汉英
土木工程大词典

下

A Chinese-English Civil Engineering Dictionary

中交第四航务工程勘察设计院有限公司

罗新华 主编

内 容 提 要

《汉英土木工程大词典》共收集词条约106万，内容涵盖房屋建筑工程、港口航道工程、道路与桥梁工程、铁路工程、岩土工程、地下与隧道工程、市政工程、水利水电工程、城市轨道交通工程、矿冶工程、工程机械、建筑材料、健康安全环保(HSE)以及数理化等相关学科专业方面的词组、短语，可供从事土木工程相关方面的广大科技人员参考使用。

图书在版编目(CIP)数据

汉英土木工程大词典：全2册 / 罗新华主编. -- 北京：人民交通出版社股份有限公司，2016.4

ISBN 978-7-114-12909-4

Ⅰ. ①汉… Ⅱ. ①罗… Ⅲ. ①土木工程 – 词典 – 汉、英 Ⅳ. ①TU-61

中国版本图书馆CIP数据核字(2016)第065039号

书　　名：汉英土木工程大词典(下)

著 作 者：罗新华

责任编辑：杜　琛　邵　江

出版发行：人民交通出版社股份有限公司

地　　址：(100011)北京市朝阳区安定门外外馆斜街3号

网　　址：http://www.ccpress.com.cn

销售电话：(010)59757973

总 经 销：人民交通出版社股份有限公司发行部

经　　销：各地新华书店

印　　刷：北京市密东印刷有限公司

开　　本：880×1230　1/16

印　　张：293.25

字　　数：24000千

版　　次：2016年4月　第1版

印　　次：2016年4月　第1次印刷

书　　号：ISBN 978-7-114-12909-4

总 定 价：998.00元(上、下册)

《汉英土木工程大词典》
编　委　会

《汉英土木工程大词典》
编　辑　组

责任编辑：杜　琛　邵　江

审　　稿：邵　江　杜　琛　王　霞　李　坤　尤晓暐　周往莲　富砚博　田克运

编　　辑：杜　琛　卢　珊　陈力维　卢俊丽　张　鑫　王景景　李学会　张　洁　韩彩君

总　　序

土木工程是一门古老的学科,人类在远古的时候就进行各种土木工程活动,如盖房子、修路、筑堤坝、烧黏土砖等。当时土木工程涉及的领域比较少,后来慢慢发展起来,到17世纪西方工业革命以后,土木工程得到高速的发展。我国也在近代派遣了许多学生到西方学习土木工程学和建筑学,他们学成归来,为我国该类学科做出了巨大的贡献。但是,在新中国成立之前,基本上没有出版过有关英汉或者汉英土木工程方面的辞书。

新中国成立以后,随着国际交流日益增多,有关出版社也出版了一些英汉土木工程方面的辞书,例如《英汉铁路词典》(1975年),这些辞书基本上是应当时国家援建项目需要而编写的。

1978年,改革开放的春风吹遍祖国的大江南北,祖国各行各业的发展蒸蒸日上,土木工程学也得到空前的发展,一大批包括工程建设公司、建设监理、咨询公司等在内的外资企业进驻中国,开展基本建设方面的业务。同时,由于国际间土木工程方面的学术交流日益增多,对英汉、汉英土木工程辞书的需求也显得更为迫切。有关出版社也出版了各行各业的英汉、汉英辞书,在一定程度上满足了当时发展的需要。但是随着科技的高速发展,以及超大型工程项目,如港珠澳大桥、上海洋山港深水港区等的建设,涉及了大量的土木工程方面问题,需要国际间土木工程专家共同商讨并解决。为了交流沟通,专业的翻译显得特别重要,工作过程中也离不开这方面的辞书。

2013年9月和10月,中国国家主席习近平在出访中亚和东南亚国家期间,先后提出共建"丝绸之路经济带"和"21世纪海上丝绸之路"的重大倡议,得到国际社会高度关注。随着国家"一带一路"战略的实施,中国将有更多的工程技术人员走向海外,参与境外土木工程建设,并成为新常态。

从事土木工程专业英语翻译的广大科技人员都有一个愿望,就是手头拥有一部综合的英汉土木工程方面词典和一部汉英土木工程词典或者这方面的电子书,这样工作起来就比较得心应手。该大词典的主编罗新华硕士和编写组大多数人员长期从事国内外土木工程项目咨询、勘察、设计、施工和监理工作,深深感到编写该辞书的重要性和必要性,经过25年的不懈努力,终于完成了《英汉土木工程大词典》(共收集词条约83万条)和《汉英土木工程大词典》(共收集词条约106万条),前者已于2014年初出版,出版后受到业界的广泛好评,也盼望后者能够早日出版。在交通运输部科技司、中国交通建设股份有限公司、人民交通出版社、中交第四航务工程勘察设计院有限公司以及其他兄弟单位的大力支持下,编写组又经过两年多的努力,《汉英土木工程大词典》如愿完成,为我国土木工程界增添了一部实用型基础工具书。

《汉英土木工程大词典》一书的最大特点是收集了海量词条(约106万条),为了满足各方面的要求,不但收集大量工程需要的新词汇,以满足目前正在进行的国内外土木工程项目的需要,而且还收集了各个时期已有的词汇,以满足在高校或者科研机构从事土木工程研究的科技人员以及广大工程技术人员的需要。同时与时俱进,为适应当今社会智能化时代的需要,词典以电子书的思路进行编写,这样用起来更方便。比如承包商、承包人、承包者这三个中文词条对应的英文都是contractor,使用者可以根据需要通过电子书进行查找选择。如果只给出中文词条其中的一个,查找时就会出现经常找不到词的情况。

我相信该大词典的出版,将为从事土木工程的科技人员提供极大方便。

中国科学院院士

邱大洪

2015年12月于大连

序

2013 年 9 月和 10 月，中国国家主席习近平在出访中亚和东南亚国家期间，先后提出共建“丝绸之路经济带”和“21 世纪海上丝绸之路”（以下简称“一带一路”）的重大倡议，得到国际社会高度关注，作为从事土木工程工作的科技工作者也受到极大的鼓舞。随着国家“一带一路”战略的实施，尤其是我国各行业海外业务的不断扩大，国内原有传统的有关汉英土木工程专业方面的辞书已经不能适应新的要求，很有必要编写一部综合性的汉英土木工程方面的辞书，以满足广大科技工作者在国内外学习和工作中的需要。

《汉英土木工程大词典》是在《汉英港湾工程大词典》（2000 年 4 月人民交通出版社出版，约 18.6 万词条）和《英汉土木工程大词典》（2014 年 2 月人民交通出版社出版，约 83 万词条）的基础上，参考了大量的国内外相关辞书、现行规范和标准术语，进一步广泛收集了与土木工程相关的专业词汇，包括房屋建筑工程、港口航道工程、道路与桥梁工程、铁路工程、岩土工程、地下与隧道工程、市政工程、水利水电工程、城市轨道交通工程、矿冶工程、工程机械、建筑材料、健康安全环保（HSE）以及数理化等相关学科专业方面的词汇编纂而成，共收集词条约 106 万条。编写组经过两年多的不懈努力，终于完成了《汉英土木工程大词典》的编纂工作，该书是迄今国内最为完整的汉英土木工程方面的辞书。

该大词典的出版，将在很大程度上满足从事土木工程方面工作的广大科技工作者和工程技术人员阅读和翻译英文的需求。

中国工程院院士

谢世楞

2015 年 9 月于天津

前　言

《汉英土木工程大词典》历经25载不懈努力，终于与读者见面了，这是一件值得庆贺的事情！20世纪80年代初的某个晚上，编者在广州中山大学图书馆晚自修，因学习需要进入外文工具书阅览室借阅有关地质方面的英汉词典。但看到的是，宽敞的书架上只放了几本不到300页(16开本)的非土木工程专业词典和两本厚厚的美国人编的韦氏词典(大16开本)。当时编者便感慨于韦氏词典编者的伟大；佩服之余，心生今后也要编出这样大部头综合性权威词典的理想。1990年，编者硕士毕业，其后一直在国内外从事岩土工程勘察类工作，经常接触到工程报告、招投标书的翻译，也经常因找不到可用的相关专业词典而苦恼，于是萌发了编写汉英土木工程方面词典的想法，并在王汝凯、麦土金等专家教授的鼓励下开始了资料搜集工作。

1997年初，编者完成了《汉英港湾工程大词典》的编撰工作，收录词汇约186000条，该书于2000年4月由人民交通出版社出版。在该词典编写过程中，编者偶然在一本有关土木工程的辞书中看到这样一句话："土木工程学科发展到今天，英汉土木工程词汇估计有三十多万条"。为了这个答案，在好奇心的驱使下，编者凭借兴趣和长期在国外及港澳地区的工作需要，尝试收集，结果发现，在实践中收集的词条远远超过这些。

2003年初，许多专家向编写组提议，仅有汉英部分不是很全面，应该配上英汉部分使用才方便。于是便把《汉英港湾工程大词典》转换为《英汉港湾工程大词典》(没有出版)继续收集编写。2014年，时值中交第四航务工程勘察设计院成立50周年，为了庆祝建院50周年，便于2014年2月先期出版了《英汉土木工程大词典》(人民交通出版社出版，收录词条约86万)，并在此基础上，继续进行《汉英土木工程大词典》的编写工作。

在该词典的编写过程中，编者参考了大量的国内外相关辞书、现行规范和标准术语，进一步广泛收集了与土木工程相关的专业词汇，包括房屋建筑工程、港口航道工程、道路与桥梁工程、铁路工程、岩土工程、地下与隧道工程、市政工程、水利水电工程、城市轨道交通工程、矿冶工程、工程机械、建筑材料、健康安全环保(HSE)以及数理化等相关学科专业方面的词汇编纂而成的，共计词条约106万。编者在本书收录词条的过程中，特别注意以下两方面：第一，高校和科研单位需要；第二，土木工程项目招投标、勘察设计以及施工过程中的需要，以期最大化满足土木工程科技工作者的实际需要。

本词典的词条首先源自近年来土木工程专业大类相关出版单位出版的土木工程类词典、专业图书、行业规范和标准术语等，并在收集时做了大量核查比对工作，力求对每个词条给出最为准确、常用和实用的解释；其次，参考了全国科学技术名词审定委员会公布的《土木工程名词》，并对其中部分内容进行了修订。此外，根据编者近二十年在国外和我国港澳地区参与国际、地区土木工程建设的经验，目前的土木工程勘察设计与施工特别注重健康、安全、环保(HSE)方面问题,因此本词典也收集了这方面的词条,方便广大科技工作者在工作中使用。

在词典编排顺序上,本词典采用计算机自动排序方法,首先按拼音排列,其次按声调顺序排列,这种方式符合当今智能化环境下一般科技工作者的行为习惯。文前列出了汉语拼音音节索引,便于使用者查询。

本词典在编写过程中,先后有一百多位国内外专家、教授、勘察设计大师、博士、硕士、英语专业毕业的大学生、在校大学生、常年奋战在工程项目第一线的工程师、技术人员参与。同时,也得到了交通运输部科技司、中国交通建设集团有限公司、人民交通出版社、中交第四航务工程勘察设计院有限公司以及其他兄弟单位的大力支持。中山大学地球科学与地质工程学院王将克教授、张珂教授更是在词典编写过程中给予不少帮助和指导。本词典的编纂可以说是凝聚了我国从事土木工程工作的广大科技人员的心血,在此表示衷心的感谢!

科技发展突飞猛进,土木工程新词汇不断出现,且随时间的推移某些词的词义也有所变化,热切期望广大读者使用过程中发现问题给予批评指正,也欢迎随时提供该词典没有收集到的相关词条,为以后再版输送新鲜血液。相信有读者的热心参与,该词典将更加完善。

主编:罗新华

2016 年 3 月 3 日于广州

总　　目

凡　　例

1. 本词典系按汉语拼音字母顺序采用计算机程序进行排列，术语单词、复合词和短语一律顺排。对于含有阿拉伯数字和西文字母的词条，以阿拉伯数字、西文字母、汉字拼音字母顺序排列；对于含有标点符号的词条，不考虑标点符号的影响，只以标点符号前后两字的顺序排列；同音异调的汉字按声调顺序排列。

2. 当中文词条有多个英文词条释义时（即一词多译情况），不同的英文释义词条分列，并用分号（;）将不同释义分隔开。例如：

框架结构 frame(d) construction;frame(d) structure;rahmen;skeleton structure;skeleton type construction

3. 括号的用法

（1）（ ）圆括号

表示可省略的汉字、字母或单词，例如：

路拌柏油（沥青）混合料路面 tar-road-mix surface

框架 frame (bent);pigsty(e);skeleton (frame)

（2）[　]方括号

① 可替换前面的中文汉字，但括号外字形为推荐字形，例如：**黏[粘]**

② 可替换前面的单词，例如：**蜡片** wax disc[disk];**缆索吊装设备** cableway erecting equipment[plant]

③ 英文缩略语，例如：**零功率铀系统** zero energy uranium system[zeus]

④ 英文缩略语解释，例如：**零功率热核装置** Zeta[zero Energy Thermonuclear Assembly]

⑤ 英文单词的复数形式，例如：**卵石饰** ovum[复 ova]

（3）< >尖括号

表示对中文词条的进一步注释或对英文释义用法的补充说明，例如：

螺旋式整平摊铺机 <**混凝土**> combination screw-screed spreader

螺旋管塞 casing head <**井钻的**>;casing cap

（4）【　】鱼尾括号

表示专业或学科，例如：**【铁】，表示铁路工程**

4. 当主词条有多个英文释义时，为节省篇幅和便于查阅进行适当合并，并用正斜杠（/）分隔，表示多个英文释义可互相替代使用，例如：

横撑 cross arm[bar/ beam/ bracing/ strut/ tie/ transom]

5. 专业或学科代号

【数】数学
【道】道路工程
【航空】航空学
【地】地质学
【气】气象学
【天文】天文学
【动】动物学
【农】农业学
【港】港口航道工程
【航海】航海学
【无】无线电子学
【矿】矿物学
【机】工程机械
【救】救捞工程
【生】生物学
【军】军事工程
【建】房屋建筑工程
【化】化学化工
【给】给排水
【水文】水文学
【疏】疏浚工程
【声】声学
【暖通】暖通工程
【物】物理
【铁】铁路工程
【岩】岩土工程
【测】测量学
【计】计算机
【植】植物学
【冶】矿冶工程

汉语拼音音节索引

yù 郁 1771
yù 狱 1771
yù 峪 1771
yù 浴 1771
yù 预 1772
yù 域 1790
yù 阈 1790
yù 寓 1791
yù 御 1791
yù 裕 1791
yù 遇 1791
yù 愈 1791
yù 誉 1791

YUAN

yuān 鸢 1791
yuān 鸳 1791
yuán 元 1791
yuán 员 1792
yuán 园 1792
yuán 垣 1792
yuán 原 1793
yuán 圆 1800
yuán 援 1810
yuán 缘 1810
yuán 源 1810
yuán 辕 1811
yuǎn 远 1811
yuàn 院 1815
yuàn 垸 1815
yuàn 愿 1815

YUE

yuē 约 1815
yuè 月 1816
yuè 岳 1818
yuè 钥 1818
yuè 悦 1818
yuè 钺 1818
yuè 阅 1818
yuè 跃 1818
yuè 越 1818

YUN

yún 云 1819
yún 匀 1820
yún 芸 1821
yún 耘 1821
yǔn 允 1821
yǔn 陨 1821
yùn 孕 1821
yùn 运 1822
yùn 晕 1832
yùn 酝 1832
yùn 韵 1832
yùn 熨 1832
yùn 蕴 1832

Z

ZA

zā 匝 1833
zá 杂 1833
zá 砸 1835

ZAI

zāi 灾 1835
zāi 甾 1835
zāi 栽 1835
zǎi 宰 1835
zài 载 1835
zài 再 1838
zài 在 1843

ZAN

zān 簪 1846
zǎn 攒 1846
zàn 暂 1846
zàn 赞 1848
zàn 錾 1848

ZANG

zāng 赃 1848
zāng 脏 1848
zàng 葬 1848
zàng 藏 1848

ZAO

zāo 遭 1848
zāo 糟 1848
záo 凿 1848
zǎo 早 1850
zǎo 枣 1851
zǎo 蚤 1851
zǎo 澡 1851
zǎo 藻 1851
zào 灶 1851
zào 皂 1851
zào 造 1852
zào 噪 1854
zào 燥 1856
zào 躁 1856

ZE

zé 择 1856
zé 泽 1856
zé 责 1856

ZENG

zēng 增 1856
zēng 憎 1861
zèng 甑 1861
zèng 赠 1861

ZHA

zhā 渣 1861
zhā 扎 1862
zhá 轧 1862
zhá 札 1864
zhá 闸 1864
zhá 铡 1866
zhǎ 眨 1866
zhà 乍 1866
zhà 诈 1866
zhà 栅 1866
zhà 炸 1866
zhà 蚱 1867
zhà 榨 1867

ZHAI

zhāi 摘 1867
zhái 宅 1867
zhái 翟 1867
zhǎi 窄 1867
zhài 债 1869
zhài 寨 1869

ZHAN

zhān 沾 1869
zhān 毡 1869
zhān 粘 1869
zhān 詹 1869
zhān 瞻 1869
zhǎn 斩 1869
zhǎn 展 1870
zhǎn 崭 1871
zhǎn 辗 1871
zhàn 占 1871
zhàn 战 1871
zhàn 栈 1872
zhàn 站 1872
zhàn 绽 1873
zhàn 蘸 1873

ZHANG

zhāng 张 1873
zhāng 章 1875
zhāng 獐 1875
zhāng 樟 1875
zhāng 蟑 1875
zhǎng 长 1875
zhǎng 涨 1875
zhǎng 掌 1876
zhàng 丈 1876
zhàng 帐 1876
zhàng 杖 1876
zhàng 胀 1876
zhàng 账 1877
zhàng 障 1877
zhàng 嶂 1878
zhàng 幛 1878
zhàng 瘴 1878

ZHAO

zhāo 招 1878
zhāo 朝 1878
zhāo 啁 1878
zháo 着 1878
zhǎo 找 1878
zhǎo 沼 1879
zhào 召 1880
zhào 兆 1880
zhào 赵 1880
zhào 照 1880
zhào 罩 1883
zhào 肇 1884

ZHE

zhē 遮 1884
zhé 折 1885
zhé 哲 1889
zhé 蛰 1889
zhé 摺 1890
zhé 辙 1890
zhě 锗 1890
zhě 赭 1890
zhě 褶 1890
zhè 这 1891
zhè 柘 1891
zhè 蔗 1891

ZHEN

zhēn 针 1891
zhēn 侦 1892
zhēn 珍 1893
zhēn 真 1893
zhēn 砧 1899
zhēn 祯 1899
zhēn 甄 1899
zhēn 榛 1899
zhēn 帧 1899
zhěn 诊 1900
zhěn 枕 1900
zhèn 阵 1900
zhèn 振 1901
zhèn 赈 1908
zhèn 镇 1908
zhèn 震 1908

ZHENG

zhēng 争 1910
zhēng 征 1910
zhēng 挣 1910
zhēng 蒸 1910
zhěng 拯 1918
zhěng 整 1918
zhèng 正 1925
zhèng 证 1940
zhèng 郑 1940
zhèng 政 1940
zhèng 症 1941

ZHI

zhī 之 1941
zhī 支 1941
zhī 汁 1947
zhī 芝 1947
zhī 吱 1947
zhī 枝 1947
zhī 知 1947
zhī 织 1948
zhī 肢 1948
zhī 脂 1948
zhī 蜘 1948
zhí 执 1949
zhí 直 1949
zhí 值 1964
zhí 职 1965
zhí 植 1966
zhí 殖 1968
zhǐ 止 1968
zhǐ 只 1970
zhǐ 纸 1970
zhǐ 咫 1971
zhǐ 指 1971
zhǐ 枳 1976
zhǐ 趾 1976
zhǐ 酯 1976
zhì 至 1976
zhì 志 1976
zhì 制 1976
zhì 治 1984
zhì 质 1984
zhì 栉 1987
zhì 秩 1987
zhì 致 1987
zhì 掷 1988
zhì 窒 1988
zhì 智 1988
zhì 滞 1989
zhì 蛭 1990
zhì 稚 1991
zhì 置 1991
zhì 雉 1991

ZHONG

zhōng 中 1991
zhōng 忠 2017
zhōng 终 2017
zhōng 钟 2020
zhōng 舯 2021
zhǒng 肿 2021
zhǒng 种 2021
zhǒng 冢 2021
zhǒng 踵 2021
zhòng 中 2021
zhòng 仲 2021
zhòng 众 2021
zhòng 种 2021
zhòng 重 2022

ZHOU

zhōu 州 2033
zhōu 舟 2033
zhōu 周 2033
zhōu 洲 2037
zhōu 粥 2037
zhóu 轴 2037
zhǒu 肘 2042
zhǒu 帚 2043
zhòu 宙 2043
zhòu 绉 2043
zhòu 昼 2043
zhòu 皱 2043
zhòu 骤 2044

ZHU

zhū 朱 2044
zhū 侏 2044
zhū 株 2044
zhū 珠 2044
zhū 猪 2044
zhū 蛛 2045
zhū 槠 2045
zhū 潴 2045
zhú 竹 2045
zhú 烛 2045
zhú 逐 2045
zhǔ 主 2047
zhǔ 煮 2061
zhù 住 2061
zhù 助 2063
zhù 苎 2063
zhù 注 2063
zhù 贮 2066
zhù 驻 2066
zhù 柱 2067
zhù 著 2071
zhù 蛀 2071
zhù 筑 2071
zhù 铸 2072

ZHUA

zhuā 抓 2074
zhuǎ 爪 2075

ZHUAN

zhuān 专 2076
zhuān 砖 2080
zhuǎn 转 2083
zhuàn 啭 2097
zhuàn 赚 2097
zhuàn 撰 2097

ZHUANG

zhuāng 庄 2097
zhuāng 桩 2097
zhuāng 装 2100
zhuàng 壮 2113
zhuàng 状 2113
zhuàng 撞 2113

ZHUI

zhuī 追 2114
zhuī 椎 2115
zhuī 锥 2115
zhuì 坠 2118
zhuì 缀 2118
zhuì 赘 2119

ZHUN

zhǔn 准 2119

ZHUO

zhuō 拙 2122
zhuō 捉 2122
zhuō 桌 2122
zhuó 卓 2122
zhuó 灼 2122
zhuó 茁 2122
zhuó 斫 2122
zhuó 浊 2122
zhuó 酌 2122
zhuó 啄 2122
zhuó 着 2122
zhuó 琢 2123

ZI

zī 兹 2124
zī 咨 2124
zī 姿 2124
zī 资 2124
zī 孳 2124
zī 滋 2127
zī 辎 2127
zī 锱 2127
zǐ 仔 2127
zǐ 籽 2127
zǐ 子 2127
zǐ 姊 2129
zǐ 梓 2129
zǐ 紫 2129
zì 字 2130
zì 自 2130
zì 恣 2173
zì 渍 2173

ZONG

zōng 枞 2173
zōng 宗 2173
zōng 综 2173
zōng 棕 2177
zōng 踪 2177
zōng 鬃 2177
zǒng 总 2177
zòng 纵 2186

ZOU

zǒu 走 2191
zòu 奏 2192

ZU

zū 租 2192
zú 足 2193
zú 卒 2193
zú 族 2193
zú 镞 2193
zǔ 阻 2193
zǔ 组 2196
zǔ 祖 2201

ZUAN

zuàn 钻 2201

ZUI

zuǐ 嘴 2209
zuì 最 2209
zuì 罪 2233
zuì 醉 2233

ZUN

zūn 尊 2233
zūn 遵 2233

ZUO

zuǒ 左 2233
zuǒ 佐 2234
zuò 作 2234
zuò 坐 2237
zuò 柞 2238
zuò 唑 2239
zuò 座 2239
zuò 做 2239

M

麻斑的 pitted

麻包针 gunny needle
麻包装皮 jute wrap(ping)
麻痹 paralysis
麻痹药 paralyzer[paralyser]
麻布 burlap; carbarsus; flax; gunny; gunny cloth; hessian; jute burlap; sack-cloth
麻布带 cambric tape
麻布袋 gunny bag
麻布(袋)拉平表面 <混凝土路面> sack-rubbed surface
麻布垫 jute burlap mat
麻布刮平抹光 <混凝土路面的> jute hessian drag finish
麻布基底上加沥青 <防潮用> anderite
麻布卷尺 linen-tape
麻布绝缘管 insulated cambric pipe
麻布拉毛 burlap finish
麻布拉平 <水泥混凝土路面施工用> burlap-drag(finish)
麻布沥青油毡 asphalt saturated and coated jute fabric
麻布抹面 sack rub
麻布饰面 burlap finish
麻布水龙带 hessian fire hose
麻布纹整饰 linen finish
麻布线头 hemp shives
麻布油毡 bituminous flax felt; hessian-based bituminous felt; hessian-based bituminous sheeting
麻车 barrow; stall
麻衬垫 jute insertion
麻粗纤维 flax tow
麻袋 burlap(bag); burlap sack; gunny (sack); hemp-bag; hemp sack; jute bag; matting; tow sack
麻袋包装 bale packing
麻袋布 sack-cloth; sacking
麻袋帆布衬垫 burlap canvas mat
麻袋帆布铺盖 burlap canvas mat
麻袋混凝土 bag concrete
麻袋夹持器 bag holder; sack grip; sack holder
麻袋墨 jute bag marking ink
麻袋片 gunny cloth; hessian cloth; jute mat
麻袋起重机 sack hoist
麻袋堰 sack dam
麻袋装沙 hessian sandbag
麻袋装卸 sack handling
麻刀【建】hair; hemp; hemp cut; oakum; hemp fiber[fibre]
麻刀打散器 hair beater
麻刀灰 fibered plaster; fibrous plaster; haired mortar; hemp-fibered plaster; lime plaster with hemp cut; staff
麻刀灰拌和钉耙 hair hook
麻刀灰打底 hemp-fibered plaster base
麻刀灰浆 fibered plaster; hair mortar
麻刀灰抹墙角 staff angle
麻刀灰抹圆形线条 staff bead
麻刀灰泥 hair-fibered cement mortar; hair-fibered plaster; hair mortar
麻刀灰墙角 staff angle
麻刀灰圆护角 staff bead
麻刀灰圆护条 staff bead
麻刀灰罩面 hemp-cut plaster finishing coat; hemp-fibered plaster finishing coat
麻刀石膏 hair gypsum
麻刀石膏灰泥 hair gypsum-lime mortar
麻刀石灰泥浆 hair-fibered lime mortar
麻刀水泥泥浆 hair-fibered cement mortar
麻刀丝 hair
麻刀纤维 hemp fiber[fibre]
麻刀油灰 grammet; gummet
麻地垫 hemp floor mat
麻点 hard spot; mechanical pitting; mottle; mottling; pitting; pock (mark); pock marked; pock marking; sesame spot; pimpling <钢材缺陷>
麻点的 <油漆不匀而产生的表面麻点> bitty
麻点腐蚀 pitting attack; pitting corrosion
麻点裂纹 catface
麻点式穿孔板 foraminate perforation plate
麻点图像 pit pattern
麻点釉 sesame spot glaze
麻点状分布的局部凹陷 <位于海底表面浅部> pock mark
麻点状腐蚀 pitting corrosion
麻点阻力 pitting resistance
麻垫料 hemp jointing; hemp packing
麻垫片 hemp gasket
麻叠层 jute lamination
麻叠片 jute lamination
麻洞 pitting due to cavitation
麻堆锥形压紧机构 conic(al) packer
麻垛 stooks
麻帆布 flax canvas; hemp canvas; tarare
麻纺厂废水 flax mill wastewater
麻风病医院 lepers hospital
麻风树 monioca
麻风树油 curcas oil
麻疯病院 laxar house; lazaret(to); leprosarium; leprosery
麻杆输送器 stalk-end transporter
麻钢混捻钢丝绳 combi-rope
麻姑观音 <瓷器名> Avalokitesvara with coiffure of Maku
麻花壁缘 twisted rope frieze
麻花辊 felt widening roll; worm roll
麻花活钻头 twist bit
麻花链环 twist-lint chain
麻花平钻 twist flat drill
麻花取土钻 twisted auger
麻花手钻 twist gimlet
麻花条纹 twisted fringe
麻花线 twisted wire
麻花形手柄 double interlaced handle
麻花形柱 salomonica; twisted column; wreathed column
麻花柱孔钻 auger twist bit
麻花状扭曲 propeller twist
麻花钻 auger; cross auger; earth auger; earth screw; fluted twist drill; milled twist drill; rotating auger; screw auger; soil boring auger; twist auger; twist drill; wood drill; worm auger
麻花钻槽铣刀 cutter for fluting twist drill; twist drill cutter
麻花钻活钻头 finger rotary detachable bit
麻花钻机 auger
麻花钻径规 twist drill ga(u)ge
麻花钻孔 auger boring
麻花钻浅层取样 shallow sampling with a soil auger
麻花钻取土器 earth boring auger
麻花钻头 auger bit; auger drill head; spiral bit; twist bit twist drill
麻花钻头刃磨机 twist drill grinding machine
麻花钻芯 twist web
麻花钻研磨器 twist drill grinder
麻花钻钻探 auger boring
麻黄式穿孔 foraminate perforation
麻黄式穿孔板 ephedroid perforation plate
麻黄型穿孔板 ephedroid perforation plate
麻加工厂 jute factory
麻夹变换器 clamp changer
麻浆纸 jute paper
麻浆纸袋纸 rope bag paper
麻蕉 abaca; abaca banana
麻筋 hemp cut; hemp fiber[fibre]
麻茎 straw
麻靠垫 jute backing
麻坑 eye hole; pit; pock mark
麻孔 air checks; air marking; air marks
麻口化 mottling
麻口镍铸铁轧辊 nickel alloy grain roll
麻口生铁 mottled(pig) iron
麻口铁 mottle cast-iron
麻口细晶粒合金铸铁轧辊 grain roll
麻口铸铁 mottled cast-iron
麻类 flex; hemp
麻类纤维 bast-fiber[fibre]; long vegetable fiber
麻类作物 crude-fiber[fibre] crop
麻栎 chittagon(y) wood; spotty oak
麻栗木 teakwood
麻栗木材 teak
麻栗树 teak
麻粒玻璃 <透射紫外线的玻璃> corning glass
麻粒的 granulitic
麻粒构造 granulose structure
麻粒结构 granule texture; granulitic texture
麻粒岩 granulite
麻粒岩化作用 granulitization
麻粒岩相 granulite facies
麻粒状 granulose
麻棉布 <窗帘用> scrim
麻面 bug hole; granular structure surface; mat(te) finish(ing); spongy surface; surface pockmark; hungry spots; pitted skin; pitted surface; pitting surface; surface pit; surface voids; rough-surfaced pavement <路面>; granular structure <塑料制品缺陷>; gas pocket
麻面玻璃纤维 roughened surface glass fiber[fibre]
麻面锤 granulated hammer; granulating hammer
麻面的 ballast-surfaced
麻面防止剂 anti-pitting agent
麻面钢丝 indented wire
麻面卵石 pitted pebble
麻面磨损 pitted wear
麻面状态 pitted surface appearance
麻姆统 <晚侏罗世>【地】Malm series
麻木[医] anaesthesia; coma; torpor
麻木的 impassible
麻泥型 loam mo(u)ld
麻泥型铸造 loam casting
麻片 <止水用> jute board
麻纱 cambric; Flax hair-cords; thread
麻纱软垫 spun yarn packing
麻纱胀堵漏剂 <用于铅麻子管> yarn swellant
麻山式磷矿床 Mashan-type phosphate deposit
麻梢 crop end
麻生电钻 <一种电钻> Mason master
麻绳 fibre rope; flax rope; hemp cable; hemp cord; hemp rope; hemp-twist; hessian rope; jute rope
麻绳厂废水 rag rope waste
麻绳处理的混凝土表面印花 ribbed rope finish
麻绳打捆 string tying
麻绳打捆机构 twine-tying mechanism
麻绳放线 string lining
麻绳放样 string lining
麻绳机 twine unit
麻绳捆扎 hemp lashing
麻绳捆扎机 string binder
麻绳捆扎机构 twine-tying mechanism
麻绳缆 hemp hawser
麻绳绳卡 Babcock manila-rope socket; wing socket
麻绳填料 hemp packing; rope packing
麻省公式 <一种设计柔性路面厚度的古典公式> Massachusetts formula
麻省理工学院 <美> Massachusetts Institute of Technology
麻省设计法 <美国一种设计柔性路面厚度的古典方法> Massachusetts method
麻省准则【道】Massachusetts's rule
麻石子 crushed granite
麻蚀 cavitation erosion
麻蚀电位 pitting potential
麻丝 flax silk; hair; hemp; hemp cut; hemp thread; jute waste
麻丝板 <建筑用的> mill board
麻丝垫环 grommet[grummet]
麻丝封填 hemp packing
麻丝沥青 bitumen-dipped hemp; bituminous flax
麻丝嵌填 hemp ca(u)lking; hemp jointing
麻丝砂浆 haired mortar; hair-fibered [fibred] mortar
麻丝石膏 hair-fibered[fibred] gypsum
麻丝石灰砂浆 haired lime mortar
麻丝水泥灰浆 haired cement mortar
麻丝水泥砂浆 hair-fibered[fibred] cement mortar
麻丝填料 hemp packing; spun yarn packing
麻丝油灰 grommet
麻塑混编袋 jute and plastics mixed bag
麻塑交织袋 plastic and gunny mixed spun bag
麻填 hemp packing
麻填料 hemp jointing
麻填密 hemp packing
麻填密活塞 hemp packed piston
麻条 ramie stripe
麻网袋 sisal netting bag
麻纤维 bast-fiber[fibre]
麻线 hemo yarn; linen thread; twine
麻线放样 string lining
麻线放样标高 string level
麻线填料 hemo packing
麻屑 hards; hemp hards; hemp tow; tow
麻屑抖落器 tow shaker
麻屑加工机 tow scutcher
麻屑清理机 tow cleaner
麻屑填料 tow packing
麻芯 hemp center[centre]; hemp core <钢缆的>
麻芯钢丝绳 hemo-core(d) wire rope
麻芯钢索 hemp cored wire rope
麻芯索条 heart yarn
麻芯线 hempen core
麻芯真空浸油器 vacuum oil penetrator for hemp core
麻型 linen look
麻絮 hemp fiber[fibre]; loose hemp fibre
麻絮丝 oakum
麻叶形纹饰 flax ornament
麻与金属丝合股绳 spring lay rope
麻织水龙带 hemp hose

麻织系带 hemp strap
麻脂填料 hemo tallow packing
麻竹 dendrocalamus affinis
麻醉室 anaesthesia induction room;anaesthetic room;anesthetizing room
麻醉台 anaesthesia table

马

马鞍 gap bridge;riding saddle;saddle;bridge piece <车床的>
马鞍凹处旋径 swing over gap
马鞍车床 geared-head gap-bed lathe
马鞍床身 gap bed
马鞍点 saddle point
马鞍点共沸体 saddle-point azeotrope
马鞍峰 saddle hill
马鞍剪床 gap shears
马鞍菌属 <拉> Helvella
马鞍壳 shell of hyperbolic paraboloid
马鞍链 bridle chain;chain bridle
马鞍石 horse saddle;saddle pebble
马鞍式车床 gap-bed lathe;gap lathe
马鞍式全齿轮车床 geared-head lathe with gap bed
马鞍式压床 gap(-frame)press
马鞍栓塞 saddle embolus
马鞍镶块 gap piece
马鞍形 shape of saddle; U-shaped; saddle type <流态等的>
马鞍形车床 gap-bed lathe
马鞍形窗间条 saddle bead
马鞍形床身 gapped bed
马鞍形分布 saddle distribution
马鞍形锅炉 saddle boiler
马鞍形壳 saddle shell
马鞍形扩孔锻造 saddle forging
马鞍形三通管 saddle tee
马鞍形塔填充物 saddle-shaped tower packing
马鞍形天线绝缘子 shell insulator
马鞍形填料 saddle packing
马鞍形柱 <固定管子的> saddle tee
马鞍压床 gap hydraulic press; gap press
马背岭 horse back;horseback mountain
马鞭草 verbena
马表 <俗名> stop watch
马波里板 <由菱苦土和沥青乳液拌制而成的一种无缝地板> Marbolith
马波维尔层【地】Maplewell bed
马勃属 <拉> Lycoperdon
马厂类型陶器 Machang type pottery
马车 horsedrawn vehicle;horse vehicle;horse wagon;wain;gharry <印度、埃及等>
马车出租站 cabstand
马车道 bridle road;bridle track;bridleway;cart road;cart way;driveway;horse track;horse way;wagon way
马车道路 horse road
马车夫 <俚语> hair pounder
马车交通 horsedrawn traffic
马车路 carriage road;cart road;cart track;cart way;wagon road;bridle way
马车旅行 staging
马车拖运 drayage
马车屋 mews
马车行 livery
马车业 cab service;staging
马车运费 cartage
马车运货费 drayage
马车运输 animal drawn traffic;cartage; carting; horsedrawn traffic; horse traffic
马车运输承包商 teaming contractor
马刺 spur
马达 electric(al)motor;motor
马达泵 motor-driven pump; motor-mounted pump;motor pump
马达变极启动器 pole-change motor starter
马达传动 motor drive
马达的主要部件 vitals of a motor
马达阀 motor valve
马达格达响声 knocking
马达过热保护 motor protection against overheat
马达和泵的布置 motor-and-pump arrangement
马达和泵的配置 motor-and-pump arrangement
马达化 motorize
马达加斯加钩 Madagascar hook
马达加斯加海盆 Madagascar basin
马达加斯加红厚壳【植】mastwood; Calophyllum;kamani
马达降速器 motor reducer
马达轮驱动 motorized wheel drive
马达轮自由轮工况随动齿轮 free-wheeling driven gear
马达排量 motor displacement
马达启停开关 <电传打字机的> motor on-off switch
马达起动器螺钉 screw for motor starter
马达驱动的 motor-driven
马达驱动的可变焦距镜头 motorized zoom lens
马达驱动的手推车 motor barrow
马达燃料 automotive fuel
马达试验台 motor test stand
马达输出转速 motor output speed
马达推进式凿岩机 motor feed drill
马达拖动泵 motor-driven pump
马达外罩 motor case
马达移动式风动起重机 air motor hoist with motor-driven trolley
马达油 motor oil
马达油掺和剂 additive for motor oils
马达油添加剂 additive for motor oils
马达支承板 rotor bearing cock
马达支座 motor maintainer
马达轴 motor-drive shaft
马达装在车轮里的马达轮 motor tire [type]
马大道 bridle path
马刀树 <即醉林> bent tree;diverted tree;tilted tree
马刀形气动锯 pneumatic sabre saw
马道 berm(e); bridle path; bridle road; bridleway; catwalk; drift way;horse path;packway;riding
马道边缘 berm(e)edge
马道修整 berm(e)trimming
马道养护 berm(e)maintenance
马德加建筑 Mudejar architecture
马德拉斯港 <印度> Port Madras
马德里 <西班牙首都> Madrid
马凳 split heads
马镫 stirrup
马镫钢 stirrup iron
马店 <设马车房的> mews
马丁钢 martensite steel; Martin's (hearth)steel
马丁刮痕硬度试验机 Martin's scratch hardness tester
马丁过滤器 Martin's filter
马丁-海因硬度 Martin-Heyn hardness
马丁胶结料 Martin's cement
马丁炉 Martin's (hearth) furnace; open-hearth furnace
马丁炉渣 Siemens-Martin steel slag
马丁密度计 Martin's density meter
马丁平炉炼钢法 Martin process
马丁散铁 martinsite
马丁式锚 close stowing anchor; Martin's anchor
马丁水泥 hard finish plaster;Martin's cement
马丁体 martensite
马丁体钢 martensitic steel
马丁直径 Martin's diameter
马钉 dog(iron);metal bridging
马尔代夫 <亚洲> Maldives
马尔古斯槽式输送机 Marcus trough conveyer[conveyor]
马尔柯夫分析 <是利用某一变量的现在状态和动向去预测该变量未来的状态及其动向的一种分析技术> Markov analysis
马尔柯夫更新过程 Markovian renewal process
马尔柯夫过程 Markovian process
马尔柯夫决策论 Markovian decision theory
马尔柯夫链 Markovian chain
马尔柯夫模型 Markovian model
马尔柯夫矢量法 Markovian vector approach
马尔柯夫随机变数 Markovian variable
马尔柯夫转移模型 Markovian transition model
马尔科夫算法 <用于事件概率的> Markovian algorithm
马尔可夫过程 Markoff process
马尔可夫矩阵 Markoff matrix
马尔可夫链 Markoff chain
马尔默港 <瑞典> Port Malmo
马尔萨斯人口论 Malthusian population theory;Malthusian thesis
马尔萨斯主义 Malthusianism
马尔特氏硬度 dynamic(al)hardness
马耳他 <欧洲> Malta
马耳他十字架 Maltese cross
马房 livery;livery stable;stable;stall
马菲吊运车 Mafi portal-lift
马粪纸 mill board
马粪纸板 compressed straw slab; strawboard
马蜂 wasp
马夫的斯板 <一种木纤维板> Maftex
马弗炉 muffle;muffle furnace;retort
马弗窑 muffle kiln
马戈格带 Magog belt
马格 <工程质量单位,=1千克的力获得每平方秒1米加速度的质量> mug
马格伯斯蒂科建筑材料 <一种菱镁土地面铺料> Magbestic
马格达连时代 <欧洲旧石器时代的最后期> Magdalenian age
马格德林教堂 church of the Magdalen
马格列夫多样性指数 Margelef diversity index
马格列夫指数 Margelef index
马格纳斯数 Magnus number
马格纳斯浴槽 Magnus' bath
马格尼特耐大制品 <商品名> Magnit
马硅砷锰矿 m(a)cgovernite
马赫 <在同一介质中物体速度与声速的比值> Mach
马赫表 Machmeter;Mach's indicator; Mach's meter
马赫波 Mach's diamond;Mach's line; Mach's wave
马赫冲波 Mach's region wave
马赫带 Mach's band
马赫单位 Mach's unit
马赫对比环 Mach's fringe
马赫反射 Mach's reflection
马赫反射区 Mach's reflection region
马赫计 Mach's meter
马赫角 Mach's angle
马赫镁铝合金 Mach's metal
马赫喷嘴 <超音速喷气脱水装置> Mach nozzle
马赫区 Mach's region
马赫扰动面 Mach's front;Mach's stem
马赫数 Mach's number
马赫数大于2 Mach 2-plus
马赫数陡度 Mach's number gradient
马赫数范围 range of Mach's numbers
马赫数分量 Mach's number component
马赫数计算机 Mach's number computer
马赫数控制器 Mach's number controller
马赫数指示器 Machmeter;Mach's indicator
马赫特含铁黄铜 Macht's metal
马赫条纹 Mach's fringe
马赫线 Mach's line
马赫效应 <反射冲击波叠加效应> Mach's stem effect
马赫型反射 March type reflection wave
马赫原理 Mach's principle
马赫-曾德耳干涉仪 Mach-Zehnder interferometer
马赫阵面 Mach's front
马赫锥 Mach's cone
马赫锥母线 edge of Mach's cone
马基诺矿 mackinawite
马脊岭 roll;swell
马技表演场 hippodrome
马家滨文化陶器 Majiabin culture pottery
马家沟灰岩 <中奥陶世> Machiakow limestone
马家窑类型陶器 Majiayiao type pottery
马架 mandrel supporter; saddle support
马架标 horsy survey beacon
马颈轭 hames
马厩 commercial stable;mews;riding stable;stable
马厩的排水坑 <在污水进入排水沟管前> horse pot
马厩改建的住房 mews
马厩隔墙外端的柱子 <内端为饲料槽旁的柱子> kicking post
马厩集装箱 horse stall container
马厩门 stable door
马具店 saddlery
马具室 harness room;tack room
马嚼式链条 curb chain
马卡拉潮湿期 Makalian wet phase
马卡路合金钢筋 Macalloy bar
马卡洛夫海盆 Makarov basin
马卡亚海槽 Macaya trough
马康那弗罗回收矿泥输送法 Marconaflo slurry transport
马康那弗罗式矿粉液化装卸法 Marconaflo slurry system
马科姆系统 Macomb strainer
马科维茨方程 Markovitz equation
马可尼磁性检波器 Marconi magnetic detector
马可尼等幅波发射机 Marconi undamped generator
马可尼定向天线 Marconi beam aerial;Marconi beam antenna
马可尼换向器 Marconi commutator
马可尼检波器 Marconi detector
马可尼金属屑检波器 Marconi coherer
马可尼天线 Marconi(type)antenna
马可尼自动调谐系统 Marconi self-tuning system
马可式旁通节流器 Macco side door choke

马可式旁通油嘴 Macco side door choke
马克板 <一种木纤维板> macaboard
马克当路 macadam
马克当路面 water-bound macadam
马克当筑路法 Macadam's construction
马克地沥青混合料 <一种冷铺沥青碎石混合料> macasphalt
马克-豪温克常数 Mark-Houwink constant
马克拉斯特 <一种木块地板冷凝乳胶> Maclast
马克莱姆线 Macrame cord
马克里奥公式 <一种设计柔性路面厚度的公式> Mcleod formula
马克隆尼气流式纤维细度测试仪 Micronaire
马克隆尼值标尺 Micronaire scale
马克斯发生器 Marx generator; surge generator
马克苏托夫反折射物镜 Maksutov catadioptric objective
马克苏托夫改正镜 Maksutov corrector
马克苏托夫望远镜 Maksutov telescope
马克苏托夫物镜 Maksutov objective
马克苏托夫系统 Maksutov system
马克苏托夫校正板 Maksutov corrector
马克西尔赖含铜奥氏体不锈钢 Maxilvry
马克西-西尔伯斯通曲线 Maxey-Silberston curve
马克样板 Maco template
马克制度 Mark system
马口生铁 mottled(pig) iron
马口铁 bright tin plate; galvanized iron; galvanized iron sheet; galvanized sheet (iron); mottled cast-iron; tin-coated steel; tinned iron; tinned iron sheet; tin(ned) plate; tin sheet iron
马口铁罐 can; tin case
马口铁卷板 black plate for tinning; cold-reduced steel
马口铁皮 black sheet(iron); tinned iron sheet; tinned steel sheet; sheet tin
马口铁皮粗锉 horse mouth rasp
马口铁器皿 tinware
马口铁条 tinplate strip
马口铁屋顶 galvanized sheet roof; zinc roof
马口铁屋面板 tin shingle
马口铁轧机 tinplate mill
马夸特瓷 Marquardt porcelain
马奎斯顿薄膜浮选机 Macquisten filmflo(a)tation machine
马阔里海岭 Macguarie ridge
马拉 horse traction
马拉博 <赤道几内亚> Malabo
马拉侧向搂草机 horsedrawn side delivery rake
马拉车 horsedrawn cart
马拉传动 horse power
马拉传动装置 horse gear; horse motor
马拉的 horsedrawn
马拉斗式铲土机 Fresno slip
马拉斗式刮土机 Fresno slip
马拉刮路机 horse scraper
马拉滚筒 horsedrawn roller
马拉机械 horsedrawn machine
马拉集草机 horserake
马拉开波湖桥 <第一座多跨斜拉桥，计5跨235米，1962年建于委内瑞拉> Maracaibo Lake bridge
马拉克赖聚酯纤维 Maracray
马拉犁 horsepower plow
马拉搂草机 horsedrawn rake
马拉搂耙 horserake
马拉路碾 horsedrawn roller
马拉耙 horse harrow
马拉奇拉纤维 <黄麻代用品> malachra fibre[fiber]
马拉扫路机 horse sweeper
马拉式多铧犁 horse gang plow
马拉式割捆机 horsedrawn binder
马拉式机动喷粉机 horsedrawn motorized duster
马拉式机具 horsedrawn
马拉式起垄犁 horsedrawn ridging plow
马拉式碎土锄 horse hack
马拉提升滚筒 horse gin
马拉维 <非洲> Malawi
马拉维储备银行 Reserve Bank of Malavi
马拉消防车 horsedrawn vehicle
马拉有轨街车 horsedrawn tramway
马拉展式钉齿耙 horsedrawn expanding harrow
马拉指轮式搂草机 horsedrawn finger-wheel rake
马来改性醇酸树脂 neolyn resin
马来改性松香树脂 amberol resin
马来酐 maleic anhydride
马来海松酸 maleopimaric acid
马来海松酸酯 maleopimarate
马来黑漆 theetsee
马来红柳桉 red seraya; seraya; seriah
马来化 maleation
马来化桐油 maleinized tung oil
马来栲树 berangan
马来克司聚乙烯 Marlex
马来漆树 Benghas gluta
马来醛 malealdehyde; maleic dialdehyde
马来乳胶 guatta-percha; gummi pertscha; gummi plasticum
马来乳胶溶液 liquor gutta percha; traumaticin
马来式关节复涨机车 Mallet articulated compound locomotive
马来树胶 gutta percha
马来松香 maleated rosin; maleic rosin
马来酸 maleic acid; pyrisuccideanol maleate; stivane toxilicacid
马来酸二苄酯 dibenzyl maleate
马来酸二丙酯 dipropyl maleate
马来酸二丁锡 dibutyltin maleate
马来酸二丁酯 dibutyl maleate
马来酸二环己酯 dicyclohexyl maleate
马来酸二巯基乙酯 dimercaptoethyl maleate
马来酸二烯丙酯 diallyl maleate
马来酸二乙酯 diethyl maleate
马来酸钙 calcium maleate
马来酸酐值 maleic anhydride value
马来酸季戊四醇树脂 pentaerythritol maleic resin
马来酸钠 sodium maleate
马来酸氢盐 bimaleate
马来酸氢酯 bimaleate
马来酸树脂 maleic(acid) resin
马来酸树脂清漆 maleic resin varnish
马来酸酰胺 maleamic acid; maleic acid monoamide
马来酸亚锡 stannous maleate
马来酸盐 maleate
马来酸酯 maleate
马来酸酯树脂 maleic acid ester resin
马来西亚和新加坡元 straits dollar
马来西亚橡木 Mempening
马来酰胺 maleamide; maleic amide
马来酰胺酸 maleamic acid; maleic acid monoamide
马来酰肼 maleic hydrazide
马来酰亚胺 maleimide
马来亚胶 gutta percha
马来亚石 malayaite
马来樟脑 Malay Camphor
马来植物亚区 Malay floristic subregion
马来酯化树脂 maleic ester resin
马来制品 <多种混凝土制品的统称> Marley
马兰变性弹力丝 Marlene
马兰戈尼效应 <因界面张力梯度引起的液流动> Marangoni effect
马兰黄土 <中国> Malan loess
马兰矿 malanite
马兰纳达风 <西班牙卡塔尼亚一种海风> marinada
马兰期 Malan stage
马兰期锡釉陶器 Manans faience
马朗科风 <意大利马乔列一带东南风> marenco
马勒-布雷斯劳原理 Muller-Breslau's principle
马勒混合粉碎机 Muller mixer
马勒量热器 Mahler calorimeter
马勒铝硅(活塞)合金 Mahler alloy
马勒模型 Muller model
马雷特合金 Mallet alloy
马雷特黄铜 Mallet alloy
马累 <马尔代夫首都> Male
马里埃塔法 Marietta process
马里埃塔连续采煤机 Marietta miner
马里板 <一种刨花板> Marlith
马里波萨组 Mariposa formation
马里抗氧化合金 Marlie's alloy
马里兰道路试验 <美国在1950年进行的刚性路面重复荷载试验> Maryland road test
马里兰古绿石 Maryland verde antique
马里纳克盐 <硫酸锡钾> Marignac salt
马里耐氧化合金 Marlie's alloy
马里亚纳俯冲带 Mariana subduction zone
马里亚纳海沟 Mariana trench
马里亚纳海盆 Mariana basin
马里亚纳海盆巨地块 Mariana ocean basin block
马里亚纳型俯冲带 Mariana-type subduction zone
马里中央银行 Banque Centrale du Mali
马力标定 horsepower rating
马力吨位 power tonnage
马力负荷 horsepower loading
马力荷载 horsepower loading
马力米制单位 pferidekraft
马力谱 horsepower spectrum
马力汽油机 horsepower petrol engine
马力牵引 horse traction
马力曲线 horsepower curve
马力输出 horsepower output
马力输入 horsepower input
马力损失 horsepower loss
马力系数 power factor
马力小时 <功率的一种旧计量单位，相当于1马力连续供给1小时单位> horsepower hour
马力与重量之比 horsepower to weight ratio
马力重量比 horsepower to weight ratio
马丽兰得栎 blackjack oak; scrub oak
马利布冲浪板 Malibu board
马利亚水泥 Malia cement
马栗树 Aesculus hippocastanum
马栗油 horse chestnut oil
马莲 iris ensata
马莲垛 <纵横叠置的木支座> cribwork
马铃薯 potato
马铃薯淀粉 farina; potato starch
马铃薯淀粉加工废水 wastewater from potato starch processing
马铃薯废水 potato waste(water)
马铃薯加工废水 wastewater from potato processing
马铃薯加工废物 potato processing waste
马铃薯渣 potato pulp
马硫铜银矿 mckinstryite
马六甲海峡 Strait of Malacca
马六甲藤 Malacca cane
马六甲锡 Malacca tin
马笼头 bridle; halter; headstall
马笼头电极 bridle electrode
马炉 muffle
马鲁古海 Maluku Sea
马路 avenue; causeway; causey; road; street
马路边沟 gutter offtake
马路边排水口 gutter offtake
马路标线漆 road marking paint; traffic paint
马路冲洗 street flushing
马路划线漆 exterior traffic paint; traffic paint
马路划线涂料 road line; road line coatings; traffic paint
马路明沟 open road ditch
马路人行横道 zebra crossing
马路弯灯 swan-neck street lamp
马路中线下排水管 centre drain
马路中心园地 terrace
马吕斯定理 Malus theorem
马吕斯定律 Malus law
马吕斯光线定律 Malus law of rays
马吕斯余弦平方定律 Malus cosine-squared law
马纶聚丙烯腈系纤维 Malon
马罗裱糊法 marouflage
马罗拉仿金属线 Malora
马罗特易熔合金 Malotth(metal)
马洛里合金 Mallory
马洛里-沙顿钛铝锆合金 Mallory-Sharton alloy
马洛里无锡高强度青铜 Mallory metal
马洛隧道窑 Marlowe kiln
马洛特合金 Malotte's metal
马洛伊迪昂铜镍锌耐蚀合金 Malloydium
马马虎虎的人 <工作等> sloven
马马特勒风 <西西里岛的西北风> mamatele; mamatili
马毛 cilice; horsehair
马毛筛绢 rapatelle
马毛绳 giller
马毛纤维 horsehair fibre
马毛织的家具罩 horsehair
马莫阶【地】Marmor
马默斯事件 Manmoth event
马姆奎斯特改正 Malmquist correction
马木斯反向极性超带 Mammoth reversed polarity superzone
马木斯反向极性亚带 Mammoth reversed polarity subzone
马木斯反向极性亚时间带 Mammoth reversed polarity subchronzone
马木斯反向事件 Mammoth event
马木斯棉 <美国佐治亚产晚熟陆地棉> Mammoth cotton
马木斯式起重机 Mammoth crane
马木斯天线 Mammoth antenna
马木斯旋曝气器 Mammoth rotor
马那瓜 <尼加拉瓜首都> Managua
马纳吉尔 al-manazil
马尼巴赫律 Manebach law
马尼克铜锰镍合金 Manic
马尼拉包装纸 Manil(l)a wrapping
马尼拉港 <菲律宾> Manil(l)a Port Manil(l)a
马尼拉海槽 Manil(l)a trough

M

马尼拉厚纸 Manil(l)a bristol
马尼拉麻 Manil(l)a hemp;quilot fibre[fiber];abaca;Cebu hemp
马尼拉麻浆砂纸 rope Manila sand paper
马尼拉麻缆索 Manil(l)a rope
马尼拉麻丝 Manil(l)a rope
马尼拉牛皮纸 Manil(l)a paper
马尼拉树脂 Manil(l)a copal;Manil(l)-a resin
马尼拉纤维 Manil(l)a fibre
马尼拉纸 <电缆绝缘纸> Manil(l)a paper
马尼拉纸板 Manil(l)a board
马尼拉纸料 Manil(l)a board
马尼拉纸绳纸 Manil(l)a rope paper
马尼拉制品 abaca
马尼拉(棕)绳 Manil(l)a rope
马涅尔布莱登千斤顶 <预应力钢筋混凝土用> Magnel-Blaton jack
马涅尔布莱登(预应力)计算图 Magnel-Blaton diagram
马涅尔布莱登(预应力)张拉系统 Magnel-Blaton system
马涅尔锚固体系 Magnel anchoring system
马涅尔预应力计算图 Magnel prestressing diagram
马涅尔预应力张拉体系 Magnel prestressing system
马涅尔预应力张拉系统 Magnel prestressing system
马涅尔张拉系统 Magnel system
马棚 stable
马皮手套 horsehide gloves
马普托 <莫桑比克首都> Maputo
马青烯 malthene
马球场 polo court;polo ground
马肉矿 horseflesh ore
马肉色木材 horseflesh
马萨诸塞 <美国州名> Massachusetts
马萨兹尔地槽 Mazatzal geosyncline
马塞德勒皮内半影仪 Macedelepinary half-shade
马塞港 <法国> Port Marseilles
马塞卢 <莱索托首都> Maseru
马塞卢斯页岩 Marcellus shale
马塞式瓦 Marseilles pattern tile
马塞瓦 Marseilles tile
马赛厄斯饱和蒸汽液体密度关系式 Mathias formula
马赛厄斯定则 Matthias rule
马赛克 ceramic mosaic;ceramic mosaic tile;earthenware mosaic;mosaic;majolica mosaic <涂有不透明釉的>
马赛克壁画 mosaic mural;mural in mosaic
马赛克玻璃生产线 production line for mosaic glass
马赛克部件 mosaic piece
马赛克瓷砖 mosaic clay tile;mosaic tile
马赛克的 inlaid
马赛克地板 mosaic fingers
马赛克地面 tessera flooring
马赛克覆盖 mosaic cover(ing)
马赛克覆盖层 mosaic clad
马赛克功能 mosaic function
马赛克教堂 mosaic church
马赛克剧院乐池 mosaic parquetry
马赛克楼板 mosaic floor(ing)
马赛克楼面 tessera floor(ing);mosaic floor(ing)
马赛克模拟地图 mosaic mimic map
马赛克磨光石 mosaic terrazzo
马赛克能手 mosaic artist
马赛克黏[粘]土地砖 mosaic clay tile
马赛克铺地 mosaic floor(ing)
马赛克铺地面层 mosaic flooring
马赛克铺地小方石 mosaic paving sett
马赛克铺面 mosaic pavement
马赛克墙砖 mosaic wall tile
马赛克穹隆 mosaic dome
马赛克饰面 mosaic facing;mosaic surface
马赛克水磨石 mosaic terrazzo
马赛克贴面 mosaic surface
马赛克外观装饰 mosaic exterior finish;mosaic external finish;mosaic outdoor finish
马赛克镶面 mosaic tiling
马赛克艺术家 mosaic artist
马赛克圆顶 mosaic cupola
马赛克砖 opus sectile;ceramic mosaic tile
马赛克装潢 mosaic decoration
马赛克装饰 mosaic decoration;mosaic enrichment
马赛克装饰性整修 mosaic decorative finish
马赛瓦 French tile
马桑尼特爆破法 <纤维分离> Masonite process
马桑尼特纤维板 masonite
马桑属 coriaria
马森图 Mason graph
马山斑岩 masanophyre
马山港 <韩国> Masan Port
马山岩 masanite
马石 horse stone
马士登方 Marsden square
马士登图 Marsden chart
马氏扁式松胀仪 Marchetti's flat dilatometer
马氏常数 Mallar's constant
马氏淬炼 marquenching
马氏管 Malpighian tube
马氏回火 martemper(ing)
马氏机构 geneva mechanism;Maltese cross
马氏间歇机构 geneva gear
马氏距离检验 Ma-distance test
马氏快凝水泥 Martin's quick setting cement
马氏漏斗型黏[粘]度计 Marsh funnel viscometer
马氏铝镁合金 Mach's metal
马氏罗盘 Maas(borehole) compass
马氏铅锑合金 <轴承合金> Mathesius metal
马氏体 martensite
马氏体板条 martensite lath
马氏体不锈钢 martensitic stainless steel
马氏体常温加工 marstraining
马氏体成分 martensitic component
马氏体淬火 marquench;marquench flame hardening;martensite quenching
马氏体的 martensitic
马氏体的形成 martensite formation
马氏体等温淬火 marquenching;martemper(ing);time quenching
马氏体范围 martensite range
马氏体分级淬火 marquenching
马氏体钢 martensite steel;martensitic steel
马氏体回火 martenaging martempering;martensite tempering
马氏体基体 martensitic matrix
马氏体结构 martensitic structure
马氏体晶格 martensite lattice
马氏体开始形成点 martensite start(ing) point
马氏体可逆转度 reverse transformation of martensite
马氏体冷作处理 martensite cold working
马氏体片 martensite plate
马氏体区域 martensite range;martensite region
马氏体时效(处理) quench ag(e)ing;martenaging;martensite ag(e)ing;mar-ag(e)ing
马氏体时效钢 <一种高强度低碳铁镍合金> mar-ag(e)ing steel
马氏体式相变 martensitic-type phase transformation
马氏体温度 martensite temperature
马氏体温度范围 martensite range
马氏体相变 martensitic phase transformation
马氏体形变处理 marstraining;temforming
马氏体形变热处理 marstressing process
马氏体形变时效 marstraining
马氏体形态学 morphology of martensite
马氏体型转变 martensite transformation
马氏体应变 marstraining
马氏体应力 marstressing
马氏体应力淬火 marstressing
马氏体硬化 martensitic hardening
马氏体中的显微裂纹 microcrack(ing) in martensite
马氏体铸铁 martensitic cast-iron
马氏体转变 martensitic transformation
马氏体转变点 martensite point
马氏体转变开始温度 martensite start(ing) point
马氏体转变区 martensite range;martensitic range
马氏体转变终止点 martensite finish(ing) point
马氏体转变终止温度 martensite finish(ing) point
马氏体组织 martensitic structure
马氏颜料 Mars pigment
马氏硬度 Martens hardness
马氏永磁体 martensitic steel
马氏预应力张拉系统 Magnel anchoring system
马氏铸铁 martensitic cast-iron
马氏钻孔偏斜测量 Maas survey
马市 horse market
马术练习场 manege
马术学校 manege;riding academy
马刷(子) body brush;horse brush
马水硅钠石 makatite
马斯河谷烟雾事件 Meuse valley incident
马斯喀特 <阿曼首都> Muscat
马斯凯特近似公式 Marschet approximation formula
马斯柯利特绝缘材料 <一种消振绝缘材料> Mascolite
马斯里奇特阶 <早白垩世>【地】 Maestrichtian
马斯塔巴皇陵 royal Mastabah tomb
马苏达-科勒尔图 Masuda-Coryell diagram
马塔角 Matterhorn
马塔拉佐财团 <巴西> Grupo Matarazzo
马塔牛斯加风 Matanuska wind
马坦普整磁用铁镍合金 Mutemp
马唐【植】crab grass
马陶-赫绍格质量分析器 Mattauch-Herzog mass analyser[analyzer]
马特基反向极性带 Mateke reversed polarity zone
马特基反向极性时 Mateke reversed polarity chron
马特基反向极性时间带 Mateke reversed polarity chronzone
马特拉赛 matelasse
马特劳摆式叉车装卸车 Matbro swing forklift
马腾斯表面划痕试验 Martens surface scratching test
马腾斯测淀积银密度计 Martens densitometer
马腾斯分光镜 Martens spectroscope
马腾斯刮痕硬度计 Martens sclerometer
马腾斯光楔 Martens wedge
马腾斯-海因硬度 Martens-Heyn hardness
马腾斯划痕硬度 Martens scratch hardness
马腾斯(耐热)试验 <塑料热弯变形> Martens test
马腾斯偏光光度计 Martens polarization photometer
马腾斯硬度试验 Martens' hardness test
马腾斯照明器 Martens illuminator
马提厄方程 Mathieu equation
马提厄函数 Mathieu function
马提厄微分方程 Mathieu differential equation
马蹄钉 wire staple
马蹄拱 horseshoe arch;mosque arch
马蹄钩 clevice;clevis
马蹄规 horseshoe ga(u)ge
马蹄环 horseshoe collar
马蹄夹 clevis;mo(u)ld clamps
马蹄礁 horse shore reef
马蹄礁油气藏趋向带 horseshoe reef pool trend
马蹄钳栓 clevis pin
马蹄式 horseshoe type
马蹄式混合机 horseshoe type mixer
马蹄式混合器 horseshoe type mixer
马蹄式搅拌机 horseshoe mixer;horseshoe stirrer
马蹄式楼梯 horseshoe stair(case)
马蹄铁 horseshoe iron;horseshoe shape magnet(iron);U-shaped magnet;cutting shoe <其中的一种>
马蹄铁管 horseshoe main
马蹄铁形磨盘 shoe nog plate
马蹄铁质量 horseshoe quality
马蹄形 U-shape
马蹄形冰川 horseshoe-shaped glacier
马蹄形衬砌 horseshoe-shaped lining
马蹄形磁铁 horseshoe shape magnet(iron);U-shaped magnet
马蹄形的 horseshoe-shaped;U-shaped
马蹄形灯丝 horseshoe filament
马蹄形镫铁 U-shaped stirrup
马蹄形底座 horseshoe base
马蹄形电磁铁 horseshoe electromagnet
马蹄形断面 horseshaped section;horseshoe(-shaped) section
马蹄形断面隧道 horseshoe tunnel
马蹄形断面隧洞 horseshoe sectional tunnel
马蹄形断面阴沟 horseshoe section sewer
马蹄形盾地 horseshoe-shaped betwixtoland
马蹄形盾构 horseshoe-shaped shield
马蹄形拱(圈) Arabic arch;horseshoe arch;Moorish arch;Moresque;Saracenic arch
马蹄形钩 clevis
马蹄形钩孔 clevis eye
马蹄形管道 horseshoe conduit;horseshoe manifold
马蹄形管渠 horseshoe conduit
马蹄形涵洞 horseshoe culvert
马蹄形河曲 horseshoe curve;horseshoe meander;mule-shoe curve

马蹄形湖 horseshoe lake
马蹄形火山口 breached crater
马蹄形火焰炉 U-flame furnace
马蹄形夹具 horseshoe holder
马蹄形礁 horseshoe reef
马蹄形截面 horseshoe section;horseshoe-shaped section
马蹄形开采法 horseshoe mining method
马蹄形块 U-shaped block
马蹄形零速度线 horseshoe-shaped curves of zero velocity
马蹄形炉膛 horseshoe type furnace
马蹄形螺栓 anchor loop;U-bolt
马蹄形铆钉机 horseshoe riveter
马蹄形排水管 horseshoe conduit;horseshoe(-shaped)sewer
马蹄形喷嘴 trailer nozzle
马蹄形切刀 horseshoe trip
马蹄形切刀的棕绳切割器 horseshoe trip knife
马蹄形区段 horseshoe-shaped section
马蹄形曲线 horseshoe curve;muleshoe curve
马蹄形沙丘 hairpin bend dune
马蹄形山谷 canoe-shaped valley;canoe valley
马蹄形山系 orocline
马蹄形输水道 horseshoe conduit
马蹄形水道 horseshoe conduit
马蹄形水管 horseshoe conduit
马蹄形隧道 horseshoe sectional tunnel;horseshoe(-shaped)tunnel
马蹄形隧洞 horseshoe(-shaped)tunnel
马蹄形搪烧窑 U-shaped enamelling furnace
马蹄形推力轴承 horseshoe type thrust bearing
马蹄形洼地 oxbowshoed depression
马蹄形弯道 hairpin bend;horseshoe bend;oxbow
马蹄形弯头 U-bend
马蹄形涡流 horseshoe vortex;U-vortex
马蹄形涡流系 horseshoe vortex system
马蹄形污水管 horseshoe-shaped sewer
马蹄形物 horseshoe
马蹄形下水道 horseshoe sewer
马蹄形线路 <在高坡地段的一种设计> horseshoe line
马蹄形旋礁 horseshoe reef
马蹄形旋涡 horseshoe vortex
马蹄形焰 horseshoe flame
马蹄形窑炉 horseshoe furnace
马蹄形阴沟 horseshoe-shaped sewer
马蹄形圆拱 mosque arch
马蹄形圆室 horseshoe apsis
马蹄形展线 horseshoe(curve)development
马蹄形支架 horseshoe set
马蹄形总管 horseshoe main
马蹄焰炉 end-fired furnace
马蹄窑 hairpin furnace
马铁 cramp
马厅 horse parlor
马桶 close-stool;closet bowl;commode;lavatory bowl;nightstool
马桶盖 closet seat lid;lavatory seat lid;toilet seat lid
马桶排出口 closet horn
马桶水箱 closet tank
马桶座 lavatory seat
马桶座圈 toilet seat;water closet seat
马头梁 false set;horsehead beam
马头门 horsehead;landing bottom
马头门横梁 horsehead girder
马头墙 corbel steps
马头丘 escarpment;scarp
马头山冰期 Matoushan(glacial)stage
马头山墙 corbie gable;crow gable;step gable
马托巴 <一种混凝土用焊接编织钢筋> Matobar
马拖车 horsedrawn cart
马拖的 horsedrawn
马尾衬 haitcloth
马尾港 Port Mawei
马尾蛤灰岩 Hippurite limestone
马尾构造 horsetail structure
马尾矿 horsetail ore
马尾毛 horse tail hair
马尾筛 hair sieve
马尾式 horsetail;pony-tail hair style
马尾树 Rhoiptelea chiliantha
马尾丝线 horsetail
马尾丝状构造 horsetail structure
马尾丝状矿脉 horsetail vein
马尾松 Chinese red pine;masson pine
马尾云 mare's tails
马尾藻海 Sargasso Sea
马尾状 horsetail shape
马尾状断层 horsetail fault
马尾状分岔 horsetail splitting
马尾状矿脉 horsetail ore
马尾状裂隙 horsetail fissure
马纬度 horse latitude
马文日照计 Marvin sunshine recorder
马西森接合 Matheson joint
马西型(球)磨机 Marcy mill
马西修斯铅-碱金属合金 Mathesius metal
马希森铜丝电阻标准 Matthiessen's standard
马希氏试验 <检砷或锑> Marsh's test
马戏场 circle;circus;hippodrome;ring
马戏场主帐篷 circus "big top"
马献德管 <一种U形氯化钙管> Marchand tube
马歇尔K指标 Marshall K
马歇尔-埃奇沃思-鲍利指数 Marshall-Edgeworth-Bowley index
马歇尔-奥尔金分布 Marshall-Olkin distribution
马歇尔阀装置 Marshall valve gear
马歇尔钢瓦挂铅青铜法 Masher process
马歇尔混合料设计法 Marshall mix design method
马歇尔混合料设计诊断法 diagnostics of Marshall mix design
马歇尔击实锤 Marshall compaction hammer
马歇尔计划 Marshall plan
马歇尔劲度 <马氏稳定度与流值之比> Marshall stiffness
马歇尔精制机 Marshall refiner
马歇尔勒纳条件 Marshall-lerner condition
马歇尔(沥青路面稳定度)试验与设计法 Marshall test and design procedure
马歇尔流动值 <沥青混凝土的> Marshall flow;Marshall flow value
马歇尔-帕尔默分布 Marshall-Palmer distribution
马歇尔群岛 Marshall Islands
马歇尔设计法 <沥青混合料按填隙密实概念取马歇尔试验数据求最佳沥青用量的设计法> Marshall-type design
马歇尔氏(沥青)混合料配合比设计法 corps of Engineers mix design method
马歇尔式设计法 Marshall-type design method
马歇尔试件 Marshall briquette
马歇尔试块 Marshall briquette
马歇尔试验 Marshall test
马歇尔试验标本 Marshall test specimen
马歇尔试验用击压锤 Marshall compaction hammer
马歇尔试验用试件 Marshall test specimen
马歇尔试验中的击压锤 Marshall compaction hammer
马歇尔稳定度 <沥青混凝土强度的一种指标> Marshall stability
马歇尔稳定度试验 Marshall stability test
马歇尔稳定度仪 Marshall stability apparatus;Marshall stabilometer
马歇尔稳定度值 Marshall stability value
马歇尔线 Marshall line
马歇尔需求函数 Marshallian demand function
马歇尔需求曲线 Marshallian demand curve
马歇尔圆柱体试验头 <沥青混凝土的> Marshall cylindric(al)testhead
马歇尔圆柱形试验 <沥青混凝土的> Marshall cylindric(al)test
马歇尔援助 Marshall aid
马歇尔指示器 Marshall's indicator
马歇特钨钢 Mushet('s)steel
马谢氏单位 <镭射气浓度单位> Mache unit
马辛寇维茨定理 Marcinkiewicz's theorem
马行道 horse road;horse track;horse way
马休函数 Massieu function
马休黄 Martius yellow
马靴 riding boots
马牙茬【建】 indenting;toothing
马牙槎 <砌体的> tusk;stubble
马牙接搓 tusking
马牙榫接【建】 combed joint;laminated joint
马雅尼制管机 Maynani pipe machine
马雅尼制瓦板机 Maynani board machine
马雅式建筑 Maya architecture
马亚克对喷式气流粉碎机 Majac jet pulverizer
马亚克磨 Majac mill
马耶斯提克轻载巴比合金 Majestic Babbitt
马依哈克声应变计 Maihak strain ga(u)ge
马扎抄取制管法 Mazza process
马扎克锌基(压铸)合金 Mazak alloy
马掌 horseshoe
马掌钉 frost nail;horseshoe nail
马掌型 style of the shoe
马朱罗环礁 Majuro atoll
马鬃 horsehair;horse mane
马鬃灯刷 horsehair lamp brush
马鬃扫帚 horsehair broom
马鬃手刷 horsehair hand brush
马鬃刷 horsehair brush

吗

吗啡(碱)morphine

玛

玛瑙贝壳 cowrie shells

玛多地块【地】Madoi block
玛尔蒂纶聚烯烃纤维 Multilene
玛钢存水弯 malleable iron trap
玛钢管 malleable iron pipe
玛钢管件 malleable cast-iron pipe fitting
玛钢卡头 galvanized malleable clip
玛格尔式后张法 Maguel post-tensioning
玛拉巴吉纳树胶 Malabar kino
玛里莫缝编机 Malimo knitting machine
玛鲁克丝管 electromalux
玛姆砂岩 malmstone
玛瑙 agate
玛瑙杯 agate cup
玛瑙碧玉 agate jasper;jaspagate
玛瑙玻璃 agate glass
玛瑙钵杵 agate pestle
玛瑙蛋白石 agate opal
玛瑙刀口 agate knife-edge
玛瑙导丝器 agate guide
玛瑙雕刻品 agate carving
玛瑙端面轴承 agate end stone jewel
玛瑙工艺品 agate artware
玛瑙尖顶 agate cap
玛瑙臼和杵 agate mortar and pestle
玛瑙面 agate plane
玛瑙抛光机 agate burnisher
玛瑙皮 manual skin
玛瑙片 agate plate
玛瑙器皿 agateware
玛瑙球 agate ball
玛瑙球形轴承 agate spheric(al)jewel
玛瑙乳钵 agate mortar
玛瑙体活字 <美> agate
玛瑙天平刀子 agate balance edge
玛瑙调刀 agate spatula
玛瑙通孔轴承 agate hole jewel
玛瑙纹饰陶器 agateware
玛瑙研钵 agate mortar
玛瑙玉雕 agate carving
玛瑙珍珠 agate pearl
玛瑙轴承 agate bearing
玛瑙珠 agate bead
玛西拉黑页岩 <中泥盆纪>【地】 Marcellus black shale
玛雅建筑 <中美洲印第安人的建筑> Maya architecture
玛雅式拱(门)Maya arch
玛雅天文学 Mayan astronomy
玛琋砂胶脂橡皮混合料 mastic-bitumen rubber
玛琋树胶 gum mastic
玛琋脂 mastic;mastic gum
玛琋脂保温 mastic insulation
玛琋脂隔离层 mastic insulation
玛琋脂隔热层 mastic insulation
玛琋脂贯入式路面 mastic penetration pavement
玛琋脂合成物 mastic compound
玛琋脂胶粘剂 mastic adhesive
玛琋脂接缝 mastic joint(ing)
玛琋脂绝热层 mastic insulation
玛琋脂康 <一种装配钢窗玻璃的玛琋脂> Masticon
玛琋脂路面 mastic pavement
玛琋脂玛蹄填料 mastic filler
玛琋脂密封 mastic seal(ing)
玛琋脂密封剂 mastic sealer
玛琋脂熔锅 mastic melting kettle
玛琋脂绳 mastic cord
玛琋脂水泥 mastic cement
玛琋脂填(塞)料 mastic filler;mastic filling
玛琋脂涂布器 mastic spreader
玛琋脂线 mastic cord

码

码 code;yard measure;yard <英美制长度单位,1码=0.9144米>

码磅度量衡法 yard pound method
码包机 bag loader;stacker
码比率 code rate
码变换 code conversion
码变换器 code converter

M

码表测时 stopwatch timing
码表测时法 stopwatch time study
码表计时法 stopwatch time method
码长度 code length;running yard
码尺<一码长的尺杆,杆上常有细分的度数> yardstick;yard measure;yard wand
码尺圆规 yardstick compass
码袋机 bag stacker
码点【计】code-point
码堆机 palletizer
码对称 code symmetry
码多项式 code polynomial
码垛 pile tally;stacking;stow
码垛不良 improper stowage
码垛车 pallet truck
码垛堆放 palletization[palletisation]
码垛堆积 palletize;palletizing
码垛高度 height of stacking
码垛机 hacking machine; stacker (crane); stacker-loader; stacking machine;yarder
码垛架 stacking rack
码垛托盘 stacking pallet
码垛系统 cubing system
码垛装置 cubing plant
码分 code division;yards per minute
码分辨率 code distinguishability
码分表 yardage clock
码分多路复用 code division multiplex
码分多址(方式)code division multiple access
码分多址蜂窝系统 code division multiple access cellular system
码分多址联结 code division multiple access
码符号 code sign;code symbol
码符号集 code alphabet
码号目录 numeric(al)catalog
码恒定性 code invariance
码基数 code base
码间干扰 intersymbol interference
码检验 code check
码距 code distance
码开关 code switch
码控无源反射器天线 coded passive reflector antenna
码块 block
码率 code check;code rate
码(耐火)砖机 setting machine
码盘【计】code(d)disc[disk]
码盘数字电子秤 encoder digital electronic balance
码坯机 setting machine
码坯系统 setting system
码群 code group
码冗余度 code redundance[redundancy]
码失真 code distortion
码矢 code vector
码树 code tree
码数 code number;fiber count;yardage
码数表 yardage recorder
码速率 bit rate
码速调整 justification
码速调整服务数字 justification service digit
码速调整率 justification rate
码速调整数字 justification digit
码条 code bar
码条代码 Codabar code
码条驱动磁铁 code bar drive magnet
码条式 code type
码条式接线器 code switch
码条式自动电话交换机 code switch automatic telephone system
码头 wharf; quay; jetty; dock quay; landing(stage); water pier; terminal;pier;staith(e)<水边有高架结构供装船的,特别是煤码头>
码头安全 wharf safety; quay safety; dock safety
码头岸壁 quay wall; wharf quay; wharf wall;dock wall
码头岸边 quayside;wharf side;dockside
码头岸肩 berth apron;apron
码头岸面构造 deck construction;wharf face structure
码头岸墙 quay wall;wharf wall
码头岸线 face-line of wharf; waterfront of wharf
码头搬运长 wharf boss;dock boss
码头搬运费 wharf dues;wharfage;stevedorage;stevedoring charges
码头搬运工 quayman;dockman;stevedore
码头搬运工头 foreman stevedore
码头搬运工资 stevedore wage
码头搬运公司 stevedoring company
码头搬运领班 foreman stevedore
码头搬运装卸费 shifting charges
码头本体设施 wharf facility
码头办公室 wharf office;dock house;dock office;head house
码头包角型护舷 rubber fender for wharf corner;rubber fender for dock corner
码头边 quayside;wharfside
码头边板 string piece
码头边的活动联系搭板 quayside movable link span;dockside movable link span
码头边的木靠把 camel
码头边的设备 dockside equipment
码头边公用设施 dockside utilities
码头边集装箱起重机 dockside container crane
码头边锯木厂 cargo mill
码头边棱 arris of quay
码头边起重机 wharfside crane;quayside crane;dockside crane
码头边水深 depth alongside
码头边线 pierhead line; quay line; wharf line
码头边缘 quay edge
码头边装卸工(人)stringpiece man
码头驳岸 bulkhead; bulkhead wall; quay(wall)
码头驳船 harbo(u)r barge;pierhead pontoon;quay pontoon
码头泊位数 number of berthes;quayage
码头仓单 dock warrant;wharfinger's warrant
码头仓库 quay shed; shipping point; terminal depot; transit godown; transit shed;wharfinger's warrant; dock warehouse
码头仓库后面的 rear-of-shed
码头仓库外墙上的和船上的联合作业起吊系统 married gear;union purchase
码头操作费 wharf handling charge
码头测隙规 wharf feeler;wharf sensor
码头长 berth master;pier master
码头长臂起重机 long boom wharf crane
码头长度 length of quay;length of wharf;quayage;quay length
码头场地 dock yard
码头车场 dock yard
码头车站<办理货物承运和交付,并无轨道> station pier
码头沉降量 settlement of wharf
码头撑柱 spur shore
码头出入证 dock pass
码头传感器 wharf feeler;wharf sensor
码头带缆工人 dockman; pier crew; pierman;quayman;wharfman
码头带缆柱 bollard;wharf post
码头带缆桩 checking bollard; combined mooring and warping bollard
码头当局 dock authority
码头挡土墙块 block quay wall
码头到船舶的装卸搬运 quay-to-ship stevedoring
码头到户 pier-to-door;pier-to-house
码头到码头 pier to pier;port-to-port; quay-to-quay
码头到码头的运输业务 pier to pier service;port-to-port service;dock-to-dock service
码头到码头运输 quay-to-quay traffic;quay-to-quay transportation
码头到门 pier-to-house
码头到内地仓库 pier-to-inland depot
码头道路 dockside road
码头堤岸防护桩 fender post
码头堤岸护舷桩 fender pile; fender post
码头底板 pier base plate
码头地板 quay floor
码头地面 quay floor; quay surface; wharf surface
码头地域 quay space
码头端部翼墙 wingwall at quay end
码头短工 dock walloper
码头堆货高度 tiering height
码头吨税 berthage;dock tonnage dues
码头吨位费 dock tonnage dues
码头墩柱 quay pier
码头趸船 captive barge;captive breasting barge; landing pontoon; pierhead pontoon;wharf boat
码头法线 base line of wharf
码头防冲设备 dock fender
码头防冲桩 fender pier
码头防冲装置 dockside fender
码头防护(桩)pier guard
码头防舷材 wharf fender
码头防撞设备 berthing fender;dock fender;fendering system;pier fender
码头防撞设施 berthing fender;dock fender;fendering system;pier fender
码头房屋 terminal budding;terminal building
码头费 quayage;wharfage;wharf dues; quay dues; keelage; keyage; pierage; pier dues;dock dues;dock dues and charges;dock tonnage dues;berthing dues
码头费和装货起运 dock dues and shipping
码头浮桥 ro/ro ramp[roll-on/roll-off ramp];pontoon
码头负责人 dockmaster
码头附属区 quay space
码头港务助理 wharf traffic assistant
码头格体 dock cell
码头工程地质勘察 engineering geologic(al)investigation of dock
码头工会 longshore union
码头工具管理员 gearman;tool house foreman
码头工人 harbo(u)r worker; longshoreman; lumper; pierman; railman; roustabout; rouster; stevedore; water sider; wharfhand; wharfier; dock labo(u)rer; dock worker;dock labo(u)r;dockman; docker;dock hand;dock walloper
码头工人罢工 docker's strike;harbo(u)r strike
码头工人手钩 docker's hook
码头工人装卸污臭货物的额外工资 dirty money
码头工作 dock work
码头公司 terminal company; marine terminal operator; maritime terminal operator; terminal operator; dock undertaking
码头公用设施 ship-terminal utilities
码头拱墩 pier impost
码头拱廊 pier arcade
码头固定系缆 fixed mooring
码头管沟 quay conduit; wharf conduit
码头管理(人)员 dockmaster; terminal master;wharfinger;wharf master;dock superintendent
码头管理办公楼 dockmaster building
码头管理机构 dock authority
码头管理人 wharfinger;wharf master
码头管理人收据 wharfinger's receipt
码头管理室 dock control house
码头管理所 administration office
码头过挡船 wharf boat
码头和城市税 dock and town dues
码头后方铁路线 backland trackage of dock
码头护栏 pier fender
码头护木 fender pile; pier fender; dock bumper; dock fender; pile boom;timber chock
码头护坡 wharf slope protection;dock slope protection
码头护舷 wharf fender;dock fender
码头护舷木 fender
码头护缘 bull rail;edge protection
码头活动 port activity
码头活动问题 port activity problem
码头或堤上斜滑道 slip
码头货棚 quay shed;transit shed
码头货运量 terminal volume
码头货运站 terminal depot
码头基点 dock base station
码头及船坞费用条款 port and dock charges(condition)clause
码头及道路冲洗喷头 flushing sprinkler for wharf and roadway
码头集装箱起重机 portainer
码头加油站 wharf fuel terminal;fuel station of wharf
码头监督员 wharf superintendent
码头检验条款 jetty clause
码头检验员 landing surveyor
码头减压平台 relieving platform
码头建设 dock construction
码头建设界线 dock line
码头建筑(物)quayside building; terminal building;waterfront structure; wharf structure;docking structure
码头建筑线 pierhead line
码头交换价(格)ex dock;ex quay;ex-wharf;free at quay;free on quay
码头交换价格条件 free at wharf
码头交换条件 ex quay term
码头交货 delivered from quay; ex dock; ex pier; export; ex quay; ex-wharf;faq;free alongside(at)quay; free dock;terminal delivery
码头交货价(格)ex dock;ex quay;ex quay duty paid;ex-wharf;ex pier; free at quay;free on quay
码头交货价格条件 free at wharf
码头绞车 dock winch
码头阶梯 quay staircase
码头结构 quayside structure;quay and pier construction;dock structure
码头进口处卡车排队车道 queuing lane for trucks at terminal entrance
码头经理 wharfinger; terminal manager
码头经营人 terminal operator
码头静电接触装置 earthing device; static grounding device
码头捐 dockage;dock due

码头靠泊费 berthage;berth charges
码头靠船构件 wharf breasting structural member
码头可调节跳板 dock level(l)er
码头空货位 quayage
码头库区 dock and depot district
码头扩充 dock expansion
码头扩建 berth extension;quay extension;wharf extension
码头廊道 quay-gallery
码头老板 wharf master
码头雷达 jetty mounted radar
码头立面 berthing face
码头利用率 wharf occupancy
码头临时工的挑选 shape up
码头路面 quay area pavement
码头路面排水坡度 drainage slope of quay area pavement
码头旅客接待员 landing clerk
码头门 ingate
码头面 quay face;berthing face;esplanade;quay surface;surface of esplanade
码头面板 wharf deck;deck construction;decking;dock slab;quay floor
码头面板下棱体稳定性 underdeck slope stability
码头面板下面净空 underdeck clearance
码头面层<现场加铺混凝土的> topping of wharf;topping of quay
码头面大梁 deck girder
码头面高程 berth level;cope level;quay level;quay surface level;wharf surface level
码头面积 area of quay;area of wharf;quayage;quay surface
码头面前沿 cope
码头面纵梁 deck stringer
码头内拆箱<集装箱> terminal devanning
码头内调运 terminal transit
码头内装箱 terminal vanning
码头爬梯 dock ladder;ladder;step ladder
码头排架桩 jetty bent pile
码头旁建筑(物) quayside building
码头棚 dock shelter
码头碰垫 wharf fender;dock fender;marine dock fender
码头碰垫桩 fender pile;pile fender
码头起重机 berth crane;building slip(way) crane;cargo crane;dock crane;dockside crane;harbo(u)r crane;pier crane;wharf crane
码头起重机轨道 crane track;crane way;dock crane track
码头起重机前轨 waterside rail;quayside rail
码头砌筑 pier bond
码头前方 transit shed
码头前方仓库 transit shed;dock shed;quay shed;wharf shed
码头前方货栈 dock shed;quay shed;wharf shed
码头前方作业地带 transit area
码头前水深 water depth in front of wharf
码头前沿 apron wharf surface;dock face;dockfront;dockside;front apron;pier apron;pier side;wharf apron;wharf-front;wharf frontage;apron<码头边至仓库墙壁之间地段>;quayside
码头前沿岸墙 quay breastwork;quay wall
码头前沿长度 frontage;quay frontage;dock frontage;length of waterfront
码头前沿地带 quay apron
码头前沿地带宽度 width of apron
码头前沿地面 quay space;surface of esplanade
码头前沿高程 wharf frontal elevation;cope level along the berth;cope elevation along wharf;elevation of wharf apron;apron elevation
码头前沿轨道 quay track;apron track
码头前沿进深 depth of approach
码头前沿纵深 depth of approach
码头前沿宽度 apron width
码头前沿廊道 apron gallery;seaside gallery
码头前沿起重机 quay edge crane;quayside crane;wharfside crane
码头前沿起重机轨道 apron crane rails
码头前沿区 apron space;quay apron space;wharf apron space
码头前沿区轨道 apron track;wharf apron track
码头前沿区铁路 apron railway
码头前沿设备 quayside appliance
码头前沿设计水深 design depth alongside wharf
码头前沿水深 depth alongside berth;depth alongside wharf;depth at quay;water depth on front of wharf;wharf-frontage depth
码头前沿水域 docking area
码头前沿铁路 marginal track;quayside railway;quayside track;apron track
码头前沿铁路轨道 quayside railway track
码头前沿铁路线 shipside and wharf apron track
码头前沿线 quay edge;wharf line;bulkhead line;dock face line;face-line of quay;face-line of wharf;quay line
码头前沿线路 apron track
码头前沿移动式起重机 dockside travel(l)ing crane;quayside travel(l)ing crane
码头前沿作业地带 apron space;quayside operation area
码头墙的锚碇 anchorage for quay wall
码头桥式起重机 quay bridge crane
码头区 port area;quayside;quay space;terminal complex;marine terminal;water terminal;dock land;quay system;docks;dockfront;dockside
码头区集装箱起重机 quayside container crane
码头区域 terminal area;wharf area;quay area;dock area
码头区照明 berth areas lighting
码头群 dock group;grouped docks
码头人员 wharfman;dockman
码头容量 terminal capacity
码头上长方形交易所 pier basilica
码头上的发货仓库 shipping room
码头上的连拱饰 pier arcading
码头上堆货处 pitch
码头上高架输送机 elevated dock conveyer
码头上交换价 free on board quay
码头上龙门起重机 pier-mounted gantry crane
码头上铁轨 on-dock rail
码头上挽缆桩 snubbing post
码头设备 port facility;quay appliance;terminal facility;wharfage;wharf equipment;wharf facility;dock(ing) accommodation;docking facility;dock installation
码头设施 quay appliance;terminal facility;wharf equipment;wharf facility;docking accommodation;docking facility
码头声呐 jetty mounted sonar
码头使用费 pier dues;quayage;wharfage;dockage;dock charge
码头使用税 quayage
码头市场 terminal market
码头试车 dock trial
码头收货单 warehouse receipt;wharf receipt;dock receipt
码头收货凭证 dock receipt
码头收据 dock receipt
码头收支单 dock warrant
码头枢纽化 terminalization
码头水手 dockman
码头水位 quay level
码头税 berthage;berth dues;dock due;jettage;pierage;pier dues;quayage;wharf dues
码头速度 approaching speed
码头所有人 wharfinger
码头坍塌 wharf collapse
码头梯 wharf ladder
码头提单 berth bill of lading
码头提货费率<不经仓库的卸货> quay rate
码头条件 berth term
码头跳板 gang plank
码头铁路(线) quay line;quayside railway track
码头停泊费 quayage
码头停泊位 quay berth
码头通道建筑物 approach structure of a wharf
码头通过能力 capacity of wharf;throughput capacity of wharf;wharf throughput
码头统计 port statistics
码头凸堤 quay apron
码头突堤 pier finger;jetty
码头拖轮 dock tug
码头外端 pier head
码头完税价 ex quay
码头舾装和安装 quay outfitting and installation
码头系船柱 bollard head;bollard
码头系缆钩 jetty hook
码头系缆桩 mooring pile;mooring post
码头系统 quay system;quayage
码头下的 under-pier
码头下游端 downstream extremity of the dock
码头线 berth(ing) line;dock line;harbo(u)r line;harbo(u)r track;quay siding;terminal line;wharf siding;pierhead line<连接码头前端的线>
码头卸货程序 terminal unloading procedure
码头卸货延滞费 wharf demurrage
码头卸货延误费 wharf demurrage
码头卸货账单 dock landing account;wharf landing account
码头信号 dock signal
码头信息控制系统 terminal information control system
码头型货运站 lighter freight station;terminal depot
码头型式 quay pattern
码头胸墙 breast wall;capping wall;coping wall;wave deflector on the quay;wharf breast wall;wharf shoulder
码头延伸 wharf extension
码头沿岸起重机 quayside crane;wharfside crane
码头沿边铁路线 apron track
码头业务 quay operation;terminal service;wharfage
码头液化天然气收集罐 jetty LNG drain drum
码头一侧 dockside
码头引道 approach to ferry
码头营运人 terminal operator
码头用长臂起重机 long boom wharf crane
码头用地 quayage;wharf surface
码头用起重机 wharf crane
码头用设备 quay furniture
码头用手动起重机 hand wharf crane
码头淤积 dock siltation
码头员工等候费 stevedore labo(u)r standby
码头栈单 dock warrant;wharfinger's warrant
码头栈桥 pier bridge;pierhead trestle;quay trestle;jetty trestle
码头栈桥上的道路 roadway on quay trestle
码头整体稳定性 global stability of the quay wall
码头正面<靠船的一面> face of wharf;berthing face
码头之间的运输 quay-to-quay transportation
码头至码头 wharf to wharf
码头滞留费 wharf demurrage
码头重量记录单 dock weight;dock weight note
码头主 wharfinger;wharf master
码头主的责任 wharf owner's liability
码头主任 wharfinger;wharf master
码头蛀虫 wharf borer
码头装卸登记表 dock sheet
码头装卸费 wharfage
码头装卸工(人) stevedore;wharfman;dockman;longshoreman;pierman;docker dockage
码头装卸工作 longshore work;longshoring work;stevedoring
码头装卸能力 stevedoring capacity;berth throughput
码头装卸桥 quay crane
码头装卸区 loading area
码头装卸设备 harbo(u)r equipment;harbo(u)r handling equipment
码头装载 block stowage;dock stevedorage;dock stowage
码头总长 quayage
码头作业 dockside operation;dockside service;terminal operation;terminal service;wharfage
码位 code bit
码效率 code efficiency
码信标 code beacon
码信号 code signal
码型变换器 code pattern converter
码型发生器 pattern generator
码序 code order
码序发生器 code sequence generator
码序列 code sequence
码延迟 code delay
码窑 setting-in a kiln
码元 code element
码元错误率 element error rate
码元流 stream of bits
码元同步 symbol synchronization
码元同步误差 symbol synchronization error
码元组合 grouping of bits
码再生 code regeneration;code rewriting
码纸机 laying machine;sheet handling machine
码重 code weight;Hamming weight
码砖机 setter
码转换 code conversion

M

码子 chip;counter;numeral;traditionally used by shopkeepers to mark prices
码字 code character;code letter;code word;codon
码字定位多项式 code word-locator polynomial
码字权重计数子 code word weight enumerator
码字首 prefix of code word
码组 code block;code character;encode block;signal code
码组差错率 block error rate
码组合 code combination;encode combination
码组校验过程 block check procedure

蚂蟥钉 steel dog;timber dog;clip;cramp iron;dog;timber dog 蚂蝗

蚂蟥螺钉 strap bolt
蚂蚁的 formic

杩槎 <一种临时性挡水建筑物> abat(t)is;weighted wooden tripod

杩槎坝 abat(t)is dike;wood tripod dam

埋安装 recessed fixture

埋版 caking
埋标石 monumentation
埋藏 bedding;burial;bury
埋藏爆破 buried blasting
埋藏变质作用 burial metamorphism
埋藏冰 buried ice
埋藏冰川 buried glacier
埋藏槽 burial tank
埋藏层 buried layer;buried stratum
埋藏场 burial ground;graveyard
埋藏冲刷面 buried erosion surface
埋藏储油罐 buried storage tank
埋藏处 disposal site
埋藏的 buried
埋藏地垒 causeway;causey
埋藏地形 buried landform
埋藏电极 implanted electrode
埋藏法 burying storage
埋藏缝合 buried suture
埋藏古夷平面 buried ancient planation surface
埋藏谷 buried valley;hidden valley
埋藏关系 buried relationship
埋藏管道 embedded pipe
埋藏海底泥土中 seabed insertion
埋藏河 buried river;buried stream
埋藏河槽 buried channel
埋藏河道 buried channel;pal(a)eochannel
埋藏河干河床 gutter
埋藏阶地 buried terrace;defended terrace
埋藏矿体 buried ore body
埋藏量 reserve
埋藏裂隙 buried fracture
埋藏露头 suboutcrop
埋藏锰结核 buried manganese nodule
埋藏侵蚀面 buried erosion surface
埋藏丘 buried hill
埋藏群落 taphocoenose
埋藏砂矿 buried placer
埋藏山 buried hill
埋藏山脊 buried ridge
埋藏深成岩体 buried pluton
埋藏深度 burial depth;buried depth;depth of burial;depth of burying;bury
埋藏式反应堆外壳 embedded reactor shell
埋藏式排水 buried drain
埋藏式砌石护坡 embedded stone pitching
埋藏式调压井 recessed surge tank
埋藏式压力管道 buried penstock
埋藏式压力水管 buried penstock;covered penstock;embedded penstock
埋藏水 concealed water;connate water;fossil water;trapped water
埋藏条件 mode of occurrence
埋藏土层 buried soil horizon
埋藏土(壤)buried soil
埋藏蜗壳 embedded spiral case
埋藏线 buried line
埋藏型喀斯特 buried karst
埋藏型岩溶 buried karst
埋藏学 para-ecology;taphonomy
埋藏盐土 concealed solonchak
埋藏夷平面 buried planation surface
埋藏异常 buried anomaly
埋藏晕 buried halo
埋层 burial layer;buried layer
埋层电缆 buried cable
埋层法 buried layer process
埋插筋 dowel(l)ing
埋沉基础 sunk foundation
埋厨房垃圾 burying garbage
埋地板 ground floor;ground plate
埋地处理 ground disposal
埋地电缆 buried cable
埋地敷设 buried underground;embedded in ground installment
埋地窿【救】embedment of holding foundation for ground tackle
埋地热油管道 buried-heated pipeline;underground heated pipeline
埋地深度 depth of burying
埋地天线 buried antenna;underground antenna
埋地线 underground line
埋钉 concealed nail(ing)
埋墩式桥台 buried abutment
埋放在混凝土中的毛石 displacer
埋封 embedding;embedment
埋封波导 imbedded(wave)guide
埋封胶 embedding compound
埋杆长度 embedded length of bar
埋沟电荷耦合器件 buried-channel charge coupled device
埋固式应力计 rigid inclusion ga(u)ge
埋刮板(链槽式)输送机 en-masse conveyer[conveyor];conveying en masse chain
埋刮链输送机 conveying en masse chain;en-masse conveyer[conveyor]
埋管 buried penstock;embedded conduit;embedded pipe;laying of pipe;penstock buried;pipe installation;tube placing
埋管驳船 burying barge
埋管程序 tubing program(me)
埋管刀板 deed blade
埋管的弯曲度 bend radius
埋管工人 pipe layer
埋管工作 pipe grout;pipe laying
埋管机 pipe layer;pipe-laying machine;pipeline burying machine
埋管接地 driven ground
埋管开挖法 open cut method
埋管扩孔块宽度 <挖掘机> bullet width
埋管冷却 embedded pipe cooling
埋管深度 cover depth
埋管时占地宽度 working width
埋管式冷却(法)embedded pipe cooling
埋管式喷灌系统 underground sprinkler system
埋管涂油工人 ragman
埋管位置探测器 pipe detector;pipe locator
埋焊 slugging
埋弧 buried arc;submerged arc
埋弧半自动焊 squirt welding
埋弧电炉 buried arc furnace;hidden arc furnace
埋弧电渣焊接 submerged slag pool welding
埋弧斗自动焊 squirt welding
埋弧焊 embedded arc welding;hidden arc welding;submerged arc welding;union-melt weld
埋弧焊机 submerged arc welding machine
埋弧焊接 submerged arc welding
埋弧焊漏斗 hopper for the union melt
埋弧炉 buried arc furnace;submerged arc furnace
埋弧自动焊 Lincoln weld;submerged arc welding;submerged automatic arc welding;union-melt welding
埋弧自动焊机 automatic submerged-arc welding machine
埋火消耗量 banking loss
埋积谷 waste-filled valley
埋浸式热电偶 immersion thermocouple
埋垃圾 burying garbage
埋缆 cover depth
埋缆刀板 deed blade
埋缆导管 <挖掘机> feed tube
埋缆的弯曲度 bend radius
埋缆机犁板转向角 blade steer angle
埋缆扩孔块宽度 <挖掘机> bullet width
埋铆钉头 sink in a rivet
埋没 entombment
埋没河道 buried channel
埋没河流 buried stream
埋没露头 buried outcrop
埋没式出水口 submerged outlet
埋木法 plug wood
埋平橱柜门闩 flush cupboard catch
埋弃场 burial ground;burying place
埋弃储槽 burial tank
埋砌的石护坡 embedded stone pitching
埋嵌条饰 sunk fillet mo(u)ld
埋嵌线条 sunk fillet
埋嵌线条饰 sunk fillet mo(u)ld
埋丘 buried hill
埋入 anchoring;bury;embed(ding);embedment;imbedding;immergence;immersion;sink;submergence
埋入部分 embedded parts;inlet part
埋入测温器 imbedded temperature detector
埋入测温器绝缘 imbedded temperature-detector insulation
埋入插销 flush bolt
埋入长度 embedded length;embedment length;length of embedment
埋入的 buried;embedded;flush;nested
埋入的管道 buried tubular conduit
埋入地下 disposal to land;land disposal;subterranean disposal;underground disposal <放射性废料>
埋入地下的 buried
埋入地下的排水管 buried drain
埋入法 embedment method;implantation
埋入钢筋 embedded steel
埋入混凝土 buried concrete
埋入混凝土内 embedded in concrete
埋入件 built-in fittings;embedment inserts;inlet parts
埋入键 recessed key
埋入金属踏步 climbing irons;stepping iron
埋入锚 embedment anchor;burial anchor
埋入砌体工程的锚杆 masonry anchor
埋入墙内的锚栓 wall anchor
埋入墙内的柱 embedded column
埋入区 <海工结构在泥面以下的> buried zone
埋入区的防蚀 protection in buried zone
埋入深度 buried depth;depth of burying;depth of embedment;depth of setting;set-depth
埋入式 flush type;immersion type
埋入式岸墩 buried abutment
埋入式板把手 flush type slab handle
埋入式测温计 embedded temperature detector
埋入式测温器 embedded temperature detector
埋入式插座 flush plug receptacle;flush type receptacle
埋入式承台桩基 embedded footing on piles
埋入式吹氧法 immersion oxygen-blowing process
埋入式挡土墙 embedded retaining wall
埋入式堤脚护坦 buried toe apron
埋入式电气盒 flush wall box
埋入式高温计 immersion pyrometer
埋入式沟槽 buried channel
埋入式管道 embedded conduit;embedded pipe
埋入式基脚 embedded footing
埋入式检温计 embedded temperature detector
埋入式结构设计 design of buried structure
埋入式结构物 buried structure
埋入式进气口 submerged inlet;submerged intake
埋入式开关 flush switch;submerged switch
埋入式路缘 kerb below ground
埋入式锚 drag embedment anchor
埋入式锚碇 deadman anchorage
埋入式锚碇物 deadman anchorage
埋入式面板 false deck
埋入式桥墩 buried abutment;buried pier
埋入式桥台 buried abutment;spill-through abutment
埋入式绕组 imbedded winding;tunnel(l)ed winding
埋入式热电偶 built-in thermocouple;immersion-type thermocouple
埋入式柔性压力涵管 buried flexible pressure conduit
埋入式生命线 buried lifeline
埋入式水池 embedded tank
埋入式探测器 embedded detector
埋入式温度计 embedded temperature detector;immersion thermometer
埋入式温度探测器 embedded temperature detector
埋入式镶板 sunk panel
埋入式压力钢管 buried penstock;embedded penstock
埋入式压力管道 buried penstock;embedded penstock
埋入式压力水管 buried penstock;embedded penstock
埋入式岩墩 buried abutment
埋入式衣橱 in-built wardrobe
埋入式应变计 embedded strain ga(u)ge

埋入式圆形涵管 buried circular conduit
埋入式圆柱壳 embedded cylindric(al)shell
埋入式运输机 buried conveyer[conveyor]
埋入式照明器具 recessed lighting fitting
埋入式振捣棒 immersion poker
埋入式振动棒 immersion needle;immersion poker
埋入式贮水池 embedded tank
埋入式锥型螺栓 cone bolt
埋入土壤 place into the soil
埋入物 embedded object
埋入压力 embedment pressure
埋入仪器 embedded instrument
埋入应变片 embedded strain ga(u)ge
埋入有机结合料的石屑 embedded organic-binder chip(ping)s
埋入装置 flush mounting
埋入作用 embedding action;imbedding action
埋烧 sawdust firing
埋设 built-in;buried laying;installation;laying
埋设暗管线的内隔墙 chase wall
埋设传力杆的缝 tied joint
埋设的机内 built-in
埋设的设备 buried tubular conduit
埋设管 buried pipe
埋设基础 sunk foundation
埋设件 built-in fittings;embedded parts
埋设结构 embedded structure
埋设里程碑 kilometre marking
埋设深度 buried depth;depth of burying;embedded depth;laying depth
埋设式浴盆 in-built(bath)tub
埋设物防护 utility protection
埋设线 buried wire
埋设线路 buried line
埋设仪器 embedded instrument
埋设在大石块内的起吊钢拉杆 lifting pin
埋设在混凝土内的锚固螺栓 swedge bolt
埋设在砌体内的螺栓 lewis bolt
埋设在砌体内的锚固螺栓 lewis anchor
埋深 depth of embedment;depth of tunnel;embedded depth;embedment depth;placing depth;setting depth
埋深比 depth ratio
埋深因数 <地基承载力,土坡> depth factor
埋深因素 <地基承载力,土坡> depth factor
埋石 displacer;mark at or below ground level;monumentation;refixation;setting monument
埋石测站 marked station;permanent station
埋石点 fixed point;monumented station;monumented(survey)point;permanent station
埋石固定点 fixed station
埋石混凝土 cyclopean concrete;rubble concrete
埋石控制点 monumented control point
埋石水准点 monumented benchmark
埋式电缆 buried cable;direct burial cable
埋式光缆 buried optic(al)cable
埋式海底光缆 buried submarine cable
埋式接缝 buried joint
埋式排水管 buried drain
埋式桥台 buried abutment;hidden abutment
埋填垃圾 sanitary landfill
埋填入土 landfill
埋铁(接)缝 metal cavity joint
埋铁条砌体 chain bond
埋头 concealed hinge;counterbore;countersunk head;flush head;sinking head
埋头带槽钉 countersunk chequered head nail
埋头道钉 dog spike
埋头的 countersunk;dormant;sunk
埋头垫圈 countersunk finishing washer;countersunk washer;cup washer;socket washer
埋头钉 dog nail;lost-head nail;secrete nailing;sinker nail
埋头钉头 countersunk head
埋头法兰连接 recessed flanged joint
埋头方颈螺栓 countersunk(head)square neck bolt
埋头机螺钉 countersunk head machine screw
埋头键 sunk key
埋头孔 counterbore;countersink
埋头孔铰刀 countersinking reamer;countersunk reamer
埋头苦干 beaver
埋头螺钉 countersunk bolt;countersunk head screw;countersunk screw;dormant screw;flush rivet;recessed head screw;sunk screw;secrete screwing
埋头螺孔 countersunk hole
埋头螺帽 counter nut;countersunk nut
埋头螺母 counter nut;countersunk nut
埋头螺栓 boxer bolt;countersink bolt;countersunk(headed)bolt;dormant bolt;flush bolt;flush headed bolt;patch bolt
埋头螺丝 countersunk screw;dormant screw;sunk screw
埋头螺丝卸扣 carvel shackle;countersunk shackle;screw pin
埋头铆钉 blind rivet;countersink and chipped rivet;countersink(head)rivet;countersinking of rivets;countersunk and chipped rivet;countersunk head rivet;countersunk rivet;flush head rivet;snug rivet;sunk heat rivet;sunk rivet
埋头铆接 flush rivet(t)ing
埋头铆接结构 flush-riveted construction
埋头铆壳 flush-riveted covering
埋头木螺丝 sunk wood screw
埋头平顶铆钉 countersunk flat head rivet;flat countersunk rivet
埋头平头钻头 countersunk flat head bit
埋头倾斜钉入 toshnailing
埋头绒头纱 dead pile yarn
埋头销 sunk pin
埋头圆顶螺钉 countersunk oval head screw
埋头圆顶铆钉 countersunk oval head rivet;round-head counter sunk rivet
埋头轴环 countersunk collar
埋头轴颈 countersunk spigot
埋头钻杆钻机 countersinker
埋头钻孔 counterbore;countersunk drilling
埋头钻(头)countersink bit;countersinker;chamfering bit;counter drill;countersink drill;countersinking;countersinking bit;countersinking drill;countersinking reamer;sinking bit;sunk drill
埋土法 mounding
埋瓦管 tile laying
埋吸胶管 embedded suction hose
埋线工业吸水胶管 steel wire embedded water suction rubber hose
埋线器 imbedding tool
埋线吸水胶管 embedded wire water suction hose
埋线吸酸碱胶管 embedded wire acid and alkali suction hose
埋线吸油胶管 embedded wire oil suction hose
埋陷 bog down
埋压机 compaction-burial machine
埋在灰浆中的管道 conduit embedded in plaster
埋在砖柱间的加固薄片 nogging strip
埋葬 entombment;inhumation;interment;sepulture
埋葬的 mortuary
埋葬地 burial place
埋葬时间测定 estimation of burial time
埋葬学 taphonomy
埋渣焊 metal buried welding
埋植 heeling in
埋置 embed(ment);imbed(ding);imbedment;nest
埋置板 embedded panel
埋置部分 embedded parts
埋置层 buried layer
埋置长度 buried length;bury length;embedded length;embedment length;bond length;grip length <钢筋在临界点以外的>
埋置长度当量 <钢筋在临界点以外的> embedment length equivalent
埋置的 embedded
埋置的钢结构 encased steelwork
埋置的供热电缆 embedded electric(al)heating cable
埋置电缆 buried cable
埋置电路学 embedded circuitry
埋置钢筋 bedded bar;embedded reinforcement;embedded steel;imbedded steel
埋置隔热 embedded insulation
埋置沟道 buried channel
埋置刮板给煤机 buried scraper coal-feeder
埋置管道 embedded conduit;embedded pipe
埋置拉杆 embedded tie
埋置冷盘管 embedded cooling coil
埋置沥青膜 buried membrane
埋置螺栓 embedded bolt
埋置膜 buried membrane
埋置入混凝土中 embedded in concrete
埋置深度 buried depth;covering of bars;depth of burying;depth of embedment;depth of laying;embedded depth;embedment depth;laying depth;depth of burial
埋置式传感器 embedded sensor
埋置式桥台 buried abutment;embedded abutment;hidden abutment
埋置式温度计 embedded temperature detector
埋置式温度探测器 embedded temperature detector
埋置式蒸发 sunken evapo(u)ration
埋置式蒸发皿 sunken evapo(u)ration pan
埋置式蒸发器 sunken evapo(u)ration pan
埋置温度传感器 embedded temperature detector
埋置温度探测 embedded temperature detector
埋置蜗壳 embedded spiral case
埋置误差 built-in error
埋置消火栓 buried fire hydrant
埋置效应 effect of embedment
埋置在混凝土中 embedded in concrete
埋置在混凝土中的 concrete-embedded
埋置在混凝土中的支架或桩 supports or piles embedded in concrete;ahset
埋种压土轮 seed press wheel
埋装插座 flush socket
埋装开关 flush switch
埋装式 flush type
埋装式插座 flush plug receptacle
埋装式活动开关 flush snap-switch
埋装式结构 flush type construction
埋装式开关 sunk-type switch
埋装式仪表 flush type instrument
埋装钻杆 flush drill pipe
埋着的岩石 encasing rock
埋钻 build-up in the(bore)hole;drilling failure;sticking of drilling rod
埋钻深度 position of covered tools

霾

霾层 haze layer

霾层顶 haze horizon
霾度 <能见度单位> coefficient of haze
霾雾 haze and fog
霾线 haze line
霾因子 haze factor

买

买办 comprador

买保险 buy insurance
买本国货 buy national commodity
买得便宜 buy at a good bargain
买得优先权 buy the refusal
买方 buyer;buying party;client;clientage;licensee;purchaser;vendee
买方安排 buying disposition
买方仓库交货价格 ex buyer's godown
买方出价 buying offer
买方出价最高者 highest bidder
买方代理人 purchasing agency
买方贷款 buyer's credit
买方对所有权得保证 guarantee of title
买方风险 buyer's risk;consumer risk
买方负全责 client full responsibility
买方负责 caveat vendor
买方和卖方直接合同 buyer and supplier contract
买方还价 counter offer
买方检查货物 inspection of goods
买方经纪人 buyer's broker
买方经理人 purchasing agency
买方开价 buying offer
买方垄断 buyer's monopoly;monopsony
买方垄断利润 monospoy profit
买方设施 buyer's facility
买方市场 buyer's market;consumer market
买方收货后付款 charges forward
买方双头垄断 duopsony
买方套购保证 buyer's hedging
买方提供的样品 buyer's sample
买方信贷 buyer's credit
买方信贷担保 buyer's credit guarantee
买方佣金 buying commission
买方有利时机 buyer's opportunity
买方远期信用证 buyer's usance letter of credit
买回 beam-covering;bear covering;repurchase

M

买汇 buying exchange
买货方 buyer
买货人选定 buyer's option
买货佣金 buy commission
买价 bid rate;buying price;purchase money; purchase price; purchasing price
买价与卖价 bids and offers
买进股票 buy in stock
买进原价 first cost
买旧货 buy secondary-hand
买廉价货 bargain hunting;buy goods at the sale
买卖 bargain; business; buying and selling;commercial act;commercial transaction;deal;trafficking
买卖公平 buy and sell at reasonable price
买卖共同条件 general conditions of sale
买卖合同 bargain;buy-sell contract; contract note; contract of sales; sales contract; agreement of sale; contract of purchase
买卖价格 basis of price
买卖契约 buying contract
买卖市场 bargain center[centre]
买卖双方 both vendor and purchaser
买卖双方直接交易 principal-to-principal transaction
买卖双方直接交易的 over-the-counter
买卖条件 terms of sale
买卖赃物的场所 fence
买卖者 trafficker
买卖中心 bargain center[centre]
买期保值 buying hedge
买入财产抵押 purchase money mortgage
买入汇率 buying rate
买入价(格)buying price;purchasing price;buying rate
买下……的全部产权 buy out
买现成服装的商店 slop shop
买新到货 buy goods on arrival
买招牌 buy out a business
买主 bargainee;bargainor;buyer;client;customer;grantee;offeree;purchaser;vendee
买主不付定金全部靠贷款的房地产交易 no money down
买主的留置权 vendee's lien
买主独家垄断 monopsony
买主过多 buyers over
买主垄断价格 oligopsony price
买主权益 buyer's interest
买主剩余 buyer's surplus
买主信用 buyer's credit
买主选择权 buyer's option; buyers option
买主意图 buyer's buying intention
买主账(户)bull account
买主资力保证 del credere

迈

迈埃金反应 Meigen's reaction
迈步式车辆 walking machine
迈步式破碎机 walker crusher;walking crusher
迈步式索斗铲 walking dragline
迈步式挖掘 walking excavation
迈步式挖掘机 walker excavator;walking dragline excavator
迈步式支架 step-type support;walking support
迈尔 <热容量单位> mayer
迈尔德 Mired
迈尔德值 Mired value
迈尔公式 Mayer's formula
迈尔合成法 Mayer's synthesis
迈尔洪水等级 Myer's flood scale
迈尔精密冲裁法 Mayer's process
迈尔率定法 Myer's rating
迈尔凝结理论 Mayer's condensation theory
迈尔-普劳斯尼茨吸附溶液理论 Myer-Prausnitx adsorbed solution theory
迈尔曲线 Mayer curve[M-curve]
迈尔溶液 <氯化银碘化锌溶液> Mayer's fluid
迈尔硬度 Meyer hardness
迈格表 megger;megohmmeter;tramegger
迈加洛波利斯的塞锡利翁 <希腊> Thersilion at Megaloplis
迈卡邦德绝缘材料 Micabond
迈克尔加成反应 Michael addition reaction
迈克尔逊-盖尔环形干涉仪 Michelson-Gale ring interferometer
迈克尔逊干涉显微镜 Michelson interference microscope
迈克尔逊干涉仪 Michelson's interferometer
迈克尔逊恒星干涉仪 Michelson stellar interferometer
迈克尔逊基准反射器 reference reflector of Michelson
迈克尔逊阶梯光栅 Michelson echelon
迈克尔逊莫雷试验 Michelson-Morley experiment
迈克尔逊日射计 Michelson actinograph
迈克尔逊双金属直接日射计 Michelson bimetallic pyrheliometer
迈克尔逊微干涉仪 Michelson microinterferometer
迈克尔逊星体干涉仪 Michelson stellar interferometer
迈克尔逊型镉灯 Michelson type cadmium lamp
迈克尔逊型镉电池灯 Michelson type cadmium lamp
迈克尔逊型激光器 Michelson type laser
迈克尔逊旋转镜 Michelson rotating mirror
迈纳尔炸药 Minol
迈尼特制铝法 Minet aluminum process
迈诺氧化物 manool oxide
迈欧特阶【地】Meotian
迈森瓷器 Meissen porcelain
迈森绿 Meissen green
迈氏硬度 Meyer hardness
迈斯纳法 Meissner method
迈斯纳效应 Meissner effect; Meissner-Ochsen field effect
迈斯纳振荡器 Meissner oscillator
迈斯维尔阶【地】Maysvillian
迈瓦西尔高真空绝缘材料 myvaseal
迈亚里 R 低合金耐热钢 Mayari R
迈因纳假设 Miner's hypothesis
迈因纳线性损伤理论 Miner's linear damage theory
迈因泽单位 <地下水渗透率单位,1 迈因泽单位 = $0.04075m^3(d \cdot m)^2$> Meinzer unit

麦

麦 <磁通量单位> Maxwell
麦波郎 <塑料专用名> Mipolam
麦迪西软瓷 Medici porcelain
麦迪逊氏废物处理过程 Madison process
麦地那 barbital sodium
麦地那石英岩 Medina quartzite
麦地那水泥 Medina cement
麦地那统 <晚志留世>【地】<美> Medinan series
麦东白 Meudon white
麦东周期【给】Metonic cycle
麦尔马克塑料 melmak
麦饭石 maifanshite;medical stone
麦麸 middling
麦杆 culm
麦杆黄(色)leghorn
麦秆焦油 wheat-straw tar
麦基奇尼高强度黄铜 Mckechnie's bronze
麦吉尔可铸可锻铝青铜 McGill metal; McGill alloy
麦吉尔铝铁青铜 McGill metal
麦钾沸石 merlinoite
麦角甾醇 ergosterol
麦角甾烷 ergostane
麦秸 straw;wheat straw
麦秸板 straw panel
麦秸的 strawy
麦秸堆 mow
麦秸骨料 wheat aggregated
麦秸画 Chinese wheat straw picture; picture in straw;straw patchwork
麦秸黄 straw yellow
麦秸集料 wheat aggregated
麦秸筐 straw basket
麦秸贴画 wheat stalk patchwork
麦秸装饰品 wheat stalk ornament
麦金道斯氏(轻便手提式)勘察工具 Mackintosh boring and prospecting tool
麦金利图版法 Mckinley type curve matching
麦金托什探测仪 Mackintosh prospector
麦金托什探头 Mackintosh probe
麦卡利地震烈度表 Mercalli(intensity)scale
麦卡利修正地震烈度表 Mercalli modified scale of earthquake intensity
麦卡利悬轴式旋回破碎机 McCully gyratory
麦卡特尼氏内桩调整夹 Macartney's collet adjuster
麦考尔腐蚀试验机 MacCaull corrosion tester
麦克风 microphone;mike
麦克康奈螺线 <用于汽车高速环行试验跑道,可使缓和曲线具有保证颠振加速度为最小的颠振运动特性,又称福特曲线> McConnell's curve
麦克科尔保护系统 McColl protective system
麦克奎德爱恩晶体粒度试验 Mc-Quaid-Ehn test
麦克莱恩堆集法 Mac-Lane system
麦克莱斯菲尔德 Macclesfield
麦克莱斯菲尔德筘号 Macclesfield reed count
麦克劳德低压计 Mac-Leod pressure ga(u)-ge
麦克劳德方程 Mac-Leod equation
麦克劳德麦氏真空规 Mac-Leod vacuum ga(u)ge
麦克劳德(压力)计 Mac-Leod ga(u)-ge
麦克劳德压强计 Mac-Leod ga(u)ge
麦克劳德真空规 Mac-Leod ga(u)ge
麦克劳德真空计 Mac-Leod vacuometer
麦克劳林定理 Maclaurin's theorem
麦克劳林公式 Maclaurin's formula
麦克劳林级数 Maclaurin's series
麦克劳林球体 Maclaurin's spheroid
麦克劳林展开 Maclaurin's expansion
麦克劳真空规 Mclead vacuum ga(u)-ge
麦克勒釉 Mackler's glaze
麦克里伦氏活塞式取土器 McClelland piston sampler
麦克利犁雪机 Macly snow-spreading car
麦克林伯格方程 Mechlenburg equation
麦克隆姆大地电流指示计 McCollum terrestrial ammeter
麦克罗斯碱性耐火制品 macros
麦克马洪填充物 MacMahon packing
麦克马洪填料 McMahon packing
麦克迈克尔黏[粘]度计 McMichael visco(si)meter
麦克米仑加速试验箱 Mchmullen accelerated cabinet
麦克米契尔黏[粘]度 MacMichael degree
麦克米契尔黏[粘]度计 MacMichael viscometer
麦克默里变流器 McMurray inverter
麦克内特镍铬耐热钢 Mackenite metal
麦克尼马尔检验 McNemar's test
麦克诺(海峡)桥 <位于美国密执安州> Mackinac Bridge
麦克佩斯照度计 Macbeth illuminometer
麦克试验 Mackey test
麦克水泥 Mack's cement
麦克斯韦 <旧磁通量单位> Mx[Maxwell]
麦克斯韦-贝蒂互换定理 Maxwell-Betti reciprocal theorem
麦克斯韦变位互等定理 Maxwell's theorem of reciprocity
麦克斯韦变位互等定律 Maxwell's law reciprocal deflection
麦克斯韦变位互等原理 Maxwell's theorem of reciprocity
麦克斯韦-玻耳兹曼方程 Maxwell-Boltzmann's equation
麦克斯韦-玻耳兹曼分布 Maxwell-Boltzmann distribution
麦克斯韦-玻耳兹曼分布律 Maxwell-Boltzmann distribution law
麦克斯韦-玻耳兹曼经典统计学 Maxwell-Boltzmann classical statistics
麦克斯韦常数 Maxwell's constant
麦克斯韦场 Maxwell's field
麦克斯韦场方程 electromagnetic field equation
麦克斯韦磁通计 Maxwellmeter
麦克斯韦等面积法则 Maxwell's equal area rule
麦克斯韦第二方程 second equation of Maxwell
麦克斯韦第一方程 first equation of Maxwell
麦克斯韦电波理论 Maxwell's electromagnetic wave theory
麦克斯韦电磁场方程 Maxwell's electromagnetic field equation
麦克斯韦电桥 Maxwell bridge
麦克斯韦定理 Maxwell's theorem
麦克斯韦定律 Maxwell's law
麦克斯韦定则 Maxwell's rule
麦克斯韦方程 Maxwell's equation
麦克斯韦分布 Maxwellian distribution;Maxwell's distribution
麦克斯韦分布(定)律 Maxwellian distribution law; Maxwell's distribution law
麦克斯韦分配 Maxwell's distribution
麦克斯韦各向同性黏[粘]性介质 Maxwell's isotropic(al)viscous medium
麦克斯韦公式 Maxwell's formula
麦克斯韦关系式 Maxwell's relation
麦克斯韦观察法 Maxwellian viewing
麦克斯韦观察系统 Maxwellian viewing system

麦克斯韦光斑 Maxwell's spot
麦克斯韦光的电磁论 Maxwell's electromagnetic theory of light
麦克斯韦光学理论 Maxwell's theory of light
麦克斯韦-哈林顿冲击试验 Maxwell's and Harrington impact test
麦克斯韦互等变位定律 Maxwell's law of reciprocal deflection; Maxwell's law of reciprocity
麦克斯韦环流 Maxwell's circulating current
麦克斯韦混色盘 Maxwell's disk
麦克斯韦基本方程 Maxwell's fundamental equation
麦克斯韦基色 Maxwell's primary
麦克斯韦计 fluxmeter;Maxwellmeter
麦克斯韦假说 Maxwell's hypothesis
麦克斯韦截面 Maxwell's cross-section
麦克斯韦扩散系数 Maxwell's coefficient of diffusion
麦克斯韦力 Maxwell's force
麦克斯韦力多边形 Maxwell polygon of forces
麦克斯韦流变模型 Maxwell's rheological model
麦克斯韦流体 < 即黏[粘]弹性流体 > Maxwell's fluid;viscoelastic fluid
麦克斯韦模型 < 表示材料流变性质的一种力学模型,弹簧与缓冲壶并联 > Maxwell's model
麦克斯韦-摩尔法 Maxwell-Mohr's Method
麦克斯韦能量分布 Maxwellian energy distribution
麦克斯韦能谱 Maxwell's spectrum
麦克斯韦黏[粘]弹性流体的应力应变关系 Maxwell's stress-strain relation for elasto-viscous fluid
麦克斯韦平衡 Maxwellian equilibrium
麦克斯韦平均自由行程 Maxwell's mean free path
麦克斯韦气体 Maxwellian gas
麦克斯韦三角形 Maxwell's triangle; X-Y chromaticity diagram
麦克斯韦色品图 Maxwell's colo(u)r triangle
麦克斯韦色三角 Maxwell's colo(u)r triangle
麦克斯韦氏挠度互等定律 Maxwell principle of reciprocal deflection
麦克斯韦松弛方程 Maxwell's relaxation equation
麦克斯韦松弛时间 Maxwell's relaxation time
麦克斯韦松驰试验 Maxwell's relaxation test
麦克斯韦速度分布 Maxwellian velocity distribution
麦克斯韦速度分布定律 Maxwell's law of velocity distribution
麦克斯韦特性 Maxwellian character
麦克斯韦体 Maxwell's body;Maxwell's substance
麦克斯韦图 Maxwell's diagram
麦克斯韦-瓦格纳机构 Maxwell's Wagner mechanism
麦克斯韦效应 Maxwell's effect
麦克斯韦妖 demon of Maxwell
麦克斯韦液体 Maxwell's body; Maxwell's liquid
麦克斯韦原理 Maxwell's theorem
麦克斯韦圆盘 Maxwell's disk
麦克斯韦匝 Maxwell-turn
麦克斯韦正交流变仪 Maxwell's orthogonal rheometer
麦克斯韦状态方程 Maxwell's equation of state
麦克唐纳道格拉斯飞机公司 McDonnel Douglas Aircraft Company
麦克唐纳式给水控制器 Macdownel feed water controller
麦克亚当辨色阈 MacAdam's chromaticness discrimination threshold
麦克亚当极限 MacAdam's limit
麦克亚当色差单位 MacAdam's unit
麦克亚当色差方程式 MacAdam colo(u)r difference equation
麦克亚当椭圆 MacAdam's ellipse
麦克亚瑟扩底桩 MacArthur pedestal pile
麦克亚瑟桩 MacArthur pile
麦克因托希分类 McIntosh classification
麦肯齐汞(合金) Mackenzie's amalgam
麦肯齐河 Mackenzie River
麦肯森三油楔动压轴承 Mackensen bearing
麦夸尔-基奥法 Mcguire-Keogh Act
麦夸里机械能力测验 Mac Quarrie test for mechanical ability
麦郎选除器 cockle separator
麦郎选除筒 cockle cylinder; recockling cylinder
麦硫锑铅矿 madocite
麦露奇型瓦斯检查器 Mcluckie gas detector
麦马克马斯公式 < 计算涵洞等出水口面积用 > McMarth's formula
麦马克马斯望远镜 McMarth telescope
麦麦克斯存储器 memex
麦芒帽 wheat awn cap
麦镁镍矿 gaspeite
麦面纱 mull
麦纳麦 < 巴林首都 > Manama
麦宁炸药 < 即苦味酸 > melinite
麦片纸 oatmeal paper
麦片粥 oat
麦羟硅钠石 magadiite
麦丘恩氏三角 Macewen's triangle;suprameatal triangle
麦丘利神塑像 Flying Mercury
麦砷钠钙石 mcnearite
麦-施-卡烈度表 MSK intensity scale
麦氏地震烈度表 Mercalli scale
麦氏平整度仪 Mays meter
麦氏压力计 Mac-Leod ga(u)ge
麦氏真空计 Mac-Leod ga(u)ge
麦斯里希特阶【地】Maastrichtian
麦斯尼定向天线 Masny directive antenna
麦松刻图片 Mason scribe
麦穗草 crested wheatgrass
麦穗状交叉 repeated Y types intersection
麦塔林钴铜铝铁合金 Mataline
麦碳铜镁石 mcguinnessite
麦头和号码 marks and numbers
麦逊法 Matson method
麦逊公司方式 < 集装箱运输 > Matson system
麦芽糖 Barley sugar;maltose
麦芽作坊地面砖 pament
麦哲伦海峡 Strait of Magallan
麦哲伦流 Magellanic stream
麦哲伦系 Magellanic system

卖出 bring to market; put on the market;selling-out

卖出的期货合同 < 美 > short position
卖出(汇)价 selling rate
卖出汇率 selling rate
卖出旧船再租回 charter back
卖出买入选择权 put and call option
卖出期权 put option
卖出实收价 net proceeds
卖出选择权 put option
卖出指示 market order
卖船合同 contract of sales
卖得高价 fetch a high price
卖底货 close out
卖掉 trade away;trade-off;unload
卖短 selling short
卖断 outright sale
卖断交易 outright sell
卖方 seller;the selling party;vendor
卖方保证条款 packer guarantee clause
卖方报价 asked price; asking price; offer price;offer rate;selling price
卖方不履行合同 seller's failure to perform
卖方仓库交货价(格) ex-godown;ex-seller's godown; ex-seller's warehouse;ex-warehouse
卖方撤回前有效的报盘 offer good until withdrawn by seller
卖方撤回有效报价 offer good until withdrawn by seller
卖方承担的风险 let the seller beware
卖方出具给买方的确认已保险信函 letter of insurance
卖方惯例 seller's usance
卖方航运文件 vendor's shipping documents
卖方价格 price of the selling party; seller's price
卖方交纳进口税的交货条件 duty-paid term
卖方决定 seller's option
卖方开价 ask price;offer rate
卖方看货后定发盘 offer subject to buyer's inspection or approval
卖方联合 bear pool
卖方联(合经)营 bear pool
卖方留置权 seller's lien;vendor's lien
卖方垄断 seller's monopoly
卖方皮重 shipping tare
卖方期货 seller's hedging
卖方确认报价 offer subject to seller's confirmation
卖方确认后有效发盘 offer subject to seller's confirmation
卖方确认有效报价 offer subject to seller's confirmation
卖方设施 seller's facility
卖方市场 seller's market
卖方索价 offer
卖方习惯包装 seller's usual packing
卖方信贷 seller's credit;supplier's credit
卖方选择(权) seller's option
卖方样品 seller's sample
卖方要价 asked
卖方盈余 seller's surplus
卖方远期信用证 seller's usance credit
卖方愿意收回抵押 owner will carry mortgage
卖方自报价 unsolicited offer
卖给你方 book your order for
卖回溢价 call premium
卖汇票预约 selling contract
卖汇票预约申请书 selling contract slip
卖货方 seller
卖家套头保值 seller's hedging
卖价 asked price;asked quotation;asking price; offer(price); offer rate; seller's price; seller's rate of exchange;selling price;selling rate
卖价汇率 offer rate
卖价极便宜的交易 dead bargain
卖价市场 offered market
卖酒处 < 夜总会酒吧间 > watering hole
卖酒柜台 bar
卖据 bills of sales;deed of purchase; deed of sales
卖空 bear drive;over-sell;over-sold; sell(ing) short
卖空补进 covering
卖空的证券交易投机商 bear seller
卖空合同 short contract
卖空交易 short selling;short sale < 指股票投机等 >
卖空浪潮 bear raid
卖空市场 bear market
卖空套头 short hedge
卖空投机 deport
卖空行为 bear operation;short selling
卖空账户 short account
卖空者 bears; short interest; short seller
卖廉价商品的商人 cheap John
卖品 article for sale
卖期保值 hedge selling;selling hedge
卖期汇率 selling rate
卖期货选择权 put option
卖气旺盛 aggressive selling
卖契 act of sale;bill of sales;deed of bargain and sale;deed of purchase; deed of sale
卖清 sell out
卖清存货 sell off
卖脱 bargain away
卖外汇 selling exchange
卖现货 spot buy
卖者过多 sellers over
卖主 bargainer [bargainor]; granter [grantor];seller;vendor
卖主标准结算程序 vendor standard settlement program(me)
卖主的开价 asked price;offered price
卖主的开叫价 asking price
卖主的最低售价 reserve price
卖主(对货物)的留置权 vendor's lien
卖主规格 vendor rating
卖主过多 sellers over
卖主检查 vendor inspection
卖主可靠性 vendor reliability
卖主控制价格 oligopoly price;seller's price
卖主控制均势 cooperative equilibrium
卖主扣押权 seller's lien;vendor's lien
卖主垄断 oligopoly
卖主垄断价格 oligopoly price
卖主评级 vendor rating
卖主剩余 seller's surplus
卖主市场 seller's market; vendor's market
卖主信贷 supplier's credit; vendor's credit
卖主选择权 seller's option
卖主债效指数 vendor performance index
卖主账(户) bear account; supplier's account;vendor's account
卖主装运配置 vendor shipping configuration
卖主装运说明书 vendor shipping instruction
卖主租约 vendor lease

脉壁 vein wall

脉壁泥 fault gouge;leaderstone
脉壁黏[粘]土【地】flucan;donk;capel [kapel]
脉壁黏[粘]土带 capel [kapel] clay zone
脉壁黏[粘]土皮 selvage
脉波 pulse wave
脉波后的 postsphygmic
脉波计 sphygmometer
脉波描记法 pulse tracing
脉波描记器 sphygmograph

M

脉波前的 presphygmic
脉波数 pulse number
脉波学 kymatology
脉搏计 pulsimeter;pulsometer;sphygmometer
脉层 streak
脉成岩类 phlebite
脉冲 impulse;kipp;outbreak;pulse;bang <回声测深仪的>;pip <荧光屏上的>
脉冲摆动 impulse hunting
脉冲半波 pulse half wave
脉冲包络 pulse envelope
脉冲包络检波器 pulsed envelope detector
脉冲包络原理 pulsed envelope principle
脉冲包线 pulse envelope
脉冲包线法 pulse envelope method
脉冲包线检波器 pulse envelope detector
脉冲包线指示器 envelope viewer
脉冲暴 impulsive burst
脉冲倍压发生器 pulse voltage multiplier
脉冲泵 governor impeller;impulse pump;pulse pump
脉冲比 impulse ratio;pulse ratio
脉冲比编码 pulse ratio encoding
脉冲比较器 pulse comparator
脉冲比调制 pulse ratio modulation
脉冲边沿 porch;pulse edge;pulse porch
脉冲边沿差 displacement of pulse porches
脉冲边沿检测网络 edge sense network;pulse edge sense network
脉冲边沿同步的D型触发器 edge clocked D-type register
脉冲边沿修整器 pulse corrector
脉冲边缘 edge of a pulse;pulse edge
脉冲边缘陡度 steepness of pulse edge
脉冲编码 impulse coding;pulse code
脉冲编码管 pulse coding tube
脉冲编码和相互关系 pulse coding and correlation
脉冲编码器 pulse coder;pulse encoder
脉冲编码调节 pulse code modulation
脉冲编码调整盘 pulse code reticle
脉冲编码调制 impulse code modulation;pulse code(d) modulation
脉冲编码调制盘 pulse code retic(u)le
脉冲编码调制系统 pulse code modulated system
脉冲编码调制遥测系统 pulse code modulation switching system
脉冲编码调制综合测量仪 pulse code modulation integrative measuring set
脉冲编码系统 coding pulse multiple
脉冲编码遥测系统 pulse code modulation switching system
脉冲编码震源 pulse coded source
脉冲编码制 pulse coding system
脉冲编码装置 coder;moder;pulse encoder;pulse moder
脉冲变换 impulse transfer;pulse transformation
脉冲变换器 pulse converter;pulse transducer
脉冲变量器 pulse transformer
脉冲变流器 pulse transformer;pulsing transformer
脉冲变狭 pulse shrinkage
脉冲变压加速器 pulsed transformer accelerator
脉冲变压器 impulse transformer;kick transformer;pulse transformer
脉冲变压器室 pulse transformer box
脉冲变窄 pulse narrowing
脉冲标记 pulse label(l)ing
脉冲标记技术 pulse label(l)ing technique
脉冲标准化 pulse normalization;pulse standardization
脉冲拨号盘 dial pulse dialer
脉冲拨号器 pulse dialer
脉冲波 impulse wave;pulse wave
脉冲波包 pulse envelope;pulse wave packet
脉冲波动 pulse ripple
脉冲波对比 pulse correlation
脉冲波多普勒超声流量计 pulsed wave Doppler ultrasonic flow meter
脉冲波峰因数 pulse crest factor
脉冲波尖 pulse spike
脉冲波列 pulse train
脉冲波前 flank;front
脉冲波前校正线路 impulse corrector;pulse corrector
脉冲波试验 front of wave test
脉冲波速试验 pulse velocity test
脉冲波纹 pulse ripple
脉冲波形 impulse wave form;pulse shape;pulse waveform
脉冲波形发生器 pulse pattern generator
脉冲波形图 timing chart
脉冲波振荡器 impulse wave generator
脉冲波至 pulse arrival
脉冲补偿加法器 pulse bucking adder
脉冲布袋除尘器 pulsed jet cloth filter
脉冲步进电动机 impulse stepping motor
脉冲步进器 pulse stepper
脉冲部件 pulse unit
脉冲参数 pulse parameter
脉冲操作 pulse operation
脉冲操作电离室 pulse operated chamber;pulse operator chamber
脉冲操作回旋加速器 pulse operated cyclotron
脉冲测高计 pulse altimeter
脉冲测距 pulse ranging
脉冲测距导航 pulse ranging navigation
脉冲测距法 method of distance measurement by pulse
脉冲测距计 pulse distance finder
脉冲测距器 impulse distance measuring instrument;tellurometer
脉冲测距系统 pulse type distance-measuring system
脉冲测距仪 impulse distance measuring instrument;tellurometer
脉冲测量 impulsive measurement
脉冲测量设备 impulse measurement facility
脉冲测量示波器 pulse measuring oscilloscope
脉冲测量仪 impulse measuring device
脉冲测试 pulse test(ing)
脉冲测试笔 pen-type pulse tester
脉冲测试机 impulse testing machine
脉冲差拍 pulse beating
脉冲产生点 pulse site
脉冲颤动 pulse jitter
脉冲长度 pulse length
脉冲长度调制 pulse length modulation
脉冲场 pulsed field
脉冲场恒陡度加速器 pulsed field constant-gradient accelerator
脉冲场恒陡度同步加速器 pulsed field constant-gradient synchrotron
脉冲场环 pulsed ring
脉冲场加速器 pulsed field accelerator
脉冲场交变陡度加速器 pulsed field alternating-gradient accelerator
脉冲超导磁场 pulsed superconducting magnetic field
脉冲超高压发生器 pulse extra high-tension generator
脉冲超声波探伤仪 soniscope
脉冲成对性 pulse pairing
脉冲成四倍 pulse quadrupling
脉冲成型线路 pulse former
脉冲成型 pulse shaping
脉冲成型级 pulse shaping stage
脉冲乘法器 impulse type multiplier
脉冲澄清池 pulsator;pulsator clarifier;pulse clarifier
脉冲持续 pulse persistence
脉冲持续时间 duration of ground;duration of impulse;impulse duration;pulse duration;pulse length;pulse length impulse;pulse width
脉冲持续时间调制 pulse duration modulation;pulse length modulation;pulse width modulation
脉冲持续时间调制调频 pulse duration modulation-frequency modulation;pulse width modulation-frequency modulation
脉冲冲击 pulse burst
脉冲重叠 pulse overlap
脉冲重发 impulse repeating
脉冲重发器 impulse repeater;pulse repeater
脉冲重复 impulse repetition;pulse recurrence interval
脉冲重复倍频 pulse rate multiplication
脉冲重复倍数 pulse rate multiplication
脉冲重复继电器 pulse repeating relay
脉冲重复间隔 pulse recurrence interval;pulse repetition interval
脉冲重复率 impulse recurrence rate;pulse recurrence rate;rate of pulse repetition
脉冲重复率半延迟 half pulse repetition-rate delay
脉冲重复频率 impulse recurrence frequency;pulse frequency repetition rate;pulse repetition frequency;pulse repetition rate;rate of repetition;recurrence frequency;recurrent frequency;impulse repetition rate
脉冲重复频率倍增 pulse rate multiplication
脉冲重复频率的分频 skip-keying
脉冲重复频率发生器 pulse repetition frequency generator
脉冲重复频率分隔 pulse rate division
脉冲重复频率分频器 pulse rate divider
脉冲重复频率极限 pulse repetition frequency limitation
脉冲重复频率开关 pulse repetition frequency switch
脉冲重复频率振荡器 pulse repetition frequency generator
脉冲重复时间 pulse recurrence time
脉冲重复性 pulse reproducibility
脉冲重复周期 pulse recurrence period;pulse repetition cycle;pulse repetition period
脉冲重合检测器 pulse coincidence detector
脉冲重合效应 pulse coincidence effect
脉冲重调 pulsed reset
脉冲初动 impulse onset;impulsive onset
脉冲除尘器 pulse dust collector
脉冲触发 pulse trigger action;pulse triggering;trigger action
脉冲触发二进元 pulse triggered binary
脉冲触发器 pulse trigger
脉冲触发双稳态 pulse triggered binary
脉冲触簧 pulse spring
脉冲穿孔 pulse punching
脉冲传播 pulse propagation
脉冲传播测定计 pulse propagation meter
脉冲传播模式 pulse transmission mode
脉冲传播时间 pulse propagation time;pulse transit-time
脉冲传播时间法 pulse transit-time method
脉冲传播试验仪 pulse transmission tester
脉冲传播速度 pulse propagation velocity
脉冲传递函数 pulsed transfer function
脉冲传感器 impulser
脉冲传声器 impulse microphone
脉冲传输 impulse transmission;pulse transmission
脉冲传输反射波法 pulse transmission-reflection method
脉冲传输函数 pulsed transfer function
脉冲传输回波法 pulse transmission-reflection method
脉冲传输技术 pulse transmission technique
脉冲传输模 pulse transmission mode
脉冲传送 pulse transmission
脉冲传送线路 pulse transmission circuit
脉冲串 burst;pulse series;pulse string;pulse train;train of impulses;train of pulses
脉冲串长度 burst length
脉冲串传输 burst transmission
脉冲串发生器 pulse series generator;pulse train generator
脉冲串方式 burst mode
脉冲串方式传输 burst transmission
脉冲串继电器 pulse train relay
脉冲串式 burst mode
脉冲串信号 burst signal
脉冲窗 pulse window
脉冲床 pulsed bed
脉冲床吸着 pulsed bed sorption
脉冲磁场功率 pulsed magnet power
脉冲磁(场)狭缝 pulsed magnetic slit
脉冲磁场装置 pulsed magnetic field apparatus
脉冲磁功率 pulsed magnet power
脉冲磁镜 pulsed magnetic mirror
脉冲磁控管 pulsed magnetron
脉冲磁快门 pulsed magnetic shutter
脉冲磁铁 impulse magnet;pulsed magnet
脉冲磁引出装置 pulsed magnetic extractor
脉冲次数继电器 notching relay
脉冲次序检测器 pulse sequence detector
脉冲猝发 burst
脉冲淬火 pulsed quenching
脉冲萃取塔 pulsed extraction column
脉冲存储器 pulse accumulator
脉冲存储时间 pulse storage time
脉冲代码 pulse code
脉冲(代)码调制 pulse code modulation
脉冲带宽 impulsive bandwidth;pulse bandwidth
脉冲带宽滤波 pulse bandwidth filtering
脉冲袋式除尘器 impulse type bag filter
脉冲单音制 pulsed individual tone system

M

脉冲导纳 impulsive admittance
脉冲导燃炸药 pulsed infusion explosive
脉冲导燃装药放炮 pulsed infusion shot-firing
脉冲倒相器 pulse inverter
脉冲到达不同接收站的时间差 time difference of arrival
脉冲到达时间差 time difference of arrival
脉冲的产生 pulsing
脉冲的成型与再成型 pulse shaping and re-shaping
脉冲的发生 pulsing
脉冲的极限重复频率 pulse limiting rate
脉冲的计算波形 computed pulse shape
脉冲的平顶 flatness of wave
脉冲的前沿 front porch
脉冲灯 pulsed lamp
脉冲等离子焊机 impulse plasma welding machine
脉冲等离子弧焊 pulsed plasma arc welding
脉冲等离子体 pulsed plasma
脉冲底 pulse base
脉冲地震法 seismic impulse method
脉冲地震记录 impulse seismogram
脉冲巅值检波器 pulse peak detector
脉冲巅值振幅 pulse spike amplitude
脉冲点焊 impulsed spot welding; pulsation spot welding; pulsed spot welding
脉冲点焊机 impulse spot welder
脉冲点火 pulse firing
脉冲点钎焊烙铁 pulse dot soldering iron
脉冲电磁动能发电机 pulse electromagnetic kinetic energy generator
脉冲电磁法 pulse electromagnetic method
脉冲电磁力 pulse electromagnetic force
脉冲电磁能机 pulse electromagnetic energy apparatus
脉冲电磁铁 impulse magnet; impulse solenoid
脉冲电磁线圈 impulse solenoid
脉冲电磁仪 pulse airborne electromagnetic instrument
脉冲电动机 impulse motor; pulse motor
脉冲电弧 pulsed arc
脉冲电弧焊(接) impulse arc welding; pulse arc welding
脉冲电解 pulse electrolysis
脉冲电缆 pulse cable
脉冲电离器 pulse ionizing device
脉冲电离室 pulsed ion chamber; pulse ionization chamber
脉冲电离箱 pulse ionization chamber
脉冲电流 impulse current; impulsive current; pulsating current; pulse current
脉冲电流半自动对焊机 impulse current semiautomatic butt welding machine
脉冲电流测量仪 impulse current meter
脉冲电流磁化 flash magnetization
脉冲电流发生器 impulse current generator
脉冲电流缝焊机 impulse current seam welding machine
脉冲电流继电器 impulse current relay
脉冲电流牵引电动机 traction motor using pulse current
脉冲电流系统 impulse current system
脉冲电流自动对焊机 impulse current automatic butt welding machine
脉冲电路 impulse(-fed) circuit; kick circuit; pulse circuit; pulsing circuit
脉冲电路理论 pulse circuit theory
脉冲电码 impulse code; pulse code
脉冲电码解调器 pulse code demodulator
脉冲电码组 pulse code group
脉冲电平 impulse level
脉冲电视热成像 pulse video thermography
脉冲电位门 impulse potential gate
脉冲电吸铁 electric(al) pulse magnet
脉冲电压 impulse voltage; pulse voltage; shock voltage; surge voltage
脉冲电压表 pulsing voltmeter
脉冲电压发生器 impulse voltage generator
脉冲电压分压器 pulse voltage divider
脉冲电压记录器 klydonograph; pulse voltage recorder
脉冲电压记录图 clydonogram[klydonogram]
脉冲电压摄测仪 clydonograph
脉冲电压示波器 impulse voltage oscilloscope
脉冲电压试验 impulse voltage test(ing)
脉冲电压特性 impulse voltage characteristics
脉冲电压显示照片 clydonogram
脉冲电源 impulsing power source; kickback type of supply; pulse power source
脉冲电晕放电 pulsed corona discharge
脉冲电子俘获检测器 pulsed electron capture detector
脉冲电子管振荡器 pulse tube oscillator
脉冲电子加速器 pulsed electron accelerator
脉冲电子静电加速器 pulsed electron electrostatic accelerator
脉冲电子碰撞离子源 pulsed electron impaction source
脉冲电阻 pulse resistance
脉冲叠加 pulse superposition
脉冲顶部 pulse top
脉冲顶部补偿 pulse top compensation
脉冲顶部倾斜 pulse droop; pulse top droop
脉冲顶部限幅器 pulse top clipper
脉冲顶点 pulse apex
脉冲顶峰 pulse crest
脉冲定标器 impulse recorder; impulse scaler; pulse counter
脉冲定深器 pinger
脉冲定时 pulse timing
脉冲定时多路传输 pulse time multiplex
脉冲定时器 impulse timer; pulse timer
脉冲定相 pulse phasing
脉冲定相装置 pulse shifter
脉冲动量 pulsed momentum
脉冲动量关系 impulse momentum relationship
脉冲抖动 pulse jitter
脉冲陡度 pulse steepness
脉冲断开 pulse switch out
脉冲断路器 impulse circuit breaker
脉冲断路信号 pulsed off signal
脉冲断路信号发生器 pulse(d) off signal generator
脉冲断续器 pulse chopper
脉冲煅烧 impulse firing
脉冲堆 pulsed reactor
脉冲堆积 pulse pile-up
脉冲堆积器 pulse stacker
脉冲堆积效应 pile-up effect
脉冲对称相位调制 pulse symmetrical phase modulation
脉冲对条信号比 pulse to bar ratio
脉冲钝化 pulse despiking
脉冲钝化电路 despiker circuit
脉冲多路编码 pulse multiplex coding
脉冲多路式测距装置 pulse multiplex distance-measuring equipment
脉冲多路通信[讯]制 impulse time division system
脉冲多频发送 multifrequency pulsing
脉冲多普勒跟踪雷达 pulsed Doppler tracker
脉冲多普勒激光雷达 pulsed Doppler lidar
脉冲多普勒监视雷达 pulse Doppler surveillance radar
脉冲多普勒雷达 pulse Doppler radar
脉冲多普勒雷达跟踪 pulsed Doppler tracking
脉冲多普勒频谱 pulse Doppler spectrum
脉冲多普勒搜索与捕捉雷达 pulse Doppler search and acquisition radar
脉冲多普勒系统 pulse Doppler system
脉冲多普勒预警雷达 pulse Doppler early warning radar
脉冲多普勒杂波 pulse Doppler clutter
脉冲多腔速调管 pulsed multi-cavity klystron
脉冲二重性 pulse ambiguity
脉冲发电机 impulse generator
脉冲发电器 surge generator
脉冲发码器 pulse coder
脉冲发射 impulse transmission; pulse emission
脉冲发射管 impulse transmitter tube; impulse transmitting tube
脉冲发射机 impact transmitter; impulse transmitter; pulse transmitter
脉冲发射器 pulse emitter
脉冲发射器负载 pulse emitter load
脉冲发生 pulse generating
脉冲发生的选择 quantization
脉冲发生电路 pulse generating circuit
脉冲发生器 discharge impulse oscillator; impulsator; impulse deviser; impulse generator; pulse generating means; pulse generator; pulse machine; pulser
脉冲发生器架 pulser rack
脉冲发生器时标 pulse generator clock
脉冲发生器系统 pulse generator system
脉冲发生器圆盘 pulse generator disc
脉冲发送 pulse sending
脉冲发送机 impulse machine; impulse sending machine; pulse transmitter
脉冲发送继电器 impulse transmitting relay
脉冲发送结束信号 end of pulsing
脉冲发送起始信号 start of pulsing
脉冲发送器 impulse machine; impulser; impulse sender; impulse transmitter; pulse transmitter
脉冲阀 impulse valve; pulse valve
脉冲法 impulse method; pulse method; pulse wise
脉冲法电波探测 impulse method electric(al) finding
脉冲法校准 pulse method calibration
脉冲反吹袋式除尘器 jet type dust collector
脉冲反控制 revertive control of impulsing
脉冲反射 impulse reflection; pulse echo; pulse reflection
脉冲反射超声图 pulse echo ultrasonogram
脉冲反射法 impulse reflection process; pulse echo method
脉冲反射器 trigger reflector
脉冲反射式故障定位器 pulse reflection fault locator
脉冲反射响应 impact-echo response
脉冲反射原理 pulse reflection principle
脉冲反应堆 pulsed reactor
脉冲反应函数 impulse response function
脉冲反应器 pulse type reactor
脉冲反褶积 pulse deconvolution; spike deconvolution; spiking deconvolution
脉冲范围 pulsating sphere; pulse regime
脉冲方程式 impulse equation
脉冲方式 pulse mode
脉冲放大 amplification of pulse; pulse amplification
脉冲放大分析仪 amplifier-pulse analyser[analyzer]
脉冲放大管 pulse amplitude
脉冲放大器 pip amplifier; pulse amplifier
脉冲放电 impulsive discharge; pulse discharge
脉冲放电灯 pulsed discharge lamp
脉冲放电管 pulsed discharge tube
脉冲放电器 surge gap
脉冲放射源 pulsed radioactive source
脉冲分辨(能)力 pulse resolution
脉冲分层全息照相术 pulsed sandwich holography
脉冲分极器 pulsator classifier
脉冲分离 pulse separation
脉冲分离器 impulse separator; pulse separator
脉冲分离装置 pulse separator; pulse separator unit
脉冲分配放大器 pulse distribution amplifier
脉冲分配器 pulse distributor; pulse divider
脉冲分频 countdown; counting down; pulse frequency dividing
脉冲分频电路 step divider circuit
脉冲分频计数器 impulse recorder; impulse scaler; pulse counter
脉冲分频器 counter-down; pulse divider; pulse frequency divider; scaler
脉冲分散 split separation of impulses
脉冲分析 pulse analysis
脉冲分析仪 impulse analyser[analyzer]; pulse analyser[analyzer]
脉冲分压器 pulsed attenuator
脉冲封口 impulse sealing
脉冲封锁 pulse inhibit
脉冲峰高 pulse peak height
脉冲峰功率 pulse peak power
脉冲峰化 peaking; pulse peaking
脉冲峰宽度 pulse peak width
脉冲峰压 peak pulse voltage
脉冲峰值 impulse peak; pulse peak
脉冲峰值安培计 surge-crest ammeter
脉冲峰值电压 peak impulse voltage; peak pulse voltage
脉冲峰值幅度 peak pulse amplitude
脉冲峰值功率 impulse peak power; peak output of pulse; peak pulse power; pulse peak power
脉冲峰值检波器 pulse peak detector
脉冲浮标 impulse buoy
脉冲符合 pulse coincidence
脉冲符合计数器 pulse coincidence detector
脉冲幅变分析 pulse amplitude analysis
脉冲幅度 impulse amplitude; pulse amplitude; pulse height

M

脉冲幅度编码调制遥测技术 pulse amplitude code modulation telemetry
脉冲幅度分布 pulse amplitude distribution
脉冲幅度分布曲线 distribution curve of pulse amplitude
脉冲幅度分析 kicksort(ing);pulse amplitude analysis
脉冲幅度分析器 kicksorter;pulse amplitude analyser[analyzer]; pulse height analyser[analyzer]
脉冲幅度鉴别器 pulse height discriminator
脉冲幅度-时间变换器 pulse amplitude-to-time converter
脉冲幅度调制 pulse amplitude modulation
脉冲幅度调制载波 pulse amplitude-modulated carrier
脉冲幅度限 cut-off pulse height
脉冲幅度选择器 pulse height selector
脉冲幅度甄别器 pulse amplitude discriminator
脉冲幅分析 pulse amplitude analysis
脉冲幅谱 pulse amplitude spectrum
脉冲幅值 cut-off amplitude; cut-off pulse height;pulse height
脉冲辐射 impulse radiation;pulse radiation
脉冲辐照 pulse irradiation
脉冲负荷 intermittent load(ing)
脉冲负压空气喷射净化器 pulse air induction reactor
脉冲负载 pulse load(ing)
脉冲复位 pulsed reset
脉冲复制器 pulse duplicator
脉冲副载波 pulse subcarrier
脉冲傅立叶转换 pulsed Fourier transform
脉冲改良限度 pulse improvement threshold
脉冲干扰 impulse interference;impulsive interference; impulsive noise; pulse interference
脉冲干扰分离器 pulse interference separator
脉冲干扰分离器和消除器 pulse interference separator and blanker
脉冲干扰机 impulse jammer
脉冲干扰限制器 spot limiter
脉冲干扰消除器 pulse interference eliminator
脉冲干扰消隐器 pulse interference blanker
脉冲干扰仪 pulse interferometer
脉冲干燥 impulse drying
脉冲感应 pulse induction
脉冲缸 impulse cylinder
脉冲高度 pulse height
脉冲高度测定 pulse height measurement
脉冲高度分辨率 pulse height resolution
脉冲高度分布 pulse height distribution
脉冲高度分析 pulse height analysis
脉冲高度分析器 kicksorter; multichannel analyser[analyzer]; pulse height analyser[analyzer]
脉冲高度分析仪 kicksorter; multichannel analyser[analyzer]; pulse height analyser[analyzer]
脉冲高度谱 pulse height spectrum
脉冲高度选择器 pulse height selector
脉冲高频机 pulse high frequency apparatus
脉冲高温分解 pulse pyrolysis
脉冲高压发生器 pulsed high-voltage generator
脉冲高压放电 pulsed high-voltage discharge
脉冲给定调节器 pulse setting controller
脉冲跟踪电路自锁 automatic block with impulse track circuits
脉冲跟踪器 pulse follower
脉冲跟踪系统 pulse tracking system
脉冲工作方式 pulsed mode
脉冲功率 impulse power;pulse power
脉冲功率击穿 pulse power breakdown
脉冲功率激励器 pulse power driver
脉冲功率输出 pulse power output
脉冲功率行波管 pulsed power travelling-wave tube
脉冲功率源 pulse power supply
脉冲供电周期 pulse powering cycle
脉冲共振 pulse resonance
脉冲共振法 pulse resonance method
脉冲共振光谱学 pulsed resonance spectroscopy
脉冲谷 pulse valley
脉冲固相输送 pulse phase conveying
脉冲故障指示器 pulse fault indicator
脉冲管 pulsatron; pulse line; pulse tube;pulse valve
脉冲管制冷机 pulse tube refrigerator
脉冲管制冷器 pulse tube refrigerator
脉冲惯性 impulse inertia
脉冲光 pulsed light
脉冲光电倍增管 pulsed photomultiplier
脉冲光电倍增器 pulsed photomultiplier
脉冲光发生器 pulsed light generator
脉冲光束云高计 pulsed light ceilometer;pulsed light cloud height indicator
脉冲光线测距仪 pulsed light range finder
脉冲光源 flashing light;light pulse generator
脉冲轨道电路 half-wave track circuit;impulse track circuit;pulsating track circuit; pulsed track circuit; pulse type track circuit; transient track circuit
脉冲轨道电路电流 impulse track circuit current
脉冲轨道继电器 code following track relay
脉冲过程 pulse process
脉冲过渡函数 impulse transition function
脉冲含角区 pulse packet
脉冲函数 Dirac beta function; Dirac delta function; impulse function; impulsive function
脉冲函数发生器 impulse function generator
脉冲焊 impulse welding; pulsation welding
脉冲焊机 pulsating welder
脉冲焊接 energy storage welding;impulse sealing;impulse welding;shot weld(ing)
脉冲焊接断续器 pulse welding interrupter
脉冲焊接计时器 pulsation welding timer;pulse welding timer
脉冲航电仪 pulse airborne electronic magnetic system
脉冲耗散 pulse dissipation
脉冲合成 pulse synthesis
脉冲合成器 pulse synthesizer
脉冲和平滑干扰 peak and flat noise
脉冲核磁共振分光计 pulsed nuclear magnetic resonance spectrometer
脉冲核磁共振分光仪 pulsed nuclear magnetic resonance spectrometer
脉冲核反应 pulsed nuclear reaction
脉冲荷电除尘器 pulse charging precipitator
脉冲荷载 impulse load(ing);impulsive load(ing); pulsating load(ing);pulse load(ing)
脉冲荷载强度 strength at pulsating load
脉冲轰击 pulsed bombardment
脉冲红宝石光学激光器 pulsed ruby optic(al)laser
脉冲红宝石激光器 pulsed ruby laser; pulsed ruby optic(al)laser
脉冲后的尖头信号 tail
脉冲后沿 back edge of impulse;lagging edge; pulse back edge; pulse trailing edge
脉冲后沿持续时间 pulse decay time
脉冲后沿时间 pulse trailing-edge time; trailing-edge pulse time
脉冲后沿斩断 pulse tail clipping
脉冲弧焊电源 pulsed arc-welding power source
脉冲弧焊整流器 pulsed arc-welding rectifier
脉冲滑翔道 pulsed glide path
脉冲滑翔道显示图形 pulse glide display
脉冲环形放电 pulsed ring discharge
脉冲簧片闭合 impulse springs make
脉冲恢复 pulse regeneration
脉冲回波 pulse echo
脉冲回波测试仪 pulse echo tester
脉冲回波超声学 pulse echo ultrasonics
脉冲回波法 pulse echo method
脉冲回波接收器 pulse echo receiver
脉冲回旋加速器 pulse(d)(operated) cyclotron
脉冲混合 pulse mixing
脉冲混合澄清器 pulsed mixer-settler
脉冲混频器 pulse mixer
脉冲击穿电压 pulse puncture voltage
脉冲击穿功率 pulse power breakdown
脉冲积分法 pulse integration
脉冲积分检波器 pulse integrating detector
脉冲积分器 pip integrator;pulse integrator
脉冲积分选择器 integrating divider; pulse integrating divider
脉冲积算器 impulse meter
脉冲基线 pulse base
脉冲畸变 pulse distortion
脉冲激发 impulse excitation;pulse excitation
脉冲激发的 pulsed
脉冲激光 pulsed laser
脉冲激光测距仪 pulsed laser rangefinder
脉冲激光放大器 impulse laser amplifier;pulsed laser amplifier
脉冲激光功率 pulsed laser power
脉冲激光光束 pulsed laser beam
脉冲激光焊机 pulsed laser welder
脉冲激光焊接 pulsed laser bonding; pulsed laser welding
脉冲激光加工 pulsed laser processing
脉冲激光器 pulsed laser;pulsing laser
脉冲激光器微功率探测仪 micropower detector for pulsed laser
脉冲激光全息术 pulsed laser holography
脉冲激光全息图 pulsed laser hologram
脉冲激光全息照相 pulsed laser hologram
脉冲激光束 pulse laser beam
脉冲激光外延 pulse laser induced epitaxy
脉冲激光系统 pulse laser system
脉冲激光信标 pulsed laser beacon
脉冲激光照明器 pulsed laser illuminator
脉冲激光振荡 pulsed laser action
脉冲激光指向标 pulsed laser beacon
脉冲激光作用 pulsed laser action
脉冲激弧器 pulsed arc-exciter
脉冲激励 impulse excitation; impulsing;pulse excitation
脉冲激励发射机 impact excited transmitter
脉冲激励器 impulse exciter
脉冲激励天线 pulse excited antenna
脉冲激励振荡器 impulse excited oscillator
脉冲激能集尘器 electrostatic precipitator with pulse energization;pulse energized electrostatic precipitator
脉冲激振 impulsiving
脉冲极化器 pulse polarizer
脉冲极谱 impulse polarography
脉冲极谱测定(法)pulse polarographic determination
脉冲极谱法 pulse polarography
脉冲极谱仪 pulse polarograph
脉冲极性 pulse polarity
脉冲极性选择器 polaraulic selector; polarity selector; pulse polaraulic selector;pulse polarity selector
脉冲急速相 impulsive hard phase
脉冲计 pulse meter; pulsimeter[pulsometer]
脉冲计时 pulse timing
脉冲计时器 impulse timer;pulse timer
脉冲计数 pulse counting;step-by-step counting
脉冲计数编码器 pulse count encoder
脉冲计数电路 pulse scaling circuit; scaling circuit
脉冲计数方法 pulse counting method
脉冲计数光度计 pulse counting photometer
脉冲计数换算器 pulse count converter
脉冲计数检测器 pulse counter detector
脉冲计数鉴别器 pulse counting discriminator
脉冲计数模件 pulse counting module
脉冲计数器 electronic counter; impulse counter; impulse register; pulse counter;scaler;impulse meter
脉冲计数式检波器 pulse counter detector
脉冲计数式鉴频器 pulse counter discriminator
脉冲计数调制 pulse count modulation
脉冲计数系统 pulse counting system
脉冲计数制 pulse counting system
脉冲计数装置 pulse counting equipment
脉冲计数组件 pulse counting module
脉冲记录 pulse recording
脉冲记录器 impulse recorder;impulse reporter
脉冲记录线路 pulse registering device
脉冲记录纸 impulse recording paper
脉冲技术 impulse technique; pulse technique
脉冲剂量 pulsed dosage
脉冲继电器 code pulse relay;code responsive relay; impulse relay; impulsing relay;pulsating relay;pulse relay;pulse train relay;stepping relay
脉冲寄存器 impulse register
脉冲加倍 pulse doubling
脉冲加宽 pulse widening
脉冲加宽电路 pulse-stretching circuit

脉冲加速 impulsive acceleration
脉冲加速器 pulsatron;pulsed accelerator
脉冲加压型喷墨头 on-demand ink gun
脉冲夹层全息图 pulsed sandwich hologram
脉冲尖峰 spike
脉冲尖峰幅度 pulse spike amplitude
脉冲尖峰信号 pulse spike signal
脉冲间的 interpulse
脉冲间的相互关系 pulse to-pulse correlation
脉冲间隔 peak separation;pulse interval;pulse separation;pulse spacing;pulse time
脉冲间隔编码 pulse spacing code;pulse spacing coding
脉冲间隔时间 intrapulse time;pulse interval time;pulse time
脉冲间隔调制 pulse interval modulation;pulse spacing modulation
脉冲间隔跳动 pulse interval jitter
脉冲间隔校核 pulse spacing check
脉冲间距 peak separate;pulse distance;pulse spacer;pulse spacing
脉冲间相关性 pulse to-pulse correlation
脉冲间歇时间 interpulse period
脉冲监控器 pulse monitor
脉冲监听序贯法 ping-listen-train method
脉冲剪切机 impulse cutting machine
脉冲检波 pulse detection
脉冲检波器 pulse detector
脉冲检测器 pulse detector
脉冲建立时间 pulse build-up time
脉冲鉴别器 pulse discriminator
脉冲鉴频器 pulse frequency discriminator
脉冲键控 pulsing key
脉冲键控器 pulse keyer
脉冲交错 pulse interlacing
脉冲交织 pulse interlacing;pulse interleave;pulse interleaving
脉冲校正 impulse correction
脉冲校正器 impulse corrector;pulse corrector
脉冲校正系数 pulsating correction factor
脉冲接插件 pulse connector;surge connector
脉冲接触焊 impulse welding;pulsating weld
脉冲接点 impulse contact
脉冲接收 pulse receiving;reception of impulse
脉冲接收机 pulse receiver
脉冲接收继电器 impulse accepting relay
脉冲接收器 pulse receiver
脉冲解码器 constant-delay discriminator;pulse code demodulator;pulse demoder
脉冲解调 pulse demodulation
脉冲解调器 pulse demodulator
脉冲解吸分析 pulse stripping analysis
脉冲介质 pulsating medium
脉冲金属蒸气激光器 pulsed metal vapo(u)r laser
脉冲近似值 impulse approximation
脉冲禁止 pulse inhibit
脉冲精密声级计 impulse precision sound level meter
脉冲静电偏转板 pulsed electrostatic deflector
脉冲矩阵隔离 pulsed matrix isolation
脉冲距离制 pulse distance system
脉冲聚变系统 pulsed fusion system
脉冲聚积 pulse pile-up
脉冲聚焦 pulse concentration
脉冲聚焦飞行时间质谱仪 impulse focused time-of-flight mass spectrometer
脉冲均衡器 pulse equalizer
脉冲均调 pulse matching
脉冲开关 currency-impulse switch;impulse breaker;impulse switch;pulse switch
脉冲开关光电路 pulse switching optic(al) circuit
脉冲可见度 pulse visibility
脉冲可见率 pulse visibility factor
脉冲刻度校准 matching of pulses;pip matching
脉冲空化 pulse cavitation
脉冲空气 pulsating air;pulse air;control air <气动调节系统>
脉冲空气清洗 pulse air cleaning
脉冲空气振动 pulse air vibration
脉冲控制 impulse control;pulse control
脉冲控制电子管振荡器 pulse controlled tube oscillator
脉冲控制调制 pulse control modulation
脉冲控制发生器 pulse controlled generator
脉冲控制键 pulsing key
脉冲控制器 impulse controller;pulse controller
脉冲控制器具 pulse control device
脉冲控制系统 pulse control system
脉冲控制振荡器 impulse governed oscillator
脉冲块 pulse packet
脉冲宽度 impulse duration;impulse width;pulse duration;pulse length;pulse persistence;pulse width
脉冲宽度编码 pulse width coding;pulse width encoding
脉冲宽度编码器 pulse duration coder;pulse width encoder
脉冲宽度波动 pulse time jitter
脉冲宽度多路调制器 pulse width multiplexer
脉冲宽度法 pulse width method
脉冲宽度鉴别器 pulse duration discriminator;pulse width discriminator
脉冲宽度键控器 pulse width keyer
脉冲宽度可控多谐振荡器 pulse width mulitivibrator
脉冲宽度控制 pulse duration control;pulse width control
脉冲宽度调整 pulse width control
脉冲宽度调制 length modulation;pulse duration modulation;pulse length modulation;pulse width modulation
脉冲宽度调制的斩波器 pulse width-modulated chopper
脉冲宽度调制功率信号 pulse width-modulated power signal
脉冲宽度调制器 pulse width modulator
脉冲宽度调制系统 variable duration impulse system
脉冲宽度调制原理 pulse width modulation principle
脉冲宽度误差 pulse length error
脉冲宽度显示器 pulse widescope
脉冲宽度选择器 pulse width selector
脉冲宽度译码器 pulse width decoder
脉冲宽度制 pulse duration system
脉冲宽度主控多谐振荡器 pulse width multivibrator
脉冲馈给 pulsed feeding
脉冲扩散 pulsed diffusion
脉冲扩展器 pulse extender;pulse stretcher
脉冲扩张电路 pulse-stretching circuit
脉冲拉伸疲劳极限 endurance limit under pulsating tension
脉冲拉伸应力范围 range for pulsating tensile stresses
脉冲雷达 pulsed radar impulse radar
脉冲雷达测距系统 pulse radar distance-measuring system
脉冲雷达发射机 impulse radar transmitter;pulse radar transmitter
脉冲雷达高度表 pulse radar altimeter
脉冲雷达设备 pulsed radar set
脉冲雷达系统 pulsed radar system;impulse radar system
脉冲累加器 pulse accumulator
脉冲累加装置 impulse accumulator
脉冲冷却控制系统 pulsed cooling temperature control system
脉冲离子回旋共振波谱仪 pulsed ion cyclotron resonance spectrometer
脉冲离子束 pulsed ionizing beam
脉冲力 impulsive force;pulsating force;surging force
脉冲连接器 pulse connector
脉冲连续波 pulsed continuous wave
脉冲涟波 pulse ripple
脉冲链 impulse train;pulse packet;pulse string;pulse train;train of pulses
脉冲链发生器 pulse train generator
脉冲链路增音机 pulse link repeater
脉冲链路转发器 pulse link repeater
脉冲链重发器 pulse link repeater
脉冲量化器 scrambler
脉冲量离散 spread in sizes
脉冲量热法 pulse calorimetry
脉冲列 pulse group;pulse train;train of pulses
脉冲裂变箱 pulse fission chamber
脉冲流 surging current;surging flow
脉冲流特性 pulse flow performance
脉冲滤波 pulse bandwidth filtering
脉冲滤波器 discontinuous filter;pulsed filter
脉冲率 pulsation rate
脉冲率表 pulse rate metre[meter]
脉冲轮计 impulse wheel meter
脉冲轮廓 pulse profile
脉冲逻辑系统 pulse logic system
脉冲螺线管 impulse solenoid
脉冲马达 impulse motor
脉冲码 impulse code
脉冲码接收器 pulse code receiver
脉冲码解调器 pulse code demodulator
脉冲码速调整 pulse justification
脉冲码调制传输测试仪 pulse code modulation transmission tester
脉冲码调制器 pulse code modulator
脉冲码系统 pulse code system
脉冲门 pulse gate
脉冲密度 impulse density;pulse density
脉冲密度调制 pulse density modulation;pulse number modulation
脉冲面 impulse front
脉冲描记法 pulse tracing
脉冲模糊性 pulse ambiguity
脉冲目标捕捉雷达 pulse target acquisition radar
脉冲目标图 impulse target diagram
脉冲能量 pulse energy
脉冲能量技术 pulsed power technology
脉冲能谱学 pulse spectrometry
脉冲凝胶电泳 pulsed field gel electrophoresis
脉冲扭转疲劳极限 endurance limit under pulsating torsion
脉冲偶极 pulsed dipole
脉冲偶极子 spike doublet
脉冲耦合 pulse coupling
脉冲排除器 pulse rejector
脉冲盘柱 pulsed plate column
脉冲喷吹袋式除尘器 pulse-jet baghouse
脉冲喷气式袋集尘器 pulse-jet filter
脉冲喷射电弧 pulsed spray arc
脉冲喷射过渡 pulsed spray transfer
脉冲喷射器 pulsed ejector
脉冲疲劳机 pulsator;pulsto-fatigue machine
脉冲疲劳试验机 pulse fatigue machine
脉冲匹配 matching of pulses
脉冲匹配法 pulse matching technique
脉冲偏压 pulsed bias
脉冲偏压法 pulsed biasing
脉冲偏压式辐帖康 pulsed biased photicon
脉冲偏压式移像光电摄像管 pulse biased photicon
脉冲偏置法 pulse biasing
脉冲偏转 pulsed deflection
脉冲偏转板 pulsed deflector
脉冲偏转磁铁 pulsed bending magnet
脉冲偏转器 pulsed inflector
脉冲偏转器电压 pulsed inflector voltage
脉冲频带 pulse band
脉冲频带宽度 impulse bandwidth;pulse bandwidth
脉冲频率 impulse frequency;pulse frequency
脉冲频率分除法 counting down
脉冲频率核磁共振流量计 pulse frequency nuclear magnetic resonance flowmeter
脉冲频率计 counting rate meter
脉冲频率控制 pulse frequency control
脉冲频率时间编码 frequency-time coding of pulse
脉冲频率特性 impulse frequency response
脉冲频率调制 pulse frequency modulating;pulse repetition-rate modulation
脉冲频率遥测 impulse frequency telemetering
脉冲频率远距离测量系统 pulse-frequency telemetering system
脉冲频率制 impulse frequency system
脉冲频谱 pulse frequency spectrum;pulse spectrum;spectrum of spike
脉冲频谱包络 pulse spectrum envelope
脉冲频谱带宽 pulse spectrum bandwidth
脉冲频闪灯 pulsed frequent flash lamp
脉冲平滑性偏移 deviation from pulse flatness
脉冲平均电路 pulse averaging circuit
脉冲平均幅度 average pulse amplitude
脉冲平均轮廓 mean pulse profile
脉冲平均时间 pulse average time
脉冲平均输出功率 average output power of pulse
脉冲谱 pulse spectrum
脉冲起动 impulse starting
脉冲起动器 impulse starter
脉冲起动转子 pulse initiating rotor
脉冲起始稳定性 pulse starting stability
脉冲气动激光器 pulsed gas dynamic(al)laser
脉冲气体断路器 pulse gas shutter
脉冲气体分析仪 pulsed gas analyser[analyzer]
脉冲气体激光器 pulsed gas laser
脉冲气体清洗 pulsed gas cleaning

M

脉冲器 pulsating column; pulsator; pulser
脉冲铅粉机 lead powder impulse filler
脉冲前沿 pulse front edge; rising edge
脉冲前沿陡度 width of transition steepness
脉冲前沿上升时间 leading edge pulse time
脉冲前沿时间 leading edge pulse time; pulse leading edge time; pulse rise time
脉冲前沿校正 pulse rise time correction; rise-time correction
脉冲强磁装置 pulse strong magnetic assembly
脉冲强度 pulsed intensity; pulse strength
脉冲强流加速器 pulsed strong-current accelerator
脉冲强迫响应 impulse-forced response
脉冲切断式信号发生器 pulsed offsignal generator
脉冲氢闸流管 pulsed hydrogen thyratron
脉冲倾斜 pulse tilt
脉冲驱动 pulsed drive
脉冲驱动的机心 impulse movement core
脉冲驱动器 pulse driver
脉冲驱动(时)钟 impulse driven clock
脉冲驱动循环反应器 pulse-driven loop reactor
脉冲取样 pulse sampling
脉冲取样和保持电路 pulse sample-and-hold circuit
脉冲取样机 pulsating sampler
脉冲取样技术 pulse sampling technique
脉冲取样示波器 pulse sampling oscilloscope
脉冲全息图 pulse hologram
脉冲群 impulse train; pulse burst; pulse group; pulse packet; pulse train; train of pulses
脉冲群波形 burst waveform
脉冲群振荡器 grouped pulse generator
脉冲燃料棒 pulse fuel rod
脉冲燃烧 pulsating combustion; pulse combustion; resonant combustion
脉冲染料激光密度测定法 pulsed dye laser densitometry
脉冲染料技术 pulsed dye technique
脉冲扰动 pulse type disturbance
脉冲热封机 impulse sealer
脉冲热焊 thermal impulse welding
脉冲热合 impulse sealing; pulse sealing
脉冲热解单元 pulsed pyrolysis unit
脉冲锐化 pulse sharpening
脉冲锐化电路 narrowing circuit
脉冲塞入 pulse stuffing
脉冲塞入技术 pulse stuffing technique
脉冲三轴试验 pulse triaxial test
脉冲散射计 pulse scatterometer
脉冲色散 pulse dispersion
脉冲筛板抽提塔 pulsed sieve-plate extraction column
脉冲筛板萃取柱 pulsed sieve-plate extraction column
脉冲筛板塔 pulsed perforated-plate tower; pulsed sieve-plate column
脉冲筛板柱 pulsed sieve-plate column
脉冲闪光气体放电灯 pulsed flash gas discharge lamp
脉冲闪络 impulse sparkover
脉冲闪络电压 impulse flashover voltage
脉冲上升边 pulse rising edge
脉冲上升率 pulse rate of rise
脉冲上升时间 pulse rise time
脉冲上升延迟 pulse rise delay
脉冲烧成 impulse firing; pulsating firing
脉冲烧结 pulse sintering
脉冲烧嘴 impulse burner
脉冲设备 pulsing unit
脉冲射电源 pulsating radio source
脉冲射解 pulse radiolysis
脉冲射解作用 pulse radiolysis
脉冲射流电弧焊 pulsed spray arc welding
脉冲射频 pulse radiation frequency
脉冲摄谱仪 pulse spectrograph
脉冲伸张管 pulse stretcher
脉冲伸张器 impulse stretcher; pulse stretcher
脉冲声 impulsive sound
脉冲声波 pulsed sound; pulsed sound wave
脉冲声方程 impulsive sound equation
脉冲声级 pulsed sound level
脉冲声级计 impulse sound level meter
脉冲声呐 pinging sonar; pulse sonar
脉冲失真 faulty selection; pulse distortion
脉冲时标发生器 pulse timing marker oscillator
脉冲时标振荡器 pulse timing marker oscillator
脉冲时差双曲线导航系统 long-range navigation system
脉冲时分多路复用 pulse time division multiplex
脉冲时间 burst length; pulse duration; pulse time
脉冲时间分布 pulse time distribution
脉冲时间计数器 pulse duration counter
脉冲时间继电器 impulse timer
脉冲时间鉴别器 pulse duration discriminator
脉冲时间解调器 pulse time demodulator
脉冲时间调制 pulse time modulation
脉冲时间调制(无线电)探空仪 pulse time-modulated radiosonde
脉冲时间系统 impulse duration system
脉冲时间裕度 impulse time margin
脉冲时码 pulse time code
脉冲时延 impulse time delay
脉冲时钟 impulse clock
脉冲时钟发生器 pulse clock generator
脉冲实验 pulse experiment
脉冲示波器 impulse oscilloscope; impulse oscilograph; impulsing oscilloscope; pulscope; pulse oscilloscope
脉冲示波术 surge oscillography
脉冲式 impulse type
脉冲式拨号 pulse dialing
脉冲式测高计 pulse type altimeter
脉冲式超声波探伤器 soniscope
脉冲式超声波探伤仪 soniscope
脉冲式低压紫外线治疗机 low-pressure pulse ultraviolet treatment unit
脉冲式电介质激光器 pulsed dielectric(al) laser
脉冲式电压调整器 impulse type voltage regulator
脉冲式电钟 pulse type electric(al) clock
脉冲式多路传输 pulse mode multiplex(ing)
脉冲式多普勒系统 pulse Doppler
脉冲式多普勒仪 pulse wave Doppler apparatus
脉冲式多普勒制 pulsed Doppler system
脉冲式放电 pulsed discharge
脉冲式红宝石激光器 pulsed ruby laser
脉冲式回旋加速器 pulsed cyclotron
脉冲式混凝土检测示波仪 impulsive concretescope
脉冲式机载雷达 pulse type airborne radar
脉冲式继电器 impulse type relay
脉冲式空气输送 pulsated air conveying
脉冲式雷达 pulse radar
脉冲式雷达测高计 pulsed radar altimeter
脉冲式雷达测高仪 pulsed radar altimeter
脉冲式列车调度电话 pulse type train dispatching telephone
脉冲式喷气发动机 aeropulse engine; aerosonator; pulse jet; pulse-jet engine
脉冲式喷头 pulse sprinkle head
脉冲式气动量仪 air pulse; air pulse ga(u)ge
脉冲式气力输送 air pulse conveying
脉冲式气流干燥器 pulsed pneumatic drier[dryer]
脉冲式燃气流 Buchi system of gas flow
脉冲式伤口清洗器 pulse wound cleaner
脉冲式手持吸铁器 pulse hand magnet
脉冲式疏水器 impulse trap; steam trap, impulse type trap
脉冲式速度计 pulse type speedometer
脉冲式调压器 impulse type voltage regulator
脉冲式通风机 impulsion fan
脉冲式涡轮增压 pulse system turbocharging
脉冲式相关器 pulse type correlator
脉冲式遥测 pulse type telemetering
脉冲式遥测计 impulse type telemeter; pulse type telemeter
脉冲式遥测系统 impulse type telemetering system
脉冲式遥测仪 impulse type telemeter
脉冲式遥测装置 impulse type telemeter
脉冲式振荡 pulsative oscillation
脉冲式蒸汽疏水器 impulse steam trap
脉冲式直流电动机 pulse direct current motor
脉冲式转速计 impulse tachometer
脉冲试验 impulsive test; pulse test(ing)
脉冲试验机 impulse testing machine; pulsator
脉冲试验仪 surge tester
脉冲室 counter chamber; pulse chamber
脉冲收发两用机 transponder
脉冲输出 pulse output
脉冲输出电压 pulse output voltage
脉冲输出放大器 impulse type output amplifier
脉冲输出功率 pulse output power
脉冲输出能量 output energy of pulse
脉冲输入 impulse input; pulse input
脉冲输入异步网络 pulse input asynchronous network
脉冲束 pulsed beam
脉冲束流 pulsed beam current
脉冲束流输送系统 pulsed beam transport system
脉冲数 pulse count
脉冲数调制 pulse number modulation
脉冲数位 pulse digit
脉冲数字间隔 pulse digit spacing
脉冲衰变 pulse decay
脉冲衰减 pulse(d) attenuation; pulse droop
脉冲衰减器 pulse attenuator; pulsed attenuator
脉冲衰减时间 pulse decay time; pulse fall time
脉冲双星 binary pulsar
脉冲水 pulsating water
脉冲水解 impulse hydrolysis
脉冲水力喷射(器) pulsed water jet
脉冲瞬变 pulse ringing
脉冲瞬间热封机 impulse heat sealer
脉冲伺服机构 pulse servo(mechanism)
脉冲伺服系统 pulse servosystem; pulse system
脉冲搜索雷达的探测范围 pulse acquisition radar coverage
脉冲速度 impulse speed; pulsating speed; pulse velocity
脉冲速度测定 pulse velocity determination; pulse velocity measurement
脉冲速度界限曲线图 impulse target diagram
脉冲速度试验 pulse velocity test
脉冲速率 pulse speed; pulsing rate
脉冲速率乘法器 pulse rate multiplier
脉冲速调管 pulsed klystorn
脉冲随动系统 pulsed servo(system); sampling servosystem
脉冲缩短 pulse shortening
脉冲锁相环 impulse phase-locked loop; pulse phase-locked loop
脉冲塔 pulse column
脉冲塔板 pulse column plate
脉冲探测 pulse detection
脉冲探测精确度 pulse finding accuracy
脉冲探测器 pulse detector
脉冲探向 pulse direction finding
脉冲探向器 pulse direction finder
脉冲特性 pulse characteristic
脉冲特性分析仪 impulse response analyser[analyzer]; pulse response analyser[analyzer]
脉冲提前 pulse advancing; pulse lead
脉冲天线 pulse antenna
脉冲填充 pulse stuffing
脉冲填料塔 pulsed packed tower
脉冲填料柱 pulsed packed column
脉冲条信号比 pulse to bar ratio
脉冲调幅 time sampling; timing sampling
脉冲调幅器 pulse amplitude modulator
脉冲调幅式高度表 pulse amplitude-modulated altimeter
脉冲调幅载波 pulsed amplitude-modulated carrier
脉冲调节 pulse modulation
脉冲调节器 impulse controller; impulse regulator
脉冲调频 pulse frequency modulation
脉冲调频信号 pulsed frequency modulated signal
脉冲调频遥测术 pulsed frequency modulation telemetry
脉冲调频载波 pulsed frequency modulated carrier
脉冲调速 chopper speed control
脉冲调相 pulse phase modulation
脉冲调相-调幅制 pulse phase modulation-amplitude modulation system
脉冲调相载波 pulsed phase-modulated carrier
脉冲调整室 sampling chamber
脉冲调制 impulse modulation; pulse modulation; pulsing; quantization of amplitude; sampling
脉冲调制波 pulse modulated wave
脉冲调制传输 pulse modulation transmission

脉冲调制磁控管 pulse modulated magnetron
脉冲调制的 pulse modulated
脉冲调制多路传输 pulse multiplex
脉冲调制多普勒雷达系统 pulsed Doppler radar
脉冲调制发射机 pulse modulation transmitter
脉冲调制法 pulse modulation method
脉冲调制放大器 modulated pulse amplifier
脉冲调制干扰 pulse modulated jamming
脉冲调制管 impulse modulator tube; pulsed modulator tube
脉冲调制红外系统 pulse modulation infrared system
脉冲调制技术 pulse modulation technique
脉冲调制接收机 pulse modulation receiver
脉冲调制控制 pulse modulated control
脉冲调制控制式轨道电路 pulse modulated-control type track circuit
脉冲调制雷达 impulse modulated radar; pulse modulated radar
脉冲调制器 impulsator; pulse modulator; quantizer; sampler
脉冲调制器雷达 pulse modulator radar
脉冲调制射束 sampled beam
脉冲调制声呐 pulse modulated sonar
脉冲调制通信[讯]系统 pulse modulated communications system
脉冲调制系统 pulse code modulating system; pulse modulation system
脉冲调制相位 sampling phase
脉冲调制信号 pulse modulated signal
脉冲调制遥测 impulse modulated telemetering
脉冲调制载波 pulsed modulated carrier
脉冲调制振荡器 pulse modulated oscillator
脉冲跳动 pulse jitter
脉冲跳增 overshoot
脉冲停止 pulse dropout
脉冲停止周期 pulse off period
脉冲通道 pulse passage
脉冲通过时间 time of flight of impulse
脉冲通信[讯] impulse signaling; pulse communication
脉冲通信[讯]制 pulse communication system
脉冲同步 pulse synchronization; pulse timing
脉冲同步加速器 pulsed synchrotron
脉冲同步示波器 pulse synchroscope
脉冲同核去耦法 impulse homonuclear decoupling method
脉冲铜蒸气激光器 pulse copper vapour laser
脉冲透平 impulse turbine
脉冲透热机 pulse diathermy apparatus
脉冲透射法 pulse transmission method
脉冲透射运转 pulse transmission mode operation
脉冲凸轮 impulse cam; pulse cam; pulsing cam
脉冲图形 pulse pattern
脉冲退磁 pulse demagnetization
脉冲退火 pulse annealing
脉冲拖尾 pulse stretching; tail of pulse
脉冲拖影 pulse smearing
脉冲脱落 pulse dropout
脉冲外相干多普勒雷达 externally coherent pulse Doppler radar
脉冲网络 pulse network
脉冲网络调制器 pulse network modulator
脉冲微波雷达 pulsed microwave radar
脉冲微分器 pulse differentiator
脉冲微扰 pulsed perturbation
脉冲尾部 pulse tail; tail of pulse
脉冲位调制 pulse displacement modulation
脉冲位置 pulse position
脉冲位置调制 pulse phase modulation; pulse position modulation
脉冲位置调制码 pulse position modulation code
脉冲位置调制器 pulse position modulator
脉冲位置调制系统 pulse position modulation system
脉冲位置指示器 pulse position indicator
脉冲温度调节控制板 pulse temperature adjusting control board
脉冲温度陡度 pulsed temperature gradient
脉冲纹波 pulse ripple
脉冲稳定性 pulse stability
脉冲稳弧焊机 welding set with a surge injector
脉冲稳弧器 pulsed arc stabilizer
脉冲涡流测量仪 pulsed eddy current instrument
脉冲钨极惰性气体保护电弧焊 pulsed tungsten inert gas arc welding
脉冲钨极气体保护焊 pulsed tungsten gas arc welding
脉冲钨极氩弧焊 pulsed tungsten argon arc weld(ing)
脉冲吸附床 pulsed adsorption bed
脉冲吸附床法 pulsed adsorption bed process
脉冲吸附床过程 pulsed adsorption bed process
脉冲吸收 digit absorption; pulse absorption
脉冲吸收的 endomomental; pulse-absorbed
脉冲吸收器 pulse absorber
脉冲吸收选择器 digit-absorbing selector
脉冲系列 impulse train; pulse packet; pulse sequence; pulse train; train of impulses
脉冲系数 impulse ratio; pulse factor
脉冲系统 dynamic(al) system; impulse signal; impulse system; pulse system; pulsing system
脉冲系统仿真 pulse system simulation
脉冲系统微分 pulse system differential
脉冲系统微分分析器 pulse system differential analyser[analyzer]
脉冲下降边 pulse trailing edge
脉冲下降时间 pulse decay time; pulse fall time
脉冲下降延迟 pulse fall delay
脉冲下降延迟时间 pulse delay fall time
脉冲氙灯 pulsed xenon lamp; pulsed xenon light; xenon flash lamp; xenon repeated pulse flash light
脉冲氙灯光源 xenon flash light source
脉冲显示 pulse display
脉冲线路 impulse circuit
脉冲线路增音机 pulse link repeater
脉冲线圈 pulsed coil
脉冲限幅器 pulse slicer
脉冲限制 pulse clipping
脉冲限制电路 clamper circuit; clamping circuit
脉冲限制级 slicer
脉冲限制率 pulse limiting rate
脉冲限制器 pulse slicer
脉冲相 pulsion phase
脉冲相加 impulse summation
脉冲相位调整 pulse phasing
脉冲相位调制 displacement modulation; pulse phase modulation
脉冲响应 impulse response; impulsive response; pulse response
脉冲响应分析 pulse response analysis
脉冲响应函数 impulse response function
脉冲响应技术 impulse response technique
脉冲响应矩阵 impulse response matrix
脉冲响应宽度 pulse response duration; pulse response width
脉冲响应模 impulse response model
脉冲响应时间 pulse response time
脉冲响应特性 pulse response characteristics
脉冲响应特性曲线 impulse response characteristic curve; pulse response curve
脉冲响应自适应 pulse response adaptive
脉冲削波 pulse cutting
脉冲削波器 pulse clipper
脉冲消除装置 pulsation damper
脉冲消失现象 pulse nulling phenomenon
脉冲协调 pulse matching
脉冲信号 impulse signal; pulse(d) signal; pulsing signal
脉冲信号发生器 impulse signal generator; pulse signal generator
脉冲信号发生器控制的调速器 pulse signal generator governor
脉冲信号群 ensemble of pulses
脉冲信号源 pulse signal source
脉冲信噪比改善阈值 pulse improvement threshold
脉冲星 pulsar
脉冲星接收机 pulsar receiver
脉冲星时标 pulsar time scale
脉冲星时间传递系统 pulsar time transfer system
脉冲星同步器 pulsar synchronizer
脉冲星信号 pulsar signal
脉冲行程 impulse stroke
脉冲形 pulse form
脉冲形成 build-up of pulse; pulse formation
脉冲形成单元脉冲形成部件 pulse forming unit
脉冲形成电路 shaper
脉冲形成电路板 pulse forming panel
脉冲形成放大器 pulse forming amplifier; shaper amplifier
脉冲形成管 impulse forming tube; pulse generating tube
脉冲形成环节 impulse forming link; pulse forming link; pulse shaping link
脉冲形成级 pulse shaping stage
脉冲形成器 impulse shaper; pulse former; pulse shaper
脉冲形成时间 pulse build-up time
脉冲形成网络 pulse forming network
脉冲形成线 pulse forming line; pulse shaping line
脉冲形成线圈 pulse forming coil
脉冲形成延迟线 pulse forming delay line
脉冲形式 impulse form; pulse form; pulse mode
脉冲形状 impulse form; pulsing form
脉冲形状分析器 pulse shape analyser[analyzer]
脉冲形状甄别器 pulse shape discriminator
脉冲型 impulse type; pulse mode
脉冲型电离箱 pulse type ionization chamber
脉冲型多路传输 pulse mode multiplex(ing)
脉冲型荷载 impulse type load(ing)
脉冲型激光二极管 pulse type laser diode
脉冲型探测器 pulse type detector
脉冲型遥测 pulse type telemetering
脉冲型遥测计 pulse type telemeter
脉冲性噪声 pulsive noise
脉冲休止间隔 inter-train pause
脉冲修尖 peaking; pulse peaking
脉冲修尖电路 peaker
脉冲序列 impulse sequence; impulse train; pulse sequence; pulse series; pulse train; train of pulses
脉冲序列发生器 pulse series generator; pulse train generator
脉冲序列分析 pulse train analysis
脉冲序列间歇 impulse train pause
脉冲序列频谱 pulse train frequency spectrum; pulse train spectrum
脉冲序批间歇式反应器 pulsed sequencing batch reactor process
脉冲选叫调度电话 pulse selective call dispatching telephone
脉冲选取 selection pulse
脉冲选通技术 gating technique
脉冲选通系统 pulse gated system
脉冲选择器 impulse selector; pulse selector
脉冲寻址多址连接 pulse address multiple access
脉冲询问 pulse interrogation
脉冲压力 pulsating pressure
脉冲压实 impulse compaction
脉冲压缩 narrowing of pulse; pulse compression
脉冲压缩比 pulse compression ratio
脉冲压缩波 pulsed compression wave
脉冲压缩法 pulse compression method
脉冲压缩技术 pulse compaction technique
脉冲压缩接收机 pulse compressive receiver
脉冲压缩雷达 chirp radar; pulse compression radar
脉冲压缩滤波器 pulse compression filter
脉冲压缩系统 pulse narrowing system
脉冲压缩应力范围 range for pulsating compressive stresses
脉冲亚毫米波激光干涉仪 pulsed submillimeter laser interferometer
脉冲延长器 pulse lengthener
脉冲延迟 impulse delay; pulse delay
脉冲延迟模 pulse delay mode
脉冲延迟器 pulse delay unit
脉冲延迟时间 impulse delay time; pulse delay time
脉冲延迟时间列线图 pulse delay nomogram
脉冲延迟网络 pulse delay network
脉冲延迟线 pulse delay line
脉冲延时器 pulse delay unit
脉冲沿 pulse edge
脉冲衍射 pulse diffraction
脉冲厌氧滤池 pulsed anaerobic filter
脉冲厌氧折流板反应器 pulsed anaerobic baffled reactor
脉冲遥测方法 pulse telemetering method
脉冲遥测计 impulse telemeter
脉冲遥测术 pulse telemetry
脉冲遥控系统 pulse remote control system

M

脉冲遗漏 missing pulse
脉冲抑制器 pulse suppressor
脉冲译码器 pulse decoder; pulse demoder
脉冲引爆器 impulse blaster
脉冲引出 pulsed extraction
脉冲引弧器 pulse arc starter; surge injector
脉冲引力波 pulsed gravity wave
脉冲引入电路 pulse inserting circuit
脉冲引线 pulse lead
脉冲应答 impulse response; pulse reply
脉冲应力 pulsating stress; pulse stress
脉冲应力极限 limit at pulsating stress
脉冲应力疲劳极限 endurance limit at pulsating stress
脉冲荧光法二氧化碳监测仪 pulsed fluorescence sulfur dioxide monitor
脉冲荧光法硫化氢监测仪 pulsed fluorescence hydrogen sulfide monitor
脉冲拥挤 impulse crowding; pulse crowding
脉冲有效宽度 effective impulse width; effective pulse width
脉冲预选 impulse preselection
脉冲阈能 pulse threshold energy
脉冲元件 impulse element
脉冲源 impulse source; impulsive source; pulser; pulse source
脉冲运行 pulsing operation
脉冲运行回旋加速器 pulse operated cyclotron
脉冲运行时间 pulse operating time
脉冲运用 pulse operation
脉冲杂波 impulsive noise
脉冲杂波信号 impulsive noise signal
脉冲载波 pulse carrier
脉冲载波系统 pulsed carrier-current system
脉冲再发器 pulse regenerator
脉冲再生 impulse regeneration; pulse regeneration
脉冲再生器 impulse regenerator; pulse regeneration unit; pulse regenerator
脉冲再生式放大器 pulse regenerative amplifier
脉冲再生振荡器 pulse regenerative oscillator
脉冲再现计时控制 impulse recurrent timing control
脉冲噪声 burst noise; impulse noise; peaked noise; pulse noise; impulsive noise
脉冲噪声测量 impulse noise measurement
脉冲噪声处理 impulse noise processing
脉冲噪声倒置器 impulse noise inverter
脉冲噪声电平 impulse noise level
脉冲噪声发生器 impulse noise generator; pulse noise generator
脉冲噪声分析 impulse noise analysis
脉冲噪声敏感性 impulse noise susceptibility
脉冲噪声频率 impulse noise frequency
脉冲噪声系数 pulse noise margin
脉冲噪声限制器 impulse noise limiter
脉冲噪声消除 impulse noise blanking
脉冲噪声消隐 impulse noise blanketing
脉冲噪声信号 impulse noise signal
脉冲增长 build-up of pulse
脉冲增量调制器 pulse delta modulator
脉冲增压 pulse pressure-charging
脉冲闸流管 pulsed thyratron
脉冲闸门 pulse gate
脉冲窄化器 pulse narrowing device
脉冲斩波器 pulse chopper
脉冲展开 pulse spreading
脉冲展宽 impulse broadening; pulse broadening; pulse spreader; pulse stretching
脉冲展宽电路 impulse spreading circuit; pulse broadening circuit; stretch circuit
脉冲展宽鉴定 pulse spreading specification
脉冲展宽器 pulse stretcher
脉冲占空比 pulse duty factor; pulse rate factor
脉冲占空系数 pulse duration ratio; pulse duty factor; pulse time ratio
脉冲占空因数 duty factor; pulse duty factor
脉冲占空因素 pulse duty factor
脉冲占空因子 pulse duty factor
脉冲张弛放大器 pulse relaxation amplifier
脉冲胀接 pulsed expansion joint
脉冲真空管电压表 pulse vacuum tube voltmeter
脉冲真空蒸汽消毒器 pulse vacuum steam disinfector
脉冲甄别器 pulse discriminator
脉冲振荡 impulse hunting
脉冲振荡雷达 pulse oscillator radar
脉冲振荡器 impulse generator; impulse oscillator; pulse generator; pulse oscillator
脉冲振荡装置 pulse generating means
脉冲振幅 impulse amplitude; pulse amplitude
脉冲振幅大小 pulse size
脉冲振幅分布 pulse height distribution
脉冲振幅分析器 kicksorter; pulse height analyser[analyzer]
脉冲振幅鉴别 pulse amplitude discrimination
脉冲振幅鉴别器 amplitude discriminator; pulse height discriminator
脉冲振幅谱 pulse amplitude spectrum
脉冲振幅清晰度 pulse height resolution
脉冲振幅时间转换器 pulse height-to-time converter
脉冲振幅调制 pulse amplitude modulation
脉冲振幅选择器 amplitude selector; diffractional pulse height discriminator; pulse height selector
脉冲振铃 pulse ringing; ringing of pulse
脉冲整流器 impulse rectifier; pulse rectifier
脉冲整形 pulse normalization; pulse reshaping; pulse shape; pulse shaping
脉冲整形电路 pulse shaper; pulse shaping circuit
脉冲整形放大器 shaper amplifier
脉冲整形分频器 pulse shaper-divider
脉冲整形和再整形 pulse shaping and re-shaping
脉冲整形器 peaker; pulse shaper
脉冲正反馈 pulse regeneration
脉冲值 pulse value
脉冲指引 pulse steering
脉冲制 dynamic(al) system; pulsing system
脉冲制导波束 pulsed guidance beam
脉冲制导激光雷达 pulse guidance laser radar
脉冲制导激光器 pulsed guidance laser
脉冲制导雷达 pulse guidance radar
脉冲质谱仪 pulse mass-spectrometer
脉冲中继器 impulse repeater
脉冲中子 pulsed neutron
脉冲中子测井 pulsed neutron log
脉冲中子堆 pulsed neutron reactor
脉冲中子发生器 pulsed neutron generator
脉冲中子法 pulsed neutron method
脉冲中子活化 pulsed neutron activation
脉冲中子技术 pulsed neutron technique
脉冲中子探询法 pulsed neutron interrogation
脉冲中子消失法 pulsed neutron die-away method
脉冲中子源法 pulsed neutron source method
脉冲终止信号 end-of-pulsing signal
脉冲周期 impulse period; pulse cycle; pulse interval; pulse period; pulse recurrence interval; sampling action
脉冲注入 pulse injection
脉冲注入器 surge injector
脉冲柱 pulsed column
脉冲转发 pulse transponding
脉冲转发机 transponder
脉冲转发器 impulse repeater; pulse repeater; transponder
脉冲转数计 impulse tachometer
脉冲转速表 pulsation tachometer
脉冲转速计 pulse tachometer
脉冲装置 pulser; pulsing device
脉冲状态 pulse condition
脉冲状态工作 pulsed operation
脉冲状态试验 impulse condition test
脉冲追踪 pulse chase
脉冲紫外激光测距仪 pulsed ultraviolet laser rangefinder
脉冲自动记录器 impulse recorder
脉冲自调制 self-pulsed modulation; self-pulsing
脉冲自调制振荡器 self-pulsed oscillator
脉冲自(动)调制发射机 self-pulsed transmitter
脉冲自动增益控制 pulsed automatic gain control
脉冲总线 pulse bus
脉冲阻隔计算器 pulse blocking counter
脉冲阻抗 impulsive impedance
脉冲阻尼二极管 pulse damping diode
脉冲阻尼器 pulsation damper; pulse damper
脉冲阻塞 pulse blocking
脉冲阻塞管 impulse blocking tube
脉冲组 pulse group
脉冲组发生器 pulse series generator
脉冲作用 impulse action; impulsive action; sampling action
脉顶倾斜 pulse droop
脉动 bounce; microseism; oscillatory motion; pulsating movement; pulsation; pulsative movement; pulsative oscillation; pulse ripple; pulsing; pumping; ripple; saltation; surging; microtremor【地】
脉动百分数 ripple percentage
脉动暴 microseismic storm
脉动泵供油系统 jerk-pump fuel system
脉动比 ripple ratio
脉动比率 pulsatory ratio
脉动变化 impulse change; impulsive variation
脉动变星 pulsating variable; pulsation variable
脉动表面 pulsating surface
脉动波 pulsating wave; pulsation wave; pulsatory oscillation
脉动波浪荷载 pulsating wave load
脉动波阻尼器 ripple eliminator
脉动不稳定带 pulsation instability strip
脉动不稳定性 pulsation instability
脉动采样器 pulsating sampler
脉动测量 microtremor measurement
脉动常数 pulsation constant
脉动场 pulsating field
脉动超导磁铁 pulsed superconducting magnet
脉动成分 ripple component
脉动澄清池 pulsator clarifier
脉动冲程 pulsion stroke
脉动抽提柱 pulse extraction column
脉动除冰设备 pulsating deicer
脉动除尘器 jet type dust collector
脉动传感器 oscillation pickup
脉动床 pulsed bed
脉动床层 mobile bed
脉动床离子交换器 pulsed bed ion exchanger
脉动床滤池 pulsed bed filter
脉动磁场 pulsating field; pulsating magnetic field
脉动淬火压床 pulse quenching press
脉动萃取塔 pulse extraction column
脉动的 intermittent; pulsating; pulsatory
脉动点 pulsation point
脉动电动势 pulsating electromotive force
脉动电弧 pulsating arc
脉动电流 fluctuating current; pulsating current; pulsatory current; pulse current; ripple current
脉动电流电动机 pulsating current motor; ripple current motor
脉动电流因数 pulsating current factor
脉动电压 pulsating pressure; pulsating voltage; ripple voltage
脉动电源 pulsafeeder; pulsating electric(al) source
脉动发射 pulsed emission
脉动发射电流 pulsed emission current
脉动发生器 flutter generator
脉动反应 impulse response; pulse response
脉动范围 oscillation limit
脉动放大器 ripple amplifier
脉动放射 pulsed emission
脉动分带 pulsative zoning
脉动分级器 pulsator classifier
脉动分量 flutter component
脉动风速 pulsation wind velocity
脉动负荷 pulsating load(ing)
脉动负载 fluctuating load; pulsating load(ing)
脉动干扰 flutter; pulsation interference
脉动给进 pulsating feed
脉动给料机 pulsafeeder
脉动功率 pulsating power
脉动功率控制 pulsating power control
脉动供给 pulsating feed
脉动观测 microseismic observation
脉动光 pulsating light; pulsed light
脉动光灯 undulating light
脉动光束 fluctuating light beam
脉动光源 pulsating light source
脉动规度装置 pulse normalizer unit
脉动过程 microtremor process
脉动过滤器 pulsed filter
脉动焊接 pulsation welding
脉动耗损 pulsation loss
脉动荷载 fluctuating load; pulsating load(ing)
脉动荷载下拉伸试验 repeated tensile stress test
脉动弧 pulsating arc
脉动琥珀色灯 <一种交通信号灯> pulsating amber light
脉动簧 impulse spring; pulse spring
脉动簧断开 impulse spring break
脉动混合沉降器 pulse mixer-settler
脉动活塞杆 pulsed ejector

脉动机 pulsating machine
脉动机构 pulsing mechanism
脉动基频 fundamental ripple frequency
脉动激光作用 pulsating laser action
脉动激励 impulse excitation
脉动激振 microtremor excited vibration
脉动极光 pulsating aurora
脉动极光带 pulsating auroral zone
脉动极光雷达回波 pulsating auroral radar echoes
脉动极化器 pulse polarizer
脉动极限 oscillation limit
脉动计数器 pulsed counter
脉动加料 pulsed feed
脉动加速 pulsation acceleration
脉动交叉光谱 pulse cross-spectrum
脉动接点 impulse spring
脉动节拍比 pulsation ratio
脉动节拍扩大比 wide pulsation ratio
脉动节制器 pulsation damper
脉动界限 oscillation limit
脉动进气孔 pulsating air intake
脉动进位二进制计数器 ripple carry binary counter
脉动进位加法器 ripple carry adder
脉动浸渗 pulsed infusion
脉动净压力 net pressure fluctuation
脉动抗弯应力疲劳强度 pulsating fatigue strength under bending stress
脉动空泡 pulsating cavity
脉动空气分级机 pulsating air classifier
脉动空蚀 pulsating cavity;pulsedcavitation
脉动空穴 pulsating cavity
脉动控制 ripple control
脉动控制器 pulsator controller;pulse controller
脉动拉伸荷载 fluctuating tensile load
脉动理论 pulsation theory
脉动力 fluctuating force;force of pulsation; intermittent force; pulsating force
脉动量 pulsating quantity;undulating quantity
脉动流 oscillating flow;pulsating flow; surging flow
脉动流测定 pulsating flow measurement
脉动流出 positive discharge
脉动流化床 pulsating fluidized bed
脉动流量 pulsating flow
脉动流量测量 pulsating flow measurement
脉动流速 fluctuating velocity;fluctuation velocity; pulsation current velocity
脉动滤波器 ripple filter
脉动滤除扼流圈 ripple-filter choke
脉动模式 pulsation mode
脉动能(源)pulsating energy
脉动喷井 gurgling well
脉动喷射 pulse jet
脉动喷射器管嘴 oscillating big gun
脉动疲劳试验机 pulse fatigue testing machine
脉动漂白 pulse bleaching
脉动频率 oscillation frequency;ripple frequency;pulsation rate
脉动谱 microseism spectrum; microtremor spectrum
脉动起动机 impulse starter
脉动气泡 pulsating bubble
脉动气体流态化 pulsed gas fluidization
脉动器 pulsation equipment;pulsator
脉动器真空管路 pulsator line;pulsator vacuum line
脉动牵引电机 pulsating traction motor
脉动球体 pulsating sphere
脉动球体声源 pulsating sphere source
脉动燃烧 intermittent burning;intermittent combustion;pulsating combustion;pulsation combustion;spasmodic burning
脉动染色 pulsator dy(e)ing
脉动染色机 pulsator dy(e)ing machine
脉动染浴 pulsating dye bath
脉动筛 pulsating screen;pulsator screen
脉动闪光 pulsating scintillation
脉动设备 pulsation equipment
脉动射流增强器 pulsed jet intensifier
脉动式磁流体发电机 pulse type magnetohydrodynamic(al)generator
脉动式萃取塔 pulsed extraction column
脉动式的 by-heads
脉动式地表热流 pulsating surface heat-flow
脉动式给矿机 pulse feeder
脉动式空气喷气发动机 aeropulse; aeroresonator; pulse duct engine; pulse-jet engine; impulse duct engine;resonant jet
脉动式排水阻气阀 impulse trap
脉动式喷出 head flow
脉动式喷气发动机 intermittent-firing duct engine; pulse duct; pulse jet; resojet
脉动式喷气发动机导弹 pulsejet missile
脉动式燃气轮机 pulsating flow gas turbine
脉动式燃烧 spasmodic burning
脉动式疏水器 impulse trap
脉动式液压马达 pulse motor
脉动试验 pulsating test
脉动室 pulsation chamber
脉动室真空记录曲线 pulsation chamber vacuum record curve
脉动输入 intermittent input
脉动数 pulse number
脉动数据 microseismic data
脉动衰减器 pulsation dampening
脉动水流 fluctuating flow; pulsating current;pulsating flow;ripple flow
脉动水舌 undulated nappe
脉动水压力 pulsatory(water)pressure
脉动说 pulsation hypothesis
脉动速度 fluctuation velocity; pulse velocity
脉动损耗 pulsation loss
脉动塔 pulsed column
脉动弹簧 impulse spring
脉动跳汰机 air pulsated jig;pulsator jig
脉动通量 oscillating flow; pulsating flux
脉动推进 pulsing feed
脉动温度 fluctuating temperature
脉动涡流 pulsed eddy
脉动无级变速器 impulse stepless gear box
脉动误差 pulsating error
脉动吸尘器 pulsator deduster
脉动吸入周期 pulsion-suction cycle
脉动系数 pulse coefficient;ripple contain factor;ripple factor;ripple ratio
脉动系统 pulsation system
脉动氙灯 pulsed xenon light
脉动现象 pulsation phenomenon
脉动削波 pulse clipping
脉动消除器 pulsation dampener
脉动效应 pulsating effect; pulsation effect
脉动卸料斗式提升机 positive-discharge bucket elevator
脉动信号 fluctuating signal; microtremor signal
脉动星 pulsating star
脉动性状 microtremor behavio(u)r
脉动修正系数 pulsating correction factor
脉动旋转磁场 pulsating rotating field
脉动学说 pulsation theory
脉动循环 pulsation cycle
脉动循环荷载 pulsating cyclic load
脉动循环遥测 rapid cycle telemetry
脉动压机 pulsating press
脉动压力 fluctuating pressure;oscillatory pressure; pulsating pressure; pulsatory pressure
脉动压力分布 fluctuating pressure distribution
脉动压缩荷载 pulsating compressive loading
脉动氧化裂解 pulsating oxidative pyrolysis
脉动仪 microtremor instrument; oscillometer
脉动抑制 ripple rejection
脉动因数 ripple contain factor
脉动引起的振动 pulsation-induced vibration
脉动应力 fluctuating stress;pulsating stress
脉动源 pulsation source
脉动噪声 microseismic noise
脉动噪声谱 microseismic noise spectrum
脉动张应力疲劳强度 fatigue strength under pulsating tensile stresses
脉动照明光源 pulsed light generator
脉动振荡 pulsative oscillation
脉动振幅 pulse amplitude
脉动直流 pulsating direct current
脉动值 pulsating quantity; ripple quantity
脉动质量 pulsation mass
脉动周期 fluctuation period; oscillation period; period of pulsation; pulsation cycle;pulsation period
脉动注射 pulse injection
脉动注水 pulsed infusion
脉动注水爆破 pulsed infusion blasting;pulsed infusion shot-firing
脉动柱 pulse column
脉动柱塞泵 pulsating plunger pump
脉动转矩 pulsating torque
脉动转数计 impulse tachometer
脉动装置 pulsating equipment
脉动阻尼 damping of pulsation;pulsation damping
脉动阻尼薄膜 pulsation damping diaphragm
脉动阻尼器 pulsation damper;ripple damper
脉动阻塞 pulse blocking
脉动作用 panting action;pulsation action
脉动作用泵 jerk pump
脉度 measurement of channels
脉短波 pulsed short wave
脉伐<可变电抗混频放大,低噪声微波放大器> mixer amplification by variable reactance
脉峰 pulse crest;pulse peak
脉浮 floating pulse
脉浮紧 floating and tense pulse
脉浮数 superficial and rapid pulse
脉幅编码调制 pulse amplitude code modulation
脉幅谱 pulse amplitude spectrum
脉幅调制 pulse amplitude modulation; pulse duration modulation; pulse length modulation; pulse width modulation
脉幅调制调频 pulse amplitude modulation-frequency modulation
脉幅调制交换机 pulse amplitudemodulation switching system
脉幅调制样值 pulse amplitude-modulated sample
脉幅选择器 pulse height selector
脉革菌属<拉> Cytidia
脉管制冷器 pulse tube refrigerator
脉管状的 vasiform
脉光灯 pulselite
脉横线 abscissa
脉后间期 postsphygmic interval
脉环 circulus venosus
脉混合岩 veinite
脉脊 vein rib
脉尖 apex of vein
脉尖编码 pip coding
脉尖放大器 pip amplifier
脉尖积分器 pip integrator
脉接合 junction of veins
脉节理黏[粘]土 floocan
脉金 lode gold
脉控振荡器 pulsed oscillator
脉宽 impulse width; pulse length; pulse width
脉宽倍增器 pulse width multiplier
脉宽编码 pulse width coding
脉宽编码器 pulse duration coder
脉宽标准化 pulse width standardization
脉宽多谐振荡器 pulse width multivibrator
脉宽记录(法)pulse width recording
脉宽记录器 pulse width recorder
脉宽鉴别 pulse duration discrimination
脉宽鉴别器 pulse width discriminator
脉宽键控器 pulse width keyer
脉宽可调起搏器 adjustable pulse width pacemaker
脉宽失真 pulse width distortion
脉宽收缩 pulse width shrinkage
脉宽调整 pulse width control
脉宽调制 impulse width modulation; pulse code modulation;pulse duration modulation; pulse position modulation; pulse width modulation;width-pulse modulation
脉宽调制变换器 pulse width-modulated inverter
脉宽调制波 pulse width-modulated wave
脉宽调制放大器 pulse width modulation amplifier
脉宽调制复用设备 pulse code modulation multiplex equipment
脉宽调制模数转换器 pulse width-modulated analog/digital converter
脉宽调制器 pulse width modulator
脉宽调制式调幅发射 amplitude modulation transmitter with PDM[pulse-duration modulation]
脉宽调制式调幅发射机 AM transmitter with pulse duration modulation
脉宽调制系统 variable duration impulse system
脉宽调制信号 pulse width signal(ling)
脉宽调制斩波器 pulse width modulation chopper
脉宽相位检波器 pulse width phase detector
脉宽译码器 pulse width decoder
脉宽周期比 pulse duration ratio;pulse time ratio
脉矿 lode
脉矿采矿用地 quartz claim
脉矿床 vein deposit
脉矿石 lode ore
脉理构造 lineage structure
脉力计 pulsimeter

M

脉沥青 albertine;vein bitumen
脉量 pulse volume
脉裂隙 vein fissure
脉流 slugging
脉流串激牵引电动机 pulsating current series traction motor
脉流牵引电动机 pulsating current traction motor
脉率计 pulse clock
脉率式遥测术 pulse rate telemetering
脉络 venation
脉码 pulse code
脉码传输 pulse code transmission
脉码多路系统 coding pulse multiple
脉码发生器 pulse pattern generator
脉码管 pulse code tube
脉码控制系统 pulse code control system
脉码调制 pulse code modulation
脉码调制变换器 pulse code modulation converter
脉码调制的量化电平 pulse codemodulated level
脉码调制电缆 pulse code modulation cable
脉码调制-调频 pulse code modulation-frequency modulation
脉码调制多路复用器 pulse code modulation multiplexer
脉码调制多路复用设备 pulse code modulation multiplex equipment
脉码调制广播 pulse code modulation broadcast(ing)
脉码调制话音 pulse code modulated vice
脉码调制记录器 pulse code modulation recorder
脉码调制交换机 pulse code modulation exchange; pulse code modulation switching system
脉码调制模拟器 pulse code modulation simulator
脉码调制器 pulse code modulator
脉码调制时分多址联结 pulse code modulation-time division multiaccess
脉码调制数字 pulse code modulation digital
脉码调制数字制彩色电视 pulse code modulated digital colo(u)r television
脉码调制系统 pulse code modulating system
脉码调制遥测系统 pulse code modulation switching system
脉码调制终端机 pulse code modulation terminal
脉码信号系统 pulse code signalling system
脉内成巷 in-seam driving
脉内矿物 lodestuff
脉内平巷 reef drift
脉内噪声 intrapulse noise
脉能测量器 energometer
脉容量测量器 volume bolometer
脉塞 maser[microwave amplifier by stimulated emission of radiation]
脉塞材料 masering material
脉塞带宽 maser bandwidth
脉塞干涉仪 maser interferometer
脉塞接收器 maser receiver
脉塞介质 maser medium
脉塞前置放大器 maser preamplifier
脉塞增益 maser gain
脉塞作用 maser action
脉梢 vein end
脉石 barren rock;burrow;dead rock; deads; gangue; gangue material; lodestone; matrix[复 matrixes/matrices];rocky impurity;veinstone
脉石膏 vein gypsum
脉石夹层 horse of barren rock
脉石矿床 root deposit
脉石矿物 gangue mineral;rocky mineral
脉石类 gangue quartz
脉石泡沫 gangue froth
脉石剔除 gangue rejection
脉石岩 gangue rock
脉石英 vein quartz
脉石元素 gangue element
脉时解调器 pulse time demodulator
脉时调制 pulse time modulation
脉时调制系统 pulse time modulation system
脉首波 papillary wave;percussion wave
脉缩 rugosity
脉体 metasome
脉调载波 pulse modulated carrier
脉铁 loded iron
脉外开拓 development works in stone
脉外溜井 rock hole
脉外天井 rock raise
脉外巷道 rock tunnel
脉网 network of vein;vein mesh
脉微 weak pulse
脉位采样器 position sampler
脉位取样器 position sampler
脉位调制 pulse position modulation
脉位调制光学跟踪仪 pulse position modulation optical tracker
脉位调制码 pulse position modulation code
脉纹 vein
脉纹大理石 veined marble
脉纹片麻岩 veined gneiss
脉锡 lode tin;vein-tin
脉系 pulse system;vein system
脉相 pulse phase
脉相调制 pulse phase modulation
脉相系统 pulse phase system
脉象仪 electropulsograph
脉序 venation
脉岩 dike rock;dikite;vein rock
脉音听诊器 sphygmophone
脉泽 maser[microwave amplifier by stimulated emission of radiation]
脉泽前置放大器 maser preamplifier
脉泽振荡器 maser oscillator
脉振速度 velocity of pulsation
脉质 ledge matter; vein filling; vein material;vein matter
脉中群 pulse group
脉状 veiny
脉状层理【地】flaser bedding
脉状层理构造 flaser bedding structure
脉状的 veined
脉状汞矿床 veined mercury deposit
脉状构造【地】vein(-type) structure; veined structure
脉状混合岩 <侵入的岩脉呈层状分布> arterite; arteritic migmatite; ve(i)nite
脉状角砾岩 vein breccia
脉状结构 vein texture
脉状矿床 lode deposit
脉状矿体 vein(-type)orebody
脉状沥青铀矿硫化物矿床 vein-type pitchblende-sulfide deposit
脉状裂隙水 vein(ed)(-type) fissure water
脉状硫铁矿床 vein(-type) pyrite deposit
脉状排水干渠 arterial drain
脉状排水系统 arterial drainage
脉状片麻岩 veined gneiss; vein-type gneiss
脉状铅锌矿床 vein lead-zinc deposit; vein-type lead-zinc deposit
脉状闪长岩 anchorite
脉状石墨 vein graphite
脉状水 vein water
脉状条纹长石 ader perthite
脉状铜矿床 vein-type copper deposit
脉状纹痕 <宽阔排列的> broad veining
脉状岩墙 vein dike[dyke]
脉状组织 vein structure

蛮

蛮石 boulder; cyclopian; displacer; land waste; nigger head; plum stone; pudding rock; pudding stone; rubble; plum <大体积混凝土用>

蛮石底基 boulder base
蛮石底盘 boulder bed
蛮石地基 boulder base
蛮石堆 cyclopean
蛮石工程 cyclopean masonry work
蛮石骨料 cyclopean aggregate
蛮石混凝土 cyclopean block; cyclopean concrete;rubble concrete
蛮石基层 boulder bed
蛮石基础 boulder base
蛮石集料 cyclopean aggregate
蛮石开采 boulder quarry
蛮石路面 boulder pavement
蛮石铺地 boulder pavement
蛮石铺面 boulder pavement
蛮石铺砌 bouldering
蛮石铺砌层 boulder setter
蛮石铺砌工 boulder setter
蛮石砌体 cyclopean masonry
蛮石砌筑坝 boulder dam
蛮石墙 boulder wall;cyclopean wall
蛮石贴面(工程) cyclopean rustication
蛮石圬工 cyclopean masonry; cyclopean rubble masonry; rubble masonry
蛮石圬工坝 cyclopean masonry dam; rubble masonry dam
蛮石峡谷 boulder stream canyon
蛮岩 land waste

馒

馒头山 dome mountain

馒头形 cabochon
馒头形顶 <印度和耆那教建筑中高塔顶部的> amalaka
馒头形饰 boultin
馒头形碗 bowl in the shape of a steamed bun
馒头碹 domed arch
馒头窑 dome kiln;kiln in the shape of steamed bun
馒形饰 ovolo;thumb
馒形线脚 ovolo mo(u)lding

鳗

鳗的洄游 migration of eels

鳗鲡迁移 eel migration; migration of eels
鳗鲡顺流洄游 downstream eel migration;downstream migration of eels
鳗梯 eel ladder
鳗鱼油 eel oil

满

满岸 bankfull

满岸流 banker
满岸水位 bankfull stage;bank stage
满版图 flush plate
满包封包机 filled bag closing machine
满包量 ladleful
满杯 cupful
满泵 full pumping
满标 full-scale
满标度 full range;full-scale
满标度值 end scale value; full-scale value
满标灵敏度 full-scale sensitivity
满标偏转 full-scale meter deflection
满标输出 full-scale output
满标误差 full-scale error
满标循环 full-scale cycle
满标值 end scale value;full-scale reading
满膘 full seeding;highly finished
满表值 <仪表> end scale value
满播 full seeding
满仓 binful
满舱 chock a block off;full and down
满舱满载货物 full and complete cargo
满槽 full bath;full coverage of bath
满槽的 bankfull
满槽对接焊 full-penetration butt weld
满槽焊 complete joint penetration
满槽河宽 bankfull width
满槽滑动 full trough gliding
满槽流 bankfull flow
满槽流量 bankfull discharge;bankfull flow;bank high flow
满槽容量 level-full capacity
满槽水流 bankfull flow
满槽水位 bankfull stage;bankfull(water)level
满潮围堰 full-tide cofferdam; whole-tide cofferdam
满潮 full sea;full tide;high tide;high water
满潮标记 flood level mark(ing)
满潮池 tide pool
满潮高 height of high tide
满潮间隙 high water interval
满潮平流 slack-water on the flood
满潮围堰 full-tide cofferdam; whaletide cofferdam
满充低缺陷能级 filled lower defect level
满充壳层 closed shell;complete shell; filled shell
满充能带 filled band
满充能级 filled level
满出 flow over;overfill
满储 full storage
满打满算 reckoning in every item of income or expenditures
满带 filled band;full filled band
满带能级 filled-band level;filled level
满袋 purseful
满到边 brim
满堤 bankfull
满底泥驳 elevator barge;elevator hopper barge;hopper barge without bottom door;non-hopper-door barge
满地图案 all-round pattern
满地釉 all-over glaze
满地照明 full earth illumination
满点 extreme point
满电压 full voltage
满丁砌合 header bond;header joint; heading bond
满丁砖行 course of headers; header course;heading course
满斗 binful;bucketful
满斗率 bucket fill degree;fillability
满斗系数 bucket factor; bucket fill factor;dipper factor
满斗阻力 filling resistance
满度偏转 full-scale deflection
满度系数 charge coefficient
满堆 stow all over
满对接焊 full penetration butt weld

满舵 full helm;full rudder;hard over
满舵回转直径 final diameter
满额 full allowance
满额荷载 full load
满额孔隙压力比 full pore-pressure ratio
满额生产 full-scale production
满二叉树 full binary tree
满帆 full sail;full spread;in full sail
满帆前进 crack on
满帆倾斜行驶 sailing on her ear
满帆顺风 full and by
满房间 roomful
满分 full marks;perfect score
满份儿 full lot
满风 bag
满负荷 capacity load; full load; full output
满负荷操作 full load operation
满负荷的 full-sized
满负荷电流 full load current
满负荷额定转速排烟特性 full load rated speed smoke characteristic
满负荷工作 full load operation;working at full capacity
满负荷功率 full load efficiency
满负荷耗量 full load need
满负荷交通 full load traffic
满负荷生产 full production;to go into full production
满负荷生产量 full capacity
满负荷试验 full load test; full-scale load(ing) test
满负荷试运转 full power trial
满负荷速度 full load speed
满负荷运行 operation at full load
满负荷运转 running at full capacity
满负载 overall loading
满负载的 full load
满负载电流 full load current
满负载时的下沉量 loaded draught
满负载条件 full load condition
满负载效率 full load efficiency
满功率 full power
满功效条件 full power condition
满管 full package
满管自停装置 full bobbin stop motion;full cap stop motion
满灌 full irrigation
满光栅 full raster
满规钻孔 full ga(u)ge drill hole
满过头 overwinding;over-wound cake
满焊 full-length welding
满荷 full charging
满荷损耗 full load loss
满火红 circulating fireman
满简 call down
满键 full key
满浆穗 laden earhead
满角焊 all fillet weld
满角焊缝 full fillet weld
满角焊接 full fillet welding
满绞自停装置 full hank stop motion
满矩阵 full matrix
满矩阵法 full matrix method
满距绕组 full-pitch winding
满卷自停装置 full lap stop motion
满开的 full gate
满开闸门 full gate
满刻度 full-scale range
满刻度测量 full-scale measurement
满刻度的 full-scale
满刻度读数 full-scale reading
满刻度范围 full-scale range
满刻度校正标准气体 < 气体分析仪 > span gas
满刻度量程 full-scale range
满刻度偏转 full-scale deflection
满刻度偏转力 full-scale deflecting force
满刻度输入(功率)full-scale input
满刻度吸收度单位 absorbance unit of full scale
满刻度响应 full-scale response
满刻度行程 full-scale travel
满刻度值 full-scale value
满控制法 full control method
满口 mouthful
满库 full reservoir
满库水面 full reservoir surface
满库水时间 period of reservoir with full water
满库水位 bankfull stage;full pool level;full reservoir surface
满宽堰 < 即指占全河渠宽的 > full-width weir
满励磁继电器 full field relay
满量程 full range
满流 bankfull flow; flowing full; full flow
满流管 full pipe
满流过滤器 full flow filter
满流浇注 decant
满流式滤油器 full flow oil filter
满流水位 bankfull stage
满满地 abrim
满满一仓 binful
满能级 occupied level
满排 set solid
满平容积 level capacity
满屏处理 full-screen processing
满屏幕 flooding
满屏显示 push-through presentation
满铺草皮 broadcast(ing) sodding
满铺地毯 wall-to-wall carpeting
满铺沥青 solid mopping
满铺砂浆垫层 full mortar bedding
满期 become due;date due;efflux of time;expiration;expire;expiry;fall due; terminate; termination; termination of term
满期保(险)费 earned premium
满期的 time-expired
满期前都附利息 interest to maturity
满期日 day of due date;day of maturity;due date;terminal of an agreement
满期日缺口暴露 maturity gap exposure
满期为止 runoff
满旗铆接 full pointed rivet
满旗绳 dressing line
满旗纵挂法 dressed full with up and down flags
满容量的 full-sized
满容量炼炉 full-size furnace
满扫描 full scan
满射【数】epimorphism;surjection
满射映射 surjective mapping
满师工人 time served worker
满师徒工 journal-man;journey-man
满十计数法 cross-ten method
满是灰尘的 dusty
满收满付制 full receipts and full payments system
满数输出 full count out
满水 full water
满水保护阀 anti-flooding valve
满水时期 replenishing period
满水水位 full supply level
满台平台 all stage scaffolding
满态 filled state
满堂草皮 broadcast sodding
满堂焊 full weld
满堂红混凝土垫层 oversite concrete
满堂红支架 full framing
满堂基础 mat foundation; raft foundation
满堂脚手架法 full staging method
满堂开挖 overall excavation
满堂式基础 mattress
满堂摊料 < 用平地机在整个车行道摊铺材料 > blade over the roadway
满堂支架 full framing
满堂支架施工法 support construction method
满天星 multistar;open herding
满填铆接 full pointed rivet
满条砌法 running bond; stretcher bond;stretching bond
满同态 epimorphism
满桶 bucketful
满筒 fullrun
满筒计数器 full package counter
满筒率 call down rate; rate of full package
满筒自停 full bobbin stop motion
满筒自停装置 full can stop motion
满涂 solid mopping
满外尺寸 out-to-out dimension; out-to-out distance;overall
满五计数法 cross-five method
满物料 full with substance
满细胞(的木材防腐)法 < 一种木材防腐的压力处理法 > full-cell process; full cell method; full-cell treatment
满线测定装置 < 铁路编组线 > fullness measuring apparatus
满线跳杆簧 stop latch spring
满相 full phase
满行方式 full line mode
满行秩 full row rank
满蓄 full storage
满蓄率 coefficient of fullness;fullness coefficient
满眼钻进 packed hole drilling
满眼钻进工艺 packed hole drilling technology
满眼钻具(组合) packed hole assembly
满窑 kiln filling
满窑工 kiln filler
满液式冷却器 flooded cooler
满液式盘管 flooded coil
满液式蒸发器 flooded evapo(u)rator
满一料斗 binful
满一周年而还未满二周年的 yearling
满意的质量水平 acceptability quality level
满意工资基础 basis of satisfactory wages
满意供水 acceptable water supply
满意解 < 优化设计用 > satisfactory solution;acceptable solution
满意效果 promising result;satisfactory result
满意证明 satisfaction proof
满意质量 satisfactory quality
满溢 brim;flood;overbrimming;overflow(ing)
满溢出口 overflow outlet
满溢船坞 flooding dock
满溢的 flooded
满溢式冷却器 flooded cooler
满溢式系统 flooded system
满溢指示器 overflow indicator
满油补充阀 prefill surge valve
满油阀 prefill valve
满油止挡 full oil stopper
满员编制 complement of personnel
满月 full moon;plenilune
满月大潮 full moon spring tide
满月脸 moon-shaped face
满月亮度 full noon brightness
满月时期 plenilune
满月照明 full moon illumination
满载 capacity load;full and down;full cargo; full loading; loaded to full capacity;overall loading
满载饱和曲线 full load saturating curve
满载比 charge ratio
满载操作 capacity operation
满载吃水 < 船舶 > deep draught; draught fully laden;full load draft; laden draft; laden draught; load draft; load draught; load (ed) draught
满载吃水标线 < 船舶的 > load line mark
满载吃水线 deep-water line;full load (water-)line;load water-line
满载吃水线标图 loading disc
满载吃水线长度 loaded waterline length
满载吃水线吃水标志 full load line mark
满载吃水线平面 full load water plan
满载吃水装载量 draft full load
满载出力 full load output
满载船 fully laden ship;fully loaded ship
满载船舶 laden hull;loaded hull
满载船吃水线 loaded water line
满载磁场 full load field
满载的 full-laden; fully loaded; fully laden
满载电流 full-load current
满载额定转速 full-rated speed
满载负荷电流 full-load current
满载干舷 loaded freeboard
满载功率 full load power
满载航速 full load speed;laden speed; loaded speed
满载货物 full and complete cargo
满载激励 full load excitation
满载阶段 full load period
满载励磁 full load excitation
满载量操作 capacity operation
满载列车 fully loaded train
满载率 full factor;full load ratio
满载扭矩 full load torque
满载排水吨位 full load displacement tonnage; loaded displacement tonnage
满载排水量 displacement fully laden; full load displacement tonnage;load(ed) displacement
满载起动 start at full load
满载起动器 full load starter
满载欠载装置 loading and underloading device
满载情况 full load condition
满载容量 full(load) capacity;heaped capacity < 矿车或铲斗的 >
满载失速 full load stall
满载时 full load
满载时的提升速度 full load lift speed
满载时的转数/分钟 full load revolutions per minute
满载试验 busy test;full load test;full-scale test;heavy test
满载试运转 full load test run
满载水线 full load line
满载水线长 length of load waterline
满载水线面 load water plane
满载水线面积 load water plane area
满载水线面面积系数 load water-line coefficient; waterplane area coefficient
满载速度 full load speed
满载损失 full load loss
满载特性 full load characteristic
满载特性曲线 full load characteristic curve
满载调整 full load adjustment
满载调整装置 full load meter adjustment
满载系数 charge ratio;full load coefficient

M

满载效率 full load efficiency
满载行程 full load trip
满载运行 full load running
满载运转 full load running
满载正常航速 commercial speed;service speed
满载重量 all-up weight;filled weight; full weight; laden weight; total weight
满载转差率 full load slip
满载转矩 full load torque
满载状态 full load condition; loaded condition
满载状态下的高度 laden height
满载租船合同 deadweight charter
满载作业 capacity operation
满增益 full gain
满帧 full frame
满支承 full bearing
满秩 full rank
满秩的 non-singular
满秩矩阵 full rank matrix;non-singular matrix
满秩平差 full rank adjustment
满秩线性变换 non-singular linear transformation
满中子通量 full neutron flux
满重荷载 full weight load
满轴列车 fully loaded train
满轴自停装置 automatic full beam stop motion
满转周期 period of complete rotation
满装 stow all over
满装容量 brimful capacity
满装式 full type
满装载的 fully laden
满装炸药 solid loading
满足比 satisfactory fraction
满足标准 satisfying criterion
满足补丁的 patchy
满足程度单位 units of satisfaction
满足的 satisfactory
满足的模型 satisfying model
满足多种需求的审计 multiple use audits
满足共同需要部分 common satisfaction of needs
满足规范要求 meet the requirement of specifications
满足化与适应性的合理决策 satisfying and adaptively rational decision
满足技术条件 meet the specification
满足静力条件的解(答) statically acceptable solution
满足上诉 meet a cassation
满足条件 meet a condition
满足需求的沟通 <表达情绪状态,解除内心紧张> consummatory communication
满足需要 answer the demand;meet a demand; meet the needs; meet the requirement; satisfaction of wants; satisfy the needs;satisfy wants
满足需要比较原理 comparative principle of satisfying needs
满足需要的产品 needs-satisfying products
满足需要可比原理 comparative principle of satisfying needs
满足需要资料 means of one's satisfaction
满足要求 answer the demand;answer the purpose; meet the challenge; meet the requirement; satisfy the demand
满足有效期 satisfy the prescriptive period
满足运动条件的解 kinematically acceptable solution
满阻尼 full damping
满钻 fully jewelled

螨

螨病 mange

曼

曼彻斯特编码 Manchester's code
曼彻斯特黄 Manchester's yellow
曼彻斯特极板 Manchester's plate
曼彻斯特经密制 Manchester's sett system
曼彻斯特棕 Manchester brown
曼代尔-克雷尔效应 Mandel-Cryer effect
曼尔法伦 melphalan
曼戈尔铝锰合金 Mangol
曼戈尼克镍锰合金 Mangonic
曼格莱特铝基活塞合金 Mangalite
曼格林瓦钴铁镍锰合金 Mangelinvar
曼古司(铜镍锌)合金 Mungoose metal
曼谷港 <泰国> Bangkok Port
曼谷银行 Bangkok Bank Ltd
曼哈顿距离 Manhattan distance
曼哈顿商业区 Manhattan
曼海姆法 Mannheim process
曼海姆(铜锌锡代金合)金 Mannheim gold
曼海姆吸收装置 Mannheim absorption system
曼海姆制盐饼炉 Mannheim salt-cake furnace
曼加内斯页岩群 Manganese Shale Group
曼加那尔高强度耐磨含镍高锰钢 Manganal
曼金镜 <消球差的折反射镜> Mangin mirror
曼金主镜 Mangin primary
曼克斯板岩层 Manx slate
曼肯德尔检验 Mann-Kendall test
曼里斯碱 mannich base
曼利-罗方程 Manley-Roew equation
曼内斯曼辊式穿孔法 Mannesmann roll-piercing process
曼内斯曼式穿孔机 Mannesmann piercer;Mannesmann piercing mill
曼内斯曼式轧管法 Mannesmann process
曼内斯曼效应 <锻件或轧件内> Mannesmann effect
曼内斯曼斜辊穿轧机 <管材> Mannesmann piercing roll
曼内斯曼斜轧机 Mannesmann mill
曼内斯曼制管法 <斜辊轧管法> Mannesmann process
曼奈普冰期【地】Menapian glacial stage
曼奈普寒冷期【地】Menapian cold epoch
曼宁糙率系数 Manning's roughness coefficient;Manning's roughness factor
曼宁常数 Manning's constant
曼宁粗糙度系数 Manning's coefficient of roughness; Manning's roughness factor; Manning's roughness coefficient
曼宁方程式 Manning's equation
曼宁公式 <计算水流速度> Manning's formula
曼宁计算图表 Manning's nomograph
曼宁摩擦系数 Manning's friction coefficient
曼宁-史崔克勒模拟 Manning-Strickler similitude
曼宁阻力系数 Manning's friction factor;Manning's roughness factor
曼钮林建筑风格 Manueline style
曼钮林建筑形式 Manueline architecture
曼诺耶施工法 <一种高烟囱施工法,由预制八边形混凝土块体分段砌筑而成> Monnoyer system
曼塞尔氏标准(土的)颜色判定法 Munsell colour
曼塞尔氏图 <用于判定土的颜色> Munsell chart
曼氏轧管法 Mannesmann process
曼斯菲尔德炼铜法 Mansfield copper process
曼斯菲尔德砂岩 Mansfield Sandstone
曼斯菲尔德油气发生器 Mansfield oil gas producer
曼斯霍尔特计划 Mansholt plan
曼斯纳格冲击试验 Mesnager impact test
曼斯纳格铰链 Mesnager hinge
曼斯纳格缺口 Mesnager notch
曼廷尼石 mantienneeite
曼托试验 Mantoux test
曼希海沟 Romanche trench

墁

墁板 surfacer
墁光表面 trowel(l)ed surface
墁灰 plaster(work)
墁平的泥灰面 set fair
墁平刮尺 nib grade
墁平机 trowel(l)ing machine
墁平(路)面 floated surface
墁平抹光 smooth by trowelling
墁平准条 ironing-screed
墁墙金属板条 metal lathing
墁饰花纹 parget
墁涂 floating coat;topping coat
墁涂层 trowel-applied coat
墁涂粉刷 trowel plaster

幔

幔 veil
幔源花岗岩 mantle-derived granite
幔源岩浆 mantle-derived magma

慢

慢焙固树脂 slow-curing resin
慢闭活门 slow closing faucet
慢编码 slow coding
慢变函数 slowly varying function
慢变化场 slowly changing field
慢变系统 slowly varying system
慢表面波 slow surface wave
慢表面态 slow surface state
慢波 slow(mode)wave
慢波比 delay ratio;slow wave ratio
慢波波长 slow wave wavelength
慢波导 slow wave waveguide
慢波结构 slow wave structure
慢波结构同轴线耦合器 slow wave structure to coaxial line coupler
慢波聚束器 slow wave buncher
慢波螺线 delay line helix
慢波系统 slow wave system
慢波线 slow wave line
慢波(线)节距 slow wave pitch
慢波线路 slow wave circuit
慢波(线)周期 period of slow wave structure
慢部分 slow component
慢长(树)木材 slow-grown wood
慢车【铁】local train; omnibus train; ordinary train; slow train; way train; stopping train; accommodation train <美铁路>
慢车出油口 slow running outlet
慢车道 climbing lane; cruising way; cycling way;non-motorized vehicle lane; slow lane; slow line; slow traffic lane;slow vehicle lane
慢车检测器 slow vehicle detector
慢车坡道 climbing lane
慢车速度 crawl speed
慢车系统 idling system
慢车线 slow line
慢车行车道 slow moving traffic
"慢车右行"标志 slow traffic keep right sign
慢车专用车道 climbing lane
慢车专用坡道 climbing lane
慢沉的(土)颗粒 slow-setting particles
慢存取 slow access
慢倒动作 reverse slow motion
慢的 slow;tardy
慢递减函数 slowly decreasing function
慢电码 slow coding
慢电压机 slow-acting press
慢电子 low-velocity electron
慢电子轰击 slow electron bombardment
慢电子扫描 low-velocity scanning
慢动的 slow acting
慢动度盘 slow-motion dial
慢动夹 slow-motion clamp
慢动压机 slow-acting press
慢动作 action slow;fine motion;slow operation;slow motion
慢动作继电器 slow-acting relay; slow-to-operate relay
慢动作起动器 slow-motion starter
慢动作凸轮 slow-acting cam
慢动作效果 slow-motion effect
慢动作影片 <超速拍摄的> ultra-rapid picture
慢冻结试验 slow-freeze test
慢读出 slow read-out
慢度 <速度的倒数> slowness
慢度分量 slowness component
慢度时间域 slowness-time domain
慢度梯度 slowness gradient
慢断路器 slow chopper
慢堆堤法 slow banking(method)
慢而持续的通货膨胀 creeping inflation
慢反应 long response
慢反应动作电位 slow response action potential
慢反应时间 long response time
慢反应物质 slow reacting[reaction/reactive] substance
慢风操作 fanning
慢盖格米勒氏管 slow Geiger-Muller tube
慢干的液体道路沥青 slow-curing paving binder
慢干剂处理的抹布 tack rag
慢干沥青 slow-curing oil
慢干树脂 slow-curing resin
慢干水泥 slow cement
慢干稀释沥青 slow-curing cutback asphalt
慢干液体沥青材料 slow-curing liquid asphaltic material
慢干油 hard oil
慢干油墨 slow drying ink; slow-setting ink
慢感光层 slow emulsion
慢拱 long arm
慢固化树脂 slow-curing resin
慢关闭油路 idle cut-off
慢光 slower ray;slow ray
慢过程 slow process
慢过程电影摄影法 time-lapse photography
慢过滤器 slow filter
慢化 slow down;slowing;slowing down
慢化本领 slowing-down power
慢化比 moderating ratio

慢化不泄漏几率 slowing-down non-leakage probability
慢化不足 undermoderation
慢化不足混合物 undermoderated mixture
慢化不足体系 undermoderated system
慢化不足栅格 undermoderated lattice
慢化材料 moderating material; slowing material
慢化长度 moderation length;slowing-down length
慢化的 moderating
慢化等离子粒团 decelerated plasmoid
慢化堆芯 moderated core
慢化法 moderative method
慢化反射层 moderating reflector
慢化反应堆 moderated reactor
慢化非漏失概率 non-leakage probability of slowing down
慢化辐射 moderated radiation; slowing-down radiation
慢化过程 moderating process
慢化函数 slowly varying function
慢化核 moderated kernel; slowing-down kernel
慢化核反应堆 moderated nuclear reactor
慢化积分方程 slowing-down integral equation
慢化剂 moderator
慢化剂俘获 moderator capture
慢化剂换热器 moderator heat-exchanger
慢化剂净化 moderator purification
慢化剂密度起伏 moderator-density fluctuations
慢化剂排放安全机构 moderator dumping safety mechanism
慢化剂砌体 moderator structure
慢化剂栅格 moderator lattice
慢化剂循环系统 moderator circulation system
慢化剂液位调节阀 control valve of moderator level
慢化剂溢流 moderator overflow
慢化剂与燃料比 moderator-to-fuel ratio
慢化剂组件 moderator assembly
慢化截面 slowing-down cross-section
慢化矩 slowing-down moment
慢化冷却剂 moderator coolant
慢化理论 slowing-down theory
慢化密度 slowing-down density
慢化面积 slowing-down area
慢化模型 slowing-down model
慢化能力 moderating power;slowing-down power
慢化能量分布 slowing-down energy distribution
慢化能谱仪 slowing-down spectrometer
慢化年龄理论 moderation age theory
慢化球 moderating sphere
慢化区 slowing-down region
慢化石灰 slow slaking lime
慢化时间 retardation time
慢化时间测谱学 slowing-down time spectrometry
慢化束 degraded beam
慢化速度 moderation velocity
慢化探测器 moderating detector
慢化系数 moderating ratio
慢化效率 moderating efficiency
慢化效应 moderating effect
慢化性能 slowing property
慢化中子 moderated neutron
慢化组 slowing-down group
慢挥发 slow vapo(u)rization
慢挥发溶剂 slow-evapo(u)rating solvent
慢挥发稀释 slow-evapo(u)rating diluent
慢回波 slow echo wave
慢火 ash fire
慢积集 slow accumulation
慢计数率 slow counting rate
慢加荷载 slowly applied load
慢加热法 slow heating technique
慢加速 gentle acceleration
慢加速波 slow accelerating wave
慢剪 slow shear
慢剪试验 consolidated-drained direct shear test;drained shear test;slow (shear) test;S-test
慢剪直剪试验 slow direct shear test
慢交换 slow exchange
慢角变量 slow angular variable
慢结合 slow combination
慢解乳浊 slow-breaking emulsion
慢进度 slow progress
慢进给 jog;slow feed
慢镜头 slow motion
慢开口阀 slow opening valve
慢空穴 slow hole
慢快中子双区反应堆 slow fast reactor
慢扩散源 slow diffusant
慢拉辊 pulldown roller
慢老化 slow aging
慢累积剂量 slowly-accumulated dose
慢冷 slow cooling
慢冷法 slow cooling method
慢冷熟料 slowly cooled clinker
慢离子 slow ion
慢裂 slow-breaking
慢裂变 slow fission
慢裂变反应 slow fission reaction
慢裂的 slow banding;slow-setting
慢裂沥青乳液 fully stable emulsion; slow-setting asphaltic emulsion; slow set emulsion
慢裂乳化地沥青 slow-breaking emulsified asphalt; slow-setting emulsified asphalt
慢裂乳液 slow-breaking emulsion; slow-setting emulsion
慢裂阳离子路用乳液 <英> slow acting cationic road emulsion
慢溜放示像 hump-slow aspect
慢溜放显示 hump-slow indication
慢流 slug flow
慢流速纸 slow flow rate paper
慢漏失 slow leakage
慢滤池 low-rate filter; slow filter; slow sand filter
慢滤(法) slow filtration
慢滤器 low-rate filter;slow filter
慢脉冲星 slow pulsar
慢慢绞 heave in easy
慢慢进行 worm
慢慢爬 snail-like movement
慢慢倾斜 shelving
慢慢倾注 decant
慢慢移动 moving slowly;shuffling
慢慢煮沸 simmer
慢门机 slow-speed device
慢弥散 slow dispersion
慢模式波 slow mode wave
慢黏[粘]接 slow sticking
慢凝 slow hardening
慢凝玻璃 slow-setting glass
慢凝导火线 slow match
慢凝的 slow-curing;slow-setting
慢凝(地)沥青 slow-curing asphalt
慢凝环氧灌浆 epoxy slow setting grout
慢凝灰浆 slow-setting mortar
慢凝混凝土 slow-setting concrete
慢凝集 slow condensation
慢凝精制地沥青 slow-curing cutback asphalt
慢凝轻制地沥青 slow-curing cutback asphalt
慢凝乳化沥青 slow-setting emulsified asphalt
慢凝乳液 slow-setting emulsion
慢凝式乳化沥青 slow-setting asphalt emulsion
慢凝水泥 setting cement;slowly taking cement; slow-setting cement; slow(-taking) cement
慢凝稀释地沥青 slow-curing cutback asphalt
慢凝液体地沥青 <可由轻制或直接蒸馏残渣取得> slow-curing liquid asphalt
慢凝液体地沥青材料 slow-curing liquid asphaltic material
慢排列 slow permutation
慢排液膜 slow draining film
慢盘齿轮 back gear
慢漂爆(发) slow drift burst
慢漂移 slow drift
慢坡 easy grade;glacis
慢坡的 acclivous
慢切削 slow cutting
慢燃 lingering
慢燃导火线 slow match
慢燃的 slow-burning
慢燃构造 slow-burning construction; slow-burning structure
慢燃黑色炸药 slow powder
慢燃结构物 slow-burning construction;slow-burning structure
慢燃烧 slow combustion
慢燃无烟火药 slow-burning smokeless powder
慢燃物 heavy fuel
慢燃线 slow-burning wire
慢燃药 slow-burning powder
慢燃引信头 slow match
慢溶物质 slowly soluble material
慢蠕变 slow creeping
慢扫描 long scan;slow scan(ning)
慢扫描示波器 slowed sweep oscilloscope
慢扫描速度 slow scanning rates
慢扫描同步 slow scan sync
慢扫描同步检测 slow scan sync detection
慢砂滤池 slow sand filter
慢砂滤床 slow sand filter bed
慢砂滤器 slow sand filter
慢射 slow fire
慢渗入试验 slow penetration test
慢渗透性 slow permeability
慢时间标度 extended time scale;slow-time scale
慢示器 bradyscope
慢释 slow release
慢衰减标准偏差 standard deviation of slow fading
慢衰落 long-term fading;slow fading
慢衰落余量 slow fading margin
漫水桥 submersible bridge
慢速标度 slow-time scale
慢速操作机器 lugging
慢速车道 slow vehicle lane
慢速齿轮 slow gear
慢速传动 inching drive; slow-speed drive
慢速存储器 slow memory;slow storage;slow store
慢速存取 slow access
慢速存取存储器 slow-access memory
慢速倒车 slow astern
慢速道 <为运输火箭或宇宙飞船而建的> crawlerway
慢速的 idling;jogging;slow
慢速点燃 slow ignition
慢速电动机 inching motor;torque motor
慢速电子枪 low-velocity gun
慢速电子束 low-velocity electron beam
慢速冻结 slow freezing
慢速度 jogging speed;low speed
慢速度混合器 slow-speed mixer
慢速断续器 slow-speed interrupter
慢速负片 slow negative film
慢速隔离阀 slow closing valve
慢速给进齿轮 slow feed gear
慢速固结压缩试验 slow consolidated compression test
慢速过滤法 slow filtration
慢速化学吸附 slow chemisorption
慢速计数管 slow counter
慢速记忆 slow memory
慢速继电器 slow relay
慢速加载试验 slow test
慢速搅拌器 slow stirrer
慢速卷扬机 low-speed winch; slow-motion hoist
慢速控制 slow-speed control
慢速连续输送机 creeper
慢速耙式搅拌器 slow rake-stirrer
慢速拍摄 time lapse
慢速刨煤机 slow-speed coal plough
慢速喷灌 aeration irrigation
慢速喷嘴 idling jet;slow jet
慢速喷嘴调节装置 idling adjustment
慢速起动装置 <快速过滤池的> slow starter
慢速器件 slow device
慢速桥式起重机 slow-speed bridge crane
慢速扫描 slow sweep
慢速砂滤 slow sand filtration
慢速砂滤层 slow sand filter
慢速砂滤池 low-speed sand filter
慢速砂滤床 slow sand filter bed
慢速摄像机 slow-motion camera
慢速摄影 memomotion;time-lapse photography
慢速摄影机 slow-motion camera
慢速摄影术 memomotion photography
慢速摄影研究 memomotion study
慢速渗滤 slow infiltration; slow rate filtration
慢速渗滤土地处理系统 slow-rate infiltration-land treatment system
慢速渗滤系统 slow land treatment system
慢速时标 slow-time scale
慢速时间比例 extended time scale; slow-time scale
慢速适应 chronic adaptation
慢速双辊破碎机 double roll slow speed crusher
慢速丝杠 slow-motion screw
慢速调节 idling adjustment;slow adjustment
慢速透镜 slow lens
慢速图像搜索 slow visual search
慢速显示 slow-speed indication
慢速显影 slow development
慢速行驶 slowly-moving
慢速移动 creeping
慢速硬化 slow hardening
慢速运动 creeping motion
慢速运动时间 slowing time
慢速运输 slow traffic
慢速运行 slow running
慢速运转 microrunning
慢速轧机 low-speed mill
慢速蒸煮 slow cooking
慢速正车 slow ahead
慢速执行 slow execution
慢速直接记录式纸带记录器 slow-

M

speed direct recording tape recorder
慢速纸带记录器 slow-speed tape recorder
慢速指示 slow-speed indication
慢速中子探测器 slow neutron detector
慢速重放 slow regeneration
慢速轴 slow axis
慢速筑堤法 slow banking(method)
慢速转动 barring
慢速自行底盘 creeper speed-propelled carrier
慢锁定 slow genlock
慢台从锁相 slow genlock
慢态 slow surface state
慢天空辐射 sky radiation
慢停车 <站外停车> deferred stop
慢停堆馏出 run down
慢通道 slow channel
慢透射密度 diffused transmission density
慢推送示像 push-slowly aspect
慢弯管 easy bend
慢弯试验 slow bend test
慢吸 slow pickup
慢吸快释继电器 slow-operate fast-release relay
慢吸收 slow trapping
慢下来 slow
慢相 slow phase
慢响应 slow response
慢响应时间 long response time
慢消化石灰 slow slaking lime
慢行 crawling; slow-speed; slow-speed running
慢行臂板 slow-speed arm
慢行标志 go-slow sign; reduce speed sign; slow sign
慢行车道 crawler lane
慢行车辆 slow moving vehicle
慢行程 idle stroke
慢行道路 <发射场等> crawlerway
慢行地段 slow section
慢行规定 slow order
慢行交通 slow moving traffic
慢行路标 go-slow sign
慢行命令 slow order
慢行牌 slow board; speed indicator; yellow slow board
慢行汽车 slow moving vehicle
慢行区域 slow drive zone
慢行弱火 creeping fire
慢行速度 velocity of creep
慢行速率 creep speed
慢行停车 slow stoppage
慢行线段 slow track
慢行信号 slow-down signal; slow drive signal; slow-speed signal; speed slackening signal
慢行信号旗 slow flag
慢型 slow type
慢性暴露 chronic exposure
慢性病 chronic disease
慢性病(医)院 chronic-disease hospital
慢性的 chronic
慢性毒性 chronic toxicity
慢性毒性试验 chronic toxicity test
慢性毒作用带 chronic toxic effect zone
慢性放射病 chronic radiation disease
慢性放射损伤 chronic radiation injury
慢性氟中毒 chronic fluorine poisoning
慢性辐射 chronic radiation
慢性辐射危害 chronic radiation hazard
慢性辐射效应 chronic radiation effect
慢性辐照 chronic irradiation
慢性高山病 chronic mountain sickness
慢性镉中毒 chronic cadmium poisoning
慢性汞中毒 chronic mercury poisoning
慢性轰击 chronic bombardment
慢性活动抑制试验 chronic immobilization test
慢性磷中毒 phosphorism
慢性钼中毒 mollybdenosis
慢性伤害 chronic damage
慢性摄入 chronic intake
慢性砷中毒 arseniasis; chronic arsenic poisoning
慢性实验 chronic experiment
慢性危害 chronic hazard
慢性污染 chronic pollution
慢性污染源 chronic pollution source
慢性效应 chronic effect
慢性锌中毒 zincalism
慢性溴中毒 chronic bromism
慢性氧化 eremacausis; slow oxidation
慢性阈剂量 chronic threshold dose
慢性阈浓度 chronic threshold concentration
慢性炸药 slow explosive
慢性照射 chronic exposure
慢性支气管炎 chronic bronchitis
慢性中毒 chronic oxygen poisoning; chronic poisoning
慢絮凝 slow coagulation
慢旋转 slow spin
慢选择转子 slow chopper
慢延迟轴 slow retardation axis
慢延火焰 slow flame
慢移低气压 slow moving depression
慢移动低(气)压 slow moving depression
慢引出 slow extraction
慢引出束 slow extracted beam
慢引出系统 slow extraction system
慢引出装置 slow extractor
慢硬玻璃 slow-setting glass
慢硬的 slow hardening
慢硬混凝土 slow-hardening concrete
慢硬水泥 slow-hardening cement
慢运转 slow running
慢增长函数 slowly growing function
慢振荡 slow oscillation
慢振方向 slow vibration direction
慢震颤 coarse tremor; tremor tardus
慢蒸发性溶剂 slow-evapo(u)rating solvent
慢制动 slow application
慢中子 slow(-speed) neutron; thermal neutron
慢中子反应堆 slow reactor
慢中子干涉仪 slow neutron interferometer
慢中子过滤器 slow neutron filter
慢中子计数器 slow neutron counter
慢中子裂变 slow neutron fission
慢中子探测室 slow neutron detecting chamber
慢中子选择器 slow neutron chopper
慢中子照射室 slow neutron exposure chamber
慢轴 slow axis
慢转 idle run
慢转变 sluggish inversion
慢转变装置 barring unit
慢转唱片 long-playing disc
慢转的 slow acting; slow-speed
慢转换 lap dissolve
慢转换器 lap dissolve shutter
慢转换装置 lap dissolve shutter
慢转控制回路 slow turning control circuit
慢转录音或放音 long playing
慢转履带 slower-moving track
慢转密纹唱片 long-playing record
慢转天体 slow rotator
慢转中子谱仪 slow chopper neutron spectrometer
慢子 tardyon
慢走刀 jog
慢作用计数管 slow counter

漫

漫布 dispersed; overspread
漫步道 foot walk; footway
漫步路 trail
漫槽河(流) overflow river; overflow stream
漫槽水位 overflow level
漫出 overflow; overrun
漫地水流 overland flow
漫顶 overtopping; overwash; topping
漫顶(流)量 overtopping discharge; quantity of overtopping; rate of overtopping
漫顶率 overtopping rate; rate of overtopping
漫顶水位 overtopping(water) stage
漫顶围堰 overtopped cofferdam
漫反射 diffuse(d) reflection; scattering reflection; scattered reflection
漫反射斑 diffusing disk
漫反射本领 diffused reflecting power
漫反射比 diffused reflectance
漫反射玻璃 diffused reflection glass
漫反射带 diffused bond
漫反射的 irreflexive
漫反射度 diffused reflectance
漫反射分光法 diffuse reflectance spectroscopy
漫反射光度计 diffused reflection photometer
漫反射光谱 diffused reflection spectrum
漫反射光谱法 diffused reflectance spectroscopy
漫反射镜 diffused reflector
漫反射率 diffused reflectance; diffused reflecting power; diffuse-reflection coefficient
漫反射目标 diffused reflecting target
漫反射能力 diffuse-reflecting power
漫反射器 diffused reflector
漫反射腔 diffuse-reflective cavity
漫反射体 diffused reflector
漫反射系数 albedo; coefficient of diffuse(d) reflection; diffused reflectance; diffused reflection coefficient; diffused reflection factor; diffuse(d) reflectivity
漫反射罩 diffused reflector
漫辐射 diffused radiation; stray radiation
漫辐射场 diffused radiation field
漫辐射形态 diffused radiation form
漫箍缩效应 diffused pinch effect
漫灌 artificial flooding; basin irrigation; broad irrigation; catchwork irrigation; copious irrigation; flooding; flood(ing) irrigation; flush irrigation; irrigation by flooding; surface flood(ing); wild-flooding irrigation
漫灌法 flooding irrigation method; inundation method; wild-flooding method
漫灌式入渗仪 flooding type infiltrometer
漫灌式下渗仪 flooding type infiltrometer
漫灌塘 flooding basin
漫灌系统 flood irrigation system; inundation irrigation system
漫光塑料板 diffusing plastic sheet
漫光塑料膜 diffusing plastic sheet
漫过坝顶 topping
漫洪 sheet flood; sheet wash
漫洪草原 water meadow
漫画法 <市场调查> balloon test
漫画化人头像装饰 mascaron
漫画家 caricaturist
漫流 overflow; overland flow; sheet flood; sheet flow; spreading(out); spread of flow; water spreading
漫流长度 length of sheet flood; length of sheet flow
漫流沉积 sheet-flood deposit
漫流法 overland flow method
漫流过程线 overland flow hydrograph
漫流河段 bayou
漫流流速 cross-flow rate
漫流式间歇泉 flooded geyser
漫流水层 nappe
漫流水层的分离 separation of water layer
漫流滩地 overflow land
漫流系统 overland flow system
漫流性 flowability
漫密西神殿 <埃及> Birth House
漫强谱 diffuse-enhanced spectrum
漫燃 spread of fire
漫散 diffuse; scattering
漫散边带 diffused sideband
漫散减光滤光片 diffuse-cutting filter
漫散球 diffusing globe
漫散射 diffused scattering
漫散射背景 diffused scattering background
漫散射玻璃 non-reflecting glass
漫散射面 diffused scattering surface
漫散射中子 diffusely scattered neutrons
漫散生长 diffused growth
漫散条纹 diffused fringe
漫散衍射 diffused diffraction
漫散着丝粒 diffused kinetochore
漫射 diffuseness; diffuse scattering; diffusion; random scatter(ing)
漫射 X 线反射 diffused X-ray reflection
漫射斑 diffused spot
漫射板 diffusing panel
漫射本底 diffusing background
漫射壁腔 diffused wall cavity
漫射表面 diffusing surface
漫射波 diffused wave
漫射玻璃 depolished glass; diffusing glass; diffusion glass
漫射玻璃屏 diffusing glass screen
漫射测标 diffused measuring mark
漫射场 diffused scattering field
漫射传输 diffused transmission
漫射窗 diffused window
漫射带 zone of diffuse(d) scattering
漫射灯 direct diffused light
漫射灯泡 diffusion lamp
漫射灯罩 diffusing fitting
漫射电辐射 diffused radio emission
漫射发射 diffusely transmitting
漫射反光镜 diffusion reflector
漫射方程 diffusion equation
漫射放大机 diffusion enlarger
漫射峰 diffused maximum
漫射辐射 diffusion radiation
漫射辐射能 stray radiant energy
漫射附件 diffusion attachment
漫射关系 irreflexive relation
漫射光 broad light; diffuse(d) light; diffusion light; stray light
漫射光玻璃 lighting glass
漫射光带吸收 diffused band absorption
漫射光度计 diffusion photometer

漫射光放大机 diffusion-type enlarger
漫射光均匀照明 diffused lighting;general diffuse(d) lighting
漫射光亮度 stray radiance
漫射光密度 diffused density
漫射光谱 diffused spectrum
漫射光天花板体系 light-diffusing ceiling system
漫射光线 general diffuse(d) ray;diffused light;diffused ray
漫射光源 diffused light source;direct diffused light
漫射光晕 diffusion halo
漫射光照明 diffused illumination;diffused lighting;direct-indirect lighting
漫射光照明器 diffuse light luminaire
漫射光照明墙(壁) diffusing wall
漫射光照明天花板 diffusing ceiling
漫射光照明装置 lighting diffuser
漫射光锥 stray light cone
漫射环 diffused ring
漫射回声 diffused echo
漫射极光 diffused aurora
漫射角 angle of diffusion;scattering angle
漫射宽容度 diffusion altitude
漫射滤光器 diffused filter
漫射媒质 diffusing media
漫射面 diffused surface;mat(te) surface
漫射面加工 diffuse finish
漫射模糊圈 diffused blur circle
漫射(能)源 diffusing source
漫射盘 diffusion disc[disk]
漫射片 diffusing disc[disk];diffusion disc[disk]
漫射屏 diffused screen
漫射谱带 diffused band
漫射器 diffuser;diffusing globe
漫射球 globe of diffusion
漫射圈 circle of diffusion;diffusing disk
漫射全息照相 diffused holography
漫射热 stray heat
漫射热辐射 diffused heat radiation
漫射日光 diffused daylight
漫射日照数据 diffused insolation data
漫射柔光器 diffusing disc
漫射声(音) diffused sound;reverberant(steady-state) sound
漫射式放大机 diffused enlarger
漫射式滤色镜 diffused filter
漫射式微密度计 diffused microdensitometer
漫射束 scattered beam
漫射特性 diffusive property
漫射体 diffuser
漫射天花板 diffuser ceiling
漫射天然光 diffused daylight
漫射透光度 diffused transmissivity
漫射透光率 diffused transmittance
漫射透光系数 diffused transmission factor
漫射透镜 diffusing lens
漫射透射比 diffused transmissivity
漫射透射率 diffused transmittance
漫射图 circle of diffusion
漫射物体 diffused object
漫射物体光束 diffused object beam
漫射误差 stray light error
漫射系 diffused series
漫射系数 diffusion coefficient;diffusion factor
漫射相干照明 diffused coherent illumination
漫射因数 diffused transmission factor;diffusion factor
漫射音 diffused sound
漫射晕 diffused halo;diffused surround
漫射照明 diffused lighting;stray lighting
漫射照明成像系统 diffused illumination imaging system
漫射照明全息图 diffused illumination hologram
漫射遮光屏 diffusing screen
漫射遮光装置 diffusing screen
漫射值 diffusion value
漫射指示量 indicatrix of diffusion
漫射转移 diffusion transfer
漫生 overgrowth
漫水 water overtopping
漫水坝 basin dam;overtopped dam
漫水草地 water meadow
漫水层 flood coat
漫水地 overflow land
漫水丁坝 submersible spur dike
漫水管桥 submerged tube bridge
漫水滚筒 submersible roller
漫水涵洞 submerged culvert
漫水阶地 first bottom
漫水路 overflow road
漫水桥 low-level bridge;overflow bridge;submergible bridge
漫水土层 submerged earth
漫水屋顶 flooded roof
漫滩 bet;land liable to flood;tidal flat
漫滩沉积(物) flood plain deposit;overbank deposit;overbank sediment
漫滩冲积泥沙 splay deposit
漫滩冲积扇 flood plain splay;overbank splay
漫滩冲积物 flood play sediment;splay deposit;splay sediment
漫滩地 flood land;landwash;valley flat;flood plain
漫滩堆积(物) flood plain deposit;flood plain sediment
漫滩见泽 backswamp
漫滩阶地 flood plain bench;flood plain terrace
漫滩流 overbank flow
漫滩流量 bankfull discharge;bankfull flow;overbank flow;overland flow
漫滩泥沙 overbank deposit
漫滩平原 overbank plain
漫滩区 overbank area
漫滩式 <指水库形式> flood plain type
漫滩水流 overbank flood plain
漫滩水位 bankfull(water) level;bankfull(water) stage;overbank stage
漫滩台地 flood plain bench
漫滩-天然堤沼泽 back-levee marsh
漫滩洼地 backswamp depression
漫滩蓄水量 overbank storage
漫滩沼泽 backswamp
漫滩沼泽地 back marsh
漫滩浊积岩 overbank turbidite
漫天空辐射 diffused sky radiation
漫天要价 price oneself out of the market
漫透射 diffused transmission
漫透射率 diffused transmissivity;diffuse transmittance
漫透射密度 diffused transmission density
漫透射系数 diffused transmission factor
漫吸收谱带 diffused absorption band
漫线系 diffused series
漫烟型 fumigation
漫溢 flood irrigation;overflow(ing);overtopping;spill
漫溢临界水深 overflow critical depth;overrun critical depth
漫溢预防 overflow-prevention
漫涌破碎波 spilling breaker
漫涌碎浪 spilling breaker
漫游 flooding;navigate;nomadism;rove
漫游电子 roaming electron
漫游硅藻 vagile diatom
漫游客票 ranger ticket
漫游生物 errantia
漫游水生动物 vagile aquatic animals
漫游藻类 errant algae
漫照明 diffused illumination
漫折射 diffused refraction

蔓 tendril;vindicareine

蔓草 sod grass;vine;weed vine
蔓草类型 twing habit
蔓草植物 gadding plant
蔓草作物 vine plant
蔓格排水系统 grapevine drainage
蔓茎状 vine
蔓毛壳属 <拉> Herpotrichia
蔓棚 pergola
蔓棚架下步道 arbor walk
蔓棚藤架荫道 covered walk
蔓生 overgrow;ramp
蔓生草 creeping bent;sod grass
蔓生植物 liana;vine crop
蔓藤浮雕 creepers
蔓藤花纹 arabesque;arabesquitic
蔓藤棚架 pergola
蔓藤棚架下步道 arbour walk
蔓形 calyculate
蔓性植物 trailing plant
蔓性种蔷薇 rambler rose
蔓延 dilatation;extend;gradation;overgrowth;overspread;propagation;sprawl;spread(ing);contagion <学说、思想、感情、谣言等的>
蔓延草 creeper
蔓延城市 spread city
蔓延的 rampant
蔓延飞火 spotting fire
蔓延分布 contagious distribution
蔓延井 extension well
蔓延前的 preinvasive
蔓延时间 <火焰的> sweep time
蔓延式发展 sporadic development
蔓延速率 rate of spread
蔓延速率计算器 rate-of-spread meter
蔓延习性 trailing habit
蔓延中的火 developing fire
蔓叶花饰 trayle;vignette
蔓叶花样 flourish
蔓叶类曲线 cissoidal curve
蔓叶线 <一种几何曲线> cissoid
蔓状丛 pampiniform plexus
蔓状的 pampiniform;racemose

镘 trowel

镘板 floating rod;floating rule;hawk;lute;mortar-board;pallet;patter;rubbing board;smoother;smoothing board;smoothing trowel
镘尺 darby
镘刀 <抹灰用> float;hand float;hand float trowel;laying on trowel;mason's float;skimmer float;steel trowel(ling);wooden float;darby;finishing trowel
镘刀抹平混凝土表面 float finish
镘刀饰面 float finish
镘刀涂光 trowel(l)ing
镘刀涂抹 trowel application
镘刀用法 trowel application
镘工 floater
镘光 trowel-applied finish(ing);trowel finishing;trowel(l)ed finish
镘光表面 trowel(l)ed surface
镘光带 <混凝土路面> smoothing belt
镘光灰泥面 set fair
镘光混凝土 floated concrete
镘光机 trowel(l)ing machine
镘光路面 smoothing pavement
镘光石 float stone
镘灰 plaster
镘灰板 hawk
镘具整修 float finish
镘抹痕迹 trowel mark
镘抹面 trowel finish
镘抹平 trowel finishing
镘抹无缝地面材料 trowel-applied seamless floor(ing) material
镘平 render-float-and-set;trowel(led) finish(ing)
镘平表面 floated surface;trowel finish
镘平机 mechanical float;trowel(l)ing machine
镘平面 trowel(l)ed face;trowel(l)ed surface
镘平作业 floating
镘涂 trowel(l)ing
镘涂层 trowel coating
镘涂稠度 trowel(l)ing consistency
镘涂灰浆 buttering
镘涂涂料 trowelable coating
镘修整 float finish
镘整地沥青(混合料) floated asphalt
镘整混凝土 floated concrete

忙等待 busy waiting

忙(蜂)音 busy-back tone;busy-buzz
忙呼叫计数表 congestion call meter
忙回信号 busy back
忙季 rush season
忙接点 busy contact
忙碌测试 busy test
忙碌等待 busy waiting
忙碌期间 busy period
忙碌试验 engage test
忙碌通道 busy channel
忙碌位 busy bit
忙碌信号 busy signal
忙碌状态测试 busy test
忙期【数】busy period
忙期分布 busy period distribution
忙时 busy hour(traffic);heavy hour;rush hours
忙时串杂音 busy hour crosstalk and noise
忙时呼叫 busy hour call
忙时业务 rush-hour traffic
忙时运量 rush-hour traffic
忙线 busy line;engaged line
忙线路 busy line
忙线信号灯 busy lamp;visual busy lamp;visual engaged lamp
忙音 busy-back signal;busy tone;busy tone signal;engaged signal;engaged tone
忙音继电器 busy-back interrupter relay
忙音连续片 busy link
忙音塞孔 busy-back jack;busy jack
忙音信号 busy signal;busy tone signal
忙音中继线 busy tone trunk
忙用时间 busy period
忙状态 busy condition

芒草 Chinese silvergrass;erianthus gianteus

芒草篮 fern basket

芒长 awn length
芒茨合金 malleable brass
芒刺清除机 bur cleaner
芒刺式散热器 barb radiator
芒垫 fern cushion
芒果胶 mango gum
芒果树 mango
芒壳油 bur oil
芒麦草 wild barley
芒坡下降曲线的巴氏函数 Baplorfstz function of depression curve for positive inclination
芒特艾萨矿业控股公司 < 澳大利亚 > MIM Holdings Ltd.
芒硝 Glauber salt; mirabilite; sal mirabile; salt cake; sodium sulphate
芒硝层 < 钙积层的上部 > chuco
芒硝锅 salt cake pan
芒硝湖 mirabilite lake
芒硝回转筛 salt cake rotary screen
芒硝炉 salt cake roaster
芒硝泡 grey blibe
芒硝侵蚀 mirabilite cutting
芒硝岩 mirabilite rock
芒屑 awn chaff
芒云母 montdorite
芒制草绳 fern plaited rope
芒状 aristiform

盲板 blank cap; blank flange; blanking plate; blankoff flange; blankoff plate; blank plate; blind plate; blind flange < 集装箱的 >

盲边框 blank jamb
盲肠 c(a)ecum[复 ceca/caeca]
盲肠的 cecal
盲肠河汊 billabong
盲穿孔机 blind perforator
盲窗 blank window; blind window; dead window; false window
盲带 dead band
盲袋 blind bag
盲道 dummy road; sidewalk for the blind
盲底板 blankoff plate
盲点 black spot; blindness; blind spot; scotoma
盲洞 blind cave; blind hole
盲度 darkness
盲端 c(a)ecum[复 ceca/caeca]; cap end; dead end
盲断层 blind fault
盲发送 blind sending
盲法兰 blank flange; blankoff flange; blind flange
盲盖板 blank cap
盲港 blind harbo(u)r
盲拱 arcature
盲拱廊 blind arcade
盲沟 < 用碎石或砾石填满的 > blind ditch; blind drainage conduit; blind subdrain; boulder ditch; catch-drain; dry stone drain; filter drain; gravel drain; infiltration ditch; mole drain; mound drain; rubble drain; sough; stone drain; underdrain; blind drain; French drain; stone-filled drain
盲沟进水口 blind inlet
盲沟排水 blind drainage; French drainage
盲沟设计 blind ditch design; blind drain design
盲沟式截留井 blind catch basin
盲沟式进水井 gravel-type gully
盲谷 hope; poljie[复 polgia]; blind valley < 灰岩区分段出现的河谷 >
盲管 blind pipe
盲管洞 blind pipe cave
盲管段 dead leg
盲管流量测定 blinded-pipe yield test
盲航法 dead reckoning
盲机 blind machine
盲键盘 blind keyboard
盲浇口 dead runner
盲角 blind angle
盲铰 blind hinge
盲接头 blind joint
盲节 blind knot
盲节理 blind joint
盲井 blind; blind shaft; bonstay; dead hole; foramen caecum; staple pit; underground shaft; winze
盲孔 blind(bore) hole; dead hole
盲孔盖 blank cap
盲孔螺母 blind nut
盲孔千分尺 intarometer
盲孔丝锥 bottoming tap
盲孔芯 blind set core
盲孔钻进技术 blind borehole technique
盲矿 blind ore
盲矿床 blind deposit
盲矿体 blind ore body
盲流人口 aimlessly drifting population
盲铆钉 blind rivet
盲目布井 Australian offset
盲目的计划 haphazard plan
盲目地起作用 blind workings of the value principle
盲目发展 unchecked development
盲目飞行条件 instrument flight rules condition
盲目飞行图 instrument chart
盲目航行 blind navigation
盲目驾驶 blind driving
盲目降落 instrument landing
盲目降落临场导航图 instrument approach chart
盲目降落临场设备 instrument landing system
盲目进场 blind approach
盲目进场信标系统 beam approach beacon system; blind approach beacon system
盲目竞争 blind competition; unbridled competition
盲目开采的 wildcatting
盲目控制 blind monitoring
盲目控制器系统 blind controller system
盲目拦截 blind interception
盲目起降跑道 instrument runway
盲目上项目 rashly launching new project
盲目生产 blind production
盲目试验 blind trial
盲目搜索 blind search
盲目投资 plunge
盲目线性化 blind linearization
盲目引进 blind introduction
盲目着陆 blind approach; blind landing; instrument landing
盲目着陆方式 instrument landing approach
盲目着陆脉冲系统 pulsed landing system
盲目着陆实验所 Blind Landing Experiment Unit
盲目着陆试验装置 blind landing experiment
盲目着陆系统 instrument landing system
盲目着陆系统信标 instrument landing system localizer
盲目钻井区 wildcat area
盲目钻探 cold nosing; wildcat drilling; wildcatting
盲炮 misfire; unexploded charge
盲区 < 接收不到信号的区域 > black-out area; blind area; blindness; blind sector; dead ground; dead space; dead spot; dead zone; fade zone; blind zone
盲区流量 dead-band current
盲区图 fade chart
盲曲 blind-loop
盲色胶片 colo(u)r blindness film
盲式集团 blind pool
盲式集资 blind pool
盲试验 blind test
盲树 blind tree
盲竖井 blind shaft
盲水道 blind channel
盲凸缘 blank flange
盲湾 < 河口泻湖湾的 > blind estuary
盲纹 blind fret
盲纹孔 blind pit
盲巷 dummy road
盲斜井 blind slope
盲芯 blind set core
盲哑院 blind and dumb asylum
盲眼 blind hole
盲叶轮 blanked off impeller
盲异常 blind anomaly
盲鱼 blind fish
盲障 blindage
盲枝 blight wood
盲蛛 daddy-longlegs

莽草油 illicium oil

莽那卡特炸药 Monarkite

猫道 catwalk

猫道钢构件 foot bridge steelwork
猫洞 < 门、墙上 > cathole
猫脸 < 抹面瑕疵 > catface
猫头链 < 拧钻杆、套管用的 > spinning line
猫头轮 capstan head
猫头绳 spinning cable; spinning line
猫头绳导向轮 catline guard
猫头绳钩 catline hook
猫头绳滑车 catline sheave
猫头石 cat's head
猫头饰线脚 < 古世纪 > cat's head mo(u)lding
猫尾树 cat-tall tree
猫眼 cat's eye; peep hole < 俗称 >
猫眼灯 automatic sensor light
猫眼反光路钉 cat's eye reflecting road stud
猫眼反射镜 cat's eye reflector
猫眼管 cathode-ray tuning indicator
猫眼光栏 cat's eye diaphragm
猫眼后向反射器 cat's eye retroreflector
猫眼石 chatoyant; harlequin opal; sunstone; cat's eye
猫眼式波形 cat's eye waveform
猫眼效应 chatoyance[chatoyancy]

毛胺染料 setamine colo(u)rs

毛把托架 handle bracket
毛白杨 Chinese white poplar
毛白蚁科 Serritermitidae
毛斑 < 油漆面的 > sleepiness
毛板 flag slab; hair plate; rough board; rough plank
毛板地板 rough lumber floor
毛板下锯法 though-and-though sawing
毛保(险)费 gross premium
毛杯菌属 < 拉 > Cookeina
毛背面胶片 mat-backed film
毛比重 bulk specific gravity
毛笔 sable writer; writing brush
毛笔腐蚀 brush etching
毛笔画 brush work
毛笔形的 penicillate; penicilliary; penicilliform
毛边 burred edge; flash; fragment; free edge; fringe selvedge; raw edge; rough edge; rough selvedge; scallop; slack side; edge as cut < 板玻璃切割后的 >; fraze < 模锻上清除飞刺后的 >
毛边的 deckle-edged; unedged
毛边的石块 dragged
毛边毛刺 burr
毛边木料 unedged timber
毛边刨削机 burr removing machine
毛边书 uncut
毛边套 trichotillomania
毛边纸 moben paper
毛辫整流 haired fairing
毛表面 hairy surface
毛冰 rime
毛病原因 source of trouble
毛玻璃 clouded glass; depolished glass; diffusing glass; etched glass; frosted glass; frosting glass; matted glass; mat(te)-surfaced glass; non-glare glass; rough cast glass; sand-blasted glass; sponge glass; visionproof glass; acid-ground glass < 用氢氟酸加工的 >; Arctic glass < 有金属丝加固的 >
毛玻璃板 rough plate glass
毛玻璃标度盘 milk-glass scale
毛玻璃测管 ground glass sounding tube
毛玻璃测距仪 ground glass rangefinder
毛玻璃窗 clouded window; obscuring window
毛玻璃灯泡 frosted lamp
毛玻璃灯罩 elen shade
毛玻璃检影 ground glass viewing
毛玻璃聚焦摄影机 camera with ground glass focusing
毛玻璃棱镜 ground glass prism
毛玻璃漫射器 ground glass diffuser
毛玻璃片 frosted glass plate
毛玻璃片对光 ground glass focusing
毛玻璃屏 ground glass screen
毛玻璃屏像 ground glass screen image
毛玻璃器皿 frosted glass plate
毛玻璃投影屏 ground glass screen
毛玻璃网目片 ground glass screen
毛玻璃寻像器 ground glass finder
毛玻璃样 ground glass appearance; ground glass-like
毛玻璃罩 frosted glass lamp shade
毛布 haircloth; whittle
毛布洗涤器 felt cleaner; felt conditioner
毛材 rough lumber; undressed lumber; unwrought timber
毛糙的 careless; coarse; crude
毛糙地面 rough floor; rough terrain
毛糙地面作业 < 起重机 > rough terrain operation
毛糙眼 rough eye
毛糙重 rough weight
毛测 full-scale
毛层 batt
毛差额 gross spread
毛产量 gross output; gross production
毛长 staple length
毛长石 rough ashlar
毛衬布 hymonette; wool linen
毛尺寸 nominal dimension; nominal measure; nominal size

毛齿菌属 <拉> Hydnochaete
毛赤铜矿 chalcotrichite; hair copper; plush copper ore
毛翅 <轧材缺陷> list edge
毛虫 caterpillar
毛虫毒毛 toxicfurs of caterpillar
毛虫状气孔 worm hole
毛出 outlet seam
毛出力 gross output
毛窗框 rough frame
毛刺 barb; burr; flash (burr); fuzzy grain; grainlifting; sentus; skin needling; spilliness <钢丝表面缺陷>; sliver <轧制缺陷>
毛刺沟痕 chip marks; score
毛刺砂光 scuffing
毛刺丝 barbed filament
毛丛 floccus[复 flocci]; penicilium
毛丛长度 staple length
毛簇 tuft; tufted setae
毛撮 coma; pencil; tuft
毛锉 bastard file; coarse file
毛刀 rough cutter
毛的 gross
毛的门框 subcasing
毛登公式 Mogden formula
毛地板 blind floor; carcass floor(ing); counterfloor; plank floor(ing); rough floor; subfloor(ing)
毛地板材料 rough flooring; subflooring
毛地板钉 plank nail
毛地板覆盖层 plank floor cover(ing)
毛地板腻子 subfloor stopping
毛地板填料 subfloor filler; subfloor stopper
毛地面 rough floor
毛点 trichobotheria
毛电能消耗量 raw energy consumption
毛垫 pulvinis
毛垫圈 black washer
毛垫毡 felt-lined
毛钉菌属 <拉> Dasyscypha
毛定额 gross rating
毛洞 gross cave; gross tunnel; unlined tunnel
毛洞口 stud opening
毛断面 gross section; rough section
毛断面尺寸 rough section size
毛吨数 gross tonnage
毛额 gross amount
毛额租金 gross lease
毛尔伸缩装置 Mauer expansion installation
毛发 curled hair
毛发电水头 nominal productive head
毛发缝 hair seam
毛发灰泥粉刷 bastard stucco
毛发蓬松 bush
毛发塞垫 hair-lock
毛发湿度计 absorption hygrometer; hair hydrograph; hair hygrometer; polymeter
毛发湿度仪表 hair hydrograph; hair hygrometer
毛发水晶 hair stone
毛发丝状裂隙 hairline crack
毛发状 hairy
毛发状的 crinite
毛发状黄铁矿 capillary pyrite
毛矾石 alunogen; ceramohalite; feather alum; hair slat
毛方 <刨切单板用> flitch
毛方基床 rubble bed
毛方木 cant
毛方排锯 sash gang saw
毛方石 rough ashlar; rubble ashlar; squared rubble
毛纺厂 woolen mill
毛纺厂废水 woolen-mill wastewater
毛沸率 gross rate
毛沸石 erionite
毛粉饰 stucco
毛粉饰型板 stucco pattern
毛粉刷 rough cast plastering; rough coating; stucco
毛粪石 bezoar; hairball; trichobezoar
毛缝 raw seam
毛干 hair shaft
毛稿 dirty manuscript
毛革盖菌 Cladoderris hairy fungus; Coriolus hirsutus
毛工饰边花边 hairpin lace
毛工作日 calendar day
毛工作水头 gross operating head
毛功率 <包括损失的> gross power
毛沟 field ditch
毛沟排水 rill drainage
毛估 gross estimate; gross estimation; interim estimate; rough estimate
毛估标价 unbalanced bid(quotation)
毛估流量收费 <给排水的> flat rate
毛估损失 apparent loss
毛估盈亏 gross margin
毛骨料 <未经筛选的天然骨料> all-in aggregate
毛管 hair canal
毛管边界层 boundary zone of capillary
毛管波 capillary wave
毛管迟滞性 capillary hysteresis
毛管持水量 capillary (moisture) capacity
毛管传导度 capillary conductivity
毛管带 capillary zone
毛管灌水 capillary watering
毛管含水量 capillary moisture capacity
毛管间孔隙 infracapillary space
毛管间隙 capillary interstice
毛管检液器 capillarimeter
毛管孔隙 capillary interstice
毛管孔隙率 capillary porosity
毛管拉力 funicular force
毛管力 capillary force
毛管力差 capillary potential gradient
毛管连系破裂含量 moisture of capillary bond disruption
毛管流(动) capillary flow
毛管黏[粘]度计 capillary visco(si)-meter
毛管凝缩作用 capillary condensation
毛管潜能 capillary potential energy; potential capillarity
毛管容量 capacity of capillarity; capillary moisture capacity
毛管上升 capillary ascension; capillary lift
毛管上升水 anastatic water
毛管上限 capillary fringe
毛管上限水 anastatic water
毛管渗透 capillary percolation; capillary penetration
毛管湿锋 capillary front
毛管势(能)梯度 capillary potential gradient
毛管水 capillary water
毛管水饱和层 zone of capillary saturation
毛管水边界层 boundary zone of capillarity; boundary zone of capillary
毛管水边缘 capillary fringe
毛管水层 capillary layer
毛管水迟滞点 lento-capillary point
毛管水分 capillary moisture
毛管水高度 capillary rise; height of capillary water
毛管水扩散 capillary migration
毛管水流 capillary flow
毛管水迁移 capillary migration
毛管水上升边缘 capillary rise fringe
毛管水上升(高度) capillary rise
毛管水上升率 rate of capillary rise
毛管水试验 capillary water test
毛管水位 capillary stage
毛管水位能 capillary potential
毛管水下降 capillary depression
毛管水压力 capillary pressure
毛管水运动 capillary movement
毛管调节 capillary adjustment
毛管吸附力 capillary adsorption
毛管吸附水 capillary absorbed water
毛管吸力 capillary suction
毛管吸升高度 capillary suction head
毛管吸水头 capillary suction head
毛管吸引力 capillary attraction
毛管现象 capillarity phenomenon
毛管移动 capillary migration
毛管张力 capillary pulling power
毛管铸造法 cast shell process
毛管作用 capillarity; capillary action
毛管作用带 boundary zone of capillary; capillary fringe
毛灌溉需水量 gross irrigation requirement
毛灌水定额 gross duty (of water); head-gate duty of water
毛灌水率 head-gate duty of water
毛龟裂性褐腐 brown cubical
毛蠓属 Psychoda
毛黑变 melanotrichia
毛厚的 bushy
毛厚度 gross thickness
毛花雀稗 Dallis grass
毛化面 frosted face
毛化整理 frosted finish
毛环不齐 ununiform loops
毛灰泥抛毛 daubing
毛回丝 wool waste
毛混凝土面 as-cast finish
毛火 gross fire
毛基质 hair matrix
毛集料 all-in aggregate
毛交叉 cruces pilorum; hair cruces
毛交通量 raw traffic capacity
毛脚 dry foot
毛校样 foul proof
毛接法 coarse joining method
毛结花纹 nep pattern
毛截面 gross cross section; gross section
毛截(面)面积 gross area of section; gross cross-sectional area; gross section(al) area; total area of cross-section
毛巾 face towel; mokador; napkin; towel; wash cloth; wash rag
毛巾布 dowlas
毛巾橱 towel cabinet
毛巾分放架 towel dispenser
毛巾干燥器 towel airer
毛巾杆 towel bar; towel rail
毛巾搁架 towel supply shelve
毛巾钩 towel hook
毛巾挂 towel bar
毛巾挂环 towel ring
毛巾挂零件 towel bar fitting
毛巾柜 towel cabinet
毛巾红显色机 para-red developing range
毛巾环 towel ring
毛巾架 towel hanger; towel holder; towel ladder; towel rack; towel rail
毛巾起圈装置 towel motion
毛巾衣物架 valet towel holder
毛金边 rough gilt edges
毛筋 hairline
毛锯木板房屋 slab house
毛刻 roughing
毛孔 pin-hole; pore; trichopore
毛孔闭塞 closure of pores
毛孔开放 opening of the pores
毛孔密度 follicle population
毛孔状 poriform
毛口 bur(r); rag
毛口磨光 burr; burring
毛口磨光机(器) burr removing machine
毛库容 gross capacity of reservoir; gross reservoir capacity; gross storage
毛块 aegagropilus
毛兰 hair orchid
毛蓝土布 blue nankeen; dyed nankeen
毛雷尔等方位曲线航用图 Maurer's nomogram
毛雷尔型螺旋压榨机 Maurer type screw press
毛雷尔组织图 Maurer constitution diagram; Maurer equilibrium diagram
毛里求斯海槽【地】Mauritius trough
毛里塔尼亚 <非洲> Mauritania
毛利 contribution margin; gross return; profit contribution; mark-up <加上商品成本上的>; gross profit
毛利百分率 gross profit percentage
毛利比率 gross margin ratio
毛利测验法 gross profit test
毛利差异 gross profit variation
毛利定价法 gross margin pricing
毛利对销货净额比率 ratio of gross profit to net sales
毛利额 gross profit margin
毛利分析 gross profit analysis
毛利菊石地理大区 Maorian ammonite region
毛利率 gross margin percentage; gross profit margin; gross profit rate; gross profit ratio; rate of gross profit; rate of margin
毛利率测试 gross profit ratio test
毛利率法 gross profit method
毛利润 gross margin; gross profit
毛利润百分比 gross margin percentage
毛利润率 gross margin percentage
毛利息 gross interest
毛荔枝 durian
毛粒 neps
毛粒构成 nep formation
毛粒状粗砂岩 trigonia grit
毛量 gross amount
毛量原料 gross raw material
毛料 challie; challis; rough lumber; rough material; run-of-quarry material; pit-run <直接取自料坑的>
毛流 flumina pilorum; hair-streams
毛流动资本 gross working capital
毛流量 gross discharge
毛柳 almond leaved willow
毛楼板 rough floor
毛楼梯梁 rough string
毛路试验台 rough road tester
毛轮 bob
毛螺栓 rough screw; stove bolt
毛曼陀罗 datura innoxia
毛毛虫状吸声板 fissured acoustic(al) tile
毛毛雨 drizzle; drizzling rain; light rain; mizzle
毛毛雨滴 drizzle drop
毛煤 runoff mine coal
毛煤总计 run-of-mine coal grade total
毛门洞 buck opening
毛门框 rough door frame
毛密度 gross density
毛棉布料 delaine
毛面 abrasive coating; blackness; mat-

M

(te)finish(ing);mat(t)plane;mat(t)surface;rough surface;torn fiber[fibre]<粗锯解成的>
毛面玻璃 matt-surface glass;obscured glass
毛面处理 roughly dressed
毛面的 quarry-faced;unfinished
毛面的玻璃砖 dull-glazed tile
毛面镀层 dull deposit
毛面浮雕 bossage
毛面钢板 dull-finish(ed)sheet
毛面光洁度 frosted finish;satin finish
毛面辊 dull roll;mat roller
毛面混凝土 rough concrete
毛面积 gross area
毛面剂 matting agent
毛面加工 mat(te)finish(ing)
毛面加工过程 matting process
毛面浸液 mat(te)kip
毛面精整 butler finish
毛面刻花 mat cutting
毛面螺栓 non-finished bolt; unfinished bolt
毛面清漆粉刷 satin finish varnish
毛面石 quarry-faced stone
毛面蚀刻 mat etching
毛面饰面构件 matt-finish structural facing unit
毛面酸蚀膏 mat-etching paste
毛面酸蚀盐 mat-etching salt
毛面砖 rustic brick
毛膜 trichilemma
毛母质 hair matrix
毛木板 rough plank
毛木地板 timber sub-floor
毛呢光面整理 clear woollen finish
毛盘 hair disc
毛泡桐 karri-tree;kiri<日本>;Princess tree
毛坯 blank;raw stock;rough coating;semi-finished product; slug; work-blank;unwrought<木材>
毛坯安装 blank mounting
毛坯板 blank flat
毛坯玻璃 raw glass
毛坯厂料机 nibbler
毛坯尺寸 blank dimension
毛坯储存 rough storage
毛坯的 rough-hewn
毛坯盖 blank cap
毛坯管 shell
毛坯加工 blank processing
毛坯检验 inspection of blank
毛坯件 blank
毛坯条 blank slug
毛坯图 blank drawing
毛坯外径 blank diameter
毛坯下料机 nibbler
毛坯直径 blank diameter
毛坯铸件 rough casting
毛皮 fell;leather goods and furs;pelt; peltry;furs
毛皮垫 fur scatter rug
毛皮动物农场 fur farm
毛皮黑 fur black
毛皮加工者 furrier
毛皮剪毛工艺 fur-cutting operation
毛皮类 furs
毛皮染料 fur dyes
毛皮上光机 fur polishing machine
毛皮兽场 fur farm
毛皮洗涤剂 fur scouring agent
毛皮颜色 fur colo(u)r
毛皮制品 fur goods
毛皮装饰 furring
毛屏 frosted screen
毛砌坡 rubble-slope(protection)
毛腔 felt chamber;frichophore
毛墙面 harie;harl
毛鞘 hairy sheath
毛青铜矿 buttgenbachite
毛氰钴矿 julienite
毛氰钴石 julienite
毛球 bulbus pili;hair bulb
毛球架 bank creel
毛球壳菌<拉> Lasiosphaeria
毛区 hair-fields;hair-scales
毛渠 distributing ditch; farm ditch; field ditch;sublateral canal
毛渠中最大水深 coverage(in field ditch)
毛取水率 gross draft rate
毛圈 looped pile;pile loop
毛圈边 loop selvedge
毛圈边线 ga(u)ging thread
毛圈布布样 terry swatch
毛圈长毛绒织物 henkelplush
毛圈导轨 pile rail
毛圈地毯 looped carpet;looped rug
毛圈府绸 terry poplin
毛圈高度控制 unlatching of pile warp
毛圈护针 pile retaining pin
毛圈花式线 frise yarn
毛圈花线 picot yarn
毛圈密度 pile density
毛圈起圈装置 loop lifting device
毛圈绒地毯 uncut loop carpet
毛圈绒头 loop pile
毛圈三角 pile cam;plush cam
毛圈式地毡 loop-pile carpet
毛圈撕裂强力 pile tear resistance
毛圈突出 sprouting
毛圈线 lock yarn;looped yarn
毛圈形成 loop formation
毛圈针片 pile point
毛圈织物 astrakhan cloth;loop cloth; looped fabric; loop raised fabric; tadpole
毛热耗率 gross heat rate
毛热效率 gross thermal efficiency
毛绒 lint;plush
毛绒成珠工艺 ratteening
毛绒处理机 hair-processing machine
毛绒疵点 whiskering
毛绒搓磨成珠机 rattening machine
毛绒电光整理 electrifying
毛绒防风条 brush weather strip
毛绒卷曲 curling
毛绒卷曲工艺 friezing
毛容量 bulk unit weight; bulk volume;gross capacity
毛乳头 hair papilla
毛色 coat colo(u)r;hair colo(u)r
毛色产生的 trichochromogenic
毛色鉴定性别法 colo(u)r sexing
毛纱 hosiery yarn;sliver;wool yarn
毛砂面 ground surface
毛生漆 Chinese lacquer
毛石 rubble(stone);rustic stone;cut stone;found stone;freestone;hewn stone;plum(stone) <指埋入混凝土中的大石块>; quarry-faced stone;quarry rubble;quarry(-run) rock; quarry(-run) stone; ratchel; rough-hewn stone;rough stone
毛石坝 rubble dam
毛石壁炉 rubble fireplace
毛石不分层砌体墙 random rubble wall
毛石挡土墙 rubble retaining wall
毛石的 rubbly
毛石底层 rubble backing
毛石垫层 rubble cushion(ing)
毛石堆 heap of rubble;rubble mound; rubble pile; rubble-slope (protection)
毛石堆防波堤 rubble breakwater; rubble-mound breakwater
毛石工 roughcast;rubble mason
毛石工程 quarry-stone bond;ragwork; ranged rubble;rubble works
毛石拱 rough arch;rubble arch
毛石骨料 rubble aggregate
毛石荷载 rubbish load;rubble load
毛石护坡 rubble mound;rubble pitching;rubble-slope(protection)
毛石回填料 rubble backfill
毛石混凝土 cyclopean block; cyclopean concrete;rubble concrete
毛石混凝土墙 rubble concrete wall
毛石基础 pierre perdue;rubble footing;rubble-mound foundation;rubble stone footing;rubble stone subbase;rubble subbase;stone footing; stone foundation
毛石基床 rubble base;rubble bed
毛石集料 rubble aggregate
毛石截水沟 rubble catchwater drain
毛石块 rough block;rough ashlar
毛石块料<混凝土用> plum
毛石块面层墙 bastard masonry
毛石料<未经筛选的> quarry(-run) rock;quarry-run stone;rock quarry run stone;untrimmed quarry stone
毛石盲沟 rubble drain
毛石排水沟 French drain;rubble drain
毛石平台 taula
毛石铺底 rough stone pitching
毛石铺面 rubble paving
毛石铺砌 quarry pavement; rubble paving
毛石铺砌层 rough stone pitching
毛石砌成的 rubbly
毛石砌谷坊 rubble masonry check dam
毛石砌合 quarry(-stone)bond
毛石砌墙 random coursed wall;rough stone masonry wall;rough walling
毛石砌体 blockage;broken range ashlar; cyclopean masonry; cyclopean rubble masonry;freestone masonry; masonry rubble;masonry work;ordinary masonry; rubble masonry; rubble works;snecking
毛石砌体坝 rubble masonry dam
毛石砌体分层砌筑法 range work
毛石砌体基础 bonder mass foundation
毛石砌筑 quarry-stone bond; rough walling;rubble works
毛石砌筑墙 rough walling
毛石墙 random rubble wall;antiquum opus;rough wall; rubble masonry; rubble wall
毛石墙边扁石 skinner
毛石墙面琢石 bastard ashlar
毛石墙琢石 bastard ashler
毛石穹顶 rubble stone vault; rubble vault
毛石砂浆 ash mortar
毛石踏步 quarry run
毛石填筑 quarry-run rockfill; rubble fill;rubble run rockfill
毛石贴面 polygonal rubble facing; ragwork
毛石贴面圬工 stone-faced rubble masonry
毛石筒拱 rubble stone vault
毛石圬工 cyclopean(rubble) masonry; freemason; freestone masonry; ordinary masonry; polygonal random rubble masonry; quarry-stone masonry;rough walling;rubble masonry
毛石圬工坝 rubble masonry dam
毛石圬工工程 rubble masonry work
毛石圬工箱制接头 clip joints of rubble masonry
毛石圬工墙 incertum
毛石圬工桥墩身 bridge pier body of rubble masonry
毛石消波设备 rubble wave absorber
毛石斜坡堤 rubble mound
毛石选块填筑 classified rubble fill
毛石整层砌 regular coursed rubble
毛石整砌层 range masonry(work)
毛饰刺绣 quill embroidery
毛收入 gross credit; gross earnings; gross income
毛收入法定分类 statutory grouping of gross income
毛收入税 gross income tax
毛收益率 gross yield
毛束 pieces
毛刷 felt finger; hair brush; camel-hair brush<涂料用>
毛刷擦亮 mop polishing
毛刷仓盖 brush housing door
毛刷仓上挡板 brush housing upper baffle
毛刷给浆辊筒 brush furnisher
毛刷过滤器 brush filter
毛刷开针舌器 brush latch opener
毛刷罗拉 bristle roll;brush roller
毛刷清洁 brush clearing
毛刷式锯齿轧花机 brush saw gin; brush stripping saw gin
毛刷式起球试验仪 brush pilling tester
毛刷梳棉 brush-comb
毛刷涂布 brush coating
毛刷修版 brush retouching
毛刷一样的 brushy
毛刷轴密封垫 brush arbor seal
毛水头 gross head
毛顺向 hair-streams
毛丝 fuzz
毛丝体 trichoblast
毛丝消除器 hairiness reducer
毛松香 barras;solidified oleoresin
毛酸浆 strawberry tomato
毛损 gross loss
毛损率 gross margin percentage
毛毯 woolen blanket
毛毯洗涤装置 felt washing device
毛毯张紧器 felt tightener
毛毯织机 blanket loom
毛体积 bulk volume;gross volume
毛体积饱和面干相对密度 bulk saturated surface dry specific gravity
毛体积比重 bulk specific gravity; bulk specific weight
毛体积干密度 dry bulk density
毛体积密度 bulk density;bulk specific gravity; bulk unit weight; gross volume density
毛体积密度试验 bulk density test
毛体积压缩系数 bulk compressibility
毛填料润滑脂 hair grease
毛条 top wool
毛条标题 catch-line
毛铁 pig iron
毛铁工 iron driver
毛铜矿<一种赤铜矿> chalcotrichite
毛头 bur(r)
毛头铰刀 bur(r)reamer
毛投资<未扣除设备折旧的> gross investment
毛团 egagropilus
毛网压紧装置 web compressor; web squeezer
毛屋顶 carcase roof(ing)
毛吸收量 gross absorption
毛细饱和法 saturating of capillary
毛细饱和(状态)capillary saturation
毛细泵 capillary pump
毛细边界带 boundary zone of capillary
毛细表面 capillary surface
毛细波 capillary wave;ripple
毛细波痕 capillary ripple

毛细常数 capillarity constant; capillary constant; surface tension constant
毛细持水度 specific capillary retention; specific retention of capillary water
毛细抽吸时间 capillary suction time
毛细抽吸时间试验 capillary suction time test
毛细带 zone of capillarity
毛细带含水量 water content in capillary zone
毛细导管 capillary vessel
毛细的 capillary; hair-shaped
毛细地下水 capillary ground water
毛细电极 capillary electrode
毛细电解 capillary electrolysis
毛细多孔体 capillary porous body
毛细分析 capillary analysis
毛细分析法 capillary analytical method
毛细改正 capillary correction
毛细隔离层 capillary cut-off
毛细管 capillary; capillary pipe; capillary tubing; capillary tube
毛细管半径 average throat width; capillary radius
毛细管饱和 capillary saturation
毛细管比色计 capillator
毛细管壁 capillary wall
毛细管边缘 capillary fringe
毛细管边缘水 fringe water
毛细管玻璃电极 capillary glass electrode
毛细管测黏[粘]度法 capillary visco(si)metry
毛细管测液器 capillarimeter
毛细管层 capillary fringe
毛细管常数 capillary constant
毛细管超导性 capillary superconductivity
毛细管出口 capillary outlet
毛细管传导度 capillary conductivity
毛细管传导率 capillary conductivity
毛细管传导性 capillary conductivity
毛细管脆性 capillary fragility
毛细管带 capillary fringe; capillary zone
毛细管导水率 capillary conductivity
毛细管导液法 capillary drainage
毛细管道 capillary channel
毛细管的灯芯作用 capillary effect
毛细管电泳 capillary electrophoresis
毛细管电泳-质谱法联用 coupling capillary electrophoresis to mass spectroscopy
毛细管定型生长 capillary action
毛细管定型生长技术 capillary action shaping technique
毛细管堵塞 capillary break
毛细管断面 capillary intersection
毛细管法 capillary method; capillary tube method
毛细管法熔点 capillary melting point
毛细管分流器 capillary splitter
毛细管分析 capillary analysis
毛细管分析法 capillary analysis method
毛细管封闭 capillary seal
毛细管干燥 capillary drying
毛细管高度 capillary height
毛细管高度法 capillary-height method
毛细管给油 capillary feed
毛细管给油器 capillary oiler
毛细管过滤 capillary filtering
毛细管过滤系数 capillary filtration coefficient
毛细管含水量 capillary moisture capacity
毛细管虹吸作用 capillary siphoning
毛细管化学 capillary chemistry
毛细管绘图笔 capillary pen
毛细管活性 capillary activity
毛细管几何形 capillary geometry
毛细管计 capillarimeter
毛细管技术 capillary technique
毛细管间的 intercapillary
毛细管间隙 capillary interstice; capillary space
毛细管检波器 capillary detector
毛细管校准法 calibrated capillary method
毛细管接头 capillary fitting
毛细管结构 capillary structure
毛细管截面效应 capillary cross section effect
毛细管浸渗 capillary dip
毛细管净气器 capillary air washer
毛细管静电计 capillary electrometer
毛细管空间 capillary space
毛细管空腔 capillary cavity
毛细管空隙 capillary void
毛细管空隙度 capillary porosity
毛细管孔(径) capillary opening; capillary pore
毛细管孔隙 capillary interstice; capillary pore; capillary porosity
毛细管孔隙度 capillary porosity
毛细管控制作用 capillary control
毛细管冷凝作用 capillary condensation
毛细管离子源 capillary ion source
毛细管力 capillary force
毛细管流变计 capillary rheometer
毛细管流变仪 capillary rheometer
毛细管流(动) capillary flow; film flow <土内水分的>
毛细管流量计 capillary flow meter
毛细管流体动力学 capillary hydrodynamics
毛细管漏孔 capillary leak
毛细管膜 capillary membrane
毛细管内层 endocapillary layer
毛细管内径 capillary bore
毛细管内聚力 capillary cohesion
毛细管能 capillary energy
毛细管黏[粘]度测定法 capillary flow measuring method
毛细管黏[粘]度计 capillary (tube type of) visco(si)meter
毛细管凝集试验 capillary tube agglutination test
毛细管凝结 capillary condensation
毛细管凝结理论 capillary condensation theory
毛细管凝结水 capillary condensed water
毛细管凝聚 capillary condensation
毛细管喷淋洗涤机 capillary cell washer
毛细管平衡状态 state of capillary equilibrium
毛细管破坏 capillary break
毛细管破裂 capillary break-up
毛细管气相层析 capillary gas chromatography
毛细管气相色谱(法) capillary gas chromatography
毛细管潜能 capillary potential
毛细管鞘 capillary sheath
毛细管切变取向 capillary-shear-induced orientation
毛细管区 capillary zone
毛细管容(水)量 capillary capacity
毛细管入口 capillary inlet
毛细管入口压力降 capillary-entrance-pressure drop
毛细管润滑 capillary lubrication
毛细管润滑剂 capillary lubricant
毛细管色谱柱 capillary chromatographic column
毛细管上升 capillary ascension; capillary ascent; capillary lift; capillary elevation; capillary rise
毛细管上升边缘 capillary rising fringe
毛细管上升法 capillary rise method
毛细管上升范围 boundary zone of capillarity
毛细管上升高度 capillary rise height; capillary rising height
毛细管上升高度试验 capillary lift test; capillary rising height test
毛细管上升速度 capillary rise velocity; capillary rising rate
毛细管上升速率 capillary rise rate; rate of capillary rise
毛细管上升限 capillary front
毛细管上升限度 boundary zone of capillarity
毛细管上升校正 capillary rise correction
毛细管上限 capillary fringe
毛细管射流 capillary jet
毛细管渗流 capillary seepage
毛细管渗漏 capillary leak; capillary seepage
毛细管渗水槽 capillary seepage trench
毛细管渗透 capillary percolation
毛细管渗透电解 capillary electrolysis
毛细管渗透性 capillary permeability
毛细管渗透作用 capillary imbibition; water of capillary; capillary penetration
毛细管湿润法 capillary wetting method
毛细管式功率表 capillary wattmeter
毛细管势 capillary potential
毛细管势梯度 capillary potential gradient
毛细管试验仪 capillometer
毛细管输送 capillary transport
毛细管输运 capillary transport
毛细管束 bundle of capillary tubes
毛细管数 capillary number
毛细管栓塞 capillary embolism
毛细管栓子 capillary embolus
毛细管水 capillary water; water of capillarity; fringe water; water of capacity water
毛细管水饱和 capillary saturation
毛细管水饱和带 zone of capillary saturation
毛细管水边缘 zone of capillarity
毛细管水边缘带 capillary fringe belt
毛细管水层 capillary layer
毛细管水带 zone of capillary
毛细管水分 capillary moisture
毛细管水上升 capillary elevation
毛细管水上升边缘 capillary fringe
毛细管水上升高度 capillary height; capillary lift
毛细管水上升高度试验 capillary lift test
毛细管水势差 capillary potential
毛细管水头 capillarity suction head; capillary head
毛细管水位 capillary stage
毛细管水下降 capillary depression
毛细管水压力 capillary pressure
毛细管水移动 capillary migration
毛细管水银灯 capillary mercury lamp
毛细管水张力 capillary tension
毛细管(水作用)带 zone of capillarity
毛细管探测器 capillary detector
毛细管特性 capillary behavio(u)r; wicking property
毛细管填料柱 capillary packed column
毛细管透过作用 capillary penetration
毛细管脱水装置 capillary dewatering unit
毛细管弯液面 capillary meniscus
毛细管网 capillary network
毛细管网眼式洗涤器 capillary cell type washer
毛细管微池 capillary microcell
毛细管位能 capillary potential
毛细管吸附时间 capillary suction time
毛细管吸附水 capillary absorbed water
毛细管吸力 capillary attraction; capillary power
毛细管吸力时间 capillary suction time
毛细管吸升高度 capillarity suction head
毛细管吸升水头 capillarity suction head
毛细管吸湿量 capillary capacity
毛细管吸湿作用 capillary action
毛细管吸收作用 absorption by capillarity; capillarity absorption; capillary absorption
毛细管吸水试验 capillary water absorption test
毛细管吸水作用 absorption by capillarity; capillary adsorption
毛细管吸引力 capillary attraction; force of capillary attraction
毛细管下降 capillary depression
毛细管显微镜 capillaroscope
毛细管显微镜检查 capillaroscopy
毛细管现象 capillarity; capillary phenomenon; capillary action; capillary attraction
毛细管效应 capillary effect
毛细管泄漏 capillary leak
毛细管性 capillarity
毛细管学 capillarity
毛细管循环迟缓 dysdiemorrhysis
毛细管压力法 capillary manometric method
毛细管压力计 capillary manometer
毛细管压头 capillary pressure head
毛细管仪 capillarimeter
毛细管移动 capillary migration
毛细管引力 capillary attraction
毛细管引流法 capillary drainage; hydrocenosis
毛细管运动 capillary motion; capillary movement
毛细管运动的 capillariomotor; capillomotor
毛细管载体 capillary support
毛细管张力 capillary tension
毛细管张力理论 capillary tension theory
毛细管整列 capillary array; multifiber array
毛细管中断 capillary break
毛细管柱 capillary column
毛细管阻抗测定器 capillary resistance measure apparatus
毛细管作用 capillarity; capillary action; capillary attraction
毛细管作用带 boundary zone of capillary; zone of capillary
毛细管作用定律 law of capillarity
毛细管作用净气器 capillary air washer
毛细管作用速率 capillary tube
毛细管作用吸湿器 capillary capacity
毛细管作用仪器 capillary apparatus
毛细灌水法 capillary watering
毛细龟裂纹 capillary fissure
毛细虹吸作用 capillary siphoning
毛细弧光离子源 capillary arc ion source
毛细活栓 capillary stopcock
毛细活性化合物 capillary active compound
毛细唧送参量 capillary pumping pa-

M

rameter
毛细唧送压头 capillary pumping head; pumping head of capillarity
毛细间隙 capillary interstice
毛细检液器 capillarimeter
毛细角 angle of capillarity
毛细校正 capillarity correction
毛细阶段 capillary stage
毛细接合 <在承插式接合中利用毛细管作用吸入熔化焊料而连接的方法> capillary joint
毛细结合 capillary bond
毛细静水压力 capillary hydrostatic pressure
毛细孔 capillary bore; capillary opening
毛细孔穿透 capillary penetration
毛细孔道 capillary channel
毛细孔结构 capillary structure
毛细孔模型 capillary-orifice model
毛细孔水压力 negative pore water pressure
毛细孔体积 capillary pore volume
毛细孔通道 capillary channel
毛细孔隙度 capillary porosity
毛细孔隙率 capillary porosity
毛细孔张力 capillary tension
毛细冷凝 capillary condensation
毛细力 capillary force
毛细力带 capillary zone
毛细裂缝 capillary crack; hairline crack(ing)
毛细裂纹 capillary crack; hairline crack(ing)
毛细流动 capillary flow
毛细漏孔 capillary leak
毛细滤器 <湿式集尘装置用> capillary filter
毛细面振动 capillary oscillation
毛细能 capillary energy
毛细黏[粘]结 capillary bond
毛细黏[粘]滞度计 capillary visco(si)-meter
毛细平衡 capillary equilibrium
毛细平衡高度 capillary equilibrium height
毛细平衡液位 capillary equilibrium level
毛细瓶 capillary flask
毛细迁移 capillary migration
毛细热管 capillary heat pipe; heat pipe of capillarity
毛细上升 capacity elevation; capillary ascent
毛细上升高度 capillary head; capillary rise
毛细上升区 area of capillary rise
毛细上升速度 velocity of capillary rise
毛细上升作用 capillary elevation
毛细渗流 capillary seepage
毛细渗入 capillary penetration
毛细渗透 capillary penetration
毛细渗透作用 capillary osmosis
毛细升高 capillary rise
毛细升高度 capillary height
毛细升高值 wicking height
毛细湿润法 capillary wetting method
毛细石英盒 capillary quartz cell
毛细势 capillary potential
毛细势差 capillary potential gradient
毛细试验 capillary test
毛细收集器 capillary collector
毛细水 capillary moisture; capillary water; water of capillarity
毛细水边缘 <地下水位以上的> capillary fringe
毛细水带 capillary water zone; zone of capillary water
毛细水顶 capillary front
毛细水分 capillary moisture; cellular moisture
毛细水分试验 capillary moisture test
毛细水高度 capillary height
毛细水隔离层 capillary cut-off
毛细水扩散 capillary diffusion
毛细水联结 capillary bond
毛细水区 capillary(fringe)zone
毛细水容度 specific capillary retention
毛细水上升 capillary lift
毛细水上升高度 height of capillary rise
毛细水上升高度试验 capillary rise test
毛细水头 capillary head
毛细水位势 capillary potential
毛细水下降 capillary depression
毛细水压力 capillary pressure
毛细水移动 capillary migration
毛细水运动 capillary movement
毛细水运动变慢的极限点 lento-capillary point
毛细水张力 capillary pull; capillary tension
毛细水作用带 boundary zone of capillary
毛细提升高度 capillary rise height; height of capillary rise
毛细途径 capillary path
毛细脱水 capillary dewatering
毛细脱水污泥 capillary dewatering sludge
毛细吸附水 capillary adsorbed water
毛细吸附作用 capillary adsorption
毛细吸管 capillary pipette; capillary syringe
毛细吸力 capillary force
毛细吸湿量 capillary capacity
毛细吸湿能力 capillary capacity
毛细吸湿(状态) capillary absorption
毛细吸引力 capillary attraction
毛细下降 capillary depression
毛细显微镜 capillary microscope
毛细显微术 capillary microscopy
毛细现象隔层 capillary break
毛细限 capillary limit
毛细效应 capillary effect
毛细泄漏 capillary leak
毛细悬挂水 suspended capillary water
毛细压低值 capillary depression
毛细压强 capillary pressure
毛细移动 capillary migration; capillary movement
毛细应力 capillary stress
毛细运动 capillary motion
毛细张力 capillary pressure
毛细滞后现象 capillary hysteresis
毛细柱色谱法 capillary column chromatography
毛细装置 device for demonstrating capillarity
毛细状裂纹 hair crack
毛细作用 capillary
毛细作用带 capillary fringe
毛细作用理论 capillary theory
毛细作用力 capillary force
毛细作用中断 capillary break
毛下颚 penicillate maxilla
毛下槛 rough sill
毛纤维 hairy roving; wool
毛纤维形态学 morphology of the wool fiber
毛线 knitting wool; woolen-knitting yarn; woolen yarn
毛小皮印痕 hair cuticle print
毛效率 gross efficiency
毛效益 gross benefit
毛屑 flue; hards; soft flocks
毛屑床垫 flock mattress
毛屑垫床 flock bed
毛蟹 Eriocheir sinensis
毛序图 setal map
毛絮 flock
毛絮构造 flocculent structure
毛絮石膏 flocculent gypsum
毛亚麻油 raw linseed oil
毛洋槐 roseacacia locust
毛样 dirty galley
毛样的 piliform
毛应力 gross stress
毛罂红 erythroglaucin
毛罂蓝 setoglaucine
毛樱桃 manchu cherry
毛用水率 gross duty(of water); headgate duty
毛雨量 gross precipitation; gross rainfall
毛运行水头 gross operating head
毛载 bulkload(ing)
毛轧板 mo(u)lder
毛轧机 mo(u)lder
毛毡 blanket(ing); curled hair; felt; hair felt(d fabric)
毛毡保温 blanket insulation
毛毡擦试器 felt wiper
毛毡衬垫 felt packing
毛毡衬里的楼面覆盖层 felt-base floor cover(ing)
毛毡衬里的亚麻油毡 felt-backed lino(leum)
毛毡衬面 felt covering
毛毡底 felt base
毛毡地面 wool felt floor cover(ing)
毛毡地毯 felt carpet; felt floor covering
毛毡垫 felt base
毛毡垫圈 felted fabric washer; felt gasket
毛毡封 felt seal
毛毡干燥法 blanket method
毛毡隔声板 felt deadener
毛毡滚筒 felt calender
毛毡过滤元件 felt element
毛毡花坛 carpet bed(ding)
毛毡缓冲垫 felt buffer
毛毡零件 felt element
毛毡滤清器 felted membrane
毛毡滤芯 felt element
毛毡滤芯滤清器 felt filter
毛毡抛光 cloth polishing
毛毡抛光机 felt lap; felt polisher
毛毡抛光圆盘 felt polishing disk[disc]
毛毡伸张器 felt stretcher
毛毡式砑光机 felt calender
毛毡苔 sundew
毛毡提花纹板 felt cardboard
毛毡填充物 felt dust packing
毛毡填料 felt packing
毛毡橡胶 hair-bonded rubber
毛毡性质 felting property
毛毡压烫机 felt press machine
毛毡油封外壳 felt retainer
毛毡油封座圈 felt retainer
毛毡油环 felt oil ring
毛毡圆筒 felt cylinder
毛毡织品 felted fabric backing
毛毡织物 felted fabric web
毛毡坐垫 felt pad
毛折 trichatrophia
毛针织品 woollen knitwear
毛针织物 woollen jersey
毛蒸发量 gross evapo(u)ration
毛枝栗 chincapin [chinkapin]; chinquapin
毛枝五针松 Wang's pine
毛织品 soft goods; stuff goods; wool; wool(en)fabrics; woollen
毛织物 antique satin; deny; drapery; wool fabric
毛织物质量标签 wool labeling
毛直径 nominal diameter
毛值 gross value
毛制厂 wool-making mill
毛制的 hairen
毛重 gross weight; apparent weight; bulk weight; full weight; gross load; invoice weight; laden weight; packed weight; rough weight; weight-gross
毛重公斤数 kilogram gross weight
毛重条件 gross weight terms
毛重与净重之比 proportion of gross to net load
毛竹 mao bamboo; moso bamboo
毛庄克层【地】Mauch Chunk bed
毛庄克统 <早石炭世>【地】Mauch Chaunk series
毛装运重量 gross shipping weight
毛状的 hairy
毛状感受器 trichoid sensillum
毛状刮纹 hair scratch
毛状痕迹 hair scratch
毛状灰尘 fluffy dust
毛状菌体 trichothallic
毛状排水器【植】trichome hydathode
毛状饰 feather edging
毛状手感 wool-like handle
毛状纹理 fuzzy grain; woolly grain
毛状云 fibratus; filosus
毛锥 pin feather
毛琢石 roughly dressed ashlar
毛鬃 curled hair; hair pencil
毛鬃油漆刷 camel-hair mop
毛总出生率 crude birth rate
毛总量 adds
毛总死亡率 crude death rate
毛作净重 gross for net

矛

矛白铁矿 spear pyrites

矛盾方程 inconsistent equation
矛盾法则 antinomy
矛盾信号监控器 hard-wire conflict monitor
矛盾虚幻示像 conflicting phantom aspect
矛式划行器 spear marker
矛式夹 <一种格栅固定夹,固定于混凝土的> spearpoint
矛式钻 spear-pointed drill
矛式钻头 center[centre] bit; digging bit; spud bit
矛头 spearhead
矛头形钻头 spearhead bit
矛头状顶死 spearhead dieback
矛形松土铲 spearhead shovel; spearpoint shovel
矛形錾 spearpoint gad
矛形指针 lance pointer; spear pointer
矛状器具 lance
矛状体 spear

牦

牦牛 yak

茅

茅草 cogongrass; quitch grass; reed; thatch; witch grass

茅草板 thatchboard
茅草干扰 grass; hash
茅草捆 yelm; yelven
茅草蜡 esparto wax
茅草棚 thatched shack; thatched shed
茅草屋 <中非> banda
茅草屋顶 reed roof; thatched roof; thatching; thatch roof(ing); yelm
茅草屋顶材料 thatching

M

茅草屋顶椽木 brotch
茅草屋面 grass thatch
茅草檐 eaves
茅草纸浆 esparto pulp
茅滴混剂 Chloropon
茅坑 dry closet;latrine pit;lavatory pit;privy vault
茅栎 grassy oak
茅舍 cot(e);grass cottage;hovel;hustiement;hut(ch);sukkah;kraal <南非>
茅舍状烟囱 hovelling
茅栗 grassy chestnut
茅亭 thatched pavilion
茅屋 cot(e);hovel;hut(ch);shack;succah[sukkah];thatched cottage;bourock <苏格兰>
茅屋顶 straw roof(ing);thatch
茅屋顶椽木 buckle
茅屋顶村舍 thatched cottage
茅屋顶的 thatchy
茅屋顶铺设辅助工 thatcher's labo(u)rer;thatcher's server
茅屋顶铺设工 thatcher
茅香木蠹 <拉> Cossus cossus
茅竹 bamboo cane

锚 anchor(nail);grapnel

锚板 anchor plate;anchor slab;anchor strap;tie plate;anchorage plate
锚板机 anchor tripper
锚板螺栓 plate anchor;plate bolt
锚杯 <预应力> anchor female cone;female cone;anchorage cup;socket of anchor
锚臂 anchor arm;anchor boom;arm;arm of anchor
锚臂连接底座 plinth for anchor arm
锚标 anchor buoy;cable buoy
锚标绳 anchor buoy rope
锚镖装饰 anchor dart
锚冰【水文】anchor ice;bottom ice;ground ice;ice anchor
锚柄 anchor shaft;anchor shank
锚泊 anchorage;anchoring;at anchor;lie at anchor;lying at anchor;mooring;ride at anchor
锚泊驳船 anchor barge
锚泊泊位 anchoring berth
锚泊冲击力 berthing impact
锚泊船只 berthed vessel
锚泊的 at anchor
锚泊灯 anchoring lamp;anchoring light;riding light
锚泊地 anchorage;anchorage area;anchorage basin;anchorage place of mooring;anchoring place;hold ground;mooring area;mooring space;mooring basin;holding ground for anchors;mooring ground
锚泊方式 mooring type at anchorage;way of anchoring
锚泊费 anchorage dues;anchor dues
锚泊浮标 anchor buoy
锚泊浮标站 anchored buoy station
锚泊浮筒 anchorage buoy
锚泊港 mooring harbo(u)r;mooring port
锚泊港池 anchorage basin
锚泊(固定)趸船 mooring pontoon
锚泊球 anchor ball
锚泊区(域)anchorage(area);mooring area;mooring basin;roadstead
锚泊设备 anchoring and mooring system;berthing accommodation;berthing plant;mooring equipment
锚泊设施 berth structure
锚泊时的荷载 load at mooring
锚泊时锚链受力不大 ride easy
锚泊时锚链受力很大 ride hard
锚泊索具 ground hold;ground tackle <锚链等之总称>
锚泊挖泥船 anchor dredge(r)
锚泊系留力 holding power of anchoring
锚泊形式 mooring pattern
锚泊原因 cause of anchorage
锚泊中波浪打进锚链孔 ride hawse full
锚泊中船随着潮水涨落而回转 tail to tide;tail up and down the stream;tail with the stream
锚泊中船尾朝着水流方向 tail with the stream
锚泊中的 riding
锚泊中浪从船首打上 over-rake
锚泊属具 ground hold;ground tackle
锚泊驻留力 holding power of anchor
锚泊装置 anchorage device;anchorage unit;ground tackle
锚泊装置设计 mooring design
锚泊装置外形 mooring geometry
锚泊作业 anchor works;mooring operation
锚捕捉器 anchor catcher
锚槽 anchor slot
锚叉 anchor stock
锚吃得住 the anchor holds
锚齿 anchor palm
锚出水 anchor awash;anchor up
锚出土 anchor broken out
锚穿链扣 anchor shackle
锚船浮筒 anchor buoy
锚船桩群 dolphin
锚床 anchor bed;anchor board;anchor shoe;billboard
锚挡尾球 nut of anchor
锚灯 anchor lamp;anchor lantern;anchor light;riding light;stay light
锚地 anchorage(area);anchorage space;anchoring basin;anchoring ground;anchoring space;anchoring station;boat basin;haven;mooring basin;mooring berth;surgidero;tying-up place;roadstead <海上敞开的>
锚地泊位 anchor(ing) berth
锚地参考资料 reference data for anchoring
锚地底质 holding ground
锚地趸船 pontoon at anchorage area
锚地浮标 anchorage buoy;roadstead buoy <海上敞开的>
锚地界限 anchorage limit
锚地界限浮标 anchorage limit buoy
锚地面积 anchorage area
锚地停泊方式 mooring type in anchorage
锚地土壤 holding ground
锚地图 anchorage chart
锚地作业工艺系统 cargo-handling system of anchorage
锚点 anchor;anchor point
锚垫板 anchor plate
锚垫圈 washer shim
锚吊杆 anchor davit
锚吊箍【船】balancing band
锚钉 holdfast;stud
锚钉撑锤 bucking tool
锚碇 <又称锚定> anchorage fixing;anchor(ing);guy anchor;holdfast;tieback;grapple;grappling
锚碇岸壁 anchor(age) bulkhead;anchored bulkhead;tied wall
锚碇板 anchor sheeting;anchor slab;anchor washer;form anchor
锚碇板挡(土)墙 anchor plate retaining wall;tieback wall;anchor slab retaining wall
锚碇板结构 anchor slab structure
锚碇板桥台 anchor slab abutment
锚碇板式挡土墙 anchorage bulkhead retaining wall;anchored bulkhead retaining wall;retaining wall with anchored bulkhead
锚碇板桩式码头 sheet pile type wharf with anchored piles
锚碇板式桥台 anchorage bulkhead abutment;anchored bulkhead abutment
锚碇板桩 anchorage sheet piling;anchored sheet piling
锚碇板桩岸壁 tied-back bulkhead;anchored sheet pile bulkhead
锚碇板桩墙 anchored sheet pile wall;anchored sheet piling wall;sheet-pile anchor wall;tied sheet pile wall
锚碇板桩式 tied sheeting type;tie sheet pile type
锚碇臂杆 mooring boom
锚碇变形 anchorage deformation
锚碇布置 anchorage layout
锚碇舱 <吊桥的> anchorage chamber;anchoring chamber
锚碇槽 anchor slot
锚碇叉桩 forked piles anchorage;raking-piles anchorage
锚碇插座 anchor socket
锚碇柴捆 anchoring fascine
锚碇沉块 mooring clump;mooring sinker
锚碇沉箱 anchor caisson
锚碇船 anchoring ship
锚碇船坞 anchored dock
锚碇带 anchor strap
锚碇单排板桩墙 anchored single-wall
锚碇挡(土)墙 anchor(ed) bulkhead;tieback wall;tied(back) retaining wall
锚碇挡土板 anchorage bulkhead;anchored bulkhead
锚碇的 anchored;wall-anchored
锚碇的墙 anchored wall
锚碇的无效长度 ineffective length of anchorage
锚碇的有效长度 effective length of anchorage
锚碇点 anchored point;anchoring point;anchor point
锚碇垫板 anchor plate
锚碇墩 anchoring abutment
锚碇墩台 anchorage abutment
锚碇方法 anchoring system;method of anchoring;bottom-pull method;method of anchorage
锚碇风速表 anchor anemometer
锚碇缝隙 anchor slot
锚碇浮标 anchored buoy
锚碇浮标系统 anchored buoy system
锚碇杆 anchor(ing) rod;anchor tie
锚碇钢板 anchor plate
锚碇钢板桩岸壁 tied-back sheet-piling wall
锚碇钢筋 anchor(age) bar;anchorage bend;anchorage steel
锚碇钢筋垫板 <浇捣混凝土时用> reinforcement plate
锚碇钢丝索 anchor strand
锚碇钢索 anchor tendon;cable anchor
锚碇钢桩 anchor steel pile
锚碇高度比 anchor level ratio
锚碇格型板桩 anchored cellular sheet pile
锚碇钩 deadman
锚碇构件 anchorage element;anchoring element
锚碇骨架 anchor frame
锚碇灌浆 anchor grout(ing)
锚碇滑动 anchorage slip
锚碇滑移 anchorage slip
锚碇环 anchor loop;anchor ring
锚碇混凝土块 anchor concrete block
锚碇混凝土桩 anchor concrete pile
锚碇基础 anchorage foundation
锚碇基团 anchor group
锚碇架 anchor frame
锚碇件 anchor element;anchor log
锚碇件的加固钢筋 anchoring accessories
锚碇角度 anchor angle
锚碇角钢 anchor angle steel
锚碇绝缘子 anchorage insulator
锚碇孔 anchor hole
锚碇跨 anchor spalling
锚碇块(体)anchor(age) block;mooring block;deadman anchorage;sinker deadman;cylindric(al) block <预应力混凝土的>;land tie
锚碇块体 fixed mooring
锚碇拉杆 land tie;mooring rod
锚碇拉力 anchorage pull;anchor pull
锚碇拉索 anchor cable;anchor rope;anchor stay
锚碇拉条 anchor tie
锚碇拉线 anchoring wire
锚碇喇叭管 anchor trumpet
锚碇缆索 anchored cable
锚碇力 anchorage force
锚碇链 <浮筒的> ground cable;ground chain
锚碇梁 anchor beam
锚碇梁的墙上锚杆 beam anchor
锚碇螺杆衬套 brake anchor pin bushing
锚碇螺栓平面布置图 plan of anchor bolt
锚碇螺杆凸轮 anchor pin cam
锚碇螺帽 anchor nut
锚碇螺栓 anchor(ed) bolt;screw anchor;tie-down bolt;fang bolt
锚碇螺栓螺母 anchor bolt and nut
锚碇螺丝 anchoring screw
锚碇螺旋 mooring screw
锚碇门 <重闸门用> anchor gate
锚碇面板 <后拉条面板> face-plate
锚碇抛锚 anchorage
锚碇器 anchor
锚碇强度 anchorage strength;anchoring strength
锚碇墙 anchor wall
锚碇墙式码头 anchored quaywall
锚碇墙支撑板桩堤岸 sheet-pile bulkhead with anchor-wall support
锚碇区 anchorage area;anchorage zone;ground tackle
锚碇设备 anchorage device;anchoring facility;mooring facility
锚碇设施 anchoring facility;land tie
锚碇失效 anchorage failure
锚碇式岸壁 anchored bulkhead
锚碇式板桩 anchored sheet piling;anchor sheeting pile
锚碇式板桩墙 anchor sheeting;anchor sheeting pile wall
锚碇式挡土结构 retaining structure with anchor
锚碇式挡土墙 anchored retaining wall;tied retaining wall
锚碇式封隔器 anchor packer
锚碇式干船坞 anchored drydock
锚碇式墙 anchored wall
锚碇式坞底板 tied drydock floor
锚碇式悬索桥 anchorage suspension bridge;anchored suspension bridge
锚碇室 <吊桥的> anchorage chamber;anchoring chamber
锚碇双排板桩墙 anchored double-wall

锚碇水文站 anchor hydrographic(al) station
锚碇索 anchor strap
锚碇塔 anchorage tower; anchor(ing) tower
锚碇体 anchor block;anchor boat
锚碇体系 anchorage system
锚碇体系块<预应力混凝土的> cylindric(al) block
锚碇铜丝 copper wire loose anchor; copper wire tie
锚碇筒 anchor cylinder
锚碇土 anchored earth
锚碇托架 anchor bracket
锚碇弯曲 anchorage bend
锚碇位挖泥船 anchor dredge(r)
锚碇位置 anchored location;anchored station;location of anchorage
锚碇坞墙 tied wall
锚碇物 anchor;deadman
锚碇吸扬式 moored suction type
锚碇系统 anchor(age) system;system of anchoring
锚碇先张法<预应力混凝土的> anchored pretensioning
锚碇销 anchor pin
锚碇效应 anchoring effect
锚碇斜拉式挡土墙 anchored bulkhead
锚碇型陶瓷板 anchored-type ceramic veneer
锚碇影响区 anchorage zone
锚碇预应力钢索 anchor tendon
锚碇在混凝土中 anchor embedded in concrete
锚碇在土中 anchored in the ground
锚碇支承板 anchor bearing plate
锚碇支架 anchoring jack
锚碇支柱 anchorage spud; anchoring spud
锚碇支座 anchorage abutment;anchoring abutment
锚碇重块 buoy stone;cast anchor
锚碇柱 anchorage mast; anchorage spud;anchoring jack;pull-off pole
锚碇桩 anchoring jack; anchoring spud;anchor log;anchor(ing) pile; anchor spike; mooring pile; pin pile;stay pile;stressing abutment; deadman;grouser<在河底固定木排或挖泥船等的>;land tie
锚碇桩观察孔 deadman's eye
锚碇装置 anchorage(device); anchorage unit; anchoring; tieback; ground anchorage<埋设在岩石或土层内的>;exterior tieback
锚碇锥(头) anchoring cone
锚碇钻塔 anchored tower
锚碇作用 anchorage; function of anchorage
锚碇座 anchor block
锚碇座板 anchor bearing plate
锚锭螺栓 lewis bolt
锚端链节 swivel shot
锚端套管 anchor sheath
锚端卸扣 bending shackle;end shackle
锚段 anchor section;tension length
锚段长度 tension length
锚段点 anchor point
锚段关节 overlap section;overlap span
锚墩 anchorage block; anchorage pier;anchor(ing) block;deadman; stressing abutment; abutment<架空雪道两端的>;anchor(ing) pier
锚轭 anchor yoke
锚方位 anchor bearing
锚缝 anchor slot
锚浮标 anchor buoy;deadhead
锚杆 anchor(arm); anchor bar; anchor bolt;anchor jack;anchor pin; anchor rod; anchor shaft; anchor shank; anchor stock; anchor tie; bolt;earth anchor;ground anchor; rock bolt;tieback;dowel bar
锚杆安装机 bolting machine; rock-bolt setter
锚杆变形 anchorage deformation
锚杆布置示意图 schematic arrangement of rock bolt
锚杆测力计 rock-bolt dynamometer
锚杆测量计 bar ga(u)ge
锚杆长度 bolt length;rock-bolt length; anchor length
锚杆承载力 bearing capacity of anchor
锚杆冲击机 rock bolter
锚杆冲凿机 rock bolter
锚杆挡墙 anchored bolt retaining wall
锚杆挡土墙 anchored retaining wall; anchor rod retaining wall
锚杆箍 anchor stock hoop
锚杆(护坡)工程 anchor works
锚杆环 balancing link;balancing ring; fish shackle
锚杆机 bolter; rock bolting jumbo; roof-bolter
锚杆基本试验 basic test of anchor
锚杆加固 bolting
锚杆间距 anchor pitch; bolt space [spacing];rock-bolt spacing
锚杆铰链 anchor hinge
锚杆金属网 bolts and wire mesh
锚杆静压桩 anchored and jacked pile
锚杆抗力 resistance of bolt
锚杆孔 anchor hole;bolt hole
锚杆孔凿岩 bolt hole drilling
锚杆孔钻车 roof-bolting jumbo
锚杆孔钻进 bolt hole drilling
锚杆拉拔器 bolting pull tester
锚杆拉拔试验 pull-test of rock bolts
锚杆拉力 bolt pulling force
锚杆类型 rock-bolt type
锚杆螺帽 anchor nut
锚杆排距 bolt row space
锚杆喷射混凝土 rock-bolt shotcrete
锚杆墙 anchor wall
锚杆容许承载力 bearing capacity of anchor
锚杆设计拉力 design tension of anchor
锚杆式挡土墙 anchorage retaining wall by tie rod;anchored retaining wall by tie rod;retaining wall with anchored tie-rod; tie-rod anchored retaining wall
锚杆收紧器 anchor winch
锚杆数 number of rock-bolt
锚杆台车 anchor rod jumbo
锚杆体 anchor shank
锚杆头 anchor head
锚杆系统 anchor system
锚杆销 forelock;linchpin
锚杆效应 effectiveness of anchor
锚杆斜拉式驳岸 anchored bulkhead
锚杆眼 square
锚杆验收试验 anchor acceptance test
锚杆应变 anchor strain
锚杆用钻孔机 anchor boring machine
锚杆圆头 anchor ball
锚杆张力 bolt tension
锚杆支撑 anchoring
锚杆支承销 anchored bolt
锚杆支护 anchorage;anchor bolt support;bolt support;rock anchorage; rock bolt(ing) support; suspension roof support;bolting
锚杆支护法 pin timbering
锚杆支护工艺参数 bolting technical parameter
锚杆支护系统【岩】anchor support system;rock bolting system
锚杆支架 bolting support
锚杆支柱 anchor prop
锚杆直径 rock-bolt diameter
锚杆轴力 axial force of rock bolt
锚杆轴力测量 measurement of tension of bolts
锚杆钻机 jumbolter
锚杆钻孔机<钻凿顶板锚杆孔用> countersink
锚钢 anchorage steel
锚更【船】anchor watch
锚工厂 anchor shop
锚工场 anchor shop
锚钩 grasp
锚钩环【船】anchor shackle; bending shackle
锚钩式行走起重机 grapnel travel(l)ing crane
锚钩索 backrope;cat back
锚箍 anchor loop
锚固 anchorage; anchor(ing); bolting;ca(u)lking
锚固安装 anchorage fixing
锚固板桩 anchor sheet piling; tied sheet piling
锚固板桩墙 anchored sheet pile wall
锚固变形 anchor(age) deformation; anchorage slip
锚固槽钢 anchor channel; anchoring rail
锚固长度 anchorage length;anchoring length; bond length; development length; embedded length; grip length
锚固齿板 anchorage blister
锚固的 anchored
锚固的板桩墙 tied bulkhead
锚固的单层板桩挡土墙 single-walled anchored sheet pile bulkhead
锚固的管接头 anchored joint
锚固点 anchorage point;anchor point
锚固吊桥 anchored suspension bridge
锚固端<预应力混凝土> anchored end
锚固段 anchorage section
锚固段黏[粘]结应力 development bond stress
锚固墩 anchorage pier
锚固方法 anchoring method; anchoring system;method of anchoring
锚固钢拱 anchored steel arch
锚固钢筋 anchorage;anchor bar;bent bar;steel anchor bar;anchoring reinforcement
锚固钢筋栈桥 anchored steel trestle
锚固构件 anchorage element; anchoring element
锚固管 anchor tube
锚固灌浆 anchor grouting
锚固后张预应力筋 bonded post-tensioning
锚固滑动引起的预应力损失 prestressing due to slip at anchorage; prestressing loss due to slip at anchorage
锚固滑移 anchorage slip;anchor slip
锚固基础 anchorage foundation; anchoring foundation
锚固剂 anchoring agent
锚固夹 anchoring clip
锚固夹具 fastening gripping device
锚固件 anchor
锚固件变形 anchorage deformation
锚固件滑动 anchorage deformation slip
锚固件滑动变形应力损失 anchorage deformation of slip;anchorage loss
锚固件滑脱 anchorage deformation
锚固件滑移变形应力损失 anchor loss
锚固件位移 anchorage deformation slip
锚固角度 anchor angle
锚固距离 anchorage distance
锚固拉力 anchor force;anchor pull
锚固缆索 bridle cable
锚固类型 anchor type
锚固力 anchorage; anchorage force; anchoring effort;anchoring force
锚固联板 anchoring bracket
锚固螺钉 anchoring screw
锚固螺杆 stay bolt
锚固螺母 anchor nut
锚固螺栓 anchor bolt;anchor screw; anchor stud; holding-down bolt; screw anchor<桥梁支座用>;rag bolt
锚固能力 anchorage capacity
锚固黏[粘]结<预应力钢丝端部> anchorage bond;anchoring bond
锚固黏[粘]结应力 anchorage bond stress
锚固黏[粘]着力 anchor grip
锚固破坏 anchorage failure; anchoring failure
锚固强度 anchorage strength; anchoring strength
锚固区 anchorage region; anchorage zone
锚固圈 anchor loop
锚固设备 anchorage device
锚固失效 anchoring failure
锚固式板桩 anchored sheet piling
锚固式驳岸 anchored bulkhead
锚固式挡土墙 anchored bulkhead
锚固式钢栈桥 anchored steel trestle
锚固式搅拌器 anchor agitator
锚固式锚杆 anchored bolt
锚固式起重机 grapple equipped crane
锚固式悬索 anchored suspension bridge
锚固竖井<吊桥> anchorage shaft
锚固损失 anchorage loss
锚固索 anchor cable
锚固套筒 sleeve anchorage
锚固体 anchorage
锚固铁件 anchor channel;precast anchor
锚固头滑动 slip at anchorage
锚固涂层 anchor coat
锚固弯钩<混凝土结构中伸过零应力点的> end anchorage
锚固系索<联结锚固缆索与滑轮的> bridle hitch
锚固系统 anchorage system; anchoring system
锚固效应 anchoring effect
锚固形式 type of anchorage
锚固性能 anchorage performance
锚固压板 holding-down clip
锚固压力 anchor pressure
锚固应力 anchorage stress
锚固圆木 anchor log
锚固在混凝土中 anchor embedded in concrete
锚固在土中 anchored in the ground
锚固支承座 anchorage bearing; anchored bearing
锚固支座 anchorage bearing;anchored bearing
锚固砖石砌体用角钢 shelf angle
锚固桩 anchor picket; anchor pile; stay pile
锚固装置 anchor fitting; anchoring device;cable anchorage
锚挂的 anchored
锚挂钢筋混凝土护坡 anchored concrete pavement with wire mesh reinforcement

锚冠 anchor crown;anchor head;head of anchor
锚冠卡环 anchor crown shackle
锚和锚链检验合格 anchors and chains proved
锚和锚链证书 anchor and chain certificate
锚喉 anchor throat
锚环 anchor loop;anchor rim;anchor ring;anchor shackle;clamp ring;club link;female cone;fixture ring;Jew's harp;torus ring;anchor and collar <闸门上的金属铰>
锚环系索 ring rope
锚环张拉法 hoop cone system
锚机持链轮 cable holder;cable lifter;cable wheel;chain grab;gypsy wheel;sprocket wheel;wild cat wheel
锚机滚筒 gipsy;gypsy
锚机滚筒接合栓 locking head
锚机间 anchor windlass room
锚机轴 windlass shaft
锚基 anchorage
锚基础 anchor foundation
锚夹预应力负弯矩钢筋 capping cables
锚甲板 foremast deck
锚架 anchor chock;cat;cathead
锚尖 anchor dart
锚尖加板 <防滑措施> shoe an anchor
锚尖装饰 anchor dart
锚肩 anchor shoulder
锚检验钢印记 anchor stamping
锚件 masonry anchor <固定门框于墙体的>;studs set;anchor(age);anchor parts【港】
锚件变形 anchorage deformation
锚件附件 anchoring accessories
锚件区 anchorage zone
锚键 anchor wedge
锚箭饰 anchor dart
锚桨式搅拌器 anchor paddle mixer
锚绞缠 anchor is foul
锚接 anchoring
锚接强度 anchorage strength;anchoring strength
锚接绳固定线夹 clamp for anchor rope;guy clamp
锚结 anchor bend;anchor knot;fisherman's bend
锚筋 anchor;anchor bar;anchor reinforcement;anchor rod;beam pitman;dowel
锚筋支座 beam pitman bearing
锚进入锚床的状态 housing
锚井 spud well
锚纠缠 anchor foul
锚具 anchorage(hardware);anchorage element;ground tackle;ground tackling;anchorage device <预应力混凝土>
锚具变形 anchorage deformation;anchorage slip
锚具变形损失 loss due to anchorage deformation
锚具滑动 slip at the anchorages
锚具温差损失 loss due to anchorage temperature difference
锚具箱 anchorage box
锚具预应力损失 anchorage loss
锚橛 anchor spike;anchor stake
锚孔 anchorage space;anchorage span;anchor eye;anchor opening;anchor space;anchor span;fixture holing
锚孔距 anchor spalling
锚控 anchoring
锚口 collar of anchorage device
锚跨 anchorage space;anchorage span;anchor space;anchor opening <索桥、悬臂桥的>
锚跨孔 anchor span
锚块 anchor(age)block;stay block
锚拉挡土墙施工法 tieback method
锚拉杆 anchor tie
锚拉基础 anchored foundation
锚拉式坞底板 tied dock floor
锚拉支座 anchorage bearing
锚缆 anchorage cable;anchor cable;anchor wire;bridle cable <与拉索成直角的>
锚缆导桩 anchor line guide spud
锚缆结 anchor rope bend
锚缆卷筒 anchor cable drum
锚雷 anchored mine
锚离底 anchor away;anchors aweigh;atrip
锚离水 anchor is clear of water
锚连接器 anchor connector
锚链 anchor(age)chain;anchorage rope;anchor cable;anchor line;anchor rope;cable chain;chain cable;Moore chain;mooring chain
锚链半绞花 an elbow
锚链标记 chain cable mark;mark the chain cable
锚链舱 cable locker;chain bin;chain lockage;chain locker;chain well
锚链舱底甲板 orlop
锚链舱内空的部位 tier
锚链缠绞 break her sheer;break sheer
锚链缠住锚杆 anchor fouled by the stock
锚链缠住锚爪 anchor fouled by the flukes
锚链长(度)length of mooring line;cable length
锚链长度标志 marking
锚链朝后 chain leading aft
锚链朝前 chain leading forward
锚链车间 chain shop
锚链掣 cable compressor;riding stopper
锚链掣动器 dolphin
锚链垂直 anchor up and down
锚链打横 leading to starboard
锚链导向器 chain cable fairleader
锚链端环 chain cable end link
锚链方向 chain direction;chain leading
锚链放松 slack way chain
锚链附件 chain fittings
锚链钩 chain hook
锚链管 chain locker pipe;chain(taker)pipe;deck pipe;monkey pipe;naval pipe
锚链管唇口 dock collar
锚链管电话型系泊浮筒 hawse pipe telephone-type mooring buoy
锚链管竖链型系泊浮筒 hawse pipe riser chain-type mooring buoy
锚链管型浮筒 hawse pipe-type buoy
锚链滚子 anchor link roller
锚链环 chain link;chain ring
锚链检验 cable inspection
锚链检验合格 chain proved
锚链检验室 chain cable proving house
锚链绞缠 foul cable;foul hawse;seamen's disgrace;chain across ship's head
锚链绞花 elbow in the hawse
锚链绞盘 gangspill
锚链节标志 shackle marks
锚链解脱器 cable reliever
锚链紧了 chain taut
锚链卡环 anchor chain shackle;anchor shackle
锚链孔 anchor hawse;cathole;hawse-(-hole);hawse pipe
锚链孔唇口 anchor bolster;hawse bolster
锚链孔防水塞 jackass
锚链孔盖 blind buckler;buckler;buckler plate;half buckler;hawse buckler;hawse flap;riding buckler
锚链孔肋板 hawse hook
锚链孔肋材 hawse piece
锚链孔塞 jackass;mouth plug;plug
锚链孔塞包 hawse bag
锚链孔所在的船首部分 hawse
锚链孔下端塞 <防浪软包> hawse bag;hawse jackass
锚链拉力表 anchor chain meter
锚链连接卸扣 chain cable shackle;connecting shackle;joining shackle
锚链轮 gypsy wheel;anchor chain wheel
锚链末端链环 end link;end ring;open link
锚链配件 chain fittings
锚链普通链环 chain cable common link
锚链强度 cable strength
锚链强度证书 cable certificate
锚链绕过船头 anchor leading abeam
锚链上加绑缆 backing a chain
锚链示数器 anchor cable indicator
锚链松出 veer cable
锚链筒 hawse pipe
锚链筒滚口 hawse pipe flange
锚链筒上端塞 hawse block;hawse plug
锚链筒凸缘 hawse pipe flange
锚链筒下端塞 <防浪软包> hawse bag;hawse boy;hawse jackass
锚链向后 leading after
锚链向前 leading ahead
锚链向右 leading to port
锚链卸扣 anchor chain shackle;cable shackle;joiner shackle
锚链性能 mooring line characteristic
锚链悬链线 chain catenary
锚链旋转环 anchor swivel;chain swivel
锚链旋转接头 anchor swivel;chain swivel
锚链与船首尾向所成的角 trend
锚链在首舷方向 chain on the bow
锚链在尾舷方向 chain on the quarter
锚链在正横 chain on the beam
锚链枕垫 bolster naval hood;hawse bolster
锚链证书 chain cable certificate
锚喷支护 shotcrete-anchorage
锚链制 lanyard stopper
锚链制动器 anchor cable stopper;cable stopper;chain stopper
锚链转环 cable mooring swivel;swivels for anchor chain
锚链转盘 anchor link roller
锚链桩 riding bitt
锚梁 <码头堤岸用> anchor(age) beam;anchor girder;beam tie;tie beam
锚令 anchor order
锚露出水面 anchor awash
锚没有抓底 anchor dragging
锚木 anchor log
锚木档铁箍 anchor hoops
锚钮 anchor button
锚喷 anchoring and guniting;shotcrete and rock bolt
锚喷网支护 shotcrete rockbolt mesh support
锚喷支护 support by rock anchoring and shotcreting
锚喷支护法 bolting and shotcreting method
锚片 germanium wafer
锚旗 anchor flags
锚扦(索)夹具 anchor grip
锚枪 cartridge powered tool
锚墙 anchor wall
锚清爽 clear anchor;anchor is clear
锚球 anchor ball;black ball
锚圈 anchor loop;anchor ring
锚圈套 anchor cup
锚塞 anchoring plug;anchor male cone;male cone
锚设备 anchor arrangement
锚设备试验 anchor trial
锚设备抓力 holding power of ground
锚绳 anchor line
锚石 bonder;moorstone
锚式挡土墙 tied retaining wall
锚式混合机 anchor mixer
锚式混合器 anchor mixer
锚式机芯 lever movement
锚式搅拌机 anchor mixer
锚式搅拌器 anchor impeller;anchor type agitator;anchor type mixer;anchor type stirrer
锚式搅动器 anchor mixer
锚式擒纵叉销 anchor pin
锚式擒纵叉爪 anchor bill
锚式擒纵机构 anchor escapement;recoil escapement
锚式套管封隔器 anchor casing packer
锚式挖沟机 anchor ditcher
锚式鞋 anchor shoe
锚式闸门 anchor gate
锚试验 anchor testing
锚饰箭头 anchor dart
锚栓 anchor bolt;anchor screw;cathead;cotter bolt;crab bolt;drift bolt;fang bolt;ground bolt;hacked bolt;holding-down bolt;stay bolt;stone bolt;truss bolt;wall screw
锚栓测点 measuring bolt
锚栓铰链 anchor hinge
锚栓孔 anchor hole
锚栓力 anchoring force
锚栓支护 rock bolting
锚栓支柱 anchor prop
锚丝 anchoring filament
锚索 anchorage cable;anchorage rope;anchoring wire;anchor line;anchor rope;anchor stay;anchor strap;anchor wire;cable anchor;contact wire mid-point anchor;hawser;mooring wire rope;stay rope;tendon;tie back cable;anchor tendon
锚索插接 cable splice;mariner's splice
锚索挡墙 anchored wall
锚索导口 anchor warp leader
锚索浮标 ball-and-line float;cable buoy
锚索环 anchor yoke
锚索拉绳 viol;voyal;voyol
锚索千斤顶试验 cable jack test
锚塔 anchor tower
锚条 anchor strip;anchor tie
锚铁 anchor iron
锚艇 anchor boat
锚头 anchorage;anchorage head;anchor cone;anchor head
锚头变形 anchorage deformation
锚头滑动 slip at the anchorages
锚头滑移标准 anchor slip criterion
锚头缆 pendant line
锚头蠕动 anchorage creep
锚头绳 cathead line
锚头套筒 socket of anchor
锚头轴 cathead shaft
锚腿式锚泊装置 anchor leg mooring
锚腿式系泊 anchor leg mooring
锚腿数 number of mooring legs
锚腕 wrist

M

锚位 anchoring space; anchor position; berth; mooring stall
锚位标 anchor buoy
锚位灯 position light
锚位浮标 anchor buoy; cable buoy
锚位浮筒 anchor buoy
锚窝 anchor pocket anchor recess
锚系岸壁 anchored bulkhead
锚系板桩 anchored sheet-pile; anchored sheet piling
锚系板桩墙 anchored sheet pile wall; tied sheeting wall
锚系布置 anchor pattern
锚系定位 anchor mooring
锚系基阵 moored array
锚系声呐浮标 moored sonobuoy
锚系总成 anchor assembly
锚线 anchor line
锚箱 anchor cell
锚销 anchor pin
锚效率 anchor efficiency
锚楔 anchor chock; anchor spike; anchor wedge
锚卸扣 anchor shackle
锚形的 anchor shaped
锚形钩 anchor hook
锚形箭头饰 anchor dart
锚形擒纵机 anchor escapement
锚形铁钩 grapnel
锚型树脂 anchoring resin
锚靴 anchorage shoe; anchor shoe
锚眼 anchor eye
锚眼工 hole digger
锚叶 anchor blade
锚已检验合格 anchors proved
锚已收妥 anchor engaged
锚已抓牢 anchor brought up
锚用浮标 anchor buoy
锚用钢丝绳 mooring wire rope
锚用绳环 anchor strop
锚缘 anchor rim
锚在拖 anchor dragging
锚爪 anchor clutch; anchor flukes; fluke; palm
锚爪垫 anchor fluke chock
锚爪尖部 anchor bill; anchor pea(k)
锚抓牢 anchor holding
锚爪鞘 anchor shoe
锚爪与锚柄夹角 fluke to shank angle
锚爪转角 fluke angle
锚掌 anchor palm
锚着 anchorage; anchoring; grapple
锚着 grapple
锚着长度 anchorage length; bond length; grip length
锚着点 anchorage point
锚着钢筋 anchor bar
锚着于地面的 ground-anchored
锚枕 anchor block
锚证书 anchor certificate
锚支闸门 anchor gate
锚支座 anchor chair
锚制动器扳手 spanner for anchor brake
锚钟 anchor bell
锚住 grappling
锚柱 anchorage mast; anchorage picket; anchor column; anchoring picket; anchor mast; anchor post; anchor stud; pull-off pole; vertical anchor
锚柱跨距 anchor span
锚柱门 anchor gate
锚抓拉索 tripping line
锚抓牢 anchor holding
锚抓力 anchor holding power; holding capacity; holding power of anchor; anchor holding capacity
锚抓力不好的底质 bad holding ground
锚砖 anchor brick
锚桩 anchorage deadman; anchorage picket; anchored peg; anchor log; anchor picket; anchor pile; anchor spike; anchor stake; pile anchor; holddown type pile; tension pile
锚桩孔 anchor post hole
锚桩拉线 anchor stay
锚状的 ancyroid
锚锥(头) anchorage cone; anchor(ing) cone; tapered die
锚作艇 anchor handling boat
锚作拖船 anchor handling tug
锚座 anchorage(bearing); anchorage shoe; anchor bed; anchor socket; anchor stock; billboard
锚座类型 anchorage type
锚座破坏 anchorage failure
锚座位置 anchorage location

卯

卯酉面 prime plane
卯酉圈 prime vertical circle
卯酉圈曲率半径 radius of curvature in prime vertical circle

岿

岿 replat; single hill; shoulder【建】

昂

昂 <指古建筑> false cantilever principal rafter

铆

铆叉 riveting handle
铆冲器 riveting punch
铆锤 ca(u)lking hammer; riveting hammer; snap hammer
铆搭接 lap riveting
铆搭接的 lap-riveted
铆钉 rivet; clincher; dolly; jointing rivet
铆钉半埋头 fillister head of rivet
铆钉半圆头 snap rivet head
铆钉背距 backpitch
铆钉边距 edge distance of rivets
铆钉表 rivet list
铆钉拆除工具 rivet removal tool
铆钉铲 buster; rivet buster
铆钉长度 rivet length
铆钉厂 rivet factory
铆钉撑锤 holding-up hammer; riveting hammer
铆钉撑锤顶住铆钉头 buck-up
铆钉撑杆 riveted staybolt
铆钉(冲)模 rivet snap
铆钉冲头 rivet punch; rivet set
铆钉锤 cup-shaped hammer; riveting (machine) hammer
铆钉锤端 hammered point
铆钉搭接 riveted lap joint; rivet(ing) lap joint
铆钉大梁 rivet girder
铆钉挡工 rivet holder
铆钉抵棒 dolly; dolly bar; hobby
铆钉抵锤 holding-up hammer
铆钉垫圈 rivet-back plate; rivet washer
铆钉垫铁 riveting horn
铆钉钉接器 riveter
铆钉顶 riveting knob
铆钉顶棒 bucking bar; holder-on
铆钉顶撑 dolly
铆钉顶锤 holder-on; holding-up hammer
铆钉顶棍 dolly bar
铆钉定位钳 rivet pitching tongs
铆钉端部 rivet point; rivet tail
铆钉锻炉 riveting forge furnace
铆钉对接 butt riveted joint; riveted butt joint
铆钉镦锻机 rivet header
铆钉镦头 bat; rivet tail; tail head
铆钉墩座 button set
铆钉飞边 collar
铆钉封合工具 rivet closing tool
铆钉符号 rivet symbol
铆钉杆 rivet bar; rivet rod; rivet shank; rivet stem; shaft of rivet; shank of rivet
铆钉杆长度 rivet length
铆钉杆面积 area of rivet shaft
铆钉杆身 shaft of rivet; shank
铆钉钢 rivet bar; rivet iron; rivet steel
铆钉钢丝 rivet wire
铆钉割炬 rivet cutting blowpipe
铆钉工 bucker-up; iron driver
铆钉工的辅助工 bumper up
铆钉工具 snap tool
铆钉菇属 <拉> Gomphidius
铆钉固定 fix with rivet
铆钉棍 dolly bar
铆钉行距 ga(u)ge distance; ga(u)ging distance; rivet pitch
铆钉行列 row of rivets
铆钉行(数) rivet row; row of rivets
铆钉行线 ga(u)ge line of rivets
铆钉行中心线 ga(u)ge line
铆钉横距 transverse pitch of rivets
铆钉机 riveter; rivet driver; rivet(ed) buster; riveting hammer; riveting machine; riveting press; rivet squeezer; squeeze riveter
铆钉机顶具 head cup
铆钉挤压器 compression rivet squeezer
铆钉加热 rivet heating
铆钉加热工(人) rivet heater
铆钉加热炉 rivet heater; rivet heating forge
铆钉加热器 rivet heater
铆钉夹 rivet grip
铆钉夹具 rivet holder; riveting clamps
铆钉夹钳 riveting clamps; riveting handle; riveting tongs
铆钉尖 rivet point; rivet tail
铆钉间隔 rivet space; rivet spacing; spacing of the rivets
铆钉间距 distance between rivets; pitch of rivets; rivet interval; rivet pitch; rivet spacing
铆钉间隙 rivet space
铆钉接缝 riveted seam
铆钉接合 riveted bond; riveted connection; riveted joint
铆钉截断器 rivet buster
铆钉截面 rivet section
铆钉茎长 grip of rivet
铆钉距 pitch of rivets; rivet pitch
铆钉距线 pitch line
铆钉孔 rivet(ed) hole; riveting hole; tack hole
铆钉孔冲 drift
铆钉孔扣除 riveted allowable; riveted hole deduction
铆钉孔扩孔钻 reaming iron
铆钉孔裕量 rivet allowance
铆钉连接 riveted connection; riveted bond; riveted joint
铆钉漏斗送料机构 rivet hopper feeder
铆钉炉 rivet(ing) furnace; rivet(ing) forge; rivet(ing) hearth
铆钉毛口 riveting burr
铆钉铆合 closing up
铆钉铆合工具 rivet ca(u)lking tool
铆钉帽 rivet nut
铆钉模 closer; rivet set; rivet stamp; set die; set punch; snap head die
铆钉排列 row of rivets
铆钉排列形式 rivet pattern
铆钉坯 rivet bar
铆钉平接 riveted butt joint
铆钉气割机 autogenous rivet cutter
铆钉器械 snap tool
铆钉钳 rivet clipper; riveting clamps; rivet tongs
铆钉枪 pneumatic hand riveter; pneumatic riveter; riveter; riveting gun; riveting hammer; rivetter; hitter
铆钉枪头 rivet snap
铆钉强度 rivet value
铆钉切断机 rivet buster
铆钉切割 rivet cutting
铆钉深入度 grip of rivet
铆钉渗水 weeping rivet
铆钉试验 <一般包体反弯和钉头锤扁试验> rivet test
铆钉试验工 rivet tester
铆钉双复板对接接头 riveted double butt strap joint
铆钉梭 rivet holder up
铆钉锁簧 riveted dog
铆钉碳钢 carbon rivet steel
铆钉套环 rivet collar
铆钉体 rivet shank; rivet stem; shaft of rivet; shank of rivet
铆钉体长度 length of rivet shank
铆钉铁 rivet iron
铆钉头 head of rivet; manufactured head; rivet (ed) head; riveting head; riveting nut; swage head; head button
铆钉头距离 grip of rivet
铆钉头压模 snap die
铆钉头压型 snap die
铆钉头圆边击平锤 snap hammer
铆钉托 heel dolly bar; holder-cup; holder up; rivet holder; riveting dolly
铆钉尾 rivet tail
铆钉窝头 rivet die
铆钉窝子 rivet snap
铆钉线 ga(u)ge line
铆钉心距 rivet pitch
铆钉旋压机 rivet spinner
铆钉用钉模 dolley set
铆钉用具 rivet(ing) set; setting punch; snap tool
铆钉圆头成型器 button set
铆钉罩锤 rivet snap hammer
铆钉直径 rivet diameter
铆钉制造机 rivet making machine
铆钉中(心)距 center-to-center of rivets; rivet centers
铆钉种类表 rivet list
铆钉轴身 shaft of rivet
铆钉准线 rivet line
铆钉阻力 resistance of rivet
铆顶 dolly(holder); hobby; rivet holder; riveting dolly
铆顶工具 holding-up tool
铆顶棍 dolly bar
铆缝 rivet(ed) seam
铆工 holder-on; holder up; rivet driver; riveting worker; rivet(t)er
铆工班 riveting gang; riveting team
铆工工作 rivet work
铆工模 rivet(ing) knob
铆工用大 U 形棒 banjo bar
铆工助手 bumper
铆工组 riveting gang; riveting party; riveting team; rivet squad
铆固 rivet clasp
铆固簇绒地毯 rivet head carpet
铆焊 plug weld(ing); rivet weld
铆焊(并用)接合 composition joint
铆焊工段 plate-work and welding section
铆焊混合接头 combined joint
铆合 assemble by welding; riveting
铆合板大梁 riveted plate girder; riveting plate girder

M

铆合板梁 riveted plate beam;riveting plate beam
铆合车盘 riveted hull
铆合储罐 riveted tank
铆合大梁 riveted girder
铆合的 riveted
铆合钢管 riveted steel pipe
铆合构造 riveted construction
铆合管 riveted pipe
铆合桁架 riveted truss
铆合互搭接头 riveted lap joint
铆合架 rivet frame
铆合结构 riveted construction;riveting structure
铆合梁 riveted beam;riveted plate girder;riveting beam
铆合凸缘 riveted flange
铆合箱 riveted tank
铆机 riveter;riveting machine
铆接 rivet(connection);riveted bond;riveted butt joint;riveted connection;riveting
铆接板梁 riveted plate girder
铆接板桅 metal mast;plated mast
铆接板桩 riveted connection sheet pile
铆接补板 hard patch
铆接槽 riveted conduit
铆接冲孔机 riveting punch
铆接大梁 riveted girder
铆接的 riveted
铆接的槽 riveted tank
铆接的罐 riveted tank
铆接的套管 riveted casing
铆接的箱 riveted tank
铆接低碳钢管 riveted mild steel pipe
铆接法 riveting
铆接缝 riveted joint;riveted seam
铆接杆件 riveted member
铆接刚度 riveted rigidity
铆接钢板 riveted sheet steel
铆接钢管 riveted steel conduit;riveted steel pipe[piping];riveted steel tube[tubing]
铆接钢桁架桥 riveted steel truss bridge
铆接钢结构 riveted construction
铆接钢梁 riveted steel girder or truss
铆接钢桥 riveted steel bridge
铆接格栅 riveted grating
铆接工具 clincher tool;riveting set;riveting tool
铆接构件 riveted element
铆接管 riveted pipe[piping];riveted tube[tubing];straight riveted pipe
铆接桁架 riveted truss;rivet girder
铆接桁架桥 riveted truss bridge
铆接机 riveter;riveting machine;riveting hammer
铆接架 riveted frame
铆接间距 rivet spacing
铆接(接)头 riveted connection;riveted joint
铆接节点 riveted joint
铆接结点的钢结构 steel framework with riveted joints
铆接结构 riveted construction;riveted structure
铆接抗剪拼(联)接板(处) riveted shear splice
铆接框架 riveted frame
铆接梁 articulated beam;riveted beam
铆接锚链环 split link
铆接模 riveting die
铆接嵌衬 rivet insert
铆接桥 riveted bridge
铆接套管 riveted casing tube
铆接铁轮箍 riveted iron tyre
铆接头 rivet joint
铆接托架 riveting jack
铆接外罩 riveted casing
铆接线 rivet line
铆接行线 ga(u)ge line of rivets
铆接压力 riveting pressure
铆接烟囱 riveted chimney
铆接用具 riveting set
铆接砧 riveting stake
铆接主梁 riveted girder
铆接装管法 stove pipe method
铆接纵缝 riveted seam
铆接作业 driving of rivets
铆结钢 riveted steel
铆紧 rivet tight
铆距 rivet pitch
铆孔留量 <计算铆接用> rivet allowance
铆牢 rivet
铆螺栓 riveted bolt
铆平铆钉 close up the rivet
铆枪 riveter;riveting gun
铆上 rivet on;rivet over
铆上销子 rivet pin
铆头模 die;rivet snap;setting punch;snap head die;snap(set)
铆头压型 snap die
铆行距 ga(u)ge of rivets
铆轧机 rivet holder up
铆砧 riveting stake
铆钻 rivet holder up

茂比乌斯变换 linear fractional transformation;Mobius transformation

茂比乌斯带 Mobius band
茂比乌斯函数 Mobius function
茂福式加热用电阻炉 muffle resistance furnace
茂基阴离子 cyclopentadienyl anion
茂金属 metallocene
茂密处 thick
茂密的森林 dense forest
茂密灌木丛 thick undergrowth
茂盛 grow very heavy
茂盛的(树林等) thick
茂树 <香叶子> myrtle
茂铁三苯硅烷 ferrocenyltriphenylsilane

冒孢子 <小孢子和花粉的总称> miospore

冒槽 overswelling
冒充的好货 shoddy
冒充货 adulterated goods
冒充真货 pass for genuine
冒出 drop;drop in;emerge
冒出的水 outgoing water
冒出水 gushing water
冒地槽【地】miogeosyncline;miomagmatic zone
冒地槽棱柱体 miogeosynclinal prism
冒地槽型沉积建造 miogeosyncline type formation
冒地斜 miogeocline
冒地斜组合 miogeoclinal association
冒顶 badly bleeding;breakaway;bump;cave-in;caving;fall of ground;puking;roof failure;roof fall;caving-in <崩落开拓时的>;collapsed face <隧洞>
冒顶钢锭 badly bleeding ingot
冒风险 risk proneness;run risk
冒估 overvaluation
冒号分类法 colon classification
冒火 belch
冒火热损失 sting-out loss
冒尖装载 loaded to pointed crest
冒浆 leakage of grout;grout oozing out
冒浆轻微 light mud pumping
冒浆情况 mud-pumping condition
冒浆严重 serious mud pumping
冒进停车信号机【铁】running past a stop signal
冒进信号 overrun a signal;overrunning of signal;passing a signal
冒进信号机【铁】overrun a signal
冒空 kiting
冒口 casting head;head metal;outgate;riser;riser-head;rising head;roof fall;roof of caving;shrinkhead;shrinking head;spread out;whistler;feed head <补缩的>
冒口保温箱 pouring box;runner box
冒口补缩区 feeding zone
冒口补缩系统 feeding system
冒口残根 pad
冒口发热剂 riser compound
冒口防缩剂 anti-piping compound;riser compound
冒口高度 feeder height;riser height
冒口回涨 bleeding of a feeder
冒口浇注 riser gating
冒口颈 feeder neck;feeding neck;riser neck
冒口颈长度 riser distance
冒口模棒 riser pin
冒口模数 feeder modulus;riser modulus
冒口切割 arm sprue cut
冒口圈 feeder bush;riser bush;riser sleeve
冒口贴边 riser padding
冒口通气芯 atmospheric riser core
冒口窝 riser base
冒口系统 riser system;runner system
冒口效率 riser efficiency
冒口压力 riser pressure
冒了顶的钢锭 badly bleeding ingot
冒落带 collapse zone
冒落带高度 height of collapse zone
冒落拱 roof pressure arch
冒落线 caving line
冒名 colo(u)r of title
冒名顶替者 imposter
冒名者 personator
冒泥 mud blasting;mud-pumping
冒牌 adulterate;imitate a trademark;imitation brand;mockery
冒牌的 wildcat
冒牌货 adulterated goods;adulteration;counterfeit;fake;pinchbeck;shoddy product
冒牌商品 counterfeit articles
冒泡 bubbling;froth-over
冒泡的喷泉 effervescent
冒泡分类法 bubble sort
冒泡泉 bubbling spring
冒泡微灌 bubbler irrigation
冒气 gassing
冒气程度 gassing level
冒气泡 effervesce
冒汽地带 steaming zone
冒汽地面 steaming ground
冒汽地面型地热含水层 steaming ground type geothermal aquifer
冒汽地面型热田 steaming ground type field
冒汽塘 steam pool
冒受汇兑风险 exposure to exchange risk
冒暑 affection caused by summerheat;heat stroke
冒暑眩晕 dizziness due to heatstroke
冒水翻沙 sand boil
冒水翻沙现象 boil(ing);boil phenomenon
冒水孔 seep hole
冒水汽 water smoking
冒水速度 issuing velocity
冒水蒸气 reek
冒头 door rail;rail
冒险 peril;run the hazard;run the risk
冒险的 risky
冒险借款 bottomry;respondentia
冒险借款债券 bottomry bond
冒险率 hazard rate
冒险事业 venture
冒险事业会计 venture accounting
冒险投资 embark;reckless investment;venture capital
冒险行事 take a fall flier
冒险性 hazard
冒险性的决策 decision in the face of risk
冒险性决策 decision-making under risk
冒险性企业 wildcat
冒险性投资 venture capital
冒险者 venturer
冒险政策 brinkmanship
冒险资本 <原指股东对入股企业的追加投资,现一般指投向高技术产业的资本> venture capital;risk capital
冒泄井 bleeder well
冒烟 aflame;belch;lunt;reek;smoke;smolder(ing)
冒烟的 fuming;smoking
冒烟的东西 smoker
冒烟公共汽车 smoky bus
冒烟列车 smoke train
冒烟燃烧 smoking combustion
冒烟指数 smoke-developed rating
冒用商标 infringement
冒油 bleed
冒渣口 relief sprue
冒涨 surge
冒涨钢 wild steel
冒涨金属 wild metal
冒着波浪 ship a sea

贸易保护法 trade safe-guarding act

贸易保护主义 trade protectionism
贸易报告 returns of trade
贸易比例矩阵 trade matrix
贸易壁垒 barriers to trade;trade barrier;trade wall
贸易表 trading schedule
贸易波动 fluctuations trade;trade fluctuation
贸易不振 bad trade
贸易部 Department of Trade <美>;Board of Trade <英>
贸易部门 commercial department
贸易财货 tradable goods
贸易采购 commercial procurement
贸易草约 trade protocol
贸易查询资料 trade reference
贸易差额 balance of trade;merchandise balance;trade balance;trade bid;trade gap
贸易差额论 theory of the balance of trade
贸易常规 commercial custom of trade;custom of trade
贸易承兑汇票 trade acceptance
贸易城市 trade city
贸易乘数 trade multiplier
贸易赤字 trade deficit
贸易出超 favo(u)rable balance of trade
贸易创造 trade creation
贸易促进 trade promotion
贸易促进会 Board of Trade;trade board
贸易促进条例 act for the encourage-

ment of trade
贸易促进中心 trade promotion center [centre]
贸易代办处 trade agency
贸易代表 trade representative
贸易代理人 commission agent
贸易代理商 commission merchant
贸易带 trade zone
贸易贷款利率 commercial loan rate
贸易的技术性壁垒 technical barrier to trade
贸易对环境的影响 trade impact on environment
贸易对象 trade partner
贸易吨位 trade burdens
贸易额 trade turnover;trade volume;value of trade
贸易发展委员会 Committee on Trade and Development
贸易法 commercial law;law of trade;trade act;trade law
贸易繁荣 brisk trade
贸易范围 trading limit
贸易方式 mode of trade;trade method;type of trading
贸易方向 direction of trade
贸易访问团 trade mission
贸易放宽 trade liberalization
贸易费用 trade charges
贸易份额 share of trade;trade share
贸易风 trade wind
贸易风暖流 trade current;trade drift
贸易风险 trade hazard;trade risks
贸易复苏 revival of trade
贸易干扰手段 trade distorting device
贸易港(口) commercial port;trading port;trade port
贸易格局 pattern of trade
贸易公司 commercial company;commercial corporation;commercial firm;trading company;trading corporation;trading firm
贸易关系 commercial relation;trade relation
贸易关系协会 Trade Relations Association
贸易官员 commercial agent
贸易管理 commercial management
贸易管制 trade control;trade restriction
贸易惯例 customs of trade;trade custom;trade practice;trade usage
贸易航路 trade route
贸易合同 business contract;commercial contract;trade contract
贸易合作组织 Organization for Trade Cooperation
贸易和发展理事会 Trade and Development Board
贸易和工业部 <英> Department of Trade and Industry
贸易和支付协定 trade-and-payment agreement
贸易和支付自由化 trade-and-payment liberalization
贸易恒等式 trade identity
贸易互惠性 reciprocity in trade
贸易环境风险指数 Business Environmental Risk Index
贸易汇兑 commercial exchange
贸易汇率 commercial rate
贸易汇票 trade bill
贸易汇票额 volume of trade bills
贸易伙伴 trade partner
贸易货栈 trade warehouse; trading warehouse
贸易机构 commercial organization;trading mechanics
贸易机会 trade access
贸易基地 commercial base
贸易畸形发展 trade distortion
贸易及发展会议 Trade and Development Committee
贸易及关税法 Trade and Tariff Act
贸易及关税总协定 General Agreement on Trade and Tariffs
贸易集团 trade block;trade group
贸易集中 concentration of trade
贸易技术性壁垒标准 standard as technical barrier to trade
贸易加权贬值 trade-weighted depreciation
贸易加权汇率 trade-weighted exchange rate
贸易加权平均数 trade-weighted average
贸易价格 trade price
贸易减让 trade concession
贸易交易会 trade fair
贸易结构的调整 re-structuralization of trade
贸易介绍人 commission broker
贸易金融 commercial finance; trade finance
贸易禁运 commercial embargo
贸易经济 trade economy
贸易经营 commercial business
贸易净差额 net balance of trade
贸易净利 net trading profit
贸易竞争 commercial competition; trade war
贸易救济服务处 Trade Remedy Assistance Office
贸易卡特尔 trade cartel
贸易开辟 trade creation
贸易开拓 trade creation
贸易科目 trade account
贸易控制程度 degree of administered-trade
贸易库存 trade inventories
贸易扩展 trade expansion
贸易扩展法 Trade Expansion Act
贸易利润 commercial profit
贸易量 trade volume
贸易量增长率减低 reduced growth in trade volume
贸易流量 trade flow
贸易路线 trade route
贸易论 discourse of trade
贸易毛利 gross trading profit
贸易美元 trade dollar
贸易秘密 trade secret
贸易摩擦 trade friction
贸易逆差 adverse balance; adverse balance of trade;adverse trade balance; passive trade balance; trade deficit;unfavo(u)rable balance of trade
贸易年度 trade year
贸易配额协定 trade-quota agreement
贸易票据 commercial bill; commercial draft
贸易平衡 balanced trade; balance of trade;trade balance;trade bid
贸易普查 census of trade
贸易歧视 trade discrimination
贸易契约 contract note of sales;trade contract
贸易前景 trade prospect
贸易清算账户 commercial clearing account
贸易清算制 clear system
贸易情报交换所 clearing house for trade information
贸易情况 state of trade;trade information
贸易区 business zone; trade area; trade zone
贸易区域 trading limit
贸易渠道 trade channel
贸易权利 trade rights
贸易融资 foreign trade finance;trade financing
贸易入超 overbalance of imports;unfavo(u)rable balance of trade
贸易商 merchant
贸易商号 trading concern
贸易商品 traded goods
贸易商行 trading house
贸易商行雇员 trade crew
贸易商行职员 trade crew;trade gang
贸易上的往来关系 connexion
贸易申诉局 Office of Trade Ombudsman
贸易实际成本条件 real cost terms of trade
贸易市场 trading market
贸易收入条件 income terms of trade
贸易收益 gain from trade
贸易收支 balance of trade;trade payment
贸易收支差额 balance of foreign trade
贸易术语 trade term
贸易术语解释的国际通则 International Rules for the Interpretation of Trade Terms
贸易衰退 depression of trade
贸易税 trade tax
贸易顺差 active trade balance; favo(u)rable balance of trade;favo(u)rable trade balance; favo(u)rable trade payment; surplus trade balance;trade surplus
贸易谈判 trade negotiation
贸易体制 trade framework
贸易条件 terms of trade
贸易条件效果 terms-of-trade effect
贸易条件指数 index of terms of trade;terms of trade indexes
贸易条款 terms of trade;trade term
贸易条例 trade regulation
贸易条约 commercial treaty
贸易调整 trade adjustment
贸易调整救济 trade adjustment assistance
贸易通货 trade currency;trading currency
贸易同盟 trade bloc
贸易投资信托公司 commercial investment trust company
贸易外汇 foreign currencies earned through trade; trade foreign exchange earnings
贸易外收支 trade invisible trade
贸易往来 commercial intercourse; trade contact
贸易危机 commercial crisis
贸易危险 trade hazard
贸易无差别曲线 trade indifference curve
贸易习惯 trade practice;trade usage
贸易线路 trade route
贸易限制 restraint of trade; restriction of trade;trade restriction
贸易限制中的联合 combination in restraint of trade
贸易项目 trade account
贸易象征 badges of trade
贸易效果 effect of trade
贸易效益 commercial efficiency
贸易协定 commercial agreement;trade agreement
贸易协定书 trade protocol
贸易协定税率 trade agreement rate of duty
贸易协会 Finance House Association; mixed chambers of commerce
贸易协会联合会 Trades Union Congress
贸易信贷 trade credit
贸易信贷公司 commercial credit corporation
贸易信贷及预付款 trade credit and advances
贸易信用 mercantile credit; trade credit
贸易信用保险 commercial credit insurance
贸易许可证 trade license
贸易许可证条例 license to trade ordinance
贸易循环 trade cycle
贸易一方 trade party
贸易依存度 dependence on foreign trade;foreign trade dependence
贸易议定书 trade protocoi
贸易异常现象 trade distortion
贸易盈余 trade surplus
贸易用商标纸 trading stamp paper
贸易用语 trade term
贸易优惠 trade preference
贸易优先权 trade preference
贸易有利条件 trading advantage
贸易与工业 trade and industry
贸易约束 restraint of trade
贸易云母 mica of commerce
贸易杂志 trade journal;trade magazine
贸易战 trade war
贸易站 post;trading post
贸易账户 trade account
贸易障碍 barrier to trade;trade barrier
贸易折扣 commercial discount
贸易者 trader
贸易者一览表 trader's schedule
贸易争吵 trade brawl
贸易争议 trade dispute
贸易值 trade value
贸易制裁 trade sanction
贸易中心 entrepot;mercantile center [centre];trade center[centre]
贸易周期 trade cycle
贸易周转额 amount of business
贸易注册 commercial registration
贸易转向 trade diversion
贸易转向效果 trade diverting effect
贸易转向援助 trade diverting aid
贸易转移 trade diversion
贸易资本 commercial capital
贸易资本流动 trade capital movement
贸易自由 freedom of trade;free trade;liberty of trading
贸易自由港 commercial free port
贸易自由化 liberalization of trade;trade liberalization
贸易总额 total volume of trade
贸易总量 total volume of trade
贸易总值 total value of trade
贸易组成 composition of trade

帽

帽瓣 <御寒用> earflaps;eartabs
帽孢锈菌属 <拉> Pileolaria
帽边扩张机 brim-stretching machine
帽材 cap[ping] piece;cap strip
帽带 bat ribbon
帽灯 cap lamp
帽钉 hat peg;rivet
帽顶事故 roof-fall accident
帽盾 hat front;hat shield
帽儿梁 lattice framing
帽盖 hood;capping <管子端头的>
帽盖的卷边 feint
帽盖泛水 cap flashing;counter flashing
帽盖泛水件 counter flashing
帽盖混凝土 cap concrete
帽盖装脚式绝缘子 cap-and-pin insu-

lator
帽钩 hat hook
帽管 <加盖管> capped pipe
帽徽 cap badge
帽架 hat rack;hat stand
帽梁 cap(ping) beam;coping beam;bent cap;pile cap
帽模 hat-block
帽木 cap[ping] piece;cap plate;header;timber cap;heading collar <导坑木支撑的>
帽坯 hat felt
帽坯机 hat-forming machine
帽墙 <堆石堤上的> crown wall
帽圈 hat band
帽绒 hatter's plush
帽塞 cap stopper;helmet
帽石 capping stone;capstone;coping stone
帽式 hat type
帽式金属底盘 <安装小型机器用> hat type foundation
帽刷 hat brush
帽套 cap sleeve
帽铜 cap copper
帽销式绝缘子 cap-and-pin insulator
帽形成 cap formation
帽形法兰 hat flange
帽形钢锭缺陷 top hat
帽形截面 hat section
帽形孔板 hat orifice
帽形孔型 hat pass
帽形密封 cap seal;hat seal
帽形密封件 L type packing
帽形起重机 helmet crane;visor crane
帽形圈 hat washer
帽形填密件 flange mo(u)lded packing
帽形止水 hat type seal
帽形轴封 hat packing
帽岩成分 composition of cap core
帽檐 visor
帽檐模型 cap model
帽缘 cap ridge
帽罩 calotte;cap piece;helmet
帽罩齿形边缘 slotted edge of cap
帽柱 bracket-like column
帽桩 capping brick
帽状阀 bonnet valve
帽状脊瓦 bonnet hip tile
帽状物 bonnet;calotte
帽状云 cloud cap
帽子 headgear;louse cage <俚语>
帽子模式 cap mode
帽子模型 <土的> cap model

没

没按直径或长度比例绘制 not to scale for diameter or length

没光漆 flatting varnish
没设计得不好的 ill-designed
没敲钉的 nailless
没人认领的物品 unclaimed freight
没烧旺的 unheated
没食子 gallnut
没食子酸 gallic acid
没完全发育 not fully developed
没销路 go begging
没用过的水 unused water
没有按期完成 behind completion
没有把握的(工作等) dubious
没有补助金的 unsubsidized
没有差别的产品 non-differentiated products
没有成本的 costless
没有出口的谷 court dock
没有达到预期效果 fall flat
没有的事物 non-exercise
没有抵押的 unencumbered
没有定义的 undefined
没有断绳保险的 dogless
没有发觉的 unaware
没有发现错误 no-fault found
没有防护的 unprotected
没有腐蚀性的水 inert water
没有负担的财产 unencumbered property
没有根据的 unauthorized;unfounded
没有挂列车的机车 light engine
没有关系 no relation
没有焊透 non-fusion
没有夯实 undertamping
没有基础的 baseless;ungrounded
没有基金的退休金计划 funded pension plan
没有计划的 undersigned
没有价值 void value
没有价值的 nugatory;valueless
没有价值的小东西 button
没有坚实路面的 unimproved
没有建筑物的荒地 undeveloped land
没有建筑物的空地 undeveloped land
没有节疤或缺陷的木材 clear lumber
没有结果的 unproductive
没有经验的 inexperienced;unfamiliar
没有经验的工作人员 green man
没有经验的顾客 unsophisticated buyer
没有坑道的蛇行 snake without tunnel
没有空气的 air-free
没有孔的管 blank pipe
没有孔隙性就没有溶提作用 no eluviation without porosity
没有快餐部客车 snack bar coach
没有拉杆或斜撑的 freestanding
没有篱笆的 unfenced
没有理由的索赔 unwarranted claim
没有利用率 unavailability
没有帘的 uncurtained
没有路的 passless
没有轮缘的 rimless
没有抹灰的墙 unsheathed wall
没有幕的 uncurtained
没有内容的 thin
没有黏[粘]性的 non-stick
没有扭曲 out-of-wind
没有排泄的 undrained
没有硼 without boron
没有偏差的 unerring
没有确定的关系 incertae sedis
没有入侵 non-intrusion
没有砂眼的 grit-free
没有声音的 noiseless
没有时间限制的 open ended
没有使用的期间 period of unavailability
没有收入来源的失业者 inactive unemployed
没有弹簧的锁紧螺栓 dead bolt
没有提单的货物 cargo without bill of lading
没有通风设备的 unventilated
没有涂焦油的麻屑 untarred tow
没有涂焦油的拖绳 untarred tow
没有完成必要的凭证手续 failure to complete required paper work
没有污染的填方 clean fill
没有屋顶的 roofless
没有限定的权利 indefinite right
没有限制的 unqualified
没有销路的 unsalable
没有销路的产品 unsalable products
没有销路的商品 unsalable products
没有写下的 unwritten
没有选择的 non-optional
没有(牙)齿的 toothless
没有严格限制条件 without qualification
没有用挡土板或其他支撑的露天开挖 open excavation
没有用的 unavailable
没有用帘幕遮住的 unscreened
没有增加的价值 unimproved value
没有障碍物 clear
没有蒸发的燃料 dribbled fuel
没有执照的 unlicensed
没有值的对策 game without a value
没有重量的 weightless
没有重要性 count for nothing
没有专门知识的人 outsider
没有装玻璃的 unglazed
没有装饰的 unadorned
没有准则的 casual
没有资产因而不能对他实行资产扣押者 judg(e)ment proof
没有资格 incapable of
没有资助的 unsubsidized
没有作出反应的投标 non-responsive bid
没熨过的窄褶 cartridge pleats
没载货的船 clean ship

枚

枚举 enumerate;enumeration

枚举定理 enumeration theorem
枚举法 enumeration method
枚举功能 enumeration function
枚举函数 enumerating function
枚举类型 enumeration type
枚举树 enumeration tree
枚举算法 enumeration algorithm
枚举文件 enumeration file

玫

玫瑰 Hedgerow rosa;rose supreme;rugosa rose

玫瑰瓣状应变片丛 rosette ga(u)ge
玫瑰窗 marigold window
玫瑰蛋白石 rose opal
玫瑰冠 rose comb
玫瑰红 rhodamine;rose Bengal;rose colo(u)r;rose red;rosiness
玫瑰红的 auroral
玫瑰红刚玉 rose A
玫瑰红染色剂 Rhodamine dye
玫瑰(红)色 rose pink
玫瑰红色的 rose colo(u)red
玫瑰花格窗 traceried rose window
玫瑰花环形成试验 rosette formation test
玫瑰花结 rosette
玫瑰花圃 rosary
玫瑰花饰 rosette
玫瑰花坛 rosary
玫瑰花图 rose pattern
玫瑰花线脚 rose mo(u)lding
玫瑰花形 rosette
玫瑰花形应变片 strain-ga(u)ge rosette
玫瑰花型 rose
玫瑰蜡 oleoptene;stearoptene
玫瑰榄球 rose olive
玫瑰绿宝石 rose beryl
玫瑰木 rose wood
玫瑰凝结物 rose concrete
玫瑰棚架 rose-arbour
玫瑰色的 roseate;rosy
玫瑰色颗粒 rose grain
玫瑰色磨料 rose grain
玫瑰砷钙石 roselite
玫瑰饰 rose;rose ornament
玫瑰水 rose water
玫瑰酸 roseolic acid
玫瑰头钉 rosehead
玫瑰图 rose (chart);rose diagram;rosette
玫瑰图案压花玻璃 rose-patterned glass
玫瑰线 curve;rose curve;rose
玫瑰线脚 rope polygon
玫瑰线脚饰 rose mo(u)lding
玫瑰形 rose pattern
玫瑰形窗 Catherine wheel window;rose window;wheel window
玫瑰形机用绞刀 rose chucking reamer
玫瑰形铣刀 rose cutter;rose reamer
玫瑰形应变片 strain rosette
玫瑰形钻头 rose bit
玫瑰眼 rose-eye
玫瑰园 rosarium;rosary;rose garden;rosery
玫瑰状断口 rosette fracture
玫瑰状环 rose ring
玫瑰状石墨 rosette graphite
玫瑰状松香甘油酯 rosette ester
玫瑰状松香甘油酯清漆 rosette ester varnish
玫瑰状铣头导向器 rose bit pilot
玫瑰状钻头 rose bit
玫瑰紫 rose purple
玫瑰紫釉 crimson purple glaze
玫红卟啉 rhodo-porphyrin
玫红初卟啉 rhodo-etio-porphyrin
玫红黄玉 rose topaz
玫红黄质 thujorhodin
玫红尖晶石 balas(ruby)
玫红品 rhodopin
玫红紫素 rhodopurpurin

眉

眉棱 <木板缺陷> brow

眉梁 safety lintel
眉毛 brow;eye brow
眉批 head notes
眉题 overline
眉条 crassula[复 crassulae]
眉纹的 superciliary
眉状结焦 eye brow

莓

莓实树【植】Strawberry tree

莓系属牧草 <一种野生的优良牧草> blue grass

梅

梅 bai;Japanese apricot

梅达尔特型矫直机 Medart straightener
梅达林铜钴铝铁合金 Metalline
梅德韦德夫地震烈度表 Medvedev intensity scale
梅德韦德夫-史邦比尔-卡米克地震烈度等级 Medvedev-Sponbeuer-Kamik scale[MSK]
梅德韦德-史邦比尔-卡米克地震烈度表 MSK scale of earthquake intensity
梅尔尼合金 Melni alloy
梅尔尼镍基耐热合金 Melni alloy
梅尔生电压调整图 Mershon diagram
梅耳罗斯泵 Melrose pump
梅格尔土壤电阻测定器 Megger earth tester
梅格派尔铁铬铝电阻丝合金 Megapyr
梅格珀姆铁镍锰高导磁率合金 Megaperm
梅红色 plum;plum colo(u)r
梅花 Flos Mume;Japanese apricot flower;mume flower
梅花扳手 box(-end) wrench;box key;box spanner;ring spanner
梅花点式 quincunx
梅花电槌 needle gun
梅花股芯钢丝绳 star wire rope
梅花联结轴 wobbler spindle

M

梅花砌砖法 Flemish bond
梅花山 plum hills
梅花式挖泥法 quincuncial dredging method
梅花试棒 cloverleaf coupon
梅花饰 cinquefoil;quinquefoil
梅花双头扳手 closed wrench; ring spanner;double offset ring spanner
梅花头 wabbler
梅花头铣床 wobbler milling machine
梅花形 plum-blossom pattern; quincunx
梅花形布置 staggered pattern
梅花形的 quincuncial
梅花形模式 staggered pattern
梅花形栽植 superposed square planting
梅花形栽植制度 quincunx system of planting
梅花玉 spotted jade
梅花针 percussopunctator;plum-blossom needle;plum-blossom pyonex;plum needle
梅花抓斗 orange-peal clamshell; orange-peel bucket
梅花桩 quincuncial piles; staggered piles
梅花桩墙 secant pile
梅花状交错桩 quincuncial piles;staggered piling
梅花钻 box bit drill; rose bit; rose drill;star drill
梅花钻头 rosette bit;star bit
梅坎顿反向极性巨带【地】Mercanton reversed polarity hyperzone
梅坎顿正向极性巨时 Mercanton normal polarity hyperchron
梅坎顿正向极性巨时间带 Mercanton normal polarity hyperchronzone
梅克尔灯 Meker burner
梅拉铝合金 Meral alloy
梅拉梅克群【地】Meramec Group; Meramecian
梅拉姆机器人 <较高级的移动式多关节双臂机器人> MELARM robot
梅兰蒂木 meranti
梅雷瓦尔页岩层 Merevale shale
梅里尔-克劳法 Merrill-Crowe process
梅里克称重计 Merrick weightometer
梅立恩公式 Merian's formula
梅林变换 Mellin transform
梅洛特铋锡铅易熔合金 Mellotte's alloy
梅纳旁压仪 <测定土和软的岩石的原位强度和变形模量用> Menard pressuremeter
梅尼振荡器 Meny oscillator
梅瓶 plum vase
梅丘陵砂岩 May Hill sandstone
梅塞尔钢板桩 Messer steel sheet piling
梅塞尔施工法 Messer sheet-pile method
梅塞尔隧道施工法 Messer tunneling method
梅赛公式 Massay formula
梅森定理 Mason's theorem
梅森公式 Mason's formula
梅氏冲击试验机 Maybach impact testing machine
梅氏逆流平洗机 Mezzera counter acting open width washing machine
梅树 plum
梅斯分度镜 May's graticule
梅斯塔油膜轴承 Mesta bearing
梅斯专利 <将球形门执手装到轴上的专利方法> Mace's patent
梅特列克斯(式空心)插销 Metlex plug
梅通周期 <18.6 年周期的潮汐变化> Metonic cycle
梅西叶号数 Messier's number
梅西叶星表 Messier's catalog
梅西叶星云星团表 Messier Catalogue
梅形茶具 plum design;tea set
梅逊定律 Mersenne's law
梅亚变换 Meiyer's transform
梅亚型船 Meiyer form ship
梅亚型船首 Meiyer form bow
梅耶棒 Meyer's bar
梅耶泵 Meyer's pump
梅耶二氧化碳吸收管 Meyer's tube
梅耶公式 Meyer's formula
梅耶合成法 Meyer's synthesis
梅耶霍夫(承载力)公式 Meyerhof's formula
梅耶霍夫法 <用屈服线理论计算水泥混凝土路面最大荷载的方法> Meyerhof method
梅耶凝结理论 Meyer's condensation theory
梅耶试验 Meyer's test
梅耶往复泵 Meyer reciprocating pump
梅耶问题 Meyer's problem
梅耶吸收器 Meyer's absorber
梅耶硬度 Meyer's hardness
梅耶指数 Meyer's index
梅耶酯化定律 Meyer's law
梅雨 bai-u;blossom shower;mo(u)ld rain;plum rain;Meiyu <我国长江以南6~7月份间的雨季>
梅雨季 bai-u season
梅雨开始 beginning of the bai-u
梅雨期 bai-u rainy period
梅雨前兆雨 prebai-u rainfall
梅制品 prune products
梅兹洛合金 Mazlo alloy
梅兹-普赖斯保护系统 Merze-Price protection system
梅子青 plum green

媒

媒合物 solvate

媒剂 mediator
媒介 agency;instrumentality;intermediary agent;medium
媒介传疾病 vector-borne disease
媒介的 intermediary;vehicular
媒介电化学氧化 mediated electrochemical oxidation
媒介动物 vector
媒介剂 catalyst agent;mixtion
媒介体 vehicle base
媒介物 carrier;intermediary;intermedium[复 intermedia]; vector; vehicle
媒介效应 intermediation effect
媒染 mordanting
媒染的 mordant
媒染红 mordant rouge
媒染黄 mordant yellow
媒染剂 colo(u)r fixing; dye mordant;mordant;mordant in dy(e)-ing
媒染剂红液 mordant rouge;red acetate;red liquor
媒染染料 mordant colo(u)r;mordant dye
媒染助剂 mordanting assistant
媒色颜料 lake
媒体 medium
媒液 vehicle
媒液漆料 oil vehicle paint
媒液油漆 oil vehicle paint
媒质 mediator;medium[复 media]
媒质石英摄谱仪 medium quartz spectrograph
媒质损耗光纤 medium-loss fiber[fibre]
媒质吸收系数 absorption coefficient of medium
媒质相干干扰 media coherence disturbances
媒质增益曲线 gain curve of medium
媒质增益系数 gain coefficient of medium

湄

湄公河 Mekong River

楣

楣 dormant tree;lintel;lintel beam; summer

楣板 cornice
楣部 supercilium
楣窗 fanlight(catch); fanlight transom(e) window; transom (window)
楣窗开关 fanlight opener
楣窗拉杆 shadbolt
楣式构造 trabeated construction
楣柱连接的小拱拱腋 <椭圆拱脚处的最小半径弧> hance

煤

煤 coal;black diamond <俚语>

煤柏油脂 coal-tar pitch
煤磅 coal scale;coal weigher
煤包裹体 coal inclusion
煤胞 cenosphere
煤崩落 coal fall
煤泵送 coal pumping
煤比 coal ratio
煤壁线 rib line
煤变质程度 metamorphic rank of coal
煤变质带 metamorphic belt of coal
煤变质阶段 metamorphic stage of coal
煤变质作用 coal metamorphism;metamorphism of coal
煤变质作用类型 type of coal metamorphic
煤饼 briquet(te); caking coal; coal briquet(te)
煤驳 coal barge;coal lighter
煤(驳)船 haulabout
煤驳码头 coal barge jetty
煤仓 bin;bunker;coal bin;coal bunker <船舶的>; coal hopper; coal pocket;coal shed;coal storage;coal silo
煤仓间 bunker bay
煤仓容量 bunkering capacity;capacity of coal bunker
煤仓式 hopper type
煤舱 bunker; coal bin; coal breaker; coal bunker; coal hold; coal hopper;coaling hatch
煤舱壁横支撑 bunker stay
煤舱孔 coal hole
煤舱口 scuttle hatch(cover)
煤舱肋骨 bunker frame
煤舱门 bunker door
煤舱平舱 trimmed in bunkers
煤舱运输机 bunker conveyer[conveyor]
煤藏 coal deposit;coal reserves
煤槽 coal chute
煤层 carbonic rock;coal bed;coal deposit;coal horizon;coal layer;coal measures; coal rake seam; coal seam
煤层包裹体 balk
煤层包体 coal seam inclusion
煤层边界 coal seam boundary
煤层变薄 squeeze
煤层剥露 coal seam uncovering
煤层出露边缘 streak
煤层垂向断面图 vertical section of coal seam
煤层垂向横断面图 vertical cross-section of coal seam
煤层垂向纵断面图 vertical longitudinal section of coal seam
煤层的不连续面 break
煤层的厚度 thickness of coal formation
煤层的密度 density of coal formation
煤层的软土顶板 clod
煤层的真厚度 true thickness of coal formation
煤层等厚图 isopachy map of coal seam
煤层底板 coal seam floor;seat earth
煤层底板的软碳质页岩 rashings
煤层底板等高线图 contour map of bottom surface of coal seam
煤层底部 bottom
煤层底界面的深度 low surface depth of coal formation
煤层底黏[粘]土 clunch;rootlet bed
煤层顶板 coal seam roof;roof
煤层顶板类型 type of coal seam roof
煤层顶板页岩 ramble
煤层顶部底板页岩 clod
煤层顶底板页岩 clod
煤层顶界面的深度 upper bound depth of coal formation
煤层顶面等高线图 contour map of top surface of coal bed; contour map of top surface of coal seam
煤层对比 coal(seam) correlation
煤层对比图 comparative map of coal seam
煤层风化带宽度 width of weathered coal zone
煤层赋存 occurrence of coal seam
煤层构造 coal seam structure
煤层含尘量 seam dustiness
煤层厚度 coal seam thickness;thickness of coal seam
煤层厚度变异系数 coefficient of coal thickness variation
煤层厚度标准差 standard deviation of coal thickness
煤层厚度差值 difference of coal seam thickness
煤层几何形态 geometry of coal bed; geometry of coal seam
煤层加倍变厚 seam doubling
煤层夹矸 coal parting
煤层甲烷 coal bed methane
煤层尖灭 balk;coal pinch out;fouls
煤层间的通道 boutgate
煤层间距 spacing of coal seam
煤层结构 coal bed texture;coal seam texture
煤层结核 bullion
煤层截面 coal face
煤层掘进机 coal ripper
煤层可(开)采厚度 minable thickness of coal
煤层累计等厚图 isopach map of total coal seam
煤层累计厚度 total thickness of coal seam
煤层立面投影图 vertical plane projection of coal bed
煤层露头 coal basset;coal outbreak;outcrop of coal bed
煤层露头线 exposure line of coal seam
煤层煤样 coal seam sample
煤层密度 density of coal seam;density of coal zone
煤层名称 name of coal seam
煤层内的黄铁矿 scruff
煤层剖面 coal bed profile;coal seam profile

煤层气 coal mining methane
煤层气抽放 coal mining methane drainage
煤层气利用 coal mining methane utilization
煤层气排放 coal mining methane emission
煤层倾角 coal bed pitch; coal seam dip angle; coal seam pitch
煤层取样平面图 sampling plan of coal seam
煤层软泥土顶板 clod
煤层石球 coal ball
煤层水平断面图 horizontal section of coal bed; horizontal section of coal seam
煤层素描图 coal seam sketch
煤层瓦斯 gas in coal
煤层瓦斯含量 coal seam gas bearing capacity; coal seam methane content
煤层瓦斯压力 coal-bed gas pressure; coal seam gas pressure
煤层伪底 coal seam false floor
煤层伪顶 coal seam false roof
煤层吸附气 absorbed gas in coal
煤层下黏[粘]土 seat earth; underclay
煤层巷道 heading
煤层型别 category of coal seam
煤层氧化带宽度 width of oxidized coal seam; width of oxidized coal zone
煤层要素 element of coal seam occurrence
煤层圆形结核 blister
煤层质量 quality of boring coal seam
煤层中岩石轧包体 ba(u)lk
煤层注水 coal seam infusion; infusion in seam
煤层总厚度 overall thickness of coal bed; overall thickness of coal seam
煤产量 coal production; coal yield; yield of coal
煤产品类型 product type of coal
煤铲 coal shovel; fire shovel
煤铲轮齿装置 shovel gear
煤厂 coaling plant
煤场 coal depot; coal stock; coal storage ground; coal (storage) yard; stock yard
煤场管理 coal yard management
煤车 coal wagon; jimmy; tramcar
煤车翻车机 coal tipper
煤车加热解冻器 hotdogs
煤尘 coal dust; dust-methane-air-mixture; fine breeze; smuts
煤尘爆炸 coal dust explosion
煤尘爆炸事故 dust explosion accident
煤尘爆炸试验仪 test instrument for coal dust explosion
煤尘爆炸危害 dust explosion hazard
煤尘爆炸性 explosibility of coal dust
煤尘爆炸指数 explosive index of coal dust
煤尘超限 excessive amount of dust
煤尘肺 anthracosis; black lung
煤尘和烟雾检测仪 dust and fume monitor
煤尘焦粉 braize
煤尘煤样 coal sample for determination of dust
煤尘黏[粘]合剂 coal dust cementing agent
煤尘排除 coal dust suppression
煤尘指数 coal dust index
煤沉积 coal deposit
煤成气 gas from coal and coal measure
煤成气量 amount of coal-derived gas; amount of coal-formed gas; amount of coal-related gas
煤储库 coal storage
煤船 coal carrier; coaler; collier
煤床【地】coal measures
煤脆性 friability of coal
煤萃 coal extract
煤袋 coalsack
煤当量 coal equivalent
煤当量吨 ton of coal equivalent
煤的 coaly
煤的 G 指数 caking index of coal
煤的 G 指数分级 caking index G graduation of coal
煤的采样和制样 sample taking and preparation of coal
煤的产品 productized
煤的产品类型 product type of coal
煤的成因类型 genetic(al) type
煤的低温灰 ash with low temperature
煤的地下气化 in-situ coal gasification; underground coal gasification
煤的发热量 calorific value of coal
煤的分类 coal classification
煤的分子结构 molecular structure of coal
煤的粉磨设备 pulverized fuel plant
煤的粉碎 coal pulverization
煤的辐射剖面 radial section of coal
煤的高温灰 ash with high temperature
煤的工业分析 proximate coal analysis
煤的工业类型 industry type
煤的工业牌号 industry brand
煤的工艺性试验 technologic(al) test of coal
煤的化学净化 chemical cleaning of coal
煤的化学组成 chemical composition of coal
煤的灰分 coal ash
煤的灰分产率 ash production rate of coal
煤的间接液化 indirect liquefaction of coal
煤的结构 structure of coal
煤的可选性 coal washability
煤的孔隙率 porosity of coal
煤的矿山价格 pit-head price
煤的框架口交货价格 pit-head price of coal
煤的粒度 coarseness of coal; size of coal
煤的裂隙 fissure of coal
煤的裂隙类型 type of coal fissure
煤的黏[粘]结指数 caking index
煤的气化 coal gasification; gasification of coal
煤的氢化 coal hydrogenation; hydrogenation of coal
煤的氢化处理法 coal hydrogenation process
煤的清洁用法 clean use of coal
煤的热分解 coal pyrolysis
煤的热稳定性 K 分级 coal thermal stability K graduation
煤的输送系统 coal delivery system
煤的炭化废水 coal carbonization wastes
煤的物理性质 physical property of coal
煤的消耗量 coal consumption
煤的液化 coal liquefaction; liquefaction of coal
煤的应用类型 uses type of coals
煤的油处理 oil-treatment of coal
煤的蕴藏量 coal reserves
煤的折射率 refringence index of coal
煤的装卸 coal handling; coal transshipment
煤的自然类型 natural type
煤的自燃 coal spontaneous combustion
煤的综合利用 coalplex
煤地层学 coal stratigraphy
煤地下气化 in-situ coal gasification
煤电钻 electric(al) coal drill
煤吊斗 coal grabbing bucket
煤吊篮 coal grabbing bucket
煤定点循环列车 coal unitrain
煤斗 coal bin; coal bucket; coal grab; coal scuttle; fuel bunker; hod; scoop; scuttle
煤斗防冻盘管 bunker coil
煤斗装置指示器 bunker position indicator
煤堆 coal dump; coal(in) pile
煤堆场 coal yard; coal heap; coal tip
煤堆排水 coal pile drainage
煤堆修整器 coal pile trimmer
煤堆自燃 spontaneous combustion of coal dump
煤对二氧化碳反应性 carboxy reactivity
煤垩 slum coal; smut
煤二次开采 secondary coal recovery
煤房 bordroom
煤房颈 room neck
煤房宽度 bordroom width
煤房运煤支巷 going bord; going headway
煤房运输机 room conveyer[conveyor]
煤房杂工 bordroom-man
煤肺病 black lung; anthracosis
煤肺症 soot lung
煤分层厚度 separating coal seam thickness
煤分级 coal grading; grade of coal
煤分类法 classification of coal; coal classification
煤分类系统 coal classification system
煤分析基准 basis of coal analysis
煤酚皂 cresol saponatus; saponated cresol
煤酚皂溶液 saponated cresol solution
煤粉 <粒径 0~0.5 毫米> coal meal; coal powder; coal slack; dust coal; finely divided coal; powder(ed) coal; pulverised[pulverized] coal; coal fines
煤粉爆炸 coal dust explosion; dust explosion
煤粉仓 coal dust bin; coal dust storage hopper; coal meal silo; pulverised[pulverized] coal-storage bin
煤粉层抑制剂 coal dust suppressant
煤粉尘 coal dust
煤粉出口 discharge
煤粉发动机 coal dust engine
煤粉分离器 coal dust classifier; pulverised[pulverized] coal classifier
煤粉风机 pulverized coal exhauster
煤粉浮沉试验结果表 float-and-sink analysis of fines result table
煤粉鼓风机 pulverised[pulverized] coal blower
煤粉管 pulverized coal pipe
煤粉管路 pulverized coal conduit
煤粉锅炉 pulverised[pulverized] coal fired boiler; pulverised[pulverized] coal firing boiler; pulverised[pulverized] fuel boiler
煤粉烘干粉磨机 <商品名> Pyrator
煤粉火焰 powdered coal flame
煤粉集中粉磨经中间仓的间接燃烧系统 bin system of coal grinding and firing
煤粉计量系统 <德国洪堡公司产品名> Pyro-Control
煤粉加热的 pulverised[pulverized] coal fired
煤粉焦粉 braize
煤粉喷燃器 coal burner
煤粉喷嘴 pulverised[pulverized] coal nozzle
煤粉清除工 gummer
煤粉燃料 powdered coal fuel; pulverised[pulverized] coal fuel
煤粉燃烧炉 powdered coal burner
煤粉燃烧器 coal burner; powdered coal burner pulverised[pulverized] coal bunker
煤粉燃烧设备 pulverised[pulverized] coal firing plant
煤粉燃烧室 pulverised[pulverized] coal combustion chamber
煤粉筛分试验 sieve analysis of powdery coal
煤粉收集器 pulverised[pulverized] coal collector
煤粉输送器 cuttings conveyer[conveyor]
煤粉系统 pulverised[pulverized] coal system
煤粉压榨机 roll mill pulverizer
煤粉制备 pulverised[pulverized] coal preparation
煤浮选 coal flo(a)tation
煤副产品 coal by-product
煤干馏 coal carbonization; coal distillation; dry distillation of coal
煤干馏工业 coal carbonization industry
煤干馏轻油 gas light oil
煤干洗 dry washing of coal
煤干洗方法 coal dry cleaning process
煤干燥器 coal dryer[drier]
煤矸 bastard coal
煤矸石 coal gangue; coal stone; coal spoil; coal waste; colliery spoils; colliery waste; culm; duns; gangue
煤矸石堆 gangue piles
煤矸石空心砌块 gangue hollow block
煤矸石利用 utilization of gangue
煤矸石棉 coal gangue wool
煤矸石砖 cliff brick; coal spoil brick
煤矸砖 gangue brick
煤港 coal harbo(u)r; coal(ing) port
煤高压加氢 high-pressure coal hydrogenation
煤膏硬脂 coumarone resin
煤镐 coal pick
煤工尘肺 coal-worker's pneumoconiosis
煤工业分析 proximate analysis of coal
煤工作面注水法 water infusion method
煤古植物学 coal paleobotany
煤含量计 coal content instrument
煤耗 coal consumption
煤耗率 coal consumption rate; rate of coal consumption
煤合成燃料 coal-based synthetic(al) fuel
煤合成物 coal complex; synthetics from coal
煤和瓦斯突出 coal gas burst
煤和瓦斯突出矿井 coal gas outburst mine
煤核 coal ball apple core
煤褐色的 fuliginous
煤黑色 coal black
煤烘干机 coal-drying drum
煤华 blossom
煤滑板 coal board slide
煤化半丝质体 coal rank semifusinite; rank semifusinite
煤化程度 degree of coalification

M

煤化粗粒体 rank macrinite
煤化轨迹 coalification track
煤化间断 coalification break
煤化沥青 nigritite
煤化木 coalified wood; coal wood
煤化丝质体 rank fusinite
煤化梯度 coalification gradient
煤化系列 coalification series
煤化形式 coalification pattern
煤化学 coal chemistry
煤化学废水 coal chemical waste water
煤化学工业 coal chemical industry
煤化学制品 chemical products from coal; coal chemicals
煤化跃变 coalification jump
煤化作用 anthracolitization; bituminization; carbonification; incoalation; coalification
煤化作用阶段 coalification stage
煤灰 cinder; coal ash; coom; crock; flying-ash; fuel ash; soot
煤灰成分分析 coal ash analysis; analysis of coal ash
煤灰处理池 ash-lagoon
煤灰船 ash boat
煤灰吹除器 ash ejector
煤灰吊车 ash hoist
煤灰堆场 cinder dump; slag dump
煤灰堆放场 ash storage area
煤灰分离器 grit separator
煤灰缸 ash tray
煤灰混凝土 ash concrete
煤灰监测仪 coal ash monitor
煤灰黏[粘]度 coal ash viscosity
煤灰圈 coal ash ring; coal dust ring
煤灰熔点 fusion point of coal ash
煤灰熔融性 coal ash fusibility
煤灰熔融性分级 coal ash fusibility graduation
煤灰纱 coal dust stained yarn
煤灰损失 ash loss
煤灰提升机 ash lift
煤灰吸收 coal ash absorption
煤灰吸收量计算 calculation of coal ash absorption
煤灰渣 coal ash and slag
煤灰砖 coal dust brick
煤火补贴 heating allowance
煤机 stoker
煤基发电 coal-based power generation
煤基质 matrix of coals
煤计量机 coal meter
煤计量器 coal weigher
煤加氢气化预处理 hydrogasification of pretreated coal
煤加湿 coal tempering
煤夹层 coal band
煤间 coal room
煤简选样 easy washing sample of coal
煤碱剂 coal-alkali reagent
煤浆 coal liquid; coal paste; coal pulp; coal slurry; slurry
煤浆管道 coal-slurry pipeline; slurry pipeline
煤浆进料机 coal feeder; slurry feeder
煤浆输送管道 coal-slurry pipeline
煤胶体 carbogel
煤胶物质 tarry matter
煤焦沥青 coke-oven coal tar
煤焦沥青产品 coal-tar product
煤焦沥青聚氨酯涂料 coal-tar urethane coating
煤焦轻油 coal-tar light oil
煤焦炭 coal coke
煤焦炭更换比 coal-to-coke replacement ratio
煤焦稀释剂 coal-tar thinner
煤焦油 coal oil; coal-tar (oil); gas tar; tar oil
煤焦油玻纤油毡 coal-tar glass felt
煤焦油产品 coal-tar product
煤焦油磁漆 coal-tar enamel
煤焦油催化加氢 catalystic hydrogenation of coal tar
煤焦油底漆 coal-tar primer
煤焦油防腐剂 coal-tar creosote
煤焦油防腐油 coal-tar creosote oil
煤焦油废水 coal-tar waste
煤焦油废物 coal-tar waste
煤焦油工业 coal-tar industry
煤焦油化学 coal-tar chemistry
煤焦油环氧涂层 coal-tar epoxy coating
煤焦油环氧涂料 coal-tar epoxide paint; coal-tar epoxy paint
煤焦油环氧树脂 coal-tar epoxy
煤焦油混合燃料 coal-tar mixture fuel
煤焦油基涂层底料 creosote primer; creosoting primer
煤焦油晶碱 hot stuff
煤焦油聚氨酯涂料 coal-tar urethane paint
煤焦油沥青 coal-tar asphalt; coke-oven (coal) tar; coal-tar pitch
煤焦油沥青黏[粘]着剂 adhesive based on coal tar
煤焦油馏出物 coal-tar distillate
煤焦油煤气 coal-tar gas
煤焦油氢化 coal-tar hydrogenation
煤焦油燃料 coal-tar fuel
煤焦油染料 coal-tar colo(u)r; coal-tar dye
煤焦油溶剂 coal-tar solvent
煤焦油乳剂 coal-tar emulsion
煤焦油软膏 coal-tar ointment
煤焦油树脂 coal-tar resin
煤焦油树脂磁漆 coal-tar resin enamel
煤焦油烃 coal-tar hydrocarbon
煤焦油涂层 coat of tar; coal tar enamel
煤焦油涂料 coal-tar paint
煤焦油性的 carbolic
煤焦油样粪 tarry stool
煤焦油油漆 coal tar paint
煤焦油油纸 coal-tar-saturated organic felt
煤焦油杂酚油 coal-tar creosote
煤焦油脂 coal pitch; coal-tar emulsion; coal-tar pitch
煤焦油脂黏[粘]结剂 bonding adhesive based on coal tar
煤焦油脂乳液 coal-tar pitch emulsion
煤焦置换比 coal-to-coke replacement
煤焦重油 coal-tar heavy oil
煤焦砖 coalite
煤角砾岩 coal breccia
煤窖 coal cellar
煤结核 coal apple; coal ball
煤经济储量 economic coal reserves
煤精 black amber; gagate; jet
煤井 coal pit; colliery
煤就地气化 coal gasification in situ
煤可采储量 recoverable coal reserves
煤可采性 workability of coal
煤可燃成分 coal combustibles
煤可燃性 coal combustibility; coal ignitability
煤可燃质 pure coal substance
煤可选性 washability of coal
煤可选性分级 washability graduation
煤坑 coal pit
煤孔隙率 coal porosity; porosity of coal
煤库 coal house; coal shed; coal stock; coal storage system; coal store
煤块 coal cinder; fuel brick; range coal
煤块尺寸 lump size
煤矿 coal mine; coal pit; colliery
煤矿安全规程 safety regulations in coal mine
煤矿安全生产监测 monitoring of coal mine safety
煤矿保证条款 colliery guarantee
煤矿爆发后毒气 afterdamp
煤矿爆炸 explosion of coal mines
煤矿采掘面 coal face
煤矿采空区 coal mine working-out section
煤矿床 coal deposit
煤矿地质学 coal mine geology
煤矿顶板分级 classification of coal mining roof
煤矿非机采型 non-mechanized coal mining
煤矿废料 colliery shale; colliery spoils
煤矿废石料 mine stone
煤矿废水 colliery waste; wastewater from coal mine
煤矿废土 colliery spoils
煤矿废物 colliery waste; waste from coal mine
煤矿废物堆 colliery waste tip
煤矿工程 colliery engineering
煤矿工人 coal miner; coal mine worker; collier; pitman
煤矿工人尘肺 coalminer's lung; coal-worker's pneumoconiosis
煤矿回采 stope of coal mines
煤矿机采型 mechanized coal mining
煤矿井下充填工 pack builder; packer; pillar man; waller
煤矿井下检查员 inspector in down coal mine
煤矿开采 coal mining
煤矿开采和加工废水 coal mining and processing wastes
煤矿开拓掘进 development and driving of coal mines
煤矿勘探 exploration of coal mines; exploratory types of coal mine
煤矿坑 coal pit
煤矿坑煤气 afterdamp
煤矿矿井废水 coal mine wastewater
煤矿矿井排水 drainage of coal pits
煤矿矿井生产流程 production process of coal pits
煤矿矿井生产能力 production capacity of coal pits
煤矿矿井压缩空气 compressed-air in coal pits
煤矿粒状废弃材料 <苏格兰地区> blaise[blaize]
煤矿露天开采 opencast coal mining
煤矿贸易 coal-ore trade
煤矿名 coal mine name
煤矿排水 coal mine drainage; drainage of coal mines
煤矿弃土 colliery spoils
煤矿区 coalfield; coal-fired
煤矿区名 coal district name
煤矿乳化炸药 coal mining emulsifying explosive
煤矿设计 design of coal mine
煤矿石 coal ore
煤矿水 coal mine water
煤矿水采 hydraulic coal mining
煤矿水胶炸药 coal mining water gel explosive
煤矿酸性废水 acid mine drainage from coal mine
煤矿委员会 <煤矿主和矿工共同推派代表组成> pit-head committee
煤矿业 coal mining
煤矿用 D 炸药 big coal D
煤矿用钢丝绳 colliery wire rope
煤矿渣 cinder
煤矿炸药 permissible explosives for coal mines
煤矿正规循环作业 normal cyclic operation of coal mines
煤矿职业性皮肤病 occupational dermatosis of mines
煤矿中的运输巷道 haulage way
煤矿主 coal master
煤矿柱 bearing block; pillar
煤矿专用线 colliery siding
煤矿装备管理员 property man
煤矿综合经营 integrated management of coal mines
煤沥青 coal pitch; coal-tar; coal-tar emulsion; coal-tar pitch; tar <为有机物,如煤、木材、页岩、石油等的蒸馏产物>
煤沥青饱和的 coal-tar-saturated
煤沥青磁漆 coal-tar-base enamel; coal-tar enamel
煤沥青分馏试验 distillation test of liquid tar
煤沥青工程 asphalt(tar) cold process construction
煤沥青基黏[粘]结剂 cementing agent based on coal tar
煤沥青胶泥 coal-tar pitch cement
煤沥青浸透的 coal-tar-impregnated
煤沥青聚氨酯漆 coal-tar urethane paint
煤沥青聚合物 tar polymer
煤沥青木材防腐剂 coal-tar creosote
煤沥青黏[粘]结剂 bonding adhesive based on coal tar
煤沥青溶液 coal-tar pitch solution
煤沥青乳液 road tar emulsion
煤沥青洒布车 tar spray car
煤沥青填料 pitch filling
煤沥青涂层 coat of tar
煤沥青屋面油毡 coal-tar-saturated roofing felt
煤沥青(系)涂料 coal-tar paint
煤沥青研究协会 <英> Coal Tar Research Association
煤沥青预拌石屑 chippings precoated with tar
煤沥青杂酚油 coal-tar creosote oil
煤沥青罩面 coal-tar finish
煤沥青脂乳液 coal-tar pitch emulsion
煤砾岩 coal conglomerate
煤粒 rice coal
煤粒沉淀 deposition of coal particle
煤裂隙型 type of coal fissure
煤溜槽 coal chute; coal drop
煤溜口 coal spout
煤流 coal stream
煤流态化 fluidization of coal
煤馏出物 coal extract
煤馏油 coal liquid; coal-tar oil; coal oil
煤(漏)斗 coal hopper
煤炉 coal range
煤炉加热式育雏器 coal-heated brooder
煤卵石 coal pebble
煤螺(旋)钻 coal auger
煤码头 coal berth; coal dock; coal terminal; coal wharf
煤面拐角 buttock
煤磨 coal grinding mill; coal pulverizer; coal pulverizing mill
煤磨机 coal pulverizing mill
煤末 coal dust; dust coal; powdered coal; pulverised [pulverized] coal; slack coal; smalls
煤末沉着病 bituminosis
煤母 dant; mother of coal
煤泥 coal mud; culm clay; slime; coal slurry; watery silt
煤泥浮沉试验 float-and-sink analysis of fines
煤泥输送管道 coal-slurry pipeline
煤泥输送装置 coal-slurry conveyer
煤黏[粘]度混合比 coal mixing rate on

viscosity
煤凝结值 agglomerating value of coal
煤盆地 coal basin
煤膨润 swelling of coal
煤品级 coal rank
煤气 carbon monoxide;coal gas;coal oven gas;gas;natural gas;oil gas
煤气安全阀 gas relief valve
煤气安全检查员 gas-testing safety man
煤气安全控制 emergency gas control
煤气安装工程 gas fitting work
煤气柏油脂 gas pitch
煤气爆炸 explosion of firedamp;gas explosion
煤气爆炸火灾 gas explosion fire
煤气焙烧炉 gas roaster
煤气闭塞装置 gas barrier
煤气标量计 gasometer
煤气表 gas meter;gas counter
煤气表房 gas meter house
煤气表过滤层 gas meter filter
煤气表过滤器 gas meter filter
煤气表盒 gas meter box
煤气表控制 gas meter control
煤气表控制阀 gas meter control valve
煤气表铅封 meter seal
煤气表罩壳 gas meter inclosure[enclosure]
煤气冰箱 gas refrigerator
煤气采暖 gas-fired heating;gas(-fired) warming
煤气采暖锅炉 gas-fired heating boiler
煤气采暖炉 gas fire
煤气采暖设备 gas-fired unit heater
煤气采暖装置 gas-fired plant;gas warming appliance;gas warming device;gas warming unit
煤气残液 coal gas waste-water
煤气产量 gas yield
煤气厂 gas house;gas plant;gas undertaking;gas works;producer gas plant;retort bench
煤气厂废水 gashouse waste;gas mill wastewater;gas plant waste(water);gasworks waste;wastewater from coal gas plant
煤气厂焦油 gasworks tar
煤气厂焦油沥青 gashouse coal-tar pitch;gasworks tar pitch
煤气厂沥青 gasworks pitch
煤气厂(煤)焦油 gashouse coal tar
煤气厂水 gasworks liquor
煤气厂蒸馏罐 retort
煤气车间 gas works
煤气沉渣室 gas slag
煤气沉渣室拱顶 <平炉的> gas slag arch
煤气成分 gas composition
煤气城边点 city gate station;town border station
煤气城边站 city gate station;town border station
煤气冲天炉 gas-fired cupola
煤气出口处 gas outlet point
煤气出口钢管 steel exit gas pipe
煤气储存罐 gas holder;gas storage tank
煤气储存站 gas-holder station
煤气储罐 gas holder
煤气储热器 gas storage heater
煤气处理 gas conditioning;gas processing;gas treatment
煤气吹管 gas blow torch
煤气吹送机 gas pump
煤气锤 gas-driven hammer
煤气存储器 gas reservoir
煤气导管 gas pipeline
煤气的种类 gas of kinds
煤气灯 gas luminaire(fixture);has lamp
煤气灯管 gas bracket
煤气灯光 gas light(ing)
煤气灯纱罩托 mantle carrier
煤气灯头钳 gas burner pliers
煤气灯(网)罩 gas mantle
煤气灯装置 gas fixture
煤气点火棒 gas poker
煤气点火的转化燃烧器 conversion gas-fired burner
煤气点火器 gas ignitor;gas lighter
煤气电力驱动 gas electric(al)drive
煤气吊灯 gas light;gaselier;gas lamp;gasolier
煤气煅烧窑 gas-fired calcining
煤气锻炉 gas forge
煤气发电厂 gas power plant
煤气发电机 gas-driven generator
煤气发动机 gas motor;gas power engine;natural gas engine;gas engine
煤气发生 gas generation
煤气发生厂 gas plant
煤气发生过程 producer gas process
煤气发生井 gas production well
煤气发生炉 coal gas generator;coal gas producer;gas furnace;gas generator;gas producer;generating furnace;producer;producer furnace
煤气发生炉的煤灰 gas generator ash;gas generator cinder
煤气发生炉的煤渣 gas generator ash;gas generator cinder
煤气发生炉隔墙 end block
煤气发生炉罐 gas producer retort
煤气发生炉褐煤焦油 producer gas brown-coal tar;producer gas lignite tar
煤气发生炉灰渣 generator ash
煤气发生炉混合装置 mixing device of gas
煤气发生炉机组 gas generator unit
煤气发生炉焦油沥青 producer gas tar pitch
煤气发生炉炉渣 generator cinder;generator clinker
煤气发生炉(煤)焦油 gas producer coal tar;producer gas tar
煤气发生炉汽车 gas producer vehicle
煤气发生炉循环 gas generator cycle
煤气发生率 productivity of gas
煤气发生器 coal gas generator;gas-generating set;gas generator;gasifier;gas producer;producer gas generator
煤气发生器用褐煤的焦油 reducer gas brown coal tar
煤气发生站 gas-generating station
煤气发生站废水 wastewater from gas generation plant
煤气发生装置 gasogene
煤气发生组 gas-generating set
煤气阀 gas lock;gas valve
煤气阀簧 gas valve spring
煤气阀座 gas seat
煤气反射炉 gas-fired air furnace
煤气防护 gas protection
煤气房 gas house
煤气房间采暖 gas room heating
煤气放气阀 gas escape valve
煤气放热器 gas radiator
煤气废水 coal gas waste;gaseous waste
煤气废物焚化炉 gas rubbish incinerator
煤气费用 gas cost
煤气分表 submeter(ing)
煤气分配 distribution of gas
煤气分配系统 distribution system of gas
煤气辐射供暖器 gas radiator
煤气辐射加热器 gas radiant heater
煤气副产品 gas by-products
煤气干管 gas main
煤气坩埚炉 gas-fired crucible furnace
煤气钢管 gas steel pipe
煤气工厂 gas house;gas producing plant;gas works;gas making plant
煤气工程 gas engineering;gas works
煤气工(人) gas fitter;gas maker;gasman;gas producer man
煤气工业 gas industry
煤气公司 distribution company;gas company;gas utility
煤气公司煤气 utility gas
煤气供暖 gas heating
煤气供暖机组 gas heater
煤气供暖器 space gas heater
煤气供热 gas-fired heating
煤气供热装置 gas-fired heating installation
煤气供应 gas supply;gas utility
煤气供应阀 gas grid
煤气供应方法 gas-distributing system
煤气供应管(道) gas feeder;gas service pipe
煤气供应管阀 gas service valve
煤气供应管升高长度 gas service riser
煤气供应系统 gas supply system
煤气供应热水器 gas circulator
煤气鼓风机 gas blowing engine
煤气关断阀 gas shut-off valve
煤气管 gas barrel;gas line pipe;gas pipe;gas tube
煤气管(扳)钳 gas wrench
煤气管道 gas conduit;gas flue;gas (pipe) line;gas piping
煤气管道安装工程 gas installation pipework
煤气管道安装工作 gas installation work
煤气管道材料 gas line material
煤气管道凝结水 gas drip
煤气管道输送 gas pipeline transportation service
煤气管道系统 gas piping system
煤气管道压缩式旋塞 gas pipeline packed cock
煤气管道装配技师 master gas fitter
煤气管阀 gas service valve
煤气管扶手 gas barrel handrail;gas-tube handrail
煤气管接头 gas fittings;gas pipe connector
煤气管连接 gas pipe connection
煤气管路 gas pipeline
煤气管螺丝扳牙和扳手 gas stock and dies
煤气管螺纹 gas pipe thread
煤气管钳 gas pipe tongs;gas pliers;gas tongs
煤气管网 gas pipe network
煤气管网干管 gas feeder
煤气管线 gas pipeline
煤气管线施工设备 gas pipeline construction rig
煤气管压力自动调节器 gas governor
煤气罐 gas cylinder;gas receiver;gas storage reservoir;gas tank
煤气罐表 gas meter;gasometer
煤气罐车 special wagon for carriage of gas
煤气柜 gas chamber;gas holder;gasometer;gas tank
煤气锅炉 fired boiler;gas(-fired) boiler
煤气锅炉烟道 gas boiler flue
煤气过滤 gas filtration
煤气过滤器 gas filter drain cock
煤气过滤器放水旋塞 gas filter drain cock
煤气过滤器壳 gas filter shell
煤气过滤器输出管 gas filter outlet pipe
煤气含硫量 sulfur content in gas
煤气耗量 gas consumption
煤气和焦炭制造业 coal gas and coke manufacture
煤气烘烤机 gas roaster
煤气烘炉 gas oven;gas stove
煤气红外线辐射供暖 gas-fired infrared heating
煤气红外线供暖系统 gas type infrared heating system
煤气弧光灯 gas arc lamp
煤气化 gasification
煤气化法 coal gasification process
煤气化废水 coal gasification wastewater
煤气化废水处理 coal gasification wastewater treatment
煤气化联合循环发电 integrated coal gasification combined cycle
煤气化联合循环工艺 combined cycle gasification process
煤气化煤气 product gas
煤气化铁炉 gas-fired cupola
煤气化系统 gasification system
煤气化型 type of coal gasification
煤气换向阀 gas reversal valve
煤气回收阀 coal gas recovery valve
煤气混合器 air and gas mixer
煤气混合箱 gas mixing tank
煤气火(焰) gas fire
煤气机 gas engine;gas motor;natural gas engine
煤气机车 gas power locomotive
煤气机电动车 gas electric(al)car
煤气机发动 gas engine drive
煤气机进气阀 gas engine inlet valve
煤气及水费 gas and water expenses
煤气集中供热 gas central heating
煤气计 gasometer
煤气计量器 gas meter;gasometer
煤气加工 gas conditioning
煤气加热 gas heating;gas-warmed
煤气加热板红外(线)烘干 infrared drying by gas heated panels
煤气加热的 gas-heated
煤气加热的板 gas-heated panel
煤气加热的空气 gas-heated air
煤气加热锅炉 gas heating boiler
煤气加热空气 gas-warmed air
煤气加热烙铁 gas soldering copper
煤气加热炉 gas-fired furnace
煤气加热器 gas-fired heater;gas heater;gas warmer
煤气加热设备 gas-fired heating unit
煤气加热式沥青贮仓 (gas heating) storage
煤气加热式育雏器 gas-heated brooder
煤气加热式蒸煮器 gas-heated steamer
煤气加热熏房 gas operated smokehouse
煤气加热装置 gas heating device;gas heating installation;gas heating unit
煤气加压风机 gas booster fan
煤气间 gas house
煤气检测锤 gas detection hammer
煤气检测器 gas detector
煤气降压站 city gate station;town border station
煤气交换器 gas reversal valve
煤气胶 gum
煤气焦化废水 gas and coke plant waste
煤气焦炭 gas coke;gasworks coke
煤气焦油(沥青) gashouse coal tar;gas tar;gasworks coal tar
煤气接口 socket outlet
煤气节流杆 gas throttle lever
煤气紧急截止阀 gas emergency trip

M

valve
煤气进口 gas inlet
煤气进气管 gas feeder
煤气井 gasser
煤气净化 gas cleaning; gas purification; gas purifying
煤气净化剂 gas scavenger
煤气净化器 gas cleaner
煤气净化塔 gas purification tower
煤气净化装置 gas cleaning device
煤气聚集管 gas collecting tube
煤气聚集器 gas collector
煤气空气混合器 gas air mixer
煤气空气混合站 gas-air mixing plant
煤气控制杆 gas control lever
煤气口 gas end
煤气口端墙 gas end
煤气库 gas holder; gas tank
煤气快速热水器 rapid gas water heater
煤气冷柜 gas refrigerator
煤气冷凝 condensation of gas
煤气冷凝液 gas liquor
煤气冷却 gas cooling
煤气冷却器 gas cooler
煤气量计 gas meter
煤气量热器 gas calorimeter
煤气流量计 gas flow indicator
煤气龙头 gas cock
煤气漏失 gas escape
煤气炉 gas-fired stove; gas furnace; gas heater; gas range; hot plate
煤气炉壁龛 gas-fired recess; gas furnace recess
煤气炉灶 gas-fired range; gas cooker
煤气炉渣混凝土 gas slag concrete
煤气炉中常燃小火 pilot light
煤气滤器 gas filter
煤气(煤)焦油脂 gasworks coal tar pitch
煤气内燃机 gas(ignition) engine
煤气内燃机驱动的热泵 heat pump driven by gas engine
煤气排管设备 gas pipeline construction rig
煤气配件 gas fittings
煤气配气管网 gas distribution piping
煤气喷出口 gas end; gas port
煤气喷灯 gas cock; gas arc lamp; gas burner
煤气喷管 gas jet tube
煤气喷口 gas jet
煤气喷枪 heat gun
煤气喷燃器 gas burner
煤气喷射 gas injection
煤气喷头 gas jet
煤气喷嘴 gas burner; gas jet
煤气瓶 gas cylinder
煤气企业 gas undertaking
煤气起动机 gas starter
煤气气流调节板 gas flow adjusting flange
煤气汽车 compressed gas truck
煤气切断安全阀 gas shut-off relief valve
煤气清洁器 gas cleaner
煤气清洗 gas cleaning
煤气驱动的钻探设备 gas rig
煤气取暖炉 gas header
煤气取暖器 gas fire
煤气燃料公共汽车 gas-fuelled bus
煤气燃烧 gas-fired
煤气燃烧壁龛 gas burning appliance recess
煤气燃烧器 gas(-fired) burner; argand burner <具有空心管子的>; batswing <排成扇形喷头的>
煤气燃烧室 gas burner
煤气燃烧水泥窑 gas-fired cement kiln
煤气燃烧装置 gas burning appliance
煤气热风采暖 gas-fired warm air heating
煤气热风供热 gas-fired warm air heating
煤气热辐射器 gas radiant tube heater
煤气热空气取暖 gas warming air heating
煤气热量单位 <在英国相当于 105B.T.U.,在美国相当于 1000 大卡> therm
煤气热水采暖器 gas-fired hot water heater
煤气热水供热装置 gas-fired water heating appliance
煤气热水供应装置 gas-fired hot water heater
煤气热水器 gas-fired water heater; gas geyser; gas water heater; water heater
煤气热水系统 gas-fired hot water system; gas hot water system
煤气热水循环采暖 gas water circulation heating
煤气热水循环供热 gas water circulation heating
煤气热值 calorific value of gas; gas caloricity
煤气容器 gas vessel
煤气融雪器 gas snow melter
煤气入口管 gas inlet pipe
煤气软管 gas flexible conduit; gas hose; gas pipe hose
煤气散热器 gas radiator
煤气砂双层滤池 anthracite-sand filter
煤气上升道 gas uptake
煤气烧毛 gas singeing
煤气烧毛机 gas flame singeing machine
煤气烧水炉 gas water heater
煤气烧嘴 gas-fired burner
煤气烧嘴口 gas port nose
煤气设备 gas-fired equipment; gas fittings; gas fixture; gas installation
煤气设备壁龛 gas appliance recess
煤气设备的烟道 flue for gas appliance
煤气设备工程师 gas appliance engineers
煤气渗碳 carbonizing of gas; carburization of gas
煤气渗碳法 gas carburizing
煤气渗碳炉 gas-fired reverberatory furnace
煤气渗碳烧结 gas carburization sintering
煤气生产 gas manufacture; gas production
煤气生产废水 gas plant waste(water)
煤气声压器 gas booster
煤气湿法除尘 wet gas cleaning
煤气石灰 gas lime
煤气试样 gas sample
煤气室 gas chamber
煤气室内采暖 gas space heating; gas space warming
煤气收费员 gasman
煤气收集器 gas collector
煤气输出量 gas output
煤气输出钻孔 gas offtake borehole
煤气输送管道 gas transmission pipeline
煤气输送管路 gas transmission line
煤气水 coal liquor; gas liquor
煤气水罐 gas liquor tank
煤气水化法 gas hydrate process
煤气水溶液 gas liquor
煤气水溶液罐 gas liquor tank
煤气炭黑颜料 gas carbon black pigment
煤气探测器 gas locator
煤气体转化器 gas reformer
煤气调压站 gas pressure-regulating station
煤气调整器 gas governor
煤气通道 gas passage
煤气桶扶手 gas barrel handrail
煤气筒汽车 liquefied gas truck
煤气头灯 gas headlight
煤气拖拉机 producer gas tractor
煤气稳压阀 gas regulator
煤气污染 gas pollution
煤气吸收式制冷机 gas absorption refrigerator
煤气矽肺 anthrao-silicosis
煤气洗涤 gas scrubbing; gas washing
煤气洗涤废水 coal gas washing wastewater; gas scrubbing waste; gas washing wastewater
煤气洗涤机 bubbling washer; gas cleaner; gas-fired washing machine; gas washing machine
煤气洗涤瓶 gas scrubbing bottle; Drexel bottle
煤气洗涤器 gas scrubber; gas washer; wet cleaner
煤气洗涤水 gas scrubbing water
煤气洗涤塔 gas scrubbing tower; gas wash tower
煤气箱 gas tank
煤气小龙头 nose cock
煤气效用 gas utility
煤气形成的化学物质 gas-forming chemical
煤气蓄热室 gas regenerator; gas regenerator chamber
煤气蓄热室(火)烟道 gas regenerator flue
煤气旋塞 gas cock; gas tap; nose cock
煤气循环器 gas circulator
煤气压力 gas pressure
煤气压力表 gas pressure indicator
煤气压力计 gas ga(u)ge
煤气压力检测点 gas pressure test point
煤气压力调节阀 gas pressure regulator
煤气压缩泵 gas compression pump
煤气压缩机 gas compression pump
煤气压缩旋塞 gas compression tap
煤气窑 gas-fired kiln
煤气液 ammonia liquor; gas liquor
煤气引燃器 gas lighter; gas pilot
煤气引入点 gas point
煤气硬煤沥青 gasworks coal tar pitch
煤气用户管道 consumer gas piping
煤气用户屋内设备 gas supply line
煤气用具 gas(-fired) appliance
煤气用煤 gas coal
煤气与汽油三通阀 gas and gasoline three-way valve
煤气运输公司 gas transmission company
煤气杂质 gaseous impurity
煤气再加热器 gas re-heater
煤气灶 gas burner; gas cooking stove; gas fire; gas kitchener; gas oven; gas range; gas ring; hot plate
煤气灶及管道辅助用具 accessory gas appliance
煤气闸 gas lock
煤气站 gas house; gas station
煤气照明 gas illumination; gas lighting
煤气照明装置 gas luminaire(fixture)
煤气止回阀 gas check valve
煤气制冷 gas refrigeration
煤气制取过程 gas manufacturing process; gas producing process
煤气制造厂 gas works
煤气制造法 gas making process
煤气制造记录簿 gas making log book
煤气质量 gas quality
煤气质量调整 gas conditioning
煤气窒息 asphyxiant carbonics
煤气中毒 carbon monoxide poisoning; gas poisoning; poisoning by charcoal fumes
煤气种类 kinds of gas
煤气转化器 gas reformer
煤气装备 gas fittings
煤气装配 gas fittings
煤气装修工 gas fitter
煤气装置 gas apparatus; gas fittings
煤气自动阀 differential pressure valve
煤气自动截止阀 automatic gas shut-off valve
煤气自动控制阀 automatic gas control valve
煤气自动切断装置 automatic gas shutoff device
煤气自停装置 automatic gas suspension device
煤气总表 main-line meter; master meter; primary meter
煤气总管 gas(collecting) main
煤气总流量表 positive total-flow gas meter
煤气组成 gas composition
煤气最大消耗量 maximum gas consumption
煤氢化装置 coal hydrogenation unit
煤氢气化 coal hydrogasification; hydrogasification of coal
煤清洗厂 coal-cleaning plant
煤球 artificial coal; briquet(te); coalite; ovoid briquette
煤球炉式气冷反应堆 pebble-bed gas cooled reactor
煤球形燃料元件 pebble fuel element
煤燃料比 coal fuel ratio
煤燃料富集 coal fuel enrichment
煤热解 pyrolysis of coal
煤容量 coaling capacity
煤洒水器 coal sprinkler
煤砂双层滤料滤池 coal-sand dual media filter
煤筛 coal screen
煤商 coaler; coal factor
煤释放能 coal delivered energy
煤输送系统 coal delivery system
煤水泵 coal-lifting pump; coal-water pump; slurry pump
煤水比 coal-water ratio
煤水车【铁】water tender; tender
煤水车铲煤板 tender shovel plate
煤水车车钩 tender coupler
煤水车挡煤板 tender coal board
煤水车挡煤板滑门 tender coal board slide
煤水车挡煤门 tender coal gate
煤水车挡煤圈 tender collar
煤水车灯 deck lamp
煤水车底架 tender frame; tender truck
煤水车地板 tender deck
煤水车渡板 tender fall plate
煤水车防擦板 tender chafing block
煤水车放水杯 tender drain cup
煤水车辅助机 auxiliary locomotive
煤水车工具箱 tender tool box
煤水车后端定位铁 tender back lug
煤水车缓冲器 tender bumper
煤水车机车 tender engine
煤水车脚蹬 tender step
煤水车煤槽 tender coal bunker
煤水车牵引杆座 drawhead; tender drawhead

M

煤水车牵引装置 tender draft gear
煤水车前罩 tender roof
煤水车身承梁 body tender bolster; tender bolster
煤水车水柜人孔罩 tender tank man hole shield
煤水车水位表 tender water level ga(u)ge
煤水车水箱 tender tank
煤水车水箱前定位铁 tender front tank lug
煤水车在前 <蒸汽机车逆行时> tender first
煤水车闸缸 tender brake cylinder
煤水车折棚 tender vestibule
煤水车制动机 tender brake
煤水车制动器 tender brake gear
煤水车中间保险链 tender intermediate safety chain
煤水车轴箱 tender journal box
煤水车注水口 funnel
煤水车转向架 tender bogie
煤水混合物 coal water mixture
煤水机车 tender locomotive
煤水浆的制备及应用 preparation and utilization of coal slurry
煤水浆燃烧技术 burning technique of coal water mixture
煤水油乳胶 coal-water-oil emulsion
煤素质 maceral unit; micropetrological unit
煤塔秤 coal tower scale
煤台 coaling stage
煤炱 black; smut; smutch; soot
煤炱属 <拉> Capnodium
煤炭 pit coal
煤炭搬运 coal haulage
煤炭泊位 coal berth
煤炭采样 coal sampling
煤炭出口港 coal export port
煤炭出口码头 coal export terminal
煤炭储藏与装袋厂 coal storage and bagging plant
煤炭储量 coal reserves
煤炭储量变动统计 statistics for coal reserves variation
煤炭地理学 coal geography
煤炭地下气化 underground coal gasification; underground gasification
煤炭地质学 coal geology
煤炭堆场 coal depot; coal storage yard; coal yard
煤炭法 <英> Coal Act
煤炭翻车机 coal car-dumper
煤炭分层装车法 layer loading
煤炭分类法 coal classification system
煤炭粉化 coal slaking
煤炭粉末 coal dust; coal fines
煤炭粉碎机 coal pulverizer
煤炭干馏副产品 coal by-product; coal carbonization by-product
煤炭工业 coal industry; coal mining industry
煤炭工业废水 coal industry waste; coal industry wastewater
煤炭工业国有化法 <英> Coal Industry Nationalization Act
煤炭工业全国咨询委员会 Coal Industry National Consultative Council
煤炭工业协会 Coal Industry Society
煤炭工业远景规划 long-range plan of coal industry
煤炭工业指标 commercial factor for coal
煤炭含矸石率 rate of coal containing waste rock
煤炭化 coal carbonization
煤炭化学工业废料 waste from coal chemical industry
煤炭化学工业废水 wastewater from coal chemical industry
煤炭化学加工废水 wastewater from coal chemical
煤炭基地 coal base
煤炭技术 colliery engineering
煤炭加工 coal processing
煤炭科学 coal science
煤炭利用化学 chemistry of coal utilization
煤炭利用理事会 Coal Utilization Council
煤炭利用率 coal utilization rate
煤炭利用研究咨询委员会 Coal Utilization Research Advisory Committee
煤炭粒度分级 granular classification of coal
煤炭露天开采 coal stripping
煤炭码头 coal pier
煤炭能量净化 cleaning of coal energy; coal energy cleaning
煤炭牌号 coal trademark
煤炭(品质)分类 grade of coal
煤炭平舱机 coal distributor; coal trimmer
煤炭企业管理 enterprise management of coal industry
煤炭起货机 coal winch
煤炭气化 coal gasification
煤炭气化动力学 kinetics of coal gasification
煤炭倾卸装置 coal tip
煤炭容量 bin space
煤炭撒布机 coal distributor
煤炭输送机 coal conveyer[conveyor]
煤炭提升机 coal hoist
煤炭田 coal basin; coalfield
煤炭突出 coal bump
煤炭突堤式码头 coal pier
煤炭脱灰 ash removal from raw coal
煤炭洗选工艺 coal preparation process
煤炭洗选加工 coal-washing and dressing
煤炭消耗 coal consumption
煤炭卸车导板 <煤车> coal discharge guiding plate
煤炭卸船泊位 coal unloading berth
煤炭卸船码头 coal unloading dock; coal unloading terminal; coal unloading wharf
煤炭卸船设施 coal unloading facility
煤炭旋转干燥器 coal rotary drier [dryer]
煤炭学 anthracology
煤炭液化 coal liquefaction
煤炭运输 coal haulage; coal transport(ation)
煤炭专用港 coal port
煤炭专用线 coaler; coal road
煤炭装船泊位 coal loading berth
煤炭装船机 coal shiploader
煤炭装船码头 coal loading dock; coal loading terminal; coal loading wharf
煤炭装船设施 coal loading facility
煤炭装卸工(人) coal heaver; coal porter
煤炭装卸设备 coal handling plant
煤炭资源 coal resources
煤炭资源管理 management of coal resources
煤炭资源勘探 coal resources exploration; exploration of coal resources
煤炭资源勘探阶段 step of coal resources exploration
煤炭资源数据库 coal resources data base
煤炭综合利用 comprehensive utilization of coal
煤炭总局 National Coal Board
煤炭作业区 coal terminal
煤特性 coal characteristic
煤提升机 coal elevator
煤田 basin; coal deposit; coal field
煤田暴露(类)型 exposed types of coalfield
煤田测井结果解释 interpretation of coal log
煤田测井资料分析 data analysis of coal log
煤田地形地质图 topographic-geologic(al) map of coalfield
煤田地震勘探 coal seismic prospecting
煤田地质 coalfield geology
煤田地质图 geologic(al) map of coalfield
煤田地质学 coal geology
煤田地质钻孔 coal geologic(al) hole
煤田翻斗提升设备 coal skip-winding plant
煤田构造 coalfield structure
煤田勘探 coal exploration
煤田名称 coalfield name
煤田浅层气 shallow gas in coal field
煤田区 coalfield area
煤田筛选长 coal screening plant
煤田筛选设备 coal screening plant
煤田水文地质图 hydrogeologic(al) map of coalfield
煤田提升机 coal lift
煤田预测 coalfield prediction; prediction of coalfield
煤田远景评价方法 evaluating method for coal-field prospect
煤田自燃 coalfield self-combustion
煤田钻探 drilling in coal
煤条带 coal band; coal bump
煤铁矿 blackband
煤烃 coaleum
煤烃化作用 coal alkylation
煤桶 coal scuttle; scuttle
煤脱硫 coal desulfurization
煤脱硫脱氮 coal desulfurization and denitrification
煤微生物学 coal microbiology
煤污病 black blight; smudge; sooty mold
煤矽肺 anthracosilicosis
煤系 coal-bearing formation; coal measures; coal series
煤系沉积硫铁矿床 sedimentary pyrite deposit in coal series
煤系单位 coal-measures unit
煤系等厚图 isopach map of coal formation
煤系地层 coal-measures strata
煤系气 coal measure gas
煤系软页岩 coaly rashings
煤系数 coal factor
煤系旋回 coal-measures unit
煤系页岩 coal-measures shale
煤系中的岩石包裹体 rock inclusion in coal series
煤系中的页岩或泥岩 blue bind
煤显微成分 maceral
煤显微结构 microstructure of coal
煤线 coal streak
煤相 coal facies
煤箱 bunker; bunker bin; coal bin
煤巷 coal drift; coal road(way); gate road
煤巷内端 gate end
煤屑 breeze; burgy; cinder; coal dust; coal screenings; coal slack; culm; duff; fine coal; nickings; slack; slack coal; small coal
煤箱 coal bin
煤屑路 cinder road
煤屑铺跑道 dirt track
煤屑砌块 slack block
煤芯采取率 percentage recovery of coal core
煤芯长度 coal core length
煤芯煤样 coal core sample
煤型 coal type
煤压碎机 coal cracker
煤烟 coal smoke; coom; fuligo; smoke black; smoke from burning coal; smut; soot
煤烟癌 soot cancer
煤烟病 dark mildew
煤烟处理设备 smoke treatment equipment; smoke treatment plant
煤烟地面浓度 ground-level concentration of smoke and soot
煤烟发生量 diesel soot emission
煤烟附着 carbon deposit
煤烟隔板 soot barrier
煤烟浓度 concentration of soot
煤烟污染 coal smoke pollution; smoke pollution; soot stain
煤烟形成 soot formation
煤烟状的 fuliginous
煤岩层对比图 comparative map of coal seam and strata
煤岩成分 composition of coal petrology
煤岩惰性组分 inertinite
煤岩可选样 washability sample of coal petrology
煤岩类型 lithotype
煤岩煤样 coal sample for petrographic(al) analysis
煤岩石学 anthracology; coal petrology; coal petrography
煤岩碎片体 micrinite
煤岩显微组分 maceral unit; micropetrological unit
煤岩相学 coal petrography
煤岩学 petrology of coal
煤研磨 coal grinding
煤氧化 oxidation of coal
煤氧化程度 degree of coal oxidation
煤氧化作用 oxidation of coal
煤样 coal sample
煤样编号 coal sample number
煤样的掺和 mixing of coal sample
煤样的破碎 size reduction of coal sample
煤样的缩分 coal sample division
煤样的制备 coal sample preparation
煤样设备 coal sample preparation; preparation of coal samples
煤样质量 quality of coal sample
煤窑 coal cell
煤业工作日 colliery working day
煤页岩 colliery shale; dank
煤液化 coal liquefaction; liquefaction of coal
煤液化速率 coal liquefaction rate; rate of coal liquefaction
煤油 kerosene oil; kerosine; American paraffin(e) oil; paraffin(e); paraffin(e) oil
煤油的标准颜色 kerosene standard colo(u)rs
煤油的储存变质 reversion of kerosene
煤油的赛波特环值 Saybolt ring number of kerosene
煤油灯 kerosene burner; kerosene lamp; oil light
煤油灯芯 kerosene wick
煤油灯罩 kerosene glass chimney
煤油点火的转化燃烧器 conversion oil-fired burner
煤油发动机 kerosene(oil) engine
煤油法 <测密度用> kerosene method

煤油法孔隙度 porosity of kerosene method
煤油浮选 kerosene flo(a)tation
煤油号数 kerosene number
煤油滑油混合物 kerosene oil mixture
煤油混合燃料 coal-oil fuel
煤油混合燃料比容 specific volume of coal-oil mixture
煤油混合燃料黏[粘]度 viscosity of coal-oil mixture
煤油混炼液化法 coal-oil mixing liquefaction method
煤油机 gas engine; kerosene oil engine
煤油沥青 kerites
煤油馏出物 burning oil; kerosine distillate
煤油馏分 kerosene distillate
煤油炉 kerosene fuel burner; kerosene stove; oil burner; oil-stove
煤油喷气燃料 kerosene propellant
煤油气 kerosene oil-gas
煤油切割器 kerosene cutting torch
煤油切割枪 kerosene cutting torch
煤油燃料 kerosene stock
煤油燃烧炉 kerosene fuel burner
煤油燃烧试验 kerosene burning test
煤油燃烧性质 kerosene burning quality
煤油色度试验 kerosene colo(u)r test
煤油闪点 kerosene flash point
煤油水混合燃料 coal-oil water fuel
煤油涂料 kerosene paint
煤油脱脂 kerosene degreasing
煤油烷基苯 kerylbenzene
煤油吸收量 <油毡的> kerosene number; kerosene value
煤油烯 kerenes
煤油稀释沥青 kerosene cutback asphalt
煤油型喷气燃料 kerosene type jet fuel
煤油型燃气轮机燃料 kerosene type turbine fuel
煤油烟 kerosene smoke
煤油颜色测定 kerosene colo(u)r test
煤油页岩 kerosene shale
煤油引擎 kerosene engine
煤油用气提塔 kerosene stripper
煤油蒸汽灯 pressure vapo(u)r lamp
煤油值 kerosine number
煤油(中)沉积物的测定 kerosene sediment test
煤油中毒 kerosene poisoning
煤油中硫的测定 kerosene sulfur[sulphate] test
煤油中硫的检测 kerosene sulfur[sulphate] test
煤有机化学 organic chemistry of coal
煤淤泥 coal sludge
煤与瓦斯突出 coal gas outburst
煤玉 gagate; gagatite; jet; jet coal
煤玉化作用 gagatization
煤玉似的 jet black
煤玉页岩【地】jet shale
煤元素分析 ultimate analysis of coal
煤运码头 coal pier
煤在油中悬浮液 coal-in-on suspension; colloidal fuel
煤渣 bean; breeze; cinder; clinker; coal cinder; coal refuse; coal slag; coal waste; pan breeze
煤渣步行道 cinder path; cinder side walk; cinder track
煤渣场 coal tip
煤渣池挡板 ashpit door
煤渣池门 ashpit door
煤渣船 ash boat; ash lighter
煤渣床 cinder bed
煤渣道渣 <轻便铁道的> cinder ballast
煤渣地面 cinder floor
煤渣垫实 cinder fill
煤渣粉碎机 cinder mill
煤渣骨料 cinder aggregate
煤渣固结砖 breeze fixing brick
煤渣回填 cinder fill
煤渣混凝土 breeze concrete; cinder concrete; coke breeze concrete; concrete cinder
煤渣混凝土砌块 cinder concrete block
煤渣混凝土砖 cinder concrete brick; coke breeze concrete brick
煤渣集料 cinder aggregate
煤渣坑 cinder fall; cinder pit
煤渣块 breeze block; lump cinder
煤渣漏斗 cinder chute
煤渣路 cinder road; coal cinder road
煤渣路面 cinder road surface; cinder surface; clinker pavement
煤渣面层 cinder road surface
煤渣拍实 tampered cinder
煤渣跑道 cinder track; dirt track <供摩托车比赛用>
煤渣铺路小方块 cinder paving sett
煤渣砌块 breeze block; cinder block
煤渣人行道 cinder path; cinder walk
煤渣筛 cinder sifter
煤渣水泥砖 breeze(cement) brick
煤渣填实 cinder fill(ing)
煤渣填土 cinder fill(ing)
煤渣小路 cinder path
煤渣研磨机 cinder mill
煤渣砖 breeze brick; cinder block; cinder brick; slag brick; coal briquet(te)
煤渣子 small piece of coal
煤闸板 coal gate
煤栈 coal depot; coal storage; coal store
煤站 coal depot
煤制备厂 coal preparation plant
煤制备装置 coal preparation plant
煤质 anthrax; coal quality; nature of coal
煤质测井 coal quality log
煤质分析 coal analysis
煤质分析结果的表示方法 show methods of coal analysis result
煤质分析试验成果表 coal analysis result table
煤质分析图 coal quality analysis plot
煤质管理 coal quality control; quality control of coal
煤质聚合物 coaly polymeric
煤质评价 coal valuation
煤质试验 coal testing
煤质型 coaly type
煤质页岩 bone coal
煤质主要指标的分级标准 grade scale of coal quality
煤中黄铁矿结核 coal brass
煤中结核 coal ball concretion
煤中矿物质 mineral matter in the coal
煤中树脂状烃 stannekite
煤注水机 water injector for coal train
煤贮库 coal storage
煤柱 coal block; coal pillar; coal wedge
煤柱爆破 pillar blast
煤柱放顶线 pillar line
煤柱分隔回采法 splitting method of pillar recovery
煤柱回收 pillar extraction; pillar mining
煤柱上打眼 pillar extraction drilling
煤柱线 pillar line
煤柱压碎 coal pillar thrust
煤柱支撑顶板法 roof control by wall-shape pillar
煤爪 coal grab
煤砖 briquet(te); coal brick; fuel briquette; full brick; mo(u)lded coal
煤砖机 briquet(te) press
煤砖专用沥青 briquetting asphalt
煤转化中产生的污水 coal conversion wastewater
煤状隔担耳 Septobasidium carbonaceum
煤自燃发火臭味 gob stink
煤自燃倾向性 coal ignitability; coal spontaneous combustion tendency
煤组构 coal constitution
煤组名称 name of coal seam group
煤钻 coal auger; coal borer; coal drill

酶法分析 enzymatic analysis

酶反应 enzyme reaction
酶解 zymo(ly)sis
酶学 zymology
酶原 zymogen
酶质 enzymatic material
酶作用 zymo(ly)sis

霉斑 blue stain

霉变表面 mildewed surface
霉臭 frowst
霉菌 mo(u)ld; mo(u)ld a form; mo(u)ld fungus
霉菌病 mycosis
霉菌传染 fungus infestation
霉菌的生长 growth of fungi
霉菌侵蚀 fungus attack
霉菌侵袭 fungus infestation
霉菌试验 mo(u)ld test
霉烂 mildew and rot; rotten
霉腥味 mo(u)ldy odour
霉抑制剂 mildew inhibited; mildewstat
霉渍 mould-stained
霉渍点 mildew-stained

每100 货物列车公里平均小时数 average hours per 100 freight train kilometers

每 200 米距离 bihectometric range
每安瓿 per ampoule
每百 per centum
每百分 pct; per cent
每百元产值工资含量 per hundred output included wage
每班 per shift
每班操作循环 <隧洞掘进的> rounds per shift
每班产量 overall output per manshift
每班掘进循环数 rounds per shift
每班每人 per man per shift
每班炮眼组数 rounds per shift
每班修理机器停工时间 in-shift repair time
每班移动机器停工时间 in shift moving time
每拌毛重 batch weight
每磅功率 <指引擎重量> power per pound
每包毛重 packed weight
每包重量 weight per package
每遍扩充工作码 each-pass own code
每搏输出量 stroke output; stroke volume
每槽电极片数 number of anode plates
每层【建】per floor; per stor(e)y
每层重新架立模板 resetting of forms
每层贯击数 blows per layer
每层两顺一丁砌砖法 flying Flemish bond; Yorkshire bond
每层砌体的层高 masonry lift
每层中的工作定额 working quota for one stratum
每钞(钟)英寸数 inches per second
每车道每小时通行车辆数 vehicles per lane per hour
每车净载量 net load per vehicle
每车辆(绿灯)延期 vehicle interval
每车每天车租费 <车辆过轨他路后> per diem charges
每车平均乘员数 < = 乘员人数/通过某点车辆数>【交】average vehicle occupancy
每车平均英里程 mileage per car
每车日净重吨公里数 net ton-kilometers per car day
每车延误 individual vehicle delay
每车装货吨数 tons loaded per car
每齿进给量 feed engagement
每齿切屑荷载 chip-load per tooth
每冲击一次套管的延深 <用锤夯入> penetration per blow
每船 per ship
每锤贯入量 <打桩的> penetration per blow
每锤击贯入度 penetration per blow
每锤桩打入尺寸 set per blow
每锤桩的贯入度 set of pile per blow
每锤桩的贯入量 set of pile per blow
每次 each time
每次冲击 per blow
每次冲击贯入深度 set per blow
每次冲击率 per blow rate
每次冲击能量 energy per blow
每次冲击下陷深度 set per blow
每次锤击的英尺磅 foot-pound per blow
每次错动量 amount of offset in each time
每次荷载时间 time of per loading
每次呼叫固定路径选择 fixed routing per call
每次呼叫固定路径选择法 fixed routing per call method
每次活动时间 duration of each movement
每次活动性质 feature of each movement
每次降水入渗系数 infiltration coefficient of precipitation each time
每次磨锐钎头的进尺率 output per resharpening
每次磨锐钻头的进尺率 output per resharpening
每次收率 yield per pass
每次所扣的税款 the tax withheld from each payment
每次行程 each run
每次行程走刀量 feed per stroke
每次噪声轰鸣时间 duration of each burst of noise
每次注射 per injection
每次装载时间 time of per loading
每寸点数 dot per inch
每吋扣数 threads of per inch
每单位 per unit
每单位量的 per unit
每单位面积 per-unit area
每单位平均值 mean per unit
每单位容量 per-unit volume
每单位时间的切筋次数钢 shearing-number per unit time
每单位体积 per-unit volume
每单位质量能量 energy per unit mass
每单位重量 per-unit weight
每单位重量的功率 power-weight ratio
每单元弯钩数 number of hooks per

unit
每道次压下量 <轧件通过轧辊的> draught per pass; reduction per pass; reduction per area
每道次延伸量 elongation for each pass
每道起始相关时间 beginning correlation time of each trace
每道压缩量 draught per pass
每道样点数 sampling number of each trace
每调车钩平均车数 number of cars per cut
每斗链节数 <链斗式多斗挖掘机的> number of links per bucket
每吨成本 cost per ton
每吨海里的运输成本 transportation cost per tonne-nautical mile
每吨货的运费 tonnage
每吨货物平均运送时间 average time in transit per ton of freight
每吨货物平均运送速度 average speed in transit per ton of freight
每吨利润 ton-profit
每吨容许载重的体积 cubic(al) capacity per permissible tonnage capacity
每吨英里汽油消耗量 gasoline consumption per ton mile
每二周 biweekly
每分输出量 minute output; output per minute
每分通气量 minute ventilation
每分有效通气量 effective minute ventilation
每分钟 per minute
每分钟变化量 variation per minute
每分钟操作次数 operations per minute
每分钟冲程 strokes per minute
每分钟冲程次数 stroke per minute
每分钟打击次数 <锻锤的> blows per-minute
每分钟倒斗次数 frequency of bucket emptying per minute
每分钟击数 blows per-minute; number of blows per minute
每分钟计数 counts per-minute
每分钟计算次数 counts per-minute
每分钟记录 counts per-minute
每分钟加仑数 gallons per minute; gal/min
每分钟进尺(量) feed per minute
每分钟进刀量 feed per minute
每分钟卡片数 cards per minute
每分钟立方英尺 cubic(al) feet per minute
每分钟流量 flow per minute
每分钟脉冲数 impulses per minute
每分钟泥浆循环加仑数 mud circulation-gallons per minute
每分钟千克数 kilograms per minute
每分钟切削行程 cutting strokes per minute
每分钟输出 output per minute
每分钟衰变数 disintegration per minute
每分钟往复次数 <柱塞的> reversal per minute
每分钟文件量 documents per minute
每分钟行程 strokes per minute
每分钟行程数 number of strokes per minute
每分钟旋转次数 number of revolutions per minute
每分钟英尺(数) feet per minute
每分钟英寸数 inches per minute
每分钟振动次数 vibrations per minute
每分钟周数 cycles per minute
每分钟转动次数 revolutions per minute
每分钟转数 number of revolution per minute; revolutions per minute; rotations per minute
每分钟字数 words per minute
每分钟走刀量 feed per minute
每分钟最大转数 maximum r.p.m.[revolution per minute]
每分钟最高行程数 maximum number of strokes per minute
每缸马力 horsepower per cylinder
每隔一定时间 at regular interval
每隔一段时间 at set interval
每隔一口井 every second well
每隔一日 every other day
每隔一天 every other day; every two days
每隔一月 every other month
每个方向客流图 passenger flow diagram for each direction
每个房间 room-by-room
每个工人的产量 productivity per worker
每个金刚石钻头(损)耗量 diamond loss per bit
每个农场种植的英亩数 number of acres grown per farm
每个破产公司的平均债务 average liabilities per failure
每个生产日筒数 barrels per stream day
每个数位误差 error per digit
每个像素比特数 bippel
每个样品用 50 秒 fifty seconds per sample
每个组合单位的贡献毛利 contribution margin per composite unit
每个钻头进尺 production per point
每个钻眼均分的工作面积 <以平方英尺计> square footage per drill hole
每根长 length per bar
每根轴上的闸瓦压力 brake-shoe pressure on each axle
每工班计划小时数 schedule hours per shift
每工作班的保养 shift maintenance
每公斤 per kilogram
每公斤日粮含量或百分率 amount or % per kilogramme ration
每公里测量中误差 mean square error per kilometer
每公里成本 cost per km
每公里孔数【岩】holes per kilometer
每公里税金 kilometric allowance
每公里线路货车数 wagons per kilometre of line
每公顷 per hectare
每公顷产量 output per hectare; yield per ha
每股 per share
每股股利与目前价格的比率 dividend yield
每股股票的账面价值 book value per share(of stock)
每股基本收益 primary earnings per share
每股基本盈利 basic earnings share
每股加权平均收益 weighted average earnings per share
每股净资产值 net assets value per share
每股目标股息 target dividend per share
每股票面价值 book value per share (of stock)
每股平均收益 earnings per share
每股清算价值 liquidation value per share
每股收益 earnings per share
每股收益率 per share earnings ratio
每股所获得现金额 cash generated shares
每股现金收益 cash earnings per share
每股盈余 per share earnings
每股账面价值 book value per share (of stock)
每股主要盈利额 primary earnings per share
每户平均人数 average size
每户以一大室作灵活分间的住宅体系 one-room system
每击贯入度 <打桩的> penetration per blow
每机车公里净重吨公里 net ton-kilometers per locomotive kilometer
每机车公里燃料吨数 tons of fuel per locomotive kilometer
每机车公里燃料费 fuel cost per locomotive kilometer
每机车公里总重吨公里 gross ton-kilometers per locomotive kilometer
每机车日净重吨公里数 net ton-kilometers per locomotive day
每级荷载 load(ing) increment
每级卸荷 load(ing) decrement
每季 quarter
每季保险费 quarterly premium
每季产量 quarterly output
每季的 quarterly
每季付款 quarterage
每季付款日 quarterage day
每季付一次的股息 quarterly dividend
每季估计 quarterly estimate
每季一次 quarterly
每加仑汽油所行英里数 miles per gallon
每加仑英里程 mileage per gallon
每加仑重 weight per gallon
每件的时间 time per piece
每节管道重量 weight/section
每节铰接车厢载容量 estimated capacity per articulated unit
每进口道平均等待时间 average wait per approaching
每开尔文 reciprocal Kelvin
每颗金刚石上的压力 diamond pressure
每克拉金刚石粒数 diamonds per carat; stones per carat
每孔(钻进)成本 cost per hole
每口井(钻进)成本 cost per well
每块构件长 length per member
每厘米吃水差平力矩 moment to change trim per centimeter
每厘米吃水差校力矩 moment to change trim one centimeter
每厘米吃水吨数 tons per centimeter immersion
每厘米浸水吨数 tons per centimeter immersion
每厘米螺纹圈数 number of turns per centimeter
每离子对能量损失 energy loss per ion pair
每历日筒数 barrels per calendar day
每立方厘米克 gram per cubic centimeter
每立方厘米的粒子数 particles per cubic(al) centimeter
每立方码散料消耗的功率 horse power per loose cubic yard
每立方米坑道炸药消耗量 explosive consumption per cubic meter
每立方米千克数 kilograms per cubic(al) meter
每立方米原岩雷管消耗量 detonation consumption per cubic entity rock
每立方米原岩炮眼消耗量 shothole consumption per cubic entity rock
每立方米造价 cost cubic(al) meter [metre]
每立方英尺造价 cost per cubic(al) foot
每立方英尺重量 weight per cubic foot
每立体角单位光通量 luminous flux per steradian
每两个会车地点间的运转时分 point-to-point timing between each pair of crossing places
每两年期 per biennium
每辆客车旅客人数 number of passengers per carriage
每辆客车平均乘客人数 average number of passengers per carriage
每列车调车钩数 number of cuts per train
每列车公里总重吨公里 gross ton-kilometers per train kilometer
每列车解体调车钩数 number of switching cuts per train
每列车平均分解间隔时间 average sorting interval per train
每列车平均分解时间 average sorting time per train
每列车厢数 number of cars per train; number of coach-units per train
每列车小时货物吨公里数 ton-kilometers per train hour
每列车小时平均净重吨公里 net ton-kilometers per train hour
每列车小时平均总重吨公里 gross ton-kilometers per train hour
每列单孔编码 single-column duodecimal coding; single-column pence coding
每列多孔卡片 ducol-punched card
每炉熔炼量 heat
每路一载波多址联结 single-channel per carrier multiple access
每马力发动机重量 engine weight per horsepower
每马力小时消耗量 consumption per horse-power-hour
每马力重量 weight per horsepower; weight per unit of power
每码 per yard
每毛吨 per gross ton[pgt]
每美元边际效用 marginal utility per dollar
每美元复利终值 accumulated amount of 1 dollar
每美元外币价 foreign currency per US dollar
每米 per meter; reciprocal meter[metre]
每米金钢石消耗 diamond loss per meter
每米坑道雷管消耗量 detonation consumption per meter
每米坑道炮眼消耗量 shothole consumption per meter
每米坑道炸药消耗量 explosive consumption per meter driftage
每米钻井成本 drilling cost per meter
每米钻头磨耗 bit wear and tear per meter
每米最大集中荷载 maximum concentrated weight per meter[metre]
每秒 per second
每秒……厘米 <一种速度单位> centimetre per second
每秒百万条指令 million instructions per second
每秒读数 readings per second
每秒公尺数 meters per second
每秒计数 counts per second

M

每秒进尺 feet per second
每秒克数 grams per second
每秒厘米(数) centimeter[centimetre] per second
每秒立方米 cubic(al) meters per second
每秒立方英尺 cubic(al) feet per second
每秒立方英尺日 cubic(al) feet per second per day;second-feet-day
每秒流量 flow rate per second; throughput
每秒脉冲数 pulses per second
每秒每秒 per second each second
每秒每秒厘米 <加速度单位> centimeter[centimetre] per second per second
每秒平均执行指令数 average instructions per second
每秒平均指令数 average instructions per second;average orders per second
每秒气耗量 gas consumption per second
每秒千克数 kilograms per second
每秒千周 kilocycles per second
每秒闪光次数 flashes per second
每秒数 cycle per second
每秒位数 bits per second[bps]
每秒旋转次数 revolutions per second
每秒循环数 number of cycles per second
每秒一跳秒表 independent seconds watch
每秒兆周 megacycles per second
每秒帧数 frame per second; number of frames per second; number of pictures per second
每秒指令数 orders per second
每秒钟 per second
每秒钟的指令数目 orders per second
每秒钟计数 counts per second
每秒钟加仑数 gallons per second
每秒钟英尺数 feet per second
每秒钟转数 revolutions per second
每秒周期数 cycles per second; number of cycles per second
每秒周数 Hertz
每秒转数 revolutions per second[rps]
每秒转速 revolution per second
每秒字符数 characters per second
每秒最大流量 maximum discharge per second
每摩尔电能量 amount of electric(al) energy per mole
每亩成本 cost per mu
每年 annuity; per annum; year after year;year by year; year in (and) year out;yearly
每年按生产率调高工资的条款 annual improvement factor
每年报税表格 annual returns
每年产量 annual output
每年成本等值 equivalent annual cost
每年乘客量 passengers per annum
每年的 annual;yearly
每年的费用标准 scale of annual fee applicable
每年的负荷变化 yearly load variation
每年等额应计法 equal annual accrual method
每年递增 an average annual increase of
每年垫付 annual advance
每年垫款 annual advance
每年度必须交纳的会费 obligatory annual contribution
每年分期偿还数 annual amortization factor
每年分摊 annual amortization
每年腐蚀深度 inches penetration per year
每年更新期 yearly renewable term
每年工作日 number of working days per year
每年工作天数 number of working days per year
每年耗用数 annual consumption
每年可拿工资的病假天数 sick leave
每年利润 profit perineum
每年利润率 annual rate of profit
每年免税额 annual allowance
每年平均速率 mean annual rate
每年平均值 per capita
每年实际死亡率 annual actual mortality
每年使用时间数 annual usage
每年收获循环 annual cropcycle
每年收益 annual earnings
每年现金投入 annual cash input
每年相等的费用 equivalent uniform annual cost
每年消耗量 consumption per year
每年续保制 yearly renewable term
每年一次的休假 annual leave
每年载荷率 yearly load factor
每年折旧免税额 annual depreciation allowance
每年整修费 annual improvement factor
每年支付给雇员之薪酬报税表 annual returns of emoluments paid to staff
每年最低支付额 minimum annual royalty
每年最高支付额 maximum annual royalty
每帕斯卡 reciprocal Pascal
每盘混凝土数量 batch quantity
每盘配料量 batch quantity
每盘重量 weight per batch
每批配料量 batch quantity
每批容量 batch
每批运价率表 <零担货物> scale of rates applied per consignment
每批重量 batch weight; weight per batch
每皮等厚的圬工 range masonry (work)
每皮二顺一丁砌合 flying bond
每皮两顺一丁砌砖法 flying Flemish bond;monk bond
每皮三顺一丁砌合 Flemish garden bond;Flemish garden wall bond
每匹 per animal
每品脱浆料的重量 slop weight
每品脱重量 weight per pint
每平方公里孔数【岩】holes per square kilometer
每平方码磅数 pounds per square yard
每平方米舱面负荷 deck load per square metre
每平方米克重 grammes per square metre
每平方米造价 cost of per-square-meter
每平方英尺磅数 pounds per square foot[lbs. per sq.ft.]
每平方英尺造价 square foot cost
每平方英寸 per square inch
每平方英寸磅数 pounds per square inch[lbs. sq.in.]
每平方英寸孔数 mesh per square inch
每平方英寸压碎荷重 crushing load per square inch
每普式耳价格 price per Bushel
每期每元复利本利和 compound amount of 1 per period
每期一元的本利和 compound amount of 1 dollar per period
每期一元现值 present value of one dollar per period
每期应付款项付讫 instalment paid
每期折旧费用 depreciation charges per period
每期折旧后财产情况 condition per cent
每起火灾造成的死亡 deaths per fire
每千 per thousand
每千立方英尺加仑数 gallons per millenary cubic feet
每千米备用钢轨 spare rails per kilometer of road
每千瓦费用 cost per kilowatt
每人 each person; every person; per capita;per head
每人典型废水量 typical per capita wastewater flow
每人工时产量 out per man
每人计算 per capita calculation
每人剂量 per capita dose
每人每班产量 output per manshift
每人每班劳动生产率 output per manshift
每人每分钟所做的工作 man minutes manit
每人每日 per capita per day
每人每日耗水量 consumption per capita per day; daily per capita consumption; per capita consumption daily;water consumption per capita per day
每人每日加仑量 gallon per capita per day
每人每日污水量 sewage per capita per day;sewage per head per day
每人每日消耗量 consumption per capita per day; daily per capita consumption; per capita consumption daily
每人每日用水量 consumption per capita per day; daily water consumption per capita; water consumption per capita per day
每人每日最大用水量 maximum water-consumption per capita per day; maximum water-consumption per head per day
每人每天 per head per day
每人每小时产量 production per man-hour
每人平均乘车次数 journeys per capita
每人平均货物吨公里数 freight ton-kilometers per capita
每人平均剂量当量 per caput dose equivalent
每人所得 per capita income
每人消耗量 consumption per capita; consumption per head; per capita consumption
每人一班产量 output per manshift
每人用水量 consumption per capita
每人有效建筑面积 floor space per person
每人装卸货物吨数 <即装卸工劳动生产率> ton of goods loaded and unloaded per man
每日 day-to-day; omn dieb; per day; per diem
每日……桶 barrels per day
每日保(险)费 daily premium
每日保养 daily maintenance
每日保养工作 daily maintenance task
每日备忘录 daily memorandum
每日变动 daily variation
每日变动系数 daily variation coefficient
每日变化 diurnal variation
每日变化系数 daily variation coefficient
每日补偿期 daily compensation period
每日材料平均支出数 average outlays of material per day
每日测产一天记录 weigh-a-day-a-day
每日产量 daily making; daily output; output per day
每日潮高差 diurnal range
每日出工工人数报告 labo(u)rers' daily report
每日处理报告 daily transaction reporting
每日存货比率 days inventory ratio
每日单价 day work rate
每日的 daily;diurnal
每日的供应 day-to-day issue
每日的例行公事 everyday routine
每日对账单 daily statement of account
每日吨数 tons per day
每日发货 make daily deliveries
每日范围 diurnal range
每日非值班工作时间 non-shift hours each day
每日费用 daily expenses; per diem expenses
每日分期付款 daily instalment
每日工程 day's work
每日工作报告单 daily work report
每日工作记录 daily sheet
每日工作进度表 calendar progress chart
每日工作量 daily work load
每日工作时间登记卡 daily time card
每日工作小时数 daily hours of operation
每日故障记录表 perturbations of daily schedule
每日规定食物量 dietarg
每日耗用量 daily consumption
每日换班 daily relay
每日获利 daily gains
每日价格变动限幅 daily limit of price changes
每日价格限幅 daily price limit
每日价位变动限幅 daily limit of price changes
每日检查 daily check; daily inspection
每日降水量 daily rainfall; precipitation per day
每日降雨量 daily rainfall; precipitation per day
每日交通量 daily traffic volume
每日津贴 daily allowance; per diem allowance
每日警惕测试 daily acknowledgment test
每日开支 daily expenses
每日历日桶数 barrels per calendar day
每日利息 daily interest
每日两次观读水位 twice-daily ga(u)ge reading
每日两次水尺读数 twice-daily ga(u)ge reading
每日量 daily amount
每日流量 daily flow
每日每舱口 per hatch per day
每日每平方分米毫克损失 milligrams loss per square decimeter per day
每日每适宜舱口 per working hatch per day
每日平均存款余额 average daily balance
每日平均值 average per day
每日情报 daily information
每日情报摘要 daily intelligence digest
每日容许摄入量 acceptable daily in-

take
每日润滑 daily lubrication
每日三班(工作)制 three-shift work day system
每日升水 daily premium
每日生产量 daily capacity
每日生活津贴 daily subsistence allowance
每日石油桶数 barrel of oil per day
每日输送量 daily flow volume
每日输送能力 daily flow capacity
每日天气图 daily climate chart; daily weather chart
每日贴水 daily premium
每日桶数 barrels daily; barrels per day
每日现金需要比率 daily cash ratio
每日消费量 daily consumption
每日消耗量 day's expenditures
每日销售总值 daily total sales value
每日循环 diurnal cycle
每日业务报告 daily service report
每日一次水尺读数 one ga(u)ge reading per day
每日营运成本 daily operating cost
每日用水量 consumption per day; daily water consumption
每日余额 daily balance
每日余额计算法 daily balance method
每日允许摄入量 acceptable daily intake
每日运转 daily run
每日运转情况 daily performance
每日账表 daily statement (of account)
每日转动 daily rotation
每日状况 daily status
每日总量负荷 daily mass loading
每日钻探记录单 driller's tour report
每日作业 daily work
每升毫克数 milligrammes per liter
每升克 gram per liter
每十五分钟测一次 a fifteen minutes frequency
每时间单位 per time unit
每四分之一加仑英里程 mileage per quart
每台备件 spare parts per unit
每台附件 accessories per unit
每台机车每日总重吨公里 <即机车日产量> gross ton-kilometers per locomotive day
每天办理(出入)车数 <车站> number of cars handled per day
每天变化量 variation per day
每天单方向断面最大客流量 maximum passengers per day per direction
每天的 day-to-day; quotidian
每天工作班数 shifts worked per day
每天工作成绩 daily performance
每天工作量 <例如编组场每天到发的重空车数> daily throughput
每天购票乘客人数 paying passengers per day
每天结算燃料消耗 day-to-day loss control
每天进尺 feet per day
每天试验 day-to-day test
每天天气图 daily weather chart; daily weather map
每天通勤乘客 daily commuter
每天通勤交通 daily commuter
每天通勤旅客 daily commuter
每天往返运行次数 <货车> trip per day
每天消耗量 consumption per day
每天行程 <日公里> daily run
每天月票乘客 daily commuter
每天走的路线 daily round
每通过一次 per pass
每通路单路载波 single carrier per channel
每桶磅数 pounds per barrel
每头牲畜所用畜舍面积 space requirement
每瓦流明 <国际光通量单位> lumen per watt
每万吨公里工资含量 per ten thousand ton-kilometre included wage
每万吨公里燃料消耗量 fuel consumption per ten thousand gross ton-kilometers
每万换算吨公里占用固定资产原价 per ten thousand converted ton-kilometer employed fixed assets cost
每万流动资金完成的换算吨公里 per ten thousand current funds finished converted ton-kilometre
每万流动资金完成的运输收入 per ten thousand current funds finished transport revenue
每万元工业产值占用固定资产原价 per ten thousand transport industry output employed fixed assets cost
每万元运输收入占用固定资产原价 per ten thousand transport revenue employed fixed assets cost
每万总重吨公里成本 cost per ten thousand-gross weight-ton-kilometer
每位 n 个磁芯的存储器 n-cores-per-bit storage
每位错误概率 probability of error per digit
每位两磁芯式存储器 two-core-per-bit storage
每位两磁芯系统 two-core per bit system
每位两个磁芯 two-core per bit
每位误差 error per digit
每位一个磁芯存储器 one-core-per-bit storage
每五年一次的 quinquennial
每项费用均衡缩减 across-the-board reduction
每小时 per hour
每小时编解车数 number of cars classified per hour; number of cars sorted per hour
每小时变化 hourly fluctuation; hourly variation
每小时变化量 variation per hour
每小时变化系数 hourly variation coefficient; per hour change coefficient
每小时波动 hourly fluctuation
每小时产量 hourly capacity; hourly output; production per hour
每小时潮高 hourly height
每小时成本 hourly cost
每小时赤纬变化量 declination change in one hour
每小时单方向 per hour per direction
每小时单方向断面客流量 section passengers per hour per direction
每小时单向客流量 passengers per hour per direction
每小时单向最大列车对数 maximum number of trains per hour in each direction
每小时的 horary; hourly
每小时定额件数 piece-per-hour rate
每小时吨数 tons per hour
每小时工资 hourly earnings; per hour wage
每小时公吨 metric(al) tons per hour
每小时公里数 kilometers per hour; kilo per hour
每小时航行里 knot
每小时耗用量 hourly consumption
每小时耗油量 per hour factor
每小时换气次数 air changes per hour
每小时加仑数 gallons per hour
每小时降水量 hourly precipitation; hourly rainfall(depth)
每小时降雨量 hourly rainfall(depth)
每小时交通量 hourly traffic volume
每小时克数 grams per hour
每小时利用率 hourly utilization ratio
每小时列车开行对数 trains per hour
每小时平均车速 hourly average speed
每小时平均工资 average hourly earnings
每小时平均速度 hourly average speed
每小时平均值 hourly mean value; hourly medium value
每小时汽耗 steam consumption per hour
每小时千卡 <冷冻能力计量单位> frigory
每小时熔量 melt per hour
每小时生产吨数 capacity in tons per hour
每小时生产量 production per hour
每小时生产率 hourly capacity; production rate per hour
每小时生产能力 hourly capacity
每小时收费率 hourly rate
每小时输出率 output rate per hour
每小时数量 hourly rate
每小时双向最大列车对数 maximum number of trains per hour in two direction
每小时损失率 hourly loss rate
每小时通行车辆数 vehicles per hour
每小时消耗量 consumption per hour; hourly consumption
每小时印数 impression per hour
每小时英里 miles per hour
每小时英里数 miles per hour
每小时用水量 consumption per hour; hourly consumption; hourly water consumption
每小时最大容量 maximum hourly capacity
每小时最大需用量 maximum hourly demand
每小时最大用水量 hourly maximum water consumption
每小时最高室外温度 maximum hourly outdoor temperature
每小时最高需要量 maximum hourly requirement
每星期休息期 weekly rest period
每行程走刀量 feed per stroke
每循环爆落实体岩石量 blast entity rock per circulation
每循环进尺 advance per attack; advance per round
每循环炮眼消耗量 shothole consumption per cycle
每延公尺线路上的荷载 load per meter of the line
每延米 per meter
每延米轨道压力 linear track load
每延英尺磅数 pounds per linear foot
每延英寸磅数 pounds per linear inch
每一百英亩的土地税 hidage
每一车道宽度 width of a traffic lane
每一次行程 <混凝土路面刷毛机> brush stroke
每一单位 per unit
每一单位积累额 accumulated funds of each enterprise
每一单位面积 per-unit area
每一单位体积 per-unit volume
每一单位重量 per-unit weight
每一单元 per unit
每一定量某种商品 every definite quantity of a commodity
每一动作制动机的吨位 tons per operative brake
每一回次 per tour
每一机车小时的净吨英里 net ton miles per engine hour
每一居室的平均居民数 average number of persons per habitable room
每一旅游者日平均花费 average outlays per tourist per day
每一毛吨 per gross ton
每一美元兑换本国货币数 national currency per US dollar
每一千读者的收费标准 cost per thousand readers
每一扫描循环分钟数 minute per scan
每一商品价值 value of every commodity
每一设备的 per unit
每一线路公里的总吨公里数 gross ton-kilometers per kilometer of line
每一英寸长的网孔数 meshes per linear inch
每一有效坡道制动机的吨位 tons per effective grade brake
每一转所需秒数 second per revolution
每英尺深锤击数 blows foot
每英尺重量 weight per foot
每英尺锥度 taper per foot
每英尺(钻进)成本 cost per foot
每英尺钻进的金刚石损耗量 diamond loss per foot drilled
每英尺钻进的钻头费用 per-foot-bit cost
每英尺钻孔 per-foot-of hole drilled
每英尺钻头成本 per-foot-bit cost
每英寸吃水差力矩 moment to alter trim one inch
每英寸吃水吨数曲线 curve of tons per inch immersion
每英寸齿数 teeth per inch; threads per inch
每英寸厚磅数 pounds per inch in thickness
每英寸螺纹圈数 number of turns per inch
每英寸螺纹数 threads per inch
每英寸排水吨数 tons per inch
每英寸倾斜力矩 inch trim moment
每英寸(筛网)孔(眼)数 mesh per inch
每英寸位数 bit per inch
每英寸行数 lines per inch; rows per inch
每英寸转数 turns per inch
每英寸字符数【计】character per inch
每英寸字节数 bytes per inch
每英寸纵倾排水量变化 change of displacement per inch of trim
每英亩产量 acreage yield
每英亩产油桶数 barrel per acre method
每英亩收获量价值 value per acre
每元本利和 amount of 1
每元的复利终值 accumulated amount of 1; compound amount of 1
每元定期复利累积值 accumulation of one per period
每元复利本利和 accumulated amount of one; compound amount of one
每元欠款到期偿还的款额 <包括复利> amount of one
每元现值每期付款数 periodic(al) payment with present value of one Yuan
每原子能量损失 energy loss per atom
每月 per mensem; per month

M

每月保险费 monthly premium
每月波动 monthly fluctuation
每月不规则劳动时数 monthly irregular hours worked
每月产量 monthly output
每月的 mensal;monthly
每月等量装运 equal quantity monthly
每月订货 monthly order
每月对账单 monthly statement of account
每月二次的 bimonthly
每月分摊数 monthly allotments
每月付款 monthly payment
每月负荷率 monthly load factor
每月个人工资计算表 monthly individual payroll sheet
每月工资 monthly earnings
每月还款的借款 monthly payment loan
每月交货量 monthly delivery
每月结账记录 monthly closing entries
每月津贴 monthly allowance
每月进度报告 monthly process report
每月企业净增数 net business formation
每月气温变化 diurnal temperature change
每月生产报告 monthly production report
每月生产能力 monthly capacity
每月数字 monthly data
每月筒数 barrels per month
每月消费量 monthly consumption
每月应纳的税款 the tax to be paid each month
每月预提所得税 monthly withholding tax
每月运转 monthly run
每月指数图表 monthly index chart
每张支票登账需加付费用的支票往来户 pay-as-you-go account
每帧显示字符数 display characters per frame
每帧行数 lines per picture
每帧有效扫描行数 number of active lines
每帧字符数 characters per frame
每种商品的产业部门总需求 total industrial demand for each commodity
每种试样 per sample
每种特殊产品 each particular kind of product
每周保险费 weekly premium
每周变化 weekly variation
每周工时分配 weekly time distribution
每周工薪单 weekly payroll
每周工作班数 shifts operated per week
每周检测 weekly measurement
每周孔数 hole numbers at every circuit
每周两次 twice a week
每周两次的 semi-weekly
每周四十小时工作制 forty hours week
每周通勤交通 weekly commuter
每周通勤旅客 weekly commuter
每周新闻简报 weekly news summary
每周圆孔行数 line numbers in one circuit
每周月票乘客 weekly commuter
每周运转 weekly run
每周中除星期日以外的日子 weekday
每轴自重 deadweight per axle
每轴最大荷载 maximum load per axle
每昼夜油井产量 size of the well
每转冲击次数 blows per turn
每转进尺量【岩】penetration per revolution
每转走刀量 feed per revolution
每字节价格 cost per byte
每字误差概率 per-word error probability

美 <音调单位> mel

美钞 Federal Reserve Note;greenback <俚语>
美蛋白石 cacholong
美的感觉 esthetic feeling
美的观念 esthetic idea
美的思想 esthetic idea
美的吸引力 esthetic charm
美的效果 esthetic effect
美的自然性 nature of beauty
美杜莎白 <美国白硅酸盐水泥> Medusa white
美吨 <美制质量单位,1 美吨 = 2000 磅或 907.2 千克> short ton; net ton
美发廊 hair salon
美发厅 hairdressing saloon
美工 art designer
美工车间 art designing department
美观程度 <色彩调配的> aesthetic measure
美观方面 esthetic aspect
美观控制 aesthetic control
美观区划 aesthetic zoning
美国 United States
美国安全工程师学会 American Society of Safety Engineers
美国八大会计师事务所 Big Eight
美国白硅酸盐水泥 atlas white
美国白腊树 white ash
美国白腊木 American ash;white ash
美国白皮松 Alpine white bark pine; creeping pine
美国白松 American white pine
美国北达科他州(柔性路面)设计法 <即圆锥本法> North Dakota design method
美国北达科他州圆锥体 <用于试验土基承载量> North Dakota cone
美国北卡罗来纳青石 North Carolina bluestone
美国北卡罗来纳州承载板(试验)法 North Carolina bearing method
美国北卡罗来纳州法 <设计柔性路面厚度的一种方法> North Carolina method
美国本土 Continental United States
美国杓兰属 lady's slipper
美国标准 American Standards;United States of American Standards;United States Standards; US specification
美国标准编码Ⅱ【计】American standard code Ⅱ
美国标准槽钢 American standard channel
美国标准产品设计 American standard product line
美国标准粗牙螺纹 national coarse (thread)
美国标准电梯规范 American Standard Elevator Codes
美国标准钢绳钻进方法 American Standard cable system
美国标准公司 American Standard Incorporation
美国标准(管路)配件 American-standard fittings
美国标准管螺纹 Brigg's standard
美国标准(规格)协会 American Standard Association
美国标准化图上的一方格 <南北 17 英里 <1 英里 = 1609.34 米>,东西 11 ~ 15 英里> quadrangle
美国标准局 American Bureau of Standards; United States Bureau of Standards
美国标准梁 American standard beam
美国标准螺纹 American national screw thread; American standard thread
美国标准试验手册 American Standards Test Manual
美国标准委员会 American Standards Associations
美国标准细牙螺纹 National fine (thread)
美国标准线规 American Standard Wire Ga(u)ge;United States Standard Ga(u)ge
美国标准型钢 American standard steel sections
美国标准学会 USA Standard Institute
美国标准直管螺纹 American Standard Straight Pipe Thread
美国标准锥管螺纹 American Standard Taper Pipe Thread
美国表面与涂装研究中心 Center for Surface and Coating Research of American
美国薄钢板规格 United States Steel Sheet Gage
美国材料标准 American material standard
美国材料试验标筛 standard sieves of American Standard of Testing Material
美国材料试验标准 American Standard of Testing Material[ASTM]
美国材料试验学会 American Society for Testing and Materials
美国材料试验学会的标准筛 standard sieves of ASTM
美国材料试验学会会刊 Proceedings of American Society for Testing and Materials
美国材料试验学会黏[粘]度-温度特性曲线图 ASTM slope
美国材料试验学会涂料试验手册 Paint Testing Manual of ASTM
美国材料试验学会颜色标准 ASTM colo(u)r standard
美国财产评估员协会会员 Fellow of American Society of Appraisers
美国(采)矿、冶(金)、石油工程师学会 American Institute of Mining, Metallurgical and Petroleum Engineers
美国采暖与通风工程师学会 The American Society of Heating and Ventilation Engineers
美国采暖、制冷与空调工程师学会 The American Society of Heating, Refrigerating and Air Conditioning Engineers
美国侧柏 giant arbor-vitae
美国钞票 American bank note
美国钞票公司 American Bank Note Company
美国车辆制造协会制定的车辆交换检查规则 Master Car Builders' Rules of Interchange
美国承包商总会 Associated General Contractors of America
美国城市公共交通协会 American Public Transit Association
美国城市规划(工作)者学会 American Institute of Planners
美国尺寸标准 American ga(u)ge
美国初期方格嵌板门 colonial panel door
美国初期建筑 colonial architecture of America
美国初期建筑风格 colonial style
美国初期门窗饰边 colonial casing
美国初期新英格兰式建筑 New English colonial
美国初期住房建筑 colonial house
美国储蓄联合会 United States League of Savings Associations
美国储蓄债券 United States saving bond
美国船舶局 American Bureau of Shipping
美国船级社 American Bureau of Shipping
美国枞 Douglas fir
美国大坝委员会 United States Committee on Large Dams
美国大陆 Continental United States
美国大学教授联合会 American Association of University Professors
美国大学联合会 Association of American Universities
美国袋装水泥 American sack
美国得克萨斯州(柔性路面)设计法 Texas design method
美国得克萨斯州三轴试验法 <设计柔性路面厚度的一种方法> Texas triaxial method
美国得克萨斯州三轴压力筒 <用于柔性路面设计中的三轴压力试验> Texas pressure cell
美国地理委员会 United States Geographic Board
美国地理学家协会 Association of American Geographers
美国地球物理联合会 American Geophysical Union
美国地球物理协会 American Geophysical Union
美国地球物理学会 American Society of Geophysics Union
美国地震学会 Seismologic(al) Society of America
美国地震学会会刊 Bulletin of the Seismologic(al) Society of America
美国地质调查局 Geologic(al) Survey
美国地质调查局水质实验室 National Water Quality Laboratory of United States Geologic(al) Survey
美国地质调查局水资源处 Water Resources Branch of United States Geologic(al) Survey
美国地质勘测局 United States Geologic Survey
美国地质学会 American Geological Institute;Geologic(al) Society of America
美国第四纪研究协会 Association of America for Quaternary Research
美国电焊学会 American Welding Society
美国电机工程师协会 American Association of Electrical Engineers
美国电机工程师学会 American Institution of Electrical Engineers
美国电气工程师学会 American Institute of Electric(al) Engineers
美国电气化铁路工程协会 American Electric(al) Railway Engineering Association
美国电气化铁路协会 American Electric(al) Railway Association
美国电子工程师学会 American Institution of Electronic Engineers
美国东部杨木 Eastern cottonwood
美国短线铁路协会 American Short Line Railroad Association
美国吨 US ton
美国多圆锥投影 polyconic(al) projection of USA

美国多脂松夹条地板 American pitch pine strip floor(ing)
美国俄勒冈松 Oregon pine
美国鹅掌楸【植】saddle tree; American whitewood; canary wood; Tulip tree; white tree
美国法氧化锌 American process zinc oxide
美国房地产估价师协会 American Institute of Real Estate Appraisal
美国房地产估价协会 American Society of Real Estate Appraisers
美国房地产及城市经济协会 American Institute of Real Estate Urban Economics Association; American Real Estate and Urban Economics Association
美国纺织化学师和着色师协会 American Association of Textile Chemists and colo(u)rists
美国分类 classification used in USA
美国枫树 American sycamore
美国枫香 alligator tree; red gum; satin walnut
美国钢结构涂装委员会 American Steel Structures painting Council
美国钢结构协会 American Institution of Steel Construction
美国钢结构研究所 American Institute of Steel Construction
美国钢铁学会 American Iron and Steel Institute
美国钢铁制造商协会 Association of American Steel Manufacturers
美国港务协会 American Association of Port Authorities
美国高级研究项目机构 Advanced Research Projects Agency
美国高速铁路公司 American High Speed Rail Corporation
美国各州公路工作者西部协会 Western Association of State Highway Officials
美国各州公路工作者西部协会的道路试验 < 于1951年在爱达荷州进行 > WASHO road test
美国各州公路工作者协会 American Association of State Highway Officials[AASHO]
美国各州公路工作者协会道路试验 < 1958～1962年在伊利诺斯州的渥太华进行 > AASHO road test
美国各州公路工作者协会方法 < 多指柔性路面和水泥混凝土路面设计方法 > AASHO method
美国各州公路工作者协会击实试验 AASHO density test
美国各州公路工作者协会柔性路面设计法 AASHO flexible pavement design method
美国各州公路工作者协会土分类(法) AASHO soil classification
美国各州公路工作者协会岩土分类法 AASHO system
美国各州公路与运输工作者协会 American Association of State Highway and Transportation Officials
美国给水工程协会 American Water Works Association
美国工程标准委员会 American Engineering Standard Committee
美国工程教育学会 American Society of Engineering Education
美国工程师及建筑师学会 American Society of Engineers and Architects
美国工程师协会 American Association of Engineers; Association of American Engineer
美国工业管理协会 American Association of Industrial Management
美国工业设计师学会 American Society of Industrial Designers
美国工业卫生协会 American Industrial Hygiene Association
美国公共交通学会 American Transit Association
美国公共交通运输学会 American Public Transit Association
美国公共汽车协会 American Bus Association
美国公共卫生协会 American Public Health Association
美国公共卫生协会色标 American Public Health Association's System
美国公路改善学会 American-Association for Highway Improvement
美国公路管理局土的分类法 < 分成 A-1 到 A-8 八种 > Public Roads Administration Classification of Soil
美国公路局 ABPR[American Bureau of Public Roads]
美国公路局土的分类系统 Bureau Public Road Classification System
美国公路局土的分类体系 Bureau of Public Road Classification System
美国供暖、制冷及空气调节工程师学会 Refrigerating and Air-conditioning Engineers
美国供暖制冷空调工程师学会指南和资料手册 Guide and Data Book of ASHRAE
美国供水工程协会学报 < 月刊 > American Water Works Association, Journal
美国股票交易所 American Stock Exchange
美国管理协会 American Management Association
美国惯用计量单位 US customary unit
美国光学会 Optic(al) Society of America
美国广播公司 American Broadcasting Company
美国规范 United States Standards; US specification
美国规划工作者协会 The American Institute of Planners
美国规划咨询学会 American Society of Consulting Planners
美国硅酸盐水泥协会 Portland Cement Association
美国硅酸盐水泥协会(刚性路面)设计法 Portland Cement Association (design) method
美国国防部五角大楼 the Pentagon
美国国际开发署 United States Agency for International Development
美国国家标准 American National Standards
美国国家标准局 United States National Bureau of Standards[NBS]
美国国家标准局分解力试验图 NBS resolution test chart
美国国家标准局色差单位 NBS unit
美国国家标准局研究杂志 Journal of Research of the National Bureau of Standards
美国国家标准学会 American National Standards Institute
美国国家防火学会 National Fire Protection Association of the USA
美国国家废料政策法令 National Waste Policy Act of United States
美国国家腐蚀工程师协会 National Association of Corrosion Engineers
美国国家腐蚀工程师协会规定的金属表面喷砂处理的第三级 commercial blast
美国国家公路安全局 National Highway Safety Bureau
美国国家规定的房屋建筑最低标准 minimum standards bylaw
美国国家航空和宇宙航行管理局 National Aeronautics and Space Administration
美国国家科学院院报 Proceedings of National Academy of Sciences of the USA
美国国家沥青培训中心 National Asphalt Training Center
美国国家水力试验所 United States Hydraulics Laboratory
美国国家水污染控制法令 National Water Pollution Control Act of United States
美国国家水质委员会 National Water Quality Commission of United States
美国国家水资源委员会 National Water Resources Committee of United States
美国国家银行 United States National Bank
美国国内的 stateside
美国国营铁路旅客运输公司 < 接收经营各私营铁路在各大城市间的长途客运业务 > National Railroad Passenger Corporation[Amtrak]
美国海岸和大地测量勘查局 United States Coast and Geodetic Survey
美国海岸警卫队 United States Coast Guard
美国海军厂署 Bureau of Yards and Docks
美国海军路面设计法 navy(design) method
美国海军水道测量部 United States Navy Hydrographic(al) Office
美国海军天文台 Navy Observatory
美国海事案例 American Maritime Cases
美国焊接学会 American Welding Society
美国航务委员会 United States Maritime Commission
美国航运总署 United States Maritime Administration
美国核废料政策法令 Nuclear Waste Policy Act of United States
美国黑果稠李 American cherry; black cherry
美国黑核桃木 American black walnut
美国黑松 lodgepole pine
美国亨利中心天然气价格 Henry hub price of USA
美国红果云杉 eastern spruce
美国厚度标准 American ga(u)ge
美国花柏 Oregon cedar; Port orford cedar
美国化学工程师学会 American Institute of Chemical Engineers
美国化学文摘 American Chemical Abstract
美国化学学会 American Chemical Society
美国环境保护局 United States Environment Protection Agency[USEPA]
美国环境保护局国家二级饮用水法令 National Secondary Drinking Water Regulations of USEPA
美国环境保护局国家排污物毒性评价中心 National Effluent Toxicity Assessment Center of USEPA
美国环境保护局国家水质监视系统 National Water Quality Surveillance System of USEPA
美国环境保护局国家一级饮用水法令 National Primary Drinking Water Regulations of USEPA
美国环境保护局国家一级饮用水修订法 National Revised Primary Drinking Water Regulations of USEPA
美国环境保护局国家一级饮用水暂行标准 National Interim Primary Drinking Water Standards of USEPA
美国环境保护局联邦水质管理局 Federal Water Quality Administration USEPA
美国环境保护局农药规划处 Office of Pesticide Program(me) of USEPA
美国环境保护局水质处 Water Quality Office of USEPA
美国环境保护局水质基准 United States Environmental Protection Agency Water Quality Criterion
美国环境法 Environmental Law of the United States
美国环境管理体制 US system of environmental management
美国环境和资源委员会 United States Environment and Resources Council; United States Environment Protection Council
美国环境政策法清洁水法国家污染物排放消除体系 National Pollutant Discharge Elimination System of USEPA Clean Water Act
美国缓和政府限制条例所赋予的自由 < 即州际商务委员会 ICC 对铁路运输业不予限制或减少限制 > Staggers Act freedom
美国黄松 Ponderosa pine
美国灰分次品硬木级别的方法 scoot
美国会计学会 American Accounting Association
美国混凝土管协会 American Concrete Pipe Association
美国混凝土路面协会 American Concrete Pavement Association
美国混凝土学会 American Concrete Institute[ACI]
美国混凝土学会材料学报 < 美国双月刊 > Material Journal
美国混凝土学会会志 < 美期刊名 > Journal of the American Concrete Institute Proceedings
美国混凝土学会结构学报 < 美国双月刊 > ACI Structural Journal
美国混凝土学会杂志 Journal of the American Concrete Institute
美国混凝土压注协会 American Concrete Pumping Association
美国机床制造(业)协会 American Machine-tool Builders Association
美国机动车工程师协会定额 Society of Automotive Engineers rating
美国机动车工程师学会 Society of Automotive Engineers
美国机械工程师学会 American Institution of Mechanical Engineers; American Society of Mechanical Engineers
美国机械师 < 半月刊 > American Machinist
美国机械协会 American Machinery Association
美国计量挖方单位 mile yard
美国计算机控制系统应用小组 Computer Control System Applications Group
美国技术工程师联合会 American Federation of Technical Engineers
美国技术情报服务处 national technical information service
美国加里福尼亚州产天然沥青 aragotite
美国加仑 United States gallon; wine gallon
美国加州原油氧化沥青 Obispo
美国家用瓷 American household china

M

美国建筑 American architecture; architecture of the United States of America
美国建筑承包商协会 American Building Contractors Association
美国建筑师协会 American Institute of Architects
美国建筑师学会 American Institute of Architects
美国交通安全服务协会 America-Traffic Safety Service Association
美国胶合板协会 American Plywood Association
美国结构工程师协会会报 <美期刊名> Proceedings of American Society of Structural Engineers
美国捷运信用卡 American Express Card
美国军事技术与国防科研文献 AD report
美国军用标准 American Military Standard
美国军用标准化局 American Military Agency for Standardization
美国军用规格 American Military Specification
美国开口沉箱 American caisson; American open caisson
美国科学促进会 American Association for the Advancement of Science
美国科学发展协会 American Association for the Advancement of Science
美国科学工作者协会 Association of America of Scientific Workers
美国科学技术研究院 American Academy of Arts and Sciences
美国科学期刊 American Journal of Science
美国科学情报研究所 American Science Information Institute
美国科学院 American Academy of Sciences
美国客运计价人员协会 American Association of Passenger Rate Men
美国肯塔基州(柔性路面)设计法 Kentucky design method
美国快递卡 American Express
美国矿业局 Bureau of Mines
美国矿业与冶金协会规定用筛 Institute of Mining and Metallurgy sieve
美国矿渣协会 National Slag Association
美国昆虫协会 Entomological Society of America
美国劳动部职业安全与卫生局 Occupational Safety and Health Administration
美国冷藏工程师学会 American Society of Refrigeration Engineers
美国历史性建筑调查 historic(al) American building survey
美国栎 bastard oak
美国联邦电力委员会 United States Federal Power Commission
美国联邦公路(管理)局 Federal Highway Administration
美国联邦规范 American Federal Specification; US Federal Specification
美国联邦规格委员会 American Federal Specification Board
美国联邦海运委员会 Federal Maritime Commission
美国联邦航空管理局 Federal Aviation Administration
美国联邦航空规程 Federal Aviation Regulation
美国联邦航空局 Federal Aviation Agency
美国联邦航空局道面设计法 pavement design method of the Federal Aviation Administration
美国联邦航空局路面设计(法) Federal Aviation Agency design
美国联邦航空局土分类法 Federal Aviation Agency soil classification
美国联邦航空局岩土分类系统 Federal Aviation Administration System
美国联邦交通部 United States Department of Transportation
美国联邦清洁水法国家污染物排放消除许可证制度 National Pollutant Discharge Elimination Permit System in Federal Clean Water Act
美国联邦铁路管理局调查研究开发处 FRA's [Federal Railroad Administration's] Office of Research and Development
美国联邦通信[讯]委员会 Federal Communications Commission
美国联邦油漆规格 American Federal Paint Specification
美国联合抵押银行协会 United Mortgage Bankers of America
美国联合国协会 United Nations Association of USA
美国联合货运公司的成组运煤列车 Conrail unit coal train
美国联合情报委员会 American Joint intelligence Committees
美国联合情报中心 American Joint Intelligence Center
美国联合铁路公司 Consolidated Rail Corporation of America[Conrail]
美国量规 American ga(u)ge
美国流体动力协会 National Fluid Power Association
美国陆海军联合规范 United States Joint Army-Navy Specifications
美国陆军工兵部队 corps of Engineers, US Army
美国陆军工兵部队(沥青)混合料配合比设计法 corps of Engineers mix design method
美国陆军工兵部队设计法 <设计路面厚度的一种方法> Corps of Engineers method, US Army
美国陆军工程兵 United States Army Corps of Engineers
美国陆军工程兵团 Army Corps of Engineers of the United States Army; United States Corps of Engineers
美国陆军工程师兵团航道试验站 Waterways Experiment Station of the US Army Corps of Engineers
美国陆军工程师团 Corps of Engineers of the United States Army; US Army Corps of Engineers
美国陆军工程师团分局 Engineer District
美国陆军工程师团区局 Engineer Division
美国旅馆瓷 American hotel china
美国旅客运输管理人员协会 American Association of Passenger Traffic Officers
美国落叶松 great western larch; western larch
美国麻省粉红色花岗岩 Milford pink granite
美国毛檫木 silky sassafras
美国煤气协会 American Gas Association
美国密苏里州(柔性路面厚度)设计法 Missouri design method
美国民航管理局法 <设计刚性路面厚度的一种方法> Civil Aeronautic Administration method
美国民用航空委员会 Civil Aeronautics Board
美国木材 American timber
美国木材标准 American Lumber Standard
美国木材防腐协会 American Wood Preserving Association
美国木豆树 Indian bean
美国木结构学会 American Institute of Timber Construction
美国内政部 United States Department of Interior
美国(内政部)地质调查局 United States Geological Survey
美国(内政部)垦务局 United States Bureau of Reclamation
美国南方红桧 red cedar
美国南太平洋型的低平板车 <能装运两层集装箱> South Pacific prototype low-loader
美国能源部 United States Department of Energy
美国能源情报局 Energy Intelligence Agency
美国农药管理委员会 Association of American Pesticide Control Officials
美国农业部 United States Department of Agriculture
美国农业部农业研究所 US Agricultural Research Service
美国农业工程师学会 American Society of Agricultural Engineers
美国农业工程师学会论文集 <双月刊> Transaction of ASAE [American Society of Agricultural Engineers]
美国农业研究服务局 United States Agricultural Research Service
美国农业研究中心 United States Agricultural Research Center
美国农作物生态学研究所 American institute of Crop Ecology
美国暖气通风工程师学会 American Society of heating and Ventilating Engineers
美国匹兹堡平板玻璃公司 Pittsburgh Plate Glass Company
美国菩提树 American linden
美国气象局 United States Weather Bureau
美国气象学会 American Meteorologist Society
美国汽车 <美期刊名> American Automobile
美国汽车工程师学会 Society of Automotive Engineers
美国汽车工程师学会编号 S. A. E. Number
美国汽车工程师学会规定功率 S.A.E. Horsepower
美国汽车工程师学会会刊 S.A.E. Journal
美国汽车工业研究协会 Motor Industry Research Association
美国汽车货运协会 American Trucking Association
美国汽车协会 American Automobile Association
美国桥梁公司 American Bridge Company
美国桥梁、隧道及收费道路协会 American Bridge, Tunnel and Turnpike Association
美国清洁法 United States Clean Water Act
美国情报科学学会 American Society of Information Science
美国全国工程研究院 National Academy of Engineering
美国全国广播公司 National Broadcasting Company
美国全国卷材涂装工作者协会 National Coil Coaters Association
美国全国油漆协会 National Paint Varnish and Lacquer Association
美国全国油漆与涂料协会 National Paint and Coatings Association
美国人口普查局 United States Bureau of the Census
美国韧性铸铁 black malleable casting
美国润滑工程师学会 American Society of Lubrication Engineering
美国三大汽车公司 Big Three
美国三里岛核电站事故 US Three Mile Island nuclear power plant incident
美国色材学会 colo(u)r Association of United States
美国色彩联络协会和国家标准局制订的色名表示方法 colo(u)r name charts of ISCC-NBS
美国森林保护协会 Forest Conservation Society of America
美国森林局 US Forest Service
美国筛系 US sieve series
美国山地白松 Idaho white pine; mountain white pine
美国山核桃 apocarya; pecan
美国山杨 American aspen
美国商务部 United States Department of commerce
美国声学及隔声材料协会 Acoustic(al) and Insulating Materials Association
美国声学学会 Acoustic(al) Society of America
美国石油地质学家协会 American Association of Petroleum Geologists
美国石油公司 American Oil Company
美国石油协会 American Petroleum Institute[API]
美国石油协会比重(度) API gravity
美国石油协会比重计 API hydrometer
美国石油协会比重计标度 API (hydrometer) scale
美国石油协会标准比重表 API gravity Scale
美国石油协会标准钻杆丝扣 API drill pipe thread
美国石油协会车用机油牌号 API Designations
美国石油协会发动机润滑油使用条件分类系统 API Engine Service Classification System
美国石油协会规格 API Specifications
美国石油协会锅炉 API Standard Boiler
美国石油协会锅炉规范 API boiler standard & specifications
美国石油协会技术规范 API specification
美国石油协会容重指标 API gravity index
美国石油协会润滑委员会 API Lubrication Committee
美国石油学会工艺过程 API Process
美国石油学会重度 API gravity
美国石油组织组合标记 API unite mark
美国市区铁路及电气化铁路雇员联合会 Association of Street and Electric(al) Railway Employees of America, Amalgamated
美国市政工程协会 American Public Works Association
美国式底质采样器 United States bed material sampler
美国式多盘过滤器 American disc [disk] filter
美国式攻锥 Yankee screwdriver

M

美国式螺丝刀 Yankee screwdriver
美国式砌合 American bond
美国式砌墙法 American bond
美国式切刀 American style cutter
美国式水轮机 American turbine
美国式隧道法 American method of tunnel driving
美国式天沟 Yankee gutter
美国式卧车 <普尔曼工厂制造的直通式坐卧两用车> Pullman car
美国式檐槽 Yankee gutter
美国式檐沟 Yankee gutter
美国试验与材料协会 American Society for Testing and Materials[ASTM]
美国试验与材料学会标准 ASTM Standards
美国试验与材料学会临时规范编号 ASTM appendix number
美国室内装饰家学会 American Institute of Decorators
美国鼠李 cascara buckthorn
美国水保持研究室 United State water Conservation Laboratory
美国水保护实验室 US Water Conservation Laboratory
美国水道实验站 Waterway Experiment Station
美国(水的)硬度 US hardness
美国水环境联合会 American Water Environment Federation
美国水轮机 American wheel
美国水文办公室 United States Hydrographic(al) Office
美国水研究理事会 American Water Research Council
美国水政策局 American Office of Water Policy
美国水质管理协会 American Water Quality Association
美国水质研究理事会 American Water Quality Research Council
美国水资源局 United State Water Resources Service
美国水资源理事会 American Water Resources Council; Water Resources Council(USA)
美国水资源实验室 Water Resources Laboratory(USA)
美国水资源协会 American Water Resource Association
美国松 bastard spruce;Douglas pine;glass pine;Puget Sound pine;Washington fir;Douglas fir
美国塑料学会 Plastics Institute of America
美国碎石协会 National Crushed Stone Association
美国碎石协会路面设计法 National Crushed Stone Association pavement design method
美国太平洋铁路 Pacific Railroads;PacRail
美国陶瓷协会 American Ceramic Society
美国特种化学品制造商联合会 Chemicals Specialty of Manufacturers Association
美国天气控制咨询委员会 United States Advisory Committee on Weather Control
美国天体方位表 <赤纬24°~70°> Blue Azimuth Table
美国天文年历 American ephemeris
美国铁道工程协会 American Railway Engineering Association
美国铁道协会车辆管理科 <管理全国私营铁路货车的组织> Car Service Division AAR[Association of American Railroad]
美国铁道协会货物损失赔偿申请及防止损坏处 AAR Freight Claim and Damage Prevention Division
美国铁道协会数据系统部 Data Systems Division AAR
美国铁道协会通信[讯]信号处 AAR Communication Signal Division
美国铁道养路学会 Roadmasters, Maintenance-of-way Association of America
美国铁路部门负责人员协会 American Association of Railroad Superintendents
美国铁路车辆学会 American Railway Car Institution
美国铁路发展协会 American Railway Development Association
美国铁路工程师协会 American Railway Engineers Association
美国铁路工程协会 American Railway Engineering Association
美国铁路货车 Amfleet
美国铁路基金会 <设在纽约> American Railroad Foundation
美国铁路客车 Amcoach
美国铁路旅客运输公司便餐车 Amtrak Care
美国铁路旅客运输公司车队 Amtrak [National Railroad Passenger Corporation] fleet
美国铁路桥梁与建筑协会 American Railway Bridge and Building Association
美国铁路售票代理商协会 American Association of Railroad Ticket Agents
美国铁路退休局 <设在纽约> US Railroad Retirement Board
美国铁路外科医师协会 American Association of Railway Surgeons
美国铁路协会 United States Railway Association; Association of American Railroad <私营铁路>
美国铁路协会安全委员会 Association of American Railroads[AAR] Security Committee
美国铁路协会标准和建议实施办法手册 AAR Manual of Standard and Recommended Practice
美国铁路协会规范 AAR specifications
美国铁路协会通信[讯]信号组 AAR Communication and Signal Section
美国铁路协会通信[讯]组 AAR Communication Section
美国铁路协会信号组 AAR Signal Section
美国铁路协会信号组推荐实施方法 Signal Section recommended practice
美国铁路养路学会 Roadmasters, Maintenance-of-way Association of America
美国铁杉 American hemlock
美国通信[讯]协会 American Communications Association
美国图书馆协会 American Library Association
美国图书馆学会 American Library Institute
美国涂料研究学会 Paint Research Institute of American
美国涂装与装饰承包商协会 Painting and Decorating Contractors of America
美国土地测量分区线 township line
美国土木工程师和建筑师学会 American Society of Civil Engineers and Architects
美国土木工程师协会 American Society of Civil Engineers
美国土木工程师协会会刊 Transaction of the American Society Journal of Civil Engineers
美国土木工程师学会 American Society of Civil Engineers
美国土木工程师学会会刊 American Society Journal of Civil Engineers; Proceedings of American Society of Civil Engineers
美国土壤保护局 United States Soil Conservation Service
美国土壤局 United States of Soil
美国土壤综合分类系统 the United States Comprehensive Soil Classification
美国卫生工程标准协会 Society of Domestic and Sanitary Engineering Standard
美国卫生工程学会 American Society of Sanitary Engineers
美国卫生署 United States Public Health Service
美国文献学会 American Documentation Institute
美国文献资料研究学会 American Documentation Institute
美国污染控制协会 American Association for Contamination Control
美国无线电公司 Radio Corporation of America
美国梧桐 buttonball;Platanus American plane(tree)
美国五针松 Eastern white pine;Weymouth pine
美国西部侧柏 Pacific red cedar; western red cedar
美国西部黄松 heavy pine;heavy pinus pondarosa
美国西部牧区铁路 granger railroad
美国西部铁杉 Western coast hemlock
美国西加云杉 great tide-land spruce; Western spruce
美国咸水办公室 United States Office of Saline Water
美国线规 American wire ga(u)ge; Brown and Sharpe ga(u)ge
美国线径标准 American wire ga(u)ge
美国线径规 American steel and wire ga(u)ge
美国香槐 American yellowwood
美国镶木地板协会标准 American Parquet Association Standard
美国新地方主义(建筑) new regionalism
美国信息处理学会联合会 American Federation of Information Processing Societies
美国信息交换标准代码 American standard code for information interchange[ASCII]
美国信息交换标准(代)码健盘 ASCII keyboard
美国信息中心 United states Information Center
美国行李运输经理协会 American Association of Baggage Traffic Managers
美国悬钩子 American dewberry
美国悬铃木 American plane
美国学会间颜色委员会 Inter-Society colo(u)r Council of American
美国压敏胶带委员会 Pressure Sensitive Tape Council of American
美国压缩空气与气体协会 Compressed Air and Gas Institute
美国亚拉巴马大理石 Alabama marble
美国亚拉巴马州(柔性路面)设计法 Alabama design method
美国岩黄松 Rocky Mountain Ponderosa pine
美国盐渍土研究室 United State Salinity Laboratory
美国药典 United States Pharmacopoeia
美国冶金学会 American Metallurgical Society
美国仪器协会 Instrument Society of American
美国移动卫星公司 American Mobile Satellite Corporation
美国邮船运输贸易 US liner trade
美国油布 American cloth
美国油松 <含脂松木> American pitch pine
美国油脂化学家学会 American Oil Chemists Society
美国预算局 United States of the Budget
美国原子能委员会 United States Atomic Energy Commission
美国运筹学学会 Operations Research Society of America
美国运货人协会联合会 American Institution for shippers Association
美国运输部 Department of Transportation
美国运输部发展高速铁路办公室 DOT[Department of Transportation of America] Office of High Speed Railroad Development
美国运输部规定的设计压力 Department of Transportation design pressure
美国运输事务管理局 Administration of American Transportation Facilities
美国运输协会 <设在美国纽约> Transportation Association of America
美国运输研究所 Transportation Research Board
美国杂草科学协会 Weed Science Society of America
美国在线 America Online
美国皂荚 Kentacky coffee tree
美国造船工程师学会 American Society of Civil Engineers
美国造的 US-built
美国噪音污染和减除噪声法令 noise Pollution and Abatement Act
美国政府出版局科技文献报告 Publication Board
美国政府方格测量系统 government rectangular survey of USA
美国职业安全与卫生管理局 Occupational Safety and Health Administration
美国职业安全与卫生条例 Occupational Safety and Health Act of American
美国制加仑 US gallon
美国制造的 American-built
美国质量管理协会 American Society for Quality Control
美国仲裁协会 American Arbitration Association
美国州际公路系统 interstate highway system
美国(州)智能车辆道路协会 Intelligent Vehicle Highway Society of America
美国(洲)智能运输系统协会 Intelligent Transportation Society of America
美国住房法 US Housing Act
美国住房顾问学会 American Institute of Housing Consultants
美国住房和城市发展部 Department of Housing and Urban Development
美国住房及城市开发部 US Depart-

M

ment of Housing and Urban Development
美国住宅建设厅 National Housing Agency
美国筑路者协会 American Road Builders Association
美国铸造学会 American Foundrymen's Society
美国专利 American Patent; United States Patent
美国专利法协会 American Patent Law Association
美国专利局 United States Patent Office
美国专业地质师学会 American Institute of Professional Geologists
美国砖 American brick
美国资产评估员协会 American Society of Appraisers
美国资源局委员会 United States Inter-Agency Committee on Water Resources
美国紫杉 Oregon yew;Pacific yew
美国紫树 tupelo
美国自来水协会 American Water Works Association
美国自由列车 American Freedom Train
美国总统水资源保护委员会 United States President's Water Resource Protection Commission
美国总统水资源政策委员会 United States President's Water Resource Policy Commission
美国总统水资源政策咨询委员会 United States President's Water Resource Policy Consulting Commission
美国最早的货币 colonial bill of credit
美哈斯特砂岩 <一种产于美国俄亥俄州的浅灰色或浅黄色或杂色石料> Amherst sandstone
美好的形式 perfect form
美化 beautification; beautify; decorating;purfle;sugar over;transfigure
美化处理 artistic treatment
美化的拱门饰 decorated archivolt
美化地区 aesthetic area
美化地区规划 aesthetic zoning
美化环境 beatify the environment; landscaping design for environmental purposes
美化市容地带 amenity area;amenity plot;amenity strip
美化市容种植地带 amenity planting strip
美化自然 landscaping
美磺胺 mesulfamide
美晶石英 Bristol stone
美景大理石 landscape marble
美景大理岩 landscape marble
美蓝 methylene blue;methylthioninium chloride
美蓝还原试验 methylene blue reduction test
美蓝树脂法 methylene blue resin method
美蓝伊红染色法 methylene blueeosin staining method
美乐女神星【天】Euphrosyne
美乐统【地】Mylor series
美丽桉 scarlet-flowered gum
美丽的视野 graceful sweep
美丽花桉 scarlet eucalyptus
美丽石英 Bristol diamonds
美丽紫苑 aster venustus
美利坚合众国 <即美国> United States of America
美联邦储备会员行存款平均额 bank credit proxy
美硫镉矿 hawleyite
美铝榴石金伯利岩 pyrope kimberlite
美莫康 <电子计算机的一种型式,容量16千字,磁鼓64千字> Memocon
美人蕉 Canna indica
美人鱼雕像 <城市雕塑> mermaid; Triton
美容剂 cosmetic
美容室 beauty salon
美容厅 beauty parlour
美容院 beauty parlour;beauty salon; beauty-spot
美生板 masonite board
美式白铁剪 American type; tinman's snips
美式沉箱 American caisson
美式地下室 American basement
美式管钳 American pattern pipe pliers;pipe wrench;rigid type tongs
美式旅馆计价制 American plan
美式砌法 American bond; common bond
美式砌合 flying bond
美式台虎钳 American type bench vice
美式无声链 Moorse chain
美式羊角锤 American type claw hammer
美饰漆 effect lacquer
美饰涂层 decorative coating
美术 fine arts
美术标准 aesthetic criterion
美术玻璃 art glass
美术玻璃器皿 artistic glassware
美术陈列馆 art gallery;gallery
美术陈列室 art gallery
美术充皮纸 art vellum
美术瓷 artistic porcelain
美术的 artistic
美术灯泡 art bulb
美术地毯 art carpet
美术电影制片厂 animation film studio
美术工艺室 art and craft room
美术顾问 artistic adviser
美术馆 art gallery; art museum; fine arts studio;pinacotheca;pinakothek
美术广告纸 art poster paper
美术广告纸板 art poster board
美术贺卡 artistic greeting card
美术家 artist
美术家的 artistic
美术教室 art room
美术解剖学 artistic anatomy
美术呢 art felt
美术品陈列室 cabinet of curiosities
美术品商店 fine arts shop
美术漆 pattern paint
美术三原色 artists' primary
美术商店 art store
美术设计 artistic design
美术设计人 graph artist
美术史 history of art
美术室 fine arts studio
美术手工艺品 artistic handicraft
美术书皮纸 art book cover paper
美术水磨石 artistic terrazzo
美术水磨石地面 artistic terrazzo flooring
美术陶瓷 art ceramics;artistic ceramics;art pottery and porcelain
美术陶器 artistic pottery
美术涂饰剂 fancy finish
美术涂装 novelty finish
美术相角 art corner
美术学校 art school
美术学院 academy of fine arts; college of fine arts
美术印刷纸 art printing paper
美术用品 artistic drawing set;art material
美术油 artists' oil
美术釉 artware glaze
美术造型艺术 <绘画> fine arts
美术纸 art paper
美术纸板 fancy embossing card
美术装潢 artistic presentation
美术装饰漆 fancy paint
美术装饰纸 art cover paper
美术字 graphic(al) arts quality character;graphic(al) character
美术字幕 art-title
美术棕色 art brown
美斯屈罗风 maestro
美索不达米亚 <位于底格里斯河与幼发拉底河之间,为古代西亚文明发祥地> Mesopotamia
美索不达米亚建筑 Mesopotamian architecture
美铁 <美国全国铁路客运公司> Amtrak
美铁道协会标准 Association of American Railroads standard
美锡尼建筑 Mycenaean architecture
美螅属 Clwia sp
美学 aesthetics;esthetics
美学标准 aesthetic criterion
美学的 aesthetic
美学的感觉 esthetic sense
美学方面 aesthetic aspect
美学概念 aesthetic concept
美学观点 aesthetic aspect; aesthetic consideration
美学观念 aesthetic concept
美学家 aesthete;esthete
美学价值 aesthetic value
美学角度 aesthetic aspect
美学教育 aesthetic education
美学理论 aesthetic theory
美学理念 aesthetic sense
美学魅力 aesthetic charm
美学设计 aesthetic design
美学设想 aesthetic concept; aesthetic idea
美学史 history of aesthetics
美学思想 aesthetic idea
美学效果 aesthetic effect
美学要求 esthetic appeal
美学质量 aesthetic quality
美学作用 aesthetic appeal; aesthetic charm
美叶桉 Marri eucalyptus
美英制 US and British system
美育 aesthetic education
美元 US dollar [US $]; American dollar;buck dollar
美元报价 dollar pry off
美援信用证 AID Credit;Credit of Agency for International Development
美制 American-built;American system
美制白木 American whitewood
美制粗牙螺纹 national coarse(thread)
美制加仑 <1 美制加仑 = 3.785 升> American gallon;US gallon
美制捞砂筒 American pump
美制量线规 <一种表示金属丝直径大小的制度,自4/0号(0.46英寸<1英寸 = 0.0254米>)至48号(0.00124英寸)> Brown and Sharpe wire gage
美制螺纹 US standard screw thread
美制细牙螺纹 national fine thread
美制硬度 US hardness
美洲 occident;the Occident
美洲白木 cotton wood
美洲板块 American plate
美洲北方动物区 American boreal faunal region
美洲侧柏 American arbor-vitae
美洲(大叶)山毛榉 American beech
美洲冬青 American holly
美洲椴 American linden;basswood
美洲椴木 baswood
美洲鹅耳枥 American hornbeam;blue beech
美洲非洲南极洲三向联结构造 American-African-Antarctic triple junction
美洲辐尾藻 Uroglena americana
美洲港务协会 American Association of Port Authorities
美洲国际城市规划学会 Inter-American Planning Society
美洲国际卫生工程协会 Inter-American Association of Sanitary Engineering
美洲国际住宅及城市规划中心 Inter-American Housing and Planning Center
美洲国家组织 The Organization of American States
美洲核桃木 tiger wood
美洲黑栎 black oak;yellow oak
美洲红树 American mangrove; red mangrove
美洲红树生态系统 mangrove ecosystem
美洲花柏 ginger pine; Lawson's cypress;matchwood;Port Orford cedar
美洲花楸 American mountain ash
美洲黄栌 American smoketree
美洲家禽品种标准 American Standard of Perfection
美洲金缕梅【植】American witch hazel
美洲栗 American chestnut
美洲落叶松 American larch; eastern larch;hackmatack
美洲毛果(芸香)油 jaborandi oil
美洲葡萄 fox grape
美洲乔木 <一种可以做材用的> banak
美洲热带高山草地 paramo
美洲熔点 American melting point
美洲山毛榉 white beech
美洲杉 sequoia
美洲深红色木材 <用于镶板> virola
美洲湿热带地区 American humid tropics
美洲水蛭 American leech
美洲铁木 hop hornbeam;ironwood; lever-wood
美洲乌木 cocuswood;Jamaica ebony
美洲梧桐 buttonwood
美洲香槐(木)yellow wood
美洲橡胶 American rubber
美洲橡胶树 American rubber tree
美洲悬铃木 buttonball;buttonwood; Platanus American plane(tree)
美洲银行 Bank of America
美洲印第安人建筑 American-Indian architecture
美洲有孔虫地理区系 American foraminifera realm
美洲榆 American elm
美洲皂荚 honey locust
美洲朱砂 <鲜红颜料> American vermilion
美洲总承包商联合会 Associated General Constructors of America
美洲钻井方法 American system of drilling
美兹三重岩芯管 Mazier triple tube
美兹三管牵引器岩芯管 Mazier triple tube retractor core barrel

镁

镁白云石 konite
镁白云石砖 magnesite dolomite brick

镁板 magnesium sheet
镁冰晶石 weberite
镁尘糊块 goop
镁川石 jimthompsonite
镁带 magnesium ribbon
镁的 magnesic
镁的回收 magnesium recovery
镁的总浓度 total concentration of Magnesium
镁底子抹子 magnesium float
镁电池 magnesium cell
镁电偶 magnesium couple
镁电气石 dravite
镁锭 magnesium ingot
镁毒石 picropharmacolite
镁方沸石 pictandcime
镁方解石 magnesian calcite; magnesium calcite
镁肥 magnesium fertilizer
镁粉 magnesite powder; magnesium dust; magnesium powder; powdered magnesium
镁斧石 magnesioaxinite
镁钙片 magnesium oxide and calcium carbonate tablets
镁钙闪石 tschermakite
镁钙盐 tachyhydrite
镁橄榄石 fayalite; white olivine; forsterite
镁橄榄石白坯陶瓷器皿 forsterite whiteware
镁橄榄石大理岩 forsterite marble
镁橄榄石耐火材料 forsterite refractory
镁橄榄石耐火材料制品 forsterite refractory product
镁橄榄石耐火砖 forsterite-based brick; forsterite refractory brick
镁橄榄石(陶)瓷 forsterite ceramics
镁橄榄石透辉石大理岩 forsterite diopside marble
镁橄榄石砖 boltonite brick; forsterite brick
镁锆砖 magnesite-zirconia brick
镁铬合金 magnesia-chrome; magnesite chrome
镁铬合金砖 magnesite chrome brick
镁铬榴石 knorringite
镁铬耐火材料 magchrome refractory; magnesia-chrome refractory
镁铬铁矿 magnesian chromite; magnesiochromite; magnochromite
镁铬铁矿矿石 magnesiochromite ore
镁铬质耐火材料 magnesite chrome refractory
镁铬质砖 magnesite chrome brick
镁铬砖 magnesite chromite brick; magnesium-chrome brick
镁光 magnesium light
镁光灯 flash bulb; magnesium (light) lamp
镁光照明弹 magnesium flare
镁硅比 Mg-Si ratio
镁硅钙石 melilite; merwinite
镁硅灰石 magnesium wollastonite
镁硅铝合金 Montegal
镁硅铝榴石 pyrope
镁硅砂 magnesium silicate
镁硅酸盐 silicate of magnesium
镁硅铀矿 magursilite
镁含量分析仪 magnesium analyser [analyzer]
镁合金 electron alloy; elektron; magnesium alloy; Mazlo alloy
镁合金薄板 magnesium alloy sheet
镁合金阳极氧化法 elomag process
镁红钠闪石 magnophorite
镁红闪石 magnesiokatophorite
镁花雕刻 through-carved work
镁化铈 cerium magneside

镁还原的 magnesium-reduced
镁还原电弧炉 magnesium reduction arc furnace
镁黄长石 akermanite[oakermanite]
镁灰统【地】Zechstein
镁灰岩 dunstone; Zechstein <上二叠纪>
镁辉石 magauigite
镁混合物 magnesium compound
镁基的 magnesia based
镁基合金 magnesium-base alloy; Magnuminium (alloy); Magnox <核燃料的覆盖材料>
镁基润滑脂 magnesium-base grease
镁剂 magnesia mixture
镁钾钙矾 krugite
镁钾钙霞石 pharaonite
镁尖晶石 magnesiospinel
镁尖晶石耐火砖 magnesia-spinel brick
镁碱沸石 ferrierite
镁焦 impregnated coke; mag coke; magnesium coke
镁角闪石 magnesiohornblende
镁校正法地热温标 magnesium revised geothermometer
镁菌素 magnesedin
镁矿(石) magnesium ore
镁蓝铁矿 baricite
镁离子 magnesian ion; magnesium ion
镁离子交换容量 Mg-cation exchange capacity
镁锂闪石 magnesioholmquistite
镁磷锰石 talktriplite
镁磷石 newberyite
镁磷铀云母 salecite
镁鳞绿泥石 klementite
镁菱锰矿 kutnahorite
镁菱铁矿 sideroplesite
镁硫酸铜整流器 magnesium-copper sulphide rectifier
镁铝矾 pickeringite
镁铝合金 duralium; electron metal; magnesium-alumin(i)um alloy
镁铝合金电极 electron pole
镁铝榴石 pyrope
镁铝钠闪石 eckermannite
镁铝耐火材料 lamagal; magnesite-alumin(i)um refractory
镁铝蛇纹石 amesite
镁铝水滑石 magnesium-alumin(i)um hydrotalcite
镁铝铁矾 idrizite
镁铝(铜)合金 magnalium
镁铝直闪石 magnesiogedrite
镁铝砖 magnesium-alumina brick
镁绿钙闪石 magnesiohastingsite
镁绿帘石 picroepidote
镁绿泥石 amesite; pictoamesite
镁绿闪石 magnesiotaramite
镁氯氧水泥 magnesium oxychloride cement; oxychloric cement
镁蒙脱石 sobotkite
镁锰电池 magnesium-manganese battery
镁锰方解石 kutnahorite
镁锰干电池 Mg-Mn dried battery
镁锰橄榄石 picroknebelite; picrotephroite; talc-knebelite
镁锰合金 magnesium-manganese alloy
镁锰榴石 magnesia blythite
镁锰铁尖晶石 magnojacobsite
镁锰锌合金 magnesium-manganese-zinc alloy
镁明矾 magnesian alum; pickeringite
镁沫岩 aphrolite
镁钠闪石 magnesioriebeckite; rhodusite; torendrikite
镁钠闪石石棉 magnesiocrocidolite
镁铌铁矿 magnocolumbite

镁镍合金 magnesium-nickel alloy
镁镍华 nickel-cabrerite
镁诺克斯合金 Magnox
镁泡石 afrodite; aphrodite
镁硼石 kotoite
镁片 magnesium sheet
镁坡缕石 mountain wood; pilolite
镁青铜 <用于窗栏杆扶手等> magnesium bronze
镁溶液 magnesium solution
镁乳 magnesia magma; milk of magnesia
镁乳浆 magma magnesiae; magnesium hydroxide mixture
镁砂 magnesia; magnesite
镁砂川闪石 magnesiosadanaguite
镁砂粉砂轮 magnesite wheel
镁砂炉底 magnesite bottom
镁砂内衬 magnesite lining
镁闪石 magnesiocummingtonite
镁砷锌锰矿 magnesium-chlorophoenicite
镁十字石 zebedassite
镁石灰 dolomilic lime
镁石棉 magnesia-asbestos
镁石水泥 magnesium oxychloride cement; Sorel's cement
镁水绿矾 jaroschite; kirovite
镁钛矿 dauphinite; geikielite
镁钛铁矿 picrocrichtonite
镁碳质耐火材料 magnesite carbon refractory
镁碳质砖 magnesite carbon brick
镁条 magnesium rod
镁铁白云石 tharandite
镁铁比 Mg-Fe ratio
镁铁矾 magnesiocopiapite
镁铁橄榄石 hortonolite
镁铁铬矿 magnoferrichromite
镁铁硅质的 mafelsic
镁铁尖晶石 ceylonite; chrisotite; pleonaste
镁铁矿 magnesioferrite; magnoferrite
镁铁榴石 majorite; pyralmandite; rhodolite
镁铁钠闪石 bababudanite
镁铁青石棉 abriachanite
镁铁闪石 cummingtonite
镁铁闪石变粒岩 cummingtonite leptynite
镁铁闪石片岩 cummingtonite schist
镁铁钛矿 armalcolite
镁铁锌类晶石 magnoferrogahnite
镁铁氧体 magnesium ferrite
镁铁云母 svitalskite
镁铁珍珠云母 calciotalc
镁铁指数 mafic index
镁铁质 mafic
镁铁质矿物 mafic mineral
镁铁质石 mafic rock
镁铁质岩 mafic rock
镁铜合金 magnesium copper alloy
镁铜铝合金 elektron; magnalium
镁烷 magnane
镁硝石 nitromagnesite
镁屑 magnesium chips
镁锌合金 magnesium-zinc alloy
镁锌铁尖晶石 magnofranklinite
镁星叶石 magnesioastrophyllite
镁亚铁钠闪石 magnesioarfvedsonite
镁盐 magnesium salt
镁盐类 magnesium salts
镁盐侵蚀 magnesium salt attack
镁盐水解物 magnesium salt hydrolysis product
镁阳极 magnesium anode
镁阳极法 magnesium anode method
镁氧 magnesia
镁氧八面体 magnesium octahedron
镁氧半水硬石灰 magnesian semi-hydraulic lime
镁氧棒 magnesia rods
镁氧化水泥 magnesium oxide cement
镁氧混合剂 magnesia mixture
镁氧尖晶石 magnesia spinel
镁氧建筑板 magnesite (building) sheet
镁氧耐火材料 magnesia refractory; magnesite refractory
镁氧耐火制件 magnesite refractories
镁氧耐火制品 magnesite refractory; magnesite refractory product
镁氧耐火砖 magnesia refractory; magnesite refractory
镁氧石灰 dolomitic lime; magnesia lime
镁氧石灰膏 magnesian lime paste
镁氧石灰岩 magnesia limestone
镁氧石棉护管套 magnesia-asbestos pipe covering
镁氧水泥 <强度很高,常用作无缝楼面的黏[粘]合剂> magnesia cement; magnesite cement
镁氧水泥混凝土 magnesia cement concrete
镁氧水泥甲板敷料 litosilo; magnasil; magnesite
镁氧陶瓷 magnesia ceramics
镁氧脱硫法 magnesia based desulfurization[desulphurization]
镁氧牙膏 magnesia tooth paste
镁氧硬度 magnesia hardness
镁氧砖 magnesia brick
镁冶金 metallurgy of magnesium
镁叶绿矾 magnesiocopiapite
镁硬度 magnesium hardness
镁硬绿泥片岩 sismondinite
镁硬绿泥石 venasquite
镁硬石膏 wathlingite
镁铀矿 magurasphyllite
镁铀云母 saleeite
镁云母 magnesia mica; magnesium mica
镁云碳酸岩 beforsite
镁云碳酸岩类 beforsite group
镁皂 magnesium soap
镁针钠钙石 magnesiopectolite; walkerite
镁珍珠云母 magnesiomargarite
镁蒸气 magnesium vapo(u)r
镁直闪石 magnesioanthophyllite
镁质白垩 magnesian chalk
镁质白云石 magnesiodolomite
镁质白云石耐火材料 magnesite dolomite refractory
镁质半水石灰 magnesian semi-hydraulic lime
镁质超镁铁岩 magnesium ultramafic rock
镁质瓷器 magnesia porcelain
镁质大理石 dolomitic marble; magnesian marble
镁质大理岩-硅质岩建造 magnesiomarble-siliceous rock formation
镁质的 magnesian
镁质灰岩层 magnesian limestone
镁质混凝土 magnesian concrete
镁质角岩 magnesiohornfels
镁质耐火材料 magnesia refractory materials; magnesite refractory
镁质耐火混凝土 magnesite refractory concrete
镁质黏[粘]土 attaclay; magnesian clay
镁质生石灰 magnesian (quick) lime
镁质石灰 magnesium lime
镁质石灰粉 pulverised [pulverized] magnesian lime
镁质石灰腻子 magnesian lime putty
镁质石灰石 magnesium limestone
镁质石灰岩 magnesian limestone;

M

magnesium limestone
镁质石灰岩浆 magnesian lime paste
镁质石灰油灰 magnesian lime putty
镁质水泥 magnesia cement; magnesite cement; magnesium cement
镁质陶瓷 magnesia porcelain
镁质无缝地板材料 Magbestic
镁质无光釉 magnesian matt(glaze)
镁质矽卡岩 magnesiosharn
镁质系数 magnesian coefficient
镁质制品 magnesian product
镁种铀云母 novacekite
镁珠 magnesium globule
镁铸(造)合金 magnesium casting alloy
镁砖 magnesia brick; magnesite brick
镁砖地板 magnesite flooring
镁砖吊式炉顶 magnesite suspended arch
镁砖炉衬 magnesite brick lining
镁砖填料 magnesite fill
镁族 magnesium group

闷气 stifle

闷热 stuffiness; sultriness
闷热的 close; sultry
闷热的房间 stuffy room
闷热极限 sultry limit
闷热气候 sweltering hot weather
闷热天气 muggy weather; oppressive weather; sultry weather
闷头 baffle; bulkhead; bulkhead gate; choke plug; closure head; pipe cap
闷头法兰 blind flange
闷头印 < 玻璃制品缺陷 > baffle mark
闷头主管 dead-end main

门安装 door system

门凹壁 gate recess
门凹凸榫 door rabbet; door rebate
门拗 knob
门把配件 knob door fitting; knob door furniture
门把球 door knob
门把手 door handle; lever handle
门把手垫板 escutcheon plate
门把手零件 door handle fittings
门把手饰板 rose
门把锁 key-in knob(door) lock
门把五金配件 knob door hardware
门把小五金 door handle fittings
门坝 wicket dam
门百页 door louver
门板 door panel; door plank; shutter; door sheet < 集装箱的 >
门板衬里 door panel-lining
门板框 door frame
门板拉条 back head brace
门板系木 doorbrand
门鼻子 strike plate
门边半槽 door rabbet
门边框 al(l)ette; door cheek; door post; side jamb
门边框踢脚板 jamb footing
门边立木 buck; door stud; door buck
门边梁 square head
门边石 mo(u)ld stone
门边锁 rim lock
门边樘 door cheek; door jamb; door post
门边梃 door cheek; door post; stile
门边梃锁 rabbeted lock
门边镶条 edging strip
门边装饰 door finish
门边装修 door finish
门柄 door knob
门玻璃 door glass
门玻璃槽 door glass channel
门跋 door knocker
门部间流量均衡方程 equation of intersector equilibrium
门采光面 door light
门操纵机构 door operating gear
门槽 gate groove; gate guide; gate recess; gate slot
门槽顶部盖板 gate camber cover
门槽口 door rabbet; door rebate
门侧 < 外墙与门窗的间的 > reveal; door reveal
门侧壁 < 古腊梅建筑构造中支撑门楣的 > postis
门侧玻璃 side panel
门侧服务员 doorman
门侧铰链柱 heel post
门侧扇 wing of door
门侧梃 side jamb
门侧翼 wing of door
门侧柱 door jamb; framed ground
门侧装修 door casing
门插销 box bolt; door latch (with lock); extension bolt; extension flush bolt; fall bar; door bolt
门厂 door factory
门衬 door gasket; door lining
门撑 portal bracing
门触点 door contact; gate contact
门触开关 door switch
门窗安装 hang
门窗暗插销 mortise bolt
门窗凹进处 kernel
门窗帮 rebval
门窗边框 jamb; side jamb
门窗边框侧墙 jamb wall
门窗边框衬板 jamb lining
门窗边框石 jamb stone
门窗边框石柱 jamb stone pillar; jamb stone post
门窗边框筒子板 jamb lining
门窗边框线脚 architrave jamb; jamb mo(u)lding
门窗边框线条 jamb mo(u)lding
门窗边框厢 jamb casing
门窗边框圆角砌块 jamb block
门窗边框圆角砖 jamb brick
门窗边框柱 jamb post
门窗边石 mo(u)ld stone
门窗边梃 outside stile
门窗扁凿 pocket chisel
门窗玻璃块 pane of glass
门窗布局 fenestration
门窗布局设计 fenestration design
门窗彩花玻璃 vitrail
门窗侧板 jamb lining
门窗侧壁 door jamb; ingo; jamb; jamb wall; reveal
门窗侧壁槽 rebate
门窗侧壁护铁 jamb guard
门窗侧壁木面 rybat(e)
门窗侧壁石 jamb stone
门窗侧壁石面 rybat(e)
门窗侧壁砖面 rybat(e)
门窗侧边 ingo; jamb(wall); reveal
门窗侧墙 ingo; jamb(wall); reveal
门窗侧石柱 jamb stone
门窗侧柱 jamb post; jamb shafts
门窗侧柱混凝土垫块 jamb block
门窗插销 Canadian latch; Suffolk latch; thumb latch
门窗长插销 Cremo(r)ne bolt
门窗衬垫 lining of door casing
门窗撑挡 stay
门窗垂饰 lambrequin
门窗搭扣 hasp
门窗挡风塞条 weather strip
门窗挡风雨条 weather strip
门窗挡线脚 rabbeted stop
门窗的额楣 casing trim
门窗的匣形竖框 boxed mullion
门窗顶部双弯曲线装饰 accolade
门窗顶葱形线饰 accolade
门窗顶泛水 head guard
门窗顶上的立砌砖拱 soldier arch
门窗顶贴脸 cap trim
门窗顶线脚 cap mo(u)ld(ing); cap trim
门窗洞边砌石 quoining up jambs
门窗洞侧墙面 scuncheon
门窗洞测墙面的拱 scuncheon arch
门窗洞口 rebval
门窗洞口边框 discharging arch
门窗洞口珠饰 casing bead
门窗洞口装饰线脚 antepagment; antepagmentum
门窗额饰 head mo(u)ld(ing)
门窗防尘 door and window dustproofing
门窗防风 door and window draughtproofing
门窗工程 works for doors and windows
门窗钩 cabin hook; door and window hook
门窗钩眼 hook ring
门窗柜木砖 framed ground; frame grounds
门窗过梁 cap; lintel of doors and windows; summer; summer beam
门窗过梁层 lintel course
门窗过梁上三角形檐饰的顶尖 pediment apex
门窗横挡 plain rail; transom bar; transom
门窗横棂 stay bar
门窗及五金 door and window hardware; door and window ironmongery
门窗及五金表 schedule of doors and windows and ironmongery
门窗间墙 wall space
门窗检修间 doors and windows repair workshop
门窗铰接框 butt stile
门窗孔 aperture
门窗口 framed opening
门窗口侧面筒子板 reveal lining
门窗口侧墙面镶衬 reveal lining
门窗口串珠状线脚 sill bead
门窗口上面的线脚 head mo(u)ld(ing)
门窗口筒子板 jamb lining
门窗口凸圆 sill bead
门窗框 casing; frame for fittings; jamb
门窗框边框 break jamb; jamb guard
门窗框边框柱 post jamb
门窗框边挺 break jamb
门窗框槽沟 scuncheon
门窗框槽刨 sash fillister
门窗框侧壁 break jamb
门窗框侧柱 jamb shafts
门窗框顶梁 head sill
门窗框合角加固棘栓 sash pin
门窗框架 plank flume
门窗框角突肘 cropping; crosette
门窗框木砖 framed ground
门窗框内屋角石 scontion
门窗框塞口 rough opening
门窗框上槛檐板 architrave cornice; back cornice
门窗框饰 chambranle
门窗框竖框 monial
门窗框条 sash bar
门窗框突角 horns
门窗框外缘饰带 backbend
门窗框缘转角线条 reprise
门窗立砌砖过梁 soldier arch
门窗立柱 jamb shafts
门窗帘 door window curtain; pelmet board; portiere
门窗帘板 pelmet board
门窗帘棍 portiere rod
门窗帘圈 curtain ring
门窗帘铜棍 brass curtain rod
门窗帘匣 pelmet(box)
门窗梁上饰物台座 acroterion
门窗绿化 door and window greening
门窗冒头 plain rail
门窗密封条 weather strip
门窗模数 window module
门窗木框 plank frame
门窗木料 wood for fittings; wood for fixture
门窗内侧呈八字形斜壁 scuncheon; scuntion
门窗内侧呈八字形斜面 scoinson
门窗内镶边 inside trim
门窗耐火等级 fire door rating
门窗披水板 weather-board(ing); weather mo(u)lding
门窗披水线条 label course
门窗漆 trim paint
门窗-墙面积比 void-solid ratio
门窗扇材料 sash stuff
门窗扇开关侧边 leading edge
门窗扇中梃 mullion
门窗上壁拱 frontage; frontal; frontispiece
门窗上部的三角顶饰 frontal
门窗上部三角饰 frontage
门窗上人字形小墙 frontispiece
门窗上缘的筋形突出部 crosette
门窗枢轴 center[centre] hinge
门窗竖框 door window mullion; monial
门窗数量表 list of quantity of doors and windows
门窗闩 latch bolt
门窗锁安锁的一边 leading edge
门窗樘子 ingoing
门窗贴脸 exterior trim; trim
门窗贴脸板 stile edging board
门窗铁纱装置 fly screen
门窗梃上闭合槽 hook rebate
门窗头的出檐线脚 hood mo(u)lding; hook-and-band hinge
门窗头泛水 head guard
门窗头框 head casing
门窗头线 head mo(u)ld(ing)
门窗头线条板 epistyle; head casing; inside trim
门窗外边梃 outside stile
门窗外部侧面墙 exterior reveal
门窗外衬饰 outside casing
门窗外框 outside casing; outside facing
门窗外十字条 counter bar
门窗线脚 accolade
门窗线条 architrave jamb
门窗线条板 hyperthyrum
门窗线条板边框 architrave jamb
门窗线条板座块 architrave block
门窗相遇的边梃 vertical meeting rail
门窗压条 stop bead
门窗一览表 schedule of doors and windows; door and window schedule
门窗用玻璃 glass for glazing
门窗用镀铜吊链 copper-plated sash chain for door and window
门窗止动条 inside stop
门窗止条 rabbeted stop
门窗中槛 transom
门窗中梃 muntin bar
门窗周围线条板 architrave
门窗装饰性短帘 lambrequin
门窗装修木料 factory and shop lum-

ber;shop lumber
门搭扣 elbow catch
门搭扣插销 <货车> door pin
门带 door strap
门单元 gate cell
门弹弓 door closer-spring
门挡 door back stop;door bumper;door shield;door stop;gate stop;floor stop <地板上的>
门挡板 door shield
门挡板上部的楣饰 apron mo(u)lding
门挡块 <装在地板上的> floor knob
门挡镶板门 framed and panel(l)ed door
门挡衣钩 bumper hook
门导轨 door guide
门到场 <集装箱的> door to container yard;door to cy
门到码头 house to pier
门到门 door-to-door;house-to-house
门到门背负式运输业务 door-to-door piggyback service
门到门背驮式运输业务 door-to-door piggyback service
门到门服务 door-to-door service
门到门货物运程 <一般指集装箱货物运输> door-to-door journey
门到门集装箱运输 door-to-door container transport
门到门交货 door-to-door delivery
门到门捷运作业 door step service
门到门取送货物流通 door-to-door circulation of goods
门到门系统 <集装箱的> door-to-door system
门到门行程时间 door-to-door travel time
门到门运输 door-to-door delivery;door-to-door conveyance;door-to-door service;door-to-door traffic;door-to-door transportation
门到门(运输)系统 door-to-door system
门到门运输用托盘 <集装箱> pallet for door-to-door transportation
门到站 <集装箱> door to container freight station;door to cfs
门道 anteport;doorway;gate head;gateway;diathyrum <古希腊住宅的>;Torana <印度庙宇的>
门道防护柱 batter post
门道桥 <澳大利亚> Gateway Bridge
门道入口 door aperture
门道一边的立柱 batter post
门德斯页岩层 Mendez shale
门的安装 door installation
门的暗式开关 concealed closer
门的保持装置 door holding device
门的布局 door system
门的电动启闭器 electric(al) shutter door operator
门的封口条 door sealing fillet
门的附件 door accessories;door fitting <不包括门锁和铰链>
门的过梁 supercilium
门的厚玻璃 dead light
门的缓冲夹钩 door holder with cushion
门的间隙 <门与地板之间> door clearance
门的接触开关 door contact
门的金属配件 door hardware
门的精细加工 door joinery
门的开启装置 door gear;door mechanism
门的扣链 chain door fastener
门的木框架 durn
门的配件 door accessories;door equipment
门的剖面 door section
门的上横档 upper door rail
门的上冒头 upper door rail
门的锁梃 shutting stile
门的贴接面 gate meeting face
门的外观 door profile
门的无界延迟 unbounded gate delay
门的五金 door fitting
门的下横挡护板 kicking plate
门的镶板设计 door panel arrangement
门的镶板中设置的百叶 punched louver
门的形状 door shape
门的修整 door finish
门的占用时间 gate occupancy time
门的中部横框 center door rail
门的中横框 middle door rail
门的中间柱 intermediate gate post
门的装配 door assembly
门的装修 door joinery
门的自动开闭装置 door operator
门的组成件 door profile
门的左右摆向 door hand;door swing
门的左右手向 hand of door
门灯 entrance lamp;gate light
门磴座 architrave block;plinth block;skirting block
门底闭门器连杆 bottom arm
门底边 bottom rail of door
门底插销 bottom bolt
门底挡雨线脚板 door strip
门底横木 <用以排除雨水和穿堂风> Adam's water bar
门底脚螺栓 foot bolt
门底净空 door clearance
门底闩 bottom bolt
门底弹簧 vertical spring-pivot hinge
门底缘材 <集装箱> door edge member
门底自动防风隔声设施 automatic threshold closer
门底自动防风隔音设施 automatic threshold closer
门第 pedigree
门点 doors
门电极 gate electrode
门电流 gate current
门电路 gate circuit;gating circuit
门电路触发二极管 gate trigger diode
门电路电子管 threshold tube
门电路断开 gate turnoff
门电路转换开关 strobe
门电子管 gate tube
门垫板 door gasket
门垫里 door panel-lining
门垫圈 door gasket
门垫塘 mat well
门吊 gantry crane;gantry type jib crane;portal jib crane;gallows frame <架桥用>;portal luffing crane;portal crane
门吊架 gantry frame
门钉 decorative nails on door leaf
门顶插销座 soffit bracket
门顶窗 transom(e);fairlight <英>
门顶拱 door arch;door niche arch
门顶关门器座 soffit bracket
门顶肩托过梁 shouldered architrave
门顶可锁室 lockable compartment
门顶框 head jamb
门顶装饰 overdoor
门定位器 door holder
门动电开关 door release
门动开关 door switch
门动力控制机构 door engine
门洞 door opening;aperture of door;door aperture
门洞尺寸 dimensions of door opening;door opening size
门洞口 door opening
门洞内的门 recessed portal
门斗 air lock;anteroom;delivery gate;door hood;storm porch
门段 gate segment
门对门运输 door-to-door service;door-to-door transportation;gate-to-gate transportation;house-to-house service;house-to-house transportation
门墩 gate block;gate step <支承铁门门枢的铁制品>
门墩柱 door pier;gate pier
门多拉白云石 <晚寒武世> Mendola dolomite
门多西诺断裂带 Mendocino fracture zone
门额 head casing
门耳【建】crossette
门阀 doorway
门反相器 gate inverter
门防火等级 door class
门房 commissionaire's room;doorkeeper's house;doorkeeper's lodge;doorkeeper's room;doorman;gate house;gatekeeper's house;gatekeeper's lodge;gate lodge;janitor's room;lodge;porter;porter's lodge
门放大器 gate amplifier
门扉 leaf
门封 door dike[dyke];door seal(ing);gate seal
门缝 appentice;door gap
门缝渗入量 door infiltration
门格尔定理 Menger's theorem
门格条 munting
门格栅 door grill(e)
门拱 arch;gate arch;gateway arch
门钩 door hanger;door holder;door hook;door stop;gate hook;gate stop;snib
门挂板 door strap
门挂钩 gudgeon
门管区 portal area
门管小叶 portal lobule
门规格表 door schedule
门轨 gate rail;outdoor rail
门滚轮 door roller
门滚轴 door roller
门过梁 brow piece;door lintel;front lintel
门合页 door hinge;gate hinge
门横档【建】door rail
门横框 door rail
门横木 door rail
门横向位置 gate lateral position
门后搭钩 back catch;back hook;catch
门后钩 back catch;gate back catch;silent door hook
门后挂钩 back catch;back hook;catch
门后夹 door catch
门后扣钩 back catch;back hook;catch
门户 portal
门户阶【建】stoop
门户节点【交】gateway
门户开放 open door
门户开放政策 open door policy
门护脚板 door protecting plate
门环 door knocker;knocker;rapper;ring door knocker;ring knocker
门缓闭器 door closer
门簧铰链 floor spring
门或窗周围线条板 architrave
门机 gate hoist
门机机构 door actuator
门机试验台 car door mechanism testing stand
门基板 plinth board
门基石 base block
门级逻辑模拟 gate level logic simulation
门级模拟 gate level simulation
门极可断开关 gate turnoff
门夹具 door jack
门架 door frame;gantry;mast;portal(frame);upright mounting <叉车>;gantry mounting <门座起重机的>
门架安装座 mast mount
门架大梁 portal girder
门架吊车 gantry travel(l)er
门架吊机 gantry crane
门架法 portal method
门架横撑 portal strut
门架后倾角 <叉车> backward inclination angle for upright mounting;backward tilting angle for upright mounting
门架建筑 portal frame block;portal frame building
门架结构 portal framed structure
门架净高度 height of portal clearance
门架净空 gantry clearance;leg clearance
门架跨距 portal span
门架框梁 portal frame
门架联结系 portal bracing
门架起重机 trestle crane
门架前倾角 <叉车> forward inclination angle for upright mounting;forward tilting angle for upright mounting
门架桥 portal bridge
门架桥轴 gantry axle
门架倾斜角 tilting angle of upright
门架伸长时高度 extended height
门架式叉车 mast lift truck
门架式工作台 portal frame type platform
门架式结构 portal framed structure;portal-frame stroke
门架式解法 <计算柱剪力分布的> portal method
门架式起重机 portal jib crane
门架收缩后高度 closed height
门架支撑(系)portal bracing
门架支柱 gantry column
门架柱 portal leg
门架组合构件 portal frame compound unit
门架作用 portal effect
门尖券尖 gentese
门检波器 gated detector
门槛 baffle sill;carpet strip;dern;door abutment piece;door cill;door saddle;door sill;door threshold;gate seat;gate sill;saddle-back board;sill;threshold plate
门槛板 sill plate
门槛处石踏步 door stone
门槛地脚螺栓 sill anchor
门槛电位 threshold level
门槛高(度)sill high
门槛混凝土砌块 sill block
门槛价格 threshold price
门槛检测法 threshold detection method
门槛密封条 threshold seal(er)
门槛砌块 sill block
门槛嵌板 saddle-back board
门槛嵌缝条 door strip;draught excluder;weather strip
门槛石 door stone;stone sill of door
门槛水深 lock significant depth;water depth above sill
门槛应力强度因子 threshold stress

intensity factor
门槛值 threshold value
门槛铸件 sill casting
门脚护板 kick(ing) plate;mop plate
门脚护条 kick strip
门铰链 door butt; door hinge; gate hinge;hingle
门铰链收进 hinge backset
门铰销 door hinge pin
门阶 door step;door stone;stoop
门接线端 gate terminal
门捷列夫周期表 Mendeleyev's chart; Mendeleyev's periodic table
门捷列夫周期律 Mendeleyev's law
门捷列夫周期系 Mendeleyev's periodic system
门金属配件 door metal furniture
门警 warder
门净空 door opening
门径 avenue
门静噪器 door silencer
门静噪声 door noise
门镜 door mirror
门鸠尾槽 door control equipment
门鸠尾榫 dovetail female
门臼石 hinge stone
门矩阵 gate matrix
门具 door gear;door mechanism
门橛石 stop stone
门开放信号 door release
门开关 door contact; door trigger; door trip;gate switch
门开关联锁装置 door interlock;gate interlock
门开启宽度 opening width of door
门龛 gate recess
门龛拱 door niche arch
门坎 door bank;door sill;door threshold
门坎铁板 door threshold plate; threshold plate
门壳 door shell
门空隙 door clearance
门孔 aperture of door
门孔全开度 full gate opening
门控 gate
门控触发器 gated flip-flop
门控放大器 valve control amplifier
门控缓冲器 gated buffer
门控开关 door contact; gate-controlled switch;gate turnoff switch; door switch
门控同核去耦法 door control homonuclear decoupling method
门控噪声测量 gated noise measurement
门控制块 gate control block
门控制屏 door control panel
门口 aperture of door;door aperture; door stead;doorway;ostium;porch; threshold;dar <印度和波斯建筑中的>
门口擦脚条 door mat
门口擦鞋垫 welcome mat
门口宽度 clear door opening; width of entrance
门口铺石 door stone
门口上方的装饰墙面 overdoor
门口踏板 door saddle
门口效应 <视觉上路幅收缩效应> gate effect
门口毡 door mat
门口照明 egress lighting
门扣 button catch;detent;door catch; door clamp;door holder;door latch (with lock);kep
门宽 gatewidth
门宽调整 gatewidth control
门框 dern; door case; door frame; door framing;gateway;port frame
门框边框 door jamb
门框边挺 adjacent plank
门框边柱 door post
门框槽沟 scuncheon
门框槽口 giblet check
门框侧板 door check
门框衬垫 frame gasket;lining of door casing;lining of door frame
门框尺寸 door opening;opening size
门框的方竖框 square door jamb
门框的固定 door frame fixing
门框的上部分 door head
门框底锚 base anchor;base clip
门框垫块 floor stilt
门框顶部横梁 headsill
门框顶梁 door case top rail
门框方竖框 square door jamb
门框防风设施 threshold draft-proofer
门框防风条 threshold draft-proofer
门框封条 door draught excluder
门框回转式起重机 gantry slewing crane
门框架 door frame (work); gate frame(work)
门框架建筑结构 gateway
门框架接缝 door frame junction
门框间隙 frame clearance
门框锚固件 door frame anchor
门框企口 door stop
门框清扫机 jamb cleaner
门框上槛 door head
门框上冒头 head rail
门框上球形把手 sash knob
门框式电热器 door frame heater
门框式起重机 portainer
门框饰 door casing;door trim
门框竖框 monial
门框条 door stile
门框贴脸 door trim
门框梃 adjacent plank;hinge jamb
门框线条板 hyperthyrum
门框楹柱线条 epistylium
门框周围衬里 paneled lining
门框周围衬砌 paneled lining
门框做成的(门)槛 cut-off stop
门窥视镜 door viewer
门窥视孔 door viewer
门廊 <有圆柱的> portico
门拉手 catch;door pull;door (pull) handle;sneck head
门拉手金属件 door knob hardware
门拉手零件 door knob fittings
门拉手小五金 door knob hardware
门拉闩锁 gated latch
门栏杆 impages
门栏密封绝缘材料 sill-sealer insulation
门廊 antium; foyer; lobby (area); narthex;porch;porticus;prostoon; stoop;galilee <哥特式教堂西端的>;lanai <上有棚盖的>;pronaos <寺庙的>
门廊壁柱之间 prostasis
门廊椽 porch rafter
门廊格构 porch lattice
门廊格栅 porch lattice
门廊栏杆 porch rail
门廊瓶饰 <早期教堂建筑的> cantharus
门廊漆 porch paint
门廊入口 entrance lobby
门廊柱 porch column
门类 ramification
门连窗 door with window
门帘 door curtain;door grill(e);door screen;portiere;spere;amphithura <希腊教堂圣壁入口的>
门帘杆 door-curtain rod;portiere rod
门联锁触点装置 <升降机的> door unit contact system
门联锁(装置) gate interlock
门链 door chain
门链栓 door chain bolt
门铃 door bell;door chime;jingle bell
门铃拉索 bell pull
门铃系统 door-bell and button system
门铃信号 door warning signal
门棂 door mullion;parting rail
门楼 gate tower
门路 access;approach;business connection;knack
门路连接器 gateway
门轮滑轨 troll(e)y track
门罗阶【地】Monroan(stage)
门罗统 <美国晚志留世>【地】Monroan series
门脉冲 gate impulse;gate pulse;gating impulse;gating pulse
门脉冲发生器 gate pulse generator; gating pulse generator;pulse generator;strobing pulse generator
门脉冲放大逆变器 gate inverter
门脉冲宽度 gatewidth
门脉系统 portal system
门脉循环 portal circulation
门帽 head jamb
门楣【建】door head; door cornice; door lintel; head still; lintel of a door
门楣饰 lintel mo(u)lding
门楣中心 tympan;tympanum[复 tympana/tympanums]
门门送货 <自发货人仓库门至收货人仓库门> store door delivery
门密封嵌条 door sealing fillet
门面 shop front; storefront; the facade of a shop
门面测距仪 ranger
门面长度 <商店> sign frontage
门面修饰 cosmetic improvement
门面招牌 facia sign
门幕 <希腊教堂圣壁入口的> amphithura
门内侧壁衬砌 reveal lining
门内侧壁销子 reveal pin
门内折叠床 in-a-door bed
门囊 <活动门的> door pocket
门尼单位 Mooney unit
门尼剪切圆盘黏[粘]度计 Mooney shearing-disk viscometer
门尼菱形镜 Mooney rhomb
门尼黏[粘]度 Mooney viscosity
门尼黏[粘]度计 Mooney visco(si)-meter
门钮形变换 door knob transition
门钮形转变 door knob transition
门钮形转换器 door knob transformer
门钮指示器 indicator button
门牌 door plate
门牌号 house number
门牌号数 street number
门跑车 coupe
门配件 door fitting;door unit
门碰 base knob;gate stop
门碰挡 door bumper
门碰联动开关 car door contact
门碰球 ball catch;bullet catch
门碰锁 door latch(with lock);latch
门碰头 door stop; floor knob; floor stop;surface(-mounted) astragal <双扇扯门前沿边安装的>
门票 admission ticket;entrance ticket;slip <集装箱>
门屏 door screen
门前(擦鞋)棕垫 door mat
门前阶梯 antium
门前露台 <美> stoop
门前铺石 door stone
门前热风幕 air curtain doorway
门前石阶 door step;step stone
门前踏步 door step
门前庭院 door yard
门前之雨篷 appentice
门嵌缝密封 door seal(ing)
门桥 portal frame
门球 ball knob;door knob
门驱动机构 door actuation mechanism
门驱动器 gate driver
门塞法 <测量岩石内应力的>【岩】doorstopper method
门塞式岩体应力测量设备 doorstopper rock stress measuring equipment
门纱 door screen
门扇 door leaf;door wing;gate leaf; leaf
门扇边梃 door stile
门扇防风刷 door sweep;sweep strip
门扇盖缝条 astragal;overlapping astragal;wrap-around astragal
门扇护条 overlapping astragal;wraparound astragal
门扇开关方向 handing of door
门扇亮子 observation panel
门扇冒头 door rail
门扇上冒头 door top rail;top rail of door
门扇上下冒头 rails
门扇枢轴端 quoin end
门扇下冒头 door bottom rail
门扇斜接端 miter[mitre] end
门扇中冒头 door middle block;door middle rail
门扇中梃 door muntin;muntin
门上安全镜 door safety mirror
门上把手护板 finger plate
门上把手锁眼处防指污的板 finger plate
门上部的 overdoor
门上部横框 top door rail
门上窗 door window; upper door sash
门上的扇形窗 fanlight
门上滴水 weather seal channel
门上附件 door accessories;semprax
门上槛 yoke
门上金属扣件 backplate
门上框 head jamb
门上球形捏手 door knob
门上热风幕 warm-air door curtain
门上输水 gate-opening conveyance
门上弹簧锁 door spring lock
门上梃 door head
门上推手板 hand plate
门上镶边木条带 edging strip
门上镶玻璃面积 door light
门上橡胶防碰垫 rubber door stopper
门上摇头窗 sectional overhead door
门上液压关门器 hydraulic overload door gear
门上雨篷 door hood
门上雨篷的支托 dorsal
门上雨罩 dorsal;dorse
门上注意牌 door notice plate
门上装饰 door finishing
门上装饰(小)五金 finish hardware for gate
门上装锁板 lock block
门舌 gate flap
门渗漏量 door leakage
门市 retail sale; sell retail over the counter
门市部 retail department;retail sales department;sale retail sales department;sales department;salesroom; selling retail department
门市贷款 street loans
门市发票 cash sale invoice;cash ticket

门市商品 goods sold over the counter
门式 portal type
门式安全装置 gate closure type safety device;gate guard
门式搬运起重机 gantry transfer crane
门式标志 overhead sign;portal framed sign
门式窗 casement window
门式窗插销 midloc
门式吊车 gantry crane;portal crane
门式吊机 gantry crane;portal crane
门式吊架 gallows frame;portal frame
门式墩 portal frame pier; transom pier
门式刚架 portal frame
门式刚架撑杆 portal strut
门式刚架结构 portal frame(d) structure
门式刚架支撑 portal strut
门式钢架 portal frame
门式拱 portal arch
门式构架撑杆 portal strut
门式刮料机 portal scraper
门式机器人 portal robot
门式剪板机 plate squaring shears
门式交通标志 overhead traffic sign
门式结构 portal structure
门式进料 gate feed
门式进料系统 gantry loading system
门式框架 portal frame
门式框架建筑 portal frame building
门式框架结构 portal frame construction;portal frame(d) structure
门式框架跨度 portal frame span
门式框架梁 portal frame beam
门式框架塔 <斜拉桥的> portal frame tower
门式缆索起重机 portal cable crane
门式炉 wicket type heater
门式螺旋卸煤机 gantry spiral coal unloader
门式平面磨床 planer-type surface grinder;plano-type grinder
门式起重机 gallows frame;gauntree;ga(u)ntry; portal bridge crane; portal crane;trestle crane
门式起重机吊架 portal crane bucket
门式起重架 gallows frame derrick
门式桥墩 portal pier;transom pier
门式桥塔 portal-type tower;transom pylon
门式取料机 portal reclaimer
门式升降机 gantry hoist
门式塔(架) portal-type pylon
门式铁塔 gantry tower
门式消防栓 gate type hydrant
门式卸卷机 gallows
门式行车 crane portal
门式行动吊车 gantry travel(l)er
门式悬臂起重机 portal jib crane
门式旋臂吊机 portal jib crane
门式旋臂起重机 portal jib crane
门式移动吊车 gantry travel(l)er
门式移动起重机 portal travel(l)ing crane
门式移动悬臂起重机 travel(l)ing portal jib crane
门式抓斗起重机 grabbing goliath
门式装船机 loading gantry
门手把 door handle
门守卫室 doorkeeper's house; doorkeeper's lodge; doorkeeper's room
门枢 gate pivot;hinge
门枢轴 door pivot
门枢轴砖 door pivot brick
门闩 bolt; crossing bar; gate bar; keeper;latch bar;locking bar;port bar;stop;ring latch <环形执手启闭的>
门闩插孔 latch jack
门闩扣 sneck
门闩拉线 latchstring
门闩圈 lock-strip gasket
门闩锁 door latch(with lock);latch
门闩铁 door bar;door bolt
门闩线路 latches;latches circuit
门闩座 door bolt bracket; door bolt keeper;shutting shoe
门栓 crossbar; door bar; door bolt; doorbrand;door latch(with lock); door leaf; keeper; locking bar; dog back; manhole dog; panic bar; strong back
门栓大螺丝 latch screw
门栓机构 door bolt mechanism
门栓控制钮把手 turning piece
门栓锁 door latch(with lock)
门栓弹簧 handle latch spring
门栓支条 dog stay
门栓指示器 <室内有无占用> indicating bolt
门栓座 shutting shoe
门锁 door lock;gate lock
门锁按钮 thumb piece
门锁按钮把手 thumb knob
门锁把手 door knob;door lock handle
门锁板 escutcheon;escutcheon plate
门锁边梃 locking stile
门锁槽 box staple;nab
门锁插销 box bolt
门锁弹簧 door latch spring; door spring lock
门锁垫盘 faced disk
门锁定位卡子 lock clip
门锁防松装置 check lock
门锁杆 door lock rod
门锁杆凸轮 lock door cam
门锁横档 lock rail
门锁环 door hook
门锁金属垫板 lock reinforcement
门锁开关 door interlock switch
门锁孔盖 finger plate
门锁控制钮把手 turn knob
门锁扣座 door lock keeper;door lock nosing
门锁螺栓防松器 check lock
门锁母 door lock keeper; door lock nosing
门锁汽缸 door lock cylinder
门锁舌片 boss strike plate;box strike plate;striking plate
门锁闩 door latch(with lock); door lock bolt
门锁闩扣座 door latch keeper
门锁闩轴 door latch spindle
门锁梃 lock stile
门锁托架 door lock rod bracket
门锁眼挡 door latch rosa
门锁装置 door locking device
门锁撞针 door lock striker
门塔 access tower
门塔式起重台架 derrick tower gantry
门弹簧 gate spring
门弹簧锁 night latch
门毯坑 mat sink
门樘 door buck
门樘边柱 door jamb
门型框架 portal frame
门樘上槛 head jamb
门樘筒子板 outer reveal
门樘外框 door buck
门樘(子) door frame
门樘子边梃 door jamb
门套 door pocket
门特罗格阶【地】Maentwrogian
门踢板 kick plate
门提升装置 gate lifting device
门贴脸 architrave; door dressing; door profile;trim
门贴脸座 architrave block;base block; plinth block;shirting block
门厅 antehall;coulisse;entrance hall; foyer;hallway;lobby(area);porch chamber;prodomus;prostas;reception hall;vestibule;hall way <美>; zaguan <西班牙建筑的>
门厅壁橱 hall closet
门厅的 vestibular
门厅镜 hall mirror
门厅、楼梯共用的公寓 apartment house of direct access
门厅门 vestibule door
门厅墙壁 foyer wall
门厅上空 upper part of lobby
门厅卧室 hall bedroom
门厅银行业务 lobby banking
门庭 <上有棚盖的> lanai
门庭若市 roaring business
门梃 door jamb;door stile;montant; portal rigid frame; principal post; stile;locking stile <装锁用>
门梃插石 heel stone
门梃钩 ajar hook
门梃臼石 heel stone
门梃石 heel quoin
门梃锁 jamb lock;rabbeted lock
门梃斜边 door bevel
门通 gate on;gate open(ing)
门筒子板 door lining
门头板 door cap;door header
门头合页 pivot hinge
门头花饰 overdoor;sopraporta
门头梁 door head
门头饰板 overdoor
门头饰画 overdoor
门头线 adjacent plank; architrave; door trim;trim
门头线墩子 architrave block
门头线上突出的檐口 hypertherium
门头线饰 coronet
门头线条板 head casing
门头线条饰 architrave of a door
门头镶板 overdoor
门头装饰 sopraporta
门推板 finger plate;hand plate;push plate;stile plate
门推手 push bar;push hardware
门腿【机】lower portal leg
门外汉 laity;layman;outsider
门外假板 <船上挡风暴的> dead door
门外摄影 outdoor camera
门外手柄 outdoor handle
门外锁 stock lock
门外遮板 dead door
门外装置的锁 stock lock
门卫 caretaker; door keeper; gatekeeper
门卫电话设备 porter system
门卫公寓 porter's apartment(unit); porter's flat
门卫管理员 gate clerk
门卫居住单元 porter unit
门卫室 gatekeeper's office; guard house;janitor's room;porter's room
门卫种植 guard planting
门卫住房 porter's dwelling; porter's lodge
门卫住所 porter's house
门五金 door ironmongery
门系板 door strap
门下导轨 door guide
门下横框 door bottom rail
门下间隙 floor clearance
门下输水 under-gate conveyance
门下通风篦子 door grill(e)
门线图 gate diagram
门限 threshold
门限电流 threshold current
门限电位器 threshold potentiometer
门限电压 threshold voltage
门限改善 improvement threshold
门限检波器 threshold detector
门限检测 threshold detection
门限结构 threshold structure
门限解码器 threshold decoder
门限扩展解调器 extended threshold demodulator; threshold extension demodulator
门限扩展微波 threshold extension microwave
门限能量 cut-off energy; threshold energy
门限频率 threshold frequency
门限信号 threshold signal
门限信号电平 threshold level
门限译码 threshold decoding
门限元素 threshold element
门限值 threshold;threshold quantity
门限制约 threshold constraint
门限钻压 threshold weight
门销 base anchor;base clip
门小叶 portal lobule
门楔 gate wedge
门芯 core stock
门芯板 replum;door panel;lay panel; panel((l)ing)
门芯板护圈 door panel retainer
门信号 gate signal
门信号多谐振荡器 gate multivibrator;gate-producing multivibrator
门信号放大器 gated amplifier
门信号宽度 gate length
门形 door shape;portal type
门形吊架 gallows frame
门形刚架 pin-ended portal frame; portal rigid frame
门形构架 gantry
门形夹铁 cramp iron
门形架 portal
门形排架 portal bent
门形起重桅 goal post
门形索塔 portal framed tower
门形铁塔 gantry tower; portal steel tower
门形支撑 dead shoring
门碹 door arch
门牙 cutter
门檐 door eaves
门眼镜框 frieze panel
门堰 gate weir
门叶 gate flap;gate leaf
门叶开度 gatage
门叶开度指示器 gatage indicator
门一览表 door schedule
门引线比 ratio of gate-to-pin
门引用 gate reference
门用零件 door fixture
门用弹簧 door spring
门用五金 door hardware;pull hardware
门用小五金(件) door furniture
门与窗的建筑 building joinery
门与门框成套单元 door and frame packaged unit
门与门框相碰的一面 narrow side
门元件 gate element;gating element
门缘饰 architrave of a door
门钥匙 door key
门簪 decorative cylinder;door nail
门扎头 door stopper;elbow catch
门闸 entrance lock
门折页 door butt;door hinge
门针比 ratio of gate-to-pin
门诊 inquiry
门诊病人 outpatient
门诊部 clinic; dispensary; outpatient department;policlinic

门诊所 ambulatorium; clinic; outpatient clinic
门枕 door bearing
门枕石 bearing stone
门阵列 gate array
门整直器 door straightener
门执手 grip handle
门执手护板 handle-plate
门执手五金配件 knob hardware
门执手与盖板 plate and knot
门止(挡)door holder
门止挡卡簧 door holder catch
门止挡卡架 door holder bracket
门止器 door fixer
门至港 house to pier
门至门 house-to-house
门制 door check; door stay; door stop
门制动器 door bumper; door check; door stop; door closer
门制止槽 door rabbet
门制止器 door bumper; door check; door holder; door stop
门中档装饰 apron mo(u)lding
门中间石柱 trumeau
门中立樘 door mullion
门中帽板装饰线脚 apron mo(u)lding
门中梃 door mullion; middle stile; muntin; munting
门中轴线 door axis
门周护条 door strip
门轴 door spindle; gudgeon
门轴调速器 gate shaft governor
门轴线 door centerline; door centre line; enfilade
门轴支座 gate trunnion
门轴柱 heel post <船坞、闸门的>; gate gudgeon; quoin post; stile of door; stile of gate <边梃的>
门轴柱支座铰 quoin hinge
门轴柱支座砌石 quoin stone
门柱【建】door pillar; door post; door cheek; door tree; gate post; gate strut; goal post; portal column
门柱铰链 harr
门柱铰链钩片 gate hook
门柱铁活装饰物 overthrow
门爪 bar claw
门砖 tweel blocks
门转轴 door pivot
门桩 goal post
门装置 door gear; door mechanism
门子板 shuttering
门阻塞器(方)法 <量测岩石应力的> doorstopper method
门组合件 doorset
门座 gate seat
门座回转式起重机 portal slewing crane
门座净空 portal clearance
门座取料机 portal reclaimer
门座石座 heel stone
门座式起重机 gantry crane; portal (slewing) crane
门座式悬臂起重机 high pedestal jib crane
门座塔式起重机 portal tower crane
门座旋臂起重机 portal jib crane
门座抓斗卸船机 portal grab ship unloader

闷

闷顶楼板 camp ceiling
闷扶梯基 close string housed string
闷盖 pipe cap
闷罐车 box car
闷罐退火 closed annealing; pot annealing
闷光玻璃 devitrified glass; obscured glass
闷光玻璃窗 obscuring window
闷光的 frosted
闷光灯泡 frosted lamp
闷坏 suffocate
闷火 smoldering fire; smo(u)lder
闷火处理 matted finish
闷料 <利用某些工业废料筑路时的备料措施> enclosing material
闷炉 banking; closed fusing
闷炮眼 spent shot; standing bobby
闷墙 blank wall; blind wall; dead wall
闷烧 smo(u)lder(ing)
闷湿的阴雨天 wet spell
闷室 dumb chamber
闷死 suffocation
闷熄 smother; suffocating; suffocate
闷心短管 blind nipple
闷压感 sense of oppression
闷眼 blind hole
闷住 smother
闷住声音 muffle

萌

萌蔽地 ground coverage
萌出 eruption
萌出期 eruption period
萌出前期 preeruption
萌地槽 embryogeosyncline
萌地槽阶段 embryogeosyncline stage
萌地台 embryoplatform
萌地台阶段 embryoplatform stage
萌发 germination
萌生林 virgulata
萌芽 bud; sprout
萌芽的 embryonic
萌芽断层 incipient fault
萌芽后 post emergence
萌芽阶段 embryonic stage
萌芽节理 incipient joint
萌芽林 sprout land
萌芽期 embryonic stage; stages in germination
萌芽体 germinate
萌芽褶皱 incipient fold
萌芽状态 coming thing

盟

盟成员(局) administration member of a union
盟员局 administration member of a union
盟约 covenant

濛

濛气差 astronomic(al) refraction

朦

朦胧状态 shadowiness

猛

猛插动作 thrust action
猛冲 onrush; storming; wallop
猛冲海岸 <指波浪> churn of wave
猛度 brisance; explosive brisance; violence
猛度测定 explosive grading
猛沸金属 wild metal
猛击 batter; ding; slam(ming); bang
猛降 pelt
猛砍 slash
猛拉 hitch; yank
猛浪 <海面浪高 12 ~ 20 英尺 1 尺 = 0.3048 米>【气】high sea
猛力传动 hard driving
猛力驾驶 hard driving
猛烈爆发 belch; burst
猛烈爆炸 heavy shot
猛烈的火力 large volume of fire
猛烈东北风 <地中海中西部及欧洲沿岸的> gregale
猛烈沸腾 wild; wildness
猛烈风暴 severe storm; violent storm
猛烈风暴警报 severe storm warning
猛烈进入缓行器 heavy buffing
猛烈雷暴 rattler; severe thunderstorm; tatter
猛烈喷出 blow wild
猛烈热带风暴 severe tropic(al) storm
猛烈射孔 heavy shot
猛烈水位 fast-falling level
猛犸树 mammoth tree
猛犸型破碎机 Mammut crusher
猛烧 blaze
猛释气体 flush liberation
猛推动作 thrust action
猛推急拉的器具 jerker
猛性氯酸盐炸药 albit
猛炸药 high explosive
猛涨 sharp rise; shot to; soaring; spate
猛涨起来的物价 boom prices
猛涨水位 fast-rising level
猛涨物价 skyrocket
猛掷 hurl
猛撞 cannon; collide; smash

蒙

蒙……特许 by courtesy of
蒙巴萨港 <肯尼亚> Mombasa Harbo(u)r; Port Mombasa
蒙巴萨鞘刃脂 Mombasa gum
蒙版 masking-out
蒙布 fabric
蒙布漆 dope
蒙布收缩性 dope textile shrink property
蒙布涂刷性 dope brushability
蒙次黄铜 Muntz
蒙大拿 <美国州名> Montana
蒙丹蜡 lignite wax; montan wax
蒙导法 mentor method
蒙导植株 mentor plant
蒙得维的亚 <乌拉圭首都> Montevideo
蒙德法 Mond process
蒙德煤气 Mond gas
蒙蒂沉箱 Montee caisson
蒙蒂盖尔合金 Montegal alloy
蒙顶板 ceiling trim
蒙盖 muffle
蒙盖结构 open construction
蒙哥反向极性亚带 Mungo reversed polarity subzone
蒙古包 ger; Mongolian yurt; yurt(a)
蒙古包式土房 yurt-type earth house
蒙古弧形构造带 Mongolia arc
蒙古栎 Mongolian oak
蒙古人民共和国 the People's Republic of Mongolia
蒙古式建筑 Mongol architecture
蒙绘 mask artwork; tracing
蒙里克夫公式 <用于设计低碳钢柱子> Moncrief's formula
蒙磷钙铵石 mundrabillaite
蒙罗维亚 <利比里亚首都> Monrovia
蒙昧 blindness; obscuration
蒙钠长石 monalbite
蒙乃尔白铜 Monel metal
蒙乃尔防雨板 Monel flashing piece
蒙乃尔高强度耐蚀合金 Monalmetal; Monel
蒙乃尔高强度耐蚀镍铜合金 Monel metal
蒙乃尔合金 Monel
蒙乃尔合金衬里 Monel-lined
蒙乃尔铜镍合金 Monel metal
蒙乃尔泻水板 Monel flashing piece
蒙尼马克斯软磁合金 Monimax
蒙皮 coating; cover(ing); envelope; skin; thin ga(u)ge skin
蒙皮材料 skin material; upholstery
蒙皮车体 stressed-skin body
蒙皮铆钉 skin rivet
蒙皮曲面 skinning surface
蒙皮热辐射探测 heated skin detection
蒙皮(受力式)结构 stressed-skin construction
蒙皮温度 skin temperature
蒙片 frisket; mask; masking film; masking sheet
蒙片图 mask artwork
蒙奇定理 Monge's theorem
蒙气差 atmospheric refraction
蒙气差改正 refraction correction
蒙瑞屋面瓦 Monray roof tile
蒙桑托共振弹性计 Monsanto resonance elastometer
蒙上水气的玻璃 weathered glass
蒙湿气 humidity blushing
蒙受损失 incur loss; sustain loss
蒙受噪声限度 noise exposure limit
蒙塔纳统 <美国晚白垩世>【地】Montanian series
蒙太奇 montage
蒙陶克大楼 Montauk building
蒙套 jacketing
蒙特公司系列数控测井仪 Mont sopris series digital logging system
蒙特卡罗抽样分布 Monte-Carlo sampling distribution
蒙特卡罗(方)法 <一种模拟技术的方法利用随机数以产生随机变量的值这些值必须与其所具有的概率分布相一致> Monte-Carlo method; Monte-Carlo approach; Monte-Carlo process; Monte-Carlo technique
蒙特卡罗分析法 Monte-Carlo analysis
蒙特卡罗计算 Monte-Carlo calculation
蒙特卡罗技术 Monte-Carlo technique
蒙特卡罗模拟 Monte-Carlo simulation
蒙特卡罗模拟法 Monte-Carlo simulation method
蒙特卡罗实验 Monte-Carlo exercise; Monte-Carlo experiment; Monte-Carlo investigation
蒙特卡罗试验 Monte-Carlo experiment
蒙特卡罗研究 Monte-Carlo study
蒙特卡罗预测 Monte-Carlo forecast
蒙特雷西班牙式建筑 Monterey Spanish architecture
蒙特雷页岩 Monterey shale
蒙特利尔港 <加拿大> Port Montreal
蒙特罗斯型雪取样器 Mount Rose snow sampler
蒙特尼格罗试验 Montenegro test
蒙特佩利厄黄 Montpellier yellow
蒙提蓝 Monthier's blue
蒙脱阶 <古新世>【地】Montian
蒙脱黏[粘]土 montmorillonite clay; montmorillonitic clay
蒙脱石 ascanite[askanite]; daunialite; montmorillonite; montmorillonoid; smectite
蒙脱石带 smectite zone
蒙脱石含量 montmorillonite content
蒙脱石结构 smectite structure
蒙脱石类 smectites
蒙脱石泥岩 montmorillonite mudstone
蒙脱石黏[粘]土 montmorillonite clay
蒙脱石土壤 nortisol
蒙脱石-伊利石混合带 mixed zone of

smectite-illite
蒙脱土 askanite clay; imvite; montmorillonite; montmorillonite clay
蒙脱岩 montmorillonoid
蒙脱质的 montmorillonitic
蒙瓦克岩 mengwacke
蒙雾表面修饰 haze surface finish
蒙雾试验 haze test
蒙雾修饰 haze finish
蒙影地带 twilight zone
蒙影光弧 twilight arch
蒙雨天气 crachin
蒙皂石 smectite
蒙纸 masking sheet

锰 manganese

锰(Ⅴ)过氧络合物 Mn(Ⅴ) peroxo complex
锰白云母 manganese muscovite; manganmuscovite
锰白云石 kutnohorite; mangandolomite
锰钡白云母 manganobarium muscovite
锰钡矿 hollandite
锰尘沉着病 manganoconiosis
锰尘肺 manganoconiosis
锰臭氧催化 manganese catalyzed ozonation
锰磁绿泥石 grovesite
锰磁铁矿 manganesian magnetite; manganmagnetite
锰催干剂 manganese drier[dryer]; manganic drier
锰催化氧化工艺 manganese catalyst oxidation process
锰胆矾 manganese vitriol
锰的 manganic
锰的树脂酸盐 resinate of manganese
锰电池 manganese cell
锰电气石 tsilaisite
锰毒害 manganese poisoning
锰矾 szmikite
锰钒钢 manganese vanadium steel
锰钒工具钢 manganese vanadium tool steel
锰钒铀云母 fritzscheite
锰方解石 calcimangite; manganocalcite
锰方硼石 chambersite; ericaite
锰肥 manganese fertilizer
锰沸石 manganese zeolite
锰粉 manganese powder
锰氟磷灰石 manganfluorapatite
锰符山石 manganvesuvianite
锰斧石 manganaxinite; tinzenite
锰钙长石 manganese-anorthite
锰钙橄榄石 manganmonticellite
锰钙黄长石 manganese-gehlenite
锰钙辉石 johannsenite
锰钙铁辉石 manganhedenbergite
锰干电池 manganese dry cell
锰干料 manganese drier[dryer]
锰橄榄石 tephroite
锰钢 ferro-manganese; ferro-manganese steel; manganese steel; Tisco manganese steel
锰钢衬板 manganese
锰钢(钢)轨 manganese steel rail
锰钢尖轨转辙器 manganese tipped switch
锰钢链板 cast manganese steel pan; manganese flight
锰钢溜槽给料器 manganese steel pan feeder
锰钢履块 manganese shoe
锰钢耙斗 manganese scraper
锰钢镶嵌辙叉 manganese steel insert crossing
锰钢制楔形棒条内衬 <球磨机> manganese wedge-bar liner
锰钢制抓卡 manganese bronze catching finger
锰铬合金结构钢 manganese-chromium structural steel
锰铬铁矿 manganochromite
锰工具钢 manganese tool steel
锰钴土 black oxide of cobalt
锰硅钒铝矿 ardennite
锰硅钢 manganese silicon steel
锰硅工具钢 manganese silicon tool steel
锰硅合金 manganese-silicon
锰硅灰石 bustamite; manganwollastonite
锰硅铝矿 ardennite
锰硅铝石 mangandisthene
锰硅弹簧钢 manganese silicon spring steel
锰硅铁 ferro-manganese-silicon
锰硅锌矿 troostite
锰海绿石 manganglauconite
锰海泡石 mangansepiolite
锰合金 manganese alloy
锰合金电阻压力计 manganin soil ga(u)ge
锰合金钢整铸辙叉 manganese alloy steel cast crossing
锰褐帘石 mangan-orthite
锰黑 manganese black
锰黑颜料 manganese black pigment
锰黑云母 manganophyllite
锰红 manganese-alumina pink; manganese red
锰红磷铁矿 manganostrengite
锰红柱石 kanonaite; manganandalusite; viridine
锰化合物 manganic compound
锰环烷酸盐 manganese naphthenate
锰黄长石 manganjustite
锰黄砷榴石 manganberzeliite
锰黄铁矿 manganpyrite
锰黄铜 manganese brass; silvel
锰辉石 kanoite
锰基 manganese group
锰基锰合金 Mangonic
锰钾矿 cryptomelane
锰钾镁矾 manganleonite; manganolangbeinite
锰钾铁矾 manganvoltaite
锰尖晶石 galaxite
锰结核 halobolite; manganese nodule
锰结核成因 cause of formation of nodule
锰结核分布图 distribution map of manganese nodules
锰结核壳调查 manganese nodules crust survey
锰结核类型 type of manganese nodule
锰结核区 manganese nodule province
锰结核性状 manganese nodule property
锰结核元素含量分布图 distribution map of element content of manganese nodules
锰结壳 manganese crust
锰结皮 manganese pavement
锰晶石 rhodochrosite
锰精砂 manganese concentrate
锰矿 manganese mineral; manganese ore
锰矿瘤 manganese nodule
锰矿石 manganese ore
锰矿岩 manganolite
锰矿渣 manganese mud
锰蓝 mineral blue
锰蓝颜料 manganese blue pigment
锰离子 manganese ion
锰锂云母 masutomilite
锰帘石 sursassite
锰磷灰石 manganapatite
锰磷锂矿 lithiophilite
锰磷锂锰矿 mangani-sicklerite
锰磷锰矿 phosphoferrite
锰菱黑稀土矿 manganosteenstrupine
锰菱铁矿 manganoan siderite; manganosiderite; oligonite
锰菱锌矿 manganese zinc spar; mangansmithsonite
锰榴石 blythite
锰榴石英岩 gondite
锰瘤 manganese nodule
锰铝矾 apjohnite
锰铝合金 mangal; manganese-aluminum(alloy); Aluflex <电缆电线用的>
锰铝榴石 spessartine
锰铝镁基合金 Magnuminium alloy
锰铝坡缕石 manganpalygorskite
锰铝蛇纹石 kellyite
锰铝石榴石 spessartine
锰铝铜强磁性合金 Heusler's alloy
锰绿 Cassel green; manganese green
锰绿矾 manganmelanterite
锰绿帘石 manganese epidote
锰绿泥石 manganchlorite; pennantite
锰绿砂 manganese green sand
锰绿铁矿 frondelite
锰帽 chapeau de mangan; manganese stain
锰媒染剂 manganese mordant
锰镁硅钙石 manganese-merwinite
锰镁合金被覆硬铝 duralplat
锰镁铝蛇纹石 baumite
锰镁闪石 tirodite
锰镁锌矾 mooreite
锰明矾 apjohnite
锰钠沸石 manganonatrolite
锰钠矿 manjiroite
锰钠磷铁矿 manganalluaudite
锰钠铁闪石 juddite
锰铌铁矿 manganocolumbite
锰镍耐磨钢 Tisco Timang steel
锰镍青铜 wyndaloy
锰镍铜合金 manganin(alloy)
锰硼合金 manganese-boron(alloy)
锰硼镁铁矿 manganludwigite
锰硼石 jimboite
锰坡缕石 yofortierite
锰铅黄铜 manganese lead brass
锰铅矿 coronadite
锰青铜 manganese bronze
锰热臭石 manganpyrosmalite
锰人造沸石 manganese permutite
锰三斜辉石 pyroxmangite
锰砂 permanganate permutite
锰闪石 manganese-amphibole; richterite
锰砷钙镁矿 manganberzeliite
锰十字石 nordmarkite
锰水铝石 mangandiaspore
锰水泥 manganese cement
锰酸 manganic acid
锰酸钡 barium manganate
锰酸钙 calcium manganate
锰酸根离子 manganese ion
锰酸钾 potassium manganate
锰酸钠 sodium manganate
锰酸铷 rubidium manganate
锰酸锌 zinc manganate
锰酸盐 manganate
锰钛合金 manganese-titanium
锰钛铁矿 manganilmenite
锰弹簧钢 manganese spring steel
锰钽矿 ixiolite; ixionolite; manganotantalite
锰碳钢 manganese carbon steel
锰铁 manganese iron; ferro-manganese; ferro-manganese iron
锰铁白云石 manganankerite
锰铁比 Mn-Fe ratio
锰铁钒铅矿 brackebuschite
锰铁橄榄石 manganese fayalite; manganesian fayalite; manganfayalite
锰铁合金 ferro-manganese alloy; manganeisen; manganese iron
锰铁尖晶石 jacobsite
锰铁矿渣 manganese iron slag
锰铁榴石 calderite; polyadelphite
锰铁铝榴石 manganalmandite
锰铁绿鳞石 knebelite
锰铁绿泥石 grangesite
锰铁闪石 danemorite
锰铁透辉石 schefferite
锰铁脱氧剂 ferro-manganese deoxidizer
锰铁钨矿 wolfram(ite)
锰铁锌矾 dietrichite
锰铁锌尖晶石 manganoferrogehnite
锰同位素 manganese isotope
锰铜 copper-manganese; isa; manganese copper; manganin(alloy)
锰铜标准电阻丝合金 Minalpha
锰铜低合金钢 Jalten
锰铜电阻线 manganese copper resistance wire
锰铜合金 copper-manganese alloy; manganese copper alloy
锰铜矿 crednerite
锰铜镍合金 manganese copper-nickel
锰铜镍线 manganin(alloy)
锰铜线 manganin wire
锰透辉石 schefferite
锰透闪石 manganotremolite
锰土 black ocher; manganomelane; ouatite; reissacherite; wad(clay); wadite
锰土微菌 Pedomicrobium manganicum
锰团块 manganese nodule
锰污染 pollution by manganese
锰系电阻材料 isabellin
锰系元素 manganides
锰细菌 manganese bacteria
锰纤闪石 manganuralite
锰榍石 groutite
锰锌辉石 <辉石的变种,含锰、锌> jeffersonite; manganese zinc pyroxene
锰锌铁氧体 manganese-zinc ferrite
锰星 manganese star
锰星泥石 ekmanite
锰星叶石 kupletskite
锰亚油酸盐 linoleate of manganese
锰盐 manganese salt
锰颜料 manganese pigment
锰阳起石 manganactinolite
锰氧化物的清除作用 scavenging action of manganese hydroxide
锰氧磷灰石 manganoxyapatite; manganvoelckerite
锰叶绿泥石 manganese pennine
锰叶泥石 ekmanite
锰硬绿泥石 ottrelite
锰铀云母 manganese-autunite
锰游合金 mu-metal
锰黝帘石 thulite
锰浴 manganese bath
锰皂 manganese soap
锰增量 manganese addition
锰针磷铁矿 mangankoninckite
锰针钠钙石 manganopectolite
锰直闪石 mangano-anthophyllite
锰质玻璃 manganese glass
锰质沉积物 manganese sediment
锰质结核 manganese concretion; manganese nodule

M

锰质石 manganolite
锰质岩 manganese rock
锰质岩类 manganese rocks
锰质皂 manganese soap
锰质紫色涂料 manganese violet pigment
锰中毒 manganese poisoning; manganismus
锰柱石 orientite
锰柱星叶石 mangan-neptunite
锰铸铁 manganese cast-iron
锰锥辉石 schefferite
锰紫 manganese violet
锰棕(色) manganese brown
锰族元素 manganese family element

蠓

蠓 oso strea

孟

孟布酮 menbutone

孟德尔定律 Mendel's law
孟德立酸 mandelic acid
孟海窑 Mendhein kiln
孟寒尔颜色系列 Munsell colo(u)r system
孟加拉红 rose Bengal
孟加拉湾 Bay of Bengal
孟留斯石灰岩 <志留纪>【地】Manlius limestone
孟买斜纹布 <印度> bombay twill
孟农加希拉统 <美国晚石炭世>【地】Monongahelan series
孟塞尔标度 Munsell scale
孟塞尔表色系统 Munsell colo(u)r system
孟塞尔表色系统彩度环 Munsell chroma circle
孟塞尔表色系统彩度值 Munsell chroma number
孟塞尔表色系统色树 Munsell colo(u)r tree
孟塞尔彩度 Munsell chroma
孟塞尔彩色标准 Munsell colo(u)r standard
孟塞尔彩色制 Munsell colo(u)r system
孟塞尔尔云母 munsell
孟塞尔明度 Munsell value
孟塞尔明度函数 Munsell value function
孟塞尔曲线 Munsell curve
孟塞尔色标 Munsell colo(u)r scale
孟塞尔色表坐标系统 Munsell colo(u)r system
孟塞尔色(彩) Munsell hue
孟塞尔色度值 Munsell value
孟塞尔色立体 Munsell colo(u)r solid
孟塞尔色品 Munsell chroma
孟塞尔色素 Munsell colo(u)r system
孟塞尔体系 Munsell system
孟塞尔新表色系统 Munsell renovation system
孟塞尔颜色表示法 Munsell colo(u)r notation
孟塞尔(颜色)色调 Munsell hue
孟塞尔颜色体系 Munsell colo(u)r system
孟塞尔值 Munsell value
孟莎屋顶 curb roof
孟特松 Montezuma pine
孟席斯型新型水力分选机 Menzies hydroseparator
孟侬巨像 <古埃及底比斯> Colossi of Memnon
孟滋合金 malleable brass
孟滋黄铜 Muntz brass; Muntz metal
孟滋(锌铜)合金 Muntz metal

梦

梦湖 <月球> lacus somniorum

梦沼 palus somnii

咪

咪唑 imidazole[iminazole]

咪唑啉 imidazoline
咪唑啉酮 imidazolinone

弥

弥补差额 make good the deficit

弥补成本后的边际报酬率 marginal rate of return over cost
弥补费用 cost of cover
弥补分歧 close difference
弥补工作 make-up job
弥补进口技术需要 offset the amount used for imported technology
弥补亏绌拨款 deficiency appropriation
弥补亏空 cover the deficit; made good the deficit
弥补亏欠 come to deficit
弥补亏损 come to deficit; cover the deficit; deficit coverage; made good the deficit
弥补损失 cover the loss; make good a loss; make-up for a loss; recover loss
弥补需要的数额 make-up the required number
弥补预算政策 compensatory budget policy
弥分散相 disperse phase
弥缝剂 sealer
弥复效应 healing effect
弥合 bridge; close
弥合差距 close up the gap; stop a gap
弥合缝隙的树胶 light gum veins
弥合裂痕 bridge the gap
弥漫 interfuse; pervade; suffusion
弥漫常数 diffusion constant
弥漫的 diffuse; suffuse; permeant
弥漫的尘土 pother
弥漫功能测定 diffused function determination
弥漫光照明 diffused lighting
弥漫函数 spread function
弥漫物质 diffused matter
弥漫型 diffused type
弥漫性 diffusibility; diffusivity
弥漫性的 diffuse
弥漫性钙化 diffused calcification
弥散 debunching; diffusion; dispersal; dispersion; dissemination
弥散斑 disc of confusion
弥散波 dispersive wave
弥散场 fringing field
弥散沉淀 intergranular precipitation
弥散带长度 length of dispersion band
弥散带宽度 width of dispersion belt
弥散点 dispersion point
弥散度 degree of dispersion; dispersion degree; dispersity
弥散分析 dispersion analysis
弥散功能 diffusion function
弥散光 diffused light
弥散光纤 diffusion optic(al) fiber
弥散光线照明法 diffused illumination
弥散计 dispersimeter
弥散剂 dispersed medium; disperse means; dispersing medium; dispersion medium
弥散胶体法 dispersion method
弥散角 angle of dispersion
弥散介质 continuous medium; dispersion medium; dispersive medium
弥散距离 distance of dispersion
弥散理论 dispersion theory
弥散力 dispersion force
弥散量 diffusing capacity
弥散率测定器 diffusiometer
弥散媒介物 dispersion medium
弥散模型 dispersion model
弥散盘 disc of confusion
弥散气体 dispensing gas
弥散器 disperser
弥散强化 dispersion-strengthening
弥散强化材料 dispersion-strengthened material
弥散强化的金属 dispersion-strengthened metal
弥散强化合金 strengthened dispersion alloy
弥散强化铅 dispersion-strengthened lead
弥散圈 figure of confusion
弥散韧化 dispersion toughening
弥散生长 diffused growth
弥散试验法渗透系数 permeability coefficient of dispersion test method
弥散试验方法 method of dispersion test
弥散试验设备 equipment of dispersion test
弥散损失 dispersion loss
弥散体 dispersoid
弥散条纹 diffused streak
弥散通量 dispersion flux
弥散网络 dispersion network
弥散物 diffusate
弥散系数 dispersion coefficient; dispersivity; longitudinal dispersion coefficient
弥散现象 diffusing phenomenon; dispersion phenomenon
弥散相 dispersed network phase
弥散效应 dispersion effect
弥散形式 dispersion pattern
弥散型 dispersion type
弥散性 diffusibility; dispersivity; disseminated
弥散性投射系统 diffused projection system
弥散因数 dispersion factor
弥散硬度 dispersion hardness
弥散硬化 dispersed phase hardening; dispersion hardening; dispersion-strengthening; precipitation hardening; strain ag(e)ing
弥散硬化不锈钢 precipitation-hardening stainless steel
弥散硬化材料 dispersion-hardened material
弥散硬化合金 dispersion-hardened alloy; dispersion-hardened material
弥散圆 air circle; blur circle
弥散增强复合材料 dispersion-strengthened composite
弥散张量 dispersion tensor
弥散障碍 impeded diffusion
弥散质 <分散于介体中的颗粒>【物】dispersed part; disperse phase
弥散质点 dispersoid; dispersoid particle
弥散质点分布 dispersoid distribution
弥散质点粒度 dispersoid particle size; dispersoid size
弥散状态 disperse state
弥散锥体 conic(al) dispersion
弥雾 atomize
弥雾发生器 mist generator
弥雾机 atomizing sprayer; micron sprayer; mister; mist sprayer; mist sprinkler
弥雾喷粉机 mist duster

迷

迷彩 baffle; camouflage coat; cryptic mimicry

迷层 stray
迷迭香【植】rosemary
迷宫 labyrinth; maze
迷宫泵 labyrinth pump
迷宫齿 labyrinth teeth
迷宫阀 labyrinth valve
迷宫法 maze method
迷宫环 labyrinth ring
迷宫活塞式压缩机 labyrinth piston compressor
迷宫间隙 labyrinth clearance
迷宫绿篱 labyrinth hedge
迷宫密封 tongue and groove labyrinth
迷宫密封螺母 labyrinth nut
迷宫密封梳齿边 labyrinth packing edge
迷宫密封压力比 labyrinth pressure ratio
迷宫密封装置 labyrinth gland
迷宫内的挡路水池 labyrinth waterstop
迷宫汽封 labyrinth gland packing; labyrinth seal
迷宫汽封疏齿 labyrinth fin
迷宫汽封体 labyrinth casing
迷宫曲径 labyrinth fret
迷宫曲径环 labyrinth
迷宫曲径回文饰 meander
迷宫式充填物 labyrinth sealing
迷宫式的 daedal(ian); mazy
迷宫式分级机 labyrinth classifier
迷宫式隔板汽封环 laby rinth diaphragm packing ring
迷宫式护油圈 labyrinth oil retainer
迷宫式活塞 labyrinth piston
迷宫式活塞压缩机 labyrinth piston compressor
迷宫式集尘器 labyrinth dust collector
迷宫式集油器 labyrinth oil retainer
迷宫式窥视装置 labyrinth viewing device
迷宫式密封 chev(e)ron seal; labyrinth gland; labyrinth seal(ing); labyrinth packing
迷宫式密封环 labyrinth collar; labyrinth ring
迷宫式密封件 labyrinth packing
迷宫式密封圈 labyrinth collar; labyrinth ring
迷宫式密封箱 labyrinth box
迷宫式密封装置 labyrinth seal gland
迷宫式汽封 labyrinth gland
迷宫式溶洞 labyrinth cave karst
迷宫式润滑脂封闭器 labyrinth-grease seal
迷宫式润滑脂密封器 labyrinth-grease seal
迷宫式疏水器 labyrinth trap
迷宫式填缝条 labyrinth waterstop
迷宫式通道 maze-like conduit
迷宫式无油润滑压缩机 oil-free labyrinth compressor
迷宫式压盖 labyrinth gland
迷宫式压缩机 labyrinth compressor
迷宫式扬声器 labyrinth loudspeaker
迷宫式障板 chev(e)ron baffle; labyrinth baffle; shell-type baffle
迷宫式止水 labyrinth sealing
迷宫式止水环 labyrinth seal ring
迷宫式轴封 labyrinth shaft seal; labyrinth sleeve; leak-off-type shaft seal
迷宫算法 labyrinth algorithm
迷宫填料 labyrinth packing
迷宫形磁畴 maze domain
迷宫型堰顶 labyrinth crest; labyrinth

sill
迷宫溢洪道 labyrinth spillway
迷宫轴承 labyrinth bearing
迷管 aberrant ductule; ductuli aberrantes
迷航 disorientation
迷惑 bewilder; maze
迷惑试验 fogging test
迷惑信号 babble signal
迷孔菌属 < 拉 > Daedalea
迷流 extraneous current; foreign current; stray current; vagabond current
迷流测试柜 stray current testing cabinet
迷路 meander; wander
迷路走线算法 maze-running algorithm
迷失方向 disorientation
迷失状态 lost condition
迷向 isotropism; isotropy
迷向面 isotropic(al) plane
迷向曲面 isotropic(al) surface
迷向曲线 isotropic(al) curve
迷向直线 isotropic(al) line
迷向锥面 isotropic(al) cone
迷向坐标 isotropic(al) coordinates
迷信行为 superstitious behavio(u)r
迷于目标的 goal obsession
迷走电流 currency from irregular sources

谜语 riddle

醚 ethyl ether

醚不溶树脂 anthracoxenite; ether-insoluble resin
醚醇 ether alcohol
醚合三氟化硼 boron trifluoride etherate
醚合物 etherate
醚化 etherify
醚化度 degree of etherification
醚化剂 etherifying agent
醚化了的尿素树脂 etherified resin
醚化作用 etherification
醚环 ether ring
醚键 ether link
醚交换 transesterification
醚裂开 ether cleavage
醚溶性浸出物 ether-soluble extractive
醚酸 ether acid
醚香料 ethereal essence
醚油 ethereal oil; ether oil
醚酯 ether ester
醚制的 ethereal
醚状液 ethereal liquid

糜化 chylification

糜烂 erosion
糜烂剂 vesicant
糜烂性毒气 blistering gas; vesicant war gas
糜滥石 mélange
糜棱构造 mylonitic structure
糜棱化 mylonitization
糜棱化散体结构 mylonitic loosen texture
糜棱化岩 mylonitized rock
糜棱结构 mylonitic texture
糜棱煤 mylonitic coal
糜棱片麻岩 mylonite gneiss
糜棱岩 miliolite; mylonite
糜棱岩带 mylonite zone
糜棱岩化方式 mylonitization way
糜棱岩化作用 mylonetization; mylonization
糜棱岩系列 mylonite series
糜棱状的 mylonitic

米 meter[metre]

米安(培) meter[metre]-ampere
米巴赫高效闪光对焊机 Miebach high efficiency flash welding machine
米-白氏原理 < 即变位线相当于影响线 > Muller-Breslau's principle
米百分值 meter percent
米波 metric(al) wave
米波拉 < 隔热用泡沫塑料的商名 > mepolam
米波拉保温材料 Mipora
米波雷达 meter wave radar; metre-wave radar
米禅勒城之狮门 < 古希腊 > Gate of Lions
米长杆尺 < 用于校正水准标尺 > rod meter
米尘 rice dust
米尺 meter ga(u)ge; meter rule; meter stick; metric(al) ga(u)ge; metric(al) scale
米尺计量 metric(al) measure
米袋机 bag bunch-sealing machine
米德富特层【地】Meadfoot bed
米德里姆灰岩 Mydrim limestone
米德里姆页岩 Mydrim shale
米迪风 vent du midi
米淀粉 rice starch
米吨秒(单位)制 meter-ton-second system; meter-ton-second unit
米吨秒计量法 meter-ton-second system
米顿板层 Mytton flags
米尔恩-肖氏地震仪 Milne-Shaw seismograph
米尔斯-内克现象【给】Mills-Reincke phenomenon
米/分 < 旧速度单位 > meter per minute
米格拉生铁 Migra iron
米格纸 < 俗称 > millimeter paper
米公斤秒安制 meter-kilogram-second ampere system
米汞柱 < 压强单位 > meter mercury column; meter of mercury head
米管 mitron
米轨距 < 窄轨距 > meter[metre] ga(u)ge
米轨铁路 meter[metre] ga(u)ge railway
米轨线路 meter[metre] ga(u)ge line; meter[metre] ga(u)ge track
米花状混凝土 popcorn concrete
米花状聚合物 popcorn polymer
米花状聚合作用 popcorn polymerization
米花状无细骨料混凝土 popcorn concrete
米级无烟煤 rice coal
米计量 metric(al) unit
米浆 rice starch paste
米胶 rice glue
米卡他绝缘板 micarta
米卡塔胶纸板 micarta board
米凯利斯 · 马斯特压力灌注桩 Michaelis Mast pressure pile
米凯利斯常数 Michaelis constant
米凯利斯菱形区 Michaelis rhomboid
米糠蜡 rice bran wax
米糠油 rice bran oil; rice oil
米克劳利特 < 一种陶瓷刀具 > Mikrolit
米克罗依 < 绝缘材料 > Mykroy
米克洛硬度试验机 Mikro-tester
米克斯粉末混合度测量仪 Mixee
米拉比来铝合金 mirabilite; sal mirabile
米拉比来铝镍合金 Mirabilite alloy
米拉合金 Miramant
米拉赖特耐蚀铝合金 Miralite
米拉耐酸蚀铜合金 Mira metal
米拉丘洛依耐高压铸造合金 Miraculoy
米拉铜基合金 Mira alloy
米拉铜铅合金 Mira
米拉硬木 < 印度产 > Milla
米兰大教堂 Milan Cathedral Church
米兰氏拱 Melan arch
米兰主教堂 Milan Cathedral Church
米勒-白司老原理 Muller-Breslau's principle
米勒比色汁 Mill's colo(u)rimeter
米勒-布雷斯劳原理 < 即变位线相当于影响线 > Muller-Breslau's principle
米厘条 floating rule
米利都布局 Milesian layout
米粒 rice grains
米粒煤 rice coal
米粒石面 grained stone facing; granulated stone facing
米粒装饰 rice grain decoration
米粒状碎裂 rice pattern fracture
米林顿混响公式 Millington reverberation formula
米洛丽蓝 < 一种铁亚氰酸盐,铁蓝的别名 > milori blue
米洛丽蓝颜料 milori blue pigment
米洛丽绿 milori green
米洛通 < 一种隔声挡板 > Melotone
米马力 metric(al) horsepower
米蒙尔木 < 一种红色硬木 > Memel timber
米/秒 < 速度单位 > meter/second; meters per second
米纳尔法铜锰镍合金 Minalpha
米纳里 < 一种木材防腐滞火材料 > Minalith
米纳斯吉拉斯型铁矿床 Minas Gerais type iron deposit
米纳型车钩 Miller coupler
米奈喷酯 minepentate
米尼表 microindicator; minimeter
米尼管 Minitron
米尼金计算法 Minikin method
米尼劳定理 theorem of Menelaus
米尼劳斯定理 Menelaus theorem
米牛阿诺风 minuano
米诺斯建筑 < 古希腊 > Minoan architecture
米诺瓦低膨胀高镍铸铁 Minovar
米诺瓦合金 Minovar metal
米胚芽油 rice germ oil
米千克 < 扭矩单位 > meter kilogram
米千克力秒单位制 meter-kilogram force second
米千克秒安制 meter-kilogram-second ampere system
米千克秒(单位)制 meter-kilogram-second system
米千克秒计量法 meter-kilogram-second system
米千克秒制 metre-kilogram-second system
米千克秒制单位 meter-kilogram-second unit
米塞斯屈服准则 Von Mises yield criterion
米赛斯屈伏面 Mises yield surface
米赛斯屈服函数 Mises yield function
米赛斯屈服准则 Mises yield criterion
米赛斯圆柱面 Mises cylinder
米赛斯准则 < 强度理论的 > Von Mises criterion
米色 beige; buff; cream; cream colo(u)r
米色布纹纸 cream wove
米色的 Beige; cream-colored; creamy; off-white
米色或蓝色帆布 bretagne
米色陶器 creamware
米色釉 cream glaze
米色直纹纸 cream laid
米筛 rice huller screen; rice sieve
米筛纹 rice sieve design
米实方/米 < 铲土量 > bank cubic meter/meter
米氏参数 Mie's scattering parameter
米氏函数 Mie's scattering function
米氏合金 pyrophoric alloy
米氏激光雷达 Mie's scattering laser radar
米氏理论 Mie's theory
米氏强度 Mie's scattering intensity
米氏散射 Mie's scattering
米氏散射系数 Mie scattering coefficient
米市 rice market
米水柱 < 压力单位 > meter water column; meter water ga(u)ge; meter of water head
米斯科合金 Misco metal
米斯科镍铬铁系耐热耐蚀合金 Misco
米斯特克建筑 Mixtec architecture
米索间冰阶【地】Messo interstade
米索间冰期【地】Messo interglacial stage
米通(花样) rice grain pattern
米托 < 一种铝漆 > metal
米线 rice noodle; rice stick
米线生产线 production line for rice noodle making
米/小时 meters per hour
米箱 rice bin
米歇尔-班克水轮机 Michell-Banki turbine
米歇尔参数 Michell's parameter
米歇尔式轴承 Michell's bearing
米歇尔推力轴承 Michell thrust bearing
米歇尔小夹 Michell's clips
米歇尔型推力轴承 Michell type thrust bearing
米歇尔轴承 Michel-type bearing
米雪 grain of ice; granular snow; snow grain
米英尺换算 meters feet conversion
米英尺换算表 meters feet conversion table
米纸 rice paper
米制 centimeter-gram-second system; metric(al) system(of units)
米制比长仪 meter comparator
米制比例尺 scale of meter; metric(al) scale
米制标度 metric(al) scale
米制测链 metric(al) chain
米制尺 metric(al) ga(u)ge
米制尺寸 metric(al) dimension
米制尺度 metre ga(u)ge; metric(al) scale
米制单位 metric(al) unit
米制的 metric(al)
米制等高线 metric(al) contour
米制度量 metric(al) measure
米制度量衡 metric(al) system of measurement
米制吨 metric(al) ton
米制格令 metric(al) grain
米制公约 metric(al) convention
米制规 metre ga(u)ge; metric(al) ga(u)ge
米制海图 metric(al) chart
米制化 metrication

米制换算 metric(al) convention
米制刻度 metric(al) graduation
米制螺距规 metric(al) screw pitch ga(u)ge
米制螺纹 metric(al) screw; metric(al) thread
米制螺纹齿轮装置 metric(al) thread gearing
米制马力 meter horsepower; metric(al) horsepower
米制模数单位 metric(al) modular unit
米制品 rice made products
米制曲线 metric(al) curve
米制数据 metric(al) data
米制水表 meter water ga(u) ge
米制水平标尺 meter rod
米烛 meter-candle
米烛光 meter-candle; lux; lux candle
米烛秒 meter-candle-second
米字交汇 umbrella-stand type intersection
米字纹 rice character design
米字形桁架 rhombic(al) truss
米字型 meter-type

脒染料 amidine dye stuffs

泌出变熔作用方式 ektexis way

泌出的 weeping
泌出混合岩化作用 ektexis
泌出水 sweating
泌出水泥浮浆 < 混凝土表面 > bleeding
泌硅生物 silica secreting organism
泌浆 < 指混凝土 > bleeding; water gain
泌浆率 bleeding rate; bleeding ratio
泌颗粒 secretory granule
泌尿化验室 urology laboratory
泌尿科 urological department
泌水 bleed water; guttation; secretion; weep(age); weeping < 水泥混凝土的 >; water gain < 未凝固混凝土的 >; segregation of water < 混凝土或砂浆 >
泌水测量 < 测泌水速率及总泌水量 > measurement of bleeding
泌水量 bleeding capacity
泌水率 bleeding capacity; bleeding rate; bleeding ratio
泌水能力 bleeding capacity
泌水速率 bleeding rate
泌水通道 bleeding channel
泌水性 bleeding
泌液 bleeding
泌油 weeping
泌脂 resinosis
泌脂原木 bled timber

觅食构造 feeding structure

觅食迹 feeding trace
觅食潜穴 feeding burrow
觅数过程 search process

秘方 nostrum

秘簧 secret spring
秘级调用锁 invocation privacy lock
秘诀 mystique; secret
秘密 privacy; secrecy
秘密保护 privacy protection
秘密采购 secret purchasing
秘密筹款 backdoor financing
秘密出价投标 closed bid
秘密代营企业 dummy company
秘密贷款 classified loan
秘密的 off-the-record; secret
秘密地牢 oubliette
秘密电报 confidential message; secrecy message
秘密反对票 black ball
秘密分类账 secret ledger
秘密合伙人 secret partner; silent partner; undisclosed partner
秘密会议 conclave; meeting in camera; private meeting
秘密交易 clandestine sale; insider dealing
秘密接头 wiretap
秘密理财 backdoor financing
秘密利润 secret profit
秘密(联系)渠道 back channel
秘密贸易 clandestine trade
秘密情报 secret message
秘密融资 backdoor financing
秘密入境 clandestine immigration
秘密审讯 hearing in camera
秘密使节 confidential envoy
秘密投标 dumb-bidding; sealed bid; sealed tender
秘密投票 secret vote
秘密协议 confidential agreement
秘密信托 secret trust
秘密以最低价格竞争 dumb-bidding
秘密盈余 hidden reserve
秘密准备(金) secret reserves
秘密资产 non-ledger assets
秘密总账 private ledger
秘色 private colo(u)r
秘色瓷 secret colo(u)r porcelain; secret colo(u)r ware
秘书长 secretary general
秘书处 secretariat(e)
秘书室 clerk's office; secretarial pool; secretary office
秘液【地】ichor

密胺甲醛 melamino-formaldehyde

密胺树脂胶 melamine resin adhesive
密斑 dense patch
密斑油 oil of mirbane
密闭 closeness; confinement; encapsulate; encapsulation; hermetic closure; obturage; tighten
密闭暗沟 closed conduit
密闭爆发器 manometric(al) bomb
密闭部件 hermetic unit
密闭操作箱 globe box
密闭槽 closed cell
密闭层 seal coat; seal course
密闭储存地 enclosed storage area
密闭储罐 closed tank
密闭带 sealing tape
密闭的 air proof; air-tight; hermetic; inclosed; leak-free; pressure-tight
密闭的安全楼梯井 enclosed exit stairwell
密闭的汽油桶 protected fuel tank
密闭的洗衣干燥小室 enclosed cabinet for drying washing
密闭电动机 closing motor
密闭电离室 closed ionization chamber
密闭度 containment; leakproofness; leak tightness
密闭发酵法 closed fermentation method
密闭防尘 dust-tightness
密闭服 full pressure suit
密闭盖 air-tight cover; hermetic cover
密闭工作室 < 沉井下部的 > air dome
密闭构造 air-tight construction
密闭管 seal pipe; seal tube
密闭管道 closed conduit
密闭和填缝 sealing and ca(u)lking
密闭呼吸气体系统 closed respiratory gas system
密闭环境 closed environment
密闭机械式 close-mechanical type
密闭集气罩 closed hood
密闭集装箱 sealed container
密闭加压进料式离心机 hermetic pressure-feed type centrifuge
密闭搅拌器 closed agitator
密闭接缝 air-tight joint
密闭接头 air-tight joint
密闭结构系统 enclosing structure system
密闭壳 containment vessel
密闭门 air-tight door; sealing door; tight-fitting door
密闭能力 sealing ability
密闭排气通风 enclosure exhaust ventilation
密闭泡沫 < 隔绝空气的 > closed cell foams
密闭膨胀器 closed expansion vessel
密闭墙 dam; stank
密闭青贮塔 tight silo
密闭群落 closed community
密闭热水器 closed water heater
密闭人工开挖式盾构 close-type manual shield
密闭容器 closed container; closed vessel; hermetically sealed container; well-closed container
密闭生态系统 closed ecological system
密闭生物过滤 enclosed trickling filter
密闭式 closed type; shell type
密闭式保险丝 enclosed fuse
密闭式表层 closed surface
密闭式给水 closed-feed water
密闭式锅炉 closed fireroom
密闭式锅炉舱 closed stokehold
密闭式呼吸设备 closed circuit breathing apparatus
密闭式混合机 internal mixer
密闭式混胶机 internal rubber mixer; rubber-internal mixer
密闭式混炼 Banbury mixing
密闭式混炼机 Banbury mixer; internal mixer
密闭式混炼器 Banbury mixer
密闭式机组 hermetically sealed unit
密闭式集装箱 < 备有装货口卧式圆筒的微小谷物集装箱 > pressurized container
密闭式冷却箱 closed cooling box
密闭式炼胶机 internal mixer
密闭式炉 closed vessel furnace
密闭式滤池 closed filter
密闭式螺杆泵 air-tight screw pump; hermetic screw pump
密闭式模具 positive mo(u)ld
密闭式喷嘴 closed-type nozzle
密闭式膨胀箱 closed expansion tank
密闭式(气用)具 balanced flued appliance
密闭式潜水服 closed diving suit
密闭式潜水钟 closed diving bell
密闭式燃气轮机发动机 closed-cycle gas turbine engine
密闭式燃(气用)具 direct vented type
密闭式杀菌器 closed sterilizer
密闭式循环冷却水系统 closed recirculating cooling water system
密闭式循环冷却系统 closed recirculating cooling system
密闭式循环潜水装置 closed circuit diving apparatus; closed recirculating diving apparatus
密闭式压缩机组 hermetically sealed compressor
密闭式叶片搅拌机 close blade pug mill
密闭式叶片搅拌器 close blade pug mill
密闭式运输带 zipper conveyer[conveyor]
密闭式轧机 closed roll mill
密闭式制冷系统 hermetically sealed refrigerating system
密闭式抓斗 enclosed type grab
密闭式自动输送 automatic custody transfer
密闭式组合开关 enclosed type combination switch
密闭输送 custody transfer
密闭双层玻璃窗 air-tight double glazing window
密闭水泥罐卡车 pressurized cement lorry
密闭填料 gasket; gasket material; gaskin
密闭条 sealing rope; sealing strip
密闭退火 box annealing; closed annealing; pot annealing
密闭系统 closed system; enclosed system
密闭系统磨细(材料) closed circuit grinding
密闭匣 closed enclosure
密闭箱 closed casing
密闭小间 zeta
密闭形燃油阀 closed-type fuel valve
密闭形轴承 closed-type bearing
密闭型 closed face
密闭型电动机 permissible motor
密闭性 air-tightness; stopping property
密闭循环 closed circuit
密闭循环空气冷却 enclosed circulating air cooling
密闭循环水冷却器 closed circuit water cooler
密闭循环系统 closed circulating system; closed circulation system; closed-cycle system
密闭压滤机 closed filter press
密闭源 sealed source
密闭闸刀开关 enclosed knife switch
密闭毡条 felt seal
密闭罩 air-tight hood; enclosed hood; exhausted enclosure; hermetic hood
密闭支撑 close timbering
密闭状态 closed state
密闭座舱 sealed cabin
密编格子 close-woven trellis
密编码 code
密变分割 density slicing
密波 condensational wave; condensation wave; wave of condensation
密播 close planting; close seeding; dense sowing; heavy seeding
密布灰泡 heavy seed
密部 compact part
密撑 close timbering
密匙 cryptographic protocol
密椽屋顶 single roof
密的 thick
密点闭合对比 closure correlation of dense point
密电译文 decipher
密斗提升机 en-masse elevator
密度 bank density; compactness; density; specific mass; thickness
密度曝光量曲线 density-exposure curve
密度比 density ratio
密度比例 density scale

密度臂 density arm
密度变化 density fluctuation;density shift
密度变化测定法 density change method
密度变化方程 density change equation
密度变化区 density transition zone
密度标度 density scale
密度标准 < 木材分类标准 > density rule
密度表 density meter
密度表面张力球 density-surface tension bob
密度表示法 densimetric representation
密度波 density wave
密度波理论 density wave theory
密度不连续面 surface of density discontinuity
密度不平衡 density imbalance
密度不足 density defect
密度参考标度 density reference scale
密度参数 density parameter
密度测定法 densi(to)metry;gravimetry
密度测定计 densimeter
密度测定瓶 density bottle
密度测定器 density ga(u)ge
密度测定仪 gravimetry instrument
密度测井 densilog;density log(ging)
密度测井读数 reading of density log
密度测井记录 density log(ging)
密度测井刻度值 density log calibration value
密度测井曲线 densilog curve
密度测井仪 densilog instrument;density logger;density tool
密度测量 densi(to)metric measurement;density measurement
密度测量分析 densi(to)metric analysis
密度测量计 gravi(to)meter
密度测量均方差 mean square error of density measurement
密度差函数 density difference function
密度差异 density variation;density difference
密度差异分层 density difference stratification
密度场 field of density
密度车速关系 density-speed relationship
密度沉降 gravity settling
密度冲量 density impulse
密度传递器 density transmitter
密度大的货物 dense cargo
密度大的泥沙 dense sediment
密度大底片 dense negative
密度低的 low density
密度-电阻率(测井)法 density-resistivity method
密度定则 density rule
密度对生长的相关性 density dependence of growth
密度法 densimetry;densi(to)metric method;density method
密度法测井 density log
密度反差 density contrast
密度反常 density anomaly
密度范围 density range;dynamic(al) range;range of density
密度分布 density distribution;density portion;density-spread;density profile < 沿板厚方向的 >
密度分布法 density distribution method
密度分布函数 density distribution function
密度分布梯度 density distribution gradient
密度分布型 density profile
密度分层作用 < 水体中密度不同引起的 > density stratification
密度分割 density slicing
密度分割法 density slicing method;density slicing process
密度分割图像 density sliced image
密度分割仪 density slicer;density slicing device
密度分级法 density step-procedure
密度分级刨花板 graded-density particle board
密度分级片 density step tablet
密度分离法 density fractionation;density separation;gravitational separation
密度分析仪 density type analyser[analyzer]
密度分选 density separation
密度风洞 density channel
密度弗劳德数 densi(to)metric Froude number
密度改善 density correction
密度干涉仪 density interferometer
密度高度 density altitude
密度光楔 density wedge
密度规则 density rule
密度-含水量关系 density moisture content relationship
密度-含水量曲线 density moisture content curve
密度函数 density function
密度换算定则 density transformation
密度换相 < 用于车辆感应信号灯 > density change
密度或直立程序 density or erectness
密度级 density level
密度计 densi(to)meter[densometer];density ga(u)ge;density indicator;density meter
密度计法颗粒分析 densitometer particle analysis
密度计分析 hydrometer analysis
密度计量学 densi(to)metry
密度计量仪 density measuring instrument
密度记录器 density recorder
密度加权平均浓度 population-weighted average concentration
密度降低 density decrease
密度校正 density correction
密度阶变图 density step tablet
密度界面 density interface
密度镜 pick glass
密度矩阵 density matrix
密度刻度器 density log calibrator
密度空速 density airspeed
密度孔隙度 density porosity
密度控制 densi(to)metric control;density control
密度控制阀 density control valve
密度控制排料 density-controlled discharge
密度控制器 density controller
密度控制装置 density control unit
密度流 current of higher density;density current;density flow;density induced flow
密度流沉积 density current bed;density current deposit
密度流底层 density current bed
密度率 degree of density
密度浓度 density concentration
密度瓶 density bottle;pycnometer[pyknometer]
密度起伏 density fluctuation;fluctuation of density
密度气压柱 density gradient column
密度区域 density district
密度曲线 density curve;densograph
密度曲线图 densogram
密度曲线自动描绘仪 densograph
密度深度坡线 < 单位深度的密度变化 > density-depth gradient
密度湿度综合探测计 combination density and moisture meter
密度数据 density data
密度算符展开 density operator expansion
密度探测器 density probe
密度探针 density probe
密度梯尺 density wedge
密度梯度 density gradient;gradient of density
密度梯度分离(法) density gradient separation
密度梯度离心(法) density gradient centrifugation
密度梯度离心分离作用 density gradient centrifugation
密度调剂 density transfer
密度调节材料 < 调节水泥浆或泥浆的密度 > density controlling material
密度调制 density modulation
密度调制光束 density-modulated beam
密度调制声道 variable-density channel;variable-density track
密度图 density chart;density map
密度无关因素 density-independent factor
密度无关因子 density-independent factor
密度误差 density error
密度系数 < 指每英里道路上的车辆数 > density factor;bulk factor;density coefficient
密度限度 limit of density
密度小的 low specific gravity
密度效应 density effect
密度修正弗鲁德数 densi(to)metric Froude number
密度修正量 density correction
密度压力计 densi-tensimeter
密度液 density fluid
密度依存 delayed density dependence
密度异常 density anomaly
密度有关因素 density-dependent factor
密度有关因子 density-dependent factor
密度源类型 type of density current
密度跃层 density transition layer;density transition zone;discontinuous layer of density;pycnocline
密度增长率 density increase rate
密度增高 increase in density
密度增加而收益下降 diminishing returns from increasing intensity
密度涨落 density fluctuation
密度指示计 density indicating meter
密度指示器 density indicator
密度指数 density index
密度制约的 density-dependent
密度制约因素 density-dependent factor
密度中线定律 law of rectilinear diameters
密度中心弗劳德数 densimetric Froude number
密度中心速度 densimetric velocity
密度中子测井法 density-neutron method
密度周期 density cycle
密度坐标图 densograph
密断统 discontinuum
密堆积 close packing
密堆积的 close-packed
密堆积点阵 close-packed lattice
密堆积结构 close-packed structure
密堆积晶体 close-packed crystal
密堆积六角结构 closed-packed hexagonal structure
密垛法 bulk stacking
密耳 < 等于千分之一英寸,1 英寸 = 0.0254米 > mil
密耳英尺 < 1 英尺 = 0.3048 米 > mil-foot
密耳英寸 milli-inch
密耳圆 circular mil
密封 air seal;air-tight seal;air-tight test;capsulation;closeness;compaction;confinement;encapsulate;encapsulation;enclose;enclosure;glands;hermetic closure;hermetization;luting;packaging;packing off;pressurization;seal(ed)-in;sealing(-off);seal off;seal up;tight seal;deep seal < 合流制排水管的 >
密封板 seal(ing) plate;seal plate;set flashing;staunching plate
密封包件 sealed pad
密封包装 hermetic package;sealed pack-age
密封保径滑动镶齿钻头 sealed ga(u)ge sliding tungsten carbide insert bit
密封保径铣齿钻头 sealed ga(u)ge milled bit
密封保径镶齿钻头 sealed ga(u)ge tungsten carbide insert bit
密封保温玻璃 sealed insulating glass
密封泵 canned pump;leakproof pump;sealed pump
密封边 hermetically sealed edge;sealing strip
密封标单 competitive sealed bid
密封标单投标 sealed-bid tender
密封表 enclosed watch
密封拨杆 sealing driving rod
密封波纹管 seal bellows
密封箔 sealing foil
密封薄膜 sealing film;sealing foil
密封薄膜养护 curing with sealing membrane
密封薄片 sealing sheeting
密封不严 undersealing
密封不足 undersealing
密封部分 hermetic unit
密封材料 encapsulant;gasket material;jointing material;packing material;sealer;seal(ing) material;sealant
密封材料抗弯试验机 stopper bending tester;stopping bending tester
密封仓 gas-tight silo;sealed compartment
密封舱 air-tight cabin;capsule;man lock;pressure-tight body
密封舱盖 insulated hatch cover;plug hatch
密封舱室 air lock
密封槽 closed cell;sealing groove
密封层 liquid sealant;sealant;sealer;sealing coat;seal(ing) course;sealing layer
密封产品 encapsulated product
密封衬板 close-boarded;close sheeting
密封层底层涂料 sealant primer
密封衬垫 gasket;packing gasket;packing gland;sealing gasket
密封衬垫接缝 sealing gasket joint
密封衬片 sealing gasket
密封衬套 gland bush;sealing bush
密封齿缘 packing edge
密封冲洗 seal flush
密封冲洗口 seal flush port

M

密封冲洗压力 seal flush pressure
密封冲洗液 seal flush liquid
密封出价 sealed bid
密封处理 sealing treatment
密封窗 air-tight sash;encapsulated window;sealed window
密封唇(口) sealing edge; seal(ing) lip;lip of seal
密封磁头 sealed head
密封催化消解法 sealing catalytic decomposed method
密封带 sealant tape; sealing strip; sealing tape;tape sealant
密封的 airproof; air-tight; aquaseal; bottle tight; canned; encapsulated; environmentally sealed; fully locked; hermetic; hermetically sealed; immersible; impermeable; leak-free; leakiness free; leakproof; leak tight; leak tree; liquid sealed; sealed; staunch; tight
密封的铲斗连杆 sealed loader linkage
密封的框架 wiping blade
密封的轻便泵 canned pump
密封的投标人名单 closed list of bidders
密封的投标书 sealed bid
密封的双层玻璃 sealed double glass
密封的文件 seal documents
密封地下爆炸 contained underground burst
密封递价 closed bid; sealed bid; sealed proposal; sealed tender
密封点 seal point
密封电磁继电器 hermetically sealed electromagnetic relay
密封电动机 canned motor
密封电缆 hermetically sealed cable
密封电路 potted circuit
密封垫 filler plate; gasket ring; gaskin; gland packing; packing plate; sealer; sealing plate; sprinkler; static seal; stuffing box
密封垫层 back bed
密封垫带 gasket
密封垫片 gasket seal;sheet gasket
密封垫圈 gasket; gasket ring; gasket seal; joint washer; packing plate; packing washer; seal packing; sheet gasket; seal(ing) washer
密封垫圈接缝 gasket joint
密封垫镶嵌玻璃 gasket glazing
密封顶板 <气压沉箱的> air deck
密封定位架 seal retainer
密封(动环)传动套 seal drive sleeve
密封度 air-tightness; containment; degree of tightness; tightness
密封度试验 leak-tested
密封舵盖 insulated hatch cover; plug hatch
密封发盘 offer under seal
密封阀 backed valve
密封法 scaling method
密封法兰 sealing flange; tongued and grooved flange
密封方法 sealing method
密封防尘车 sealed dust-proof van
密封防火墙 firedamp
密封防石击涂料 anti-stone-bumping sealing paint
密封防水 waterproofing sheet
密封防水舷窗 scuttle
密封防水油膏 water sealant
密封放电器 sealed spark gap
密封分离器 hermetic separator
密封粉料 stopper powder
密封风机 seal fan
密封封口 pressurizing window
密封缝 sealed joint
密封敷层 seal coating
密封服 encapsulating suit
密封盖 air-tight cover; gland cover; joint cap; sealed cap; sealed cover; sealing cap; sealing cover; gland bonnet <轴端>
密封盖板 close-fitting cover
密封盖环 cover ring
密封盖折卸工具 seal replacer
密封干运转 seal running dry
密封干燥空气层 blanket of dry air
密封钢丝绳 locked coil wire rope; sheathed wire rope
密封膏 mastic sealer; sealant
密封膏背衬材料 sealant backing
密封膏浆 slurry seal
密封膏条 bedding
密封隔热玻璃 sealed insulating glass
密封隔热玻璃制品 sealed insulating glass unit
密封给水加热器 closed-feed water heater
密封工程 sealing engineering
密封供应压力 seal supply pressure
密封管 sealed capsule; sealed pipe; sealed tube
密封管分解 decomposition by sealed tubes
密封灌浆 seal grouting
密封罐 sealable tank; seal pot
密封光阀 sealed-off light valve
密封滚动铣齿钻头 sealed roller milled bit
密封焊道 pressure-tight weld; sealing bead; seal weld
密封焊缝 seal joint; seal weld
密封焊接 ca(u)lking weld; sealing bead; seal(ing) weld(ing); tight weld
密封耗损泄漏 seal drain leakage
密封盒 gland pocket; packing box; packing case; seal box; stuffing box
密封护板 skin casing
密封护圈 seal retainer
密封护脂圈拆卸器 grease retainer remover
密封戽斗 sealing bucket
密封滑板 <带式烧结机的> drop bar
密封滑动铣齿钻头 sealed sliding milled bit
密封滑块 <带式烧结机的> drop bar
密封环 annular seal; anti-leak ring; compression ring; gland ring; junk ring; packing ring; packing washer; ring seal; seal cup; sealing collar; sealing ring; lantern ring <水泵的>
密封环盖 sealing ring cap
密封环境 sealed environment
密封换能器 closed transducer
密封黄油 packing grease
密封回热式空气预热器 sealing regenerative-type air preheater
密封混合料 sealing mix(ture)
密封混凝土 sealed concrete
密封货柜 right container
密封机 sealer; sealing machine
密封机罩 hermetically sealed casing
密封集装箱 closed container
密封技术 hermetic sealing technique; sealing engineering
密封剂 encapsulant; jointing compound; liquid sealant; sealant; sealer; sealing agent; sealing compound; aquaseal <电缆绝缘涂敷用>
密封剂填料 bedding compound
密封坚固舱 sealed cabin
密封检查 leakage check; leakiness check
密封件 air-tight packing; seal; sealing component; sealing element; sealing member; staunching piece; seal agent
密封胶 dope; fluid sealant; gasket cement; joint sealant; liquid packing; sealant; sealing compound; seal(ing) gum
密封胶卷 pack film
密封胶泥 lute(in); luting
密封胶皮 stripper rubber
密封胶圈 O-ring seal
密封胶水 liquid sealant
密封胶条 sealing joint strip
密封焦油 aquatard
密封绞刀 shrouded screw conveyer [conveyor]
密封接触型 sealed contact type
密封接缝 sealing joint
密封接合 sealing compound
密封接合器 packing maker
密封接合抓斗 lip seals on grab
密封接头 air-tight joint; closed joint; gland joint; seal fitting; seal(ing) joint; seal nipple; seal point; hermetic seal
密封节(波导) sealing section
密封结 sealed junction
密封结构 hermetically sealed construction
密封结合器 packing maker
密封解调器 encapsulated demodulator
密封介质 sealing medium
密封金属容器 sealed metal container
密封紧力 sealing load
密封进料联动装置 sealed charging linkage
密封静环 seal seat; stationary seal ring
密封静环护圈 seal seat retainer; stationary seal ring retainer
密封绝热养护 mass curing
密封抗压容器 sealed expansion vessel
密封壳 hermetic case
密封壳体 pressure-tight body; seal casing; seal housing
密封可靠 reliable seal
密封可靠性 sealing reliability
密封空气 sealing air
密封空气管 seal-air pipe
密封空腔谐振器 sealed cavity
密封孔 closed hole
密封块 sealing mass
密封蜡 sealing cement; sealing wax
密封肋 sealing rib
密封冷冻压缩机 sealed refrigeration compressor
密封离心机 bermetic centrifuge
密封锂电池 sealed lithium battery
密封力 sealing force load
密封连接 packing joint; seal point; tight coupling; tight fitting
密封料 joint sealant; joint sealing compound; seal; sealing compound
密封料垫层 <嵌玻璃或墙板槽中> back bed
密封漏泄 packing leakage
密封铝皮 alumin(i)um seal(ing) sheeting
密封螺钉 seal(ing) screw
密封螺母 packing nut
密封螺栓 packing bolt
密封螺纹管接头 sealing nipple
密封玛瑞脂 sealing mastic
密封脉冲变压器 hermetically sealed pulse transformer
密封帽 sealed cap
密封门 air-tight door; hermetically sealed door; hermetic door
密封迷路 labyrinth seal
密封面 facing; facing surface; packing surface; sealant profile; seal contact face; sealing face; sealing surface; tight surface; trim tight surface
密封面层 facing surface
密封面尺寸 size of seal contact face
密封面积 sealing area
密封面贴合 seating
密封面泄漏 sealing face leakage
密封模内养护 mo(u)ld-encased curing
密封膜 diaphragm seal
密封膜片容积式液位传感器 diaphragm sealed displacement transmitter
密封摩擦 seal friction
密封能力 plugging ability; sealability; sealing ability
密封腻子 sealant
密封黏[粘]胶 airtack cement
密封黏[粘]结剂 sealing cement
密封浓缩物 sealing concentrate
密封排放口 seal drain port
密封排灰阀 air-seal dust-valve
密封盘管 sealing coil
密封培养 sealed cabinet
密封配电盘 cubicle
密封配件 seal fitting
密封喷头 <取暖装置> sealed jet
密封喷雾式冷却器 enclosed spray type cooler
密封棚子 closed frame
密封皮碗 cup leather; fullering cup; leather package; primary cup; sealing cup
密封皮碗套 sealing cup body
密封片 diaphragm seal; gasket ring; sealing strip
密封屏蔽 closed shield
密封破坏 seal break-off
密封鳍状物 sealing fin
密封企口 seal groove
密封气垫 seal-in air cushion
密封砌筑巨石圬工 megalithic masonry
密封器 packing maker; sealer
密封铅酸蓄电池 sealed lead-acid battery
密封腔 seal cavity
密封腔压力上升 seal cavity pressure rise
密封球 ball sealer
密封区 seal point
密封曲轴箱 pressurized crankcase
密封取芯工具 dealed coring tool
密封圈 rubber seal ring <橡皮式>; air seal ring <空气式>; ca(u)lking ring; joint ring; locking ring; O-ring seal; packer ring; packing ring; sealing lip; seal(ing) ring; seal packing ring; stuffing box; grease seal <滚动轴箱>
密封圈弹簧 packing spring
密封圈的保护圈 backup ring
密封圈接头机 gasket splicer
密封圈座 packing holder
密封绕组 encapsulated winding
密封刃 sealing edge
密封容器 air-tight container; closed container; pressurized reserve; sealed container
密封软木橡胶 cork rubber
密封润滑履带 sealed and lubricated
密封润滑器 sealed lubricator
密封润滑式履带 sealed and lubricated track
密封润滑脂 packing grease
密封塞(子) sealing stopper; closing plug
密封砂浆 seal(ing) mortar
密封上釉单元 sealed glazing unit
密封烧结 sealed sintering

密封设计 encapsulation
密封绳 rope sealing
密封绳索 sealing rope
密封失效 sealing failure
密封时间 seal up time
密封式板撑 tight sheathing
密封式保险开关 safety enclosed switch
密封式变压器 hermetically sealed type transformer
密封式测量仪器 hermetically sealed instrument
密封式的 hermetically sealed
密封式电动泵 canned motor pump
密封式电动机 enclosed motor; hermetic motor; impervious machine; hermetically sealed motor
密封式电机 gas-tight machine; hermetic machine; impervious machine
密封式电路 packaged circuit
密封式电热塞 sealed plug sheathed plug
密封式电容器 potted capacitor
密封式发动机 canned motor
密封式翻车深护结构 enclosed roll-over protection structure
密封式罐车 <装运水泥等> hermetically sealed tank car
密封式光束灯泡 sealed beam lamp
密封式滚柱轴承 sealed roller bearing
密封式呼吸器 drying breather
密封式活动支座 encapsulated expansion bearing
密封式机器 air-tight machine
密封式积分陀螺仪 hermetic integrating gyroscope
密封式记录纸驱动电动机 sealed chart-drive motor
密封式继电器 hermetically sealed relay; sealed relay
密封式晶体管 packaged transister [transistor]
密封式开关 secret switch
密封式开关箱 hermetically sealed switchbox
密封式冷凝机 hermetically sealed condensing unit
密封式离心机 hermetic centrifuge; sealed centrifuge
密封式炼塑机 closed plastic refining machine
密封式履带链轨节 sealed track
密封式螺旋输送机 shrouded screw conveyer[conveyor]
密封式前大灯 sealed beam lamp
密封式人孔盖板 tight manhole cover
密封式入井盖板 tight manhole cover
密封式水冷系统 permanent seal cooling
密封式叶片搅拌机 close blade pug mill
密封式闸刀开关 safety enclosed switch
密封式整流阀 sealed rectifier
密封式制冷压缩机 sealed refrigeration compressor
密封式重水和轻水反应堆 pressurized heavy and light water reactor
密封试验 pressure test(ing)
密封室 air-tight chamber; gland body; sealing chamber
密封室的端子 sealed chamber terminal
密封寿命 sealing life
密封双层玻璃 hermetically sealed double glazing
密封双层玻璃窗 factory-sealed double-glazing unit; sealed double-glazing unit; sealed double glazing window
密封双层玻璃单元 sealed double-glazed unit
密封水 sealed water; sealing water
密封缩颈 sealing constriction
密封弹簧压圈 seal spring compression ring
密封套 gland; gland cover; packing gland; seal gland cartridge; seal(ing) jacket; sleeve gasket; solid enclosure
密封套管轴承 enclosed tubular bearing
密封套夹 boot clamp
密封体 seal
密封体引线头 hermetic enclosure header
密封填充物 sealing filler
密封填缝石灰 joint sealing lime
密封填料 fitting of stuffing; gasket; gland packing; packing material; sealer; sealing bush; seal packing; sheet packing; toe bead
密封填料环 gland ring
密封填片 sheet gasket
密封条 air lock strip; band seal; closure strip; draught excluder strip; gland strip; sealant strip; sealing bead; sealing rod; sealing rope; sealing strip; sealing tape; stamping steel ribbon; weather strip
密封条三维伸缩装置 three-dimension expansion installation with sealing strip
密封投标 sealed bid(ding); sealed proposal; sealed tender
密封投标书 sealed bid
密封凸缘 packing flange; sealing flange
密封涂层 air-tight coating; gasket coating; seal coat(ing); stopping coat
密封涂层混凝土 seal-coat concrete
密封涂层乳胶 seal-coat emulsion
密封涂层乳液 seal-coat emulsion
密封涂料 primer-sealer; sealing paint
密封土样 sealed soil sample
密封外壳 can; containment shell; sealed outer housing
密封外壳发动机 canned motor
密封外壳法灌浆 containment grouting
密封外壳燃油蒸发排出物确定试验 sealed housing evaporative emission determination test
密封碗 obturating cup; sealing cup
密封微波溶出 hermetic microwave digestion
密封微孔泡沫 closed cell foam
密封维修备件 service seal
密封物 encapsulant; sealer
密封物质 sealing mass
密封系数 gasket factor
密封系统 sealed system; sealing arrangement; sealing system
密封线 potted line
密封箱 air-tight container; sealing box; stuffing box; water-tight chest
密封箱烧结 sintering in sealed box
密封橡胶 sealing rubber
密封橡胶圈 lute; rubber-ring packing; sealing rubber ring
密封橡皮圈 lute
密封销 link block
密封效率 leakage efficiency
密封斜插板阀 hermetic deflection damper
密封泄漏 sealing leak
密封型 closed type
密封型电动机 hermetically sealed motor; hermetic motor
密封型热敏电阻器 enveloped thermistor
密封型式 seal style; type of seal
密封性 air-tightness; imperviousness; leakproofness; leak tightness; tightness
密封性测定仪 tightness measuring instrument
密封性测试器 leakage tester
密封性分析 hermeticity analysis
密封性检验 leak(age) test
密封性检验仪 leak-testing apparatus
密封性能 sealing property; stopping property
密封性试验 leak(age) test
密封悬架 seal hanger
密封压盖 gland; gland cover; gland plug; packing gland; seal cover; seal end plate; stuffing box gland; stuffing gland; sealing gland
密封压盖随动件 gland follower
密封压花机 closure-sealing embossed machine
密封压力 sealing load; sealing pressure
密封压缩 sealing constriction
密封压缩机 hermetic compressor
密封烟道 sealed flue
密封样品 sealed sample
密封液出口 seal fluid outlet
密封液接口 seal fluid connection
密封液入口 sealing fluid inlet
密封液(体) sealing fluid; sealing liquid
密封液压系统 sealed hydraulic system
密封仪器 hermetically sealed instrument
密封引线 sealing wire
密封用衬垫 tightening flap; tire band
密封用料 ramming mix(ture)
密封用气体喷烧器 sealing-in burner
密封用铅条 drawn lead trap
密封用软木橡胶 cork rubber
密封用物质 sealing medium
密封用油 seal oil
密封用轴套 shaft installing sleeve
密封邮件 first-class mail
密封油 blocked oil
密封油膏 adhesive seal; mastic seal(ing); sealing cement; sealing compound
密封油灰 sealing cement
密封油漆 sealing paint
密封油漆面饰 sealing paint finish
密封油任 ground seat union
密封油箱 leakproof fuel cell
密封元件 potted component; sealing member
密封源 sealed source
密封云母电容器 hermetically sealed mica capacitor
密封云母环 sealed mica ring
密封增压救生圈 hyperbaric lifeboat
密封增压座舱 manometric(al) capsule
密封窄条 sealing fillet
密封毡 sealing felt(ed fabric)
密封涨圈 sealing ring
密封胀圈 packing flange
密封罩 enclosed fitting; seal cover; sealed cowl(ing)
密封锗探测器 encapsulated Ge detector
密封蒸汽 sealing steam
密封支承环 backup ring
密封脂 stop-leak compound
密封止水 hermetic seal
密封轴承 sealed bearing; stuffing box bearing
密封轴封 gland bush
密封皱纹管 seal bellows
密封注胶 seal pouring
密封转子泵 canned-rotor pump
密封装置 containment; dust stop; gland; gland seal; hermetic unit; obturator; packaged plant; packing assembly; seal; sealable equipment; sealed unit; sealing arrangement; sealing device; sealing gland; sealing installation <盾构等>
密封装置反馈试验 seal oil back up test
密封装置与封接 seals and sealing
密封锥面 sealing cone
密封组件 block box; seal assembly; seal set
密封座 seal retainer
密缝 joint close; neat seam; tight joint; closed joint
密缝工事 sealing works
密缝胶 close-contact glue
密缝接合 close(d) joint; hooked joint
密缝接头 close(d) joint; hooked joint
密缝圬工 tightly jointed masonry
密缝性 impermeability of joints; imperviousness of joints
密缝錾 ca(u)lking iron; ca(u)lking tool
密缝凿 ca(u)lker; ca(u)lking; ca(u)lking iron; ca(u)lking tool; clincher iron
密高岭土 lithomarge; terratolite
密格街道网 close-meshed street network
密根 heavy roots
密沟灌溉法 corrugated furrow irrigation
密固 tightness
密固接缝 tight-strong seam
密管系统 dense tubular system
密灌丛 scrub; thicket
密灌木丛 scrub
密合 driving fit
密合层 close binder(course)
密合挡板围栏 close-boarded fencing
密合的盖子 close lid
密合间隙 close clearance
密合接缝 coped joint
密合接头 closed joint
密合铺板 close-boarded
密合丝扣 close fit
密合屋脊 close-cut hip
密合装填 closed packing
密黑氧化 tight black oxide
密烘铸铁 <一种孕育铸铁> meehanite cast-iron; meehanite(metal)
密厚云 spissatus
密花石柯 tanbark oak
密化粉末 densified powder
密化器 densifier
密环菌属 <拉> Armillariella
密积冰泥 slob
密级 classification category; degree of classification
密级代码 security code
密级配 dense gradation
密级配柏油路面 dense tar surfacing
密级配柏油面层 dense tar surfacing
密级配拌和料 densely graded mixture
密级配表面处治碎石路 dense-coated macadam
密级配的 close(ly) graded; dense(ly) graded
密级配底基层 densely graded subbase
密级配骨料 close-graded aggregate; dense-graded aggregate; densely graded aggregate
密级配黄砂 closely graded sand
密级配混合料 closed mix; dense-gra-

M

ded mix
密级配集料 close(ly)-graded aggregate;dense(ly) graded aggregate
密级配集料沥青混凝土 close-graded aggregate asphalt(ic) concrete
密级配焦油沥青路面 dense tar surfacing
密级配焦油沥青面层 dense tar surfacing
密级配焦油沥青碎石路 dense tar macadam
密级配结合层 close binder(course)
密级配矿物骨料 dense-graded mineral aggregate
密级配矿物集料 dense-graded mineral aggregate
密级配沥青混凝土 close-graded bituminous concrete; dense-graded asphalt concrete; dense-graded bituminous concrete
密级配沥青面层 close-graded bituminous surfacing
密级配沥青碎石(路) dense bitumen macadam
密级配路面 close-graded pavement; dense-graded pavement
密级配煤沥青表面处治碎石路 dense-coated macadam
密级配磨耗层 dense friction course
密级配砂 close sand
密级信息 security information
密集 close;close-packed;close packing;close up;compaction;compression; concentrated; congestion; deepening;massing
密集杯 cluster cup
密集冰 close ice;close pack;compact ice; conglomerated ice; ice nip; packed ice
密集冰的下风边缘 compacted ice lee edge
密集冰块 pack ice
密集病害流行 close epiphytotic disease
密集波 density wave
密集波分复用系统 density wave distribution multiplexer
密集补强 concentrated reinforcement
密集层 dense layer
密集车头时距 close headway
密集抽样 intensive sampling
密集的 close; closely spaced; condensed;conglomerate;thick;dense; intensive
密集的灌木篱笆 hedgerow
密集的晶面 closest-packed crystal plane
密集的信号显示 conglomeration of signal aspects
密集低小微波型 numerous minute echoes; numerous small and low echoes
密集点阵 close-over lattice; close-packed lattice
密集度 closeness;concentration; crowding level;intensity
密集堆冰 compact pack ice
密集二进码 dense binary code
密集放牧 bunched up herding;close herding
密集浮冰群 close pack ice
密集共生 intimate intergrowth
密集厚冰 consolidated ice
密集环形图案 concentric(al) ring pattern
密集混凝土 compacted concrete
密集建筑 density district
密集建筑规划 compact planning
密集建筑物 abutting building
密集节理式劈理 close-joint cleavage
密集晶格 close-over lattice; close-packed lattice
密集晶体 densely packed crystal
密集井框支架 solid crib timbering
密集居住区 compact settlement
密集矩阵 dense matrix
密集菌 heavy bacteria
密集连续图形 dense map continuous
密集裂缝 cluster cracking
密集林场 close woodland
密集林分 dense stand
密集林冠 close canopy
密集流冰 very close pack ice
密集流冰群 close drift ice;close ice
密集流水 close ice;close pack
密集六方晶系 closed-packed hexagonal system
密集六角晶格 close-over hexagonal lattice
密集龙头 close hydrant
密集排桩防波堤 row-of-piles breakwater
密集配焦油沥青磨耗层 dense tar surfacing
密集配焦油沥青碎石路 dense tar macadam
密集配沥青碎石路 dense bitumen macadam
密集喷灌系统 short-distant sprinkler system
密集气孔 porosity
密集区 compact district;dense area
密集区域 density district
密集曲线 tight curve
密集取样 close sampling
密集绕组 concentrated winding
密集射流 fire stream
密集生长 matted growth
密集式书库 compact storage
密集饲料箱 close feedbox
密集损伤 heavy break;heavy injury; heavy lesion
密集索引 dense index
密集台阵 dense array
密集体 conglomerate
密集团聚体 dense aggregate
密集微震台网 dense microtremor network
密集问题 congested problem;congestion problem
密集物 thicket
密集雾滴 density drop
密集系统 congestion system
密集小林 wood
密集小气泡 heavy seed
密集小群 pod
密集效应 density effect
密集型联合 conglomeration
密集性 compactability
密集样本 cluster sampling
密集仪器台阵 intensive instrumentation arrays
密集云层 dense sky cover
密集运输 mass transport
密集再生草 close silage aftermath
密集整枝法 intensive system
密集支架 close-standing; intensive support
密集支柱 prop wall
密集桩排防波堤 row-of-piles breakwater
密集装车 concentrated load(ing)
密集状分布 concentrated distribution
密集资本 deepening of capital
密集组 modal group
密集钻进 close drilling;multiple drilling
密集钻眼 close drilling
密加索尔圆锥 <测定泥浆稠度的仪器> Mecasol cone
密间隔 closed pitching;closed spacing
密间隔气相沉积 closed space vapo(u)r deposition
密间距灌水垄沟 fine ridge
密件 confidential; confidential documents
密接 air-tight joint; fit; joint seal(ing);knit;water joint
密接变速比 closed ratio
密接的挡土板 close boarding
密接点 osculation point;point of osculation
密接缝 bonded joint
密接金属套 <混凝土路面胀缝传力杆的> close-fitting metal cap
密接双星 contact binary
密结的 tightlock
密结合层 closed binder
密结合料 close binder(course)
密结式车钩 tightlock coupler
密经条纹 crammed stripe
密井网 close well spacing
密聚 conglomeration
密聚波 condensation wave
密聚体 conglomerate
密卷螺簧 close coiled helical spring
密卷云 cirrus densus;cirrus spissatus
密孔剂 beaumontage
密孔砌块 honeycombed block
密孔筛 close-meshed screen
密孔砖 terra-cotta lumber
密拉聚酯薄膜 Mylar plastic
密拦污栅 fine trash rack
密勒发射天线 Miller radiator
密勒符号 Miller's symbol
密勒积分器 Miller integrator
密勒锯齿波产生器 Miller sawtooth generator
密勒码 Miller code
密勒平方码 Miller square code
密勒全自动辘轳机 Miller machine
密勒效应 Miller effect
密勒岩石分类 Miller classification of rock
密勒指数 crystal index;Miller index
密肋 ribbed
密肋板 multiribbed plate;multiribbed slab
密肋板式基础 waffle footing
密肋空心砖楼板 filler joist floor
密肋梁 slab-and-beam rib
密肋梁式楼板结构 pan construction
密肋楼板 ribbed(beam) floor;ribbed slab;waffle slab
密肋楼盖 beam and girder floor;ribbed floor;ribbed slab
密肋楼面 ribbed floor;ribbed slab
密肋抹灰构造 close studding
密肋穹顶 waffle
密肋式楼板 cellular-type of floor; waffle floor
密肋式楼盖 cellular-type of floor
密肋式楼面 cellular-type of floor
密篱 thick-set
密立根静电计 Millikan electrometer
密立根油滴实验 Millikan oil-drop experiment
密砾砂混合物 dense gravel-sand mixture
密炼 banburying
密炼机 Banbury mixer;internal mixer
密林 closed forest;dense crop;dense stand;jungle;midwood
密林带 large forest region
密林地区 weald
密林区 densely-wooded area;density-wooded area
密螺纹接套 close nipple
密螺旋体 treponema
密码室 coding office
密码子 codon
密铆 close riveting
密面表 hunting case watch
密苗 close stand
密谋策划 conspire
密木材 pycnoxylic wood
密木纹 close grain
密年轮木材 narrow-ringed timber
密涅瓦神庙 Temple of Minerva
密耦 close couple;tight coupling
密耦泵 close-coupled pump
密耦合 close coupling;tight coupling
密耦水箱坐式大便器 close-coupled tank and bowl
密排 close packing;solid matter
密排板墙筋 close studding
密排板桩 close sheeting pile
密排舱底板 close ceiling
密排插板 forepiling plate
密排法 close-spaced method
密排灌注桩 contiguous bored pile
密排技术 close-spaced technique
密排晶格 close-over lattice
密排框架 closely spaced frame
密排六方 close-packed hexagonal
密排六方结构 close-packed hexagonal structure
密排六方晶格 close-packed hexagonal lattice
密排模板隔墙 close-boarded screen
密排木版围篱 close-boarded fencing
密排深水口站头 simulated insert bit
密排天线阵 closely spaced array
密排形式挡板 <土方开挖中> close sheeting
密排形式支撑 <土方开挖中> close timbering
密排桩 close pile
密判图 cup seal
密配合 close fit;drive fit;fay;snape; snug fit
密配合的 leakproof fit
密配合螺栓 driving fit bolt;reamed bolt;reamer bolt
密铺板屋面 close-boarded(battened) roofing
密铺衬板 tight sheathing
密铺挡板 tight sheathing
密铺挡土板 tight sheathing
密铺面板 closed deck
密铺木板屋顶 close-boarded roof
密铺排木格栅地板 plank-on-edge floor
密铺望板 tight sheathing
密铺屋面板 close-sheeted roofing
密切【数】osculate;osculation
密切插值法 osculating interpolation
密切的 intimate;osculatory
密切点 osculating point;point of osculation
密切二次曲面 osculating quadric
密切关系 close correlation
密切轨道 osculating orbit
密切轨道根数 osculating element
密切轨道椭圆 osculating orbital ellipse
密切轨道元素 osculating orbit element
密切合作 hand-in-glove
密切接触 intimate contact
密切结合的 married
密切抛物线 osculating parabola
密切平面 osculating plane
密切球面 osculating sphere
密切三次曲线 osculating cubic curve
密切圆【数】osculating circle
密切坐标 osculating coordinates
密圈螺旋弹簧 closed coiled helical spring
密圈弹簧 closed coil spring
密绕螺旋弹簧 close-coiled spring

密伞圆锥花序 panicled thyrsoid cyme
密桑奈特炸药 methanite
密砂 compact sand
密砂布 closed coat
密砂砾混合物 dense sand-gravel mixture
密砂纸 closed coat
密筛孔状 non-clathrate
密栅 fine rack
密栅云纹法 Moiré method
密上胶层 closed coat
密烧的 dense-burned
密生灌丛 mogote
密生苗 more population
密生群落 closed community
密实 densify
密实表面 close(ly)knit surface
密实冰 compact ice
密实部件 compact part;compact unit
密实材料 compact material
密实层 close bed;dense bed;dense layer
密实程度 compaction rate
密实充填 solid packing;solid stowing
密实的 close-grained;compact;void-free;voidless;dense
密实的表面 well-knit surface
密实的石灰石 compact limestone
密实的原土 tight bank
密实地壤 close-settled soil
密实度 degree of compaction;degree of density;compactibility;compactness;consistence[consistency]
密实度比 solidity ratio
密实度控制 <控制现场浇筑混凝土的密实度> density control;compaction control
密实度曲线 compaction curve
密实度试验 density test
密实堆积 <木料等> tight stacking
密实方法 compaction method
密实缝 close joint
密实干重 dry compacted weight
密实钢筋混凝土 reinforced dense concrete
密实骨料 dense aggregate
密实焊缝 ca(u)lking weld;composite weld
密实化 densification
密实化胶合木 densified laminated wood
密实混合料 closed mix(ture);compressed mix(ture);dense mix(ture)
密实混凝土 dense(aggregate)concrete;air-free concrete;air-tight concrete;concrete of low porosity
密实混凝土块(体) dense concrete block
密实混凝土砌块 dense concrete block
密实混凝土墙 dense concrete wall
密实积雪 closely packed snow
密实级配的 close-grained
密实级配结构 close-grained structure
密实级配沥青路面 dense-graded bituminous surfacing
密实级配沥青面层 dense-graded bituminous surfacing
密实集料 dense aggregate
密实剂 densifier;sealant
密实浇注法 end cast method
密实胶合板 super-pressed plywood
密实结 closed binder course
密实结构 close texture;compact structure;dense structure;dense texture
密实结合层 close binder(course)
密实颗粒 compact grain
密实颗粒的 compact grained
密实矿物骨料 dense mineral aggregate
密实矿物集料 dense mineral aggregate
密实理论 compaction theory
密实流 internal pressure current
密实面层 closed surface;close(ly)knit surface;compacted surface layer;skin <多孔材料的>
密实磨耗层 dense friction course
密实木材 densified wood;super-pressed wood
密实泥炭 <形成泥炭沼泽地底基> baken peat
密实黏[粘]土砂岩 dauk[dawk]
密实强固焊缝 tight-strong weld
密实墙 compact wall
密实砂 <又称密实沙> closed sand;compact sand;packsand;dense sand
密实砂层 tight sand
密实设备 compaction equipment
密实射流 solid jet
密实湿度 compaction moisture
密实石灰岩 mountain limestone
密实弹性体压缩密封垫 dense elastomeric compression seal gasket
密实套筒 solid sleeve
密实铁 close-grained iron
密实图 compact;dense
密实土(壤) close-settled soil;compact soil;hard compact soil;solid ground;tight soil
密实纹理木材 close-grained wood
密实物料 close substance
密实物质 close substance
密实系数 compacting factor;compaction factor;packing coefficient
密实纤维 void-free fibre
密实纤维板 solid fiberboard
密实橡胶 dense rubber
密实效果 compacting effect
密实型路堤 compact embankment
密实性 compactibility;compactness;denseness;solidity
密实雪 compact snow
密实岩块 closely jointed rock mass
密实岩石 compact rock
密实因数 compacting factor
密实因素 compacting factor
密实轧机 compaction roll
密实指数 compaction index
密实制品 full density product
密实桩 compaction pile
密实状态 dense condition;dense state
密实组织的 close textured
密史脱柱风 mistral
密式结合层 close binder(course)
密室 back room;conclave;sanctum;adytum <古时庙宇中的>
密水网的地区 dense waterway net region
密丝组织 plectenchyme
密斯风格【建】Miesian style
密斯风格建筑 Miesian architecture
密苏里河 Missouri River
密苏里统【地】Missourian
密苏里州大理石 Missouri marble
密苏里州红色花岗岩 Missouri red granite
密苏里州柔性路面设计法 <美> Missouri design method
密索 multiple stays;multistays
密索体系 multicable system
密索体系斜拉桥(动系统) cable-stayed bridge with multiple stay system
密索斜拉桥 multicable stayed bridge
密锁自动耦合器 tightlock coupler;tight-lock coupler
密特勒恩 <计量信息的单位> metron
密体 dense body
密填 compact fill
密填沥青材料的仰拱 paved invert
密条播 dense sowing in line;solid planting
密贴 close
密贴层 osculatory
密贴道岔 closed switch
密贴活塞环 conformable ring
密贴尖轨 closed point;closed point rail;closed tongue rail;securely closed tongue
密贴浇筑(法) <后张预应力装配式预制节段的一种施工方法> match-(ing)casting
密贴浇筑接缝 <依次分段预制浇筑的梁段之间的接缝> match-cast joint
密贴调整杆【铁】adjustable switch operating rod
密贴压扁试验 flatten(ing)close test
密铜铁矿 mahogany ore
密陀僧 lead monoxide;lead oxide;litharge;lithargite;Lithargyrum
密陀僧中毒 lithargysmus
密妥耳 metol
密网 detail network
密网布置钻井 close-spaced wells
密网格穹隆 geodesic dome
密网街道网 close-meshed street network
密网筛 close-mesh;fine structure mesh
密网栅极 fine-mesh grid
密网眼的 close-meshed
密网钻进 close drilling;dense drilling
密位 mil unit of angular measure
密位尺 milrule
密位公式 mil formula
密位刻度尺 milscale
密文 cipher text
密文剩余类 cryptogram residue class
密纹 dense grain;fine groove;microgroove;minigroove
密纹唱片 fine-grooved disc;long play;microgroove;microgroove record
密纹的 close(d)-grained
密纹理 close grain
密纹螺线 closely spaced spiral
密纹木(材) close-grained timber;close-grained wood;dense wood;narrow-ringed timber;slow-grown timber;closely ringed timber;comb-grained wood
密纹木料 narrow-ringed
密纹组织 compact-grain structure
密雾 density fog
密西西比层(或系) <美国石炭纪>【地】Mississippi(an)system
密西西比河 <美> Mississippi River
密西西比河谷式铅锌矿床 Mississippi-valley-type lead-zinc deposit
密西西比河里程数 mileage number
密西西比河墨西哥湾人工水道 Mississippi river gulf outlet
密西西比河型浮标 Mississippi river-type buoy
密西西比纪 <美国,约相当于欧洲的早石炭世>【地】Mississippian period
密西西比三角洲 Mississippi delta
密西西比深海扇 Mississippi abyssal fan
密西西比系 <属美国石灰纪>【地】Mississippi system
密西西比(下石灰)纪砂岩【地】Mississippian-age sandstone
密线凿石面 comb chiselled finish
密相 dense phase
密相流化床 dense-phase fluidized bed
密相气力输送 dense-phase pneumatic conveying
密相(气流)输送系统 dense-phase transporting system
密相气升输送法 hyperflow conveying method
密镶的 <金刚石> heavy set
密歇尔螺旋锚(杆) Mitchell's screw anchor
密歇尔推力轴承 Mitchell's thrust bearing
密歇尔止推轴承 Mitchell's thrust bearing
密屑体 densinite
密信 confidential message;secrecy message
密押 cipher;code number;test key
密眼筛 hair sieve
密眼围网 minnow seine
密叶 heavy foliage
密叶饰 stiff leaf
密叶饰柱顶 stiff-leaf capital
密叶饰柱帽 stiff-leaf capital
密钥 cipher code;key
密钥产生器 key generator
密钥加密【计】key cryptograph
密钥组件 key module
密跃层 pycnocline
密云 dense cloud;thick cloud
密云天空 very cloudy sky
密蒸器 autoclave;pressure boiler
密枝木 branchy wood
密执安式平行龟裂陶槽 Michigan cut
密执安州公路处柔性路面设计法 Michigan State Highway Department method
密植 close planting;close seeding;compact crop;compact planting;condensed planting;high plant population;solid-planted;thick planting
密植度 tangled vegetation
密植距 close spacing
密植作物 close growing crop
密植作物轮作 alternate close-grown crop
密质 compact substance;substantial compacta
密致材料 dense material
密致结构 fine texture
密致零件 dense part
密致青铜 dense bronze
密置钢筋 dense reinforcement
密置天线阵 closely spaced array
密株 more seedlings
密柱式 <古希腊、古罗马神庙的,柱间为1.5米柱径的形式> pyknostylos[picnostyle]
密装 close package
密着剂 adherence promoter
密着检验 <橡胶> adhesion testing
密着力试验 friction(al)pull test
密着强度试验 <橡胶> adherence strength test(ing);adhesion strength testing
密着性 bondability
密族的【地】flocculent
密族结构【地】flocculent structure

幂 index;mathematic(al)power;power;degree

幂乘积定理 power product theorem
幂次加速度 cresceleration
幂等 idempotent
幂等变换 idempotent transformation
幂等定律 idempotent law
幂等矩阵 idempotent matrix
幂等性 idempotence
幂等性质 idempotent property

幂等因子 idem factor
幂等元 idempotent element
幂定律 power law
幂法 power method
幂规律 power rule
幂函数 exponential function; power function
幂函数关系式 power law relation
幂函数模型 power function model
幂函数曲线型 curve type of power function
幂函数校正 power function correction
幂和对称函数 power-sum symmetric(al) function
幂积分 exponential integral
幂级 power level
幂级数 positive series; power series; series of powers
幂级数的对数 power-series logarithm
幂级数的收敛性 convergence of generating function; convergence of power series
幂级数的松弛 relaxation of power series
幂级数解法 power-series solution
幂级数展开 expansion into power series; power series expansion
幂集(合) power set
幂刻度尺 power scale
幂零的【数】nilpotent
幂零矩阵 nilpotent matrix
幂零流形 nilmanifold
幂零群 nilpotent group
幂零算子 nilpotent operator
幂零阵 nilpotent matrix
幂律 power law
幂律廓线 power law profile
幂律流(体) power law fluid
幂律失真 power law distortion
幂律指数 power law index
幂频谱 power spectrum
幂平均值 power mean
幂群计数定理 power group enumeration theorem
幂剩余 residue of the power
幂数 exponent; power of number
幂特性 power characteristic
幂线 radical axis
幂因数 power factor
幂因子 power factor
幂指数 power exponent; power index; power series
幂指数风廓线 power law wind profile; wind speed power law
幂指数符号 exponent sign

嘧啶 pyramine

嘧啶磷 pyimithate
嘧菌醇 triarimol

蜜胺 cyanuramide; melamine

蜜胺层压板 melamine laminate
蜜胺合成树脂胶 melamine resin glue
蜜胺甲醛胶 melamine-formaldehyde glue
蜜胺甲醛黏[粘]胶剂 melamine-formaldehyde adhesive
蜜胺甲醛树脂 melamine-formaldehyde resin
蜜胺胶 melamine glue
蜜胺脲醛树脂 melamine-urea resin
蜜胺树脂 melamine; melamine-formaldehyde; melamine resin
蜜胺树脂层压板 laminated melamine resin board
蜜胺树脂清漆 melamine resin varnish
蜜胺塑料 melamine plastics; melaminoplast
蜜苯胺 melaniline
蜜环菌 honey fungus; shoestring fungus
蜜黄长石 meliphane; meliphanite
蜜黄色 honey-gold
蜜饯 confection
蜜距 nectariferous spur
蜜蜡 beeswax; bee wax; ceromel; wax
蜜蜡黄 beeswax yellow
蜜蜡黄玉 beeswax jade
蜜蜡石 honeystone; mellite
蜜蜡提取器 wax extractor
蜜流季节 honey flow
蜜露 honeydew
蜜醛塑料 melamac
蜜糖 syrup
蜜陀僧 <氧化铅> lead monoxide; yellow lead(oxide)
蜜味桉 yellowbox eucalyptus

绵绸 bourette

绵火药 pyroxylin(e)
绵矿 floss
绵纶 caprone
绵马油树脂 oleoresin aspidium
绵皮孔菌属 <拉> Spongipellis
绵雨 continuous rain
绵织品 cotton manufactured goods
绵纸 tissue paper

棉白杨 cotton wood

棉包橡皮包布和油包布 cotton floater
棉包用麻布 cotton bagging
棉被 quilt
棉被形采暖器 quit radiator
棉编胶管 rubber hose braided with cotton wire
棉布 cotton cloth; fabric cotton; muslin-delaine
棉布隔膜袋 muslin bag
棉擦布 cotton rag
棉尘 cotton dust
棉尘沉着病 byssinosis
棉绸 noil cloth; noil poplin
棉垫 <养护混凝土用> cotton mat
棉短绒 linter
棉缎 sateen
棉缎织物 sateen-weave fabric
棉法兰绒 flannelet(te)
棉帆布 cotton canvas
棉帆布盖布 cotton duck tarpaulin
棉帆线 cotton twine
棉纺厂废水 textile-mill waste(water)
棉纺业用水水质标准 quality standard of water for cotton industry
棉纺(织)厂 cotton mill
棉腐卧孔菌 Poria raporaria
棉花 cotton; lint
棉花带气候 cotton belt climate
棉花火药 cotton powder; nitrocellulose
棉花基地 cotton base
棉花收割机 cotton picker
棉花胎 bats
棉花土 cotton soil
棉花栽培 culture of cotton
棉浆黑液 cotton pulp black liquor
棉浆液 cotton pulp
棉胶 celloidin
棉胶干板 collodion dry plate
棉胶湿板 wet collodion plate
棉胶湿片法 collodion process
棉胶纸 celloidin paper
棉绞线 cotton sewing thread in hank
棉结测试仪 nep meter
棉结检验机 nep testing machine
棉结检验仪 nep tester
棉经马鬃纬衬里 baline
棉经纸纬地毯 fiber[fibre] rug
棉卷 cotton roll
棉卷辊 lap roller
棉卷均匀度试验机 lap evenness tester
棉蜡 cotton wax
棉蜡绳 cotton paraffined rope
棉蓝染液 cotton blue staining solution
棉料绝缘板 Elephantide pressboard
棉隆 dazomet; mylone
棉麻帆布 jute canvas
棉麻混合织物 lined and cotton mixture
棉麻交织物 cotton warp linen
棉麻绳 cotton(and) hemp rope
棉麻织物 ramie cotton fabric
棉毛 linter
棉毛混纺地毯 drugget
棉毛混纺地毯纱 bi-fibre
棉毛混纺织物 cotton and wool mixture; cottonette
棉毛交织物 half-wool
棉球 cotton ball
棉球固定镊 tampon forceps
棉球缨 <球形边饰> cotton ball tassel
棉球云 cotton ball clouds
棉区 cotton region
棉圈球状边饰 cotton loop ball fringe
棉染蓝 cotton blue
棉染料 cotton dye
棉绒 cotton flock; cotton linter; cotton velvet; cotton wool; lint; velour; velveteen
棉绒布 winsey
棉绒除去器 linter
棉绒浆 linter pulp
棉绒胶乳涂料 linted latex paint
棉绒毯 pile cotton blanket
棉绒填料 cotton flock filler
棉绒纤维 linters
棉绒纸 lint paper
棉塞 cotton plug; cotton-wool tampon; tampon
棉塞弹簧垫 innerspring cotton-filled mattress
棉塞套管 tampon cannula
棉塞支托法 columning; columnization
棉纱 cotton; cotton yarn; rag; spun cotton; thread
棉纱包电缆 cable in cotton
棉纱擦帚 cotton swab
棉纱绝缘 cotton insulation
棉纱绝缘铅包配线电缆 lead-sheathed cotton-insulated distributing cable
棉纱撇缆 cotton heaving line
棉纱人造丝混合织物 cotton and artificial silk mixture
棉纱人造丝镶边带 cotton and rayon gimp
棉纱绳 cotton cord; cotton rope
棉纱手套 cotton gloves
棉纱头 cotton waste
棉胎 wadding
棉填料 cotton packing
棉条 strip wool
棉纤胎油毡 rag felt
棉纤维强力试验机 cotton fiber strength tester
棉线 cotton; cotton hair; cotton string; cotton thread
棉箱顶盖 basket lid
棉屑沉着病 byssinosis
棉屑肺 byssinosis
棉心 cotton core
棉絮 batting; cotton batting; cotton wool; flocs
棉絮沉着病 byssinosis
棉絮抛光圆盘 cotton polishing disk
棉絮状 flocculence
棉絮状渗出点 cotton-wool spots
棉硬脂 cotton-seed oil stearin
棉油 cotton(-seed) oil; oleum gossypii seminis
棉毡 blanket; felt
棉织薄膜 cotton fabric membrane
棉织衬垫 cotton filler
棉织带 cotton tape
棉织帆布 cotton duck
棉织品 cotton; cotton fabric; cotton goods; cotton textile
棉织品供应过剩 glut of cotton goods
棉织物 cotton fabric; fabric cotton
棉织斜纹布 cotton drill
棉质碎布 cotton rag
棉籽 cottonseed
棉籽壳堵漏材料 cotton-seed hulls
棉籽清选机 cotton-seed cleaner; cotton-seed cleaning machine
棉籽绒 linter
棉籽油 cotton-seed oil
棉籽油硬脂精 cotton-seed oil stearin
棉籽油脂肪酸 cotton seed oil fatty acid

免办登记证券 exempt securities

免办年度所得税申报 exempt from filing annual income tax returns
免保养的 maintenance-free
免测沉淀法 immunoprecipitation
免拆验 exemption from customs examination
免偿债务 acquittance of a debt; debt relief; exemption of debt; forgive a debt; release of debt; remit a debt
免除 absolution; absolve; acquittance; dispensation; exempt; exemption; exoneration; free; immunity; obviate
免除保养 maintenance prevention
免除部分责任 partially exempt obligation
免除偿还的背书 endorsement without recourse
免除担保付款的背书 endorsement without recourse
免除的 immune
免除吨位 tonnage exemption
免除罚金 relief against forfeiture
免除罚款 remission of penalty
免除干扰 immune from interference
免除工商税 exempt from commercial and industrial taxes
免除故障 fail-safety
免除和开敞部位吨位 tonnage of exempted and open spaces
免除汇票上的责任 exchange of liability on the bill
免除捐税 immune from taxation
免除清偿 discharge of repayment
免除使用费 dispense with royalties
免除双重税收 double taxation relief
免除所得 exempt income
免除所得税 exemption from income tax
免除条件 exemption clause
免除条款 escape clause
免除刑罚 abatement and exemption from penalty
免除义务 relief from obligation
免除责任 dissolution of responsibility; exemption from liability

免除责任背书 endorsement without recourse
免除责任条款的删除 deletion of exclusion
免除债券 exempt bonds
免除债务 abatement of debts; acquittance of a debt; debt relief; discharge of debt; forgive debt; forgiveness of debts; forgiveness of liabilities; release from debt; remission of debts
免除债务证书 quittance
免除者 remitter
免除证书 exemption certificate
免除重税 take-off a heavy tax
免除作成拒绝书 protest waived
免错 fault-avoidance
免逗点码 comma free code
免毒试验 avoidance test
免罚 impunity
免罚条款 exception clause; exemption clause; exoneration clause; escape clause <贸易等的>
免罚协议 indemnity agreement
免费 exempted from charges; franco; free; free of charges; gratis; gratuity; toll-free; without charges
免费搬运 free haul
免费保管期 free storage period; period of free storage; period of free time
免费保险 free insurance
免费保修期 free maintenance
免费保养服务 after-sale
免费驳运装卸 free lighterage
免费乘车 free riding
免费乘车人 deadhead
免费乘(公交)车政策 zero-fare policy
免费乘客 non-revenue passenger
免费储存时间 free time for storage
免费存放物品 free storage of articles
免费存车场 free car park
免费搭车 hopping
免费搭乘 hopping
免费代理程序 free agent
免费道路 toll-free road
免费的 cost-free; no charges; gratuitous; no cost; not charged; on the house; uncharged
免费的示范录像带 free demonstration videotape
免费递送 free delivery; free dispatch
免费电话 free telephone call; toll-free call
免费堆存期 free storage period; free time
免费额<重量、价费等> free allowance
免费发送 free dispatch
免费奉送 free of cost
免费服务 free service; gratis service; gratuitous service
免费服务和修理 free service and repair
免费服务区 base rate area
免费公共交通 free public transportation
免费公路 toll-free highway
免费伙食与住宿 freeboard and room
免费货样 sample of no value
免费检验 inspection free
免费交货 free delivery
免费交货地点 point of free delivery
免费留存期限 period of free storage
免费旅行 free travel; gratuity of travel
免费率(额)条款 franchise clause
免费年龄 fare free age
免费期间 free period
免费桥 toll-free bridge
免费热线 toll-free hotline
免费入场 admission free
免费入场名单 free list
免费入场券 free admission
免费软件 free-ware
免费商品 free goods; gratuitous goods
免费时间 free time
免费时间的截止 expiration of free time
免费试用 free trial
免费送货 free delivery
免费随带行李 free baggage allowance
免费所得 franked income
免费停车场 free car park
免费通话 non-registered call; "no charge" call
免费通行 frank
免费土方运输 free-haul yardage
免费物品 free goods
免费享用公共货物者 free rider
免费卸载 free discharge
免费行李 baggage allowance; free baggage allowance
免费宣传 free publicity
免费样品 free sample
免费医疗 free health service; free medical care; free medical service; free medical treatment
免费用户电报 free telex call
免费优待用户电报 franking privilege telex call
免费邮递 free delivery
免费邮寄 frank
免费运距 free haul; free-haul distance
免费运输 free haul; free-haul traffic; free transit; traffic conveyed free
免费运送 carriage free; free haul
免费运送到家 free house
免费赠品 bonus; lagniappe
免费治疗 free treatment
免费注销 flat cancellation
免费资源 free resources
免付关税 free of customs; free of duty
免付利息 no interest; waiving interest
免付给租佣金 free of address
免付速遣费 free despatch[dispatch]
免付所得税 free income tax
免付委托佣金 free of address
免付印花税 free of stamp
免付佣金的 no-load
免付邮资 postage free
免付运费 carriage free; carriage paid
免给 exempt from customs examination
免耕 no-till(age)
免耕法 no-till cultural treatment; no-tillage system; zero till(age)
免耕农业 no-till agriculture; till-less agriculture
免耕制 no-tillage system
免关税 exempt from customs duty
免毫釉 hare's fur
免检 exemption; testing exemption
免检的 exempted from inspection
免检压力容器 pressure vessels exempted from inspection
免交保险费 waiver of premium
免缴石油收益税 exempt from petroleum revenue tax
免缴税入息额 tax threshold
免缴所得税界限 exemption limit
免接费 terminal pickup allowance
免结汇 no-exchange surrender(ed)
免扣预提税证 deduction exemption certificate
免纳个人所得税 be exempted from individual income tax
免纳所得税 free of income tax
免赔 abatement
免赔额 deductible; franchise <美>
免赔额条款 franchise clause
免赔率 franchise
免赔(率)条款 franchise clause
免赔事项 exceptions form liability
免疲热处理 immunizing
免票 free of charges; free pass; free ticket; deadhead <搭车、看戏的>
免票乘车(的人) deadhead
免票搭车<看戏等> deadheading
免去海损估价 free of all average
免去债务 cancellation of indebtedness
免润滑密封履带 non-lub sealed track
免试准则 no test criterion
免收关税地区 free zone
免收航空附加费的航空函件 undercharged airmail correspondence
免收进(出)口税的货物单 free list
免收利息 free of interest
免收税款 waive duty
免收邮费的 franco
免收运费 freight free
免收运费的 franco
免受惩罚 impunity
免受断裂 safety against rupture
免受开裂 safety against cracking
免受倾倒 safe against overturning
免受损失 indemnification
免受弯曲 safe against buckling
免受限制的职工 exempt employees
免税 duty exemption; exemption from tax(ation); free from duty; free of charges; free of duty; free of tax; immunity from taxation; net of tax; remission of tax; tax avoidance; tax exclusion; tax exemption; tax retirement; toll-free; zero rate of duty
免税报单 free entry
免税标准 level of tax avoidance
免税材料 duty free material
免税财产 exempt property
免税财产交换 non-taxable exchanges of property
免税仓库 duty free storage
免税出口货物 tax exempt exports; tax-free export
免税出入港 port franco
免税储蓄证书 all savers' certificate
免税待遇的承诺 zero-duty binding
免税单 bill of sufferance; duty free slips; exemption certificate
免税的 duty free; tax-exempt; tax-free; toll-free; untaxed
免税的所得 income exempted from tax
免税地产 exempt property
免税地带 duty free zone
免税地区 tax haven
免税递送 frank
免税点 exemption point
免税店 duty-free shop
免税额 allowance; zero bracket amount
免税法<尤指部分所得税> exemption
免税方案 tax credit scheme
免税方法 exemption method
免税房地产 tax-exempt property
免税放行 exemption of duty; release without payment of duty
免税分配额 tax-free distribution
免税服务 free service
免税服务费 toll-free
免税港口 tax haven
免税公路 toll-free highway
免税公债 tax-exempt bond
免税规定 requirement for tax exemption
免税好处 advantages of certain tax benefits
免税货单 free list
免税货名表 free list
免税货物 commodities exempt from taxes; exempt goods; free articles; free cargo; free commodities; free from particular average; from taxes; zero-rated goods
免税货物(进口)报单 entry for free goods
免税货物(明细)表 free list
免税机制 exemption mechanism
免税及无配额限制 duty and quota free
免税奖学金 tax-exempt scholarship
免税交换 tax-free exchange
免税交易 exempted exchange; tax-free exchange
免税津贴 tax-free allowance
免税进口 duty free entry; duty free import; duty free importation; enter duty free; free import; free importation; free of duty; tax-exempt import; tax-free import; temporary admission
免税进口货 free imports
免税进口货单 free list
免税进口货物 duty free goods; exempted goods; free goods; tax-exempt import; tax-free import
免税进口货物报单 entry for free goods
免税进口申报单 free entry
免税进口许可证 permission for duty free importation
免税进入市场 duty free access to the market
免税口岸 free port
免税宽减额 tax-free allowance
免税利润 tax-free profit
免税利息 tax-exempt interest
免税利息收入 non-taxable interest income
免税利益 tax-free interest
免税贸易区 tax-free trade zone
免税排水 exempted discharge
免税票证 tax-exempt bond
免税品 duty free goods; free commodities; free entries; free goods
免税品进口报单 entry for free goods
免税品进口报关单 transhipment free entry
免税品起货批准单 lading order for free goods
免税凭证 duty free certificate; tax exemption certificate
免税期 exemption period; free time; tax holiday <口语>
免税期满 the expiration of the period for exemption
免税期限 limitation of the period for tax exemption
免税清算 tax-free liquidation
免税区 duty free territory; exempt zone; free trade area; free zone
免税人 exempt
免税商店 duty free shop
免税商品 articles free; free commodities
免税商品表 free list
免税商品目录 free list
免税声明 tax affidavit
免税收入 exempt income; tax-exempt income
免税收益 exemption of income; franked income; tax-exempt income
免税售出 tax selling
免税输出 tax-exempt export tax-free export
免税输入 tax-exempt import tax-free import
免税输入品 duty free import
免税数 freedom from taxation; free pay; toll-free number

免税所得 income exempted; tax-free income
免税所得额 exempt income
免税条款 exemption clause
免税通过 free of duty entry
免税通行证 <法语> passavant
免税投资 tax-free investment
免税物品 duty free articles
免税限额表 duty free quota list
免税项目 allowance; tax-exempt item; tax-exempt items allowance
免税信贷 tax sparing credit
免税研究奖金 tax-exempt fellowship
免税邮寄 frank
免税运距 free haul; free-haul distance
免税债票 tax exempt
免税债券 non-taxable securities; tax-exempt bond; tax-free bond
免税证明 tax exemption certification
免税证明书 tax exemption certificate; duty free certificate
免税证券 tax-exempt securities
免税证书 tax exemption certificate; duty free certificate; exemption certification
免税支出 allowable expenses
免税执照 duty free certificate; tax exemption certificate
免税制度 exemption system; tax exemption system
免税周围地带 free perimeter
免税周围区 free perimeter
免税转让 non-taxable transference
免税装 bill of sufferance
免税准备基金 tax-free reserve funds
免税资本利得 capital gain free of taxation
免税资产 exempt property
免税资产交换 tax-free exchange
免税资产交换或转让 tax-free exchange or transfer
免税资产转让 tax-free transfer
免税组织 tax-exempt organization
免送费 terminal delivery allowance
免损力 damage resistance
免所得税 free of income tax
免提式 hands free
免调 adjustment-free
免调的烧油系统 adjustment-free fuel system
免调整的 a adjustment-free
免调整的系统 adjustment-free system
免贴额条款 free of franchise clause
免投抑制 immunosuppression
免息 free of interest
免息贷款基金 interest free loan fund
免息借券 loaned flat
免消费税输入申报单 free consumption entry
免修极限尺寸 no-repair limit
免压带 pressure-release zone
免压圈 loosen zone; Trompeter zone
免验 exempt from customs examination
免验放行 pass without examination
免验证 laissez-passer
免疫的 immune
免疫放射测定 radioimmunoassay
免疫化学 immunochemistry
免疫力 immunity
免疫学 immunology
免疫证书 bill of health
免印花税 free of stamp
免油轴承 oilless bearing
免于处罚 remit
免于公诉 immunity from prosecution
免于核监督 exemptions from safeguards
免于扣押和执行的财产 property exempt from attachment and execution
免于纳税的收入 exempt income
免于起诉 exemption from prosecution
免于验关 exempt from customs examination
免予灭鼠证书 deratization exemption certificate
免予熏舱证书 deratization exemption certificate
免予征税 be exempted from taxation
免予追索 without recourse
免运费 freight free
免运费的 franco
免运费回扣 <指租船> free of address
免遭洪水泛滥的能力 flood immunity
免遭损失 indemnification
免责 exemption; impunity; relief
免责理由 reasons of impunity
免责事项 exception from liability
免责条款 escape clause; exception clause; exceptions; exemption clause; exoneration clause; hedge clause
免责约定 hold harmless agreement
免债 remit a debt
免债条款 negligence clause
免征地方所得税 exemption from the local income tax
免征额 zero bracket amount
免征个人所得税 be exempted from individual income tax
免征关税 exempt from the levying of tariff
免征进口税的货物 free goods
免征所得税 be exempted from income tax
免征消费税 exempt from consumption tax
免租 remit rent
免租金的 rent-free
免租期 rent-free period

勉强过半数 bare majority

勉强合格的能力 marginal ability
勉强合用的材料 marginal material
勉强可感地震 scarcely noticeable

冕玻璃 crown glass

冕洞 coronal hole
冕珥 coronal prominence
冕火石玻璃 crown flint glass
冕牌玻操作过程 crown process
冕牌玻璃 alkali-rich glass; crown; crown glass
冕牌玻璃棱镜 crown-glass prism
冕牌玻璃透镜 crown-glass lens
冕牌玻璃制作法 crown process
冕牌光学玻璃 crown optic(al) glass
冕牌萤石玻璃 fluor crown
冕式透气 <存水弯上面通气用> crown vent(ing)
冕透镜 crown lens
冕云 coronal cloud
冕状叉 crown fork
冕状齿联轴节 crown gear coupling
冕状齿轮 contrate wheel; crown gear; crown wheel
冕状顶部 crown-like top
冕状轮擒纵机构 crown wheel escapement; verge escapement
冕状曲线 crowning curve
冕状爪式擒纵机构 verge dub-footed escapement

缅甸本色粗布 hypin

缅甸大理石 <一种粗纹理呈粉紫中灰色调的> Deer Isle
缅甸方柱石 Burmese scapolite
缅甸佛堂 payawut
缅甸红色黄檀 Burma tulipwood; tamalin
缅甸黄檀 Burma blackwood
缅甸建筑 Burma architecture
缅甸-马来亚地槽【地】Burma-Malaya geosyncline
缅甸枪木 Burma lancewood
缅甸式(建筑) Burmese style
缅甸寺院的塔 ceti
缅甸塔 payasat
缅甸桃花心木 Burma mahogany; thitka
缅甸天然漆 Burmese lacquer
缅甸硬琥珀 burmite
缅甸玉 Burma jade; Burmese jade
缅甸郁金香木 Burma tulipwood; tamalin
缅甸紫檀 amboina rosewood; Andaman rosewood; Burmacoast padauk; Burma padauk; Burmese rosewood
缅漆酚 thitsiol
缅因州黑云母花岗岩 <产于美国缅因州> Somes Sound granite
缅状岩 roe stone

面板 boarding; cover plate; cover slab; deck (slab); decking; face panel <L 形码头、岸壁或加筋土等结构的>; face-plate; face slab; facing; facing slab; front panel; panel; shell skin plate; skin plate; skin plating; surface skin; skin <夹板门的>; face-sheet <夹层结构的>

面板安装 panel mounting
面板安装可变光衰减器 panel mount variable optic(al) attenuator
面板暴露边 shingle butt
面板编号 panel number
面板插口 panel jack
面板垫层区 face-supporting zone
面板堆石坝 decked rockfill dam
面板服务 panel service
面板钢筋 deck reinforcement; slab reinforcement
面板荷载 deck load
面板混凝土 facing element concrete
面板接缝 deck joint; facing joint; slab joint; face joint <混凝土面板堆石坝的>
面板接缝件 panel divider
面板结构的坡度 inclination of deck structure
面板开关 deck switch; panel switch
面板开口图 panel cut out drawing
面板控制 front panel control
面板离缝 face checking
面板两侧作互锁缝的门 lock seam door
面板螺钉 face-plate screw
面板螺栓 panel bolt
面板模板 <钢筋混凝土> shell formwork
面板平面上的变形 face-plane deformation
面板上活荷载 live load on deck
面板上填碎石的码头岸壁 platform wharf wall
面板设备 face equipment
面板伸缩缝 facing expansion joint
面板式开关板 panel type switchboard
面板式控制器 board-mounted controller
面板式控制台 board-type console
面板式仪表 panel meter; surface type meter
面板贴敷 veneering
面板无接线的配电盘 dead front switchboard
面板显示 panel display
面板型电流表 panel type ammeter
面板型自动交换机 panel type automatic switch board
面板仪表 panel instrument
面板应变 face strain
面板用仪表 panel meter; surface type meter
面板与机架接头 rack and panel connector
面板照明灯 front panel illuminator
面板支撑 face support
面板支承梁 deck girder
面板支承区 <面板坝的> face-supporting zone
面板中断屏蔽 panel interrupt mask on
面板座 panel bed
面包厂 bakery
面包车 light bus; minibus; van
面包店 bake shop; tommy-shop
面包房 bake house; bakery room
面包皮状构造 bread-crusted structure
面包铺 pastry
面包师傅 baker
面包树 bread fruit tree
面包砖 bloated brick
面北的立面 <建筑物> north-facing facade
面泵浦 face-pumping
面冰 sheet ice
面波 plane wave; surface wave
面波法 area method
面波法波速测试 surface wave velocity test
面波勘探仪 surface wave hydrophone
面波震级 surface wave magnitude
面部 face
面部保护用具 face protector
面部畸形 facial deformity
面部距离 facial dimension
面部轮廓描记器 profilograph
面部平面 facial plane
面部三角区 facial triangle
面部相片 mug
面部纵裂 longitudinal facial crack
面材 face bar; facing material
面彩色全息图 surface colo(u)r hologram
面层 armo(u)red layer; coat; cover(age); crust; face layer; facing; floor finish; mat coat; skin coat; surface coat(ing); surface course; surface dressing; surface layer; surface finish; surfacing(course); top course; top coat; topping; veneer <常指面层的一种薄木板>
面层板 topping slab
面层保护物质 protective finish mass
面层保护罩 face saver
面层剥离 removal of coat
面层材料 floor finishing; surface material; surfacing; surfacing material; finished flooring
面层操作 topcoat operation
面层草皮 top sod
面层测厚仪 covermeter
面层单板 face veneer
面层导热率 film conductance
面层叠缝 topping joint
面层翻铺法 repaving process
面层粉刷石膏 finish plaster; ga(u)ging plaster
面层符号 face mark
面层光弹试验法 photoelastic coating method

面层龟裂 <混凝土> surface distress
面层厚度的标定粉刷 <俗称出拓饼> ga(u)ging plaster for finish(ing)
面层混合料 surface course mix; surface mixture; face mixture
面层混凝土 facing concrete; top concrete; pavement concrete
面层胶混合料 topping compound
面层接缝 facing expansion joint; facing joint
面层裂纹 surface check
面层流线 top flow line
面层密度 surface density
面层磨平机 surface planing machine
面层抹灰 finish coat; surface coat
面层配合料 topping mix(ture)
面层铺路工 facing pavio(u)r
面层铺路机 facing pavio(u)r
面层曲线 friction(al) curve
面层伸缩缝 facing expansion joint
面层石屑 gritty coverstone
面层损坏 surface distress
面层涂层 back coat(ing)
面层涂料 investment precoat; surface paint; top coat; coating material
面层土(壤) surface soil
面层为水磨石的梯级 terrazzo faced tread
面层未涂柏油的粗油毡 uncoated tar(red) rag felt
面层未涂沥青的粗油毡 uncoated asphalt rag felt
面层下层 lower coat; base course <英>
面层下的毛抹灰 topping coat
面层修补 resurfacing; surface repair
面层修整 topping finish
面层修筑 surface coating operation
面层压实机 surface compactor
面层养护 maintenance of surface; surface maintenance
面层用清漆 body varnish
面层油漆 exterior paint; surface paint
面层油毡 cap sheet
面层再生(利用) surface recycle
面层凿平器 pavement cuter
面层正规性 surface regularity
面层装饰 decorative finish
面层做法 surface finish
面铲 face shovel
面长 face length
面朝上的 face up
面朝上焊接(法) face-up bonding
面朝下的 face down
面朝下焊接(法) face-down bonding
面衬 furring
面窗 face-window
面吹 top blast
面吹法 surface-blowing
面磁荷 surface magnet charge
面存储密度 areal packing density areal storage density
面的关系 relation of plane
面的几何力矩 geometric(al) moment of area
面的计数 cover count
面的矩心 centroid of area
面的轮廓度 profile of any surface; profile tolerance of a surface
面的内角【数】interior angle
面的倾斜与扭曲 ramp and twist
面的提升与扭曲 ramp and twist
面电导 sheet conductance
面电荷 surface charge
面钉 face nailing
面顶点 vertex of surface
面对 confront; face
面对称 mirror symmetry; plane symmetry
面对大海的屋顶阳台 widow's walk
面对的 facing
面对方向 facing direction
面对角线 face diagonal
面对面 face-to-face; nose-to-nose; tete-a-tete; vis-à-vis
面对面的会议 face off
面对面地 face-to-face
面对面耦合 face-to-face coupling
面对面人兽图案雕饰 affronte(d)
面对严峻挑战 meet the serious challenge
面额 face amount; face value; nominal amount
面发射发光二极管 surface-emitting light emitting diode
面发射体 surface emitter
面放大率 area magnification
面非点污染源 areal non-point source
面非点污染源流域环境响应模拟模型 areal non-point source watershed environment response simulation model
面分布力 force per unit area
面分析 area analysis
面粉 flour
面粉厂 flour mill
面粉尘 flour dust
面粉磨 flour mill
面粉黏[粘]结剂 cereal binder; flour adhesive
面粉全险条款 flour all risks clause
面粉筛 flour bolt; flour sieve
面粉装袋机 flour packer; flour sacker
面粉状氧化铝 floury alumina
面粉自动包装机组 automatically flour banding plant
面缝宽度 bed-and-joint width
面浮车 surface skimmer
面浮船 surface skimmer
面辐射强度 radiance
面干饱和的 saturated-surface-dried
面干饱和重量 saturated-surface-dry weight
面干饱和状态 saturated-surface-dry condition
面干骨料 surface-dry aggregate
面干含水量 dry-face water content
面干集料 surface-dry aggregate
面干铸型 skin dried mo(u)ld
面干状态 <骨料或集料表面干燥、内部水饱和的状态> surface-dry condition
面割理 face cleat
面弓 face-bow
面贡献量 surface contribution
面垢 dirty complexion
面估计方差 section term of estimation variance
面光 ceiling spotlight; front lighting; limelight flap; spotlight flap
面光胶合板 good one side
面光源 area lighting source; surface light source
面海防波堤 seaward breakwater
面荷载 area load(ing); surface loading
面横裂 transverse facial cleft
面灰 <嵌门窗玻璃的> back putty; face puttying; front putty
面积 area; dimension; superficial content
面积比例尺 square-measure scale
面积比例规 planimegraph
面积比(率) area ratio; area scale; ratio of area; square-measure scale; surface ratio
面积变化 areal deformation
面积变形 alternation of area; area distortion
面积变形比 area-distortion ratio; area-distribution ratio
面积变形理论 theory of distortion of areas
面积变形系数 area-distortion coefficient
面积测定 area measurement
面积测量 areal metric; measurement of area
面积测速计 area meter
面积乘积 product of areas
面积尺 square chain
面积抽样(法) area(l) sampling
面积单位 area unit; superficial unit; unit of area; square measure
面积的 areal; superficial
面积的粗略估算 area take-off
面积的二次矩 second moment of area
面积的再分配 area redistribution
面积定额 building area quota
面积定律 area law; law of area
面积定则 area rule
面积度规 areal metric
面积法 area method; method of determining yield by area; planimetry
面积放大(率) area magnification
面积分 surface integral
面积分布测定法 measurement of area distribution
面积分布曲线 area-distribution curve
面积分法 surface integration method
面积分配(关系)曲线 area-distribution curve
面积分区 <土地利用规划的> area district
面积分区制 <城市规划的> area zoning
面积分析 areal analysis
面积高程分配曲线 area-elevation distribution curve
面积-高程曲线 area-elevation curve
面积-高程曲线图 area-elevation graph
面积高度曲线 area-altitude curve
面积估计方差 estimation variance of surface
面积估计误差 estimation error of surface
面积估价法 area method; area method of estimating cost
面积惯矩 second moment of area
面积归一法 measurement of area distribution
面积归一化法 area normalization method
面积函数 area function
面积航空摄影 area aerial photography
面积和线的形心 centroids of areas and lines
面积荷载 area load(ing)
面积划分 division of area
面积回声测深仪 area echograph
面积回声曲线图 area echograph
面积集中 areal concentration
面积计 planimeter
面积计量 superficial measurement
面积计量器 area meter
面积计量仪 area meter
面积计算 area computation
面积加权平均分辨率 area weighted average resolution
面积减少法 area-reduction method
面积校正 area correction
面积矩 area moment; moment of area
面积矩心 centroid of area
面积矩定理 theorem of area moment
面积距离法 area distance method
面积距离曲线 area distance curve
面积库容曲线 area-capacity curve
面积夸大 area exaggeration
面积扩大 <在受压试件中,最大横断面积与原来横断面积的差值,常用百分比表示> expansion of area
面积力 facial force
面积力矩法 area moment method
面积利用系数 area utilization factor
面积量算 measure of area
面积令 area age
面积流量计 area flowmeter
面积流量仪 area flowmeter
面积流速测流法 area-velocity ga(u)ging method
面积流速法 area velocity method
面积律 area rule
面积律概念 area rule concept
面积律辖域 area rule scope
面积率 specific surface area
面积目标 area target
面积平差 area adjustment; block adjustment
面积平衡波形 area-balanced waveform
面积平衡电流 area-balanced current
面积平衡法 balanced area method
面积平均法 area averaging method
面积平均压力 areal average pressure
面积普查 area reconnaissance
面积曲线 area curve
面积热流量 areal heat flow rate
面积容积曲线 area-capacity curve; area-volume curve
面积三角测量 area triangulation
面积扫油 area sweep
面积深度分布曲线 area-depth distribution curve
面积深度分配曲线 area-depth distribution curve
面积深度关系 depth-area relationship
面积深度降雨曲线 area-depth curve for rainfall
面积深度曲线 area-depth curve
面积失配损失 area mismatch loss
面积式流量计 area flowmeter
面积收缩 contraction in area; contraction of area; shrinkage of area
面积收缩率 percentage reduction of area
面积数据 area data
面积水准测量 areal level(l)ing; grid level(l)ing
面积速度 area(l) velocity
面积速度(测流)法 area velocity method
面积速率 areal rate
面积缩小 area reduction; reduction of area
面积缩小率 percentage reduction in area
面积体积比 area-volume ratio; surface-volume ratio
面积条形图 area bar-chart
面积调节法 area regulation
面积调制声道 variable area track
面积图 area chart; area graph; planimetric(al) map
面积弯矩 area moment
面积为 100 米 × 100 米 an area of 100m x 100m
面积稳定性 area stability
面积误差 area error; error in area
面积系数 area coefficient; area factor; coefficient of area
面积细测 area fine measurement
面积效率 area efficiency
面积效应 area effect
面积形心 centroid of area
面积型流量计 area type meter
面积要求 area requirement
面积要素 areal feature
面积一次矩 first moment of area
面积仪 area meter; planimeter [pla-

M

nometer]; platometer
面积仪方向导杆 planimetric(al) arm
面积仪平面控制基线 planimetric(al) base
面积因数 area factor
面积元(素) area(l) element; cell area; differential of area; element of area; surface element
面积增长 area increment
面积增长法 area increment method
面积张力 interfacial tension
面积丈量 square measure
面积直方图 area histogram
面积中心 center[centre] of area
面积中心二次矩 central second moment of area
面积重力测量 area gravity survey
面积重量比 area/weight ratio
面积注水 pattern flood
面积总值法 area summation method
面积组合 area array
面积坐标 area coordinates
面基的 basifacial
面畸变 area distortion
面极化 surface polarization
面极化系数 coefficient of surface polarization
面际的 interfacial
面际张力 interfacial tension
面颊 cheek
面价值的提高 write up
面架后的加强壁骨 stud behind the face frame
面架后的加强支柱 stud behind the face frame
面间表面张力 interfacial surface tension
面间层 <两种液体或固体接触面上的一层> interfacial layer
面间电阻 interface resistance
面间腐蚀 interfacial corrosion
面间角 interfacial angle
面间角守恒定律 law of constancy of interfacial angle; law of constant angles
面间接触(状况) interfacial contact
面间距 identity distance
面间膜 interfacial film
面降水量 areal precipitation
面交角 interfacial angle
面交通控制 area traffic control
面角 face angle; facial angle; plane angle
面角造型 cant
面接 face bond
面接触 face contact; surface contact
面接触二极管 surface contact diode
面接触钢丝绳 facial-contracted wire rope; plane contact lay wire rope
面接触系统 surface contact system
面接对偶 lower pair
面接合(法) face bonding
面接合型光电晶体管 junction photo transistor; junction phototransistor
面接合型晶体三极管 junction transistor; junction triode
面接角 surface contact angle
面结 junction
面结构 face structure
面结型场效应 junction field effect
面结型晶体二极管 junction diode
面结型晶体管 junction type transistor
面界限长度标准 end standard of length
面巾纸生产线 face tissue production line
面金属量 areal productivity
面筋 gluten
面矩法 area moment method
面具 face guard; face mask; face-piece; mask
面具式传声器 mask microphone
面壳 face-piece
面空化 face cavitation
面控制 area control
面控制交通系统 area traffic control system
面盔 visor
面阔 building width; width
面蜡 mercolized wax
面垒型半导体探测器 surface barrier semiconductor detector
面垒型探测器 surface barrier detector
面理【地】foliation
面理的产状 attitude of foliation
面理的世代 generation of foliation
面理的研究 study of foliation
面理类型 type of foliation
面理图解 diagram of foliations
面理与主构造的关系 relationship between foliation and main structure
面理褶皱 foliation fold
面力 surface force
面立体角乘积 area-solid-angle product
面连续性系数 coefficient of planar continuity
面料 facing; facing material; plus material; lining <俗称来令>【气】
面料层 precoat
面裂 facial cleft
面裂隙率 fissure ratio on plane
面临 face
面临的危机 coming crisis
面临海的场所 sea front
面临海滩的 beachfront
面临水域 <船厂、码头的> water frontage
面临挑战 face the challenge
面临危机 face a crisis
面临危险 faced with danger
面临一个问题 face a problem
面临有限资源 faced with limited resources; facing the limited resources
面临种种可能的 in the face of all odds
面灵敏度 area sensitivity
面流 surface current
面流速 surface velocity
面流消能 dissipation of surface flow; energy dissipation of surface regime
面流形态 surface flow pattern
面露骨料纹理 exposed aggregate texture
面露集料纹理 exposed aggregate texture
面律 plane law
面貌 feature
面密度 areal density; planar density
面密封 face seal
面面接触 face-face contact
面模 face(unit) mo(u)ld
面摩擦试验 skin friction test
面磨削表面 ground surface; land surface; terrain surface; terrestrial surface
面目标 area target
面内云纹法 in-plane moiré method
面内振动 in-plane vibration
面盆出水管 lavatory waste
面膨胀 surface expansion
面偏振波 plane polarization wave; plane polarized wave
面偏振光 plane polarized light
面平均降水量 areal mean precipitation; areal precipitation; average area precipitation; mean areal precipitation
面平均降水深度 mean areal depth of precipitation
面平均雨量 areal average rainfall; areal mean rainfall; areal rainfall; mean areal precipitation
面平行玻璃板 plane parallel glass plate
面平整度 surface evenness
面坡 surface slope; batter of facing <墙面或坝面的>
面坡椽 <四坡屋顶的> angle ridge; jack rafter; pien(d) rafter; valley jack rafter
面漆 clear top(coating); face coat(ing); facing coat; finish(ing) coat(ing); finishing paint; surface finish; surface paint; top(coating); top paint
面漆的色泽深度 depth of finish
面漆厚度 depth of finish
面漆配套性 overcoatability
面漆涂层 overcoating
面漆下涂层 undercoat
面漆消光处理 mat(te) finish(ing)
面起翘校正机 cover bending machine
面洽 face-to-face negotiation
面墙 fascia wall; face wall; head wall <洞道进出口>; leaf <空心墙的>
面曲率 curvature of face
面全息图 surface hologram
面缺陷 face defect; planar defect
面色 face colo(u)r; top colo(u)r
面砂 facing sand
面上安装的灯座 surface mounting lampholder
面上的堆焊 surface welding
面上分布 areal distribution; distribution in area; surface distribution
面上和装满 face and fill
面上结冰 freeze over
面上水流 surface current
面上置放 surface fixing
面深关系 depth-area relationship
面深曲线 depth-area curve
面声源 area source of sound
面施药 area spray
面石 face stone; facing stone
面石修饰锤 patent hammer
面石修饰的 patent-hammered
面试 interviewing
面饰 facing
面饰砌块 faced block
面束 pencil of planes
面水准测量 area(l) level(l) ing
面速度 <空气流入设备的> face velocity
面损伤 surface damage
面缩 contraction of area; reduction of area
面谈(调查) personal interview
面谈室 interview room
面条生产线 noodle producing line
面条式(建筑) noodle style
面图投影 orthography
面图像传感器 area image sensor
面涂 top coat
面涂抹面 finish coat floating
面土滑动 detritus slide
面外的 out-of-plane
面外振动 out-of-oriented vibration
面弯试验 face-bend test
面弯试样 face-bend test specimen
面网 net; plane group
面网符号 symbol of net
面网间距 interplanar spacing
面网间距 d 值 interplanar spacing d value
面网密度 planar net density
面网密度理论 surface network density theory
面位显示 pan
面污染 areal pollution
面污染特征 area source pollution characteristic
面污染源 area(l) pollution source; area(l) source; non-point pollution source; plan pollution source
面污染源调查 area source survey
面西的窗 west-facing window
面铣 facing cut; surface milling
面相关 area correlation
面向 face; facing
面向比特规程 bit-oriented procedure
面向材料的规范 material-oriented specification
面向车辆系统 vehicle-oriented system
面向程序的语言【计】program(me)-oriented language
面向出口的工业 export-oriented industry
面向出口的工业化 export-led industrialization
面向磁盘系统 disk oriented system
面向存储器系统 memory oriented system
面向存取的 access-oriented
面向代数的语言 algebra-oriented language
面向地区需要的 region-oriented
面向对象程序 object-oriented program(me)
面向对象程序设计 object oriented programming
面向对象的 object-oriented
面向对象的结构 object-oriented construction
面向对象模型 object-oriented model
面向对象(目标)数据库 object oriented database
面向对象(目标)的程序设计方法 object-oriented program(me)
面向方便的 convenience-oriented
面向非机动车的 non-auto-oriented
面向概念 idea-oriented
面向工程的 engineering-oriented
面向公路运输的工业联合企业 highway-oriented industrial complex
面向功能需要模型 ability requirement-oriented model
面向观念 idea-oriented
面向规则的 rule-oriented
面向规则的系统 rule-oriented system
面向规则示例 rule-oriented paradigm
面向过程 procedure-oriented
面向过程的语言 procedural language; procedure-oriented language
面向海港的一边 harbo(u)r side
面向环境的项目 environment-geared project
面向活动的模拟 activity-directed simulation
面向机动车的 auto-oriented
面向机器的程序系统 machine-oriented program(m)ing system
面向机器的语言 <第二代程序>【计】machine-oriented language
面向计划的 plan oriented
面向计算机的 computer-oriented
面向计算机语言 computer-oriented language
面向记录的传输 record-oriented transmission
面向记录的设备 record-oriented device
面向记录的数据 record-oriented data
面向阶段的 stage-oriented
面向节约的政策 conservation-orientated policy
面向空间 space-oriented

M

面向控制的功能 control-oriented function
面向控制的微型计算机 control-oriented microcomputer
面向块的 block-oriented
面向块的相联处理机 block-oriented associative processor
面向框架法 frame-oriented approach
面向扩充系统 expansion-oriented system
面向扩展的环境 growth-oriented environment
面向连接的网络层协议 connection oriented network layer protocol
面向连接协议 connection oriented protocol
面向(某一特定)计算机的语言 computer-dependent language
面向目标的 goal-oriented;object-oriented
面向目标的方法 goal-oriented approach;object-oriented approach
面向目标的分析程序 goal-oriented parser
面向目标的识别程序 goal-oriented recognizer
面向目标的推理 goal-oriented inference
面向目标的系统 object-oriented system
面向目标示例 object-oriented paradigm
面向汽车的城市 auto-oriented city
面向人的语言 human-oriented language
面向任务的网络 mission-oriented network
面向软件 software-oriented
面向上 face up;upward facing
面向设备的 device oriented
面向时间的顺序控制 time-oriented sequential control
面向市场的经济结构 market-oriental structure
面向市场的生产 market-oriented production
面向事件的 event-oriented
面向事件的模拟 event-directed simulation
面向事件模拟法 event-directed simulation;event-oriented simulation
面向事项处理应用 transaction-oriented application
面向受雇者的计划 employee-oriented program(me)
面向数据的检验 data-oriented testing
面向思想 idea-oriented
面向铁路的工业 rail-oriented industry
面向通信[讯]的生产信息及控制系统 communications oriented production information and control system
面向图像的 image-oriented
面向网络的计算机系统 network-oriented computer system
面向文件的程序设计 file-oriented programming
面向文件的系统 file-oriented system
面向问题的 problem oriented
面向问题的软件 problem-oriented software
面向问题的语言 problem-oriented language
面向系统的 system-oriented
面向系统的计算机 system-oriented computer
面向系统的硬件 system-oriented hardware
面向下 downward facing;face down
面向下的 prone
面向线路系统 wayside-oriented system
面向消息的 message-oriented
面向效率的 effect-oriented
面向应用的 application-oriented
面向应用的协议 application-oriented protocol
面向应用的语言 application-oriented language
面向应用系统 application-oriented system
面向用户的语言 user-oriented language
面向用户系统 user-oriented system
面向语法的编译程序 syntax-oriented compiler
面向语法的识别算法 syntax-directed recognition algorithm
面向语言的编译程序 syntax-directed compiler
面向原料产业 resource-oriented industry
面向原文的数据库 text-oriented data base
面向政策模型 policy-oriented model
面向中断的系统 interrupt-oriented system
面向终端网络 terminal-oriented network
面向字符的数据流 character-oriented data stream
面向作业的语言 job-oriented language
面向作业终端 job-oriented terminal
面斜裂 oblique facial cleft;prosopoanoschisis
面谐函数 surface harmonic function;surface harmonics
面心 center of area;center of figure;centroid
面心长方点阵 centered rectangular lattice
面心单斜晶系 face-centered monocline system
面心的 face-centered[centred]
面心点阵 face-centered lattice
面心格子 face-centered lattice
面心晶格 face-centered lattice
面心晶体 face-centered crystal
面心立方的 face-centered cubic(al)
面心立方堆积 face-centered cubic(al) packing
面心立方格子 face-centered cubic(al) lattice
面心立方结构 face-centered cubic(al) structure
面心立方晶格 face-centered cubic(al) lattice
面心立方晶格的排列 face-centered cubic(al) array
面心立方体 face-centered cube
面心立方铜金合金 face-centered cubic(al) copper-gold alloy
面心网格 face-centered grating
面心正交 face-centered orthorhombic
面心正交晶格 face-centered orthorhombic lattice
面心最紧密填充 face-centered closest packing
面形 surface shape
面形测定器 profilometer
面形描记器 profilograph
面修复术 facial restoration
面选择信号 side select signal
面穴 surface cavity
面询 interview
面延烧性 surface spread of flame
面岩溶率 karst factor counted on plane
面样模板 face mo(u)ld
面页错误 page fault
面油灰 front putty;face putty <嵌门窗玻璃的>
面釉 cover coat enamel;overglaze
面釉装饰 overglaze decoration
面雨量 areal rainfall;rainfall on area
面元大小 surface element size
面元法 surface element method
面元素 elements of a surface;surface element
面元网的列数 columns of area element mesh
面元网的行数 rows of area element mesh
面元质量彩色显示 bin quality colo(u)r display
面元中心点坐标 central coordinate of area element
面源 area(l) source;planar source
面源污染 areal source pollution;non-point pollution
面源污染特征 area source pollution characteristic
面砟【铁】upper layer of ballast;top ballast
面轧染色 nip dyeing
面罩 face guard;face mask;face-piece;face shield;mask;respirator;veil
面罩层 surface coating
面罩加压呼吸 face-masked pressure respiration
面震源 area(l) source
面征 facial sign
面正交系统 orthotomic system
面支承 surface bearing
面直径测量器 faceometer
面值 denominational value;face amount;face value;par(value)
面值重量 nominal weight
面指数 facial index
面砖 ashlar brick;brick slip;face brick;facing (clay) brick;facing tile;flagstone;furring brick;lining brick;wall tile
面砖剥落 chippage
面砖粗糙背纹 plaster-base finish tile
面砖的抗压强度 tile compression strength
面砖加釉 tile glazed coat(ing)
面砖铺设机 tile laying machine
面砖铺贴 setting of wall tiles;wall tile setting
面砖砌合 face brick bond
面砖切割机 tile cutting machine
面砖色泽 tile colo(u)r;tile tint
面砖砂轮机 tile grinding machine
面砖饰面 tile(d) finish
面砖贴面 tile facing
面砖土 face brick clay
面砖脱皮 chippage
面砖压机 wall tile press
面砖压制机 tile press
面砖用灰浆 tiling plaster
面砖制造厂 tile making factory
面装金属电线管 surface metal raceway
面装圈带 surface-mounted astragal
面状 surface form
面状沉降 areal settlement
面状出水点 out flow pattern
面状地物符号 <依比例尺的> plan symbol
面状断裂法 planar fracture method
面状分布 plane distribution
面状风化壳 sheet residuum
面状符号 areal symbol;face symbol
面状构造【地】planar structure
面状均匀补给 uniform recharge over a plane
面状模式 area pattern
面状暖异常 planar warm anomalies
面状侵蚀 sheet erosion
面状入侵 intrusion on an area
面状要素 areal feature;planar element
面状组构 planar fabric
面锥 face cone
面坐标 areal coordinates

苗

苗床 nursery;nursery bed;plant bed;seedbed;seeding bed;seed plot;seminary
苗床保护框 seedbed frame
苗床除草机 seedbed weeder
苗床格子 latorex
苗床支柱 seedbed stakes
苗宽 width of seedling shoot
苗龄 seeding age
苗木 nurs(e)ling;nursery-grown plant;nursery stock;seedling
苗木培育圃 plant school
苗木石 naegite
苗木箱 plant box
苗圃 free nursery;nursery;nursery bed;nursery garden;nursery school;plant school;seedling nursery;seed plot;seminary;tree nursery
苗圃播种机 nursery drill;nursery planter
苗圃草皮 nursery sod
苗圃工作者 nurseryman
苗圃供热 plant nursery heating
苗圃移植铲 scoop for nursery planting
苗圃与试验地播种机 plot seeder
苗数 seedling population
苗头 incipient tendency
苗榆 Japanese hop hornbeam
苗榆属 hop hornbeam;Ostrya <拉>
苗株和成株 seeding and adult
苗族建筑 architecture of the Miao nationality

描

描笔 pen;tracer
描笔式记录器 pen recorder
描笔式示波器 pen oscillograph
描笔中心调整 pen centering
描波器 kymograph
描好的图 traced drawing
描花工 flowerer
描花纹机 pattern tracer
描画 limn
描画器 delineator
描画误差 drawing error
描画针 stylus
描绘 artwork;copy;delineate;delineation;depiction;describe;draw;plot;portray;portrayal;scaling;trace;tracing;tracking
描绘板 plotting board;plotting table
描绘出图形花样的玻璃条 glass strip for outlining patterns
描绘单元 delineation unit
描绘等高线 contour drafting;contour drawing;contour sketching;generate contour;generating contour;plot contour;trace contour;tracing of contour lines
描绘等高线程序 contour-hypsocline generator
描绘地貌 topographic(al) plotting
描绘地物 planimetric(al) plotting
描绘点 tracing point
描绘断面 profiling

描绘法 method of portrayal; method of presentation; portrayal method
描绘工具 tracer
描绘拱块 traversing
描绘畸变 tracing distortion
描绘轮廓 delineate; outlining
描绘器 plotter; tracer
描绘曲线笔尖 curve following stylus
描绘设备 plotting unit
描绘射束 scan-off beam
描绘手段 means of representation
描绘台 drawing desk; tracing table
描绘铁路简图 railway sketching
描绘外形 delineation
描绘像片 plot the photograph
描绘制版法 block out method
描绘装置 tracing device
描迹笔 tracing pen
描迹笔尖 tracing pen
描迹臂 tracing arm
描迹点 tracing point
描迹函数 describing function
描迹轮 tracer wheel; tracing wheel
描迹器 hodoscope; tracer
描迹曲线 tracing curve
描迹图 trace diagram
描迹仪 hodoscope
描迹针 tracing needle; tracing point
描记笔 recording stylus
描记法 graphic(al) method; tracing
描记杠杆 writing lever
描记气鼓 recording tambour
描记器 tracer
描记式蒸发计 atmidometrograph
描界器 diagraph
描金 decoration with liquid gold; gild
描金胶 gold size
描金烤清漆 gold stoving varnish
描金漆器 lacquer with gold design
描略图 outline
描轮廓线 key drawing
描摹复制 duplicating
描青 blue drawing
描述表 description list
描述布 glacitex
描述词 descriptor
描述地质学 geognosy
描述法 method of description; scenario method
描述分类 interpretive classification
描述符 descriptor
描述符方法 descriptor approach
描述符语言 descriptor language
描述函数 described function; describing function
描述函数法 describing function method
描述环境 describe environment
描述技术 description technique
描述模型 description model
描述器 describer
描述设备 describe-equipment
描述水文学 descriptive hydrology
描述系统工作的综合方程 mathematic(al) model
描述项 description entry; discriptive item
描述信息 descriptor
描述性程序设计 descriptive programming
描述性抽样调查 descriptive sampling survey
描述性地层学 descriptive stratigraphy
描述性地震学 descriptive seismography
描述性调查 descriptive survey
描述性定义 descriptive definition
描述性工艺过程 descriptive process
描述性古生物学 descriptive palaeontology
描述性规定 descriptive provision
描述性过程 descriptive procedure
描述性海洋学 descriptive oceanography
描述性晶体学 descriptive crystallography
描述性决策 descriptive decision making
描述性决策理论 descriptive decision theory
描述性矿物学 descriptive mineralogy
描述性模拟模型 descriptive simulation model
描述性模型 descriptive model
描述性气候学 descriptive climatology
描述性气象学 aerography; descriptive meteorology
描述性生态学 descriptive ecology
描述性生物地理学 areography
描述性数列 descriptive series
描述性天文学 descriptive astronomy
描述性统计 descriptive statistic
描述性统计法 descriptive statistical method
描述性统计学【数】descriptive statistics
描述性形态学 descriptive morphology
描述性岩石学 descriptive petrology
描述性研究 descriptive research
描述性指数 descriptive index
描述岩石学 petrography
描图 counterdraw; plotting; trace; tracing; tracing of drawing
描图笔 plot pen
描图布 claratex; drawing cloth; plantex; tracing cloth; tracing linen
描图附件 tracing attachment
描图杆 tracing bar
描图工具 tracer
描图机 tracing machine
描图控制程序 drafter control program(me)
描图麻布 tracing linen
描图膜 drafting film
描图器 drafter; draughter; tracer
描图室 tracing house; tracing room
描图仪(器) tracing instrument
描图员 draftsman; draughtsman; tracer
描图者 delineator
描图针 tracing needle
描图纸 detail paper; light-tracing paper; traceable paper; trace paper; tracing cloth; tracing paper; tracing sheet
描图桌 tracing table
描外形 delineate
描纹畸变 tracing distortion
描纹失真 tracing distortion
描纹损耗 tracing loss
描纹误差 tracing error
描线 banding
描线规 lineograph
描线轮 tracing wheel
描像器 camera lucida
描写 limn; portray
描写者 delineator
描形针 tracer needle
描印 overprint
描影(法) shading
描准器 foresight

瞄得准 aim true

瞄得准的 well-placed
瞄视误差 boresighting error
瞄线法 sighting line method
瞄直法 sighting line method
瞄准 acquiring; aim(ing); collimate; collimation; home on; homing on; laying; pointing; ranging into line; sighting; take sight; target; training; boning in < 检测二点间水平差 >
瞄准板【测】vane
瞄准标尺 leaf sight
瞄准标杆 aiming post
瞄准标志 sighting mark
瞄准捕捞 controlled-directed fishing
瞄准尺 aiming rule; sight rule
瞄准锤 sighting pendant
瞄准点 aiming point; boresight; point of sight; sighting point
瞄准点提前量 aim-off
瞄准点修正 holding-off
瞄准方位角连测 geodetic connection of aiming azimuth
瞄准放大器 sight amplifier
瞄准符 aiming symbol
瞄准杆 sighting bar; sighting mast
瞄准环 aiming circle
瞄准机构 pointing device
瞄准基准距离 sight base distance
瞄准极限 sight reach
瞄准几何学 line-of-sight geometry
瞄准技术 point technique
瞄准架 sight bracket
瞄准件 sight unit
瞄准角 angle of collimation; angle of sight; sighting angle
瞄准校正器 aiming corrector
瞄准精度 pointing accuracy; pointing precision
瞄准镜 finder adapter; telescopic(al) sight; aiming rule sight < 表尺的 >
瞄准镜插座 claw mount; telescope claw mounts
瞄准镜视差 visual tube parallax
瞄准镜水准器 telescope level
瞄准镜筒 sight tube
瞄准镜头 traversing head
瞄准镜装置 sight mechanism
瞄准具 collimating sight; gunsight
瞄准具调整环 reticle adjusting ring
瞄准具分划 sight graduation
瞄准具光轴 sight axis
瞄准具盒 sight case
瞄准具计算机 gunsight computer
瞄准具校靶 sighting harmonization
瞄准具校正 zero-in
瞄准具卡座 sight bracket
瞄准具刻度 sight graduation
瞄准具雷达距离指示灯 sight radar range light
瞄准具灵敏度 sight sensitivity
瞄准具炉额垫 sight headrest
瞄准具升降器 sight extension
瞄准具十字线 sight reticule
瞄准具头部 sight head
瞄准具支架 sight mount
瞄准具轴线 sight axis
瞄准具装定 sight adjustment
瞄准距离 sighting range
瞄准刻线 sighting string
瞄准孔 backsight; peep hole; sight hole; sight vane
瞄准口 backsight
瞄准面 plane of direction
瞄准器 aiming device; alignment clamp; breech-sight; cross wires; diopter [dioptre]; finder; guidance unit; gunsight; hairline pointer; hairline sight; open sight; peep-sight; sight gun; sight vane
瞄准器标高 elevation of sight
瞄准器校孔 < 测斜仪 > hole for alignment
瞄准器刻度 sight scale
瞄准器孔径 finder aperture
瞄准器小孔 viewfinder eye
瞄准器游丝 diopter hair
瞄准设备 sighting device
瞄准摄影机 boresight camera
瞄准手用望远镜 pointer's telescope
瞄准速度 pointing velocity
瞄准塔 collimation tower; sighting mast; sighting tower
瞄准台 sighting platform
瞄准提前量 sighting offset
瞄准调节螺钉 aiming adjust screw; aiming screw
瞄准调整 point harmonization
瞄准透镜 finder lens; scanning lens
瞄准图 sighting diagram for bombing
瞄准拖网捕捞 aimed trawling
瞄准望远镜 finder telescope; optic(al) sight; sniperscope; sighting telescope; aiming telescope < 水准仪上的 >
瞄准位置 aiming position
瞄准问题 pointing problem
瞄准误差 aiming error; collimating fault; collimation error; error of sighting; pointing error; sighting error
瞄准系统 pointing system; sighting system
瞄准显微镜 sight microscope
瞄准线 aiming line; cross hairs; boresight; collimation line; diopter thread; hair cross; hairline; light of sight; line of aim; line of sight; range line; sighting line
瞄准线半自动控制 semi-automatic command to line of sight
瞄准线检验 boresighting test
瞄准线零位调整 boresight adjustment
瞄准线稳定 line-of-sight stabilization
瞄准线与目标重合 sight alignment
瞄准销售技术 rifle technique
瞄准斜率 boresight slope
瞄准修正 aiming off
瞄准仪 aiming sight; pointing instrumentation; sighting rule
瞄准仪校正器 aim corrector
瞄准用计算装置 computing sight
瞄准用准星 sight head
瞄准站 sighting station
瞄准照相枪 gunsight aiming point camera
瞄准中层拖网捕捞 aimed midwater trawling
瞄准重合线 colineated line of sight
瞄准轴 axis of sighting; guidance axis
瞄准轴线 axis of sighting
瞄准装定器 range setter
瞄准装置 aiming mechanism; finder; gunsight device; sighting instrument
瞄准装置孔径 finder aperture
瞄准装置透镜 finder lens
瞄准装置遮光罩 finder hood

秒摆 second pendulum

秒摆时针 second pendulum clock
秒表 chronograph; game watch; second chronograph; second counter; stop watch; timer; job watch < 代替天文钟测天用 >
秒表读数 stopwatch reading
秒表循环时间 stopwatch cycle time
秒差电雷管 second delay electric(al) detonator
秒差距 < 表示天体距离的单位,视差为一秒的距离相当于 3.259 光年 > parsec; parallax second
秒差雷管 second delay blasting cap;

M

second delay detonator
秒齿轴 second pinion
秒锤杆定位垫圈 second hammer maintaining washer
秒锤杆簧位钉 second hammer spring foot
秒锤杆制动器 second hammer damping stop
秒倒数 reciprocal sec
秒分指针 split-second-hand
秒复位装置 second reset device
秒回零杆锤头 second hammer
秒棘爪轮 second pawl wheel
秒交通量 traffic volume per second
秒控制 second control
秒控制器 second controller
秒立方尺 cubic(al)feet per second
秒立方米 cubic(al) meters per second
秒立方英尺 cusec;second cubic(al) foot
秒立方英尺/天 second cubic(al)foot perday
秒流 second flow
秒流量 flow rate per second
秒流速 flow rate per second
秒轮 second wheel
秒轮传动齿轴 second wheel driving pinion
秒轮附加锁杆 second wheel additional lock
秒轮夹板 second wheel bridge
秒轮摩擦簧 second wheel friction spring
秒轮锁杆 second wheel lock
秒轮锁杆簧 second wheel lock spring
秒脉冲 pulse per second
秒/米/厘米 s/m/cm
秒欧 <电感的旧单位,≈1 赫兹> secohm
秒拍 second beat
秒数计数器 second counter
秒桃轮 second heart
秒桃轮簧 second heart spring
秒显示 second display
秒延迟电雷管 second delay electric(al)detonator
秒延迟雷管 second delay blasting cap;second delay detonator
秒英尺 second-foot
秒英尺日 second-feet-day;second-foot-day
秒针 second-hand
秒钟 second clock

渺

渺羟萘磺酸 croceic acid
渺位缩合环 kata-condensed ring

妙

妙想 bright idea

庙

庙 joss house;mausoleum[复 mausoleums/mausoloa];shrine
庙的内殿 adytum
庙舍 hieron
庙塔 temple tower
庙堂内室 adytum
庙堂前栅栏围墙 transenna
庙宇 temple;mortuary temple <祭帝王伟人的>
庙宇城 temple city
庙宇地平面图 temple ground plan
庙宇坟场 temple tomb
庙宇复合建筑 temple complex
庙宇建筑 temple building;temple construction;euthynteria <希腊>
庙宇林荫路堤 temple terrace
庙宇门道 temple gateway
庙宇门廊 temple portico
庙宇内殿 secos[sekos]
庙宇平面图 temple plan
庙宇前部 temple front
庙宇前庭 temple forecourt
庙宇入口 temple gateway
庙宇寺堂占地 abbeystead
庙宇庭院 temple court
庙宇围墙 enceinte wall;peribolos;temple enclosure
庙宇辖区 temple precinct
庙宇阳台 temple terrace
庙宇院子 hypaethros
庙宇正面 temple facade
庙宇中心 temple nucleus
庙宇柱廊 temple portico

灭

灭波器 ripple eliminator
灭草 weed eradication;weed killing
灭草机 weed destroyer
灭草剂 herbicide
灭草剂配方 herbicide formula
灭草灵 swep
灭草隆 <一种持久性除莠剂> monuron
灭草喷雾机 weed sprayer
灭草喷雾器 weed spraying gear
灭草器 weed destroyer
灭茬机 stubble cleaner;stubble skim plough
灭茬圆盘耙 one-way disk harrow;single-row disk harrow
灭尘 lay the dust
灭尘剂 dust palliative
灭尘系统 dust suppression system
灭尘油 dust-binding oil;dust laying oil
灭尘装置 dust suppression device
灭虫法 disinfestation;disinsection;disinsectization
灭虫库 <客车> disinfecting shed
灭磁 de-excitation;de-magnetization;excitation suppression;field discharge
灭磁电压 dropout voltage
灭磁开关 field circuit breaker;field suppressing switch;magnetic blow-out switch
灭磁器 de-exciter
灭磁绕组 killer winding
灭磁时间常数 de-excitation time-constant
灭磁系统 de-excitation system
灭磁滞 free hysteresis;hysteresis free;hysteresis suppression
灭灯 lighting-off;light out
灭灯报警 lamp failure alarm
灭灯继电器 light-out relay
灭点 perspective center[centre];vanishing point
灭点控制 vanishing point control
灭电弧 arc suppression
灭毒剂 viricide
灭洪 flood abatement
灭弧 arc blow-out;arc control;arc extinction;arc-extinguishing;arc suppressing;blowout;quench
灭弧变压器 arc suppressing transformer;quenching transformer
灭弧触点 arcing contact
灭弧磁体 blowout magnet
灭弧电抗器 arc suppressing reactor
灭弧电路 arc-suppression circuit
灭弧电压 blowout voltage;extinction voltage;extinguishing voltage;reseal voltage
灭弧二极管 field quenching diode
灭弧工作 rupturing duty
灭弧沟 arc chute
灭弧角开关 horn-break switch
灭弧接点 blowout contact
灭弧介质 arc-extinguishing medium
灭弧能力 arc-rupturing capacity
灭弧器 arc deflector;arc extinguish device;arc suppressor;extinguisher;quencher
灭弧腔 arcing chamber;spark chamber
灭弧栅 arc chute
灭弧时间 quenching time
灭弧时刻 quenching moment
灭弧室 arc chute;arc extinguish chamber;arcing chamber
灭弧室开关 quenching pot
灭弧双电路 quenching circuit
灭弧系统 blowout system
灭弧线圈 arc-extinguishing coil;arc suppressing coil;arc suppression coil;blowout coil
灭弧箱 explosion chamber;quenching pot
灭弧性塑料 arc-extinguishing plastics
灭弧罩 arc chute
灭弧装置 arc-control device;arc-extinguishing equipment;arc-suppressing apparatus
灭活率 inactivated ratio
灭活污泥 inactivated sludge
灭活效率 inactivated efficiency
灭活作用 deactivation;inactivation
灭火 dowse;extinguishing;extinguishment;extinguish the fire;fire control;fire-extinguishing;fire suppression;outfire;quench;smother
灭火安全 fire-fighting safety
灭火按钮 fire button;fire committal button
灭火保护 flame failure protection
灭火泵 fire engine;fire-extinguishing pump;fire pump
灭火材料 extinguishment material
灭火弹 fire-extinguishing bomb;fire-extinguishing bullet;fire-fighting bomb;grenade
灭火电极 pick-off electrode
灭火阀 fire valve
灭火(方)法 extinguishing method;extinguishment method;fire-fighting method;fire-extinguishing method;fire-fighting procedure
灭火飞机 air tanker
灭火风机 fire-extinguishing fan
灭火干粉 fire-fighting powder
灭火工程 extinguishing engineering;fire extinction engineering
灭火管 fire-extinguishing tube
灭火规定 fire-fighting order
灭火花板 spark blow-out plate;spark plate
灭火花的 spark-extinguishing
灭火花电路 spark-extinguishing circuit
灭火花电容器 spark condenser
灭火花器 spark extinguisher
灭火花用电容器 spark capacitor
灭火环形主管 fire-fighting ring main
灭火混合物 fire-extinguishing composition
灭火机 extinguisher;fire-extinguisher;hand flame cutter;fire extinction
灭火机车 fireless locomotive
灭火集尘器 spark arrester[arrestor];spark catcher
灭火计算尺 preparedness meter
灭火技术 fire-extinguishing engineering;fire-fighting technique
灭火剂 extinguishant;fire-extinguishing agent;fire-extinguishing chemical;fire-retardant;flame retardant;suppressant
灭火剂钢瓶 fire-extinguishing bottle
灭火剂控制阀 flame failure control valve
灭火剂浓度 agent concentration
灭火剂装充率 charging ratio of fire-extinguishing agent
灭火救援 fire rescue
灭火控制阀 fire suppression control valve;flame failure control
灭火口 fire access
灭火力量 fire-fighting force
灭火料 extinguish material
灭火龙头 fire hydrant;hydrant
灭火路线 extinguishing line
灭火沫 fire foam
灭火能力 extinguishing ability;flame extinguishing ability
灭火浓度 flame extinguishing concentration
灭火炮 fire cannon
灭火泡沫 extinguishing foam;fire-fighting foam
灭火泡沫材料 fire-extinguishing foam material
灭火泡沫发生器 fire foam-producing machine
灭火泡沫剂 fire foam
灭火泡沫液 fire-fighting foam agent
灭火喷射器 fire pump and sprayer
灭火喷水阀 fire-protection sprinkler valve;fire-protection valve
灭火喷水管 sprinkler main
灭火喷头阀门 sprinkler valve
灭火皮带管站 fire hose station
灭火瓶 fire-extinguisher bottle;fire suppression bottle
灭火气氛 extinctive atmosphere
灭火气体 fire-extinguishing gas;fire suppressant gas;fire-smothering gas
灭火器 annihilator;extinguisher;fire annihilator;fire-extinguisher;fire-extinguisher bottle;fire-extinguishing apparatus;fire suppression bottle;flame arrester[arrestor];flame damper;quencher;quenching unit
灭火器材 fire apparatus
灭火器材储藏所 fire-tool cache
灭火器橱 fire-extinguisher cabinet;fire hose and extinguishing cabinet
灭火器的配置 distribution of fire-extinguisher
灭火器符号 fire-extinguisher symbol
灭火器干粉 fire-extinguisher dry chemical
灭火器柜 fire-extinguisher cabinet;fire hose and extinguishing cabinet
灭火器架 fire-extinguisher bracket
灭火器件 extinguishing device
灭火器开关 fire-extinguisher cock
灭火器灭火 fire-fighting with fire-extinguisher
灭火器泡沫的给水栓 foam hydrant
灭火器喷射开关 fire-extinguisher switch
灭火器设备 fire-extinguisher equipment
灭火器手柄 fire-extinguisher handle
灭火器推车 fire-extinguisher trolley
灭火器箱 fire-extinguisher box;fire-extinguisher cabinet
灭火器行业协会 Fire Extinguishing Trade Association
灭火器液 fire-extinguisher fluid
灭火器用灭火剂 fire-extinguisher agent

M

灭火器转换开关 fire-extinguisher transfer switch
灭火前环境状况的调查 prefire situation inspection
灭火枪 extinguish gun;fire gun
灭火圈 fire suppression ring
灭火洒水喷头 sprinkler head
灭火洒水系统 fire-sprinkling system
灭火塞 fire plug
灭火砂 fire-extinguishing sand
灭火砂箱 fire sand box
灭火设备 extinguishing equipment;extinguishing installation;fire extinction equipment;fire-extinguishing apparatus;fire-extinguishing appliance;fire-extinguishing device;fire-extinguishing equipment;fire-fighting apparatus;fire-fighting equipment;fire-fighting facility
灭火设施 fire-extinguishing installation
灭火射流 extinguishing jet;extinguishment stream
灭火时间 flame extinction time
灭火试验 extinguishing run;fire-extinguishing test
灭火手雷 extinguishing bomber;extinguishing grenade;fire-fighting grenade
灭火手推车 fire-extinguishing cart
灭火栓 fire plug
灭火栓箱 fire hydrant chamber
灭火水泵 fire-extinguishing pump
灭火水泵驱动 fire pump drive
灭火水带线 fire attack line
灭火水带线路 fire-fighting line
灭火水龙带 fire hose
灭火水龙带柜 fire hose cabinet
灭火水龙带接头 fire hose connection
灭火水龙带支架 fire hose rack
灭火水枪 fire hose nozzle;fire nozzle
灭火水压力 fire pressure;fire water pressure
灭火速度 blow-off velocity
灭火毯 fire(-smothering) blanket
灭火通道大厅 fire-fighting access lobby
灭火系统 extinguishing system;fire-extinguisher system;fire-extinguishing system;fire suppression system
灭火效率 fire-fighting efficiency
灭火效能 fire-extinguishing effectiveness;fire-extinguishing efficiency
灭火行为 fire-fighting behavio(u)r
灭火性能 extinguishment characteristic;fire-extinguishing performance;fire-extinguishing property
灭火演习 fire-extinguishing drill
灭火焰的 flameless
灭火药剂 extinguishing agent;extinguishing medium;extinguishing product
灭火药沫 foamite
灭火要求 extinguishing requirement
灭火液 fire-extinguisher fluid;fire-extinguishing fluid
灭火用哈龙 fire-fighting halon
灭火用泡沫材料 extinguishing foam
灭火战斗 fire fighting
灭火战略 fire attack strategy;fire-fighting strategy
灭火战术 fire-fighting tactics;fire tactics
灭火蒸气 fire-extinguishing vapo(u)r
灭火蒸汽 extinguish steam;steam for fire-extinguishing
灭火直升飞机 helitanker
灭火帚 fire broom
灭火主管道 fire main(pipe)
灭火装置 extinguishing installation;fire-extinguishing apparatus;fire-extinguishing equipment;fire-extinguishing installation;fire-extinguishing plant;fire-fighting equipment;fire prevention equipment;fire-protection equipment;fire unit
灭火资源 fire-fighting resources
灭火总指挥 fire boss
灭火组织 fire suppression organization
灭火作战计划 fire control plan(ning);prefire planning
灭迹 obliteration
灭迹元素 extinct element
灭绝 die out;eradication;extinction;extirpate;uproot
灭绝动物 extinct animal
灭菌 disinfect;pasteurize;sterilizing
灭菌处理 biocidal treatment;germicidal treatment
灭菌灯 sterilamp
灭菌法 sterilization
灭菌废水 sterilized wastewater
灭菌粉 sterilized powder
灭菌硅 Ceresan universal trockenbeize;methoxyethylmercury silicate
灭菌滑石 sterilized talc
灭菌混合机 sterilizing mixer
灭菌混悬液 sterilized suspension
灭菌机 sterilizer
灭菌计 sterilometer
灭菌剂 bactericidal agent;disinfectant;sterilizing agent
灭菌接触控制 contact tank
灭菌冷冻干燥猪真皮 lyophilized porcine dermis irradiated
灭菌磷 Plondrel
灭菌器 sterilizer
灭菌溶液 sterile solution
灭菌软膏剂 sterilized ointment
灭菌散剂 sterilized powder
灭菌设备 sterilizer instrument;sterilizing installation
灭菌射线 bactericidal rays
灭菌室 sterilizing chamber
灭菌水 aqua sterilisa
灭菌铜 Omazene
灭菌外加剂 fungicidal admixture
灭菌消毒 bacteria disinfection
灭菌效应 sterilizing effect
灭菌药剂 sterilized pharmaceutics
灭菌液循环泵 sterilizer circulator
灭菌指示器 sterility detector
灭菌质 anti-microbin
灭菌作用 disinfecting action;sterilizing effect;disinfection
灭昆虫 deinsectization
灭雷器具 mine neutralization vehicle
灭励 de-excitation
灭能剂 inactivator
灭声器 muffle(r)
灭声器爆声 muffler explosion
灭声设备 sound-damping arrangement
灭失或不灭失条款 lost of not lost clause
灭失记录 <航运> loss experience
灭失条款 lost clause
灭虱 delousing
灭虱剂 lousicide
灭虱药 pediculicide
灭鼠 deratization;rat destruction
灭鼠剂 rodenticide
灭鼠灵 warfarin
灭鼠器 mouse killer;mouse-trap
灭鼠特 thiosemicarbazide
灭鼠药 disinfectant;rodent poison bait
灭鼠药分析 analysis of rodent poison bait
灭鼠优 pyrinuron;ratkiller;vacor
灭鼠证书 deratization certificate
灭鼠舟 promurit
灭酸 fenamic acid
灭梭威 methiocarb
灭亡系数 extinction coefficient
灭蚊弹 mosquito bomb
灭蚊的 culicidal
灭蚊剂 mosquitocide
灭蚊器 mosquito killer
灭蚊药 culicide;mosquitocide
灭焰的 flameless
灭焰器 flame damper
灭音 muffle
灭蝇剂 fly spray
灭蝇磷 Nexa fly-trap
灭蝇制剂 fly destroying preparation
灭蚤 depulization
灭藻 algae removal
灭藻剂 algicide
灭蟑螂药 cockroach killer
灭酯灵 hepronicate

篾

篾框 gabion

篾筐 <筑堤用的> pannier

蠛

limpet

民

民办地图制图业 private cartography

民办公助 run by the local people and subsidized by the state
民变险 civil commotion
民兵(部队) militia
民德冰期【地】Mindel glacial epoch;Mindel glaciation
民德阶【地】Mindel(stage)
民德-里斯冰期【地】Mindel-Riss ice period
民德-里斯间冰期【地】Mindel-Riss interglacial stage
民堤 local dike[dyke]
民法 civil law
民法(法)典 civil code
民法规范 norm of the civil law
民法事实 juristic(al) fact in civil law
民防标志 civil defence sign
民防构筑 civil defense construction
民防系统 civil defence system
民房建筑 domestic architecture
民工 civilian worker
民工建勤 civilian laborers working on public project;peasants voluntary labo(u)r;public works duties
民航 public navigation
民航的 civil aeronautic
民航飞机利用率 utilization ratio of civil aviation aircraft
民航飞行事故 civil accident
民航机 civil aeroplane;civil aircraft
民航机标记 civil air ensign
民航接送车 airport limousine
民航经济学 civil economics
民航条例 civil air regulations
民航运输组织 organization of civil aviation transport
民间的 civil;civilian;private
民间访问 civilian visit
民间工程 deposit(e) works
民间工艺品 folk arts and crafts
民间会计 private accounting
民间集体企业 private collective enterprise
民间纠纷 dispute among the people
民间救助 civil salvage
民间贸易 non-governmental trade;people-to-people trading activity
民间贸易代表团 non-government trade mission
民间贸易协定 non-governmental trade agreement
民间企业 private enterprise
民间契约 civil contract
民间青花 folk blue and white
民间社团 voluntary association
民间审计 non-governmental audit;private audit
民间团体 private Association
民间习俗 folk custom
民间信用 non-governmental credit
民间艺术 folk art
民间准则 private regulation
民井简易抽水试验 simple pumping test from domestic well
民居 folk house;vernacular dwelling;vernacular housing
民居厨房 dwelling kitchen
民力 financial resources of the people
民权 civil right
民生 people's livelihood
民生经济 livehood economy
民事案件 civil case;suit at law
民事案件证据 civil evidence
民事保释 civil bail
民事被告 civil defendant
民事补偿 civil remedy
民事补救(办)法 civil remedy
民事不法行为 civil wrong behavio(u)r
民事财产关系 civil property relations
民事裁判权 civil jurisdiction
民事处罚 civil penalty
民事处分 civil sanction
民事的侵权行为 <不包括违背契约> tort
民事登记体系 system of civil registration
民事第一审程序 civil procedure of first instance
民事罚款 civil penalty
民事法律关系 civil legal relation(ship)
民事法律关系的客体 object of civil legal relationship
民事法律行为 civil juristic act
民事法庭 civil court;civil court of sessions;civil division
民事法院 civil court
民事犯 civil prisoner
民事方面的共谋 civil conspiracy
民事诽谤罪 civil libel
民事关系 civil relation
民事管辖豁免 exemption from civil jurisdiction
民事管辖(权) civil jurisdiction
民事过失 civil negligence
民事纠纷 civil dispute;dispute among the people
民事客体 civil object
民事立法 civil legislation
民事判决 civil judg(e)ment
民事赔偿 civil compensation
民事起诉书 memorandum of complaint
民事强制措施 coercive measure in civil suits
民事侵权法 law of tort
民事侵权行为 civil trespass
民事权利 right in civil affairs;right relating to civil law
民事(上的)罚款 civil forfeiture
民事(上的)没收 civil forfeiture
民事上诉 civil appeal
民事上诉法庭 court of common pleas
民事审判 civil judg(e)ment;civil justice;civil trial
民事审判管辖 civil jurisdiction

民事审判体制 system of civil justice
民事审判庭 civil adjudication; civil court; court of tribunal
民事时效 civil prescription
民事(司法)统计 civil judicial statistics
民事诉讼 civil action; civil case; civil lawsuit; civil-use civilian use
民事损害 civil injury
民事损害赔偿 civil damages
民事调解 civil mediation
民事投诉 civil complaint
民事违法行为 civil offence
民事刑事诉讼 civil suit; criminal suit
民事行政 civil administration
民事义务 civil obligation; duty in civil affairs
民事油污染事故责任公约 convention on civil liability for oil pollution damage
民事原告 civil plaintiff
民事责任 civil liability; civil responsibility
民事责任保险 civil liability insurance; civil responsibility insurance
民事责任赔偿 civil remedy
民事债务 civil debt; civil obligation
民事指控 civil charge
民事制裁 civil sanction
民事主体 civil subject
民事住所 civil domicile
民俗 folk custom; folkways
民俗旅游 ethnic tour; ethnic tourism
民俗史 history of folk custom
民俗学 folklore
民选机构 publicly elected body
民窑 civil ware; folk kiln
民意 public opinion
民意测验 opinion poll; poll; public opinion poll; public opinion survey; test of opinion
民意测验者 pollster; polltacker
民意调查 public opinion poll
民意调查者 pollster; polltacker
民意对策 public opinion game
民意听取会 public hearing meeting
民意征询 popular consultation
民营岔线 private siding; private track
民营的 private
民营企业 private enterprise
民营铁路 private line; privately owned line; private(ly owned) railway
民营性质的公有制形式 public ownership with a popularly-run nature
民用 civilian application; civil-use civilian use; for civil use
民用闭环太阳能热水器 closed loop domestic water heater
民用波段 citizen-band
民用波段无线电通信[讯] citizens band radio
民用补给品 civilian supplies
民用柴油 domestic diesel oil
民用产品 civilian industry product
民用车辆 commercial vehicle
民用晨昏蒙影 civil twilight
民用储备金 civil contingency fund
民用船 civil ship; merchant ship
民用的 civil; domestic
民用地图 civilian map
民用电路 commercial circuit
民用电视 home television
民用电台频道 citizens band channels
民用电子产品 consuming electronic product
民用防空 civil defense against air raids
民用防空洞 civil defense shelter door
民用防空洞通风装置 civil defense shelter ventilation
民用防空工程 civil air defense engineering
民用防(空)建筑 civil defense structure
民用房屋 civic building; dwelling house
民用房屋漆 house paint
民用飞机 civil aircraft
民用飞机场 air port; civil airfield construction
民用负荷 appliance load; domestic load
民用高技术 high technology for civilian use
民用给水 domestic water supply
民用工程 civil engineering; civil works; utility
民用工业 civilian industry; civil industry
民用供热 domestic heating
民用管 commercial tube
民用规格 domestic grade
民用航(空) civil aviation
民用航空的 civil aeronautic
民用航空管理制度 system of civil aviation management
民用航空技术管理 technical management of civil aviation
民用航空局 <美> Civil Aeronautics Board
民用航空摄影学 civilian air photography
民用航空线 civil aviation lines
民用航空运输 civil air transportation
民用航空运输协定 Air Service Agreements
民用航空组织 civil aviation organization
民用耗水量 domestic consumption
民用合成洗涤剂 household synthetic detergent
民用核设施安全许可制度 safety licence system of civil nuclear device
民用环境 commercial environment
民用机场 civil aerodrome; civil airfield; civil airport
民用机场管理暂行规定 Provisional Rules on Administration of Civil Airports
民用基本建设经济学 civil economics of capital construction
民用技术 civilian technology
民用间防护 civil protection
民用建筑 building of civic; civil architecture; civil building; civil construction; civilian construction; domestic architecture; residential construction
民用建筑材料 building materials for civilian use
民用建筑师 civil architect
民用建筑物 civil building; civilian structure
民用建筑专业 residential construction industry
民用焦炭 domestic coke
民用接收机 commercial receiver
民用结构 secular structure
民用经济 civilian economy
民用开环太阳能热水器 open loop domestic water heater
民用垃圾处理系统 household disposal system
民用垃圾粉碎机 household garbage grinder
民用劳动力 civilian labor force
民用滤水器 household water filter
民用铝 commercial aluminum
民用码头 commercial dock
民用煤气 domestic gas; household fuel gas; town gas
民用煤油 domestic kerosene
民用能源要求 domestic energy requirement
民用年 calendar year; civil year; equinoctial year; natural year; solar year; tropic(al) year
民用排水系统 domestic sewerage
民用频带 citizens' band
民用频段 citizens' radio band
民用品 civilian goods
民用漆 house paint
民用企业 civil enterprise
民用气垫船 hovergem
民用汽油 domestic gasoline
民用取暖油 domestic heating oil
民用燃料 domestic fuel; household fuel
民用燃料气 domestic fuel gas; household fuel gas
民用燃料油 domestic fuel oil
民用热水 domestic hot water
民用热水系统 domestic hot-water system
民用日 calendar day; civil day
民用日出日落 civil sunrise and sunset
民用日历 civil calendar
民用日期 civil date
民用乳胶涂料 latex house paint
民用设备的可靠性 commercial reliability
民用生产 civilian production; domestic production
民用时 civil mean time; civil time
民用室外型焚化炉 domestic outside-type incinerator
民用收音机 people's receiver
民用曙光时 civil twilight time
民用曙暮光 civil twilight
民用水 domestic use of water
民用水表 domestic water meter
民用水水源 domestic water supply
民用水预热储存 domestic water pre-heat storage
民用太阳能热水器 domestic solar water heater
民用无线电频带 citizens' radio band
民用无线电台频带 citizens' radio band
民用无线电业务 citizens' radio service
民用物资 civilian goods
民用洗涤槽 housemaid's sink
民用洗涤剂 detergent for household use
民用下水道(系统) domestic sewage
民用型 civil version
民用型煤 coal briquet(te) for civilian use
民用烟火 civilian pyrotechnics
民用月 calendar month
民用载重车 civil truck
民用栅栏 privacy fence
民用炸药 commercial explosive
民用蒸汽锅炉 steam boiler for domestic heating
民用支出 civil expenditures
民用住房 civil housing; civilian construction
民用住宅 residential houses
民用自动火灾报警系统 privately operated automatic alarm system
民用自卫武装船 private armed vessel
民有经济 burgher economy
民约论 social contract
民泽单位【地】Meizer unit
民政 home affairs
民政部门 civil administration department
民政当局 civilian authority
民政管理 civil administration
民政机关 civil administration
民政局 Bureau of Civil Administration
民政人员 civilian personnel
民政事务 civil affairs
民众参与 popular participation; public participation
民众骚乱 civil commotion
民主国 commonwealth
民主集中制 democratic centralism
民族本体 national identity
民族传统 national tradition
民族地理学 ethnogeography
民族地区 regions inhabited by ethnic groups
民族地毯 nationality carpet
民族风格 nationalistic style; national style
民族风情瓶 vase of national amorous feeling
民族工业 national industry
民族工业保护 protection of home industries
民族宫 Palace of Nations
民族画 national drawing
民族集居地区 national area
民族纪念碑 national monument
民族建筑形式 national architecture
民族建筑学 national architecture
民族经济学 national economics
民族经济综合体 national economic complex
民族考古学 national archaeology
民族款式 folklore; folkstyle
民族浪漫风格 national Romantic style
民族贸易企业 nationality products enterprises
民族区域自治 regional national autonomy
民族商业 minority nationalities commerce
民族生态学 ethnoecology; national ecology
民族史 history of nationality; national history
民族市场 national market
民族事务 nationalities affairs
民族嗜好 national preferences
民族特色 national feature
民族特色建筑 national features of architecture
民族特性 national identity
民族图案 national ornamental
民族娃娃 doll in national costumes
民族委员会 Nationalities Committee
民族文化宫 the Cultural Palace of the Nationalities
民族乡 nationality township(s)
民族学领域考察项目 ethnographic program(me)
民族学院 institute for nationalities
民族遗产 national inheritance
民族意识 national awareness; national consciousness
民族用品生产基地 base of producing goods for minority nationalities
民族装饰 national ornamental
民族资本 national capital
民族资源 national resources
民族自治区 national autonomous areas
民族组成 national composition

皿

头方颈螺栓 plough bolt

皿头方形螺栓 plough bolt
皿头螺钉 flush bolt
皿头铆钉 flush head rivet
皿形螺钉 pan head screw
皿形铆 pan rivet head

皿形头 pan head
皿形钟 gong

抿

抿墙工人 dauber

敏

敏拜楼 <清真寺内的讲坛> minbar

敏度 acuity;acuteness
敏感 sensitization;subtlety
敏感 X 线片 diaphax
敏感比 <土壤试验> sensitivity ratio
敏感边缘 sensitive edge
敏感变送器 sensor transmitter
敏感标度 sensitivity scale
敏感表面 sensing surface
敏感部件 sensing unit
敏感材料 sensitive material; sensitized material
敏感层 sensitive layer
敏感产品 sensitive product
敏感成分 sensitive composition
敏感单元 sensing unit
敏感的 responsive;sensible;sensitive;susceptible;susceptive
敏感地区 sensitive area
敏感点 sensitive spot
敏感点法 sensitivity points method
敏感电路 sensing circuit
敏感电阻器 sensitive resistor
敏感度 quick reaction capability;sensitiveness; sensitivity; sensitivity level;susceptibility
敏感度分析 sensitivity analysis
敏感度降低 decrease of sensitivity
敏感度控制 sensitivity control
敏感度试验 sensitivity test(ing)
敏感度特性曲线 sensitivity characteristic
敏感度系统 sensory system
敏感段 sensitive segment
敏感分析 sensibility analysis
敏感过程 sensitizing
敏感过度的 super-sensitive
敏感化 sensitization
敏感继电器 sensitive relay
敏感价格 sensitive price
敏感监测器 sensitive monitor
敏感检测器 sensitive monitor
敏感件 position-sensing unit
敏感金属 sensitive metal
敏感晶体 sensing crystal
敏感面 sensitive area
敏感黏[粘]土 sensible clay;sensitive clay
敏感起爆 sensitive priming
敏感气压计 baroscope
敏感器件 sensing device;sensitive device
敏感区 sensitizing range
敏感区技术 sensing zone technique
敏感热丝 sensing filament
敏感色 sensitive colo(u)r
敏感栅 sensitive grid
敏感商品价格 sensitive goods price
敏感生物测定 sensitive bioassay
敏感试验 sensitiveness test;sensitization test
敏感水体 sensitive water body;sensitive waters
敏感探头 sensing probe
敏感特性 sensitivity characteristic
敏感体积 sensitive volume
敏感头 sensing head
敏感温度计 sensitive thermometer
敏感问题 tender subject
敏感物质 sensitized material
敏感系数 sensitivity coefficient
敏感线圈 sensor coil
敏感项目 sensitive item
敏感效应 sensitizing effect
敏感形式 sensitive form
敏感型 sensitive form
敏感性 receptance;sensibility;sensitivity;susceptibility;susceptiveness
敏感性产品 sensitive product
敏感性当地材料 sensitive local materials
敏感性地层 sensitive formation
敏感性分析 sensitive analysis;sensitivity analysis;susceptivity analysis
敏感性股票 cyclic(al) stock
敏感性价格 sensitive price
敏感性检验 sensitive test;sensitivity test(ing)
敏感性能 sensitive property
敏感性商品 sensitive goods;sensitive market
敏感性市场 sensitive market
敏感性试验 sensitive test;sensitivity analysis;sensitivity test(ing)
敏感性问题 sensitive question
敏感性物价指数 index number of sensitive prices
敏感性物种 sensitive species
敏感性页岩 sensitive shale
敏感性炸药 sensitive explosive
敏感性植物 sensitive plant
敏感性自动调整电阻器 sensitive resistor
敏感液 sensing solution
敏感仪表 sensing instrument
敏感元件 end instrument; pick-off; sensing device; sensing element; sensing instrument; sensing unit; sensitive element; sensitive pick-up;sensor;susceptor element
敏感元件数 sensing component number
敏感元件系统 sensor system
敏感元件信息 sensory information
敏感增强机制 sensitization-invigoration mechanism
敏感中心 sensitivity centre
敏感种 intolerant species; sensitive species
敏感装置 senser;sensing device;sensor
敏化 activation;sensibilization;sensitization;sensitize
敏化材料 sensitized material
敏化层 sensitizing layer
敏化处理 sensitizing treatment
敏化发光 sensitization luminescence;sensitized luminescence
敏化法 sensitizing
敏化分解 sensilized decomposition
敏化腐蚀 sensitization corrosion
敏化光电管 sensitized photocell
敏化光解 sensitized photodecomposition
敏化过程 activation process; sensitizing
敏化剂 sensibilizer;sensitizer;sensitizing agent;sensitizing compound
敏化剂离子 sensitizer ion
敏化立方 sensitized cube
敏化了的 sensitized
敏化磷光 sensitized phosphorescence
敏化脉冲 sensitizing pulse
敏化面 sensitized surface
敏化潜力 sensitizing potential
敏化强度 sensitizing intensity
敏化区 sensitized zone
敏化区腐蚀 <焊接后> weld decay
敏化染料 sensitizing dyestuff
敏化筛网 sensitized screen
敏化时间 sensitization time
敏化物 sensitizer
敏化效应 sensitizing effect
敏化液 sensitizing solution
敏化阴极 sensitized cathode
敏化荧光 sensitized fluorescence
敏化织物 sensitized fabric
敏化纸 sensitized paper
敏化中心 sensitivity speck;sensitizing center[centre]
敏化作用 sensibilization;sensitization
敏捷的 expeditious;quick
敏捷性测定 dexterity test
敏觉杆 sensitive lever
敏勒炸药 minesite
敏锐 acuteness
敏锐的 subtle
敏锐度 acuity;sharpness
敏锐度光度计 acuity photometer
敏锐性 acuity
敏锐指数 sharpness index
敏瓦尔低膨胀系数合金铸铁 Minvar
敏压卸载阀 pressure-sensing unloading valve
敏于迷路的 maze-bright
敏载制动器 load brake

名

名标 identifier

名表目 name entry
名册 beadroll;list;panel;register;roll
名册单位 listing unit
名产 famous product;speciality goods;special product;staple
名常数 name constant
名称登录【计】name entry
名称、地址和住址卡片 name, address and residence card
名称混乱 poikilonymy
名称卡 designation card
名称量 denominate quantity
名称码 name code
名称牌 badge plate
名称上的 nominal
名称术语 identification terminology
名称学 onomasiology;onomastics
名称转换 name resolution
名词汇编 glossary;nomenclature
名词术语 vocabulary of terms
名词索引 glossarial index
名词学 terminology
名词组 noun group
名词组角色 noun group case
名代码 name code
名单 beadroll;list;panel
名单抽样 list sampling
名单框抽样 list frame sampling
名单上列名者 panelist
名地址服务器 name address server
名服务程序 name server
名符其实 live up to one's name
名贵 famous and precious
名贵的 rare
名画 famous painting;great picture
名家 eminence;master
名空间 name space
名空间的名 name in a name space
名块 name block
名利 wealth and fame
名列第一 take the first place
名录服务【计】directory service
名目 denomination;item
名目所得 nominal income
名目作业 dummy activity
名牌 designation strip;famous brand;placard;name plate <门上刻写姓名的牌子>
名牌产品 brandname products; famous brand products
名牌大学的古老建筑物 <美> halls of ivy
名牌货 famous brand; goods of a well-known brand;name brand
名牌货商店 speciality shop
名牌商品 name brand; name merchandise
名牌上的额定值 name-plate rating
名牌油 branded oil
名片 business card;calling card;name card;visiting card
名片袋 card pocket
名片盒 cardcase
名片机 card printer
名片框 <门上的> card plate;card frame
名区 point of interest
名人词典 biographic(al) dictionary
名人巨像 colossus
名入口 name entry
名声不佳 objectionable character
名胜 place of interest;point of interest;show place
名胜地 scenic spot;show place
名胜地点 place of interest
名胜地区的共有公寓房屋 resort condominium
名胜古迹 places of historic interest and scenic beauty; popular resort and historic spot; scenic spots and historic(al) sites
名胜古迹区 scenic spot and historic resort
名数 concrete number; denominate number【数】
名数制 denominational number system
名义报价 nominal quotation
名义被告 nominal defendant
名义层高 nominal height
名义产量 nominal yield
名义长度 nominal length
名义车道 notional lane
名义车速 crest speed;nominal speed
名义成本 nominal cost
名义成分 nominal composition
名义尺寸 nominal dimension;nominal measure;nominal size;specified size
名义尺寸范围 nominal dimension range
名义尺度 nominal mean power;nominal measure
名义存款 nominal deposit
名义代理 ostensible agency
名义代理权 ostensible authority
名义代理人 ostensible agent
名义的 nominal;ostensible
名义电压 nominal voltage
名义董事 hono(u)rary director
名义动力 nominal power
名义断裂强度 nominal breaking strength
名义堆装容量 nominal heaped capacity
名义吨位 rated pressure
名义费率 apparent rate
名义费用 nominal cost
名义幅度 nominal width
名义负债 nominal liability
名义刚度 nominal rigidity
名义刚性 nominal rigidity
名义高度 nominal elevation;nominal height
名义工时 nominal hours
名义工资 nominal wages
名义工资额 nominal wage bill
名义工资率 rate of nominal wages
名义工资指数 index number of nominal wages;nominal wages indexes
名义工资总值 nominal wage bill

名义工作时间 nominal hours; nominal working hours
名义工作周时间 nominal work week
名义公差 published tolerance
名义功率 nominal horsepower; nominal rating
名义供热输出量 nominal heating output
名义股东 nominal partner
名义关税 nominal customs duties; nominal tariff
名义关税率 nominal tariff rate
名义滚动半径 nominal rolling radius
名义合伙人 nominal partner; ostensible partner
名义合同价格 nominal contract price
名义荷载 nominal load
名义厚度 nominal thickness
名义汇价 nominal rate; nominal rate of exchange
名义汇率 nominal exchange rate; nominal rate; nominal rate of exchange
名义汇票 nominal exchange
名义获利 nominal yield
名义价格 nominal price
名义价值 nominal value
名义交易 nominal transaction
名义焦面 nominal focal surface
名义截面 nominal cross-section
名义金额 nominal amount
名义劲度 nominal stiffness
名义抗扭强度 nominal torsional strength
名义颗粒度 nominal grain size
名义可变因素 nominal variables
名义孔径 nominal aperture
名义宽度 nominal width
名义拉(伸)应力 nominal tensile stress
名义利率 explicit interest rate; face rate of interest; nominal interest rate; nominal rate
名义利润 nominal profit
名义利息 nominal interest
名义流量 nominal flow capacity
名义马力 nominal horsepower
名义美元 nominal dollar
名义密度 nominal density
名义密封线 nominal sealing line
名义面积 apparent area
名义能力 nominal capacity
名义年利率 nominal annual rate
名义牌价 nominal quotation
名义配合比 nominal mix
名义偏向 nominal deviation
名义票面价格 nominal par
名义票面值 nominal par
名义牵入转矩 nominal pull-in torque
名义强度 nominal strength
名义屈服应力 nominal yield stress
名义群体技术 nominal group technique
名义容量 nominal capacity
名义上持股 nominal holding; nominee holding
名义上的当事人 nominal party
名义上的负债 nominal liability
名义上的股东 ostensible partner
名义上的汇兑 nominal exchange
名义上的税收 nominal tax
名义上的所有者 nominal owner; straw man
名义上的营业支出 nominal expenditures
名义上赔偿 nominal damage
名义上下限截止频率 nominal upper and lower cutoff frequencies
名义深度 nominal depth
名义审计 nominal audit
名义实际汇率 nominal effective exchange rate
名义收入 nominal income
名义收益 nominal earning
名义收益率 nominal yield
名义寿命期 nominal lifetime
名义输出功率 nominal heating output
名义水灰比 nominal water-cement ratio
名义水力停留时间 nominal hydraulic residue; nominal hydraulic retention time
名义水平 nominally horizontal
名义税 apparent tax
名义税率 apparent tax rate; crest speed; nominal rate
名义速率 <指在道路的一定区段中，多数驾驶人在没有交通干扰的情况下所能达到的行驶速率> crest speed; nominal speed
名义损失 nominal damage
名义所得 nominal income
名义条款 nominal term
名义贴现率 nominal discount rate
名义停留时 nominal retention period
名义托运人 nominal shipper; notional shipper
名义危险区 nominal risk area
名义误差 nominal error
名义显著性水平 nominal significance level
名义现金余额 nominal cash balance
名义型数据 nominal data
名义需求 notional demand
名义压力 nominal pressure
名义摇臂比 nom rock ratio
名义阴极 dummy
名义应变 apparent strain; conventional strain; nominal strain
名义应力 nominal stress
名义有效汇率 nominal effective exchange rate
名义余额 nominal balances
名义约定价格 nominal contract price
名义载重量 nominal load (bearing) capacity
名义账户 nominal account
名义账面额 nominal amount
名义账目 nominal account
名义账项总账 nominal ledger
名义整坡曲线 nominal grading curve
名义支出 nominal expenditures
名义支付 payment for honour
名义直径 basic size; nominal diameter; normal diameter
名义值 nominal value
名义转数 nominal revolution
名义转折角 nominal deflection
名义资本 nominal capital
名义资产 immaterial assets; nominal assets
名义总体 nominal population
名义最大速限 <较决定的略小，用以事先警戒驾驶人> prima facies speed limit
名义作价 nominal allowed price
名优产品 famous brand high-quality products; high-quality well-known goods
名誉保险单 hono(u)r policy
名誉博士 hono(u)rary doctor
名誉不佳的 of bad repute
名誉董事长 hono(u)rary chairman
名誉的 hono(u)rary
名誉顾问 hono(u)rary advisor
名誉会员 hono(u)rary member
名誉教授 hono(u)rary professor; emeritus professor
名誉契约 hono(u)r agreement; hono(u)r contract
名誉损害赔偿 indemnity for defamation
名誉团体 hono(u)rary
名誉校长 chancellor
名誉职位 hono(u)rary post
名誉职务 hono(u)rary office
名字表 name table

明暗(表现) light and shade

明暗层次级数 gradation series
明暗灯标 intermittent light; occulting light
明暗灯光组 group occulting light
明暗等高线 illuminated contour; illuminated contour method
明暗度 brightness
明暗对比 contrast between light and shade; light and shade contrast
明暗对比法 chiaro(o)scuro
明暗对照法 chiaro(o)scuro
明暗法 shading
明暗反差 contrast in black-white transitions
明暗光 occulting light
明暗互光 alternating light
明暗急闪光 occulting quick flashing light
明暗交替(灯)光 alternating light; intermittent light
明暗界线 terminator
明暗面 light and shade face
明暗配合 chiaro(o)scuro
明暗融合 shading and blending
明暗相间灯 occulting light
明暗相间闪光灯光 occulting quick flashing light
明白指定(资金)用途 earmark
明板铰链 flap hinge
明孢臭属 <拉> Armatella
明保险丝 open fuse
明报 telegram in plain language
明冰 clear ice
明槽 open canal; open channel; open flume
明槽冲击波 shock wave in open channel
明槽流 channel flow
明槽式水轮机 open flume turbine
明槽式装置 open flume setting
明槽水面线 flow profile in open channel
明槽装配 <涡轮机叶片的> open flume setting
明插销 rim latch; surface latch
明察的可见度 see-through visibility
明察的视野 see-through visibility
明场 light field
明场图像 bright field image
明椽 open rafter; show rafter
明德林理论 Mindlin theory
明钉法 exposed nailing
明洞 cut-and-cover tunnel; gallery; open cut tunnel; tunnel without cover; open tunnel
明硐 open tunnel
明度 brightness
明度标 brightness scale; lightness scale
明度纯度 colo(u)rimetric purity
明度对比 brightness contrast
明度范围 brightness range
明度视角变异 brightness flop
明度试验 brightness test
明度梯级 brightness scale
明度谐调 value harmony
明度属性 brightness attribute
明对接 open butt joint
明墩墩头 pier head
明矾 alaum; allum(e); alumbre; potash alum; potassium-alumin(i)um sulfate; white alum
明矾板岩 alum slate
明矾板岩混凝土 alum slate concrete
明矾玻璃 alum glass
明矾沉淀作用 alum precipitation
明矾处理 alum treatment
明矾粉 alum powder
明矾化 alunitization
明矾回收 alum recovery
明矾卡红 alum-carmine
明矾块 alum cake; cake of alum
明矾矿石 alum ore
明矾媒染剂 alum mordant
明矾煤 alum coal
明矾黏[粘]土 alum clay
明矾片岩 alum schist
明矾片岩混凝土 alum schist concrete
明矾(溶)液 alum solution
明矾鞣 alum tannage
明矾鞣革 alum dressed leather; alumed leather
明矾石 alumite; alum rock; alum stone; alunite; newtonite
明矾石板岩 alum slate
明矾石高强水泥 alunite high strength cement
明矾石化 alunitization
明矾石混凝土膨胀剂 alunite expansion agent for concrete
明矾石矿床 alunite deposit
明矾石膨胀水泥 alunite expansive cement
明矾石片岩 alum schist
明矾石水泥 alunite cement
明矾水 alum solution
明矾水浴 alum bath
明矾苏木精 alum hematoxylin
明矾苏木精染液 muchematein
明矾投加器 alum feeder
明矾土 alum earth; alumite; alum soil
明矾污泥 alum sludge
明矾洗涤液 alum cleaning bath
明矾絮凝 alum flocculation
明矾絮凝处理 alum coagulation treatment
明矾胭脂红染剂 alum-carmine stain
明矾岩 alum rock; alunite rock
明矾页岩 alum schist; alum shale
明矾页岩混凝土 alum shale concrete
明矾液贮存罐 alum storage tank
明反衬 bright contrast
明反应 light reaction
明缝 exposed joint; face joint; gap butt; open joint; top-stitched seam
明缝管 open-seam tube
明缝接头 face joint; open joint
明缝燕尾形榫接 box dovetail
明扶壁岸壁 buttress wall
明扶壁坝 buttress dam
明扶壁挡土墙 buttress wall
明敷电线 surface cable
明敷管 conduit on wall and ceiling
明敷线 exposed wiring
明腹板钢材 open-web steel
明杆 rising stem
明杆内螺纹 inside screw and rising stem
明杆平行式双闸滑阀 rising stem double-disk parallel slide valve
明杆式闸阀 rising stem gate valve
明格栅 exposed joist; open joist
明给虹吸润滑器 sight feed siphon lubricator
明给加油器 sight feed oiler
明给润滑器 <俗称牛眼或儿眼给油器> sight feed lubricator
明给油滴指示器 sight feed indicator
明拱 open arch

M

明沟 area drain; cut drain; ditch; gutter; land drain; open channel; open conduit; open cut drain; open ditch; open drain; open trench; small trough; surface ditch; surface drain; water furrow
明沟节制闸 ditch check
明沟排水 ditch drainage; drainage in open; drained joint; gutter cleaner; gutter drainage; open channel drainage; open cut drainage; open ditch drainage; open drainage; open sewer; surface drain; surface drainage; trench drain(age)
明沟生态系统 ditch ecosystem
明沟疏干法 open trough drainage
明沟系统 open drain system
明沟斜度 bank of ditch
明故宫遗址 ruins of the Ming palace
明管 exposed carcassing; exposed conduit; exposed penstock; exposed pipe; exposed piping; open conduit
明管布设法 exposed installation
明管道 open conduit
明管渠(道) open conduit
明光玻璃 clear glass
明光视觉 photopic vision
明轨 surface track
明涵(洞) culvert without top-fill; exposed culvert; open culvert
明合页 back-flap hinge; butt hinge; edge hinge; flap hinge
明河床 open channel
明弧 apparent arc; free-burning arc; open arc; visible arc
明弧灯 open arc lamp
明弧焊(接) open arc welding
明弧炉 open arc furnace
明火 direct fire; luminous fire; naked flame; naked light; open fire; open flame
明火标灯 cresset
明火操作 visible flame operation
明火灯 naked-flame lamp; open light
明火灯架 torchere
明火灯矿井 naked-flame mine
明火放炮 bunch blasting; cap and fuse blasting
明火加热 direct fire heating; open firing
明火加热炉 direct firing furnace
明火加热设备 firing equipment
明火炉 direct-fired heater
明火烧成 direct firing
明火试验法 open flame method
明火试验反应 response to open flame exposure
明火头 naked fire
明火直烧干燥器 direct-fired drier [dryer]
明火作业 working with naked fire
明基床 rubble bedding foundation
明间 case bay; center opening; central bay
明间旁的一间 side bay
明浇 casting-in open
明浇砂型 open sand mo(u)ld
明浇铸件 open sand casting
明浇铸型 open mo(u)ld
明浇铸造 open sand casting
明胶 celluloid; colla taurina; gel; gelatin(e); gelatin(e) glue
明胶板 <照相制板> artotype; gelatin(e) plate
明胶版复印 gelatin(e) duplicating
明胶版复印纸 gelatin(e) duplicating process paper
明胶版印刷 gelatin(e) printing
明胶包埋法 gelatin(e) embedding
明胶包衣 gelatin(e) glaze
明胶薄膜 gelatin(e) foil
明胶层 gelatin(e) layer
明胶处理 gelatin(e) treatment
明胶代用品 gelatin(e) substitute
明胶蛋白 glutin
明胶底漆 clairecolle; clearcole
明胶废水 gelatin(e) wastewater
明胶分解的 gelatinolytic
明胶粉 jelly powder
明胶浮雕 gelatin(e) relief
明胶复印 gelatin(e) printing
明胶干版 gelatin(e) dry plate
明胶干照相底版 gelatin(e) dry photographic(al) plate
明胶感光胶印法 aquatone
明胶海绵 absorbable gelatin sponge; gelatin(e) foam; gelatin(e) sponge
明胶合物 gelatinate
明胶化 gelatinisation [gelatinization]; gelation
明胶剂 gelatina
明胶甲醛 glutol
明胶浆 gelatin(e) size
明胶滤光片 gelatin(e) filter
明胶滤色片 colo(u)red gelatin filter; gelatin(e) dry filter
明胶滤色器 gelatin(e) filter
明胶氯化银相纸 gelatino-chloride paper
明胶模铸线脚 gelatin(e) mo(u)lding
明胶膜 gelatin(e) coating; gelatin(e) film; gelatin(e) layer
明胶墨辊 composition roller
明胶培养 gelatin(e) culture
明胶培养基 gelatin(e) culture medium
明胶片 gelatin(e) foil
明胶溶化保温器 gel solubilizing warmer
明胶乳剂 gelatin(e) emulsion
明胶朊 glutin
明胶软片 gelfilm
明胶石膏模 gelatin(e) mo(u)ld
明胶凸版制版法 gelatin(e) relief process
明胶涂层 gelatin(e) coating; gelatin(e) overcoat
明胶涂层纸 gelatin(e)-coated paper
明胶网格模型 gelatin(e) network model
明胶相片 gelatin(e) image; gelating relief image
明胶效应 gelatin(e) effect
明胶溴化银干板 gelatino-bromide plate
明胶溴化银相纸 gelatino-bromide paper
明胶样的 gelatin(e) form; gelatinoid
明胶液化 gelatin(e) liquefaction
明胶液化试验 gelatin(e) liquefaction test
明胶印刷版 gelatin(e) printing plate
明胶硬化 hardening of gelatin(e)
明胶炸药 dynamite; gelatin(e) dynamite; gelatin(e) explosive; gelatin(eous) blasting explosive; gelatin(e) powder; nitrogelation
明胶炸药卷 stick of gelatin(e) dynamite
明胶照相干版 gelatin(e) plate
明胶纸 gelatin(e) paper
明胶纸板 gelatin(e) board
明胶制版法 gelatin(e) process
明胶制模 gelatin(e) mo(u)lding
明礁 bare rock; exposed rock; rock uncovered; uncovered rock
明铰链 back-flap hinge; edge hinge
明接管 open-jointed pipe
明接合 exposed joint
明接接头 exposed joint
明接头 exposed joint; open joint
明进给嘴 sight feed nozzle
明净的玻璃 clear glass
明鸠尾榫 open dovetail
明鸠尾榫接合 open dovetailing
明开挖 open cut(ting); open excavation
明可夫斯基不等式 Minkovski inequality
明可夫斯基电动力学 Minkovski electrodynamics
明可夫斯基度规 Minkovski metrics
明可夫斯基范数 Minkovski norm
明可夫斯基几何学 Minkovski geometry
明可夫斯基矩阵 Minkovski matrix
明可夫斯基-列昂节夫矩阵 Minkovski-Leontief matrix
明可夫斯基时空 Minkovski space-time
明可夫斯基氏法 Minkovski method
明可夫斯基世界 Minkovski world
明坑 open reservoir
明框玻璃幕墙 exposed framing glass curtain wall
明朗点 clear point
明亮部分 highlight area
明亮的背景 light ground
明亮的琉璃瓦 bright enamel tile
明亮度 brightness; lightness
明流 barely flow; flow in open air; free flow
明流槽 free flowing channel
明流管道 open conduit
明流进水(口) exposed intake
明流渠(道) free flowing channel
明流水位流量关系 open water rating
明流隧洞 free flow tunnel; free level tunnel
明流隧洞(引水)式电站 free flow tunnel development
明流堰 clear overflow weir; overflow weir
明流溢水 clear overflow
明龙骨吊顶系统 exposed ceiling grid system
明垄沟 open furrow
明楼梯 open stairway
明楼梯梁 cut string; stepped string; mitered-and-cut string <梯级竖板与搁板相斜接的>
明楼梯斜梁 <楼梯竖板与搁板斜接的> cut-and-mitered[mitred] string
明露承重柱 exposed post
明露灯 bare lamp
明露格栅 open joist
明露管道 open conduit
明露楼梯斜梁 front string
明露元件 exposed unit
明炉 open oven
明路沟 open road ditch
明路堑 through cut
明路声学流量计 open channel acoustic flowmeter
明轮 paddle wheel
明轮板 paddle
明轮船 paddle boat; paddler; paddle steamer; paddle vessel; paddle wheeler; paddle wheel steamer
明轮发动机 paddle engine
明轮架 paddle beam
明轮壳 paddle box
明轮客轮 paddle passenger steamer
明轮梁 paddle beam
明轮汽船 paddle wheel steamer
明轮推进器 paddle wheel propeller
明轮推进器承梁 paddle beam
明轮推进器罩 paddle box
明轮罩 paddle wheel house; wheelhouse
明轮轴 paddle shaft
明码 listed price; ordinary telegraph code; plain code
明码电报 code telegram; plain code telegram
明码实价 clearly marked prices
明码售货 put goods on sale with the prices clearly marked; sell at market prices
明码通信[讯] clear text
明码系统 code-transparent system
明码坐标 plain coordinates
明买明卖 open transaction
明满过渡流 flow transition
明冒口 open riser; open-top(feeder); open-top riser
明门锁 rim lock
明灭相间灯 occulting light
明纳金特铜镍合金 Minargent alloy
明尼苏达操纵速度试验 Minnesota rate of manipulation test
明尼苏达大学圣安东尼瀑布水工试验所 <美> St. Anthony Falls Hydraulic Laboratory, the University of Minnesota
明尼苏达工程类比测试 Minnesota engineering analogies test
明尼苏达黑色花岗石 <美> Minnesota black
明尼苏达花岗岩 <美> Minnesota stone
明尼苏达曼卡托石灰岩 <美> Minnesota Mankato
明诺福合金 Minofor Alloy
明排管道 ditch conduit
明排水 sumping
明排水接头 <混凝土护墙之间的> open drained joint
明配套 open piping
明铺管道 open conduit
明堑 open cut(ting); open excavation
明桥面 bridge open floor; open bridge floor; open floor
明清度 brilliance
明区 area pellucida; light region
明曲线 open curve
明渠 free channel; free surface flow channel; open canal; open channel; open cut drain; open ditch; open trench; uncovered canal; uncovered drain
明渠坝段 open channel dam section
明渠导流 channel diversion; open channel diversion
明渠导流模型 model test of open channel diversion
明渠道流 free surface flow; open channel flow
明渠动床演算 movable-bed routing in open channel
明渠分水 channel diversion
明渠干舷 freeboard of channel
明渠均匀流 stream-flow in open channel
明渠开发 open channel development
明渠流 channel flow; open channel flow
明渠流量计 open channel meter
明渠流渠道 free surface flow channel
明渠流速分布 open channel velocity distribution
明渠流体动力学 open channel hydrodynamics
明渠落差 open-conduit drop
明渠排水 canal drainage; ditch drainage; drainage by open channel; gutter drainage; open channel drainage; open cut drainage; open ditch drainage
明渠排水法 open trench system
明渠渠道 free flowing channel

明渠式进水渠道 channel-type race
明渠收缩段 open channel constriction
明渠水动力学 open channel hydrodynamics
明渠水力学 channel hydraulics; hydraulics of open channel; open channel hydraulics
明渠水流 open channel flow
明渠水流更迭段 alternate stages of open channel flow
明渠水流要素 hydraulic element
明渠紊流 turbulent open channel flow
明渠污水流 open channel sewer flow
明渠泄水道 open sluiceway
明渠演算 open channel routing
明渠涌浪 hydraulic bore
明渠障碍 open channel constriction
明渠中回水 backwater in open channel; backwater in reservoir
明确承诺 definite undertaking
明确的证据 tangible proof
明确定义的 well-defined
明确的方法 deterministic approach
明确废除 express abrogation
明确分工 clear-cut division of labo(u)r; cleat-cut division of work
明确熔点 sharp melting point
明确收益方案 defined-benefit plans
明确证实 be unequivocally established
明熔丝断路器 open fuse cut-out
明熔丝片 open link fuse
明锐度 acutance; sharpness
明锐性 sharpness
明三彩器 Ming three-colo(u)red ware
明散热器 bare radiator; exposed radiator
明设 exposed installation
明设电线管 tubular conduits on the surface
明升暗降 kicked upstair(case)
明石海峡大桥 < 日本 > Akashi Kaikyo Bridge
明示保释 express bailment
明示保证 expressed warranty
明示承诺 express acceptance
明示承认 express recognition
明示担保 express guarantee(ship); express warranty
明示放弃 express waiver
明示合同 express contract
明示和解 express arrangement
明示或默示 express or implied
明示交付 express delivery
明示拒绝履行 express renunciation
明示诺言 express promise
明示弃权 express waiver
明示弃权书 express waiver
明示契约 express deed
明示授权 express authority
明示条件 express condition
明示条款 express term
明示同意 express consent
明示协议 express agreement
明示信托 express trust
明示要约与承诺 express offer and acceptance
明示异议 express objection
明示引渡 express extradition
明示约因 express consideration
明示转让 express assignment
明式基床 bedding of rubble mound type
明式家具 Ming-style furniture of hardwood
明视场 bright field
明视持久度 duration of photopic vision
明视地物点 clearly definable natural point; outstanding point; uniquely definable point
明视度 legibility
明视度曲线 visibility curve
明视距离 distance of distinct vision; normal reading distance; visibility distance; vision distance
明视觉 daylight vision; photonic vision; photopia vision
明视觉光谱光视效率 relative luminous efficiency
明视式实体 visible mass
明视式支座 visible bearing seat
明视适应性 light adaptation
明视野 bright field
明视远点 far point
明视最短距离 least distance of distinct vision
明视最小距离 least distance of distinct vision
明室 daylight room
明室暗室制版照相机 two-room camera
明室照相机 two-room camera
明室装暗盒 daylight loading cartridge
明室装片暗盒 daylight loading cassette
明适应 bright adaptation; light(ness) adaptation < 由暗处到明处视力恢复的过程 >
明收缩缝 butt contraction joint
明水道 open channel
明榫 open mortise[mortice]
明榫槽接合 open mortise and tenon joint
明榫纳接 shouldered face housing
明缩管 open pipe
明缩孔 open cavity; open shrinkage
明锁 surface latch
明特立体声系统 minter stereo system
明特洛甫波 Mintrop wave
明梯井式楼梯 staircase of open-well type
明体 phaneroplasm
明条纹 bright fringe
明挖 open cut; open dredging; open excavation; through cut; opencast; ditch
明挖爆破 open face blasting
明挖采矿 opencast mining
明挖挡墙支柱 strut in trench work
明挖法 < 隧道 > cut-and-cover method; open dredging process; open surface method; open excavation
明挖法施工 cut-and-cover construction method; open cut method < 开挖隧道的 >
明挖方量 excavation in open cut; open cut volume
明挖改建 daylighting
明挖管道基槽 open-cut trenching
明挖回填 cut-and-cover; open excavation and backfill
明挖回填地段 cut-and-cover section
明挖回填法 cut-and-cover method
明挖回填隧道 cut-and-cover tunnel
明挖基础 open cut foundation; open dug foundation; shallow spread foundation
明挖路堑 through cut
明挖排水 open cut drainage
明挖隧道 cut-and-cover tunnel; open cut tunnel; open trench tunnel
明挖隧道工程 open cut tunnel(l)ing
明挖隧道结构钢筋 structural reinforcement in open cut tunnel
明挖隧洞 cut-and-cover tunnel; open cut tunnel; open trench tunnel
明挖尾水渠 open cut tailrace
明挖引道 open cut approach
明挖支撑 bracing in open cut
明文 clear text; in the clear; plaintext
明文电报 telegram in plain language
明文公告 express proclamation
明文规定 expressly agreed terms; express provision; express term
明文条款 expressed covenant; express provision
明屋谷 open valley
明息 contractual interest; explicit interest
明晰 distinctness
明晰的 clear; distinct
明晰的平面图 freely articulated (ground) plan
明晰度 limpidity; perspectivity; transparency[transparence]
明晰器 clarifier
明晰清楚原则 principle of clarity
明晰圈 circle of least confusion
明晰视距 clear sight distance; clear vision distance
明晰视觉 distinct vision
明晰听觉 distinct hearing
明晰性 pellucidity
明细表 breakdown; comprehensive list; descriptive schedule; detailed list; detailed schedule; detailed statement; detail list; itemized schedule; particular sheet
明细单 bill of particulars
明细档案 detail file
明细地图 comprehensive map
明细对比 detail contrast
明细分类账 detail(ed) account; detail(ed) ledger; subsidiary ledger
明细分类账核对 proving the subsidiary ledger
明细分类账户 subsidiary ledger account
明细分类账审计 detail ledger audit
明细附表 detailed supporting statement
明细规范 specified criterion[复 criteria]
明细计划 explicit program(me)
明细记录 detail record; itemized record; subsidiary record
明细进度表 detailed schedule
明细科目 classification item
明细栏 item list
明细流程图 detail flowchart
明细图 detail drawing; detailed map
明细预算 detail budgeting
明细账 detail(ed) account; detail(ed) ledger; itemized account
明细账簿 subsidiary book
明细账单 account stated
明细账户 controlled accounts
明细账科目余额试算表 trial balance of subsidiary ledger
明细支出预算 line item budget
明霞缎 min xia satin brocade
明纤维 light fiber[fibre]
明显 evidence
明显边界 distinct boundary
明显边缘 distinct edge
明显标出的 well-marked
明显标志 distinguishing mark; obvious mark; visible marker
明显表示的 ostensive
明显差异 apparent disparity
明显成本 explicit cost
明显的 apparent; evidently; overt; pronounced; sharp; transparent; unambiguous; visible
明显的分界面 sharp interface
明显的局部不匀度 remarkable local irregularity
明显的模型 explicit model
明显的缺陷 patent defect
明显的证据 clear evidence
明显低于正常 be considerably subnormal
明显地层缺失 apparent stratigraphic(al) gap
明显地物 clear topographic(al) feature; well-defined feature
明显地物点 outstanding point; uniquely definable point; clearly definable natural point
明显地震 larger observed
明显点 definite point; well-defined point
明显度 < 目标对于背景的 > discreteness
明显方向变化 clear directive variation
明显废止 express repeal
明显分层的黏[粘]土 varved clay
明显缝 rusticated joint
明显高于 apparently higher than
明显故障 obvious fault
明显故障状态 fail-obvious condition
明显好转 be clearly better; improve markedly
明显痕迹 sharp trace
明显接触 sharp contact
明显节点 pronounced node
明显解 trivial solution
明显解劈理 easy cleavage
明显聚沉 apparent coagulation; clear coagulation
明显聚焦图像 sharply focused image
明显开裂 visible crack
明显磨损 noticeable wear
明显屈服 sharp yield
明显屈服点 sharp yield point
明显缺陷 open defect; patent defect
明显溶蚀 clear dissolution
明显色渍 pronounced staining
明显损失制 call lost system
明显特征 distinguishing characteristic
明显物标 obvious target
明显下降 dramatic decline
明显协变性 manifest covariance
明显性 distinctiveness
明显絮状沉淀法 distinct flocculence
明显增加 be markedly increased
明显债务 explicit debt
明线 aerial conductor; air wire; bright line; exposed carcassing; naked cable; object line; open line; open wiring; surface-mounted wire
明线保险器 open fuse
明线布线 exposed wiring; front wiring; open wiring; surface wiring
明线布置 exposed wiring; front wiring; open wiring; surface wiring
明线参数 open wire parameter
明线传输线 open wire transmission line
明线的 open wire
明线电路 open wire circuit
明线干线 open main
明线光谱 bright line spectrum
明线话频线路 open wire voice frequency circuit
明线回路 open wire loop
明线交叉 open wire transposition
明线开关 surface wiring switch
明线谱 bright line spectrum
明线图像 clear-line image
明线线路 open wire circuit; open-wire(pole) line
明线载波电话 open wire carrier telephony
明线载波电话电路 open wire carrier telephone channel
明线载波机 overhead wire carrier e-

M

quipment
明线载波通信[讯] open wire carrier communication
明线载波通信[讯]系统 open wire carrier communication system
明线载波系统 open wire carrier system
明线载波线路 open wire carrier line
明箱 camera lucida
明孝陵 <南京> Tomb of Emperor Zhu Yuanzhang;Tomb of First Emperor of Ming Dynasty
明信片 postcard;postal card <美>
明信片软件 postcard-wale
明信片质询(法)post card questionnaire
明延岩 akenobeite
明焰炉 open flame furnace
明焰烧嘴 open flame burner
明焰窑 direct-fired furnace; direct-fired kiln;open flame kiln
明溢洪道 open spillway
明雨水管 exposed downpipe
明语电报 plain language message
明约 express contract
明置地物 clear topographic(al) feature;well-defined feature
明置点 definite point
明置基础 surface footing; surface foundation
明置天然地物 prominent natural feature
明置像点 easily identification image point
明桩码头 open jetty
明桩栈桥 open jetty
明装 surface-mounted
明装白炽灯 surface incandescent lamp
明装搬把开关 surface wiring switch
明装插销 surface bolt
明装的金属电缆管道 metal mo(u)lding;surface metal raceway
明装阀门 open valve
明装管道 exposed pipe;open piping
明装碰锁 surface latch
明装小五金 surface-mounted hardware
明装荧光灯 surface fluorescent fixture
明装荧光灯带 surface continuous row fluorescent fixture
明装照明设备 surface-mounted luminaire
明子 light wood; resinous wood; stumpwood
明子干馏 resinous wood distillation
明子林 light wood
明子松节油 stump turpentine

鸣

鸣笛标 blow whistle mark;whistle mark; whistle post; whistling mark
鸣笛浮标 whistling buoy
鸣笛牌 whistle post
鸣笛信号 whistle signal
鸣笛(预告)标志 whistle sign
鸣管 syrinx
鸣号 sound horn
鸣叫 chirping
鸣铃 ring
鸣谱 phonatome
鸣汽笛 whistle board
鸣沙 <又称鸣砂> booming sand; singing sand;whistling sand
鸣沙沙丘 booming dune
鸣声 wooliness
鸣声汽笛 bell chime steam whistle
鸣声器 squealer
鸣响警报浮标 sound warning buoy
鸣振现象 singing phenomenon
鸣震 ringing
鸣钟 ring

茗

茗荷儿 lepadidae;lepas;ship barnacle

冥

冥王星【天】Pluto

铭

铭记点 engraved point
铭刻 inscription
铭牌 name board; name plate; data plate; designation strip; marking plate; placard; rate plate; rating plate;escutcheon <标志船名处>
铭牌出力 manufacturer's rating;name-plate rating
铭牌电流 name-plate current
铭牌额定 name-plate rating
铭牌额定容量 nominal rated capacity
铭牌额定值 name-plate capacity
铭牌负荷 nominal load
铭牌规定的 nominal
铭牌名义容量 nominal rated capacity
铭牌容量 name-plate capacity
铭牌数据 name-plate information
铭牌位置 name-plate set
铭文 epigraph;inscription;legend
铭文学 epigraphy
铭文艺术 epigraphy

螟

螟害 borer injury

命

命令变更 change of order
命令付款 order to pay
命令授权 command authority
命令作业 command job
命脉 life line
命名 denomination; designation; entitle;naming;nominate
命名表达式 designational expression
命名常数 named constant
命名程序模块 named program(me) module
命名的 named;nomenclative
命名段落 named paragraph
命名法 nomenclature;terminology
命名方法 designation; nomenclature system
命名符 designator
命名公用块 named common
命名公用区 named common area
命名实体 named entity
命名式 name form
命名输出文件 named output file
命名文件 named file
命名系统 named system
命名学 glossology;onomatology
命名用户程序 named user program(me)
命名原则 nomenclature principle
命名源文件 named source file
命名约定 naming convention
命名者 denominator
命名主义 nominalism
命数法 numeration
命数系统 numeral system; numeration system
命题 proposition;statement
命题变量 propositional variable
命题常数 propositional constant
命题代数 algebra of propositions; propositional algebra
命题定律 propositional law
命题法则 statement law
命题公式 propositional formula;statement formula
命题函数 propositional function;statement function
命题函数的置换实例 substitution instance of the statement function
命题矩阵 proposition matrix
命题逻辑 propositional logic; statement logic
命题逻辑布尔代数 Boolean algebra of propositional logic
命题逻辑式 well-formed formula
命题树 proposition tree
命题演算 calculus of proposition; propositional calculus
命题姿态 propositional attitude
命题字母 proposition letter
命中概率 hit probability
命中率 hit rate;hit ratio;occurrence of hits
命中率数 number of hitting
命中面 impact area
命中摄影术 strike photography
命中文件 hit file
命中型控制器 deadbeat controller
命中中心 center of impact

谬

谬拉滤池 Miura filter
谬误推导 fallacious derivation
谬误推理 fallacious derivation; fallacious inference; fall derivation (inference)
谬误之推理 golden rule fallacy

摸

摸彩袋 grab bag
摸到水底 touch bottom
摸绕酸 morolic acid
摸索反射 groping-reflex

摹

摹本缎 mopen
摹仿 take after
摹仿本 imitation copy
摹绘 isography
摹绘图 draft copy;facsimile map
摹描石印 grained stone lithography
摹拟的化学战剂 simulated agent
摹图材料 tracing material
摹写 copy; facsimile; fax; imitate; tracing
摹写传输业务 fax service
摹写登记 facsimile posting
摹写品 replica
摹写通信[讯] direct recording facsimile;tape facsimile
摹写者 tracer
摹真 facsimile
摹真印花 facsimile printing

模

模 2 和 binary sum; modulo-two-sum
模 2 和运算 modulo-two-sum
模 9 检验 modulo-nine-check
模 9 校验 cast-out-9-check
模 m modulo-m
模 n 的余数 modulo-n residue
模 n 校验 modulo n check
模拔钢绞线 deform-strand
模版 stencil(ing)
模版彩色印刷 stencil colo(u)red print
模版复印机 stencil duplicator
模版活性 template activity
模版印刷 stencil printing
模版印刷油墨 stencil ink
模版原版 stencil master
模版纸 template paper
模版纸板 template board
模半径 mode radius
模半序线性空间 modular semi-ordered linear spaces
模包迹 mode envelope
模壁 die wall
模壁光洁度 die wall finish
模壁面 die side wall;die wall surface
模壁磨擦 die wall friction
模壁润滑 die wall lubrication
模壁润滑剂 die wall lubricant
模壁寿命 die wall life
模壁效应 die wall effect;wall effect
模变 moding
模变换 mode transformation;modular substitution
模变换电路 mode transducer circuit
模变换器 mode converter; mode transducer;mode transformer
模表示 modular representation
模柄 die shank;punch shank;stalk
模不变式 modular invariant
模不等式 modular inequality
模槽 cavity; die cavity; die hole; die impression;die opening;die space
模槽材料 impression material
模槽机 formgrader
模常数 modular constant
模场 mode field
模场同轴度 mode field concentricity
模场直径 mode field diameter
模巢 mo(u)ld cavity
模衬 die insert; die liner; mo(u)ld casing;mo(u)ld lining
模冲 plunger;stamp(ing)
模代数 modular algebra
模袋混凝土 fabriform concrete
模的 moding;modular
模的分离 mode separation
模的完全性 completeness of modes
模的正交性 orthogonality of modes
模底板 baffle plate;die shoe;stool
模底线印 bottom plate seam
模电极 membrane electrode
模电压 mode voltage
模垫 bolster;die cushion
模垫缓冲器 die cushion
模叠加 addition of modes
模度坯料 die block;dies blank
模锻 contour forging; drop forge; forming; impact forging; press forming;stamp forging;stamp out; swaging
模锻不足 underpressing
模锻车间 stamping room
模锻齿轮 stamped gear
模锻锤 die hammer; drop hammer; stamp (ing) hammer; swaging hammer
模锻飞边 flash
模锻工 puncher
模锻焊 die weld(ing)
模锻活扳手 drop monkey
模锻机 die forging machine
模锻机油 punching oil
模锻件 die forging; drop forging; stamp work
模锻件活板车 drop monkey
模锻空气锤 die forging air hammer
模锻螺栓 swedge bolt
模锻模 closed die; impression die; stamping die
模锻摩擦压力机 Vincent press
模锻配件 swaged fitting
模锻设计 stamping design

M

模锻水压机 hydraulic die press
模锻斜度 draft angle;leave
模锻卸扣 drop forged shackle
模锻压力机 drop press; mechanical forging press;stamping press
模锻锥套 swaged socket
模对象 module object
模发射 mode launching
模泛函 modular functional
模范 exemplar;pacesetter;prototype
模范工业城 model industrial city
模方【数】norm
模方位角定位系统 modular azimuth positioning system
模仿 imitation;mimesis;mimicry;simulate;simulation
模仿本能 instinct of imitation
模仿的 mimic;mock
模仿定价法 imitative pricing
模仿复制品 replica
模仿机器人 copying robot
模仿价格 imitative pricing
模仿器 emulator;imitator
模仿色 imitative colo(u)r
模仿现象 echo-phenomenon
模仿效应 imitation effect
模仿者 copier;simulator
模分解 modular resolution
模分配噪声 mode partition noise
模缝 die slot; joint mark; mo(u)ld seam
模缝痕 seam mark
模缝脊 fin
模缝裂纹 seam check
模缝条痕 mo(u)ld mark
模缝线 back mo(u)ld joint;bad mo(u)ld joint; joint line; mo(u)ld mark;parting line;seam line
模缝线偏移 <玻璃制品缺陷> offset seam
模符号 modulo symbol
模杆 tringle
模格 modular lattice; modular reference grid
模股 template strand
模光纤 mode fiber[fibre]
模函数 mode function;modular function
模函数论 modular function theory
模盒 case mo(u)ld;mo(u)ld box
模盒边框挡板 edge form
模盒标尺板条 <混凝土路坡> grade strip
模盒紧固拉杆 form(work)tie
模盒式伸缩器 bellows expansion joint
模盒压力计 bellow ga(u)ge
模痕 die line;mo(u)ld mark
模后收缩 die shrinkage
模糊 blur; confusion; fogging; fuzziness; indeterminacy; indistinctness; lack of definition; lack of sharpness;obscuration
模糊斑点 blur spot
模糊闭集 fuzzy closed set
模糊编码 ambiguity encoding;ambiguous encoding
模糊变换 blurring mapping;fuzzy mapping
模糊变量 fuzzy variable
模糊表达式 fuzzy expression
模糊不明的 nubilous
模糊不清 blur
模糊不清的 ambiguous;dumb
模糊不清的复制本 vague copy
模糊不清区域 fuzzy region
模糊部分图 fuzzy partial graph
模糊测度 fuzzy measure
模糊簇聚 fuzzy clustering
模糊簇聚算法 fuzzy clustering algorithm
模糊错误 ambiguity error;fuzzy error
模糊代数 fuzzy algebra
模糊导数 fuzzifying derivation;fuzzy derivation
模糊的 bleary;blurred;cloudy;dull;fuzzy; indistinct; misty; washed-out;blurring;obscure
模糊的木纹 woolly grain
模糊的声音 obscure sound
模糊等高线 faint contour
模糊笛卡儿积 fuzzy Cartesian product
模糊地平线 misty horizon
模糊点 ambiguity;blob;blurred spot;fuzzy point
模糊动态规划 fuzzy dynamic(al) programming
模糊动态系统 fuzzify(ing) dynamic(al) system
模糊度 ambiguity degree;haze;haziness;equivocation
模糊度函数 ambiguity function
模糊对策 fuzzy game
模糊对策值 value of a fuzzy game
模糊多准则建模 fuzzy multicriteria modelling
模糊反馈控制系统 fuzzy feedback control system
模糊范畴 fuzzy category
模糊方程 fuzzifying equation; fuzzy equation
模糊概率【数】fuzzy probability
模糊概率场 fuzzy probability field
模糊概率分布 fuzzy probability distribution
模糊概率回归 fuzzy probability regression
模糊概念 fuzzy concept
模糊干扰 fuzzy noise
模糊公式 blurring formula;fuzzy formula
模糊关系 ambiguity relation fuzzy relation;fuzzy relation
模糊关系的定义域 domain of a fuzzy relation
模糊关系的合成 composition of fuzzy relations
模糊关系的截口 section of a fuzzy relation
模糊关系的自反性 reflexivity of a fuzzy relation
模糊关系方程 equation of a fuzzy relation;fuzzy relation equation
模糊关系系统 fuzzy relation system
模糊光点 soft spot
模糊规划 fuzzy programming
模糊轨道 fuzzy orbit;fuzzy trajectory
模糊过程 fuzzy process
模糊函数 ambiguity function; fuzzy function
模糊函数的导数 derivation of a fuzzifying function
模糊函数的分类 classification of fuzzy functions
模糊函数的分析 analysis of a fuzzy function
模糊函数的极值 extremum of a fuzzy functions
模糊函数的静态冒险 static hazard of fuzzy functions
模糊函数的综合 synthesis of fuzzy functions
模糊函数分析 analysis of fuzzy functions
模糊恒等函数 fuzzy identity function
模糊恒等映照 fuzzy identity morphism
模糊化过程 fuzzification process
模糊化函数 fuzzification function
模糊化函数因子 fuzzification functor
模糊划分 fuzzy partition
模糊划分矩阵 fuzzy partition matrix
模糊环境 fuzzy environment;inexact environment
模糊积分 fuzzifying integral fuzzy integral
模糊基数 fuzzy cardinality
模糊级 blur level
模糊极值 fuzzify(ing)extremum
模糊集的补集 complementation of a fuzzy set
模糊集的高 height of a fuzzy set
模糊集的基数 cardinality of a fuzzy set
模糊集的交 intersection of fuzzy sets
模糊集法 fuzzy set method
模糊集范畴 category of fuzzy sets
模糊集(合)fuzzy set
模糊集(合)论【数】fuzzy set theory
模糊集合系统 fuzzy set system
模糊集基数 cardinality of fuzzy sets
模糊集理论 fuzzy set theory
模糊集论域 universe of fuzzy sets
模糊集相等 equality of fuzzy sets
模糊计算机 blurring computer;fuzzy computer
模糊假言推理 fuzzy modus ponens
模糊简单析取分解 fuzzy simple disjunctive decomposition
模糊奖金 anonymous bonus
模糊结构 fuzzy structure
模糊界限 fuzzy boundary
模糊矩阵 fuzzy matrix
模糊距离 fuzzy distance
模糊聚类 fuzzy cluster
模糊聚类分析 fuzzy cluster analysis
模糊决策 fuzzifying decision; fuzzy decision;non-clear decision
模糊决策树 fuzzy decision-tree
模糊开关函数 fuzzy switching function
模糊开关函数图 graph of a fuzzy switching function
模糊开集 fuzzy open set
模糊科学 fuzzy science
模糊空间划分 fuzzy space partition
模糊控制 fuzzy control
模糊控制器 fuzzy controller
模糊理论【数】fuzzy theory
模糊隶属函数 fuzzy membership functions
模糊连接词 fuzzy connective
模糊连续 fuzzy continuity
模糊连续的 fuzzy continuous
模糊联盟 fuzzy coalition
模糊量 fuzzy quantity
模糊列扩张 fuzzy column extension
模糊列凝聚 fuzzy column condensation
模糊裂缝模型 smeared cracking model
模糊流程图 fuzzy flow chart
模糊滤波器 fog filter;fuzzy filter
模糊轮廓 blurred contour
模糊逻辑【数】fuzzy logic
模糊逻辑公式的有效性 validity of formula in fuzzy logic
模糊逻辑函数 fuzzy logic(al) function
模糊逻辑控制 fuzzy logic(al) control
模糊逻辑控制器 fuzzy logic(al) controller
模糊逻辑器 fuzzy logic(al) controller
模糊逻辑学 fuzzing logic
模糊命题 fuzzy proposition
模糊模型 fuzzy model
模糊目标 blurred target;fuzzy object
模糊评估 fuzzy evaluation
模糊评估法 fuzzy assessment method
模糊评判 fuzzy evaluation
模糊谱线 diffused line
模糊区 confusion region
模糊区间 fuzzy interval
模糊区域 blurred region
模糊曲面 equivocal surface
模糊圈 blur circle;circle of confusion
模糊权【数】fuzzy weight
模糊群 fuzzy group
模糊熵 fuzzy entropy
模糊上界 fuzzy upperbound
模糊上限 fuzzy upperbound
模糊射线 obscure ray
模糊神经网络 fuzzy neural network
模糊生成 fuzzy production
模糊识别 fuzzy diagnosis
模糊事件【数】fuzzy event
模糊事件的独立性 independence of a fuzzy event
模糊事件的可能性 possibility of a fuzzy event
模糊事件均值 mean of a fuzzy event
模糊输出变换 fuzzy output transformation
模糊输出映射 fuzzy output map
模糊数 fuzzy number
模糊数除法 fuzzy number division
模糊数加法 fuzzy number addition
模糊数绝对值 fuzzy number absolute value
模糊数熵 fuzzy number entropy
模糊数学 ambiguity mathematics;fuzzing mathematics; fuzzy mathematics;indescribable mathematics
模糊数学规划 fuzzy mathematical programming
模糊数指数 fuzzy number exponential
模糊素隐含 fuzzy prime implication
模糊算法 fuzzy algorithmic approach
模糊随机变量 fuzzy random variable
模糊特性 fuzzy behaviour
模糊特征 fuzzy characteristics
模糊凸包 fuzzy convex hull
模糊凸集合 fuzzy convex set
模糊图 ambiguity diagram;fuzzy graph
模糊图像 blurred image; broad image;degraded image; fuzzy image;fuzzy piece; hazy picture; non-distinct image;washed-out picture
模糊推断 fuzzy assertion
模糊推理 fuzzy inference
模糊拓扑 fuzzy topology
模糊拓扑空间 fuzzy topological space
模糊文字 fuzzy literal
模糊稳定的 fuzzy stable
模糊物体 fuzzy objective
模糊误差 ambiguity error
模糊系统 fuzzy system
模糊系统映射 fuzzy system mapping
模糊现象 blooming
模糊线 unsharp line
模糊线性规划 fuzzy linear programming
模糊线性回归 fuzzy linear regression
模糊线性序 fuzzy linear ordering
模糊限制 fuzzy restriction
模糊相关 fuzzy correlation
模糊响应 fuzzy response
模糊向量 fuzzy vector
模糊像 vague image
模糊效应 blurring effect; smearing effect
模糊效用 fuzzy utility
模糊协方差 fuzzy covariance
模糊信号 blurred signal
模糊信息 fuzzy information
模糊信息机 fuzzy processor
模糊形象 blurred image
模糊性 fuzzification;fuzziness[fuzzy-

M

ness]
模糊性测度 measure of fuzziness
模糊性指数 index of fuzziness
模糊延迟 ambiguity delay
模糊样品 fuzzy prototype
模糊一致 fuzzy consensus
模糊仪 <测水泥细度> obscurometer
模糊因数 blur factor
模糊音装置 fuzzbox
模糊隐含 fuzzy implication
模糊印记 illegible mark
模糊印刷 muddy print
模糊印样 bad copy; burred image; fuzzy image
模糊影像 blurred image; blurred picture; fuzzy image; indistinct image
模糊映射 fuzzy mapping
模糊映照 fuzzy morphism
模糊优化 fuzzy optimum
模糊有限状态自动机 fuzzy finite state automaton
模糊预测 fuzzy forecasting
模糊域 fuzzy field
模糊圆 circle of least confusion
模糊约束 fuzzy constraint; fuzzy restriction
模糊蕴涵 fuzzy implication
模糊噪声 fuzzy noise
模糊照相法 smear photography
模糊真值 fuzzy truth
模糊终端调整器 fuzzy terminal regulator
模糊柱形扩张 fuzzy column extension
模糊柱形凝聚 fuzzy column condensation
模糊转变函数 blurring transition function; fuzzy transition function
模糊转移矩阵 fuzzy transition matrix
模糊转移矢量 fuzzy transition vector
模糊状 cloudiness
模糊状态 fuzzy state
模糊状态整流子 fuzzifying state regulator; fuzzy state regulator
模糊状态转移矩阵 fuzzy state transition matrix
模糊状态转移树 fuzzy state transition tree
模糊子集 fuzzy subset
模糊子集域 fuzzy field of subset
模糊子图 fuzzy subgraph
模糊自动机 fuzzy automata
模糊综合评估 fuzzy comprehensive evaluation; fuzzy synthetic evaluation
模糊组合法则 fuzzy composition law
模糊最佳控制 fuzzy optimal control
模化程度 modularity
模化流道 model flume
模化流体 model liquid
模化水槽 model flume
模环 die ring; modular ring
模绘板 stencil
模激发 mode excitation
模夹 die clamp; mo(u)ld-holder
模夹变形丝 forming between profile
模架 die carrier
模架油井钻台 template oil-drilling platform
模间隔 mode spacing
模间畸变 intermodal distortion
模间间隔频率 intermode spacing frequency
模间耦合 mode coupling
模间耦合器 intermodal coupler
模间拍 intermode beats
模件 modular unit; modulus[复 moduli]; moudle
模件板 prototype board
模件包装 modular package
模件插入件 mo(u)ld insert
模件宽度 module width
模件扩展板 module extender board
模件联锁系统 modular interlocking system
模件联锁制 modular interlocking system
模件培训法 module training
模件式结构 modular construction
模件式联锁 modular interlocking
模件试验装置 module test set
模件系统 modular system
模件信号系统 modular signaling system
模件信号制 modular signaling system
模件性 modularity
模件原则 modular principle
模件制 building block system; modular system
模角 modular angle
模阶 mode step
模接合 <把集成电路的小片焊接到同一衬底上> die bonding
模接头 die adapter
模结构 mode structure
模截面 mode cross section
模九检验 modulo-nine's checking; nines check
模矩阵 modular matrix
模抉数据 simulated data
模抉系统 simulated system
模烤压切机 wire-cut cake machine
模壳 blinding; form; mo(u)ld box; mo(u)ld case; shuttering
模壳板 form board(ing); form panel; sheathing
模壳板条 shutter lath; shutter slat
模壳边撑 form clamp
模壳拆除 form removal
模壳衬板 form sheathing
模壳衬垫 shuttering lining
模壳承包商 formwork contractor
模壳尺寸 shuttering dimension
模壳捣实混凝土 packing concrete in forms
模壳反力模量 formwork reaction modulus
模壳刚度 rigidity of modulus
模壳膏 form paste
模壳工 form fixer; form setter
模壳工程安装车 <隧道、坑道施工> forms handler
模壳工程规范 formwork specifications
模壳工作 formwork
模壳横档 shuttering tie
模壳蜡 shuttering wax
模壳栏杆 shuttering tie
模壳楼板 mo(u)lded waffle slab
模壳密封料 shuttering sealer
模壳强度 form strength
模壳设置 form placing
模壳受到的压力 pressure on shuttering
模壳图(样) formwork drawing
模壳涂料 form coating; shuttering paint
模壳系统 forming system
模壳小五金 form hardware
模壳压力 pressure on forms
模壳元件 forming element
模壳振捣器 shutter vibrator
模壳振动 form vibration; shutter vibration
模壳振动器 form vibrator
模壳作业 forming job
模空间 modular space
模孔 die hole; die opening; die orifice
模孔变形锥 <模孔第二部分> die approach angle
模孔出口锥 dieback; die exit angle; die relief angle
模孔光学检查仪 profiloscope
模孔喇叭角 generating angle
模孔入口锥角 die approach angle
模孔修磨 reconditioning
模孔压缩锥 die reduction angle
模孔针磨机 needle die grinding machine; needle die polishing machine; needle grinding machine
模孔直径 die throat diameter
模孔铸型 cavity-molding
模口 die orifice
模口半径 die radius
模口部分 die bearing
模口挤出膨胀 extrusion die swell
模口角度 die angle
模口膨胀比 die swell ratio
模口支承面 die bearing
模块 block box; die block; die piece; module
模块板 module board
模块程序(定位)设计 modular program(me) position
模块处理系统 modular processing system
模块的 modular
模块电路板 module circuit board
模块度 modularity
模块对 pair of module
模块分解 modular decomposition
模块复合库 module complex library
模块功耗 module dissipation
模块构造 module structure
模块互连 module interconnection
模块化 blocking; modularity; modularization; modularize
模块化程序 modularized program(me)
模块化程序设计 modular programming
模块化的 modular; modularized
模块化电路 modularized circuit
模块化电气设备 modular electrical equipment
模块化方法 modular approach
模块化方式软件的结构运行和试验 modular approach software construction operation and test
模块化光纤光学测试仪 modular optical test instrument for fiber[fibre]
模块化机器 module machine
模块化计算机 modular computer
模块化结构 modular structure
模块化巨型机体系结构 modular supercomputer architecture
模块化巨型(计算)机 modular supercomputer
模块化冷水机组 modular liquid chillers
模块化脉冲信号发生器 modular pulse generator
模块化设计 building block design; modular design
模块化生产 module production
模块化实验室计算机 modular laboratory computer
模块化微程序 structured microprogram(me)
模块化微程序机 modular micro-program(me) machine
模块化系统 modular system
模块化硬件 modularized hardware
模块化转换器 modular converter
模块化组装式给水装置 packaged modularised feed unit
模块技术 building block technique
模块间通信[讯] intermodule communication
模块检查 module check
模块交互作用 module interaction
模块接日 intermodular interface
模块结构 modular architecture; modular construction
模块结构式微型计算机 modular structure microcomputer
模块开发系统 prototype development system
模块控制 modular control
模块扩充 modular expansion
模块扩充性 modular expansibility
模块名 module name
模块内容 module content
模块耙头 modular draghead
模块冗余度 modular redundancy
模块设计 modular design
模块式多层采样器 modular multilevel sampler
模块式概念 modular concept
模块式工具系统 modular tooling system
模块式监视雷达 modular surveillance radar
模块式结构 modular organization
模块式雷达 modular radar
模块式卫星 modular satellite
模块式综合公用事业系统 modular integrated utility system
模块说明 module declaration
模块体 module body
模块性【电】modularity
模块指明 module specification
模块属性 module attribute
模块自同步 module self synchronization
模宽 mo(u)lded breadth, beam mo(u)lded
模拉 die drawing
模蜡 pattern wax
模来石 mullite
模棱两可 amphibology; amphiboly
模棱两可的 ambiguous; equivocal
模理论 modular theory; module theory; modulus theory
模理想 modular ideal
模梁 beam
模量 modulus[复 moduli]
模量比 modular ratio; modulus ratio
模量比法设计 m-design; modular-ratio design
模量参考制 modular reference system
模量密度比 modulus-to-density ratio
模量温度曲线 modulus(-versus)-temperature curve
模量效应 modular effect
模量制 modularized system
模量重量比 modulus-weight ratio
模料储存期 storage life of mo(u)lding compound
模鳞 mo(u)ld scale
模漏 mode ship
模律 modular law
模密度 density of modes; mode density
模面 die surface; modular surface; mo(u)ld surface
模面切割造粒 die-face pelletizing
模膜直径比 blow-up ratio
模内 intramode
模内箔装饰 in-mo(u)ld decorating with foils
模内捣实混凝土 packing concrete in forms
模内粉末涂覆 powder in mo(u)ld coating
模内畸变 intramodal distortion
模内可见度计算机 analog(ue) visibility computer
模内面 inner die surface
模内喷涂装饰 in-mo(u)ld decorating

with coatings
模内强度调制 analog-intensity modulation
模内失真 intramodal distortion
模内收缩 mo(u)ld shrinkage
模内涂覆 in-mo(u)ld coating
模拟 analog(ue);emulation;imitate;imitation;mimesis;model(l)ing;simulate;simulating;simulation
模拟摆 false bob;mock pendulum
模拟板 analog(ue) board;breadboard;dummy panel
模拟板电路 breadboarded circuit
模拟板区域 breadboard area
模拟板设计 breadboard design
模拟板装置 breadboard set up
模拟保护(装置) analog(ue) protection
模拟备份 analog(ue) backup
模拟比较器 analog(ue) comparator
模拟编码器 analog(ue) encoder
模拟编译程序 analog(ue) compiler
模拟编译程序系统 analog(ue) compiler system
模拟变换式记录 record of analog(ue) transform
模拟变量 analog(ue) variable
模拟变量赋值 analog(ue) variable assignment
模拟变量指定 analog(ue) assignment of variables;analog(ue) variable assignment
模拟变数字的 analog(ue)-to-digital
模拟表示(法) analog(ue) representation
模拟冰雪天的路面 winterized pavement
模拟波谱分析器 analog(ue) wave spectrum analyser[analyzer]
模拟博奕 simulation game
模拟布线 simulation wire
模拟布线表 simulation wire list
模拟部件 simulated part
模拟采样 analog(ue) sampling
模拟参数 simulation parameter
模拟操纵板 mimic panel
模拟操纵台 mimic panel
模拟操作程序 simulated operating procedure
模拟槽 simulated slot
模拟测试 simulation test
模拟测试器 analog(ue) tester
模拟测试组 simulation test deck;test deck
模拟测图 analog(ue) photogrammetric plotting
模拟测图仪 analog(ue) plotting instrument
模拟沉降物 simulated fall-out
模拟成型性试验 simulative formability test
模拟乘法 analog(ue) multiplication
模拟乘法器 analog(ue) multiplier;analog(ue) multiplier unit
模拟乘法器部件 analog(ue) multiplier unit
模拟程序 imitator;model(l)ing process;simulator;simulator program(me)
模拟程序包 simulation package
模拟程序磁带 analog(ue) program tape
模拟程序的编译程序 simulator compiler
模拟程序的调试应用 simulator debug utility
模拟程序连续系统 model(l)ing program(me) continuous system
模拟程序软件 simulator software
模拟程序软件包 simulator software package
模拟程序设计语言 simulation programming language
模拟程序调试 simulator debug
模拟冲击实验器 simulated impact tester
模拟冲击试验 simulated impact test
模拟冲击试验机 simulated impact tester
模拟抽样 simulated sampling
模拟除法器 analog(ue)-divider
模拟处理 analog(ue) processing;analog(ue) treatment;simulation manipulation
模拟处理机 analog(ue) processor
模拟处理机控制器 analog(ue) processor controller
模拟处理器 analog(ue) processor
模拟处理设备 analog(ue) processing equipment
模拟传感器 analog(ue) sensor
模拟传输 analog(ue) transmission
模拟传输设备 analog(ue) transmission facility
模拟传输线 mimic transmission line
模拟传输线段 nominal section
模拟船舶模型 vessel simulator model
模拟磁带记录地震仪 analog(ue) tape seismograph
模拟存储 analog(ue) record(ing);analog(ue) storage
模拟存储扩张机构 analog(ue) storage expander
模拟存储模块 analog(ue) storage module
模拟存储器 analog(ue) memory(device)
模拟代码加密器 analog(ue) code encryption unit
模拟大理石花纹的装饰 marbling
模拟带 analogy tape
模拟单晶衍射计 analog(ue) single-crystal diffractometer
模拟道路 simulated roadway
模拟道路试验 road test simulation;simulated road test
模拟道路条件 simulated road condition
模拟的 analog(ue);analogous;mimetic;mimic;mock;simulative;stimulant
模拟的方位 simulated-azimuth
模拟的空间环境 simulated space environment
模拟的外界条件 simulated environment
模拟的再入大气层环境 simulated reentry environment
模拟地层 simulated formation
模拟地图 analog(ue) map;mimic map
模拟地震 analog(ue) earth quake;simulated earthquake;simulation earthquake
模拟地震动试验 simulated ground motion test
模拟地震反演 model(l)ing seismic inverse
模拟地震运动 simulation earthquake motion
模拟地震振动台 earthquake simulating shaking table;earthquake simulator;simulated earthquake vibration stand
模拟点 analog(ue) point;model point
模拟电动机 simulating motor
模拟电路 analog(ue) circuit;artificial circuit;circuit cheater;dummy load;mimic bus;mimic channel;simulator
模拟电路板设计 breadboard design
模拟电路法 simulative circuit method
模拟电路分析 analog(ue) circuit analysis
模拟电路技术 analogic(al) circuit technique
模拟电路接线 phantom connection
模拟电容器 artificial capacitor
模拟电视电话 analog(ue);picture-phone
模拟电压 analog(ue) voltage
模拟电压信号 analog(ue) voltage signal
模拟电子计算机 analog(ue) electronic computer
模拟定标 analog(ue) scaling
模拟定时脉冲 analog(ue) timing pulse
模拟定位机 simulated locator
模拟定子 model stator
模拟读出 analog(ue) readout
模拟读出装置 analog(ue) readout device
模拟断路开关 mimic-disconnecting switch
模拟断路器 mimic-disconnecting switch
模拟断面记录仪 analog(ue) profiler
模拟对策技术 simulation-game technique
模拟对讲机 analog interphone
模拟多路复用器 analog(ue) multiplexer
模拟多路复用设备 analog(ue) multiplex equipment
模拟多路调制器 analog(ue) multiplexer
模拟多路转换器 analog(ue) multiplexer
模拟发射 simulated emission
模拟发射井 simulated silo
模拟发送器 analog(ue) transmitter;simulative generator
模拟法 analog(ue) method;analog(ue) procedure;analogy approach;analogy method;model method;simulation method
模拟法测图 analog(ue) method of photogrammetric mapping
模拟法空中三角测量 analog(ue) aerial triangulation;analog(ue) aerotriangulation
模拟法求解 analog(ue) approach
模拟法水位改正 correction of water level by analog(ue) method
模拟反馈系统 analog(ue) feedback system
模拟反应堆 mock-up reactor
模拟反应计算机 analogy response computer
模拟范围 simulation context
模拟方程 simulation equation
模拟方程解算器 analog(ue) equation solver
模拟方法 analogy method;analogy procedure;simulation method
模拟方法论 simulation methodology
模拟方法学 simulation methodology
模拟放大器 analog(ue) amplifier
模拟放射 simulated emission
模拟放射效应 radiomimetic effect
模拟飞行 simulated flight
模拟飞行测试 flight simulation test
模拟飞行计划 simulated flight plan
模拟飞行试验设备 simulated flight test facility
模拟分布 analog(ue)-distribution
模拟分光光度计 analog(ue) spectrophotometer
模拟分配器 analog(ue) commutator;analog(ue)-distributor
模拟分权组织 simulated decentralization organization
模拟分析 analogy analysis;simulated analysis;simulation analysis;model analysis
模拟风压 simulated wind pressure
模拟伏特计 analog(ue) voltmeter
模拟负载 artificial load
模拟复飞 simulated missed approach
模拟复印 analog(ue) copying
模拟复印机 analog(ue) copier
模拟概率评估 probability assessment with simulation
模拟干扰 simulated countermeasure
模拟感觉 simulated feel
模拟高度 simulated altitude
模拟高度环境 simulated altitude environment
模拟高温计 analog(ue) pyrometer
模拟格式器 analog(ue) formatter
模拟耕作试验 model tillage test
模拟工况 simulated condition
模拟工艺 analog(ue) technique;analog(ue) technology
模拟工作记录表 simulation worksheet
模拟功能 analog(ue) function
模拟功能转换开关 analog(ue) function switch
模拟攻击 simulated strike
模拟拱廊 mock arcade
模拟故障 simulated failure
模拟关系 simulative relation
模拟管理作业 simulated management operation
模拟光点记录地震仪 analog(ue) light spot seismograph
模拟光合 mimic photosynthesis
模拟光偏转器 analog(ue) light deflector
模拟光谱分析器 analog(ue) spectrum analyser[analyzer]
模拟光学距离测试器 simulated optic(al) range tester
模拟过程 simulation process
模拟过程的物理量 analog(ue) quantity
模拟函数发生器 analog(ue) function generator
模拟航空母舰甲板 simulated carrier deck
模拟航线 analog(ue) strip
模拟航向 false course
模拟和动态程序编制 simulation and dynamic(al) programming
模拟河槽 model channel
模拟黑色信号 artificial black signal
模拟后援 analog(ue) backup
模拟滑翔道 false glide path
模拟环回 analog(ue) loopback
模拟环回测试法 analog(ue) loopback testing
模拟环境 simulated environment;simulation environment
模拟环境销售 environmental selling
模拟缓冲器 analog(ue) buffer
模拟换向器 analog(ue) commutator
模拟回放系统 analog(ue) playback system;analogy playback system
模拟回声测深图形记录仪 analog(ue) graphic(al) echo sounding recorder
模拟汇编程序 simulate assembler
模拟绘图机 analog(ue) drawing machine;analog(ue) plotter
模拟绘图仪 analog(ue) plotter
模拟活动 simulation game
模拟活动控制中心 simulation operations control center[centre]

模拟活动中心 simulation operation center[centre]
模拟火花点火发动机 simulated spark-ignition engine
模拟火箭 simulated rocket
模拟火球条件 simulated fire ball condition
模拟机 analog(ue) machine; analogy machine; simulator
模拟机动 simulated maneuver
模拟机器标引 simulated machine indexing
模拟机器人学 robotics
模拟积分 analog(ue) integration
模拟积分器 analog(ue) integrator
模拟积分仪 analog(ue) integrator
模拟级配 model(l)(ing) gradation
模拟极 analogous pole
模拟集成电路 analog(ue) integrated circuit
模拟计划 simulation game
模拟计数器 simulative counter
模拟计算 analog(ue) calculation; analog(ue) computation; analogy calculation; analogy computation; model(l)ing calculation
模拟计算机 analogy computer; simulator
模拟计算机仿真 analog(ue) computer simulation
模拟计算机控制系统 analog(ue) computer control system
模拟计算机系统 analog(ue) computer system
模拟计算器 analog(ue) calculator; analogy calculator
模拟计算系统 analog(ue) computing system
模拟计算装置 analog(ue) computing system; analogy computing device
模拟记录 analog(ue) record(ing)
模拟记录测井仪 analog(ue) logging system
模拟记录器 analog(ue) recorder
模拟记录仪 analog(ue) recorder
模拟记录组 analog(ue) recording group; model file
模拟技术 analog(ue) technique; model(l)ing technique; simulation technique
模拟加法器 analog(ue) adder
模拟加速器 analog(ue) accelerator
模拟加载 analog(ue) loading
模拟驾驶舱 simulator cockpit
模拟监督 simulation monitoring
模拟监控 simulation monitoring
模拟建筑 mock architecture
模拟降雨试验 simulated rain trails
模拟交通试验 simulated traffic test
模拟接地 analog(ue) ground
模拟接收机 analog(ue) receiver
模拟结构 model configuration
模拟结果 simulation result; analog(ue) result
模拟截击 simulated interception
模拟解 analog(ue) solution
模拟解码器 analog(ue) decoder
模拟介质的类型 kind of analogy medium
模拟进场 simulated approach
模拟井筒 simulation wellbore hole
模拟纠正 analog(ue) rectification; simulating rectification
模拟矩阵寻址 analog(ue) matrix addressing
模拟距离 simulated range
模拟开关 analog(ue) switch
模拟科学 simulation science
模拟可见度计算机 analog(ue) visibility computer
模拟空间环境条件 simulated space condition
模拟控制 analog(ue) control; analogy control; simulating control
模拟控制机 cybernetic model
模拟控制器 analog(ue) controller
模拟控制设备 analog(ue) control equipment
模拟控制中心 simulation control center[centre]
模拟扩展 analog(ue) extension
模拟拉紧 fictitious tensioning
模拟雷达图像 analog(ue) radar image; simulated radar image
模拟雷达吸收装置 analog(ue) radar absorber
模拟类型 analog(ue) type
模拟理论 analog(ue) theory; theory of models
模拟例行程序 simulated routine
模拟量 analog(ue) quantity; analog(ue) variable
模拟量表示 analog(ue) representation
模拟量多工器 analog(ue) multiplexer
模拟量分配器 analog(ue)-distributor
模拟量输出通道 analog(ue) output channel
模拟量输入 read analog(ue) input
模拟量输入通道 analog(ue) input channel
模拟量/数字转换 analog(ue)/digit conversion
模拟量限值验算 analog(ue) limit check
模拟裂变中子谱 mock fission neutron spectrum
模拟流动 simulated flow
模拟滤波器 analog(ue) filter; analogy filter
模拟逻辑 analog(ue) logic; model logic
模拟脉冲【计】analog(ue) pulse
模拟脉冲传输 simulation pulse transmission
模拟脉冲功率 analog(ue) pulse power
模拟门 analog(ue) gate
模拟面试 simulated interview
模拟面谈 simulated interview
模拟命令 simulation command
模拟模式 analog(ue) mode; analogy mode; simulation mode
模拟模型 analog(ue) model(ing); analogy model; simulation model
模拟模型参数估计 simulation model parameter estimation
模拟模型数据流程图 flowchart of data by simulation model
模拟模型研究 analog(ue) model study
模拟母线 analog(ue) bus; dummy bus; mimic bus
模拟目标 simulated target
模拟耐候试验 resistance to simulated weathering test
模拟黏[粘]性项 simulated viscosity term
模拟盘 mimic board
模拟盘驱动器 mimic board driver
模拟碰撞警告 simulated collision warning
模拟疲劳试验机 simulated service testing machine
模拟频率变换器 analog(ue) to frequency converter
模拟屏 mimic board; mimic panel
模拟屏接口 mimic board interface; mimic panel interface
模拟剖面 simulated section
模拟谱分析仪 analog(ue) spectrum analyser[analyzer]
模拟期 simulation period
模拟气氛 simulated atmosphere
模拟气候室 climatic chamber
模拟气候条件 simulated climatic condition
模拟器 imitator; simulator
模拟器件 analog(ue) device
模拟器座舱 simulator cabin
模拟牵引特性 simulating tractive performance
模拟潜水 simulated diving
模拟潜水装置 diving simulator; simulated diving device
模拟求解法 analog(ue) approach
模拟区域运价 simulated regional freight rate
模拟曲线描绘器 analog(ue) curve plotter
模拟燃料块 simulated fuel slug
模拟热平衡 simulate thermal equilibrium
模拟人造卫星的观察与研究气球 satellorb
模拟容器 simulation chamber
模拟软件 simulation software
模拟软件程序 simulation software program(me); simulator software program(me)
模拟扫描 analog(ue) sweep
模拟扫描变换器 analog(ue) scan converter
模拟色 mimic colo(u)ring
模拟色谱图 analog(ue) chromatogram
模拟设备 analog(ue) device; analog(ue) machine; imitator; simulated equipment; simulation device; simulation equipment; simulator
模拟设计 breadboard design
模拟射击 simulated gunnery
模拟摄影 analog(ue) projection; simulated photography
模拟摄影测量 analog(ue) photogrammetry
模拟摄影测量制图 analog(ue) photogrammetric mapping; analog(ue) photogrammetric plotting
模拟深度 simulated depth
模拟深度记录仪 analog(ue) depth recorder
模拟生料 model raw meal
模拟声波测井 simulated sonic log
模拟声频系统 simulating audio system
模拟失重 simulated weightlessness
模拟失重条件 simulated weightlessness condition
模拟石英表 quartz analog(ue) watch
模拟石英钟 quartz analog(ue) clock
模拟时间 simulated time
模拟时间段 segment of simulated time
模拟时间数字转换 analog(ue) to time to digital
模拟时钟 simulated clock simulation clock
模拟识别算法 pattern recognition algorithm
模拟实际工作条件 simulated working condition
模拟实物 simulating reality
模拟实验 analog(ue) experiment; experimental model test; mock-up experiment; simplex; simulation experiment
模拟实验法 experimental analogic method; experimental simulation method; simulation test method
模拟实验室 simulation laboratory
模拟使用 < 在试验时进行的 > simulated use
模拟使用试验 simulated service test
模拟使用寿命试验 simulated life test
模拟使用条件 service-simulated condition
模拟市场调节 simulated market regulation
模拟式 analogous; analog(ue) type
模拟式导航设备 analogous navigation set
模拟式电摄影技术 analog(ue) electrophotographic technology
模拟式仿真 analog(ue) simulation
模拟式干扰 simulation jamming
模拟式计算机 analog(ue) computer
模拟式计算装置 analog(ue) computing device
模拟式立体测图仪 analog(ue) plotter; simulated stereoautograph
模拟式模拟 analog(ue) simulation
模拟式扫描 analog(ue) scanning
模拟式设备 analog(ue) equipment
模拟式铁氧体移相器 analog(ue) type ferrite shifter
模拟式微波中继系统 analog(ue) microwave relay system
模拟式仪表 analog(ue) instrument; analog(ue) meter
模拟试验 analog(ue) experiment; analog(ue) test; mock-up test; model experiment; model test; preevaluation test; prevalution test; scale down test; simulated test; simulating test; simulation experiment; simulation test
模拟试验程序 simulation program(me); simulator program(me); simulated program(me)
模拟试验的种类 type of analogy tests
模拟试验技术 model(l)ing test technique
模拟试验结果 analog(ue) test result
模拟试验台 simulator stand
模拟试验仪器 instrument of modeling test
模拟室 simulation chamber
模拟手 artificial hand
模拟寿命 agree life
模拟输出 analog(ue) out(put)
模拟输出程序 simulated output program(me)
模拟输出分组件 analog(ue) output submodule
模拟输出扫描器 analog(ue) output scanner
模拟输出通道 analog(ue) output channel
模拟输电线 mimic transmission line
模拟输入 analog(ue) input; simulated input
模拟输入操作 analog(ue) input operation; simulated input operation
模拟输入传感器 analog(ue) input sensor; simulated input sensor
模拟输入电压 analog(ue) input voltage; simulated input voltage
模拟输入扩展器 analog(ue) input expander
模拟输入模件 analog(ue) input module
模拟输入设备 simulation input device
模拟输入输出装置 analog(ue) input/output unit
模拟输入条件 simulated input condition
模拟输入通道 analog(ue) input channel
模拟输入信号 mimic imput signal;

simulator input
模拟输入组件 analog(ue) input module
模拟熟料 model clinker
模拟数据 analog(ue) data; simulated data; simulation data
模拟数据传输 analog(ue) data transmission
模拟数据计算机 analog(ue) data computer
模拟数据记录器 analog(ue) data recorder
模拟数据交换 analog-digital data interconversion
模拟数据接转器 analog(ue) data sink
模拟数据数字化转换器 analog(ue) data digitizer
模拟数据信道 analog(ue) data channel
模拟数字变换器 analog(ue)(-to)-digital converter[convertor]; digitalyer
模拟数字变换器检验程序 analog(ue)(-to)-digital converter check program(me)
模拟数字变换器误差 analog(ue)(-to)-digital converter error
模拟数字程序控制 analog(ue)(-to)-digital programmed control
模拟数字传感 analog(ue)(-to)-digital sensing
模拟数字的 analog(ue)(-to)-digital
模拟数字混合仿真 hybrid analog(ue)(-to)-digital simulation
模拟数字混合计算机 combined analog(ue)(-to)-digital computer
模拟数字混合模拟 hybrid analog(ue)(-to)-digital simulation
模拟数字计算机 analog(ue)(-to)-digital computer
模拟数字计算系统 analog(ue)-digital computing system
模拟数字记录器 analog(ue)(-to)-digital recorder
模拟数字记录设备 analog(ue)(-to)-digital recording equipment
模拟数字控制系统 analog(ue)(-to)-digital control system
模拟数字模拟转换器 analog(ue)-digital-analog(ue) converter
模拟数字模拟转换系统 analog(ue)-digital-analog(ue) converter system
模拟数字式读出 analog(ue)(-to)-digital sensing
模拟数字适配器 analog(ue)(-to)-digital adapter
模拟数字数据记录设备 analog(ue)(-to)-digital data recording device
模拟数字随动系统 analog(ue)(-to)-digital control system
模拟数字系统 analog(ue)(-to)-digital system
模拟数字元件 analog(ue)(-to)-digital element
模拟数字转换 analog(ue)(-to)-digital conversion
模拟数字转换程序 analog(ue)(-to)-digital converter[convertor]
模拟数字转换脉冲 analog(ue)(-to)-digital conversion pulse
模拟数字转换器 analog(ue)(-to)-digital converter[convertor]; digitizer
模拟数字转换速度 analog(ue)(-to)-digital conversion rate
模拟数字转换装置 analog(ue)(-to)-digital commutator
模拟数字转换准确度 analog(ue)(-to)-digital conversion accuracy
模拟数字综合转换器 analog(ue)(-to)-digital integrating converter; analog(ue)(-to)-digital integrating translator
模拟水流 simulated flow
模拟水平偏转 analog(ue) horizontal deflection
模拟伺服机构 servo-simulator
模拟伺服系统 analog(ue) servo system
模拟随机过程试验 analog(ue) random process test
模拟桃花心木状 mahoganize
模拟天气图 analog(ue) weather map
模拟条件 simulated condition; simulation condition
模拟电阻 artificial resistance
模拟调制 analog(ue) modulation
模拟调制方式 analog(ue) modulation system
模拟调制解调器 analog(ue) modem
模拟调制系统 analog(ue) modulation system
模拟调制装置 analog(ue) modulation system
模拟听力损失 simulated hearing loss
模拟停车着陆 simulated flameout landing
模拟通道 analog(ue) channel
模拟通信[讯] analog(ue) communication
模拟通信[讯]量 artificial traffic
模拟通信[讯]系统 analog(ue) communication system
模拟投放 simulated drop
模拟凸轮 copying cam
模拟突变体 mimic mutant
模拟图 mimic diagram; simulated diagram
模拟图像存储器 analog(ue) pattern memory
模拟图像扫描器 analog(ue) image scanner
模拟图像显示 analog(ue) graphic(al) display
模拟图形 analog(ue) image
模拟推断 game
模拟外界环境(条件) simulated environment
模拟网络 analog(ue) network; artificial network; dummy; model network; simulative network
模拟微波接力通信[讯]系统 analog(ue) microwave relay system
模拟微分分析机 analog(ue)-differential analyser[analyzer]
模拟微分分析器 analog(ue)-differential analyser[analyzer]
模拟微析机 analog(ue)-differential analyser[analyzer]
模拟污垢 model soil
模拟无线电极光 artificial radio aurora; radio aurora
模拟物 simulacrum; stand in
模拟误差统计 simulation error statistic
模拟熄火 simulated flame out
模拟系统 analog(ue); analog(ue) system; copying system; simulated system; simulation system
模拟系统板 mimic system panel
模拟系统构架 frame of model(l)ing system
模拟系统图 sight-reading chart
模拟显示 analog(ue) display; analogy display
模拟显示部件 analog(ue)-display unit
模拟显示屏 <路线现状、信号、车流活动等> mimic panel
模拟显示装置 analog(ue)-display unit
模拟现场法 simulated in-situ method
模拟现场图 mimic diagram
模拟线 artificial line; simulated line
模拟线路激励器 analog(ue) line driver
模拟线路图 mimic diagram
模拟线圈 former-wound coil
模拟相关函数分析仪 analog(ue) correlator
模拟相关器 analog(ue) correlator
模拟像处理 analog(ue) image processing
模拟斜率检测 analog(ue) slope detection
模拟斜率检测器 analog(ue) slope detector
模拟泄漏 simulated leakage
模拟信道 analog(ue) channel
模拟信号 analog(ue) signal; dummy signal; simulated signal; simulating signal
模拟信号泵 analog(ue)-responsive pump
模拟信号处理转发器 analog(ue) signal processing transponder
模拟信号发生器 analog(ue) signal generator
模拟信号机【铁】analog(ue) signal; dummy signal; fictive signal; simulating signal
模拟信号加密器 analog(ue) signal encryption scrambler
模拟信号解调设备 analog(ue) signal demodulation equipment
模拟信号设备 analog(ue) signal(l)ing device
模拟信令 analog(ue) signal(l)ing
模拟信息 analog(ue) information
模拟信息滤波器 analog(ue) filter
模拟形式 analog(ue) form
模拟形网屏 chequerboard screen
模拟型 analog(ue) type
模拟型化 simulation modelling
模拟型试验 simulative type test
模拟性能 simulated performance
模拟序列 simulated series; simulation series
模拟旋转重力 rotational pseudo gravity
模拟学 analogy
模拟训练 simulated training
模拟训练电视系统 television system for simulation training
模拟训练设备 simulation training system
模拟压力 simulated pressure
模拟压力指示器 analog(ue) pressure indicator
模拟岩层移动 simulating strata movement
模拟研究 analog(ue) study; simulation study; simulator investigation; simulator study
模拟研究报告 simulation report
模拟延时装置 analog(ue) delay unit
模拟演示工作站 workstation for simulation and demonstration
模拟验证 verification of model(ing)
模拟样板 breadboard
模拟样机程序 simulation model program(me)
模拟样机研制 simulator prototype development
模拟遥测计 analog(ue) telemeter
模拟遥测仪 analog(ue) telemeter
模拟液体 model fluid
模拟仪 analog(ue) meter
模拟仪器 analog(ue) instrument
模拟仪器系统 analog(ue) instrumentation system
模拟移位寄存器 analog(ue) shift register
模拟译码 analogy decode
模拟译码器 analog(ue) translator; analogy decoder
模拟引起注意中断 simulated attention breaking off
模拟应变 simulated strain
模拟应答 analog(ue) answer
模拟应用 simulation application
模拟应用试验 simulated service test
模拟用户模块 analog(ue) subscriber module
模拟优选 simulation optimization
模拟有拱的长形房屋 mock arcade
模拟诱惑 imitative deception
模拟语言 analogous language; simulation language
模拟元件 analog(ue) element
模拟原始条件 pseudo-primeval condition
模拟运动 simulated motion; simulated movement
模拟运算步骤 simulated operating procedure
模拟运算部件 analog(ue) operational unit
模拟运行 simulation run
模拟运转条件的技术 environmental engineering
模拟造型 simulated model(l)ing
模拟噪声 modal noise
模拟增音机 analog(ue) repeater
模拟张拉 fictitious stretching
模拟振荡器 simulative generator
模拟震动 simulated ground motion
模拟执行 simulation executive
模拟值 analog(ue) value; simulated value; value of simulation
模拟指令舱 simulated command module
模拟指示器 analog(ue) indicator
模拟质量 analogous quality
模拟置乱器 analog(ue) scrambler
模拟中断条件 simulated interrupt condition
模拟中继 analog(ue) trunk
模拟中继模块 analog(ue) trunk module
模拟中继器 analog(ue) repeater
模拟重结晶法 simulated annealing
模拟重力 simulated gravity
模拟昼夜时钟 simulated clock
模拟注册 simulated log-on
模拟注意信号 simulated attention
模拟柱 analogous column; analog(ue) column
模拟转换 analog(ue) conversion
模拟转换开关 analog(ue) multiplexer
模拟转换器 simulated converter
模拟转接中心 simulated transit center[centre]
模拟装配夹具 simulated installation fixture
模拟装置 analog(ue) device; simulant; simulator(rig)
模拟装置工作时间 simulator time
模拟装置模型 simulator model
模拟准则 model(l)ing criterion; model(l)ing rule
模拟资料 analog(ue) data
模拟子集 analog(ue) subset
模拟紫外分光光度计 analog(ue) ultra-violet spectrophotometer
模拟自动驾驶仪 analog(ue) auto pilot
模拟总线 emulation bus
模拟总线驱动器 analog(ue) line driver
模拟走廊设备 model corridor facility

M

模拟阻抗 simulating impedance
模拟组合装置 analog(ue) nest unit
模拟作业 simulation job
模拟作业试验 simulated-operations testing
模拟作业研究设计 simulation research design
模拍频 mode beating
模盘 mo(u)ld
模配式啮合 modulated engagement
模配组合装置 Modulok
模匹配透镜 mode matching lens
模片 template;templet
模片安装 templet laydown
模片法 templet method
模片隔开 die separation
模片固定 die attachment
模片键合 die bonding;die mo(u)lding
模片切缝器 machine for slotting templet;slot cutter;slotted templet cutter;templet cutter
模片组合 templet assembly
模平衡理论 modal balancing theory
模谱 mode spectrum
模腔 die cavity;die opening;die space;mo(u)ld cavity;mo(u)ld form;mo(u)lding chamber
模腔衬环 container ring
模腔衬圈 container liner
模强函数 modular majorant
模切 die cut(ting)
模切纸盒 die-cut carton
模群 modular group
模绕法 former winding;form winding
模绕线圈 formed coil;former-wound coil
模绕线圈试验装置 form-wound motorette
模绕组 form(er) winding
模容量 mode volume
模塞 force piston;force plug;formpiston;male plug
模塞成型 plug forming
模塞助压成型 plug forming
模色散 mode dispersion
模砂 sand
模筛样板 mo(u)ld;running mo(u)ld
模深 mo(u)lded depth
模失真 mode distortion
模时延差 differential modal delay
模氏Ⅱ型 Mobitz type Ⅱ heart block
模式 model((l)ing);mode pattern;mode shape;module;pattern;schema[复 schemata];type
模式被输截入缩合 input-truncated condensation
模式变化 mode change
模式变换 mode change;mode conversion
模式辨 pattern recognition
模式辨认 model recognition;pattern discrimination
模式辨识 pattern identification
模式标本 type specimen
模式标识符 mode identifier
模式表 pattern list
模式操作编辑序列 pattern operation editing sequence
模式操作数 pattern operand
模式操作(字)符 pattern operator
模式产地 type locality
模式处理语句 pattern handling statement
模式传递函数 mode transfer function
模式串 pattern string
模式纯度 mode purity
模式存储器 mode memory
模式大气 model atmosphere
模式的解释 interpretation of scheme
模式地点 type locality
模式地区 type area
模式定义 mode-definition
模式多谱段扫描仪 modular multispectral scanner
模式反应堆 prototype reactor
模式范畴 category;schema category
模式分隔 mode separation
模式分类 pattern classification
模式分类器 pattern classifier
模式分析 modal analysis;pattern analysis
模式符号 mode symbol
模式函数 mode function;pattern function
模式核对语句 pattern matching statement
模式核对运算 pattern matching operation
模式花样 mode pattern
模式化 hipping
模式化记号 hip token
模式化家具 modular furniture
模式化施工法 modular construction method
模式混合器 mode mixer
模式基元 pattern primitive
模式激励器 mode exciter
模式集 set of patterns
模式简并 mode degeneracy
模式鉴别干涉仪 mode-discriminating interferometer
模式校验 modulo check
模式校准 model calibration
模式结构 mode configuration
模式解释 interpretation of scheme
模式竞争 mode competition
模式空间中的半平面 half plane in the pattern space
模式控制 mode control
模式类别 pattern class
模式理论 pattern theory
模式列举 pattern enumeration
模式林 normal forest
模式令级分配 normal age-class
模式令级序列 normal series of age gradations
模式滤除器 mode filter
模式轮廓 mode envelope
模式弥散 modal dispersion
模式描述 pattern description
模式描述语言 pattern description language;schema description language
模式敏感故障 pattern-sensitive fault
模式名 schema name
模式模拟器 mode simulator
模式母树枝 normal seed stand
模式年龄 model date
模式年龄计算方法 calculating methods of model ages
模式耦合 mode coupling
模式耦合激光器 mode coupled laser
模式耦合器 mode coupler
模式培养 type culture
模式匹配 pattern matching
模式偏差 model deviation
模式频谱 mode spectrum
模式剖面 type profile;type section
模式牵引 mode pulling
模式牵引效应 mode pulling effect
模式驱动子程序 pattern-driven subroutine
模式色散 mode dispersion
模式生长量 normal increment
模式生成 pattern generation
模式生态系统 model ecosystem
模式识别 pattern recognition
模式识别程序 pattern recognition program(me)
模式识别处理 pattern recognition processing
模式识别的鉴别法 discriminant approach to pattern recognition
模式识别的判定理论法 decision-theoretic approach to pattern recognition
模式识别法 pattern recognition method
模式识别机 pattern recognizer
模式识别类型 pattern recognition type
模式识别器 pattern recognizer
模式识别系统 pattern recognition system
模式收获表 normal yield table
模式收获量 normal yield
模式输出分析 model output diagnosis
模式输出统计 model output statistics
模式数据描述语言 schema data description language
模式数值 model value
模式说明 mode declaration pattern specification
模式搜索 pattern search
模式搜索法 pattern search method
模式损耗 modal loss
模式锁定 mode locking
模式跳变 mode jump
模式跳越 mode hopping
模式同步组列 mode-locked train
模式图 ideograph;mode chart;schema chart;schematic diagram
模式图样 mode pattern
模式微扰 mode perturbation
模式位 mode bit
模式系数 mode factor
模式相位弥散 modal phase dispersion
模式项(目) schema entry
模式信息 pattern information
模式信息处理 pattern information processing
模式信息处理系统 pattern information processing system
模式蓄积 normal growing stock
模式选择程式 model select program(me)
模式选择器 mode selector
模式研究法 model methodology
模式移动 pattern move
模式遗迹 type site;type statian
模式抑制 mode rejection
模式展开法 model-expansion method
模式振荡 mode oscillation
模式指示 mode indication
模式制作系统 pattern making system
模式中的活动 activity in the model
模式中心 mode top
模式种 type species
模式重复循环 pattern repeat cycle
模式自适应写入 pattern adaptive writing
模式综合法 mode synthesis method
模式组 type series
模饰样板 running mo(u)ld
模数 mode number;module;modulo;modulus[复 moduli]
模数 n 计数器 modulo-n counter
模数 n 加法器 modular-n adder
模数比 modular proportion;modulus ratio
模数比尺方格 planning grid
模数编码器 analog(ue)-to-digital encoder
模数变换器 analog(ue)(-to)-digital converter[convertor]
模数表示 modular representation
模数参考制 modular reference system
模数测定仪 modulus tester
模数插口 modular jack
模数程序设计 modular programming
模数尺寸 modular dimension;modular size
模数尺寸箱式结构建筑 modular box system building
模数尺寸砖 modular brick
模数传感 analog(ue)-to-digital sensing
模数大小 modular size
模数代数 modular algebra
模数带 modular zone
模数单元 modular component
模数的 modular
模数叠合式墙体(系统) modular-stack-type wall system
模数法 modular method;module method
模数法设计 modular design;modular design method
模数方格 modular grid
模数分类法 modular systematics
模数分类学 modular systematics
模数构件 modular member
模数规划格 modular planning grid;planning grid
模数函数 modular function
模数化 modularization;modulization
模数化部件 modular component;modular element
模数化尺寸 modulated dimension;modular size
模数化方格 modular grid
模数化构件 modular construction unit;modular unit
模数化构造 modular construction
模数化集装箱自动装卸 modular automated container handling
模数化结构 unit construction
模数化控制 modulated control
模数化楼层高 modular story height
模数化平面 modular plane
模数化起重机 modular crane
模数化砌块 modular(building) unit
模数化施工(构造) modular construction
模数化自动化集装箱装卸 modular automatic container handling
模数化自动化集装箱装卸桥 modular automatic container handling portainer
模数化组件 modular automatic container handling module
模数混合计算机【计】hybrid computer
模数混合模拟 hybrid analog(ue)-digital simulation
模数计算机 analog(ue)-digital computer
模数检验 modular check
模数校验 modular check
模数结构 modular construction;modular structure
模数空间 modular space
模数量纲 modular dimension
模数螺纹螺距范围 range of module thread cut
模数螺纹头数 number of module threads cut
模数模转换器 analog(ue)-digital-analog(ue) converter
模数配合 modular coordination
模数平面 modular plane
模数砌块 modular masonry unit
模数砌筑单位 modular masonry unit
模数三维方格 modular space grid
模数设计 modular design
模数设计法 modular design method
模数体系 modular system;fuzzy system

模数体制 modular system
模数网格 modular grid
模数铣刀 module-milling cutter
模数系统 modular system
模数协调 modular coordinating; modular coordination
模数协调法 modular coordinating
模数与数模转换 analog(ue)-digital and digital-analog(ue) conversion
模数制 dimensional framework; modular coordination; modularized system; modular measure system; modular system; modulus system
模数制房屋 modular house; modular housing
模数制结构 modular structure
模数制设计 design by modulus system
模数制图系统 modular CAD system
模数制住房 modular house; modular housing
模数砖 modular brick
模数转换 analog(-to)-digital conversion
模数转换器 analog(-to)-digital converter[convertor]
模数转换器性能 performance of analog(ue)-to-digital convertor
模衰减差 differential modal attenuation
模双曲方程 normal hyperbolic equation
模似 mimicry
模似负载 fictitious load
模塑 form; mo(u)ld
模塑材料 mo(u)lding material
模塑成型机 mo(u)lding machine
模塑成型压力 mo(u)lding pressure
模塑的 mo(u)lded
模塑地图 relief map
模塑法 method of mo(u)lding
模塑粉合料 mo(u)lding powder blends
模塑粉料 mo(u)lding powder
模塑合成物 mo(u)lded composition
模塑化合物 mo(u)lded plastic compound
模塑机 mo(u)ld machine
模塑胶合板 plymold
模塑料 mo(u)lding compound
模塑料收缩性 shrinkage of mo(u)lding compound
模塑料压缩比 bulk factor of mo(u)lding compound
模塑硫化 mo(u)ld cure; mo(u)ld vulcanization
模塑品 mo(u)lded work
模塑品外观疵点 visual defect in mo(u)ldings
模塑缺陷 mo(u)lding defect
模塑收缩 mo(u)ld shrinkage
模塑条件 condition of mo(u)lding
模塑物 mo(u)lding
模塑效率 mo(u)ld efficiency
模塑压机 mo(u)lding press
模塑元件 mo(u)lded element
模塑造型 die mo(u)lding
模塑者 mo(u)lder
模塑指数 mo(u)lding index
模塑制品 mo(u)lded product; mo(u)lding article
模塑周期 mo(u)lding cycle
模算子 modular operator
模索经验 search for experience
模锁定 mode locking
模态 modality
模态参数 modal parameter
模态参数识别 mode-parameter identification
模态的 modal
模态叠加 modal superposition; mode superposition
模态叠加法 method of modal superposition
模态反应 modal response
模态分析 modal analysis
模态分析程序 modal analysis program(me)
模态分析软件 modal analysis software
模态刚度 modal stiffness
模态互斥 mode repulsion
模态矩阵 modal matrix
模态控制 modal control
模态列 modal column
模态逻辑 logic of modality; modal logic
模态逻辑学 modal logic
模态耦合 mode coupling
模态平衡 modal balancing
模态矢量 modal vector
模态系统 modal system; model system
模态运算子 modal operator
模态展式 model expansion
模态综合法 modal expansion method; modal synthesis method
模态阻尼比 modal damping ratio
模态组合 modal combination
模膛 die opening; impression
模膛的布排 location of impressions
模套 die sleeve; external form; chase
模特儿似的 modelly
模特儿职业 model(l)ing
模特性 module feature
模体 body mo(u)ld; die body
模体积 mode volume
模跳跃 mode hopping
模同步 self-mode-locking
模同态 module homomorphism
模头 die head
模头加热区 die heating zone
模凸轮 pattern cam
模图样 mode pattern
模推离 mode pushing
模托 die shoe
模瓦 die shoe
模网 lay wire
模纹种植 mosaic culture
模系 mode group
模隙 die gap
模下压板 die slide
模线 mo(u)ld line
模线缝 match mark
模限制 mode confinement
模箱 mo(u)ld(ing) box
模写 facsimile
模芯 core rod; mo(u)ld core; mo(u)ld kernel; kernel
模芯支架 spindre
模形式 modular form
模型 model(set); pattern; prototype; cast; former; matrix [复 matrixes/matrices]; shape maker; test piece model; form matter <制作塑料立体地图的>
模型板 breadboard model; form board(ing); mo(u)ld blade; mo(u)ld plate; prototype board; stamping board
模型爆破 model blasting
模型比(例)尺 model scale
模型比例尺确定 fixing model scale
模型边界 model boundary
模型边线 model boundary
模型变换 model(l)ing transformation
模型变率 distortion ratio of model
模型变态 model distortion
模型变形 model deformation
模型变形公差 pattern distortion allowance
模型变形校正 correction of model deformation
模型标本 archetype
模型标定 model calibration
模型标准 model standard
模型表达 model formulation
模型表面 mo(u)ld surface
模型表述 model formulation
模型材料 cast material; form matter; model material; mo(u)lding material; pattern-making material
模型参考自适应控制 model reference adaptive control
模型参数 model parameter
模型测定 model estimation
模型测量 model measurement
模型测试 model measurement
模型长度 length of model
模型厂 model plant
模型车 model car
模型车间 model room; pattern shop
模型池塘 model pond
模型尺寸 mo(u)lded dimension
模型尺度 model dimension
模型尺度比 scale ratio
模型抽样 model sampling
模型磁铁 model magnet
模型大地构造学 model tectonics
模型大气分析 model atmosphere analysis
模型弹 mock bomb
模型捣动 ramaway
模型的 breach board
模型的表示法 model notation
模型的估计量性质 property of estimator of a model
模型的记法 model notation
模型的阶 model order
模型的结构 structure of the model
模型的期望性质 desirable property of a model
模型的数学完备性 mathematic(al) completeness of model
模型的一部分 mo(u)ld portion
模型堤坝 model fill
模型底板 pattern board
模型地裂运动 model earthquake motion
模型地(形)图 relief map
模型地震学 model seismology
模型地震运动 model earthquake motion
模型电动机 model motor
模型顶视图 model top view
模型定缝销钉及套 peg-and-cup dowel
模型定律 model law
模型定式 model specification
模型动物 animal pattern
模型锻造 closed-die forging; die forging
模型堆场 mo(u)lds stockyard
模型对原型的比例 model-to-prototype scale
模型多项式方程 polynomial model equation
模型法 model method
模型范畴 category of model
模型范围 model area; model range; model scope; range of models
模型方程 model equation
模型方法 model(l)ing approach
模型房 model room
模型飞边 mo(u)ld mark
模型飞机 model aircraft
模型分割 model split
模型分类 category of model
模型分析 model analysis
模型符号 model symbol
模型辐射带 model radiation belt
模型复杂性 model complexity
模型改进 model refinement
模型高 model height
模型工 model maker; pattern maker
模型工厂 pattern shop
模型工场 pattern shop
模型工车床 pattern-maker's lathe
模型工用尺 pattern-maker's rule
模型工用锤 pattern-maker's hammer
模型公差 pattern allowance
模型构成 model formulation
模型构造 construction of model; mo(u)ld construction
模型构造地质学 model tectonics
模型估计 model estimation
模型光带 model strip light
模型光核反应 model photonuclear reaction
模型过滤器 prototype filter
模型函数 pattern function
模型号数 pattern number
模型合理化论证 validation of the model
模型合理性论证 validation of the model
模型合销 pattern dowel
模型和真型 model and prototype
模型河槽 model channel
模型河流 model river; model stream
模型宏语句 prototype macrostatement
模型互搭接头 mo(u)lded chime lap joint
模型划线 pattern scribing
模型化 model(l)ing
模型化的进展 evolution of model(l)ing
模型化合物 model compound
模型黄铜 matrix brass
模型混合料 mo(u)ld compound
模型机 prototype
模型基础 mo(u)lded base
模型基础试验 model footing test
模型基线 model base(line); mo(u)lded base
模型基准自适应控制系统 model reference adaptive control system
模型及图样 pattern and drawing
模型计划 model plan
模型计算机 normatron
模型计算结果的稳定性 stability of model calculation
模型技巧摄影 model shot
模型技术 model(ling) technique
模型加工 pattern forming
模型加强筋 pattern reinforcing rib
模型架 model support
模型假曲率 counter-camber
模型检核 model verification
模型检验 model testing; pattern checking
模型建立 model building; model formulation; model(l)ing
模型建立的方法论 model(l)ing methodology
模型建立的过程 model(l)ing process
模型建立阶段 model(l)ing phase
模型建立与改进 model development and refinement
模型建造 model building
模型建筑 mo(u)ld construction
模型浇口 mo(u)ld opening
模型铰链销 mo(u)ld-hinge pin
模型校验 model checking
模型校正器 model corrector
模型校正装置 model device for the model
模型校准 model calibration
模型接合缝 mo(u)ld mark
模型结构 model building; model

M

structure
模型结构分析 model structural analysis
模型金属 model metal;pattern metal
模型紧固夹 mo(u)ld clamps
模型精确度 model accuracy
模型静力研究 static model investigation
模型镜头 model lens
模型距离 model distance;model range;model spacing
模型卷尺 model scale
模型卡紧 seizing of mo(u)ld
模型可靠度 model reliability
模型可靠性 model reliability
模型可再生性 molds reuseability
模型空间 model space
模型控制 model control
模型控制盘 diagrammatic control board
模型库 model library;pattern storage;pattern store
模型框图 model framework chart
模型雷诺数 model Reynolds number
模型类型 model type;type of models
模型冷却 mo(u)ld cooling
模型理论 model theory
模型连接 bridging of models;model connection
模型量测 measurement of models;measurements of the stereo-model
模型列车 miniature train
模型列举 pattern enumeration
模型灵敏度 pattern sensitivity
模型流动 model flow;scale model flow
模型留量 pattern allowance
模型硫化 mo(u)ld cure
模型律 model law
模型率定 model calibration
模型率定曲线 model calibration curve;model rating curve
模型论 model theory
模型面积 model area
模型模拟 model simulation
模型目标 model objective
模型目的 model objective;purpose of model
模型内的变数 variable in the model
模型内腔 mo(u)ld form;mo(u)ld impression
模型拟合 model fitting
模型黏[粘]土 model(l)er's clay
模型扭曲 model distortion;model warpage;warping of model
模型排列 model listing
模型盘 model board
模型配件 mo(u)ld parts
模型膨胀 pattern swelling;swelling of the pattern
模型匹配 model matching;pattern matching
模型拼接 model marriage
模型频率特性 model response
模型平面 model plane
模型评定 pattern evaluation
模型前池 model head-pond
模型桥墩 model pile
模型趋近 model approximation
模型绕组 diamond winding;pattern winding;spool winding
模型人格 model personality
模型认证 model verification
模型润滑剂 mo(u)ld lubricant;mo(u)ld release;parting agent
模型扫描 model scanning
模型砂 model material;model sand
模型上漆 pattern varnish
模型设备 simulator
模型设定 model specification;specification of model
模型设定不完善 imperfect specification of model
模型设计 model design;pattern design;pattern plan
模型设计法 model design method
模型设计图 pattern layout
模型生态系统 model ecosystem
模型石膏 model plaster;mo(u)ld plaster
模型石膏废料 waste mo(u)ld gypsum
模型石膏粉刷 gypsum mo(u)lding plaster
模型识别 model discrimination
模型视觉 model-based vision
模型试验 mock-up test;model experiment;model(l)ing test(ing);scaled model test;scale down test;model study
模型试验操作技术 model-operating technique
模型试验池 model tank;model testing basin;towing tank
模型试验法 model experiment method;model test method
模型试验技术 model(l)ing technique
模型试验设备 model test equipment
模型试验塔 rig for model test
模型试验台 model basing;rig for model test
模型试验网 experimental model net
模型试验研究 model investigation;model study
模型试验验证 verification by model test
模型试验用加砂器 bed-load feeder
模型试验装置 rig for model test
模型室 model room
模型收缩公差 pattern(-maker's) shrinkage
模型寿命 mo(u)ld life
模型术 model(l)ing technique
模型数据 model data;model information
模型数学形式的设定 specification of the mathematical form of model
模型水槽 model flume
模型水陆生态系统 aquatic model ecosystem
模型水轮机 model turbine;model water turbine
模型说明(书) model specification;specification of model
模型丝 model filament
模型搜索 pattern search
模型搜索法 pattern search method
模型塑料 model plastic
模型算法 model algorithm
模型缩尺 model scale
模型缩尺效应 model scale effect
模型缩放 scaling of model
模型所代表的实体 antitype
模型填塞【电】pattern-filling
模型调整 model adjustment
模型调整法 model adjustment technique
模型铁路 model railway
模型图 illustration of model
模型托板 follow board
模型完备性 completeness of a model
模型完全化 model completion
模型完全性 model complete
模型网络 prototype network
模型违例 model violation
模型文件 model file
模型稳定性 model stability
模型问题 model problem
模型物质 model substance
模型误差 model error
模型铣床 pattern mill(er);pattern milling machine
模型系统 model system
模型纤维 model fiber[fibre]
模型线 mo(u)ld line;pattern line
模型线圈 model coil
模型限制 model constraint
模型相似律 model similarity law
模型相似性 model similarity
模型箱 model casing
模型箱闸刀开关 mo(u)ld case fused knife switch
模型效率 model efficiency
模型信息 model information
模型行为 model behavio(u)r
模型形成 pattern generation
模型型 mo(u)ld form
模型型腔 mo(u)ld form;mo(u)ld impression
模型性 modularity
模型修改方式 model modifying mode
模型修整 cast trimming
模型修整器 model trimmer
模型修正 adjustment of model
模型选择 model select;pattern selection
模型选择程序 model select program(me)
模型选择问题 problem of model selection
模型选择准则 model selection criterion
模型研究 scale model investigation;model study
模型眼 reduced eye;schematic eye
模型验收 model acceptance
模型验收试验 model acceptance test
模型验证 model conformation;model(l)ing verification;verification of model(ing)
模型验证试验 verification by model test
模型一致的 model consistent
模型因子 factor of a model
模型硬化剂 model hardener
模型用浆 model(l)ing paste
模型用木材 pattern wood
模型用润滑剂 mo(u)ld lubricant
模型用石膏 mo(u)ld plaster
模型油 dope;mo(u)ld oil;mo(u)ld release oil
模型油漆 mo(u)ld paint
模型余量 pattern allowance
模型与原型比较试验 model-prototype comparison test
模型与原型关系 model-prototype relation(ship)
模型语句 model statement;prototype statement
模型预测法 model prediction
模型预测能力检验 test of predictive power of a model
模型预测效能 forecast performance of model
模型预示法 predictive method by model
模型元件 model element
模型园 model garden
模型原模 prototype
模型约束 model constraint
模型照片 model photograph
模型振动器 mo(u)ld vibrator
模型制品 mo(u)lded articles;mo(u)lding
模型制造 model building;model(l)ing;mo(u)lding of model;mo(u)ld manufacture
模型制造车间 model shop
模型制造者 pattern maker
模型制作 model building;model construction;model fabrication;model(l)ing
模型制作技术 model fabrication technique;model(l)ing technique
模型置平 horizontalizing model;level(l)ing(of) model
模型重建 restitution of model
模型住房 model home
模型助绘地图 model-assisted mapping
模型转轮 model runner
模型桩 model pile
模型桩群 model pile group
模型装配 model rigging
模型锥度 conicity of model;tape of mo(u)ld;taper of model
模型准备室 model preparation room
模型资料 model data
模型自适应控制 model adaptive control
模型组 model set
模型组成 model formulation
模型组成部分 model component
模型坐标 coordinates for model;model coordinates
模型坐标系统 model coordinate system
模选择 mode selection
模选择器 model selector
模选择性 mode selectivity
模穴 die cavity;mo(u)ld impression
模穴套板 nest plate
模压 blanking;coining;contour forging;die stamping;die work;extrude;mo(u)lding;mo(u)ld pressing;press forming;punching;stem pressing;swaging
模压板 embossed sheet;mo(u)lded board;stamped sheet
模压棒 mo(u)lded rod
模压玻璃 mo(u)lded glass
模压部件 mo(u)lded part
模压材 mo(u)lded timber
模压成型 compression mo(u)lding;die forming;mo(u)ld pressing;press forming;press mo(u)lding
模压成型法 stamping process
模压成型件 die formed parts
模压成型绕组 former winding
模压成型温度 mo(u)lding temperature
模压窗台 mo(u)lded window cill[sill]
模压锤 swaging hammer
模压淬火 die quenching;press quenching
模压的 die-formed;shaped
模压电路 die-stamped circuit
模压锻 die forging
模压法 die pressing
模压钢钩头链 pressed steel hook chain
模压钢开沟器 pressed steel boot
模压工 pressman;stamper
模压回火 press tempering
模压混凝土型材 extruded concrete profile;extruded concrete section;extruded concrete shape
模压机 block press;dieing machine;mo(u)lding machine;mo(u)lding press;stamping press;stem machine;swaging hammer
模压机头 punching head
模压机压气具 air compressor for mo(u)lding machine
模压加工 embossed work;embossing;embossment
模压加工性 mo(u)ldability
模压件 forming
模压胶合板 formed plywood;mo(u)-

lded plywood
模压金属门 pressed metal door
模压梁 pressed girder
模压轮廓 pressed profile
模压螺母 stamped nut
模压螺纹 stamped thread
模压锚碇头 swage head
模压铆钉 swage rivet head
模压铆钉头 swage head;swage rivet head
模压捻线机 forming twister
模压泡沫塑料 mo(u)lded foam plastics
模压器皿 press mo(u)lding ware
模压缺陷 bedding fault;mo(u)lding fault
模压绳索 rope mo(u)lding
模压时间 clamp time
模压示镜玻璃 mo(u)lded spectacle glass
模压收缩 mo(u)ld shrinkage
模压收缩率 die shrinkage
模压塑料 mo(u)lded plastic;mo(u)lding plastics
模压塑料流动性 flowability of mo(u)lding compound
模压塑料照明灯具 mo(u)lded plastic light fitting
模压填料环 rings of die-molded packing
模压铁(粉)芯 mo(u)lded core
模压筒管 profiled tube
模压透镜 mo(u)lded lens
模压透镜毛坯 mo(u)lded lens blank
模压涂层 die coating
模压系数 mo(u)lding index
模压箱 mo(u)lded suitcase
模压橡胶内套 extruded liner;mo(u)lded liner
模压性能 mo(u)ldability
模压阴极 mo(u)lded cathode
模压云母板 mo(u)lding micanite
模压织物 mo(u)lded fabric
模压制板 mo(u)lded board
模压制模法 die hobbing
模压制模压力机 die hobbing press
模压制品 mo(u)lded part;mo(u)lded piece;pressing(piece)
模压铸造 casting-forging method;extrusion casting;liquid(metal)forging;squeeze casting
模移 mode shift
模抑制 mode suppression
模抑制技术 mode suppression technique
模印 pressing back;stamping;stamping back
模印浮雕装饰法 mo(u)lded decoration
模油 mo(u)lding oil
模域 modular field
模域光纤传感器 modal-domain fiber optic(al)sensor
模原理 modulus principle
模运算 modular arithmetic;modulo arithmetic
模造材 sham wood
模造大理石 ston(e)y
模造大理石门 ston(e)y gate
模造黑药 mo(u)lded powder
模造物 dummy
模造性能 mo(u)ldability
模振幅 mode amplitude
模帧 module frame
模制 forming;mo(u)lding
模制棒 mo(u)lded rod
模制(标准)试块(水泥) briquet(te) test specimen
模制玻璃 mo(u)lded glass;corning-steuben <一种供建筑和装饰用的专利产品>
模制玻璃块 mo(u)lded glass block
模制玻璃组件 <一种专利产品> Cristol glass
模制材料 mo(u)lding material
模制层压材料 mo(u)lded laminate
模制层压管材 mo(u)lded laminated tube
模制导板 template guide
模制倒角 mo(u)lded chamfer
模制的 mo(u)lded
模制的管道保温材料 mo(u)lded pipe insulation
模制底板 mo(u)lded base
模制电木盒 mo(u)lded bakelite case
模制电阻器 mo(u)ld resistor
模制法 mechanography
模制粉末铁氧体 mo(u)lded ferrite
模制隔热材料 mo(u)lded insulation;mo(u)lded insulator
模制隔热制品 mo(u)lded insulation
模制工序 mo(u)lding process
模制公差 mo(u)lding tolerance
模制构件 mo(u)lded articles
模制固体催化剂 mo(u)lded solid catalyst
模制管 mo(u)lded tube
模制管工设备 mo(u)lded plumbing unit
模制盒型断流器 mo(u)lded-case circuit breaker
模制盒式断路器 mo(u)lded case circuit breaker
模制混凝土 mo(u)lded concrete
模制混凝土板 mo(u)lded concrete block
模制混凝土块 mo(u)lded concrete ashlar;mo(u)lded concrete block
模制机 mo(u)lding press
模制基础 mo(u)lded base
模制继电器 mo(u)lded relay
模制件 mo(u)lded articles;mo(u)lded part;mo(u)lded piece
模制件缺陷 mo(u)lding fault
模制件收缩量 mo(u)lding shrinkage
模制结构 mo(u)ld construction
模制聚氯乙烯衬垫污水管配件 mo(u)lded PVC gasketed sewer fitting
模制聚乙烯泡沫浮选 mo(u)lded polythylene foam flo(a)tation
模制绝缘 mo(u)ld insulation
模制绝缘材料 mo(u)lded insulating material;mo(u)lded insulation
模制绝缘子 mo(u)lded insulator
模制颗粒燃料球 mo(u)lded fuel sphere
模制颗粒燃料元件 mo(u)lded fuel element
模制壳 mo(u)ld case
模制块状颗粒燃料元件 mo(u)lded block fuel element
模制棱角 mo(u)lded chamfer
模制链节 profile chain link
模制零件 mo(u)lded articles;mo(u)lding
模制楼板 mo(u)lded floor
模制煤 mo(u)lded coal
模制门框石 mo(u)ld stone
模制内胎 mo(u)lded tube
模制配件 stamping
模制膨胀珍珠岩砌块 mo(u)lded expanded pearlite block
模制片 mo(u)lded tablet
模制品 mechanograph;mo(u)lded goods;mo(u)lded product;mo(u)lded work
模制瓶 mo(u)ld formed bottle
模制人造橡胶闸缸皮碗 mo(u)lded synthetic rubber piston seal
模制石棉 mo(u)lded asbestos
模制式 mo(u)lded-on type
模制试件 mo(u)lded specimen
模制试块 briquet(te)
模制试样 mo(u)lded specimen
模制树脂 mo(u)lded resin
模制塑料 mo(u)lded plastic
模制塑料趸船 molded plastic float pontoon
模制塑料绝缘 mo(u)lded insulation
模制塑料天窗 mo(u)lded plastic skylight
模制天沟 mo(u)lded gutter
模制挑檐 mo(u)lded cornice
模制透镜 mo(u)lded lens
模制圬工 mo(u)lded masonry
模制纤维板 mo(u)lding fiber[fibre] board
模制纤维轴承 mo(u)lded-fabric bearing
模制小五金 template hardware
模制斜面 mo(u)lded chamfer
模制压力 mo(u)lding pressure
模制檐槽 mo(u)lded gutter
模制阴极 mo(u)lded cathode;mo(u)ld matrix cathode
模制油漆 formwork paint
模制云母 mo(u)lded mica
模制云母纸 mo(u)lding mica sheet
模制炸药 mo(u)ld explosive
模制纸 mo(u)lding paper;mo(u)ld-made paper
模制纸板 mo(u)lded board
模制砖 mo(u)lded brick;smooth finish tile
模制桩 premo(u)lded pile
模制装饰耐热玻璃 lalique
模制组成 mo(u)lding composition
模中涂布系统 in-mo(u)ld coating system
模中装饰系统 in-mo(u)ld decorating system
模轴 mo(u)ld shaft
模轴式涡轮机 horizontal turbine
模筑混凝土 cast concrete;cast-in-place concrete;cast-in-situ concrete
模铸 die cast;mo(u)lded casting
模铸材料 die casting
模铸法 die casting
模铸钢 mo(u)lded steel
模铸钢锭 static ingot
模铸合金 alloy for die casting
模铸机 die-casting machine
模铸架 mo(u)lding frame
模铸件 die casting
模铸金属 die-casting material;die metal
模铸空铅 die cast furniture
模铸铝合金 alumin(i)um die cast alloy
模铸树脂 mo(u)lded resin
模铸橡胶 mo(u)lded rubber
模铸元件 die-casting element
模铸轴承 die cast bearing
模砖 modular brick
模转闭合高度 die height
模转化干扰 mode change-over disturbance
模转换 mode conversion
模转换干涉 mode conversion interference
模转换器 mode switch
模转寿命 life of the die
模装 mock-up
模装填比 die fill ratio
模组 <熔模铸造用> cluster;die set;mo(u)ld train;pattern assembly
模组分 mo(u)ld split
模座 die bed;die block;die holder;die seat;master block

膜 membrane

膜板 diaphragm;lamina membrane
膜板阀 diaphragm valve
膜板塞门 diaphragm cock
膜板式夹头 diaphragm chuck
膜板压块 diaphragm follower
膜瓣收集器 petal catcher
膜本征函数 membrane eigenfunction
膜本征值 membrane eigenvalue
膜本征值问题 membrane eigenvalue problem
膜壁 membrane wall;membranous wall
膜壁厚度 wall thickness
膜壁屈服 membrane yield
膜剥裂试验 film stripping test
膜部 membranous part
膜材料 membrane material
膜测压力计 membrane manometer
膜层 membranous layer
膜层测定仪 layer thickness meter
膜层测厚仪 layer thickness ga(u)ge
膜层的化学稳定性 chemical durability of layers
膜层的机械牢固度 machinable durability of layers
膜层附着力 film adhesion
膜层空气界面 film-air interface
膜层强度测定仪 film intensity measuring device
膜层散热系数 convection coefficient;film coefficient
膜层温度 film temperature
膜层应力 stress in thin film
膜层阻力 film resistance
膜池 membrane cisterna
膜传导理论 membrane theory of conduction
膜传热系数 film heat transfer coefficient
膜刀 hymenotome
膜导管 membrane duct
膜导热系数 film heat conductance coefficient
膜的不对称性 membrane asymmetry
膜的分离效率 membrane separation efficiency
膜的构成 film building
膜的强度 film strength
膜的渗透性 membrane permeability
膜的形成 filming
膜的形成过程 film process
膜的选择性分离 membrane-based selective separation
膜的有效厚度 effective film thickness
膜的状态 filminess
膜点 film spot
膜电导 membrane conductance
膜电感 membrane inductance
膜电荷 membrane charge
膜电极 membrane electrode
膜电解 membrane electrolysis
膜电流 membrane current;membrane electric(al)current
膜电容器 membrane capacitance
膜电势 membrane potential
膜电位 film potential;membrane potential
膜电泳 membrane electrophoresis
膜电阻 membrane resistance
膜动控制阀 diaphragm-operated control valve
膜堆 membrane stack
膜法 <离子交换> membrane method;membrane process
膜法脱盐法 membrane demineralization process
膜反应器 membrane reactor

膜反应曲线 membrane reaction curve
膜泛水 membrane flashing
膜方程 membrane equation
膜防水 membrane waterproofing
膜放热 film heat transfer
膜分离 barrier separation; membrane separation; separation
膜分离法 membrane separating process; membrane separation process
膜分离工艺 membrane separation process
膜分离活性污泥法 membrane separation activated sludge process
膜分离技术 membrane separation technique
膜分离器 membrane separator
膜分离生物反应器 membrane separation bioreactor
膜分离系统 membrane separation system
膜分离装置 membrane separation device
膜分馏物 fractional distillation product of membrane
膜封闭 membrane seal
膜封闭的 membrane sealed
膜辅助电解 membrane-assisted electrolysis
膜共振 membrane resonance
膜盒 aneroid; bellows; capsule (capstan); diaphragm capsule; sylphon
膜盒阀 aneroid valve
膜盒感压膜片 aneroid diaphragm
膜盒高度表 aneroid altimeter
膜盒高度计 aneroid altimeter
膜盒记录器 aneroidograph
膜盒加速度计 diaphragm accelerometer
膜盒控制阀 bellows operated pilot valve
膜盒连接 bellows joint
膜盒量热计 aneroid calorimeter
膜盒流量计 aneroid flowmeter
膜盒密封 bellows seal
膜盒气压表 aneroid barometer
膜盒气压测量计 aneroid surveying barometer
膜盒气压计 aneroid barometer; aneroidograph
膜盒气压记录器 barograph; barometrograph
膜盒气压曲线 aeroidogram
膜盒式 bellows-type
膜盒式差动流量计 bellows differential flowmeter
膜盒式差压流量计 bellows differential flowmeter
膜盒式混合比调节器 aneroid mixture controller
膜盒式加速表 diaphragm type accelerometer
膜盒式量热计 aneroid calorimeter
膜盒式料位控制器 diaphragm box level controller
膜盒式流量计 bellows flowmeter
膜盒式煤气表 diaphragm gas meter
膜盒式气量计 bellow gas meter
膜盒式气体流量计 bellows-type gas flowmeter; diaphragm gas meter
膜盒式气压计 aneroid barometer
膜盒式气压记录器 aeroidograph; aneroid barograph; barograph
膜盒式气压记录仪 aneroid barograph
膜盒式压差计 bellows manometer
膜盒式压力表 capsule ga(u)ge; capsule pressure ga(u)ge; diaphragm manometer; diaphragm type pressure ga(u)ge
膜盒式压力计 bellows ga(u)ge; bellows manometer; capsule-type manometer; diaphragm type pressure ga(u)ge; diaphragm ga(u)ge
膜盒式液位计 diaphragm capsule type level ga(u)ge
膜盒式仪表 bellows instrument; bellows-type meter
膜盒式真空计 capsule-type vacuum ga(u)ge
膜盒形接点 bellow form contact
膜盒压差计 bellows differential ga(u)ge
膜盒压力表 diaphragm barometer
膜盒压力传感器 aneroid capsule; aneroid sensor; bellow
膜盒压力计 bellows pressure ga(u)ge; membrane-case manometer
膜盒液面计 aneroid liquid-level meter
膜盒真空计 capsule ga(u)ge; diaphragm ga(u)ge
膜盒组 capsule stack
膜盒组件 bellows; capsule stack
膜厚 film thickness; build
膜厚测定法 film thickness measuring
膜厚测定仪 elcometer; film thickness measuring device; film thickness tester
膜厚测量 film thickness measuring; measurement of film thickness
膜厚度 film thickness
膜厚度不均 non-uniform film thickness
膜厚计 film thickness ga(u)ge
膜厚监测仪 film thickness monitor
膜厚均匀性 film thickness uniformity
膜厚指示器 film thickness indicator
膜化 membranization
膜化学 membrane chemistry
膜回弹性 film resilience
膜混合集成电路 film hybrid integrated circuit
膜技术 membrane technology
膜接触器 membrane contactor
膜结构 membrane structure
膜界层 membrane boundary layer
膜静电位 resting membrane potential
膜均匀性 film uniformity
膜科学 membrane science
膜孔 iris opening
膜孔电积酌 electrostenolysis
膜孔耦合波长计 iris-coupled wavemeter
膜孔耦合截止衰减器 iris-coupled cut-off attenuator
膜孔耦合滤波器 iris-coupled filter
膜孔苔虫 Menbranipora
膜孔透镜 aperture lens; pin-hold lens
膜孔型腐蚀 pitting corrosion
膜肋加固的内气压系统 <充气建筑> inside pressure pneumatic system reinforced by membrane ribs
膜冷式发动机 film-cooled engine
膜粒 peplomer
膜连 symphysis
膜料 coating material
膜流体性 membrane fluidity
膜滤 membrane filtration
膜滤法 membrane filter technique
膜滤活性污泥法 membrane filtration activated sludge process
膜滤技术 membrane filtration technique
膜滤器 membrane filter; molecular filter
膜滤器标准板式计数 membrane filter standard plate count
膜滤器隔室 membrane filter chamber
膜滤器计数 membrane filter count
膜滤器技术 membrane filter technique
膜滤器室 membrane filter chamber
膜滤器指标 membrane filter index
膜滤特性 membrane filtration characteristic
膜滤指标 membrane filtration index
膜螺旋板 membranous spiral lamina
膜密封式机械泵 membrane seal mechanical pump
膜面 face; membrane surface
膜模型 membrane model; model membrane
膜黏[粘]度 film viscosity
膜耦合活性污泥法 membrane coupled activated sludge process
膜耦合活性污泥系统 membrane coupled activated sludge system
膜泡传输 membrane vesicle transport
膜泡塑模冷却 bubble mo(u)ld cooling
膜皮 hymeniderm
膜片 capsule; film; iris diaphragm; membrane; diaphragm
膜片泵 diaphragm pump
膜片泵喷雾机 diaphragm sprayer
膜片避雷器 film lightning arrester
膜片操作控制阀 diaphragm-operated control valve
膜片电位 membrane potential
膜片堆 diaphragm pile; diaphragm stack
膜片阀 diaphragm valve; membrane valve
膜片阀控制电动机 diaphragm motor
膜片刚度 membrane tension
膜片激励器 diaphragm actuator
膜片加载波导 diaphragmatic waveguide
膜片胶粘剂 supported film adhesive
膜片孔径 iris aperture
膜片喇叭 diaphragm horn
膜片密封渗漏损失 diaphragm packing leakage loss
膜片耦合式波长计 iris-coupled wavemeter
膜片喷射放大器 diaphragm-ejector amplifier
膜片气动有效容量 effective pneumatic capacitor
膜片式 diaphragm type
膜片式差压计 diaphragm type differential pressure ga(u)ge
膜片式储能器 diaphragm type accumulator
膜片式动力制动器 diaphragm type power brake
膜片式堵管装置 iris stop equipment
膜片式风箱 membrane
膜片式风压表 diaphragm blast ga(u)ge; diaphragm draught ga(u)ge; diaphragm type wind pressure ga(u)ge
膜片式计量泵 diaphragm type metering pump
膜片式计量器 diaphragm meter
膜片式继电器 diaphragm relay
膜片式加速表 diaphragm type accelerometer
膜片式静压计 static plate manometer
膜片式连续液位检测器 continuous diaphragm type level detector
膜片式逻辑 membrane logic
膜片式煤气表 diaphragm type gas meter
膜片式密封 diaphragm seal
膜片式密封流量计 diaphragm seal flowmeter
膜片式泥浆泵 diaphragm type sludge pump
膜片式耦合滤波器 iris-coupled filter
膜片式喷漆器 diaphragm type paint sprayer
膜片式气量计 diaphragm type gas meter
膜片式气压计 diaphragm barometer
膜片式燃料泵 diaphragm fuel pump
膜片式(燃)油泵 diaphragm type fuel pump
膜片式热交换器 plate heat exchanger
膜片式微压计 diaphragm type micromanometer
膜片式蓄能器 diaphragm accumulator
膜片式压力表 diaphragm manometer
膜片式压力传感开关 diaphragm pressure switch
膜片式压力传感器 diaphragm type pressure sensor
膜片式压力计 diaphragm type pressure ga(u)ge
膜片式压力平衡检测器 diaphragm type pressure balance detector
膜片式压缩机 diaphragm type compressor
膜片式液位计 diaphragm level ga(u)ge; diaphragm type level meter
膜片式液压保险 hydraulic fuse
膜片式应变计 diaphragm strain ga(u)ge
膜片式真空计 diaphragm vacuum ga(u)ge
膜片弹簧 diaphragm spring
膜片弹簧(式)离合器 diaphragm spring clutch
膜片调节阀 diaphragm regulating valve
膜片调节器 diaphragm regulator
膜片型元件 moving diaphragm element
膜片压力表 plate manometer
膜片压力计 plate manometer
膜片压模 pressure unit
膜片压圈 diaphragm ring
膜片摇臂室 rotachamber
膜片运动 motion of membrane
膜片张力 membrane tension
膜片胀缩接合 diaphragm expansion joint
膜片振动 diaphragm oscillation
膜片振动音 diaphragm-transmitted sound
膜片致动器 diaphragm actuator
膜片作用 iris action
膜平衡 membrane equilibrium
膜破裂 film rupture
膜谱学 membrane spectroscopy
膜曝气 membrane aeration
膜曝气生物反应器 membrane aeration bioreactor
膜曝气生物膜反应器 membrane aerated biofilm reactor
膜气提法 membrane air stripping
膜强度 film toughness
膜清洗 membrane cleaning
膜热阻 film resistance
膜容量 membrane capacity
膜融合 membrane fusion
膜入水轮 nebula covering the pupil
膜色谱 membrane chromatography
膜渗平衡 membrane balance; membrane equilibrium
膜生物反应器 membrane biologic(al) reactor; membrane bioreactor
膜生物反应器污着 membrane biologic(al) reactor fouling; membrane bioreactor fouling
膜生物反应器系统 membrane biologic(al) reactor system; membrane bioreactor system
膜生物技术 membrane biologic(al) technology

M

膜生物膜反应器 membrane biofilm reactor
膜生物学 membrane biology
膜式泵 membrane pump;surge pump
膜式舱柜 membrane tank
膜式操作 film-wise operation
膜式传感器 diaphragm transducer
膜式弹簧 diaphragm type spring
膜式电阻器 film resistor
膜式阀 diaphragm valve
膜式反应器 film reactor
膜式过滤器 membrane filter
膜式和活塞式水泵 diaphragm and piston pump
膜式磺化器 film sulfonator
膜式计量泵 diaphragm metering pump
膜式加速度计 diaphragm accelerometer
膜式冷凝 film type condensation
膜式冷却器 Baudelot type cooler; film type cooler
膜式气化炉 film type boiler
膜式施工用泵 diaphragm contractor's pump
膜式水泵 diaphragm pump
膜式水冷壁 fin panel casing; membrane wall
膜式水冷壁燃烧器 film-cooled combustion chamber
膜式水力测压器 hydraulic capsule
膜式探测器 membrane detector
膜式填料 film packing
膜式通风计 diaphragm draught ga(u)ge
膜式吸收器 membrane type absorber
膜式吸收塔 membrane type absorbing tower
膜式压力计 membrane manometer; membrane type manometer
膜式压敏电阻器 film varistor
膜式织物支撑 membrane type fabric support
膜式转换器 diaphragm transducer
膜试验 film test
膜受体 membrane receptor
膜酸 film acid
膜态沸腾 film boiling
膜态沸腾区 film-boiling range
膜态冷凝 film condensation
膜态流动 film flow
膜弹性 film elasticity
膜天平 film balance
膜通量 membrane flux
膜通透性 membrane permeability
膜透率 membrane permeability
膜退化 membrane degradation
膜脱盐法 membrane desalting
膜位移 membrane displacement
膜位置 position of membrane
膜稳定作用 membrane stabilization; membrane stabilizing action
膜稳态通量 membrane steady-state flux
膜污染 membrane pollution
膜污着 membrane fouling
膜污着层 membrane fouling layer
膜污着成因 cause of membrane fouling
膜污着机理 membrane fouling mechanism
膜污着控制 membrane fouling control
膜污着物 membrane foulant
膜污着阻力 membrane fouling resistance
膜吸收 membrane absorption
膜吸收系统 membrane absorption system
膜吸液滤器 membrane syringe filter
膜系数 film coefficient
膜下腐蚀 under-film corrosion
膜下锈蚀 underrusting
膜下致密层 hypolemmal compact layer
膜相结构 meinbranous structure
膜效率 membrane efficiency
膜形态学 leptonomorphology
膜型 membranous type
膜修复 film healing
膜序批间歇式反应器 membrane anaerobic reactor
膜选择性 membrane selectivity
膜学 hymenology
膜学说 membrane theory
膜压 surface pressure
膜压壳型机 diaphragm shell mo(u)lding machine
膜压力 film pressure
膜压造型 diaphragm mo(u)lding
膜压造型机 diaphragm mo(u)lding machine
膜厌氧反应器 membrane anaerobic reactor
膜厌氧反应器系统 membrane anaerobic reactor system
膜样的 membraniform;membranoid
膜一体化技术 membrane integration technique
膜印片 mo(u)lded tablet
膜应变 membrane strain
膜应力 diaphragm stress; membrane stress
膜元(件) membrane element
膜运载 membrane transport
膜再循环假说 membrane recycling hypothesis
膜振动本征模式 proper mode of membrane vibration
膜蒸发 pervapo(u)rization
膜蒸馏 membrane distillation
膜脂质相变 membrane lipid phase transition
膜制备 film preparation
膜质的 membranous
膜中毒 membrane poisoning
膜重测定 coating weight
膜状的 membraniform;membranous
膜状沸腾 film boiling
膜状过程 film process
膜状胶粘剂 film adhesive
膜状冷凝 film condensation; filmwise condensation
膜状凝结 film condensation
膜状色 film colo(u)r
膜状物 membranoid substance
膜阻力 membrane resistance
膜组件 membrane module

摩

摩必尔转化法 Mobire transformation method

摩擦 attrite;attrition;chafe;dragging; fractionating; frictionate; frictionize;grating;rubbing
摩擦板 chafing plate; friction(al) plate;bulldog plate <防止搭接木料移动>
摩擦板插 chafing iron pocket
摩擦拌流 <船舶航行时的> friction(al) wake
摩擦保护层 friction(al) coat
摩擦背轮 friction(al) back gear
摩擦倍率 friction(al) multiplier
摩擦泵 friction(al) pump; Wesco pump
摩擦比降 friction(al) slope
摩擦臂测定法 friction(al) arm method
摩擦边界层 friction(al) boundary layer
摩擦变速钻床 friction(al) drilling machine
摩擦变形 friction(al) texturizing
摩擦变形量 <摩擦焊的> burn-off length
摩擦变形速度 burn-off rate
摩擦变质作用 friction(al) metamorphism
摩擦表面 friction(al) surface
摩擦补偿 friction(al) compensation
摩擦材料 derodo; ferrodo; friction(al) material
摩擦材料试验机 friction(al) material test machine
摩擦操纵转向机构 friction-controlled steering mechanism
摩擦测功器 friction(al) dynamometer;Prony brake
摩擦测力计 friction(al) brake; friction(al) dynamometer
摩擦层 friction(al) layer; wearing course
摩擦插片 ferrodo
摩擦差 friction(al) drop
摩擦差微效应 differential friction(al) effect
摩擦差异 differential friction
摩擦(产生的)水头损失 loss of head in friction
摩擦常数 constant of friction;friction(al) constant
摩擦衬块 lining pad
摩擦衬面 lining of friction
摩擦衬片 friction(al) facing;friction(al) lining
摩擦衬片表面 lining surface
摩擦衬片磨合 lining break-in
摩擦衬片装合工具 relining tool
摩擦衬套 friction(al) slip
摩擦齿轮 friction(al) gear
摩擦充电 triboelectric(al) charging
摩擦触点 rubbing contact
摩擦传动 friction(al) transmission; slip gearing;friction(al) drive
摩擦传动的筒子 friction(al) driven spools
摩擦传动辊 friction(al) roll
摩擦传动绞车 friction(al) geared winch;friction(al) winch
摩擦传动轮 friction(al) gearing wheel;idle wheel
摩擦传动装置 friction(al) gear(ing)
摩擦窗撑 friction(al) stay
摩擦锤 friction(al) hammer; jump hammer
摩擦搭扣 friction(al) catch
摩擦打捞器 friction(al) socket
摩擦打捞筒 friction(al) socket
摩擦带 black tape;friction(al) band; friction(al) belt;friction(al) tape
摩擦带电 friction(al) electrification; triboelectric(al) charging
摩擦带电成像 triboelectric(al) imaging
摩擦带电方法 triboelectric(al) means
摩擦带电关系 triboelectric(al) relationship
摩擦带电接触 triboelectric(al) contact
摩擦带电特性 triboelectric(al) characteristic
摩擦带电效应 triboelectric(al) effect
摩擦带电性能 triboelectric(al) behavio(u)r; triboelectric(al) property
摩擦带电作用 triboelectric(al) action
摩擦带离合器 friction(al) band clutch
摩擦带式提升机 belt friction(al) elevator
摩擦挡 friction(al) catch
摩擦挡布 chafing path
摩擦导布辊 pinch roller
摩擦导卫板 friction(al) guide
摩擦的 friction(al); fricative; grinding
摩擦等离子体 triboplasma
摩擦点 friction(al) point
摩擦点火 friction(al) ignition
摩擦点火管 friction(al) primer
摩擦点火器 friction(al) igniter
摩擦电 franklinic electricity;friction(al) electricity;triboelectricity
摩擦电荷 triboelectric(al) charge
摩擦电机 friction(al) electrical machine
摩擦电键 triboelectric(al) bond
摩擦电流 friction(al) working current
摩擦电偶 tribocouple
摩擦电亲和力 triboelectric affinity
摩擦电双层 triboelectric(al) double layer
摩擦电吸引力 triboelectric(al) attraction
摩擦电序列 triboelectric(al) series
摩擦垫圈 friction(al) washer
摩擦吊锚架 friction(al) cathead
摩擦掉色 crocking
摩擦定律 law of friction
摩擦动力系数 kinetic coefficient of friction
摩擦端承桩 friction(al) end bearing pile
摩擦发电 triboelectricity
摩擦发光 triboluminescence
摩擦发光的 triboluminescent
摩擦发光发射 triboemission
摩擦发火 ignition by friction
摩擦发热点 hot point;hot spot
摩擦发音 stridulate;stridulation
摩擦翻木机 friction(al) nigger
摩擦粉料 friction-type mix
摩擦风 antitriptic wind
摩擦缝 friction(al) joint
摩擦腐蚀 friction(al) corrosion; frottage (corrosion); chafing corrosion;fretting corrosion;wear oxidation
摩擦腐蚀压痕 false brinelling
摩擦杆 friction(al) lever
摩擦杠杆式 friction(al) lever type
摩擦功 friction(al) work; rubbing work
摩擦功率 friction(al) horsepower; friction(al) power
摩擦功率计 friction(al) dynamometer
摩擦构件 friction(al) member
摩擦轨 friction(al) rail
摩擦滚动舱口盖 friction(al) rolling hatch cover
摩擦滚轮 friction(al) roller
摩擦滚筒 friction(al) roller
摩擦滚轴 friction(al) roller
摩擦滚柱 friction(al) roller
摩擦滚柱传动(装置) friction(al) roller drive
摩擦海流 friction(al) current
摩擦焊(接) friction(al) welding; spin welding
摩擦焊接机 friction(al) welding machine
摩擦荷载 friction(al) load(ing) <船舶与靠泊面之间的>;rubbing load <船舶和护舷之间的>
摩擦痕 fricative track
摩擦痕迹 abrasion mark
摩擦滑动式离合器 friction(al) slip coupling

摩擦滑动作用 friction(al) sliding
摩擦滑轮 friction(al) pulley
摩擦化学 tribo-chemistry
摩擦环 balk ring; drag ring; friction-(al) ring; rubbing ring
摩擦缓冲弹簧 friction(al) draft spring
摩擦缓冲器 friction(al) draft gear
摩擦火花 friction-sparking
摩擦机构 friction(al) mechanism
摩擦积分器 ball-and-disk integrator
摩擦极限 limit of friction
摩擦极限角 limiting angle of friction
摩擦极限设计 friction(al) limit design
摩擦棘轮止动器 friction-ratchet stop
摩擦计 friction(al) ga(u)ge; tribometer
摩擦剂 abradant
摩擦加捻 friction(al) twisting
摩擦加捻拉伸变形机 friction-twist draw-texturing machine
摩擦加热 friction(al) heating
摩擦加速度反应谱 friction(al) acceleration response spectrum
摩擦夹板 friction(al) clamping plate
摩擦夹紧装置 friction(al) grip
摩擦夹盘 friction(al) chuck
摩擦夹片 friction(al) jaws
摩擦夹钳 friction(al) clamp
摩擦减振器 snubber
摩擦减震 damping by friction
摩擦减震器 friction(al) damper; friction(al) disk shock absorber; snubbed
摩擦剪节理 friction(al) shear joint
摩擦剪力 shear friction
摩擦角 angle of friction; friction(al) angle
摩擦角砾岩 friction(al) breccia
摩擦铰链 friction(al) hinge
摩擦搅拌 attrition mixing
摩擦接触 friction(al) contact; rubbing contact
摩擦接点 wipe contact; wiping contact
摩擦接头 friction(al) joint
摩擦紧固螺栓 friction(al) grip bolt
摩擦静电粉末喷涂 triboelectrostatic powder spray
摩擦静电喷枪 triboelectrostatic gun
摩擦锯 friction(al) saw
摩擦锯床 friction(al) saw machine
摩擦锯切 friction(al) sawing
摩擦卷筒 friction(al) drum; friction-(al) roll
摩擦卷扬器 friction(al) windlass
摩擦抗力 friction(al) resistance
摩擦块 friction(al) block; rubbing block
摩擦扩散 friction(al) dissipation
摩擦离合器 friction(al) clutch
摩擦离合器副轴 countershaft clutch type; friction(al) countershaft
摩擦离合器外盘 friction(al) clutch cup
摩擦离合器总成 friction(al) clutch assembly
摩擦离合制动 friction(al) clutch
摩擦力 friction; force of friction; friction(al) force; confriction
摩擦力测定仪 tribometer
摩擦力测量术 tribometry
摩擦力测量仪 tribometer
摩擦力方向 direction of friction
摩擦力界 friction(al) field
摩擦力矩 friction(al) moment; moment of friction
摩擦力矩测试仪 friction(al) torque testing instrument
摩擦力控制 friction(al) pressure control
摩擦力调节器 friction(al) adjuster
摩擦力止动的窗撑 woodlock
摩擦连接 friction(al) bond; friction-(al) joining
摩擦联结器 friction(al) clutch
摩擦联轴节 friction(al) coupling
摩擦裂缝 friction(al) crack
摩擦裂纹 rubbing crack
摩擦裂隙 friction(al) crack
摩擦流(动) friction(al) flow
摩擦流速 friction(al) velocity
摩擦轮 friction(al) pulley; friction-(al) wheel; Koepe pulley; sheave of friction
摩擦轮传动 friction(al) wheel drive
摩擦轮传动的混凝土搅拌机 friction-(al) wheel-drive concrete mixer
摩擦轮垫 surging pad
摩擦轮和钢丝直径比 sheave-to-wire ratio
摩擦轮积分器 friction(al) wheel integrator
摩擦轮离合器 friction(al) pulley clutch
摩擦轮驱动 friction(al) wheel drive
摩擦轮驱动的混凝土搅拌机 friction-(al) wheel-drive concrete mixer
摩擦轮式积分器 wheel-and-disc integrator
摩擦轮式转速计 friction(al) wheel tachometer
摩擦轮送纸器 friction(al) wheel feeder
摩擦轮提升机 friction(al) winder
摩擦轮转速计数器 friction(al) wheel speed counter
摩擦螺旋压机 friction(al) screw press
摩擦马力 friction(al) horsepower
摩擦锚夹 friction(al) grip
摩擦锚夹栓接(头) friction(al) grip bolt joint
摩擦面 rubbed surface; rubbing surface; surface of contact; surface of friction; friction surface
摩擦面表面釉光化 lining glaze
摩擦面的粗糙度 minute projections
摩擦面焊合(点) interfacial weld
摩擦面积 friction(al) area; rubbing area
摩擦敏感度 sensitiveness to friction
摩擦黏[粘]附分选 tribo-adhesion separation
摩擦扭矩 friction(al) torque
摩擦盘式变速器【机】friction(al) disk type transmission
摩擦抛光 burnishing
摩擦抛光机 friction(al) glazed machine
摩擦抛光轮 burnishing wheel
摩擦抛光器 burnisher
摩擦片 ferodo; friction(al) disk[disc]; friction(al) lining; friction(al) wafer; friction (al) plate; wearing piece; wearing plate
摩擦片表面釉光化 lining glaze
摩擦片离合器 friction(al) plate clutch
摩擦片调节 wearing plate adjustment
摩擦坡(降) <水流的> friction(al) slope
摩擦坡角 friction(al) slope
摩擦起步 <汽车的> friction(al) start
摩擦起电 triboelectrification
摩擦起电器 friction(al) electrical machine
摩擦起弧 scratch start
摩擦起火 produce fire by friction
摩擦牵引系数 friction(al) drag coefficient
摩擦强度 friction(al) strength
摩擦侵入 fretting corrosion
摩擦清除表面脏物 abrasive wear-(ing)
摩擦清洁表面法 abrasive cleaning method
摩擦球 friction(al) ball
摩擦去皮机 abrasive peeler
摩擦圈 friction(al) ring
摩擦热 friction(al) heat; heat due to friction; heat of friction
摩擦热能 friction(al) heat energy
摩擦热损失 friction(al) heat loss
摩擦刹车 friction(al) brake
摩擦刹车片 friction(al) shoe
摩擦深度 friction(al) depth
摩擦生热 excite heat by friction
摩擦声 grating
摩擦式安全离合器 friction-type safety clutch
摩擦式安全器 friction(al) trip
摩擦式弹簧锁 friction(al) snap latch
摩擦式缓冲装置 friction(al) buffer
摩擦式加捻体系 friction(al) based twisting system
摩擦式假捻器 friction(al) twister
摩擦式减振器 friction(al)-type shock absorber
摩擦式绞车 friction(al) hoist; friction-(al) windlass
摩擦式金属支架 friction(al) metal chock
摩擦式离合器 slip clutch
摩擦式连接 <螺栓连接> friction-type connection
摩擦式锚定 friction-type anchorage
摩擦式锚杆 friction(al) bolt
摩擦式锚基 friction-type anchorage
摩擦式锚具 friction-type anchorage
摩擦式清洗器 friction(al) cleaner
摩擦式清选机 friction(al) separator
摩擦式释放器 friction(al) release
摩擦式提升 friction(al) hoisting; friction(al) winding; Koepe hoisting
摩擦式调节器 friction(al) governor
摩擦式推钢机 friction(al) pusher
摩擦式喂纱装置 friction(al) yarn feed installation
摩擦式扬声器 friction(al) driven loudspeaker
摩擦式应变计 friction(al) strain ga-(u)ge
摩擦式毡垫 <翼锭细纱机> friction-type felt cartridge
摩擦式真空规 friction(al) type vacuum ga(u)ge
摩擦式真空计 friction(al) type vacuum ga(u)ge
摩擦式织针 friction-type knitting needle
摩擦式转数表 friction(al) revolution counter
摩擦式转速计 air braking tachometer; friction(al) tachometer
摩擦式自动给纸装置 friction(al) feeder
摩擦式阻尼器 friction(al) damper
摩擦试验 friction(al) test; rub test
摩擦试验机 friction(al) testing machine
摩擦衰减 friction(al) attenuation
摩擦水头 friction(al) head
摩擦水头损失 friction(al) head loss
摩擦松脱(安全)器 friction(al) release
摩擦速度 <流水沿壁面的> friction-(al) velocity; friction(al) speed; rubbing speed; rubbing velocity
摩擦速率 rubbing speed
摩擦损耗 friction(al) loss; friction-caused wear; rubbing wear
摩擦损耗系数 friction(al) loss factor
摩擦损伤 friction(al) damage
摩擦损失 friction(al) lass; friction-(al) loss; loss due to friction; loss of friction
摩擦(损失的)压头 friction(al) head
摩擦损失水头 friction(al) loss of (water) head; friction head
摩擦损失系数 friction(al) loss factor
摩擦弹簧 friction(al) spring
摩擦套 friction(al) coat; friction(al) sleeve
摩擦提升(法) friction(al) winding
摩擦提升机 friction(al) (drive) hoist; friction(al) winder
摩擦筒提升机 friction(al) drum winder
摩擦涡度 friction(al) vorticity
摩擦涡轮机 drag turbine
摩擦物理学 tribophysics
摩擦误差 friction(al) error
摩擦系数 coefficient of friction; constant of friction; friction(al) coefficient; friction(al) factor
摩擦系数测定器 friction(o) meter
摩擦系数测定仪 friction(al) meter; friction(al) tester; friction(o) meter
摩擦橡胶缓冲器 rubber friction draft gear
摩擦消耗马力 friction horsepower
摩擦效应 friction(al) effect
摩擦楔 bolster control wedge; control wedge; friction(al) casting
摩擦型 friction-type
摩擦型连接 friction-type connection
摩擦型锚杆 friction(al) rock bolt; friction-type bolt
摩擦型装置 friction-type unit
摩擦性间接费用 friction(al) overhead
摩擦性能 friction(al) behavio(u)r; nature of friction
摩擦性失业 <劳动力市场职能上缺陷所造成的临时性失业> friction-(al) unemployment
摩擦性质 friction(al) property; nature of friction
摩擦选矿 friction(al) separation
摩擦选矿法 friction(al) separation method
摩擦学 tribology
摩擦压床 friction(al) press
摩擦压光机 friction(al) press
摩擦压光纸板 friction(al) glazed paper board
摩擦压降 friction(al) pressure drop
摩擦压力 friction(al) pressure
摩擦压力机 friction(al) (screw) press
摩擦压力降 friction(al) drop
摩擦压力控制 friction(al) pressure control
摩擦压力损失 friction(al) pressure loss
摩擦压制 friction(al) calendaring
摩擦压砖机 friction(al) press
摩擦堰 friction(al) weir
摩擦曳力 friction(al) drag
摩擦仪表 friction(al) meter
摩擦因数 rubbing factor
摩擦因素 friction(al) factor
摩擦因子 friction(al) factor
摩擦音 crunch
摩擦引起的漆膜失光 sweating
摩擦引起的预应力损失 loss of pre-

stress friction;prestressing loss due to friction
摩擦应力 friction(al) stress; stresses due to friction
摩擦影响 friction(al) influence
摩擦影响深度 depth of friction(al) influence
摩擦预应力损失 friction(al) pressure loss
摩擦圆 friction(al) circle
摩擦圆法 friction(al) circle method
摩擦圆分析法 friction(al) circle analysis
摩擦圆盘 friction(al) circle; friction(al) disk[disc]
摩擦圆盘锯 friction(al) disk saw
摩擦圆盘离合器 friction(al) circle clutch;friction(al) disk clutch
摩擦圆盘座 friction(al) disk seat
摩擦造成的粗糙面 galling
摩擦噪声 friction(al) noise
摩擦轧点 friction(al) nip
摩擦轧光 friction(al) calendaring; friction(al) glazing
摩擦轧光机 friction(al) calendar; glazing calender
摩擦轧光细布 glazed calico
摩擦轧光艳丽花布 glazed chintz
摩擦轧光整理 chintz finish; glazed chintz finish;graged finish
摩擦闸 friction(al) lock
摩擦闸箍 friction(al) band brake
摩擦闸瓦 friction(al) brake; friction(al) dogs
摩擦闸摇柄 friction(al) brake handle
摩擦支承 friction(al) bearing
摩擦支柱 friction(al) prop
摩擦支座 friction(al) support
摩擦指数 friction(al) index; index of friction;rubbing index
摩擦制动鼓 friction(al) brake drum
摩擦制动机 friction(al) brake
摩擦制动(器) friction(al) brake
摩擦制动系统 friction(al) brake control system
摩擦制动旋转 friction(al) braked rotation
摩擦制动装置 friction(al) catch; friction(al) clamp;friction(al) skidding device;friction(al) stopping device
摩擦制品 friction(al) goods
摩擦致热量 heat of friction
摩擦滞后 friction(al) lag
摩擦中心 center[centre] of friction
摩擦轴 friction(al) axis; friction(al) mandrel;friction(al) shaft
摩擦轴承 friction(al) bearing
摩擦轴承接触表面 friction-bearing surfaces
摩擦柱帽 friction(al) cap
摩擦爪 friction(al) dogs; friction(al) pawl
摩擦桩 friction(al) pile;buoyant pile; floating pile
摩擦桩基(础) floating pile foundation;friction(al) (pile) foundation
摩擦锥 cone of friction
摩擦锥轮 friction(al) cone
摩擦着的 abradant
摩擦阻抗 friction(al) drag
摩擦阻力 friction(al) drag; initial resistance; resistance of friction; resistance to friction; friction(al) resistance
摩擦阻力矩 friction(al) resistance moment
摩擦阻力头曲线 graph of friction(al) resistance curve
摩擦阻力因数 friction(al) factor
摩擦阻尼 friction(al) damping

摩擦作用 friction(al) action; rubbing effect
摩电式测车器 triboelectric(al) detector
摩尔 mol;mole;molecule
摩尔百分数 mole per cent
摩尔扳手 Mole wrench
摩尔包络曲线 intrinsic(al) curve
摩尔比 molar ratio;mol ratio
摩尔比(率)法 mole ratio method
摩尔比热 molal specific heat; molar specific heat; molecular heat; molecular specific heat;mole heat
摩尔表面能 molar surface energy
摩尔冰点下降常数 molar depression constant
摩尔布里登陆港 < 二战时期盟军登陆港 > Mulberry invasion harbo(u)r
摩尔布里港 < 用预制块筑成的人工港 > Mulberry harbo(u)r
摩尔臭气度 molar olfactometry
摩尔臭味单位 molar olfactory
摩尔磁化率 molar susceptibility
摩尔单位 molar unit
摩尔的 molecular
摩尔等张比容 molar parachor
摩尔等张体积 parachor
摩尔电导 molar conductance
摩尔电导率 molar conductivity
摩尔电荷 molar charge
摩尔电化学能 molar electrochemical energy
摩尔电化学势能 molar electrochemical potential energy
摩尔沸点上升 molar elevation of boiling point
摩尔沸点升高 molar elevation of boiling
摩尔沸点升高常数 molar elevation constant
摩尔分数 molar fraction; mole fraction;molfraction
摩尔丰度 molar abundance
摩尔干格尼柱 Morgagni's columns
摩尔构型熵 molal configurational entropy
摩尔光吸收 molar absorbance; molar optic(al) absorbance
摩尔焓 molar enthalpy
摩尔(焓熵)图 Mollier chart; Mollier diagram
摩尔豪斯研磨机 Morehouse mill
摩尔化学计量反应 mol stoichiometric(al) reaction
摩尔吉布斯能量 molar Gibbs energy
摩尔极化度 molar polarization
摩尔建筑式(装饰) Moorish
摩尔量 gram molecular weight; molal weight; molar amount; molar weight
摩尔量热函数 molar heat content
摩尔内聚能 molar cohesion energy
摩尔内能 molar internal energy
摩尔能量 energy per mole
摩尔凝固点降低 molar depression of freezing point
摩尔浓度 molar concentration; molarity;molar ratio
摩尔平均沸点 molar average boiling point
摩尔破坏(圆)图解 Mohr's rupture diagram
摩尔气体常数 molar gas constant; mole gas constant
摩尔汽化热 molar heat of vapo(u)rization
摩尔热电偶 Mole thermopile; Moll thermopile
摩尔热函 molar heat content; molecular heat capacity

摩尔热函数 molar heat content
摩尔热力学能量 molar thermodynamic(al) energy
摩尔热容 molar heat capacity; molecular heat capacity; moll heat capacity
摩尔溶解度 molar solubility
摩尔溶液 gram molecular solution; molar solution
摩尔色散 molar dispersion
摩尔熵 molar entropy
摩尔渗透压浓度 Morie osmolarity
摩尔式阿拉伯花饰 Moorish arabesque
摩尔式多叶形拱 Moorish multifoiled arch
摩尔式风格的大型公共建筑 < 中古时代北非或西班牙的 > Moorish palace
摩尔式宫殿 Moorish palace
摩尔式拱 Moorish arch
摩尔式建筑 Moorish architecture; Moorish style architecture;Moresque
摩尔式马蹄形发券 Moorish horseshoe arch
摩尔式马蹄形拱 Moorish horseshoe arch
摩尔式穹顶 Moorish cupola
摩尔式穹隆 Moorish dome
摩尔式圆顶 imperial cupola
摩尔式圆屋顶 Moorish cupola
摩尔式柱顶 Moorish capital
摩尔式柱头 Moorish capital
摩尔数 mole number
摩尔数比 mole ratio
摩尔数法 molar number method
摩尔数量 molal quantity
摩尔体积 molal volume; molar volume; molecular volume; mole volume
摩尔条纹 Morie fringe
摩尔条纹技术 Morie fringes technique
摩尔维特投影 Mollweide's projection
摩尔温差电堆 Moll thermopile
摩尔吸光系数 molar absorbancy index; molar absorption coefficient; molar absorptivity
摩尔吸收率 molar absorptivity
摩尔吸收系数 molar absorptivity
摩尔系数 mole coefficient
摩尔响应 mole-basis response
摩尔消光系数 molar extinction coefficient
摩尔性质 molar property
摩尔旋光度 molar rotation
摩尔选择性系数 < 离子交换 > molal selectivity coefficient
摩尔质量 molar mass
摩尔自由能量 molar free energy
摩尔族建筑 Moorish style architecture
摩根 < 荷兰等国面积单位,1 摩根 = 8565.18 平方米 > morgen
摩根斯坦边坡稳定性分析法 Morgenstern method of slope stability analysis
摩耗 friction(al) loss
摩耗层 < 道路 > top wearing surface
摩加迪沙 < 索马里首都 > Mogadishu
摩羯座 Capricorn;Capricornus
摩拉瓦学校 Morava school
摩勒姆式体系建筑 Mowlem
摩立登湾豆树 Moreton Bay chestnut
摩林氏表 < 列有各种静止角和摩擦系数 > Morin's tables
摩伦迪锚 < 意大利莫兰第 > Morandi anchorage
摩洛哥面砖 Morocco faced tile

摩洛哥式柱头 capital of Moorish column
摩洛哥针头玻璃 pinhead Morocco
摩-马公司试验法 Morgan-Marshall test
摩面联结器 friction(al) coupling
摩纳哥市 < 摩纳哥首都 > Monaco-Ville
摩莎拉布式建筑 < 九世纪以后建的西班牙建筑 > Mozarabic architecture
摩斯雾信号 Morse code fog signal
摩损极限 wear limit
摩损期限 wear-life
摩损寿命长的 long wearing
摩天办公大楼 office skyscraper
摩天大楼 skyscraper; super-skyscraper
摩天大楼公寓 skyscraper block
摩天大楼林立的城市 skyscraper city
摩天大厦 skyscraper
摩天的 cloud-kissing
摩天楼 gratte-ciel; tower building; rock pile < 俚语 >
摩天楼城市 vertical city
摩托车 autobike; motor bike; motorcar; motor bicycle; autobicycle; motorcycle;autocycle < 美 >
摩托车比赛场 motorcycle-racing arena;velodrome < 室内的 >
摩托车边车 side car body
摩托车场 cycle stand
摩托车车道加边 cycle path edging
摩托车的边车 bathtub
摩托车的跨斗 side-car
摩托车后座 pillion
摩托车驾驶员的全套衣帽装备 harness
摩托车赛跑道 speed way
摩托车手 motor cyclist
摩托车司机 motor man
摩托车胎 cycle tyre
摩托车停放处 cycle park
摩托车无线电设备 motorcycle radio
摩托车用铅酸蓄电池 motorbicycle lead-acid storage battery
摩托车越野赛 moto-cross
摩托车组 motor unit
摩托车组制动机 railcar brake
摩托化 motorization; motorize
摩托化比率 motorization rate
摩托化扫雪机 motorized snowplow
摩托化水准测量 motorized level(l)ing
摩托客车 motor coach
摩托快艇 skimming dish
摩托罗拉电子定位系统 Motorola electronic positioning system
摩托棚车 motor van
摩托艇 autoboat; cabin cruiser; motor boat; motor dory; motor lorry; power-boat
摩托艇码头 marina
摩托雪橇 snow mobile
摩托凿岩机 motor drill
摩托自行车 bis-motor
摩阻 friction(al) resistance
摩阻板 friction(al) slab
摩阻比 friction(al) ratio; friction-cone resistance ratio; sleeve-cone resistance ratio
摩阻变化 friction(al) change; friction(al) variation
摩阻层 friction course
摩阻传动器 friction(al) clutch
摩阻挡 friction(al) catch
摩阻副流 friction(al) secondary flow
摩阻固定 anchorage by friction
摩阻夹紧器(件) < 如高强螺栓 > friction(al) grip fastener
摩阻力 friction resistance

M

摩阻力锚固 anchorage by friction
摩阻联结器 friction(al) coupling
摩阻量水堰 friction(al) measuring weir
摩阻流速 friction(al) velocity
摩阻锚固 anchoring by friction; friction(al) anchorage
摩阻面层 friction(al) course
摩阻强度 friction(al) strength
摩阻刹车 friction(al) brake
摩阻式螺栓接头 friction(al) type bolted joint
摩阻数 friction(al) number
摩阻水头 friction(al) head; resistance head
摩阻水头损失 loss of friction head
摩阻损失 friction(al) lass; friction(al) loss
摩阻梯度 gradient of friction
摩阻系数 coefficient of friction((al) resistance); friction(al) coefficient; friction(al) factor; friction(al) wobble coefficient <管道的>
摩阻性土(壤) friction(al) soil
摩阻续纸装置 friction(al) feeder
摩阻应力 friction(al) stress
摩阻应力损失 stress loss due to friction
摩阻圆 friction(al) circle
摩阻圆法 friction(al) circle method
摩阻圆锥 cone of friction
摩阻制动器 friction(al) brake
摩阻桩 friction(al) pile

磨

磨 grind; hogger; polish; rub; skive; wear; whet; skive <宝石表面的>

磨板 nog plate
磨板机用瓷球 porcelain ball for graining machine
磨板拉铁 form tie
磨版机 graining machine; plate grinder; plate grinding machine
磨版砂 graining sand
磨版刷 polishing brush
磨版细度 fineness of grind
磨棒 burnisher
磨棒装入口 rodding hole
磨边 edge grinding; edging
磨边残留 shiner
磨边残屑 edge-arrissing residue
磨边倒角机 chamfering machine
磨边工序 edging operation
磨边机 edge grinding machine; edge smoothing machine; glass edging machine; machine for edging
磨边机器 edger
磨边镜玻璃 bevel(l)ed mirror glass
磨边镜子 bevel-edge mirror
磨边锯 ground off saw
磨边磨角开槽机 chamfering machine
磨边设备 edger unit
磨边石 edging stone
磨边装置 edge deleting device
磨冰机 ice mill
磨饼 mill cake
磨玻璃或宝石的油灰粉 glass putty
磨薄 wear down
磨薄边缘 feather edging
磨薄机 drill drift
磨槽 millcourse; mill race
磨槽机 notch grinder
磨衬 mill lining
磨成凹痕 worn into ruts
磨成粉 flour; grind down; powdering
磨成粉的 ground-up
磨成粉状的 powdered; pulverize
磨成凸轮形 cam grinding
磨成细粉状高炉熔渣 ground granulated blast furnace
磨成小球 pebbling
磨成锥形 worn on a taper
磨橙 milling orange
磨齿 gear grinding; grinding teeth; mill teeth; mola; toothing
磨齿滚刀 ground hob
磨齿机 gear grinding machine
磨齿铰刀 ground reamer
磨出面 rubbed finish
磨穿的 worn-out
磨床 bench grinder; grinder; grinding machine; grinding mill; lapper; rubbing bed; sharpener; sharpening machine
磨床床头 grinding carriage; wheel carriage
磨床床头滑架 wheel slide
磨床刀架 grinding rest
磨床进料器 grinder feeder
磨床平卡盘 runner back
磨床头 grinding machine head(stock)
磨床托盘 runner back
磨床用卡规 grinding ga(u)ge
磨床组件 grinding element
磨床座架 grinder carrier
磨刺辊机 licker-in grinder
磨搓作用 abrasion action
磨挫法 draw filing
磨锉 burring
磨锉工具 roughening tool
磨带 abrasive belt; grinding belt; sand belt; strip
磨带锯机 band sharpening machine
磨刀 mill knife; sharpening; stoning
磨刀车间 tool-grinding department
磨刀附加装置 knife-grinding attachment
磨刀工(人) knife grinder
磨刀机 blade sharpening machine; knife grinder; knife grinding machine; knife sharpener; nail cutter grinder; tool-grinding machine; tool sharpener
磨刀夹 blade holder
磨刀夹刀板 grinder jaw
磨刀架 knife board
磨刀间 tool grindery
磨刀皮带 razor strop
磨刀片器 blade stropper
磨刀器 knife machine; knife slicker; sharpener
磨刀砂 sharp sand
磨刀砂轮 blade grinder; knife grinder
磨刀砂岩 Farewell rock
磨刀石 blade grinder; burr; grinding stone; hone(stone); knife grinder; knife stone; oilslip; rubstone; sickle grinder; stone; whetstone; grindstone; sharpening stone; slip stone
磨刀石隔块 hone spacing block
磨刀石架 whetstone holder
磨刀石水盘 grindstone-trough
磨刀小油石 oilstone slip; slip stone; stone
磨刀斜角导板 protractor tool guide
磨刀油 honing oil
磨刀油石条 oilstone slip
磨刀凿用的油石 grinding slip
磨刀装置 knife grinder
磨刀锥形砂轮 bevel(l)ed sickle wheel
磨得过早 premature sharpening
磨的 grinding
磨底 head of mill; grinding of base <玻璃、陶瓷器皿的>
磨掉 abrade; cutting down; grind off; wear off
磨掉的面层 lost surfacing
磨掉光泽 flatting down
磨端面 face grinding
磨端中空轴 mill end trunnion
磨端轴颈 mill end trunnion
磨钝 dulling
磨钝的金刚石 blunt edge stone; blunt stones
磨钝的钎头 blunt drill; smooth bit; worn bit
磨钝的钻头 blunt drill; smooth bit; worn bit
磨钝寿命 dulling life
磨阀 grinding valve; valve grinding
磨阀面 valve facing
磨阀面机 valve-refacer machine; valve refacing machine
磨阀砂 valve grinding sand
磨阀物 valve grinding compound
磨坊 grindery; grinding mill(ing); gristmill; mill
磨坊安装工 millwright
磨坊出水槽 mill tail
磨坊储水池 mill pond
磨坊的立柱屋架 mill bent
磨坊放水槽 mill tail
磨坊进水道 mill race
磨坊式厂房 mill building
磨坊水坝 mill dam
磨坊水轮 mill wheel
磨坊水渠 leat
磨坊土 miller
磨坊引水槽 mill race
磨坊用电动机 mill motor
磨坊主 miller
磨坊(贮)水池 mill pond
磨房 mill building
磨粉 abradant; powdering
磨粉布 crocus cloth
磨粉机 attritor; comminuter; flour mill; flour milling machine; granulator; pulverizer
磨粉浆 abrasive slurry
磨粉品质 milling characters
磨粉人 triturator
磨粉损耗 milling loss
磨粉系统 scratch system
磨粉细度 fineness of grinding
磨钢板机 slab grinder
磨钢球砂轮 grinding wheel for steel balls
磨钢丝 grinding
磨钢丝车工人 stripper and grinder
磨革玻璃块 sleeking glass
磨革机 buffing machine
磨工 branner; furbisher; grinder; grinding machine operator
磨工车间 tool-grinding department; grindery
磨工尘肺 grinders pneumoconiosis
磨工肺坏疽 grinder's rot
磨工工人 grinders
磨工工作 grinding work
磨工具的设备 tool sharpener
磨光 grinding; grinder finish; ground finish; abrading; abrasion finishing; barrel finishing; brighten; buffing; burnish(ing); cutting down; final grind; finish grinding; flatting down; hone out; levigate; levitation; mill finish; planish; polished dressing; polish(ing); politure; rubbed finish; glassed <对大理石或花岗岩表面>
磨光板 polished plate; rubbed slab
磨光棒 polishing stick
磨光标本 polished specimen
磨光表面 polished work
磨光表面的粗糙度 roughness of polished surface
磨光玻璃 abraded glass; abrased glass; mat glass; milled glass; polished glass; polished plate; sleeking glass; slicking glass; smoothed glass
磨光玻璃表面 glassed surface
磨光玻璃片 polished glass plate
磨光薄板 bright-polished sheet
磨光薄膜 milled thin section; polished thin section
磨光薄片 polished thin section
磨光材料 polishing material
磨光层 flush coat
磨光车床 scratch lathe
磨光带 finishing strip
磨光的 bright finished; dead bright; glassed; milled; polished; sanded
磨光的混凝土路面 polished concrete pavement
磨光的晶体 grinding crystal
磨光的石料 glassed stone
磨光的石棉水泥板 polished asbestos-cement board
磨光的钻杆 polish rod
磨光地板 finished floor
磨光地面 finished floor
磨光段 polished section
磨光断层 polish fault
磨光废砂 burgee
磨光粉 burnishing powder; grinding powder; grit
磨光钢丝 ground steel wire
磨光工具 smoothing tool; burnishing tool
磨光工作 millwork
磨光骨料 polished aggregate
磨光辊 burnisher; burnishing brush; burnishing roll
磨光滚筒 finished roller; trembling barrel
磨光混凝土 enamel(l)ed concrete; glazed concrete; rubbed concrete
磨光混凝土表面层的工艺 glazeraise
磨光机 abrader; abrading tool; branner; brilliant cutter; buffing lathe; buffing machine; finishing machine; flint glazing machine; glassing jack; glassing machine; glazing machine; grater; grinder; polishing lathe; polishing machine; rasping machine; refacer; sander; sanding machine; smoothing machine; trowel(l)ing machine
磨光集料 polished aggregate
磨光剂 buffing compound; grinding material; polish; polishing agent; polishing material
磨光夹丝玻璃 polished wire(d) glass
磨光接合 ground joint
磨光接头 ground joint
磨光金 best gold; burnished gold; burnishing gold; gold scouring; polished gold
磨光金属 burnished metal
磨光卷筒 scouring box
磨光刻花法 brilliant cut(ting)
磨光块 rub brick
磨光棱角 roughing the angles
磨光棱面 surface burnishing facet
磨光梁 finishing screed; smoothing screed
磨光了的 dressed
磨光卵石 faceted pebble
磨光轮 buff wheel; burnishing wheel; polishing wheel; rag wheel
磨光面 bright finish; burnishing surface; fine-rubbed finish; grinding finish; honed; polished(sur)face; rubbed surface; smooth finish; sand rubbed; rubbed finish
磨光敏感性 polish susceptibility
磨光木地板 flogging

磨光盘 buffing disc[disk]
磨光盆 sleeking tub
磨光皮 buff
磨光皮革 boarded leather
磨光片 polished section
磨光平板玻璃 polished plate;polished plate glass
磨光漆 polishing varnish;rubbing varnish;sanding lacquer
磨光漆面 rubbed finish
磨光器 burnisher;polisher;sleeker;slicker
磨光墙面板 sanded-face shingle
磨光清漆 body varnish;rubbing varnish
磨光砂纸 buffing paper
磨光生产线 polishing line
磨光石 abraded stone;polishing stone
磨光石料 ground stone;polished stone;rubbing stone
磨光石面 polished face
磨光试验 polishing test
磨光试样 polished sample
磨光饰面 rubbed finish
磨光书边 burnished edge
磨光术 polishing technique
磨光刷 polishing brush
磨光特性 polishing characteristics
磨光条 burnishing stick
磨光涂银玻璃 polished silvered plate glass
磨光物 polishing material
磨光系数 polished coefficient;polishing coefficient
磨光橡胶轮 rubber polishing wheel
磨光性(能)polishing characteristic
磨光压板 polished press plate
磨光银 burnished silver;burnish(ing)silver
磨光用铁 rub iron
磨光用砖 rubbing brick;rub brick
磨光油 buffing oil
磨光织物 buffing fabric
磨光止水肋条 machined rib
磨光止水条 machined sealing strip
磨光止条 machined sealing strip
磨光纸 buffing paper;friction(al)glazed paper;grinding paper
磨光轴杆 <深井泵> slick rod;slip stick
磨光砖 dressed brick;rubber
磨光作用 burnishing action
磨轨 rail grinding
磨轨车 rail grinding coach;rail grinding wagon
磨轨机 rail grinding machine
磨轨饰面 rubbed finish
磨辊 grinding roller
磨辊悬置系统 roller suspension system
磨辊用弹簧加压的辊磨 roller mill with spring-loaded rollers
磨过的 ground;lapped
磨耗 abrade;abrasion;abrasive wear(ing);attrition loss;defacement;detrition;fret;fretting corrosion;friction(al)wear;grinding wheel;tear and wear;wear;wear away;wear off;wear out
磨耗百分率 percentage of wear
磨耗板 friction(al)plate;wear(ing)plate
磨耗比 wear ratio
磨耗测定 abrasion test
磨耗层 wearing course;wearing layer;wearing carpet;wearing coat;wearing surface;abrasion surface;road-mix course road crust;abrasive surface <路面>;carpet(coat);
磨耗层骨料 wearing course aggregate
磨耗层集料 wearing course aggregate
磨耗车轮 worn wheel
磨耗程度 degree of abrasion;rate of wear(ing)
磨耗的 abrasive;torn-up;worn
磨耗的帆布缝线 soft seams
磨耗的面层 lost surfacing
磨耗等级 wear rating
磨耗掉 grinding off
磨耗度 abrasion degree;abrasiveness;degree of abrasion;degree of wear
磨耗度磨耗层 <一种由至少两种磨耗度不同石料拌制而成的磨耗层> delugrip
磨耗故障 wear-out failure
磨耗规 wear and tear ga(u)ge
磨耗和撕裂 wear and tear
磨耗机 abrasion device
磨耗机试验 Los Angeles test;rattler test
磨耗及损伤 wear and tear
磨耗极限 limit of wear
磨耗减低因数 wear reduction factor
磨耗减量 wear loss
磨耗件 wearing parts
磨耗量 abrasion loss;abrasion value;abrasion volume;abrasion wear;amount of wear;wear and tear volume
磨耗率 rate of wear(ing);specific abrasion;rattler loss <磨耗试验的>
磨耗轮缘 worn flange
磨耗面 wearing surface
磨耗器 wearometer
磨耗强烈程度 degree of wearing intensity
磨耗时间 wearing time
磨耗试验 abrasion test;attrition test;wear(ing)test(ing)
磨耗试验机 abrasion machine;abrasion tester;abrasion test(ing)machine;attrition testing machine;rattler
磨耗试验器 abrading device
磨耗速度 rate of wear(ing)
磨耗损失量 abrasion loss
磨耗条款 metal(l)ing clause
磨耗图纹 wear pattern
磨耗系数 abrasion coefficient;abrasiveness factor;coefficient of waste;coefficient of wear
磨耗型 wear profile
磨耗型车轮踏面 profiled wheel tread;worn wheel profile
磨耗型踏面 worn tread
磨耗性 abradability;abrasiveness
磨耗性能 wearing quality
磨耗硬度 abrasion hardness;abrasive hardness
磨耗增强板 wear plate
磨耗值 abrasion value;attrition value;wear(ing)value
磨耗指示器 wear indicator
磨耗指数 abrasion index;abrasion resistance index;wear index
磨耗装置 abrasion device
磨耗阻力 abrasion resistance;abrasive resistance;resistance to abrasion;resistance to wear
磨耗作用 abrasive action;attrition
磨合 backfit;bedding-in;bread-in;breaking-in;burnish;grind in;lapping;regrinding;run(ning)-in;wear-in
磨合表面 broken-in surface
磨合了的钻头 broken in bit
磨合期 breaking-in period;run-in time;running-in period
磨合期磨损 run-in wear
磨合前 prebreak-in;preburnish
磨合前制动检查试验 preburnish check
磨合时间 run(ning)-in time;seating time
磨合时间表 running-in schedule
磨合试验台 run-in stand
磨合性 running-in ability
磨合性能 running-in characteristic
磨合样机 running-in machine
磨合用(润滑)油 running-in oil;breaking-in oil
磨合油 break-in oil
磨合运转 break-in run;run-in
磨合运转时期偏移力 deflection force for run-in period
磨合轴颈 worn-in journal
磨合装置 running-in machine
磨痕 buffing mark;grinding crack;polishing scratch
磨滑机 abrading tool
磨滑器 burnisher
磨坏的钢板 renewable steel
磨毁作用 blasting
磨机 branner
磨机产出物 mill product
磨机产量 milling capacity
磨机衬板 mill liner
磨机衬板磨损 mill liner wear
磨机衬里 mill liner
磨机传动 mill drive
磨机的电声控制 electroacoustic(al)mill control
磨机负荷 mill load(ing)
磨机负荷的单电耳控制 single electric(al)ear control for mill loading
磨机负荷控制系统 mill load control system
磨机负荷声音控制 acoustic(al)mill load control
磨机盖板 hardwall of mill
磨机隔仓板 mill diaphragm partition;mill partition
磨机给料 mill feed
磨机给料仓 mill feed bin
磨机给料器 mill feeder
磨机工作转速 working speed of mill
磨机滚筒筛 mill trammel
磨机进料口 intake of mill;mill inlet
磨机矿浆 mill slurries
磨机立碾轮 edgestone
磨机料位控制 mill level control
磨机临界转速 critical speed of mill
磨机流量 mill stream
磨机磨头仓 grinding mill feed bin
磨机内径 mill diameter inside liners
磨机排出的余风 exhaust from mill
磨机排料 mill discharge
磨机配料 mill batch
磨机气流 mill atmosphere
磨机生产能力 milling capacity
磨机双传动 dual drive of mill
磨机填充率 filling ratio of mill;percentage loading of mill
磨机调整 tune-up the mill
磨机通风 mill vent air;mill ventilation
磨机筒体 mill cylinder
磨机外壳 mill casing;mill shell
磨机喂料仓 feed bin of mill
磨机喂料斗 mill feed hopper
磨机喂料机 mill feeder
磨机喂料监控 mill charge monitoring
磨机橡胶衬里 rubber lining of mill
磨机卸料口 mill outlet
磨机需用功率 mill power draft
磨机循环 mill circuit
磨机研磨体填充系数 percentage of loading of mill
磨机用润滑脂 mill grease
磨机中空轴内螺旋喂料装置 drum-and-scoop feeder
磨机周边排料 peripheral discharge
磨机转速 mill speed
磨机装量 grinding charge
磨机装料 mill charge
磨机装料声控装置 audio mill loading device
磨机最佳工作转速 optimum speed of mill
磨剂 grinding composition;grinding compound
磨加法 mill addition method
磨加工 stoning
磨加工留量 grinding allowance
磨加工裕量 allow for grinding
磨尖 sharpen(ing)
磨尖的 sharpened
磨尖机 pipe pointer
磨浆 defibrination
磨浆机 paste mill;paste roller;pulping grinder;pulping machine;refining mill;stuff grinder
磨浆石 pulpstone
磨角 angle lap(ping)
磨角机 corner grinding machine
磨角架 bevel(l)ing post
磨角染色法 angle lap stain method
磨角染色分析 grinding angle and colo(u)ring analysis
磨角样品 angle lapped specimen
磨具 abrasive tool;grinding apparatus;rubber;sharpener;sharper
磨具结构 structure of abrasive tool
磨具结合剂 binder for abrasive
磨具刻画线条 thread grinding
磨具硬度 grade of abrasive tool
磨具硬度计 grade tester
磨锯齿 saw gumming
磨锯齿的砂轮 saw gumming wheel
磨锯齿机 saw doctor;saw sharpener;saw sharpening machine
磨锯锉 saw file
磨锯虎钳 saw vice
磨锯机 saw grinder
磨刻 cutting;engraving
磨刻槽纹 fluting
磨刻车间 cutting shop
磨刻面 flat
磨坑 grinding dish
磨孔器 hone
磨口玻璃 ground glass
磨口玻璃接头 ground glass joint
磨口玻璃瓶 ground glass stoppered bottle;ground glass stoppered flask
磨口玻璃塞 ground glass stopper
磨口玻璃仪器 ground glass apparatus
磨口的 ground-in
磨口堵头 blind ground joint
磨口接头 ground connection;ground in joint;ground joint
磨口蛇形管蒸馏器 distilling apparatus with ground on coil condenser
磨口锥塞 blind taper joint
磨块 abrasive brick;grinding;rubbing block;sharpen;whet(ting)
磨快 sharpening
磨快的 sharpened
磨快器 sharpener
磨快圆锯齿 saw gumming
磨矿 grinding ore;milling grinding
磨矿板 buck-plate
磨矿车间 grinding building
磨矿段 grinding section
磨矿法 milling method
磨矿分级 grind grading
磨矿回路 grinding circuit
磨矿机 grinder;kominuter;milling pit;ore mill
磨矿机衬里 mill liner;mill lining

M

M

磨矿机滚筒 mill barrel
磨矿机卸料端提升机 mill head elevator
磨矿机械 grinding mill
磨矿介质 grinding media; milling medium
磨矿精选装置 grinding-concentration unit
磨矿流程图 grinding flowsheet
磨矿设备 grinding attachment
磨矿石厂 ore mill
磨矿速度 grinding rate
磨矿效率 grinding efficiency
磨拉石(层)【地】molasse
磨拉石建造 molasse formation
磨拉石砂岩 molasses sandstone
磨拉石碎屑盆地 molasse clastic basin
磨拉石相 molasse facies
磨拉石型沉积 molasse-type sediment
磨拉石组合 molasses association
磨利的 sharpened
磨砾 <砾磨机用> grinding pebble; mill pebble
磨砾层 molasse formation
磨砾层分子内聚力 molasse molecular cohesion
磨粒 abrasive grain; abrasive particle
磨粒变钝 glazing
磨粒的硬度 hardness of grain
磨粒磨损试验机 grain-abrasion testing machine
磨炼机 mill
磨亮 rubbing down
磨亮的 brighten; burnish
磨料 abradant; abradant material; abraser; abrasive; abrasive material; cutting agent; grinding aid; grinding compound; grinding material; grinding media; polishing material; rubbing compound
磨料表层 abrasive coating
磨料堆放棚 grits storage hangar; gritting material hangar
磨料堆栈 gritting material store
磨料飞散 grit emission
磨料分级 classification of abrasive
磨料粉 abrasive powder
磨料粉尘 abrasive dust
磨料粉砂沉积池 grade waste pond; sand pond; silt field
磨料钢球 abrasive steel shot
磨料钢珠 abrasive steel shot
磨料过滤器 abrasives filter
磨料盒 grinding material box
磨料级金刚石 abrasive diamond
磨料锯 mud saw
磨料颗粒 abrasive particle
磨料粒度 abrasive grain; abrasive grain size; abrasive particle; grit size
磨料粒度数 grit number
磨料粒径 abrasive grain size
磨料磨擦面 abrasive surface
磨料磨损 grinding abrasion
磨料磨损试验机 abrasive wear testing machine
磨料黏[粘]结剂 abrasive cement
磨料喷射清理 abrasive blasting
磨料砂 abrasive sand
磨料筛目数 grit number
磨料伤痕 plate mark; runner cut
磨料细度 fineness of grinding
磨料研磨 abrasive lapping
磨料研磨机 abrasive lapping machine
磨料研磨液 lapping fluid
磨料硬度 abrasive hardness
磨料制备槽 slough
磨料制品 abrasive product; product of abrasive
磨料制品液压机 abrasive molding hydraulic press
磨料种类 type of abrasive
磨料贮仓 gritting material bin
磨轮 abrasion wheel; abrasive blade; abrasive disc[disk]; abrasive grinding wheel; brush wheel; buzzer; edge runner; grinding wheel; sanding disc[disk]
磨轮表面修饰 crush dressing
磨轮的整修 dressing of grinding wheel
磨轮机工 millwright
磨轮夹紧法 clamping of the grinding wheel
磨轮架 wheel stand
磨轮接触面 grinding wheel contact surface
磨轮黏[粘]结剂 binder in abrasive wheels
磨轮式碾机 edge runner mill
磨轮试验机 grinding wheel testing machine
磨轮修整器 tool dresser
磨轮压刮整型机 crush-form dresser
磨轮压刮整型器 crush-form dresser
磨轮眼镜 grinding wheel spectacles
磨轮整型器 abrasive wheel dresser
磨毛机 roughening machine
磨煤机 coal grinding mill
磨煤机负荷率 mill load ratio
磨煤设备耗电量 power to pulverize coal
磨面 mill flour; surfacing
磨面革 buffed leather
磨面工作 rubbed work
磨面混凝土 rubbed concrete
磨面机 face-milling machine; surface grinder; surfacer; surface sander; fluffing machine <磨皮革等的>
磨面巨砾 faceted boulder
磨面砌体 rubbed work
磨面绒革 buff leather
磨面石 rubbing stone; stone surfacer
磨面砖 dressed brick; rub(bed) brick
磨灭 deface
磨木锭子机 skewer pointer
磨木机 wood grinder
磨木浆 groundwood pulp
磨木浆浓缩机 groundwood decker
磨木浆纤维 groundwood fiber
磨木浆芯层纸板 groundwood filled board
磨木浆印刷纸 groundwood printing paper
磨木浆制纸板 mechanical pulp board
磨木木素 milled wood lignin
磨木石 pulpstone
磨木纸浆废水 groundwood wastewater
磨内荷载 mill charge
磨内结块 agglomeration in mill
磨内结圈 ring formation in mill
磨内通风 ventilation of mill; ventilation within mill
磨内物料与钢球总量 charge
磨内圆 grinding out
磨内装球表面积 ball-charged surface in mill
磨泥机 clay grinding machine
磨泥盘 clay-grinding pan
磨盘 abrasive disc[disk]; buhrstone; circular stone; grinder; grinding base; grinding bowl; grinding disk [disc]; grinding pan; grinding stone holder; grinding table; polisher; polishing disk; rotating table; sand disc [disk]; stone grinder
磨盘粉磨机 burr mill
磨盘沟纹 stone furrow
磨盘肋 nog; runner bar
磨盘上的刻槽 millstone cutter
磨盘石 millstone
磨盘式磨粉机 attrition mill; burring grinder; burring mill; plate mill
磨盘式磨碎机 burring grinder; burring mill
磨盘速度 speed of grinding disc
磨盘钻探 rotary drilling
磨盘钻探用的复合润滑油 running-in compound
磨片 abrasive disc [disk]; microsection; microspecimen; section; thin section
磨片分析 thin-section analysis
磨片机 lapping machine
磨片组织图 macrosection
磨偏 worn out of place
磨平 planish; rubbing down; rubbing flat; rub down; wear down; wear flat; worn flat; flatting down <油漆作业>
磨平材料 abrasive material
磨平打光 polish grind
磨平刀 finishing knife
磨平的端面 face machined flat
磨平机 planing mill
磨平机和抛光机 grinder and polisher
磨平块 <混凝土> rub brick
磨平路面 cold milling
磨平面 flat surface grinding
磨平石 smoothing stone
磨平头的钉 grinding point
磨平修整 rubbed finish
磨平凿 span chisel
磨破 fray
磨气门机 valve refacer
磨气门座机 valve seat grinder
磨汽门机 valve grinder
磨钎工 bit dresser; grinder
磨钎机 bit dresser; bit grinder; bit grinding machine; bit sharpener; drill grinder; drill sharpener; drill sharpening machine; grinder; mechanical sharpener; shank grinder; sharpener; steel sharpening machine
磨钎刃 edge grinding
磨钎装置 grinding attachment
磨铅心机 lead sharpener
磨墙机 wall grinding machine
磨墙面机 face-milling machine
磨球 ball; grinding mill ball; grinding pebble
磨球崩落 avalanching of grinding ball
磨球初装量 initial ball charge
磨球的回转周期 travel(l)ing period of grinding ball
磨球分级 graduation of grinding ball
磨球分选 ball grading
磨球滚落轨迹 rolling path of grinding bodies
磨球机 ball crusher; ball grinder(machine)
磨球路线 ball path
磨球面 contouring
磨球磨损 ball loss; grinding ball loss
磨球配量 ball rationing
磨球平均球径 average diameter of grinding ball
磨球瀑落 cascading; cataracting of grinding media
磨球损耗 ball loss
磨球消耗量 ball consumption
磨球泻落 avalanching
磨球运动路线 ball path
磨去 abrade; buff away; grind off; thinning
磨刃 sharpening
磨刃机 sharpener
磨刃夹具 honing ga(u)ge
磨刃器 sharpening tool
磨锐 grinding; sharpening
磨锐刀具 sharpening tool
磨锐工具 whetter
磨锐角 taper angle
磨锐式钻头 sharpable bit
磨锐性 regrindability
磨砂 dull polish; frost; frosted finish
磨砂白炽灯 frosted incandescent lamp; incandescent lamp
磨砂边显微镜载玻片 ground edge; microscope slide
磨砂表面 sand surfacing
磨砂表面处理 mat finish
磨砂表面涂料 mat finish paint
磨砂玻璃 depolished glass; etched glass; frosted glass; ground glass; mat(ted) glass; mat(te)-surfaced glass; sand-blasted glass
磨砂玻璃窗 frosted glass window
磨砂玻璃塞 ground glass stopper
磨砂玻璃装饰 glister[glistre]
磨砂处理 grit finish
磨砂的 feathered washboard; frosted
磨砂灯泡 dim lamp; frosted (lamp) bulb; frosted lamp globe; opal bulb
磨砂花纹玻璃 frosted glass with muslin pattern
磨砂机 frosting machine; sand grinding mill
磨砂轮 abrasive wheel
磨砂面 frosting; frosting surface
磨砂面层 rubbed finish
磨砂球形灯泡 frosted lamp globe
磨砂显微镜载玻片 frosted
磨伤 abrasion damage; fretting
磨声喂料控制 acoustic(al) mill feed control; acoustic(al) mill load control
磨失光泽 abrasion loss of gloss
磨石 abrader; burr (stone); edgestone; grinding wheel; holystone; millstone(grit); oil stone; pan mill; quern-stone; rotten stone; rubber down; rubberstone; rubstone; runner; sharpening stone; strickle; whetstone
磨石板机 beveler
磨石(粗)砾砂 millstone grit
磨石粉 rotten stone powder
磨石工 floatsman; rubbing-bed hand; stone grinder; stone polisher; stone rubber
磨石规格 grinding stone ga(u)ge
磨石机 stone grinder; stone mill
磨石井 grindstone pit
磨石刻槽器 grindstone dresser
磨石磨孔法 honing
磨石盘 grinding lap
磨石片 polishing slate
磨石砂砾 buhrstone; millstone grit
磨石砂轮 grinding stone wheel
磨石铁 <用以磨砖上粗面痕迹> float stone
磨石屑 grindstone swarf
磨石子 terrazzo
磨石子地面 granolithic concrete surface; terrazzo floor(ing)
磨石子地坪混凝土 grano concrete; granolithic concrete
磨蚀 abrade; abrasion; abrasive wear(ing); wearing away
磨蚀处理 scarification
磨蚀的 abrasive
磨蚀度 abrasivity
磨蚀粉 abrading powder
磨蚀腐蚀 erosion corrosion
磨蚀膏 abrasive paste
磨蚀沟 abrasion groove
磨蚀海湾 abrasion embayment

磨蚀河湾 abrasion embayment
磨蚀痕迹 abrasion mark
磨蚀机 abrader
磨蚀基准面 base level of abrasion
磨蚀剂 abradant;abrasive;abrasive agent
磨蚀抗力 abrasion resistance; abrasive resistance
磨蚀刻痕 abrasion mark
磨蚀率 rate of abrasion;specific abrasion
磨蚀面 abraded plane;abrasion plane;abrasive surface
磨蚀喷射钻井 abrasive jet drilling
磨蚀疲劳 chafing fatigue;fretting fatigue
磨蚀平原 plain of abrasion
磨蚀器 abrading apparatus
磨蚀钎头岩 ga(u)ge-wearing rock
磨蚀浅滩宽度 width of abrasion shoal patch
磨蚀浅滩坡角 slope angle of abrasion shoal patch
磨蚀强度 abrasive resistance
磨蚀深度 depth of wear
磨蚀式回旋钻进 abrasion drilling
磨蚀试验 abraded test;abrasion test
磨蚀试验机 abraded tester;abrader;abrasion machine;abrasion tester
磨蚀数 abrasion number
磨蚀微粒 abrasion particle
磨蚀性 abradability;erosivity;abrasiveness
磨蚀性侵蚀 abrasive erosion
磨蚀性岩层 abrasive formation
磨蚀修边 abrasive deflashing
磨蚀液 abrasive suspension
磨蚀硬度 abrasion hardness;abrasive hardness
磨蚀油 abrasive belt grinding lubricant
磨蚀裕量 allowance for abrasion
磨蚀原因 erosion caused by
磨蚀皂 abrasive soap
磨蚀造成的损伤 damage wear
磨蚀指数 abrasion index
磨蚀周期 abrasion cycle
磨蚀阻力 resistance to abrasion
磨蚀钻井 abrasive cutting drilling
磨蚀作用 ablation;fretting corrosion;abraded action; rubbing-abrading action
磨碎 breakdown by grinding; confrication; disintegration; granulate; grinding; mash; milling pulverization; pulverize; size degradation; size reduction;triturate;trituration
磨碎拌和机 blender grinder
磨碎并过筛机 rasper
磨碎玻璃纤维 milled glass fiber[fibre]
磨碎程度 reduction ratio
磨碎处理 grinding treatment
磨碎的 ground;mashed;pulverized
磨碎的材料 pulverized material
磨碎的谷壳 ground grain hulls
磨碎的固体垃圾 milled refuse
磨碎的石灰石 pulverized limestone
磨碎砥石 grinding plate
磨碎动力学 kinetics of comminution
磨碎段 reduction stage
磨碎混合 attrition mixing
磨碎机 attrition mill; comminutor; communicator; grinder; grinding drum; reducing machine; grinding mill
磨碎机进料器 grinding mill feeder
磨碎机筒 mill barrel
磨碎碱性炉渣 ground basic slag
磨碎浸取过程 grind-leach process
磨碎垃圾 ground garbage
磨碎了的 milled
磨碎炉渣 ground slag
磨碎面 mill surface
磨碎能力 grinding property
磨碎燃烧浸取过程 grind-burn-leach process
磨碎设备 grinding plant; milling equipment
磨碎时的加入物 interground addition
磨碎速度 grinding rate
磨碎添加剂 interground additive
磨碎物 triturate
磨碎细度 fineness of grinding
磨碎细度调节器 fineness regulator
磨碎装置 grinding mechanism
磨碎作用 milling action
磨损 abrasion; abrasive wear(ing); wear and tear; wear away; wear down; wearing; wear out; attrition (loss); attrition wear; foreworn; fray out;frazzle;fretting corrosion; friction (al) wear; friction-caused wear; loss due to friction; obliteration; rattler loss; rubbing wear; scour;scuff;tear and wear;frayed <指装货物的编织袋磨损>
磨损百分率 percent of wear
磨损报警传感器 wear alarm transmitter
磨损变细 drawing small
磨损标志 signs of wear
磨损表 wear tables
磨损表面 wear surface; worn-out surface
磨损补偿 compensate for wear;wear-compensating
磨损部分 wearing parts
磨损部件 wearing parts;wearing terrain;wear segment
磨损部位 abrading section; wearing terrain
磨损残坏硬币 defaced coins
磨损测量的基本尺寸 wear reference point
磨损层 wear(ing) course;wear(ing) carpet;wear(ing) coat;wear(ing) layer
磨损产物 wear debris
磨损程度 attrition rate; degree of wear;rate of wear(ing)
磨损处 fray;fret;gall;scuff
磨损带 wear land
磨损的 abrasive; battered; worn; worn-out
磨损的材料 lost material
磨损的导套 worn bushing
磨损的导柱 worn guide pin
磨损的固定资产 worn-out fixed assets
磨损的金刚石 abraded diamond
磨损的颗粒 wear particle
磨损的路面 torn-up surface
磨损的螺栓 skinned bolt
磨损的面层 lost surface
磨损的模具 dull die
磨损的末端或边缘 frazzle
磨损的绳索 grind
磨损的支承 burnt bearing
磨损的钻杆 lost drill pipe
磨损地面 worn-out pavement
磨损度 abradability;abrasiveness;resistance to wear; wearing quality; wear intensity
磨损帆布 frayed canvas
磨损范围 wearing area
磨损腐蚀 fretting erosion
磨损公差 wear allowance
磨损公差带 wear allowance zone
磨损故障 deterioration failure;wear-out failure
磨损规律 wear pattern
磨损轨 worn-out rail
磨损过程 progress of wear; wear process
磨损痕迹 wear trace;wear track
磨损厚度 depth of wear; wearing depth;wearing thickness
磨损环 wear ring
磨损机理 abrasion mechanism
磨损极限 limit of wear;wearing limit
磨损计 wearometer
磨损检查 abrasion inspection
磨损检验 examination of wear;examine for wear and tear
磨损件 wear-out parts;worn-out articles;worn-out parts
磨损均匀性 even wear
磨损控制 wearing control
磨损类型 wear pattern
磨损力学 wear mechanics
磨损量 abraded quantity; abrasion loss; abrasiveness rating; amount of wear;wear extent;wearing capacity
磨损量测 wearing measurement
磨损量测仪 wear and tear ga(u)ge
磨损量规 wearing ga(u)ge
磨损量极限 limit of wear
磨损量块 wear block
磨损了的设备 run-down equipment
磨损了的绳子 fretted rope
磨损零件 wear and tear parts;wearing parts
磨损留量 tear-and-wear allowance; wear allowance
磨损留量极限 limit of allowance for wear
磨损路面 worn-out pavement; worn pavement
磨损率 attrition rate; rate of wear(ing); specific abrasion; specific wearability;wear rate
磨损面 abrasion; abrasive surface; torn surface; wear course; wear flat;wear(ing) surface
磨损面层 lost surfacing; wearing course
磨损面积 wearing area
磨损模式 wear pattern
磨损模型 wear model
磨损疲劳 abrasion fatigue;fretting fatigue
磨损破坏 wear-out failure
磨损期限 wearing life;wear-out period
磨损汽缸重镗 reboring of worn cylinder
磨损倾向 liability fraying
磨损情况 wear phase
磨损曲线 wear curve;wear history
磨损圈 wear ring
磨损容差 tear-and-wear allowance; wear allowance
磨损容量 wear-out allowance
磨损深度 depth of wear; wearing depth
磨损生成物 wear products
磨损失效 wear-out failure
磨损失效周期 wear-out-failure period
磨损矢径 wear-out of ga(u)ge
磨损试验 abrasion test;attrition test; grinding test;rattler test;reciprocal friction test;wear(ing) test(ing)
磨损试验机 abrasion machine;abrasion tester;abrasive tester;attrition tester; attrition testing machine; rattler;wear tester;abrader
磨损试验器 abrasion tester
磨损试验转数 abrasion cycle
磨损寿命 wear-life
磨损速度 abrasiveness rating
磨损特性 wearing characteristic;wearing quality
磨损调整 adjustment for wear
磨损物件 worn-out article
磨损系数 abrasiveness factor;coefficient of waste;coefficient of wear;mortality factor
磨损限度 wearing limit
磨损限值 wear allowance
磨损性 abradability; abrasiveness; polishing machine; resistance to wear;wearability;wearing quality;wear-life
磨损性剥蚀 abrasive erosion
磨损性载荷 abrasive duty
磨损性质 abrasive nature
磨损学 tribology
磨损压力 abrasion pressure
磨损氧化 fretting oxidation;wear oxidation
磨损硬币以盗取金银粉末 sweating
磨损硬度 abrasion hardness;abrasive hardness;wear(ing) hardness
磨损值 abrasion value; amount of wear;attrition value;wearing value
磨损指示器 wear detector
磨损指数 abrasion index; abrasion wear index
磨损指数分析仪 wear index analyser [analyzer]
磨损状态 wear phase
磨损阻力 abrasion resistance; wear(ing) resistance; wear-resisting property
磨损作用 abrasive action;wearing action
磨台 breaking-in
磨碳化物和钢用金刚石品种 diamond for carbide and steel
磨添加物 mill addition
磨条 abrasive stick
磨头 bistrique;end wall;grinding head; grinding unit;grinding wheel head; head of mill; mill end; mill head; mounted wheel;wheel-head
磨头仓 mill feed bin; mill feed hopper;mill primary hopper;prebin
磨头衬板 end lining; end wall liner; headliner
磨头电动机 grinding head motor
磨头盖板 end plate
磨头磨尾机 top and end grinding machine
磨头退出槽 honing clearance grooves
磨头纵向移动的磨床 travel(l)ing head grinder
磨涂料机 paint mill
磨退 flatting down;rubbing
磨退面 rubbed finish
磨退效果 rubbed effect
磨褪 satin;satin finishing
磨褪处理 satinizing
磨尾 exit wall
磨纹 abrasion pattern
磨细 fine grinding; finish grinding; levigating; levigation; size degradation;size reduction
磨细白垩粉 whiting
磨细材料 ground material
磨细掺和剂 interground addition
磨细掺加剂 interground addition
磨细的白土 milled clay
磨细的物料 ground material
磨细度 grinding rate
磨细高炉矿渣 ground blast furnace slag
磨细硅质材料<具有火山性质的> alfesil
磨细火山灰 ground pozzolana
磨细矿渣 ground slag
磨细粒化高炉渣 finely pulverized

M

blast furnace slag
磨细粒状高炉熔渣 ground granulated blast furnace slag
磨细轮胎粉 powdered tire[tyre]
磨细砂 ground sand
磨细筛目 mog
磨细生石灰 burnt and ground lime; ground quick lime; processed quick-lime
磨细石膏灰泥 sized gypsum
磨细石料 ground stone
磨细碳酸钙 ground calcium carbon-ate
磨细铁矿高炉渣 ground iron blast-furnace slag
磨细岩石 ground rock
磨细作用 degradation in size
磨削 abrasive planing; burnish; grind in; grinding; skiving
磨削比 grinding ratio
磨削变质层 grinding skin
磨削表面 grinding surface
磨削操作 grinding action; grinding operation
磨削层 surface layer
磨削长度 grinding length
磨削到标准尺寸 grinding to ga(u)ge
磨削附件 grinding attachment
磨削工具 grinding tool
磨削滚筒 polishing drum
磨削环 milling ring
磨削机 milling machine
磨削技术 grinding technique
磨削加工 abrasive machining; ground finish
磨削加工留量 grinding allowance
磨削加工余量 grinding allowance
磨削交叉花纹 overlapping curve
磨削接触 grinding contact
磨削靠模 grind master
磨削空刀槽 grinding undercut
磨削宽度 polishing width
磨削力 abrasive force
磨削裂纹 grinding crack
磨削裂纹敏感性 grinding sensitivity
磨削留量 grinding stock; grinding tol-erance
磨削面 grinding(sur)face
磨削抛光机 grinder-polisher
磨削容量 grinding capacity
磨削润滑冷却液 grinding fluid cool-ant
磨削砂带 grinding belt
磨削伤 block rack; cullet cut; rake
磨削烧伤 grinding burn
磨削时高温回火 grinding temper
磨削损伤 grinding damage
磨削位置 grinding position
磨削屑 grindstone dust
磨削性 grindability
磨削循环 grinding cycle
磨削液 grinding fluid
磨削应力 grinding stress
磨削用冷却剂 grinding paste
磨削油 grinding oil
磨削裕度 grinding tolerance
磨削越程槽 grinding undercut
磨削针 abrasive point
磨削砖 ga(u)ged brick
磨削装置 grinding attachment
磨削灼伤条痕 grinding burnt line
磨削钻头 drag bit
磨削作用 ablation; grinding action
磨斜边 bevel(l)ing
磨斜边机 bevel(l)er
磨斜角 bevel grinding
磨斜棱 bevel(l)ing
磨斜面 bevel grinding
磨斜盘 bevel grinding
磨斜切边 bevel grinding
磨鞋 drag shoe; milling shoe
磨屑 abraded parts; abrasive dust; grindings
磨牙丝锥 ground tap
磨洋工 scamping
磨样 grind away
磨样法 ground sample process
磨油机 oil grinding machine
磨油石 abrasive grinding wheel
磨余面 grinding lap
磨圆边角 round-off corner
磨圆的 rounded
磨圆度 psephicity; roundness
磨圆角 cavetto
磨圆角工具 <用于新浇混凝土> ar-rissing tool
磨圆金刚石 tumbled diamond
磨圆颗粒 rounded grain
磨圆卵石 rounded pebble
磨圆面 rounded face
磨圆盘砂轮 disk grinder
磨圆器 <新浇混凝土边缘> arrissing tool
磨圆作用 rounding
磨掌 runner bar
磨制表面 honed finish
磨制矿浆用磨石 pulpstone
磨制木浆 defibrination
磨制暖气片丝锥 ground tap for radi-ator
磨制物 polish
磨制楔形拱砖的盒 cutting box
磨制直柄麻花钻 ground straight-shank twist drill
磨砖 brick rubbing; cutter and rub-ber; rubbed brick; rubbing brick
磨砖对缝 rubbed bricks with tight joints; rubbed joint
磨砖对缝拱 ga(u)ged arch
磨砖对缝砌筑 rubbed brickwork
磨砖对缝墙 smoothened-brick wall
磨砖工人 brick trimmer
磨砖勾缝 rubbed bricks with pointed joints
磨砖机 rattler
磨砖密缝拱 bonded brick arch
磨砖密缝券 bonded brick arch
磨砖石 float stone
磨砖试验 rattler test
磨琢 filing
磨琢金刚石 cut diamond
磨琢面 rubbed dressing
磨钻 abrasive drilling
磨钻机 sharpener; sharpening machine; bits grinders
磨钻头机 drill sharpening machine

蘑菇 agaric; round mushroom

蘑菇冰 mushroom ice
蘑菇钉 mushroom nail
蘑菇钉路 <汽车可靠性强化试验路的一种> mushroom road
蘑菇工人肺 mushroom workers' lung
蘑菇片 sliced mushroom
蘑菇石 mushroom rock; pedestal rock; cheesewring <美国科尼什产的一种浅灰色花岗岩>
蘑菇式 mushroom hair style
蘑菇式穿孔机 cone-roll piercing mill
蘑菇式结构 mushroom construction
蘑菇式绝缘子 mushroom insulator
蘑菇亭 mushroom pavilion
蘑菇头 <用于固定集装箱> cone; stud
蘑菇头方颈螺栓 mushroom-head square neck bolt
蘑菇头形螺栓 mushroom-head bolt
蘑菇香精 lenthionine
蘑菇形的 mushroomed
蘑菇形叠加褶皱 mushroom-shaped superposed folds
蘑菇形阀 mushroom valve
蘑菇形结构 mushroom construction
蘑菇形截面 mushroom cross section
蘑菇形开挖法 mushroom excavation; mushroom-type tunnel(l)ing meth-od
蘑菇形壳(体) mushroom-type shell
蘑菇形楼板建筑 mushroom slab con-struction
蘑菇形螺钉头 mushroom head
蘑菇形锚 mushroom anchor
蘑菇形模型 mushroom mo(u)ld
蘑菇形桥 mushroom-type bridge
蘑菇形天线 mushroom antenna
蘑菇形通风筒 mushroom ventilator
蘑菇形先拱后墙法 mushroom-type tunnel(l)ing method
蘑菇形轧辊的穿孔机 rotary rolling mill
蘑菇形柱 mushroom-head column
蘑菇形柱头 mushroom head
蘑菇型柱头的无梁楼板建筑 mush-room system of flat-slab construc-tion
蘑菇栽培室 mushroom hothouse
蘑菇属 <拉> Agaricus
蘑菇状的 mushroom
蘑菇状的矮石柱 staddle stone
蘑菇状积雪 snow mushroom
蘑菇状矿物 agaric mineral
蘑菇状物 mushroom
蘑菇状(烟)云 mushroom cloud
蘑菇状岩石 cheesewring

魔镖 magic dart

魔方 magic square
魔镜 witch mirror
魔力热袋 magic heat pack
魔日 devil date

抹 plaster; swab(bing); wipe

抹八字角 filleting
抹边镘刀 margin trowel
抹边抹子 margin trowel
抹玻璃器皿布 glass towel
抹布 cleaning cloth; cotton waste for cleaning; dish cloth; duster; duster cloth; napery; rag; scouring cloth; swabbing cloth; tack rag
抹除 erase
抹除电子枪 erasing gun
抹除符号组 ignore character block
抹除区 wipe out area
抹除振荡器 erase oscillator
抹粗灰浆 parg(et)ing
抹粗灰泥墙 harl
抹带 collar
抹刀 buttering trowel; spattle; spatu-la; trowel
抹刀形的 spatulate
抹底灰 brown-out
抹第二层灰 <面层下的一层> browning coat
抹掉 blot; blur out; erase; erasion; ex-punction
抹掉信息的读出 destructive reading; destructive readout
抹掉(字)符 delete character
抹工 dauber
抹刮油灰 face puttying; puttying
抹光 finishing; scouring; trowel-applied finish(ing); trowel(l)ing
抹光板横向摆幅 smoothing plate am-plitude
抹光不足 under-trowelling
抹光布轮 cloth finishing mop
抹光带 smoothing strip
抹光带横向摆幅 smoothing strip am-plitude
抹光的混凝土路面 polished concrete pavement
抹光过度 over-trowelling
抹光机 trowel(l)ing machine
抹光烙铁 <沥青路面接缝用> smoothing iron
抹光镘刀 finish trowel
抹光面 trowel finish; trowel(l)ed (sur)face
抹光面层 float finish; ground finish
抹光性能 trowel(l)ability
抹光用玛瑞脂 trowel-mastic
抹过灰的板条 rendered lath
抹过腻子的木料 plugged lumber
抹灰 application of mortar; buttering; plastering; trowel(l)ed stucco; trowel(l)ing; wet plaster
抹灰凹圆角 plaster cove
抹灰板 floating rule
抹灰板墙隔热垫层 lathing insulating mat
抹灰板条 batoon; lathing for stucco; plasterer's lath (ing); rendering lath; wood(en) lath
抹灰板条锤 plasterer lath hammer
抹灰板条墙 battered wall
抹灰板条上抹灰 plaster on wood-(en) lath(ing)
抹灰拌和 plaster mix(ture)
抹灰拌料 rendering mix(ture)
抹灰爆裂 blowing(of plaster); pop-ping
抹灰边角的保护钢条 plaster bead
抹灰边抹子 angle trowel
抹灰剥落 addling
抹灰材料 plastering material; setting stuff; rendering material
抹灰层 backing coat; float(ed) coat; plaster coat
抹灰层底漆 plaster primer
抹灰层底涂料 plaster primer
抹灰层间的楔子 mechanical bond
抹灰层面层 face coat
抹灰承包人 plaster contractor
抹灰冲筋 plaster screed; screed strip
抹灰出现松散细粉的缺陷 chalking
抹灰次序 plaster system
抹灰打底 key rendering; pricking-up (coat); primary coat(ing); prime coat(ing); rendering
抹灰打底材料 <一般使用灰泥> coarse stuff
抹灰打底层 scratch course
抹灰挡板 plaster stop
抹灰导点 inner bead
抹灰的 plastered
抹灰的板条 lathing
抹灰的爆裂缺陷 popping
抹灰的底 base for plaster(ing)
抹灰的底层 key floating
抹灰的拉毛工具 scratch tool
抹灰的门券 plaster arch
抹灰的网工具 small tool
抹灰底层 key floating; plaster base; pricking up; primary coat(ing); render coat; rough coat; roughing-in; scratch coat
抹灰底层板 plaster baseboard
抹灰底面 background for plastering
抹灰底涂层 rendering coat
抹灰底子结构 rendering background
抹灰垫层 browing
抹灰垫圈 plaster ring
抹灰顶棚 plaster(ed) ceiling(panel);

plaster(ed) ceiling(slab); stucco ceiling
抹灰定厚木条 plaster ground
抹灰方法 plaster method
抹灰方式 plaster system
抹灰分隔条 screed
抹灰辅助工 plasterer's labo(u)rer
抹灰钢刮板 base screed
抹灰钢丝网 integral lath(ing); lathing mesh
抹灰格间 floating bay
抹灰工 plasterer; trowel hand; trowel man; fixer <现场的>
抹灰工程 floated work; plaster work; rendering work
抹灰工工长 foreman plasterer
抹灰工刮尺 plasterer's darby
抹灰工灰板 plasterer's hawk
抹灰工具 floater; hair hook; hand float; plasterer's tool; plastering tool
抹灰工抹子 plasterer's float
抹灰工人 plaster
抹灰工细灰膏 plasterer's putty
抹灰工用脚手架 plasterer scaffold(ing)
抹灰工作 float(ed) work; plastering; plaster work
抹灰工作头道打底 first coat in plastering work
抹灰光面 parg(et)ing
抹灰厚度 depth of plastering; thickness of coating; thickness of plastering
抹灰护角条 plaster head; plaster staff
抹灰滑动橡胶样板 running mo(u)ld
抹灰机 mechanical float; mechanical trowel; pasting machine; plastering machine
抹灰基层 base coat; plaster base
抹灰基层修整 dubbing out
抹灰基底刻槽接合 mechanical key
抹灰技术 rendering technique
抹灰简图 plaster scheme
抹灰浆 buttering
抹灰结构 plastering background
抹灰金属网 metal lath
抹灰靠尺 screed rail; screed strip; slipper
抹灰空鼓 gaul
抹灰空心砖 furring tile
抹灰篱笆墙 wattle and da(u)b
抹灰立面 rendered facade
抹灰麻刀 plaster(er)'s hair
抹灰镘刀 plasterer's trowel; plastering trowel
抹灰面层 finishing of coat of plaster; finish plaster
抹灰面疵病 plastering defect
抹灰面发裂 crazing of plaster
抹灰面房屋 rendered building
抹灰面刮尺 floating rule
抹灰面划痕 scratching
抹灰面积 floating area; floating bay; plaster area
抹灰面交叉划痕 cross scratching
抹灰面开裂 cracking of plaster; crazing of plaster
抹灰面用嵌线条 floating rule
抹灰面油漆 paint on plaster
抹灰抹子 plaster's float; plastering trowel
抹灰木板条 wooden lathing
抹灰木纤维护壁板 wood fiber[fibre] plaster baseboard
抹灰内装暗线 line installed in plaster
抹灰泥 marling; plaster
抹灰泥刀 plastering trowel
抹灰泥的 plastered
抹灰泥的机械抹子 power trowel
抹灰泥的基底 lathing
抹灰泥的外墙正面 external plaster facade
抹灰腻子 plasterer's putty
抹灰黏[粘]结剂 <一种水泥砂浆> liqualino
抹灰黏[粘]结力的提高 improvement of plaster bond
抹灰盘 pallet
抹灰喷浆器 air trowel
抹灰喷射机 plaster-throwing machine
抹灰平顶 flat plaster ceiling
抹灰砌石结构 solid background
抹灰嵌缝 putty in plastering
抹灰墙 plaster(ed) wall(ing); wet wall
抹灰撒黏[粘]石子、卵石或壳墙面 rock dash
抹灰砂 plastering sand
抹灰砂浆 rendering mortar
抹灰饰面 float(ed) finish
抹灰饰面建筑 stucco finished building
抹灰天花板 plaster ceiling; plaster panel; plaster slab
抹灰挑檐 plaster cornice
抹灰铁丝网 iron lath(ing)
抹灰凸包 bulb
抹灰涂底油灰 plasterer's putty
抹灰脱层 adding
抹灰脱落 falling plaster
抹灰污斑 plaster stain
抹灰线脚 plaster mo(u)lding
抹灰压光面 hard finish; putty coat; trowel finish in plastering
抹灰样板 screed rail
抹灰以后安装的部件 second fixing
抹灰印痕 brooming
抹灰用板 plaster's hawk
抹灰用板条 plaster lath(ing)
抹灰用板条锤 plasterer's lath hammer
抹灰用锤 plasterer's hammer
抹灰用钢丝网 steel lathing; wire lath
抹灰用隔热纤维板条 insulating fiber [fibre] board lath
抹灰用划痕器 scratcher
抹灰用灰板 plasterer's hawk
抹灰用建筑材料 plastering material
抹灰用靠尺 plaster ground
抹灰用镘刀 plasterer's darby
抹灰用抹子 plasterer's float
抹灰用砂 plastering sand; rendering sand
抹灰用水泥 <基层为白色硬水泥> Astroplax
抹灰用硬板条 rock lath(ing)
抹灰用油灰 plaster putty
抹灰增强网 plaster reinforcing mesh
抹灰找平 dubbing
抹灰找准木条 plaster ground
抹灰罩面 finish plaster
抹灰罩面层 setting coat (plaster); skimming coat
抹灰整平板 base screed
抹灰柱条 furring nails
抹灰柱头 furring
抹灰砖工 plastered brickwork
抹灰砖砌体 plastered brickwork
抹灰准尺 running rule; slipper; slipper guide
抹灰准带 running screed
抹灰准木 plaster ground
抹灰准条 floating screed; running screed
抹灰作业 floated work; plaster work
抹迹 erase
抹迹电路 erase circuit
抹迹射束 play-off beam
抹迹速率 erasing speed
抹浆 browning plaster; plastering; grout pour
抹浆厚度 grout lift
抹浆砌筑 buttered masonry
抹胶 daub
抹胶刮刀 trowel tool
抹角的 splayed
抹角梁 cornered beam
抹角镘 arrissing tool
抹角泥刀 corner trowel
抹镜水 lens cleaner
抹镜纸 lens tissue
抹棱工具 arrissing tool
抹里衬 back mortaring; backplastering
抹里皮 back mortaring
抹两层灰 two-coat work
抹密封剂 buttering
抹面 finishing; float work; stuke; surface trowel(l)ing; trowel(led) finish
抹面剥落 addling; disintegration of plaster
抹面层 float coat; skimming
抹面产生瑕疵污点 catface
抹面法 texturing
抹面工作 face work; plastering
抹面刮糙器 scarifier
抹面刮粗 score
抹面灰浆 finishing mortar; float coat; plaster
抹面混凝土 floated concrete
抹面机 finisher; finishing machine; mechanical trowel
抹面腻子 front putty
抹面起鼓 blistering in plaster
抹面砂浆 decorative mortar; mortar for coating; plastering mortar
抹面水纹 water crack
抹面土层 earth lining
抹面细灰浆 fine stuff
抹面油灰 front putty
抹面砖墙 plastered brickwork
抹泥 claying; daub
抹泥刀 claying knife; plaster(ing) trowel
抹泥修墙 spackling
抹腻白坯 filled
抹盘 trowel(l)ing plate
抹平 fettle; floating; screeding; strike off; trowel-applied finish (ing); trowel(l)ing
抹平的表面 floated surface
抹平的混凝土 floated concrete
抹平的涂面 trowel coating
抹平地面 float finish floor
抹平混合料 levelling compound
抹平机 floating machine; trowel(l)ing machine
抹平料 filling compound; grouting filling; screeding compound
抹平面 floated coat; floated surface
抹平试验 <现场测定混凝土工作度的> trowel test
抹平填料 floated filler
抹平子 floater
抹墙充筋 wall furring
抹去 blotting; efface; erase; strike off
抹去磁头 erase head
抹去读数 destructive read
抹去能力 erasing ability
抹去时间 erasing time
抹去树 erasing of tree
抹去数据 dataout
抹去速度 erasing speed
抹去信息读出 destructive readout; destructive reading
抹去信息读数【计】destructive read
抹去账目 wipe out an account
抹去账项 strike-off an entry; write off an entry
抹去字符 erase character
抹杀 efface
抹砂浆层 parg(et)ing
抹砂浆镘刀 buttering trowel
抹头 rail; window stool
抹涂 brush coating; trowel(l)ing
抹涂层 trowel coating
抹纤维性灰泥工作 rag and stick work
抹香鲸 sperm whale
抹香鲸酸 physeteric acid
抹香鲸烯酸 palmitoleic acid; physetoleic acid
抹音(磁)头 erase head; erasing head
抹音联锁装置 erase interlock
抹音器 eraser
抹圆角边 bull-nose edge; thumb edge
抹圆角线脚 thumb mo(u)lding
抹纸筋灰 daub
抹子 darby; finishing tool; hand float (trowel); mason's float; paddle; pallet; plane; rubbing board; strike off; trowel
抹子缝 trowel(led) joint
抹子拉毛 trowel(l)ed stucco
抹子修整混凝土表面 floated finish
抹子压光 trowel finish

末遍 last pass

末遍扩充工作码 last pass own code
末波分量 meter component
末层包布 final wrap
末层涂漆 finish varnishing
末茶 dust tea
末次读数 final reading
末次付款 final payment
末次会晤 last schedule
末次会议 closing meeting
末次冷却器 after-cooler
末次碾压 final rolling
末次碰撞 last collision
末档齿轮 top gear
末档速度 top gear speed
末道 extreme trace
末道表面粉刷 final coat exterior plaster
末道并条机 finishing drawing frame
末道瓷漆 enamel finish
末道粗纱 dandy roving
末道粗纱机 dandy finisher; dandy rover; jack frame
末道底漆 finished primer
末道反应 terminal reaction
末道粉墙泥灰料 final stucco stuff
末道粉刷 fining-off
末道粉刷层 finish(ing) coat (plaster); setting coat
末道工序 finish
末道灰泥 brown coat; finish plaster
末道混合涂料 final coat mixed plaster
末道抹灰 final rendering; finish plaster
末道抹灰材料 final rendering stuff
末道抹灰层 skim coat plaster; face coat
末道抹灰混合料 final rendering mix(ture)
末道漆 final coating; final lacquer; finish
末道清漆 finishing varnish
末道饰面层 finish coat
末道涂层 final coat; finishing coat
末道涂层混合料 final coat external plaster stuff
末道研磨机 final grinder
末道油漆 final coat paint
末道渣 crinkle finishing
末电压 <蓄电池放电终了电压> final voltage
末端 dead end; distal end; end(ing);

M

extreme; extremity; fag-end; tail (end); terminal; termination; terminal section <防波堤等的>
末端氨基 terminal amino group
末端板加油器 end plate oiler
末端棒眼 terminal bar
末端本征值 end eigenvalue
末端泵站 terminal pump station
末端标记 end mark
末端部 terminal part
末端部件 terminal member
末端测试 tag end test
末端掺入 terminal incorporation
末端产物抑制 terminal product inhibition
末端朝前供给 endwise feed
末端朝前输送 endwise feed
末端朝上的 endway; endwise
末端程序 extremity routine
末端传动齿轮 final drive gear
末端传动齿轮箱 final drive gear box
末端搭叠 end lap
末端搭接 end lap
末端的 distal; extremital; terminal
末端电池 milking cell
末端电池式 end cell system
末端电池整流器 end cell rectifier
末端电池转换开关 end cell switch
末端电线杆 end pole
末端电阻材料 end-resistance material
末端队列 final queue
末端对接 end-on
末端发光光纤 end glow fiber[fibre]
末端反射损失 end reflection loss
末端分析 end-group analysis; terminal analysis
末端分支 end branch
末端改正 end correction
末端干管 dead-end main
末端杆 end pole
末端钩环 loop break
末端荷载 end load; terminal load
末端护罩 end shield
末端滑轮 tail block
末端基 end group
末端畸变 end distortion
末端交叉 terminal crossing
末端节点 end node; end point node
末端结点 end node
末端截面 tip section
末端井框 end crib
末端距 end-to-end distance
末端卡扣 end shackle
末端开间 end bay
末端控制 terminal control
末端控制装置 control terminal
末端跨 end bay
末端框架 end frame
末端扩大 enlarged end
末端扩大的岩石锚碇装置 under-reamed anchorage
末端拉杆 end tie-bar
末端冷却 end cooling
末端链节 end shot
末端链路 end link
末端脉动式继电器 final impulse operating relay
末端片 end piece; end wafer
末端屏蔽 end shield(ing)
末端汽缸 end cylinder
末端砌固 end built-in
末端砌入 end built-in
末端嵌固 end built-in
末端亲水基 terminal hydrophilic group
末端情况 end condition
末端缺失 terminal deletion
末端设备 end instrument
末端设备的 end equipment
末端收缩 end contraction
末端双键 end double bonds
末端速度 terminal velocity
末端羧基 terminal carboxyl group; terminal hydroxyl
末端套管 end sleeve
末端条件 end condition
末端调压电池 end voltage-regulating cell
末端温差 terminal temperature difference
末端吸入式泵 end suction pump
末端吸入消防泵 end suction fire pump
末端线圈 end coil
末端相接的 abutting
末端向前地 endway; endwise
末端效应 end effect
末端效应带 end effect zone
末端信息 final word
末端压力 terminal pressure
末端氧化 terminal oxidation
末端用户 end-user
末端约束 end restraint
末端约束效应 end restraint effect
末端再热系统 terminal reheat system
末端镇墩 end anchorage
末端整体空调器 packaged terminal air conditioner
末端支承 end bearing
末端执行器 end effector
末端装置 air terminal device; end equipment; terminal device
末段 end portion; end section; latter end
末段导引 terminal guidance
末段引导精度 terminal accuracy
末段制导 final guidance; terminal guidance
末付款额 amount in arrear
末级 final stage; last stage; output stage; upstage
末级沉淀池 final settling tank
末级除氧器 final stage deaerator
末级传动 final drive
末级传动防护器 final drive guard
末级传动磨损防护器 final drive wear guard
末级传动牙齿箱 final drive transmission
末级电路 final circuit
末级发动机 final stage (rocket) engine
末级放大 final magnification
末级放大器 final amplifier; last amplifier; output amplifier
末级功率放大器 final power amplifier
末级过热器 final superheater; finishing superheater
末级缓行器 final rail brake; final retarder; last retarder
末级激励器 final driver
末级集电极 final collector
末级加热器 top heater
末级减速器 final rail brake; final retarder; last retarder
末级检波器 last detector
末级控制元件 final control element
末级冷却器 after-cooler
末级清洗槽浓度 final rinse tank concentration
末级施控元件 final controlling element
末级视频放大器 final video-amplifier
末级输出 final output
末级阳极电压 final-anode voltage
末级叶轮 impeller
末级叶片 exhaust stage blade; last stage blade
末级羽片 ultimate pinna
末级再热器 final reheater
末级蒸发器 final evapo(u)rator
末级中继线 final trunk
末极放大器 final amplifier
末开垦土壤 virgin soil
末控元件 final controlling element
末煤 small coal
末期 final stage; terminal phase
末期当量【物】late-stage equivalence
末期干燥 terminal drying
末期裂缝率 final cracking rate
末前级 <倒数第二级> penultimate stage
末梢 ending; tip end
末梢的 acroteric; distal
末速度 end speed; end velocity; final velocity; terminal speed; terminal velocity; final speed
末涂层 skimming coat
末尾 end(ing); fag-end; tag end; tail; closure
末尾的 extreme
末尾页面 end page
末位 final bit
末位有效数字 last significant figure
末项 last term
末项的 terminal
末行 footline
末站 final station
末站掺入 terminal incorporation
末站设备 terminal outfit
末站月台 arrival platform
末桩 <指在一排中最后打入的桩> closing pile

没落 ruination; sinking

没落时代 twilight
没食子酸盐 gallate
没收 confiscate; confiscation; dispossession; expropriate; expropriation; foreclosure; impound (ing); seize; seizure; sequester; sequestrate; sequestration
没收财产 confiscation of property; expropriation
没收财物 escheat
没收担保品 foreclose
没收担保品的销售 foreclosure sale
没收担保品价值 foreclosure value
没收的财产 forfeiture
没收抵押品 forfeited securities; forfeited securities arising from loans
没收抵押品价值 foreclosure value
没收地 escheat
没收定金收益 income from forfeited deposits
没收法例 Forfeiture Act
没收公告 declaration of forfeiture
没收股款 forfeited stock subscription
没收货(物) confiscated goods
没收品变卖 sale of confiscated goods
没收汽车执照 forfeit motor licence [license]
没收人 confiscator
没收条款 forfeiture clause
没收无遗嘱的财产税 doctrine of intestate escheat
没收物 forfeit(ure)
没收物的处置 disposal of forfeits
没收物品(罚金)的补偿 recovery of forfeitures
没收性赋税 confiscatory taxation
没收性税金 confiscatory taxation
没收押品 forfeited securities arising from loans
没收者 confiscator
没收执照 forfeit
没收走私货物 seize smuggled goods; seizure of smuggled goods
没影点 <透视画中的线条会聚点> vanishing point
没影点方法 vanishing point method
没影点条件 vanishing point condition
没影面 vanishing plane
没影线 vanishing line
没影轴 vanishing axis

茉莉【植】jessamin(e)

秣仓 hayloft

莫艾型井 Maui-type well

莫邦材料 Moppon
莫尔包迹线 Mohr's envelope
莫尔包络面 Mohr's envelope
莫尔包(络)线 Mohr's envelope
莫尔比重天平 Mohr's balance
莫尔变曲线 Mohr's bending curve
莫尔道熔融石 moldavite; pseudo-chrysolite; vitavite
莫尔灯 Moore lamp
莫尔滴定法 Mohr's titration
莫尔地形测量学 Moiré topography
莫尔法 <用硝酸银滴定测定卤素离子的方法> Mohr's method; Moiré technique
莫尔干涉条纹 Moiré fringe
莫尔干涉条纹测量系统 Moiré fringe measuring system
莫尔干涉条纹图样 Moiré pattern
莫尔干涉效应 Moiré effect
莫尔格子 Moiré grille
莫尔格子砖 Moll checker
莫尔过滤器 Moore filter
莫尔毫升 Mohr's cubic centimeter
莫尔机 Moore machine
莫尔技术 Moiré technique
莫尔校正法 Mohr's correction method
莫尔夹 Mohr's clamp
莫尔-卡柯脱包络线 <在弹性界限时莫尔圆的包络线> Mohr-Caqout envelope
莫尔-库仑定律 Mohr-Coulomb law
莫尔-库仑方程 Mohr-Coulomb equation
莫尔-库仑滑落包络线 Mohr-Coulomb failure envelope
莫尔-库仑剪切强度参数 Mohr-Coulomb shear strength parameter
莫尔-库仑理论 Mohr-Coulomb theory
莫尔-库仑模型 Mohr-Coulomb model
莫尔-库仑破坏准则 Mohr-Coulomb failure criterion
莫尔-库仑强度包线 <三轴压力试验> Mohr-Coulomb strength envelope
莫尔-库仑屈服面 Mohr-Coulomb yield surface
莫尔-库仑屈服函数 Mohr-Coulomb yield function
莫尔-库仑屈服准则 Mohr-Coulomb yield criterion
莫尔-库仑土模型 Mohr-Coulomb soil model
莫尔-库仑准则 <强度理论的> Mohr-Coulomb criterion
莫尔冷凝器 Mohr's condenser
莫尔理论 Mohr's theory
莫尔立方厘米 Mohr's cubic centimeter
莫尔练漂法 Mohr's bleaching process
莫尔滤机 Moor filter
莫尔曼指数 Mohlaman index
莫尔蒙砂岩 Mormon sandstone
莫尔模型 Moore model

莫尔-彭罗斯逆变换 Moore-Penrose inverse
莫尔破坏包络线 Mohr's failure envelope;Mohr's rupture envelope
莫尔破坏理论 Mohr's theory of failure
莫尔破坏图解 Mohr's failure diagram;Mohr's rupture diagram
莫尔破裂包(络)线 Mohr's rupture envelope
莫尔强度理论 Mohr's strength theory;Mohr's theory of strength
莫尔桥 <支承在过梁上> Moore's bridge
莫尔屈服准则 Mohr yield criterion
莫尔全息术 Moiré holography
莫尔全息照相法 Moiré-holography method
莫尔升 <约为 1.002 公升> Mohr's litre[liter]
莫尔斯标准 Morse standard
莫尔斯电报机 Morse set;Morse telegraph
莫尔斯电报技术 dot and dash technique
莫尔斯电报键 Morse key
莫尔斯电码 dot-and-dash code;Morse code
莫尔斯多工电报制 Morse multiplex system
莫尔斯方程式 Morse equation
莫尔斯分布 Morse distribution
莫尔斯符号 Morse code
莫尔斯继电器 Morse relay
莫尔斯收报 Morse reception
莫尔斯收报机 Morse receiver
莫尔斯双工机 Morse duplex
莫尔斯信号灯 Morse code light;Morse signal light
莫尔斯音响器 Morse sound
莫尔斯锥柄键槽铣刀 Morse taper shank keyway cutter
莫尔斯锥柄立铣刀 Morse taper shank slotting cutter
莫尔斯锥度 Morse taper
莫尔斯钻 Morse drill
莫尔弹簧夹 Mohr's pinchcock
莫尔条纹 Moiré fringe
莫尔条纹测量 Moiré topography
莫尔条纹法 Moiré fringe technique
莫尔条纹图 Moiré topograph
莫尔图 Mohr's circle;Mohr's diagram
莫尔图解法 Mohr's graphical solution
莫尔图示法 Mohr's graphic representation
莫尔图形 Moiré pattern
莫尔弯曲曲线 Mohr's bending curve
莫尔型多叶真空过滤器 Moore filter
莫尔应变圆 Mohr's strain circle
莫尔应力圆 Mohr's circle;Mohr's circle of stress;Mohr's stress circle;Mohr's stress diagram
莫尔应力圆图 Mohr's circle diagram
莫尔硬度计 Mohr's scale
莫尔硬度仪 Mohr's scale
莫尔圆 circle of Mohr;Mohr's circle
莫尔圆的适应范围 Mohr's circle applicability
莫尔圆分析法 Mohr's circle analysis method
莫尔圆绘制 construction of Mohr's circle
莫尔圆图 Mohr's diagram
莫尔兹比港 <巴布亚新几内亚首都> port Moresby
莫根森筛 Mogenson screen
莫古灯头 mogul base
莫霍不连续面【地】Moho's discontinuity
莫霍间断面 Moho discontinuity;Mohorovicic discontinuity
莫霍克巴比特合金 Mohawk babbitt alloy
莫霍罗维奇(不连续)面 Moho;Mohorovicic discontinuity
莫霍面【地】Moho surface;Moho;Mohorovicic discontinuity
莫霍面边界 Moho boundary
莫霍面等深线 Moho discontinuity
莫霍面等深线图 Moho contour map
莫霍面深度【地】depth of Moho-discontinuity;depth contour;Moho depth
莫霍面形态【地】form of Moho-discontinuity
莫霍面钻探 Mohorovicic discontinuity hole;Mohole
莫霍钻 Mohole;Mohorovicic discontinuity hole
莫-坎-西烈度表 M-C-S intensity scale
莫-肯二氏电隧道窑 Moore-Camhbell kiln
莫拉 <杂色硬木,产于圭亚那> Mora
莫来石 mullite;porcelainite;porzite;keramite <莫来石的别名>
莫来石瓷 mullite white ware
莫来石大砖 mullite block
莫来石化 millitisation[mullitization]
莫来石基复合材料 mullite-based composite
莫来石-堇青石耐火材料 mullite-cordierite refractory
莫来石砌块 mullite block
莫来石陶瓷 mullite ceramics
莫来石纤维 mullite fiber[fibre]
莫来石质瓷 mullite porcelain
莫来石质耐火材料 mullite refractory
莫来石质耐火砖 mullite fire brick
莫来石砖 mullite brick
莫勒均化法 Moller homogenizing process
莫雷自由下落取土器 <用于深水下的> Moore free-fall corer
莫里森方程 <计算波浪力的公式> Morison's equation
莫里森公式 <计算波浪力的公式> Morison's formula
莫里森-努森发动机 <美> Morrison-Knudson engine
莫里斯式安乐椅 Morris chair
莫里亚图 <压焓图> Moilier chart
莫利特精密水准尺 <带有圆水准器和温度计,最小分划 2 毫米> Molitor precise leveling rod
莫罗尼 <科摩罗首都> Moroni
莫洛金斯基法 Molodensky method
莫洛金斯基理论 Molodensky theory;theory of Molodensky
莫洛凯破裂带 Moloka fracture zone
莫梅埃油脂不饱和试验 Maumene test
莫姆 <力迁移单位> mohm
莫诺泵 Mono pump
莫诺方程 Monod equation
莫诺菲尔特丙纶机织土建布 Monofilter
莫诺公式 Monod formula
莫诺赖特栗红 monolite maroon
莫诺特龙刻痕硬度试验 monotron indentation hardness test
莫诺特龙速调管 monotron
莫诺特龙硬度检验仪 monotron
莫诺特龙硬度试验 monotron hardness test
莫诺硬度计 monotron
莫诺铸排机 Monotype
莫诺自动排字铸印机 monotype
莫塞莱定律 Moseley law
莫桑比克海流 <非洲> Mozambique current
莫桑比克海盆 Mozambique basin
莫桑比克海峡 Mozambique channel
莫桑电炉 Moissan furnace
莫砷硒铜矿 mgriite
莫氏标度 Mohs' scale
莫氏标度值 Mohr's scale number
莫氏标准锥度 Morse standard taper
莫氏标准锥度(塞)规 Morse standard taper plug ga(u)ge
莫氏电码 Morse code
莫氏分级 Mohs' scale
莫氏缓冲器 Murray friction draft gear
莫氏刻痕硬度试验 <用以确定产生一定刻痕深度所需的荷载> monotron indentation hardness test
莫氏量规 Morse ga(u)ge
莫氏麻花钻 Morse twist drill
莫氏耐磨性指数 Modell number
莫氏强度理论 Mohs' strength theory
莫氏强度值 Mohs' scale number
莫氏天平 Mohs' balance
莫氏相对硬度标 Mohs' relative hardness scale;Mohs' hardness scale
莫氏硬度 <矿物硬度标度> Mohs' scale;Mohs' hardness
莫氏硬度标 Mohs' scale of hardness
莫氏硬度表 Mohs' scale of hardness
莫氏硬度分度法 Mohs' scale of hardness
莫氏硬度分级 Mohs' hardness range
莫氏硬度计 Mohs' hardness ga(u)ge;Mohs' hardness scale;Mohs' scale of hardness
莫氏硬度值 Mohr's scale of hardness;Mohs'(scale) number
莫氏圆锥 Morse's cone
莫氏圆锥铰刀 Morse conical reamer
莫氏圆锥钻 drill with Morse cone
莫氏蒸汽图 Morse's chart
莫氏锥柄麻花钻头 Morse taper shank twist drill
莫氏锥柄钻 Morse taper shank drill
莫氏锥度 Mohs' taper;Morse taper
莫氏锥度的钻 drill with Morse taper
莫氏锥度铰刀 Morse taper reamer
莫氏锥度量规 Morse taper ga(u)ge
莫氏锥度套筒 Morse taper sleeve
莫氏锥度钻套 Morse taper drill sleeve
莫氏锥形粗加工铰刀 Morse taper roughing reamer
莫氏锥形铰刀 Morse taper reamer
莫氏锥形精加工铰刀 Morse taper finishing reamer
莫氏锥形孔 Morse tapered hole
莫水硅钙钡石 macdonaldite
莫斯-布尔斯坦效应 Moss-Burstein effect
莫斯金干法 Muskingum method
莫斯金干洪水演进法 Muskingum routing method
莫斯凯特法 Muskat method
莫斯凯特截距 Muskat intercept
莫斯科维茨氏试验 Moschcowitz's test
莫斯利定律 Moslay's law
莫斯氏加强混凝土粗筋 Moss bar
莫斯西基电容器 Moscicki condenser
莫斯硬度 Moss hardness
莫卧儿帝国建筑 <印度> architecture of the Mogul Empire
莫卧儿建筑 <印度> Mogul architecture
莫泽方程 Moser equation

漠

漠境土 yermosol
漠视 disregard
漠土 desert soil

墨

墨 China ink
墨斑 ink speck
墨泵 ink pump
墨彩 <陶瓷装饰法> china-ink painting;grisaille painting
墨层过厚 monk
墨辰砂 metacinnabarite
墨床 ink rest
墨带 ink ribbon;ribbon
墨滴 ink droplet
墨地白花 white-and-black;white-on-black
墨地三彩 tricolo(u)r with china-ink ground
墨点 ink dot
墨锭 ink stick
墨斗 carpenter's ink number;duct;ink fountain;inking pot
墨斗刮刀 duct blade;duct knife
墨斗轨 ink rail
墨斗辊 doctor roller;duck roller;ink duct roller
墨斗辊摆动杆 rocking-lever for ink duct roller
墨斗键 ink adjustment key
墨斗搅拌器 ink fountain agitator
墨斗调节螺丝 duct adjusting screw
墨斗线 black mark-line;carpenter's ink box and line
墨斗装置 ink system
墨度计 nigrometer
墨尔本港 <澳大利亚> Melbourne Port
墨尔本岩 Melbourn rock
墨粉 dry toner;ink powder;powdered toner
墨粉不足传感器 toner empty sensor
墨粉层电位 toner layer potential
墨粉分配传动机构 toner dispense gear
墨粉调节辊 toner metering roller
墨粉载体配合用材料 material for the toner-carrier combination
墨辊 inking roller
墨辊摆动中断器 ink vibrator interrupter
墨辊冲洗机 inker wash-up machine
墨辊冲洗设备 roller washing
墨辊杠子 roller mark
墨辊浇铸机 inking roller casting machine;roller casting apparatus
墨辊膨胀 roller swelling
墨辊清洗装置 ink roller cleaning device
墨辊调节 roller setting
墨辊脱墨 stripping
墨辊洗涤槽 roller trough
墨辊转移印刷 pad-transfer printing
墨滚 inker
墨盒 ink box
墨盒和镇尺 ink case and paper-weight
墨黑 inkiness
墨黑的 coal black;swarthy
墨黑色 ink black;raven black
墨化剂 graphitizer;graphitizing medium
墨灰色 raven gray[grey]
墨迹 blot;inkblot
墨迹测验 inkblot test;Rorschach test
墨迹反射 ink reflectance
墨迹均匀性 ink uniformity
墨迹式画图 inking
墨角藻 focus vesiculosus;fucus;seat angle
墨角藻状砂岩 fucoid(al) sandstone
墨结球 black sphere
墨金刚石 carbonado
墨晶 black crystal;black quartz;o-

M

paque crystal;smoky quartz
墨晶石雕 smoky quartz carving
墨镜 sun visor
墨卡托方位 Mercator direction;rhumb direction
墨卡托方位角 Mercator(al)bearing;rhumb bearing
墨卡托方向 Mercator direction;rhumb direction
墨卡托海里 Mercator's mile
墨卡托海图 Mercator chart
墨卡托航迹计算法 Mercator sailing;rhumb-line sailing
墨卡托航线 Mercator track
墨卡托航向 Mercator course;rhumb course
墨卡托航行法 Mercator sailing
墨卡托航用图 Mercator chart
墨卡托投影 Mercator('s)projection
墨卡托投影地图 Mercator chart;Mercator's projection chart;Mercator's projection map
墨卡托投影海图 equatorial cylindric(al)orthomorphic chart;Mercator's projection chart
墨卡托型图 Mercator-type graph
墨累河 Murray River
墨流印刷 marbling print
墨绿 blackish green;dark-green;green black;greenish black
墨绿环氧烘干黑板漆 blackish-green epoxy baking paint for blackboard
墨绿色 atrovirens;invisible green;jasper
墨绿色布 hunter green cloth
墨绿色的 greenish black
墨绿砷铜矿 cornwallite
墨绿砷铜石 erinite
墨洛温王朝建筑 Merovingian architecture
墨色 mass colo(u)r;overtone;top tone
墨色不匀 bad colo(u)r
墨色不足 friar
墨色反射特性 ink reflectance
墨色均匀 good colo(u)r
墨色浓度 print density
墨色样本 colo(u)r specimen
墨守成规 scholasticism
墨水 atrament;writing ink
墨水笔 ink pen;wet pen
墨水笔绘记录器 ink writing recorder
墨水笔绘图机 pen plotter
墨水带 ink ribbon
墨水点 inkspot
墨水反射率 ink reflectance
墨水盒 ink cartridge
墨水痕 iron mo(u)ld
墨水画 ink drawing
墨水绘图机 ink plotter
墨水挤出 ink squeeze out
墨水记录仪 ink recorder
墨水精 ink essence
墨水均匀性 ink uniformity
墨水蓝 ink blue
墨水描迹器 ink plotter
墨水黏[粘]度计 inkometer
墨水浓度 ink density
墨水喷射打印机 ink-jet printer
墨水飘浮试验法 ink flo(a)tation test
墨水瓶 ink bottle;inkpot
墨水瓶架 inkstandish
墨水瓶状孔 ink-bottle pore
墨水容器 ink tank
墨水色的 atramental
墨水色料 ink dyestuff
墨水渗出 ink bleed
墨水渗迹 ink bleed
墨水水池 ink well
墨水台 inkstand;standish
墨色调整 colo(u)r adjusting
墨水涂染 inked
墨水味 ink taste
墨水污迹 ink smudge
墨水雾化打印机 ink mist printer
墨水雾式打印机 ink fog printer
墨水雾印刷机 mist printer
墨铜矿 valleriite
墨污 blot
墨西哥柏油 Mexico asphalt
墨西哥产巴西果油 Mexican oiticica oil
墨西哥城 <墨西哥首都> Mexico City
墨西哥大铁锤 Mexican crusher
墨西哥地槽 Mexican geosyncline
墨西哥缟玛瑙 Mexican onyx
墨西哥海流 Gulf stream
墨西哥建筑 Mexican architecture
墨西哥沥青 Chapapate;Mexico asphalt
墨西哥暖流 Gulf stream
墨西哥钳子 Mexican speed wrench
墨西哥式人造草坪 chinampa
墨西哥式十字镐 Mexican back-hoe
墨西哥湾 Gulf of Mexico
墨西哥湾岸盆地 Mexico Gulf Coastal basin
墨西哥湾地槽 Mexican Gulf Coast geosyncline
墨西哥湾各港 Gulf ports
墨西哥湾海岸 gulf coast
墨西哥湾流 Gulf stream
墨西哥沿岸航路 Gulf intracoastal waterway
墨西哥沿岸运河 Gulf intracoastal canal
墨西哥油 chia oil
墨西拿海峡桥 <意大利> Messina Strait crossing
墨西拿造山旋回 Messina cycle
墨线 carpenter's line;ink line
墨线闩 <木工用> ga(u)ge
墨线图 ink drawing
墨油胶 black factice
墨玉 black diamond
墨云片岩 graphite mica schist
墨汁 Indian ink;liquid ink

默

默奥斯奎地毯 Mousquet

默冬章 lunar cycle
默硅镁 merwinite
默硅镁钙石 merwinite
默罕默德建筑 Mahometan architecture
默科(铅)青铜 Merco bronze
默里环路定位试验 Murray's loop test
默里环路试验 Murry loop test
默里环线试验法 Murray's circuit testing method
默里破裂带 Murray's fracture zone
默契 privity;tacit understanding
默契的接受 tacit acceptance
默契的协议 tacit agreement
默契的转期 tacit renewal
默契值 default value
默认 implied recognition;tacit acquiescence;tacit recognition
默认保证 implied warranty
默认承诺 tacit acceptance
默认的驱动器 default drive
默认合伙 implied partnership
默认合同 accepted contract
默认合同续订 tacit renewal
默认契约 implicit contract;implied contract
默认驱动(器)default drive
默认(缺省)打印机【计】default printer
默认容许浓度 mum permissible concentration
默认条件 implied condition;implied term
默认协议 implied agreement;tacit agreement
默认续订 <合同的> tacit renew
默认值 default value
默认作废 repeal by implication
默示保证 implied warranty
默示承担风险 implied acceptance of risk
默示承诺 implied promise;implied recognition;tacit acceptance
默示承认 implied recognition
默示代理 implied representation
默示担保 implied warranty
默示的授权 implied authority
默示放弃 implied waiver
默示管辖权 implied jurisdiction
默示合同 implied contract;implied deed
默示加入 tacit adherence
默示批准 implied ratification
默示契约 implied deed
默示取消 implied revocation
默示权力 implied power
默示权益保留 implied reservation
默示税收 implicit tax
默示条件 implied term
默示条款 implied term
默示同意 implied consent;tacit consent
默示协定 implied agreement
默示信托 implied trust
默示义务 implied undertaking
默示异议 implied objection
默脱面 <复杂三棱镜的胶合面> Merte surface
默许 implied consent;sufferance;tacit consent
默许代理权 implied authority
默许订价 price fixing
默许上市委托 implied listing
默许条件 implied condition
默许信托 implied trust
默许租赁 tenancy at sufferance
默扎珀地毯 mirzapur

牟

牟取暴利者 profiteer

谋

谋取暴利 profiteering

谋士 idea man

某

某地纬度 latitude of a place

某点海拔高度 spot elevation
某点平行轴面的厚度 thickness parallel to axial plane at given point
某点正交厚度 orthogonal thickness at given point
某个动作所需的平均时间 required time
某个来源的排污量 discharge of a pollutant
某国商船(的总称)<只用单数> marine
某航向上的航程 miles on course
某几类产品的平均关税 average duty on specified category of product
某间隔的钎杆 equi-spaced steel
某些观察 some observation on
某些土壤性质 some soil property
某些土壤与作物关系 some soil plant relationships
某些微生物 certain micro-organisms
某些植物 certain plants
某一波长区域 wavelength region
某一点的交通量调查(或计数)spot count
某一订货人 the order of
某一类型船避航区 area to avoid by ships of certain classes
某种程度的 partial

模

模板 formwork board;formwork plate;boxing;casting box;die block;die plate;falsework;follow board;match board;match plate;mitre templet;moldboard;mother plate;mo(u)ld(ing) board;mo(u)ld(ing) plate;oddside board;panel shuttering;pattern plate;profile board;sheathing;sheathing board;sheathing plate;shutter-board;shuttering;shuttering form(work);shuttering panel;spandril;stencil(plate);template;templet;wood planking;joint mo(u)ld <预制灰质饰件的>

模板安放 form layering;form laying
模板安装 form erection;form placing;form setting
模板安装工 formwork setter;shuttering erector
模板安装器 shuttering erector
模板疤 form scabbing
模板板条 shutter lath;shutter slat
模板保温 form insulation
模板保温材料 form insulation material
模板比较 template matching
模板边撑 form clamp
模板表面 forms facing
模板不变的 form-retentive
模板不变性 form-retentiveness
模板布置 form layout;formwork planning
模板布置图 form panel layout
模板部件 formwork element;shuttering element
模板材 form lumber
模板材料 template material
模板侧压力 form lateral pressure;form side pressure;side pressure of formwork
模板拆除 form removable;form stripping
模板拆卸 dismantling of shuttering
模板拆装工作 form handling
模板车升降机 hoist for mo(u)ld board carrier
模板衬板 form sheathing
模板衬垫 form lining;shuttering lining
模板衬里 forms lining;formwork liner;mo(u)ld liner
模板衬料 form lining
模板衬砌 form lining
模板撑挡 form spreader;spreader
模板撑杆 form spreader
模板承包商 formwork contractor
模板挡 form stop
模板的侧板 formwork sideboard
模板的侧木 timbering sideboard
模板的反拉钢丝 wire back-tie
模板的活络搭扣 wrecking strip
模板的绝热 form insulation
模板的木板 shuttering board
模板垫衬 form liner;shutter liner
模板垫条 form liner;shutter liner
模板吊架 form hanger;form traveler
模板钉 form nail
模板钉结 form-tying
模板定距标尺 form spacer
模板堵头 form stop;stunt end

模板端头 form stop
模板反力模量 formwork reaction modulus
模板防黏[粘]涂料 form lacquer
模板费 form cost
模板封闭剂 forms sealer
模板膏 form paste
模板格栅 formwork joist
模板隔离剂 form coating
模板工 form fixer;form setter
模板工场 template shop
模板工程 former;form project;formwork (engineering); shuttering works
模板工程安装车 < 隧道、坑道施工 > forms handler;formwork handler
模板工程的底板 formwork bottom
模板工程工作 formwork work
模板工程规范 formwork specifications
模板工程压力 pressure on formwork
模板工锤子 formwork setter's hammer
模板工作 < 浇灌混凝土之前的 > formwork
模板构件 formwork member
模板箍 < 钢筋混凝土柱的 > column clamp
模板箍圈 formwork yoke
模板股 template strand
模板骨架 form cage
模板固定件 formwork fixture
模板挂钩 form hanger
模板管理人 shuttering handler
模板规划 formwork plan
模板夯实机 < 道路 > form tamper
模板痕(迹) < 混凝土表面上的 > shuttering mark
模板横撑 ranger;spelled waler;waler
模板横档 shuttering tie
模板机理 template mechanism
模板机械制图 mechanical template plot
模板夹具 form clamp
模板间隔物 form spacer
模板间隙 die clearance
模板件 prototype board
模板接缝 form joint;formwork joint
模板节缝 form joint
模板紧固螺栓 dies plate tap bolt
模板绝热材料 form insulation
模板卡箍 formwork yoke
模板框 pattern plate bolster;vibrator frame
模板框架横木 frame timber
模板拉杆 form (work) tie; shuttering tie
模板拉条 form brace
模板拉铁 formwork tie
模板蜡 shuttering wax
模板里衬 formwork lining;formwork liner
模板立柱 soldier
模板檩条 ledger
模板螺栓 sheave bolt
模板脉冲控制 template impulse control
模板锚定板 form anchor
模板锚定螺丝 formwork anchored screw
模板锚定器 form anchor
模板锚定物 form anchor
模板锚杆 form anchor
模板锚固装置 form anchor
模板密封材料 form sealer
模板密封料 shuttering sealer
模板面 form face
模板面涂油 face of oiling
模板磨耗 abrasion of form
模板内侧的三角形嵌缝木条 angle fillet
模板内衬 form(work)lining
模板内面 face of form
模板牛油 formwork grease
模板喷花 stenciling
模板匹配 template matching;templet matching
模板铺设 form spread
模板切割法 template cutting
模板切割器 template cutter
模板清除机 formwork cleaner
模板清除器 formwork cleaner
模板清洁器 formwork board cleaning machine; shuttering board cleaning machine
模板润滑膏 formwork grease
模板润滑油 form lube;formwork oil
模板上支撑木块 kicker
模板设备 < 建筑物外伸部分的 > stepping formwork equipment
模板设计 formwork plan
模板升高 formwork movement
模板施工 mo(u)ld construction
模板式钻模 template jig
模板刷滑 stenciling
模板梭动机构 pattern shuttle
模板台车 formwork jumbo
模板填孔料 filler
模板贴料 form liner
模板图(样)form(work)drawing
模板涂层 form coating;form(work) lining
模板涂料 form coating; form oil; shuttering paint
模板涂油 form coating
模板挖掘机 template excavate
模板外部振捣器 form (work) vibrator
模板外壳 forms facing
模板弯曲 formwork deflection
模板纹混凝土 board-marked concrete
模板吸水衬里 absorptive form lining
模板洗涤机 formwork board cleaning machine; shuttering board cleaning machine
模板系杆 forms tie
模板系统 shuttering system
模板系统的立柱 spreader
模板型的 stencil-like
模板学说 template theory
模板压力 form(work)pressure;pressure developed by concrete on formwork
模板移动 formwork movement
模板移动车架 travel(l)er
模板移动机构 pattern movement mechanism
模板移动装置 form traveler
模板印刷机 stenciler
模板印纹 board-marked texture
模板用胶合板 form plywood; formwork plywood
模板用木板 form lagging; shuttering board
模板用油 form oil
模板油 concrete oil;form oil;mo(u)ld oil
模板油喷雾器 formwork oil atomizing device
模板(与钢筋之间的)垫块 formwork spacer
模板运输车 formwork transport wagon
模板造模 pattern plate mo(u)lding
模板造型 plate mo(u)lding
模板罩 platen cover
模板振捣 formwork vibration
模板振捣混凝土 < 与插入振捣混凝土不同 > form-vibrated concrete
模板振捣器 external vibrator;shutter(ing) vibrator
模板振动 shutter vibration
模板振动器 form vibrator;mo (u) ld vibrator
模板整型 mo(u)ld board plough
模板整型犁 mo(u)ld board plough
模板支撑 form brace[bracing]; form hanger; form tie; formwork support;shuttering;soldier(pile)
模板支撑装配 form tie assembly
模板支撑构件 form tie assembly
模板支撑件 shuttering aid
模板支撑装配 form-tying
模板支架 falsework;subpurlin(e)
模板支柱 stud
模板重复使用 reuse of forms;reuse of shuttering
模板周转 go round of form;repeated use of form;turnover of form
模板周转率 form cycling rate;rate of formwork turnover
模板主柱 soldier
模板铸件 planchet casting
模板铸模 match plate dies
模板装配 form assembly; formwork erection
模板装配工 formwork erector
模板装配工作 shuttering assembly work
模板装卸装置 shuttering handler
模板作用 template action
模板座 dies holder
模具 die;die arrangement;die equipment;die set;former;mo(u)ld;patrix and matrix
模具把 handle for die block
模具闭合量 die closure
模具布 mo(u)ld cloth
模具拆除装置 swage-setting
模具厂 die shop
模具车间 die making shop
模具成型部分 bearing of die
模具淬火 die quenching
模具存放架 die storage rack
模具锉刀 die sinker's file
模具的安装 die setting
模具的成型部分 bearing of die
模具的气垫 die cushion
模具的叶状模槽 die button
模具的再研磨 regrinding of die
模具的最大负荷 peak die load
模具等级 die caste
模具垫板 adapter plate; die backing plate;die tie-plate
模具定位块 die spacer block;mo(u)ld spacer block;spacer block
模具定心装置 die center[centre]
模具对准 mo(u)ld alignment
模具非常规热处理工艺 process of unconventional heat treatment of mo(u)ld
模具复合抛光机 complex die polisher
模具改良 die modifications
模具钢 die steel;mo(u)ld steel
模具高度调节量 die set height adjustment
模具工具磨床 die tool grinding machine
模具工作部分 working portion of die
模具灌铅检验法 < 检查模腔精度用 > die proof
模具痕迹 mo(u)ld mark
模具滑移装置 die slide
模具划痕 die line
模具缓冲装置 die cushion
模具加工机 die processing unit
模具间隙 die clearance
模具精修 die finishing
模具宽度 die width
模具面刃口 die face land
模具磨损 die wear
模具配对 mo(u)ld alignment
模具坯料 die blank;die block
模具拼块 die piece
模具嵌件 mo(u)ld insert
模具清洁剂 mo(u)ld cleaner
模具润滑剂 die lubricant
模具生产工人 pattern maker
模具寿命 die life;rupture life of die;service life of die
模具套环 die adaptor
模具调整不良 die misalignment
模具涂料 die lubricant
模具铣床 die sinker; mo (u) ld and die milling machine
模具铣刀 die mill
模具镶块 die edge;die insert
模具镶套 button dies;die bushing
模具锌合金 kirksite
模具型面刨削 stamp slotting
模具型腔的加工 die sinking
模具悬杆 mo(u)ld hanger
模具压板 clamping cap
模具银块 die insert
模具用砂轮 die grinder
模具油漆 forms paint
模具预热炉 die preheating furnace
模具振动 mo(u)ld vibration
模具制造(工)厂 die making shop
模具周转率 mo(u)ld turnover rate
模具装夹机构 clamping device
模具装配压力机 die setting press
模具总成 die assembly
模具组合 die assembly
模样 mode;model
模样工缩尺 pattern-maker's rule
模样花式 mode pattern
模样花坛 carpet bed(ding)
模样上的芯头 core print
模样芯头(突)core print
模子 die;former;matrix[复 matrixes/matrices];mo (u) ld (a form) ; pattern;stamper
模子闭合高度 die height
模子成型 mo(u)ld formation
模子定心盘 die center[centre]
模子量规 mo(u)ld ga(u)ge
模子上油 dopping;swabbing
模子寿命 life of the die

母摆 master pendulum

母斑 birthmark;mother's mark
母板 base metal;mother blank;mother board; parent metal; parent plate;platter
母版 master mask; metal positive; mother set
母本 female parent;maternal plant
母本类型 maternal form
母本品系 female parent line
母本实生苗 maternal seedling
母本园 maternal plant plantation
母本植株 maternal plant
母表 matrix[复 matrixes/matrices]
母材 base metal; basic metal; maternal material; mother metal; parent material;parent metal
母材尖钉状突起 rouge peak
母材晶格 parent lattice
母材料 parent material
母材熔合区 base metal fusion zone
母材试件 base metal test specimen; parent metal test specimen
母插件 mother board
母差保护屏 differential relaying panel
母城 < 城市规划用语 > mother city; mother town;parent city
母程序钟 master program(me)clock
母虫 queen

M

母船 lighter aboard ship;mother ship;mother vessel;parent ship
母船式渔业 mother-type fishery
母地槽 mother geosyncline
母点 generatrix[复 generatrices]
母点面 generator
母点线 generator
母点阵 parent lattice
母电话局 parent exchange
母电流 mother current
母电钟 control clock;controlling electric(al)clock
母段 parent segment
母分 mother stock
母峰 parent peak
母公司 controlling company;holding company;parent company;parent corporation;parent firm
母公司报表 parent statement
母公司拨入资金 funds from parent;funds from parent company
母公司对子公司的股本转移 split-off
母公司监督作用 oversight function of a parent company
母公司投资人 investor from a parent company;parent-investor
母公司直接收益 direct earnings of parent company
母管制 piping-main scheme
母管制系统 header system
母光栅 prototype grating
母国 <指航运船只所属国> home country
母函数 determining function;generating function
母函数的收敛性 convergence of generating function
母合金 foundry alloy;hardener;key metal;master alloy;mother alloy
母核 mother nuclide;original nucleus;parent nucleus
母机 launching aircraft;machine tool;mother aircraft;parent aircraft
母集合 superclass
母集团 parent population
母舰 depot ship;mother ship;parent ship
母礁 mother reef
母接头 box joint;tool joint box <锁链接的>
母接头台肩 shoulder of box
母金属 mother metal
母晶(体) mother crystal;parent crystal
母聚焦 parent focusing
母扣 box thread
母扣端 box end
母扣端面 box face
母扣连接 female connection
母扣面 box face
母扣弯头 female ell
母扣型钻头 box type bit
母离子 parent ion
母离子稳定性 parent ion stability
母联 busbar interconnection switch;busbar tie
母联断路器 busbar tie circuit breaker;bus-bite breaker
母联开关柜 busbar tie cubicle
母链 fundamental chain
母料 base material;base metal;parent material;raw and processed materials;raw material
母料粉 powder concentrate
母料配料 master batching
母料着色 concentrate colo(u)ring;master batching
母螺纹 box thread
母螺纹端 box end
母马 mare
母脉 mother lode
母面 generatrix[复 generatrices]
母模 case mo(u)ld;face unit mo(u)ld;master pattern;mother plate;pattern master;pattern mo(u)ld
母模用石膏 case mo(u)ld plaster
母曲线 generating curve
母时钟 master clock;mother clock
母式 matrix[复 matrixes/matrices]
母树 maternal plant;seed-bear
母题题材 subject for a motif
母体 general population;generatrix[复 generatrices];mother substance;parent;parent body;parent matrix;parent population;parent substance;population;precursor
母体玻璃 matrix glass;mother glass
母体部分 parent fraction
母体材料 fertile material;parent material
母体毒性 maternal toxicity
母体方差 population variance
母体放射性同位素 parent radioisotope
母体分布 parent distribution
母体分子 parent molecule
母体峰 parent peak
母体感应 maternal impression
母体化合物 index compound;parent compound
母体环境 maternal environment;parent environment
母体金属 <粉末冶金的> matrix metal;base metal;key metal;parent metal
母体矩 parent moment
母体聚合物 matrix polymer;precursor polymer
母体类型 maternal form
母体离子 parent ion
母体链 parent chain
母体量 size of population
母体目录 parent directory
母体设备制造用涂料 original equipment manufacture coating
母体适应 maternal adaptation
母体统计单位 parent statistical unit
母体物质 parent material
母体吸收 stem absorption
母体纤维 precursor fiber[fibre]
母体硝胺 parent nitramine
母体效应 maternal effect
母体信使 maternal messenger
母体性决定 maternal sex determination
母体性状 maternal character
母体选择 maternal-line selection
母体印痕 maternal impression
母体影响 maternal effect;maternal influence
母体元素 matrix element;meta-element;parent element
母体原料 parent stock
母体植株 maternal plant
母体转化法 precursor conversion
母体阻滞 maternal handicap
母同位素 mother
母凸轮 master cam
母图 main map;master map;mother map
母椭圆 generating ellipse
母物 parent substance
母物件 parent member
母线 bus;busbar;busbar wire;bus line;bus wire;collecting main;generating line;generatrix[复 generatrices];main lead;omnibus bar;omnibus rod;trunk;lead wire <爆破用>
母线安装 installation of bus line
母线棒 bus rod
母线保护 busbar protection
母线保护装置 busbar protective devices
母线布置 busbar arrangement
母线槽 bus duct;busway
母线穿墙隔板 diaphragm for penetration of bus
母线单元 busbar unit
母线导管 batch duct;busbar duct
母线导纳矩阵 bus admittance matrix
母线的腰面 central plane of a generator
母线电流变压器 busbar current transformer
母线电压 busbar voltage
母线电压互感器 busbar potential transformer
母线电压损失 bus voltage loss
母线电压调整器 bus voltage regulator
母线分段断路器 busbar section circuit breaker;bus section circuit breaker
母线分段开关 busbar sectionalizing switch
母线分离 bus separation
母线分配器 bus allocator
母线隔板 busbar partition
母线隔断电抗器 bus sectionalizing reactor
母线故障 busbar fault
母线管道 busbar channel;bus duct;busway
母线管道工程 bus duct work
母线盒 bus chamber
母线夹 busbar clamp
母线架 busbar frame
母线间隔 busbar partition
母线检查廊道 busbar corridor
母线接地 busbar grounding
母线接合 busbar coupling;bus-coupling
母线接线端子 busbar terminal
母线截面 busbar cross section
母线进线管道 busbar duct
母线绝缘器 busbar insulator
母线绝缘子 busbar insulator
母线廊道 busbar gallery;busbar tunnel
母线连接布置 busbar connection arrangement
母线连接器 busbar coupler
母线联结 busbar connection
母线联络断路器 busbar coupler circuit breaker;busbar tie circuit breaker
母线联络间隔 barbus tie cell
母线联络开关 busbar coupler;busbar tie switch;bus coupler
母线联络开关柜 busbar coupler switch panel
母线路 <指正线、梯线、基线等> parent track
母线耦合 busbar coupling;bus-coupling
母线耦合器 bus coupler
母线桥【电】bus gallery
母线驱动器 bus driver
母线伸缩接头 busbar expansion joint;bus compensator
母线式穿墙套管 busbar wall bushing
母线式电流互感器 bus-type current transformer
母线室 busbar chamber;bus compartment;bus room
母线条 bus rod
母线通道 busbar channel;busway;electric(al)busway
母线通信[讯]设备 bus communication equipment
母线系统 busbar system;bus system
母线箱 busbar chamber
母线支持绝缘子 busbar supporting insulator
母线支架 bus structure
母型 master model;master mo(u)ld;master pattern
母型船 basis ship;parent ship;type ship
母型反应堆 mother reactor
母压缩波 parent compressional wave
母岩 country rock;matrix[复 matrixes/matrices];mother rock;native rock;original rock;pan;parent rock;preexisting rock;source bed;source rock;underlying rock
母岩沉积 source sediment
母岩的成分和构造 composition and structure of parent rocks
母岩干酪根中无潜力的碳 non-potential carbon of kerogen in source rock
母岩干酪根中有成烃潜力的碳 potential carbon of kerogen in source rock
母岩厚度 thickness of source rock
母岩浆 mother magma;original magma;parental magma
母岩类型 type of parent rock
母岩碎屑 fragment of mother rock
母岩物质 parent material
母岩中干酪根碳 carbon of kerogen in source rock
母液 liquor;mother;pregnant antimony solution;pregnant liquor;pregnant solution;mother liquid;mother liquor
母液包藏 mother liquor occlusion
母液槽 mother liquor tank
母液罐 mother liquor tank
母液瓮 mother vat
母圆 generating circle
母质 matrix;mother materials;parent material;parent substance
母质层 C-horizon;horizon C;parent material horizon;lithic contact <土壤岩石接触层>
母质地形组合 parent material landform association
母钟 clock synchronizer;control clock;main clock;mother clock;primary clock
母锥 bell socket;box bell;die collar;female coupling tap;generating cone
母锥开口端 bell end
母子船坞 pontoon aboard floating dock
母子单元 mother and baby unit
母子公司关系 parent subsidiary relationship
母子关系 mother-child relationship
母子候车室 waiting room for mothers with children
母子起重机 <码头用的> combined type wharf crane
母子卫星 hitchhiker satellite
母子钟系统 master clock system
母组织 rib weave

亩

亩 <中国土地丈量单位,单数复数同,1 亩 = 667 平方米或 0.067 公顷> mu

牡

牡丹 Paeonia suffruticosa;tree peony
牡丹草亭 leontine

牡丹绸 peony brocade
牡丹红 peony
牡丹(红)色 peony red
牡丹纹 peony design; tree peony figure
牡桂 cassia flowertree
牡荆属 Chastetree
牡蛎 Concha Ostreae; oyster; oyster shell
牡蛎采集船 oyster dredge(r)
牡蛎层 oyster bed
牡蛎叉 oyster fork
牡蛎场 oyster bed
牡蛎筏 oyster raft
牡蛎干 dried oyster
牡蛎灰 oyster grey[gray]
牡蛎铗 oyster tongs
牡蛎礁 oyster reef
牡蛎礁海岸 oyster reef coast
牡蛎介壳虫 oyster shell scale
牡蛎蚧 oyster shell scale
牡蛎壳 oyster shell
牡蛎壳粉 ground oyster shell
牡蛎壳混凝土 oyster shell concrete
牡蛎壳石灰 oyster shell lime
牡蛎苗 spat of oyster
牡蛎耙 oyster rake
牡蛎耙网 oyster dredge(r)
牡蛎砂 oyster sand
牡蛎属 Ostrea
牡蛎属头巾式饰结 Turk's head
牡蛎养殖 farming of bivalve; oyster farming
牡蛎养殖场 oyster bed; oyster farm; oyster land
牡蛎养殖架 oyster cloister
牡蛎幼体 larvae of oyster
牡蛎中毒 ostreotoxismus
牡瓦 capping tile; concave tile; convex tile

姆巴巴纳 <斯威士兰首都> Mbabane

姆欧 <电传导率单位>【物】reciprocal of ohm; reciprocal ohm; mho
姆欧计 mhometer

拇拧螺(丝)钉 thumb screw

拇指 thumb
拇指操纵杆 thumb lever
拇指甲 thumbnail sketch
拇指开关 thumb switch
拇指孔 thumb hole
拇指控调扳手 thumb control adjustable wrench
拇指轮 thumbwheel
拇指模 thumb mo(u)lding
拇指饰【建】thumb mo(u)lding
拇指套 thumbstall
拇指纹 thumb-mark; thumb print
拇指旋转式控制器 thumbwheel controller
拇指圆(花)饰 echinus; conge

木爱奥尼亚式柱 wooden Ionic column

木扒钉 timber dog
木把手 wooden handle
木坝 log dam; timber dam; wooden dam
木白蚁科 <拉> Kalotermitidae
木百叶窗 timber shutter(ing); timber window shutter; wooden louver [louvre]; wooden shutter; wood louver window
木柏油 wood tar
木摆杆 wood pendulum rod
木板 board(ing); plank; timber board; wooden board; wooden slab; wood laminate; wood plank; wood strip
木板暗沟 batten underdrain
木板坝 flitch dam
木板百叶窗 board shuttering
木板搬运叉 lumber fork
木板搬运器 plank carrier
木板壁 timber wall; wood board; wood siding
木板边缘孔洞缺陷 edge void
木板车道 plank roadway
木板衬板固定木条 backing
木板衬垫 shingle lining
木板衬里 wooden board lining
木板床 plank bed
木板道 duck board <泥泞道上铺的>; boardwalk <尤指海滨>
木板地板 wood board flooring; constratum <古罗马>
木板地面 plank floor(ing)
木板钉 plank nail
木板渡槽 plank flume; wood(ed) stave flume
木板对接缝 board butt joint
木板房 barrack
木板房屋 frame house
木板覆盖 wooden board covering
木板盖面 wooden board covering
木板隔墙 board(ed) partition
木板隔热隔声 board insulation
木板工房 <临时的> wooden barrack
木板工作台 board platform
木板谷坊 plank check dam
木板刮(路器) plank drag
木板管 stave pipe
木板规整化 regularizing
木板过水渡槽 wood-stave flume
木板和板条组合 board and batten
木板桁架 plank truss; wooden plate girder
木板护壁 plank sheathing
木板画 wood engraving; xylograph
木板基层 plank base
木板及装饰材料 lumber and decorative materials
木板夹 clip with wood(en) board
木板夹钢板的组合板 flitch plate
木板夹钢板的组合大梁 flitch girder
木板夹钢板的组合梁 flitch beam
木板夹具 board grapple
木板假顶 false timber root; gob floor; plank floor(ing)
木板交叉道 plank crossing
木板交叉口 plank crossing
木板胶合 veneering
木板胶粘剂 board adhesive
木板接合【建】batten and button; board joint
木板井壁 plank tubbing
木板卷百叶窗 wooden slatted roller blind
木板框架 plank frame[framing]
木板框子 plank trim
木板拦砂坝 plank check dam
木板栏栅 plank hurdle works
木板耮 plank drag
木板篱笆 board fence
木板梁 timber plate girder; wooden plain(web) girder; wooden plateau; wooden plate girder; wood plate girder
木板梁桥 wooden plate girder bridge
木板量尺 board rule
木板路 lumber plank road; plank road; plank track
木板路刮 plank drag
木板路面 <多用于地下工程施工中> plank pavement
木板门 boarded door
木板面积 face measure
木板模 timber shutter(ing)
木板摩擦挤胶拼缝 rubbed joint
木板抹子 wood float
木板捻缝锤 reeming beetle
木板刨平 regularizing
木板平台 plank platform
木板铺屋顶 shingle roof covering
木板铺屋面 shingle roofing
木板嵌镶工 boarded parquetry
木板嵌镶作业 boarded parquetry
木板墙 close timbering; shingle wall; wall of planks; wood siding wall
木板墙衬垫 shingle wall lining
木板墙筋 timber stud
木板墙面 shingle siding
木板桥 plank bridge; plank floored bridge
木板人行道 board walk
木板软百叶帘 wooden slatted Venetian blind
木板饰面 panel(l)ing; wooden facing
木板水槽 plank flume; wooded stave flume
木板水渠 timber ditch
木板梯 chicken ladder
木板挑檐 board eaves
木板条 batten; bearer; lath; plank lath(ing); strapping; strip lumber; timber batten; timber boarding; wooden slat; wooden stave; wood lath; wood slat; wood strip; wooden batten; lacing board
木板条编织的板 woven board
木板条吊顶 suspended wood(en) lath ceiling
木板条管 wood(en) stave pipe
木板条机箍水管 wooden stave machine-banded pipe
木板条拼成的穹隆 wooden lamella dome
木板条拼成的圆顶 wooden lamella cupola
木板条水管 stave pipe
木板条围栏 woven wood fences
木板贴面 timber cladding
木板瓦 shake; wooden shingle
木板瓦墙 vertical shingling
木板瓦天沟 shingle valley
木板纹混凝土饰面 board-marked concrete finish
木板纹饰面粉刷 board finish plaster
木板屋顶 board deck; flat timber roof; flat wood(en) roof; plank roof; shingle
木板屋盖 board roof
木板屋面 shingle roof covering
木板屋面瓦 shingle tile
木板无榫缝 carve joint
木板镶槽端 hood end
木板镶嵌工作 boarded parquetry
木板镶嵌细作 boarded parquetry
木板镶嵌作业 boarded parquetry
木板心胶合板 battenwood
木板修饰钻头 veneer trimming bit
木板堰 timber(deck) weir
木板运输船 deals ship
木板栅栏 board fence; fence of boards
木板支撑 <沟槽的> pinchers
木板制品 panel product
木板桩 pile plank; plank pile; timber piling; timber sheeting; timber sheet pile; timber sheet piling; wooden sheet pile; wooden sheet piling
木板桩岸壁 curtain of timber; timber bulkhead
木板桩丁坝 timber sheet-pile groin; timber sheet-pile groyne
木板桩护岸 timber sheet-pile bulkhead
木板桩围堰 wooden sheet pile cofferdam; timber sheet pile cofferdam
木板桩堰 wooden sheet piling weir
木板装卸叉 lumber fork
木板走道 board runway
木板走廊 plank track
木版画 wood block; woodcut; xylograph
木版刻 wood engraving
木版印画 xylography
木版印刷 block printing
木棒 barling; timber bar
木包裹层 timber surround
木包角 wood trim
木包装 wooden packing
木背板 plank lagging; timber lagging
木本观赏植物 woody ornamentals
木本花卉 woody flowering plant
木本群落 lignose; woodland
木本油料树 oil-bearing tree
木本沼泽 swamp
木本植被 lignosa
木本植物 ligneous plant; woody plant; xylophyta; xypohyta
木壁板 board forms; board formwork; board shuttering; wooden panel(l)ing
木壁柱 wood stud
木壁柱构造 wood stud construction
木边刨 shooting
木标尺 wooden staff
木标桩 peg stake; timber bollard
木柄 wooden handle
木柄薄螺(丝)钉起子 thin screwdriver with wood(en) handle
木柄穿心螺丝刀 screwdriver with through tang wood(en) handle
木柄钢锯架 flush-cutting saw with wood(en) handle; hacksaw frame with wood(en) handle
木柄鸡尾锯 compass saw with wood(en) handle
木柄棘轮式螺(丝)钉起子 ratchet screwdriver with wood(en) handle; wooden handle ratchet screwdriver
木柄螺丝刀 screwdriver with wood(en) handle
木柄螺(丝)钉起子 wooden handle screwdriver
木柄摩擦落锤 board drop hammer
木柄木螺钉刀 stubby driver
木柄平头锯 flush-cutting saw with wood(en) handle
木柄十字尖头锤 wood handle cross peen hammer
木柄式固定钢锯架 solid hacksaw frame with wood(en) handle
木柄手把 brace head; brace key
木柄手锯 back saw with wood(en) handle; hand saw with wood(en) handle
木柄套箍木凿 firmer chisel; wooden handle with ferrule
木柄套锯 nest of saws with wood(en) handle; wooden casing saw; wooden handle nest of saws
木柄套装螺丝刀 screwdriver set with wood(en) handle
木柄弯刨 spokeshave with wood(en) handle
木柄橡胶锤 rubber mallet wood(en) handle
木柄一字螺丝刀 slotted screwdriver with wood(en) handle

M

木玻璃条 wooden patent glazing bar
木菠萝 jackfruit
木薄壁组织 wood parenchyma; xylem parenchyma
木薄壳 wooden shell
木薄壳屋顶 wooden shell roof
木薄壳圆顶 wooden shell cupola
木擦条 wood rubbing strip
木擦洗板 timber scrub board; wooden scrub board
木材 forest product; log; lumber; timber(ing); wood; lignum
木材暗节 enclosed knot
木材搬运叉 log and lumber fork
木材搬运车 go-devil
木材搬运橇 go-devil
木材搬运用拖车 timber trailer
木材板院气干 yard seasoning
木材板制品 <胶合板、木屑板等> panel product
木材包商 wood contractor
木材包扎法 <防腐处理> method of bandage
木材保存 wood preservation; wood preserving
木材报价单 stock note
木材本色 natural wood colo(u)r
木材比重计 xylometer
木材边纹 edge grain
木材变色 sap stain; stain; timber stain
木材变色菌 wood-staining fungus[复 fungi]
木材变质 wood deterioration
木材标售 auction of timber
木材标准 timber and lumber standard
木材标准单位 standard unit of timber
木材标准化 timber and lumber standardization
木材表 cutting list
木材表面处理 superficial treatment of timber; surfacing of lumber; wood superficial treatment
木材表面节疤 margin knot
木材表面炭化处理 superficial charring
木材表面炭化(法) superficial charring; surface carbonizing
木材表面炭化防腐 charring
木材病虫害 timber pests
木材波形跑锯 snake
木材驳船 log barge
木材薄片 ply
木材材积 timber measurement; wood volume
木材材积计量 board measure
木材采伐 logging
木材采伐权 common of estovers
木材采伐证 timber licence
木材采运 harvest; logging
木材采运工业 logging industry; lumbering
木材采运作业 logging operation
木材仓库 timber storage shed; timber warehouse
木材测容器 xylometer
木材层积计量 <以考得为单位> Cord measure
木材产量 timber yield
木材产品工业 wood product industry
木材产品加工 timber product processing; wood product processing
木材厂 timber mill
木材厂废料 hardboard mill waste; lumber waste
木材厂废水 hardboard mill wastewater; lumber wastewater; wastewater from wood preparation
木材场 lumber yard
木材场地 timber ground
木材车 lumber car; pole wagon; timber wagon
木材车削 wood turning
木材车转向架 logging truck
木材成材等级 grade of lumber
木材尺寸 timber dimension; timber size
木材尺寸检查证 <美> certificate of measurement
木材虫 wood warm
木材抽出物 wood extractive
木材出口贸易 timber export trade
木材处理 wood treatment
木材处理厂 timber treating plant; treating plant; wood treating plant
木材处理车间 timber treating plant
木材处理加压筒 treating cylinder
木材传声性 acoustic(al) conductivity of wood
木材船 timber carrier; lumber carrier
木材串集 trail
木材垂直叠合法 vertical lamination
木材疵病 defects in timber
木材粗纹 coarse grain
木材脆折性 brashness
木材搭接 scarfed joint
木材打光 polishing of wood
木材打印 timber marking
木材大头 butt end
木材丹宁 wood tannin
木材淡水载重线 lumber freshwater load line
木材导管 trachea[复 tracheae/tracheas]
木材盗伐 timber trespass
木材的 ligneous
木材的薄壁组织 parenchyma of wood
木材的不均匀纹理 uneven grain
木材的风蚀 weathering of timber
木材的刚性 stiffness of wood
木材的结合水 absorbed water in wood
木材的抗弯极限强度 strength in static bending
木材的连皮直径 diameter over bark
木材的劈裂性能 splitting property of wood
木材的清漆罩面 wood clear finish
木材的去皮直径 diameter under bark
木材的韧性 swelling of wood; toughness of wood
木材的同心圈 concentric(al) rings of wood
木材的应力分级 stress grading of timber
木材的蒸汽处理 reconditioning of timber
木材的自由水 free water in wood
木材等级 board-foot measure; grading of timber; lumber grade; lumber grading; lumber scale; timber grade
木材底板 timber base
木材底层用调合漆 primer mixed paint
木材底涂料 wood sealer
木材电处理防腐法 Noden-bretteuneau
木材淀粉 wood starch
木材雕刻 wood
木材雕刻的墙面装饰 carved wood(en) wall panel(l)ing
木材吊具 timber grab
木材吊平顶 timber counter ceiling
木材冬季北大西洋载重线 lumber winter North Atlantic load line
木材冬季载重线 lumber winter load line
木材堵塞 log jam
木材端裂 end check; end shake; end split
木材端面 end surface grain
木材端部表面裂纹 end check
木材端面裂纹 end checking
木材断裂 upset
木材堆场 timber yard
木材堆场机械系统 cable logging
木材堆存场 timber dock
木材堆单位 cord
木材堆放场 lumber yard
木材堆放稳定木条 ba(u)lk
木材堆棚 lumber storage shed
木材堆置场 log yard; lumber yard; timber storage yard; wood yard; timber yard
木材堆置处 lumber yard
木材发光性 luminescence of wood
木材发运税 timber delivery tax
木材筏道 log chute
木材筏运工 rafter
木材反跳 <锯木时> back lash
木材贩卖 timber selling
木材防腐 preservation of timber; preserving timber; timber preservation; wood preservation; wood preserving
木材防腐保护层 timber protection coat
木材防腐厂 timber treating plant; timber treatment plant; wood preserving plant
木材防腐处理 preservative treatment of timber; wood preservation; wood preserving process; wood treatment
木材防腐打底油 creosote primer
木材防腐废液 wood preserving waste
木材防腐工业 wood preserving industry
木材防腐工艺 tanalith process
木材防腐剂 timber preservative(agent); wood preservative(agent); antiseptics for wood; pamatol; antifungin; parmatol <四氯酚和五氯酚混合的有机溶液>; Triolth <一种含氟化钠、重铬酸钾及二硝基酚>; Minalith <一种兼作滞火剂用,含磷酸铵、硼酸、硼酸钠、磷酸氢二铵的木材防腐剂>; murolineum <主要成分为硅氟化锌>
木材防腐加压法 pressure process of timber preservation
木材防腐气溶胶涂料 rot-repellent aerosol coating
木材防腐水剂 aqueous wood preservative
木材防腐涂料 wood preservative coating
木材防腐盐 wood preservative salt
木材防腐油 coal-tar creosote solution; creosote oil; wood preservative creosote; wood preserving oil
木材防腐油浸渍 impregnate with creosote; impregnation with creosote
木材防火防腐剂 minolith
木材防水剂 aqueous wood preservative
木材放流槽 logway
木材废料 wood waste
木材废料制品 wood waste
木材分割机 wood splitter
木材分级 grading of timber; sorting of wood
木材分级准则 lumber grading rules
木材分类 timber assortment
木材分类场 timber sorting yard
木材风干 seasoning
木材封闭底漆 wash coat; wood filler
木材浮护舷 floating camel fender
木材浮运 lumber floating; timber floating
木材浮运分类场 timber floating sorting yard
木材浮运水槽 chute raft
木材辐裂 checking of wood; check in wood; end check; end shake; end split; wind shake
木材腐节 dead knot
木材腐烂 decay of wood; timber rot
木材腐朽 fox(e)y; foxiness; rot of wood; timber rot
木材腐朽菌 wood decaying fungus; wood decomposing fungi; wood rotting fungi
木材复合塑料板 wood plastics composite
木材干腐 cellar fungus
木材干枯 cellar fungus
木材干裂 honeycombing; rift crack; season crack; ware crack
木材干馏 destructive distillation of wood; dry distillation of wood; wood distillation; wood pyrolysis
木材干馏厂 wood distillation plant
木材干曲 case harden
木材干湿交变 movement
木材干缩翘曲 cupping
木材干缩性 shrinkage of wood; wood shrinkage
木材干舷 <运木材时的干舷高度> lumber freeboard; timber freeboard
木材干燥 wood seasoning
木材干燥厂 timber drying plant
木材干燥法 desiccation of wood; drying of wood; lumber drying; seasoning of timber; seasoning of wood; wood drying
木材干燥机 timber drier[dryer]
木材干燥翘曲 cupping
木材干燥设备 timber drier[dryer]
木材干燥窑 lumber(drying) kiln; timber(drying) kiln
木材干燥装置 timber drier[dryer]
木材赶羊流送 river driving
木材钢板夹层梁 sandwiched girder
木材钢铆钉法 timber-steel-rivet method
木材港 lumber harbo(u)r; lumber port; timber harbo(u)r; timber port
木材港池 timber basin; timber pond
木材割脂伤 pitch defect
木材根端直径 butt diameter
木材根墩 butt end
木材工厂 timber work
木材工场 lumber mill
木材工程 timber engineering
木材工人 timberer; timberman
木材工业 logging industry; lumber industry; timber industry
木材工艺学 wood technology
木材工作 timber work
木材构件厂 wooden structure factory
木材构造 structure of wood; wood structure
木材构造学 xylology
木材光泽 luster of wood
木材归楞工 decker
木材龟裂 honeycombing; honeycombing of wood
木材规格 timber and lumber standard
木材规格表 cutting list
木材害虫 timber worm
木材含水量 moisture content of wood; water content of timber; wood moisture content
木材含水率 moisture content of wood; water content of timber; wood moisture content
木材合缝修整 jointing
木材横截面 butt end
木材横裂 cross shake

木材横纹切割 flat grain
木材烘干炉 timber drier[dryer]
木材烘干设备 timber drier[dryer]; timber drying plant
木材烘干窑 lumber kiln
木材厚度 timber thickness
木材厚度规 lumber ga(u)ge
木材护面的 timber faced
木材滑道 log slip; skid road; sledge-way; timber chute; timber slide
木材化石 xyltile
木材化学 wood chemistry
木材化学防腐 chemical wood preservation
木材化学防腐法 chemical wood preservation
木材化学工艺 chemical technology of wood
木材环裂 circular shake; cup shake; ring gall; shake; wind shake; wood shake
木材环形裂口 cup shake
木材混凝土组合梁 wooden concrete composite beam
木材货 timber cargo
木材及木船 pickling
木材计量标准单位 <165 立方英尺> timber standard
木材加工 lumbering; lumber sawing; timber processing; timber work; woodcraft; wood processing; wood working
木材加工残留物 plant residue
木材加工厂 fabrication plant of timber; timber processing plant; timber work; wood mill works; wood working factory; lumber mill
木材加工场 planning mill
木材加工废料 wood curing waste
木材加工废水 timber curing wastewater; wood curing wastewater
木材加工废物 wood curing waste
木材加工工业 wood industry
木材加工机 timber working machine; wood working machine
木材加工机械 wood working machinery
木材加工面 working face
木材加工刨床 wood working planer
木材加工业 wood processing industry; wood working industry
木材加压防腐的 pressure-preservative treated
木材加压灌油防腐 bethelizing
木材加压灌油防腐法 Bethel's process
木材加压注入杂酚油防腐 bethell process
木材夹钳 span dogs
木材甲板货 timber deck cargo
木材剪切试验机 wood shear testing machine
木材检查所 lumber inspection bureau
木材检尺 culls; timber measurement
木材检尺记录牌 <约 2 英寸×10 英寸×1 英寸, 1 英寸 = 0.0254 米> tally board
木材检尺员 grader; tallyman
木材检验员 marker
木材建筑 wooden construction
木材建筑物 timber work; wooden building
木材降等 dockage
木材交织纹 double cross grain
木材胶合板 laminboard
木材胶粘剂 wood adhesive
木材接合 timber connection
木材接合法 timber jointing method
木材接合件 timber connector; timber joint connector
木材节疤 brittle heart; cat eye; knur; twisted growth
木材结构 structure of wood; texture of wood
木材截面 wood's butt end section
木材截片术 xylotomy
木材截头 docking
木材解剖术 xylotomy
木材解剖学 xylotomy
木材金属平台肩架 carpenter's bracket scaffold
木材金属托架平台脚手架 carpenter's bracket scaffold
木材浸硫酸铜处理 copper sulfate[sulphate] treatment
木材浸水处理法 water-seasoning
木材浸水两周后风干法 water-seasoning
木材浸渍 <一种防腐处理> timber pickling; pickling of timber
木材浸渍处理 impregnation of timber
木材浸渍处理法 dipping method of timber treatment
木材浸渍瓮废水 wood soaking vat wastewater
木材径面纹理 quarter-sawn grain
木材局部片状腐朽 pecky
木材锯方工 squarer
木材锯后尺寸 ripping size
木材锯解 woodcut; wood cutting
木材锯切机 timber dapper
木材菌霉防腐剂 wood rotting-fungi preventative
木材开裂性 splitting property of wood
木材抗疲劳性 fatigue of wood
木材抗弯强度 bending strength of wood
木材科学 timber science
木材可裂性 cleavability of wood
木材跨运车 log straddle carrier; lumber straddle carrier
木材捆运链 logging chain
木材拉坡机 alligator
木材滥采 timber mining
木材离心面 <木表> outside surface; outer surface
木材连接件 timber connection pieces; Maf
木材两边翘 cupping
木材两端各切一半 half-and-half lap scarf
木材量度 feet board measure
木材量积 wood measurement
木材裂缝 lag
木材裂开 checking
木材裂纹 check of wood
木材磷光(现象) phosphorescence of wood
木材溜放道 log(way) chute
木材流变学 rheology of wood
木材流放 rafting; river driving
木材流放坝 log dam; logging dam
木材流放槽 raft chute
木材流放工(人) log driver; raftman [复 raftmen]
木材流放渠 rafting canal; raft pass; raft path; raft sluice
木材流放堰 log weir
木材流放闸 raft lock
木材流送 floatage; river driving
木材流送工人 driver
木材流运 river driving
木材露天慢腐 eremacausis; orema causis
木材码垛 log yard; timber yard
木材码头 timber terminal
木材猫脸斑 catface
木材毛料 rough lumber
木材贸易 lumber trade; timber trade
木材密实度 density of wood substance
木材面油漆 paint on timber
木材内环裂 internal shake
木材耐潮 methylol urea
木材耐腐期 time of duration of wood
木材耐久性 durability of wood
木材耐磨性 abrasion resistance of wood
木材腻子 paste wood filler; wood filler
木材年轮 annual ring(of timber)
木材年轮裂 through shake
木材黏[粘]合剂 wood adhesive
木材捻缝工人 wood ca(u)lker
木材扭纹斑 epins
木材扭转纹 twisted growth
木材漂运 floating of timber
木材拼接(法) timber splicing
木材品种 wood species
木材平衡含水量 equilibrium moisture content of timber
木材平衡含水率 equilibrium moisture content of timber; equilibrium moisture content of wood
木材平面拼接 timber plain splice
木材破坏 wood destruction; wood failure
木材破坏率 wood failure ratio
木材铺盖面积 timbered area
木材铺面 timber floor(ing); wooden pavement; wood paving
木材铺面工作 timber paving work
木材起重臂 wood derrick
木材气干 seasoning
木材气息 odo(u)r of wood
木材汽蒸 steaming of wood
木材嵌填料 wood filler
木材戗丝 grainlifting
木材强度 strength of wood; timber strength; wood strength; strength of timber
木材翘曲 warp in wood
木材切片机 timber slicer
木材切削 wood cutting
木材青面纤维层 cambium layer
木材清漆 wood varnish
木材缺棱 wane
木材缺陷 defects in timber
木材染料 wood dye
木材染色 wood staining
木材热带淡水载重线 lumber tropical fresh water load line
木材热带载重线 lumber tropical load line
木材热解 thermolysis of wood; wood pyrolysis
木材容重 volume weight of wood
木材肉眼识别检索表 lens key
木材软腐朽 soft rot
木材软纹 open grain
木材商(人) lumberman; timber dealer; timberman; timberer
木材商业 lumber trade
木材上垛装置 lumberjack
木材上松节疤 loose knot
木材梢径 top diameter
木材烧焦 scorching of wood
木材渗出的树脂 resin streak
木材湿朽 wet rot
木材湿胀性 swelling of wood
木材识别 wood identification
木材市场 log market; lumber market; timber market
木材收获 timber yield
木材收缩 lumber shrinkage; wood shrinkage
木材树脂 wood resin
木材树脂产品 wood naval stores
木材树脂道 resin duct
木材竖积堆 pile
木材水解 wood hydrolysis
木材水解产物污水 wood hydrolysate effluent
木材水解厂 timber hydrolysis plant
木材水解物 wood hydrolyzate
木材水力采伐 liquid cutting of wood
木材水力切割 liquid cutting of wood
木材水泥屋顶 wood cement roof
木材水运 wood transport by water
木材松节油 wood terebene oil
木材髓线 ray
木材碎料 particle
木材弹性极限 elastic limit of wood
木材弹性模量 modulus of elasticity of wood
木材炭化 carbonization of wood; wood dry distillation
木材碳水化合物 wood carbohydrate
木材体积 timber volume; wood volume; log scale
木材体积测量计 cordometer
木材体积单位 <等于 0.71 立方米> serch
木材天然干燥法 conditioning
木材填缝膏 paste wood filler; wood paste filler
木材填缝腻子 wood paste filler
木材填孔剂 wood filler; wood sealer
木材填孔用油灰 wood filling
木材填料缝 wood filler
木材透声性 acoustic(al) permeability of wood
木材透水护坡 timber permeable revetment
木材涂装 wood finish
木材拖车 timber carriage
木材拖链 loading chain
木材拖索 timber jack chain
木材拖运 log haul; log slip
木材拖运车 timber hauling vehicle
木材拖运机 log haul-up
木材外露疵病 open defects in wood
木材弯曲 wood-bending
木材万能试验机 wood universal testing machine
木材危机 wood crisis
木材纹理 grain of wood; wood grain
木材无压防腐处理 non-pressure treatments for timber
木材物理学 timber physics
木材物质 wood substance
木材吸水性 water-absorbing capacity of wood
木材习居菌 wood inhabiting fungus
木材细胞纹孔 pit of wood cell
木材细裂纹 checking of wood
木材细纹 close grain
木材下河 watering
木材夏季载重线 lumber summer load line
木材先浸水后风干法 water seasoning
木材纤维 wood fiber[fibre]
木材纤维板 wood fiber[fibre] board
木材纤维饱和点 wood fiber[fibre] saturation point
木材纤维素 wood cellulose
木材纤维细胞 wood element
木材线 pith ray
木材向心面 inside surface
木材小节疤 pin knot
木材削片 chipped wood
木材削片机 chipper
木材消毒 sterilization of timber
木材消耗量 wood consumption
木材心材 duramen
木材心材面 inner surface
木材心裂 growth shake; rift
木材芯板 lumber core
木材朽腐 wood deterioration
木材朽节 decayed knot; unsound knot

木材蓄积量 stocking
木材学 timber science
木材压力防腐工艺 Rütger's process
木材压缩测定仪 wood compressometer
木材研究发展协会 Timber Research and Development Association
木材堰 log boom
木材药剂拌土处理 pudding treatment
木材业 lumbering
木材隐节 intergrown knot
木材隐伤 invisible defect
木材应力 timber stress
木材应力等级 stress grade of lumber
木材硬度 hardness of wood
木材硬化法 hardness process(ing) of wood
木材拥塞 jam of logs
木材拥塞积水湖 raft lake
木材用杀虫剂 insecticide for timber
木材油灰 wood putty
木材油饰工序 wood finishing
木材预制建筑 timber prefabricated construction
木材运输车 timber carriage; timber cart
木材运输撬 go-devil
木材杂酚油 creosoting of wood
木材载重水线 timber load water-line
木材载重线 <装运木材的满载吃水线> lumber load line; timber load line
木材在大气中慢性腐烂 eremacausis
木材在年轮间开裂 through shake
木材糟朽 boat; decay of lumber; dote; rot of wood
木材丈量 timber measure
木材真空干燥法 vacuum drying of timber
木材真空注油防腐 bethelizing
木材蒸干 steaming of wood; wood steaming
木材蒸炼厂 wood preserving plant
木材蒸馏厂 wood distilling plant
木材蒸馏(法) wood distillation
木材蒸馏废液 spent liquid form wood cooking
木材蒸馏设备 wood distilling plant
木材蒸汽处理 steam-remedial treatment of lumber
木材蒸汽干燥 steaming of wood
木材正常使用年限 normal timber life
木材支护的井筒 wood lined shaft
木材支架 wood support
木材制品 timber product; timber work
木材滞燃处理 wood fire-retardant treatment
木材中的开裂节疤 open knot in wood
木材中的小孔缺陷 pin-hole defect in lumber
木材中的自由水分 absorbed water in wood
木材(中)渗水组织 conducting tissue of wood
木材种类 timber assortment
木材注油 <用杂酚油或煤油馏灌入木材> bethelizing
木材蛀虫 lyctidae
木材抓斗 timber grab
木材抓钩 log grapple; timber grapple
木材抓具 timber grab
木材抓起机 span dogs
木材抓取机 span dogs
木材抓取器 timber grab
木材装车场 brow; log landing
木材装饰 wood finishing
木材装卸叉 log and lumber fork; lumber fork
木材装卸港 lumber port; raft port
木材装载 loading of timber
木材装载叉 lumber fork
木材装载机 timber loader; wood loader
木材装载机的鹅颈型吊臂 gooseneck boom
木材着色剂 wood stain
木材着色料 wood stain
木材资源 timber resources
木材综合利用 comprehensive use of timber; integrated logging; integrated utilization of wood
木材综合(利用)采运 integrated logging
木材总量 <以考得为单位的> cordage
木材总数 <以 128 立方英尺为单位测量的,1 立方英尺 = 0.02832 立方米> cordage
木材阻燃处理 fire-retarding treatment of timber; wood fire-retardant treatment
木材阻塞 log gorge
木材组织学 timber histology
木仓 timber bin
木舱壁 wooden bulkhead
木舱盖端护铁 hatch end protection
木舱口盖 wooden hatch board
木槽 timber flume; trow
木槽板 wooden sanitary cove
木槽板条布线法 wood casing system
木草铲 timber spud
木侧板 wood board; wooden board lining
木层孔菌 <拉> Phellinus
木插口 cup bearing
木插销 cat bar; cat bar bolt
木插座 wood receptacle
木柴 fire wood
木柴堆的体积单位 <等于 4 英尺 × 4 英尺 × 1 英尺,即 16 立方英尺> cord-foot
木柴含能量 wood energy
木柴块 billet
木柴液体燃料转换 conversion of wood to liquid fuel
木觇标 wooden tower
木产品 wood product
木场曳引机 yarder
木车工 bodger; wood turner
木尘 wood dust
木沉井 timber box caisson
木沉排 lumber mattress; pole mattress
木沉箱 timber box caisson; timber caisson; wood(en) caisson
木沉箱板 cleading
木衬板 timber board sheathing; timber lining; wooden(board) lining
木衬垫 wooden dunnage; wooden insert; wooden packing
木衬管 wood lined pipe
木衬护 timber lagging
木衬砌 timber lining; wood lining
木撑 timbering; timber strut; timber support; wood strut
木撑板 horizontal sheeting
木撑杆 timber stanchion; wooden reinforcement
木撑架桥 timber strut framed bridge
木撑隧道 timbered tunnel
木成型机 shaper
木承辊 timber runner
木承重结构 timber bearing structure; timber weight-carrying structure
木尺 wooden ruler; wood rule
木储仓 wooden bin
木船 junk; wooden boat; wooden vessel
木船舱盖板内侧肋骨支架 rider
木船船底漆 wood boat bottom paint
木船船底升高 dead rising
木船船底用沥青防污漆 toxic chian
木船的桅座板 mast bed
木船壳 wooden hull
木船模型 junk and sail-boat model
木船首部抗冰强度板 trebling
木船首尾部翘起弧度 sny
木船舷侧厚板 bends
木椽 timber rafter
木椽子 wooden rafter
木窗 timber window; wood(en) window
木窗百叶 wooden window shutter
木窗玻璃镶条 wooden astragal
木窗搭闩 wooden truss button
木窗挡板 timber window shutter
木窗格 wooden grid
木窗框油灰 wood sash putty
木窗框裕量 wood allowance
木窗棂 wooden astragal; wooden sash bar
木窗扇 wooden sash; wood sash
木窗扇变形 distortion of wood(en) sash
木窗芯子 <镶嵌玻璃> wooden glazing bar; wooden sash bar
木窗遮板 dead light
木槌 beetle; mallet; wooden hammer; wood(en) mallet; wooden maul
木锤 mallet; steel beetle; steel mall; wooden hammer; wooden mallet; wooden maul
木醇 methanol; methyl alcohol; wood alcohol; wood spirit
木醇油 wood alcohol oil
木疵 callus; rind gall
木醋酸 pyroligneous acid
木簇桩 wood pile dolphin
木锉(刀) cut file; rasp; rasp-cut file; rasping file; wood file; wood rasp; babbitt metal file
木溚 wood tar
木大梁 timber girder; wood(en) girder
木大(闸)门 wooden gate
木单板 <制胶合板用> wood veneer
木担 wooden arm
木弹 wooden round
木蛋白石 wood opal
木挡 wooden bar
木挡板 timber apron; timber baffle; timber sheeting; wooden fender; wooden screen
木挡土板 timber runner
木导管 trachea; wood conduit; wood vessel
木导架 timber head; timber lead
木导墙 timber guide wall
木捣 wooden punner
木道尺 wooden centre rail ga(u)ge; wooden centre track ga(u)ge; wooden rail ga(u)ge
木的 ligneous; wooden
木的纹理 grain of wood
木底 wooden sole
木底板 plank floor(ing); timber apron; timber base plate; timber floor(ing); wooden base
木底板部分更换 partial replacement plank floor
木底板去污 clearing wood(en) floor
木底层地板 timber sub-floor; wooden ground floor
木底架 cribbing and matting; timber underframe; wooden under frame
木底框 wooden under frame
木底座 floor sill; timber seat
木地板 ground plate; planch; plank floor(ing); timber boarding; timber floor(ing); wood covering; wood decking; wood decking board; wooden floor(ing)
木地板承包者 wood flooring contractor
木地板覆盖层 timber floor cover(ing)
木地板固定条 floor clip
木地板花式 parquetry
木地板面层 wooden floor cover(ing)
木地板木键 counter battens; counter wood(en) ground
木地板刨平机 floor machine
木地板铺装 wood block floor-installation
木地板条 strip flooring; wooden floor chip; wood floor strip
木地撑 wooden ground bracing
木地面 wood flooring
木垫板 layer of wood(en) blocks; timber foot block; underlayment; wooden bed plate; wooden cushion pad; wooden insert; wooden-packing plate; wood stilt
木垫块 spacer; timber blocking; wood block filler; wooden crusher block; wooden cushion pad
木垫梁 wood filler
木垫排 wooden matting
木垫圈 wooden washer; wood washer
木垫桩 <打桩机的> timber dolly
木雕 carving on wood; sculpture in wood; wood sculpture
木雕橱 carved wood(en) cabinet
木雕船 carved boat
木雕船模 solid-block model
木雕垂饰 carved wood(en) pendant
木雕灯 carved wood(en) lamp; wooden carving lantern
木雕凳 carved wood(en) stool
木雕果盘 carved wood(en) fruit plate
木雕画 wooden engraving picture
木雕火炉凳 carved wood(en) fire bench
木雕家具 carved wood(en) furniture; wooden carved furniture
木雕咖啡桌 carved wood(en) coffee table
木雕炕几 carved wood(en) low table
木雕刻 wooden sculpture
木雕刻刀 sloyd knife; wood carving tool
木雕刻工 bodger
木雕刻画 wood engraving
木雕刻品 wood carving
木雕屏风 carved wood(en) screen
木雕人物坠 carved wood(en) figure shaped pendant
木雕首饰 carved wood(en) jewellery
木雕首饰盒 carved wood(en) jewellery box
木雕术 xylography
木雕箱 carved wood(en) chest
木雕项链 carved wood(en) necklace
木雕像 wood figurine
木雕小套箱 carved wood(en) fancy chest
木雕写字台 carved wood(en) writing desk
木雕椅 carved wood(en) chair
木雕制品 wood carving wares
木雕镯子 carved wood(en) bracelet
木吊顶棚 wooden counter ceiling
木吊扣塔 wooden hoist tower
木吊盘 wooden stage
木叠合肋 built-up rib
木叠(合)梁 timber compound beam; wooden double beam
木丁坝 timber groin; timber groyne

木丁头 timber header
木钉 brace block; carpentry; dowel pin; nog; peep stake; peg (stake); peg table; timber peg; tre(e) nail; trunnel; wooden nail; wooden peg; wooden pin; wood nog; wood plug
木钉板肋拱桥 nailed wooden rib arch bridge
木钉板梁 nailed plate girder; nailed wooden girder
木钉板梁桥 nailed wooden girder bridge
木钉结合 wood pegging
木钉结合靴 wood pegged boot
木钉裤 pegged leggers
木钉块 wooden nailing plug
木钉梯 peg ladder
木钉条 wooden nailing plug
木顶梁 wooden header; timber roof bar <采矿>
木堵头 tre(e) nail
木渡槽 plank flume; stave flume; timber flume; wood flume
木蠹 chelura spengler; japonica; limnoria
木蠹蛾 carpenter moth; goat moth; wood beetle
木蠹蛾科 <拉> Cossidae
木段 juggle
木堆 lumps of wood; stack of wood
木堆栈 timber store
木墩 pigsty(e); plancon; timber pier
木垛 cog; crib; pigsty(e); steer; stick; timber grillage; timber pack; wood cell; wooden chock; wooden crib
木垛坝 timber crib dam
木垛填石挡土墙 retaining crib wall
木垛通气道 chimney
木垛支护 crib protection; pigsty timbering
木垛支架 chock; nog; timber crib
木垛支座 cribbing
木垛砖层 layers of wood(en) blocks
木舵后边板 angle strap
木舵后边缘 bow piece
木耳 Jew's ear; Auricularia <拉>
木筏 log raft; raft; raft log; timber float; wooden raft
木筏道 log pass; timber pass
木筏(浮)桥 raft bridge
木筏港 raft(er) harbo(u)r
木筏工人 log driver; raftman[复 raftmen]
木筏流放 rafting
木筏上的陋室 wanigan
木筏输送工作 rafting operation
木筏拖船 crab
木筏转向设备 log deflector
木帆船 <中国式的> junk
木防冲桩 timber fender pile
木防护板 wooden fender
木房(屋) log house; timber building; timber house; wood house
木分隔墙 timber partition(wall)
木粉 wood flour; wood meal
木粉填粉 wood meal
木粉填料 wood meal filler
木蜂 carpenter bee
木佛堂 timber ciborium
木扶手 timber handrail; wooden hand rail; wooden rail
木芙蓉 confederate rosa; cotton rose [rosa]
木浮标 boom; deadhead
木浮栅 log boom
木浮筑地板 wooden suspended floor
木浮子 wood float
木腐病 wood rot
木腐菌 conk; wood decaying fungus; wood-destroying fungus[复 fungi]; wood rotting fungi; wood rotting fungus; timber fungus
木腐生物 wood-destroying organisms
木腹板 boarded web
木覆板 batten ceiling; wood covering; wood sheathing
木覆面 wood covering
木盖板 timber sheathing
木盖面积 timbered area
木杆 pole; raddle; timber bar; timber post; timber stick; wood(en) pole; wood(en) post
木杆绑接 <制脚手架时> marrying of wooded pole
木杆道 corduroy; pole road; skipper road
木杆钩 cant dog; cant hook
木杆轨道 pole tram-way
木杆脚手架 pole scaffold(ing)
木杆埋深程度 depth of pole setting
木杆排水沟 pole drain
木杆上加固线担的槽 gaining
木杆围栏 wooden post and rail fences
木杆线路 wooden pole line
木刚架 wooden rigid frame
木刚性结构 timber rigid frame
木刚性框架 timber rigid frame
木杠 darby; darby float
木杠杆 cant hook
木杠刮平 darbying
木格板隔墙 wood panelled partition
木格板回转式升运器 wood turnover elevator
木格床 timber grillage
木格床基础 wooden grillage footing; wooden grillage foundation
木格构穹隆 wooden latticed cupola
木格构式梁 wooden latticed beam
木格构圆顶 wooden latticed cupola
木格架底脚 timber grillage footing
木格框 lumber grillage; wooden grillage
木格笼 cribwork
木格排 timber grillage; wooden grillage
木格排基础 wooden grillage footing; wooden grillage foundation
木格栅 timber grid; timber grille; timber joist(ing); wooden grid; wooden joist; wood joist
木格栅的尾端伸入墙内固定 tail in
木格栅底脚 timber grid footing
木格栅地板 timber joist floor; wooden joist floor
木格栅楼板 wooden joist floor
木格栅上层盖板 wooden joist upper floor
木格栅上层楼板 single upper timber floor; timber joist upper floor
木格栅屋顶地板 wooden joist roof floor
木格栅屋顶楼板 timber joist roof floor
木格填料 wood grid packing
木格条 timber bar
木格箱 skeleton case
木格选别机 slab grating
木格子 wooden grid
木格子窗 timber grille
木格子基础 wooden grid footing
木格子细工 wood fret
木格子转鼓 wooden slatted drum
木隔板 timber boarding; timber partition (wall); timber separator; wooden lagging; wooden lining; wooden screen; wood(en) separator
木隔断 wooden partition(wall); wood partition
木隔断的压顶木 head piece
木隔断墙 stud partition
木隔离物 wooden spacer
木隔墙 framed partition(wall); timber partition(wall); trussed partition (wall); wooden partition (wall); wood partition
木隔墙下槛 sole; sole plate
木铬胶 ligno-chrome gel
木工 square man; squarer; timberman; wood worker; wright; carpentry (work); joinery; woodwork <指工作>
木工安全刨床 safety lumber milling machine
木工暗铲口 concealed routing
木工凹形铣刀 concave milling cutter for wood working
木工板刨 rabbet plane
木工扁钻头 power wood bit
木工操作 carpentering; carpenter undertaking
木工长刨 joiner's plane
木工厂 carpenter's shop; carpenter's workshop; carpentry shop; timber mill
木工厂制品 planing mill products
木工场 carpenter's shop; carpenter's yard; carpenters yard; joiner's shop
木工车床 pattern-maker's lathe; wood working lathe; wood turning lathe
木工车刀 wood turning tool
木工车间 carpenter's shop; joiner's shop; wood shop; wood working (machine) shop
木工成型机 wood working shaper
木工尺 carpenter's rule
木工锤 carpenter's hammer; joiner's hammer
木工粗刨 carpenter's jack plane; carpenter's trying plane
木工打孔钻头 shell bit
木工打眼机 wood working driller
木工带锯 band saw for wood; carpenter's band saw; wood working band saw
木工带锯机 wood working band saw; wood working band sawing machine
木工刀头 wood working cutting head
木工的 wood working
木工的安装工作 finish carpentry
木工电动机 motor for wood working
木工电刨 electric(al) planer for wood working
木工雕合 scribe
木工雕刻锉 woodcarver's file; woodcarver's rasp
木工钉冲 carpenter's punch
木工方锤 carpenter's square hammer
木工斧 bench ax(e); carpenter's axe; carpentry tongue; woodman's axe
木工钢角尺 carpenter's steel square
木工工长 foreman carpenter and joiner
木工工程 woodwork
木工工具 carpenter's tool; wood working instrument; wood working tool
木工工具箱 carpenter's tool kit
木工工人 carpenter; joiner; wood worker
木工工业 woodwork industry
木工工艺 carpenter's art
木工工艺学 carpentry technology
木工工作 woodwork
木工工作裤 <俚语> dashboard running jeans
木工工作台 carpenter's bench; joiner's belt; joiner's bench
木工工作台挡头木 bench hook
木工刮刀 carpenter's scraper
木工规 hand trammel
木工规尺 joiner's ga(u)ge
木工行业 carpenter trade
木工虎钳 wood working vise[vice]
木工划线尺 marking ga(u)ge
木工划线刀 marking knife
木工划线规 marking ga(u)ge
木工划线盘 carpenter's ga(u)ge; marking ga(u)ge
木工画线规 butt ga(u)ge
木工画线机 wood working marking machine
木工画线盘 butt ga(u)ge
木工绘图圆规 carpenter's drawing compass
木工活 woodwork
木工机床 timber working machine tool; wood working machine; wood working machine tool
木工机器刨刀 wood machine planing knife
木工机器人 wood working robot
木工机械 carpenter machinery; wood working machinery
木工机械师 wood working mechanist
木工技术 woodcraft
木工技术行业 joinery
木工技艺 carpenter art
木工加工机械 machinery for processing woodwork; wood working machine
木工夹 carpenter's clamp; sliding head steel bar clamp; wood worker's clamp
木工夹具 claw bar
木工间 carpenter's room
木工监工 foreman carpenter
木工件的基准边 work edge
木工件的基准端 work end
木工件的基准面 work face
木工胶 carpenter's glue; joiner's glue
木工胶夹 glue press
木工角尺 carpenter's square; framing square; L-square; sliding bevel
木工角夹 corner clamp
木工校正刨 adjusting plane
木工接点 carpentry joint
木工接缝 carpentry joint
木工接合器 wood joiner
木工截切机 bench trimmer
木工锯 carpenter's saw
木工锯床 carpenter's sawing machine
木工锯机 carpenter's sawing machine; wood working sawing machine
木工锯架 bow saw frame; wood working saw frame
木工锯条 wood working blade
木工空心钻 core drill for wood working
木工拉锯锯条 cross-cut arch saw blade
木工领班 master carpenter
木工螺钉 hand screw
木工螺栓 hanger bolt
木工模板 woodwork form
木工磨床 wood tool grinder; wood working grinder
木工墨斗 carpenter's ink box
木工墨线 carpenter's line
木工刨 carpenter's plane; jack plane
木工刨床 carpenter's planer; wood shaping machine; wood working planing machine; wood planer
木工刨刀 wood working plane edge
木工刨刀片 cutting iron
木工刨可调动柄 adjusting lever
木工刨榫机 wood milling-slotting ma-

chine
木工平面规 carpenter's ga(u)ge
木工平面刨床 wood planning machine; wood working surface planer
木工平刨 carpenter's flat plane
木工平刨床 surface planer
木工平錾 carpenter's chipping-off chisel
木工企业 carpenter undertaking
木工铅笔 lumber crayon
木工嵌合 tabling
木工师傅 master carpenter
木工手锤 claw hammer
木工手艺 carpenter trade
木工水平尺 carpenter's (water) level rule
木工水平仪 carpenter's level
木工台 carpenter's bench; joiner's belt; joiner's bench
木工台挡头 bench stop
木工台刀挡头 bench knife
木工台虎钳 bench vice for wood working; table vice for wood working; wood working anvil vice
木工台夹钳 bench screw
木工台刨 bench plane
木工台钳 woodwork's vice[vise]
木工梯 builder's ladder; pole ladder
木工挑出式脚手架 carpenter's bracket scaffold
木工托臂脚手架 carpenter's bracket scaffold
木工铣床 wood working milling machine
木工细刨 carpenter's smooth plane; smoothing plane
木工(小)括刀 cabinet scraper
木工斜角尺 mitre square
木工修整机 bench trimmer
木工学徒 carpenter's apprentice
木工业 carpentering; carpentry
木工用的矩尺 roofing square
木工用钉 carpenter's nail
木工用斧 bench ax(e); hacket
木工用钢曲尺 roofing square
木工用刮刀 draw knife; draw shave
木工用弧口凿 firmer gouge
木工用胡桃钳 carpenter's pincers
木工用夹紧装置 carriage clamp
木工用墨线庳 joiner's ga(u)ge
木工圆锯机 wood working round sawing machine
木工凿 carpenter's flat chisel; carpentry tongue; joiner's chisel; sharp bit; sharp chisel; wood chisel
木工凿子 firmer chisel
木工直角组接 corner locking
木工直锯 wood working rip saw
木工职业 carpenter's trade
木工制品 joinery
木工制造及安装 joinery work
木工助手 joiner's labo(u)rer
木工装修 carpenter's finish; joinery
木工装修作业 finish carpentry
木工锥子 awl; bradawl
木工钻 carpenter's auger; carpenter's brace; gimlet; solid center[centre] eyed auger; wood drill; wood working drill
木工钻床 wood(working) boring machine; wood working drilling machine
木工钻孔机 wood working drilling machine
木工钻孔器 carpenter's(bit) brace
木工钻头 solid center bit
木工最后安装作业 <安装门窗、圈梁、踢脚板、门窗线条板等> finish carpentry
木工最后三道整修 <房屋的> third fixing
木工作架 timber falsework
木工作业 timber work
木工作业工具 firmer tools
木拱 timber arch; wood arch; wooden arch
木拱大梁 timber arched girder
木拱底面 wooden soffit
木拱顶 timber vault; wooden vault; wood vault
木拱腹 timber soffit
木拱架 timber centring[centering]; wood centring[centering]
木拱脚悬臂托梁 hammer beam
木拱脚悬臂托梁屋顶 Ardand type polygonal roof
木拱桥 timber arch bridge; wooden arch bridge
木拱楔块 arch block
木拱形圆顶 wood vault
木拱膺架 wood centering
木构涵洞 timber culvert
木构架 gallows timber; timber frame; timber framing
木构架包砖结构 timber-framed brick construction
木构架地板 timber-framed floor; wood-framed floor
木构架房屋 frame house; timber-framed house; wood frame house
木构架构件 carcassing timber
木构架构造 timber frame construction
木构架间填砖作业 brick nog(ging)
木构架建筑(物) frame wood construction; wooden framed building; timber-framed building; timber framed construction
木构架结构 pole-frame construction; timber frame construction; timber frame structure; wood frame construction
木构架斜撑 gallows bracket
木构架与填充墙的组合结构 needlework
木构件 first fixings; timber component; timber member; wood element; wood member
木构件侧接 side joint of timber members
木构件的键结合 timber member connected by hard wood key
木构件接合件 timber joint connector
木构件连接 joinery
木构件用开环和螺栓结合 timber member connected by split ring and bolt
木构件用螺栓和钉结合 timber member connected with bolts and nails
木构教堂 <斯堪的纳维亚> stave church
木构楼板 timber-framed floor
木构造 timber construction; wood construction
木构造固定构件 wood-framed in-built units
木构造类型 wooden construction type
木构造立面 wood-framed facade
木构造嵌入构件 wood-framed in-built units
木构造型式 timber construction type
木构造正面 wood-framed facade
木构造柱 wood constructional column
木骨板壁 <壁井内壁支撑> cribbing
木骨架 timber frame; timber framework; timber framing; wood skeleton
木骨架的底木条 sill plate
木骨架顶板 head plate
木骨架顶木条 head plate
木骨架房屋 timber framework building
木骨架构造 timber framing
木骨架建筑 pole-frame construction; half-timbered building <伊丽莎白时代的建筑式样>
木骨架结构 half-timber construction
木骨架平顶镶板 horizontal nogging piece
木骨架支柱 stor(e)y pole
木刮板 plank drag; wooden scraping straight edge; wooden screed; wooden separator; wood flight; xylometer drag
木刮刀 wood spatula
木挂瓦条 timber batten; wood batten
木管 wooden pipe; wood pipe
木管乐器 wood wind instruments; wood winds
木罐车 wooden tank car
木轨 wooden runner
木轨撑 wooden check
木轨道 wooden track
木轨枕 timber tie
木滚筒 wood roll
木过街楼 wooden gatehouse
木过梁 timber lintel; wooden lintel; wood lintel
木涵洞 timber culvert; wood culvert
木夯 beetle; heavy mallet; steel beetle; steel mall; wooded rammer; wooden punner; wooden rammer; wooden tamper
木和混凝土组合的结构构件 timber/concrete composite structural elements
木桁 ribband
木桁船首向外张的最顶部的厚外板 harpin(g)s
木桁构大梁 wooden trussed girder
木桁构框架 wooden truss frame
木桁架 plank truss; timber truss; wooden truss
木桁架构架 timber truss frame
木桁架桥 plank truss bridge; timber truss bridge; wooden truss bridge
木桁架桥横梁 transverse beam for timber truss bridge
木桁架式大梁 timber trussed girder
木桁梁 timber truss girder
木桁条 ribbon; ribbon board; ribbon strip
木横臂 wooden cross arm
木横担 wooden cross arm
木横挡 <门窗的> timber transom; wooden rail; wood waling; wooden transom
木横理抗拉强度 tensile strength of wood perpendicular to grain
木横梁 wooden header
木横纹破坏强度 cross-breaking strength(of wood)
木瑚菌属 <拉> Lentaria
木槲树 oriental white oak
木护板 timber lining
木护板通道 timber-lined passage
木护壁 plank sheathing
木护栏 timber fender; wooden rail
木护栏柱 wooden fence post
木护栏桩 <码头> wood fender pile
木护墙 wainscot; wall paneling; wood paneling of interior wall
木护墙板 wooden panel(l)ing; wooden wainscoting; wood panel(l)ing; wood sheathing
木护坦 timber apron; wood apron; wood protection
木护套 timber sheathing
木护条 guard bead
木护舷 timber fender; wood fender
木护舷防摩擦条 timber rub strip
木护舷桩 wood fender pile; timber fender pile
木花板 xylolite[xylolith]
木花格梁 timber lattice beam
木花盆 wood core box
木花瓶 wooden flower-vase
木花墙纸 wood chip wallpaper
木花制品 wood chip products
木华盖 timber ciborium; wooden ciborium
木滑板 timber sliding panel
木滑车 wooden block
木滑道 log chute; timber chute; timber pass; timber runner; timber runway; timber slide; wooden ground slipway; wood slip
木化石 dendrolite; fossil wood; petrified wood; wood opal; woodstone
木化纤维 lighted fiber[fibre]
木画 wood drawing
木缓冲器 wood buffer
木缓衝器 wood buffer
木灰 wood ash
木灰板条 wooden firring
木灰条 wood lath
木回纹花饰 wood fret
木混合桩 wooden composite pile
木混凝土板 wood concrete slab
木混凝土地板垫块 wood concrete soffit floor filler block
木混凝土合成梁 wood concrete composite beam
木混凝土合成桥(梁) composite timber concrete bridge; composite wood concrete bridge
木混凝土混合梁 wood concrete composite beam
木混凝土混合梁桥 wooden concrete composite beam bridge
木混凝土结合桥(梁) composite timber concrete bridge; composite wood concrete bridge
木混凝土块 wood concrete block; wood concrete tile
木混凝土填料 wood concrete filler
木混凝土屋面板 wood concrete roofing slab
木混凝土组合梁 wood concrete composite beam
木混凝土组合梁桥 wooded concrete composite beam bridge
木混凝土组合桥面 composite wood concrete deck
木或金属板舱壁 joiner bulkhead
木基材料 <指纤维板和碎料板等人造板> wood base materials
木基础 <纵横搭接逐层排列的> timber footing
木基底 wooden ground
木基刨花板 wood base particle panel
木基塑料 wood base plastics
木基纤维板 wood base fiber[fibre] board
木基纤维板料 wood base fiber[fibre] panel materials
木基制品 wood-based products
木加固 timbering
木夹 wood clamp
木夹板 clamp timber
木夹板固定术 wooden splintage
木夹板或斜张甲板 plywood diagonal deck plating
木夹层地板覆盖层 timber floating floor cover(ing)
木夹层地板终饰 timber floating floor(ing) finish
木夹件 clamp timber

木夹盘 wood chuck
木甲板 wooden deck
木甲板边(缘)板 boundary plank; margin plank; margin plate; nibbing plank
木甲板缝 devil
木甲板牵板 tie plate
木甲板上的边沟 chimb; chine
木甲板室直木柱 studding
木甲板条 deck plank(ing)
木假顶 timber mat
木架 wooded form; wooden stand
木架钉齿耙 wooden framed spiked harrow
木架房屋 stick-built
木架隔层横木 ledger board
木架构基础 wood foundation
木架构造 wood frame construction
木架间水平短撑 nogging
木架空楼板 timber suspended floor
木架梁 nog
木架楼板 timber-framed floor
木架填砖隔墙 brick nog(ging); brick-and-stud work
木架填砖砌墙建筑 brick nogging building
木架土屋 earth lodge
木架箱 skeleton case
木架砖壁 brick-and-stud; nogging; brick nog(ging)
木架砖壁作业 nogging
木架砖梁 nogging
木架砖墙 brick-and-stud work; brick-nogged wall
木架砖墙房屋 <即立贴式房屋> brick nogging building
木尖端装饰 timber spire
木尖塔 timber spire; wooden broach; wooden spire; wooden steeple; timber steeple <教堂上的>
木尖塔屋顶 wooden roof spire
木间壁 <不承重的> common partition
木间二酚 xylorcinol
木间隔承重结构 timber space load-bearing structure
木间木栓 interxylary cork
木间柱 wood studding
木建部分 <房屋内部的> woodwork
木建筑 timber architecture; timber building; timber construction; wood construction
木建筑块材模数 wooden building block module
木建筑砌块模数 timber building block module
木键 brace block; treenail; wood(en) key <叠合梁的>
木键合梁 dowel(l)ed beam; keyed girder
木键接合 keyed joint
木槛 timber beam; timber sill; wooden door threshold
木浆 lignosol; pulp; wood pulp
木浆材搬钩 pulp hook
木浆材料 pulpwood; wood pulp material
木浆粗纤维纸 wood chip paper
木浆废料 wood pulp waste
木浆化学 wood pulp chemistry
木浆纱 pulp yarn
木浆水解厂 wood pulp hydrolysis plant
木浆卫生纸 wood pulp wadding paper
木浆纤维 wood pulp cellulose
木浆压制板 presspahn
木浆研磨机 pulp grinder
木浆纸 wood pulp paper; groundwood paper
木浆纸板 wood pulp board
木浆贮仓 pulp bin
木讲坛 timber pulpit
木讲堂 wooden pulpit
木匠 carpenter; square man
木匠储藏室 carpenter's store
木匠工长 master carpenter
木匠木凳刀 carpenter's bench knife
木匠水平尺 carpenter's level
木匠外套 carpenter's overalls
木匠用的木槌 carpenter's mallet
木匠证书 carpenter's certificate
木匠执照 carpenter's certificate
木匠助手 carpenter's mate
木胶 wood glue; wood gum
木焦油 gum; pine tar; wood tar
木焦油醇 <作防腐剂用的> creosol
木焦油防腐油 wood tar creosote
木焦油酚 creosol
木焦油沥青 wood tar; wood tar pitch
木焦油馏分 wood tar fraction
木焦油酸 lignoceric acid; tetracosanoic acid
木焦油杂酚油 wood(tar) creosote
木脚手架 timber falsework; timber scaffold(ing); timber staging; wooden falsework
木脚手架工 timber scaffolder
木教堂 wooden church
木阶梯 wooden step
木阶梯式建筑(物) timber staircase builder
木接合 timber connection; wood connection; wood joint
木接合方法 wood joining method
木接合件 timber fastener
木接头 wood connector
木接腿 wooden reinforcement
木节 knot; timber knot; burl; cat's eye; curly grain; gnarl; knag; knar(i); knaur; wart
木节包装纸 screenings wrapping
木节封闭 knots sealing
木节封闭剂 knot sealer
木节封闭清漆 knotting varnish
木节刮油 <木节眼做刮油处理,以便油漆> knotting
木节孔 knot hole
木节黏[粘]土 kibushi clay; knar clay
木节水虱 Asellus aquatics
木节纸浆 screenings pulp
木结疤 knar(i)
木结点 wood connection
木结构 carpentry; timber construction; timbering; timber structure; timber work; wood construction; wooden construction; wooden structure; wood structure; woodwork construction
木结构车库 wooden garage
木结构的齿环连接件 spike grid
木结构的高强横梁 poitrail
木结构的最下部分 groundsel
木结构房屋 timber structure house
木结构房屋的横梁 petrail
木结构杆件 wood structural member
木结构工程 timber engineering
木结构构件 wood structural member
木结构技术 timber structural technique; wood structural technique
木结构建筑 building of wood(en) construction; timber building; timber-framed building; wood frame construction
木结构脚手架 timber scaffold
木结构节点连接件 timber joint connector
木结构结合件 timber connector
木结构框架 timber framework
木结构类型 timber type of construction
木结构连接 wooden connection
木结构连接件 structural timber connector; timber connector; structural wood fastener
木结构体系 timber structural system; wood(en) structural system
木结构突堤 timber jetty
木结构图 timber structure drawing
木结构制作 structural carpentry
木结构中的防冲块 bumping block
木结构住宅 timber house
木结构最下部分 ground sill
木结节 knar(i)
木筋隔断 wooden stud partition (wall)
木筋隔墙 quarter(ed) partition; stud partition; wooden stud partition (wall)
木筋混凝土 ligno-concrete
木筋墙 wood stud wall
木紧固件 timber fastener; wooden fastener
木槿 shrubalth(a)ea; shrubby alth(a)ea
木槿根粉 hibiscus root powder; powdered hibiscus
木槿属 majagua
木精 methyl alcohol; wood alcohol; wood spirit
木井架 timber derrick; wooden tower
木井框 timber set; wooden crib
木井框支架 curbing
木警卫室 timber gatehouse
木酒精 methanol
木锯 bucksaw; timber saw; wood saw
木锯架 deal frame
木聚糖【化】xylan(e)
木卷百叶窗 wooden roller shutter
木橛 peg(stake); stake; treenail
木菌 wood fungus
木菌甲虫 ambrosin beetle
木菌素 dermadine
木开口滑车 wooden snatch block
木龛室 wooden ciborium
木颗粒骨料 wood particle aggregate
木壳 timber shell
木壳板 planking
木壳船造船厂 wooden shipyard
木壳粗刨 wooden jack plane
木壳穹隆 timber shell dome
木壳屋顶 timber shell roof
木壳销 plate lock
木壳圆屋顶 timber shell cupola
木刻 wood cutting; wood(en) engraving; xylograph
木刻版 wood engraving
木刻版本 block book
木刻版工 xylographer
木刻版印刷(术) block printing; hand blocking; wood block printing
木刻材料 carving wood; wood engraving material
木刻刀 woodcut knife; wood-cutting blade
木刻动物 carved wood(en) animal
木刻工具 wood carving tool
木刻工艺 wood carving
木刻画 woodcut
木刻活字 wood letters
木刻家 woodcutter
木刻师 xylographer
木刻术 woodcraft; wood engraving; xyloglyphy[xylography]
木刻水印 wood block printing
木刻印版 engraved block
木刻制品 wood carving articles
木空间承重结构 timber spatial weight-carrying structure; timber three-dimensional load-bearing structure
木空间框架 timber space frame
木空心地板填料 timber hollow floor filler
木扣件 timber fastener; wooden fastener
木块 block(wood); chump; cut log; lumps of wood; rough ground; skids; timber block; timber brick; wood block; wood brick; wooden block; fag(g)ot wood <路面用>
木块地面 wood block flooring
木块堵塞 log jam
木块骨料混凝土地面 cement-wood floor
木块集料混凝土地面 cement-wood floor
木块胶合实心板 solid wood staved core door
木块茎 lignotuber
木块砾石路面 wood gravel pavement
木块楼面 wood block flooring
木块路面 pavement wood block; wood block pavement; wood block paving; wooden-block pavement; wood pavement; wood paving
木块路面铺设工人 wood pavior
木块锚标 deadhead
木块拼花地板 wood block flooring; wood block parquetry flooring
木块拼镶实心板 solid wood staved core door
木块铺楼面 wood block floor(ing)
木块铺面 block wood pavement; pavement wood block; wood block pavement; wood block paving; wooden-block pavement; wood pavement; wood paving
木块铺砌 wood block paving; wood paving
木块铺砌层 wood pavement
木块铺砌路面 block wood pavement; wooden-block pavement; wood pavement
木块铺屋面 wood block flooring
木块砂砾路面 wood gravel pavement
木块设计测验 block design
木块湿度控制器 wood block humidity controller
木块芯细木工板 block board
木筐 coffin
木框 case frame; crib; timber crib
木框坝 crib dam; timber crib dam; timber dam
木框架 deal frame; wooden framing; wooden frame; plank frame <厚木板作非承重的隔墙和墙体的>; sway frame; timber framing
木框架底部横档 grundsill
木框架房屋 post and girt; wooden framed house; wood frame house
木框架建筑(物) timber framed building; wood frame construction; wooden framework building
木框架结构 wood(en) frame(d) construction
木框架墙 timber-framed wall; timber framework wall; wooden framed wall
木框架式坝 rafter-and-strut framed dam
木框架外贴附柔性板围堰 flexible sheeting on timber framing cofferdam
木框架堰 frame weir
木框架正面 timber-framed facade
木框架砖或混凝土潜栏台坎 box sill
木框胶合板 wood-framed plywood
木框镜(子) timber-framed mirror; wooden framed mirror

M

木框螺栓夹板 hand screw
木框门窗 plank frame
木框石心基础 foundation by timber casing with stone filling;foundation filling
木框式沉排 framed mattress
木框填充墙板 timber-framed infilling panel
木框镶板墙 wood frame panel wall
木框压榨器 rack presser
木框堰 crib weir
木框支撑式廊道 timbering gallery
木框转角凹凸榫上联珠线脚 return bead and rebate joint
木框子 chess
木扩孔器 timber broach
木拉桩 timber anchor pile
木蜡笔 timber crayon
木蜡树 wax-tree
木兰 lily magnolia
木兰属植物 magnolia
木栏导流堤 log training wall
木栏底下的长石板 gravel board; gravel plank
木栏杆 timber handrail; timber rail; wooden baluster
木栏杆(小)柱 wooden baluster
木栏栅 paling
木栏着地长板 gravel plank
木缆柱 timber bollard
木缆桩 timber bollard;timber head
木类构造 wooden type of construction
木棱柱壳屋顶 wooden prismatic shell roof
木棱柱形薄壳屋顶 timber prismatic shell roof
木楞式接缝 batten seam
木楞式金属面 batten seam metal
木楞条 wooden fillet
木篱笆 wood fence
木理纹 grain pattern;wood grain design
木立筋 timber stud;wooden stud
木立筋隔断 timber-stud partition (wall);wood stud partition(wall)
木立筋隔墙 wood stud partition (wall)
木立筋构造 wood stud construction
木立筋结构 wood stud construction
木立筋锚件 nailing anchor; wood stud anchor
木立柱隔墙 stud partition
木立柱结构 wood stud construction
木沥青 wood asphalt;wood pitch
木连接 timber joint
木连接器 wood connector
木帘滞流装置 curtain-pole retarder
木联结格栅 timber binding joist
木联结梁 timber binding beam;wooden binder
木联系格栅 wooden binding joist
木联系(小)梁 wooden binder
木梁 nailing strip; summer tree; timber beam;wooden beam
木梁底面 wooden soffit
木梁地板 timber beam floor
木梁搁接法 corking
木梁格栅 grating of timber
木梁基 wood beam-base
木梁楼板 wooden beam floor
木梁桥 timber beam bridge
木梁式桥 wooden beam bridge
木料 forest product;lumber;timber; wood;planch <作木地板用的>
木料迸裂 popping
木料表面修整 timber surfacing
木料裁角机 trimming machine
木料场 lumber yard;timber yard
木料尺 board foot
木料虫 timber worm
木料储存棚 timber shed
木料垂直于木纹拉力 tension perpendicular to grain of timber member
木料垂直于木纹压力 compression perpendicular to grain of timber member
木料大头 stub end
木料带锯机 log band saw mill
木料的表面处理 lumber surfacing; surfacing of lumber
木料的腐节 dead knot
木料的临时工程 timbering
木料的气味 odo(u)r of wood
木料堆 rick
木料防腐 wood preservation
木料防腐处理车间 timber preservation plant
木料防腐剂 wood preservative (agent)
木料粉红打底漆 pink priming paint of timber
木料根头 stub end
木料横截机 log cross-cutting machine
木料横锯机 log cross-cutting machine
木料环裂 shake in timber
木料夹叉 logging grabble
木料夹钳 lumber tongs
木料胶结剂 marine glue
木料接长 lengthening of timber
木料结合器 timber connector
木料框锯机 log frame sawing machine
木料面上记号 surfacing of lumber
木料碾碎机 defibrater[defibrator]
木料刨花 paring
木料铺面 timber surfacing
木料清单 cutting list
木料去湿法 boulton process
木料缺陷 defect in timber
木料砂处理机 wood sanding machine
木料上的斜纹 cross grain
木料升降机 timber elevator
木料输送机 log conveyer[conveyor]
木料顺纹拉力 tension parallel to grain of timber member
木料顺纹压力 compression parallel to grain of timber member
木料死节 unsound knot
木料炭厂 wood carbonization plant
木料拖车 log trailer
木料斜角铣刀 wood angle cutter
木料心材 duramen
木料悬桥面 timber suspended floor
木料养护 log curing
木料英尺 board foot;board measure
木料正面 marked face
木料支撑的 timbered
木料蛀虫 timber borer
木料组合梁 beam girder
木裂缝 timber crack
木临时支撑 timber falsework
木檩(条) wooden purlin(e)
木馏甲醇 wood distillation methanol
木馏油 creosote(oil);kreosote;wood tar creosote
木瘤 canker growth of wood; curly grain;gnarl;knag;knar(i)
木龙骨 rough ground;wood joist
木龙骨间填砖作业 brick-and-stud work
木龙骨楼板 wood joist floor
木龙桥台 logged crib abutment
木笼 basket crib; crib; log-crib; timber cage; timber crib; wood cell; wooden crib
木笼岸壁 cribwork wall
木笼坝 crib dam; log-crib dam; timber crib dam
木笼挡土墙 crib retaining wall;cribwork retaining wall
木笼丁坝 crib groin;crib groyne;crib spur
木笼防波堤 box breakwater; crib breakwater; timber-crib breakwater
木笼工程 cribwork
木笼构筑物 <填石> cribwork;timber cribwork
木笼谷坊 crib check dam
木笼护岸 log-crib revetment
木笼护墩框架 <混凝土桥墩的> basket curb
木笼护坡 log-crib pitching; log-crib revetment
木笼基础 crib foundation
木笼基础铺放的木料 crib foundation log
木笼结构的面层木料 face log of crib structure
木笼井架混凝土桥墩 cribwork
木笼拦沙坝 crib check dam
木笼码头 crib wharf;timber wharf
木笼码头岸壁 cribwork quay wall
木笼木桩堤 crib and pile dike[dyke]
木笼排 woven wooden mattress
木笼桥墩 crib pier
木笼桥台 log(ged) crib abutment
木笼式岸墩 log(ged) crib abutment
木笼式防波堤 timber crib breakwater
木笼式桥台 logged crib abutment; logged crib pier
木笼式填石堰 rock-filled crib weir
木笼式突堤码头 limber crib pier
木笼式围堰 timber crib type cofferdam;crib type cofferdam
木笼填石 cribwork; rock-fill timber crib
木笼填石坝 beaver-type timber dam; crib dam; crib dam filled with stone;rock-crib dam;rock-fill(ed) timber crib dam;timber crib dam
木笼填石丁坝 crib groin;crib groyne; stone-filled timber crib groin
木笼填石堰 weir of dry stone and timber work
木笼围堰 crib cofferdam;timber cage cofferdam; wooden cage cofferdam;timber crib(work) cofferdam
木笼围堰桥墩 cribwork
木笼堰 crib weir
木笼组件 crib member
木笼作业 cribwork
木楼板 wooden floor;tabulatum <古罗马>
木楼板地面 boarded floor
木楼板格栅撑 floor strutting
木楼板龙骨 floor batten;floor joist; floor sleeper;timber floor joist
木楼板龙骨夹 floor(ing) clip
木楼板面层 wooden floor cover(ing)
木楼梯 timber stair (case); wood(ed) stair(case)
木楼梯的厚踏步板 stepping wood
木楼梯栏杆 timber banister
木楼梯露明小梁 cut string
木楼梯露明斜梁 open stringer
木楼梯平台丁头梁 platform header
木楼梯施工人员 wooden stair(case) builder
木楼梯踏步板 wooden tread
木楼梯斜梁 timber string; wooden string(er)
木楼梯中间斜梁 rough string
木路面 wooden pavement; wooden road; wooden runway; wooden track
木辘砖地面 wood block floor
木轮车 bummer
木轮滑车 wooden sheave block
木螺钉 grub screw;lag wood screw; log screw; Parker-Kalon; screw; screw nail; screw nail of wood; screw spike;wood screw
木螺(丝)钉车床 wood screw lathe
木螺钉锤 screw spike driver
木螺钉垫圈 screw cup
木螺钉螺纹 wood screw thread
木螺栓 blunt bolt; lag bolt; timber bolt
木螺丝 grub screw;log screw;screw nail;wood screw
木螺(丝)钉成套设备 complete equipment for making wood screw
木螺旋 wood screw
木螺旋泵 wood screw pump
木螺旋桨 wooden propeller; wooden screw
木螺旋形体 timber spire
木螺旋桩 wooden screw pile
木螺锥 auger bit
木螺钻 auger bit
木落砖 built-in nailing block; fixing block;fixing brick;wood brick
木落砖固端梁 built-in beam
木落砖孔 <砖墙上用的> break-in
木麻花钻 wood screw drill
木麻黄 beach she-oak;beef wood
木麻黄栲胶 beefwood extract
木麻黄属 beef wood;bull oak;Casuarina <拉>
木马 gym-horse;saw block;saw horse; wooden horse
木马架 horse scaffold
木码头 timber pier;wood jetty
木码头护木 bull rail
木镘 patter
木镘板 wood(en) float; wood(en) trowel
木镘刀 wood float
木镘粉平的 wood-floated
木镘抹光面 <混凝土路面的> wood float finish
木镘修 <混凝土表面的> wood float finish
木镘修整 <混凝土路面的> wood float finish
木毛 excelsior;wood wool
木毛地板 wooden sub-floor
木锚杆 wood(en)(rock) bolt
木锚钎 wood bolt
木锚桩 timber anchor pile
木煤气 wood gas
木门 all-timber door;all-wood door; timber door;wood door
木门窗镶边 wood trim
木门房 timber gatehouse; wooden gatehouse
木门槛 timber door threshold;timber threshold;wood sill
木门框 plank frame; timber door frame;wooden door frame
木门闩 bobbin latch;wooden latch
木门樘 wooden door frame
木门芯 core stock
木棉 bombax cotton;capoc;Ceiba fibre; floss; kapok; kapok ceiba; mockmain; silk cotton; vegetable down;vegetable silk
木棉板 woodwool slab
木棉布 capoc[capok] cloth
木棉垫 kapok cushion
木棉救生衣 kapok lifejacket
木棉树 cotton tree; silktree; wood cotton tree
木棉纤维 kawo kawo fibre
木棉油 baobab oil
木面 face side;wood surface
木面板 plank facing; timber deck; timber planking

木面板堰 timber deck weir
木面层 timber finish
木面积仪 planimeter
木面加湿 sissing
木面经度【天】zenographic longitude
木面料 board facing
木面起毛 fluffy
木面纬度 zenographic latitude
木面学 zenography
木面坐标 zenographic coordinate
木模 model;timbering;timber mould; wooden; wooden former; wooden pattern; wood former; wood mo(u)lding;wood pattern
木模板 board forms; board formwork; board shuttering; form lumber; plank sheathing; timber form(work); timber shutter(ing); wood(en) form(work); form board(ing); shuttering board; timber falsework; timber shuttering; wooden shuttering
木模仓库 pattern store;wood pattern store
木模车间 model(l)ing shop; wood modelling shop;wood pattern shop
木模工 patterner;wood pattern maker
木模工程 timber form(work);wood formwork
木模活块 loose piece
木模基 wooden base
木模胶 pattern glue; wood pattern glue
木模壳 timber form(work); timber shutter(ing)
木模框边框保护条 crush plate
木模毛坯料 pattern stock
木模漆 pattern coating
木模图 form drawing; wood form drawing
木模涂层 pattern coating
木模涂色 pattern colo(u)r; wood pattern colo(u)r
木模型 wooden model; wooden mo(u)ld
木模支撑 timbering
木摩擦条 timber rubbing
木抹搓平 scouring
木抹抹平 smoothing with float
木抹修整 float finish;wood float finish
木抹子 float; hand float; patter; wooden float;wood float
木抹子粉平的 wood-floated
木乃伊 mummy
木泥煤 wood peat
木黏[粘]合剂 timber binder
木碾轮 wooden runner
木牛头刨 wood shaper;wood shaping machine
木牛腿 wooden bracket
木偶 puppet;wooden image
木偶剧场 marionette;puppet theatre; toy theater[theatre]
木偶剧院 marionette;puppet theatre; toy theater[theatre]
木偶戏 puppet play
木偶状结核 <黄土中钙结核> puppet
木爬梯 wooden ladder
木耙 strike board;wooden rake
木排 float board; log raft; raft; raft tow; timber bundle; timber float; timber matting; timber raft; weaving-pole
木排坝 rafter dam
木排道(路)corduroy road
木排浮坝 <河口或港口的> log boom
木排浮运 bundle floating; timber floating
木排工(人)raftman[复 raftmen]
木排护岸 fascine revetment
木排护堤 fascine dike[dyke]
木排架 timber bent; timber trestle; wooden trestle
木排架桥 timber trestle bridge;wooden trestle bridge
木排拦河埂 log boom
木排流放 downstream floating of log raft
木排流送材 raft-wood
木排路 brush corduroy; corduroy road <美>
木排起拖 commencement of raft towing
木排(圆木)铺面 corduroy mat
木排运河 rafting canal
木盘车工 dish turner
木刨 wood(en) plane;wood working plane
木刨床 wood planning machine
木刨花 wood shavings;wood wool
木刨花过滤器 wood-shaving filter
木刨花建筑板 wood excelsior building slab
木刨花焦油 wood-shaving tar; wood splitting tar
木刨花吸声板 wood excelsior acoustic(al) sheet
木刨压板 chip cap
木配电板 timber panel(ling)
木盆 tub
木棚 timber set;timber shed;wooden shed;wood shed
木棚屋 timber hut;wooden hut
木棚洗涤器 wood hurdle scrubber
木碰垫 wooden fender
木披叠板 wood siding
木皮 sap wood;wood veneer
木片 chip; chump; piece wood; spill; wood chip;shingle
木片板 scale wood
木片破碎机 chip-breaker;chip crusher
木片切削机 wood chipper
木片清洁机 chip cleaner
木片瓦 wooden shake;wooden shingle
木片屋顶 shingle roof
木片屋面 splinter roof
木片楔子 chock
木片压碎器 chip crusher
木片再碎机 rechipper
木片振动筛 vibratory chip screen
木片装料机 chip packer
木平底船 <沉箱下水船只> timber packing
木平开窗 wooden casement window
木平面门 wood flush door
木平台 wood deck
木平屋顶 timber flat roof; wooden flat roof;wooden roof floor
木坡道 timber ramp
木破(损)wood failure
木铺板 batten ceiling; wood(en) planking
木铺面 timber floor(ing); wooden paving
木铺片瓦屋顶 wooden shingle roof
木铺砌工人 wood pavior
木铺砌工作 wooden paving work
木铺砌块 wooden paving block
木普通腹板大梁 timber plain web (bed) girder
木普通腹板梁 timber plain web (bed) beam
木企口楼板 rebated wood(en) floor(ing)
木企口楼面 rebated wood(en) floor(ing)
木起重臂 wood derrick
木起重机塔 wooden hoist tower
木气窗 timber transom
木砌块 wood block filler; wooden block
木砌块模数 timber unitized unit
木器 carpentry;wooden articles;woodenware;woodware
木器厂 furniture factory
木器等表面的光泽 patina
木器加工 wood working
木器家具仿古装饰 distressing
木器品 wood product
木器漆 wood lacquer
木器漆塌渗 lack of filling power
木器水性填孔剂 water filler
木器涂胶层 glue sizing
木器涂料 wood lacquer
木器涂装 woodwork coating
木器用颜料填孔着色剂 pigment wiping filler stain
木扦绑接 marrying
木签子 leggatt
木嵌块 wood block filler
木嵌条 common ground;timber fillet; wooden fillet
木戗脊折板屋顶 wooden hipped-plate roof
木墙 wooden wall
木墙板 wood board;wood siding
木墙筋 wooden stud
木墙裙 wood dado
木墙雨淋板 wooden siding shake
木桥 timber bridge; wooded bridge; wooden bridge; wooden trestle bridge;sanga(r) <原始的>
木桥墩 timber bridge pier
木桥面 plank floor(ing);timber floor of bridge;timber of bridge
木桥面板 floor plank for timber bridge
木桥面底层 timber sub-floor
木桥面栏杆 railing for timber bridge deck
木桥上部结构 timber bridge superstructure
木桥上的护轮木 wheel guards for timber bridge floor
木桥枕 wooden bridge sleeper
木桥桩 timber bridge pier
木倾板屋顶 wooden tilted-slab roof
木穹隆 timber dome;wooden dome
木球 bowl;wood ball
木球草地 bowling green
木球车床 wood ball turning lathe
木曲角钉 wooden brad
木曲头钉 wooden brad
木圈梁 chain timbers;girt(h)
木群桩 wood pile cluster
木染料 wood stain
木刃脚 <沉箱的> timber cutting curb
木绒 wood flock
木蠕虫 woodworm
木萨特 <一种轻质骨料> Lignacite
木塞 bung; cork stopper; dook; tie plug;wooden plug;wooden stopper
木塞穿孔器 cork borer
木塞法 cork method
木塞块 dutchman;wooden chock
木塞压紧机 corker
木塞压紧器 cork pressor
木塞钻孔器 cork drill
木三角板 wooden triangle
木三铰拱(框架)boomerang
木砂皮纸 coarse glass paper
木砂箱 timber moulding box; wood flask
木砂纸 abrasive paper for wood; wood flask paper
木砂纸用磨料 flint abrasive
木栅 paling;palisade;timber fender; wooden fender;zare(e)ba
木栅板 <浴室地面用> wood grating
木栅格 lumber grillage;timber grill
木栅谷坊 palisade check dam
木栅门 batten door
木栅填料 wooden grid packing
木栅网 barrier net
木栅围篱 post-and-paling
木栅堰 needle weir
木梢 wane
木哨所 timber gatehouse
木射线 medullary ray;pitch ray;ray; wood ray;xylem ray
木射线薄壁组织 wood ray parenchyma
木神龛 timber ciborium
木石结构 <木构架中填石及灰泥> black and white work
木石界面 wood stone interface
木石刻品 wood & stone carving
木石脑油 wood naphtha
木实心腹板梁 timber solid web(bed) beam
木饰面 wooden surfacing
木饰面板 veneer;wood veneer
木饰面蜡克 wood finishing lacquer
木饰面清漆 wood finishing lacquer
木饰线条 wooden mo(u)lding
木饰线条机 wood mo(u)lding machine
木梳 wooden comb
木薯淀粉 tapioca starch
木薯粉 cassava;cassava starch
木薯粉浆糊 cassava glue
木薯粉胶 cassava glue
木薯粉胶结剂 cassava adhesive
木薯酒精废液 cassava alcohol wastewater
木薯(属植物)cassava
木束梁 timber-pack beam
木闩 bobbin latch
木栓 cork; nailing block; nog; peg; phellem; timber peg; tre(e) nail; trunnel;wooden peg;wooden plug; wood nog;wood plug;suber【植】
木栓板 cork board
木栓层 phellem layer;suberous layer
木栓醇 friedelinol
木栓化 suberification;suberization
木栓化壁 suberized wall
木栓化部分 brown wood
木栓化作用 suberinisation
木栓壳 cork crust
木栓离层 abscission phelloid
木栓瘤 cork wart
木栓膜 cork film
木栓皮层 cork cortex
木栓深度 hub-deep
木栓酸 suberic acid
木栓酮 suberone
木栓烷 friedelane
木栓形成层 cork cambium;phellogen
木栓榆 rock elm
木栓脂 suberin
木栓质 suberin
木栓质层 suberin lamella
木栓质次结构体 suberinite-posttelinite
木栓质结构体 suberinite-telinite
木栓质类 suberinite
木栓质煤 suberain
木栓质煤素质 suberinite
木栓质似无结构体 suberinite-precollinite
木栓质体 suberinite
木栓质无结构体 suberinite-collinite
木栓状的 suberose

木栓组织 cork tissue
木水槽 wood(en) cistern;wood(en) flume
木水管 wood-stave pipe
木水落管 square shoot;wood downspout
木水平尺 wooden level
木水塔 timber cistern;timber water tower;wooden water tower
木水准仪 wooden spirit level
木丝 excelsior;wood fiber[fibre];wood wool;woody fiber[fibre]
木丝板 cellulose fiber tile;fiber[fibre] building board;wood excelsior slab;wood fiber board;wood fiber[fibre] slab;woodwool board;woodwool building slab;woodwool slab;excelsior board;fiber[fibre] board
木丝板吊顶 woodwool slab ceiling
木丝板隔墙 wood excelsior partition;woodwool slab partition(wall)
木丝板皮料 woodwool slab
木丝板平顶 woodwool slab ceiling
木丝保护绳 excelsior covering rope
木丝混凝土 excelsior concrete;wood excelsior concrete;wood fiber[fibre] concrete;woodwool concrete
木丝混凝土板 wood excelsior concrete slab
木丝混凝土平板 excelsior concrete slab
木丝混凝土屋顶板 wood concrete roof slab
木丝机 excelsior chuting machine
木丝建筑平板 excelsior building slab
木丝绝缘 excelsior insulation;wood excelsior insulation;woodwool insulation
木丝空隙填充料 excelsior hollow filler
木丝空心垫衬 wood excelsior hollow filler
木丝滤器 excelsior filter
木丝木屑绳 woodwool covering rope
木丝刨花绳 woodwool covering rope
木丝平板 excelsior slab
木丝平板隔墙 excelsior slab partition
木丝墙板 plastergon
木丝声控板 excelsior sound-control board
木丝水泥 woodwool cement
木丝水泥板 wood cement board;woodwool cement slab
木丝填空料 woodwool hollow filler
木丝填料 woodwool filler
木丝调音板 excelsior sound-control board
木丝吸声板 excelsior acoustic(al) board;wood excelsior absorbent board
木丝吸收板 excelsior absorbent board
木丝吸音板 wood excelsior absorbent sheet
木丝永久性模板 woodwool permanent formwork
木丝制造机 woodwool making machine
木松节油 sulfate wood turpentine;wood turpentine(oil)
木素 lignin(e);xylogen
木素苯酚甲醛树脂 lignin(e) phenol formaldehyde resin
木素腐殖质复合体 lignin-humus complex
木素含量 lignin(e) content
木素化 lignifying
木素磺化盐 lignosulfonate
木素磺酸 lignin(e) sulphonic acid;lignosulfonic[lignosulphonic] acid
木素磺酸钙 calcium lignin(e) sulphonate;lignosite
木素磺酸铬 <泥浆添加剂> chrome lignin(e) sulphonate;chrome lignosulfonate
木素磺酸盐 lignin(e) sulfonate
木素磺酸酯 lignin(e) sulfonate
木素结构单元 lignin(e) building unit
木素酸 lignin(e) acid
木素脱除 delignification
木素纤维素 lignocellulose
木塑窗 wood plastic window
木塑料复合材料 wood plastic composite
木塑制品 wood plastic product
木髓 core;pith;medulla <拉>
木髓斑 pitch flock;pit flock
木髓球验电器 pith-ball electroscope
木髓条纹 pith fleck
木髓物 pith
木髓线 pith ray;medullary ray
木碎料 wood slashings
木榫 timber dowel;wood nog;wood plug
木榫接合 cross corking
木榫头 wood joint
木塔(楼) timber tower;wooden tower;wood pagoda
木踏板 <楼梯等的> timber tread
木踏板绳梯 jack ladder;Jacob's ladder;accommodation net <大的有横木的>
木踏板舷梯 accommodation net
木踏步 timber step;wooden step;wooden tread
木胎钻生 sealing bare wood
木台 timber staging;wood stand
木台架 bunning
木台阶 timber step
木弹簧门 wooden swing door
木弹簧式护木 wood spring-type fender
木弹簧式护舷 wood spring-type fender
木炭 burned wood;burnt wood;char;charcoal;chark;vegetable charcoal;wood char;wood charcoal;woodcoal;wooden coal;xylanthrax
木炭笔 charcoal stick
木炭车 charcoal burning vehicle
木炭床过滤 charcoal bed filtration
木炭斗式输送机 char bucket conveyer[conveyor]
木炭发动机 charcoal burning engine
木炭粉 finely ground charcoal
木炭粉饼炉 charcoal dust cake stove
木炭腐病 charcoal rot
木炭高炉 charcoal blast furnace
木炭过滤法 charcoal filtration
木炭过滤器 charcoal filter
木炭和木材能 charcoal and firewood energy
木炭黑 charcoal black;Frankfort black;wood ash black
木炭化泥炭 carbonized peat
木炭画 charcoal drawing
木炭灰 ash wood
木炭加热筒 charcoal cartridge
木炭精炼炉 finery
木炭精炼铁 charcoal knobbled iron
木炭扩散层 charcoal diffuser bed
木炭炉 charcoal burner
木炭滤池 charcoal filter
木炭抛光 charcoal finishing
木炭气 wood gas
木炭燃气 gasogene[gasogene]
木炭生铁 charcoal pig iron
木炭石油能源开发过程 char-oil energy development process
木炭铁 wood charcoal iron
木炭吸附测定法 charcoal(adsorption) test method
木炭吸附法 charcoal adsorption process
木炭吸附过程 charcoal adsorption process
木炭铣铁 charcoal pig iron
木炭消耗 charcoal consumption
木炭液氧炸药 charcoal liquid oxygen explosive
木炭熨斗 charcoal iron
木炭纸 charcoal paper
木糖 wood sugar;xylose
木糖醇 xylitol
木糖和木糖醇制药废水 xylose and xylitol pharmaceutical wastewater
木糖胶 xylan(e)
木梯级 wooden step
木梯梁 wooden string(er)
木梯上踏脚横档 rounds
木梯(子) wooden ladder
木踢脚 timber floor base;timber shirting;wood shirting
木踢脚板 timber mopboard;timber skirting(board);timber washboard;wood base(board);wooded washboard;wooden base plate;wooden mopboard;wooden sanitary cove;wooden scrub board;wooden skirting shake;wood skirting board
木踢脚板龙骨 wood shirting rough ground
木体系 wooden system
木体系构造 wooden system construction
木天沟 wood(en) gutter;wooden rainwater gutter;wooden valley gutter
木天沟雨水管 timber valley gutter
木填缝料 wooden filler
木填块 wood filler block;wood packing
木填料 wood filling
木填条 blocking
木条 accouplement;covering fillet;wooden bar;wooden strip;wood strip
木条板墙 strapped wall
木条板踏步 ga(u)ge boarding
木条板屋顶 split roof
木条板箱 crate
木条板闸门振动 shutter vibration
木条地板 batten floor;strip floor cover(ing);strip wooden flooring;wood strip flooring;wood floor
木条地板斜钉 toe-nailing wood-strip flooring
木条垫 <放在澡堂、洗衣间地板上和门口的> plank foundation platform
木条固定的玻璃 ribbed glass
木条刮路器 split-log drag
木条或极轻型槽钢 furring strip
木条加固咬口 <金属薄板表面> batten seam
木条胶接缝加热 strip heating
木条卷帘 pinoleum blind
木条楼板 wood strip flooring
木条路刮 split-log drag
木条碰垫 rod fender
木条嵌花 wooden mosaic
木条墙面 wood lath facing
木条踏脚 <放在澡堂、洗衣间地板上和门口的> plank foundation platform
木条跳板 duck board;gang boarding
木条箱 wooden crate
木条镶边 banding;lipping
木条镶花地板 parquet flooring
木条子 batten
木跳板 <用于摊铺混凝土的> plank run(way)
木贴面 wooden facing;wooden lining
木贴接面 meeting faces timber
木贴面通道 wood-lined passage
木酮糖 xylulose
木桶 barrel;cannikin;cask;kit;tub;vat;wooden casks;tun <容量 252 加仑,1 英制加仑 = 4.546 升,1 美制加仑 = 3.785 升>
木桶浮标 keg buoy
木桶箍 quarter hoop
木桶箍紧原理 wooden barrel principle
木桶味 wood taste
木桶业 cooperage
木桶制造业者 cooperage
木头 log
木突堤 timber pier
木推手板 wooden finger plate
木托 bearer
木托板 wooden pallet
木托架 timber underframe
木托梁 timber joist;wood(en) joist
木托瓦屋顶 timber-and-tile roof
木托座 wooden bracket
木拖车 timber trailer
木砣 hand lead
木瓦 perfections;shake;shide;shingle tile;tile of wood;wood tile
木瓦板 shingle;wood shingle
木瓦板屋面 shingle roofing
木瓦搭接 head lap;shingle lap
木瓦钉 tile pin
木瓦短圆材 shingle bolt
木瓦覆盖屋顶 shingle roof cladding
木瓦结构 shingle construction
木瓦斯 wood gas
木瓦屋顶 timber-and-tile roof
木瓦屋面 splinter roof
木瓦用材 wood for shingles
木瓦用钉 shingle nail
木望板 board sheathing;plank sheathing;sheathing;timber board sheathing;wooden board sheathing
木围板 wooded coaming
木围框导缆口 coaming chock
木围栏 timber rail;wooden crib;wooden fence
木围篱 wooden palisade fence
木围囹 wood waling
木围墙 timber fender;wooden fender
木围绕物 timber surround
木围堰 log cofferdam;timber cofferdam
木桅顶方榫 head tenon
木桅(杆) wooden mast
木尾防腐 butt-end treatment
木纹 chord;medullary ray;timber grain;vein;wood grain
木纹板 grain board;wood card
木纹板面 board-marked
木纹波纹绸 Moiré ronde
木纹处理 substrate preparation
木纹垂直截面 end surface grain
木纹发白 whitening in the grain
木纹方向 grain direction
木纹合成装饰板 wood grain synthetic(al) decorative board
木纹黄石 serpeggiante
木纹加工 wood grain finish
木纹交叉的胶合板 crossbanded lumber veneered board
木纹交叉黏[粘]结 cross band
木纹角 grain slope
木纹角度 slope of grain
木纹开裂 grain cracking
木纹理罩面处理(家具) oystering
木纹隆起 grain raising;raised grain
木纹面板 veneer

木纹描绘用具 graining tool
木纹膜胶粘剂 adhesive for wood grain film
木纹拼花 book matching;herringbone matching
木纹平整着色剂 non-grain-raising stain
木纹平整着色料 non-grain-raising stain
木纹漆 graining liquid;graining paint;wood grain paint
木纹漆刷 overgrainer
木纹(倾)斜度 grain slope
木纹色 graining colo(u)r
木纹色衬 engrain lining paper
木纹色墙纸 engrain wallpaper
木纹石 xylotile
木纹饰面 <拼缝锯板做模板> A finish;graining;type A finish
木纹刷 graining brush
木纹突起 grain raising;raised grain
木纹突起处理 burned finish
木纹涂装法 graining
木纹细密的 close-grained
木纹显露 pick-up of grains
木纹斜度 angle of the grain;slope of grain;slope of timber grain;grain slope
木纹研光机 Moiré calender
木纹印花 wood grain print
木纹印刷 wood grain printing
木纹凿 veiner
木纹纸 grained paper;paper overlay
木纹纸板 grained board
木纹装饰 graining
木纹状的 grained;grainy
木纹状断口 woody fracture
木纹状结构 woody structure
木纹状微裂纹 grain checking
木纹纵裂 thunder shake
木纹纵向抗拉强度 tensile strength parallel to the grain
木蜗杆 timber worm
木屋 block house;frame house;log cabin;lumber cabin;timber crib;timber-framed;wooden house;chalet <瑞士山中倾斜屋顶的>
木屋顶 plank roof;timber roof;wood(en) roof
木屋顶材料 timber roofing
木屋顶地板 timber roof floor
木屋顶覆盖层 timber roof cover(ing)
木屋顶盖板 timber roof sheathing
木屋顶桁架 timber roof truss
木屋顶桁条 timber purlin(e)
木屋顶架 wooden roof truss
木屋顶尖端 timber roof spire
木屋顶尖塔 wooden roof spire
木屋顶桁 wooden roofed
木屋顶系梁 tigna
木屋盖 carpenter's roofing
木屋架 plank roof truss;plank truss;roof woodwork;wooden truss;wood roof truss
木屋架拉梁 tigna
木屋架上承托檩条的三角木 timber cleat
木屋角柱 teazel post
木屋教堂 timber church
木屋面 timber roofing;wooden roofing
木屋面板 wooden board sheathing;wood sheathing
木屋面覆盖物 timber roof cladding
木屋面盖板 wooden roof cladding;wooden roof sheathing
木屋棚户区 <巴西> favela
木屋区 squatter area
木犀【植】sweet-scented osmanthus;sweet osier
木犀草属【植】mignonette
木犀属 <拉> osmanlhus
木锡矿 wood tin
木洗衣板 timber washboard
木铣床 shaper
木系梁 wooden binding beam
木下部结构 timber supporting structure
木纤腻子 wood dough
木纤维 lignified fiber[fibre];lignin(e);xylogen;xylon
木纤维板 beaver board;Huntonit;vegetable fiber[fibre] board;wood fiber[fibre] pane;wood fiber[fibre] board
木纤维板顶棚 wood fiber[fibre] board ceiling
木纤维板吸声顶棚 wood fiber[fibre] board absorbent ceiling
木纤维保温 woody fiber[fibre] insulation
木纤维保温材料 wood fiber[fibre] insulation material
木纤维材料 wood-fibered material
木纤维断裂 upset
木纤维粉刷 wood fiber[fibre] plaster
木纤维敷料 alignimum
木纤维隔热(材料) woody fiber[fibre] insulation
木纤维隔声板 wood(y) fiber[fibre] sound-deadening board
木纤维隔音板 celotex
木纤维骨料 wood fiber[fibre] aggregate;woody fiber[fibre] aggregate
木纤维硅化 wood fiber[fibre] silicification
木纤维灰浆 wood fiber[fibre] plaster
木纤维灰泥 wood fiber[fibred] plaster
木纤维混凝土 wood fiber[fibre] concrete
木纤维集料 wood fiber[fibre] aggregate;wood(y) fiber[fibre] aggregate
木纤维绝热 <材料> woody fiber[fibre] insulation
木纤维抹灰底板 wood fiber[fibre] plaster baseboard
木纤维墙板 Insulite;Insulwood
木纤维热绝缘墙板 ankar;beaver board
木纤维石膏板 gypsum wood-fibered plaster
木纤维水泥 <木纤维水泥屋顶用的> dressing compound for Hausler type roof(ing)
木纤维水泥板 cemented wood fiber[fibre] board
木纤维素 lignose
木纤维素纤维 cellulosic fiber[fibre]
木纤维吸声板 sound-absorbent wood fiber[fibre] board
木纤维吸音天花板 wood fiber[fibre] board acoustic(al) tiled ceiling;wood fiber[fibre] board sound absorptive ceiling
木纤维增强水泥复合材料 woody fiber[fibre]-cement composite
木纤维质 lignone
木纤油灰 wood dough
木线槽 boxing;wooden wiring casing
木线担 wooden arm
木线脚(饰) wood(ened) mo(u)lding)
木香花 banks rose
木箱 coffin;timber bin;wooden bin;wooden case;wooden crate
木箱包装 wooden box package
木箱货 case cargo
木箱内衬铝箔纸 lined with aluminium foil in the wood(en) case
木箱式暗沟 wooden box drain
木箱式进水闸 wooden box headgate
木箱式排水沟 wooden box drain
木镶板 mosaic parquet panel;timber panel(ling);wood(en) panel(ling);wooden wainscoting;wood mosaic
木镶板隔墙 wood panelled partition
木镶边 wooden surround
木镶块 wood mosaic
木镶框 wooden surround
木镶面 wood lining
木镶面板 timber facing
木镶嵌 mosaic parquet panel
木镶嵌饰 intarsia;wooden inlay
木销 dowel
木销钉 coak;dowel;timber dowel;trunnel;wooden dovetail;wooden dowel
木小舍 wooden hut
木楔 chock;timber cleat;wedge;wooden chock;wooden insert;wooden key;wooden plug;wooden quoin;wooden wedge;wood wedge
木楔形墙板 clapboard
木斜板屋顶 wooden tilted-slab roof
木斜槽 timber chute
木斜杆 wooden strut
木斜沟槽 timber valley gutter;wooden valley gutter
木鞋 sabot
木屑 chip;dook;hog fuel;saw dust;spill;swarf;timbering residue;wood chip;wood dust;wood flour;wood meal;wood refuse
木屑板 chipboard;particle board;sawdust board;wafer board;wooden chipboard;wooden particle board;wood particle board;xylolite[xylolith] <(slab)由水泥和锯屑制成>;flakeboard
木屑材料 wood particle material
木屑铲斗 wood chip bucket
木屑储存 wood chip storage
木屑电木制品 wood plastic article
木屑分离器 wood scraps separator
木屑骨料 sawdust aggregate
木屑合成纤维 defibering
木屑回收 wood chip recovery
木屑混凝土 sawdust concrete;wood waste concrete
木屑混凝土板 wood excelsior concrete slab
木屑集料 sawdust aggregate
木屑漏斗车 wood chip hopper car
木屑黏[粘]土炮泥 sawdust-and-clay stem
木屑刨花板 flake board
木屑片 sawdust chip;sawdust cleat
木屑墙纸 wood chip wallpaper
木屑砂浆板 wood cement concrete slab
木屑石 pictite
木屑水泥混凝土 wood cement concrete
木屑塑料窗 wood plastic window
木屑塑料拼花地板 wood plastic parquetry
木屑塑料制品 wood plastic article
木屑吸声顶棚板 wood chip absorbent ceiling board
木屑纤维板 particle board
木屑制品 wood chip products
木屑抓斗 wood chip grab
木蟹 wooden float
木蟹找平 floated screening
木芯 wooden core;core of wood;duramen
木芯辐裂 heart check;heart shake
木芯腐烂 heart rot
木芯环裂 growth shake;heart check;heart shake;internal annular shake;ring shake
木芯裂 ring shake
木芯髓线 ray;wood ray
木芯坐标 zenocentric coordinate
木星 Jupiter
木星卫星地质学 geology of Jupiter's satellites
木行 lumber yard
木行人桥 wooden foot bridge
木型箱 wood flask
木性的 xyloid
木雄榫 carpentry tongue
木蓄水池 wooden cistern
木旋床 wood lathe;wood turning lathe
木旋转门 timber revolving door;wooden revolving door
木压缝条 wood cover fillet;wood cover strip
木压杆 wooden strut
木压条 bead;timber bead;wooden batten;wood trim;applied mo(u)lding <门窗、家具的>
木檐 board eaves;wood gutter
木檐沟 timber eaves gutter;wooden eave(s) gutter;wood(en)(roof) gutter
木堰 log weir;wooden splash-dam;wooden weir
木堰堤 wooden dam
木堰口 timber weir
木样板 wooden screed
木腰窗 wooden transom
木摇头窗 timber transom
木叶纹 tree leaf design
木叶纹饰 leaf pattern decoration
木叶装饰的 foglie
木蚁 carpenter ant
木鹰架 timber falsework
木营房 timber hut
木用填料 wood filler
木油 wood oil
木油树 Aleurites montana;wood oil tree
木油树酸 alsnritolic acid
木油树油 Aleurites montana oil
木油桐 wood oil tree
木鱼 wooden fish;wooden knocker
木鱼鳞板 wooden roof shake;wooden shake;wood weather-boarding
木与金属连接 timber-to-metal connection
木与木连接 timber-to-timber connection
木雨水檐沟 wooden rainwater gutter
木浴盆 timber bath(tub)
木浴桶 wooden bathtub
木浴澡盆 wooden bathtub
木杂酚油 wood tar creosote
木凿 wood chisel;bevel chisel;pocket chisel
木凿柄 wooden handle for chisel
木造的 timbered
木造农舍 chalet
木造小房 shack
木贼泥炭 equisetum peat
木渣 wood wool
木闸 log stop;sasse
木闸板 wood weir block
木闸块 wooden brake block
木闸门 cleading
木闸瓦 wooden brake block;wooden brake shoe
木闸瓦的条状制动带 jointed brake
木栅拦沙坝 palisade check dam
木栅栏 abat(t)is;boarded fence;pali-

sade; post-and-paling; timber fence; wooden(palisade) fence; woodhenge; board fence; picket fence
木栅栏丁坝 transverse board fence dike
木栅栏护岸 fencing revetment
木栅栏桩 timber fence post
木窄板 wood batten
木栈道 timber trestle; wooden trestle
木栈桥 timber jetty; timber trestle; timber trestle bridge; wooden trestle; wooden trestle bridge; wood trestle(bridge)
木栈桥导航堤 timber jetty
木栈桥码头 timber jetty; timber pier
木折板屋顶 timber folded plate roof; timber folded slab roof; wooden folded plate roof; wooden prismatic shell roof
木折板屋盖 timber folded plate roof; timber folded slab roof; wooden folded plate roof; wooden prismatic shell roof
木折尺 folding wood(en) rule; pocket rule; wooden folding rule
木折叠门 timber folding door; wooden folding door
木折椅 wooden folding chair; wood folding chair
木摺【建】fillet
木针 fixing fillet; fixing slip; ground; timber needle
木砧 anchor brick; timber fillet
木枕 timber sleeper; wood(ed) block(ing); wooden sleeper; wooden tie
木枕代用物 substitute timber sleeper
木枕防腐 preservation of wooden tie
木枕防腐处理 conserving timber sleeper; creosoting timber sleeper
木枕锯 sleeper saw
木枕刻痕 incising of wooden tie; notching
木枕刻字标记 date nail
木枕无砟轨道 wooden tie ballastless track
木枕线路 timber-sleepered track
木枕预钻孔 preboring of spike holes; preboring of wooden tie
木蒸松香 wood rosin
木支撑 lignum shotcrete; sheeting with timber; timbering; timber strut; wooden backing; wooden brace; wooden strut; wooden support; timbering set <隧道内的>
木支撑尺寸 timber size
木支撑的导坑 timber heading
木支撑的英式方法 English system of timbering
木支撑的英式体系 English system of timbering
木支撑工程 timbering work
木支撑及顶撑 timbering and shoring
木支撑架 racked timbering
木支撑用金属紧扣件 nail spike
木支承 timber support
木支承结构 timber supporting structure
木支墩 wooden pillar
木支护 wooden support
木支护扒矿平巷 timber scram
木支护的对角工作面 timbered rill
木支护的水平分层充填采矿法 timbered-horizontal cut and fill stoping
木支护顶板 timbered back
木支护井底车场 timbered bottom
木支护平巷 timbered drift
木支护竖井 timbered shaft
木支护天井 timbered raise
木支架 timbering; timber lining; timber set; timber support; wooden timbering; wooden support
木支架构件接头 timber joint
木支架加工车间 timber framing shop
木支架喷浆 timber support guniting
木支架起火 adustion of timbering
木支柱 peg-leg; post; timber mast; timber prop; timber stanchion; timber strut; wooden strut; wooden stanchion
木支柱式支撑 timbering
木支座 timber saddle
木肢起重杆 timber stiff leg derrick
木执手 wooden handle
木止水条 log seal; timber seal
木纸浆 wood pulp
木纸浆厂废水 spent liquid form wood cooking; wood pulp waste
木制矮垫脚凳 cricket
木制安全销 wood break pin
木制便桶盖 timber sanitary cove
木制标尺 wooden rod
木制玻璃格条 wooden astragal
木制槽板 wooden raceway
木制蝉滑板 wooden dark slide
木制巢窝式储粮箱 nested wood(en) silo
木制车 wooden car
木制储水器 timber cistern
木制穿线板 wooden raceway
木制窗芯子 wooden astragal
木制大拱 wood arch
木制大门 timber gate
木制道路 timber road
木制的 ligneous; timbered; wooden; woody; xyloid
木制地下排水管 wooden underdrain(age) pipe
木制堵漏板 wooded patch
木制二拼滑车 wooden double block
木制翻斗车 wood skip; wood tip(ping) wagon
木制风桥 wood overcast
木制格子板 wood grating
木制隔空条 wood rule
木制沟堤板 timber ditch dike
木制管 wood conduit; wood pipe; wood tube; wood tubing; wood piping
木制管道闸门 wooden pipe penstock
木制(轨道)水平尺 wooden track
木制轨距尺 wooden center[centre] rail ga(u)ge; wooden center[centre] track ga(u)ge
木制混凝土模板 wooden concrete form
木制或金属烛台架 hearse
木制集装箱 wood container; wooden container
木制家具 wooden furniture; woodware
木制件 ligneous piece; woodwork
木制建筑嵌板 wood building panel
木制讲坛 wooden pulpit
木制卷帘百叶窗 wooden roller shutter
木制卷升百叶窗 timber roller shutter; timber rolling shutter
木制绝缘子 wood stick break
木制框板 wooden frame
木制临时营房 wooden hut
木制轮辐 wooden spoke
木制门槛 wooden door threshold
木制排气管 wooden stave
木制排水槽 wooden drainage
木制棚车车体 wooden boxcar body
木制品 wooden articles; wooden products; woodenware; woodware; woodwork
木制品工业 woodwork industry; wood working industry
木制品加工业 wood working industry
木制品装饰 wood finishing
木制品装修 wood finishing
木制砌块 wooden block
木制嵌件 wooden insert
木制人体模型 dressmaker's dummy
木制人字形闸门 wooden mitering [mitring] gate
木制容器 wooden container
木制三角架 wooden tripod
木制实腹梁 wooden plain(web) girder
木制守车 <俚语> way car
木制竖铰链窗 wooden casement window
木制双开式弹簧门 timber swing door; wooden swing door
木制水轮机 wooden turbine
木制通风筒 wooden ventilator
木制透水护坡 timber permeable revetment
木制托盘 wood pallet
木制拖行玩具 toddler toy
木制脱粒机 wooden thresher
木制玩具 wooden toy
木制往复式马铃薯分选机 wooden reciprocating potato sorter
木制威尼斯软百叶窗 wooden Venetian blind
木制卫生凹圆形板 timber sanitary cove
木制卧铺上部隔板的加强板 head board plate
木制屋盖 roofed in wood
木制下部结构 wooden substructure
木制线槽敷设工程 wooden raceway work
木制线担 wooden crossarm
木制线脚装饰 wooden mo(u)lding
木制箱 wooden-packing case
木制箱式梁 timber box girder
木制箱形涵洞 wooden box culvert
木制漩涡形柱头装饰 wooden roll
木制压力水管 wooden pipe penstock
木制叶片结构 wooden blade construction
木制溢流导管 wooden pump logs
木制引鞋 wooden guide shoe
木制有扶手的高背长靠椅 box settle
木制圆屋顶 wooden dome
木制闸门 wooden gate
木制折叠箱形托盘 wooden collapsible box pallet
木制支架 lazy board
木制直角尺 wooden square
木制纸 wooden paper
木制轴承 wooden bearing
木质 woodiness; xylogen; xylon
木质靶机 wooden target drone
木质板 wood-based panel
木质板拼成管 stave pipe
木质部【植】hadromestome; xylem
木质部射线 wood ray; xylem ray
木质部蛀虫 wood borer
木质材料 woody material
木质层 layer of wood
木质层压板 wood veneer laminate
木质层压材料 compregnated wood
木质车体客车 coach with wooden body
木质成套设备 timber unitized unit
木质船体 wooden hull
木质道路 timber road
木质的 ligneous; wooden; xyloid; woody; xylary
木质顶劈接 woody terminal cleft grafting
木质反应 lignin(e) reaction
木质防尘罩 wooden dust shield
木质防火门 fire-proof wood(en) door
木质放水门 wooden wicket
木质飞机 wooden airplane
木质酚油 wood creosote
木质风筒 wood ventilation pipe
木质浮船坞 timber floating dock
木质感混凝土 wood(chip) concrete; woodcrete
木质感水泥 wood cement
木质梗 woody stem
木质骨料 wood(fiber) aggregate
木质骨牌 wooden domino
木质光敏素 xylochrome
木质涵洞 timber culvert; wood culvert
木质褐煤 woody brown coal; woody lignite
木质花纹饰面 <混凝土> marked finish
木质滑板 slider
木质化 lignification; lignifying
木质化程度 degree of lignifications
木质化木栓 lignified cork
木质化组织 lignified tissue
木质化作用 lignification
木质缓冲器 wooden bumper
木质磺酸 lignin(e) sulphonic acid
木质磺酸钙 calcium lignosulfonate
木质磺酸铁铬 ferro-chrome salts
木质货车 timber wagon
木质机身 wooden fuselage
木质集料 wood aggregate
木质建筑玩具 wood construction set
木质胶 collose
木质接缝填料 wood joint filler
木质结构腐木质体 texto-ulminite
木质结构镜质体 xylotelinite
木质结构体 textinite
木质井架 wood(y) headframe
木质镜煤 xylovitrain
木质镜煤半丝炭体 xyloritrosemifusinite
木质镜丝炭 xylovitrofusinite
木质镜质丝质体 xylovitrofusinite
木质救生筏 wooden lifecraft
木质瘤 xyloma
木质煤 board coal; xyloid coal
木质蒙皮 wooden envelope
木质模板 wooden form
木质泥炭 wooden peat; woody peat
木质排水管 wooden drain(pipe)
木质刨花芯板 wood flake core panel
木质破坏率 wood failure ratio
木质器皿 woodware
木质人造板 man-made wood board; wood base fibre and particle panel material
木质人造板材 wood base panel material
木质人字起重机 timber derrick
木质溶解菌 lignin(e)-dissolving fungi
木质水封 timber seal
木质水管 stave pipe
木质水泥混凝土 wood cement concrete
木质丝煤 xylofusinite
木质丝炭 xylofusinite
木质素 lignin(e); lignose; wood element
木质素柏油脂 lignin(e) tar pitch
木质素酚醛树脂 lignin(e) phenol formaldehyde resin
木质素腐殖质 ligno-humus
木质素铬盐 chrome lignin(e)
木质素和铬、木纤维稳定土法 lignin and chrome-lignin process
木质素磺酸 lignosulfonic acid

木质素磺酸钙 calcium lignosulphonate
木质素磺酸钠 sodium lignin(e) sulfonate;sodium lignosulphonate
木质素磺酸盐 calcium lignin(e) sulphonate;lignin(e) sulphonate;lignosulphonate <一种混凝土缓凝剂>
木质素磺酸盐泥浆 lignosulphonate mud
木质素磺盐酸钠 Polyfon
木质素减水剂 lignin(e)-type water-reducing admixture
木质素胶 lignin(e) paste
木质素胶合板 lignified wood
木质素焦油(沥青) lignin(e) tar;lignin tar
木质素焦油脂 lignin(e) tar pitch
木质素黏[粘]胶剂 lignin(e) adhesive
木质素树脂 lignin(e) resin
木质素树脂固着生物膜厌氧流化床 lignin(e) resin fixed biofilm anaerobic fluidized bed
木质素塑料 lignin(e) plastics
木质素稳定 lignin stabilization
木质素絮凝剂 lignin(e) flocculant
木质素硬沥青 lignin(e) tar pitch
木质素有机物 lignin(e) organic matter
木质塑料 lignin plastic
木质酸 xylonic acid
木质填料 wood filler
木质贴接面 timber meeting face
木质通用构件 timber unitized unit
木质托座 bita[byatt]
木质屋顶 wood deck
木质系船柱 deadhead;wooden bitt
木质纤维 boony fiber;lignified fiber[fibre];wood(y) fiber[fibre]
木质纤维板 wood fiber slab
木质纤维壁板 Sundeala
木质纤维的 lignocellulosic
木质纤维隔声吊顶板 acoustic(al) celotex board;acoustic(al) celotex (tile)
木质纤维合成板 cellulose sheet
木质纤维素 bastose;lignocellulose
木质纤维塑料 lignin(e) plastics
木质纤维无烟煤 lignocellulosic anthracite
木质酰胺 xylonamide
木质镶嵌工艺品 wood mosaic
木质小门 wooden wicket
木质型 woody type
木质翼梁 wooden spar
木质止水条 timber seal
木质制品 timber product
木质状结构 woody structure
木质组 xylinoid group
木质组织 lignum
木质钻塔 wooden derrick
木肘材 bosom;bosom knee;knee piece;knee timber;timber knee
木主杆 timber principal post
木主柱 wooden principal post
木柱 timber column;timber pillar;timber post;wood column;wooden column;wooden pillar;wooden post;wood post;timber prop <采矿>
木柱承板 foot block
木柱隔墙 stud partition;timber-stud partition(wall)
木柱架 timber trestle
木柱筋结构 wood stud construction
木柱橛 wooden peg
木柱墙壁 timber-stud wall
木柱式 wooden column order;wooden order
木柱铸铁帽 post cap
木柱桩 timber pillar
木柱座 foot block
木蛀虫 timber worm
木砖 anchor brick;dowel brick;grounds;nailing block;nog;soldier;timber brick;wood block;wood box;wood brick;wooden brick;wood mosaic;rough ground <钉罩面板的>;anjan <印度,有黑色条纹,颜色由红到暗棕色>;shelf nog <支搁板的>
木砖地面 block wood flooring;wooden brick floor
木砖孔 break-in
木桩 earth pile;peep stake;peg;peg stake;peg table;picket;spile;spiling;surveyor's stake;timber peg;timber pile;timber post;wooden peg;wooden pile;wooden post;wood pile;stump <障碍物>
木桩坝 piled dyke[dike]
木桩丛桩型系船桩 wood-pile dolphin
木桩的过夯 over-driving of wood pile
木桩堤 timber pile dike[dyke]
木桩堤坝 pile and stone dyke[dike]
木桩丁坝 timber pile dike[dyke];timber pile groin
木桩顶的保护 protection wood pile head
木桩顶钢箍 timber pile top ring
木桩顶铁箍 timber pile top ring
木桩顶头钢箍 driving band
木桩堆石堤 pile and stone dyke[dike]
木桩腐烂 deterioration of wood pile
木桩高 picker level
木桩工作平台 wooden pile staging
木桩工作台 wooden pile stage
木桩横撑丁坝 pile and waling groin
木桩横撑透水丁坝 pile and waling permeable groin
木桩护岸法 raw-bank system
木桩校正法【测】peg adjustment;peg method;peg test
木桩接长 splicing of timber piles
木桩接头 wooden pile joint
木桩靠船墩 wood pile dolphin
木桩码头 timber piled wharf
木桩锚碇 timber pile anchorage
木桩排架 timber pile bent
木桩排架桥台 timber pile bent abutment
木桩排架桥台挡土板 bulkhead plank of timber pile bent abutment
木桩群 wood pile cluster
木桩式码头 timber pile(d) jetty
木桩式突堤 timber pile(d) jetty
木桩碎裂 brooming
木桩台 timber pile-staging
木桩头部桩箍 driving band
木桩突码头 timber piled jetty
木桩围栏 hell fence
木桩靴 shoe of timber pile
木桩栅栏 palisade;picket fence
木桩支护竖井 timber shaft
木桩锥度 taper of timber pile
木装饰面 timber facing
木装修 joiner's finish;joiner's work;joinery work;wooden decoration
木装修板 boarding
木锥 gimlet
木准尺 guiding rule
木准条 wooden screed
木纵梁 longitudinal timber;timber longitudinal beam
木纵梁间的交叉撑 cross bracing of timber longitudinal beam
木组合梁 Clark beam;wood combined beam;wooden compound beam
木组织 wood tissue
木钻 auger;wood borer
木钻床 wood borer;wood driller;wood drilling machine
木钻机 wood drilling machine
木钻架 wooden drilling derrick
木钻井架 timber drilling derrick
木钻塔 timber derrick;wood derrick
木钻头 wooden broach
木钻钻头夹头 wood drill bit chuck
木作 carpenter's work;carpentry;woodwork
木作表面 face side;work face
木作尺 butt ga(u)ge
木作工程 woodwork construction
木作接头 carpentry joint
木作企业 carpenter's undertaking
木座位 timber seat;wooden seat

目标 aim(ing);destination;eyemark;feature;goal;mark;object;objective;standpoint;target

目标百分数 percent of goal
目标暴露量 target exposure
目标背景 target background
目标逼真度 object fidelity;target fidelity
目标变量 object variable;target variable
目标变量的允许变化范围 allowable variation range of objective variable
目标变数 target variable
目标标示系统 target-designating system
目标标志器 target marker
目标表 object table
目标捕获 target acquisition
目标捕捉 object snap
目标操作 object run
目标测位 target position finding
目标层次 level of objective
目标长度 target length
目标成本 object cost;target cost
目标成本测定 target cost determination
目标成本分解 target cost decomposing
目标成本管理 target cost management
目标成本合同 target cost contract
目标成本控制 target cost control
目标成本落实 target cost implement
目标成本设计 design to cost goal
目标成就的衡量标准 criterion for measuring objective achievement
目标成像 target imaging
目标程序 objective program(me);object program(me);object routine;target program(me);target routine
目标程序编制 object program(me) development
目标程序表 object program(me) list
目标程序长度 object program(me) size
目标程序穿孔卡片叠 object pack
目标程序的输出 output of the object program(me)
目标程序地址 object program(me) address
目标程序方面 object phase;run phase
目标程序机 target machine
目标程序计算机 object computer;target computer
目标程序结构 object configuration
目标程序卡片组 object deck
目标程序开发 object program(me) development
目标程序库 goal program(me) library;object base;objective program(me) library;object library;object program(me) library;text library
目标程序码的生成 object program(me) code generation
目标程序模块 object program(me) module
目标程序配置 object configuration;target configuration
目标程序设计 goal programming
目标程序时间 object time
目标程序优化 object program(me) optimization
目标程序语言 object program(me) language
目标程序执行阶段 object phase
目标程序执行时间 object time
目标程序指令 objective program(me) instruction
目标程序状态 object phase;run phase
目标程序准备 object program(me) preparation
目标尺寸 target size
目标穿孔卡片叠 object pack
目标词 target word
目标存储器 target memory
目标达成分 goal-achievement analysis
目标达到情况 goal attainment
目标代码 destination code;object code
目标代码表 object code listing
目标代码程序 object code program(me)
目标代码兼容性 object code compatibility
目标代码模块 object code module
目标带 object tape
目标带格式 object tape format
目标档案 target dossier
目标导向行为 goal-directed behavio(u)r
目标的不确定性 uncertainty of objective
目标的产生 generation of objective
目标的发展 evolution of objective
目标的可识别性 target identifiability
目标的特征 clarification of objective
目标的现时位置 target present position
目标的修正 modification of goals;modification of objectives
目标的演变 evolution of objective
目标的一致性 compatibility of goals
目标灯 target lamp
目标等效反射面 radar cross-section
目标点 object point
目标点速度 <溜放车辆通过缓行器后的> target point speed
目标叠 object deck
目标定点 pinpoint;target pinpoint
目标定时 target timing
目标定位 localization of target;target location
目标定位镜 index mirror
目标定位数据 localization data of target;target positioning data
目标定位系统 object locating system
目标定向程序设计 object oriented programming
目标定义 object definition
目标反射 object reflection
目标反射率 target reflectivity
目标反射器 <测距仪用> target reflector
目标反射信号 target echo
目标反应 goal response

目标范围 target coverage;target range
目标方位 target bearing
目标方向 target direction
目标分布 target distribution
目标分解 target decomposition;target deconstruction
目标分析 target analysis
目标符号 aiming symbol
目标覆盖 plank covering
目标高度 object height;target elevation
目标跟踪 target following;target tracking
目标跟踪激光雷达 target-tracking laser radar
目标跟踪校正 target track correction
目标跟踪雷达 target-tracking radar
目标跟踪滤光器 target-tracking filter
目标跟踪头 target seeker
目标跟踪仪 target-sighting apparatus
目标管理 goal management;management by objectives;management by results;managing for results;objective management;target management
目标规定过程 goal setting process
目标规划 goal plan(ning);goal programming;objective planning;object planning;target programming
目标函数 object function;objective function;result function
目标函数的约束条件 bond for objective function
目标函数方程 objective function equation
目标回波 target echo
目标机 target machine
目标级配 target grading
目标急速移动 scintillation
目标集(合) goal set;object set
目标集中 target concentration;target tightening
目标计划 objective planning
目标计算机 object machine
目标(计算机)配置 object configuration;target configuration
目标记录 object record
目标继电器 target relay
目标价格 target price
目标(价)合同 target contract
目标检测 object detection;target detect
目标简况 target brief
目标鉴别 target discrimination
目标鉴别雷达 decoy discrimination radar
目标鉴别器 target discriminator
目标鉴别声呐 classification sonar
目标角色 object case
目标阶段 object phase;target phase
目标结构密度 density of object structure;density of target structure
目标距离 object distance;objective distance;objective range
目标距离及方位计算机 target range and bearing computer
目标控制 target control
目标棱镜 target prism
目标利润 profit target;target profit
目标利润定价 target profit pricing
目标利润率 target rate of return
目标利润原理 target returns principle
目标利润原则 target returns principle
目标利润转移价格 target profit transfer price
目标码 target code
目标码符号组 object code field
目标密度 target concentration
目标模件 object module
目标模块 object module
目标模块(程序)库 object module library
目标模块卡片 object module card
目标模块数据集 object module data set
目标模拟 target simulation
目标黏[粘]滞度 target viscosity
目标喷雾 target spray
目标平均强度 target mean strength
目标评价 target assessment
目标破坏概率 target failure probability
目标强度 target intensity;target strength
目标区 objective zone;region of target;target zone
目标区地形测量 target area survey
目标区域 target area
目标群 target complex
目标任务法 objective-and-task method
目标日期 <指工程完工> target date
目标容量 target capacity
目标色 aim colo(u)r
目标上的 on-target
目标设计 object design
目标摄影 point reconnaissance;spot photography
目标深度测量声呐 depth scan(ning) sonar
目标失效概率 target failure probability
目标失踪 contact lost;target lost
目标识别 object identification;target discrimination;target identification;target recognition
目标识别器 target marker
目标识别设备 target identification equipment
目标识别声呐 classification sonar
目标识别系统 target identification system
目标示例 object instance
目标事物 goal object
目标收益率 target rate
目标收益率价格设定 target-rate-of-return pricing
目标寿命 target lifetime
目标树 target tree
目标水深 target depth
目标水质基准 target water quality criterion
目标搜索 acquisition of target;target search;target seeking
目标搜索法 goal-seeking approach
目标搜索雷达 target acquisition radar
目标搜索器 target-seeking unit
目标搜索装置 target-seeking device
目标搜寻 target acquisition;target homing
目标搜寻环 goal-seeking loop
目标速度计算程序 target speed calculation program(me)
目标缩影 aspect of approach
目标探测 acquisition of target;target acquisition;target detection
目标探测技术 target acquisition technique;target detection technique
目标探测器 target detecting device;target detector
目标探测系统 acquisition system;target detecting system
目标探测装置 target detection unit
目标特性 target property
目标特征 target signature
目标特征识别 target characteristic identification
目标特征图 target signature
目标体积 target volume
目标体系 objective set;target complex;target system
目标替换 target swop
目标投影比 aspect of approach
目标图 object map;target chart
目标图像 target image
目标位置图 target position map
目标文件 destination file;file destination;object file
目标吸附质 target adsorbate
目标显示度 <为交通标志设计质量评定指标之一> target value
目标线 line of position
目标相对于背景的明显度 discreteness
目标相容性模式 goal consistency model
目标像函数 object-image function
目标像片 pinpoint photo(graph);spot photo(graph);target photo(graph)
目标协调 goal coordination
目标协调制度 goal congruent system
目标信号 echo signal;target signal;video signal
目标信号减弱 target fade
目标信号检测门 target acquisition gate
目标信号消失 target fade
目标信息 target information;target intelligence
目标信息源 source of target information
目标行动 goal activity
目标行为 target behavio(u)r
目标行销 target marketing
目标选择分析 target selection analysis
目标选择器 target selector
目标仰角 target elevation
目标一致 congruence;goal congruence
目标映象 object map
目标有效截面 target cross section
目标语句 object statement
目标语言 object language;target language
目标语言程序 object language program(me);object routine
目标预测位置 target predicted position
目标预算 target estimate
目标预算法 budgeting to objective
目标噪声源 orifice of target noise
目标增长率 target growth rate;target rate of growth
目标照度 object illumination;target illumination
目标照明光源 target illuminating source
目标照明激光雷达 target illuminating laser radar
目标照准 on-target
目标照准器 object vane;target vane
目标照准指示器 on-target detector
目标真实运动显示雷达 target true motion radar
目标真运动指示器 true motion indicator
目标整定 target setting
目标值 required value;target value
目标指令 target instruction target word
目标指示 target designation
目标指示精度 pointing accuracy
目标指示雷达 acquisition radar
目标指示瞄准具 target-indicating sight
目标指示器 designation indicator;target designator;target indicator
目标指示系统 target designation system;target-indicating system
目标指示站 target director post
目标指向 goal-oriented
目标制导功能调用 goal-directed function invocation
目标制定 goal setting
目标制动 target brake
目标质量测定 target mass determination
目标属性 target attribute
目标转移 aim transference
目标状态 problem status
目标资料图 target material chart;target material graph
目标子句 goal clause
目标子模式 object subschema
目标字 target word
目标字段 aiming field
目标组成 object configuration
目标组匣 object module
目测 by sight;estimate by eye;eye estimation;eye examination;eye measure(ment);eye observation;eye peruse;eye survey;eye work;field sketching;measuring by sight;ocular estimation;ocular measurement;perusal;sketching;sketch survey;visual estimate;visual estimation;visual examination;visual measurement;visualize approximate;surveying
目测比色计 visual colo(u)rimeter
目测标定法 method by naked eye
目测波高 visual wave height
目测波浪 visual wave observation
目测波浪无线电测向仪 visual radio direction finder
目测波浪资料 visual wave data
目测草图 eye draft;eye sketch
目测滴定法 visual titration
目测定位法 visual positioning method
目测定线 visually align
目测法 eye-survey method;ocular estimation method;visual method;ocular estimate;visual observation
目测范围 visual range
目测方位 estimated bearing by eye;visual bearing
目测放大镜 eye ga(u)ge
目测感光计 visual sensitometer
目测高温计 visual pyrometer
目测估计值 eye estimate
目测光度计 visual photometer
目测光学爆裂记录法 eye-observation optic(al) decrepitation record method
目测航向指示器 visual course indicator
目测航行 eyesight navigation;naked eye navigation
目测滑油流量计 visible oil flow ga(u)ge
目测监测器 visual monitor
目测检查 visual inspection;visually inspect
目测检验 visualize
目测距离 estimated distance by eye;judge distance;visual distance
目测均匀性 homogeneity to the eye
目测品位 visual estimated grade
目测评分 visual score
目测气象条件 visual meteorological condition
目测气液比测温法 visual thermometry by estimating gas/liquid ratio
目测清洁度 visual cleanliness
目测示踪剂 visual tracer
目测式无线电测向仪 visual radio direction finder
目测随手方法 visual freehand projec-

tion
目测天体亮度 visual astrophotometry
目测调谐 visual tuning
目测图 thumbnail sketch
目测外观检查 visual inspection visually inspect
目测微计 ocular micrometer
目测位置 visual position
目测位置线 visual line of position
目测星等 visual magnitude
目测选种 visual selection
目测油标 sight oil indicator
目测油表 sight oil ga(u)ge
目测指示 visual indication
目测指示器 sight indicator; viewing indicator; visual indicator
目测准直法 alignment by sight
目测资料 visual data
目次 table of contents
目的 aim(ing); destination goal; goal; motive; object; objective; purpose
目的包 destination packet
目的操作数 destination operand
目的层保振幅子波处理 conservation amplitude wavelet processing of desired layer
目的层深 target stratum depth
目的层位 target stratum
目的程序 object routine; object deck <在计算机语言中一堆穿孔组成的一种程序>【计】
目的程序穿孔卡片套 object deck
目的程序计算机 object computer
目的程序库 object program(me) library
目的程序模块 object module
目的程序设计 goal programming
目的程序时间 object time
目的程序优化 object program(me) optimization
目的程序语言【计】object language
目的程序运行阶段 target phase
目的抽样 purposive sampling
目的存储器 destination memory
目的存储器单元 destination memory location
目的贷款 purpose loan
目的单纯的工厂 simple purpose plant
目的地 arrival point; bourn(e); destination; journey's end; place of destination; point of destination
目的地编码 destination code
目的地标志 destination mark
目的地表示器 destination indicator
目的地代码 destination code
目的地定位 outer fix
目的地队列 destination queue
目的地管制声明 destination control statement
目的地国 country of destination
目的地国旗 destination flag
目的地寄存器 destination register
目的地交货 free delivered; free on board destination
目的地交货合同 free contract
目的地交货价(格) free delivered; free on board destination; freight on board destination; price of delivery to destination
目的地交通量调查 destination traffic survey
目的地角色 destination case
目的地控制表 destination control table
目的地码头交货 ex dock; ex-terminal; ex-wharf
目的地买方指定地点交货价 franco rendu
目的地契约 arrival contract; destination contract
目的地提单 destination bill
目的地网络 destination network
目的地元字段 destination element field
目的地址 destination address
目的地址寄存器 destination address register
目的地址主字段 destination address field prime
目的地址字段 destination address field
目的地址字段加小撇 destination address field prime
目的地指示标志 destination sign
目的地指示代码 destination code
目的地指示掩码 destination mask
目的地字段 destination field
目的点 point of destination
目的定址方式 destination addressing mode
目的范畴 destination category
目的分析 goal analysis
目的港 destination port; final destination; final port; port of debarkation; port of destination
目的港岸上交货价 landed terms
目的港变更 alteration of destination
目的港标志 destination mark; port mark
目的港船边交货 free on side
目的港船上交货 delivered ex-ship; ex-ship
目的港船上交货价(格) free overside; ex lighter port of arrival; ex-ship
目的港到岸价 coast, insurance and freight port of entry
目的港码头交货价 ex-wharf; franco quay; duties on buyer's account; duty paid; ex quay; delivered ex-quay <关税已付价>
目的港名称 named port of destination
目的工作区寄存器 destination workspace register
目的规划 goal programming
目的计算机 target computer
目的卡片组 object deck
目的口岸 port of destination
目的论 purposivism; teleology
目的模件 object module
目的区 destination zone
目的-手段分析 ends-means analysis
目的性 purposiveness
目的在于 with the view of
目的站 destination station
目的指向行为 goal-directed behavio(u)r
目的制动【铁】objective braking; target point braking; target braking; target shoot
目的制动车辆缓行器 final rail brake
目的制动车辆减速器 final rail brake
目的制动缓行器 final retarder; last retarder
目的制动减速器 final retarder; last retarder
目的主机 destination host
目读 eye reading
目估 eye estimate; eye estimation; ocular estimate; ocular estimation; ocular lens; visual estimate
目估法 eye estimate method; eye estimating method; ocular estimate method
目估高度 visible height
目估冠幅 visible diameter
目估树高 visible tree height
目光 eye-beam
目光远的 far-seeing
目击 vision; witness
目击试验 witness test
目击者 eye-witness; observer; ocular witness; witness
目基 interocular distance
目检 visual inspection
目检标准 visual criterion
目检方法 visual procedure
目捷法 eye and key method
目镜 eyeglass; eye lens; eyepiece; helioscope eyepiece; ocular glass; ocular lens; ocular piece; sighting piece
目镜安全罩 eye guard
目镜标记 eyepiece mark
目镜测定 visual photometry
目镜测微计 eyepiece micrometer; facies micrometer; ocular micrometer
目镜测微器 eyepiece micrometer; ocular micrometer
目镜挡光片 eye shade; eye shield
目镜的接目镜 eye lens of the eyepiece
目镜的像差 eyepiece aberration
目镜灯 eyepiece lamp
目镜读数 eyepiece reading
目镜对光 eyepiece focusing; ocular focus(s)ing
目镜放大率 eyepiece magnification; magnification of eyepiece; magnifying power of eye-piece
目镜分辨率 visual resolution
目镜分光镜 eyepiece spectroscope; ocular spectroscope
目镜分划板 eyepiece graticule
目镜盖 eyepiece cap
目镜管 draw of eyepiece; eyepiece assembly; eyepiece tube
目镜光栏 eyepiece stop
目镜光阑 eyepiece diaphragm; ocular blind
目镜盒 ocular box
目镜滑轨 eye lens slide
目镜滑筒 eyepiece slide
目镜环 eyecup; eyepiece ring
目镜焦距 eyepiece focal length
目镜聚光透镜 eyepiece collective
目镜框 eyepiece frame
目镜棱镜 eyepiece prism; ocular prism
目镜量角器 eyepiece goniometer
目镜滤光片 eye filter; eyepiece filter
目镜螺(丝)钉【测】eyepiece screw
目镜螺纹 ocular thread
目镜螺旋 eyepiece screw
目镜帽 eyepiece cap
目镜目测(法) ocular lens
目镜千分尺 eyepiece micrometer; ocular micrometer
目镜散斑 subjective speckle
目镜色片 dark eye piece; eye shade; eye shield
目镜十字线 eyepiece cross-hair
目镜调焦 eyepiece focusing
目镜调焦环 eyepiece focusing ring
目镜调节 ocular accommodation
目镜调节范围 range of ocular accommodation
目镜筒 eyepiece barrel; eyepiece sleeve; eyepiece stalk; eyepiece tube
目镜头 eyepiece head; ocular head
目镜透镜 eyepiece lens
目镜网格测量颗粒面移法 eyepiece network measure grain areal method
目镜微尺测量法 eyepiece micrometer measure method
目镜微尺累积测量法 eyepiece micrometer integrated measurement method
目镜箱 eyepiece housing; housing of eyepiece
目镜选择杆 eyepiece selector lever
目镜照相机 camera with eyepiece
目镜蛛丝 ocular thread
目镜转向棱镜 eyecap
目镜装置 eye guard
目镜总成 eyepiece assembly
目镜座 eyepiece mount
目力 eyesight; vision
目力飞行 contact flying
目力分辨率 visual resolution
目力估测法 rule-of-thumb method
目力监测系统 visual observation system
目力检查 visual inspection
目录 beadroll; caption; catalog(ue); content; directory; table of contents
目录保护 directory maintenance
目录表 catalog(ue)(listing); list; directory list【计】
目录表管理 directory management
目录册 catalog(ue)
目录查找 directory search
目录成员项 directory member entry
目录存储器 catalog(ue) memory
目录单 booklet; catalog(ue) listing
目录档案 directory file
目录的层次结构 hierarchic(al) structure of directory
目录登记项 directory entry
目录服务 directory service
目录工具书 bibliographic(al) tool
目录功能 catalog(ue) function
目录关键字 directory key
目录管理程序 contents supervisor; directory maintenance; directory manager
目录函数 catalog(ue) function
目录号 catalog(ue) number
目录恢复区 catalog(ue) recovery area
目录记录 catalog(ue) record; directory record
目录价(格) catalog(ue) price; list price
目录检索 catalog(ue) search; contents retrieval; directory retrieval
目录检索系统 catalog(ue) retrieval system
目录结构 bibliographic(al) structure; directory structure
目录卡 card catalog(ue); card file; card index; unit card
目录卡片 catalog(ue) cards
目录控制项 directory control entry
目录名 directory name
目录区 directory area
目录容量 directory capacity
目录商品展室 catalog(ue) appliance showrooms
目录上没有 uncatalog
目录式路径选择 directory routing
目录式路由选择 directory routing
目录式数据集 cataloged data set
目录书 index book
目录数据集 directory data set
目录数据库 catalog(ue) data base
目录搜索 directory search
目录索引 catalog(ue) directory
目录文件 catalog(ue) file; directory file
目录项 catalog entry; catentry
目录学 bibliography
目录学家 bibliographer
目录字典 catalog(ue) directory; contents dictionary; contents directory
目前的工资率 existing rate
目前地震平静区 now-quiescent area of earthquake
目前工艺水平 state of art; state-of-

M

the-art;the state of the art
目前技术水平 state-of-the-art
目前利润 immediate profit
目前趋势 current trend
目前生产的产品 current production
目前税项与价格之比率 present incidence
目前土地使用情况的地图 existing land-use map
目前文件 current file
目前先进水平 present advanced stage
目前需水量 present water demand
目前压力下原油体积系数 volume factor of crude at the present pressure
目前业务 current operation
目前运量 existing traffic
目前状态 current state
目视 unaided eye; visual; visual observation
目视报警器 visual alarm unit
目视曝光计 instoscope
目视比较法 visual comparison method
目视比色法 optic(al) colo(u) rimetry;visual colo(u)rimetry
目视比色计 visual colo(u)rimeter
目视边缘匹配坐标仪 visual edge match comparator
目视操纵 visual control
目视操作控制台 visual operation console
目视测光 subjective photometry;visual photometry
目视测向仪 visual direction finder
目视导航 visual navigation
目视导航辅助设备 visual aid to navigation
目视的 macroscopic;ocular
目视的敏锐性 visual acuity
目视地标飞行图 visual chart
目视定位 visual fix
目视定向 empiric(al) orientation;visual orientation
目视读出 visual readout
目视读数 eye reading
目视读数面板 digital visual panel
目视对光 visual focusing
目视发光度 visual luminosity
目视法 visual method
目视反馈控制 visible feedback control
目视飞行 contact flying
目视飞行规则天气 visual flight rule weather
目视飞行条件 visual flight condition
目视分辨率 resolution of the eye;visual acuity; visual resolution; visual resolving power
目视分光光度计 subjective spectrophotometer;visual spectrophotometer
目视分光镜 ocular spectroscope
目视分析 visual analysis
目视跟踪 visual trace;visual tracking
目视工作 visual task
目视估计 eye estimate;eye estimation
目视观测 eye observation; visual observation
目视观察 visualization; visual observation
目视光变曲线 visual light curve
目视光度测量(法) visual photometry
目视光度计 subjective photometer; visual photometer
目视光度学 visual photometry
目视光学高温计 visual optic(al) pyrometer
目视光学仪器 visual optic(al) instrument
目视航向 line-of-sight course
目视极点 visual end point
目视计数法 counting by naked eye
目视记录 visual record
目视监督控制器 visual supervisory control
目视检查 sight check; visual check; visual examination; visual inspection;visual test
目视检核 check by sight
目视检验 apparent survey; sight check;sighting survey;visual check
目视渐进 visual approach
目视渐进坡度指示系统 visual approach slope indicator system
目视降落方向指示装置 visual landing direction indicator
目视解译 visual interpretation
目视解译方法 visual interpretation method
目视距离 visual distance
目视可见 visual access
目视孔 eyesight
目视控制 sight control
目视量 visual magnitude
目视敏度 visual efficiency
目视敏锐度 visual acuity
目视模型 eye pattern
目视拟合 fitting by eye
目视判别 visual identification
目视判定 visual confirmation
目视判读 visual interpretation
目视判读性 visual interpretability
目视判断 visual appraisal
目视判译 visual interpretation
目视跑道 visual runway
目视偏振计 visual polarimeter
目视气象条件 visual meteorological condition
目视散斑干涉量度仪 visual speckle interferometer
目视色度测量 visual colo(u)rimetry
目视色度计 visual colo(u)rimeter
目视识别 visual acquisition
目视式 visual type
目视数字显示 visual digital display
目视双星 visual binaries;visual doubles
目视水准瞄准器 eye-level viewfinder
目视搜索时间 visual search time
目视探向仪 visual direction finder
目视天顶筒 visual zenith telescope
目视天顶仪 visual zenith telescope
目视调谐 visual tuning
目视调谐指示器 visual tuning indicator
目视调整 visual alignment
目视筒 eyepiece slide
目视图 eye pattern
目视图像 visual picture
目视网格 eyepiece grid
目视望远镜 visual telescope
目视位势 sighting potential
目视位置线 visual lien of position
目视系统 see-and-be-seen system; visual system
目视显示 visual display
目视星等 visual magnitude
目视引导 visual guide
目视指示器 visual indicator
目视中心 center[centre] of perspective;center[centre] of view;center[centre] of vision; visual center[centre]
目视助航设备 visual aids
目态 problem mode;problem status; user mode
目眩 dazzle

沐池 needle bath

沐浴 ablution;bathing;lave
沐浴更衣室 bath closet

牧斑 spot grazing

牧草 fodder grass;forage grass;grass fodder;grazing;hay crop;pasture;pasture plant;red top
牧草地 grassland for grazing; grass plot; meadow; meadowy land; sward;grazing land
牧草良种繁殖场 forage seed breeding farm
牧草轮作 alternate meadow
牧草人工干燥设备 barn hay drying plant
牧草收割机 grass harvester
牧草中毒 forage poisoning
牧草种植地 hayland
牧场 glazing land;farm;feedlot;grassland; grazing; grazing land; ground pastplain land; hayfield; home range; ley; lea; leasow; meadow; outfit;pascuum;pasturage;pasture; pasture ground; pasture land; pasture pastplain; prairie; rangeland; swale
牧场保持 range conservation
牧场草皮 field sod
牧场草坪 meadowy land
牧场调查 range survey
牧场阀 alfalfa valve
牧场房屋 ranch house
牧场废物 waste from stockfarming
牧场改良 pasture improvement
牧场更新 pasture renovation
牧场工人 rancher;ranchman
牧场管理 pasture management; rang management
牧场经营学 range science
牧场面积 grazing acreage
牧场平房 rambler
牧场式平房住宅 <通常建有车库> ranch house
牧场土地保护 rang land conservation
牧场员工 stockman
牧场主 farmer;rachero;rancher;ranchman;stockman
牧场主住宅 rancher
牧场住房 ranch house
牧场住宅型窗 ranch style window
牧地 lea;rangeland
牧地单位 pasture unit
牧地管理 pasture management;range management
牧豆树 mesquite
牧民 nomadic people
牧牛工 rancher
牧(农)场主住宅 ranch house
牧区 grazing district; pastoral area; pasturing area;ranching area
牧区的春秋羊场 spring-fall range
牧区畜牧业 pastoral industry
牧区植被 range plant cover
牧群 herd
牧人小屋 herdman's hut
牧师唱诗班 priest's choir
牧师俸地 parsonage building
牧师跪祷桌 litany desk
牧师会教堂 chapter house
牧师会组织的小教堂 collegiate chapel
牧师住宅 manse; parsonage; parsonage building; priest's house; vicarage
牧畜围栏 stock yard
牧羊场 sheepwalk
牧羊场主 <澳洲> squatter
牧羊小道 sheep-tracks;terracettes
牧羊小屋 sheal(ing)
牧养群 breeding herd
牧业 livestock
牧业区 stock farming district
牧业生产 livestock production
牧业图 grazing service map;pedologic(al) map
牧者 herder

苜蓿 clover;Medicago sativa

苜蓿草 alfalfa;bur clover;lucern(e)
苜蓿尘 alfalfa dust
苜蓿干燥机 alfalfa dehydrator
苜蓿形立体交叉 cloverleaf interchange
苜蓿叶 cloverleaf
苜蓿叶式 cloverleaf type
苜蓿叶式环形匝道 cloverleaf loop
苜蓿叶式交叉 cloverleaf junction; cloverleaf-leaf junction
苜蓿叶式交叉布置 cloverleaf layout
苜蓿叶式立交 cloverleaf flyover junction
苜蓿叶式立体交叉 cloverleaf crossing; cloverleaf intersection; cloverleaf grade separation; cloverleaf interchange
苜蓿叶式匝道 cloverleaf loop
苜蓿叶形 cloverleaf;cloverleaf type
苜蓿叶形格体 cloverleaf cell
苜蓿叶形格型钢板桩 cloverleaf cell
苜蓿叶形互通式交叉 cloverleaf crossing; cloverleaf grade separation;cloverleaf interchange
苜蓿叶形立体交叉 full clover-leaf interchange
苜蓿叶形天线 cloverleaf antenna

钼靶 molybdenum target

钼靶 X 线摄影 radiography with molybdenum target tube
钼板 molybdenum plate;molybdenum sheet
钼棒 molybdenum bar; molybdenum rod
钼棒炉 molybdenum bar furnace
钼铋矿 koechlinite
钼波马合金 Mo-perm alloy
钼波民瓦尔合金 Mo-perminvar alloy
钼玻璃 molybdenum glass
钼箔 molybdenum foil
钼箔薄膜压力计 molybdenum-foil diaphragm ga(u)ge
钼材 molybdenum materials
钼衬里 molybdenum-liner
钼承料网 molybdenum supporting grid
钼橙红颜料 molybdate orange pigment
钼单晶 molybdenum single crystal
钼的 molybdenic
钼的多因复成矿床 polygenetic compound molybdenum deposit
钼电极 molybdenum electrode
钼电极保护 protection of molybdenum electrode
钼电阻加热元件 stratit element
钼垫板 Mo-mat
钼钒钢 molybdenum-vanadium steel
钼反射器 molybdenum reflector
钼肥 molybdenum fertilizer
钼粉 molybdenum powder
钼钙矿 powellite
钼钙铀矿 calcurmolite
钼杆 molybdenum rod
钼钢 ferromolybdenum; molybdenum steel
钼高速钢 molybdenum high speed steel
钼隔料网 molybdenum feed retainer

M

钼铬橙 chrome vermil(l)ion;molybdate orange;moly orange
钼铬钢 molybdenum-chrome steel
钼铬合金钢 molybdenum-chrome alloy steel
钼铬红 molybdate red;molybdenum red
钼铬红环氧底漆 molybdate red epoxy primer
钼铬红颜料 molybdate red pigment
钼光阑 molybdenum aperture
钼含量分析仪 molybdenum analyser [analyzer]
钼焊接玻璃 molybdenum solder glass
钼合金 molybdenum alloy
钼合金电阻炉 molybdenum alloy resistance furnace
钼合金铸铁 molybdenum alloy iron
钼黑 moly-blacks
钼华 molybdine
钼加热器 molybdenum heater
钼精矿 molybdenum concentrate
钼桔红 molybdate orange;molybdate red
钼橘红颜料 molybdate orange pigment
钼矿化探 geochemical exploration for molybdenum
钼矿石 molybdenum ore
钼矿物 molybdenum mineral
钼矿异常 anomaly of molybdenum ore
钼蓝 molybdenum blue
钼蓝比色法 molybdenum blue colo(u)rimetry
钼蓝导数分光光度法 molybdenum blue-derivative spectrophotometry
钼蓝反应 molybdenum blue reaction
钼蓝凝聚-有机溶剂溶解分光光度法 molybdenum blue condensation-organic solvent dissociation spectrophotometry
钼镁铀矿 cousinite
钼锰法陶瓷金属封接 ceramics-to-metal seal by Mo-Mn process
钼锰克混剂 Granox P-e-M
钼镍铁导磁合金 molybdenum-permalloy
钼镍铁(高导磁)合金 Monimax
钼屏 molybdenum shield
钼铅矿 wulfenite;yellow lead ore
钼色淀 molybdenum lake
钼砂 molybdenum ore
钼砷铜铅石 molybdoformacite
钼丝 molybdenum filament;molybdenum wire
钼丝炉 molybdenum wire furnace;molybdenum wound furnace
钼酸 molybdenic acid;molybdic acid
钼酸铵 ammonium molybdate
钼酸铵分光光度法 ammonium molybdate spectrophotometry
钼酸钡 barium molybdate
钼酸铋 bismuth molybdate
钼酸钆 gadolinium molybdate
钼酸钙 calcium molybdate
钼酸钴 cobalt molybdate
钼酸钾 potassium molybdate
钼酸钠 sodium molybdate
钼酸镍 nickel molybdate
钼酸钕 neodymium molybdate
钼酸铅 lead molybdate
钼酸锶 strontium molybdal
钼酸锌 zinc molybdate
钼酸盐 molybdate
钼酸盐玻璃 molybdate glass
钼酸盐活化磷 molybdate-reactive phosphorus
钼碳钢 carbon molybdenum steel
钼锑抗分光光度法 molybdoantimony anti-spectrophotometry
钼锑抗钼蓝法 molybdoantimony anti-molybdenum blue method
钼铁合金 ferromolybdenum alloy
钼铁矿 kamiokite
钼铜比 Mo-Cu ratio
钼铜合金 molybdenum-copper
钼铜矿 lindgrenite
钼铜矿石 Mo-bearing copper ore
钼钍金属陶瓷 thoria-molybdenum cermet
钼污染 molybdenum pollution;pollution by molybdenum
钼钨钙矿 powellite
钼钨铅矿 chillagite
钼线 molybdenum wire
钼锌酸钡 barium zinc tantate
钼铀矿 umohoite
钼质模具 molybdenum die
钼组玻璃 molybdenum-group glass

募捐 collection;purse

墓 cenotaph;tomb

墓碑 cenotaph;funerary monument;grave monument;mortuary monument;sepulchral monument;sepulchral stone;tombstone
墓碑石块料 tombstone in block;tombstone in slab
墓碑艺术 funeral art
墓壁龛 tomb niche
墓地 burial chamber;burial ground;burial place;cemetery;God's acre;God's area;grave surround;graveyard;in-place grave;sepulcher;necropolis <大型的>
墓地教堂 cemetery church
墓地廊道 lich[lych]-gate
墓地礼拜堂 graveyard chapel
墓地门道 lych-gate
墓地寺庙 sepulchral temple
墓柜 tomb-chest
墓华盖 baldacohino
墓架 hearse
墓门 <教堂的> corpse gate
墓石 grave slab;gravestone;head stone;sepulchral stone;tombstone
墓石碑 gravestone
墓室 burial chamber;coffin chamber;sepulchral chamber;tomb chamber;chamber tomb;passage grave <由巨石长道引向用土盖起的>;tjandi <印度 8~14 世纪的>
墓屋 tabernacle work
墓像 sepulchral effigy
墓穴 mortuary block;sepulchral chamber;sepulchral pit;sepulchre;sepulture;tomb chamber
墓穴板 funerary slab
墓穴盖石 grave slab
墓园 park cemetery
墓葬 grave
墓志铭 epitaph
墓桌 sepulcrum of a mensa

幕 curtain;shroud;tabernacle;tentorium[复 tentoria]

幕被拉起的 uncurtained
幕壁计数管 screen wall counter
幕拱 curtain arch
幕轨 curtain rail;Huntland rail
幕后 background
幕后参谋主管 staff executive
幕后操纵者 behind-the-scene master
幕后策划 behind-the-scheming
幕后的 off-stage
幕后董事 shadow director
幕后合伙人 dormant partner
幕后交易 backstage deal;insider dealing
幕后投影 back projection
幕间小节目 carpenter's scene
幕间休息 interval
幕帘顶端 heading
幕帘快门 focal shutter
幕帘涂清漆法 lacquer curtain coating (method)
幕帘涂装 curtain painting
幕门 curtain door
幕前执行主管 line executive
幕墙 curtain masonry wall;curtain wall;panel wall;wall curtain
幕墙板 curtain-wall panel
幕墙玻璃 cladding glass
幕墙不透明玻璃 spandrel glass
幕墙材料 skin
幕墙建筑 panel wall building
幕墙结构 curtain-wall construction;curtain-wall structure
幕墙面板 front panel curtain wall
幕墙墙板 panel-for-panel curtain wall
幕墙式防波堤 curtain-wall type breakwater
幕墙外皮 outer skin of curtain walling
幕墙小室 dressing cubicle
幕式灌浆 <防渗漏> curtain grouting
幕式集尘器 screen collector
幕式淋涂 curtain coating
幕式淋涂机 curtain flow coater
幕涂涂层 curtain-coated finish
幕涂涂料 curtain-coated finish
幕效应 curtain effect
幕状链 curtain chain
幕状黏[粘]连 curtain-like adhesion
幕状射流 curtain jet

睦邻政策 good neighbo(u)r policy

暮光 evening twilight

暮辉 evening glow

穆宾体制 <指钢结构和混凝土结构的大生产方式> Mopin system

穆德尔阶【地】Murderian
穆罕默德建筑 Mohammedan[Muhammadan] architecture
穆雷铅焊料 Mouray's solder
穆磷铝铀矿 mundite
穆硫铁铜钾矿 murunskite
穆硫锡铜矿 mohite
穆水钒钠石 munirite
穆斯堡尔参数 Mossbauer parameter
穆斯堡尔光谱 Mossbauer spectroscopy
穆斯堡尔核 Mossbauer nuclear
穆斯堡尔(能)谱 Mossbauer spectrum
穆斯堡尔谱法 Mossbauer spectroscopy
穆斯堡尔谱分析 Mossbauer spectroscopy
穆斯堡尔谱图 Mossbauer spectrogram
穆斯堡尔谱仪 Mossbauer spectrometer;Mossbauer spectroscope
穆斯堡尔效应 Mossbauer effect
穆斯堡尔重力计 Mossbauer gravity meter
穆斯鲍尔效应 <一种广泛应用于研究理论应用科学中的光谱技术> Mssbauer effect
穆斯林布道坛 mimbar
穆斯林法院议事厅 diwan
穆斯林建筑 Moslem[Muslim] architecture
穆斯林历 Moslem calendar
穆斯林墓地 Moslem tomb
穆斯林圣堂 <麦加大寺院中的> Caada;Kaaba;Kabah
穆斯林头巾式刷子 Turk's head
穆斯林学校 Muslim school
穆斯琼测力计 <美国产的一种充气电阻应变片式测力计> Mustran cell
穆西利特殊型锡铋汞合金 Musily silver

N

拿到 take possession of

拿高工资 get good wages
拿破仑绿 <1000 度高温色料> Napoleon green
拿破仑维尔阶【地】Napoleonville (stage)
拿起 take-up
拿骚 <巴哈马首都> Nassau
拿拖布拖 mop

镎 neptunium

那波里银行 <意大利> Banco di Napoli

那不勒斯港 <意大利> Port Naples
那不勒斯黄 Naples yellow
那达砂堆模拟法 <求塑性状态扭转剪应力的方法，薄膜模拟法的延伸> Nadal's sandheap analogy
那丹哈达海槽 Nadanhada marine trough
那丹哈达优地槽褶皱带 Nadanhada Eu geosynclinal fold belt
那丹哈达早中生代俯冲带 Nadanhada Early Mesozoic subduction zone
那里兹阶【地】Narizian(stage)
那络剂 <一种硅酸，用作水泥的硬化剂和防水剂> Naylorite
那摩尔试验路 <国际道路会议常设委员会建筑在那摩尔的混凝土试验路> Namur test road
那慕尔阶 <早晚石炭世>【地】Namurian
那一面 far side

纳 nano; noy <感觉噪声度单位>

纳安计 nanoammeter
纳巴罗-赫林蠕变 Nabarro-Herring creep
纳本 obligatory presentation copy
纳程序 nanoprogram(me)
纳达 <一种抗变色铜基合金> Nada
纳达铜合金 Nada alloy
纳地方税人 <英> rate payer
纳尔 <长度单位> nail
纳尔旁风 narbonnais
纳尔逊电池 Nelson's cell
纳尔逊隔膜电解槽 Nelson's diaphragm cell
纳尔逊滚筒系统 Nelson's rollers system
纳尔逊河 Nelson River
纳法 millimicrofarad; nanofarad
纳费 toll
纳费贷款 premium loan
纳费通知 premium note
纳费业务标识 paid service indication
纳格拉式水轮机 Nagler turbine
纳骨处 ossuary
纳痕量分析 nanotrace analysis
纳亨 nanohenry
纳霍德卡港 <俄罗斯> Nakhodka Port
纳加那 <非洲锥虫病> nagana
纳加那红 nagana red; nagarot
纳里钙质层 nari
纳丽特(铝青铜)合金 Narite(alloy)
纳滤 nanofiltration
纳滤表面电荷 nanofiltration membrane surface charge
纳滤工艺 nanofiltration process
纳滤膜 nanofiltration membrane
纳滤膜分离 nanofiltration membrane separation
纳滤膜加宽 nanofiltration membrane broaden
纳滤陶瓷膜 nanofiltration ceramic membrane
纳米 nanometer; bicron; micromillimeter[micromillimetre]; millimicron
纳米比亚 <非洲> Namibia
纳米材料 nanomaterial; nanophase material
纳米多孔聚合物 nanoporous polymer
纳米二氧化硅-聚乙烯醇复合超滤膜 nanosized silica-polyvinyl alcohol composite ultrafiltration membrane
纳米二氧化钛 nano titanium dioxide
纳米二氧化钛催化剂 nano titanium dioxide catalysis
纳米二氧化钛光催化剂 nano titanium dioxide photocatalyst
纳米二氧化钛光催化氧化 nano titanium dioxide photocatalytic oxidation
纳米二氧化钛光电催化降解 nanometer titanium dioxide photoelectrocatalytic degradation
纳米沸石分子筛 nano zeolite molecular sieve
纳米沸石分子筛吸附 nano zeolite molecular sieve adsorption
纳米分辨率 nanometer resolution
纳米复合材料 nanocomposite material
纳米复合膜 nanocomposite membrane
纳米改性混凝剂 nanometer-size modification coagulant
纳米高岭土 nano-kaolin
纳米光催化氧化 nanophotocatalytic oxidation
纳米棍 nanometer club
纳米海绵状吸附介质 nanoponge adsorbent media
纳米合成物 nanocomposite
纳米化学 nanochemistry
纳米级复合材料 nanocomposite
纳米级功能材料 nanometer scale function material
纳米级晶体 nanometer scale crystal
纳米结构 nano-structure
纳米结晶 nanocrystalline
纳米晶复合材料 nanocrystalline composite material
纳米晶陶瓷 nanocrystalline ceramics
纳米晶体 nanocrystal
纳米晶体的 nanocrystalline
纳米颗粒 nano-particle
纳米零价铁 nanoscale zero-valent iron
纳米锐钛型二氧化钛 nano-anatase TiO_2
纳米陶瓷 nanoceramics; nanophase ceramics
纳米铜铋钒矿 namibite
纳米吸附剂 nano adsorbent; nano-adsorption material
纳米相 nanophase
纳米氧化铝陶瓷膜 nanometric alumin(i)um oxide ceramic membrane
纳米氧化锌 nanometer zinc oxide
纳米液滴 nano droplet
纳秒 millimicrosecond; nanosecond
纳秒脉冲 nanosecond pulse
纳秒上升时间 nanosecond rise time
纳秒时间分辨率光谱学 nanosecond time-resolved spectroscopy
纳秒示波器 nanoscope
纳姆合成树脂黏[粘]接剂 Narm tape
纳诺技术 nanotechnology
纳皮尔差动螺旋操舵装置 Napier's differential screw steering gear
纳皮尔对偶式 Napier's analogy
纳皮尔对数 Napierian logarithm
纳皮尔对数底 Napierian base
纳皮尔法则 Napier's rules; Napierian rule
纳皮尔类比 Napier's analogy
纳皮尔罗经自差表 Napier's Deviation table
纳皮尔三角式双冲程柴油机 Napier's deltaic two stroke engine
纳皮尔氏公式 Napier's equation
纳皮尔相似式 Napier's analogy
纳皮尔自差曲线图 Napier diagram
纳普底压计 Knapp bottom pressure ga(u)ge
纳普姆 <材料透气率单位，等于 10^{-9} 普姆> nanoperm
纳热水域 heat receiving water region
纳入接 socketing
纳塞 <导电玻璃> Nasa
纳纱制品 petit-point articles
纳什泵 Nash pump
纳氏泵 Nash pump
纳氏底 Napierian base
纳氏对数 Napier's log(arithm); Napierian logarithm
纳氏粒铁直接冶炼法 Nesbitt method
纳税 pay duty; payment of duties; pay tax; tax payment; toll
纳税保证金 security of tax payment
纳税差别的非收入基准 non-income base for tax discrimination
纳税大众 tax-paying public
纳税单位 tax unit; unit of taxation
纳税单位分列计税制 separate entity system
纳税的价值 rateable value
纳税的支付能力原则 ability-to-pay principle of taxation
纳税登记工作 tax registration
纳税地点 place of tax payment; tax payment place
纳税对象 object of taxation
纳税额 ratal
纳税辅导 taxpayers consultant
纳税负担 burden of paying tax; incidence of taxation
纳税公平 equitable tax; tax equity
纳税公平原则 equity principle of taxation
纳税横向均等 horizontal equity
纳税后个人可用收入 disposable income; disposable personal income
纳税后价格 price after tax
纳税后净收入 net income after taxes; net of tax income
纳税后扣除 after-tax deduction
纳税后利润 after-tax profit; post-taxation profit; profit after tax
纳税后利润留存比率 tax retention ratio
纳税后利润率 after-tax profit rate
纳税后收入 after-tax income
纳税后收益 return after all taxes
纳税后所得 after-tax income
纳税基础 base of taxation
纳税价值 duty-paying value
纳税检查 tax inspection
纳税鉴定 tax appraisal
纳税均等 tax equalization
纳税扣除 tax deduction
纳税留置 lien for taxes
纳税能力 ability (of the subject) to pay; taxability; taxable capacity
纳税年度 tax year
纳税抛售 tax selling
纳税平等 equality of taxation; equalization of tax payment
纳税凭证 duty receipt; tax payment receipt
纳税期限 term of tax
纳税起征线 taxable limit
纳税前 before tax
纳税前的投资利息收入 gross interest
纳税前净利润 net profit before taxes
纳税前净收益 <纳税指纳所得税> net income before taxes
纳税前利润 profit before taxes
纳税前收益额 earnings before taxes
纳税清册 tax roll
纳税饶让 tax sparing
纳税人 tax bearer; taxpayer; rate payer <英国地方税>; contributor
纳税人分级 class of taxpayer
纳税人抗税 taxpayer revolt
纳税人申报的税额 taxes assessed by taxpayers' report
纳税人身份 status of taxpayers
纳税日 tax day
纳税日期 date of tax levied; tax payment date
纳税申报 tax declaration
纳税申报表(格) tax forms
纳税申报单 tax returns
纳税时不能扣除的支出项目 expenditure not deductible
纳税时的股息扣除或抵免 dividend credit
纳税时的养老金基金扣除额 pension fund deduction
纳税时的医疗费用扣除 medical expense deduction
纳税时各项费用的冲销 tax write-off for all the expenditures
纳税时可扣除的项目 tax deduction
纳税时亏损结转 tax-loss carry-forward
纳税时亏损转回 tax-loss carry-back
纳税时四舍五入计去的金额 rounding-off dollars
纳税收据 duty receipt; tax payment receipt
纳税通知书 notice of tax payment; tax paper
纳税义务 obligation to pay tax; rateability; tax liability; tax obligation
纳税义务人 taxpayer
纳税隐藏所 tax haven; tax paradise; tax shelter
纳税者 taxpayer
纳税证明书 certificate of tax payment; certification on tax payment
纳税住所纳税准备存款 deposit(e) for tax payments
纳税准备存款 deposit(e) for tax payments
纳税准备金 earmarked deposits for taxes; tax reserves
纳税准备证书 tax reserve certificates
tax home
纳斯列多夫石 nasledovite
纳瓦 <等于 10^{-9} 瓦特> nanowatt
纳瓦霍砂岩 Navajo sandstone
纳瓦霍小地毯 Navajo rug
纳瓦罗阶【地】Navarroan(stage)
纳维尔定理 Navier's theorem
纳维尔定则 Navier's law
纳维尔方程 Navier equation
纳维尔假定 Navier's hypothesis
纳维尔假设 Navier's hypothesis
纳维尔-斯托克斯方程(式) <不可压缩黏[粘]性流体运动方程> Navier-Stokes equation

纳维尔-斯托克斯激波结构 Navier-Stokes shock structure
纳污河流 receiving river; receiving stream
纳污水体 receiving water body; receiving waters
纳污水体生态系统 receiving water ecosystem
纳污水体水质 receiving water quality
纳污水体水质模型 receiving water quality model
纳污水体污染物浓度 receiving water pollutant concentration
纳污水体污水 wastewater in receiving water
纳污水体影响 receiving water impact
纳污水域 receiving water body; receiving waters
纳希定理 Nash's theorem
纳希瞬时单位过程线 Nashs instantaneous unit hydrograph
纳希-威廉公式 Nash-William's formula
纳希液封型真空泵 Nash pump
纳新卡板块 Nazaca plate
纳质火成岩型霞石正长岩 agpaitic nepheline syenite

钠

钠 D 线 sodium D-line

钠铵矾 lecontite
钠板石 allevardite; rectorite; tablite
钠饱和黏[粘]土 sodium-saturated clay
钠钡玻璃 soda-baryta glass
钠钡长石 banalsite
钠泵 sodium pump
钠泵功能障碍 dysfunction of sodium pump
钠泵衰竭 sodium pump failure
钠冰长石 sodium adularia
钠玻璃 soda(ash)glass; soft glass
钠差 alkaline error; sodium error
钠长斑岩 albitophyre
钠长粗面岩 albite trachyte
钠长玢岩 albite porphyrite; albitite
钠长花岗岩 albite granite; sodaclase granite
钠长辉绿岩 albite-diabase
钠长角闪辉绿岩 minverite
钠长粒玄岩 albite-dolerite
钠长流纹岩 albite rhyolite
钠长绿帘阳起角 albite-epidote actinolite hornfels
钠长绿泥片岩 albite-chlorite schist
钠长绿泥千枚岩 albite-chlorite phyllite
钠长千枚岩 albite phyllite
钠长浅粒岩 albite leuco granoblastite
钠长闪长岩 albite-diorite
钠长石 albite; anorthite; felspar; olafite; sodaclase; soda feldspar; sodium feldspar; tetartine; white feldspar; white schorl
钠长石斑岩 albitite
钠长石钙长石系列 albite-anorthite series
钠长石化【地】albitization
钠长石卡斯巴双晶律 albite-Carlsbad twin law
钠长石律 albite law
钠长石绿帘石角岩相 albite-epidote hornfels facies
钠长石青铝闪石岩 albite-crossite rock
钠长石双晶 albite twin
钠长石双晶律 albite twin law
钠长石岩 albitite
钠长微斜正长岩 bigwoodite
钠长榍伟晶岩 varnsingite
钠长阳起片岩 albite-actinolite schist
钠长英板岩 adinole; adinolite slate
钠长英辉正长斑岩 albite-akerite
钠长正长岩 albite-sienite[syenite]
钠沉积 sodium deposition
钠传导陶瓷固体电解质 sodium-conducting ceramic solid electrolyte
钠粗面安山岩 doreite
钠代丙二酸酯 sodio-malonic ester
钠代甲基丙二酸盐 sodio-methylmalonate
钠代甲基丙二酸酯 sodio-methylmalonic ester
钠代氰基乙酸酯 sodium cyanoacetic ester
钠代酮酸酯 sodio-ketoester
钠代烷基丙二酸酯 sodio-alkylmalonic ester
钠代乙基丙二酸酯 sodio-ethylmalonic ester
钠代乙酰乙酸酯 sodio-acetoacetic ester
钠蛋白石 natroopal
钠道 sodium channel
钠灯 sodium discharge lamp; sodium lamp; sodium vapo(u)r lamp
钠灯光 sodium light
钠电导 sodium conductance
钠电流 sodium current
钠矾 sodium alum
钠矾硝石 darapskite; nitroglauberite
钠方解石 natrocalcite
钠方硼解石 ulexite
钠沸石 fargite; lehuntite; mesotype; mooraboolite; natrolite; radiolite; savite; sodalite; soda mesotype; soda zeolite; sodium zeolite
钠负平衡 negative balance of sodium
钠钙玻璃 soda-lime glass
钠钙长石 soda-lime feldspar
钠钙锆石 lavenite
钠钙硅玻璃 soda-lime silica glass
钠钙砷铀矿 natrium uranospinite
钠钙霞石 natroncancrinite
钠钙铀云母 natroautunite
钠钙柱石 mizzonite
钠锆石 catapleiite
钠汞合金 sodium amalgam
钠汞齐 natrium amalgam; sodium amalgam
钠汞齐萃取(法) sodium amalgam extraction
钠汞齐化 sodium amalgamation
钠汞齐氧电池 sodium amalgam-oxygen cell
钠光 sodium light
钠光灯 high-intensity discharge lamp; natrium lamp; sodium lamp; sodium vapo(u)r lamp
钠光电池 Na-photocell
钠硅氧玻璃 sodium-silica glass
钠含量 sodium content
钠黑云母 natronbiotite
钠黑蛭石 tabergite
钠红沸石 barrerite
钠弧灯 sodium lamp
钠花岗岩 soda-granite
钠滑石 achlusite
钠化合物 sodium compound
钠辉绿岩 sodadiabase
钠辉石 soda pyroxene
钠辉微晶花岗岩 sodic pyroxene microganite
钠辉细岗岩 rockallite
钠辉叶石 eggletonite
钠回路 sodium loop
钠基 sodium base
钠基膨润土 sodium bentonite
钠基润滑脂 sodium-base grease; sodium(-soap)grease
钠基脂 soda grease; sodium-base grease
钠钾泵 sodium-potassium pump
钠钾玻璃 soda-pearl ash glass; soda-potash glass
钠钾地热温标 sodium-potassium
钠钾钙地热温标 sodium-potassium-calcium
钠钾共晶合金 Nak
钠钾合金 Na-K alloy
钠钾冷却堆 Na-K-cooled reactor
钠钾锂伟晶岩 soda-potassic-lithium pegmatite
钠钾锂细晶岩 sodic-potassic-lithium aplite
钠钾芒硝 glaserite
钠钾霞石 natrodavyne; natrodavynite
钠钾云母 euphyllite
钠钾质长石斑岩 soda-potassic feldspar porphyry
钠钾质微晶正长岩 soda-porphyry microsyenite
钠钾质伟晶岩 soda-potassic pegmatite
钠钾质细晶岩 sodic-potassic aplite
钠钾质霞石微晶正长岩 soda-potassic nepheline microsyenite
钠钾质霞石正长岩 soda-potassic nepeline syenite
钠钾质正长岩 sodic-potassic syenite
钠间隙 sodium space
钠碱 soda
钠碱粉 soda ash
钠碱灰 soda ash
钠角闪石 soda amphibole
钠结渣 sodium dross
钠金云母 sodium phlogopite
钠聚丙烯腈 sodium polyacrylonitrile
钠冷堆 sodium-cooled reactor
钠冷反应堆 sodium-cooled reactor
钠冷排气阀 sodium-cooled valve; sodium-filled exhaust valve
钠冷却 sodium cooling
钠冷却气门 natrium cooled valve; sodium-cooled valve
钠离子 sodion; sodium ion
钠离子导体 sodium ion conductor
钠离子化黏[粘]土 sodium-ionized clay
钠离子交换 sodium ion exchange
钠离子交换器 Na-ion exchanger; sodium cation exchanger; sodium exchanger; sodium ion exchanger
钠离子交换容量 Na-cation exchange capacity
钠离子/氯离子当量比 equivalent weight ratio of sodium ion to chloride ion
钠离子氯离子/硫酸根离子当量比 equivalent weight ratio of sodium ion-chloride ion to sulfate ion
钠离子内流 sodium ion inflow
钠离子选择电极 sodium selective electrode
钠锂大隅石 sugilite
钠锂钾伟晶岩 soda-lithium-potassic pegmatite
钠锂钾细晶岩 sodic-lithium-potassic aplite
钠锂铯钾伟晶岩 soda-lithium-caesium-potassic pegmatite
钠锂铯钾细晶岩 sodic-lithium-caesium-potassic aplite
钠磷锂铝石 fremontite; natramblygonite; natromontebrasite
钠磷铝铅矿 natrohitchcockite
钠磷铝石 harbortite
钠磷锰矿 lemnasite
钠磷锰铁矿 arrojadite; soda triphylite
钠菱沸石 gmelinite; sodium chabazite
钠流纹岩 soda liparite; soda rhyolite
钠硫电池 sodium-sulfur[sulphur] battery; sodium-sulphur cell
钠榴石 soda-garnet
钠滤色镜 sodium filter
钠铝硅酸盐 sodium alumin(i)um silicate; sodium aluminosilicate
钠铝辉石 percivalite
钠绿鳞高铁石 natrodufrenite
钠镁大隅石 elfelite
钠镁矾 loew(e)ite; vanthoffite
钠蒙脱石 sodium montmorillonite
钠锰电气石 tsilaisite
钠锰辉石 blanfordite
钠敏化热离子检测器 sodium-sensitized thermoionic detector
钠明矾 mendozite; soda alum; sodalumite; sodium alum
钠明矾石 natroalunite
钠铌矿 lueshite; natroninobite
钠镍矾 nickelbloedite
钠硼矾石 alunite
钠硼钙石 tincalcite; tiza
钠平衡 sodium balance
钠气灯 sodium discharge lamp; sodium vapo(u)r lamp < 一种现代化道路照明灯,发黄光,多用于立交层桥隧道进口 >
钠铅合金 Hydrone; sodium lead alloy
钠青铜 sodium bronze
钠球云母 ephesite
钠缺乏 sodium deficit
钠热还原 reduction by sodium; sodium reduction
钠砂 sodium sand
钠闪花岗细晶岩 riebeckite granite-aplite
钠闪花岗岩 lindinosite; riebeckite granophyre
钠闪辉长岩 mafraite
钠闪碱流岩 comendite
钠闪石 osannite; riebeckite
钠闪石霏细岩 riebeckite felsite
钠闪石石棉 blue asbestos; crocidolite
钠闪微岗岩 ailsyte; paisanite
钠闪微晶花岗岩 sodic amphibole microganite
钠闪细晶花岗斑岩 riebeckite aplite granite porphyry
钠闪细晶花岗岩 riebeckite aplite granite
钠闪细晶岩 riebeckite aplite
钠闪正长岩 lakarpite
钠砷铀云母 sodium uramospinite
钠石灰 soda lime
钠试验 sodium test
钠双线 sodium doublet lines
钠水玻璃 soda soluble glass; water soda ash glass
钠水潴留 retention of sodium and water
钠丝压制器 sodium press
钠羧甲基纤维 sodium carboxy methyl cellulose
钠钛闪石 anophorite
钠钽矿 natrotantite
钠碳石 natrite
钠体系 sodium system
钠铁矾 natrojarosite
钠铁非石 aenigmatite
钠铁坡缕石 tuperssuatsiaite
钠铁闪石 arfvedsonite; soda hornblende
钠通道 sodium channel
钠铜矾 natrochalcite
钠铜锌矾 nanuwite
钠透长石 sodium sanidine
钠透闪石 imernite; richterite; soda-tremolite

N

钠微斜长石 soda microcline; sodium microcline
钠钨青铜 sodium tungsten bronze
钠误差 alkaline error; sodium error
钠吸附比 sodium adsorption ratio
钠吸收比 sodium absorption ratio
钠系统 sodium system
钠细晶石 natrobistantite
钠霞正长石 canadite; mariupolite
钠霞正长伟晶岩 canadite pegmatite
钠霞正长岩 canadite; mariupolite
钠线折射率 sodium line index
钠橡胶 sodium rubber
钠硝矾 darapskite
钠硝石 Chile slatpeter[slatpetre]; natron saltpeter; nitratine; nitratite; soda saltpeter; soda niter <即智利硝石>
钠销 saltier
钠烟减退 sodium-smoke abatement
钠盐 sodium salt
钠盐水 natrium brine
钠衍生物 sodio-derivative
钠伊利石 brammallite; sodium illite
钠硬硅钙石 miserite; natroxonotlite
钠油酸 sodium oleic acid
钠铀云母 sodiumautunite
钠黝帘石化 saussuritization
钠云 sodium cloud
钠云母 paragonite; soda mica; sodium mica
钠云母片岩 paragonite schist
钠云霞正长岩 litchfieldite
钠载体 sodium carrier
钠皂 hard soap; soda soap; sodium soap
钠皂基润滑脂 soda-soap-base grease; sodium-soap grease; sponge grease
钠皂润滑脂 sodium-base grease; sodium-soap grease
钠沾染物 sodium contaminant
钠蒸气灯 sodium vapo(u)r lamp
钠正长石 barbierite; soda orthoclase; sodium orthoclase; sodium orthoclase
钠正长岩 soda-syenite
钠脂 soda-base grease
钠质长石斑岩 sodic feldspar porphyry
钠质粗面岩 sodic trachyte
钠质霏细岩 soda felsite
钠质辉石粗面斑岩 sodic pyroxene trachyte porphyry
钠质辉石粗面岩 sodic pyroxene trachyte
钠质辉石花岗岩 sodic pyroxene granite
钠质辉石流纹岩 soda pyroxene rhyolite
钠质辉石微晶正长岩 sodic pyroxene microsyenite
钠质辉石正长岩 sodic pyroxene syenite
钠质火成岩类 agpaite
钠质交代作用 sodium metasomatism
钠质角闪石粗面斑岩 sodic amphibole trachyte porphyry
钠质角闪石粗面岩 sodic amphibole trachyte
钠质角闪石花岗岩 sodic amphibole granite
钠质角闪石流纹岩 soda amphibole rhyolite
钠质角闪石微晶正长岩 sodic amphibole microsyenite
钠质角闪石正长岩 sodic amphibole syenite
钠质黏[粘]粒 sodium clay
钠质膨润土矿石 sodic bentonite ore
钠质碳酸岩熔岩 soda carbonatite lava
钠质土 sodic soil; sodium soil
钠质危害 sodium hazard
钠质微晶正长岩 sodic microsyenite
钠质霞石微晶正长岩 soda-nepheline microsyenite
钠质霞石正长岩 soda nepheline syenite
钠质正长岩 sodic akenite
钠珠云母 ephesite; soda-margarite
钠柱晶石 kornerupine
钠柱石 marialite
钠柱石化 marialitization

乃

乃奎斯特不稳定性 Nyquist's instability
乃奎斯特采样定理 Nyquist's sampling theorem
乃奎斯特定理 Nyquist's theorem
乃奎斯特轨迹 Nyquist's locus
乃奎斯特极限 Nyquist's limit
乃奎斯特间隔 Nyquist's interval
乃奎斯特截面 Nyquist's flank
乃奎斯特解调器 Nyquist's demodulator
乃奎斯特理论 Nyquist's theory
乃奎斯特滤波器 Nyquist's filter
乃奎斯特率 Nyquist's rate
乃奎斯特判据 Nyquist's criterion[复 criteria]
乃奎斯特频率 Nyquist's frequency
乃奎斯特前沿带宽 Nyquist's interval bandwidth
乃奎斯特取样定理 Nyquist's sampling theorem
乃奎斯特取样间隔 Nyquist's sampling interval
乃奎斯特取样频率 Nyquist's sampling rate
乃奎斯特速率 <理想信道极限传输的> Nyquist rate
乃奎斯特图 Nyquist's diagram; Nyquist's plot
乃奎斯特围线 Nyquist's contour
乃奎斯特稳定度判据 Nyquist's stability criterion
乃奎斯特稳定度准则 Nyquist's stability criterion
乃奎斯特稳定性定理 Nyquist's stability theorem
乃奎斯特噪声 Nyquist's noise
乃奎斯特噪声定理 Nyquist's theorem of noise
乃奎斯特准则 Nyquist's criterion

奶

奶白釉 creamy white glaze
奶泵 milk pump
奶场 dairy farm; lactary
奶黄色 cream
奶黄色不透明玻璃 custard glass
奶酪废水 creamy waste
奶酪生产车间 dairy building
奶牛 dairy cattle; dairy stock; milch cattle; milch cow; milk cattle
奶牛场淤泥 dairy-farm slurry
奶品厂废物 dairy plant waste
奶品工业 dairy industry
奶品制造业 dairying
奶瓶清洗机 milk bottle washer
奶瓶形筒子 milk-bottle bobbin
奶色石 Bath stone
奶山羊 milk goat
奶桶吊挂架 racking for milk churns and buckets
奶桶卡车 churn lorry
奶桶输送机 milk can conveyer[conveyor]
奶桶压力补偿器 can equalizer
奶头状物 pap
奶油 cream
奶油黄色 butter yellow
奶油沥青 bog butter
奶油色 creamy colo(u)r
奶油色白云石 <美国约克郡的> anston
奶油色大理石 caroline
奶油色的 creamy
奶油色石灰石 <产于英国威尔特郡的> box ground
奶油色陶器 queen's ware
奶油制造厂 creamery
奶子榔头 ball peen hammer

氖

氖测电笔 neon tester
氖灯 inditron; neon lamp; neon light
氖灯方向指示器 neon-light regulator
氖灯泡 neon bulb
氖灯脉冲触发器 neon pulse trigger
氖灯信号 neon sign
氖灯指示器 neon indicator
氖电压指示器 neon voltage indicator
氖二极管计数管 neon diode counter
氖伏特计 neon voltmeter
氖管 neon(filled) tube; neon(glim) lamp; neon pipe
氖管变压器 neon-tube transformer
氖管灯 neon-tube lamp; neon-tube light
氖管电极 neon-tube electrode
氖管峰值电压表 neon-tube peak-voltmeter
氖管光电导体 neon-photoconductor
氖管广告牌装置 neon-sign installation
氖管火花试验器 neon spark tester
氖管稳定器 neon stabilizer
氖管稳压器 neon voltage regulator
氖管照明 neon lighting
氖管振荡器 neon oscillator
氖管整流器 neon-tube rectifier
氖管指示器 neon indicator
氖管装置 neon-tube installation
氖光灯 neon bulb; neon light; neon tube
氖光管 neon(filled) tube
氖光式火花塞试验器 neon type spark plug tester
氖氮辉光灯 osglim lamp
氖弧(光)灯 neon arc lamp
氖辉光 neon glow
氖辉光灯管 neon glow lamp
氖气 neon gas
氖气辉光灯 neon glow lamp
氖气计算元件 neon computing element
氖气闪光管 neon flash tube
氖氢气泡室 neon-hydrogen bubble chamber
氖栅屏 neon-grid screen
氖示谐管 neon tuning-indicator
氖氙气体激光器 neon-xenon gas laser
氖氩发光灯 neon argon luminous tube
氖氩发光管灯 neon argon luminous tube light

奈

奈 <电信传输单位, 1 奈 = 8.686 分贝> napier
奈恩陆核 Nairn Nucleus
奈恩砂岩 Nairn sandstone
奈尔皮克采样器 Neyrpic sediment sampler
奈尔皮克流速仪 Neyrpic current meter
奈耳壁 Neel wall
奈耳点 Neel point
奈耳理论 Neel's theory
奈耳温度 <反铁磁物质最高温度> Neel temperature
奈奎斯特环路 Nyquist loop
奈奎斯特(频率)稳定度判据 Nyquist stability criterion
奈奎斯特图 Nyquist diagram
奈奎斯特噪声 Nyquist noise
奈良法隆寺 Horyuji
奈迈 <负有效质量放大器与振荡器> Nemag
奈培 <电信传输单,一种衰耗单位, 1 奈培 = 8.686 分贝>【无】nepier[np]
奈培表 nepermeter
奈培计 nepermeter
奈硼钠石 nasinite
奈塞 <透明导电模> nesa
奈塞玻璃 <一种透明导电薄膜半导体玻璃> nesa glass
奈氏比色管 Nessler tube
奈斯伯(底沙)采样器 Nesper sampler
奈斯勒比色法 Nesslerization
奈斯勒比色管【化】Nessler jar; Nessler tube
奈斯勒氏比色管 Nessler's colo(u)r comparison tube
奈斯勒试剂 Nessler's reagent
奈斯勒试剂比色法 Nessler's reagent
奈斯勒试剂分光光度法 Nessler's reagent spectrophotometry
奈斯勒试剂光度法 Nessler's reagent photometry
奈特 <一种度量信息的单位, 1 奈特 = 1.443 节或位> nat
奈特里尔纤维 nytril fibre
奈特龙炸药 Nitrone
奈伊投影地图 Ney's chart

耐

耐氨聚合物胶结料 ammonium salt resisting polymerize binder
耐氨砂浆 ammonia resisting mortar
耐拔性 resistance to pull-off
耐白垩化合成树脂涂料 non-chalking synthetic(al) resin paint
耐爆应变 bursting strain
耐崩裂性 chipping resistance
耐变压器油性 resistance to former oils; transformer oil resistance
耐波浪的船 sea boat
耐波性 seakeeping(ability); seaworthiness
耐剥落性 flaking resistance; spalling resistance
耐擦 scratchproof
耐擦伤性 mar proof; mar resistance; resistance to scuffing; scratch resistance
耐擦洗试验 scrubbing resistance test
耐擦洗性 scrub resistance
耐擦性 abrasion resistance; resistance to scrubbing; scuff resistance
耐擦油墨 non-scratch ink
耐踩踏的面饰 walked-on finish
耐超低温胶粘剂 ultra-low temperature resistant adhesive
耐超速能力 over-speed capability
耐潮 resistance to moisture
耐潮的 insusceptible to moisture
耐潮胶粘剂 moisture-resistant adhesive
耐潮湿试验 humidity test
耐潮性 moisture resistance
耐潮性的 damp resistant; moisture-resistant
耐尘的 dust-fast

耐尘度 fastness to dust
耐尘性 dust fastness
耐冲击丙烯酸模塑料 high impact acrylic mo(u)lding powder
耐冲击的 shock-proof
耐冲击钢 notch ductile steel
耐冲击破坏性 resistance to impact fracture
耐冲击试验 impact endurance test
耐冲击塑料 high impact plastics;impact-resistant plastics;rubber-resin alloy
耐冲击性 impact resistance;resistance to shock;shock resistance;resistance to impact
耐冲蚀的 erosion resistant
耐冲蚀性 abrasion resistance
耐冲试验 impact endurance test
耐冲刷能力 resistance to washing
耐冲刷性 abrasion resistance;erosion resistance;washout resistance
耐冲洗性 washability
耐冲性聚苯乙烯 high impact polystyrene
耐臭氧龟裂性 resistance to ozone cracking
耐臭氧性(能) ozone resistance;ozone proof;resistance to ozone
耐储藏的 long keeping
耐储存性 storage stability;storage property
耐储性 storable property
耐储性试验 storage stability test
耐处理化合物 refractory compound
耐处理物质 refractory substance
耐穿透性 penetration resistance
耐穿性 wearing value
耐穿着牢度 fastness to wear
耐醇性 alcohol fastness;alcohol-proof;alcohol resistance;fastness to alcohol
耐大气腐蚀钢 weathering steel
耐大气牢度 fastness to atmospheric gases
耐淡碱渍牢度 fastness to acid spotting;fastness to alkali spotting
耐低 low overload withstand capacity
耐低温试验 cold test
耐低温性 low-temperature resistance
耐低温油毡 low-temperature resistant roofing felt
耐地压的拱 geostatic arch
耐地震的 quake-proof;aseismatic;aseismic;earthquake-proof;earthquake resistant;earthquake-resisting
耐电弧性 arc resistance
耐电流 withstand current
耐电强度 withstand electric(al) strength
耐电强度试验 withstand electric(al) strength test
耐电压 proof voltage;voltage withstand;withstand voltage
耐电压测试器 dielectric(al) strength tester
耐电压试验 disruptive test;withstand voltage test
耐电晕放电 corona resistant
耐电晕放电(击穿)能力 corona resistance
耐电晕性 corona resistance;resistance to corona
耐冬性 cold resistance;freezing resistance;frost resistance;winter hardiness
耐冬油 winter oil
耐冬植物 winter resistant plant
耐动态压痕性 resistance to dynamic(al) indentation
耐冻材料 frost-proof material;frost-resisting material
耐冻垂直孔黏[粘]土砖 frost-proof vertical coring(clay) brick;frost-resistant vertical coring(clay) brick
耐冻的 freeze-proof;frost-proof;frost-resisting;non-freezing
耐冻多孔砖 frost-proof porous brick
耐冻硅酸盐砖 frost-proof calcium silicate brick;frost-proof sand-lime brick;frost-resistant calcium silicate brick
耐冻灰砂砖 frost-proof lime-sand brick;frost-proof sand-lime brick
耐冻混凝土 frost-resistant concrete
耐冻耐熔 freeze-thaw stable
耐冻砌筑工作 frost-resistant masonry work
耐冻融性 freezing-melting resistance
耐冻润滑油 winter oil
耐冻实心砖 frost-proof solid brick
耐冻水平孔黏[粘]土砖 frost-proof horizontal coring(clay) brick;frost-resistant vertical coring(clay) brick
耐冻性 cold resistance;cold resisting property;freeze resistance;freezing resistance;frost hardiness;frost resistance;frost-resisting property
耐冻砖 frost-proof brick;frost-resistant brick
耐毒极限 toxic limit
耐毒下限 lower toxic limit
耐毒限度 toxic limit
耐毒性 resistance to poison;toxic tolerance
耐短路的 short-circuit proof
耐尔蓝 Nile blue
耐二氧化硫气体牢度 fastness to stoving
耐反复挠曲性 resistance to flexing
耐放射性污染物 resistance to radioactive contaminants
耐沸水试验 boiling water test
耐沸水性 boiling water resistance;resistance to boiling water
耐沸腾胶结的 boil-proof-glued
耐粉化 chalk resistance
耐粉化性 chalk(ing) resistance
耐风暴的 stormproof
耐风抽性 resistance to wind flap
耐风化 weather resistance;weather-resistant
耐风化的 weather resisting
耐风化胶合板 weatherproof plywood
耐风化密封剂 weather sealant
耐风化密封胶 weather sealant
耐风化强度 resistance to weathering
耐风化外层 weather skin
耐风化性 resistance to efflorescence;weatherability;weather resistance;weather exposure capacity <暴露于大气中的>
耐风化性能 resistance to weathering
耐风火柴 fusee
耐风加固电杆 storm-guyed pole
耐风揭性 uplift resistance;wind uplift resistance
耐风浪的 seaworthy
耐风浪性 seaworthiness
耐风蚀 weatherproof
耐风蚀测试机 weatherometer
耐风蚀钢 weathering steel
耐风蚀矿物 resistant mineral
耐风蚀强度 resistance to weathering
耐风蚀性(能) resistance to weathering;weathering quality
耐风树木 wind-enduring tree
耐风压玻璃 anti-storm glazing;storm-resistant glazing;wind-resisting glazing
耐风雨 weatherproofing
耐风雨的 all weather;weather fast
耐风雨的包装 weatherproof dressing
耐风雨的化合物 weatherproofing compound
耐风雨的外壳 weatherproof enclosure
耐风雨发动机 weatherproof engine
耐风雨绝缘 weather-resistant insulation
耐风雨侵蚀能力 weather resistance
耐风雨侵袭能力 weatherproofness
耐风雨绳索 weatherproof rope
耐风雨涂层 weather coat
耐风雨涂层的 weather-coated
耐风雨涂层的更新 weather-coating renewal
耐风雨涂料 weather-coating material
耐风雨外层 weather skin
耐风雨线 weatherproof wire
耐风雨性 weatherproofness
耐风载性 wind load resistance
耐风植物 wind-enduring plant
耐辐射玻璃 radiation-resistant glass
耐辐射玻璃棉 radiation-resistant glass wool
耐辐射玻璃纤维 radiation-resistant fiber[fibre]
耐辐射材料 radio-resistant material
耐辐射齿轮油 radiation resistance gear oil
耐辐射电工玻璃纤维 radiation-resistant electric(al) glass fiber[fibre]
耐辐射电绝缘玻璃纤维 irradiation-resistant electric(al) glass fiber[fibre]
耐辐射光学玻璃 irradiation-resistant optic(al) glass
耐辐射容器 radiation-proof container
耐辐射损伤性 radiation-damage stability
耐辐射陶瓷 radiation-resistant ceramic
耐辐射涂料 radio-resistant coating
耐辐射污染性 resistance to radiative contaminants
耐辐射性 radiation stability;radioresistance;radio tolerance
耐辐射植物 radio-resistant plant
耐辐照度 radiation stabile
耐腐蚀 anti-corrosion;corrosion-proof
耐腐蚀材料 corrosion-resistant material;resistant material
耐腐蚀处理 immunizing
耐腐蚀的 corrosion-proof;inoxidable;resistant to corrosion;resistant to rust;rust-proof(ing);unattackable;corrosion-resistant;corrosion-resisting
耐腐蚀低合金钢 corrosion-resistant low alloy steel
耐腐蚀钢 corrosion-proof steel;corrosion-resistant steel
耐腐蚀硅钢 corrosion
耐腐蚀合金 regulus of Venus
耐腐蚀混凝土 corrosion-resistant concrete
耐腐蚀绝缘 erosion-resisting insulation
耐腐蚀离心泵 corrosion-resisting centrifugal pump
耐腐蚀铝合金 corrosion-resisting alumin(i)um alloy
耐腐蚀试验 anti-corrosion test;corrosion test;prolonged corrosion test
耐腐蚀套管 corrosion-resistant casing
耐腐蚀铜合金 Albrac
耐腐蚀性 corrosion resistance;corrosion resistivity;erosion resistance;inoxidability;inoxidizability;noncorrodibility;resistance to corrosion
耐腐蚀性测定 corrosion resistance measurement
耐腐蚀阳极涂层 corrosion-resistant anodic coating
耐腐蚀轴承 corrosion-resisting bearing
耐腐性 decay resistance;resistance to corrosion;rot-proofness;wearability
耐钙表面活性剂 lime-resistant surfactant
耐钙性 lime resistance
耐干热牢度 fastness to dryheat
耐干湿交替性 humid-dry cycling resistance
耐干湿循环性 endurance to alternate wetting and drying
耐干燥性 dry strength
耐钢比 ratio of refractory to crude steel
耐高渗透压的 osmophilic
耐高温玻璃 hard glass;pyroceram
耐高温材料 chamot(te);heat-resisting material;high-temperature resistant material
耐高温瓷制品 high-temperature porcelain
耐高温的 boil-proof;burn-out proof;fire-resistant;heat-resisting;resistant to elevated temperatures
耐高温分解性 pyrolytic stability
耐高温坩埚 high-temperature crucible
耐高温钢 high-temperature steel
耐高温骨料 refractory aggregate
耐高温合金 super-alloy
耐高温黄油 high-temperature grease
耐高温混凝土 <由高矾土水泥和耐火集料如耐火碎砖制成> high-temperature concrete;refractory concrete
耐高温集料 refractory aggregate
耐高温金属 refractory metal
耐高温聚合物 heat-resistant polymer;high-temperature resistant polymer
耐高温绝缘涂层 high-temperature insulation coating
耐高温沥青漆 furnace black;stove black
耐高温密封 refractory seal
耐高温棉 refractory mineral wool
耐高温(耐火)混凝土 high-temperature resisting refractory concrete
耐高温黏[粘]土砖 high heat duty fireclay brick
耐高温漆 heat-resistance paint;heat-resistant paint;high-temperature resistant paint
耐高温砌合 boil-proof bond
耐高温软化性 resistance to softening at high temperature
耐高温试验 high-temperature test
耐高温水泥 high-temperature cement
耐高温陶瓷 pyroceram
耐高温陶瓷黏[粘]合剂 pyroceram
耐高温涂层 high-temperature coating
耐高温涂料 high-temperature coating
耐高温纤维 high-temperature resistant fiber[fibre]
耐高温性 heat-resisting quality;high-temperature resistance;resistance to high temperature;resistance to severe heat
耐高温性能 resistance to elevated temperatures
耐高温氧化物纤维 refractory oxide fiber[fibre]

耐高温银灰漆 high-temperature heat-resisting aluminum paint
耐高温硬质合金 high-temperature cemented carbide
耐工业腐蚀性 resistance to industrial corrosion
耐工业溶剂性 resistance industrial solvent
耐汞的 mercury-resistant
耐汞菌 mercury-resistant bacteria
耐拱楼盖 brick arch floor
耐固性试验 soundness test
耐刮 scratchproof
耐刮伤性 scratch resistance
耐光 fast light;light-resistant
耐光材料 lightfast material
耐光的 light resistance; light-resistant;photostable;light-proof
耐光度 fastness rate;fastness to light;light-fastness
耐光色牢度 colo(u)r fastness to light
耐光物质 light stabilizing(agent)
耐光性 fastness to light; light-fastness; light-proofness; light resistance;light stability;photostability
耐光性试验 light exposure test;radiation proofing test
耐光照的 light fastness
耐光泽的 glossproof
耐龟裂性 checking resistance
耐海水的 resistance to seawater;seawater-resistant
耐海水低合金钢 seawater-resistance low alloyed steel
耐海水钢材 marine steel
耐海水金属 admiralty metal
耐海水浸蚀性 resistance to corrosion from seawater
耐海水牢度 fastness to sea water
耐海水油漆 marine paint
耐海水性 seawater resistance
耐寒的 cold proof; frost-resisting; non-freezing
耐寒度 cold endurance
耐寒力 cold endurance; cold hardiness;frost-resisting power
耐寒强度 resistance to cold
耐寒式电动机 frost-proof motor
耐寒试验 low-temperature test
耐寒性 cold endurance; cold resistance; cold tolerance; freeze resistance; freezing resistance; frost-resisting property; hardiness; low-temperature resistance; resistance to cold;winter hardiness
耐寒性润滑油 Everlube; low setting point oil
耐寒性试验 freezing test
耐寒一年生植物 hardy annual plant
耐寒砧木 hardy stock
耐寒植物 hardy plant
耐汗度 fastness to perspiration
耐汗性 perspiration resistance;resistance to sweat
耐旱的 tolerant to drought
耐旱灌木 low suckering shrub
耐旱品种 drought-tolerant species
耐旱性 drought hardiness;drought resistance;drought tolerance
耐旱植物 <在沙漠铁路种植> drought-resistant plant;drought-enduring plant
耐旱作物 drought-enduring crop
耐航包装 seaworthy packing
耐航力 seaworthiness
耐烘烤性试验 resistance-to-bake test
耐红丝腐蚀性 filifor corrosion resistance
耐候钢 weathering steel;weatherproof steel;weather-resistant steel;weather-resisting steel;weather withstand steel
耐候化性能 resistance to elements;resistance to weathering
耐候胶 weatherproof glue; weatherproof sealant
耐候密封膏 weatherproofing sealant;weather sealant
耐候试验 weathering test
耐候试验机 weatherometer; weather tester
耐候涂层 weather coat
耐候性 fastness to weathering;resistance to weather; weatherability; weather(ing) resistance; weather resisting property
耐候性胶合板 weatherproof plywood
耐候性胶粘剂 weatherproof adhesive
耐候性试验 atmospheric exposure test;weather exposure test
耐弧的 arc tight
耐滑移性 resistance to sliding
耐化学玻璃纤维 chemically resistant glass fiber
耐化学腐蚀 resistance to chemical attack
耐化学品圬工 chemical-resistant masonry
耐化学漆 chemical-resistant paint
耐化学侵蚀的砌体单元 chemical resistance masonry unit
耐化学侵蚀系数 coefficient of chemical resistant
耐化学侵蚀性 chemical resistant
耐化学酸碱侵蚀能力 chemical resistance
耐化学涂层 chemical-resistant coating
耐化学洗涤牢度 fastness to chemical washing
耐化学性 chemoresistance
耐化学药剂性 chemical-proof
耐化学药品的陶土制品 chemicals resistant clayware
耐化学药品涂层 chemical-resistant coating
耐化学药品涂料 chemical-resistant paint
耐化学药品性 chemical resistance;resistant to chemicals
耐化学制品性 resistant to chemicals
耐划 scratchproof
耐划痕 mar proof
耐划痕性 mar resistance
耐划试验 lead pencil scratch test
耐环境(条件)龟裂性 resistance to environment(al)cracking
耐环境条件性 resistance environment
耐环境应力抗裂性 environmental stress crack resistance
耐灰性 dirt resistance
耐挥发性 fixedness; fixity; volatility resistance
耐火 refractory proof;withstand fire
耐火鞍形棒 <装窑用> saddle
耐火百叶窗 fire-resistant shutter;fire-resisting shutter;fire shutter
耐火板 fire-resistant board;refractory slab
耐火包装 fire-resisting casing
耐火保护层 refractory protection
耐火保险箱 Salamander
耐火玻璃 fire-resisting glass; fire-retarding glazing; flame-resistant glass; hard borosilicate glass; hard glass;Pyrex glass;refractory glass;fire-resisting glazing
耐火玻璃纤维 refractory glass fiber [fibre]
耐火薄膜结构 non-combustible membrane structure
耐火补炉料 refractory repairing mass
耐火材 fire-proofed wood;fire-proofing wood
耐火材的 slow-burning
耐火材料 refractory(material);fireproof(ing) material; fire-resistant material; fire-resisting material; fire-resistive material; fire-retarding material; fire safe material; flameproof material;gan(n)ister;grog;incombustible material;resistive material;silicon liner;Corhart
耐火材料焙烧 firing of refractories
耐火材料剥落 shelling of refractory;spalling of refractory materials
耐火材料车间 refractory shop
耐火材料衬里 refractory liner
耐火材料衬里的 refractory lined
耐火材料衬里的铜炉体 refractory-lined copper bowl
耐火材料衬砌的炉膛 refractory furnace
耐火材料的内衬导管 flare bed
耐火材料的侵蚀现象 erosion of refractory
耐火材料的氧热修补 oxythermal repair of refractory
耐火材料底板 <窑具> thimble bat
耐火材料锻膨胀 burning expansion
耐火材料坩埚 refractory crucible
耐火材料工艺 refractory technology
耐火材料结石 refractory stone
耐火材料绝缘 refractory insulation
耐火材料壳型 refractory shell mo(u)ld
耐火材料矿产 refractory raw material commodities
耐火材料锚固件 refractory anchor
耐火材料黏[粘]结剂 fire bond
耐火材料喷管 refractory nozzle
耐火材料侵蚀 refractory corrosion
耐火材料熔渣 slagging of refractory
耐火材料散裂 spalling of refractory
耐火材料上加衬 lining over refractory
耐火材料伸缩缝 refractory expansion joint
耐火材料制件 refractory product
耐火材料制品间 refractory clay product room
耐火舱壁 fire-proof bulkhead;fire-resisting bulkhead
耐火槽 debiteuse
耐火层 flame-retardant coating; refractory coating
耐火衬垫 silicon liner
耐火衬砌 refractory lining
耐火衬套 fire-proof casing;refractory lining
耐火处理 fire-proof;fire-proof treatment; fire-retardant treatment; incombustible transaction
耐火窗 lock light
耐火传送带 flame-resistant belt
耐火瓷 refractory porcelain
耐火瓷漆 refractory enamel
耐火导线 fire-resistant wire
耐火捣筑料 mo(u)ldable material
耐火的 fire-proof(ing); fire-resistant; fire-resistive; flameproofing; flame-resisting; high melt(ing); non-ignitable;refractory;fire-resisting;apyrous;calcitrant
耐火的补炉料 refractory patching mixture
耐火的档案室 fire-resistive file room
耐火的浸渍化合物 <一种纤维素接枝剂> methylol urea
耐火的铁矿石 refractory iron ore
耐火等级 fire endurance rating; fire grading; fire-protection rating; fire rating; fire-resistance class; fire-resistance rating; fire-resistant [resistance] grading <结构件的>
耐火等级测定 fire-resistive rating; fire-retardant rating
耐火等级隔墙 rated partition
耐火等级期限 fire-resistance grading period
耐火等级墙 rated wall
耐火等级试验 <建筑材料> flame spread classification test;fire rating test
耐火底环 refractory bottom ring
耐火地板 fire-resisting floor; smith floor;King floor <其中的一种>
耐火地板墙 fire-resisting floor wall
耐火地面 fire-resisting floor; fire-resistive flooring
耐火电缆 fire-resistant cable;fire-resisting cable;flame-resistant cable
耐火电缆电线 fire-proof cable and wire
耐火垫 refractory block
耐火垫板 refractory support
耐火吊顶 fire-resistive suspended ceiling
耐火顶板 fire-resisting ceiling;fire-resistive ceiling
耐火顶层楼面 fire-proof roof floor
耐火顶棚 fire-resisting ceiling;fire-resistive ceiling
耐火度 degree of fireproof; flame resistance; pyrometric cone equivalent;refractoriness
耐火度测温锥当量值 pyrometric cone equivalent
耐火额定值 fire-resistance rating
耐火房屋 fire-proof building
耐火分级 fire rate
耐火分级的 fire-rated
耐火粉料 powdered refractory; refractory powder
耐火风门 fire damper; fire-resisting damper
耐火封口材料 brasq(ue)
耐火盖板 refractory cover
耐火盖布 flame-resistant tarpaulin
耐火盖层 fire-proof lining
耐火坩埚 fireclay crucible
耐火高铝水泥 refractory alumina cement
耐火隔板 fire-rated partition;fire-resistant shield
耐火隔壁 fire-proof bulkhead
耐火隔块 printer's bit
耐火隔离片 <彩烧装窑用> printer's bit
耐火隔墙 fire partition;fire-resisting partition
耐火隔热混凝土 refractory insulating concrete
耐火隔热砖 insulating firebrick; refractory and insulating fire brick
耐火铬矿石 refractory chrome ore
耐火工业材料 refractory industry
耐火构件 fire-resistant member;fire-retardant member;fire-retardant unit; fire-retarding component; fire-retarding unit
耐火构造 fire-proof construction; fire-resisting construction; fire-resistive construction; incombustible construction;mill construction
耐火骨料 fire-proof aggregate;refractory aggregate
耐火固体 refractory solid
耐火管 refractory tube
耐火管材 flue lining
耐火管道 fire-resisting duct; fire-re-

sisting trunking
耐火硅质岩 fire stone
耐火焊条 annealing welds
耐火合金 refractory alloy
耐火盒 fire-resisting casing
耐火花电压 sparking voltage resistance
耐火化合物 refractory compound
耐火化学材料 fire-resistant chemicals;fire-retardant chemicals
耐火灰浆 fire-proofing plaster; fire-retardant(mixed)plaster;fire-retarding (mixed) plaster; refractory mortar
耐火混合物 compo
耐火混凝土 castable;castable refractory concrete; fireclay concrete; fire-proof concrete; fire-resistant concrete; fire-resisting concrete; high-temperature concrete; refractory concrete
耐火混凝土衬里 poured-refractory backing
耐火混凝土的热前强度 cold strength
耐火混凝土骨料 refractory concrete aggregate
耐火混凝土集料 refractory concrete aggregate
耐火混凝土结构 refractory concrete structure
耐火混凝土砌块 refractory concrete block
耐火机壳 fire-resistant housing
耐火基础 fire-resistant foundation
耐火极限 duration of fire resistance; fire endurance;fire-resistance duration
耐火极限计算 fire endurance calculation
耐火集料 fire-proof aggregate;refractory aggregate
耐火剂 fire-proofing chemical; fire-resistant chemical; fire-retardant chemicals;flameproofing agent
耐火夹芯板 label(l)ed panel
耐火建筑材料 non-combustible constructional material; refractory building material; refractory construction material
耐火建筑构件 fire-retarding unit
耐火建筑(物) fire-proof building; fire-proof construction; protected non-combustible construction; refractory building; slow-burning construction; fire-resisting construction
耐火浆料 plastic refractory
耐火浇灌料 castable refractory; refractory castable;refractory casting
耐火浇注料 castable refractory; refractory castable;refractory casting
耐火胶合板 fire-proofing plywood
耐火胶结材料 fire cement;fire-proof cement; fire-resisting cement; refractory cement;thermolith
耐火胶泥 refractory mortar
耐火结构 fire-proof construction; fire-proof structure; fire-resistant construction; fire-resistant structure;fire-resisting structure;fire-retardant construction; incombustible construction
耐火结构材料 fire-proof construction(al) material; refractory structural material
耐火结构地区 fire-protection structural zone
耐火结构房屋 full-mill
耐火金属 refractory metal
耐火救生带 fire-proof life-belt
耐火救生绳 fire-proof life-line
耐火救生艇 fire-proof lifeboat
耐火聚酯树脂 flame-resistant polyester resin(a)
耐火卷帘门 fire-resistant shutter;fire-resisting shutter
耐火绝热混凝土 refractory insulating concrete
耐火绝缘层 slow-burning insulation
耐火绝缘导线 fire-resistant wire
耐火绝缘体 refractory insulator; refractory thermal insulator
耐火绝缘线 flameproof wire; slow-burning wire
耐火卡箍 fire-resistant clamp
耐火壳 fire-resisting casing
耐火空心球制品 refractory bubble product
耐火空心砖 refractory "bubble"; refractory hollow brick
耐火类型 fire-resistive type
耐火梁 fire-resisting beam
耐火料 refractory matter
耐火菱镁矿矿石 refractory magnesite ore
耐火楼板 fire-proof(slab)floor;fire-resisting floor;fire-resistive floor
耐火楼面 fire-resistant floor
耐火楼梯 fire-proof stair(case);fire-resistant stair(case);fire-resisting stair(case)
耐火露天楼面 fire-proof uncovered floor
耐火炉 refractory furnace;refractory-lined oven
耐火炉壁 refractory furnace wall
耐火炉衬 refractory lining
耐火炉衬混合料 refractory lining mixture
耐火炉衬燃烧室 refractory chamber
耐火铝水泥 refractory alumina cement
耐火率 fire endurance rating;fire rating; refractory quotient; refractory value
耐火门 fire(-proof)door;fire-resistant door; fire-resisting door; fire-retardant door
耐火棉 refractory mineral wool
耐火面 refractory surface
耐火面层 fire-proofing coat;refractory coat
耐火模板 fire-rated form board;fire-rated form(work)
耐火木材 fire-proof timber;fire proof wood;fire-resisting timber;fire-resisting wood; refractory timber; fire-retardant wood
耐火木结构 heavy timber; mill construction
耐火木框架建筑 protected wood frame construction
耐火泥 castable refractory;fire clay; refractory clay; refractory mix(ture);seat clay;structural clay
耐火泥厂 grog mill
耐火泥衬 fireclay lining
耐火(泥)灰浆 fireclay mortar
耐火泥混凝土 chamot(te)concrete
耐火泥脊瓦 fired clay ridge tile
耐火泥浆 bat wash;refractory mortar
耐火泥排水落水器 fireclay sink unit
耐火泥轻质骨料 fired clay light (weight)aggregate
耐火泥轻质集料 fired clay light (weight)aggregate
耐火泥曲瓦屋顶 fired clay curved tile roof
耐火泥商品 fireclay goods
耐火泥熟料 fireclay grog refractory
耐火泥条 <封匣钵口用> wad; refractory wad
耐火泥桶 fireclay ware
耐火泥土 saggar
耐火泥屋脊瓦 fired clay hip tile
耐火泥屋面曲瓦 fired clay curved roof(ing)tile
耐火泥浴盆 fireclay bath tub
耐火泥制件 fireclay ware
耐火泥制品 fired clay product
耐火黏[粘]合剂 fire-proofing adhesive
耐火黏[粘]土 refractory clay;clunch; coal clay; daugh; fire stone; grog; moler; sagger; seat clay; seggar clay;fire clay
耐火黏[粘]土衬里 fireclay lining
耐火黏[粘]土过梁 <负荷楼板用的> dawnay's lintel
耐火黏[粘]土厚层 thill
耐火黏[粘]土浆 slurry
耐火黏[粘]土矿材 fireclay mineral
耐火黏[粘]土矿床 fireclay deposit
耐火黏[粘]土矿石 refractory clay ore
耐火黏[粘]土矿物 fireclay mineral
耐火黏[粘]土模 fireclay mo(u)ld
耐火黏[粘]土破碎机 grog mill
耐火黏[粘]土砌块 refractory fireclay block
耐火黏[粘]土砂浆 grog fireclay mortar
耐火黏[粘]土熟料 chamot(te);fireclay chamotte;fireclay smog
耐火黏[粘]土水泥 pysuma
耐火黏[粘]土陶瓷 chamot(te)ceramics
耐火黏[粘]土卫生器 fireclay sanitary ware
耐火黏[粘]土砖 fireclay brick; high heat duty firebrick
耐火喷射料 gunning refractory
耐火喷涂材料 refractory gunning material
耐火喷嘴 fireclay nozzle
耐火屏蔽 fire-resistant shield
耐火期 fire-resistance period
耐火漆 fire-coat paint; fire-proof paint;fire-resistant paint;fire-resisting paint; flameproof paint; non-flammable paint
耐火汽油桶 fire-proof petrol tank
耐火砌块 fire block;refractory block
耐火器材 refractory ware
耐火器皿 flame ware
耐火强度 <混凝土燃烧后的抗压、挠曲强度> fired strength
耐火墙 fire-cutting partition; fire-proof wall;fire-resisting wall;fire-resistive wall; flameproofing wall; flame shield;refractory wall
耐火墙板 fire-resistant wall board; Kimoloboard <一种由矿物(硅藻土)组成的>
耐火切割系统 hot face cutting system
耐火热风炉栅 refractory hot-gas grate
耐火润滑油 fire-proof grease;fire-resistant grease
耐火润滑脂 fire-proof grease;fire-resistant grease
耐火塞 refractory stopper
耐火砂 fire sand;refractory sand
耐火砂浆 fireclay mortar; fire-coat mortar; fire mortar; fire-proofing mortar; fire-resistance mortar; refractory mortar
耐火烧针 refractoriness pins
耐火绳 fire-proof line
耐火石 fire stone; ovenstone; quartzite-schist;refractory pebbles
耐火石膏板 fire fighter gypsum board
耐火石灰 refractory lime
耐火时间 fire duration; fire endurance; fire-resistance hour; fire-resistance period
耐火时限 fire endurance
耐火试验 fire experiment;fire-resistance test
耐火饰面 fire-resisting finish;fire-retardant finish
耐火熟料 grog;grog refractory
耐火熟料砖 grog brick
耐火竖井 fire-resisting shaft
耐火水泥 fire cement; fire-coat cement;fire-proof(ing)cement; fire-resisting cement; high-temperature cement;refractory cement;thermolith; Kestner cement <耐高温达1300℃>
耐火水泥熔融水泥料 refractory fused cement
耐火塑料 fire-resisting plastics;flame-resistant plastics; plastic refractory;refractory plastics
耐火塑料布 Saran
耐火燧石土 flint fireclay
耐火陶瓷 refractory ceramics
耐火套 fire-resisting casing
耐火特性 fire-resistant property
耐火天花板 fire-resisting ceiling;fire-resisting roof
耐火填缝灰浆 jointing refractory cement
耐火填缝灰泥 jointing refractory cement
耐火填料 fire-resistant (in) filling; fire-resistant packing
耐火通道口 protected opening
耐火涂层 fire-coat;fire-proofing plaster;fire-retardant coating;refractory coating
耐火涂料 fire-proof coating;fire-proof dope; fire-proof paint; fire-resisting dope;fire-retardant paint;refractory coating; refractory dress-(ing); refractory paint
耐火涂料涂层 fire-proof paint coat(ing)
耐火涂抹(材料)refractory wash
耐火土 stove clay
耐火土杯 fireclay cup
耐火土块 ball;bloom
耐火土盘 fireclay disk[dish]
耐火土熟料 burned fireclay
耐火瓦 burner tile
耐火完整性 fire integrity
耐火稳定性 fire stability
耐火圬工 refractory masonry
耐火屋顶 fire-resistant roof; fire-resisting roof
耐火物 fire-proofing;refractory body
耐火席子 fire-proof mat
耐火匣体 saggar[sagger]
耐火纤维 refractory fiber[fibre]
耐火纤维湿毡 moist felt
耐火纤维增强塑料 refractory fibre reinforced plastic
耐火纤维制品 refractory fiber product
耐火线 fire wire
耐火箱 fire-resisting casing
耐火效果 fire action
耐火芯板 fire core
耐火型电缆 refractory cable
耐火性 fire durability;fire-proofness; fire-resistance; flame resistance; non-ignition; refractoriness; resistance to fire
耐火性分级 fire-resistance classification;fire-resistance grading

N

耐火性能 fire performance; fire-protecting performance; fire-resisting property
耐火性试验 fire(-proof) test; fire-resistance test; flame-resistance test
耐火悬件 <指帘帷等> fire-proofing hanging
耐火岩 refractory rock
耐火岩类 refractory rocks
耐火岩石 fire stone
耐火焰处理 flame-retardant
耐火焰侵蚀性 resistance to flame erosion
耐火氧化皮 fire scale
耐火氧化物 refractory oxide
耐火氧化物坩埚 refractory oxide crucible
耐火窑衬 refractory lining of kiln
耐火液体涂层材料 fire-retarding liquid coating material
耐火液压油 fire-resistant hydraulic liquid
耐火硬金属 refractory hard metal
耐火用橄榄岩 refractory peridotite
耐火油漆 fire-proofing paint; fire-resistant[resisting] paint; fire-retardant[retarding] paint
耐火油漆涂料 fire-resisting finish
耐火油液 fire-resistant fluid
耐火釉 fire-resisting glazing
耐火原料 refractory raw material
耐火原料分级 classification of refractory raw material
耐火运输带 fire-resistance belt(ing); fire-resistant belt(ing)
耐火罩 fire-resisting casing
耐火罩面 fire-resistive coating; fire-retardant coating
耐火支架 incombustible lining
耐火织物 flame-resistant fabric
耐火值 fire-resistance rating
耐火植物 pyophyte
耐火纸 incombustible paper
耐火纸面石膏板 fire-resistance gypsum plaster board
耐火制品 refractory composition; refractory product
耐火制品外观缺陷 apparent defect of refractory product
耐火中间楼面 fire-proof intermediate floor
耐火铸石 fire-proof cast stone
耐火砖 brick fireproofing; chamot(te) brick; fire brick; fireclay refractory material; fire-proof brick; firing brick; floor tile; furnace brick; grog refractory; hard stock brick; neutral brick; refractory brick
耐火砖衬 firebrick lining
耐火砖衬里 firebrick lining; lining of fire brick
耐火砖衬砌 firebrick lined
耐火砖成型机 Jolly
耐火砖挡圈 refractory dam ring
耐火砖盖 firebrick lid
耐火砖拱 firebrick arch; flue bridge
耐火砖骨料 firebrick aggregate
耐火砖集料 firebrick aggregate
耐火砖结构 firebrick structure; refractory brickwork
耐火砖绝缘 clay brick insulation
耐火砖炉墙 fire brick wall
耐火砖内衬烟囱 firebrick lined chimney
耐火砖坯 fire bat(t)
耐火砖墙 refractory wall
耐火砖散裂试验 spalling test of fireclay brick
耐火砖套 shroud
耐火砖瓦 fire-proofing tile
耐火砖载体 firebrick support
耐火装置 fire-proofing
耐击穿试验 breaking down test
耐击穿性 resistance to sparking
耐激冷激热性 spalling resistance
耐激冷热性 resistance to thermal shocks
耐急冷急热玻璃 thermal glass
耐急冷能力 chilling resistance
耐碱玻璃 alkali-proof glass
耐碱玻璃纤维 alkali-proof glass fibre; alkali-resistant glass fiber[fibre]
耐碱的 alkalifast; alkaline resisting; alkali-proof; alkali-resistant; limefast
耐碱底油 alkali-resistant primer
耐碱度 alkali fastness; alkali resistance; alkali tolerance; fastness to alkali
耐碱封闭剂 alkaline resistant sealer
耐碱钢 alkali-proof steel
耐碱骨料 alkaline resisting aggregate; alkali-resisting aggregate
耐碱混凝土 alkali fast concrete; alkali-proof concrete
耐碱混凝土地面 alkali-resistant cement concrete flooring
耐碱集料 alkaline resisting aggregate; alkali-resisting aggregate
耐碱矿(物)棉 alkali-resistant mineral wool
耐碱矿(物)棉增强水泥 alkali-resistant mineral wool reinforced cement
耐碱楼地面 alkaline resistant flooring; alkali-resistant flooring
耐碱率 alkali resistance rate
耐碱玛琋脂 alkaline resistant mastic
耐碱耐火材料 alkali-resistant refractory
耐碱黏[粘]土砖 alkali-resistant fired-clay brick
耐碱漆 alkali-proof paint; alkali-resisting paint; alkali-resisting resistant paint; cement paint
耐碱清漆 alkali-proof varnish; alkali-resisting varnish
耐碱砂浆 alkaline-resistant mortar; alkali(ne) resisting mortar
耐碱水泥 alkali fast cement; alkali(ne) resisting cement
耐碱水泥砂浆楼地面 alkaline resistant cement mortar flooring
耐碱涂料 alkali-proof paint; alkali-resisting paint; alkali-resisting resistant paint
耐碱性 alkali proofness; alkali(ne) resistance; alkali tolerance; fastness to alkali; resistance to alkali
耐碱性鉴定 be evaluated for salt tolerance
耐碱性能 alkaline resistance property
耐碱性试验 alkali resistance test; alkali-proof test
耐碱颜料 limefast pigment
耐碱植被 alkali-resistant vegetation
耐碱植物 alkali soil plant
耐碱煮牢度 fastness to soda boiling
耐碱铸铁 alkali-resisting cast iron
耐交变力学应力性 resistance to alternating mechanical stress
耐交染牢度 fastness to cross-dy(e)-ing
耐胶合牢度 fastness to trubenizing
耐金属划痕性 metal making resistance; metal ring resistance
耐近海气候性 coastal climate resistance
耐浸型 flood-proof type
耐浸渍性 immersion resistance
耐静态压痕性 resistance to static indentation
耐久 duration; lasting; lastingness
耐久比 durability ratio; endurance ratio
耐久材 durable wood
耐久程度 endurance degree
耐久存的食品 lasting food
耐久的 durable; indestructible; long-lasting; long lived; resistant; resisting; time-proof
耐久的物品 durables
耐久定形 <俗称永久定形> permanent press
耐久定型 durable press
耐久度 degree of durability; degree of resistance
耐久高强混凝土 durable high-strength concrete
耐久混凝土 durable concrete
耐久极限 endurance limit; threshold strength
耐久极限试验 fatigue limit test
耐久极限应力 endurance limit stress
耐久结构 durable structure
耐久力 durability; endurance
耐久路面 long-lasting pavement
耐久率 durability; durability index
耐久面漆 long life top coat
耐久年限 maintenance-free life
耐久破坏 endurance failure
耐久嵌缝 durable seal; lasting seal
耐久强度 <无限重复荷载后水泥混凝土路面的> endurance strength
耐久色 weatherproof colo(u)r
耐久渗透试验 permeability test
耐久试验 endurance trial; fatigue test(ing); long duration test; long-time test
耐久寿命 endurance life
耐久水膜 enduring water-proof membrane
耐久弹簧 resistant spring
耐久涂层 long lived coating
耐久系数 endurance coefficient; endurance ratio; fatigue coefficient
耐久现象 endurance phenomenon
耐久限度 endurance limit; endurance strength
耐久效果 lasting effect
耐久性 ag(e)ing resistance; durability; endurance; keeping property; keeping quality; perdurability; permanence; permanent stability; persistence; robustness; ruggedness; service durability; stability; staying power; viability
耐久性材料 durable material
耐久性参数 durability parameter
耐久性道路试验 endurance road testing
耐久性等级 durability rating
耐久性拷花整理 permanent embossed finish
耐久性快速试验 accelerated durability test
耐久性路线 <汽车的> endurance road test
耐久性能 endurance behavio(u)r; lasting property; lasting quality
耐久性曲线 durability line
耐久性上光整理 permanent lustred finish
耐久性设计 design for durability
耐久性试验 durability test; endurance test; life test(ing); duration test
耐久性试验台 endurance rig; endurance test bed
耐久性系数 coefficient of durability; life factor; durability factor
耐久性要求 life requirement
耐久性因素 durability factor; life factor
耐久性有形资产 permanent tangible property
耐久性运行 endurance running
耐久性运转 endurance running
耐久性指数 durability index
耐久压烫 durable press; permanent press
耐久阴极 life-boost cathode
耐久褶裥加工 permanent pleating
耐酒精性 fastness to spirit; spirit fastness
耐酒精性能 resistance to spirit
耐巨浪的船体 holsom; wholesome
耐开裂性 cracking resistance
耐烤玻璃器皿 ovenware
耐空气腐蚀性 resistance to atmospheric corrosion
耐空气污染牢度 fastness to air pollution
耐拉钢 high-tensile steel
耐拉合金 high-tensile alloy
耐拉绝缘器 strain insulator
耐拉绝缘子 dead-end insulator; strain insulator; stretching insulator; tensioning insulator
耐拉伸的 stretch-proof
耐拉铁塔 strain tower
耐拉线夹 strain clamp
耐劳的 hardy
耐劳幅度 fatigue range
耐劳津贴 fatigue allowance
耐劳强度 fatigue resistance
耐劳橡胶 fatigue-proof rubber
耐老化 weather resistance
耐老化钢 weathering steel
耐老化黏[粘]结剂 weathering resistant adhesive
耐老化润滑脂 ag(e)ing-resistant grease
耐老化涂料 weather-reflective paint
耐老化性 resistance to ag(e)ing; weathering resistance; weathering quality
耐老化状态 ag(e)ing behavio(u)r
耐涝的 flood resistant
耐涝性 resistance to overhead flooding injury
耐冷的 cold proof
耐冷度 cold resistance; fastness to cold
耐冷性 bending brittle point
耐冷性能 cold resistance; cool-resistance; resistance to cold
耐冷油 cold test oil
耐力 proof; stamina; staying power
耐力测验 tolerance test
耐力程度 degree of resistance
耐力焊接 stress weld
耐力铆钉 stress rivet
耐力试验 endurance test
耐力限度 endurance limit
耐量 tolerance
耐量试验 tolerance test
耐裂纹性 resistance to crazing
耐磷 tolerate P-levels
耐硫酸的 sulfate[sulphate]-resisting; sulfate[sulphate]-resistant
耐硫酸黏[粘]结剂 sulfur acid-resisting binder
耐硫酸盐 sulphate resistance
耐硫酸盐的 sulfate-resistant
耐硫酸盐水泥 sulphate-resistant cement; sulphate-resisting cement
耐氯化牢度 fastness to chlorination
耐氯牢度 fastness to chlorine
耐氯漂牢度 fastness to chlorine-bleach-

ing
耐氯水牢度 fastness to chlorinated water
耐纶 nylon
耐轮椅磨损的 resistant to wheelchairs
耐霉(菌)性 fungus resistance
耐泌水度 fastness to bleeding
耐摩 anti-friction
耐摩板 anti-friction plate
耐摩擦程度 rubbing fastness
耐摩擦的 friction-resistant
耐摩擦牢度 fastness to rubbing
耐摩擦脱色牢度 fastness to crocking
耐摩材料 anti-friction material
耐摩镀层 friction(al) coat
耐摩涂层 friction(al) coat
耐磨 anti-friction;long wearing
耐磨板 impact block;wear(ing) plate;wear-resisting plate
耐磨泵 < 抽汲有磨损作用的物料的吸料泵 > wear pump
耐磨表面 anti-frictional surface;wearing coat;wearing surface
耐磨部件 wear parts;wear segment
耐磨擦性 rub resistance
耐磨材料 abrasion resistance material;abrasive material;anti-frictional material; anti-friction material; high-abrasive material;wear material;wear-resistant material;wear-resisting material
耐磨层 abrasion-resistant coating;wearing coat;wearing course;wearing layer;wearing surface;wear layer
耐磨插芯喷嘴 hard center nozzle
耐磨衬板 wear-resisting lining board
耐磨程度 rub fastness;rub proofness
耐磨带 wearing strip
耐磨刀片 wearing blade
耐磨道路水泥 wear-resisting road cement
耐磨的 abrasion-proof;abrasion-resistant; anti-abrasion; attrition resistant;hardwearing;tear proof;wearable; wear-proof; wear-resistant;anti-wear;toughlay < 绳索 >
耐磨地板 wearing floor;wear-resisting floor
耐磨地面面层 hardwearing floor cover(ing)
耐磨地面涂料 deck paint
耐磨电刷 non-dusting brush
耐磨垫 wearing carpet
耐磨垫圈 wear washer
耐磨度 abradability;abrasion resistance; abrasive resistance; endurance;fastness to rubbing;resistance to abrasion;wearing strength
耐磨镀层 scuff-resistant coating;wear-resistant coating
耐磨端板 wear cap
耐磨堆焊 hard-facing
耐磨堆焊层 wear-resisting overlay
耐磨堆焊焊条 hard-facing electrode
耐磨房屋 wearing area
耐磨钢 abrasion-resistant steel;wear-resistant steel;wear-resisting alloy;wear-resisting steel
耐磨钢轨 anvil faced rail;wear resisting[resistant] rail
耐磨骨料 abrasive aggregate;non-polishing aggregate
耐磨骨料砂浆 mortar with metallic aggregate
耐磨管道 wear resistant pipe
耐磨轨 wear resisting rail
耐磨耗的 abrasion-resistant
耐磨耗性 abrasion performance;erosion resistance
耐磨合金 abrasion alloy;abrasive resistant alloy; anti-friction alloy; wearing-proof alloy;wear-resistant alloy;wear-resisting alloy
耐磨合金钢 abrasion-resistant steel alloy;wear-resisting alloy steel
耐磨花纹 cut resistant tread
耐磨环 wear(ing) ring
耐磨混凝土 < 用坚硬骨料拌制的 > granolithic concrete; wear resistance concrete
耐磨机械装置 wear mechanism
耐磨极限 endurance limit
耐磨集料 abrasive aggregate;non-polishing aggregate
耐磨剂 anti-wear agent;hardener
耐磨件 wearing piece;wear-resistant parts
耐磨金属 anti-friction metal;wear-resistant metal;wear-resisting metal
耐磨锯条 < 用于割断混凝土接头 > abrasive blade
耐磨肋骨 wear rib
耐磨力 tear resistance
耐磨粒料 abrasive aggregate
耐磨零件 anti-abrasion part
耐磨面 wear-resistant surface
耐磨面层 hardwearing coat;scuff-resistant coating;wear-proof coating;wear-resistant coat
耐磨耐蚀铝青铜 dynamobronze
耐磨能力 abradability; wearing capacity;wear resistance;abrasive resistance
耐磨强度 abrasive resistance;resistance to abrasion; scuff resistance;wear rate;wear resistance
耐磨清漆 rubbing varnish
耐磨扇形衬板 wear segment
耐磨蚀材料 ant-friction material;resistant material
耐磨蚀度 abrasion hardness;abrasion test
耐磨蚀轴承 corrosion-resisting bearing
耐磨试验 abrasion test;attrition test;wear test(ing)
耐磨试验机 abrasion wear test machine;wear-testing machine
耐磨试验仪 abraser;abrasimeter
耐磨寿命 wearing life
耐磨损 anti-attrition;stand wear and tear
耐磨损材料 anti-abrasive material
耐磨损的 anti-abrasive
耐磨损牢度 fastness to abrasion
耐磨损性 mar proof;resistance to wearing; scuff resistance; wear-proof;wearability
耐磨特性 wearing characteristic;wearing quality;wear-resisting property
耐磨填料 abrasion-resistant filler
耐磨条 wearing strip;wear rib
耐磨铁板 wear plate
耐磨涂层 abrasion-resistant coating;anti-abrasion coating; wear-resistant coating
耐磨涂料 wear-resistant paint(ing)
耐磨系数 wearing coefficient
耐磨限度 endurance limit
耐磨橡胶 abrasive rubber
耐磨性 abradability; abrasion performance;abrasion resistance;abrasive hardness; abrasive resistance;anti-friction property; attrition resistance;friction(al) resistance;resistance to attrition; resistance to wear; wear hardness; wearing capacity; wearing property; wearing quality;wear(ing) resistance;wearlessness;wear-resistant
耐磨性测定 falling sand test
耐磨性的 wear-resisting
耐磨性绝缘 abrasion insulation
耐磨性能 wear-resistant property;wear-resisting property
耐磨性试验 resistance-to-abrasion test;wearing test(ing)
耐磨仪 jet abrader
耐磨硬度 abrasion hardness;abrasive hardness;attrition hardness;passive hardness
耐磨硬度试验 abrasive hardness test
耐磨硬度试验仪 wear hardness testing apparatus
耐磨罩面涂层 abrasion concealing coating
耐磨指数 abrasion resistance index
耐磨轴承 anti-friction(al) bearing
耐磨铸铁 wear-resistant cast iron;wear-resisting casting iron
耐磨砖 abrasive brick; anti-friction brick
耐磨阻力 abrasive resistance
耐内压试验机 increment internal pressure tester
耐挠曲性 flexing life
耐农药性 resistance to pesticide
耐欧来特 < 耐磨合成橡胶化合物 > Neolite
耐喷烧钢 torch-resistant steel
耐膨胀度 swelling fastness
耐膨胀性 fastness to swelling; swelling resistance
耐碰击性 resistance to impact
耐碰撞试验 shock-resistance test
耐碰撞性能 crashworthiness
耐疲劳 anti-fatigue
耐疲劳度 endurance; fatigue endurance
耐疲劳度试验 endurance test
耐疲劳极限 endurance fatigue limit;fatigue endurance limit
耐疲劳能力 fatigue capability;fatigue life capability
耐疲劳强度 fatigue strength
耐疲劳特性 fatigue durability
耐疲劳限度 fatigue capability
耐疲劳性 fatigue resistance;long fatigue life
耐片落性 flaking resistance
耐漂白牢度 fastness to bleaching
耐破度试验 burst test
耐破坏特性 resistance against breakage
耐起粉性 chalking resistance
耐起泡性 blistering resistance
耐起皮性 scaling resistance
耐起霜性 fastness to blooming
耐气候钢 weatherproof steel;weather-resistant steel
耐气候牢度 fastness to weathering
耐气候老化性 resistance to deterioration on weathering
耐气候性 fastness to weather;resistance to weather; weatherability;weather fastness;weathering resistance
耐气候性检验 weathering test
耐气候性试验 weathering test
耐气候整理 weatherproof finish
耐气流性 airflow resistance
耐气流阻率 airflow resistivity
耐汽油的 gasoline-proof
耐汽油软管 benzine resisting hose
耐汽油涂料 gasoline-resistant coating
耐汽油性 gasoline resistance; petrol resistance
耐汽油性试验 gasoline resisting test
耐汽蒸牢度 fastness to steaming
耐侵蚀的 erosion resistant
耐侵蚀性 erosion resistance;erosion resisting;resistance to erosion
耐燃的 flameproofing;flame-resisting
耐燃工厂建筑 slow-burning mill construction
耐燃构造 flame-retardant construction;slow-burning construction
耐燃剂 flame-retardant
耐燃建筑(物) slow-burning construction;slow-burning structure
耐燃绝缘 slow-burning insulation
耐燃料油性 fuel resistance
耐燃木材 fire proof wood
耐燃木建筑 slow-burning timber construction
耐燃漆 flame-retardant paint
耐燃烧性 burning resistance
耐燃试验 ignition resistance test
耐燃型阻火器 fire-retardant type flame arrester
耐燃性 burning resistance;fire durability;flame resistance
耐燃液体 fire-resistant fluid
耐燃液压油 fire-resistant hydraulic fluid
耐染污性 stain resistance
耐热 resist heat
耐热包皮 fire-resisting covering
耐热包装 tropicalized packing
耐热保护层 refractory protection;refractory coating
耐热变形性 resistance to heat distortion
耐热玻璃 flameproof glass;flame-retarding glass;hard glass;heat-proof glass; heat-resistant glass; heat-resisting glass; ovenproof glass; Pyrex; Pyrex glass; pyroceram; vycor glass
耐热玻璃炊具 oven-top ware;stovetop ware
耐热玻璃瓶 heat-resistant bulb;Pyrex bulb
耐热玻璃器皿 flame ware
耐热玻璃设备 heat-resistant unit;Pyrex unit
耐热不起皮钢 oxidation-resistant steel
耐热布 heat-resisting fabric
耐热材料 heat-proof material; heat-resistant material; heat-resisting material;heat stable material
耐热餐具 oven-to-table ware
耐热衬垫 heat-resisting gasket
耐热持久性 continuous heat resistance;heat-resisting durability
耐热冲击 heat-shock resistance
耐热冲击性 heat impact resistance;resistance to thermal shocks
耐热瓷 heat-resisting porcelain
耐热瓷漆 heat(ing)-resistant enamel paint
耐热瓷器 heat-proof porcelain; refractory porcelain
耐热的 flameproofing;heat-resistant;heat stable;high heat;temperature-resistant; thermoduric; thermostable; thermotolerant; heat-proof;heat-resisting;tolerant to heat
耐热等级 resistance to heat class;temperature classification; thermal rating
耐热电动机 heat-resistant motor
耐热电机 heat-tight machine
耐热垫片 super-gasket
耐热度 fastness to heat;heat durability; heat endurance; refractoriness;thermal stability;heat resistance

耐热防锈钢 heat and corrosion resistant steel
耐热分解性 resistance to thermal decomposition
耐热风机 heat-resisting exhauster
耐热敷层 temperature-resistant coating
耐热幅射性 resistance to heat emission
耐热坩锅 heat-resistant crucible
耐热钢 heat-resistant steel; heat-resisting steel; high-temperature steel; refractory steel; resisting steel
耐热钢焊条 heat-resistant steel electrode
耐热钢链条 heat-resistant steel chain
耐热高压润滑剂 extreme pressure lubricant
耐热骨料 refractory aggregate
耐热合金 heat-resistant alloy; heat-resisting alloy; high-temperature alloy; super-alloy; nichrome (alloy); Calorite
耐热合金钢 heat-resisting alloy steel
耐热和耐紫外辐射试验 bake and ultraviolet irradiation test
耐热滑脂 heat-resisting grease
耐热混凝土 heat-resistance concrete; heat-resistant concrete; heat-resisting concrete; refractory concrete
耐热混凝土板 heat-resisting cement plate
耐热集料 heat-resisting aggregate; refractory aggregate
耐热夹布胶管 heat-proof laminated rubber hose
耐热金属 heating resisting metal
耐热聚合物 heat-resistant polymer
耐热绝缘 high-temperature insulation
耐热蜡油墨 hot wax ink
耐热牢度 fastness to heat
耐热力 heat hardiness
耐热裂解性 resistance to thermal decomposition
耐热流速计 heat-resistance current meter
耐热流速仪 heat-resistance current meter
耐热炉 heat-resistant furnace
耐热铝粉漆 heat-resisting alumin(i)um paint
耐热铝钢 Sichromal
耐热铝合金 heat-resisting alumin(i)um alloy
耐热轮胎 heat-resistant; heat-resisting tire[tyre]
耐热面层 high heat coat(ing)
耐热面漆 heat-resistant finish
耐热敏层 temperature-resistant
耐热耐蚀合金 thermalloy
耐热耐蚀铝青铜 Crotorite
耐热耐酸铝铸铁 Alsiron
耐热能力 thermal capacity
耐热镍铬铁合金 Incoloy; Xite
耐热镍基合金 Chlorimet
耐热漆 heat-resistance paint; thermostat varnish; Berlin black < 漆火炉的 >
耐热器 heat-resistor
耐热器皿 flameproof ware
耐热润滑剂 heat-resistant lubricant
耐热渗铝法 alumincoat
耐热试验 heating test; heat-resistance test; heat-resisting test; heat test(ing); oven test
耐热树脂基体 heat-resistant resin matrix
耐热水测定 hot-water resistance test
耐热水牢度 fastness to hot water
耐热水泥板 heat-resistance cement plate; heat-resisting cement plate
耐热水平 thermal level
耐热塑料 heat-resistant plastic
耐热塑料喷管 heat-resistant plastic nozzle
耐热搪瓷 heat-resistant enamel; heat-resisting enamel
耐热陶瓷 heat-proof porcelain
耐热特性 heat-resistant quality; heat-resisting quality
耐热贴面 heat-resisting tile
耐热铁铬合金 pyrocast
耐热铁铬镍铝合金 Fahralloy
耐热涂层 heat-resisting coating
耐热涂料 heat-resistant paint; heat-resisting paint
耐热瓦斯油 refractory gas oil
耐热微晶玻璃涂层 heat-resisting glass-ceramic coating
耐热温度 refractory temperature
耐热稳定试验 oven stability test
耐热物质 heat resistant substance
耐热细菌 thermoduric bacteria; thermophilic bacteria; thermotolerant bacteria
耐热橡胶 heat-resistant rubber; heat-resisting rubber
耐热芯子 heat-resistant core
耐热性 heat endurance; heat fastness; heat-resistant quality; heat-resisting quality; heat-resistivity; resistance to heat; resistance to the effects of heat; temperature resistance; thermal endurance; thermal resistance; thermal stability; thermostability; thermotolerance; heat resistance
耐热性能 fire-resisting property; heat-resisting property; resistance to effect of heat
耐热性能试验 resistance-to-high-temperature test
耐热性试验 thermal test(ing)
耐热压烫牢度 fastness to hot pressing
耐热颜料 heat-resistant paint; heat-resistant pigment
耐热阴极 refractory cathode
耐热油泥 manganesite putty
耐热油品 heat-resistant oil
耐热油漆 heat-resistant paint; heat-resisting paint
耐热有机玻璃 heat-resistant organic glass; plexidur; plexiglass; plexigum
耐热阈 heat tolerance threshold
耐热炸药 heat-resistant explosive
耐热震炻器 thermal shock resisting stoneware
耐热震性 resistance to heat shocks; resistance to thermal shocks; spalling resistance; thermal shock resistance
耐热织物 heat-resistant fabric
耐热值 heating resistance
耐热植物 heat-resistant plant; heat-resisting plant; heat stable plant; thermophyte
耐热铸管 refractory casting tube
耐热铸件 heat-resisting casting
耐热铸铁 heat-resistant cast iron; heat-resisting cast iron; heat-resisting iron
耐热铸铁模 heat-resistant cast iron mo(u)ld
耐人造光牢度 fastness to artificial light
耐日光牢度 fastness to light
耐日光晒裂 anti-sun checking
耐日晒的 resistant to sunlight
耐日晒色牢度 colo(u)r fastness daylight exposure
耐溶剂溶胀性 resistance to swelling by solvent
耐溶剂涂料 solvent resistant coating
耐溶剂性 resistance to solvent; solvent resistance
耐溶性 cold resistance
耐溶胀性 resistance to swelling; swelling resistance
耐熔材料 refractory material
耐熔的 refractory
耐熔度 refractoriness
耐熔钴铬碳合金 refractory cobalt chromium carbon alloy
耐熔金属 refractory metal
耐熔铝青铜 dynamobronze
耐熔面 hot face
耐熔物质 refractory
耐熔性 refractoriness
耐熔质 refractory
耐揉搓性 scuff resistance
耐揉度 folding strength
耐蠕变钢 creep-resistant steel
耐蠕变合金 creep-resistant alloy
耐蠕变试验 creep test
耐润滑油性 grease resistance; lubricating oil resistance; resistance to lubricating oil
耐砂浆性 mortar resistance
耐砂磨损试验 sand abrasion test
耐晒不褪色的(颜料或油漆) lightfast
耐晒橙 hansa orange
耐晒的 sun-proof
耐晒红 light red
耐晒黄 Hansa yellow
耐晒裂 anti-sun checking
耐晒漆 sun-proof paint
耐晒试验器 fadeometer
耐晒塑料 sun-proof plastics; sun-resistant plastics
耐晒涂料 sun-proof paint
耐晒猩红 hansa scarlet
耐晒性 light-fastness
耐烧时间 combustion duration; duration of combustion
耐烧蚀性 ablation resistance
耐烧性 burning resistance
耐渗漏度 fastness to bleeding
耐升华牢度 fastness to sublimation
耐生根性 root resistance
耐湿的 moisture-proof; moisture-resistant; vapo(u)r-resistant
耐湿环氧树脂 moisture-tolerant epoxy resin
耐湿绝缘 moisture-proof insulation
耐湿老化性 resistance to humid ag(e)ing exposure
耐湿磨性 wet abrasion resistance
耐湿气老化性 humid ag(e)ing resistance
耐湿强度 wet strength
耐湿热性 humid heat-resistance; resistance to heat and humidity
耐湿石膏板 green board
耐湿性 dampproofness; humidity resistance; moisture-proof; moisture proofness; moisture resistance; wet fastness
耐湿性抗裂指数 index of anti-moist cracking resistance
耐湿植物 damp tolerant plant
耐石灰的 limefast
耐石灰性 fastness to lime
耐石灰颜料 limefast pigment
耐蚀 anti-corrosion; corrosion resistance
耐蚀泵 corrosion free pump; corrosion-resisting pump
耐蚀薄膜 anti-corrosion film
耐蚀材料 corrosion-resisting material; rust-resisting material
耐蚀成分 resistant component
耐蚀的 anti-rust; corrosion-proof; corrosion-resistant; corrosion-resisting
耐蚀底基层 non-erodible subbase
耐蚀钢 corrosion-resisting steel
耐蚀钢板 corrosion-resistant plate; corrosion-resisting steel plate
耐蚀高镍铸铁 nickel resist; nickel-resist cast iron; Nimol; Ni-resist
耐蚀高强度铜合金 super-ston
耐蚀高强黄铜 delta metal
耐蚀硅钢 corrosiron
耐蚀硅铜合金 Silcurdur
耐蚀硅砖 Resistal
耐蚀合金 anti-corrosion alloy; corrosion-resistant alloy; corrosion-resisting alloy
耐蚀黄铜 inhibited brass; naval brass
耐蚀金属 Chlorimet; corrosion-proof metal; corrosion-resistant metal; corrosion-resisting metal; resistant metal
耐蚀力 corrosion strength
耐蚀铝硅合金 permite alumin(i)um alloy
耐蚀铝合金 Alcoa
耐蚀铝基合金 Pantal alloy
耐蚀铝青铜 dynamobronze
耐蚀玛琋脂 corrosion-resistant mastic
耐蚀面层 corrosion-proof coating
耐蚀耐热合金钢 corrosion-proof and heat-proof alloy; Era
耐蚀耐热铜合金 Ampco(loy)
耐蚀镍的 nickel resist
耐蚀镍铬合金 Pioneer metal
耐蚀镍钼铁合金 Corronel
耐蚀漆 hold paint
耐蚀铅合金 Roofloy
耐蚀(铅)锡黄铜 hydraulic bronze
耐蚀强度 corrosion strength
耐蚀青铜 Hercules bronze
耐蚀试验 corrosion-resisting test; Kesternich test; prolonged corrosion test
耐蚀铜镍合金 Niconmetal; tempaloy
耐蚀系数 coefficient of chemical resistance
耐蚀性 corrosion resistivity; corrosion stability; corrosion strength; corrosive strength; stain resistance; corrosion resistance
耐蚀性能 corrosion-resisting property
耐蚀性试验 resistance to chemical attack
耐蚀压铸铝合金 Ruselite
耐蚀岩石 resistant rock
耐蚀增强塑料设备 reinforced plastic corrosion-resistance equipment
耐蚀铸铁 corrosion-resistant cast iron; corrosion-resist cast iron; sublimation
耐室外曝晒老化性 resistance deterioration on weathering
耐室外气候老化性 outdoor ag(e)ing characteristics
耐受 tolerance
耐受极限 tolerance limit
耐受剂量 tolerance dose
耐受天气自然作用的能力 resistance to weather
耐受物种 tolerance species
耐受限 tolerance limit
耐受性 survivability; tolerance; toleration
耐受性嵌合体 tolerance chimera
耐受原 toleragen
耐暑 tolerance to heat
耐鼠咬性 rodent resistance
耐刷能力 brushability
耐刷洗牢度 fastness to brushing

耐刷洗试验 scrubbing resistance test
耐刷洗性 scrubbing resistance
耐霜的 frost-resistant
耐霜冻 resistance to frost action
耐霜性 frost resistance
耐水 waterproof
耐水玻璃 waterproof glass
耐水玻璃性 fastness to soluble glass; fastness to water glass
耐水冲的 jetproof
耐水的 water-resistant; water-resisting
耐水度 water fastness; water tolerance
耐水化性 slaking resistance
耐水剂 waterproofing agent
耐水胶 marine glue; waterproof glue; water-resistant glue; wet-use adhesive
耐水胶合产品 water-borne adhesive product
耐水胶粘剂 moisture-resistant adhesive
耐水浸牢度 fastness to wear
耐水泥胶结度 fastness to cement
耐水黏[粘]合剂 moisture-resistant adhesive
耐水泡胀性 resistance to swelling by water
耐水漆 water-resistant paint
耐水清漆 spar varnish; waterproof varnish; water-resistant varnish
耐水砂纸 waterproof sand paper
耐水石膏板 water-resistant gypsum board
耐水石膏衬板 water-resistant gypsum backing board
耐水试验 water-resisting test
耐水添加剂 water-resisting additive
耐水涂料 water-resistant paint
耐水外用清漆 spar varnish
耐水桅杆清漆 water-resisting spar varnish
耐水线 waterproof wire
耐水芯 water-resistant core
耐水性 resistance to water; resistivity against water; water fastness; water resistance; water tolerance
耐水性胶 water-resistant adhesive
耐水性黏[粘]结剂 waterproof adhesive
耐水性试验 water tolerance test
耐水压的 water-tight
耐水压力试验方法 hydraulic test method
耐水压性 resistance to water pressure
耐水蒸气 water vapo(u)r resistance
耐水蒸气的 water vapo(u)r resistant
耐水纸面石膏板 water-resistant gypsum plaster board
耐水渍牢度 fastness to water spotting
耐丝光牢度 fastness to mercerizing
耐丝状腐蚀性 filiform corrosion resistance
耐撕裂性 resistance to tearing
耐四氧化二氮涂料 nitrogen tetroxide resistant coating
耐松节油试验 turpentine oil-resisting test
耐酸 acid heat; acid protection
耐酸泵 acid pump; acid-resisting pump
耐酸标准 acid-resisting standard
耐酸玻璃蛇管 acid-resisting glass coil
耐酸材料 acid-proof material; acid-resistant material; acid-resisting material
耐酸衬里 acid-proof lining; acid-proof trap; acid-resisting lining
耐酸衬料 acid-proof lining; acid-proof trap; acid-resisting lining
耐酸衬砌 acid lining
耐酸瓷漆 acid-proof enamel; acid-resisting enamel
耐酸瓷砖 acid-resisting ceramic tile
耐酸存水湾 acid-proof trap
耐酸的 acid-fast; acidophil(e); acidoresistant; acid-resisting; refractory; acid-proof; acid-resistant
耐酸地板 acid-proof floor; acid-resistant floor(ing)
耐酸地面 acid-proof floor; acid-resistant floor(ing)
耐酸电池箱 acid-proof battery box
耐酸电动机 acid-resistant motor
耐酸度 acid-fastness; acid-resistance
耐酸阀 acid-proof valve
耐酸凡立水 acid-fast varnish
耐酸防腐层 acid-proof coating
耐酸杆菌 acid-fast bacillus
耐酸缸瓷器 acid-proof stoneware
耐酸缸器 acid-proof stoneware
耐酸缸瓦器 acid-resisting stoneware
耐酸缸砖地面 acid-resistant tile flooring
耐酸钢 acid-resistant steel; acid-resisting steel
耐酸勾缝 acid-resistant gull(e)y
耐酸骨料 acid-resisting aggregate
耐酸管 acid-resistant pipe; line for acids
耐酸合金 acid-proof alloy; acid-resistance metal; acid-resisting alloy
耐酸黄铜 acid brass; admiralty brass
耐酸灰泥 acid-resisting mortar
耐酸挥发清漆 acid-proof spirit varnish
耐酸混凝土 acid fast concrete; acid-proof concrete; acid-resisting concrete; fast-fast concrete
耐酸集料 acid-resisting aggregate
耐酸碱的 acid and alkali-resistant
耐酸碱泥浆 acid and alkali-resistant grout
耐酸碱砂浆 acid and alkali-resistant mortar
耐酸碱水泥浆 acid and alkali-resistant grout
耐酸碱水泥砂浆 acid and alkali-resistant mortar
耐酸碱性 resistance to acid(attack)
耐酸浇注料 acid-resistant castable
耐酸胶合料 acid-fast mastic; acid-resistant mastic
耐酸胶结料 acid-resisting binder
耐酸胶泥 acid-proof cement
耐酸接缝 acid-fast joint
耐酸接头 acid-fast joint
耐酸金属 acid metal
耐酸浸蚀性 acid spot resistance
耐酸绝缘 acid-resistant insulation
耐酸菌 aciduric bacteria
耐酸沥青 acid-resisting asphalt
耐酸沥青黑清漆 acid-resisting black varnish
耐酸沥青面砖 acid-fast asphalt(ic) tile; acid-proof asphalt(ic) tile
耐酸沥青清漆 acid-proof asphalt varnish; acid-resisting bituminous varnish
耐酸硫磺砂浆 acid-proof sulphuric mortar
耐酸铝 alumite
耐酸率 acid-resistance rate
耐酸玛瑞脂 acid-fast mastic; acid-proof mastic; acid-resistant mastic
耐酸面层 acid-proof coating
耐酸面砖 acid-resisting tile
耐酸内衬 acid-proof lining
耐酸耐火材料 acid-proof refractory material
耐酸耐火混凝土 acid-fast refractory concrete; acid-resistant refractory concrete
耐酸耐火砖 acid-resistant refractory brick
耐酸耐温砖 acid-resistant thermotolerant ceramic brick
耐酸品种 acid tolerant species
耐酸漆 acid-fast paint; acid-proof paint; acid-resistant paint; acid-resisting paint; etching ink
耐酸铅 acid lead
耐酸青铜 acid bronze
耐酸清漆 acid-fast varnish; acid-proof varnish
耐酸软管 acid-proof hose
耐酸砂浆 acid-proof mortar; acid-resistant mortar; acid-resisting mortar
耐酸设备 acid-resistant system
耐酸炻器 acid-proof stoneware; chemical stoneware
耐酸试验 acid-proof test; acid-resistant test
耐酸试验后搪瓷表面侵蚀程度试验 blurring highlight test of vitreous enamel after acid-resistant test
耐酸试验器 acid tester
耐酸手套 acid-proof gloves
耐酸水玻璃砂浆 acid-proof water glass mortar
耐酸水泥 acid-proof cement; acid-resisting cement; anti-acid cement
耐酸水泥砂浆 acid-resistant cement mortar
耐酸水泥刷面 brush-on acid-resisting cement finish
耐酸缩绒牢度 fastness to planking
耐酸搪瓷 acid-proof enamel
耐酸陶瓷板 acid-resisting ceramic veneer
耐酸陶瓷泵 acid-resistant ceramic pump; acid-resistant stoneware pump
耐酸陶瓷阀门 acid-resistant stoneware valve
耐酸陶瓷鼓风机 acid-resistant stoneware blower
耐酸陶瓷管 acid-proof ceramic pipe; acid-resistant ceramic pipe
耐酸陶瓷管道 acid-resistant ceramic pipeline
耐酸陶瓷机械密封 anti-acid ceramic mechanical sealing
耐酸陶瓷矩鞍形填料 acid-resistant ceramic saddle packing
耐酸陶瓷喷射泵 acid-resistant ceramic jet pump
耐酸陶瓷容器 acid-resistant stoneware container
耐酸陶瓷砂浆泵 acid-resistant ceramic slurry pump
耐酸陶瓷塔 acid-resistant stoneware tower
耐酸陶瓷制品 acid-proof ceramic products; acid-resisting ceramic ware
耐酸陶管 acid-proofing ceramic pipe
耐酸陶器 acid-proof stoneware; acid-resisting stoneware; chemical stoneware
耐酸铜合金 acid-resistant copper alloy
耐酸涂层 acid-resistant coating; anti-acid coat(ing)
耐酸涂料 acid-fast paint; acid-proof coating; acid-proof paint; acid-resistant paint; acid-resisting paint
耐酸圬工 acid-fast masonry(work)
耐酸无碱玻璃纤维 alkali free chemical resistant glass fiber
耐酸细菌 acid-fast bacteria; acidophil(e) bacteria; acid-resisting bacteria; aciduric bacteria
耐酸纤维 oxytalan
耐酸纤维溶解 oxytalanolysis
耐酸性 acid-fastness; acid-resistance; acid resistivity; resistance to acid (attack)
耐酸性的 acid-fast
耐酸性能 acid-resistance property
耐酸性去污剂纤维 acid detergent fiber
耐酸性试验 acid-fastness test; acid-resistance test
耐酸烟囱 acid-fast chimney
耐酸印剂 acid-proof ink
耐酸油墨 acid-proof ink
耐酸油漆 anti-acidic paint
耐酸釉 acid-resistant glaze
耐酸釉面砖 acid-resisting glazed tile
耐酸纸 acid-proof paper
耐酸铸件 acid-proof casting; acid-resistance casting; acid-resisting casting
耐酸铸铁 acid-proof cast iron; Ariron
耐酸砖 acid brick; acid-proof brick; acid refractory brick; acid-resistant brick; acid-resisting brick; chemical brick
耐酸砖石工 acid-fast masonry(work)
耐碎片性 chipping resistance
耐损度 damage fastness; damage resistance
耐损害 refractory damage
耐损伤性 mar resistance
耐损试验 non-destructive test(ing)
耐损性 damage fastness; damage resistance
耐缩的 anti-shrink
耐炭化牢度 fastness to carbonizing
耐特高压润滑剂 hypoid lubricant
耐特压添加剂 extreme pressure additive
耐天气性 weather fastness; weather resistance
耐天然光牢度 fastness to daylight
耐土埋牢度 fastness to soil burial
耐脱胶牢度 fastness to degumming
耐唾沫性 resistance to spittle
耐外延性 external pressure resistance
耐弯曲性 flexing life; flexing resistance; resistance to bend(ing)
耐温 heat-resistant
耐温差性 resistance to temperature variation; temperature difference resistance
耐温度变化能力 resistance to temperature change
耐温度变化性 temperature change resistance
耐温度变化性试验 resistance to temperature change test
耐温极限 <材料的> thermal limit
耐温水泥 insulating cement
耐温性 temperature resistance
耐温性抗裂指数 index of anti-thermal cracking resistance
耐污能力 durability against pollution; tolerance pollution capacity
耐污染 fouling resistant
耐污染生物 pollution tolerant organism
耐污染物种 pollution tolerant species; tolerance species; tolerant species
耐污染性 stain resistance
耐污染植物 pollution tolerant plant
耐污生态 tolerance ecology

N

耐污型散热器 trash-resistant radiator
耐污性 dirt resistance; resistance to soiling
耐雾的 fog-type
耐雾绝缘子 fog-type insulator
耐洗的 washable; wash-fast; wash-proof; wash-wear
耐洗的水漆 calcarium
耐洗涤的白漆 washable white
耐洗涤的水性漆 washable water paint
耐洗涤剂腐蚀 detergent resistance
耐洗涤剂性 detergent resistance
耐洗涤性 resistance to washing
耐洗坚牢度 washing fastness
耐洗蓝 washing blue
耐洗牢度 fastness to washing
耐洗牢度试验仪 launderometer
耐洗刷试验机 wet abrasion tester
耐洗刷性 scrab scrub resistance
耐洗性 fastness to washing; washability; washing fastness
耐洗织物 wash goods
耐消蚀性 ablation resistance
耐硝基漆性 nitrocellulose lacquer resistance; resistance to nitrocellulose lacquer
耐性 proofness; tolerance
耐性等级 scale of tolerance
耐性定律 law of tolerance
耐性界限 limit of tolerance
耐性生态学 toleration ecology
耐锈钢 corrosion-resisting steel
耐锈力 rust resistance
耐锈(蚀)的 anti-rust
耐锈蚀钢 anti-rust steel
耐锈性 rust resistance
耐旋光性 light resistance
耐压 pressurization; withstanding pressure
耐压保护壳 pressure-proof protecting housing
耐压舱 pressure chamber
耐压舱壁 holding bulkhead; pressure bulkhead; strength bulkhead
耐压测试器 dielectric (al) strength tester
耐压层 pressure-proof layer
耐压的 overpressure resistant; pressure-proof
耐压电视摄影机 pressure-proof television camera
耐压垫块 crusher block
耐压拱 geostatic arch
耐压管 pressure-proof pipe
耐压管路 pressure line; pressure piping(-line)
耐压管线 pressure pipeline; pressure piping(-line)
耐压环 pressure ring
耐压壳(体) inner hull; inner shell; pressure hull
耐压力 resistance to pressure
耐压力斑点性 pressure mottling resistance
耐压密封的 pressure-tight
耐压木材 stress grade of timber
耐压瓶 pressure bottle
耐压器皿 pressure ware
耐压强度 breakdown field strength; compression strength; compressive resistance; compressive strength; crushing strength; resistance to compression
耐压容器 pressure container; pressure tank; pressure vessel
耐压软管 pressure hose
耐压烧瓶 pressure flask
耐压升降装置 pressure hoisting gear
耐压试验 air-tight pressure test; breakdown test; compression test; disruption test; pressure test(ing); pressure-tight test; puncture test; withstand test; high-voltage holding test【电】
耐压试验变压器 transformer for withstand voltage test
耐压试验电压 withstand test voltage
耐压试验机 compression tester
耐压水舱 pressure tank
耐压碎性 resistance to crushing
耐压艇体结构 pressure-hull structure
耐压外壳 pressure casing; pressure-proof housing
耐压吸引胶管 pressure and suction hose
耐压系统 pressurized system
耐压细菌 baroduric bacteria
耐压橡胶管 pressure rubber pipe; pressure rubber piping; pressure rubber tubing
耐压性 pressure resistance; resistance to pressure
耐压油箱 pressure-proof tank
耐烟气牢度 fastness to gas fumes
耐烟树 smoke enduring plant
耐烟头点燃性 cigarette ignition resistance
耐烟雾瓷漆 fume-proof enamel
耐烟雾漆 fume-proof paint
耐烟雾性 fume resistance
耐盐的 haloduric; salt resisting
耐盐度 salt endurance; salt tolerance; tolerance of salinity
耐盐碱植物 saline-alkaline tolerant plant
耐盐绝缘子 salt-resistive insulator
耐盐蚀楼面 salt-resistant flooring
耐盐水试验 salt-water resistant test
耐盐水性 salt-water resistance
耐盐酸镍基合金 Hastelloy
耐盐污泥 slat-tolerant sludge
耐盐雾性 salt fog resistance
耐盐性 resistance to salt; salinity tolerance; salt endurance; salt tolerance
耐盐性的 salt-tolerant
耐盐性能 salt-resistant property
耐盐植被 salt-resistant vegetation
耐盐植物 salt-tolerant plant
耐盐种 salt-avoiding species; salt-tolerant species
耐焰性 flame proofness
耐阳光的 sunlight-proof; sunlight resistant
耐阳光降解性 resistance degradation by sunlight
耐氧的 aerotolerant; oxytolerant
耐氧化氮牢度 fastness to nitrogen oxides
耐氧化能力 oxidation resistance
耐氧化碳碳复合材料 oxidation-resistant carbon-carbon composite
耐氧化油 passivated oil
耐氧菌 aerotorelant bacteria
耐氧量 oxygen tolerance
耐氧细菌 aerotolerant bacteria
耐药量 medicinal tolerance; tolerated dose
耐药量试验 tolerance test
耐药性 drug-fastness; drug resistance; drug tolerance
耐药中浓度 median tolerant limit
耐液压冲击性 resistance to hydraulic shock
耐阴性 rain fastness; shade bears; shade endurance; shade tolerance
耐荫树 tolerant tree
耐荫性 tolerance of shade
耐荫植物 shade plant; shade-tolerant plant; tolerant plant
耐印痕性 print resistance
耐印力 pressrun
耐应力碎裂性 resistance to stress crazing
耐用的 durable; long-lasting; long lived; robust; serviceable; time-proof; heavy-duty
耐用灯 rough service lamp
耐用度 abrasion resistance; endurance; incorruptibility
耐用堆焊 hard-facing welding
耐用机械 long life machine
耐用极限 endurance limit
耐用极限状态 limit state of durability
耐用胶粘剂 boil resistant adhesive
耐用路面 long-lasting pavement
耐用率 serviceability ratio
耐用年数 longevity
耐用年限 durability; durable years; endurance life; period of depreciation; required length of life; serviceable life; service life
耐用年限表 life table
耐用品 durable goods; durables
耐用品价格 durable goods price
耐用品新订货量 new orders for durable goods
耐用期 durability; durability period; life duration
耐用商品 durability goods; hard goods
耐用设备 durable facility; long lived facility
耐用生产资料 producers' durable goods
耐用寿命特性 endurance life characteristic
耐用条款 durable clause
耐用涂料 maintenance finish
耐用消费品 consumer durable goods; consumer durables; durable consumer goods; hard goods
耐用性 durability; durableness; endurance; maintainability; robustness; ruggedness; serviceability; service durability; viability; wearing property
耐用性评定 serviceability rating
耐用性趋向 <即耐用性与旋荷次数的曲线> serviceability trend
耐用性试验 endurance test; life test(ing); serviceability test
耐用性指数 durability index; serviceability index
耐用油 passivated oil
耐油 grease-proofness; oil-proofing; oil-proofness
耐油的 fuel oil-proof; fuel oil-resistant; oil-proof; oil-resistant; oil-resisting
耐油地面 oil-protecting floor
耐油垫圈 oil seat gasket
耐油混凝土 oil-proof concrete; oil-resisting concrete
耐油砂浆 oil-proof mortar
耐油石棉橡胶接合垫片 oil-proof asbestos rubber joint sheet
耐油(食品)包装纸涂料 grease-proof paper coating
耐油试验 oil-proof test; oil-resisting test
耐油酸手套 oil acid free plastic gloves
耐油搪瓷 oil-proof enamel
耐油涂层 oil-resistant coating
耐油橡胶 oil-proof rubber; oil-resistant rubber
耐油橡胶管 oil-resistant rubber hose
耐油性 fastness to oil; oil fastness; oil resistance; oil resistivity
耐油性涂料 grease-proof paint
耐油性油墨和涂料 grease-proof ink and coating
耐油脂涂料 grease resistant coating
耐油脂性(能) grease resistance; oil and grease resistance
耐油纸 glassine; pergamyn paper
耐有机溶剂牢度 fastness to organic solvent
耐有机溶剂性 resistance to organic solvent
耐雨淋牢度 fastness to rain
耐雨性 rain fastness
耐再涂性 fastness to overvanishing
耐脏 soil and oil repellent
耐脏的 dirtproof
耐皂化性 saponification resistance
耐皂化作用的 resistant to saponification
耐皂洗牢度 fastness to soaping
耐增塑剂性 plasticizer resistance
耐张绝缘子 dead-end insulator; strain insulator; tension insulator
耐张力铁塔 pylon tower
耐张塔杆 strain pole
耐张线夹 dead-end clamp; strain clamp; strain cleat
耐张支持物 anchor support
耐胀绝缘子 dead-end insulator; strain insulator
耐胀线夹 strain clamp
耐折处理剂 crease resistant finish
耐折度 folding strength
耐折强度 folding strength
耐真菌的 funginert
耐真空性 vacuum resistance
耐振的 shakeproof; vibration-proof
耐振灯泡 shock-proof lamp; vibration service lamp
耐振地基 vibration-proof foundation
耐振电气设备 shock-proof electric(al) apparatus
耐振荡 resistance to shock
耐振机器 ruggedized machine
耐震 seismic restraint
耐震玻璃 safety glass; shatter-proof glass
耐震玻璃门 shatter-proof glass door
耐震玻璃片 shatter-proof sheet glass
耐震窗玻璃 shatter-proof window glass
耐震的 aseismic; earthquake resistant; shatter-proof; shock-proof; tremostable
耐震灯泡 mill-type lamp; rough service lamp
耐震电灯 rough service lamp
耐震电气设备 shock-proof electric(al) apparatus
耐震电子管 mobile tube
耐震隔热玻璃 shatter-proof insulating glass
耐震管 ruggedized tube
耐震计算机 ruggedized computer; ruggedized machine
耐震建筑 earthquake resistant construction
耐震建筑物 hazard-resistant building
耐震接头 earthquake-proof joint
耐震结构 aseismatic structure; earthquake resistant structure; quake-proof structure; ruggedized construction
耐震平板玻璃 shatter-proof plate glass
耐震强度 shock resistance
耐震设计 aseismatic design; aseismic design; earthquake-proof design
耐震设计规范 earthquake resistant design code
耐震水准仪 shock-proof level
耐震台座 shock-proof mount(ing)

耐震性 aseismicity;resistance to shock
耐震性能 earthquake resistant behavio(u)r
耐震性试验 vibration test(ing)
耐蒸度 fastness to decatizing
耐蒸汽的 vapo(u)r-resistant
耐植物生根性 resistance to plant roots
耐重力 anti-gravitation;anti-gravity
耐周期位移性 resistance to cycle movement
耐周围条件性 resistance to environment
耐皱 anti-wrinkling
耐皱加工牢度 fastness to anti-crease processing
耐皱涂层 crease resistant finish
耐煮的 boil-proof
耐煮度 fastness to boiling
耐煮沸的 coctostabile
耐煮性 boiling fastness
耐贮藏性 keeping quality
耐贮性 bin stability
耐转移性 migration characteristics
耐撞性 impact strength
耐紫外线变色性 resistance to colo(u)r change by ultraviolet;resistance to ultraviolet discoloration
耐紫外线性 resistance to ultraviolet
耐自燃腐朽性 natural decay resistance

萘【化】albocarbon;naphthale(i)ne

萘胺 naphthylamine
萘胺橙 naphthylamine orange
萘胺黑 naphthylamine black
萘胺蓝 naphthamine blue
萘胺亮枣红 naphthamine brilliant Bordeaux
萘胺浅蓝 naphthamine light blue
萘胺正蓝 naphthamine pure blue
萘胺中毒 naphthylamine poisoning
萘二胺 naphthalene diamine
萘分解菌 naphthalene oxidizing bacteria
萘酚 naphthol;naphtholum
萘酚铋 bismuth naphtholate
萘酚黄 naphthol yellow
萘酚蓝黑 naphthol blue black
萘酚类颜料 naphthols pigment
萘酚染料 naphthol dyestuff
萘酚色淀颜料 naphthol lake colo(u)r
萘酚盐 naphtholate
萘酚中毒 naphtholism
萘铬绿 naphthochrome green
萘含量 naphthalene content
萘黑 naphthalene black
萘黄酮 naphthoflavon
萘磺酸 naphthalene sulfonic acid
萘磺酸盐 naphthalene sulfonate
萘基 naphthyl
萘甲酰苯胺颜料 naphthanilide pigment
萘醌 naphthoquinone
萘球 naphthalene ball
萘醛树脂 naphthalene resin
萘酸 naphthoic acid
萘羧酸 naphthalene carboxylic acid
萘丸 naphthalene ball
萘烷 decahydronaphthalene
萘烷醇 decalol
萘衍生物 naphthalene derivative
萘乙酸 naphthlacetic acid
萘乙酸胺 naphthalene acetamide
萘油 naphthalene oil
萘状断口 naphthalene fracture

男舱室 berth for male
男厕所 gentlemen's room;male rest room;male toilet;men toilet
男更衣室 dressing room for man;men's changing room
男盥洗室 <公共建筑中的> men's room;men toilet
男会员 male member
男人卧室壁橱 bedroom closet for men
男卫生间 men's room
男像柱 atlantes;Persian;telamon[复 telamones]
男性女像柱 male caryatid
男性人头狮身像 androsphinx
男休息室 male rest room
男修道院 friary
男修道院宿舍 dorter[dortour]
男衣帽间 men's changing room
男用房间 <古希腊> andron
男子公寓 andron

南奥克尼海槽【地】south Orkney trough

南澳大利亚海盆 south Australian basin
南澳洲 South Australia
南半球 southern hemisphere;water hemisphere
南半球春季 austral spring;southern spring
南半球冬季 austral winter;southern winter
南半球冬季季节区带 southern winter seasonal zone
南半球副热带无风带 calms of Capricorn;calm zone of Capricorn
南半球高纬度地区 high southern latitudes
南半球植物地理区系 south Hemispheric floral realm
南半天球 southern celestial hemisphere
南北不对称性 north-south asymmetry;south-north asymmetry
南北地洼区 north-south Diwa region
南北地震带 the south-north seismotectonic zone
南北分量 north-south component
南北构造带 north-south structural zone
南北构造系 north-south tectonic system
南北航线 south-north route
南北回归线之间地带内的任何地区 intertropics
南北贸易 north-south trade;vertical trade
南北谈判 north-south negotiation
南北线 meridian line
南冰洋 the Antarctic Ocean
南部潮湿亚热-热带 southern China humid subtropical to tropical zone
南部大洋中脊 Mid oceanic ridge of Atlantic Ocean
南部地区的山毛榉 southland beech
南部非洲 Southern Africa
南部分等(货运) <美> Southern classification
南部海域 Southern sea area
南部台坳 southern platform depression
南侧廊 south aisle
南朝鲜海槽 south Korean marine trough
南朝鲜沿海平原 south Korean coastal plain
南赤道海流 equatorial current;south equatorial current
南窗 austral window;southern window
南磁极 south magnetic pole
南磁极位置 south magnetic pole position
南磁极性 southern polarity
南磁倾角极 south magnetic dip pole
南达科他 <美国州名> South Dakota
南大巴台褶带 Southern Daba platform folded belt
南大风 souther
南大距 southern elongation
南大西洋 South Atlantic ocean
南大西洋高压 South Atlantic high (pressure)
南大西洋海流 South Atlantic current
南大西洋航行警告 navigational warning for South Atlantic
南大西洋间断分布 South Atlantic disjunction
南大西洋深层和底层水 South Atlantic deep and bottom water
南大西洋双壳类地理亚区 South Atlantic bivalve subprovince
南大西洋西伯利亚波系 South Atlantic-Siberia crustal-wave system
南大西洋中脊 southern mid-Atlantic ridge
南大洋 Great Southern Ocean;southern ocean
南丹海沟 Nandan trench
南地(理磁)极 south geographic(al) pole
南点 south point
南东 southeast[SE]
南东东 east southeast
南东南 south southeast
南方 south
南方白扁柏 southern white cedar
南方贝壳杉 Cowrie
南方采光的锯齿形屋顶 <南半球> south-light roof
南方赤栎 Southern red oak
南方的 austral;meridional;southward;southern
南方阀动装置 southern valve gear
南方黑钙土 southern chernozem
南方黄松 southern yellow pine
南方假山毛榉 southern falsebeech
南方区 austral region
南方区系 Alvinokaffric realm
南方散白蚁 Reticulitermes virginicus
南方生物带 austral
南方双壳类地理区 austral bivalve province
南方涛动 southern oscillation
南方腕足动物地理区系 Malvinoakffric brachiopod realm
南方位标 south cardinal mark
南方西洋海流 South Atlantic current
南方香脂冷杉 southern balsam fir
南方型半干旱亚热带 semi-arid subtropical zone of southern china type
南方(油)松 Southern(pitch)pine
南方针叶林带 Austroriparian life zone
南非尺 cape foot
南非航线 South-Africa shipping line
南非红宝石 cape ruby
南非金刚石 cape diamond
南非罗汉松 bestard yellow wood
南非玫瑰木 African rosewood;South African rose-wood
南非石棉 cape asbestos;cape blue asbestos
南非之星 Star of South Africa
南非洲建筑 South Africa architecture
南风 auster;southerly;south wind
南格南 <一种青铜> Nongran
南拱廊 <古希腊式> South stoa
南瓜 China squash;pumpkin;Spanish gourd
南海 South China Sea
南海沉降海盆 <中国> South China Sea subsiding basin
南海地台 <中国> South China Sea platform
南海海槽 <中国> South China sea trough
南海海盆张裂系 Tensile fracture zone in South China Sea Basin
南海中央海盆隆起地带 Central sea basin uplift of South China Sea
南寒带 Antarctic zone;south frigid zone
南寒带动物区 anti-boreal faunal region
南寒风 <澳大利亚东南沿岸夏天的偏南阵风> southerly burster;burster
南赫布里底海沟 south Hebrides trench
南回归线 the Tropic(al)of Capricorn
南回归线无风带 calm of Capricorn
南极 Antarctic Pole;South Pole
南极冰源 Antarctic ice sheet
南极臭氧层空间 Antarctic ozone hole
南极春季臭氧消耗 Antarctic spring time ozone depletion
南极大陆 Antarctica;Antarctic Continent
南极大陆冰川 Antarctic inlandsis
南极带 south frigid zone
南极地带 Antarctic zone;Antarctica
南极地带的 Antarctic
南极地区 Antarctic region;Antarctica;Antarctogea
南极冬季平流层绕极涡旋 winter Antarctic stratospheric circumpolar vortex
南极动物区 Antarctic faunal region
南极反气旋 Antarctic anti-cyclone
南极锋 Antarctic front
南极辐合带 Antarctic convergence;Southern Polar Front
南极光 aurora australis;australis aurora;southern lights
南极光带 southern auroral zone
南极汇聚 Antarctic convergence
南极距 south polar distance
南极考察船 south pole exploration ship
南极冷流 Antarctic west wind drift
南极平流层绕极涡旋 Antarctic stratospheric circumpolar vortex
南极企鹅 Antarctic penguin
南极区 Antarctic zone;southern polar region
南极圈 Antarctica;Antarctic Circle;south polar circle
南极(圈)气团 Antarctic air
南极绕极(海)流 Antarctic circum polar current;west wind drift
南极乳白天空 Antarctic whiteout
南极生态系统 Antarctic ecosystem
南极石 antarcticite
南极水域 Antarctic waters
南极探险船 Antarctic exploration ship
南极条约 Antarctic treaty
南极西风洋流 Antarctic west wind drift
南极星座 Octans
南极性 southern polarity
南极研究科学委员会 Scientific Committee on Antarctic Research
南极印度洋海盆 Indian Antarctic basin
南极源水 Antarctic source water
南极植物地理区系 Antarctic floral realm
南极植物区 Antarctic region
南极制图 Antarctic mapping
南极中层水 Antarctic intermediate

water
南极洲 Antarctica;Antarctic Continent
南极洲板块 Antarctic plate
南极洲地台 Antarctic platform
南极洲界 Antarctic realm
南交叉甬道<教堂的> southern transept
南郊 southern suburbs
南界标 south cardinal mark
南界的 Notogaean
南界限 southern limit
南进 southing
南京长江大桥 Yangtze River Bridge, Nanjing
南京椴 miquel linden
南京剪纸 Nanjing paper cuts
南京土布 Nankeen
南京云锦 Nanjing brocade
南卡罗来纳<美国州名> South Carolina
南立面 south elevation
南岭楝树 ceylan cedar
南岭石 nanlingite
南流河 south-going river;south-going stream
南鲈 Nandus nandus
南美板块 South American plate
南美草<产于南美,有丝光,可用于装饰的> campas grass
南美大草原<尤指阿根廷的> pampa
南美大陆与非洲大陆分离 South America-Africa separation
南美地台 south American platform
南美地震学区域中心 regional center [centre] of seismology for South America
南美航线 South-America shipping line
南美红木 tulipwood
南美槐属 balsam wood
南美杉 Chile pine
南美杉属 podocarpus pine;Araucaria<拉>
南美亚马逊河口暴涨潮 Amazon Bore;pororoca
南美洲大草原 lano
南美洲建筑 South American architecture
南门廊 southern portio;south porch
南蒙古深断裂系 Southern Mongolian deep fracture zone
南面的 austral
南南东 south southeast
南南合作 South-South cooperation
南南会议 South-South conference
南南西 south southwest
南欧海松 clusted pine
南欧黑松 Austrian pine;larch pine
南欧紫荆 Judas tree
南偏东 south by east
南偏西 south by west
南热带扰动 south tropical disturbance
南森采水瓶 Nansen bottle
南森集水器 Nansen bottle
南森炼锌法 Nansen zinc process
南森瓶 Nansen bottle;Petterson-Nansen water bottle;reversing water bottle
南闪石岩 hornblendite
南设德兰海槽 south Shetland trough
南水北调 south-north water transfer;south water to north
南斯拉夫船舶登记局 Judgement Jugoslav Resister of Shipping
南斯拉夫船级社 Jugoslav Resister;Yugoslavian Ship Classification Society
南斯拉夫建筑 Yugoslav architecture;Yugoslavian architecture
南太平洋 South Pacific Ocean
南太平洋常设委员会 Permanent Committee for the South Pacific
南太平洋海流 South Pacific current
南太平洋海盆 South Pacific basin
南太平洋航线 South Pacific shipping line
南太平洋红木 red lauan;South Pacific mahogany
南太平洋间断分布 South Pacific disjunction
南太平洋区域环境方案 South Pacific Regional Environmental Program(me)
南唐二陵 Tombs of the Emperors of Southern Tang dynasty
南涛指数 Southern oscillation index
南特港<法国> Port Nantes
南特提斯海槽 south Tethyan trough
南天参考星 southern reference star
南天恒星 Southern Star
南天极 south celestial pole
南天星系 southern galaxy
南天竹 Nandina domestica;sacred bamboo;heavenly bamboo
南通港 Port of Nantong
南图廓(线) lower border;bottom border;bottom edge;bottom margin;lower margin;southerly limit
南纬 latitude south;southern latitude;south latitude
南纬差 southing
南温带 south temperate zone;south temperature zone
南温带的 south temperate
南温带双壳类地理区系 south temperate bivalve realm
南西南 south southwest
南喜马拉雅地洼区 Southern Himalayan Diwa region
南下河流 south-going river;south-going stream
南向 south orientation
南向采光锯齿形天窗 south-light roof
南向窗 austral window
南向门柱<古建筑的> antium
南向坡 southern slope
南向阳台 south-facing balcony
南星 Southern Star;Star of the South
南行的 south bound
南行航程 southing
南行列车 southbound train
南亚和远东电缆 South Asia and Far East cable
南亚松 Merkus pine
南洋材 south sea timber
南洋柳安 bagtikan
南洋杉 Araucaria;monkey puzzle;Norfolk-island-pine
南洋香椿 Spanish cedar
南印度风格 South Indian style
南印度洋海流 South Indian current
南印度洋海盆 south Australian basin
南中【天】culmination
南中国海 South China Sea
南中国海海盆张裂区【地】Extensional fracture region of central South China Sea basin
南中国海盆 South China basin
南中天 southing
南重 south heavy
南轴极 austral axis pole
南柱廊<古希腊式> South stoa
南走廊 south aisle;south porch

难保无误 undercorrection

难变形区 stagnant zone
难操纵 unhandy;unwieldy
难操纵船 clumsy ship
难超越的 insurmountable
难沉颗粒 difficult-to-settle particle
难冲刷的浅滩段 shoal stretch hard to be scoured
难抽 hard-pumped
难除掉杂草 troublesome weed
难除物 refractory
难除杂质 troublesome impurity
难处理的 unhandy;unmanageable
难处理的非线性 intractable nonlinearity
难处理地基 difficult foundation
难处理流沙 treacherous quicksand
难处理土 troublesome soil
难船抽水 pumping-out water from a distressed vessel
难船固位 positioning a stranded vessel to prevent shifting
难船救助船 salvor
难船救助人员 salvor
难船抢险 beaching of a distressed vessel
难磁化方向 hard direction
难达到的 unattainable;ungetatable
难得数据 hard-to-get data
难得文献 hard to get titles
难得文献资料 hard-to-obtain materials
难点 bottleneck;crux[复 cruxes/cruces];hard-point;knotty point
难读地图 faint map;recondite map
难度 degree of difficulty;difficulty
难度测验 power test
难分解有机物 refractory organics
难干集料 hard to dry aggregate
难根除的杂草 troublesome weed
难估量的 inestimable
难关 bottleneck
难灌(筑的)混凝土 unworkable concrete
难混汞的金 rusty gold
难混熔金属 immiscible metal
难加工 intractable
难加工材料 difficult-to-machine material;hard-to-cut material
难加工的木材 refractory timber
难加工钢材 low machinability steel
难加工合金 difficult-to-cut alloy
难加工性 unworkability;unworkableness
难驾驭的自然环境 unmanaged environment
难驾驭 impracticability
难降解 refractory degradation
难降解废水 refractory wastewater
难降解化工有机废水 chemical industrial organic refractory wastewater
难降解化合物 recalcitrant compound
难降解挥发性固体 refractory volatile solid
难降解挥发性有机物 refractory volatile organic matter
难降解离子化合物 refractory ionic compound
难降解污染物 refractory pollutant
难降解物质 non-degradable substance
难降解有机废水 hard-degradable organic wastewater;refractory organic wastewater
难降解有机废水处理 refractory organic wastewater treatment
难降解有机化合物 hard-degradable organic compound
难降解有机微污染物 refractory organic micro-pollutant
难降解有机污泥 refractory organic silt
难降解有机污染物 persistent organic pollutant
难降解有机物 refractory organic matter;refractory organics
难浇筑混凝土 unworkable concrete
难接近 inaccessibility
难接近的 inapproachable
难接近地区 inaccessible region
难接受的 unacceptable
难解 hardness;knotty
难解的 abstruse;obscure;transcendental
难解性 intractability
难开采的油井 bear cat
难刻符号 hard-to-scribble symbol
难控性 difficulty of control
难控制的 intractable;uncontrollable;ungovernable;unruly;unwieldy
难控制性 difficulty to control;resistance to control
难理解的 unintelligible
难冒落顶板 difficultly falling roof
难民 displaced persons
难民和灾民 refugee and evacuee
难民收容所 collecting station;house of refugee;refugee shelter
难民营 refugee camp
难明故障 gremlin
难凝炸药 low-freezing explosive
难弄的 tricky
难劈向 hardway
难破面 hardway
难起动 hard-to-start
难迁移的元素 elements of difficult migration
难切断 hard-to-break
难切削钢材 low machinability steel
难区别的 indistinguishable
难确定的洪水灾害 intangible flood damage
难燃材料 hard inflammable material;non-flammable material
难燃处理 fire-retardant treatment
难燃的 flame-resisting
难燃建筑 slow-burning construction
难燃垃圾 hard-to-burn-refuse;refuse difficult to burn
难燃烧的 incombustible
难燃烧体 hard-combustible component;non-flammable material
难燃物 incombustible
难溶白云石 high-sintering dolomite
难溶的 indissolvable;insoluble
难溶建筑材料 infusible material
难溶解 difficult soluble
难溶(解)的 indissoluble
难溶素 dyslysin
难溶性化合物 difficult soluble compound
难溶盐含量 content of bad-soluble salt;content of difficultly soluble salts
难溶盐试验 bad-soluble salt test;slightly soluble salt test
难熔玻璃 high melting glass
难熔的 high heat;high melt(ing);infusible;refractory
难熔合金 refractory alloy
难熔化合物 infusible compound;refractory compound
难熔灰分 difficult-fusible ash
难熔金属 refractory metal
难熔金属电接触器材 refractory metal contacts
难熔金属复合材料 refractory metal base composite material
难熔金属合金 refractory metal alloy
难熔金属烧结法 sintered refractory metal process
难熔金属碳化物 refractory carbide;refractory metal carbide
难熔矿石 refractory mineral;refractory ore
难熔矿渣 refractory slag

难熔沥青 wetherilite
难熔稀有金属 rare refractory metal
难熔性 infusibility;refractory behavio(u)r;refractory quality
难熔元素 refractory element
难筛材料 difficult-to-screen material
难筛颗粒 near mesh particle
难筛粒 difficult particle of screening
难烧结 hard-to-sinter
难烧生料 hard burning mix
难生化降解有机废水 refractory biochemical degradation organic wastewater
难生物降解物质 recalcitrant substance
难生物降解洗涤剂 biologically hard detergent
难生物降解有机污染物 refractory biodegradation organic pollutant
难使用的 unwieldy
难事 crux[复 cruxes/cruces]
难塑造的 inductile
难题 crux[复 cruxes/cruces]; crux problem; knotty problem; poser; puzzle; quiz; thorny problem; thorny subject;tickler;troublesome problem
难题地段 problem section
难题地区 problem area
难调节的 ungovernable
难通行地区 difficult country;difficult ground
难闻的 frowy
难闻的气味 disagreeable odo(u)r; nasty smell;niff;natal smell
难下定义 baffle definition
难相信的 unbelievable
难消化 indigestibility
难销货物 drug
难信任的 unreliable
难行车 bad runner;poor runner;slow car rolling car; slow runner; hard rolling car;poor rolling car【铁】
难行路 poor road
难行路段 difficult section; difficult stretch of road
难行线 <编组场> hard rolling track; hard running track; poor rolling track;poor running track;bad running track
难选 hard washing
难选金 refractory gold
难选矿石 refractory ore
难选硫化物 refractory sulfide
难压实的 incompactible
难氧化的 dysoxidizable
难移的 unalterable
难以操作的 unworkable
难以达到的 hard-to-reach
难以防除杂草 tolerant grasses
难以负担 ill afford
难以改正的 irremediable
难以估量的 inestimable
难以估量的价值 inestimable value
难以管理的 unmanageable
难以归类的 nondescript
难以过筛的材料 hard-to-screen material
难以加工的 harsh working; unmanageable
难以接近 out of(the)reach
难以纠正的 incorrigible
难以觉察的 imperceptible
难以控制的 unmanageable
难以理解的 impalpable;inapprehensible
难以取得的数据 hard-to-get data
难以确定的风险 borderline risk
难以确定的因素 intangible factor
难以数清的 incalculable
难以提供 ill afford
难以通过的障碍物 blank-out wall
难以通行的道路 heavy-going road
难以维持温饱生活的工资 starvation wages
难以行得通的 unfeasible
难以形容 baffle description
难以制造的机械 hard-to-manufacture machinery
难以转为现金的资产 inliquidity
难以追索的产权变迁 wild interest
难应付的 unmanageable
难釉 hard axis
难于查点的人群 hard to enumerate group
难于处理的 unworkable
难于机械加工的 hard to machine
难于凝聚的气体 fixed gas
难于疏干的矿床 ore deposits of difficult drainage
难于通信[讯] difficult communication
难于通行的 heavy-going
难于通行的道路 heavy-going road
难于通行的地带 heavy ground
难凿岩石 hard-to-cut rock
难轴 hard axis
难捉摸异常 subtle anomaly
难着色 chromophobe
难钻地层 bad ground
难钻进的硬岩石 hard-to-drill
难钻岩石 resistant rock

楠木 nanmu

楠木油 Machilus oil

囊包式活动支座 encapsulated expansion bearing

囊孢壳属 <拉> Physalospora
囊袋 sac
囊腐 <木材缺陷> pocket rot
囊护泥浆 encapsulating mud
囊孔菌属 <拉> Hirschioporus
囊式防喷器 annular preventer
囊式空气弹簧 bellows-type air spring
囊式密度计 balloon densi(to)meter
囊式体积仪 balloon volumeter
囊式压力计 capsule-type pressure ga(u)ge
囊脱石 nontronite
囊形膜板 bellowphragm;bellows diaphragm
囊状 scrotiform
囊状点蚀 encapsulated pitting
囊状风化壳 sack-like residuum
囊状腐朽 white pocket
囊状构造 sack-like structure
囊状空气弹簧 bladder type pneumatic spring
囊状矿体 pockety ore body;sack-like body
囊状矿物 chambered vein
囊状黏[粘]土 pocket clay
囊状体 cystidium[复 cystidia]
囊状物 bladder
囊状云 sack-cloud;sack of coals

挠棒 tommy bar

挠臂锤式振打机 tumbling hammer type rapper
挠点线 torsal line
挠度 amount of deflection; deflection【物】; arching; buckling; degree of flexibility; flexivity; hogging; sag; swag
挠度测定 deflection measurement
挠度测量 deflection survey
挠度测量仪 deflection indicator
挠度传感器 deflection transducer
挠度等值 deflection contour
挠度等值线 deflection contour
挠度法 flexibility method; method of deflection
挠度分量 component of deflection
挠度感应器 deflection sensor
挠度公式 deflection formula
挠度观测 deflection observation
挠度规定 deflection criterion
挠度过大 undue deflection
挠度横向分布 transverse distribution of deflection
挠度计 deflection ga(u)ge;deflection indicator; fleximeter [flexometer]; flexure meter;deflectometer <测量梁受弯时挠曲量的装置>
挠度矩阵 flexibility matrix
挠度控制 control of deflection
挠度理论 deflection theory
挠度器 flexometer
挠度曲线 deflection curve; deflexion curve;sag curve
挠度试验 deflection test; flexibility test;transverse test
挠度试样 transverse test-piece
挠度数 flexibility number
挠度调节 sag adjustment
挠度图 deflection diagram; deflectogram
挠度系数 coefficient of deflection; flexibility factor
挠度线 line of deflection
挠度协调方程式 equation of compatibility of deflection
挠度延性系数 deflection ductility (factor)
挠度仪 flexometer;flexural meter
挠度应力 flexural stress
挠度与时间关系曲线 load-time deflection curve
挠度增量 deflection increment
挠杆式天线 turnstile antenna
挠钢 bend bar
挠管 bend pipe
挠矩 moment flexure;moment of deflection
挠矩面积 area moment;moment area
挠矩曲线 moment curve
挠矩试验 moment test
挠矩图 bending diagram;bending moment diagram
挠矩轴 moment axis
挠裂 flex crack(ing)
挠率 torsion
挠率半径 radius of torsion
挠模 torsion module
挠偏转 bending deflection
挠平行性 parataxy
挠切点 torsal point;torsional point
挠切面 torsal plane
挠曲 bewel;buckle;deflection;deflexion; flexure; torture; twist; warp(ing)
挠曲 O 形圈注射配件 flexible O-ring injection fittings
挠曲半径 bending radius
挠曲比 deflection ratio
挠曲变位 bending deflection
挠曲变形 flexural deflection; flexural deformation;hogg(ing)deformation
挠曲测量计 deflection meter
挠曲测量仪 deflection ga(u)ge
挠曲常数 flexural constant
挠曲沉降 distortion settlement
挠曲导致的变形性 deformability due to bending
挠曲的 flexural;thrawn
挠曲的混凝土板 flexing concrete slab
挠曲的指针 sagitta
挠曲地震 warping earthquake
挠曲点 bending point; point of rupture
挠曲度 bending deflection; flexibility factor;warpage
挠曲缝 flexural gap
挠曲附加力 additional flexural force; additional flexural load
挠曲刚度 flexible rigidity;warping rigidity
挠曲钢键 deflected tendon
挠曲钢筋 bent bar; bent reinforcement;bent steel
挠曲钢筋束工艺 raised-cable technique
挠曲公式 flexing formula;flexion formula;flexure formula
挠曲荷载 flection load;flexural load
挠曲滑动 flexural slip
挠曲滑动褶皱【地】flexure-slip fold
挠曲毁坏 bending failure
挠曲机 bender;flexing machine
挠曲极限 flection limit;flexing limit; flexure limit; ultimate flexural capacity
挠曲计 flexion spring; flexometer; flexure meter
挠曲尖轨转辙器 points with flexible tongue
挠曲角 angle of bend;bending angle
挠曲劲度 flex stiffness;flexural stiffness
挠曲理论 deflection theory;theory of flexure
挠曲力 bending force;flexural force; force of flexion
挠曲力矩 bending moment; flexural moment; hog(ging) moment; moment of deflection;moment of flexure
挠曲力矩图 bending moment diagram
挠曲梁 flexural beam
挠曲量 amount of bend(ing)
挠曲裂缝 flex crack(ing); flexural cracking
挠曲流动褶皱【地】flexure-flow fold
挠曲面 deflection surface; plane of flexure; torsional plane; warped surface
挠曲面积 <原位与挠曲线之间的面积> deflected area;deflective area
挠曲模量 flexural modulus; modulus of rupture
挠曲磨损 flex abrasion
挠曲木材 bent wood
挠曲扭转 flexure torsion
挠曲疲劳 bending fatigue;endurance in flexure;fatigue in flexure;flex fatigue; flexural endurance; flexural fatigue
挠曲偏斜 bending deflection
挠曲破坏 bending failure
挠曲破坏理论 theory on failure by bending
挠曲强度 cross-breaking strength;flexing strength;flexural rigidity;flexural strength; modulus of rupture; strength in bending; transverse strength
挠曲曲线半径 deflexion curve radius
挠曲试验 bend(ing)test;flexing test; flexion test; flexural test; flexure test
挠曲试验机 flexer;flexing machine; flexure meter; flexure test machine
挠曲寿命 flexible life;flex(ing)life
挠曲受力筋 moment flexure;moment

of deflection
挠曲说 buckling hypothesis
挠曲速率 rate of deflection
挠曲弹性 bending elasticity;elasticity of flexure;flexing elasticity
挠曲温度 deflection temperature
挠曲稳定 deflection stability
挠曲稳定性 bending stability
挠曲握裹力 flexural bond
挠曲系数 bending coefficient;flexibility coefficient; flexibility factor; torsion coefficient
挠曲纤维应力 flexural fiber[fibre] stress
挠曲线 bending line; flexing curve; flexion curve;flexure curve;line of deflection; skew curve; twisted curve;twisting curve
挠曲线活塞刮油环 flexline oil piston ring
挠曲限度 flexible limit
挠曲性 flexibility
挠曲性联轴节 resilient connector
挠曲循环 flex cycle
挠曲仪 deflection indicator
挠曲应变 bending strain; flexural strain;strain of flexure
挠曲应力 bending stress; buckling stress;flexing stress;flexion stress; transverse stress; warping stress; flexural stress;flexure stress
挠曲预应力钢索 deflected tendon
挠曲折断两用机 bender and cutter
挠曲褶皱作用 flexure fold(ing)
挠曲振动 bending vibration; flexural vibration
挠曲中心 center of flexure; flexural center[centre]
挠曲周期 cycle of bending;flexing cycle
挠曲阻力 flexing resistance
挠曲作用 bending;flexing action
挠群 torsion group
挠软封袋法 flexible packaging
挠式振动器 flexible type vibrator
挠四边形 skew quadrilateral
挠弹性的 flexible resilient
挠性 flexibleness;pliability;pliancy
挠性安全护栅 flexible safety barrier
挠性板 flexible plate;flex plate
挠性板超音速喷管 flexible plate supersonic nozzle
挠性板联轴节 flexible plate coupling
挠性保护管套 flexible protective casing for pipe
挠性扁钢钎 flexible flat drill steel; flexible steel in flat section
挠性变速杆 flexible gear-shift lever
挠性波导管 flexible waveguide
挠性波纹管 flexible bellows
挠性波纹管连接 flexible bellows connection
挠性不锈钢密封盒 flexible stainless-steel bellows
挠性操纵 flexible control;flexible operation
挠性草捆收集器 flexible bale collector
挠性车箱 flexible bodywork
挠性衬套 flexible bush
挠性撑条 flexible stay
挠性尺寸 flexible dimension
挠性传动 flexible drive;flexible transmission
挠性传动机构 flexible gear
挠性传动张紧装置 take-up screw
挠性传动装置 flexible gearing
挠性从动齿轮 flexible driven gear
挠性锉刀 flexible file
挠性带状钢钎 flexible ribbon steel
挠性袋模塑 flexible membrane mo(u)lding
挠性刀具 flexible knife
挠性(导)管 flexible conduit;flexible pipe
挠性导线管 flexible conduit
挠性导像光缆 flexible imaging bundle
挠性地垫 track mat
挠性电缆 flexible cable
挠性垫 flexible mat
挠性防波堤 flexible breakwater
挠性封闭屏 flexible closed screen
挠性浮式防波堤 flexible floating breakwater
挠性腹板 flexible wall
挠性刚度 flexural rigidity
挠性钢带 flexible ribbon steel
挠性钢钎 flexible drill steel;flexible steel
挠性构件 flexural member; flexure member
挠性挂接装置 flexible hitch
挠性管 flexible tube;hose pipe
挠性管道清洁器 flexible pipeline pig
挠性管阀 flexible tube valve
挠性管路 flexible tubing
挠性管式螺旋输送器 flexible tube auger conveyer[conveyor]
挠性罐道 flexible cage guide;flexible guide
挠性光导 flexible lightguide
挠性滚柱轴承 flexible roller bearing; flexible type of roller bearing;Hyatt roller bearing
挠性滚子 flexible roller
挠性滚子保持架 flexible roller cage
挠性褐煤 dysodile
挠性护墩桩 flexible dolphin
挠性回复装置 flexible return
挠性火棉胶 flexible collodion
挠性尖轨 flexible switch
挠性件 flexible member;tension element
挠性剑杆 flexible rapier
挠性接合 flexible joint
挠性接合器 flexible jumper
挠性接头 elastic joint;flexible connection;flexible connector;flexible joint; kidney joint; pigtail; woggle joint
挠性接头部件 flexible extension
挠性节 flexible hinge
挠性结构减震系统 damping system of flexible structure
挠性结合 flexonics
挠性金属管 flexible metal hose;flexible metallic hose
挠性金属密封套 flexible metal shoe
挠性金属膜盒 flexible metal bellows
挠性聚氨基甲酸酯接合剂 flexible polyurethane bonding
挠性聚氨酯清管器 flexible urethane pig
挠性空腔波导管 flexible hollow waveguide
挠性控制 flexible control
挠性控制索 flexible control cable
挠性矿物 flexible mineral
挠性框架 flexible frame
挠性连接 flexible connection flexible joint;woggle joint
挠性连接传动 flexible gear(ing)
挠性连接器 flexible connector
挠性联管节 flexible pipe union
挠性联结 bent coupling link;flexible connection
挠性联结器 flexible coupling
挠性联轴节 flexible coupling;flexible joint;vernier coupling
挠性联轴器 elastic clutch
挠性链 flexible chain
挠性链齿耙 flexible harrow
挠性履带式拖拉机 flexible crawler tractor
挠性轮距 flexible wheel base
挠性螺旋 flexible auger
挠性螺旋输送器 flexible auger conveyer[conveyor]; flexible screw conveyer[conveyor]
挠性门 flexible door
挠性密封 flexible seal
挠性模数 modulus of flexibility
挠性幕罩 <气垫车> flexible curtain
挠性盘式联轴节 cushion disk coupling
挠性喷嘴 flexible nozzle
挠性膨胀节 flexible expansion piece
挠性气升吸泥管 flexible air lift mud pipe
挠性墙纸 flexible wall covering
挠性曲线样板 flexible curve
挠性屈曲 flexural buckling
挠性燃料桶 flexible fuel tank
挠性容器 flexible container
挠性容器增压 flexible tank pressurization
挠性软管 flexible hose
挠性软管泵 flexible hose pump
挠性润滑器 flexible lubricator
挠性砂岩 flexible sandstone
挠性设备连接器 flexible appliance connector
挠性石棉 mountain flesh
挠性试验 flexural test;flexure test
挠性试验机 flexure test machine
挠性树脂 flexible resin
挠性水运油囊 flexible oil barge
挠性松土覆盖镇压机 flexible roller-mulcher-tiller
挠性塑料 flexiplast(ics)
挠性索 flexible rope
挠性天线反射器 flexible reflector curtain
挠性铁塔 flexible tower
挠性铜线 flexible copper cord
挠性涂层 flexible coating
挠性托轮 flexible idler
挠性陀螺(仪) flexible gyroscope
挠性弯头 flexible bend
挠性万向节 flexible universal joint
挠性无缝管 flexible seamless tubing
挠性吸泥管 flexible suction tube
挠性系船柱 flexible dolphin
挠性纤维 flexible fiber[fibre]
挠性线 flexible wire
挠性橡胶清管器 flexible rubber pig
挠性星轮碎土器 flexible sprocket-wheel pulverizer
挠性型 flexible type
挠性压力 flexible pressure
挠性叶轮泵 flexible-impeller(rotary) pump
挠性油环 flexible oil piston ring;flexvent oil ring
挠性元件 flexural member
挠性云母 flexible mica
挠性运输机 flexible conveyer[conveyor]
挠性折叠翼 flexible folding wing
挠性振捣器 flexible vibrator
挠性振动 flexural vibration
挠性支撑 flexible strut
挠性支持 flexible support
挠性指杆 flexible finger
挠性中间散料容器 flexible intermediate bulk container
挠性轴 flexible axle; flexible shaft; quill shaft
挠性轴承 flexible bearing
挠性轴传动 flexible shaft transmission;flexible shaft drive
挠性轴的水准控制器 flexible shaft level controller;flexible shaft level indicator
挠性轴机床 flexible shaft machine
挠性轴机械传动 flexible shaft mechanical transmission
挠性轴式 flexible shaft type
挠性轴套管 <机械传动用> flexible casing
挠性轴系 flexible shafting
挠性柱 flexible dolphin
挠性转角螺旋钻 angle auger
挠性转向架 flexible truck
挠性转子 flexible rotor
挠性桩 flexible dolphin;flexible pile
挠性钻杆 flexible drill stem
挠性钻杆储藏架 flexible drill stem storage basket
挠性钻杆接头 flexible pipe joint
挠性钻杆卷筒 flexible drill stem storage reel
挠性钻钢 bendable steel;flexible drill steel
挠应变 bending strain
挠应力 bending stress
挠折带 flexural zone
挠折模量 modulus of rupture;rupture modulus
挠折收缩量 <受压弹簧的> buckling deflection
挠褶滑动 bedding plane slip;flexural slip
挠振现象 panting
挠振作用 panting action
挠转能量 bending-rotation capacity
挠子 scratcher

脑

脑椽 brain rafter;upper rafter

脑壳 skull
脑力 mentality
脑力工作 intellectual work; white-collar jobs
脑力激荡计划 brainstorming program(me)
脑力激荡术 <多数人在一起能收到思想上互相激励、创造新观念的效果> brainstorm(ing)
脑力劳动 brain work;headworks;intellectual labo(u)r; intellectual work;mental labo(u)r
脑力劳动产物 brain child
脑力劳动者 brain worker; mental worker; white-collar worker (or employee)
脑损伤 brain damage
脑震荡 concussion

瑙

瑙加海德 <乙烯树脂涂面织物,作为家具和墙布> Naugahyde

瑙鲁 <在西太平洋> Nauru
瑙鲁磷酸盐公司 Nauru Phosphate Corp.
瑙云母 naujakasite

闹

闹铃 vibrating bell

闹市 busy street;hive
闹市区 commercial center[centre]; crowded downtown area; downtown(area);rumble strip
闹市区街道 downtown street
闹速动轮 alarm rocket wheel
闹指示器 alarm indicator
闹钟 alarm clock;alarum

内阿米巴 entamoeba

内鞍 inner saddle
内岸 inner bank
内岸纵主梁 < 减荷台式码头的 > inshore longitudinal girder
内凹槽车刀 inside recessing tool
内凹角 reentrant corner
内凹轮廓 curved profile
内凹穹顶 domical vault
内凹式老虎窗 < 斜屋顶的 > internal dormer
内凹手把 socket holder
内凹踢脚板 coved skirting
内凹砖 internal bull-nose brick
内凹钻 cove bit
内凹座 inside recess
内八角砖 internal octagon brick
内八字脚 pigeon-toed;toe(d)-in
内把手 inside handle
内靶 internal probe target; internal target
内靶辐照 internal target irradiation
内摆角 < 钻车的 > inward divergence
内摆线 hypocycloid;inner cycloid
内板 back head;internal plate
内半径 inside radius
内拌和翼 internal mixing blade
内瓣 inner valve
内包络断面 inner enveloping profile
内包络轮廓 < 外形的 > inner enveloping profile
内包络线转子工作面 < 转子发动机的 > inner envelope rotor flank
内包叶 inner envelope
内包装 inner packaging; inside packing
内保温 internal insulation
内保险计时器 internal failsafe timer
内暴露量 internal exposure
内爆 cavitation damage;implosion
内爆波 implosion wave
内爆压碎钻进 implosion drilling
内爆凿岩 implosion drilling
内爆炸 implode
内背板 entodorsum;entotergum
内倍性 endoploidy
内泵壳 inner pump housing
内比 internal ratio
内闭括号 internal closing bracket
内庇三角式双冲程柴油机 Napier deltaic two stroke engine
内壁 inner side;inner surface inwall; inner wall(ing);inwall
内壁板 internal wall(ing) panel; lining board
内壁间距 inner wall spacing
内壁检验器 introscope
内壁涂层 inside coating
内壁涂瓷釉的管道 glass-lined pipe
内边 inner edge;inner side
内边界层 internal boundary layer
内边界条件 internal boundary condition
内边界(线)internal boundary
内边缘 inward flange
内变晶 endoblast
内变量 internal variable
内变形 < 金属轧断时的 > internal flow
内变形的【地】enterolithic
内变形损耗 inner deforming loss
内变质带 endomorphic zone
内变质作用 endometamorphism; endomorphism
内标 interior label;internal mark;internal standard
内标尺式温度计 double tube thermometer;enclosed scale thermometer
内标度 < 阴极射线管的 > internal graticule
内标法 internal method; internal standard;internal standard method
内标归一化法 internal normalization; internal standard normalization
内标记 internal labeling
内标记寄存器 inner flag register
内标率 < 核磁共振的 > internal reference
内标面积 internal standard area
内标试样 internal standard sample
内标物 internal standard
内标线 internal standard line
内标元素 internal standard element
内标准法 < 磁共振的 > internal reference method
内表 interior finish
内表层 inner skin;internal skin
内表面 inner face; inner surface; inside(sur)face;internal(sur)face
内表面过滤器 internal surface filter
内表面换热系数 heat-exchange coefficient of interior surface
内表面积 inner surface area; inside surface area;internal surface area
内表面加工 internal operation;internal work
内表面检查 intersurface inspection
内表面闪光灯泡 inside-frosted lamp
内表面散热系数 inside film coefficient
内表面温度 inside surface temperature
内表面修装 internal facing; internal surfacing
内表面研磨 internal grinding
内表示镜 backlight blinder;backlight shutter;back spectacle;signal back light blinder
内滨 inshore zone;shoreface
内滨带 inshore area;inshore zone
内滨流 inshore current
内滨区 inshore area
内冰碛【地】englacial moraine;englacial till
内禀 intrinsic(al)
内禀传导性 intrinsic(al)conductivity
内禀磁感应强度 intensity of magnetization;intrinsic(al) flux density;intrinsic(al) induction; magnetic polarization
内禀磁矩 intrinsic(al) magnetic moment
内禀导纳 intrinsic(al)admittance
内禀法拉第旋转 internal Faraday rotation
内禀感应 intrinsic(al)induction
内禀共振 intrinsic(al)resonance
内禀光电发射 intrinsic(al) photoemission
内禀光度 intrinsic(al)luminosity
内禀红移 intrinsic(al)redshift
内禀矩 eigen-moment
内禀亮度 intrinsic(al)brightness
内禀弥散度 intrinsic(al)dispersion
内禀能 intrinsic(al)energy
内禀能量 self energy
内禀黏[粘]滞性 intrinsic(al)viscosity
内禀时间 intrinsic(al)time
内禀吸收 intrinsic(al)absorption
内禀效率 intrinsic(al)efficiency
内禀性质 intrinsic(al)property
内禀耀度 intrinsic(al)brilliancy
内禀宇称 intrinsic(al)parity
内禀噪声温度 intrinsic(al)noise temperature
内禀增长率 intrinsic(al) rate of increase
内禀周期 intrinsic(al)period
内禀自然增长率 intrinsic(al) rate of natural increase
内禀阻抗 intrinsic(al)impedance
内波 interfacial wave;internal wave; subsurface wave;internal wave
内波效应 effect of internal wave
内波运动 internal wave motion
内玻璃门 inside glass door
内剥离物 inner stripping material
内伯石 neighborite
内薄集 internally thin set
内补偿运算放大器 internally compensated operational amplifier
内捕俘 internal trap
内布拉斯加冰期【地】Nebraskan glacial epoch;Nebraskan glacial stage
内布拉斯加冰碛 Nebraskan drift
内布拉斯加层 Nebraska bed
内布拉斯加州工程学会 < 美 > Nebraska Engineering Society
内部 inner portion; inside; interior portion;inward
内部安装 internal installation;interior glazed;internal glazing < 玻璃的 >
内部安装玻璃法 inside glazing
内部安装的 capsulate(d)
内部安装的设备 built-in comfort
内部安装工作 internal installation work
内部包装 interior packing
内部保留数准备 reserve for interdepartmental encumbrance
内部报酬率 internal rate of return
内部报告 internal reporting; interwork report
内部报告会计 accounting for internal reporting
内部比率 internal ratio
内部壁骨 internal stud
内部边材 internal sapwood
内部编号 internal number
内部编码 in-line coding
内部变换 inner transformation
内部变形 internal deformation
内部标高 internal elevation
内部标号 interior label;internal label
内部标记 interior label;internal label
内部标志 interior label;internal label
内部标准 internal standard
内部标准比率 internal standard ratio
内部表 internal table
内部表示法 internal representation
内部表征 internal attribute
内部玻璃安装 internal glazing
内部玻璃门 inner glass door;internal glass door
内部补偿 internal compensation
内部不经济性 internal diseconomy
内部布线 interior wiring; internal wiring
内部布置 inner layout;inside arrangement;interior collocation
内部布置图 internal layout
内部财务 internal financing
内部财务收益率 internal financial rate of return
内部财源 internal sources of revenue
内部采光 internal lighting
内部操纵指令 internal manipulation instruction
内部操作 built-in function
内部操作比 internal operation ratio
内部侧廊 inner side aisle
内部侧卸式起重机 internal skip-hoist
内部侧院 interior side yard
内部层裂 internal lamination
内部查找 internal searcher; internal searching
内部差异 internal diversity
内部产热 internal heat generation
内部长廊 interior gallery
内部长途中继线 intertoll trunk
内部常数 internal constant
内部超静定的 interior statically indeterminate
内部潮气 inner moisture
内部车库 in-built garage
内部车库标志 interim garage sign;interior garage sign
内部沉降量测 internal settlement measurement
内部沉陷量测 internal settlement measurement
内部衬垫 inner liner
内部衬砌 inner lining;interior lining; internal lining;inside lining
内部成本 internal cost
内部乘法逻辑 built-in multiplication logic
内部尺寸 inner dimension;inside dimension;internal dimension
内部冲刷 internal scour
内部冲洗 internal wash
内部筹款 internal financing; self-finance
内部除法逻辑 built-in division logic
内部厨房 built-in kitchen;built kitchen cabinet;in-built kitchen
内部处理 in-line processing;internal treatment
内部处理机错误 internal processor error
内部传感器 internal sensor
内部传热膜层系数 inside film coefficient
内部窗 inward window
内部窗侧 inside reveal
内部窗(花)格 interior tracery
内部磁记录 internal magnetic recording
内部刺激机制 internal incentive mechanism
内部存储残片 internal fragmentation
内部存(储)程序 internally stored program(me)
内部存储器 internal memory;internal storage
内部存储碎片 internal fragmentation
内部错误 internal error
内部错误检查 built-in error checking
内部错误校正 built-in error correction
内部搭配 interior collocation
内部大气层区 internal atmospheric zone
内部代码 internal code
内部挡块 interior stop
内部导体 inner conductor; internal conductor
内部的 build in;built-in;in-house;inline;inside;internal;intramural;intrinsic(al);inner;interior
内部的报告制度 internal report system
内部的不经济 internal diseconomy
内部的节约 internal economy
内部等时线 internal isochron
内部抵抗 internal resistance
内部地带 zone of interior
内部地址 home address;internal address
内部地址分辨表 build address resolution table
内部电磁干扰 inherent interference
内部电话 house telephone; intercall telephone;intercommunication telephone;internal(tele)phone;interphone; intertelephone; talk-back telephone
内部电话机 intercall telephone

内部电话交换机 interphone control box

内部电话交换系统 house telephone system; intercommunicating system; interphone system

内部电话设备 intercom[intercommunication system]

内部电话系统 internal telephone system

内部电话装置 service telephone plant

内部电极位置控制 internal electrode position control

内部电缆 inside cable

内部电路 domestic circuit; interior circuit

内部电路测试 in circuit testing

内部电路仿真器 in-circuit emulator

内部电容 internal capacitance

内部电位振动 internal potential vibration

内部电压降 internal voltage drop

内部电源 internal electric(al) source

内部吊索 internal rigging

内部调拨 internal transference

内部调拨定价 transfer pricing

内部调拨价格 transfer price

内部调查 internal research

内部定位 inner orientation

内部定位条 interior stop

内部定相 internal phasing

内部定向 inner orientation; internal orientation; interior orientation

内部董事 inside director

内部动水应力 < 打混凝土管桩的 > internal hydrodynamic stress

内部对称群 internal symmetry group

内部对称原理 internal symmetry principle

内部对话装置 intercommunicating set

内部对讲电话 intercommunicating (tele) phone

内部对讲电话系统 intercommunicating system; interphone system

内部对讲电话增音机 interphone amplifier

内部对讲电路 talk-back circuit

内部对象 internal object

内部二进计数制 binary internal number base; number base

内部发券【建】inner arch

内部发热 inner heat

内部发行 privacy issue

内部阀 interior valve

内部反常电压 internal abnormal voltage

内部反合 internal return

内部反射 internal reflection

内部反射层 inner reflector

内部反射的散射光 internally reflected scattered light

内部反射照明 internal-indirect lighting

内部方差 internal variance

内部防波堤 inner breakwater; inner waterbreak

内部防火 interior fire protection

内部房间 penetralia

内部非承重墙 interior non-bearing walls

内部废物处理系统 internal waste disposal system

内部分程序 internal block

内部分隔 inner partitioning; inside partitioning; internal partitioning

内部分隔重新布置 rearrangement of interior partitioning

内部分机通话 home extension intercom

内部分级 < 磨机的 > internal classification

内部分类 internal sorting

内部分类阶段 internal sort phase

内部分析 internal analysis

内部分线交换机 private branch exchange

内部封闭 inner seal(ing); inside seal(ing)

内部封接 inner seal(ing)

内部扶手 interior handrail

内部浮力装置 internal buoyancy appliance

内部浮体 internal buoyancy

内部符号 insymbol; internal symbol

内部符合 inner consistency [consistence]; internal agreement; internal consistency[consistence]

内部辐射剂量 internal radiation dose

内部腐蚀 interior corrosion; internal corrosion; internal erosion

内部腐蚀电池 corrosion cells within concrete

内部腐朽 inside rot; internal decay

内部负债 interior liabilities

内部复制 internal reproduction

内部干斑 internal dry patch

内部干裂 internal check

内部干扰 internal interference

内部干燥试验 internal drying test

内部杆件 internal member

内部感光点 internal sensitivity speck

内部感受器 internal receptor

内部刚度 internal rigidity

内部港 close port

内部高程 internal elevation

内部格式 internal form

内部隔板货车 internal-partition wagon

内部隔断 interior partition

内部隔墙 inside cellar wall

内部隔声 inner insulation; inside insulation

内部跟踪表 internal trace table

内部跟踪结构 built-in tracing structure

内部工程 inner work

内部工程用材料 material for interior work

内部工会 inside union

内部工资差别 internal wage differentials

内部工作 inner work

内部功率 internal power

内部功能 built-in function

内部功能寄存器 internal function register

内部供能的机车 internally powered locomotive

内部供暖 inner heat; inside heat

内部供水式钻机 internal water-feed machine

内部供应和销售收入 internal supply and sale income

内部供应和销售支出 internal supply and sale expenses

内部沟 internal access

内部构形 internal configuration

内部构造 internal structure

内部构造(底)槛 interformational cill [sill]

内部估价 internal valuation

内部故障 internal fault

内部故障成本 internal failure cost

内部顾问师 internal consultant

内部观测 internal observation

内部管道 internal piping

内部管道系统 inner piping(system)

内部管理 inner management

内部管理报告 internal report for management

内部管理控制 internal administrative control

内部管理制度 internal control system

内部管线 interior piping; internal pipeline

内部管涌 internal piping

内部光阑 internal stop

内部规定指标 internally specified index

内部规划图 internal layout

内部过程 in-line procedure; internal procedure; intrinsic(al) procedure

内部过电压 internal overvoltage

内部含有水分的 enhydrous

内部函数 built-in function; intrinsic(al) function

内部函数的分配 allocation built in function

内部函数的配置 allocation built in function

内部焊接 internal welding

内部合成 internal composition

内部核查 internal check

内部核心 internal core

内部核心贸易 internal core trade

内部荷载 internal load

内部恒温器 inner thermostat; internal thermostat

内部红斑 < 木材 > inside red spot

内部宏指令 inner macroinstruction; internal macro instruction

内部后视镜 < 车身的 > inside rear view mirror; interior mirror

内部厚度 interior thickness

内部湖震 internal seiche

内部互联放大器 interconnected amplifier

内部互联网 intranet

内部互通电话机 intercommunicating (telephone) set; intercommunication telephone set

内部护板 interior lining panel

内部护墙板 internal panel

内部环境 internal environment

内部环裂 internal shake

内部缓冲区 internal buffer

内部灰 fixed ash

内部回弹能 internal resilience

内部回动装置 internal reversing gear

内部回路 home loop; local loop

内部回路操作 home loop operation

内部回收率 internal rate of return; rate of internal return

内部回收率法 internal rate of return method

内部回转半径 inside turning radius

内部汇票 house bill; house paper

内部会计 internal accounting

内部会计控制 internal accounting control

内部会计事项 internal transactions

内部混合机 internal mixer

内部混合喷嘴 inside mix nozzle

内部货币 inside money

内部积累 internal accumulation

内部积水 internal stagnant water

内部基础结构 internal infrastructure

内部稽核 internal audit

内部几何条件 internal geometry

内部计时 internal clocking

内部计数制 internal number system

内部计算 internal arithmetic

内部计算机 inner computer

内部寄存器 internal register

内部寄生菌 endophyte

内部加工车间 closed job-shop

内部加厚的钻管 internal upset drill pipe

内部加热式干燥机 internal heat drier

内部价格 inside price

内部间断 internal discontinuity

内部间隔货车 internal-partition wagon

内部间接照明 internal-indirect lighting

内部间隙 < 滚动轴承等的 > internal clearance

内部减热器 internal desuperheater; submerged type desuperheater

内部减压 inner pressure relief

内部检查 internal examination; internal survey; internal inspection

内部检修 internal check

内部检验 built-in check; internal examination; internal inspection

内部建筑板 inner building board; inside building board; internal building board

内部建筑薄板 internal building sheet

内部奖励 internal rewards

内部交叉 internal crossing

内部交叉拉条 internal cross bracing

内部交通 internal traffic

内部交易 intratransaction

内部角隅 internal corner

内部搅拌叶片 internal mixing blade

内部教育 internal education

内部校验 built-in check

内部校准 internal calibration

内部接线图 cutaway view; internal wiring diagram

内部街区 back block

内部节点 interior joint

内部结构 innards; inner structure; interior structure; internal structure

内部结垢 internal encrustation

内部结露 < 围护结构 > condensation within structure

内部结算价格 internal settlement price

内部捷传 forwarding

内部介电损耗 internal dielectric(al) loss

内部禁运 civil embargo

内部经济报酬率 internal economic rate of return

内部经济核算 internal economic accounting

内部经济回收率 internal economic rate of return

内部经济收益率 internal economic rate of return

内部经济性 internal economy

内部晶化 interflorescence

内部净空比 inside clearance ratio

内部净空断面形状 inside shape

内部聚值集 inner cluster set

内部卷扬机 internal hoist

内部决策 internal decision making

内部绝热 inner insulation; inside insulation

内部绝缘 interior insulation

内部均压型热动膨胀阀 internal equalizing thermal expansion valve

内部均压装置 internal equalizer

内部开关 internal switch

内部开裂 internal cracking

内部刊物 house organ; in-house literature; house journal < 侧重技术内容 >; house magazine < 侧重于综合报导 >

内部刊物目录 house magazine directory

内部勘测 internal survey

内部可操作性 interoperability

内部空间 inner space; interior space; internal space; space-enclosed

内部空气 internal air

内部空气冷却 internal air cooling

内部空气冷却器 inner air cooler; internal air cooler

内部空气温度 inside air temperature
内部空隙度 internal porosity
内部空隙分率 internal void fraction
内部空芯墙 inner cavity wall
内部孔穴 internal cavity
内部控制 built-in control; internal control
内部控制报告 internal control report
内部控制标准 internal control standard
内部控制点 internal control point
内部控制调查表 internal control questionnaire
内部控制系统 internal control system
内部控制制度审计 audit of internal control system
内部控制组织图 organization charts for internal control
内部跨度 interior span
内部拉条 internal bracing
内部蓝变 interior blue stain
内部廊道 internal gallery
内部劳动规则 internal rules of labo(u)r
内部劳动力市场 internal labor market
内部冷却 internal cooling
内部冷却的 inner-cooled; internal cooled
内部立管 internal riser pipe
内部利(润)率法 internal rate of return method
内部利润所得税 income tax on intercompany profit
内部连接 internal connection
内部联系 interconnection;internal relation
内部链接 internal linking
内部梁 built-in beam
内部裂缝 heart check;internal crack;internal fissure;internal flaw;structural cracking
内部裂纹 internal crack;internal fissure;structural cracking;underbead covering
内部裂隙 pit hole
内部淋浴间 in-built shower stall
内部零件 inner body;internal unit
内部流动特性 internal flow characteristics
内部流失 internal drain
内部楼面面积 interior area; interior floor area
内部楼梯 back stair(case)
内部漏泄 internal leakage
内部路径 inner track
内部路线 interior route
内部氯化处理的橡胶漆 interior chlorinate rubber paint
内部码 internal code; internal machine code
内部贸易 intratrade
内部门窗侧壁砌块 interior door jamb block
内部门窗侧壁砌体 interior door jamb block
内部门窗镶嵌用玻璃 inner glazing
内部门扉 internal leaf
内部密封箱 inner sealed box
内部面饰 internal facing
内部灭火系统 internal fire-extinguishing system
内部名字 internal name
内部命令 internal command
内部模板工程 internal frame(work)
内部模块 internal module
内部模拟环路 internal analog(ue) loop
内部模式 internal schema
内部摩擦 internal friction
内部磨板 internal wearing plate
内部磨损 internal wear
内部抹灰 internal plastering
内部木器工 interior woodwork
内部木条百叶窗 inner slatted blind; inside slatted blind
内部目标 internal object
内部能量平衡 internal energy balance
内部能源 internal power source
内部黏[粘]结剂 interior bonding agent
内部黏[粘]结力 internal bond
内部黏[粘]结强度 internal bond strength
内部黏[粘]着力 interior adhesive
内部排气再循环 internal exhaust gas recirculation
内部排水 interior drainage
内部排水层 internal drainage layer
内部排水管 internal drain
内部排水系统 internal drainage system
内部排土场 inner dump
内部排序 internal sort
内部配件 internal fittings
内部配置 interior collocation
内部喷嘴 inwardly projecting orifice
内部喷嘴全收缩注孔 inwardly projecting orifice
内部偏移 internal bias
内部偏振调制 internal polarization modulation
内部漂流 internal drift current
内部平差 internal adjustment
内部平衡 internal equilibrium
内部平均信息量 internal entropy
内部屏蔽 internal screening; internal shield
内部破裂试验 internal fracture test
内部破碎 internal fracture
内部破损 contents broken
内部气封 internal packing
内部气孔 deep-seated blowhole
内部气泡 trapped air
内部气驱 dissolved gas drive;gas depletion drive;internal gas drive;solution gas drive
内部汽提段 internal stripping section
内部砌壁 <隧道> permanent supporting system
内部牵引力 internal traction force
内部牵制 internal check
内部牵制制度 internal check system; system of internal check
内部潜热 internal latent heat
内部切削加工 internal machining operation
内部清算 internal clearing
内部情报 inside info; internal information
内部求导 inner derivation
内部球面几何学 intrinsic(al) spheric(al) geometry
内部区段 interior zone;internal zone;zone of interior
内部区域 interior zone;internal zone;zone of interior
内部取暖 internal heat gain
内部权 internal weight
内部缺水量 internal water deficit
内部缺陷 <铸件橡胶制品等的> inherent vice;internal defect
内部缺陷引起的损坏 inherent weakness failure
内部燃烧 interior combustion; internal ignition
内部燃烧炉 internally fired furnace
内部热传导 internal thermal conductance
内部热交换 internal heat exchange
内部热量 inner heat;internal heat
内部热增量 internal heat gain
内部容积 <船舶的> internal cubic(al) capacity
内部软件延迟 built-in software delay
内部润滑法 internal lubrication
内部塞绳 rear cord
内部散射 scattering-in
内部色彩 interior stain
内部商店 capitive shop;tommy-shop
内部上光清漆 inner gloss(clear) varnish;inside gloss(clear) varnish;internal gloss(clear) varnish
内部上釉的 interior glazed
内部烧蚀 internal ablation
内部设备 internal equipment;internal fittings; internal plant; internal unit;in-house facility <企业的>
内部设计 internal design
内部设计师 internal designer
内部审计 internal audit(ing); management audit
内部审计标准 internal auditing standard
内部审计处 internal audit service
内部审计规程 internal auditing manual
内部审计师 certified internal auditor
内部审计员 internal auditor
内部审计制度 internal audit system
内部审计组织 organization of internal audit
内部生物遗迹 internal lebensspuren
内部湿度 inner humidity;inner moisture;inner primer;internal humidity;internal moisture
内部时基稳定度 internal time base stability
内部时滞 inside lag
内部时钟 internal clock(ing)
内部时钟脉冲 internal clock pulse
内部时钟脉冲信号 internal clock pulse signal
内部时钟源 internal clock source
内部食堂 mess hall;mess room
内部使用 internal use
内部市场 inside market
内部示踪物 intrinsic(al) tracer
内部视图 inside view
内部饰面 interior finish
内部适配器 built-in adapter
内部收缩 internal shrinkage;intrinsic(al) shrinkage
内部收益率 internal rate of return
内部收益率计算法 internal rate of return method;yield method
内部输出电路 internal output circuit
内部输入电路 internal input circuit
内部输送指令 internal transport instruction
内部数 internal number
内部数据 internal data
内部数据与语言传输 interpolated data and speech
内部数组 inarray
内部水 <指地下饱和层以下部分的> internal water;inner water
内部水流 internal flow
内部水落管 internal downpipe
内部水暖安装 internal plumbing
内部水下区 internal submerged zone
内部水循环 internal water circulation
内部水域 internal waters
内部水质 internal water quality
内部顺序号 internal sequence number
内部损坏 internal breakdown; low-temperature break-down
内部损伤 internal defect;internal lesion
内部损失 internal loss
内部缩孔 internal shrinkage
内部锁闭 inside locking
内部淘洗 internal scour
内部特性 internal characteristic
内部特性曲线 internal characteristic curve
内部天井 internal court
内部天线 inside antenna;internal antenna
内部填缝 inner seal(ing);inside seal(ing);internal sealing
内部填塞式活塞泵 inside-packed type piston pump
内部条件 internal condition
内部贴现率 internal rate of return
内部铁路(线) internal rail
内部庭园 internal court
内部通道 inner passage;internal passage
内部通风沟 internal air duct
内部通话机 interphone
内部通话开关 intercommunication switch
内部通话设备 interphone equipment
内部通话系统 intercommunicating system
内部通话扬声器 intercom loudspeaker;intercommunication speaker
内部通话制 intercom[intercommunication system]
内部通话主机 intercommunication master set
内部通话装置 intercommunication installation
内部通信[讯]插塞式交换机 intercommunication plug switchboard
内部通信[讯]电路 intercommunication circuit
内部通信[讯]机 intercommunicator
内部通信[讯]联络系统 <轮船、飞机等用的> intercom[intercommunication system]
内部通信[讯]联络系统模件 intercom module
内部通信[讯](联系) intercommunication; interior communication; internal communication
内部通信[讯]设备 intercommunication set;intercom set;internal plant
内部通信[讯]系统 intercommunication system; interior communication system; internal communication system
内部通信[讯]线路 intercommunicating line; internal communication circuit
内部通信[讯]小交换机 private branch exchange switchboard
内部通信[讯]制 intercommunication system;interior communication system;internal communication system
内部通信[讯]装置 inside plant;intercommunication installation; interphone
内部透射度 internal transmittance
内部图 interior view
内部涂层 internal coating
内部涂料 interior finish;interior paint coat;internal paint
内部涂有沥青的管子 loricated pipe
内部拓扑 inner topology
内部外抱 inner reveal;inside reveal; internal reveal
内部弯矩 interior bending moment
内部网 intranet
内部网络 in-house network
内部网络协议 internal network protocol
内部往来 intrabranch

N

内部围砌分隔墙 inside masonry dividing
内部围岩 interlayer rock
内部卫生管道工程 internal plumbing
内部位移量测设备 internal movement measuring device
内部温度 interior temperature; internal temperature
内部温度控(制) inner temperature control; internal temperature control
内部温度调节 internal temperature control
内部文件 house document
内部文件关系 interfile relationship
内部稳定性 internal stability
内部圬工分隔墙 internal masonry dividing wall
内部污水管 private sewer
内部无线电话 interphone radio
内部无影照明 internal-indirect lighting
内部舞台 inner proscenium
内部误差校正 built-in error correction
内部吸收 intrinsic(al) absorption
内部系统 built-in system
内部系统结果【计】internal system result
内部细木工作 interior joinery; internal joinery
内部瑕疵 internal flaw
内部纤维板装饰 inner fiber[fibre] board finish; inside fiber[fibre] board finish; interior fiberboard finish; internal fiber[fibre] board finish
内部纤维板装修 inner fiber[fibre] board finish; inside fiber[fibre] board finish; interior fiberboard finish; internal fiber[fibre] board finish
内部线路 domestic circuit; in-house line; internal wiring
内部线性变换 internal linear transformation
内部相对含水量 interior relative moisture
内部相容性 internal consistency[consistence]
内部相移 internal phase shift
内部镶条 inside trim
内部向量 intrinsic(al) vector
内部消防系统 internal fire-extinguishing system
内部消光 internal delustring
内部消耗 internal consumption; internal drain <指排水>
内部消息 inside information
内部销紧机构 internal locking mechanism
内部小交换机通信[讯] internal intercommunication
内部小淋浴间 built-in shower
内部效益 internal benefit
内部效益率 internal rate of return
内部协调 internal coordination
内部协议 internal arrangement
内部泄漏 inner leak; interior leak; internal leak
内部泄漏量 inner leakage; interior leakage; internal leakage; passing unregistered
内部泄漏损失 internal leakage loss
内部信号 internal signal
内部信息 internal information
内部信息处理系统 internal information processing system
内部行驶阻力 internal motion resistance
内部形式 internal form
内部型芯 interior core
内部性态 internal performance
内部性状 internal behavio(u)r
内部修饰 interior finishing; interior trim(ming)
内部锈蚀 internal corrosion
内部悬挂式脚手架 interior hung scaffold
内部循环 internal recycling
内部压力 internal pressure; intrinsic(al) pressure
内部压气机 internal compressor
内部压缩爆破 <岩石中> blasting for internal compression(in rock)
内部压条 interior stop
内部烟囱 interior chimney
内部研究 in-house research; internal research
内部杨氏模量 intrinsic(al) Young's modulus
内部氧化物 subscale
内部业务 internal service
内部一致性 internal consistency[consistence]
内部一致性指数 internal consistency index
内部一致性准则 internal consistency criterion
内部仪表装配 in dash installation
内部移民 internal migration
内部引线接合法 inner lead bonding
内部引线接合器 inner lead bonder
内部应变 built-in strain
内部硬质纤维板装修 internal hard-board finish
内部优先级 internal priority
内部优先数 internal priority
内部有突齿的匣体 nibbed saggar[sagger]
内部预(加)应力 internal prestress
内部预加应力的 internally prestressed
内部元素 inner element
内部源程序 internal source program(me)
内部源程序指示字 internal source program(me) pointer
内部运输 internal transport(ation)
内部运输系统 internal transportation system
内部运算 internal arithmetic
内部再聚束 internal rebunching
内部再生区 inner blanket; internal breeder
内部再生系数 internal breeding ratio
内部造型 <汽车车身的> interior styling; internal styling
内部噪声 internal noise
内部闸 interior gate; internal gate
内部闸阀 internal gate valve; interior gate valve
内部债券 internal bond
内部丈量 internal measurement
内部照明的牌照板 inwardly illuminated license plate
内部照明度 inner illumination; internal illumination; inside illumination; interior lighting
内部照明式标志板 internal illumination sign plate
内部罩面层 inner skin; inside skin
内部真空度 internal vacuum
内部真空室 inner vacuum vessel
内部振荡 inherent oscillation
内部振捣 <混凝土的> pervibration; internal vibration
内部振捣器 pervibrator
内部振动 internal vibration
内部振动器 internal vibrator
内部争端 domestic dispute
内部蒸汽加热式转动薄膜干燥机 rotary steam-heated film drier
内部整修 inside trim
内部政策 policy for internal observance
内部支撑 internal bracing
内部支承 interior support
内部支柱 interior support
内部直接调度电话 internal direct dispatching telephone
内部指令【计】built-in command
内部指示灯 interior lamp
内部指示剂 internal indicator
内部质量 internal soundness
内部秩序 internal order
内部中断 internal interrupt; internal interruption
内部重排 internal rearrangement
内部重新装饰 inner redecoration; inside redecoration; internal redecoration
内部重新装修 inner redecoration; inside redecoration; internal redecoration
内部周期 intercycle
内部周转 internal turnaround
内部属性 built-in attribute; internal attribute
内部贮存 internal storage
内部转拨(款项) interior transfer
内部转换 internal conversion
内部转换程序 internal conversion routine
内部转让价格 internal transfer price; intratransfer price
内部转让利润 internal transfer profit
内部装璜 interior decorating; interior decoration; interior furnishing; interior trim(ming); internal decorating; internal decoration
内部装配 internal rigging
内部装饰 interior decorating; interior decoration; interior furnishing; interior trim(ming); internal decorating; internal decoration
内部装饰板 liner panel
内部装饰及装备的设计者 interior designer
内部装饰品 interior fixtures
内部装饰色彩 interior trim colo(u)r
内部装饰条材 trim for interior work
内部装饰物 interior fixtures
内部装修 inside decorating; interior decoration; interior trim(ming); inside finish; interior finish
内部装修保护板 drop cloth
内部装修材料 finish
内部装修工程 interior finish work
内部装修胶合板 interior plywood
内部装修用漆 internal finish(ing) paint
内部装药(爆破) internal charge
内部装有许多拉紧钢绳的钻塔 spider-wed the rig
内部装置 interior arrangement; internal arrangement; internal fittings; internal fixture
内部准备对净利润比率 ratio of internal reserves to net profits
内部着色 integral colo(u)ring
内部资金 <收益留存加折旧> internal finance
内部资金提供 internal financing
内部资金效应 cash flow effect
内部资料 inside information; internal data; restricted data
内部子程序 built-in subroutine
内部自测试 built-in self test
内部自动电话交换机 unit automatic exchange
内部自动电压调节 inherent voltage regulation
内部自动审计制度 system of automatic internal audit
内部自动调节 inherent regulation
内部自治 home rule
内部阻力 internal drag
内部阻尼 internal damping
内部阻尼器 inside damper
内部组成 bulk composition
内部组织 interior tissue
内部组织经济学 economics of internal organization
内部最小平方法 internal least square
内部作用 internal action
内部坐标 intrinsic(al) coordinates
内擦准法 hole lapping
内材面 inner surface
内参比电极 internal reference electrode
内参比线 internal reference line
内参考电极 internal reference electrode
内参数 internal parameter
内藏电机式振动器 motor-in vibrator
内藏裂纹 endokinetic fissure
内藏式安全踏脚板 concealed safety step
内藏式冷凝器 internal-mounted condenser
内藏式踏脚板 concealed running board; concealed safety step
内操作寄存器 internal function register
内操作指令 internal manipulation instruction
内槽 inside groove; internal groove; internal slot
内槽壁 internal groove sidewall
内槽径 groove-bottom diameter
内槽镗杆 recessing bar
内槽柱 hypostyle column
内侧 inner flank; inner side; inside
内侧坝趾 inside embankment toe
内侧保径金刚石 inside-ga(u)ge stone
内侧壁 inside reveal; inner reveal <门窗框的>; madial wall
内侧边坡 inslope
内侧部 medial part
内侧舱室 inboard cabin; inside cabin
内侧车道 fast lane; inner lane; liner lane; centre line lane <公路弯道的>; inner wheelpath <靠近道路中线的车行道>
内侧车轮的回转角 inner steered angle
内侧尺寸 inside clearance; inside measurement
内侧船首三角帆 inner jib
内侧船闸 land lock
内侧唇 labium mediale; medial lip
内侧粗劣砌块 internal quality block
内侧粗劣砖 internal quality brick
内侧刀齿 inside blade
内侧道 inner loop
内侧的 inboard; medial
内侧堤坝坡脚 inside embankment toe
内侧底段 medial basal segment
内侧底支 medial basal branch
内侧地块 inside lot
内侧端 medial extremity
内侧段 medial segment
内侧堆积曲流 scroll meander
内侧舵 inrudder
内侧发动机 inboard engine
内侧沟 inside ditch
内侧刮铲 inside dozer
内侧护轨 inner check rail; inner guard rail
内侧护梁 inside check beam; outside

guard beam

内侧护面 lee armo(u)r

内侧角 medial angle

内侧脚 medial leg

内侧径规 inside cal(l)ipers

内侧开口运动 inside shedding

内侧可见的光滑(面) smooth visible under-face; smooth visible underside

内侧力 inside force

内侧连接 back connection

内侧联锁送气表 purge meter interlock

内侧隆起 medial eminence

内侧轮 <双轮的> near-side wheel

内侧轮胎 inside tire

内侧轮胎最小转弯半径 minimum turning inner radius

内侧螺旋桨 inboard screw

内侧面 facies medialis; medial surface

内侧墙板 side lining

内侧墙砖 inside jamb block

内侧人行道 inner footpath

内侧刃保径金刚石 <钻头的> inner-ga(u)ge stone

内侧刃金刚石 <钻头的> inner stone; inside kicker; inside reamer; inside stone

内侧踏盘 inside tappet

内侧头 medial head

内侧推土机 inside dozer

内侧推土机刮铲 inside dozer blade

内侧镶板 interior panelling

内侧斜坡 <水库、水池的> inside slope

内侧信道 inboard channel

内侧悬挂制动 inside hung brake

内侧硬质板饰面 inner hardboard finish; inside hardboard finish

内侧鱼尾板 inner fish-plate

内侧缘 medial margin

内侧支 medial branch

内侧轴 inboard shaft; inner shafts

内侧转向半径 <叉车> radius of inside steering

内测度 inner measure; interior measure

内测量用量规 male ga(u)ge

内测千分卡尺 inside micrometer cal(l)ipers

内测微计 hole ga(u)ge; internal micrometer

内层 inner coat; inner layer; inside coat; inside layer; interior layer; internal leaf <空芯墙的>

内层安全壳 primary containment

内层百叶遮阳 inner slatted blind; inside slatted blind

内层表面络合物 inner sphere surface complex

内层薄板 cross banding veneer

内层窗樘 inner window frame

内层带 core band

内层单板 cross banding veneer

内层电子 inner-shell electron

内层缸 inner casing

内层轨道 inner orbit

内层交叉单板 cross band; crossing

内层壳板 inner skin

内层空间 inner space

内层络合物 inner sphere complex

内层(木壳)板 inner planking

内层幕墙 inner curtain(wall); inner enceinte(wall)

内层墙 internal skin

内层线圈 inner coil

内叉拱 interlacing arches

内叉拱廊 interlacing arcade

内插 interpolate; interpolating; interpolation

内插边缘 built-in edge

内插表 interpolation; interpolation table

内插等高线 interpolated contour

内插等深线 interpolated contour

内插点 interpolated point

内插多项式 interpolation polynomial

内插法 interpolation process

内插法相对分析 interpolative relation analysis

内插高程点 interpolated point between contours

内插公式 interpolation formula

内插函数 interpolation function

内插基点【数】basic point of interpolation

内插及外插程序 interpolation and extrapolation program(me)

内插间距 interpolated interval

内插梁 built-in beam; encastre beam

内插器 interpolator

内插曲面 surface for fitting

内插式【数】interpolant

内插式电势计 interpolating potentiometer

内插式电位计 interpolating potentiometer

内插式滑门 pocket sliding door

内插式设计 interpolation design

内插式振捣器 internal vibrator

内插数据与语言传输 interpolated data and speech transmission

内插同步观测 interpolated synthetic-(al) observation

内插图解法 graphic(al) interpolation

内插图像编码 interpolative picture coding

内插误差 interpolated error; interpolation error

内插误差方程式 error equation for interpolating

内插误差滤波器 interpolation error-filter

内插信号 interpolated signal

内插行 interpolation line

内插样条函数 interpolating splines

内插因子 interpolation factor; proportional parts

内插圆半径 radius of interpolation circle

内插振荡器 interpolating oscillator; interpolation oscillator

内插振实作用 immersion vibration

内插值 interpolated value

内插(值)法 interpolation method; method of interpolation

内插重力值 interpolation gravity

内插钻井 offset drilling

内差式针形阀 interior differential needle valve

内差速补偿小行星齿轮 internal differential compensating pinion

内缠绕层 inner warp(ing)

内长尺 inside cal(l)ipers

内长的 endogenous

内场 internal field

内场推动 internal vertical drive

内超过 internal trans

内潮 internal tide

内潮汐三角洲 inner tidal delta

内沉积作用【地】internal sedimentation

内衬 inner lining; inner(sur) facing; inside facing; interior lining; interline; internal liner; internal lining; internal surfacing; inwall; lining; lining-up; secondary lining; soffit

内衬板 inner lining board; inside lining board; inside welt; side lining

内衬部件 lining component

内衬层 air impervious liner; air resisting liner; air retaining wall; liner

内衬防潮层 furring

内衬拱 back arch

内衬过梁 back lintel

内衬砌 inside lining

内衬砌喷浆混凝土 inner ring shotcrete

内衬墙 inner wall(ing); inwall

内衬套 neck bush

内衬橡胶帆布管 rubber-lined canvas hose

内衬支护 <井巷道> crib

内撑 internal stay

内撑裙 inner skirt

内成 endogenesis; endogeny

内成变形 endogenetic deformation

内成沉积 endogenetic sediment

内成的 endogenetic; endogenous

内成角砾岩 endolitihic breccia

内成角砾作用 endolithic brecciation

内成节理 endokinetic joint

内成结构 endogenetic texture

内成力 endogenetic force; endotenous force

内成裂缝【地】endokinetic fissure; endogenetic fissure

内成裂隙 endogenetic fissure; endokinetic fissure

内成隆起 endogenous dome

内成喷出物 endogenous ejecta

内成穹隆 endogenous dome

内成碎屑灰岩 intraclast

内成型 internal shaping

内成型变 endogenetic deformation

内成岩 endogenetic rock; endogenic rock; ingenite

内成岩浆热液分导作用 endomagmatic hydrothermal differentiation

内成岩类 endogenetic rocks

内成岩墙 endodyke

内成因水库诱发地震 endogenous reservoir-induced earthquake

内成铀矿 endogenous uranium ore

内成作用【地】endogenetic action; endogenic action

内成作用过程 endogenic process

内呈鲕粒状的 entoolitic

内承重墙 internal load bearing wall; spine wall <平行于建筑物主轴的>

内城区 <美> inner city

内城衰落 inner city decay

内乘法 inner multiplication

内程序 internal program(me)

内程序计算机 internally program(me)ed computer

内尺寸 clear dimension; inside dimension; inside measurement; size in the clear

内齿 internal tooth

内齿层 barbule; endoperistome

内齿根 <套式斗齿的> tooth base

内齿环 annular gear; annular wheel

内齿离合器 internal gear clutch

内齿轮 annular gear; annular wheel; inner gear(ing); interior gear; internal gear; internal-teeth gear

内齿轮传动(装置) internal-teeth spur gearing; internal gear drive

内齿轮回转泵 internal gear rotary pump

内齿轮检查仪 internal gear tester

内齿轮磨床 internal gear grinder

内齿轮水泵 internal gear pump

内齿轮油泵 gerotor

内齿摩擦片 internal toothed friction plate

内齿啮合 internal gearing

内齿圈轮边减速式驱动桥 internal geared axle

内齿锁紧垫圈 internal tooth lock washer

内齿型锁紧垫圈 internal-tooth washer

内冲头 inner punch

内初切 interior ingress

内厨房 inside kitchen

内储存器 built-in storage; internal memory; internal storage

内处理 in-line processing

内传动的 internally driven

内传动接合 inner drive joint

内传力法预加应力 <有握裹力的预加应力> prestressing with bond

内传力钢丝 compressor wire

内窗 inner window; inside window; internal window

内窗板 inside window panel

内窗层【地】inlier

内窗(花)格 inner tracery; inside tracery; internal tracery

内窗槛 inner window cill[sill]; inside window cill[sill]; interior window cill[sill]

内窗框(架) inner window frame; interior window frame; internal window frame

内窗台 inside window cill[sill]; internal window cill[sill]; stool; window stool

内垂直轴 inner vertical shaft

内唇 inner lip; internal lip

内唇面 inside lip surface

内磁层 inner magnetosphere

内磁场 internal magnetic field

内磁道 inner track

内磁轭 inner yoke

内磁感应强度 intrinsic(al) inductance

内磁铁 internal magnet

内粗外细砖砌体 backing tier

内催化剂床 internal catalyst bed

内存保留区 save memory

内存常驻区 core memory resident

内存程序 stored program(me); stored routine

内存(程序)段 inclusive segment

内存储 physical memory

内存储程序 internal stored program(me)

内存储逻辑 logic(al) in memory

内存储器 built-in storage; inner memory; inner storage; internal memory; internal storage

内存储器容量 internal storage capacity

内存储系统 internal storage system

内存储信息位置图示 topogram

内存储装置 internal storage system

内存存储文件 core file

内存大小 memory size

内存单元 internal storage location

内存地址寄存器 internal memory-address register

内存分配 memory allocation

内存复归程序 core memory reentrant routine

内存管理单元 memory management unit

内存化 memorize

内存缓冲器 core buffer

内存结构 internal storage structure

内存联系指示位 presence bit

内存区 memory field

内存数据 stored data

内存位置 core position

内存信息转储 memory dump

内存映像 core image

内存映像程序库 core image library

内存映像转储 core image dump

内存转储 core dump

内错角 alternate interior angle

内搭接 inside lap; interlap
内大门 inner gate
内带 inner zone
内带滑车 internal band block; internal bound block; internal strapped block
内带夹角 angle between foliation and kink plane inside
内带式制动 expanding band brake
内带外张式制动器 inner band brake
内带闸 internal band brake
内单向阀 <油罐的> internal check valve
内担 liner
内挡泥板 inner fender
内挡圈 inner thrust collar
内档 home record
内导 internal conduction
内导风板 internal air baffle
内导杆 inner guide
内导环 inner pivot ring
内导流堤 back levee
内导曲轨 inside curved lead rail
内导水堤 back levee
内导条 <垂拉窗> guide bead
内导线 inner conductor; inside conductor
内岛 lagoon island
内倒角镀铬活塞环 inside bevel(led) chromium piston ring
内倒角锥形活塞环 inside bevel(led) taper-face piston ring
内倒转术 internal version
内堤 inland dam; inland dike[dyke]; inner dike[dyke]
内底 inner bottom; innersole; invert; tank top; invert level <管道、沟渠等的>
内底板 inner bottom plating; tank top plate
内底标高 <管道的> invert level
内底部最低点标高 invert level
内底高程 grade
内底漆 inner primer; internal primer
内底式建筑 atrium architecture
内地 hinterland; inland; inland native country; interior; midland; outback; up-country; back of beyond <英>
内地保税货站 inland clearance depot
内地槽【地】internal geosyncline; intrageosyncline
内地城市 inland city; inland town; interior city
内地的 fresh water; midland
内地调运站 inland distribution depot
内地段 interior lot
内地港(口) inland harbo(u)r; inland port
内地工业 industry in the interior; inland orientated industry
内地国家 landlocked country
内地海关 inland customs(house)
内地河道 inland waterway
内地火力发电站 inland thermal power plant
内地货物 inland commodity
内地货物集散地 inland clearance depot
内地货运 inland freight haulage
内地货运站 inland depot; inland freight depot
内地集装箱货场 inland container depot
内地集装箱装卸站(栈) inland container depot
内地经济 interior economy
内地壳层 infracrust
内地幔【地】inner mantle
内地偏僻地区 outback
内地沙漠 inland desert
内地省份 inland province
内地市场 market in the interior
内地水道 inland waterway
内地水域 inland waters
内地水运 inland water-borne transport
内地税 inland duty; internal revenue tax
内地向斜 intrageosyncline
内地验关站 inland clearance depot
内地运费 inland forwarding expenses
内地运输 inland freight; inland transport(ation)
内地运输业 <办理港口与内地间的输出、输入货运> inland carrier
内点 inner point; interior point
内点变换 internal transformation
内碘值 inner iodine value
内电池盒 inner battery housing
内电解法 inner electrolysis; interior electrolysis; internal electrolysis
内电解厌氧好氧工艺 interior electrolytic-anaerobic-oxic process
内电抗 internal reactance
内电路 in-circuit
内电势 internal voltage
内电位 inner potential
内电压 internal voltage
内电重量法 internal electrogravimetry
内电阻 internal resistance; source resistance
内电阻熔融法 internal resistance electric(al) melting
内电阻斜率 internal slope resistance
内店铺 <下巷内的> alley house
内殿 cella; inner hall; sanctum; innermost part <寺庙、教堂的>
内殿窗 cella window
内殿后部房间 epinaos
内殿门 cella door
内殿内墙立面 cella facade
内殿墙 cella wall
内吊斗提升 internal skip-hoist
内迭代 inner iteration
内叠关系 inlaid relationship
内叠河漫滩 inlaid flood plain
内叠阶地 inlaid terrace
内叠式洪积扇 inner-lapping pluvial fan
内钉法 inside fastening
内顶板 ceiling board; ceiling sheet; head lining; inside ceiling panel; roof lining; inside ceiling <冷藏车>
内顶板端板 ceiling partition panel
内顶板压条 ceiling mo(u)lding
内顶盖 inner head cover; inner top cover; internal head cover; runner inner lid
内定位条 interior stop
内定向元素 interior element
内定心环 inside centering ring
内定用人名单 slate
内定子 inner stator
内动力地质作用 endogenic geologic(al) process
内动力型的 endodynamorphic
内动力型土(壤) endodynamic(al) soil; endodynamomorphic soil
内动力学 internal dynamics
内堵 <空芯墙的> internal leaf
内端 inner end
内端子 inner terminal
内对称性 internal symmetry
内对搭接板 inside butt strap
内对光 interior focusing; internal focusing
内对光透镜 interior focusing lens; internal focusing lens
内对光望远镜 inner focusing telescope; interior focusing telescope; internal focusing telescope
内对光系统 interior focusing system
内对角 inner opposite angle; interior opposite angle
内多倍体 endopolyploid
内多倍体化 endopolyploidization
内多倍性 endopolyploidy
内耳 inner ear; internal ear
内发动机 intrinsic(al) motivation
内发热 internal heat generation
内发射线 internal emission line
内阀(门) interior valve; internal valve
内法兰 inner flange; inside flange
内法线 inner normal
内翻 introversion; inversion
内反光器 subsurface illuminator
内反馈 inherent feedback; internal feedback; self feedback
内反馈式磁力放大器 amplistat
内反射 internal reflection
内反射反应堆 inner reflected reactor
内反射观察法 internal reflection observation method
内反射率 internal reflectivity
内反射色 internal reflection colo(u)r
内反向齿轮 internal reversing gear
内方位 inner orientation; interior orientation; internal orientation
内方位误差 error of inner orientation; error of interior orientation; error of internal orientation
内方位元素 element of inner orientation; element of interior orientation; element of internal orientation; plate constant
内防波堤 inner jetty; interior breakwater
内防水 inner waterproofing; inside waterproofing
内防水套 internal applied tanking; internal waterproofing lining of basement
内放射源 internal source
内非正则点 internal irregular point; internally irregular point
内分比【数】internal ratio
内分布型 internal distribution pattern
内分程序 internal block
内分点 internal point of division; plus point
内分点法 method of interior point
内分(割) internal division
内分角线 internal bisector
内分解化合物 endolytic compound
内分类 internal sort; internal sorting
内分离 internal separation
内分泌(腺)的 endocrine
内粉饰浆 interior plaster
内粉刷 internal surfacing
内封补偿 sealed pressure balance; Westland-Irving balance
内封相 enclosed phase
内峰 internal peak
内缝 inseam
内夫尼尔罩 Nipher screen; Nipher shield
内夫氏锤 <电流启闭锤> Neef's hammer
内扶手 <栏杆的> inside handrail; internal handrail; inner handrail
内浮盖式换热器 internal floating-head exchanger
内浮力 interior buoyancy; internal buoyancy
内浮坞门座 inner seat
内符号 insymbol; internal symbol
内符号形式 internal symbol form
内幅板 <带箍车轮> back face plate
内辐射带 inner radiation belt; inner radiation zone
内辐照 internal irradiation
内辐照剂量学 internal irradiation dosimetry
内腐 endosepsis; hollow heart
内腐蚀 etch back; internal corrosion
内负载电路 internal load circuit
内覆盖 interlap
内盖 inner cap; internal shroud
内干扰 internal interference
内甘丁 nergandin
内杆 <坑道内伸缩顶杆的> upper prop
内钢套 inner steel bushing
内港 inner harbo(u)r; interior harbo(u)r; interior port
内港池 inner basin; inner harbo(u)r basin; interior basin
内港防波堤 wave breaker
内港界线 inner harbo(u)r line; inner port line
内港小型防波堤 inner breakwater; inner waterbreak
内杠杆 inner lever
内杠杆套 inner lever bushing
内高 inside height
内阁 cabinet <美>; ministry
内阁总理 premier
内格 internal grid
内格点 internal grid point
内格勒式水轮机 Nagler type turbine
内格累氏规律 delivery date rule; Nagele's rule
内格累氏倾斜 Nagele's obliquity
内格罗河 Negro River
内隔板 inner casing; inner curtain (wall)
内隔壁 endoseptum
内隔环 inner split ring
内隔墙 interior partition; lining wall
内隔墙用的空心砖 partition tile
内隔圈 cone spacer; inner ring spacer
内隔热导流管 guide pipe with inner insulation
内隔声 internal sound insulation
内给水管 internal feed pipe
内公切面 internal common tangent plane
内公切线 inner common tangent; inner tangent common; internal common tangent
内功 internal work
内功函数 inner work function
内攻丝 internal tapping
内供电式加热器元件 internal electrically energizable heater element
内拱 back arch; inner arch; inside arch; internal arch; rere arch
内拱弧梅花雕饰的小双尖拱 two-cusped arch
内拱砌块 <坝的承重墙> subintradosal block
内拱圈 intrados
内拱圈半径 radius of intrados; radius of soffit
内共生 endosymbiosis
内构架转向架 inside-frame bogie
内毂 inner hub
内股公司 <股票不对外公开> close corporation
内骨骼 internal skeleton
内骨架 inner frame; skeleton core
内鼓 inner drum
内鼓摩擦片 inner clutch plate
内鼓式过滤器 inside-drum filter
内固 internal stability
内固数 internal stability number
内刮板 inside strake
内挂板 internal cladding

内挂钩 back hook
内挂脚手架 interior hung scaffold
内关节 intrinsic(al) articulation
内管 inner pipe; inner tube; inside tube; inner core tube <指岩芯管>
内管扳手 inner-tube wrench
内管壁 inside tube wall
内管导向套<岩芯管> inner-tube adapter[adaptor]
内管道 interior conduit
内管短节 inner extension tube
内管管鞋 inner-tube shoe
内管壶<瓷器名> ewer with inner tube
内管径 internal pipe diameter
内管拼合式双层岩芯管 split inner-tube core barrel
内管凸出体 inner-tube projects
内管稳定器 inner-tube stabilizer
内管岩芯卡取器 inner-tube core lifter
内管直径 inside tube diameter
内管注水泥 inner string cementing
内管组件 inner-tube assembly
内管钻头<双层岩芯管突出的> projected inner bit
内光电效应 inner photoeffect; internal photoelectric(al) effect
内光电效应光电管 photoconducting cell; photoresistance cell
内光路 internal optic(al) path; internal short path
内光密度 internal optic(al) density
内光楔调整 internal wedge adjustment
内光学参量振荡 internal optic(al) parametric oscillation
内光学参量振荡器 internal optic(al) parametric oscillator
内光学密度 internal optic(al) density
内光源 flashing light supply
内硅铝地槽【地】ensialic geosyncline
内硅镁地槽 ensimatic geosyncline
内轨 inner rail; inside track; low track
内轨道 inner rail
内轨线 inner line of rail
内轨型 inner-orbital configuration
内滚珠框架轴承 inner track roller frame bearing
内过程 internal procedure
内过梁 internal lintel
内海 closed sea <有岬角环抱的>; coastal sea; enclosed sea; inland waters; inland sea <英>; inner sea; internal sea; island sea; land-locked
内海潮 internal tide
内海港线 inner harbo(u)r line
内海航运线 barge line
内海石油蕴藏量 onshore oil reserves
内海引航员 inland sea pilot
内含报酬率 internal revenue rate
内含边材 double sapwood; included sapwood; inside sapwood; internal sapwood
内含成本 implicit cost
内含成员 include member
内含程序 inclusive routine
内含的 inset
内含反应 inclusive reaction
内含过程 inclusive process
内含校验 built-in check
内含节 enclose
内含颗粒 inclusion granule
内含利率 implied rate of interest
内含利息 implicit interest
内含利息的开支 implicit interest charges
内含利息收入 implicit interest revenue
内含量 intensive amount; intensive magnitude
内含谱 inclusive spectrum
内含气体<缝隙中的> included gas
内含韧皮部 included phloem; internal phloem; interxylary phloem
内含示踪剂 intrinsic(al) tracer
内含体 inclusion body
内含文件 include file
内含纹孔口 included pit aperture
内含物 inclusion
内含物不多 lack content
内含误差 inherent error
内含增长率公式 implied growth rate equation
内含政策 implicit policy
内含子 introne
内函数 inner function; intrinsic(al) function
内涵 connotation; intension; interior extent
内涵的 intensive
内涵法推断的温度 temperature of interpolation
内涵价格 hedonic price
内涵纹孔口 included aperture
内涵意义 connotative meaning
内焊缝 inside weld
内行操作 hands-on operation
内航道 intracoastal waterway; interior channel
内壕 escarp; scarp
内壕悬崖 escarp
内耗 inner friction; internal friction
内耗不等式 internal dissipation inequality
内耗热 internal heat rate
内耗系数 decay coefficient
内河 inland river; inland sea; inland stream; inland waters; internal waters; national waters; inland waterway
内河驳船 fluvial barge; inland barge
内河驳船队 river tow
内河船(舶) inland boat; inland craft; inland fleet; inland ship; inland vessel; inland waterway vessel; river boat; river craft; river fleet
内河船舶保险 river hull insurance
内河船队 inland fleet; inland waterway fleet; river fleet
内河船坞 inland (dry) dock; river dock
内河船闸 river lock; waterway lock; inland waterway lock
内河导航雷达 river navigation radar
内河导航系统 river navigation system
内河到岸价格 cost, insurance, freight inland waterway
内河的 river-bound
内河服务艇 river service launch
内河浮标 river buoy
内河港(埠) inland port; inland waterway port; river port; waterway harbo(u)r
内河港口 barge port; barging port; inland harbo(u)r; inland port; river harbo(u)r; river port; inland waterway port
内河港口疏浚 dredged inland port
内河工作船 river utility craft
内河航标 aids-to-navigation on inland waterway
内河航标等级 classification of inland aids to navigation
内河航标制式 aids-to-navigation system on inland waterway
内河航道 inland channel; inland navigation channel; inland navigation fairway; inland waterway; interior waterway; navigable inland channel; river channel
内河航道分级标准 inland waterway classification standard
内河航道公报 inland waterway bulletin
内河航道网 inland waterway network
内河航道系统 inland waterway network
内河航行 internal navigation; river navigation; inland (waterway) navigation
内河航行规章 inland navigation regulation; inland rule
内河航行权 inland navigation rights; right of inland navigation
内河航行图 inland waterway chart
内河航行系统分析 inland navigation system analysis
内河航运 inland (waterway) navigation; internal navigation; river navigation; river shipping; river traffic
内河航运港 inland navigation harbo(u)r
内河航运工程 inland navigation project
内河航运提单 inland waterway bill of lading
内河货物运输 carriage of goods by inland river
内河货运量 river traffic
内河集散系统 river feeder system
内河交通 river traffic
内河军用艇 riverine warfare craft
内河可用船舶数 number of inland vessels available
内河领航雷达 river navigation radar
内河码头 domestic pier; inland terminal; inland terminal depot; river quay; riverside jetty; river wharf
内河平底驳船 lighter
内河三角洲 indelta
内河深水航道<可航远洋船的> seaway
内河石油库 river terminal
内河疏运系统 river feeder system
内河水道 inland waterway; interior waterway
内河水道网 inland waterway network
内河水域 inland waters
内河水运 inland water transportation
内河水闸 inland lock
内河税 river dues
内河提单 river bill of lading
内河通航标准 navigation standard of inland waterway
内河通航水域 navigable internal waterways in use
内河推轮 river pusher; stream pusher
内河拖带 river towing
内河拖轮 river tug
内河挖泥船 river dredge(r)
内河吸泥船 river suction dredge(r)
内河小船 river boat
内河小艇 river launch
内河巡逻警察 river patrol
内河巡逻艇 river patrol craft
内河引航员 river pilot
内河渔业 inland fishery; river and lake fishery
内河运费 river freight
内河运煤船 griper; inland water gripper; inland waterway gripper
内河运输 river transport; inland water(way) transport(ation); river traffic; transport over inland waterways
内河运输保险 river transportation insurance
内河运输费 river freight
内河造船厂 river yard
内河治安警察 river police
内河治安警察分局 river police station
内河助航标志 aids-to-navigation on inland waterway
内河装舱挖泥船 river hopper dredge(r)
内核 inner core; kernel
内核层 inner nuclear layer
内核技术 kernel technology
内核仁 endonucleolus
内核外核界面 inner-outer core boundary
内核栈无效失败 kernel stack not valid abort
内核质量 mass of inner core
内荷载<集装箱等的> internal loading
内横隔板 interior diaphragm
内喉板 inside throat sheet
内后四分体 inner posterior quadrant
内弧 inner arc; knock-knee
内弧断层 fault of intrados
内弧盆地 inner-arc basin
内弧曲率 intrados curvature
内弧揉皱带 crumple zone of intrados
内湖 enclosed lake
内护板 inner casing
内护道 inner berm
内护轨 inner guard rail; inside guard rail
内护舷纵材 inwale
内花键 female spline; internal spline
内花键量规 inside spline ga(u)ge
内花园 inner garden
内华达<美国州名> Nevada
内华达造山运动 Nevadian orogeny; Nevadic orogeny
内滑 inside slip
内滑道 inner slide
内化作用 internalization
内环 inner belt; inner circle; inner race; inner ring; internal ring; minor loop
内环层 inner circular layer
内环的 endocyclic
内环骨板 internal circumferential lamella
内环境 internal milieu
内环境平衡 homeostasis
内环弹簧<缓冲器的> inner ring
内环境稳定 homeostasis
内环裂<木材的> internal shake
内环路 inner belt; inner circle; inner circumference highway; inner loop; inner ring road
内缓冲器 buffer; internal inner ring
内缓冲区 internal buffer
内回采工作面 inner stope
内回归 internal regression
内回流 inner reflux; internal reflux
内回流比 internal reflux ratio
内回路 inner loop
内回授 self feedback
内回授放大器 self-feedback amplifier
内回转 internal rotation; rifle-bar rotation <凿岩机>
内回转凿岩机 self-rotating rock drill
内混合 internal mixing
内混合喷枪 internal mixing spray gun
内混合式烧嘴 inner mixing air type burner
内混合型多组分喷嘴 internal mixing multiple-component spray nozzle
内混合岩化方式 endomigmatization way
内混雾化器 internal mix atomizer
内混浊 internal haze

N

内活动接头 < 万向接头的 > inner casing of joint
内活塞 inner carrier; inner piston
内火山构造 endovolcanic structure
内火山锥 volcanello
内火室锅炉 internal firebox boiler
内火箱 inner fire box
内货变糖浆 contents becoming molasses
内货不详 contents unknown
内货短少 contents short
内货发芽 contents sprouting
内货漏出 contents leaking out
内货气味外溢 contents smelling out
内货全无 contents empty
内货融化 contents melted
内货脱出 contents running out
内货外露 contents exposed
内货沾污 contents stained
内货重量不足 contents short weight
内机座 <汽轮发电机的> inner cage
内积 internal product; scalar product; inner product
内积反馈控制 inner product feedback control
内积空间 inner product space
内积控制 inner product control
内基线测距仪 self-contained-base range finder
内基准发生器 internal reference generator
内基准抑制 internal reference muting
内级存储器 inner level memory
内级光带 inner auroral zone
内极点 inpolar
内极二次曲线 inpolar conic
内极式交流发电机 inner-pole type alternator
内极式同步发电机 internal field alternator
内极位置 pole inside figure
内棘轮 < 底开门车 > handle ratchet wheel
内集 inner set
内集汽管 internal steam pipe
内计时器 internal timer
内计数 inside counting
内计数管 internal counter
内剂量 internal dose
内寄生物 endoparasite; endosite; entorganism
内祭坛天盖 peristerium
内加工指令 internal manipulation instruction
内加工装置 internal attachment
内加厚 inside upset; internal upset(ting)
内加厚油管 tubing with internal upset ends
内加厚钻杆 internal upset drill pipe
内加热 inner heating; internal heating
内加热法破岩 internal heating method for rock fragmentation
内加热高压釜 internally heated pressure vessel
内加热面 inner heating surface
内加热气体装置 internally heated gas apparatus
内加压隔间 pressurized compartment
内加压司机室 pressurized compartment
内夹角 internal angle
内夹棉纱的橡胶板 cotton cloth inserted rubber sheet
内架 inner tower < 觇标的 >; skeleton core < 空心门的 >
内假潮 internal seiche
内尖角 square corner
内尖角嵌接 square-corner halving
内间距比 < 取土器的 > inner clearance ratio; inside clearance ratio; ratio of internal interval
内间隙比 inner clearance ratio; inside clearance ratio; ratio of internal interval
内肩 internal shoulder
内肩车削 internal shoulder turning
内肩节 < 平旋桥 > inner hip
内肩镗孔 internal shoulder boring
内剪强度包线 intrinsic(al) shear strength curve
内检 internal control analysis
内检分析 internal control analysis
内检实验室 laboratory for internal examination
内检相对误差 internal examining relative errors
内检样品化学分析 chemical analysis for internal examination
内检样品数 number of samples for internal examination
内建场 built-in field
内建电路 built-in channel
内建电压 built-in voltage
内建功能 built-in function
内建管道 built-in channel
内建偏压 built-in bias
内建势 built-in potential
内键 internal key
内键槽 internal keyway
内键盘 internal keyboard
内浆 entoplasm; entosarc
内浇道 gate; ingate
内浇道厚度 gate thickness
内浇道宽度 gatewidth
内浇口 flow gate; ingate; ledge
内浇口心 ingate core
内胶结强度 internal bond
内焦点 interior focal point
内角 inner angle; inside angle; interior angle; internal angle; internal corner; reentrant corner; reentrant part
内角撑杆 inside corner brace
内角导线 interior angle traverse
内角拱 squinch
内角焊 inside fillet weld
内角焊缝 inside fillet
内角焊接 inside fillet welding; inside corner weld(ing)
内角加工 corner cut
内角加强用三角条 corner block
内角抹子 inside-angle tool
内角石属的化石 endoceratite
内角线条 inside corner mo(u)lding
内角镶条 inside corner mo(u)lding
内角削斜的线脚 sprung mo(u)lding
内矫顽磁力 intrinsic(al) coercivity
内脚手架 internal scaffold(ing)
内脚手砌墙法 overhand work
内铰合板 inner hinge plate
内铰孔 internal reaming
内校验测试 internal verification testing
内校样 reader's set
内校准 internal calibration
内校准源 internal calibration source
内接 inscribe
内接齿轮 annulus[复 annuli/annuluses]
内接触 interior contact; internal contact
内接的 inscribed
内接多边形 inpolygon; inscribed polygon
内接多面体 inpolyhedron; inscribed polyhedron
内接符 inconnector
内接管 interconnecting pipe
内接活管接 socket union
内接角 inscribed angle
内接控制区 inscribed control region
内接棱柱(体) inscribed prism
内接棱锥 inscribed pyramid
内接三角形 inscribed triangle
内接三棱形 inscribed prism
内接四边形 inscribed quadrilateral; inscribed square
内接头 nipple; collar bushing < 深井泵的 >
内接图形 inscribed figure
内接小齿轮 internal pinion
内接圆 incircle; inscribed circle
内接圆心 incenter
内接圆锥 inscribed cone
内节 internal segment
内节点 interior nodal point; internal nodal point; interior node; internal node
内结合能 internal bonding energy
内结晶 interior crystalline
内结晶水 intercrystalline water; intracrystalline water
内截面 internal cross-section
内解 interior solution
内介壳 endostracum
内界面 interface
内界面泡沫浮选 pulp-body froth flo(a)tation
内界膜 internal limiting membrane
内进汽 < 蒸汽机的 > inside admission; internal admission
内进汽汽阀 inside admission valve
内进站信号【铁】inner home signal
内经两脚规 inside cal(l)ipers
内经千分表 inside micrometer
内颈轴 inside journal axle; within inside journals axle
内净径 inner clearance
内净径比 inner clearance ratio; inside clearance ratio
内净空比 inner clearance ratio
内径 bore(size); clear mesh; clear width; core diameter; inner diameter; inradius; inside diameter; inside radius; internal diameter; width in the clear; cylinder < 数据库用 >
内径边刃金刚石 inner diameter ga(u)ge stone
内径标准尺 inner diameter ga(u)ge
内径表面检查仪 boroscope
内径测微表示器 inside dial indicator
内径测微计 inside micrometer
内径测微器 inside micrometer; telescope cal(l)ipers
内径测微指示计 inside indicator
内径代号 inside diameter character
内径法 internal diameter method
内径杆规 end measuring ga(u)ge; end measuring rod
内径公差 inner diameter tolerance
内径规 hole ga(u)ge; inside ga(u)ge; internal ga(u)ge; male ga(u)ge; plug ga(u)ge; inside calipers
内径弧长 length of inner diameter arc
内径极限规 internal limit ga(u)ge
内径精测仪 indicating plug ga(u)ge; indicating snap ga(u)ge; passimeter
内径锯 inside diameter saw[ID saw]; inner diameter saw; tubular saw
内径卡尺 inside cal(l)ipers; internal caliber ga(u)ge
内径卡规 inside cal(l)ipers; internal cal(l)iper ga(u)ge
内径量测仪器 internal measuring instrument
内径量规 inside ga(u)ge; straight cal(l)ipers
内径配合 minor diameter fit
内径千分表 inside dial indicator; internal dial ga(u)ge; internal micrometer
内径千分(卡)尺 inside cal(l)ipers ga(u)ge; inside dial indicator; inside micrometer; inside micrometer cal(l)ipers; micrometer for inside measuring; internal micrometer
内径掏槽刃 inside diameter kicker
内径隙规 inside clearance
内径应力 bore stress
内径指示规 passimeter[passometer]
内镜筒 inner lens cone
内矩 internal moment
内距 internal spur
内聚 linkage
内聚功 cohesional work; cohesive work
内聚机制 cohesion mechanism
内聚焦望远镜 interior focusing telescope; internal focusing telescope
内聚结构 coherent structure
内聚力 adhesive power; coherence; cohesion; cohesive force; force of cohesion; power of cohesion
内聚力断裂 cohesive fracture
内聚力摩擦力应变试验 cohesion-friction-strain test
内聚力强度 cohesive strength
内聚力系数 coefficient of cohesion; cohesional coefficient
内聚力种类 cohesion type
内聚能 cohesive energy
内聚能密度 cohesive energy density
内聚黏[粘]结 cohesive bonding
内聚破坏 adhesive failure; cohesion failure
内聚强度 cohesive strength
内聚束 internal bunching
内聚束力 binding force
内聚体 interpolymer
内聚性 cohesion; cohesiveness
内聚压力 cohesion pressure; cohesive pressure
内卷 involution
内卷的 involute
内绝热建筑 inside insulating building
内绝缘 inner insulation; inside insulation; internal insulation
内均质结 built-in homojunction
内菌幕 inner veil
内开 inward opening; opening in < 门、窗的 >
内开边梃铰接的 in-swing side hinged
内开槽 internal recessing
内开窗 inward opening window
内开窗扇 in-swinging casement
内开门 in-opening door; inwardly opened door; inward opening door
内开扇 casement opening in; inwardly opened casement
内开上悬窗 top-hung window opening inwards
内开式窗 in-opening window; inswinging casement window; inwardly opened window
内开式窗扇 casement opening in; inwardly opened casement; swinging-in casement
内开锁 < 内用把手外用钥匙开的锁 > drawback lock; drawbolt lock
内开下旋窗扇 hopper frame
内开下旋气窗 hopper light; hopper lite
内开旋窗 top-hinged in swinging window
内抗矩 internal resisting moment; intrinsic(al) resisting moment
内科 internal medicine
内科病房 medical ward

内科学 medicine
内科治疗 medicine
内颗粒 endoparticle
内颗粒层 internal granular layer
内壳 inner casing;inner hull;pressure hull;endoconch【地】;inner lining <垂直扯窗窗框内重锤箱>
内壳层 hypostratum; internal shell; inner shell
内壳电离 inner-shell ionization
内壳缝 inner fissure
内壳构造 infrastructure
内空隙 internal pore;internal void
内空闲时间 internal idle time
内孔 bore
内孔表面检查仪 boroscope
内孔的 female
内孔翻边镦粗模 coining dimpling die
内孔规 hole ga(u)ge
内孔检视仪 borescope;introscope
内孔铰刀 internal reamer
内孔窥视仪 introscope
内孔拉削 burnish broaching
内孔连接器 female connector
内孔隙度 internal porosity
内孔隙率 internal porosity
内控 internally piloting
内控标准 inner quality standard
内控阀 internally piloted valve
内口板 endostoma
内口倒角 corner-reversal cutting
内口式 entognathous type
内扣扳机 inside trigger
内扣式接口 hermaphroditic coupling
内跨度 interior span; inner span; internal span
内跨梁 interior span
内宽 inner width;inside width
内宽外窄的开口漏斗状斜面墙 embrasure
内框 inside casing;inside lining;inside trim;interior casing
内框架 inner frame
内框架结构 bearing wall and frame structure;framed structure without exterior columns
内窥镜 endoscope
内窥镜灯泡 bulb for endoscope
内窥镜检查 endoscopy; splanchnoscopy
内窥图像 endoscopic picture
内窥仪 endoscope
内扩散 inner diffusion;interior diffusion;internal diffusion
内扩散结 internally diffuse(d) junction
内拉床 internal broacher; internal broaching machine
内拉刀 inside broach;internal broach
内拉幅装置 internal stretcher
内拉杆 draw-in bar;inside link
内拉格朗日点 inner Lagrangian point
内拉簧夹盘 draw-in chuck
内拉簧卡盘 draw-in chuck
内拉簧套圈 draw-in collet
内拉夹套 pull-in collet
内拉孔 internal broaching
内拉削 internal broaching
内廊 central corridor; centre corridor; interior corridor; internal corridor;middle corridor
内廊道式公寓建筑 gallery apartment building; inner [internal/ interior] gallery apartment building
内廊式 double-loaded corridor type; middle corridor type
内廊式房屋 interior-corridor type building
内廊式住宅 middle corridor type dwelling house
内廊式住宅建筑 central-corridor residential building
内涝 inland inundation; land flood(ing);water-logging
内冷 inner-cooling
内冷电机 inner-cooled machine
内冷定子线圈 inner-cooled stator coil
内冷发电机 internal cooling generator
内冷发电机转子 hollow-conductor-cooled rotor
内冷凝器 inner condenser
内冷却 internal cooling
内冷却磨削 internal cooling grinding
内冷式吹管 internal cooling blow pipe
内冷式的 internal cooled
内冷水 internal liquid cooling
内冷铁 densener; inner chill; internal chill
内冷增压 internally cooled supercharging
内离合齿轮 internal clutch gear
内里 inwardly
内里层 endonexine
内里克斯炸药 Nerex explosive
内力 endogen force; hypogene; inner force;internal force
内力包络图 enveloping curve of internal force
内力地质作用 endogenic force of geologic(al) function
内力分布 internal force distribution
内力构造的 endotectonic
内力矩 interior moment; internal moment;moment of resistance
内力偶 internal couple
内力偶臂 arm of internal force couple
内力偶法 internal couple method
内力强度 intensity of internal force
内力素 elements of internal force
内力图 internal force diagram
内力协调方法 <结构力学上的> compatibility method
内力作用【地】endogenetic action; endogenetic process; endogenic action;hypogene action
内力作用过程 endogenic process
内立管 internal riser pipe
内粒层 internal granular layer
内粒层纹 stria Baillargeri internal
内连杆 inside connecting rod
内连(接) interconnection; intraconnection
内联企业 domestically associated enterprise
内链接 internal chaining
内梁 inner beam;interior beam
内量规 caliber ga(u)ge; caliper ga(u)ge
内量子数 inner quantum number
内列板 inner strake; inside strake; sunken strake
内裂 honeycombing; implode; implosion;interior check;internal break; internal check;internal shake
内裂缝 clinking;honeycomb
内磷负荷 internal phosphorus loading
内流 inflow; influx; interior flow; internal flow
内流的 in-streaming
内流电流 inflow current
内流动性 internal mobility
内流湖 basinal lake; endor(h)eic lake;inland lake
内流空气动力学 internal aerodynamics
内流流域 basin of internal drainage; closed basin;closed drainage basin
内流盆地 basin of internal drainage; cut-off basin; endor(h)eic basin; inland basin
内流区 endor(h)eic region
内流熔岩【地】interfluent
内流熔岩流 interfluent lava flow
内流式水轮机 centripetal turbine; inward flow turbine
内流式透平 centripetal turbine; inward flow turbine
内流式涡轮机 centripetal turbine; inward flow turbine
内流水系 blind drainage; closed drainage;endor(h)eic system
内流线型接头 <冲洗液孔呈锥形> stream-flow coupling; counterbored coupling
内六角扳手 inner hexagon spanner; socket screw hexagon wrench; hexagon socket screw key
内六角导向螺母 socket pilot nut
内六角螺钉 cap screw; socket head screw
内六角螺(丝)钉扳手 allen wrench; socket screw wrench
内六角头固定螺钉 hollow head setscrew
内六角头机螺钉扳头 hexagonal socket cap screw wrench
内六角头螺(丝)钉 Allen screw; socket head screw
内龙骨 inner keel;keel batten;keelso
内龙骨护板 false keelson
内龙骨角钢 keelson angle
内龙骨翼板 futtock plate; limber strake;keelson plate
内龙骨与肋板连接的短角钢 keelson lug
内楼面 interior flooring
内楼梯 internal stair(case)
内楼梯基 rough string
内楼梯拦杆 inner stair(case) rail; inside stair(case) rail; internal stair(case) rail
内楼梯斜梁 inner string
内露层【地】inlier
内炉离子源 internal furnace ion source
内炉膛 internal furnace
内陆 continental interior; hinterland; inland;interior continent
内陆冰(川) inland ice
内陆常绿阔叶林 hammock forest
内陆城市 inland city
内陆单旋回盆地 interior single cycle basin
内陆的 intracoastal; landlocked; midland;outback
内陆低地 interior lowland; intracontinental lowland
内陆地槽 intracontinental geosyncline
内陆地区 endor(h)eic region; inland country; inland region; landlocked country
内陆地震 inland earthquake
内陆复合盆地 intracontinental composite basin
内陆钢铁公司 <美> Inland Steel
内陆港(口) inland harbo(u)r; inland port;landlocked harbo(u)r
内陆高原 continental plateau
内陆公共点 overland common point
内陆共同点 overland common point
内陆国(家) landlocked country; landlocked state
内陆海 landlocked sea; inland sea <英>; continent(al) sea; epicontinental sea;island sea;interior sea
内陆海堤 inland sea dike[dyke]
内陆航道 thoroughfare
内陆航行运河 inland navigation canal
内陆河(流) continental river; continental stream; fluvial river; inland river;inland stream
内陆河(流)流域 endor(h)eic drainage;endor(h)eic drainage basin
内陆弧 inland arc
内陆湖 astatic lake;closed lake;endor(h)eic lake; inland lake; interior lake;landlocked lake;static lake
内陆湖泊 enclosed lake
内陆集装箱 inland container
内陆架 inner continental shelf
内陆交货价格 inland delivery price
内陆流域 continental basin; enclosed basin; endor(h)eic basin; endorheism
内陆流域盆地 inland drainage basin
内陆排水干线 <堤坝> main-line of inner drainage
内陆排水系 inland drainage; interior drainage
内陆盆地 continental basin; enclosed basin; endor(h)eic basin; inland basin;interior basin
内陆盆地红层组合 inner continental basin red beds association
内陆盆地型 continental basin type
内陆拼装站 inland consolidation depot
内陆平原 inland plain;interior plain
内陆气候 inland climate
内陆萨巴哈沉积 inland Sabkha deposit
内陆萨巴哈沉积模式 inland Sabkha sedimentation model
内陆萨布哈相 inland Sabkha facies
内陆三角洲 indelta;inland delta;interior delta
内陆沙漠 inland desert
内陆沙丘 inland dune
内陆砂 inland sand
内陆山间盆地型 continental intermountainous basin type
内陆商港 <英> close port
内陆枢纽站 inland terminal depot
内陆水 inland water; inner water; internal water
内陆水道 inland watercourse; inland waterway; interior waterway; intracoastal waterway
内陆水道港口 inland waterway port
内陆水路系统 inland waterway
内陆水上运输 inland water transport
内陆水上运输保险 inland marine insurance
内陆水体 inland waters
内陆水体富营养化 eutrophication of inland waters
内陆水文测量 inland water survey
内陆水系 closed drainage; endor(h)-eic drainage; inland waters; internal drainage;landlocked drainage
内陆水域 inland waters; inner waters;inside waters;internal waters; landlocked waters;national waters
内陆水域环境的无害管理 environmental innocuity management of inland waters
内陆水域污染 inland waters pollution
内陆水域渔业 inland waters fishery
内陆水运 inland water-borne transportation; inland (waterway) navigation
内陆水运(保)险 inland marine insurance
内陆水运提单 inland waterway bill of lading
内陆税收 inland revenue
内陆损害 <海上货物保险附加险> country damage
内陆提单 inland bill of lading; inland BL

N

内陆提单条款 inland bill of lading clause
内陆通关基地 inland clearance depot
内陆洼地 depression of land
内陆洼地电站 inland depression plant;land depression plant
内陆雾 inland fog
内陆卸货地点 inland place of discharge
内陆移动沙丘 blowing dune;inland-moving dune
内陆用集装箱 Binnen container;Maxicadre container
内陆渔业 inland fishery
内陆运费 inland freight
内陆运费用 inland forwarding expenses;inland haulage;inland transportation charges
内陆运河 barge canal;inland canal;inland navigation canal
内陆运输 inland freight;inland shipment;inland traffic;inland transport(ation);overland transportation
内陆沼泽 continental swamp;boli
内陆沼泽土 continental swamp soil
内陆种(类)landlocked species
内陆装卸站 inland depot
内滤鼓式过滤机 internal drum filter
内滤式过滤器 inside-drum filter;internal surface filter
内路诊断 internal diagnostics
内履 inside shoe
内乱 civil commotion
内轮迹带 inner wheelpath
内轮迹线 inner wheelpath
内轮廓锯法 internal contour sawing
内轮裂 <木材的> internal shake
内轮胎补缀 tube patch
内罗毕 <肯尼亚首都> Nairobi
内罗温泉 Thermae of Nero
内螺车丝机 pipe tap drill
内螺模 male die
内螺(丝)(齿纹)internal thread
内螺纹 box thread;female thread;inside thread;internal screw thread;internal thread
内螺纹半锁接箍 box coupling
内螺纹半锁接头 box joint
内螺纹部分 box member
内螺纹车刀 internal screw cutting tool;internal threading tool
内螺纹车丝锥 pipe drill;pipe tap
内螺纹车削 internal threading
内螺纹导程 inside lead
内螺纹导程(螺距)仪 inside lead ga(u)ge
内螺纹管 ribbed pipe;ribbed piping;ribbed tube;ribbed tubing;rifled pipe;rifled tube
内螺纹管接头 female connector;female fitting
内螺纹管接头配件 female fitting
内螺纹管口 female end of pipe
内螺纹规 internal screw ga(u)ge;internal thread ga(u)ge
内螺纹过渡管接头 female adapter
内螺纹过渡接头密封环 female support ring
内螺纹接头 female adapter
内螺纹截止阀 internal screw thread stop valve
内螺纹卡尺 thread inside cal(l)ipers
内螺纹连接 female connection
内螺纹连接管 female coupling
内螺纹联管节 female union
内螺纹量规 female thread ga(u)ge
内螺纹螺栓 female screw
内螺纹面 box face
内螺纹明杆 inside screw rising stem
内螺纹磨床 internal thread grinder;internal thread grinding machine
内螺纹抛光 internal screw finish;internal thread finish
内螺纹配件 female fitting
内螺纹瓶口 internal screw finish;internal screw thread finish
内螺纹塞规 internal screw ga(u)ge
内螺纹三通 female branch tee[T]
内螺纹三通管接 female branch tee[T]
内螺纹梳刀 inside chaser
内螺纹外径 root diameter
内螺纹弯管接头 female elbow
内螺纹弯头 female ell
内螺纹旋塞阀 internal screw thread cock valve
内螺纹终端接头 socket end fitting
内螺纹钻杆 internal threaded drill rod
内螺纹钻头 box type bit;internal threaded bit
内螺旋 female screw;inside spin;internal screw;internal spiral
内螺旋泵 internal screw pump
内螺旋联管节(双向)female union
内螺旋束 inner spiral bundle
内螺旋线 <旋流集尘器> inner vortex
内络合物 inner complex
内络盐 inner complex salt
内落水 interior drainage
内落水管 built-in gutter
内埋件 insert
内埋裂纹 buried crack
内埋缺陷 buried flaw
内埋式采暖板 embedded heating panel
内曼抽样 Neyman sampling
内曼模型 Neyman model
内曼配置 Neyman allocation
内曼-皮尔逊理论 Neyman-Pearson theory
内曼-斯科特模型 Neyman-Scott model
内毛细水 inner capillary water
内锚 shore anchor
内锚泊地 inner roadstead
内门 inner door;inside door;interior door;internal door
内门侧柱 interior door jamb
内门窗框 inner casing
内门窗梃 inner stud;inside stud
内门架 inner mast【港】;inner upright mounting <叉车>
内门坎板 <车身> inner sill panel
内门框 door lining
内门框装修 framed lining
内门廊 <在古庙神坛尾部的> opisthodomos;posticum
内闷光(毛玻璃)灯泡 inside-frosted lamp
内醚 inner ether
内密封 internal sealing
内冕 inner corona
内面 inner face;medial surface;soffit;underside
内面带法兰盘的丘宾筒 inner face tubing
内面的 interior
内面积比 internal area ratio
内面校正 internal adjustment
内面磨砂 internal frosting
内面磨削机 internal grinder
内面墙 interior wall
内面铣刀 internal milling cutter
内面阻 interface layer resistance
内模 center[centre] form(work);inner mo(u)ld;internal frame(work);internal mo(u)ld
内模板 inner formwork;internal shuttering
内模挤压(制管)法 internal moulding process
内模壳 internal shuttering
内模式 internal schema
内模型 inner model
内膜 inner capsule
内膜系数 inner film coefficient
内摩擦 intrinsic(al)friction;viscosity
内摩擦加捻原理 internal friction twist principle
内摩擦角 angle of internal friction;internal angle of friction;internal friction angle
内摩擦力 anelasticity;internal frictional force;inner friction;internal friction
内摩擦损失 internal friction(al)loss
内摩擦系数 coefficient of friction;coefficient of internal friction;internal friction coefficient
内摩(擦)阻力大的土(壤)friction(al)soil
内摩阻力 internal friction
内磨附件 internal grinding attachment
内磨砂玻璃灯 inside-frosted lamp
内磨砂灯泡 inside-frosted bulb;internally frosted bulb
内磨削夹具 internal grinding fixture
内磨作用 internal grinding action
内末端曲线 inner terminal curve
内木栓层 internal cork
内目录 in-list
内幕交易 insider dealing;insider trading
内能 inner energy;internal energy;intrinsic(al)energy
内能除霜 internal defrosting
内能模型 internal-energy model
内能热 internal heat
内能消耗率 rate of internal energy combustion dissipation
内黏[粘]聚力 <土或材料的> internal cohesion
内黏[粘]聚力破坏 cohesive failure
内黏[粘]滞性 internal viscosity
内黏[粘]滞阻尼 internal viscous damping;intrinsic(al)viscous damping
内啮合 inner gearing;inside gearing;internal gearing;internal toothing
内啮合齿轮 inside engaged gear;internal gear
内啮合齿轮泵 crescent gear pump;gear-within-gear pump;internal gear pump
内啮合齿轮沥青泵 inside gear asphalt pump
内啮合齿轮马达 crescent gear motor;gear-within-gear motor
内啮合回转轮盘 inner mesh slewing rim
内凝聚力 internal cohesion
内扭矩 internal torque
内扭转 intorsion;intort
内爬塔式起重机 internal self-climbing tower crane
内排齿 inner row teeth
内排气管 inside exhaust pipe
内排水 indirect drainage;inside drainage
内排水系统 indirect drainage system;inside drainage system;interior storm system
内盘管蒸发器 internal coil evapo(u)rator
内盘旋 inside turn
内旁瓣 near-in sidelobe
内配流径向柱塞泵 radial piston pump with interior admission
内配位层 inner coordination sphere
内皮 inside skin;under bark
内皮尔对数 Napierian log
内皮漆 interior varnish
内平板龙骨 inner flat keel;inner keel
内平的 internal flush
内平管 <钻杆> flush-joint pipe
内平衡 inner equilibrium
内平衡环 inner gimbal;inner gimbal ring
内平衡架 inner gimbal
内平接头连接套管 flush-joint liner
内平均洪流量 mean annual flood
内平均径流量 mean annual runoff
内平均流量 mean annual discharge
内平均温度 mean annual temperature
内平扣 internal flush thread
内平油管 tubing with plain ends
内平钻杆 internal flush drill pipe
内平钻杆接头 internal flush jointed coupling;internal flush tool joint
内屏蔽 inner shield;internal screening;internal shield
内屏蔽圈 internal shield ring
内坡 back slope;inner slope;inside bank;inside slope;inslope;inward slope;landside slope
内坡道 <立体交叉的> internal ramp
内坡机 back sloper
内坡屋顶 double lean-to roof;V-roof
内破裂 implosion;internal broken
内剖面 interior profile
内栖动物 infauna
内栖生物 endobiont
内气体计数管 internal gas counter
内气压 internal gas pressure
内汽缸 inner casing;inside cylinder
内汽缸机车 inside connected locomotive
内砌砖 inbond brick
内碛 internal moraine
内卡板管接头 socketed grip
内卡尺 inside calipers
内卡规 caliber ga(u)ge;caliper ga(u)ge;inside cal(l)ipers
内卡钳 inner cal(l)ipers;inside cal(l)ipers;internal cal(l)ipers
内卡钳千分表 inside cal(l)ipers micrometer
内迁移 internal migration
内前四分体 inner anterior quadrant
内潜能 internal latent heat
内嵌 encapsulation
内嵌板 infill panel
内戗道 inner berm
内腔 bore;cave
内腔Q开关 internal Q-switch
内腔加工 chambering
内腔检视仪 introscope
内腔镜 endoscope
内腔镜检查 endoscopy
内腔镜摄片投影仪 endoscopic film projector
内腔容积测定法 endometry
内腔式 intracavity
内腔式变像管 intracavity image converter
内腔式电光调制器 intracavity electrooptic modulator
内腔式共振器 intracavity resonator
内腔探视仪 introscope
内墙 inside wall;internal wall(ing)
内墙板 inner lining;inner plate;inside lining;interior panel;interior wall lining;lining;lining board
内墙板垫木 inside lining stud
内墙板间柱 lining stud
内墙板条 lining strip
内墙表面 inside wall surface
内墙干饰面 drywall finish
内墙干饰面材料 drywall material

内墙基石 inside wall sill
内墙角 interior corner
内墙角加固件 interior corner reinforcement
内墙角加强筋 interior corner reinforcement
内墙角增强筋 interior corner reinforcement
内墙金属龙骨 interior metal stud
内墙框架 inside wall frame
内墙面 inner surface of wall;internal wall surface; surface of internal wall
内墙抹面 parget
内墙饰面 interior facing
内墙饰面底涂 interior primer
内墙(镶)板 internal wall(ing) panel
内墙小窗 hagioscope
内墙用金属立筋 interior metal stud
内墙装饰板 gypsum drywall
内墙装饰板锯 drywall saw
内墙装饰板切割刀 drywall knife
内墙装饰板砂眼缺陷 drywall blister
内墙装饰板用钉 drywall nail
内墙装饰覆盖物 wall covering
内桥连接 internal bridge connection
内桥砖 inside bridge wall
内切【数】internally tangent; interior contact
内切刀齿 inside cutting blade
内切断 internal cutting off
内切拱 interior arch
内切河曲 ingrown meander
内切河湾 ingrown meander
内切倾岸壁 quay with battered face; quay with sloping face
内切球 inscribed sphere
内切球心 <四面体的> incenter
内切形【数】inscribed figure
内切圆 circumferential circle; incircle; inscribed circle; internally tangent circle
内切圆圆心 incenter[incentre]
内切圆柱 inscribed cylinder
内侵蚀 internal erosion
内勤工程师 office engineer
内勤人员 back office force
内倾 bank sided; fall home; fall(ing)-in; introversion; slant inward; tumble home; tumble house; tumble in; tumbling-in
内倾岸壁码头 quay with sloping face
内倾的 inward
内倾轨道 laterally angled track
内倾角 angle of toe-in
内倾水平梯田 gradient terrace
内倾舷墙 tumble-home bulwark
内倾(斜)斜度 inward batter
内倾柱 A 形塔架 A-shaped pylon with inward leaning legs
内倾砖层 tumbling course
内倾桩 batter peg
内情通报 <股票等> tip sheet
内情向量 dope vector
内球 endosphere
内区域 inner region
内曲 incurvation; incurve
内曲柄 inside crank
内曲的 reentrant
内曲合拢轨 inside curved lead rail
内曲率 incurvature
内曲面 negative camber
内曲球 incurve
内曲线 inner curve; inside curve
内曲线液压马达 internal curve hydraulic motor
内曲轴 inside crankshaft
内驱力 drive
内屈 introflexion
内屈服压力 <深钻工艺> internal yield pressure
内取芯管 inner barrel; inner tube
内圈 inner circle; inner race; inner ring <滚动轴承的>
内全反射 inner total reflection
内燃 internal combustion
内燃泵 internal combustion pump
内燃叉车 diesel fork lift truck
内燃叉车起重机 engine fork lift
内燃车组 diesel multiple unit
内燃打桩机 diesel pile hammer
内燃捣固机 diesel tamping machine
内燃电车 diesel tramcar
内燃电动车 diesel-electric(al) bus; oil-electric(al) bus
内燃电动传动机车 diesel-electric(al) locomotive
内燃电动的 oil-electric(al)
内燃电动公共汽车 diesel-electric(al) bus
内燃电力车 diesel-electric(al) bus; oil-electric(al) bus
内燃电力传动动车 diesel-electric(al) railcar
内燃电力传动机车 diesel-electric(al) locomotive
内燃电力传动起重机 diesel-electric-(al) transmission crane
内燃电力浮式起重机 diesel-electric-(al) floating crane
内燃电力汽车 oil-electric(al) car
内燃动车 diesel railcar
内燃动车列车 diesel railcar train
内燃动车组 diesel multiple unit; self-propelled diesel train
内燃动车组组合列车 diesel multiple unit train
内燃发电机 oil-electric(al) engine
内燃发电机组 diesel generator set
内燃发电站 diesel power station
内燃发动机 explosive motor
内燃浮式起重机 diesel floating crane
内燃轨行汽车 diesel railbus
内燃锅炉 internally fired boiler
内燃夯实机 internal combustion compactor; internal combustion rammer
内燃和内燃电力机务段 diesel and diesel electric (al) locomotive terminal
内燃化 dieselization[dieselisation]
内燃化线路公里里程 dieseline kilometrage
内燃化线路里程 dieselized kilometrage
内燃活塞式发动机 internal combustion piston engine
内燃机 combustion engine; combustion motor; diesel; diesel engine; gas engine; internal combustion engine; motor; oil engine
内燃机泵 unipump; unit pump
内燃机厂 internal combustion engine plant
内燃机车 diesel; diesel locomotive; internal combustion loco(motive); railway motor car
内燃机车的前节 lead-unit of diesel locomotive
内燃机车间 diesel shop
内燃机车牵引 internal combustion locomotive haulage
内燃机车牵引的 diesel-hauled
内燃机船 motor ship; motor vessel
内燃机的汽缸数 <俚语> banger
内燃机的消音装置 quieter
内燃机动的 petro-engined
内燃机发爆噪声 pink noise of internal combustion engine
内燃机发电驱动 oil-electric(al) drive
内燃机发电站 internal combustion power station
内燃机发动的振动器 engine type vibrator
内燃机废气排放 exhaust emission
内燃机负荷特性 load characteristics of internal combustion engine
内燃机工作容积 swept volume of internal combustion engine
内燃机公司 Combustion Engineering Inc.
内燃机轨道车 internal combustion rail-car; motor trolley; trackmobile
内燃机化油器 carburet(t)or
内燃机火花塞 plug
内燃机火星捕捉器 internal combustion engine spark arrester
内燃机机务段 diesel locomotive terminal
内燃机空转替代继电器 diesel idling speed substitution relay
内燃机冷却系统 cooling system of internal combustion engine
内燃机列车 diesel train
内燃机喷流荷载 internal blast loading
内燃机起动系统 starting system of internal combustion engine
内燃机气化器 gas generator
内燃机牵引 diesel traction
内燃机牵引车 diesel tractor; motor truck
内燃机牵引的 motor-driven
内燃机驱动的 oil-engine driven
内燃机驱动的发电机 petrol-electric-(al) generating set
内燃机驱动的热泵 heat pump driven by combustion engine
内燃机驱动的钻机 power rig
内燃机驱动设备 oil engine driven plant
内燃机驱动装置 oil engine driven device
内燃机燃料 motor fuel
内燃机润滑系统 lubrication system of internal combustion engine
内燃机润滑油 mobile oil
内燃机润滑油防锈试验 rust test for engine oil
内燃(机式)打夯机 explosion ram
内燃机特性 characteristics of internal combustion engine
内燃机调节器 diesel motor regulator
内燃机凸轮挡 catch of internal combustion engine
内燃机拖动的发电机 internal combustion engine driven generator
内燃机效率 efficiency of internal combustion engine
内燃机械传动动车 diesel mechanical rail-car
内燃机械传动机车 diesel mechanical locomotive
内燃机械传动起重机 diesel mechanical transmission crane
内燃机型拖拉机 internal combustion tractor
内燃机性能指标 performance index of internal combustion engine
内燃机修理船 internal combustion engine repair ship
内燃机油船 motor tanker
内燃机总厂 general internal combustion engine plant
内燃炉 internally fired furnace
内燃轮胎起重机 diesel tyred crane
内燃喷嘴 internal combustion burner
内燃牵引车 gas tractor
内燃牵引干扰 diesel traction interference
内燃牵引机组 diesel traction unit
内燃牵引列车 diesel train
内燃牵引铁路 diesel traction railway
内燃烧 internal combustion
内燃烧炉 internal fired furnace
内燃烧室 fire pot; inner combustion chamber
内燃烧砖法 internal firing of brick
内燃式打夯机 power rammer
内燃式地面抹光机 internal combustion trowelling machine
内燃式发动机 internal combustion engine
内燃式夯实整平样板 <一种用内燃机震动的混凝土路面夯击修整样板> petrol compacting and finishing screed
内燃式夯样机 petrol compacting and finishing screed
内燃式卷扬机 internal combustion engine winch
内燃式破碎机 motor breaker
内燃式燃气轮机 internal combustion gas turbine
内燃式涂料喷射机 internal combustion engine driven paint sprayer
内燃式挖土机 gasoline shovel
内燃式振捣器 gas internal vibrator
内燃水泵 Humphrey gas pump
内燃铁路起重机 diesel locomotive crane
内燃拖拉机 diesel tractor
内燃液力传动机车 diesel hydraulic transmission locomotive
内燃液力传动起重机 diesel hydraulic transmission crane
内燃液力机械传动机车 diesel hydro-mechanical locomotive
内燃液力机械传动有轨车 diesel hydro-mechanical rail-car
内燃凿岩机 diesel engine rock drill; gasoline-powered drill; internal combustion rock drill; motor drill; petrol-driven rock drill; motor jack hammer
内燃砖 brick fired with combustible additives
内燃装载机 diesel loader
内热 internal thermal; intrinsic (al) heat
内热的 endothermal; endothermic
内热电偶 internal thermocouple
内热动物 endothermic animal
内热法 internal heating method; internal resistance electric(al) melting
内热釜 internal gas heated retort
内热生成 internal heat generation
内热式烤箱 direct-fired oven; directly heated oven; internally heated oven
内热源 endogenous pyrogen; internal heat source
内热阻 internal thermal resistance
内刃角 inner cutting angle
内刃脚 inner shoe
内容 content; subject matter
内容暴露 contents exposed
内容编址存储器 content-addressable memory; content-addressed storage
内容不明 contents unknown
内容不详条款 contents unknown clause
内容充实的 full-bodied
内容定址 content addressing
内容定址存储器 annex(e) storage; associative memory; associative storage; parallel search storage; searching storage; search memory
内容丢失 contents lost
内容独立地址 content-independent address
内容度 inner content
内容发霉 contents mildewed
内容分析 content analysis

内容腐烂 contents rotten
内容级功能 content level function
内容监督器 contents supervisor
内容监控器 contents supervisor
内容检索 content retrieval
内容介绍 prospectus
内容据发货人申报 said by shipper to contain
内容可更换的存储器 changeable storage
内容控制 parental control
内容矿物<包含于他种矿物内者> endomorph
内容量 inner capacity
内容明细表 specification of contents
内容目录 contents directory
内容贫乏 vacuity
内容器 inner container; inner jar; interior container
内容提要 indicative abstract; informative abstract
内容体 endomorph
内容同步 content synchronization
内容显示器 content indicator
内容修订 revise for content
内容要素 information content
内容一览表 content directory; table of contents
内容指示器 content indicator
内容属性 contents attribute
内容转换 contents conversion
内乳胶漆 inner emulsion paint
内锐角钢 sharp backed angle
内润滑钢丝绳 internally lubricated wire rope
内润滑增强热塑料 internally lubricated reinforced thermoplastics
内润湿剂 built-in wetting
内塞盖 inner plug
内塞雾化铁粉生产法 Naeser process
内散焦 internal defocusing
内散射 inscattering
内散射校正 inscattering correction
内扫描 interscan
内扫描激光器 internally scanned laser
内色 integral colo(u)r
内色散 internal dispersion
内沙坝 inner bar
内沙洲 inner bar
内刹车 inner brake; internal brake
内刹车毂 inner brake hub
内栅极 inner grid
内煽风干燥窑 internal fan kiln
内扇 inner fan
内扇沉积 inner fan deposit
内熵 internal entropy
内舌片 internal tongue
内蛇管 inner coil
内设的厨房 in-built kitchen
内设荧光灯 internal fluorescent lighting
内射包络 injective envelope
内射的 injective
内射的函数 injective function
内射分解 injective resolution
内射流 internal jet
内射模 injective module
内射上链复形 injective cochain complex
内射束分离层 internal beamsplitting layer
内射同态 injective homomorphism
内射维数 injective dimension
内射性 injectivity
内射映射 injective mapping
内射自同态 injective endomorphism
内伸臂 inboard extension
内渗 endomose; endosmosis
内渗当量 endosmotic equivalent
内渗(透)计 endosmometer
内渗透率 specific permeability
内渗透现象 endomose; endosmosis; endosmosis-mose
内渗透性 intrinsic(al) permeability; specific permeability
内渗现象 endosmose
内渗压测定器 endosmometer
内渗仪 endosmometer
内生 endogen; endogeny
内生包体 endogenous enclosure; endogenous inclusion; endogenous spore
内生变动 endogenous change
内生变量 endogenous variable; endogenous variate
内生变数 endogenous variable
内生变质作用 endomorphism
内生成矿建造【地】endogenetic metallogenic formation
内生成矿作用 endogenic mineralization
内生的 endogenetic; endogenous
内生动物群生物 infaunal organism
内生洞穴 endogenous cave
内生过程 endogenous process
内生河曲 ingrown meander
内生河湾 ingrown meander
内生环 endogenous cycle
内生活动 endogenous activity
内生寄生植物 endobiophyta
内生解释变量 endogenous explanatory variable
内生矿床 endogenetic deposit; hypogene deposit
内生裂缝 cleat; endokinetic fissure
内生裂隙 cleat; endokinetic fissure
内生流体 endogenous fluid
内生硫细菌 endothiobacteria
内生脉冲源 endogenous pulsed source; inherent pulsed source
内生能 endogenetic energy; endogenic energy
内生曲流 ingrown meander
内生韧皮部 internal phloem
内生砂 endogenetic sand
内生生长 endogenous growth
内生生物 endophyte
内生时间信号 internally generated time signal
内生收入假设 endogenous income hypothesis
内生树 endogen tree
内生投资 endogenous investment
内生紊乱 internal irregularity
内生岩 endogenic rock
内生异常 endogenic anomaly
内生源 endogenous origin
内生运动 endogenic movement
内生晕 endogenic halo; endogenous halo
内生藻类 endophytic algae
内生滞后变量 endogenous lagged variable
内生中柱鞘 inner pericycle
内生转矩 internal torque
内生作用 endogenic action; hypogene action
内生作用过程 endogenic process
内绳 inner rope; inside cord
内绳扣 inside clinch
内施胶 internal sizing
内施凝固剂 internal coagulant
内湿度 internal humidity
内石板百叶窗 internal slatted blind
内时标 internal clock
内时理论 endochronic theory
内时循环 internal time loop
内时钟 internal clock
内时钟发生器 internal clock generator
内史密斯打桩机 Nasmyth pile-driver
内始式 end arch
内氏焦点 Nasmyth focus
内势垒 internal barrier
内视图 inside view; interior view
内室 inner chamber; internal cell; thalamus<古希腊建筑的>; spence<苏格兰语>
内适性 endo-adaptation
内收墙身 wall offset
内收缩 internal contraction
内收转角缝 setback corner joint
内受热面 internal heating surface
内束 internal beam
内束流 internal beam current
内束能量 internal beam energy
内树皮 inner bark
内数据段 internal data field
内数据结构 internal data structure
内数据说明符 interior label
内数据域 internal data field
内双峰 internal double peak
内双键 internal double bond
内双曲面 inner hyperboloid
内水 internal water; local runoff; national waters; on-site runoff
内水分欠缺 internal water deficit
内水分循环 internal water circulation
内水合层 inner hydration sphere
内水口 inner nozzle
内水利用 on-site water use
内水落管 inside leader; interior downpipe
内水压力 internal water pressure
内水域 internal waters
内水源 interior supply water
内丝扣 internal thread
内丝扣接头 female coupling
内丝扣套管 female coupling
内死点 inner dead point
内碎屑 intraclast
内碎屑硅质岩 intraclast siliceous rock
内碎屑灰岩 intraclastic limestone
内碎屑结构 intraclastic texture
内碎屑亮晶灰岩 intrasparite
内碎屑亮晶砾屑灰岩 intrasparudite
内碎屑亮晶砂屑灰岩 intraspararenite
内碎屑磷块岩 intraclastic phosphorite
内碎屑铝质岩 intraclast aluminous rock
内碎屑泥晶灰岩 intramicrite
内碎屑泥晶砾屑灰岩 intramicarenite; intramicrudite
内碎屑泥状铝质岩 intraclast pelitomorphic aluminous rock
内碎屑生物球粒泥晶灰岩 intrabiopelmicrite
内碎屑微晶灰岩 intraclastic-micritic limestone; intramicrite
内碎屑微晶砾屑灰岩 intramicrudite
内损耗系数 internal loss coefficient
内损失 internal loss
内损转矩 internal loss torque
内缩合作用 internal condensation
内缩量 neck-in
内缩排水孔 reentrant orifice
内缩松 internal porosity
内缩速度 retraction speed
内缩酮 keto-lactol
内缩位置 retracted position
内缩行程 retraction stroke; retract stroke
内锁信号 internal lock signal
内塔 inner tower
内踏盘运动<指开口运动> inside treadle motion
内胎 air bag; air tube; inner tire [tyre]; inner tube; inner-tube of tire [tyre]; inner tire [tyre]; tire [tyre] tube; tire[tyre] tube
内胎补缀 tube patch
内胎衬带 chafing patch
内胎放气 inner-tube deflation
内胎胶补丁 repair patch
内胎硫化热压器 tube press
内胎模 tube mould
内胎气门嘴 inner-tube valve; tire valve
内胎气门嘴垫 valve pad
内胎气门嘴帽 inner-tube valve cap
内胎热补机 inner tube vulcanizer
内台阶设计 inside step design
内太阳系 inner solar system
内态变量 internal state variable
内滩 inner bar; in shore
内弹簧 inner spring
内弹簧安全阀 internal spring safety relief valve
内弹簧垫子 interior spring
内弹性膜 internal elastic membrane
内探针 internal probe
内堂<教堂的> presbyterium
内膛腐蚀 erosion in the bore
内膛挤压硬化法 autofrettage
内套管 inner casing; inside covering; internal sleeve
内套件 internal member
内套圈 inside race
内套式泥泵 double wall dredge pump
内套筒 inner sleeve
内特斯通层 Nettlestone bed
内梯 inside ladder
内梯玻璃 inner stepped glass
内梯透镜 inner step(ped) lens; inside step(ped) lens
内提升臂及挺杆总成 inner lift arm and tappet assembly
内蹄式制动器 inside shoe brake
内蹄外张式制动器 internal block brake; internal expanding shoe brake
内蹄制动器 internal brake
内体积 inner volume; internal volume
内天窗 internal dormer
内天井 indoor court-yard; interior court; internal court; piazza
内天线 internal antenna
内填砂沉箱 sand-filled caisson
内填作用 intussusception
内条纹 intra-striate
内调和测度 inner harmonic measure
内调和函数 internal harmonics
内调焦 inner-adjustable focus; interior focusing; internal focusing
内调(焦)平行光管 inner-adjustable focus collimator
内调(焦)望远镜 interior focusing telescope
内调节速率 inherent regulation rate
内调式瞄准镜 internal adjustment scope
内调望远镜 internal focusing telescope
内调荧光 intermodulated fluorescence
内调整补偿器 internal adjustment compensator
内调制 intermodulation; internal modulation
内调制法 intermodulation method
内调制器 internal modulator
内贴袋 sandwich pocket
内贴脸<门窗的> inside casing
内贴面单板 ceiling veneer
内铁式变压器 core type transformer
内铁芯电抗器 core type reactor
内厅 ben
内庭 atrium[复 atria/triums]; central court(yard); interior court; internal court
内庭式建筑 atrium architecture
内庭院<一边有土地界线的> inner

lot line court
内通风式电动机 internally ventilated motor
内通路长度 internal path length
内通气孔 inner vent
内同步 inter-sync
内同步方式 internal synchronization mode
内同步振荡器 interlocked oscillator
内同构 inner isomorphism
内筒 dip pipe; inner core; inner couette
内投作用 introjection
内透射比 internal transmittance
内透射率 internal transmission factor
内透射系数 internal transmittance
内透视中心 interior perspective center[centre]
内凸轮环 reaction ring
内凸面离合器 inner cam clutch
内凸通风口 project in vent
内凸缘 inward flange
内突堤 inner jetty
内突环式连接管 lip union
内图廓【测】edge-frame; inner border; inner edge
内图廓分度线 neatline gradation
内图廓线 geographic (al) limit; neat line
内图廓注记 neatline mark
内涂层 undercoat(ing)
内涂阴极 internally-coated cathode
内推 interpolate
内推法 interpolation
内推下旋式窗扇 hopper light
内退解 internal unwinding
内吞作用 endocytosis
内托板 inner shoe
内脱模剂 internal release agent
内瓦 inner tile; Neva <聚酰胺纤维>
内外摆线轮齿 epicycloidal and hypocycloidal gear tooth
内外表面都略呈扁平形 flattened both on the outer surface
内外拨号制 inward-outward dialing system
内外层错 intrinsic-extrinsic fault
内外插法 interior extrapolation method
内外差异费率 in-and-out rates
内外齿离合器 gear clutch
内外齿轮离合器 internal external gear clutch
内外齿轮装置 internal external gear arrangement
内外传送 radial transfer
内外搭接板 in-and-out plating
内外倒转术 combined version
内外叠板 in-and-out strakes
内外叠板法 in-and-out plating; in-and-out system; raised and sunken plating; sunken and raised plating; system plating
内外共同形成的 endectoplastic
内外构型 endo-exo configuration
内外管<岩芯管> inner and outer tube
内外函喷气发动机 ducted fan; turbo-fan
内外荷载 internal and external load(ing)
内外机座型电机 double casing machine
内外机座型发电机 double-framed generator
内外加厚<钻杆> external and internal thickening
内外加厚的 internal external upset
内外加厚钻杆 internal external-upset drill pipe
内外夹 curb pins
内外键槽连接 slip joint
内外接活管接 street union
内外接三通管接 service tee[T]
内外接头<管子的> street ell
内外接弯头 street bend; street elbow; street ell
内外接弯头活管接 street bend union; street elbow union; union ell male and female
内外接弯头联管节 street bend union
内外径比 boss ratio; diameter ratio
内外径规 internal and external ga(u)ge; internal and external snap ga(u)ge
内外径卡钳 combination cal(l)ipers
内外径啮合链传动装置 over-and-under chain gear
内外卡钳 hermaphrodite cal(l)ipers; outside-and-inside cal(l)ipers; scribing cal(l)ipers
内外空间互贯 interpenetration of internal and external space
内外两面承受水压力的中间坞门 bilaterally-loaded inner gate
内外两用卡钳 double cal(l)ipers
内外螺母 bushing
内外螺纹管接头 extension piece
内外螺纹弯头 male and female bend
内外贸同船同码头 domestic cargo and foreign trade cargo loaded on the same ship and handled on the same terminal
内外磨削 inside and outside grinding
内外排列 collateral arrangement
内外配合直径 outside-inside mating diameter
内外平衡 internal and external equilibrium
内外墙间隔带 intervallum
内外墙中层 intermural
内外三心拱 convex arch
内外双开门框 double rebated
内外水槽 inside or outside water channels; inside or outside water ways
内外四心桃花拱 keel arch
内外四心桃尖拱 ogee arch; ogival arch
内外图廓间坐标注记 grid reference box
内外图廓坐标记 grid reference box
内外效应 inside-outside effect
内外压力 internal and external pressure
内外研磨法 inside and outside grinding
内外异构化 endo-exogenetic isomerization
内外因的 endexoteric
内外因演替 endo-exogenetic succession
内外重叠 in-and-out
内外转子泵 internal external rotary pump
内外装修 interior and exterior finishes; interior and exterior decoration
内外自动电话系统 inward-outward dialing system
内弯 incurvation; incurve; introversion; inward curve
内弯矩 inner bending moment; internal bending moment; internal moment
内弯肋骨 concave bend frame
内万向悬挂支架 inner gimbal
内网 inner-mesh
内网器 Golgi apparatus; internal reticular apparatus
内网织层 inner plexiform layer
内韦季摆筛 Newage vibrating screen
内围层 inlier
内围栏 inner rail
内围砌分隔墙 inner masonry dividing wall
内围墙 inner enceinte(wall)
内位加成 internal addition
内位移坐标 internal displacement coordinate
内稳定状态 homeostasis
内务部 Department of Interior
内务操作 bookkeeping (operation); housekeeping (operation); overhead operation; red-tape operation
内务操作程序 housekeeping routine
内务操作方式 overhead method
内务程序 housekeeping program(me); house-wife program(me)
内务处理 housekeeping
内务处理程序 housekeeper; housekeeping program(me); housekeeping routine
内务的 overhead
内务府 imperial palace
内务辅助信息 housekeeping information
内务工作 housekeeping
内务例行程序 housekeeping routine
内务命令 built-in command
内务软件 housekeeping software
内务指令 housekeeping instruction
内吸的 systemic
内吸附 internal adsorption
内吸光率 internal absorptance
内吸剂 systemics
内吸磷 demeton
内吸磷混合物 demeton mixed
内吸杀虫剂 systemic insecticide
内吸杀菌剂 systemic fungicide
内吸收 inner absorption; internal absorption
内吸收比 internal absorptance
内吸收法 internal absorbent method
内吸收率 internal absorption factor
内吸收系数 internal absorptance
内吸收性除草剂 systemic herbicide
内吸收因数 internal absorption factor
内吸性农药 systemic pesticide
内吸转移的除草剂 translocate herbicide
内吸作用 systemic action
内矽卡岩 endoskarn
内烯烃 internal olefin
内洗 wash-in
内细齿 internal serration
内狭泄水孔 reentrant orifice
内酰胺 lactam
内酰亚胺 lactim
内舷 inside
内舷墙 inner bulwark
内线 domestic wiring; indoor wiring; interior wiring; internal wiring
内线电话 house phone
内线符号 house line
内线航道 inside passage
内线交易 insider trading
内线条 inner bead
内线性化 inner linearization
内线性模型 intrinsically linear model
内线自动电话机 interphone; talk-back circuit
内陷 invagination; wash-in
内陷气孔 sunken stomata
内相【化】internal phase
内相比 internal phase ratio
内相位角 internal phase angle
内镶边 interior trim
内镶玻璃 inside glazing; internal glazing
内镶(嵌)板 infill panel
内向 introversion; inward orientation; toe-in
内向爆破 implode; implosion
内向爆炸 implode; implosion
内向变差 inward variation
内向侧翼窗 window with wings opening inwards
内向冲程 instroke
内向代收 inward collection
内向电流 inward electric(al) current
内向对应 inward correspondence
内向法线 inward normal
内向构型 endo-configuration
内向光 inward-bound light
内向机会变差 inward chance variation
内向径流 radial-inward flow
内向开发政策 inward-looking development policy
内向流 indraft
内向漏失 inward leakage
内向人格 introverted personality
内向渗流 inward osmosis
内向实质变差 inward substantial variation
内向水流 inward flow
内向弯曲 introversion
内向弯曲的 incurvate
内向削边窗 inward chamfered window
内向旋转 inboard turning; inward turning
内向压力 inward pressure
内向运转 inward running
内向折边 inward flange
内项 inner term; mean term; middle term
内消旋体 mesomer(e)
内消旋作用 internal compensation
内销商品 commodities for the home market
内效率 internal efficiency
内楔形炮孔<掏槽用> inner wedge
内斜撑 interior leg brace
内斜的 intraclinal
内斜面活塞环 inside bevel(led) piston ring
内斜榫 internal miter[mitre]
内泄油 internal drainage
内卸 internal discharge
内卸斗式提升机 internal discharge bucket elevator
内心 entad; incenter[incentre]; inner center[centre]; interior center[centre]
内心投影 internal projection
内芯 inside core
内芯高压法灌浆 containment grouting
内芯型 belly core
内行程 inner stroke
内行推动 internal horizontal drive
内行星 inferior planet; inner planet; interior planet
内行星齿轮 inner planet(ary) gear
内型 inner block; former
内旋 inner side of a vertical surfaces; internal rotation
内旋的 involute
内旋结 inner clinch; inside clinch
内旋式 inboard turning; inward turning
内旋式入口 involute inlet
内旋双车 inturning screws
内旋调制盘叶片 involute reticle blade
内旋转 internal rotation
内选通 internal gating
内靴<双层岩心钻的> inner cutting

shoe
内循环 internal circulation; internal recycle; internal recycling; inner loop
内循环反应器 internal circulation reactor
内循环流化床 inner loop fluidized bed
内循环三相生物流化床反应器 interior circulation three-phase biological fluidized bed reactor
内循环水解酸化 internal cycle hydrolytic acidification
内循环厌氧反应器 internal circulation anaerobic reactor
内循环移动床生物膜反应器 internal circulation moving bed biofilm reactor
内压防爆型设备 gas-filled explosion proof-type equilibrium
内压盖 internal gland
内压降 internal drop
内压力 internal pressure; interpressure; intrinsic(al) pressure
内压流 internal pressure current
内压强 internal pressure; interpressure; intrinsic(al) pressure
内压强度 built-in strength; internal pressure strength
内压强度试验 internal pressure strength test
内压式发动机 inward compression engine
内压试验 inner pressure test; internal pressure test
内压试验机 internal pressure tester
内压条 inside casing; inside stop
内芽 internal bud
内亚波利顿黄 <锑酸铅黄色彩料> Neapolitan yellow
内烟道 inner flue
内烟道锅炉 Cornish boiler
内烟管 internal flue
内烟筒 inside stack; smoke stack extension
内延 interior extent
内岩基带 endobatholite zone
内岩基的 endobatholithic
内岩浆矿床 intramagmatic deposit
内岩芯管 inner barrel; inner tube
内岩芯管接长管 inner-tube extension
内岩芯管连接管 inner-tube adapter [adaptor]
内岩芯管岩芯提断器 inner-tube core lifter
内岩芯筒 inner barrel
内岩芯筒长度 inner core barrel length
内岩芯筒尺寸 inner core barrel size
内岩芯筒靴 inner-tube shoe
内研磨杆 internal lap
内檐斗拱 inner-eaves corbel bracket
内檐线条 cornice
内檐装修 interior finish work
内焰 inner cone of flame; inner flame
内焰管锅炉 internal flue boiler
内焰(锥部)高度指数 cone height index
内阳台 inset balcony
内氧化 internal oxidation
内业 indoor work; office operation; office work
内业编制的 office-compiled
内业处理 <测图成果> office analysis
内业处理测量成果 plot a survey
内业工程师 desk engineer; office engineer
内业工作 office work
内业计算工作 calculating office work
内业记录 office record
内业加密仪器 bridging instrument
内业检核 office control
内业胶片 indoor film
内业控制 office control
内业判读 office-identification
内业装备 office equipment
内叶 internal lobe
内叶轮 inner impeller
内叶轮泵 intra-vane pump
内叶片 intra vane
内叶片泵 intra-vane pump
内叶片热泵 heat intra vane pump
内液冷 internal liquid cooling
内衣壳 inner capsid
内衣口袋 vest-pocket
内移 ingression
内移铰链轴补偿器 inset hinge compensator
内移入口 displaced threshold
内移位 internal shift
内异构体 internal position isomer
内抑制 internal inhibition
内翼泵 intra-vane pump
内因 immanent cause; internal agent; internal cause
内因变星 intrinsic(al) variable star
内因沉降 endogenetic subsidence; endogenic subsidence
内因动态演替 endodynamic(al) succession
内因故障 primary failure
内因论 theory of the internal origin
内因特网【计】intranet
内因演替 endogenetic succession
内因运动 endogenous movement
内因子 intrinsic(al) factor
内阴极保护 internal cathodic protection
内隐反应 covert response; implicit response
内隐行为 covert behavio(u)r; implicit behavio(u)r
内应变 exterior strain; inner strain; internal strain; intrinsic(al) strain
内应变计 internal strain-ga(u)ge unit
内应力 body stress; inner stress; interior stress; internal stress; intrinsic(al) stress; locked-in stress; locked-up stress
内应力脆性 tension brittleness
内应力裂缝 season crack
内应力松弛 internal stress relaxation
内应力消除过程 stress-relieving process
内应力增大 strain-raise
内营力 endogenetic force; endogenous force
内营力的 endogenic; endogenous; endogenetic
内颖 inner glume
内映象 interior mapping
内涌浪 internal surge
内用瓷漆 interior enamel
内用胶合板 interior plywood; interior-type plywood
内用抗静电剂 internal anti-stat
内用末道清漆 interior finishing varnish
内用漆 indoor paint; interior paint
内用清漆 interior varnish
内用罩光清漆 interior finishing varnish
内油槽 inner oil sump
内油醇硫酸酯 internal oleyl sulfate
内油挡 inner grease retainer
内油封 inner oil seal
内油封环 inner oil seal ring
内有杂质 inherent impurity
内余面 inside lap
内余隙 inside clearance
内鱼尾棒 inner splice bar
内隅角 reentrant corner
内浴室 inside bath(room)
内预告信号(机) inner distant signal
内预应力 inner prestress
内域 internal area
内渊 intradeep
内园 inner garden; interior garden
内原性的 endogenetic; endogenic; entogenous
内圆车刀 internal turning tool
内圆滚线【数】hypocycloid
内圆滚线的 hypocycloidal
内圆珩床 internal honing machine
内圆珩磨 internal honing
内圆加工 internal work
内圆角 fillet; filleted corner; sunk fillet
内圆角半径 fillet radius
内圆角与外圆角【机】fillet and round
内圆角子 <修型工具> inside corner
内圆锯 inside diameter saw
内圆磨床 internal grinder; internal grinding machine
内圆磨头 internal grinding head
内圆磨削 internal grinding
内圆磨削附件 internal grinding attachment
内圆磨削砂轮轴 internal wheel spindle
内圆盘 <接合器的> internal disc [disk]
内圆刨 concave round plane
内圆切片机 inside diameter slicer
内圆筒 inner cylinder
内圆无心磨床 internal centerless grinder; internal centerless grinding machine
内圆牙螺纹 internal rope thread
内圆直径 internal diameter
内圆周 inside circumference
内圆锥 female cone
内圆锥管螺纹牙高测高仪 inside taper ga(u)ge
内缘 inner edge; inner margin
内缘翻边 plunging
内缘钻石 <金刚石岩心钻头的> inner-ga(u)ge stone; inside kicker
内源 endogenesis
内源包体 endogenic inclusion; endogenous enclosure; endogenous inclusion
内源变量 endogenous variable
内源场 field of the internal source
内源代谢作用 <细菌的> endogenous metabolism
内源的 autogenic; autogenous
内源河 autogenic river; autogenic stream; autogenous river
内源呼吸 endogenous respiration
内源节律 endogenous rhythm
内源谱仪 internal source spectrometer
内源生长期 endogenous phase of growth
内源污染物质量 endogenous pollutant mass
内源相 endogenous phase
内源性的 endogenous
内源性生物修复 intrinsic(al) bioremediation
内源性调节器 endogenous modulator
内源性微生物修复 intrinsic(al) micro-bioremediation
内源性污染 intrinsic(al) pollution
内源因素 Castle's intrinsic(al) factor; intrinsic(al) factor
内源正比计算器 internal proportional counter
内源周期性 endogenous periodicity
内院 garth; inner court; inside court; interior court; interior yard; internal court; patio
内院宽度 inner court width
内院式住宅 atrium house; court house; patio house
内运过境 inward transit
内运流 interior convection; internal convection
内蕴对 intrinsic(al) parity
内蕴方程 intrinsic(al) equation
内蕴函数 intrinsic(al) function
内蕴几何学 intrinsic(al) geometry
内蕴假设 intrinsic(al) hypothesis
内蕴面积 intrinsic(al) area
内蕴能 intrinsic(al) energy
内蕴曲线方程 intrinsic(al) equations of a curve
内蕴同调 intrinsic(al) homology
内匝道 <立体交叉的> internal ramp
内载波伴音 intercarrier sound
内载波伴音系统 intercarrier sound system
内载波电视伴音接收机 intercarrier sound receiver
内载波接收法 intercarrier sound reception
内载波接收方式 intercarrier system
内载波接收系统 intercarrier receiving system
内载波信号 intercarrier signal
内载波噪声抑制 intercarrier noise suppression
内载负荷 <集装箱的> internal loading
内在保安设备 built-in safeguard
内在保证 implied warranty
内在报酬 intrinsic(al) reward
内在报酬率 internal rate of return
内在不对称性 intrinsic(al) dissymmetry
内在不均匀性 intrinsic(al) inhomogeneity
内在不稳定性 inherent instability
内在层错 intrinsic(al) fault
内在陈旧性 built-in obsolescence
内在成本 implicit cost
内在磁感应 intrinsic(al) induction
内在错误 inherent error
内在代谢活动 intrinsic(al) metabolic activity
内在导纳 intrinsic(al) admittance
内在导数 intrinsic(al) derivative
内在的 immanent; inherent; inner; internal; intrinsic(al)
内在的不燃性 intrinsic(al) non-flammability
内在底栖生物 endobenthos; endobiosis
内在电平 intrinsic(al) level
内在动机 intrinsic(al) motivation
内在动能 internal kinetic energy
内在动态 intrinsic(al) dynamics
内在动因 internal agent
内在断裂能 intrinsic(al) breaking energy
内在防爆安全 intrinsic(al) safety
内在分辨率 intrinsic(al) resolution
内在付现成本 implicit cash cost; implied cash cost
内在各向异性 inherent anisotropy; intrinsic(al) anisotropy
内在功能 built-in function
内在关系 internal relation
内在管理界限 inner control limits
内在规律 inherent law
内在过程 intrinsic(al) procedure
内在函数 intrinsic(al) function
内在函数名 intrinsic(al) function name

内在函数引用 intrinsic(al) function reference
内在化 internalization
内在化为关系 internalized as relations
内在环境 internal environment
内在灰分 fixed ash; inherent ash; inherent moisture
内在混凝土热量 inherent concrete heat
内在活动 internal activity
内在活性 intrinsic(al) activity
内在机械性能 intrinsic(al) mechanical property
内在夹杂物 endogenous enclosure; endogenous inclusion
内在价值 immanent value; inmost value; inner value; intrinsic(al) value; intrinsic(al) worth
内在剪切强度 intrinsic(al) shear strength
内在校验 built-in check
内在精(确)度 intrinsic(al) accuracy; intrinsic(al) precision
内在抗力 inherent resistance
内在可靠性 inherent reliability; intrinsic(al) reliability
内在孔隙 inherent porosity
内在矿物质 inherent mineral mater
内在利息 implicit interest
内在利益 intrinsic(al) interest
内在联系 inner link; internal logic(al); internal relation; interrelationship
内在联系模型 interconnection model
内在流量 internal liquidity
内在矛盾 immanent contradiction; inherent contradictions; inner contradictions
内在敏感性 endogenous susceptibility
内在(内装的)特征判析设备 built-in diagnostic equipment
内在能力 endogenous capacity
内在黏[粘]度 inner viscosity
内在评价 built-in evaluation
内在强度 intrinsic(al) strength
内在强度曲线 intrinsic(al) strength curve
内在曲线方程 intrinsic(al) equations of a curve
内在缺点 inherent vice
内在缺陷 inherent defect; inherent vice; latent defect
内在缺陷故障 inherent weakness failure
内在热 natural heat
内在伸缩性 built-in flexibility
内在渗透性 intrinsic(al) permeability
内在湿度 inherent moisture
内在时间差滞 inside lag
内在收缩 chemical shrinkage; intrinsic(al) shrinkage
内在水分 inherent moisture; internal moisture
内在特性 intrinsic(al) characteristic; intrinsic(al) property
内在特征 intrinsic(al) property
内在调节 inherent regulation
内在透水性 intrinsic(al) permeability
内在凸齿的匣钵 nibbed saggar [sagger]
内在危险 internal hazard
内在微分 intrinsic(al) differentiation
内在微分几何 intrinsic(al) differential geometry
内在稳定器 built-in stabilizer
内在稳定性 intrinsic(al) stability
内在稳定作用 built-in stabilization
内在问题 intrinsic(al) problem
内在误差 inherent error; intrinsic(al) error
内在瑕疵 hidden defect
内在效用 intrinsic(al) utility
内在信托 implied trust
内在形变 inner deformation
内在性 inherence [inherency]; internality
内在性能 intrinsic(al) property
内在压力 intrinsic(al) pressure
内在要素汇集 convergence of inner demands
内在抑制作用 intrinsic(al) inhibition
内在因素 intrinsic(al) factor
内在因子 internal factor
内在应变 built-in strain
内在应力 inherent strain; inherent stress
内在原因 immanent cause; internal cause
内在杂质 intrinsic(al) contaminant
内在增长率 intrinsic(al) growth rate; intrinsic(al) rate of growth
内在质量 inherent quality; inner quality
内在转换 inner conversion
内在自励磁电机 machine with inherent self-excitation
内在自然增长率 intrinsic(al) rate of natural increase
内在阻抗 intrinsic(al) impedance
内在阻力 inherent resistance
内在阻尼 internal damping
内噪声 internal noise; self-noise
内噪声带宽 internal noise bandwidth
内噪声源 internal noise source
内增结核 excretion concretion
内增塑作用 internal plasticization
内增压器 internal supercharger
内增殖 internal breeding
内增殖比 internal breeding ratio
内闸 inner brake; inside brake; internal brake
内闸门 inner gate
内窄外宽的水槽 expanding waterway
内债 domestic borrowing; domestic debt; domestic loan; internal debt; internal loan
内战 civil war
内站 inside plant
内张力 internal tension
内张式离合器 inner expanding clutch
内涨 internal expanding
内涨式摩擦离合器 expansion clutch
内胀 internal expanding
内胀式车轮制动器 internal wheel brake
内胀式管接头 internal expanded coupling
内胀式制动器 expanding inside brake; inner expanding brake; inside expanding brake; internal expanding brake
内照明地球仪 illuminated globe
内照射 internal exposure; internal irradiation; internal radiation
内照射防护 internal irradiation protection; protection of inner radiation
内照射危险 internal irradiation hazard
内罩 inner casing; inner cover
内折 infolding
内褶缘型【地】intraplicate
内褶皱面 intrados
内针门 inner door
内真空制管法 inner suction pipe process
内真空制管机 inner suction pipe machine
内诊镜 endoscope
内阵列扩散 intra-array diffusion
内振荡器 internal oscillator
内振捣 < 混凝土作业 > internal vibration
内振动 internal vibration
内蒸阱 flash chamber
内蒸器 flash chamber
内正齿轮 internal spur gear
内正方形 female square; internal square
内正则集 inner regular set
内正则性 inner regularity
内政部 < 美 > Department of Interior
内政的 municipal
内支 internal branch
内支撑 internal bracing; strut
内支撑片 inside spider
内支承块 inner shoe
内支式挡土墙 counterforted type retaining
内支式桥墩 counterforted abutment
内直和 internal direct sum
内直角尺 female square; internal square
内直角光子 inside square-corner slick
内直径 inside diameter; internal diameter
内直线车削 internal straight turning
内直线镗孔 internal straight boring
内植 interplantation
内止点 inner dead center[centre]; inner dead point
内止回阀导座 inner check valve guide
内指标点 < 机场 > inner marker
内指点标 inner marker
内指示剂 internal indicator
内酯 inner ester; internal ester; lactone; lactonic ketone
内酯法 lactone process
内酯规则 lactone rule
内酯环 lactonic ring
内酯染料 lactone colo(u)ring matter
内酯酸 lactonic acid
内酯异构现象 lactone isomerism
内制动 inside brake
内制动臂 inner brake arm
内制动杆 inner brake arm
内制动器 inner brake; inside brake; internal brake
内制动器毂 inner brake hub
内制动蹄 internal brake shoe
内制动瓦 inner brake shoe
内质 inwardness
内质混合岩化作用 endomigmatization
内质应力 isotropic(al) stress
内滞水 internal stagnant water
内置泵 < 与发动机安装在一起的 > built-on pump
内置的 built-in
内置故障防止效能 built-in fail-safe behavio(u)r
内置注解【计】in-place comments
内中断 internal interrupt
内中堂 < 古希腊神殿 > nanos
内中心钻孔 internal center drilling
内钟 entocodon
内重复 endoreduplication
内周水压 internal hydraulic pressure
内轴 inner spindle; internal shaft
内轴承 inner bearing
内轴承盖 inside cap
内轴套 inner bushing
内轴箱 inside axle box
内肘板 inner bracket
内皱 implexed
内主应力 inherent principal stress
内驻波 internal standing wave; standing internal wave
内柱 hypostyle column; inner column; inside column; interior column; internal column
内柱插 inside stake pocket
内柱式集装箱 inner post type container; interior post type container; smooth panel container
内爪卡盘 inside jaw chuck
内砖 inside tile; internal tile
内转 involution
内转的 self-rotated
内转换比 internal conversion ratio
内转换谱仪 internal conversion spectrometer
内转换系数 internal conversion coefficient
内转换因子 internal conversion factor
内转换源 internal conversion source
内转换中子探测器 internal conversion neutron detector
内转结 inside rolling hitch
内转弯 inside turn
内转弯半径 inside turning radius
内转向轮转向角 inside wheel turning angle
内转移 internal trans
内转匝道 < 立体交叉中的 > inner loop
内转子 inner rotor
内装 in-built
内装板 interior board
内装曝光表 built-in exposure meter; built-in meter
内装壁炉 engaged chimney; engaged fire-place
内装玻璃窗 interior glazed window
内装玻璃的 interior glazed
内装测标 built-in measuring mark
内装测试设备 built-in test equipment
内装储柜的床 box couch
内装窗玻璃镶条 interior stop
内装单向阀 integral check valve
内装的 built-in; inset; nested
内装的车库 in-built garage
内装的诊断系统 built-in diagnosis
内装对流器 built-in converter
内装风扇式干燥 internal fan kiln
内装缓冲器 built-in cushion
内装祭礼用塑像的壁龛 niche containing the cult statue
内装减速器的电动机 internally geared motor
内装空调器 built-in air-conditioner
内装冷气机组 built-in cooling unit
内装流路 built-in flow circuit
内装配玻璃的钢窗油灰 steel sash putty of the interior glazing type
内装热交换装置 integral heat-exchange unit
内装润滑系统 build-in lubricating system
内装散热器 built-in radiator
内装设施的顶棚 ceiling incorporation services
内装式 built-in
内装式泵 built-in pump
内装式怠速浓度调程限制器 internal type idle limiter
内装式单向阀 integral check valve
内装式电动机 integral motor
内装式防溅板 built-in antisplash
内装式灰盆 built-in ashtray
内装式计算机 built-in computer
内装式起重器 in-built jack
内装式消声器 integral sound suppressor
内装式选粉机 internal classifier
内装式支承环 built-in crutch
内装式支承件 built-in crutch
内装式支柱 built-in crutch

N

内装饰 inner finish;interior finish(ing)
内装饰涂料 inner finish(ing) paint
内装天线 built-in antenna
内装填料的容器 packing vessel
内装物 content
内装显微镜 built-in microscope
内装修 inner decorating;interior finishing
内装修木材 wood interior
内装修用灰泥 interior stucco
内装修作业 finished interior
内装烟道 built-in chimney
内装颜色转换系统 built-in colo(u)r changing system
内装遥控传感器的恒温阀门头 thermostatic valve head with remote sensor
内装支柱 built-in crutch
内状态 internal state
内锥 endocone;inner cone
内锥齿轮 internal bevel gear
内锥度 internal taper
内锥度量规 inside taper ga(u)ge
内锥面扩孔器 taper-wall core shell
内锥面钻头 taper-wall bit
内锥形车削 internal taper turning
内锥形孔 internal tapered hole
内锥形镗孔 internal taper boring
内锥形折射 internal conical refraction
内子程序 in-line subroutine
内自聚焦 internal self-focusing
内自同构 inner automorphism
内自同构群 inner automorphism group
内自陷 internal trap
内自由度 internal degree of freedom
内纵的 inboard
内纵梁 inner stringer;interim beam;interior beam;interior stringer
内纵墙 spine wall
内走廊 central corridor;inner corridor;inner gallery;inside gallery;internal corridor
内走廊房屋 central-corridor residential building
内走廊式公寓建筑 inside gallery apartment building
内走廊型建筑物 interior-corridor type building
内阻抗 internal impedance
内阻力 inherent resistance;internal resistance;internal drag;internal friction
内阻力曲线 internal drag curve;internal resistance curve
内阻尼 internal damping
内阻尼曲线 internal damping curve
内阻尼损失 internal damping losses
内阻尼系数 internal damping coefficient
内阻压降 potential drop of internal resistance
内钻孔 internal drilling
内坐标 internal coordinate
内座壳属<拉> Endothia
内座圈 inner race;internal race

嫩

嫩黄 bright yellow;canary yellow
嫩煎锅 brat pan
嫩柳覆盖的沙洲 towhead
嫩绿 tender green
嫩树皮 peel
嫩叶 browse;leaflet
嫩炸盘 brat pan
嫩枝 browse;sprig;sprout

能

能把货物迅速变成现金的 self-liquidating
能保持水分的土壤 retentive soil
能爆炸的 detonable;detonatable
能操作的 serviceable
能层【天】ergosphere
能层效应 ergosphere effect
能拆卸的 knock down
能产生颜色的 colo(u)rific
能撤回的 withdrawable
能沉淀的 precipitable
能成比例的 commensurable
能成型性 figurability
能承压的 pressure-tight
能除数 aliquot number
能处理的 manageable
能存活的细菌 viable bacteria
能达到的可靠性 achieved reliability
能达到性 reachability
能达点 accessible point
能带 band;energy band
能带边缘 band edge;energy band edge
能带发射 band emission
能带间距 band separation
能带间隙 energy band gap
能带结构 energy band structure
能带宽度 bandwidth;energy gap
能带理论 band theory;zone theory
能带模型 band model
能带图 energy band diagram
能带图式 band scheme
能带尾 bandtail
能带尾伸 band-edge tailing
能带隙 band gap
能带学说 energy band theory
能当量 energy equivalent;equivalent energy
能倒出 in-tap
能得到的资料 available information
能的储量 energy content
能的递降 degradation of energy
能的量子化 quantization of energy
能的退降 degradation of energy
能的消耗 waste of energy
能的需要量 power demand
能的转化 transformation of energy
能动的 kinetic;motile
能动断层 capable fault
能动关系 active relation
能动光学 active optics
能动力 motility
能动论 activism
能动论心理学 activist psychology
能动投资 active investment
能动性 activism;activity;dynamic(al) role;initiative;motility;movability;theory of conscious activity
能动元件 active device;active element
能动资本 active capital
能独立运行的 stand-alone
能多次结果的 sychnocarpous
能阀 energy threshold
能反光的路缘石 reflectorizing curbstone
能反向吊机臂 reversible jib
能防腐的 imputrescible
能飞距离 range ability
能峰 energy barrier
能浮的 buoyant
能浮起面<矿物> water-repellent surface
能复制的 replicable;reproducible
能改正的 amendable
能干的 able-minded;high-powered
能感光的 photosensitive
能高线 energy head line;line of energy head
能高线原理 principle of the line of energy head
能构造性 constructibility
能够承受的车辆荷载 capable of bearing wheel loads
能够维持生活的工资 living wage
能谷 energy-valley
能观测性 observability
能贯穿 penetrative
能贯穿的 penetrable
能含量 intrinsic(al) energy
能耗 energy consumption
能耗降低率 decreasing rate of energy source consumption
能耗制动 dynamic(al) braking
能横过的 traversable
能恢复的 non-expendable
能恢复原状 resilient
能挥发的 volatilizable
能回潮的 re-wetable
能回用的 recycable
能汇 energy sink
能毁坏的 destructible
能获得的 obtainable
能积曲线 energy-product curve
能级 level
能级不变 constance [constancy] of level
能级参数 energy level parameter
能级差 energy head;level difference
能级的自然宽度 natural width of energy level
能级多重性 energy level multiplicity
能级分布 energy level distribution
能级分裂 energy level splitting;splitter of levels;splitting of energy levels
能级分配图 energy level diagram
能级耗尽 energy level depletion
能级记录器 level recorder
能级记录仪 energy level recorder
能级间的跃迁 transition between energy level
能级间隔 energy level spacing;level spacing
能级简并 degeneracy of energy level
能级交叉 energy level-crossing
能级交叉法 energy level-crossing method
能级交叉光谱学 level-crossing spectroscopy
能级结构 energy level structure
能级距离 energy[level] gap
能级宽度 energy level width;level width
能级粒减少 energy level depopulation
能级粒子数 energy level population
能级密度 level density
能级强度 strength of level
能级去激活 level deexcitation
能级寿命 level life;lifetime of energy level
能级数 population of levels
能级图 energy level diagram;energy level scheme;level scheme
能级跃迁 energy level transition
能级指示器 level indicator
能计算的 calculable
能加工成薄片的 laminable
能见的 visible
能见度 visibility (limit);visibility range;distance visibility;distance vision;range of visibility;conspicuity
能见度半径 radius of visibility
能见度变坏 obstruction to vision
能见度不定 visibility poor;visibility variable
能见度不好 visibility poor
能见度不好的天气 thick weather
能见度不良 low visibility;poor visibility
能见度不良时的声号 sound signals in restricted visibility
能见度测定表 nephelometer
能见度测定计 nephelometer;visibility meter
能见度测定器 nephelometer;visibility meter
能见度测量仪 hazemeter;transmittance meter;visibility measuring set
能见度测试仪 visibility detector
能见度等级 scale of visibility;visibility scale
能见度等于零的浓度 zero fog
能见度低 bad visibility
能见度电码 visibility code
能见度范围 limit of visibility
能见度好 visibility good
能见度很好 visibility very good
能见度极好 ceiling and visibility unlimited;visibility unlimited
能见度极限 limit of velocity;limit of visibility
能见度减低 reduction in visibility
能见度检定器 visibility detector
能见度界限 limit of velocity;limit of visibility
能见度距离 range of visibility
能见度曲线 visibility curve
能见度试验 visibility test
能见度受限制 restricted visibility
能见度条件 visibility condition
能见度图表 visibility chart
能见度为零的 zero-zero
能见度限制 restriction of visibility
能见度一般 visibility moderate
能见度仪表 visibility meter
能见度异常好 visibility exceptional
能见度因数 visibility factor
能见度指示器 visibility indicator
能见范围 visible area;visual range;area of visibility;range of visibility;range of vision;visibility distance;visibility range
能见光弧 arc of visibility;sector of light
能见函数 visibility function
能见弧度 arc of visibility
能见极限 visibility limit
能见检索 visible search
能见角 visibility angle
能见距离 sight distance;visibility distance;visibility range;visible distance;visible range;vision distance;visual distance
能见情况 visibility condition
能见区地图 visible area map
能见显示 visual display
能见限度 limit of velocity
能见信标 visual beacon
能见锥区 cone of visibility
能降低成本的 cost-effective
能接受的最低收益率 cut-off rate of return;lowest acceptable rate of return
能解决的 solvable
能进人冷藏室 walk-in
能进人水箱 walk-in
能进入的 accessible;walk-in
能浸透的 saturable
能经受恶劣条件的 heavy-duty
能阱 energy sink;energy trap(ping)
能看到表情 two-way communication
能控制的 manageable
能扩建的建筑 building adaptable to extension
能垒 energy barrier
能力 ability;brain power;capability;capacity;competence;potency;power
能力保护构件 capacity protected member
能力比降 energy slope
能力测验 ability test;aptitude test

能力差异 capacity variance
能力储备 reserve capacity; maneuver margin
能力储备系数 coefficient of reserve capacity
能力工资制 wage system based on ability
能力过剩 overcapacity
能力建设 capacity building
能力降低系数 <结构设计强度的> capacity reduction factor
能力交换 exchange of abilities
能力较低者 inferior
能力开发 development of faculty
能力利用 utilization of capacity
能力利用系数 capacity utilization factor
能力判决书 letter of capacity
能力任务相互关系 ability-task interaction
能力任务相互作用 ability-task interaction
能力设计 capacity design
能力水平 level of competency
能力消散 dissipation of energy
能力型测验 power test
能力研究 capacity study
能力与负荷分析 analysis of capacity and load
能力与收入 ability and earnings
能力裕度 capability margin
能力证书 certificate of competence [competency]
能链 energy chain
能量 capacity; energy; energy capacity; quantity of energy
能量保持 conservation of energy; energy conservation
能量保护 energy preservation
能量倍乘数 crusher capacity multiplier
能量倍增器 energy doubler
能量倍增因子 energy multiplication factor
能量比 duty ratio; energy ratio
能量比耗 specific energy consumption
能量比降 energy slope
能量变化 fluctuation of energy
能量变化灵敏的 energy-sensitive
能量变化率 energy gradient
能量变化顺序检测 energy-variant sequential detection
能量变换 energy conversion; power conversion
能量变形曲线 energy-deflection curve
能量标度 energy scale
能量表 energy meter
能量波动 energy hunting
能量补偿 energy bucking
能量不变 energy conservation
能量不等式 energy inequality
能量不灵敏探测器 energy-insensitive detector
能量不灭 conservation of energy; energy conservation; persistence of energy
能量不灭定律 law of conservation of energy
能量不守恒 non-conservation of energy
能量不足 energy deficiency; undercapacity
能量测定 energometry
能量测定器 energometer
能量测量 energy measurement
能量差 energy difference; energy head
能量差异 capacity variance
能量产量 energy yield
能量产生 energy production
能量常数 energy constant
能量沉积 energy deposition
能量沉积事件 energy deposition event
能量（乘）积曲线 energy-product curve
能量储备 energy reserves; energy storage; margin energy; margin of energy
能量储备降低系数 reserve energy reduction factor
能量储备能力 reserve energy of capacity
能量储备系统 energy conservation system
能量储存 energy storage; margin of energy
能量储存材料 energy storage material
能量处理能力 energy-handling capability
能量传递 energy transfer; energy transmission; transmission of energy
能量传递边界 energy-transmitting boundary
能量传递系数 energy transfer coefficient
能量传输 energy transfer; energy transport; transmission of energy
能量传送的 energy-delivering
能量窗 energy window
能量脆性转变温度 energy transition temperature
能量存储 energy storage
能量存留 energy retention
能量代谢 energy metabolism
能量代谢率 relative metabolic rate
能量带 energy band
能量带模型 energy band model
能量带通 energy band pass
能量单色性 energy monochromaticity
能量单位 energy unit; power unit
能量当量能谱 energy equivalence
能量的对数变化 logarithmic energy change
能量的惯性 inertia of energy
能量的回收 recovery of energy
能量的利用 energy use
能量的受控释放 controlled release of energy
能量的最重要来源 the most important sources of energy
能量递降 energy degradation
能量定标因数 energy scaling factor
能量定理 energy theorem
能量定律 energy law
能量定值 power demand
能量动量法 energy-momentum approach
能量动量关系 energy-momentum relationship
能量动量矩阵 energy-impulse matrix
能量动量守恒 energy-momentum conservation
能量动量赝张量 energy-momentum pseudotensor
能量动量张量 energy-momentum tensor
能量对流 convection of energy
能量发散 energy divergence [divergency]; energy spread
能量发射 energy emission
能量法 energy approach; energy method
能量法则 energy theorem
能量反馈 energy feedback
能量反射积累因数 energy reflection build-up factor
能量反射率 energy reflectivity
能量反射系数 energy reflection coefficient; energy reflectivity
能量反照率 energy albedo
能量范围 energy range; energy region
能量方程式 energy equation
能量放大 energy amplification
能量分辨本领 energy resolving power
能量分辨力 energy resolution
能量分辨（率）energy resolution
能量分布 distribution of energy; energy distribution; energy spectrum; partition of energy
能量分布函数 energy distribution function
能量分布曲线 energy distribution curve
能量分层 quantization of energy
能量分解 energy resolution
能量分配 energy distribution; energy sharing
能量分配曲线 energy distribution curve
能量分散 energy dispersion; energy dissipation; energy spread; energy straggling
能量分散 X 射线分析 energy-dispersive X-ray analysis
能量分散光谱法 energy-dispersive spectroscopy
能量分散微观分析 energy-dispersive microanalysis
能量分析 energy analysis
能量分析磁铁 energy analyzing magnet
能量分析法 energy analysis method
能量分析显微镜 energy analyzing microscope
能量峰面 energy front
能量辐射 energy beaming; energy radiation; radiation of energy
能量辐射率 energy radiation rate
能量辐射损失 energy loss by radiation
能量负荷 demand of energy
能量高【铁】energy head
能量高点 energy hill
能量供给 delivery of energy
能量供应 energy supply; power supply
能量共振转移 resonance energy transfer
能量估算 energy budget
能量估算法 energy budget method
能量关系 energy dependence
能量管理 energy management
能量管理系统 energy management system
能量函数 energy function
能量耗散 dissipation of energy; energy dissipation
能量耗散部件 energy-dissipating element
能量和天然驱动类型 energy and natural driving type
能量核 energy kernel
能量核算 energy budget
能量互易定理 reciprocal energy theorem
能量恢复 energy recovery; recovery of energy
能量恢复系统 energy recovery system
能量回收 energy recovery
能量回收器 recuperator
能量回收时间 energy payback time
能量回收式制动 regenerative braking
能量积 energy product
能量积存 energy deposition
能量积分 energy integral; integral of energy
能量积聚 energy storage
能量积聚制动 energy storage braking
能量积累 energy accumulation
能量积累因数 energy built-up factor
能量积累因子 energy accumulation factor; energy built-up factor
能量基准因素 energy guideline factor
能量级 energy stage; energy level
能量极限 energy limit
能量极限值 energy bound
能量集中 energy concentration
能量价格 energy cost
能量间隔 energy bite; energy interval; energy separation; energy spacing
能量减到最小法 energy minimization
能量减量 energy decrement
能量减缩器 capacity reducer
能量降低 energy degradation
能量降级 energy degradation
能量交换 energy exchange; interchange of energy
能量交换时间 energy exchange time
能量交换通量 energy exchange flux
能量校正系数 energy correction factor
能量校准标准 energy calibration standard
能量接收度 energy acceptance
能量节约 energy economy
能量结算 energy balance
能量金字塔 energy pyramid
能量经济学 economics of energy
能量聚集 energy converge; focusing of energy
能量聚焦 energy focusing
能量决策 capacity decision
能量均分 energy equipartition; equipartition of energy
能量均分时间 energy equipartition time
能量均分原理 principle of the equipartition of energy
能量均衡线 energy grade line
能量均配 equipartition of energy
能量均匀性 energy uniformity
能量抗扰度 energy noise immunity
能量刻度 energy calibration
能量控制 capacity control; energy management
能量控制系统 energy control system
能量枯竭 depletion of energy
能量块 energy dispersion block
能量扩充投资 capacity expansion investment
能量扩散 energy diffusion; energy dispersal
能量扩散器 energy disperser
能量扩散特性 energy dispersal characteristics
能量扩散信号 energy dispersal signal
能量扩展度 energy spread
能量浪费 energy dissipation
能量累积 energy accumulation
能量累积曲线 mass energy curve
能量离散 spread in energy
能量利用 energy utilization; utilization of energy
能量利用差异 capacity utilization variance
能量利用率 capacity usage ratio
能量利用效率 power efficiency
能量利用效室 energy utilization efficiency
能量连续区 energy continuum
能量灵敏度 energy sensitivity
能量流（动）energy flow
能量流量 energy fluence
能量流量率 energy fluence rate

N

能量流率 energy flux rate
能量率 specific energy
能量脉冲 energy impulse
能量密度 energy density
能量密度函数 energy density function
能量密度谱 energy density spectrum
能量密度图 energy density spectrum
能量密集的 energy intensive
能量密集型工业 energy intensive industry
能量面土反馈 energy ground feedback
能量模型 energy model
能量-挠度关系曲线 <码头护弦木的> energy-deflection curve
能量耦合 energy coupling
能量判据 energy criterion
能量匹配 energy matching
能量漂移 energy jitter
能量频谱 energy distribution
能量平衡 energy balance; energy budget; energy equilibrium
能量平衡表 energy balance sheet
能量平衡法 energy balanced approach; energy balance method; energy budget method
能量平衡方程 energy balance equation; equation of energy balance
能量平衡气候学 energy-balance climatology
能量坡度线 energy grade line
能量坡降 energy slope
能量坡降线 energy grade line
能量谱 energy spectrum; power spectrum
能量谱密度 energy spectrum density
能量歧离 energy straggling
能量起始阈值 beginning threshold value of energy
能量迁移 energy migration; energy transfer
能量强度 energy intensity
能量区 energy range; energy region
能量曲线 energy curve
能量曲线中断处 energy gap
能量缺乏 energy deficiency; energy deficit
能量群 energy group
能量散射的球面系数 global coefficient of energy scattering
能量散逸 dissipation of energy; energy dissipation
能量散逸微观分析 energy-dispersive microanalysis
能量色散 energy dispersion
能量色散 X 射线分析 energy dispersion X-ray analysis
能量色散 X 射线探测仪 energy-dispersive X-ray detector
能量色散 X 射线荧光 energy-dispersive X-ray fluorescence
能量色散法 energy dispersion method
能量色散分析 energy dispersion analysis
能量射束 energy beam
能量生产成本 cost of power production
能量实验中心 experimental energy center[centre]
能量试验台 power test rig
能量试验装置 power test rig
能量释出 energy release
能量释放 energy liberation; energy release
能量释放速率 energy release rate
能量释放中心 centre of energy release
能量收支 energy budget
能量守恒 conservation of energy; energy conservation; persistence of energy; preservation of energy
能量守恒定理 energy conservation principle; principle of construction of energy; principle of energy conservation
能量守恒定律 conservation of energy law; energy conservation law; law of conservation of energy; law of construction of energy; law of energy conservation
能量守恒定则 energy conservation principle
能量守恒方程 energy conservation equation
能量守恒和转换定律 law of conservation and conversion of energy
能量守恒原理 conservation of energy principle; principle of conservation of energy; principle of construction of energy; principle of energy conservation
能量寿命 energy life time
能量输出 energy output; energy yield; power export
能量输入 energy import; energy input
能量输入率 rate of energy input
能量输送 energy transport; transmission of energy
能量输运 energy transport
能量数据 energy datum
能量衰减 attenuation of energy; energy attenuation; energy decay
能量衰减函数 energy decay function
能量水头 energy head
能量算符 energy operator
能量损耗 energy dissipation; energy loss; energy waste; expenditure of energy
能量损耗率 rate of energy loss; specific energy loss
能量损耗系数 figure of loss
能量损失 energy degradation; energy dissipation; energy loss; loss in energy; lost energy; loss of energy
能量损失机理 energy loss mechanism
能量损失控制 energy loss control
能量损失率 rate of energy loss
能量损失谱 energy loss spectroscopy
能量损失强度 energy thickness
能量损失系数 coefficient of energy loss; energy loss factor
能量塔 energy pyramid
能量弹性 energy elasticity
能量探测器 energy-probe
能量特性 energy characteristic; energy response
能量梯度 energy gradient; energy slope
能量梯度线 energy grade line; energy gradient line
能量体系 energy system
能量调节 capacity modulation; energy regulation
能量调节阀 capacity regulating valve; energy regulator
能量调节器 capacity controller; energy regulator
能量调节装置 capacity regulating device
能量调制 energy modulation
能量跳变 energy jump
能量通量 energy flow; flux of energy
能量头 energy head
能量投影算符 energy projection operators
能量透射积累因子 energy transmission build-up factor
能量图(解) energy diagram; energygram; energy scheme
能量途径 energy path(way)
能量退降律 law of degradation of energy
能量危机 energy crisis
能量(位)垒 energy barrier
能量温度曲线 energy-temperature curve
能量稳定 energy stabilization
能量稳定度 energy stability
能量无规则化 randomization of energy
能量误差 energy error
能量吸收 absorption of energy; energy absorption; energy deposition; power consumption; energy-absorbing
能量吸收材料 energy absorbing material
能量吸收法 energy absorbing method
能量吸收管 energy absorbing tube
能量吸收过程 energy absorption process
能量吸收截面 energy absorption cross-section
能量吸收能力 energy absorption capability; energy absorption capacity
能量吸收器 energy absorber
能量吸收前端 <汽车的> energy absorbing front end
能量吸收设备 energy absorber
能量吸收式保险杠系统 energy absorbing bumper system
能量吸收式测功机 absorption dynamometer
能量吸收式车架 energy absorbing frame
能量吸收式路边防撞护栅 energy absorbing roadside crash barrier
能量吸收式转向柱 energy absorbing steering column
能量吸收系数 coefficient of energy absorb; energy absorption coefficient
能量吸收系统 <防撞安全设施> energy absorption system; energy absorbing system
能量吸收型转向装置 energy absorbing steering assembly
能量吸收制动器 energy consumption brake
能量系数 coefficient of energy; energy coefficient; energy efficiency
能量系统 energy system
能量线 energy line
能量陷阱 energy trap(ping)
能量相关通量 energy-dependent flux
能量相关性 energy dependence
能量响应 energy response
能量响应函数 energy response function
能量项 energy term
能量消费 energy expenditures
能量消耗 consumption of energy (power); energy consumption; energy cost; energy dissipation; energy expenditures; expenditure of energy; power consumption; waste of energy
能量消耗监控 energy management
能量消耗率 rate of energy consumption; rate of energy expenditures
能量消耗值 energy consumption value
能量消散 energy dissipation
能量消散率 energy dissipator; energy killer
能量消散器 energy dissipator
能量消散微观分析 energy-dispersive microanalysis
能量效率 energy efficiency
能量效(率)比 energy efficiency ratio
能量效应 energy effect
能量楔 energy-dissipating wedge
能量需求 energy demand
能量需要 demand of energy; energy demand; energy requirement
能量选择 energy selection
能量选择缝 energy-selective slit
能量学 energetics
能量循环 energy cycle
能量研究 capacity study
能量一致 energy coincidence
能量依赖性 energy dependence
能量已调制束 energy-modulated beams
能量意外费用规划 energy contingency planning
能量因数 energy factor
能量因素 capacity factor; energy factor
能量因子 energy factor
能量引出装置 probe
能量预算 energy budget
能量预算法 energy budget method
能量阈 energy cut-off
能量原理 energy principle
能量约束时间 energy confinement time
能量跃迁 energy jump
能量再生 energy regeneration
能量-噪声抗扰度 energy noise immunity
能量增量 energy increment
能量增益 energy gain
能量展布波形 energy spreading waveform
能量张量 energy tensor
能量甄别法 energy discrimination method
能量振荡 energy oscillation
能量-震级 energy-magnitude
能量-震级关系 energy-magnitude relationship
能量直接输入 direct energy input
能量直接转换 direct energy conversion
能量值 amount of energy; energy content; energy value
能量指数 energy index
能量质量当量 energy mass equivalence
能量置换 energy replacement
能量置换时间 energy-loss-time; energy-replacement time
能量注量 energy fluence
能量转变 conversion of energy; energy conversion
能量转化 conversion of energy; energy conversion; energy transformation
能量转化者 energy transformer
能量转换 conversion of energy; energy conversion; energy transfer; energy transformation; energy transmission
能量转换工程 energy conversion engineering
能量转换机理 energy transfer mechanism
能量转换技术 energy conversion technique
能量转换率 energy conversion rate; rate of energy transformation
能量转换器 energy absorber; energy converter
能量转换途经 path of energy transformation
能量转换系数 energy conversion factor
能量转换系统 energy conversion system

能量转换效率 energy conversion efficiency
能量转换装置 energy conversion device
能量转移 energy transfer
能量转移过程 energy transfer process
能量转移函数 energy transfer function
能量转移矩阵 energy transfer matrix
能量转移时间 energy transfer time
能量子 energy quantum
能量阻尼器 energy damper
能流 energy current;energy flux;energy stream;flow of energy;power flow
能流标度校准 energy scale calibration
能流反应 energy reaction
能流分布 power-flow distribution
能流改向 change of energy flow
能流回级 energy recovery power
能流率 rate of energy flow
能流密度 energy flux density
能流密度矢量 Poynting vector
能流模式 energy flow mode
能流模型 energy flow model
能流速度 rate of energy flow
能流速率 energy flow rate
能流图 energy flow chart; energy flow diagram
能路 energy circuit
能率 energy rate
能率比 duty ratio
能率促进 <机械等的> speed-up
能率试验 capacity test
能霉素 capacidin
能面 energy surface
能耐的 tolerant
能黏[粘]结的 cohesible
能黏[粘]聚的 cohesible
能爬升的坡度 <汽车> climbable gradient
能漂白的 decolo(u)rant
能坡 energy gradient;energy slope
能坡线 <管路水流的> energy grade line;energy gradient line
能谱 power spectrum
能谱编录值 spectrum documentary value
能谱的软化 spectrum degradation
能谱分布 spectral energy distribution
能谱分析 energy spectrum analysis
能谱分析法 energy spectrometry
能谱函数 energy spectrum function
能谱宽度 energy spectrum width
能谱密度 energy spectral density
能谱取样换算系数 spectrum sampling converted coefficient
能谱软化 spectrum degradation
能谱外的 extraspectral
能谱仪 energy spectrometer
能谱仪换算系数值 conversion coefficient value of spectrometer
能谱仪净计数 net count of spectrometer
能谱仪型号 spectrometer model
能谱异常三源图 three-resource figure of spectrum anomaly
能清偿的 solvent
能区 ergoregion
能取得航行通告 availability of notice to mariners
能溶和的 miscible
能溶解的 dissoluble;solvable
能溶于苯乙醇 benzene-ethanol soluble
能溶于二硫化碳 carbon bisulfide soluble
能溶于干性油中的干燥剂 drier soluble in drying oil
能溶于酒精的 spirit-soluble
能溶于石油醚 petroleum ether soluble
能溶于酸的碱 acid-soluble alkaline
能熔化的 liquefiable
能散度 energy divergence[divergency]
能散分析 energy-dispersive analysis
能杀菌的 bacteriolytic
能杀菌素 capacidin
能上楼梯的轮椅 stairwalking hand truck
能上楼梯的手推小车 stairwalking hand truck
能伸缩 flexible
能伸缩的 telescopic(al)
能伸缩的吊杆起重机 telescopic(al) boom crane
能伸缩活动铁架 iron expansion shield
能伸缩耙 expanding harrow
能伸展 tensible
能升坡度 climbable gradient
能生长植物的浅水区 <湖泊> phytal zone
能生存的 viable
能胜任的工人 competent worker
能胜任的人 competent person
能胜任的人员 competent personnel
能识性原则 identifiability criterion[复 criteria]
能使荷载扩散的 load-shedding
能使用两种语言的 bilingual
能收缩的 contractible;contractile
能手 adept;artist;consummator;master;proficient
能受钉的 nailable
能水合的 hydratable
能斯脱-爱因斯坦关系式 <扩散定律> Nernst-Einstein relation
能斯脱定理 Nernst's theorem
能斯脱方程 Nernst's equation
能斯脱分布定律 Nernst's distribution law
能斯脱分配定律 Nernst's distribution law
能斯脱高温发热体 Nernst's body
能斯脱公式 Nernst's equation
能斯脱近似公式 Nernst's approximation formula
能斯脱理论 Nernst's theorem; Nernst's theory
能斯脱-林德曼热量计 Nernst-Lindemann calorimeter
能斯脱流量单位 Nernst's unit
能斯脱-普朗克方程 Nernst-Planck equation
能斯脱燃料电池 Nernst's fuel cell
能斯脱热定理 Nernst's heat theorem
能斯脱-汤姆逊规则 Nernst-Thomson rule
能斯脱响应 Nernstian response
能斯脱效应 Nernst's effect
能斯脱学说 Nernst's theory
能松开和冲洗液循环的打捞筒 releasing and circulation overshot
能速 velocity of energy
能缩回的 retractile
能态 energy eigenstate;energy state;quantum state;stationary state
能态密度 density of energy state;energy state density
能态寿命 lifetime of energy state;lifetime of the state
能提取的 withdrawable
能调整角度的刨 chamfer plane
能听度 audibility
能听范围 area of audibility
能听极限 limit of audibility
能听区 zone of audibility
能通量 energy fluence;energy flux
能通量密度 energy fluence rate;energy flux density
能通行管沟 passable conduit
能通行情况 trafficability condition
能头损失 loss of energy head
能头损失机理 energy loss mechanism
能透光化线的 diactinic
能透过的 pervial;pervious
能脱色的 decolo(u)rant
能网 energy network
能稳定性 stabilizability
能隙 energy gap
能线图 energy profile
能陷振动模式 energy trap vibration mode
能效率 energy efficiency
能效率比 energy efficiency ratio
能效系数 energy-efficiency factor
能行性 effectiveness
能蓄热的 heat retaining
能迅速变现的资产 quick assets
能迅速还本生利的投资 self-liquidating investment
能延期的 deferrable
能延伸的建筑 building adaptable to extension
能液化的 liquefiable
能移开的 withdrawable
能用金刚石 usable diamond
能用水稀释的 water thinnable
能育性 fertility
能预防的 precludable
能域 energy gap
能域分析 energy gap analysis
能源 energy supply; power source; source of energy;source of power; energy resources
能源安全 energy security
能源保护 energy conservation
能源保护计划 energy conservation program(me)
能源部 <美> Department of Energy
能源材 energy wood
能源材料 energy material; material for energy application
能源策略 energy strategy
能源产量(收益) energy yield
能源产品 energy product
能源成本 energy cost
能源成本系数 energy cost ratio
能源储备 energy conservation;energy reserves
能源储存设备 energy storage device
能源储存系统 energy storage system
能源储量 energy reserves
能源当量 energy equivalent
能源的持久使用 sustainable use of energy
能源的经济节约 economy of energy
能源的需求与供给 demand and supply of energy
能源地理 geography of energy
能源定价 energy pricing
能源短缺 energy shortage
能源法规 laws and regulations of energy
能源费用 power cost;energy cost
能源分布 energy distribution;power-flow distribution
能源分布图 map of energy
能源分配系统 energy distribution system
能源丰富国家 energy rich
能源港 energy harbor
能源革命 energy revolution
能源工程学报 <美国土木工程学会不定期刊> Journal of Energy Engineering
能源工业 energy source industry
能源工艺领域 energy technology sphere
能源功效 energy efficiency
能源供需平衡 balanced supply and demand of energy
能源供应 energy source feed;energy source supply
能源管道与系统 energy pipelines and systems
能源管理 energy control;energy management
能源管理合同 energy management contract
能源管理技术员 energy technician
能源管理局 supply authority
能源规划 energy plan(ning);power scheme
能源合理利用 energy conservation
能源合作 cooperation in the field of energy
能源环境 energy environment
能源环境政策 energy environmental policy
能源回收 energy recovery
能源回收进程 energy recovery process
能源回收系统 energy recovery system
能源基地 base of energy resources
能源计划 energy plan(ning);energy program(me)
能源技术 energy technic;energy technique
能源技术经济学 technologic(al) economics of energy
能源价格研究 energy pricing study
能源建设 energy source building
能源交通基金 energy and transportation fund
能源交易选择权 energy option
能源节约 conservation of energy;energy conservation;energy-saving
能源结构 energy structure
能源紧张的 energy intensive
能源经济 economy of energy
能源经济学 economics of energy;energy economics
能源经济学家 energy economist
能源开发 energy development
能源勘察 energy resource survey
能源勘探 energy exploration;energy resource survey
能源控制 power-operated control
能源控制装置 energy source controller
能源枯竭 depletion of energy
能源库 energy depot
能源矿产 commodities for energy source
能源浪费 energy dissipation
能源类别 classification of energy source
能源利用 energy use;energy utilization
能源利用技术 energy utilization technology
能源利用率 energy utilization rate
能源利用效果 utilization results of energy
能源利用效率 energy utilization efficiency
能源利用者 energy user
能源流优化模型 energy flow optimization model
能源码头 energy quay;energy terminal;energy wharf
能源密集的 energy intensive
能源密集的技术 energy intensive technology
能源贫乏 energy poor
能源贫乏国家 energy poor

能源品质 energy quality
能源平衡 energy balance
能源评议 energy source assessment
能源坡降 energy gradient
能源坡降线 energy-gradient line
能源期货 energy future
能源设备实验 energy systems laboratory
能源设施 energy facility
能源审计 energy auditing
能源生产和使用工艺技术领域 energy technology sphere
能源生产结构 energy production structure
能源生产能力 energy production capacity
能源生产弹性系数 energy production elasticity
能源生产与消耗量 energy production and consumption
能源使用管理 energy utility management
能源数据系统 energy resources data system
能源弹性系数 energy elasticity coefficient
能源弹性值 energy elasticity
能源梯级利用 graded use of energy
能源统计 energy statistics
能源危机 depletion of energy;energy crisis;energy dilemma;energy disruption
能源未来 energy future
能源问题 energy problem
能源吸收法 energy-absorbing means
能源消费 energy consumption;energy use
能源消费结构 energy consumption structure;structure of energy consumption
能源消费弹性系数 energy consumption elasticity coefficient
能源消费总量 total energy consumed
能源消耗 energy dissipation;expenditure of energy
能源消耗定额管理 quota control of energy consumption
能源消耗管理 management of energy consumption
能源消耗管理系统 energy management system
能源消耗奖惩制 reward and penalty system of energy consumption
能源消耗量 energy consumption
能源消耗增长系数 energy consumption growth coefficient
能源效率 energy efficiency
能源效率比 energy efficiency ratio
能源效率标准 energy performance standard
能源效应 energy efficiency
能源信息管理局 Energy Information Administration
能源需求 energy need;energy demand
能源需求量 energy demand
能源需求趋势 energy demand trend
能源需要 energy need
能源研究和发展管理局 Energy Research and Development Administration
能源诱发的全球气候变化 energy induced global climate change
能源预测 energy forecasting
能源预算 energy budget
能源噪声 noise in sources
能源增长率 energy growth rate
能源政策 energy(sources)policy;power policy
能源中心站 central energy station
能源转换 energy conversion;energy transfer
能源转换器 energy converter
能源资源 energy resources;energy sources
能源资源保护 conservation of energy
能源作物 energy crop
能再现的 reproducible
能再现的示踪 reproducible tracing
能再现数据 reproducible data
能在任何情况下着陆的 omniphibious
能增生的铀 fertile uranium
能障 energy barrier
能折入箱中的百叶窗扇 boxed shutter
能证明 verifiability
能支承的 supportable
能直接在传真感光纸上记录的地震仪 facsimile seismograph
能重复利用的 recycable
能蛀木的 xylotomous
能自动换片的唱机 record autochanger
能自给的 self-feed
能自燃货物 spontaneously combustible cargo
能自行灌水的 self-flooding
能钻木的 xylotomous

尼

尼阿格黄铜 <一种含铅黄铜> Niag

尼奥博拉拉灰岩 Niobrara limestone
尼奥卡尔 <阴离子渗透剂> Neokal
尼泊尔乌木 Nepal ebony
尼丑西拉尔镍硅铁 nitrosilal
尼尔抛物线 Neil's parabola
尼尔气压分析法 Neil method
尼尔森桥 <一种斜吊杆系杆拱桥> Nielsen bridge
尼尔瓦合金 nilvar
尼尔-约翰逊质量分析器 Nier-Johnson mass analyser[analyzer]
尼尔质谱计 Nier's mass spectrometer
尼钙石 nifontovite
尼姑庵 Buddhist nunnery;nunnery(church)
尼古丁 nicotine
尼古拉兹砂粒糙率 Nikuradse sand roughness
尼加费尔防缩法 Negafel process
尼加拉瓜 <北美洲> Nicaragua
尼加罗炼镍法 Nicaro process
尼卡洛伊合金 <一种高导磁铁镍合金> Nicalloy
尼科巴热 Nicobar fever
尼科波尔锰矿床 Nikopor manganese deposit
尼科尔斯辐射计 Nichols radiometer
尼科尔斯曲线 Nichols curve
尼科尔淤泥焚化炉 Nichols sludge incinerator
尼科耳辐射计 Nicol's radiometer
尼科耳棱镜 Nicol;Nicol's prism
尼科耳偏振棱镜 Nicol polarizing prism
尼科耳乳化油漆 Nicoloid
尼科罗斯(合金)nicorros(alloy)
尼科牌液压往复式给矿机 Nico hydrostroke feeder
尼科铜镍合金 Nico metal
尼科西亚 <塞浦路斯首都> Nicosia
尼可尔斯图 Nichols diagram
尼克拉尔 <一种铝合金> Nicral
尼克拉黄铜 <一种铅黄铜> Nicla
尼克劳合金 Nickeloy
尼克劳斯型锅炉 Niclausse boiler
尼克利特镍铬合金 <一种镍铬耐热合金> Nicrite
尼克林高阻合金 Nickelin
尼克洛西 <一种高硅镍合金> Nichrosi
尼克洛依镍铜 Nicloy
尼克森色度计 Nickerson colorimeter
尼阔林铜镍合金 <一种铜镍合金> Nickoline
尼勒克斯 <一种镍铁合金> Nilex
尼龙 nylon;polyamide
尼龙-11 粉末涂料 nylon-11 coating;Rilsan coating
尼龙-6 <纤维叫锦纶-6,卡普纶> polycaprolactam
尼龙-β poly-beta-alanine
尼龙坝 fabridam;nylon dam
尼龙薄膜 nylon film
尼龙布 nylon cloth
尼龙布袋 nylon bag
尼龙衬套 nylon bush
尼龙齿轮 nylon gear
尼龙绸 ninon
尼龙锤 hammer nylon
尼龙单丝 nylon monofilament
尼龙底绝缘板 nylon-base insulator
尼龙地毯 nylon carpet
尼龙垫圈 polythene washer
尼龙吊带 nylon sling
尼龙吊具 nylon sling
尼龙吊索 nylon sling
尼龙钉 nylon staple
尼龙废料 nylon waste
尼龙酚醛树脂 nylon phenolic resin
尼龙酚醛塑料 nylon phenolics
尼龙覆面钢 nylon-coated metal
尼龙格栅 nylon grill(e)
尼龙管 nylon tube
尼龙轨下胶垫 nylon rail pads
尼龙滚子泵 nylon-roller pump
尼龙和涤纶合成的吊索 nylon and terylene lifting sling
尼龙和聚丙烯黏[粘]胶纤维 bonded fibers[fibres] of nylon and polypropylene
尼龙环氧 nylon epoxy
尼龙环氧黏[粘]合剂 nylon-epoxy adhesives
尼龙加固带 nylon-reinforced belt
尼龙加筋橡胶板 nylon-reinforced rubber sheet
尼龙加筋橡胶片 nylon-reinforced rubber sheet
尼龙夹层皮带 nylon belt
尼龙结构 nylon fabric
尼龙缆(索)nylon rope;nylon line
尼龙链钩 nylon sling
尼龙滤层防冲系统 nylon filter antiscour system
尼龙清理刷 nylon cleaning
尼龙燃料箱 nylon-fuel cell
尼龙沙袋防波堤 sand-sausage breakwater
尼龙纱 nylon yarn
尼龙筛 nylon sieve
尼龙筛网 nylon screen
尼龙绳 nylon cord;nylon rope;nylon wire
尼龙树脂 nylon resin
尼龙刷 <刷毛混凝土路面用> nylon brush
尼龙栓 nylon plug
尼龙丝 nylon wire;nylon yarn
尼龙塑料 nylon plastics
尼龙索 nylon cord
尼龙套 nylon bush
尼龙填料 nylon gasket
尼龙填圈 nylon gasket
尼龙涂层 nylon coating
尼龙涂料 nylon paint
尼龙网 nylon net
尼龙网坝 nylon net dam
尼龙纬线 nylon-weft yarn
尼龙细绳 nylon marline
尼龙纤维 nylon fiber[fibre]
尼龙纤维增强水泥 polyamide fiber[fibre] reinforced cement
尼龙线 nylon wire;nylon yarn
尼龙线织轮胎 nylon cord tyre
尼龙织品 nylon fabric
尼龙织物 nylon fabric;nylon woven fabric
尼龙织物护岸 fabriform protection
尼龙轴承 nylon bearing
尼龙轴套 nylon bushing
尼龙柱销联轴器 nylon pin coupling
尼龙铸粉 nylon mo(u)lding powder
尼龙铸造齿轮 nylon mo(u)lded gear
尼罗河 Nile River
尼罗河泥砖 Nile mud brick
尼罗河水位仪 nilometer;niloscope
尼罗蓝 <稍带绿色的> Nile blue
尼罗绿 <浅淡的蓝绿色> Nile green
尼洛斯皮带接头夹 Nilos belt fastening nip
尼洛斯皮带扣 Nilos belt fastening nip
尼曼三角 Niemann's triangle
尼曼试验 Niemann's test
尼曼振荡器 Nieman's oscillator
尼孟合金 <一种耐热镍基变形合金> nimonic alloy
尼莫尔(铸铁)Nimol
尼莫尼克(合金)Nimonic
尼哦油 niobe oil
尼欧克姆阶 <早白垩世早期>【地】Neocomian(stage)
尼帕极性超带 Nepa polarity superzone
尼帕极性超时 Nepa polarity superchron
尼帕极性超时间带 Nepa polarity superchronzone
尼硼钙石 nifontovite
尼普科夫圆盘 Nipkow disk
尼日尔港 Port Niger
尼日尔河 Niger River
尼日尔三角洲 Niger delta
尼日尔三向联结构造 Niger triple junction
尼日尔深海扇 Niger abyssal fan
尼日利亚椴木 movingue
尼日利亚红木 Nigerian mahogany
尼日利亚石 nigerite
尼日利亚桃花心木 Nigerian mahogany
尼赛尔管 <一种充气的指示管>【物】Nixie
尼森式小(活动)屋 Nissen hut
尼斯盘铁镍合金 nickel-Span alloy
尼塔玻璃 Nitaline
尼特 <等于 1.44 位的信息单位>【计】nit
尼特计 Nitometer
尼特拉蒙炸药 nitramon
尼沃内耳耐蚀合金 Nionel
尼亚加拉河 Niagara River
尼亚加拉统【地】Niagaran
尼亚美 <尼日尔首都> Niamey

呢

呢绒 cloth;drapery;stuff goods;woolen cloth

泥

泥 slob

泥巴试验 experiment with clay model;model experiment with clay
泥笆墙 wattle;wattle and da(u)b
泥坝 mud bar
泥板 derby
泥板波 mud wave
泥板条 clay lath;clay strip
泥板岩 argillite;clay lath;clay slate;

mudstone
泥板岩振动器 shale shaker
泥拌砾石混合料【道】hoggin
泥包 mud drum;sludging up
泥包的 sludge bound
泥包卡钻 bit bailing
泥包钻头 balling of the bit
泥崩 mud avalanche; mudflow avalanche
泥泵 mud pump; slime pump; tubing drill; dredge pump; dredging pump【疏】
泥泵衬板 lining board of dredge pump
泥泵电磁离合器 electromagnetic clutch for dredge pump
泥泵阀 mud pump valve
泥泵工况 working condition of dredge pump
泥泵功率 power of dredge pump
泥泵构造 construction of dredge pump
泥泵联轴节 dredge pump coupling
泥泵联轴器 dredge pump coupling
泥泵链斗挖泥船 bucket dredge(r) with sand pump
泥泵磨耗 abrasion of dredge pump; dredge pump wearing
泥泵磨耗量 abrasion value of dredge pump
泥泵磨耗试验 abrasion test of dredge pump
泥泵磨损检查 inspection for dredge pump wearing
泥泵排量 displacement of dredge pump
泥泵排泄压力 dredge pump discharge pressure
泥泵清孔 bailing
泥泵驱动 dredge pump drive
泥泵驱动功率 driving power for dredge pump
泥泵取样 thief sampling; pump sampling
泥泵设计 design of dredge pump
泥泵绳轮 sand reel
泥泵剩余扬程 surplus head of dredge pump
泥泵试验 dredge pump test
泥泵试验结果 test results of dredge pump
泥泵试样 thief sample
泥泵水封腔 water seal chamber of dredge pump
泥泵特性曲线 characteristic curve of sand pump;dredge pump characteristic curve
泥泵脱开 disengagement of dredge pump
泥泵脱流 off-flow of dredge pump
泥泵相似定律 law of similarity for dredge pump
泥泵效率 dredge pump efficiency
泥泵压力表 pressure for dredge pump
泥泵扬程 discharge head of dredge pump;head of dredge pump;lift of dredge pump
泥泵叶轮 impeller of dredge pump
泥泵叶片 blade of dredge pump
泥泵真空 vacuum of dredge pump
泥泵真空表 vacuum ga(u)ge for dredge pump
泥泵主机 dredge pump engine
泥泵装置 dredge pump unit
泥泵自动控制器 automatic pump controller
泥泵总扬程 gross head of dredge pump;total head of dredge pump
泥蔽 mud sheet;mud work
泥饼 cake;flake of sludge;muck cake;mud cake
泥饼产率 muck cake production; sludge cake production rate
泥饼沉积 <孔壁上的> mud bridge
泥饼电阻率 mud cake resistivity; resistivity of mud cake
泥饼厚度 cake thickness;thickness of mud cake
泥饼结构 filter cake texture
泥饼密度 mud cake density
泥饼摩擦系数 cake friction factor
泥饼黏[粘]附卡钻 cake sticking
泥饼影响校正 mud cake correction
泥波 mud wave
泥波层理 laser bedding
泥驳 dredging barge; dump barge; dumping barge; hopper barge; mud barge; mud boat; mud lighter; mud scow; spoil barge
泥驳船 mud scow
泥驳轮换 alternation of mud barges
泥驳容量 hopper capacity
泥驳吸卸泵 barge sucker; suction dredge(r) for emptying barge
泥驳需要量计算 calculation of mud barge required
泥驳装载量 loading capacity of hopper barge
泥补裂缝 mud daub
泥舱 hopper; mud hold; spoil hold; spoil hopper
泥舱船 hopper barge
泥舱计量【疏】hopper measurement; measurement in the hopper; hopper loading measurement
泥舱门 dump door;hopper door
泥舱密度计 hopper well densimeter
泥舱容量 hopper capacity
泥舱容量曲线 curve of hopper capacity;hopper capacity curve
泥舱深装舱系统 deep loading system of hopper
泥舱式挖泥船 hopper dredge(r)
泥舱吸泥管 hopper suction pipe
泥舱稀释装置 diluting installation for hopper
泥舱溢流门 hopper overflow gate; hopper overflow weir
泥舱溢流堰 hopper overflow weir
泥舱溢流装置 hopper overflow device
泥舱装载量 hopper load
泥层厚度 height of cut
泥层渗透性 slip permeability
泥铲 grafting tool;mud shovel;spoon
泥铲吊索 <挖泥船上的> spud rope
泥铲压抹修整 spaded concrete
泥车 muck car
泥沉积带 mud belt
泥沉积物压实阶段 compaction stage of argillaceous sediment
泥池 earthen pond; mud basin; mud sump
泥处理法 activated sludge process
泥穿刺 mud diapire
泥船 central hopper
泥刺穿圈闭 mud diapir trap
泥刀 laying on trowel; pallet; slicer; sword;trowel
泥道拱 axial bracket arm
泥的包壳作用 coating of mud
泥的絮凝体 mud flock
泥底(锚地) muddy bottom
泥底辟 mud diapir
泥底辟遮挡 mud diapir barrier
泥底层 muddy bottom
泥地 earth floor;lake and river mud; mud basin;muddy land;mud-field; mudflat
泥地高通过性车辆 mud performer
泥地行驶能力 mud ability
泥点 slime spot
泥斗 bagger;bucket;dredge(r) bucket;sludge hopper
泥斗充泥系数 coefficient of bucket filling
泥斗容量 bucket capacity
泥斗销 bucket pin
泥斗运转计数器 bucket running counter
泥斗运转速度 bucket running speed
泥斗转速指示仪 bucket speed indicator
泥堆 mudbank
泥多边形土 mud crack polygon;mud polygon
泥阀 mud valve
泥肥 peaty fertilizer;sludge
泥肥洒布器 slurry spreader
泥封 luting; mudding; mud off; mud-up
泥封保温 scove
泥封爆破(法) mud blasting
泥封的 mudded off
泥封窑 Scove kiln
泥封窑烧的砖 Scove kiln-burnt brick
泥缝 mud seam
泥敷剂 cataplasm;poultice
泥盖 mudcap(ping);mud drape
泥盖爆破法 mudcapping method; adobe blasting;mud blasting
泥盖法 <将炸药放在岩石表面,盖泥后再行爆破的方法> mudcapping method
泥岗 mud mound
泥膏剂 pastas
泥膏岩 muckle-gypsum rock
泥膏样品 sludge sample
泥工 brick layer;mud scraper
泥工线 bricklayer's line;mason's line
泥工行业用熟石灰 trowel trades hydrated lime
泥垢 sludge
泥海星 mud-star
泥河 madspate;mudflow;mud stream
泥河弯沉降 <更新世> Nihowan sediment
泥糊堆雕 paste-on-paste
泥滑 mudslide;mud wave
泥化 sloughing
泥化带 argillation zone
泥化夹层 argillaceous intercalated bed; argillic intercalated layer; argillized seam; clay gouged intercalation;siltized intercalation
泥化夹层分带 zoning if intercalated soft layer
泥化作用 argillization
泥环 mud collar
泥灰板圬塥 marl loam
泥灰板岩 margode;marl slate
泥灰板岩层 marl slate stratum
泥灰方解石 argillo-calcite
泥灰粉末 marl pellet
泥灰灰岩 marly limestone
泥灰混浊度 marl turbidity
泥灰结碎石/砾石 marl-bond macadam/gravel
泥灰结碎石路面 clay-lime bound macadam
泥灰沥青 marl asphalt
泥灰沥青质页岩 marly bituminous shale
泥灰抹子 pallet
泥灰泥 chalky clay
泥灰球条 marl pellet
泥灰砂浆 mud mortar;soil mortar
泥灰砂质壤土 marly sandy loam
泥灰石 caliche;marlstone
泥灰石板瓦 marl slate
泥灰石灰石 marly limestone
泥灰土 lake marl; marl; marl soil; marly earth;marly soil
泥灰土壤 marl ground
泥灰岩 chalky clay;clay marl;malm; marlite;marl
泥灰岩采掘场 marl pit
泥灰岩的 malmy;marlaceous
泥灰岩过渡组合 marl transitional association
泥灰岩矿 marl pit
泥灰岩矿床 marl deposit
泥灰岩相 marlaceous facies
泥灰(岩)质石灰岩 marly limestone
泥灰岩砖 gault brick;marl brick
泥灰页岩 marl shale
泥灰质白垩 argillo-calcareous chalk
泥灰质的 argillo-calcareous; marlaceous;marly
泥灰质黄土 marl loess
泥灰质泥 marly mud
泥灰质黏[粘]土 gault(clay); marly clay;ga(u)ging clay
泥灰质壤土 marl loam
泥灰质砂黏[粘]土 marl sand clay
泥灰质砂岩 marly sandstone
泥灰质土(壤) marly soil
泥灰质团粒 marl pellet
泥灰质岩 marlite;marlstone
泥灰质页岩 marlaceous shale; marly shale
泥火山 hervidero; mud lump; mud volcano;salse
泥火山底辟背斜聚集带 accumulation zone of mudlump and diapir fold
泥火山活动 mud-volcanic activity
泥火山气 mud-volcanic gas
泥火山锥 mud cone;puff cone
泥基泥浆 shale laden mud
泥夹层 mud seam
泥槛 mud sill
泥浆 batter; clay grout; clay mortar; clay slip; clay slurry; clay slush; clay suspension; moil; mud; mud fluid; mud slurry; mud solution; ooze;pulp;silt slurry;slime sludge; slop;sludge;slurry;slush;mud laden fluid <比重 1.2 以上>
泥浆 pH 值计 pH value meter of mud
泥浆般的黏[粘]土 slushy clay
泥浆拌和 clay mortar mix
泥浆拌和机 ground batching plant
泥浆杯 mud cup
泥浆泵 mud pump; bailer[bailor]; dredging pump; excavating pump; floating pump; mash; mud hog pump; pump dredge(r); scum pump;sheet pump;shoe shell;slime pump; sludge pump; slurry pump; slush fitted pump; solids handling pump;dredge pump;slush pump <钻井、钻孔用>
泥浆泵安全阀 mud relief valve
泥浆泵表 slush pump ga(u)ge
泥浆泵参数 coefficient of mud pump
泥浆泵阀 mud pump valve
泥浆泵功率 power of mud pump
泥浆泵活塞 mud piston
泥浆泵流量 capacity of mud pump
泥浆泵排出阀 mud valve
泥浆泵软管 mud pump hose
泥浆泵输入功率 input horsepower of mud pump
泥浆泵数量 numbers of mud pumps
泥浆泵送车间 slurry pumping plant
泥浆泵送极限 pumping limit of mud
泥浆泵送装置 slurry pumping plant
泥浆泵拖车 mud pump trailer
泥浆泵往复次数 reciprocating number of mud pump

N

泥浆泵吸池 pump suction pit
泥浆泵旋轴 bailer swivel
泥浆泵压力波动 mud pump pulse
泥浆泵液力端 mud end of the pump
泥浆泵振动压力 mud pump shock pressure
泥浆泵最大压力 maximum pressure of mud pump
泥浆比重 mud weight
泥浆比重测定 test for specific gravity of slip
泥浆比重计 mud scale; mud water hydrometer; mud weight indicator
泥浆比重计秤 mud weight balance
泥浆比重瓶 gravity bottle of mud
泥浆材料 mud material
泥浆槽 canal ditch; conductor box; ditch; grout trough; mud flume; mud port; mud return ditch; mud sluice; mud trough; slurry tank
泥浆槽壁 slurry wall
泥浆槽法 <地基防渗处理的> slurry trench method
泥浆槽盖 channel cover
泥浆槽黏[粘]土截水墙 slurry trench cutoff
泥浆槽墙 slurry(trenched) wall
泥浆槽取样 ditch sampling; mud ditch sampling
泥浆槽岩屑样品 ditch sample
泥浆侧流槽 mud flume
泥浆测井 mud log(ging)
泥浆层 mud layer; slush layer
泥浆厂 mud plant
泥浆沉淀 slurry settling
泥浆沉淀槽 mud launder
泥浆沉淀池 catch pit; mud settling pit; mud settling pump; mud sump; settling vessel of mud
泥浆沉淀系统 settling system of mud
泥浆沉淀装置 mixture sedimentation device
泥浆沉积物 mud deposit
泥浆沉降 slurry sedimentation
泥浆沉降器 mud settler; mud still
泥浆成木 mud cost
泥浆澄清 sludge clarification; slurry clarification
泥浆池 mud chamber; mud ditch; mud flume; mud pit; mud settling pit; mud sump; mud trough; reservoir for mud; slime dam; slurry pond; slurry pool; slurry tank; slush pit; slush pond; spoil pool; suction pit
泥浆尺 mud ga(u)ge
泥浆充填法 slurry packing method
泥浆冲洗 mud flush
泥浆冲洗法 mud flush system; mud system; sludging process
泥浆冲洗钻进 hydraulic drilling; mud drilling; mud flush boring; mud flush drilling
泥浆冲洗钻井法 mud flush method of sinking
泥浆抽筒 <冲击钻进用> mud barrel; mud socket
泥浆抽吸车间 slurry pumping plant
泥浆抽吸装置 slurry pumping plant
泥浆稠度 slip consistency
泥浆稠化 live ring; thickening of mud
泥浆稠化机 slurry thickener
泥浆出口 circulating hole
泥浆除砂器 mud desander
泥浆储槽 sludge storage tank
泥浆储存池 slurry storage tank
泥浆处理 mud conditioning; mud control; sludge handling; slurry handling; slurry treatment; treatment of mud
泥浆处理剂 mud conditioner
泥浆处理设备 mud treating equipment; reconditioning plant
泥浆触变 slurry rheopexy; slurry thixotropy
泥浆传动涡轮机 mud propelled turbine
泥浆船 sludge boat; slurry carrier
泥浆挡板 mud baffle
泥浆道砟 slurried ballast
泥浆的 oozy
泥浆的化学处理 chemical treatment of mud
泥浆的固体干重 dry solids
泥浆的流动特性 flow characteristics of mud
泥浆底子 mud base
泥浆地 slob-land
泥浆电阻率 mud resistivity; resistivity of mud
泥浆电阻率测定器 mud resistivity tester
泥浆电阻率测井 mud log(ging); mud resistivity log
泥浆电阻率测井曲线 mud resistivity log curve
泥浆电阻率测井曲线图 mud resistivity log plot
泥浆动切力计 dynamic(al) shearometer of mud
泥浆堵漏 mudding
泥浆堵钻 mudded up
泥浆堆花浮雕 pate-sur-pate
泥浆堆积 accumulation of mud
泥浆盾构 slurry shield
泥浆翻腾 mud boil
泥浆反应器 slurry reactor
泥浆返回 return of drilling mud
泥浆防溅盒 mudsaver bucket
泥浆防喷 mud lubrication
泥浆防喷盒 spray arrester
泥浆防渗墙 slurry cutoff
泥浆房 mud house
泥浆肥料泵 liquid sludge pump
泥浆费 mud cost
泥浆分离 slime separation
泥浆分离器 mud separator
泥浆分析测井 mud analysis logging
泥浆分析记录 mud analysis log
泥浆封层 slurry coat; slurry seal (coat)
泥浆封面处理 slurry seal surface treatment
泥浆覆盖的 muddy
泥浆覆盖的道砟浆 muddy ballast
泥浆干燥机 slurry drier[dryer]
泥浆高位槽 sludge head tank
泥浆工程师 mud engineer
泥浆工段 slip house
泥浆沟 mud trough
泥浆沟槽 mud ditch; slurry trench
泥浆固壁 slurry stable wall; slurry wall stabilizing
泥浆固壁法施工 slurry trench construction
泥浆固结法 slurry consolidation method
泥浆固相 mud solid phase
泥浆管 sludge pipe; sludge tube
泥浆管道 mud line; slurry pipeline
泥浆管路 mud line; slurry pipeline
泥浆管线 mud line; slurry pipeline
泥浆管线设备 mud manifolding
泥浆管涌 mud boil
泥浆灌浆 slush-grout
泥浆灌浆作业 mud grouting
泥浆灌注 mud grouting; mud injection; slush grouting
泥浆罐 mud pot; mud tank; sludge tank
泥浆规格 mud program(me)
泥浆过滤 sludge filtration
泥浆过滤器 slurry filter
泥浆过滤前预热器 slurry filter preheater
泥浆含砂量 sand content of mud
泥浆含砂量计 sand content ga(u)ge
泥浆含水量 water content of mud
泥浆护壁法 hole protection method with mud
泥浆护壁基槽 slurry trench
泥浆护壁钻孔法 slurry drilling method; slurry hole-boring method
泥浆护罩 mudsaver bucket
泥浆化 sliming; sludging
泥浆化学师 mud chemist
泥浆画 trailed decoration
泥浆挥流 slurry return
泥浆回转接头 mud swivel
泥浆回转钻进法 mud rotary drilling method
泥浆绘制图案 trailing
泥浆混合泵 slurry agitator pump
泥浆混合器 mud mixing appliance
泥浆机 pump dredge(r)
泥浆基 mud base
泥浆基质 mud body
泥浆集流沟 mud gully(trap)
泥浆挤出机 slip trailer
泥浆挤出器 bulb trailer; slip trailer
泥浆记录 mud record
泥浆加稠 live ring; thickening of mud
泥浆加压盾构(法) mud pressure shield; slurry pressure shield
泥浆加压式盾构 bentonite shield
泥浆加重剂 weighing material of mud
泥浆交叉 mud cross
泥浆浇注 slip casting; slurry casting
泥浆浇铸砖 slip cast brick
泥浆搅拌 blunging
泥浆搅拌池 slip agitating tank
泥浆搅拌机 grout agitator; mud mixer; slurry agitator; mixer for stabilizing solution <用于地下连续墙施工>
泥浆搅拌器 grout agitator; slurry agitator; slurry mixer
泥浆搅拌输送螺旋 slurry auger
泥浆搅拌装置 mud gun
泥浆校正 mud correction
泥浆结构黏[粘]度 rigidity of mud
泥浆结团 livering of slurry
泥浆进料 slurry feed
泥浆井 drop chute
泥浆净化设备 mud cleaning device
泥浆绝对含水率 absolute water content of slip
泥浆掘进机 slurry mole
泥浆均压器 mud lubricator
泥浆坑 mud pit; sludge pit; sludge pond; slurry trench; slush pit
泥浆孔 mud port
泥浆口 mud port
泥浆库 slurry silo; slurry storage tank
泥浆类型 mud type
泥浆冷却塔 <地热钻进时用> mud cooling tower
泥浆离心机 sludge centrifuge
泥浆离心机脱水 centrate
泥浆流变学 mud rheology; slip rheology
泥浆流动速度 velocity of mud circulation
泥浆流动性 slip fluidity
泥浆流量 mud volume
泥浆流速指示仪 mixture velocity indicator
泥浆漏斗 marsh funnel; mud hopper; mud scow <取海泥的>
泥浆漏失 lose return; lose water; lost circulation; mud loss
泥浆漏失测定器 mud loss instrument
泥浆滤液电阻率 resistivity of mud filtrate
泥浆录井 mud(dy)log(ging)
泥浆滤饼 slurry filter-cake
泥浆滤液 mud filtrate
泥浆滤液电阻率 mud filtrate resistivity
泥浆路基 slurry base
泥浆密度 mud density
泥浆密度计 mud balance
泥浆名 name of mud
泥浆磨 slurry grinding mill
泥浆黏[粘]度 mud viscosity; slurry viscosity; viscosity of mud
泥浆黏[粘]度计 visco(si)meter of mud
泥浆凝胶强度和剪切力测定仪 eykometer
泥浆浓度 sediment concentration; slurry concentration
泥浆浓度计 mud meter
泥浆浓缩 thickening of slurry
泥浆浓缩器 slurry thickener
泥浆排放泵 sludge draw-off pump
泥浆排气阀 mud valve
泥浆排卸管道 mud discharge line
泥浆配制员 mud chemist
泥浆喷枪 slurry gun
泥浆喷射机 air mortar gun
泥浆喷射器 mud diffuser
泥浆棚屋 mud hut
泥浆铺撒器 grout spreader
泥浆铺设器 slurry spreader
泥浆气侵 cutting of mud by gas
泥浆枪 mud gun
泥浆切力计 shearometer of mud
泥浆清孔底 mud slip
泥浆区面积 mud tank area
泥浆取样杆 thief rod
泥浆取样筒 slurry sampler
泥浆去砂器 mud desander
泥浆圈 mud ring; slurry ring
泥浆燃烧法 slurry-burning process
泥浆软管 flexible mud hose; mud hose
泥浆润滑套 <沉井> clay slurry jacket
泥浆筛 mud desander; mud screen; mud shaker
泥浆筛余物 knocking
泥浆设计 mud design
泥浆渗透 mud percolation
泥浆渗透性 slip permeability
泥浆失水量计 water-lossmeter of mud
泥浆失水量特性 filtration loss quality of mud
泥浆使用说明书 mud program(me)
泥浆输送管道 slurry pipe; slurry pipeline
泥浆输送管路 slurry charge line
泥浆输送管站 slurry charge line
泥浆水 earth(y) water; mud(dy) water; slime water
泥浆四通阀 mud cross
泥浆送料机 slurry feeder
泥浆损失 slurry loss
泥浆(套)法 slurry coat method
泥浆套法下沉沉井 sinking open caisson by slurry coating
泥浆体 mud laden fluid
泥浆体积浓度 volumetric(al) density of slurry
泥浆天平 mud balance
泥浆添料 mud additive
泥浆填充法 slurry packed method
泥浆调节 mud conditioning
泥浆调质剂 mud conditioner
泥浆通道 mud flue
泥浆桶 mash tun; mud tank

泥浆突然涌入 mud run
泥浆涂绘 slip trailing
泥浆土壤 mud soil
泥浆脱水 slurry dewatering
泥浆挖槽法 slurry trench excavation; slurry trench method
泥浆挖掘机 slurry modulus
泥浆喂料 slurry feed
泥浆温泉 mud pot
泥浆稳定性 mud stability
泥浆吸入软管 mud suction hose
泥浆稀释剂 mud thinner; slurry thinner
泥浆洗井 clay flushing
泥浆洗孔钻进 hydraulic drilling; mud flush boring; mud flush drilling
泥浆相生物处理 slurry phase biological treatment
泥浆箱 mud box; mud tank; sludge tank
泥浆性能处理 mud property control
泥浆性能调整 mud conditioning
泥浆性能指数 mud property ratio
泥浆性质调整 adjustment of slip
泥浆序批间歇反应器活性污泥法 soil slurry-sequencing batch reactor activated sludge process
泥浆絮凝 slurry flocculation
泥浆循环 <旋钻技术中的> circulation of mud; mud circulation; sludge circulation
泥浆循环损失 loss of circulation
泥浆循环系统 mud circulating system; mud flush system; mud system
泥浆压井器 mud lubricator
泥浆压力 mud pressure; slurry pressure
泥浆压力计 mud pressure indicator
泥浆压滤机 mud press
泥浆压缩系数 compression coefficient of mud
泥浆盐浸 salt mix in mud
泥浆液流 mud stream
泥浆液相 mud liquid phase
泥浆液压运输 slurry convey
泥浆移动 movement of mud
泥浆釉 slip glaze
泥浆淤积 accumulation of mud; mud filling
泥浆运输船 sludge carrier
泥浆再生 mud reclamation
泥浆造壁 mudding; wall building of mud
泥浆造壁性能测验仪 wall building tester of mud
泥浆造壁作用 mudding action; wall building property of mud
泥浆增稠剂 mud thickener; slurry thickener
泥浆增强的覆盖层 slurry-reinforced overburden
泥浆站 mud plant
泥浆真空处理机 slip vacuum treatment machine
泥浆振动筛 shale screen; shale shaker; vibrating mud screen
泥浆支架 slurry support
泥浆制备 slurrying; slurry preparation
泥浆制备厂 clay slurrying plant
泥浆置换 slurry displacement
泥浆中固相含量 solids content in mud
泥浆中杂质 overburden
泥浆重量 mud weight
泥浆注浆 mud injection
泥浆贮藏箱 mud bin
泥浆柱 <钻孔> mud column; column of mud
泥浆桩 mud pile
泥浆装饰 trailed decoration
泥浆装饰的 slip-decorated
泥浆状的 slimy
泥浆状滑塌 slurry slump
泥浆状土 muddy soil
泥浆钻进 mud(flush) drilling
泥浆钻进法 well drilling with bentonite
泥浆钻进用水龙头 mud swivel
泥浆钻孔 mud hole; mud well; slurry drilling
泥浆钻孔法 slurry drilling method
泥浆钻孔桩 slurry drilled pile
泥浆钻探 mud flush drilling
泥胶 earth rubber
泥结的 clay-bound
泥结砾石 clay-bound gravel
泥结碎石 clay-bound macadam
泥结碎石路 clay-bound macadam road
泥结碎石路面 clay-bound macadam pavement
泥结碎石铺面 clay-bound macadam pavement
泥结稳定路面 stabilized soil-bound surface
泥结稳定面层 stabilized soil-bound surface
泥结稳定土路 stabilized soil-bound surface
泥晶化作用 micritilization
泥晶灰岩 micrite; micritic limestone; pelsparite
泥晶加大作用 micrite enlargement
泥晶套 micrite envelope
泥晶套组构 micrite envelope fabric
泥晶组构 micrite fabric
泥壳 mud shell; slurry cake
泥克点 <河床的侵蚀交叉点> nick-point
泥坑 mire; slough; slunk
泥孔 mud hole
泥块 clod; clot; grog; lumps of clay
泥块煤 clod
泥框 gabion
泥涝 water-logging
泥砾 mud boulder
泥砾层 boulder clay; clay boulder; muddy gravel
泥砾沉积 muddy gravel
泥砾混杂岩 olistostromic melange
泥砾土 boulder clay; cimolite
泥砾岩 <杂色块状砾石> gompholite; nagelfluh
泥砾岩层 nagelfluh; olistostrome
泥粒石灰岩 packstone
泥粒状灰岩 packstone
泥料 pug
泥料可塑性 mud plasticity
泥料真空处理 de-airing of clay
泥料蒸汽加热处理 steam-heating of clay
泥料组分 batch composition
泥裂 desiccation crack; shrinkage crack(ing); sun crack
泥裂缝 mud crack
泥裂痕 mud crack
泥龄 sludge age
泥流 creepwash; earth flow; eruptive tuff; flowing slope; flowing soil; flow slide <顺坡滑动的>; mudspate; mudflow avalanche; mud-spate; mud stream; sandbree; soil flow; solifluction flow; mud current
泥流搬运砾石 kneaded gravel
泥流沉积（物） lahar deposit; mud flowage deposit; mudflow deposit; solifluction sediment
泥流堤 mudflow levee
泥流覆盖 soil-flow cover
泥流构造痕迹火山泥流 mudflow
泥流黄土 solifluction loess
泥流火山碎屑 mud lava
泥流角砾岩 cenuglomerite
泥流阶地 mudflow terrace; solifluction terrace
泥流砾 kneaded gravel
泥流砾岩 cenuglomerate
泥流蠕动 mudflow creep
泥流梯度 mudflow gradient
泥流土 mudflow soil
泥流物 mudflow
泥流席 solifluction sheet
泥流型滑坡 flow slide
泥流岩 aqueous lava
泥流作用 gelifluction; solifluction [solifluxion]
泥路强固 improving dirt road
泥镘板条 Staub clay lath(ing)
泥猫 <一种环保型吸扬挖泥船> Mudcat
泥锚位 mud berth
泥煤 boghead(ite) coal; earth coal; humic coal; laminated coal; moor peat; mush; peat; peat coal; sapropelic coal; sapropelite; slime peat; sooty coal; turf; turf peat
泥煤柏油 peat tar
泥煤爆破法 peat blasting
泥煤层 peat layer
泥煤尘 peat dust
泥煤的 peaty
泥煤地 peat hag
泥煤堆 peat bank
泥煤粉 peat dust; peat meal
泥煤焦炭 peat charcoal; peat coke
泥煤焦油 peat tar
泥煤焦油沥青 peat-tar pitch
泥煤焦油酸 peat-tar acid
泥煤焦油脂 peat-tar pitch
泥煤矿区 peaty ground
泥煤蜡 peat wax
泥煤沥青 peat-tar pitch
泥煤炉渣 peat cinder
泥煤气体 peat gas
泥煤区爆炸筑路法 road construction by peat blasting
泥煤燃烧室 peat combustion chamber
泥煤石蜡 peat paraffin
泥煤田 peatland
泥煤挖掘机 peat excavator
泥煤烟 peat reek
泥煤硬沥青 peat-tar pitch
泥煤沼 peat bog
泥煤砖 peat brick
泥门 bottom door; dumping door; hopper bottom door
泥门启闭装置 bottom door control device; hopper door control device
泥门液压柱塞 hopper door hydraulic ram
泥面控制器 blanket-height controller
泥面上混凝土垫层 mud slab
泥模 loam mo(u)ld
泥漠 takyr
泥泞 miriness; muddiness; silting; sloppiness; slough
泥泞冰 grease ice; ice fat; lard ice
泥泞道路 greasy road; muddy road
泥泞的 dirty; founderous; miry; muddy; quaggy; quagmiry; slimy; sloppy; sloughy; sludgy; swamped
泥泞的废物 slimy waste material
泥泞地 Fen; muddy ground; quagmire; slimy ground; slob
泥泞地带 muddy terrain
泥泞地面 heavy-ground
泥泞地区 boggy country
泥泞地上的木板道 duck board
泥泞路 dirt road; earthen road
泥泞路和雪路用轮胎 mud and snow tire[tyre]
泥泞沼泽 swing moor
泥偶像 clay idol
泥耙 mud rake
泥炮 clay gun; notch gun; spray gun; tap-hole gun
泥喷泉 mud geyser; mud pot
泥盆纪【地】Devonian period
泥盆纪冰期 Devonian glacial stage
泥盆纪火成岩 Devonian igneous rocks
泥盆纪石灰岩 Devonian limestone
泥盆系 Devonian system
泥皮 clay coating; mud cake <钻孔壁上的>; filter cake; mud cake
泥皮电阻率 mud cake resistivity
泥皮形成 <孔壁上的> mud cake growth
泥片 slice
泥片麻岩 pelite-gneiss
泥片黏[粘]结成型 slabbing
泥片岩 fissile shale
泥片印花法 batt printing
泥坪 mudflat
泥坪沉积 mudflat deposit
泥坪海岸 mudflat coast
泥铺裂缝 mud daub
泥铺跑道 dirt track
泥前滨 mud of foreshore
泥枪 mud gun
泥墙 dirt wall
泥丘 mud cone; mud lump <三角洲地区的>; mud mound
泥球 ball clay; mud ball; tuff ball
泥区 mud-field
泥圈 mud collar
泥泉 earth spring; mud spring
泥熔岩 aqueous lava; moya; mud lava
泥蕊打出机 core knockout
泥塞 bat; bod; bot; bott; clay plug; plug of clay
泥沙 agger arenal; alluvium[复 alluvia/alluviums]; atteration; erosion material; silt; stream-borne material
泥沙坝 bank of muddy sand
泥沙百分率 percentage of sediment
泥沙搬运 silt transport; transportation of sediments; transport of silt
泥沙饱和的液体 saturated liquid of mud-sand
泥沙比值 ration of mud/sand
泥沙补给 sediment supply
泥沙捕集装置 sediment trap; sludge trap
泥沙采集器 sand catcher; sediment catcher
泥沙采样瓶 bottle silt-sampler
泥沙采样器 sediment bulk; sediment sampler; sediment sampling equipment; silt-sampler
泥沙采样设备 sediment sampling equipment
泥沙槽蓄 sediment storage
泥沙测定 sediment determination
泥沙测量 sediment survey; sediment measurement
泥沙测量方法 method of sediment measurement
泥沙测量仪 silt meter
泥沙测验 sediment measurement; silt-determination measure
泥沙测验仪器 instrument of sediment measurement
泥沙层 layer of sediments
泥沙层理 stratification of sediments
泥沙超载系数 sediment overload factor
泥沙沉淀动力学 dynamics of silt sedimentation
泥沙沉淀监测 sedimentation monitor

泥沙沉淀物 sediment
泥沙沉积 deposition of silt; sediment deposit; sediment deposition; silt deposit; silt logging; entrapment of sediment
泥沙沉积机理 mechanism of silt deposit; sedimentary mechanism
泥沙沉积物示踪剂 sediment tracer
泥沙沉降 settling of sediment
泥沙沉降速度 fall velocity of sediment; sedimentation velocity; settling velocity of sediment
泥沙沉速 fall velocity of sediment; sedimentation velocity; settling velocity of sediment
泥沙池 atteration tank
泥沙冲积层 inwashed sediment
泥沙冲泻质 wash load(ing)
泥沙冲淤平衡 scour sedimentation equilibrium
泥沙传递系数 sediment transfer coefficient
泥沙垂向分布 sediment profile
泥沙垂直于岸线的输移 cross-shore sediment transport
泥沙纯度 purity of sand
泥沙存蓄 sediment storage
泥沙的垂直分布 sediment profile; vertical distribution of sediment
泥沙的横向分布 lateral distribution of sediment
泥沙的级配特性 grading characteristic of sediment
泥沙的几何平均粒径 geometric(al) mean-diameter of silt
泥沙的算术平均粒径 mathematics mean-diameter of silt
泥沙的自由流动 free running of sand
泥沙地 slob
泥沙调查 sediment survey
泥沙堆积 sediment encroachment; silting
泥沙过多 sediment overload
泥沙放射性示踪试验 radioactive sediment tracer test
泥沙分布 sediment distribution
泥沙分布范围 sediment range
泥沙分层 stratification of sediments
泥沙分流设施 sediment diverter
泥沙分流隧道 tunnel-type sediment diverter
泥沙分流隧洞 tunnel-type sediment diverter
泥沙分析 sediment analysis
泥沙分析仪 siltometer
泥沙分选 differentiation of sediment; sorting of sediment
泥沙分选机理 sediment sorting mechanism
泥沙负荷 sediment load
泥沙干容重 dry bulk density of sediment; unit dry weight of sediment
泥沙干燥皿 sediment dish
泥沙工程学 sediment(ation) engineering
泥沙观测站 sediment station
泥沙含量 percentage of sediment; sediment charge; sediment concentration; sediment content; silt carrying capacity; silt charge; silt concentration; silt content
泥沙含量百分率 percentage of sediment
泥沙含量分析 test for silt
泥沙河槽 dirty channel
泥沙河床 sediment bed
泥沙荷载 burden
泥沙横向交换 transverse interchange of sediment
泥沙横向运动 transverse movement of material
泥沙混合滩 sand and mud foreshore
泥沙混合物 sediment mixture
泥沙积聚 sand accretion; sediment accumulation; silt accumulation; upbuilding
泥沙级配 grading of sediment
泥沙级配曲线 sediment grading curve
泥沙监测器 silt monitor
泥沙鉴别 differentiation of sediment
泥沙阶地 <河道的> bench of silt
泥沙径流 flow of solid matter; sediment runoff
泥沙颗粒 sand grain; sedimentary particles; sediment grain; sediment particle
泥沙颗粒的跃移长度 saltation step of sediment
泥沙颗粒的跃移高度 saltation height of sediment
泥沙颗粒分级标准 sediment grade scale
泥沙颗粒分析 particle-size analysis; sediment grain analysis; sediment granulometry; sediment particle analysis; sediment size analysis
泥沙颗粒分析器 sand-size analyser [analyzer]; sediment grain analyser [analyzer]; sediment particle analyser[analyzer]; sediment size analyser[analyzer]
泥沙颗粒分析设备 sediment size analysis equipment
泥沙颗粒跟踪 grain tracking
泥沙颗粒级配 sediment grading
泥沙颗粒级配系统 packing coefficient
泥沙颗粒剪应力 grain shear stress
泥沙颗粒结构 sediment texture
泥沙控制 sediment control; silt control
泥沙控制设施 sediment control structure; sediment control works
泥沙扩散 sediment diffusion
泥沙来源 sediment source; sediment supply; source of sediments; sediment resource
泥沙拦截装置 sediment catcher
泥沙粒度 <颗粒平均直径> sediment size; silt grade
泥沙粒级 sediment size fraction
泥沙粒级尺寸 sediment grade size
泥沙粒径 particle diameter of sediment; sediment grain size; sediment particle size
泥沙粒径分析 grain-size analysis of sediment
泥沙粒径组成 grain-size fraction; silt combination
泥沙量 quantity of sediment; sediment yield
泥沙临界流速 critical tractive force
泥沙流 silt flow; suspension current
泥沙流饱和度 saturation of silt flow
泥沙流不饱和 silt flow unsaturation
泥沙流过饱和 silt flow supersaturation
泥沙流量 sediment discharge
泥沙流量关系 load-discharge relationship
泥沙流量过程线 sediment hydrograph
泥沙流量曲线 sediment discharge curve
泥沙流入流出计算 sediment budget
泥沙率定曲线 sediment-discharge rating curve; sediment-rating curve
泥沙密度 sediment density
泥沙模型 sediment model
泥沙模型试验 sand model test; sediment model test
泥沙年沉积量 annual accumulation of sediment
泥沙年淤积量 annual accumulation of sediment
泥沙浓度 sediment concentration; silt concentration
泥沙旁通设施 by passing plant
泥沙喷射泵 silt ejector
泥沙漂移 sediment drift
泥沙平均粒度 mean sediment size
泥沙平均浓度 mean sediment concentration
泥沙起动 incipient sediment motion; initiation of sediment transport; sediment entrainment; sediment incipient motion; sediment pickup
泥沙起动剪切力比尺 scale of shear force of sediment initiation motion
泥沙起动流速 critical friction velocity; sediment-moving incipient velocity; threshold velocity
泥沙起动速度 starting velocity of sediment
泥沙潜量 sediment potential
泥沙侵入 sediment encroachment; sediment intrusion
泥沙侵蚀 sediment erosion
泥沙清除 evacuation of sediment
泥沙球度 sediment sphericity
泥沙球面比 sediment sphericity
泥沙取样 load sampling; sampling sediment; sediment(-load) sampling
泥沙取样器 sediment catcher; sediment sampler; silt-sampler
泥沙容重比例 scale of sediment of specific weight; specific weight scale of sediment
泥沙示踪剂 sediment tracer
泥沙试样 sediment-load sample; sediment sample
泥沙疏导 diversion of sediment; training for sediment; training of sediment
泥沙输出量 sediment yield
泥沙输送 load transport; sediment transmission; sediment transport; transport of silt; mud transport
泥沙输送公式 sediment-transport equation
泥沙输送量测量 sediment transport measurement
泥沙输送设备 silt supply apparatus
泥沙输送系数 coefficient of silt transfer; sediment transfer coefficient
泥沙输移 sediment bypassing; sediment delivery; sediment transport; sediment transportation; silt transportation; solid transport; transportation of debris; transport of silt
泥沙输移比 sediment delivery ratio
泥沙输移过程 sediment delivery process; sediment transport process; transportation process
泥沙输移量计算 sediment transport budget
泥沙输移率 rate of sediment transport; sediment transport rate
泥沙输移能力 sediment transport capacity; sediment transport competence; sediment transport competency
泥沙输移曲线 sediment transport curve
泥沙竖向浓度分布 vertical sediment concentration distribution
泥沙水力学 sediment hydraulics
泥沙顺岸输移 longshore sediment transport
泥沙体积 sediment volume
泥沙填塞 blocking up; deposition of sediments
泥沙推算 sediment routing
泥沙推移 travel of sand
泥沙推移力 sediment dislodging force
泥沙问题 sediment-oriented problem
泥沙污染 silt pollution
泥沙系数 silt factor
泥沙相对含量 relative sediment charge
泥沙相对粒径 relative sediment size
泥沙携带的 sediment-borne
泥沙携带能力 sediment-carrying capacity; silt carrying capacity
泥沙絮凝作用 sediment coagulation; sediment flocculation
泥沙悬浮物质 silt suspension
泥沙悬移 sediment suspension
泥沙学 sedimentology
泥沙压力 mud-sand pressure; silt pressure
泥沙岩与层速度关系曲线 relation curve between mud sandstone and interval velocity
泥沙演进计算 sediment routing
泥沙样品 sampling sediment
泥沙移动 sediment transport; sediment transportation; silt displacement
泥沙移动状态 phase of sediment transportation; phase of transport of sediment
泥沙移运 transport of silt
泥沙异重流 silt density current
泥沙因子 silt factor
泥沙引起的 sediment-induced
泥沙涌流 mud surge
泥沙游移 shifting of earth
泥沙淤积 filling up; sand accretion; sediment accumulation; sediment deposit; silt accumulation; silting deposit; upbuilding
泥沙淤积垫底库容 dead storage of sedimentation
泥沙淤积量 accumulation of sediment; sediment yield
泥沙淤积物 inwashed sediment
泥沙淤积形成的死库容 dead storage of sedimentation
泥沙淤积纵剖面图 sediment profile
泥沙淤塞 blocking up with silt; deposition of silt; silt plug
泥沙源地 sediment source
泥沙源区 sediment-source area
泥沙跃移 rippling
泥沙运动 movement of sediment; movement of silt; sediment motion; sediment movement
泥沙运动的冲刷 scour with sediment motion
泥沙运动力学 sedimentation mechanics
泥沙运动临界速度 critical velocity of sediment movement
泥沙运动模型 sediment motion model; sediment movement model
泥沙运动相似 similarity of sediment motion; similarity of sediment movement
泥沙蕴藏量 sediment potential
泥沙站 sediment station
泥沙直径 sand diameter
泥沙指数 sediment index
泥沙质的 argillo-arenaceous
泥沙质混杂体 argillo-arenaceous melange
泥沙中值粒度 median sediment size
泥沙中值粒径 median sediment size
泥沙重率 sedimentary specific weight

泥沙转移 sediment transfer;sediment diversion
泥沙转移设施 sediment diverter
泥沙资料 sedimentary data
泥沙总量 total sediment;total sediment load;total volume of silt
泥沙总重 total sediment weight
泥沙纵向运动 longshore movement of material
泥砂 earthy material
泥砂泵 sand pump;shell pump;sludge pump
泥砂的跳跃前进 saltation
泥砂地 argillo-arenaceous ground
泥砂泛积土地 balagh
泥砂浆 compo;mud mortar;soil mortar
泥砂砾 kneaded gravel
泥砂路面 sand clay road surface
泥砂轮 sand wheel
泥砂水力分选程序 sorting operation
泥砂土石 earth material
泥砂挟带标准 entrainment criterion
泥砂岩 siltite;siltstone
泥砂淤积 siltation
泥砂质的 argillo-arenaceous;argillo-calcareous
泥砂质石灰岩 argillo-calcite
泥砂洲 bar of muddy sand
泥筛出料箱 discharge box for the mud screen
泥石 mud rock;mudstone;stone clay
泥石崩坍 detritus avalanche
泥石沉淀室 detritus chamber
泥石法 soil plague method
泥石混杂物 block clay
泥石筐垒成的坝 gabionade
泥石筐垒成的堤 gabionade
泥石流 mudflow;mud-rock flow;mudstone flow;mud and stone flow;mud avalanche;avalanche of sand and stone;block glacier;debris avalanche;debris flow;debris-laden flow;debris-laden stream;debris stream;earth flow;lahar;rock flow;rock glacier;rubble flow;detritus stream
泥石流沉积 debris flow deposit;lahar deposit <火山型的>
泥石流沉积物 debris flow sediment
泥石流冲沟 mudflow gully
泥石流冲击力 impulsive force of mudflow
泥石流冲积堆 dry flood
泥石流冲积扇 mudflow fan
泥石流稠度 consistency state of mud-stone flow
泥石流粗糙系数 roughness index of mud-stone flow
泥石流的灾害 hazard of debris flow
泥石流调查 mudstone flow investigation
泥石流动态观测 observation of movement of mud-rock flow
泥石流渡槽 mud avalanche aqueduct
泥石流堆积区 end accumulation area of debris flow
泥石流发生地点 happened place of debris flow
泥石流发生时间 happened time of debris flow
泥石流防治 mudflow protection
泥石流防治措施 measures of mud-stone flow treatment
泥石流规模 dimensions of debris flow
泥石流厚度 thickness of mud-stone flow
泥石流拦挡坝 mud avalanche retaining dike[dyke]
泥石流流量 discharge of mud-stone flow
泥石流流通区 passage channel of debris flow
泥石流流速 velocity of mud-stone flow
泥石流明洞 mud avalanche cave
泥石流排导沟 mud avalanche ditch
泥石流前锋 mudflow front
泥石流容重 unit weight of debris flow;unit weight of mud-stone flow body
泥石流体 mudflow body
泥石流体长 length of mud-stone flow body
泥石流体横断面积 cross area of mud-stone flow body
泥石流停淤场 mud avalanche retarding field
泥石流形成区 material source of debris flow
泥石流种类 type of mud-stone flow
泥石芯 stone-clay core
泥石硬石膏矿石 pelitic ore
泥蚀变 argillic alteration
泥水泵 sand pump;slurry pump
泥水泵送挖泥法 hydraulic dredge(r)
泥水测井记录 mud log(ging)
泥水处理 slurry treatment
泥水处理厂 slurry treatment plant
泥水处理系统 slurry treatment system
泥水盾构 bentonite tunneling machine;bentonite-type shield;hydroshield
泥水盾构掘进机 slurry shield tunneling machine
泥水分界线 mndline
泥水分离 slurry separation
泥水工 bricklayer;dauber;mason;masonry;plasterer;tiler
泥水工作 plastering
泥水沟槽 slurry trench
泥水混合物 silt-water mixture
泥水混和箱 <水力冲填的> hog box
泥水加压盾构 mud shield;pressurized slurry;shield machine;slurry shield
泥水加压盾构旋工 construction according to slurry shield
泥水间歇泉 mud geyser
泥水匠 bricklayer;plasterer;tiler
泥水匠工具 plasterer's tool
泥水匠助手 hawk-boy
泥水坑 puddle
泥水块 sludge cake
泥水浓度调整槽 slurry density control tank
泥水喷发 mud eruption
泥水泉 earthy spring
泥水塘 bog pool
泥水稳定液施工法 <防护钻孔壁面用的> liquid method
泥塑 clay figure modelling;clay sculpture;loam mo(u)ld
泥塑模型 clay model
泥塑像 clay figure
泥酸 mud acid
泥滩 gumbo bank;mudbank;mudflat;mud foreshore;mud island;mud lump;mud bar;mud beach <河海口的>
泥滩群落 ochthium;pelochthium
泥潭 bog pool;lair;mire;mire bog;mud puddle
泥炭 peat;peat(char)coal;peat cube;bog muck;muck;sedge peat;turf(peat)
泥炭包体 inclusion of peat
泥炭爆破 blasting of peat
泥炭爆破法 peat blasting
泥炭爆炸 bog blasting
泥炭采后的废坑 moss hag
泥炭采掘地 peat cutting
泥炭草甸 peaty meadow
泥炭层 moss layer;peat bed;peat deposit;peat stratum;peaty deposit;turf bed
泥炭产地 peatery
泥炭产生地 turf development
泥炭铲 peat spade
泥炭沉积 peat deposit
泥炭成岩作用 peat diagenesis
泥炭冲洗 peak scours
泥炭岛 peat island
泥炭的 boggy
泥炭的形成 peat formation
泥炭地 bog;peat bed;peatland;peaty spoil;Spagnum moss
泥炭地基 muck foundation;peat foundation
泥炭堆(积物)peat
泥炭肥料 peat fertilizer
泥炭粉 peat dust;peat powder
泥炭腐植酸 humic acid in peat
泥炭腐殖质育苗钵 peat plant block
泥炭腹泻 peat scour
泥炭隔热板 peat insulating board
泥炭含量 peat content
泥炭化阶段 peatification stage
泥炭化作用 peatification
泥炭黄腐酸 peat fulvic acid
泥炭灰壤 peat podzol;peat podzol soil
泥炭灰化土 peat-gley soil
泥炭混凝土 peat concrete
泥炭级腐殖煤 humocoll
泥炭夹层 inclusion of peat
泥炭建筑板材 peat building board
泥炭焦油 peat tar
泥炭焦油脂 peat-tar pitch
泥炭角砾 peat breccia
泥炭结构 peat structure
泥炭矿工 peatman
泥炭矿区 peaty ground
泥炭沥青 peat tar
泥炭炉渣 peat cinder
泥炭末 peat dust
泥炭黏[粘]土 peaty clay
泥炭剖面 peat profile
泥炭潜育土 heat gley soil;peat-gley soil
泥炭切割刀 peat cutter
泥炭切取机 peat cutting machine
泥炭丘 turf mount
泥炭丘沼泽 palsabog
泥炭球 peat ball
泥炭熔块 peat clinker
泥炭生成史 turf development
泥炭熟料 peat clinker
泥炭水 peat water
泥炭似的 peaty
泥炭碎屑 peat litter
泥炭碎屑简易厕所 peat-litter privy
泥炭苔(藓)peat moss
泥炭苔藓草本群落 sphagniherbosa
泥炭田 clob;peat bed;peat bog;peat-land;turfary
泥炭填充料 peat filler
泥炭填絮 peat wadding
泥炭土 peaty earth;peaty soil;slime peat soil;dal soil;histosol
泥炭土的发展 development of peat land
泥炭吸附 peat adsorption
泥炭系 Carboniferous system
泥炭纤维 peat fiber[fibre]
泥炭纤维绳索 peat fibre cord
泥炭藓草本群落 sphagniprata
泥炭压榨机 peat press
泥炭压制播种饼 peat seeding pellet;peat seeding starter
泥炭烟 peat-reek
泥炭硬沥青 peat pitch
泥炭硬沥青煤 peaty pitch coal
泥炭沼 moss;peatery;Spagnum moss
泥炭沼地 black bog;moorland
泥炭沼泽 bog;moor;moss-land;peat-hag(g);peaty moor;quagmire;turfary;turf swamp;muskeg;peat bed;peat bog;peat moor;peat moss;turf moor
泥炭沼泽沉积 bog deposit
泥炭沼泽群落的 oxodic
泥炭沼泽相 bog facies
泥炭沼泽形成过程 moor forming process
泥炭质腐泥 peat-sapropel;sapropel-peat
泥炭质黏[粘]土 peat(y)clay
泥炭质土 cumulosol;peaty soil
泥炭质硬沥青 peat pitch
泥炭属性 peat attribute
泥炭砖 peat brick
泥炭状的 peat
泥炭状土壤 peat(y)soil
泥炭钻 peat auger
泥炭钻机 bog drill
泥碳钠钙石 nyererite
泥塘 containment area;bog;dystrophic lake;mire;morass;pulk
泥塘排水 containment area drainage
泥塘围埝 containment area enclosure
泥塘性 quagginess
泥条 clay column;clay lathing;wad
泥条盘筑成型法 clay-strip forming technique
泥条软度计 Hoimester
泥条筑成法 clay-strip building method
泥铁矿 blackband;gubbin;ironstone
泥铁石 clay band
泥铁岩 clay ironstone
泥铜 cement copper
泥土 argilla;clay;earth(material);earthy material;soil
泥土层 dirt bed
泥土处理方法 soil disposal method
泥土的 earthen;earthy
泥土地面 dirt floor(ing)
泥土垫层 mud mat
泥土堵塞 earth plug
泥土翻腾 mud boil
泥土覆盖 earth blanket
泥土干裂 desiccation crack
泥土干缩裂缝 mud crack
泥土灌浆 soil grouting
泥土灰浆 mud mortar
泥土回填 clay backfill
泥土混入 clay intrusion
泥土建筑 pise;terre pise
泥土搅拌机 pugmill mixer
泥土流 soil flow
泥土路 dirt road
泥土内的大圆石 turtle stone
泥土拌和机 pugmill mixer
泥土膨胀性 dilatancy of soil
泥土铺地坪 dirt floor(ing)
泥土气味 argillaceous odo(u)r;argillaceous smelt
泥土戗台 earth berm
泥土取样器 mud sampler
泥土台 grass table
泥土填充 clay fill(ing)
泥土污物 dirt
泥土压力 silt pressure
泥土移动 shifting of earth
泥土质结构 argillaceous texture
泥土住房 <爱斯基摩人住的> barrabora

泥土抓斗 clamshell type grab; mud grab
泥土砖 mud brick
泥土钻 earth(boring) auger; earth drill; soil drill
泥土钻铲 clay auger
泥团 balling
泥瓦刀 slicer
泥瓦工 mason; plasterer; tiler; bricklayer
泥瓦工工头 bricklayer's charge
泥瓦工具 bricklayer's tool
泥瓦工抹子 mason's float
泥瓦工平头凿 mason's flat-ended chisel
泥瓦工砌砖技巧 pick and dip
泥瓦工水泥 masonry cement
泥瓦工水准尺 mason's level
泥瓦工水准器 mason's level
泥瓦工水准仪 mason's level
泥瓦工梯 bricklayer's ladder
泥瓦工跳板 mason's runway
泥瓦工用尺 mason's rule
泥瓦工用砂 mason sand
泥瓦工用水平器 mason's level
泥瓦工用水准器 mason's level
泥瓦工用型板 mason's mo(u)ld
泥瓦匠用水准器 mason's level
泥湾 liman
泥位指示器 sludge level detector
泥污 muddiness
泥污道砟 slurried ballast
泥污砟床 foul ballast; slurried ballast
泥溪 muddy creek
泥虾 Laomedia astacina
泥线 mud line
泥线悬挂系统 mudline suspension system
泥箱 dirt pocket; dust pocket; mud box; mud drum
泥箱管道开关 mud plug
泥箱土 saggar[sagger] clay
泥鞋室 mud room
泥屑 mud clast; spare
泥屑白云岩 dololutite
泥屑灰岩 calcilutite; calcilutyte; mudstone
泥屑结构 mud clastic texture
泥屑石灰岩 calcipelite
泥芯 centering core; loam core; sand core
泥芯坝 earth-fill puddle core dam
泥芯撑 core chaplet
泥芯打出机 core knockout
泥芯干燥炉 core-baking oven
泥芯骨 core grid; core iron
泥芯盒 core box
泥芯架 lantern
泥芯件 expanding mandrel
泥芯孔螺纹堵塞 screwed core plug
泥芯块 sand cake
泥芯落砂机 core jarring machine
泥芯气孔 core blow
泥芯头 core print
泥芯油 core oil
泥型 chamot(te) mo(u)ld; loam mo(u)ld
泥型铸造 loam mo(u)lding
泥雪地轮胎 mud-snow tire[tyre]
泥雪碎屑堆 dirt cone
泥压 silt pressure
泥岩 clay rock; moya; mud rock; ruffite
泥岩残积层 eluvium of moya
泥岩层 turf bed
泥岩的密度 density shale
泥岩类 argilloid
泥岩蚀变 argillic alteration
泥窑 stall
泥窑焙烧 stall roasting
泥页岩 argillutite shale; clay shale; mud shale
泥荫鱼 mud minnow
泥涌 mud boil; mud spoil
泥釉 clay glaze; slip glaze
泥釉黑陶 black clay glaze pottery; black pottery with clay glaze
泥釉缕 threadlike surface flaw
泥釉黏[粘]土 slip clay
泥釉铸塑技术 slip casting
泥淤 silting
泥鱼 mudfish
泥雨 mud rain
泥浴 bog bath; mud bath
泥泽土 muskeg
泥渣 body refuse; mud residue; slime sludge; sludge(silt)
泥渣捕集器 mud box
泥渣层 sludge blanket; slurry blanket
泥渣层过滤 blanket filtration; slurry blanket
泥渣沉积 subsidence settling
泥渣沉积槽 subsidence tank
泥渣抽出器 sewage solid extractor
泥渣堆积 sludge accumulation
泥渣分离器 mud separator; slime separator; sludge separator
泥渣回流 return of slurry
泥渣孔 salt door
泥渣浓缩池 slurry concentrator
泥渣浓缩器 slurry concentrator
泥渣脱水 slurry dewatering
泥毡 sludge blanket
泥毡层 mud blanket
泥沼 lair; mire; morass; quagmire; quake ooze; slew; slime peat; slough; slue
泥沼爆炸 bog blasting
泥沼的 miry; quaggy; sloughy
泥沼地 bog; marshy area; marsh(y) land; muskeg swamp; quagmire; swampy area; turfary; turf moor
泥沼海湾 bayou
泥沼河道 sloughy channel
泥沼河口 bayou
泥沼湖口 bayou
泥沼坑 bog hole
泥沼煤 boghead(ite) coal
泥沼区 swampy area
泥沼水 bog water
泥沼土 moor peat
泥沼质土 peat(y) soil
泥沼状的 swamped
泥沼状态 bogginess
泥支撑 mud-supported
泥支撑组构 mud-supported fabric
泥质白云岩 argillaceous dolomite
泥质板岩 argillaceous slate; argillite
泥质板岩地面 clay slate ground
泥质板岩针 slate needle
泥质材料 argillaceous material; earthy material
泥质层凝灰岩 argillolith
泥质产品 slime product
泥质沉淀 mud precipitation
泥质沉积(物) argillaceous sediment; muddy sediment
泥质充填 clayey filling
泥质的 argillaceous; argillic; argilliferous; argillious; clayey; lutaceous; pelitic; clayish
泥质的含氢指数 hydrogen index of shale
泥质的视密度孔隙度 apparent density porosity of shale
泥质的视声波孔隙度 apparent sonic porosity of shale
泥质的视中子孔隙度 apparent neutron porosity of shale
泥质的相对体积 fractional volume of shale
泥质的中子测井孔隙度 neutron porosity of shale
泥质地层 sludge formation
泥质多边形土 mud polygon
泥质粉砂 clayey silt
泥质粉砂岩 argillaceous siltstone; pelitic siltstone
泥质海滩 muddy beach; bank of muddy sand
泥质含量 bulk volume fraction of shale; shale content
泥质化 argillation; argillization
泥质灰岩 argillaceous limestone; pelitic limestone
泥质混杂物 argillaceous ingredient
泥质夹层 argillaceous intercalation; muddy intercalation
泥质胶结 pelitic cement
泥质胶结物 argillaceous cement; clayey cement
泥质角砾岩 argillaceous breccia
泥质角页岩 pelitic hornfels
泥质结构 argillaceous texture; pelitic texture
泥质矿物 clay mineral
泥质砾 clayey gravel
泥质砾石 argillaceous gravel
泥质砾岩 argillaceous conglomerate
泥质流 current of higher density
泥质铝矾土 argillaceous bauxite
泥质煤 carbargilite
泥质泥灰岩 argillaceous marl
泥质凝灰岩 pelitic tuff
泥质片麻岩 pelite-gneiss; pelitic gneiss
泥质片岩 argillaceous schist
泥质浅滩 mud shoal; mudbank
泥质沙滩 bank of muddy sand
泥质砂 <又称泥质沙> silty sand; argillaceous sand; clayey sand
泥质砂层 dirty sand
泥质砂地 argillaceous sand ground
泥质砂岩 argillaceous sandstone; dauk[dawk]; pelitic sandstone
泥质砂岩中纯砂岩夹层的地层因数 formation factor of sand laminate in shaly sand
泥质声波时差 sonic wave interval transit time of shale
泥质石膏 argillaceous gypsum; pelitic gypsum
泥质石膏-硬石膏矿石 pelitic gypsum-anhydrite ore
泥质石灰岩 argillaceous limestone
泥质陶 clay pottery
泥质陶器 argillaceous earthenware
泥质铁矿 argillaceous iron ore; argillaceous ironstone; iron clay
泥质微晶灰岩 pelmicrite
泥质物质 muddy material
泥质岩 argillaceous rock; argillite; lutite[lutyte]; pelite[pelyte]
泥质岩包裹体 argillaceous inclusion
泥质岩结构 texture of argillaceous rocks
泥质岩石 argillaceous rock; clayey rock
泥质岩相 argillaceous facies; argillaceous rock facies
泥质页岩 argillaceous shale; argillite shale; bat; blue metal; dauk[dawk]
泥质页岩建造 argillaceous shale formation
泥质应变带 pelitic strain bands
泥质硬砂岩 calley stone
泥质硬石膏-石膏矿石 pelitic anhydrite-gypsum ore
泥质组分 argillaceous fracture
泥洲 mudbank
泥铸 loam casting
泥砖 loam brick; mud brick
泥砖建筑 sun-dried brick construction
泥砖墙 sun-dried brick masonry wall
泥状废水 muddy wastewater
泥状料 gunk
泥状铝质岩 pelitomorphic aluminous rock
泥状物质 gunk
泥状岩 mud rock
泥锥 mud cone
泥子 putty
泥钻 ground auger

铌

95 辐射 radiation of 95Nb
铌铋锌陶瓷 Nb_2O_2-Bi_2O_3-ZnO ceramics
铌的 niobic
铌钙矿 fersmite
铌钙石 fersmite
铌钙钛矿 dysanalyte; latrappite
铌基质超导材料 niobium-base superconducting material
铌钶铁矿含量 ferrocolumbite
铌矿 niobium ores
铌镁锆钛酸铅陶瓷 lead niobiummagnesium zirconate titanate ceramics; lead niobiumzine zirconate titanate ceramics
铌镁矿 magnoniobite
铌镁酸铅 lead magnesio-niobate
铌镁酸铅陶瓷 lead magnesio-niobate ceramics
铌锰矿 manganocolumbite
铌镍酸铅陶瓷 lead niobate-nickelate ceramics
铌石 nioboxide
铌铈钇钙矿 polymignite
铌酸 niobic acid
铌酸钡钠 barium sodium niobate
铌酸钡钠晶体 barium sodium niobate crystal
铌酸钡钠陶瓷 sodium barium niobate ceramics
铌酸钡锶 barium strontium niobate
铌酸钙 calcium niobate
铌酸镉 cadmium niobate
铌酸光栅 potassium tantallum niobate grating
铌酸钾 potassium niobate
铌酸钾锂 lithium potassium niobate
铌酸钾钠 potassium-sodium niobate
铌酸钾钠压电陶瓷 potassium sodium niobate piezoelectric(al) ceramics
铌酸钾锶 strontium potassium niobate
铌酸锂 lithium niobate
铌酸钠 sodium niobate
铌酸铅 lead niobate
铌酸锶钡 strontium barium niobate
铌酸钛 titanium niobate
铌酸钽钾 potassium tantallum niobate
铌酸锌 zinc niobate
铌酸盐 niobate
铌酸盐晶体 niobate crystal
铌酸盐陶瓷 niobate ceramics
铌酸盐系陶瓷 niobate system ceramics
铌酸盐系压电陶瓷 niobate system piezoelectric(al) ceramics
铌酸应力计 lithium niobate stress ga(u)ge
铌钛锰石 manganbelyankinite
铌钛铀矿 betafite; hatchetttolite
铌钽交代蚀变花岗岩矿 niobium and tantalum deposit in altered granite
铌钽矿(石) niobium-tantalum ore[Ni-Ta ore]

铌钽酸钾 potassium niobate-tantalate
铌钽酸锂 lithium niobate-tantalate
铌钽铁矿(含量)columbite-tantalite
铌钽铁铀矿 ishikawaite
铌钽伟晶岩矿床 niobium and tantalum-bearing pegmatite deposit
铌钽铀矿 liandratite
铌锑矿 stibiocolumbite
铌铁 ferroniobium
铌铁矿 columbite; ferrocolumbite; greenlandite; niobite
铌铁矿矿石 niobite ore
铌铁矿-钽铁矿矿石 niobite-tantalite ore
铌铁锰矿 columbite
铌铁铀矿 petscheckite
铌星叶石 niobophyllite
铌氧基 niobyl
铌叶石 niobophyllite
铌钇矿 nuevite; samarskite
铌钇矿含量 samarskite
铌易解石 nioboaeschynite
铌易解石矿石 Nb-eschynite ore
铌制模具 <热压用> niobium die

霓磁斑岩 aegirinolite

霓橄粗面岩 kenyte
霓虹灯 festoon lighting; neon bulb; neon(filled) tube; neon lamp; neon light; neon strip lighting; neon illumination
霓虹灯标 neon beacon
霓虹灯广告 neon; neon sign
霓虹灯街 neon-street
霓虹灯式波高计 neon-tube type wave meter
霓虹管灯 neon-tube light
霓虹散射 rainbow scattering
霓虹招牌 neon sign
霓辉石 acmite-augite; aegirine augite
霓角斑岩 lahnporphyry
霓磷灰石 aegiapite
霓石 <为锥辉石的同义词,有时指含钙、镁或铝等的不纯锥辉石> aegirite; aegirine
霓石白云(石)碳酸岩 aegirine rauhaugite
霓石方解石碳酸岩 aegirine alvikite
霓石霏细岩 aegirine felsite
霓石黑云碳酸岩 aegirine sovite
霓石化 aegirinzation
霓石镁云碳酸岩 aegirine beforsite
霓细斑岩 felsite porphyry
霓细花岗岩 grorudite
霓细晶岩 aegirine aplite
霓霞钠辉岩 melteigite
霓霞石 ijolite
霓霞岩的分类图 classification of ijolites
霓霞岩类 ijolite group

你方结存 balance due you from us; balance in your favour; balance to your credit

你方结欠 balance to your debit
你方结余 balance due you from us; balance in your favour; balance to your credit
你方(受益的)余额 balance in your favour
你方账户 <国际汇兑> due to balance; vostro account

拟凹的效用函数 quasi-concave utility function

拟凹规划 quasi-concave programming
拟凹函数 quasi-concave function
拟凹和凸的生产函数 quasi-concave and convex production function
拟凹性 quasi-concavity
拟板法 quasi-slab method; slab analogy
拟半局部环 quasi-semi-local ring
拟包被 pseudo-peridium
拟保角映射 quasi-conformal mapping
拟保形 quasi-conformality
拟倍数 quasi-multiple
拟本原的 quasi-primitive
拟比率尺度 quasi-ratio scale
拟变量 quasi-variable
拟变数 dummy variable
拟遍性假设 quasi-ergodic hypothesis
拟标准化分布 quasi-standardized distribution
拟表型 phenocopy
拟并行处理 quasi-parallel processing
拟补 quasi-complement
拟补格 quasi-complemented lattice
拟不变测度 quasi-invariant measure
拟不变测度空间 quasi-invariant measure space
拟不可分解的 quasi-indecomposable
拟草案 protocol
拟测地线 quasi-geodesic
拟测井记录 image log
拟测井技术 pseudo-logging well technique
拟层孔菌属 <拉> Fomitopsis
拟长度 quasi-length
拟乘法 quasi-multiplication
拟程序 programming
拟赤杨【植】Chinabells
拟充分性 quasi-sufficiency
拟出 strike out
拟纯粹局部环 quasi-unmixed local ring
拟大纲 block in; blockout
拟代数 quasi-algebra
拟代数闭域 quasi-algebraically closed field
拟代数体的 quasi-algebroidal
拟单的 quasi-simple
拟单调的 quasi-monotone; quasi-monotonic
拟单调函数 quasi-monotone function
拟单环 quasi-simple ring
拟蛋白 albuminoid
拟倒品字锈菌属 <拉> Triphragmiopsis
拟等价类 quasi-equivalence classes
拟等价单表示 quasi-equivalent unitary representation
拟等位性 pseudo-allelism
拟等值 quasi-equality
拟等值函数 quasi-equality function
拟电日期 date of preparation
拟顶极植物群落 quasi-climax
拟订 draw up; map out; work out
拟订机器和设备技术条件 specifying machines and equipment
拟订投标人名单 restricted list of bidders
拟订投标人清单 restricted list of bidders
拟订相同价格的行动 conscious parallel action
拟订预算 drafted budget
拟订总预算 drafted general budget
拟定 sketch out
拟定报告 report development
拟定尺寸 dimensioning
拟定初步方案 roughcast
拟定的计划 adopted plan
拟定合理的项目 generate sound project
拟定监测项目 proposed monitoring item
拟定阶段 concept phase
拟定试验程序 advanced development
拟定义 quasi-definite
拟动力法 pseudo-dynamic(al) method
拟动力试验 pseudo-dynamic(al) test
拟动态模型 quasi-dynamic(al) model
拟度量的 quasi-metric
拟对角的 quasi-diagonal
拟对角(线)矩阵 quasi-diagonal matrix
拟对偶空间 quasi-dual space
拟多孢锈菌属 <拉> Xenodochus
拟多孔菌属 <拉> Polyporellus
拟多项式 quasi-polynomial
拟多项式函数 polynomial-like function
拟范数 quasi-norm; semi-norm
拟仿埃尔米特的 quasi-parahermitian
拟仿复的 quasi-paracomplex
拟仿射代数簇 quasi-affine algebraic variety
拟分量 quasi-component
拟分裂 quasi-split
拟分裂代数群 quasi-split algebraic group
拟分析方法 quasi-analytic method
拟分支空间 quasi-component space
拟辐射 radiomimesis
拟辐射化合物 radiomimetic chemical
拟辐射活性 radiomimetic activity
拟辐射剂 radiomimetic agent
拟辐射物质 radiomimetic substance
拟辐射效应 radiomimetic effect
拟负算子 quasi-negative operator
拟复格 quasi-complex lattice
拟复流形 quasi-complex manifold
拟复兴整修购置房产 acquisition-with rehabilitation
拟复性 quasi-complexity
拟赋范线性空间 quasi-normed linear space
拟概率 quasi-probability
拟革盖菌属 <拉> Coriolopsis
拟各态历经系统 quasi-ergodic system
拟根环 quasi-radical ring
拟共振 pseudo-resonance
拟构等高线 approximate contour; form line; inaccurate contour; landform line; sketching contour
拟构实线等高线 continuous form line
拟构造岩 mimetic tectonite
拟古典主义 pseudo-classicism
拟观测值 quasi-observation
拟函数 quasi-function
拟合常数 fitting constant
拟合的非参数检验 distribution-free test of fit
拟合的回归直线 fitted regression line
拟合的线性回归直线 fitted linear regression line
拟合多重回归方程 fitted multiple regression equation
拟合(方)法 fitting method
拟合假设 fit hypothesis
拟合检验 fitting check; test of fit
拟合良好性 goodness of fit
拟合良好性检定【数】goodness of fit test(ing)
拟合良好性检验 goodness of fit test(ing); testing goodness of fit
拟合劣度 badness of fit
拟合模型 fitted model
拟合抛物线 parabola of fit
拟合平滑法 adaptive smoothing method
拟合曲线 fitted curve; fitting a curve; fitting curve
拟合取等值母式 quasi-conjunctive equality matrix
拟合适当 adequacy of fit
拟合数据 fitting data
拟合双曲线 hyperbola of fit
拟合条件 fitting condition
拟合同 draw up a contract
拟合误差 error of fitting
拟合系数 fitting coefficient
拟合优度【数】goodness of fit
拟合优度度量 measure of goodness of fit
拟合优度检测 goodness of fit test(ing)
拟合优度检定 goodness of fit test(ing)
拟合优度检验 goodness of fit test(ing); testing goodness of fit; test of goodness of fit
拟合优良度 goodness of fit
拟合直线 fitting a straight line
拟合值 fitted value
拟合指数曲线 exponential curve of fit
拟合准则 fitting criterion
拟核 nucleoid
拟核映射 quasi-nuclear mapping
拟恒等式 quasi-identities
拟化学方法 quasi-chemical method
拟环 quasi-ring
拟积分的 quasi-integral
拟基本解 parametrix
拟基点霉属 <拉> Phomopsis
拟极大似然估计 quasi-maximum likelihood estimation
拟极小集 quasi-minimal set
拟极值紧统 quasi-extremal compactum
拟计划 crayon
拟季节波动 quasi-seasonal fluctuation
拟寄生 parasitoid
拟加速度 pseudo-acceleration
拟加速度谱 pseudo-acceleration spectrum
拟简波 quasi-simple wave
拟简谐系统 quasi-harmonic system
拟建岸壁 proposed quay wall
拟建道路 future road
拟建第二线【铁】proposed second(ary) track
拟建房屋 proposed building
拟建房屋的土地 lot zone
拟建工程 proposed project
拟建公路里程 draft feeder highway kilometerage
拟建公路路基 proposed highway subgrade
拟建架空索道长度 draft telpherage meterage
拟建架空索道规格 planning telpherage standards
拟建码头 proposed dock; proposed quay; proposed terminal; proposed wharf
拟建入口道路 proposed access road
拟建输电线长度 draft transmission line meterage
拟建输送管道长度 draft transport pipeline meterage
拟建输送管道规格 planning transport pipeline meterage
拟建支线里程 draft feeder railway kilometerage
拟渐近稳定 quasi-asymptotical stability
拟渐近线 quasi-asymptote
拟渐近线法 quasi-asymptote method
拟解析函数 quasi-analytic function
拟解析性 quasi-analyticity
拟金茅 false golden cogongrass
拟紧集 quasi-compact set
拟紧聚集 quasi-compact cluster
拟紧空间 quasi-compact space

N

拟紧性 quasi-compactness
拟晶的 mimetic
拟晶结晶【地】mimetic crystallization
拟晶体 crystalloid
拟静力 pseudo-static force
拟静力触探试验 quasi-static penetration test
拟静力地震分析 pseudo-static earthquake analysis; pseudo-static seismic analysis
拟静力法 pseudo-static method
拟静力过程 pseudo-static process; quasi-static process
拟静力破坏机制 pseudo-static failure mechanism
拟静力试验 pseudo-static test
拟静力位移 pseudo-static displacement
拟静力影响系数 pseudo-static influence coefficient; pseudo-static influence factor
拟静态传递 pseudo-static transmission
拟静态过程 pseudo-static process; quasi-static process
拟静态试验 qualification test
拟静态问题 quasi-static problem
拟静自然电位 pseudo-static spontaneous potential
拟局部环 quasi-local ring
拟局部算子 quasi-local operator
拟菊海鞘(属)<拉> Botrylloides
拟矩 quasi-moment
拟距离 quasi-distance
拟绝热膨胀 pseudo-adiabatic expansion
拟掘建航道里程 draft channel kilometerage
拟均匀分布 quasi-uniform distribution
拟均匀收敛 quasi-uniform convergence
拟开变换 quasi-open transformation
拟开映射 quasi-open mapping
拟可公理化的 quasi-axiomatizable
拟可逆元 quasi-invertible element
拟拉格朗日极小化问题 quasi-Lagrangean minimization problem
拟离散谱 quasi-discrete spectrum
拟连锁 pseudo-linkage
拟连续测度空间 quasi-continuous measure space
拟连续泛函 quasi-continuous functional
拟连续过程 quasi-continuous process
拟连续函数 quasi-continuous function
拟连续性 quasi-continuity
拟连续性原理 quasi-continuity principle
拟裂变 deep inelastic transfer; incomplete fusion; quasi-fission; relaxed peak process; strongly damped collision
拟零群 quasi-nil group
拟滤子化的 quasi-filtered
拟滤子化范畴 quasi-filtered category
拟螺线 quasi-spiral
拟马尔可夫链 quasi-Markov chain
拟蒙特·卡罗法 quasi-Monte-Carlo method
拟迷孔菌属<拉> Daedaleopsis
拟幂等 quasi-idempotency
拟幂等阵 quasi-idempotent matrices
拟幂级数 quasi-power series
拟幂零环 quasi-nilpotent ring
拟幂零算子 quasi-nilpotent operator
拟幂零元 quasi-nilpotent element
拟内点 quasi-interior point
拟逆元 quasi-inverse element
拟逆阵 quasi-inverse matrix
拟黏[粘]滞流 pseudo-viscous flow
拟凝聚层 quasi-coherent sheaves
拟牛顿法 quasi-Newton's method
拟派股利 proposed dividend
拟平面 quasi-plane
拟平面曲线 quasi-plane curve
拟平稳过程 quasi-stationary process
拟平稳现象 quasi-stationary phenomenon
拟平移 quasi-translation
拟期望值 quasi-expectation
拟奇异顶点 quasi-singular vertex
拟气管 pseudo-tracheae
拟强对角线 quasi-dominant diagonal
拟强连通图 quasi-strongly connected graph
拟侵填体 tylosoid
拟球壳霉属 <拉> Sphaeropsis
拟球体 globoid
拟球心阑透镜组 hypergon
拟群 quasi-group
拟人 personification
拟人机器人 genera factory
拟人型常压潜水服 anthropomorphic diving suit
拟三角矩阵 almost triangular matrix
拟三维计算 quasi-three dimensional computation
拟色 assimilation colo(u)r; mimic colo(u)ration
拟设输气管长度 draft gas pipeline meterage
拟射影簇 quasi-projective variety
拟射影代数簇 quasi-projective algebraic variety
拟射影的 quasi-projective
拟射影概型 quasi-projective scheme
拟剩余的 quasi-residual
拟实验分析的有效性 validity of quasi-experimental analysis
拟实验设计 quasi-experimental design
拟收敛 quasi-convergence
拟收敛的 quasi-convergent
拟收缩半群 quasi-contraction semigroup
拟四元数的 quasi-quaternionic
拟似变形 affine deformation
拟速度 pseudo-velocity
拟速度谱 pseudo-velocity spectrum; spectral pseudovelocity
拟速率常数 pseudo-rate constant
拟算子 quasi-operator
拟随机抽样 quasi-random sampling
拟随机的 quasi-random; quasi-stochastic
拟随机点 quasi-random point
拟随机二值信号 pseudo-random binding signal
拟随机过程 pseudo-random process
拟随机模式 quasi-random pattern
拟随机数量 pseudo-random number; quasi-random number
拟随机数目 pseudo-random number; quasi-random number
拟随机数序列 pseudo-random number sequence; pseudo-random sequence of numbers; quasi-random sequence of numbers
拟态 mimicry; simulation
拟态贝 Mimella
拟态的 mimetic; mimic
拟态相似 analog(ue) simulation
拟态性 pseudo-pisenmatic character
拟态重结晶 mimetic recrystallization
拟弹性近似(法) pseudo-elastic approximation
拟特提斯 <古地中海边缘区> Parathethys
拟特征标 quasi-character
拟特征函数 quasi-characteristic function
拟凸泛函 quasi-convex functional
拟凸函数 quasi-convex function
拟凸性 quasi-convexity
拟凸序列 quasi-convex sequence
拟椭圆函数 quasi-elliptic(al) function
拟椭圆几何学 quasi-elliptic(al) geometry
拟椭圆空间 quasi-elliptic(al) space
拟挖总方量 total volume to be dredged
拟完备局部凸空间 quasi-complete locally convex space
拟完备空间 quasi-complete space
拟完全映射 quasi-perfect mapping
拟微分 quasi-differential
拟位势算子 quasi-potential operator
拟位势映射 quasi-potential mapping
拟稳的流动准稳流 quasi-stationary flow
拟稳定分布 quasi-stable distribution
拟稳定分布函数 quasi-stable distribution function
拟稳定分布律 quasi-stable distribution law; quasi-stable law
拟稳定流 pseudo-stationary flow
拟稳定状态压力分布 pseudo-steady-state pressure distribution
拟稳平差 quasi-stable adjustment
拟稳态 quasi-stable state
拟稳状态 quasi-stationary state
拟析取等值母式 quasi-disjunctive equality matrix
拟析取式 quasi-disjunction
拟析因设计 quasi-factorial design
拟线性的 quasi-linear
拟线性泛函 quasi-linear functional
拟线性方程 quasi-linear equation
拟线性方程组 quasi-linear system of equations
拟线性化 quasi-linearization
拟线性化法 quasi-linearization method
拟线性化规划 quasi-linear planning; quasi-linear programming
拟线性偏微分方程 quasi-linear partial differential equation
拟线性双曲型方程 quasi-linear hyperbolic equation
拟线性双曲型方程组 quasi-linear hyperbolic systems
拟线性双曲型组 quasi-linear hyperbolic systems
拟线性算子 quasi-linear operator
拟线性椭圆型方程 quasi-linear elliptic(al) equation
拟线性微分方程 quasi-linear differential equation
拟线性系统 quasi-linear system
拟线性性 quasi-linearity
拟相对速度谱 pseudo-relative velocity spectrum
拟小数 quasi-decimal
拟需求表 pseudo-demand schedule
拟序 quasi-ordering
拟旋转 quasi-rotation
拟循环群 quasi-cyclic(al) group
拟压面 surface to be pressed
拟样本 quasi-sample
拟样本容量 quasi-sample size
拟液态密度 pseudo-liquid density
拟一致 quasi-uniform
拟一致分布 quasi-uniform distribution
拟一致空间 quasi-uniform space
拟一致连续的 quasi-uniformly continuous
拟一致性 quasi-uniformity
拟议中项目 pipeline project
拟诣零环 quasi-nil ring
拟因子 quasi-divisor; quasi-factor
拟用建筑物 proposed structure
拟用路线 proposed alignment
拟用年限 proposed life
拟用坡度 proposed grade
拟酉的 quasi-unitary
拟域 near field; quasi-field
拟圆柱 quasi-cylinder
拟圆柱面【数】cylindroid
拟圆柱面的 cylindroid
拟圆柱投影法 pseudo-cylindric(al) projection
拟圆锥投影法 pseudo-conic(al) projection
拟张量 quasi-tensor
拟整格序半群 quasi-integral lattice-ordered semigroup
拟正规的 quasi-regular
拟正规算子 quasi-normal operator
拟正规性 quasi-normality
拟正规族 quasi-normal family
拟正交多项式 quasi-orthogonal polynomial
拟正交异性板法 quasi-orthotropic plate method; quasi-orthotropic slab method
拟正则点 quasi-regular point
拟正则性 quasi-regularity
拟正则序列 quasi-regular sequence
拟正则元素 quasi-regular element
拟正则左理想 quasi-regular left ideal
拟直和 quasi-direct sum
拟直线应力图 quasi-rectilinear stress circle
拟植色 phytoscopic
拟指令 pseudo-instruction; quasi-instruction
拟指令方式 quasi-instruction form
拟指令形式 quasi-instruction form
拟制流程图 flowcharting
拟重合 quasi-coincidence
拟重入 quasi-reentrant
拟周期 quasi-periodicity
拟周期变化 quasi-periodic(al) variation
拟周期波动 quasi-periodic(al) oscillation
拟周期轨道 quasi-periodic(al) orbit
拟周期函数 quasi-periodic(al) function
拟周期解 quasi-periodic(al) solution
拟周期运动 quasi-periodic(al) motion
拟周期振荡 quasi-periodic(al) oscillation
拟柱公式 prismoidal formula
拟自共轭扩张 quasi-self-conjugate extension
拟自相关 quasi-autocorrelation
拟自旋群 quasi-spin group
拟最优的 quasi-optimal
拟左连续性 quasi-left-continuity

逆

逆半对数回归 inverse semilog regression
逆包化 depacketize
逆饱和分析 reverse saturation analysis
逆保序映射 inverse isotone mapping
逆爆破 suction blast
逆本征值 reciprocal eigenvalue
逆比降 adverse grade; adverse slope
逆比时限 inverse time limit
逆变 contravariance
逆变单元 inverter unit
逆变的 contravariant
逆变分 inverse variation

逆变分量 contravariant component
逆变换 inverse transform; inverse transformation; retransformation; reverse transformation
逆变换器 decommutator
逆变流 inversion
逆变器 DC-to-AC converter[inverter]; inverter[invertor]
逆变器保护单元 inverter protection unit
逆变器本身 inverter proper
逆变器调换钮 inverter knob
逆变器晶(体)闸(流)管 inverter thyristor
逆变器控制的列车 inverter controlled train
逆变器用变压器 inverter transformer
逆变器组 inverter group
逆变条件 inversion condition
逆变向量 contravariant vector
逆变形 inverse metamorphism
逆变运行 inverse operation; inversion operation
逆变张量 contravariant tensor
逆变整流器 <由直流变为交流> inverted rectifier
逆变指标 contravariant index
逆变质作用【地】diaphthoresis; retrograde metamorphism
逆变作用 reversion reaction
逆变坐标 contravariant coordinates
逆表层硬化 reverse case hardening
逆表示 reciprocal representation
逆波 up wave
逆波摆动 backward-wave oscillation
逆波痕 anti-dune
逆布雷顿循环 dense air refrigeration cycle; reverse Brayton cycle
逆步 contragradience
逆步的 contragradient
逆侧风 cross wind
逆插法 inverse interpolation
逆插值法 inverse interpolation
逆差 adverse balance; adverse trade balance; excess of imports; import surplus; negative balance; overbalance of imports; passive balance; unfavo(u)rable balance; unfavo(u)rable balance of deficit; unfavo(u)rable balance of trade
逆差异 unfavo(u)rable variance
逆潮(流) counter tide; foul tide; head tide; interbed tide; inverted tide; opposite tide; reversed tide; opposing tide
逆车流方向 contraflow
逆衬砌法 <隧道施工时,先进行顶部挖掘与衬砌,再进行两侧挖掘与衬砌,适用于地质不良与大跨径隧道> inverted lining
逆城市化 counterurbanization; deurbanization
逆程 back trace; blanking; retrace; return stroke; return trace
逆程率 retrace ratio
逆程期 return period
逆程扫描 retrace
逆程损耗 return loss
逆程特性 retrace characteristic
逆程消隐 flyback blanking; retrace blanking
逆程消隐混合器 blanking mixer
逆程重影 return ghost
逆充电 reversed charge
逆冲断层 thrust fault
逆冲断层面【地】thrust plane
逆冲推覆体 thrust block; thrust nappe; thrust plate; thrust sheet; thrust slice
逆冲岩席 thrust sheet
逆抽样 inverse sampling
逆触变性 dilatancy
逆传递闭包 inverse transitive closure
逆吹 backwash; blowback
逆吹风 traverse wind
逆垂趾曲线 contrapedal curve
逆磁玻璃 anti-magnetic glass
逆磁场 counter field; counter-magnetic field
逆磁场制动 contra-field braking
逆磁畴 reverse domain
逆磁化 magnetic reversal
逆磁矩 diamagnetic moment
逆磁性 diamagnetism
逆磁性的 diamagnetic
逆磁性矿物 diamagnetism mineral
逆磁影响 diamagnetic contribution
逆刺 reversed barb
逆大陵变量 antalgol
逆代法 back substitution
逆代换 inverse substitution
逆代码 inverse code
逆带宽 inverse bandwidth; inverted bandwidth
逆导磁体 backleg
逆导可控硅 reverse conducting triode thyristor
逆的 inverse; inverted
逆狄利克雷分布 inverted Dirichlet distribution
逆地貌 obsequent landform
逆地下水布置的井 upstream well
逆地址解析协议 reverse address resolution protocol
逆点法【测】method by inversion
逆点阵 reciprocal lattice
逆电动势 back voltage; counter-electromotive force; counter voltage
逆电感矩阵 inverse inductance matrix
逆电流 counter-current; inverse current; reversal of current; reverse current
逆电流保护 reverse-current protection
逆电流保护设备 reverse-current protecting equipment
逆电流断路器 reverse-current circuit breaker
逆电流继电器 reverse-current relay; reversed relay
逆电流开关 reverse-current switch
逆电流释放器 reverse-current release
逆电流制动 counter-current braking
逆电流自动切断 reverse-current trip
逆电渗析工艺 electrodialysis reversal process
逆电势期 inverse period
逆电压 back voltage; counter voltage; inverse voltage; opposing voltage
逆迭代(法) inverse iteration
逆顶风 stormy head wind
逆定理 converse theorem; inverse theorem; reciprocal theorem
逆动 reserve motion; retroaction; retrogressive movement; reverse movement
逆动式 back action
逆动作 reverse acting
逆断层【地】reverse fault; abnormal fault; jump-up; overfault; pressure fault; ramp; thruput; thrust fault; thrust slip fault; upcast fault; upleap; upthrow fault
逆断层圈闭 reverse fault trap
逆断线崖 obsequent fault-line scarp
逆对应 inverse correspondence
逆多项式抽样 inverse multinomial sampling
逆帆 aback; lay aback
逆反价格体系 adverse pricing structure; adverse pricing system
逆反摄影测量 reverse photogrammetry
逆反式 contrapositive
逆反式规则 contrapositive rule
逆反效应 snob effect
逆反演 anti-inversion
逆反应 back reaction; backward reaction; counter reaction; inverse reaction; reverse reaction
逆反应速率常数 reverse rate constant
逆方位 reverse bearing
逆方向转动防止器 runback preventer
逆方阵 inverse square matrix
逆放大器 inverse amplifier
逆分布 inverse distribution
逆分量 inverse component
逆分析 back analysis
逆风 adverse wind; against wind; bating wind; contrary wind; counter-blast; counterwind; cross wind; dead-wind; foul wind; head-on wind; head wind; opposing wind; teeth of the gale
逆风边 upwind leg
逆风潮 weather tide
逆风的 upwind
逆风的帆 backwinding
逆风顶浪 against weather
逆风航程 upwind course; upwind range
逆风航行 sail into wind; steer against wind; stem
逆风滑行 upwind taxiing
逆风火 draft fire
逆风浪航行 thrash
逆风流 inverted air current
逆风流速 upwind velocity
逆风跑道入口 upwind threshold
逆风起飞 upwind takeoff
逆风前进 thrash[thresh]
逆风驶帆 working to windward
逆风污染物 upwind wind source of pollution
逆风用火 back-burn
逆风转弯 upwind turn
逆风着陆 upwind landing
逆辐射 back radiation; counter radiation
逆复背斜层 abnormal anticlinorium
逆复向斜 abnormal synclinorium
逆傅立叶变换 inverse Fourier transform(ation)
逆改正 inversion reduction
逆概率 inverse probability
逆高斯分布 inverse Gaussian distribution
逆功率 reverse power
逆功率保护 converse power protection; reverse-power protection
逆功率保护装置 converse power protection unit; reverse-power protector
逆功率动作 reverse-power tripping
逆功率断路器 reverse-power circuit breaker
逆功率继电器 reverse-power relay
逆功率脱扣装置 reverse-power tripping device
逆共振 inverse resonance
逆估计 inverse estimation
逆关系 converse of relation; inverse relation
逆光 backlighting
逆光的 backlighted
逆光角 shielding angle
逆光摄影 backlighted shot; shadowgraph
逆光条件 adverse lighting condition
逆光线 backlight; counter light
逆光性 reverse photoproduction
逆光照明 backlighting
逆辊涂布 reverse roll coating
逆过程 inverse process
逆过渡机动 <垂直起落飞机的> reconversion maneuver
逆过载继电器 reverse over-current relay
逆函数 inverse function
逆合成 retrosynthesis
逆合成分析 retrosynthetic analysis
逆河口 inverse estuary
逆核 reciprocal kernel
逆虹吸作用 backsiphon(age)
逆弧 back-fire; flash back; inverse arc; reignite; arc-back
逆滑动 <滑动自坡脚开始> retrogressive slide
逆滑断层 reverse slip fault
逆化皂 invert soap
逆环流 counter-current circulation; indirect circulation; recirculate
逆环流圈 reverse cell
逆换流器 inverted converter
逆回归 inverse regression
逆汇 adverse exchange; bill by negotiation; draft by negotiation; reverse remittance
逆汇编程序 inverse assembler
逆汇兑 redraft
逆火 after burning; back-fire; back-kick; flare back; flash back
逆火止回阀 back-fire check valve
逆积分器 inverse integrator
逆级配 inverse grading
逆极限谱 inverse limit spectrum
逆兼容性 reverse compatibility
逆剪切 upshear
逆检索 retrospective search
逆检索法 retrospective retrieval method; retrospective search method
逆检系统 retrospective search system
逆进供应 regressive supply
逆矩阵 inverse matrix; inverse of a matrix; inverting matrix; matrix inversion; reciprocal matrix; reverse matrix
逆矩阵系数表 inverse matrix table
逆聚合 reverse polymerization
逆卷积 inverse convolution
逆卡诺循环 reverse Carnot cycle
逆康普顿效应 inverse Compton effect
逆亏损 converse deficiency
逆扩散 counter-diffusion
逆扩散长度 inverse diffusion length
逆廓影 reversed silhouette
逆拉普拉斯变换 inverse Laplace's transform(action); inverse Laplace transform
逆朗道阻尼 inverse Landau damping
逆浪 countersea; head sea
逆棱 inverse edge
逆冷凝 retrograde condensation
逆立摆 invert pendulum
逆利率 negative interest
逆联系 inverse relation
逆良序集 inversely well-ordered set
逆流 adverse current; against stream; anarrhea; backflow; backset; backset current; back streaming; backward flow; conflicting stream; contraflow; counter-current flow; counter-flow; counter tide; cross current; flow reversal; inverse current; inverse flow; opposed current; opposing current; opposite-flow; reflow; refluence; reflux; return current; return flow; reverse;

N

reverse current; reverse direction(al) flow; reverse flow; reversing current; underset; upflow
逆流泵 reversing pump
逆流泵站 upstream pumping unit
逆流波痕 regressive ripple
逆流操作 counter-current operation; counter-flow operation
逆流测量 reverse-current metering
逆流层 backset bed
逆流层理 backset bedding
逆流车道 contrafloe lane; contraflow lane
逆流衬垫 counter-flow liner
逆流澄析 counter-current elutriation
逆流冲洗 backflush; backwash
逆流船 upstream vessel
逆流床 counter bed
逆流萃取 counter-current extraction
逆流的 opposite-flow; reverse; upstream
逆流等离子体 counterstreaming plasma
逆流地 counter-current-wise
逆流断电器 reverse-current cut-out
逆流断路器 reverse-current breaker
逆流发生气体过程 reverse flow gas process
逆流阀 reflux valve; reverse flow valve; reversing valve
逆流方法 counter-current method
逆流防止门 backwater gate
逆流分布 counter-current distribution; counter-flow distribution
逆流分级 counter-current classification; counter-current fraction
逆流分级机 counter-current classifier
逆流分粒机 counter-current sizer
逆流分配法 counter-current distribution method
逆流干燥 counter-current drying; counter-flow drying
逆流干燥器 inverse dryer; reversed current drier[dryer]
逆流管壳式热交换器 counter-current tube and shell heat exchanger
逆流锅炉 counter-current boiler
逆流过程 counter-current process; counter-flow process
逆流过滤器 reverse-current filter
逆流航行 running against the seas; sail against current; steer against current
逆流河 counter-flow river; counter-flow stream; inverted river; inverted stream; reversed stream
逆流烘干 drying in counter-current
逆流烘干机 reversed current drier [dryer]
逆流换气 counter-flow scavenging
逆流换热 counter-current flow
逆流换热器 counter-current heat exchange
逆流混合 counter-flow mixing
逆流混料机 counter-flow mixer
逆流级 back current step; counter-flow stage
逆流级联 counter-current cascade
逆流技术 reversed flow technique
逆流继电器 reversed current relay
逆流搅拌 counter-current agitation
逆流截门 water check
逆流浸提 counter-current leaching
逆流空气分级机 reversed current air classifier
逆流空气分离器 reversed air classifier
逆流快速拌和机 counter-flow rapid action mixer
逆流冷凝器 counter-current condenser; counter flow condenser
逆流冷凝水 counter-flow condensate
逆流冷却 counter-current cooling
逆流冷却机 counter-current cooler; cross-current cooler
逆流冷却器 counter-current cooler; counter-flow cooler; cross-current cooler
逆流冷却水 counter-flow cooling water
逆流冷却水系统 counter-flow cooling water system; reverse flow cooling system
逆流离心法 counter-current centrifuging
逆流离心机 counter-current centrifuge
逆流离子电泳法 counter-current ionphoresis
逆流离子交换 counter-current ion exchange
逆流立式冷却机 counter-current vertical cooler
逆流立式冷却器 counter-current vertical cooler
逆流冷凝器 counter-current jet condenser
逆流炉 counter-current furnace
逆流螺旋热交换器 counter-current spiral heat exchanger
逆流泥炭过滤 counter-current peat filtration
逆流凝结 counter-flow condensation
逆流凝汽器 counter-flow condenser
逆流盘式混合机 Lancaster mixer
逆流喷淋塔 counter-current column
逆流喷射 upstream injection
逆流喷雾干燥器 counter-current spray dryer
逆流平衡 counter-current balance
逆流坡度 non-sustaining slope
逆流气化 counter-current gasification
逆流前进 advancing against current; advancing against stream
逆流倾析 counter-current decantation
逆流倾析洗涤法 counter-current decantation method
逆流倾斜呈叠瓦状 imbricated tilt against current
逆流区 zone of reverse flow
逆流热交换器 counter-current heat exchanger; counter-flow regenerator
逆流色谱法 counter-current chromatography
逆流沙嘴 opposing spit
逆流施工【疏】upbound dredging
逆流式拌和(法) counter-flow mixing; counter mixing
逆流式拌和机 counter-current mixer
逆流式浮选机 counter-current machine
逆流式干燥器 counter-flow drier [dryer]
逆流式干燥系统 counter-current drying system; counter-flow drying system
逆流式辊合机 counter-flow mixer
逆流式锅炉 counter-flow boiler
逆流式烘干机 counter-current drier [dryer]; counter-flow drier[dryer]
逆流式混合机 counter-current mixer
逆流式加热系统 counter-flow system of heating
逆流式搅拌(法) counter-flow mixing
逆流式搅拌机 counter-current mixer
逆流式快速搅拌机 counter-flow rapid action mixer; counter rapid action mixer
逆流式冷凝器 contraflow condenser
逆流式冷却塔 counter-current cooling tower; counter-flow cooling tower
逆流式立窑 counter-flow shaft kiln
逆流式凉水塔 counter-flow cooling tower
逆流式炉 counter-flow furnace
逆流式黏[粘]度计 reversed-flow type viscometer
逆流式凝气器 reversed-flow condenser
逆流式燃烧器 counter-flow combustor; reversed-flow type combustion chamber
逆流式燃烧室 counter-flow combustor
逆流式热(水)交换器 contraflow heat exchanger; contraflow regenerator; counter-flow heat exchanger
逆流式省煤器 upstream economizer
逆流式通风机 <隧道用> contra fan
逆流式洗涤器 counter-current scrubber
逆流式洗砾机 contraflow gravel washer
逆流式选粉机 counter-current classifier
逆流式压缩机 return flow compressor
逆流式烟道采暖炉 downflow furnace
逆流式转筒烘干机 counter-current rotary drier[dryer]
逆流水管锅炉 counter-flow water tube boiler
逆流送料法 backward feed
逆流碎波 overfalls
逆流塔 counter-current tower; counter-flow tower
逆流淘洗 counter-current elutriation
逆流填充料层过滤器 counter-current packed-bed filter
逆流填充料层吸尘器 counter-current packed-bed filter
逆流通风 back draft; back draught
逆流挖泥 dredging against current
逆流吸附 counter-current absorption
逆流洗涤 backflush; counter-current washing; counter-flow washing
逆流洗砾(石)机 contraflow gravel washer
逆流洗衣机 contraflow washer
逆流系统 counter-current system
逆流效率 counter-current efficiency
逆流效应 backsetting effect; backward effect
逆流悬浮预热器 counter-current suspension preheater
逆流旋流器 reverse flow cyclone
逆流循环 counter-current circuit; counter-current circulation; counter-current recycling
逆流延时继电器 reverse-current time-lag relay
逆流窑 counter-current fired kiln
逆流与单流两用干燥筒 combined contra-flow uniflow drying drum
逆流与单流两用烘缸 combined contra-flow uniflow drying drum
逆流原理 counter-current action
逆流运动 retrogression
逆流运行 reverse flow operation
逆流再生 contraflow regeneration; counter-current regeneration; counter-flow regeneration; upflow regeneration
逆流蒸汽机 counter-flow engine
逆流制动 reverse-current braking
逆流制动开关 plugging switch
逆流注水冷凝器 counter-flow jet condenser
逆流自动断路开关 discriminating breaker
逆流自断器 automatic reverse current cutout
逆留 retrograde stationary
逆滤波 inverse filtering
逆滤波器 inverse filter
逆螺纹 minus thread
逆螺线操纵试验 reversed spiral test
逆幂法 inverse power method
逆幂法则 inverse power law
逆幂校正 inverse power correction
逆幂型流体 inverse power fluid
逆命题【数】contrary proposition; converse; inverse proposition
逆木纹 cross grain
逆奈奎斯特图 inverse Nyquist diagram
逆奈奎斯特阵列法 inverse Nyquist array method
逆捻 ordinary lay; regular lay
逆捻钢丝绳 reverse laid rope
逆凝析油池 retrogradation condensate pool
逆喷 upstream spray pattern
逆匹配 inverse matching
逆偏析 inverse segregation
逆偏移 reversed migration
逆频散 inverse dispersion
逆平行 anti-parallel
逆平行畴 anti-parallel domain
逆平移断层 reverse-wrench fault
逆坡 adverse grade; adverse slope; reverse gradient; reverse slope
逆坡倾斜隔水底板 reversed inclined impervious bottom bed
逆谱 reciprocal spectrum
逆气流反吹袋式除尘器 reverse air baghouse
逆牵引 antidrag; reversed drag; dip reversal
逆牵引背斜 reversed drag anticline
逆牵引构造 reversed drag structure
逆牵引构造宽度 width of reverse drag structure
逆倾斜 <地层的> reversal dip; reverse caster; reversed dip; up-dip
逆倾斜工作面 raise stope
逆倾斜掘进 driving up the pitch
逆曲线 inverse curve
逆屈光系统 contracurrent system; katoptric system
逆燃 back-fire; flame flash-back
逆热 backheating
逆蠕动 anti-creep
逆蠕动波 anti-peristaltic wave
逆蠕动的 anti-peristaltic
逆乳液 inverted emulsion
逆乳浊液 inverse emulsion
逆塞曼效应 inverse Zeeman effect
逆三角函数 inverse trigonometric(al) function
逆散射 back-scatter
逆散射变换 inverse scattering transform
逆沙丘 anti-dune
逆沙丘运动 anti-dune motion; anti-dune movement
逆扇形褶皱【地】abnormal fan-shaped fold
逆扇状褶皱【地】abnormal fan-shaped fold
逆摄影 backlighted shot
逆渗透 reverse osmosis
逆渗透分离 reverse osmosis separation
逆渗透净化 reverse osmosis purification
逆渗透膜 reverse osmosis membrane
逆渗透渗漏计 reverse osmosis permeator

逆渗透渗漏装置 reverse osmosis device;reverse osmosis unit
逆时定时限继电器 inverse time definite time limit relay
逆时定时限特性 inverse definite time limit characteristic
逆时断路器 inverse time circuit breaker
逆时计 inverse hour
逆时间偏移 reverse-time migration
逆时偏振电磁波 counter-clockwise polarized electromagnetic wave
逆时延迟 inverse time-delay
逆时针 inhour;inverted hour
逆时针多边形 counter-clockwise polygon
逆时针方向 anti-clockwise direction; counter; counter-clockwise direction
逆时针方向编号 numbering in anti-clockwise direction
逆时针方向的 contraclockwise;counter-clockwise
逆时针方向旋转 anti-clockwise rotation;counter-clockwise rotation
逆时针方向运动 anti-clockwise motion;counter-clockwise motion
逆时针角 counter-clockwise angle
逆时针满舵 fully anti-clockwise
逆时针翘倾摆动 sinistrally tilting swing
逆时针旋转 anti-clockwise rotation; left-handed rotation;reverse spin
逆时针旋转的 contraclockwise
逆时针转的 anti-clockwise
逆时钟方向的 counter-clockwise
逆矢量轨迹 inverse vector locus
逆视杜布雷斯棱镜 retrograde-vision Daubresse prism
逆视四面棱镜 retrograde-vision tetrahedral prism
逆视四面体棱镜 retrograde-vision Daubresse prism
逆收益差额 reverse yield gap
逆数 inverse number
逆双工电路 inverse duplex circuit
逆顺向河 reversed consequent stream
逆斯塔克效应 inverse Stark effect
逆四周地层的倾斜方向而下降的 anaclinal
逆送风 inverse blow
逆算符 inverse operator
逆算子 inverse operator
逆台阶工作面 flat-back stope
逆台阶式掘进法 <先挖底半部,后挖顶半部> bottom heading method
逆台阶式开挖法(先挖底半部,后挖顶半部)top-cut method
逆梯度 anti-gradient
逆梯度发展 inverting stratified development
逆调节效应 counter-regulatory effect
逆铁淦氧 inverse ferrite
逆通风 back draft;back draught
逆同步加速吸收 inverse synchrotron absorption
逆同伦 inverse homotopy
逆同态 inverse homomorphism
逆同态函数 inverse homomorphic function
逆同位素交换法 reverse isotopic exchange method
逆同位素稀释 inverse isotopic dilution;reverse isotopic dilution
逆同位素稀释法 inverse isotopic dilution method; reverse isotopic dilution method
逆同位素稀释分析 reverse isotope dilution analysis
逆图形 inverse figure
逆推力 reverse thrust
逆网络 inverse network
逆微波激射 inverse maser
逆微波激射效应 inverse maser effect
逆位 inversion
逆温 air temperature inversion;inversion(of) temperature; temperature inversion;thermal inversion
逆温层 inversion layer;inversion layer of temperature; temperature inversion layer;thermal layer
逆温层底 inversion base
逆温层顶 inversion lid
逆温层高度 inversion height
逆温层形成 inversion layer forming
逆温的 katothermal
逆温湖 <温度随深度增高> katothermal lake
逆温霾 inversion haze
逆温破坏 inversion break-up
逆温雾 inversion log
逆温下沉型烟羽 inversion fumigation plume
逆温下气层 subinversion layer
逆温云 inversion cloud
逆温状况 inversion condition
逆纹 against the grain; cross grain; perpendicular to grain
逆纹的 cross-grained
逆纹理 cross grain
逆纹镘 cross-grained float
逆纹木材 cross-grained timber
逆问题 inverse problem
逆涡 counter vortex
逆涡流 back eddy
逆稀释法 reverse dilution method
逆洗 back washing
逆洗泵 back wash pump
逆洗阀 backwash valve
逆铣 conventional milling;up-cut;up milling
逆系数矩阵 inverse coefficient matrix
逆掀块山 obsequent tilt block mountain
逆显影 reverse development
逆线性回归 inverse linear regression
逆线性最优控制 inverse linear optimal control
逆相 anti-phase
逆相变 reverse transformation
逆相关 inverse correlation; inversely related;retrocorrelation
逆向 contraflow;inverse direction;inversion;negative direction
逆向安匝数 back ampere-turn
逆向爆炸 back shooting;reverse shooting
逆向变换成本 inversely varying cost
逆向变换器 flyback converter
逆向潮 opposing tide
逆向车流 contraflow;counter-flow
逆向成本法 reversal cost method
逆向冲采 counter efflux
逆向抽水设备 back-pumping equipment
逆向储层 retrograde reservoir
逆向传导 anti-dromic conduction
逆向传导阻滞 retrograde conduction block
逆向磁化 reversed magnetization
逆向错位断层 reverse-separation fault
逆向导叶 contravane
逆向道岔锁闭控制器 facing point locking controller
逆向道岔锁闭器电路控制器 facing point lock circuit controller
逆向的 anti-dromic; backhand(ed); backrun;sense reversing;obsequent <与地层倾斜反向的> reverse
逆向递推 backward recursion
逆向电极 reversible electrode
逆向电码轨道电路 reversed code track circuit; reversing code track circuit
逆向电容 reciprocal capacitance
逆向电阻 back resistance
逆向反射 retrodirective reflection; retro-reflection
逆向分带 inverse zoning;reverse zoning
逆向分段焊接 step-back welding
逆向赋压层 negative confining bed
逆向杠杆作用 reverse leakage
逆向谷 obsequent valley
逆向辊 reverse roll
逆向辊式涂布机 contra coater
逆向辊涂 reverse roller coating
逆向辊涂机 reverse roll coater
逆向滚削 convectional hobbing
逆向焊 backhand welding; backstep welding
逆向河口 inverse estuary
逆向河(流) obsequent river; obsequent stream; anti-consequent river;anti-consequent stream
逆向滑(动)断层 reverse(-dip)-slip fault
逆向彗尾 anti-tail
逆向混合 backmixing
逆向极限 inverse limit
逆向极限空间 inverse limit space
逆向计数 counting in reverse
逆向剪刀差 negative price scissors
逆向交叉【铁】reverse intersecting
逆向拉制 reverse drawing
逆向离合器 reverse clutch
逆向量 inverse vector
逆向流动 counter-current flow;obsequent flow
逆向流斜板沉淀 oppositing plate precipitator
逆向流褶皱 reverse-flowage fold
逆向码 reverse code
逆向弥散 counter-current dispersion
逆向磨削 up-cut grinding
逆向排水 backward drainage
逆向排水井 inverted drainage well
逆向喷射 contrainjection
逆向偏压 reverse bias
逆向偏移 updrift
逆向漂流 counter drift
逆向漂移 updrift
逆向坡 obsequent slope
逆向气流 inverted draft
逆向侵蚀 regressive erosion
逆向侵蚀断层崖 reversed erosion fault scarp
逆向倾销 reverse dumping
逆向射流 backward jet
逆向渗透 <从污水盐水提取淡水的方法> reverse osmosis
逆向输送 inverted transport
逆向数位排序法 reverse digit sorting method
逆向水流 cutback stream
逆向水系 backward drainage
逆向调节器 backward acting regulator
逆向推移【地】updrift
逆向系统 inverse system
逆向销售 back-selling
逆向小河 obsequent stream
逆向行的 anti-dromic
逆向行驶 facing movement
逆向旋转 retrograde rotation
逆向应力 reverse stress
逆向运动 counter motion
逆向运行 anti-kinesis; reverse running;reversible working;running-in reverse
逆向匝 back turn
逆向占有 notorious possession
逆向支流 barbed tributary;hook valley
逆向转动 counter-rotating
逆像 inverse image
逆效率 reverse efficiency
逆效应 adverse effect;converse effect
逆斜杆 counter diagonal
逆斜河 <河道坡降与地层倾向相反> anaclinal river' anaclinal stream
逆谐振 inverse resonance
逆信道 inverse channel
逆行 backrun(ning); backset; backstroke; flyback; regress; retrace; retrogradation; retrograde motion; retrogression
逆行变态 retrogressive metamorphosis
逆行波 retrogressive wave; reverse wave
逆行波放大器 backward-wave amplifier
逆行波痕 regressive ripple mark
逆行程 backstroke;reverse drive
逆行冲动 anti-dromic impulse
逆行的 anti-dromic;retrograde;retrogressive
逆行叠绕组 reversed loop winding
逆行段 return run
逆行返回 retrograde return
逆行分化 regressive differentiation
逆行轨道 retrograde orbit;retrograde trajectory
逆行海流 countersea
逆行滑动 retrogressive slide
逆行扩散 reverse diffusion
逆行流痕 regressive ripple
逆行流滑动 retrogressive flowslide
逆行排放 anti-dromic volley
逆行圈 loop of retrogression
逆行沙波 regressive sand wave
逆行沙痕 regressive sand ripple
逆行沙丘 anti-dune;reversing dune
逆行沙丘交错层理构造 anti-dune cross bedding structure
逆行射线 retrograde ray
逆行栓塞 retrograde embolism
逆行栓子 retrograde embolus
逆行水流 negatively progressive flow
逆行停止 back stop
逆行纤维 anti-dromic fibers
逆行性变性 ascending degeneration; retrograde degeneration
逆行性传导 retrograde conduction
逆行演化 regressive evolution;retrograde evolution
逆行演替 regression;retrogression of succession;retrogressive succession
逆行再入 retrograde reentry
逆行制气 backrun
逆行制气法 backrun process
逆行转移 paradoxical metastasis;retrograde metastasis
逆形式 inverse form
逆性洗涤剂 reverse detergent
逆序 backward sequence;inverted sequence; negative sequence; reversed order;reverse manner
逆序操作 back-out(of)
逆序场 inverse field
逆序换算 last in
逆序计数 countdown
逆序粒层 inverse graded bedding
逆序列 opposite sequence
逆序列相关 inverse serial correlation
逆选择 adverse selection
逆循环 cross-over circulation;inverse cycle;reverse cycle;reverse circu-

lation
逆循环除霜 reverse cycle defrosting
逆循环法 reverse circulation method
逆循环供热 reverse cycle heating
逆循环加热 reverse cycle heating
逆循环轻便钻机 reverse circulation portable drill
逆循环钻进岩芯管 reverse circulation core barrel
逆循环钻孔 reverse circulation boring
逆压电效应 converse piezoelectric(al) effect
逆压法 retroclusion
逆芽接 reversed budding
逆掩层 overthrust sheet
逆掩大断层 overthrust fault
逆掩断层【地】reversed fault;overlap fault;overthrust;overthrust fault;reversed thrust;shove fault;thrust fault;upthrust fault
逆掩断层背斜 overthrust anticline
逆掩断层面 overthrust plane
逆掩断层圈闭 overthrust trap
逆掩断层山 overthrust mountain
逆掩断层推覆体 overthrust block;overthrust nappe
逆掩断距 overthrust distance
逆掩断块 overthrust block
逆掩断裂褶皱聚集带 accumulation zone of thrust-fold
逆掩盘 thrust sheet
逆掩推覆体 overthrust nappe
逆掩岩席 overthrust sheet
逆掩褶皱【地】overthrust fold
逆掩褶皱油气田 thrust fold oil-gas field
逆掩作用 thrusting
逆应变椭圆 reciprocal strain ellipse
逆应力 inverse stress
逆映射 inverse mapping
逆映射系 inverse mapping system
逆映象 inverse mapping
逆涌 backward surging;head swell
逆优惠 reverse preference
逆游标 retrograde vernier
逆游标尺 retrograde vernier
逆元(素) inverse element
逆运算 inverse operation
逆增 inversion
逆增轨迹 contrail
逆增烟云消散 inversion break-up fumigation
逆增益干扰 inverse gain jamming
逆折射图 inverse refraction diagram
逆褶断层【地】reversed fold-fault
逆褶积【地】inverse convolution
逆褶皱【地】reversed fold
逆真方位 reverted true bearing
逆蒸发 retrograde evapo(u)ration
逆整流器 <直流变交流> inverted rectifier
逆正则表示 inverse regular representation
逆止吊门 <排水口的> flap trap
逆止阀 backflow pressure valve;backflow preventer;backflow valve;backpressure valve;backwater gate;backwater valve;check valve;chuck valve;clack valve;downstep check;flap trap;inverted valve;non-return valve;one-way valve;reflex valve;reflux valve;rising stem valve;suction valve <吸泥管或吸水管的>
逆止器 hold-back;hold-back device;reverse stop device
逆止水阀 check sluice
逆止水跃 check jump
逆止水闸 check sluice
逆止闸门 flap gate
逆指标 inverse index;inverse indicator
逆制套 ratchet coupling
逆置换 inverse permutation
逆中介 disintermediation
逆重购协议 reverse repurchase agreement
逆周期 conversion period
逆筑法 top-down construction method
逆转 backing;backkick;back-off;back turn;cutback;inversion;kick back;reversal;reverse motion;reverse spin;reversing;reversion;revert;setback
逆转层 inversion layer
逆转成带现象 inversion of zonation
逆转磁化 reversed magnetization
逆转的 reversed
逆转点 inversion point
逆转点法 reversal point method
逆转翻笼 kickback dump
逆转风 backing wind
逆转改正 Rudzki (inversion) reduction
逆转过程 Umklapp process
逆转横轴 slewing gear intermediate shaft
逆转换 inversion transformation;shift transmission
逆转换器 inverse converter
逆转机构 reverser;reversing gear;reversing gearbox;tumbler
逆转离合器 reversing clutch
逆转票据 redraft
逆转期 reversed epoch
逆转速试验 reverse rotation speed test
逆转条件 condition of inversion
逆转显影 reversal development
逆转相序 reversed phase sequence
逆转效应 reversed effect;Umkehr effect
逆转循环 reverse circulation
逆转移 countertransference
逆转运行 reverse speed operation
逆转轴 countershaft;transmission countershaft
逆转株 revertant
逆转装置 change-over gear;inversion set;reverse gear;reversing gear
逆转作用 oversteepening
逆桩号而上 up-station
逆子午线 anti-meridian
逆阻抗 inverse impedance
逆阻尼 reversed damping
逆作法 downward construction method;top-down method
逆作施工 upside-down construction

匿

匿报 concealment
匿报所得额 concealing the amount of income
匿名的 innominate
匿名服务器 anonymous server
匿名股东 dormant partner;silent partner;sleeping partner
匿名合伙人 dormant partner
匿名合伙营业 dormant partnership
匿名合作伙伴 silent partner
匿名控制股票 warehousing
匿名块 anonymous block
匿名文件传输服务器 anonymous file transfer protocol server
匿名者 anonym
匿影 blackout;blanketing

溺

溺岸 coast of submergence;drowned coast;submerged coast;submergence coast
溺谷 drowned valley;liman;submerged valley
溺谷海岸 liman coast
溺谷海岸礁 drowned valley coast reef
溺谷名称 name of drowned valley
溺河 drowned river;ria
溺河海岸 drowned coast;ria coast
溺壶色 chamber pot colo(u)r
溺礁 drowned reef;submerged reef;sunk reef
溺流 drowned flow
溺泉 drowned spring
溺水 drowning
溺水假死 apparently drowned
溺湾 ria
溺堰 drowned weir

腻

腻感 soapy feeling
腻料 lute;mastic
腻子 badigeon;chinking;luting;paint filler;paste filler;putty;sal ammoniac;smoothing cement;stopping;surfacer
腻子稠度计 putty consistency apparatus
腻子胶 ca(u)lking compound
腻子料 spackle;sparkling
腻子泥刀 mastic trowel
腻子涂层 stopper coat

年

年百分利率 annual percentage rate
年百分率 percent per annum
年保险费 annual premium
年保证工资 guaranteed annual wage
年报 annual report;year book;yearly report
年变层 <沉积的>【地】vary
年变程 annual range
年变幅 annual amplitude;annual range
年变化 annual change;annual variation;yearly change;yearly variation
年变量 yearly variation
年变率 annual variation rate
年变异 annual change;annual variation;yearly variation
年表 annals;chronology
年冰 one-year ice
年波动 yearly fluctuation
年补偿 annual payment
年不等 annual inequality
年不均衡性 annual inequality
年不平衡 annual unbalance
年层 annual layer
年差 annual change;annual variation
年差异量 annual inequality
年产出 yearly output
年产电量 yearly output
年产量 annual capacity;annual output;annual production;annual yield;cumulative production to year end;yearly capacity;yearly production
年产沙量 annual sediment yield
年长 length of the year;seniority
年沉积量 annual sediment deposition
年沉积速率 annual rate of deposition
年成本 annual cost
年成本比较法 annual cost method of comparison
年承 year-bearer
年冲刷量 annual erosion
年抽水量 annual volume of water abstraction
年出口量 annual export
年出力 annual power output;yearly output
年初 beginning of year
年储存成本 annual holding cost
年磁变 annual magnetic variation
年磁差 magnetic annual change;magnetic annual variation
年代 age;chron;years
年代比 age ratio
年代表 chronological table
年代不明的 Dateless
年代测定 age dating;annual age determination;dating
年代带 chronostratigraphic(al) zone;chronozone
年代地层单位 chronolith;chronolithologic unit;chronostratigraphic(al) unit;time-stratigraphic(al) unit
年代地层单位名称 name of chronostratigraphic(al) unit
年代地层对比表 chronostratigraphic(al) correlation chart
年代地层相 chronostratigraphic(al) facies
年代地层学 chronostratigraphy;time-stratigraphy
年代记 chronicle
年代时序表 chronological table;chronological time scale
年代效应 age effect
年代学 chronology
年代学的 chronological
年代因数 age(d) factor
年代因素 age(d) factor
年贷款偿还率 annual load constant
年等值比较法 equivalent annual worth comparisons
年底 close of the year
年底收入 yearly income
年地温差 difference of yearly earth temperature
年地震积累数 annual accumulation of earthquake
年地震极限 annual earthquake extreme
年地震累积数 annual accumulation of earthquake
年第 30 位最大小时交通量 thirtieth highest annual hourly volume
年电能输出 annual energy output
年订货成本 annual ordering cost
年冻层 annual frost zone
年度 annum
年度保险费 annual premium
年度保养 annual upkeep;yearly maintenance
年度保养维修 annual maintenance
年度报表 annual statement
年度报告 annual report;yearly report
年度拨款 annual appropriation
年度拨款计划 annual funding program(me)
年度拨款支付书 appropriation warrant
年度财务报表 annual financial statement
年度财务决算报告 annual financial report
年度财务决算表 annual financial statement
年度成本 annual cost
年度成本法 annual cost method
年度大检查 annual general inspection
年度贷款计划 annual borrowing program(me)
年度的 annual

年度定额 annual quota;annual standard
年度费用 annual charges;annual cost
年度概算 annual estimate
年度工作规划 annual working program(me)
年度工作计划 annual program(me) of work;annual working program(me)
年度工作量 yearly working volume
年度汇编 annual summary
年度货物运输计划 yearly goods transport plan
年度基准 annual basis
年度计划 annual plan;annual program(me);yearly plan;yearly program(me)
年度计划利润 planned annual profit
年度技术进展报告 annual technical progress report
年度假 annual leave
年度检查 annual inspection;yearly inspection
年度检修 annual inspection;annual maintenance;annual overhaul;yearly maintenance
年度检验 yearly test
年度建设计划 annual construction plan
年度结算 annual account;annual closing;annual reckoning
年度结账 annual closing
年度进展报告 annual progress report
年度经费 annual appropriation
年度经营费用 annual business expenses
年度决算 annual account;annual closing;annual reckoning
年度决算表 annual financial statement
年度决算书 annual balance sheet;annual financial statement
年度矿量变动 variation of annual ore reserves
年度亏损 annual deficit;annual loss;losses in a year
年度免税额 annual allowance
年度目标 annual objectives;annual target
年度配额 annual quota
年度全面检查 annual general inspection
年度审查 annual review
年度审计 annual audit
年度施工组织设计 annual construction organization design
年度收入 annual receipt;annual revenue
年度损失 annual loss
年度缩减 annual decrement
年度索引 annual index
年度讨论会 annual workshop
年度统计表 annual returns;annual statistic(al) table
年度推销费 annual salable expenditures
年度维护 annual maintenance;yearly annual maintenance;yearly maintenance
年度维修 annual maintenance;annual repair;annual upkeep;yearly annual maintenance;yearly maintenance
年度维修成本 annual maintenance cost
年度维修费用 annual cost of upkeep;annual maintenance charges;yearly maintenance charges
年度现金流量 annual cash flow
年度消耗量 annual consumption;yearly consumption
年度性趋向 yearly trench
年度修理 annual overhaul
年度需要量 annual requirement
年度养护 annual maintenance;yearly maintenance
年度养护费 annual maintenance cost;yearly maintenance cost
年度英里程 annual mil(e)age
年度盈利率 annual rate of return
年度预算 annual budget;yearly budget
年度原始成本 annual first cost
年度运量 annual volume of traffic
年度运输计划 annual transport plan
年度增长 annual increment;yearly increment
年度增加额 annual increment;yearly increment
年度账 annual account
年度折旧 annual depreciation
年度支出 annual expenditures
年度支出预算 budget for annual expenditures
年度支付最高限额 annual cash ceiling
年度指标 annual target;annual achievement
年度终了 year ending
年度主要成本 annual first cost
年度住房调查 annual housing survey
年度资产负债表 annual balance sheet;revenue balance sheet
年度资料 annual data
年度总(决算)报告 general annual reports
年度最高小时(交通)量 highest annual hourly volume;maximum annual hourly volume
年短缺量 annual short
年堆存量 annual storage
年吨 metric(al)ton per year
年额 annual amount
年发电量 annual energy output;annual energy production;annual power generation;annual power output;annual power production;capacity of electric(al) production
年伐量 annual cut;annual yield
年伐面积 annual cutting
年泛滥湿地 toich
年飞机起落架次 annual aircraft movements
年废水排放量 annual waste water discharge
年费 yearly payment
年费用 annual charges;annual cost;annual expenses
年费用函数 annual cost function
年分 fraction of the year
年峰荷 annual peak load
年幅度 annual range
年负荷 annual load
年负荷历时曲线 annual load duration curve
年负荷率 yearly load factor
年负荷曲线 annual load curve;yearly loaded curve
年负荷曲线图 annual load diagram
年负荷系数 annual load factor
年负荷因数 annual load factor;yearly load factor
年负荷因子 yearly load factor
年负载率 annual load factor;yearly load factor
年负载曲线 annual load curve;yearly loaded curve
年负载因数 annual load factor
年富余量 annual residual
年工作小时数 yearly working hours
年固定费用 annual fixed charges;annual fixes cost
年固定支出 annual fixed charges;annual fixes cost
年过程 annual march
年耗尽率 rate of annual depletion
年耗亏减率 rate of annual depletion
年耗水率 rate of annual depletion
年河流入流 annual river inflow
年荷载变化 yearly load variation
年荷载率 yearly load factor
年荷载曲线 yearly load curve
年洪峰系列 annual flood peak series
年洪峰序列 annual flood peak series
年洪水量 annual flood;annual output
年洪水序列法 <洪水频率的> annual-flood-series method
年还本付息 annual debt service
年环 annual ring(of timber)
年环境容量 annual environmental capacity
年会 annual conference;annual convention;annual meeting;annual session;annual symposium;annual workshop
年货运量 annual cargo tonnage;annual tonnage(capacity)
年货运周转量 annual cargo turnover
年货运总量 total annual goods transport
年积累环境承载能力 annual cumulative environmental carrying capacity
年积累环境承载容量 annual cumulative environmental carrying capacity
年极大(值) annual maximum
年极小(值) annual minimum
年计成本 annual cost
年纪 eon
年际变化 interannual change;interannual variation;variation between years
年际变率 interannual variability
年际冲淤变化 interannual variation of scour and fill
年际关系 interannual correlation
年际气压差 year-to-year pressure difference
年际相关 interannual correlation
年剂量 annual dose
年剂量率 annual dose-rate
年鉴 almanac yearbook;annual;year book
年降水变化率 variability of annual precipitation;variability of annual rainfall
年降水量 annual precipitation;annual rainfall;yearly precipitation;yearly rainfall
年降水量因素 annual precipitation factor;annual rainfall factor
年降水入渗深度 infiltration depth of annual precipitation
年降水入渗系数 infiltration coefficient of annual precipitation
年降雪量 annual snowfall
年降雪线 annual snowline
年降雨变化率 variability of annual precipitation;variability of annual rainfall
年降雨量 annual precipitation;annual rainfall;yearly rainfall;yearly precipitation
年降雨量因素 annual precipitation factor;annual rainfall factor
年交通量 annual traffic
年交通量变化 variation in yearly traffic
年交通顺位图 <以图或表按递减次序排列,显示全年每小时的交通量> yearly traffic pattern
年较差 annual amplitude;annual range
年界法 age-boundary method
年金 annuity;benefit;pension;reprise
年金保险 annuity;endowment assurance
年金表 annuity table
年金的复利值 compound value of an annuity
年金的现值 present value of an annuity
年金递减土地估价法 declining
年金法 annuity method
年金复利系数 compounding factor for per annum
年金基金 annuity funds
年金契约 annuity agreement
年金授予 pension vesting
年金现值 present value of annuity;present worth of an annuity
年金现值系数 present worth of an annuity factor
年金享受权 annuity
年金信托账 annuity trust account
年金折换 commutation of annuity
年金折旧法 compound interest method;depreciation-annuity method;depreciation-equal-annual payment method
年金终值 final value of annuity
年进尺 year footage
年进度 annual advance
年进口量 annual import
年经营费 annual operating charges;annual operating cost;yearly operating cost
年净堆积量 <冰川的> net annual accumulation
年净利 annual net profit
年净输沙率 annual net transport rate
年净消融量 <冰川的> net annual ablation
年净效(收)益 annual net benefit
年净增长 net annual growth
年净值 net annual value
年径流分配 annual distribution of runoff
年径流量 annual flow;annual runoff;yearly runoff
年径流率 annual ratio of runoff
年径流深度 annual runoff depth
年久而产生的柔和色彩 patina
年久失修的 beat up
年距平 annual anomaly
年均日照百分率 yearly man sunshine percentage
年均增长量 average annual growth
年均值 annual mean value
年开采量 annual mining output
年开采下降速度 mining downing velocity of year
年刊 annals;annual;year book
年客运总量 total annual person travel
年亏耗量 <地下水> annual depletion(rate)
年老衰退 age involution
年老退休 superannuation
年累积气温 yearly accumulated air temperature
年累积增长率 cumulative annual rate of growth
年历 almanac;year book
年利率 annual interest rate;rate of interest per annual
年利润 annual profit
年利息率 annual percentage rate
年利用率 annual utilization rate
年利用系数 annual capacity factor
年利用小时 utilization hours in a year
年量大洪水量 annual flood

N

年龄比 age ratio
年龄变化 ag(e)ing change
年龄标记 age indicator
年龄标志 age mark
年龄测定 age dating; age determination
年龄测定方法 method of age determination
年龄测量方法种类 the kinds of measuring geologic(al) ages
年龄单位 the usual units of ages
年龄等级 age class
年龄发生率 age incidence
年龄范围 age-bracket
年龄分布函数类型 type of age distribution function
年龄分布图 age distribution
年龄分配函数 age distribution function
年龄分组 age cohort
年龄构成 age composition; age structure
年龄和资历条件 age and seniority condition
年龄换算系数 age-conversion factor
年龄级 <动物出生年> age class; year class
年龄级周转 turnover
年龄鉴别 age determination
年龄鉴定 age assessment
年龄校正基准 age-corrected basis
年龄校正死亡率 age-adjusted death rate; age-adjusted mortality rate
年龄结构 age structure
年龄金字塔(图) age pyramid; human pyramid
年龄理论 age theory
年龄频率分布图 histogram of age frequency distribution
年龄普查 age census
年龄谱 age spectrum
年龄期 age period
年龄群 age-group
年龄群组 age cohort
年龄生长 age growth
年龄省图 map of age provinces
年龄死亡率 age-specific death rate
年龄特征 age characteristic
年龄统计角锥状图 age pyramid
年龄系列 <土壤的> chronosequence
年龄系列表 age-table
年龄相近的人们 age grade
年龄效率 age-effectiveness
年龄性别出生率 age-specific birth rate
年龄性别构成 age-sex composition
年龄直方图 age pyramid
年龄锥体 age pyramid
年龄组 age-group
年龄组成 age composition
年流量 annual discharge; annual flow
年流量变动率 coefficient of variation of annual discharge
年流量率 annual ratio of runoff
年旅客吞吐量 annual passenger movements
年率 annual rate; per annum rate; rate per annum
年率法 annual rate method
年轮 <树木的> annual growth; annual ring; growth layer; ring layer; secondary ring; growth ring; annual growth ring; annual zone; seasonal ring; tree ring; yearly ring; year ring
年轮层 annual growth layer; annual layer; yearly layer of wood
年轮界 growth ring boundary
年轮宽度 <树木> width of annual ring
年轮裂 ring shake
年轮密度 annual ring density
年轮气候学 dendroclimatology; tree ring climatology
年轮水文学 dendrohydrology
年轮学 dendrochronology
年轮状水系 annular drainage
年落淤量 annual sediment deposition
年毛收入 gross revenue earning
年末法 end of year method
年内变化 variation within a year
年内不均衡性 annual inequality
年内分配 annual distribution
年内减少数 decrease during the fiscal year
年泥沙淤积量 annual accumulation of sediment; annual sediment deposition
年逆转 annual turnover
年年 annually; year after year; year by year; yearly
年年变化 year-to-year variation
年年不断 year in(and) year out
年年结果树 annual bearer
年排出量 annual output
年排沙量 annual sediment discharge
年排水量 annual drainage
年平均 annual mean; average annual; average year; mean annual
年平均产量 average annual output
年平均潮汐周期 annual mean tidal cycle
年平均沉积量 annual average sediment yield
年平均发电量 average annual energy output; average annual output
年平均发震率 average annual rate of earthquake occurrence
年平均费用 annual average cost
年平均负荷(容)量 average yearly loading capacity; mean yearly loading; mean yearly loading capacity
年平均负荷因素 annual load factor
年平均工作量 average annual working capacity
年平均海平面 annual mean sea level; yearly mean sea-level; yearly sea-level
年平均含沙量 average annual sediment concentration; yearly mean sediment concentration
年平均洪水流量 average annual flood; mean annual flood
年平均降水量 average annual precipitation; mean annual precipitation; average annual rainfall
年平均降雨量 mean annual rainfall
年平均交通量 annual average daily traffic
年平均较差 mean annual range
年平均径流量 average annual flow; yearly mean runoff; average annual runoff(volume); mean annual runoff
年平均历时曲线 average annual duration curve
年平均量 annual average; annual mean
年平均流量 annual average flow; average annual discharge; average annual flow; average yearly flow; mean annual discharge; mean yearly discharge
年平均浓度 annual average concentration; mean annual concentration
年平均排水量 annual mean(water) discharge
年平均气候温度 mean annual climatic temperature
年平均气温 mean annual temperature; yearly mean air temperature
年平均日车流量 annual mean daily flow
年平均日交通量 annual average daily traffic; annual average daily traffic volume; annual mean daily flow; average annual daily volume
年平均设备利用率 average annual capacity factor
年平均生产能力 annual average productive capacity
年平均湿度 mean annual humidity
年平均疏伐量 average annual stand depletion
年平均输沙量 annual mean sediment discharge; average annual sediment discharge; yearly mean sediment discharge
年平均水流量 annual mean(water) discharge
年平均水位 annual mean water level; mean annual water level; mean yearly water level
年平均温差 annual mean temperature difference
年平均温度 annual mean temperature; mean annual temperature
年平均淤积量 annual average sediment yield
年平均雨量 average annual rainfall
年平均运行时间 average annual use hours
年平均增长率 average annual growth rate; mean annual growth rate
年平均蒸发量 mean annual evapo(u)ration
年平均值 annual mean
年平均总磷浓度 average yearly concentration of total phosphorus
年平均最大洪峰 mean annual maximum flood
年平均最大洪水量 average annual flood
年期望损失率 annual expected loss ratio
年起伏 yearly fluctuation
年气候 year-climate
年气温变幅 difference of annual air temperature variation
年侵蚀量 annual erosion
年青的 juvenile; young
年青地台 young platform
年青地台阶段 young platform stage
年青裂谷 young rift
年青泥炭 young peat
年青山地 young mountain
年青褶皱带 young folding zone
年轻断层崖 young fault scarp
年轻恒星 young star
年轻水混入模型 mixed models of modern water
年趋势 annual trend
年全距【天】annual range
年热能消耗量 annual heat consumption
年人均国民生产总值 annual per capita gross national product
年日照时数 yearly sunshine time
年闰余【天】annual epact
年散发量 transpiration depth
年沙暴日数 days of yearly sandstorm
年摄入量限值 annular intake limit; annular limit of intake
年升降变化 yearly fluctuation
年生产能力 annual capacity; yearly capacity; annual throughput capacity
年生长层 annual growth layer; annual layer; yearly layer of wood
年生长量 annual growth; annual increment; annual production; year increment
年生长率 annual growth rate
年释放能量 annual energy displacement
年收成 annual crop
年收获量 annual crop; annual cut; annual harvest; annual yield
年收获率 annual yield; annular yield
年收入 annual income
年收益 annual earnings; annual profit; annual revenue; purchase
年收益函数 annual-benefit function
年收益增长率 rate of growth of benefits
年输出功率 annual power output
年输沙量 annual load; annual sediment discharge; annual sediment transport; annual volume of carried silt; yearly sediment discharge; yearly sediment transport
年数总和折旧法 sum-of-the-year digits depreciation method
年衰减深度 annual damping depth
年霜带 annual frost zone
年水量平衡 annual hydrologic balance
年水量指数 index of wetness
年水土流失 annual soil loss
年水文平衡 annual hydrologic balance; yearly hydrologic balance
年税额 annual tax
年死亡率 annual mortality rate
年岁差 annual precession
年损 annual loss
年调节 annual regulation; yearly regulation
年调节库容 annual storage; seasonal storage
年调节(库容)电站 annual storage plant
年调节量 annual storage
年调节水库 annual artificial lake; annual balancing reservoir; annual impounding reservoir; annual regulating reservoir; annual-storage reservoir
年调节水库电厂 annual storage plant
年调节水库电站 annual-storage station
年通过能力 annual throughput capacity; annual tonnage capacity; annual traffic ability; annual traffic capacity
年头 the beginning of a year
年投资回收额 annual payment
年土壤流失量 annual soil loss
年土壤流失率 annual erosion rate
年推进度 annual advance
年推进量 annual advance
年退水量 annual depletion rate; depletion rate of a year
年吞吐量 annual throughput; annual tonnage
年危险性 annual risk
年维持费用 annual maintenance cost
年维护费用 annual maintenance cost
年维修成本 annual maintenance cost
年维修费用 annual maintenance cost
年位移量 year displacement
年温差 annual range of temperature
年温度变化 annual temperature variation
年温度差 annual temperature range; year temperature difference
年污染负荷 yearly pollution load
年息 annual interest; annum running; interest per annum
年系列 annual series
年限 defined in years

年限总额折旧法 sum-of-the-year digits depreciation method
年限总和法 sum-of-the-year digits
年相关 yearly correlation
年消耗量 annual consumption
年消退率 annual depletion rate
年效率 annual efficiency;yearly efficiency
年薪 annual earnings;annual pay
年型 model year
年修 annual overhaul
年序堆积层 diachronous;time transgressive
年蓄能量 annual storage;yearly storage
年蓄水量 annual storage;yearly storage
年雪线 annual snowline
年循环 annual cycle
年循环能源系统 annual cycle energy system
年窑 Nian ware
年翼管 inner finned tube
年引水量 annual abstraction volume; annual volume of water abstraction
年营业额 annual sale volume;annual turnover
年营运天数【港】annual operation days(of port)
年用量 annual consumption
年优良品种 the best varieties for
年有效黏[粘]度 annual effective viscosity
年有效温度 annual effective temperature
年幼的 junior
年淤积量 annual accumulation of sediment;annual sediment deposition; annual sediment yield; annual siltation
年淤积率 annual rate of deposition
年淤量 annual accumulation of sediment
年雨量 annual precipitation; annual rainfall
年雨量指数 index of wetness
年月日次序 chronological order
年运输成本 annual transport cost
年运输量 annual traffic
年运输能力 yearly transportation capacity
年运行报告 annual operating report
年运行费 annual operating charges; annual operating cost
年运转小时 annual operating hours
年增长量 annual increase;annual increment;year increment
年增长率 annual growth rate;annual increase rate;annual percent of increase;annual rates of increases
年增长速度 annual growth rate
年增长系数 annual improvement coefficient; annual improvement factor
年债务偿还常数 annual debt constant
年涨落 yearly fluctuation
年折旧 annual depreciation
年折旧费用 annual depreciation cost
年折旧回收 annual depreciation reserve
年折旧率 annual depreciation rate; yearly depreciation
年折算费用 annular reduced cost
年振幅 annual range
年蒸发量 annual evapo(u)ration; yearly evapo(u)ration discharge
年正常径流 annual normal runoff
年正常径流量 annual normal flow
年支出 annual charges
年支付 annual payment
年值 annual value
年值法 <经济分析中的一种方法> annual cost method; annual worth method
年中人口 mid-year population
年终 year-end
年终报告 annual report;year-end report
年终股利 year-end dividend
年终股息 year-end dividend
年终汇总 year-end summarization
年终奖金 December bonus;year-end bonus
年终结算 annual closing
年终结余 annual balance; year-end balance
年终结账 year-end closing
年终决算表 annual balance sheet
年终决算书 annual balance sheet
年终库存 year-end stock
年终审计 year-end audit
年终生产能力 year-end capacity
年终调整 year-end adjustment
年终账单 yearly account
年终资产负债表 year-end balance sheet
年钟 year clock
年周风 anniversary wind
年周期 annual cycle;annual period
年周期潮 annual tide
年周转(额) annual turnover
年周转量 annual turnover
年装卸能力 annual handling capacity
年资 seniority
年资工资制 seniority order wage system;wage-by-age system
年自行 annual proper motion
年综合折旧率 annual compositive depreciation rate
年总产量 gross annual output;gross annual production; total annual yield;total yearly output
年总产水量 revenue-producing water
年总成本 annual capital cost
年总堆积量 <冰川的> gross annual accumulation
年总辐射量 yearly mean sunshine percentage
年总收入 gross annual income;gross revenue earning
年总输沙率 annual gross transport rate
年总消融量 <冰川的> gross annual ablation
年租 annual rental
年最大地震 annual maximum earthquake
年最大负荷 annual peak load
年最大洪水量 maximum annual flood
年最大洪水流量 annual maximum flood flow; maximum annual flood flow
年最大径流 annual maximum runoff
年最大流量 annual flood; annual flood flow;annual maximum flow; maximum annual flow; maximum yearly peak
年最大容许剂量当量 annual maximum permissible dose equivalent
年最大事件 annular largest event
年最大水位 annual maximum level; annual maximum stage
年最大系列 annual maximum series
年最大小时交通量 annual maximum hourly traffic (volume); maximum annual hourly traffic volume;maximum annual hourly volume
年最大(值) annual maximum;yearly maximum
年最高额 annual maximum
年最高负载 yearly maximum load
年最高量 annual maximum
年最高小时交通量 highest annual hourly traffic volume
年最冷月份 annual coldest month
年最热月份 annual hottest month
年最小系列 annual minimum series
年最小(值) annual minimum; yearly minimum

黏

黏[粘]板 haft-plate;plywood

黏[粘]板岩 adhesive slate;clay slate; slate;slate clay
黏[粘]闭度 puddlability
黏[粘]闭土 puddled land
黏[粘]闭土壤 puddled soil
黏[粘]闭现象 puddling phenomenon
黏[粘]闭性 puddlability
黏[粘]表层 mazaedium
黏[粘]补剂 sticker
黏[粘]层 track coat
黏[粘]层成球法 coating process
黏[粘]插图纸条 guards
黏[粘]冲 sticking
黏[粘]虫 armyworm; oriental army worm
黏[粘]稠的 gummy appearance;ropy; sticky;viscous
黏[粘]稠的沥青 mineral rubber
黏[粘]稠地沥青 <不含矿粉,美国> asphalt cement
黏[粘]稠度 sliminess;viscosity
黏[粘]稠灰浆 heavy mortar
黏[粘]稠进给料 sticky feed
黏[粘]稠沥青 penetration-grade asphalt; penetration-grade bitumen <英>
黏[粘]稠密封剂 non-sag sealant
黏[粘]稠泥炭沼泽 ropy peat-bog
黏[粘]稠性 mucosity; stickiness; viscidity
黏[粘]稠亚麻子油 bodied linseed oil
黏[粘]稠液 viscous fluid
黏[粘]稠液状的 semi-liquid
黏[粘]稠质的 heavy-bodied
黏[粘]刺槐 clammy locust
黏[粘]带机 belt building machine
黏[粘]单 allonge
黏[粘]的 glutenous;gummy;mucose; sticky
黏[粘]底板钢锭 stool sticker
黏[粘]点 liquid limit;sticky point
黏[粘]点试验仪 sticky point tester
黏[粘]度 viscosity;viscosity grade
黏[粘]度保持性 <滑油的> viscosity retention;retention of viscosity
黏[粘]度杯 cup visco(si)meter;viscosity cup
黏[粘]度杯法 efflux cup method
黏[粘]度比 ratio of viscosities;viscosity ratio
黏[粘]度比重常数 viscosity-gravity constant
黏[粘]度比重图 viscosity-gravity chart
黏[粘]度变化范围 spread of viscosity;range of viscosity
黏[粘]度变化斜率 viscosity slope
黏[粘]度标准油 viscosity standard oil
黏[粘]度表 visco(si)meter
黏[粘]度测定 viscosity test
黏[粘]度测定法 visco(si)metry
黏[粘]度测定分析 visco(si)metric analysis
黏[粘]度测定计 visco(si)meter
黏[粘]度测定术 visco(si)metry
黏[粘]度测定数据 viscometric data
黏[粘]度测定学 viscometry
黏[粘]度测量 viscosity measurement
黏[粘]度测量法 visco(si)metry
黏[粘]度掺和线图 viscosity blending chart
黏[粘]度常数 viscosity constant
黏[粘]度稠节剂 viscosity regulator
黏[粘]度粗估仪 viscoscope
黏[粘]度单位 viscosity unit
黏[粘]度倒数 reciprocal viscosity
黏[粘]度滴定 visco(si)metric titration
黏[粘]度定律 viscosity law
黏[粘]度动力系数 kinematic(al) coefficient of viscosity
黏[粘]度法 visco(si)metry
黏[粘]度反常性 viscosity abnormality
黏[粘]度方程 viscosity equation
黏[粘]度分析器 viscosity analyser[analyzer]
黏[粘]度改进剂 viscosity modifier
黏[粘]度估算 estimating of viscosity
黏[粘]度管 viscosity tube
黏[粘]度函数 viscosity function
黏[粘]度换算 viscosity conversion
黏[粘]度换算表 viscosity conversion table
黏[粘]度极 viscosity pole
黏[粘]度极高度 viscosity pole height
黏[粘]度极线 viscosity pole line
黏[粘]度计 viscometer;viscosity meter;adhesive meter;caplastometer; cohesiometer;fluidimeter;fluid(o)-meter
黏[粘]度计比率 visco(si)meter proportions
黏[粘]度计标定 calibration of viscosity
黏[粘]度计不同直径毛细管组 series of viscosimeter tips
黏[粘]度计尖头 visco(si)meter tip
黏[粘]度计式真空计 viscometer ga(u)ge
黏[粘]度记录控制器 viscosity recording controller
黏[粘]度记录器 viscosity recorder
黏[粘]度监视器 viscosity monitor
黏[粘]度减低 viscosity breaking
黏[粘]度降低 viscosity loss;viscosity reducing
黏[粘]度降落 fall of viscosity
黏[粘]度校正 viscosity correction
黏[粘]度结果报表 report table of viscosity result
黏[粘]度警报记录器 viscosity alarm recorder
黏[粘]度控制 viscosity control
黏[粘]度控制剂 viscosity control agent
黏[粘]度控制器 viscosity controller
黏[粘]度流出时间 flow time of viscosity
黏[粘]度/密度比 viscosity-density ratio
黏[粘]度平均分子量 viscosity average molecular weight
黏[粘]度平均值 viscosity average
黏[粘]度倾点 viscosity pour point
黏[粘]度曲线 <通常指黏[粘]温曲线> viscograph;viscosity curve
黏[粘]度曲线图 viscosity diagram
黏[粘]度收率曲线 viscosity-yield curve
黏[粘]度数 viscosity number
黏[粘]度探测器 viscosity sensor
黏[粘]度梯度 viscosity gradient
黏[粘]度调节 viscosity control
黏[粘]度调节剂 viscosity-controlling agent;viscosity modifier
黏[粘]度调整器 viscosity adjuster
黏[粘]度温度关系特性 viscosity-tem-

perature characteristic
黏[粘]度-温度曲线 viscosity-temperature curve
黏[粘]度温度曲线图 viscosity-temperature chart
黏[粘]度温度系数 viscosity-temperature coefficient
黏[粘]度温度值 viscosity-temperature number
黏[粘]度稳定剂 viscosity stabiliser [stabilizer]
黏[粘]度稳定性 viscosity stability
黏[粘]度系数 coefficient of viscosity; viscosity coefficient; viscosity factor
黏[粘]度效应 viscosity effect
黏[粘]度学说 viscosity theory
黏[粘]度压力计 viscosity manometer
黏[粘]度压力系数 pressure coefficient of viscosity
黏[粘]度因数 viscosity factor
黏[粘]度因素 viscosity factor
黏[粘]度增加速度 bodying speed
黏[粘]度值 viscosity value; viscosity number
黏[粘]度指示控制器 viscosity indicating controller
黏[粘]度指示器 viscoscope; viscosity indicator
黏[粘]度指数 viscosity index; viscosity index figure
黏[粘]度指数改进剂 viscosity index improver
黏[粘]度指数改善剂 viscosity index improver
黏[粘]度指数扩展 viscosity index extension
黏[粘]度指数调合值 viscosity index blending value
黏[粘]度指数线圈 viscosity index chart
黏[粘]度指数延伸 viscosity index extension
黏[粘]度指数延伸值 viscosity index extention
黏[粘]度指数组分 viscosity index constituents
黏[粘]度重度常数 viscosity-gravity constant
黏[粘]度重力数 viscosity-gravity number
黏[粘]度转换 viscosity conversion
黏[粘]度自动调节器 viscosimat
黏[粘]度自记仪 viscosity recorder
黏[粘]分生子团 pionnate
黏[粘]粉土沉积平原 adobe flat
黏[粘]缝带 joint tape
黏[粘]附 adhere; adherence; adherency; bonding; clog seize; cohesion; conglutinate; seizing; stick(ing)
黏[粘]附保护膜 adhesion-preventing film
黏[粘]附层 adhesion layer; adhesive coating
黏[粘]附尘埃 stuck on dust; stuck-on particles
黏[粘]附尘埃的玻璃 adhered glass
黏[粘]附的 adherent; adhesive
黏[粘]附的化合物 anchored compound
黏[粘]附度 adhesiveness
黏[粘]附机理 mechanism of adhesion
黏[粘]附计 adherometer
黏[粘]附剂 adhesion agent; adhesion promoting agent
黏[粘]附件 adherend
黏[粘]附抗剪强度 shearing adhesion
黏[粘]附颗粒 attached particle
黏[粘]附力 adhesion(quality); adhesion strength; adhesion stress; adhesive bond; adhesive force; adhesive power; adhesive strength; adhesive ability
黏[粘]附力计 adhesion meter
黏[粘]附料 adhesion agent
黏[粘]附率 adhesion ratio
黏[粘]附面 surface of adherence
黏[粘]附面积 area of adhesion
黏[粘]附膜 adhesive film
黏[粘]附能 adhesional energy
黏[粘]附能力 adhesive capacity; adhesive power; adhesivity
黏[粘]附破坏 adhesion failure
黏[粘]附气泡 attached bubble
黏[粘]附牵引 adhesion traction
黏[粘]附强度 adhesion strength; adhesive strength
黏[粘]附强度表 adhesion meter
黏[粘]附强度增强剂 adhesion promoting agent
黏[粘]附乳液 adhesive emulsion
黏[粘]附伤痕 seizing mark
黏[粘]附式过滤器 viscous impingement type filter
黏[粘]附式空气净化设备 viscous type air cleaner
黏[粘]附试验 adherence test(ing)
黏[粘]附水 adhesion water; adhesive water
黏[粘]附水分 adhesive moisture
黏[粘]附特性 adhesion characteristic; adhesion property
黏[粘]附体 adherend
黏[粘]附体破坏 adherend failure
黏[粘]附物 adherend; bur; conglutination
黏[粘]附系数 adhesion coefficient; adhesion factor; adhesive coefficient; adhesive factor; factor of adhesion; sticking coefficient
黏[粘]附现象 adhesion phenomenon
黏[粘]附型陶瓷面砖 adhesion-type ceramic veneer
黏[粘]附型陶瓷饰面板 adhesion-type ceramic veneer
黏[粘]附性 adhesion; adhesive capacity; adhesiveness; adhesive property
黏[粘]附性能 adhesion property; blocking property; bonding performance; bonding property
黏[粘]附性试验 adhesiveness test
黏[粘]附选矿法 coherent separation; surface-adhesion separation
黏[粘]附溢流 adherent nappe
黏[粘]附溢流水舌 adherent nappe
黏[粘]附在叶片上 adhering to blade
黏[粘]附张力 adhesion tension; adhesive tension
黏[粘]附状态 coherent condition
黏[粘]附作用 adhesive effect
黏[粘]钙土 calcisol
黏[粘]垢 fouling; slime
黏[粘]固 adhere
黏[粘]固板 sticking board
黏[粘]固粉 cement
黏[粘]固粉玻璃板 cement glass plate
黏[粘]固粉充填器 cement plugger
黏[粘]固粉液 cement liquid
黏[粘]管 collophore
黏[粘]辊 roll banding; roll coating
黏[粘]合 cementing; agglutination; bind(ing); bonding; conglutinate; conglutination; gluing; pasting; seizing; slur; stick together
黏[粘]合斑 adhesion spot
黏[粘]合玻璃 cleaved glass
黏[粘]合布 bonded fabric
黏[粘]合材料 binding material; bonding material; matrix [复 matrixes/matrices]
黏[粘]合侧面 joint edge
黏[粘]合层 adhesive coat(ing); binding course; binding layer; bond; bond coat; bonding course; bonding layer; bond line
黏[粘]合层厚度 bondline thickness
黏[粘]合衬 bonded fabric
黏[粘]合疵点 adhesive failure
黏[粘]合带 adhesive strip; adhesive tape; adhesive zone; binding tape
黏[粘]合的 adhesive; binding; bonded; conglutinant
黏[粘]合底漆 binding primer
黏[粘]合叠层木 adhesive-laminated wood
黏[粘]合度 adhesiveness
黏[粘]合构件 bonded member
黏[粘]合混凝土 bond concrete
黏[粘]合火山角砾岩 binding volcanic breccia
黏[粘]合集块岩 agglutinate; binding agglomerate
黏[粘]合剂 adhesion agent; adhesive; adhesive material; binder; binding agent; binding material; binding medium; blinding material; bonding adhesive; bonding agent; bonding medium; bunch; cemedin(e); cementing agent; parent matrix; setting glue
黏[粘]合剂层 adhesive phase; bond line
黏[粘]合剂分散剂 binder dispersion
黏[粘]合剂含量 binder content; binder portion
黏[粘]合剂含量测定 binder content determination
黏[粘]合剂含量确定 binder portion determination
黏[粘]合剂回收 binder recovery
黏[粘]合剂混合料 binder aggregate mix(ture); binder mix(ture)
黏[粘]合剂计量 binder measuring; binder metering
黏[粘]合剂计量泵 binder metering pump
黏[粘]合剂沥青加热装置 binder asphalt heating installation
黏[粘]合剂膜 binder film
黏[粘]合剂黏[粘]固期 conditioning period
黏[粘]合剂配量泵 binder measuring pump
黏[粘]合剂配量器 binder batcher
黏[粘]合剂配料 binder batching; binder proportioning
黏[粘]合剂配料泵 binder batching pump; binder proportioning pump
黏[粘]合剂配料计量器 binder batcher
黏[粘]合剂喷枪 binder gun
黏[粘]合剂乳化液 binder emulsion
黏[粘]合剂撒布 binder application; binder spreading
黏[粘]合剂撒布器 adhesive sprayer
黏[粘]合剂桶 binder barrel
黏[粘]合剂涂层 adhesive coating
黏[粘]合剂悬浊液 binder suspension
黏[粘]合剂用喷枪 glue gun
黏[粘]合剂注入 binder injection
黏[粘]合剂注入装置 binder injecting device
黏[粘]合夹具 bonding fixture
黏[粘]合胶 adhesive-glue
黏[粘]合接头 adhint; bonded joint
黏[粘]合结构 adhesive system
黏[粘]合介质 bonding medium
黏[粘]合界面 adhesive bond
黏[粘]合金属板 bonded metal
黏[粘]合抗力 bond resistance
黏[粘]合力 adhesion; adhesive power; binding energy; binding force; binding power; bonding power
黏[粘]合力试验机 bond testing machine
黏[粘]合力损失 loss of adhesion
黏[粘]合料 binder; binding material
黏[粘]合料按比例混合用的泵 binder measuring pump
黏[粘]合料按比例配料 binder batching
黏[粘]合料按计量配料 binder metering
黏[粘]合料薄膜 binder film
黏[粘]合料成分 binder content
黏[粘]合料的扩展 spreading of binders
黏[粘]合料的撒布 spreading of binders
黏[粘]合料的涂刮 spreading of binders
黏[粘]合料灌注器 binder injecting device
黏[粘]合料混合 binder mix(ture)
黏[粘]合料计量泵 binder metering pump
黏[粘]合料加热装置 binder heating installation
黏[粘]合料颗粒 binder particle
黏[粘]合料配量泵 binder measuring pump
黏[粘]合料配料用的泵 binder metering pump
黏[粘]合料熔锅 binder cooker; binder heater
黏[粘]合料散装贮藏 binder bulk storage
黏[粘]合锚固件 adhesive anchor; adhesive tie
黏[粘]合面积 bond area
黏[粘]合膜 adhesive film
黏[粘]合耐久性 endurance of bond
黏[粘]合能 adhesional energy; bonding energy; energy of adhesion
黏[粘]合能力 bounding capacity
黏[粘]合黏[粘]土 bond clay
黏[粘]合疲劳 endurance of bond
黏[粘]合期 joint ag(e)ing time
黏[粘]合漆包线 bonded wire
黏[粘]合强度 adhesion strength; adhesive strength; binding strength; bonding action; bonding strength
黏[粘]合强度试验机 adhesive bond strength tester
黏[粘]合强度试验器 bond tester
黏[粘]合墙粉 bonding plaster
黏[粘]合清漆 adhesive varnish; binding varnish
黏[粘]合热 adhesion heat
黏[粘]合失效 application failure
黏[粘]合湿润 adhesional wetting
黏[粘]合时间 closure time
黏[粘]合试验 adhesive test; gumming test
黏[粘]合树脂 bonding resin
黏[粘]合素 conglutinin
黏[粘]合碳化钨 cemented tungsten carbide
黏[粘]合体 adherend
黏[粘]合体系 bonding system
黏[粘]合土 bond soil; bond clay <砌土坯墙用>
黏[粘]合温度 bonding temperature
黏[粘]合物 agglutination
黏[粘]合物层 adhering zone
黏[粘]合物镜 cemented objective
黏[粘]合物质 adhesion substance
黏[粘]合纤维 bonded fiber[fibre]
黏[粘]合纤维网 bonded joint
黏[粘]合线 cementing line
黏[粘]合橡胶 adhesive rubber

黏[粘]合橡胶(软)垫 bonded rubber cushion(ing)
黏[粘]合小带 adhesion tape
黏[粘]合效应 binding effect
黏[粘]合型炸药 slurry explosive
黏[粘]合性 adhesiveness;adhesivity;bonding characteristic
黏[粘]合性能 blocking property;bond behavio(u)r;bonding performance;bonding property
黏[粘]合岩 coherent rock
黏[粘]合液 liquid adhesive;slurry
黏[粘]合应变计 bonded wire strain ga(u)ge
黏[粘]合应变片 bonded wire strain ga(u)ge
黏[粘]合毡 bonded mat
黏[粘]合张力 adhesion tension;adhesive tension
黏[粘]合织物 bonded fiber fabric;bonded mat
黏[粘]合指数 bondability index
黏[粘]合质 cement in
黏[粘]合作用 bonding action
黏[粘]痕 sticker mark;sticking mark
黏[粘]花装饰 prunt decoration
黏[粘]滑成因 stick-slip genesis of earthquake
黏[粘]滑地震 stick-slip earthquake
黏[粘]滑断层 stick-slip fault
黏[粘]滑断裂 stick-slip fault
黏[粘]滑方式 stick-slip behavio(u)r
黏[粘]滑机制 stick-slip mechanism
黏[粘]滑力学 stick-slip mechanics
黏[粘]滑速率 rate of stick-slip
黏[粘]滑性 mucosity
黏[粘]滑运动 stick slip
黏[粘]化 sliming
黏[粘]化层 argillic horizon
黏[粘]化旱成土 argid
黏[粘]浆 mucilage
黏[粘]浆处理法 viscous processing
黏[粘]胶 adhesive paste;mucilage;sticking;viscose glue
黏[粘]胶泵 viscose pump
黏[粘]胶标签 stick-on labels
黏[粘]胶薄膜 viscose film
黏[粘]胶布 adhesive tape
黏[粘]胶层 cementing layer
黏[粘]胶长丝 viscose filament yarn
黏[粘]胶带 adhesive tape;cellulose tape;Scotch tape
黏[粘]胶短纤维 rayon staple;viscose staple fiber[fibre]
黏[粘]胶法 viscose process
黏[粘]胶废水 viscose wastewater
黏[粘]胶过滤性 filterability of viscose
黏[粘]胶海绵 viscose sponge
黏[粘]胶合剂 adhesive composition;adhesive compound
黏[粘]胶剂 adhesive cement;cementing medium
黏[粘]胶接合 adhesive joint
黏[粘]胶帘布 viscose cord fabric
黏[粘]胶窿 viscose cellar
黏[粘]胶模型 plasticine model
黏[粘]胶黏[粘]合剂 viscose binder
黏[粘]胶清除剂 adhesive remover
黏[粘]胶人造丝 viscose artificial thread;viscose rayon yarn;viscose silk
黏[粘]胶人造纤维 viscose rayon fibre
黏[粘]胶溶液 viscose solution
黏[粘]胶纱 viscose yarn
黏[粘]胶熟成 viscose ripening
黏[粘]胶熟成桶 viscose ripening tank
黏[粘]胶丝 viscose rayon;vissilk
黏[粘]胶丝束 rayon tow;viscose rayon tow
黏[粘]胶桶 viscose tank
黏[粘]胶纤维 rayon;viscose;viscose fiber[fibre]
黏[粘]胶纤维基碳纤维 rayon-based carbon fiber
黏[粘]胶纤维素 prezenta;viscose cellulose
黏[粘]胶研磨机 viscose grinder
黏[粘]胶液 viscose
黏[粘]胶杂质分离器 viscose trap
黏[粘]胶纸 viscose paper
黏[粘]胶质 baregin;glairin;pectin
黏[粘]接 adhesive bond;bond;cementation;luting;splice;sticking(up)
黏[粘]接材料 adhesive;bonding material
黏[粘]接层 bonding course;bonding layer
黏[粘]接磁带 splicing tape
黏[粘]接点 sticking point
黏[粘]接点焊 weld-bonding
黏[粘]接法 adhesive jointing
黏[粘]接缝 adhesive-bonded joint
黏[粘]接合缝 adhesive joint
黏[粘]接剂 binder;bonder;bonding adhesive;bonding cement
黏[粘]接胶泥 bonding cement
黏[粘]接胶粘剂 bonding adhesive
黏[粘]接接合 adhesive-bonded joint;adhesive joint;bonded joint
黏[粘]接接头 adhesive-bonded joint;adhesive joint;bonded joint
黏[粘]接结合面 adhesive-bonded joint;adhesive joint;bonded joint
黏[粘]接金属 bonding metal
黏[粘]接块 stick block
黏[粘]接能力 adhesive capacity
黏[粘]接强度 adhesive strength;bond strength
黏[粘]接乳剂 adhesive emulsion
黏[粘]接失效 adhesive failure;bonding failure;fail in bond
黏[粘]接试验 adhesive test
黏[粘]接松脱 contact failure
黏[粘]接温度 sticking temperature
黏[粘]接系统 system of jointing
黏[粘]接型锚杆 full bonded bolt
黏[粘]接型陶瓷板 adhesion-type ceramic veneer
黏[粘]接修理 adhesive bonding repair
黏[粘]接应力 bond stress
黏[粘]接用泥 mending clay
黏[粘]结 binding;bonding;cake;caking;caking cementation;cementing;sticking together
黏[粘]结变形砖 crozzle
黏[粘]结不良 imperfect adhesion;imperfect bonding;poor bond
黏[粘]结材料 adhesive material;binding material;bonding material;cementing material;cohesive material;cementitious material
黏[粘]结层 binder coat;binder course;binding course;bonding course;bonding layer;tack coat
黏[粘]结层混合料 binder course mix(ture)
黏[粘]结层之间黏[粘]结力 ply adhesion
黏[粘]结长度 bond length
黏[粘]结常数 adhesion constant
黏[粘]结成球 ball up
黏[粘]结处理 adhesive treatment;bonding treatment
黏[粘]结促进剂 bonding additive
黏[粘]结带 cementing zone
黏[粘]结的 astringent;cementatory;cementitious;coking;glomerate
黏[粘]结的砾石屋顶 bonded gravel roof
黏[粘]结的石砾 cemented gravel
黏[粘]结的氧化皮疤 cinder patch
黏[粘]结底层 adhesive primer
黏[粘]结点 spotting
黏[粘]结丁砖 bonder brick
黏[粘]结度 cementitious value
黏[粘]结度试验 cohesiometer test
黏[粘]结度值 cohesiometer value
黏[粘]结法 bonding;bonding process;cement fixing method;mull technique
黏[粘]结范围 bonding area
黏[粘]结防水 bonded water-proofing
黏[粘]结分隔材料 bond breaker
黏[粘]结钢筋 bonded reinforcement
黏[粘]结钢筋束 bonded tendon
黏[粘]结钢丝束 bonded tendon
黏[粘]结膏 cement plaster
黏[粘]结工艺技术 bonding technique
黏[粘]结构件 bonded member
黏[粘]结固结法 glue fixing method
黏[粘]结管材 cemented tube
黏[粘]结焊剂 bond flux
黏[粘]结耗损 loss of bond
黏[粘]结合金 bonding alloy
黏[粘]结后张(法) bonded post-tensioning
黏[粘]结滑动 bond slippage
黏[粘]结滑移 bond slip
黏[粘]结灰 caked ash
黏[粘]结灰膏 bond plaster
黏[粘]结灰浆 adhesive mortar
黏[粘]结灰泥 adhesive mortar;bond(ing) plaster
黏[粘]结灰岩 bound stone
黏[粘]结灰质砾石 limy gravel
黏[粘]结混合料 bonding compound
黏[粘]结混合物 bonding compound
黏[粘]结混凝土 bonding concrete
黏[粘]结机理 bond mechanism
黏[粘]结剂 adhesion additive;agglomerant;agglomerator;binder;binding agent;binding element;binding material;binding admixture;bonding agent;bonding cement;bonding compound;bonding medium;bridging agent;cementation material;cementing agent;cementing medium;cementitious agent;coherent material;cohesive material;coupling agent;grouting agent;linking agent;plastering agent
黏[粘]结剂的悬浮 medium suspension
黏[粘]结剂法 glued method
黏[粘]结剂固化炉 binder curing oven
黏[粘]结剂管 binder pipe
黏[粘]结剂搅拌机 glue mixer
黏[粘]结剂金属 cementing metal
黏[粘]结剂黏[粘]合 bond between binder and aggregate
黏[粘]结剂喷涂 binder spray
黏[粘]结剂施加段 binder application section
黏[粘]结剂施加装置 binder depositor
黏[粘]结剂摊铺机 glue spreader
黏[粘]结剂体系 mastic system
黏[粘]结剂填充混合料 binder-filler mix(ture)
黏[粘]结加入物 binding admixture
黏[粘]结浆糊 bonding paste
黏[粘]结胶 assembly glue;sealing mastic glue;viscose
黏[粘]结胶带 sealing tape
黏[粘]结胶泥 bonding putty
黏[粘]结结合 adhesive bonding;adhesive joint
黏[粘]结介质 bonding medium;cementing agent;cementing medium
黏[粘]结金属 binding metal
黏[粘]结块 lumped mass
黏[粘]结矿物纤维芯板 felted mineral core
黏[粘]结力 adhesion;adhesive power;application bond;binding power;bond force;bonding capacity;bonding force;cementing bond;cementing capacity;cementing power;cement value;cohesive force;cohesive force;cohesiveness;cohesion;bond(ing) stress < 混凝土与钢筋的 >
黏[粘]结力长度 < 图解法中的 > bond length
黏[粘]结力破坏 bond failure
黏[粘]结力强的材料 tenacious material
黏[粘]结(沥青)混合料 bonding compound
黏[粘]结沥青煤 byerite
黏[粘]结砾石 cement gravel
黏[粘]结粒 binder
黏[粘]结良好的 well-bonded
黏[粘]结料 binder;binding agent;binding medium;bonding;cementing material;cementitious material
黏[粘]结料按比例配料 binder proportioning
黏[粘]结料按比例配料用的泵 binder proportioning pump
黏[粘]结料的按重量配料器 binder weighing batcher
黏[粘]结料抗滑阻力 liquid friction
黏[粘]结料配料用泵 binder pump
黏[粘]结料喷射泵 binder spray(ing) bar
黏[粘]结料喷射机(器) binder spray(ing) machine
黏[粘]结料温度计 binder thermometer
黏[粘]结料贮存装置 binder storage installation
黏[粘]结路面加铺层 bonded pavement overlay
黏[粘]结锚固 bond anchorage
黏[粘]结煤 agglomerating coal;baking coal;binding coal
黏[粘]结面 adhesive surface;bond face
黏[粘]结面积 bond(ing) area
黏[粘]结膜 binding film
黏[粘]结木块 glue block
黏[粘]结耐久性 bond durability
黏[粘]结能 binding energy;binding power
黏[粘]结能力 adhesive capacity;bonding power;caking capacity;caking power;cementation power;cementing power;cementitiousness;power of cementation
黏[粘]结劈裂破坏 bond splitting failure
黏[粘]结疲劳 < 混凝土 > fatigue of bond
黏[粘]结破坏 bond-breaking;bond(ing) failure
黏[粘]结气泡 bond blister
黏[粘]结强度 adhesion strength;adhesive strength;binding strength;bonding strength;cementing strength;cohesive strength;interface strength;application bond < 两个黏[粘]贴面之间的 >
黏[粘]结强度试验 adherence test(ing)
黏[粘]结柔性材料 coherent flexible material
黏[粘]结乳剂 bonding emulsion
黏[粘]结软木砖 agglomerated cork brick;baked cork brick
黏[粘]结砂浆 adhesive mortar;bond(ing) mortar;bonding grout < 混凝

N

土罩面用的 >
黏[粘]结失效 cohesive failure
黏[粘]结石膏 bond(ing) plaster
黏[粘]结试验 adhesion test; bonded test; bond(ing) test
黏[粘]结试验器 bond tester
黏[粘]结树脂 bonding resin
黏[粘]结水 adhesive water; pellicular water
黏[粘]结水磨石 bonded terrazzo
黏[粘]结水泥 < 仓库内的 > stock-house set
黏[粘]结松脱 contact failure
黏[粘]结损坏 bonding failure
黏[粘]结碳酸岩 bound stone
黏[粘]结条痕 sticker break
黏[粘]结涂层 bond(ing) coat; glue coating
黏[粘]结外部粉刷 bonding stucco
黏[粘]结外加剂 adhesion additive
黏[粘]结温度 bonding temperature
黏[粘]结物 attachment
黏[粘]结物质 bonding compound
黏[粘]结系数 coefficient of cohesion
黏[粘]结相 bonding phase
黏[粘]结型施工缝 bonded type construction joint
黏[粘]结性 binding quality; bonding characteristic; bounding; caking property; cementing property; cementing quality; cohesiveness; cok(e) ability; cementation; coherence
黏[粘]结性的 coherent
黏[粘]结性底漆 bonding primer
黏[粘]结性煤 binding coal; caking coal
黏[粘]结性能 blocking property; bond(ing) performance; bond(ing) property
黏[粘]结性石膏灰浆 bonding plaster
黏[粘]结性试验 cementation test; cohesiveness test
黏[粘]结性土壤 coherent soil; cohesive soil
黏[粘]结性岩层 sticky formation
黏[粘]结性指数 cementation index; cohesive index; sticking index
黏[粘]结性质 bonding property
黏[粘]结悬浮液 bonding suspension
黏[粘]结悬胶液 adhesive suspension
黏[粘]结岩 bindstone
黏[粘]结研究 bond study
黏[粘]结窑皮 incrustation
黏[粘]结应力 adhesion stress; bond stress
黏[粘]结用泥浆 jointing slip
黏[粘]结用石膏灰 gypsum bond(ing) plaster
黏[粘]结油灰 bonding putty
黏[粘]结预应力钢(丝)束 bonded prestressed tendon
黏[粘]结预应力筋 bonded prestressed tendon
黏[粘]结原理 bond principle
黏[粘]结增进剂 adhesion promoter
黏[粘]结毡 bonded mat
黏[粘]结值 adhesion value; cementing value; cementitious value
黏[粘]结指数 agglutinating index; caking index; cementation index; cementing index
黏[粘]结质量 bond(ing) quality; cementing quality
黏[粘]结住的 bonded-on
黏[粘]结装配 adhesive-assembly
黏[粘]结阻力 cohesive resistance
黏[粘]结作用 adhesion; adhesive effect; binding function; bonding; cementation; cementing action; cohesive action
黏[粘]精 mucosin
黏[粘]聚 adhesive aggregation
黏[粘]聚的 cohesive
黏[粘]聚混凝土 cohesive concrete
黏[粘]聚抗力 cohesional resistance; cohesive resistance
黏[粘]聚力 cohesion; cohesive force; cohesive strength; force of cohesion
黏[粘]聚力计 cohesiometer
黏[粘]聚力截距 cohesion intercept
黏[粘]聚力矩 cohesion moment
黏[粘]聚力试验 cohesiometer test
黏[粘]聚力仪 cohesiometer
黏[粘]聚力值 cohesiometer value
黏[粘]聚氯乙烯薄膜钢板 < 商品名 > Artbond
黏[粘]聚强度 cohesive resistance
黏[粘]聚试验 cohesion test
黏[粘]聚水 cohesive water
黏[粘]聚系数 coefficient of cohesion; cohesional coefficient
黏[粘]聚性 coherence; cohesion; cohesiveness
黏[粘]聚性材料 cohesive material
黏[粘]聚性物质 cohesive matter
黏[粘]聚阻力 cohesional resistance; cohesive resistance
黏[粘]聚作用 cohesive action
黏[粘]菌 myxomycete; slime fungi; slime mold
黏[粘]菌生长 slime growth
黏[粘]壳孢属 < 拉 > Gloeodes
黏[粘]块 gob
黏[粘]蜡 adhesive wax; stick wax; sticky wax
黏[粘]力 adhesive force
黏[粘]力计 cohesiometer
黏[粘]沥青 viscous bitumen
黏[粘]粒 clay grain; clay particle
黏[粘]粒部分 clay fraction
黏[粘]粒成分 clay fraction
黏[粘]粒葱皮黏[粘]结 clay onion-skin bond
黏[粘]粒复合体 clay complex
黏[粘]粒灌浆 clay grouting
黏[粘]粒含量 clay content
黏[粘]粒胶膜 clay coating
黏[粘]粒胶体 clay colloid
黏[粘]粒粒级 clay fraction
黏[粘]粒粒组 clay fraction
黏[粘]粒酸 clay acid
黏[粘]粒絮凝 clay flocculation
黏[粘]粒移动 clay movement
黏[粘]粒组 clay fraction; clay grain grade
黏[粘]连 accretion; adhesion; blocking; sticking
黏[粘]连钢筋束 bonded tendon
黏[粘]连间断层 bond breaker course
黏[粘]连现象 adhesion phenomenon
黏[粘]连性 adhesive
黏[粘]料 binder; cementing material; sticking(of a charge)
黏[粘]流 coherent flow
黏[粘]流态 plastic state; viscous state
黏[粘]流体 rheid
黏[粘]氯乙烯薄膜钢板 arbond
黏[粘]霉菌类 slime mold
黏[粘]膜 adhesive in film form; mo(u)ld sticking; mucous membrane; slime; slimes; sticking
黏[粘]膜脱落 sloughing
黏[粘]磨 high finish
黏[粘]泥 slime; viscous mud
黏[粘]泥板岩 sticky sale
黏[粘]泥剥离 slimes dispersant
黏[粘]泥层 slime
黏[粘]泥堆放场 slime bank site
黏[粘]泥分离器 slime separator
黏[粘]泥浆 clay slurry
黏[粘]泥浆盾构 slime shield
黏[粘]泥菌 slime fungi
黏[粘]泥涂抹 slime coating
黏[粘]泥形成细菌 slime-forming bacteria
黏[粘]泥悬浮质 clay suspension; slime suspension
黏[粘]泥指数 slime index
黏[粘]腻污物 gunk
黏[粘]鸟胶 birdlime
黏[粘]凝 gum set
黏[粘]盘孢属 < 拉 > Myxosporella
黏[粘]磐 clay pan
黏[粘]磐黑钙土 clay pan chernozem
黏[粘]磐土 clay loam; clay pan soil; planosol
黏[粘]品装填机 viscous fillers
黏[粘]球 slimeball
黏[粘]壤土 clay-loam soil
黏[粘]韧冰碛 gumbotil
黏[粘]韧点 sticky point
黏[粘]韧结持度 sticky consistency
黏[粘]溶液 viscous solution
黏[粘]熔相 viscous melt-phase
黏[粘]熔岩流 coulee
黏[粘]乳 slimy milk
黏[粘]润 adhesion wetting
黏[粘]润作用 adhesional wetting
黏[粘]砂 < 又称黏[粘]沙 > gummy sand; adhering sand; burning-on of sand; burnt-on sand; fat sand; sand fusion; sand-gritting; sand penetration
黏[粘]砂刚纸 floor sanding paper
黏[粘]砂沥青油毡 sanded bitumen felt
黏[粘]砂土 clayey sand
黏[粘]砂岩 < 玛姆砂岩 > malm rock
黏[粘]上 affixture; stick-on
黏[粘]湿 adhesional wetting
黏[粘]湿表面分馏塔 wetted surface column
黏[粘]石粉刷 kellstone
黏[粘]土石 leck
黏[粘]刷性 ropiness
黏[粘]丝 haircuts; viscin thread
黏[粘]丝法 spun-bonded process
黏[粘]丝体 viscoid
黏[粘]塑流 viscoplastic flow
黏[粘]塑弹性 viscoplastoelastic
黏[粘]塑体的 viscoplastic
黏[粘]塑性 viscoplasticity
黏[粘]塑性变形 viscoplastic deformation
黏[粘]塑性材料 viscoplastic material
黏[粘]塑性的 viscoplastic
黏[粘]塑性固体 viscoplastic solid
黏[粘]塑性理论 viscoplasticity theory
黏[粘]塑性流动 plastic-viscous flow; viscoplastic flow
黏[粘]塑性流体 viscoplastic fluid
黏[粘]塑性能 viscoplastic property
黏[粘]塑性土 viscoplastic soil
黏[粘]塑性土壤 plastic cohesive soil
黏[粘]塑主应变差 viscoplastic principal strain difference
黏[粘]酸 glactaric acid; mucic acid
黏[粘]损 pick-up
黏[粘]弹固体 viscoelastic solid
黏[粘]弹计 viscoelastometer
黏[粘]弹流体 Maxwell fluid; Maxwell's fluid
黏[粘]弹区 viscoelastic region
黏[粘]弹塑性的 viscoelastoplastic
黏[粘]弹塑性介质 viscoelastic-plastic medium
黏[粘]弹塑性土 viscoelastoplastic soil
黏[粘]弹特性 viscoelastic behavio(u)r; viscoelastic property; viscous-elastic behavio(u)r
黏[粘]弹体 viscoelastic body
黏[粘]弹响应 viscoelastic response
黏[粘]弹性 viscoelasticity
黏[粘]弹性板 viscoelastic plate
黏[粘]弹性半空间 viscoelastic half-space
黏[粘]弹性变形 viscoelastic deformation
黏[粘]弹性材料 viscoelastic material; viscous-elastic material
黏[粘]弹性层 viscoelastic layer
黏[粘]弹性层状体系 viscoelastic layer system; viscoelastic multi-layer system
黏[粘]弹性的 viscoelastic
黏[粘]弹性地面 viscoelastic ground
黏[粘]弹性对应原理 viscoelastic correspondence principle
黏[粘]弹性反应 viscoelastic response
黏[粘]弹性反应分析 viscoelastic response analysis
黏[粘]弹性分散体 viscoelastic dispersion
黏[粘]弹性分析 viscoelastic analysis
黏[粘]弹性横向效应 viscoelastic cross effect
黏[粘]弹性结构 viscoelastic structure
黏[粘]弹性介质 viscoelastic medium
黏[粘]弹性聚合物 viscoelastic polymer
黏[粘]弹性理论 viscoelastic theory
黏[粘]弹性流动 viscoelastic flow
黏[粘]弹性流体 viscoelastic fluid
黏[粘]弹性流体系统 viscoelastic fluid system
黏[粘]弹性模量 viscoelastic modulus
黏[粘]弹性模型 viscoelastic model
黏[粘]弹性能 viscoelastic behavio(u)r; viscoelastic property
黏[粘]弹性蠕变 viscoelastic creep
黏[粘]弹性设计 viscoelastic design
黏[粘]弹性数分析 viscoelastic numerical analysis
黏[粘]弹性特征 viscoelastic behavio(u)r
黏[粘]弹性特征时间 natural time of viscoelasticity
黏[粘]弹性体 viscoelastic body
黏[粘]弹性体系 viscoelastic system
黏[粘]弹性土 viscoelastic soil; viscoscope soil
黏[粘]弹性物质 viscoelastic material; viscoelastic substance
黏[粘]弹性系数 viscoelastic modulus
黏[粘]弹性性态 viscoelastic behavio(u)r
黏[粘]弹性性质 viscoelastic behavio(u)r; viscoelastic property
黏[粘]弹性性状 viscoelastic behavio(u)r
黏[粘]弹性岩石圈 viscoelastic lithosphere
黏[粘]弹性应变 viscoelastic strain
黏[粘]弹性应力 viscoelastic stress; viscoscope stress
黏[粘]弹性指标 criterion[复 criteria] of viscoelasticity
黏[粘]弹性阻尼 viscoelastic damping
黏[粘]弹性阻尼器 viscoelastic damper
黏[粘]贴 paste; pasting
黏[粘]贴壁画画布 marouflage
黏[粘]贴标签 sticking label
黏[粘]贴电阻丝应变计 bonded wire strain ga(u)ge
黏[粘]贴电阻丝应变片 bonded wire strain ga(u)ge
黏[粘]贴浮雕玻璃图样 springing
黏[粘]贴钢板 fastened steel sheet
黏[粘]贴胶合板工作 veneering work

黏[粘]贴金箔用的透明涂料 clairecolle;clearcole
黏[粘]贴墙面砖的快凝灰泥 casting plaster
黏[粘]贴软木沥青 cork setting asphalt
黏[粘]贴石膏 gypsum bond(ing) plaster
黏[粘]贴式应变计 bended strain ga(u)ge;bonded type strain ga(u)ge
黏[粘]贴水磨石 bonded terrazzo
黏[粘]贴线脚 stuck mo(u)ld
黏[粘]贴线条 stuck mo(u)ld
黏[粘]贴型陶瓷面板 adhesion-type ceramic veneer
黏[粘]贴应变片的装置 ga(u)ge installation equipment
黏[粘]贴用灰泥 application mortar
黏[粘]贴纸板 fastened cardboard
黏[粘]贴纸袋 pasted-end sack
黏[粘]土 clay; argillaceous earth; binder soil; mota; potter's clay; sticky soil; tenacious clay; weald-clay
黏[粘]土坝 puddle dyke[dike]
黏[粘]土百分率 clay fraction
黏[粘]土板 clay slab
黏[粘]土板条 clay lathing
黏[粘]土板岩 slate clay;clay slate
黏[粘]土拌和机 blunder; clay mill; clay mixer; kneader; malaxator; mud mixer;pugmill
黏[粘]土拌和器 blunder; clay mill; clay mixer; kneader; malaxator; mud mixer;pugmill;loam mixer
黏[粘]土包壳 clay coating
黏[粘]土崩解 slaking of clay
黏[粘]土扁平体 clay lenses
黏[粘]土表面 clay surface
黏[粘]土薄层 clay laminae
黏[粘]土薄片 argillaceous gall
黏[粘]土部分 clay fraction
黏[粘]土采掘 clay winning;getting of clay
黏[粘]土采矿 clay mining
黏[粘]土层 clay bed; clay(ey) stratum[复 strata]; clay layer; clay stratification;feu;layer of clay
黏[粘]土层隧道 tunnel in clayey stratum
黏[粘]土层钻头 mud bit
黏[粘]土掺料 clay mortar mix
黏[粘]土铲 clay digger;clay spade
黏[粘]土沉淀 clay deposit
黏[粘]土沉积 clay deposit
黏[粘]土沉积平原 clay outwash plain
黏[粘]土沉降 clay settlement
黏[粘]土衬层 clay lining
黏[粘]土衬砌 clay lining
黏[粘]土成分 clay fraction
黏[粘]土冲击钻头 <钻机的> clay cutter
黏[粘]土稠度 consistency of clay
黏[粘]土储藏 clay storage
黏[粘]土处理收率 clay yield
黏[粘]土纯化 wash clay
黏[粘]土催化剂 clay catalyst
黏[粘]土带 clay belt
黏[粘]土袋穴法 <盾构施工中采用的> clay-pocketing method
黏[粘]土单元 clay element
黏[粘]土捣塑 pugging
黏[粘]土、稻草和卵石混合的土墙 cob wall
黏[粘]土的 argilliferous;bolar
黏[粘]土的成团作用 aggregation of clay
黏[粘]土的电化学固结 electrochemical solidification of clay
黏[粘]土的聚集作用 aggregation of clay
黏[粘]土的开采 mining of clay
黏[粘]土的可塑性 plasticity of clay
黏[粘]土的灵敏度 sensitivity of clay
黏[粘]土的挖掘 mining of clay
黏[粘]土堤 clay bank; puddle dyke[dike]
黏[粘]土底板 clay bottom
黏[粘]土地 loamy soil
黏[粘]土地板方砖 quarry tile
黏[粘]土地基 argillaceous bottom; clay bed;clay foundation
黏[粘]土地面 clay floor;earth floor
黏[粘]土地砖 clay tile
黏[粘]土电化固结 electrohardening of clay
黏[粘]土叠层 clay laminae
黏[粘]土堵塞 clay plug
黏[粘]土断层 clay fault;clay slip
黏[粘]土煅烧 clay burning
黏[粘]土煅烧曲线 clay burning curve
黏[粘]土煅烧窑 clay burner; clay burning unit
黏[粘]土煅制 clay burning
黏[粘]土堆场 heap of clay
黏[粘]土堆棕色 clay bank
黏[粘]土防渗槽 cut-off trench of puddle clay
黏[粘]土防渗层 clay membrane;earth membrane
黏[粘]土防渗面层 <堆石坝的> earth blanket
黏[粘]土防渗墙 clay diaphragm;clay membrane;earth membrane
黏[粘]土防渗芯墙 clay core
黏[粘]土防水层 clay seal
黏[粘]土肥料 clay fertilizer
黏[粘]土分散剂 dispersion of clay
黏[粘]土粉 dry mud
黏[粘]土粉末 clay powder
黏[粘]土粉砂质砂岩 clay silty sandstone
黏[粘]土封层 clay blanket
黏[粘]土封孔 clay sealing hole
黏[粘]土封口 bod
黏[粘]土封水层 clay seal
黏[粘]土浮雕花样 sprig
黏[粘]土附着力 adhesion of clay
黏[粘]土覆盖层 clay overburden;clod top;clay blanket <压气隧道>
黏[粘]土盖层 clay coating
黏[粘]土干燥机 clay dryer[drier]
黏[粘]土坩埚 clay crucible;clay pot
黏[粘]土缸砖铺路面 clay block paving
黏[粘]土膏 clay puddle
黏[粘]土膏衬层 clay puddle lining
黏[粘]土膏浆填封 clay puddle seal
黏[粘]土膏浆填塞 clay packing
黏[粘]土膏芯墙 clay puddle core wall
黏[粘]土镐 clay pick
黏[粘]土隔层 clay barrier
黏[粘]土隔墙 clay diaphragm
黏[粘]土隔水层 clayey aquitard
黏[粘]土隔水桩 pug pile
黏[粘]土工业 clay industry
黏[粘]土骨架 clay skeleton
黏[粘]土骨料混凝土 clay aggregate concrete
黏[粘]土固沙 fixed sand by bedding clay
黏[粘]土管 clay pipe
黏[粘]土管瓦 clay drainage tile
黏[粘]土灌浆 clay grouting; mud grouting;mud injection
黏[粘]土灌浆法 injection process of clay
黏[粘]土滚筒 clay roll
黏[粘]土含量 clay content
黏[粘]土夯实 cob construction;puddling
黏[粘]土夯筑围堰 stock ramming
黏[粘]土和砂混合机 clay and sand mixer
黏[粘]土和水搅拌器 blunder
黏[粘]土烘干 clay drying
黏[粘]土糊 clay slip
黏[粘]土护层 clay blanket
黏[粘]土滑动 clay slide;clay slip
黏[粘]土滑坡 clay slide;clay slip
黏[粘]土化 argillation; argillization; clayization
黏[粘]土化带 argillaceous zone
黏[粘]土化夹层 clay gouged intercalation
黏[粘]土荒漠 argillaceous desert;clay desert
黏[粘]土黄褐色 clay drab
黏[粘]土灰浆 clay mortar
黏[粘]土灰浆混合料 <磨细的> clay mortar mix(ture)
黏[粘]土回填 back-stuffing with clay; clay backfill
黏[粘]土混合物 <由黏[粘]土砂砾石混合碾压而成的> clay paddle
黏[粘]土混凝土 clay concrete;loam concrete;soil concrete
黏[粘]土混凝土铺盖(层) clay concrete blanket
黏[粘]土活性 activity of clay
黏[粘]土基 clay base
黏[粘]土基础 clay foundation
黏[粘]土基床 clay bed
黏[粘]土基耐火混凝土 fireclay grog castable
黏[粘]土基座 clay base
黏[粘]土集料混凝土 clay aggregate concrete
黏[粘]土剂 clay bond
黏[粘]土加工 clay getting;clay working;working of clay
黏[粘]土加工处理 clay processing
黏[粘]土加工机(械) clay-working machine
黏[粘]土夹层 clay band;clay gouge; clay parting;clay seam;clay course
黏[粘]土夹层地基 foundation with interbedded argillous soil
黏[粘]土夹缝 clay seam
黏[粘]土(夹)砾石 clay gravel
黏[粘]土夹心墙 clay-core wall of dam
黏[粘]土夹心式防波堤 clay core type embankment
黏[粘]土建筑砖 building brick
黏[粘]土浆 argillaceous mud; clay grout; clay puddle; clay slip; clay slurry;clay suspension;wash clay; puddle clay <含水率近于液限,用于防水涂层>
黏[粘]土浆制备厂 clay slurrying plant
黏[粘]土胶结料 clay binder;clay cement
黏[粘]土胶结物 clay binder;clay cement
黏[粘]土胶泥 <含水率在液限与塑限之间,用于防水涂层> clay puddle
黏[粘]土胶体活动性 colloidal activity of clay
黏[粘]土胶体溶液 clay solution
黏[粘]土胶性 colloid(al) properties of clay
黏[粘]土铰刀 clay cutter
黏[粘]土搅拌机 clay kneading machine
黏[粘]土结构 clay structure
黏[粘]土结构地面砖 structural clay floor tile
黏[粘]土结构空心砖 structural clay tile
黏[粘]土结构面砖 structural clay facing tile
黏[粘]土结合的硅砖 clay-bonded silica brick
黏[粘]土结合剂 clay binder
黏[粘]土结合浇注耐火材料 clay-bonded castable refractory
黏[粘]土结合料 clay binder
黏[粘]土结核 <岩石中的> stone gall
黏[粘]土截水墙 clay cut-off wall
黏[粘]土介质 clay medium
黏[粘]土精炼 clay refining
黏[粘]土净化器 clay cleaner;clay purifier
黏[粘]土菌 slime fungi
黏[粘]土开采 clay quarry;clay working
黏[粘]土开采场 clay quarry
黏[粘]土颗粒 argillaceous grain;argillaceous particle; clay grain; clay particle
黏[粘]土坑 clay pit
黏[粘]土空心砌块 double clay pot; hollow clay block
黏[粘]土空心砖 hollow clay brick; hollow clay tile; partition tile; structural clay tile
黏[粘]土空心砖过梁 hollow clay tile lintel
黏[粘]土空心砖填充物 clay tile filler
黏[粘]土块 ball clay;clay block;clay chunks;clay mass
黏[粘]土矿 clay pit
黏[粘]土矿床 clay deposit
黏[粘]土矿床上的废土与废石 callow
黏[粘]土矿石 clay ore
黏[粘]土矿物 <即胶体矿物> clay mineral;sialite
黏[粘]土矿物材料 clay mineral materials
黏[粘]土矿物成分分析 clay mineral composition analysis
黏[粘]土矿物分布 distribution of clay;distribution of clay mineral
黏[粘]土矿物含量 content of clay mineral
黏[粘]土矿物和土壤 clay minerals and soil
黏[粘]土矿物结构 clay mineral structure
黏[粘]土矿物颗粒 clay mineral particle
黏[粘]土矿物亲水的程度 degree of hydrogen-ion capacity
黏[粘]土矿物脱水 clay mineral dehydration
黏[粘]土矿物性质 clay mineral property
黏[粘]土矿物学 clay mineralogy
黏[粘]土矿物组合 clay mineral composition
黏[粘]土老化 ag(e)ing of clay
黏[粘]土类土 clayey soil
黏[粘]土砾石 clayey gravel
黏[粘]土粒 lutum
黏[粘]土粒度范围 clay range
黏[粘]土粒级 clay fraction
黏[粘]土粒径 clay size
黏[粘]土隆起 clay swelling
黏[粘]土垆坶 clay loam
黏[粘]土路 clay road
黏[粘]土路床 clay bed
黏[粘]土路面 clay surface
黏[粘]土螺旋钻 clay auger
黏[粘]土脉 clay vein;dirt slip
黏[粘]土脉壁 clay course
黏[粘]土锚碇 clay anchor
黏[粘]土霉菌 slime mold
黏[粘]土镁絮凝物 clay-magnesium floc

N

黏[粘]土密封 clay sealing
黏[粘]土密封墙 wax wall
黏[粘]土面砖 clay file
黏[粘]土模型 clay mo(u)lding
黏[粘]土模型制作者 model(l)er
黏[粘]土膜 clay film
黏[粘]土耐火材料 clay refractory product
黏[粘]土耐火砖 fire brick;refractory fireclay block
黏[粘]土泥 cob walling
黏[粘]土泥灰岩 argillaceous marl; clay marl
黏[粘]土泥浆 argillaceous mud;clay mortar;clay-water mix(ture);clay mud
黏[粘]土泥塞 botting
黏[粘]土泥岩石 clay mud rock
黏[粘]土黏[粘]合 clay bond
黏[粘]土黏[粘]合剂 ceramic bond
黏[粘]土黏[粘]合料 clay binder
黏[粘]土黏[粘]结料 ceramic bond
黏[粘]土耙对地压力 scarifier pressure
黏[粘]土盘 clay pan
黏[粘]土膨胀 clay swell(ing)
黏[粘]土坯 clay masonry unit;tapia
黏[粘]土片 clay gall
黏[粘]土片晶 clay platelet
黏[粘]土漂洗 wash clay
黏[粘]土平台 clay terrace
黏[粘]土平瓦 Broseley tile;flat clay roof(ing) tile
黏[粘]土平瓦屋顶 flat clay tile roof
黏[粘]土平原 aftout;clay plain
黏[粘]土破碎机 clay crusher;clay disintegrator
黏[粘]土铺盖 clay blanket;clay packing
黏[粘]土铺盖衬砌 day blanket lining
黏[粘]土气泡 clay bubble
黏[粘]土砌块 clay masonry unit
黏[粘]土器皿 clayey vessel
黏[粘]土强度各向同性 clay strength isotropy
黏[粘]土强度各向异性 clay strength anisotropy
黏[粘]土墙 clay diaphragm;clay wall
黏[粘]土切割 clay cutting
黏[粘]土切割机 clay cutter;clay-cutting machine
黏[粘]土切割器 clay cutter
黏[粘]土切碎机 clay-cutting machine
黏[粘]土切削器 clay cutter
黏[粘]土穹顶 clay cupola;clay dome
黏[粘]土球 clay ball
黏[粘]土球粒 clay gall
黏[粘]土取样器 clay sampler
黏[粘]土容器 clay vessel
黏[粘]土揉混槽 kneading trough
黏[粘]土蠕变 clay creep
黏[粘]土乳液 clay emulsion
黏[粘]土润滑剂 clay grease
黏[粘]土润滑脂 clay grease
黏[粘]土塞 clay plug
黏[粘]土塞料 clay tamping
黏[粘]土色 clayey colo(u)r
黏[粘]土砂浆 clay mortar;earth mortar;clay puddle
黏[粘]土砂浆混合物 clay mortar mix
黏[粘]土砂砾 clay-grit
黏[粘]土、砂、砾石混合料 clay-sand-gravel mix(ture)
黏[粘]土砂泥 loam
黏[粘]土砂泥捏和机 loam kneader;loam mill
黏[粘]土砂泥型 loam(sand)mo(u)ld
黏[粘]土砂岩 clayey sandstone;clay sandstone;malm rock;malmstone
黏[粘]土砂障 <又称黏[粘]土沙障> clay bank sand-break
黏[粘]土砂质粉砂岩 clay-sandy siltstone
黏[粘]土烧成 burning of clay
黏[粘]土烧结 vitrified bond
黏[粘]土烧结的砂轮 vitrified bond (grinding) wheel
黏[粘]土烧结方法 clay sintering process
黏[粘]土湿润时间指数 slaking value of clay
黏[粘]土石 claystone;leek
黏[粘]土石灰混合料 clay-lime mixture
黏[粘]土石墨坩埚 clay-graphite crucible
黏[粘]土石墨搪料 clay-graphite mixture
黏[粘]土时间固结关系 time-consolidation relation for clay
黏[粘]土熟料 calcined clay;chamot(te);grog
黏[粘]土栓 clay plug
黏[粘]土双曲瓦 clay pantile
黏[粘]土水化作用 clay hydration
黏[粘]土水泥 argillaceous cement
黏[粘]土水泥灌浆 clay-cement grouting;clay-cement injection
黏[粘]土水泥混合料 clay-cement mixture
黏[粘]土水泥混凝土 clay containing concrete
黏[粘]土水泥浆 clay-cement grout
黏[粘]土水泥砂浆 clay-cement mortar
黏[粘]土似的 argillaceous;argilliferous;argillious;clayey
黏[粘]土塑造的 fictile
黏[粘]土碎屑 shearclay
黏[粘]土陶粒 <黏[粘]土加一定数量的外加料,烧成球状体,表面光滑,经过破碎后,成为较细的级配集料> leca;expanded clay
黏[粘]土陶粒和陶砂 coarse and fine aggregate of expanded clay
黏[粘]土陶粒块 leca block
黏[粘]土陶粒轻集料 lightweight expanded clay aggregate
黏[粘]土淘洗 clay wash
黏[粘]土提纯器 clay purifier
黏[粘]土体 clay mass
黏[粘]土填封 <炮眼用> clay tamping
黏[粘]土填缝 clay sealing
黏[粘]土填料 <即级配混合料中的黏[粘]土成分> clay-filler;clay fill(ing)
黏[粘]土填塞 clay tamping
黏[粘]土填实的防渗隔离槽 <土坝> cut-off trench of paddled clay
黏[粘]土填筑的截水墙 clay-filled cut-off
黏[粘]土填筑体 clay fill(ing)
黏[粘]土条 <压力机压出的> clay lath;clay column
黏[粘]土调料 clay preparation
黏[粘]土铁絮凝物 clay-iron floc
黏[粘]土铁质岩 clay ironstone
黏[粘]土筒 clay cutter
黏[粘]土筒仓 clay silo
黏[粘]土透镜 lenticle of clay
黏[粘]土透镜体 clay lens;lens of clay
黏[粘]土涂层 clay coat(ing);clay film
黏[粘]土涂覆 clay coating
黏[粘]土涂料 coating clay
黏[粘]土涂面 clay-coated finish
黏[粘]土涂铺机 clay coating machine
黏[粘]土土壤 clay soil
黏[粘]土团(块) clay lump;argillaceous gall;clay ball
黏[粘]土挖掘铲斗 clay bucket
黏[粘]土瓦 clay shingle;clay tile;earthen tile
黏[粘]土瓦工厂 clay tile factory
黏[粘]土瓦管 clay tile
黏[粘]土瓦污水管 crock
黏[粘]土瓦屋顶 clay tile roof
黏[粘]土瓦屋面 clay tile roof cladding
黏[粘]土完全软化程度 fully softened strength of clay
黏[粘]土围堰 clay cofferdam;puddle cofferdam;puddle cofferdam weir
黏[粘]土帷幕 clay blanket
黏[粘]土稳定 clay stabilization
黏[粘]土稳定的 clay-stabilized
黏[粘]土稳定砂 clay-stabilized material
黏[粘]土窝 clay pocket
黏[粘]土污水管 clay sewer pipe
黏[粘]土屋脊瓦 clay book tile
黏[粘]土屋面瓦 clay roofing tile
黏[粘]土吸附 clay absorption
黏[粘]土洗涤剂 clay wash
黏[粘]土细工 clay work
黏[粘]土下卧层 bed of clay
黏[粘]土小球架 clay cross
黏[粘]土(小)团(球) clay key;clay pellet
黏[粘]土小洼地 gilgai
黏[粘]土斜坡 clay slope
黏[粘]土斜坡铺布器 clay slope revetment spreader
黏[粘]土斜墙 clay-filled inclined wall;clay sloping core
黏[粘]土斜墙土石坝 sloping core earth-rock dam
黏[粘]土斜心墙(土)坝 earth dam with inclined clay core
黏[粘]土屑 clay chip(ping)s
黏[粘]土芯墙 clay core wall;puddle core wall;puddle wall
黏[粘]土芯墙坝 earth dam with clay core
黏[粘]土芯墙堆石坝 rock-fill dam with clay core;earth core rockfill dam
黏[粘]土芯墙土石坝 clay core earth-rock dam
黏[粘]土形成 clay formation
黏[粘]土型 loam mo(u)ld
黏[粘]土型芯 loam core
黏[粘]土絮凝剂 flocculant of clay
黏[粘]土悬浮剂 deflocculant;deflocculating agent;deflocculation agent
黏[粘]土悬浮体 clay suspension
黏[粘]土悬浮液 clay suspended water;clay suspension;dispersion of clay
黏[粘]土穴 clay pocket
黏[粘]土压实 clay packing
黏[粘]土压重柴捆 clay weighed fascine
黏[粘]土压重梢捆 clay weighed fascine
黏[粘]土岩 argillaceous rock;bass;clayey rock;clay rock;claystone;pelite
黏[粘]土岩类型 clay rock type
黏[粘]土岩脉 clay dike[dyke]
黏[粘]土岩柱 clay pillar
黏[粘]土页岩 bury;clayey shale;clay shale;shaking shale
黏[粘]土页岩的冻胀 heaving shale
黏[粘]土硬层 clay pan
黏[粘]土硬磐 clay hardpan
黏[粘]土、油页岩或天然石块墙面 ashlar facing
黏[粘]土釉 earthen glaze
黏[粘]土与硅酸钙砌块铺面 clay and calcium silicate paving
黏[粘]土再活化 clay reactivation
黏[粘]土造浆率 yield of clay
黏[粘]土造型 loam mo(u)lding
黏[粘]土植物群落 spiladophytia
黏[粘]土止水 clay seal;water sealing with clay
黏[粘]土制备 clay preparation
黏[粘]土制备机 clay preparation machine
黏[粘]土制模 clay modeling
黏[粘]土制品 <烧制的> clay article;clayware
黏[粘]土制品融化及冷冻试验 clay product thawing and freezing test
黏[粘]土制品硬度试验 clay product hardness test
黏[粘]土制型法 loam casting
黏[粘]土质 heavy soil;douke
黏[粘]土质白云岩 clayey dolomite
黏[粘]土质板岩 clay slate
黏[粘]土质材料 argillaceous material;clay product
黏[粘]土质沉积物 clayey sediment
黏[粘]土质的 argillaceous;argillic;clayey;clayish;lutaceous
黏[粘]土质底土 clay bottom
黏[粘]土质粉砂 clayey silt
黏[粘]土质风化岩 clayey stone;claystone
黏[粘]土质灰岩 argillaceous limestone;clayey limestone
黏[粘]土质混合物 argillaceous ingredient
黏[粘]土质建材制品 heavy clay ware
黏[粘]土质结构 argillaceous texture
黏[粘]土质矿石洗矿 puddling
黏[粘]土质砾石 clayey gravel
黏[粘]土质垆坶 clayey loam
黏[粘]土质面具 clay mask
黏[粘]土质母岩 clay matrix
黏[粘]土质耐火材料 fireclay refractory
黏[粘]土质耐火浇注料 clay castable
黏[粘]土质耐火砖 schamot(te) brick
黏[粘]土质泥灰岩 clay(ey) marl
黏[粘]土质泥沙 clayey sediment
黏[粘]土质黏[粘]结剂 bonding clay
黏[粘]土质黏[粘]结料 clay bond
黏[粘]土质片岩 blue metal
黏[粘]土质砂(土) argillaceous sand;clayey sand
黏[粘]土质砂岩 argillaceous sandstone;dauk[dawk]
黏[粘]土质石灰石 argillaceous limestone;clayey limestone
黏[粘]土质陶器 clay earthenware
黏[粘]土质铁矿 argillaceous iron ore
黏[粘]土质铁矿石 argillaceous ironstone
黏[粘]土质土 argillaceous soil;clayey soil
黏[粘]土质物体 clay object;clay substance
黏[粘]土质物质 clay substance
黏[粘]土质岩 argillaceous rock
黏[粘]土质岩脉 clayey vein
黏[粘]土质岩石 clod;soil clod
黏[粘]土质页岩 argillaceous shale;clay shale;dauk[dawk];dunn bass;batt
黏[粘]土质淤泥 clayey mud
黏[粘]土质原料 argillaceous raw material
黏[粘]土质装料 clay batch
黏[粘]土中薄壁取土器 clay cutter
黏[粘]土中硫化铁团块 kidney stone
黏[粘]土柱 column of clay
黏[粘]土铸法 loam casting

黏[粘]土铸造(物)loam mo(u)ld
黏[粘]土砖 chamot(te);chamot(te) brick;clay brick;earthen brick;earthenware tile;fireclay brick;loam brick;structural clay tile
黏[粘]土砖板地面 clay slate ground
黏[粘]土砖层 undercloak
黏[粘]土砖承重横墙 weight-carrying clay brick cross-wall
黏[粘]土砖窗花格 clay brick tracery
黏[粘]土砖的包装 pack of clay brick
黏[粘]土砖地面 clay brick floor
黏[粘]土砖发券 clay brick wall arch
黏[粘]土砖粉尘 clay brick dust
黏[粘]土砖格子窗 clay brick grille
黏[粘]土砖隔墙 clay brick cross-wall
黏[粘]土砖基础 clay brick foundation
黏[粘]土砖集料 clay brick aggregate
黏[粘]土砖集料混凝土 clay brick aggregate concrete
黏[粘]土砖集料混凝土砌块 clay brick aggregate concrete block
黏[粘]土砖建大教堂 clay brick cathedral
黏[粘]土砖建汽车库 clay brick-built garage
黏[粘]土砖结构 clay brick construction
黏[粘]土砖梁 clay brick beam
黏[粘]土砖梁托 clay brick corbel
黏[粘]土砖坯 dobie
黏[粘]土砖铺地面 clay tile floor cover(ing)
黏[粘]土砖砌空芯墙 clay brick cavity wall
黏[粘]土砖砌轮式窗 clay brick wheel window
黏[粘]土砖砌隧道拱顶 clay brick tunnel vault
黏[粘]土砖砌台阶 clay brick step
黏[粘]土砖砌体 clay masonry
黏[粘]土砖砌体壁炉 clay brickwork fireplace
黏[粘]土砖砌筑 clay brick setting
黏[粘]土砖切割器 clay cutter
黏[粘]土砖竖井 clay brick shaft
黏[粘]土砖体 tile clay body
黏[粘]土砖外墙 clay brick exterior wall
黏[粘]土砖柱基 clay brick plinth
黏[粘]土状的 clayey;clayly;gumbo <美>;clay-like
黏[粘]土状细粉 clay-like fines
黏[粘]土状细粒料 <岩石断层或滑动面区> clay gouge
黏[粘]土钻头 clay auger;clay bit
黏[粘]脱下来 felting down
黏[粘]温关系 viscosity-temperature dependency
黏[粘]温关系图 viscosity-temperature graph
黏[粘]温模数 viscosity-temperature modulus
黏[粘]温曲线 viscosity-temperature chart
黏[粘]温系数 viscosity-temperature coefficient
黏[粘]温线图 viscosity-temperature chart
黏[粘]温性质计算尺 visco(sity-temperature) calculator
黏[粘]污 <电极的> pick-up
黏[粘]吸力 adhesive attraction
黏[粘]吸作用 adhesive attraction
黏[粘]锡 bottom tin
黏[粘]细菌 myxobacteria;slime bacteria;slime-forming bacteria
黏[粘]纤维素 muco-cellulose
黏[粘]限 sticky limit
黏[粘]性 adhesiveness;bond(ing) property;glutinosity;gummosity;mucosity;plasticity(of clay);stickness;viscidity;viscous property
黏[粘]性板岩 adhesive slate
黏[粘]性泵 visco pump
黏[粘]性比 viscosity ratio
黏[粘]性变形 viscous deformation;viscous yielding
黏[粘]性波动 viscosity fluctuation
黏[粘]性薄膜 adhesive in film form
黏[粘]性材料 cohesive material;viscous material
黏[粘]性产碱杆菌 Bacillus lactis viscosus
黏[粘]性沉淀 viscous precipitate
黏[粘]性沉积物 cohesive deposit;cohesive sediment
黏[粘]性冲击收尘器 <取尘样用> viscous impingement filter
黏[粘]性处理纸 release paper
黏[粘]性触击式过滤器 viscous impingement type filter
黏[粘]性磁滞 viscous hysteresis
黏[粘]性的 cohesive;gummy;limy;mucilaginous;pasty;pitchy;ropy;sticky;viscid;viscous
黏[粘]性底层 viscous sublayer
黏[粘]性发热 viscous heating
黏[粘]性反常 viscosity anomaly
黏[粘]性非弹性流体 viscoinelastic fluid
黏[粘]性分散 viscous dissipation
黏[粘]性粉尘 sticky dust
黏[粘]性高温计 viscosity pyrometer
黏[粘]性过程 viscose process
黏[粘]性过滤器 viscous filter
黏[粘]性耗散函数 viscous dissipation function
黏[粘]性河床 cohesive bed;cohesive riverbed
黏[粘]性河床底泥 cohesive bed sediment
黏[粘]性河底 cohesive river bottom
黏[粘]性黑土 karail
黏[粘]性回弹 viscous recoil
黏[粘]性计的牛顿流动方程 Newtonian flow equation for tackmeter
黏[粘]性计的塑性黏[粘]度方程 plastic viscosity equation on tackmeter
黏[粘]性价格 sticky price
黏[粘]性剪力 viscous shear
黏[粘]性减震器 viscous damper
黏[粘]性键固 viscous keying
黏[粘]性胶粒 clay-water micelle
黏[粘]性胶体 viscolloid
黏[粘]性介质 resisting medium
黏[粘]性空气过滤器 viscous air filter
黏[粘]性力 force due of viscosity;force due to viscosity;force of cohesion;viscosity force;viscous force
黏[粘]性料浆 viscous slurry
黏[粘]性料块粉碎机 cob mill
黏[粘]性流 shear flow;slip flow;stream of viscousness
黏[粘]性流变仪 visco corder
黏[粘]性流层 viscous flow region
黏[粘]性流动 viscous flow
黏[粘]性流动传质机理 material transfer mechanism
黏[粘]性流动模量 viscous flow modulus
黏[粘]性流动漏孔 viscous flow leak hole
黏[粘]性流渗漏孔 viscous flow leak
黏[粘]性流体 viscous fluid
黏[粘]性流体方程 Navier-Stokes equation
黏[粘]性流体力学 mechanics of viscous fluids
黏[粘]性流体系统 viscous fluid system
黏[粘]性流体运动方程 viscous equation of motion
黏[粘]性流状态 viscous regime of flow
黏[粘]性膜 adhesive film
黏[粘]性摩擦(力) viscous friction;fluid friction
黏[粘]性摩擦应力 shear stress;viscous friction stress
黏[粘]性末端 cohesive end;sticky end
黏[粘]性耐火材料 viscous refractory
黏[粘]性泥沙 cohesive deposit;cohesive material;cohesive sediment
黏[粘]性泥石流 structure type mud flow;viscous debris flow;viscous mud-flow;viscous mud-stone flow
黏[粘]性泥石流沉积 cohesive-debris-flow deposit
黏[粘]性黏[粘]土悬浮液 viscous clay suspension
黏[粘]性凝胶 viscogel
黏[粘]性破坏 viscous fracture
黏[粘]性牵伸系数 coefficient of viscous traction
黏[粘]性青铜 plastic bronze
黏[粘]性屈服 viscous yielding
黏[粘]性热 viscous heat
黏[粘]性热弹材料 viscous thermoelastic material
黏[粘]性溶液 plastic solution
黏[粘]性乳化 viscous emulsion
黏[粘]性砂浆 fat mortar
黏[粘]性试验 adhesive test
黏[粘]性衰减 viscous damping
黏[粘]性水泥 adhesive putty;sticky cement
黏[粘]性水泥浆液 viscous cement grout
黏[粘]性体 viscoid;viscous body
黏[粘]性体系的结构学子体系 structural subsystem of viscous system
黏[粘]性填土 cohesive backfill
黏[粘]性填土的主动土压力 active earth pressure of cohesive backfill
黏[粘]性土 cohesive soil;binder soil;clay-bearing soil;clayed ground;clay(ey) soil;heavy soil;heavy-textured soil
黏[粘]性土层 adhesive layer;cohesive bed;cohesive clay;cohesive stratum
黏[粘]性土稠度 consistency of cohesive soil
黏[粘]性土的贯入阻力 penetration resistance of cohesive soil
黏[粘]性土的活动度 activity of cohesive soil
黏[粘]性土的活动性等级 grade of activity of clay mineral
黏[粘]性土的抗剪强度 shear strength of cohesive soil
黏[粘]性土的可塑性与稠度 plasticity and consistency of cohesive soil
黏[粘]性土的灵敏度 sensitivity of cohesive soil
黏[粘]性土的膨胀性与崩解性 swelling-shrinkage and disintegration of cohesive soil
黏[粘]性土的土压力 earth pressure of cohesion soil
黏[粘]性土的稳定 stability of cohesive soil
黏[粘]性土河岸 cohesive bank
黏[粘]性土滑坡 cohesive soil landslide
黏[粘]性土开挖 excavation in cohesive soil
黏[粘]性土类 clay system
黏[粘]性土破裂 ravefing
黏[粘]性土壤 clay soil;cohesive soil;impervious soil
黏[粘]性土压缩层 cohesive compressible soil stratum
黏[粘]性土中打入式开口管桩摩阻力设计(λ)法 lambda method of pile design
黏[粘]性温度关系 viscosity-temperature relation
黏[粘]性涡流 viscous vortex
黏[粘]性物质 stickum
黏[粘]性系数 coefficient of viscosity;viscosity coefficient;viscosity factor
黏[粘]性限 cohesive limit;cohesive ultimate
黏[粘]性项 viscosity term
黏[粘]性效果 quasi-viscous effect
黏[粘]性效应 cohesive effect
黏[粘]性悬浮物 adhesive suspension
黏[粘]性岩石 gummy formation
黏[粘]性液流动 viscous fluid flow;viscous liquid flow
黏[粘]性液体 viscous liquid
黏[粘]性液体流 friction(al) flow
黏[粘]性液体试样 viscous liquid sample
黏[粘]性因素 viscosity factor
黏[粘]性应变 viscous strain
黏[粘]性油 viscous oil
黏[粘]性油状液体 viscous oily liquid
黏[粘]性原油废水 viscous crude oil wastewater
黏[粘]性渣 viscous slag
黏[粘]性炸弹 sticky bomb;sticky charge;sticky grenade
黏[粘]性指数 viscosity index
黏[粘]性粥状 viscous gruel state
黏[粘]性转数计 viscosity tach(e)ometer
黏[粘]性状态 viscous state
黏[粘]性阻力 viscosity resistance;viscous drag
黏[粘]性阻尼 viscous damping
黏[粘]性阻尼器 viscous damper;viscous type damper
黏[粘]性阻尼系数 coefficient of viscous damping;viscous damping coefficient
黏[粘]絮除泥剂 slimicide
黏[粘]雪 clog snow
黏[粘]压结合 squeezed joint
黏[粘]页岩 adhesive shale
黏[粘]液 mucago;mucosa;mucus;phlegma;slime;slime flux
黏[粘]液比重计 areopycnometer[areopyknometer]
黏[粘]因 <黏[粘]滞度单位> rein
黏[粘]蝇纸 flypaper
黏[粘]油罐车 viscous oil tank car
黏[粘]釉 bonded glaze;glaze sticking
黏[粘]在一起 sticking together
黏[粘]脏 set-off
黏[粘]褶菌属 <拉> Gloeophyllum
黏[粘]织在一起 felting down
黏[粘]质 cement;viscidity
黏[粘]质暗色土 grum(m)usol
黏[粘]质的 muculent
黏[粘]质底泥 clayey sediment
黏[粘]质底土 argillaceous bottom
黏[粘]质粉砂 fine sandy loam
黏[粘]质粉土 clayey silt;clay silt
黏[粘]质海底 stiff bottom
黏[粘]质矿物 clay mineral
黏[粘]质沥青 steep pitch
黏[粘]质垆坶 clayey loam;clay loam
黏[粘]质泥灰岩 clay marl
黏[粘]质壤土 clay(ey) loam

黏[粘]质砂土 clay(ey) sand
黏[粘]质砂岩 clay sandstone
黏[粘]质土(壤) clayey soil;cohesive soil;heavy(-textured) soil;sea silt
黏[粘]质物 mucilage;slime
黏[粘]质细砂(土) clayey fine sand
黏[粘]质纤维 gelatinous fiber[fibre];mucilaginous fiber[fibre]
黏[粘]质纤维素 muco-cellulose
黏[粘]质页岩 clay shale
黏[粘]滞边界 viscosity boundary;viscous boundary
黏[粘]滞边界层 viscous boundary layer
黏[粘]滞变形 viscosity deformation;viscous deformation
黏[粘]滞波 viscosity wave
黏[粘]滞层 viscous layer
黏[粘]滞常数 viscosity constant
黏[粘]滞稠度 viscid consistency;viscous consistency
黏[粘]滞传导 viscous conductance
黏[粘]滞磁化 viscous magnetization
黏[粘]滞次层 viscous sublayer
黏[粘]滞的 viscid;viscous
黏[粘]滞底层 viscid sublayer;viscous sublayer
黏[粘]滞度 glutinousness;viscosity;viscosity grade;viscousness
黏[粘]滞度计 visco(si)meter
黏[粘]滞度降低 lowering of viscosity
黏[粘]滞度试验 viscosity test
黏[粘]滞度数 viscosity number
黏[粘]滞度仪 visco(si)meter
黏[粘]滞发酵 viscous fermentation
黏[粘]滞过滤器 viscous filter
黏[粘]滞耗散 viscous dissipation
黏[粘]滞后缘涡流 viscous trailing vortex
黏[粘]滞滑动 stick slip
黏[粘]滞计 visco(si)meter
黏[粘]滞剂 viscosity agent
黏[粘]滞减震器 viscous dashpot
黏[粘]滞结持度 viscous consistency
黏[粘]滞空气过滤器 viscous air filter
黏[粘]滞力 viscosity force;viscous force
黏[粘]滞流(动) plastic flow;viscous flow;slug flow;sluggish flow;viscosity flow
黏[粘]滞流(动)传质机理 viscous flow material transfer mechanism
黏[粘]滞流(动)方程 viscous flow equation
黏[粘]滞滤气器 viscous air filter
黏[粘]滞率 kinematic(al) viscosity
黏[粘]滞模量 viscosity modulus
黏[粘]滞摩擦 viscosity friction;viscous friction
黏[粘]滞摩擦损失 viscous friction loss
黏[粘]滞摩擦系数 viscous friction coefficient
黏[粘]滞黏[粘]质 argillaceous material;binder material
黏[粘]滞耦合地震计 viscous coupled seismometer
黏[粘]滞倾点 viscous pour point
黏[粘]滞区 stagnant zone
黏[粘]滞蠕变 viscous creep
黏[粘]滞润滑 viscous lubrication
黏[粘]滞润滑油 sluggish lubricant
黏[粘]滞剩余磁化强度 viscous remanent magnetization
黏[粘]滞式空气滤池 cohesive air filter
黏[粘]滞衰减 viscous damping;viscous decay
黏[粘]滞水 held water
黏[粘]滞塑性变形 visco-plastic deformation
黏[粘]滞损失 viscous loss
黏[粘]滞弹性 viscoelasticity
黏[粘]滞体 viscous body
黏[粘]滞位移 viscous displacement
黏[粘]滞系数 coefficient of viscosity;visco(si)metry coefficient;visco(si)metry factor;viscosity coefficient;viscosity factor;viscous factor
黏[粘]滞系数绝对值 absolute viscosity coefficient
黏[粘]滞限度 sticky limit
黏[粘]滞相似性 viscous similarity
黏[粘]滞性 adhesive retention;astringency;viscidity;viscosity
黏[粘]滞性副层 viscosity sublayer
黏[粘]滞性过高 hyperviscosity
黏[粘]滞性理论 viscosity theory
黏[粘]滞性流体 viscosity fluid;viscous fluid;viscous liquid
黏[粘]滞性试验 visco(si)metry test-(ing);viscosity test
黏[粘]滞性退磁 viscous demagnetization
黏[粘]滞性效应 viscosity effect;viscous effect
黏[粘]滞性亚层 viscosity sublayer
黏[粘]滞性液流渗透率 viscous flow permeability
黏[粘]滞性质 viscous behavio(u)r
黏[粘]滞压力 viscous pressure
黏[粘]滞曳力 viscous drag
黏[粘]滞液体 viscous liquid
黏[粘]滞因数 viscosity factor;viscous factor
黏[粘]滞因素 viscosity factor;viscous factor
黏[粘]滞应变 viscous strain
黏[粘]滞应力 viscous stress
黏[粘]滞应力张量 viscous stress tensor
黏[粘]滞运动 viscous motion
黏[粘]滞运动方程 viscous equation of motion
黏[粘]滞真空规 sticky ga(u)ge
黏[粘]滞真空计黏[粘]度仪 viscosity ga(u)ge
黏[粘]滞撞击滤尘器 viscous impingement filter
黏[粘]滞阻力 viscosity resistance;viscous resistance
黏[粘]滞阻力气体密度测量器 viscous-drag gas-density meter
黏[粘]滞阻尼 viscous damping
黏[粘]滞阻尼常数 viscous damping constant
黏[粘]滞阻尼系数 viscous damping factor
黏[粘]滞作用 viscous effect
黏[粘]中性油 viscous neutral oil
黏[粘]重常数 viscosity-density constant
黏[粘]重土(壤) heavier textured soil;heavy soil
黏[粘]粥冰 snow sludge
黏[粘]住 cling;clogging;conglutinate;sticking together
黏[粘]住的 stamp
黏[粘]状式过滤器 viscous impingement type filter
黏[粘]着 adhere;adherence;cling;gumminess;sticking
黏[粘]着材料 sticky material
黏[粘]着长度 coherence length
黏[粘]着的 adherent;bonded;mucid;sticky;frozen <矿石和围岩等>
黏[粘]着的水泥 sticky cement
黏[粘]着点 blocking point
黏[粘]着电压 sticking voltage
黏[粘]着度 bond value;sticky limit
黏[粘]着段落 adhesion division
黏[粘]着方式 adhesion system
黏[粘]着钢筋 bonded reinforcement
黏[粘]着钢丝 bonded steel wire
黏[粘]着高度 cohesion height
黏[粘]着和齿轨合用铁路 combined adhesion and rack railway
黏[粘]着剂 adhesion additive;adhesion agent;adhesive(agent);bonding agent;sticker;sticking agent
黏[粘]着键 adhesive bond
黏[粘]着结持度 sticky consistency[consistence]
黏[粘]着结构 bond structure
黏[粘]着界限 coherent limit
黏[粘]着抗力 bond(ed) resistance
黏[粘]着控制 adhesion control
黏[粘]着块 adhesion briquet(te)
黏[粘]着蜡 adhesive wax
黏[粘]着力 adhesion(capacity);adhesion force;adhesion strength;adhesion stress;adhesive ability;adhesive attraction;adhesive bond;adhesive force;adhesive strength;aggregation force;binding force;bond force;bond value;bounding capacity;coherence;stick force
黏[粘]着力计 adhesive meter
黏[粘]着力筋 bonded tendon
黏[粘]着力破坏 bond failure
黏[粘]着力强度极限 adhesion limit
黏[粘]着力试验 adhesion test
黏[粘]着力损失检测 adhesion loss detection
黏[粘]着力损失检验系统 adhesion loss detection system
黏[粘]着力特性 adhesion characteristic;adhesive characteristic
黏[粘]着力线 adhesion line
黏[粘]着力消失 adhesion loss
黏[粘]着力值 bond value
黏[粘]着率 adhesion ratio
黏[粘]着面 adherent surface;bonding plane
黏[粘]着面积 area of cohesion
黏[粘]着膜 adhesive film
黏[粘]着摩擦 sticking friction
黏[粘]着摩托车 adhesion motor rail coach
黏[粘]着能力 adhesion capability;adhesion capacity;adhesive capacity;stickability;adhesive power
黏[粘]着破坏 bond failure;fail in bond;loss of adhesion
黏[粘]着牵引力 adhesion traction;adhesion tractive effort
黏[粘]着强度 adhesion strength;adhesive strength;adhesive tension;bond resistance;bond strength
黏[粘]着区 stick zone;zone of zero slip
黏[粘]着色 adhesive colo(u)r
黏[粘]着失效 bond failure
黏[粘]着时间 closure time
黏[粘]着式空气过滤装置 viscous type air cleaner
黏[粘]着水 adhesive water;viscous water
黏[粘]着水分 adhesive moisture
黏[粘]着铁路 adhesion railway
黏[粘]着温度 blocking point
黏[粘]着污染物 adhered pollutant
黏[粘]着物 adhesion
黏[粘]着系数 adhesion coefficient;adhesion factor;adhesion of drivers;adhesive coefficient;coefficient of adhesion;factor of adhesion;specific adhesion
黏[粘]着线 bond line
黏[粘]着性 adhesion capacity;adhesiveness;adhesivity;cohesiveness;ropiness;stickability;stickiness
黏[粘]着性能 cling property
黏[粘]着性试验 adhesiveness test
黏[粘]着性质 adherent property
黏[粘]着徐变 bond creep
黏[粘]着应力 bond(ing) stress
黏[粘]着植物 haptophyte
黏[粘]着制动 stuck brake
黏[粘]着重量 adhesion weight;adhesive weight
黏[粘]着阻力 resistance to bond
黏[粘]着作用 adhesion;adhesive effect

捻

捻 entwist;pinch;stranding;twine;twirl;twist

捻比系数 <钢丝绳> ratio of lay
捻边装置 twist fringing device
捻长 length of lay
捻成的 twined
捻度 twist;twisting number
捻度不稳定性 twist liveliness
捻度不匀 twist irregularity
捻度测试仪 twist tester
捻度常数 twist constant
捻度齿轮 twist gear
捻度平衡 twist balance
捻度平衡指数 twist balance index
捻度试验机 twist tester;twist testing machine
捻度试验仪 twist-and-contraction meter
捻度系数 twist factor;twist multiplier
捻度转换区 twist change over region
捻度转移 twist migration
捻发音 crepitation
捻缝边 ca(u)lking side
捻缝槽 ca(u)lking pocket
捻缝锤 ca(u)lking hammer;driving mallet
捻缝工 ca(u)lker
捻缝工具 ca(u)lking tool
捻缝工具箱 ca(u)lker box
捻缝钩 reeling hook
捻缝尖刀 butt iron;sharp iron
捻缝胶 marine glue;seam composition
捻缝口 ca(u)lking;fullering
捻缝料 blare
捻缝麻絮 pledget;thread of oakum
捻缝棉条 ca(u)lking cotton;cotton wicking
捻缝木槌 ca(u)lking mallet;hawsing mallet;making mallet;making maul;setting mallet
捻缝填料 ca(u)lkage;ca(u)lking stuff
捻缝条 ca(u)lking strip
捻缝隙 ca(u)lking groove
捻缝削刀 deck iron;dumb iron
捻缝油灰 luting
捻缝凿 ca(u)lking chisel;ca(u)lking iron;common iron;creasing iron;hawsing iron;horsing iron;making iron
捻股 stranding
捻股机 strander;stranding machine
捻管生头 twist tube threading
捻合 pigtail splice
捻回传递 twist propagation
捻级 twist level
捻角 <钢丝绳的> spiral angle
捻接 blind splice;splicing wire
捻接法 twisting-in
捻结织网机 twisted knotless net mak-

ing machine
捻矩 twisting moment;twisting torque
捻距 length of lay
捻距长度 length of lay
捻距范围 range of lay
捻口 ca(u)lking
捻口接头 ca(u)lked joint;ca(u)lking joint
捻料 ca(u)lking compound
捻裂法原纤化薄膜条 twist-fibrillated tape
捻密度 turns per inch
捻密缝 ca(u)lk seam
捻弄 twirl
捻绳机 closer
捻数 twisting count
捻丝 thrown silk
捻丝机 twisting frame
捻缩 twist contraction;twist retraction;twist shrinkage;twist takeup
捻缩率 twist contraction rate
捻系数 twist factor
捻线 twisted thread;twist(ing)
捻线钢领 twister ring
捻线机 twister;twisting frame
捻向 direction of lay
捻旋作用 screw action
捻织 twisted weave
捻织网 twist wire weave
捻皱缩 twist-lively
捻转角 angle of twist

辇

辇道 <中国> ancient imperial road(China)

碾

碾 crush;roller

碾出法 rolling-out process
碾的 rolling
碾坊 grain mill
碾杆 roll bar
碾钢板 rolled plate
碾工 roller man
碾谷滚筒 scourer cylinder
碾光 calendering;rolling
碾光机 calender
碾痕 roll mark
碾环机 ring rolls
碾灰机 ash crusher machine
碾结碎石(路) roller-bound macadam
碾料间 mill room
碾轮 muller;runner wheel
碾轮式混砂机 sand roller mill;Simpson mill
碾轮用刮板 wheel scraper
碾米厂 grain-polishing mill;rice mill
碾米机 paddy-pounder;rice mill;rice scourer;rice sheller
碾磨 attrition grinding;milling;mulling;pan grinding
碾磨材料 mill material
碾磨厂 rolling mill
碾磨打光 polish grind
碾磨机 attrition mill;attritor;attritor mill;edge runner;mill;mortar mill;muller
碾磨介质 grinding media
碾磨盘 pan mill
碾磨细度 fineness of grinding
碾磨效率 grinding efficiency
碾泥机 malaxator;mud mixer;pugmill
碾盘 edge runner pan;grinding base
碾盘式粉碎机 edge runner mill
碾盘式(碾)磨机 pan mill
碾盘式破碎机 pan crusher
碾皮机 bark mill
碾平 flattening out;crimp;planish(ing);roll
碾枪 muller
碾砂机 kollergang;sand mill;wet pan
碾实 compaction by rolling
碾实光面混凝土路面 rolled concrete
碾实厚度 compacted thickness
碾实混凝土 rolled concrete
碾实面层 compacted surface
碾实填土 rolled earth fill
碾碎 bray;breakdown by grinding;pulverizer;scrunch
碾碎板 crumbing shoe
碾碎材料 <未过筛的> crusher-run product
碾碎的 crushed
碾碎的玻璃 crushed glass
碾碎的材料 crushed material
碾碎辊 cracker
碾碎机 bucker;chaser mill;edge mill;edge runner;edge runner mill;end-runner mill;pan mill;powdering machine
碾碎了的 ground
碾碎时的添加料 interground addition
碾条 muck bar
碾铁 rolled iron
碾铁板 rolled plate
碾土机 clay mill;mangle
碾压 compacting;compaction (by rolling);roller compaction;rolling (wheel compaction)
碾压坝 rolled(fill) dam;roller dam
碾压板 mill board
碾压遍数 number of roller passes;number of rolling coverage
碾压不匀 uneven rolling
碾压步骤 rolling process
碾压的 rolled(-on)
碾压的堆石 rolled rockfill
碾压的路面 rolled pavement
碾压的铺面 rolled pavement
碾压底土 roller subsoil
碾压断面 rolled section
碾压堆石 compacted rockfill
碾压堆石坝 rolled rockfill dam
碾压法 rolled-on method;roller compaction;rolling compaction
碾压负荷 crushing load
碾压工作 rolling work
碾压滚筒 tamping drum
碾压滚轴光泽机 mangle
碾压过的 rolled
碾压过度的 overrolling
碾压混凝土 rollcrete;rolled concrete;roller-compacted concrete
碾压混凝土坝 <特指日本的碾压混凝土坝技术> rolled-compacted concrete dam
碾压混凝土路 roller-compacted concrete pavement
碾压混凝土路面 rolling compacted concrete pavement
碾压混凝土土坝 roll-compacted concrete dam
碾压混凝土围堰 roller-compacted concrete cofferdam
碾压混凝土质量控制系统 roller-compacted concrete quality control system
碾压机 bucker;chaser;compaction wheel;mangle;roller(mill);rolling mill
碾压机械 rolling machinery
碾压基础地基 rolled foundation subsoil
碾压接缝 rolling a joint
碾压卡车 rolling truck
碾压宽度 rolling width
碾压沥青 rolled asphalt
碾压裂纹 roller check
碾压路面 rolled surfacing
碾压轮迹 roller mark;rolling groove
碾压轮碾脚尖 compactor tips
碾压模 rolling jig
碾压贫混凝土 lean rolled concrete;rolled lean concrete
碾压铅板 milled lead;rolled sheet lead
碾压铅皮 milled lead
碾压强度 crushing strength
碾压切削 mangling
碾压设备 compacting plant;compaction equipment;compaction plant
碾压式地沥青 <英> rolled asphalt
碾压式地沥青混合料 rolled asphalt mixture
碾压式堆石 rolled rockfill
碾压式干硬性混凝土 dry compacted concrete
碾压式沥青路 rolled asphalt pavement
碾压式沥青磨耗层 rolled asphalt wearing course
碾压式水泥混凝土 rolled cement concrete
碾压式填筑坝 rolled fill dam
碾压式土坝 rolled(earth) dam;rolled fill dam
碾压式土石坝 rolled earth-rock fill dam
碾压试验 flow test;roller compaction test;rolling test;rolling trial;test embankment;trial embankment
碾压试验填土 test fill
碾压试验曲线 rolling trial curve
碾压水 water of compaction
碾压速度 rolling speed
碾压碎石 rolled broken stone;rolled stone
碾压碎石基层路面 rolled-stone base pavement
碾压碎石路 traffic-bound macadam
碾压碎石路面 roller-bound surface
碾压填土 rolled earth fill;rolled fill
碾压填土坝 rolled earthfill dam
碾压填土墙芯 rolled earthen core
碾压填筑法 <路堤的> rolled-embankment method
碾压铁板 rolled sheet iron
碾压土坝 rolled fill dam
碾压土堤 rolled earth fill
碾压温度 rolling temperature
碾压因数 rolling factor
碾压因素 rolling factor
碾压整直车间 straightening-roller shop
碾压筑堤 rolled-fill embankment
碾轧 becking;expansion;rolling;roll-off
碾轧操作 crushing operation
碾轧产品 crushed product
碾轧厂 becking mill
碾轧成的粒状炉渣 crushed granulated cinder
碾轧成的石灰石砂 crushed limestone sand
碾轧过程 crushing process
碾轧机 rolling mill
碾制的 rolled
碾轧钢 rolled steel section
碾制(金属)板 rolled sheet metal
碾制凸缘 expanded flange
碾子 edge mill;edge runner;fair leader;roller
碾子锥头 roller bit

廿

廿六烷基 ceryl

念

念珠 paternoster

念珠状的 miniliform;toruloid;torulose
念珠状负荷模 torise load zasts

酿

酿酒厂 beverage distillery;brewery;brewhouse;winery

酿酒场 still
酿酒废水 brewing proems wastewater
酿酒工业 brewery industry
酿酶 zymase
酿造 brewing
酿造厂 brewery
酿造废水 brewage wastewater;brewing process wastewater
酿造锅 brew kettle
酿造和蒸馏废水 brewery and distillery wastewater
酿造排放液 malting effluent
酿造学 zymurgy
酿造业 brewing industry
酿造渣 brewery residues

鸟

鸟巢 bird('s) nest

鸟巢菌属 bird's nest fungus;nidulariaceous fungus
鸟池 birdbath
鸟岛 bird island
鸟粪 bird droppings;bird manure
鸟粪胺 guanamine
鸟粪胺树脂 guanamine resin
鸟粪层 guano
鸟粪磷钙土 guano phosphorite
鸟粪磷矿石 struvite ore
鸟粪石 guanite;guano;struvite
鸟粪石型矿床 guano-type deposit
鸟冠 pileum
鸟桕 tallow tree
鸟瞰 aerial perspective;air view;through view
鸟瞰测量 bird's eye survey
鸟瞰的 bird's eye
鸟瞰地形图 morphographic map
鸟瞰透视图 aerial perspective;bird's eye perspective
鸟瞰图 aerial view;aeroview;bird's eye perspective;bird's eye view (drawing);aeroplane view;airscape;air view map;bird's eye view map;bird view;perspective view(ing);top view
鸟瞰图照相机 kite camera
鸟瞰照片 bird's eye photograph
鸟类保护区 bird sanctuary
鸟类和哺乳类时期【地】age of birds and mammals
鸟类环志 bird banding
鸟类纪念碑 monument for birds
鸟类生态学 bird ecology
鸟类学 ornithology
鸟笼式 birdcage
鸟笼式换热器 tubular cage recuperator
鸟笼形避雷器 bird cage lightning
鸟禽学 ornithology
鸟舍 aviary;nestle box
鸟兽纹 bird and beast design;bird and beast pattern
鸟头饰 cat's head
鸟头饰线脚 <中世纪> cat's head mo(u)lding
鸟头线脚装饰 <中世纪的> bird's-head mo(u)lding
鸟纹 bird pattern
鸟屋 aviary
鸟形杯 bird-shaped cup
鸟形盖罐 bird-shaped jar with lid
鸟形嘴 beak
鸟眼 <木材缺陷> bird's eye

N

鸟眼构造 birds eye structure
鸟眼花纹 <木材缺陷> bird's eye; bird's eye pattern
鸟眼花纹织物 bird's eye
鸟眼灰岩 bird's eye limestone; loferite
鸟眼孔隙 bird's eye porosity
鸟眼木纹 bird's eye in wood
鸟眼槭树 bird's eye maple
鸟眼纹 bird's eye
鸟眼纹材 bird's eye wood
鸟眼纹理 <木材> bird's eye figure; bird's eye grain; bird's eye pattern
鸟眼状长烟煤 curley cannel
鸟眼状煤 eye coal
鸟翼上的翼镜 mirror
鸟爪形裂纹 crow foot crack
鸟爪状三角洲 bird's foot delta; bird's-foot margin; digitate margin
鸟啄斑纹 bird peck
鸟啄纹 bird peck
鸟啄形线脚 quirk mo(u)lding
鸟啄装饰线脚 hawksbeak
鸟足形三角洲 bird's foot delta
鸟足状半裂的 pedately cleft
鸟足状地 pedately
鸟足状裂的 pedatifid
鸟足状脉的 pedately veined; pedatinerved
鸟足状浅裂的 pedatilobed
鸟足状全裂的 pedatisect
鸟足状三角洲 bird foot delta
鸟足状深裂的 pedatipartite
鸟足状水系 digitate drainage pattern
鸟足状叶 pedate leaf
鸟嘴刨 snipe's-bill
鸟嘴式线脚 bird's beak; bird's beak mo(u)lding
鸟嘴式线饰 bird's beak mo(u)lding; bird's beak ornament
鸟嘴式装饰 bird's beak ornament
鸟嘴饰 beak(head) mo(u)lding
鸟嘴线脚 beak(head) mo(u)lding
鸟嘴铁 beakiron
鸟嘴头饰 <诺尔曼门道上的富丽装饰> beakhead ornament; beakhead mo(u)lding
鸟嘴头饰线脚 beakhead mo(u)lding
鸟嘴物 beak
鸟嘴线脚 beakhead mo(u)lding
鸟嘴形 beak; bird's-mouth
鸟嘴形承接 birdsmouthing
鸟嘴砧 anvil with an arm; beakiron; beckern
鸟嘴装饰线脚 hawksbeak

袅

袅混线脚 cima reversal

尿

尿布 napkin

尿池 urinary
尿环石 uricite
尿醛树脂 urea resin
尿素 urea
尿素除蜡 dewaxing with urea
尿素合成树脂胶 urea synthetic resin adhesive
尿素甲醛 urea formaldehyde
尿素甲醛塑料 bexoid
尿素络合 urea adduction
尿素酶 urea
尿素清漆树脂 urea varnish resin
尿素三聚氰胺甲醛树脂 urea melamine formaldehyde resin
尿素石 urea
尿素树脂 urea resin
尿素树脂层压塑料 urea resin laminated plastics
尿素树脂胶 beetle cement; urea resin glue
尿素树脂胶粘剂 carbamide resin adhesive; urea resin adhesive
尿素树脂接合剂 kaurit
尿素树脂黏[粘]合剂 urea adhesive; urea resin adhesive
尿素树脂黏[粘]结剂 carbamide resin adhesive; urea resin adhesive
尿素树脂泡沫(材料) urea resin foam
尿素树脂清漆 urea resin varnish
尿素树脂造型合成材料 urea resin mo(u)lding compound
尿素装置 urea plant; urea unit
尿烷泡沫 urethane foam
尿烷清漆 urethane lacquer
尿烷树脂 urethane resin
尿烷涂料 urethane coating
尿烷亚麻籽油 urethane linseed oil
尿烷油漆 urethane paint

脲

脲【化】urea

脲基甲醇 monomethylolurea
脲基甲酸盐 allophanate
脲基甲酰乙酸 malonuric acid
脲甲醛树脂 carbamide formaldehyde resin
脲醛 urea formaldehyde
脲醛胶 urea formaldehyde adhesive; urea formaldehyde glue
脲醛胶粘剂 urea formaldehyde adhesive
脲醛黏[粘]合剂 urea formaldehyde adhesive
脲醛泡沫 urea formaldehyde foam
脲醛热固树脂 plaskon
脲醛树脂 urea formaldehyde resin
脲醛树脂胶 urea formaldehyde resin adhesive
脲醛树脂木材黏[粘]合剂 urea resin wood adhesive
脲醛树脂黏[粘]合剂 beetle cement; urac
脲醛树脂清漆 urea resin varnish
脲醛塑料 mouldrite; urea formaldehyde plastics; urea plastics
脲羰基乙酸 malonuric acid
脲烷泡沫填缝料 urethane foam filler
脲烷弹性纤维生产废水 urethane elastic fibre[fiber] production wastewater
脲脂黏[粘]胶 urea resin glue
脲脂黏[粘]结剂 urea resin adhesive

捏

捏拌机 pug(mill) mixer <窑泥等>; pugmill(type mixer)

捏拌机拌和 pugmill mixing
捏拌机拌和的 pugmill-mixed
捏拌机搅拌 pugmill mixing
捏拌机搅拌的 pugmill-mixed
捏雕 applied relief
捏合 kneading; tempering
捏合机 dough mixer; kneader; kneading machine; kneading mill
捏合挤出分批混合机 kneader extruder batch mixer
捏合挤压机 kneading extruder
捏合结构 kneaded structure
捏合器 malaxator
捏合物质 kneaded mass
捏合型搅拌机 kneader type mixer
捏和 kneading; pug(ging)
捏和机 kneader; pugging mill
捏和结构 kneaded structure
捏和碾磨机 pugging mill
捏紧 clutch
捏扣 masticate
捏练 pugging mullering
捏练过的黏[粘]土 pugged clay
捏练机 kneader; kneading machine
捏炼混合机 kneading and mixing machinery
捏磨机 pugging mill
捏泥 batter
捏泥机 clay kneading machine; wedging mill
捏泥盘 tempering tub
捏黏[粘]土 temper
捏揉 kneed
捏手 handle knob
捏手螺母 star knob nut
捏塑 kneading model; sculpture
捏塑体 dough
捏塑造型 dough mo(u)lding
捏缩效应 rheostriction
捏土机 kneading machine; pugmill mixer
捏造 fabricate
捏造报告 cook up a report
捏造物 fabrication
捏制 kneading

涅

涅昂纳尔炸药 Neonal

涅昂纳留姆 <一种铝合金> Neonalium
涅昂纳铝铜合金 Neonalium
涅墨西斯神庙 Temple of Nemesis
涅瓦河 Neva River
涅瓦斯通 <厨房设备的商品名称，如洗涤盆等> Nevastone

啮

啮齿类 rodent

啮出齿轮 recess-action gear
啮出轨迹 recess path
啮出接触 recess contact
啮出相位 recess phase
啮合 gearing in; gear into; in gear; engagement; engager; engaging; falling-in; mesh(ing); meshing engagement; put into gear; toothing
啮合部分 mate
啮合采泥器 bottom grab
啮合长度 length of action
啮合齿廓 contacting profile
啮合齿轮 meshing gear
啮合齿轮副 pair of(meshed) gears
啮合齿面 drive side of tooth
啮合齿条 pinion rack
啮合触面 mating surface
啮合搭接 joggled lap joint
啮合单面搭接 joggled single lap joint
啮合的 joggled; meshed
啮合的齿轮 meshed gears
啮合点 meshing point; point of engagement
啮合电磁铁 clutch magnet
啮合对接 joggled butt joint
啮合范围 engagement range
啮合方式 engagement system
啮合干涉 meshing interference
啮合杠杆 engaging lever
啮合拐角 lock corner
啮合轨迹 path of action
啮合过程 engagement process
啮合痕迹 meshing trace
啮合弧 arc of contact; contact arc
啮合机构 engaging mechanism
啮合机接件 joggle
啮合间隙 back lash
啮合检查结构 meshing frame
啮合检查仪 meshing tester
啮合件 joggle(d) piece
啮合角 angle of engagement; contact angle; engaged angle; engaging angle; generating angle; locking angle
啮合校正 mesh adjustment
啮合接(榫) joggle(d) joint
啮合接头 crossette; joggle(d) joint
啮合精度 accuracy of mesh
啮合颗粒 interlocking particle
啮合扣 joggle
啮合扣接榫 joggle tenon
啮合梁 joggle(d) beam
啮合面 field of conjugate action
啮合摩擦 engaging friction
啮合木 joggled timber
啮合木柱 joggle post
啮合盘 toothed disc
啮合器 engage switch
啮合前状态 preengagement condition
啮合区 meshing zone
啮合伞齿轮 meshing bevel gear
啮合深度 depth of engagement; meshing depth
啮合时限 gear-meshing time limit
啮合榫 joggle
啮合锁 toothed lock
啮合弹簧 engage spring
啮合条件 meshing condition
啮合调整 mesh adjustment
啮合调正 mesh adjustment
啮合系数 contact ratio; coupling coefficient
啮合线 <齿轮传动> contacting line; line of contact; line of engagement; line of pressure; path of contact; pressure line; mesh lines
啮合镶接 joggle joint
啮合(小)过梁 joggled lintel
啮合效果 tooth effect
啮合隅角 angle bond
啮合原理 theory of engagement
啮合周期 mesh cycle
啮合柱 joggled piece; joggled post
啮合作用 tooth effect
啮角 nip angle
啮接 birdsmouth joint; bridge joint; bridle joint
啮侵蚀 bite
啮入轨迹 approach path
啮入间隙 entry gap
啮入接触 approach contact
啮入相位 approach phase
啮蚀状的 erose
啮食迹 grazing trace
啮咬 nibble
啮轧机 crimping machine

镊

镊子 forceps; nippers; pinch; tweezers

镍

镍 nickel

镍白口铁 nickel white iron
镍白铜 Alpaka
镍包钢 nickel-clad steel
镍包石墨粉 nickel-coated graphite powder
镍包氧化铝粉 nickel-coated alumin(i)um oxide powder
镍饱和溶液 nickel-saturated liquor
镍贝塔蜡蛇纹石 nickel-β-kerolite
镍钡合金 nickel-barium alloy
镍被膜 <搪瓷> nickel dipping; nickel flashing; nickel pickling
镍冰铜 nickel matte
镍箔 nickel foil
镍卟林 nickel porphyrin
镍层 nickel layer; nickel dam <轴瓦上巴氏合金和铅青铜之间的>

镍产物 nickel product
镍吹炼 nickel converting
镍磁铁矿 nickel-magnetite;trevorite
镍催化剂 nickel catalyst;nickel catalyzator;Raney nickel
镍催化异构化法 nickel-isomerization
镍带 nickel strap
镍带测辐射热计 nickel-strap bolometer
镍当量 nickel equivalent
镍导线 nickel wire line
镍的 nickelous
镍的电处理 nickelizing
镍电极 nickel electrode
镍电解液净化 nickel electrolyte purification
镍电阻温度计 nickel resistance thermometer
镍镀合金 nickel-beryllium alloy
镍矾 retgersite
镍矾石 nickelalumite
镍钒钢 nickel-vanadium steel
镍钒铸铁 nickel-vanadium cast iron
镍方钴矿 nickel-skutterudite
镍粉 nickel powder
镍辐射屏 nickel radiation screen
镍覆盖层 nickel coating
镍橄榄石 liebenbergite
镍钢 nickel steel
镍钢气焊条 nickel steel gas welding rod
镍钢线 nickel steel wire
镍镉电池 nickel-cadmium battery; nickel-cadmium cell;nickel cell
镍镉蓄电池 cadmium-nickel storage cell; nickel-cadmium [Ni-Cd] accumulator;nickel-cadmium battery
镍铬 nickel-chromium triangle
镍铬薄膜 nickel-chromium thin film
镍铬不锈钢 nickel-chromium stainless steel
镍铬不锈钢钢罐 Inconel charge can
镍铬的电镀涂层 electrophoretic coating of nickel plus chromium
镍铬低膨胀系数合金 nilo
镍铬电阻 nichrome resistance
镍铬电阻合金 Jellif
镍铬电阻器 nickel-chromium resistor
镍铬电阻线 nichrome resistance wire
镍铬发热线圈 nichrome heating spiral
镍铬发热元件 nichrome heat element
镍铬钢 nichrome steel;nickel-chrome steel;nickel-chromium steel
镍铬钴低膨胀合金 Actanium
镍铬钴高导磁率合金 permivar
镍铬钴耐热蚀合金 Sirius
镍铬硅合金 Nichrosi
镍铬硅铁磁合金 Rhometal
镍铬硅铁合金 nicrosilal
镍铬硅铸铁 nichrosilal;nicrosilal
镍铬焊料合金 Nicrobraz
镍铬合金 nichrome (alloy); nickel-chrome alloy; nickel-chromium alloy; Chromel; Illium; nicochrome; pyrolic alloy;Q alloy
镍铬合金保险丝 nichrome fuse
镍铬合金镀层 nichrome coating
镍铬合金粉末 nichrome powder
镍铬合金钢 nickel-chrome alloy steel;nicochrome steel
镍铬合金片 nichrome film
镍铬合金三角【化】nichrome triangle
镍铬合金丝 < 一种高阻合金线 > nichrome wire
镍铬合金珠 nichrome bead
镍铬恒弹性钢 Elinvar
镍铬康铜热电偶 chromel-constantan thermocouple
镍铬冷硬铸铁 Niharl
镍铬铝奥氏体耐热钢 Calmet
镍铬铝铜热电偶 nichrome alumino-copper thermocouple
镍铬锰钢 nickel-chromium manganese steel
镍铬钼钢 nickel-chromium molybdenum steels
镍铬钼钛钢 < 透平叶片用的 > Discaloy
镍铬耐磨白口铁 Ni-hard
镍铬耐热钢 nichrome steel
镍铬耐热合金 < 因科镍 > cronite; kuromore;Incochrome nickel
镍铬耐蚀可锻钢 staybrite
镍铬耐酸钢 utiloy
镍铬镍硅热电偶 nickel-chromium-nickel silicon thermocouple
镍铬镍铝热电偶 nichrome alumino-nickel couple; nichrome alumino-nickel thermocouple
镍铬镍热电偶 nickel-chromium-nickel thermocouple
镍铬青铜 nickel-chrome bronze
镍铬丝 chromel-filament;nickel-chrome wire
镍铬丝加热线圈 nichrome heating coil
镍铬钛合金 Nimonic
镍铬钛铁定弹性系数合金 Ni-Span (alloy)
镍铬铁 nickel-chromium-iron
镍铬铁电热合金 calomic
镍铬铁防锈合金 Ascoloy
镍铬铁合金 Economet; nickel-chromium-iron alloy
镍铬铁矿 nichromite
镍铬铁锰合金 firearmor
镍铬铁耐热合金 Accoloy;Alray;Camloy;tophet
镍铬铁耐蚀合金 Misco metal
镍铬铜低合金钢 Skhl steel
镍铬铜合金铸铁 Causul metal
镍铬系精密级电阻材料 Karma
镍铬线 chrome wire
镍铬预热片 nickel-chrome glow strip
镍铬铸铁 Minvar; nickel-chromium cast iron
镍汞合金 nickel amalgam
镍钴比 nickel-to-cobalt ratio
镍钴合金 Konal;nickel-cobalt(alloy)
镍钴锍 nickel-cobalt matte
镍钴铁流 nickel-cobalt-iron matte
镍钴土 nickel asbolane
镍刮 nickel scratch
镍光泽彩 nickel lustre[luster]
镍硅 nisiloy
镍硅镀层 Nicasil
镍硅合金镀覆层 Einisil coating;nickel silicon alloy coating
镍硅碳化物 nickel silicon carbide
镍海泡石 falcondite
镍焊丝 nickel bare welding filler
镍合金 nickel alloy
镍合金钢 alloyed nickel steel;nickel alloy steel
镍合金罐 nickel alloy pot
镍合金线 < 一种高阻合金线 > nickeline wire
镍褐煤 kerzinte
镍黑 nickel black
镍华 anabergite;nickel arsenite;nickel bloom;nickel ocher
镍滑石 villiersite;willemseiye
镍还原高压釜 nickel reduction autoclave
镍黄 nickel yellow
镍黄铁矿 capillary pyrite; hair pyrites;millerite;nickel pyrites;nicopyrite;pentahydrite;pentlandite
镍黄铜 nickel brass; nickel oreide; Neogen < 其中的一种 >
镍黄铜合金 Benedict's metal
镍辉钴矿 dzhulukulite
镍基超耐热合金 nickel-base heat-resisting superalloy
镍基底 nickel base
镍基合金 nickel-base alloy
镍基合金焊丝 nickel-base alloy bare welding filler metal
镍基合金焊条 nickel-base alloy covered electrode
镍基耐热合金 refractoloy
镍基热电偶合金 Alumel alloy
镍结碳化钛硬质合金 nickel-cemented titanium carbide
镍结碳化钽硬质合金 nickel-cemented tantalum carbide
镍结碳化钨硬质合金 nickel-cemented tungsten carbide
镍金属陶瓷 nickel cermet
镍金属陶瓷镀层 nickel cermet coating
镍精炼 nickel refining
镍可铁 nicofer
镍克林电阻 nickeline resistance
镍克罗 nichrome(alloy)
镍孔雀石 glaukosphaerite
镍块 nickel block
镍矿 nickel minerals;nickel ore
镍矿石 nickel ore
镍劳特合金 < 一种铜镍耐蚀合金 > nickeloid
镍冷硬磨球 Ni-hard grinding ball
镍离子 nickel ion
镍利蛇纹石 nepouite
镍粒 nickel shot
镍磷青铜 nickel phosphor bronze
镍硫钴矿 nickel linnaeite
镍硫铁矿 horbachite;inverarite
镍锍 nickel matte;nis matte
镍锍吹炼 nickel matte converting
镍锍精炼 nickel matte refining
镍滤网 nickel filter net
镍铝镀层 nickel-alumin(i)um coating
镍铝矾 carrboydite
镍铝铬铁合金 Calite
镍铝合金 nickel alumin(i)um alloy
镍铝尖晶石 nickel aluminate
镍铝锰硅合金 Alumel alloy
镍铝镍铬合金 Alumel chromel
镍铝镍铬热电偶 Alumel chromel thermocouple; Alumel chromel thermoelement
镍铝青铜 nickel-alumin(i)um bronze
镍铝蛇纹石 brindleyite
镍铝旋磁铁氧体 Ni-Al gyromagnetic ferrite
镍绿 nickel green
镍绿泥石 garnierite;nepouite;nickel-chlorite;nimite
镍络偶氮黄 nickel azo yellow
镍锰电阻合金 magno
镍锰钢 konik(e)
镍锰合金 nickel-manganese; nickel-manganese alloy
镍锰黄铜 nickel-manganese brass
镍锰青铜 nickel-manganese bronze
镍锰铁高导磁合金 Megaperm
镍锰铸钢 nickel-manganese cast steel
镍钼钢 nickel-molybdenum steel
镍钼合金 nickel-molybdenum alloy
镍钼热电偶 nickel-molybdenum thermocouple
镍钼铁 nickel-molybdenum iron
镍钼铁超导磁合金 malloy
镍钼铁合金 < 一种磁畴定向的软磁性合金 > Dynamax
镍钼铁弹簧合金 vibralloy
镍钼温差电偶 nickel-molybdenum thermocouple
镍黏[粘]结硬质合金 nickel-cemented carbide
镍配合基体系 nickel-ligand system
镍屏 nickel screen
镍铅铜 nickel-leaded bronze
镍青铜 nickel-bronze
镍青铜合金 nickel-bronze alloy
镍氢蓄电池 nickel-hydrogen battery; nickel hydrogen storage cell
镍砷钴矿 nickel-smaltite
镍试剂 nickel reagent
镍束纹石 maufite
镍水镁石 nibrucite
镍水蛇纹石 genthite
镍酸盐 nickelate
镍钛硬质合金 nickel-cemented titanium carbide
镍钽合金 nickel tantalum alloy
镍碳化硅镀盖缸壁 nickel silicon carbide coated housing
镍碳化硅敷层 nickel silicon carbide coating
镍碳化硅缸壁型面 nikel silicon carbide surface
镍碳化钨敷层 nickel tungsten-carbide coating
镍碳化钨涂层 nickel tungsten-carbide coating
镍碳热电偶 nickel-carbon couple
镍碳铁矿 cohenite
镍锑辉矿 nickel antimony glance
镍条刮擦 nickel scratch
镍铁 ferro-nickel (iron); nickel ferrite;nickel-iron
镍铁层 nifesphere
镍铁磁合金 Rhometal
镍铁磁芯材料 Orthonik
镍铁带 nife
镍铁地核假说 NiFe hypothesis
镍铁电池 nickel-iron cell;NiFe cell
镍铁电池组 Edison battery; nickel-iron battery
镍铁锭 ferro-nickel ingot
镍铁高磁导率合金 mu-metal
镍铁钴钒合金 velinvar
镍铁钴高导磁率合金 permivar
镍铁硅镁带 nifesima
镍铁合金 Dilver (alloy); nickalloy; nickel-iron alloy;Nickeloy
镍铁碱性蓄电池 nickel-iron alkaline cell; nickel-iron alkaline storage battery
镍铁矿 josephinite
镍铁锍 nickel-iron matte
镍铁钼超导磁合金 super-malloy
镍铁双金属 nickel-iron bimetal
镍铁体 nickel ferrite
镍铁蓄电池 nickel-iron storage battery;Nife accumulator
镍铁陨石 ataxite
镍铁陨星 catarinite
镍铜比 nickel-copper ratio
镍铜铬耐蚀铸铁 Ni-resist
镍铜铬铸铁 nickel-copper-chrome cast iron; nickel-copper chromium cast iron
镍铜合金 constantan; cupro-nickel; konstantan; nickel-copper (alloy); thermalloy
镍铜锰铁合金 Monel(metal)
镍铜钎焊 nickel brazing
镍铜丝 constantan wire
镍铜铁氧体 nickel-copper ferrite
镍铜锌合金 nickel brass
镍铜锌合金电阻丝 platinoid
镍铜锌系合金 Gallimore
镍丸 nickel rondelle;nickel shot
镍网 nickel screen
镍纹石 taenite
镍污染 nickel contamination; pollu-

N

tion by nickel
镍钨合金 nickel tungsten
镍钨硬质合金 nickel-cemented tungsten carbide
镍锡合金 nickeltin
镍纤蛇纹石 pecoraite
镍心 nickel core
镍锌电池 nickel-zinc cell
镍锌合金 coronite
镍锌水滑石 eardleyite
镍锌铁氧体 nickel-zinc; nickel-zinc ferrite
镍锌铜合金 German silver
镍延迟线 nickel delay line
镍盐 nickel salt
镍阳极 nickel anode
镍氧化铝敷层 nickel-alumin(i)um oxide coating
镍药皮焊条 nickel covered welding electrode
镍冶金 metallurgy of nickel
镍叶绿泥石 rottisite
镍衣铝粉粒 nickel alumide
镍阴极 nickel cathode
镍银 electrum;nickel silver;Silveroid; silviroid
镍银铋合金 proplatinum
镍银合金 baza(a)r metal;nickel silver alloy
镍印法 nickel print
镍营养 nickel nutrition
镍浴槽 nickel-plating bath
镍纸 nickel paper
镍制连枷状搅拌器 nickel flail stirrer
镍制配件 nickel fittings
镍滞延线 nickel delay line
镍中毒 nickel poisoning
镍珠 nickel bead;nickel pellet
镍铸铁 nickel cast iron;nickel-tensilorin
镍铸铁轧辊 nickel-iron roll

宁

宁静磁层 quiet-time magnetosphere
宁静分量 quiet component
宁静辐射电平 base level
宁静光谱 quiescent spectrum
宁静极光 quiet-form aurora
宁静逆温污染 calm inversion pollution
宁静区 calm zone
宁静热发射 quiet thermal emission
宁静日变化 quiet-day variation
宁静日珥 quiescent prominence
宁静日冕 quiet corona
宁静石 tranquillityite
宁静太阳 quiet sun
宁静太阳风 quiet-time solar wind
宁静太阳射电 quiet solar radio radiation
宁静太阳条件 quiet sun condition
宁静太阳噪声 quiet sun noise
宁静态光度 quiescent luminosity
宁静烟雾 calm smog

拧

拧出 screw off;screw out;unscrew
拧到头 screw home
拧得过紧 overtighten
拧断螺纹 overturn the thread
拧盖 screw capping
拧干 wring out
拧管机 breakout gun;breakout table
拧管链条 spin-up chain
拧管扭矩指示器 tong torque controller;tong torque indicator
拧管时把丝扣拧坏 take a chance on wrenching bits out of round
拧管子用旋转扳手 pipe hook
拧绞 <线材或轧件的> coil buckling;kinky
拧接 screw together
拧接螺栓 extension bolt
拧接式钻杆 sectional rod
拧接钻杆 extension rod
拧结机构 twisting mechanism
拧结器 twister
拧紧 bolting;coarctation;screw home; screw in;screw on;screw up;tighten;turn home;buck-up <管接头>
拧紧扳手 impact wrench
拧紧不够 undertighten
拧紧的接头 screwed coupling
拧紧的连接管子 screwed coupling pipe
拧紧的耦联器 screwed coupling
拧紧管接头 handling tight
拧紧铰接卡钳 firm-joint cal(l)ipers
拧紧接头 tightening coupling
拧紧力矩 tightening torque
拧紧螺母 nut-running
拧紧螺(丝)钉 thread mounting
拧紧扭矩 tightening torque
拧紧配合 wringing fit
拧紧器 tightener
拧紧头 tappet head
拧开 <钻头或钻杆> ring off;screw off;twist-off;unscrew
拧开螺栓 unbolt
拧开瓶盖 twist-off closure
拧龙头 turn a tap
拧螺钉杆 wrenching bar
拧螺帽 nutting
拧螺母机 nut driving machine; nut fastening machine
拧螺栓工人 wrencher
拧螺柱器 stud remover
拧埋头栓工具 countersinking tool
拧钎头器 bit breaker
拧入 screw in
拧入式滤光片环 screw-in filter cell
拧入式滤清器 spin-on filter
拧上 screw on;tie on;wrench up
拧上的管子 screwed-on pipe
拧上的铰链 screwed-on hinge
拧上的锁 screwed-on lock
拧上螺钉的 screwing on
拧上丝扣 thread up
拧松 back-out(of); screw off; unscrew
拧松螺(丝)钉 unscrew
拧索 cable-laid rope
拧下[螺(丝)钉] screw off;unscrew
拧卸工具 making-up or breaking-out tools
拧卸及摆管 tailing-out rods
拧卸钻杆 breaking down the pipe
拧转 retort(ion)
拧转式放炮器 twist machine

柠

柠康酸 citraconic(al)acid
柠康酸酐 citraconic(al)anhydride
柠檬 Citrus limon;lemon;limo
柠檬桉 Eucalyptus citriodora;lemon eucalyptus;lemon gum;lemon scented gum
柠檬苍白色 lemon pale
柠檬虫漆 lemon shellac
柠檬铬黄 chromo-citronine; citron yellow;lemon chrome yellow;primrose(chrome)yellow
柠檬铬黄颜料 chrome lemon yellow
柠檬褐腐病 lemon brown rot
柠檬黄级紫胶 lemon shellac
柠檬黄(色) citrine; citron; lemon chrome yellow;lemon yellow
柠檬黄颜料 lemon yellow pig
柠檬胶 lemon lac
柠檬木 lemon wood
柠檬皮条片 lemon grating
柠檬塞缝片 lemon spline
柠檬色 citrine;citron
柠檬树 lemon
柠檬素钠 sodium citrate
柠檬酸 citrate acid;citric acid
柠檬酸铵 ammonium citrate
柠檬酸丁酯 butylcitrate;tributyl citrate
柠檬酸法烟气脱硫 flue gas desulfurization with sodium citrate
柠檬酸废水 citric acid wastewater
柠檬酸废水污泥 citric acid wastewater sludge
柠檬酸钙 calcium citrate
柠檬酸工业废水 citric acid industrial wastewater
柠檬酸钾 potassium citrate
柠檬酸锂 lithium citrate
柠檬酸连二硫酸盐重碳酸盐 dithionite critrate-bicarbonate
柠檬酸镁 magnesium citrate
柠檬酸钠 natrium citricum; sodium citrate
柠檬酸氢钠 natrium hydrocitricum
柠檬酸三丁酯 tributyl citrate
柠檬酸三乙酯 triethyl citrate
柠檬酸石膏 citric acid gypsum
柠檬酸铁铵 ferric citrate;iron citrate
柠檬酸铜 copper citrate
柠檬酸戊酯 amyl citrate
柠檬酸循环 citric acid cycle
柠檬酸盐 citrate;citrated copper salt
柠檬酸盐利用试验 citrate utilization test
柠檬酸盐溶性磷 citrate soluble phosphate
柠檬酸盐物种 citrate species
柠檬酸银 silver citrate
柠檬酸酯 citric acid ester
柠檬酸酯增塑剂 citrate plasticizer
柠檬铁铵 ferric ammonium citrate
柠檬盐 lemon salt;salacetos;sal limonis
柠檬油 lemon oil
柠檬子油 lemon pips oil;lemon seed oil

凝

凝冰器 cryophorus;ice condenser
凝成 cement together
凝点 set point;solidifying point
凝点安定性 pour stability
凝点记录仪 eutectometer
凝定 briquet(te); briquetting; set; setting
凝定点 setting point
凝定时间 setting time
凝固 clotting; coagulate; coagulation; concretion;concretize;congeal;congealation;congealing;curdling;freeze-(in);harden(ing);set hard;solidify-(ing);development of rigidity <水泥浆、水泥砂浆、混凝土的>
凝固不足 under-ag(e)ing
凝固程度 freezing level
凝固催速剂 setting accelerator
凝固萃取 liquid solid extraction
凝固带 zone of consolidation
凝固的 cak(e)y;concretionary;consolidated;fixed;hardened;hard set; freezing
凝固的锭料 solidified ingot
凝固的混凝土 set concrete
凝固点 chill point;congealing point; congelation point;freezing point; freezing temperature;point of solidification;set(ting)point;solidification point;solidification temperature;solidification value;solidifying point;solid point;zero pour;condensation point
凝固点测定计 kryoscope
凝固点测定装置 measuring apparatus for solidification point
凝固点降低 depression of freezing point;freezing-point depression
凝固点降低常数 freezing constant
凝固点降低法 cryoscopic method
凝固点曲线 freezing-point curve
凝固点试验 setting-point test
凝固点以上温度 temperature above freezing
凝固点以下温度 temperature below freezing
凝固锭料 frozen ingot
凝固法 freeze-out method; freezing method;set method
凝固反应 coagulation reaction;coaguloreaction;freezing reaction
凝固分界面 freezing interface
凝固高度【气】 freezing level
凝固过程 process of consolidation; process of setting; setting procedure
凝固核 freezing nucleus
凝固后构造 post-solidification structure
凝固混凝土 hardened concrete
凝固机理 freezing mechanism; mechanic of set(ting)
凝固剂 coagulant; coagulating(re)agent;coagulator;peptizer;setting(-up)agent
凝固胶 room temperature setting adhesive
凝固结构 consolidated structure;solidification structure
凝固介质 setting medium
凝固界面 freezing interface
凝固金属 frozen metal
凝固进程 progress of consolidation
凝固控制 set control
凝固力 power of hardening
凝固锍 solidified matte
凝固煤油 solidified kerosene
凝固面 solidifying front
凝固泥浆 solidified slip
凝固皮带 endless belt;setting belt
凝固平衡 liquid solid equilibrium
凝固期 period of hardening; solid stage
凝固期试验针 Gillmore needle
凝固汽油 gelatinized gasoline;gelled gasoline; incinderjell; jellied gasoline;solidified gasoline
凝固汽油弹 petrol bomb
凝固汽油剂 napalm
凝固汽油燃烧弹 jelly bomb
凝固汽油箱 napalm tank
凝固潜热 latent heat of solidification
凝固强度 setting strength;strength of set(ting)
凝固区 zone of consolidation
凝固区间 freezing range
凝固曲线 freezing curve;setting curve
凝固燃烧剂 incinderjell
凝固热 heat of freezing;heat of hardening;heat of solidification;setting heat;solidification heat
凝固乳胶 coagulum
凝固石油 solidified petroleum
凝固时间 clotting time; coagulation time;freezing time;set time;setting

time;time of setting
凝固时间测定仪 setting-time apparatus
凝固试验 hardening test;setting test
凝固试验机 setting tester
凝固试验仪(器)setting testing apparatus
凝固收缩 liquid solid contraction;setting shrinkage;solidification shrinkage
凝固树脂 cured resin;hardened resin
凝固素 coaguin
凝固速度 freezing rate;rate of coagulation;setting rate
凝固速率 rate of cure
凝固特性 setting behavio(u)r
凝固梯度 solidification gradient
凝固图像 frozen picture
凝固温度 condensation temperature;freezing temperature;setting temperature;solidification temperature;temperature of solidification
凝固温度范围 solidification range
凝固物 coagulum
凝固雾 Snowcrete
凝固线 line of solidification
凝固相 solidifying phase
凝固效应 setting effect
凝固性 coagulability;fixedness;fixity;setting quality
凝固性过低 hypocoagulability
凝固性过高 hypercoagulability
凝固性过高的 hypercoagulable
凝固性质 setting property
凝固雪 Snowcrete
凝固因子 coagulation factor
凝固因子缺乏 coagulation factor deficiency
凝固硬化点 congealing point
凝固釉 solidified glaze
凝固浴 spinning bath
凝固浴管 bath-fed tube
凝固渣 solidified slag
凝固障碍 coagulation disorder
凝固状况 setting condition
凝固状态 freeze mode
凝固作用 freezing action;solidification
凝花【化】flore
凝华核 sublimation nucleus
凝灰辉绿岩 ash-bed diabase
凝灰火山 tuff volcano
凝灰火山岩 tuff volcanic rock
凝灰火山锥 ash cone;cinder cone
凝灰集块熔岩 tuff agglomerate;tuff agglomerate-lava
凝灰角砾岩 tuff breccia
凝灰熔岩 tuff lava
凝灰石 travertine
凝灰岩 ash rock;ashstone;eruptive tuff;tufa;tuff;tuffaceous limestone;tuff rock
凝灰岩层 ash bed
凝灰岩的 tuffaceous
凝灰岩壤土 tuff loam
凝灰岩水泥 tuff(aceous)cement
凝灰岩(水泥)混凝土 tuffcrete
凝灰岩碎屑 clastic tuff
凝灰岩相 tuffaceous facies
凝灰岩锥 tuff cone
凝灰质白云岩 tuffaceous dolomite
凝灰质板岩 ash slate;tuffaceous slate
凝灰质粗砾岩 tuffaceous cobblestone
凝灰质粗砂岩 tuffaceous coarse sandstone
凝灰质的 tuffaceous
凝灰质粉砂岩 tuffaceous siltstone
凝灰质硅质岩 tuffaceous siliceous rock
凝灰质化学岩 tuffaceous chemical sedimentary rock
凝灰质集块岩 tuffaceous agglomerate
凝灰质角砾岩 tuffaceous breccia
凝灰质巨砾岩 tuffaceous boulder
凝灰质砾岩 tuffaceous conglomerate
凝灰质泥灰岩 tuffaceous marl
凝灰质泥岩 tuffaceous mudstone
凝灰质黏[粘]土 gault(clay)
凝灰质砂岩 tuff sandstone
凝灰质生物碎屑灰岩 tuffaceous bioclastic limestone
凝灰质石膏岩 tuffaceous gyprock
凝灰质石灰岩 tuffaceous limestone
凝灰质细砾岩 tuffaceous granulestone
凝灰质细砂岩 tuffaceous fine sandstone
凝灰质盐岩 tuffaceous halite rock
凝灰质页岩 ashy shale;tuffaceous shale
凝灰锥 tuff cone
凝灰浊流岩 tuff-turbidite
凝集 cohesion bond;congeal
凝集测验 agglutination
凝集簇 agglutinophore
凝集的 agglutinative
凝集反应 agglutination reaction
凝集反应镜 agglutinoscope
凝集反应镜检查 agglutinoscopy
凝集反应器 agglutometer
凝集反应试管 agglutination reaction test tube
凝集极限 cohesion limit
凝集剂 agglomerant;agglutinant;coagulator;settle accelerator
凝集价 agglutination titer
凝集价测定 determination of agglutination titer
凝集检查镜 glutoscope
凝集能 agglutinability
凝集能密度 cohesive energy density
凝集溶解试验 agglutination lysis test
凝集试验 agglutination test
凝集素 agglutinin
凝集素吸收 agglutinin absorption
凝集素吸收试验 agglutinin absorption test
凝集为绒毛状沉淀 flocculate
凝集物 agglutinator
凝集吸收作用 agglutinative absorption
凝集现象 agglutination phenomenon
凝集性 compendency
凝集原 agglutinogen;agglutogen
凝集作用 agglutination
凝胶 gelatin(e);gelatum;jell(y)
凝胶表面积 gel surface area
凝胶层 gelatin(e)layer
凝胶层析 gel chromatography
凝胶沉淀概率 gel precipitation probability
凝胶沉淀过程 gel precipitation process
凝胶沉积 gel deposition
凝胶稠度 gel consistency;jelly consistency
凝胶处理 gel treatment
凝胶的 gelatinous
凝胶底层 gel sub
凝胶电聚焦 gel electrofocusing
凝胶电泳 gel electrophoresis
凝胶法 <测孔斜的> gelating method
凝胶纺丝 gel spinning
凝胶分段分离器 gel fractionator
凝胶构造 gelatin(e)structure
凝胶骨架 gel skeleton
凝胶过滤 gel filtration
凝胶过滤层析 gel filtration chromatography;molecular exclusion chromatography;molecular sieve chromatography
凝胶过滤色谱(法)gel filtration chromatography;molecular exclusion chromatography;molecular sieve chromatography
凝胶海绵 gel sponge
凝胶化 gelatination;gelating;gelatinisation[gelatinization];gelatinize;gelation;curdling
凝胶化度 jelly grade
凝胶化法 gelling technique
凝胶化浑圆体 gelified circleinite
凝胶化基质体 gelified groudmassinite
凝胶化菌类体 gelified sclerotinite
凝胶化时间 gel time
凝胶化组 gelinite group
凝胶化作用 gelatification;gelling
凝胶剂 gelata;gels;jellies
凝胶加添过程 gel addition process
凝胶鉴定法 gelodiagnosis
凝胶结构 gel structure
凝胶空间比 gel-space ratio;gel-void ratio
凝胶空隙 gel pore
凝胶孔水 gel pore water
凝胶孔隙 gel pore
凝胶孔隙比 gel-space ratio
凝胶孔隙率 gel porosity
凝胶块 gel particle
凝胶扩散 gel diffusion
凝胶扩散沉淀试验 gel diffusion precipitation;gel precipitin test
凝胶扩散法 gel diffusion method
凝胶类 gelinite
凝胶罗盘 <测孔斜的> gelatin(e)-compass
凝胶煤素质 collinite
凝胶内聚力 gel coherence
凝胶排阻色谱(法)gel exclusion chromatography
凝胶盘状电泳 gel disc electrophoresis
凝胶喷射器 gel coat sprayer
凝胶漆 gel coat(ing);gel lacquer;gel paint;jelly paint
凝胶气孔 gel pore
凝胶强度 gel strength
凝胶强度试验器 Alexander tester
凝胶染色法 gel dy(e)ing method
凝胶溶胀度 gel swelling
凝胶润滑剂 lubricating jelly
凝胶色谱(法)exclusion chromatography;gel chromatography;molecular exclusion chromatography;steric exclusion chromatography
凝胶闪烁体 gel scintillator
凝胶渗透层析 gel permeation chromatography
凝胶渗透色层(分离)法 gel permeation chromatography
凝胶渗透色谱(法)gel permeation chromatography
凝胶渗透液相层析 gel permeation liquid chromatography
凝胶渗透液相色谱(法)gel permeation liquid chromatography
凝胶渗压计 gel osmometer
凝胶生长法 gel growth method
凝胶时间测定计 gelemeter
凝胶试验 gel(ation)test
凝胶水 gel water
凝胶丝 gelatin(e)silk
凝胶丝炭化作用 gelefusainization
凝胶态离子交换树脂 gel type ion exchange resin
凝胶体 gel;gelinite;jell
凝胶体积 gel pore volume
凝胶涂料 gel coat(ing)
凝胶涂刷面 gel coated surface
凝胶网络 gel network
凝胶微粒 microgel particle
凝胶物质 gelling material
凝胶析出 gel precipitation
凝胶纤维 gelatinous fiber[fibre];gel(led)fiber[fibre]
凝胶橡胶 gel rubber
凝胶效应 gel effect
凝胶形成 gel formation
凝胶型 gel type
凝胶型防锈添加剂 gelling type rust preventive
凝胶型固体 gel type solid
凝胶型基料 gel type binder
凝胶型树脂 gel type resin
凝胶型涂料 gel type lacquer
凝胶型阳离子树脂 gel type ion exchange resin
凝胶液纺丝 gelatin(e)solution spinning
凝胶液相色谱仪 gel permeation chromatography
凝胶柱 gel column
凝胶状沉淀 gelatinous precipitate
凝胶状的 gelatinous;gel-like compound
凝胶状化合物 gel-like compound
凝胶状乳凝块 gelatinous curd
凝胶状水泥 jelled cement
凝胶状态 gel condition;gel state
凝胶自旋 gel spinning
凝胶作用 jellification
凝结 clot;coagulate;coagulating(re)-agent;compaction;condense;condensation <气体变为液体的>;congeal;conglomeration;curdle;curdling;gelation;jelly;precipitate;setting up;clotting;coagulation;concretion;congealment;congelation;setting <液体变为固体的>
凝结槽 coagulating bath
凝结层 condensation layer
凝结沉淀池 coagulating basin;condensation precipitation tank
凝结沉淀现象 condensation precipitation phenomenon
凝结沉降作用 coagulating sedimentation;condensation precipitation
凝结成的 concrete
凝结池 coagulation basin
凝结迟延 retardation of set(ting)
凝结处理 coagulation treatment
凝结的 coagulative;concretionary
凝结的水气(或雨、露等)precipitate
凝结点 coagulating point;condensation temperature;coagulation point;congelation point;set(ting)point
凝结动力学 kinetics of condensation
凝结法 condensation method
凝结反应 reaction of set(ting);set(ting)reaction
凝结防湿 condensation dampproofing
凝结放热量 heat of hardening
凝结放热系数 condensing coefficient
凝结干扰 disturbance of set(ting)
凝结干燥 drying by condensation
凝结高度 condensation level;freezing level
凝结构造 coagulation structure
凝结管 condenser tube
凝结过程 process of setting;setting process
凝结核 condensation nucleus;embryonic droplet;nucleus[复 nuclei] of condensation
凝结核大小分光计 condensation nucleus size spectrometer
凝结核计数器 condensation nucleus counter
凝结机理 setting mechanism
凝结激波 condensation shock wave
凝结剂 agglomerating agent;coagu-

N

lant; coagulating (re) agent; coagulator; setting agent
凝结加速剂 setting accelerator
凝结降水量 condensation precipitation; condensation rainfall
凝结降雨量 condensation precipitation; condensation rainfall
凝结结构 cemented formation
凝结介质 condensating medium
凝结金属 frozen iron
凝结绝热 condensation adiabat
凝结块 coagulum
凝结理论 condensation theory
凝结力 coagulability; coagulating power; setting power
凝结流 condensing flow
凝结率 setting rate; setting value
凝结面 condensation level
凝结能 energy of set(ting)
凝结能力 capacity of set(ting); coagulability
凝结能量 setting energy
凝结胚 embryonic droplet
凝结膨胀 setting expansion
凝结期 setting period
凝结气驱 condensing gas drive
凝结气压 adiabatic condensation pressure; adiabatic saturation pressure; condensation pressure
凝结器 coagulator; condensator; condenser; distiller
凝结强度 set(ting) strength
凝结区 condensing zone
凝结曲线 curve of set(ting)
凝结趋向 condensation tendency
凝结扰动 setting disturbance
凝结热 heat of condensation; heat of setting; set heat
凝结乳脂 bitty cream
凝结时间 jelling time; pot life; set(ting)(-up) time; time of setting
凝结时间测定 set test
凝结时间测定仪 setting(time) tester; setting-time apparatus
凝结时间可调整的水泥 regulated-set cement
凝结时间试验 setting-time test
凝结时间试验器 setting-time apparatus
凝结时间调节剂 set control agent
凝结时石膏膨胀 gypsum expansion on setting
凝结实验 setting test
凝结试验 set test; setting-time test
凝结试验仪 set testing apparatus
凝结室 coagulation chamber
凝结收缩 set(ting) shrinkage
凝结收缩裂缝 setting shrinkage crack
凝结水 condensate; condensation water; condensed water; condensing water; precipitation water; set(ting) water; water of condensation; water of setting
凝结水背压力 back-pressure of steam trap
凝结水泵 condensate pump; hot well pump
凝结水补给 recharge of condensation water
凝结水槽 condensation channel; condensation groove; condensation gutter; condensation trough
凝结水处理 condensate treatment
凝结水阀 condensate valve
凝结水分 congealed moisture; precipitation moisture
凝结水管(道) condensate pipe; condensing water conduit; condensing water pipe; condensing water piping
凝结水管路 condensate line
凝结水管线 condensate circuit
凝结水过冷(却) condensate depression; hot well depression
凝结水过滤器 condensate filter; condensation water filter
凝结水回流 condensate return
凝结水回流管 condensate return pipe
凝结水回流集水池 condensate return collecting tank
凝结水回路 condensate circuit
凝结水回收率 condensate recovery percentage
凝结水回收箱 condensate return tank
凝结水井 catch pot; hot well
凝结水控制 condensate control
凝结水冷却器 condensation cooler
凝结水利用装置 condensate scavenging installation
凝结水量 condensation water quantity
凝结水滤网 condensate strainer
凝结水泥 set cement
凝结水泥的楔紧作用 wedging-in action of setting cement
凝结水排水泵 condensate extraction pump
凝结水盘 condensate drain pan
凝结水升压泵 condensate booster pump
凝结水系统 condensate circuit; condensate system
凝结水箱 condensate tank; condensation tank; hot well
凝结水循环管 condensate circulating water pipe
凝结水溢流管 condensate spill over pipe
凝结水引出管 condensate outlet
凝结水再循环管路 condensate return piping
凝结速度 rate of condensation; rate of set; set(ting) rate
凝结速率 rate of setting
凝结损耗 loss due to condensation
凝结损失 condensation loss
凝结尾迹 condensation trail; contrail
凝结尾迹形成图 contrail-formation graph
凝结尾流 contrail
凝结温度 adiabatic saturation temperature; condensation temperature; setting temperature; temperature of set(ting)
凝结温度图 condensation temperature diagram
凝结物 coagulate; coagulum; concrete; concretionary; condensate
凝结现象 phenomenon of coagulation
凝结效果 <瓷釉的> curdle effect
凝结效率 condensation efficiency
凝结性 coagulability; condensability; condensation; condensibility
凝结性的 concretive
凝结性能 set behavio(u)r
凝结絮化 coagulation-flocculation
凝结岩层 cemented formation
凝结液 condensing liquor
凝结液储槽 reservoir for condensation
凝结硬化 setting and hardening
凝结淤泥 coagulated silt
凝结增温 heating by condensation
凝结值 coagulation value; condensation value
凝结质量 quality of set(ting)
凝结终止 end of set(ting)
凝结状态 set behavio(u)r
凝结作用 coagulation; desublimation
凝晶质 crystalloid
凝聚 agglutinate; coagulate; coalescence; coalescing; condensation; condensing; conglomeration
凝聚波 condensational wave; wave of condensation
凝聚层 coacervate course; coacervate
凝聚沉淀 coagulation sediment; coagulation sedimentation; coagulative precipitation
凝聚沉淀池 coagulating sedimentation tank; coagulating settling tank; coagulative tank
凝聚沉淀法 coagulation and flocculation process
凝聚沉淀装置 cyclator
凝聚成团 conglomerate
凝聚程度 stage of aggregation
凝聚池 coagulating tank; coagulation tank; coagulating basin; coagulation basin
凝聚粗颗粒 seed
凝聚点 accumulation point; congealing point
凝聚反应 polycondensation reaction
凝聚反应速度 coagulate reaction
凝聚分布 degenerate distribution
凝聚浮冰 <群> conglomerated pack
凝聚过程 aggregation process; coagulative process
凝聚过滤器 coalescent filter; coalescer
凝聚函数 coherency function
凝聚极限 coagulation threshold
凝聚剂 agglomerant; agglomerator; coagulant; coagulating (re) agent; coagulator; coalescer; congelation agent; flocculating agent
凝聚胶 coagel
凝聚胶体 flocculated colloid
凝聚结构 coherence structure
凝聚块 condensation nucleus
凝聚力 coagulating power; coherence; cohesion; cohesive affinity; cohesive force; cohesiveness; force of cohesion
凝聚滤床处理 granular packed bed coalescer
凝聚面积 area of cohesion
凝聚能力 coagulability
凝聚破坏 cohesive failure
凝聚器 coagulator; knockout
凝聚区 condensation region
凝聚热传递 condensing heat transfer
凝聚设备 coagulative device
凝聚室 coagulating chamber; coagulation chamber
凝聚态 condensed state
凝聚态物质 condensed matter
凝聚态物质理论【物】condensed-matter theory
凝聚体 aggregate; aggregation body; coacervate
凝聚体系 coacervated system; condensed system
凝聚物 coagulated matter; coagulum; condensation polymer; condensation product
凝聚物系 condensed system
凝聚物质 condensed substance
凝聚系数 coefficient of cohesion; coefficient of concentration; condensation coefficient
凝聚系统 condensed system
凝聚系相律 condensed system phase rule
凝聚相 coacervated phase; condensed phase
凝聚箱 coagulation box; coagulation tank
凝聚效应 agglomeration effect
凝聚星 condensed star
凝聚型气溶胶 coagulated aerosol
凝聚性 coagulability; coherence [coherency]; cohesion; cohesiveness; cohesive property
凝聚性冲积层 coherent alluvium
凝聚异常 coagulate anomaly; condensation anomaly
凝聚增长 agglomeration; coagulation
凝聚者 agglomerator
凝聚值 coagulating value; coagulation value
凝聚阻力 cohesional resistance
凝聚作用 coacervation; coagulation; coagulation action; condensation action; flocculation; polycondensation reaction
凝壳 ice; kish; scull[skull]
凝壳炉 skull crucible; skull furnace
凝壳熔炼法 skull melting
凝壳铸造 slush casting
凝块 clot; concretion; gelosis
凝块的 clotty
凝块密度 bunch density
凝块石 catagraph
凝离 segregate
凝露检测器 moist detector
凝膜 haptogen
凝气瓣 gas trap
凝气管 catch tank
凝气器 capacitor; vapo(u)r condenser
凝汽 condensed steam
凝汽发动机 condensing engine
凝汽阀 condenser valve; steam trap
凝汽阀体 condenser valve body
凝汽罐 drainage receiver
凝汽机组 condensing engine set
凝汽盘管 condenser [condensator] coil; condensing coil
凝汽器 condenser
凝汽器安全阀 condenser relief valve
凝汽器安全装置 condenser safety device
凝汽器抽管距离 clearance for pulling condenser tubes; condenser tube pulling space
凝汽器垫密片 condenser gasket
凝汽器管道 condenser piping
凝汽器管道摩阻 condenser tube friction
凝汽器颈部法兰 condenser flange
凝汽器热负荷 condenser duty
凝汽器热交换 condenser heat transfer
凝汽器水室 condenser water box
凝汽器水位调节阀 condenser level control valve
凝汽设备 condensing equipment; condensing plant
凝汽式抽汽涡轮机 condensing bleeder turbine
凝汽式电厂 condensing power station
凝汽式发电机 condensing power plant
凝汽式机车 condensing locomotive
凝汽式汽轮机 condensing steam turbine
凝汽式透平 condensing steam turbine
凝汽式涡轮机 condensing steam turbine
凝汽筒 trap
凝汽油剂 napalm
凝汽装置 condensing plant
凝前期 presetting period; resetting period
凝溶试验 agglutination lysis test
凝乳 <沥青凝聚时的产物> curd; curdled milk
凝乳剂 milk coagulant
凝乳结块不良 gelatinous curd deject
凝乳状沉淀物 curdy precipitate

凝入 freezing in
凝入反应 freezing in reaction
凝入杂质 < 结晶等 > freeze-in impurity
凝砂块 < 砂丘砂和碳酸钙胶凝砂的混合料 > kurkar
凝霜 efflorescence;glazed frost
凝水 condensation;fogging;glare ice
凝水密封 condensate seal
凝水排除器 steam trap
凝水自动排除器 automatic steam trap
凝酸 saltpetering
凝缩器 densener
凝缩性 condensability
凝土拌和车 truck mixer
凝土路面铺筑机械 concrete pavement machine
凝土四脚锥体(块) < 防波堤用 > concrete tetrapod
凝土锥形桩 concrete taper pile
凝析气 condensate gas
凝析气藏分布区 distribution area of condensated gas pool
凝析气藏平面图 planimetric(al) map of condensate gas pool
凝析气藏剖面图 sectional drawing of condensate gas pool
凝析气井 gas condensate well
凝析气田 gas condensate field
凝析油 gas distillate;gasol;oil liquor
凝析油伴生气 gas associated with condensate oil
凝析油带 condensed oil gas zone
凝析油气藏 condensed oil gas pool
凝絮试验 floc test
凝液 lime set
凝液井 condensate well
凝液收集袋 drip pocket
凝液收集管线 condensate collection line
凝液收集器 condensate collector
凝液收集匣 drip pocket
凝溢 syneresis
凝硬时间 < 混凝土的 > setting-up time
凝硬性的 pozz(u)olanic
凝硬作用 < 水泥的 > pozzolanic action
凝渣 crust block
凝脂润滑油 solid grease
凝脂油 grease;grease oil

牛

牛鼻钳 bull-holder
牛鼻形不取芯钻头 bull-nose bit
牛鼻形砖 bull-nose brick
牛鼻砖 jamb brick
牛场废水 cowshed waste
牛车 bullock cart
牛顿 Newton;large dyne
牛顿八分之三法则 Newton's three-eighths rule
牛顿材料 Newtonian material
牛顿参考坐标系 Newtonian reference frame
牛顿插值 Newton's interpolation
牛顿插值公式 Newton's interpolation formula
牛顿的 Newtonian
牛顿的运动方程 Newton's equation of motion
牛顿第二定律 Newton's second law
牛顿第三定律 Newton's third law
牛顿第一定律 Newton's first law;first law of motion
牛顿迭代法 Newton's iteration method
牛顿定理冷却 Newtonian cooling
牛顿定律 Newton's law
牛顿法 Newton's method
牛顿反射式望远镜 Newtonian reflector
牛顿方程 Newton's equation
牛顿概念 Newton's concept
牛顿公式 Newton's formula
牛顿关系式 Newton's relation
牛顿合金 < 一种低熔点合金 > Newton's alloy;Newton's metal
牛顿恒等式 Newton's identity
牛顿环 Newton's ring
牛顿活化分析 Newton activation analysis
牛顿极限黏[粘]度 Newtonian limiting viscosity
牛顿假说 Newton's hypothesis
牛顿剪切黏[粘]度 Newtonian shear viscosity
牛顿焦点 Newtonian focus
牛顿近似法 Newton's approximation
牛顿近似计算法 Newton's method of approximation
牛顿-卡塞格仑望远镜 Newtonian-Cassegrain telescope
牛顿-柯特斯公式 Newton-Cotes formula
牛顿-拉夫逊迭代法 Newton-Laphson iteration
牛顿-拉夫逊法 < 潮流计算 > Newton-Laphson method
牛顿-拉夫逊公式 Newton-Laphson formula
牛顿-拉夫逊算法 Newton-Laphson algorithm
牛顿冷却定律 Newton's cooling law;Newton's law of cooling;Newtonian cooling law
牛顿力学 Newtonian mechanics
牛顿流动 Newtonian flow
牛顿流动性 Newtonian flow property
牛顿流体 < 剪切率与剪应力成正比时的流体 > Newtonian fluid
牛顿流体冷却 Newtonian cooling
牛顿米 Newton's meter[metre]
牛顿米扭矩单位 Newton's meter torque unit
牛顿/秒 Newton per second
牛顿模型 Newton's model
牛顿摩擦定律 friction(al) law
牛顿内容量 Newtonian inner capacity
牛顿黏[粘]性流动定律 Newton's law of viscous flow;Newton flow of viscous flow
牛顿黏[粘]滞度 Newtonian viscosity;Newton's viscosity
牛顿黏[粘]滞体 Newton's viscous body
牛顿碰撞理论 Newton's theory of collision
牛顿平方根法 Newton's square-root method
牛顿平行四边形 Newton's parallelogram
牛顿前向插值公式 Newton's forward interpolation formula
牛顿区 Newtonian region
牛顿圈 Newton's ring
牛顿容量 Newtonian capacity
牛顿润滑层 Newtonian lubricating layer
牛顿三叉线 trident of Newton
牛顿蛇形线 Newton's serpentine
牛顿升力理论 Newton's theory of lift
牛顿声速 Newtonian speed of sound
牛顿时(间) Newtonian time
牛顿(氏)色盘 < 七色盘 > Newton's disk[disc]
牛顿式反射望远镜 Newton's reflecting telescope
牛顿式望远镜 Newton's telescope
牛顿式指示器 Newton's viewfinder
牛顿数 Newton's number
牛顿速度 Newtonian velocity
牛顿特性 Newton's behavio(u)r
牛顿透镜公式 Newton's lens equation
牛顿图解 Newton's diagram
牛顿外容量 Newtonian outer capacity
牛顿万有引力常数 Newtonian constant of gravitation
牛顿万有引力定律 Newton('s) law;Newton's law of gravitation;Newton's law of gravity;Newton's law of universal gravitation
牛顿望远镜 Newtonian telescope
牛顿位势 Newtonian potential
牛顿位势函数 Newtonian potential function
牛顿向后插值公式 Newton's backward interpolation formula
牛顿向前插值公式 Newton's forward interpolation formula
牛顿行为 Newtonian behavio(u)r
牛顿性 Newtonianism
牛顿液体 Newton's liquid;Newtonian liquid;simple liquid
牛顿液体模型 Newton's liquid model
牛顿易熔合金 Newton's alloy
牛顿引力 Newtonian attraction
牛顿引力常数 Newtonian gravitational constant
牛顿引力理论 Newton's theory of gravitation
牛顿有限差分法 Newton's finite difference method
牛顿宇宙论 Newtonian cosmology
牛顿运动定律 Newton's laws of motion
牛顿折射 Newton's refraction
牛顿制 < 黏[粘]度单位 > Newton's system
牛顿阻力定律 Newton's law of resistance
牛轭湖 abandoned channel;abandoned meander;banc(o);bayou(lake);by-water lake;crescentic lake;cut-off;cut-off lake;cut-off meander;loop lake;lunate lake;meander lake;moat;mortlake;oxbow lake
牛轭湖沉积 mortlake deposit
牛轭湖相 mortlake facies
牛轭湖形弯道 oxbowu
牛轭湖沼泽 oxbow swamp
牛轭洼地 oxbowshaped depression
牛轭形的 oxbow
牛轭形弯道 oxbow
牛肝菌属 < 拉 > Boletus
牛骨胶 bone glue;cow glue
牛角匙 horn spoon
牛角浇口 horn gate
牛角线 cornoid
牛津吹张器 Oxford inflator
牛津单位 Oxford unit
牛津阶 < 晚侏罗世 >【地】Oxfordian;Divesian
牛津黏[粘]土 Oxford clay
牛津统【地】Oxford series
牛栏 byre;cattle pen;kraal;oxer;penfold;stanchion;timber crib
牛轮 bull wheel
牛洛铜镍合金 < 一种耐蚀铜镍合金 > Newloy
牛毛 cow hair;cattle hair
牛毛毡 cattle hair felt
牛奶 milk
牛奶厂 milk plant
牛奶场 dairy(farm)
牛奶场废水 dairy waste;dairy wastewater
牛奶场建筑 dairy building
牛奶场排污 dairy-farm slurry
牛奶车 milk car
牛奶房 dairy;milk house
牛奶壶 milk jug
牛奶加工厂 milk-processing waste
牛奶加工设备 milk production equipment
牛奶搅拌器 milk agitator
牛奶冷冻 cooling milk
牛奶冷却器 milk cooler
牛奶冷却设备 milk cooling equipment
牛奶冷却装置 milk cooling unit
牛奶流量计 milk flow meter
牛奶棚 dairy
牛奶铺 dairy;milk bar
牛奶与奶油泵 milk and cream pump
牛奶运输车 milk lorry;wagon for the transport of milk
牛奶装瓶机 milk dispenser
牛排餐厅 steak house
牛排菌属 < 拉 > Fistulina
牛棚 bull pen;cowshed;neat house;shippen
牛皮 cow hide
牛皮包装纸 kraft wrapping
牛皮锤 cowhide hammer
牛皮袋纸 kraft bag paper
牛皮筏 bullboat
牛皮浆漂白 kraft bleaching
牛皮胶 carpenter's glue;oxhide glue
牛皮胶粉 glue powder
牛皮卡纸 kraft liner
牛皮靠把 leather boat-body-rest pads
牛皮块 cow's lip
牛皮抛光轮 disk leather wheel;leather polishing wheel
牛皮浅水船 bullboat
牛皮止水 water sealing with cattle hide
牛皮纸 brown packing paper;craft paper;kraft(paper)
牛皮纸板 draft board;kraft liner
牛皮纸包的玻璃纤维或矿棉毯 < 用于木墙筋和平顶格栅之间的隔热材料 > batt insulation
牛皮纸箔 kraft paper foil
牛皮纸厂废水 kraft mill effluent
牛皮纸袋 kraft bag
牛皮纸袋包装 kraft paper bag package
牛皮纸废水 kraft mill waste
牛皮纸浆 kraft pulp
牛皮纸浆厂 kraft pulp mill
牛皮纸浆厂污水 kraft pulp-mill wastewater
牛皮纸浆工厂 kraft mill
牛皮纸浆制法 kraft process
牛皮纸漂白厂废水 kraft bleach plant effluent
牛皮纸漂白厂污水 kraft bleach plant effluent;kraft bleach wastewater
牛皮纸漂白厂污水生色团 kraft bleach plant effluent chromophores
牛皮纸漂白废水 kraft bleach effluent;kraft bleach wastewater
牛皮纸漂白污水 kraft bleach effluent
牛皮纸贴面建筑防潮板 kraft-faced building insulation
牛皮纸箱纸板 kraft test liner
牛皮纸液 kraft liquor
牛皮纸造纸机 kraft machine
牛肉 beef
牛乳场 dairy
牛乳房 milk bar
牛舌饰 calf's[复 calves'] tongue mo(u)lding
牛舌线脚 calf's[复 calves'] tongue mo(u)lding
牛舍 cowhouse;oxstall;shippen

牛舍废水 cowshed waste
牛市 bull market
牛栓枷 stanchion
牛特 <英国运动黏[粘]度单位> Newt
牛蹄油 neatsfoot oil
牛头雕饰 bucranium
牛头骨饰 oxscull[oxskull]
牛头骨状雕带 bucrane frieze; bucranium frieze
牛头骨状雕饰 bucranium mo(u)lding; bucrane <在古罗马爱奥尼亚和柯林斯柱式上的中楣内>
牛头骨状雕饰中楣 bucrane frieze
牛头骨状饰 <罗马爱奥尼亚和考林辛柱型上的中楣内> bucranium
牛头刨 bull-nose plane
牛头刨床 shaper; shaping machine
牛头刨的刨头 ram of shaper
牛头刨虎钳 shaper chuck; shaping chuck
牛头刨夹具 shaper chuck; shaping chuck
牛头刨零件 shaper part
牛头饰 oxhead; oxscull[oxskull]
牛头尊 oxhead
牛腿【建】ancon(e); bragger; cantilever bracket; fixing bracket; pannier; angle table; bracket; console; corbel; cut bracket <边缘有线脚装饰的>; builder's jack <墙上支承脚手架>; prothyride <砖石砌体上突挑的>
牛腿窗 bull's eye window; ox-eye
牛腿连接 bracket connection
牛腿梁 dapped-end beam
牛腿式雕饰 <楼梯踏步及梁端部的> step bracket
牛腿式装饰 step bracket
牛腿托砖 corbel table
牛腿支架 bracket support
牛腿支座 bracket support
牛腿柱冠 <早期西班牙式的> bracket capital
牛形底桩 button bottom pile
牛眼玻璃 bull's eye glass
牛眼窗 bullion; oculus; oeil de boeuf; oxeye(window); shed dormer
牛眼灯 bull's eye lamp
牛眼灯玻璃 bulleye lamp glass
牛眼环 bull's eye ring; heart shape thimble; pear-shaped thimble
牛眼木饼 bull's eye wooden cake
牛眼透镜 bull's eye lens
牛眼形拱 bull's eye arch
牛眼状 <黄铁矿结核心> bull's eye
牛羊圈养 stock raising in livestock shed
牛油 grease; tallow
牛油杯【机】grease cup
牛油环润滑器 Stauffer lubricator
牛油枪 grease gun; hydraulic grease gun
牛油枪喷嘴 grease nipple
牛油容器 grease horn
牛油软脂 tallow oleine
牛油树脂 shea butter
牛油填料 grease packing
牛栅 cowhouse
牛脂 beef tallow; tallow
牛脂胺 tallow amine
牛脂油 tallow oil

扭

S 形块体 dino(saur) block
扭摆 torsional pendulum; torsion oscillation
扭摆地震计 torsional seismometer
扭摆缓冲器 torsional oscillation damper
扭摆试验 torsional pendulum method
扭变 torsional deflection
扭变测定计 troptometer
扭变点 distortional point
扭变模型 distorted model
扭辫 torsional braid
扭辫分析 torsional braid analysis
扭波 torsional wave
扭波导 twist(ed) waveguide
扭成对 twinning
扭秤 torsional balance; torsional gravimeter; torsional scale
扭秤臂 torsional balance beam
扭秤横杆 variometer bar
扭秤头 torsional head
扭船式 twist boat
扭大 turn up
扭带分析 torsional braid analysis
扭的 torsional; twisted
扭点 pinch-point; twisting point
扭动 twisting motion; wriggle; wring
扭动的 aswivel
扭动构造体系 shear structural system; shear tectonic system
扭断 twist-off
扭断层 wrench fault
扭断强度 torsional strength
扭方阵 skew square matrix
扭杆 torsion(al) bar torsion(al) lever; torsion(al) rod; twist(ed) bar; twisting rod
扭杆连接 torque rod joint
扭杆式带负荷限制器 torsional bar belt-force-limiter
扭杆弹簧 torsional bar spring; torsion bar; torsion bar spring
扭杆弹簧固定塞 torsional bar spring anchor plug
扭杆弹簧固定塞盖 torsional bar spring anchor plug cover
扭杆弹簧扭转 torsional bar spring tension
扭杆弹簧式独立悬挂 torsional bar independent suspension
扭杆悬架 torsional bar suspension
扭杆座 torsional bar seat
扭钢 torsional steel
扭工字块体 Dolos
扭鼓常数 torsional drum coefficient
扭鼓改正 torsional drum correction
扭鼓改正值 torsional drum correction value
扭合接头 torsional joint
扭荷载 twisting load(ing)
扭花环饰 torsel
扭花环装饰 torse
扭坏 torsional break; torsional failure
扭环 torsional circle; torsional ring
扭簧 torsional spring
扭簧比较仪 reed-type comparator
扭回 retortion
扭剪(切) torsion(al) shear
扭剪(切)螺栓 torshear bolt
扭剪(切)模量 modulus of torsion-(al) shear
扭剪(切)破坏 wrench-shear fracture
扭剪(切)破裂 wrench-shear fracture
扭剪(切)强度 torsional shear strength
扭剪(切)试验 torsion shear test
扭剪(切)试验装置 torsional shear apparatus
扭角 angle of distortion; angle of torsion; angle of twist; torsional angle; twist angle
扭角协调方程 equation of compatibility of torsion angle
扭角仪 troptometer
扭绞八芯电缆 quad pair cable
扭绞磁场 twisted magnetic field
扭绞导条 twisting bar
扭绞电缆 strand(ed) cable; strand fiber[fibre] cable
扭绞二股线 twisted pair wiring
扭绞二线电缆 twisted pair cables
扭绞二线馈线 twisted-pair feeder
扭绞钢丝 twisted wire
扭绞机 strander
扭绞连接 twist joint
扭绞双股电缆 twisted pair cables
扭绞双线电缆 twisted pair cables
扭绞四芯电话电缆 twist pair type telephone cable
扭绞四芯电缆 multiple twin cable; quadded cable; quarded cable; spiral quad
扭绞系数 lay ratio
扭绞线 twisted line
扭绞线对 twisted pair
扭绞运动 twisting motion
扭接 dry joint; kinking; solderless joint; twisted joint; twisting joint
扭接连杆 torque link
扭接套管接头 twisted sleeve joint
扭接头 twist(ed) joint
扭节 <电缆的> knuckling
扭节理 torsional joint
扭节链 twist coil chain
扭节式胎链 twist-link type chain; twist-link type tyre chain
扭结 kinky; knot; snarl; kinking <钢丝绳>; coil buckling <线材或轧件的>
扭结带 joint drag; kink band; knick zone
扭结的双股钢筋 twin-twisted bar reinforcement
扭结钢筋 twin-twisted reinforcement
扭结链 kinked chain
扭结器 kinker; twister
扭结器轴 kinker shaft
扭结小齿轮 twister pinion
扭结轴 <打结器的> kinker
扭紧 lashing
扭紧螺(丝)钉 furbuckles
扭矩 moment of torque; moment of torsion; torque(moment); torsional moment; twist(ing) moment
扭矩扳手 torque spanner; torque wrench
扭矩比 torque ratio
扭矩臂 torque arm
扭矩变换 torque conversion
扭矩变换器 torque converter
扭矩表 rotational torque meter; torque chart; torque ga(u)ge; torquemeter; torquer; torsional ga(u)ge; torsional meter
扭矩波动图 torque fluctuation diagram
扭矩补偿 torque compensation
扭矩补偿器 torque compensator
扭矩不中断换挡的变速器 continuous drive transmission
扭矩测定 torque test
扭矩测定器 torque detector
扭矩测力计 torque dynamometer
扭矩测量 torque measurement
扭矩测量曲线图 torquemeter chart
扭矩测量台 torque-reaction stand
扭矩测量仪 torductor; torquemeter; torsionmeter
扭矩测台 torque stand
扭矩差动型常四轮驱动 torque-biasing full time four-wheel drive
扭矩秤 torsional balance
扭矩传递链 torque chain
扭矩传递同步器 torque synchro
扭矩传递装置 torque transmitter
扭矩传感联轴节 torque sensing coupling
扭矩传感器 torque sensor; torque transducer
扭矩的偏差范围 <紧固件等的拧紧> torque scatter
扭矩反向 torque reversal
扭矩反作用 torque reaction
扭矩放大器 torque amplifier
扭矩分布 torsion distribution; twisting moment distribution
扭矩分解 torque split
扭矩分流传动 split torque transmission
扭矩分配 torque transfer
扭矩负载 torque load(ing)
扭矩负载特性曲线 torque-load characteristic(curve)
扭矩杆 torque rod
扭矩杆的固定零件 torque rod mounting parts
扭矩杆端销 torque rod end pin
扭矩杆上销钉 torque rod upper pin
扭矩杆托架 torque rod bracker
扭矩杆下销钉 torque rod lower pin
扭矩杆橡胶衬套 torque rod rubber bushing
扭矩杆悬挂 torque rod suspension
扭矩管 torque tube
扭矩管式流量计 torque-tube flowmeter
扭矩管万向节传动 torque-tube drive
扭矩-惯矩比 torque to inertia ratio
扭矩换算 torque conversion
扭矩计 rotational torque meter; torque ga(u)ge; torquemeter; torquer; torsiometer; torsional ga(u)ge; torsional meter; torsionmeter
扭矩记录器 torque recorder
扭矩加捻曲线 torque-twist curve
扭矩减少 torque reduction
扭矩降 torque drop
扭矩降低器 torque reducer
扭矩抗力 torque resistance
扭矩控制 torque control
扭矩控制法 torque control method
扭矩控制系统 torque control system
扭矩马达 torque motor
扭矩黏[粘]度 torque viscosity
扭矩扭转角曲线图 <零件的> torque-twist diagram
扭矩平衡 torque balance
扭矩平衡变换器 torque-balance converter
扭矩平衡法 torque-balance system
扭矩平衡装置 torque-balance device
扭矩黏[粘]度计 torque viscometer
扭矩器 torquemeter
扭矩曲线(图) torque curve
扭矩屈服限 torque yield
扭矩容量 torque capacity
扭矩式黏[粘]度计 torque(-type) visco(si)meter
扭矩试验机 machine for testing torsion
扭矩试验仪 torque tester
扭矩释放离合器 torque release clutch
扭矩弹簧 torque spring
扭矩特性 torque performance
扭矩特性曲线图 torque diagram
扭矩调节 torque control
扭矩调节器 torque controller
扭矩同步机 synchronous link
扭矩图 torque chart; torque diagram; torsiogram; twisting moment diagram
扭矩系数 torque coefficient
扭矩限度 torque limitation
扭矩限制离合器 torque-limiter clutch

扭矩限制器 torque limiter
扭矩效率 torque efficiency
扭矩选数器 torque selector
扭矩要求 torque requirement
扭矩液动机 torque actuator
扭矩仪 torque ga(u)ge;torque indicator; torque level; torquemeter; torsiograph;tong torque ga(u)ge <机械大钳用>
扭矩与速度关系特性 torque-speed characteristic
扭矩与转数特性曲线 speed-torque characteristic curve
扭矩增大 multiplication; torque amplification; torque multiplying; torque rise
扭矩增大比 torque multiplication
扭矩增殖 multiplication;torque rise
扭矩张力试验 torque tension test
扭矩指示计 torsional indicator
扭矩重量比 ratio of torque to weight
扭矩轴 torque axis
扭矩转换器 torque converter
扭矩转角曲线 torsion(al)curve
扭距 pitch of strand
扭锯器 saw wrest
扭开 turn-on
扭抗盒式结构 torsion-box structure
扭壳 hypar shell;hyperbolic paraboloidal shell
扭壳体 hyperboloidal shell
扭亏增盈 turn from deficits to profits
扭亏增盈运输利润指标 transport target from loss to profit achieved
扭力 torque capacity; torque force; torsion; torsion(al)force; twisting effort;twisting force
扭力扳手 tension wrench; torque spanner;torque wrench
扭力搬手 torque tension wrench
扭力棒 torsional bar
扭力臂 toggle;torque arm
扭力表 torque ga(u)ge; torquemeter; torsional ga(u)ge; torsional meter
扭力波 torsional wave
扭力测定仪 torquemeter;torquer
扭力测功率计 torque-type power meter;torsional dynamometer
扭力测功器 torque-type power meter;torsion(al)dynamometer
扭力测试 torsional test
扭力常数 torsional constant
扭力秤 torque balance;torsional balance;torsion balance
扭力冲击试验 torsional impact test
扭力传递装置 torque transfer
扭力传感器 torque transducer
扭力传输路径 torque flow
扭力带分析仪 torsion-braid analyser [analyzer]
扭力带试验机 torsional braid tester
扭力分析 torsional analysis
扭力辐 torque spider
扭力杆 torque rod; torsion(al)bar; torsion shaft
扭力杆拆卸工具 torsional bar replacer
扭力杆端部塞子 torsional bar end plug
扭力杆固定螺(丝)钉 torsional bar retaining cap screw
扭力杆后挡块 torsional bar rear retainer
扭力杆内套筒 torsional bar housing inner
扭力杆前挡块 torsional bar front retainer
扭力杆式悬架 torque rod suspension
扭力杆外套筒 torsional bar housing outer
扭力杆稳定器 torsional bar stabilizer
扭力杆限制块 torsional bar retainer
扭力杆支座 torsional bar anchor;torsional bar support
扭力钢筋 torsion(al)reinforcement
扭力功率计 torsional dynamometer; torsional meter
扭力管 torque pipe;torque tube
扭力管传动 torque-tube drive
扭力荷载 torsional load(ing)
扭力计 torque balance;torque ga(u)ge; torque level; torquemeter; torque pickup; torsiograph; torsiometer; torsional dynamometer; torsion(al)indicator;torsional meter;torsionmeter;troptometer
扭力记录仪 torsiograph
扭力减震器 torsional damping arrangement
扭力减震装置 torsional damping arrangement
扭力矩 torque(moment);torsion(al) moment;twisting moment
扭力控制阀 torque control valve
扭力拉杆 torsion(al) bracing
扭力拉条 torsional bracing
扭力模数 torsional modulus
扭力黏[粘]度计 torsional visco(si)meter
扭力偶 torsional couple;turning couple;twisting couple
扭力盘形弹簧 torsional coil spring
扭力疲劳试验 endurance torsion test
扭力平衡器 torsional balancer
扭力破坏 torsional break
扭力破裂 torsional fracture
扭力强度试验 torsional strength test
扭力柔度 torsional compliance
扭力式功率计 torque-type power meter
扭力试验 torque test;twisting test
扭力试验机 torsional testing machine
扭力试验器 twisting tester
扭力试验仪 torsion tester
扭力输出量 torque output
扭力丝 torsional wire
扭力松弛 torsional relaxation; twist stress relaxation
扭力弹齿式 cultivator with torsion(al)spring tines
扭力弹簧 torque spring; torsional spring;torsion spring
扭力特性 torque characteristic
扭力天平 torque balance;torsion(al) balance
扭力杆调平式悬架 torsional level(l)ing suspension
扭力杆调整套筒 torsional bar spacer
扭力图 torsiogram
扭力系数 torque coefficient
扭力消震器 torsional vibration damper
扭力仪 torquemeter;torquer;torsional meter
扭力仪扳手 torquemeter wrench
扭力圆锥稠度计 torsion-cone consistometer
扭力支柱 torsional prop
扭力指示器 torsion(al)indicator
扭力轴 torsion(al)shaft
扭力自动记录仪 torsiograph
扭力作用 torque reaction
扭裂缝泥巴实验 experiment with clay on shear fractures
扭裂面 plane of shear fracture
扭裂模量 modulus of rupture in torsion
扭裂运动【地】rhaegmageny
扭麻花构造 figure S structure
扭模 twisting die
扭疲劳试验 torsional fatigue test
扭切模量 modulus of torsion(al) shear
扭曲 distortion; contort(ion); curling; flexural torsion; hogging; torsional buckling; torsional bending; tortuosity;twist(ing)(buckling); warp(ing)
扭曲板 twisted board
扭曲板压机 buckle plate press
扭曲板桩 buckle plate sheet piling
扭曲边 distorted bevel;twisted bevel
扭曲变形 distorted deformation; distortional deformation; torsional deformation
扭曲变形分布图 <车身的> torsional distribution graph
扭曲波导管 twisted waveguide
扭曲波痕 curved ripple mark
扭曲波理论 distorted wave theory
扭曲薄壳 skewed shell
扭曲不稳定性 kinking instability; twisting instability;wriggling instability
扭曲层【地】contorted bed
扭曲层理【地】contorted bedding;distorted bedding
扭曲沉降 distortion settlement
扭曲程度 twisting degree
扭曲次数 number of twists
扭曲带钢 twisted strip
扭曲的 kinked;tortile;tortuous
扭曲的轿厢闸门 collapsing car gate
扭曲地层 contorted stratum; twisted stratum
扭曲点 distortion point
扭曲动作 twisting action
扭曲度 degree of twisting; torsional resistance;tortuosity
扭曲断层 contortion fault; torsional fault
扭曲刚度 twisting rigidity
扭曲钢筋使其偏移原来中心线 offset bend
扭曲格网 distorted grid
扭曲河湾 distorted bend
扭曲荷载 twisting buckling load
扭曲环 hog ring
扭曲剪应力 torsional shear stress
扭曲劲度 twisting stiffness
扭曲梁的直段 wreath piece
扭曲量 warpage
扭曲量增量 increment of face advance
扭曲裂缝 contorted fissure; contortion crack;contortion fissure
扭曲路 twisty road
扭曲面 skew surface; torse; warped surface;warp surface
扭曲模 kink mode
扭曲模件 torsional modulus
扭曲模量 modulus of torsion
扭曲模型 distorted model
扭曲木材 wavy grown timber
扭曲木纹 interlocked grain; twisted grain
扭曲能 energy of distortion
扭曲喷管 canted nozzle
扭曲疲劳试验 repeated torsion(al) test
扭曲平衡轨道 kinked equilibrium orbit
扭曲破坏模量 modulus of rupture in torsion
扭曲破碎 breaking by curling
扭曲强度 twisting strength; warp strength;torsion strength
扭曲式活塞环 twist type piston ring
扭曲试验 torsional test;twisting test
扭曲试验机 torsion testing machine
扭曲试验跑道 twist course
扭曲微细裂纹 strain line
扭曲稳定性 stability under torsion
扭曲系数 buckling factor;coefficient of torsion;twisting coefficient
扭曲纤维 twisted fiber[fibre]
扭曲线型梁 initially twisted beam
扭曲相思树 twisted acacia
扭曲效应 twisted effect
扭曲斜边 crooked bevel
扭曲形扶手 wreathed handrail(ing)
扭曲形翼墙 warped wing wall
扭曲形柱 torso
扭曲型方格路网 warped grid network
扭曲型间界 twist boundary
扭曲悬挂 torsion flex suspension
扭曲压屈 distortional buckling
扭曲叶片 twisted blade;warped blade
扭曲因数 buckling factor
扭曲应变 strain due to torsion
扭曲应力 buckling stress; curling stress;torsional stress
扭曲张量 torsional tensor
扭曲者 twister
扭曲褶皱 contorted fold; contortion fold
扭曲振荡 torsional oscillation
扭曲振动 torsional vibration
扭曲中心 center[centre] of twist
扭曲状水系模式 contorted mode
扭屈 torsional buckling;twisting;warping
扭蠕变 torsional creep
扭塞 petcock
扭伤 sprain;wrench
扭式杆式悬架 torsional suspension
扭丝 torsional ribbon
扭丝的 cross-grained
扭丝式黏[粘]度计 torsion(al)(wire) visco(si)meter
扭丝仪 <测试橡胶的> torsional wire apparatus
扭松螺(丝)钉 unscrew
扭损 torsion(al)failure
扭索 hawser twist
扭索饰【建】guilloche
扭锁 swing lock;twist lock
扭锁销 twist lock span
扭筒法黏[粘]度计 visco(si)meter for cylinder torsion(al)method
扭歪 contort(ion);distortion;strain
扭弯 bending;contort(ion);crumple; fold back;twist;wreath
扭弯的 tortile
扭弯扶手 wreathed handrail(ing)
扭弯力矩 torsion-bending moment
扭弯模量 modulus of torsion
扭弯内缘 neck out
扭弯双力矩 torsion-bending bimoment
扭王字块体 <一种防波堤护面异形块体> accropode
扭王字块体护面 <一种防波堤护面异形块体> accropode armo(u)r
扭纹钢 <一种钢筋牌号> Twisteel
扭纹木材 cross-fibered [fibered] wood;cross-grained wood
扭熄 turn-off
扭线 torsional wire
扭线电流计 torsion-string galvanometer
扭线换位 twisted-lead transposition
扭向卸扣 reverse key shackle
扭销 twist lock pin;twist pin
扭斜板 skew plate
扭心 center[centre] of twist
扭形均衡器 twist equalizer
扭形柱 wreathed column

扭性结构面 shear structural plane
扭悬 torsional suspension
扭旋波导管 twist waveguide
扭压性结构面 shearing compressive plane
扭腰式转向 articulated steering
扭腰转向拖拉机 frame-steered tractor
扭叶高加索冷杉 twistleaf nordmann fir
扭叶片 twisted blade
扭叶松 <北美西部> western jack-pine
扭椅式 twist chair
扭应变 twisting strain
扭应力 torsion(al) stress; twisting stress
扭张力 torsional capacity
扭张性结构面 shearing tensile plane
扭折 kink(y)
扭折不稳定性 kinking instability
扭折带 kind band; joint drag
扭折的缆索 cranky
扭振 torsional oscillation
扭振激发装置 torsional exciter system
扭振减振器 torsional vibration damper
扭振减振阻尼器 torsional damper
扭振减震器 torque-vibration damper; torsional balancer
扭振模数 torsional modulus
扭振频率 torsional frequency
扭振平衡器 torsional vibration balancer; torsional vibration damper
扭振图 torsiogram
扭振橡胶弹簧 torsional gum
扭振形式 torsional mode
扭振自(动)记(录)仪 torsiograph
扭振自记器 torsiograph
扭振阻尼器 torsional damper; torsional vibration damper
扭震防冲器 torsional shock absorber
扭枝银枞 twisted silver fir
扭轴 axis of torsion
扭住 cinch; running-in
扭转 torsion twist; twist rotation; torsional stiffness graph <车身的>
扭转摆 torsional pendulum
扭转摆动衰减器 torsional oscillation damper
扭转半径 radius of torsion
扭转比拟 torsional analogy
扭转变形 torsional deflection; torsional deformation
扭转变形钢筋 twisted deformed bar
扭转变形模数 torsional modulus
扭转波 torsional wave
扭转层状位移 torsional laminar displacement
扭转颤振 torsional flutter
扭转常数 torsion(al) constant
扭转车叶 twisted blade
扭转程度 twisting degree
扭转持久极限 torsional endurance limit
扭转磁致伸缩传感器 torsional magnetostriction pickup
扭转错位 torsiversion malposition
扭转打滑 torque creep
扭转导板 twisting box
扭转的 torsional; tortile
扭转的钢丝绳 twisting wire
扭转地面运动 torsional ground motion
扭转地震计 torsional seismometer
扭转地震仪 torsional seismograph; torsional seismometer
扭转电流计 torsional galvanometer
扭转度 degree of torsion; torsion degree; twist
扭转断层【地】pivotal fault; basculating fault; torsional fault; wrench fault <走向滑距近于垂直的断层>
扭转断口试验 split torsion(al) test
扭转断裂 torsional fracture; twist-off
扭转断裂模量 modulus of rupture in torsion
扭转发散 torsional divergence
扭转法 torsional method
扭转反应谱 torsional spectrum
扭转防震器 torsion balancer
扭转放大器 twist amplifier
扭转分布 twist distribution
扭转幅度 torsional amplitude
扭转负荷 torsional load(ing); twisting load(ing)
扭转复原试验 torsional recovery test
扭转杆 torque rod; torsion bar
扭转刚度 torsional rigidity; torsional stiffness
扭转钢筋 twisted bar; twisted steel
扭转钢筋网 twisted square bar fabric
扭转公式 torsion(al) formula
扭转共振柱试验仪 torsional resonant column apparatus
扭转共振柱仪 torsional resonant column apparatus
扭转荷载 torsion(al) load(ing); twisting load(ing); twist load
扭转环剪仪 torsion ring shear apparatus
扭转回能 torsional resilience
扭转活塞环 torsional piston ring
扭转加捻卷曲 reel-twist crimping; torsional twist crimping
扭转减震器 torsional vibration damper
扭转剪力 torsional shear
扭转剪力试验 torsional shear test
扭转剪切 torsional shear
扭转剪(切)试验 torsional shear test
扭转剪(切)应力 torsional shear(ing) stress
扭转角 angle of distortion; angle of torque; angle of torsion; angle of twist; angle torsion; torsion(al) angle
扭转矫正术 detorsion
扭转节理 torsion joint
扭转晶界 twist boundary
扭转静电计 torsional electrometer
扭转卷曲法 torsional crimping process
扭转开关 turn switch
扭转开裂 torsional crack
扭转孔型 twisting pass
扭转亏损 carry-back of losses
扭转理论 theory of torsion
扭转力 torque force; torsion; torsional force; twisting resistance
扭转力矩 moment of torque; moment of torsion; moment of twist(ing); torque moment; torsional couple; torsional moment; turning moment; twisting couple; twisting moment; leverage
扭转力偶 twisting couple
扭转连杆 twisted connecting rod
扭转裂缝 torsional cracking
扭转临界转速 torsional critical speed
扭转流动 torsional flow
扭转模量 modulus of torsion; torsion(al) modulus; twisting modulus
扭转模态 torsional mode
扭转挠曲 torsional-flexural buckling; twist warp
扭转挠性 torsional flexibility
扭转黏[粘]度计 torque-type visco(si)meter; torsional viscometer
扭转扭转的 torsionally rigid
扭转耦联建筑物 torsionally coupled building
扭转配对 torsional pairing
扭转疲劳 torsional fatigue
扭转疲劳极限 fatigue limit for torsion; torsional endurance limit; torsional fatigue limit
扭转疲劳破坏 torsional fatigue failure
扭转疲劳强度 torsional fatigue strength
扭转疲劳试验 torsional fatigue test
扭转疲劳试验机 reverse torsion(al) machine
扭转疲劳限度 torsional endurance
扭转偏心 torsional eccentricity
扭转频率 torsional frequency
扭转破坏 torque failure; torsion(al) failure; twisting failure
扭转破坏模量 modulus of rupture in torsion
扭转谱 torsional spectrum
扭转潜伏能 potential energy of twist
扭转强度 torsion(al) strength; twisting strength
扭转翘曲 twist warp
扭转屈服试验 torsion yield test
扭转屈曲 torsional buckling
扭转圈数 revolutions of twisting
扭转蠕变 torsional creep
扭转蠕变裂纹 torsional creeping crack
扭转伸缩性 torsional flexibility
扭转失效 torsional failure
扭转湿度计 torsional hygrometer
扭转式测力计 torsional dynamometer
扭转式风速计 torsional anemometer
扭转式检流计 torsional galvanometer
扭转势能 potential energy of twist
扭转试验 torsion(al) test; twisting test
扭转试验机 torsional tester; torsional testing machine
扭转试验台 torsional rig
扭转试验图 torsional test diagram
扭转双层元件 twister bimorph
扭转弹簧 torsional spring
扭转弹性 elasticity of torsion; torsional elasticity; torsional resilience; torsional spring
扭转弹性变形 torsional resilience
扭转弹性联轴节 torsionally flexible coupling
扭转弹性模量 modulus of torsion(al) elasticity
扭转弹性模数 modulus of torsion
扭转特性 torque characteristic; torque property
扭转弯曲 torsional bending; torsion buckling
扭转位移 torsional displacement
扭转纹 torsion-fibre[fiber]; twisted fiber[fibre]; twisted growth
扭转纹的 twisted-fibred; twisted-grained
扭转纹理 <木材的> spiral grain; twisted grain
扭转稳定性 torsional stability
扭转问题 torsion(al) problem
扭转系数 coefficient of torsion; coefficient of twisting; torsional coefficient
扭转响应 torsional response
扭转效果 effect of torsion
扭转效应 torsional effect
扭转信号开关 turn signal switch
扭转形变 torsional deformation
扭转型延迟线 torsional mode delay line
扭转性能 torque characteristic; torque property
扭转悬置 torsion-bar suspension
扭转压法 torsoclusion
扭转叶片 twist blades
扭转仪 torsionmeter; troptometer
扭转应变 strain due to torsion; twisting strain; torsional strain
扭转应力 torsional stress; twisting stress; distorting stress
扭转映射 twist mapping
扭转约束 torsional restraint
扭转运动 twist motion; wrench motion; wrench movement
扭转载荷 torsion load
扭转振摆 torsional shake
扭转振荡 torsion(al) oscillation; yawing oscillation
扭转振动 torsional mode; torsional vibration; twisting vibration
扭转振动传感器 torsional vibration pick-up
扭转振动法 torsional oscillation method
扭转振(动)记录仪 torsiograph
扭转振动减震器 torsional vibration damper
扭转振动频率 torsional frequency
扭转振动图 torsiogram
扭转振动振幅 amplitude of rotary oscillation
扭转振动阻尼器 torque-vibration damper; torsional balancer
扭转振型 torsional mode
扭转支撑的 torsionally braced
扭转指示器 torsion(al) indicator
扭转中心 center[centre] of torsion; center[centre] of twist; twist center[centre]
扭转重力仪 torsional gravimeter
扭转周期荷载三轴仪 torsional cyclic(al) load triaxial apparatus
扭转轴 torsion(al) shaft
扭转轴承 torsional bearing
扭转轴弹簧 torsion-bar spring
扭转自由振动试验 free torsion(al) vibration test
扭转阻力 torsional resistance
扭转作用 twisting action
扭阻力 twisting resistance

纽

纽比特 <硬质合金> NewBide
纽伯恩砂质泥灰岩 Newbourn crag
纽伯格白垩 Neuberg chalk
纽伯格蓝 Neuberg blue
纽带 vinculum[复 vinculums/vincula]
纽带系统 cordonnier system
纽点 knot
纽芬兰 Newfoundland
纽结 chinckle; kink; knot
纽结表 knot table
纽结等价 knot equivalence
纽结多项式 knot polynomial; K-polynomial
纽结环面 knot torus
纽结理论 knot theory; K-theory
纽结曲面 knot sphere
纽结射影 knot projection
纽结图 knot diagram
纽结问题 knot problem
纽结型 knot type
纽卡斯尔层【地】Newcastle bed
纽卡斯尔煤 Newcastle coal
纽卡斯尔窑 Newcastle kiln
纽康理论 Newcomb theory
纽康算符 Newcomb operator
纽里花岗石 <产于爱尔兰纽里地区> Newry
纽林标准零点 <英> ordnance datum Newlyn
纽林零点 Newlyn datum

纽络<一种硬质合金> Nuloy

纽马克感应图<求土基中任意点的垂直压应力用> Newmark's influence chart

纽马克图 Newmark's chart

纽曼表<各种土壤的安全抗阻力表> Newman's tables

纽曼灰岩 Newman limestone

纽曼六位规则 Newman's rule of six

纽曼-皮尔逊准则 Neyman-Pearson criterion

纽恰尔牌活性炭 Nuchar

纽森钻井法 Newson's boring method

纽绳 bandage

纽索饰 plait-band

纽瓦克系【地】Newark series

纽维德尔绿 Neuwieder green

纽沃尔制<公差配合基孔制> Newall system

纽形剖线 loop cut;retrosection

纽约<美国州名> New York

纽约测杆<画有细线的水准杆> New York rod

纽约港 New York Port;Port of New York

纽约港务局 Port of New York Authority

纽约水准尺<美国的一种有活动觇板的水准标尺> New York leveling rod

纽约哑铃形平面(住房) dumbbell tenement

纽约(原木)计量法<用于桃花心木和香椿木原木的材积计量> Constantine measure

纽约证券交易所 Big Board

纽转纳 neodrenal

钮板式卸载输送机 chain-and-slate unloader

钮板式卸载输送器 chain-and-slate unloader

钮钩触点 buttonhook contact

钮孔 buttonhole

钮扣传声器 button microphone

钮扣钩 buttonhook

钮扣式冲模 button dies

钮扣式电容器 button capacitor

钮扣形底桩 button-bottom pile

钮扣型电池 button cell

钮扣状过滤管 button-screen pipe

钮扣自选机 button sizer

钮式引流法 button drainage

钮头弹簧 button-head spring

钮形底桩 button bottom pile

钮形孔距规 toolmaker's button

钮形物 button

钮形心柱 button stem

钮状(虫)胶 button lac

钮状紫胶 button lac

钮子 button

钮子板 buttonboard

农仓 barn

农仓板墙 barn siding

农仓空场 barn yard

农产品 agricultural materials;agricultural products; agriculture commodity;farm product;yielder

农产品仓库 produce warehouse

农产品集散中心市场 terminal market

农产品加工废物 farm products processing waste

农产品加工(工)业 agroindustry; farming industry

农产品加工建筑 agricultural building

农产品加工深度和转化率 the degree of processing and the conversion rate of agriculture products

农产品加工学 agrotechny

农产品预购合同 contract of ordering agricultural products

农场 farm(yard);grange

农场出售价 farm gate price

农场的农田 farmstead

农场抵押 farm mortgage

农场电气化 farm electrification

农场房屋 ranch house

农场肥料 farmyard manure

农场废水 farmyard wastewater

农场废物 farm waste

农场工程 farmstead engineering

农场工人 farmhand

农场供水效率<作物灌溉耗水量与供水量之比> farm delivery efficiency

农场固定资产 farm capital

农场管理 farm management

农场灌溉车 farm-irrigator

农场灌溉率 farm duty(of water)

农场及其建筑物 farmstead

农场给水工程 farm water supply engineering

农场建筑(物) farm building; farmstead

农场交货 explantation

农场进水口 farm inlet

农场经营人员 farm operator

农场空地 farm yard

农场绿化 plantation planting

农场内部的 on-farm

农场内部水道 farm waterway

农场内部水管理 on-farm water management

农场内建筑物 steading

农场喷灌系统 farm sprinkler system

农场平房 rambler

农场企业 agricultural industrial enterprise

农场起居室 farm house place

农场渗透损失 farm percolation loss

农场生产费用 farm production cost

农场式 ranch shape

农场式林地 farm woodland; farm woodlot

农场式林业 farm forestry

农场式森林 farm woods

农场式住宅 ranch style house

农场水塘 farm pond

农场所有权人 homesteader

农场投资 farm investment

农场蓄水池 farm reservoir

农场用带篷挂车 farm wagon

农场用冻结装置 farm freezer

农场用机器 farm machine

农场与集市间交通 farm to market traffic

农场种植 plantation planting

农场主 farmer

农场住房 ranch house

农村 countryside;rural district

农村暴雨水污染 rural stormwater pollution

农村病 rural disease

农村城市化 rural urbanization

农村城镇 agrotown

农村大气环境 rural atmosphere

农村道路 agricultural road;farm road; hay road;rustic road

农村的 campestral; outbound city; out-city;rural

农村地带 agricultural belt; country belt;rural belt;rural zone

农村地区 rural area

农村地区城镇 agroindustrial town

农村电话网 rural telephone network

农村电话线 rural line

农村电话线路增音机 rural line repeater

农村电气化 rural electrification

农村都市化 rural urbanization

农村繁荣 rural prosperity

农村房舍 rustic home

农村房屋 farm dwelling;rural building;rural home;rustic home

农村辅助建筑物 farm service building

农村副业 rural subsidiary occupations

农村给水 rural water supply

农村工业 agroindustry

农村工业化 industrialization of rural areas

农村公路 rural highway

农村供水 rural water supply

农村构筑物 farm structure

农村雇工的房屋 farm employee housing

农村规划 country planning; rural planning

农村规划目标 rural planning object(ive)

农村化 villagization

农村环境 rural environment

农村环境规划 rural environmental planning

农村基本设施 rural infrastructure

农村基础设施 rural infrastructure

农村集合村 rural conglomeration

农村集市道路 farm market road

农村集约化 intensive agriculture

农村建设 rural construction

农村建筑 farm building;rural architecture;rural construction

农村建筑贷款 rural housing loan

农村建筑工程 farm building construction

农村教堂 field church

农村教育 rural education

农村结构 rural structure

农村经济 rural economy

农村经济的分散化 decentralization of rural economy

农村经济学 rural economics

农村景色 farmscape;rurality

农村景象 farmscape

农村就业机会 rural job creation

农村开发 rural development

农村旅客列车<日本> rural passenger train

农村排水工程 rural sewage;rural sewerage

农村排污工程 rural sewerage

农村排污系统 rural sewerage

农村桥梁 farm bridge

农村人口 agrarian population; rural population

农村人口外流 rural exodus

农村生活废水 rural domestic waste (water)

农村生活污水 rural domestic waste (water)

农村生活用水 rural domestic water

农村式布置 village-type arrangement;village-type grouping

农村式组合 village-type grouping

农村收入 rural income

农村特征 rurality

农村通信[讯] farmer communication

农村土地利用对策 rural land use strategy

农村卫生 rural hygiene

农村污水 rural sewerage

农村污水处理 rural sewage disposal; rural sewage treatment

农村污水处置 rural sewage disposal

农村污水工程 rural sewerage

农村污水灌溉田 rural sewage farm

农村污水系统 rural sewage system

农村乡镇 rural community

农村向城市移民 rural-urban migration

农村小教堂 field chapel

农村小路 rural penetration

农村医院 village hospital

农村移民 rural resettler

农村移民安置 rural resettlement

农村饮用水 rural drinking water

农村饮用水泵站 rural drinking water pump station

农村用地 rural land use

农村有线广播网 rural wire broadcasting network

农村运货大车 farm wagon

农村住房 dwelling;dwelling house

农村住宅 country house;grange;rural residence

农村自动电话局 village automatic exchange

农村综合发展 integrated rural development

农地 cultivated area;cultivated soil

农地林业 agroforestry;farm forestry

农地小坝 farm dam

农副产品 native and subsidiary products

农耕方式 farming practice

农耕效率 farming efficiency

农工联合企业 agro-based-industrial complex;agroindustrial complex

农工企业 agroindustrial enterprise

农工商联合企业 agroindustrial-commercial enterprise; farm-industry-commerce-enterprise

农沟 farm ditch;field lateral

农户 farmholding

农户劳动收入 farmer's family labo(u)r return

农话电缆 rural cable

农话网 rural telephone network

农会 agricultural association

农活 farm work

农机 dead stock

农机设备 agricultural equipment

农机制造厂 farm equipment factory

农机自动底盘 tool frame

农基工业 agrobased industry

农家 farmholding

农家场院废物 farmyard waste

农家肥料 farm manure;farmyard manure;self-supplied manure

农家空地 farm yard

农家庭院 farm yard

农家养禽场地 base-court;basse-court

农家庄院 farm yard

农具 agricultural equipment; agricultural implement; dead stock; farm implement;implement

农具改革 reform of farmtools

农具棚 implement shed

农具脱钩 unhitch

农具下降速度调节 implement lowering control

农垦区 agricultural settlement

农矿产品 commodity

农历 lunar calendar

农历年 lunar year

农林部 Ministry of Agriculture and Forestry

农林化工 agriculture and forest chemical industry

农林徽 farming and forestry badge

农林间作 alternation of agricultural and forest crops

农林牧结合 coordination of farming forestry and animal husbandry

农林牧系统 agro-sylvo-pastoral sys-

N

tem
农林学 agroforestry
农林作业 agrisulviculture
农忙期 rush period
农贸市场 farm product market; free market
农民 farmer; tiller
农民的 agrarian
农民肺 farmer's lung
农民聚落 rural settlement
农民用陶器 peasant pottery
农牧轮作 rotation pasture
农(牧)业区 grass roots
农渠 distributary minor; distributing ditch; farm ditch; farm lateral; feeder canal; field lateral; quarternary canal
农商联合企业 agribusiness
农舍 cottage; farm house; grange; ranch house
农舍式房屋 chalet
农事季节 farming season
农事年 farmer's year
农田 agricultural plot; arable land; cultivated land; farm(field); farmland; agricultural land; cropland
农田边界 field boundary
农田测量 farm survey
农田池塘 farm pond
农田底土排水 subsoil agricultural drain
农田调查 farm survey
农田防护林 agricultural protection forest; farmland shelter forest; shelter forest on farmland
农田废水 farm wastewater
农田废物 farm waste
农田覆盖不宜露采 not opencast for farmland cap
农田杆泵 spindle drag pump
农田供水 agriculture water supply
农田供水量 farm delivery requirement
农田供水水文地质调查 hydrogeologic(al) survey of agriculture water supply
农田供水水文地质图 hydrogeologic(al) map on agricultural water supply
农田供水要求 farm delivery requirement
农田灌溉 irrigation of farmland
农田灌溉布置 farm irrigation layout
农田灌溉单元 irrigating unit
农田灌溉率 farm duty(of water)
农田灌溉水质标准 irrigation water quality standard
农田灌水 watering of farmland
农田规划 farm planning
农田净灌溉需水量 farm delivery requirement
农田径流 agricultural runoff; farm runoff
农田林网 green belt around farmland
农田排水 agricultural drainage; farm drainage; field drain(age)
农田排水沟 agricultural drain; land drain
农田排水管(道) agricultural drain pipe; agricultural pipe drain; field drain pipe; land-drain pipe
农田排水渠 agricultural drain
农田排水瓦管 farm drain tile; field tile
农田喷水灌(溉) agricultural sprinkler irrigation
农田破坏 surface and trespass damage
农田青苗补偿 crop loss compensation
农田取水口 farm turnout
农田壤土 agricultural soil
农田设施 on-farm facility
农田渗透损失 farm percolation loss
农田生态系统 farmland ecosystem
农田受损 surface and trespass damage
农田水沟 agricultural drain
农田水利 water conservancy of agriculture land
农田水利工程 agricultural hydraulic engineering; farm field; farm plot; grain field; irrigation and drainage engineering; water conservation measures in agriculture
农田水利规划方案 program(me) of water conservancy of farmland
农田水利化程度 degree of bringing all farmland under irrigation
农田水情 field water condition
农田水土流失率 rural erosion rate
农田投资 farm investment
农田土地协会 Farm and Land Institute
农田土壤 agricultural soil
农田瓦管 farm tile
农田污染 farm pollution
农田小气候 agricultural microclimate
农田蚁类 harvester ants
农田用滚筒 agricultural roller
农田用(排灌)瓦管 farm drain tile
农田用水季节 irrigation season
农田栅栏 <分隔公路与农田用> farm fence
农田丈量 farm survey
农闲季节 farm slack season; non-agricultural season; slack farming season
农闲作物 catch crop
农学 agriculture; agronomy
农学家 agriculturist; agronomist
农学杂志 Agronomy Journal
农药 agricultural chemicals; agriculture chemicals; agrochemicals; biocide; farm insecticide; insecticide; pesticide
农药安全使用法 good agricultural practice
农药安全使用规定 good agricultural regulation; regulations for safe use of pesticide
农药安全使用制度 system of save use of pesticide
农药安全预防方案 pesticide safety precaution scheme
农药残毒 residual toxicity; toxicity of pesticide residue
农药残留 agricultural chemical pesticide; agricultural chemical residuum; pesticide residue
农药残留标准 trace standard of agricultural chemical; trace standard of agricultural chemistry; trace standard of pesticides
农药残留量 pesticide residue
农药残留量分析 pesticide residue analysis
农药残留容限量 pesticide residue tolerance
农药残留容许量 pesticide residue tolerance
农药残留危害性 pesticide residue hazard
农药残留物 pesticide residue
农药残留允许量 pesticide residue tolerance
农药残效 residual effect of pesticide
农药残液 remaining liquid of pesticide
农药残余 pesticide residue; residue of pesticide
农药残余危害 pesticide residue hazard
农药厂 pesticide factory; pesticide processing plant
农药厂废水 wastewater from pesticide factory
农药沉降 fallout of pesticide
农药持久性 persistence of pesticides
农药代谢 pesticide metabolism
农药的长期留存 persistence of pesticides
农药的沉降 fallout of pesticide
农药的代谢 metabolism of pesticide
农药的附加效应 additive effect of pesticide
农药的滥用 indiscriminate use of pesticide
农药的微生物降解 microbial breakdown of pesticide
农药定量分析 quantitative analysis of agricultural chemicals
农药毒理学 pesticide toxicology
农药毒性 pesticide toxicity; toxicity of pesticide
农药毒性等级 classes of pesticide toxicity
农药肥料 pesticide-added fertilizer
农药废料 pesticide waste
农药废水 pesticide wastewater
农药废水处理 pesticide wastewater treatment
农药分解模型 pesticide analytical model
农药分析 analysis of agricultural drugs; pesticide analysis
农药辅助剂 pesticide adjuvant
农药公害 pesticide pollution
农药管理法 Agricultural Chemicals Regulation Law
农药化学 pesticide chemistry
农药环境模拟室 pesticide environmental chamber
农药挥发 volatilization of pesticide
农药混合物 pesticide combination
农药剂型 pesticide formulation
农药监测 pesticide monitoring
农药检测 pesticide detection
农药降解 degradation of pesticide; pesticide degradation
农药解毒作用 detoxification of pesticide
农药解吸 desorption of pesticide
农药径流模拟器 pesticide runoff simulator
农药控制 control of pesticide; pesticide control
农药联合毒性 combined toxicity of pesticide
农药流失 pesticide runoff
农药耐久性 persistence of pesticides
农药耐量 pesticide tolerance
农药年度 pesticide year
农药浓度 pesticide concentration
农药喷射器 crop dusting
农药飘失 pesticide drift
农药清除 elimination of pesticide
农药容许量 pesticide tolerance
农药溶化估算与化学模型 leaching estimation and chemistry model-pesticide
农药生产废水 pesticide production wastewater
农药生物测定 bioassay of pesticide
农药施用 application of pesticide
农药施用方法 application ways of pesticide
农药室内毒力测定 toxicity experiment of pesticide
农药输移 pesticide transport
农药输移和径流模型 pesticide transport and runoff model
农药水污染 aquatic pesticide contamination
农药污染 contamination by pesticides; pesticide contamination; pesticide pollution; pollution by farm chemicals; pollution by pesticides
农药污染势 pesticide contamination potential
农药行径和归宿 pesticide pathway and fate
农药药害 pesticide phytotoxicity
农药药剂 pesticidal chemical; pesticide formulation
农药药效 pesticide effectiveness
农药影响 effect of pesticide; pesticide effect
农药用乳化剂 pesticide use emulsifier
农药有效成分 effective ingredient of pesticide
农药允许残留量 pesticide residue tolerance
农药增效剂 synergist of pesticide
农药脂溶度 fat-solubility of pesticide
农药中毒 being poisoned by agricultural chemicals; pesticide poisoning; poisoning by agricultural chemicals
农药助剂 inert ingredient
农药注册 pesticide registration
农药综合治理 integrated pesticide management
农业 agriculture; geoponics; husbandry; terra culture
农业保护纲要 agricultural conservation program(me)
农业保水措施 water conservation measures in agriculture
农业变化 agricultural change
农业病虫害 agricultural pest
农业病虫害防治 control of agricultural pest
农业补偿政策 agricultural compensation policy
农业布局 distribution of agriculture
农业部 <美> Department of Agriculture
农业部门 agricultural sector
农业残余物 agriculture residue
农业残渣 agriculture residue
农业产值 agricultural output value
农业场地 farm yard
农业承包工程 agricultural contract works
农业城市 agricultural city
农业城镇 agrotown
农业储藏 agricultural storage
农业处理 agricultural disposal
农业措施 agricultural method; agronomic practice
农业贷款 agricultural credit; agricultural loan
农业的 agric; agricultural; geoponic; rural
农业地产管理 management of agricultural estates
农业地带 agricultural belt; agricultural district; country belt; grass roots; rural belt
农业地理 agricultural geography
农业地理学 agrogeography
农业地区 agricultural district
农业地质图 agricultural map; agrogeological map
农业地质学 agricultural geology; agrigeology; agrogeology
农业电气化 farm electrification
农业调查 agricultural research
农业动力 farm power
农业毒物 agricultural poison

农业多种经营 diversified agricultural activities
农业发展 agricultural development
农业发展顾问局 Agricultural Development Advisory Service
农业发展区 agricultural development zone
农业发展项目 agricultural development project
农业法 agricultural act; agricultural law;farmer's law
农业方法 agricultural method
农业防治 agricultural control;agricultural prevention and treatment
农业放牧 agricultural pasture
农业非点污染源 agricultural nonpoint source pollution;sources of agricultural nonpoint pollution
农业非点污染源水质模型 agricultural nonpoint source water quality model
农业非点污染源污染 agricultural nonpoint source pollution
农业非点污染源污染控制 agricultural nonpoint source pollution control
农业非点污染源污染模拟 agricultural nonpoint source pollution simulation
农业非点污染源污染模型 agricultural nonpoint source pollution model
农业非点源污染 agricultural nonpoint source pollution
农业肥料 agricultural fertilizer
农业废(弃)物 agricultural waste
农业废水 agricultural effluent; agricultural wastewater
农业废物处置 agricultural waste disposal
农业分区 agriculture division
农业改造 rural transformation
农业改组 agricultural regrouping
农业革命 green revolution
农业工程 agricultural engineering
农业工程师 agricultural engineer
农业工具 agricultural implement
农业工人 farmhand;field hand
农业工人宿舍 farm labo(u)rer's quarters
农业工人宿舍区 farm labo(u)r camp
农业工业 agricultural industry;agro-industrial-industry;agroindustry
农业工业的 agroindustrial
农业工业结合体 agricultural industrial complex;agroindustrial complex
农业固体废物 agricultural solid waste
农业规划 agricultural planning
农业害虫 agricultural insects
农业害虫的防治 agricultural insect management
农业害物 agricultural pest
农业航空 agricultural aviation
农业合作化 agricultural cooperation
农业化肥 agrochemicals
农业化肥的使用 use of agrochemical
农业化学 agricultural chemistry; agrochemistry
农业化学的 agrochemicals
农业化学加工 chemurgy
农业化学品管理条例 Agricultural Chemicals Regulation Law
农业化学土壤图 agricultural map;agrochemical soil map
农业化学(药)品 agricultural chemicals;agrochemicals
农业环境 agricultural environment
农业环境保护 agricultural environmental protection
农业环境管理 agricultural environmental management
农业环境监测 agricultural environmental monitoring
农业环境破坏 agricultural environmental destroy
农业环境学 agricultural environmental science
农业回流污染 agricultural backwater pollution
农业回水污染 agricultural backwater pollution
农业会计 agricultural accounting
农业活动 agricultural activity; rural activity
农业机具 agricultural equipment
农业机具的管理 management of tools and equipment
农业机具和设备 agricultural tools and equipment
农业机械 agricultural machinery; agricultural operation equipment; farm equipment;farm machinery
农业机械的 agricultural and mechanical
农业机械工业 farm machinery industry
农业机械化 farm mechanization
农业机械化水平 level of farming mechanization
农业机械及农具 farm machinery and implements
农业机械设计 agricultural machinery design
农业机械生产 farm machinery production
农业机械通用性 versatility of farm tools
农业机械用润滑油 harvester oil
农业集水区 agricultural watershed
农业集约化 agriculture integration
农业计划 agricultural plan;agricultural program(me)
农业计划方案 program(me) for agriculture
农业计划分析 analysis of program(me)
农业技术 agricultural technique;agriculture technique; agronomic practice;agrotechnique
农业技术改革 technologic(al) change in agriculture
农业技术人员 professional of agriculture
农业技术水平 agrotechnical level
农业技术推广 popularizing agricultural technique
农业技术推广站 station for popularizing agricultural technique
农业技术员 agrotechnician; agricultural technician
农业季节工人 migrant worker; migrant labo(u)r;migratory labo(u)r <美>
农业加工废物 agricultural processing waste
农业加工建筑 agricultural processing building
农业界 agricultural sector
农业经济 agricultural economy;farming economy
农业经济局<美> Bureau of Agricultural Economics
农业经济学 agrarian economics;agricultural economics; agricultural farm economics
农业景观 farming landscape
农业就业 farm employment
农业聚落 farming settlement
农业开发 agricultural development
农业矿物 agricultural mineral
农业矿渣 pit silo
农业昆虫学 agricultural entomology
农业立法 agricultural legislation
农业利用 agricultural use
农业联合配料器 agrocement batcher
农业林学 agroforestry
农业流域 agricultural catchment
农业留置权 agricultural lien
农业面源污染 agricultural area source pollution
农业年度 crop year
农业排水 agricultural drainage
农业排水沟 agricultural drain
农业排水管 agricultural drain
农业排水系统 agricultural drainage system
农业品散市场 agricultural terminal market
农业普查 census of agriculture
农业企业 agricultural enterprise
农业气候带 agroclimatic zone
农业气候的 agroclimatic
农业气候分界 agroclimatic delimitation
农业气候分区 agroclimatic delimitation
农业气候区 agroclimate zone; agroclimatic zone
农业气候区划 agroclimatic delimitation
农业气候区域 agroclimatic region
农业气候图 agriclimatic map;agri(o)-climatic map
农业气候学 agricultural climatology; agroclimatology
农业气候资源 agroclimatological resources
农业气象委员会 Commission of Agricultural Meteorology
农业气象学 agricultural meteorology;agrometeorology
农业气象预报 agrometeorological forecast
农业气象站 agrometeorological station
农业区 agricultural area; agricultural region; agriculture area; grass roots;rural area
农业区划 agricultural division; agricultural regionalization;agricultural zoning
农业区铁路事业 grass roots railroading
农业区域环境 agricultural regional environment
农业人口 agricultural population; peasant population
农业森林的 agro-forestal
农业杀虫剂 agricultural insecticide
农业杀螨剂 agricultural miticide
农业商品 agricultural commodity
农业社会资源 agrosocial resources
农业生产 agricultural production
农业生产建筑 farm production building
农业生产率低 low productivity in agriculture
农业生产值 agricultural output value
农业生态恶性循环 vicious cycle of agroecosystem
农业生态工程 agroeco-engineering
农业生态环境 agro-ecological environment
农业生态经济系统 agroeco-economic system
农业生态良性循环 good cycle of agroecosystem
农业生态系(统) agricultural ecosystem;agroecosystem
农业生态系统研究 agroecosystem research
农业生态型 agroecotype
农业生态学 agrobiology;agroecology
农业生态预测 agricultural ecosystem forecasting
农业生物技术 agricultural biotechnology
农业生物气象学规划 agricultural biometeorology program(me)
农业生物学 agrobiology
农业生物学分析 agrobiology analysis
农业生物学研究 study on agrobiology
农业剩余物 agricultural waste
农业师 agronomist
农业示范带 agricultural demonstration strip
农业试验站 agricultural experiment station;farm research station
农业收入 agricultural income
农业数据 agricultural data
农业数据分析 analysis of agricultural data
农业水利工程 cultural hydraulic engineering
农业水文学 agricultural hydrology; agrohydrology
农业水资源 agricultural water resources
农业税 tax on agriculture
农业损害 agricultural damage
农业损伤 agricultural injury
农业天气预报 agricultural weather forecast
农业投资 farm investment
农业土地 agricultural land;farmland
农业土地管理 farmland management
农业土壤 agricultural soil
农业土壤发生 genesis of agricultural soil
农业土壤改良 agricultural amelioration;agroamelioration
农业土壤类型 agro-type
农业土壤流失 agricultural soil loss
农业土壤流失率 agricultural soil loss rate
农业土壤退化 agropedic degradation
农业土壤学 agrology;agropedology
农业土壤用水 farmland water
农业拖拉机 agricultural traction engine;agricultural tractor;agrimotor
农业拖拉机燃料 farm tractor fuel
农业微生物学 agromicrobiology
农业温标 thermometric scale for agriculture
农业污染 agricultural contamination; agricultural pollution
农业污染处理 agricultural pollution treatment
农业污染物 agricultural pollutant
农业污染源 agricultural pollution sources;agricultural source;sources of agricultural pollution
农业污水 agricultural wastewater
农业物候 agricultural microclimate
农业物理性质 agrophysical property
农业系统 agricultural system
农业系统的起源 origin of agricultural system
农业系统工程 agricultural systems engineering
农业细菌学 agricultural bacteriology
农业现代化 modernization of agriculture;modernization of farming
农业兴旺 prosperity in farming
农业型 agro-type
农业性状 agronomical traits
农业迅速增长 rapid agricultural growth
农业研究 agricultural research;farming research
农业研究课题 agricultural research project

N

农业研究所 agricultural research service
农业遥感 agricultural remote sensing;remote-sensing for agriculture
农业移民 displaced farmer
农业移民区 agricultural settlement
农业因素<土地利用的> agronomic factor
农业用变压器 agricultural service transformer
农业用地 agricultural land;agricultural plot;agricultural soil;farmland
农业用工作装置 agricultural attachment
农业用机械 agricultural machine
农业用水 agricultural water; water for agriculture;agriculture water
农业用水管理 agricultural water management
农业用水评价等级 grade of water in agriculture
农业用水水质标准 quality standard of agricultural water
农业用水水质要求 water quality requirement for agricultural water
农业用途 agricultural purpose
农业用拖拉机 agrimotor;farm tractor
农业用贮水池 farm pond
农业运输 agricultural traffic
农业政策 agricultural policy;agropolitics
农业植保员 peasant plant protection
农业主要投资 main input into agriculture
农业贮仓 pit silo
农业专家 agronomist
农业资本 agricultural capital
农业自然资源 agronatural resources
农业综合发展 comprehensive development of agriculture
农业综合防治 agricultural comprehensive prevention and treatment
农业综合公司<美> agricorporation
农业综合企业 agribusiness
农业总产量 total agricultural output
农业总产值 total output value of agriculture
农艺 agriculture;terra culture
农艺的 agricultural
农艺分类 agricultural classification
农艺化学 chemurgy
农艺师 agriculturist;agronomist
农艺学 agronomy
农艺学的 agronomic
农用扳钳 agricultural wrench
农用扳手 agricultural wrench
农用长柄杓 hoe dipper
农用车辆 agricultural vehicle
农用粗硫粉 agricultural sulphur
农用道口 farming crossing
农用地下排水沟 subsoil drain
农用电动机 farm motor
农用动力 farm power
农用动力机 agrimotor
农用斗门 farm head gate
农用发动机 agricultural engine;farm engine
农用房屋 agricultural building
农用飞机 agricultural aircraft
农用非饮用水 non-potable water for agricultural purposes
农用钢窗 steel window for agricultural use
农用钢丝网水泥船 ferro-cement agricultural boat
农用工具 agricultural implement
农用工业 agroindustry
农用挂车 farm trailer;farm wagon
农用航空 agricultural aviation
农用化学品 agrichemicals
农用化学污染 agricultural chemical contamination
农用化学物 agricultural chemicals; agrochemicals
农用化学物定量分析 quantitative analysis of agricultural chemicals
农用化学物废水 agricultural chemicals wastewater
农用化学物溶化评估法 leaching evaluation of agricultural chemical methodology
农用机动车<指拖拉机、汽车等> agrimotor
农用机具 farm implement
农用机械 farm machinery
农用建筑 agricultural building
农用抗菌素 agricultural antibiotic
农用排水沟 agricultural drain
农用排水管 agricultural drain pipe
农用喷灌泵 agricultural spray pump
农用喷雾机 agrosprayer
农用喷雾器 agroatomizer
农用平地机 farm leveller
农用起重机 farm crane
农用汽车 agricultural automobile;agrimotor
农用牵引车 farm tractor
农用杀虫剂 agricultural insecticide
农用杀螨剂 agricultural miticide
农用石灰 aglime; agricultural lime; agstone
农用水库 farm reservoir
农用塑料 agriplast;agriplastics
农用拖车 farm cart;farm wagon
农用拖拉机 agricultural traction engine; agricultural tractor; agrimotor;farm tractor
农用拖拉机轮胎 farm tractor tyre
农用小(土)坝 farm dam
农用压路机 farm roller
农用药剂 agricultural chemicals
农用运输汽车 agricultural truck
农用载货汽车 farm truck
农用载重汽车 farm truck
农用装载机 agricultural loader;farm loader
农用装载设备 farm loading equipment
农庄 farm(stead);grange;steading
农作季节 agricultural season
农作物 agricultural crop; agronomic crop;crop;farm crop;field crop; tillage
农作物病害 murrain
农作物产量 crop yield
农作物疾病 disease of agricultural plants
农作物类型 agro-type
农作物轮作 crop rotation
农作物名称 cultivar name
农作物群落 agrophytocoenosium
农作物受害 hazard to crops
农作物损失 crop loss
农作物图 crop map
农作物栽培 arable farming
农作物种植机 crop planting machine
农作学 geoponics
农作制度 farming system

浓

浓氨溶液 liquor ammoniae fortis; strong ammonia solution
浓氨水 strong aqua;stronger ammonia water
浓拌和 fat mix(ture)
浓茶色的 umber
浓差电池 concentration cell
浓差电流 concentration current
浓差电位 concentration potential
浓差极化 concentration polarization
浓差扩散 concentration diffusion
浓差渗析 concentration dialysis
浓差压 osmosic pressure; osmotic pressure
浓尘 smother
浓稠的 heavy-bodied
浓稠度 heavy-bodied degree
浓萃 quintessence
浓淡 dark-and-light
浓淡计算机图形 shaded computer graphics
浓淡曲面 shaded curve surface
浓淡曲线 shade curve
浓淡色调法 half-tone method
浓淡色度 value
浓淡设计 shade design
浓淡图 shaded picture
浓的 concentrated; dense; strong; thick
浓碘溶液 Lugol's solution;strong iodine solution
浓度 concentration; consistence[consistency]; deepness; degree of consistency [consistence]; degree of density;loading;strength;thickness
浓度百分率 percentage concentration;percent concentration
浓度比 concentration ratio; density ratio
浓度变化相对速度 relative rates of change in concentration
浓度标准 concentration standard
浓度测定 concentration determination
浓度测流速法 dilution ga(u)ging;dilution method
浓度差 concentration difference;differential concentration
浓度差能 salinity gradient energy
浓度常数 concentration constant
浓度淬火 concentration quenching
浓度单位 unit of concentration
浓度单位等值线 concentration unit contour
浓度等级 grade of concentration
浓度等值线 concentration contour
浓度断面图 concentration cross-section;concentration profile
浓度反应曲线 concentration response curve
浓度范围 concentration range;range of concentration
浓度分布 concentration distribution
浓度分布曲线 concentration distribution curve; concentration distribution profile
浓度分布图 concentration profile
浓度活度关系 concentration-activity relationship
浓度极限 concentration threshold; threshold concentration
浓度计 concentration meter
浓度界限 threshold concentration
浓度克拉克值 Clarke of concentration
浓度控制 concentration control
浓度控制装置 density control unit
浓度扩散 concentration diffusion
浓度历时曲线 concentration-duration curve
浓度敏感型检测器 concentration sensitive detector
浓度内带 concentration inner zone
浓度平衡 concentration equilibrium
浓度剖视图 concentration profile
浓度曲线 concentration curve
浓度容度积 concentration solubility product
浓度时间乘积 product of concentration and time
浓度水平 concentration level
浓度探测器 density probe
浓度梯度 concentration gradient
浓度外带 concentration outer zone
浓度系数 concentration coefficient
浓度限度 concentration limit;limit of concentration
浓度效应 concentration effect
浓度效应律 law of concentration effect
浓度指示器 concentration indicator
浓度指数 concentration index
浓度中带 concentration intermediate zone
浓度周期 cycles of concentration
浓度组分 density component
浓肥料 dense manure
浓肥料处理 dense manure treatment
浓废水 strong wastewater
浓膏 stiff paste
浓过磷酸钙 concentrated superphosphate
浓黑 dense black
浓黑烟 dark-smoke
浓红银矿 pyrargyrite
浓厚拌和 rich mix(ture)
浓厚的 inspissate;rich;thick;viscid
浓化时间 thickening time
浓黄绿色 sap green
浓黄土 sienna
浓黄土矿床 sienna deposit
浓灰浆 fat mortar;heavy mortar;rich mortar
浓混合气 rich mix(ture)
浓混合气调节 rich metering
浓积云 cumulus congestus cloud
浓集 compaction
浓集产物供料 enriched feed
浓集法 concentration method
浓集反应气体 enriched reactant gas
浓集克拉克值 concentration Clark value
浓集势 concentration potential
浓集污水处理厂 upgrading wastewater treatment plant
浓集系数 concentration factor
浓集中心 concentration center[centre]
浓加工颜料 pigment concentrate
浓碱式醋酸铅溶液 liquor plumbi subacetatis fortis
浓浆 jatex;underflow
浓浆泵 underflow pump
浓浆法 thick slurry process
浓菌膜 dense bacterial membrane
浓料浆 concentrated slurry
浓烈的气味 tang
浓硫酸 concentrated sulfuric acid; concentrated sulphuric acid; oil of vitriol;vitriol
浓硫酸硫酸钾硫酸铜消解蒸馏等浓比色法 strong sulfuric acid-potassium sulfate-copper sulfate digestion-distillation-nesslerization
浓绿 invisible green
浓密冰泥 dense sludge;slob;slob ice
浓密产品 thickened product
浓密的 bushy
浓密底流 thickened underflow
浓密高层云 altostratus densus
浓密灌丛<班图> msitu
浓密机层 thickener tray
浓密机澄清区 thickener clear zone
浓密机底流 thickener underflow
浓密机泥浆区 thickener sludge zone
浓密机排砂坑道 thickener tunnel
浓密剂 thickening agent
浓密精矿 thickened concentrate
浓密脱泥矿浆 thickened deslimed

pulp
浓密旋流器 thickening cyclone
浓密(云) densus
浓泥 thickened underflow;underflow
浓泥浆 thick mud;thick slurry
浓配合 rich mix(ture)
浓配合混凝土 rich mixed concrete
浓喷雾液 spray concentrate
浓氰化物电镀液 cyanide-concentrated plating solution
浓溶剂槽 tight flux bath
浓溶液 rich solution; strong liquor; strong solution
浓色 intense colo(u)r;rich colo(u)r
浓色的 rich
浓色河水 highly colo(u) red river water
浓色母料 colo(u)r concentrate
浓色深度 depth of mass tone
浓色相 rich shade
浓色效应 hyperchromic effect
浓砂浆 fat mortar;rich mortar
浓生石灰 white quicklime
浓湿雾 camanchaca;garua;harrua
浓石灰 fat lime
浓石灰砂浆 rich lime mortar
浓霜 depth hoar; heavy frost; sugar snow
浓水剂 aqueous concentrate
浓水流 concentrated stream
浓水泥浆 coarse mix(ture)
浓酸 concentrated acid
浓酸浸出 concentrated acid leaching; high acid leaching
浓缩 boil off; bring down; condense; enrich(ment); evaporate; solidification;thickening;upgrade
浓缩靶 enriched target
浓缩倍数 concentration multiple; cycles of concentration
浓缩泵 concentrate pump
浓缩比 concentration ratio
浓缩材料 enriched material
浓缩残油 short residuum
浓缩槽 thickening tank
浓缩层 enriched layer; enrichment layer
浓缩产品 condensation product
浓缩厂 condensery
浓缩沉淀池 thickening settling tank
浓缩池 concentration basin; concentration tank; concentrator; thickener;thickening tank
浓缩池扩大 scale-up of thickener
浓缩处理 thickening treatment
浓缩粗柴油 enriched gas oil
浓缩猝灭 concentration quenching
浓缩催干剂 concentrated drier[dryer]
浓缩导槽 concentration guide
浓缩的 concentrated
浓缩的活性污泥 concentrated activated sludge
浓缩底流 thickened underflow
浓缩度 condensation rate; grade of concentration
浓缩度计 enrichment meter
浓缩度控制 enrichment control
浓缩段 enriching section
浓缩法 concentration method; enrichment process; method by condensation
浓缩反应 polycondensation reaction
浓缩反应堆 enriched reactor
浓缩肥料 concentrated fertilizer; concentrating fertilizer; high analysis fertilizer
浓缩废活性污泥 thickened waste-activated sludge
浓缩粉状垃圾 enriched pulverized refuse
浓缩工厂 enrichment plant
浓缩功能 concentrating function
浓缩管 evaporating pipe
浓缩罐 concentrating pan
浓缩锅 concentrating pan
浓缩海水 enriched seawater
浓缩核燃料 enriched nuclear fuel
浓缩化合物 enriched compound
浓缩回收 enriching recovery
浓缩混合物 enriched mixture
浓缩活性污泥 thick activated sludge
浓缩机 concentrator;thickener
浓缩机底槽 < 圆网 > thickener vat
浓缩机底流 thickener underflow
浓缩技术 concentration technique
浓缩剂 concentrating agent; densifier;thickener
浓缩检验 concentration test
浓缩胶乳 concentrated latex
浓缩硫化乳胶 revultex
浓缩馏分 enriched fraction
浓缩卤水 brine blowdown
浓缩滤器 thickening filter
浓缩率 enrichment factor
浓缩螺滤脱水 concentration-spiral filtration dehydration
浓缩模 enrichment mode
浓缩能力 condensability
浓缩泥浆 thickened slurry
浓缩汽油 enriched gas
浓缩器 column evapo(u) rator; concentrator; densifier; evaporator; thickener
浓缩情报 < 指经过整理和压缩篇幅的情报资料 > enriched information
浓缩燃料 enriched fuel
浓缩燃料堆 enriched-fuel reactor
浓缩燃料堆芯曲率 enriched buckling
浓缩溶液 concentrated solution
浓缩石油 inspissated oil
浓缩时间 concentration time; time of concentration
浓缩食物 emergency ration
浓缩试验 concentration test
浓缩室 enriched chamber
浓缩树脂 polycondensation resin
浓缩水 concentrated water
浓缩酸 concentrated acid
浓缩塔 concentrating tower
浓缩碳化铀 enriched uranium carbide
浓缩碳化铀粒子 enriched uranium carbide particle
浓缩添加剂 additive concentrate
浓缩污泥 concentrated sludge; concentrating sludge; thickened sludge; thick sludge
浓缩物 concentrate
浓缩物质 enriched material
浓缩稀释试验 concentration dilution test
浓缩系数 concentrated coefficient; concentration coefficient; concentration factor; enrichment coefficient;enrichment factor
浓缩系数法 concentration coefficient method
浓缩橡浆 jatex
浓缩硝酸盐废水 concentrated nitrate waste
浓缩效率 concentration efficiency
浓缩悬浮液 concentrated suspension
浓缩旋流器 cyclone thickener
浓缩亚硫酸路用结合料 < 浓缩到 50% 固体含量 > sulphite road binder concentrate
浓缩氧化铀 enriched uranium oxide
浓缩液 concentrated liquor; concentrated solution
浓缩液舱柜 concentrate tank
浓缩液体硅烟 < 一种提高混凝土抗渗性的掺和剂 > condensed liquid silica fume
浓缩因素 concentrating factor
浓缩因子 enrichment factor
浓缩油 enriched oil
浓缩油藏 inspissated deposit
浓缩铀 enriched uranium; uranium concentrate
浓缩铀点燃棒 enriched uranium booster
浓缩铀堆 enriched uranium reactor
浓缩铀反应堆 enriched uranium reactor
浓缩铀固体均匀反应堆 enriched uranium solid homogeneous reactor
浓缩铀裂变室 enriched uranium fission chamber
浓缩铀轻水慢化反应堆 enriched uranium light water moderated reactor
浓缩铀石墨慢化反应堆 enriched uranium graphite moderated reactor
浓缩铀水池式反应堆 enriched uranium swimming pool reactor
浓缩铀水均匀反应堆 enriched uranium aqueous homogeneous reactor
浓缩铀重水慢化反应堆 enriched uranium heavy water moderated reactor
浓缩装置 enrichment facility
浓缩作用 concentration; inspissation; polycondensation reaction
浓污泥 thickened sludge
浓污水 strong sewage;strong waste
浓雾 gross fog; heavy fog; pea soup fog; smog; smother; solid fog; soup < 俚语 >; thick fog < 能见度 50 ~ 200 米 >; dense fog < 能见度小于 50 米 >;fog bank
浓雾报警器 megafog
浓雾剂 atomising concentrate
浓雾笼罩 befog
浓雾笼罩的海岸 fogbound coast
浓雾密团 fog bank
浓雾信号 fog signal
浓咸水 brine
浓相 dense phase
浓硝酸 aqua fortis; concentrated nitric acid
浓烟 dense smoke; smeech; smother; smudge
浓盐水 concentrated brine;strong brine
浓盐水溶液 brine; concentrated salting liquor
浓盐酸 concentrated hydrochloric acid
浓盐液 brine
浓颜料分散体 concentrated pigment dispersion
浓颜料制备物 colo(u)r concentrate
浓艳色彩 rich colo(u)r
浓焰 rich flame
浓叶时期 heavy foliage period
浓液 dope
浓液性 thick fluidity
浓乙酸 spirit acid
浓荫的 umbrageous
浓荫树 shade tolerant tree;shade tree
浓油斑 < 试验沥青时的 > heavy stain
浓油剂 oil miscible concentrate
浓云 congestus;dense cloud
浓纸浆 thickened pulp
浓重的 miasma [复 miasmata/miasmas]

脓

脓肿 abscess

弄

弄【建】alley

弄暗 obscure
弄凹 pit
弄薄 weaken(ing)
弄成粉 pulverize
弄成崎岖不平 roughen
弄成畦 stitch
弄成衰弱 crock
弄成台地 terrace
弄成条纹 stripe
弄成小尖塔形 pinnacle
弄成小块 wad
弄成斜面 splay
弄成杂色 variegate
弄单纯 simplify
弄淡 subdue
弄得晕头转向 disorient
弄短 shorten(ing)
弄钝 blunt
弄红 ruddy
弄滑 sleeking
弄尖 cut to a point;nib;taper
弄简单 simplify
弄结实 knit
弄空 deplenish;vacate
弄乱 clutter; foul-up; muss up; snarl; tangle
弄明白 ravel
弄模糊 blur;blur out
弄平 plane; planing; planish; scrape down;dub out < 木板、瓦等 >
弄平的 platten
弄平整 decurl
弄破 crumple
弄清楚 clarify
弄清洁 cleanse
弄缺 nick
弄软 soften
弄上污点 spot
弄湿 dabble;humidification;moisten
弄湿的 moistened
弄湿了的地方 slop
弄碎 cob;crumble;pulverize
弄堂 drong
弄歪 sway
弄歪曲 misshape
弄弯 buckle;buckling;curve
弄污 bitch
弄细 attenuate
弄香 perfume
弄斜 canting;splay
弄圆 conglobate;radiusing;round off
弄圆边角 edge rounding
弄脏 begrime; defile; fouling-up; maculate;muss up;pollute;spot;taint
弄脏的 contaminative
弄糟 box-up;muss up
弄折(书页的)纸角 dog ear
弄整洁 snug
弄整齐 slick
弄直 straighten;unbend
弄皱 crumple; ruck; ruckle; ruffle; shrink

奴

奴隶 bondsman

奴役 bond service
奴役现象 helotism;slavery

驽

驽马 skate

努

努阿内齐反向极性带 Nuanetsi reversed polarity zone

努阿内齐反向极性时 Nuanetsi reversed polarity chron
努阿内齐反向极性时间带 Nuanetsi reversed polarity chronzone
努比亚砂岩 Nubian Sandstone
努德森(准则)数 Knudsen number

N

N

努封保径滑动铣齿钻头 sealed ga(u)-ge sliding milled bit
努库阿洛法<汤加首都> Nukualofa
努拉铝合金 Nural
努拉岩<一种石棉沥青> Nuralite
努里斯坦地块 Nuristan massif
努力做 labo(u)r at
努尼瓦克事件 Nunivak event
努尼瓦克正向极性亚带【地】Nunivak normal polarity subzone
努尼瓦克正向极性亚时【地】Nunivak normal polarity subchron
努尼瓦克正向极性亚时间带【地】Nunivak normal polarity subchron-zone
努普法 Knoop's method
努普和维克斯硬度计 knoop and Vickers indenter
努普刻痕微硬度试验 Knoop's indentation microhardness test
努普微硬度 Knoop's microhardness
努普微硬度试验 Knoop's microhardness test
努普学说 Knoop's theory
努普压痕 Knoop's indentation
努普压痕试验 Knoop's indentation test
努普压头<硬度试验用> Knoop's indenter
努普硬度标度 Knoop's hardness scale
努普硬度计 Knoop's hardness tester
努普硬度仪 Knoop's indenter
努普硬度值 Knoop's hardness number;Knoop's number
努森表 Knudsen's table
努森采样器 Knudsen sampler
努森层 Knudsen layer
努森方程 Knudsen's equation
努森公式 Knudsen formula
努森室 Knudsen cell
努森余弦定律 Knudsen cosine law
努森真空计 Knudsen(vacuum)ga(u)-ge
努碳镍石 nullaginite
努瓦克肖特<毛里塔尼亚首都> Nouakchott
努希尔<一种沥青防水剂> Nuseal

怒潮 eager[eagre];eager pororoca;tidal bore;acker;bore

怒潮高度 bore height
怒潮形式 bore shaped pattern
怒号 roar
怒涛<八级风浪> mountainous sea;angry wave;precipitous sea;confused or phenomenal sea

女搬运工 portress

女宾休息室 powder room
女舱室 berth for female
女厕(所)ladies' room;ladies' toilet;ladies' water-closet[WC];ladies wash closet;women toilet
女乘务员 stewardess
女创造者 creatress
女创作者 creatress
女大使 ambassadress
女导演 directress
女店主 tradesivoman
女雕刻家 sculptress
女雕塑家 sculptress
女雕像柱 canephorae;caryatid
女董事 directress
女董事长 chairwoman
女儿平均(值)法 daughter average
女儿墙【建】dwarf wall;parapet (wall);blocking course;breast work;domali
女儿墙凹 breast of a window
女儿墙壁脚板 parapet skirting
女儿墙衬砌 parapet lining
女儿墙防水 parapet water proofing
女儿墙高度 parapet wall height
女儿墙格栅 parapet grill(e)
女儿墙建筑构件 parapet building component
女儿墙脚泛水 parapet skirting
女儿墙节间 parapet panel
女儿墙勒脚 parapet skirting;upstand at parapet
女儿墙勒脚泛水 base flashing of parapet shirting
女儿墙黏[粘]土砖 parapet clay brick
女儿墙排水沟 parapet gutter
女儿墙砌块 parapet wall block
女儿墙墙板 parapet panel
女儿墙墙裙 parapet skirting
女儿墙上排水孔 weep hole
女儿墙上泄水洞口 outlet
女儿墙饰面 parapet facing
女儿墙踢脚板 parapet skirting
女儿墙天沟 parapet gutter
女儿墙涂底 parapet lining
女儿墙位移 parapet displacement
女儿墙斜天沟<不等宽的> tapered parapet gutter
女发明家 inventress
女房地产经纪人协会 Women's Council of Realtors
女飞机驾驶员 aviatress;aviatrix
女飞行员 aviatress;aviatrix
女服务员 portress;stewardess
女更衣室 dressing room for woman;female changing room;ladies' changing room
女工 factory girl;female worker;workwomam
女工长 forelady;forewoman
女工头 forelady;forewoman
女工作者 workwomam
女管理员 administratrix[复 administratrices/administratrixes]
女盥洗室 ladies' room;ladies' toilet;powder room;women toilet
女继承人 heiress
女驾驶员<汽车的> chauffeurette
女监工员 forewoman
女教师 instructress
女经理 manageress
女警察 policewoman
女看门人 portress
女客厅 ladies' drawing room
女矿工 balmaiden
女郎头顶祭盘的装饰品 canephora
女理发店 female hairdressing shop;ladies' hair dressing shop
女理事长 chairwoman
女立遗嘱人 testatrix
女列车员 portress;conductress<美>
女零售商 tradesivoman
女领班 forelady;forewoman
女领工员 forelady
女帽毡 millinery felt
女陪审员 jurywoman
女清洁工 portress
女人卧室壁橱 bedroom closet for women
女商人 tradesivoman
女施工员 buildress
女士的私室 boudoir
女式服装 costume;toilet
女室 female room
女售票员<公共汽车、电车等> conductress
女司机<美> chauffeurette
女所有人 proprietress
女委员长 chairwoman
女卫生间 ladies' room;powder room
女向导 conductress
女像柱 caryatid
女像柱门廊 caryatid porch
女新闻记者 newswoman
女休息室 female drawing room;female rest room;ladies' drawing room
女修道会 convent
女修道院 convent;minchery;nunnery;nunnery church;mynchery<老盎格鲁-撒克逊语>
女修道院侧厅 conventual limb
女修道院厨房 conventual kitchen
女修道院房屋 conventual building
女修道院建筑 convent architecture;conventual architecture
女修道院教堂 conventual church
女修道院客厅 conventual parlour
女修道院社区 conventual community
女业主 proprietress
女佣用的水槽 housemaid's sink
女用鞍 side saddle
女用盥洗盆 vanity basin
女用露指长手套 mitten
女杂工 char;portress
女贞【植】privet;glossy privet
女执行人 executrix[复 executrixes/executrices]
女职工 workmistress
女指导人 directress
女指导员 instructress
女指导者 conductress;directress
女指挥 conductress;directress
女主席 chairwoman;madam president
女助手 coadjutress
女子更衣室 maid's changing room
女子健身房 girl's gymnasium
女子理发店 hairdresser's parlo(u)r
女子理发室 ladies' hair dressing shop

钕玻璃 Nd(-in)glass;neodymium glass

钕玻璃激光器 neodymium glass laser
钕钒玻璃 neodymium-vanadium glass
钕铬玻璃 neodymium-chromium glass
钕激光辐射 neodymium laser radiation
钕激光器 Nd laser;neodymium laser
钕激光照明器 neodymium laser illuminator
钕激光振荡器 neodymium laser oscillator
钕晶体激光器 neodymium crystal laser
钕矿 neodymium ore
钕镨 didymium
钕镨玻璃 didymium glass
钕镨玻璃标准 didymium glass standard
钕镨混合物 didymium
钕锶地幔向量 neodymium-strontium mantle vector
钕锶同位素相关性 neodymium-strontium isotope correlation
钕铜玻璃 neodymium-copper glass

疟疾 malaria;ever and ague;mosquito fever;paludism

疟疾的 paludal;paludine
疟蚊 anopheles
疟蚊属 anopheles
疟原虫 malarial parasite

暖白光 warm white

暖棒 warm rod
暖冰川 temperate glacier;warm glacier
暖床 hot bed
暖床器 warming pan
暖带林 warm temperate forest
暖带雨林 warm temperate rain forest
暖带植物区 warm temperate district
暖等温水层 epilimnion(layer)
暖低压【气】warm low
暖调白色 warm white
暖洞 warm cave
暖房 conservatory;glasshouse;greenhouse;hot house;warmhouse;warming-house;warm room;calefactory<寺院内的>
暖房玻璃 greenhouse glass
暖房腐蚀 greenhouse rot
暖房效应 glasshouse effect;hothouse effect
暖房育鸡房 brooder
暖房栽培 glass culture
暖风 warm winds
暖风干燥机 warm-air drier[dryer]
暖风干燥器 warm-air drier[dryer]
暖风机 forced convection air heater;heater unit;hot-air heat;unit heater
暖风机组 unit air heater
暖锋【气】thermal front;warm(-air)front
暖锋波动 warm-front wave
暖锋锢囚【气】warm-front occlusion
暖锋降水 warm-front precipitation
暖锋雷暴 warm-front thunderstorm
暖锋面 anaphalanx;warm-front surface
暖锋雾 warm-front fog
暖锋雨 warm-front rain
暖高压【气】warm anticyclone;warm high
暖高压脊 warm ridge
暖管 warming pipe
暖管旁通阀 warm-up by-pass(ing)
暖锅 chafing dish
暖和晴朗天气 open weather
暖核 warm core
暖湖 warm lake
暖环境 warm water environment
暖黄土 warm loess
暖机 heating of turbine;warming
暖机阀 warm-up valve
暖机管 warming pipe
暖机过程 warming-up process
暖机时间 warm-up period
暖机特性 warm-up characteristics
暖机因数 warm-up factor
暖机预备 hot reserve
暖机运转 warm-operation
暖脊 warm ridge
暖季 warm season
暖季牧草 warm-season grass
暖季植物 warm-season plant
暖季作物 warm-season crop
暖脚器 foot warmer
暖空气舌 tongue of warm air
暖浪 warm wave
暖流 warm current;warm flow;warm water stream
暖流带 warm current zone
暖流低压槽 warm trough
暖昧石 griphite
暖面器 surface heater
暖膜风速表 hot film anemometer;warm-film anemometer
暖盘 warming plate
暖棚育鸡房 brooder
暖瓶架 pot-rack

暖期 warm period
暖气 central heating; heating; heating air
暖气布置图 heating plan
暖气厂 heating plant
暖气除霜器 hot-air heater-defroster
暖气阀盖垫片 heater valve cover gasket
暖气阀轴 heat control valve shaft
暖气分区 zone heating
暖气工程师 heating engineer
暖气供给 warm-air feed
暖气管 caliduct; flue; heating pipe; radiator pipe
暖气管道 column of radiator; heating duct; hot-air duct; warm-air pipe
暖气管道竖井 air duct riser; warm-air duct riser
暖气管沟 heating trench
暖气管柱 column of radiator
暖气火炉 warm-air stove
暖气机 warming machine
暖气加热系统 warm-air heating system
暖气流 warm air stream
暖气炉 hot-air heater
暖气炉片 heating radiator
暖气盘管 heating coil
暖气盘旋管 heating spiral
暖气片 finned radiator; heating element; hot-water steam radiator; radiated flange; radiating fin; radiator section
暖气片的防护罩 radiator guard
暖气片的回水装置 radiator return fitting
暖气片的排出孔 radiator bleeder tap
暖气热量 heating heat
暖气软管 steam heater hose
暖气软管接头 heating hose coupling
暖气软管联结器 steam hose coupler
暖气设备 heater; heating equipment; heating installation; heating plant; warm-air heating plant
暖气设备室 heating plant room
暖气设备罩 heater casing
暖气设计 heating design
暖气室 heating chamber
暖气输出量 heated air output
暖气竖管 air duct riser; warm-air duct riser
暖气通道 heating passage
暖气团 warm-air mass
暖气系统 heating system
暖气压力表 steam heat ga(u)ge
暖气要求 heating heat requirement
暖气与通风 heating and ventilating
暖气罩 heating mantle; radiator box
暖气装置 heating apparatus; heating installation; heating plant; installation of heating; radiator
暖气装置挡板 heating hood
暖气装置防护套 heating hood
暖气装置外壳 heating hood
暖气装置外套 heating jacket
暖气总管 heating main
暖汽端阀 steam end valve
暖汽管 heater pipe; heating pipe; main steam pipe; steam-heating pipe
暖汽管端阀 end train-pipe valve; train line end valve
暖汽管联结 heating hose coupling
暖汽管罩 heater pipe casing
暖汽减压阀 steam heat reducing valve
暖汽连接器上接头 end pipe connection
暖汽联结器卡环 clamp lock
暖汽联结器卡铁 coupler latch
暖汽炉水箱 heater tank
暖汽软管 heater hose
暖汽调整阀 vapo(u)r regulating valve; vapo(u)r regulator
暖汽调整阀垫圈 deflector ring
暖汽箱 heater box
暖汽照明组合车 heating and lighting wagon
暖汽主管 heater train pipe
暖器 heater
暖墙 <设采暖烟道的墙壁> hot wall; oven wall
暖区 warm section; warm sector; warm zone
暖泉 warm spring
暖燃导火线 slow-burning fuse
暖热量 heat-rejection load
暖色 <有暖感的颜色,指红黄橙等> warm colo(u)r; advancing colo(u)r
暖色调 warm tone
暖色调的 warm-toned
暖色调像片 paper with warm image tone
暖舌 warm tongue
暖湿空气 warm moist air
暖湿亚带 warm humid subzone
暖室 hot house; stove; warmhouse
暖水表 warm water meter
暖水层 warm water sphere
暖水池 warm pool
暖水带 <黑潮的高温水带> warm belt
暖水管道 caliduct
暖水灌溉 warm water irrigation
暖水壶 heating kettle
暖水流速仪 warm meter
暖水区 <指南北极圈以外的海洋> warm sea; warm water area; warm waters
暖水上涌 <深层的> warm upwelling
暖水团 warm water mass
暖水性海鲜鱼类 warn-ocean fish
暖水有孔虫丰度 abundance of foraminifera warm water
暖通 heating and ventilating; heating and ventilation
暖通工程师 heating and ventilating engineer
暖通工程师学会期刊 <英> Journal of Institute of Heating Ventilating Engineer
暖通设计 heating and ventilation design
暖通调控 heating, ventilation and air-conditioning
暖通专业 heating and ventilating discipline
暖卫 heating and plumbing
暖温带 warm temperate zone
暖温带混交林黄褐土带 warm temperate zone mix forest yellow cinnamon soil zone
暖温带亚带 warm temperate subzone
暖温-亚热带亚带 warm temperate to subtropical subzone
暖雾 warm fog
暖西南风 Chinook wind
暖箱 incubator
暖心扰动 warm-cored disturbance
暖型锢囚锋 warm-occluded front
暖性反气旋 warm anticyclone; warm high
暖性气旋 warm cyclone; warm low
暖性针叶林 warm needle-leaf forest
暖雨 warm rain
暖云 warm cloud
暖云雨 rain from warm clouds

挪后桩 <船靠码头> shift aft

挪前桩 <船靠码头> shift forward
挪威 <欧洲> Norway
挪威白石英 Norwegian quartz
挪威船级社 Det norske Veritas
挪威海 Norwegian Sea
挪威海盆 Norwegian basin
挪威海事管理局 Norwegian Maritime Directorate
挪威红松 Norway pine
挪威建筑 Norwegian architecture
挪威黏[粘]土 Norwegian clay
挪威槭树 Norway maple; plane maple
挪威杉木 white deal
挪威式板条教堂 Norwegian stave church
挪威式掏槽 Norwegian cut
挪威水工试验所 Norwegian Hydrotechnical Laboratory
挪威土工研究所 Norske Geotechnical [Geotekinske] Institute
挪威岩土工程研究所 Norwegian Geotechnical Institute
挪威云杉 Norway spruce; Picea abies; spruce fir
挪用 misappropriation; peculate; stealing
挪用偿债基金 raid the sinking fund
挪用度 appropriation
挪作他用 misappropriate

诺阿卫星 NOAA satellite

诺埃尔木块地板 Noel
诺埃曼边界条件 Neumann's boundary condition
诺埃曼层状组织 Neumann band
诺埃曼带 Neumann band
诺埃曼公式 Neumann's formula
诺埃曼函数 Neumann's function
诺埃曼井函数 Neumann's well function
诺埃曼-柯普法则 Neumann-Kopp rule
诺埃曼问题 Neumann's problem
诺埃曼原理 Neumann's principle
诺贝尔检验器 Nobel blastometer
诺贝尔奖金 Nobel Prize
诺贝尔奖金获得者 Nobelist; Nobelman
诺贝尔淘析器 <粒度分析用> Nobel elutriator
诺贝尔油 Nobel oil
诺布尔法 noble method
诺丁汉沥青试验机 <诺丁汉大学用于评定沥青混合料力学性能的一种试验机> Nottingham asphalt tester
诺顿齿轮 Norton's gear
诺顿等效 Norton's equivalent
诺顿定理 Norton's theorem
诺顿管(深)井 <用桩锤将一尖头多孔的井管打入含水层形成的井> Norton's tube well
诺顿理论 Norton's theory
诺尔阶 <晚三叠世>【地】Norian
诺尔曼式建筑装饰 nebule
诺尔曼式曲折线脚 churn mo(u)lding
诺尔斯畅流 <一种管子接头商品名> Knowle's free flow
诺尔斯电解池 Knowles's cell
诺尔斯多臂装置 Knowles' dobby motion; Knowles' head; Knowles' positive dobby
诺尔制动机 Knorr brake
诺福克插销 Norfolk latch; thumb latch
诺福克南洋杉 Norfolk-island-pine
诺福克锁 Norfolk latch
诺福沥青 novophalt
诺伽重力仪 Norgaard gravimeter
诺克斯 <弱光照度单位> nox
诺克斯-奥克斯本焙烧炉 Knox's and Oxborne furnace
诺克斯裂化法 <气相热裂化法> Knox cracking
诺克斯过程 Knox's process
诺克斯气相裂化过程 Knox's true vapour phase process
诺克斯装置 Knox's unit
诺拉斯托 <一种防锈漆> Norusto
诺利斯航海表 Nories' nautical tables
诺硫铁铜矿 nukundamite
诺马尔斯基显微镜 Nomarski microscope
诺曼玻璃块 Norman slab
诺曼底大桥 <主跨 856 米,世界上建成的最大跨径斜拉桥之一> Normandie Bridge
诺曼底式建筑中拱的线脚 double cone mo(u)lding
诺曼风格 Norman style
诺曼哥特风格 Norman Gothic style
诺曼拱顶 Norman vault
诺曼建筑风格 Norman style
诺曼连拱式圆屋顶 Norman vault
诺曼式 Norman style
诺曼式窗(户) Norman window
诺曼式地穴 Norman crypt
诺曼式拱 Norman arch
诺曼式建筑 Norman(style) architecture
诺曼式屋顶瓦 Norman roofing tile
诺曼屋面瓦 Norman roofing tile
诺曼值 Naumene number
诺曼砖 <规格为 2.5412 英寸,1 英寸 = 0.0254 米> Norman brick
诺模图 abac; nomographic(al) chart; nomogram; nomograph (alignment chart); alignment diagram; alignment line; alinement chart
诺模图法 nomography
诺硼钙石 nobleite
诺三水铝石 nordstrandite
诺思希尔式锚 Northill anchor
诺斯特漆 <一种金属防锈油漆> Nust
诺塔(固体)混合机 Nauta mixer
诺塔混合器 Nauta mixer
诺特格罗夫易切岩 Notgrove freestone
诺特环 Noetherian ring
诺瓦硅基粉末 <一种含硅石的防水、油、酸的粉末材料> Novoid
诺瓦勒克司(玻璃透镜) Novalux
诺瓦里特 <一种铜铝合金> Novalite
诺威奇砂质泥灰岩 Norwich Crag
诺维特 <硬质合金> Novite
诺维特炸药 novit
诺伊尔青铜 Noil
诺伊曼问题 Neumann problem
诺依曼计算机 Neumann computer

O

欧 <电阻单位> ohm

欧安计 ohm ammeter
欧勃波 Erb's wave
欧勃点 Erb's point
欧代联轴节 Oldham's coupling
欧椴树 lime tree
欧多克斯球 spheres of Eudoxus
欧非道路 Euro-African roads
欧非共同体 Eurafrica
欧吉齐紫树 sour tupelo
欧几里得测度 Euclidean measure
欧几里得单纯复形 Euclidean simplicial complex
欧几里得单形 Euclidean simplex
欧几里得度量 Euclidean meter
欧几里得多面体 Euclidean polyhedron
欧几里得范数 Euclidean norm; Euclid norm
欧几里得复形 Euclidean complex
欧几里得刚体 Euclidean rigid body
欧几里得公理 Euclidean axiom
欧几里得公设 Euclid's postulate; Euclidean postulate
欧几里得固体 Euclidean solid
欧几里得环 Euclidean ring
欧几里得几何码 Euclidean geometry code
欧几里得几何(学) Euclidean geometry
欧几里得距离 Euclidean distance
欧几里得空间 Euclidean space
欧几里得联络 Euclidean connection
欧几里得模量 Euclidean norm
欧几里得平面 Euclidean plane
欧几里得区 Euclidean domain
欧几里得算法 Euclid's algorithm; Euclidean algorithm
欧几里得体 Euclidean body; Euclidsolid
欧几里得物体 <在任何压力下都不产生变形的理想固体> Euclidean solid
欧几里得向量空间 Euclidean vector space
欧几里得域 Euclidean field
欧几里得运动 Euclidean motion
欧几里得运动群 Euclidean group of motions
欧几里得辗转相除法 Euclidean algorithm
欧几里得整环 Euclidean domain; Euclidean integral domain
欧几里得直线 Euclidean straight line
欧鲫 crucian carp
欧卡拉石灰石 ocala
欧空局遥感卫星 European space administration remote sensing satellite
欧拉变换 Euler's transformation; Eulerian change
欧拉-伯努利理论 Euler-Bernoulli theory
欧拉-伯努利梁理论 Euler-Bernoulli beam theory
欧拉参数 Euler's number
欧拉禀性方程 Eulerian intrinsic(al) equation
欧拉长柱公式 Euler's formula for long columns
欧拉常数 Euler's constant
欧拉定理 Euler's theorem
欧拉定律 Euler's law
欧拉断裂应力 Euler's fracture stress
欧拉多项式 Euler's polynomial
欧拉法 Eulerian method
欧拉方 Euler's square; Eulerian square
欧拉方程(式) Euler's equation; Eulerian equation
欧拉方法 Euler's method
欧拉风 Eulerian wind
欧拉公式 Euler's formula
欧拉规则 Euler's rule
欧拉轨道 Eulerian trajectory
欧拉函数 Euler's function
欧拉荷载 Euler's load; Eulerian load
欧拉恒等式 Euler's identity
欧拉迹 Eulerian trace
欧拉积 Euler's product
欧拉积分 Euler's integral
欧拉角 Euler's angle Eulerian angles; Eulerian angle
欧拉角法 Euler's angle method
欧拉静水力学定律 Euler's hydrostatical law
欧拉柯西法 Euler-Cauchy method
欧拉可分割空间公式 Euler's polyhedron formula
欧拉-拉格朗日方程 Euler-Lagrange equation
欧拉理论 Euler's theory
欧拉力 Euler's force
欧拉临界荷载 Euler's buckling load
欧拉临界(断裂)应力 <纵向弯曲> Euler's crippling stress
欧拉路径 Eulerian path
欧拉罗德里格参数 Euler-Rodriguse parameter
欧拉螺旋曲线 Euler spiral
欧拉拟阵 Eulerian matroid
欧拉判别准则 Euler's criterion
欧拉-庞加莱方程 Euler-Poincare equation
欧拉平均流 Eulerian mean flow
欧拉求和公式 Euler's summation formula
欧拉求解法 Euler's approach
欧拉曲线 <压应力与长细比的关系曲线> Euler's curve
欧拉屈曲 Euler's buckling
欧拉屈曲长度 Euler's buckling length
欧拉屈曲公式 Euler's buckling formula
欧拉屈曲荷载 Euler's buckling load
欧拉屈曲应力 Eulerian buckling stress
欧拉三角形 Euler's triangle
欧拉剩余判别准则 Euler's criterion for residues
欧拉示性数 Euler's characteristic
欧拉数 Eulerian number; Euler's number
欧拉数值法 Eulerian numerical method
欧拉双曲线 Euler's hyperbola
欧拉水动力定律 Euler's hydrodynamical law
欧拉撕裂应力 Eulerian crippling stress
欧拉图 Euler's diagram; Euler's graph
欧拉弯曲应力 Euler's buckling stress
欧拉网孔数公式 Euler's mesh formula
欧拉微分方程 Euler's differential equation
欧拉涡轮方程 Euler's turbine equation
欧拉无穷积表示 Euler's infinite product representation
欧拉线 Euler's line
欧拉线路 Eulerian circuit
欧拉相关 Eulerian correlation; synoptic(al) correlation
欧拉消去法 Euler's method of elimination
欧拉循环 Euler's cycle; Euler's loop
欧拉压头 Euler's head
欧拉应力 Euler's stress
欧拉压屈荷载 <按照欧拉公式计算的> Euler's buckling load
欧拉游程 Eulerian tour
欧拉圆 Euler's circle
欧拉运动 Eulerian motion
欧拉运动方程 Eulerian equation of motion
欧拉展开 Euler's expansion
欧拉展开式 Euler's expansion formula
欧拉指数 Eulerian index
欧拉周期 Euler's period; Eulerian period
欧拉柱公式 Euler's column formula; Eulerian column formula
欧拉自由周期 Eulerian free period
欧拉坐标 Euler's coordinates; Eulerian coordinates
欧罗阿克里尔 <聚丙烯腈系纤维> Euroacryl
欧罗巴瓷 European porcelain
欧罗剪切牢度试验仪 Euroflock shearing tester
欧罗磨损试验仪 Euroflock abrasion tester
欧马德 <旧的电阻单位> ohmad
欧美 occident; the Occident
欧美大陆 Euramerica
欧美人 occidental
欧美双壳类地理大区 Euramerican bivalve region
欧美植物地理区系 Euramerican floral realm
欧盟关于危险物质限定的指令 Restrictions on Hazardous Substances
欧米伽 Omega
欧米伽导航系统 Omega navigation system
欧米伽导航仪 Omega navigator
欧米伽电子定位系统 Omega electronic positioning system
欧米伽定位 Omega fixing
欧米伽定位设备 Omega positioning and locating equipment
欧米伽发射台 Omega transmitting station
欧米伽建筑 <一种预制钢筋混凝土现场拼接的建筑型式> omega construction
欧米伽接收机 Omega receiver
欧米伽介子 Omega meson
欧米伽器 omegatron
欧米伽双曲线网格坐标系 Omega hyperbolic grid system
欧米伽双曲线族 Omega hyperbola family
欧米伽台名 designation of Omega station
欧米伽系统 Omega system
欧米伽巷识别 Omega lane identification
欧米伽信号格式 Omega signal format
欧姆 ohm
欧姆安培计 ohm ammeter
欧姆表 ohm ga(u)ge; ohmmeter
欧姆的 ohmic
欧姆的倒数 <电导或导纳的单位> reciprocal of ohm
欧姆电极 ohmic electrode
欧姆电阻 ohmage; ohmic resistance; plain resistance; true resistance
欧姆电阻器 ohmic resistor
欧姆电阻试验 ohmic resistance test
欧姆电阻损耗 ohmic resistance loss
欧姆定律 Ohm's law
欧姆定律轮 Ohm's law wheel
欧姆极化 ohmic polarization
欧姆集电极电阻 ohmic collector resistance
欧姆计 ohm ga(u)ge; ohmmeter
欧姆继电器 ohm relay
欧姆加热 ohmic heating
欧姆接触 ohmic contact
欧姆量程 ohm range
欧姆 · 米/厘米 ohm · m per cm
欧姆损耗 in-phase loss; ohmic loss; wattful loss
欧姆损失 ohmic loss
欧姆线圈 resistance winding
欧姆型电抗继电器 ohm type reactance relay
欧姆值 ohmage; ohmic value
欧姆阻抗 ohmage
欧佩克 Oil Producing and Exporting Countries
欧石楠属 <拉> Erica
欧氏极 Eulerian pole
欧氏几何码 Euclidean geometry code
欧氏空间 Euclidean space
欧斯马铝锰合金 Osmayal
欧特里阶 <白垩世>【地】Hauterivian
欧铁联盟研究所 Office of Research, Experiment
欧卫矛【植】prickwood
欧文尘埃计 Owen's dust counter
欧文粉尘计数器 Owen's jet dust counter
欧文破裂带 Owen fracture zone
欧文器 Owen's organ
欧西-北极有孔虫地理区系 Eurasia-Arctic foraminifera realm
欧亚板块 Eurasian plate
欧亚大陆 Eurasia
欧亚大陆桥 Eurasian continental bridge
欧亚古陆 Eurasian paleocontinent
欧亚海盆 Eurasia basin
欧亚区系 Eurasia realm
欧亚山字形构造体系【地】Eurasian epsilon structural system
欧亚珊瑚地理区系 Eurasia-Asian coral realm
欧夜鹰 puck
欧元 Eurodollar
欧扎克统 <早奥陶世>【地】Ozarkian series
欧洲 occident; the Occident
欧洲安全标准 European safety code
欧洲安全规范 European safety code
欧洲-澳大利亚航线 Europe-Australia route
欧洲白桦 droop birch; European white birch; white birch
欧洲白蜡树 common ash; European ash; Fraxinus excelsior
欧洲白冷杉 Abies pectinata; European fir
欧洲白杨 European white poplar
欧洲-北美鹦鹉螺地理大区 European-north American nautiloid region
欧洲标准 European standard
欧洲标准船舶 Europe class ship
欧洲标准化委员会 European Committee for Standardization
欧洲驳船货运系统 European barge carrier system
欧洲驳船运输方式 European barge carrier system
欧洲赤松 Norway fir; Pinus Sylvestris; redwood; Scots pine
欧洲传统式洒水喷头 European conventional style sprinkler
欧洲刺柏【植】common juniper
欧洲枞木 Corsican fir

欧洲大规模森林损害调查 large-scale forest damage survey in Europe
欧洲大陆产柳属 sallow
欧洲大西洋部 European Atlantic province
欧洲道路管理系统 road management system for Europe
欧洲道路交通安全设施(计划)<欧共体联合研究智能交通系统的一项计划> Dedicated Road Infrastructure for Vehicle Safety in Europe
欧洲道路运输信息技术实施协调组织 European Road Transport Telematics Implementation Coordination Organization
欧洲地球物理勘探工作者协会 European Association of Exploration Geophysicists
欧洲地球物理学会 European Geophysics Society
欧洲地震工程协会 European Association of Earthquake Engineering
欧洲地震学委员会 European Seismological Commission
欧洲地中海部 European Mediterranean province
欧洲地中海地震学中心 European-Mediterranean Seismologic(al) Centre
欧洲电视网 Eurovision
欧洲椴 linden
欧洲椴木 European lime;Tilia vulgaris
欧洲法郎 Eurofranc
欧洲风格 Europeanism
欧洲枫木 aul
欧洲钢结构会议 European Convention for Constructional Steelwork
欧洲钢铁标准化委员会 European Committee for Iron and Steel Standardization
欧洲港口 European port
欧洲高效安全交通计划 Program(me) for European Traffic with Highest Efficiency and Unprecedented Safety
欧洲工程机械委员会 the Committee on European Construction Equipment
欧洲共同市场<欧洲经济共同体的俗称> Euromarket; European Common Market; European Economic(al) Community
欧洲共同市场科技数据通信[讯]网络 Euronet
欧洲共同市场科技网 Euronet
欧洲共同体 European Communities
欧洲共同体理事会 Council of the European Community
欧洲共同体情报服务机构 European community information services
欧洲共同体委员会 Commission of the European Communities
欧洲共同体委员会环境领域计划 Environmental Realm Plan of the European Communities
欧洲古陆 Europe paleocontinent
欧洲规范 Eurocode
欧洲规范第二部分<关于混凝土结构设计> Eurocode 2[EC2]
欧洲规范第八部分<关于结构抗震设计> Eurocode 8[EC8]
欧洲国际承包商协会 European International Contractors
欧洲国际银行 European Banks of International Company
欧洲国家通用插头 Europlug
欧洲合叶 meadow sweet
欧洲核桃木 European walnut
欧洲红端木 gaiter
欧洲红瑞木 red dogwood
欧洲红松 Scots pine
欧洲(互联)网【计】EUnet
欧洲花楸 European mountain ash; rowan;Rowan tree
欧洲花楸树 mountain ash
欧洲桦(木) European birch; Betula alba;Common birch
欧洲环境机构 European Environmental Agency
欧洲环境监测和资料网络 European Environmental Monitoring and Information Network
欧洲环境节 European Environmental Festival
欧洲混凝土委员会 European Committee for Concrete
欧洲火棘 evergreen thorn
欧洲货币 Eurocurrency;Euromoney
欧洲货币单位 Eurocurrency unit;European currency unit
欧洲货币联盟 European Monetary Union
欧洲货币市场 Eurocurrency market
欧洲货币体系 European Monetary System
欧洲计算机厂家协会 European Computer Manufacturer's Association
欧洲计算机网络 European Computer Network
欧洲计算机制造商学会 European Computer Manufacturer's Association
欧洲鲫鱼 crucian
欧洲建筑设备委员会 Committee of European Construction Equipment
欧洲交换网络 European switching network
欧洲交通部部长会议 European conference of Minister's of Transport
欧洲接骨木 European red elder
欧洲结构力学实验室协会 European Association of Structural Mechanics Laboratories
欧洲结构评定实验室 European Laboratory for Structural Assessment
欧洲经济共同体 European Economic(al) Community; European Economic Communities
欧洲经济共同体的官员 Eurocrat
欧洲经济合作组织 Organization for European Economic Cooperation
欧洲经济委员会<联合国> Economic Commission for Europe
欧洲经济学会 association Economic for Europe
欧洲卷材涂装工业协会 European Coil Coating Association
欧洲卡梣 common ash
欧洲科学基金会 European Science Fund
欧洲空间局 European Space Agency
欧洲空间数据中心 European Space Data Center
欧洲空间组织情报检索服务中心 European Space Agency Information Retrieval Service
欧洲空气压缩机真空泵和风动工具制造商委员会 the European Committee of Manufacturers of compressors, Vacuum Pumps and Pneumatic Tools
欧洲冷杉 European silver fir
欧洲梨 common pear
欧洲沥青固化设备 Eurobitum
欧洲沥青协会 European Bitumen Association[Eurobitume]
欧洲栎 common oak;English oak;European oak; Quercus robur; truffle oak
欧洲栗(木) Europe(an) chestnut; sweet chestnut; Italian chestnut; Spanish chestnut
欧洲联合经济委员会 United Economic Commission for Europe
欧洲联盟 European Federation
欧洲林业委员会 European Forestry Commission
欧洲落叶松 common larch;Corsican larch;European larch;Larix decidua;Scotch larch
欧洲煤炭钢铁工业 European coal and steel industry
欧洲煤炭和钢铁共同体 European Coal and Steel Community
欧洲美元 Eurodollar
欧洲牡丹 Paeonia peregina
欧洲牡蛎 European oyster
欧洲内部的 intra-European
欧洲七叶树 Aesculus hippocastanum; European horsechestnut
欧洲桤木 aar;common alder;European alder
欧洲汽车 European car
欧洲情报网 European Information Network
欧洲山毛榉 common beech;European beech;Fagus sylvatica
欧洲山杨 aps;European aspen;populus tremula;trembling poplar
欧洲商品编号 European Article Number
欧洲石油地质家协会 European Association of Petroleum Geologists
欧洲式导向架 European lead
欧洲式高效曝气(法) European highrate aeration
欧洲式型号 Europe type
欧洲式样 Europeanism
欧洲水泥协会 European Cement Association
欧洲水青冈 Fagus sylvatica
欧洲穗子榆 European hophorn beam
欧洲甜樱桃 mazzard cherry; sweet cherry
欧洲铁路研究所 European railway research institute
欧洲铁木 European hophorn beam; Ostrya carpinifolia
欧洲拖轮船东协会 European Tugowners Association
欧洲网络 European network
欧洲委员会 European Commission
欧洲卫矛 European spindle-tree
欧洲文献翻译中心 European Translation Center[Centre]
欧洲污染物长途运输方案 program(me) on long-range transport of pollutants over Europe
欧洲西伯利亚植物亚区 European-Siberian floristic subregion
欧洲系统 European system
欧洲先进集成公路系统 advanced integrated motorway system for Europe
欧洲信号网 Euro-Signal Network
欧洲信息网 European Information Network
欧洲型自航驳船 Europe self-propelled barge;Europe ship
欧洲型自航船 Europe self-propelled ship
欧洲银行 Eurobank
欧洲引水员协会 European Marine Pilots Association
欧洲英镑 Eurosterling
欧洲鹦鹉螺地理大区 European nautiloid region
欧洲有孔虫地理大区 European foraminifera region
欧洲月桂树 baytree
欧洲云杉 common spruce;European spruce
欧洲运输部长会议 European conference of Minister's of Transport
欧洲运输组织 European Transport Organization
欧洲载驳船 Europe barge carrier
欧洲造船商协会 Association of West European Shipbuilders
欧洲债券 Eurobond
欧洲榛 hazel tree;nut-tree
欧洲植被图 European vegetation map
欧洲质量管理机构 European Organization for Quality Control
欧洲中北部大陆冰盖 glacial sheet of north-mid-Europe
欧洲主要港口 European main ports
欧洲紫杉 European yew
欧洲自然保护资料中心 European Information Center for Nature Conservation
欧洲自由贸易联盟 European Free Trade Association

殴

殴打 buffet
殴斗 injury accident from hitting each other

瓯

瓯穴 evorsion hollow; giant's kettle;pot-hole;sline
瓯窑 Ou kiln;Ou ware

呕

呕吐 emesis;sickness;vomit

偶

偶 duplet;pair
偶 A 同位素 even-A isotope
偶 N 同位素 even-N isotope
偶倍数 even multiple
偶层 double layer
偶差 accident(al) error
偶成湖泊 accidental lake
偶成气泡 accidental air;entrapped air
偶次的 even-order
偶次谐波 even harmonic;even-order harmonic
偶次谐波工作 even harmonic operation
偶次谐波失真 even-order harmonic distortion
偶粗线 zygo-pachynema
偶代换 even substitution
偶单力组【物】wrench
偶氮 azo
偶氮苯 azobenzene;azobenzol;diphenyldiimide;phenylazobenzene
偶氮二异丁腈 azobisisobutyronitrile
偶氮芳基醚 azoaryl ether
偶氮腐殖酸 azo-humic acid
偶氮化合物 azocompound
偶氮黄 azo yellow
偶氮磺酰胺 neoprontosil
偶氮甲碱颜料 azomethine pigment
偶氮甲烷 azomethane
偶氮卡红 azocarmine
偶氮蓝 azo blue
偶氮品(红) azofuchsine
偶氮染料 azo dyestuff
偶氮染料废水 azo-dye wastewater
偶氮色素 azopigment
偶氮缩合颜料 disazo condensation pigment
偶氮涂料 azopigment
偶氮酰胺 azoamide

O

偶氮型染料 azoform colo(u)r
偶氮亚胺 azoimide
偶氮胭脂红 azocarmine
偶氮颜料 azopigment
偶氮衍生物 azo derivative
偶氮异丁基腈 azoisobutyl cyanide
偶氮玉红 Azorubin
偶氮紫 azo violet
偶得利润 casual profit
偶点 pair of points;pair-point
偶电层 electric(al) double layer
偶对 couple
偶对称 even symmetry
偶尔 off-and-on
偶发的 adventitious
偶发电流 accidental current
偶发故障 random failure
偶发旱情 episodic drought events
偶发畸变 fortuitous distortion;jitter
偶发记录器 accident data recorder
偶发脉冲 accidental preliminary impulse
偶发事故条款 contingency clause
偶发事件 a chance occurrence;incident
偶发事件自动侦测系统 automatic incident detecting system
偶发误差 accident error
偶发原因 chance cause
偶发作用 incidental effect
偶分拆 even partition
偶广义函数 even generalized function
偶函数 even function
偶函数与奇函数 even and odd functions
偶行扫描 even line interlacing
偶合 accouple
偶合的 linked
偶合断开 bond open
偶合红 para red
偶合品红 pair magenta;parafuchsin
偶合染料 para-dye
偶合事件 sporadic event
偶和 even summation
偶环流 bicirculation
偶环流向量 bicirculation vector
偶极部位 dipolar site
偶极测深 dipole-dipole sounding
偶极测深法 dipole-dipole sounding method
偶极层 dipole layer
偶极场中半无限板状体的综合参数 synthetic(al) parameter of half-infinite sheet in dipole field
偶极场中无限板状体的综合参数 synthetic(al) parameter of infinite sheet in dipole field
偶极磁场 dipole magnetic field
偶极的 dipolar
偶极电磁剖面法 profiling method with electric(al) dipole
偶极法 double distribution method
偶极分布 doublet distribution
偶极分子 dipole molecule
偶极加宽 dipole broadening
偶极键 dipolar bond
偶极晶格 dipole lattice
偶极矩 dipolar moment;dipole moment
偶极矩阵元 dipole matrix element
偶极馈源 dipole feed
偶极离子 dipolar ion;zwitterion
偶极力 dipole
偶极剖面法 dipole profiling method
偶极剖面曲线 dipole-dipole profiling curve
偶极剖面装置 array for dipole-dipole profiling
偶极损耗 dipolar loss
偶极弹性损耗 dipole-elastic loss
偶极天线 dipole antenna;double(t) antenna
偶极天线阵 dipole array
偶极调制 dipole modulation
偶极位错 dipole dislocation
偶极引力 dipole attraction
偶极与辐射器 dipole feed
偶极圆盘馈电 dipole disk feed
偶极跃迁 dipole transition
偶极子 doublet;electric(al) doublet;dipole
偶极子场 dipole field
偶极子弛豫 dipole relaxation
偶极子辐射 dipole radiation
偶极子辐射器 dipole radiator
偶极子辐射图样 dipole radiation pattern
偶极子极化 dipole polarization
偶极子极化率 dipolar polarizability
偶极子矩 moment of dipole
偶极子馈电 dipole feed
偶极子流动 doublet flow
偶极子耦合 dipole linkage
偶极子声场 dipole sound field
偶极子天线 <带 1/4 波长匹配线的> Q-aerial;Q-antenna;bipole antenna
偶极子天线的摆动 flutter
偶极子吸收 dipole absorption
偶极子源 dipole source
偶极作用场 dipole effect field
偶件 matching parts;mating part
偶键 duplet bond
偶角圆线条 cut-and-mitered[mitred] bead
偶校验 even-parity check
偶接 joint coupling
偶接器 coupling unit
偶矩 moment couple
偶聚焦 even focusing
偶力的分解 resolution of couple
偶力矩 couple(d) moment
偶联管 pipe coupling
偶联剂 resin acceptor
偶量子数 even quantum number
偶螺纹 even thread
偶脉冲 even pulse
偶脉冲响应值 even pulse response
偶偶核 even-even nucleus
偶偶元素 even-even element
偶耦合 even coupling
偶排列 even permutation
偶栖宿主 accidental host
偶奇定则 even-odd rule
偶奇函数 even and odd functions
偶奇核 even-odd nucleus[复 nucle]
偶奇元素 even-odd element
偶然暴雨 occasional storm
偶然变动 accidental fluctuation;random fluctuation
偶然变化 accidental change;accidental variation
偶然变量 chance variable;random variable
偶然变异 accidental variation
偶然波动 random fluctuation
偶然成本 accidental cost
偶然成功 chance success
偶然抽查 random check
偶然动作 random operation
偶然冻死 random water-killing
偶然发火 haphazard ignition
偶然发生 heterogenesis
偶然废料 non-recurrent waste
偶然费用 incidental expenses
偶然分布 casual distribution
偶然分布带 zone of accidental distribution
偶然分布区 zone of accidental distribution
偶然浮游生物 tychoplankton
偶然符合 accidental coincidence;random coincidence;stray coincidence
偶然概率 accident probability
偶然干扰 accidental jamming;random disturbance;random noise
偶然故障 accidental failure;chance failure;occasional fault;random failure;sadden failure
偶然故障期 random failure period
偶然故障周期 random failure period
偶然过程 contingency procedure
偶然荷载 accidental load(ing)
偶然基点 accidental base point
偶然畸变 fortuitous distortion
偶然极端值 occasional extreme value
偶然寄生物 incidental parasite
偶然交易 casual business
偶然接地 accidental ground
偶然进气 air accidental
偶然进入 drop in
偶然聚积 accidental aggregation
偶然聚集 accidental aggregation
偶然来访 drop in
偶然雷击火 lightning-caused fire
偶然利益 casual profit
偶然漏油 accidental spillage
偶然扭转 accidental torsion
偶然抛出物 accidental ejecta
偶然喷出物 accidental ejecta
偶然偏差 eventual deviation
偶然偏心 accidental eccentricity
偶然漂移 erratic drift
偶然破坏 accidental damage;accidental failure
偶然迁移 accidental dispersal
偶然缺点 accident fault;incidental defect
偶然缺陷 accident fault;incidental defect
偶然升降 random fluctuation
偶然失效 random failure
偶然失效率 constant failure rate
偶然失效期 random failure period
偶然事故 chance failure;contingence[contingency];contingent
偶然事故的损失 loss from accident
偶然事件 accident;chance event;contingence;fortuity;happenchance;happenstance;random event;random occurrence
偶然事件拥挤 non-reciprocal nonrecurrent congestion
偶然收敛 accidental convergence
偶然收入 occasional income
偶然收益 incidental revenue
偶然损耗 random depletion
偶然损坏 random failure
偶然损益 non-recurring profit and loss
偶然污染 accidental contamination;accidental pollution
偶然误差 accident(al) error;chance error;data noise;erratic error;irregular error;non-systematic(al) error;random error
偶然误差定律 law of accidental errors
偶然误差曲线 curve of random error
偶然吸合 random operation
偶然现象 accidental phenomenon;fortuitous phenomenon;random event;stochastic event
偶然相关 chance correlation
偶然信号 random signal
偶然信息 incidental information
偶然行动 chance-medley
偶然性 chance-medley;change;contingence[contingency];eventuality;fortuity;law of chance;randomness;haphazard
偶然性变化 random variation
偶然性故障 random failure
偶然性规划 contingency planning
偶然性中断 contingency interrupt
偶然野火 accident wildfire
偶然一致 chance coincidence
偶然因素 accidental factor
偶然原因 chance cause;occasional cause
偶然再生 occasional regeneration
偶然早期放牧 random early bite
偶然噪声脉冲 occasional noise pulse
偶然中断 involuntary interrupt;occasional interruption
偶然重合 chance coincidence
偶然状况 accidental situation
偶然阻力 incidental resistance
偶然作用 accidental action
偶生成本 non-recurring cost
偶生费用 non-recurrent cost;non-recurrent expenses;non-recurring cost
偶生利益 non-recurring gains
偶生收益 non-recurring income
偶生损益 non-recurring profit and loss
偶生盈利 non-recurring gains
偶适地性生物 geoxene
偶数比特 even bit
偶数步 even number step
偶数层 even level;even number of piles
偶数差错 even-numbered error
偶数场 even field
偶数齿数 even number of teeth
偶数存储单元 even location
偶数道次 even-numbered pass
偶数的 even number
偶数对 even-even
偶数级 even level
偶数继电器 even number relay
偶数桨叶螺旋桨 even bladed propeller
偶数校验 even check
偶数阶 even-order
偶数跨 even span
偶数列车 <上行列车> even number train
偶数螺距 even pitch
偶数螺纹 even number of threads
偶数脉冲 even number impulse;even number pulse
偶数幂 even power
偶数模 even mode
偶数奇偶校验 even parity;even-parity check
偶数奇偶性 even parity
偶数通路 even channel;even-numbered channels
偶数同位 even parity
偶数同位校验 even-parity check
偶数信道 even-numbered channels
偶数行 even-numbered line
偶数行穿孔 normal-stage punch(ing)
偶数行数 even number of lines
偶数元素 even elements
偶态 even state;gerade
偶碳数优势 even carbon number predominance
偶图 bigraph;bipartite graph
偶完全数 even perfect number
偶向河 insequent river
偶项 even term
偶像 graven image;idol;joss
偶因论 occasionalism
偶姻【化】acyloin
偶用灯 occasional light
偶有的 stray
偶宇称态 even-parity state
偶宇称振动 even-parity vibration
偶遇暴雨 occasional storm
偶遇假说 encounter hypothesis
偶支 even component

偶质量 even-mass
偶质量数同位素 even-mass isotope
偶置换 even permutation
偶子图 even subgraph

耦

耦对的 symplectic

耦合 catenation; couple; interdependence; linkage; linking
耦合百分率 coupling percentage
耦合摆 coupled pendulum
耦合变量 coupling variable
耦合变压器 coupling transformer
耦合波 coupled wave
耦合不足 undercoupling
耦合部分 coupling unit
耦合部件 coupling unit
耦合参数 coupling parameter
耦合层 coupling layer
耦合常数 coupled constant; coupling constant
耦合场矢量 coupled field vector
耦合场向量 coupled field vector
耦合存储器分配 linked memory allocation
耦合带 strap
耦合的 coupled; coupling
耦合的反射能力 coupled reflectance
耦合的反射系数 coupled reflectance
耦合电感 coupling inductance
耦合电感器 coupling inductor
耦合电抗 coupling reactance
耦合电抗器 mutual reactor
耦合电缆 coupling cable
耦合电路 coupled circuit; coupling circuit; interlinked circuit
耦合电容 coupling capacitance
耦合电容器 coupling capacitor; coupling condenser; duct capacitor; duct condenser
耦合电子 coupled electron
耦合电阻 coupling resistance
耦合度 degree of coupling
耦合度测试器 coupling meter
耦合短截线 coupling stub
耦合短线 coupling stub
耦合断开 bond open
耦合对 coupled pair
耦合反应 coupled[coupling] reaction
耦合方案 coupling scheme
耦合方程 coupled equation; coupled wave equation
耦合方式 coupled mode
耦合放大器 coupling amplifier
耦合分裂 coupling split
耦合缝 coupling slot
耦合负载 coupled load
耦合刚度 coupling rigidity
耦合膏 coupling paste
耦合工艺 coupled process
耦合功率 coupled power
耦合共振 coupling resonance
耦合管 coupling tube
耦合光栏 coupling iris
耦合光谱 coupling spectrum
耦合光纤 tail optic(al) fiber[fibre]
耦合光谐振腔 coupled optic(al) resonators
耦合荷载 coupled load
耦合环(线) coupling loop
耦合回路 coupling loop; coupling network
耦合机构 coupling mechanism
耦合剂 couplant; coupler; coupling agent; coupling medium; coupling paste
耦合结构 coupled structure
耦合介质 couplant; coupling medium
耦合镜 coupling mirror
耦合矩阵 coupled matrix
耦合开关 linked switch
耦合空穴 coupling hole
耦合孔 coupling aperture; coupling hole; coupling slot
耦合控制 coupling control
耦合理论 coupled wave theory
耦合滤波器 coupling filter
耦合面 coupling plane
耦合模型 coupled model
耦合膜片 coupling iris
耦合频率 coupling frequency
耦合器 bonder; coupled equation; coupler; coupling(mechanism); jigger
耦合器插座 coupler socket
耦合器环舌 coupling buckle
耦合器件 coupled apparatus
耦合器壳 coupling housing
耦合腔 couple resonator
耦合腔技术 coupled cavity technique
耦合强度 stiffness of coupling
耦合区 coupled zone
耦合曲线 connecting curve
耦合圈 pick-up loop
耦合溶剂 coupling solvent
耦合输出 coupling-out
耦合衰减 coupling attenuation
耦合损耗 coupler loss
耦合损失 coupling loss
耦合态 coupled state
耦合探针 coupling probe
耦合套筒 coupling sleeve
耦合天线 coupled antenna
耦合调整 coupling adjustment
耦合调制 coupling modulation
耦合透射比 coupled transmittance
耦合网络 coupling network
耦合位 coupling bit
耦合位置 coupling position
耦合系数 coefficient of coupling; coupled factor; coupling coefficient; coupling factor; index of coupling
耦合系统 coupled system
耦合隙缝 coupling gap
耦合线【电】parallel coupled lines
耦合线圈 coupling coil; coupling inductor; pick-up coil
耦合项 coupling term
耦合效率 coupling efficiency
耦合效应 coupling effect
耦合谐振腔 coupled resonator
耦合谐振子系 coupled harmonic oscillator
耦合因数 coupling factor
耦合因子 coupling factor
耦合应力 coupling stress
耦合用胺 coupling amine
耦合元件 coupler; coupling element; coupling unit; matching plug
耦合跃迁 coupled transition
耦合运动 coupled motion
耦合振荡模 coupled mode
耦合振荡器 coupled oscillator
耦合振动 coupled vibration
耦合振动器 coupled vibrator
耦合振动系统 coupled vibration system
耦合振幅 coupled amplitude
耦合振幅近似 coupled amplitude approximation
耦合振子 coupled oscillator
耦合指令 link order
耦合指数 index of coupling
耦合置换 coupled substitution
耦合装置 coupled system; coupling device
耦合自耦变压器 auto-leak transformer
耦合阻抗 coupled impedance; coupling impedance; mutual impedance
耦合作用 coupled action; coupled behavio(u)r; coupling function; linked reaction
耦接器 coupling unit
耦联 couple
耦联部位 coupling site
耦联反应 coupling reaction
耦联管 coupling
耦联机理 coupling mechanism
耦联剂 coupling agent
耦联间期 coupling interval
耦联晶片 bimorph crystal
耦联聚合 coupling polymerization
耦联抗剪墙 coupled shear wall
耦联器 coupler; coupling
耦联体 couplet
耦联体系 coupled system
耦联氧化絮凝剂 coupling oxidation flocculant
耦联因子 coupling factor
耦腔式磁控管 strapped magnetron; strip type magnetron

藕

藕节形石香肠 lotus-rootlike boudin

藕节状河道 lotus-root-shaped channel
藕色 pale pinkish grey
藕状的 lotus-root like; lotus-root-shaped

沤

沤肥 water-logged compost; wet compost

沤烂 retting
沤麻 flax retting
沤麻地 retting clam
沤麻期 season of flax retting
沤麻渍 water retting
沤软 ret

P

杷

杷椰动物地理区 Palang faunal province

杷荏油 beou oil

爬

爬车器 creeper;height compensator

爬虫 creeper

爬地野草 trailer

爬电电流 creepage current

爬电电阻 creepage resistance;resistance to tracking

爬电距离 creepage distance;creepage length

爬动 creeping motion

爬动式牵引车 creeper type tractor

爬陡坡 climb very steep slopes

爬伏椽 creeping rafter

爬杆 climbing pole

爬杆脚扣 climbing irons;grab;grappler;pole climbers

爬杆器 climbing equipment

爬杆式工作台架 mast climbing work platforms

爬杆效应 Weissenberg effect

爬高 climb(ing);run-up;altitude gain;ascent;swash height

爬高法 ascending method

爬高工具 climbing equipment

爬高角 ascending angle

爬高速度 <飞机> rate of climb;run-up speed

爬高用绳索 climbing rope

爬高用靴刺 climbing spurs

爬罐 raise climber

爬溅岸 <海岸的上部> swash bank

爬径 line of creep

爬径长度 <指渗流> creep path length

爬距 creep distance;creeping length

爬跨检查器 mount detector

爬犁 jumper;sleight

爬犁脚 runner

爬犁脚链 runner chain

爬犁起重架 jammer

爬路 lift of creep

爬模 climbing form

爬坡 climb(ing);grade climbing

爬坡车道 climbing lane;crawler lane;crawling lane

爬坡车道标志 climbing lane sign

爬坡道岔 climbing switch

爬坡法 hill-climbing(method)

爬坡角 ascending angle;ramp angle;sliding-up angle

爬坡界限 limit of uprush

爬坡轮牙 <车辆> cog

爬坡能力 ability to climb gradients;climbing ability;climbing capacity;climbing power;gradeability;grade climbing ability;grade climbing capacity;hill climbing ability

爬坡让车道 crawler lane

爬坡时车体荷重阻力点 weight-resistance point

爬坡时车体荷重作用点 weight-effective grade point

爬坡试验 climbing test

爬坡速率 crawl speed;creep rate;creep speed

爬坡性能 gradeability performance

爬坡性能试验 climbing ability test

爬坡自动制动器 automatic hill holder

爬坡总阻力 total climbing resistance;effective grade

爬墙虎【植】Boston ivy

爬墙梯 escalade

爬墙植物 walling plant

爬山比喻法 hill-climbing metaphor

爬山车 trail bike

爬山法 hill climbing method <控制论>;climbing method【数】

爬山过程 hill-climbing process

爬山虎【植】Boston ivy;Japanese ivy

爬山缆车 mountain rope hoist

爬山廊 climbing corridor

爬山者 climber

爬上轨面 climbing on rail

爬升 climb(ing)

爬升比 ratio of climbing

爬升波痕 climbing ripple

爬升波痕纹层 climbing ripple lamination

爬升高度 <异重流> aspiration height

爬升机构 climbing mechanism

爬升加料机 inclined skip hoist

爬升角 angle of ascent;angle of climb

爬升率指示器 rate-of-descent indicator;variometer;vertical speed indicator

爬升模板 climbing form(work);jump form <美>

爬升起重机 climber crane

爬升时间 time-to-climb

爬升式脚手架 climbing scaffold(ing)

爬升式搅拌机 climbing mixer

爬升式起重机 creeper travel(l)er;climbing crane

爬升式塔吊 climbing tower crane

爬升速度 climbing speed

爬升速度表 rate-of-climb meter

爬升速率指示器 rate-of-climb indicator

爬升塔式起重机 climbing tower crane

爬升套架 climbing frame

爬升限制重量 climb-limited weight

爬升装置 climbing device

爬生器 climber

爬式翻斗加料桶 side-dump skip bucket

爬式加料机 inclined skip charger;inclined skip hoist

爬梯 access ladder;accommodation ladder;cat ladder;ladder;lifting rack;safety ladder

爬梯安全装置 ladder guard

爬梯槽 ladder recess

爬梯的休息梯级 resting rung

爬梯式自动起落机构 rack lift

爬梯竖杆 ladder rails

爬梯梯级 ladder rung

爬卧 sprawl

爬芯 drop core;wing core

爬行 climbing;crabwise motion;crawl(ing);creep(ing);worm;creeping of track【铁】

爬行板 chicken ladder;creepies <铺屋面瓦片时用的>;crawling board <有刻痕的木板>

爬行变速齿轮 crawling gear

爬行长度 <渗流的> creep path length

爬行车 caterpillar;creeper travel(l)er

爬行打滑 creeping

爬行倒相器 line crawl inverter

爬行道 creeper land

爬行的 reptile;snailish;snail-paced;snail-show

爬行(低顶)通道 crawlerway

爬行吊车 crawler crane;creeper derrick;creep traveler

爬行吊机 crawler crane

爬行动物 reptile

爬行动物时期【地】age of reptiles

爬行固定汇率 crawling peg exchange rate;sliding peg exchange rate

爬行管道 crawlway;creeping trench <人可以进去里面进行检查>

爬行缓冲器 line crawl buffer

爬行迹 crawling trace

爬行甲虫 ground beetle

爬行距离 creepage distance

爬行空间 crawl space;basementless space <地板下铺设管道设备所留出供人爬行的空间>

爬行力 creeping force

爬行清除 line crawl cancel

爬行情况 creep behavio(u)r

爬行式的汇率调整 crawling peg

爬行式吊车 creeper crane

爬行式吊机 mule travel(l)er

爬行式起重机 <悬臂架桥的> creeper crane

爬行式升降台 raise climber

爬行式探测仪 creeping weasel

爬行式转向 four-wheel crab steering

爬行速度 crawl speed;creep speed;rate of creep;snail's pace;velocity of creep

爬行速率弯沉 creep-speed deflection

爬行特性 creep behavio(u)r;creep characteristic

爬行通道 crawlway

爬行物 crawler;creeper

爬行纤维 climbing fiber[fibre]

爬行曳引车 crawler

爬行者 crawler;creeper

爬移沼泽地 climbing bog

爬越道岔 climbing turnout;continuous rail points

爬越式道岔 run over type turnout

爬越式钢轨 elevated rail

爬云梯 escalade

爬子钩 extractor

耙

耙笔石属 <拉> Rastrites

耙臂 drag arm

耙臂耳轴 trunnion of drag arm

耙臂绞车自动控制器 suction tube automatic winch controller

耙臂位置监控器 suction tube position monitor

耙冰机 ice rake

耙柄 rabble arm

耙不松的 non-rippable

耙齿 cultivator tooth;rake tooth;rastellus;ripper tooth;spike

耙齿柄 scarifier shank

耙齿插座 tine adapter

耙齿工作宽度 shank working width

耙齿贯入深度 tooth penetration depth

耙齿尖 scarifier tip

耙齿伸出 pitch out

耙齿收拢 pitch in

耙齿头 ripper tip

耙齿座杆 <松土机的> tooth bar

耙出 rake-off

耙出物和筛余物 rakings and screenings

耙除 raking

耙除物 rakings

耙的刮刀 scraper

耙底层 harrow sole

耙地 harrowing

耙地松土压土器 cultimulcher

耙斗后壁 scraper tail gate

耙斗机 scraper unit

耙斗式装岩机 scraper type loader

耙斗提升机 scraper hoist

耙斗运载机 scraper loader

耙斗装料机 scraper loader;slide loader

耙粪器 dropping board scraper

耙痕 harrow mark

耙集式抓斗 trim type bucket

耙集式装料机 gathering loader

耙焦机 coke-drawing machine

耙角 drag angle;drag rake

耙掘 dragging;harrowing

耙菌属 <拉> irpex

耙开 prong

耙犁 plough

耙路板 pavement scarifier machine

耙路工(人) raker

耙路机 grade ripper;mechanical rake;pavement scarifier machine;raker;ripper;road harrow;road ripper;scarifier;harrow

耙泥船 scraper dredge(r)

耙平 drag;hack;rake

耙平损失 rack loss

耙平阻力 rack resistance

耙石机 rock rake

耙石爪 rock rake

耙式分级机 drag classifier;rake classifier

耙式分级器 rake classifier

耙式分料器 trough classifier

耙式分选机 rake classifier

耙式刮路机 rake-type drag

耙式混合器 rack mixer;rake mixer

耙式搅拌机 rake mixer

耙式路刮 rack-type-drag;rake-type drag

耙式筛 drag screen

耙式输送机 rake conveyer[conveyor];rake-type conveyer[conveyor]

耙式挖泥船 scraper dredger

耙式洗毛机 rake machine

耙式运输机 rake-type conveyer[conveyor]

耙式整坡器 currycomb

耙式装岩机 scraper type rock loader

耙式钻头 bit with teeth

耙松 spike up

耙松特性 rippability

耙松岩层 rock ripper

耙松岩石 rock loosening

耙田 harrowing

耙头 <耙吸式挖泥船的> drag head;trailing head

耙头吊架 draghead gantry

耙头架 draghead ladder

耙头架起落装置 hoisting gear for draghead ladder

耙头绞车自动控制器 automatic draghead winch controller

耙头接头管 draghead adapter

耙头磨损 abrasion of draghead;draghead wearing

耙头罩 draghead visor

耙头罩控制器 draghead visor controller

耙头遮板控制器 draghead visor controller

耙土机 harrower;towed scarifier

耙土器升降装置 scarifier lift device

耙土器液压缸 <平地机> scarifier cylinder

耙土装置 rake attachment(for graders);scarifier device

耙吸 trailer suction

耙吸式挖泥船 drag-suction dredge(r);trailing hopper suction dredge(r);trailing suction hopper dredge(r);scoop-type suction dredge(r)

耙吸式挖泥机 drag-suction dredge(r);scoop-type suction dredge(r)

耙吸式挖泥机掘头 drag head

耙吸式挖泥阻力 resistance of trailing dredging

耙吸(挖泥)法 <挖泥的> drag-suction method

耙吸装置 drag and suction device

耙形牙 rake teeth

耙形支撑 raking support
耙寻 comb
耙岩机 rock rake
耙中【疏】intermediate joint
耙爪式装载机 gathering arm loader
耙状礁 rake-form reef
耙子 raker

帕埃斯图姆的波赛冬神庙 Temple of Poseidonat Paestum

帕埃斯图姆的德墨忒耳神庙 Temple of Demeter at Paestum
帕-鲍氏计量槽 Polmer-Bowlus flume
帕德矛斯法 Padnos method
帕多瓦圆形大教堂 <意大利> Arena Chapel at Padua
帕尔巴蜡 palba wax
帕尔量热器 Parr calorimeter
帕尔默-巴伯公式 <设计柔性路面厚度的一种公式> Palmer and Barber formula
帕尔默-巴伯公式板 Palmer and Barber formula panel;panel board
帕尔默干旱严重(程度)指数 Palmer drought severity index
帕尔默桁架 Palmer truss
帕尔默石 <一种产于美国缅因州带浅红色的花岗岩> Palmer granite
帕尔纳特 <一种单线螺纹锁紧螺母> palnut
帕尔帖(反热电偶)效应 Peltier effect
帕尔希风 puelche
帕尔浊度计 Parr turbidimeter
帕耳 <固体振动的无量纲单位> pal
帕耳计 palmeter
帕柯勒斯 <一种专利防水剂> Percolex
帕克尔法 Parker's process
帕克尔桁架 <上弦呈多边形> Parker's truss
帕克尔镍铬黄铜 Parker's alloy
帕克尔水泥 Parker's cement
帕克式桁架 Parker's truss
帕克水泥 <一种天然水泥> Parker's cement
帕克斯隔热板 Poxboard
帕克斯隔热毡板 Pox felt
帕奎尔经验扩散公式 Pasquill's practical diffusion formula; Pasquill's empirical diffusion formula
帕奎尔实用扩散公式 Pasquill's practical diffusion formula; Pasquill's empirical diffusion formula
帕拉第奥古典式建筑风格 Palladian classicism
帕拉第奥建筑构图特色 Motif Palladio
帕拉第奥建筑形式 Palladianism
帕拉第奥建筑主义 Palladianism
帕拉第奥式窗 Palladian window
帕拉第奥式风格 Palladian motive
帕拉第奥式 Palladian
帕拉第奥式建筑 Palladian architecture;Palladian motive
帕拉第奥式建筑处理手法 Palladian motif
帕拉第奥式建筑的复兴 Palladian revival
帕拉第奥式建筑特色 Palladian motif
帕拉第奥式建筑特色窗 Palladian motif window
帕拉第奥式水磨石 Palladian terrazzo
帕拉第奥学派 Palladianism
帕拉尔黄褐色硬木 <产于缅甸、印度> Paral
帕拉弗雷克斯钢丝绳 Paraflex wire rope
帕拉格阶 Pragian
帕拉格模型 Prager model
帕拉拉号【船】Paralla
帕拉马里博 <苏里南首府> Paramaribo
帕拉(橡)胶 Para rubber
帕兰蒂铝模铸造法 Parlanti casting process
帕雷兹液压锥尖及侧阻触探仪 Parez hydraulic cone/friction sleeve
帕累托标准 Pareto Criterion
帕累托乘子 Pareto multiplier
帕累托非最优状态 Pareto nonoptimality
帕累托改进 Pareto improvement
帕累托合理性 Pareto rationality
帕累托极小 Pareto minimum
帕累托排列图 Pareto diagram
帕累托优边界 Pareto optimal boundary
帕累托指数 Pareto index
帕里安奈沥青 parianite
帕里西维拉海盆 Parece Vela basin
帕利拉底壳(基础托换用小直径钻孔)桩 Pali Radice Pile
帕利塞德变动 Palisade's disturbance
帕利塞德岩床 Palisade's sill
帕卢紫褐色硬木 <印度产> Palu
帕罗塞尘暴 palouser
帕洛玛天图 Palomar Sky Survey
帕秒 <动力黏[粘]度单位> Pascal second
帕末朗丘克冷却 Pomerronchuk cooling
帕姆函数 Palm function
帕姆佩罗风 pampero
帕帕加约风 papagayo
帕秋卡槽 Pachuca tank
帕秋卡搅拌槽 Pachuca agitator
帕萨迪运动 Pasadenan orogeny
帕萨尔加德的赛勒斯大帝宫 <伊朗> Palace of Cyrus the Great at Pasargadae
帕森汽轮机级 Parson-stage steam turbine
帕森数 Parson's number
帕森斯-邓肯铸锭法 Parsons Duncan process
帕水硅铝钙石 partheite
帕斯板 <一种绝缘板> Paxboard
帕斯顿沟槽 <用来收集暖房或温室内外潮气的> Paxton gutter
帕斯卡 <压力单位> Pascal
帕斯卡秒 <动态黏[粘]滞度单位> Pascal second
帕斯卡定理 Pascal's theorem
帕斯卡定律 Pascal's law
帕斯卡分布【数】Pascal's distribution
帕斯卡/开 Pascal per Kelvin
帕斯卡流体 Pascal's liquid; Pascalian fluid
帕斯卡/秒 Pascal per second
帕斯卡三角形 Pascal's triangle
帕斯卡液体 <理想的稠度为零的非黏[粘]性液体> Pascalian liquid
帕斯卡液压定律 Pascal's law of fluid pressure;Pascal's law of pressure
帕斯卡原理 Pascal's principle;Pascal's theory
帕斯卡滞后分布 Pascal's lag distribution
帕斯卡滞后模型 Pascal's lag model
帕斯柯岸壁集装箱装卸桥 Paceco portainer
帕斯科-肯尼威克桥 <密索混凝土斜拉桥,主跨 300 米,1978 年建于美国哥伦比亚河上> Pasco-Kennewick bridge
帕斯奎尔-吉福德扩散参数 Pasquill-Gifford diffusion parameter
帕斯奎尔-特纳尔稳定度分类法 Pasquill-Turner stability category
帕塔普斯岩层 Patapsco formation
帕太克斯浸油探伤法 Partex penetrant process
帕特拉法 Patera process
帕特孙投影(法)Patterson projection
帕特孙综合法 Patterson synthesis
帕提农神庙 Great Temple of Parthenon
帕提亚建筑 Parthian architecture
帕廷森铅白 Pattinson's white lead
帕邢定律 Paschen's law
帕歇尔量水槽 Parshan measuring flume
帕亚里钻孔测斜仪 Pajari apparatus

怕地段衰落而引起的竞卖 panic selling

怕光 keep in dark place
怕冷 protect against cold
怕热 protect against heat
怕压 <不可装在重货之下> not to be stowed under heavy cargo; not to be stowed below other cargo

拍岸大浪 heavy surf

拍岸巨浪 land swell
拍岸浪 beach comber; breaker; climb;land swell;surf(beat)
拍岸浪带 breaker zone;surf zone
拍岸浪的折射 refraction of swash
拍岸浪花 surf breaker
拍岸浪区 surf zone
拍岸碎浪 surf
拍岸涌 land swell
拍板 clapper
拍板成交 knock down; strike a bargain
拍板式衔铁 clapper-type armature
拍板座 clapper box
拍打 beat;clap;flap;spank;whip
拍打成型 paddle and anvil;thwacking
拍打机构 beater mechanism
拍打器 beater
拍打饰面 sparrow peck
拍捣 beating
拍动 flap
拍幅 amplitude of beat
拍杆 rapping bar;rapping iron
拍合式继电器 clapper relay
拍火器 fire beater
拍击 clap;rattle;rattling
拍击试验 panting test
拍击音 flapping sound
拍击应力 panting stress
拍击作用 slap
拍节电路 cycling circuit
拍叩 slapping percussion
拍快照 snapshot
拍快照者 snapper
拍隆 <太阳辐射强度单位> pyron
拍卖 auction sale; open sale; outcry; public auction; public sale; sale by auction; selling off goods at reduced prices
拍卖场 auction market;bidding block; salesroom
拍卖成交人 successful bidder
拍卖(大)厅 auction hall
拍卖底价 upset price
拍卖地价 auction land price
拍卖掉 auction off
拍卖费用 auction charges;auctioneer's fee
拍卖黄金 gold auction
拍卖价(格)auction price
拍卖买主 auctioneer vendee
拍卖目录 auction catalogue
拍卖前交易 preauction trading
拍卖钱 lot money
拍卖清单 statement of auction
拍卖人 auctioneer;public saler
拍卖人的小槌 auctioneer's hammer
拍卖人佣金 auctioneer's commission
拍卖商 auctioneer;public saler
拍卖商的小槌 auctioneer's gavel
拍卖市场 auction market
拍卖所 auction hall
拍卖条件 conditions of auction sale
拍卖通告 auction sale notice
拍卖行 auction house; auction room; hastarium
拍卖佣金 auctioneer's commission;lot commission;lot money
拍卖中出价 bid at auction
拍门 beat gate;pin gate
拍泥 thwacking mud
拍频 beat frequency
拍频波 beating wave
拍频波图形 beat pattern
拍频波形图 beat pattern
拍频测试器 beat frequency meter
拍频差率 beat rate
拍频倒相器 clocked inverter
拍频放大器 beat frequency amplifier
拍频辐射声 beat radiation sound
拍频干扰 beat interference
拍频干扰现象 double super phenomenon
拍频干扰效应 double super effect
拍频检波器 beat frequency detector
拍频接收机 beat frequency receiver; beat receiver
拍频率 beat frequency
拍频起放大器 beat frequency amplifier
拍频式接收机 beat frequency receiver
拍频效应 beat effect
拍频信号输出 beat output
拍频音调 pitch of beat(note)
拍频原理 beat principle
拍频振荡器 beat frequency oscillator;beating oscillator
拍频振幅 amplitude of beat
拍频指示器 beat frequency indicator
拍频质量 beat mass
拍频周期 beat period
拍入 beating-in
拍摄 mutograph;shoot
拍摄电视记录片 kinescope
拍摄范围 coverage;covering power
拍摄镜头 taking lens
拍摄距离 focusing distance
拍摄头部 head shot
拍摄位置 taking position
拍摄物反差 subject contrast
拍实 compaction
拍实土 compacted earth
拍手 clap
拍水声 splashing sound
拍同 beating-in
拍现象 beat phenomenon
拍效应 beat effect
拍摇筛 impact screen
拍音 beat(tone)
拍音干涉滤波器 beat interference filter
拍照 photography
拍振筛 impact type screen
拍纸簿 jotter; pad; scribbing-block; tablet
拍周期 beat period
拍子 tempo
拍子浸入深度 dip of the bait

徘徊 cruise;wander

排班表 duty roster

排板式空气过滤器 panel type air cleaner

排板式空气冷却器 panel type air cooler
排版 composition; imposition; make-up; typesetting
排版尺 composing rule
排版隔条 gutter
排版花饰 flower
排版架 chase
排版者 composer
排笔 combined pen brush
排边沟 gutter
排便管 hopper tube
排便器 <客车> hopper
排冰道 deicing sluice; floating ice sluice
排冰路 ice sluice
排冰门 ice gate
排冰闸(门) deicing sluice; ice gate
排不容量 displaced volume
排尘 dust discharge; dust exhausting
排尘风扇 dust exhausting fan
排尘浮动门 exhaust damper
排尘管 dust chimney; dust-extractor duct
排尘机 exhauster
排尘器 dust ejector; dust evacuator; dust exhauster[exhaustor]; dust exhausting device; dust exhausting hood
排尘设备 dust exhausting equipment
排尘雾设备 fume extractor
排尘系统 dust exhaust system
排尘装置 dust ejector; dust exhaust apparatus
排成梯队 echelon; in echelon
排成梯形 echelon; in echelon
排成行 alignment; lineage
排成行的 lined
排成形的 in alignment
排成一串 stringing
排成一行 line up
排成一直线 lineage
排成直线的 in-line
排程序合理化 optimum coding
排斥 bar off; drive out; exclude; lock-out
排斥的 rejective; repellant
排斥工人 discard
排斥极 repeller
排斥力 repellant; repelling force; repulsion; repulsive force
排斥能力 repellent capacity
排斥能量 repulsive energy
排斥任选 exclusive option
排斥势 repulsive potential
排斥水平 rejection level
排斥问题 exclusive problem
排斥效应 exclusion effect; repulsive effect
排斥性能 repellent property
排斥性条款 exclusion clause
排斥压力 disjoining pressure
排斥属性 exclusive attribute
排斥转移 exclusive branch
排斥作用 repulsive interaction
排冲 gang punch
排冲压机 gang punch
排臭 odo(u)r release
排臭气烟囱 odo(u)r discharging chimney
排臭器 hopper deflector; hopper ventilator
排出 bleed-off; blow-off; discard; discharging; drain(age) off; draw-off; eject; exhaust(ing); expel; expulsion; extraction; flushoff; let down; off-take; outgush; perspiration; perspire; pumping out; snap-down; snap-out; transpiration; venting; withdrawal
排出泵 discharge pump; excavating pump
排出侧阱 exhaust side trap
排出侧筒形接头 discharge side cartridge
排出冲程 discharge stroke
排出氮氧化物少的技术 low-NO technology
排出道 drip pipe; outlet duct
排出的废气 exhaust fume; exhaust gas
排出的空气 outgoing air
排出的气体 burned gas; discharge gas; effluent air; exhaust air; exhaust gas; expellant gas
排出的燃烧产物 exhaust product
排出的热 reject(ed) heat
排出的试样 drain sample
排出的水 drain water
排出的污水 drainage
排出的烟 exhaust smoke
排出的烟量 exhaust smoke level
排出动能 kinetic energy rejection
排出端 outlet side
排出端泵盖 discharge cover
排出阀 bleed valve; blowout valve; delivery clack; discharge service valve; discharge valve; exhaust valve; release valve
排出废气 combustion gas
排出废气系统 exhaust system
排出废水 drain water
排出废物量 quantity of refuse
排出废液 effluent discharge
排出粉尘 exit dust
排出风量 draft capacity
排出管 bleeder pipe; blow-off pipe; building drain; discharge conduit; discharge leader; discharge pipe; discharge tube; mouth; off-take; outlet pipe; scavenger pipe; scavenger tube
排出管道的压力 manifold pressure
排出管汇 exhaust manifold
排出管路 bleed-off line; escape route; vent(pipe) line; discharge line
排出管线 bleed-off line; discharge line; vent(pipe) line
排出滚轮 trip roller
排出黄油 tap grease
排出或截留洪水 draining or intercepting flood water
排出节流 exhaust choke
排出静压头 static discharge head
排出开关 waste cock
排出空气 air exhaust; disinflate; exhaust air; exit air; leaving air; outgoing air
排出空气通路 passage of outgoing air
排出孔 bleed hole; discharge hole; discharge orifice; ejection opening; exhaust port; outage; outlet opening; outlet point; spur; tap(ping arrangement)
排出控制 blow-off control
排出口 delivery outlet; discharge orifice; draining point; escape hole; escapement; exhaust port; exhaust slot; mouth; outfall works; outlet; outlet port; spur; tap hole
排出口节流 throttling discharge
排出口小室 outlet compartment
排出口直径 outlet diameter
排出冷凝器 eductor condenser
排出力 expulsion force
排出量 discharge capacity; discharge rate; discharging rate; displacement discharge rate; outage; output; quantity discharged; rate of discharge; withdrawal; displacement
排出流 discharge current; discharging current; wash
排出流量 discharge
排出流量数 pumping capacity number; throughput number
排出流束 outflow jet
排出流速度 discharge velocity; outflow velocity
排出率 discharge rate; ejection rate; excretion rate
排出凝结水 drainage
排出喷嘴 discharge cone; discharge nozzle
排出期 ejection period; expulsive stage
排出气 exit gas; relief gas
排出气流 exhaust stream
排出气体 disinflate; effluent gas; fume off; vent gas
排出器 ejector; eliminator; evacuator
排出器板 ejector plate
排出曲线 exhaust line
排出塞 draw-off plug
排出剩余灌溉水的水沟 a ditch to carry extra irrigation water
排出时间 draining time; efflux time
排出室 exhaust chamber
排出水 discharge(d) water; displaced water; drainage(water); flow water; outlet water
排出水处置 effluent disposal
排出水回收 drainage recovery
排出水数据 effluent data
排出速度 discharge velocity; efflux velocity; outlet velocity; velocity of discharge; velocity of exhaust
排出速率 rate of discharge
排出损失 exhaust loss
排出套管 drain sleeve
排出通道 escape route
排出危害 elimination of damage
排出温度 delivery temperature; discharge temperature
排出污染的空气 remove foul air
排出污染物 emission; release of pollutant
排出污水 effluent sewerage; outfall sewage
排出污浊空气管道 foul air duct
排出物 discharge; displacer; drained product; effluvium[复 effluvia]; ejection
排出物沉淀池 effluent settling chamber
排出物沉降池 effluent settling chamber
排出物处理 effluent treatment
排出物分离系统 effluent segregation system
排出物分流系数 effluent segregation coefficient
排出物收集器 drained product collector
排出物数据 effluent data
排出物税收 emission tax
排出系数 efflux coefficient
排出下水管 outfall sewage
排出箱 outlet case
排出压力 bleed-off pressure; delivery pressure; discharge pressure; exhaust pressure; head pressure
排出压力表 delivery ga(u)ge; discharge ga(u)ge; high-pressure ga(u)ge
排出烟雾 exhaust smoke
排出岩屑 <炮眼中> chip removal
排出扬程 discharge head
排出叶片 discharge blade
排出液 discharge liquid; discharge liquor; effluent; exhausted liquid; released liquor; waste liquid
排出液体的总体积 total volume of expulsive liquid
排出油气中的有机质 organic matter of expulsive oil and gas
排出油气中碳 carbon of expulsive oil and gas
排出油脂 tap grease
排出淤泥量 silt discharge
排出闸门 discharge gate
排出蒸汽 discharge steam; exhaust steam; extraction of steam
排出支管 exhaust outlet
排出装置 discharger; distributor; eduction gear
排出总管 discharge header
排除 clear away; clear off; clear out; disposal; drain(age) off; draw-off; eliminate; eliminating; elimination; exclusion; foreclose; freeing; lock-out; preclusion; purge; reject(ion); removal; remove; banishment; breed out; cast aside
排除部分 exclusive segment
排除程序中的错误【计】 debug
排除穿堂风 draft exclusion; draught exclusion
排除错误 debug(ging)
排除的 eliminant; evacuant; expellent; preclusive
排除地表水 drainage surface water
排除地下水 drainage ground water
排除电路 lockout circuit
排除段 exclusive segment
排除发火装置 disarm
排除阀 bleed valve; blowdown valve; drain; release valve
排除法 elimination process; exclusive method
排除法判读标志 elimination key
排除法判读样片 elimination key
排除废气 scavenger
排除废气的噪音 exhaust noise
排除干扰 suppress interference
排除故障 clearing of a fault; debugging; fault clearing; obstacle avoidance; remedy of a trouble; removal of faults; trouble clearing; trouble-shoot(ing)
排除故障程序 debugging routine
排除故障工人 trouble-shooter
排除故障阶段 debugging phase
排除故障装置 clearing device
排除管 drainage pipe; purger
排除和根除对比 exclusion and eradication versus
排除积涝 disposal of excess water
排除积水泵 sump pump
排除计算机故障 debug(ging)
排除技术 elimination technique
排除剂 eliminant; excluder
排除继电器 lockout relay
排除镜面反射的反射度测定 specular reflectance excluded
排除可燃废气的燃烧室 after burner
排除落崩用设备 avalanche brake
排除气体系统 degassing system
排除器 clearer; drain; ejector; eliminator; excluder; expeller; releaser
排除容量 removal capacity
排除色谱法 exclusion chromatography
排除速率常数 elimination rate constant
排除速率系数 depuration rate coefficient
排除通风 exclusion of draught
排除挖泥气体系统 degassing system
排除挖泥中气体 degasification
排除危险 obviate
排除污染 decontamination
排除物 displacer
排除细泥的大型圆锥分级机 sloughing-off cone
排除性采购 preclusive buying
排除岩粉 <从孔内> removal of cuttings

排除因素 rejection factor
排除余气 outgas
排除雨水的檐槽 gutter
排除杂质 despumation
排除在外 exclude
排除障碍 abatement of nuisance; clearance; smooth the way
排除装置 remover
排错 misarrangement
排代泵 displacement pump
排代次序 series of potentials
排代滴定法 displacement titration
排代电镀 displacement plating
排代反应 displacement reaction; substitution reaction
排代剂 displacer
排代流量计 positive-displacement meter
排代容积 displacement volume
排代色谱法 displacement chromatography
排代时间 displacement time
排代式水表 displacement(water)meter
排代速度 displacement velocity
排代投配机 displacement feeder
排代周期 displacement period
排代作用 displacement
排挡杆 joy stick
排挡间距 meros
排挡间饰 metope
排挡数 gear
排刀程序 gang tool operation
排刀切削 rack cutting
排灯 bank light
排叠 piling
排钉 chain-riveting
排定 schedule
排定的工程时间 schedule engineering time
排定时序的 time-sequenced
排锭器 ingot ejector
排毒力与管理对策 toxicity elimination and management strategy
排毒药 expellent
排队 queue(-up); queuing; waiting line
排队长度 queue length
排队长度加权数 queue weight
排队长度限制 queue constraint
排队超长损失(费用) penalty for excess queue
排队程序 queuing routine
排队存取 queued access
排队存取法 queued access method
排队动作 queuing behavio(u)r
排队法 waiting line approach
排队方式 queuing system
排队分析 queuing analysis
排队管理 queue management control
排队管理(控制)技术 queue management technique
排队规则 queue discipline; queue rule; queuing discipline; queuing rule
排队过程 queuing process
排队后背 back of queue
排队候车棚 queue shelter
排队缓冲器 queuing buffer
排队检测器 queue detector
排队距离 queue distance; queue space
排队空间 queue space
排队控制 queue(management)control
排队控制部件 queue control block
排队控制块 queue control block
排队控制模式 queue control mode
排队控制组件 queue control block
排队理论 queen theory; waiting line theory; queuing theory; theory of queues; theory of queuing
排队量 queue size
排队论 waiting line theory; queue theory; queuing theory and waiting time problems <一种随机服务系统理论>
排队论和等待时间课题 queuing theory and waiting time problems
排队描述 queuing description
排队名单 waiting list
排队模型 queuing model
排队期望人数 expected number in the queue
排队器 queue equipment
排队请求 queue(d)request
排队区 queuing area
排队人群(等候公共汽车) bus-queue
排队时间【数】 queuing time
排队式自动机 push-down automaton
排队顺序存取 queued sequential access
排队顺序存取法 queued sequential access method
排队通信[讯]量 queue traffic
排队网络 queuing network
排队文件 queue file
排队问题 queuing problem; wait line problem
排队系统 queuing system
排队系统中顾客期望滞留时间 expected customer time in the queue system
排队系统中期望顾客人数 expected number of customers in the queue system; queue system
排队线长度受限制 limited queue length
排队延时 queuing delay
排队溢出 queue overflow
排队用栅栏 queue barrier
排队原则 queuing discipline; queuing principle
排方程式组 equation set-up
排房 row house; terrace; town house
排放 blow-off; discharge; discharging; release
排放泵 discharge pump; emptying pump
排放比 discharge ratio; emission rate
排放标准 emission criterion; emission level; emission standard; discharge standard; effluent standard <污水>
排放标准极限值 limited value of discharge standard
排放表 emission inventory
排放不足 discharge deficiency
排放材料速度 mass rate of emission
排放槽 let-down tank
排放测量 discharge measurement
排放场 drain field
排放出喷嘴接管 discharge nozzle
排放刺激物 releaser stimulus
排放大户 major discharge
排放带 discharging zone
排放点 discharge point; emission point
排放阀 bleeding valve; blowdown valve; draining valve; escape cock; exhaust valve
排放阀滤网 drainage valve screen
排放方法 drainage(method)
排放废气风道 vitiated air floor flue
排放废气格栅 vitiated air grid
排放废气孔 foul air hole; foul air opening
排放废气炉箅 vitiated air grate
排放废气竖井 foul air shaft
排放废物 effluent discharge
排放费用 discharge fee; effluent charges
排放分离器 blowdown separator
排放负荷 emission load
排放改变 flow variation
排放干线 main drain
排放高度 discharge height; height of release
排放管 bleeder; bleeding pipe; discharge pipe; discharger; discharging tube; eductor; relief tube
排放管出口 pipe-away exhaust
排放管道 discharge pipeline; drainage pipeline
排放管系 discharge pipe system; discharge pipework
排放管线 discharge pipe line
排放规定 effluent control
排放规范 effluent specification
排放基准 discharging criterion
排放极限 limit of release
排放计量 emission measurement
排放记录 discharge record; emission inventory; record of discharge
排放监测 discharge monitoring; emission monitoring
排放交易 emission trading
排放胶管 emptying hose
排放经过处理污水的流水渠 effluent trough
排放警报 discharge alarm
排放孔 discharge orifice; draining aperture; outlet hole
排放控制 blow-off control; effluent control; emission control
排放控制标准 emission control standard
排放控制极限 emission control deadline
排放控制设备 emission control equipment
排放口 discharge outlet; discharge port; drain; relief outlet; tap hole
排放口扩散管 outfall diffuser pipe
排放量 discharge amount; discharge capacity; outflow volume
排放流量 discharge flow rate; effluent flow
排放流速 <水工建筑物下游的> retreat velocity; velocity of retreat
排放漏气 blow-by gas
排放明沟 open drain
排放能力 delivery capacity; discharge capacity; drainability
排放泥浆 mud removal
排放浓度 emission concentration; emission strength
排放浓度限制 effluent concentration limit
排放喷嘴 discharge nozzle
排放期 discharge period
排放气(体) exhaust gas; effluent gas
排放器 discharge device; escaper
排放情况 discharge condition
排放情形 emission behavio(u)r
排放区边界 boundaries of discharge area
排放区中心 center[centre] of discharge area
排放权 emission rights
排放入海 discharge into the sea
排放软管 drainage hose
排放软管组 discharge hose assembly
排放塞 drainage plug
排放塞门 draw-off cock
排放设备 bleeding device; draw-off
排放设施 outlet structure; outlet works
排放时间 flow time
排放时间特性 emission-time pattern
排放时期 discharge period; displacement period
排放试验 drain testing; emission testing
排放数据库 emission inventory; emission database
排放栓 blow-off cock
排放水 discharge water
排放水管 tapping conduit
排放水平 emission level
排放水质 drainage water quality
排放水质标准 drainage water quality standard
排放税 discharge fee
排放速度 discharge velocity; velocity of discharge
排放速率 discharge rate; rate of emission; emission rate
排放特性 emission characteristic; emission performance
排放条件 discharge condition
排放通道 discharge carrier; discharge channel
排放瓦斯 fire(damp) drainage; methane drainage
排放瓦斯(用的重型气动)钻机 firedamp drainage drill
排放污染物 discharging pollutant
排放污水管 discharge sewer
排放物 discharge matter; effluent; emission
排放物测量 emission measurement
排放物取样 emission sampling
排放物收集器 drainage collector
排放物温度 temperature of effluent
排放物质 emission substance
排放系数 discharge coefficient; discharge factor; drainage coefficient; emission factor
排放系统 drainage
排放限度 emission limit
排放限制 effluent limitation
排放箱 <筛余渣> runoff box
排放消减单位 emission reduction unit
排放消减权 emission reduction right
排放性能 emission behavio(u)r
排放性质 discharge property
排放许可 discharge permit
排放许可证 emission certification; licence to discharge; permit to discharge
排放旋塞 bleeding cock; discharge cock
排放旋栓 blow cock
排放压力 blowdown pressure
排放压载 deballast
排放烟囱 ventilating stack
排放烟道 discharge tunnel
排放液 effluent
排放因数 emission factor
排放因素 emission factor
排放因子 emission factor
排放预防 discharge prevention
排放源 emission source
排放源采样 emission source sampling
排放噪声 displacement noise
排放闸门 waste gate; waste sluice
排放者 discharger
排放质量标准 effluent-quality standard
排放中间产品提升机 secondary reject elevator
排放装置 blowdown apparatus; eductor
排放装置末端 discharge end
排放状况 emission status
排放总量 total discharge; total release
排肥机构 fertilizer mechanism
排肥量调节杆 fertilizer quantity lever
排肥量调节器 fertilizer quantity regulator
排肥盘 fertilizer apparatus; fertilizer disk; fertilizer feed
排肥器 fertilizer apparatus
排肥器传动装置 fertilizer feed drive; fertilizer gearing
排肥器离合器 fertilizer clutch
排肥箱 fertilizer can
排肥闸门 fertilizer gate
排肥装置 fertilizer distributor; fertil-

izer unit
排废 waste discharge
排废气格栅 foul air grate
排废气管 foul air pipe
排废气烟囱 foul air chimney
排废气噪声 exhaust noise
排废水定额 wastewater norm
排废线 exhaust line
排粉 cutting removal
排粉槽 chip way
排粉机 mill exhauster; mill fan; pulverizer exhauster
排粉间隙 clearance of clearing dust; chipway space < 钻头 >
排粉能力 < 冲洗液的 > cutting-carrying capacity
排粉盘 dust discharging plate
排粪管 soil branch
排风 air exhaust; blowdown; discharge air; exhaust; ventilation
排风阀 relief damper
排风干燥 fan-drying
排风格栅 extraction grille
排风管 air stack; exhaust duct
排风管道 exhaust air duct; exhaust airway
排风柜 flue hood; fume hood
排风过滤器 exhaust air filter
排风机 air exhaust ventilator; discharge fan; exhaust blower; exhaust fan; fan; vent burner; vent(ilating) fan
排风机平台栏杆 handrail for fan deck
排风机室 exhaust fan room
排风机组 extract ventilation unit
排风井 removing shaft
排风静压室 exhaust air plenum
排风孔 air exit hole
排风口 air exhausting vent; air exit; exhaust outlet; exhaust port; exit; extraction grille
排风口堵 exhaust plug
排风量 draft capacity; exhaust air rate; exhaust air volume
排风气窗 air exhaust ventilator
排风器 draught excluder
排风前置过滤器 exhaust air pre-filter
排风切换杆 shifting lever for exhaust air
排风入口 exhaust inlet
排风扇 fan(blower); air-ejecting fan; blower fan; exhaust fan; flow fan
排风扇鼓风机 ventilating fan blower
排风扇气流折射板 fan blast deflector
排风设备 exhaust equipment
排风式冷却塔 induced-draft water-cooling tower
排风式凉水器 induced draught water cooler
排风室 air discharge compartment; exhaust air box
排风竖井 blowing-out shaft; exhaust shaft
排风速度 outlet velocity
排风速度压力 fan velocity pressure
排风塔 exhaust chimney; exhaust stack
排风筒 discharge stack
排风温度 draft temperature; temperature of outgoing air
排风系统 extract system; ventilation exhaust system
排风罩 exhaust hood; hood
排浮冰闸 floating ice sluice
排干 drain(ing); unwater(ing)
排干的沼泽地 inning
排干田 drained field
排杆 tier pole
排矸场 mine dump
排矸提升机 primary reject elevator
排沟深度 depth of draining

排故障程序 debugging-aid routine
排管 run of pipe; tube bank
排管费 cost of laying pipe
排管式 calabash
排管式加热件 calandria
排管式加热器 calandria
排管体 calandria
排灌 drainage and irrigation
排灌动力机械 power-driven irrigation and drainage equipment
排灌机械 drainage and irrigation machinery
排灌犁沟 water furrow
排灌设备 drainage and irrigation equipment
排灌水沟 deep-furrow
排灌网 irrigation and drainage network; irrigation net(work)
排灌站 drainage and irrigation station; irrigation and drainage pumping station
排海口污水扩散 ocean outfall dispersion
排汗 perspiration
排汗冷却效率 cooling efficiency of sweating
排夯 ramming row
排洪道 flood drainage way
排洪沟 tidal channel
排洪能力 flood carrying capacity
排洪区工程地质勘察 engineering geologic(al) investigation of draining flooded fields
排洪渠(道) flood discharging channel; flood relief channel
排洪闸 lock-gate hatch
排灰 blowout
排灰仓 ash discharge hopper
排灰斗 ash discharge hopper
排灰阀 ash valve; dust remove valve
排灰口 dust extraction port
排灰器 ash ejector; ash exhauster; ash gun
排灰锁斗 ash lock
排灰装置 ash exhauster
排机锯 gang-sawing machine
排挤 supplant
排架 bank frame; bent; bent frame; frame(d) bent; bay < 集装箱船的 >
排架墩座 pile pier
排架盖梁 bent cap
排架工程 trestle works
排架横撑 sash brace
排架结构 bent structure; card-house structure; trestle structure
排架结构鹰架 bent centering
排架桥 trestle bridge; trestle stand
排架式脚手架 trestle scaffold(ing)
排架式满布木拱架 full-span wooden bent centering[centring]
排架引桥 trestle approach
排架支柱钢连接件 clip angle
排架轴线 axis of bent
排架柱木 trestle tree
排架桩墩 bent pile pier; pile bent pier
排架桩基础 pile bent foundation
排架桩柔性墩 bent pile flexible pier
排架座木 grating beam; grating cill [sill]; mud sill; timber bent sill
排间距 row spacing
排间距离 distance between rows; row spacing; row-to-row distance
排间绝缘 bar insulation
排浆 drainage of slip; slip exhausting
排浆泵 slurry pump
排胶装置 batching out unit
排节 crib(bing)
排净空气 exhaust air
排距 array pitch; distance between rows; row spacing

排锯 frame saw; gate saw; gang saw
排锯的石块表面 gang-sawn
排锯机 gang saw
排锯切割技术 gang-sawing technique
排锯制材厂 gang mill; gang sawmill
排空 air-out; delivery of empty cars; drain(ing); emptying; evacuate; evacuation
排空阀 blow-off valve; washout valve
排空列车 deadhead train
排空漏斗 cone of exhaustion
排空时间 emptying time
排空式油槽车 gull(e)y emptier
排空试验 depletion test; dry test
排空数 emptying number; number of delivered empty cars【交】
排空旋塞 emptying cock
排空压舱水的浮船坞 emptied floating dry dock; pumped-out dock
排孔 holes in pattern
排孔爆破 row shooting
排孔节距 pitch of holes in pattern
排孔削弱系数 weakening factor of holes in pattern
排孔轴向削弱系数 weakening factor of axial holes in pattern
排矿槽 discharge trough
排矿斗 discharge cone
排矿口 out end
排矿箱 discharge box
排矿圆锥 discharge cone
排矿渣渠 tailrace
排蜡 dewaxing; wax removal
排蜡烧结炉 dewaxing-sintering furnace
排缆装置 spooling gear
排涝 draining water-logged
排涝标准 criterion[复 criteria] of water log control
排涝沟 surface drain
排涝模数 modulus of drainage
排涝桥 flood relief bridge; relief bridge
排练 rehearsal
排练室 rehearsal room
排练厅 hall choir; hall quire
排链 range of cable; tier a cable
排链待检 cable ranging
排梁 ceiling girder
排量 discharge; displacement volume; flow rate; fluid volume; output volume; swept capacity; swept volume
排量泵 displacement pump
排量表 < 安装在油罐上 > outage ga(u)ge
排量测定 delivery determination
排量风缸 displacement reservoir
排量厚度 displacement thickness
排量均衡器 < 泵的 > suction flow equalizer
排量膜板 displacement diaphragm
排量式压气机 displacement compressor
排量试验 capacity test
排量系数 coefficient of flow rate
排量油泵 displacement oil pump
排料 blowdown; discharge
排料按钮 discharge button
排料槽 discharge duct; reject chute
排料端 exhaust end
排料端排渣室 discharge-end refuse extraction chamber
排料阀 blow-off valve
排料接口 adaptor for product discharge
排料接受槽 dump tank
排料口 discharge gate; material outlet; spigot discharge
排料口间隙 crusher setting; crusher gap; crusher interval

排料口宽度 breath of discharge opening
排料口调定开度 crusher setting opening
排料口张开时的宽度 open-side setting
排料螺旋 load displacing screw
排料门 discharge door
排料门底座 door support
排料盘 feed disk[disc]
排料启闭器 door latch(with lock)
排料启闭器汽缸 latch cylinder
排料前端 discharge front end
排料箱 draw-off box
排列 alignment; apportion; arrange(in order); array; collocation; configurate; configuration; dispose; line up; listing; marshal(ling); ordination; permute; rank(ing); windrow; queuing【计】; permutation【数】
排列表 permutation table
排列长度 spread's length
排列成层 tiering
排列成行 align(ment); line in; range into line
排列成行的淋浴装置 gang showers
排列次序 ordering; rank; sequence
排列的第二号品种 number two ranking variety
排列的阶 order of a permutation
排列的型式 type of array
排列法 arrangement
排列方向 orient; orientation
排列符号 permutation symbol
排列规则 arrangement rule; queuing discipline
排列进路 route setting; setting of route
排列矩阵 permutation matrix
排列类似【地】homotaxis
排列类型 spread's type
排列了的进路 lined(-up) route
排列密度 packing density
排列匹配 ranked matching
排列偏离 ordering bias
排列平面图 disposition plan
排列区 alignment area
排列取向 oriented
排列缺陷 stacking fault
排列群 permutation group
排列时间 setting time; time of setting
排列式砌合 ranging bond
排列式砌筑 ranging bond; ranging bridge
排列顺序 chronological order
排列索引 permutation index
排列相似性【地】homotaxis
排列项 line item
排列因数 array factor
排列余量 justify margin
排列与组合 permutation and combination
排列在中心上的 spaced on centers [centres]
排列整齐 marshaling
排列支架 line timber
排列着的盥洗盆 basins in range
排列着的洗手盆 basins in range
排流 current drainage
排流坝 groin(e); groyne
排流鼻坎 flip bucket
排流变压器 draining transformer
排流电缆 current drainage cable
排流端子箱 current drainage terminal box
排流轨 current drainage rail
排流柜 current drainage cabinet
排流角 efflux angle
排流接线柜 current drainage connect-

ing cabinet
排流流速 velocity of retreat
排流能力 discharge capacity
排流器 drainage equipment; drainage system; electric(al) drainage【电】
排流速度 velocity of retreat
排流条 discharge bar
排流网 current drainage net
排流峡谷 gorge
排流线圈 bleeder coil; drainage coil
排流柱 drainage post
排硫杆菌 thiobacillus thioparus
排路完毕 route completion
排路缘石 curbing
排锚链 tiering
排锚链的钩 hooks for piling of anchor's chain
排锚链人 tierer
排煤渣灰器 ash gun
排棉台 picking table
排沫沟 scum gutter
排沫管 scum pipe
排沫旋塞 scum cock
排木 putlock; putlog
排泥 disposal of spoil; mud cock; sludge evacuation; spoil discharge; spoiling
排泥半径 dumping radius
排泥泵 dredge pump; sludge pump
排泥驳船 dumb barge
排泥船 pump-out(hopper) dredge(r)
排泥阀 hydrostatic valve; mud valve
排泥放水龙头 blow-off hydrant
排泥浮管 floating pipeline for spoils
排泥管 blow-off pipe; delivery pipe line; discharge pipe; dredge pipe-(line); extraction conduit; hydraulic fill pipe; mud drum; mud piping; silt ejector; sludge discharge pipe; sludge draw-off conduit; sludge draw-off pipe; sludge draw-off tube; sludge extraction pipe; sludge extraction tube
排泥管出口 outlet of discharge pipeline
排泥管道 mud pipe
排泥管堵塞 pipeline blockage
排泥管浮筒 discharge pipeline float; pipeline float; pontoon for pipeline
排泥管活动接头 movable joint of discharge pipeline
排泥管架 bracket mount for discharge pipeline; discharge pipeline supporting frame
排泥管接岸装置 shore connecting plant for pipeline
排泥管内沉淀 sedimentation in discharge pipeline
排泥管内摩擦阻力 friction(al) resistance in discharge pipeline
排泥管挠性接头 flexible joint of discharge pipeline
排泥管抛泥 hydraulic pipeline placement
排泥管特征曲线 characteristic curve of discharge pipeline
排泥管线【疏】 discharge(pipe) line; hydraulic fill pipeline; dredging pipe(line); discharging piping
排泥管线摩擦损失 friction(al) loss of discharge pipeline
排泥管线磨损检查 wearing inspection of discharge pipeline
排泥管线水头损失 head loss of discharge pipeline
排泥机构 discharge mechanism
排泥井 blow-off chamber; sludge discharge well
排泥坑 dredging pit
排泥孔 mud hole
排泥口 mud outlet; silt orifice; sludge outlet
排泥螺钉 screw for mud draining
排泥器 silt ejector
排泥区 deposit(e) area; deposit(e) ground; disposal area; disposal region; disposal site; dumping ground; dumping site; dumping space
排泥软管 sludge discharge hose
排泥三通 sludge discharging tee
排泥上岸装置 pump ashore unit
排泥室 mud chamber
排泥塘 dredged spoil basin
排泥筒 mud drum
排泥周期 period of blowdown
排尿槽 urinal channel
排刨 gang planning
排漂通道 floating debris pass
排齐 lining-up
排齐的引线 in-line pin
排气 air-out; air release; aerofluxus; air bleed; air elimination; air exhausting; air relief; air removal; back pressure steam; bled steam; bleed-off; blow-off; deaerate; deaeration; deaering; discharge gas; exhaust air; exhaust steam; exit air; exit gas; extracted air; fume off; gas blow off; off-take; vent; ventilation
排气暗管 exhaust manifold
排气板 exhaustion plate
排气瓣 exhaust clack; exhaust flap
排气爆震 exhaust detonation
排气背压(力) exhaust gas counter pressure; exhaust back pressure
排气泵 air-displacement pump; air exhausting pump; air pump; discharge pump; displacement pump; exhaust pump; extraction pump; off-gas pump; return pump
排气闭路阀 exhaust cut-out
排气箅子 exhaust grill(e)
排气补燃锅炉联合循环 exhaust-fired-boiler combined cycle
排气补燃器 exhaust gas afterburner
排气布 vent cloth
排气操作 exhaust operation; bleeding
排气侧 exhaust side
排气(侧)余面 exhaust lap
排气测定 exhaust measurement
排气层 venting layer
排气成分 exhaust gas composition
排气池 air separating tank
排气冲程 ejection stroke; outstroke; exhaust stroke
排气出口 air escape; exhaust outlet; exit air opening
排气储筒 exhaust reservoir
排气处理 exhaust gas disposal
排气传动 exhaust-driven
排气窗 exhaust grill(e); outlet ventilator
排气催化反应器 exhaust gas catalytic reactor
排气催化净化器 catalytic exhaust pure
排气催化系统 exhaust gas catalytic system
排气催化转化器 exhaust gas catalytic converter
排气存水弯 gas trap
排气打开 exhaust open
排气导程 exhaust lead
排气导管 exhaust guide; exhaust guidance
排气导流扳 exhaust splitter
排气导流叶片 exhaust stator blade
排气道 air channel; air chute; air drain; air passage; exhaust vent; gas withdrawal; vapo(u)r chimney; whistler; local vent <便池间的>
排气道用砖 extract venting tile
排气的 deaerated; ventilating
排气笛 exhaust whistle
排气端 exhaust end
排气短管 exhausting stub tube
排气阀 air discharge valve; air bleeder; air bleed valve; air escape cock; air escape valve; air evacuation valve; air outlet valve; air purge valve; air release valve; air relief valve; air (vent) valve; bleeder valve; blowdown valve; blow-off cock; blow-off valve; delivery valve; discharge service valve; discharge valve; draw-off valve; emptying valve; escape valve; exhaust flap; exhaust pipe; exhaust valve; negative pressure valve; outlet valve; petcock; release valve; snifter valve; snifting valve; vent valve
排气阀导管 exhaust valve guide
排气阀调整 exhaust valve regulation
排气阀盖 exhaust valve cap
排气阀杆 exhaust valve stem
排气阀机构 exhaust valve mechanism
排气阀间隙 exhaust valve clearance
排气阀升降装置 exhaust valve lifting gear
排气阀弹簧 exhaust valve spring
排气阀挺杆 exhaust valve lifter; exhaust valve tappet
排气阀头部 exhaust valve head
排气阀凸轮 exhaust valve cam
排气阀镶座 exhaust valve insert
排气阀销 exhaust valve pin
排气阀轴 exhaust valve spindle
排气反压力 exhaust back pressure
排气分流片 exhaust splitter
排气分析 exhaust gas analysis
排气分析器 exhaust analyser[analyzer]; exhaust emission analyser[analyzer]; exhaust gas analyser[analyzer]
排气分析系统 exhaust gas analysis system
排气分析仪 exhaust gas analyser[analyzer]
排气分析仪表 exhaust gas instrumentation
排气风道 discharge duct; extract venting duct
排气风阀 automatic vent damper
排气风管 exhaust flue
排气风机 air exhauster; exhaust(air) fan; fan exhauster; suction fan
排气风机的保护 protection of exhaust fan
排气风门 discharge damper; exhaust damper
排气风扇 air exhausting ventilator; exhaust air fan; exhaust blower; exhaust(er) draught fan; exit air fan; extraction fan; extractor fan; scavenger fan; ventilating fan; ventilator; discharge fan; exhaust fan
排气风筒 exhaust duct; exhaust gas duct
排气风箱 exhaust blower; exhaust gas duct
排气辅助器 booster; ejector; extractor
排气副储筒 auxiliary exhaust reservoir
排气干管 arterial vent
排气缸 exhaust cylinder
排气高温计 exhaust pyrometer
排气格孔 exhaust grill(e)
排气格栅 air discharge grill(e); exhaust air grill(e); exhaust grill(e)
排气格眼 exhaust grill(e)
排气格子窗 exhaust grill(e)
排气供热 exhaust steam heating
排气供热开口循环 exhaust-heated open cycle
排气沟挖掘机 land drainer
排气管 blow(-off) pipe; air bleeder; air exhauster; air outlet pipe; air vent pipe; blast pipe; discharge channel; downcomer; eduction pipe; eduction tube; effluent pipe; escape pipe; exhaust pipe; exhaust stacking; exhaust tube; exhaust tubulation; exit gas pipe; extract duct; extract venting pipe; freeing pipe; gas escape tube; off-take; outlet air pipe; puff pipe; purge pipe; purger; release pipe; relief pipe; vent flue; vent(ilating) pipe; vent line; vent tube; waste pipe
排气管爆燃 exhaust explosion
排气管挡水盖板 flashing of a vent (pipe)
排气管道 air exhaust duct; discharge pipe; discharge duct; evacuation line; exhaust air duct; exhaust duct; exhaust line; exhaust piping; exit air duct
排气管道装置 exhaust plumbing
排气管点火废气净化系统 exhaust port ignition cleaner
排气管法兰 discharge flange
排气管防雨盖板 flashing of a vent (pipe); rain cover of a vent(pipe)
排气管防雨帽 exhaust rain cap
排气管"放炮" exhaust explosion
排气管隔板 exhaust partition wall
排气管护板 exhaust blanking plate
排气管夹 exhaust pipe clamp
排气管接头 exhaust fitting
排气管截面积 releasing sectional area
排气管口盖 exhaust cover
排气管扩板 exhaust blanking plate
排气管理规则 regulation of exhaust gas
排气管路 discharge line; exhaust line; exit gas pipeline; gas exhaust piping
排气管内噪声 backshot
排气管喷嘴 exhaust pipe nozzle
排气管软接头 flexible exhaust fitting
排气管送风器喷嘴 exhaust pipe blower nozzle
排气管损失 hood loss
排气管套 exhaust shroud; exit gas socket
排气管填密物 exhaust pipe packing
排气管通风帽 vent cowl
排气管凸缘密封片 exhaust pipe flange gasket
排气管凸缘气封 exhaust pipe flange seal
排气管托架 exhaust pipe bracket
排气管弯头 bend in exhaust pipe
排气管网 air exhaust duct network; exhausting ductwork
排气管消焰器 exhaust flame damper
排气管消音器 blown-down silencer
排气管延伸 exhaust pipe extension
排气管罩 exhaust pipe shield; vent cap
排气管支架 exhaust pipe support
排气管阻焰器 exhaust flame damper
排气管座 air vent seat; exit gas socket
排气罐 exhaust tank
排气规范 exhaust schedule
排气柜 exhaust gas cabinet; fume cupboard; ventilated case
排气滚净筒 exhaust tumbling barrel; exhaust tumbling mill
排气锅炉 exhaust gas boiler

P

排气过程 exhaust process
排气过滤器 discharge filter; exhaust gas filter
排气和通气联合系统 combination waste-and-vent system
排气黑烟滤清器 exhaust gas smoke cleaner
排气恒温调节器 discharge gas thermostat
排气烘烤 exhaust bake-out
排气后燃(净化)器 exhaust after-burner
排气回流阀 exhaust gas recirculator valve; recycling valve
排气回收装置<制冷机冷媒> bleeding recovery
排气活门 air escape valve; exhaust clack; exhaust valve
排气活塞 exhaust piston
排气活塞栓 blow-off plug cock
排气火焰 exhaust flame
排气机 air exhauster; air extractor; air pump; evacuating machine; exhaust blower; exhauster; exhaustor; gas exhauster; gas expeller; exhaust fan
排气唧筒 air pump
排气及进气歧管【机】exhaust and intake manifold
排气集管 exhaust header; exhaust manifold
排气集合管 discharge manifold; exhaust manifold
排气集合环 exhaust collection ring; exhaust collector ring
排气集气管 exhaust collector
排气技术 exhaust technique
排气加热 exhaust heating; heating by exhaust gases
排气加热喷雾器 exhaust-heated atomizer
排气加热器 exhaust feed heater; exhaust heater; exhaust jacket
排气监测器 exhaust air monitor
排气减音器 detuner
排气胶管 exhaust hose
排气角管 exhaust horn
排气搅拌机<混凝土> deaerating mixer
排气搅拌器<混凝土> deaerating mixer
排气接管 vent connection
排气接口 exhaust port
排气接头 bleed connection
排气截止阀 discharge shutoff valve
排气进口<气波增压器> exhaust gas intake port
排气井 blast pit; exhaust shaft; extract shaft
排气井道 extract venting shaft
排气警报器 exhaust alarm
排气(净化)催化处理 catalytic exhaust treatment
排气净化目标 exhaust emission target
排气净化器 exhaust(gas) purifier
排气净化系统 emission control system
排气净化箱 exhaust conditioning box
排气净化装置 emission control equipment
排气开度 exhaust opening
排气开关 air discharge cock; exhaust close; exhaust cut-out
排气坑 exhaust pit
排气坑道 upcast
排气孔 vent hole[holing]; air drain; air escape; air outlet; air vent; deflation opening; discharge orifice; escape opening; exhaust hole; exhaust opening; exhaust orifice; exhaust outlet; exhaust vent; gas nozzle; gas vent; kicker port; loop vent; outage; scavenging port; ventilation hole
排气孔阀 air vent valve
排气孔塞 vent peg; vent plug
排气控制 blow-off control
排气口 air outlet opening; air escape; air exhaust opening; air gate; air outlet slit; air vent; blow vent; discharge outlet; escape hole; exhaust opening; exhaust orifice; exhaust outlet; exhaust port; gas outlet; gas vent; off-take; outcome; outlet-vent; relief opening; air exhaust
排气口堵 exhaust plug
排气口积炭 exhaust port deposition
排气口接头 gas vent connector
排气口控制 exhaust nozzle control
排气口圈 ring of exhaust port
排气口温度 exit temperature
排气口镶套 exhaust insert
排气口罩 exhaust cover
排气口直径 diameter of outlet
排气扩压器 exhaust diffuser
排气冷凝器 vent condenser
排气冷却法 exhaust cooling method
排气冷却器 vent gas cooler
排气量 air discharge; air displacement; air output; delivery; displacement; exhaust volume; external throughput; quantity discharged
排气量测定 discharge measurement
排气流 exhaust gas stream; exhaust steam jet
排气流量 extraction flow
排气流速 exhaust gas flow rate; exhaust gas velocity
排气龙头 draw-off cock
排气滤器 exhaust filter
排气滤网 exhaust screen
排气路线 extract venting line
排气轮机增压器 exhaust turbo-blower
排气螺杆 vented screw
排气脉冲空气喷射<抽吸式> exhaust pulse air injection
排气门 blast gate; exhaust port
排气门调节器 exhaust valve regulator
排气面积 leaving area; venting area
排气能 exhaust energy
排气旁通阀 exhaust by-pass valve
排气旁通阀控制线转环 exhaust by-pass valve control wire switch
排气喷管 discharge nozzle; ejector exhaust pipe; exhaust nozzle
排气喷孔 jet orifice
排气喷口 jet exhaust; jet orifice
排气喷射器 exhaust steam injector; exhaust suction pipe
排气喷射引水泵 exhaust ejector primer pump
排气喷泄损失 exhaust blow down loss
排气喷嘴 exhaust nozzle; final nozzle; gas-discharge nozzle
排气膨胀 exhaust expansion
排气偏导装置 exhaust deflector
排气偏转环 exhaust deflecting ring
排气偏转控制 exhaust deflection control
排气歧管 delivery manifold; exhaust collector pipe; exhaust gas manifold; exhaust manifold; gas exhaust manifold
排气歧管反应器 exhaust manifold reactor
排气歧管管套 exhaust manifold jacket
排气歧管加热阀 exhaust manifold heat valve
排气歧管连合凸缘 exhaust manifold companion flange
排气歧管密封片 exhaust manifold gasket
排气歧管压力 exhaust manifold pressure
排气气口 exhaust riser
排气汽轮式压缩机 exhaust turbo-compressor
排气器 air ejector; air eliminator; air exhauster; evacuator; exhauster; exhaust stack; exhaust ventilator; exsufflator; gas exhauster; purge unit; vent fan
排气器滤网 breather screen
排气器帽 breather cap
排气器通风风扇 exhauster draft fan
排气强度 discharge intensity; discharge strength
排气曲线<示功图的> exhaust curve
排气驱动增压器 exhaust-driven supercharger
排气取样 exhaust gas sampling
排气去污系统 exhaust air decontamination system
排气圈 exhaust ring
排气热 exhaust heat
排气热损失 exhaust heat loss; waste heat rejection
排气容积 delivery space; exhaust volume
排气软管 exhaust hose; vent hose
排气塞 core box vent; core vent
排气三通 air exhausting tee; blow-off tee
排气刹车 exhaust brake
排气扇 outlet ventilator; vent fan
排气设备 air discharge equipment; air relief installation; pumping equipment; air drainage equipment
排气射流 exhaust jet(stream)
排气声音 exhaust sound
排气湿度 exhaust wetness
排气时间 evacuation time; exhaust time
排气式 exhaust; exhaust-driven; exhaust system
排气式打手 exhaust beater
排气式燃料元件 vented fuel element
排气式调节器 bleed type controller
排气式凿岩机 vented-type drill
排气式钻机 vented-type drill
排气室 blow-off chamber; discharge (air) chamber; exhaust chamber; exhaust room; exhaust space
排气收集器 exhaust collector
排气受阻 exhaust braking
排气竖风道 exhaust shaft
排气竖管 exhaust vertical pipe; gas vent; venting stack; exhaust shaft
排气竖井 exhaust shaft; extract venting shaft; air relief shaft
排气栓 blow(-off) cock; draw cock; petcock
排气速度 delivery speed; discharge velocity; efflux velocity; exhaust velocity; velocity of exhaust
排气塑模 degassing mo(u)ld
排气损耗 breathing loss
排气损失 discharge loss; exhaust loss
排气塔 de-airing tower; exhaust tower
排气特性 discharge characteristic
排气提前【机】exhaust advance
排气体系 exhaust system
排气调节环 exhaust regulating ring
排气调节器 exhaust regulator
排气铁花格 exhaust grill(e)
排气停截阀 exhaust cut-out
排气停止 exhaust close
排气通道 exhaust duct; extract venting; exhaust passage
排气通道砌块 extract venting block
排气通风道 discharge flue
排气通风(法) aspiration ventilation; bleed venting; blow-off ventilation; exhaust(duct) ventilation
排气通风管 air exhaust vent-pipe; exhaust ventilating duct
排气通风机 air-ejecting fan; air exhaust ventilator; suction fan
排气通风筒 exhaust ventilator
排气通风系统 exhaust ventilation system; extraction ventilation system
排气通风装置 exhaust ventilator
排气通路 exhaust passage
排气筒 discharge stack
排气头 exhaust head
排气透平机 blowdown turbine
排气凸轮 exhaust(valve lifting) cam
排气凸轮轴 exhaust cam shaft
排气推力 exhaust thrust
排气脱硫 desulfurization from exhaust gas; exhaust gas desulfurization
排气弯管 exhaust elbow
排气尾管 tail pipe
排气温度 delivery temperature; discharge temperature; downstream exhaust temperature; exhaust (gas) temperature; exit gas temperature; outlet gas temperature
排气温度过高切断器 high discharge temperature cut-out
排气温度计 exhaust temperature ga(u)ge
排气温度计接管 nozzle for thermometer of exhaust gas
排气温度曲线 temperature curve of exhaust
排气涡轮 exhaust-driven gas turbine
排气涡轮传动增压器 exhaust turbine driven supercharge
排气涡轮式增压器 exhaust gas turbine supercharger
排气蜗壳 discharge volute
排气污染 exhaust pollution; exhaust emission
排气污染标准 exhaust air pollution standard
排气污染成分分析仪 emission analyser
排气污染程度 emission level
排气污染法规 emission regulation
排气污染鉴定<汽车发动机的> emission certification
排气污染控制 exhaust emission control; exhaust pollution control
排气污染控制装置 emission control device
排气污染试验 emission test
排气污染特征 exhaust emission characteristics
排气污染物 exhaust contaminant; exhaust pollutant
排气污染物的形成 emission formation
排气污染物浓度 emission concentration
排气污染总量测定 mass emission measurement
排气吸气冲程 exhaust suction stroke
排气系数 exhaust coefficient; exit coefficient; outgassing coefficient
排气系统 discharge system; emptying system; exhaust gas system; exhaust system; exit gas system; gas withdrawal system; scavenger system; vent(ing) system
排气线 exhaust line
排气线路 exit gas line
排气相 exhaust phase
排气箱 exhaust box; exhaust chest;

exhauster chamber
排气消火器 exhaust flame suppressor
排气消声器 exhaust(gas) muffler;exhaust silencer;exhaust snubber
排气消声器开关 exhaust muffler cut-out
排气消音器 exhaust muffler;exhaust silencer
排气效应 gas expelling effect
排气行程 exhaust stroke;instroke
排气旋塞 air cock;drain cock;draw-off cock;drip cock;relief cock
排气穴 exhaust cavity
排气循环 exhaust cycle
排气循环法 exhaust-recirculation method
排气压比 exhaust pressure ratio passage
排气压力 abandonment pressure;blow-off pressure;delivery pressure;discharge pressure;exhaust pressure;gas surging pressure;pressure at expulsion
排气压力表 discharge pressure ga(u)ge
排气压力测潮仪 gas-purging pressure tide ga(u)ge
排气压力检查孔 inspection hole for exhaust gas pressure
排气压力控制型废气再循环阀 exhaust pressure modulated exhaust gas recirculation valve
排气压力验潮仪 gas surging pressure tide ga(u)ge
排气压头 discharge head
排气烟囱 exhaust chimney;exhaust duct;exit air chimney;exit gas installation;extract venting chimney;oven stack;venting stack;exhaust stack
排气烟道 draft flue;draught flue
排气摇杆 exhaust rocker
排气叶片 exhaust blading;nozzle vane
排气音调 exhaust note
排气有害成分控制仪 emission exhaust control device
排气有害物流动试验室 mobile emissions test laboratory
排气余面 exhaust lap;inside lap
排气余气 remainder of exhaust gases
排气预留孔 weep hole
排气缘 exhaust edge
排气再热 < 用于排气净化 > exhaust reheat(ing)
排气再循环 < 一种降低排气中一氧化氮含量方法 > exhaust gas recharging
排气(再)循环控制器 exhaust gas recirculation controller
排气再循环系统 exhaust gas recycling system
排气再循环装置 exhaust gas recirculation device
排气噪声 exhaust roar
排气闸 exhaust brake
排气闸门 exhaust air port
排气站 exit air station
排气罩 draft hood;exhaust canopy;exhaust hood;ventilating cowl
排气罩壳 suction cap
排气罩种类 hood type
排气真空泵 vacuum pump for vent
排气蒸汽机 non-condensing engine
排气支管 blow off branch pipe;bypass vent;exhaust branch(pipe);gas exhaust manifold
排气制动 exhaust brake
排气制动指示灯 exhaust brake indicator lamp
排气滞后 exhaust lag
排气中氨测定法 method for determination of ammonia in exhaust gas
排气中氮氧化物测定法 method for determination of oxides of nitrogen in exhaust gas
排气中氯测定法 method for determination of chlorine in exhaust gas
排气中氰氢酸测定法 method for determination of hydrogen cyanide in exhaust gas
排气中一氧化碳测定法 method for determination of carbon dioxide in exhaust gas
排气周期 exhaust period
排气转化器 exhaust gas converter
排气装置 air eliminator;air exhaust;air exhaust device;air relief installation;breathing apparatus;disengaging device;exhaust gas ducting system;exhaust gear;exhaust installation;exhaust unit;releaser;gas barrier < 废物处理场的 >
排气状态 exhaust condition
排气着火 exhaust flaming
排气总管 discharge manifold;exhaust belt;exhaust collector;exhaust main;exhaust trunk;exhaust manifold
排气阻力 exhaust resistance
排气嘴 air escape cock;venting nipple
排汽 dump steam;exhaust steam;extraction of steam;steam dumping
排汽道 steam exhausting way
排汽点 exhaust point;point of release
排汽端 exhaust steam end
排汽阀 draw-off valve;exhaust steam valve
排汽阀拉杆 exhaust rod
排汽缸 exhaust casing;exhaust hood;exhaust steam casing;outlet casing
排汽缸损失 hood loss
排汽供给管 exhaust steam supply pipe
排汽管 eduction pipe;steam exhaust pipe
排汽管道 blow-off line
排汽管路 dump line
排汽级 exhaust stage
排汽给水加热器 exhaust steam feed heater
排汽进口 exhaust steam inlet
排汽孔 exhaust steam port
排汽口 steam exhaust port
排汽扩压器 exhaust diffuser
排汽能 exhaust energy
排汽喷射器 exhaust steam injector
排汽喷嘴 < 机车的 > exhaust steam nozzle
排汽器 steam-jet exhauster
排汽湿度 exhaust steam moisture
排汽式涡轮 exhaust type steam turbine
排汽室【机】exhaust hood;exhaust room
排汽头 < 蒸汽机的 > exhaust head
排汽压力 exhaust steam pressure
排汽余面 steam lap
排汽蒸汽机 non-condensing engine
排汽支管 blow-off branch
排汽总管 exhaust steam main
排砌形式 laying pattern
排球场 volleyball court
排驱效率 expulsion efficiency
排驱压力 replacement pressure
排去 draining
排(燃)油塞 fuel drain plug
排扰线 drain wire
排热 abstraction of heat;carry-off heat;egress of heat;heat abstraction;heat egress;heat extraction;heat rejection;heat removal;heat withdrawal;removal of heat
排热法 elimination of heat;heat elimination
排热回路 heat-rejection circuit
排热机理 heat-removal mechanism
排热剂 heat-removing agent
排热率 heat extraction rate
排热设备 heat-removal equipment;heat-removal mechanism
排热设施 heat venting facility
排热系统 heat-extraction system;heat-removal system
排日程计划 scheduling program(me)
排如锯形 arris-wise
排入 discharged into;immission
排入大气 atmospheric dilution
排入地表水体 drainage to surface water
排入地下 underground disposal
排入海洋 ocean disposal
排入海中 sea disposal
排入环境 environmental release
排入环境中 discharge into the environment
排入进水口 drainage inlet
排入水中 discharge into the water
排沙 < 又称排砂 > desilt(ing);diversion of sediment;sediment ejection;sediment outflow;sand removal;sediment removal
排沙泵 hydraulic ejector
排沙槽 wasteway channel
排沙道 sand escape
排沙底孔 sediment bottom sluice
排沙阀 sand flush valve;scour valve
排沙构筑物 sand removal structure
排沙管 discharge pipe;grit blow-off
排沙孔 clearance hole;sediment flushing outlet;sediment sluice;sand sluice
排沙孔口 silt orifice
排沙口 sand outlet;silt orifice
排沙量 sediment outflow
排沙漏斗 desilting funnel
排沙率 sediment delivery percentage;sediment delivery rate
排沙能力 sediment discharge capacity
排沙清淤 sediment removal
排沙渠(道) desilting canal;discharge canal;discharge channel
排沙设备 by passing plant;desilter;sediment ejector
排沙设施 desilting work;sand by-passing;sediment ejector
排沙隧洞 sluice tunnel
排沙闸(门) desilting sluice;flushing sluice;sand escape;sand gate;sand sluice;scouring sluice
排沙装置 by passing plant;sediment ejector
排渗特性 draining characteristic
排渗特征 drainage characteristic
排渗帷幕 curtain drain
排绳 leg along
排绳传送带 rope belt
排绳器 rope-arranging device;rope guider;rope guiding device
排湿气管 wet vent
排湿气孔 wet vent
排式 format
排式冲床 gang punch
排式发动机 row engine
排式接头 in-line coupling
排式铆钉 gang riveting
排式水尺 multiple tide staff;multiple water ga(u)ge
排式铣床 gang mill
排式铣削 gang milling
排式压床 gang press
排式钻床 gang drill(er)
排式钻机 gang drill(er)
排数影响 effect of rows
排水 discharging;drainage water;abstraction of water;bailing;discharge of opening;draining off the water;flow of catchment;pumping out;unwater(ing);dewatering < 基坑的 >;water diversion;water drainage;water exclusion;water release;water removal
排水暗沟 blind drain(age);covered drain(age);dewatering culvert;drainage fill;drainage tunnel;gravel fill;infiltration ditch;mole drain;trench drain(age)
排水暗沟进水口 blind inlet
排水暗管 buried drain;closed drain;buried drainage pipe
排水暗管桥 drainage bridge
排水暗管清洗机 subsurface drain cleaner
排水暗渠 drain conduit
排水暗(瓦)管垫层 bedding of drain tile
排水板 drain(age) board;drain plate
排水板底 tip of drain
排水板顶 top of drain
排水板法 cardboard drain method;geodrain method;sheet drainage;wick drain
排水半径 drainage radius
排水伴流 displacement wake
排水泵 dewatering pump;discharge pump;drainage pump;drain(ing) pump;outlet pump;sump pump;unwatering pump;wet pit pump
排水泵房 sewerage pumping house
排水泵站 drainage pumping station
排水比降 flow gradient
排水箅 drainage grating;draining grate
排水箅栅 drainage grate
排水边沟 drainage gutter;gutter
排水边坡 drain gutter
排水边线 drainage divide(line)
排水标准 quality standard of discharge water
排水表面 drainage(sur)face
排水不畅流域 poorly drained stream basin
排水不良 impeded drainage;poor drainage
排水不良的 imperfectly drained
排水不良土壤 poorly drained soil
排水布置 drainage arrangement;draining arrangement
排水布置方式 drainage arrangement pattern
排水布置模式 drainage pattern
排水槽 by-wash channel;catch gutter;discharge channel;discharge gutter;discharge sump;discharge trough;drainage channel;drainage tray;drain flute;outlet trough;overflow trough;rhone;water shoot < 屋檐 >
排水槽铁面板 canting strip
排水槽支承板 layer board;lear board
排水测量 discharge measurement
排水层 drainage blanket < 美 >;drainage course;drainage layer;drainage stratum;pervious blanket
排水铲斗 drainage bucket
排水沉污井 drainage tray;drainage well
排水池 discharge sump;drain(age) basin;drain(age) tank;drip chamber;sump tank

P

排水冲水廊道 emptying and filling conduit; emptying and filling culvert; emptying and filling gallery
排水出口 discharge outlet; drain(age) opening; drainage outlet
排水出口工程 drainage outlet works
排水出路 discharge outlet; drainage outlet
排水处 off-take
排水处理装置 treatment equipment for the wastewater
排水次干管 submain
排水导洞 drainage heading
排水导水沟 pilot ditch
排水道 culvert; dewatering way; dike [dyke]; discharge channel; drainage passage; drainage path; drainage way; emissary; outlet; outlet drain; discharge manifold <闸室的>
排水的 drained
排水的行政管理区 drainage district
排水底 dewatering conduit
排水底基层 <混凝土路面下的> draining subbase
排水底孔 bottom door; dewatering conduit
排水地埂 drainage terrace
排水地基 drained ground base
排水地区 drainage
排水地下水管系(统) drainage pipe system
排水点 draining point; off-take point
排水垫层 drainage blanket; drainage cushion; pervious blanket
排水吊桶 bailer; dewatering bucket
排水定额 drainage quota; drainage requirement; wastewater flow norm; wastewater flow quota
排水动力消耗 power consumption of drainage
排水洞 adit for draining
排水斗 drain bin
排水陡槽 drain chute
排水度 degree of drainage; freeness
排水度数 freeness number
排水吨 displacement ton
排水吨数 displacement tonnage
排水吨位 <船舶> displacement tonnage; cubic(al) displacement; tonnage displacement; vessel tonnage
排水阀(门) bale valve; discharge valve; discharging valve; drainage valve; drain(ing) valve; emptying valve; water drain apparatus; draw-off valve; blowdown valve; blow-off valve; scupper valve; sewage water valve
排水法 displacement method <浮游生物定量方法>; drainage; draining
排水反复直剪试验 drained repeated direct shear test
排水方法 drainage; unwatering method
排水方式 drainage pattern
排水防气瓣 drainage tray; drain trap; intercepting trap
排水费 cost of draining; drainage cost
排水分界线 drainage divide(line)
排水风车 windmill for drainage
排水干沟 arterial drain; main drain
排水干管 arterial drainage; arterial drainage pipeline; drainage collector; drainage trunk; drain collector; drain trunk; main drain; main sewer; off-take main; outlet header; trunk main
排水干渠 arterial drainage canal; main drain
排水干渠系统 arterial drainage(system)
排水干线 drainage line
排水纲 drainage scheme
排水钢管 steel drain(age) pipe; steel drain water pipe
排水格栅 drainage grate [grating]; drainage grid
排水工 drainer
排水工程 dewatering excavation; drainage project; outfall works; sewerage; sewerage and sewage treatment; water drainage works; drainage works
排水工程队 drainage gang; draining gang
排水工程设计实践 sewerage design practice
排水工程学 drainage engineering; sewerage engineering; wastewater engineering
排水工具 drainage tool
排水工时消耗 man-hour consumption of drainage
排水沟 drain(age) trench; drain(ing) channel; drain(ing) ditch; draining gutter; aphodus; berm(e) ditch; by-wash(channel); catch-drain; catch-water drain; channel gull(e)y; culvert; cut drain; delivery conduit; discharge ditch; discharge gutter; disposal ditch; flushing canal; foul drain; gull(e)y(drainage); gutter way; off-take; outfall ditch; outlet channel; outlet conduit; relief ditch; sluice; sough; storm drain; swamp ditch; trench drain(age); trough gutter; water-diversion ditch; water drain; water exhaust; water furrow; water shoot; water trough; weeper drain; yard gutter <堆场的>; tide channel <潮汐排水用>
排水沟边坡 gutter slope
排水沟铲 drain-trench spade
排水沟出口 outlet of drainage ditch
排水沟出口处 drainage exit
排水沟道 drainage line; draining line
排水沟断面 discharge section
排水沟机械 drainage machinery; Kjellmann-Franki machine
排水沟级配反滤层 graded filter of drainage
排水沟检查盖板 access gulley of drainage
排水沟建造工具 gutter tool
排水沟口 off-take
排水沟犁 draining plough
排水沟落差 drainage drop
排水沟模型 ground mo(u)ld
排水沟配件 gutter fitting; channel fitting
排水沟坡降 gutter gradient
排水沟清除垃圾 access gully
排水沟清理器 drain cleaner
排水沟渠 drainage channel; drainage ditch; off-take; waterway; escape canal
排水沟深度 drain depth
排水沟疏浚机 discharge ditch sweeper
排水沟填土 drainage fill
排水沟头 head of drain
排水沟挖掘机 drain(age) digger; drainage trench digger; land drainer
排水沟弯段 channel bend
排水沟网 drainage pipework; drainage piping; network of drains; waterway net(work); drainage net(work)
排水沟要求 requirement of drain
排水沟有孔盖板 channel grating
排水沟预制块 block channel
排水沟栅 drain grating
排水沟整修器 gutter tool
排水沟支承板 lear board
排水沟终端 gutter end stop; channel stop-end
排水沟终端瓦 gutter end tile
排水沟主渠与支渠连接点 channel junction
排水沟锥形连接段 taper channel
排水构筑物 drainage structure; drainage work; outlet structure
排水固结法 drain-consolidation method
排水骨料 drainage aggregate
排水管 blow-off pipe; by-wash; collector drain; cut-off; delivery conduit; delivery pipe; drainage duct; drain conduit; draining pipe; exhaust pipe; exhaust tube; exhaust water pipe; fall tube; foreyn; foul drain; foul sewer; freeing pipe; leader; leading; off-take; pass way of water; pipe drainage; porous pipe; relief drain; sewer; sewer pipe (drain); ware pipe; waste pipe; water drain pipe; water shoot; weep drain; pipe weep <混凝土砌体背后排水用>; service drain(age) <用户通至街道污水管的>
排水管道 dale; dewatering conduit; discharge conduit; discharge duct; discharge pipe; drainage conduit; drain(age) pipe; drainage pipeline; drain(age) tube; drip hole; off-take pipe [piping]; overflow pipe; pipe drain; scupper pipe; unwatering pipe; water discharge pipe [piping]; water drainage pipe [piping]; water drainage tube [tubing]; discharge pipe line; draining pipeline
排水管道布置 layout of sewerage system
排水管道布置图 plan of sewerage system
排水管道沟槽 sewer line trench
排水管道建筑 sewer line construction
排水管道连接 sewer line connection
排水管道漏水试验 asphxiator
排水管道施工 sewer line construction
排水管道系统 drainage piping; draining pipe system
排水管的帽管 capped pipe
排水管垫层 bedding of drain; discharge duct; drainage piping
排水管端 head of drain
排水管法兰 discharge flange
排水管腐蚀 corrosion of sewer
排水管沟 by-wash; drainage basin
排水管管顶腐蚀 crown corrosion of sewer
排水管管理区 waste and stormwater collection section
排水管裹料 drainage envelope
排水管户线 house drain
排水管间隔 drain spacing
排水管接头 drainage fitting; drain connection; drain connector
排水管净化剂 drainage cleaner
排水管坑道 drainage pipe gallery
排水管廊 drainage pipe gallery
排水管理区 drainage district
排水管理系统最优化设计 optimization design of sewer pipe system
排水管理线 drain line
排水管路 discharge pipe line; drainage piping; drainage system; drain line
排水管铺管机 drain(age)(draw) layer
排水管铺设机 drain(age)(draw) layer
排水管清扫器 badger
排水管清洗机 drain(age) cleaner
排水管清洗器 badger
排水管渠 sewer; water drainage conduit
排水管深度 drain depth
排水管试验 drain testing
排水管疏通软杆 drain rod
排水管栓 drain plug; drain stopper
排水管水头 leader head
排水管探测器 drain detector
排水管通沟室 cleaning chamber
排水管外压试验 external loading test for drain pipe
排水管弯头 draining pipe elbow
排水管网 drainage net(work); drainage pipe system; drainage pipework; drainage piping; network of drains
排水管系 drainage piping; pipe drainage; pipe system
排水管系统 drainage system
排水管线 discharge(pipe) line; drainage pipeline; drain tile line; flowing line; sewer line
排水管(压头)损失 discharge pipe loss
排水管淤积 drain silting
排水管支管进口 back-inlet gull(e)y
排水管制品 sewer goods; sewer product
排水灌浆 displacement grouting
排水灌毛沟 water furrow
排水涵洞 discharge culvert; drainage culvert
排水涵管 drainage culvert
排水和倒滤层 drainage and filter
排水河槽 discharge channel
排水荷载循环 drained load cycle
排水湖 drainage lake; exorheic lake
排水花管 perforated drain pipe
排水化学板 chemical board drain; drained chemical board
排水汇流 convergence of drainage
排水机 draft engine; drainage machine; draught-engine
排水机械 drainage machinery
排水箕斗 bail(ing) skip
排水及通气合用系统 combination waste-and-vent system
排水极劣 very poorly drained
排水集管 drain(age) header
排水集料 drainage aggregate
排水集水坑 drainage sump
排水计划 drainage project; scheme for irrigation and drainage
排水技术经济指标 technical-economic index of drainage
排水加荷 drainage loading; drained load
排水间 pumpway
排水间隔 spacing of drain
排水间距 drain spacing; spacing of drain
排水减压 drain to reduce pressure
排水减压式干船坞 drainage drydock
排水剪力试验 drained shear test
排水剪力特性 drained shear characteristic
排水剪切强度 drained shear strength
排水剪切试验 drained shear test
排水建筑 drainage works; outfall head works
排水建筑物 drainage structure
排水交叉建筑物 drainage crossing
排水胶管 discharge hose
排水阶地 drainage terrace
排水接触式乙炔发生器 water displacement contact type generator
排水接缝 drained joint
排水接户支管 exhaust connecting branch
排水接头 drainage connection; drained joint

排水结构(物) drainage structure;outfall structure
排水井 bleeder well;blind catch basin;catch pit;dewatering sump;dewatering tank;drainage pit;drainage sump;draining well;dry well;gull(e)y pot;negative well;offset well;pumping shaft;relief well;sump well;well drain
排水井点 dewatering hole;well-drain point
排水井口防臭设备 gull(e)y trap
排水井系统 well system
排水开挖 drainage excavation
排水抗剪强度 drained shear strength
排水坑 dewatering pit;discharge sump;drainage pit;drainage sump;drain pit;drain sump;sump pit;unwatering pit;unwatering sump
排水坑道 adit for draining;adit opening;drainage drift;drain tunnel
排水空腔 drained cavity
排水孔 bleed hole;dewatering hole;dewatering orifice;discharge hole;discharge orifice;drainage hole;drainage opening;drain(ed) hole;draining hole;drilled drainage hole;drip hole;escape hole;exhaust outlet;opening for drainage;scupper <船舶或坞门的>;sinker;spiracle;wash port;weeper;weep hole
排水孔带铰链的盖 port flap
排水孔管 <混凝土砌体背后排水用> weeping conduit;weeping piping;weeping tube;weeping tubing;weeping pipe
排水孔口 head of drain;emptying port
排水孔隙率 drainable porosity
排水孔折合盖 port flap
排水孔钻进 drainhole drilling
排水口 discharge opening;discharge point;discharge port;drainage hole;drain port;escape hole;exhaust outlet;flowing mouth;free port;gargoyle;outfall;outlet;rain spout;scupper;taping spout;freeing port;wash port <船舷的>
排水口阀门 scupper valve
排水口格子盖 scupper grating
排水口翼盖 port flap;port lid;port sash
排水口止回板 port lip
排水口止回阀 storm valve
排水快剪试验 consolidated quick shear test
排水廊道 dewatering gallery;drainage adit;drainage gallery;drain(age) tunnel;unwatering gallery
排水棱体 drainage prism
排水棱柱体平台 drainage berm
排水冷却器 drain cooler
排水历时图 histograph;histograph of drainage
排水立管 drainage riser;soil stack
排水利用 water recovery
排水沥青 drainage asphalt
排水连接管 discharging connection;fixture drain;drain(age) connection
排水良好的 excessively drained;well-drained
排水良好的地块 excessively drained field
排水良好的流域 well-drained stream basin
排水两千吨 displacement two thousand tons
排水量 delivery capacity;discharge of water;displacement (of water);displacement tonnage;drain yield;submerged displacement;water displacement;withdrawal;water discharge
排水量标尺 displacement scale
排水量船长比 displacement length ratio
排水量船长系数 displacement length coefficient
排水量吨 <船舰的> displacement ton
排水量计算 discharge calculation
排水量计算表 displacement sheet
排水量平衡 drainage equilibrium
排水量曲线 curve of displacement;displacement curve
排水量系数 displacement coefficient;water discharge coefficient
排水量与给水量之比 proportion of water supply reaching sewers
排水流 discharge flow
排水流量 drain flux;drain(age) discharge
排水流域 drainage area;drainage region
排水龙头 discharge cock
排水垄沟 furrow drain
排水垄沟形式 ridge and furrow type of drainage
排水漏斗 drainage funnel;water drain funnel
排水滤层 drainage filter
排水滤体 drainage filter
排水滤网 draining screen
排水路径 drainage path
排水率 degree of drainage;discharge rate
排水落差 drain drop
排水(慢速)加荷载 drained loading
排水盲沟 blind ditch;ditch drainage;weeper drain;blind drainage
排水毛沟 furrow drain
排水门 freeing scuttle
排水门盖 port lid
排水密度 drainage density
排水面层 draining blanket
排水面积 drain(age) area;draining area;watershed
排水明沟 ditch;drain;open drain;drainage ditch
排水模量 drainage modulus;draining modulus;modulus of drainage
排水模数 drainage modulus;draining modulus;modulus of drainage
排水幕 drainage curtain
排水内摩擦角 drained angle of internal friction
排水能力 drainability;drainage ability;drainage capacity;flow capacity
排水盘 drain pan
排水盆地 drainage basin;watershed
排水平峒 drainage gallery;off-take drift
排水平洞 draining adit;sough
排水平衡 drainage balance;drainage equilibrium
排水平巷 drainage level
排水坡度 drainage slope;slope for drainage
排水铺盖 drainage blanket
排水铺面 drainage blanket
排水气隙比 drain air space ratio
排水器 delf;drainer;drain trap;hydathode;water drain apparatus
排水铅槽 <平屋顶,通过墙身的> lead shoot
排水铅管 lead draining pipe
排水浅洼 drainage dip
排水强度 discharge intensity;drainage intensity
排水区 drainage area;drainage basin
排水区内最高点 drainage head
排水区内最远点 drainage head
排水区域 drainage area;drainage district
排水渠(道) discharge canal;discharge channel;discharge conduit;drainage canal;drainage channel;drainage conduit;draining canal;draining channel;draining conduit;draining outfall;escape canal;gull(e)y drain;outfall ditch;swamp ditch;conduit drain;culvert;discharge ditch;drainage ditch;draining ditch;off-take;outfall channel;storm drain;take-out channel;watering;waterway
排水渠落差 head difference of drain
排水渠桥 conduit bridge
排水渠首 head of drain
排水渠水头 head of drain
排水渠弯道 channel bend
排水权 drainage(water) rights
排水容积 volume of displacement
排水容积计 volumenometer
排水容量 volume disposal;volume of displacement
排水容器 drainage vessel;water displacement vessel
排水蠕变 drained creep
排水入口 drainage inlet
排水褥垫 drainage blanket
排水软管 drain hose;flexible exhaust pipe;scupper hose【船】
排水塞 draining plug;drain stopper
排水塞门 draining cock
排水三轴试验 drained triaxial test
排水砂垫层 drainage sand mat;sand mat of subgrade
排水砂井 displacement well;drainage wick;filter well;sand drain;vertical sand drain;drainage well
排水砂井顶部透水覆盖层 blanket connection for sand drains
排水砂井网 grid of drainage wicks
排水砂桩 drain(age) pile;sand drain;sand pile
排水舌 sill
排水设备 dewatering equipment;drain(age) equipment;drainage facility;drainage plant;drain device;emptying device;pump drain equipment;water-freedom arrangement
排水设备能力 drainage equipment capacity
排水设备折旧摊销及大修费 pumping equipment depreciation apportion and overhaul charges
排水设计 drainage design
排水设计标准 drainage design criterion
排水设计重现期 design return period of drainage;design frequency of drainage
排水设计准则 drainage design criterion
排水设施 drainage appliance;drainage facility;outlet works;sewerage;sewerage installations
排水设施通行权 drainage right-of-way;drainage easement
排水渗流井 <井中填石的> dry well
排水施工法 dewatering method
排水十万吨 displacement one hundred thousand tons
排水十五万吨 displacement one hundred and fifty thousand tons
排水时间 discharge time;displacement time;drainage time;draining time;emptying time
排水式土壤蒸发仪 drainage-type lysimeter
排水室 drip chamber
排水收费 drainage rate
排水竖井 drainage shaft;draining shaft
排水栓 scupper plug
排水水头 drainage head
排水水质 quality of sewage
排水水质标准 effluent-quality standard
排水税率 drainage rate
排水速度 discharge velocity;velocity of discharge
排水速率 rate of drainage
排水塑料板 dewatering plastic sheet;geodrain
排水塑料管 drain tile
排水隧道 drainage tunnel;sewerage culvert;sewerage tunnel;sewer tunnel;water discharge tunnel
排水隧洞 drainage tunnel;sewerage culvert;sewerage tunnel;water tunnel
排水塔 intake tower;outlet tower
排水陶管 drain tile;stoneware drain
排水陶土管 clay sewer pipe
排水提升泵站 sewage hoisting pumping station
排水体 displacement hull
排水体积 volume of displacement
排水体制 sewerage system
排水条件 condition of drainage;drainage condition
排水通道 drainage adit;drainage way
排水筒 adjutage;bailer;scupper shoot
排水土层 open layer
排水(土工)布 drain fabric
排水瓦管 bleeder(line of) tile;clay drainage tile;clay sewer pipe;drain(age) tile;tile culvert;tile drain
排水瓦管出口 outlet of tile drain
排水洼地 drainage swale
排水弯管 drain elbow
排水网 anastomosing drainage;drainage system;draining system
排水网密度 drainage density
排水帷幕 curtain drain;drainage curtain
排水五万吨 displacement fifty thousand tons
排水吸附器 draining sucker
排水系人孔 catch pit
排水系数 <以毫米/天计的排水深度> drainage factor;drainage coefficient
排水系统 collecting system;discharge system;double main system;drainage(work);emptying system;hydraulic emptying system;natural system;pumping-out system;sewerage(system);sewer system;unwatering system;water release system;drainage system <英>
排水系统布置 sewerage system layout
排水系统布置图 layout of sewerage system
排水系统方案 drainage scheme
排水系统管理机构 drainage district
排水系统检漏试验 leak test of drainage system
排水系统类型 drainage pattern;drainage type
排水系统区 drainage area;drainage basin
排水系统设计 drainage system design
排水系统竖向通风口 circuit vent
排水系统水管 drainage piping
排水系统泄漏 exfiltration
排水系统用的铸铁制品 cast-iron products for sewerage systems
排水系统用地范围 drainage right of ways
排水系统中的最高点 head of drain
排水线 discharge line
排水箱 discharge casing;drain box;dump tank

排水巷道 drainage gallery
排水小沟 field ditch
排水斜管 drainage chute;drain chute
排水信息系统 drainage information system
排水型船 displacement boat;displacement ship
排水型客轮 passenger displacement ship
排水型艇 displacement boat;displacement ship
排水性 drainability;drainage
排水性材料 drainable material
排水性沥青 draining asphalt
排水性能 drainage characteristic
排水性能良好的 free draining
排水性能良好的粒状材料 free-draining granular material
排水蓄水槽 drainage cistern
排水蓄水池 drainage cistern
排水旋塞 bleeding plug; discharge cock;drainage cock
排水旋塞阀 discharge cock
排水窨井 inlet well
排水压力 discharge pressure; drainage pressure;sewerage force
排水压力干管 sewerage force main
排水要求 drainage requirement
排水叶轮 water impeller
排水液 water-displacing liquid
排水翼盖 port flap;port lid
排水因数 drainage factor
排水因素 drainage factor
排水阴沟 drainage culvert;drain sewer;lode
排水窨井 inlet well
排水用 U 形管 horseshoe drain
排水用铲 draining spade
排水用的小方石 drain blockage
排水用地 drainage right of ways
排水与降水 drainage and dewatering
排水原理 principles of drainage
排水源头 drainage head
排水运河 drainage canal
排水再利用 drainage water reuse
排水闸 sluice
排水闸门 drain(age) gate; drainage sluice;outlet gate;outlet sluice
排水栅网 drainage screen
排水站 drainage station
排水障碍 blocking of drainage;obstacle to drainage
排水沼泽 fenland
排水支沟 horizontal branch; subsidiary drain
排水支管 branch drain; branch sewer; discharge branch; horizontal branch;tributary drain
排水支管沟 by-wash
排水支渠 horizontal branch
排水纸板 cardboard drain; paper drain
排水指示器 drainage indicator
排水制度 sewer system
排水制品 drainage goods; drainage product;drainage ware
排水中等的 moderately well drained
排水中心 center[centre] of displacement
排水重量 displaced weight
排水周期 drain period; period of drainage
排水主管 drainage collector
排水铸铁管 soil pipe
排水装备 drainage fitting
排水装置 discharger;drain(age) device;drain(age) system;outlet device;pumping equipment
排水状 <气垫船等的> hull-borne
排水状态水线 displacement water line
排水锥管 delivery cone
排水总管 drain main;drain manifold; main drain; main sewer; drainage-water main
排水走向 the routing of water drainage
排水阻碍 blocking of drainage
排水阻滞 retardation of discharge
排水钻孔 boring for drainage;dewatering boring;drain boring
排水作业 draining(operation)
排水作业用长柄勺 hoe dipper for drainage work
排水作业用长柄铲 hoe dipper for drainage work
排送机 exhauster
排送静压头 static discharge head
排送距离 discharge distance; discharge length
排送量 delivery capacity
排送能力 dischargeable capacity
排送损失 pumping loss
排塑 plastics removal
排他的补救 exclusive remedies
排他调度方式 exclusive dispatching facility
排他法 exclusion method
排他性的管辖条款 exclusive jurisdiction
排他性的私有财产形式 form of exclusive private property
排他性调用 exclusive call
排他性合同 exclusive contract
排他性区域规划 exclusionary zoning
排他性区域合作 closed regional cooperation
排他性条款 exclusivity clause
排他性销售协议 exclusive distribution agreement
排他性许可证 sole license[licence]
排他性许可证合同 sole license[licence] contract
排他性引用标准 exclusive reference to standards
排他原则 exclusion principle
排他运算 exclusion operation
排他属性 exclusive attribute
排题板【计】 patchboard
排烃系数 expulsive hydrocarbon coefficient
排烃效率 expulsive efficiency of hydrocarbon
排筒机构 empty case ejector mechanism
排头 crab
排土 casting;dumping
排土场 dump pit
排土场下沉系数 subsidence factor of dump
排土改良 improvement by soil removing
排土犁 draw plough;dumping plough
排土量 discharge earth volume
排土率 mucking ratio
排土桥 conveyor bridge;overburden
排土桩 displacement pile
排外主义 exclusionism;xenophobia
排位泵 displacement pump
排位次序 rank order
排位置字符 layout character
排污 bleed; blowdown <锅炉的>; blow through; exhaust pollution; pollution discharge;unloading
排污杯 force cup
排污泵 drain pump;dredge pump;effluent pump; sewage pump; trash pump; residual sea water pump【船】
排污泵站 sewage pumping station
排污比 discharge ratio
排污标准 effluent-quality standard; effluent standard; emission standard; pollutant discharge standard; pollution exhaust criterion; sewage drainage standard; standards for discharge of pollutant; wastewater discharge standard
排污层深度 depth of waste disposal stratum
排污层时代 age of waste disposal stratum
排污层岩性 lithologic(al) characters of waste disposal stratum
排污池 blow-off tank;sewage tank
排污船 sewage boat;sewage vessel
排污道 sewage channel;trashway
排污地下水 discharge of pollution to underground water
排污点 discharge location; discharge point;discharge site;drainage location;drainage point;effluent point; point of discharge; release point; waste discharge point;waste drainage point
排污短管 blow-off nozzle
排污阀 block down valve;blowdown valve; blow-off valve; dirt trap; drainage valve; drain valve; foul valve;sewage water valve;sludge valve
排污费(用) charges for disposing pollutants; discharge fee; drainage cost;effluent fee;pollution charges
排污负荷优化配置 optimal allocation for pollution load discharged
排污工程 drainage work;engineering of waste disposal
排污功能 drain function
排污沟 sewage drain
排污沟渠 drainage sewage canal
排污管 bleeder; blowdown (pipe); blowing tube; blow-off (pipe); drain;drainage pipe;draw-off pipe; off-take; outlet sewer; sewage drain;soil drain
排污管道 blow-off line;effluent sewerage
排污管接头 blowdown connection; blow-off connection
排污管域边界 sewershed boundary
排污管域范围 sewershed range
排污管域水文学 sewershed hydrology
排污罐 blowdown drum
排污河 recipient river
排污河道底泥 sewage river sediment
排污回收 blowdown return
排污监测 emission monitoring;monitoring of discharge
排污监测报告 discharge monitoring report
排污交易 emission trading
排污接管 blowdown connection; blow-off connection
排污井 gull(e)y trap;well for waste disposal
排污开关 scum cock
排污孔 drain hole;drain port
排污控制 effluent control; emission control
排污控制标准 emission control standard
排污控制设备 emission control equipment
排污控制总量 total discharge control of pollutant;total emission control
排污控制总量规划 total amount of pollution emission control planning
排污口 blow-off; drain; outfall; sewage outfall;sewage outlet
排污口监测 outlet monitoring
排污口隧洞 sewage outfall tunnel
排污冷却器 blowdown cooler
排污离心泵 trash-handling centrifugal pump
排污量 discharge capacity
排污量观测 water disposal observation
排污龙头 cleaning cock
排污率 discharge rate of effluent; drainage rate; emission rate; rating of blowdown
排污螺塞 drain plug
排污泥 sludge discharge
排污泥开关 sludge cock
排污泥量 emission amount of sludge
排污泥龙头 sludge cock
排污泥器 de-sludger
排污浓度 effluent concentration
排污浓度曲线 effluent concentration curve;effluent concentration history
排污浓度限度 effluent concentration limit;emission concentration limit
排污喷射器 sew(er)age ejector
排污权 emission rights
排污三通 blow-off tee
排污设备 blow-off equipment; discharger
排污设计 effluent design
排污申报登记 discharge reporting and registering
排污申报登记制度 system of reporting and registering pollutant emission
排污时间 blowing-off time
排污收费 pollution charges; effluent charges
排污收费制度 system of effluent charges
排污水 blowdown water; blow-off water;sewer;bleed water;blowout water
排污水处理 blowdown water treatment
排污水井 disposal well
排污水库 drainage sewage reservoir
排污水平 emission level
排污水渠 effluent channel
排污水软化 blowdown softening
排污水系统 drainage system; sink drainage
排污税 effluent levy;effluent tax;emission tax
排污速率常数 emission rate constant
排污通道 drain outlet
排污物 emission
排污系数 pollution discharge coefficient
排污系统 effluent system
排污系统基础设施 wastewater infrastructure
排污限度 discharge limit;effluent limitation;emission limit
排污限额 discharge ration
排污箱 blowdown tank;blown down tank
排污消减 emission mitigation
排污斜槽 trash chute
排污许可证交易 trade in pollution permit
排污许可证实行程序 procedure of permit system of pollutant discharged
排污许可证市场 market of discharge permit
排污许可证制度 permit system for discharging pollutants; permit system of pollutant discharged
排污悬浮固体浓度 effluent suspended solid concentration
排污旋塞 blow-off cock; cleaning cock;delivery cock;foam cock

排污烟缕 effluent plume
排污遥测 remote measurement of discharge
排污要求 emission requirement
排污因子 emission factor
排污源 discharge source (of wastewater);emission source
排污源采样 emission source sampling
排污闸 trash sluice
排污闸门 waste gate
排污栅 drain grating
排污支管 blowdown branch
排污指标 effluent guideline; effluent index
排污周期 blowing-off period
排污装置 blowdown apparatus;blowdown plant
排污浊空气的烟囱 foul air chimney
排污浊空气格栅 foul air grate
排污浊空气管 foul air pipe
排污浊空气孔 foul air hole;foul air opening
排污总管 outfall sewer
排污总量控制 total quantity control of pollutant discharge
排屋 terrace;terraced dwelling
排屋居住单元 terrace dwelling unit
排物井 removing shaft
排吸线圈 pumping loop
排铣 gang milling
排铣机 gang mill
排系统示意图 draining system diagram
排线 winding displacement
排线架 creel stand
排线器排幅 throw of traverse
排线装置 traversing device
排泄 dejection; egest; evacuate; voidance; digestion; excretion【生】; bypass to waste; discharging; drainage; draining <指水、气等>
排泄泵 discharge pump; drainage pump; draining pump; sewage pump
排泄槽 drain grating
排泄产物 excretory product
排泄道 effluent channel
排泄点 release point
排泄阀 bleeder valve; blow-off valve; draining valve; escape cock; outlet valve
排泄方式 discharge pattern
排泄沟 drain grating;drain way
排泄管(道) bleeder pipe; blowdown pipe; blowdown stack; discharge conduit; eduction column; eduction pipe; eductor; exhaust line; scavenge pipe; drain connection; discharge pipe; release pipe; waste conduit;waste pipe
排泄开关 petcock
排泄空气 evacuated air
排泄孔 bleeder hole;drain grating; eduction port; escape orifice; opening for drainage;outlet hole
排泄口 discharge outlet;draining point; outfall;outlet;outlet port;scavenge port;scupper <排除平屋顶或楼板积水在墙上开的>
排泄口流量 outlet discharge
排泄量 discharge rate;quantity of discharge
排泄龙头 blow-off cock
排泄门 emptying gate
排泄能力 discharge capacity
排泄器 eductor
排泄区 discharge area;drain district; region of outflow
排泄区位置 location of discharge area
排泄塞 drainage plug;drain(ing)plug
排泄时间 time for escape
排泄试验 dry test
排泄栓 blow-off cock
排泄水管 effluent discharge conduit
排泄隧洞 discharge tunnel
排泄调节器 escape regulator
排泄途径 escape route
排泄物【生】 dejection; egesta; excrement; excreta; excretion product; feces; fecula; merde; ordure; rejectamenta; excrete
排泄物监测 excretion monitoring
排泄物渗漏 faeces seepage
排泄系统 excretory system; systems excretion
排泄线 dropout line
排泄旋塞 escape cock
排泄用的 emunctory
排泄障碍 discharge obstacle
排泄周期 period of blowing
排泄浊气 vitiated expired air
排泄阻力 bleeder resistance
排卸 disassembly
排卸机 discharger
排屑 chip removal; cutting removal; escape of chip(ping)s
排屑槽 chip area;chip flute;chip room; chip space;cutting chute;flute
排屑孔径 clearance hole
排屑器 chip cleaner
排序 collate;collating;collation;ordering;sequencing
排序程序 collate program(me);collator;sequencer;sequencing routine
排序方法 ranking method
排序机 collator
排序论 scheduling theory
排序模式 sequencing model
排序判据 ranking criterion
排序偏差 ordering bias
排序算法 sort algorithm
排序文件 sort file
排序问题 sequencing problem
排序序列 collating sequence
排序语句【计】 ordering statement
排序装置 collating unit;collator
排雪槽 snow gutter
排雪沟 snow gutter
排雪机 mechanical snow plough; mechanical snow plow;snow plough
排雪具 plough
排雪犁 snow plough
排雪器 plough
排衙石 guard stone
排烟 exhaust fume; smoke control; smoke disposal; smoke emission; smoke evacuation; smoke exhaust; smoke removal
排烟舱口 smoke scuttles
排烟出口 outlet port
排烟窗 exhaust smoke window; vent window
排烟窗口 smoke scuttles
排烟道 smoke duct; smoke outlet; smoke pipe;smoke vent
排烟阀 smoke exhaust damper
排烟分隔 smoke exhaustion compartment
排烟风机 smoke discharge fan;smoke exhaust fan;smoke extract fan
排烟风扇 draft fan
排烟管(道) discharge flue; exit flue; smoke duct; smoke pipe; smoke vent;fume pipe
排烟管外管 smoke pipe casing
排烟柜 fume cupboard
排烟活门 smoke door;smoke hatch
排烟机 induced-draft fan; smoke extractor
排烟机房 smoke exhaust machine room
排烟极限 smoke limit
排烟孔道 fire vent
排烟口 exhaust port;fire vent;smoke and fire vent; smoke door; smoke extract;smoke hatch;smoke outlet; smoke vent
排烟口高度 height of smoke outlet
排烟扩散理论 dispersion theory of smoke
排烟门 smoke door
排烟密度 smoke density
排烟能力 stack capacity
排烟浓度 exhaust smoke density; smoke level
排烟浓度测量 measurement of exhaust smoke concentration
排烟喷射器 exhaust smoke ejector
排烟器 smoke exhauster;smoke ventilator
排烟腔 <舞台防火幕滑槽中的> smoke pocket
排烟热损失 waste heat loss
排烟软管 fumes disposal hose
排烟设备 fume extractor; smoke equipment
排烟设施 smoke discharge facility
排烟式折流板 draught diverter
排烟竖井 smoke shaft
排烟速度 flue gas velocity
排烟损失 flue gas loss;stack loss
排烟塔 smoke(-proof) tower
排烟塔窗 smoke tower window
排烟通道 smoke uptake; discharge flue
排烟脱氮 exhaust smoke denitrogenation
排烟脱硫 exhaust gas desulfurization; stack desulfurization
排烟温度 outlet gas temperature
排烟系统 smoke evacuation system; smoke exhaustion system
排烟消防车 smoke-eliminating car
排烟小炉 exhaust port
排烟循环器 vented circulator
排烟烟道 exit flue
排烟罩 enclosure;exhaust hood;fume hood; hood; plume trap; smoke hood;smoke jack
排烟罩挡板 smoke hood check damper
排烟罩壳 suction cap
排烟装置 fume extractor
排岩屑槽 sludge groove
排演 rehearsal
排演场 rehearsal room
排演厅 rehearsal hall;rehearsal studio
排氧装置 oxygen exhauster
排样 black layout;stock layout
排液 liquid discharge
排液槽 sump pit
排液池 draining tank
排液阀 liquid-out valve
排液分离器 drain separator
排液管 downcomer
排液灌浆泵 positive-displacement grout pump
排液接管 drainage connection
排液接头 drainage connection
排液孔 outage
排液口 leakage fluid dram
排液量 displacement
排液腔 exhaust chamber
排液线 drainage thread
排液旋塞 drain cock
排液装置 pumping equipment
排液总管 liquid header
排溢铸造 displacement casting
排印 typography
排印上的 typographic
排油 drainage; fuel expulsion; oil drain(age);oil outlet;oil vent;reservoir sweep
排油半径 drainage radius
排油泵 delivery pump;oil drain pump; oil scavenging pump
排油槽 oil scupper
排油阀 emptying valve;residual cake valve; residual slag valve; vent valve
排油方法 drainage method
排油管 drainage pipe;oil exit pipe
排油管龙头 oil drain cock
排油管套 drainage yoke
排油罐 oil drain drum
排油结构 displacement configuration
排油孔 outgate;outlet
排油口 oil drain out
排油量 oil outlet value
排油龙头 oil disposal hydrant
排油歧管 discharge manifold
排油栓 oil disposal hydrant
排油系统 emptying system; drainage system
排油烟风扇 vapo(u)r exhaust fan
排油烟罩 oil drainage hood
排油锥形筒 oil drainage cone
排油总管 oil drainage main
排淤管 scour(ing)pipe
排淤渠(道) desilting canal
排余气体 remainder of exhaust gases
排雨槽 channel stone
排雨剂 rain repellent
排雨量 rain discharge
排雨水管 rainwater sewer
排雨水渠道 storm-water channel
排(雨)水用制品 drainage article
排运 rafting
排杂辊 trash roll
排杂器壳 trash chamber
排杂型散热器 trash-resistant radiator
排渣 clinkering;deslagging;slag flow; slagging;slag-off
排渣勺 scummer
排渣阀 residual cake valve
排渣机 slag extractor
排渣及放气 drain;waste and vent
排渣冒口 runoff feeder;runoff riser
排渣器 ash gun;refuse ejector
排渣渠 <矿渣> tailrace
排渣设备 residue extraction mechanism
排渣式燃烧 slagging incineration
排渣式燃烧室 slagging combustion chamber
排渣室 refuse extraction chamber
排张拉锚具 capped anchorage
排障 fault removing
排障底板 pilot base
排障杆 pilot bar
排障宏指令 debug macroinstruction
排障架 rail guard
排障器 fender;life guard;obstruction-guard; pilot; stone sweeper; track-clearer;rail guard <铁路机车的>
排障器杆 pilot bar
排障器构架 pilot frame
排障器角铁 pilot angle
排障器脚蹬 pilot step
排障器拉条 pilot brace
排障器梁 pilot beam
排障器条 pilot bar
排障器雪犁 pilot shield
排障器支架 pilot bracket
排障器嘴 pilot nose
排障桩 pilot pile
排障装置 <机车前的> cowcatcher
排蒸汽管 exhaust steam pipe; steam discharge pipe; steam discharge tube
排中律 law of excluded middle
排钟 bells;campana

P

排柱 organ timbering;row of supports
排柱岸壁 campshed
排柱寺院 monopteral temple
排柱圆庙 monopteral temple
排柱圆屋 monopteron[复 monoptera]
排柱中填土 campshed
排砖立砌 soldier course
排砖立砌层 soldier course
排砖竖砌层 soldier course
排桩 line of piles;pile bent;pile row;piling;row of piles;soldier pile
排桩岸壁 < 两排桩中填土 > campshed
排桩丁坝 pile and waling groin;piled dike[dyke];pile groin;pile groyne
排桩填石堤 rock-filled dike[dyke];rock-filled pile dike[dyke]
排桩填石堤岸 rock-filled dike[dyke];rock-filled pile dike[dyke]
排桩透水坝 stake dam
排桩透水堤 pervious pile dike[dyke]
排桩透水丁坝 pervious pile dike[dyke]
排桩围堰 pile cofferdam
排桩栈道 pile trestle
排桩栈桥 pile trestle
排桩滞流 pile retard
排桩滞流透水坝 stake dam
排字车间 composing room
排走 venting
排阻 exclusion
排阻极限 exclusion limit
排阻色谱法 exclusion chromatography
排钻 gang drill;multidrill;multispindle drilling machine
排作业时间表 job scheduling

牌

牌匾 fa(s)cia board;tablet
牌坊 memorial archway;memorial gateway
牌盖 flapper
牌号 brand;chop;designation;grade;mark;shop sign;tally;trade mark
牌号标准化 brand standardization
牌号名称 brand name
牌号选择 brand choice
牌价 initial price;listed price;list of quotations;market price;market quotation;posted price;quotation;quotation of prices;quoted price;state price
牌架式升降机 frame hoist
牌楼 archway;decorated gateway;pailoo[pailou] < 中国 >
牌名 trade name
牌名货 branded goods;branded product
牌名政策 brand policy
牌照 license[licence](plate)
牌照板 license[licence] plate
牌照板支架 number-plate support;number stay
牌照持有人 holder of license[licence]
牌照灯 license[licence] plate lamp;number-plate lamp;number-plate light;registration mark light
牌照费 fee of permit;license fee
牌照架 license[licence] holder;license[licence] plate support
牌照税 fee for permit;license[licence] tax
牌桌 card table
牌子 brand;docket;tablet;tally;trade mark
牌子产品份额 brand share
牌子产品经理 brand manager
牌子改用模型 brand-switching model
牌子意识 brand awareness
牌子影响范围 brand franchise

哌

哌啶盐 piperidinium salt
哌嗪 piperazine

派

派出人员 expatriate personnel
派出所 local police station;police post;police substation
派代表 deputation
派道 send
派定 allot
派定泊位 allotment of berth
派定的工作 stint
派恩卡多硬木 < 一种产于印度的红褐色至纯红色的硬木 > pyinkado
派恩玛硬木 < 一种产于印度的淡红至红褐色的硬木 > pyinma
派尔默-勒斯波槽 < 下水道窨井内量水用 > Palmer-Bowlus flume
派给工作 assignment
派工簿 work order book
派工单 job order;job sheet;work card;work(ing) order;work notice;work ticket
派工卡 operation job card
派购 purchase by state quotas
派购价格 apportioned purchase price
派款 collection of compulsory contributions
派来恩方法 < 一种制造泡沫砂浆的方法 > Pyrene process
派朗海绵铁粉 Pyron iron powder
派勒克斯玻璃 < 商标名称 > Pyrex
派勒克斯玻璃冷却套 Pyrex cooling jacket
派勒克斯耐热玻璃 Pyrex glass
派洛克 < 一种轻质砂浆,可用喷枪喷涂 > Pyroc
派洛迈克 < 一种镍铬耐热合金 > pyromic
派洛石膏混凝土 Pyrofill gypsum concrete
派洛斯 < 一种耐热镍合金 > Pyros
派尼大梁 Peine girder
派尼钢板桩 Peine(sheet)pile
派尼截面 Peine section
派遣 despatch[dispatch];detach;send out
派遣的代表团 contingent
派遣费 dispatching money
派遣国 accrediting state
派遣台 dispatch station
派遣中国专家 sending of Chinese specialist abroad
派人看守平交道口 staffing of level crossing
派任项目经理 appointed project manager
派森黄铜 Parson's brass
派生 derivation;derive;spin-off;variant
派生变量 concomitant variable
派生标引 derivative indexing
派生波 derived wave
派生潮 derived tide
派生词 derivative word
派生的 derivative;parasitic(al)
派生的统计数字 derived statistics
派生地图 derivative map;interpretive map
派生地下水 allochthonic groundwater
派生放射性同位素 daughter radioisotope
派生构造 derivative structure
派生价值 derived value
派生结构 derivative structure
派生利息 derivative interest
派生码 generated code
派生收益 derivative revenue
派生数据 derived data
派生所得 derivation income
派生文献 derivative document
派生系列 subseries
派生小侵入体内外接触 inner and outer contact of associated small intrusion
派生形式 derivative form;subform
派生需求 derivative demand;derived demand
派生指数 derivative index
派西菲克型直接成条机 Pacific converter
派息率 dividend payout ratio
派息日 < 股票 > date payable
派系 clique;coterie
派形板 double slab

蒎

蒎烯 pinene

潘

潘得罗弹簧扣件 Pandrol flexible fastenings
潘得罗扣件 Pandrol fastenings
潘恩木材防火法 Payne's process
潘姜港 < 印度尼西亚 > Port Panjiang
潘马氏(闭杯法)闪火点 < 试验轻质油类及沥青材料用 > flash point of Pensky-Martens
潘马氏闪火点 Presky-Martens flash-point
潘宁放电 < 磁场内的电弧放电 > penning discharge;pig discharge
潘帕草原 pampa
潘塔尔铝合金 Pantal alloy

攀

攀比速度 pursuitting speed to match others
攀登 clamber;climb(ing);scale
攀登道岔 climbing switch
攀登杆 climbing rod
攀登架 climbing apparatus;climbing frame
攀登空间 climbing space
攀登铁爬梯 hand iron
攀登柱 climbing shaft
攀钩 gaff
攀墙植物 wall plant
攀升式潮后周期层序 climbing tidal-cyclic(al) sequence
攀升式层序 climbing sequence
攀索 guess rope;guest rope;guess warp;guest warp
攀踏结 harness hitch
攀藤拱架 trellis(ed) arch
攀梯 < 坡度大于 75° > rung ladder
攀岩技术 rock climbing skill
攀移 climbing
攀移过程 climbing process
攀移式起重机 climbing crane
攀移运动 climb motion
攀缘灌木 climbing shrub;scandent shrub
攀缘茎 winding stem
攀缘能力 climbing capacity
攀缘蔷薇 rambler rose;climbing rose
攀缘式起重机 climbing crane
攀缘一年生植物 climbing therophyte
攀缘植物 climber;climbing plant
攀折花木 pick flowers

盘

盘斑灰泥岩 dismicrite
盘草菌属 < 拉 > Aleurodiscus
盘长孢属 < 拉 > Gloeosporium
盘肠构造 enterolithic structure
盘肠管 pipe coil
盘肠状结构的【地】 enterolithic
盘肠状褶皱的【地】 enterolithic
盘车 barring;jigger;jolley;hand-operated rotation
盘车电动机 barring motor
盘车机 jacking engine;turning engine;turning gear
盘车孔 barring hole
盘车拉坯 < 制陶器 > jigger(ing)
盘车装置 barring device;barring gear;rolling gear;rotor turning gear;shaft turning gear;spin gearing;turning gear
盘秤 hang scoop scale;pan scale
盘齿式镇压器 disk spider roller
盘存 inventory survey;inventory taking;stock taking;take stock;taking inventory
盘存报表 inventory sheet
盘存标签 inventory tag
盘存表 inventory table
盘存成本 cost of inventory
盘存等式 inventory equation
盘存点货通知 circular of stock-taking
盘存点料单 inventory count slip
盘存短耗 inventory shortage
盘存法 inventory method
盘存方法对照 inventory methods contrasted
盘存公式 inventory equation
盘存估价准备 reserve to reduce inventory value to market
盘存管理 inventory control
盘存会计 accounting for inventories
盘存货(物) goods on hand;inventory stock
盘存计价折旧法 inventory depreciation method
盘存计数单 tally sheet
盘存计数卡 tally card
盘存记录 inventory record
盘存截止凭单 inventory cut-off voucher
盘存截止期 inventory cut-off date
盘存清单 inventory count slip;inventory list
盘存缺溢 inventory short and over
盘存日期 inventory date
盘存损益 inventory short and over
盘存通知 circular of stock-taking
盘存盈余 inventory profit
盘存账户 inventory account
盘存折旧法 depreciation-inventory method
盘存制度 inventory system
盘存准备金 inventory reserve
盘单毛孢 < 拉 > Monochaetia
盘刀式切布机 rag cutter with disc knives
盘刀式切碎器 radial-knife cutterhead
盘道 winding road
盘底桩 disc[disk] pile
盘点 inventory;pipe coil;stock taking
盘点存料单 inventory count slip
盘点单 inventory sheet
盘电泳 disc electrophoresis
盘碟底部不稳定检测器 whirler
盘碟(类) flatware
盘东【测】 level east
盘动法 spooling method
盘动装置 shaft turning gear
盘多毛孢属 < 拉 > Pestalotia
盘二孢属 < 拉 > Marssonina
盘阀 disc valve
盘阀捞砂筒 disk valve bailer
盘法炭黑 disc block
盘封 disk seal;loop seal
盘封管 disk-seal tube;lighthouse tube;megatron
盘封三极管 lighthouse triode

盘杆阀 poppet valve
盘钢筋 bundled bar
盘革耳属 <拉> Eichleriella
盘根 gland packing;packing
盘根槽 packing chamber; packing groove
盘根盒 packing gland
盘根盒式防喷器 blowout preventer of stuffing box type
盘根盒套管头 stuffing box casing head
盘根盒压套 stuffing box gland
盘根压紧垫圈 packing washer
盘谷 central;zungenbecken
盘鼓形汽轮机 combination turbine; disk-and-drum turbine
盘管 coiled pipe;pipe coil;serpentine pipe; spiral coil; spiral pipe; spiral tube; tube coil; worm conduit; worm pipe; worm piping; worm tube;worm tubing;pigtail
盘管泵 coil pump
盘管壁 coil wall
盘管长度 coil length
盘管虫属 Hydroides
盘管抽头 coil tap
盘管的热器 coil heat exchanger
盘管地位 coil position
盘管盖板 coil deck
盘管高度 coil height
盘管供暖 coil heating
盘管供热 coil heating
盘管锅炉 coil boiler
盘管过滤介质 spiral-coil filter medium
盘管过滤器 coil filter;spiral-coil filter
盘管簧机 coiling machine
盘管机 coiler
盘管加热 coil heating
盘管加热器 heater coil
盘管加热装置 heater coil
盘管节距 coil pitch
盘管进入接头 coil in
盘管空间 coil space
盘管冷凝器 coil(ed) condenser
盘管冷却 coil cooling
盘管冷却器 coil(ed)(pipe) cooler; serpentine cooler
盘管喷涂 coil coating
盘管器 coil former
盘管取暖板 <埋入墙内的> heating panel
盘管散热器 coiled radiator
盘管设计 coil design
盘管深度 coil depth
盘管省煤器 continuous loop(-type) economizer;loop economizer
盘管式冰 ice-on-coil
盘管式加热器 coil heater;spiral heater
盘管式散热器 coiled radiator
盘管水冷水箱 water tank with cooling coil
盘管涂刷 coil coating
盘管引出接头 coil-out
盘管迎风面积 coil face area
盘管迂回运行循环 coil run around cycle
盘管装置 coil installation
盘管状态曲线 coil condition curve
盘管组件 coil pack
盘辊磨 bar race mill;peters mill
盘号 reel number
盘荷波导 diaphragmatic waveguide
盘黑 disc[disk] black
盘后布线 back-of-board wiring;back-of-panel wiring
盘后接线 back connection; rear connection
盘后接线图 back connection diagram; rear connection diagram
盘后配线 back-of-board wiring
盘花 disc[disk] floret
盘花群类 Disciflorae
盘环形折流板 disk-doughnut baffle
盘簧 coil(ed)(mechanical) spring; disc spring; recoil spring; spiral spring
盘簧凹形弹簧片 dish spring
盘簧承座 coil spring retainer
盘簧底圈 coil spring retainer
盘簧隔振器 coil spring type vibration isolator
盘簧机 helical spring
盘簧减震器 spring damper coil
盘簧悬置 spiral spring suspension
盘簧制动器 coil brake
盘回皱纹【地】crumpling
盘或排列锚链 range the cable
盘货 stock taking;take stock
盘机孔 barring hole
盘架 plate rack;reel;tray frame;deck truss
盘架干燥 shelf drying
盘架干燥炉 shelf drier[dryer]; tray drier[dryer]
盘架干燥器 shelf drier[dryer]; tray drier[dryer]
盘架炉 shelf burner; shelf furnace; shelf kiln;shelf oven;tray burner
盘架燃烧 shelf burning
盘脚桩 disc-footed pile; disc[disk] pile;footed pile
盘径 disk track
盘锯 circular saw;disk saw
盘卷 coil down;pony roll
盘卷货 coil cargo
盘卷机 curler
盘卷绳索 coil rope
盘刻 facet(te);panel-cut
盘口面 fa(s)cia[复 fa(s)ciae/fa(s)cias]
盘库 make an inventory of goods in a warehouse
盘类制品 shallow article
盘类自动成型线 automatic plate-forming line
盘里炒干的 pan-dried
盘联轴器 disk coupling
盘料 preform
盘料坯 stock
盘滤机 disc[disk] filter
盘轮 disc[disk] wheel; plate wheel; straight-type wheel
盘轮式积分器 disc[disk]-and-wheel integrator
盘面 panel
盘面比 disc[disk] ratio
盘面接线 live front
盘面接线式配电盘 live front switchboard
盘面设计 panel layout
盘模 pan form(work)
盘模楼梯 pan form stair(case)
盘模踏步 pan steps
盘磨 disc[disk] mill;disc[disk] attrition mill;pan grinder
盘磨机 buhr(stone) mill; disc[disk] grinder;disc[disk] refiner
盘木 <船坞用> keel block
盘木顶块 cap piece of keel block
盘木及侧盘木 keel-and-bilge block
盘木台 keel block tier
盘木线脚 brace mo(u)lding
盘泥条成型 coiling
盘碾 pan grinder
盘碾砂机 pan mill
盘前接线 disc[disk] front connection
盘区 panel
盘区边界 panel boundary
盘区车场 panel landing
盘区开采 panel(l)ing
盘区开采年限【矿】panel life
盘曲管 coil(pipe);sinuous coil
盘曲皱纹 crumpling
盘圈 coil
盘绕 circumvolution; coil; convolution; tortile; twine; voluminosity; voluminous
盘绕的 tortile
盘绕法 coil method
盘绕加工 swirl finish
盘绕楼梯 wreathed stair(case)
盘绕式水泵 coil pump
盘绕弹簧 coil mechanical spring
盘绕线圈式灯丝 coiled-coil filament
盘绕柱身的饰带 frilals for wreathed column
盘塞 disc[disk] plug
盘筛 disc[disk] screen;riddle
盘山道路 lacet;lacet road
盘山路 switchback
盘山线 winding line
盘山展线 spiral track
盘扇 disc[disk] fan
盘上安装 panel mounting
盘声记录 canned;canned record
盘绳机 coiling machine
盘绳栓 kevel
盘式拌和机 mixer pan; pan mixer; open-top mixer
盘式避雷器 disc[disk] lightning arrester
盘式便器 pan closet
盘式采样器 pan-type sampler; tray-type sampler
盘式车架 tray-type frame
盘式车架结构 tray frame construction
盘式车轮 disc[disk] wheel
盘式成球机 disc[disk] granulator;disc[dish] type nodulizer
盘式除氧器 tray shape deaerator
盘式穿孔机 disc[disk] piercer
盘式磁选机 disc[disk]-type magnetic separator;tray separator
盘式存储器 jukebox storage
盘式打磨 disc-sanding
盘式打磨机 disc[disk]-type cutter
盘式刀架 <铣刀杆上的> arbor flange
盘式的 reel-to-reel
盘式灯头 disc[disk] base
盘式底沙采样器 pan-type bed-load sampler
盘式电泳法 disc[disk] electrophoresis
盘式定位附着装置 <用以固定桥梁测试仪器> locating disc[disk] attachment system
盘式发电机 disc[disk] generator
盘式发动机 disc[disk] engine
盘式阀 disc[disk] valve
盘式放电器 disc[disk] discharger
盘式飞轮 disc[disk] flywheel; web-type flywheel
盘式分离机 disc[disk] separator
盘式分选器 tray-type separator
盘式粉碎机 pan breaker;pan crusher
盘式干燥 tray drying
盘式干燥机 disc[disk] drier[dryer]
盘式干燥器 pan drier[dryer]; tray drier[dryer]
盘式格栅 disc[disk] screen;pan grid
盘式格栅地板 pan joist floor
盘式给矿机 pan feeder;pan ore feeder
盘式给料机 cradle feeder;feed table; pan feeder;revolving plate feeder
盘式给料机构 disc[disk] feeder
盘式给料器 feed table;pan feeder;revolving plate feeder
盘式辊磨机 bowl-type mill
盘式过滤机 disc[disk] filter;pan filter
盘式过滤器 disc[disk] filter;pan filter
盘式河底推移质采样器 tray-type bed-load sampler
盘式河底质采样器 pan-type bed-material sampler
盘式河砂采样器 pan-type sampler
盘式河砂取样器 pan-type sampler
盘式烘干机 pan drier[dryer]
盘式烘干器 pan drier[dryer]
盘式烘燥机 tray drier[dryer]
盘式混合机 horizontal pan mixer; pan-type mixer
盘式混合器 pan mixer
盘式混料机 mixing pan; pan-type mixer
盘式混凝土搅拌机 pan-type concrete mixer
盘式激动器 disc[disk] impeller
盘式集装箱 panel container
盘式计算机 disc[disk] calculator
盘式记录器 disc[disk] recorder
盘式加料机 disc[disk] feeder;revolving plate feeder
盘式加料器 disc[disk] feeder
盘式加湿器 pan(-type) humidifier
盘式减压阀 disk-type reducing valve
盘式溅水器 dish sprayer
盘式搅拌机 bowl mixer; disc[disk] mixer;pan mixer
盘式截煤机 disc[disk] coal cutter
盘式井底封隔器 disc[disk] bottom hole packer
盘式绝缘子 disc[disk] insulator
盘式绝缘子半自动成型机 semi-automatic machine for disc[disk] insulator forming
盘式开关 dial switch;wafer switch
盘式快门 disc[disk] shutter
盘式扩散器 disc[disk] diffuser
盘式冷却器 flange cooler
盘式离合器 disc[disk] clutch; plate clutch
盘式离心法 disc[disk] centrifugal process
盘式离心机 disc[disk] centrifuge
盘式联轴节 disc[disk] coupling
盘式淋灰机 tray-type lime watering treater
盘式流量计 disc[disk] flowmeter;disc[disk] meter; nutating-disc[disk] flowmeter
盘式楼梯 pan-type stair(case)
盘式滤池 disc[disk] filter
盘式轮 web wheel
盘式螺旋钻 disc[disk] auger
盘式模板 pan form(work);pan masterplate
盘式模型 tray mo(u)ld
盘式摩擦离合器 disk-type friction clutch
盘式磨粉机 disc[disk] mill
盘式磨耗(测定)法 abrasion disc[disk] method
盘式磨矿机 pan grinder
盘式黏[粘]度计 pan visco(si)meter; disc[disk] visco(si)meter
盘式碾矿机 pan grinder
盘式碾磨机 rim discharge mill
盘式碾碎搅拌机 mortar mill
盘式捏练机 tempering pan
盘式频闪观察仪 disc[disk] stroboscope
盘式平面磨床 disc[disk] grinding machine
盘式破碎机 disc[disk] crusher
盘式曝气器 dish sprayer;tray aerator
盘式气体洗涤器 tray-type gas scrubber
盘式切片机 disc[disk] slicer
盘式球机 disk-type nodulizer
盘式绕组 disc[disk] winding; pie

winding; sandwich winding
盘式色度计 disc[disk] colo(u)rimeter
盘式色谱 disc[disk] chromatography
盘式砂光机 dish sander
盘式筛 disc[disk] screen
盘式筛砂机 tray sand screen
盘式烧结机 Greenawalt type sintering machine
盘式输送机 disc[disk] conveyer[conveyor]; pan conveyer[conveyor]; tray conveyer[conveyor]
盘式数片机 disc[disk] tablet counter
盘式水表 disc[disk] water meter; disc[disk] meter
盘式水上浮油回收装置 disc[disk] skimmer
盘式水银传感温度计 dial mercury sensing thermometer
盘式送风口 pan outlet
盘式送料机 table feeder
盘式塔 tray column; tray tower
盘式踏板 pan-type tread
盘式弹簧秤 spring scale with stabilized pan
盘式跳汰机 pan jig
盘式推移质采样器 pan-type bed-load sampler; tray-type bed-load sampler
盘式喂料机 cradle feeder; disc[disk] feeder; feed disc[disk]; rotary disc[disk] feeder; table feeder
盘式喂料器 cradle feeder; feed disc[disk]; rotary disc[disk] feeder; table feeder
盘式洗涤器 disc[disk] washer; tray scrubber
盘式铣刀 disc[disk] mill
盘式线圈 sandwich wound coil
盘式卸料篦子 dish type discharge grate
盘式卸载车 pan-type car; pan-type vehicle
盘式蓄电池 tray accumulator
盘式压碎机 disc[disk] crusher
盘式研磨机 disc[disk] grinder
盘式叶轮搅拌机 vaned disc[disk] agitator
盘式叶片风扇 disc[disk] fan
盘式仪表 disc[disk] meter
盘式凿岩机 plate-shaped drill
盘式造粒机 pan-type pelletizer
盘式造球混料机 <烧结机,二次混料机的一种> disc[disk] pelletizer mixer
盘式增湿器 pan-type humidifier
盘式轧碎机 disc[disk] breaker; disc[disk] crusher
盘式真空过滤机 disc[disk] type suction filter
盘式真空过滤器 disc[disk] type suction filter
盘式振捣器 disc[disk] vibrator
盘式蒸发 pan evapo(u)ration
盘式蒸发计 pan evapo(u)rimeter
盘式蒸发仪 pan evapo(u)rimeter
盘式蒸汽压力温度计 dial vapo(u)r-pressure thermometer
盘式制动器 caliper(disc) brake; disc[disk] brake
盘式制动运输机 disk(-type) retarder
盘式制粒机 pan pelletizer
盘式转子 disc[disk] rotor
盘式钻头 rotary disc[disk] bit
盘首纹和盘尾纹 spiral lead-in and spiral throw-out
盘水车轮 disc[disk] type wheel
盘丝 parcel; wire coil
盘丝车床 scroll lathe
盘斯里铝硅合金 Panseri alloy
盘索 bight; fake
盘梯 screw stair(case); spiral stair(case); corkscrew stair(case); dancing winders; helical stair(case); round ladder; stair winder; winder
盘梯级 scroll steps
盘条 bull rod; coil rod; coil stock; pencil rod; rod bundle; rod coil; rod iron; rolled steel wire; roll of wire; wire bar; wire coil; wire rod; rolling steel rod <卷成盘状的细钢筋>
盘条打捆机 revolving bundle holder
盘条翻转台 tilting chair
盘条挂送装置 bundle buster
盘条紧捆机 rod coil compressor
盘条卷取导管 reel pipe; run pipe
盘条收集机 capstan
盘条卸卷机 gallows
盘条形填料 coil packing
盘条轧机 rod mill
盘条轧制控制冷却法 Stelmor process
盘铜 rosette copper
盘头螺钉 pan head screw
盘头螺栓 pan head bolt
盘头铆钉 pan head rivet
盘头桩 disc[disk] pile
盘头锥颈铆钉 swell-neck pan head rivet
盘托 pan arrest
盘陀道路 winding road
盘弯头 flange bend
盘碗壁架 plate rail
盘碗架 plate rack
盘碗离心机 disk-bowl centrifuge
盘尾纹 lead-out groove; throw-out spiral
盘文件 disc[disk] file
盘问 challenge; cross-examination; cross question
盘涡形 volute
盘西【测】 level west
盘隙发射机 disc[disk] gap transmitter
盘香管 pancake coil; tube coil
盘香管灯 scroll lamp
盘香管容器 scroll tank
盘香式芯撑 radiator chaplet
盘香状热子 coiled-coil heater
盘形凹地 pan
盘形瓣膜 discoidal valve
盘形避雷器 disc[disk] type lightning arrester
盘形插齿刀 disc[disk] slotting cutter
盘形车轮 disc[disk] wheel
盘形衬片 disc[disk] facing
盘形齿轮 bevel gear; disc[disk] gear
盘形齿轮刀具 disc[disk] gear cutter
盘形齿轮铣刀 disc[disk]-type gear milling cutter
盘形猝熄火花放电器 lepel quenched spark-gap
盘形挡板 dish baffle
盘形刀具 disc[disk] cutter
盘形刀片 desk blade
盘形的 discoid; dished; disked
盘形底 dished end
盘形电枢 disc[disk] armature
盘形垫圈 dished washer
盘形顶棚 tray ceiling
盘形阀 disc[disk] valve; plate valve
盘形飞轮 disc[disk] flywheel
盘形封接管 direct-seal valve
盘形封口的 disc[disk] seal
盘形管膨胀补偿器 expansion loop
盘形辊 disk-shaped roll
盘形壶 dish-shaped ewer
盘形混凝土块 pancake concrete
盘形活塞 disc[disk] piston
盘形活塞鼓风机 disc[disk] piston blower
盘形继电器 disc[disk](type) relay
盘形浇口 diaphragm grate
盘形铰刀 block-type reamer; disk-type reamer
盘形搅拌机 pan-type mixer
盘形节风器 pan register
盘形结构 pan construction
盘形井壁封隔器 disk-wall packer
盘形孔板 plate orifice
盘形离合器 disc[disk] clutch
盘形联轴节 disc[disk] coupling
盘形滤池 disc[disk] filter
盘形轮 disc[disk] wheel; straight wheel
盘形螺母 disc[disk] nut
盘形模板 pan form
盘形磨擦轮 disc[disk] friction wheel
盘形腔 disc-shaped cavity
盘形曲柄 wheel crank
盘形曲拐 disc[disk] crank
盘形绕组 disc[disk] winding; pancake winding; pie; sandwich winding; slab winding
盘形砂轮 disc[disk] wheel
盘形筛 disc[disk] sieve
盘形水量计 disc[disk] water meter
盘形送风口 pan-type air outlet
盘形弹簧 Belleville spring; cup spring; disc[disk] spring
盘形剃齿刀 disc[disk] shaving cutter; rotary gear shaving cutter; rotary type cutter
盘形天线 disc[disk] antenna
盘形头 pan head
盘形头螺(丝)钉 pan head screw
盘形头螺栓 pan head bolt
盘形头铆钉 pan head rivet
盘形凸轮 disc[disk] cam; periphery cam
盘形挖器 discoid excavator
盘形外浇口 pouring dish
盘形物 dish
盘形铣齿刀 disc[disk] gear cutter
盘形铣刀 disc[disk] cutter; disc[disk] milling cutter
盘形线 slab line
盘形线圈 disc[disk] coil; pancake coil; slab coil
盘形穴 pan head hole; panhole
盘形阳极 anode disc[disk]
盘形叶轮 disc[disk] rotor
盘形仪表 panel type instrument
盘形灶面板 dished hotplate
盘形增湿器 disc[disk] humidifier
盘形振动器 shaker pan
盘形制动 disc[disk] brake; rotor brake
盘形制动机 disc[disk] brake
盘形制动器 disc[disk] brake
盘形转子 disk-type rotor
盘形风栅 pan grille
盘形砂轮 saucer wheel
盘旋 convolution; convolve; nutation; serpentuate; spiral; spiry
盘旋场 <雷达> nutation field
盘旋抽样 zigzag sampling
盘旋道路 twisty road
盘旋的 convolute; serpentine
盘旋飞行 spiral flight
盘旋扶梯 newel stair(case)
盘旋公路 serpentine highway
盘旋共轴电热器 coiled coaxial electric(al) heater
盘旋箍筋 wire spiral
盘旋管 pipe coil; sinuous coil
盘旋角 nutation angle
盘旋阶梯 scroll steps
盘旋进场区 circling approach area
盘旋馈电 nutating feed
盘旋馈入天线 nutating antenna
盘旋馈入装置 nutator
盘旋力矩 yawing moment
盘旋楼梯 circular stair(case); cockle stair(case); corkscrew stair(case); elliptic(al) stair(case); geometric(al) stair(case); spiral stair(case); winding stair(case)
盘旋楼梯斜梁 wreathed stringer
盘旋路 loop line
盘旋面 convolute
盘旋排列 serpentining
盘旋膨胀管 coiled expansion pipe
盘旋气泡 serpiginous hollow
盘旋砂眼 serpiginous hollow
盘旋式楼梯 screw stair(case); winding stair(case)
盘旋式斜坡道 helicline; spiral ramp
盘旋梯 screw stair(case)
盘旋梯级 scroll steps
盘旋梯阶 scroll steps
盘旋线 zigzag
盘旋形 convolute
盘旋形摆动拉幅机 serpentine jigging tenter
盘旋形的 convolute
盘旋引导灯 <机场> circling guidance light
盘旋展线法 zigzag development
盘旋柱 acclivous column; coil column; winding stair(case)
盘窑 kiln jacking
盘用电压表 panel type voltmeter
盘右【测】 face right; circle right; face right position (of telescope); inverted position of telescope; reversed position of telescope; right circle
盘圆 wire rod; mill coil <细钢筋或钢丝>
盘圆钢 coil steel
盘噪声电平 rumble level
盘账 audit account; check account; examine the account
盘针孢属 <拉> Libertella
盘枝叶饰柱 corollithic column
盘制石膏 gypsum heated in pan
盘珠 pan bead
盘珠饰 fusorole
盘筑泥条 coiling
盘转磨机 grinding pan
盘装壳孢属 <拉> Discella
盘状 circular; discoidal; disc[disk]
盘状半月板 discoid meniscus
盘状拌和机 pan mixer
盘状冰碛石 discoid morine stone
盘状的 discal; disciform; discoid
盘状电极 circular electrode; welding wheel
盘状电泳 disc[disk] electrophoresis
盘状阀 plate valve
盘状感应指示器 puck cursor
盘状构造 dish structure
盘状灌溉 cup irrigation
盘状核粒【地】 discolith
盘状花的 discifloral
盘状花托 cotyloid receptacle
盘状假星象 spurious disc[disk]
盘状结构 disklike structure
盘状结核 discoidal nodule
盘状模制品 biscuit
盘状闪烁体 scintillating disc[disk]
盘状弹簧 saucer spring
盘锥形天线 discone antenna
盘桌 tray-top table
盘着的导线 bundled conductor
盘子 dish; tray
盘子搁架 tray rack
盘左【测】 circle left; direct position of telescope; face left position (of telescope); left circle; normal position of telescope

磐斑灰泥岩 dismicrite

磐石 block of stone;monolith
磐梯岩 bandaite

蹒跚 lurch;stagger

蟠花柱 wreathed column

蟠龙柱 dragon-wreathed column

判辩法 interpretation technique

判标 award
判标纪要 process-verbal of awarding
判别 differentiate;discrimination
判别比 discrimination ratio
判别边界 boundary distribution;decision boundary
判别变量数 amount of discriminatory variant
判别参数 discriminant parameter
判别得分 discriminatory score
判别方程式 discriminant equation
判别分界值 discriminatory critical value
判别分析 discriminatory analysis
判别分析法 discriminant analysis; techniques of discriminant analysis
判别分析模型 discriminate analysis model
判别符号 distinguished symbol
判别归组 grouping of discriminant
判别函数 criterion function;discriminating[discrimination] function
判别计量 discriminating score
判别路径表 criterion path scheduling
判别器 arbiter;discriminator
判别区域 critical region
判别式 criterion;evaluation formula; discriminant
判别式分析 discriminant analysis
判别式函数 discriminant function
判别试验 crucial test
判别条件 criterion
判别推理方法 decision-theoretic approach
判别显现 distinguish
判别序列 distinguished sequence;distinguishing sequence
判别阈 discriminant threshold
判别值 discriminant score; discriminating value
判别指标 discriminatory index
判别指令 decision instruction; discrimination instruction;discriminating order
判层深度 depth of seam decision
判定 decision-making;determine;discrimination;finding;judge;judg(e)ment;predicate
判定边界 decision boundary
判定标准 standard of criterion[复 criteria]
判定表 critical table;decision table
判定表编译程序 decision table compiler
判定表处理程序 decision table processor
判定表逻辑 decision table logic
判定表语言 decision table language
判定表预处理程序 decision table preprocessor
判定表转换 decision table conversion
判定程序 decision procedure
判定抽样 judg(e)ment sample
判定电路 decision element; decision(-making) circuit;voting circuit
判定法 criterion;decision method
判定反馈 decision feedback
判定反馈系统 decision feedback system
判定方案 decision plan;decision schedule;decision scheme
判定方法 decision method; decision procedure
判定规则 decision rule
判定过程 decision procedure; decision process
判定函数 critical function; decision function
判定机 determinate machine
判定机构 decision mechanism
判定积分器 decision integrator
判定计算 decision accounting
判定精度 precision of definition
判定空间 decision space
判定块 decision block
判定框图 decision box
判定(理)论 decision theory
判定逻辑电路 decision logic circuitry
判定面 decision surface
判定模型 decision model
判定模型库 decision model base
判定树 decision tree
判定瞬间 decision instant
判定算法 decision algorithm
判定通路 decision path
判定图 critical diagram
判定土体承载力用触探仪 soil assessment cone penetrometer
判定网格表 decision grid chart
判定系数 coefficient of determination
判定系统 decision-making system
判定性的 deterministic
判定性检索 deterministic retrieval
判定验证 decision verification
判定液位 fluid level determination
判定应付款项 judg(e)ment payable
判定域 decision space
判定阈 decision thresholding
判定元件 decision element
判定元素 decision element
判定指令 decision instruction; discriminating order;discrimination instruction
判定装置 decision-maker
判定准则 standard of criterion
判读 interpretation;reading
判读标志 interpretation key;photokey
判读法 method of interpretation
判读方法 interpretation method
判读技术 interpretation technique
判读略图 identification map sketch
判读器 decipherer;decoder
判读时间 legibility time
判读误差 reading error
判读性 interpretability
判读样片 interpretation key;photokey
判读仪 interpretoscope
判读员 interpreter
判读(照)片 identified photograph; interpreted photograph
判读组 interpretation group;interpretation unit
判断 decide; diagnosis; discretion; judg(e)ment
判断标识信号 identification burst
判断标准 criterion
判断抽选 judg(e)ment selection
判断抽样 judg(e)ment sampling
判断错误 error of judg(e)ment;misjudge;misjudg(e)ment
判断错误的概率 probability of misjudg(e)ment
判断概率 judg(e)mental probability
判断公正 equilibrium[复 equilibria/equilibriums] fairness
判断过程 deterministic process
判断函数 discriminant function
判断合格值 acceptance value
判断力下降 decline in judg(e)ment
判断失当的 ill-judged
判断时间 judg(e)ment time
判断市场机会 assess market opportunities
判断误差 error in judg(e)ment
判断样本 judg(e)ment sample
判断预测 judg(e)ment forecast
判断指令 decision instruction
判给 award
判据 criterion;standard of criterion
判决 adjudication; decree; deliverance; finding; judg(e)ment; sentence
判决产权归属 quiet title
判决的豁免 immunity from judg(e)ment
判决的执行 enforcement of judg(e)ments
判决抵押 judg(e)ment lien
判决方案 decision scheme
判决费率 judg(e)ment rate
判决费用 cost of decisions
判决规则 decision rule
判决过程 decision process
判决记录 judg(e)ment record
判决空间 decision space
判决理论 judg(e)ment theory
判决留置财产 judg(e)ment lien
判决留置权 judg(e)ment lien
判决没收 condemnation
判决某人败诉 decide against somebody
判决某人胜诉 decide in favor of somebody
判决确定的费率 judg(e)ment rate
判决确定的债务 judg(e)ment debt
判决确定债权人 judg(e)ment creditor
判决确定债务人 judg(e)ment debtor
判决试验 crucial test
判决收益人 judg(e)ment creditor
判决书 letter of award
判决通货条款 judg(e)ment currency clause
判决无效 nullity of judg(e)ment
判决无罪 acquittal
判决性试验 decisive experiment
判决应付款项 judg(e)ment payable
判决应收款项 judg(e)ment receivable
判决摘要 abstract of judg(e)ment
判决值 decision value
判空函数 empty function
判例 jurisprudence;precedent;prejudication
判例案件 test case
判例法 case law;judge-made law
判识地质过程数 identify processes number
判释法 method of interpretation
判释方法 interpretation technique
判小蠹属 <拉> Scolytoplatypus
判译 interpretation
判优程序 arbiter
判优法 arbitration;method for arbitration
判优逻辑 arbitration logic
判优速度 arbiter speed
判域 critical region

滂胺偶氮猩红 pontamine diazo scarlet

滂胺染料 pontamine colo(u)r
滂铬 pontachrome
滂阡树脂 pontianak
滂阡树脂胶 pontianak gum
滂梭红 ponsol red
滂梭黄 ponsol yellow
滂梭染料 ponsol dye

庞贝的潘萨住宅 house of Pansa at Pompeil

庞贝坟墓街 Street of the Tombs at Pompeii
庞贝红 Pompeian red
庞贝建筑 Pompeian architecture
庞大根系 heavy root system
庞大计划 vast scheme
庞大建筑物 megastructure
庞加莱猜想 Poincare conjecture
庞加莱根数 Poincare elements
庞加莱群【地】Poincare group
庞加莱椭球体 Poincare spheroid
庞加利法郎 Poincare franc
庞加偏振球 Poincare sphere
庞然大物 colossus;jumbo;mammoth
庞式桁架 Pong truss
庞斯莱水轮机 <弯曲轮叶无冲击的下冲式水轮机> Poncelet wheel
庞斯列图解法 <土压力> Poncelet graphical construction
庞特里雅金极大原理 Pontryagin's maximum principle
庞特斯福德统【地】Pontesfordian Series
庞兹劳炻器 Bunzlau ware

旁暗阴极加热器 dark heater

旁板 side plate;side plating
旁瓣 minor lobe; secondary lobe; sideband;side lobe
旁瓣电平 sidelobe level
旁瓣对消 sidelobe cancellation
旁瓣回波 sidelobe echo
旁瓣消隐 sidelobe blanking
旁瓣抑制 sidelobe suppression
旁边给水的尾钎 shank rod for separate flushing
旁边透气管 side vent
旁波瓣 minor lobe; secondary lobe; side lobe
旁侧导坑法 <隧道开挖的> core-leaving method; corduroy-leaving method
旁侧地界线 side property line
旁侧供水凿岩机 external water-feed machine
旁侧供水装置 side water supplying device
旁侧给水冲洗 separate flushing
旁侧间距 side spacing
旁侧建筑翼部入口 side transeptal portal
旁侧教堂十字形耳堂门 side transeptal portal
旁侧进口人孔 side entrance manhole
旁侧进水口 side intake
旁侧开式舱盖 side opening hatchcover
旁侧控制极 side gate
旁侧来水 lateral inflow;side inflow
旁侧礼拜堂 lateral chapel
旁侧片 adfrontal
旁侧入流 lateral inflow;side inflow
旁侧(扫描)声呐 side scan sonar
旁侧(扫描)声呐勘测 side scan sonar prospecting
旁侧(扫描)声呐勘探 side scan sonar prospecting
旁侧(扫描)声呐探测 side scan sonar

sounding
旁侧水道 side channel
旁侧(土地)区划线 side plot line;side layout line
旁侧延拓 lateral continuation
旁侧溢洪道 by(e) channel; lateral spillway channel;side spillway
旁测点 side shot
旁测声呐测量 side scan sonar survey
旁插棒 side contact spike
旁承 side bearing;topside bearing
旁承垫铁 side bearing brace
旁承盖 side bearing cover
旁承拱架 <三轴转向架> bolster bridge
旁承间隙 side bearing clearance
旁承支重 side bearing loading
旁承支重转向架 balanced side bearing truck;side bearing truck
旁触传递 ephaptic transmission
旁带效应 sideband effect
旁导 by-passing
旁导洞 side-pilot tunnel
旁道 by-path;by-road
旁道比【机】by-pass ratio
旁道控制 by-pass control
旁道水表 by-path meter
旁道系数 <旁道空气量与总空气量之比> by-pass factor
旁点观测 <导线测量中观测导线两旁的点,这些目标点需要绘在地图上但不作控制点> side observation; side sight;side shot
旁阀 side valve
旁放射 side emission
旁风 side wind
旁锋后让角 side relief angle
旁锋余隙角 side clearance angle
旁锋缘角 side cutting edge angle
旁锋正刀面角 normal side rake angle
旁锋正后让角 normal side relief angle
旁锋正留隙角 normal side clearance angle
旁盖 side cap
旁杆 side bar
旁沟 by-pass canal
旁观 side looking
旁观者 beholder; looker-on[复 lookers-on]; observer; spectator; stander-by
旁轨 side-track
旁滑 side skid
旁滑式模 lateral slide mo(u)ld
旁击 side blow
旁剪钳 side cutting pliers
旁交点 side point;wing point
旁角 side bottom
旁脚 side bottom
旁接滑轮 non-axial trolley
旁街 by-street;side street
旁近路线 pass-course
旁进式入孔 side entrance manhole
旁进式窨井 side entrance manhole
旁开式吊卡 sidedoor elevator
旁开式提引器 sidedoor elevator
旁孔 side hole
旁跨 side space;side span
旁馈 parafeed
旁馈给 sideways feed
旁馈耦合 parafeed coupling
旁廊 side corridor
旁廊式客车 side-corridor coach
旁列板 side strake
旁流 bypass flow;flow by-pass
旁流法 sidestream process
旁流管 by-pass line
旁流管线 sidestream(pipe)line
旁流过滤 sidestream filtration
旁流过滤器 sidestream filter
旁流控制 by-pass control
旁流水 side stream
旁流水处理 sidestream treatment
旁流调节 by-pass control
旁漏 bleeder
旁漏电流 bleeder current
旁漏网络 bleeder network
旁滤 side filtration
旁路 by line;by-passage;by-pass highway;by-passing;by-path;canal by-pass(ing); in bridge; pass-by circuit; shunt (circuit); side road; sideway;siding road
旁路比 by-pass ratio
旁路避雷器 by-pass arrester
旁路处理 by-pass procedure
旁路挡板 by-pass damper
旁路电容 shunt capacity
旁路电容器 by-pass capacitor; by-pass condenser;shunted capacitor
旁路电阻器 by-pass resistor
旁路阀 by-pass valve;side valve;stage valve
旁路放风阀 by-pass valve
旁路放风系统 by-pass system
旁路供电线 by-pass feeder
旁路管 shunt valve
旁路管道 by-pass duct
旁路管线 by-pass line
旁路过程 by-pass procedure
旁路过滤器 by-pass strainer
旁路过滤设施 bypass filter
旁路级 by-pass stage
旁路接地 by-pass to ground
旁路接续器 by-pass set
旁路开关 by-pass switch
旁路开关柜 by-pass panel
旁路馈电线 by-pass feeder
旁路连接 shunt connection
旁路门 by-passing door
旁路母线 hospital busbar;interbus
旁路汽轮机 pass-out steam engine
旁路去耦 by-pass decoupling
旁路三通 by-pass tee
旁路式系统 line-by-pass system
旁路式旋风吸尘器 by-pass cyclone (vacuum cleaner)
旁路式装置 line-by-pass device
旁路调节 by-pass control
旁路调节管 shunt regulator tube
旁路调温 by-pass temperature control
旁路系数 by-pass factor
旁路系统 by-path system; by-pass system
旁路线 by-pass line
旁路线圈 by-pass coil
旁路信道 by-pass channel
旁路信息组 by-pass block
旁路整流器 by-pass rectifier
旁路指示信号 bypass indicator
旁路制 by-pass system;by-path system
旁路中继 by-pass relay
旁路转流术 by-pass operation
旁掠射线 grazing incidence ray
旁门 by-passing door
旁门楼梯 service stair(case)
旁面三角台 prismatoid
旁模 septate mode
旁内龙骨 bottom side girder; side keelson
旁频带 sideband
旁频防卫度 carrier to side component ratio
旁频率 side frequency
旁切 escribe
旁切球 escribed sphere
旁切圆 escribed circle;excircle
旁切圆心 escenter
旁渠 by-pass canal
旁热 indirect heat
旁热管 separate heater tube
旁热式 heater type
旁热式电子管 cathode-heater tube; heater cathode type tube; heater tube;indirectly heated tube
旁热式二极管 alternate current heated diode
旁热式热敏电阻(器)heater-type thermistor;indirectly heated thermistor
旁热式阴极 equipotential cathode; heated cathode;indirectly cathode; separately cathode
旁热式真空管 separate heater valve
旁热型 heater type
旁热阴极 separately heated cathode
旁绒球 paraflocculus
旁入口 lateral entrance
旁入式搅拌器 side entering type agitator
旁山沟 sidehill canal;sidehill ditch
旁射光 accidental light
旁施 side dressing
旁施法 side dressing method
旁施时间 time of side dressing
旁蚀 lateral abrasion;lateral erosion
旁矢状面的 parasagittal
旁示信道 commentary channel
旁视 <前后视以外的水准视线> intermediate sight
旁视雷达 side-looking radar
旁视声呐 lateral sonar; sidelocking sonar
旁室 side chamber
旁输送 sideways feed
旁送口 run-out
旁听席 public gallery
旁通 by-passing;by-path;pass-by
旁通安全阀 by-pass relief valve
旁通比【机】by-pass ratio
旁通变风量系统 by-pass VAV[variable air volume] system
旁通打捞筒 sidedoor basket
旁通挡板 by-pass baffles
旁通导管 by-pass conduit;loop
旁通到地 by-pass to ground
旁通道 by-pass conduit;by-pass route; cross passage
旁通电阀 by-pass electrovalve
旁通阀 bleed-off valve;by-pass damper;by-pass valve;by-path valve;relief valve;shunt valve
旁通阀盖 by-pass valve cap
旁通阀活塞 by-pass valve piston
旁通阀活塞端盖 by-pass piston cap
旁通阀活塞杆 by-pass valve piston rod
旁通阀活塞涨圈 by-pass piston ring
旁通阀汽管 by-pass valve steam pipe
旁通阀弹簧 by-pass valve spring
旁通阀体 by-pass valve body
旁通阀橡胶垫 by-pass valve rubber seat
旁通阀泄管 by-pass valve escape pipe
旁通法 by-pass method
旁通风道 by-pass air duct
旁通风阀 by-pass air valve
旁通风门 by-pass damper
旁通管(道)by-pass conduit;by-passing; by-pass (pipe) line; by-pass pipe; by-pass tube; down pipe; jumper tube;path-by;shunt
旁通管件 cross-over
旁通管节 side port nipple
旁通管流量 by-pass rate
旁通过滤器 by-pass filter
旁通涵洞 by-pass culvert
旁通活塞 by-pass piston
旁通减压阀 by-pass relief valve
旁通截断阀 by-pass stop valve
旁通开关 by-pass switch
旁通空气导管 by-pass air duct
旁通孔 by-pass opening;by-pass port
旁通控制 by-pass control; by-pass governing
旁通控制的喷油泵 controlled by-path injection pump
旁通控制阀 by-pass control valve
旁通口 by-pass port
旁通连接 by-pass connection
旁通流 by-pass flow
旁通路 bye-pass;bye-path
旁通气门 sidedoor choke
旁通球阀 <滤清器的> release valve ball
旁通渠 by-channel;by-pass channel; by-pass tunnel
旁通塞 by-pass plug
旁通筛道 by-pass screen
旁通施工【疏】by-pass dredging
旁通式过热器 by-pass superheater
旁通输沙道 sand by-pass; sediment by-pass
旁通输沙设施 sand by-passing plant
旁通输沙系统 sand by-passing system
旁通隧洞 by-pass tunnel
旁通套管 tapping sleeve
旁通调节 by-pass governing
旁通系统 by-pass system
旁通线 by-pass line
旁通旋塞 by-pass cock
旁通烟道 by-pass flue
旁通堰 by-pass weir
旁通液压油滤清器 by-pass hydraulic(al)filter
旁通溢流阀 vent relief valve
旁通因数 by-pass factor
旁通油过滤器 by-pass oil filter
旁通油路 bleed-off circuit; by-pass circuit
旁通运行 by-pass operation
旁通闸门 by-pass gate;filler gate
旁通蒸汽 pass-out steam
旁通装置 by-pass collar
旁透气管 side vent
旁推调车 poling of cars
旁拖 tow abreast;tow alongside
旁弯 side sway;sidewise bending
旁系 collateral series; indirect descent;offshoot
旁系的 collateral
旁系继承人 collateral heir
旁系亲属 collateral consanguinity
旁隙 side clearance
旁线 side line
旁线快速修理 rapid repairs on sidings
旁向航摄范围 ground gained sideways
旁向控制点跨度 bridging distance of control points cross strips; control point interval cross strips
旁向偏差 lateral deviation;lateral declination
旁向倾角 lateral tilt;tip angle;roll angle <飞机的>
旁向倾斜 lateral tilt;side tilt
旁向倾斜控制 rotating control;rotational control
旁向倾斜摄影 lateral-oblique photograph
旁向形变 lateral deformation
旁向折射 lateral refraction
旁向重叠 lateral (over) lap; side (over) lap
旁向重叠宽度 sidelap width
旁向重叠中线 central line of lateral overlap
旁泄 by-passing leakage
旁泄塞 by-pass plug
旁卸的 side-dump
旁卸式 side-dump type
旁心 escenter[escentre]; excenter[excentre]

旁心分角线定位法 outside fix
旁压极限压力 pressuremeter limit pressure
旁压力 lateral pressure;side pressure
旁压模量 pressuremeter modulus
旁压强 lateral pressure
旁压曲线 lateral compression curve
旁压曲线外推法 extrapolated method of lateral pressure curve
旁压试验 pressuremeter test;lateral pressure test;side pressure test
旁压试验加压等级 loading grade of lateral pressure test
旁压试验数 number of lateral pressure test
旁压试验稳定标准 stability criterion of lateral pressure test
旁压系数 lateral compression coefficient
旁压仪 lateral pressure apparatus;pressiometer;pressure meter
旁压仪极限压力 pressuremeter limit pressure
旁压仪类型 pressuremeter type
旁压仪模量 pressuremeter modulus
旁压仪探头种类 type of probe of pressuremeter
旁压钻孔 lateral compression boring
旁夷作用 lateral planation
旁移 side sway
旁移力矩 sidesway moment
旁溢洪道式坝 side-spillway dam
旁因 tributary cause
旁应变 lateral strain
旁院 side yard
旁折光 lateral refraction
旁证 circumstantial evidence;collateral evidence;side witness
旁支 offshoot
旁支导流堤 offshoot jetty
旁支管 lateral branch
旁支渠道 lateral canal
旁支渠系统 lateral system
旁支溢洪道 lateral spillway channel
旁支运河 lateral canal
旁枝 outgrowth
旁置接触轨 side conductor rail
旁轴常数 paraxial constant
旁轴的 paraxial
旁轴电子束 paraxial beam
旁轴光束 paraxial beam
旁轴光线 paraxial ray
旁轴焦点 paraxial focus
旁轴射束 paraxial beam
旁轴射线 paraxial ray
旁注 marginal note;sidenote
旁注线 leader line
旁柱 side post
旁纵桁 side girder;wing girder

彷

彷徨变异测试 fluctuation test

螃

螃蟹 crab

胖

胖边 bulb edge

胖裥 box pleating
"胖"柱 stocky column

抛

抛补 cover(ing)

抛补空头 covered bear;protected bear
抛草杆 dump rod
抛草机外壳 throwing chamber
抛撤飞行 final run
抛出 eject;knockout;shoot;throw-out
抛出船外 drop overboard;pitch overboard
抛出尾纹 throw-out tail
抛出纹 throw-out spiral
抛串联锚 back(ing) an anchor
抛堆毛石防波堤 rubble-mound breakwater
抛堆土石料 depositing fill;placing filling;random dumping
抛方块防波堤 block-mound breakwater
抛放系统 jettisoning system
抛负荷 load rejection;load thrown off
抛光 barrel finish;burnishing;colo(u)ring;dead bright;glazing;gloss finish;graze;grind in;ground finish;hone out;lapping;polish(ing);politure;satin finishing;surface finish
抛光板 caul;polished slab
抛光边加工 polished edge arising
抛光表面 polished surface
抛光玻璃 polished glass
抛光薄板 luster[lustre] sheet
抛光不足 directional polish;short finish
抛光布 polishing cloth
抛光布轮 cloth finishing mop
抛光材料 abradant material;abrasive material;polishing material
抛光车床 burnishing lathe;polishing lathe
抛光车间 polishing department;polishing plant;polishing section
抛光处理 clear finish;polished finish;polishing treatment
抛光窗框 glazed frame
抛光锤 polishing hammer;sleeking hammer
抛光锉 polishing file
抛光打磨机 polishing sander
抛光带 fine grinding belt;finishing belt;polishing belt;polishing the tape;sand belt
抛光到镜面 looking-glass finish
抛光到消失火花 spark out
抛光的 bright finished;satin finished;wrought
抛光的粗底层 polished rough coating
抛光的钢条 polished steel bar
抛光的金刚石 polished diamond
抛光的抹灰层 glazed finish
抛光的木材 wrought stuff
抛光的石面 polished stone
抛光的岩芯钻头 polished core bit
抛光动力头 buff unit
抛光方法 finishing method
抛光粉 abrasive grain;burnishing powder;crocus;crocus martis;polishing powder
抛光氟硅酸盐 polishing fluosilicate
抛光钢丝 polished steel wire
抛光膏 anti-scuffing paste;polishing compound;polishing cream;polishing paste;rubbing compound;rubbing paste
抛光工 furbisher;polisher
抛光工厂 polishing plant
抛光工段 polishing department
抛光工具 polishing tool
抛光工作 polished work
抛光鼓涂布 polished drum coating
抛光辊 burnishing brush;burnishing roll
抛光滚筒 burnishing barrel;polishing drum
抛光黄铜 polished brass
抛光(混合)剂 polishing compound
抛光活塞 polished piston
抛光机 abrasive finishing machine;branner;buffing lathe;buffing machine;burnishing machine;glazing machine;polisher;polishing machine;sander;sanding machine;blocking machine <平板玻璃的>
抛光剂 brightener;brilliant polish;buffing compound;lapping agent;polisher;polishing agent;polishing composition;polishing medium;politure;rumbling compound
抛光夹具 polishing clamp
抛光件 polished part
抛光浆 rubbing paste
抛光介质 polishing medium
抛光金 burnished gold
抛光金属 polishing metal
抛光金属箔 polished foil
抛光浸蚀 polish attack
抛光壳体 burnishing shell
抛光框格 glazed frame
抛光拉伸装置 buffing and draw gear
抛光蜡 polishing-type wax
抛光铝饰面 polished alumin(i)um finish
抛光轮 buff(ing) wheel;burnishing wheel;glazer;glaze wheel;mop;polishing wheel;rag wheel
抛光轮架 buffing head
抛光轮轮毂 centre buff
抛光螺栓 polished bolt
抛光棉轮 cotton polish wheel
抛光面 polished finish;polished surface;bright finish
抛光模 polished die
抛光磨痕 polishing marks
抛光木纹 chipped grain
抛光盘 flat plate;polisher;polishing disk;polishing runner
抛光盘刻痕 plate mark;runner cut
抛光漆 polishing varnish
抛光器 burnisher;polisher
抛光清漆 flatting varnish;polishing varnish
抛光散热器 coil radiator
抛光砂轮 buffing wheel;finishing wheel;polishing wheel
抛光设备 polishing apparatus;polishing equipment;polishing unit
抛光生产线 polishing line
抛光石 rotten stone
抛光石料 polished rock
抛光试件 polished specimen
抛光试验 polishing test
抛光饰面 polished finish;sand finish
抛光刷 polish(ing) brush
抛光速度 buffing speed
抛光台 polishing block
抛光特性 polishing characteristics
抛光条 burnishing stick
抛光铁丹 polishing rouge
抛光铜质玻璃割刀 polished brass glass cutter
抛光筒 polishing barrel;polishing drum
抛光头 buff unit
抛光线 polishing line
抛光屑 buffings
抛光性 polishability
抛光研磨器 polishing stake
抛光氧化铝粉 abradum
抛光样品 polished sample
抛光液 liquid polishing agent;polishing fluid;polishing solution
抛光银 burnished silver;polished silver
抛光用红丹粉 polishing rouge
抛光用料 polishing material
抛光用毛毡 feltless polishing bob
抛光用呢 polishing cloth
抛光用铁丹粉 polishing rouge
抛光用毡 polishing felt
抛光油 buffing oil;burnishing oil;polishing compound
抛光圆棒 polished-barrel rod
抛光圆盘 sand disc[disk]
抛光皂 clear boiled soap
抛光(轧)辊 polishing roll
抛光毡 polishing felt
抛光毡轮 felt buff wheel;felt polishing disk[disc]
抛光纸 polishing paper
抛光纸片 polishing paper disc[disk]
抛光至镜面光泽 minute finish
抛光质量 quality of finish
抛光中心线 center[centre] buffing
抛光装置 buffing attachment;buffing gear
抛光锥 buffing cone;polishing cone
抛光灼晕 buff-burned pattern
抛光钻头 polished rod head
抛海 tipping at sea
抛混凝土型块斜坡堤 ripraped concrete block breakwater
抛货 <船遇险时为减载> jettison
抛近锚 anchor short stay;short stay
抛开 ditching
抛壳钩杆 ejector rod
抛壳机构 ejection case mechanism;extractor
抛空 empty
抛捆器 sheaf discharger
抛缆 heaving line;throwing line
抛缆器 hawser apparatus;line thrower
抛缆筒火药 impulse ammunition
抛粒 throwing
抛粒护板 grit guard
抛亮 glossing
抛料带式输送机 thrower belt conveyor;thrower belt unit
抛料机 extractor;shedder;thrower
抛落线 line of departure
抛毛石基床 rubble mound
抛锚 anchor(age);anchoring berthing;breakdown on the way;cast anchor;drop anchor;let go anchor;running the anchors;setting anchor
抛锚掣住船的运动 snubbing
抛锚船 anchor boat;anchor handling vessel
抛锚地 anchorage
抛锚掉头 turning short round by anchor
抛锚杆 anchor boom
抛锚杆绞车 anchor boom winch
抛锚杆绞车速度控制器 anchor boom winch speed controller
抛锚起锚 anchor handling
抛锚驶靠 going alongside with an anchor down
抛锚信号球 anchor ball
抛锚作业 anchor operation;anchor works
抛煤机 spreader feeder
抛煤机加煤锅炉 spreader-stoker-fired furnace
抛煤机炉 spreader stoker boiler
抛煤机炉排 spreader stoker;sprinkler stoker
抛煤面 stoker surface
抛磨 rubbing down
抛磨膏 polish paste
抛泥【疏】 spoil disposal;disposal of spoil;spoiling;disposal of dredged material
抛泥槽 disposal chute
抛泥场 dumping ground;dumping place;dumping site;spoil(ing) ground;spoil(ing) area
抛泥场围埝 dam of spoil area
抛泥池 sump rehandler

抛泥堆 spoil bank
抛泥区【疏】 disposal area; disposal ground; disposal site; dumping ground; dumping site; spoil ground; spoil site
抛泥区浮标 spoil ground buoy
抛起锚口令 anchoring orders
抛弃 abandon(ment); blunder away; cast aside; cast off; castaway; dereliction; dump; memory dump; reject(ion); repudiate; throwaway; throw off; jettison < 船在遇难时减轻船重的一种措施 >
抛弃舱面货物 jettison of cargo
抛弃错误 drop-error
抛弃的 rejective
抛弃法烟气脱硫 throwaway process of fume gas desulfurization
抛弃废河道 abandoned channel
抛弃和浪打落海货物 jettison and washing overboard
抛弃浚挖物质 disposal of dredged material
抛弃块体 roughly set block
抛弃垃圾入海 disposal at sea
抛弃能量 dump energy
抛弃式电导温深探测器 expendable conductivity temperature depth sonde
抛弃式海水温度法 expendable bathythermography
抛弃式海水温度计 expendable bathythermograph
抛弃物 derelict; discard
抛前后锚时船为一链所阻不能旋转 girded; girt(h)
抛人工块体防波堤 block-mound breakwater
抛撒轮 spreading reel
抛撒输送器 spreading conveyer[conveyor]
抛撒装置 scattering mechanism
抛散距离 pulvation distance
抛砂 ramming; slinging; sand fill
抛砂驳 spreader barge
抛砂机 sand projection machine; sand slinger; sand thrower; sand throwing machine; slinger
抛砂机壳体 impeller casing
抛砂机型砂捣击锤 sand rammer
抛砂器 ramming head
抛砂速度 ramming speed
抛砂头 impeller head; slinger head
抛砂头横臂 ramming arm
抛砂造型 impeller ramming; slinger mo(u)lding
抛砂造型机 sand slinger
抛射 catapult; slinging
抛射机 flinger
抛射剂 propellant[propellent]
抛射距离 range of trajectory; jetting distance < 喷射运输机 >
抛射粒子 projectile particle
抛射器 flinger; thrower
抛射式取样机 projectile sampler
抛射速度 ejection velocity
抛射体 projectile
抛射体发射室 projectile expelling chamber
抛射体运动 projectile motion
抛射物 trajectile
抛射线 trajectory
抛射线法 trajectory method
抛射作用 heaving action
抛绳 rocket line; whip line
抛绳枪 line throwing gun; lyle gun
抛绳设备 line throwing appliance; mortar and rocket apparatus
抛石 drop fill rock; loose rock dump; placed riprap; rippling stone; riprap(ping); riprap stone; rock riprap; rubble discharging; rubble disposition; rubble tipping; stone riprap; clean-dumped rockfill【港】; rock fill; rubble fill; tipped rubble; stone fill
抛石坝 dumped rock embankment; riprap dike[dyke]; rock dike[dyke]; rock-fill dam; rock weir
抛石驳 dump barge; dump scow; rock-dumping barge
抛石层 rubble layer
抛石充填 loose rock-fill
抛石船 dumping barge; stone dumper; rubble dumping boat
抛石堤 riprap dike; riprap mound; rubble mound
抛石堤脚 rubble toe
抛石堤破坏 mound failure
抛石堆 rock windrow; rubble mound
抛石堆填 bulk rockfill
抛石防波堤 mound breakwater; riprap breakwater; rock-mound breakwater; rubble(-mound) breakwater
抛石防潮海堤 rubble-mound seawall
抛石防护 riprap protection
抛石防坡堤 riprap breakwater; rubble-mound breakwater; stone riprap; rubble filling
抛石工程 random(-placed) riprap; riprap works
抛石工事 riprap works
抛石灌浆 grouted riprap
抛石海漫 riprap apron
抛石护岸 dumped riprap; enrockment; rippling revetment; riprap bank protection; riprap revetment; rock revetment; stone filling
抛石护面 natural stone facing; pavement of riprap
抛石护坡 dumped riprap; natural stone facing; rippling protection; riprap protection of slope; riprap slope protection; rock revetment; rubble pitching; stone riprap; random riprap; riprap on slope
抛石护坦 riprap apron; stone apron; stone riprap; falling apron < 在出现冲刷后 >; rubble apron
抛石基 pierre perdue
抛石基础 riprap foundation; rubble-mound foundation; rubble(-mounted) foundation
抛石基床 enrockment; pierre perdue; rubble base; rubble bed; rubble(-mound) foundation; stone bedding; rubble mattress; stone mattress; riprap mound; rubble mound; tipped rubble; stone bedding
抛石建筑物 rubble-mound structure
抛石结构物 rubble-mound structure
抛石抗冲刷保护 rubble scour protection
抛石抗冲刷护坦 rubble anti-scour apron
抛石棱体 riprap prism; rock-fill toe bund; rock-fill toe mound; rock mound prism; rubble-mound (for foot protection); rubble-mound prism; rock berm; rock riprap; rubble backing; tipped rubble; rubble toe
抛石理坡 grading of rock mound; slope grading of rock mound
抛石路堤 riprapped embankment; dumped rock embankment
抛石路基 riprapped embankment
抛石潜坝 rock sill; submerged riprap dam
抛石桥台 rockfill abutment
抛石体 bulk rockfill; rock-filling; stone riprap; dumped riprap; enrockment
抛石填充 loose rock-fill; rock-fill
抛石围堰 riprap cofferdam; rockfill cofferdam
抛石斜坡导流堤 rubble mound jetty
抛石斜坡堤 rubble-mound breakwater
抛石斜坡丁坝 rubble-mounded groyne
抛石堰 rock(fill) weir
抛石整理护岸 dumped riprap placement
抛石整平 leveling of a rubble bed
抛石止水材料 riprap sealing compound
抛实 impeller ramming
抛首尾锚 anchor head and stern; anchoring by the head and stern
抛售货物 sell large quantities of goods; undersell; dump materials
抛售物资 sell large quantities of goods; undersell; dump materials
抛双锚 moor with two anchors
抛水砣 cast the lead
抛送机 catapult
抛坍爆破 collapse blasting
抛体运动 projectile motion
抛填 dumping fill
抛填的 dumped
抛填堆石(体) bulk rockfill; tipped rockfill; dumped rockfill; loose rock dump
抛填方材料 deposit(e) fill material
抛填风化砂 dumped weathered sand
抛填块石 tipping stone rubble; tipped rubble; tipped stone
抛填料 drop fill material; dumped fill
抛填乱石 random rubble fill; riprap stone rubble; tipped stone rubble
抛填毛石 tipping stone rubble
抛填片石 tipping stone rubble
抛填石料 drop fill rock
抛填土 dumped earth fill; dumped fill; tipped fill
抛头 ramming head; throwing wheel
抛投强度 < 河道截流时的 > dumping intensity
抛土 spoiling
抛土溜槽 spoil chute
抛丸 ball blast
抛丸冲击清理 impact cleaning
抛丸除锈 blast cleaning
抛丸除锈法 shot peening
抛丸滚筒 tumblast
抛丸机 shot-blasting machine
抛丸喷射清理 impact cleaning
抛丸器 impeller head
抛丸清理 throw shot cleaning
抛丸清理滚筒 shot-blast cleaning barrel; shot tumblast
抛丸清理机 abrator; airless blast cleaner; arbiter
抛丸清理转台 swing table abrator
抛丸清理装置 wheel abrator
抛丸头 impeller head
抛丸硬化法 shot peening
抛丸装置 blaster
抛物蚌线 parabolic(al) conchoid
抛物波电流 parabolic(al) current
抛物波电压 parabolic(al) voltage
抛物层 parabolic(al) layer
抛物插值法 parabolic(al) interpolation
抛物超柱 parabolic(al) hypercylinder
抛物点 parabolic(al) point
抛物度量几何 parabolic(al) metric geometry
抛物度量群 parabolic(al) metric group
抛物分布 parabolic(al) distribution
抛物盒天线 pill-box antenna
抛物环面天线 parabolic(al) torus antenna
抛物回归 parabolic(al) regression
抛物几何 parabolic(al) geometry
抛物尖点 parabolic(al) cusp
抛物渐近线 parabolic(al) asymptote
抛物镜面天线 parabolic(al) radiator; waveguide feed
抛物空间 < 即欧几里德空间 > parabolic(al) space
抛物螺(旋)线 Fermat's spiral; parabolic(al) spiral
抛物面 paraboloid
抛物面传声器 parabolic(al) microphone
抛物面的 parabolic(al); paraboloidal
抛物面发射天线 transmitting paraboloid
抛物面法 paraboloid method
抛物面反光罩 parabolic(al) reflector; paraboloidal reflector; paracyl reflector
抛物面反射传声器 parabolic(al) microphone
抛物面反射镜 parabolic(al) mirror; parabolic(al) reflector; paraboloidal reflector
抛物面反射镜式传声器 parabolic(al) reflector microphone
抛物面反射器 dish; parabola; parabolic(al) dish; parabolic(al) mirror; parabolic(al) reflector; paraboloid; paraboloid reflector
抛物面反射器式传声器 parabolic(al) microphone
抛物面反射式传声器 parabolic(al) reflector microphone
抛物面反射体 parabolic(al) reflector
抛物面辐射器 parabolic(al) radiator
抛物面集热器 parabolic(al) collector
抛物面接收天线 receiving parabola; receiving paraboloid
抛物面镜 parabolic(al) mirror; parabolic(al) reflector; paraboloidal mirror
抛物面聚光灯 parabolic(al) projector
抛物面聚光镜 paraboloid condenser
抛物面聚光器 parabolic(al) concentrator; parabolic(al) condenser; paraboloid condenser
抛物面聚焦集热器 parabolic(al) focussing collector
抛物面喇叭 parabolic(al) horn
抛物面盘 parabolic(al) dish
抛物面壳(体) paraboloid shell
抛物面伞形屋顶 paraboloid umbrella roof
抛物面扫描 parabolic(al) scanning
抛物面扫描器 parabscan
抛物面射电望远镜 parabolic(al) radio telescope
抛物面天线 dish antenna; parabolic(al) aerial; parabolic(al) antenna; paraboloid; paraboloid dish(antenna)
抛物面透镜 paraboloid lens
抛物面屋顶 paraboloid roof
抛物面屋面 paraboloid roof
抛物面形薄壳 paraboloidal shell
抛物面形伞状薄壳 paraboloidal umbrella shell roof
抛物面形伞状的折板薄壳屋顶 paraboloidal umbrella roof of folded shell
抛物面形天线 parabolic(al) antenna
抛物面形屋顶 paraboloidal roof
抛物面型壳 paraboloid shell
抛物面坐标 paraboloidal coordinates

抛物挠线 cubic(al) parabola
抛物平面 parabolic(al) plane
抛物曲面拱顶 paraboloid calotte
抛物曲线 gravity curve; parabolic(al) curve
抛物扇形域 parabolic(al) sector
抛物体运动 projectile motion
抛物天线 parabolic(al) antenna
抛物线 parabola; paraboloid; paracurve
抛物线逼近法 parabolic(al) approximation
抛物线波形 parabolic(al) waveform
抛物线薄壳 parabolic(al) shell
抛物线插补器 parabolic(al) interpolator
抛物线大梁 parabolic(al) girder
抛物线的 parabolic(al)
抛物线的变换 transformation of parabola
抛物线的顶点 vertex of a parabola
抛物线的一段 segment of a parabola
抛物线的正焦弦 latus rectum of parabolas
抛物线递增分配方式 parabolic(al) increase allocation method
抛物线定律 parabolic(al) law
抛物线段 parabolic(al) segment
抛物线断面 parabolic(al) section
抛物线法 parabolic(al) method
抛物线反射器 parabolic(al) reflector
抛物线方程(式) parabolic(al) equation
抛物线飞行 parabolic(al) flight
抛物线分布 parabolic(al) distribution
抛物线刚构 parabolic(al) frame
抛物线公式 <设计柱用的> parabolic(al) formula
抛物线拱 parabolic(al) arch
抛物线拱坝 parabolic(al) arch dam
抛物线拱顶 parabola vault; parabolic(al) vault
抛物线拱断面 parabolic(al) arch section
抛物线拱肋 parabolic(al) rib
抛物线拱桥 parabolic(al) arch bridge
抛物线拱支架 parabolic(al) arch support
抛物线构架 parabolic(al) frame
抛物线关系 parabolic(al) relation
抛物线轨道 parabolic(al) orbit
抛物线轨迹 parabolic(al) path
抛物线函数 parabolic(al) function
抛物线函数乘法器 quarter-square multiplier
抛物线桁架 parabolic(al) truss
抛物线弧 parabolic(al) arc
抛物线缓和曲线 parabolic(al) transition curve
抛物线基线校正 parabolic(al) baseline correction
抛物线加荷 parabolic(al) loading
抛物线加腋板 parabolic(al) haunched slab
抛物线加腋梁 parabolic(al) haunched girder
抛物线加载 parabolic(al) loading
抛物线检波 parabolic(al) detection
抛物线截角锥体 frustum of parabola; parabola frustum
抛物线解法 parabolic(al) solution
抛物线控制电位器 parabola control potentiometer
抛物线梁 parabolic(al) beam; parabolic(al) girder
抛物线流量特性 parabolic(al) flow characteristic
抛物线路拱 parabolic(al) crown
抛物线模板 parabolic(al) form
抛物线模型 parabolic(al) model
抛物线内插法 parabolic(al) interpolation
抛物线起拱大梁 parabolic(al) arched girder
抛物线穹顶 parabolic(al) vault
抛物线穹隆 parabolic(al) dome
抛物线穹形屋顶 parabola cupola; parabola dome
抛物线沙丘 parabolic(al) dune
抛物线生长 parabola-growth
抛物线式拱肋 parabolic(al) rib
抛物线速度 parabolic(al) velocity
抛物线速度分布 parabolic(al) velocity distribution
抛物线特性 parabolic(al) characteristic
抛物线体 parabolic(al) body; paraboloid
抛物线调节器 parabolic(al) governor
抛物线筒柱 parabolic(al) cylinder
抛物线透镜 parabolic(al) lens
抛物线屋顶桁架 parabolic(al) roof truss
抛物线屋面桁架 parabola(roof) truss
抛物线弦 parabolic(al) chord
抛物线形边缘 parabolic(al) edge
抛物线形波 parabolic(al) wave
抛物线形插销 parabolic(al) plug
抛物线形大灯 parabolic(al) headlamp
抛物线形大梁 parabolic(al) girder
抛物线形反射器 parabolic(al) reflector
抛物线形分布荷载 parabolic(al) distribution load; parabolic(al) load
抛物线形钢丝束 parabolic(al) cable; parabolic(al) wire
抛物线形荷载 parabolic(al) load
抛物线形黑斑 parabolic(al) shading
抛物线形桁架 parabolic(al) truss
抛物线形(横)断面 parabolic(al) cross-section
抛物线形激波 paraboloidal shock
抛物线形镜子 parabolic(al) mirror
抛物线形喇叭 parabolic(al) horn
抛物线形料斗 parabolic(al) hopper
抛物线形裂缝 parabolic(al) crack
抛物线形流量密度模型 parabolic(al) flow-concentration model
抛物线形螺旋曲线 parabolic(al) spiral curve
抛物线形喷嘴 German nozzle; parabolic(al) nozzle
抛物线形坡度 parabolic(al) grade
抛物线形曲线 parabolic(al) curve
抛物线形沙丘 parabolic(al) dune
抛物线形式 parabolic(al) form
抛物线形竖曲线 parabolic(al) vertical curve
抛物线形四片式板簧器 parabolic(al) 4-leaf spring
抛物线形弹簧 parabolic(al) spring
抛物线形头部 parabolic(al) nose
抛物线形喂料仓 parabolic(al) kettle feed bin
抛物线形旋转面 paraboloid revolution
抛物线形堰 parabolic(al) measuring weir; parabolic(al) weir
抛物线形预应力筋 parabolic(al) tendon
抛物线(形折弦)桁架 parabolic(al) chord truss
抛物线形帧信号 vertical parabola
抛物线形柱面 parabolic(al) cylinder
抛物线型 parabolic(al) type
抛物线型变化 variation of parabolic(al) type
抛物线型钢构架 parabolic(al) frame
抛物线型沙丘 hairpin bend dune
抛物线堰 parabolic(al) weir
抛物线圆拱 parabolic(al) conoid
抛物线圆拱壳体 parabolic(al) conoid shell
抛物线圆穹壳体 parabolic(al) conoid shell
抛物线圆屋顶 parabolic(al) cupola
抛物线质谱 parabola mass spectrum
抛物线质谱法 parabola mass spectrography
抛物线质谱仪 parabola mass-spectrograph; parabola mass-spectrometer
抛物线状 parabolic(al) shape
抛物线纵断面 parabolic(al) profile
抛物线坐标 parabolic(al) coordinates
抛物形变换 parabolic(al) transformation
抛物形二次超曲面 parabolic(al) quadric hypersurface
抛物形二次曲面 parabolic(al) quadratic surface
抛物形反射镜 parabolic(al) reflector
抛物形方程 parabolic(al) equation
抛物形轨道 parabolic(al) trajectory
抛物形渐近线 parabolic(al) asymptote
抛物形喇叭筒 parabolic(al) horn
抛物形黎曼曲面 parabolic(al) Riemann surface
抛物形偏微分方程 parabolic(al) partial differential equation
抛物形延迟失真 parabolic(al) delay distortion
抛物形秩 parabolic(al) rank
抛物形纵面图 parabolic(al) profile
抛物性射影 parabolic(al) projectivity
抛物性透射 parabolic(al) homology
抛物性直射 parabolic(al) collineation
抛物悬链线 parabolic(al) catenary
抛物叶形线【数】parabolic(al) folium
抛物运动 projectile motion
抛物柱面 parabolic(al) cylindrical surface; parabolic(al) dish
抛物柱面反射镜 parabolic(al) reflector
抛物柱面函数 parabolic(al) cylinder function
抛物柱面镜 parabolic(al) mirror
抛物柱面镜控制 parabolic(al) mirror control
抛物柱面曲面 paraboloidal surface
抛物柱面天线 parabolic(al) cylinder antenna; pill-box antenna
抛物柱面形方程 parabolic(al) cylindrical equation
抛物柱面坐标 parabolic(al) cylinder coordinates
抛物柱体 parabolic(al) cylinder
抛物柱体曲面 paraboloidal surface
抛物子代数 parabolic(al) sub-algebra
抛物子群 parabolic(al) sub-group
抛下物 droppings
抛小锚移船 kedge; kedging
抛小石粗面 pebble dashing
抛小石粗面加工 pebble dashing
抛卸高度 dumping height
抛卸器 ejector
抛卸装置 jettison device
抛雪装置 snow casting device
抛压载装置 ballast jettisoning arrangement; ballast jettisoning device
抛油环 flinger; oil ring; oil slinger; oil thrower; slinger
抛油环式密封 slinger seal
抛油环式润滑 slinger type lubrication
抛油环套筒 slinger sleeve
抛油器 oil thrower
抛油圈 flinger; oil slinger; oil thrower; oil throw ring
抛油圈垫 oil slinger gasket
抛掷 cast; throw
抛掷爆破 blasting for throwing rock; explosive casting; pinpoint blasting; throw(ing) blasting
抛掷充填 sling stowing
抛掷充填机 centrifugal stower
抛掷堆高度 ejecta crest height
抛掷法试验 drop test
抛掷分散器 dispersion sling
抛掷杆 discharge arm; dump rod
抛掷机构 ejection mechanism; tripping mechanism
抛掷机皮带 flight belt
抛掷器 kicher; knockout; spinner; sling
抛掷式充填机 flight belt filler
抛掷式充填机皮带 throwering belt
抛掷式分级机 ballistic separator
抛掷式分选机 ballistic separator
抛掷式胶带填充机 beltstower
抛掷式清扫车 abandon sweeper
抛掷式输送器 jet conveyer[conveyor]; throw-out conveyer[conveyor]
抛掷式填充机 flight belt
抛掷物 thrower
抛掷药包 pinpoint charge
抛掷炸药 propellant explosive; propelling explosive
抛掷装置 slinger; swivel piler
抛掷装置减荷器 slinger-unloader
抛掷装置卸载机 slinger-unloader
抛筑 pell-mell placing; random placing; riprapping
抛筑的 pell-mell placed; random-placed
抛筑的混凝土块体 concrete block at pell-mell; pell-mell-placed concrete block
抛筑的块体 block at pell-mell; concrete block pell-mell; pell-mell block; random block
抛筑构造 pell-mell construction
抛筑块体 random block; roughly set block
抛筑块体防波堤 block-mound breakwater
抛筑块体工程【港】query-random blockwork
抛左/右锚/双锚 let go port/starboard/both anchors

咆

咆哮 bluster; roar

刨

刨洞 <小断面开挖> small drift

刨根锄 grub hoe
刨去 shoot off
刨土机 power shovel; shale planer; stripper

袍

袍树 gland-bearing oak

跑

跑边 edges deviation

跑表 stop watch; timer
跑步 double time
跑车 roadster; sporting sedan; sport roadster
跑车时间 travel time
跑车速度 travel(l)ing speed
跑车装置 runaway device
跑出轨道 run-out track
跑道 landing runway; landing strip; racecourse; race track; runway (strip); buckled track【铁】; buckling of rail; track buckling <轨道侧涨>

P

跑道板 run plank;runway board <机场>
跑道边界 runway boundary
跑道边界灯 runway edge light
跑道边界照明 runway edge lighting
跑道边线灯 runway edge light
跑道边指挥车 runway control van
跑道编号体系 runway numbering system
跑道标志 band display;runway marking
跑道标志灯 runway light
跑道布置 runway configuration
跑道布置方案 runway layout
跑道承载能力 runway bearing capacity
跑道导路 <飞机场跑道前段> worm-up apron
跑道导向标志 taxiway guidance sign
跑道道肩 runway shoulder
跑道道面 runway pavement;runway surfacing
跑道道面结冰探测器 runway surface ice[icing] detector
跑道的侧安全道 runway safety shoulder
跑道灯 runway light
跑道定位标发射机 landing beam transmitter
跑道定位信标 runway localizing beacon
跑道定线 runway align
跑道端识别灯光 runway end identification lights
跑道泛光灯 runway floodlight
跑道方向 runway heading
跑道分布图 runway pattern
跑道钢轨 <高架起重机> runway rail
跑道构型 runway configuration
跑道管理 runway control
跑道环境 runway environment
跑道基层 runway base course
跑道基础 runway base
跑道加固 runway strengthening
跑道检查 runway check
跑道界限 <机场> runway threshold
跑道界限指示灯 runway threshold light
跑道进近端 approach end runway
跑道进口 runway threshold
跑道距离标志 runway distance marker
跑道可用系数 usability factor of runway
跑道控制门 runway gate
跑道拦阻网 runway arrester net
跑道拦阻装置 runway arrester;runway arresting gear;runway barrier
跑道路面 runway pavement
跑道目视距离 runway visual range
跑道能见度 runway visibility
跑道配置 runway arrangement
跑道铺面 runway surfacing
跑道铺砌层 runway pavement
跑道起飞端 take-off end of the runway
跑道气象观测 runway observation
跑道浅槽 <防止降雨时飞机滑移危险> runway grooving
跑道清洁器 runway cleaner
跑道清扫机 runway sweeper
跑道容量 runway capacity
跑道入口 runway threshold
跑道入口标志 runway threshold marking
跑道入口灯 runway threshold light
跑道上降落 runway landing
跑道上离机 runway escape
跑道上弹射 runway ejection
跑道上着陆滑跑操纵 runway roll-out guidance
跑道设备 runway facility;runway gear
跑道剩余长度 remaining runway (length)
跑道剩余段 remaining runway (length)
跑道视程 runway visual range
跑道视距 runway visual
跑道视距测定装置 runway visual range equipment
跑道弹射座椅 runway ejection seat
跑道通行能力 runway capacity
跑道头保险道 runway overrun area
跑道头标志 runway end marking
跑道头警告灯 runway caution light
跑道托架 <机场> runway bracket
跑道位置指示系统 runway locating system
跑道温度 runway temperature
跑道稳定性 racetrack stability
跑道形电磁分离器 race track
跑道形放电管 race track
跑道型轨道 racetrack orbit
跑道型加速器的环形室 racetrack torus
跑道型同步加速器 racetrack synchrotron
跑道型微波加速器 racetrack microtron
跑道延长段 runway extension
跑道引导 runway guidance
跑道有效坡度 effective gradient of runway
跑道障碍 runway obstruction
跑道照明 <机场> runway lighting
跑道照明灯 runway light;runway surface lights
跑道照明设备 runway lighting equipment;runway lighting installation
跑道指示灯 course light;runway light
跑道中线灯 center line runway lights
跑道中心线 runway centerline[centreline]
跑道中心线信号灯系统 runway centerline lights system
跑道终端 end of runway
跑道轴线 runway axis
跑道状况 runway condition
跑道状况报告 runway condition report
跑道最大容量 maximum capacity of runway
跑电 <蓄电池> shelf depreciation
跑动干扰信号 running rabbit
跑钢 badly bleeding;bleed-out;break-out;run-out
跑狗场 canidrome club
跑光 edge light
跑过 overrun
跑合 break(ing)-in;run(ning)-in
跑合对中 running alignment
跑合面 running surface
跑合期 running-in period
跑合期磨损 run-in wear
跑合时间 seating time
跑合试车 run-in test
跑合试验 run-in test
跑合速度 running-in speed
跑合性能 running-in characteristic
跑合用油 break-in oil
跑合运转 break-in run
跑合作业 brunishing-in
跑火 exudation
跑浆 void of paste
跑街 <携带样品推销员> demonstrator
跑锯材 miscut lumber
跑料 leak;running;run-out
跑马 horse-riding
跑马场 racecourse
跑马场大看台 racecourse grand stand
跑马场看台 racecourse stand
跑马道 bridle path;bridle road;riding trail
跑米 meter[metre] run
跑偏 off-tracking
跑偏监视器 off-line running monitor
跑偏控制器 off-track controller
跑漆 cissing[sissing]
跑气 <气压盾构> blow in
跑生料 meal push
跑水 metal break out;metal runout
跑速 running speed
跑索 <架空索道> track rope
跑腿活 legwork
跑油事故 oil leak accident
跑釉 aired ware
跑釉制品 aired ware
跑纸 paper throw
跑纸符号 paper throw character
跑钻【岩】 runaway of (drilling) bit;run-down of drill string

泡 blister;vesicle

泡包 steam drum
泡铋矿 bismutite
泡菜坛 pickle jar
泡畴 bubble domain
泡畴材料 magnetic bubble material
泡畴破裂器 bubble domain collapser
泡点 bubble point
泡点曲线 bubble point curve
泡点温度 bubble point temperature
泡点压力 bubble point pressure
泡发射极工艺 washed emitter technology
泡筏 bubble raft
泡沸 <油漆等遇热产生的> intumescence
泡沸电气石 aphrizite
泡沸剂 intumescent agent
泡沸井 bubbling well
泡沸面 effervescence level
泡沸黏[粘]土 effervescing clay
泡沸泉 bubbling spring
泡沸石 zeolite
泡钢 <由熟铁渗碳而成的钢> blister steel
泡海 mare spumans
泡核沸腾 nucleate boiling
泡痕 dobying
泡化 alveolation
泡环塑料筒垫 foam plastic cushion
泡货 <体大质轻> bulky cargo
泡碱 native soda;natron
泡碱湖 natron lake
泡界线 scum line
泡浸 dip
泡浸足 immersion foot
泡孔 abscess
泡溃灭 bubble collapse
泡立水 alcoholic varnish;polish;polituture;French polish
泡利不相容原理 Pauli's exclusion principle
泡利反常矩项 Pauli's anomalous moment term
泡利方程 Pauli's equation
泡利矩阵 Pauli's matrices
泡利梁 Pauli's girder
泡利式电渗析器 Pauli-type electrodialyzer
泡利原理 exclusion principle
泡利重排定理 Pauli's rearrangement theorem
泡利自旋磁化率 Pauli's spin susceptibility
泡利自旋空间 Pauli's spin space
泡流 burble;froth flow
泡流分离 burble separation
泡流分离角 burble angle;burble point
泡瘤状纹 blister figure
泡帽 bell cap;bubble cap
泡锰铅矿 cesarolite
泡面钢 blister steel
泡膜 bubble film
泡膜剂 membrane-foaming compound
泡沫 foam;bubble;froth(flo(a)tation);lather;scum;spume;suds
泡沫板 foam(ed) board
泡沫保持能力 foam retention
泡沫保温 foam insulation
泡沫保温材料 cellular insulant;foamed insulating material
泡沫保温管套 foam pipe lagging
泡沫背衬 foam-back
泡沫倍数 foam multiple;foam factor
泡沫苯乙烯混凝土 gas-forming styrol concrete
泡沫泵 foam pump
泡沫比例器 foam proportioner
泡沫玻璃 bubble glass;cellular glass;foam(ed) glass;frothed glass;glass foam;multicellular glass;perforated glass;sound-absorbent glass;sponge glass
泡沫玻璃板 expanded glass
泡沫玻璃骨料 foamed glass aggregate
泡沫玻璃块 foamed glass block
泡沫玻璃砖 foam glass block;foam glass brick
泡沫玻璃砖块 perforated glass block
泡沫捕尘 foam arrested dust
泡沫捕集器 foam catcher
泡沫材料 cellular material;foam compound;foam (ed) material;foam mat;sponge
泡沫材料黏[粘]合 foam bonding
泡沫测定仪 foamer meter
泡沫层 foaming;foam layer;layer of foam;layer of scum
泡沫层织物 foam-laminated fabric
泡沫掺加剂 foam-entraining admixture
泡沫产品 froth pulp
泡沫产品回收率 froth recovery
泡沫沉落度 <衡量泡沫质量的一项指标> foam slump
泡沫沉陷距 foam slump
泡沫衬垫硬木地板 foam-cushioned hardwood flooring
泡沫成布法 foam-to-fabric process
泡沫持久性 foam persistence
泡沫充填护舷 foam-filled fender
泡沫充填混凝土块体 foam filled concrete block
泡沫充填料 foam fillings
泡沫冲洗 foam flushing
泡沫冲洗钻进 foam drilling
泡沫除尘器 bubble dust collector;bubble dust scrubber;foam dust separator
泡沫促进剂 foam booster;foam-promoting builder;froth promoter;lather booster
泡沫带饰 foam ribbon
泡沫的 spumous;spumy
泡沫的粉煤灰硅酸盐混凝土 foam-ash-silicate concrete
泡沫的消散 foamy meltdown
泡沫的应用 foam application
泡沫底垫 <隧道用> foam base mat
泡沫地沥青 foam asphalt
泡沫地沥青装置 foam asphalt unit
泡沫垫层 foam carpet
泡沫垫层法 foam cushion method
泡沫堵塞 froth buildup

泡沫发生器 foam-generating unit;foam generator;foam maker;frother
泡沫发生室 foam mixing chamber
泡沫发生塔 foam column;froth tower
泡沫法 foaming method; foaming process
泡沫法探漏 bubble method leak detection
泡沫翻滚的波浪 white horses
泡沫矾土砖 alumina bubble brick
泡沫防尘 foam dust suppression
泡沫分布型 foam profile
泡沫分级分离 foam fractionation
泡沫分级(器) foam fractionation
泡沫分类 bubble sort
泡沫分类法 bubble sort method
泡沫分离 foam separation
泡沫分离法 foamet;foam separation process
泡沫分离器 foam separator;skimmer
泡沫分裂 foam fractionation
泡沫分析 foam analysis
泡沫分选 froth separation
泡沫粉末 insulating powder
泡沫敷层 porous coating
泡沫浮层 foam floating roof
泡沫浮选 foam flo(a)tation;froth flo(a)tation;pneumatic flo(a)tation
泡沫浮选槽 bubble cell
泡沫浮选法 froth flo(a)tation method;froth flo(a)tation process
泡沫浮选分离 froth flo(a)tation separation
泡沫浮选水转分选机 froth flo(a)tation hydrotator
泡沫浮子 foam float
泡沫覆盖层 foam cover
泡沫覆盖灭火 foam blanketing
泡沫干燥 foam drying
泡沫刚玉砖 foamed corundum brick
泡沫高度 foam height
泡沫(高炉)炉渣 foamed blast-furnace slag
泡沫隔热层 foamed insulation
泡沫隔声层 foamed(sound) insulation
泡沫给进<钻眼防尘用> foam feed
泡沫供给器 aerofloat reagent feeder
泡沫构造 pumiceous structure
泡沫刮板 froth paddle
泡沫管 bubbler tube;foam pipe
泡沫管接头 foam connection; foam couple
泡沫硅酸钙制品 aerated calcium silicate
泡沫硅酸盐 foam-silicate
泡沫硅酸盐混凝土 foam-silicate concrete
泡沫硅酸盐混凝土板 foam-silicate concrete slab
泡沫过滤器 foam filter
泡沫合成树脂 foam synthetic(al) resin
泡沫和表面活化剂 foam and surface activating agent
泡沫痕 foam impression;foam mark
泡沫花纹 blister figure
泡沫灰浆 foamed mortar
泡沫灰泥 foam mortar
泡沫混凝土 aerated concrete; aero(con) crete; blown-out concrete; bubble concrete; cellular concrete; concrete aerated with foam; concrete foam; foam; foam(ed) concrete; foaming concrete; porous concrete
泡沫混凝土板 foamed concrete slab
泡沫混凝土地面 foamed concrete ground surface
泡沫混凝土刮尺 foamed concrete screed
泡沫混凝土建筑砖 foamed concrete (building) tile
泡沫混凝土搅拌机 foamed concrete mixer
泡沫混凝土砌块 foamed concrete (building) block; cellular concrete block
泡沫混凝土墙板 foamed concrete wall slab
泡沫混凝土墙块 foamed concrete wall block
泡沫混凝土填料 foamed concrete filler
泡沫混凝土瓦 foamed concrete tile
泡沫混凝土砖 foamed concrete brick; foamed concrete block
泡沫混凝土砖块 gas concrete block
泡沫火山岩类 bubble-rocks
泡沫唧筒车<消防用> foam pump car
泡沫集尘钻眼法 foam drilling
泡沫计 frothmeter
泡沫剂 foamer; foam-forming admixture; foaming agent; frother; frothing agent; gas foamer; gas-foaming admixture; gas-forming admixture
泡沫加筋混凝土 reinforced cellular concrete
泡沫加气钢筋混凝土 reinforced concrete aerated with foam
泡沫加气混凝土 foam-gas concrete
泡沫加强混凝土 reinforced cellular concrete
泡沫胶 foamed glue
泡沫胶布 foam coated fabric; foam fabric
泡沫胶合 foam gluing
泡沫胶截断机 splitter
泡沫胶硫化模 foam sponge mo(u)ld
泡沫胶耐压试验机 foam sponge indentation tester
泡沫胶乳 foamed latex;frothed latex; latex foam
泡沫胶粘剂 cellular adhesive;foam(ed) adhesive;foam(ed) glue
泡沫焦性石墨 pyrofoam
泡沫搅拌室 foam mixing chamber
泡沫接触式冷却器 froth cooler
泡沫结构 cellulation; foaming structure;foamy structure
泡沫结构破坏 foam collapse
泡沫界线 foam line
泡沫金属 foam(ed) metal
泡沫进给 foam feed
泡沫经济 bubble economy
泡沫精矿 froth concentrate
泡沫精练 foam degumming
泡沫精练机 foam degumming apparatus;froth boiling apparatus
泡沫净化器 foamed purifier
泡沫聚氨酯板 polyurethan(e) foam board;polyurethan(e) foam sheet
泡沫聚氨酯条 polyurethan(e) foam strip
泡沫聚苯乙烯 expanded polystyrene; foamed polystyrene; polystyrene foam;styrene foam;styrofoam
泡沫聚苯乙烯隔热板板 foamed polystyrene board for thermal insulation purposes
泡沫聚苯乙烯绝缘板 styrofoam insulation
泡沫聚苯乙烯绝缘材料 styrofoam insulation material
泡沫聚苯乙烯模 expanded polystyrene pattern
泡沫聚苯乙烯砖 polystyrene foam tile
泡沫聚苯乙烯砖瓦 expanded polystyrene tile
泡沫聚丙烯 expanded polypropylene
泡沫聚氯乙烯 foamed polyvinyl chloride;polyvinyl chloride foam;vinyl foam
泡沫聚亚氨酯 foamed polyurethane
泡沫聚亚氨酯条 foamed polyurethane strip
泡沫聚乙烯 cellular polyethylene; foamed polyethylene
泡沫聚乙烯绝缘 foamed polyethylene insulation
泡沫绝热 foam insulation
泡沫绝热材料 cellular thermal insulation;foam-thermal insulation
泡沫绝缘 cellular insulation; foam(ed) insulation
泡沫绝缘电缆 foamed insulation cable
泡沫空化 foam cavitation
泡沫空间 foaming space
泡沫控制 foam control;froth-control
泡沫控制粉剂 foam control agent powder
泡沫控制硅酸盐混凝土 foam-slag silicate concrete
泡沫控制剂 foam control agent
泡沫控制系统 foam control system; froth-control system
泡沫矿渣 foamed slag
泡沫矿渣混凝土 foamed slag concrete
泡沫矿渣混凝土板 foamed slag concrete slab
泡沫矿渣混凝土厚板 foamed slag concrete plank
泡沫矿渣混凝土空心砌块 foamed slag concrete hollow block
泡沫矿渣混凝土套管 foamed slag concrete casing
泡沫沥青 fortlay bitumen;frothy bitumen
泡沫溜槽 foam chute
泡沫流 foam flow
泡沫炉渣 foamed slag
泡沫炉渣混凝土 foamed slag concrete
泡沫炉渣混凝土砌块 hollow expanded cinder concrete block
泡沫铝 foamed alumin(i)um
泡沫氯丁(二烯)橡胶 foam(ed) neoprene
泡沫泌水量 bleeding of foam
泡沫密度 foam density
泡沫密封层 foamy sealant
泡沫密封垫 foam gasket
泡沫灭火 extinction by foam;fire extinction by foam
泡沫灭火法 foamite extinguishing method
泡沫灭火机 bubble extinguisher;foam fire extinguisher; foam fire extinguishing system; froth fire extinguisher
泡沫灭火剂 control foam; fire-extinguishing foam; fire foam; foam extinguishing agent;foam-forming fire retardant agent;foamite
泡沫灭火剂容器 froth liquid vessel
泡沫灭火器 bubble extinguisher;foam annihilator;foam can;foam extinction; foam extinguisher; foam fire extinguisher; foam fire extinguishing system;forth fire extinguisher
泡沫灭火器消防栓 foam extinguisher hydrant
泡沫灭火系统 foam extinguishing system; foamite system; froth fire extinguishing system
泡沫灭火装置 fixed froth installation; foam installation
泡沫泥浆 foamed mud
泡沫黏[粘]度 bubble viscosity
泡沫黏[粘]度计 bubble visco(si) meter
泡沫黏[粘]合剂 foamed adhesive; foam glue
泡沫黏[粘]土 foam clay
泡沫脲甲醛 foamed urea-formadehyde
泡沫排除区 froth removal zone
泡沫喷枪 foam lance
泡沫喷洒器 foam sprinkler
泡沫喷射器 froth nozzle
泡沫喷嘴 foam jet
泡沫膨胀材料 cellular-expanded material
泡沫破裂 lather collapse
泡沫破碎盘管 foam breaking coil
泡沫气孔 foam cell;foam porosity
泡沫气泡 foam bubble
泡沫强化剂 froth stiffener
泡沫轻质混凝土 foamed lightweight concrete
泡沫清管器 foam go-devil;foam pig; foam scraper
泡沫去除 foam removal
泡沫染色 foam dy(e)ing
泡沫溶液 foam solution
泡沫熔岩 foamed lava; foamy lava; scoriaceous lava
泡沫熔岩混凝土 foamed lava concrete;scoriaceous lava concrete
泡沫熔岩混凝土墙板 foamed lava concrete wall slab;scoriaceous lava concrete wall slab
泡沫熔渣 foamed slag
泡沫熔渣混凝土 foamed slag concrete
泡沫熔渣混凝土板 foamed slag concrete slab
泡沫熔渣混凝土厚板 foamed slag concrete plank
泡沫熔渣混凝土套管 foamed slag concrete casing
泡沫乳胶 foam(ed) latex;latex foam; plastic foam
泡沫色谱 froth chromatography
泡沫砂浆 cellular-expanded mortar; cellular mortar; foam(ed) mortar; mortar aerated with foam
泡沫生成 foam formation
泡沫生成管 foam-making duct
泡沫生成外加剂 foam-forming admixture
泡沫湿矿渣混凝土 foamed slag concrete
泡沫石膏基底 foam plaster base
泡沫石膏造型 foamed plaster mo(u)lding
泡沫石棉 asbestos foam
泡沫石英玻璃 foamsil
泡沫石油沥青 foamed asphalt
泡沫式灭火器 dry chemical(fire) extinguisher;foam type
泡沫试验 foam test
泡沫试验器 foam-tester
泡沫室 bubble chamber; foam chamber
泡沫室自记潮汐计 bubbler tide ga(u)ge
泡沫收集器 foam catcher;foam trap
泡沫寿命 foam life
泡沫束带层 foam ribbon
泡沫树脂 cellular resin;foamed resin
泡沫数值 foam value
泡沫水泥地面 foam cement screed
泡沫水泥混凝土 foam cement concrete
泡沫水泥砂浆 air-entrained mortar
泡沫水泥制品 bubblestone
泡沫水泥砖 foam cementitious brick
泡沫水平探测器 froth level detector

泡沫似的 foamy
泡沫塑料 aerated plastics;cellular plastics;cellular resin;expanded plastics;expanded polystyrene;foam(ed)plastics;laminac;plastic foam;polyfoam;porous plastics;sponge plastics
泡沫塑料板 foamed plastic board;plastic foam board;rigid foam
泡沫塑料板材 foam board
泡沫塑料保温板 insulating plastic foam board
泡沫塑料薄板 expanded plastic sheet;foamed plastic sheet;foamed sheet;plastic foam sheet
泡沫塑料层压机 foam laminating machine
泡沫塑料层压黏[粘]合 foam laminating
泡沫塑料层压织物 foam laminate fabric
泡沫塑料衬垫 foamed plastic cushioning
泡沫塑料衬里 foam backing
泡沫塑料衬里织物 foam-backed textile
泡沫塑料成型机 foam slab machine
泡沫塑料隔热材料 plastic foam insulation
泡沫塑料隔热舱 foamed plastic insulated hold
泡沫塑料隔热层 plastic foam insulation
泡沫塑料隔声层 plastic foam insulation
泡沫塑料滚筒 foamed plastic roller
泡沫塑料滚压装置 foamed plastic roller
泡沫塑料焊接 plastic foam seal(ing)
泡沫塑料灰泥底板 plastic foam plaster baseboard
泡沫塑料灰泥踢脚板 plastic foam plaster baseboard
泡沫塑料混凝土 foamed plastic concrete
泡沫塑料夹层结构 foamed plastic sandwich construction
泡沫塑料绝缘 expanded plastic insulation;foamed plastic insulation;plastic foam insulant
泡沫塑料滤器 sponge filter
泡沫塑料冒口 foamed polystyrene feeder
泡沫塑料密封 expanded plastic seal(ing);foamed plastic seal;foamed seal;plastic foam seal(ing)
泡沫塑料密封条 foamed plastic sealing strip
泡沫塑料面织物 foam-back fabric;foam-laminated fabric
泡沫塑料模 foamed plastic pattern;foam mo(u)ld
泡沫塑料模型 foamed polystyrene pattern
泡沫塑料片材切割机 die press
泡沫塑料轻质混凝土 plastic foam lightweight concrete
泡沫塑料穹隆 foam dome;foamed plastic dome;plastic foam dome
泡沫塑料球 foamed plastic ball
泡沫塑料热熔黏[粘]贴 foam flame bonding
泡沫塑料石膏底板 foamed plastic plaster baseboard
泡沫塑料水密封 foamed waterstop
泡沫塑料填料 foamed-in-place filler
泡沫塑料芯板 foam core
泡沫塑料圆筒 foamed plastic cylinder
泡沫塑料圆屋顶 foamed cupola;foamed plastic cupola;plastic foam cupola
泡沫塑料止水剂 foamed waterstop
泡沫塑料止水器 foamed waterstop
泡沫塑料止水条 foamed plastic waterbar;foamed water bar
泡沫塔 foam column;foam tower
泡沫塔除尘器 foam tower scrubber
泡沫碳 foam carbon
泡沫陶瓷 expanded ceramic;foamed ceramics
泡沫陶瓷过滤器 ceramic foam filter
泡沫陶器 foam ceramics
泡沫特性 foam characteristic
泡沫体 rigid foam
泡沫体积 foam volume;lather volume
泡沫天然橡胶 foamed natural rubber
泡沫填充缓冲浮筒 foam-filled spring buoy
泡沫调节剂 foam modifier
泡沫涂层 foamed coating
泡沫涂层织物 foam coated fabric
泡沫涂层装置 foaming equipment
泡沫涂料 foamed coating
泡沫脱胶 froth degumming
泡沫稳定剂 foam stabilizer
泡沫稳定性 foam stability
泡沫稳定性试验 foaming stability test
泡沫吸收器 bubble absorber;foam scrubber
泡沫洗涤器 foam scrubber
泡沫现象 frothing
泡沫线 <波浪破碎后到达海滩形成的> foam(ing)line
泡沫线痕迹 foam mark
泡沫相分离法 foam phase separation
泡沫橡胶 air foam rubber;cellular rubber;foam(ed)rubber;froth rubber;rubber foam;scum rubber;sponge rubber;latex foam
泡沫橡胶底衬 foam rubber backing
泡沫橡胶浆 foamed rubber latex
泡沫橡胶轮胎 foam rubber tire[tyre]
泡沫橡胶切割器 foam rubber cutter
泡沫橡胶乳汁 foamed rubber latex
泡沫橡胶软垫 foam rubber cushion
泡沫橡胶制品 foam rubber product
泡沫橡胶座 foam rubber seat
泡沫橡皮 cellular rubber
泡沫消除剂 froth destroyer;froth killer
泡沫消除器 froth killer
泡沫消防龙头 foam hydrant
泡沫消防炮 foam water cannon
泡沫消防系统 fire-fighting foam system
泡沫消声器 cellular silencer
泡沫消失率 disappearance rate of foam
泡沫携带 foam over
泡沫芯层 foam core
泡沫芯夹层板 foam core sandwich panel
泡沫形成 physallization
泡沫型灭火器 foam type fire extinguisher
泡沫型灭火系统 foam type fire extinguishing system
泡沫旋塞 foam cock
泡沫循环 foam recycle
泡沫循环液 foam circulation fluid
泡沫岩 aphrolite;pumice;pumice stone
泡沫液 foam concentrate
泡沫液体排出(岩粉)速率 film drainage rate
泡沫仪 foam meter;Latherometer
泡沫乙烯树脂 foamed vinyl resin
泡沫抑尘 foam dust suppression
泡沫抑制剂 foam breaker;foam inhibitor;foam suppressor;froth destroyer
泡沫溢出 froth overflow
泡沫引气法 entraining method of foaming
泡沫渣 foam(ing)slag
泡沫珍珠岩 foamed pearlite
泡沫政策 bubbles policy
泡沫值 foam number;foam value;lather value
泡沫指数 foam index
泡沫制品 foam article
泡沫终点 foam end point;lather end point
泡沫绉 foam crinkle crepe
泡沫注射 foam injection
泡沫贮槽 froth reservoir
泡沫贮液模塑工艺 foam reservoir mo(u)lding
泡沫铸塑 foam casting
泡沫装置 foamite system
泡沫状的 frothy;pumiceous;spumescence;spumous;spumy
泡沫状鼓风炉渣 foamed blast-furnace slag
泡沫状结构 frothy texture
泡沫状结壳 foam crust
泡沫状聚氨基甲酸乙脂 polyurethan(e)foam
泡沫状聚氨酯绝缘体 cellular polyurethane insulation
泡沫状聚苯乙烯绝缘体 cellular polystyrene insulation
泡沫状聚乙烯电介质 foamed polyethylene dielectric
泡沫状空化 foam cavitation
泡沫状组织 vesicular tissue
泡沫总管 foam main
泡沫阻止剂 <酸洗附加剂> anti-foaming agent
泡沫钻进 foam drilling
泡囊 vesicle
泡泡玻璃 bubble glass
泡泡纱 crimp cloth;seersucker
泡泡绉 crinkle crepe
泡砂石 quartzific sandstone;sand rock
泡砂岩 quartzific sandstone;sand rock
泡石 afrodite
泡石英 water quartz
泡式图 bubble diagram
泡式微型车 <有透明圆罩的微型汽车> bubble car
泡水 <指水文现象> boiling water;ponding;boil(ing);river boil
泡松 bull pine
泡酸量 amount of acid
泡塔 bubble column
泡腾 effervesce;froth-over
泡腾成粒 effervescent granulation
泡腾的 effervescent
泡腾合剂 effervescent mixture
泡腾混合物 effervescent mixture
泡腾磷酸盐 effervescent phosphate
泡腾片剂 effervescent tablets
泡腾现象 effervescence phenomenon
泡腾盐 effervescent salt
泡腾浴台 effervescent bath tables
泡梯 bubble ladder
泡田 steeping field
泡桐 Fortunes Poulownia;Paulownia imperials
泡桐属 Paulownia
泡铜 blister;blister copper;crude copper;raw copper
泡铜精炼 blister refining
泡铜块 blister cake
泡铜熔炼炉 blister furnace
泡霰石 conchite
泡形水平仪 spirit level-bubble
泡形罩 blister
泡漩 boils-vertical eddy;boil vortex;whirlpools and eddies
泡漩水流 boil-eddy flow;boil-vortex flow
泡漩险滩 boil-eddy-caused danger passage;boil-eddy rapids;hazardous passage of boil-eddy type
泡油量 amount of oiling
泡在盐水或醋中 pickle
泡渣砖 foamed slag brick
泡胀 swell(ing)
泡胀度 degree of swelling
泡胀剂 swelling agent
泡罩 bubble cap;bubbling hood
泡罩板 bubble-cap plate
泡罩板式塔 bubble cap plated tower
泡罩层塔 bubble plate tower
泡罩层蒸溜塔 bubble plate tower
泡罩齿缝长度与宽度 slot length and width of bubble cap
泡罩分馏塔 bubble cap fractionating column
泡罩分馏柱 bubble cap fractionating column
泡罩高度 height of bubble cap
泡罩式分布器 bubblecaps
泡罩塔 bubble-cap plate column;bubble-cap plate tower;bubble-cap tray tower
泡罩塔盘 bubble-cap tray;bubble deck
泡罩外径 outside diameter of bubble cap
泡罩与塔板间隙 clearance between bubble cap and tray
泡罩蒸馏塔 bubble-cap tower
泡制 infusion
泡柱式浮选机 bubble-column machine
泡状包裹体 bubble inclusion;bubble cover
泡状花纹 quilted figure
泡状结构 bubble structure
泡状流 bubble flow
泡状碳 carbon foam
泡状纹理 blister grain
泡状压力线 bulb pressure

炮

炮兵 artillery
炮兵测地 artillery survey
炮兵测地勤务 artillery survey service
炮兵地图 artillery map
炮兵罗盘 artillery director
炮兵照相写景图 artillery panorama picture
炮车 gun carrier
炮弹 cannonball;projectile
炮的仰角 gun elevation
炮点 shot point
炮点低速度带厚度 weathering thickness at shot point
炮点高程 shot point elevation
炮点位置 shot point site
炮点下水深度 water depth under shot
炮点增量值 shot point increment
炮队 artillery
炮队镜 battery commander's telescope
炮队镜量角器 azimuth worn knob
炮耳座 trunnion bed;trunnion seat
炮杆 stemmer
炮工 powderman;shooter;shotfirer
炮管 barrel;gun barrel;gun tube
炮管钻头 boring drill;gun drill;gun drilling bit
炮辊 tamper bar
炮棍 beater;stemmer;tamping bar;tamping pole;tamping stick
炮合金 gun metal
炮后风锥形区 backblast area
炮击 bombardment
炮记录道数 shot traces;traces of a shot
炮架 gun carriage;gun carrier;gun cradle;gunmetals mount;gun mount

炮架型车架 cradle frame
炮检距调整 offset adjustment
炮舰 gunboat;gunnery ship
炮脚碾压机 tamping foot roller
炮井编号 sequence of well
炮井地震测井 uphole shooting
炮井深度 borehole depth
炮孔 cell;drill hole;shothole
炮孔并联 parallel hole connection
炮孔成组起爆 shot-firing in rounds
炮孔串联 series hole connection
炮孔定位 emplacement of blast-hole; emplacement of borehole
炮孔定向器 burden ga(u)ge
炮孔扩底 borehole springing
炮孔排 row of holes
炮孔排列(方式)drilling pattern
炮孔深度 borehole depth
炮孔套管 shot casing
炮孔凿岩机 shothole drill
炮孔炸药 borehole loading
炮孔直径 borehole diameter;hole diameter
炮孔装药 borehole loading;hole loading
炮孔组爆破 round throw
炮孔钻 blast hole drill
炮孔钻机 < 采石场 > quarry blast-hole rig
炮孔钻针 blast hole drill
炮口 embouchure;muzzle
炮口塞 tampion
炮口速度 muzzle velocity
炮口烟光遮蔽 smoke defilade
炮楼 defensive tower
炮泥 clay tamping; stemming plug; stemming stick;tamping plug
炮泥材料 stemming material
炮泥枪 mud gun
炮泥纸袋 stem bag
炮排方位角 azimuth of shoot array
炮钎钢 hollow drill steel
炮塞衬垫 spaser
炮身 cannon
炮手 shooter
炮术 artillery
炮索 breeching
炮塔 gun turret;turret
炮台 barbette;fort(ress);presidio
炮台甲板 gundeck
炮铁 gun iron
炮艇 gunboat;gunship
炮铜 gun brass;gun bronze;gun metal;gunmetal bronze
炮铜合金 government bronze
炮铜轴衬 gunmetal bush
炮铜轴承 gunmetal bearing
炮筒 gun barrel;gun metal < 铜锡锌合金,有时加入铅和镍 >
炮筒来复线槽拉刀 rifle barrel groove cutting broach
炮筒离心铸造 watertown process
炮尾 breech
炮窝子 dead hole
炮学 artillery
炮烟 blackdamp;blast fume
炮眼 arrow loop; blasthole; bore (hole); embrasure; kernel; mine chamber; port (hole); shothole; dead hole < 爆破后的 >
炮眼爆破 shotfiring
炮眼爆破次序 firing sequence
炮眼爆破法 shothole blasting method
炮眼壁探缝器 breakfinder
炮眼布置 blast geometry; blast-hole arrangement; hole layout; hole placement; hole placing; hole setting;point of holes
炮眼布置方案 pattern layout
炮眼布置图 drilling pattern
炮眼(布置)形式 drilling pattern
炮眼参数 shothole parameter
炮眼长度 hole length
炮眼城垛 crenel(le);kernel
炮眼导向架 burden ga(u)ge;hole director
炮眼底 bootleg;bottom of the hole; hole bottom;end of drill-hole
炮眼点火放炮 fire a hole
炮眼法伽马取样 gamma sampling by uphole method
炮眼分置 layout of round
炮眼封泥 stemming
炮眼封泥棒 stemming rod;tamping rod
炮眼刮杓 sludger
炮眼间距 spacing; spacing between holes
炮眼交错布置 staggered arrangement of shotholes
炮眼角度 hole angle
炮眼进(英)尺 foot-of-hole
炮眼开孔 starting the borehole
炮眼口 heel of a shot;hole collar
炮眼扩大 blast-hole springing
炮眼扩孔 chambering;springing
炮眼离装药最远部分 heel of a shot
炮眼利用率 capacity ratio of shothole;percentage of shot hole(depth in blasting); utilization factor of shot hole
炮眼名称 shothole nomenclature
炮眼内炸药过早起爆 premature burning
炮眼内炸药与自由间距 drill hole burden
炮眼排列 boring pattern
炮眼排列方式 pattern of a round
炮眼倾斜 grip of hole
炮眼清除刷 swabstick
炮眼塞 tamping plug
炮眼扇形布置 fanning
炮眼扇形排列花样 fan pattern
炮眼深度 borehole depth;hole depth
炮眼数 number of shot-holes
炮眼掏壶 chambering
炮眼套管 shothole casing
炮眼斜度测量 surveying of borehole
炮眼用塑料水袋 water ampo(u)e
炮眼针 needle pricker
炮眼直径 borehole diameter;diameter of shot-hole;hole diameter;hole size
炮眼装药 blast-hole charge;borehole charge;borehole loading;hole loading;loading of hole
炮眼装药量 explosives-loading weight of hole
炮眼装药器 blast-hole charger
炮眼装药系数 explosives-loading factor of hole
炮眼组 blasting round; drill round; round;round of holes;set of holes
炮眼组布置 layout of round; round layout
炮眼组布置和放炮程序图 boring pattern and procedure
炮眼组的排列 pattern of round
炮眼组的正确布置 right face of the round
炮眼组环形排列花样 ring-pattern
炮眼组深度 depth of round
炮眼钻 blast hole bit;shothole drill
炮眼钻机 blast-hole drill; blast-hole machine
炮眼钻进 blast-hole drilling
炮眼钻头 blast-hole bit
炮眼钻针 blast-hole drill
炮焰 flame of shot
炮用金属 gun metal
炮竹 fire-cracker
炮座 barbette;gun platform

疱状突起修整 blister repair

胚胎 embryo

胚胎的 embryonic
胚胎矿【地】protore
胚胎期 embryonic stage
胚芽 germ
胚芽期 embryo

陪伴 companion

陪伴旅客的服务员 courier
陪伴物 accompaniment
陪都 second capital
陪集 coset
陪集代表 coset leader
陪集码 coset code
陪集权 coset weight
陪集首 coset leader

培长石 bytownite

培尔顿(水斗式)水轮机 Pelton's(water) wheel (turbine); Pelton's turbine
培尔顿水轮机外壳 Pelton wheel case
培肥灌溉 manuring irrigation
培根燃料电池 Bacon fuel cell
培柯铜盐脱硫 Perco copper sweetening
培雷式火山的 pelean
培垄铲 ridged sweep
培氏浊度计 Baylis turbidimeter
培特煤层耐火黏[粘]土 better bed fireclay
培提反变位理论 < 马克思威尔理论的推演 > Bettis reciprocal theorem
培提氏交互定理 Bettis reciprocal theorem
培土 earth backing; earthing; earth up;hill up
培土铲 earthing blade;ridged sweep
培土锄铲 hiller hoe
培土机 banking machine
培土犁 earthing-up plow;ridge plow; ridging plough
培土犁体 banking body; hilling bottom
培土器 coverer; covering body; earthing-up plow;hiller;ridger
培土浅种法 surface planting
培修堤防 embankment repair
培训 training
培训班 training course
培训部 educational service
培训车间 instructional(work)shop
培训程序 training program(me)
培训厨房 training kitchen
培训船 training ship
培训方案 training scheme
培训费(用)training cost;training expenses
培训工场 training workshop
培训工作人员 training officer
培训顾问 training adviser[advisor]
培训规范 training specification
培训过度 overtraining
培训回路 training loop
培训计划 educational program(me); training plan(ning)
培训间 training room
培训教室 training room
培训教员 trainer
培训科目 training course
培训期满的学徒 journeyman
培训区 training bay
培训设计 training design
培训时间 training time
培训手册 training manual
培训提升 upgrading
培训效果检验 validation of training
培训协议 learnership
培训学员 trainee
培训业绩指标 training performance indicator
培训职能 training function
培训中心 training center[centre]
培养 cultivate;culture;education;patronage;training
培养本国人才 training of local personnel
培养池 cultivation tank
培养的 cultural
培养碟【化】culture dish
培养法 cultivation; cultivation method;culture method
培养缸 aquarium
培养关税 educational duty; educational tariff
培养黄油 culture butter
培养基 culture medium; incubation medium; medium; nutrient; nutrient medium;substratum[复 substrata]
培养基烧瓶 culture flash
培养技术 culture technique
培养介体【化】culture medium
培养皿【化】 double dish; culture dish;culture plate;Petri dish
培养目标 training objective
培养瓶 culture bottle;culture flask
培养期 incubation period
培养器 culture dish
培养容器 culture vessel
培养时间 culture time
培养试管 culture test tube
培养水产的池塘 farm
培养特性 cultural character; cultural trait
培养物 culture
培养细菌 culture of bacteria
培养箱 incubater[incubator]
培养性质 cultural property
培养液 culture fluid; culture solution; inoculums[复 inocula];nutrient fluid
培养液瓶 culture jar
培育 cultivation
培育措施 cultural measure; cultural operation
培育法 cultivation method
培育期 rearing stage
培育箱 incubater[incubator]
培育新的经济增长点 foster new growth areas in economy
培植 implant
培植钵 culture pan
培植接种 inoculate

赔本 lose one's capital; lose one's outlay;run a business at a loss; sustain losses in business

赔本价(钱)junk price
赔偿 atone(ment); indemnification; indemnity; make satisfaction for; payback;quittance;recoupment;redress;reimburse(ment);remuneration;repayment;solatium
赔偿保险 compensation insurance
赔偿保证书 back(ward) letter;bond of indemnity;letter of indemnity
赔偿标准 measure of indemnity
赔偿程度 measures of indemnity
赔偿处 claims department; loss department
赔偿代理人 claim settling agent

P

赔偿的 compensatory;reparative
赔偿的减少 mitigation of damages
赔偿的义务 obligation to compensation
赔偿额 amount of claim;reexchange
赔偿法 law of restitution
赔偿范围 extent of compensation
赔偿方式 mode of satisfaction
赔偿费(用) reimbursable expenses;amortization fund;damages;indemnification;penal sum;compensation
赔偿付讫 claims paid
赔偿负担费用 pay for
赔偿估算 claims estimated
赔偿合同 contract indemnity;contract of indemnity;indemnity contract
赔偿毁损 indemnity for damage or loss
赔偿基金 compensation fund
赔偿金 compensation money;compensatory payment;damages liquidated;indemnification;smart money;solatium;indemnity;recompense
赔偿金额 claim amount
赔偿金限额 monetary limit of liability
赔偿率 reparation duty
赔偿名誉 indemnity for defamation
赔偿期限 temporal limit of liability
赔偿契约 deed of indemnity;indemnity bond;indemnity contract
赔偿清算 claims settlement
赔偿权保险 claim right insurance
赔偿人 compensator;indemnitor
赔偿申请办事员 claim clerk
赔偿申请查询单 claim tracer
赔偿申请的仲裁 arbitration of claims
赔偿申请通知 notice of claims
赔偿实物 indemnity in kind
赔偿损害 make a reparation for an injury;repair the injury
赔偿损害要求 claim for damages;damage claim
赔偿损毁 recover damage
赔偿损失 compensate for loss;compensation for a loss;compensation for loss;indemnity for damages;make good a loss;make restitution;pay for a loss;redress damage;respond in damages;make amend for;reparation
赔偿损失的加倍付给 double value compensation
赔偿损失的要求 claim for damages
赔偿损失的义务 obligation of compensation for losses
赔偿损失时的加倍付给 double value
赔偿所受损失 indemnify for the loss incurred
赔偿条款 indemnity clause;penalty clause
赔偿通知书 notice for accident indemnity
赔偿物 indemnification;indemnity;satisfaction
赔偿限度 measure of indemnity
赔偿限额 limit of compensation;maximum amount of accident indemnity
赔偿协定 amending agreement;indemnity agreement;reparation's agreement
赔偿协议 indemnity agreement;reparation agreement
赔偿要求 appeal for compensation;claim for compensation of damages;compensation claim;reexchange
赔偿要求书 request for accident indemnity
赔偿预算 claims estimated
赔偿责任 liability for damages;liability to pay compensation
赔偿责任承保人 liability insurer
赔偿责任的免责事项 exceptions form liability
赔偿责任的时效和地效 temporal and geographical limits of liability
赔偿责任制度 regime(n) of liability
赔偿责任转移 hold harmless
赔偿者 compensator
赔偿证书 deed of indemnity
赔付率 loss ratio
赔付损失 settlement of loss
赔款 damages;indemnity;make compensation;pay reparations;reparations
赔款成本 burning cost
赔款后保险金额复原条款 loss reinstatement clauses
赔款率 loss ratio
赔款率分保合同 stop-loss treaty
赔款条款 penalty clause
赔款选择保险 settlement option
赔款支付 reparation payments
赔款支付命令 loss order
赔款准备金 loss reserve
赔钱 be out of pocket
赔钱的 submarginal
赔钱货 distress commodity;distress merchandise
赔账 pay for the loss of cash or goods entrusted to one

锫矾 zircosulphate

裴甲的 steel-clad

裴斯莱石 peisleyite

佩带式小型话筒 lapel microphone

佩丹黄红色硬木 <缅甸产> Petthan
佩尔蒂埃绿 Pelletier's green
佩加玛雕带 <小亚细亚的遗迹> Pergamum frieze
佩克莱数 Peclet number
佩肯活钩 Peliocan hook
佩利果特蓝 Peligot blue
佩纶 <无纺织物> Pellon
佩斯波黎斯宫 <波斯古建筑> Palace of Persepolis
佩斯利涡旋纹花呢 <苏格兰制> paisley
佩碳钡铈矿 cordylite
佩特罗夫摩擦系数公式 Petroff equation
佩特罗式高速锤 petro-forging machine
佩特诺斯特泵 Paternoster pump
佩希伯克尔(转式)炉 Pershbecker furnace
佩兹伐和 Petzval sum
佩兹伐曲率 Petzval curvature
佩兹伐曲面 Petzval surface
佩兹伐条件 Petzval condition
佩兹伐透镜 Petzval lens

配板 matching

配拌骨料 prepare aggregate
配拌集料 prepare aggregate
配备 accouterment;allocate;deploy;dispose;fit out;furnish;outfit(ting);provide;rig
配备船舶人员 commission a ship
配备船员 man the ship
配备的 equipping
配备品 outfit
配备品保险 outfit insurance
配备齐全的 well-appointed;full manned <指人员>
配备人员 man(ning);staffing;to provide staff
配备容量 equipped capacitor
配备水下呼吸器的潜水员 scuba diver
配备拖拉机 allocate tractor
配比 match(ing);mixture ratio;proportion of mixture
配比挡板 proportioning damper
配比公式 proportioning formula
配比和搅拌 proportioning and mixing
配比计量机 proportioning machine
配比燃烧器 proportional burner
配比探头 proportioning probe
配比限度 proportionality limit
配比研究 matching study
配比组 matched group
配边 cobordism
配冰网 ice distribution network
配菜室 vegetable preparation room
配菜桌 dresser
配餐室 butlery;pantry;cuddy <船上>
配餐台 service table
配车 car distribution
配车股 rolling stock distribution section
配车命令 distribution order
配车数【铁】 quantity of allocation trains
配车员 car distributor;distributor
配称建筑 jawab
配出器 dispenser
配出线 feeder conductor
配船 allocation of ships;distribution of ship
配船计划 cargo planning;sailing schedule
配锤 bobweight
配错的 mismatched
配搭极佳 well assorted
配搭误差 matching error
配达台 ticket distributing position
配带电机型号 complete motor type
配电 current distribution;distribution;load distribution;power distribution;switching;electric(al) distribution
配电板 panel;panel switchboard;cabinet panel;distributing board;distribution block;distribution board;distribution panel;distribution switchboard;electric(al) panel;electric(al) panelboard;instrument board;keysets;power bay;power board;power panel;switchboard(panel)
配电板电压 board voltage
配电板汇流条 omnibus bar
配电板控制盘 control board
配电变电站 distributing substation;distribution transformer station
配电变压器 distribution transformer
配电变压器保险器 catch holder
配电部件 power distribution unit
配电场 switch yard
配电成本 distribution cost
配电电缆 distributing cable;distribution cable
配电电压 distribution voltage
配电段 distribution section
配电断流器 distribution cutout
配电房 power distribution building;power distribution room;switching house
配电费用 distribution cost
配电分站 distributing substation;distribution substation
配电干线 distributing main;distribution main;distribution line
配电干线管道 distributor duct
配电和控制电缆 power and control cable
配电和控制设备 switchgear and control equipment
配电盒 block terminal;distributing box;distributing cabinet;distribution cabinet
配电架 distributor shaft
配电间 distributing room;switchgear room;switch yard
配电柜 electric(al) panelboard
配电开关板 cross board
配电控制室 transmission line control room
配电馈线 distribution feeder
配电联动器 switch gear
配电母线通道 feeder busway
配电盘 board;cabinet panel;control board;control panel;disposition board;dissipator;distribution board [disk/ panel/ switchboard];distributor disk[disc];distributor(plate);electric(al) panel;panel(board);panel box;patch bay;patchboard;patch panel;power board;power distributing panel;power panel;power strip;service panel;switchboard;switch panel
配电盘保险丝盒 distribution fuse block
配电盘及故障点标定装置 switchboard and fault locator
配电盘室 electric(al) panel room;panel room;switchboard room
配电盘线路箱 panel board box
配电盘用仪表 back connected instrument
配电屏 distribution board;distribution panel;switchboard
配电器 distributing switch;sparger;distributor
配电器触点 distributor point
配电器电容器 distributor condenser
配电器电刷 distributor brush
配电器电转子 distributor rotor
配电器断电臂 distributor breaker arm
配电器控制臂 distributor control arm
配电器提早发火指针 distributor advance pointer
配电器轴 distributor shaft
配电曲线 distribution curve;load distribution curve;load distribution line
配电设备 switch gear
配电室 distributing room;distribution room;electric(al) control room;switch(board) room;switch(board) house
配电所 distribution substation;load-dispatching office
配电通道 busway;electric(al) busway
配电网 distributing net(work);distribution circuits;distribution grid;distribution net(work);electric(al) distribution circuit;electric(al) distribution network;electricity grid
配电网导线 electric(al) distribution circuit conductors
配电系统 distributed system;distributing system;distribution system;electric(al) distribution system;power distribution system
配电线 distribution wire
配电线路 distribution line
配电箱 block terminal;conduit box;cross board;distributing box;distribution box;distributor box;electric-

P

(al) cabinet; electric (al) distribution box; electric (al) switch box; panel box; pot head; power distributing panel; switch box
配电与保险丝盘 distribution and fuse board
配电与熔丝盘 distribution and fuse board
配电员 load dispatcher
配电站 distributing center [centre]; distributing station; distribution center[centre]; distribution station; power distribution station; substation
配电中心 center[centre] of distribution; distributing center [centre]; distribution center[centre]; electric (al) distribution center[centre]
配电柱 distribution pillar
配电装置 distribution apparatus; distribution installation; electric (al) distribution unit; power distribution; power distribution unit; switch gear
配电装置的栅 cubicle
配电装置间隔 cubicle
配殿 side hall
配定标准液 fixanal
配对 association; conjugate; conjugation; geminate; matched pair; mate; mating partner; pairing
配对比较(法) pairwise comparison; comparison of matched pairs
配对标志 match marking
配对操作 matching operation
配对差检验 paired difference test
配对齿廓 counterpart profile; mating profile
配对齿轮 mating gear
配对抽样 paired sampling
配对穿孔卡 matching punch cards
配对的 handed; mating; paired
配对滴定管 paired buret(te)
配对调查 matched study; matched survey
配对法兰 companion flange; counterflange
配对法兰盖 companion blind flange
配对管 pair tube
配对函数 pairing function
配对结合 pair bond
配对晶体管 matched pair transistor
配对零件 mate
配对区 collochore
配对取样 paired sampling
配对设计 matched pairs design; paired design
配对试验 paired experiment
配对物 mate; counterpart
配对像片 matched photo; matched picture; matched print
配对小齿轮 mating pinion
配对小区 paired plot
配对选取非比例抽样 disproportionate sampling with paired selections
配对照片 <配立体像对用> matched print
配对资料 paired data
配额 quota; ration(ing)
配额抽样 quota sampling
配额分记 quota allocation
配额供应本 ration book
配额管理 quota administered
配额期限 quota periods
配额权利 quota rights
配额市场 allocation market
配额数量 quota quantity
配额调整 adjustment of quota
配额外汇 rationing exchange
配额以上 above quota
配额制 quota system
配发用料空地 shipping material space
配方 batch formula; composition; compounding; dispensation; dispensing; formula [复 formulae/formulas]; formulation; recipe
配方比率 formula ratio
配方变更 formula change
配方差异 compounding variation; formula variation
配方改进 compositional refinement
配方计算 formula calculation
配方设计 formula design; formulating; formulating of recipe
配方设计师 formulator
配方式规范 recipe-type specification
配方手册 formulary
配方调整 formula adjustment
配方用料 formula ingredient
配分比 partition ratio
配分函数 partition function; sum of states; sum over states
配分矩阵 partitioned matrix
配分系数 partition coefficient
配分子 complex molecule
配风(空气调节)器 <锅炉的> air register
配风室 distribution plenum
配复筋的 doubly reinforced
配复筋混凝土 doubly reinforced concrete
配赋法 compensation method
配钢筋的 reinforced
配工 allotment of labo(u)r
配箍筋的混凝土 hooped concrete
配箍筋的桩 hooped pile
配箍(筋)桩 hooped pile
配箍率 stirrup ratio
配骨料 <一种较细的> leca
配刮 facing up
配管 pipe arrangement; piping; tubing
配管布置 pipe arrangement
配管部件 tube parts
配管单元 piping unit
配管地区 piping tract
配管设计 piping design
配管图 piping diagram; piping drawing
配管图例 piping symbol
配管自控流程图 piping and instrumentation diagram
配光 candle power distribution; light distribution
配光曲线 contour map of light; distribution curve flux; luminous intensity distribution curve
配好的 made-up
配好的骨料 matched aggregate
配好混合料 proportioned mix
配好料浆 finished slurry
配合 accouplement; adap(ta)tion; bedding-in; commensurate; compounding; conjunction; coordination; couple; coupling; fit(ting-in); matching; match together; match up; mate; pair; proportioning <按比例定量>; team work
配合白 broken white
配合泵 proportioning pump
配合比 mixing proportioning; mix(ing) proportion(ratio); mix(ing) ratio; mixture ratio; ratio of mixture
配合比理论 <混凝土的> theory of proportioning
配合比例 <混凝土等的> proportion of mixture
配合比设计 design mixture; design of mix; design of mix proportion; mix design; proportioning
配合比设计的计算机程序 computer programming of mix design
配合比试验 mix design test
配合比调整 adjustment of mix(ture)
配合比细度模量法 fineness modulus method of proportioning
配合比选择 proportioning
配合比选择器 mixing selector
配合边缘 matched edge
配合标记 mating mark
配合标志 match marking
配合表面 mating surface
配合不当 mismatch; mismate
配合不良 misfit; poor fit
配合仓 blending bunker
配合操纵器 <水力透平机> combinator
配合铲装法 conjugate shovel-run method
配合长度 length of fit
配合厂 proportioning plant
配合成分 grad(u)ation composition
配合齿轮对 match gearing
配合触面 mating surface
配合单位 unit of fit
配合的 combinative; combinatorial
配合的材料 matched material
配合的骨料 matched aggregate
配合的集料 matched aggregate
配合的木材 matched timber; matched wood
配合的贴面板 matched veneer
配合的屋顶板 matched shingle
配合等级 fit quality; grade of fit; quality of fit
配合地板 matched floor
配合地槽 yoked basin; zeugogeosyncline
配合点 match point
配合点的降深值 drawdown value of matching point
配合点的井函数值 well function values of matching point
配合点的井函数自变量值 argument value of well function at matching point
配合点的时间值 time vale of matching point
配合点坐标 coordinate of matching point
配合垫板 fit strip
配合动作 team work
配合度 degree of adaptability; grade of fit
配合度检定 goodness of fit test(ing)
配合端面 counterface
配合对应原则 principle of matching
配合舵栓 fitted pintle
配合法 matching method; method of adjustment; method of proportioning
配合方式 system of fit
配合防撞设备的雷达 radar with anti-collision equipment
配合废料 proportioning waste
配合符号 match mark(ing)
配合工序 compounding operation
配合公差 allowance; fitting allowance; fit(ting) tolerance; tolerance of fit; tolerance on fit
配合公差等级 class of fit
配合观念 matching concept
配合惯例 <费用与收益相> matching convention
配合换位 coordinated transposition
配合基 ligand
配合级别 class of fit
配合极限 fitting limit
配合计算机 couple computer
配合记号 match mark(ing)
配合技术 compounding technique
配合剂 additive agent; compounding ingredient; synergist
配合加料(漏)斗 proportioner
配合间隙 fit clearance; tolerance clearance
配合检验 goodness of fit test(ing)
配合件 fitting parts; fitting piece; matching parts; mating member
配合键 fit key
配合浇制法 match casting method
配合孔 mating hole
配合类别 class of fit
配合良好 no-float
配合良好的 well-proportioned
配合良好的混凝土 well-proportioned concrete
配合量 use level
配合料 batch material; mixed batch; raw batch; raw mixture
配合料薄层 batch blanket
配合料层 batch layer
配合料车 batch car
配合料成分设计 mix design
配合料成团 batch agglomeration
配合料储存 batch handling
配合料斗 batch bucket
配合料飞散 batch carry-over
配合料分层 batch segregation; demixing
配合料粉尘 batch dust
配合料罐 batch can
配合料挥发量 volatile loss from batch
配合料挥发率 rate of fusion loss; rate of melting loss
配合料混合 batch mixing
配合料结块 batch agglomeration; batch caking
配合料结石 batch stone
配合料块 batch briquet; batch lump
配合料粒化 batch pelletizing; granulating of the batch
配合料料仓 batch bucket; batch silo
配合料料车 batch barrow
配合料料斗 batch hopper
配合料料堆 batch pile
配合料料方 batch formula
配合料料粉 batch powder
配合料料块 batch lump
配合料料粒 batch pellet
配合料料团 batch pile
配合料气体 batch gas
配合料球粒 batch pellet
配合料容器 batch container
配合料熔尽时间 batch-free time
配合料润湿 batch wetting
配合料压块 batch briquetting
配合料预熔 fritting of the batch
配合料运输 batch handling
配合料造粒 batch pelletizing
配合料制备 preparation of batch
配合料装卸 batch handling
配合料组成 batch composition
配合料组成变化 batch change
配合料组分 batch component; batch ingredient
配合轮 match wheel
配合螺母 attaching nut
配合螺栓 fit(ted) bolt
配合门 marriage-gate
配合密封面 matching trim
配合面 fitting surface; matching surface
配合模 mating die
配合盆地 yoked basin
配合器 dispenser; ducon
配合染料 complex dye
配合色 broken colo(u)r
配合色产品 broken-colo(u)r work
配合设计 mix preparation
配合使用 fitment

配合试验 compatibility test
配合适度 goodness of fit
配合手轮 fly
配合饲料 compound feed;mixed feed
配合体 ligand
配合条 setting strip
配合条件 matching requirement
配合物 complex
配合误差 mismatch error
配合橡胶 compound rubber
配合销 adjusting pin
配合研磨法 equalizing lapping
配合运输 coordinated transportation
配合制 system of fit
配合制度 fit system
配合质量 fit quality
配合轴承 fitted bearing
配合装置 adapting device
配合准备 mix preparation
配合准地槽 yoked basin; zeugogeosyncline
配合作用 mating reaction
配和力 combing ability
配衡 tare
配衡的 tared
配衡滤器 tarred filter
配衡烧瓶(已称过容器皮重的烧瓶) tared flask
配衡体 tare
配环筋混凝土 hooped concrete
配换齿轮 change wheel; selective gear
配货 allocate cargo;stowage
配货单 invoice;order blank
配货楼层 distributing stor(e)y; distribution floor
配货人 shipper
配机线 idle locomotive track
配级 grading
配极 polarity
配极变换 polarization
配极的自共轭元素 self-conjugate element of a polarity
配极论 polar theory
配极曲面 polar reciprocal surface
配极曲线 polar curve
配极系 polar system
配极运算子 polarizing operator
配极锥面 reciprocal cone
配给 allocate; allot; allotment; distribution;ration
配给程序 rationing procedures
配给代替交换 substitution of rationing for exchange
配给的份额 quota allowed to somebody
配给价格 ration price
配给名单 ration roll
配给品 allowance
配给器 allocator
配给系统 distribution system
配给制 ration system
配价 coordinate;coordination
配价化合物【化】 coordinate compound
配价键(合) coordinate link(age); coordinate bond;dative bond
配价键型吸附 coordinate type adsorption
配价金属配合物 coordinated metal complex
配价络盐 coordinate complex salt
配价体 ligand
配件 accessories; accessory parts; adjunct; appurtenance; carcase; complement; duplicate parts; fitment; fitting metal; fitting piece; fittings; matching parts; matching unit; mountings; parts; repair parts; repair piece; replacement parts; service parts;subassembly
配件表 parts list
配件工厂 fittings factory
配件号 parts number
配件加工车间 parts and processing shop
配件库 spare parts depot
配件细部图 reinforcement detail drawing
配浆比率 paper formulation
配浆机 slurry compounding machine
配铰 align reaming
配接玻璃 adapter glass
配接法兰 adapting flange
配接环 adapter ring
配接凸缘 adapter flange; adapting flange
配接线图 wiring diagram
配筋 arrangement of bars; distributed steel; distribution reinforcement; fabrication; placing of steel bars; re-bar; reinforcement; reinforcement placement;reinforcing bar
配筋百分率 steel ratio
配筋比例平衡的梁 proportioning beam
配筋表 bar list;bending schedule;cutting list
配筋不足的 under-reinforced
配筋不足的混凝土 under-reinforced concrete
配筋不足断面 under-reinforced section
配筋布置 bar arrangement
配筋大样 reinforcement detail
配筋分散性系数 dispersion coefficient of reinforcement
配筋灌浆砖砌体 reinforced hollow unit masonry; reinforced grouted brick masonry
配筋灌浆砌石砌体 reinforced grouted masonry
配筋过多的 over-reinforced
配筋过多的混凝土 over-reinforced concrete
配筋混凝土砌块结构 reinforced concrete masonry construction
配筋混凝土砌体 reinforced concrete masonry
配筋空心砌块砌体 reinforced hollow unit masonry
配筋率 percentage of reinforcement; ratio of reinforcement; reinforcement percentage; reinforcement ratio; steel area ratio; steel percentage; steel ratio; stirrup ratio of reinforcement
配筋密度<以千克/平方米计> reinforcement density
配筋明细表 bar list
配筋耐火砖 reinforced refractory brick
配筋砌块工程 reinforcement blockwork
配筋砌体 reinforced masonry
配筋墙 reinforced wall
配筋砂浆带 reinforced mortar band
配筋说明 reinforcement notes
配筋图 arrangement of bars; arrangement of reinforcement; bar arrangement drawing; detail drawing of reinforcement; reinforcement drawing
配筋圬工 reinforced masonry
配筋限度 reinforcement limitation
配筋形式 form of reinforcement
配筋指数 reinforcement index
配筋砖 reinforced brick
配筋砖过梁 brick beam
配筋砖结构 reinforced brick construction
配筋砖砌体 reinforced brick masonry;reinforced brick work
配筋砖墙 reinforced brickwork wall
配筋砖圈梁 reinforced brick ring beam
配景 foil; objective view; perspective;entourage【建】
配景的 perspective
配克<英国量名,粒状物容量单位,合 1/4 蒲式耳,1 配克 =9.09 升或 2 加仑> peck
配矿 ingredient ore
配矿方法 method of ingredient
配框(架) enframe
配离子 complex ion
配离子生成法 complex formation method
配力钢筋 distribution bar; distribution bar reinforcement; distribution rod; distribution steel; lacing; lacing bar
配力轴 distribution shaft
配凉台高平房<印度供旅游者住宿的> dawk
配量 dosing;proportioning;ration
配量泵 proportioning pump
配量比 dosing ratio
配量斗 quantifier
配量阀 metering valve; proportional valve;proportioning valve
配量阀组 metering block
配量计 proportioning meter; quantifier
配量给料器 ratio feeder
配量螺旋 metering auger
配量器 dispenser
配量油流 metered flow of oil
配量装置 dosing mechanism; metering device
配料 batch(feeder); batching <混凝土的>; blend(ing); burden(ing); charge material; charge mixture; compounding; dosage; dosing; feed proportioning; furnish; ingredient; make-up of charge; mixture(making);proportioning
配料板 distributor plate
配料拌和时间 concrete mixing time
配料报告 dispersing report
配料泵 dispenser; dispensing pump; dispersing pump
配料比 burden; charge ratio; mixture ratio; ratio of components; ratio of mixture
配料比计算 mix calculation
配料表 batch table; burden sheet; tabulation of mixture
配料部分 proportioning section
配料部门 batching department; batching section
配料仓 batch bin; blending bin; distributing bin; proportioning bin; stock bin
配料舱 batch bin; blending bin; distributing bin
配料操作 batch operation
配料槽 dosage bunker; measuring pocket
配料槽式输送机 proportioning trough conveyer[conveyor]
配料槽输送机 batching trough conveyer[conveyor]
配料厂 batch plant;proportioning plant
配料场 batch(ing)plant
配料车 batch cart; batch lorry; batch truck; burden charging carriage
配料车间 batch house; batch plant; batch preparation plant; furnishing department;proportioning plant
配料称 proportion scale
配料称量车 proportioning cart
配料称量传动装置 proportioning scale
配料称量单元 proportioning unit
配料称量斗 proportioning bin
配料称量鼓筒 proportioning drum
配料称量盒 proportioning box
配料称量机组 proportioning unit
配料称量架 proportioning frame
配料称量螺杆输送机 proportioning worm conveyer[conveyor]
配料称量螺旋输送机 proportioning screw conveyer[conveyor]
配料称量器 batcher;batch weigher
配料称量容器 proportioning container
配料称量输送带秤 proportioning conveyor belt scale
配料称量箱 proportioning box; proportioning frame
配料称量循环 proportioning cycle
配料成分 batch composition
配料成分控制 raw mix composition control
配料程序分选器 proportioning selector
配料秤 batcher scale; batch weigher; batch weighing scale; proportioning scale; weigh batcher <工地用>
配料池 dosing chamber
配料储仓 batch holding bin
配料传送槽 measuring trough conveyer[conveyor]
配料传送带 measuring trough conveyer[conveyor]
配料单 batch sheet; charge(r) sheet; charging sheet;mixing ratio sheet
配料的断续拌和 batch mixing
配料的砂浆空隙法 mortar-void method of proportioning materials
配料的制备 preparation of batch
配料的组合 composition of batch
配料斗 batch bin; batch hopper; distributing bucket; ga(u)ging box; weighbatching hopper <按重量的>
配料断续拌和厂 batch mixing plant
配料断续拌和时间 batch mixing time
配料断续时间 batch mixing time
配料阀门 proportioning lock
配料方法 distribution
配料房 batch house;mixing room
配料给料装置 aggregate feeder unit
配料工厂 batch(ing)plant
配料公式 batch formula
配料供应品 ingredient supply
配料灌区 dispersing depot; dispersing tank farm
配料灌容量 dispersing tankage
配料规范 mixture specification
配料规范控制图 "standard given" control chart
配料后直接入磨 direct dosing
配料混合试验法 line blend
配料机 proportioner
配料及搅拌设备 batching and mixing plant
配料计 proportioning meter
配料计量器 batch counter; batcher; batching machine
配料计数器 batch counter
配料计算 batch calculation; burden calculation; charge calculation; mix proportion(ing) calculation; mixture calculation;mixture making
配料计重器 weigh batcher
配料记录 batch counting; batch record
配料记录本 charge book
配料记录器 batch counter; batch recorder

配料加热器 < 混凝土骨料 > batch heater
配料架 batching frame
配料间 weighing area
配料渐减 batch decrescence
配料桨式拌和机 batch paddle mixer
配料绞刀 proportioning screw conveyer [conveyor]; proportioning screw feeder; proportioning worm conveyer[conveyor]
配料搅拌塔 batching and mixing tower
配料卡车间隔 batch lorry compartment
配料控制 blending control
配料跨 stockyard bay
配料量 dosage
配料量斗 batch box; batcher
配料料斗 batch hopper
配料螺旋计量输送机 batching screw conveyer[conveyor]
配料螺旋计量运输机 batching screw conveyer[conveyor]
配料螺旋输送机 proportioning screw conveyer[conveyor]
配料螺旋喂料机 proportioning screw feeder; proportioning worm conveyer[conveyor]
配料皮带 proportioning belt
配料皮带秤 weigh belt feeder
配料品 batch box
配料平路法 < 用平地机或推土机 > blading method of proportioning
配料起重机 make-up crane
配料器 batcher; batch unit; dispenser; measuring hopper; mix selector
配料器装置 batcher installation
配料轻便车 batching cart
配料区 dispensing depot; dispensing station; dispersing bulk plant
配料区段 proportioning section
配料容器 batch(ing) container
配料乳胶的制备 latex compounding
配料砂 lending sand
配料设备 batching equipment; dispensing equipment; dosing appliance; dosing device; dosing instrument; proportioning device; proportioning installation; proportioning plant
配料设计 design mix; design of mixture; mix design; mixture design
配料升降机 batch elevator
配料湿度表 batching hygrometer
配料湿度计 proportioning hygrometer
配料试验 batch test; dispensing test
配料缩减 batch reduction
配料锁定器 batching lock
配料塔 proportioning tower
配料塔楼 batching tower
配料台 batching desk
配料体积 batch volume
配料筒仓 batching silo; proportioning silo
配料喂料斗 batching hopper
配料误差 mixing error
配料系统 dosing system
配料箱 batch bin; batcher; batch hopper; batch(ing) box < 拌和楼的 >; batching frame; measuring frame
配料斜槽 distributing chute; distribution chute
配料星号 batching star
配料星行轮 batching star
配料星形架 proportioning star
配料选择器 batching selector
配料要求 burden requirement
配料业务 dispersing service
配料用胶带秤 proportioning belt balance; proportioning belt scale
配料圆筒 batching drum
配料运货卡车 batch lorry
配料运输带 batch(ing) conveyor belt
配料运送带 batch(ing) conveyor belt
配料闸门 proportioning lock
配料站 batch(ing) plant
配料者 mix selector
配料蒸锅 < 蒸馏 > feed make-up boiler
配料周期 batching cycle
配料专利 batch patenting
配料装料机提升装置 batch loader unit
配料装料装置 batch loader unit
配料装置 batching installation; batching apparatus; batching set-up; batching unit; distributor; proportioning installation; proportioning plant
配料作业 batch operation
配列方法 seriation method
配列控制 configuration control
配流阀 flat valve
配流盘 port plate; valve plate
配流盘表面 plane valving surface
配流因子 orificing factor
配流轴 pintle
配楼 accessory block
配螺旋钢筋的混凝土柱 spirally reinforced column
配螺旋形钢箍的混凝土 spirally bound concrete
配煤 coal blending; mixing coal
配煤比 coal blending rate
配煤厂 coal mixing plant
配煤设备 coal mixing plant
配煤系统 coal proportioning system
配模更换法 module replacement
配磨 mated pair
配磨自动定尺寸 matching sizing
配磨自动定时磨削 match grinding
配偶 partner
配偶(部)件 mating part
配偶电子 paired electrons
配偶津贴 spouse allowance
配平 balancing
配平电阻 balancing resistor
配平角 angle of trim(ming)
配平调整片 trimming tab
配漆 let down
配漆机 colo matching system; colo(u)r matching system
配漆用漆料 mixing varnish
配齐 assort
配齐工具 truing tool
配气 distribution of gas
配气板 pneumatic panel
配气吊顶 air-distributing ceiling
配气顶棚 air-distributing ceiling
配气定时 valve timing
配气阀 air-distributing valve; air distributor; air-thrown valve; compression valve; distributing valve
配气阀形式 form of distribution value
配气管道 gas distribution pipe line
配气机构 valve gear
配气器 air distributor; gas distributor
配气吸声顶棚 air-distributing acoustic(al) ceiling
配气系统 air regulating system; distribution system; gas distribution system
配气相位 phase of valve distribution
配气相位图 valve timing diagram
配气箱 air chest
配汽杆系 linkwork
配汽主管 vapo(u)r distribution line
配球台 marble distribution station
配入量 loading; dosage
配色 blendent; blending colo(u)r; colo(u)r combination; colo(u)r ring; colo(u)r match(ing); match the colo(u)r
配色表 colo(u)r system
配色程序 colo(u)r matching program(me)
配色法 colo(u)r matching method; colo(u)r scheme
配色方案 scheme of colo(u)r
配色函数 colo(u)r matching function
配色基釉 base glaze for colo(u)ration
配色计算机 colo(u)rant mixture computer; colo(u)r computer; colo(u)r matching computer
配色间 colo(u)r mixing room; colo(u)r shop
配色浆 tint base
配色镜 chromatoscope
配色钮扣 matching button
配色人员 colo(u)rist; colo(u)r matcher
配色三角形 chromaticity diagram; colo(u)r triangle
配色色位 colo(u)r way
配色师 colo(u)rist
配色调和喷漆 mixing lacquers
配色线缝 matching seam; matching stitching
配色性能试验 colo(u)r matching aptitude test
配砂工段 sand-conditioning plant
配砂箱 sand proportioning box
配膳厨房 serving kitchen
配膳间 pantry room; serving room
配膳时间 serving time
配膳室 buttery; pantry
配膳室与食堂之间的窗口 buttery hatch
配声 dub
配时 timing
配时计划 < 自动及联动信号系统用 > timing plan
配时键 timing key
配受压筋的梁 double reinforced beam
配竖筋砖砌体 brick masonry with vertical bars; quetta bond < 巴基斯坦砌墙法 >
配水 water-distribution; water proportioning
配水泵 water service pump
配水泵调整轮 water pump idle roller
配水泵性能 water pump performance
配水槽 distribution box; distribution channel; distribution header
配水池 distributing reservoir; distribution basin; distribution reservoir; distribution tank; service basin; service reservoir
配水道 distribution conduit
配水点 distribution point; point of water distribution
配水阀 delivery valve; distributing valve; distribution valve
配水法 water distance
配水干管 distributing main; distribution main; service main
配水干管系统 arterial system
配水干渠 distributing main; distribution main
配水干线 distributing main; distribution main
配水干线系统 arterial system
配水工程 distribution works
配水沟 distributing ditch
配水沟槽 distributing gutter; distribution gutter
配水沟支流 distributary
配水管(道) distributing pipe; distributing water pipe; water distribution piping; distributary; distribution header; distribution pipe; sparge pipe; spiral distributor; water-distributing pipe
配水管渠网 distributing net(work)
配水管图 water service pipe plan
配水管网 distributing net(work); distribution net(work); water distribution network; distribution grid; distribution network piping system; distribution system (of water supply); pipe[piping] system; separating pipe; water distribution pipe; distribution system of water
配水管网系统 water distribution system
配水管系统 water service system; main system
配水管系图 water service plan
配水管线 distributing(pipe) line; distribution(pipe) line; water distribution piping
配水井 conduit box; distributing well; distribution box; distribution well
配水孔 distribution orifice
配水孔板 honeycomb plate
配水库 service reservoir
配水量 distribution capacity; water distribution capacity
配水量水箱 delivery and measuring box
配水龙头 distribution cock
配水母管 distribution manifold
配水盘装置 paring disc device
配水歧管 distribution manifold
配水器 distributor; water distributor
配水区 distribution zone
配水区域 distributing area; distribution area
配水渠 distributing channel; distributing ditch; distribution canal; distribution channel; distribution ditch; distributor
配水设备 distribution equipment; water distribution installation
配水试验 water supply test
配水水库 distribution reservoir
配水水头 distributing head
配水损失 distribution loss; transmission loss
配水塔 distribution tower
配水瓦管 distribution tile
配水网 distribution system; water distribution system
配水系统 distributing system; distribution system; system of distribution; water distribution system
配水系统的法向压力 normal dynamic(al) pressure of distribution system
配水系统生物稳定性 distribution system biostability
配水系统水质 distribution system water quality
配水系统网络 water distribution system network
配水系统压力试验 pressure test of distribution system
配水系统中的储水设施 distribution reservoir
配水系统组成部分 distribution system component
配水箱 delivery box; distribution box; distribution reservoir; service reservoir; sprinkler basin; water make-up tank
配水消防栓 distribution hydrant
配水压力 distribution pressure
配水闸(门) check gate; delivery gate; distribution gate; distribution structure; division box
配水支管 branch distributing pipe; distributing branch

配水装置 distribution apparatus; flow distributing device; water distribution installation; water service installation
配水总管 distribution main; distribution manifold; distribution line
配水总渠 distribution main
配水总线 distribution main
配速齿轮 selective speed gear
配速活塞 speed setting piston
配酸 complex acid
配套 assort; complement; corollary; fit; form a complete set; match(ing); match together; mating; set-up; unitization
配套百分率 percentage of complement
配套材料 associated material
配套程度 level of integrity
配套出售 systems selling
配套储藏的消防工具 fire-tool cache
配套措施 coordinated sets of measures; supporting measure
配套单据 aligned series
配套的 associated; assorted; self-contained; self-contained complete
配套的经济体制改革 complete set of economic structural reform
配套的水下呼吸器 scuba[self-contained underwater breathing apparatus]
配套的卫生设备 self-contained sanitation
配套的消防探测装置 self-contained fire detector unit
配套的小型钢铁联合企业 small self-contained iron and steel complex
配套电机型号 type of mating electric (al) machine
配套法兰盘 companion flange
配套服务 adequate and systematic service
配套工程 auxiliary project; parts and accessories for imported equipment
配套挂图 flip chart
配套机 building machine
配套机械 supporting machinery; auxiliary machinery
配套件 parts and auxiliary equipment; parts and components
配套建筑物 attached building
配套扩充插件 extended card kit
配套零件 kit
配套能力 ability to provide the auxiliary items
配套齐全的排水系统 self-contained drainage system
配套器材 necessary accessories
配套认可书 parts release notification
配套商品 closed stock
配套设备 associated equipment; corollary equipment; matching equipment; supplementary equipment; supporting facility
配套设计的 functionally designed
配套设施 associated facility; corollary facility
配套生产 form a complete production network
配套使用 fitment
配套无线电零件 multiple component units
配套系统 complete system; self-contained system
配套项目 support item
配套样机部件 prototyping kit
配套硬模型 matched die mo(u)lding
配套元件 kit
配套运输 transportation in assembled state
配套装置 rig
配套资金 counterpart finance; counterpart fund; supporting fund
配套组合 form a complete set of
配调整用平衡重块 adjustable counterweight
配位【化】 coordination; configuration; coordinate
配位场理论 ligand field theory
配位场稳定化能 ligand field stabilization energy
配位催化 complex catalysis; coordinating catalysis
配位催化剂 coordinating catalyst
配位滴定(法) complexometric titration; complexometry; coordinate titration
配位度 complex formability
配位多边形 coordination polygon
配位多面体 coordination polyhedron
配位多面体规则 coordinated polyhedral rule
配位多面体类型 type of coordination polyhedron
配位多面体连接方式 connection type of coordination polyhedron
配位反应 complexation reaction
配位高聚物 coordination polymer
配位共价键 coordinate covalent bond
配位化合物 coordinated complex; coordination complex; coordination compound; Werner complex; complex compound
配位化学 coordination chemistry
配位活性 coordinative activity
配位基 dentate; ligand
配位基吸附 dentate adsorption
配位基型 coordinate motif(pattern)
配位价 coordinate bond; coordinate [coordination] valence
配位价力 coordinate valence force
配位键 coordinate bond; coordination link(age)
配位结构 coordination structure
配位晶格 coordinate crystal; coordination lattice
配位晶体 coordinate crystal
配位聚合 coordination polymerization
配位聚合物 coordinate polymer; coordination polymer
配位离子 coordinating ion
配位理论 coordination theory
配位络盐 coordinate complex salt
配位羟离子 complex hydroxy ion
配位色谱法 ligand chromatography
配位式 locate mode
配位数 coordinate number; coordination number
配位水 coordinated water; water of coordination
配位体 ligand
配位体份额 ligand share
配位体交换 ligand exchange
配位体交换层析 ligand exchange chromatography
配位体交换色谱法 ligand exchange chromatography
配位体聚合物 ligand polymer
配位体离子 ligand ion
配位体浓度 ligand concentration
配位体溶剂 ligand solvent
配位体色谱(法) ligand chromatography
配位体数 ligand number
配位仪 coordinator
配位仪程序 coordinator routine
配位异构 coordination isomerism
配位异构体 coordination isomer
配位原子 coordinating atom
配伍组试验 randomized block experiment
配线 conductor arrangement; conductor configuration; curve fit(ting); wiring; siding【铁】
配线板 control panel; patchboard; plugboard; wiring board
配线表 jumper list
配线程序 externally stored program(me)
配线程序计算机 wired program(me) computer
配线点 distribution point
配线电缆 distributing cable; distribution cable; intermediate cable
配线电缆管 distribution channel
配线法 superimposed line method
配线符号 wiring symbol
配线工工作 wiring work
配线管(道) distribution channel; subsidiary conduit
配线盒 distributing box; outlet box
配线盒进线孔 pot head tail
配线护墙板 wiring clapboard; plugmold
配线架 connecting rack; cross connecting field; distributing frame; distribution frame; line distributor
配线架的接线端子 distribution terminal
配线绝缘子 distributed insulator; distributing insulator
配线盘 control panel; distributing board; distribution block; distribution board; patch bay; patch(ing) panel; plugboard
配线情况 wiring condition
配线区 distributing area; distribution area; distribution district
配线人孔 distribution chamber
配线熔丝盘 distributing fuse board; distributing fuse panel
配线熔线盘 distribution fuse panel
配线设备 commutation equipment
配线踢脚板 plugmold
配线图 allocation scheme; layout drawing; scheme of wiring; wiring scheme
配线装置 bus compartment; switch gear
配谐 chimb[chime]
配研 face up; facing up
配盐 complex salt
配阳离子 complex cation
配药 dispensation; dosage; pharmacy
配药方 compounding
配药者 dispenser
配页 collating; gathering
配页机 collator
配页装钉联动机 gather-stitcher
配阴离子 complex anion
配音 dub
配音复制 dubbing
配音机 dubbing machine
配用车<装运伸臂起重机时,需要另挂一辆容纳伸臂的车辆> match wagon
配用功率过大(的驱动装置) overpower
配用马达 attached motor
配油泵 distributor pump
配油器 oil distributor
配油站 bulk station; petroleum bulk station
配有车轮的 mounted on wheels
配有堵头的 blanked off
配有盖板的 blanked off
配有箍筋的柱 column steel hooping; column with lateral tie reinforcement
配有计算机的拌和厂 computer-aided batcher plant
配有计算机的配料厂 computer-aided batcher plant
配有家具的厨房单元 unit furniture kitchen
配有立柱的平车 flat car with stanchions
配有螺旋筋的柱 column with spiral hooping
配有人造光的植物灌溉 plant irrigation with artificial light
配有弹簧制动器的踏面制动单元 tread brake unit with spring loaded brake
配员 crew
配载【船】 stowage
配载货物 berth cargo
配载计划 cargo planning; prestowage planning; stowage planning
配载计划员 stowage planner
配载气 make-up carrier gas
配载清单 boat note
配载设备 corollary equipment
配载图 cargo plan; loading plan; prestowage plan; stowage plan
配载要求 stowage requirement
配载仪 loading instrument
配渣计算 slag calculation
配制 configurate; make-up; making; processing
配制拌和的材料 batched material
配制拌和的骨料 batched aggregate
配制拌和的集料 batched aggregate
配制拌合[和]物 batching
配制产品 formulated products
配制工作 make-up job
配制供使用的涂料 paint prepared for use
配制供使用的油漆 paint prepared for use
配制胶 prepared glue
配制室 dosing chamber
配制水 make-up water
配制油 make-up oil
配质 genin
配置 allocate; allocation; arrangement; configuration; decentralization; decentralize; disposal; dispose; disposition; distribution; gadget; ordonnance; positioning; set(ting) up; siting; staging; structure
配置标识 configuration identification
配置程序 configurator
配置磁铁 distributed magnet
配置错误的线圈 misplaced winding
配置点 collocation point
配置法<结构力学中的> collocation method
配置钢筋 embedded reinforcement
配置过程 layout procedure
配置和布局 layout and location
配置键 grouping key
配置晶体管的脉冲计数器 transistor scaler
配置开关 configure switch
配置控制 configuration control
配置控制寄存器 configuration control register
配置排水管 disposal drain
配置平面图 disposition plan
配置器 configurator
配置审查 configuration audit
配置受压钢筋的梁 beam with compression steel
配置图 allocation plan; arrangement drawing; arrangement of plan; arrangement plan; layout(plan); set-up diagram; storage plan
配置误差 error of allocation

配置线 layout line
配置项目 configuration item
配置效率 allocative effect; allocative efficiency
配置选择 configuration option
配置阴极射线管的加法器 cathode-ray tube adder
配置于产品上 put part on products
配置于一线 dead in line
配置圆 layout circle
配置职能 allocation function
配置指令 configuration command
配置状态报告表 configuration status accounting
配置资金 deploy funds
配重 additional weight; back balance; balance(d) weight; balancer; balancing weight; ballast (weight); bob-weight; compensating weight; counter-balance; counter-balance weight; counterweight; dummy mass; equipoise; mass balance; trimming ballast; weight counterbalance
配重拨火门 counterpoised poke door
配重滑车 counterweight pulley
配重均衡制动 counter-balanced brake
配重块 counterweight
配重块底座 setting base for counter weight
配重块伸出后的内间隙 < 吊机 > under extended counterweight clearance
配重块座架 counterweight mount
配重平衡 counterweigh
配重平衡块 balance
配重平衡器 weight equalizer
配重铁块 kentledge
配重箱 ballast tank; weight basket; weight box; weight tray
配重用混凝土 ballast concrete
配重运输机带 counterpoised conveyor belt
配重闸 deadweight brake
配重制动器 counterweight brake
配属车 allocated car
配属机车【铁】allocated locomotive
配属枢纽 home terminal
配属站 < 列车或特种车辆 > home station
配属终点站 home terminal
配装号 number for matched loading
配准 matching; registering; registration; registry
配准不佳 lack of registration
配准不良 misregistration; misregistry
配准测试卡 registration chart
配准的显像管 registered tube
配准度 registration
配准精度 registration accuracy
配准控制 registration control
配准漂移 registration drift
配准缺陷 registration fault
配准误差 registration error
配准系统 registration arrangement
配准线 matching line
配子 gamete
配子体 gametophyte

喷

喷氨处理 injection of ammonia
喷白浆两道 sprayed twice with white wash
喷摆动杆 nozzle oscillating lever
喷斑 spray mottle
喷笔 air brush; air compress ink spray; atomizer; hand spray
喷笔系统 spray system
喷玻璃丸清表面 glass bead cleaning
喷播 seed spraying
喷补 spray repair
喷补的内衬 gunned lining
喷补机 gun
喷补料 gunning mix; gunning refractory
喷补炉衬 gunning lining
喷补盛钢桶 gunned ladle
喷补用耐火材料 gunned refractory
喷布机 distributor; spray bar; sprayer
喷布机喷管 distributor nozzle
喷布器 priming nozzle
喷布砂浆 spread mortar
喷彩 aerograph; air brush; spray decoration
喷层 spray-up
喷层厚度 depth of shotcrete
喷成雾 pulverize
喷成雾状的燃油 atomized oil
喷程 carry of spray; throw
喷出 belching; blowing; blow-off; blowout; boiling; break forth; effuse; eject (ion); extrude; gush; gust; jetting; outgush; spout; squ(i)rt; upwelling; extrusion < 熔岩等的 >
喷出层 eruptive sheet
喷出冲程 ejection stroke
喷出的 anogene; effusive; eruptive; extrusive; jetted
喷出的地热流体 outcoming geothermal fluid
喷出的堆积物 eruptive stock
喷出的水 squirt water
喷出火成岩 extrusive igneous rock
喷出角砾岩 effusive breccia
喷出口 ejection opening; issuing jet; jet orifice; outlet port < 小炉的 >
喷出口顶 port roof
喷出口端墙 port end
喷出口拱顶 port arch
喷出流 discharging jet
喷出率 emission rate
喷出气体 emission gas
喷出试验 spout test
喷出水 atomized water
喷出速度 spouting velocity
喷出物 effusion; ejecta; ejected matter; eruption; gusher; spew
喷出物面积 area of volcanic products
喷出物体积 volume of volcanic products
喷出物质 ejected matter
喷出相 extrusive facies
喷出压力 flowing pressure
喷出岩 effusive rock; ejected rock; eruptive rock; extrusive rock
喷出岩体 extrusive rock body
喷出岩席 extruded sheet
喷出蛛网状漆丝 cobwebbing
喷吹 blowing; jetting
喷吹玻璃棉 blowing wool; blown wool
喷吹出 blowing
喷吹法 blowing process
喷吹精炼 spray refining
喷吹率 injection rate
喷吹棉 blowing wool
喷吹清灰的袋式吸尘器 jet filter
喷吹清理 blast cleaning; power cleaning
喷吹燃烧器 blowing burner
喷吹设备 < 玻璃纤维 > bushing blower
喷吹施工玻璃棉 fiberglass blown wool
喷瓷 ceramic spraying
喷大白 whitewashing
喷的铁丸 blasting shot
喷灯 air burner; bench torch; blast burner; blast lamp; blow lamp; blow torch; brazing torch; Bunsen-type burner; burner; flame gun; flush burner; heating lamp; jet burner; plumber's furnace; smokeless burner; soldering lamp; spray burner; torch (burner); torch lamp; torch pipe; sealing-in burner < 封闭玻管用 >
喷灯除旧漆 torching
喷灯除漆 torching
喷灯焊接处 blown joint
喷灯环 burner ring
喷灯机 torch machine
喷灯口 burner nozzle; burner orifice; burnt orifice
喷灯喷嘴 nipple
喷灯铺毡 torch-applied method
喷灯射口 burner head; burner jet; burner nozzle
喷灯施工油毡 torch-applied asphalt felt; burner construction linoleum
喷灯铜焊 torch brazing
喷灯头 torch head
喷灯油 torch oil
喷灯嘴 burner nozzle; nipple
喷电电势 spray potential
喷电电源 spray supply
喷电极 jet electrode
喷电针尖 spray point
喷电针支撑杆 needle bar
喷动床 spouted bed
喷动床干燥 spouted bed drying
喷动床干燥器 spouted bed drier[dryer]
喷动流化床 spouted fluidized bed
喷镀 cladding; deposition; jet plating; metallization; metallizing; plating; sherardizing; spatter; spraying; spray plating
喷镀薄膜 sputtered film
喷镀层 deposited metal
喷镀的金属 sprayed metal
喷镀堆焊 spraying overlay
喷镀法 metallikon; metal spraying; spray-on process
喷镀范围 area of deposition
喷镀工人 plater
喷镀管道 coating pipe
喷镀金属 metallic cementation; metallikon; metal(l)ing
喷镀金属层 spray metal coating
喷镀金属的灯泡 metalized lamp bulb
喷镀金属法 metallization; metallizing; spray metal coating
喷镀金属(膜)玻璃 metallized glass
喷镀金属器 metallizer
喷镀金属陶瓷 metalloceramics
喷镀面积 area of deposition
喷镀枪 refractory gun; spray gun; spray torch
喷镀区 area of deposition
喷镀软钎焊 spray soldering
喷镀锌 zinc spraying
喷镀用的金属丝 spray wire
喷发 blasting; effusion; ejection; erupt(ion)【地】; outburst
喷发沉积 eruptive deposit
喷发次数 times of eruption
喷发道 vent of eruption
喷发的 effusive
喷发方式 eruption type
喷发管道【地】eruption canal; eruption pipe
喷发环境 environment of eruption
喷发火山渣 ejected scoria
喷发角砾岩 eruptive breccia
喷发口 vent of eruption
喷发矿床 eruptive mineral deposit
喷发类型 eruption type
喷发裂缝 eruption fissure
喷发裂隙 eruption fissure
喷发模式 eruption mode
喷发年代【地】time of eruption
喷发凝灰岩 eruptive tuff
喷发期 effusive period; effusive stage
喷发前兆 eruption symptom
喷发强度 eruption intensity
喷发体 effusive mass
喷发物质 eruptive material
喷发相 eruptive facies
喷发形式 eruption type
喷发岩 effusive rock; eruptive rock; explosive rock
喷发岩盖 eruption laccolith; eruptive laccolith
喷发岩相 eruptive facies
喷发韵律 eruption rhythm
喷发征兆 eruption symptom
喷发锥 cone of eruption
喷发作用 eruptive process
喷放管道 blow line
喷放密度 discharge density
喷放形式 discharge pattern
喷放装置 discharge device
喷粉 disperse; dusting; powder spraying
喷粉复印 smoke printing
喷粉管 discharge tube
喷粉机 duster sprayer; dusting machine; powder applying machine; powder gun; powdering machine
喷粉黏[粘]土 dusting clay
喷粉器 aeroduster; blow gun; disperser; dry sprayer; duster; dusting beak; powder blower; powder sprayer
喷粉式加料器 spout feeder
喷粉压送系统 < 喷粉机的 > spout delivery system
喷粉桩 jet powder pile
喷粉嘴 dust distributor
喷风口 air nozzle
喷敷 sprayed coating
喷敷管 pistol
喷敷枪 pistol
喷敷石棉层 sprayed asbestos
喷干釉 dusting
喷杆 injection rod; nozzle mast; spray lance
喷杆延长杆 boom extension
喷杆组 < 喷雾器的 > boom section
喷割器 cutting blowpipe
喷管 atomizer device; atomizing pipe; blast pipe; effuser; injection cock; jet exit; jet nozzle; jet piping; nozzle; nozzle tube; spray bar; spray tube
喷管摆角 angle of cant
喷管边缘 lip of the jet
喷管衬垫 nozzle liner
喷管衬套 nozzle liner
喷管尺寸 size of the jet
喷管出口 nozzle exit
喷管出口动量 nozzle exit momentum
喷管出口截面 nozzle exit section; tip of nozzle
喷管出口截面面积 exit area; jet area
喷管出口面 nozzle-exhaust plane; nozzle exit plane
喷管出口速度 nozzle velocity
喷管传热率 nozzle heat transfer rate
喷管的收敛部分 convergent section
喷管电磁阀 nozzle solenoid
喷管端部 nozzle-end
喷管轭 nozzle tube yoke
喷管阀门 nozzle valve; orifice tap
喷管法兰 nozzle flange
喷管隔片 nozzle blade; nozzle vane

喷管隔热层 jet pipe shroud
喷管根部 nozzle-end
喷管箍环 nozzle ring
喷管管夹 nozzle block
喷管喉部 nozzle throat
喷管喉部面积 nozzle throat area; throat opening area
喷管喉口 nozzle throat
喷管环 nozzle ring
喷管火焰 nozzle flame
喷管架 nozzle tube yoke
喷管进口的气体参数 nozzle inlet condition
喷管进口气流马赫数 nozzle inlet Mach number
喷管进口温度 nozzle inlet temperature
喷管开口比 nozzle opening ratio
喷管口 nozzle opening
喷管扩张损失因数 nozzle divergence loss factor
喷管冷却 nozzle cooling
喷管临界截面 exit nozzle throat; throat plane
喷管临界截面积 throat area
喷管临界截面压力 nozzle throat pressure
喷管流量特性 metering characteristics of nozzle
喷管帽 blast pipe cap
喷管面积比 exit area ratio
喷管内壁形状 nozzle interior contour
喷管排气测温计 jet pipe thermometer
喷管喷嘴 jet pipe tip
喷管膨胀比 nozzle expansion ratio
喷管切口 nozzle edge
喷管倾斜 cant of the jet
喷管曲率 nozzle curvature
喷管入口 nozzle entry
喷管入口段 nozzle entrance section
喷管实际流速 actual exhaust velocity
喷管式流量计 nozzle flow meter
喷管式喷雾器 boom sprayer
喷管收敛系数 jet area contraction coefficient
喷管水头损失 nozzle head loss
喷管调节 nozzle control governing; nozzle cut-out governing; nozzle governed
喷管调节螺钉 nozzle adjusting screw
喷管通道最小截面 minimum nozzle area
喷管通路面积 nozzle area
喷管头子 nozzle tip
喷管凸缘 nozzle flange
喷管推力系数 nozzle thrust coefficient
喷管外形 nozzle contour
喷管温度调节器 jet pipe temperature controller
喷管系统 sprinkler pipe system
喷管形状 nozzle shape
喷管形式 nozzle shape
喷管压力 nozzle exit pressure
喷管压力比 nozzle pressure ratio
喷管液压控制器 jet pipe oil-operated controller
喷管噪声 nozzle noise
喷管胀圈 nozzle filler block
喷管罩 nozzle shroud
喷管直径 jet size; size of the jet; width of nozzle
喷管锥体 jet tail cone
喷灌 irrigation by sprinkling; overhead irrigation; projecting spray irrigation; shower; spouting; spray irrigation; sprinkler irrigation; sprinkling (irrigation); squirting irrigation
喷灌草地 sprinkling lawn
喷灌场 sprinkling bed
喷灌防霜冻 frost protection by sprinkler irrigation
喷灌覆盖面 sprinkler coverage
喷灌管道 spray line
喷灌管道系统 spray pipe system; sprinkler pipe system
喷灌花圃 sprinkling bed
喷灌混凝土 guncreting
喷灌机 irrigator; rainger; sprinkler; sprinkling machine; watering machine; air placer; air placing machine <混凝土>
喷灌机控制箱 sprinkler box
喷灌机喷嘴 sprinkler nozzle
喷灌机器 water sprinkler
喷灌剂 injection agent; injection aid
喷灌滤池 spraying irrigation filter
喷灌配水设备 water distribution equipment
喷灌器 injection agent; sprinkler apparatus
喷灌枪 sprinkler gun
喷灌强度 intensity of spray irrigation
喷灌设备 rainmaker
喷灌试验小区 sprinkled trial plot
喷灌头 sprinkler gun
喷灌系统 spraying irrigation system; sprinkler system
喷灌系统中的出水口 outlet for water in a sprinkler system
喷灌形式 sprinkler pattern
喷灌用软管 squirt hose
喷灌油灰 gun(ned) putty
喷灌雨 delivery of rain
喷灌支管 sprinkler lateral
喷灌装置 irrigation rig; rainer; sprinkler plant
喷灌装置摆动机构 sprinkler oscillating mechanism
喷灌装置的喷管 sprinkler jet
喷灌装置管道移动器 pipe-mover
喷灌装置管接头 sprinkler coupler; sprinkler coupling
喷灌装置调节器 sprinkler control
喷柜 spray booth
喷焊 surfacing
喷焊器 blow pipe; spray welding unit; welding blowpipe
喷焊器混合管 mixing head
喷壶 spray can; sprinkler; sprinkling can; watering pot
喷花 spray decoration
喷画笔 aerograph
喷灰浆 Tyrolean plaster
喷灰泥 torching
喷混凝土 shotcrete; shotcreting; sprayed concrete; spraying-on concrete
喷混凝土衬砌设计 shotcrete support design
喷混凝土的施工缝 board butt joint
喷混凝土方法 shotcrete method
喷混凝土盖面 flash coat
喷混凝土回弹 shotcrete rebound
喷混凝土机械手 spray robot
喷混凝土空隙 void in shotcrete
喷混凝土肋 shotcrete rib
喷混凝土喷射机 shotcrete sprayer
喷混凝土枪 shotcrete gun
喷混凝土应用 shotcrete application; shotcrete usage
喷混凝土质量控制 shotcrete quality control
喷火 breaking-out; flaming; spread out; sting-out
喷火管架 blowpipe stand
喷火孔 eye
喷火口 bocca; crater; port mouth; port opening
喷火口砖 burner block; quarl block
喷火口状的 crateriform
喷火器 flame projector; flame thrower; liquid fire gun
喷火筒 gallery burner
喷火焰 application of flame
喷火凿槽 flame gouging
喷火自流井 blow-well
喷火嘴 burner; nipple
喷溅 dry spray; expulsion; flashing; jet; overspray; spatter; splash; sputtering; surface flash; spitting
喷溅腐蚀装置 spray pickling unit
喷溅麻点 spatting
喷溅喷嘴 spray-making valve
喷溅区 splash zone
喷溅润滑 splash lubrication; spray lubrication
喷溅式间歇泉 fountain geyser
喷溅水 splash water
喷溅涂覆法 sputtering
喷溅物 ejection; splashings
喷浆 cement mortar blowing; grout injection; guncreting; gunite mortar; gunite-shooting; gunned castable; gun spraying; jetcrete; refractory gun mix; shotcreting; spout of stock
喷浆保护层 gunite coat(ing); gunite covering; gunite layer
喷浆表层 gunite coat(ing); gunite layer
喷浆材料 gunite material; gunning material
喷浆层 gunite
喷浆车间 gunite plant
喷浆衬砌 gunite lining
喷浆吊顶 shotcrete ceiling
喷浆法 <混凝土、水泥浆等> gunite method; gunite process; mortar gunite method; shotcrete system; guniting; air placing
喷浆粉尘 air float fines
喷浆粉末 air float fines
喷浆粉刷 sprayed-on plaster
喷浆辅助器 guniting aid
喷浆盖层 flash coat
喷浆工 gunman
喷浆工作 gunite work
喷浆罐衬 gunite lining of tank
喷浆厚度 guniting thickness
喷浆环 gunite ring
喷浆混凝土 gunite(d) concrete; jetcrete; shotcrete
喷浆混凝土板 gunite slab
喷浆混凝土衬砌 shotcrete lining
喷浆机 cement gun; cement throwing jet; concrete injector; guniting machine; painter; patching machine; shotcrete equipment; shotcrete machine; throwing jet
喷浆孔 grout hole
喷浆面层 gunite coat(ing); gunite covering
喷浆黏[粘]合层 gun mix bond
喷浆黏[粘]合料 gun mix bond
喷浆黏[粘]结剂 gun mix bond
喷浆器 gunite gun; guniting machine; mortar sprayer; spray gun; white washing sprayer
喷浆枪 batch gun; gunning rig; refractory gun
喷浆设备 gunning rig; shotcrete equipment
喷浆式曝气池 fountain aerator
喷浆水泥 injection cement
喷浆涂层 gunite coat(ing); gunite covering
喷浆挖掘法 jet cutter method
喷浆挖泥法 <冲水打井,用泵吸泥> jet cutter method
喷浆系统 injection system
喷浆橡胶管 grout spray rubber hose
喷浆修理 gunned repair; spray finishing
喷浆用灰浆 pneumatically applied mortar
喷浆用水泥 gun consistence [consistency]
喷浆罩面 gun finish; gunite covering
喷浆整理 spray finishing
喷浆支护 gunite lining
喷浆直径 gunning diameter
喷浆装置 gunite plant; refractory gunning equipment
喷浇技术 gunning technique
喷胶机 adhesive sprayer
喷胶器 binder gun
喷金属粒(除锈) shot blasting
喷净 shot blasting
喷净法 blast cleaning; shot blasting
喷咀流量系数 nozzle flow coefficient
喷咀数量 numbers of nozzle
喷咀直径 nozzle size
喷咀阻力系数 nozzle friction coefficient
喷孔器 widow maker
喷口 nozzle outlet; bocca; discharge spout; ejector nozzle; jet exit; jet orifice; jetting tip; snoot; spraying jet
喷口被废气烧蚀 nozzle erosion
喷口尺寸 jet size
喷口端 nozzle-end
喷口端点火 nozzle-end ignition
喷口盖 nozzle cap; nozzle closure
喷口盖板 nozzle plate
喷口隔板 nozzle closure
喷口管 jet pipe
喷口接合 nozzle coupling
喷口截面 nozzle exit
喷口截面中心 center [centre] of jet exit
喷口控制器 nozzle register
喷口块 port block
喷口扩张 nozzle divergence
喷口面积 area of injection orifice; nozzle exit area; throat area
喷口内锥 cone in the jet
喷口砌块 port block
喷口送风 nozzle outlet air supply
喷口送风系统 air jet system; ejector system
喷口速度 muzzle velocity
喷口调节盘 nozzle orifice disk
喷口调节片 efflux door
喷口调节通路 throttle channel
喷口调节锥 exhaust cone; throat bullet
喷口通道塞 jet passage plug
喷口外部火焰 off-port flame
喷口形状 nozzle configuration
喷口形状修正 nozzle-contour correction
喷口整流锥 bullet
喷口锥形体 bullet
喷口组 nozzle cluster
喷拉丝涂覆法 cobwebbing
喷粒处理 grit blast(ing)
喷裂 spit-out; spitting
喷淋 spray(ing); spray pump; spray shower; sprinkling; water spray
喷淋表面冷却器 Baudelot surfacing cooler
喷淋车 sprinkler truck
喷淋池 spray pond
喷淋池冷却 spray-pond cooling
喷淋除尘器 shower scrubber
喷淋冻结 spray(ing) freezing

喷淋阀 spray valve
喷淋分离器 spray separator
喷淋腐蚀装置 spray pickling unit
喷淋固化(过程) spray solidification
喷淋管 shower (pipe); spray line; spray pipe
喷淋罐 sprayed tank
喷淋辊 spray roll
喷淋机 spraying machine
喷淋机械 spraying machinery
喷淋加湿器 spray humidifier
喷淋降温屋顶 spray-pond roof
喷淋净化器 spray eliminator
喷淋绝缘 sprayed insulation
喷淋控制阀 sprinkler control valve
喷淋拉伸 spray drawing
喷淋冷却 shower cooling; spray cooling
喷淋冷却器 shower cooler; sprayed cooler
喷淋冷却塔 spray cooling tower
喷淋立管 sprinkle riser
喷淋流程 spray drift
喷淋密度 sprinkle density
喷淋黏[粘]膜法 spray bonding method
喷淋喷雾器 spray atomizer
喷淋皮带管 spray hose
喷淋器 spray jet; spray thrower; sprinkling machine sparger; water sprinkler
喷淋器阀 sprinkler valve
喷淋枪 spray pistol
喷淋软管 spray hose
喷淋润滑 spray lubrication
喷淋润滑器 spray oiler; sprinkling can
喷淋杀菌剂 spraying fungicide
喷淋筛 spraying screen
喷淋式抽提塔 spray-type extraction column
喷淋式冻结法 spray-freezing method
喷淋式冻结间 spray freezer
喷淋式冻结装置 spray freezer
喷淋式干燥器 spray-type drier[dryer]
喷淋式空气干燥器 spray drier[dryer]
喷淋式空气冷却器 spray-type air cooler
喷淋式空气洗涤器 spray-type air washer
喷淋式冷却 spray-type cooling
喷淋式冷却器 spray-type cooler; water drip cooler
喷淋式气体吸收器 spray gas scrubber
喷淋式气体洗涤塔 spray column; Glover tower
喷淋式热水加热器 cascade heater
喷淋式蛇管热交换器 spray-type coil heat exchanger
喷淋式浴缸 jet tub
喷淋式蒸发器 spray-type evapo(u)rator
喷淋试验 raining test; spray(ing) test
喷淋室 spray chamber; sprinkling chamber
喷淋水 shower water; spray(ing) water
喷淋水槽 spray shower recess
喷淋水管 spray header
喷淋水和去染剂系统 spray water and defoamant system
喷淋水接受器 spray shower receptor
喷淋塔 spray tower
喷淋头 spray header; spray rose
喷淋脱脂 spray degreasing
喷淋网 spraying screen
喷淋帷幕 spray shower curtain
喷淋屋面 sprayed roof
喷淋雾化器 spray atomizer
喷淋吸收器 spray absorber
喷淋洗涤 spray cleaning
喷淋洗涤器 sprayer-washer; spray scrubber
喷淋洗涤塔 spray scrubber
喷淋系统 drencher system; spray system; sprinkler system; sprinkling system
喷淋系统的设计密度 design density of sprinkler system
喷淋系统的设计面积 design area of sprinkler system
喷淋型热交换器 spray-type heat exchanger
喷淋压力 spraying pressure
喷淋养护法 sprayed-on method of curing
喷淋油 spray oil
喷淋浴 jet shower
喷淋折板 spray deflector
喷淋装置 spray system; sprinkler system
喷流 cascade; exhausted jet stream; jet current; jet efflux; plume; spout(ing)
喷流动量 momentum of the jet
喷流方向 jet direction
喷流管阀 jet pipe valve
喷流结构 jet structure
喷流静压力 jet static pressure; static jet pressure
喷流幕 jet curtain; spray sheet
喷流偏斜控制系统 jet system
喷流偏转 jet deflection; jet deflexion
喷流偏转器 jetavator
喷流偏转系统 jet deflection system
喷流速度 jet speed; jet velocity
喷流调节 jet control
喷流推进装置 jet-reaction unit
喷流温度计 jet temperature indicator
喷流温度指示器 jet temperature indicator
喷流星系 jet galaxy
喷流型 spray pattern
喷流诱导的环流控制 jet-induced circulation control
喷流噪声 jet noise
喷流重热温度控制 jet reheat temperature control
喷流状结构 jet-like structure
喷流阻尼 jet damping
喷硫(现象) sulfur blooming
喷滤池 sprinkling filter
喷路波动 spitting spray
喷路搭接流平时间 lapping time
喷铝 sprayed alumin(i)um
喷铝涤纶薄膜 aluminized polyester film
喷锚 bolting and shotcreting
喷锚衬砌 shotcrete and bolt lining
喷锚构筑法 shotcrete-bolt construction method
喷锚联合支护 combined support with bolting and shotcreting; combined bolting and shotcreting
喷锚支撑 combined bolting and shotcrete
喷锚支护 anchorage and gunite support; combined bolting and shotcrete; rock bolt support with shotcrete; shotcrete and rock bolts protection; shotcrete and rock bolts support; spout anchor supporting; shotcrete-anchorage support
喷锚支护法 rockbolts and shotcrete supporting method
喷煤(油)管 burner pipe; coal dust burner pipe
喷煤嘴 burner tip; coal burner; pulverised[pulverized] coal nozzle
喷灭火器系统 sprinkler system
喷膜养护 spray film maintenance
喷磨机 aeropulverizer
喷沫 barbotage
喷墨打印机 ink-jet printer; jet printer
喷墨法 ink-jet method
喷墨绘图机 ink-jet plotter
喷墨绘图仪 ink-jet plotter
喷墨记录 ink mist recording
喷墨记录法 ink vapo(u)r recording
喷墨记录器 ink vapo(u)r recorder
喷墨式绘图仪 spray plotter
喷墨式控制台打字机 ink jet console
喷墨图像 ink-jet image
喷墨印刷机 ink-jet printer
喷墨印刷油墨 jet printing ink
喷泥 splashing
喷泥机 <刚性路面的> blow-hole
喷泥浆 mud spout
喷泥孔 <刚性路面的> blow-hole
喷泥现象 <刚性路面的> blowing phenomenon
喷泥作用 <刚性路面的> blowing action
喷凝水泥 <起源于美国生产的一种早强水泥> jet set cement
喷铺砂浆 gun-applied mortar
喷漆 aerographing paint; aerosol paint; lacker; lacquer(ing); painting spray(ing); spray(ing) lacquer; spray(ing) paint(ing)
喷漆薄膜电容器 lacquer film capacitor
喷漆层 doped coating; gunned coat of paint; mist coat
喷漆场 spraying shop
喷漆车间 painting workshop; paint spraying shop; spray(ing) shop
喷漆橱 booth; paint spray; spray booth
喷漆唇形罩 spraying lip mask
喷漆挡板 spray plate
喷漆的 spray-painted
喷漆底层用油灰 lacquer putty
喷漆发生器 air-painting equipment
喷漆法 lacquer curtain coating(method); paint spraying system; spray method
喷漆房 paint spray booth
喷漆废水 painting waste
喷漆粉尘 spray dust
喷漆工具 paint spraying outfit
喷漆工用面罩 spray mask
喷漆工作 spray lacquering; spray painting
喷漆光滑法 spray finishing system
喷漆机 lacquering machine; paint sprayer; paint spraying machine; spray booth; spraying machine
喷漆机器人 spraying robot
喷漆间 dope room; paint spray booth; spray booth; spraying shop
喷漆器 air brush; air-painter; compressed-air painting gun; lacquer and paint sprayer; paint blower; paint sprayer; paint-spray(ing) gun
喷漆枪 aerograph; airgun; compressed-air painting gun; marking gun; oil gun; paint blower; paint-(ing) gun; paint spraying gun; paint spraying pistol; spray(ing) gun; spray(ing) pistol; varnish spray gun
喷漆球 paint sprayer-bulb
喷漆设备 air-painter; air-painting equipment; paint spraying apparatus; paint spraying system; spray-painting equipment
喷漆室 paint spray booth; spray booth; spray chamber
喷漆涂料 lacquer paint
喷漆温度 paint temperature
喷漆稀料 lacquer thinner
喷漆橡胶管 paint spray rubber hose
喷漆用面具 paint spray mask
喷漆用清漆 lacquer varnish
喷漆用油 lacquer oil
喷漆中的结晶物 crystallization in painting
喷漆装置 paint spraying apparatus; spray-painting plant
喷漆嘴 painting nozzle
喷漆作业 paint-spraying
喷气 air blowout; air injection; blast; blowing; emanation exhalation; exhale; gas injection; jet blast
喷气泵 jet-air pump
喷气变形 air jet bulking; air jet texturing
喷气部分 jet part
喷气操纵 jet steering
喷气操纵的飞机 reaction control aircraft
喷气侧壁共振器哨 jet-edge resonator whistle
喷气沉积矿床 exhalation-sedimentary mineral deposit
喷气抽风 induced draft
喷气除垢 air blast
喷气除雪机 snow blower; snow blowing machine
喷气传动 jet drive
喷气船 <发动机驱动的> jet boat
喷气单板干燥机 jet veneer drier[dryer]
喷气单板干燥器 jet veneer drier[dryer]
喷气导流控制片 jet vane
喷气灯 blow torch
喷气动力 jet power
喷气动力装置 air-breathing power unit; jet power unit
喷气短管 augmenter tube
喷气舵 jet vane
喷气发电机组 air-breathing power unit
喷气发动机 aeration jet; jet motor; jet propulsion; propulsive duct; propulsor; reaction engine; reaction motor
喷气发动机零件 jet engine parts
喷气发动机燃料 jet fuel(oil); jet propellant
喷气发动机燃油 jet engine fuel
喷气发动机润滑剂 jet engine lubricant
喷气发动机射流涡轮机 blast turbine
喷气发动机试车间 jet engine test cell
喷气发动机推动 jet power
喷气发动机推动的 jet-propelled
喷气发动机推力 jet thrust
喷气发动机推力反向器 jet thrust reverser
喷气发动机消声器 jet noise suppressor; jet silencer
喷气发动机压气机 jet engine compressor
喷气发动机叶片 jet blade
喷气发动机组件 jet pack
喷气发射器 jet ejector
喷气发生系统 jet-generating system
喷气发声 snort(en)ing
喷气阀 snifting valve
喷气防护板 blast plate
喷气飞机航空站 jet port
喷气飞机用的燃料 jet aircraft fuel
喷气分散(法) <土颗粒的> air jet dispersion
喷气粉磨机 aeropulverizer
喷气干燥炉 jet drier
喷气钢 spray steel
喷气功率单位 air-breathing power unit

喷气功率环 jet power ring
喷气鼓风机 jet blower
喷气管 air blowpipe
喷气管嘴 <溶炉的> tuyere
喷气烘燥式浆纱机 jet dryers sizing machine
喷气环 nozzle ring
喷气火山 gas volcano
喷气机场 jet port
喷气机机场 jet airfield
喷气机用的跑道 jet runway
喷气机着陆拦阻装置 jet landing barrier
喷气技术进展 jet progress
喷气进行表面清洁 abrasive jet cleaning
喷气井 gasser
喷气净化法 air blast
喷气净化器 cleaning blower; cleansing blower
喷气孔 gas orifice; gas vent; gas volcano; fumarole【地】
喷气孔阀 air vent valve
喷气孔活动 fumarolic activity
喷气孔田 fumarolic field
喷气口 air jet; gas vent; jet exhaust; puff port; gas maar【地】
喷气快艇 jet boat
喷气矿床 exhalation deposit
喷气扩散器 sparger diffuser
喷气冷却 jet cooling
喷气炼钢 pneumatic steel
喷气流 gaseous blast; jet flow; jet stream
喷气流断路器 air blast breaker
喷气流偏转器 jetevator
喷气螺旋桨发动机 jet-propeller engine
喷气脉冲吸尘器 jet pulse filter
喷气泥锥 gassing mud cone
喷气起动机 jet starter
喷气起飞助推器 jet assisted take-off unit
喷气器 air fountain; air jet; air sparger; blast; air blast
喷气器通风花格窗 ejector grille
喷气器透平 sparger turbine
喷气清理 liquid honing; vapo(u)r blasting; vapo(u)r honing
喷气清洗 jet cleaning
喷气驱除油气法 steam ejector gas freeing
喷气燃料 jet engine fuel; jet fuel (oil); propellant[propellent]
喷气燃料注入 propellant injection
喷气燃气透平 jet gas turbine
喷气燃烧炉 jet ignition
喷气刹车 reverse jato
喷气声 <蒸汽机> snort
喷气式班机 jet airliner; jetliner
喷气式除雪机 air-jet snow remover
喷气式单板干燥器 jet single-panel drier
喷气式动力装置 reaction power plant
喷气式防波堤 pneumatic breakwater
喷气式发动机 jet engine; jet motor; turbojet
喷气式飞机 aerojet; jet aeroplane; jet aircraft; jet airplane; jet plane; jet-propelled aeroplane; rocket plane; squirt
喷气式飞机场 jet port
喷气式飞机航空图 jet-propelled aircraft chart
喷气式飞机航路 jet route
喷气式飞机航线 jet route
喷气式飞机跑道 jetway; jet runway
喷气式飞机用的加油车 jet refueling vehicle
喷气式飞机用燃料 jet aircraft fuel; jet propulsion fuel
喷气式飞机噪声 rocket plane noise
喷气式干燥窑 steam spray kiln
喷气式客机 jet airliner; jetliner
喷气式空气泵 ejector air pump
喷气式螺旋桨 jet propeller
喷气式喷雾器 jet-propelled sprayer; jet sprayer
喷气式汽车 jet-powered car
喷气式燃料中萘系烃含量 naphthalene hydrocarbons in jet fuel
喷气式燃气轮机 jet gas turbine
喷气式水泵 ejector air pump
喷气式透平油 jet turbine oil
喷气式推进器 jet propeller
喷气式消波设备 pneumatic breakwater
喷气式旋翼 rotorjet
喷气式运输机 cargojet
喷气式运载工具 jet-propelled carrier
喷气式直升机 jetocopter; reaction-powered helicopter
喷气速度 efflux velocity; jet velocity
喷气碎屑沉积物 gasoclastic sediment
喷气提升 air jet lift
喷气提升机 lift(ing) jet
喷气通道 jet passage
喷气头 jet head
喷气涂布 air knife coating
喷气推动 jet drive
喷气推动的 jet-propelled
喷气推动功率 jet power
喷气推进 duct propulsion; gas jet propulsion; jet drive; jet propulsion; rocket propulsion
喷气推进发动器 jet propulsion engine
喷气推进牵引车 jet-propelled carriage
喷气推进实验室 jet propulsion laboratory
喷气推进式船舶 jet-propelled vessel
喷气推进水翼艇 jet hydrofoil
喷气推进艇 jet-propelled boat
喷气推进效率 jet propulsive efficiency
喷气消波防波堤 air breakwater
喷气岩 exhalite
喷气引擎鼓风机 jet engine blower
喷气引射泵 ejector pump
喷气噪声 air jet noise; jet noise
喷气指示灯 gas-discharge strobe light
喷气注水器 inspirator
喷气阻拦屏 jet barrier
喷气嘴 air nozzle; blast nozzle
喷气作用 blast effect; exhalative process
喷气作用滤池 jet action filter
喷汽泵 steam-jet pump; water-jet pump
喷汽鼓风机 steam-jet blower
喷汽管 blow(er) pipe
喷汽活动 fumarole activity
喷汽机 steam ejector
喷汽搅拌槽 steam-jet agitator
喷汽孔沉淀物 fumarole deposit
喷汽孔洞 steam jet
喷汽孔景观 landscape of fumarole
喷汽孔类型 type of fumarole
喷汽孔凝结水 fumarole condensate
喷汽孔排气 fumarole discharge
喷汽孔喷发 fumarole eruption
喷汽孔气体 fumarole gas
喷汽孔式气体喷射 fumarolic-type gas emission
喷汽冷却系统 steam-jet cooling system
喷汽器 perforated steam spray; steam jet
喷汽声 snort
喷汽式燃烧器 steam-injector type burner
喷汽式热水器 steam-jet water heater
喷汽水 water-jet
喷汽送风机 steam-jet blower
喷汽制冷设备 steam-jet refrigerating equipment
喷器 spray apparatus
喷枪 airgun; air placer; blow gun; burner; doper; gun(ite); gunjet; hand lance; injection gun; injection lance; jetcrete; jet(ting) lance; lance; paint-spray gun; pistol; shotcrete equipment; spray(ing) gun; spray(ing) lance; spray(ing) pistol; air brush <喷漆用的>
喷枪操作工 nozzleman
喷枪杆 jetting cutter rod
喷枪级粉刷 gun-grade rendering
喷枪控制器 spray gun controller
喷枪控制系统 spray gun control system
喷枪拉丝 combwebbing of spray gun
喷枪轮箍 tyre of spray gun
喷枪喷路颤振 spitting spray
喷枪喷洒 gun spraying
喷枪喷射水泥浆 jetcrete
喷枪喷涂 gun spraying
喷枪喷雾 gun spraying
喷枪喷嘴 air nozzle; spray tip
喷枪软管 gun hose
喷枪式喷雾机 gun sprayer
喷枪式喷嘴 gun jet nozzle
喷枪式燃烧器 gun-type burner; lance type burner
喷枪手柄 pistol grip
喷枪套 gunite jacket
喷枪涂敷 spray application
喷枪修补 spray gun repair
喷枪修整 <混凝土表面的> gun finish
喷枪用混合料 spray gun mix
喷清漆 gunned varnish; lacquering
喷泉 eruptive fountain; eruptive spring; gushing spring; jet dean; jet fountain; keld; spouting fountain
喷泉场地 fountain plaza; fountain site
喷泉池 fountain basin
喷泉管 fountain pipe
喷泉广场 fountain plaza
喷泉井 blowing well
喷泉口 gusher hole
喷泉流 fountain flow
喷泉模型 fountain model
喷泉射流 fountain jet
喷泉式冷凝器 spray-pond condenser
喷泉式凝结器 spray-pond condenser
喷泉式曝气池 fountain aerator
喷泉式洗涤槽 <为水磨石环槽,同时可供六人使用> ablution fountain
喷泉式引水龙头 drinking fountain
喷泉式饮水机 drinking-fountain
喷泉式饮水器 bubbler; bubbler fountain; drinking fountain
喷泉水流 fountain flow
喷泉水头 fountain head
喷燃管内缩口 throat inside burner
喷燃剂切割 flux injection cutting
喷燃器 blow lamp; burner; pressure atomizer; pressure atomizing burner
喷燃器操纵盘 burner control panel
喷燃器风箱 burner wind box
喷燃器壳 burner box
喷燃器壳体 burner housing
喷燃器耐火块 burner firing block
喷燃器燃料管 burner manifold
喷燃器热能量 burner energy; burner loading
喷燃器特性 burner characteristic
喷燃器组 bench of burners
喷燃式燃烧器 blow burner
喷染 spray painting
喷染器 aerograph
喷染术 aerography
喷绒 flock spraying
喷熔 spray fusing
喷入射流 stream penetrating
喷入物 injectant
喷软粒 grit blasting
喷软木屑处理 corking
喷软砂 grit blasting
喷洒 sparge; spatter dash; spray(ing); sprinkle; sprinkling
喷洒柏油 tar spraying
喷洒保温层 spraying isolation
喷洒侧向 sprinkler lateral
喷洒车 spraying vehicle; spraying wagon; spray truck; sprinkler lorry; sprinkler truck; tank sprayer
喷洒出水口 spray outlet
喷洒除草机 weed killer; weed killing machine
喷洒除霜设备 water spray defrost system
喷洒处理 sprinkle treatment
喷洒萃取塔 spray-type extraction column
喷洒的沥青 gunned asphalt
喷洒滴滤床 sprinkling bed
喷洒法 sprinkler method
喷洒方法 spraying method
喷洒防火材 sprayed fireproofing
喷洒分布率 spreading rate
喷洒干燥法 spray drying
喷洒干燥器 spray drier[dryer]
喷洒高度 spraying altitude
喷洒隔热层 spraying isolation
喷洒管 spray line
喷洒管嘴 sprinkler nozzle
喷洒灌溉(法) spray irrigation
喷洒后降雨 rainfall subsequent to spraying
喷洒混合灰泥 spray mixed plaster
喷洒机 distributor; flush coater; liquid distributor; spraying distributor; spraying machine; spraying tanker; sprinkling machine; tank sprayer; water sprayer; water sprinkler tank
喷洒技术 spray application
喷洒剂 spray reagent; monkey blood
喷洒焦油沥青 tar spraying
喷洒绝缘层 <防火、隔热或隔音用> spray on insulation
喷洒绝缘矿物棉 sprayed mineral fiber
喷洒均匀度 emission uniformity
喷洒冷却装置 trickling cooling plant
喷洒沥青 gun-grade asphalt
喷洒沥青的小型罐车 oil pot
喷洒沥青覆盖 asphalt spray mulch
喷洒龙头 spraying cock
喷洒滤池 sprinkler filter; sprinkling filter
喷洒率 emission rate
喷洒面积 sprinkler area
喷洒灭火系统 fire sprinkler system; fire-sprinkling system; sprinkler fire extinguishing system; sprinkler system
喷洒黏[粘]合法 spray bonding
喷洒农药 spray insecticide
喷洒喷嘴 spreader nozzle
喷洒曝气器 spray aerator
喷洒器 distributor; hydroconion; nebulizer; sparge pipe; sparger; sprayer; spraying device; spray jet; sprinkler; thrower
喷洒器横向侧管 sprinkler lateral
喷洒器配水管 sprinkler distribution pipe
喷洒器喷淋头 sprinkler head
喷洒器强度 sprinkler intensity; sprin-

kler strength
喷洒器式集尘器 spray-type collector
喷洒器械 spraying device
喷洒器压力 sprinkler pressure
喷洒器站 sprinkler station
喷洒器支管 sprinkler arm; sprinkler branch
喷洒强度 sprinkling intensity; sprinkling strength
喷洒人造石铺面楼地板 granolithic sprinkle finish floor(ing)
喷洒软管 spray hose
喷洒设备 spray appliance; spray rig; sprinkling device
喷洒施肥 spring application
喷洒施工 spray-applied
喷洒石灰干燥法 lime spray drying
喷洒石棉砂浆 sprayed asbestos
喷洒式吸收器 spray absorber
喷洒水器 sprinkler
喷洒塑料饰面 spray plastic finish
喷洒塔 spray column
喷洒填料床生物膜反应器 sparged packed-bed biofilm reactor
喷洒通气器 spray aerator
喷洒头 spray head; sprinkler head
喷洒吸收器 spray absorber
喷洒系数 coefficient of sprinkling
喷洒系统 spraying system; sprinkling system
喷洒盐水(养护)法 brine-spray method
喷洒养护装备 spray-curing rig
喷洒养护装置 spray-curing rig
喷洒主支管 water works main and branch pipes
喷洒装备 spray rig
喷洒装置 flusher; spraying device; spraying gun; sprinkling machine
喷洒嘴 sprinkler nozzle
喷塞 ejector plug
喷散 spreading
喷散泵 disintegrator pump
喷散器 spray gun
喷散纤维混合料 sprayed fibrous mix
喷色浆两道 sprayed twice with colo(u)red
喷砂 < 又称喷沙 > blasting; blow sand; grit blast; sand-blast; sand blow(ing); sand boiling; sand eruption; sand spraying; spray sand
喷砂斑 blasted patch
喷砂表面处理 sand-blast finish
喷砂玻璃 sand-blasted glass
喷砂测试 sand-blasting test
喷砂车间 blasting shop; sand-blast shop
喷砂冲(毛) sand-blast
喷砂冲刷 sand-blasting
喷砂除垢 blast cleaning
喷砂除污 sand-blasting
喷砂除锈 cleaning rust by sand blasting; grit blasting; sand-blast cleaning; sand-blasting; abrasive blasting
喷砂除锈到出白 white blast
喷砂除锈到出白级的金属 white metal
喷砂处理 blast finishing; blast sanding; gritting blasting; sand-blast(ing); sand jet; grit blasted; grit blasting
喷砂处理过的板 sand-blasted panel
喷砂打光 sand blowing; shot blasting
喷砂打磨表面 abrasive blasting
喷砂打磨(法) grit blasting; abrasive blasting
喷砂的 sand-blasted
喷砂雕刻 grave sand dust; modelled sandblast
喷砂法 blasting; grit blasting; sand-blasting; sand-jetting(method) <沉管施工>
喷砂法除锈 derusting by sandblast
喷砂法处理 shot blasting
喷砂工 nozzleman
喷砂工具 jetting tool
喷砂滚筒 sand-blast barrel mill
喷砂过度 overblasting
喷砂壶 blasting kettle
喷砂机 air sand blower; blaster; blasting machine; peening machine; sand-blast apparatus; sand blower; sand ejector; sander; sanding machine; sand sprayer; sand spraying device; sand spraying machine
喷砂加工 abrasion blasting; shot-blast
喷砂间 shot-blast cabinet
喷砂浆 sprayed mortar
喷砂浆机 mortar spraying machine; sand-blaster
喷砂胶管 sand-blast hose
喷砂净化 blast cleaning
喷砂口 sand-blasting nozzle; sand vent
喷砂冒水 boils and mud spouts; handbills and waterspouts; sands and waterspouts; sand spraying and water oozing; sand boil
喷砂灭草机 sand-blast mower
喷砂磨齿机 airdent
喷砂磨光 sand-blast(ed) finish
喷砂磨光设备 abrasive blast equipment; sand-blast equipment
喷砂磨耗试验 jet type abrasion test
喷砂磨料 sand-blasting abrasive
喷砂磨蚀 abrasive jet wear testing
喷砂磨损试验 abrasive jet wear testing
喷砂盘 sanding disc[disk]
喷砂喷枪 sand spraying gun
喷砂器 air sand blower; clean(s) ing blower; sand-blast apparatus; sand-blaster; sand blower; sand ejector; sander; sanding; sanding gear; sand spraying device; sand jet
喷砂枪 sand-blasting gun
喷砂强化 stress peening
喷砂强化效应 peening effect
喷砂轻磨 touch sanding
喷砂清除法 grit blasting; sand-blast
喷砂清洁法 abrasive clearing
喷砂清理 abrasive blast cleaning; sand-blast cleaning
喷砂清理法 abrasive blasting; abrasive cleaning; blast cleaning; sand-blasting method; sand washing; grit blasting method
喷砂清理机 abrator; sand-blasting machine
喷砂清理器 sand-blast cleaner
喷砂清理室 sand-blast cleaning room
喷砂清洗 sand-bath; sand washing
喷砂丘 sand core
喷砂软管 sand-blast(ing) hose
喷砂沙子 sand-blast sand
喷砂设备 blaster; blasting equipment; sand-blast apparatus; sand-blast device; grit blasting equipment
喷砂蚀面 sand-blasting finish; sand-blast obscuring; sand etch
喷砂使表面粗糙 sand-blasting
喷砂使混凝土露出骨料 grit blasting
喷砂式清洁器 sand cleaner
喷砂试验 falling sand test; sand-blast test
喷砂饰面 sand-blast(ed) finish
喷砂饰面混凝土 sand-blasted concrete
喷砂室 blast room
喷砂碎面(法) grit blasting
喷砂筒 sand-blasting drum
喷砂头 abrator head
喷砂箱 sand-blasting box
喷砂橡胶管 sand blast rubber hose
喷砂消除法 grit blasting
喷砂修整 blast finishing; sand-blast finish
喷砂用的砂子 sand-blast sand
喷砂用砂 blast sand
喷砂装饰磨光 sand-blast decorative finish
喷砂装置 blasting equipment; blasting set; sand-blast apparatus; sand-blasting unit; sand blower; sander; sanding gear; sand spraying gear; blaster
喷砂锥 sand emitting cone
喷砂嘴 blast nozzle; sand-blast(ing) nozzle; sand blowing nozzle
喷上 throwing on
喷烧 torch firing
喷烧穿孔机 jet-piercing drill
喷烧管 burner tube
喷烧器 burner; pulverizing jet; spray burner
喷射 counterjet; ejecting; ejection; flashing; gun application; gunning; gun spraying; inject(ion); jetting; sparging; spouting; spray; spray-up; spurt; squirt; shooting
喷射板 sprinkler plate
喷射棒 injection rod
喷射爆破缆 projected charge
喷射泵 eductor pump; injector; injector pump; jet-air pump; jet ejector; jetting pump; sprayed pump; squirt pump; water-jet pump
喷射泵操纵杆 injection pump lever
喷射泵排水 jet drainage
喷射泵启动 ejector priming
喷射泵前级 ejector fore-stage
喷射泵输送 jetting pump transfer
喷射泵体 injection pump housing
喷射泵挖泥船 jet lift dredge(r); jet pump dredge(r)
喷射泵文丘里涤气器 ejector Venturi scrubber
喷射泵柱塞 injection pump plunger
喷射泵装燃料入口 charging port
喷射泵钻粒钻井 jet pump pellet drilling
喷射避雷器 spray arrester
喷射材料 gunning material
喷射舱 jet flap rudder
喷射插口 injection socket
喷射衬砌 sprayed lining
喷射成型 gunning process; injection mo(u)lding; jet moulding; reaction injection mo(u)lding; spray forming; spray-up; spray-up process
喷射成型法 spray-up method
喷射成型零件 injection mo(u)lded part
喷射成型塑料 injection mo(u)lded plastics
喷射迟后 injection lag
喷射持续时间 injection duration
喷射冲击式磨机 jet impact mill
喷射冲洗 jet douche; jet wash; spray washing
喷射冲洗器 jet washer
喷射抽气泵 ejector air pump
喷射抽气机 jet exhauster
喷射抽气器 water-jet suction apparatus
喷射出 jet out
喷射出口速度 spouting velocity
喷射处理 spray treatment
喷射(处理)混凝土 radcrete
喷射床 spouted bed
喷射淬火 flash quenching; jet hardening; jet quenching; spray quenching
喷射大爆破缆 projected charge demolition kit
喷射带 jet band
喷射导管 spraying conduit; schedule air pipe
喷射的 sprayed
喷射的外层粉刷打底 sprayed exterior rendering
喷射的外层粉刷抹灰 sprayed exterior plastering
喷射的外墙粉刷打底 sprayed exterior rendering
喷射的外墙粉刷抹灰 sprayed exterior plastering
喷射涤气器 jet scrubber
喷射底板 bottom ejector plate
喷射点火系统 jet ignition system
喷射电镀 jet electro-plating
喷射电荷 spray charge
喷射电弧 spray arc
喷射电流反馈 spray current feedback
喷射顶角 jet apex angle
喷射定时 injection timing
喷射冻结 jet-freezing
喷射舵 jet flap rudder
喷射发泡器 dispenser
喷射阀 injection valve; jet valve; shooting valve
喷射法 air placing; gunite; guniting; injection method; injection process; jetting process; spray(ing) method; spray(ing) process
喷射法成孔 jet hole
喷射法除锈 shot rust removing
喷射法磷化处 spray bonderizing
喷射法抹灰 projection plastering
喷射法膨化变形 jet bulking
喷射法输送 jetting(pump) transfer
喷射法挖泥 jet drive
喷射方法 spraying method
喷射方式 spray regime
喷射方向 jet direction
喷射分馏塔盘 jet tray
喷射分散 injection spread; jet spread
喷射分散角 jet-spread angle
喷射粉磨 jet milling; jet pulverizing
喷射粉刷 spray plastering
喷射粉碎机 jet pulverizer
喷射腐蚀装置 spray pickling unit
喷射干燥 jet drying; spray drying
喷射干燥机 jet drier[dryer]
喷射干燥器 ejector drier[dryer]; jet drier[dryer]
喷射杆 jetting rod
喷射钢筋混凝土 ferro-shotcrete
喷射钢粒钻进 jetted particle drilling
喷射钢丝网水泥 ferrogunite
喷射钢纤维混凝土 steel fiber reinforced concrete
喷射高度 jet(ting) height
喷射工 nozzleman
喷射工作风压 concrete-spouting air-pressure
喷射工作水压 concrete-spouting hydraulic pressure
喷射功 injection work
喷射鼓风机 jet fan
喷射管 adjutage; air blowpipe; air ejector; beam tube; injection pipe; intrusion pipe; jet(ting) nozzle; jet(ting) pipe; nozzle pipe; shower pipe; torch pipe
喷射管道 jet passage
喷射管管口 mouthpiece of a jet pipe
喷射管管嘴 mouthpiece of a jet pipe
喷射管嘴 injection nozzle
喷射灌浆 beton project; jet grouting; gunite
喷射灌浆法 jet grouting method; guniting
喷射灌浆混凝土 guncreting

P

喷射灌浆围堰 jet grouted cofferdam
喷射过渡 projected transfer; spray transfer
喷射厚度 thickness of spraying
喷射化学抛光 jet chemical polishing
喷射化油器 pressure carburet(t)or
喷射环 sparge ring
喷射灰浆 injection mortar;projection plaster
喷射灰泥 projection plaster
喷射混合 jet mixing
喷射混合灰泥 spray mixed plaster
喷射混合料 gunning mix
喷射混合流 jet mixing flow
喷射混合器 injector mixer;jet agitator;jet mixer
喷射混合系统 jet mixing system
喷射混凝土 air-blown concrete; blocrete; gun-applied concrete; guncreting; jetcrete; pneumatically applied concrete; pneumatically placed concrete; projected concrete; shooting concrete; shotcrete; shotcreting; spray concrete; spraycrete; sprayed(-on) concrete
喷射混凝土板式平接施工缝 board butt joint
喷射混凝土包裹的铁管 concrete-spraying surfaced iron pipe
喷射混凝土表面 concrete-spraying surface
喷射混凝土表面改进剂 concrete-spraying surface improver
喷射混凝土表面缓凝剂 concrete-spraying surface retardant
喷射混凝土表面接缝密封合成物 concrete-spraying surfacing joint sealing compound
喷射混凝土表面修饰 concrete-spraying surface finish
喷射混凝土表面硬化剂 concrete-spraying surface hardener
喷射混凝土表面优化剂 concrete-spraying surface improver
喷射混凝土薄层 <弥补混凝土表面缺陷的> flash coat
喷射混凝土操作 shotcrete operation
喷射混凝土槽板 concrete-spraying trough slab
喷射混凝土层 layer of sprayed concrete
喷射混凝土衬砌 poured concrete lining; shot concrete lining; shotcrete lining
喷射混凝土稠度 concrete-spraying workability
喷射混凝土储罐 concrete-spraying tank
喷射混凝土窗槛 concrete-spraying window sill
喷射混凝土的供料及调节装置 concrete-spaying feed wheel
喷射混凝土的抗拉强度 concrete-spraying tension strength
喷射混凝土的孔隙 void in shotcrete
喷射混凝土的拉伸 concrete-spraying tension
喷射混凝土的压缩空气供料器 concrete-spaying pneumatic feeder
喷射混凝土的应变 concrete-spraying strain
喷射混凝土的应力 concrete-spraying stress
喷射混凝土底层地板 concrete-spraying sub-floor
喷射混凝土电视塔 concrete-spraying television tower
喷射混凝土顶盖 concrete-spraying umbrella
喷射混凝土定线及定坡度的钢丝 ground wire of concrete-spraying for positioning line and slope
喷射混凝土发生坍落 slough
喷射混凝土防水 concrete-spraying waterproofing
喷射混凝土防水粉 concrete-spraying waterproofing powder
喷射混凝土防水剂 concrete-spraying waterproofer
喷射混凝土废水管 concrete-spraying waste pipe
喷射混凝土粉尘 shotcrete dust
喷射混凝土腹板 concrete-spraying web
喷射混凝土工 gunman;nozzleman
喷射混凝土工场进水沟 concrete-spraying yard inlet gulley
喷射混凝土工程 concrete-spraying work
喷射混凝土工艺参数 concrete-spouting technical parameter; shotcrete technical parameter
喷射混凝土工艺师 concrete-spraying technologist
喷射混凝土工艺学 concrete-spraying technology
喷射混凝土灌筑 placing of shotcrete
喷射混凝土过多 concrete-spraying surcharge
喷射混凝土和易性 concrete-spraying workability
喷射混凝土厚度 concrete-spraying thickness
喷射混凝土回弹 rebound of shotcrete
喷射混凝土机进料器 feed wheel
喷射混凝土技师 concrete-spraying technician
喷射混凝土加料提升塔 concrete-spraying tower
喷射混凝土减水剂 concrete-spraying water-reducing agent
喷射混凝土建筑类型 concrete-spraying type of construction
喷射混凝土搅拌机 injector mixer
喷射混凝土街道进水口 concrete-spraying street(inlet) gulley
喷射混凝土结构用板 concrete-spraying structural slab
喷射混凝土井圈 concrete-spraying well ring
喷射混凝土壳体 shotcrete shell
喷射混凝土拉伸区 concrete-spraying tensile zone
喷射混凝土类型 concrete-spraying type
喷射混凝土立方试块 concrete-spraying test cube
喷射混凝土梁腹 concrete-spraying web
喷射混凝土楼梯 concrete-spraying stair(case)
喷射混凝土楼梯水塔楼 concrete-spraying staircase tower
喷射混凝土楼梯踏步 concrete-spraying(staircase) step
喷射混凝土楼梯梯级 concrete-spraying(staircase) step
喷射混凝土路面 concrete-spraying surfacing
喷射混凝土锚杆支护 shotcrete and rock bolts protection; shotcrete and rock bolts support
喷射混凝土门槛 concrete-spraying threshold
喷射混凝土面层 concrete-spraying topping
喷射混凝土磨损表面 concrete-spraying wearing surface
喷射混凝土强度 concrete-spraying strength
喷射混凝土墙 concrete-spraying wall
喷射混凝土墙洞 concrete-spraying wall pot
喷射混凝土墙块 concrete-spraying wall block
喷射混凝土穹隆 concrete-spraying vault
喷射混凝土软管 shotcrete hose
喷射混凝土设备 shotcrete equipment
喷射混凝土施工队 gunning crew
喷射混凝土湿密度 concrete-spraying wet density
喷射混凝土试验 concrete-spraying test
喷射混凝土试验锤 <回弹仪> concrete-spraying test hammer
喷射混凝土试验方法 concrete-spraying testing method
喷射混凝土试验机 concrete-spraying testing machine
喷射混凝土试验梁 concrete-spraying test beam
喷射混凝土饰面 gun finish; shotcrete facing; shotcrete surfacing
喷射混凝土受拉开裂 concrete-spraying tension crack
喷射混凝土(输送)槽 concrete-spraying trough
喷射混凝土输送设备 positive displacement device of concrete
喷射混凝土束带层 concrete-spraying string course
喷射混凝土水箱 concrete-spraying water tank
喷射混凝土塔式建筑 concrete-spraying tower
喷射混凝土台架 concrete-spraying stand; concrete-spraying table
喷射混凝土体系 shotcreting system
喷射混凝土填缝 guniting of the joints
喷射混凝土涂层 gun finish
喷射混凝土托梁板 concrete-spraying trimmer plank
喷射混凝土瓦 concrete-spraying tile
喷射混凝土瓦厂 concrete-spraying tile works
喷射混凝土瓦的压制机 concrete-spraying tile press
喷射混凝土瓦屋顶铺设 concrete-spraying tile roof cover(ing)
喷射混凝土外加剂 shotcrete admixture
喷射混凝土温度 concrete-spraying temperature
喷射混凝土污水防止器 concrete-spraying water waste preventer
喷射混凝土污水管 concrete-spraying waste pipe
喷射混凝土楔 concrete-spraying wedge
喷射混凝土应用的单位 concrete-spraying unitized unit
喷射混凝土用稠度剂 concrete-spraying workability agent
喷射混凝土用地基纸 concrete-spraying subgrade paper
喷射混凝土用工作度(剂) concrete-spraying workability
喷射混凝土用和易性剂 concrete-spraying workability agent
喷射混凝土游泳池 concrete-spraying swimming pool
喷射混凝土圆形穹隆 concrete-spraying wagon vault
喷射混凝土圆柱体试件 concrete-spraying test cylinder
喷射混凝土整平刮板 cutting screed
喷射混凝土支承介质 concrete-spraying supporting medium
喷射混凝土支护 shotcrete support
喷射混凝土支护的井筒 shotcrete-lined shaft
喷射混凝土支护法 shotcrete system
喷射混凝土制品 concrete-spraying ware
喷射混凝土制品厂 concrete-spraying ware factory
喷射混凝土中的钢筋 concrete-spraying steel
喷射混凝土砖圬工墙 concrete-spraying tile masonry wall
喷射混凝土装饰面 gun finish
喷射混凝土组织 concrete-spraying texture
喷射机 gun; injection machine; jetting machine; shotcrete machine; shotcrete sprayer; spraying machine
喷射机构 injection equipment; injection mechanism
喷射机台车 jumbo for shotcrete sprayer
喷射机械 spraying machinery
喷射挤压机 ejecting press
喷射技术 spray application technique
喷射加工硬化法 peening(method)
喷射加料 splash loading
喷射加湿 jet humidification
喷射加湿器 injection humidity; jet (type) humidifier
喷射加速 jet assist
喷射浇灌 gunited concreting; pneumatic placing
喷射浇灌混凝土 pneumatically placed concrete; gunited concrete
喷射浇模 gunning pattern
喷射浇水 spray shower
喷射浇水单元 spray shower unit
喷射浇水阀(门) spray shower valve
喷射浇水架 spray shower tray
喷射浇水控制 spray shower control
喷射浇水控制室 spray shower cubicle
喷射浇水控制台 spray shower cubicle
喷射浇水莲蓬头 spray shower head
喷射浇水喷嘴 spray shower rose
喷射浇水软管 spray shower hose
喷射浇水设备 spray shower set
喷射浇水水龙带 spray shower hose
喷射浇水塔 spray shower column
喷射浇水柱 spray shower column
喷射浇水装置 spray shower installation
喷射浇筑混凝土 pneumatic applied concrete
喷射角度 angle of spray; injection angle; jetting angle; spray angle
喷射搅拌 jet agitation
喷射搅拌法 dry jet mixing method
喷射搅拌机 injector mixer; jet agitator
喷射金属法结合 sprayed-metal bonding
喷射进料器 injection feeder
喷射浸出 jet leaching
喷射井 injection well
喷射井点 eductor(well point); ejector; ejector(type) well point
喷射井点降水 ejector type well-point method
喷射井点排水系统 eductor system
喷射井点系统 eductor well point system; ejector well point system
喷射净化器 jet purifier
喷射距离 jetting distance; shooting distance
喷射颗粒 jet particle
喷射空气 blast air; injection air
喷射空气泵 ejector air pump
喷射孔 injection hole; injection orifice; jet hole; jet orifice; spray-hole

喷射控制 jet probing;jetting control
喷射口 injection orifice; jet burner; jet orifice;squirt
喷射扩散 jet diffusion
喷射扩散泵 jet diffusion pump
喷射拉毛粉刷 spray stucco
喷射拉毛粉刷工程 spray stuccowork
喷射冷凝 jet condensation
喷射冷凝器 injection condenser; injector condenser
喷射冷却 cooling spray;effusion cooling;jet cooling;jet quenching;shower cooling;water spray cooling
喷射冷却柜 spray tank
喷射冷却器 quenching nozzle; shower cooler
喷射力 jet power
喷射沥青 air blow asphalt;gun-grade asphalt
喷射粒子 jet particle
喷射量 emitted dose
喷射淋浴 spray shower bath
喷射流 reaction jet; spraying jet; jet flow
喷射流扩展锥 jet cone
喷射流收敛 jet contraction
喷射流收缩 jet contraction
喷射龙头 spray tap
喷射率 injection rate;spraying rate
喷射落后 injection lag
喷射密度 injection density
喷射模法 jet moulding
喷射模拟器 sprayer simulator
喷射模塑法 jet mo(u)lding;spray-up mo(u)lding
喷射模塑喷嘴 jet mo(u)lding nozzle
喷射模型 jetting model
喷射模铸机 injection mo(u)lding machine
喷射磨 jet mill;jet pulverizer
喷射磨粉机 jet pulverizer
喷射磨碎 jet pulverizing
喷射磨损 injection wear
喷射耐火混凝土 refractory gunite
喷射泥浆 mud jacking
喷射黏[粘]度计 jet visco(si)meter
喷射碾机 jet mill
喷射碾磨机 jet pulverizer
喷射凝结 condensation by injection
喷射凝汽器 ejector condenser; jet condenser
喷射凝铸 spray casting
喷射排泥管 elephant trunk
喷射抛出 ejectment
喷射抛光 jet polishing
喷射泡沫塑料打捞法 salvage by injecting plastic foam
喷射喷雾干燥机 jet spray dryer
喷射皮带管 spray hose
喷射曝气 jet aeration
喷射曝气器 injection aerator; jet aerator
喷射曝气器设计 injection aerator design
喷射曝气装置 injection aerator
喷射期 injection period
喷射气流 air jet; exhausted jet stream;jet-stream wind
喷射气流速度 jet speed
喷射汽缸 injection cylinder
喷射器 eductor; ejector(nozzle); ejector priming;gun;injector;inspirator;jet apparatus;jet eductor;jet ejector; kicher; pulverizer; spray(ing)gun;squirt(gun);jet
喷射器导向滚轴 ejector guide roller
喷射器的变速阀 ejector speed change valve
喷射器滴油总管 injector dribble gallery pipe
喷射器断流阀 injector stop valve
喷射器阀 ejector valve
喷射器放油管 injection drain pipe
喷射器缝口 injector slot
喷射器负重轮 ejector carrier roller
喷射器管 gun hose
喷射器管道 ejector line
喷射器控制阀 ejector control valve
喷射器控制杆 ejector control lever
喷射器连接阀 ejector clutch
喷射器联合嘴 injector combining nozzle
喷射器喷管 jet injection pipe
喷射器喷头 injector spray tip
喷射器喷嘴 injector nozzle
喷射器气流噪声 jet noise
喷射器浅盘溢流防护器 ejector pan overflow guard
喷射器绳索 ejector cable
喷射器式压气机 ejector type air compressor
喷射器试验 injector testing
喷射器输出管 injector delivery
喷射器缩回装置 atomizer retraction gear
喷射器体 injector body
喷射器通风栅门 jet ventilation grid
喷射器系列阀 ejector sequence valve
喷射器橡胶管 squirt hose
喷射器械 educator
喷射器性能试验装置 nozzle tester
喷射器压缩机 thermal compressor
喷射器烟囱 ejector chimney
喷射器溢油管 injector overflow pipe
喷射器圆柱 ejector jack
喷射器止动装置 ejector stop
喷射器制动阀 ejector brake valve
喷射器注入管 ejector filler
喷射器自动选速阀 ejector automatic speed selector valve
喷射器嘴 injector nozzle
喷射器嘴座 injector nozzle holder
喷射嵌缝膏 gun sealant
喷射枪 applicator;injection lance;injection rod; pneumatic gun; spray gun;air-water jet <混凝土水平工作缝>;patch gun
喷射切管器 jet casing cutter
喷射切削 jet cutting
喷射清舱泵 jet strip pump
喷射清舱系统 jet strip system
喷射清釜装置 jet cleaner
喷射清理 blast cleaning;grit blast;jet cleaning;shot blasting
喷射清理机 seed blast
喷射清洗 jet cleaning
喷射驱动 hydraulic propulsion; jet propulsion
喷射泉 eruptive fountain
喷射燃料 burner oil
喷射燃烧器 atomizer burner; spraying burner
喷射染色机 jet dy(e)ing machine
喷射日珥 spray prominence
喷射软管 delivery hose; injection hose;jet(ting)hose;spray hose
喷射软钎焊 spray soldering
喷射润滑 injection lubrication;jet lubrication;splash lubrication
喷射砂浆 pneumatically applied mortar;sprayed mortar;spread mortar;air-blown mortar
喷射砂浆支护 sand-cement coating support
喷射舌片 jet tab
喷射设备 injection plant; splashing device;spraying equipment
喷射深度 spray penetration
喷射施工 spray-applied
喷射施工缝 board butt joint
喷射石油沥青 air-blown asphalt
喷射时间 discharge time;injection time
喷射式泵 jet pump
喷射式充气器 jet aerater
喷射式除尘器 spray scrubber
喷射式除气器 jet type deaerator
喷射式除氧加热器 jet type deaerating heater
喷射式穿孔器 jet perforator
喷射式穿孔药包 jet perforator charge
喷射式吹管 injector blow pipe
喷射式电镀法 jet electrolysis-plating method
喷射式定边装置 jet deckle
喷射式发动机 injection type engine
喷射式反循环钻具 jet type reverse circulation tool
喷射式分析器 jet analyser[analyzer]
喷射式风机 ejector type ventilator
喷射式格栅 ejector type grille
喷射式鼓风机 jet blower
喷射式过油管下井仪 ejector type through-tubing tool
喷射式焊炬 injector blow pipe;injector torch
喷射式化油器 injection carburet(t)or;spraying carburet(t)or
喷射式换热器 spray-type heat exchanger
喷射式回收炉 spray-type recovery furnace
喷射式混合 ejector mixing
喷射式混合流体机 jet mixer
喷射式混合气体燃烧器 nozzle-mix gas burner
喷射式混合器 ejector mixer
喷射式机 ejector machine
喷射式集尘器 jet collector
喷射式计尘器 jet dust counter
喷射式加料器 spout feeder
喷射式加湿器 jet type humidifier
喷射式降温仪 spurt type cooler
喷射式搅拌机 jet mixer
喷射式搅拌器 ejector type agitator; hydraulic jet mixer;jet agitator
喷射式孔底反循环取芯钻具 bottom-hole partial jet reverse circulation core tool; partial jet hole bottom reverse circulation core barrel
喷射式冷凝器 eductor condenser; injection cooler; ejector condenser; jet condenser
喷射式离心泵 jet centrifugal pump
喷射式硫黄燃烧炉 spray-type sulfur burner
喷射式磨机 fluid energy mill; injector-type mill; jet mill; micronizer [microniser]
喷射式碾磨机 jet mill
喷射式排风机 jet air exhauster
喷射式排泥管 hydraulic ejector
喷射式排气歧管 ejector type exhaust manifold
喷射式喷雾机 jet sprayer
喷射式汽化器 injection carburet(t)or;jet type carbureter
喷射式潜入污水泵 sewage ejector submersible pump
喷射式清洗机 jet type washer;spray washer
喷射式燃烧器 after burner;gun-type burner;injection burner
喷射式润滑 spray lubrication
喷射式三牙轮钻头 jet type tricone bit
喷射式射孔器 jet perforator
喷射式深井泵 jet deep-well pump; jet-well pump
喷射式深水泵 jet-well pump
喷射式绳状染槽 jet beck
喷射式水泵 ejector(air)pump; jet pump
喷射式水泥筒仓过滤器 jet-cement silo filter
喷射式通风机 jet blower
喷射式挖泥船 jet dredge(r);jet lift dredge(r)
喷射式微粉磨机 jet micronizer;jet-O-Mizer
喷射式涡轮机 jet turbine
喷射式牙轮钻头 jet nozzled rock bit
喷射式油燃烧器 gun-type oil burner
喷射式凿井机 spray rig
喷射式凿岩机 spray rig
喷射式真空泵 jet vacuum pump
喷射式煮浆锅 jet cooker
喷射式钻头 abrasive jet;hydraulic HP bit;jet bit;jet type bit
喷射式钻头水眼 jet type water course
喷射式钻头钻进 jet bit drilling
喷射事故 discharge accident
喷射试验 ejection test;spraying test
喷射输送器 jet conveyer[conveyor]
喷射水 injection water;jet(ting)water; jet of water; sparge water; spraying water; spring water; water-jet
喷射水泵 water ejector; water-jet pump
喷射水流 air jet;jet stream;water jet
喷射水泥 guniting; gunned castable; jet set cement;one-hour cement
喷射水泥衬里 poured-refractory backing
喷射水泥混凝土 cement gun concrete
喷射水泥砂浆 gunite; jetcrete; shotcrete; pneumatic applied mortar; air-blown mortar
喷射水舌 discharge jet
喷射水束 discharge jet
喷射送料 splash feed;spray feed
喷射送料器 squirt feeder
喷射速度 inlet velocity;jet velocity; spouting speed;spouting velocity
喷射碎屑 spraying muck
喷射损失 injection loss
喷射探测 jet probing
喷射探测器 jet probe
喷射探查 jet probing
喷射提前 injection advance
喷射替续器 jet relay
喷射条件 injection condition
喷射调节 injection regulation
喷射调节器 jet regulator
喷射铜焊 spray brazing
喷射筒排放装置 cartouch(e) release mechanism
喷射头 injector head; shower head; splash head;jet head
喷射头部表面 injector face
喷射图 injection scheme
喷射涂布装置 jet foundation applicator
喷射推进 jet propulsion
喷射推进船 hydrojet propelled ship
喷射推进气体 propulsive gas; reaction gas
喷射推进器 jet propeller
喷射推进装置 jet propulsion unit
喷射推力 jet thrust
喷射挖掘法 <冲水打井、用泵吸泥> jet cutter method
喷射温度 injection temperation;injection temperature
喷射涡流 jet-stream whirl
喷射雾化器 blast atomizer;jet atomizer
喷射雾锥 spray-cone
喷射吸泥泵 jet drive pump;jet eductor pump

喷射洗涤 jet cleaning;jet washing
喷射洗涤除尘 jet cleaning precipitation
喷射洗涤机 jet scrubber;jet washer
喷射洗涤器 jet scrubber;jet washer
喷射洗井 < 用清泥浆 > blow a well clean
喷射洗石法 water-jet method
喷射系数 coefficient of injection;jet coefficient
喷射系统 injection system; injector system;sparge system
喷射现象 jet phenomenon
喷射箱 distributing and hood jet
喷射消防系统 fire-protection sprinkler system
喷射效率 jet efficiency
喷射效应 blowing-off effect
喷射卸货系统 sprinkling system
喷射卸料槽 kerne tipple
喷射形塔盘 jet tray
喷射形状 spray configuration
喷射型除尘器 jet type dust collector
喷射型井点 jetting type well-point
喷射性滴答声 ejection click
喷射旋流式浮选机 jet-cyclo flo(a)tation cell;jet-cyclo flotator
喷射循环 injector circulation
喷射压力 expulsion pressure; injection pressure; jet pressure; nozzle pressure;spraying pressure
喷射压气机 jet compressor
喷射压缩 jet compression
喷射延迟 injection delay
喷射研磨 jet grinding;jet milling;jet pulverizing
喷射盐雾 sprayed salt mist
喷射溢流染色机 jet overflow dy(e)ing machine
喷射音 ejection sound
喷射印花装置 jet printing system
喷射印刷 jet printing
喷射应用 spray application
喷射硬化 shot peening
喷射用灰浆 pneumatic mortar
喷射用砂浆 pneumatically applied mortar;pneumatic mortar
喷射油泵 high-pressure injection pump;jet oil pump
喷射油滴 spray-droplet
喷射油壶 squirt can
喷射浴 douche bath
喷射源 injection source
喷射运输器 < 短距离运输器械 > jet conveyer[conveyor]
喷射再生式潜水装置 injecto-regenerative diving outfit
喷射凿井机 spraying rig
喷射凿岩机 spraying rig
喷射造粒 air prilling
喷射噪声 injection noise;jet noise
喷射噪音 jet noise
喷射增压泵 ejector booster pump
喷射真空泵 ejector vacuum pump;jet exhauster
喷射蒸汽加湿器 injection steam humidity
喷射蒸汽器 steam-jet blower
喷射制动 jet brake
喷射制品 jet-formed product
喷射滞后 injection delay
喷射滞后时间 delayed action time of discharge
喷射蛭石 spraying vermiculite
喷射注浆 jet grouting
喷射注入法 < 水泵启动 > ejector priming
喷射注入混合 jet mixing
喷射注入器 ejector filler
喷射铸造(模型)法 injection mo(u)lding

喷射桩 < 即旋喷桩 > churning pile; chemical churning pile
喷射装置 ejector device;injection apparatus;injection gear;injection installation; injection plant;jet apparatus; jet bubbler; jetting device; sprayer;spray rig
喷射状 spurting
喷射锥角 spray-cone angle
喷射钻机 jet bit;jetting drill
喷射钻进 jetting drilling;wash boring
喷射钻进泥浆 jet drilling mud
喷射钻井设备 jet drilling rig
喷射钻头 jet bit
喷射嘴 ejector nozzle;injection nozzle;jet blower;jet nozzle
喷射作用 ejecting action;spray application;jet action
喷湿 squirt
喷湿剂 humidification spray;spraying reagent;streak reagent
喷湿器 humidification spray
喷湿砂法 < 混凝土接头施工方法 > wet sand blasting process
喷湿砂洗模 wet blast mo(u)ld cleaning
喷湿试剂 spray reagent
喷石枪 rock gun
喷石屑 grit blinding
喷束 spray fan
喷束角 spray angle
喷束图形 spray pattern
喷束直径 spray diameter pattern
喷束锥 spray-cone
喷刷沥青防水层 brush coating
喷刷路搭接 overlap
喷刷器 air brush
喷霜 blooming
喷水 atomization of water; drench; sprinkle with water; water ejection;water eruption;watering;water injection;water-jet;water spouting; water spray(ing); water sprinkling
喷水泵 water injection pump;water-jet pump
喷水冰层 ice blister
喷水采泥船 jet dredge(r)
喷水沉桩法 jetting piles into place
喷水池 font;fountain(basin);fountain pool;spray cooler;spray fountain; sprinkling basin; sprinkling pool;water fountain;phiale < 教堂前 > ;spray pond
喷水池台座 tazza
喷水池屋顶 spray-pond roof
喷水冲洗器 water spray scrubber
喷水抽气泵 < 一种机械真空泵 > water ejector pump
喷水除尘器 hydrojet cleaner
喷水除尘装置 wet scrubber
喷水除鳞 water-jet descaling
喷水处理 spray disposal;water blasting;water spray treatment
喷水处理法 spray disposal method; water spray treatment method
喷水船 hydrojet craft;water-jet(propelled) boat
喷水淬火 flush quenching;jet hardening;spray quenching;stream hardening
喷水的 artesian
喷水段 spray chamber;spray-type air washer section
喷水阀组合 sprinkling valve package
喷水法 spraying
喷水干管 < 灭火 > sprinkler main
喷水干燥窑 water spray kiln
喷水高度 jet height
喷水工 nozzleman

喷水管 ajutage;fountain pipe;shower pipe; sparge conduit; sparge pipe; sparge tube; spout; spraying pipe; spraying tube; spray pipe; spray water pipe;spray water tube;sprinkler pipe;sprinkling bar
喷水管布置 piping schema
喷水管道系统 spray pipe system
喷水管系 water spray piping
喷水灌溉 overhead irrigation; spray irrigation; sprinkler irrigation; sprinkling irrigation
喷水壶 blow can;sprinkling;watering can
喷水环 water ring
喷水火花捕集器 wet cap
喷水机 spray distributor
喷水集管 spray manifold
喷水减温器 direct contact desuperheater; spray attemperator; spray manifold;spray-type desuperheater
喷水减温箱 spray manifold
喷水降温 spray desuperheating
喷水降温器 spray tank
喷水井水压头 artesian pressure
喷水空气泵 ejector water air pump; water-jet air ejector; water-jet air pump
喷水孔 blow-hole; jet hole[holing]; spout hole
喷水控制阀门 deluge valve
喷水口 gargoyle;jet teau;spray jet; sprinkling outlet; water injector; water-jet;water spout
喷水冷凝 jet condensation
喷水冷凝泵 jet condenser pump
喷水冷凝器 jet condenser;spray condenser;water-jet condenser
喷水冷却 injection water cooling; pipe quenching
喷水冷却池 spray cooling pond;spray(ing) pond
喷水冷却机 spray cooler
喷水冷却器 spray cooler
喷水冷却式发动机 water injection cooled engine
喷水冷却塔 spray tank;spray tower
喷水冷却系统 cooling spray system
喷水莲蓬头 jet rose
喷水量 delivery of rain;working water-jet capacity
喷水流 jet flow;jet stream
喷水滤器 sprinkling filter
喷水冒沙 boils and mud spouts;sand boil
喷水灭火器 sprinkling tank
喷水灭火设备报警器 sprinkler alarm
喷水灭火水箱 sprinkler tank
喷水灭火系统 deluge system; fire-protection sprinkler system; fire-sprinkling system; sprinkler system;drencher system < 指一种手动洒水装置 >
喷水灭火装置 sprinkler system
喷水泥 gunite
喷水泥枪 gunite gun
喷水凝气器 injection condenser
喷水凝气式泵 jet condenser pump
喷水凝汽器 jet condenser
喷水排列 bank
喷水盘管 spraying bank
喷水炮 scooter
喷水喷头 sprinkler head
喷水器 feed-water injector;fountain; gunjet; hand lance; jet elevator; lance; moistener; projector; spraying lance;spray jet;spray water unit; sprinkler; sprinkling machine; water eductor; water-jet; water spray(er);water sprinkler;water vessel

喷水器间隙 jet clearance
喷水枪 monitor;spray gun;water gun
喷水清理 hydropeening; water horning
喷水清洗 hydropeening;jet cleaning
喷水清洗筛 rinsing-spraying screen
喷水区域 sprinkling zone
喷水泉 fountain
喷水日光计 water-jet actinometer
喷水日光能量测定仪 water-jet actinometer
喷水筛 spraying screen
喷水设备 sprinkler
喷水湿润器 jet humidifier
喷水时间延迟试验 discharge time delay test
喷水式避雷器 spray arrester
喷水式除尘器 water spray separator
喷水式防波堤 hydraulic breakwater; jetting wave absorber
喷水式减温器 spray-type desuperheater
喷水式冷却 water-sprinkled cooling
喷水式炉腹 water-sprinkled bosh
喷水式喷漆室 water wash spray booth
喷水式清底螺旋土样钻 bottom cleaning jet auger
喷水式调节器 spray-type controller
喷水式洗手器 wash fountain
喷水式饮水口 bubbler
喷水式饮水器 bubbler fountain
喷水室 sprinkler chamber
喷水刷 fountain brush
喷水水淬法 jet granulation
喷水水墩 spray dome
喷水水束挖掘法 hydraulicking
喷水水塔 rain chamber
喷水伺服机构 water injection servo
喷水通风器 ventilator with water spray;water-jet ventilator
喷水通气器 fountain aerator
喷水头 fountain head;quenching head; spray water head;sprinkler head
喷水推动装置 jet unit
喷水推进 hydraulic jet propulsion
喷水推进机 hydraulic propeller
喷水推进器 hydraulic propeller; jet propeller
喷水推进艇 hydrojet propelled ship; jet boat
喷水推进装置 hydraulic propulsion unit
喷水推力 jet thrust
喷水挖泥船 water injection dredge(r)
喷水雾 water spray
喷水雾化 atomization of water
喷水吸尘室 rain chamber
喷水洗涤 washing by jet
喷水洗涤器 water-jet scrubber
喷水系统 drencher system; sprinkler system;sprinkling system;water injection system; water spray system;water-spreading system
喷水消尘 water spray for damping dust
喷水消防栓 sprinkling hydrant
喷水形成的粒状矿渣砂 water spray granulated slag sand
喷水养护 curing by sprinkling;spray curing;sprinkler curing;wet curing
喷水抑尘法 dust-control with water-jetting; water-jet method of dust-control
喷水用给水管 spray water pipe; spray water tube
喷水增湿器 jet humidifier
喷水装置 moistener;sprinkling installation; water ejector; water injec-

tor; water sprinkler; water sprinkling device; waterworks
喷水嘴 water nozzle; water spray nozzle
喷丝板 extrusion nozzle
喷丝板沉积物 jet crater
喷丝板细孔 spinneret orifice
喷丝板组合件 spinneret assembly
喷丝板座 spinneret holder
喷丝帽浆环 jet ring
喷丝枪 wire pistol
喷丝头 spinneret; spinning jet; spinning nozzle
喷丝头拉伸 spinneret draft
喷丝头室 jet holder; jet room
喷丝涂料 webbing finish paint
喷丝涂装法 veiling; webbing
喷丝嘴 spinneret
喷送器 pneumatic transport placer
喷速 nozzle velocity
喷塑 spray plastic
喷塑彩色饰面 glazed finish
喷塑机 spray plastic machine
喷酸液机 acid blast machine
喷酸作业 acid jetting
喷搪 spray enamelling
喷腾层 spouted bed
喷腾式分解炉 spouted calciner
喷铁粉火焰清理 powder scarfing
喷铁砂 cloudburst
喷筒活塞 ejector ram
喷头 cap piece; injection head; injector; injector head; injector spray tip; nozzle tip; spray head; spray tip; spread head; sprinkler; sprinkler head; sprinkler nozzle
喷头间歇转位式喷灌器 wave-type sprinkler
喷头孔板 tip orifice
喷头拉伸 jet stretch
喷头拉伸比 jet stretch ratio
喷头装置系统 sprinkler head system
喷投成型法 gunning and slinger process
喷涂 extrusion coating; flame plating; gun application; gunite; gunning; spray coating; spray-down; spraying; spray-on; spray painting; sputtering; curtain coating
喷涂保护层 sprayed protective coating
喷涂保温层 sprayed-on insulation
喷涂薄膜 gunned film; sprayed-on film; sprayed-on membrane
喷涂布线 spray wiring
喷涂材料 gunited material; sprayed(-on) material
喷涂层 gunned coat; mist coat
喷涂衬垫 sprayed(-on) lining
喷涂成型 spray(-up) mo(u)lding
喷涂处理 spray(ing) treatment
喷涂挡水板 spray eliminator
喷涂的 sprayed; spray-painted
喷涂的加热炉绝热材料 gun-applied furnace insulation
喷涂地沥青 sprayed-on asphalt
喷涂法 curtain coating method; spraying method; spraying painting; spraying process; spray method
喷涂防滑料 spray grip
喷涂防火 sprayed fireproofing
喷涂防火材料 sprayed fireproofing material
喷涂分界线 break line of spraying
喷涂粉末 dusty spray; spray dust
喷涂粉刷打底层 sprayed-on rendering
喷涂封层 sprayed-on seal(ing)
喷涂隔热 sprayed insulation
喷涂隔热层 sprayed-on insulation
喷涂隔热隔声材料 gunned insulation
喷涂隔声吊顶 suspended sprayed acoustic(al) ceiling
喷涂隔声天花板 sprayed acoustic(al) ceiling
喷涂隔音灰泥 sprayed acoustic(al)-plaster
喷涂工艺 spray-coating technique
喷涂工作间 spray booth
喷涂管道 coating pipe
喷涂化学制品 spray chemical
喷涂灰浆的分隔墙 thrown-on plaster partition(wall)
喷涂灰泥 projection plaster
喷涂混合物 sprayed-on compound
喷涂混凝土 gunning concrete
喷涂机 painter; paint sprayer; paint spraying machine; sprayer; spray-painting equipment
喷涂机械 spraying machinery
喷涂积层法 spray lay-up
喷涂技术 gunning technique
喷涂间 spray booth
喷涂金属 metallization; metallizing; sprayed-on metal; spray metal
喷涂金属粉 metallisation
喷涂金属膜 metallic spray(ed) coat(ing)
喷涂金属涂层 sprayed-on metal coating
喷涂聚氨酯泡沫 sprayed polyurethane foam
喷涂聚氨酯涂料 sprayed polyurethane coating
喷涂绝热材料 sprayed insulation material; sprayed thermal insulant
喷涂绝缘 sprayed insulation; spray-on insulation
喷涂绝缘材料 spray-on insulation material
喷涂绝缘材料的混凝土衬垫 sprayed insulating concrete lining
喷涂控制器 spray controller
喷涂矿棉 sprayed mineral wool
喷涂矿物绝缘层 sprayed mineral insulation
喷涂矿物纤维绝热材料 spray-applied mineral fiber[fibre] insulation
喷涂冷却法 splat cooling
喷涂沥青 gun asphalt; gunned mastic
喷涂量 quantity for spray
喷涂料 paint-spraying; spraying-on material
喷涂玛瑞脂 gunned mastic; spray(ed-on) mastic
喷涂面 coating surface
喷涂面层 mist coat; sprayed coat; sprayed-on facing
喷涂面积 spray area
喷涂抹灰 projection plastering; shot-plastering
喷涂耐火材料 sprayed fire-resistive material; sprayed-on fireproofing
喷涂能力 capacity of spray coating
喷涂泡沫 sprayed-on foam
喷涂泡沫材料 spray foam material; gun-grade form
喷涂皮带管 spray hose
喷涂器 spray coater; sprayer
喷涂枪 spray gun
喷涂清漆 lacquer varnish; spraying varnish
喷涂容量 capacity of spray coating
喷涂软管 spray hose
喷涂软木(屑) gun-grade cork; sprayed-on cork
喷涂软木屑涂料 corking
喷涂砂浆 gunite mortar; sprayed-on mortar
喷涂设备 sprayer; spraying apparatus; spraying equipment; spraying plant
喷涂石膏 projection plaster
喷涂石灰 spray lime
喷涂石棉 limpet asbestos; sprayed(-on) asbestos
喷涂石棉保温层 sprayed-on asbestos insulation
喷涂石棉层 sprayed asbestos
喷涂石棉的涂层 sprayed asbestos insulation
喷涂石棉粉刷 sprayed-on asbestos plaster
喷涂石棉隔热层 sprayed-on asbestos insulation
喷涂石棉隔声材料 sprayed asbestos insulation material
喷涂石棉拉毛粉刷 sprayed-on asbestos stucco
喷涂石棉抹灰 sprayed-on asbestos plaster
喷涂石棉外墙粉刷打底 sprayed-on asbestos external rendering
喷涂石棉终饰 sprayed-on asbestos finish
喷涂时喷枪发生飞弧 whipping during spraying
喷涂时喷丝 cobwebbing during spraying; cocoon spraying
喷涂时形成的网纹 cobwebbing during spraying
喷涂式应变仪 strain-ga(u)ge installed by spraying process
喷涂试验 spraying test
喷涂饰面 coating; spray texture
喷涂室 spray chamber
喷涂水化石灰 spray lime
喷涂水泥 gun cement; sprayed-on cement
喷涂水泥砂浆 gunite
喷涂水泥外墙粉刷打底 sprayed-on cement external rendering
喷涂塑料 gunned plastics; sprayed-on plastics; spray plastic
喷涂塑料地面 sprayed plastic flooring; spray-on plastic flooring
喷涂塑料饰面 sprayed plastic finishing; spray-on plastic finishing
喷涂透明清漆 sprayed-on clear varnish
喷涂涂层 sprayed-on(paint) coat
喷涂涂饰剂 spray finish
喷涂涂锌 spray galvanizing
喷涂土薄膜 sprayed-on geomembrane
喷涂脱模剂 sprayed-on release agent
喷涂屋顶薄膜 sprayed-on roof(ing) membrane
喷涂屋面 sprayed roof
喷涂物 spraying substance
喷涂物质 sprayed-on mass
喷涂吸声材料 sprayed-on acoustical material
喷涂吸声顶棚 sprayed-on absorbing ceiling; sprayed-on acoustical ceiling; sprayed-on sound absorbent ceiling
喷涂吸声粉刷 sprayed acoustic(al) plaster
喷涂吸声抹灰层 sprayed acoustic(al) plaster
喷涂吸音粉刷 absorptive sprayed on plaster
喷涂细片冷却法 splat cooling
喷涂锌 zinc spraying
喷涂性 sprayability
喷涂压力 spraying pressure
喷涂压料罐 spraying pressure container
喷涂养护剂 spray-on membrane
喷涂阴极 sprayed cathode
喷涂油灰 sprayed-on putty
喷涂油漆 spray painting
喷涂油漆工 spray painter
喷涂蛭石 sprayed vermiculite; gun-grade vermiculite
喷涂蛭石砂浆 spray(ed)(-on) vermiculite plaster
喷涂装饰 projection plastwork
喷涂装饰瓷 jet-enamelled ware
喷涂装置 spraying plant
喷涂着色 spray colo(u)ration
喷涂作业 gun-applied; paint-spraying
喷丸 ball blast; blast grit; blasting shot; cloudburst; grit blasting; peeling; rotoblast; shot peen
喷丸不足 short shot
喷丸残渣 shot-blast debris
喷丸车间 shit blasting shop
喷丸成型 cloud burst treatment forming; contour peening; shot peening forming
喷丸除鳞 rotoblasting
喷丸除锈 shot blast(ing)
喷丸除锈车间 shot-blasting shop
喷丸除锈机 shot-blaster; shot-blasting machine
喷丸除锈设备 shot blasting equipment
喷丸处理 ball(shot) peening; ball spray treatment; blast finishing; grit blasting; shot-blast(ing)(processing); shot peening
喷丸处理表面 peened surface; shot-blasting surface; shot-peened surface
喷丸处理法 cloudburst process
喷丸打磨 wheelabrating
喷丸范围 shot peening coverage
喷丸滚筒 barrel-type shot blasting machine
喷丸机 compressed-air shot-blasting machine; peener; peening machine
喷丸加工 shot-blast; shot peening
喷丸加工轧辊 shot-blasted roll
喷丸净化 blast cleaning
喷丸冷加工(处理) shot peening
喷丸喷头 abrator head
喷丸器 atomizer; fog spray; grit blast; spray can; sprayer; spray jet; wheel abrator
喷丸强化 shot peening strengthening; stress peening
喷丸强化效应 peening effect
喷丸清理 ball blasting; ball-shooting; blast cleaning; blasting; grit blast; impact cleaning; peening; shot blast(ing); shot-cleaning; wheel abrating
喷丸清理滚筒 rotoblast barrel
喷丸清理机 airless shot blasting machine; pneumatic shot blasting machine; wheel abrator; abrator
喷丸清理室 shot-blasting cabinet; shot-blast chamber
喷丸清理转台 table blast
喷丸清理装置 abrator; wheel abrator
喷丸清面 shot blasting
喷丸清砂装置 shot cleaning unit
喷丸室 shot-blast cabinet; shot-blast room; wheel abrator cabinet
喷丸硬度试验 cloud burst hardness test
喷丸硬化 peen hardening
喷丸硬化处理 shot cold hardening; shot peening; cloud burst treatment
喷丸硬化处理试验 cloud burst test
喷丸硬化法 cloud burst process; peening
喷丸装置 blaster; peener; shot-blast unit

喷纬嘴 pick-blowing nozzle
喷雾 atomise[atomize]; atomizing; dewing; fog spray(ing); mist propagation; mist spray(ing); nebulize; pulverization; sparge; spray; spraying; spray-up; water sprinkling
喷雾泵 atomizing pump
喷雾薄膜蒸发器 spray film evapo(u)rator
喷雾槽 spray cistern
喷雾车 spraying car
喷雾沉残渣 spray residue
喷雾除草 weed spray
喷雾除尘器 aerosolator; spray scrubber
喷雾除气器 spray deaerator
喷雾处理 shot-cleaning; spray treatment
喷雾吹风电动机 mist fan motor
喷雾淬火 fog quenching
喷雾的 atomizing; sprayed atomized
喷雾冻凝法 congealing spray
喷雾洞室 chamber mist
喷雾锻造 spray method
喷雾阀 spray valve
喷雾阀盖 spray valve cover
喷雾阀簧 spray valve spring
喷雾法 atomization method; spraying method; spraying process; spray-on process; spray gun process
喷雾法上釉 glazing by atomization; glazing by spraying
喷雾法铁粉 Rz-powder
喷雾飞逸 spray loss
喷雾沸腾干燥器 spray-type fluidized bed drier[dryer]
喷雾分散度测定仪 sprayograph
喷雾风机 spray fan
喷雾风扇 aerosolizing fan; air-douche unit with water atomization; mechanical humidifier with fan; spray fan; sprinkling fan
喷雾腐蚀机 spray etcher
喷雾附加装置 sprayer attachment
喷雾附着 spray catching; spray deposit
喷雾干燥 spray drying
喷雾干燥法 spray drying method
喷雾干燥法烟气脱硫 spray drying process of fume gas desulfurization
喷雾干燥过程 spray drying process
喷雾干燥器 nubilose; spraying drier [dryer]
喷雾干燥室 spray drying chamber
喷雾干燥塔 spray drying tower
喷雾干燥洗涤剂 spray-dried detergent
喷雾杆 spray lance
喷雾给湿机 mist spray damping machine
喷雾给湿器 spray damping machine
喷雾管 atomizing jet; injection spray pipe; spray line; spray pipe
喷雾管路 atomizer passage; spray pipe passage
喷雾管嘴 fog-cured; fog nozzle; spray nozzle
喷雾罐 aerosol spray-can
喷雾罐中的喷雾剂 spray-can propellant
喷雾罐装涂料 aerosol paint
喷雾罐装油漆 aerosol paint
喷雾烘干机 atomizing drier[dryer]; spray drier[dryer]
喷雾护面罩 spray mask
喷雾环 perforated ring
喷雾机 atomizing machine; damping machine; dewing machine; pulverizer; spray(ing) machine
喷雾机械 spraying machinery
喷雾机压力控制器 spray-pressure controller
喷雾计 spraymeter
喷雾剂 nebula; pressurized container spray; pressurized spray; spraying agent
喷雾剂瓶 aerosol container
喷雾加工法 spray processing
喷雾加湿器 atomizing humidifier; humidification spray; spray damping machine
喷雾降温 mist cooling
喷雾胶管 spraying hose
喷雾角度 spray angle; spreading of spray; spreader of spray
喷雾结晶 spray crystallization
喷雾聚合 spray polymerization
喷雾颗粒尺寸 spray particle size
喷雾刻蚀机 spray etcher
喷雾口 spray jet
喷雾冷凝 spray congealing
喷雾冷凝法 spray-freezing method
喷雾冷凝器 spray condenser
喷雾冷却 mist cooling; spray cooling
喷雾冷却法 spray cooling process
喷雾冷却风机 mist-spraying cooling fan
喷雾冷却器 spray cooler
喷雾冷却塔 spray cooling tower
喷雾淋盘式除氧器 spray tray(type) deaerator
喷雾媒介 atomizing medium
喷雾面罩 spray helmet
喷雾苗床 mist bed
喷雾灭尘 water spray for damping dust
喷雾黏[粘]着剂 spray sticker
喷雾沤 mist retting
喷雾盘 spray disc[disk]
喷雾喷粉机 sprayer duster
喷雾喷粉器 sprayer and duster
喷雾喷灌器 mist sprinkler; spray head sprinkler
喷雾皮带管 spray hose
喷雾曝气器 spray aerator
喷雾器 airgun; air sprayer; atomizer; atomizing device; blow gun; diffuser; dissipater[dissipator]; fog sprayer; gun; hydroconion; mist atomizer; mist blower; mist sprayer; nebulizer; pulverizer; rugosa; sparger; spray atomizer; spray bar; spray can; spray device; sprayer; spraying apparatus; spraying equipment; spraying gun; spraying unit; spray jet; spray lance; spray producer; vapo(u)rizer
喷雾器拱形喷杆 jet arch
喷雾器管道 injector manifold
喷雾器弧形喷管 sprayer jet arc
喷雾器环 spray ring
喷雾器集流腔 injector manifold
喷雾器喇叭口 atomizer cone
喷雾器喷头 mister head
喷雾器喷嘴 <莲蓬式> rose
喷雾器头 nozzle tip
喷雾浅盘式除氧器 spray and tray tape deaerator
喷雾枪 spray pistol
喷雾清理机 spray cleaner; jenny
喷雾圈 atomizing cup
喷雾燃烧 spray burning; spray combustion
喷雾燃烧器 atomizing burner; atomizing oil burner; spray burner
喷雾热分解 spray pyrolysis
喷雾熔接 spray welding
喷雾熔烧 spray roasting
喷雾软管 spray hose
喷雾润滑 mist lubrication
喷雾润湿器 spray damper
喷雾撒粉机 mist duster
喷雾洒水 sprinkle; water sprinkling
喷雾洒水设备 water spray system
喷雾上浆 spray sizing
喷雾上浆机 spray starching machine
喷雾设备 spraying device; spraying equipment; spraying plant
喷雾射程 carry of spray; spray penetration
喷雾射流 spraying jet
喷雾式除尘器 spray dust scrubber; water spray separator
喷雾式除气器 atomizing deaerator
喷雾式除氧器 spray-type deaerator
喷雾式电熨斗 steam iron
喷雾式干燥机 spray drier[dryer]
喷雾式化油器 jet carburet(t)or; pulverization carburet(t)or
喷雾式加湿器 atomizing-type humidifier
喷雾式减湿器 spray-type dehumidifier
喷雾式进气口 atomizing air intake
喷雾式冷却 spray-type cooling
喷雾式冷却器 spray-type cooler; spray-type fire extinguisher
喷雾式冷却塔 spray-type tower
喷雾式灭火器 spray-type fire extinguisher; water spray (fire) extinguisher
喷雾式汽化器 atomizing carburet(t)er; jet carburet(t)or; spray carburet(t)or
喷雾式脱氧器 atomizing deaerator
喷雾式洗片机 jet-spray processor
喷雾式油燃烧器 atomizing-type oil burner
喷雾试验 spray(ing) test(ing)
喷雾室 air scrubber; air-washer room; fog room; jet chamber; spray(ing) chamber; sprinkling chamber
喷雾刷色器 air brush
喷雾水 spraying water
喷雾水枪 diffuse branch; diffuser spray nozzle
喷雾酸洗 spray acid cleaning
喷雾损失 spray loss
喷雾塔 atomizing column; spray tower
喷雾探头 spray probe
喷雾调节温度 spray attemperation
喷雾调湿 spray damping
喷雾筒唧筒 spray pump
喷雾头 fog-spray nozzle; spray head
喷雾涂布机 Bracewell coater
喷雾涂层 fog coat; spray painting
喷雾涂料 aerosol paint
喷雾涂搪 spray enamelling
喷雾脱脂剂 spray degreaser
喷雾温度 vapo(u)rizing temperature
喷雾涡轮机 sparge turbine
喷雾吸入器 aerohaler
喷雾吸收塔 spray tower for absorption
喷雾洗涤汽器 spraying reagent equipment
喷雾匣 spraying cabin
喷雾消尘 water spray for damping dust
喷雾效率 nebulization efficiency
喷雾形状 spray pattern
喷雾性 sprayability
喷雾性能 pulverability
喷雾压力 spraying pressure
喷雾压力控制器 spray-pressure controller
喷雾厌氧培养皿 spray anaerobic dish
喷雾养护 fog cure; fog curing; fog spraying curing
喷雾养护的 fog-cured
喷雾氧化法 spray oxidizing process
喷雾增湿 spraying humidification
喷雾增温塔 spray-type humidifier
喷雾针 spray needle
喷雾蒸发 spray evapo(u)ration
喷雾蒸发器 spray evapo(u)rator
喷雾质量测定仪 sprayograph
喷雾柱 spray column
喷雾装置 atomizing device; spray device; sprayer unit; spray system
喷雾状物 spray
喷雾锥 atomizer cone; atomizing cone; spray(ing) cone
喷雾锥角 spray angle
喷雾灼烧 spray calcining
喷雾嘴 air cup; air nozzle; atomizer aperture; atomizing jet; fog(-type) nozzle; spray(ing) nipple; spray(ing) nozzle; spray(ing) jet
喷雾作用 atomization; nebulization
喷吸泵 ejection pump; ejector; jet pump; pump nozzle
喷吸气泵 ejection air pump; liquid jet air pump
喷吸器带条 ejector strap
喷吸嘴 pump nozzle
喷吸作用 blow torch action; ejector effect
喷洗 blow wash; hydropeening; spray cleaning; spraying; spray rinsing; spray-wash
喷洗机 injection rinsing machine; spray washer
喷洗皮带管 spray-down hose
喷洗器 jet purifier; jetter; spray(ing) cleaner; spray(ing) washer
喷洗枪 cleaning gun
喷洗软管 spray-down hose
喷洗设备 spray-down equipment
喷洗室 spray chamber; rodding chamber <车辆的>
喷洗水龙带 spray-down hose
喷洗装置 jetter; jetter assembly
喷显剂 spraying(re) agent; streak reagent
喷箱 spray box
喷屑 fragmental ejecta
喷锌 zinc spraying
喷锌层 sprayed zinc coating
喷絮植绒 flock spraying
喷压提升机 lift(ing) jet
喷烟 aerosol spraying; smoke jet; puff
喷烟测漏器 <管道的> smoke rocket; rocket tester
喷烟管 smoke nozzle
喷烟喷雾两用机 fogging and spraying machine
喷烟器 fogging machine
喷盐腐蚀试验 salt spray test
喷盐(雾)试验 <快速耐腐蚀试验> salt spray test
喷眼 key holing
喷焰 bright eruption; flare; puff
喷焰除草机 flame cultivator; flame weeder
喷焰除鳞 flame descaling
喷焰连续退火 flame strand annealing
喷焰偏转器 blast deflector
喷焰钻机 jet-pierce machine
喷焰钻进 jet-piercing
喷焰钻孔 <岩石内> jet drilling
喷焰钻孔机 jet drilling rig
喷焰钻燃烧室 jet-piercing burner
喷窑爆炸 spouting explosion
喷窑灰 insufflation dust
喷药残渣 spray residue
喷药伤害 spray injury
喷液 hydrojet
喷液比重计 spray(o)meter
喷液车 liquid distributor
喷液淬火 spray quenching
喷液挡板 flow catch(er)

喷液涤气器 spray scrubber
喷液管 sparge pipe
喷液机 liquid distributor;liquid-spreader
喷液冷却 diluent cooling
喷液器 spreader
喷逸 dry spray;overspray
喷溢 outpouring
喷溢口 vent of eruption
喷溢型 effusive type
喷银 spray silvering
喷印 spray printing
喷涌 gush
喷用胶浆 spray-type cement
喷用砂 sand-blast sand
喷油 fuel oil injection;oil injection;oil throw
喷油泵 fuel injection pump;injection pump
喷油泵试验台 fuel injection pump tester
喷油泵试验装置 fuel injection pump testing instrument
喷油泵套筒 injection pump barrel
喷油泵凸轮轴 fuel injection camshaft
喷油泵外壳 housing of fuel injection pump;injection housing
喷油泵柱塞 injection pump plunger
喷油泵柱塞齿轮 plunger gear
喷油泵柱塞行程 injection pump stroke
喷油泵总成 injection pump assembly
喷油点火装置 fuel injector igniter
喷油定时 injection timing
喷油发动机 injection oil engine
喷油阀 injection valve;spray valve
喷油阀关闭压力 injection-valve closing pressure
喷油阀开启压力 injection-valve opening pressure
喷油阀嘴 spray valve nozzle
喷油翻土机 <可以翻土喷油,就地处理土壤> suboiler
喷油缸 ejection cylinder
喷油管 oil atomizer;spray bar
喷油管道 injection(pipe)line
喷油管路 injection(pipe)line
喷油罐 spout pouring pot
喷油机 oil distributor
喷油剂量系统 dosing system
喷油间隔 injection interval
喷油浆粉刷 plaster sprayed with oil paint
喷油井 gusher;spouter
喷油孔 nozzle opening; oil spray hole;squirt hole
喷油控制齿条 fuel control rack
喷油冷却 spray oil cooling;spread oil cooling
喷油量 <即沥青洒布机生产率> distributive value
喷油率 fuel injection rate
喷油灭弧室 oil-blast explosion chamber
喷油能力 <即沥青洒布机生产率> distributive ability
喷油漆 paint-spraying
喷油起点 fuel injection beginning
喷油器 ejector;fuel injector;fuel oil injector;fuel valve;injector;oil atomizer;oil ejector;oil sprayer
喷油器垫 injector pad
喷油器冷却泵 injector cooling pump
喷油器溢流孔 injector overflow
喷油器针阀 oil fuel injector needle-valve
喷油器组 <包括燃油泵与喷嘴> unit injector
喷油燃烧器 spraying burner
喷油润滑 oil jet lubrication;oil spray lubrication
喷油润滑系统 grease spray lubrication system
喷油射流 oil jet
喷油时限 injection timing
喷油式料斗送料器 oil jet hopper feeder
喷油提前 early admission of fuel;injection advance
喷油提前装置 injection advance device
喷油调节 injection regulation
喷油调时装置 injection timer
喷油通路 oil jet path
喷油头 nozzle tip
喷油凸轮 fuel injection cam
喷油凸轮轴 fuel injection camshaft
喷油拖车 trailer oil distributor
喷油雾 oil spray
喷油(雾化)燃烧炉 atomizing oil burner
喷油雾润滑法 atomized lubrication
喷油雾养护装备 atomized lubrication
喷油熄弧式脉冲断路器 oil blow-up type impulse breaker
喷油系统 injection system
喷油小孔 oil spit hole
喷油针 fuel injection needle;jet needle;nozzle needle
喷油滞后 injection lag
喷油终点 fuel injection end
喷油周期 oiling period
喷油柱前锋 jet front;jet head
喷油柱外层 jet envelope;jet mantle
喷油柱形状 jet-form
喷油装置 fueling injection equipment
喷油嘴 fuel injection nozzle; fuel sprayer;fuel spray nozzle;injector;nozzle(tip);oil atomizer nozzle;oil injection nozzle; oil nozzle;spraying nozzle
喷油嘴安装 <喷射器的> injector setting
喷油嘴阀体 nozzle body
喷油嘴固定螺柱 injector stud
喷油嘴紧帽 nozzle cap nut
喷油嘴孔 injector orifice
喷油嘴倾角 nozzle inclination
喷油嘴试验器 injection nozzle tester
喷油嘴溢油管 dribble pipe
喷油嘴针阀 injection nozzle valve;injector valve;nozzle needle
喷油嘴总成 injector assembly
喷釉 glazing by atomization;glazing by spraying;spray glaze;spraying glazing;sprinkling
喷釉波纹 spray sagging
喷釉橱 spray booth
喷釉法 spray glazing
喷釉柜 glaze spray booth
喷釉流痕 spray sagging
喷釉枪 glazing spray gun
喷釉室 glaze spray booth;spray booth
喷雨枪 irrigation grease; rotary rain gun
喷皂粉 spraying soap powder
喷针杆 needle stem
喷针接力器 spear servomotor
喷蒸干燥窑 steam-jet kiln
喷制的石棉 sprayed asbestos
喷制土 sprayed earth
喷制橡胶 sprayed rubber
喷珠法 shot blast
喷珠硬化法 peening
喷注 insufflation;jetting;spout(ing)
喷注边棱发声器 jet-edge generator
喷注材料 air-placed material;injection material
喷注插口 injection socket
喷注的雾化 jet dispersion
喷注冻结法 jet-freezing
喷注发生器 jet generator
喷注混合系统 jet mixing system
喷注混凝土 air placed concrete
喷注孔 jet hole
喷注棱超声发生器 jet-edge ultrasonic generator
喷注器 ejector filler;injector;inspirator
喷注砂浆 injection mortar
喷注式超声波发生器 jet ultrasonic generator
喷注式发生器 jet generator
喷注水管 jetting hose
喷注水泥浆 gunite
喷注替续器 jet relay
喷注效应 fountain effect
喷注压力 injection pressure;jet pressure
喷注噪声 injection noise;jet noise
喷锥 atomizing cone
喷嘴 air jet nozzle;air spout;air spray;atomizer;beam tube;bib(b);burner nozzle; discharge jet; discharge nozzle;downspout;ejection nozzle;ejection opening;ejector pump;gas orifice; injecting nozzle; injection cock; injection orifice; injector; jet apparatus;jet atomizer;jet orifice;jet(ting)nozzle;nose(piece);nozzle(exit);placing nozzle;reducing nozzle; spout; spraying head; spray(ing) nozzle; sprinkler [sprinkling] nozzle;nozzle bushing <钻头的>
喷嘴扳手 nozzle wrench
喷嘴壁 nozzle wall
喷嘴残滴 dribble from jet
喷嘴测流计 nozzle meter
喷嘴测流仪 nozzle meter
喷嘴插芯 nozzle core
喷嘴出口 jet exit;jet expansion;nozzle mouth
喷嘴出口截面 jet area
喷嘴出口面积 exit area of nozzle
喷嘴出气角 nozzle outlet gas angle
喷嘴挡板 nozzle flapper;nozzle screen
喷嘴挡板机构 nozzle-baffle mechanism;nozzle-flapper mechanism
喷嘴挡板系统 nozzle flap
喷嘴导向叶片 nozzle guide vanes
喷嘴导叶 nozzle guide vanes
喷嘴的喷孔圆盘 nozzle orifice disk
喷嘴的锐孔 nozzle orifice
喷嘴的压降 injection differential pressure
喷嘴的锥形雾化喷头 tapered atomizer cup
喷嘴垫圈 nozzle orifice disk; spray disc[disk]
喷嘴堵塞 nozzle clogging
喷嘴阀 nozzle group valve; nozzle valve
喷嘴阀操作盘 nozzle valve operating panel
喷嘴阀杆 nozzle valve rod
喷嘴阀开闭调速 nozzle cut-out
喷嘴法 nozzle process
喷嘴反作用 reaction of nozzle
喷嘴分离 jet separation
喷嘴负荷调整范围 burner load range
喷嘴盖板 nozzle plate
喷嘴杆 jet stem
喷嘴格栅 fueling injection grate
喷嘴供液管路 jet line
喷嘴共振 nozzle resonance
喷嘴管 nozzle pipe;nozzle tube;blast tube
喷嘴管路滤器 nozzle line strainer
喷嘴管体 nozzle body
喷嘴过滤器 nozzle filter
喷嘴焊钳 tip holder
喷嘴横截面 nozzle throat area
喷嘴喉部 nozzle throat
喷嘴弧段 nozzle segment
喷嘴环 nozzle cascade; nozzle diaphragm;nozzle ring
喷嘴环隙 nozzle clearance
喷嘴环叶片 nozzle blade
喷嘴混合器 nozzle mixer
喷嘴火焰 nozzle flame
喷嘴集管 spray header
喷嘴夹 nozzle clamp
喷嘴架 injection nozzle carrier;spray bar
喷嘴尖 nozzle tip
喷嘴间距 injector spacing
喷嘴间隙 jet clearance;nozzle clearance
喷嘴检查盖 inspection cover for nozzle
喷嘴角 nozzle angle
喷嘴接合 bean joint
喷嘴接头 bean joint;nozzle stub
喷嘴节距 nozzle pitch
喷嘴节流调速 nozzle control governing
喷嘴截面 nozzle cross-section
喷嘴截面积 sectional area of nozzle
喷嘴截面面积 cross-section area of nozzle
喷嘴进口 nozzle inlet
喷嘴颈 nozzle neck
喷嘴绝对等值流 equivalent absolute nozzle flow
喷嘴开启压力 nozzle opening pressure
喷嘴壳体 nozzle body
喷嘴空气分级机 jet classifier
喷嘴孔 injection orifice;nozzle bore;nozzle hole
喷嘴孔的通过截面面积 injector hole area
喷嘴孔径 nozzle hole diameter
喷嘴孔口 nozzle opening
喷嘴孔砌体 nozzle masonry
喷嘴(控制)阀 nozzle control valve
喷嘴口 nozzle exit;nozzle orifice
喷嘴口环 nozzle mouth-ring
喷嘴扩散角 nozzle angle
喷嘴扩张角 nozzle divergence angle
喷嘴喇叭口 nozzle throat
喷嘴临界截面 critical throat section
喷嘴临界面积 critical throat area
喷嘴流量 nozzle flow
喷嘴流量系数 discharge coefficient of flow nozzle;nozzle discharge coefficient
喷嘴漏斗 funnel with nozzle
喷嘴滤网 nozzle strainer
喷嘴滤油网 nozzle screen
喷嘴帽 nozzle cone
喷嘴密度 spray-nozzle density;spray-type air water section
喷嘴面积 jet area;nozzle area
喷嘴面积比 nozzle area ratio
喷嘴面积控制系统 nozzle area control system
喷嘴面积膨胀率 nozzle expansion ratio
喷嘴面积收缩率 nozzle-contraction-area ratio
喷嘴面积系数 nozzle area coefficient
喷嘴内衬 nozzle liner
喷嘴内径 nozzle inside diameter
喷嘴内腔 nozzle chamber
喷嘴排布 injector configuration
喷嘴排出空气 nozzle air bleed
喷嘴抛头丸 abrator head
喷嘴配件 nozzle fittings

P

喷嘴配置 injector staging
喷嘴喷气分离 nozzle jet separation
喷嘴喷射方向 nozzle spray direction
喷嘴喷射角度 nozzle spray angle
喷嘴喷射压力 nozzle injection pressure
喷嘴喷雾器 nozzle atomizer
喷嘴剖面 nozzle section
喷嘴曝气器 nozzle aerator
喷嘴气流角 nozzle-stream angle
喷嘴气流离体 nozzle separation
喷嘴气体 orifice gas
喷嘴器输出管 injector delivery
喷嘴前空间 jet space
喷嘴腔 nozzle chamber
喷嘴切口 nozzle edge
喷嘴清洁器 < 清洁喷嘴的细丝 > jet cleaner
喷嘴清洗器 jet cleaner
喷嘴燃烧器 jet burner; spray-nozzle burner
喷嘴热风干燥器 nozzle jet drier
喷嘴入口端 upstream side of the injector
喷嘴梢 spray tip
喷嘴射流 nozzle jet; injection stream
喷嘴射流速度 nozzle velocity
喷嘴式分离器 jet orifice separator
喷嘴式加料器 spout feeder
喷嘴式净化器 nozzle type purifier
喷嘴式沥青乳化机 nozzle bitumen emulsifying machine
喷嘴式流量计 nozzle flow meter
喷嘴式流速计 nozzle meter
喷嘴式喷雾干燥器 nozzle type spray drier[dryer]
喷嘴式水轮机 nozzle type turbine; nozzle type water wheel
喷嘴式送风口 nozzle outlet
喷嘴式网前箱 nozzle headbox
喷嘴式岩粉捕集器 < 空气钻进时 > nozzle sludge trap
喷嘴式堰板 jet slice
喷嘴试验泵 injector testing pump
喷嘴试验器 nozzle testing stand
喷嘴试验台 nozzle testing stand
喷嘴室 nozzle box; nozzle chest; nozzle supply
喷嘴束流 contraction of nozzle
喷嘴(水头)损失 nozzle loss
喷嘴速度 nozzle velocity
喷嘴损失系数 nozzle loss factor
喷嘴套 nozzle cage
喷嘴体 nozzle assembly
喷嘴调节 cut-out governing; nozzle governed; nozzle governing; nozzle regulation
喷嘴调节阀 nozzle control valve; nozzle regulator
喷嘴调节螺旋 jet adjuster
喷嘴调节汽轮机 multivalve turbine
喷嘴调节器 jet adjuster
喷嘴调节特性 nozzle metering characteristic
喷嘴调节装置 jet adjuster; nozzle control governor gear
喷嘴通道 nozzle passage
喷嘴通道摩擦 nozzle passage friction
喷嘴通气器 nozzle aerator; spray aerator
喷嘴通针 tip cleaner
喷嘴头 burner cup; head of nozzle; injection cup; injector cup; injector spray tip; nozzle head; spray cup; spray head
喷嘴头部面积 injector face
喷嘴突缘 nozzle flange
喷嘴托 nozzle holder
喷嘴弯头 elbow for nozzle
喷嘴网 injector grid
喷嘴尾流共振 nozzle wake resonance
喷嘴雾化 nozzle atomization; nozzle atomizing
喷嘴雾化型喷雾干燥器 nozzle type spray drier[dryer]
喷嘴吸风机 nozzle blower
喷嘴系统 jet assembly
喷嘴箱 nozzle box; nozzle chamber; nozzle chest
喷嘴箱部套 nozzle box assembly
喷嘴效率 nozzle efficiency
喷嘴效应 nozzle effect
喷嘴形状 nozzle form
喷嘴压降 injection drop; nozzle drop
喷嘴压力 nozzle exit pressure; nozzle pressure
喷嘴叶片 nozzle blade; nozzle bucket; nozzle division plate; partition plate
喷嘴叶栅 nozzle blade cascade
喷嘴针阀 nozzle needle
喷嘴针阀塞杆 nozzle valve rod
喷嘴支承环 nozzle seat ring
喷嘴砖 burner block; burner tile; quarl block
喷嘴装配体 nozzle assembly
喷嘴装置 nozzle diaphragm; nozzle device
喷嘴锥体 nozzle cone
喷嘴组 descaling spray; injector set; nozzle block; nozzle grouping; nozzle segment; set of nozzles
喷嘴组件 jet assembly
喷嘴组调节 nozzle group control; segmental nozzle group control
喷嘴座 nozzle frame; nozzle carrier; nozzle holder

盆板 buckle plate

盆槽 basin
盆的无缝成型法 seamless forming of bowl
盆底敞车"bathtub" gondola
盆地 basin; bottom land; catchment basin; charco; court dock; flat bottom land; hoya; valley
盆地边界性质 boundary property of basin
盆地边缘沉积 basin margin deposit
盆地边缘相 basin facies
盆地冰川 basin glacier
盆地补给 recharge of basin
盆地侧下降断层 down-to-basin fault
盆地充填物性质 filling property of basin
盆地充填序列 basin-fill sequence
盆地充填样式 basin-fill pattern
盆地出口 basin mouth
盆地的边界 boundary of basin
盆地的基底 basement of basin
盆地的形态 basin shape
盆地底部地形 topography of basin bottom
盆地地层格架 stratigraphic(al) framework of basin
盆地分带 zonation of basin
盆地分类 basin classification
盆地构造 basin structure
盆地构造格架 structure framework of basin
盆地构造演化 structural evolution of basin
盆地古流体系 pal(a)eocurrent system of basin
盆地谷 cirque
盆地湖 basinal lake
盆地花坛 sunken flower bed
盆地基本数据 basin basic data
盆地几何形态 geometry of basin
盆地类型 type of basin
盆地模式 basin model
盆地平原 basin plain
盆地群 group of basin
盆地热特征 thermal property of basin
盆地山脉 basin range
盆地山脉构造 basin(-and)-range structure
盆地山脉省 basin and range province
盆地特征 basin characteristic
盆地土壤 basin soil
盆地土壤栽培 basin soil cultivation
盆地相 basin facies
盆地相的空间配置 spatial configuration of basin facies
盆地形成作用 basining
盆地形状 basin shape
盆地形式 style of basin
盆地园 sunken garden
盆地沼泽 basin bog
盆地褶曲 basin fold
盆地周围钙质沉积 rimstone
盆地组合 set of basin
盆碟室 plate closet
盆架 pot shelving; sunk shelf
盆景 bonsai; dish garden; miniature landscape; miniature trees and rockery; potted landscape; potted plant
盆景园 garden of pot plant; potted landscape garden
盆岭构造 basin range structure
盆岭省 basin and range province
盆岭(相间)地貌 basin range landform
盆内岩石 intrabasinal rock
盆式便器 pan closet
盆式多格桥面(体系) dishing-shaped multi-cellular deck
盆式炉栅 basin grid
盆式天线 dishpan antenna
盆式橡胶支座 pot neoprene bearing; pot rubber bearing
盆式支座 pot bearing
盆饰挑出线脚 sunk shelf
盆外河流 extrabasinal river
盆外岩 extrabasinal rock
盆形凹地 pan
盆形船闸 basin lock
盆形钢板 hang plate
盆形构造的 basin-structured
盆形轮 crown gear
盆形褶皱 basin fold
盆形支座 pot bearing
盆浴 tub bath
盆栽 pot culture
盆栽灌水系统 pot watering system
盆栽花草 potted plant
盆栽花坛 basined flower bed; potted flower bed
盆栽试验 pot experiment
盆栽树木 bonsai
盆栽植物 bonsai; pot-grown plant; pot(ted) plant
盆状凹地 crater; pan
盆状层【地】 basin-shaped strata
盆状架栅 trough grate
盆状排列构造 bowl arrangement structure
盆状山谷 basin valley
盆状物 pan
盆状向斜 pericline
盆子牙齿 ring gear

砰的声音 slam

砰的一声 fut
砰地放下 slam
砰地关上 slam
砰击【船】 slamming
砰砰声 ping
砰然的打击 thud
砰然声 thud; thump
砰然作声 clap

烹饪场所 cooking place

烹饪的 culinary
烹饪(法) cuisine; cooking
烹饪废物 garbage
烹饪器具 cooking apparatus
烹饪器皿 cooking apparatus; cooking ware; flame ware
烹饪用玻璃 oven glass
烹调器皿 ovenware
烹调用器皿 cooking vessel

彭佰顿水准尺 Pemberton level(l)ing rod

彭布罗克折面桌 Pembroke table
彭代里孔山的大理石 < 希腊 > Pentelic marble
彭-法心轴机器护罩 Pen-Farrel guard of mandrel machine
彭克混凝土 < 一种憎水水泥 > Pectacrete
彭林淡灰色花岗岩 < 英国康沃尔产 > Penryn
彭罗斯过程 Penrose process
彭曼法 < 计算耗水量即蒸发能力的 > Penman's method
彭木 < 一种产于印度安达曼群岛的红褐色硬木 > Poon
彭奈恩山脉 Pennine Range
彭奈恩山石灰 Pennine lime
彭尼科克屋面装配玻璃 Pennycock
彭宁放电泵 Penning pump
彭宁冷阴极电离真空计 Penning ga(u)ge

棚暗箱 booth

棚板 bat; decking; setter; shelf; slab
棚厂 hangar
棚厂操作轨 docking rail
棚厂甲板 hangar deck
棚车 box car; box freight car; caravan; closed car; closed wagon; covered car; covered truck; house-car; shed car; van; covered van < 小型的 >; box wagon < 英 >【铁】
棚车端部 box car end
棚车端门 lumber door
棚车端墙 box car end
棚车门 box car door
棚车内腰带 girt(h)
棚灯 ceiling lamp
棚灯导线 ceiling lamp wiring
棚店 stall
棚顶 shed roof
棚顶板 poling board
棚顶车身 convertible body
棚顶窗 shed dormer
棚顶梁 set collar
棚洞 hangar tunnel; shed for falling stone; tunnel shed
棚房 shanty
棚盖市场 covered market(place)
棚干材 shed drying stock
棚户 shack-dweller; slumdweller
棚户区 preliberation; shack area; shanty town; slum area
棚户区改善 slum improvement; slum upgrading
棚户区改造 slum clearance; slum rec-

lamation
棚户住区 squatter settlement
棚加车 rack-type car
棚架 arbour;ladder;trellis
棚架车 shelf car
棚架顶梁 cap
棚架构件 set member
棚架式盾构 multideck shield
棚架式拱道 trellis
棚架式平台 shelf-type platform
棚架装窑 pocket setting
棚间隔梁 studdle
棚梁中柱 studdle
棚料 bridging;scaffold
棚面板 weather-board(ing)
棚舍 barn;booth;shed;succah;sukkah
棚式 <内燃电力机车车体> body type
棚式储仓 shed storage
棚式盾构 shelf-type shield
棚式熟料库 shed-type clinker storage
棚屋 benab;bothy;cabana;hulk;hustiement;hut(ch);shack;shake-cabin;shanty;wigwam;whare <新西兰毛利人的>;basha <印度>;linhay <英>;hovel
棚屋护板 camp sheathing;camp shedding
棚屋区 squatter area
棚屋群 hutment
棚屋支柱 lodgepole
棚下浇筑混凝土 concreting under shelter
棚型焚化炉 wigwam-type incinerator
棚压滞后 grid-voltage lag
棚枝果树 dwarf fruit
棚子 hovel;shack;shed;lookum <覆盖卷扬机的>
棚子支柱 leg
棚座 <新西兰毛利人的> whare

硼

硼铵石 larderellite
硼玻璃 borax glass;boron glass
硼玻璃料 boron frit
硼掺杂层 boron-doped layer
硼氮聚合物 boron-nitrogen polymer
硼的 boric
硼碘水 boron-iodine water
硼电离室 boron chamber
硼反常 boric oxide anomaly
硼肥 boron fertilizer
硼酚醛树脂 boron-phenolic resin
硼符山石 wiluite
硼钙石 bechilite;borocalcite;colemanite;roweitwe;strontioborite
硼钙锡矿 nordenskioldine
硼钢 borax steel;boron steel
硼钢砂 boron shot
硼钢制的链轮 boralloy sprocket
硼硅钡铅矿 hyalotekite
硼硅钡钇矿 cappelenite
硼硅玻璃 borosilicate glass
硼硅玻璃冷却套 Pyrex cooling jacket
硼硅钒钡石 nagashimalite
硼硅钙镁石 harkerite
硼硅铬黄 borosilicate crown
硼硅铈矿 tritomite
硼硅酸玻璃 <一种耐热玻璃> Pyrex glass
硼硅酸锂 lithium borosilicate
硼硅酸铝玻璃 alumina-borosilicate glass
硼硅酸镁玻璃 magnesia borosilicate glass
硼硅酸钠 borsal;sodium borosilicate
硼硅酸耐热玻璃 vycol glass;vycor glass;Vycol <商品名>
硼硅酸铅 lead borosilicate
硼硅酸铅釉 lead borosilicate glaze
硼硅酸石灰玻璃 lime-borosilicate glass
硼硅酸盐 boron silicate;borosilicate
硼硅酸盐玻璃 borosilicate glass
硼硅酸盐铬黄 borosilicate crown
硼硅酸盐冕玻璃 borosilicate crown glass
硼硅铁合金 brososil
硼硅钇钙石 hellandite
硼硅钇矿 tritomite-(Y)
硼合金 boron alloy
硼化铥 thulium boride
硼化钒 vanadium boride
硼化钙 calcium boride
硼化锆 zirconium boride
硼化锆陶瓷 zirconium boride ceramic
硼化铬 chromium boride
硼化硅 silicon boride
硼化铪 hafnium boride
硼化合物 boride
硼化钪 scandium boride
硼化镧 lanthanum boride
硼化镧陶瓷 lanthanum boride ceramics
硼化铝 alumin(i)um boride
硼化密实混凝土 boron-loaded concrete
硼化钼 molybdenum boride
硼化镎 neptunium boride
硼化铌 niobium boride
硼化镍 nickel boride
硼化钕 neodymium boride
硼化铍 beryllium boride
硼化钐 samarium boride
硼化铈 cerium boride
硼化锶 strontium boride
硼化塑料 boron plastic
硼化钛 titanium boride
硼化钽 tantalum boride
硼化铽 terbium boride
硼化钍 thorium boride
硼化钨 tungsten boride
硼化物 boride
硼化物基金属陶瓷 boride base cermet
硼化物陶瓷 boride ceramics
硼化物陶瓷材料 boride
硼化物涂层 boride coating
硼化钇 yttrium boride
硼化铀 uranium boride
硼甲酸钠 sodium boroformate
硼钾镁石 kaliborite
硼金属陶瓷 boride cermet
硼聚合物 boron polymer
硼矿 boron ores
硼矿指示植物 indicator plant of boron
硼扩散 boron diffusion
硼磷镁石 luneneburgite
硼磷酸钠液体混合物 abopon
硼磷酸盐 borophosphate
硼硫酸钠 borol
硼铝钙石 johachidolite
硼铝合金 boron-aluminum alloy
硼铝镁石 sinhalite
硼铝石 jeremejewite
硼镁瓷 ascharite porcelain
硼镁矾 sulfoborite;sulphoborite
硼镁钙石 kurchatovite
硼镁锰矿 pinakiolite
硼镁石 luenburgite;szaibelyite
硼镁铁矿 ludwigite
硼镁铁石 ludwigite
硼镁铁钛矿 azoproite
硼锰矿 sussexite
硼锰锌石 roweite
硼棉 boron wool
硼钠长石 reedmergnerite
硼钠方解石 ulexite
硼钠钙石 boronatrocalcite;ulexite
硼钠镁石 aristarainite
硼镍铁矿 bonaccordite
硼铍铝铯石 rhodizite
硼铍石 hambergite
硼嗪 borazine;borazole
硼氢化合物 boron hydride;borane
硼氢化合物中毒 boron hydride poisoning
硼氢化钾 potassium borohydride
硼氢化钾溶液 potassium borohydride solution
硼氢化锂 lithium borohydride
硼氢化铝 alumin(i)um borohydride
硼氢化钠 sodium borohydride
硼氢化物 borane;hydroboron
硼氢化作用 hydroboration
硼氢基 borine radical
硼铯铷矿 rhodizite
硼砂 baroscu;borax;natrium biboricum;pyroborate;sal sedatirum;sodium(tetra) borate;tincal;tinkalite;zala
硼砂玻璃 borax glass
硼砂湖 alkaline lake;borax lake
硼砂混凝土 borax concrete
硼砂卡红染剂 borax-carmine stain
硼砂熔融 fusion with borax
硼砂熔珠 borax bead
硼砂原矿 tincal
硼砂珠 borax bead
硼砂珠技术 borax-lead technique
硼砂珠试验 borax bead test
硼石 szaiblyite
硼-石蜡准直仪 borax paraffin collimator
硼锶钙石 braitschite
硼树脂 boron resin
硼水 boron water
硼丝 boron filament
硼酸 anti-fungin;boracic acid;boric acid;ortho-boric acid;sal sedative;sodium borate
硼酸铵 ammonium borate
硼酸钡 barium borate
硼酸苯汞 phenylmercury borate
硼酸处理过的 borated
硼酸方解石 borocalcite
硼酸钙 calcium borate
硼酸钙玻璃 lime borate glass
硼酸甘油酯 glyceroborate
硼酸缓冲溶液 boric acid buffered potassium iodide
硼酸火石玻璃 boracic acid flint glass
硼酸矿物 borate mineral
硼酸锂 lithium borate
硼酸铝 alumin(i)um borate
硼酸镁 anti-fungin;magnesium borate
硼酸锰 manganese borate
硼酸钠 borax;sodium borate
硼酸镍 nickel(ous) borate
硼酸喷气孔 soffione
硼酸漂白粉 mixture of boric acid and bleaching powder
硼酸气体 boric acid gas
硼酸铅 lead borate
硼酸铅釉 lead borate glaze
硼酸三丙酯 tripropylborate
硼酸三丁酯 tributyl borate
硼酸三甲酯 trimethyl borate
硼酸三戊酯 triamyl borate
硼酸锶 strontium borate
硼酸铜包覆二氧化硅 copper borate-coated silica
硼酸戊酯 amyl borate
硼酸洗液 boric acid lotion
硼酸锌 zinc borate
硼酸岩矿床 borate deposit
硼酸盐 borate
硼酸盐玻璃 borate glass
硼酸盐石灰 borate of lime
硼酸乙酯 borogen;boron triethoxide;triethylic borate
硼酸油膏 boric ointment
硼酸酯 borate;boric acid ester
硼酸质熔块 boracic frit
硼钛镁石 warwickite
硼钛镁铁矿 warwickite
硼钽石 behierite
硼碳电阻器 borocarbon resister[resistor]
硼碳镁石 canavesite
硼铁钙矾 sturmanite
硼铁矿(石) paigeite;vonsenite
硼烷 borane
硼污染 boron pollution
硼锡钙石 nordenskioldine
硼锡锰石 tusionite
硼纤维 boric filament;boron fiber;boron filament <增强用>
硼纤维增强复合材料 boron fiber reinforced composites
硼纤维增强环氧复合材料 boron-epoxy composite
硼纤维增强铝 boron filament reinforced alumin(i)um
硼纤维增强塑料 boric fiber[fibre] reinforced plastics;boron fiber[fibre] reinforced plastics
硼氧烷 boroxane
硼硬化剂 needler
硼中毒 borism
硼族元素 boron group elements

蓬

蓬拜城之赫尔库兰门 <古罗马> Gate of Herculaneum at Pompeli
蓬蒂地槽 Ponti geosyncline
蓬蒂阶 <晚上新世>【地】Pontian stage
蓬蒂统 <晚上新世>【地】Pontian series
蓬莪术油 Zedoary oil
蓬乱的 woolly
蓬松度值 bulking value

篷

篷 awning;covers <养护混凝土用>
篷布 canvas;paulin;roof paulin;tarp(aulin);wagon sheet
篷布支杆 sheet supporting bar
篷布调度员 wagon sheet controller
篷布回送计划 wagon sheet dispatching plan
篷布孔环 tarpaulin eye;tie ring
篷布支架 sheet trestle
篷车 box freight car;fold(ing) top car;van;wagon
篷车顶 cape top
篷车旅店 caravansary;khan
篷车旅馆 caravan hotel;cho(u)ltry
篷车拖车和活动房屋 caravan;trailers and mobile homes
篷顶 tilt roof
篷顶操纵开关 rain switch
篷顶小客车 convertible sedan
篷盖 sun blind;tilting
篷夹冲模 steeling inserting die
篷裂桩头 broomed pile head;broom head of pile
篷幕紧绳器 euphroe
篷式百叶窗 awning blind
篷式窗 awning window
篷式挂车 van trailer
篷式汽车 cabriolet
篷式天窗 awning light;awning sash window;awning type window

篷式遮帘 awning blind
篷索滑车 halyard block

膨

膨大 enlargement

膨大的 inflated
膨化 swelling
膨化不均度 bulk variability
膨化度测试器 bulk tester
膨化谷物 puffed grain
膨化机 bulking machine
膨化均匀性 bulk uniformity
膨化黏[粘]土 swelling clay
膨化现象 swelling phenomenon
膨痂锈菌属 <拉> pucciniastrum
膨径螺栓 upset bolt
膨径铆钉 upset rivet
膨壳式锚杆 expansion shell bolt
膨隆 bulge;knuckle
膨起 loft
膨润度 swelling capacity
膨润剂 swelling agent
膨润黏[粘]土 bentonite clay;bentonitic clay
膨润土 amargosite;bentonite;bentonite clay;bentonitic clay;bentonitic mud;Alta-mud <商品名>
膨润土柴油 bentonite diesel oil
膨润土掺水泥乳液 <沉井外围防水灌浆用> cement bentonite milk
膨润土处理 <蓄水池防止漏水用> bentonite treatment
膨润土法施工 slurry process
膨润土粉 bentonite powder
膨润土灌浆 bentonite grouting
膨润土互磨 intergrinding of bentonite
膨润土滑润脂 bentonite grease
膨润土混凝土 bentonite concrete
膨润土浆 bentonite slurry
膨润土胶泥 bentonite mud
膨润土胶体 bentonite gel
膨润土胶液密度 density of bentonite solution
膨润土矿床 bentonite deposit
膨润土矿石 bentonite ore
膨润土泥灌浆(填孔用) bentonite grout
膨润土泥浆 <减少推进或下沉时的摩阻力> bentonite slurry;bentonite mud;bentonitic mud;water-bentonite slurry
膨润土水泥拌和团粒 bentonite-cement pellet
膨润土凝胶体 bentonite gel
膨润土球 bentonite pellet
膨润土润滑脂 bentonite grease
膨润土砂 bentonite bonded sand
膨润土施工法 bentonite process
膨润土水泥 bentonite cement
膨润土水泥粒料 bentonite cement pellets
膨润土水泥乳浆 cement bentonite milk
膨润土水溶液 bentonite-water solution
膨润土稳定液 bentonite stabilizing fluid
膨润土稳定浆液隧道掘进机 bentonite tunnelling machine
膨润土消耗量 <槽孔固壁> bentonite consumption
膨润土絮凝作用 bentonite flocculation
膨润土悬浮液 bentonite suspension
膨润土页岩 bentonite shale
膨润压力 swelling pressure
膨润作用 swelling action;imbibition
膨松变形纱 bulked yarn
膨松不匀度 bulk variability
膨松处理 relaxing bulking treatment
膨松度 bulkiness
膨松度测试仪 bulk tester
膨松度试验仪 bulkmeter
膨松度值 bulking value
膨松活性 puffing activity
膨松剂 swelling agent
膨松均匀度 bulk uniformity
膨松纱 bulky yarn;textured yarn
膨松性 bulking power
膨缩构造 pinch and swell structure
膨缩特性 swell-shrink characteristic
膨体变形能力 bulking power
膨体混凝土 Ribmet
膨体纱 bulked yarn
膨突管 puff pipe
膨土岩 amargosite;bentonite
膨压 tougor pressure;turgescence;turgor
膨胀 expansion;dilatation;inflation;bloating bulking;bulge;bulging;inflate;negative shrinkage;swell(ing);upswell;bloating <土块或砖加热至缸化时体积的膨胀>
膨胀 U 形管 expansion U-bend
膨胀安全阀 expansion relief valve
膨胀凹槽 <供弯管伸缩的> expansion bay
膨胀板 expansion plate;expansion sheet
膨胀板石轻骨料 slate light aggregate;Solite <商品名>
膨胀板石轻骨料混凝土 slate concrete
膨胀板石轻集料 slate light aggregate;Solite <商品名>
膨胀板石轻集料混凝土 slate concrete
膨胀板岩 bloated slate;expanded slate
膨胀板岩工厂 bloated slate factory;bloating slate factory
膨胀板岩骨料 expanded slate aggregate
膨胀板岩混凝土 bloated slate concrete;bloating slate concrete;expanded slate concrete
膨胀板岩集料 expanded slate aggregate
膨胀倍数 expanded ratio;expansion multiple
膨胀比率 dilatation;expansion pressure ratio;ratio of expansion;swell-(ing) ratio;expansion ratio;degree of expansion
膨胀臂 expanding arm
膨胀变化 dilatometric change
膨胀变位 expansion deflection
膨胀变形 dilatancy;swelling strain;expansion deformation
膨胀饼 expansion diaphragm
膨胀饼挡钩 diaphragm hook
膨胀饼壳 <暖气用> lower casing
膨胀饼托 diaphragm rest;diaphragm spider
膨胀饼托座 diaphragm spider seat
膨胀饼压板 diaphragm ring
膨胀饼座 diaphragm bottom seat;diaphragm cover seat
膨胀饼座托板 diaphragm seat plate
膨胀波 dilatation(al) wave;expansion fan;expansion wave;rarefaction wave
膨胀玻璃 expanded glass
膨胀玻璃球 expanded glass sphere
膨胀补偿 expansion compensation
膨胀补偿节 compensator piece
膨胀补偿器 expansion appliance;expansion bellows;expansion bend;expansion compensator;expansion cushion;expansion piece
膨胀不足 underexpansion
膨胀材料 <如珍珠岩、蛭石等> expanding material
膨胀舱 expansion tank;expansion trunk
膨胀操作 expansive working
膨胀槽 expansion bath;expansion slot;expansion trough
膨胀测定法 dilatometry
膨胀测定术 dilatometry
膨胀测量法 dilatometry
膨胀测量术 dilatometry
膨胀叉 expansion fork
膨胀掺和剂 expansion admixture
膨胀常量 dilatation constant
膨胀常数 dilatation constant
膨胀场 swelled field
膨胀衬垫 expansion insert
膨胀衬砌 expanded lining
膨胀衬套 expansion bushing
膨胀成分 <膨胀水泥的> expansive component
膨胀冲程 firing stroke;power stroke;expansion stroke
膨胀(冲程终了时的)温度 expansion temperature
膨胀床 expanded bed
膨胀床层 expanding bed
膨胀床反应器 expanded-bed reactor
膨胀床高度 expanded-bed height
膨胀床活性炭接触器 expanded-bed activated carbon contactor
膨胀床吸附 expanded-bed adsorption
膨胀带闸 expanding band brake
膨胀导承 expansion guide
膨胀导阀 expansion slide valve
膨胀的 torose
膨胀的合成软木 expanded composition cork
膨胀的耦合弹性 coupled elasticity for swelling
膨胀的天然橡胶 expanded natural rubber
膨胀的填料 tensional gasket
膨胀的预算 inflated budget
膨胀的珠光体 bloated pe(a)rlite
膨胀地基 swelling ground
膨胀垫 expanded cushion;expansion block;expansion pad
膨胀垫圈 expansion washer
膨胀顶 expansion roof
膨胀顶储罐 expansion roof tanks
膨胀度 bulking intensity;degree of expansion;degree of swelling;dila(ta)tion;expandability
膨胀度计 expansimeter
膨胀度试验 dilatometer test
膨胀端 expansion end
膨胀断路器 expansion circuit breaker
膨胀多孔混凝土砌块 cellular-expanded concrete(building) block
膨胀发泡活性污泥处理设备 bulking and foaming activated sludge plant
膨胀阀 expanding valve
膨胀阀动装置 expansion valve gear
膨胀阀过热度 expansion valve superheat
膨胀阀能力 expansion valve capacity
膨胀阀容量 expansion valve capacity
膨胀法 expansion method;swelling process
膨胀反射 distention reflex
膨胀反应 expansion reaction
膨胀方式 swelling mode
膨胀缝 dilatation joint;expansion crack(ing);expansion gap;expansion space;head space;expansion joint
膨胀缝填料 joint filler
膨胀缝条 isolation strip
膨胀浮力 expansion buoyancy
膨胀附加剂 expansion admixture
膨胀盖 expanding lid
膨胀干燥 puffing drying
膨胀杆 <锅炉> expansion link
膨胀杆导承 expansion link guide
膨胀杆试验法 expansive bar test
膨胀刚性聚氯乙烯 expanded rigid polyvinyl chloride
膨胀钢筋 expansion reinforcement
膨胀钢筋混凝土 reinforced expanded concrete
膨胀高炉矿渣 expanded blast-furnace slag
膨胀高炉矿渣混凝土 expanded blast-furnace slag concrete
膨胀高炉矿渣骨料 expanded blast-furnace slag aggregate
膨胀高炉熔渣集料 expanded blast-furnace slag aggregate
膨胀高温计 expansion pyrometer
膨胀功 expansion work
膨胀构造 expansion structure
膨胀骨料 expanded aggregate
膨胀管 bulged tube;compensating pipe;expanding pipe;expanding tube;expansion pipe;expansion tube
膨胀罐 expansion drum
膨胀柜 expansion tank
膨胀过程 expansion line;expansion process;swelling process
膨胀过度 hyperdistention;overdistension;super-distention
膨胀含水量 swelling water content;water content after swelling
膨胀合金 expanded alloy;expanded metal;expanding metal;expansion alloy
膨胀盒 expansion chamber
膨胀滑阀 cut-off slide;cut slide valve
膨胀化学剂 expanding chemical
膨胀环 expansion loop
膨胀环箍 expansion hoop
膨胀混凝土 expanded concrete;expansion concrete;expansive concrete;puffed up concrete;self-stressed concrete;expanding concrete
膨胀混凝土板材 expanded concrete slab
膨胀混凝土产品 expanded concrete product
膨胀混凝土工场 expanded concrete plant
膨胀混凝土骨料 expanded concrete aggregate
膨胀混凝土罐筒 expanded concrete pot
膨胀混凝土集料 bloated concrete aggregate;bloating concrete aggregate;expanded concrete aggregate
膨胀混凝土绝热板材 expanded concrete insulation grade slab
膨胀混凝土绝热砖 expanded concrete insulating brick
膨胀混凝土空心砌块 expanded concrete cavity block;expanded concrete hollow block
膨胀混凝土砌块 expanded concrete block
膨胀混凝土特殊砌块 expanded concrete special block
膨胀混凝土瓦 expanded concrete tile
膨胀混凝土专用砌块 expanded concrete purpose-made block
膨胀混凝土专用砖 expanded concrete purpose-made brick
膨胀火山玻璃体 expanded volcanic glass
膨胀机 decompressor;expander;expansion engine;expansion machine
膨胀机理 expansion mechanism

膨胀机增压压缩机 expander-booster compressor
膨胀激波 expansion shock
膨胀极限 swelling limit; limit of expansion
膨胀集料 expanded aggregate
膨胀计 expansimeter; expansion ga(u)ge; shell expansion indicator; swellmeter; dilatometer <测量体积变化的仪器>
膨胀计测定 dilatometer measurement
膨胀计试验 dilatometer test
膨胀剂 bloater; bloating agent; bulking agent; expanding admixture; expanding agent; expansion agent; expansion producing admixture; expansive agent; plumbing agent; swelling agent
膨胀间隙 bulge clearance; clearance for expansion; expansion clearance
膨胀件 expansion piece
膨胀胶 swelling gel
膨胀角 <喷嘴的> divergence cone angle
膨胀阶段 expansion stage
膨胀接合 expansible joint
膨胀接合的密封口 expansion joint sealing
膨胀接合屋脊顶梁 expansion joint ridge capping
膨胀接合止水带 expansion joint waterstop
膨胀接口 dilatation joint
膨胀接头 expansion coupling; expansion joint; swing joint
膨胀接头的填料压盖 expansion gland
膨胀节 bellows; compensator; expansion piece; expansion pipe
膨胀节理【地】 expansion fissure; expansion joint
膨胀节水密封 expansion joint waterstop
膨胀结疤 expansion scab
膨胀结构 expansion texture
膨胀结合 expansion joint
膨胀金属 expanding metal; expansive metal
膨胀井 expansion trunk
膨胀井舱口 expansion hatch(way); expansion trunk hatchway
膨胀聚氨基甲酸酯板 expanded polyurethane board
膨胀聚氨基甲酸酯薄板 expanded polyurethane sheet
膨胀聚氨基甲酸酯条带 expanded polyurethane strip
膨胀聚苯乙烯 expandable polystyrene; expanded polystyrene
膨胀聚氯乙烯 expanded polyvinyl chloride[PVC]
膨胀聚乙烯缝隙密封 expanded polyethylene draught seal
膨胀卡头 expansion clamp
膨胀开裂 expansive crack(ing)
膨胀颗粒污泥床反应器 expanded granular sludge bed reactor; expansile granular sludge bed reactor
膨胀壳式加热器 expansion shell type heater
膨胀壳(体) expanding shell
膨胀空间 expansion space
膨胀空腔 expanding cavity
膨胀空隙 expansion space; vacuity
膨胀口 expansion hatch(way)
膨胀块状纯软木 expanded pure agglomerated cork
膨胀矿石 expanded mineral
膨胀矿渣 expanded slag
膨胀矿渣粉末 expanded slag powder
膨胀矿渣骨料 expanded slag aggregate
膨胀矿渣混凝土 expanded slag concrete
膨胀矿渣混凝土板 expanded slag concrete plank
膨胀矿渣混凝土砌块 expanded slag concrete block
膨胀矿渣混凝土砖瓦 expanded slag concrete tile
膨胀矿渣集料 expanded slag aggregate
膨胀矿渣建筑砖瓦 expanded slag building tile
膨胀矿渣砌块 expanded slag block
膨胀矿渣瓦 expanded slag tile
膨胀扩散模型 dilatancy diffusion model
膨胀拉力运动 expanding tensional motion
膨胀劳动力资源 expanded human sources
膨胀冷却 expansion(al) cooling
膨胀力 bulging force; expansible force; expansion force; expansive force; expansive power; swelling force; tension
膨胀力试验 swelling force test; expansibility test
膨胀粒化软木 expanded granulated cork
膨胀粒化软木板 expanded granulated cork slab
膨胀连接器 expansion coupling
膨胀量 degree of swelling; swell increment; swelling amount; swelling capacity
膨胀量试验 swelling capacity test
膨胀裂缝 dilatation fissure; expansion crack(ing); expansion fissure; expansive crack(ing)
膨胀裂纹 expansion crack(ing)
膨胀裂隙 dila(ta)tion fissure; expansion fissure
膨胀流(动) expanded flow
膨胀隆起 swelling heave
膨胀炉渣 expanded cinder; expanded slag
膨胀炉渣板材 expanded cinder slab
膨胀炉渣骨料 expanded cinder aggregate
膨胀炉渣混凝土 expanded cinder concrete
膨胀炉渣混凝土罐 expanded cinder concrete pot
膨胀炉渣混凝土空心砌块 expanded cinder concrete cavity block
膨胀炉渣混凝土平板 expanded cinder concrete slab
膨胀炉渣混凝土墙板 expanded cinder concrete wall slab
膨胀炉渣集料 expanded cinder aggregate
膨胀率 dilatability; expanded rate; expansibility; expansibleness; expansion efficiency; expansion factor; expansion rate; rate of expansion; rate of swelling; ratio of expansion; specific expansion; swelling number; swelling rate; swelling ratio; percentage of bed-expansion【给】
膨胀率的测量 swell measurement
膨胀率试验 swelling rate test
膨胀氯丁橡胶 expanded neoprene
膨胀螺钉 powder-set fastener
膨胀螺栓 blind fastener; cinch bolt; expanded anchor; expanding bolt; expansion bolt; expansion shield; plug bolt; star expansion bolt
膨胀螺栓座 <埋在塑料及混凝土中的> molly
膨胀螺旋 inflationary spiral
膨胀脉动 expansion pulsation
膨胀锚碇 expansion anchor
膨胀锚杆 expansion anchor
膨胀锚固 expansion anchor
膨胀锚固螺栓 expanding anchor
膨胀铆钉 expansion rivet; expansive rivet; rivet plug; upset rivet
膨胀帽 <俗称> expansion cap
膨胀煤 swelling coal
膨胀密封膏 intumescent sealant
膨胀密封胶 intumescent sealant
膨胀模 bulging die
膨胀模量 modulus of dilatation
膨胀能 expansion energy
膨胀能力 expansion ability; expansivity; swelling capacity; swelling power
膨胀能量回收 recovering expansion energy
膨胀泥板岩 expanded shale
膨胀黏[粘]度 dilatational viscosity
膨胀黏[粘]土 bloated clay; bloating clay; dilatancy clay; effervescing clay; expandable clay; expanded clay; swelling clay
膨胀黏[粘]土的压实 compaction of expansive clay
膨胀黏[粘]土骨料 aglite; expanded clay aggregate; serlite
膨胀黏[粘]土混凝土 bloated clay concrete; bloating clay concrete
膨胀黏[粘]土混凝土砌块 bloated clay concrete solid block; expanded clay concrete block
膨胀黏[粘]土混凝土墙板 bloated clay (concrete) wall slab; bloating clay concrete wall slab; expanded clay concrete wall slab
膨胀黏[粘]土混凝土砖 bloating clay concrete solid tile
膨胀黏[粘]土集料 bloated clay aggregate; bloating clay aggregate; expended clay aggregate; serlite
膨胀黏[粘]土轻骨料 lightweight expanded clay aggregate
膨胀黏[粘]土轻集料 lightweight expanded clay aggregate
膨胀黏[粘]土体 bloating of the stone
膨胀黏[粘]土团粒滤池 expanded clay aggregate filter
膨胀黏[粘]性土的干密度 dry density of expansive clay
膨胀尿素甲醛 expanded urea-formaldehyde
膨胀凝固 freezing by expansion
膨胀盘 expanding disc[disk]
膨胀盘管 expansion coil
膨胀配合 expansion fit
膨胀喷嘴 expanding nozzle
膨胀片 expansion piece
膨胀偏心轮 expansion eccentric wheel
膨胀破坏 breaking-in bulking; failure by bulging
膨胀漆 intumescent paint
膨胀起来的部分 puff
膨胀气 expanding gas; expansion gas
膨胀气泡 expanding bubble
膨胀气驱 expansion gas drive
膨胀汽 expansion of vapo(u)r
膨胀器 bloater; expander; extender
膨胀前的饱和 pre-expansion saturation
膨胀前的径迹 pre-expansion track
膨胀潜能 swelling potential
膨胀强度 expansion strength
膨胀穹 expansion dome
膨胀区 breathing space; expansion area; zone of swelling
膨胀曲壁 swell wall
膨胀曲率 expansive curvature
膨胀曲线 cut-off curve; dilatometric curve; expansion curve; swelling curve
膨胀驱 expansion drive
膨胀趋势 swelling tendency
膨胀圈 expansion loop; expansion ring
膨胀缺口 inflationary gap
膨胀扰动 dilatational disturbance
膨胀热 swelling heat
膨胀容积 expanded volume
膨胀容器 expansion vessel
膨胀熔融黏[粘]结 expansive fusion caking
膨胀熔岩 inflated lava
膨胀如袋的 baggy
膨胀如球 balloon
膨胀软木 expanded cork
膨胀软木板 expanded cork slab
膨胀软木片 expanded cork plate; expanded cork sheet
膨胀塞 expanding plug; expanding stopple; expansion plug; expansion stopple
膨胀塞卡环 expansion plug snap ring
膨胀塞子 expanded cork; expanding stopper; expansion plug
膨胀蛇形管 expansion coil
膨胀设备 bloating plant; expanding plant
膨胀伸缩机构 dilating device
膨胀伸缩式波纹管 expansion bellows
膨胀石板工厂 bloated slate factory
膨胀石灰混凝土 expanded lime concrete
膨胀石灰混凝土砖瓦 expanded lime concrete tile
膨胀石墨 expanded graphite
膨胀时间 expansion time
膨胀式除水器 expansion trap
膨胀式除氧器 expansion deaerator
膨胀式沸水器 expansion boiler
膨胀式封隔器 expanded packer; inflatable packer
膨胀式管形封隔器 expanding type tubing packer
膨胀式化油器 expanding carburet(t)or
膨胀式建筑 inflatable structure
膨胀式锚杆 expansion-type bolt
膨胀式密封 inflatable seal
膨胀式碰垫 inflatable fender
膨胀式疏水器 steam trap, expansion type
膨胀式温度计 expansion thermometer
膨胀式岩石锚杆 expansion rock bolt
膨胀式仪表 hot-wire instrument
膨胀式蒸汽疏水阀 expansion type steam trap
膨胀式制动器 expansion brake
膨胀式致冷发动机 expansion engine
膨胀势 potential swell; swell(ing) potential
膨胀试剂 bloat agent; swelling agent
膨胀试验 bloating test; dilatancy test; dila(ta)tion test; expanding test; expansion test; floating test; swell(ing) test
膨胀试验仪 expansion pressure testing apparatus
膨胀室 expansion chamber; expansion opening
膨胀室集尘器 expansion chamber collector
膨胀室消声器 expansion chamber muffler
膨胀收缩机理 mechanism of heave or shrinkage
膨胀疏体纱 bulky yarn
膨胀栓 expansion plug
膨胀水 water of dilatation
膨胀水槽 expansion vessel

P

膨胀水管 expansion water pipe
膨胀水泥 expansive cement; self-stressing cement; expanded cement; expanding cement; expansion cement; self-stressing cement; sulfoaluminate cement
膨胀水泥灌浆 expanding grout
膨胀水泥灌浆材料 expansive-cement grout
膨胀水泥灰浆 expansive cement mortar
膨胀水泥混凝土 expanding cement concrete; expansive cement concrete
膨胀水泥混凝土路面 expansive cement concrete pavement
膨胀水泥接口 expanding cement joint
膨胀水泥砂浆 expansive-cement mortar
膨胀水箱 closed expansion tank; expansion vessel; expansion (water) tank
膨胀速度 swelling speed; swelling velocity
膨胀速率 swelling rate
膨胀塑料 expanded plastics
膨胀塑料板 expanded plastic board
膨胀塑料混凝土 expanded plastic concrete
膨胀塑料绝缘 expanded plastic insulation
膨胀塑料颗粒骨粉 expanded plastic particle aggregate
膨胀塑料轻质混凝土 expanded plastic lightweight concrete
膨胀塑料穹隆 expanded plastic dome
膨胀塑料圆屋顶 expanded plastic cupola
膨胀塑料珠骨粉 expanded plastic bead aggregate
膨胀塑性石膏踢脚板 expanded plastic plaster baseboard
膨胀塑性土 dilatable plastic soil
膨胀弹簧 expansion spring
膨胀陶粒 serlite
膨胀陶粒混凝土 <一种前苏联产品> keramzite concrete
膨胀套管 expansion sleeve
膨胀套壳 <杆柱> expandable shell
膨胀套筒 expansion sleeve
膨胀体积 expanding volume; volume of expansion
膨胀填料函 expansion stuffing box
膨胀填料盒 expansion stuffing box
膨胀调节阀 expansion regulating valve
膨胀调节器 expansion regulator
膨胀筒 expansion cylinder; expansion trunk
膨胀凸出 swelling-up
膨胀凸轮 expansion cam
膨胀土 dilatable soil; dilative soil; expanding soil; expansion soil; expansive soil; swelled ground; swell(ing) soil; swollen soil
膨胀土的鉴别 identification of swelling soil
膨胀土的胀缩性分级 swelling and shrinkage grade of swelling soil
膨胀土地基的防治 protection and treatment of swelling foundation soil
膨胀土地基的胀缩量 value of swelling shrinkage of foundation soil
膨胀土活动层 active zone for expansive soils
膨胀土收缩 shrinkage of swelling soil
膨胀土缩限 shrinkage limit of swelling soil
膨胀外加剂 expansion producing admixture
膨胀弯管 expansible bend; expansion bend; loop expansion bend
膨胀弯管法 loop expansion bend
膨胀弯头 loop expansion bend
膨胀网 expanded mesh
膨胀网银泡沫石膏 cellular-expanded gypsum
膨胀危险期 critical period of expansion
膨胀围壁 expansion trunk
膨胀围岩压力 expansion pressure of surroundings
膨胀稳定期 stable stage of expansion
膨胀涡轮 expansion turbine
膨胀污泥 bulking sludge; expanded sludge
膨胀物 dilatant
膨胀雾 expansion fog
膨胀系数 coefficient of dila(ta)tion; coefficient of expansion; coefficient of swelling; dilatation coefficient; expansion coefficient; expansion efficiency; expansion factor; expansivity; factor of expansion; swell factor; thermal expansion coefficient
膨胀系数绝对值 absolute expansion coefficient
膨胀系数匹配 matching of expansion coefficient
膨胀下陷 oil-canning
膨胀现象 swelling phenomenon
膨胀线 expansion line
膨胀相 expansion phase
膨胀箱 dilator; expansion box; expansion tank; expansion vessel; flash box; flash tank
膨胀橡胶 expanded rubber; expansive rubber; swelling rubber
膨胀消弧 expansion arc suppressing
膨胀效应 bulking effect; dilatancy effect
膨胀心轴 built-up mandrel
膨胀心轴法 <打长钢管桩> expanding mandrel method
膨胀行程 expansion stroke
膨胀型封隔器 inflatable packer
膨胀型聚苯乙烯材料 expansion styropor; styropor
膨胀型锚杆 expansion-type bolt
膨胀型土压力 swelling pressure
膨胀性 dilatability; distensibility; expandability; expansibility; expansibleness; expansivity; extensibility; turgesency; turgidity; dilatancy
膨胀性材料 expandable material; expansive material
膨胀性掺和料 expansion producing admixture
膨胀性成分 expansive component
膨胀性稠度 dilatant consistency
膨胀性的 dilatant; dilatational; expansile; expansive
膨胀性地层 expansive formation; expansive ground; swelling formation; swelling ground
膨胀性地基 expansive foundation; expansive ground; swelling foundation; swelling ground
膨胀性分类 <膨胀土的> classification of swelling property
膨胀性封隔器 pressure packer
膨胀性高价 inflationary high price
膨胀性高炉炉渣 expanded blast-furnace slag
膨胀性骨料混凝土 expanded aggregate concrete
膨胀性固定件 expansion fastener
膨胀性灌浆 expanding grout
膨胀性混凝土 expansive concrete
膨胀性集料混凝土 <一种轻集料的混凝土,所用集料如膨胀性黏[粘]土、胀性页岩、胀性矿渣、陶粒、珍珠岩等> expanded aggregate concrete
膨胀性浆液 expanding grout
膨胀性晶格黏[粘]粒 expanding lattice clay
膨胀性聚氨酯 expanded polyurethane
膨胀性聚苯乙烯 expanded polystyrene
膨胀性矿渣 expanded slag
膨胀性流体 dilatant fluid
膨胀性能 expansion character
膨胀性黏[粘]土 dilatation clay; expansive clay; heaving clay; swelling clay; expanding clay; expanded clay
膨胀性砂浆 expanding mortar; expansive grout
膨胀性生长 expansive growth
膨胀性水泥 expansive cement
膨胀性水泥砂浆 expanding cement mortar; expansive cement grout
膨胀性塑料骨料 expanded plastics particle aggregate
膨胀性酸 swelling acid
膨胀性土(壤) dilatation soil; dilative soil; expansive soil
膨胀性土压力 swelling earth pressure
膨胀性围岩 expanding rock
膨胀性无收缩水泥 expanding and non-contracting cement
膨胀性岩石 expansive rock; swelling rock
膨胀性盐 swelling salt
膨胀性页岩 expanded shale; heaving shale
膨胀性页岩粉屑 expanded shale fine
膨胀性增塑糊 dilatant plastisol
膨胀性增塑溶胶 dilatant plastisol
膨胀性指标 swelling index
膨胀性质 swelling property
膨胀压(力) bulb of pressure; dilation pressure; expansion [expansive] pressure; swelling pressure; disjoining pressure <对土壤而言>
膨胀压(力)比 expansion pressure ratio
膨胀压力测定 expansion pressure measurement
膨胀压力机 swelling press
膨胀压力试验 expansion pressure test; swelling pressure test; test for expansive pressure
膨胀压力试验装置 expansion pressure device
膨胀压缩波 compressionally dilatational wave
膨胀页岩 bloated shale; bloating shale; expanding shale; heaving shale
膨胀页岩粗骨料 bloating shale coarse aggregate; expanded shale coarse aggregate
膨胀页岩粗集料 bloating shale coarse aggregate; expanded shale coarse aggregate
膨胀页岩骨料 expanded shale aggregate
膨胀页岩混凝土 bloated shale concrete; bloating shale concrete
膨胀页岩集料 expanded shale aggregate
膨胀页岩黏[粘]土 bloated shale clay; expanded shale clay
膨胀页岩黏[粘]土混凝土 expanded shale clay concrete
膨胀页岩轻骨料混凝土 shale concrete
膨胀页岩轻集料混凝土 shale concrete
膨胀仪 dilatometer; expansion apparatus; swelling apparatus
膨胀仪检验 dilatometric test
膨胀因数 swell factor
膨胀因素 expansion factor; swelling factor
膨胀引导 expanding pilot
膨胀应变 expansion strain
膨胀应变计 dilatation strain meter
膨胀应变能 dilatational strain energy
膨胀应变能量梯度 dilatational strain energy gradient
膨胀应变指数 swelling strain index
膨胀应力 swelling stress
膨胀应力系数 dilatation stress factor
膨胀硬橡胶 expanded ebonite
膨胀硬橡皮 expanded hard rubber
膨胀油井水泥 expansive oil well cement
膨胀(油)页岩 expanded shale
膨胀余量 expansion space
膨胀余位 ullage of expansion
膨胀余位表 ullage table
膨胀余位测板 ullage board; ullage stick
膨胀余位测尺 ullage foot; ullage ga(u)ge; ullage scale
膨胀余位测孔 ullage hole; ullage port
膨胀余位测孔盖板 ullage plate
膨胀余位测孔栓塞 ullage plug
膨胀余位尺 ullage board; ullage stick
膨胀余隙 expansion clearance
膨胀与收缩系数 coefficient of expansion and contraction
膨胀浴处理 expansion bath treatment
膨胀源 dilatation source
膨胀云母灰泥 exfoliated vermiculite mortar
膨胀云室 expansion cloud chamber
膨胀增稠器 dilatant thickener
膨胀增大 swell
膨胀褶皱 expansion fold; swelling fold
膨胀珍珠岩 bloated pe(a)rlite; bloating pe(a)rlite; expanded pe(a)rlite
膨胀珍珠岩厂 pearlite expansion plant
膨胀珍珠岩粉 ground perlite
膨胀珍珠岩骨料 expanded pe(a)rlite aggregate
膨胀珍珠岩混凝土 expanded pe(a)rlite concrete
膨胀珍珠岩集料 expanded pe(a)rlite aggregate
膨胀珍珠岩砂 expanded pe(a)rlite sand
膨胀珍珠岩砂浆 expanded pe(a)rlite mortar
膨胀珍珠岩水泥制品 expanded pe(a)rlite cement product
膨胀珍珠岩碎石 expanded pe(a)rlite coarse aggregate
膨胀珍珠岩吸声板 expanded pe(a)rlite acoustic(al) tile
膨胀珍珠岩芯板 expanded pe(a)rlite core; perlite core
膨胀珍珠岩制品 expanded pe(a)rlite product
膨胀政策 reflation policy
膨胀支座 expansion bearing; expansion pedestal
膨胀直径 swell diameter
膨胀值 expanding value; expansion value; swell(ing) value
膨胀指示器 expansion indicator
膨胀指数 expansion index; exponent of expansion; swell(ing) index; swell(ing) number
膨胀指数率 swelling number
膨胀制品 <多指膨胀矿渣> expanding product
膨胀致冷 expansion refrigeration
膨胀蛭石 exfoliated vermiculite; ex-

panded vermiculite
膨胀蛭石骨料 exfoliated vermiculite aggregate;expanded vermiculite aggregate
膨胀蛭石灰浆 exfoliated vermiculite plaster; expanded vermiculite plaster;vermiculite mortar
膨胀蛭石集料 exfoliated vermiculite aggregate;expanded vermiculite aggregate
膨胀蛭石砂浆 expanded vermiculite mortar
膨胀蛭石制品 expanded vermiculite product
膨胀周期 expansion period
膨胀轴 axis of dilatation;axis of expansion;expanding shaft
膨胀轴承 expansion bearing
膨胀珠光体 expanded pe(a)rlite
膨胀砖 expanded brick;lammie
膨胀装置 dilatation device;distension [distention]; expanding device; expansion device;expansion gear
膨胀状态 expansional phase; expansional state
膨胀阻燃剂 intumescent flame retardant
膨胀组分 expansion component; expansive component;expansive constituent
膨胀钻孔桩 expanded bore pile
膨胀作用 dilatation; expanding action; expansive action; distension [distention]
膨胀作用制动器 expanding brake
膨珠 expanded slag pellet

碰冲 impaction

碰船 afoul of a ship;fall foul of a ship
碰船货损条款 freight collision clause
碰到 come actress;hit(up)on hit
碰到恶劣天气 make heavy weather
碰到好天气 make good weather
碰到坏天气 make bad weather
碰垫 bumper block; bumping block; fendering; mat fender; pudding fender
碰垫活结 open fender hitching
碰垫球 grommet fender;roll fender
碰垫栓 fender bolt
碰垫系统 fendering system
碰返机构 breakback mechanism
碰关 slam
碰关声 slam hang
碰焊 butt-weld(ing)
碰焊机 butt welder
碰痕 bruise mark; percussion mark; percussion scar
碰击指示器 hit indicator
碰磕掉釉 knocking
碰裂反应 impact disintegration
碰伤 bruise
碰碎 crash
碰损破碎险 clash and breakage risks; risk of clashing and breakage
碰损险 clash(ing) risks; risk of clashing
碰锁 instant lock; latch(ing); snap lock;spring(latch)lock
碰锁门梃 slamming stile
碰锁门梃板条 slamming strip
碰锁门梃嵌条 slamming strip
碰锁弹簧 latch spring
碰锁钥匙 latch key;passkey
碰停鼓 stop drum
碰停块 stop dog
碰停转筒 stop roll
碰停装置 sizing device
碰头边梃 meeting stiles
碰头挡 check rail
碰头航向 collision course
碰头横挡 <窗的> meeting rail
碰头门梃 meeting post
碰头站 colliding station
碰线 <电线路> swinging crossing; wires crossing
碰响试验 <估计水泥土等相对硬度用> click test
碰炸导火索 impact fuse
碰炸帽 percussion cap
碰炸起爆药包 percussion primer
碰炸引信 impact fuze
碰珠 bale's catch; ball catch; roller catch
碰珠门锁 bullet catch
碰撞 bombardment; bump; collide; collision;crash;hit;impact(ment); impinge(ment);knock;percussion
碰撞半径 impact radius
碰撞保险 collision insurance
碰撞爆破 collision blasting
碰撞崩溃 crash
碰撞边界 collision boundary
碰撞变宽 Lorentz broadening
碰撞标志 collision blip
碰撞参数 collision parameter; impact parameter
碰撞除尘器 impingement dust collector
碰撞船 colliding vessel
碰撞脆性 impact brittleness
碰撞带 collision belt;impact zone
碰撞点 impingement point; point of impingement
碰撞电离 ionization by collision
碰撞电离开关 impact ionization switch
碰撞电离探测器 impact ionization detector
碰撞调车 impact switching
碰撞定律 impact law
碰撞发响 clash
碰撞方位 collision bearing
碰撞防护 collision prevention
碰撞分析图 collision diagram
碰撞缝合 collision suture
碰撞辐射 collision radiation; impact radiation
碰撞辐射复合 collision-radiative recombination
碰撞负载 impact load(ing)
碰撞复合 impact recombination
碰撞概率 collision probability; impact probability
碰撞干扰 hit
碰撞感生不稳定性 collision induced instability
碰撞感生谱 collision induced spectrum
碰撞谷盆地 collision basin
碰撞轨道 collision orbit
碰撞过程 collisional process
碰撞航向 collision course
碰撞荷载 collision load; impact load(ing)
碰撞护栏 crash barrier
碰撞环境 impact environment
碰撞缓冲器 collision bumper
碰撞机 collider
碰撞激发 collision excitation; impact excitation
碰撞激励 impact agitation; impact excitation; repulse excitation; shock-excitation
碰撞极化 impact polarization
碰撞加热 collisional heating; gyrorelaxation heating
碰撞加速 impact acceleration
碰撞加速过程 impact acceleration process
碰撞加速计 impact accelerometer
碰撞假说 collision hypothesis
碰撞减活作用 collisional deactivation
碰撞检测 collision detection
碰撞检测器 impact detector
碰撞角 angle of import; collision angle;impingement angle
碰撞截面 collision cross-section
碰撞警报 collision warning
碰撞警报系统 collision warning system
碰撞警报指示器 collision warning indicator
碰撞警告 collision warning; conflict alert
碰撞矩阵 impact matrix
碰撞坑 impact pit
碰撞扩散 collisional diffusion
碰撞类型 collision pattern
碰撞理论 collision theory
碰撞力 collision force;impact force
碰撞粒子 impingement particle
碰撞量 impact momentum
碰撞裂纹 bruise check
碰撞率 collision efficiency; collision rate;impingement rate
碰撞密度 collision density
碰撞面积 impact area
碰撞磨损 gouging abrasion
碰撞能 impact energy
碰撞能量传递 collision transfer of energy
碰撞凝结 perikinetic aggregation; perikinetic flocculation
碰撞频率 collision frequency;encounter frequency; frequency of collision
碰撞破坏 impact wreckage
碰撞破碎 impact crushing
碰撞破损 impact failure
碰撞器 impactor
碰撞强度 collision strength; impact strength
碰撞区 impact zone
碰撞取样器 impinger
碰撞闪光传感器 impact(-light) flash sensor
碰撞射电源 collisional radio source
碰撞声 impact sound
碰撞时间 collision time;time to collision
碰撞式开关 impact switch
碰撞式破碎机 impact breaker
碰撞式悬浮微粒分离器 <空气除尘净化用的> impingement separator
碰撞式摇床 bumping table; percussion table
碰撞事故 collision accident; running-up accident
碰撞事故分析 collision analysis; running-up accident analysis
碰撞事故分析图 collision diagram; collision graph
碰撞试验 bump test; crash test; impact test(ing)
碰撞双方均有过失条款 both to blame collision clauses
碰撞速度 collision speed; collision velocity;contact speed;impact velocity
碰撞速率 collision rate;impact speed
碰撞损害条款 damage done clause
碰撞损失 collision damage; collision loss; damage in collision; impact-(ion)loss
碰撞他船的船 colliding ship
碰撞探测器 crash sensor
碰撞弹簧锁 warded rim lock
碰撞条款 collision clause
碰撞湍动 collisional turbulence
碰撞蜕变 impact disintegration
碰撞蜕变反应 impact disintegration reaction
碰撞危险 danger of collision; risk of collision
碰撞位移 knocking-out;knock-on displacement
碰撞温度 impact temperature
碰撞系数 collision coefficient
碰撞效率 collision efficiency
碰撞效率因素 collision efficiency factor
碰撞型造山作用 collision-type orogeny
碰撞雪崩渡越时间二极管 time(IMPATT)diode
碰撞压力 impact pressure
碰撞压强 impact pressure
碰撞岩 impactite
碰撞因子 collision factor
碰撞荧光 impact fluorescence
碰撞与条件观测图 collision and condition diagram
碰撞噪声 impact noise; impulsive noise
碰撞责任【交】collision liability
碰撞责任条款 running down liability clause
碰撞肇祸后开快车逃走 hit-and-run driving
碰撞致宽 collisional broadening; impact broadening
碰撞致窄 collisional narrowing
碰撞状态 manner of collision
碰撞自停装置 knock off device
碰撞阻力 resistance to impact
碰撞阻尼 collisional damping
碰撞作用 impact effect; percussion action

批变异 batch variation

批处理【计】batch processing;batch
批处理程序 batch program(me)
批处理法 batch method
批处理技术 batch process(ing) technic
批处理文件 batch file
批处理系统 batch process(ing) system
批处理终端 batch process(ing) terminal
批次 batch
批次产品完工 completion of jobs
批次号 batch number
批次间偏倚 between batch bias
批次式搅拌机 batch mixer
批次式配料秤 batch weigher
批单 endorsement;indorsement
批地 grant of land
批地条件 conditions of grant
批发 bulk sale;combination sale;sale by bulk;sell by wholesale;trade sale
批发部 wholesale department
批发采购 buy up wholesale
批发成本 wholesale cost
批发承受商 wholesale receiver
批发出售 sell wholesale
批发的 wholesale
批发店 jobbing house;wholesale firm; wholesale house
批发货栈 wholesale warehouse(building)
批发价(格) at wholesale price; combination price;inside price;trade price; wholesale price;dealer's price
批发价格曲线 wholesale price curve
批发交易 wholesaler trade
批发进货渠道 channels for purchasing goods wholesale
批发款项 wholesale money

P

批发量 wholesale quantity
批发零售商店 wholesale-retailing warehouse(building)
批发贸易 distribution trade; wholesale commerce; wholesale trade
批发贸易额 value of wholesale trade
批发企业管理 management of wholesale enterprises
批发起点 starting point of wholesale
批发商 desk jobber; distributor; jobber; merchant whole sale; stapler; wholesale dealer; wholesaler
批发商店 outlet; wholesale shop
批发商品价格水平 level of commodity price at wholesale; level of prices at wholesale
批发商品中心 jobbing center[centre]
批发商业费用 wholesale commerce expenses
批发商业库存 wholesale inventory
批发商业网 wholesale network
批发食品市场 wholesale food market
批发市场 terminal market; wholesale market
批发市价 wholesale market value
批发式 shipper style
批发售出 sell by wholesale
批发物价指数 index of wholesale prices; wholesale price index
批发销售 wholesale sale
批发销售额指数 index of wholesales
批发销售商 wholesale distributor
批发业务 wholesale business; wholesale establishment; wholesale trade
批发油库 bulk plant; bulk terminal
批发站 distributing center[centre]; distribution center[centre]; relaying station; wholesale center[centre]
批发账户 wholesale account
批发折扣 distributor discount; trade discount
批发执照者 licensor
批发中间商 jobber; wholesale middleman; wholesaling middlemen
批改条款 endorsement; rider
批公里 consignment kilometers
批号 batch number; block number; charge number; lot identification mark; lot number
批号卡 lot card
批间标准偏差 batch-to-batch standard deviation
批件 instrument ratification; parcel
批焦 coke charge
批接受概率 acceptance probability per batch
批例推定量 ratio estimate
批量 batch; batch quantity; batch size; batch size bulk; lot; lot size; wholesale
批量保付代理 bulk factoring
批量产品容许失效率 failure rate of lot tolerance
批量处理 batch processing
批量处理方式 batch process(ing) mode
批量处理系统 batch process(ing) system
批量大小 lot size
批量分析 batch quantity analysis
批量供水 bulk water supply
批量供应 bulk supplies; lump supplies
批量共计 batch total
批量货单管理内插技术 lot-size inventory management interpolation technique
批量加煤机 batch stoker
批量搅拌机 batch kneader
批量接受抽样 lot-acceptance sampling
批量控制 batch control; job lot control
批量控制采样 batch control sampling
批量平均直径 lot mean diameter
批量生产 batch-(like) process; batch production; lot production; mass-produce; mass production; quantity production; series production; volume produce; volume production
批量生产的烧结制品 mass-produced sintered ware
批量生产的预制混凝土加工品 mass-produced precast concrete ware
批量生产法 bulk method
批量试样 batch sample
批量通信[讯] batched communication
批量喂料 batch feed
批量喂料斗 batching hopper
批量问题 lot-size problem
批量系统 batch system
批量验收 acceptance of a batch
批量印刷 batch printing
批量用水 batch water
批量允许不良率 lot-tolerance percent defective
批量直径的变化值 lot diameter variation
批量制法 batch process
批量制造 mass manufacture
批量质量 quality of lot
批料 batch; batch material; charging material
批料混合系统 batch-mixing system
批料计数器 batch counter
批料加热器 batch heater
批料漏斗 batch hopper
批料重量 batch weight
批料装载机 batch loader
批命令 batch command
批内容许故障率 lot-tolerance failure rate
批内允许次品率 lot-tolerance percent defective
批内允许故障率 lot-tolerance failure rate
批内允许破损率 lot-tolerance failure rate
批式自热高温好氧消化工艺 patch-typed auto-thermal thermophilic aerobic digestion process
批试法 trial-batch method
批数 lot number
批文件 batch file
批样 lot sample
批允许废品率 lot-tolerance percent defective
批运料车 <装卸干混合料的> batch truck
批指令 batch command
批重 batch weight
批重容差 lot weight tolerance
批准 affirmation; approbation; approval; approve; clearance; fiat; ratification; ratify; sanction; validate; validation; warrant(y)
批准变更的装车数占总装车数百分率 percent of sanctioned alternations in wagon loadings to total planned
批准程序 procedure of approval
批准代理人签订的合同 ratification of agent's contract
批准单位 organization for approving
批准的 approved
批准的拨款 authorized appropriation
批准的部件清单 approved parts list
批准的代案 approved equal
批准的贷款项目表 statement of loans approved
批准的发展计划 permitted development
批准的供应人清单 approved vendor list
批准的供应商清单 approved vendor list
批准的经费 authorized appropriation
批准的设计图 design approval drawing
批准的图纸 approved plan
批准的型号 approved pattern
批准的专利 patent granted
批准的最大供水量 authorized full supply
批准发行的抵押债券 authorized mortgage bonds
批准发行的债券 authorized bonds; bonds authorized
批准方案 approval of plans
批准方式 approved method
批准房屋入住证书 occupancy permit
批准规划 approval of plans; planning approval
批准合同 affirmation of contract
批准合同之日 date of approval of the contract
批准机关 approval authority
批准集装箱 approval container
批准计划 approval of plans
批准阶段 approval stage
批准会计师 approved accountant
批准签署人 approved signatory
批准前 preauthorization
批准权 approval authority
批准人住证书 certificate of occupancy
批准(认可)标志 confirmatory sign
批准设计 approved design
批准申请 hono(u)r an application
批准施工 approved for construction
批准使用证明 established use certificate
批准书 act of ratification; confirmation form; instrument of certification; instrument of ratification; instrument ratification; letter of ratification
批准特许会计师 approved accountant
批准通知 notification of approval
批准图 approval drawing; final plat
批准图纸 approval of drawings
批准文件 document of approval
批准箱型集装箱 type-series container
批准泄流 discharge consent
批准行为 act of ratification
批准延期缴纳的税款 authorized deferred payment
批准样品 approved sample
批准者 person for approving
批准证书 certificate of approval; instrument of ratification
批准中断 interrupt acknowledge
批准专利 patent
批租 lease in batches
批租地产 batch lease real estate; batch-leasing landed property
批作业【计】 batch

坯

坯板 palette; pallet board
坯爆 body peel off
坯布 gray fabric; gray goods; loom-state fabric
坯布初洗 grey washing
坯材疤皮 shell
坯的实验式 empiric(al) formula of body
坯锭加热炉 billet furnace
坯段 billet
坯段钢 billet steel
坯房头 head of the clay room
坯钢 bloom steel
坯架 saddle; stilliards
坯件 blank
坯件布置 blank layout
坯晶 crystal blank
坯块 briquet(te); compact
坯料 blank; body; ground; ingot; primary; stock
坯料剥皮机 peeler
坯料仓库 billet storage
坯料定位窝 blank nest; nest type blank locator
坯料分配器 billet switch
坯料厚度 stock thickness
坯料剪切机 billet shears; stock cutter
坯料库 stock bank
坯料排样 blank layout
坯料去皮车床 peeler
坯料渗碳 blank carburizing
坯料压紧 blank holding
坯料轧机 billet mill
坯料制备 body preparation
坯裂 checking; crack in body; fire check
坯炉 biscuit furnace; biscuit kiln
坯面施釉重量 application weight
坯模 blank mo(u)ld
坯品 made-up article
坯式 empiric(al) formula of body
坯胎 metal body
坯体 billet; body; green body
坯体表面滚压入彩色黏[粘]土 rolled inlay
坯体成型 body formation
坯体泥浆筛上料 knotting
坯体缺陷 body defect
坯体收缩失重记录仪 Barelattograph
坯体塌软 flabbiness
坯体制备 body preparation
坯铁 pig iron
坯托印痕 stilt marks
坯斜槽 blank channel
坯修整工段 conditioning yard
坯选冶综合回收率 total rate of mining dressing and melting recovery
坯窑 biscuit kiln
坯用泥浆 body slip
坯用熔剂 body flux
坯用色料 body stain
坯釉均裂 crack both on body and glaze
坯釉配方 batching of body and glaze
坯釉适应性 body-glaze fit; glaze-body fit
坯釉中呈链状气泡 air chain
坯釉中间层 body-glaze intermediate layer; glaze-body interface
坯与釉适应性 glaze fit test
坯砖 encallow

披

披铂 platinization
披铂石棉 platinized asbestos
披铂炭 platinized charcoal
披叠 shiplap lagging
披叠板【建】 siding; rustic siding; weather-board(ing)
披叠板壁 bevel siding
披叠板外壁 drop siding
披叠木板壁 wooden bevel siding
披叠墙板 lap siding; rabbeted siding
披缝 joint flash
披盖 drape; draping
披盖沉积 draping deposit
披盖模式 drape mode
披盖褶皱 drape fold
披肩 cape; scarf
披巾状云 scarf cloud
披露事实 discovery of facts
披麻捉灰 <中国在木结构上刷油漆的传统做法> gunny-and-paste pre-

treatment for painting
披水 cover flashing; eaves flashing; upstand; watershed; weathering
披水板 ballast fin; cover flashing; flashing; gutter bed; soaker; throating plate; apron flashing
披水槽 raglet; raglin; raggle
披水槽砌块 raglet; raggle block
披水沟 flashing valley
披水面 canting strip; water-table
披水黏[粘]结剂 flashing cement
披水石 label
披水石端饰 label stop
披水饰 hood mo(u)ld
披水条 cant(ing) strip <外墙角的>; check rail; eaves pole; skew fillet; tilting fillet; weather-board(ing)
披水屋檐 water drip
披水线脚 <方形拱的> label mo(u)lding; hood mo(u)lding
披屋 lean-to roof; lookum; pentee; penthouse; pentice; tofall
披屋顶 penthouse roof
披屋居住单元 penthouse living unit
披屋入口 penthouse entrance; entrance roof overhang
披屋套间 penthouse flat
披压柱 <凿岩机> hydraulic leg
披檐 pent(house)
披针叶桦 green ash

砒霜 rude arsenic; white arsenic

铍-10 测年法 10 Be dating method

铍窗 beryllium window
铍法 beryllium method
铍方钠石 beryllosodalite; tugtupite
铍肺(病) berylliosis
铍钙大隅石 milarite
铍硅钠石 lovdarite
铍黄长石 aminoffite
铍矿(石) beryllium ore
铍砾 pebble of beryllium
铍榴石 danalite
铍镁晶石 taaffeite
铍镍膜电阻器 beryllium-nickel film resistor
铍青铜 beryllium bronze
铍青铜滚动轴承 beryllium bronze rolling journal bearing
铍砷矿 bearsite; beryllium arsenic ore
铍石 bromellite
铍酸钾 potassium beryllate
铍酸钠 sodium beryllate
铍酸盐 beryllate
铍铜合金 Beallon; beryllium-copper alloy
铍污染 pollution by beryllium
铍中毒 berylliosis
铍柱石 harstigite

劈板斧 froe; riving knife

劈材 hewn timber; split stuff; split timber
劈材工 chopper
劈成 wedge out
劈成轨枕 split tie
劈锤 chop hammer; cleaving hammer; riving hammer; splitting hammer
劈刀 brick set; froe; frow; riving knife
劈顶芽接 terminal budding
劈度 fissility
劈度表 scale of fissility
劈度等级 scale of fissility
劈枋 hewn square
劈分线对 split pair
劈斧 broad axe; chip ax(e)
劈痕 kerf[复 kerve]; saw kerf
劈接 cleft grafting; cleft-grafting stub
劈接法 cleft graft
劈开 breaking up; chap; cleat; cleavage; cleaving; rend; rift; rip; rive; splinter(ing); splitting up; straddling; wedge; wedging; cleave
劈开薄板 breakout plate
劈开的 cleft; spalt; split; rendered
劈开的板 riven slate
劈开的板条 riven lath
劈开的冰山 ice doubling
劈开的灰板条 rent lath
劈开的木材 cleaved wood
劈开的窄木条 rent pale
劈开加工的燧石 knapped flint
劈开力 breakout force
劈开木材 split wood
劈开岩石 rock splitting
劈拉强度 split strength; split tensile strength
劈拉试验 split-ring-tensile test; splitting tensile test
劈离机 splitter
劈离砖 split tile
劈离砖块 split block
劈理【地】 cleavage; cleat
劈理带 cleavage belt
劈理构造 cleavage structure
劈理观测点 observation point of cleavage
劈理痕迹 cleavage trace
劈理夹层 cleavage banding
劈理裂缝 cleavage crack
劈理面 cleavage plane; divisional plane
劈理片 gleithretter
劈理扇 cleavage fan
劈理式窗棱 cleavage-mullion
劈理线 line of cleavage
劈理域 cleavage domain
劈理域平均间隔 average spacing of cleavage domain
劈理域相对宽度比 ratio of relative width of cleavage domain
劈理域形态 shape of cleavage domain
劈理褶皱 cleavage fold
劈理组构特征 fabric feature of cleavage
劈裂 cleavage crack; laceration; popouts; split; splitting; through check
劈裂板岩 sculping
劈裂剥离强度 cleavage peel-strength
劈裂材 cloven timber
劈裂法 Brazilian tensiling method
劈裂法抗拉试验 Brazilian tensile test
劈裂法试验 splitting tensile strength test; tensile spilling strength test
劈裂法圆盘试验 splitting disk test
劈裂方向 splitting direction
劈裂缝 cleavage cracking; split crack
劈裂管 split pipe
劈裂灌浆 fracture grouting; fracturing grouting
劈裂机 splitter
劈裂抗拉强度 splitting tensile strength; splitting tension strength; tensile spilling test; tensile splitting strength
劈裂抗拉强度试验 <即巴西法试验> splitting tensile strength test; tensile spilling strength test
劈裂抗拉试验 diametral compression test; splitting tensile test; split-ring-tensile test <非直接的抗拉试验>
劈裂可能性 rippability
劈裂拉伸强度试验 splitting tensile strength test
劈裂裂缝 splitting crack
劈裂面 cleavage plane; plane of cleavage; split face
劈裂面混凝土砌块 split concrete block
劈裂模量 cleavage modulus
劈裂木板 split shakes; splitting wood
劈裂木瓦 split shakes; splitting shingles
劈裂黏[粘]结强度 bond splitting resistance
劈裂破坏 cleavage failure; splitting failure; splitting rupture
劈裂强度 cleavage strength; splitting strength
劈裂强度比 split strength ratio
劈裂实验 cleavage splitting test; split test
劈裂试验 <测抗拉强度的> Brazilian test; cleavage test; indirect tension test; split(ting)(ring) test
劈裂丝 spelly wire
劈裂线 line of cleavage
劈裂效应 cleavage effect
劈裂楔 splitting wedge
劈裂性 cleavability; rift; rippability; cleavage【地】; fissi(bi)lity <木材的>
劈裂性能 splitting property
劈裂岩石 rock splitting
劈裂注浆 fracture grouting
劈裂注浆法 fracture grouting method
劈裂砖 split brick; split tile
劈裂作用 slabbing action; splitting action
劈流器 jet splitter
劈模铸造 multiple casting
劈木机 splitter; wood splitting machine
劈木器 wood cleaver
劈木瓦刀 riving knife
劈木楔 splitting wedge
劈啪声 <空化发生时的> snap(ping sound)
劈劈拍拍地响 pop
劈劈拍拍声音 pop
劈片 cleft
劈片光度计 wedge photometer
劈片摄谱仪 wedge spectrograph
劈破 cleave
劈石 rock breaking
劈石斧 hack hammer; hammer hack; slate axe
劈石工 splitting mason
劈石机 chomper
劈石面 split-stone finish
劈石器 rock splitter
劈石凿 splitting chisel
劈水(设备) upstream pier nosing
劈碎 wedging down
劈头 split end; splitter
劈楔 clink; froe; frow
劈形的 wedge-shaped
劈形膜 wedge film
劈形算符 nabla
劈形吸声体 wedge-shaped absorber
劈样机 splitter
劈样器 sample cutter
劈因子格式 splitting scheme
劈植 notching
劈制板条 riven lath
劈制材 cleft timber
劈制的木材 billet
劈制的薪材 billet
劈竹形窑 split-bamboo kiln
劈砖斧 brick ax(e)
劈砖枕垫 brick set
劈锥曲面 conoid
劈锥(曲面)壳 conoid shell
劈锥曲面穹顶 conoidal dome

噼啪地响 crackle

噼啪声 bang; crackling; crepitation
噼啪响电弧 crackling arc
噼啪噪声 crackling noise
噼啪作响 sputter
噼拍噪声 scratching noise
噼噼啪啪声 cracking

霹雷 thunderbolt

霹雳 thunderbolt

皮安计 picoammeter

皮安(培) picoampere
皮鞍座 leather seat
皮奥里间冰期【地】 Peorian
皮包公司 bubble company; incorporated pocket-books; long firm; speculation company
皮包节 encased knot
皮包商 baggage trader; bagman
皮包线【电】 covered electric(al) conductor
皮层 bark mantle; cortex[复 cortices/cortexes]; cortical layer
皮查方程式 Pitzer equation
皮衬 leather gasket; leather sheet
皮衬垫 leather gasket; leather washer
皮尺 band tape; builder's tape; linen-tape; measuring tape; ribbon tape; tape line; tape measure; tape ruler
皮尺平差 adjustment of tapes
皮尺数 brick ga(u)ge
皮传送带 belt line
皮带 belt; hide rope; leather belt; leather strap; strap; thong
皮带包角 belt contact
皮带变速装置 belt speeder
皮带拨杆 striking gear
皮带擦拭器 belt cleaner; belt scraper; belt wiper
皮带材料 strapping
皮带槽轮 belt sheave
皮带测长台 belt bench
皮带车床 belt driven lathe
皮带衬料 belt material
皮带称量计 belt balance
皮带称重装置 conveyer belt scale
皮带秤 belt scale; belt weight; belt weight meter; belt weightometer; conveyer scale; weightometer
皮带秤量计 belt balancer
皮带秤量器 belt weigher
皮带秤喂料机 weigh belt feeder
皮带秤卸料仓 weigh belt bin
皮带尺 ribbon tape
皮带齿轮减速型 belt-gear reduction type
皮带冲孔机 belt punch
皮带冲孔器 belt punch
皮带冲头 belt punch
皮带冲压机 belt punch
皮带传动 belt drive; belt driving; belt gear(ing); belt transmission; beltwork; pulley drive
皮带传动泵 belt driven pump
皮带传动的 belt driven; belt conveyed
皮带传动电机 belt drive motor
皮带传动电梯 belt driven elevator; belt driven lift
皮带传动鼓风机 belt driven blower
皮带传动滑轮 belt pulley
皮带传动混合器 belt driven mixer
皮带传动机构 belt gear(ing); belt grinder; tape-drive mechanism

皮带传动绞车 belt drive winch
皮带传动龙门刨床 belt drive double housing planer
皮带传动铆接机 belt driven riveting machine
皮带传动升降机 belt driven elevator; belt driven lift
皮带传动生产作业 belt production
皮带传动压气机 belt driven compressor
皮带传动转速计 belt driven tach(e)-ometer
皮带传动装置 belt drive unit; belt gear(ing)
皮带传动作业机 belt drive implement
皮带传送 belt conveyance; belt transmission; tape transport
皮带锤 belt drop hammer; belt hammer
皮带从动边 slack side
皮带打滑 belt creep; belt slip(page); bootlegging; creep of belt; slipping of belt
皮带带扣 belt fastener
皮带导槽 belt guidance
皮带导向轮 belt guide pulley
皮带的毛面 hair side of belt
皮带电子秤 belt electronic weigher; belt scale
皮带定长器 belt bench
皮带斗式提升机 band-type bucket elevator; belt and bucket elevator; belt(-type) bucket elevator
皮带斗式提升加载机 belt-type bucket elevator loader
皮带堆料机 stacker; stacking conveyer[conveyor]
皮带防护罩 belt guard
皮带防滑装置 belt slip monitoring system
皮带分级机 belt grader
皮带分条搓条机 leather tape condenser
皮带分选机 belt grader
皮带给料机 belt feeder; feed belt
皮带给料器 belt feeder
皮带功率 belt horsepower
皮带钩 belt hook; alligator
皮带刮板 belt scraper
皮带刮土机 belt wiper
皮带挂杆 deflecting bar
皮带管 hose; water hose
皮带辊筒 belt roller
皮带护挡 belt guard
皮带滑车 belt roller
皮带滑动 creep of belt; slipping of belt
皮带滑轮 belt pulley flywheel
皮带滑轮传动 belt pulley drive
皮带回动机构 belt reversing drive
皮带回动装置 belt reversing gear
皮带机 belt conveyer
皮带机除尘器 belt cleaner
皮带机的连续变速传动 rubber-belt CVT [continuously variable transmission]
皮带机的张紧滚轮 tension drum of conveyer[conveyor]
皮带机排泥 disposal of spoil with belt conveyer[conveyor]
皮带机拖曳式砂粒分选机 drag classifier
皮带机旋转角 conveyer swing angle
皮带及链条传动的钻机 belt-and-chain driven drill
皮带计量机 belt weigher
皮带加料器 feed belt
皮带夹 belt clamp; belt clip
皮带胶接 cemented belt joint
皮带接头 belt joint; belt lace; belt lacer; belt lacing
皮带接头固紧件 belt fastener; belt lace
皮带接头机 belt lacing machine
皮带结构 belt composition
皮带结合 belt joint; belt lacing
皮带结头 belt lacing
皮带截面 section of the belt
皮带紧边 advancing side of belt; tension side of belt; tight side of belt
皮带紧轮 belt idler; belt stretching roller; stretching pulley
皮带紧张轮 idle pulley
皮带绝缘电缆 belt insulated cable
皮带卡子 belt lacer; belt lacing
皮带空转轮 belt idler
皮带扣 belt clamp; belt fastener; belt hook; belt lacer; belt lacing
皮带扣件钢丝 belt fastener wire
皮带扣钳压器 belt hook clipper
皮带拉合 belt joint
皮带拉紧 belt tension
皮带拉紧轮 belt tightener
皮带拉力 belt tension
皮带拉力调节器 belt tension regulator
皮带蜡 belt wax
皮带连接器 belt fastener
皮带连接锁扣 claw belt fastener
皮带联结卡 belt clamp
皮带溜槽 vanner
皮带轮 band wheel; belt pulley; belt (pulley) wheel; belt (stretching) roller; leather belt wheel; pulley; pulley wheel; sheave
皮带轮安全装置 pulley guard
皮带轮槽 pulley groove
皮带轮车床 pulley lathe
皮带轮传动 pulley drive
皮带轮传动装置 pulley gear
皮带轮防护罩 guard for belt pulley
皮带轮刮板 pulley scraper
皮带轮键 pulley key
皮带轮离合器 belt pulley clutch
皮带轮螺母 pulley nut
皮带轮螺丝攻 pulley tap
皮带轮丝锥 pulley tap
皮带轮托架 pulley holder
皮带轮缘 rim of pulley
皮带轮缘滑动 belt creep
皮带轮造型机 pulley mo(u)lding machine
皮带轮罩 pulley cover; pulley shell
皮带轮支架 pulley support
皮带轮中心距 pulley center[centre] spacing
皮带轮铸坯 pulley casting
皮带轮转速 pulley speed
皮带落锤 belt(drop) hammer
皮带铆钉 belt rivet; strap rivet
皮带捻 belt lace
皮带爬行 belt creep
皮带盘 belt pulley
皮带盘垫套 packing carrier
皮带抛光 belt polishing
皮带抛光机 belt polishing machine
皮带跑偏 wandering of belt
皮带起重机 belt hoist
皮带起重装置 belt lifting arrangement
皮带牵引 drawing by belt
皮带桥 belt bridge
皮带倾斜提升机 belt slope lifter
皮带驱动的 belt driven
皮带圈 endless belt
皮带润滑剂 belt composition
皮带润滑脂 belt grease
皮带刹车 belt brake
皮带筛 belt screen
皮带上料 belt filled
皮带上煤机 belt feeder
皮带伸张器 belt stretcher
皮带升降机 belt hoist
皮带式 belt-type
皮带式称量器 belt scale
皮带式称量喂料机 belt weigh feeder
皮带式分级机 belt grader; belt sorter; diverging belt sorter; diversion belt sizer
皮带式分配装置 belt-type proportioning unit
皮带式干燥机 apron conveyor drier
皮带式供料器 feed apron
皮带式拉力计 belt tension ga(u)ge
皮带式量料装置 belt-type measuring unit
皮带式龙头 hose cock; hose connection
皮带式配料器 belt-type batching unit
皮带式取料装载机 excavating belt loader
皮带式取样器 conveyer-type sample
皮带式筛子 belt screen
皮带式输送机 belt conveyer
皮带式卸车机 conveyor type wagon unloader
皮带式卸料机 throw-off belt
皮带式悬挂磁选机 belt-type suspended magnetic separator
皮带式装载机 conveyer loader
皮带输送 belt conveyance
皮带输送秤 belt conveyor scale
皮带输送带 band conveyer; endless belt conveyer; ribbon conveyer[conveyor]
皮带输送机 band conveyer; drawing plate apron conveyer; endless belt conveyer; feeding belt conveyer; ribbon conveyer; rubber belt conveyer; travel(l)ing belt conveyer [conveyor]
皮带输送机秤 conveyor weigh meter
皮带输送机防止倒转的自动安全装置 holdback
皮带输送机桥 belt bridge
皮带输送机托辊 conveyer idler
皮带输送机系统 belt conveyor stacker
皮带输送机自动张紧装置 automatic type belt tensioning device
皮带输送式掘沟机 belt conveyer type trencher
皮带松边 loose side; receding side of belt; slack side; slack side of belt
皮带松紧度调节 belt adjustment
皮带松紧杆 belt tension release lever
皮带松紧调整器 belt tensioner
皮带送料 ribbon feeding
皮带送料带 endless belt conveyer; feed(ing) belt conveyer[conveyor]
皮带送料机 endless belt conveyer; feed(ing) belt conveyer[conveyor]
皮带损坏 belt failure
皮带锁扣 strap snap
皮带塔轮 belt cone; stepped cone
皮带提升机 belt elevator
皮带调整装置 belt aligning device
皮带涂料 belt dressing
皮带退出侧 receding side of belt
皮带托辊 belt idler
皮带拖平<水泥混凝土路面施工时的> belt finish(ing)
皮带位置校正 belt training
皮带喂料 belt feed(ing)
皮带喂料机 belt feeder; feeding belt conveyer[conveyor]
皮带系统 belt system
皮带斜切机 strap bevel(l)ing machine
皮带循环运输机 endless belt conveyer[conveyor]
皮带移动机构 belt shifting mechanism
皮带移动器 strap fork
皮带移动手把 belt shift lever
皮带移动装置 belt shifter
皮带油 belt dressing; belt filler; belt lubricant
皮带运输 belt conveyance; flexible transport
皮带运输带 band conveyer[conveyor]; belt conveyer; conveyer belt(ing); rubber conveyor belt
皮带运输的混凝土 belt-conveyed concrete
皮带运输的物料 conveying medium
皮带运输机 apron conveyer; band conveyer; belt conveyer; belt roller; conveyer[conveyor] belt(ing); rubber conveyor belt
皮带运输机道 beltway
皮带运输机的上(面传送)带 carrier side
皮带运输机的卸料机 belt tripper
皮带运输机的卸料器 belt tripper
皮带运输机收紧器 belt conveyer takeup
皮带运输机外罩 belt cover
皮带运输机装载站 belt loading station
皮带运输浇灌混凝土塔 belt concreting tower
皮带运输廊 conveyer gallery
皮带运输桥 belt bridge
皮带展幅机 belt stretching machine
皮带张紧侧 advancing side
皮带张紧度 belt tightening
皮带张紧滑轮 belt tensioning pulley
皮带张紧轮 belt idler; belting tightener; belt stretching roller; idler pulley; tightener sheave
皮带张紧器 belting tightener
皮带张紧式离合器 belt tension clutch
皮带张紧装置 belt stretcher; belt tightener
皮带张力 belt pull; belt tension
皮带罩 belt guard; belt house; belt housing
皮带直角转弯传动 belt quarter-turn drive; quarter-turn drive
皮带止水条 leather sealing strip
皮带制动器 belt brake
皮带轴 belt roller
皮带主动边 tension side
皮带助卷机 wrapping machine
皮带注油口 belt filler
皮带转运机 conveyer transfer
皮带转载 transfer by belt conveyer [conveyor]
皮带装料铲运机 ladder scraper
皮带装料机 belt conveyor stacker; belt loader
皮带装卸车 belt car unit
皮带装载机 loader conveyer[conveyor]
皮带装置 belt dressing; belting
皮带自动张紧机构 automatic type belt tensioning device
皮带走偏 belt off-centre; wandering of belt
皮垫 leather packing
皮垫圈 leather collar; leather gasket; leather packing collar; leather washer
皮吊带 belt sling
皮蒽酮 pyranthrone
皮尔巴拉古陆【地】 Pilbara old land
皮尔巴拉陆核【地】 Pilbara nucleus
皮尔彻木材防腐剂 Pilcher's stop rot
皮尔德发电机保护装置 Beard protective system
皮尔格式冷轧管 rocked pipe
皮尔格式无线钢管轧机 Pilger seamless-tube mill
皮尔格式轧管机 Pilger mill
皮尔格缩管法 Pilger tube-reducing

process
皮尔斯不稳定性 Pierce instability
皮尔斯透镜组 Pierce lenses
皮尔斯型引出器 Pierce-type extractor
皮尔斯振荡器 <控制晶体接在栅极和屏极间的自激振荡器> Pierce oscillator
皮尔塔 <英国北部和苏格兰的一种避难用的小塔或房子> Pele-tower
皮尔逊分布 Pearson's distribution
皮尔逊分布曲线 Pearson's distribution curve
皮尔逊公式 <计算斜面风压> Pearson's formula
皮尔逊环比法 Pearson's linked relative method
皮尔逊积矩相关系数 Pearson's product-moment correlation coefficient
皮尔逊均方列联系数 Pearson's coefficient of mean square contingency
皮尔逊空气颗粒分级器 Pearson's air elutriator
皮尔逊-莫斯科维奇波谱 Pierson-Moskowitz wave spectrum
皮尔逊拟合优度准则 Pearson's goodness-of-fit criterion
皮尔逊-纽曼-詹姆斯波浪推算法 Pearson-Neuman-James method
皮尔逊偏态系数 Pearson's coefficient of skewness
皮尔逊频率分布 Pearson's frequency distribution
皮尔逊频率曲线参数 Pearsonian parameter of frequency curve
皮尔逊曲线 <水文统计用> Pearson's curve;Pearson curve
皮尔逊三型分布 Pearson's type III distribution
皮尔逊三型（偏态）分布曲线 Pearson's type III distribution curve
皮尔逊-斯皮尔曼相关系数 Pearson and Spearman correlation coefficient
皮尔逊四形正态分布曲线 Pearson's type IV normal distribution curve
皮尔逊四型分布 Pearson's type IV distribution
皮尔逊系数 Pearson's coefficient
皮尔逊相关系数 Pearson's correlation coefficient
皮尔逊一型分布 Pearson's type I distribution
皮尔逊准则 Pearson's criterion
皮筏 canoe;kayak;skin raft;umia(c)k
皮法 micromicrofarad;picofarad
皮风箱 leather bellows
皮封的 bark-bound
皮肤 skin;derma;dermis;fell
皮肤病 dermatopathy; dermatosis; skin disease
皮肤的 cutaneous
皮肤接触温度 skin contact temperature
皮肤科 dermatological department; dermatology
皮肤碰伤 <俚语> boo-boo
皮肤湿润度 skin wittedness
皮肤视觉 skin vision
皮肤损伤 skin injury
皮肤温度 skin temperature
皮肤炎 dermatitis
皮肤药膏 barrier cream
皮肤诊所 skin clinic
皮肤中毒 dermal toxicity
皮革 hide;leather;skin hide
皮革保护油 dubbing
皮革厂 tannery
皮革厂出水 tannery effluent
皮革厂废水 tannery effluent;tannery wastewater
皮革厂流出水 tannery effluent
皮革厂污泥 tannery sludge
皮革厂污水 tannery sewage
皮革衬条 leather sealing strip
皮革打底色 daub of leather
皮革的软化 <鸟粪鞣化> bating of leather
皮革的石灰处理 liming of leather
皮革的调理 leather conditioning
皮革底色 daub
皮革垫圈 leather washer
皮革粉末 powdered leather
皮革铬鞣法 chrome tannage of leather
皮革工业 leather industry
皮革挂帘 leather-hanging
皮革加脂法 leather stuffing
皮革胶 leather glue
皮革焦桔酚鞣法 pyrogallol tannage of leather
皮革焦油 leather tar
皮革门 leather door
皮革面漆 hide finish
皮革黏[粘]合剂 leather adhesive
皮革喷漆 leather lacquer
皮革漆 leather dope
皮革清漆 leather varnish
皮革染料 leather dye
皮革染色 leather colo(u)ring
皮革上光油 fat liquor
皮革上油 leather fat-liquoring
皮革手垫 leather hand pad
皮革熟皮脱灰废水 bating process waste
皮革填充物 leather packing
皮革铁鞣法 iron tannage of leather
皮革涂饰剂 hide finish
皮革正面 <原长毛的面> grain side of leather;hair side of leather
皮革织物 leather fabric
皮革植物鞣法 vegetable tannage of leather
皮革纸浆板 leatherpulp board
皮革制品 leather;leather goods
皮工作手套 leather palm working gloves
皮古效应 Pigou effect
皮辊 belt roller
皮滚筒 belt roller
皮滚轴 belt roller
皮滚柱 belt roller
皮痕量分析 picotrace analysis
皮猴 anorak
皮虎石砌体 polygonal masonry
皮货 peltry;thong
皮夹子 wallet;notecase <英>
皮件胶合料 leather cement
皮胶 hide glue;skin glue
皮焦法 superficial charring(method)
皮卷尺 leather measuring tape;linen-tape <俗称>
皮卡方法 Picard's method
皮科克钙碱系数 calc-alkalic coefficient of Peacock
皮壳 rind
皮壳矿石 crust ore
皮壳状 crustose
皮壳状构造 crusty structure
皮可 micromicro;pico
皮(可)法拉 <电容量单位，常用于无线电> picofarad
皮(可)米 micromicron
皮(可)秒 picosecond
皮拉脱风 <阿拉伯南海岸的一种北风或西北风> Belat
皮拉藻类体 Pila-alginite
皮老虎 bellows; cleaning blower; hand bellows; handle air blower; leather bellows; hand blower <锻工用>
皮勒尔型沼气检验灯 Pieler's lamp
皮料薄膜 leather diaphragm
皮裂 shell shake
皮硫铋铜铅矿 pekoite
皮硫锡锌银矿 pirquiasite
皮龙扳手 hose spanner
皮旅行包 kit bag
皮毛 fur
皮毛类制品 fur goods
皮毛石 fir-like spatter dash
皮密封衬垫片 leather gasket
皮密封垫(片) leather gasket
皮密封垫圈 leather seal; leather washer
皮密封圈 leather sealing ring
皮棉打包机 cotton-press
皮面门板 drum panel(l)ing
皮秒 picosecond
皮膜校正轴试验 membrane correction shaft test
皮囊 peltry
皮囊壶 bagging pot
皮囊取样器 <悬移质泥沙测验的> collapsible bag sampler
皮囊式液压蓄能器 rubber bag type hydraulic accumulator
皮腔 bellows attachment;leather bellows
皮腔式密封 bellows seal
皮腔式膨胀接合件 bellows expansion piece
皮腔延伸 bellows extension
皮鞘刀 sheath knife
皮圈 leather collar;washer
皮软管 leather hose
皮砂轮 emery buff
皮绳 hide rope
皮手套 gloves-leather;leather gloves
皮数杆【测】 height board; height pole;profile;keeping ga(u)ge;stor(e)y rod
皮斯电火花闪点测定器 Pease's electric(al) tester
皮斯塔除砂井 Pista desilting trap
皮斯塔混凝过程 Pista coagulation process
皮斯特沉砂池 Pista grit trap
皮炭 charred leather
皮炭可 <一种绝缘材料> Pitanco
皮套 leather case
皮套袖 leather sleeves
皮特发动机润滑油试验 Petter motor oil test
皮特里盘 Petri dish
皮特里网 Petri net
皮特流速计 Pitot ga(u)ge
皮特式闸门截止阀 Peet's valve
皮特专利 <指门拉手与心轴的连接> Pitt's patent
皮填料 leather packing
皮条 raw hide;thong
皮托测压管 Pitot pressure ga(u)ge
皮托管 face tube; impact pressure tube; impact tube; Pitot; Pitot probe; Pitometer (survey); Pitot tube <测流速用>
皮托管测量 <测流速用> Pitometer survey;Pitot tube measurement
皮托管测流法 Pitot tube method of measuring flow
皮托管测压器 Pitometer
皮托管风速测量计 Pitot tube velocity survey
皮托管(进水)口 Pitot tube mouth
皮托管静力管 Pitot-static tube
皮托管静压测定 Pitot-static traverse
皮托管壳 Pitot tube cover
皮托管量测 Pitometer measurement
皮托管排 Pitot traverse
皮托管式风速计 Pitot tube anemometer
皮托管式计程仪 Pitometer log;Pitot log
皮托管式流量计 Pitot tube flowmeter
皮托管水压计程仪 Pitot tube water-pressure log;staulog
皮托管误差修正 Pitot error correction
皮托管行程 Pitot tube traverse
皮托管压差计 <测量流速的> Pitometer
皮托管压力 Pitot pressure
皮托计 Pitot ga(u)ge
皮托计测量 Pitotmeter measurement
皮托计调查 Pitotmeter survey
皮托计式计程仪 Pitometer log
皮托静压差 Pitot-static difference
皮托静压管 Pitot-static pressure tube;Pitot-static tube
皮托静压管沟 Pitot-static tubing
皮托静压管系数 coefficient of Pitot static pressure tube
皮托静压组合管 combined pitotstatic tube
皮托流速测定管 Pitot ga(u)ge;Pitot tube
皮托流速计 Pitotmeter;Pitot tube
皮托球(体) Pitot sphere
皮托损失 Pitot loss
皮托-文丘里测流元件 Pitot-Venturi flow element
皮托-文丘里管 Pitot-Venturi tube
皮托压差计 pitometer survey; Pitot ga(u)ge;Pitotmeter
皮托压力计 Pitot pressure ga(u)ge
皮托柱 Pitot cylinder
皮瓦 pico-watt
皮碗【机】 cuff;cup leather;cup ring; leather cup; leather bowl; leather bucket; leather packing; packing cup;packing leather;rubber cup
皮碗泵 cup pump;plunger
皮碗的环形螺帽 cup ring nut
皮碗垫圈 leather cup washer
皮碗阀 cup valve
皮碗活塞 leather bucket
皮碗密封 cup leather packing
皮碗式密封 cup packing
皮碗弹簧圈 expanding ring; packing expander;piston packing expander
皮碗填密 cup leather packing; cup packing
皮碗压板 follower
皮碗压板栽螺丝 follower stud
皮网箱 leather bellows
皮乌拉盆地 Piura basin
皮下的 subcutaneous
皮下盘菌属 <拉> Hypoderma
皮下氧化 subscale
皮线 covered wire; flex (cord); rubber-insulated wire; flexible cord; flexible wire【电】; electric(al) flex【电】
皮箱 luggage
皮鞋鞋跟 outsole
皮亚诺曲线 Peano curve
皮炎 dermatitis
皮油 Chinese(vegetable) tallow; haze tallow;haze wax
皮胀圈 leather cup
皮制板 leather board
皮制防渗帷幕 leather apron
皮制公事包 skin-making portfolio
皮制零星工具袋 wallet
皮制密封填料 leathering
皮重 bare weight; dead load; empty weight; tare; tare weight; weight empty
皮重和添头 tare and tret
皮重及重量损耗折扣 tare and draft

皮重率 percentage tare
皮重与总重比 ratio of tare weight to gross
皮重证明书 certificate of tare weight
皮舟式山谷 canoe-shaped valley

枇把(树) loquat

毗拱角距 apsidal angle

毗连 border(up) on; flank; juxtaposition; verge
毗连层次 adjacent layers; adjoining layers; associated layers
毗连车道 abutting lane
毗连船闸 adjacent locks
毗连的 abutting; adjacent; closely spaced; neighbo(u)ring
毗连地区 adjacent zone; contiguous zone
毗连房屋 abutting building; adjoining blocks
毗连国 contiguous state
毗连建筑物 abutting building; adjacent structure
毗连接头 abutting joint
毗连区间 interval abutting one another
毗连区间法 method of successive interval
毗连式汽车房 attached garage
毗连式住宅 terrace house
毗连水域 adjacent waters
毗连土地 adjoining land; adjoining lot
毗连线圈 adjacent coil
毗连渔区 continuous fishing zone
毗连支路 adjoint branch
毗连住房 attached dwelling
毗连住宅 attached house
毗邻的墙 adjoining wall
毗邻的停车场地 associated parking area
毗邻地产 adjoining premises
毗邻地块 adjacent land
毗邻地区 contiguous area; contiguous zone
毗邻房地产业主 abutting owner
毗邻海域 adjacent sea; adjacent waters
毗邻河流 contiguous river
毗邻环境评价及判断系统 neighbo(u)rhood environmental evaluation and decision system
毗邻环境噪声 neighbo(u)rhood noise
毗邻孔 adjoining hole
毗邻流域 adjacent catchment
毗邻碾压带 adjacent strip
毗邻区 adjacent area
毗邻生境 adjacent habitat
毗邻式汽车房 attached garage
毗邻水域 adjacent waters
毗邻停车场 associated parking area
毗邻岩层 adjoining stratum
毗邻展开数 adjacent spread's number
毗邻住房 attached dwelling
毗邻作业 contiguous operation; contiguous work

疲乏裂缝 fatigue crack; fatigue fracture

疲乏试验机 fatigue testing machine
疲劳 dark burn; distress; fatigue
疲劳(比)率 <物体疲劳强度和其静态抗拉强度的比率> fatigue ratio
疲劳变形 fatigue deformation
疲劳剥落 fatigue flake; spalling fatigue
疲劳剥蚀 fatigue pitting
疲劳补偿 fatigue allowance
疲劳部位 seat of fatigue
疲劳测定 fatigue measurement
疲劳产物 fatigue product
疲劳程度 fatigue degree
疲劳承载能力 fatigue capacity
疲劳持久极限 fatigue endurance limit
疲劳冲击强度 fatigue impact strength
疲劳导致破坏 fatigue-induced damage
疲劳抵抗力 fatigue resistance
疲劳点 fatigue point
疲劳点蚀 pit corrosion due to fatigue
疲劳电磁测定法 induflux method
疲劳毒素 fatigue toxin; kinotoxin
疲劳度 endurance degree; fatigue
疲劳度测定仪 fatigue meter
疲劳度指示器 fatigue indicator
疲劳断裂 endurance crack; endurance fracture; fatigue break(down); fatigue crack; fatigue failure; fatigue fracture; repeated stress failure <交变应力引起的>
疲劳断裂机理 fatigue fracture mechanism
疲劳反应 fatigue response
疲劳范围 fatigue range
疲劳分析 analysis of fatigue; fatigue analysis
疲劳腐蚀 fatigue corrosion
疲劳负荷(下使用)寿命 <试验时应力重复的次数> fatigue life
疲劳负载 endurance load
疲劳故障 wear-out failure
疲劳龟裂 fatigue crack(ing)
疲劳规律 regularity of fatigue
疲劳过度 overfatigue; over-strain
疲劳核 fatigue nucleus
疲劳荷载 fatigue load(ing); repeated load(ing)
疲劳荷载谱 fatigue load spectrum
疲劳毁损 <金属> fatigue failure
疲劳机理 fatigue mechanism
疲劳极限 endurance limit; fatigue limit; fatigue point; fatigue value; limit of endurance; limit of fatigue; safe range of stress
疲劳极限点 fatigue point
疲劳极限力矩 fatigue ultimate moment
疲劳极限状态 fatigue limit state; limit state of fatigue
疲劳记录计 ergograph
疲劳加载 fatigue load(ing)
疲劳剪切试验 fatigue shear test
疲劳剪切仪 fatigue shear apparatus
疲劳检查 fatigue test(ing)
疲劳界限 endurance limit; fatigue limit; fatigue point; limit of fatigue
疲劳开裂 fatigue crack(ing)
疲劳抗力 fatigue resistance
疲劳可靠度 fatigue reliability
疲劳累积损伤定律 cumulative fatigue damage law
疲劳裂缝 endurance crack; fatigue crack(ing); fatigue fracture; fracture by fatigue; tension crack
疲劳裂口 fatigue break
疲劳裂纹 endurance crack; fatigue crack; flex crack(ing)
疲劳裂纹尖端 fatigue crack tip
疲劳裂纹扩展试验 fatigue crack propagation test
疲劳裂纹扩张 fatigue crack propagation
疲劳裂纹增长 fatigue crack growth
疲劳描记器 ponograph
疲劳磨损 fatigue wear
疲劳耐久极限 fatigue endurance limit
疲劳耐久试验 fatigue endurance test
疲劳耐久性 fatigue endurance
疲劳耐限 endurance limit; fatigue limit
疲劳能力 fatigue capability
疲劳黏[粘]结强度 fatigue bond strength
疲劳破断 fatigue rupture
疲劳破坏 endurance failure; endurance fatigue; failure due to fatigue; fatigue breakdown; fatigue failure; fatigue rupture
疲劳破坏表面 surface of fatigue break
疲劳破坏的循环次数 cycles to failure
疲劳破坏概率 probability of fatigue failure
疲劳破坏前以循环次数表示的寿命 cycle life; cyclic(al) life
疲劳破坏区域 fatigue area
疲劳破坏应力 punish(ing) stress
疲劳破裂 endurance rupture; fatigue fracture; fracture by fatigue
疲劳破损 endurance failure; fatigue failure
疲劳期 fatigue life
疲劳起始 fatigue initiation
疲劳强度 endurance strength; fatigue limit; fatigue resistance; fatigue strength; limit of endurance; limit of fatigue; punish(ing) stress; strength depending on shape
疲劳强度比(率) fatigue ratio; endurance ratio
疲劳强度变化图 stress-endurance diagram
疲劳强度极限 fatigue endurance limit
疲劳强度极限与抗断强度极限之比 endurance ratio
疲劳强度计 fatigue meter
疲劳强度降低系数 fatigue strength reduction factor
疲劳强度曲线 Wohler curve
疲劳强度试验 Stromeyer test
疲劳强度衰减系数 <无应力集中时的疲劳强度和有应力集中时的疲劳强度之比> fatigue strength reduction factor
疲劳强度缩小系数 fatigue strength reduction factor
疲劳强度系数 fatigue strength coefficient
疲劳强度折减因子 fatigue strength reduction factor
疲劳曲线 curve of fatigue; fatigue curve; stress-cycle diagram; stress-endurance curve
疲劳容许度 fatigue allowance
疲劳设计 fatigue design
疲劳剩余寿命 fatigue remaining life
疲劳失效 fatigue failure
疲劳试验 durability test; endurance test; fatigue test(ing); grueling test; protracted test; repeated stress test; stress duration test
疲劳试验机 endurance testing machine; fatigue machine; fatigue tester; fatigue testing machine; protracted test machine
疲劳试验曲线图 fatigue testing diagram; S-N diagram
疲劳试验设备 fatigue rig; fatigue testing rig
疲劳试验数据 fatigue data
疲劳试验应力比 stress ratio from fatigue test
疲劳试验应力类型 fatigue testing types of stress
疲劳试验装置 fatigue rig
疲劳试样 fatigue(test(ing)) specimen
疲劳寿命 endurance life; fatigue endurance; fatigue lifetime
疲劳寿命片 fatigue life ga(u)ge
疲劳寿命循环数 life cycles; lift cycles
疲劳损耗 fatigue loss
疲劳损坏 fatigue breakdown; fatigue damage; repeated stress failure <交变应力引起的>
疲劳损伤指示器 fatigue damage indicator
疲劳损失 fatigue damage
疲劳损失累加 fatigue damage accumulation
疲劳特性 endurance character; fatigue behavio(u)r; fatigue characteristic; fatigue property
疲劳特征 endurance characteristic; fatigue characteristic
疲劳弯曲毁坏 endurance bending failure
疲劳弯曲破裂 endurance bending failure
疲劳弯曲试验 fatigue bend test
疲劳弯曲寿命 flexible life
疲劳弯折试验 endurance bending test
疲劳系数 coefficient of fatigue; endurance coefficient; endurance ratio; fatigue coefficient; fatigue factor; fatigue ratio
疲劳现象 appearance of fatigue; endurance phenomenon; fatigue effect; fatigue phenomenon
疲劳线 striation
疲劳限度 endurance limit; fatigue allowance; fatigue limit
疲劳限度范围 fatigue range
疲劳限渐增现象 coaxing
疲劳响应 fatigue response
疲劳效率减退界限 fatigue decreased proficiency boundary
疲劳效应 fatigue effect
疲劳性故障 fatigue failure
疲劳性能 fatigue performance; fatigue property
疲劳性损坏 fatigue failure
疲劳性质 endurance behavio(u)r
疲劳循环数 fatigue life
疲劳研究 fatigue study
疲劳因素 fatigue factor
疲劳引发 fatigue initiation
疲劳应变 repeated strain
疲劳应力 fatigue stress
疲劳应力比(值) fatigue stress ratio
疲劳应力集中系数 fatigue stress concentration factor
疲劳硬化 fatigue hardening
疲劳有效温度 fatigue effective temperature
疲劳预测(值) fatigue prediction
疲劳预裂 fatigue precracking
疲劳阈值 fatigue threshold
疲劳折减系数 fatigue reduction factor
疲劳折减因子 fatigue reduction factor
疲劳作用 endurance action; fatigue action
疲软的 workout
疲软的市场 easy market; soft market; weak market

啤酒厂 brewery; brewhouse

啤酒厂残渣 brewery residues
啤酒厂废料 brewery waste
啤酒厂废水 brewery effluent; brewery waste; brewery wastewater
啤酒厂废物 brewery waste
啤酒厂混合废水 mixed brewery

wastewater
啤酒车 beer wagon
啤酒地窖 beer cellar
啤酒店 beer hall;beer shop;brasserie;beer house <英>
啤酒店花园 beer garden
啤酒馆前面长凳 ale bench
啤酒冷却器 beer cooler
啤酒酿造 beer brewing
啤酒酿造废水 brewery process liquor;brewing process wastewater
啤酒桶涂料用沥青 brewer's pitch

琵

琵桶 hops

琵琶头 <绳结> eye splice

匹

匹配 accordance; accouplement; adaptation; commensurate; couple;coupling;match(ing);mate

匹配板 matching disc
匹配棒 matching pillar;matching post
匹配包层光纤 matched cladding fibre [fiber]
匹配比较 matched comparison
匹配比滤波器 matched filter
匹配变压器 matching transformer
匹配变阻器 matching transformer
匹配部件 matching parts;mating part
匹配测量计 adaptometer
匹配插头 matching plug
匹配齿轮 mating gear
匹配穿孔卡 matching punch cards
匹配传输线 matched transmission line
匹配窗 matching window
匹配窗口片 matching diaphragm
匹配磁铁 matching magnet
匹配的传输线 matched transmission line
匹配的存储体系 matching storage hierarchy
匹配点 matching point
匹配电流 matching current
匹配电路 building-out circuit;match(ing) circuit
匹配电阻 build-out resistor;matched load resistance
匹配短截线 matching stub;pill transformer
匹配短线 matching stub
匹配段 matching section
匹配放大器 adapter amplifier;matching amplifier
匹配封接 matched seal;match sealing
匹配封装 envelope match
匹配辐热检波器 matched bolometer detector
匹配负载 matched load;matching load
匹配负载电阻 matched load resistance
匹配负载功率 available power;matched load power
匹配负载线 match-terminated line
匹配罐 matching can
匹配光学 matching optics
匹配过渡 matching transition
匹配焊钢 <焊条钢料抗拉强度不小于母体钢材> matching weld steel
匹配焊接面积 matching bond area
匹配荷载 matched load
匹配盒 matching box;matching cell
匹配弧 match arc
匹配机 matcher
匹配记录指示器 matching record indicator
匹配技术 matching technique
匹配渐近展开式 matched asymptotic expansions
匹配阶层 matching hierarchy
匹配接收天线 matched receiving antenna
匹配节 matching section
匹配可变光阑 matching iris
匹配理论【数】matching theory
匹配连接 accordant connection; matched junction
匹配连接时的损耗 structural return loss
匹配滤波 matched filtering;matching filtering
匹配滤波器 matched filter;matching filter
匹配滤光片探测 matched filter detection
匹配脉冲 matching pulse
匹配脉冲拦截 matched pulse intercepting
匹配模锻斜度 matching draft
匹配膜片 <波导的> matching plate
匹配能力 adaptability
匹配器 adapter; coupler; matcher; matching box;matching unit
匹配三合透镜 matching triplet
匹配时间 match time
匹配式热照射检测器 matched bolometric(al) detector
匹配试管 matched test tube
匹配输出 matched output
匹配束 matched beam
匹配数据 matched data
匹配衰减器 matched attenuator
匹配体积 matched volume
匹配条件 match(ed) condition;matching condition
匹配同轴电缆 matching coaxial cable
匹配透镜 matched lens
匹配外壳 matching can
匹配网络 matching network
匹配问题 marriage problem;matching problem
匹配误差 matching error
匹配系数 matching coefficient
匹配显示 <有辅助调时机件的信号控制机> parent phase;match display
匹配线 matched line
匹配线圈 matched coil
匹配箱 match(ing) box
匹配信息组 matching field
匹配选择器 matcher-selector-connector
匹配延迟线 matching delay-line
匹配用电阻网络 matching resistance network
匹配用阻抗 matching impedance
匹配元件 matching element;matching unit
匹配元件特性 matching components characteristic
匹配原理 matching principle
匹配圆锥线法 matched conic(al) method
匹配运算 matching operation
匹配振荡器 matched generator
匹配终端 reflexless terminal
匹配终端负载 non-reflecting termination
匹配终端器 matched termination
匹配柱 matched column
匹配装置 co-alignment;matching device;matching unit
匹配准确度 matching accuracy
匹配准则 matching criterion
匹配自耦变压器 matching autotransformer
匹配阻抗 matched impedance
匹配阻抗天线 matched impedance antenna
匹染 piece dy(e)ing
匹染织物 piece-dyed fabric
匹头货 piece goods
匹兹堡加价法 Pittsburgh plus pricing

苉

苉 picene

辟

辟雍砚 ink-stone with water annulus

癖

癖好 hobby;predilection

癖性 aptitude

僻

僻地 nook

僻径 by-path;by-walk;byway
僻静的住处 hermitage
僻静娱乐处 hide-away
僻巷 by-lane

偏

偏辊轧机 leaning 8-rolls mill

偏摆 beat
偏摆角 angle of drift
偏般曲线 deviation curve
偏般阻尼器 directional damper
偏苯三酸 trimellitic acid
偏苯三酸酐 trimellitic anhydride
偏苯三酸盐 trimellitate
偏辟的 withdrawn
偏变换式 partial transform
偏柄螺纹梳刀架 offset shank holder for chasers
偏薄板 shaving board;shim
偏不安全误差 unsafe error
偏不等系数 partial inequality coefficient
偏测定系数 coefficient of partial determination
偏差 aberration; affect departure; bending; bias; crab【船】; declination; deflection; deviate; disagreement; discrepancy [discrepance]; divergence; drift; inaccuracy; partial increment; variation; residual; residuary【数】; bias(s)ed error <偏离于常规正值或常规负值的误差>
偏差倍增器 deflection multiplier
偏差比的方差 variance of deviation ratio
偏差标准 deviation standard
偏差补偿器 deviation compensator
偏差参数 straggling parameter
偏差测定计 <测量目标实际位置的偏差> train meter
偏差测定器 deviation detector
偏差测量 deviation measurement
偏差测量法 deflection method of measurement
偏差测量仪 drift meter
偏差查寻系统 error system
偏差常数 deviation constant
偏差点 deviation point
偏差度 degree of deviation
偏差法 method of deviation
偏差法试射 magnitude method
偏差范围 deviation range; extent of the error
偏差分 partial difference
偏差分方程 partial difference equation
偏差分量 offset component
偏差分析 variance analysis
偏差积分 variance integration
偏差计 deflectometer; deviometer; drift ga(u)ge;drift meter
偏差计算机 deviation computer;drift computer
偏差计算器 deviation calculator
偏差记录仪 deviation recorder
偏差检测装置 error measuring means
偏差角度 angle of deviation; deflection;deviation angle
偏差角方向 direction of angle of deviation
偏差校正 drift correction;offset correction
偏差校正器 parallax correction setter
偏差校准曲线 deviation calibration curve
偏差均衡器 design deviation equalizer;deviation equalizer
偏差控制 deviation control;tolerance control
偏差量 amount of deviation;amount of dispersion
偏差灵敏度 deviation sensitivity
偏差率 rate of deviation change
偏差器 declinator;declinometer;deviatoric
偏差曲线 aberration curve;departure curve
偏差上限 upper deviation
偏差失真 deviation distortion
偏差数 deviation number
偏差数值 amount of deflection
偏差调整 deviation adjustment
偏差调整法 offset method
偏差调整楔 <X 射线带钢测厚仪部件> deviation wedge
偏差调制 deflection modulation
偏差系数 change factor;coefficient of asymmetry; coefficient of deflection;coefficient of deviation;coefficient of skewness;deviation factor
偏差下限 lower limit of variation
偏差显示 deviation display
偏差修正 deviation correction
偏差压力 deviator stress
偏差异常 deflection anomaly
偏差应力 deviatoric stress;differential stress
偏差源 aberrant source
偏差值 amount of deflection;amount of deviation;deviant
偏差(值)比 discrepancy ratio
偏差指示器 deviation indicator;drift marker
偏差指向标 beacon for compass adjustment;deviation beacon
偏差钻孔 drift(bore) hole
偏出航向 fall off;off-course
偏穿孔 off-punch
偏垂的 hading
偏垂角 hade
偏磁 bias;magnetic biasing
偏磁场 bias field
偏磁串扰 bias crosstalk
偏磁技术 biasing technique
偏磁震荡 bias oscillator
偏大近似值 excessive approximate value
偏荡 slew;slue;yawing
偏荡波道放大器 yawing amplifier
偏荡幅度 yawing amplitude
偏荡计 yawmeter
偏荡角 yawing angle
偏荡力矩 yawing moment
偏荡平衡 yawing balance
偏刀 offset tool;side tool
偏导法 partial differentiation
偏导角 deflector angle
偏导矩阵 matrix of partial derivatives
偏导器【机】deflector
偏导数 local derivative; partial derivation; partial derivative; partial

P

quotient
偏导装置 deflector
偏地 eccentrically
偏等性 aeolotropism;aeolotropy
偏低 downward bias
偏低场 biased low-strength field
偏低的估计 conservative estimate
偏低设计 <安全系数不足的> under-design
偏底 flanged bottom
偏点 by point
偏电压 biasing voltage
偏东南的 southeasterly
偏动电铃 bias bell;biased ringer
偏动簧片 biasing spring
偏动态 klinokinesis
偏动弹簧 biasing spring
偏动装置 bias gear
偏度 bias;skewness
偏度表 deflection indicator
偏度测量 measure of skewness
偏度峰度检验 test of skewness and kurtosis
偏度计 deflection indicator;deflectometer
偏度系数 asymmetry coefficient;bias coefficient;coefficient of skewness
偏对称迁移 asymmetric(al) migration
偏多重相关系数 partial multiple correlation coefficient
偏二氟乙烯 vinylidene fluoride
偏二氯乙烯 vinylidene chloride;vinylidene
偏二氯乙烯树脂 polyvinylidene resin;vinylidene resin
偏二氯乙烯塑料 vinylidene chloride plastic
偏钒酸 metavanadic acid
偏钒酸铵 ammonium metavanadate
偏钒酸钠 sodium metavanadate
偏反光两用显微镜 polarized-reflected light microscope
偏方 folk prescription
偏方差 partial variance
偏方二十四面(晶)体 diploid
偏方复十二面体 diploid
偏方三八面体 trapezohedron
偏方四分面体 trapezohedral tetartohedron
偏方四分面像 trapezohedral tetartohedry
偏房 outhouse
偏分母 partial denominator
偏分析 partial analysis
偏锋齿 briar tooth;gullet tooth
偏幅 beat
偏杆摇把 crank handle with offset lever
偏高场 biased strength field
偏高碘酸 metaperiodic acid
偏高碘酸铵 ammonium metaperiodate
偏高碘酸钾 potassium metaperiodate
偏高碘酸钠 sodium metaperiodate
偏高岭石 metakaolinite
偏高岭土 metakaolin
偏铬橄榄棕 metachrome olive brown
偏铬红 metachrome red
偏铬黄 metachrome yellow
偏铬蓝 metachrome blue
偏铬酸 metachromic acid
偏铬枣红 metachrome bordeaux
偏共晶 monotectic
偏共晶合金 monotectic alloy
偏共晶平衡图 monotectic equilibrium diagram
偏共振去耦法 polarizing resonance decoupling method
偏沟 gutter
偏估计量 bias(s)ed estimator
偏光 light bias;polarization;polarized light;polar light
偏光比色计 polarization colo(u)rimeter
偏光玻璃 deflecting glass;light deflecting unit;polarized glass;spread glass
偏光部分 light deflecting part
偏光测试 polariscope test
偏光反射镜 polarizing mirror;polarizing reflector
偏光放大镜 polarization-magnifying lens
偏光干涉 polarizing interference
偏光干涉显微镜 polarization interference microscope
偏光计 polarimeter;polariscope
偏光镜 deflecting lens;deflecting roundel;Nicol's prism;polarizer;polarizing glass;polarizing screen;polarizing spectacle;polaroid glass;spreadlight lens;spreadlight roundel;stereo-visor
偏光镜应力仪 polarimeter
偏光滤光器 polarized light filter
偏光面 polarization plane
偏光片 diffuser;polarizer
偏光器 polarimeter;polariscope;polarizer
偏光色 polarization colo(u)r
偏光弹性试验 polar photoelastic test
偏光弹性学 polarization photoelasticity
偏光透镜 deflecting lens;hot spot;hot spot lens;light deflecting unit;spreadlight lens
偏光图 polarization figure
偏光显微镜 microphotometer;petrographic(al) microscope;polarization microscope;polarizing microscope
偏光显微镜技术 petrographic(al) microscopy
偏光显微镜检光板 test plate
偏光显微镜鉴定 polarizing microscope identification
偏光显微(镜)术 polarized light microscopy;polarizing microscopy
偏光显微镜速 polarized microscope
偏光性 polarity;polarization character
偏光眼镜 polaroid glasses
偏光因子 polarization factor
偏光(应力)仪 polariscope
偏光荧光计 polarization fluorometer
偏硅酸 metasilicic acid
偏硅酸钙 calcium metasilicate
偏硅酸钴 cobalt metasilicate
偏硅酸锂 lithium metasilicate
偏硅酸钠 sodium metasilicate
偏硅酸铅 lead metasilicate
偏硅酸铁 ferric metasilicate
偏硅酸锌 zinc metasilicate
偏硅酸盐 bisilicate;metasilicate
偏硅酸盐类 bisilicate;metasilicate
偏轨箱形梁 bias rail box girder
偏航 yaw(ing);yawing motion;bias;crab;drift(age);off-course
偏航安定性 direction stability;yaw stability
偏航报告 deviation report
偏航不稳定性 yawing instability
偏航操纵 yaw control
偏航操纵机构 yawer
偏航程 amount of drift
偏航传感器 yaw detector;yawhead
偏航动作 yaw maneuver
偏航惯性矩 moment of inertia in yaw
偏航滚转动作 yaw-roll maneuver
偏航滚转陀螺仪 yaw-roll gyroscope
偏航航迹 off-course track
偏航回路 yaw loop
偏航机动性 yawing maneuverability
偏航计 drift meter;yawmeter
偏航加速度 acceleration in yaw;yaw acceleration
偏航加速度自记仪 yaw-sensing accelerograph
偏航角 amount of yaw;angle of drift;angle of yaw;crab angle;drift angle;yaw(ing) angle;leeway
偏航角传感器 yaw-angle pickoff;yaw-angle sensor
偏航角控制 drift angle control
偏航角速度 yaw rate
偏航角速度控制 yaw rate control
偏航角指示器 drift angle indicator;yaw indicator
偏航校正 drift correction
偏航校正仪 yaw calibrator
偏航警报器 off-course alarm
偏航控制 yaw control
偏航控制器 yawer
偏航控制通道 yaw control channel
偏航力矩 drifting moment;moment in yaw;yawing;yawing moment
偏航灵敏度 angular deviation sensitivity;lateral sensitivity;yaw sensitivity
偏航率陀螺仪 yaw rate gyroscope
偏航敏感度 angular deviation sensitivity;lateral sensitivity
偏航模拟机 yaw simulator
偏航频率 yaw frequency
偏航平面 plane of yaw
偏航时间 drift time
偏航条款 deviation clause
偏航同步机 yaw synchro
偏航同步器 yaw synchronizer
偏航陀螺仪 yaw gyroscope
偏航物体 yawed body
偏航误差 yaw error
偏航信号 off-course signal
偏航修正 drift correction;off-course correction
偏航翼 yaw vane
偏航引起的力矩 moment due to yawing
偏航与俯仰 yaw and pitch
偏航与航速指示器 drift and speed indicator
偏航运动 yawing rotation
偏航(运动)加速表 yaw-sensing accelerometer
偏航指示 off-course indication;off-course sensing
偏航指示器 course deviation indicator;deflection indicator;deviometer;drift indicator;yawmeter
偏航指示区 clearance sector
偏航轴 yaw(ing) axis
偏航姿态 yaw-position
偏航自由度 yaw freedom
偏航阻尼器 yaw damper
偏好的顺序 ordering of preference
偏好曲线 preference curve
偏回归方程 partial regression equation
偏回归系数 partial regression coefficient
偏火 deflected burning
偏火自动控制系统 automatic deflected burning control system
偏积分 partial integration
偏畸变 bias distortion
偏畸变计 bias meter
偏极继电器 biased relay;polar-biased relay;pole biased relay
偏极天线 bipole antenna
偏寄生营养 paratrophy
偏碱性 partial alkalinity
偏见 bias;prejudice
偏胶质 metacolloid
偏角 angle of deflection;angle of deviation;angle off;angle of horizontal swing;angular deviation;cutting edge angle;declination;deflection;depression angle;deviation angle;edge angle;fore-and-aft tilt;offset angle;swing angle;angle of avertence
偏角测量 deviational survey
偏角磁强计 deflection magnetometer
偏角导线 <观测前一条边的延长线与下一条边之间的水平角> deflection angle traverse
偏角法 deflection angle method;method of deflection angles
偏角计 declinator;declinometer
偏角曲线 deflection angle curve
偏角调节 angling
偏角误差 declination error
偏角仪 declinometer;deflection ga(u)ge
偏角增大 drift angle build up
偏近点角 eccentric anomaly
偏晶的 monotectic
偏距 deflection distance;offset distance;setover
偏距尺 <测量用短杆> offset staff
偏距法 method of deflection distance;offset method
偏距光学直角头 offset optic(al) square
偏距螺钉 setover screw
偏跨 lean-to
偏浪航行 steaming with the sea on the bow
偏离 deflection;departure;deviate;deviation;disagreement;divergence;drift;out of alignment;straggling;swerve
偏离爆心投影点 offset ground zero
偏离标志 deviation beacon
偏离标准的 deviant
偏离不定度 bias uncertainty
偏离磁道 offset track;offtrack
偏离磁迹 offtrack
偏离的 deflective;off-line;off-normal
偏离等航线 departure from isochronism
偏离点 point of deviation
偏离度 degree of deviation;irrelevance
偏离对称 departure from symmetry
偏离额定值 excursion
偏离法 offset method
偏离规定路线 deviated line;deviated traffic
偏离轨道 offset track;offtrack
偏离轨道误差 off-track error
偏离航程 deviation from voyage route
偏离航道 off channel
偏离航迹误差 cross-track error
偏离航路 off-airway
偏离航线 off-course
偏离航向 off-course;offset direction;off the course
偏离航向指示 off-course indication
偏离化学计量 non-stoichiometry
偏离计 diasporometer
偏离角 angle of alternation;angle of departure;deflection angle;divergence angle;slip angle;take-off angle
偏离角方向 direction of angle of deviation
偏离进度表 departure from schedule
偏离临界的量 off-critical amount
偏离率 bias ratio

偏离配平位置的 mistrimmed
偏离平均值 departure from mean value
偏离铅锤量 variations from plumb
偏离曲线 deflection curve; departure curve
偏离设计工况 off-design
偏离设计条件 off-design condition
偏离设计条件的 off-design
偏离失真 bias distortion
偏离市场价格 deviation of market price
偏离数据 bias data
偏离图 slip chart
偏离误差 biased error
偏离吸收 deviation absorption; off-course absorption
偏离行态 offset behavio(u)r
偏离性能 offset behavio(u)r
偏离许可 deviation permit
偏离预定位置 out-of-position
偏离载波 offset carrier
偏离正常顺序的服务 out-of-sequence service
偏离值 stray value; variance
偏离指示 departure indication; deviation beacon
偏离指示器 deviation indicator; departure indicator
偏离中心 decentralization; off-center[centre]; out of center[centre]
偏离中心的 off-centered
偏离中心偶极子 off-centered dipole
偏离中心水眼 off-center waterway
偏离中心线 disalignment
偏离中心线的挖掘 offset digging
偏离重心位置的载重 out-of-balance weight
偏离轴心 disalignment
偏离轴心力 eccentric axial force
偏离主题 branch out
偏利共栖 commensalism
偏量 deviator; deviatoric
偏磷酸钠 sodium metaphosphate
偏磷酸盐 metaphosphate
偏岭石 pianlinite
偏菱形波痕 overhanging ripple
偏流 bias current
偏流板 deflector plate
偏流表 bias meter
偏流测视器 drift sight
偏流差 drift error
偏流的冲击作用 impulse of deviated flow
偏流反力 deflecting force
偏流观测 drift observation
偏流航角 crab
偏流计 bias meter; drift indicator; drift meter; drift sight; yawmeter
偏流检波 bias detection
偏流角 angle of crab; angle of yaw; crab; crab angle; drift angle
偏流角指针 crab index
偏流校正 drift correction
偏流空气 deflected air
偏流喷射式 deflector jet
偏流气闸 deflecting damper
偏流器 deflector gear; flow diverter; jet deviator; water deflector
偏流系数 drift coefficient
偏流线修正 exline correction
偏流消能设备 baffle
偏流修正 drift correction
偏流依存关系 dependence on bias current
偏流针形喷嘴 deflecting needle nozzle
偏铝酸钙 calcium meta-aluminate
偏铝酸盐 meta-aluminate
偏氯纶树脂 Saran
偏氯乙烯树脂 permalon
偏码 <集装箱> offset stacking
偏锰酸矿 vernadite
偏摩尔体积 partial molal volume
偏摩尔自由能 partial molal free energy
偏磨 eccentric wear
偏南的 meridional
偏扭 deflection eclipse; deflection skew
偏硼石 metaborite
偏硼酸 metaboric acid
偏硼酸钡 barium metaborate
偏硼酸锂熔融 fusion with $LiBO_2$
偏僻处 byplace; seclusion
偏僻村镇 <距交通线较远的、人口稀少的居民区> back country
偏僻的 obscure; out-of-the-way
偏僻的农村地区 back country
偏僻地带 nook
偏僻地点 remote point
偏僻地方 byplace; nook
偏僻地区 back block; inaccessible area; isolated area; out-of-the-way district; out-of-the-way region; remote area; remote district; remote location; remote region; tall timber
偏僻地区罐瓶站 remote filling point
偏僻地区水库 remote reservoir
偏僻港附加费 minor port surcharge; outport surcharge
偏僻港增额 minor port surcharge
偏僻小城镇 podunk
偏僻小吃店 hide-away
偏僻小路 byway
偏僻之处 inaccessible area
偏僻住宅 back residence block; back residential block; back residential building
偏偏 must needs
偏频 offset frequency
偏平面 deviatoric plane
偏企口接合 eccentric table joint
偏牵引 <与运输方向偏移的> offset pull
偏曲 lean
偏曲凸轮 deflecting cam
偏绕角 fleet angle
偏三角面体 scalenohedron
偏色 colo(u)r cast
偏商 partial quotient
偏上生长 epinasty
偏射 deflection
偏湿压实 <相对于最优含水率的> wet compaction
偏蚀 partial eclipse
偏食 <日、月> partial eclipse
偏态 skewness
偏态度量 measure of skewness
偏态分布 distribution of skewness; skew(ed) distribution
偏态频率分布 skew frequency distribution
偏态频率曲线 skew frequency curve
偏态曲线 skew curve
偏态系数 coefficient of skewness; skewness coefficient
偏态因数 skew factor
偏态因素 skew factor
偏态直方图 skewed histogram
偏钛酸钙 calcium metatitanate
偏钛酸盐 metatitanate
偏梯形扣接箍 buttress thread coupling
偏梯形扣套管 buttress thread casing
偏提取方法 partial extraction method
偏贴向岸作用 bank suction
偏头螺丝刀 offset screwdriver
偏途顶极群落 disclimax
偏途演替顶级 plagioclimax
偏途演替系列 plagiosere
偏危险误差 unsafe error
偏微分 partial differential; partial differentiate
偏微分法 partial differentiation
偏微分方程式【数】 partial differential equation
偏微分方程语言 partial differential equation language
偏微分矩阵 partial derivative matrix
偏微分算子 del
偏微商 local derivative; partial derivative
偏位 asymmetry; de-symmetry; deviation; off-normal; offsetting
偏位传感器 position error probe
偏位错 partial dislocation
偏位角 angle of displacement; axial rake <轴的>; displacement angle
偏位警报 off-location alarm
偏位框 error box
偏位裂解 meta-cleavage
偏位灵敏度 deviation sensitivity
偏位明沟 offset ditch
偏位明渠 offset ditch
偏位屈服强度 offset yield strength
偏位酸【化】 meta-acid
偏位线圈 misplaced winding
偏位指示器 bearing indicator
偏位桩 deflected pile
偏屋 lean-to
偏屋脊坡屋顶 broken gable roof
偏无烟煤 meta-anthracite
偏西距离 westing
偏析 segregate; segregation
偏析的 monotectic
偏析点 segregation spot
偏狭的 parochial
偏相关 partial correlation
偏相关比 partial correlation ratio
偏相关分析 partial correlation analysis
偏相关系数 coefficient of partial correlation; partial correlation coefficient
偏向 deflection; deflectivity; deflexion; deviate; deviation; incline; lean; swerve
偏向飞行 declination flying
偏向分离 preferential segregation
偏向估计 biased estimate
偏向光楔 <测距仪> deviating wedge
偏向角 angle of deflection; angle of deviation; deflection angle; offset angle; declination angle
偏向力 angular force; deflecting force; deflective force; deviating force
偏向力矩 yawing moment
偏向流动 deviated flow
偏向螺丝起子 offset screw
偏向配对 preferential pairing
偏向器 deflector
偏向曲线 declination curve
偏向调制 deflection modulation
偏向误差 biased error
偏向吸收 deviation absorption
偏向一边 sway
偏向一侧 amesiality
偏向运动 <汽车绕平面轴线的> yaw
偏向针形管嘴 deflecting needle nozzle
偏向针形喷嘴 deflecting needle nozzle
偏向装置 deviator; deviatoric device
偏小近似值 lower approximate value; lower approximation value
偏斜 angularity; crab; declination; declining; deflection; deviate; fining-away; inclination; joggle; run-out; skew; squint; tilt; going off <指钻孔钻斜>
偏斜边界条件 skewed boundary condition
偏斜齿轮系 skew axis gear
偏斜出错检测 skew error detection
偏斜的 deflective; divergent; oblique
偏斜地块【地】 tilted block
偏斜度 degree of inclination; degree of skewness; measure of skewness; skewness
偏斜对位 <集装箱的> angular spotting
偏斜方位 direction of deflection
偏斜方向 direction of deflection
偏斜分布 skewed distribution
偏斜改正 <测量断面> correction for angularity
偏斜概率曲线 skew probability curve
偏斜杆 partial diagonal
偏斜钢丝束 harped tendons
偏斜光线 skew ray
偏斜回归 skew regression
偏斜计 deflectometer
偏斜角 angle of declination; angle of deviation; angle of rake; crab angle; deflection angle; deflexion angle; oblique angle; offset angle; outset angle
偏斜校正 angularity correction; skew correction
偏斜井 crooked well; deflected well; deflecting well; deviating well; deviation hole
偏斜卡环 inclined ring
偏斜孔 deflected borehole; deviating hole; drifted borehole
偏斜孔岩芯 deflecting core
偏斜孔钻进 deflected drilling; inclination drilling
偏斜密度函数 skewed density function
偏斜频率曲线 skew frequency curve
偏斜器 deflecting wedge; whipstock
偏斜倾向断层 semi-transverse fault
偏斜曲线 skew curve
偏斜曲轴 offset crankshaft
偏斜容限 <试验行车偏离指定横向地位的容许限度> bias(s)ed tolerance
偏斜失真 skew distortion
偏斜式动臂 offset boom
偏斜水流 deviated flow; inclined flow
偏斜误差 skewed error
偏斜系数 coefficient of skewness; skewness coefficient
偏斜现象 skewness
偏斜限制器 skew limiter
偏斜相关 skew correlation
偏斜像片 wing photograph
偏斜楔玫瑰形钻头 wedge rose bit
偏斜靴【岩】 deflecting shoe
偏斜岩块【地】 tilted block
偏斜(直)线【数】 skew line
偏斜指示器 deflection indicator; inclination indicator; skew indicator
偏斜走向断层 semi-longitudinal fault
偏斜钻进 deviated drilling; side-tracking
偏斜钻孔 gone off hole
偏斜钻头 deflecting bit
偏心 eccentric center[centre]; decentration; mass eccentricity; off-center[centre]; off centering; offset; out of center[centre]; wandering heart <木材髓心偏离>
偏心安置 eccentric set-up
偏心扒杆式输送螺旋 disappearing finger auger
偏心半径 eccentric radius

偏心棒 eccentric rod
偏心泵 eccentric driven pump; eccentric rotary pump
偏心比率 eccentric ratio
偏心闭锁机 eccentric-screw breech-block
偏心臂 eccentering arm; eccentric arm
偏心变向器 eccentric rebel tool
偏心标志 eccentric signal
偏心测角 angle measurement an excentre; eccentric angle measurement
偏心测站 eccentric surveying point; eccentric station
偏心差 centering error; eccentric error; equation of center[centre]; error of eccentricity
偏心觇标 eccentric sighting mark; eccentric signal
偏心场 off-centering field
偏心车床 eccentric lathe
偏心车削 eccentric turning
偏心衬垫 eccentric bushing
偏心衬筒 eccentric tensioner
偏心承载基础 eccentric loaded foundation
偏心承载柱 eccentric loaded column
偏心齿轮 eccentric gear
偏心齿轮传动 eccentric gear drive
偏心冲床 eccentric shaft press
偏心传动 eccentric drive; eccentric gearing
偏心传动筛 eccentric drive screen
偏心传动装置 eccentric gear; radial gear
偏心锤 eccentric weight
偏心磁极 eccentric pole
偏心大小头 eccentric reducer
偏心带边异径管箍 beaded eccentric reducing socket
偏心刀架 eccentric cutter holder
偏心的 decentered; eccentric; non-central
偏心电路 off-center circuit
偏心吊筛 gyratory riddle
偏心动作 eccentric action
偏心度 amount of eccentricity; degree of eccentricity; misalignment; eccentricity
偏心度公差 eccentricity tolerance
偏心度系数 eccentricity factor
偏心颚式破碎机 eccentric jaw crusher
偏心鲕 eccentric ooid
偏心阀 eccentric valve
偏心阀座磨床 eccentric valve seat grinder
偏心方向 eccentric direction
偏心分量 eccentric compound
偏心负荷 eccentric load
偏心负载 eccentric load(ing); non-central load; off-center load(ing)
偏心改正 correction for eccentricity; eccentric reduction
偏心杆 eccentric arm; eccentric bar; eccentric lever; eccentric mandrel; eccentric rod
偏心杆销 eccentric rod pin
偏心杆座 eccentric rod seat
偏心缸回转油泵 eccentric cylinder rotary oil pump
偏心钢筋束 eccentric tendon; draped tendon
偏心杠杆 eccentric lever
偏心固定夹 eccentric clamp; eccentric strap
偏心观测 eccentric observation
偏心观察 eccentric observation
偏心管 eccentric pipe; offset pipe; offset piping
偏心管接头 eccentric fitting
偏心管钳 eccentric tongs
偏心轨道 eccentric orbit
偏心滚动拌和机 eccentric tumbling mixer
偏心滚轮 eccentric roller
偏心滚柱 eccentric roller
偏心滚子弯轨器 eccentric roller rail bender
偏心荷载 eccentric load(ing); non-central load; off-center load(ing)
偏心荷载的 eccentrically loaded
偏心荷载基础 eccentrically loaded footing
偏心荷载柱【建】eccentrically loaded column
偏心荷载转轴 eccentrically loaded rotating shaft
偏心荷载桩 eccentrically loaded pile
偏心荷重 non-central load; off-center load(ing)
偏心滑轮 eccentric pulley
偏心滑轮制动螺钉 eccentric sheave set
偏心滑叶泵 eccentric sliding vane pump
偏心环 eccentric band; eccentric ring; eccentric strap
偏心环衬 eccentric-strap liner
偏心环空 eccentric annulus
偏心环清除塞 eccentric-strap cleaning plug
偏心环式危急保安器 eccentric ring emergency governor
偏心环形山构造 eccentric ring structure
偏心环油杯 eccentric-strap oil cup
偏心环油槽 eccentric-strap oil pocket
偏心活塞环 eccentric piston ring
偏心机构 eccentric gear; eccentric mechanism
偏心基础 eccentric foundation; offset footing; offset foundation
偏心畸变 decentering distortion
偏心集中荷载 single non-central load
偏心计 eccentricity indicator
偏心记录仪 eccentricity recorder
偏心加固的 eccentrically stiffened
偏心加劲杆 eccentric stiffener
偏心加载 eccentric loading; off-center load(ing)
偏心夹 cam-lock
偏心夹环 eccentric clip
偏心夹具 eccentric fixture; lever clamp
偏心剪切力 eccentric shear force
偏心渐变器 eccentric reducer
偏心渐缩管 eccentric reducer
偏心角 angle of eccentricity; angular eccentricity; eccentric angle
偏心矫正装置 eccentricity correction device
偏心校正 correction for eccentricity; eccentric correction
偏心铰弧形闸门 radial gate with eccentric pivot
偏心接箍 eccentric yokes
偏心接合 eccentric connection; eccentric joint
偏心结合 eccentric connection
偏心进给运动 eccentric feed motion
偏心径向扫描 off-centered radial sweep
偏心距 eccentric distance; linear eccentricity; offsetting; eccentricity
偏心距检查仪 eccentricity tester
偏心距离 eccentric distance; offsetting distance; throw
偏心距增大系数 magnifying coefficient of eccentricity
偏心卡盘 eccentric chuck
偏心开槽式圆筒 cam drum
偏心孔板 eccentric orifice
偏心块 eccentric block
偏心拉力 eccentric tension
偏心拦阻减速 off-center arrestment
偏心力 eccentric force
偏心力矩 eccentric moment; eccentric torque; moment of eccentricity
偏心连杆 eccentric link
偏心联结 eccentric connection
偏心梁柱连接 eccentric beam-column connection
偏心律 linear eccentricity
偏心率 degree of eccentricity; eccentricity; factor of eccentricity; malalignment; misalignment; numeric(al) eccentricity; run-out
偏心率比 eccentricity ratio
偏心率改正 eccentricity correction
偏心轮 cam; eccentric; eccentric sheave; eccentric wheel; pivoting eccentric; wobbler
偏心轮泵 eccentric pump
偏心轮叉 eccentric fork
偏心轮叉子 eccentric gab
偏心轮衬面 eccentric liner
偏心轮传动 eccentric drive
偏心轮紧固柱螺栓 eccentric fastening stud
偏心轮螺栓 eccentric bolt
偏心轮配气 gab-motion
偏心轮偏心度 throw of eccentric
偏心轮式泵 eccentric pump
偏心轮销 eccentric key
偏心轮行程 eccentric throw; throw of eccentricity
偏心轮与连杆传动装置 eccentric-and-pitman drive
偏心轮运动 wobbler action
偏心轮执行机构 tapped actuator
偏心轮轴 camshaft; eccentric shaft
偏心螺杆泵 eccentered screw pump
偏心螺母 eccentric nut
偏心铆钉 off-center rivet
偏心铆接 eccentric riveted joint
偏心门 di-axis door
偏心磨损 eccentric wear
偏心目标 eccentric object
偏心啮合 overcenter engagement
偏心偶极 eccentric dipole
偏心偶极子 off-centered dipole
偏心盘 cam disc[disk]; cam gear
偏心配产器 eccentric production mandrel
偏心配件 eccentric fittings
偏心配水器 eccentric injection mandrel; eccentric water distributor
偏心配置 off-center arrangement
偏心配重 eccentric balance-weight
偏心喷燃管 eccentric burner; off-center burner
偏心喷溢 eccentric eruption
偏心碰撞 eccentric impact
偏心皮带轮 eccentric pulley; eccentric sheave
偏心皮带(轮键槽) eccentric sheave key way
偏心平衡锤 eccentric balance-weight
偏心平面位置显示器 off-center plan(e) position indicator
偏心平面显示 off-center plan(e) display
偏心起振机 rotating mass shaker
偏心气举阀 eccentric gas lift valve
偏心器 eccentraliser[eccentralizer]; excentralizer
偏心器的 eccentric
偏心铅垂线 eccentric plumb line
偏心钳 eccentric tongs
偏心曲柄 eccentric crank; offset crank
偏心曲柄机构 eccentric crank mechanism; offset crank mechanism
偏心曲柄压力机 eccentric shaft press
偏心曲柄轴 eccentric crankshaft; offset crankshaft
偏心曲拐 drag link; eccentric crank; return crank
偏心曲梁 curved beam with eccentric boundaries
偏心曲线梁 eccentrically curved beam
偏心圈出 eccentric circle-out
偏心圈入 eccentric circle-in
偏心燃烧器 eccentric burner; off-center burner
偏心绕组 off-centering winding
偏心热动力疏水器 eccentric thermodynamic(al) type steam trap
偏心筛 eccentric screen
偏心扇形齿轮 eccentric sector; eccentric segment
偏心射孔器 eccentered gun
偏心施力 off-center application of force
偏心式电极 offset electrode
偏心式焊条 offset electrode
偏心式捡拾器 eccentric-type pick up
偏心式平面位置显示器 off-centered plan position indicator
偏心式倾翻出料搅拌机 offset tumbling mixer
偏心式曲柄压力机 full eccentric type crank press
偏心式压力机 eccentric press
偏心式振动器 eccentric-type vibrator
偏心式振动筛 eccentric-type vibrating screen; link-belt PD screen
偏心受拉 eccentric tension
偏心受拉杆件 eccentrically tensile member
偏心受拉构件 eccentric tension member
偏心受压 eccentric compression
偏心受压的 eccentrically compressed
偏心受压法 eccentric compression method
偏心受压杆件 eccentrically compressed member
偏心受压构件 eccentric compression member
偏心受压柱 eccentrically compressed column
偏心闩 cam-lock
偏心水准 eccentric level
偏心套管 eccentric socket
偏心套环座 eccentric-strap seat
偏心套筒 eccentric adjusting sleeve
偏心调节套 eccentric adjusting sleeve
偏心调节装置 eccentric adjuster
偏心调整机构 eccentric adjustment
偏心凸轮 circular cam; offset cam; eccentric cam
偏心推出器 eccentricity throw-out
偏心推力 eccentricity of thrust; eccentric thrust
偏心挖掘 <挖掘机> offset digging
偏心外轮 eccentric strap
偏心弯矩 eccentric bending moment
偏心弯曲 oblique bending
偏心望远镜 eccentric telescope; extra central telescope
偏心纬度 eccentric latitude
偏心位置 eccentric position; off-center position
偏心纹 eccentric groove
偏心稳定器 eccentric stabilizer
偏心误差 eccentric error
偏心系船浮筒 eccentric mooring buoy
偏心系数 eccentric(ity) coefficient

偏心显示 off centered display
偏心线圈 off-centered coil; off-centering yoke
偏心限度 eccentricity limit
偏心像差 eccentric anomaly
偏心销(钉) cam pin; eccentric pin
偏心楔 deflecting wedge
偏心卸料 eccentric discharge
偏心信号 eccentric signal
偏心性 eccentricity
偏心悬挂式反铲挖掘机 overcenter back-hoe
偏心旋 eccentrically rotating
偏心旋转镜 eccentrically rotating mirror
偏心旋转式激振器 eccentric rotary shaker
偏心压板 cam actuated clamp; eccentric clamp
偏心压床 cam press; eccentric press
偏心压机 eccentric press
偏心压力 eccentric compression; eccentric(ity) pressure
偏心压力机 open-end press
偏心压缩 eccentric compression
偏心研磨机 eccentric grinder
偏心叶片马达 cam-vane motor
偏心异常 eccentric anomaly
偏心异径管 eccentric reducer
偏心异径管接 eccentric reducing coupling; eccentric reducing socket
偏心异径三通 eccentric tee
偏心应力 eccentric stress
偏心预压混凝土预制桩 eccentrically precompressed precast-concrete-pile
偏心预应力钢丝束 eccentric tendon
偏心预应力钢索 eccentric tendon
偏心圆 eccentric; eccentric circle
偏心圆锯 wobble saw
偏心圆盘 eccentric disk[disc]
偏心圆筒式流变仪 eccentric cylinder rheometer
偏心运动地 eccentrically
偏心晕 eccentric halo
偏心载重 eccentric load(ing); non-central load
偏心轧碎机 eccentric breaker
偏心轧头 cam-lock
偏心站 eccentric station; false station
偏心振动 eccentric vibration
偏心振动落砂机 eccentric knockout grid; vibrating shake-out with eccentric drive
偏心振动磨 eccentric vibrating mill
偏心振动器 eccentric vibrator; unbalanced weight vibrator
偏心振动筛 eccentric drive screen; eccentric vibrating screen; gyratory riddle; vibrating screen with eccentric drive
偏心振动台 eccentric shaker table; gyratory shaker table
偏心振筛 gyratory screen
偏心振子 off-centered dipole
偏心支撑 eccentric brace
偏心支撑的 eccentrically braced
偏心支销 eccentric anchor pin
偏心制动器 eccentric brake
偏心制动(装置) eccentric catch
偏心质量 eccentric mass
偏心质量式激振器 eccentric mass-type vibration generator
偏心质量式振动发生器 eccentric mass-type vibration generator
偏心质量振荡器 eccentric mass shaker
偏心重(块) eccentric weight
偏心重式振动器 eccentric weight type vibrator
偏心重轴 eccentric weight shaft
偏心轴 eccentric axis; excentric shaft; tumbling shaft; wobbler shaft
偏心轴衬 eccentric bush(ing)
偏心轴齿轮传动 camshaft gear drive
偏心轴颈 rotor journal
偏心轴链条传动 camshaft chain and sprocket drive
偏心轴套 eccentric bushing
偏心轴向荷载 eccentric axial load
偏心轴压机 eccentric shaft press
偏心轴柱销 eccentric shaft plunger
偏心注油器 eccentric oiler
偏心柱螺栓 eccentric stud
偏心爪 cam finger
偏心转动活塞 eccentric rotary piston
偏心转筒式拌和机 eccentric tumbling mixer
偏心转筒式拌和器 eccentric tumbling mixer
偏心转子 eccentric runner
偏心转子泵 rotary eccentric piston pump
偏心转子发动机 eccentric rotor engine
偏心转子滑叶式压缩机 eccentric rotor sliding vane compressor
偏心装置 eccentric; wobbler
偏心锥形管 eccentric reducer
偏心锥形筒体 eccentric conic(al) shell
偏心自锁常开式离合器 overcenter [overcentre] type clutch
偏心自锁搭扣 eccentric clamp
偏心自锁机构 overcenter mechanism
偏心钻 oscillating bit
偏心钻杆 eccentrically bored spindle
偏心钻铤 eccentric collar
偏心钻头 eccentric(drill) bit; off-balance bit
偏心钻头荷载 eccentric bit load
偏心钻压 eccentric load
偏心作用 eccentric action; wobbler action
偏性 bias
偏性嫌气菌 obligatory anaerobic bacteria
偏性嫌氧性细菌 obligatory anaerobic bacteria; obligate anaerobic bacteria
偏序 partial order; semi-order
偏序列相关系数 partial serial correlation coefficient
偏旋光镜 polariscope
偏旋式破碎机 Telsmith breaker; Telsmith crusher
偏压 bias(sing); oblique load(ing); side pressure; unsymmetric(al) load(ing); biasing voltage【电】
偏压点火 bias ignition; priming
偏压电池 bias cell; biasing battery
偏压电池组 bias battery
偏压电流 bias current
偏压电阻 bias resistance; bias resistor
偏压干压实 <相对于最优含水率的> bias dry compaction
偏压继电器 biased relay
偏压检波器 bias detector
偏压溅射 bias sputtering
偏压控制 bias(-voltage) control
偏压控制电路 bias control circuit
偏压脉冲 bias pulse
偏压漂移 bias drift
偏压湿压实 bias wet compaction
偏压隧道 deflective pressure tunnel; slope tunnel; unsymmetric(al) loading tunnel
偏压调整 bias(-voltage) control
偏压误差 bias(s)ed error
偏压线圈 bias winding
偏压整流器 bias rectifier
偏压阻抗 biasing impedance
偏亚磷酸盐 metaphosphite
偏亚砷酸钠 sodium meta-arsenite
偏亚砷酸铜 cupric meta-arsenite
偏亚砷酸锌处理的木材 zinc meta-arsenite treated timber
偏亚砷酸锌剂 <木材防腐> zinc-meta-arsenite
偏沿子 feather edge board
偏盐 metasalt
偏要 must needs
偏移 back lash; deflection; departure; deviate; deviation; drift; excursion; offset(ting); shift(ing)
偏移边界时间 edge time of migration
偏移标度 deflection scale
偏移波 offset wave
偏移彩色编码速度剖面 migrating colo(u)r code interval velocity section
偏移测量仪 drift meter
偏移差分格式参数 difference pattern parameters of migration
偏移常数 deviation constant
偏移的 off-lying; out-of-position
偏移的精度和校正 accuracy and correction of deviation
偏移叠加 migration stack
偏移叠加剖面 migration stack section
偏移叠卡法 offset stacking
偏移度 drift rate; throw
偏移度盘 offset dial
偏移对准 offset alignment
偏移法 deflection method; offset method
偏移范围 deviation range
偏移方向 offset direction
偏移滚迹 skew running mark
偏移弧 offset arc
偏移畸变 bias distortion
偏移计 deflection indicator
偏移计算 calculation of offset
偏移角 angle of crab; deflect(ion) angle; deviation angle
偏移接卡箱 offset stacker
偏移开关 offset switch; slope switch
偏移刻度 deviation scale
偏移控制 shift(ing) control
偏移量 sideplay amount
偏移量检验 slippage test
偏移率 drift rate
偏移脉冲 shift pulse
偏移面 offset surface
偏移末级前置放大器 shift driver
偏移目镜 offset eyepiece
偏移频率 deviation frequency; offset frequency
偏移起始点号 beginning point number of migration
偏移起始时间 beginning time of migration
偏移射束 offset beam
偏移深度剖面 migration depth section
偏移失真 bias distortion
偏移式速调管 drift klystron
偏移数据 offset data
偏移栓 drift key
偏移速度 migration velocity
偏移速率 deflection rate
偏移特性 offset behavio(u)r
偏移调整 shift control
偏移调制 deflection modulation
偏移位 offset bit
偏移位置 offset position
偏移误差 bias error; displacement error; offset error
偏移吸收 deviation absorption
偏移系数 deviation ratio; offset coefficient
偏移限定 offset qualification
偏移相位水平切片 horizontal slice of migration phase
偏移心轴 drift mandrel
偏移信号 shifted signal
偏移性能 offset behavio(u)r
偏移修正值 bias correction
偏移载波频率 offset carrier frequency
偏移振幅水平切片 horizontal slice of migration amplitude
偏移振型 vibration deflection mode
偏移值 deviant
偏移终了点号 end point number of migration
偏移终了时间 end time of migration
偏移轴 offset axis
偏移准则 deflection criterion
偏倚 bias
偏倚比例 bias proportion
偏倚临界域 biased critical region
偏倚误差 bias error; error due to bias
偏因 bias factor
偏应变 deviator(ic) strain
偏应变能 deviatoric strain energy
偏应力 deviation stress; deviator(ic) stress
偏应力历史 deviatoric stress history
偏应力蠕变试验 creep test under deviated stress
偏应力张量 deviatoric stress tensor
偏应力状态 deviatoric state of stress
偏右 deviation to right
偏于安全的误差 error on safe side
偏于保守的估价 loose estimate
偏于一方 one-sided
偏远的 by(e)
偏远的森林地区 backwoods
偏远地区 back of beyond
偏载 oblique load(ing)
偏载系数 eccentric load factor
偏增量 partial difference; partial increment
偏增量变差 partial increment deviation
偏张量【数】 deviatoric tensor
偏折 deviation
偏折棱镜 deviating prism; deviation prism
偏振 excursion; fall off; polarization
偏振板观察系统 polarized plate viewing system
偏振保持光纤 polarization maintaining optic(al) fiber[fibre]
偏振暴 polarized burst
偏振波 polarized wave
偏振波束 polarized beam
偏振参量 polarization parameter
偏振测定 polarimetry
偏振测定法 polarimetry; polariscopy
偏振测量 polarization measurement
偏振测量仪 fall-off meter
偏振层板 sheet polarizer
偏振成分 polarized component
偏振单色仪 polarizing monochromator
偏振等倾线 polarization isocline
偏振度 degree of polarization; polarized degree
偏振反射光 polarized-reflected light
偏振方向 polarized direction
偏振分光光度计 polarizing spectrophotometer
偏振分集 polarization diversity
偏振分量 polarized component
偏振辐射 polarized radiation
偏振复用 palarization multiplexing
偏振干涉滤光器 polarization interference filter
偏振干涉仪 polarization interferometer
偏振观察 polarized observation; polarized viewing
偏振光 polarised[polarized] light

偏振光玻璃网板 polarized glass screen
偏振光测高温计 polarizing optic(al) pyrometer
偏振光度计 polarization photometer
偏振光光测高温计 polarizing pyrometer
偏振光角 angle of polarization
偏振光镜 polariscope; polarizing screen; polaroid polarizer
偏振光镜管 polariscope tube
偏振光立体观察法 vectographic method
偏振光立体立体影像 vectographic image
偏振光零件 polarizing optics
偏振光面旋转效应 torque
偏振光模拟 polarized light analog
偏振光偏振面转动 optic(al) rotation
偏振光谱 polarized spectrum
偏振光体视镜 vectograph
偏振光椭圆率测量仪 ellipsometer
偏振光物镜 polarization objective
偏振光显微镜 micropolariscope; polarising [polarizing] light microscope; polaised [polarized] light microscope
偏振光显微术 polarized light microscopy
偏振化 polarize
偏振化方向 polarization direction
偏振化作用 polarization
偏振回波 polar echo
偏振计 polarimeter; polaristrobometer
偏振交叉 polarization cross
偏振角 Brewster angle; polarization angle; polarizing angle
偏振接收机 polarization receiver
偏振镜 polariscope; polarized disc; polarizer; polarizing spectacle
偏振镜的 polariscopic
偏振镜检查 polariscopy
偏振开关 polarization switch
偏振棱镜 polarizing prism
偏振立体摄影术 vectography
偏振立体图 stereovectograph; vectograph
偏振立体照片 vectograph
偏振滤光镜 polarizing filter; polaroid filter
偏振滤光片 polarizing filter; polaroid filter
偏振面 plane of polarization; polarization plane
偏振面的旋光 rotation of the plane of polarization
偏振目镜 polarizing eyepiece; polarizing glass; polaroid glass
偏振片 polarizing disc [disk]; polarizing film; polaroid
偏振片检影器 polarized pattern viewer
偏振片调节 polarization modulation
偏振片调置钮 polaroid adjustment knob
偏振频谱仪 spectropolarimeter
偏振平面 plane of polarization
偏振谱 polarization spectrum
偏振器 polarization apparatus; polarizer
偏振曲线 polarization curve
偏振全息摄影 polarization holography
偏振式传感器 polarimetric sensor
偏振态 polarization state
偏振天线 polarized antenna
偏振调制 polarization modulation
偏振椭圆 polarization ellipse
偏振像片 vectograph
偏振效应 polarization effect
偏振旋转 polarization rotation
偏振选择性 polarization selectivity
偏振仪 Senamont polarimeter
偏振荧光 polarized fluorescence
偏振转向 optic(al) inversion
偏振状态 state of polarization
偏振子 polariton
偏正桩 deflected pile
偏酯 partial ester
偏制 compile
偏制管 offset pipe
偏置 bias; bias(s)ing; offset; polarization; setover; misgather < 集装箱 >
偏置爆炸点 offset shotpoint
偏置部件 offset unit
偏置差动保护装置 biased differential protective system
偏置尺 offset scale
偏置单元 offset unit
偏置刀架 offset toolholder
偏置导星 offset guiding
偏置导星装置 offset guiding device
偏置的 bias(s)ed; off-centered
偏置的履带板 offset grouser shoe
偏置电流补偿 bias-current compensation
偏置电路 biasing circuit
偏置舵 offset rudder
偏置阀 eccentrically arranged valve
偏置法 offset method; offsetting; setover method
偏置杆 offset staff
偏置构件 offset unit
偏置估计 biased estimator
偏置管 offset pipe; pipe offset
偏置管道上的喷嘴 offset conduit nipple
偏置管道上的乳头 offset conduit nipple
偏置光 bias lighting
偏置畸变 bias distortion
偏置棘轮的螺钉起子 offset ratchet screwdriver
偏置技术 biasing technique
偏置溅射 bias sputtering
偏置角 offset angle
偏置截止 bias off
偏置卡塞格伦天线 offset Cassegrain antenna
偏置控制 bias control
偏置框架 offset frame
偏置馈料 offset feed
偏置连接 offset joint
偏置链节 offset link
偏置梁 offset beam
偏置六角形开口扳手 offset hexagon open-end wrench
偏置六角形螺钉起子 offset hexagon screwdriver
偏置逻辑 biasing logic
偏置螺丝刀 offset screwdriver
偏置脉冲 bias pulse
偏置姆欧继电器 offset-mho relay
偏置喷管 offset nozzle
偏置汽缸 offset cylinder
偏置器 deviator
偏置器件 offset unit
偏置曲柄机构 eccentric crank mechanism
偏置曲柄轴 offset crankshaft
偏置绕组 bias winding
偏置栅极 offset gate
偏置式安装 offset mounting
偏置式发动机 offset type engine
偏置式曲轴 offset crankshaft
偏置式圆盘耙 offset disc [disk] harrow
偏置手闸 overcenter handbrake
偏置数据 bias data
偏置探测器 biased detector
偏置套筒扳手 offset box wrench
偏置特性 biasing characteristic
偏置调定频率 bias-set frequency
偏置挺杆 offset tappet
偏置头锥 offset nose cone
偏置凸轮 offset cam
偏置误差 biased error
偏置谐波电流 harmonic bias
偏置信号 offset signal
偏置悬挂 offset mounting
偏置悬挂犁 offset plow
偏置旋凿 offset screwdriver
偏置抑制杂波 bias mute
偏置载频 offset carrier
偏置振荡器 bias oscillator
偏置指数 biased exponent
偏置轴销 offset pivot
偏置阻抗继电器 offset impedance relay
偏置阻气阀 offset choke valve
偏置钻眼 offset drilling
偏中心管件 eccentric fitting
偏中心运用 off-center operation
偏重 imbalance; lopside; overemphasis; poise error; sway
偏重的 lopsided
偏重驱动摇动筛 eccentric weight drive shaking screen
偏重输送机 unbalanced weight conveyer [conveyor]
偏重轴 eccentric weight shaft
偏轴 skew shaft
偏轴齿轮系 skew axis gear [复 skew axes gears]
偏轴弧齿近平面齿轮 spiroid
偏轴溅射 off-axis sputtering
偏轴鉴别 off-beam discrimination
偏轴角 off-axis angle
偏轴立体坐标仪 wind-tipped stereocomparator
偏轴面齿轮 planoid gear
偏轴平行光管 off-axis collimator
偏轴伞齿轮 hypoid bevel gear; hypoid gear
偏轴伞齿轮传动 hypoid bevel
偏轴色差 off-axis chromatic aberration
偏轴透镜 decentered lens
偏轴斜齿伞齿轮 skew bevel wheel; skew wheel
偏轴旋转叶片 rotating off-axis blade
偏轴英国式装置 off-axis English mounting
偏转 avertence; deflect(ing); deflexion; eccentric rotation; yawing【船】
偏转鞍座 < 桥梁修复或加强时所用预应力筋的 > deviation saddle
偏转板 deflecting plate; deflection plate; deflector plate; inflector plate
偏转标度 deflection scale
偏转部分 deflection chassis
偏转场 deflecting field; deflection field
偏转磁场 deflection magnetic field
偏转的 deflective
偏转点 deflection point
偏转电流 deflecting current; deflection current
偏转电路 deflecting circuit; deflection circuit
偏转电压 deflection voltage
偏转电子束 deflected electron beam
偏转电子束管 beam deflection valve
偏转度 deflection; degree of deflection; degree of inclination
偏转度计 deflection ga(u)ge
偏转法 deflecting method; deflection method
偏转范围 range of deflection
偏转放大器 deflection amplifier
偏转分离器 deflection separator
偏转风速表 deflection anemometer
偏转峰 deflection front
偏转钢缆 deflected cable
偏转管 deflection valve; deflector tube
偏转管嘴 deflection nozzle
偏转光 deflection light
偏转光束 deflected beam
偏转过度 overswing
偏转河 deflected river
偏转后加速电极 post-deflection accelerating electrode
偏转极 deflecting electrode
偏转角 angle of avertence; angle of declination; angle of deflection; angle of yaw; deflection angle; deflexion angle; inflection angle
偏转聚焦 deflection focusing
偏转力 deflecting force; deflection force; deflective force; deviating force; force of deviation
偏转力矩 bias torque; deflecting torque; deflection moment; deflection torque; flexure moment; yawing moment
偏转力偶 deflecting couple
偏转灵敏度 deflection factor; deflection sensitivity; sensitivity of deflection
偏转流 refracted flow
偏转喷嘴 deflection nozzle
偏转平面 deflection plane
偏转器【化】 deflector
偏转散焦 deflection defocusing
偏转升降舵产生的力矩 moment due to elevator deflection
偏转失真 deflection distortion
偏转失真校正电路 deflection linearity circuit
偏转时间 deflection time
偏转式 gauche form; case of avertence
偏转式破碎机 Telsmith breaker; Telsmith crusher
偏转式折射计 deflection refractometer
偏转索鞍 < 悬带桥的 > deviation saddle
偏转同步 deflection synchronization
偏转图 deflection diagram
偏转位置 inflection point; inflecture point
偏转系统 deflecting yoke
偏转线圈 deflected coil; deflecting coil; deflecting yoke; deflection coil; deflection winding; deflection yoke; sweep coil; yoke
偏转线圈电流 yoke current
偏转陷波电路 deflector wave trap
偏转效率 deflection efficiency
偏转信号 deflection signal
偏转仪 deflector; yawmeter
偏转翼式风速计 deflecting vane type anemometer
偏转因数 deflection factor
偏转之前 predeflection
偏转指令 deflection command
偏转中心 deflection center [centre]
偏转轴 axis of deflection
偏转轴线 yawing axis
偏转装置 deflecting device; deflection system; deflector; inflector assembly
偏转阻尼器 yaw damper
偏装法 offset method
偏装置的 offset-mounted
偏自相关 partial autocorrelation
偏最小二乘 partial least squares
偏左 deviation to left

便宜货 bargain

便宜级 economy grade

片暗销 plate dowel

片坝 vane dike[dyke]
片斑 film mottle
片板式散热器 finned plate radiator
片板状流动构造 linear flow structure; planar flow structure; platy flow structure
片帮【岩】 side pressed out; sidewall scaling; sidewall slabbing; sidewall spalling; spalling
片边号码 edge number; key number
片边缺口机 film notcher
片扁壳理论 shallow shell theory
片冰 flake ice; lolly
片冰厂 flaked ice plant
片冰机 flake ice maker
片冰设备 flaked ice plant
片玻璃 sheet glass
片材 flake; sheet
片材成型 sheet forming
片材饰面 sheet facing
片材屋面防水 sheet water-proofing
片材造型 die pressing
片材造型机 punch press
片层 lamella; ply
片层分裂 flakiness
片层结构 lamellar structure
片层膜 lamellar membrane
片层体 lamellar body; lamellasome
片窗 gate
片搭接 oblique cut grafting
片带 film strip
片带输送机 horizontal belt conveyer [conveyor]
片地沥青 <混合料> sheet asphalt
片地沥青路面 sheet asphalt pavement; sheet pavement
片地沥青砂 sand sheet asphalt
片地沥青石 <即细粒式地沥青混凝土> stone-filled sheet asphalt
片端 film end
片段 extract; fragment; part; piecemeal; segment(ing)
片段变换 segment transformation
片段的 piecewise
片断 fragment; snip
片断生产线 segment production line
片堆结构 bookhouse texture
片阀 flapper; plate valve; slatch
片筏基础 mat foundation; raft foundation; raft of foundation
片沸石 heulandite
片粉状树脂 flaky resin
片估计方差 slice term of estimation variance
片管式散热器 fin and tube(type) radiator
片硅碱钙石 delhayelite
片硅铝石 donbassite
片焊接 chip bonding
片痕 scar
片洪 sheet flood
片滑 slab slide
片簧 bearing spring; blade spring; laminated spring; spring leaf
片簧式悬置阀 reed-type suspension valve
片机 mascerating machine
片基 back; film base
片基尺寸稳定的胶片 stable base film
片基量规 jacket ga(u)ge
片基密度 base density
片级 chip level
片剂 pellet; tablet; tabloid
片剂斑点 tablet mottle
片剂崩解测定器 tablet disintegration tester
片剂硬度 tablet hardness
片剂硬度计 tablet hardness tester
片剂重量差异 tablet weight variation
片夹 film jacket
片夹面 jacket face
片架状组构 <土的结构构造的一种型式,指土粒的亚微观构造> card-house fabric; card-house structure
片间电压 <整流子的> bar-to-bar voltage
片间距离 distance between commutator segments
片间绝缘 insulation between plates
片间黏[粘]结 intersheet bonding
片间试验 bar-to-bar test
片胶 flake shellac
片礁 patch reef
片接合 chip bonding
片接头 strip terminal
片结 chip junction
片金 flake gold
片晶 lamella
片晶变形 lamellar deformation
片晶结构 lamellar structure
片晶晶格 lamellar lattice
片晶位移 lamellar displacement
片晶增厚 lamellar thickening
片距 pitch of fins; segment pitch
片锯 blade saw; web saw
片刻 an instant; a short while; for a space; tick
片孔效应 sprocket-hole effect
片肋 laminated rib
片理 foliation; schistosity
片理带 schistosity belts
片理观测点 observation point of schistosity
片理化带 schistosity zone
片理化岩 schistose rock
片理面【地】 foliation plane; plane of schistosisty; schistosity plane
片利共生 synoeciosis
片粒状 schistose granular
片链 plate dowel
片料覆盖油毡 flake surfaced asphalt felt
片裂 splinter
片裂板岩 sculp
片裂的 splintery
片裂树皮 fissure bark
片流 lamellar flow; laminar flow; laminar motion; sheet flood; sheet wash
片流层 laminar film
片流底层 laminar sublayer
片流地质作用 geologic(al) process of sheet flow
片流模型 laminar model
片流侵蚀 sheet erosion
片流区 laminar zone
片流运动 laminar motion
片流(运动)边界层 laminar boundary layer
片铝石 kayserite
片落 flake off; flaking; flaking off; scale off; scaling
片落试验 chipping test
片麻花岗岩 gneiss granite
片麻火成岩 gneiss igneous rock
片麻结构 gneissic structure
片麻理 gneissic schistosity; gneissosity
片麻岩 bastard granite; gneiss; mottled schist; trade granite
片麻岩的 gneissic; gneissoid; gneissose
片麻岩化 gneissification
片麻岩混合岩中含金石英脉 gold-quartz vein in gneiss-migmatite
片麻岩期 gneiss period
片麻岩区 gneissic regions
片麻岩砂 gneissic sand
片麻岩弯窿 gneiss dome
片麻岩系 gneiss system
片麻状的 gneissic; gneissoid; gneissose
片麻状构造 gneissic structure; gneissosity
片麻状花岗岩 bastard granite; gneissose granite; gneissic granite
片麻状节理 gneissosity
片麻状结构 gneissic texture; gneissose texture
片麻状云母片岩 gneiss-mica schist
片面的 one-sided; unilateral
片面的观点 one-sided view
片面废止 unilateral repudiation
片面解除 unilateral withdrawal
片面行为 unilateral act
片面性 one-sidedness
片面移转 unilateral transfer
片面之词 one-side argument
片面转移 unilateral transfer
片膜层 lamellar membrane
片内褶皱 intrafolial fold(ing); intraformational fold(ing)
片黏[粘]土 adamic earth
片牛角 sheet cows
片区 sector
片设计 chip layout
片石 mammock; rubble; scabbling; slabstone; flagstone; colly Weston slate <用于铺盖屋顶>
片石材料 schistose material; schistous material
片石粗琢 boaster chisel
片石底层 rubble backing
片石护坡 rubble pitching
片石混凝土 ragstone concrete; rubble concrete
片石基础 rubble foundation; rubble stone footing; stone footing
片石阔凿 boasting chisel
片石路面 quarry pavement
片石铺砌 crazy pavement; crazy paving; slab pavement; slab paving
片石砌筑 ragwork
片石墙 rag rubble walling; ragwork
片石圬工 rubble masonry
片石英 bauerite
片蚀 sheet erosion; sheet flood; sheet wash
片式变压器 lamellar transformer
片式处理器 slice processor
片式串联安装 tandem mounting
片式串联连接 sectional mounting; section-type mounting
片式单钩 laminated hook
片式电阻器 chip resistor
片式供暖管 finned heating tube
片式供暖盘管 finned type heating coil
片式锅炉 sectional boiler
片式加热器 strip heater
片式空气滤清器 panel type air; panel type air cleaner
片式离合器 lamella clutch; plate coupling
片式(沥青砂)面层 surface of sheet type
片式连接的 sectional; stack-mounted
片式流水线系统 slice pipeline system
片式滤清器 gap-type filter; lamellar filter; plate filter
片式摩擦离合器 plate clutch
片式热风机 lamellar air-heater
片式热风器 lamellar air-heater
片式热水锅炉 sectional hot water boiler
片式散热器 gilled radiator; plate radiator
片式矢量 lamellar vector; laminar vector
片式闩锁屏蔽网络 slice latch network; slice mask network
片式弹簧钢板 flat spring steel plate
片式填料 plate fill; plate packing
片式微机 slice microcomputer
片式消声器 splitter plate type sound absorber
片式压力表 plate manometer
片式振捣器 spade-type vibrator; spade vibrator
片式蒸汽锅炉 sectional steam boiler
片式铸铁锅炉 iron casting sectional boiler; sectional cast-iron boiler
片栓 plate dowel
片霜 sheet frost
片撕裂 lamellar tearing
片锁 mortice lock
片滩 patch
片弹簧 leaf spring; plate spring; flat spring
片弹簧悬挂 leaf spring suspension
片碳镁石 coalingite
片条 film strip
片通 <一种氯化聚醚塑料> penton
片头 film end; film head
片头铆钉 set-head rivet
片外标志 off-chip marker
片尾 trailer
片匣 cassette
片销 plate dowel
片屑 <木材加工剩余的> coarse residue
片屑绞碎器 chipswringer
片榍石 guarinite; hiortdahlite
片形电容器 chip capacitor
片形电阻器 chip resistor
片形铺砌 sheet asphalt pavement; sheet pavement
片型范性流变 laminar plastic flow
片型衰减器 flap attenuator; vane attenuator
片型塑性流动 laminar plastic flow
片型芯 slab core
片选 chip enable; chip select
片岩 callys; killas; laminated rock; schist
片岩板 schistosity sheet
片岩的 schistic; schistose; schistous
片岩基础 schistose subbase
片岩夹层 schist band
片岩剪切夹层 sheared schist band
片岩面 plane of schistosity
片岩石灰石 schistoid limestone
片岩质黏[粘]土 schistose clay
片岩状黏[粘]土 schistose clay
片研磨 slice lapping
片页式过滤器 leaf filter
片页岩 leaf shale
片终信号 end-of-copy signal
片柱钙石 scawtite
片状 flake; flakiness; foliated; plate; schistic; tabular
片状保险器 strip fuse
片状保险丝 strip fuse
片状避雷器 plate protector
片状边界层 laminar boundary layer
片状变质煌斑岩 lamproschist
片状冰 sheet ice; foliated ice; ice in sheet; ice wedge
片状波导 slab guide
片状玻璃 foliated glass
片状玻璃加强 flake glass reinforcement
片状剥落 exfoliation; flaking; sheeting
片状剥落的蛭石混凝土 exfoliated vermiculite concrete

P

片状剥落的蛭石石膏灰泥 exfoliated vermiculite-gypsum plaster
片状剥落腐蚀 exfoliation corrosion
片状材料 schistose material; schistous material
片状插座 subedge connector
片状成分 flaky constituent
片状冲蚀 sheet wash(ing)
片状冲刷 sheet wash
片状粗砂岩 foliated grit(stone)
片状单晶 platy-monocrystal
片状的 fissile; lamellar; lamellate; schistose; schistous; slabby; flaky
片状地沥青 sheet asphalt
片状电介质 sheet dielectric
片状电容器 chip capacitor
片状阀 disc[disk] valve; plate valve; flap valve
片状分层电容器 interdigitated capacitor
片状粉块结构 play-pulverulent structure
片状粉末 flaked powder; flake-like powder; leafed powder
片状粉末石棉 flaked asbestos
片状腐蚀 layer corrosion
片状腐蚀产物 corrosion scale
片状覆盖体 laminated cover(ing)
片状刚玉 alumina tabular
片状高岭土 delaminated kaolin
片状隔膜 plate diaphragm
片状构造 laminate structure; sheeting structure; slabby shape; schistose structure【地】
片状骨料 flaky aggregate
片状硅栅 chip silicon gate
片状合金钻头 carbide insert blade type bit
片状花岗岩 bastard granite; foliated granite
片状滑石 foliated talc
片状辉长岩 schistose gabbro
片状积沙 laminar sand-flood
片状集料 flaky aggregate
片状件边缘曲线饰 escalloped
片状胶粘剂 sheet adhesive; sheet bonding adhesive; sheet cementing agent
片状角岩 leptynolite
片状角页岩 schistose hornfels
片状结构 flake texture; laminated structure; laminated texture; layered structure; schistose texture; sheet structure; sheet texture; tubular texture
片状结晶 plate crystal; tabular crystal
片状结晶岩 schistose crystalline rock
片状金粉 flake gold powder; flitter
片状金属粉 flake metal powder
片状金属颜料 metallic flake pigment
片状晶体 crystal plate; crystal ribbon; flat crystal; lamellar crystal; plate crystal; tabular crystal
片状径流溶蚀 flake runoff corrosion
片状聚合物 sheet polymer
片状颗粒 fiat piece; flake-shaped particle; flaky grain; flaky particle; flat particle; plate-like grain; platy grain
片状颗粒材料 flaky grain material
片状空化 sheet cavitation
片状空蚀 sheet cavitation
片状孔 sheet-like pore
片状矿物 platy mineral
片状蜡 scale wax
片状冷却器 fin shield
片状沥青 sheet asphalt
片状沥青路面 sheet asphalt pavement
片状粒料 schistose particle
片状粒子 plate-like particle
片状流 sheet flow
片状铝粉 alumin(i)um flake; flake alumin(i)um; leafed alumin(i)um powder
片状铝粒 laminar alumin(i)um particle
片状氯化钙 flake calcium chloride
片状马氏体 lamellar martensite; massive martensite; plate-type martensite
片状煤 foliated coal
片状模塑料 sheet mo(u)lding compound
片状模塑料蒙皮 sheet mo(u)lding compound skin
片状膜 lamellar membrane; laminar film
片状黏[粘]连 patchy adhesion
片状黏[粘]土 foliated clay; laminated clay; plate-like clay
片状配子体 lamellar gametophyte
片状劈理 schistose cleavage; schistosity cleavage
片状片麻岩 schist gneiss
片状铺的沥青屋面 asphalt sheet roofing
片状汽蚀 sheet cavitation
片状器件 sheet device
片状侵蚀 sheet erosion
片状侵蚀作用 action of sheet erosion
片状热交换器 lamellar heat exchange
片状熔断器 plate fuse
片状熔丝 fuse strip
片状砂砾石 foliated grit(stone); foliated sandstone
片状砂埋 sand cover of slice shape
片状砂屑岩 clasmoschist
片状砂岩 foliated sandstone; schistous sandstone
片状闪电 sheet lightning; summer lightning
片状闪石 schistous amphibolite
片状烧碱 caustic in flakes; flake caustic
片状生长 lamellar growth
片状石膏 foliated gypsum
片状石蜡 scale paraffin
片状石棉 sheet asbestos
片状石墨 flake graphite; flaky graphite; graphite in flake form; lamellar graphite
片状石墨粉粒 graphite flake
片状树皮 flake-bark
片状水流 sheet flow; flat plate flow
片状碎屑岩 clasmoschist
片状胎座式 lamellate placentation
片状炭黑 flaky black carbon; smut
片状体 plate
片状填料 laminal filter
片状铁粉 flake iron powder
片状铜粉 flake copper
片状退变岩 diaphthorite
片状脱落 flaking; slabbiness
片状瓦 cleaving tile; split floor tile
片状微型组件 pellet module
片状物 platelet
片状物料 slabby
片状物质 flaky material
片状形态 lamellar morphology
片状悬垂物 flap
片状雪 flaky snow
片状岩 cleaving stone
片状岩石 slabby rock
片状颜料 flake pigment; lamellar pigment; leafed pigment
片状氧化铝 tabular alumina
片状氧化铁 leafed iron oxide
片状氧化物 oxide platelet
片状硬合金钻头 slice tungsten carbide bit
片状元件 chip component
片状原纤 lamellar fibril
片状云 pannus
片状云母 sheet mica
片状运动 sheet motion; sheet movement
片状栽植 patch planting
片状炸药 flake(d) powder; sheet explosive
片状止回阀 clack valve; clapper valve
片状指数 flakiness index
片状珠光体 lamellar pearlite
片紫胶 flake shellac
片组簧 spring set

剽

剽窃者 copyist; cribber

漂

漂 swim

漂白 blanch(ing); bleach; chemicking; decolo(u)r; decolo(u)ration; decolo(u)rization; whiten; decolo(u)rize
漂白斑 bleach spot; deoxidation sphere
漂白层 bleached bed; bleached horizon; tongues of albic horizon
漂白厂 bleaching works
漂白厂污水 bleachery effluent; bleachery waste(water)
漂白车间 bleaching plant
漂白虫胶 bleached(shel)lac; white lac
漂白的 bleaching
漂白法 bleach out process
漂白反像法 etch-bleach
漂白反应 bleaching action
漂白废水 bleaching wastewater
漂白粉 bleaching powder; calcium bleach; calcium hypochlorite; calcium hypochlorite lime; chloride; chloride of lime; chlorinated lime; hypochlorite of lime
漂白粉水 Javelle water
漂白粉消毒 disinfection by bleaching powder
漂白粉液 bleaching fluid
漂白辅助剂 bleaching assistant
漂白工 bleacher
漂白化学热力学纸浆 bleached chemithermomechanical pulp
漂白黄蜡 beeswax
漂白灰化作用 bleaching podzolisation
漂白机 bleacher; bleach hollander; bleaching machine; potcher
漂白剂 bleacher; bleaching agent; chloride; decolo(u)rant; decolo(u)riser[decolo(u)rizer]; discharger; discharging agent
漂白间 bleachery
漂白胶 bleached glue
漂白蜡 white wax
漂白麻刀 bleached fiber
漂白能力 bleaching power
漂白泥 camstone
漂白黏[粘]土 bleaching clay
漂白牛皮纸厂污水 bleached kraftmill effluent
漂白牛皮纸浆 bleached kraft pulp
漂白器 bleacher
漂白强度 bleaching intensity
漂白染色污水 bleaching and dyeing wastewater
漂白软土 alboll
漂白砂 bleached sand
漂白试验 discolo(u)ration test
漂白水 bleaching water
漂白土 active earth; albic; bleached earth; bleaching clay; bleaching earth; bleaching soil; discolo(u)ring clay; floridin; Flrida clay; Fuller's clay; fulling clay; greda; milled clay; Fuller's earth
漂白土矿床 Fuller's earth deposit
漂白土上层 cledge
漂白污水 bleaching effluent
漂白效应 bleaching effect
漂白性 bleachability
漂白亚硫酸盐纸浆 bleached sulfite pulp
漂白液 bleach(ing) bath; bleaching liquid; bleaching lye; bleaching solution; bleach liquor; sodium hypochlorite solution
漂白液沉淀池 bleach liquor settling tank
漂白油 bleach(ed) oil
漂白纸 bleached paper
漂白纸浆 bleached pulp
漂白纸浆污水 bleached pulp sewage
漂白紫胶 bleached(shel)lac; bone dry bleached shellac; white lac
漂白作用 bleaching action; decolo(u)rization
漂变 drift
漂冰 drift ice; floating ice; ice-drift(ing); ice run; pack ice; run of ice
漂冰沟痕 iceberg furrow mark
漂冰界 limit of drift ice; limit of pack-ice
漂泊 decolo(u)r; discolo(u)ration; rove
漂泊过程 bleaching process
漂泊能力 bleachability
漂泊信天翁 wandering albatross
漂泊种 fugitive species
漂布工 fuller
漂布土 malthacite
漂尘 air-borne dust
漂尘污染 floating dust pollution
漂程 driftage; drift way
漂打机 hollander
漂动 drifting; wandering
漂斗 drifting grab
漂粉 bleaching powder
漂粉机 bleaching powder machine
漂粉精片 calcium hypochlorite tablet
漂粉与漂液工段 bleaching powder and liquor section
漂浮 adrift; afloat; buoy; drifting; floating; leafing; levitate; rafting
漂浮残骸 flotsam
漂浮常数 flo(a)tation constant
漂浮超导环 levitated superconducting ring
漂浮称重式测渗仪 weighing float-type lysimeter
漂浮称重式土壤蒸发器 weighing float-type soil evaporator
漂浮持久性 leafing retention
漂浮充气器 floating aerator
漂浮导体 floating conductor
漂浮的 floating; levitate; natant; supernatant
漂浮的沉船残骸 drifting wreck
漂浮的船 floatage
漂浮的遇难船舶残骸 flotsam
漂浮电势 floating potential
漂浮动力学 kinetics of flo(a)tation
漂浮动物 floating animal
漂浮法 floating method; flo(a)tation
漂浮废物 floating trash
漂浮分析 flo(a)tation analysis
漂浮感 levitation
漂浮固体 floating solid
漂浮观测站 drift station
漂浮管道 floating pipeline
漂浮(海)藻 drift weed
漂浮环 floating ring
漂浮货柜 floatainer; floating container

漂浮剂 leafing agent
漂浮结构 floating structure
漂浮介质滤池 floating media filter
漂浮卡片 floating card
漂浮空间模拟 floating space simulation
漂浮控制器 floating controller
漂浮矿渣 floating slag
漂浮垃圾 flotsam; floating debris; floating debris
漂浮砾石灰岩 float stone
漂浮面 plane of buoyancy
漂浮面积 flo(a)tation area
漂浮木材 floating log
漂浮泥炭 floating peat
漂浮泥渣 supernatant sludge
漂浮浓集法 flo(a)tation concentration method
漂浮浓聚法 flo(a)tation concentration method
漂浮期 drift epoch
漂浮器 levitron
漂浮球形器 levitated spherator
漂浮森林 floating forest
漂浮栅极 floating grid
漂浮深度 depth of flo(a)tation
漂浮生物 plankton organism
漂浮生物的 neustic
漂浮石墨 kish;kish graphite
漂浮式采矿装置 floating mining plant
漂浮式防波堤港 floating harbo(u)r
漂浮式海洋气象自动观测站 floating oceanographic(al) and meteorologic(al) automatic station
漂浮式流速计 float current meter
漂浮式流速仪 float current meter
漂浮式土壤蒸发器 floating lysimeter
漂浮式洗矿机 floating washer
漂浮式斜拉桥 floating cable-stayed bridge
漂浮式蒸发计 afloat evapo(u)rimeter;floating evapo(u)rimeter
漂浮式蒸发量测装置 floating device for measuring evapo(u)ration
漂浮式蒸发皿 floating evapo(u)ration pan
漂浮式指示器 drake device
漂浮式自记(海洋)观测站 drifting recording station
漂浮试验 float test
漂浮水龙 floating primrose willow
漂浮水生植物 floating aquatic;hydrophyte natantia;natant hydrophyte
漂浮速度 floating velocity
漂浮碎屑 floating debris
漂浮特性 flo(a)tation characteristic
漂浮体系 floatable system
漂浮庭园 floating garden
漂浮艇 float boat
漂浮(微)生物 neuston
漂浮位置 floating position
漂浮稳定性 floating stability; leafing stability;stability of flo(a)tation
漂浮污泥 floating sludge;rising sludge
漂浮污物 floating pollutant
漂浮物 debris; driftage; floatable; flo(a)tation matter; floater; floating debris;floating drift; rejectamenta; floatage;flotsam <船舶失事后的>
漂浮物冲击力 impact force of floater
漂浮物过道 floating debris pass
漂浮物截留设备 debris trap
漂浮物拦栅 <河道中拦漂浮木料设施> drift barrier
漂浮物料 floating material
漂浮物排污道 floating debris pass
漂浮物清除器 drift eliminator
漂浮物体 floating body
漂浮物栅栏 drift barrier
漂浮物占有权 floatage

漂浮物质 floating matter
漂浮吸收剂 floating absorbent
漂浮系数 flo(a)tation coefficient
漂浮细粒 floating fine particle
漂浮现象 hydroplaning
漂浮响应 float(ing)response
漂浮橡胶护舷 floating rubber fender; foam rubber fender
漂浮效应 levitation effect
漂浮型铝粉 leafing-type alumin(i)um powder
漂浮型轮胎 floating tire[tyre]; flotation tire[tyre]
漂浮性 floatability;flo(a)tation
漂浮性海绵胶 flo(a)tation sponge
漂浮圆木 floating log
漂浮蒸发器 afloat pan;levitation evapo(u)rator
漂浮蒸发站 evaporation station on water surface; floating evapo(u)ration station
漂浮值 leafing value
漂浮植被 errantia
漂浮植物 floating plant;floating vegetation
漂浮组织 floating tissue
漂浮作用 floating action
漂航时间 drift time
漂迹层 stratified drift
漂积 drifting
漂积层【地】drift sheet;erratic form; stratified drift
漂积矿体 erratic orebody
漂积黏[粘]土 drift clay
漂积土 till
漂积物 drift
漂积物形状 erratic form
漂块 erratic block(of rock); erratic boulder
漂砾 boulder; drift boulder; erratic block(of rock);erratic boulder
漂砾层 boulder gravel;erratic formation
漂砾沉积 erratic deposit
漂砾带 boulder belt
漂砾道面 boulder pavement
漂砾河槽 boulder channel
漂砾河床 boulder bed
漂砾井 boulder well
漂砾勘探 boulder prospecting
漂砾列 boulder train
漂砾泥 clay till;drift clay
漂砾黏[粘]土 boulder clay;cimolite
漂砾墙 boulder wall
漂砾扇 boulder fan
漂砾土 erratic soil; boulder clay; glacial till
漂砾土冰川泥 boulder clay
漂砾土结构 erratic soil structure
漂砾推移质 boulder shingle
漂砾原群落 petrodium
漂砾指数 erratic block index;erratic index
漂砾质基(底层)土 erratic subsoil
漂凌 drift ice;run of ice;sailing ice
漂流 clubbing; drift(age); drift current; drifting; wash; wind drift; wind-driven current;lay by(e);lying to <船不抛锚>
漂流冰 drift(ing)ice
漂流测示器 drift meter
漂流沉积 drifting sediment; winnowed sediment
漂流船 derelict
漂流的 adrift;water-borne
漂流浮标 drifting buoy
漂流浮标站报告 drifting buoy(station)report
漂流浮标站合作小组 drifting buoy cooperation panel

漂流杆 drift pole
漂流海藻 floating seaweed
漂流航线 sweepstakes route
漂流货物 drift goods
漂流角 angle of drift; angle of leeway; crab angle; drift leeway; leeway(angle); leeway drift; drift angle
漂流矿石 float ore;shoad;shoad stone; shode
漂流理论 drift theory
漂流路径 drift path
漂流路线 drift trajectory
漂流木(材)floodwood;drift wood
漂流木泥炭 driftwood peat
漂流瓶 bottle post; drift bottle; floater; messenger bottle; route-indicating bottle;surface drift bottle;current bottle
漂流矢量 drift vector
漂流水雷 drifting mine
漂流速度 drift velocity; floating velocity
漂流碎屑 water-borne detritus
漂流物 derelict; driftage; drifter; drift way; floating debris; waif; water-borne body
漂流物沉淀池 debris barrier; debris basin
漂流物导向设施 debris deflector
漂流物撞击力 impact load of drift
漂流岩屑 water-borne detritus
漂流者 drifter
漂流植物 sudd
漂流指示锤 drift lead
漂流指示器 drift indicator
漂流着的 water-borne
漂流自动无线电气象系统 drifting automatic radio meteorologic(al) system
漂流自动无线电气象站 drifting automatic radio meteorologic(al) station
漂面 plane of flo(a)tation
漂描测量 drift-scan measurement
漂描曲线 drift curve
漂木 drift log;drift wood
漂木存放水区 log pond
漂木屏障 drift barrier
漂泥 draft mud
漂鸟 wandering bird
漂瓶 bottle
漂瓶资料 bottle paper
漂起 levitating
漂碛层 stratified drift
漂清 rinse;rinsing
漂清车 rinse truck
漂清周期 rinse cycle;rinsing cycle
漂染(工)厂 bleaching and dy(e)ing mill
漂散货物 floating cargo
漂散木 drift
漂沙<又称漂砂> blown sand;drift-(ing) sand; placer sand; quick sand;sand drift
漂沙采集瓶 delft bottle
漂沙的上游面 updrift side
漂沙改向带 nodal zone
漂沙观测 sand drift observation
漂沙快滤池 drifting-sand filter
漂沙矿床 alluvial deposit
漂沙栏 sand drift fence;sand fence
漂沙来源 supply source of littoral drift
漂沙率 rate of littoral transport
漂沙上游方向的丁坝 updrift groin [groyne]
漂沙损失 loss of littoral drift
漂失 drift
漂石【地】boulder; drift boulder; erratic block(of rock); float stone; erratic
漂石爆破作业 blockholing
漂石地基 boulder foundation
漂石击碎器 boulder buster
漂石基床 boulder bed
漂石基底 boulder base
漂石块石粒组 boulder-massive grain grade
漂石类的 bouldery
漂石裸爆覆盖层 mudcapping
漂石黏[粘]土 boulder clay
漂石铺面 boulder paving
漂石铺砌层 boulder setter
漂石墙 boulder wall
漂石原 boulder-field
漂水现象 water splashing
漂水植物 drifting biomaterial;floating plant
漂跳 dap
漂网 drift net;fly net
漂网上小浮标 bowl
漂网鱼船 drifter;drift boat;drift fisher;drover
漂网鱼船队 drifter fleet
漂物测速法 Dutchman's log
漂洗 poach;rinse;rinsing
漂洗槽 potcher
漂洗层 bleached horizon
漂洗池 rinse tank
漂洗废料 kier waste
漂洗废水 kier wastewater
漂洗回收 reclaim rinse
漂洗机 fulling mill
漂洗泥 Fuller's earth
漂洗盆 rinsing tub
漂洗全息摄影栅 bleached holographic(al)grating
漂洗筛 rinsing screen
漂洗设备 rinse device;rinsing device
漂洗水 rinse water
漂洗土 bleached earth;bleached soil
漂洗液 kier liquor
漂洗装置 rinsing equipment
漂下来 fall down
漂向岸边 drift ashore
漂向下风 crab to leeward
漂心 center[centre] of flo(a)tation; tripping center[centre]
漂心距船中距离 distance from midship to center of flo(a)tation
漂心曲线 curve of center[centre] of gravity of water plane
漂心轴 axis of flo(a)tation
漂芯 core flo(a)tation;core raised
漂悬焙烧 shower roasting
漂悬焙烧法 flash roast; suspension roast
漂悬焙烧炉 flash roaster
漂液 bleaching liquor
漂液残渣 bleach(ing)sludge
漂液澄清池 bleaching liquor settling tank
漂液处理 chemicking
漂液贮槽 bleaching liquor storage tank
漂移 deviation dispersion;dispersion; drift(age); drifting; excursion; leeway; shift(ing); shunt running; wander(ing)
漂移板块【地】drift plate
漂移倍数 multiple of drift
漂移变换 conversion by drift
漂移波 drift wave
漂移补偿 drift compensating; drift compensation;drift stabilization
漂移补偿电路 drift compensating circuit
漂移补偿放大器 drift-corrected amplifier

漂移补偿模型 model for drift compensation
漂移不稳定性 drift instability;universal instability
漂移长度 drift length
漂移常数 drift constant
漂移次脉冲 drift subpulse;marching subpulse
漂移大陆 drifting continent
漂移的 adrift
漂移电场 drift field;sweeping field
漂移电导 hopping conduction
漂移电弧 turbulent arc
漂移电流 drift current
漂移电平 drift level
漂移电位 floating potential
漂移电压 drift voltage
漂移度 driftance
漂移法 <连调管> drift method
漂移方向的逆转 backing of drift direction
漂移改正 drift correction
漂移沟道 drift channel
漂移管 drift pipe;drift tube
漂移管质谱分析 drift tube mass spectrometry
漂移轨道 drift orbit
漂移轨迹 drifting locus
漂移耗散不稳定性 drift-dissipative instability
漂移回旋波 drift cyclotron wave
漂移基板 drift base
漂移极限 drift limitation
漂移计 drift meter
漂移记录器 drift recorder
漂移角 drift angle;creep angle <前轮的>
漂移角测定器 drift sight
漂移角度 drift angle;angular shift
漂移校正 drift correction; wander correction
漂移校正放大器 drift-corrected amplifier
漂移晶体管集总模型 drift-transistor lumped model
漂移距离 drift distance
漂移空间 drift space
漂移矿石 float ore
漂移力 drift force
漂移量 amount of drift
漂移灵敏度 deviation sensitivity
漂移率 drift mobility;drift rate
漂移率测量法 rate of drift method
漂移面 drift surface
漂移频率 drift frequency
漂移迁移率 drift mobility;mobility
漂移区 drift region
漂移曲线 drift curve
漂移扫描 drift scanning
漂移沙丘 wandering dune
漂移石屑 drift detritus
漂移时间 drift time
漂移式钻岩机 drifter rig
漂移收敛 drifting convergence
漂移束不稳定性 drift beam instability
漂移数学模型 drift mathematical model
漂移水道 shifting channel
漂移速度 drift rate;drift velocity;migration velocity;shift velocity
漂移速调管 drift klystron; floating drift tube klystron
漂移特征曲线 erratic curve
漂移湍动 drift turbulence
漂移物 drifted material;erratic
漂移误差 drift error
漂移系数 coefficient of deviation
漂移线 drift line
漂移消除 drift cancellation
漂移信号 shifted signal
漂移型晶体管 drift transistor
漂移修正 drift correction
漂移学说 drift theory; allochthonous theory
漂移因素 drift factor
漂移运动 drift motion
漂移载流子 drift carrier
漂移振荡 slow drift oscillation
漂移振荡器 shifting oscillator
漂移值 value of drift
漂移轴 drift axis
漂移自动平衡电路 automatic drifting balanced circuit
漂移作用 drift action
漂油 floating oil;oil spill
漂油回收船 oil-spill collect vessel
漂油清除 clean-up of oil spills
漂油栅 floating oil barrier
漂游 excursion;floating
漂游的 erratic
漂游电位 floating potential
漂游过程 erratic process
漂游生物 plankton organism
漂游植物阶段 floating stage
漂越效应 spillover effect
漂云母 bravaisite
漂运 alligation;drifting
漂煮 kiering
漂煮废料 kier waste
漂煮锅 boiling kier;kier
漂煮锅废水 kier wastewater
漂煮(锅废)液 kier liquor
漂子门 ball valve

飘

飘尘 drifting dust;floating dust;fly-ash;suspending dust
飘尘控制 fly-ash control
飘尘浓度 concentration of floating dust
飘尘污染 air-borne dust pollution; floating dust pollution
飘带 ribbon;streamer;tab
飘带法 <水工试验研究流向的> method of streamers
飘带或长卷形花纹 band(e)rol;bannerol
飘滴 drift
飘滴损失 drift loss
飘浮固体 floating solid
飘浮机 hovering craft
飘浮式偿付款 balloon repayment
飘浮式回收机 flo(a)tation save-all
飘浮式货款 balloon loan
飘浮状 float
飘滑现象 aquaplaning
飘来雨 spillover
飘沙 blow(n)sand
飘失 drift
飘水损失率 rate of drift
飘送 waft(age)
飘损 drift loss
飘雾喷射 drift spraying
飘悬焙烧 flash roasting; suspension roasting
飘悬焙烧炉 flash roaster
飘悬干燥机 flash drier[dryer]; flash drying machine
飘雪 drift snow;nativik
飘移 migration;wandering
飘移电弧 turbulent arc
飘移调节 floating control
飘移土 drift soil

瓢

瓢虫 lady-beetle; lady-bird; ladybug;goldenknop <英>
瓢曲 <板材缺漏> buckle
瓢水 floating water loss
瓢状轮叶 spoon-shaped blade

票

票背签字 endorsement; indorsement
票插 card rack;ticket holder
票额分配计划 planned distribution of passenger tickets
票房 box office;ticket office
票房机 ticket office machine
票房售票机 booking office machine
票根 counterfoil;counterfoil stub of a checkbook;stub
票汇 remittance by banker's demand draft
票汇汇率 draft exchange rate;draft rate
票汇及电汇 draft and telegraphic(al) transfer
票货不符 non-coincidence of shipping documents and luggage or parcel
票货分离 separation of invoice and goods; separation of waybill from shipment
票货分离的货物 astray freight
票价 fare
票价变动 fare change
票价表 fare table
票价补贴 fare subsidy
票价率 rate of passenger fare
票价水平 fare level
票架 ticket rack
票兼售票机 ticket printing and issuing machine
票剪 nippers;puncher;ticket clipper
票据 bill; instrument; negotiable instrument;note;paper
票据保证 guarantee of a bill;guarantee of bills
票据背书附条 allonge
票据簿 bill book;notebook
票据承兑 acceptance of bills;acceptance of checks;bill acceptance;bill accepted
票据承兑(利)率 acceptance rate
票据承兑市场 acceptance market
票据承兑所 acceptance house
票据承兑所协会 Accepting Houses Committee
票据承兑贴现率 acceptance rate
票据承兑信贷 acceptance credit
票据承兑业 acceptance business
票据承兑银行 acceptance bank
票据承受人 endorsee
票据持有人 holder
票据处理机 billing machine
票据传送设备 classification list conveyor system
票据代换所 clearing agency
票据贷款 bill loan
票据担保人 guarantor
票据当事人 parties to a bill
票据到期 bill to fall due;bill to mature;fall due
票据到期日 due date
票据到期通知书 note notice
票据的流回 circuity of action
票据的黏[粘]度 allonge
票据的有效性 validity of the bill
票据登记簿 bills book;note register
票据等后面的签名 endorsement
票据抵补 cover for draft
票据抵押贷款 loan on notes;loans on bill
票据发行便利 note insurance facility
票据法 checking law;law of bill;law of negotiable checks;law of negotiable instruments;negotiable instrument law
票据分录簿 note journal
票据分配员 delivery clerk
票据付款人 drawee
票据付款通知 advice of bill collected;advice of bill paid
票据付款行 drawee bank
票据附页 allonge
票据副页 bill duplicate
票据高升水 shaving
票据管理审计 note management audit
票据管理员 bill clerk
票据合同 contract of bill
票据回收账簿 collection book for bills
票据汇兑 exchange by bills
票据机 billing machine
票据夹 bill fold
票据交换 bills in process of clearing; check clearing; clearance; clearing (of bills);clearing of checks
票据交换差额 clearing balance;clearing house balance
票据交换贷方传票 clearing house credit ticket
票据交换额 outclearing
票据交换工 clearing labor
票据交换后银行贷差 credit bank balance
票据交换后银行借差 debit bank balance
票据交换日期 date of clearance
票据交换市场 market for bill exchange
票据交换所 banker's clearing house; clearing house;settlement house
票据交换所报表 clearing house statement
票据交换所贷借决算表 clearing sheet
票据交换所贷款证券 clearing house loan certificate
票据交换所会议 clearing house conference
票据交换所会员银行 clearing member
票据交换所决算表 clearing house proof
票据交换所清单 clearing house statement
票据交换所清算表 clearing house settlement sheet
票据交换所资金 clearing house funds
票据交换协会 clearing house association
票据交换业务 clearing business;clearing services
票据交换轧差银行 debtor bank
票据交换中心 cedel
票据交货所 clearing house
票据交易 bill business; exchange by bills
票据交易所 clearing house
票据经纪人 bill broker
票据拒付 notice of dishono(u)r
票据拒付证书 protest of bill
票据签证 endorsement
票据清算 bill clearing
票据商 bill merchant
票据上的通融人 accommodation party
票据市场 bill market
票据贴现 discounting a bill
票据贴现率 acceptance rate;bill rate
票据贴现人员 discount clerk
票据贴现押金 bill discount deposit
票据托收 bill collection; instrument collection
票据委托书 draft advice;drawing advice
票据账目 bill account
票据支付期 time of maturity

票据转期 notes renew
票口 ticket bezel
票面的 rated
票面额 par;par value
票面价格 face par;nominal par
票面价值 denomination(al) value;denominator;face value;paper value;par value;nominal value
票面金额 face amount;face value;par
票面值 face value;par value
票面值法 par value method
票期未到 bill undue
票数相等 draw of votes;tie of votes
票外承兑 extrinsic acceptance
票务室 ticket affairs office;ticket office
票箱 ticket case
票样 specimen ticket
票证 coupon
票证单插 billing card holder

氕核 proton

撇除 decantation;scraping;skim

撇浮渣勺 scummer
撇缆 hauling line;heaving and hauling line;heaving line;rocket line;whip line
撇缆带缆 line handling
撇缆活结 heaving line slip knot
撇缆接结 heaving line bend;heaving line knot
撇缆头 sand lead
撇缆头结 monkey fist
撇离法 decantation
撇流 skimming flow
撇沫 skimming
撇沫板 <用于污水池中撇开泡沫悬渣,能升降的隔板> floating scumboard;skimmer
撇沫过滤器 skimmer filter
撇沫机 skimmer;skimming machine
撇沫机制 skimmer mechanism
撇沫器 grease-skimming tank;skimmer;skimming dish;skimming machine
撇沫堰 skimming weir
撇奶油式订价法 skim-the-cream pricing
撇奶油式价格 skimming price
撇清 skimming
撇取 skimming;skim(off)
撇去 skim(off)
撇去浮质 skimming
撇熔砂勺 slag scummer
撇乳器 skimmer
撇水板 skimming plate
撇水堰 skimming weir
撇滩裁弯 chute cut-off
撇弯切滩 chute cut-off
撇液板 skimming baffle
撇油板 skimming baffle
撇油驳船 oil skimming barge
撇油槽 skimming tank
撇油池 deoiling skimmer;fat collector;grease-skimming tank;skimming pit;skimming pond;skimming tank
撇油船 oil skimmer barge
撇油管线 skimming(pipe) line
撇油罐 oil skimming tank
撇油器 grease skimmer;oil skimmer;oil skimmer vessel;skimming device
撇油室 skimming chamber
撇渣 dross(ing);dross run;dross trap;raking off the slag;scumming;skim(ming);slag-off
撇渣板 skimming panel;skimming plate
撇渣棒 skimmer bar;slag skimmer
撇渣沉沙池 skimming detritus tank
撇渣池 skimming tank
撇渣杆 skim bar
撇渣机 scum skimmer;skimming machine
撇渣口 skin gate
撇渣面层 skim coat
撇渣泥芯 skim core
撇渣器 dam;dross extractor;scum skimmer;skimmer;skimming tool;slag separator;slag skimmer
撇渣勺 skimming ladle
撇渣芯 skimming core
撇蒸 skimming
撇蒸厂 skimming plant
撇蒸过程 skimming processing
撇脂池 grease skimmer
撇脂器 grease skimmer
撇资 devest

拼 V 形缝的单边 arrissed corner;arrissed edge

拼板 flap;splice
拼板地面 raft-slab floor
拼板防挠压条 <固定在背面与板缝垂直> counter battens
拼板工作 match boarding;matching
拼板建筑 stave construction
拼板接合 stave jointing
拼板接合机 stave-jointing machine
拼板门 match board door;barred door;batten door;ledged door <不带框的>;framed and ledged door <带框的>
拼板门的横档 door ledge
拼板木模 close timbering
拼板木支撑 close timbering
拼板天花 matched ceiling
拼板斜撑门 framed-and-ledged-and-braced door <门扇带框>;ledged and braced door <门扇不带框>
拼板心胶合板 batten board
拼版工人 stone hand
拼版晒版 key flat
拼版台 register table
拼版印刷 multiple printing
拼边刨 match plane
拼车货 less-than-carload;less-than-truckload
拼成的旋压模 collapsible spinning block
拼成图案的地面覆盖材料 patterned floor cover(ing) tile
拼成图案的砖工 pattern brick masonry work
拼凑 patch;rig
拼凑的合伙经营 bobtail pooling
拼凑式 mosaic type
拼凑式控制台 mosaic-type control panel
拼凑物 patch work
拼钉 double-headed nail;dowel pin;dual head nail
拼分 ingredient
拼缝 edge joint;wood joint
拼缝凹陷 sunken joint
拼缝地板 joint floor(ing)
拼缝机 joint applicator;splicer
拼缝间隙 gap joint
拼缝内层单板 jointed inner plys
拼缝刨 shooting plane
拼缝器 joint applicator
拼缝芯板 jointed core
拼缝织物 patch work
拼高消费水平 encouraging a high level of consumption
拼好版的打样 page proof
拼合 amalgamation;cocking;synthesis
拼合 S 形转弯 split S-turn
拼合扳牙 split-die
拼合板 match(ed) board
拼合表面 mating surface
拼合衬垫 splice pad;split bush
拼合衬套 bushing split;split housing;split bushing
拼合齿轮 split gear
拼合齿轮系驱动 split-train drive
拼合处理 parting handling
拼合带 splicing tape
拼合刀架 split holder
拼合的 split
拼合的瓦屋顶 splitter tiled roof
拼合底座 split base
拼合地板 matched floor
拼合地毯 carpet tile
拼合隔板 split diaphragm
拼合拱墩 banded impost
拼合毂 split hub
拼合管 split pipe;splitter;split tube
拼合管打入式取土器 split-tube drive sampler
拼合管取土器 split-tube sampler
拼合滑车 parting pulley;splitting pulley
拼合滑轮 parting pulley;splitting pulley
拼合环 split collar
拼合环座圈 split collar retainer
拼合记号 match marking
拼合夹紧套 split clamping bearing
拼合夹具 split clamp
拼合接头 splice joint
拼合截面 fabricated section
拼合镜面望远镜 segmented mirror telescope
拼合壳 split housing
拼合口 split
拼合链节 split link
拼合梁 joggle beam;keyed beam;pieced beam
拼合轮 parting pulley;segmental wheel;split wheel
拼合轮毂式皮带轮 split-hub pulley
拼合轮辋 split rim
拼合螺母 split nut
拼合铆钉 split rivet;stitch rivet
拼合面 parting face
拼合模 assembled die;composite die;multiple segment die;sectional model;segment die;split-die
拼合模块 die piece;die section;die segment
拼合摩擦锥轮 split friction cone
拼合木料 pieced timber
拼合木模 split pattern
拼合木模制型机 split pattern mo(u)lding machine
拼合泥驳 split barge;split-bottom barge;split dump barge
拼合扭转传动 split torque drive transmission
拼合皮带轮 split pulley
拼合石 matching gem
拼合式 split type
拼合式拐臂 split lever
拼合式滑动轴承 split box
拼合式夹紧支座 split clamping bearing
拼合式联轴器 split coupling
拼合式链节 two-piece link
拼合式链轮 split sprocket
拼合式螺栓锻模 open die
拼合式模具 assembled die;segmental die
拼合式木支持轮 split wooden carrying roller
拼合式曲柄箱 split type crank case
拼合式取土器 split spoon
拼合式凸模压板 split punch pad
拼合式外壳 two-piece housing
拼合式岩芯管 split barrel;split core barrel
拼合式转轮 split runner
拼合式自航挖泥船 split hull hopper dredge(r)
拼合竖框 split mullion
拼合条 splicing tape
拼合外壳 split casing
拼合外压圈滚柱轴承 split outer race bearing
拼合桅 armed mast
拼合文字 monogram
拼合线 split line
拼合楔 split wedge
拼合芯棒 split core rod
拼合型 sectional mo(u)ld
拼合型板 clincher pattern plate
拼合型芯 sectional core
拼合用销钉 dowel pin
拼合直棂 split mullion
拼合轴衬 split bearing shell
拼合轴承 split bearing;two-piece bearing
拼合轴瓦 slip bearing bushing;split bushing
拼合柱 pieced column
拼合铸模 split mo(u)ld
拼合锥形衬套 split tapered bushing
拼花 inlay
拼花板 match(ed) board
拼花板门 panel door
拼花边缘 parquet border
拼花玻璃 cathedral glass;pattern glass
拼花玻璃板 cathedral glass sheet
拼花玻璃装配 pattern-glazed
拼花的 tongue and groove matched
拼花地板 block floor(ing);matched floor;mosaic pavement;parquet floor(ing);wood block flooring
拼花地板块 parquet block
拼花地板磨光机 parquet sander
拼花地板木 parquet wood
拼花地板铺设 parquet flooring
拼花地板砂光机 parquet sanding machine
拼花地板条 parquet strip;parquet strip flooring;wood strip parquet
拼花地板镶边 parquet border
拼花地板砖 patterned floor(ing)(finish) tile
拼花地面 asarotum[复 asaroria];mosaic pavement;sectile opus
拼花路面 asarotum[复 asaroria];tessellated path
拼花马赛克 parquet mosaic
拼花面 mosaic surface
拼花木板 matched floor;parquet
拼花木地板块 wood mosaic
拼花木块 parquet block
拼花木作 parquetry;parquet work
拼花铺面 mosaic pavement
拼花贴面【建】mosaic surface;oystering
拼花图案 mosaic
拼花镶面板 patterned veneer
拼花镶嵌 inlay;intarsia
拼花小板条 parquet strip
拼花硬木地板 checker parquet flooring
拼花制品 tarsia
拼混染料 mixed dye
拼件 mosaic building block
拼件率 rate of combined package
拼槛 sill splicing

P

拼铰接 splice fished joint
拼接 blind splice; joint; marriage; montage; scab; splice; splice grafting; splicing wire; split joint
拼接板 butt cover plate; butt strip; joint tie; scab; splice (bar); splice board; splice piece; splice plate
拼接板式接头 spliced plate type joint
拼接草图 combined sketch
拼接场 joining yard
拼接衬垫 splice pad
拼接处 splice
拼接带 splice tape; splicing tape
拼接的 scabbed
拼接的指针 splicing needle
拼接垫板 splice pad
拼接盖板 splice cover
拼接钢筋 splice bar
拼接工场 joining shop
拼接工作 joggle work
拼接桁架 joggle truss; spliced truss
拼接技术 butting technique
拼接架 joggle truss
拼接件 splicer
拼接胶 splicing cement; splicing glue
拼接胶粘剂 splice adhesive
拼接角钢 spliced angle
拼接接合 spliced connection
拼接接头 joggled joint; splice joint
拼接块 splice block
拼接梁 splicing sill; joggle beam
拼接螺栓 splice bolt
拼接木材 pieced timber; piecing
拼接木材的穿钉 drift bolt
拼接木料 built-up beam; built-up timber
拼接黏[粘]接剂 splice cement
拼接砌合 joggle work
拼接清洁剂 splice cleaner
拼接镶嵌图 scale-ratio mosaic
拼接小过梁 joggled lintel
拼接修整 jointing
拼接桩 segmental pile; spliced pile
拼接装置 splicing apparatus
拼接籽晶 joining seed
拼卷作业线 coil build-up line
拼块梁梁段 block beam
拼块芯木门 wood blocks core door
拼栏 upmake
拼连的两所房屋 double house; twin house
拼连接 splice fished joint
拼梁 joggle beam; keyed beam; sill splicing
拼料成分 mixing ingredient
拼木板芯 stave core
拼木地板 wood block flooring
拼木块砖地板 block flooring
拼木芯板结构 lumber-core construction
拼配 matching
拼晒 make-up copying
拼设备 deploy equipment to utmost capacity; overuse equipment
拼生【植】 symphysis
拼石环氧灌浆 rock-matching epoxy grout
拼条地板 inlaid-strip floor
拼条地面铺饰 inlaid-strip floor cover(ing)
拼贴 collage; marriage; montage
拼贴底图 compilation sheet; composite sheet
拼贴画 collage
拼贴技术 <装饰工艺之一的> collage
拼箱 groupage
拼箱货 less than container load
拼箱集装箱 less than container load
拼箱提单 groupage bill of lading
拼箱运输 container transport by loading more than one consignments of goods in one container; groupage traffic
拼箱制 <航运> groupage system
拼箱装载 less than contained load
拼写 transliteration
拼写错误 cacography
拼写检查程序 spell checker
拼修 cannibalise[cannibalize]
拼制木柱 built spar
拼桩 pile splice
拼装 fitted assembly; splice
拼装场 consolidation shed
拼装承运商 consolidator
拼装承运商提单 consolidator's bill of lading
拼装大梁 section girder
拼装的 built-up
拼装点 assembly jig
拼装构件 built-up member; segmental member
拼装构造 section construction
拼装轨道 erector track
拼装机 erector
拼装阶段的测量 surveys during placement
拼装结构 section construction
拼装梁 built-up beam; compound beam
拼装轮胎 dual tires[tyres]
拼装式 unit-built type
拼装式衬砌 precast lining; prefabricated lining
拼装式船舶 built-up boat; dismountable boat; dismountable ship; dismountable vessel
拼装式挡土墙 built-up retaining wall
拼装式房屋 portable building
拼装式浮船坞 self-docking dock
拼装式钢护舷 collapsible steel fender
拼装式钢筋混凝土盖板涵 assemble reinforcement concrete cover plated culvert
拼装式钢模板 collapsible steel form
拼装式活动钢模板 collapsible steel shuttering
拼装式结构 section construction
拼装式模板 assembling form(work); collapsible form(work)
拼装式墙体 dry wall(ing)
拼装式桥墩 spliced pier
拼装式跳舞地面 portable dance floor
拼装水箱 sectional tank
拼装台 assembly jig; jib
拼装套管 split thimble
拼装性 modularity
拼装油箱 sectional tank
拼装站 consolidation depot; consolidation station
拼装制 building block system
拼装柱 built-up column
拼装作业 consolidation service
拼缀 spell

贫

贫拌和料 lean mix(ture); poor mix(ture)
贫拌合[和]物 poor mix(ture)
贫乏 leanness; paucity; poverty; tenuity
贫腐水性的 oligosaprobic
贫雇农 poor peasant and farm labo(u)rers
贫化 depletion; impoverishment
贫化材料 depleted material
贫化燃料 depleted fuel; impoverished fuel
贫化液 stripped solution
贫化铀 depleted uranium
贫灰混合料 lean mix(ture)
贫灰混凝土 concrete of lean mix; lean concrete; poor (-mix) concrete; weak concrete
贫灰浆 lean mortar
贫灰泥 lean mix(ture)
贫混合料 poor mix(ture); rare mixture
贫混合物 poor mix(ture)
贫混凝土 lean(cement) concrete; lean mix concrete; poor concrete
贫混凝土拌合[和]物 rare mixture
贫混凝土底层 lean concrete base
贫混凝土垫层 blinding; lean concrete pad; lean concrete underlayer
贫混凝土封层 blinding
贫混凝土混合料 rare concrete mixture; weak concrete mixture
贫混凝土基层 lean concrete base
贫混凝土基底 lean concrete base
贫混凝土试块 lean concrete test cube
贫混凝土试验用圆柱体 lean concrete test cylinder
贫混凝土外包层 <管道的> surround
贫瘠 hungry
贫瘠的 arid; barren; infertile; lean; meager[meagre]
贫瘠的土地 infertile soil
贫瘠的土壤 impoverished soil
贫瘠地 arid land; barren; impoverished soil; poor ground; poor soil
贫瘠地区 hungry spots
贫瘠海区 barren waters
贫瘠化 impoverishment
贫瘠黏[粘]土 meager[meagre] clay; poor clay
贫瘠土 impoverished soil; oligotrophic soil; poor soil
贫瘠土地 marginal land
贫瘠土壤 worn-out soil
贫瘠的 meager; meagre
贫瘠棕色土 oligotrophic brown soil
贫胶区 adhesive starved area
贫搅拌 lean mixing; lean mix(ture)
贫金属卤水 metal-poor brine
贫矿脉 blowout; coose; low vein
贫矿(石) halvans; low-grade ore; poor ore; lean ore; mining sterile; barren ore; halvings; lean material; poor quality of the ores
贫矿石英 dead quartz
贫困城市 distressed city
贫困地带 pocket of poverty
贫困地区 distressed area; poor area; poverty-stricken zone
贫困国家 under-privileged country
贫困劳工 pauper labor
贫困区域 pocket of poverty
贫料 lean
贫硫沥青 tscherwinskite
贫毛类环虫 <生于土中,使土壤空气流通,有利于植物生长> oligochaete
贫煤 blind coal; dry burning coal; lean coal; meager[meagre] coal
贫民救济院 poorhouse
贫民窟 back slum; rookery; shanty town; slum
贫民窟的拆除 land clearance
贫民窟的清理 rookery clearance; slum clearance
贫民窟居民 slumdweller
贫民窟清除地区 clearance slum area; cleared slum area
贫民窟问题 slum problem
贫民窟现象 slumism
贫民区 purlieu; slum; slum area; slum district; slum land; slum site; slum zone
贫民区改造 slum clearance
贫民区改造地带 slum-cleared zone
贫民所 almonry house
贫民院 almshouse; poorhouse
贫民住房 poorhouse
贫黏[粘]土 lean clay; poor clay
贫碾压混凝土 poor mixing ratio rolled concrete
贫配合比 lean mixture ratio
贫配合(比)混凝土 poor-mix concrete
贫气 lean gas
贫气提升管 lean gas riser
贫铅矿石 keckle meckle
贫穷地带 pocket of poverty
贫穷地区 impoverished area
贫燃料混合气 lean fuel mixture
贫燃料混合物最大功率 lean mixture maximum power
贫燃料气 poor gas
贫燃料燃烧 fuel-lean combustion
贫溶液 lean solution
贫砂拌和料 undersanded mix(ture)
贫砂混凝土 undersanded concrete
贫砂浆 lean mortar
贫砂沥青混合料 lean sand asphalt
贫石灰 lean lime; meager [meagre] lime; poor lime
贫石灰膏 lean lime
贫瘦煤 meager lean coal
贫树脂区 resin-starved resin
贫水泥 lean mix(ture)
贫水泥浆 lean mortar
贫水硼砂 kernite
贫水溢出带 poor-water zone
贫吸收油 lean absorption oil
贫锡矿 leap ore
贫血症 amaemia
贫养湿源 high moor
贫氧层 oxygen-poor layer
贫氧酸沼地 high moor
贫氧态 anaerobic
贫液 barren liquor; lean solution
贫液内蒸槽 lean solution flash drum
贫营养的 oligotrophic
贫营养湖 dystrophic lake; oligotrophic lake
贫营养湖型水体 oligotrophic lake type waters
贫营养湖型水域 oligotrophic lake type waters
贫营养化 oligotrophication
贫营养林沼 oligotrophic swamp
贫营养泥沼 oligotrophic mire
贫营养生境 nutrient-poor habitat; oligotrophic habitat
贫营养水 oligotrophic water
贫营养水体 oligotrophic water body
贫营养水域 oligotrophic water body
贫营养酸沼 oligotrophic bog
贫营养条件 oligotrophic condition
贫油 leaning; lean oil
贫油断路器 oil-poor circuit-breaker
贫油混合气 lean mix(ture)
贫油混合物抗爆性评定法 lean mixture rating method
贫油极限 lean-limit
贫油井 stripper
贫油沥青混合料 lean bituminous mixes
贫油调节 lean metering
贫油页岩 lean oil shale
贫渣 lean slag
贫装料 lean burden
贫滋育湖泊 oligotrophic lake

频

频爆式排土装置 repeated explosion device of soil displacement
频爆式推土机 blasting bulldozer

频爆式推土装置 repeated explosion device of soil displacement
频比 frequency ratio
频变沉区 frequency sinking zone;sink
频标 frequency scale
频标振荡器 frequency marker oscillator
频波传导 band conduction
频差 frequency deviation
频差共振 frequency difference resonance
频差检波器 frequency difference detector
频颤检验器 frequency flutter checker
频程 frequency interval;sound interval
频窗范围 window-range
频带 frequency band;wave band
频带边缘 limit of (the) band; band limit
频带补偿 band compensation
频带参差 frequency staggering
频带倒置 frequency inversion
频带分割 band sharing;band splitting
频带分割器 band splitter
频带分割同时制彩色电视系统 band shared simultaneous colour system
频带分隔滤波器 separation filter
频带分隔器 band separator
频带分离网络 band-separation network
频带分裂设备 band splitting equipment
频带分析器 band analyser[analyzer]
频带改变 band shift
频带共用 band sharing
频带级 band level; frequency band level
频带空段 suckout
频带宽度 bandwidth; broadness of band;frequency bandwidth
频带扩展 band spread(ing)
频带扩展比 bandwidth expansion ratio
频带扩展微调电感器 bandspreader
频带扩展微调电容器 bandspreader
频带模 band modal
频带内失真 inband distortion
频带偏移 band bending
频带清晰度 band articulation
频带伸展放大器 expander amplifier
频带声功率级 sound band power level
频带声压级 sound band pressure level
频带特别宽的 ultrabandwidth
频带调整 band adjustment
频带外的 out of band
频带下限 greatest lower band
频带形成电路 band shaping circuit
频带压力级 band pressure level
频带压强级 band pressure level;pressure band level
频带压缩 bandwidth compression; frequency band compression
频带压缩放大器 band compressor amplifier;compressor amplifier
频带压缩信号系统 suppressed frequency signalling system
频带展开调谐控制 bandspread tuning control
频带展开制 bandspread system
频带展宽 bandspread; band spread(ing)
频带展宽比 bandwidth expansion ratio
频带展宽系数 band expansion factor
频带指示器 band indicator
频带中心频率 band center[centre] frequency;mid-band frequency
频带重叠 band overlap
频带转换 band conversion
频道 channel;frequency channel
频道定义格式 channel definition format
频道方案 channel plan
频道间隔 channel spacing
频道宽度 channel size;channel width
频道容量 channel capacity
频道设计 channel plan
频道数 channel number; number of channels
频道选择器 tuner
频道选择器转鼓 tuner drum
频道转换电路 channel switcher
频道转换开关 channel selector;channel switcher
频低 low frequency
频度 frequency
频度累积率 accumulation rate
频度特性分析 frequency response analysis; frequency characteristic analysis
频度特性分析器 frequency response analyser[analyzer]
频度震级 frequency magnitude
频度震级参数 frequency magnitude parameter
频度震级系数 frequency magnitude coefficient
频段 frequency band;frequency channel;frequency range;wave band
频段分配 frequency allocation
频段扩展 bandspread;frequency band spreading
频段扩展(精调电容)器 bandspreader
频段名称 frequency spectrum designation
频段转换 band switching
频繁罢工 frequent strikes
频繁变换工作的人 job hopper
频繁产业调查 frequent industrial inquiry
频繁使用 congested use
频繁使用的脚手架 heavy-duty scaffold
频繁停车操作 stop-and-go driving
频繁挖掘 frequent hoeing
频分多路复用 frequency division multiplex(ing)
频分多路复用线路 frequency division multiple circuit
频分多路话频通信[讯] frequency division multiplex voice communication
频分多路设备 frequency division multiplex equipment
频分多路调制 frequency division multiplex modulation
频分多路系统 multichannel FDM[frequency-division multiple] system
频分多址接入 frequency division multiple access
频分多址(联结) frequency division multiple access
频分信道 frequency derived channel
频分与时分数据线路 frequency and time-division data link
频积 total frequency
频宽比 duty cycle;off-duty factor
频率 beat;cycles per second;frequency;freak <俚语>
频率摆动 frequency swing; warble; wobbulation
频率倍减 frequency demultiplication
频率倍减器 scaler;scale unit;scaling unit
频率倍增 frequency multiplication
频率倍增电路 frequency multiplier circuit
频率被调制的 frequency-modulated
频率比 frequency ratio
频率比较器 frequency comparator
频率比较器电路 frequency comparator circuit
频率编码制 frequency coding system
频率变换 frequency conversion; frequency frogging; frequency transformation;frequency transmission
频率变换电路 frequency-changing circuit;frequency conversion circuit
频率变换器 frequency changer
频率变换装置 frequency transforming arrangement
频率标准 frequency standard
频率表 frequency indicator;frequency list;frequency meter
频率波动 frequency jitter
频率波数分析 frequency wave number analysis
频率补偿 frequency compensation
频率捕捉 frequency capture
频率不连续性 frequency discontinuity
频率不稳定 swinging
频率不稳定度 frequency instability
频率不稳定性 frequency instability
频率参差 frequency staggering
频率参数 frequency parameter
频率测定仪 frequency meter
频率测量电桥 frequency measuring bridge
频率测量设备 frequency measuring equipment
频率测深法 frequency sounding method
频率测深曲线图 curve of frequency sounding
频率测深仪 frequency sounding instrument
频率测试台 frequency measuring station
频率插入传送法 frequency interlace technique
频率差 frequency difference
频率差动继电器 frequency differential relay
频率产生器 frequency generator
频率颤动 flutter; frequency wow; wow
频率颤动测试器 flutter meter
频率常量 frequency constant
频率常数 frequency constant
频率场函数 frequency field function
频率陈化 frequency aging
频率成分 frequency component; frequency composition; frequency content
频率程序设计 frequency programming
频率程序系统 frequency programming system
频率池 frequency pool
频率抽样定理 frequency sampling theorem
频率畴 frequency domain
频率传递函数 frequency transfer function
频率传输 frequency transmission
频率传输曲线 frequency transmission curve
频率次数 frequency times
频率带 frequency range
频率倒置串音 inverted crosstalk
频率倒置的言语 inverted speech
频率电码 frequency code
频率电压变换器 frequency to voltage converter
频率抖动 frequency jitter
频率对比 frequency comparison
频率多边形 frequency polygon
频率多路传输 <按传输频率划分为多路通道的多信息传输方法> frequency multiplexing
频率二等分 frequency halving
频率发射(转换)机 frequency translator
频率发生器 frequency generator
频率反演 frequency inversion
频率反应 frequency response
频率反应函数 frequency response function
频率反应曲线 frequency response curve
频率反应特性 frequency response characteristic
频率范围 frequency coverage; frequency domain method; frequency range
频率方程式 frequency equation
频率分辨率 frequency resolution
频率分辨(能力) frequency resolution
频率分布 frequency allocation;probability density distribution; relative frequency distribution; frequency distribution <按数值排列的观察系统,统计用>
频率分布测定 measurement of frequency distribution
频率分布的图示法 graphic(al) presentation of frequency distribution
频率分布分析 analysis of frequency distribution
频率分布记录 frequency histogram record
频率分布类型 type of frequency distribution
频率分布器 frequency histogram
频率分布曲线 curve of frequency distribution; frequency distribution curve
频率分布图 frequency distribution diagram;histogram;map of frequency distribution
频率分布直方图 frequency distribution histogram
频率分割 frequency division;frequency splitter
频率分割多路传输 frequency division multiplex(ing)
频率分割多路电报终端机 frequency division multiplex telegraph terminal
频率分割技术 frequency division technique
频率分割制 frequency division system
频率分割制交换机 frequency division switching system
频率分隔 frequency diversity; frequency division; frequency separation;frequency splitting
频率分集(制) frequency diversity
频率分解 frequency decomposition; frequency resolution
频率分离器 frequency separator
频率分量 frequency component
频率分裂 frequency splitter
频率分配 allocation of frequency; allowance of frequency; frequency allocation; frequency assignment; frequency planning
频率分配表 <载波> frequency assignment chart
频率分配的国际规程 international procedure of frequency assignment
频率分配委员会 frequency allocation committee
频率分配中心 frequency allocation center
频率分析 frequency analysis

频率分析法压缩 frequency analysis compaction
频率分析技术 frequency analysis technique
频率分析精简法 compaction; frequency analysis compaction
频率分析器 frequency analyser[analyzer]
频率分析仪 frequency analyser[analyzer]
频率复用 channel(l)ing
频率跟踪器 frequency tracker
频率共用 frequency sharing
频率固定 frequency fixing
频率关系式 frequency dependence; frequency correlation
频率管理 frequency government
频率函数 frequency function
频率合成器 frequency synthesizer
频率合成器电路 frequency synthesizer circuit
频率划分表 table of frequency allocations
频率划分多路传输制 frequency division multiplex(ing)
频率划分制电路 frequency sharing scheme
频率缓变 slow frequency drift
频率混淆 frequency alias
频率畸变 frequency distortion
频率及计时标准 frequency and time standard
频率级 frequency class;frequency level
频率级数 frequency series
频率极限 frequency limit(ation)
频率急降 frequency breakdown
频率计 cycle counter; cycle rate counter; cymometer; frequency indicator; frequency meter; frequency teller; ondometer
频率计划 frequency plan
频率计数器 frequency counter
频率计算的洪水 frequency-basis flood
频率继电器 frequency relay; frequency responsive relay
频率加强滤波 frequency emphasis filtering
频率加权函数 frequency-dependent weighting function
频率加权网络 frequency weighting network
频率间隔 frequency interval; frequency space; frequency span
频率间求差 between frequency difference
频率监测器 frequency monitor
频率监视器 frequency monitor
频率减半 frequency halving
频率减慢 frequency deceleration
频率检波 frequency discrimination
频率检波器 frequency discriminator
频率检校 frequency calibration
频率鉴别 frequency discrimination
频率键控 frequency keying
频率键控法 frequency keying method
频率键控器 frequency shift keyer
频率降低 frequency step down
频率交叉 frequency crossing; frequency frogging
频率交错 frequency interlace; frequency interlacing; frequency interleave
频率交错法 frequency interleaving
频率交错技术 frequency interlace technique
频率交错制 frequency interleaving system
频率交换 frequency frogging
频率交换信号传输 frequency exchange signaling
频率交联 cross banding
频率校正 frequency compensation; frequency correction
频率校正电路 accentuator; compensating circuit; emphasizer; frequency correction circuit
频率校正线路 deaccentuator
频率校准 frequency calibration
频率阶跃 frequency step
频率捷变 frequency agility
频率截止 frequency cutoff
频率解调 frequency demodulation
频率界限 frequency limit
频率卷积定理 frequency convolution theorem
频率均衡器 frequency equalizer
频率开关 frequency switching
频率可变振荡器 variable frequency oscillator
频率刻度 frequency scale
频率空间 frequency space
频率控制 frequency control
频率控制传动装置 frequency controlled actuator
频率控制回路 frequency control loop
频率控制激光器 frequency-controlled laser
频率控制器 frequency regulator
频率控制系统 frequency control system
频率快变 agility of frequency
频率来回变动 frequency swing
频率累计曲线 total frequency curve
频率滤波 frequency filtering
频率滤波器 frequency filter
频率码 frequency code
频率脉冲 frequency(im)pulse
频率密度 frequency density
频率密码 frequency coding
频率面 frequency surface
频率敏感器件 frequency device
频率能量分布 frequency energy distribution
频率拟合 frequency matching
频率配置 frequency allocation
频率偏差 frequency offset
频率偏差容许度 frequency tolerance
频率偏移 frequency deviation; frequency offset
频率偏移包迹 frequency-deviation envelope
频率漂移 drift of frequency; frequency departure; frequency drift; frequency deviation
频率漂移速率 rate of frequency drift
频率平滑 frequency smoothing
频率谱 frequency spectrum
频率起伏 frequency fluctuation
频率强化 accentuation
频率曲线 frequency curve
频率曲线正态分布 <统计分析的> normal distribution of frequency curve
频率容限 frequency tolerance
频率扫描 frequency scan; frequency sweep
频率扫描试验 frequency sweep test
频率上限 upper-frequency limit
频率失真 frequency distortion
频率时间谱密度函数 frequency time spectral density function
频率识别装置 frequency-identification unit
频率式轨道电路 frequency track circuit
频率疏散 frequency diversity
频率数 frequency number
频率锁定 frequency lock
频率锁相 frequency phase lock
频率特性 frequency character(istic); frequency response
频率特性的陡度 frequency characteristic gradient
频率特性的曲率调节 curvilinear regulation of frequency
频率特性校平 smoothing of frequency characteristics
频率特性曲线 response curve
频率特性曲线的折断点 break frequency
频率特性曲线上升 high boost
频率特性曲线衰减 frequency attenuation
频率特性试验 frequency characteristic run
频率特性相等的波道中的起伏噪声 flat-channel noise
频率特性响应曲线 frequency response curve
频率特征 frequency character(istic)
频率调节 frequency modulation; frequency regulation
频率调节法 frequency modulation method[FM-method]
频率调节器 frequency regulator
频率调谐带通滤波器 frequency tuned bandpass filter
频率调谐范围 frequency tuning range
频率调整 frequency adjustment
频率调整器 frequency regulator; tuner
频率调制 frequency modulation
频率调制器 frequency modulator
频率调制系统 frequency-modulated system
频率通道 frequency channel
频率同步 frequency synchronism
频率同步范围 frequency look-in range
频率同步指示器 frequency lock indicator
频率统计曲线图 frequency statistic histogram
频率突变 frequency discontinuity
频率突增 glitch
频率图(解) frequency diagram; frequency chart; frequency graph; frequency plot; interference diagram
频率推移 frequency pushing
频率推移系数 frequency pushing factor
频率微变 fractional frequency change
频率微调 frequency fine tuning; frequency trim
频率位移 frequency shifting
频率温度特性 frequency temperature characteristic
频率稳定 frequency fixing
频率稳定度 frequency stability
频率稳定性 constance[constancy] of frequency; frequency stability
频率误差 frequency error
频率系数 frequency factor
频率下降 decrease of frequency
频率下限 lower frequency limit
频率限制 frequency restriction
频率陷波器 frequency trap
频率相关 frequency dependence
频率相依 frequency dependence
频率响应 frequency response
频率响应逼真度 fidelity of frequency response
频率响应法 frequency response method
频率响应法求解 frequency response approach
频率响应分析器 frequency response analyser[analyzer]
频率响应函数 frequency response function
频率响应计算法 frequency response approach
频率响应曲线 frequency response characteristic; frequency response curve
频率响应式速度调节器 frequency response speed governor
频率响应显示设备 frequency response display set
频率向量 frequency vector
频率效应 frequency effect
频率斜率调制 frequency-slope modulation
频率信号 frequency code
频率选择 frequency selection
频率选择波导 frequency selective waveguide
频率选择电路 frequency selective network
频率选择钮 frequency selector
频率选择性衰落 frequency selective fading
频率压缩解调器 frequency compression demodulator
频率仪 frequency recorder
频率移动 frequency shifting
频率因数 frequency factor
频率因素 frequency factor
频率因子 frequency factor
频率影响 <对仪表读数准确度> frequency influence
频率诱导 entrainment of frequency
频率预测 frequency prediction
频率预测图 frequency prediction chart
频率预矫 predistortion
频率预置器 frequency presetting device
频率域 frequency domain; frequency field
频率域磁性界面反演 magnetic interface inversion in frequency domain
频率域法 frequency domain method
频率域反褶积 frequency domain deconvolution
频率域分析(法) frequency domain analysis
频率域激电仪 induced polarization instrument in frequency domain
频率域宽度 width of frequency
频率域滤波 frequency domain filter
频率阈 frequency threshold
频率跃变 frequency jumping
频率再用 frequency reuse
频率增益曲线 frequency gain curve
频率展开 frequency spread
频率占据 frequency capture
频率振幅曲线 frequency amplitude curve
频率震级二次关系式 quadratic frequency magnitude relation
频率直方图 frequency histogram; rectangular frequency diagram
频率指标 frequency index
频率指配通知 frequency assignment notice
频率指示器 frequency-indicating device; frequency indicator
频率指数 frequency index
频率滞后 frequency hysteresis
频率周期 frequency period
频率转换 frequency conversion
频率转换器 frequency converter[convertor]; frequency transformer
频率准确度 frequency accuracy
频率自动控制 automatic frequency control
频率自动控制系统 automatic-frequency control system
频率自动控制运行 automatic frequency control operation

频率综合技术 frequency synthesis technique
频率组成 frequency content
频脉 increased pulse
频敏检波器 frequency-sensitive detector
频哪酮 pinacolone
频偏 drift of frequency;frequency deviation;frequency offset;frequency shift;frequency bias
频偏范围 frequency-deviation region
频偏改正 correction for frequency deviation
频偏后的载频 displaced carrier
频偏计 frequency-deviation meter
频偏能力 deviation capability
频偏误差 frequency offset error
频偏值 frequency shift value
频漂 creep of frequency; frequency drift;frequency shifting
频漂误差 frequency drift error
频谱 frequency spectrum
频谱波长 spectral wavelength
频谱测定法 spectrometry
频谱窗 spectral window
频谱带 spectral band
频谱带宽 spectral bandwidth
频谱的 spectral
频谱段 wavelength coverage
频谱发射率 spectral emissivity
频谱法 spectral method;spectroscopic(al) technique
频谱范围 frequency coverage; frequency range;spectral range
频谱放大 spectral amplification
频谱分布 spectral distribution; spectrum distribution
频谱分布函数 spectral distribution function; spectrum distribution function
频谱分布特性曲线 spectral distribution characteristic curve
频谱分析 decomposition of spectrum;frequency spectrum analysis; spectral analysis;spectral decomposition;spectrum analysis
频谱分析图 diagram of spectrum analysis
频谱分析仪 analytic(al) spectrometer; frequency spectrum analyser [analyzer];spectral analyser[analyzer];spectrometer;spectrum analyser
频谱幅度 spectral amplitude
频谱辐射强度 spectral intensity
频谱复用技术 spectrum reuse technique
频谱激电仪系统 spectral induced polarization system
频谱技术 spectral technique
频谱交错 frequency interlacing
频谱宽度 spectrum space
频谱密度 frequency spectrum intensity;spectral density
频谱模型 spectral model
频谱强度 spectral intensity;spectrum intensity
频谱扫调指示的 panoramic
频谱扫描指示 panoramic indication
频谱上幅度平均分配的起伏噪声 flat-channel noise;flat noise
频谱上限 upper cut-off frequency
频谱试验 spectrum test
频谱图 spectrogram
频谱下限 lower cut-off frequency
频谱线 spectrum line
频谱限制 spectrum limitation
频谱相对分布 spectral relative distribution
频谱响应 spectral response
频谱谐波分析 frequency harmonic analysis
频谱形状 spectral shape; spectrum shape
频谱形状函数 spectral shape function
频谱选择器 spectrum selector
频谱选择性 spectral selectivity
频谱学 spectroscopy
频谱压缩 frequency spectrum compression
频谱仪 frequency spectrograph; frequency spectrometer; spectrograph;spectrometer
频谱因数 spectrum factor
频谱展宽器 spectrum expander
频谱展示术 < 荧光屏上频谱的展开 > panoramic technique
频谱振幅 spectral amplitude
频谱振幅法 spectral amplitude technique
频谱直视法 frequency spectrum direct vision method
频谱指数 spectrum index
频谱重正化 spectrum renormalization
频群 frequency group
频散 dispersion;frequency dispersion
频散波 dispersion wave
频散带 dispersion zone
频散电导率 frequency astigmatism conductivity
频散电阻率 frequency astigmatism resistivity
频散方程 dispersion equation
频散函数 dispersive function
频散介质 dispersion medium; dispersive medium
频散率 percent frequency effect
频散曲线 dispersion curve;dispersive curve
频散瑞利波 dispersive Rayleigh wave
频散效应 dispersion effect
频闪测距 flash ranging
频闪测速计 strobotach
频闪测速器 strobotach
频闪测速仪 stroboscope;stroboscopic instrument
频闪测向器 stroboscopic direction finder
频闪的 stroboscopic
频闪灯光 strobe light
频闪放电管 speed flash;speed light;strobotron
频闪观测管 < 其中的一种 > neostron
频闪观测盘 stroboscopic disc
频闪观测器 stroboscope
频闪观测器照相术 stroboscope photography
频闪观测仪 stroboscope
频闪观测原理 stroboscopic principle
频闪观察 < 高速机床试验 > stroboscopic observation
频闪管 stroboscopic tube;strobotron
频闪光 stroboscopic light(ing)
频闪校验法 stroboscopic checking
频闪器闪光 strobe flash
频闪闪光装置 cinestrobe
频闪摄影术 flash photography;strobe photography; stroboscopic photography
频闪式测向仪 stroboscopic direction finder
频闪式仪表 stroboscopic instrument
频闪图案轮 stroboscopic pattern wheel
频闪氙灯 stroboscopic xenon lamp
频闪效应 stroboscopic effect
频闪仪式检测仪表 stroboscopic instrumentation
频闪仪式偏振计 stroboscopic polarimeter
频闪仪式转速计 stroboscopic tach(e)ometer
频闪照明 stroboscopic illumination
频闪指示器 stroboscopic indicator
频闪转速表 strobotac
频闪转速计 strobotach
频闪转速仪 stroboscopic tach(e)ometer
频闪装置 blinking device; flickering device
频闪作用 stroboscopic action
频射无线电工程 radio-frequency engineering
频数 frequency number
频数比 frequency ratio
频数表 frequency table
频数次数计算 counting the frequencies
频数分布 frequency distribution
频数分布表 frequency distribution table
频数分布多边形 frequency distribution polygon
频数分布曲线图 histogram
频数分配表 frequency distribution table;table of frequency distribution
频数函数 frequency function
频数距 frequency moment
频数配合 frequency matching
频数曲线 frequency curve
频数数列 frequency series
频数图 frequency chart
频调抽运 frequency-modulated pumping
频调声 frequency-modulated sound
频稳度 frequency stability
频细震颤 fine tremor
频响跌落特性 roll-off characteristic
频响均衡 frequency response equalization
频响失真 frequency response distortion
频响预校网络 preemphasis network
频压法 alternating pressure method
频移 frequency excursion; frequency shift
频移键控 < 数据传输用的频率调制 > frequency shift keying
频移失真 deviation distortion
频移调制 frequency shift keying
频移吸收 deviation absorption
频移系统 frequency shift system
频移限制 audio deviation limiting
频应复元 de-emphasis
频域分析 frequency domain analysis
频域分析器 frequency domain analyser[analyzer]
频域解 frequency domain solution
频域均衡器 frequency domain equalizer
频域模型 frequency domain model
频域判据 frequency domain criterion
频域时变反褶积 time-variant deconvolution in frequency domain
频域时间数列 frequency domain time series

品

品标 brand mark
品度值 performance value
品红 aniline red;fuchsin(e);roseine;rubin;solferino;magenta;rosaniline
品红接触网 fuchsin(e)contact screen;magenta contact screen
品红醛试剂 fuchsin(e)-aldehyde reagent
品红色 fuchsin(e)red;magenta
品红色层(感绿层)magenta layer
品红色接触网目板 magenta contact screen
品红试法 fuchsin(e)test
品红试剂 Schiff's reagent
品红试验 < 陶瓷吸水性试验 > ink test
品红酮 fuchsone
品红网目板 magenta screen
品红亚硫酸法 fuchsin(e)-sulphurous acid method
品红亚硫酸试剂 fuchsin(e)-sulphurous acid reagent
品级 grade;scale
品级率 percentage of product
品级试验 rank test
品蓝(色)reddish blue;royal blue
品类 category;class
品绿 light green;malachite green
品名 name of article; name of commodity;name of part;trade name
品名表 nomenclature
品目 item
品牌晴雨报告 brand barometer report
品脱 < 容量名,1 品脱 = 1/8 加仑 > pint[pt]
品位 grade;purity;quality;tenor【地】
品位变化曲线图 variation chart in grade
品位变化系数 variation coefficient of grade
品位分级 ranking of grade
品位价值 esteem value
品位降低 down grading
品位控制 grade control
品位下限 cut-off grade
品位因数 quality factor
品位中和 quality neutralization; tensor neutralization
品物架 shelving
品系 line;strain;strain of breed
品系试验系统 strain testing system
品行 morality
品质 assortment; brand; caliber size; quality;sort;train;variety
品质按买方样品 quality as per buyer's sample
品质保证 quality assurance; quality guarantee
品质变量模型 qualitative variables model
品质标记 quality symbol
品质标志 attributive indicant;attributive indication;quality mark
品质标准 quality level;quality standard
品质不符 different quality
品质不合格 off quality
品质不匀的 unequal;uneven
品质测试 attribute test(ing)
品质差幅 quality latitude
品质差价 quality price differential
品质成本 cost of quality
品质抽样方案 sampling plan by attributes
品质担保条款 quality warranty clause
品质的 attributive
品质低劣 inferior quality
品质低劣的 adulterate
品质调查 quality survey
品质方程 quality equation
品质分类 attribute classification; qualitative classification
品质分组 qualitative grouping
品质复验 checking of quality
品质公差 quality tolerance
品质管理 qualitative control; quality control
品质管理过程 quality control process
品质管理圈 quality control circle
品质管理循环 quality control cycle

P

品质管制 quality restriction
品质规格 specifications of quality
品质号数 quality number; spinning number
品质或数量证明书 certificate of quality or quantity
品质机动幅度 quality latitude
品质监督 quality monitoring
品质检查 quality inspection
品质检验 inspection by attributes; quality inspection
品质检验(证)书 inspection certificate of quality
品质鉴定 appraisal of quality; characterization
品质鉴定证(明)书 survey report on quality
品质奖金 quality bonus
品质降低极限 quality deterioration limit
品质较差 inferior quality
品质可靠性消耗数据 qualitative reliability consumption data
品质控制 quality monitoring
品质控制装置 quality control package
品质领先 quality leadership
品质免赔限度 quality franchise
品质欠均匀 lack of uniformity
品质(上)的变质 substantial deterioration
品质试验证明书 certificate of quality test
品质数列 quality series
品质水准 quality level
品质索赔 quality claim
品质特性 qualitative characteristic; quality characteristic
品质条件 term of quality
品质条款 quality clause
品质统计 attribute statistics; statistics of attributes
品质稳定 stay in grade
品质系数 factor of merit
品质相关 correlation of attributes
品质信用限制 qualitative credit restriction
品质选择模型 model of qualitative choice
品质样品 quality sample
品质移植 quality transplants
品质意识 quality mind
品质因数 quality factor; energy factor; figure of merit; goodness; Q-factor; quality value; Q-value
品质因数表 Q-meter; quality meter
品质因数计 Q-meter; quality meter
品质因素 quality factor; factor of merit; figure of merit; goodness; Q-factor
品质因子 quality factor
品质应变量 qualitative dependent variable
品质优等 best in quality; superior quality
品质优良 best in quality; fine quality
品质优值 figure of merit
品质有缺点 defective quality
品质与报价不符 variation from quality offered
品质与订单不符 variation from quality ordered
品质与样品不符 quality variation from sample
品质证明 hall mark
品质证(明)书 certificate of quality; quality certificate
品质指标 criterion of control quality; index of quality
品质指数 quality index(number)
品质属性 qualitative attribute; qualitative property; quality characteristic
品质资料 attribute data
品种 brand; breed; race; strain; variety
品种减少 variety reduction
品种检定 variety certification
品种检索表 key to variety
品种简化 variety reduction
品种简介标签 descriptive labeling
品种鉴定 approbation; identification of variety
品种鉴定说明 key to varieties
品种结构 breed structure
品种竞争 variety competition
品种控制 variety control
品种灭绝 extinction of species
品种齐全 great variety of goods
品种区 variety plot
品种试验 variety test; variety trial
品种试验报告 report for variety test
品种试验方案 program(me) of variety test
品种试验小区 variety test plot
品种试验小区图 design of variety test plot
品种收集 variety collection
品种说明 key to variety
品种索引表 key to varieties
品种特性 varietal characteristic
品种特征 breed characteristic
品种协会 breeders association
品种证明书 species certificate
品种指标 indicator of product variety
品种资源库 species banks
品字形 trefoil

榅

榅桲 quince

聘

聘后作业 post employment activity
聘前作业 pre-employment activity
聘请 employ; engage; invite
聘书 contract; engagement letter; letter of appointment
聘用 retain
聘用及解雇成本 hiring and layoff cost
聘用条件 conditions for employment
聘用职员 staffing
聘约 contract of employment

乒

乒乓法 ping-pong procedure
乒乓开关 toggle switch
乒乓球比赛场 table tennis court
乒乓球室 ping-pong room; table tennis room
乒乓球厅 table tennis hall

平

平安保险 insurance FPA [free of/from particular average]
平安航行 plain sailing
平安时间 alcyone days; halcyon days
平安时期 alcyone days; halcyon days
平安无事 without mishap
平安险 free from particular average; free of particular average
平安险条款 FPA [free of particular average] clause
平鞍形键 flat saddle key
平岸的 bankfull
平岸流 bankfull flow
平岸流量 bankfull discharge
平岸水位 bankfull stage
平凹 plano-concave
平凹板 plano-concave plate
平凹版 deep-etch plate
平凹版腐蚀 deep etch
平凹版印刷 deep-etch printing; deep-set printing
平凹版用基漆 deep-etch lacquer
平凹的 concavo-plane; piano-concave
平凹镜片 plano-concave lens
平凹透镜 plane concave lens; plano-concave lens
平凹形 concavo-plane
平凹形的 <一面平、一面凹的> plano-concave
平凹形天线 plane concave antenna
平扳手 end wrench
平板 flag; flat board; flat bed; flat plate; flat sheet; flat slab; mo(u)ld board; panel; patten; plain board; plain sheet; planch; plane plate; plane table; platen; slab; surface plate; table
平板坝 deck dam; flat slab(buttress) dam; flat slab deck dam; slab dam
平板半挂车 flat-bed trailer; platform semitrailer; straight frame trailer
平板边缘 slab edge
平板玻璃 flat glass; glassed surface; patent glass; plate glass; sheet glass
平板玻璃半成品 rough plate blank
平板玻璃表面裂纹、皱纹 crizzle
平板玻璃池窑 plate glass furnace; plate glass tank; sheet glass furnace; sheet glass tank
平板玻璃垂直引上法 up-drawing sheet glass process
平板玻璃拉管生产法 cylinder process
平板玻璃拉制 sheet glass drawing
平板玻璃拉制法 sheet glass drawing process
平板玻璃拉制机 sheet glass drawing machine
平板玻璃毛坯 rough plate blank
平板玻璃门 plate glass door
平板玻璃切裁尺寸 cut size plate; lite
平板玻璃水平拉制法 Glavebel process
平板玻璃瓦 plain roller glass
平板玻璃小缺口 undercut
平板玻璃窑炉 flat glass furnace
平板玻璃引上法 sheet drawing process
平板玻璃折算系数 conversion factor of sheet glass
平板玻璃制造法 sheet glass process
平板材 flat-sheet material
平板测绘 plane-tabling
平板测绘仪 plane table
平板测量的 plane-tabling
平板测量器 alidade
平板测量仪 plane table
平板测数 plate count
平板测图板 plane-table board
平板插栅式车厢 platform-and-stake body
平板车 bolster wagon; dilly; flat-bed trailer; flat-bottom cart; flat carriage; flat(deck) car; flat-form lorry; flat-form truck; flat lorry; hand truck; platform car; platform lorry; platform truck; retriever; rock body truck; stake body trailer; flat wagon
平板车身 platform body; stake body
平板车厢 flat bed; platform body
平板车装运集装箱 container on flat car
平板衬砌 plate facing
平板承载试验 plate bearing test
平板承重结构 plate load-bearing structure
平板冲孔机 plate-punching machine
平板除尘器 plate precipitator
平板船首柱 plate stem
平板窗格 plate tracery
平板窗花格 plate tracery
平板锤 dinging hammer
平板粗纸板 platen-pressed chipboard
平板打夯机 flat beater
平板大货车 platform truck
平板大卡车 platform truck
平板大门 plane gate
平板带接头 apron bolt joint
平板导体 plane conductor
平板底脚 mat footing; slab footing
平板底面 slab soffit
平板底座 flat base
平板电极 plate electrode
平板电极管 flat plate lamp
平板垫圈 plate washer
平板吊车 platform crane car
平板吊架 platform sling
平板冻结装置 plate freezer
平板端部 slab end
平板断面 slab cross-section
平板堆 slab reactor
平板舵 flat plate rudder; single plate rudder
平板阀(门) disk valve; flap valve; plate valve; flat valve
平板法 flat band method
平板法剥落试验 panel spalling test
平板法导热系数测定仪 thermal conductivity measuring apparatus by guarded plate(method)
平板反射器 plane sheet reflector
平板枋 wood-plate lintel
平板分解器 slab resolver
平板分析 plate analysis; slab analysis
平板分选机 plate cleaner
平板风速计 plate anemometer
平板封头 flat plate closure
平板辐射计 flat plate radiometer
平板负载试验 navy method
平板覆盖 plate cover(ing)
平板盖 flush plate
平板刚度 slab rigidity; slab stiffness
平板钢管片 flat steel segment
平板钢筋 slab reinforcement
平板钢筋定位架 slab spacer
平板钢筋支架 slab bar bolster; slab bolster; slab spacer
平板格排 plate grillage
平板格网 plate screen
平板给料机 table feeder
平板给料器 table feeder
平板拱 plate arch
平板拱梁 plate arched girder
平板构造 slab(bed) construction; slab structure
平板挂车 flat bed; flat trailer; platform trailer <搬运行李、包裹>
平板挂车汽车运输连 flat-bed company
平板涵洞 slab culvert
平板夯捣机 <混凝土块生产中的> plate-tamping machine
平板夯的刚性底板 rigid base plate
平板荷载 plate load; slab load
平板荷载试验 plate bearing test; plate load(ing) test
平板厚度 plate depth; plate thickness; slab depth
平板滑动支座 plane sliding bearing
平板滑阀 linear action disk valve
平板划线法 plate streak
平板绘图机 flat plotter
平板绘图器 planchette; plane plate
平板绘图仪 plane table; flat-bed plotter
平板混汞法 plate amalgamation
平板货车 flat car; platform car; platewagon <铁路>
平板货柜 flat rack container; flat bed container

平板货台尺寸 platform dimensions
平板机 flat links and links knitting machine
平板基础 flat plate foundation; slab footing; slab foundation
平板集热器 flat plate collector
平板集装箱 flat rack
平板几何条件 slab geometry
平板计算 plate analysis; slab calculation
平板加工 plate finish
平板加热器 panel heater
平板加热器加热 panel heating
平板加载试验 plate bearing test; plate load(ing) test
平板架 pallet
平板剪切试验 plate-shear test
平板鉴定 identification of plates
平板胶印 offset-lith-printing
平板角铁柱 plate and angle column
平板接触冻结装置 contact plate freezer
平板接触式冷冻器 plate contact freezer
平板接缝 board joint
平板接合 plate joint
平板接头 flat connector
平板结构 flat plate structure; slab construction; slab structure
平板卡车 flat-bed lorry; flat(-bed) truck; flatted truck; plate truck; flat-bottom truck
平板开关 flush switch
平板抗碎强度 slab crushing strength
平板抗压强度 slab compressive strength
平板跨度 plate span; slab span
平板宽度 slab width
平板扩散器 flat plate diffuser
平板扩压器 flat plate diffuser
平板肋板 plate floor; solid floor; solid frame
平板冷却器 panel cooler
平板理论 slab theory
平板力矩 slab moment
平板立柱式坝 slab and column dam
平板梁 slab-and-beam
平板梁楼面 slab and girder
平板料 slab
平板列车 plate train
平板硫化机 platen press
平板龙骨 flat bar keel; flat keel; flat plate keel(son); plate keel
平板楼板 flat floor; slab floor
平板楼盖 flat plate floor
平板楼面 plate floor
平板炉 flattening oven
平板路面 slab pavement
平板罗盘 card compass; compass card
平板门 one-panel door; plain batten door; slab door; flap gate
平板密封套 plane seal housing
平板面层 topping slab
平板模板 <浇筑混凝土用> slab form(work)
平板模数 plate modulus
平板内龙骨 flat plate keel(son)
平板挠度 plate deflection
平板排水沟 plate gutter
平板抛光机 plate glazing calender
平板刨花板 platen-pressed chipboard
平板刨切机 horizontal veneer slicer
平板培养 plate culture
平板膨胀仪 flat dilatometer
平板破坏 plate failure; slab failure
平板铺的盥洗器 flat slab lavatory (basin)
平板铺面 slab pavement
平板铺砌的屋顶散步道 promenade slab roof(ing)
平板汽车 platform car
平板千斤顶 flat jack
平板强度 slab strength
平板绕流 flat plate flow
平板热压机 flat-bed press
平板日光收集器 flat plate solar collector
平板三角测量 plane-table triangulation
平板散热盘 panel coil
平板散热器 flat radiator
平板色谱 flat-bed chromatogram
平板筛 flat screen; plate screen
平板筛浆机 flat strainer
平板筛网印花机 flat screen printing machine
平板石磨 flat stone mill
平板式 flat design
平板式半挂(拖)车 platform semi-trailer
平板式变阻器 face-plate rheostat
平板式打夯机 flat beater
平板式挡水板坝 flat slab deck dam
平板式挡水面板 flat slab deck
平板式导水墙 slabs training wall
平板式电集尘器 Elex precipitator
平板式冻结法 refrigerated plate freezing method
平板式冻结机 plate freezer
平板式阀 gate valve; sluice valve
平板式扶垛坝 flat slab buttress dam
平板式轨道 slab track
平板式涵管 slab culvert
平板式换热器 flat plate heat exchanger
平板式绘图仪 flat-bed plotter
平板式混凝土振动器 plate concrete vibrator
平板式加料器 table feeder
平板式夹具 plate jig
平板式扩散器 plaque diffuser
平板式楼板 flat slab floor; pan floor
平板式抹光机 disk-type trowelling machine
平板式曝气池 plate aerator
平板式热变换器 plate-type exchanger
平板式热风机 plate-type air heater
平板式溶解器 slab dissolver
平板式上光机 flat-bed glazer
平板式太阳能集热器 flat plate solar collector
平板式拖车 flat-bed trailer; straight frame trailer
平板式蓄电池搬运车 electric(al) platform truck
平板式印刷电路 plated printed circuit
平板式运货汽车 flat-bed truck
平板式载重汽车 flat-bed truck; platform truck
平板式增压阀 plate-type delivery valve
平板式振捣器 flat plate vibrator; plate-type vibrator; plate-vibrating compactor; screed board vibrator; vibroplate
平板式振动夯压器 plate vibratory tamper
平板式振动器 flat plate vibrator; plate-type vibrator; plate-vibrating compactor; screed board vibrator; vibroplate
平板式振动压实器 vibrating plate compactor
平板式振动压土机 vibratory plate compactor
平板式振动压毡机 flat harder
平板式支墩坝 flat slab buttress dam; slab buttress dam; slab and buttress dam
平板试验 tread mill test
平板饰面 plate facing; slab facing
平板艏 flat bow
平板输送带 plate conveyer[conveyor]
平板输送机 pallet conveyer[conveyor]; pan conveyer[conveyor]; slat type conveyer[conveyor]
平板送料机 table feeder
平板碎纸胶合板 platen-pressed chipboard
平板太阳能集热器 solar collector panel
平板太阳能收集器 solar flat plate collector
平板提升机 apron elevator
平板体 platysome
平板体系 slab system
平板天线 plate aerial; plate antenna
平板条 flat bar; flat slat; slat
平板条屋顶板 plain strip shingle
平板条屋顶木瓦 plain strip shingle
平板凸轮 plate cam; cam plate
平板图纸 plane-table paper; plane-table sheet
平板推套 tappet
平板托盘 flat pallet; platform pallet
平板拖车 athey wagon; deck trailer; platform trailer; flatbed trailer
平板拖挂车 deck trailer
平板瓦 plane tile; plate tile; shingle
平板弯曲 slab bending
平板网印 flat-bed screen printing
平板位移 plate displacement; slab displacement
平板问题 plate problem; slab problem
平板屋顶 slab roof
平板吸水箱 flat suction box
平板显示 flat panel display; panel display
平板显示器 flat plate display
平板箱形孔型 bull head
平板镶衬 plate lining
平板橡胶支座 laminated rubber bearing
平板效率 plating efficiency
平板效应 plate action; slab effect
平板斜交 slab skew
平板卸底拖车 athey wagon
平板形薄管板 flat thin tube sheet
平板形管片 flat type segment
平板形凸轮 cam plate
平板型板桩 flat type pile; straight-web sheet pile
平板型车身 <卡车、拖车的> platform unit
平板型的直线电机 flat linear electric(al) motor
平板型钢板桩 flat web steel sheet pile; straight-web steel sheet pile
平板型夯土机 plate vibrator
平板型卡车 platform truck
平板型空气过滤器 panel type air filter
平板型曝气器 turbine aerator
平板型筛 plate screen
平板型收集器 flat plate collector
平板型太阳能收集器 solar flat plate collector
平板型透析器 pack type dialyzer
平板型叶栅 cascade of straightline profile
平板型正比计数管 flat counter tube
平板旋转 slab rotation
平板选煤器 plate cleaner
平板压光 plate glazing
平板压烫机 flat plate pressing machine
平板压延 taper
平板压延辊 tapered roller
平板压制 flat platen pressed
平板压制机 slab press
平板研光机 plate calender; plate glazing calender; skim coat calender
平板檐槽 plate gutter
平板扬声器 flat plane speaker
平板叶栅 flat plate cascade
平板仪【测】 planchette; plane plate; plane table; surveying panel; surveying plane table; surveyor's table
平板仪侧方交会 side intersection by plane-table
平板仪测量 plane-table measurement; plane-tabling; plane-table operation; plane-table survey(ing)
平板仪测量定点 plane-table fixing
平板仪测量法 plane-table method; stadia method
平板仪测量图 plane-table map
平板仪测图 plane-tabling; topographic(al) mapping with plane table
平板仪测图板 plane-table board; plane-table plate
平板仪测图法 plane-table method
平板仪测站 plane-table station
平板仪导线 plane-table traverse
平板仪地形测量 plane-table topographic(al) survey; topographic(al) plane-table survey
平板仪定位 plane-table fixing
平板仪定向 orientation of plane-table
平板仪后方交会 resection by plane-table
平板仪接头 <平板与脚架的接头> plane-table head
平板仪罗盘 plane-table compass; plate-table compass
平板仪前方交会 forward intersection by plane-table
平板仪三角测量 plane-table triangulation
平板仪三脚架 plane-table tripod
平板仪摄影测量学 plane-table photogrammetry
平板仪视距测量 plane-table tacheometry; plane-table tachymetric(al) survey; stadia plane-table survey
平板仪视距法 plane-table tacheometry[tachymetry]
平板仪水准管 plane-table level tube
平板仪水准器 plane-table level tube
平板仪外业原图 plate-table sheet
平板仪作业 plane-table operation
平板仪作业员 plane-table operator; planetabler
平板阴极 planar cathode; plane cathode
平板印花 flat plate printing
平板印刷 offset printing
平板印刷机 lithographic(al) press
平板印刷线圈 flat printed coil
平板印刷用玻璃 glass for lithography
平板圆锯 plate saw
平板运输车 flat bed; flat-bed truck
平板运输机 crocodile; flat-top apron conveyer[conveyor]; hinged apron; plate conveyer[conveyor]; slat type conveyer[conveyor]
平板载荷试验 plate load(ing) test
平板载货汽车 platform lorry
平板轧机 plate rolling mill
平板闸门 bulkhead gate; draft tube bulkhead gate; plain gate; plane gate
平板找正 plate alignment
平板照准仪 plane-table alidade; surveyor's alidade
平板遮盖阀 shear-seal type valve
平板振捣器 pan vibrator; plate vibrator; surface vibrator; vibrating board
平板振动机 float vibrator; plate vibrator; vibrating board; vibroplate
平板振动器 float vibrator; plate vi-

brator;vibrating board
平板振动筛 flat vibrating screen; shaker pan
平板振动压实机 plate-vibrating compactor;vibrating plate compactor
平板震动器 plate vibrator
平板整流器 slab rectifier
平板正面 slab facade
平板支撑 box sheeting
平板支撑坝 fixed deck dam
平板支承介质 slab supporting medium
平板支墩坝 Ambursen(-type)dam;flat slab deck dam;flat deck buttress dam
平板支墩蜂窝式芯墙坝 cellular-wedge core wall Ambursen dam
平板支墩堰 Ambursen-type weir
平板支座 plane surface support;plate bearing
平板纸压光机 sheet breaker; sheet calender
平板周边 plate circumference
平板柱构造 flat slab column construction
平板抓斗 plate grab
平板砖 plate block;rectangular block
平板转动 plate rotation
平板转移印花机 flat-sheet transfer unit
平板桩 flat type piling bar;slab pile
平板桩基式码头 relieving platform pile type wharf
平板桩靴 flat plate(pile)shoe;flat type pile shoe;slab pile shoe
平板状的 flat;tabular;tabulate
平板状垫片 panel washer
平板状垫圈 panel washer
平板阻力 flat plate drag
平板作用 plate action;slab effect
平版 lithographic(al)plate;surface plate
平版复印油画术 oleography
平版工艺 lith process
平版红棕油墨 litho-bronze deep red
平版印刷 lithograph;plane print;planographic(al)printing
平版印刷术 lithography;planography
平版用清漆 lithographic(al)oil;lithographic(al)varnish;litho oil
平版制版胶片 lithographic(al)film
平半圆混合拱 Carnarvon arch
平背车 flat-back car
平背拱 floor arch
平背式冰箱 flat-back type refrigerator
平壁 blank wall;planomural
平壁导热 heat-transfer through panel wall
平壁龛 flat niche
平箅式进水口 gutter grating inlet
平箅式雨水口 inlets with horizontal gratings
平箅添煤器 horizontal grate stoker
平臂起重机 level-luffing crane
平边 plain edge;platband
平边车身 flush-sided body
平边锉 flat edge file
平边的 flush-sided
平边对接焊 plain butt weld
平边条 fillet
平边凿 fillet chisel
平编 flat sennit
平编带 flat braid;plain braid
平编宽紧带 flat-braided elastic
平扁壳 shallow shell
平扁块拣选筛 flat picker
平扁轮胎 flat-built tire[tyre]
平扁小(块)石 pennystone
平扁蛛网形线圈 spider-web coil
平变位 horizontal deflection
平表面抹灰 pargetry
平波电抗器 smoothing reactor
平玻璃瓦 plain rolled glass
平玻璃屋面瓦 plain glass roof(ing) tile
平播 flat planting
平薄膜生物反应器 flat-sheet membrane bioreactor
平薄物体 planar body
平薄叶板 flat skin plate
平薄砖 fine brick
平布 parallel fabric
平仓(船舱)trimming
平舱 level(-trimming); plowing in the hold;trim
平舱费 level(l)ing charges;trimming charges
平舱费用在内的船上交货价 free on board trimmed
平舱工人 cargo trimmer;trimmer
平舱机 mechanical trimmer;thrower;trimmer
平舱机械 trimming machine
平舱口盖 flush cover
平舱器 trimmer
平舱压载 trim ballast
平舱作业 trimming operation
平槽 flatlander
平槽板 flush plate
平槽电枢 smooth core armature
平槽轨 flat grooved rail
平槽滤板 flush filter plate
平槽刨 dado plane
平槽水流 inbank flow
平槽头 flat slotted head
平槽头螺栓 flat slotted head bolt
平槽泄水能力 inbank capacity
平槽压滤板 flush filter plate
平槽压滤机 flush plate filter press
平槽应变解除法 flat-slot method
平侧面框架 plane side frame
平侧石 curb and gutter
平层 flat bed;level(l)ing
平层的 flat layer
平层堆料 areal stockpiling
平层法 flat-layer technique
平层技术 flat-layer technique
平层节理 bed joint
平层料堆 <集料的> flat layered pile
平层调整区 <电梯> level(l)ing zone
平叉 pallet fork
平插销的退进 flush bolt backset
平茬 stump
平茬苗 stump plant
平差【测】adjust(ing);adjustment; adjustment of errors; balancing(a survey); compensation of errors; error compensation
平差参数 adjustment parameter
平差残差 adjustment residual
平差程序 adjustment program(me)
平差电路 neutrodyne circuit
平差方程 adjustment equation
平差(方)法 method of adjustment; adjustment method; compensation method;consummation of errors
平差改正 adjustment correction
平差高程【测】adjusted elevation; standard elevation
平差公式 adjustment formula
平差后的位置 adjusted position
平差计算 adjustment computation; compensating computation
平差角【测】adjust(ed)angle
平差校正 adjustment correction
平差量 adjusted quantity
平差器 adjuster
平差图 adjusting diagram;adjustment diagram
平差位置 adjusted position
平差因子 adjustment factor
平差值 adjusted quantity;adjusted value
平铲 blade;flat spade;grafter;grafting tool;sweep blade
平铲闭锁装置 blade lock
平铲控制 blade control
平铲提升臂 blade lift arm
平铲提升机械装置 blade lift mechanism
平铲提升控制齿轮 blade lift control pinion
平铲提升控制箱 blade lift control housing
平铲提升连杆 blade lift post
平铲中耕机 blade cultivator
平常暴雨水污染物 conventional stormwater pollutant
平常大潮低潮位 low water of ordinary spring tides
平常大潮高潮位 high water of ordinary spring tides
平常的 mediocre
平常结账 ordinary closing of accounts
平常型式 conventional type
平场 flat field
平场改正 flat field correction
平场透镜 field flattening lens
平场物镜 flat field objective
平场照相机 flat field camera
平敞车 flat car
平潮 slack tide;slack water;stand of tide;still tide;tidal stand;tide stand
平潮航行 slack-water navigation
平潮期 slack-water time
平潮区 area of slack water;slack-water area
平潮时 flushing time;slack-water time
平车 flat bogie wagon;platform car; platform wagon
平车装运 trailer on flat car
平车装运载有集装箱的挂车 trailer on flat car
平车装运载有集装箱的挂车及平车装载集装箱的列车 trailer on flat car/container on flat car train
平衬板 flat liner;plane liner
平吃水 even keel;on even keel;trim on even keel
平吃水装载 on level trim
平尺 level(l)ing instrument;level(l)ing rule;straight edge
平齿轮 horizontal gear;plain gear
平齿型迷宫气封 plain labyrinth packing
平齿压路机 <齿比羊足压路机短> padfoot roller
平赤道 mean equator
平冲断层 horizontal thrust fault
平冲沟 talat
平冲角 angle of balance
平冲圈 balance ring;balancing ring
平冲式冲断层 horizontal thrust
平冲头 flat punch
平出滩 foreshore flat
平椽 orle
平窗采光 lay light
平窗棂 lay bar
平床 level bed
平床混合炉 flat hearth type mixer
平床货柜 flat-bed container
平床色谱法 flat-bed chromatography
平床式集装箱 flat-bed container
平床式拖车 flat-bed trailer;platform semitrailer;straight frame trailer
平吹旋转炉 Graef rotor
平锤 ca(u)lker;face(d)hammer; flattener; flat (ter) hammer; hammer flattener;setting hammer; slogging hammer;smoothing hammer
平春分点 mean equinox of date
平粗齿木锉 flat rasp
平锉 flat file;hand file;warding file
平错 heave
平错齿饰 flat billet
平搭接盖板 butt strap
平搭接接头 plain lap joint
平搭瓦 interlocking tile; single-lap tile;Spanish tile
平打绳 flat-braided cord
平带 flat ribbon;flat rubber belting
平带点 flat band point
平带电压 flat band voltage
平带拉紧装置 belt spanner
平带式谷物输送器 flat-belt grain conveyer[conveyor]
平带式驱动 flat belt drive
平带输送机 flat-belt conveyer[conveyor]
平带运输机 horizontal belt conveyer [conveyor]
平淡的 cool
平淡光 flat light
平淡图像 picture without contrast
平淡照明 flat-lighting
平挡圈 loose rib
平导板 flat bearing
平导轨 level ga(u)ge
平导针片 plain slider
平倒塌 pancake collapse
平道 level line;level track
平道坡道可调式制动机 level-gradient changeover brake
平道坡道制动机调整器 level-gradient device
平道坡道转换装置 level-gradient changeover device
平道线路 level line
平的 aflat;even;flat;planar;planus
平的半圆木条 flush bead
平的半圆线脚 flush bead
平的串珠线脚【建】flush bead;astragal
平的浮船 flat pontoon
平的浮码头 flat pontoon;flat floating wharf
平的圆石磙 flat circular stone
平灯座 bayonet flush socket
平等 equality
平等待遇 equal treatment; integration; parity of treatment; reciprocal treatment
平等待遇原则 principle of equality of treatment
平等的法人身份 equal status of legal persons
平等对待投标者 equal treatment of bidders
平等互利 equality and mutual benefit
平等互利原则 principle of equality and mutual benefit
平等交换贸易 give-and-take trade
平等竞争 compete on an equal basis; compete on an equal footing;equal competition
平等贸易 fair dealing
平等条款 parity clause
平等效力条例 <当事人对代办人的书面委托> equal dignities rule
平等协商 consultation on an equal footing;consultation on the basis of equality
平等运送旋臂起重机 level-luffing crane
平等租借交换 lease-lend;lend-lease
平低阔峰 platy kurtosis
平底 flat base;flat bottom;flat floor; flat keel;uncambered bottom;plane bottom
平底 8 英尺 <1 英尺 = 0.3048 米> 大挖深度 <挖掘机> depth of cut for

8 ft level bottom
平底板 <车身> flat underside
平底驳船 flat-bottom barge;pontoon barge
平底驳船管理人 scow banker
平底驳船员 scowman
平底部分 flat of bottom
平底仓 flat-bottom bin;flat-bottomed silo
平底槽 flat-bottom slot
平底槽刨 dado plane
平底侧面配合 flat foot side fit
平底测流槽 flat-bottomed flume
平底车 flat-bottom car
平底沉淀池 flat-bottomed clarifier
平底澄清池 flat-bottomed clarifier
平底池 flat-bottom slot; flat-bottom tank
平底冲头 flat-bottom punch;flat edge die
平底船 ark;barge;bateau;canal barge; cargo-box barge; deck scow; flat boat;flat-bottomed boat[ship/ vessel];flatty;gondola;hoy;johnboat; keel; pontoon; pram; propulsion pontoon;punt;scow
平底船船员 punter;puntist
平底船浮吊 pontoon crane
平底船计量 <水下挖方的一种计量方法> scow measurement
平底船列 barge train
平底船煤 keel
平底船上进气管 barge sucker
平底船上进油管 barge sucker
平底船式起重机 crane barge; pontoon crane
平底船系数 <有效装货体积与总体积之比> scow factor
平底船卸货用升降机 barge elevator
平底船载的 barge-loaded
平底船抓斗 grab pontoon
平底船抓斗挖泥船 grab pontoon dredge(r)
平底船钻探技术 scow drilling
平底锤 plane bottom hammer
平底大玻璃杯 tumbler
平底的 broad-bottomed;flat-bottomed
平底垫板 flat-bottom tie plate
平底渡船 bac;ferry flat
平底方驳 scow
平底浮船 bridge boat
平底浮筒 flat pontoon
平底浮筒平台 flat-bottomed pontoon
平底干燥器 hot floor
平底钢轨 flange rail;flat-bottom(ed) rail
平底高边敞车 flat-bottom gondola
平底沟 flat-bottom ditch; level bottom trench
平底谷 flat-bottomed valley; strath; U-shaped valley
平底管 flat-ended tube
平底罐 flat-bottom tank
平底轨 flat-bottom rail;girder rail
平底锅 sauce pan;sauce pot
平底河船 <旧时苏格兰内河用的> gabbart
平底河谷 broad valley;glen;strath
平底环境 level bottom environment
平底混凝土料车 flat-bottomed skip
平底混凝土料斗 flat-bottomed skip
平底锪孔 countering
平底锪孔刀具 counterboring tool
平底锪钻 counterbore
平底货物 flat-bottom goods
平底脚 flat foot
平底铰刀 bottoming reamer
平底晶洞构造 stromatactis
平底孔 flat-bottom hole
平底孔型 flat pass
平底框架 flat section
平底扩孔桩 counterbore
平底扩孔钻 counterbore
平底扩孔钻床 counterboring drill press
平底扩孔钻头 counterbore cutter
平底拉延值 flat-bottom draw value
平底漏斗 Buchner(')funnel
平底炉 hearth furnace
平底轮胎 flat base tire[tyre]
平底轮辋 flat base rim;straight base rim
平底轮缘 flat base rim
平底螺丝纹 bottoming tap
平底麻花钻 flat twist drill
平底煤驳 keel
平底泥驳 mud scow
平底盘 flat chassis
平底器皿下沉变形 dishing
平底浅盆 keeler
平底橇 <即石橇> stoneboat
平底乳钵 muller
平底舢舨 johnboat
平底烧杯 Bunsen beaker;flat-bottom beaker
平底烧瓶 Bunsen beaker;flat bottom flask
平底烧瓶化 florence flask
平底水陆两用车 alligator
平底丝锥 bottom tap
平底四角锥体 quadripod
平底挺杆 flat-bottom tappet
平底通道 flat-bottom channel
平底拖车 flat-bed trailer; low bed trailer
平底拖船 pullboat
平底挖掘深度 <挖掘机> flat-bottom digging depth
平底卧式圆筒形罐 flat-ended horizontal cylindrical drum
平底无脚玻璃杯 tumbler
平底吸泥船 scow sucker
平底洗瓶 flat-bottom washing bottle
平底狭长小船 gondola
平底箱 flat-bottomed bin
平底小船 bateau;bugeye
平底型防波堤 pontoon breakwater
平底型工作船 pontoon type work boat
平底雪橇 toboggan
平底压力盒 boundary pressure cell
平底渔船 coble
平底圆形栏杆扶手 mopstick handrail
平底运泥船 mud scow
平底锥形黏[粘]度计 cone-and-plate viscosimeter
平底钻 bottoming drill
平底钻头 square-nose bit
平底最低值 flat minimum
平底座 flat bed
平地 blading;flat country;flat ground; flat land;flat terrain;land grading; land level(l)ing landing; level ground;level terrain
平地板 flat bottom;flush deck
平地槽 flat-bottom tank
平地铲运机 scraper for land leveling
平地刀板 level(l)ing blade
平地滚压器 land roller
平地花园 flat garden
平地机 blade grader;blade machine; blader;bullgrader;carryall;grading machine;land grader;land level(l)-er;land level(l)ing machine;land scraper;land smoothing machine; level(l)er;mechanical subgrader; motor(ized)grader;plow;grader
平地机刀片 grader blade
平地机刀片操纵轮 blade control wheel
平地机导向板 curb shoe
平地机的多余行程 <即无效行程> extra round of the grader
平地机工作 <包括平路、刮路、整型、整平等工作> blading work
平地机刮刀 cutting blade; grader blade
平地机驾驶室 cockpit
平地机驾驶员 blademan
平地机上加装的松土耙 rake attachment(for graders)
平地机手 blademan
平地机作业 blade grading; blading operation
平地开堑 level cutting
平地犁 mo(u)ld board plough
平地面电波衰减 plane earth attenuation
平地模型 flat earth model
平地耙 smoothing harrow
平地器 land level(l)er
平地泉 land spring
平地山区制动机转换装置 plain-mountain device
平地升送机 elevation grader
平地升运机 elevation grade
平地时的压实 blading compaction
平地拖铲 level(l)ing drag scraper
平地拖曳刮土机 level(l)ing drag scraper
平地园林 flat-land garden
平地栽培植物 flat-growing vegetables
平地载运 flat haul
平地沼(泽)flat bog
平地作业 grading;grading operation
平点 flat-spotting;planar point
平点剖面 flat point section
平电 ordinary telegram
平电极管 flat plate lamp
平垫 plain cushion;plain washer
平垫板 flat tie plate
平垫片 flat shim
平垫片密封 seal with flat gasket
平垫圈【机】flat washer;flat gasket; machine washer; plain washer; plate washer;snug washer
平雕 flat carving
平雕刻工艺 flat carving
平吊碹 flat suspended main crown
平钉 back nail
平顶 flat ceiling;flat crest;flat roof; flat top(ping); insulex; plateau[复 plateaus/ plateaux];straight top;tabletop
平顶暗礁 guyot
平顶暗装式 ceiling built-in type
平顶板 ceiling board;ceiling floor
平顶板涵洞 slab-top culvert
平顶冰川 mountain ice-cap
平顶冰块 glacier table;ice table
平顶冰山 barrier iceberg; tableberg; table iceberg
平顶波 flat-top(ped) wave;flat wave; wave with horizontal crest
平顶波峰 horizontal crest
平顶波工作 flat-top operation
平顶波形 flat-topped waveform
平顶驳船 flat-top barge
平顶布置图 ceiling plan
平顶部件 ceiling element
平顶铲 tractor sweep
平顶长方形建筑物 flat-topped basilica
平顶窗 lay light
平顶打磨机 ceiling sander; ceiling sanding machine
平顶打磨器 ceiling sander; ceiling sanding machine
平顶的 flat-topped;straight head;flat-roofed <建筑物等>
平顶(的大)冰山 tabular iceberg
平顶灯 lay light
平顶电压脉冲 flat-topped voltage pulse
平顶吊筋 ceiling hanger
平顶短护刃器 brush lipless guard
平顶方头螺栓 flattened square head bolt
平顶房屋 flat-topped block; flat-topped house
平顶飞檐式线脚 ceiling cornice
平顶峰 flat peak;flat-top peak
平顶辐射板采暖 panel ceiling heating
平顶附属房屋 flat-topped annexe
平顶盖 flat deck roof
平顶盖板 ceiling boarding
平顶高原 pats
平顶格栅 ceiling joist;raglan
平顶格栅中间支撑 ceiling binder
平顶拱 floor vault
平顶拱石 stepped voussoirs
平顶构件 flat-topped unit
平顶孤丘 mesa-butte
平顶孤山 tafelkop;mesa-butte
平顶光束 flat-top beam
平顶锅炉 straight boiler;straight top boiler
平顶海丘 table knoll
平顶海山 guyot;table knoll;tablemount;table mountain
平顶海山名称 name of guyot
平顶涵洞 fiat-topped culvert
平顶荷载 ceiling load
平顶活塞 flat-head(ed) piston; flat-top piston
平顶火箱锅炉 Belpaire boiler
平顶畸变 flat-top distortion
平顶脊 flat-top ridge
平顶脊地形 bankveld
平顶建筑构件 flat-topped unit
平顶建筑物 flat-top(ped) building; flat-top(ped) structure
平顶礁 table reef;platform reef
平顶角焊缝 miter weld
平顶角焊缝焊接 flat-faced fillet weld
平顶接头 <暗沟的> flat-top junction
平顶结构 ceiling floor;deck construction; flat-topped structure; tabular structure
平顶金字塔 truncated pyramid
平顶开槽 ceiling groove
平顶梁 ceiling beam
平顶量水堰 flat-crested measuring weir
平顶淋浴器 ceiling shower
平顶龙骨 ceiling joist
平顶螺母 flush nut
平顶螺纹 square thread
平顶埋头螺钉 flat countersunk head screw
平顶埋头螺栓 flat-headed countersunk bolt
平顶脉冲 flat pulse;flat-topped pulse; square-topped pulse
平顶铆钉 pan head rivet
平顶模 flat-top die
平顶(内装)风扇 ceiling fan
平顶砌块 soffit block
平顶千斤顶 ceiling jack
平顶球面耙片 crimped-center[centre] disk
平顶曲线 flat-topped curve
平顶圈条器 flat-top coiler
平顶容量 struck capacity
平顶软木板 corkboard for ceiling
平顶山 even-crested ridge; flat-top hill; flat-top mountain; mesa; table mountain
平顶山脊 flat-topped crest;flat-topped ridge
平顶山口 yoke pass
平顶山丘 table knoll
平顶式单元采暖器 ceiling-type unit

heater
平顶式溢洪道 flat-crested spillway
平顶输送机 flat-top conveyer [conveyor]
平顶树 flat crown; flat tree
平顶弹簧片 ceiling reed
平顶特性 flat characteristic
平顶天窗 flat roof skylight
平顶天线 coat-top antenna; flat-top antenna; sheet antenna
平顶通道 ceiling duct
平顶凸轮 flat-topped cam
平顶涂层 ceiling plaster
平顶土堤 terreplein
平顶系板 ceiling strap
平顶下部 floor soffit
平顶线脚 ceiling cornice; ceiling mo(u)lding
平顶箱 ceiling box
平顶箱涵 flat-topped culvert
平顶镶板【建】ceiling panel; coffer
平顶镶板装饰 coffer
平顶镶边 ceiling trimming
平顶响应 flat-top response
平顶效应 ceiling effect
平顶斜角板条 Scotch bracketing
平顶斜坡坟墓 < 古埃及 > terraced mastaba(h); mastabah tomb
平顶谐振器 flat roof resonator
平顶信号 flat-topped wave
平顶修饰 ceiling trimming
平顶岩石 flat-topped rock
平顶堰 flat-crested weir; flat-topped weir
平顶圆头螺栓 cheese head bolt
平顶载重 ceiling load
平顶藻井 ceiling caisson
平顶照明 ceiling light; counter(sky) light
平顶照明装置 ceiling lighting fitting
平顶正峰 flat positive peak
平顶装配 ceiling fitting
平顶阻力 translatory resistance
平顶最高值 flat maximum
平订 flat stitching
平定 put down; still
平定管 Q tube
平动 libration; translation(al) motion
平动波 translatory wave
平动滑移 translational slip
平动连杆 guide link
平动能量 translational energy
平动配分函数 translational partition function
平动矢量 translation vector
平动损失 translation loss
平动态 translational state
平动阻力 translatory resistance
平峒 adit; day level; drift way; mine adit
平峒洞口 adit collar
平峒工作面 adit end
平峒-井筒联合开拓 adit-shaft development
平峒掘进 gallery driving
平峒开拓 adit development; tunnel development
平峒口 adit opening; adit portal
平峒偏差 drift of tunnel
平峒数 number of adit
平峒水平 adit level
平峒总长度 total length of adit
平洞 cross head(ing); day level; drift; gallery; horizontal opening
平洞掘进工程 shield tunnel(l)ing; tunnel(l)ing; tunnel works
平洞水平 gallery level
平洞推进 driftage
平硐 adit
平硐掘迷 tunnel driving method
平硐推进 drift-in
平硐展视图 geological log of adit
平斗的 levelled off
平斗斗容 level(l)ed bucket capacity
平斗容积 struck volume
平斗容量 struck capacity
平堵 horizontal closure
平堵法截流(法) horizontal closure (method)
平堵截流法 frontal dumping closure method; horizontal closure (method); level fill closure method; level-tipped closure method; transverse dumping closure
平端 flat-end; flush end
平端材 plan end lumber
平端定位螺钉 flat point set screw
平端钢管 plain-end steel pipe
平端弓形桁架 truncated bowstring truss
平端固定螺钉 flat end set screw
平端管 plain-end tube
平端管材 plain-end pipe
平端管头 plain shoe
平端渐缩管 plain-end reducer
平端接缝 butt-end joint
平端螺钉 plain-end screw; pointless screw
平端面 flat face
平端面金刚石钻头 flat-bottom diamond crown; flat-faced bit; flat-nose bit
平端面钻头 flat-bottom crown; flat-faced bit; flat-faced crown; flat-nose bit
平端头 butt end
平端转辙器 stub switch
平断层 flat fault
平锻机 bulldozer; upsetter; upsetting machine
平锻钳 flat forge tongs
平堆 < 板材水平堆积 > flat piling
平堆法 flat stacking
平堆费 level(l)ing charges
平堆机 windrow(er) evener
平对焊 butt-welding
平对接 square butt joint
平垛法 flat stacking
平发送 flat transmission
平阀 flat valve
平阀框 valve yoke
平阀框前杆 valve yoke guide
平阀体 plain valve head
平阀头 plain valve head
平筏 flat raft
平法兰连接 flat flange joint
平凡层 trivial sheaf
平凡丛 trivial bundle
平凡的 trite; unimportant; workaday
平凡的工业家 prosaic industrialist
平凡赋值 trivial valuation
平凡化 trivialization; trivialize
平凡解 trivial solution
平凡空间 trivial space
平凡邻近 trivial proximity
平凡纽结 trivial knot
平凡双图 trivial digraph
平凡图 trivial graph
平凡微丛 trivial microbundle
平凡位 trivial place
平凡线丛 trivial line bundle
平凡向量丛 trivial vector bundle
平凡性 triviality
平凡一致性 trivial uniformity
平凡子空间 trivial subspace
平凡子群 trivial subgroup
平凡族 trivial family
平反光镜 flat reflector
平方 second power; square
平方标度 square-law scale
平方标度电容器 square-law condenser
平方测链 square chain
平方差 square deviation
平方的 quadrate; quadratic
平方的积分 integrated square
平方电光效应 quadratic electro-optic(al) effect
平方电路【计】squarer
平方电位计 quadratic potentiometer
平方钉 box nail
平方度 square degree; square grade
平方多项式 second-order polynomial
平方反比(例)定律 inverse square law
平方分量 < 误差的 > quadrantal component
平方分米 square decimeter
平方根 square root
平方根变换 quadratic root transformation; square root transformation; transformation of square root
平方根倒数 inverse square root
平方根的 subduplicate
平方根电位计 square root potentiometer
平方根(定)律 square root law
平方根法 root-squaring method; square root method
平方根法则 square root rule
平方根分式 quadrant formula
平方根浮点运算 square root-floating
平方根面积仪 square root planimeter
平方根纸 square root paper
平方公尺 square meter[metre]
平方公寸 square decimeter [decimetre]
平方公分 square centimeter[centimetre]
平方公里 square kilometer
平方光电效应 quadratic electro-optic(al) effect
平方规划 quadratic planning; quadratic programming
平方毫米 square millimeter
平方和 quadratic sum; sum of squares
平方和的开方法 square root of the sum of square method
平方和的平方根法 square root of sum of squares method; square root of square sum
平方和方根法 square root of sum square method; square root the sum of the squares rule
平方和开方法 square of sum of squares
平方积分误差 integral square error
平方积分相互关系 integral-square-cross-correlation
平方可积的 square-integrable
平方可积函数 quadratically integrable function; square-integrable function
平方刻度电流表 current square meter
平方厘米 centimeter square; square centimeter[centimetre]
平方离差 square deviation
平方哩尺 < 水力学常数, 等于一平方哩面积, 一尺深的水量 > square-mile foot
平方链 < 面积单位, 为 0.1 英亩或 404.7 平方米 > square chain
平方列联 square contingency [contingence]
平方律 square law
平方律检波 square-law detection; square-law rectification
平方律检波器 square-law demodulator; square-law detector
平方律全波检波器 full-wave square-law detector
平方律调制 square-law modulation
平方律效应 quadratic law effect
平方码 < 英美制面积单位, 1 平方码 = 0.836 平方米 > square yard
平方码数 yardage
平方米 cent(i)are; square meter[metre]
平方米预算价 estimate on square metre basis
平方密耳 < 1/1000 平方英寸 > square mil
平方面积 quadrature
平方面积制 square measure
平方模数 squared absolute value
平方偏差 square deviation
平方器 square-law function generator; squarer
平方千米 square kilometer
平方取中法 middle square method; mid-square method
平方取中生成程序 mid-square generator
平方数 square number; roofing square < 指屋面材料 >
平方网络 squaring network
平方误差 square error
平方误差积分 integral square error
平方误差渐近有效估计量 squared error asymptotically efficient estimator
平方误差相容性 square error consistency
平方项 quadratic component; quadratic term
平方英尺 square-foot; superficial foot < 木板计量单位 >
平方英尺比价法 square foot method
平方英尺法 < 牧场植被测定的 > square foot method
平方英寸 square-inch
平方英寸流量 < 美国西部使用的一种流量单位 > miner's inch
平方英里 square-mile
平方英里英尺 square-mile foot
平方英码 square yard
平方余割包线型射束 cosecant-squared beam
平方余割反射器 cosecant-squared dish
平方余割天线 cosecant-squared antenna
平方余割形方向图 cosecant-squared diagram
平方子程序 quadrature subroutine
平方总和 total sum of squares
平房 bungalow; flat building; ground-floor building; one-floor house; one-stor(e)y house; single-stor(e)y building; single-stor(e)y house
平房壁板 bungalow siding
平房别墅 bungalow villas
平房仓库 flat warehouse; one-stor(e)y warehouse
平房工厂 flatted factory
平房谷仓 flat warehouse
平房建筑 ranch house
平房凉台挡板 bungalow siding
平房式的 bungaloid
平房庭院 bungalow; bungalow court
平放 flatwise; keep flat
平放 S 形瓦 pantile
平放储存 horizontal storage
平放大梁的矫正 straightening of the girder lying flat-wise
平放锭料 horizontal charge
平放水平仪 block spirit level
平放水准器 block level
平放直角拐肘 side crank
平分 bisect(ion); divide equally; halve
平分布 flat distribution

平分点 bisecting point
平分电路 halving circuit
平分法 mean division method
平分规 bisecting compasses
平分寄存器 halving register
平分经度 middle longitude
平分流 bisection flow
平分式供暖系统 upfeed system of heating
平分丝 halving line
平分误差 bisection error
平分线 bisection line; bisector; bisectrix; halving line; mean line
平封环 flat seal ring
平封头 flat head
平峰期 ordinary hour; ordinary duration
平峰小时最小行车间隔 minimum running interval on ordinary hours
平缝 bed joint; flat joint; flat seam; flat sewing; flush (ed) point; joint pointing; plain work; hick joint <圬工的>
平缝刀 rough tool
平缝法 flat seaming
平缝帆针 flat seam needle
平缝工 feller
平缝辊筒 butt roller
平缝砌 hick joint
平缝砌合 flat-joint jointed
平缝屋面 flat seam roofing
平伏 flat-lying
平伏层 flat-lying bed; flat measures
平伏的 flatly; flat-lying
平伏键 equatorial bond
平伏矿层 flat-lying seam
平伏矿床 blanket deposit; flat-lying deposit; flat mass
平伏矿脉 blanket vein
平伏砂层 blanket sand; sheet sand
平伏砂岩层 blanket sandstone
平扶手断面 flat handrail section
平浮 even keel; on even keel
平浮雕 coelanaglyphic relief; flat relief
平浮环 cardan
平幅 opening width
平幅焙烘 flat curing
平幅捶布 flat beetling
平幅烘燥机 level(l)ing dryer[drier]
平幅真空吸水干燥机 open width suction drier
平幅煮练机 open scourer
平敷黏[粘]合剂 lay-flat adhesive
平复激电动机 level-compound excited motor
平复激发电机 flat-compound generator
平复激励 level-compound excitation
平复励 flat-compounding; level-compound excitation
平复励磁 flat-compound excitation
平复励的 flat-compounded
平复励发电机 flat-compound dynamo; flat-compound generator
平复励特性 flat-compound characteristic
平复绕 flat composition; flat compound
平腹板型钢板桩 straight-web sheet pile
平腹板桩 flat web piling
平腹板桩墙 flat web sheet piling wall; straight-web sheet-piling wall
平腹工字梁 plain web beam
平腹拱形大梁 plain web(bed) arched girder
平腹矿床 blanket deposit
平腹脉 blanket vein
平腹三铰拱架 plain web(bed) three-hinge(d) arched girder
平盖开孔 opening in flat head
平盖木灵芝 shelf fungus
平盖手孔 handhole with flat cover
平钢板 flat steel plate; plain plate; smooth plate
平钢带 flat strip steel
平钢化玻璃 flat tempered glass
平钢模 flat die
平高点【测】horizontal and elevation (picture) control point
平高控制点 horizontal and vertical control point
平格槽缩样器 flat riffle
平格构工作 flat trelliswork
平格筛 flat bar screen; level grizzly; level screen of bars
平格栅 flat bar screen
平格振动筛 shaker pan
平隔板 flat partition board
平跟鞋 flattie; flatty
平耕 flatbreaking; flatbusting
平弓 flat bow
平公局 <负责审核财产税收及投诉地方政府评估房地产税的机构> Equalization Board
平拱 arch-flat; camber arch; depressed arch; flat arch; flattened arch; floor arch; plain arch; stor(e)y arch; straight arch; jack arch
平拱顶 jack vault
平拱过梁 flat arch lintel; floor arch lintel; straight arch lintel
平拱梁 flat arched girder
平拱梁屋顶 flat arched girder roof
平拱门洞 square-headed
平拱铺地瓦 flat arch floor tile
平拱桥 elliptic(al) arch bridge; flatted arch bridge
平拱形顶棚 cambered ceiling
平拱形天花板 cambered ceiling
平勾缝 flat pointing; flush-joint pointing; hick joint pointing
平沟铲斗 ditch grading bucket
平沟机 trench filler
平钩 plain hook
平钩环 plate shackle
平箍钢 flat hoop iron
平谷 flat valley
平刮 strickle
平刮(灰)缝 plain-cut joint
平挂链 flat chain system
平挂石板瓦 slate listing
平管口 plain end
平管蒸发器 horizontal tube evaporator
平罐 horizontal retort
平光 dead flat; lusterless; mat gloss
平光表面 plain face
平光锤 smoothing hammer
平光醇酸瓷漆 flat alkyd enamel
平光光泽 sheen gloss
平光光泽计 sheen gloss meter
平光剂 flattening agent
平光铆 flush rivet(t)ing
平光铆接 flush rivet(t)ing
平光面漆 flat finish
平光漆 flat paint; mat(t) paint; smoothing lacquer
平光清漆 flat varnish; mat(te) varnish; none gloss varnish
平光涂层 flat coat
平光涂料 flat paint
平光镶板 flush panel
平光眼镜 plain glass spectacles
平光油灰 flatting putty
平光油漆 mat gloss paint
平光油性漆 flat oil paint
平轨 flat rail; plate rail
平轨道 flat track
平轨机 rail level(l)er
平辊 plain-barrel(l)ed roll
平辊破碎机 plain roll crusher
平辊身 straight barrel
平辊型 parallel roll profile
平滚式舱盖装置 end or side rolling hatchcover
平滚筒 flat roller
平锅 pan
平过道【铁】cross-track passage; foot pass between platform; foot pass between tracks <跨越轨道的>
平海底 low coast; plain
平焊的 downhand
平焊法兰 slip-on flange
平焊法兰人孔 manhole with common welded flange
平焊缝 downhand weld; flush bead; flush weld
平焊钢法兰 flat position welding flange
平焊钢凸缘 flat position welding flange
平焊焊条 downhand electrode
平焊(接) butt welding; downhand welding; flat position welding; flat weld (ing); horizontal (position) weld (ing); slip-on weld (ing); flush(ing) weld(ing)
平焊桶体接缝 butt-welded body seam
平焊位置 downhand position; flat position
平夯 straight tamper
平合缝 flat seam
平合嵌接 flat scarf
平河床 flat bed; flat riverbed
平恒星时 mean sidereal time; uniform sidereal time
平桁架 plane frame work
平衡 averaging out; balance; balance out; compensate; compromise; counteract; equalization; equalize; equibalance; equilibrate; equilibrium[复 equilibria/equilibriums]; equiponderance; equiponderate; neutralization; neutralization process; neutralize; neutralizing; poise; reconciliation; stand-off
平衡 pH equilibrium pH
平衡摆 balance wheel; compensation balance; compensation wheel
平衡板 balancing plate; balance plate; outrigger base; outrigger plate
平衡板组 balance plate unit
平衡半径 equilibrium radius
平衡棒 halter
平衡饱和度 equilibrium saturation
平衡泵 ballast pump
平衡比 equilibrium ratio
平衡比降 balanced grade; equilibrium slope; slope of equilibrium
平衡比值法 ratio method of balancing
平衡闭合电路 balanced termination
平衡臂 arm of balance; balance arm; balance bob; balancing arm; counter jib; equalizer beam
平衡边界层 equilibrium boundary layer
平衡变换器 balance converter; balun unit
平衡变换装置 bazooka
平衡变压器 balancer-transformer; balancing transformer
平衡变阻器 balancing rheostat
平衡变阻箱 balancing rheostat
平衡标记 balance mark
平衡标志 balance mark
平衡表 balance sheet
平衡表法 balance method
平衡表面结构 equilibrium surface structure
平衡表面张力 equilibrium surface tension
平衡表征 equilibrium specification
平衡拨盘 Clement's driver
平衡薄板 balancing sheet
平衡补助金 equalization grants
平衡不定全区组设计 balanced incomplete block design
平衡不定全区组试验 balanced incomplete block experiment
平衡不稳 disequilibration; disequilibrium
平衡不足 underbalance
平衡布局 balanced distribution
平衡仓 balance bunker
平衡舱【船】balanced rudder; equalizing compartment; equalizing rudder
平衡操纵面 balanced surface
平衡槽 balancing groove; compensating groove; counter-balanced chute
平衡测 balancing side
平衡测力器 balancing dynamometer
平衡测量 balancing a survey; traverse adjustment
平衡策略 equilibrium strategy
平衡差 balanced differences
平衡差度 disbalance; unbalance
平衡常量 equilibrium constant
平衡常数 equilibrium constant
平衡场地 balanced field
平衡超高 equilibrium superelevation
平衡超高度 balanced superelevation; equilibrium superelevation
平衡潮理论 equilibrium theory of tide
平衡潮(汐) equilibrium tide
平衡潮汐论 equilibrium argument
平衡潮学说 equilibrium theory of tide
平衡车 balance car
平衡车道 lane balance
平衡衬套 balance bush
平衡成长率 equilibrium rate of growth
平衡成分 equilibrium composition
平衡乘积混频器 balanced product mixer
平衡池 balancing reservoir; equalizer; equalizing basin
平衡齿轮 balance gear
平衡充电 equalizing charge
平衡冲刷 equilibrium scour
平衡出水 equalization of discharge
平衡储槽 balancing reservoir; balancing tank; equalising [equalizing] reservoir
平衡储存 balanced stock
平衡储水池 balance storage
平衡处理 equalization treatment
平衡传输线 balanced line; balanced transmission line
平衡窗 counter-balanced window; suspended sash
平衡窗扇 balance sash
平衡锤 balancing weight; back balance; balance bob; balance(d) box; barney; bobweight; compensating counterweight; counter-balance; counter-balance weight; counter scale; counterweight; counterweight weight; damper weight; help weight; pull-off; tension weight; counterpoise
平衡锤臂 counterweight arm; hammerhead jib
平衡锤杆 balance weight arm; balance weight lever; counter-balance lever; counter bar; counterweight arm; side lever; weight bar
平衡锤杆枢轴 balance weight lever pivot
平衡锤固定板 weight setting plate
平衡锤盒 weight box; weight pocket; window box

平衡锤栏木 counter-balance barrier
平衡锤式车库翻门 balance-weight-type overhead garage door
平衡锤提升机 counterweight winder
平衡锤调节绳索下降 weight a brake
平衡锤系统 counterweight system
平衡锤装卸方式 counterweight system
平衡磁极 weight equalizer
平衡次斜槽 counter-balanced sub-chute
平衡催化剂 equilibrium catalyst
平衡锉 equalizing file
平衡单元 balancing unit
平衡弹 equilibrium bomb
平衡导沟 counter-balanced spout
平衡导锁 counter lock
平衡导线 equalizing conductor
平衡到不平衡的变换装置 bazooka
平衡的 balanced; balancing; compensating; equilibrious; proportional; well-balanced
平衡的边际效用 balancing margin
平衡的经济发展 balanced economic development
平衡的全面发展规划 balanced overall development planning
平衡的溶剂 balanced solvent
平衡的稳定条件 stability conditions of equilibrium
平衡的预算 balanced budget
平衡等离子体 equilibrium plasma
平衡等离子体温度 equilibrium plasma temperature
平衡低共熔物 equilibrium eutectic
平衡底板基础 balanced base foundation
平衡点 balance point; balancing point; equilibrium point; rest point
平衡点测定器 null instrument
平衡点测量 equilibrium point measurement
平衡点检测器 balance point detector; equilibrium point detector
平衡电动机电阻 balancing motor resistance
平衡电缆 balanced cable
平衡电流 balanced current; symmetric(al) alternating current
平衡电路 balanced circuit; electric(al) balance
平衡电桥 balanced bridge
平衡电桥电路 balanced bridge circuit; null-type bridge circuit
平衡电容器 balancing capacitor; neutrodon
平衡电势 balanced potential; equilibrium potential
平衡电网路 balanced network
平衡电位 equilibrium potential
平衡电压 balanced voltage
平衡电压测试法 balanced cell method
平衡电阻 balancing resistance; ballast resistance
平衡电阻器 balancing resistor; stabilizing resistor
平衡垫圈 balancing washer; equalizing washer
平衡吊架 balance hanger
平衡掉 balance out
平衡顶进法 balanced jacking method
平衡动力学 equilibrium kinetics
平衡动量 balance momentum; equilibrium momentum
平衡度 degree of balance; quality of balance; slope of equilibrium; trim
平衡端接 balanced termination
平衡短链 bridle; bridle chain
平衡断面 balanced section; equilibrium profile
平衡堆芯 equilibrium core
平衡(对称)波形 balancing waveform
平衡多边形 equilibrium polygon
平衡多层增稠器 balanced tray thickener
平衡多数空穴密度 equilibrium majority-hole density
平衡多态现象 balanced polymorphism
平衡舵 balanced rudder; equipoise rudder; rudder balance
平衡舵角 equilibrium rudder angle
平衡轭导承 equalizing yoke guide
平衡二氧化碳 equilibrium carbon dioxide
平衡二元聚合物 equibinary polymer
平衡发电机 balancer; balancing dynamo; balancing generator; equalizing dynamo
平衡阀 balance(d) valve; balancing valve; compensating valve; compensation valve; counter-balance valve; equalizing valve; equilibrium valve; holding valve; level(l)ing valve; lock shield valve; neutralizer valve; sustaining valve
平衡阀压力调节器 balance valve pressure regulator
平衡法 balanced method; null reading; zero method; balancing
平衡法的投资 balanced fund
平衡法抵押 equitable mortgage
平衡法码 counterpoise
平衡法上毁坏 equitable waste
平衡法上利益 equitable interests
平衡法上赎回权 equity of redemption
平衡法上遗产 equitable assets
平衡法上债务让与 equitable assignment of debt
平衡砝码 sliding weight
平衡反差 balance contrast
平衡反应 balanced reaction
平衡反应舵 balanced reaction rudder
平衡反应计数器 flat-response counter
平衡反应温度 equilibrium-reaction temperature
平衡方程【数】balance equation
平衡方程式 balanced equation; equation of equilibrium; equilibrium equation
平衡方法 balance[balancing] method; equilibrium method
平衡防护装置 balanced guard
平衡放大器 balanced amplifier; equalizing amplifier
平衡放射性 equilibrium activity
平衡沸点 equilibrium boiling point
平衡费用计划 level charge plan
平衡分布 equilibrium distribution
平衡分布系数【数】equilibrium distribution coefficient
平衡分割 equilibrium division
平衡分类 balanced sort
平衡分类法 balanced sorting
平衡分离系数 equilibrium separation coefficient
平衡分凝系数 equilibrium segregation coefficient
平衡分析 equilibrium analysis
平衡风门 equalizing damper
平衡浮子 displacer
平衡辐射 equilibrium radiation
平衡负荷 balanced load; integrated demand
平衡负荷继电器 load-level(l)ing relay
平衡负载 balanced load; balancing load; equilibrium load
平衡负载电度表 balanced load meter
平衡负载仪表 balanced load meter
平衡概率 equilibrium probability
平衡杆 balance arm; balance beam; compensating bar; compensating beam; equalizer; equalizer bar; equalizing bar; equalizing beam; fall lever; Hoosier pole; inverted lever; operating arm; rocker lever; rocking arm; stabilizator rod; sway beam; working beam
平衡杆冲击柱 <钢绳冲击钻机的> stout post
平衡杆件 balance bar; balance beam
平衡杆式继电器 beam relay
平衡杆枢 equalizer bar pivot
平衡感 equilibrium sense
平衡感觉障碍 paraequilibrium
平衡感受器 statoreceptor
平衡缸 compensating cylinder
平衡钢锭 balanced ingot
平衡钢筋 balance(d) reinforcement
平衡钢筋比 balanced steel ratio
平衡钢绳 balance rope
平衡杠杆 balance[balancing] lever; jack back; equalizing lever
平衡杠杆升降桥 balanced-lever lift bridge; semi-lift bridge
平衡杠杆运动 movement of equilibrium lever
平衡高度 balance height
平衡工具 poising tool
平衡工业发展布局 even out the distribution of industry
平衡工作点 matching point
平衡工作时间 balancing time
平衡工作温度 equilibrium operating temperature
平衡工作站 balancing work station
平衡公式 balance equation; equilibrium formula
平衡功率级 equilibrium power level
平衡功率计 balanced dynamometer
平衡拱 balance arm; balanced arch; ground arch
平衡共聚合作用 equilibrium copolymerization
平衡构架 balanced rigid frame
平衡构形 configuration of equilibrium
平衡构造 balanced construction
平衡购入点 equilibrium purchase point
平衡鼓 balancing drum
平衡固溶度 equilibrium solid solubility
平衡故障率 equilibrium failure rate
平衡关税 parity duty
平衡关系 equilibrium relationship
平衡管 balance pipe; compensating pipe; equalizer[equalizing] pipe
平衡管道 equalizing main
平衡管道系统 balance line system; balancing line system
平衡管道蒸气回收系统 balance vapo(u)r recovery system
平衡管网 balanced network
平衡罐 balance tank; balancing tank; compensator; surge drum; surge tank
平衡光学混频器 balanced optic(al) mixer
平衡归并 balanced merge
平衡轨道 equilibrium orbit; stable orbit
平衡轨道半径 equilibrium orbit radius
平衡过程 balance[balancing] process; equilibrate[equilibrium] process
平衡海底地形 shore equilibrium profile
平衡含水量 balance[balancing] water (content); equilibrium water
平衡含水率 balance water content; equilibrium moisture content; equilibrium water
平衡合并 balanced merge
平衡河流 poised river; poised stream; regime(n) river
平衡核间距 equilibrium internuclear distance
平衡荷载 balanced load; balancing load; equilibrium load
平衡荷载法 balance load method
平衡恒等式【数】balanced identity
平衡横断面 <厚边混凝土路面> balanced cross-section
平衡横梁 <履带底盘半刚性悬架的> pivoted equalizer bar
平衡厚度 equilibrium thickness
平衡弧 balance arc
平衡滑车(轮) equalizer sheave; equalizing sheave
平衡滑阀 balanced slide valve; equilibrium slide valve
平衡滑轮 compensating pulley
平衡滑轮补偿法 means of compensation with compensating pulley
平衡滑翔距离 equilibrium range
平衡化 equilibration
平衡化学 equilibrium chemistry
平衡化学表征 equilibrium chemical specification
平衡环 balanced-loop antenna; balancing link; balancing ring; cardan; dummy ring; equilibrium ring; fish shackle; gimbal; gimbal ring
平衡环境 balanced environment
平衡环境房间型量热计 balanced ambient room type calorimeter
平衡环形调制器 balanced ring modulator
平衡环轴 axis of the gimbal
平衡缓冲器 compensating buffer; equalizing buffer
平衡换能器 balance converter
平衡簧套 counter-balance spring casing
平衡恢复 restoration of balance
平衡回潮率 equilibrium moisture regain
平衡回流 equilibrium reflux
平衡回流沸点 equilibrium reflux boiling point
平衡回路 balanced-loop antenna
平衡回转窑 balanced rotary kiln
平衡汇率 equilibrium rate
平衡混合物 equilibrium mixture
平衡混频器 balanced mixer
平衡活动桥 balance bridge
平衡活度 equilibrium activity
平衡活塞 balance piston; balancing drum; balancing piston; dummy piston; relief piston; supported piston
平衡活塞杆接头 joint for balance piston
平衡活塞密封圈 equalizing piston packing ring
平衡活塞皮垫 leather packing for balance piston
平衡活塞皮碗 cup leather for balance piston
平衡活塞橡胶环 rubber follower for balance piston
平衡活塞与护圈 dummy piston and retainer
平衡活闸板 counter shutter
平衡机 equilibrator
平衡机构 balance mechanism; equalizing mechanism
平衡机构调节器 governor balance gear
平衡机架 balance frame

P

平衡机件 dummy
平衡机组 balancer set
平衡基床 <管道的> equalizing bed
平衡基金 equalization fund
平衡级 equilibrium level; equilibrium stage
平衡极限 limit of equilibrium
平衡计 equilibristat
平衡计划进度表 balanced schedule
平衡计算 balance calculation; balancing calculation
平衡技术 balancing technique
平衡剂 poising agent
平衡剂量常数 equilibrium dose constant
平衡继电器 balanced relay; balancing relay
平衡加捻 balanced twist
平衡加速电压 equilibrium accelerating voltage
平衡价格 equilibrium price; parity price
平衡架 bogey; compensated bogie; equalizer[equalizing] bar; equalizing frame; gimbal; load-dividing dolly <牵引车和拖车连接处的>
平衡架轴 gimbal axis
平衡假设 equilibrium hypothesis
平衡减压阀 balance pressure reducing valve
平衡剪切模量 equilibrium shear modulus
平衡剪切柔量 equilibrium shear compliance
平衡检波器 balance(d) detector
平衡检查 balance check; equilibrium test
平衡检验 balance test
平衡检验式 balance check mode
平衡检验型 balance check mode
平衡鉴别器 balanced discriminator
平衡交错套设计 balance staggered nested design
平衡交通运输 balanced traffic
平衡角 angle of equilibrium; equilibrium angle
平衡校验 balance check
平衡校验型 balance check mode
平衡接触级 equilibrium contact stage
平衡接触角 equilibrium contact angle
平衡接合常数 equilibrium binding constant
平衡接头 equalizing joint; equalizing sub
平衡结 junction at equilibrium
平衡结构 balanced construction; equilibrium structure
平衡结构胶合板 balanced construction plywood
平衡结晶作用 equilibrium crystallization
平衡截面 balanced section
平衡解 equilibrium solution
平衡解题法 equilibrium approach
平衡解调器 balanced demodulator
平衡界面张力 equilibrium interfacial tension
平衡(近似)计算法 equilibrium approach
平衡井公式 <地下水力学中的> equilibrium well formula
平衡井斜 equilibrium hole inclination
平衡静噪电路 noise balancing circuit
平衡卷筒补偿法 means of compensation with compensating drum
平衡觉 equilibratory sensation
平衡卡规 poising cal(l)ipers
平衡卡头 equalized clamp
平衡开关 balance cock
平衡开挖 balanced excavation
平衡空气蒸馏 equilibrium air distillation
平衡孔 balance hole; balancing hole; equalizing hole
平衡控制 balanced control; balancing control
平衡控制反应 equilibrium-controlled reaction
平衡口【机】 equalizing port
平衡库容 balance storage; balancing storage
平衡块 back balance; balanced weight; counter-balance; dummy mass; equalizing block; poise; weight
平衡块控制液压缸 cylinder of counterweight
平衡块伸出长度 counter-balance radius
平衡块伸距 counter-balance radius
平衡块式推力轴承 balancing sector thrust bearing
平衡块支架 counterweight frame
平衡块止推轴承 balancing sector thrust bearing
平衡馈线 balanced feeder
平衡拉力 balanced tension
平衡缆式起重机 balanced cable crane
平衡缆索 counterweight cable
平衡棱镜 equilibrium prism
平衡冷凝 equilibrium condensation
平衡冷凝曲线 equilibrium condensation curve
平衡离子 counter ion; equilibrium ion
平衡犁 swing plough
平衡理论 equilibrium theory; theory of balances; theory of equilibrium
平衡力 balance force; balancing force; counter-balance (force); counterpoise; equilibrant; equilibrium force; equipoise
平衡力矩 balanced moment; balancing moment; counter(-balance) moment; equilibrium moment; trim(ming) moment
平衡力系 equilibrant system
平衡利率 equilibrium rate of interest
平衡连杆 balanced connecting rod; stabilizer link
平衡连接 equalizing connection
平衡链 balancing chain
平衡梁 balance beam; compensating beam; equalizer; equalizer bar; equalizer beam; equalizing beam; rocker beam; walking beam; weigh-beam
平衡梁鞍座 equalizer saddle
平衡梁鞍座垫 pad of equalizer bar
平衡梁垫块 equalizer pad
平衡梁弹簧 equalizer spring
平衡梁悬挂装置 walking beam suspension
平衡梁悬架【机】 equalizer bar suspension
平衡亮度 equilibrium brightness
平衡量 equilibrium amount
平衡料仓 surge bin
平衡料轮 gob hopper
平衡磷浓度 equilibrium phosphorus concentration
平衡灵敏度 balanced sensitivity
平衡流 equilibrium flow
平衡流量 balancing flow; regime(n) discharge; regime(n) flow
平衡露点 equilibrium dew point
平衡滤失速度 equilibrium filtration rate
平衡路 neutrodyne circuit
平衡路线法 line of balance technique
平衡率 balance ratio
平衡轮 balance wheel; equalizing pulley
平衡轮组 balanced gear cluster
平衡螺钉 mass screw
平衡螺杆 <钢绳钻探用> temper screw
平衡螺栓 equalizing bolt
平衡落球法黏[粘]度计 counter-balanced falling ball viscometer; drown sphere viscometer
平衡码 balanced code
平衡毛细管压力 equilibrium capillary pressure
平衡门 balanced door; balanced gate; counter(-balanced) gate
平衡密度 equilibrium density
平衡密度泥浆法 balanced density slurry technique
平衡密度溶剂 balanced density solvent
平衡面 balanced surface
平衡面积 balanced area
平衡明线 balanced open wire
平衡模量 equilibrium modulus
平衡模式 equilibrium mode
平衡模型 equilibrium model
平衡膜厚度 equilibrium film thickness
平衡磨球装量 equilibrium ball charge
平衡木 balance beam; balancing beam
平衡能力 balanced capacity
平衡黏[粘]滞度 equilibrium viscosity
平衡凝固 equilibrium freezing
平衡凝入 equilibrium freezing-in
平衡扭转 equilibrium torsion
平衡浓度 equilibrium concentration
平衡浓缩系数 equilibrium enrichment coefficient; equilibrium enrichment factor
平衡排风管 balanced exhaust duct
平衡盘 balancing drum; dummy
平衡盘式增稠器 balanced tray thickener
平衡盘套筒 balance bush
平衡判据 equilibrium criterion
平衡跑道 balanced runway
平衡配板 balance matching
平衡配筋 balanced reinforcement
平衡配筋梁 balanced reinforcement beam
平衡配筋率 balanced steel ratio
平衡配平片操纵 trim-tab control
平衡配气阀 balanced distribution valve
平衡配重 weight balance
平衡膨胀 equilibrium swelling
平衡膨胀率 equilibrium swelling ratio
平衡偏转 balanced deflection
平衡品质 balanced strain
平衡品质 balance quality
平衡屏幕框架 counter-balanced screen frame
平衡坡度 balanced grade; balanced grading; equilibrium slope; slope of equilibrium
平衡坡降 equilibrium gradient
平衡破坏 balanced failure; disturbance of balance
平衡剖面 balanced cross-section; balanced profile; profile of equilibrium; equilibrium profile
平衡谱 equilibrium spectrum
平衡起重机 balance(d) hoist
平衡气 equilibrium gas
平衡气阀 compensating air valve
平衡气柜 relief holder
平衡气球 balanced balloon
平衡气驱 equilibrium gas drive
平衡气体饱和率 equilibrium gas saturation
平衡汽缸 balance cylinder; counter-balance cylinder
平衡汽缸盖 balance cylinder cover
平衡汽化 equilibrium evapo(u)ration
平衡汽化比 equilibrium vapo(u)rization ratio
平衡汽化釜 equilibrium flash vapo(u)rization still
平衡汽油 balanced gasoline
平衡器 balance bob; balancer; balancing device; counterpoise; equalizer; equilibrator; evener; neutralizer; stabilizer; weight equalizer
平衡器安全环 equalizer safety strap
平衡器齿轮 balancer gear
平衡器阀 equalizer valve
平衡器-分析器组合 equalizer-analyser[analyzer] combination
平衡器功能块 equalizer block
平衡器架 equalizer frame
平衡器弹簧 equalizer spring; equilibrator spring
平衡器旋钮 zero sharpening control
平衡器支杆 equalizer supporting bar
平衡钳位电路 balanced clamper
平衡桥 balance bridge
平衡切削 balanced cutting
平衡球阀 equilibrium ball valve
平衡球面 equilibrium spheroid
平衡球状体 equilibrium spheroid
平衡区 conditioning section
平衡区段设计(法) balanced block design
平衡曲柄 balanced crank
平衡曲线 equilibrium curve; equilibrium line
平衡驱动 balance drive
平衡趋势 equilibrium tendency
平衡取样门 symmetric(al) sampling gate
平衡圈 balance ring
平衡群落 equilibrium population
平衡扰乱 disturbance of balance
平衡绕组 equalizer winding; stabilizing winding
平衡容量 balancing capacity
平衡容器 balancing vessel
平衡溶解草酸盐浓度 equilibrium dissolved oxalate concentration
平衡溶解度 equilibrium solubility
平衡溶解氧浓度 equilibrium dissolved oxygen concentration
平衡溶液 balanced solution; equilibrium solution
平衡溶液成分 equilibrium solution composition
平衡溶胀 equilibrium swelling
平衡柔量 equilibrium compliance
平衡润湿 equilibrium wetting
平衡三线制 balanced threewire system
平衡三相电路 balanced three phase circuit
平衡三相制 balanced three-phase system
平衡沙量 equilibrium load
平衡纱 balanced yarn
平衡砂堤 equilibrium bank
平衡闪蒸(法) equilibrium flash vapo(u)rization; simple continuous distillation
平衡闪蒸锅 equilibrium flash still
平衡闪蒸器 equilibrium flash vapo(u)rizer
平衡闪蒸曲线 equilibrium flash curve
平衡闪蒸设备 equilibrium flash vapo(u)rizer
平衡膳食 balanced diet
平衡设计 balance(d) design
平衡深度 equilibrium depth
平衡渗析法 equilibrium dialysis

P

平衡升降机 balanced hoist
平衡绳 balance rope;neutral rope
平衡施肥 balanced fertilization
平衡施工 balanced construction
平衡湿度 equilibrium humidity
平衡湿含量 equilibrium moisture content
平衡石 lithite
平衡时间 balanced time;equilibrium time
平衡实验 equilibrium experiment
平衡式 C 形四端网络 balanced C four-pole network
平衡式百叶窗 balanced shutter
平衡式测功器 cradle dynamometer
平衡式船坞 balance dock
平衡式窗扇 balanced sash
平衡式窗扇托臂 balance arm
平衡式吊桥 counterpoise bridge
平衡式多层增稠器 balanced tray thickener
平衡式多谐振荡器 balanced multivibrator
平衡式发动机 balanced engine
平衡式浮坞 balance floating dock
平衡式滚锥轴承 balanced taper roller bearing
平衡式平滑阀 match-box-type of flat slide valve
平衡式起重机 balance crane
平衡式汽化器 balanced type carburet(t)or
平衡式上下推拉窗 counter-balanced sash
平衡式上下推拉门 balanced door
平衡式舌瓣闸门 automatic flap gate
平衡式舌簧单元 balanced armature unit
平衡式收益表 account form of income statement
平衡式双端面机械密封 balanced double mechanical seal
平衡式提水工具 counterpoise lift
平衡式通风 balanced draught
平衡式推拉门 counter-balanced sash
平衡式温度计 balanced thermometer
平衡式衔铁拾声器 balanced armature pick-up
平衡式限噪器 balanced noise limiter
平衡式溢流阀 compensated relief valve
平衡式载波电话 equilibrium carrying telephone
平衡试验 balance[balancing] test
平衡试验机 balance[balancing] tester;balancer;balancing machine
平衡试验装置 balancing rig
平衡室 balance cylinder;balancing chamber
平衡收支 revenues and expenditures balanced
平衡受力 balanced load
平衡输出 balanced output
平衡束 equilibrium beam
平衡束流尺寸 equilibrium beam dimensions
平衡束流大小 equilibrium beam size
平衡树 balanced tree
平衡数 equilibrium number
平衡数据 equilibrium criterion
平衡水 equilibrium water
平衡水舱 trim tank;water ballast chamber
平衡水池 balancing reservoir;balancing tank;counter reservoir;equalizing reservoir;make-up reservoir
平衡水袋【地】compensating[compensation] sac;compensatrix
平衡水分 balancing moisture content;equilibrium moisture content;equilibrium water content
平衡水柜 balanced tank;balancing reservoir;balancing tank;equalising [equalizing] reservoir
平衡水库 balancing reservoir;balancing tank;compensating reservoir;equalising [equalizing] reservoir;make-up reservoir;reregulating reservoir
平衡水库对置蓄水池 counter reservoir
平衡水流 regime(n) flow
平衡水面 equilibrium water surface
平衡水平 equilibrium level
平衡水汽含量 equilibrium moisture content
平衡水深 equilibrium depth
平衡水通量 equilibrium water flux
平衡水头 counter head
平衡水箱 balanced reservoir;make-up tank;balancing tank
平衡水闸 balanced sluice
平衡水准 equilibrium level
平衡水准面 neutralization level
平衡税 countervailing duty
平衡送风 balanced draft
平衡速度 balance speed;balancing speed;equilibrium speed;free running speed;neutral speed
平衡速率 balancing rate;balancing speed;equilibrium speed;neutral speed
平衡酸碱度 equilibrium pH
平衡损耗 balancing loss
平衡损益 balance the profit and loss
平衡索 compensating rope
平衡锁口 counter lock
平衡锁扣模 counter locked die
平衡塔 <安装缆索桥用的> compensation tower
平衡台 balancer
平衡太阳潮 equilibrium solar tide
平衡摊还抵押 <每月等额偿付的抵押> level-payment mortgage
平衡弹簧 balance spring;balancing spring;compensating spring;compensator spring
平衡弹簧悬架【机】equalizer spring suspension
平衡套管 balancing sleeve;sleeve for balance splicing
平衡特性 equilibrium response
平衡提升机 balanced hoist
平衡天线 balanced antenna;balancing aerial
平衡填挖方 balance(d) cuts and fills
平衡条件 balance condition;condition of balance;condition of equilibrium;equilibrium condition
平衡条(汽室) balance strip
平衡调节 balance adjustment
平衡调整 balanced control;balancing adjustment;balancing control
平衡调制 balanced modulation
平衡调制器 balanced modulator
平衡贴现率 equalizing discount rate
平衡通风 balanced draft;balance(d) ventilation;equilibrium draft
平衡通风系统 balanced ventilation system
平衡通量 equilibrium flux
平衡统计 balance statistics
平衡统计力学 equilibrium statistical mechanics
平衡筒 counterweight;surge drum
平衡投影 compensation projection
平衡投影器 equalizing projector
平衡透析 equilibrium dialysis
平衡图 balance chart;constitution(al) diagram;stable diagram;state diagram;structural diagram
平衡土方 balancing quantities
平衡土方工程 balanced earthwork
平衡土方量 balanced earthwork
平衡推进 <相对方向绿波波宽相同的车辆波状推进> balanced progression
平衡推进器给进 <冲击钻进时> temper the jar
平衡陀螺仪 balanced gyroscope;balancing gyroscope
平衡挖填 balanced cuts and fills
平衡弯矩 balanced moment;balancing moment
平衡弯矩法 balancing moment
平衡弯液面 equilibrium meniscus
平衡网络 balancing network
平衡危险性 balanced risk
平衡卫板 balanced guard
平衡位置 equilibrium position;position of equilibrium
平衡温度 equilibrium temperature
平衡稳定时间 equilibration time
平衡问题 equilibrium problem
平衡物 counterpoise;equilibrant;equilibrator;equipoise;equiponderant
平衡误差 balanced error;balancing fault;poise error
平衡吸附 equilibrium adsorption
平衡吸附关系 equilibrium adsorption relationship
平衡吸附量 equilibrium adsorption amount
平衡吸附模型 equilibrium adsorption model
平衡吸气取样 balanced draft sampling
平衡吸收 equilibrium absorption
平衡铣 balance mill
平衡系数 balancing coefficient;balancing factor;coefficient of balance;equilibrium coefficient
平衡系统 equilibrium system
平衡衔接 balanced linkage
平衡衔铁 balanced armature
平衡线 balance(d) line;line of equilibrium
平衡线络 balance line
平衡线圈 balance coil;balancing coil;ballast coil
平衡线图 line of balance
平衡相 equilibrium phase
平衡相变 equilibrium phase change
平衡相对湿度 equilibrium relative humidity
平衡相角 equilibrium phase angle
平衡相态 equilibrium phase behavio(u)r
平衡相图 equilibrium phase diagram;phase diagram of equilibrium
平衡箱 balance box;equalization box;equilibrium box;surge tank
平衡项目 balancing item;item of balance
平衡销 balancing plug
平衡小车 balancing trolley
平衡效应 balancing effect
平衡芯头 balanced print
平衡形态 equilibrium figure
平衡形状 equilibrium configuration
平衡型封头曲面 isotensoid head contour
平衡型结构 balanced construction
平衡型链路接入规程 link access procedure balanced
平衡型土压盒 balance type pressure cell
平衡型心 balanced core
平衡修正值 equilibrium corrected value
平衡蓄水池 balance storage
平衡悬臂法 balanced cantilever method
平衡悬臂分段施工 balance cantilever segmental construction
平衡悬臂架设(法) balanced cantilever erection
平衡悬臂浇筑法 cast-in-place balancing cantilever method
平衡悬臂施工(法) balanced cantilever construction;balanced cantilever erection
平衡悬浇法 cast-in-place balancing cantilever method
平衡悬拼法 precast balancing cantilever method
平衡旋塞 balancing plug cock
平衡旋塞屏蔽阀 <便于暖气片拆修> balancing valve
平衡旋涡式燃烧室 balanced vortex type combustion chamber
平衡循环 equilibrium cycle
平衡压力 balance(d) pressure;counterpressure;equalizing pressure;equilibrium of pressure;equilibrium pressure
平衡压力的通道 pressure equalizing passage
平衡压力罐 balancing pressurizing tank
平衡压力溢流支管(道) by-pass pipeline for equalizing pressure
平衡压实 equilibrium compaction
平衡压缩 equilibrium compression
平衡(压)重填土 counterweight fill
平衡烟道 balanced flue
平衡烟道出口 balanced flue nozzle
平衡烟道式采暖炉 balanced flued gas heater
平衡延迟 balancing delay;equilibrium delay
平衡延误 balancing delay
平衡岩芯管 balanced core barrel
平衡研究 equilibrium study
平衡盐对 reciprocal salt pair
平衡盐液 balanced salt solution
平衡氧化还原表征 equilibrium redox specification
平衡样本【数】balanced sample
平衡要求 equilibrium requirement
平衡叶片泵 balance vane pump
平衡叶片式继电器 balanced vane relay
平衡液 equilibrium liquid
平衡液压机 <液压千斤顶的> accumulator
平衡液柱 balancing liquid column
平衡仪 balancing apparatus
平衡因子 balance divisor
平衡油 equilibrium oil
平衡油槽 balancing tank
平衡油罐 balancing tank
平衡油箱 make-up tank
平衡预算 balance a budget;balanced budget
平衡预算乘数 balanced budget multiplier
平衡预算规则 balanced budget rule
平衡元件 balancing components;equalizing components
平衡原理 equilibrium principle;principle of equilibrium
平衡运动 balance exercise
平衡运动的 statokinetic
平衡运费到货价 freight equalization
平衡运输 balanced traffic
平衡运转 equilibrium running
平衡载荷 balancing load
平衡载流子 equilibrium carriers
平衡噪声限制器 balanced noise limiter
平衡闸门 balanced gate;counter(-bal-

anced) gate
平衡闸装置 compensation brake rigging; equalization brake gear
平衡账户 balancing account
平衡针阀 balanced needle valve
平衡振荡器 balanced oscillator
平衡振幅 equilibrium amplitude
平衡蒸发 equilibrium vapo(u)rization
平衡蒸馏 equilibrium distillation
平衡蒸馏锅 equilibrium still
平衡蒸馏曲线 equilibrium distillation curve
平衡蒸气相 equilibrium vapo(u)r phase
平衡蒸气压 equilibrium vapo(u)r pressure
平衡直升装置 balanced vertical lift
平衡值 equilibrium value; trimmed value
平衡指示器 balance indicator
平衡指数 equilibrium index
平衡制动器 balanced brake; brake equalize
平衡制动装置 equalization brake gear
平衡质量 balance mass; quality of balance
平衡质量分布 equilibrium mass distribution
平衡质子 equilibrium proton
平衡中心 center[centre] of equilibrium
平衡终接 balanced termination
平衡重 back balance; balanced weight; balancing weight; bobweight; brake counterweight; compensating weight; counter-balance weight; counter scale; donkey; kentledge; mass balance; mass balance weight; tension weight
平衡重安全装置 counterweight safety device
平衡重臂 <塔吊的> counterjib; counterweight jib
平衡重边 counterweight side
平衡重穿心轴 counterweight rod
平衡重锤 balance weight
平衡重导轨 balance weight guide; counterweight guide
平衡重的一部分 segment of counterweight
平衡重杠杆 balance weight lever; counter-balance lever
平衡重滑车 counterweight sheave
平衡重滑道框 box casing
平衡重滑轮 counterweight block
平衡重架子 counterweight
平衡重净空 counterweight clearance
平衡重举船机 shiplift with counterweight
平衡重卷索鼓轮 spiral counterweight drum
平衡重块 balance weight; ballast weight
平衡重块安全阀 deadweight safety valve
平衡重块绷紧 balance weight tensioning; deadweight tensioning
平衡重块制动器 deadweight brake
平衡重量 balance[balancing] weight; counter-balance; counterpoise; counterweight
平衡重笼 counter-balanced cage; counterweighted cage
平衡重设备 bullfrog
平衡重绳索 counterweight cord; counterweight line
平衡重式叉车 counter-balanced fork lift truck
平衡重式垂直升船机 vertical ship lift with counterweight
平衡重式挡土墙 compensated retaining wall
平衡重式吊臂 counterweight arm
平衡重式举船机 <由低水面部分一边举船过闸至高水一边> lift with counterweight
平衡重式桥台 weight-balanced abutment
平衡重式升船机 balanced vertical ship-lift; counterweight ship-lift; ship-lifter with counter-weight; vertical ship-lift with suspension gear
平衡重式升降机 counterweight lift
平衡重式水闸启闭机 counterweight hoist
平衡重室 ballast chamber
平衡重锁 counterweight catch
平衡重提升器 counterweight lift
平衡重物 weight equalizer
平衡重系统 counter-balance system; counterweight system
平衡重下锚 balance anchor
平衡重闸门 counter-balanced shutter door
平衡重组 counterweight set
平衡重座架 counterweight mount
平衡轴 balance shaft; staff
平衡轴承 countershaft bearing; equalizer bearing
平衡轴冲头 staff punch
平衡轴线 axis of equilibrium
平衡轴向力 balancing axial thrust
平衡柱塞垫圈 washer for balance piston
平衡转电线圈 balancing repeating coil
平衡转化 equilibrium conversion
平衡转换器 <平衡不平衡转换器> balun [balanced to unbalanced transformer]
平衡转子 balancing rotor
平衡装球量 equilibrium ball charge; equilibrium ball load
平衡装置 balancer; balancing device; bogie[bog(e)y]; compensating device; equalizing device; equalizing gear; equilibrator; equilibrium unit; trim system
平衡状况 equilibrium condition
平衡状态 balance condition; balanced state; condition of balance; equipondérant; state of equalization; state of equilibrium; equilibrium state
平衡(状态)含水量 equilibrium moisture content
平衡状态图 equilibrium state diagram
平衡锥 <用于稠度试验的> balance cone
平衡锥式液限仪 balance cone
平衡准则 equilibrium criterion
平衡资金 balanced fund
平衡纵断面 balanced profile
平衡纵剖面 equilibrium profile; profile of equilibrium
平衡阻碍 hysteresis
平衡阻抗 balanced impedance
平衡组分 equilibrium composition
平衡组件 balanced component
平衡组态 configuration of equilibrium
平衡作业线 line of balance
平衡作用 balanced action; balancing action; equalizing effect; equilibrating action; equilibration; equilibrium activity; poising action; trimming function
平衡座 balance seat
平后桥驱动装置 rear drive
平弧拱 schematization arch; scheme arch; skene arch
平弧屋顶 flat vaulted ceiling
平弧形天花 flat vaulted ceiling
平虎口钳 parallel-jaw pliers
平护栅断面 flat guard rail section
平滑 even; plain; smoothing
平滑板 flat run board; flat runner
平滑板链 plate link chain
平滑比 flatness ratio
平滑边坡 smooth slope
平滑边缘 slickenside
平滑表面 slick-surface; smooth surface
平滑表面的 slick-surfaced
平滑冰 level ice
平滑材面 smooth grain
平滑常数 smoothing constant; smoothing current
平滑车流 smooth flow
平滑承座传动 smooth bearing transmission
平滑单板 plain veneer
平滑的 flowing; level and smooth; slick; smooth; stepless
平滑的表面涂料 glazement
平滑的水面 <有一层油膜的> slick
平滑地毯 smooth carpet
平滑电抗器 smoothing reactor
平滑电路 smoothing circuit
平滑电容器 smoothing capacitor; smoothing condenser
平滑电枢 smooth core armature; surface-wound armature
平滑电阻 buffer resistance
平滑电阻器 smoothing reactor
平滑动作的 smooth-acting
平滑度 accuracy of levels; evenness; smoothness; flat seal <阀的>
平滑度检查 smoothness check
平滑度试验器 smoothness tester
平滑断口 even fracture
平滑断裂 smooth break
平滑扼流圈 ripple-filter choke; series reactor; smoothing choke(coil)
平滑阀 flat slide valve
平滑反应谱 smoothed response spectrum
平滑非线性 smoothed non-linearity
平滑功率谱密度函数 smooth power spectral density function
平滑功能 smoothing function
平滑管嘴 <内径渐变的> smooth nozzle
平滑辊式破碎机 smooth roll crusher
平滑过程 smoothing process
平滑过渡 smooth transfer
平滑函数 smooth function
平滑核 smoothing kernel
平滑化分布 smoothed distribution
平滑减速度 smooth deceleration
平滑接合 smooth engagement
平滑控制 slide control
平滑离合器 loose socket
平滑流 slipping stream
平滑流动 streamline flow
平滑流形 smooth manifold
平滑柳 red willow
平滑陆架 smooth shelf
平滑滤波 smoothing filtering
平滑滤波抗流圈 smooth choke
平滑滤波器 brate force filter; hum filter; rectifier filter; ripple filter; smoother; smoothing filter
平滑轮 horizontal sliding wheel
平滑轮起重机 jenny wheel
平滑面 even surface; slick(-surface); straight surface of slipping; straight slip surface <土坡滑坍的>
平滑面砌砖工程 fair-faced brickwork
平滑模板 flat run sheet; wrought shuttering
平滑模壳 wrought shuttering
平滑磨削表面 smooth ground surface
平滑黏[粘]附的电解淀积 reguline
平滑刨床 surface planer; surface planing machine
平滑喷管 smooth nozzle
平滑喷嘴 smooth nozzle
平滑劈理 smooth cleavage
平滑匹配装置 hoghorn
平滑平均声场 smooth-averaged sound field
平滑蒲公英 smooth crabgrass
平滑谱密度 smoothed spectral density
平滑器 deaccentuator; slick
平滑曲率 smooth curvature
平滑曲线 flat curve; smooth curve
平滑曲线法 method of smooth curve
平滑绕组 choking winding
平滑设计谱 smooth design spectrum
平滑石楠 smooth oriental photinia
平滑时窗长 smooth window length
平滑时限特性距离保护装置 continuous curve distance-time protection
平滑水面 greyslick; slick
平滑塑料管 smooth plastic pipe
平滑胎面 smooth tread
平滑台 smoothing board
平滑调节 slide control; stepless regulation
平滑调节变阻器 continuous rheostat
平滑调整的 infinitely variable
平滑调整电感线圈 continuously adjustable inductor
平滑系数 smooth(ing) coefficient; smooth(ing) factor
平滑下降型 smooth descend type
平滑线(路) smooth line; slick line
平滑线圈 choking winding
平滑镶板 flat run panel
平滑斜降型 smooth oblique descend type
平滑形 smooth
平滑性 flatness; gliding property; slipperiness; smooth texture
平滑性假设 smoothness assumption
平滑岩石 slick-rock
平滑仪 flatscope
平滑油箱 plain tank
平滑圆角 smoothly radiused edge
平滑运动 smooth motion
平滑运转 easy running
平滑噪声发生器 flat noise generator
平滑转换 cross fade
平滑转矩 smooth torque
平滑转折点 break-even point
平滑装置 smoothing device
平滑作用 smoothing effect
平化 flatting
平环链 Gall's chain
平缓 <指地形> gentleness
平缓背斜 arrested anticline
平缓背斜带 gentle anticline zone
平缓背斜聚集带 accumulation zone of slightly anticline
平缓比降 flat grade; flat gradient; flat slope; gentle slope; mild slope
平缓边坡 conservative side slope; conservative slope; gentle side slope; mild side slope; mild slope
平缓波 flat wave
平缓不冲刷斜坡岸 slip-off slope bank
平缓沉积岩层区 flat-lying sedimentary rock region
平缓大半径河湾 flat long-radius bend
平缓的 flat-lying; gentle; tranquil
平缓的坡度 gentle incline
平缓地 gently
平缓峰顶 horizontal crest
平缓罐 surge tank
平缓河道 mild channel
平缓河段 <河床比降变化不大的>

graded reach
平缓河流 graded stream
平缓块状岩层型 type of gently dipping bed and massive rocks
平缓流出 smooth outflow
平缓坡 conservative slope
平缓坡道 easy grade; easy gradient; flat grade; flat gradient; flat slope; gentle slope; good gradient; low gradient
平缓坡度 easy grade; easy gradient; favo(u)rable gradient; flat grade; flat gradient; flat slope; gentle slope; good gradient; gradual slope; low grade; low gradient; mild slope; retarding gradient; slight grade
平缓坡面海堤 mild-sloped seawall
平缓起伏地形 gently rolling topography
平缓倾斜 gentle dip; gentle incline
平缓倾斜的 gently dipping
平缓丘陵地 gently rolling country
平缓曲线 easy curve; flat curve; gentle curve; shallow curve; smooth curve; wide curve
平缓山 subdued mountain
平缓衰减 gradual attenuation
平缓速度 easy speed
平缓梯度 easy gradient; low gradient
平缓弹簧 easy spring
平缓弯道 easy bend
平缓弯段 easy curve; flat bend
平缓弯头 easy bend
平缓弯子 easy bend
平缓下降的 gently dipping
平缓仰拱 flat invert
平缓圆顶 easy dome
平缓圆屋顶 easy dome
平缓褶皱 gentle fold
平黄赤交角 mean obliquity of ecliptic
平黄道 mean ecliptic
平黄道太阳 mean ecliptic sun
平黄纬 mean latitude
平灰缝 flat joint; flush(ed) joint; rough-cut joint; hick joint <圬工的>
平回归年 mean tropic(al) year
平混凝土屋面瓦 plain concrete roofing tile
平货位堆货场 level storage yard
平积层 contourite
平积云 flat cumulus cloud
平极 mean pole
平挤薄膜 cast film
平挤薄膜挤塑 cast film extrusion
平脊瓦 under-ridge gable tile; under-ridge tile
平夹板 flat joint bar
平夹板结合 plain fishplate joint
平夹层玻璃 flat laminated glass
平夹式扩孔钻头 flat reaming head
平甲板 floor; flush deck
平甲板驳船 flush deck barge
平甲板船 flush decker; flush deck vessel
平甲板的板材 floor plate
平价 at par; flat price; flat rate; par(ity); par value
平价比率 parity ratio
平价差幅 deviation from par
平价兑换 conversion at par; exchange at par
平价发行(法) par issue; emission at par
平价方案 parity program(me)
平价汇率 par exchange rate; par of exchange
平价价格 parity price
平价交易 flat trades
平价商店 fair-price shop
平价商品 commodities at par
平价市场 bargain center[centre]
平价条款 parity clause
平价网络 parity grid
平价系数 parity coefficient
平价项目 offer item
平价销售 sales at state fixed prices
平价以上 above par value
平价栅 parity grid
平价债券 par bond
平价政策 parity price policy
平价指数 parity index; parity rate
平价制度 par regime; par value system
平架集装箱 flat rack folding container
平尖 flat point
平肩机 shoulder grader
平减指数 deflator
平剪 straight snips
平键 flat key; parallel key; slay key
平浆 paddle
平交 level crossing; at grade junction
平交叉道口 road crossing
平交道 highway grade crossing
平交道横越冲撞 crossing collision
平交道横越撞车 crossing collision
平交道口 accommodation crossing; intersection; level crossing <英>; grade crossing <美>; railroad crossing; railway crossing; track crossing; railroad grade crossing【铁】
平交道口边门 level-crossing side gate; level-crossing wicket gate
平交道口的远程管理 remote supervision of level crossings
平交道口防护 crossing protection; grade crossing protection; level-crossing protection
平交道口轨枕 level-crossing sleeper
平交道口交叉警告标 cross-buck sign
平交道口警铃 level-crossing bell
平交道口看守房 level-crossing keeper's house
平交道口栏木 level-crossing barrier; level-crossing boom; level-crossing gate
平交道口栏木电动机 level-crossing gate motor
平交道口路面 highway grade crossing surface
平交道口信号 level-crossing(road) signal; railroad crossing signal
平交道口栅门 railroad crossing gate
平交道口撞车 level-crossing collision
平交道拦 level-crossing gate
平交道拦路门(栅) level-crossing gate
平交道拦路木 level-crossing gate
平交防护 crossing protection
平交公路 at-grade highway
平交口 grade crossing; grade intersection
平交快速干道 at-grade expressway
平交路口 level junction
平角 flat angle; straight angle
平角交分道岔转辙器跟部的特种支撑 special brace for heel of switches with flat angle of slips; special tie for heel of switches with flat angle of slips
平角平差 angular adjustment
平铰拱桥 arch bridge with flat hinges
平铰接 butt hinge
平铰链 butt hinge; flat hinge; Hurl-hinge
平接 bell-and-plain end joint; butting; butt joint; end(-to-end) joint; even joint; flush joint; joining on butt; joint on butt; plain joint; square joint; straight joint
平接板 board joint; butt strap; flat seam; flush plating
平接壁板 butt lagging; butt laying
平接瓷砖 straight joint tile
平接的 butt jointed
平接底面 flush visible under-face
平接对焊 butt weld in the down hand position
平接缝 abutment joint; abut(ting) joint; carvel joint; flat seam; plain butt joint; end joint; edge joint <两块板在同一平面内的>
平接缝暗加固片 concealed cleat
平接盖板 butt strap
平接盖板宽度 width of butt strap
平接合 butt
平接灰浆缝 flush joint
平接角钢 butt(ing) angle
平接铰链 butt hinge
平接接合 <带盖缝条的> bead and butt
平接接头 butt joint; butt splice; flush coupling joint
平接口 batt joint
平接联结器 band coupling
平接门框 butted frame
平接模 flat-back pattern
平接内贴板 inside butt strap
平接片 butt strip
平接企口墙板 flush siding
平接式套管 flush-joint casing
平接套管 flush-coupled casing; flush-jointed casing
平接头 butt junction; opposite joint
平接瓦 straight joint tile
平接线片 spade lug
平接压焊夹具 apparatus for butt welding
平接压力环 plain pressure ring
平接钻具 flush-coupled type drill string
平节理 flat joint; flat-lying joint
平结 flat knot; reef knot; sail-making-knot; square(reef) knot; thief knot
平结构和高结构 flat structure and tall structure
平截面 horizontal section; plane section
平截面假定 plane cross-section assumption
平截面假设 assumption of plane cross-section
平截四面体 truncated tetrahedron
平截筒柱 truncated cylinder
平截头墩 frustum
平截头棱锥公式 prismoidal formula
平截头棱锥体 frustum of pyramid; prismoid
平截头棱锥体的 prismoidal
平截头台 frustum
平截头体 frustum
平截头圆锥体 frustum of a cone
平截锥 truncated cone
平界面 planar interface
平筋平接缝 plain butt joint
平近点角 mean anomaly
平经 azimuth
平经度 mean longitude
平经误差 azimuth error
平晶 optic(al) flat
平井 horizontal well
平静 calm; equability; peace; quietude; undisturbance
平静的 surgeless; tranquil; windless
平静海面 ash breeze; calm sea; smooth sea
平静海面区域 smooth sea area; smooth water area
平静浇铸 quiet pouring
平静均匀弧 quiet homogeneous arc
平静空气 smooth air
平静期 quiet period
平静区 quiescent area; zone of silence
平静日 calm day; quiet day
平静水面 undisturbed water level
平距【测】 distance reduced to the horizontal; horizontal distance
平距图像 ground range image
平锯 flat cutting; flatting
平锯材 flat-sawn timber
平锯的 plain-sawed
平锯法 flat sawing; plain-sawing
平锯木 plain-sawed; plain-sawn
平锯木材 flat-sawn lumber; plain-sawed lumber
平锯木纹 slash grain
平锯台 plain saw bench
平掘开挖法 breasting method
平均 mean; on an average; par
平均 p 叶函数 mean p-valent function
平均凹凸高度 average asperity height
平均凹凸密度 average asperity density
平均白噪声谱 average white noise spectrum
平均百分离差 average percentage deviation
平均百分误差 mean percent error
平均半潮面 mean half-tide level
平均半潮水位 average tide level; mean tide level; half-tide level
平均半径 average radius; mean radius
平均半球面亮度 mean hemi-spheric(al) intensity
平均半球面烛光 mean hemi-spheric(al) candle power
平均包装重量 average tare
平均饱和车流量 average saturation flow
平均饱和度 average saturation degree
平均饱和流量 average saturation flow
平均保(险)费 average premium; level of premium; level premium
平均保险费率 average rate
平均保养间隔期 mean time between maintenance
平均保养间隔时间 mean time between maintenance
平均保养时间 average time between maintenance
平均保证出力 average firm output power
平均暴露水平 average exposure level
平均暴雨输沙率 average storm sediment discharge; mean storm sediment discharge
平均泵压 average pump pressure
平均比标系数 averaged proportional scale coefficient
平均比降 average gradient; average slope; mean grade; mean gradient; mean slope
平均比例尺 average scale; mean scale
平均比率 average gradient; average of ratio; average rate; mean gradient
平均比热 average specific heat; mean specific heat
平均比收支 average specific budget; mean specific budget
平均比压 mean specific pressure
平均比值 average ratio; mean ratio
平均比重 mean specific gravity
平均闭合差 average closure; mean closure
平均壁厚 mean wall thickness
平均避碰角 mean avoiding angle
平均边长 mean side length
平均边界层厚度 mean boundary-layer thickness
平均边缘线 average-edge line

平均变差函数 mean variogram
平均变动成本下降 declining average variable cost
平均变幅 average range;average shift
平均变化 average shift
平均变化幅度 mean range
平均变化率 average variability;mean variability
平均变率 average variability
平均变形 average deformation
平均变形模量 average modulus of deformation;mean modulus of deformation
平均变异性 average variability;mean variability
平均遍历半群 mean ergodic semi-group
平均遍历的 mean ergodic
平均遍历定理 mean ergodic theorem
平均遍历矩阵 mean ergodic matrix
平均标准耳 average normal ear
平均标准偏差 mean standard deviation
平均表面直径 average surface diameter
平均表土流失量 average topsoil loss
平均表现 mean performance
平均冰冻深度 average frost penetration
平均冰冻指数 mean freezing index
平均冰界 mean ice edge
平均并合样品 average composite sample
平均波 average wave;mean wave
平均波长 mean wavelength
平均波动模数 <交通量> mean modulus of error
平均波高 average wave height;mean wave height
平均波浪 mean wave
平均波浪漂移力 mean wave drift
平均波(浪)周期 average wave period;mean wave period
平均波向 mean wave direction
平均剥采比 mean stripping ratio;overall stripping ratio
平均不变成本 average constant cost;average fixed cost
平均不工作时间 mean downtime
平均不准确度 average inaccuracy
平均布井法 method of well uniform distribution
平均采剥比 average stripping ratio
平均参考压力 mean referred pressure
平均参数 average parameter;mean parameter
平均残差 mean residual;residual mean
平均残留期 mean residence time
平均测试 average test
平均层厚 average thickness of the layer
平均层数 average story number
平均查全率 recall level average
平均查找时间 average search time;average seeking time;mean seeking time
平均差不匀率 coefficient of mean deviation
平均差参数 mean difference parameter
平均差额 average balance
平均差分 mean difference
平均差幅 deviation from par
平均差(值) average difference;average error;average range;mean deviation;mean difference;mean range
平均产出 average output;average product
平均产量 average output;average product;average yield;mean output;mean yield
平均产品 average product
平均产烃率 average hydrocarbon productivity
平均长度 average length;mean length
平均常水位 average of normal water levels;mean of normal water levels
平均场地条件 average site condition
平均场论 mean-field theory
平均潮差 mean range;mean range of tide;mean tidal range;mean tide range
平均潮高 mean height of the tide;mean of tide
平均潮候时差 mean establishment
平均潮流 average tidal flow;mean tidal current;mean tidal flow
平均潮面 half tide level;mean tide level;ordinary tidal level;tide level
平均潮升 average rise of tide;mean rise of tide
平均潮升间隙 mean rise interval
平均潮位 half tide level;mean tidal level;mean tide level;ordinary tidal level
平均潮位海岸线长度 length of mean-tide coastline
平均潮位曲线 average tidal curve;mean tidal curve
平均潮位升高 average rise;mean rise;mean tide level rise
平均潮(汐) mean tide
平均潮(汐)水位 mean tide level
平均车间(时)距 average headway
平均车辆调查法 average car study technique
平均车日行程 average kilometerage per vehicle-day
平均车时距 average headway
平均车速 mean speed
平均车头间距 average space headway
平均车头时距 average time headway
平均沉淀 average precipitation
平均沉降量 average settlement;average value of subsidence
平均沉降速度 <泥沙的> average settlement velocity;mean fall velocity;mean settlement velocity
平均沉降速率 average subsidence rate
平均沉降影响系数 affected coefficient of mean settlement
平均沉落速度 average fall velocity;mean fall velocity
平均成本 average cost
平均成本的行为 behavio(u)r of average cost
平均成本定价 price making according to average cost
平均成本定价法 average cost pricing
平均成本法 average cost method
平均成本计算 average costing
平均成本加成 average mark-on
平均成本利润率 average cost profit rate
平均成本曲线 average cost curve
平均成本最低 least average cost
平均成分 average assay
平均承雨量 <雨量器的> average rain catch
平均承载应力 average bearing stress;mean bearing stress
平均乘距 average length of passenger journey;average riding distance
平均吃水(深度) average draft;average draught;mean draft;mean draught
平均尺寸 average dimension;intermediate size
平均尺度 average dimension;average size;mean dimension;mean size
平均赤道惯性矩 average equatorial moment;mean equatorial moment
平均赤字 average deficit;mean deficit
平均充气 uniform charge
平均冲角 mean incidence
平均冲刷深度 average erosion(al) depth;mean erosion(al) depth
平均冲刷速度 average erosion(al) velocity;mean erosion(al) velocity
平均抽查数 average sample number
平均抽查质量 average outgoing quality
平均抽检质量 average outgoing quality
平均抽检质量界限 average outgoing quality limit
平均抽吸压头 mean pressure suction head
平均抽样检验个数 average sample size
平均抽样数函数 average sampling number function
平均抽样(数)量 average sample number
平均抽样数量曲线 average sample number curve
平均稠度 average consistency
平均出厂品质等级 average outgoing quality level
平均出厂质量 average outgoing quality
平均出厂质量极限 average outgoing quality limit
平均出厂质量水平 average outgoing quality level
平均出错率 average error rate
平均出力 average output;mean output
平均出力曲线 average power curve
平均出行次数 average trip numbers
平均初次出故障时间 mean time to failure
平均初始加速度 average initial acceleration
平均初雪(日)期 mean date of first snow cover
平均储备金账户 equalization reserve account
平均储备天数 average days of stocking
平均传导率 average transfer rate;average transmission rate
平均传热系数 mean thermal transmittance
平均传输率 average data transfer rate;average transfer rate;average transmission rate
平均垂直压力 average vertical pressure
平均垂直应力 mean normal stress
平均春分大潮 mean equinoctial springs
平均纯利润 average net profit
平均磁化曲线 mean magnetizing curve
平均磁极位置 mean pole position
平均磁阱 average magnetic well
平均次数 averaging time
平均粗糙密度 average asperity density
平均存活时间 mean survival time
平均存货 average inventory;average stock
平均存货期 days of average inventory
平均存货水平 average inventory level
平均存(款)余额 deposit(e)line
平均存料 average stock
平均存取时间 average access time
平均大潮 mean springs;mean spring tide
平均大潮差 mean spring range;spring range
平均大潮潮升 mean spring rise
平均大潮低潮面 mean low water springs;mean spring low water
平均大潮低潮升 mean spring rise of low tide
平均大潮低潮位 mean low water spring tide
平均大潮低低潮面 harmonic tide plane;Indian spring low water;Indian tide plane;mean lower low-water springs
平均大潮高潮面 mean high water springs
平均大潮高潮升 mean spring rise of high tide
平均大潮高潮位 mean high water springs;mean high water spring tide
平均大潮高度 mean spring rise
平均大潮高高潮面 mean higher high-water springs
平均大潮极低潮 mean extreme low water
平均大潮极高潮 mean extreme high water
平均大潮极高潮面 mean extreme high water springs
平均大潮升 mean spring rise
平均大气差 mean refraction
平均大气折光差 mean refraction
平均大气折射 average atmospheric(al) refraction;mean atmospheric(al) refraction
平均大小 mean size
平均单价 average unit cost;average unit price
平均单位成本 average unit cost
平均单位过程线 average unit graph
平均单位水文过程线 average unit hydrograph;mean unit hydrograph
平均单位应力 average unit stress
平均单位支付价格指数 index of average unit price paid
平均单元 averaging unit
平均导程 mean lead
平均到达率 average arrival rate;mean arrival rate
平均到期日 average due date;average maturity;equated maturity
平均道路车速 average highway speed
平均的 average;averaging;mean
平均的平均 mean of mean
平均等待时间 average waiting time
平均等高线面积法 average contour area method
平均等级 average rank;mean rank
平均等级系数 average graded coefficient
平均等值线 average isopleth;mean isopleth;normal isopleth
平均低潮 mean low tide;mean low water
平均低潮间隙 mean low water full and change;mean low water interval
平均低潮面 average low water(level);mean low tide;mean low water(level)
平均低潮升 mean low water springs;mean rise of low tide;mean low water rise
平均低潮水位 mean low water mark
平均低潮位 mean low tide;mean low water;mean low water level
平均低潮月潮间隙 mean low water lunitidal interval
平均低低潮面 mean lower low tide;mean lower low water

平均低低潮位 mean lower low water
平均低高潮(位) mean lower high water
平均低水流量 average low water flow;mean nine-month flow
平均低水位 average low water;average low water level;mean low water;mean low water level;mean of low stages
平均地表面 average sphere level;mean sphere level
平均地表温度 average surface temperature
平均地极 average terrestrial pole;mean terrestrial pole
平均地面高程 average ground elevation;average ground level;average ground surface;mean ground elevation;mean ground level;mean ground surface;average grade
平均地球椭球 average earth ellipsoid;mean earth ellipsoid
平均地温梯度 mean geothermal gradient
平均地震反应谱曲线 average response spectrum curve to earthquake
平均地租 equalization of land rent
平均递增率 average increasing rate;average rate of increase;average rate of progressive increase
平均点 midpoint
平均电荷 average charge
平均电荷密度起伏 average charge density fluctuation
平均电离 mean ionization
平均电离能 mean ionization energy
平均电流 average current;mean current
平均电流脉冲响应 average current pulse responses
平均电路 averaging circuit
平均电码长度 mean code length
平均电平 mean level
平均电平检测系统 mean level detection system
平均电平式自动增益控制 mean level auto-gain control
平均电压 average voltage
平均电轴 average electric(al)axis
平均电子浓度 mean electron concentration
平均叠合度 mean degree of polymerization
平均订约利率 average interest rate on total contracted
平均定额 average norm;average standard
平均动力放大系数 mean dynamic(al)amplification factor
平均动能 mean kinetic energy
平均动压头 mean dynamic(al)head
平均冻结指数<由平均温度求得的冰冻指数> mean freezing index
平均读数 average reading
平均度量的 mean-metric(al)
平均端面积公式 mean end area formula
平均端面积(计算)法 average end area method;mean end area method
平均端面积(计算)公式<计算土方量用的> average end area formula
平均断裂荷载 mean fracture load
平均堆存期 average storage time
平均对数差余(值) mean log residual
平均对数能量损失 mean logarithmic energy loss
平均对数偏差 average log deviation
平均对数衰减率 mean logarithmic decrement
平均吨位 average tonnage;mean tonnage
平均额外缺陷界限 average extra defective limit
平均二级差 average difference of second order
平均发散度 average divergence
平均发生率 mean occurrence rate
平均发行价格 average issuing price
平均发生频率 average frequency of occurrence
平均发展速度 average growth speed;average speed of development
平均罚款 average penalty
平均阀 application pilot valve
平均法 averaging;method of average;statistic(al)method
平均法经验公式 empiric(al)formula by method of average
平均法向变形 mean normal deformation
平均法向形变 mean normal deformation
平均法向应力 mean normal stress
平均反射率 average reflectance;mean reflectivity
平均反射能力 mean reflectance
平均反射系数 average reflectance
平均反应谱曲线 mean response spectrum curve
平均范围 average range
平均方位角 mean azimuth
平均方向 average direction
平均方向的比较 comparison of mean direction
平均放电电压 average discharge voltage
平均放电功率 average discharge power
平均放射性测量计 mean radioactivity meter
平均沸点 average boiling point;mid-boiling point
平均费用 average cost
平均费用曲线 average cost curve;mean-cost curve
平均分辨率 average definition
平均分布 average distribution;even distribution;hypodispersion
平均分点 mean equinox of date
平均分隔 equipartition
平均分配 equal distribution;equal division
平均分配方式 equalitarian distribution method
平均分期付款 equation of payments
平均分期摊销法 straight-line method of amortization
平均分散度 average divergence
平均分摊 share out equally
平均分析 average analysis
平均分子 mean molecule
平均分子量 average molecular weight;mean molecular weight
平均丰度 average abundance
平均风 averaging wind
平均风速 average wind speed;average wind velocity;mean wind speed;mean wind velocity
平均风速的高度分布 wind profile
平均风速计 mean wind speed meter
平均风险 average risks
平均封冻日期 average freeze-up date;mean freeze-up date
平均服务率 average service rate;mean service rate
平均服务时间 average service time
平均幅度 mean range
平均辐射强度 mean radiation intensity
平均辐射温度 average radiant temperature;mean radiant temperature
平均付款期 equaled time of payment
平均负担 equalization of incidence;equal sacrifice
平均负担税 share alike
平均负担租 equal sacrifice;share alike
平均负荷 average load;mean load
平均负坡度 average negative gradient
平均负载 average load;mean load
平均概率 average probability
平均概率偏差 mean difference;mean probable error
平均概率寿命 expectation of life
平均概率误差 mean probable error
平均干围 mean girth
平均高 mean height
平均高差 average height difference
平均高潮 mean high tide
平均高潮差 mean high water range
平均高潮间隙 mean high water interval
平均高潮面 mean high water(level)
平均高潮升 mean rise of high tide
平均高潮水位 mean high water tide
平均高潮位 mean high tide;mean high water(level)
平均高潮线 mean high water line
平均高潮(月潮)间隙 mean high water lunitidal interval
平均高程 average elevation;mean elevation;mean level
平均高程差 mean difference of elevation;mean height difference
平均高低潮(位) mean higher low water
平均高度 average depth;average height;center[centre] line average height;mean height;mean level
平均高度法 centre line average method
平均高度图 mean height map;average height map
平均高高潮 average higher high water tide
平均高高潮面 average higher high water tide
平均高高潮位 average higher high water tide;mean higher high water tide
平均高水位 mean high water(level);mean of high stages
平均高水位线 mean high water line
平均隔声量 mean transmission loss
平均给水温度 mean supply water temperature
平均工时数 average hour of work
平均工资 average wage
平均工资等级系数 average wage scale coefficient
平均工资计划 average wage plan
平均工资率 average earnings per unit of time;average labo(u)r rate;average wage rate
平均工资收入 average wage earnings
平均工资养老金方案 average salary pension scheme
平均工资指数 average wage index
平均工作等级因素 average rating factor
平均工作负载 average work load
平均工作时间 average operation time
平均工作寿命 average length of working life
平均工作数量 average number of job
平均工作周 average workweek
平均功率 average power;mean power
平均功率谱法 average power spectrum method
平均功率曲线 average power curve
平均功率输出 average power output
平均功率与最大功率之比 duty ratio
平均供水量 average water supply;mean water consumption
平均供水温度 mean supply water temperature
平均供应间隔天数 average days of supplying interval
平均共轭曲线 mean conjugate curve
平均共轭网 mean conjugate net
平均估计方差 average estimated variance
平均古地磁场强度 average paleointensity
平均股息率 average rate of dividend
平均固定成本 average fixed cost
平均固结度 average degree of consolidation
平均故障间隔期 mean downtime;mean time between failures;mean time to failure
平均故障间隔时间 mean time between failures;mean time to failure
平均故障检测时间 mean time to detection
平均故障历时 mean duration of failure
平均故障率 average failure rate;failure rate;mean failure rate
平均故障时间 mean downtime;mean duration of failure;mean time to failure
平均故障数 average failure number
平均故障行程 mean distance between failures
平均故障修复时间 average time for repair of breakdowns;mean time to repair
平均故障修理时间 mean time to repair;mean time to restore
平均观察法 method of averaging observation
平均官能度 average functionality
平均灌溉率 average rate of irrigation;mean rate of irrigation
平均光度曲线 average light curve;mean light curve
平均光强 average luminous intensity
平均规避法 mean avoiding method
平均规避速度 mean avoiding speed
平均规律 law of the mean
平均规模 average size
平均轨道 average orbit;mean orbit;mean trajectory
平均轨道半径 mean orbit radius
平均轨道根数 median orbital elements
平均海拔高度 above mean sea level;average altitude;mean altitude
平均海冰界线 average limit of sea ice
平均海底 mean ocean floor
平均海流 average current
平均海面 mean level;mean level of the sea;Ordnance Datum<英国陆军测量局地图所采用的基准>
平均海面测定仪 medimarimeter
平均海面传递法 synchro-measurement;transmittal method by mean sea-level
平均海面记录器 medimarimeter
平均海面记录仪 medimarimeter
平均海面季节改正 seasonal correction of mean sea level
平均海面之上 above mean sea level
平均海平面 mean level of the sea;mean sea level;zero water level
平均海平面大气压 sea level pressure
平均海平面订正 correction to average sea level;correction to mean sea level;reduction to average sea level;reduction to mean sea level
平均海平面高程 average sea level;

mean sea level
平均海平面基点 mean sea level datum
平均海平面气压 average sea level pressure;mean sea level pressure
平均海平面温度订正 reduction of temperature to mean sea level
平均海平面以上 above mean sea level
平均海深 average sea depth; mean sea depth
平均海水面 mean sea level;mean sea level surface
平均海水面基点 mean sea level datum
平均海损担保函 average guarantee
平均含沙量 <指油、水、空气中含沙量> average sediment concentration; mean sediment concentration;mean silt charge
平均含砂量 average sand content
平均含水量 average moisture content
平均含碳量 mean carbon content
平均含盐量 mean salinity
平均旱流污水量 average dry weather sewage flow
平均旱天流量 average dry weather flow;mean dry weather flow
平均航次周转次 ships turnover rate
平均航速 average speed
平均耗费的劳动时间 average labor-time expended in
平均耗散 mean dissipation
平均耗水量 average water consumption;mean consumption
平均河川流量 average river flow;average stream flow; mean river flow;mean stream flow
平均河床高程 mean channel height
平均河床高度 mean channel height
平均河底高程 average bed level; mean bed level
平均河流流量 average river flow;average stream flow; mean river flow;mean stream flow
平均河面 mean river level
平均河面高程 average river level; mean river level
平均河水位 mean river level
平均荷载 average load;mean load
平均恒星日 mean sidereal day
平均恒星时 mean sidereal time
平均横断面 average transverse section
平均洪水位 average of high water stages;mean of high water stages
平均喉道宽度 medium throat width
平均厚度 average depth; average thickness;mean thickness
平均厚度法 method of mean thickness
平均呼叫时间(间隔) mean time between calls
平均湖面 average lake level; mean lake level
平均湖深 average lake depth; mean lake depth
平均滑动方向 average slip direction
平均化 flattening
平均划线法 plate streak
平均环境容量 average environmental capacity
平均环境水流流速矢量 mean ambient flow vector
平均环路高度 mean circuit height
平均缓发时间 mean delay time
平均缓解 uniform release
平均换料率 mean refuelling rate
平均黄赤交角 mean obliquity
平均回波平衡返回损耗 mean echo balance return loss
平均回答率 average response rate
平均回风温度 mean return air temperature
平均回归潮差 mean tropic(al)range
平均回归值 average regressed value
平均回收率 average recovery rate
平均回水温度 mean return water temperature
平均回转时间 mean recurrence time
平均汇率 midpoint rate
平均活度 mean activity
平均活度乘积 mean activity product
平均活塞速度 mean piston speed
平均火成岩 average igneous rock; mean igneous rock
平均或最大风速 mean or maximum wind velocity
平均货币工资率 average money wage rate
平均货币总收入 average gross money wages
平均机 average device
平均机气泡六分仪 bubble sextant with averaging gear
平均机械钻速 average penetration rate
平均积厚 mean accumulation
平均积累效果系数 coefficient of average accumulation efficiency
平均积雪厚度 average thickness of cumulative snow
平均级配 average grading
平均级配方法 mean grading method
平均级数 average grading
平均极差 mean range
平均极大值 mean maximum
平均极限 mean limit
平均极小值 mean minimum
平均极值 mean extreme value
平均几何尺度 geometric(al)mean size
平均计价法 average pricing method
平均计算 take the average
平均计算操作 average calculating operation
平均技术熟练水平 average technical and vocational level
平均技术速度 average technical speed
平均剂量 average dosage;mean dosage
平均季节性河流 average seasonal river;mean seasonal river
平均季节性径流 average seasonal runoff;mean seasonal runoff
平均加速度 average acceleration; mean acceleration
平均加速度反应谱 average acceleration response spectrum
平均加速度率 average acceleration rate
平均价(格)average price;mean price; midpoint rate
平均间隔 equi-spaced
平均减速度 average deceleration
平均剪切模量 average shear modulus
平均剪应变 average shear strain; mean shear strain
平均剪应力 average shear stress;mean shear stress
平均检查量 average sample number
平均检出限 mean detection limit
平均检出质量 average outgoing quality
平均检修时间 mean repair time
平均检验数 average amount of inspection;average total inspection
平均建房工期 average construction period
平均降低率 average rate of decrease
平均降水量 average precipitation;average rainfall; mean precipitation; mean quantity of precipitation; mean rainfall
平均降雨量 average precipitation;average rainfall; mean precipitation; mean rainfall
平均降雨强度 average precipitation intensity;average rainfall intensity; mean precipitation intensity; mean rainfall intensity
平均交通(流)量 average traffic(volume)
平均交通密度 average traffic concentration;average traffic density
平均角 mean angle
平均角离差 mean angular deviation
平均校正坡度 average rectified slope
平均较低低潮位 mean lower low water
平均较低低水位 mean lower low water
平均较高高潮面 mean higher high water
平均较高高水位 mean higher high water
平均截流量 average inflow capacity
平均截面 average cross section
平均解冻日期 mean breakup date
平均近点角 mean anomaly
平均近点运动 average anomalistic(al) motion; mean anomalistic motion
平均近似 approximation on the average; average approximation; mean approximation
平均进尺 average footage
平均进度 average advance
平均经度 mean longitude
平均经向环流 mean meridional circulation
平均井径 average diameter of well
平均净动力负载 net average kinetic load
平均净静力负荷 net average static load
平均净利润 average net profit
平均净水头 average net head; mean net head
平均径流 average flow
平均径流量 average runoff; mean runoff
平均径流率 average runoff rate; mean runoff rate
平均径流深度 average runoff depth
平均径向误差 mean radial error
平均居住数 average number of inhabitants per building
平均矩阵 mean matrix
平均距 mean departure
平均距差法 average range method
平均距离 average distance;mean distance;mean range
平均距离误差 mean longitudinal error
平均距平 average departure
平均聚合度 average degree of polymerization
平均聚合作用 mean degree of polymerization
平均绝对百分误差 mean absolute percent error
平均绝对偏差 mean absolute deviation
平均绝对偏离 mean absolute deviation
平均绝对偏向 mean absolute deviation
平均绝对湿度差 mean absolute humidity difference
平均绝对误差 <误差绝对值的算术平均> mean absolute error; average absolute error
平均绝对星等 mean absolute magnitude
平均绝对值 mean absolute value
平均卡路里 mean calorie
平均开动钻机数 average amount of operating rigs
平均开盘利率 average initial offering yield
平均开挖半径 average excavation round
平均抗拉强度 average tensile strength
平均抗压强度 average compressive strength; mean compression [compressive] strength
平均颗粒粒径 average particle diameter;mean grain size; mean particle diameter
平均颗粒体积 average volume of grain
平均颗粒直径 average particle diameter;mean grain size; mean particle diameter
平均可变成本 average variable cost
平均可变成本曲线 average variable cost curve
平均可工作时间 mean up time
平均可靠出力 average firm output power
平均可用流量 average available discharge;mean available discharge
平均可用性 mean availability
平均课税 average taxation;averaging taxation
平均空间异常 average free-air anomaly;mean free-air anomaly
平均空气温度 mean air temperature
平均空闲时间 mean downtime
平均孔径 average pore diameter;average pore size;mean pore size
平均孔隙度 average porosity
平均孔隙水速度 average velocity of pore water
平均孔隙压力 mean pore pressure
平均枯水流量 average low flow;average low water flow; mean low flow; mean low water flow; mean minimum flow
平均枯水年 mean dry year
平均枯水位 average of low water stages;mean of low water stages
平均库存 average stock; averaging stock
平均库存量 average inventory(level)
平均库存天数 days of average inventory on hand
平均库存余额 average outstanding balance of deposits
平均库容压力 average reservoir pressure
平均块长 average block length
平均块度 average blockness
平均宽度 mean breadth;mean width
平均宽度比 mean width ratio
平均矿床数 mean deposit number
平均亏空 average deficit;mean deficit
平均扩散度 average divergence;mean divergence
平均拉力强度 average tensile strength
平均累计最大需量 average integrated demand
平均累计最大需量计 average integrated demand meter; integrated demand meter
平均离差 average deviation;mean deviation;mean difference
平均离散度 mean dispersion
平均里程 average mileage; average trip length; average trip mileage; mean mileage
平均历时曲线 average duration curve; mean duration curve
平均历元 average epoch;mean epoch
平均利率 average interest rate
平均利润 average profit;mean profit
平均利润法则 law of average profit
平均利润率 average profit rate;average rate of profit;mean profit rate
平均利润率规律 law of average rate

P

of profit
平均利用率 <无故障工作周期> mean availability
平均粒度 average grain size; mean grain size; mean particle size; particle-size average
平均粒度组成 average grading
平均粒径 average diameter; average grain diameter; average (grain) size; mean (grain) diameter; mean grain size; mean particle diameter; median size
平均粒径级配 average grading; mean grading
平均亮度 average bright; average brightness; average luminance; mean brightness; mean luminance
平均量 average amount; average magnitude; mean amount
平均量测值 average measured value; mean measured value
平均列车净吨 net tons per train kilometer
平均列车净重 net tons per train kilometer
平均列车总重 gross tons per train kilometer
平均裂缝宽度 mean crack width
平均裂隙长度 average fissure length
平均裂隙宽度 average fissure width
平均临界剪应力 average threshold shear stress; mean threshold shear stress
平均流 average current; mean current
平均流程时间 average flow time
平均流动法 moving average method
平均流动温度 mean flow temperature
平均流量 average discharge; average flow; mean discharge; mean flow; mean water flow
平均流量比 mean discharge ratio
平均流量历时曲线 average flow duration curve; mean flow duration curve
平均流量模量 average discharge modulus; mean discharge modulus
平均流量模数 average discharge modulus; mean discharge modulus
平均流率 average flow rate; mean flow rate
平均流速 average flow velocity; average stream velocity; average velocity; mean flow velocity; mean stream velocity; mean velocity; velocity of approach <水流接近量水堰的>
平均流速点 average velocity point; mean velocity point; mean velocity position
平均流速点的(水深)位置 mean velocity position
平均流速公式 mean velocity formula
平均流速曲线 average velocity curve; mean velocity curve
平均流线 center line of fluid flow
平均流域宽 mean width of basin
平均龙口宽度 mean closure-gap width
平均滤波器 averaging filter
平均路段速度 <一定路段、一定时间内所有车辆车速的平均值> average overall speed
平均旅程 average trip length; mean trip length
平均律 law of average; law of mean
平均螺距 mean pitch
平均络合常数 mean complexity constant
平均落潮历时 mean duration of ebb
平均落潮流量 average ebb discharge
平均脉冲时间 mean pulse time
平均脉冲指示器 mean impulse indicator
平均满期日 average life(period)
平均漫射光穿透率 average diffuse light transmission
平均忙时(电话)呼叫 equated busy-hour call
平均毛额 gross average
平均毛利率 ratio of average gross profit
平均毛水头 mean gross head
平均每车延误(时间) average individual vehicle delay
平均每户建筑面积 average floor area per household
平均每户人口 average number of persons per household
平均每年走行公里 average annual kilometrage
平均每人居住面积 average dwelling area per capita; average living floor area per capita
平均每人每日用水量 average water consumption per capita per day
平均每人投资 investment per capita
平均每人消费量 average per capita consumption
平均每人消耗量 average per capita consumption
平均每人用水量 average per capita consumption of water
平均每日应缴运输进款 average transport revenue payable per day
平均每日用水量 average water assumption per day
平均每日装车数 average carloadings per day
平均每套建筑面积 average floor area per suite; average floor area per unit
平均每天堆存货物吨数 average daily tonnage of cargo in storage
平均每小时收入 average hourly earnings
平均蒙气差 mean refraction
平均密度 average density; mean density; particle number density
平均面积降雨量 average precipitation over area
平均明暗度 average shading; mean shading
平均模量 average modulus
平均摩尔数量 mean molar quantity
平均磨损率 average wear rate
平均内部收益率 average internal rate of return
平均耐受限 median tolerance limit
平均能级 average level
平均能量 average energy; median energy
平均能量标准 average capacity standard
平均能量损失 average loss of energy
平均泥沙浓度 average sediment concentration; mean sediment concentration
平均年 average year; mean year; median year; normal year
平均年变幅 mean annual range
平均年沉积量 average annual accumulation
平均年沉积率 average annual sedimentation rate; mean annual sedimentation rate
平均年冲刷量 average annual erosion; mean annual erosion
平均年抽水量 average annual abstraction; mean annual abstraction
平均年地租 average annual ground-rent
平均年度变化范围 mean annual range
平均年度成本 average annual cost
平均年度费用 average annual cost
平均年洪水(流量) average annual flood; mean annual flood
平均年降水量 average annual precipitation; average yearly rainfall; mean annual precipitation; mean annual rainfall; mean yearly rainfall
平均年降雨量 average annual rainfall; average yearly rainfall
平均年金折旧法 equal annual depreciation payment method
平均年径流 mean annual runoff
平均年利率 mean annual rate of interest
平均年龄 average age
平均年流量 average annual discharge; mean annual discharge
平均年泥沙沉积量 average annual sediment yield; mean annual sediment yield
平均年生长量 mean annual increment; mean increment
平均年疏伐度 average annual stand depletion
平均年疏浚量 average annual dredging quantity
平均年输沙量 average annual sediment discharge; mean annual sediment discharge
平均年水位 mean annual water level
平均年土壤流失量 average annual soil loss; mean annual soil loss
平均年温差 mean annual temperature range
平均年限 average life(period); composite life; mean life
平均年限法 composite life method
平均年限收益率 yield to average life
平均年限折旧法 depreciation-composite life method
平均年淤积量 average annual accumulation; average annual sediment deposition
平均年雨量 average annual rainfall
平均年灾害损失 average annual equivalent damage
平均年增长量 periodic annual increment
平均年增长率 annual average rate of increase
平均年增值 mean annual increment
平均黏[粘]度 average viscosity
平均黏[粘]滑周期 average periodicity of stick-slip
平均黏[粘]结应力 average bond stress
平均黏[粘]着应力 <混凝土与钢筋的> average bond stress
平均浓度 average concentration; mean concentration
平均排出量 average discharge
平均排队超长 average excess
平均判别值 average discriminatory value
平均碰撞频率 mean collisional frequency
平均碰撞时间 mean collision time
平均碰撞数 mean number of collisions
平均皮重 average tare
平均偏差 average deviation; mean deviation; average departure; mean departure; mean variation; deviation from average
平均偏差系数 coefficient of average deviation
平均偏航角 average yaw; mean yaw
平均偏好 average preference
平均偏角 mean declination
平均偏移 average shift; mean deviation
平均偏转 parallel avertence
平均漂移 average drift; mean deviation; mean drift
平均漂移速度 average drift velocity
平均拼合样品 average composite sample; mean composite sample
平均频率 mean frequency
平均频率剖面 average frequency section
平均频谱 average spectrum
平均品位 average grade of ore
平均平方 mean square
平均平方数离差 root mean square deviation
平均平滑谱 average smoothed spectrum; mean smoothed spectrum
平均平面光强 mean horizontal intensity
平均平面烛光 average horizontal candle power; mean horizontal candle power
平均坡度 average gradient; average slope; mean gradient; mean inclination; mean slope; natural profile
平均坡降 average gradient; average slope; mean gradient; mean inclination; mean slope
平均坡面漫流长度 mean length of overland flow
平均破损数 <制品或试样在试验中的> average failure number
平均谱 average spectrum
平均起动加速度 average starting acceleration
平均气垫长度 average cushion length; mean cushion length
平均气动中心 mean aerodynamic(al) center[centre]
平均气温 average air temperature
平均气隙磁密 specific magnetic loading
平均器 averager; averaging device
平均牵引(阻)力 average draft
平均欠额 average shortfall
平均强度 average strength; mean intensity; mean strength
平均倾角 mean inclination
平均清晰度 average definition; mean definition
平均球度 average sphericity
平均球径 average diameter of grinding media
平均球面烛光 average spheric(al) candle power; mean spheric(al) candle power
平均区域速度 average zone velocity
平均区域异常 average block anomaly; average region anomaly; mean block anomaly; mean region anomaly
平均曲率 average curvature; mean curvature
平均曲率半径 Gaussian radius; mean radius of curvature
平均曲线 average curve
平均取集代表性样品 average grab sample
平均取样 average sampling; mean sampling
平均取样数 average sample number
平均燃料消化率 means specific propellant consumption
平均燃料转换比 mean fuel conversion ratio
平均热导率 mean thermal conductivity
平均热辐射强度 average heat radia-

tion intensity
平均热流密度 mean heat flow density
平均热容量 mean heat capacity
平均热值 mean calorie
平均日 mean diurnal
平均日变差 average daily variation; mean variation of daily rate; mean variation of rates
平均日变化 average daily change; mean daily change; mean interdiurnal variation
平均日潮低潮不等 mean diurnal low-water inequality
平均日低潮不等(值) average diurnal low water inequality; mean diurnal low-water inequality
平均日高潮不等(值) average diurnal high water inequality; mean diurnal high-water inequality
平均日供水量 average daily output; average daily water supply output
平均日耗量 average daily consumption; average day consumption; mean daily consumption
平均日耗水量 average daily water consumption; mean daily water consumption
平均日际变化 mean interdiurnal change; mean interdiurnal variation
平均日交通量 average daily traffic; average daily volume; mean daily traffic
平均日流量 average daily discharge; average daily flow; mean daily discharge; mean daily flow; mean day flow
平均日偏差 mean deviation in daily rate; mean deviation of daily rate
平均日十个最重车轮荷载 average daily ten heaviest wheel load
平均日输沙量 mean daily sediment discharge
平均日温差 average daily range; mean daily range
平均日污水量 average sewage rate per day
平均日消费量 average daily consumption
平均日用水量 average daily consumption of water; average day consumption
平均日运动 mean daily motion; mean diurnal motion
平均日增重 average daily gain
平均容许极限量 tolerance limit median
平均容重 average density; mean density; weighted mean of unit weight
平均熔点 average melting point
平均融化指数 <由平均温度求得的融化指数> mean thawing index
平均入渗能力 average infiltration capacity
平均入渗速率 average infiltration rate
平均散度 average divergence; mean divergence
平均骚动速度 mean velocity agitation
平均色散 mean dispersion
平均筛目 mean mesh
平均上限 average upper limit; mean upper limit
平均上涨 averaging up
平均深度 average depth; mean depth
平均渗入量 average infiltration; average permeability; mean infiltration
平均渗入率 average infiltration; average permeability; mean infiltration
平均渗透量 average infiltration; average permeability; mean infiltration
平均渗透率 average infiltration; average permeability; mean infiltration; W-index
平均渗透系数 average permeability coefficient
平均升举力 average lift (force); mean lift; mean lift force
平均生产成本 average production cost
平均生产价格 average price of production
平均生产量 average production; average production capacity
平均生产率 average output; average performance; average productivity; efficiency average; Value to cost ratio
平均生产能力 average productive capacity
平均生产期 average period of production; average production period
平均生长量 average increment
平均生长率 average life rate
平均生存期 mean survival time
平均生活标准 average standard of living
平均生命期 life span
平均生息 average yield
平均声级 average sound level
平均声能流密度 sound energy flux density
平均声速 average sounding velocity; mean sounding velocity
平均声吸系数 mean sound absorption coefficient
平均声压 average sound pressure; mean sound pressure
平均声压级 average sound pressure level; mean sound pressure level
平均剩余耐用年限 <建筑物> average remaining durable years
平均失效间隔 mean time between failures
平均施工工期 average construction period
平均湿度 median humidity
平均时纪 mean epoch
平均时间 average time; averaging time; mean time
平均时间测定 time-average measurement
平均时间测量 time-average measurement
平均时间照相 time average photography
平均时流量 average hourly flow
平均实交资本 average paid-in capital
平均实行系数 average-practice coefficient
平均使用率 average service rate
平均使用年限 average life (period); average serviceable life; average serviceable years; mean life; mean service life
平均使用期限 average life time
平均使用时间 average service time
平均使用寿命 mean life(time)
平均视差 mean parallax
平均视浪高 mean height of apparent sea
平均试验车行车 average test-car run
平均试验车行程 average test-car run
平均试样 average sample; running sample
平均室内条件 <温度25℃,相对湿度40%> average room condition
平均收款期 average collection period
平均收敛 convergence in mean; mean convergence
平均收敛速度 average rate of convergence
平均收率 average yield
平均收入 average earnings; average income; average return; average revenue; mean yield
平均收入产品 average revenue product
平均收入率 average revenue rate
平均收入率法 average revenue rate method
平均收入曲线 average revenue curve
平均收益 average revenue
平均收益率 average yield
平均收账期间 average collection period
平均寿命 average length of life; average life (period); average lifetime; failure rate; mean length of life
平均寿命长度 average duration of life
平均寿命率 mean life rate
平均寿命预期值 average life expectancy
平均受拉强度 mean tensile strength
平均受压强度 mean compression [compressive] strength
平均输出功率 mean power output
平均输沙率 average rate of sediment discharge; mean sediment discharge
平均输送率 average transfer rate; average transmission rate
平均数 average; average number; average value; averaging; mean value; typical value
平均数变动率 variation in average
平均数标准差 standard deviation of mean
平均数标准误差 standard error of the mean
平均数差数 difference between means
平均数的平均 average of averages
平均数的选择 selection of averages
平均数功用 function of averages
平均数据 average data; mean data
平均数据传递率 average data transfer rate
平均数据传输速度 average data transfer rate
平均数趋势调整法 trend adjustment for average method
平均数指数 average number index
平均衰减系数求取 evaluation of average attenuation coefficients
平均水力半径 hydraulic average radius; hydraulic mean radius; mean hydraulic radius
平均水力深度 hydraulic average depth; hydraulic mean depth; mean hydraulic depth
平均水力蕴藏量 average potential water power; mean potential water power
平均水(平)面 average level; mean (water) level; ground table <局部地区的>
平均水平曲率 average horizontal curvature
平均水平声速 mean horizontal sound speed
平均水平烛光 mean horizontal candle power
平均水深 average depth; hydraulic mean depth; mean depth; mean water depth
平均水头 average head; average water head; man head
平均水位 arithmetic(al) mean of water level; average ordinary water level; average stage; average water (level); mean ordinary water level; mean water level; mean water stage
平均水位恢复速度 average recovery velocity of water level
平均水位治导 mean water training
平均水温 average water temperature
平均税率 average rate of tax; average tax rate
平均瞬时频率 mean instantaneous frequency
平均瞬时速度梯度 average temporal velocity gradient; mean temporal velocity gradient
平均朔望 mean establishment
平均朔望潮 average establishment; mean synodic lunar month
平均朔望潮高潮间隙 average establishment; common establishment; high water full and change; mean establishment; vulgar establishment
平均朔望低潮间隙时间 low water full and change
平均朔望高潮间隙 establishment of the port
平均朔望高潮间隙时间 high-water full and change
平均送风温度 mean supply air temperature
平均速度 average rate; average speed; average velocity; mean speed; mean velocity; moderate speed
平均速度差 <一定时间内,某一道路上车辆的车速差的平均值> average speed difference
平均速度点 mean velocity position
平均速度反应谱 average velocity response spectrum
平均速度分布 flat velocity distribution
平均速度分布图 flat velocity profile
平均速度曲线 mean velocity curve
平均速度梯度 <等于加速度干扰/平均速度> mean velocity gradient
平均速率 average rate; average speed; mean speed
平均损失 average loss
平均损失比 mean loss ratio
平均塔板高度 average plate height
平均太阳半日潮 mean solar semidiurnal tide
平均太阳半日分潮 mean solar semidiurnal constituent
平均太阳年 mean solar year
平均太阳日 mean solar day
平均太阳时 mean solar time; mean sun
平均太阴半日潮 mean lunar semidiurnal tide
平均梯度 mean gradient
平均体积 average volume
平均体积膨胀系数 average coefficient of cubic expansion
平均体积弹性模数 average bulk modulus
平均体重 average volume weight
平均天空 average sky; mean sky
平均天文北极 average astronomic(al) north pole; mean astronomical north pole
平均天文台 average observatory; mean observatory
平均调节阀组合特性曲线 mean of valve loop curve
平均调整速度的极微偏差 speed drift
平均调制速率 mean modulation rate
平均贴现率 average discount rate
平均听力损失 average bearing loss
平均停车延误 <一段时间内,进口道上各停车延误的总和除以驶离进口道的车数,秒/辆> average stopped-time delay
平均停机间隔时间 mean time be-

P

tween stops
平均通航期 average navigation period;mean navigation period
平均通勤行程距离 average length of commuters' journey
平均同步的 mesochronous
平均同时潮 mean cotidal hour
平均投资标准 average investment criterion
平均投资额 average investment
平均投资数 average investment
平均透光强度 average transmitted light intensity
平均图示压力 mean indicated pressure
平均涂布率 average spreading capacity
平均土壤流失量 average soil loss; mean soil loss
平均土温 mean soil temperature
平均推进率 average excavation speed
平均退水曲线 average recession curve;mean recession curve
平均托收期 average collection period
平均托收期比率 average collection period ratio
平均托收余额 average collected balance
平均外部体积 average external volume
平均微粒子体积 mean corpuscular volume
平均维护时间间隔 average time between maintenance; mean time between maintenance
平均维修时间 average time to repair
平均维修时间间隔 average time between maintenance; mean time between maintenance
平均纬度 mean latitude
平均位错 average dislocation
平均位能 average potential energy
平均位移 average displacement; average shift; mean displacement; mean shift
平均位移反应谱 average displacement response spectrum
平均位置 mean place;mean position
平均位置作用 average position action
平均温差 mean temperature difference
平均温度 average temperature; integral temperature; mean temperature; temperature average; temperature mean
平均温度表 chronothermometer
平均温度年差 mean annual range of temperature
平均温度误差 middle-temperature error
平均温度值 average temperature value
平均纹理深度 average asperity height; mean texture depth
平均稳定时间 mean time between failures
平均握裹应力 average bond stress
平均污水流量 average wastewater flow
平均无故障次数 mean cycles between failures
平均无故障工作时间 mean time of no failure operation; mean time to double failure
平均无故障距离 mean distance between failures
平均无故障时间 mean free error time; mean time to failure; mean time between failures
平均无偏估计量 mean unbiased estimator
平均误差 average error; balanced error; error mean; error of mean square; mean error; quadratic mean deviation;standard deviation
平均误差律 normal law of errors
平均吸声率 average sound absorptivity
平均吸声系数 average sound absorption coefficient; mean sound absorption coefficient
平均吸收系数 mean absorption coefficient
平均系数法 average factor method
平均细度 average fineness
平均下渗能力 average infiltration capacity
平均下限 average lower limit; mean lower limit
平均夏雨值线 isothermobrose
平均先进定额 advanced average quota;average advanced norm
平均现金余额 average cash balance
平均线 mean line
平均线膨胀系数 average coefficient of linear expansion
平均线性连续差 mean linear successive difference
平均线性直径 mean linear diameter
平均相对波动 mean relative fluctuation
平均相对价格 average relative price
平均相对离差绝对值 average absolute relative deviation
平均相对密度 mean specific gravity
平均相对偏差 average relative error; mean relative deviation; mean relative fluctuation
平均相对湿度 mean relative humidity;average relative humidity
平均相对误差绝对值 average absolute relative error
平均相干 average coherence
平均消费量 average consumption; mean consumption
平均消费倾向 average propensity to consume
平均小潮 average neap(tide); mean neap(tide)
平均小潮差 mean neap range
平均小潮低潮面 low-water neaps; mean low water neap tides; mean low water of ordinary neap tides; neap low water; mean low water neaps
平均小潮低潮升 mean neap rise of low tide
平均小潮低潮位 mean low water neap tides; mean low water of ordinary neap tide; mean low water neaps
平均小潮高潮高 mean water neaps; neap high water
平均小潮高潮面 mean high water neaps; mean high water of ordinary neap tide; neap high water
平均小潮高潮升 mean neap rise of high tide
平均小潮高潮位 mean high water neaps; mean high water of ordinary neap tide
平均小潮升 mean neap rise
平均小时 mean hours
平均小时变化 average hour variation
平均小时工资 average hourly wage
平均小时收益 average hourly earnings; hourly earnings average
平均效率 average efficiency; average output; mean efficiency; mean output
平均效率比率 average efficient ratio
平均效率因素 mean effectiveness factor
平均效率指数 average efficiency index
平均效应 average effect
平均效应区间中值商法 mean effect range median quotient method
平均效用 average utility
平均协方差函数 mean covariance function
平均协议 <按平均值简化结算车辆延期费> average agreement
平均信息量 average amount of information; average information content; entropy; mean information content
平均信息率 average information rate; mean information rate
平均信息速率 average information rate
平均行车速度 average driving speed
平均行程 average distance;average trip
平均行程车速 <路程距离除以通过该路程全部车辆行程时间的平均值> average travel speed
平均行程时间 average travel time
平均行程速度 average journey speed; average travel speed
平均行程钻速 average trip bit speed
平均行度 mean motion
平均行驶车速 <路程距离除以通过该路程全部车辆行驶时间的平均值> average running speed
平均行驶时间 average running time
平均型 mean type
平均修复时间 mean time to repair; mean repair time
平均修复时间的观测值 observed mean repair time
平均修理时间 mean repair time; mean time to repair
平均修理时间间隔 mean time between repairs
平均需电量 average power demand
平均需气量 integrated demand
平均需要量 average demand; mean demand
平均旋度 mean rotation
平均雪线高度 average snowline height
平均寻道【计】 average seek
平均寻(找磁)道时间 average seek time
平均循环时间 mean circulation time
平均循环系充盈压 mean circulatory filling pressure
平均压力 average pressure; mean pressure
平均压力梯度 average pressure gradient;mean pressure gradient
平均压力梯度线 average pressure grade line; mean pressure grade line
平均压力吸引高度 mean pressure suction head
平均压实度 average degree of consolidation
平均延伸量 mean elongation
平均延伸系数 mean coefficient of elongation
平均延停时间 average duration
平均岩芯采取率 average percentage recovery of core
平均岩芯收获率 average core recovery
平均盐碱度 mean salinity
平均氧化数 average oxidation number
平均样本 average sample; bulk sampling;mean sample;running sample
平均样本数 average sample number
平均样品 average sample; bulk sampling;mean sample;running sample
平均样品数 average sample number
平均夜流量 average night flow; mean night flow
平均一般成本 average general cost
平均一日卸车数 average number of cars unloaded per day
平均仪 averaging device
平均移位 average shift
平均以下 below the average
平均异常推估 prediction of mean anomalies
平均引出角 mean angle of emergence
平均应变 average strain;mean strain
平均应力 average stress;mean stress
平均应力值 average stress value
平均应收账款日期 day's receivables
平均英里 <1 英里 = 1609.34 米> average mileage
平均用水量 average consumption; mean consumption
平均铀含量 average uranium content
平均有酬假日天数 average length of paid holidays
平均有效功率 mean effective horsepower;mean effective power
平均有效厚度 average effective thickness
平均有效孔隙度 average effective porosity
平均有效流量 average available discharge
平均有效螺距 average effective pitch;mean effective pitch
平均有效螺纹 average effective pitch;mean effective pitch
平均有效马力 mean effective horsepower
平均有效黏[粘]度 mean effective viscosity
平均有效容重 average effective weight
平均有效使用年限 average useful life
平均有效寿命 average useful life
平均有效温度 mean effective temperature
平均有效系数 mean efficiency factor
平均有效压力 average effective pressure; brake mean pressure; effective mean pressure; mean effective pressure
平均有效应力 average effective stress;mean effective stress
平均有效直径 mean effective diameter
平均有效值 average effective value; mean effect;mean effective value
平均有效制动压力 brake mean effective pressure
平均有用寿命 average useful life
平均余额 average balance
平均雨量 average precipitation; average rainfall
平均雨量年份 year of average rainfall
平均雨量强度 average rainfall intensity
平均语言功率 average speech power
平均预期收入 average of expected incomes
平均预期寿命 average future life; average future time; average life expectancy; average life span; mean future time
平均预期销售额 average sale expectancy
平均圆度 average roundness; mean roundness
平均月潮半日分潮 mean lunar semidiurnal constituent
平均月潮低潮间隙 mean low water

lunitidal interval
平均月潮低低潮间隙 mean lower low-water lunitidal interval
平均月潮高潮间隙 corrected establishment;mean high water lunitidal interval
平均月潮高高潮间隙 mean higher high-water lunitidal interval
平均月机械钻 every monthly penetration rate
平均月库存 average monthly inventory
平均月末余额 average monthly balance
平均月球中心 average centre of the moon; mean center[centre] of the moon
平均运程<土石方的> average haul (distance);average length of haul
平均运动 average motion; average movement; mean motion; mean movement
平均运价率 average rate
平均运距 average distance; average haul(distance); mean haul; mean haul distance
平均运输距离<从采土场到卸土场的> average haul(distance)
平均运输量 transportation average
平均运算 average calculating operation
平均运算时间 average operation time; mean operation time
平均运行速度 average operating speed
平均匝数 mean turn
平均载波 mean carrier
平均载波频率 mean carrier frequency
平均载荷 mean load
平均载气线速 mean carrier velocity
平均载雨量<雨量器的> average rain catch
平均载重(吨数)average load
平均再充气 uniform recharge
平均在途资金 average fund in transit
平均噪声 average noise
平均增长率 average growth rate;average rate of growth;average rate of increase
平均增长率法 average factor method
平均增长速度 average increment speed;average speed of growth
平均增长因素模型 average-growth-factor model
平均增减趋势法 average increase-decrease trend method
平均增量 average increment
平均占用的流动资金 average amount of current funds possessed;average occupied amount of current capital
平均占用流动资金额 average occupied amount of current capital
平均站点法 mean station method
平均站间距 average station interval
平均站间距离 average distance between stations
平均张力 mean tension
平均涨潮间隙 mean rise interval
平均涨潮历时 mean duration of flood
平均涨落功率值 mean fluctuation power
平均照(明)度 average illumination
平均折旧 depreciation straight line
平均折旧法 composite depreciation method
平均折射 mean refraction
平均折射率 mean refractive index
平均振幅 mean amplitude
平均振幅倍数 times of average amplitude
平均震级 average magnitude of earthquake
平均蒸发率 average evapo(u)ration
平均正常运行时间 mean up time
平均正常重力 average normal gravity;mean normal gravity
平均正交反射率 mean orthogonal reflectivity
平均正坡度 average positive gradient
平均正应力 mean normal stress
平均支付期 average due date
平均直接成本 average prime cost
平均直径 arithmetic(al) mean diameter;average diameter;mean diameter;mid-diameter;angle diameter <螺纹的>
平均直线射程 mean linear range
平均值 average(value); mean(value);medium value
平均值表达式 averaged expression
平均值抽样 average value sample
平均值抽样法 average value sampling
平均值抽样数 average sample number
平均值的标准偏差 standard error of the mean
平均值的概差 probable error of mean
平均值的可加性 additivity of mean
平均值的平均值 mean of mean
平均值的误差曲线 error curve of mean
平均值等深线图 average chart;average isopleth map;mean chart;mean isopleth map
平均值定理 average value theorem; law of the mean;mean value theorem;theorem of mean
平均值定律 law of the mean
平均值读数表 average reading meter
平均值法 average value method; average value process; mean value method;mean value process
平均值函数 mean function;mean value function
平均值极差控制图 X-R control chart
平均值间之不等式 inequalities between means
平均值检波器 average reading detector
平均值聚类分析 means clustering analysis
平均值调节器 mean value control system
平均值原理 mean value theorem
平均值之间的变差 between-the-mean variation
平均值指示器 average value indicator
平均指标 average index;average indicatrix
平均指标数 average index number
平均指标指数 index number of average indicatrix
平均指令执行时间 average instruction execution time
平均指示器 mean indicator
平均指示压力 mean indicated pressure
平均指示有效压力 indicated mean effective pressure
平均指数 average index number
平均制动率 average braking rate
平均制动马力 brake mean pressure
平均制动性能 equalization characteristics of braking
平均质量 average quality;mean quality
平均质量保护 average quality protection
平均质量检查最低限 average outgoing quality limit
平均质量射程 mean mass range
平均质量因素 average quality factor
平均秩法 average tank method;mean tank method
平均滞留期 average residence time
平均滞留时间 average residence time
平均终雪(日)期 mean date of last snow cover
平均钟 average clock
平均重力 mean gravity
平均重力异常 mean gravity anomaly
平均重量 average weight;mean weight; weight in average
平均重量统计法 average weight statistics method
平均重现间距 average recurrence interval
平均重现期 average return period
平均重现时间 mean recurrence time; mean return time
平均周工资 average weekly wage
平均周工作时数 average hours worked per week
平均周年变化 mean annual change
平均周期 average period;median period
平均周期函数 mean periodic functions
平均周日变化 average diurnal change; average diurnal variation;mean diurnal change; mean diurnal variation
平均周日运动 average diurnal motion;mean diurnal motion
平均周转天数 average number of days to turnover
平均轴 mean axis
平均轴重 average axle-weight
平均烛光 average candle;mean candle
平均主义 equalitarianism
平均主应力 mean principal stress
平均住院日数 average duration hospitalization
平均柱压 mean column pressure
平均转动力矩 mean turning moment
平均转数 average revolutions
平均转速 average rotary speed
平均转效点 break-even point
平均装填密度 mean load density
平均装载重量 average shipping weight; mean shipping weight
平均状况 long-run average
平均准备账户 equalization reserve account
平均准确度 average accuracy
平均资本 average capital
平均资本比率(法)average capital ratio method
平均资金利润率 average profit margin; average profit rate on funds; average rate of profit on investment
平均子午线 mean meridian
平均自差 mean deviation
平均自由程 average free path; mean free path
平均自由碰撞时间 mean free collision time
平均自由时 mean free time
平均自由时间 mean free time
平均自由通路 mean free path
平均自由行程 mean free path
平均综合成本 average combined cost
平均总成本 average total cost
平均总速率 average overall rate
平均总效率 average overall efficiency
平均纵度 average longitudinal gradient
平均纵距 average ordinate
平均纵坡 average grade;average gradient
平均纵向电阻率法 mean longitudinal resistivity method
平均纵坐标 average ordinate
平均阻力 mean resistance
平均组成 average composition; mean composition
平均钻速 mean penetration rate
平均钻头进尺 average bit footage
平均钻压 average bit weight
平均最大波高 average maximum wave height
平均最大风速 mean maximum air velocity
平均最大流量 mean highest discharge
平均最大木高 mean top wood height
平均最大需要量 average maximum demand
平均最大值 mean maximum
平均最低潮面 average lowest water level;mean lowest water level
平均最低潮位 average lowest water level;mean lowest water level
平均最低水位 average lowest water level;mean lowest water level
平均最低温度 mean lowest temperature
平均最高高潮面 average highest high tide; average highest high water; mean highest high water
平均最高高水位 average highest high tide; average highest high water; mean highest high water
平均最高水位 mean highest water level
平均最高温度 mean highest temperature
平均最小尺寸<骨料/集料的> average least dimension
平均最小流量 average lowest discharge;mean lowest discharge
平均最小值 mean minimum
平均作业时间 average operation time
平均作业循环时间 average of operating cycle time
平均作用力 mean effort
平开窗 casement window; side-hung window;window casement
平开窗插销 casement window bolt
平开窗铰链 butt casement hinge
平开窗密合式铰链 close up casement hinge
平开窗扇 casement
平开窗台 slip sill
平开窗通风门 casement ventilator
平开隔帘 casement screen
平开门 hinged leaf gate; side-hung door;vertical hinged door
平开桥 pivot bridge; swing bridge; swivel bridge;turnable bridge;turn(ing) bridge
平开纱窗 casement screen
平开式<窗的> side-hung
平康石 heikolite
平孔 horizontal hole
平口车刀 shovel-nose tool
平口成型刀具 flat-form tool
平口刀具 flat(ted) pointed tool; flat tool
平口对接 square butt joint
平口沟管 plain-end sewer pipe
平口焊接器 grass hopper
平口虎钳 parallel(-jaw) vise[vice]
平口接箍 external flush-jointed coupling
平口接头 square joint;straight joint
平口螺纹 Acme(screw) thread
平口钳 chipping tool; flat bit tongs; flat mouth tongs;flat(-nose) pliers; flat tongs; gad tongs; gripping tongs;plain vice[vise];square-nosed

pliers
平口手套 plain top glove
平口台钳 parallel-jaw vice
平口装 flush mounting
平块 plain block
平拉窗 horizontally sliding sash; horizontally sliding window
平拉法 <玻璃> Colburn process
平拉环 flush ring
平拉门 folding sliding door; sliding flush door; sliding plug type door; plug door
平拉门曲拐 door crank
平拉门锁杆 door locking bar
平拉坞门 sliding dock gate
平老虎窗 flat dormer(window)
平肋板 flat floor
平肋板条 flat rib lath(ing)
平楞 open piling
平连接的 flat jointed
平联角钢 angle for horizontal connection
平联锁瓦 flat interlocking(clay)tile
平链 band chain
平链节 plain link
平梁 flat-topped ridge
平梁板桥 flat slab bridge
平列断层 parallel faults
平列发动机簇 parallel cluster
平列鸡笼 flat deck poultry cage
平列双排发动机 flat twin(type) engine
平列纤维束强力试验 flat bundle test
平裂 plain fracture; plane fracture
平裂形气孔 paracytic type of stomata
平流 laminar current; laminar flow; slack current; slack tide; slack water; stand of tide; subcritical flow; tranquil flow; advection <水平向运动引起的转移>
平流 V 形箱分级机 sloughing-off box
平流测量 measurement of advection
平流层 <指大气> advection layer; stratosphere
平流层尘埃浓度 stratospheric(al) dust load
平流层臭氧 stratospheric(al) ozone
平流层臭氧损耗 stratospheric(al) ozone loss
平流层臭氧稳态 stratospheric(al) ozone steady state
平流层的 stratospheric(al)
平流层顶 stratopause
平流层和中层气象学小组 panel on meteorology of the stratospheric(al) and mesosphere
平流层化学 stratospheric(al) chemistry
平流层化学动力学 stratospheric(al) chemical dynamics
平流层回降 stratospheric(al) fallout
平流层临边红外监测 limb infrared monitoring of the stratosphere
平流层氯 stratospheric(al) chlorine
平流层气溶胶测定装置 stratospheric(al) aerosol measurement device
平流层气溶胶和气体实验 stratospheric(al) aerosol and gas experiment
平流层上部 high stratosphere
平流层微尘 stratospheric(al) dust
平流层污染 stratospheric(al) pollution
平流层污染物 stratospheric(al) pollutant
平流层下部 low stratosphere
平流层增温警戒信息 stratospheric(al) warming alert message
平流层振荡 stratospheric(al) oscillation
平流层中部 middle stratosphere
平流层中间层 stratomesophere
平流沉淀 horizontal sedimentation
平流沉淀池 horizontal flow sedimentation basin; horizontal flow settling tank; horizontal sedimentation tank
平流沉降槽 horizontal flow tank; horizontal sedimentation tank
平流池 horizontal flow basin; horizontal flow tank
平流粗滤 horizontal flow roughing filtration
平流辐射雾 advection radiation fog
平流干燥器 parallel flow drier[dryer]
平流隔板式反应池 horizontal flow baffled reaction basin
平流冷却 advective cooling
平流逆温 advection inversion
平流区 advective region; slack water zone
平流热效 advection of heat
平流融雪 advection snowmelt
平流砂滤池 horizontal flow sand filter
平流式沉淀池 horizontal flow sedimentation tank; horizontal settling tank
平流式沉砂池 horizontal flow grit chamber; horizontal grit settling tank
平流式粗滤池 horizontal flow roughing filter
平流式隔油池 isolating-oil pool of horizontal flow
平流式流液筒 straight throat
平流式滤池 <其中的一种> Bohna filter
平流式排列 smooth plain packing
平流式厌氧滤池 horizontal anaerobic filter
平流输送(水、气等) advect
平流水槽 horizontal flow tank
平流雾 advection fog
平流消融 advective ablation
平流性雷暴 advective thunderstorm
平龙骨 flat keel
平垄 level ridge
平垄机 ridge level(l)er
平垄器 ridge level(l)er; row knocker
平楼板 deck floor
平楼盖 deck floor
平楼面 deck floor
平漏 flat leakage
平炉 Martin furnace; open-hearth (furnace)
平炉箅 flat grate
平炉车间 open-hearth plant; Siemens-Martin plant
平炉床 flat hearth generator
平炉床发生器 flat hearth generator
平炉胆 plain furnace
平炉的气窗 gas end
平炉底 flat floor
平炉底炉膛 flat-bottom furnace; horizontal bottom furnace
平炉法 Martin process
平炉钢 Martin steel; open-hearth steel; Siemens-Martin steel
平炉钢锭 open-hearth steel ingot
平炉钢轨 open-hearth rail
平炉钢渣 open-hearth cinder; open-hearth slag; Siemens-Martin steel slag
平炉构架 open-hearth frame
平炉观察孔 wicket hole
平炉换向器 open-hearth block
平炉利用系数 capacity factor of open-hearth furnace; utilization coefficient of open-hearth furnace
平炉炼钢 open-hearth furnace steelmaking
平炉炼钢法 open-hearth (furnace) process; Siemens-Martin process; Siemens open hearth process
平炉炉底打结料 <一种合成物> martenite
平炉内坡 scarp
平炉排 plain grate
平炉喷口 open-hearth port
平炉喷嘴 doghouse
平炉冶炼法 open-hearth process
平炉渣氧化作用 open-hearth slag oxidation
平炉栅 flat grate
平路 blading; level road; level track
平路拱 flat composition crown; flat crown
平路机 blade grader; blade machine; blade of grader; blader (grader); bullgrader; grader; level(l)er; level(l)ing machine; motor grader; patrol grader; planer; profile planer; road grader; roadpacker; road planning machine; road scraper; smoother; road finisher
平路机操作 <平路、刮路、整型、整平等工作> blading operation
平路机铲刀 blade of grader
平路机刀片操纵轮 blade control wheel
平路机的蟹壳斗 road grader clamshell shovel
平路机工作 blading operation; blading work
平路机汽缸 blade cylinder
平路面机 road planer
平路噪声 flat-channel noise
平掠回桨 feather an oar
平轮 flat wheel
平轮辋 flat rim
平轮压路机 flat steel roller; flat-wheel roller; smooth-wheel roller
平轮缘 flat base rim
平螺钉 plain screw
平螺纹 flat thread
平螺旋线 flat spiral
平螺旋状 pancake
平螺旋钻 low-helix drill; slow-spiral drill
平落 flat fading
平落式槽口排种盘 flat drop plate
平落式圆孔排种盘 flat drop round-hole plate
平埋侧石 flush kerb
平埋混凝土路缘 flush concrete curb
平埋路缘 flush curb; flush kerb
平埋头铆钉 flat countersunk head rivet
平埋头钻 flat countersink
平埋缘石 flush curb; shoulder curb
平埋砖 inlaid brick
平脉冲 flat pulse; normal pulse
平镘(刀) flat trowel
平毛石墙 squared rubble
平铆钉头 flattened rivet head
平铆(接) flush rivet(t)ing
平铆接头 butt rivet joint
平煤杆 level(l)ing bar
平门 hospital door
平面 at-grade; flat (sur) face; level; plain face; plane; plane surface; planum; tabular surface
平面 X 线照相术 plane radiography
平面凹陷 depression of surface level
平面把【数】 bundle of planes
平面靶 flat target
平面摆动筛 plain shaking screen
平面板 face-plate
平面板式管 flat-faced tube
平面板式集装箱 smooth panel container
平面半径 plan radius
平面包络蜗轮蜗杆传动 planer worm gear
平面保形几何学 plane conformal geometry
平面比尺 horizontal scale
平面比例(尺) horizontal scale; planimetric(al) scale
平面边长 side length in plane
平面边界层 plan boundary layer
平面边线镶饰 mo(u)lded and flat panel
平面边线修饰 mo(u)lded and flat panel
平面编组场【铁】 shunting yard; flat yard; marshalling yard
平面变形 plane deformation
平面变形剪切图 plane strain shear diagram
平面变形状态 plane deformation state; plane distortion state; state of plane deformation
平面表格绘图 plane-table plotting
平面波 plane wave
平面波玻恩近似 plane wave Born approximation
平面波传播 plane wave propagation
平面波导 slab guide
平面波倒易校准 plane wave reciprocity calibration
平面波的分解 plane wave resolution
平面波发生器 planar [plane] wave generator
平面波反射系数 plane wave reflection coefficient
平面波解 plane wave solution
平面波量子 plane wave quantum
平面波起爆 plane wave initiation
平面波散射系数 plane wave scattering coefficient
平面波纹法兰 plain-face corrugated flange
平面波系 planar wave system; plane wave system
平面波相互干扰 plane wave interaction
平面波相互作用 plane wave interaction
平面波信号 plane wave signal
平面波阵面 plane front
平面玻璃 flat glass
平面玻璃屋面瓦 plane glass roof(ing) tile
平面泊肃叶流 plane Poiseuille flow
平面薄膜 flat film
平面薄片 plane lamina
平面不连续群 planar discontinuous groups
平面布景图 floor plan
平面布线槽 surface raceway
平面布置 flat-sheet horizontal layout; floor plan layout; ichnograph; plan(e) arrangement; plane layout; planimetric(al) map; plan layout; planning; plant design; sitting
平面布置比较方案 alternate layout
平面布置设计 layout plan
平面布置图 layout plan; layout chart; layout drawing; floor chart; floor plan; ground plan; location plan; site plan; situation plan of site
平面布置总图 general layout; general plan; key plan; layout plan
平面部件 surfacing unit
平面操纵台 flat type console
平面草图 rough plan; schematic plan; sketch plan
平面草图设计网络 planning grid
平面侧面图 plan-profile sheet
平面测角计 plane goniometer
平面测量 plane survey(ing); planim-

etering;planimetric(al) survey;stadia traversing
平面测量的 planimetric(al)
平面测量法 plane-surveying;planimetric(al) method;planimetry
平面测量概算 plane survey preliminary calculation
平面测量图 planimetric(al) mapping
平面测量学 plane geodesy;planimetry
平面测量仪 face measuring instrument
平面层 plane layer
平面层移 planar lamination
平面岔道 switch of plane
平面缠绕 planar winding
平面缠绕封头曲面 in-plane head shape
平面场 flat field;plane field
平面场磁铁 flat field magnet
平面场发生器 flat field generator
平面场均一性 flat field uniformity
平面朝下 flat down;flat way
平面衬底 plane substrate
平面呈多边形的穹隆 cupola
平面呈多角形的穹隆 cupola
平面承重结构 area covering structural element; planar load-bearing structure; planar weight-carrying structure
平面承重结构建筑物 area covering structure
平面程序 planar process
平面尺度 horizontal scale
平面齿轮 crown gear; face gear; planoid gear
平面充水镇压器 plain water-filled roller
平面出入口 hatchway
平面传递带 plane belt
平面传感器 flat-surfaced probe; flat surface sensor
平面传送带 plane belt
平面吹气器 plane diffuser
平面锤 faced hammer;flatter
平面磁场 flat magnetic field
平面磁流体波 plane hydromagnetic wave
平面磁轴 flat magnetic axis
平面大冰块 ice table
平面大地探测基础 horizontal geodetic base
平面大梁 plane girder
平面代数曲线 plane algebraic curve
平面带输送 plate belt
平面单位尺寸 measurement unit of plan
平面导板 flat guide plate
平面导杆 planimetric(al) arm
平面导轨 flat guide
平面导航 horizontal navigation
平面导向 plano-guidance desk
平面导向台 piano-guidance table;plano-guidance
平面捣固器 flat-faced tamper
平面道路交叉口 single-level road junction
平面的 aflat; at-grade; complanate; planar; planiform; planimetric(al); tabulate;two-dimensional
平面的最佳极点 best pole of plane
平面等距抽样 plane systematic sampling
平面等值线图 plane isometric line map
平面底板 flat floor slab
平面底图 planimetric(al) base map
平面地 plane earth
平面地图 planimetric(al) map
平面地位指示器 plane position indicator
平面地物的地质下沉 geologic(al) dip of planar feature
平面地形图 ground plot
平面点集 plane point set
平面电磁波 plane electromagnetic wave
平面电动机 planar motor
平面电机绘图机 linear-motor table
平面电极 plane electrode;flat electrode
平面电极管 planar electrode tube
平面垫圈 faced washer
平面吊杆 flat suspension rod
平面吊架 flat hanger
平面调车 flat shunting; shunting on level tracks
平面调车场【铁】 flat classification yard;flat marshalling yard;flat yard
平面调车电气集中【铁】 all-relay interlocking for shunting area
平面定线 horizontal alignment
平面定子电动机 planar stator motor
平面度 degree of planeness; evenness;flatness;planeness;planarity
平面度测量装置 flatness inspection devices
平面度公差 surface planeness tolerance
平面度误差 flatness error
平面断层 plane fault
平面断层照相法 plane tomography; planigraphy
平面断裂 plane fracture
平面对称流动 plane symmetry flow
平面对称异构 plane symmetric(al) isomerism
平面对称异构现象 plane symmetric(al) isomerism
平面对等砧装置 flat-faced opposite anvil
平面镦粗 plane upsetting
平面多边形 plane polygon
平面颚板 smooth jaw plate
平面二极管 planar diode
平面阀 disc valve
平面法兰 face flange;flat-faced flange; plain flange
平面反光镜 plane mirror
平面反射 reflection at plane mirror
平面反射器 flat reflector; planar reflector
平面反射器天线 plane reflector antenna
平面反射器同相多振子天线 planar reflector billboard array
平面反射因数 regular reflection factor
平面反演 planar inversion
平面方案 schematic plan
平面方位角 grid azimuth
平面方向角 grid bearing
平面防撞装置 plain bumper
平面仿射变换 plane affine transformation
平面放射形钻孔钻进 ring drilling
平面分层介质 plane layered medium
平面分界面 plane interface
平面分离度 degree of plane separation
平面分子 planar molecule
平面封头曲面 balanced-in-plane contour
平面浮雕 cavo-relievo;intaglio relievo
平面浮射流 plane buoyant jet
平面盖度 flat cover-degree
平面杆 plane link
平面杆单元 plane rod element
平面杆系 plane member system
平面刚架 plane rigid frame
平面钢板网 plane expanded metal
平面钢鱼尾板 steel plain fish-plate
平面高程加密 horizontal-vertical bridging
平面镐 flat pick
平面格网 plane network
平面构架 plane frame work
平面构架构件 planar frame element; plane frame element
平面构架结构 planar frame structure
平面构图 plane configuration
平面构型 plane configuration
平面构造应力场 planar tectonic stress field
平面股钢丝绳 flattened strand wire rope
平面骨架构件 planar frame element
平面骨架结构 planar frame structure
平面固定膜 flat fixed film
平面刮刀 facing set;flat scraper
平面光阑 planar iris
平面光栅 flat grate;plane grating
平面光栅摄谱仪 plane-grating spectrograph
平面光栅制造器 flat grate producer
平面规 planometer;surface ga(u)ge; surface indicator
平面规划 plane layout
平面辊 straight roll
平面辊式破碎机 smooth roll crusher
平面过渡点 horizontal pass point
平面过滤机 filter table; flat filter; horizontal table filter;table filter
平面过滤器 filter table;flat filter
平面海图 plane chart
平面焊 level weld
平面焊接 face bonding
平面航海法 plane sailing
平面航迹计算法 plane sailing
平面航空摄影测量 planimetric(al) air survey
平面痕迹 trace of the plane
平面桁架 plane frame; plane girder; plane truss
平面桁架工程 plane truss works
平面后方交会(法) horizontal resection;plane resection
平面花砖 plane cement tile
平面滑动 plane slide; plane sliding; plane slip;planar slide
平面滑坍 <土体的> plane failure
平面化 complanation;planarization
平面环形电枢发电机 flat ring dynamo
平面簧 flat spring
平面汇交力系 plane concurrent force system;system of plane concurrent forces
平面绘图器 plane table; surveyor's table
平面绘图仪 plane plate
平面混凝土瓦 plane concrete tile
平面火焰喷燃器 flat-flame burner
平面机电耦合系数 planar electromechanical coupling factor
平面基准 horizontal datum
平面畸变 plane distortion
平面及立面图 plan and profile
平面极化 plane polarization linear polarization
平面极化波 plane polarized wave
平面极坐标 plane polar coordinate
平面集水布置 horizontal water-collecting layout
平面几何形状 plane geometry
平面几何学 plane geometry;planimetry;geometry of plane
平面技术 planar technique
平面加工 facing
平面加密 horizontal bridging
平面夹角 plane included angle
平面间 interplanar
平面间距 interplanar distance; interplanar spacing
平面监测网 horizontal monitoring control network
平面剪切 planar shear
平面剪切变形 planar shear deformation
平面剪切破坏 plane shear failure
平面检验方式 balance check mode
平面简谐方程 plane harmonic equation
平面渐开线 plane involute
平面渐屈线 plane evolute
平面渐伸线 plane involute
平面交叉 crossing ar grade; grade cross-section;grade intersection;intersection at angle; intersection at grade; level crossing; non-separated crossing; plain intersection; grade crossing <美>; level crossing <英>;at-grade intersection
平面交叉口 at-grade intersection; road crossing
平面交叉信号 level-crossing signal
平面交错层理 planar cross-bedding
平面交错层组 planar cross set
平面交换机 flat switchboard
平面交会的处理 handling of plane surface
平面交会法 plane crossing method
平面交通分隔 horizontal traffic segregation
平面浇注 pouring on flat
平面角 <法定单位为弧度> plane angle;horizontal angle
平面铰接车架转向 horizontal articulation steering
平面教堂 longitudinal church
平面接触 area contact
平面接缝 plane(d) joint
平面结 planar junction
平面结构 planar construction; planar structure;plane structure
平面解 planar solution
平面解析几何 analytic(al) geometry of plane
平面金刚石钻头 flat-bottom crown; flat-faced bit; flat-nose bit; square-nose bit
平面金属丝筛布 flat-top wire cloth
平面近似 plane approximation
平面晶体 flat crystal
平面晶体分光计 flat-crystal spectrometer
平面晶体管 planar transistor
平面精加工机床 flat surface finishing machine
平面精压 flat coining
平面精研机 surface lapping machine
平面径向流 planar-radial flow
平面镜 plane mirror
平面距离 plan range
平面锯齿结构 planar zigzag structure
平面卡盘 cement chuck; disk chuck; draw-in gear; drawing gear; face chuck;face-plate
平面抗剪试验 in-plane shear test
平面可变光阑 planar iris
平面课题 plane problem
平面空蚀 sheet cavitation
平面空中三角测量 horizontal aerotriangulation
平面控制【测】 horizontal control
平面控制标石 horizontal control monument
平面控制测量 horizontal control survey
平面控制测量方法 method of horizontal control survey
平面控制测量精度 accuracy for horizontal control

P

平面控制测量网 horizontal control survey net(work)
平面控制地形摄影 horizontal control photography
平面控制点 horizontal control station;plane control point;planimetric(al) control point;planimetric-(al) point;horizontal control point
平面控制点坐标计算员 horizontal control operator
平面控制高程加密 horizontal and vertical extension
平面控制基线 horizontal control base;planimetric(al) base
平面控制加密 horizontal bridging operation;horizontal extension
平面控制界碑 horizontal control monument
平面控制精度 accuracy of horizontal control;horizontal control accuracy;planimetric(al) accuracy
平面控制栅极 planar control grid
平面控制数据 horizontal control data
平面控制台 flat type console
平面控制网 horizontal control network
平面控制网布设形式 mode of establishing horizontal control
平面框架 flat frame;planar frame;plane frame
平面框架杆件 flat frame element
平面框架构件 planar frame element
平面框架结构 flat frame structure;planar frame structure;plane frame structure;plane frame work
平面扩散晶体管 planar-diffused transistor
平面扩散器 plane diffuser
平面扩压器 plane diffuser
平面拉床 surface broaching machine
平面拉刀 surface broach
平面拉刀磨床 flat broach grinder
平面拉削 surface broaching
平面力系 planar force system;plane force system;plane system of forces
平面力系的合成 composition of forces in plane
平面立体 plane solid
平面利用效果 plan utility
平面连接器 planar junction
平面裂缝 plane fracture
平面流(动) planar flow;plane flow
平面流网图 flow net plan
平面炉箅 plane grate
平面滤波 plane filtering
平面履带板 flat track shoe;flat shoe
平面略图 planimetric(al) sketch
平面轮 face wheel;flat-faced sheave;plain wheel
平面轮廓 face profile
平面螺管 flat spiral
平面螺栓塞 disk plug
平面螺旋 snail
平面螺旋缠绕 planar helix winding
平面螺旋线凸轮 snail
平面门 plain door;slab door <可装配玻璃的>;flush door <冷藏车的>
平面面积 area of plane;flat area;plain area;plan area
平面膜 planar film
平面磨床 face grinder;face grinding machine;flat grinder;parallel plane grinding machine;plain surface grinder;plain surface grinding machine;plane grinder;straight grinder;surface grinder;surface grinding machine
平面磨轮 face grinding wheel;straight wheel
平面磨蚀 flat abrasion
平面磨削 flat surface grinding;flatting operation;plane grinding;surface grinding
平面磨削过程 flatting process
平面内 in plane
平面内分布 distribution in plane
平面内刚度 in-plane rigidity
平面内荷载 in-plane loading
平面内剪力裂缝 in-plane shear crack
平面内力 in-plane force
平面内压曲 buckling in plane
平面内压屈 buckling in plane
平面内振动 in-plane vibration
平面扭曲状态 state of plane distortion
平面排列 flat arrangement
平面排水 flat drainage
平面刨床 surface planer;surfacer
平面培养计数琼脂 plate count agar
平面配合模 straight die
平面配流面 plane valving surface
平面皮带轮 flat-faced pulley
平面偏光镜 plane polarizer
平面偏光仪 plane polariscope
平面偏振 plane polarization
平面偏振波 plane polarized wave
平面偏振的 plane polarized
平面偏振的辐射束 plane polarized beam
平面偏振电磁波 plane polarized electromagnetic wave
平面偏振光 linearly polarized light;plane polarization light;plane polarized light
平面偏振光镜 plane polariscope
平面偏振射束 plane polarized beam
平面偏振天线 plane polarized antenna
平面平差 planimetric(al) adjustment
平面平均值法 method of plane mean value
平面平行大气压 plane parallel atmosphere
平面平行的 plano-parallel
平面平行流 two-dimensional parallel flow
平面平行系统 flat parallel system;planar parallel system
平面破裂 plane fracture
平面剖面图 plan-profile sheet
平面剖视图 sectional plan
平面铺地砖 plain brick paver;plain paving brick
平面气隙发电机 planar air-gap generator
平面器件 planar device
平面牵出线 flat shunting neck
平面前进波 plane progressive wave;progressive plane wave
平面墙 planar wall
平面切刀 undercutting tool
平面切削 face cutting;flush cut;surface;surfacing
平面求积仪 planimeter[planometer] platometer
平面球体图 planisphere
平面球形图 planisphere;planosphere
平面曲杆 plane-curved bar
平面曲线 parallel curve;plane curve;two-dimensional curve;horizontal curve
平面曲线测设 plane curve location
平面曲线桥 plane curve bridge
平面曲线束(筋) plane-curved tendon
平面取向数 plane orientation number
平面热源测定法 plane heat-source method
平面日晷 plane sundial
平面三角形 plane triangle
平面三角学 plane trigonometry
平面扫描 flat scanning
平面扫描法 flat plane scanning method
平面扫描头 planizer
平面扫描仪 sweep optic(al) square
平面色谱法 planar chromatography
平面砂轮 flat-faced wheel;straight wheel
平面筛 horizontal screen;plansifter
平面筛网 flat-surfaced screen
平面上亮度的量度 <以烛光计> luminance
平面上为 L 形的外墙角砌块 quoin block
平面上组合 composition on the plane
平面设计 design of plane;layout design;plane design;planning
平面设计网络 planning grid
平面摄影测量网 plane photogrammetric network
平面摄影测量学 plane photogrammetry
平面石磨 flat stone mill
平面示位图 plane position indicator
平面示意图 outline plan;plan(e) sketch;rough plan;schematic plan
平面式 flat type;plane formula
平面式直线感应钠泵 flat linear induction sodium pump
平面势函 plane potential function
平面试验水槽 model test basin
平面视野计 campimeter
平面视野计检查法 campimetry
平面疏解【铁】 level untying;plane untwining
平面束 axial pencil
平面数字管 planitron
平面双层底舱 flat tank
平面水泥瓦 plane cement tile
平面四杆机构 plane four bar mechanism
平面塑性变形 plane plastic deformation
平面碎部点 planimetric(al) detail
平面弹性理论 planar theory of elasticity
平面弹性力学 plane elasticity
平面弹性系统 plane elastic system
平面探头 flat-surfaced probe
平面特性 flatness of the response
平面体 plane body
平面体系 plane system;two-dimensional system
平面天体图 planisphere
平面天线 flash antenna;flat plane antenna
平面天线阵 linear array
平面填角焊 flat fillet weld
平面通量密度 planar flux density
平面同步扫描线 horizontal-scan-(ning) line
平面投影 plane projection;projection on a plane
平面投影法 plane projection method
平面透明度 flat transparency
平面凸轮 edge cam;face cam;plane cam
平面凸缘 flat-faced flange
平面图 drawing of site;ground plan;horizontal plan;ichnograph;iconography;layout;layout sheet;orthographic(al) plan;plain view drawing;plan;planar graph;plan drawing;plane diagram;plane figure;planform;planimetric(al) map;plan view;plot
平面图案装饰 diaper ornament
平面图边缘 margins of drawing
平面图测量 planimetric(al) survey
平面图尺寸 plan dimension
平面图的 planimetric(al)
平面图法 ichnography
平面图方格 grid of drawing
平面图幅 flat sheet;plan sheet
平面图和剖面图 plan and section
平面图绘(制)法 ichnography;iconography
平面图显示器 plain indicator
平面图像的中心投影【测】 central projection of plane figure
平面图像放大器 flat-image amplifier
平面图像投影 planometric
平面图形 planar graph;plane figure
平面图原图 original plan
平面涂装法 flat finishing
平面推力球轴承 thrust ball bearing with flat seat
平面椭圆弧 plane elliptic arc
平面外 out-of-plane
平面外延法 planar epitaxial method
平面弯曲 kink;plain bending;plane bending;plane flexure;uniphanar bending;uniplane bend
平面弯曲的建筑 curved building
平面弯曲试验 <板材> camber test
平面网 plane net
平面网板 flat netted board
平面网格结构 plane grid structure
平面网络 planar network
平面微弹性 plane microelasticity
平面为多边形的穹隆 dome of polygonal plan
平面位移 horizontal displacement;planimetric(al) displacement
平面位置 planimetric(al) position;planimetric(al) representation;plan location;plan position
平面位置表示法 planimetric(al) representation
平面位置测量 horizontal measurement;plan position measurement
平面位置辅助指示器 plane repeater indicator
平面位置固定 plan position fixing
平面位置精度 horizontal position accuracy;planimetric(al) accuracy;plan position accuracy
平面位置勘测 planimetric(al) reconnaissance
平面位置雷达显示器 plan position radar indicator scope
平面位置雷达指示器 plan position radar indicator
平面位置数字化 planimetric(al) digitizing
平面位置天线 plan position antenna
平面位置投影显示器 projection plan position indicator
平面位置图 general location map;plan position map
平面位置误差 planimetric(al) error
平面位置显示 plan display;plan position indication;plan position display
平面位置显示法 plan position indication method
平面位置显示器 plan(e) position indicator;plan position indicator scope;plan position repeater;plan position set
平面位置显示器预测 plan position indication prediction
平面位置显示器中继站 plan position indicator repeater
平面位置显示扫描 plan position indicator scan
平面位置信息 plan position information
平面位置遥示器 plan position repeater;remote plan(position) indicator
平面位置指示进场 plan position indication approach

平面位置指示雷达 plan position indicator radar
平面位置指示声呐 plan position indication sonar
平面位置中误差 mean square error of horizontal position
平面文件 flat file
平面纹饰 flat ornament
平面问题 plan(ar) problem; two-dimensional problem
平面涡卷形装饰 < 楼梯扶手终端的 > lateral scroll
平面雾化器 plane diffuser
平面吸收源 plan sink
平面铣 plan milling
平面铣床 plain miller (machine); plain milling machine; plan miller; plan milling machine; slabbing machine; surface miller; surface milling machine
平面铣刀 face cutter; face mill; facing cutter; facing mill; plain milling cutter; slab cutter
平面铣刀锥 face mill driver
平面铣削 slabbing; slab milling
平面系 two-dimensional system
平面细部 plan detail
平面细节 plan detail
平面显示 flat panel display
平面显示器 two-dimensional display
平面线 planimetric(al) line; horizontal curve
平面线横净距 lateral clear distance of horizontal curve
平面线路 horizontal alignment
平面线形 horizontal alignment; plane alignment
平面相控阵 planar phased array
平面相似变换 plane similarity transformation
平面像场 flat-image field
平面像片三角测量 instrument photo triangulation
平面胁变 plane strain
平面斜底 plane slanted bottom
平面斜接 plain scarf
平面斜平行线投影 planometric projection
平面谐波 plane harmonic wave
平面谐波运动 plane harmonic motion
平面信号 planed signal
平面星图 celestial planisphere
平面行波 plane progressive wave; progressive plane wave
平面形态 plain-form; planform geometry
平面形心 centroid of a plane area
平面形荧光屏 flat-faced screen
平面形状 flat shape; plain-form; planar configuration; planform geometry; section shape
平面型二极管 planar diode
平面性 flatness; planarity
平面修整机 flat finisher
平面虚像 virtual image; virtual planar image
平面旋回筛 gyrating screen; gyratory screen; revolving screen
平面旋转 revolution of plane
平面旋转式 case of horizontal swing
平面旋转系统 flat rotation system
平面旋转向量 rotating plane-vector
平面岩层 plane bed; plane rockbed
平面研磨 face grinding; plane lapping
平面研磨机 face lapping mill; plane grinder
平面研磨盘 flat tool
平面衍射光栅 plane diffraction grating
平面氧化 planox
平面叶片 flat blade
平面仪 planimeter
平面翼 flat plane
平面阴极 planar cathode; plane cathode
平面印刷线路板 flush printed wiring board
平面应变 plane strain
平面应变不稳定性 plane strain instability
平面应变断裂 plane strain fracture
平面应变断裂韧度 plane strain fracture toughness
平面应变断裂韧性 plane strain fracture toughness
平面应变加载 plane strain loading
平面应变拉伸试验 plane strain extension; plane strain extension test
平面应变伸长试验 plane strain extension test
平面应变条件 plane strain condition
平面应变型椭球体 plane strain ellipsoid
平面应变压缩试验 plane strain compression test
平面应变仪 plane strain apparatus
平面应变应力奇点 plane strain-stress singularity
平面应变状态 plane state of strain; plane strain state
平面应力 biaxial stress; plane stress
平面应力断裂 plane stress fracture
平面应力分布 plane stress distribution
平面应力分析 plane stress analysis
平面应力问题 flat stress problem; planar stress problem
平面应力状态 flat state of stress; planar state of stress; plane state of stress; plane stress state; state of plane stress
平面映射 plane reflection
平面预测 planar prediction
平面源 planar source
平面元 flat element; planar element; plane element
平面运动 plane motion; plane movement; planimetric (al) motion; planimetric(al) movement
平面运动机构 plain motion mechanism
平面运输 horizontal transport
平面轧辊 flat roll; plain roll
平面闸门 plain gate; plane gate
平面闸门启闭机 lifter of plain gate
平面栅状应变仪 flat-grid strain ga(u)ge
平面障碍 impediment in plane
平面障板 plane baffle
平面照片三角测量 plane phototriangulation
平面折流器 plain deflector
平面针盘 surface ga(u)ge
平面阵 planar array; plane array
平面阵列雷达 planar array radar
平面阵列雷达天线 planar array radar antenna
平面振动 plane vibration
平面振动压实器 vibrating plate compactor
平面振实器 surface vibrator
平面震源 plane source of earthquake
平面支承 even bearing; flat bearing; plane bearing
平面支承结构 flat supporting structure; planar bearing structure; planar supporting structure
平面支座 flat bearing; plane bearing; surface bearing
平面支座结构 flat bearing structure
平面直角三角形计算工具 trigonometer
平面直角坐标 grid coordinates; plane rectangular coordinates; plane right coordinates; planimetric(al) rectangular coordinates; rectangular coordinates; rectangular plane coordinates
平面直角坐标系 planimetric(al) rectangular coordinate system
平面直角坐标系统 plane rectangular coordinate system
平面直线透视 plane linear perspective
平面中心 centroid of a plane area
平面种植 fiat planting
平面轴承 plane bearing; surface bearing
平面轴套 flat bush
平面注量 planar fluence
平面铸铁轧辊 plain cast iron roll
平面转动系统 planar rotational system
平面装配图 setting plan; assembly layout < 室内的 >
平面装卸 horizontal handling
平面装药 slab charge
平面锥锉 flat taper file
平面琢石 plane ashlar
平面子午线收敛角 grid convergence
平面组装 planar package
平面组装插件 planar module
平面钻进线 < 采矿 > level advance line
平面坐标 horizontal coordinate; horizontal position; phase coordinates; plane coordinates; planimetric(al) coordinates
平面坐标方位角 plane coordinate azimuth
平面坐标平差 adjustment of planimetry
平面坐标三角形 plane coordinate triangle
平面坐标网 plane network
平面坐标系(统) horizontal coordinate system; plane coordinate system; two-dimensional coordinate system
平面座阀 flat valve
平苗床器 seedbed leveler
平民 civilian; commonalty
平民居民 plainsman
平民人口 civilian population
平皿计数 plate count
平皿计数技术 plate count technique
平皿培养 plate culture
平模滚轧螺纹法 flat die thread rolling
平模流水法 flat-form process
平膜片 flat diaphragm
平磨 flat mill; plain grinding; plane abrasion
平磨机 flat-grinding machine; Hoover muller
平磨试验 flat abrasion test
平抹子 floater; flat trowel
平木板屋顶 flat timber roof
平木锉 flat rasp file; horse rasp file
平幕 flat screen
平泥板 screed board
平泥地 vloer
平年 < 非闰年 > common year; average year; mean year
平捻绳 plain laid rope
平碾 drum roller; flat-wheel roller; smooth drum roller; smooth-wheel roller
平耙头 flat draghead
平排 flat raft
平排多嘴包装机 multispout in-line packer
平盘 bench floor; flat pan; plate; platform
平盘管 pancake coil
平盘菌属 < 拉 > Discina
平盘烧针装窑法 dottling
平盘绳 Flemish coil; Flemish down; Flemish fake
平盘旋坯机 plate jiggering machine
平盘装窑法 dottling
平刨 bench rabbet plane; flat plane; plane; single face planer; surface planer; truing plane; try(ing) plane; cabinet scraper < 橱柜用的 >
平刨刨口槽 throating
平刨刨刃 flat plane iron
平炮眼 flat hole
平劈面 horizontal clearance plane; horizontal cleavage plane
平皮带 flat belt
平皮带传动 flat-belt drive
平皮带传送机 flat-belt conveyer[conveyor]
平皮带轮 flat-belt pulley
平皮带式传动 flat-belt drive
平片 plain film
平片解算电势计 flat-card resolving potentiometer
平片介质波导 slab dielectric (al) waveguide
平片介质光波导 slab dielectric (al) optical waveguide
平片式解算电位器 flat-card resolving potentiometer
平片式解算器 flat-card resolver
平坡 evenness; flat slope; gentle slope; gradual slope; horizontal slope; level grade; mild slope; near-level grade; zero slope
平坡道口 flat approaching grade
平坡度 level grade
平坡机 < 装有刮坡刀的平地机 > slope grader
平坡进口道 flat approaching grade
平坡隧道 level tunnel
平坡梯田 bench-type terracing
平坡线 depression of contour line
平坡线路 level line
平剖 flat cutting
平剖面 profile in plan; section plan
平剖面图 profile in plan
平剖面组合 composition on the plane-section
平剖图 profile in elevation; profile in plan
平铺 carvel built; flush
平铺船壳板 carvel planking
平铺的 procumbent
平铺缸砖路面 flat brick pavement
平铺管喷灌 lay-flat tube irrigation
平铺混凝土 spread concrete
平铺料堆 flat layered pile
平铺硫化 flat cure
平铺路轨 track set in paving
平铺砌层 pegtop paving
平铺式 carvel system; flush system
平铺式船壳板 carvel planking; smooth planking
平铺式舢板 carvel built boat
平铺熨烫 flat pressing
平铺砖 brick laid on flat
平谱 flat spectrum
平谱源 flat-spectrum source
平齐 abreast
平齐安装 flush mounting
平齐层 flush coat
平齐的 flush
平齐底面 < 三角形的截面梯级 > flush soffit; flush visible under-face
平齐地貌 accordant morphology

平齐峰顶线 accordant summit(level)
平齐汇流 accordant junction
平齐接缝 cash joint
平齐路缘 flush curb
平齐式分隔带 <路面分隔标线> flush stripe
平齐式接管 flat nozzle
平齐头针锉 equaling needle-handle file
平齐中心带 <与路面铺砌齐平的中心分车带> flush median strip
平齐装粉 flush-level powder fill
平齐装料 flush filling
平畦植法 flat planting
平砌 flat laying;surface
平砌层 <方块> horizontal course
平砌多孔砖 horizontal cell tile
平砌砖 brick laid on flat; brick on bed;brick on plate;inlaid brick
平砌砖过梁 flat brick lintel
平牵引力 horizontal driving force
平钳 flat nippers
平嵌的 flush-filled
平嵌灰缝 flush-filled joint
平嵌接 flat scarf
平嵌接缝 flush-filled joint
平嵌饰线【建】back fillet;reglet[riglet]
平嵌线 back fillet
平嵌砖缝 flat-joint point(ing)
平墙镘 darby
平锹 grafter;grafting tool
平桥 level bridge
平切 flat cutting;truncated
平切单板 sliced veneer
平切法 truncation method
平切面 side grain
平切丘顶 bevel(l)ed hill top
平揿钉 flat thumbtack
平倾断层 low-angle fault
平穹顶 straight vault
平曲柄 flat crank
平曲面 plane surface
平曲线 horizontal curve;plane curve
平曲线半径 horizontal radius;radius of horizontal curve;radius of plane curve
平曲线加宽 curve widening
平曲线起点 beginning of curve
平曲线桥面的超高 super-elevation of the floor surface of a bridge on a horizontal curve
平曲线曲率 plan curvature
平曲线位置 seat of plane curve
平曲线终点 end of curve
平驱动力 horizontal driving force
平壤 <朝鲜民主主义人民共和国首都> Pyongyang
平壤海槽 Pyongyang marine trough
平绕粗纱 collimated roving
平刃刀片 plain knife
平刃动刀片 plain section
平日 mean day;weekday
平日负荷曲线 <星期日除外> week-day load curve
平日小教堂 week-day chapel
平绒 panne velvet
平绒地毯 level cut carpet
平三角锉 ridged-back file
平扫描场透镜 flat field lens
平筛 diaphragm screen; flat screen; flat sieve;jog strainer;plansifter
平栅炉 flat grate
平扇块磁铁 flat sector magnet
平勺 heart and square;spatula
平射喷嘴 tee-jet nozzle
平身科斗拱 intermediate corbel-bracket set
平石【道】gutter apron
平石磨 stone mill
平时 civil time;mean time;non-peak hours
平时编制 peace footing; permanent basis
平时功率 continuous power;firm power
平时式时号 mean time type time signal
平时水位 ordinary water level;ordinary water-stage
平时钟 mean time clock
平式安全岛 pavement type island
平式缝接 overhand
平式焊接 butt-weld(ing)
平式接缝 flat-fell seam
平式空气轴承 flat air bearing
平式炉顶 flat arch
平式锁簧床 horizontal locking box
平式洗涤机 panel washing machine
平式缘石 pavement edge curb;shoulder curb
平视 orthophoria
平视差 mean parallax
平视显示器 head-up display
平手锉 flat hand file
平竖轧制法 flat and edge method of rolling; flat-and-edging method of rolling
平衰减 flat fading
平栓 level(l)ing peg
平水 mean water;par【建】;slack-water
平水池 equalizing pond;equalizing pool
平水涵洞 equalizing culvert
平水航行 still water navigation; slack-water navigation
平水河床 minor river bed
平水年 average water year; median flow year; moderate-flow year; normal year
平水期水位 water-level in common season
平水器 spirit level
平水设备 <自记水位计观测井筒> flushing device
平水塔 equalizing tank
平水位 ordinary water level
平水运河 ditch canal;level canal
平顺 easement;fair;smooth-going
平顺底边 easy bilge;slack bilge
平顺河段 smooth river stretch;well-aligned reach
平顺护岸 continuous revetment;smooth continuous revetment
平顺胶合 flatting
平顺流 smooth flow
平顺门把 flush handle
平顺坡度 easy grade;easy gradient; smooth slope
平顺曲线 easy curve;fair curve;gentle curve; shallow curve; smooth curve;wide curve;flat curve <大半径曲线>
平顺山坡 smooth hillock;smooth hillside
平朔望月 mean synodic lunar month
平锁 horizontal lock
平塔顶 plate-form roof; truncated roof
平踏脚板 plain run board
平台 berm(e);block;deck;landing; level table; pace; platform; surface plate;terrace
平台安全岛 pavement type island
平台安装 platform erection
平台板 flat pallet;landing slab;platform pallet
平台边的踏步或基石 lysis
平台标高 floor level
平台层顶 platform roof;terraced roof
平台长度 floor length
平台车 dolly;lift truck;platform car; platform truck
平台车身 platform body
平台承载能力 platform loading capacity
平台承重链式输送机 flat-top chain conveyer[conveyor]
平台尺度 dimensions of the platform
平台触点 floor contact
平台存粮装载机 loader for loading flat stored grain
平台带锯 table band saw
平台带栏条车体 platform-and-stake body
平台挡 table dog
平台导槽 guidance of table
平台导承 flat table guide;level table guide
平台的下降 platform drop
平台地质平面图 geologic(al) plan of mining bench
平台顶板格栅 landing ceiling joist
平台断层 platform fault;tabular fault
平台饭店 terrace restaurant
平台防护 terrace prevent
平台防水 deck waterproofing;terrace waterproofing
平台盖 platform door
平台钢垫板 <楼梯、车站等地方用的> steel landing mat
平台高架移动起重机 platform gantry
平台格栅 landing carriage; landing joist(of stairs)
平台挂车 flat trailer truck
平台基准面 face-plate
平台基座 quadra
平台集装箱 platform container
平台甲板 deck floor;half deck;platform deck
平台甲板平底船 decked barge
平台架 staging
平台胶印机 flat-bed offset(printing) machine
平台礁 platform reef;tabular reef
平台校准 platform alignment
平台净高 headroom of landing
平台卷扬机 platform hoist
平台开关 floor switch
平台框架 platform frame; platform framing
平台栏车 flat-form and stake racks truck
平台栏杆 platform railing
平台联结格栅 landing binding joist
平台梁 landing beam;landing girder
平台楼梯 platform stairway
平台罗经 platform compass
平台面板 deck of platform
平台模板 deck form
平台木门 wood terrace door
平台排水 terrace drainage
平台期 plateau[复 plateaus/plateaux]
平台期电位 plateau potential
平台起升范围 lifting and lowering range
平台起升高度 platform height
平台起重机 platform hoist
平台千斤顶 table jack
平台扫描仪 flat-bed scanner
平台升降机 balcony lift; platform hoist
平台升降台 lifting plate
平台式 flat bed; platform type; flat type
平台式拌和设备 flat type mixing plant
平台式并条机 rail type drawing frame
平台式驳船升降机 mechanical platform barge lift
平台式堤脚护坦 berm toe apron
平台式干燥机 platform drier[dryer]
平台式(钢)模板 deck form
平台式工作架 island serving shelf
平台式构筑 platform framing
平台式绘图机 flat-bed coordinatograph;flat-bed plotter
平台式货物提升机 freight platform hoist
平台式机械 deck machinery
平台式礁 platform reef
平台式搅拌设备 flat type mixing plant
平台式进水口结构 balcony-like intake structure
平台式厩肥小推车 platform-type manure barrow
平台式框架 western framing
平台式龙门架 platform gantry
平台式楼梯 platform stair(case)
平台式码头 platform quay; platform wharf
平台式模板 deck form; deck type form
平台式木框架 western frame
平台式起重车 platform truck
平台式起重机 pedestal(mounted) crane; platform crane; platform hoist
平台式数控绘图机 flat-bed digital plotter
平台式踏步 landing step;terrace step
平台式台秤 platform bench scale
平台式梯 platform ladder
平台式提升机 platform hoist; platform elevator
平台式屋顶 cut roof;deck roof;flat-bed roof;terrace rood
平台式移动拌和机 flat type travel(l)-ing mixer
平台式移动搅拌机 flat type travel(l)-ing mixer
平台式运输车 platform trunk
平台式载货汽车 flat-bed truck
平台式载重汽车 platform lorry
平台式振动压实器 vibrating plate compactor
平台式逐稿器 single-wide shaker
平台式自动绘图机 flat-bed plotter
平台输送机 platform conveyer[conveyor]
平台四轮车 lorry
平台伺服机构 platform servo
平台缩小雕刻机 flat table pantograph engraving machine
平台踏步 landing step
平台弹簧 platform spring
平台梯级 terrace step
平台梯架 platform ladder
平台调准 platform erection
平台推车 platform trunk
平台托梁 stemple
平台拖车 dolly bar; dolly car; platform trailer
平台位置 position of platform
平台屋顶 terrace(d) roof
平台系泊 platform mooring
平台系船柱 platform bollard
平台销售 gondola sales
平台型底架 platform frame
平台型作业船 platform-type working craft
平台印刷机 flat-bed press
平台有效载重 platform payload
平台有效载重量 platform capacity
平台园 terrace garden
平台运输器 platform conveyer[conveyor]
平台载货提升机 platform hoist
平台遮阳篷 terrace awning

平台振捣器 vibrating table; vibroplatform; vibrostand
平台振动器 platform vibrator
平台支架 platform mounting
平台支柱 studdle
平台装置式曝气器 platform-mounted aerator
平太阳 astronomic(al) sun; mean sun
平太阳赤经 right ascension of mean sun
平太阳秒 mean solar second
平太阳年 mean solar year
平太阳日 mean solar day; sea day
平太阳时 mean solar time
平太阳时迟滞差 retardation of mean solar time
平太阳时角 hour of mean solar time
平太阳小时 mean solar hour
平太阴时迟滞差 retardation of mean lunar time
平态特性 flat characteristic
平摊资金和物资 contribute the same amounts of funds and materials
平滩 flat; flat shoal
平滩流量 bankfull discharge; bank high flow; sideflat-overtopping discharge
平滩水位 bankfull stage; flow land line water level; sideflat-overtopping stage
平弹簧 coach spring
平坦 even; level; plane
平坦冰 flat ice
平坦层理 flat-bedding
平坦程度 evenness
平坦的 complanate; flat; non-sloping; plain
平坦的表面 champ
平坦的场地 flat space
平坦的耳形手柄 lug handle
平坦的频率响应 flat frequency response
平坦的坡度 even pitch
平坦的水准表面 level plane surface
平坦的速度分布线 flat speed profile
平坦地层隧道 tunnels under flat ground
平坦地带 flat terrain; flat topography; level terrain; smooth terrain
平坦地段 flat site
平坦地面 level surface
平坦地区 even ground; flat-bottom zone; flat country; flat ground; flat topography; level country; level terrain; normal country; region of no relief
平坦地势 smooth relief
平坦地形 even ground; flat terrain; flat topography; level terrain; smooth terrain; subdued relief
平坦点 flat spot
平坦定伸 flat modulus
平坦度 flatness
平坦度仪 flatscope
平坦断口 even fracture
平坦耳形手柄 lug handle
平坦分布源 flat source
平坦功率 flat power
平坦拱 < 拱心在起拱线下的 > blunt arch; depressed arch
平坦谷地 charco
平坦海岸 flat coast; flat shore; low coast
平坦海冰 flat ice; level ice
平坦海床 flat seabed
平坦海底 flat seabed
平坦海底带 level bottom zone
平坦函数 flat function
平坦河床 flat(river) bed; plane(river) bed
平坦河漫滩 flat flood plain
平坦化 equalization; planarization
平坦环境 flat space
平坦基床 plane bed
平坦基线 flat baseline
平坦加权 flat weighting
平坦空间 flat space
平坦连通 flat connection
平坦漏过功率 flat leakage power
平坦露头 gentle outcrop
平坦路拱 flat crown; sweet camber
平坦路基 plain background
平坦路面 level(l)ing-up course; level surface; level-up course
平坦面 flat face
平坦模 flat module
平坦模型 flat mode
平坦跑道 level(l)ing-up course; level-up course
平坦频率响应曲线 flat frequency response
平坦坡度 flat slope
平坦浅滩 flat shoal
平坦球对 flat sphere pair
平坦区 flat region
平坦曲线 flat profile
平坦瑞利衰落 flat Rayleigh fading
平坦射 flat morphism
平坦衰落 flat fading
平坦衰落信道 flat fading channel
平坦双模 flat bimodule
平坦铁路线 level line; level railroad line
平坦通量 flat flux
平坦通量反应堆 flat-flux reactor
平坦温度分布曲线 flat-temperature curve; flat-temperature profile
平坦线路 flat line
平坦响应 flat response
平坦响应曲线 flat response curve
平坦斜坡 flat slope
平坦型光发射器 flat-geometry light emitter
平坦型光源 flat-geometry light source
平坦型器件 flat device
平坦增益 flat gain
平坦增益控制 flat gain control
平坦子流形 flat submanifold
平坦增益调节电容器 flat gain capacitor; flat gain control capacitor
平探头 flat probe
平陶瓷管 planar ceramic tube
平陶瓦 plain clay roofing tile
平踢脚板 flush base; flush floor base
平天窗 flat skylight
平天窗采光片 roof-light sheet
平天花板底面 flush soffit
平天文纬度 mean astronomic latitude
平天文子午面 mean astronomic meridian plane
平天线 plane antenna
平添纱 plain plating
平填角焊缝 flat fillet weld; smooth fillet weld
平挑檐 flat corbel-table(frieze)
平条 riglet
平条埋嵌线饰 sunk fillet mo(u)ld
平条线脚 fillet mo(u)lding
平调 flat gain regulation; flat regulation; indiscriminate transfer of resources; unpaid transfer of resources
平调电容器 flat gain capacitor; flat gain control capacitor
平调放大器 flat regulating amplifier
平调/斜调控制盘 flat/slope control panel
平调增益主控器 flat gain master control
平调主控制器 flat gain master controller
平贴盖板 flush plate
平贴接合 flush joint
平贴门锁 straight lock
平贴现率 flat discount rate
平铁 black iron
平铁皮 flat-sheet metal; plain sheet; plain sheet iron
平铁皮的折缝 welt
平铁片 galvanized iron plain sheet
平通带 flat pass-band
平通管路 flat passage
平头 cheese head; flat head; tacky; truncate; pan head
平头 T 形接合 square-tee joint
平头摆杆摇柱 flat rocker column
平头摆杆摇座 flat rocker column
平头搬钩 cant hook
平头部 butt end
平头槽螺栓 flat-head grooved bolt
平头插销 flush bolt
平头插销座 flush bolt keeper
平头铲 butt chisel
平头铲刀 butt chisel; chipping chisel
平头窗框提手 flush sash lift
平头锤 butt rammer; coppering hammer; tack hammer
平头大铁钉 holdfast
平头捣锤 butt rammer; flask rammer
平头的 blunt-nosed; flat-headed
平头电极 flat tip; flat tip electrode
平头丁字焊接 plain tee joint
平头丁字接头 plain tee joint; square-tee joint
平头钉 brad; clout nail; hobnail; tack (nail)
平头钉拔除器 tack claw
平头钉锤 tack hammer
平头钉子 flat nail
平头钉(自动)敲打机 tack driver
平头端焊接 plain-end joint
平头对焊接 square butt joint
平头对接 unchamfered butt
平头对接焊 square butt welding
平头对接焊缝 plain butt weld; square butt weld
平头对接接头 plain butt joint; square butt joint
平头墩 blunt nosed pier
平头方尾圆底小艇 pram dinghy
平头缝 butt joint
平头改锥 flat screwdriver
平头钢轨 flat champignon rail; flat-headed rail
平头镐 flat pick
平头管子 plain-end pipe
平头滚针 flat end needle rollers
平头焊接 butt weld; joint weld; jump weld; tack weld(ing)
平头焊接钢筋网 tack-welded reinforcing mesh
平头绘图机 flat-head plotter
平头剪 snips; tin snips
平头键 recessed key
平头角焊接 plain corner joint
平头角接头 square groove corner joint
平头接合 even joint; flush cut joint; flush joint; square joint
平头接头 even joint; flush cut joint; flush joint; square joint
平头开口对接焊 open square butt joint
平头炉撑 flush staybolt
平头螺钉 butt-end bolt; flat-head screw; flat screw; grub screw; screw with flat head
平头螺母 flat nut
平头螺栓 flat-headed bolt; flush bolt
平头螺栓卸扣 countersink screw pin shackle; flush head screw shackle
平头螺丝 grub screw
平头螺丝刀 flat screwdriver
平头埋入键 recessed key
平头埋头螺钉 flat head
平头铆钉 flat-head rivet; flatten(ed) rivet; flush(head) rivet; pan head rivet; set-head rivet; tack rivet
平头铆接 butt rivet joint
平头铆头 disc rivet
平头坡口 < 焊体 > square groove
平头气门 flat-head valve
平头汽车 flat fronted vehicle
平头铅条 lead tack
平头前倾船首 pram bow; scow bow; swim bow; transom bow
平头钳 flat-head pliers; flat-nose pliers
平头切割 square cutting
平头瑞 butt end
平头式 flat-head type; cabover type
平头式载重汽车 cab over engine type
平头手把 flush handle
平头枢轴 flat pivot
平头双接 unchamfered butt
平头筒形屋面瓦 pan and roll roofing tile
平头弯纱三角 flat-nosed stitch cam
平头消防栓 flush(type) hydrant
平头消火栓 < 出口低于路面或与路面齐平 > flush hydrant
平头小锤 tack hammer
平头小钉 face nail
平头型 forward type
平头型钉 brad
平头型卡车 cab-forward type vehicle; cab over engine truck < 驾驶室在发动机上方 >
平头压实齿 foot pad
平头檐 flush eaves
平头圆铁钉 flat-head wire nail
平头錾 chipping chisel; spike iron
平头錾锤 straight-pane hammer
平头凿 butt chisel; chipping chisel
平头直颈铆钉 flat-head straight neck rivet
平头钻头 counterbore; countersinking bit; flat-head bit
平凸的 piano-convex
平凸颚板式破碎机 straight and bellied jaw crusher
平凸颚板式碎石机 straight and bellied jaw crusher
平凸颚板式轧碎机 straight and bellied jaw crusher
平凸镜片 plano-convex lens
平凸轮 periphery cam
平凸面夯击器 flatly cambered rammer
平凸透镜 plane-convex lens; plano-convex lens
平凸形 convex(o)-plane
平凸形的 < 一面平、一面凸的 > plano-convex
平凸型聚光灯 plano-convex spotlight
平凸缘连接 flat flange joint
平土 grading
平土铲 scraper blade
平土铲翻转油缸 blade tip cylinder
平土铲加长件 blade extension
平土铲托架 blade bracket
平土铲稳定器 blade stabilizer
平土铲自动控制 automatic blade control
平土铲自动控制装置 automatic blade controller
平土刀板 level(l)ing blade
平土工具 grading tool
平土工作 grading job; grading operation; grading work
平土机 blade equipment; blade grader; bull-clam shovel; bullgrader; grading machine; land level(l)er;

P

level(l)er; level(l)ing machine; scraper
平土机具 grading outfit
平土机拉铲 <俚语> clod buster
平土机手 grader man
平土机械 grader
平土宽度 graded width
平土器 earth level(l)er
平土升送机 elevating grader
平土升运机 elevating grader; grader elevator
平土提升机 grader elevator
平土作业 grading job; grading operation; grading work
平推端螺钉 flat point screw
平推断层 blatt(er)
平推法 flat stacking
平推防火门 horizontal sliding fire door
平推管井 <将管井水平地推入含水层或湖河底形成的井> push well
平推拉窗 horizontal slider; horizontal sliding window
平推(拉)门 horizontal sliding door
平推流 plug flow
平推流反应堆 plug-flow reactor
平推流反应器 plug-flow reactor
平挖法 level cutting
平挖面 front bank
平瓦 flat roofing tile; nibbed tile; pantile; plain tile
平瓦出檐 creasing
平瓦房顶 flat tile roof
平瓦片 flat tile
平瓦屋顶 flat tile roof; plain tile roof
平瓦屋面层 plain tile roof cladding
平瓦屋面覆盖 plain tile roof cladding
平弯曲和压缩的综合作用 combined flat bending and compression; combined planar bending and compression
平弯试验 <焊件 180°> flat-bend test
平网氧化器 flat gauze oxidizer
平围板 flush coaming
平纬 mean latitude
平位焊 flat position welding
平位焊缝 positioned weld
平位膨胀 flat-seated swell
平位置 mean place; mean position
平纹 plain weave; tabby
平纹板 screw die
平纹编组织 taffeta weave
平纹玻璃 brush line
平纹带 plain tape
平纹的 tabby
平纹顶棚 muslin ceiling
平纹纺织地毯 plain weave carpet
平纹(横)镶板 lying panel
平纹厚亚麻布 butcher linen
平纹开口踏盘 plain shedding tappet
平纹理 flat grain
平纹密度 plain quality
平纹木材 flat-grained timber
平纹人造丝织物 butcher rayon
平纹石 druid stone
平纹踏盘 plain tappet
平纹屋顶板 flat-grained shingle
平纹细布 sheeting
平纹细棉布 muslin
平纹镶板 lay panel
平纹织物 tabby
平稳 even; stationary
平稳爆破 smooth blasting
平稳遍历过程 stationary ergodic process
平稳泊松过程 stationary Poisson process
平稳操作 quiet run
平稳船 stable ship
平稳单自由度系统 stationary single degree of freedom system
平稳的 pacific; smooth; smooth and steady; surgeless; tranquil; stable
平稳的各态历经随机过程 stationary ergodic random process
平稳的回转 gentle turn
平稳的提升动作 cushioned shifting
平稳灯 amperite; ballast lamp
平稳点 stationary point
平稳电弧 tranquil arc
平稳独立增量 stationary independent increment
平稳度 smoothness
平稳二项随机过程 stationary binomial random process
平稳反应 smooth reaction
平稳反应计数器 flat-response counter
平稳方程 plateau equation
平稳飞行 smooth flight
平稳分布 stationary distribution
平稳分布过程 stationary distribution process
平稳分布函数 stable distribution function
平稳分布距离 equilibrium distance
平稳分配法 steady allocation method
平稳高斯过程 stationary Gaussian process
平稳各态历程 stationary ergodic process
平稳工作 smooth working
平稳过程 homogeneous process; stationary process
平稳过渡 smooth transition
平稳化假设 stationary assumption
平稳集 stationary ensemble
平稳加速 smooth acceleration
平稳价格 flation price
平稳接合 smooth engagement
平稳进场着陆 smooth approach
平稳开动 gentle start
平稳控制 trim control
平稳力 stable force; stationary force
平稳流 stationary flow; stationary stream
平稳流出 smooth outflow
平稳流动 even flow; smooth flow
平稳流化床 quiescent fluidized bed; smoothly fluidized bed
平稳流态化床 steady fluidized bed
平稳模型 stationary model
平稳目标 even-keel objective
平稳偏历噪声 stationary ergodic noise
平稳坡度 steady gradient
平稳期 stationary phase
平稳起步 gentle start
平稳起动 <又称平稳启动> smooth start(ing); soft start(ing)
平稳起动器 soft starter
平稳起落装置 delayed action lifting device
平稳器 ballast
平稳区 meadow
平稳曲线族 family of stationary curves
平稳燃烧 smooth combustion
平稳上升 steadily lift up
平稳上涨 steadily rise
平稳时间序列 stationary stochastic series; stationary time series
平稳时期 flation; plateau[复 plateaus/plateaux]
平稳速度 stable speed
平稳随机分布 stationary random distribution
平稳随机过程 stationary random process; stationary stochastic process
平稳随机过程谱分类 spectral classification of stationary random process
平稳随机过程谱距 spectral moments of stationary random process
平稳随机过程谱性质 spectral property of stationary random process
平稳随机函数 stationary random function
平稳随机输入 stationary random input
平稳随机序列 stationary random sequence
平稳随机振动 stationary stochastic vibration
平稳随机作用 stationary random action
平稳调整 smooth regulation
平稳退绕 smooth run-off
平稳下降 delayed drop; steadily drop down
平稳消息源 stationary message source
平稳效用值 stationary utility value
平稳信道 stationary channel
平稳行车 ease driving; smooth riding
平稳行车轨道 smooth-riding track
平稳行车路面 smooth-riding surface
平稳行驶 smooth-ride; smooth running
平稳性 smoothness; stationarity; steadiness
平稳性高斯过程 stationarity Gaussian process
平稳性计算【港】 trim/stability computation
平稳循环 steady-state cycle
平稳运动 smooth motion; steady motion
平稳运行 even running; silent running; smooth running
平稳运转 quiet run; smooth operation; smooth running; steady running
平稳噪声 stationary noise
平稳值 plateau value; stationary value
平稳制动 smooth braking
平稳转动 smooth running
平稳转换概率 stationary transition probability
平稳状态 plateau [复 plateaus/plateaux]; steady state; tranquil regime
平卧地层 flat-lying formation
平卧锅炉式气闸 boiler lock
平卧焊 flat weld(ing)
平卧气闸 horizontal lock
平卧位 horizontal position
平卧褶皱 lying fold; reclined fold; recumbent fold
平卧状矿床 manto
平屋顶 dead level roof(ing); deck roof; flat roof; plate-form roof; platform roof; terrace roof
平屋顶保暖 flat roof insulation
平屋顶保暖板 flat roof insulating slab
平屋顶保暖材料 flat roof insulating material
平屋顶保暖物 flat roof insulating compound
平屋顶保温 flat roof insulation
平屋顶抽气通风管道 flat roof extract ventilation duct
平屋顶窗 flat dormer(window)
平屋顶泛水 deck cant
平屋顶防水 flat roof waterproofing
平屋顶房屋 flat-roofed block; flat-roofed house
平屋顶风机 flat roof fan
平屋顶附属房屋 flat-roofed annexe
平屋顶构件 flat-roofed unit
平屋顶构造 flat construction
平屋顶桁架 flat truss
平屋顶建筑 flat-roofed building
平屋顶建筑物 flat-roofed building
平屋顶绝热组合物 flat roofing insulating compound
平屋顶绝缘 flat roof insulation
平屋顶绝缘板 flat roof insulating slab
平屋顶绝缘材料 flat roof insulating material
平屋顶绝缘物 flat roof insulating compound
平屋顶老虎窗 deck dormer
平屋顶梁垫木 furring piece
平屋顶面板 flat roof slab
平屋顶模板 flat roof form
平屋顶排水槽 flat roof gutter
平屋顶排水口 flat roof outlet
平屋顶排水设备 flat roof drainage device
平屋顶配房 flat-roofed annexe
平屋顶气窗 flat roof dormer; flat-roofed dormer-window
平屋顶墙角排水口 scupper drain
平屋顶晒台 glazed terrace
平屋顶上保护地沥青的瓦 solcheck
平屋顶上的小屋 pendice[pentice]
平屋顶体系 flat roof system
平屋顶天窗 flat roof hatch
平屋顶通风机 flat roof ventilator
平屋顶凸缘 deck curb
平屋顶涂沥青 sprinkle mopping
平屋顶无窗 deck dormer
平屋顶镶玻璃 flat roof glazing
平屋顶阻挡条 slag strip
平屋脊镜 flat roof mirror
平屋面 flat roof deck
平屋面顶层涂沥青 top mop
平屋面排水 flat roof drainage
平屋面涂沥青 sprinkle mopping
平屋面瓦 plain roofing tile
平午 mean noon
平吸式煤气发生炉 cross draft gas producer
平息 quench; quietus; subsidence
平洗机 open soaper
平铣 plain milling
平铣刀 plain cutter; plain milling cutter
平狭高地 bench land
平弦 horizontal chord
平弦桁架 flat-chord truss
平弦再分桁架 Baltimore truss; Howe truss
平线 horizontal line
平线花纹轮胎 plain tire[tyre]
平线脚 band; facet; fillet; flash mo(u)ld; flat mo(u)lding; flush bead; listel; weld <金属的>
平线脚装饰 flat mo(u)lding
平线圈 almacantar; flat coil; slab coil
平线条 flush bead
平镶 carvel built
平镶板 flush panel
平镶板门 flush paneled door
平镶接合 carvel joint
平镶式 flush system
平镶阵 flush-mounted array
平响应曲线 flatness of the response
平向挤压 horizontal extrusion
平向开合运动 horizontal opening closing movement
平向流斜板式沉淀池 inclined plank settling tank for horizontal-direction flow
平向压力 horizontal pressure
平巷 <隧道、坑道施工的> approach adit; drift way; gallery; heading; level gallery; mine gallery; mine tunnel; roadway
平巷边帮 roadway side
平巷的工作面 roadhead
平巷顶板背板 astel
平巷端 heading
平巷护顶板 astel

平巷交叉道口 roadway crossing;roadway junction
平巷掘进 drifting; heading advance; level cutting;roadway excavation
平巷掘进法 drifting method
平巷掘进工作面 drift face
平巷掘进机 drifting machine; earth boring machine; header; heading machine;ripper
平巷掘进机械 road-making plant
平巷掘进运输 tunnel(l)ing transport
平巷掘进作业 drifting operation
平巷靠壁纵向棚子 wall plate
平巷口 heading collar
平巷联合掘进机 mole
平巷配电箱 gate end panel
平巷输送机 gate conveyer[conveyor]
平巷维护 roadway maintenance
平巷压力拱 roadway pressure arch
平巷用装载机 gate end loader
平巷凿岩 drilling drift
平巷照明 roadway lighting
平巷支护 roadway support
平巷转载运输带 gate end conveyer [conveyor]
平巷转载运输机 gate end conveyer [conveyor]
平巷装车机 entry loader
平巷装载机 gangway ladder;gate end loader
平巷钻进(法)drift boring
平巷钻进机 tunnel borer
平像物镜 field flattener
平削断层 planed fault
平销钉 plain pin
平销锚栓 flat-stock anchor
平销锁 straight lock
平斜端砖 end skew on flat
平斜接缝 plain mitre joint
平斜轮 plane bevel wheel
平鞋形耙头 flat-shoe-shaped draghead
平信徒食堂 < 寺院中 > refectory for lay brethren
平行 collateral;parallel
平行安装 parallel application;parallel mount
平行凹槽间光滑表面 orlo
平行板 parallel plate
平行板波导 parallel plate waveguide
平行板电极 parallel plate electrode
平行板电极计数器 parallel plate counter
平行板电离室 parallel plate chamber; plane parallel plate ionization chamber
平行板电容器 parallel plate condenser;plate capacitor
平行板干涉仪 parallel plate interferometer
平行板隔油池 parallel plate oil interceptor
平行板画图器 clinograph
平行板加载 parallel plate loading
平行板拦截器 parallel plate interceptor
平行板模型 parallel plate model
平行板黏[粘]度计 parallel plate visco(si)meter
平行板式塑性仪 parallel plate plastometer
平行板式油(水)分离器 parallel plate oil interceptor
平行板塑性计 parallel plate plastometer
平行板条 parallel strip
平行板线 parallel plate line
平行板振荡器 parallel plate oscillator
平行棒条筛 parallel-rod screen
平行棒阵列 parallel rod array
平行棒振荡器 parallel-rod oscillator
平行本位 parallel standard
平行本位制 parallel standard system
平行辟分 parallel chorisis
平行壁座舱 parallel cabin
平行臂式独立悬架 parallel-arm suspension
平行边 parallel edge
平行边坡法 parallel banks method
平行边线 parallel border
平行变换 parallel transformation
平行变位 parallel displacement
平行变异 parallel variation
平行标线法 parallel index technique
平行标注 aligned dimensioning
平行并排的 parallel-sided
平行波 parallel waves
平行波痕 parallel ripple mark
平行玻璃板 parallel plate;plane parallel plate
平行玻璃板测微器 parallel plate micrometer;plane plate micrometer
平行不整合【地】para-unconformity; accordant unconformity; disconformity;parallel unconformity
平行布置的人造凸起障碍物 parallel bumps
平行布置进风巷道 parallel entry
平行布置系统 parallel system
平行操作 parallel operation;parallel running;simultaneous operation
平行槽 parallel slot
平行槽线夹 parallel-groove clamp
平行测定 parallel determination;replicate determination
平行测量 horizontal survey
平行测试 parallel test(ing)
平行层积 paralled laminate
平行层理 concordant bedding;evenly bedding;parallel bedding
平行层理错动 lateral movement
平行层理构造 evenly bedding structure
平行层面 concordant bedding
平行层压的 parallel-laminated
平行长度 length of parallelism
平行场 parallel field
平行成层的 parallel laminated
平行成型 parallel build
平行成型卷装 parallel build package
平行承发包 parallel contracting
平行尺 drafting machine;parallel motion protractor;parallel rule(r)
平行齿半圆锉 hook-tooth file
平行齿轮 parallel gears
平行齿轮传动 parallel-gear drive
平行冲洪积扇轴 parallel with axis of alluvial-pluvial fan
平行抽汲 parallel pumping
平行抽送 parallel pumping
平行处理 parallel processing
平行传递 simultaneous transmission
平行传输 parallel transmission
平行传输线 ribbon feeder
平行传送 simultaneous transmission
平行船中体 dead flat body; parallel body
平行垂直面 parallel vertical plane
平行存储器 parallel storage
平行带 parallel band;parallel strip
平行贷款 parallel loan
平行单色光 combined monochromatic light
平行单色图像 by-pass monochrome image
平行导杆装置 parallel guidance mechanism
平行导轨 closed slide;parallel guide
平行导坑 parallel adit; parallel connecting by-pass tunnel; parallel heading
平行导料装置 parallel stock guide
平行导流堤 parallel jetties; parallel training wall
平行的 cocurrent;paralleled
平行的半圆波纹装饰 reeded
平行的导坑方法 parallel drift method
平行的矩形框架 parallel rectangular frame
平行的指挥系统 parallel chain of command
平行登记法 parallel record method
平行堤 parallel dike[dyke]
平行电场 parallel electric(al)fields
平行电流 cocurrent
平行垫铁 parallel block; parallel clamp
平行吊索的加劲系统 restraining system for parallel suspension cable
平行叠层板 parallel laminate
平行叠置双芯电缆 twin cable
平行顶盖 parallel capping
平行定价法 parallel pricing
平行陡崖区 scarp land
平行度 depth of parallelism;parallelism
平行度公差 parallelism tolerance;tolerance of parallelism
平行度试验仪 parallelization tester
平行度调准 parallel alignment
平行渡线【铁】parallel crossover
平行段 parallel section
平行断层 parallel faults
平行断面 parallel section
平行断面法 method of parallel sections;parallel section method
平行对接 parallel docking
平行对准照相机 boresight camera
平行多面体 parallelohedron
平行多面体分类 parallelepiped classification
平行多叶阀 parallel multi-blade damper
平行二线隧道 duplicate tunnel
平行二线隧道导坑 duplicate pilot tunnel
平行反馈 parallel feedback
平行反应 parallel reaction
平行方锉 blunt square file
平行方向 parallel direction
平行方向法 parallel method orientation
平行仿射 parallel-affine
平行风路 parallel ventilation branch
平行缝干涉仪 parallel-slit interferometer
平行缝隙 parallel fissure;parallel slit
平行敷设 equal lay
平行服务系统 parallel service system
平行浮选回路 parallel flo(a)tation circuit
平行辐条状调制盘 parallel spoked reticle
平行盖顶 parallel coping
平行杆 parallel bars;parallel rod
平行杆调谐 parallel-rod tuning
平行杆筛 parallel-rod screen
平行钢绞线拉索 parallel strand stand stay cable
平行钢铰线 parallel strand
平行钢丝 parallel wire bundle
平行钢丝绳 parallel wire cable
平行钢丝束 bundled parallel wires; parallel wire strand; parallel wire unit
平行钢丝(束)张拉设备 < 预加应力用 > parallel wire unit
平行钢丝索 parallel strand; parallel wire cable
平行杠杆式独立悬架 parallel link suspension
平行格构大梁 parallel lattice girder
平行格构的穹隆 parallel lattice dome
平行格构的圆顶 parallel lattice cupola
平行格构穹顶 parallel lattice dome
平行隔板式消声器 parallel baffle muffler
平行隔距(离)parallel separation
平行隔水边界 parallel impervious boundary
平行工序 parallel process
平行工作 multiple operation
平行公理 parallel axiom; axiom of parallels
平行沟(渠)系(统)parallel ditch system
平行沟通 < 同阶层人员间的信函来往的 > horizontal communication
平行构造 parallel structure;plan-parallel structure
平行管 parallel pipe
平行管首道 parallel pipe
平行管线 parallel pipelines
平行管蒸发器 raceway coil
平行光 parallel light;collimated light
平行光变像器 collimated image converter
平行光度分析 parallel photometric(al)analysis
平行光观察 orthoscopic observation
平行光管 collimating device;collimator
平行光管法 parallel collimator method
平行光镜 collimating mirror
平行光密度 parallel light density
平行光束 collimated beam; parallel beam;parallel light beam;pencil of parallel rays
平行光束光学装置 beam collimation optics
平行光线 parallel rays
平行光学 collimating optics
平行光源 parallel lamp; source of parallel light
平行光栅 parallel grating;plane grating
平行规 parallel ga(u)ge; parallel rule(r)
平行滚柱 parallel roller
平行滚柱轴承 parallel roller bearing
平行海岸波 edge wave
平行海进 parallel transgression
平行海滩 parallel beach
平行航向 parallel course
平行河谷 river valley in parallel
平行河流 river in parallel
平行桁架 parallel truss
平行弧 parallel arc
平行虎钳 parallel vice
平行互生 parallel intergrowth
平行滑尺 parallel slide
平行滑动 translational slide
平行滑距 trace slip
平行滑翔 parallel gliding
平行滑移 parallel gliding;parallel slip
平行化 parallelization
平行划线尺 parallel marking ga(u)ge
平行划线器 tosecan
平行环 parallel ring
平行环式寄存器 parallel ring type register
平行簧片 parallel spring
平行汇率 parallel rate of exchange
平行活动 parallel activity
平行机用锉 parallel machine file
平行迹线 parallel trace of lines
平行级配法 < 试样制备的 > parallel grading method
平行极板电容器 plane parallel capacitor

P

平行极板计数管 parallel plate counter
平行极化 parallel polarization
平行计算机 parallel computer
平行记账 keep accounts in parallel; parallel bookkeeping
平行加工 parallel processing
平行夹头 parallel carrier; parallel clamp
平行架空索道 twin-cable ropeway; twin-cable tramway
平行间隙焊接 parallel gap welding
平行间隙回流焊接 parallel gap reflow soldering
平行剪 parallel shears
平行检测 parallelism detection
平行检验 parallel test(ing)
平行渐变群 parallel cline
平行键 parallel key
平行交互生长 parallel intergrowth
平行交易 parallel deal
平行角 parallel angle
平行铰刀 parallel reamer
平行校靶 parallel boresighting
平行校正 collimate; collimation
平行校正误差 collimating fault
平行阶梯状断层 echelon fault
平行接近 parallelism approach
平行接近法 constant bearing course; constant bearing navigation
平行接近距离 parallel exposure
平行结构 parallel construction
平行截割 parallel cut; face parallel cut <晶体的>
平行截口 parallel section
平行截面 parallel section
平行解理走向的节理面【地】 back joint
平行进刀 parallel feed
平行进风平巷 parallel heading
平行进风巷 parallel intake entry
平行进给 parallel feed
平行进口 parallel import
平行进料法 parallel feed method
平行进路 by-pass route; simultaneously possible route; compatible route; parallel route
平行径路 <货物> parallel route
平行镜 parallelоscope
平行矩形带调制盘 parallel-rectangular-bar reticle
平行矩形框架 parallel rectangular frame
平行矩形条调制盘叶片 parallel-rectangular reticle blade
平行卷绕 parallel wind(ing)
平行卷绕筒子 parallel wound bobbin
平行卷筒式牵引轮 parallel drum type capstan
平行龟裂掏槽 parallel-chap cut
平行坑道 parallel adit; parallel connecting by-pass tunnel
平行空炮孔掏槽爆破 parallel-cut blasting
平行空炮眼掏槽 parallel burn cut; Michigan cut
平行空炮眼掏槽法 Michigan cut method
平行孔模型 parallel pore model
平行孔掏槽爆破 parallel hole blasting
平行孔凿岩 parallel hole drilling
平行孔钻进 parallel hole drilling
平行控制 parallel control
平行控制器 parallel run controller
平行馈电线 parallel feeder
平行拉索 parallel stay cable
平行拉线 parallel stays
平行类 parallel classes
平行力 parallel forces
平行力管定律 law of parallel solenoids
平行力系的合成 composition of parallel forces
平行力系中心 centre of parallel forces
平行连杆 parallel rod
平行连杆机构 parallel linkage
平行连接 parallel linkage
平行连接杆 parallel connecting bar
平行连生 parallel growth
平行联合 parasyndesis
平行链晶 parallel group
平行梁 parallel beam; parallel girder
平行梁桥 parallel girder bridge
平行料堆 parallel stockpile
平行列车运行图 parallel train graph
平行溜放 parallel humping
平行流程 parallel circuit; parallel flow; parallel procedure
平行流(动) concurrent flow; parallel flow; split flow
平行流干燥器 parallel flow drier[dryer]
平行流喷管 parallel flow nozzle
平行流式热交换器 parallel flow heat exchanger
平行流水池 parallel flow basin
平行流涡流 parallel flow eddy
平行流蒸发器 parallel flow evapo(u)rator
平行六边形 cuboid; parallelepiped
平行六面体 parallelepipedon; parallelopiped
平行六面体分类 parallelepiped classification
平行六面体和极大似然分类程序 parallelepiped and maximum likelihood classification program(me)
平行六面体石块的砖石工程 masonry work of parallelepipedal cut stone; parallelepiped stone masonry
平行楼梯 parallel stair(case)
平行路 parallel road
平行路边停车 parallel curb parking
平行螺纹 parallel thread
平行螺旋 parallel spiral
平行螺旋齿轮 parallel helical gears
平行螺旋线 parallel spires
平行落差 parallel throw
平行铆钉 chain-riveting; parallel rivet
平行铆接 chain-riveting
平行面规 parallel-faced type of ga(u)ge
平行面夹钳 parallel cheek pliers
平行面间距 face-to-face distance
平行面切割 face parallel cut
平行面体 parallelehedra; parallel hedra
平行瞄准具 collimating sight
平行木板接缝 straight joint
平行木纹胶合板 parallel-laminated veneer
平行捻 parallel lay
平行捻钢丝绳 equal laid wire rope; lang-lay rope; parallel laid wire rope; rope of parallel wire
平行捻缆绳 parallel lay rope
平行啮合 harp mesh
平行扭接 pigtail splice
平行纽结 parallel knot
平行偶极测深曲线 curve of parallel dipole-dipole sounding
平行排放系统 parallel drainage(system)
平行排列 homogeneous alignment; parallel arrangement
平行排列结构 homogeneous texture
平行排水系统 parallel drainage
平行盘式可塑仪 parallel plastometer
平行刨床 parallel planer; parallel planing machine
平行炮孔 parallel hole
平行炮孔分段回采 parallel hole method of sublevel stoping
平行炮孔掏槽 parallel hole cut
平行炮眼 parallel hole
平行跑道 parallel runway
平行跑道间距 parallel-runway separation; separation of parallel runways; spacing between parallel runways
平行跑道进场着陆 parallel-runway approach
平行配合 parasyndesis
平行喷嘴 parallel nozzle
平行劈理 parallel cleavages
平行片状结构 parallel laminated texture
平行偏光 parallel polarized
平行漂移带 parallel drifting bands
平行平板电极 parallel planar electrode; parallel plate electrode
平行平板腔激光器 parallel plate laser
平行平错 parallel heave
平行平面 parallel planes
平行平面层 plane parallel layer
平行平面的 plane parallel
平行坡道 parallel ramp
平行剖面 parallel section
平行剖面法 parallel profile method
平行铺置纤维网 parallel laid web
平行起爆线 parallel series
平行钳口 parallel jaw
平行堑壕 parallel trench
平行切割 parallel cut
平行切线 parallel tangents
平行切线法 parallel tangent method
平行切线算法 partan algorithm
平行曲柄 parallel crank
平行曲柄机构 parallel crank mechanism
平行曲面 parallel curvature; parallel surfaces
平行曲线肋拱 parallel curve rib arch
平行曲线肋拱桥 parallel curved rib arch bridge
平行渠道 parallel canal
平行圈 parallel circle
平行圈尺度 scale in parallel
平行圈弧度测量 grade parallel measurement; parallel arc measurement
平行圈曲率 curvature of parallel
平行圈曲率半径 transverse radius of curvature
平行圈曲率校正 correction for parallel curvature
平行刃口剪切机 squaring shears
平行融资 parallel financing
平行三灯丝法 parallel triple filament method
平行扫描 parallel scan(ning)
平行沙垅 lateral dune
平行砂轮 parallel grinding wheel
平行设计 parallel design
平行射线 infinite rays
平行射线束 parallel beam
平行射向 normal sheaf; parallel pointing
平行射影 parallel projection
平行射影的比例 ratio of a parallel projection
平行摄影 parallel photographing; parallel photography
平行升降台 lifting platform; lifting table
平行声束 collimated beam
平行失活 parallel deactivation
平行驶靠 parallel approach
平行市场 parallel market
平行式 <斜拉桥拉索的布置形式> harp arrangement
平行式布置 parallel system
平行式导料装置 parallel stock guide
平行式堆垛机 parallel stacker
平行式多叶阀 parallel multi-blade damper
平行式剪切 frame cutting
平行式剪切机 frame shears
平行式码头 quay(in parallel)
平行式停车 park parallel; parallel parking
平行式自动扶梯 parallel escalators
平行试验 duplicate test; parallel test(ing); reproducibility of tests; side-by-side test
平行试验的符合性 congruity of parallel test
平行收缩 parasystole
平行输出光束 collimated output beam
平行束 collimated beam; parallel beam; parallel packet
平行束状 parallel sheaf pattern
平行双导流堤 parallel jetties
平行双电缆 twin-lead cable
平行双耳鞍子 conductor suspension clamp with hook
平行双管线 parallel pipe
平行双晶 parallel twin
平行双馈线 twin-lead type feeder
平行双螺杆挤出机 parallel twin-screw extruder
平行双面 pinacoid
平行双纬引送 parallel double pick insertion
平行双线 parallel wire
平行双线有线电路 parallel wire line
平行双芯线 twin conductor
平行水系 parallel drainage
平行水系模式 parallel drainage pattern; parallel drainage mode
平行顺次移动方式 parallel sequential moving method
平行四边式松土机 parallelogram ripper
平行四边形 crossed parallelogram; rhomboid
平行四边形的板 parallelogram plate
平行四边形调车场 parallelogram yard
平行四边形定律 law of parallelogram; parallel law
平行四边形法则 parallelogram law
平行四边形恒等式 parallelogram identity
平行四边形机构 parallelogram mechanism
平行四边形畸变 parallel distortion; parallelogram distortion
平行四边形检测仪 parallelometer
平行四边形结构 parallelogram structure
平行四边形截面 parallelogram section
平行四边形框架 parallelogram lever
平行四边形连杆结构 parallelogram type linkage
平行四边形裂土器 parallelogram ripper
平行四边形(楼)板 parallelogram slab
平行四边形偏心机构 parallelogram eccentric mechanism
平行四边形起落机构 parallel lift
平行四边形试验台 parallelogram stand
平行四边形松土机 parallelogram type ripper; parallelogram type scarifier
平行四边形松土器 parallelogram ripper; parallelogram type ripper
平行四边形缩放仪 parallel pantograph
平行四边形屋面材料 parallelogram roofing material; rexangle; rextite
平行四边形型 parallelogram type
平行四边形悬挂装置 parallel lift;

parallel lift linkage; parallel motion linkage; parallelogram linkage
平行四边形砖 rhomb brick
平行四边形转向杆系 parallelogram steering linkage
平行四边形组合臂架 double link jib in parallelogram
平行四杆回转链系 parallel crank four bar linkage
平行四杆悬挂装置 parallel link lift
平行四连杆悬挂装置 parallel link hitch
平行四线式馈线 parallel four wire feeder
平行搜索 parallel search
平行随纹机构 parallel tracking
平行碎料板 platen-pressed particle board
平行隧道 tunnel duplication
平行台 parallel block
平行台用虎钳 parallel bench vice
平行滩列 parallel road
平行掏槽 burn cut
平行梯级 flyer
平行梯线 parallel ladder track
平行天沟 parallel gutter
平行天球 parallel sphere
平行天线 parallel antenna
平行条带状【地】onyx
平行条款 parallel clause
平行条状线 parallel-strip line
平行调节器 parallel regulator
平行停车 longitudinal parking; parallel parking
平行停靠 longitudinal parking; parallel parking
平行通路 paralleled path
平行同形 parallel homeomorphy
平行筒拱的互贯 interpenetration of two parallel barrel vaults
平行筒子 parallel cheese
平行投影 orthographic (al) projection; parallel projection
平行透视 parallel perspective
平行透水边界 parallel pervious boundary
平行透水隔水边界 parallel pervious and impervious boundary
平行推进 parallel advance
平行推土法 <两台推土机> blade to blade dozing
平行推压法 <挖掘机> parallel crowd
平行托架 parallel carrier
平行挖掘 parallel cut
平行网络定律 law of the parallel solenoids
平行位移 parallel displacement
平行位移测条件屈服强度方法 offset method
平行纹理 parallel grain
平行纹理构造 parallel lamellar structure
平行纹理抗拉强度 <木材的> tensile strength parallel to grain
平行纹理压力 compression parallel to grain
平行吸光 parallel extinction
平行吸声板 parallel absorbent baffle
平行铣刀 parallel mill
平行系 collateral series
平行系统 parallel system
平行细槽形面层 stretching face
平行细锉 blunt mill file
平行下沉 parallel sinkage
平行弦 parallel chords
平行弦法 parallel chord method
平行弦杆桁架 parallel boom truss
平行弦桁架 parallel-chord truss; Pratt truss; truss (ed) beam; trussed girder; truss with parallel chords
平行弦式布置 harp arrangement
平行弦式斜拉桥 harp type cable-stayed bridge
平行弦式斜缆桥 harp type cable-stayed bridge
平行弦再分式桁架 Baltimore truss
平行线 parallels; by line; collateral lines; paratactic (al) lines
平行线变形 distortion of parallel lines
平行线测定 parallel line assay
平行线尺【测】marquois scale; parallel ruler
平行线传输线 twin line
平行线电缆 parallel wire cable
平行线法 parallel attack; parallel method
平行线分辨力 parallel line resolution
平行线分辨率 parallel line resolution
平行线规 isometrography
平行线路 parallel circuit
平行线系统 parallel wire system
平行线谐振器 parallel wire resonator
平行线性组合 linear-parallel to dominant line array
平行线凿石工艺 parallel line tooling
平行线振荡器 parallel line oscillator
平行线支票 crossed check
平行线族搜索 radial parallel search
平行向量 parallel vector
平行消光 parallel-axial extinction; parallel-axis extinction; straight extinction
平行销 parallel pin
平行小路 parallel alley
平行信息处理机 parallel information processor
平行型 parallel pattern
平行型加速车道 parallel lane type acceleration lane; parallel-type acceleration lane
平行性 parallelism
平行性拉线 parallel stays
平行性偏差 parallel misalignment
平行性破坏 decollimation
平行循迹 parallel tracking
平行训练法 parallel training
平行压顶 parallel coping
平行演化 parallel evolution
平行样 duplicate sample
平行摇摆振动 rocking vibration
平行咬夹 parallel-jawed clamp
平行仪 parallelometer
平行移动 parallel shift; parallel translation
平行移动方式 parallel mobile mode
平行移动式缆道 parallel-travel (l) ing cableway
平行移动式缆机 parallelly travel (l)-ing cable crane
平行移动式索道 parallel-travel (l) ing cableway
平行移动装置 parallel motion device
平行移位寄存器 parallel shift register
平行移轴定理 parallel-axis theorem
平行移轴公式 formula for translation of axis
平行易货 parallel barter
平行翼缘梁 parallel flange beam; parallel flanged girder
平行翼状墙 parallel wing wall
平行翼状物 parallel wing
平行诱导 parallel induction
平行于轨面 parallel to cross track level
平行于岸的结构 shore-parallel structure
平行于岸防波堤 shore-parallel breakwater
平行于干管的支管 parallel branch
平行于海岸的 shore-parallel
平行于海岸的离岸沙槛 shore-parallel offshore sill
平行于基准线的力 chord force
平行于木纹的板面 side grain
平行于铁路干线的支线 by line
平行于纹理 <木材的> parallel to grain
平行于纹理的压缩 compression parallel to grain
平行于圆周方向的 circumferential
平行于轴线 parallel to the axis
平行雨水槽 box gutter
平行雨水檐沟 parallel gutter
平行圆弧 arc of parallel; parallel arc
平行圆弧长表 table of parallel arcs
平行缘桁架 parallel flange truss
平行缘梁 parallel flanged girder
平行运动 parallel motion; parallel movement; translation
平行运动分度规 drafting machine; parallel motion protractor
平行运动杆 parallel motion bar
平行运动机构 parallel motion mechanism
平行运送旋臂起重机 level-luffing crane
平行运算器 parallel arithmetic unit
平行运行图 parallel train working diagram
平行运行图的区间通过能力 carrying capacity in the section computed on the basis of parallel train working diagram
平行运转 parallel running
平行晕线 single hatching
平行匝道 parallel ramp
平行闸阀 parallel slide valve
平行张量场 parallel tensor field
平行折页 parallel folding
平行褶皱 parallel fold
平行振荡 parallel oscillation
平行整合【地】plano-conformity
平行直角定规 precision flat try square
平行直梁式车架 parallel straight frame
平行直线 parallel (straight) lines
平行直线法 method of parallel straight curve
平行指向线 parallel index line
平行中体 parallel middle body
平行舯体 parallel body; raft body
平行轴 countershaft; parallel axes; parallel shaft
平行轴齿轮 (系) parallel-axis gears; parallel-axes gears
平行轴传动 countershaft transmission
平行轴定理 parallel-axis theorem
平行轴减速器 parallel shaft (speed) reducer
平行轴结构 parallel shaft structure
平行轴面厚度参数 thickness parameter parallel to axial plane
平行轴双绞筒提升机 tandem hoist
平行轴剃齿法 parallel-axes shaving
平行主干线的复线 alternative side street
平行转鼓 parallel drum
平行转向 parallel steer
平行状流域 parallel basin
平行状水系 parallel drainage pattern; parallel drainage system
平行状态 parallelism
平行琢石面 tooled finish of stone
平行子空间 parallel subspaces
平行自旋 parallel spin
平行鬃岗式河漫滩 parallel hogback flood plain
平行走向河 adjusted river
平行钻孔 parallel hole; cylinder cut <中孔不装药,沿孔全长平均装药>
平行钻孔爆裂 burn cut
平行作业 parallel operation; parallel working; simultaneous work (ing); working in parallel
平行作业布置 simultaneous working arrangement
平行作用 parallel action
平行坐标 parallel coordinates
平形 level shape
平形垫圈 equalizing washer
平形工作物 flat stock
平形磨瓦 flat abrasive tile
平形砂轮 straight grinding wheel
平形四边形缩放仪 parallelogram pantograph
平型 flat pattern
平型传动带 flat-toothed belt; flat transmission belt
平型带式输送机 flat-belt conveyer [conveyor]
平型回复机 flat purl machine
平型炉 low incinerator
平型木纹 flat grain
平型十字头导框 plain crosshead guide
平型橡胶制品 flat rubber product
平型运输带 flat conveyor belt
平旋 swivel
平旋的壳 planispiral
平旋阀 swing (ing) valve
平旋壳 planispiral shell
平旋跨 swing space; swing spalling; swing span
平旋轮推进器 Voith-Schneider propeller
平旋门 balance gate
平旋盘 facing head
平旋桥 pivot bridge; swing bridge; swivel bridge; turnable bridge; turn (ing) bridge
平旋桥防护信号 (机) swing bridge signal; swivel bridge signal
平旋桥护座 draw rest
平旋桥中墩护石 pivot pier protection
平旋桥中墩护桩 pivot pier protection
平旋桥中心支墩 pivot pier
平旋推进船 Voith-Schneider ship
平旋推进器 cycloidal propeller; rotating blade propeller
平旋推进器船 cycloidal propeller ship
平旋止回阀 swing check valve
平碹 flat arch; jack arch
平靴形耙头 California drag-head
平雪机 snow grader
平雪机区 <一种平整雪地的机器> snowplane
平压板 flat clamp; flat clamping plate; flat-press board
平压管 equalizing pipe; standpipe
平压涵洞 equalizing culvert
平压机推料车 spreader lorry; spreader truck
平压颗粒板 flat-press board; platen-pressed particle board
平压硫化机 press vulcanizer
平压密封 flat press seal
平压密封环 flat sealing ring
平压刨花板 flat-platen-pressed particle board; flat-press board; platen-pressed particle board
平压片 plain presser
平压切断冲模 dinking die
平压切断机 beat cutter; dinking machine
平压切形模 dinking die
平压竖管 surge pipe
平压碎料板 flat-press board; platen-pressed particle board
平压铁 flat clamp
平压印 blind impression
平压印刷机 platen machine; platen press
平压装置 surge suppressor
平岩 flat rock

P

平檐沟 level gutter
平檐雨水沟 flat(rainwater) gutter
平焰 flat flame
平焰烧嘴 flat-flame burner
平焰式坩埚窑 horizontal flame pot furnace
平焰式燃烧器 flat-flame burner
平腰线 plain flat
平咬合缝屋面 flat seam roofing
平叶风扇 flat-blade fan
平叶片 paddle blade
平叶片送风机 plate blower
平叶片通风机 plate fan
平叶涡轮 flat-blade turbine
平液管 level(l)ing tube
平液球管 level(l)ing bulb
平液容器 level(ling) vessel
平液水准瓶 level(l)ing bottle; level(ling) flask
平移 parallel translation; translation(al) motion; translatory motion
平移波 translation(al) wave; translatory wave; wave of translation; waves of distortion
平移不变度量 translational invariant metric
平移不变函数代数 translation-invariant function algebras
平移不变式 translational invariant
平移参数 translational parameter
平移传感器 rectilinear transducer
平移窗 sliding window
平移的 translational; translatory
平移地震仪 translational seismograph
平移点阵 translational lattice
平移定理 shifting theorem
平移断层【地】 displacement fault; heave fault; horizontal fault; lateral fault; off-hap; parallel displacement fault; shift fault; strike slip fault; transcurrent fault; translatory fault; wrench fault; offlap
平移断层和伴生构造 wrench-fault and associated structures
平移对称 translational symmetry
平移对称元 translational symmetry element
平移反应 translational response
平移分布【数】 translated distribution
平移分布距离 equilibrium distance
平移分量 translational component
平移负指数分布【数】 translated negative exponential distribution
平移滑动 translational slide
平移滑动泥门【疏】 sliding hopper door
平移加速度 acceleration of translation
平移角尺 drafting machine; parallel motion protractor
平移晶格 translational lattice
平移靠码头 broadside on berthing
平移壳 translational shell
平移控制 translational control
平移缆索起重机 parallelly travel(l)ing cable crane
平移流 translational flow; translatory flow
平移幂函数 translated power function
平移面 surface of translation; translational plane
平移面壳 translational surface shell
平移能 translational energy
平移频率 translational frequency
平移曲面 translational surface
平移群 translational group
平移韧性剪切带 ductile wrench-shear zone
平移时滞 translational time lag
平移矢量 translational vector
平移驶靠【船】 broadside berthing; parallel approach
平移式荷载 moving load
平移式滑坡 translational landslide
平移式泥门 sliding bottom door
平移双曲壳体 doubly curved translational shell
平移双曲线壳 hyperbolic shell
平移速度 point-to-point speed; translational velocity; translatory velocity
平移算法 translational algorithm
平移算子 translational operator
平移损失 translational ioss
平移弹性刚度 translational elastic stiffness
平移探空仪 transosonde
平移凸轮 translational cam
平移弯筋 offset bend
平移位能 translational energy
平移系数 translational coefficient
平移系统 translational system
平移型道岔 sliding switch
平移要素 translational element
平移因子 shift factor
平移原理 translational principle
平移圆形弯道 smooth circular bend
平移运动 motion of translation; strikeship movement; translational motion; translational movement; translatory motion
平移运行 translational operation
平移振型 translational mode
平移正弦函数 translation sine function
平移轴 translational axis
平移阻力 translatory resistance
平移坐标 translation(al) coordinate
平抑价格 controlling the price; keep down price; stabilize price
平易驾驶 ease driving
平庸(低劣)的成绩 mediocre performance
平油漆 flat gloss oil paint
平鱼尾板结合 plain fishplate joint
平鱼(鱼)尾板 flat fishplate
平原 blair; champaign; dol; flat topography
平原不平地区 level cross country
平原成因类型 genetic(al) type of plain
平原城市 town in a plain
平原地 <平均坡度每英里 1 ~ 5 英尺, 1 英里 = 1609.34 米> flat land
平原地槽【地】 autogeosyncline; intracratonic basin
平原地带 flat-bottomed land
平原地带试验 flat-country test
平原地貌 plain landform
平原地貌单元 geomorphic(al) unit of plain
平原地貌调查 survey of plain morphology
平原地区 easy terrain; plain area; plain country; plain region; plain zone; flat country; region of no relief
平原地区铁路 flat-land railway
平原地区选线 location in plain region; plain location
平原段 plan section; plan tract
平原高程分类 altitude classification of plain
平原海岸 plain coast
平原航道 plain channel; plain waterway
平原河段 alluvial tract; flat-land section; plain tract
平原河流 lowland river; plain river; plain stream; river of flat lands
平原洪流 plain flood
平原湖 lowland lake
平原居民 plainsman
平原绿化 plain greening
平原气候 plain climate
平原区 flat region; plain area; plain region; plain terrain
平原区地下水动态 groundwater regime in plain area
平原铁路 flat line
平原线 plain line
平原游荡性河流 plain meandering river; plain meandering stream
平圆贝超群【地】 Discinacea
平圆方角接(合) bead, butt and square joint
平圆附柱 semi-column
平圆钢 <有别于带纹钢> smooth bar
平圆拱 blunt arch; diminished arch; drop arch; Georgian arch; hance arch; imperfect arch; segmental arch; skeen arch <小于半圆的>
平圆角锉 fork file
平圆接合 bead(and) butt joint
平圆两用钳 double pick-up tongs
平圆墁刀 bead tool
平圆面 facet(te)
平圆盘 flat disk[disc]
平圆盘的 discal
平圆三心拱 cambered arch
平圆式钢链 flat and round steel-chain
平圆线脚 flush bead; flush bead mo(u)lding
平圆形的 discoid
平圆形物 discoid
平圆柱头螺栓 deck bolt
平缘 fillet; listel
平缘材 flush coaming
平缘石 flush curb[kerb]
平月亮 mean moon
平匀衰落 flat fading
平载 even keel
平载量 <铲斗或翻斗车等的> struck capacity
平錾 chipping chisel; cold chisel
平凿 broad chisel; chipping chisel; drove; drove chisel; flat chisel
平增益调整 flat gain regulation
平渣器 ballast level(l)ing device
平扎绳 flat seizing
平轧道次 flat pass
平轧的 flat rolled
平轧法 beam roughing method
平轧钢板 boiler plate
平轧立轧孔型 slab and edging pass
平展 S 面 S-plane of flattening
平展法 development method
平展纹理 flat grain
平折缝 flat folded seam
平折性 lay-flat
平折页 butt hinge
平砧 flat anvil; flat die; flat platen
平振法 plane polarized method
平整 evenness; flattening; level(l)ing
平整岸线 bank level(l)ing; bank regulation; level(l)ing and smoothing the bank line
平整板 screed
平整板材 temper rolling
平整爆破 <按预定尺寸的> smooth blasting
平整标桩 grade stake
平整表面 flat surface
平整薄板 planished sheet
平整层 level(l)ing blanket; level(l)ing course; level-up course
平整层状的 even bedded
平整场地 grading level(l)ing; ground-level(l)ing; level(l)ing of ground; site level(l)ing
平整成层的 even bedded
平整带 temper rolling
平整道次 pinch pass; planishing pass; skin pass; temper pass
平整道砟机 ballast bed finisher; ballast strip finisher
平整的 smooth
平整的表面 even surface
平整的地面 level area
平整的路面 even running surface
平整的硬地面 hard level ground
平整地面 planned surface
平整定形 flat set
平整度 accuracy of level(s); degree of planeness; planeness; smoothness
平整度测定仪 profilograph; profilometer; ride meter; rough(o)meter; rideograph <路面>
平整度偏差 deviation in level
平整度试验设备 evenness test device
平整度试验装备 evenness test device
平整法 smoothing technique
平整方石 smooth ashlar
平整工作 grading job
平整刮板机 spreader ditcher
平整光滑的 smooth
平整辊机 equalizing rolling mill
平整河床 flat bed; plane bed
平整(后)地坪标高 grade level
平整机 finisher; level(l)er; level(l)ing machine; pinch pass mill; planishing mill; planning machine
平整机推料车 spreader lorry; spreader truck
平整机座 dressing stand
平整接 plain joint
平整井壁 smooth lining
平整冷轧 skin pass rolling
平整路机 subgrading machine
平整路基 subgrading
平整路面 smooth(-riding) surface
平整木面 trueing-up
平整器 smoother
平整式 flush type
平整水泥垫 level concrete pad
平整台 level(l)ing bench
平整土地 grading; land grading; land level(l)ing landing; land level(l)ing operation; level soil; preparation of land
平整土地测量 survey for land grading; survey for land preparation
平整土方 grading; grading work
平整推土机 trimming dozer
平整性 smoothness
平整腰线 plain flat strip mo(u)lding
平整用铲斗 grading bucket
平整用推土铲 trimming dozer
平整指数 flattening index
平正锤 dead flat hammer
平正面 flat facade; out-of-wind
平正午 mean noon
平支板 breast board
平支洞口 adit collar
平织网 plain net
平直表面 flat surface
平直滨线 straight shoreline
平直程度 flatness; straightness
平直翅片 plain fin; straight fin
平直传送 flat transmission
平直粗刨刀 straight rough shaping tool
平直的 translatory
平直度 evenness; straightness; planarity <地质结构面的>
平直度测量仪 flatness measuring instrument
平直度公差 straightness tolerance
平直放大器 flat amplifier
平直幅频起伏噪声 flat-channel noise

平直幅频特性放大器 flat-channel amplifier
平直钢锭模壁 flat wall
平直海岸 regular coast;smooth coast
平直海岸线 straight coastline;straight shoreline
平直海墙 regular coast;smooth coast
平直河 straight river
平直机 straightening press
平直接合对结 flush joint
平直结构面【地】straight discontinuity;flat structural plane
平直进气口 flush air intake
平直流 smooth flow
平直模板 lined formwork
平直模壁 flat wall
平直末端 plain tail
平直频率成分 flat frequency content
平直墙 plane wall
平直区 flat region
平直曲线 flat curve
平直时空 flat space-time
平直收尾 <墙的> stop(ped) end
平直束 flat beam
平直水流 forward flow;smooth flow
平直套板 straight jacket
平直调谐 flat-tuning
平直特性放大器 flat-channel amplifier
平直特性曲线 flat characteristic curve
平直通路放大器 flat-channel amplifier
平直弯曲 <波导管> flatwise bend
平直尾喷管 plain tail pipe
平直纹理 flat grain
平直线路 level tangent track;tangent level track
平直压强响应传声器 flat-pressure-response microphone
平直崖壁 straightening cliff
平制动鼓 plain brake drum
平制动鼓轴 plain brake drum shaft
平置 bed down
平置阀 horizontal valve
平置继电器 horizontal relay
平置石 bedding stone
平置式发动机 flat engine
平周壁 periclinal wall
平周的 periclinal
平周分裂 periclinal division
平周期 average period;mean period
平轴颈 plain journal
平轴盘式碎矿机 horizontal shaft disc crusher
平轴式气孔 paracytic type
平轴瓦 plain insert
平皱 roller flattening;roller-smoothing
平皱剪卷联合机 combined spinning trimming and curling machine
平柱式凿岩机 drifter;drifter drill
平砖 plain edge tile;plain tile;scone brick
平砖拱 flat brick arch;soldier arch
平砖拱的辐射接合 sommering
平砖路面 flat brick paving
平转 flat turn
平转跨 swing space;swing span
平转桥 pivot bridge;swing bridge;swivel bridge
平转塔车床 flat-turret lathe
平转塔六角车床 flat-turret lathe
平转弯 plate turn
平转阻力 resistance to bascule action
平装 level loading
平装本 paper back;paper-bound edition;paper-bound volume;paper copy
平装玻璃 flush glazing
平装插座 flush socket
平装抽出式继电器 flush-mounted drawout relay
平装斗容量 struck;struck capacity
平装开关 flush switch;panel switch
平装开关面板 flush plate
平装门框 butted frame
平装书 paper back
平装书陈列架 paper-backed book rack
平装型 flush type
平装型仪表 flush type instrument
平锥尖 flat dog point
平锥头铆钉窝模 pan head snap
平锥形的 planoconic
平坠降落 <飞机> pancake landing
平准表 balance sheet;statement of assets and liabilities
平准基金 buffer fund;equalization fund;stabilization fund
平准螺丝 level(l)ing screw
平准溶剂 level(l)ing solvent
平准资金 buffer fund
平子午线 mean meridian
平子夜 mean midnight
平纵断面图 plane-profile map
平纵面图 plan-profile map
平足 foot pad;flatfoot
平钻 flat bit;flat bush;flat drill;flat jewel
平嘴镊 flat pliers
平座 flat seat
平座阀(门) flat-seated valve;flat (slide) valve
平座圈 flat seat washer

评

评比会场 show-ring
评比框架图 assessment framework
评比试验 competitive test
评标 bid evaluation;evaluation of bids;evaluation of tender;tender evaluation
评标报告 reporting on tenders
评标(结果)通知书 notification of award
评定 assessment;estimation;evaluate;evaluation;pass judg(e)ment on;qualification
评定报告 appraisal report
评定变色用灰色样卡 grey scale for assessing change in colo(u)r
评定标准 criterion for evaluation
评定部分 rating unit
评定(贷款)条件 in assessing terms
评定的 rated
评定的优先次序 priority rating
评定等级 rating
评定方法 appraisal method;method for assessment
评定功绩 merit rating
评定合格 assessment of conformity
评定价值 appraised value
评定金额 amount appraised
评定框架表 assessment framework
评定流程 rate the process
评定赔偿要求 adjust claim
评定平均运费率 average freight rate assessment
评定燃料 evaluate the fuels
评定人 appraiser
评定生产 level(l)ing production
评定试验 qualification test;rating test
评定税额 assess a tax
评定铁的不足 evaluation of iron deficiency
评定误差 evaluated error
评定因素 level(l)ing factor
评定沾色用灰色样卡 grey scale for assessing staining of colo(u)r
评定指标 deliberated index
评定制度 system of rating
评定质量试验 evolution test
评分法 method of point rating;point system;score system
评分函数(法) scoring function
评工法 job evaluation
评工记分 evaluation of work and allotment of points
评估 assessment;economic appraisal;estimate;evaluation
评估报告 narrative appraisal (report);appraisal report
评估报告书 appraisal report
评估标准 evaluation criterion[复 criteria]
评估表 evaluation sheet
评估地产位置归档法 geographic(al) sequence
评估方法 appraisal method
评估方式 evaluation system
评估复审员 review appraiser
评估公司 appraisal company
评估机构 appraisal institute
评估计算 calculation of assessment
评估价 extended price
评估价值 warranted value
评估阶段 projection period
评估人 appraiser;estimator
评估试验 test for grading
评估损失 assessment of loss
评估图 evaluation chart
评估小组 assessment panel
评估准则 evaluation criterion
评估资本 appraisal capital
评估资产 appraise assets;appraising property
评核 scrutiny
评核管制与反馈 controlling and feedback with evaluation
评核加速资料(技)术 rapid information technique for evaluation
评核研究 evaluation study
评级 grade;rank score;rating;rating number
评级方法 ranking method
评级函数 ranking function
评级试验 rating test;test for grading
评价 appraise(ment);assessment;estimating;estimation;evaluation of bids;expertise;perusal;rating;reckon;vaiorisation;valuation;value
评价报告 evaluation report
评价标准 assessed standard;assessment yardstick;evaluation criterion
评价步骤 evaluation procedure
评价程序 assessment procedure;assessment process;evaluation process;evaluation program(me);valuation process
评价出售的协议 agreement to sell at valuation
评价单元 <一段路况均匀的路段> rating segment
评价的视野 scope of evaluate
评价等级 assessment category;evaluation grade;opinion rating;order of evaluation
评价法 rating method
评价法则 evaluation rule
评价范围 range of values
评价方法 assessment method;evaluating method;evaluation method;methods of evaluation
评价费用 appraisal cost
评价光化学法 evaluate photochemical method
评价过程 assessment process
评价过的核数据汇编 evaluated nuclear data file
评价过的数据 evaluated data
评价过低 undervaluation;undervalue
评价过高 overrate;over-value
评价函数 evaluation function;objective function
评价基础 assessment basis
评价基础数据 basic data of evaluation
评价计算方法 evaluation calculating method
评价技术 evaluating technology
评价阶段 stage of evaluation
评价结论 conclusion of evaluation
评价井 test well
评价决策报告书 evaluation report
评价勘查 appraisal survey
评价矿床 evaluate a deposit
评价买卖 sale at a valuation
评价模件 evaluation module
评价模件接口 evaluation module interface
评价模型 evaluation model
评价期 evaluation period
评价强度 evaluation strength
评价区 region of interest
评价人 appraiser
评价日期 date of evaluation
评价矢量 pricing vector
评价试验 evaluation test;ranking test
评价书 appraisal certificate
评价水质 evaluating water quality
评价图 evaluation map
评价维修规划 evaluated maintenance programming
评价系统 evaluation system
评价项目 item of evaluation
评价效率 assess effectiveness
评价型质量监督 quality supervision for approval
评价性详探阶段 detail prospecting stage of evaluation
评价研究 evaluation study
评价依据 assessment basis
评价依据资料 dependent information of evaluation
评价因素 element of assessment;factor of evaluation
评价因子 assessed cost element
评价账户 qualifying reserve account;valuation account
评价者 valuator;valuer
评价者报酬 valuer's fee
评价值 evaluation of estimate
评价指标 criterion of evaluation;evaluating index;evaluating indicator;evaluation index
评价指标体系 assessment index system
评价指数 evaluation number
评价终点 assessment endpoint
评价准备金 valuation reserves
评价准则 criterion
评价总结 evaluation and summarization
评价钻探 appraisal drilling
评奖 bonuses issued on the basis of a general voicing;decide on awards through discussion
评奖人 juror
评可证税 excise
评论 comment;criticism;criticize;review
评论家 critic
评论研究 review research
评论员 commentator;observer
评判人 referee;umpire
评判厅 <西班牙阿尔罕伯拉宫> Hall of Judg(e)ment
评色 colo(u)r appraisal
评审 appraise and examine

评(审)标价 evaluated bid price
评审单位 appraising unit
评审级别 appraising grade
评审价 evaluated tender price
评审时间 appraising date
评审小组 adjudicating panel; evaluation group
评薪 discuss and determine wage-grade
评序 ranking
评序程序 rank program(me)
评议 counsel; jointly assess
评议案例 deliberate a case
评议等级 class of assessment
评议会 appraisal meeting; consultation; convocation
评议商定 jointly assess
评阅者 reader
评值程序 valuation process
评质图 <根据图上线段范围来评定材料性质是否符合要求> evaluation chart

P

凭

保证开发的信用证 guaranteed letter of credit

凭标准买卖 sale by standard
凭标准销售 sale by standard
凭触觉 by touch
凭单 voucher; indenture
凭单的发给 certification
凭单登记簿 voucher register
凭单付款 cash against documents; payment against documents; presented for payment
凭单付款信用证 payment against documentary letter of credit
凭单记录 voucher record
凭单夹 voucher jacket
凭单据 against documents
凭单据付款 cash against documents
凭单据付款信用证 payment against documents credit
凭单扣款 voucher deduction
凭单日记账 voucher journal
凭单审核 voucher audit
凭单索引 voucher index
凭单托收付款 payment against documents through collection
凭单账户 voucher account
凭单证付现金 cash against documents
凭单支票 voucher check; voucher cheque
凭单支票制 voucher check system
凭单制度 voucher system
凭抵押品贷款 lend money on security
凭发票付款 payment on invoice
凭付款交付 hand-over against payment of the price
凭感觉(或眼力)的试验法 thumb rule
凭规格报盘 offer by description
凭规格售货 sale by specification
凭规格销售 sale by grade
凭规模买卖 sales as per specification
凭汇票付款的信用证 payment against draft credit; payment draft credit
凭汇票付款条件 sales on draft terms
凭汇票付款信用证 payment against draft credit
凭汇票取得贷款 get an advance against the bill
凭检视买卖 sale as seen
凭借 resort
凭借风势 leaning against the wind
凭经验的 a posterior; experiential
凭经验的方法 rule-of-thumb method
凭经验的估计方法 rule-of-thumb
凭经验进行设计 rule-of-thumb design
凭经验总结的大致做法 rule-of-thumb
凭买方样品交货 quality as per buyer's sample
凭买方样品质量交货 quality as per buyer's sample
凭卖方样品 quality as seller's sample
凭卖方样品质量(交货) quality as per seller's sample
凭票付款票据 note to bearer
凭票供应 coupon-based supply
凭票即付持票人 draft or note payable to bearer; money order payable to bearer; note order payable to bearer
凭票即付持票人的汇票 draft or note payable to bearer
凭票证供应的商品 commodities used to be rationed
凭商标售货 sale by trade mark
凭时效取得财产权 usucapion
凭适销品质买卖 sale on good merchantable quality
凭收据付款 payment on a receipt
凭收条付款信用证 payment on receipt credit
凭说明书买卖 sale by specification
凭说明书售货 sale by description
凭说明销售 sale on description
凭提单交货 delivery against bill of lading
凭委托书 per procuration
凭文字说明买卖 sale by description
凭信押汇汇票 documentary bill with letter of credit
凭信用借款 draw on one's credit
凭信用证交货 delivery against letter of credit; letter of credit payment
凭信誉承兑 acceptance for honour
凭型号售货 sale by type
凭样(品)买卖 as found; sale by sample
凭样(品)销售 as found; sale by sample
凭要求付款的保证书 on-demand bond
凭运单付款 payment against presentation of shipping documents
凭账单付款 payment on statement
凭证 document; evidence; receipt; supporting voucher; token; voucher; certificate
凭证闭塞站 token station
凭证闭塞制 token block system
凭证操作法 token operation
凭证查检统计抽样法 statistic(al) sampling in voucher examination
凭证的传递 transfer of vouchers
凭证的审核 scruting of vouchers; scrutiny of vouchers
凭证递交机 token deliverer
凭证递送盒 token instrument portable magazine
凭证订正 revision of documents
凭证放牧地 community allotment
凭证付款 payment against documents
凭证附后 vouchers attached
凭证复核 revision of documents
凭证供应 document-based supply
凭证机锁 token instrument lock
凭证记录 evidence record
凭证检查 inspection of voucher
凭证件才能让渡的财产 in grant
凭证交换 token exchanging
凭证接受机 token receiver
凭证调整 balancing of tokens
凭证控制 token operation
凭证平衡 balancing of tokens
凭证式闭塞行车 token operation
凭证授受 token exchanging
凭证授受机 token exchanger
凭证文件 evidential document
凭证制度 voucher schemes
凭支票取款的银行存款 bank deposits subject to withdrawal by check
凭指示付给 pay to order

坪

<辐射计数管计数率对电压的特性曲线的平直部分> plateau[复 plateaus/plateaux]

坪长 plateau length
坪年龄 plateau age
坪曲线 plateau curve
坪特性(曲线) counting rate voltage characteristic; plateau characteristic
坪问题 plateau problem
坪斜 plateau slope
坪值 plateau[复 plateaus/plateaux]

苹

果计算机公司 Apple

苹果绿 apple green
苹果木 applewood
苹果树 apple tree
苹果酸 malate; malic acid
苹果酸靛蓝 malic acid anil
苹果酸缩苯胺 malic acid anil
苹果酸盐 malate
苹果酰胺 malic amide
苹果园 apple orchard
苹果属 crabapple

屏

板 curtain wall; screen board; shroud

屏板电导 plate conductance
屏蔽 containment shell; curtain; masking; screening
屏蔽矮墙 screen wall
屏蔽板 barricade; shielding slab
屏蔽保护 screen protection
屏蔽保护范围 screen protection range; shadow
屏蔽保护器 screening protector
屏蔽泵 canned pump; canyon-type pump; shielded pump
屏蔽变量 obscured variable
屏蔽变压器 <接在通信[讯]电缆线路上防止强电影响> reduction transformer; shielded transformer
屏蔽玻璃 shield(ing) glass
屏蔽布置 shielding configuration
屏蔽材料 screening material; shielding material
屏蔽参数 shielding parameter
屏蔽测井 guard log
屏蔽层 barrier layer; shielding layer
屏蔽插头 shielded plug
屏蔽长度 shielding length
屏蔽常数 screening constant; shielding constant
屏蔽池 shielded pond
屏蔽传输线 strip line
屏蔽窗 shielded window; shielding window
屏蔽磁铁 shielded magnet
屏蔽带 shelter belt
屏蔽导体 shield(ed) conductor
屏蔽导线 shield(ed) conductor
屏蔽导线束 shielded wire bundle; shielding harness
屏蔽的 conductively closed; protective; shielded
屏蔽电弧 shield arc
屏蔽电极 bucking electrodes; guard electrode
屏蔽电缆 bonding cable; H cable; screened cable; shield(ed) cable; shielded-conductor cable
屏蔽电缆用插头 phonoplug
屏蔽电流 screen current
屏蔽电位 screen potential
屏蔽电压 mask voltage
屏蔽法 screen method
屏蔽房屋 screen building; shielding building
屏蔽分割 split screen
屏蔽辐射源 radioactive source shielding
屏蔽港 sheltered harbo(u)r
屏蔽工程 shielded engineering
屏蔽管 shielded plate tube
屏蔽罐 coffin; shielded case; shielding tank
屏蔽核 shielded nucleus
屏蔽盒 screening box; shielded box; shielding box; shielding case
屏蔽宏观照相机 shielded macro camera
屏蔽环 grading ring; shading ring; shielded ring
屏蔽环形天线 shielded loop
屏蔽混凝土 radiation-shielding concrete; shielding concrete
屏蔽火花塞 screened sparking plug; shielded plug
屏蔽火焰 sheathed flame; shielded flame
屏蔽及景观设计 screening and landscaping
屏蔽极 consequent pole; shaded pole
屏蔽极电动机 shielded pole motor
屏蔽极仪表 shielded pole instrument
屏蔽几何形状 shielding geometry
屏蔽计数管 screen wall counter
屏蔽计数器 guarding counter; shielded counter; shielded relay
屏蔽记忆 screen memory
屏蔽寄存器 mask register
屏蔽角 screening angle; shielded angle; shielding angle
屏蔽绞合电缆 shielded twisted cable
屏蔽接地 bonding; screen earth
屏蔽接地(导)线 bonding conductor
屏蔽接头 shielded joint
屏蔽矩阵电路 masking matrix circuit
屏蔽距 mask pitch
屏蔽壳箱 shielding case
屏蔽可编程序 mask programmable
屏蔽坑 shielded pit
屏蔽孔 shielded opening
屏蔽块 shield block; shielding slab
屏蔽馈线 screened feeder
屏蔽栏 mask field
屏蔽冷却管道 shield-cooling duct
屏蔽冷却系统 shielded cooling system; shielding cooling system
屏蔽连接 screen connection
屏蔽六芯电缆 shield three pair multi-cable
屏蔽笼 screening cage
屏蔽码 mask off code
屏蔽门 platform screen door; shield(ed) door; shielding door
屏蔽门包 package of platform screen door
屏蔽门控制室 platform screen door control room
屏蔽门系统 platform screen door system
屏蔽面 face of the screen; ground plane
屏蔽面积 shielding surface
屏蔽膜 barrier film; screened film
屏蔽排管 shield-row tube
屏蔽盘 blanking disc[disk]
屏蔽气 sheath gas
屏蔽砌块 shielding block

屏蔽墙 barricade;barrier shield;shielding wall
屏蔽墙壁 screen wall
屏蔽区 shadow zone
屏蔽热电偶 shield thermocouple
屏蔽日光灯 screened fluorescent tube
屏蔽容器 shielded container;shielded flask; shielding castle; shielding container;shielding flask
屏蔽软电缆 shielded flexible cable
屏蔽塞 shielding plug
屏蔽栅 shield grid
屏蔽设备 shielding-barrier equipment;shielding device
屏蔽设计 shielding design
屏蔽设施 shielded facility
屏蔽生成程序 mask generator
屏蔽式 protected type
屏蔽式电弧焊 shielded arc welding
屏蔽式电热塞 sealed plug; sheathed plug
屏蔽式阀 masked valve
屏蔽式检流计 shielded galvanometer
屏蔽式潜水电泵 canned motor pump
屏蔽式热电偶 shielded thermocouple
屏蔽式振荡器 shielded oscillator
屏蔽试验堆 shielded test reactor
屏蔽试验装置 shielded test apparatus;shielded test facility
屏蔽室 cave; protective enclosure; screen(ed) room; shielded enclosure;shielded room
屏蔽数 screening number
屏蔽双绞线 shielded twisted pair cable
屏蔽双扭线【电】 shielded twisted wire pair
屏蔽双线 screening doublet
屏蔽双线馈线 twinax
屏蔽水 shielded water
屏蔽水池 shielding pond
屏蔽水箱 water shielding tank
屏蔽损耗 shadow loss
屏蔽探针 shielded probe
屏蔽碳弧焊 shielded carbon arc welding
屏蔽套 housing
屏蔽特性 screening characteristic
屏蔽天线 screened aerial; screened antenna;shielded antenna
屏蔽同轴电缆 screened coaxial cable
屏蔽头盔 shielding helmet
屏蔽涂层 curtain coating
屏蔽涂料 barrier coating
屏蔽外壳 screening can
屏蔽位 mask bit
屏蔽五极管 screened pentode
屏蔽物 shield(ing)
屏蔽吸收系数 screen absorptance
屏蔽系数 screen-factor;screening coefficient; screening number; shielded factor;shielding factor
屏蔽系数法 method of shielding factor
屏蔽系统 shielding harness;shielding system
屏蔽线 screening wire;shielded line; shielded wire; shield(ing) line; shrouding wire
屏蔽线对 shielded pair
屏蔽线对电缆 shielded pair cable
屏蔽线路 line with shielding;shielded line
屏蔽线圈 coil; potted coil; screened coil; shading coil; shielded coil; shielding coil
屏蔽线圈电动机 screen coil motor
屏蔽箱 screening box;shielded box; shielded case;shielding box
屏蔽效果 shielding effect
屏蔽效率 shielded effectiveness;shielding efficiency
屏蔽效应 safe(ty) action; screening effect; shadow effect; shielding action;shielding effect
屏蔽楔 shading wedge
屏蔽信号发生器 mask signal generator
屏蔽型泵 shielding can-type pump
屏蔽型测量仪表 shielded measuring instrument
屏蔽型离心机 shielding can-type centrifuge
屏蔽性能 barrier property; shielding property
屏蔽选择 mask option
屏蔽眼镜 shielding spectacles
屏蔽因数 shielding factor
屏蔽因子 shielding factor
屏蔽阴极 shielded cathode
屏蔽引线 shielded lead
屏蔽圆辊闸门 shield roller gate
屏蔽造价 shielded cost
屏蔽闸 shutter;lock shield valve <便于暖气片拆修>
屏蔽闸门 shield gate
屏蔽栅栏 screening fence
屏蔽照相机 screened camera
屏蔽罩 screening can; shielding can; shielding case
屏蔽值 masking value
屏蔽置位 masked set
屏蔽砖 shielding tile;swinging brick
屏蔽装甲 spaced armor
屏蔽装甲板 skirting plate
屏蔽装置 shielded assembly;shielding device;shielding facility
屏蔽状态 masked state
屏蔽总线 masked bus
屏蔽阻抗 shielding impedance
屏蔽组织 shielding system
屏蔽作用 shielding action; shielding effect
屏壁 screen bulkhead
屏除 exclusion
屏挡墙 screen wall
屏的光化效率 screen actinic efficiency
屏风【建】 screen; accordion shades; shoji screen
屏风分隔 screen separation
屏风隔断 screen partition
屏风过道 <中世纪客厅布置形式> screen(ing)s passage
屏风式打桩法 group driving
屏隔舱壁 screen bulkhead
屏隔式立面 <一种掩饰房屋形状或大小的立面> screen facade
屏耗 plate consumption
屏后布线 back-of-panel wiring
屏厚度 screen thickness
屏厚系数 shield coefficient
屏级电压指示 plate-voltage indication
屏极 plate electrode
屏极变压器 anode transformer
屏极槽路 plate tank
屏极电池 anode battery;plate battery
屏极电流 plate current
屏极电路 plate circuit
屏极电路检波器 plate circuit detector
屏极电路调谐振荡器 tuned-plate oscillator
屏极电容 plate capacitance
屏极电容器 plate capacitor
屏极电位稳定 anode potential stabilization
屏极电压 B plus voltage;plate voltage
屏极电压降 anode voltage drop
屏极电源 B power supply;B-source; plate supply
屏极电源电压 plate supply voltage
屏极扼流圈 plate choke coil
屏极反电压 plate inverse voltage
屏极负载 anode load;plate load
屏极负载阻抗 anode-load impedance
屏极功率 plate power
屏极过载 plate overload
屏极回路调谐的 tuned-anode
屏极检波 anode detection;anode rectification;plate detection;plate rectification
屏极检波电子管伏特计 plate-detection voltmeter
屏极检波器 bias detector
屏极帘栅极调制 plate-screen modulation
屏极帽 plate cap
屏极旁路电容器 plate by-pass capacitor
屏极-栅极特性(曲线) grid plate characteristic
屏极输出功率 plate power output
屏极输出器 anode follower
屏极输入功率 plate power input
屏极损耗 anode dissipation;plate dissipation
屏极特性 anode characteristic
屏极特性曲线 plate characteristic curve
屏极调制 plate control;plate modulation
屏极调制振荡器 anode-modulated oscillator
屏极效率 plate efficiency
屏极-阴极电压 plate-to-cathode voltage
屏极引出头 plate cap
屏极振荡回路 plate tank
屏极中和 plate neutralization
屏极阻抗 anode impedance
屏框 mask
屏路扼流圈 plate inductor
屏路效率 plate circuit efficiency
屏门 screen door
屏面 face of the screen;raster
屏面对角线 picture diagonal
屏面格式 screen format
屏幕 parclose screen;screen;shroud
屏幕安装条 screen spline
屏幕保护程序 screen server
屏幕编辑 on-screen editing
屏幕变形 hooking
屏幕捕获 screen capture
屏幕菜单 on-screen menu;screen menu
屏幕尺寸 screen size
屏幕的开孔 perforation of screen
屏幕地址 screen address
屏幕对角 screen diagonal
屏幕方式 screen mode
屏幕高度 screen height
屏幕宽度 screen width
屏幕亮度 screen intensity;screen luminance
屏幕录像机 kinescope recorder;photographic(al) recorder; kinescope recording equipment
屏幕面积 screen area
屏幕输出【计】 screen output
屏幕输入输出 screen input/output
屏幕透射比 screen transmittance
屏幕图像 screen picture
屏幕涂层 curtain coating
屏幕吸收比 screen absorptance
屏幕吸收系数 screen absorptance
屏幕显示 screen display
屏幕显示器 visual display unit
屏幕斜度 screen inclination
屏幕"雪花"干扰 snow storm
屏幕照度 screen illumination
屏幕转换器 screen changer
屏幕状态区 screen status area
屏幕坐标 screen coordinates
屏气 breath holding
屏气潜水 breath-hold diving
屏气潜水作业 breath-hold diving operation
屏曲率 panel curvature
屏曲线 screen curve
屏扫气缸 screen-wiper cylinder
屏色谱特性 screen-colo(u)r characteristic
屏栅 screen-grid
屏栅电路 screen-grid circuit
屏栅调制 screen-grid modulation
屏栅耗散 screen-grid dissipation
屏栅特性 mutual characteristic
屏栅闸流管 shield-grid thyratron
屏栅真空管 screen-grid tube
屏上安装 panel mounting
屏石 screen stone
屏式操纵台 panel type board
屏式的 platen-type
屏式管墙 curtain wall
屏式过热器 pendant superheater; plate(n) superheater
屏式控制台 panel type board
屏式凝渣管 wing wall
屏式热交换器 platen heat exchanger
屏式受热面 curtain wall
屏式中间再热器 platen reheater
屏饰 antependium
屏刷 screen wiper
屏条 screen bar
屏调整 plate control
屏调制 plate modulated
屏显风挡玻璃 head-up display windscreen
屏型仪表 panel type instrument
屏压整流器 anode rectifier
屏移法 curtain method;screen method;travel(l)ing screen method
屏荧光粉 screen phosphor
屏障【建】 jube;parclose;abri;barrier (layer); barrier wall; curtain; guard screen; screen; shelter; clo-sures du choeur <教堂唱诗班的>
屏障材料 barrier material
屏障岛 island barrier
屏障机制 barrier mechanism
屏障计数器 shielded counter
屏障记忆 screen memory
屏障滤光片 barrier filter
屏障螺杆 barrier screw
屏障墙 barricade
屏障衰减 attenuation by barrier
屏障小柱 cancelli
屏障效果 screen effect
屏障效应 barrier effect
屏障休闲地 coulisse fallow
屏障栽植 screen planting
屏障作用 barrier action
屏罩 encapsulation
屏罩应变计 capsulated ga(u)ge;encapsulated strain ga(u)ge
屏罩应变片 capsulated ga(u)ge;encapsulated strain ga(u)ge
屏帚 screen wiper
屏状滤光片 barrier filter
屏锥封接机 glass envelope sealing machine

瓶鼻状滴水槽 bottle-nose drip

瓶壁热应力集中带 murgatroyed belt
瓶测洋流海图 bottle chart
瓶底 base;punt
瓶底标记 punt code
瓶底厂家密码 base code
瓶底凸出 rocker bottom
瓶底歪斜 <玻璃制品缺陷> offset punt

瓶底下沉 dropped bottom; dropped punt
瓶底印记 lettering on bottom
瓶地窖 bottle cellar
瓶点法 bottle point method
瓶点试验 bottle point experiment
瓶阀式启闭机 screwing and unscrewing machine
瓶盖 bottle cap; capsule
瓶盖机 bottle capper
瓶盖座 cap seat
瓶罐玻璃 bottle glass; container glass; hollow glass
瓶罐玻璃池窑 bottle tank
瓶罐模制 can mo(u)lding
瓶罐碎玻璃 bottle cullet
瓶架 pot-rack
瓶肩 shoulder
瓶肩过薄 light shoulder
瓶接 bottle graft
瓶结 builder's knot
瓶颈 bottleneck; necking
瓶颈表面裂纹 smear
瓶颈分析 bottleneck analysis
瓶颈根部 root of neck
瓶颈谷 bottleneck valley; hour-glass valley
瓶颈过薄 hollow neck
瓶颈过厚 slug in neck
瓶颈过窄 choke; choked bore; choked neck
瓶颈环 ring collar
瓶颈加工工具 finish tool; mouth tool
瓶颈口 bottleneck
瓶颈路段 bottleneck road
瓶颈模 neck mo(u)ld(ing); ring mo(u)ld
瓶颈确定问题 bottleneck assignment problem
瓶颈上游专用车道 <高速干道上一种优先控制车辆的道路> exclusive lanes upstream of bottlenecks
瓶颈式 bottleneck
瓶颈式道路 bottleneck road
瓶颈式千斤顶 bottle jack
瓶颈式通货膨胀 bottleneck inflation
瓶颈式窑 flare kiln
瓶颈问题 bottleneck problem
瓶颈下部 base of bottleneck
瓶颈状态 bottleneck state
瓶颈作用 bottleneck effect
瓶口 bottle-mouth
瓶口锭模 semi-closed top mo(u)ld
瓶口结 bottleneck hitch
瓶口开裂 broken finish
瓶口歪斜 cocked finish
瓶口下液滴 droplet under finish
瓶口泄漏 leakyring
瓶料玻璃 green glass
瓶绿色 bottle green
瓶帽 capsule
瓶密接头 bottle-tight joint
瓶内黏[粘]丝 <玻璃制品缺陷> birdcage; bird swing
瓶盘菌属 <拉> Urnula
瓶塞 bottle cork; bung; stopper
瓶塞开启工具 opener
瓶塞起子 corkscrew
瓶塞钻 corkscrew
瓶式采样器 <泥沙测验的> bottle sampler
瓶式储气罐 bottle holder
瓶式浮标 bottom bottle; drift bottle
瓶式过滤器 bottle(-shaped) filter
瓶式检验 jar test
瓶式泥沙采样器 bottle silt-sampler
瓶式千斤顶 bottle jack
瓶式曲管 bottle trap
瓶式竖窑 bottle kiln
瓶式陶瓷窑大口下面的拱 glut arch
瓶式陶瓷窑加煤口 Barratt-Halsall fire mouth
瓶式悬移质取样器 bottle type suspension load sampler
瓶式窑 bottle oven
瓶式饮水冷却器 bottle type drinking water cooier
瓶式淤泥取样器 bottle silt-sampler
瓶饰【建】 vase; tazza
瓶饰地毯 vase carpet
瓶饰涂饰 vase-painting
瓶头阀 bottle valve
瓶箱托 bottlecase supporter
瓶形存水弯 bottle trap
瓶形的 flask-shaped; lageniform
瓶形搅拌机 bottle mixer
瓶形螺栓支座 bottle-shaped screw base support
瓶形竖窑 bottle kiln
瓶形物 bomb; bottle; flask; vase
瓶形柱墩 bottle-shaped pillar
瓶用冷却柜 bottle cooler
瓶用冷却售货柜 bottle vendor
瓶装的 bottled; in bottles; bottling
瓶装和软饮料厂 bottling and soft drink plant
瓶装煤气 bottled gas
瓶装喷压密封膏 bulk compound
瓶装气(体) bomb gas; bottle air; portable gas
瓶装清凉饮料厂 soft-drink bottling plant
瓶装清凉饮料厂废水 soft-drink bottling plant waste
瓶装水 bottled water
瓶装水泥 bottle(d) cement
瓶装氧气 bottle(d) oxygen
瓶装液化(石油)气 bottled gas; bugas; liquid petroleum gas[LP-gas]
瓶装液体煤气 bottle gas; calor gas
瓶状激波系 shock bottle
瓶状千斤顶 bottle jack
瓶状竖窑 bottle kiln
瓶状物 bottle
瓶状心 flask-shaped heart
瓶状叶 pitcher
瓶锥体试验法 <一种用砂量测土体积的方法> jar-cone method
瓶子 bottle
瓶子卡车 bottle truck
瓶子窑 hovel kiln
瓶子运输器 bottle conveyer[conveyor]

钋

钋210的置换率 replacement rate of 210Po
钋含量等值图 contour map of polonium content
钋含量平剖图 profile on plane of polonium content
钋离子 polonium ion

坡

坡岸 sloping bank
坡板 slope board
坡标线 gradient board
坡长 inclined length; slope length
坡长限制 grade length limitation; grade limit(ation); grade line limit
坡窗 batement light
坡道 gradient; ramp; slade; slide slope; slideway; slope ladder; slope ramp; slopeway
坡道边缘 slope edge
坡道标示 exit ramp marking
坡道表面处理 slope dressing
坡道部分 slope portion; slope section
坡道冲蚀 slope erosion
坡道单位阻力 specific gradient resistance
坡道道岔 ramp turnout
坡道的斜面 ramp incline
坡道点 ramp point
坡道段 gradient section
坡道附加阻力 additional resistance for gradient; ascent resistance; grade resistance; inclination resistance; gradient resistance
坡道化直 grade rectification
坡道击实机 slope compactor
坡道临界长度 critical length of grade
坡道隆起 slope crown
坡道面 ramp surface
坡道磨损 slope erosion
坡道抹面机 slope finisher
坡道排水 slope drainage
坡道坡度 ramp slope
坡道桥 ramp bridge
坡道倾斜角 slope inclination
坡道区段 slope portion; slope section
坡道设计 slope design
坡道施工 slope construction
坡道施工设备 slope construction equipment
坡道式停车场 ramp park
坡道式(停)车库 ramp-type garage; sloping floor parking garage
坡道凸面 slope crown
坡道线 slope line
坡道斜度 slope inclination
坡道型式 <高速干道的> ramp type
坡道压路机 slope compactor
坡道养护 slope maintenance
坡道整面机 slope finisher
坡道整平 slope flattening
坡道整平压实机 slope level(l)ing and compacting machine
坡道整修 slope dressing
坡道值 gradient ratio
坡道指示牌 gradient board
坡道制动 grade braking
坡道驻车性能 grade parking performance
坡道纵剖面图 profile of slope
坡道阻力 ascent resistance; grade resistance; inclination resistance; slope resistance
坡的高端 high point of the slope
坡底 base of slope; slope toe
坡底冲刷 hillwash; slope erosion; slope wash
坡底护坦 toe apron
坡底柔性接缝 flexible facing joint
坡底塌陷 base failure
坡底线 bench toe
坡底圆 base circle
坡地 contoured land; hillside field; rolling land; slope land; sloping field
坡地测站 slope station
坡地沉积 slope deposit
坡地成因 slope genesis
坡地冲刷 slope wash
坡地冲刷作用 hillwash
坡地地貌 hillslope landform
坡地黄土 slope loess
坡地径流损失 runoff losses from slope
坡地犁 upland plow
坡地绿化 slope planting
坡地排水沟犁 hill drainage plough
坡地切割器 bank bar
坡地侵蚀 slope erosion
坡地泉 slope spring
坡地上起重机的平衡 crane balance on slope
坡地上铁路挖方 railway cutting on sloping ground
坡地坍滑 hillwash
坡地梯田化 slope terracing; terracing of the land on the slopes
坡地拖拉机 hillside tractor
坡地形态 hillslope shape
坡地型地基 slope subgrade
坡地栽培 hill culture
坡地造林 slope planting
坡地植草 sowing a slope down to grass
坡地住宅 stepped hillside house
坡地抓 sidehill lug
坡顶 apex of grade; crest; crest of slope; slope top; top of grade; top of slope
坡顶高程 altitude of slope top
坡顶排水 slope apex drainage
坡顶屋架 pitched roof truss; pitched truss
坡顶线 bench edge
坡度 angle of slope; declivity; degree of slope; elimination; falling gradient; gradient; inclination; one-to-one slope; pendence; percent of slope; pitch; rate of slope; steepness; versant; batter <坝的背水面>
坡度1比2 one-to-two of slope
坡度凹变点 sag of the grade line
坡度凹陷 sag in grade
坡度百分比 percentage slope
坡度百分率 percentage inclination; percentage of grade
坡度板 batter board; pitch board
坡度比 ratio of slope; slope ratio
坡度变点 change point of gradient; point of change of gradient
坡度变更点 breadthwise in grade; break in grade
坡度变化 change of declivity; change of slope; slope variance
坡度变化率 rate of grade
坡度变换 break in grade
坡度变换点 point of vertical intersection
坡度变异 slope variance
坡度标 grade indicator; gradient indicator; gradient sign; slope sign
坡度标尺 scale of slope
坡度标钉 grade spad
坡度标桩【测】 bank plug; grade post; gradient post; grade stake; slope stake; blue top
坡度标准桩 blue top
坡度表示法 slope-value method
坡度测定器 gradiometer
坡度测定仪 gradient meter[metre]; grading instrument; gradiometer
坡度测量 grade measurement; gradiometry; slope level(l)ing
坡度测量仪 grading instrument; gradiometer
坡度测设 grade location
坡度层 grade course
坡度差 algebraic(al) difference between adjacent gradients; difference in gradients
坡度尺 adjustable triangle; grade rod; gradient scale; scale of horizontal equivalent; slope diagram <地形图上的>
坡度传感器 grade sensor
坡度单位 slope unit
坡度的低点 low point of the slope
坡度等级 grade of slope
坡度等级图 slope category; slope class map
坡度点 grade point
坡度电位法 graded potential system
坡度陡的 high-pitched

坡度断面图 grading plan
坡度法 slope method
坡度分析图 slope analysis map
坡度改进 grade easement
坡度改善 gradient improvement
坡度改正 correction for grade; correction for inclination of tape; grade correction; slope correction
坡度感测器 grade sensor
坡度规 tilt ga(u)ge
坡度过度升降 excessive rise and fall
坡度很小的道路 level road; level track
坡度缓和 grade easement
坡度基区 graded base
坡度基准 grade reference
坡度极限 gradability limit
坡度计 slope ga(u)ge; slope level; slope meter
坡度计算 grade calculation; grade computation
坡度记录 grade record
坡度减缓 grade elimination
坡度减小 grade reduction
坡度交会 intersection of grades
坡度角 angle of elevation; angle of gradient; angle of pitch(ing); angle of slope; gradient of slope; pitch angle; slope angle
坡度校正 correction for grade; grade correction; grade rectification; slope correction
坡度浚挖 slope dredging
坡度控制 grade control; roll-position control
坡度控制结构物 grade-control structure
坡度控制线 grade wire
坡度控制装置 slope controller
坡度流 gradient current; slope current; slope flow
坡度率 grade rate; percentage inclination
坡度摩阻 grade resistance
坡度挠度方程 slop-deflection equation
坡度排水 grading drain
坡度牌 gradient board
坡度平缓的河流 graded river
坡度平坦化 slope flattening
坡度剖面图 <道路等> grade profile
坡度千分率 gradient ratio; kilogradient ratio
坡度牵出线【铁】 draw-out track at grade; switching lead at grade
坡度倾角 angle of gradient
坡度曲线 clinographic(al) curve
坡度上车速 speed on grade
坡度设计 grade design; grade location
坡度(升高)百分比 percent grade; grade percentage
坡度施工平面图 grading plan
坡度说明 graded description
坡度损失 loss in gradient
坡度调整 grade adjustment; slope adjustment
坡度凸变点 summit of the grade line
坡度突变 knuckle
坡度突然改变 abrupt change of slope
坡度图 slope category map; slope diagram; slope map
坡度线 grade line; slope line
坡度线标桩 grade peg
坡度线高程 grade elevation(of road surface); grade line elevation
坡度限制 grade limit(ation)
坡度削减 grade reduction
坡度消除 grade elimination
坡度修正系数 gradient coefficient
坡度样板 slope board
坡度仪 declinator; declinometer; slope indicator
坡度折点 gradient break
坡度折减 compensation of gradient; grade reduction; reduction of gradient
坡度折减率 <曲线路段的> curve compensation
坡度整平机 grade trimmer
坡度指示板 grade lath
坡度指示器 slope indicator
坡度助力 grade assistance
坡度转折 gradient break
坡度转折点 gradient break point; point of change of gradient
坡度桩【测】 blue top; grade peg; gradient peg; slope stake
坡度纵断面 <驼峰调车区> gradient profile
坡度纵剖面 grade profile
坡度阻力 gradation resistance; grade resistance; gradient resistance; negotiating resistance; slope resistance
坡段 grade section
坡段长度 gradient length; length of grade section; length of gradient; length of slope element
坡风 hang wind
坡改梯 turn hillside into terraced fields
坡高 slope height
坡耕地 hillside cultivated
坡管口 bevel end
坡规 battering rule
坡后退 slope recession; slope retreat
坡画演化 hillslope evolution
坡积 slide rock
坡积层 diluvial layer; scree slope; talus material; drift bed
坡积的 diluvial
坡积矿床 talus deposit
坡积泉 talus spring
坡积裙 talus apron; talus fan
坡积砂矿 diluvial placer
坡积铁矿床 slope wash iron deposit
坡积土 clinosol; slope wash; slope wash soil; talus
坡积物 cliff debris; deluvium; slope material; slope wash; talus material
坡积相 slop wasp facies
坡基层 sloped substrate
坡尖【地】 spur
坡尖泊 spur notch
坡肩 shoulder of slope
坡肩截流沟 shoulder ditch
坡肩截水沟 shoulder ditch
坡降 fall; gradient; gradient ratio; hydraulic slope
坡降比尺 slope scale ratio
坡降产生的水流 discharge due to slope
坡降陡度 steepness of slope
坡降流 gradient current
坡降流量曲线 slope discharge curve
坡降流量图 slope discharge diagram
坡降面积测流法 slope-area discharge measurement
坡降面积法 slope-area method
坡降水位流量关系 fall-stage-discharge relation(ship)
坡降水位流量关系曲线 fall-stage-discharge relation curve
坡降线 hydraulic axis
坡角度 angle of slope; grading angle; slope angle
坡脚 foot of slope; tail of slope; toe; toe of(side) slope
坡脚保护 foot protection; toe protection
坡脚保护工程 toe protection works
坡脚冲蚀 etching of slope toe
坡脚挡土墙 toe wall
坡脚挡土桩 breast pile
坡脚底部 base of slope
坡脚底宽 base of slope
坡脚堆石 projection of talus
坡脚(反)滤层 toe filter
坡脚防护 slope protection; toe protection of slope
坡脚防冲刷地梁 toe beam
坡脚防冲刷棱体 berm(e) toe apron
坡脚防护工程 foot protection work
坡脚护坦 toe apron
坡脚滑动 toe failure
坡脚滑坍 <滑动面穿过坡脚的> toe-failure
坡脚截水墙 toe cut-off wall
坡脚开挖 toe excavation
坡脚块体 <防波堤> toe armo(u)r
坡脚拦墙 toe cut-off wall
坡脚临时压重(物) temporary toe weight
坡脚排水 toe drain
坡脚排水沟 sough
坡脚排水设施 toe drain
坡脚破坏 toe damage; toe failure
坡脚砌层 toe course
坡脚砌块 toe block
坡脚墙 dwarf wall; toe wall
坡脚坍塌 base failure of slope; toe failure
坡脚稳定 toe hold
坡脚稳定性 toe hold
坡脚线 tip grade; toe line
坡脚压载 toe loading
坡脚压重 toe weight
坡脚应力与强度对比法 stress and strength contrast method of slope foot
坡脚圆 toe circle
坡脚柱 batter peg
坡脚桩 batter peg; breast pile
坡接轨 ramp rail
坡口 bevel; divided edge; end bevel; groove; V-groove; bevel groove <焊件预开的>
坡口半径 root radius
坡口边 bevel(l)ed edge; tapered edge
坡口表面 groove face
坡口端 bevel end; groove end
坡口钝边 root face
坡口高度 groove depth
坡口焊缝 groove welding seam
坡口焊(接) bevel weld; grooving; groove weld(ing)
坡口加工 bevel(l)ing of the edge; chamfering; edge preparation
坡口加工端 bevel end
坡口加工面 prepared edge
坡口检查 inspection of edge preparation
坡口角(度) angle of bevel; angle of vee; bevel angle; groove angle; included angle; vee-angle
坡口面 fusion face
坡口面角度 angle of preparation; bevel angle
坡口切边 feather edging; feathering
坡口深度 groove depth
坡口铣刀 bevel cutter
坡口斜角 angle of chamfer
坡框式砌块 mo(u)ld blocks for slope protection
坡拉帕斯 <一种木材防腐剂> Pollapas
坡立谷 hojo; poljie[复 polgia]; polya [polye]
坡路 ramp
坡麓 foot slope
坡缕石 mountain cork; mountain leather; paligorskite [palygorskite]; rock leather
坡缕石化 paligorskitization
坡率 rate of grade; ratio of slope; slope ratio
坡面 dome; face of slope; grade surface; open face; sloping surface; hillslope【地】
坡面被破坏 destroyed slope surface
坡面不合理加载 unreasonable loading on wall face
坡面冲刷 slope erosion; slope wash
坡面道岔 switch of inclined plane
坡面的 domatic
坡面防护 slope surface protection
坡面高度桩 blue top
坡面过程 hillslope process
坡面夯实 slope compaction
坡面截水堆 ridge terrace
坡面经纬仪 slope theodolite; slope transit
坡面景观 slope landscape
坡面径流 hillslope runoff; sheet flow; slope runoff
坡面宽度 width of slope surface
坡面流 gradient current; overland runoff
坡面漫流 overland flow; sheet flow; surface flow
坡面漫流速度 velocity of overland flow
坡面排水 slope surface drainage
坡面平地机 slope grader
坡面铺砌 lining of slope
坡面侵蚀 the erosion caused by sloping surfaces
坡面渗流 effluent seepage
坡面特性 slope characteristic
坡面特征 slope feature
坡面图 slope diagram
坡面稳定性 slope stability
坡面修整 <融合地形景色的> transitional grading
坡面压路机 slope compactor
坡面压实 slope compaction
坡面圆 slope circle
坡面桩 blue top
坡莫高导磁率合金 peamafy
坡莫合金补偿 permalloy compensation
坡莫合金传感器 permalloy strip
坡莫合金"汉字"型检测器 permalloy "Chinese letter" detector
坡莫(镍铁)合金 <强磁性铁镍合金> permafrost permalloy
坡莫特 <一种铜镍钴永磁合金> Permet
坡目费树脂 <软水用的一种离子交换树脂> permufit
坡平距 horizontal equivalent
坡栖巨砾 perched boulder
坡栖漂砾 perched block
坡栖盆地 perched basin
坡栖岩块 balanced rock; perched block; perched boulder
坡桥 bridge on slope; inclined bridge; ramp bridge
坡身不稳 slope failure
坡式高架线 gradient elevated track
坡梯道 corded way
坡田 gradient terrace; sloping field
坡尾 tail of slope
坡屋 appentice; pent; pentice
坡屋边缘 canopy lip
坡屋顶 abat-vent; pitched roof; slanting roof; sloping roof
坡屋顶脚手托架 roof clippy
坡屋顶面积 pitched roof area
坡下堆积物 talus accumulation
坡下阶地 slope-covered terrace

P

坡线 contour gradient;grade line;line of slope
坡向 aspect of slope;exposal;exposure;slope aspect
坡向明沟 slope downwards to open drain
坡斜度 gradient of slope
坡形 slope shape
坡形板边 tapered edge
坡形地板 raked floor;ramped floor
坡形顶棚 camp ceiling
坡形路缘口 dropped curb
坡腰 mid-slope
坡腰平台 dike ledge
坡折【地】gradient break
坡折点 break of slope;culminating point;knick point;nick-point;slope break
坡折裂点 break-in slope
坡趾 toe of slope
坡趾破坏 toe failure
坡趾挖方 toe excavation
坡趾圆 toe circle
坡桩 grade stake;slope peg

泼出 slop

泼溅 bubbling
泼溅加油装置 splasher
泼溅润滑 splash lubrication
泼墨山水 splashed-ink landscape
泼散器 splasher
泼水 water sprinkling
泼水铁铜矾 poitevinite
泼涂层 dash coat
泼涂打底层 dash-bond coat
泼涂黏[粘]结层 dash-bond coat
泼涂饰面 dash finish
泼油 oil application
泼油管 sprinkling bar
泼油机 oil plasher
泼油器 oil plasher
泼油土 <用铺路油类或沥青处理过的土> oiled earth
泼油温度 <沥青材料> application temperature
泼釉 glazing by splashing;splashing

颇大的 sizable

颇强震 rather strong shock

婆罗科努复背斜带【地】Borohoro anticlinorium belt

婆罗科努构造段【地】Borohoro tectonic segment
婆罗门式(建筑)Brahman style
婆罗门寺院上的塔 vimana
婆罗洲玫瑰木 <产于东南亚一带> Rengas
婆罗洲樟木 Mahabon teak

鄱阳冰期【地】Poyang glacial epoch

鄱阳-大姑间冰期【地】Poyang-Dagu interglacial epoch

迫出 expel

迫降场 emergency landing field;emergency landing site
迫近 close at hand;close in
迫近的危险 imminent danger
迫卖价值 forced sale value
迫切率 urgency rate
迫使资方让步的变相怠工 work to rule
迫售价格价值 forced sale value
迫位化合物 peri-compound
迫位抗力系数 coefficient specific resistance

珀硅钛铈铁矿 perrierite

珀卡煤 <一种烧铆钉用煤> Pocabontas coal
珀拉玲 pyralin
珀林黏[粘]度杯 Perlin viscosity cup
珀莫利特 <一种木材防腐剂> Permolite
珀莫釉 <一种透明防水的墙壁涂料> Permoglaze
珀苏兹硬度 Persoz hardness

破包 bale broken

破边 broken selvedge;edge damage
破冰 bucking;ice breaking
破冰爆破 icebreaking blasting
破冰船 iceboat;ice breaker;icebreaker ship
破冰船队 icebreaking fleet
破冰船型船头 Baltic tow;icebreaking tow
破冰电缆修理船 icebreaking cable repair ship
破冰墩 icebreaking pier;sharpened pier
破冰斧 ice ax(e)
破冰工具 icebreaking tool
破冰机 breaker;glacier mill;ice breaker
破冰接接点 frost-cutting contact
破冰开河 ice break-up
破冰拦 ice breaker
破冰雷 ice mine
破冰鳍 ice fin
破冰器 ice breaker
破冰前进 slewing
破冰区 broken belt;onshore zone
破冰设备 iceboat;ice breaker;ice guard;sterling
破冰体 ice apron
破冰拖轮 icebreaking tug;ice-strengthened tug
破冰型船首 Baltic bow
破冰型船头 Baltic bow
破冰型货船 cargo icebreaker
破冰油船 icebreaking tanker
破波 breaker;breaking wave
破波波高 breaker height;breaking wave height;wave height on breaking
破波带 breaker zone;surf zone
破波带拍 surf beat
破波级 scale of surf
破波类型 type of breaker;ultimate of breakers
破波拍岸 surf beat
破波深度 depth of breaking
破波线 breaker line
破波压力 breaking wave pressure;pressure due to breaking wave
破波指数 breaker index
破布 cloth rag;clout;rag;shred;tatter;wiping rag
破布拌浆机 rag breaker
破布尘屑 rag dust
破布除尘机 rag duster;willowing machine rag duster
破布除尘器 rag willow
破布拣除器 deragger
破布拣选辊 rag sorting roll
破布浆 rag pulp
破布敲打机 rag willow
破布切断机 rag cutter
破布松散机 rag thrasher
破布条 tatter
破布条切割机 rag slitter
破布纤维 rag fiber[fibre]
破布蒸煮器 rag boiler
破布纸浆废水 rag pulping waste
破舱试算水线 trial trim line
破舱稳性 damaged stability;stability in damaged condition
破舱稳性计算书 calculation report of floodability;damaged stability calculation report
破拆工具 entry tool
破产 bankruptcy;cracker;crash;failing;failure;insolvency;liquidate;liquidation;out of business;play smash;ruination
破产案产业管理人 receiver
破产财产 bankrupt's estates
破产财产管理人 accountant in bankruptcy;bankruptcy administrator;receiver in bankruptcy;trustee in bankruptcy
破产财产扣押令 warrant in bankruptcy
破产产业的接管 receivership
破产程序 bankruptcy proceeding;procedure in bankruptcy;procedure of bankruptcy
破产程序中清偿债权 discharge of claims in bankruptcy
破产担保基金 insolvency guaranteed fund
破产的 bankrupt
破产的效力 effect of bankruptcy
破产地主 bankruptcy landlord
破产法 act of bankruptcy;bankruptcy act;bankruptcy law;insolvent laws;law of bankruptcy
破产法庭 bankruptcy court;court of bankruptcy
破产风险 clean risk of liquidation
破产概率 probability of ruin;ruin probability
破产公断人 referee in bankruptcy
破产公告 bankruptcy notice
破产管理人 administrator in a bankrupt estate;trustee in bankruptcy
破产核算员 accountant in bankruptcy
破产户之债权人 creditor of bankruptcy
破产家园豁免 homestead exemption
破产价格 ruinous price
破产连锁反应 chain-reaction bankruptcy
破产判决 sentence of bankruptcy
破产企业财产清理价值 break-up value
破产清理 administration in bankruptcy
破产清理价值 break-up value
破产清理中的保证债权 secured creditors' right in bankruptcy proceedings
破产清算人 assignee in insolvency
破产清算书 statement of liquidation
破产人 bankrupt;brokee
破产申请 bankruptcy petition;petition in bankruptcy
破产审理法庭 bankruptcy court
破产时可认债务 debts-provable in bankruptcy
破产时认可的债务 debts-provable in bankruptcy
破产事务官 official receiver
破产收益率 yield to crash
破产受托人 trustee in bankruptcy
破产诉讼 liquidation or bankruptcy proceedings
破产条款 insolvency clause
破产通知(书)bankruptcy notice
破产信托人 trustee in bankruptcy
破产行为 act of bankruptcy
破产银行 failed bank
破产优先偿付 preferential payment in bankruptcy
破产原因 cause of bankruptcy
破产债权 credit of bankruptcy
破产债权人 creditor in a bankrupt estate;creditors of a bankrupt estate
破产债务人 bankruptcy debtor
破产账 account of bankruptcy;bankrupt account
破产者 bankrupt;insolvent
破产者的管财人 syndic
破产终结判决 decree of completion of the bankruptcy proceeding
破产资本家 crashee
破产资产 bankrupt's assets
破尺丈量 broken tape measurement
破船 hulk
破堤 levee break
破底清挖道砟机械 undercutting machine
破断 breakdown;failure;fracture
破断负载 breaking load
破断截面 broken-out section
破断截面线 <制图的> broken section line
破断力 breaking force
破断联杆 <水轮机> breaking link
破断裂现象 fracture phenomenon
破断面 broken-out section;cutting plane <制图学上的>
破断面视图 broken-out section view
破断面线 cutting plane line
破断模型 fractographic(al)pattern
破断片 <材料> breakage
破断强度 breaking strength
破断试验 breaking test;rupture test
破断系数 breaking factor
破断销 breaking pin
破房 slum
破封 break a seal(ing)
破斧木 quebracho
破格提拔 by-pass convention to promote someone to leading post
破埂器 debanker;splitter
破坏罢工 strike-breaking
破坏罢工者 finks
破坏半径 damage radius;radius of destruction;radius of rupture
破坏包(络)线 envelope of failure;failure envelope;Mohr's circle
破坏比 failure ratio
破坏变形 failure deformation;failure strain
破坏变形谱 damage deformation spectrum
破坏变质 catabolism
破坏变质作用 destructive metamorphism
破坏表面 failed surface
破坏参数 damage parameter
破坏层 disrupted bed
破坏车辆之抢修工作 recovery work
破坏承载应力 failure bearing stress
破坏程度 damage level;degree of breaking;degree of damage;destructiveness
破坏代价系数 damage cost factor
破坏带 destruction belt;destruction zone
破坏的产物 product of destruction
破坏的混凝土 demolished concrete
破坏的行为 sabotage
破坏等值线图 damage contour map
破坏堤 water-break;wave breaker
破坏点 breakdown point;breaking point;fail(ure)point;point of failure

破坏电压 breakdown voltage;disintegration voltage;disruptive voltage;failure voltage;plague voltage;rupturing voltage
破坏断面 failure cross-section
破坏阀限地震 damage threshold earthquake
破坏阀限地震烈度 damage threshold earthquake intensity
破坏法结构设计 collapse method of structural design
破坏范围 failed area;failure zone
破坏方式 failure mode;mode of failure
破坏分布 failure distribution
破坏分析 failure analysis
破坏风景 visual pollution
破坏负荷 failing load
破坏负载 critical load(ing)
破坏概率 failure probability;probability of damage;probability of failure
破坏概率分布函数 damage probability function
破坏概率矩阵 damage probability matrix
破坏概率限值 bound to failure probability
破坏工作用铲斗 demolition bucket
破坏规律 failure law
破坏轨迹 failure locus
破坏过程 failure process
破坏函数 failure function
破坏合同 break(of)contract
破坏荷载 breaking load(ing);breaking weight;charge of rupture;collapse [collapsing] load;cracking load;crippling load;crush(ing)load(ing);destroy load;destructive load;failing load;failure load;fracture(d) load;fracturing load;load at failure;load at rupture;load of breakage;load to collapse;rupture load
破坏荷载系数 collapse load factor
破坏荷重 breaking weight;failing load
破坏后果 failure mode
破坏后阶段 stage after failure stage
破坏弧 <土坡的> failure arc;failure circle
破坏滑动面 slip surface of failure
破坏环境的经济活动 environmentally damaging activities
破坏环境责任 environmentally damaging liability
破坏活动 sabotage
破坏机构 breakdown mechanism;collapse mechanism;failure mechanism;mechanism of fracture
破坏机理 breakdown mechanism;collapse mechanism;failure mechanism;mechanism of fracture
破坏机制 breakdown mechanism;collapse mechanism;failure mechanism;mechanism of fracture
破坏级别 failure rank
破坏极限 failure limit;limit of rupture;rupture limit
破坏极限状态 failure limit state;limit state of rupture
破坏纪律 breach of discipline
破坏价值因素 damage cost factor
破坏假说 failure hypothesis
破坏监测系统 failure detection system;failure monitoring system
破坏检查 destructive test
破坏检验 destructive inspection
破坏角 angle of break(ing up);angle of rupture
破坏阶段 failure stage
破坏阶段法 load factor method
破坏阶段理论 yield line theory
破坏阶段设计 collapse design
破坏阶段设计法 collapse design method;plastic stage design method
破坏截面 breaking section
破坏进行过程 collapse mechanism
破坏经济秩序罪 crime against economic order
破坏矿产资源罪 crime of destroying mineral resources
破坏类别 category of damage
破坏棱体 failure wedge
破坏理论 failure theory;theory of breaking;theory of failure
破坏力 collapsing force;destructibility;destructive force;destructive power
破坏力矩 breaking moment;destabilizing moment;failure moment;moment of rupture;failure moment
破坏力学 damage mechanics
破坏连贯性 disjoint
破坏路面 destruction surface
破坏路面的交通 mutilative traffic
破坏率 damage rate;damage ratio;failure rate
破坏率曲线 rate-of-failure curve
破坏密封 break seal
破坏面 breaking plane;failure surface;rupture plane;sliding surface;surface of fracture;surface of rupture
破坏面积 failed area
破坏名胜古迹罪 crime of destroying scenic spot
破坏模式 failure pattern;mode of failure;failure mode
破坏能 damage energy
破坏能力 breaking capacity;damage capability;damage potential;destructive capacity;failure capacity
破坏能量 breaking energy;energy of rupture;failure energy
破坏扭矩 breakdown torque
破坏判别标准 failure criterion[复 criteria]
破坏判别准则 failure criterion[复 criteria]
破坏判据 failure criterion
破坏平衡 destruction of balance;disruption in the balance;rhexistasy
破坏平面 breaking plane;failure plane;plane of failure
破坏平原 destruction plain
破坏评定标准 failure criterion
破坏期 breakdown time
破坏器 destructor
破坏前的 prerupture
破坏前流动 prerupture flow
破坏前响应 prerupture response
破坏潜势 damage potential
破坏强度 breakdown strength;breaking resistance;breaking strength;cleavage strength;collapse strength;collapsing strength;crushing strength;disruptive strength;failure strength;intensity of damage
破坏强度设计法 ultimate strength design method
破坏强度试验 breaking strength test
破坏情况 state of failure
破坏球 demolition ball
破坏区 destruction region;failure area;failure zone
破坏区的厚度 thickness of failure zone
破坏区域 failure zone;rupture zone
破坏韧性 fracture toughness
破坏乳化液 breakdown of emulsion
破坏伸长度 elongation at failure
破坏伸长率 elongation at failure
破坏生态罪 ecologic(al)crime
破坏生物的 biolytic
破坏石料的生物 stone-destroying animal organism
破坏时变形 deformation at failure
破坏时间 time-to-failure
破坏识别 damage identification
破坏式加法 destructive addition
破坏试验 breakdown test;breaking down test;breaking test;crash test;crippling test;test to destruction;failure test
破坏试验方法 destructive testing method
破坏试验压力 burst pressure
破坏水力梯度 failure hydraulic gradient
破坏速度 breakdown speed
破坏特点 damage feature
破坏条件 breaking condition;collapse condition;failure condition
破坏图 circle of failure
破坏团体者 job spoiler
破坏弯角 failure bending angle
破坏弯矩 failing moment;rupture bending moment
破坏危险 risk of failure
破坏温度 fail temperature
破坏系数 coefficient of destruction
破坏现场调查 damage survey
破坏线 failure line;line of cleavage;line of rupture;rupture line
破坏效力 brisance
破坏效应 damage effect;destructive effect
破坏楔体 failure wedge
破坏协定 break an agreement
破坏协议 break agreement
破坏信息读出 destructive readout
破坏信息读出数 destructive reading
破坏信用 breach of confidence
破坏行动 sabotage
破坏行规 job stealer
破坏型式 damage pattern;failure mode;failure type;mode(l)of failure
破坏性 destructibility;destructive effect;destructiveness
破坏性板块边缘 destructive plate margin;destructive plate boundary
破坏性波浪 destructive wave
破坏性冲击 damaging impact
破坏性存储器 destructive storage
破坏性代谢作用 destructive metabolism
破坏性的 damaging;disruptive;subversive;destructive
破坏性的火 destructive fire
破坏性地震 damage earthquake;destructive earthquake;devastating earthquake
破坏性地震区 destructive earthquake area;ruin earthquake area
破坏性读出 destructive read(out)
破坏性读数 destructive reading
破坏性放电 disruptive discharge
破坏性风暴 blowdown
破坏性负载 collapse load;collapsing load
破坏性干扰 destructive interference
破坏性故障 destructive malfunction
破坏性光标 destructive cursor
破坏性货币流动 disruptive currency flow
破坏性货物 damaging goods
破坏性击穿 destructive breakdown
破坏性加法 destructive addition
破坏性检查 destructive examination;destructive inspection
破坏性检验法 destructive testing method
破坏性交通 mutilative traffic
破坏性结构设计法 collapse method of structural design
破坏性景观 destructional landscape
破坏性竞争 destructive competition
破坏性开采 irrational extraction of resources
破坏性力量 maelstrom
破坏性膨胀 destructive expansion
破坏性热负载 burn-out heat flux
破坏性三角洲 destructive delta
破坏性试验 breakdown test;overload test;puncture test;destruction test;destructive experiment;destructive test(ing);failure test;rupture test;fracture test
破坏性试验(跑)道 torture track
破坏性物质 destroying substance
破坏性压力 collapse pressure;collapsing pressure;damage pressure;failing stress
破坏性应变 strain failure
破坏性运动 destructive movement
破坏性杂草生长 destroy weed growth
破坏性震动 destructive shock;destructive vibration
破坏性质 failure property
破坏性资本流动 disruptive flow of capital
破坏性作用 disruptive effects
破坏压力 breakdown pressure;rupture pressure;collapse pressure
破坏压榨机 demolition press
破坏样式 failure pattern;failure mode
破坏应变 breaking strain;failure strain;rupture strain;strain at failure
破坏应力 breakdown stress;breaking stress;collapsing stress;crippling stress;damaging stress;detrimental stress;failing stress;failure stress;rupture stress
破坏应力条件 breaking stress condition
破坏应力与应变 failure stress and strain
破坏应力圆 breaking stress circle
破坏应力状态 failure stress condition
破坏用抓斗 demolition grapple
破坏有机质 destruction of organic material
破坏预测 failure prediction
破坏圆 circle of rupture
破坏运动 destructive motion
破坏载荷 breaking load;charge of rupture;collapse load;collapsing load;crushing load;load at failure
破坏张力 breaking strain
破坏者 destructor
破坏真空 vacuum break(ing)
破坏真空浮子 vacuum breaker float
破坏蒸馏 destructive distillation
破坏指令 command destruction
破坏指数 failure index
破坏终点 failure end point
破坏重锤 <起重机> skull cracker
破坏装置 breaking plant;destructor
破坏状况 fracture condition
破坏状态 collapse behavio(u)r;collapse state;damage state
破坏准则 criterion of failure;criterion of rupture;damage criterion
破坏钻头 demolition bit
破坏作用 damage effect;damaging effect;destructive effect;katogene【地】
破毁面 failure surface

P

破火山口 caldera
破火山口类型 type of calderas
破记录的 record-breaking
破纪录的长度 record length
破解理 fault-slip cleavage; fracture cleavage
破旧的 beaten up
破旧的汽车 crock
破旧的住房 rookery
破旧房屋拆迁地带 slum-cleared zone
破旧飞机 <俚语> crate
破旧汽车 crate
破旧衣服 cotton rag
破旧住宅 deteriorated dwelling house
破旧住宅区 deteriorated residential quarter
破开 rift cut
破开算子 operator compact implicit
破开算子法 operator-splitting method
破壳机 rejuvenator
破口 breach
破口总水压力 total hydraulic head on damaged part
破廓图 dog ear
破烂 frazzle; tumble down
破烂车辆 deadline vehicle
破烂的 rotten; tumble-down <如房子等>
破烂的帆布 ragged canvas
破烂货 junk
破烂建筑物 dilapidated building
破浪 breaker; breaking of waves; heavy sea; surf; swash of wave
破浪冲蚀沟 channel formed by breakers
破浪带 breaker zone; roller of breaking waves; surf zone
破浪堤 breadthwise water; breaker dike [dyke]; water-break; wave breaker
破浪墩 wave breaker
破浪阶地 breaker terrace
破浪前进 breast the sea; hack the sea; plough the sea; plow the ocean
破浪墙 breaker wall
破浪石砌体 <桥墩式护岸> break-water glacis
破浪线 <海滨的> line of wave breakers
破浪运动特征 motion characteristic of wave
破浪作用力 force of breaking wave
破例的 unconventional
破裂 breakaway; breaking; break off; break open; break out; break-up; chink; cracking; disrupture; fracture; give way; lacerate; outbreak; rift; rupture; splintering; splitting; tear; brashness <指木材受震动时木纹的破裂>
破裂板 fracture plate
破裂包 bale burst
破裂包(络)线 rupture envelope; envelope of rupture
破裂变位 deflection at rupture
破裂变形 failure by rupture
破裂标准 fracture criterion
破裂表面 rupture surface
破裂波 breaking wave
破裂长度 rupture length
破裂成块的 ragged
破裂尺寸 fracture dimension
破裂传播 rupture propagation
破裂传播裂隙 fracture-propagating flaw
破裂带【地】 fracture(d) zone; rift zone; rupture zone; zone of fracture
破裂带高度 height of fractured zone
破裂带名称 name of fracture zone
破裂带钻井记录 fracture log
破裂的 broken; disruptive; ruptured
破裂地层 broken ground; shelly ground
破裂点 break point; failure point; fracture point; point of rupture
破裂电导 disruptive conduction
破裂断面 breaking cross-section; broken-out section; fracture section
破裂范围 rupture area; rupture range
破裂方式 failure mode
破裂放电 disruptive discharge
破裂构件 fractured member; rupture member
破裂构造 fractural structure
破裂过程 failure process; rupture process
破裂荷载 crack loading
破裂机构 fracture mechanism
破裂机制 fracture mechanism
破裂极限 breaking limit; limit of rupture
破裂极限圆 breaking limit circle
破裂角 angle of break(ing up); angle of rupture; fracture angle
破裂孔 fracture bore
破裂理论 theory of rupture
破裂力 breakout force
破裂密度函数 failure density function
破裂面 breaking face; breaking plane; fracture(d) surface; fracture plane; lithoclase; plane of fracture; plane of rupture; surface of fracture; surface of rupture; failure plane <土体的>; plane of break <岩石的>
破裂面半径 radius of rupture
破裂面边缘 edge of fracture
破裂模量 rupture modulus
破裂模式 fracture pattern
破裂木纹 rupture(d) grain
破裂能力 breaking capacity
破裂能量 energy of rupture
破裂盘 rupture disc[disk]
破裂片 rupture disc[disk]
破裂频率 fracture frequency
破裂平面 degradation level
破裂破坏 failure by rupture
破裂强度 breakdown strength; breaking down strength; bursting strength; cracking strength; disruptive strength; fracture intensity; rupture strength
破裂区 ruptured zone
破裂声 clap
破裂试验 bursting pressure test; fracture test; rupture test; test to destruction
破裂寿命 rupture life
破裂瞬间 instant of failure
破裂速度 rupture speed; rupture velocity
破裂速率 rupture rate
破裂碎片 breakage
破裂停止温度 crack arrest temperature
破裂图 rupture diagram
破裂外观 fracture appearance
破裂弯矩 failure bending moment
破裂系 fracture system
破裂系数 coefficient of rupture; modulus of rupture
破裂线 break line; cutting plane line; fracture line; line of fracture; line of rupture; rupture line
破裂效应 rending effect
破裂楔件 failure wedge
破裂楔体 <土体的> failure wedge
破裂形状 break pattern; fracture pattern
破裂型态 fracture pattern
破裂性 broken condition; disruptiveness
破裂性地壳运动 fractured earth movement
破裂性张拉力 bursting tension
破裂性状 fracture behavio(u)r
破裂压力 breakdown pressure; bursting pressure; disruptive pressure
破裂岩 cataclasite
破裂岩层 shelly formation
破裂岩石 faulted rock; fracture(d) rock
破裂岩石区 area of faulted rock
破裂音 crackling
破裂应变 falling strain; rupture strain
破裂应力 breakaway stress; breaking stress; falling stress; rupture stress; splitting stress; disruptive stress
破裂应力与应变 rupture stress and strain
破裂应力圆 breaking limit circle; circle of rupture; rupture circle; rupture diagram
破裂缘 fracture edge
破裂运动 rupture motion
破裂转速 burst speed
破裂锥形面 fracture cone
破裂准则 fracture criterion
破裂阻力 resistance to rupture
破裂阻力系数 breaking factor
破裂作用 disruption; regmagenesis【地】
破鳞轧制 spellerizing
破漏包件必须移至安全地点 <用于易燃液体> leaking package must be removed to a safe place
破路抓斗 demolition grapple
破轮检测器 <装设在峰顶或到达场入口处> broken wheel detector
破轮缘检测指 broken flange detector finger
破落 deterioration
破落街区 deteriorating neighborhood
破落区 blighted area
破落住房 deteriorating housing
破密(码)机 code breaker
破沫板 anti-foam plate
破沫设施 defoaming device
破沫装置 anti-foam package
破泡剂 foam breaker; foamicide; foam killing agent
破劈理【地】 fracture cleavage
破劈理带 fracture cleavage belts
破片 flinders; fragment; scrap; shatter; shiver; snatch
破片飞散 fragment emission
破片飞散试验 fragmentation emission test
破片砂岩 sparagmite
破乳 breakdown of emulsions; demulsification; emulsion break(ing); splitting of emulsions
破乳点 breaking point
破乳电压 emulsion-breaking voltage
破乳化作用 de-emulsification
破乳剂 demulsifier; demulsifying agent; emulsion breaker; emulsion splitter
破乳率 emulsion breaking rate
破乳器 emulsion breaker
破乳浊 breaking of emulsion
破砂机 sand aerator
破伤风 tetanus
破伤指数 laceration index
破石锤 cutting hammer
破石工 sledger
破石凿 boasting chisel
破碎 breakage; breaking; breaking up; cataclase; comminution; fracture; fragmenting; granulation; shattering; smash
破碎板 breaker plate; crushing member; crush(ing) plate
破碎泵 disintegrator pump
破碎比 degree of size breakage; ratio of crushing; ratio of reduction; ratio of size reduction; reduction ratio; size reduction ratio
破碎变形 clastic deformation
破碎变质【地】 catachosis
破碎变质带 katamorphic zone
破碎变质现象 katamorphism
破碎变质岩 katamorphic rock
破碎表面 surface of fracture
破碎冰 ice flake
破碎波 breaker; broken wave; breaking wave
破碎波带 breaker zone; surf zone
破碎波高 breaking wave height; breaker height
破碎波荷载 broken wave load
破碎波涛 breaking sea; breaking wave
破碎波挟带空气的冲击 ventilated shock
破碎波线 breaker line
破碎波压力 breaking wave pressure
破碎玻璃 broken glass
破碎步骤 broken step
破碎材料 broken material
破碎仓 crushing compartment
破碎操作 breaking operation
破碎槽 crushing cavity
破碎厂 crushing mill; crushing plant
破碎车间 crusher department
破碎成为粗粒 coarse break-up
破碎程度 degree of crushing; rate of decay
破碎冲击器 breaking impactor
破碎锤 breaking hammer; quartering hammer; wreck ball
破碎粗块的拖板 smoother bar
破碎带 broken ground; cataclastic zone; cracked zone; crushed belt; crushed zone; crushing zone; fractured zone; shatter zone; shuttered zone; troublesome zone; zone of crush; zone of fracture
破碎带变质岩 kata-rock
破碎带埋深 buried depth of breaking zone
破碎带线性体 fracture zone lineament
破碎带走向 strike of breaking zone
破碎的 crumbling; ragged; kataclastic <岩石>
破碎的冰 broken ice
破碎的混凝土 broken concrete
破碎的垃圾 shredded refuse
破碎的铁 crumbling iron
破碎的玄武岩 broken basalt
破碎地层 broken formation; broken ground; crushed stratum; friable formation; friable ground
破碎地形 broken terrain
破碎点 shatter point; breaking point; plunge point <波浪的>
破碎顶板 breaking roof
破碎动力学 kinetics of comminution
破碎动作 crushing movement
破碎度 degree of fragmentation; fragmentation; fragmentation of blasted rock; reduction range
破碎颚 crushing member
破碎颚板 crushing jaw
破碎方法 crushing of method; fracture method
破碎粉磨机械 crushing and grinding machinery
破碎浮渣 breaking scum
破碎工厂 breaking installation; break-

ing plant
破碎工具 breaker tool
破碎工艺 crushing process
破碎功 size reduction work
破碎骨料<由碎石机轧出而未经过筛选的> crusher-run aggregate
破碎管载重 crushing pipe of proof load
破碎辊 crusher roll; crushing roll; crush roller
破碎过程 crushing process; shattering process
破碎海岸 broken coast
破碎海岸线 broken coastline; broken shoreline
破碎和裂隙岩石 blocky and seamy rock
破碎和未破碎砾石的混合骨料 partially crushed gravel
破碎核 fragmentation nucleus
破碎荷载 crush load
破碎后波高 broken wave height; post-breaking wave height
破碎后能见度试验 visibility test after fracture
破碎混凝土 concrete cutting
破碎混凝土用砂 broken concreting sand
破碎机 breaker; breaking machine; bucker; chipper; chips breaker; chips crusher; cracker (mill); crusher; crushing engine [machine/mill/ plant/ press]; destroyer; devil; disintegrating mill; disintegrator; kibbler; knapper; knapping machine; nuggetizer; shredder
破碎机产品 crusher-run product
破碎机衬板 breaker plate
破碎机传动方式 crusher drive
破碎机单元 breaker unit
破碎机的系列支撑柱 breaker prop row
破碎机的支撑柱 breaker props
破碎机颚板 crusher jaw
破碎机颚板冲程 throw of the crusher
破碎机房 crusher chamber
破碎机滚筒 crushing roll; crusher roll
破碎机滚筒护套 crushing roll shell
破碎机可动圆锥 mantle cone
破碎机类型 type of breakers
破碎机能量系数 crusher capacity multiplier
破碎机排料口闭合时的间隙宽度 closed-side setting
破碎机石块 crusher-run rock
破碎机通风装置 pulverizer aerator
破碎机头中轴<圆锥破碎机> head-center[centre]
破碎机喂料 crusher feed
破碎机械 breaker machinery
破碎机轧辊 breaker roll
破碎机中间平台 crusher intermediate floor
破碎机主动轴 crusher drive shaft
破碎机锥辊 conic(al) roll
破碎机锥头 crushing head
破碎机最大接受粒度 crusher size
破碎级 size fractions of crushed coal
破碎级数 crushing steps
破碎集料 crushed aggregate
破碎剂 fracturing agent
破碎阶段 crushing section
破碎截面 fractured face
破碎金刚石 crushing bo(a)rt; crushing bortz
破碎金属块 breaking of pigs
破碎壳 crushing bowl
破碎坑 broken pits
破碎块度 breaking size; broken blockness
破碎块裂体结构 block fracture texture
破碎矿石 ore breaking
破碎矿渣砂 broken slag sand
破碎理论 theory of comminution
破碎力 shattering force
破碎力学 crushing mechanics
破碎砾石的粗骨料 crushed gravel
破碎砾石的细骨料 crushed gravel fines
破碎粒度 crushing granularity
破碎量 crushing quantity
破碎料 crusher-run aggregate
破碎裂缝 shatter crack
破碎流程 crusher circuit; crushing circuit
破碎路面钻 paving breaker drill
破碎率 crushing rate; percentage of damage; rate of breakage; rate of reduction; shatter index
破碎面 chipped surface; fractured (sur) face; free face; plane of disruption
破碎抛掷机 crusher-thrower
破碎泡沫矿渣 broken foamed (blast furnace) slag
破碎膨胀性矿渣 broken expanded cinder
破碎片 flinders
破碎剖面 broken section
破碎器 bucker; destroyer; knapper; mastax; shredder
破碎腔 crushing cavity; crushing space
破碎强度 breaking strength; crushing strength
破碎清除法<混凝土修复工艺的> break-up and clean out
破碎球 headache ball; skull cracker; wrecking ball
破碎区 crushing zone
破碎取样联合机 crusher-sampler
破碎砂 broken sand
破碎筛分 crushing-screening; size reduction sizing
破碎筛分厂 crushing and screening plant
破碎筛分设备 crushing and screening plant
破碎筛分塔 crushing and screening tower
破碎设备 breaking installation; crushing appliance; crushing equipment; crushing plant; detritus equipment
破碎石块 broken stone; crushed stone; knobbling
破碎时拣出的废石 run-of-crusher stone
破碎试验 crushing test
破碎试样分析 battery assay
破碎室 breaking chamber; crushing cavity; crushing chamber
破碎水深 depth of breaking
破碎送料斗 breaker feed hopper
破碎特征 appearance of fracture
破碎头 breaking head; crushing member
破碎图像 ragged picture
破碎外观 appearance of fracture
破碎位置 breaker position
破碎物 crumble; trash
破碎物质 broken material
破碎系数<岩石的> grindability index
破碎险 breakage risks; risks of breakage
破碎险条款 breakage clause
破碎现象 appearance of fracture; rupture appearance
破碎效应<爆破时的> shattering effect
破碎型式 crushing type
破碎性 breakability; crushability
破碎需要功 work required of size reduction
破碎压力 breakdown pressure; crushing pressure
破碎压实机 pulverizer compactor
破碎岩 fissured rock
破碎岩石 bad rock; shattered rock; kata-rock; cataclastic rock
破碎岩石层 broken rock
破碎岩石的凿子 rock-breaking chisel
破碎岩石滑坡 broken rock landslide
破碎岩体 crushed rock
破碎页岩 broken slate
破碎应变 crushing strain
破碎应力 crushing stress; fracture stress
破碎圆锥<破碎机的> crushing head; crushing cone
破碎杂料 interground addition
破碎整合【地】fractoconformity
破碎值 crushing value
破碎指数 shatter index
破碎装置 breaking plant; crushing device; crushing installation; crushing mechanism; shredding equipment
破碎锥体的锰钢壳 crushing mantle
破碎锥头 crushing cone
破碎作业 crushing operation
破碎作用 catabasis; crushing action; fragmentation; percussion action; percussive action
破损 disrepair; failure; out-of-repair; outworn; shatter; tear; wastage
破损安全 fail safe; fail-safety
破损安全试验 fail-safe test
破损安全装置 fail-safe equipment
破损保安准则 fail-safe concept
破损保养费 breakdown maintenance
破损边缘<屋顶的> breaking edge
破损不能修理(的程度) damaged beyond repair
破损车卡片插 defect card holder
破损车辆停留线 bad order track
破损成本 cost of breaks
破损导接线 damaged bond; defective bond
破损的 non-serviceable; outworn
破损的石板 shilf
破损低压 breakdown low pressure
破损点 breakdown point; breaking (down) point; point of failure
破损定额 allowance for breakage
破损分析 failure analysis; threat analysis
破损概率 failure probability
破损荷载 breaking load; load at failure
破损货 damaged cargo
破损货车 breakdown wagon
破损货物 damaged goods
破损货物估价 assessment of damaged goods
破损机构 collapse mechanism
破损机理 distress mechanism; failure mechanism; mechanism of failure
破损迹象 failure sign
破损阶段 fractural stage; rupture stage
破损阶段法 load factor method
破损阶段设计法 fractural stage design method
破损力 spalling force
破损力矩 moment of rupture; ultimate moment
破损率 break rate
破损轮车辋 shattered rim
破损赔偿 compensation for damage
破损赔偿额 compensation for breakage
破损强 breaking down point
破损强度 breakdown point; breaking point; breaking strength
破损缺陷 open defect
破损燃料 burst slug
破损燃料元件 failed fuel element
破损设计 damage tolerant design
破损试验 breadthwise down test; breakdown test; crash test; destructive test(ing)
破损水准仪 destruction level
破损条款 break-up clause
破损通知书 advice of damage
破损维修 breakdown service
破损系数 damage coefficient
破损险 risk of breakage
破损险条款 breakage clause
破损线 damage line
破损象征 distress manifestation
破损性试验 failure test
破损修理 breakdown repair
破损压力 breaking pressure; bursting pressure
破损严重 badly broken
破损应力 damaging stress
破损预测 failure prediction
破损折扣 breakage allowance
破损支票 mutilated check
破损指数 damage index; index of damage
破图廓 bleeding edge
破图廓部分 blister; border break; broken parcel
破图廓地区 broken parcel
破图廓图 broken border diagram
破土典礼 ground-breaking ceremony
破土(动工) break earth; break ground; ground breaking; breaking of ground
破土剂 soil-fracturing agent
破土开工 break ground
破土犁 breaker
破瓦 rubble
破相 disfigurement
破镶面的 brick-veneered
破晓 dawn
破斜纹 broken twill
破屑机 chip-breaker
破心下料 sawing through the pith
破型时间 time of fracture
破岩破坏两用抓斗 rock and demolition grapple
破岩抓斗 demolition-rock grapple
破译 cryptanalysis
破译密码 decipher
破渣机 slag breaker
破绽 rip
破折线 broken line

剖

剖半砖 split brick
剖层机 splitting machine
剖分 dissection; subdivision
剖分带锯机 band resaw
剖分式齿轮 split gear
剖分式导管 split guide
剖分式阀导管 split valve guide
剖分式紧定套 split adapter sleeve
剖分式外壳 two-piece housing
剖分式轴承座 split housing
剖分数目 number of divisions
剖分为三角 triangulate
剖分轴衬 split bush
剖分轴承 split bearing
剖截 parting
剖锯 deep-sawing
剖开 dissect(ion)
剖开的木材 cleaved wood
剖开断面 ripping-up

剖开管 split pipe
剖开检验 examination by sectioning
剖开立体图 cutaway drawing; cutaway view
剖开透镜 split lens
剖壳式泵 split-casing pump
剖口 bevel; groove
剖料 breakdown
剖料锯 head saw; rift saw
剖裂锤 splitting hammer
剖露地形 exhumed topography
剖面 cross-section; section plane; broken-out section; cutaway section; cut plane; lateral section; orthography; plane section; right section; section; sectional detail
剖面爆破法 profile shooting (method)
剖面剥蚀的 truncated
剖面采样 profile sampling
剖面测量 cross-sectioning; profile survey
剖面测量法 profile surveying
剖面长度 length of profile; profile length
剖面尺寸 section size
剖面磁方位 magnetic azimuth of profile
剖面的 cutaway; sectional
剖面点 cross-section point; profile point
剖面法 cross-section method; exploration method by profile
剖面方位 orientation of profile
剖面方向 profile direction
剖面分析法 section topography
剖面符号 section symbol
剖面高度 profile height
剖面构造应力场 tectonic stress field at profile
剖面回转半径 radius of gyration of section
剖面计算机 profile computer
剖面计算器 profile calculator
剖面记录本 profile book
剖面记录表 record table of profile
剖面记录器 cross-section recorder; profile recorder
剖面间隔 cross-section interval; profile interval
剖面交点 cross point of profile
剖面阶梯型排序 echelon sections
剖面距离 profile distance
剖面立视图 sectional elevation
剖面裂缝 section crack
剖面流网图 flow net profile
剖面面积 area of section; sectional area
剖面描述 description of profile
剖面模板 profile board
剖面模量 modulus of section; profile modulus; section modulus
剖面模数 modulus of section; profile modulus; section modulus
剖面模型 profile model
剖面平视图 section plan
剖面起始号 number of starting point of the section
剖面上组合 composition on the section
剖面示意图 constructed profile
剖面数据 cross-sectional data; profile data
剖面水头分布图 profile of water head distribution
剖面缩图 contracted cross-section
剖面条数 number of profiles
剖面透视图 sectional perspective
剖面图 profile; sectional profile; sectional view; sectional diagram; sectional drawing; cutaway drawing <切去一角露出内部结构的>; relief diagram
剖面图程序 profile program(me)
剖面图法 profile method
剖面位置 location of profile; profile position
剖面系数 sectional coefficient; section modulus
剖面线 cross hatch; cross-section(al) line; hatching; line of section plane; profiling line; section line; thalweg
剖面线法 section lining
剖面线及编号【岩】geological section line and numbed; profile line and no.
剖面线与岩层倾向夹角 angle between section line and bed dip
剖面线与岩层走向夹角 angle between section line and bed strike
剖面选择器 cross-section selector
剖面样板 profile board
剖面样条 bisect
剖面仪 profiler
剖面终止号 number of end point of the section
剖面重力测量 profile gravity survey
剖面资料 profile date
剖面自动记录仪 profilograph
剖面总长度 total profile length
剖平障壁砂坝圈闭 bevel(l)ed-barrier bar trap
剖切平面 sectional plane
剖石机 split-face machine
剖视面 sectional plane
剖视图 cross-section view; cutaway illustration; cutaway view; cut-open view; exploded drawing; exploded view; sectional elevation; section-(al) view
剖土机 knifer
剖析 analyse; dissect; explode
剖析表 parse list
剖析模型 anatomic(al) model
剖线边沿 bank of a cut
剖线的正边 positive edge of cross cut

仆

仆人居住部分 servant's quarter

仆人室 servant's room
仆人浴盆 servant's bath
仆人浴室 servant's bathroom

扑

扑火 attack a fire; fire suppression

扑火大本营 base camp
扑火队员 fire fighter
扑火进度图 fire progress map
扑火力量 strength of attack
扑火器 flail
扑火外站 stand-by crew
扑火野外指挥部 base camp
扑火营部 base camp
扑火总指挥 fire boss; line boss
扑灭 extinguish; smash; stifle
扑灭方法 putting-down method
扑灭火 suppress fire
扑灭火焰 darken
扑拍门扇 flop gate

铺

铺柏油的<路面> tar-paved

铺板 berth bent; boarding; decking; floor plate; panking; plank; plate setting; slabbing
铺板道口 timbered crossing
铺板工作 planking
铺板条 slatting
铺板托梁 fish plate
铺板甬道 boarded gangway
铺扁铁条 lacing of flat bars
铺薄沥青表层 bituminous decking
铺薄铁板的人工集水区<为了增加径流的> iron-clad catchment
铺舱作业 fitting for grain
铺草 sward
铺草的花圃 sod bed
铺草的花坛 sod bed
铺草的区域 sodded area
铺草的中央分隔带<多车道的> grassed central reserve
铺草地带 grassed area; grass-surfaced area
铺草地面 grassed surface
铺草路 grass walk
铺草路堤 grass embankment
铺草排水沟 turf drain
铺草皮 sod(d)ing; turf(ing)
铺草皮的边坡 turfed slope
铺草皮的草地 turfing by sodding
铺草皮的泄水沟 sodded "leak-off"
铺草皮地带 sodded strip
铺草皮地区 grassed area; grassed region
铺草皮块草坪 sodding lawn
铺草皮路肩 grass shoulder
铺草皮路缘 sod curb
铺草皮面积 grassed area; sodding area; turfed area; turfing area
铺草皮坡 sodded slope
铺草皮水渠 sodded channel
铺草皮于 sod
铺草皮作业 sodding work; terracing with sod
铺草人行道 grass walk
铺草屋顶芦苇垫底 fleaking
铺草席固沙 fixed sand by mat shape grass land
铺草运动场 straw-yard; barnyard
铺侧石的步行道 kerbed footway
铺侧石机 curber
铺层 layup; overlap
铺层边缘的标识 paved shoulder mark(ing)
铺层成型 layup mo(u)lding
铺层翻修 pavement recycling
铺层厚度 layer thickness
铺层器 spreader
铺层清除铲斗 pavement removal bucket
铺层设计 lamina design
铺撒料 interspersed matter
铺撒填缝石屑 blinding
铺床 bed-making
铺瓷砖 tiling
铺瓷砖的 tiled
铺瓷砖工用的镘刀 floor tiler's trowel
铺瓷砖用锤 tile hammer
铺粗地面 rough flooring
铺大瓦用短柄斧 shingling hatchet
铺带机 tape-placement machine
铺导航跑道路面 pilot surfacing lane
铺道渣 coffering
铺道床 build road-bed
铺道木 skidding
铺道砟 ballasting; covered with gravel
铺道砟的 ballasted
铺道砟轨道 ballasted track
铺道砟面板 ballasted deck
铺道砟桥面 ballast bridge floor; ballasted deck; ballasted floor
铺底 bottoming; underlay
铺底层 bedding course
铺底大块石 soling
铺底混凝土 subbase concrete
铺底料 grate layer material
铺底料的运输带和漏斗 bedding conveyer and hopper
铺底流动资金 bottoming circulating funds
铺地板 boarding; floor board(ing); flooring
铺地板仓库 floored warehouse
铺地板长方小木块 sett
铺地板的漆布 floor cloth
铺地板工人 floorer
铺地板厚漆布 floor cloth
铺地板机 floor machine
铺地板(木)块 flooring block
铺地板用工程砖 engineering brick floor(ing)
铺地板油毛毡 flooring felt
铺地板(錾紧)凿 floor chisel
铺地材料 flooring material; pavement
铺地草本植物 carpet herb
铺地瓷砖 flooring finish tile; floor(ing) tile; porcelain tile
铺地大理石 flooring marble
铺地辐射板 flooring radiant panel
铺地缸砖 floor quarry
铺地建筑砖 floor brick
铺地空心块 end pot
铺地空心陶砖 hollow clay tile
铺地块材 paving block
铺地沥青 tarmac
铺地沥青水泥 paving asphalt cement
铺地马赛克 floor(ing) mosaic
铺地面 flooring
铺地面和墙面砖工的助手 floorlayer's labo(u)rer
铺地面软木 flooring softtimber; flooring softwood
铺地面织物 floor cloth
铺地面砖 floor tile; paving brick; paving tile
铺地面砖和墙面砖工人 tile slabber
铺地黏[粘]土砖 flooring clay brick
铺地砌块 pavio(u)r
铺地熔渣 paving clinker
铺地石板 flooring slate; floor slab
铺地塑料 flooring plastics
铺地毯 carpet
铺地陶砖 earthenware paving block; earthenware paving tile
铺地镶嵌锦砖 flooring mosaic
铺地用沥青 tarmac; tar macadam
铺地用漆布 jaspe
铺地用毯 flooring felt
铺地园 paved garden
铺地织物 floor cloth
铺地植被 matted vegetation
铺地植物 mat plant; mattae[mattao]; paving plant
铺地砖 floor(ing) brick; floor quarry; porcelain tile; quarry tile
铺地砖的响声试验 paving brick ratti-er test
铺地砖机 travel(l)ing block machine
铺地砖土 paving brick clay
铺第二线 laying of second track
铺电缆船 cable laying vessel; cable layout vessel
铺垫 bedding; foreshadowing
铺垫层框架 bedding frame
铺垫肩砂浆 shouldering
铺叠 layup
铺钉石棉板 asbestos boarding
铺顶沥青 roofing asphalt
铺方边地板 square-edge floor boarding
铺方块石 paving flag
铺方块石路 block stone road
铺防水面层 lay waterproof skin
铺放 laying end; laying-up; laying yard
铺放工程船 laying vessel; layout vessel
铺放轨枕的道砟面层 track bed
铺放机 layer

铺放块体护岸 placed-block revetment
铺放绒布 bench cloth; laying cloth; setting cloth
铺放水龙带 hose-lay
铺放速度 rate of placing
铺放者 layer
铺放整修机 laying-and-finishing machine
铺复轨 rerailing
铺复线 laying of second track
铺覆盖层 mulching
铺覆盖料 mulching
铺覆性 drapabiltiy
铺盖 blanketing; overlay; pavement
铺盖材料 blanketing material
铺盖灌浆 area grouting; blanket grouting
铺盖排水 blanket drain
铺盖砂 gritting sand
铺盖式固结灌浆 blanket grouting
铺盖式护岸 blanket revetment
铺盖式护坡 blanket revetment
铺钢轨的底板 railed floor
铺钢轨用辊子 permanent way roller
铺格条 riffling
铺骨料的路肩 aggregate-surfaced shoulder
铺刮沥青工具 asphalt raker
铺刮沥青工人 asphalt raker
铺管 laying of pipe; pipe laying; piping
铺管班 laying crew; laying gang; laying party
铺管驳(船) laying barge; pipe-laying barge
铺管驳施工方法 lay barge method
铺管长度 laying length
铺管车 laying vehicle
铺管船 laying barge; laying ship; pipe-laying vessel; laying vessel; layout vessel; pipe-layout vessel
铺管垫床 equalizing bed
铺管浮筒 laying buoy
铺管工 pipe layer; pipe liner; piping layer
铺管工程 piping work
铺管工地 laying site; placing site
铺管工具 laying tool
铺管工作 piping work
铺管机 pipe handling machine; pipe layer; pipe-laying crane; pipe-laying machine; pipe-laying tractor; pipe liner; pipelining machine; pipelayer; piping lining machine
铺管机械 pipe-laying equipment
铺管计划 laying plan
铺管绞车 laying winch
铺管进度表 progress chart for pipe-laying
铺管进度图表 pipe-laying progress chart
铺管履带拖拉机 laying cat; laying crawler tractor
铺管牵引车 pipe-laying tractor
铺管设备 pipe-laying gear
铺管速度 placing rate; rate of laying; rate of placing
铺管图 pipe-laying drawing
铺管托架 <铺管船上的> stinger
铺管现场 laying site; pipe-laying site; placing site
铺管业务 pipelining practice
铺管作业 pipe-laying operation; laying operation; laying work
铺轨 laying rail; track construction; tracking; track laying
铺轨班 rail shift
铺轨长度 length of track laid out
铺轨车 crawler-type vehicle; rail-laying car; track-laying car; track-laying vehicle
铺轨锤 track layer's hammer
铺轨吊车 rail-laying crane; track-laying crane
铺轨队 rail-laying gang; track-laying gang
铺轨工 groundman
铺轨工班 steel gang
铺轨工人 track layer
铺轨工作 rail-laying work; railroad track-work
铺轨机 rail layer; track layer (machine); track-laying crane; track-laying gantry; track-laying machine
铺轨机械 rail-laying equipment; track-laying machinery
铺轨基标 surveying base point for track laying
铺轨基地 track construction base; track construction depot; track-laying depot
铺轨架桥机 track-laying and girder erecting machine
铺轨列车 track construction train; track-laying train
铺轨龙门架 track-laying gantry
铺轨起重机 track-laying crane
铺轨千斤顶 tracking jack
铺轨牵引车 track-laying tractor
铺轨式拖拉机 track layer tractor
铺轨拖拉机 track-laying tractor
铺黑色路面的郊区 blacktopped purlieus
铺黑色面层 blacktopping
铺红地毯的 red carpet
铺护岸 beached bank
铺画列车运行图 diagram(m)ing
铺画列车运行线 plot the paths of trains
铺灰浆 application of mortar
铺灰镘 buttering trowel
铺混凝土 spread concrete
铺混凝土机 spreader
铺基层用碎石 crushed rock base material
铺基础底层 planting base course
铺集料的路肩 aggregate-surfaced shoulder
铺浆面 top face
铺浆砌砖法 western method
铺浇多层焦油沥青 multiple tarring
铺胶镘刀 application spreader
铺胶器 mastic spreader
铺金属片屋面材料 metal sheet roof-(ing)
铺金属屋面 metal roofing
铺开 unfold; unroll
铺开的 unrolled
铺块 paving block
铺块厚度 depth face
铺块路面 block pavement
铺沥青 blacktopping
铺沥青工程 asphalt work
铺沥青工人 asphaltic-bitumen layer
铺沥青工作 asphalt work
铺沥青路面加热器 devil
铺砾石 gravel(l)ing
铺砾石的进水口 blind inlet
铺砾石地区 gravel(l)ed area
铺料 placing and spreading
铺料机 pav(i)er; spreading machine
铺料料斗 spreading hopper
铺料式卷线机 laying reel
铺料箱 spreader box
铺楼板 flooring
铺楼面软木 flooring softtimber; flooring softwood
铺楼面用板条 flooring batten
铺楼面用石板 flooring slate
铺路 flagging; pave; paving; road laying
铺路安山石毛石 andesite paving sett
铺路板材 pavement slab; paving slab
铺路拌和机 paving mixer
铺路玻璃 paving glass
铺路薄片石 paving flag; paving stone
铺路材料 paving material; pavio(u)r; tarmac <沥青与焦油、石子等混合面而成的>
铺路承包人 paving contractor
铺路承包商 paving contractor
铺路方块 cube
铺路方石 pitcher
铺路方石块 cube paving sett
铺路方砖 square paving brick
铺路缸砖 vitrified deck floor surfaced with road pavement; vitrified paving brick
铺路钢丝垫 steel landing mat
铺路工的整面锤 paver's hammer
铺路工(人) pavio(u)r [pav(i)er]; plate layer; pavio(u)r labo(u)r; street mason
铺路骨料 paving aggregate
铺路规划 paving project
铺路夯 pavement ram; paving rammer; paving tamper
铺路夯锤 paving rammer
铺路夯具 paving beetle
铺路混合料 paving mixture
铺路混凝土 paving concrete
铺路混凝土板 paving flag
铺路混凝土砌块 concrete paving block
铺路机 laying machine; pav(i)er; paving machine; pavio(u)r; road surfacer; spreader
铺路机动列车 paving train
铺路机械列车 paving train; surfacing train
铺路机用水准仪 pavio(u)r's level
铺路机组 paving train
铺路及敷设管道工作队 bull gang
铺路集料 paving aggregate
铺路计划 paving project
铺路块 paving unit
铺路块料 pattern display; paving block
铺路矿渣 slag paving stone
铺路矿渣块料 slag paving sett
铺路沥青 road building bitumen; road oil
铺路沥青的扩展 spreading of binders
铺路沥青的撒布 spreading of binders
铺路沥青的涂刷 spreading of binders
铺路沥青(间)空隙 bituminous tubes; bituminous pipes
铺路面 paving; road surfacing; surfacing
铺路面材料的泵送 pumping of pavement
铺路面材料的工人 spreader
铺路面底基层 underlay
铺路面方石 cube paving sett
铺路面前的路基准备 conditioning of road bed
铺路面碎石 road metal
铺路面用筑路机 all-in-one pav(i)er
铺路面砖 paver[paving] tile
铺路乳化沥青 road emulsion
铺路洒油机 road oiler
铺路砂 paving sand
铺路设备 paving equipment
铺路设备列车 paving plant train
铺路伸缩缝 paving expansion joint
铺路石 cobble; flagstone; pavestone; paving cobble; paving stone; slab-stone; stone flag; pitcher <英>
铺路石板 dalle; flag(stone); paving flag; paving sett; rag stone
铺路石块 paving sett; pitching block; paving stone
铺路石块加热 sett burning
铺路石料 road stone
铺路时的密实度 placement density
铺路熟料砖 paving clinker brick
铺路水泥 paving cement
铺路税 pavage
铺路碎石 metal; slag paving stone; metal(l)ing
铺路碎石料 <英> metal
铺路填缝用的细石屑和砂 blinding
铺路小方块石 stone paving sett
铺路小方石 paving sett; sett
铺路小方石测试仪 paving sett tester
铺路小方石润滑脂 sett grease
铺路硬砖 Dutch clinker
铺路用的 paving
铺路用地沥青 paving asphalt
铺路用矿渣碎石 slag paving stone
铺路用沥青 paving asphalt
铺路用沥青水泥 paving asphalt cement
铺路用沥青碎石 bitumen macadam
铺路用六角形混凝土块 concrete paving hexacrete
铺路用卵石 cobble boulder
铺路用石板 flagging stone
铺路用油 dust-binding oil
铺路油 asphalt(um) oil; dust-binding oil; road oil; mas(o)ut <重质石油蒸馏残油>
铺路油处治 <土路的> oil treatment
铺路油处治路面 oil-treated surface
铺路油处治面层 oil-treated surface
铺路缘石机 curber
铺路振动器 paving vibrator
铺路铸铁块 cast-iron road paving block
铺路砖 clinker brick; Ipro brick; paving brick; pavio(u)r
铺卵石(路面) pebbling
铺卵石水沟 cobble gutter
铺马赛克 laying of mosaics; mosaic laying
铺茅草顶 thatching
铺面 facing; matting; pavement; paving; revetment; shop front; surfacing; non-structural top screed <混凝土>; storefront <商店的>
铺面板 decking; flooring board; pavement slab; slab; surfacing board; surfacing sheet
铺面板材 paving slab
铺面表面容许弯沉 allowable rebound deflection of surface pavement
铺面玻璃 pavement glass
铺面玻璃砖 deck glass
铺面材料 paver; paving; paving material; surface covering material; surfacing material
铺面层 surfacing course
铺面橱窗 shop window
铺面瓷砖 vitrified paving brick
铺面大石板 broad stone
铺面道路 paved road; paving road; surfaced road
铺面的底层 base course of pavement
铺面方块 cube
铺面缸砖 promenade tile
铺面工人 pavio(u)r
铺面夯具 paving beetle
铺面厚度 depth of pavement
铺面黄色硬砖 Flemish brick
铺面混合料 paving mixture; surfacing mixture
铺面混凝土 pavement concrete; surface concrete; topping concrete
铺面混凝土板 surfacing concrete slab
铺面机 pav(i)er; paving machine
铺面基层 pavement base
铺面交叉口 paved crossing

铺面结构 pavement structure
铺面块 paving block
铺面块体 floor board(ing); paving block; paving flag
铺面宽度系数 factor of pavement width
铺面栏板 stallboard
铺面沥青 paving asphalt
铺面料场 paved storage yard
铺面路 improved road; surfaced road
铺面螺钉 flooring nail
铺面码头前沿 paved apron
铺面木块 paving wood; paving wood block
铺面跑道 paved runway; paving course
铺面破碎机 paving breaker
铺面墙裙板 paved apron
铺面渠道 paved channel
铺面人行道 paved footway
铺面伸胀缝 paving expansion joint
铺面石 paving stone
铺面石板 paving sett; paving stone
铺面石屑 surface dressing chip(ping)s
铺面铁路道口 paved railway crossing
铺面停机坪 paved apron
铺面系统 paving system
铺面修整机 paving finisher
铺面砖 pavement brick; paving brick; paving tile
铺面砖嵌条 diagonal tile
铺木 wood block
铺木板 boarding; wood planking
铺木纤维板吸声顶棚 wood fiber[fibre] board acoustic(al) tiled ceiling
铺排水管 moling
铺平 spreading
铺平道路 smooth the way
铺平木 pie plate
铺平台的铝制件 alumin(i)um deck unit
铺坡砖 <大坡度时铺路用的炼砖> hillside brick
铺企口板 match boarding
铺砌 laying; layup; paving; pitching; set in
铺砌层 coverage; pavement; paving course; racking course
铺砌车道 <货场> paved driveway
铺砌锤 pavio(u)r's hammer
铺砌道路边沟 paved gutter
铺砌的 paving
铺砌的谷道 paving notch
铺砌的路肩 paved shoulder
铺砌的跑道 paved runway
铺砌的泄水沟 paved "leak-off"; paved drain
铺砌地板 paved floor
铺砌地面 paved flooring
铺砌法 pitching method
铺砌工程 paving
铺砌工(人) pavio(u)r; pav(i)er
铺砌工用的水平仪 paver's level
铺砌沟 paved ditch; paved gutter
铺砌过水路面 paved ford
铺砌夯具 paving rammer
铺砌护岸 riprap placement
铺砌护面 facing of pitching
铺砌混凝土 laid concrete
铺砌混凝土块护坡工程 block pitching
铺砌机 paving machine
铺砌基础 pitched foundation
铺砌技术 laying technique
铺砌街沟 paved ditch; paved gutter
铺砌进水口 paved inlet
铺砌块 paving block; paving tile
铺砌块面 block pavement
铺砌沥青 spraying of asphalt; spraying of bitumen
铺砌路面 block stone pavement
铺砌卵石 paving pebble
铺砌马路边沟 paved gutter
铺砌毛石 coursed rubble
铺砌面 pavement
铺砌木块 wood paving
铺砌黏[粘]结剂 bonding medium for laying
铺砌片石 pitching stone
铺砌桥面 paved floor
铺砌渠道 paved channel
铺砌石 paving stone
铺砌石板 flagging; flagstone
铺砌石块工作 pitched stone work
铺砌石块路面的方法 Kleinpflaster
铺砌式道沟 paved gutter
铺砌式道口 paved crossing
铺砌式交叉 paved intersection
铺砌式路肩 paved shoulder
铺砌速度 laying rate
铺砌条石 paving slab
铺砌图案 paving pattern
铺砌小石板 <不规则地铺砌> scantle slating
铺砌斜坡 pitched slope
铺砌仰拱 paved invert
铺砌窑底 bottom pavement; bottom paving
铺砌用模型板 laying pattern
铺砌用黏[粘]结剂 bonding adhesive for laying; bonding agent for laying; bonding medium for laying
铺砌整平机 laying-and-finishing machine
铺桥面板 flooring
铺人行道石板 stone footpath paving flag
铺软木地面 cork flooring
铺撒 scatter
铺撒砂砾 <路面> grit spreading
铺撒石屑 <路面> grit spreading
铺撒物 interspersed matter
铺洒碾压机 laydown machine; laying and finishing machine
铺洒小粒石面层 mineral surface
铺散的 diffuse
铺砂 sand dressing; sanding(-up); sand spreading; sand up; strewing sand
铺砂层 sand bed(ding); sand carpeting
铺砂的 sand cloth; sanded
铺砂法 <测路面纹理深度用> sand patch method
铺砂(法)试验 sand patch test
铺砂服务 sanding service
铺砂工程 gritting
铺砂工作 gritting
铺砂机 gritter; gritting machine; sand distributor; sand dressing machine; sand gritter; sanding machine; sand processing machine; sand spreading machine; frost-gritting machine <防止路面冰滑的>
铺砂浆 application of mortar
铺砂浆垫层 mortar bedding
铺砂浆技术 rendering technique
铺砂沥青毡 sand bitumen felt
铺砂砾 gritting
铺砂砾路 gravel(l)ed path
铺砂跑道 sand race track
铺砂器 sanding gear
铺砂设备 sand dressing machine; sanding apparatus; sanding equipment; sanding plant
铺砂石 blinding
铺砂小路 sand path
铺砂装置 sanding gear
铺杉木皮的屋面 cryptomeria bark roofing
铺上层桥面 upper-decking
铺上屋脊处的屋面材料 crown course
铺舌榫地板 tongued floor cover(ing); tongued floor(ing)
铺设 laying; pavage; placing
铺设波形钢瓦 steel pantiling
铺设薄沥青表面 bituminous decking
铺设材料 laying material
铺设侧石机械 kerb machine
铺设长度 laying length; laid length
铺设厂内管线 plant piping
铺设导线 laying guide; laying line
铺设道岔【铁】 lay a switch
铺设的 laid
铺设的混凝土面层 <地下室沥青地面上> loading concrete
铺设地板 flooring; planching
铺设地板的仓库 floored warehouse
铺设地板块 flooring block
铺设地点 laying point
铺设第二线 doubling of the track
铺设方法 laying technique
铺设防扩散覆盖层 cap placement
铺设费用 laying cost
铺设格式 laying pattern
铺设工 shingle applicator
铺设挂瓦条 ground work
铺设管道 laying pipe; pipe laying; run a line; tubing
铺设管道附件 laying attachment
铺设过渡段路面 approach pavement to the road decking
铺设海底电缆 marine cable laying
铺设豪斯勒式屋面用纸 paper for Hausler type roof(ing)
铺设厚度 laydown thickness
铺设厚度控制杆 thickness control handle
铺设机 <铺设沥青混凝土用> laydown machine
铺设机械 laying machine
铺设基础底层 laying subbase
铺设计划 laying plan
铺设技术 laying technique
铺设绞车 <管道> laying winch
铺设金属格栅地板 metal grid floor(ing)
铺设金属格子地板 metal grid floor(ing)
铺设金属网格地板 metal grid floor(ing)
铺设锦砖地面 mosaic floor(ing)
铺设宽度 laying width
铺设路缘石 curbing
铺设路缘石机械 kerb machine
铺设码垛 palletization[palletisation]
铺设面积 laying area
铺设木地板 wood flooring
铺设拼花地板 parquetry floor cover(ing)
铺设权 charter
铺设水带线 hose layout
铺设速度 laying rate
铺设铁路 laydown railway
铺设图 laying drawing
铺设温度 laying temperature
铺设线 laying line
铺设镶木地板 parquetry floor cover(ing)
铺设整平器 laying screed
铺湿砂养护混凝土 wet sand cure of concrete
铺石 stone laying
铺石板 slab; slate; slating work; slatting
铺石板工程 slater's work
铺石板工用的修琢工具 slater's iron
铺石板工作 slater's work
铺石板路 flags; paving flag
铺石板瓦 roof slating; weather slating; slating
铺石板瓦规 slating ga(u)ge
铺石板瓦屋面 slate roofing
铺石板屋面 slater's roofing
铺石锤 pavio(u)r's hammer
铺石的 ston(e)y
铺石扶壁 stone counterfort
铺石格条流洗槽 rock riffle sluice
铺石固沙 fixed sand by bedding stone
铺石护岸 beached bank
铺石花坛 paved bed
铺石块的边坡 sett-paved bank
铺石块的河岸 sett-paved bank
铺石块的简易机场 sett-paved strip
铺石块的流槽 sett-paved gutter
铺石块的明沟 sett-paved gutter
铺石路 stone road
铺石路面 stone pavement; flags; sett pavement
铺石水渠 pitched channel
铺石瓦准尺 slating ga(u)ge
铺石屑 gritting
铺石屑的 grit-blinded; gritted
铺室内地板用料 flooring
铺碎石 macadamize
铺碎石的路 macadamized road
铺碎石机 chippings machine; chippings spreader
铺碎石路面 metal surfacing
铺碎石渣桥面 ballasted deck
铺摊拌和机 paver-mixer
铺摊混凝土 spreading of concrete
铺摊修整两用机 spreader-finisher
铺摊子 rashly putting up establishments
铺填混凝土 blind concrete
铺填加固带的土 soil with reinforcement strips installed
铺填砂井 installing column of sand
铺条滑槽 windrow chute
铺条机 swather; swathmaker
铺条装置 swathing device
铺贴基层 setting bed
铺贴面砖胶粘剂 tile mastic
铺贴面砖玛瑞脂 tile mastic
铺贴油毡 felt laying
铺铁轨 rail
铺庭院小径的砾石 gravel for garden paths
铺图机 <列车运行图> grapher
铺土 earth placement; spreading process
铺土厚度 lift thickness
铺瓦 tile laying; tile paving; tile roof(ing); tiling
铺瓦的 tiled
铺瓦垫片 tilting piece
铺瓦工 roof tiler; tiler
铺瓦工作 tilework; tiling(work)
铺瓦圬工 tile masonry; tile masonry work
铺瓦作业 tiler work
铺完石渣 ballasting up
铺网 laying-up
铺网孔筋 mesh-rod placement
铺位 berth; bunk
铺位号码 berth number
铺位舷侧木板 bunk board
铺屋顶铝材 roofing alumin(i)um
铺屋顶硬沥青 roof pitch
铺屋顶用干油毛毡 dry felted fabric for roofing
铺屋顶用马粪纸 dry felted fabric for roofing
铺屋顶用硬沥青 roofer pitch
铺屋面 healing; placing the roofing
铺屋面材料 surfacing material
铺屋面金属板 roof cladding metal sheet
铺屋面石板 roofing slate
铺屋面石屑 roof chip(ping)s

铺屋面用钉 slate nail
铺屋面用干油毡 dry roofing felt
铺狭条地板 strip flooring
铺镶木地板 parquet
铺小方石块的 sett-paved
铺小路的砖 path tile
铺新路面 resurface
铺压料 cover material
铺油毡 lino(leum)-laying
铺有多层橡胶面的沥青碎石路面 rubercrete
铺有栎木镶板 paneled in oak
铺有两条对向尽头岔线的会让站 <高坡地区的一种设计> two single ended siding
铺有路面的厚度 surfaced width
铺有路面的宽度 surfaced width
铺有橡木镶板 paneled in oak
铺圆石水沟 cobble gutter
铺在地基上的混凝土垫层 oversite concrete
铺在楼梯上的地毯 stair carpet
铺在堑壕底的板道 duck board
铺在下面 underlay
铺渣 spreading ballast
铺渣车 ballast car
铺渣底板 ballasted floor
铺渣工程 cindering work
铺渣机 ballasting machine; ballast spreader; machine for ballasting
铺渣轨道构造 ballasted track structure
铺渣机 ballasting machine; ballast spreader; machine for ballasting
铺渣路面 ballast(ed) bridge floor
铺渣桥面 ballast(ed) bridge floor; solid bridge floor; ballasted deck <铁路桥>
铺渣(上承)桥 ballasted deck bridge
铺渣上承式桥 ballasted deck bridge
铺栅 skid bed
铺毡层 blanket; carpet
铺毡机 felt layer
铺展 spread(ing); spread out
铺展破坏 failure by spreading
铺展润湿 spreading wetting
铺展系数 spreading coefficient
铺展型 spreading form
铺展性 spreadability
铺张浪费 conspicuous consumption; extravagance and waste
铺枕机 tie machine
铺整两用机 spreader-finisher
铺纸机 paper applying machine; papering(interleaving) machine
铺种草皮 sodding
铺竹屋顶 bamboo roof
铺筑 layering; pavage; placement
铺筑大圆石 bouldering
铺筑道路 laying
铺筑的游戏场地 paved play area
铺筑的走道 paved walk
铺筑底层的沥青混凝土 asphalt underseal-work
铺筑地面混合料 compocrete
铺筑法 method of laying
铺筑封层 sealing
铺筑封层工作 seal coat work
铺筑工组 placing crew
铺筑规范 specifications for laying
铺筑过的场地 paved area
铺筑含水量 placement moisture
铺筑黑色面层 blacktopping
铺筑厚度 laying depth
铺筑护坡混凝土 benching
铺筑混凝土护面 concrete paving
铺筑混凝土路面 concrete paving
铺筑混凝土面层 concrete paving
铺筑机列 <道路> train of paving plants

铺筑简易路面 light surfacing
铺筑巨砾 bouldering
铺筑沥青路面层 blacktopping
铺筑路面 road surfacing; surfacing; topping of road
铺筑路面厚度 height of lift(ing)
铺筑路面设备 paving plant
铺筑面层 covering; road surfacing
铺筑设备 paving plant
铺筑石质基层 soling
铺筑石质基础 soling
铺筑时的含水量 placement water content
铺筑时的密实度 placement density
铺筑时含水量 placement moisture content
铺筑碎石路 macadamize; road metalling
铺筑温度 <道路涂面的> laying temperature
铺筑斜坡混凝土 benching
铺筑硬质路面 placing hard-surface
铺筑用方砖 square paving brick
铺筑毡层 carpeting work; carpet treatment
铺砖 brick paving; tile paving; tilework; tiling
铺砖地面加工 tile floor(ing) finish
铺砖路面 brick pavement
铺砖人行道 brick sidewalk
铺砖瓦介质 tile setting medium
铺砖圬工 tile masonry
铺装 pavement; paving; placement
铺装层 paving layer
铺装方式 installation system
铺装工程 paving work
铺装公路 light-duty road
铺装机 forming machine
铺装路 improved road
铺装路面 coursed pavement; road carriageway
铺装路面标线 pavement mark(ing)
铺装路面界限 pavement border
铺装式沥青道床 built-up asphalt ballast
铺装终止标志 pavement ends sign
铺组合屋面 built-up roof(ing)

匍匐覆盖植物 creeping covers

匍匐沟槽 creep trench
匍匐桧 creeping juniper
匍匐松 creeping pine
匍匐在地上的 decumbent
匍匐植物 creeper; mattao
匍匐砖 creeper

菩提树 pipal tree; pipul; basswood; bodhi(c); botree; lime tree; linden; poplar leaved fig(tree); pupiltree; sacred fig (tree); teil (tree)

菩伊吞地沥青 <一种天然地沥青> Boeton asphalt

葡萄 wine grape

葡氏贯入阻力 standard penetration resistance
葡萄孢属 <拉> Botrytis
葡萄串 <截流的> string of blocks
葡萄串形装饰 <英国古代用的> trayle
葡萄黑颜料 vine black pigment
葡萄架 grape trellis
葡萄酒厂废水 wastewater from grapewine factory

葡萄酒渣炭黑 lees black
葡萄壳属 <拉> Xylobotryum
葡萄链 steel short link chain
葡萄棚 gridiron
葡萄色 vine black
葡萄酒色痣 port wine stain
葡萄石 prehnite; spheric(al) stalactites
葡萄石团块 grapestone lump
葡萄石岩 prehnitite; prehnite rock
葡萄石-绿纤石相【地】 prehnite-pumpellyite facies
葡萄饰【建】 pampre; vignette
葡萄树 grapevine; vine
葡萄树叶形饰 vine leaf
葡萄糖 amylaceum; dextrose; glucose; grape sugar
葡萄糖酸内酯 gluconolactone
葡萄藤 grapevine; vine
葡萄藤黑 vine black
葡萄藤状排水系统 grapevine drainage
葡萄藤状水系 grapevine drainage
葡萄温室 grapery; vinery
葡萄牙橙 Portugal orange
葡萄牙地毯 Portuguese carpet
葡萄牙国家土木工程实验研究所 Labrat' orio Nacional de Engenharia Civil Portugal
葡萄牙建筑 Portuguese architecture
葡萄园 grapery; vinery; vineyard
葡萄园撑杆 vineyard pole
葡萄种植 viticulture
葡萄状的 aciniform; botryoid(al)
葡萄状花纹 vinaceous figure
葡萄状灰岩 grapestone
葡萄状结构 botryoidal structure; cluster structure
葡萄状结核 botryoidal nodule
葡萄状熟料 grape like clinker
葡萄子油 grape seed oil
葡萄座腔菌属 <拉> Botryosphaeria

蒲包 matting

蒲登(岛上产的天然)沥青 Asbuton
蒲丰投针问题【数】 Buffon needle problem
蒲福风级 Beaufort(wind) scale; wind in Beaufort scale; wind scale of Beaufort
蒲福风级表 Beaufort's wind scale
蒲福风力 Beaufort wind force
蒲福风力等级 Beaufort scale for wind
蒲福海(浪) Beaufort sea
蒲福零级风 calm
蒲福零级浪 glassy calm(sea)
蒲福数 Beaufort number
蒲福天气符号 Beaufort notation
蒲公英 dandelion
蒲公英赛烷 taraxerane
蒲葵 Chinese fan-palm
蒲式耳 <一种容量单位,英国 = 36.37 升,美国 = 35.24 升> bushel
蒲桃 rose apple
蒲团 hassock
蒲原黏[粘]土 kanbara clay

朴树子油 hackberry tree seed oil

朴斯茅茨球阀 Portsmouth ball valve
朴素 austerity; rusticity
朴素的圆浮雕饰 orb
朴素凸圆饰 plain boltel
朴素修饰【建】 plain dressing
朴属 hackberry; nettle tree

浦耳 <英制容积单位> bushel

浦耳孙电弧式发生器 Paulson arc generator
浦肯雅效应 <对可见光谱的视觉灵敏度> Purkinji effect

普贝克层 <晚侏罗世>【地】 Purbeckian bed

普贝克大理石 Purbeckian marble
普贝克阶 <晚侏罗世>【地】 Purbeckian stage
普遍暴雨 general storm
普遍背景 universal background
普遍采用 in general use
普遍参加条款 general participation clause
普遍冲刷 general scour(ing)
普遍磁场 general magnetic field
普遍存在 ubiquitousness
普遍存在的 ubiquitous
普遍的 all round all
普遍地 universally
普遍发作 general attacking
普遍风暴 general storm
普遍风浪谱 generalized wind wave spectrum
普遍浮动 generalized floating
普遍腐蚀 general corrosion
普遍规律 universal law
普遍化 generalization; generalizing; universalize
普遍回返 general inversion
普遍价格 prevailing price
普遍检查 general inspection
普遍利益 universal interest
普遍垄作 common ridge planting
普遍密植 general close planting
普遍命运 general destiny
普遍喷粉 common spraying
普遍起动 <泥沙的> general movement
普遍气体定律 general gas law
普遍汽车化 general motorization
普遍日工 unskilled day labo(u)r
普遍散度 generalized divergence
普遍摄动 general perturbation
普遍适用 universal relevance
普遍性 generality; universalism; universality
普遍休闲 general fallow
普遍压力范围 popular pressure range
普遍优惠 generalized preference
普遍优惠制 generalized system of preferences
普遍优惠制待遇 general preferential treatment
普遍有效性 universal validity
普遍原理 general principle
普遍造林 general afforestation
普遍增加工资 across-the-board increase
普遍增加资本 general capital increase
普遍真理 universal truth
普遍振荡 general oscillation
普遍种 cosmopolitan
普遍自动化 universal automation
普查 extensive survey; general census; general investigation; general survey; mass survey; overall reconnaissance; reconnaissance survey
普查办公室 census office
普查报告 census report; prospecting report
普查表 census form; census schedule
普查的 broad survey
普查地段 census tract

普查地区类型 type of prospecting area
普查地质报告 reconnaissance geologic(al) report
普查范围线 boundary line of general survey
普查(方)法 broad survey method; prospecting method; census method
普查方法论 census methodology
普查分析 survey analysis
普查辐射仪 survey monitor
普查后估计 post-censual estimates
普查后核查 post-enumeration check; post-enumeration test
普查剂量计 survey monitor
普查检 screening
普查阶段 stage of reconnaissance survey; survey stage
普查井 reconnaissance well
普查孔 scout hole
普查兰水泥 pozzolanic cement
普查年 census year
普查用仪器 reconnaissance instrument
普查员用表 enumerator's schedule
普查钻孔 prospecting hole
普查钻探 apply boring test; prospect drilling; wildcat drilling
普查最终 prospecting for construct mine
普得洛 <一种粉状防水剂> Pudlo
普德村林膛 puddling basin
普蒂再分式桁架 Pettit truss
普度法 Purde method
普尔加风 Purga
普尔曼二轴转向架 Pullman four wheel truck
普尔门式火车卧车 Pullman
普尔门式列车 Pullman train
普尔门式弹簧 Pullman balance
普尔门卧车公司 <设在美国芝加哥> Pullman Company
普尔萨特机 <南非洲矿山用以采金刚石的一种工具> Pulsator
普尔瓦 Pool's tile
普惠制 generalized system of preferences
普及 diffuse; dissemination; overreach; pervade; propagation; universalize; widespread
普及版 popular edition; trade edition
普及本 paper back; popular edition
普及的 omnibus
普及环保知识 popularize environmental protection knowledge
普及教育 generalization of access to education
普克尔盒 Pockel's cell
普克尔效应 Pockel's effect
普拉塞-陶依尔 Plasser-Theurer
普拉斯法尔特 <一种用废糖浆和燃料油混合而成土壤稳定剂> Plasmofalt
普拉斯基页岩和砂岩 Pulaski shales and sandstones
普拉斯特黄铜 Plaster brass
普拉特·海福特均衡理论 Pratt-Hayford theory of isostasy
普拉特桁架(梁) Pratt truss
普拉特机制 Pratt's mechanism
普拉特假说 Pratt's hypothesis
普拉特均衡理论 Pratt theory of isostasy
普拉特纳姆镍铜合金 platnam
普拉特平桁架 flat Pratt truss
普拉特式桁架 Pratt truss
普腊桁架 <交叉斜杆的> Petite truss
普腊氏桁架 Pratt truss
普腊亚 <佛得角群岛首府> Praia
普莱玻璃 plexidur; plexigum
普莱德克 <一种塑性地板材料> Plydek
普莱桑斯阶【地】Plaisancian
普莱希尔 <一种永久性胶合板组成的模板> Plysyl
普赖斯海流计 Price current meter
普赖斯流速仪 Price current meter
普赖斯旋转掘进机 Price rotary excavator
普赖斯旋转式流速仪 Price rotary current meter
普赖异常 Prey anomaly
普兰德耳双线应力图 bilinear diagram
普兰德运动量传送理论 <用于流体> Prandtl's momentum transfer theory
普蓝铁矾 planeferrite
普郎克常数 Planck's constant
普朗巴南寺 <九世纪爪哇岛的> Chandi Prambanan
普朗克 <作用量单位> planck
普朗克常数 <辐射频率与能量之间的关系常数> Planck's constant
普郎克定律 Planck's law
普朗克分布 Planck's distribution
普郎特尔边界层理论 Prandtl boundary layer theory
普朗特尔掺混长度 Prandtl mixing length
普郎特尔承载力理论 Prandtl bearing capacity theory
普郎特尔管 Prandtl tube
普郎特尔混合长理论 Prandtl's theory of mixing length
普朗特尔解 Prandtl solution
普朗特尔-罗斯方程 Prandtl-Reuss equation
普朗特尔塑性平衡理论 Prandtl plastic equilibrium theory
普郎特尔微压计 Prandtl micro-manometer
普郎特尔(准)数 Prandtl number
普劳恩石灰石 <德国> Plauen limestone
普勒瓦通风管道 Plewa vent(ing) duct
普雷克斯玻璃 Plexiglass
普雷迈磨 Premier mill
普雷派克锥 <测定泥浆稠度> Prepakt cone
普雷赛绿 Plessy's green
普雷斯特克(现场灌注)桩 Prestcore pile
普雷斯-尤因垂直向地震计 Press-Ewing vertical seismometer
普雷斯-尤因地震计 Press-Ewing seismometer
普雷斯-尤因地震仪 Press-Ewing seismograph
普里恩的波利阿斯雅典娜神庙 Temple of Athena Polias at Priene
普里克斯法 Purex process
普里克特分析法 <地下水力学> Prickett analysis
普里库特 <一种冷铺的黑色碎石机黑色石屑> Precote
普里里灰褐色硬木 <即新西兰柚木> Puriri
普里纶医用黏[粘]胶短纤维 Purilon
普里马克斯 <一种耐火水泥> Purimachos
普里斯 <电阻率单位,=1013 欧姆> Preece
普里斯钢丝镀锌层的硫酸铜浸蚀试验 Preece test
普里斯特打桩法 <一种打桩方法适用于有限净空高度和防震> Prestcore
普里索密特 <一种涂刷铸铁的沥青油漆> presomet
普里索提恩 <一种木材装饰防腐剂> presotim
普里希利页岩 <产于南威尔士> precelly
普利 <每单位长度的质量单位,约等于 17.8580 千克/米或 1 磅/英寸> pli
普列恩斯巴赫阶 <早侏罗世>【地】Pliensbachian
普列利群落土壤 prairie soil
普列利群落植被 prairie vegetation
普列斯德层压式板材 Presdwood
普林柏木板 <一种建筑用木板> Plimber(ite)
普林士黄铜 princes-metal
普硫锑铅矿 playfairite
普鲁东桥 <主跨 320 米,1977 年建于法国> Brotonne Bridge
普鲁克式换向器 Pollock type commutator
普鲁曼奇矿 plumangite
普鲁士红 mummy
普鲁士蓝 <主要成分亚铁氰化铁> iron blue; Berlin blue; Prussian blue (colo(u)r)
普鲁士蓝拱顶 Prussian cap vault
普鲁士蓝颜料 Prussian blue pigment
普鲁士绿 Prussian green
普鲁维克斯 <一种沥青防水层> Pluvex
普吕克坐标 Plucker coordinates
普罗多莱 <一种防酸、耐火或沥青敷面水泥> Prodorite
普罗克托动力试验 Proctor dynamic(al) test
普罗克托方法 Proctor method
普罗克托贯入曲线 Proctor penetration curve
普罗克托贯入试验 Proctor penetration test
普罗克托贯入针 Proctor resistance needle
普罗克托贯入阻力 Proctor penetration resistance
普罗克托击实曲线 Proctor compaction curve
普罗克托击实试验 Proctor compaction test
普罗克托击实筒 Proctor cylinder
普罗克托密实度测定针 Proctor penetration needle
普罗克托密实度试验 Proctor density test
普罗克托试验 Proctor test
普罗克托塑性计 Proctor plasticity needle
普罗克托压实法 Proctor method
普罗克托压实效力 Proctor compactive effort
普罗克托硬度 Proctor hardness
普罗克托硬度试验 Proctor hardness test
普罗克托针测含水量试验 Proctor needle moisture test
普罗克托针刺阻力 standard penetration resistance
普罗克托最大干密度 Proctor maximum dry density
普罗皮纶 <聚丙烯纤维> propylon
普罗特斯 <一种防潮乳剂> Protex
普罗梯乌木属 <拉> Protium
普罗托吉雅可诺夫数 Protodyakonov's number
普洛彻莱颜色系统 Plochere colo(u)r system
普洛谢尔式地热电站 Plowshare geothermal plant
普平线圈 pupin coil
普染 plain tint
普染面积 shaded area; solid area
普染色 flat colo(u)r; solid colo(u)r
普染要素清绘原图 discrete-area draught
普塞尔气孔率测定法 Purcell method
普色拉 <一种防火油漆> Porcella
普生种 cosmopolitan species
普氏动力试验 Proctor dynamic(al) test
普氏法 <击实试验的> Proctor method
普氏方法 Proctor method
普氏干容重 Proctor dry unit weight
普氏贯入曲线 Proctor penetration curve
普氏贯入试验 Proctor penetration test
普氏贯入仪 Proctor penetration needle
普氏贯入针 <击实试验的> Proctor penetration needle; Proctor (resistance) needle
普氏贯入阻力 Proctor penetration resistance
普氏击实曲线 Proctor compaction curve
普氏击实曲线的干燥段 dry branch of Proctor's curve
普氏击实试验 Proctor compaction test
普氏击实筒 Proctor cylinder
普氏密实度测定针 Proctor needle; Proctor penetration needle
普氏密实度计 Proctor-type compaction tester
普氏密实度试验 Proctor density test
普氏曲线 Proctor curve
普氏试验 Proctor test
普氏塑性测定锤 Proctor plasticity needle
普氏塑性贯入锤 Proctor plasticity needle
普氏塑性计 Proctor plasticity needle
普氏塑性针 <土力学试验的> Proctor plasticity needle
普氏塑性指针 Proctor plasticity needle
普氏(土)密度试验 Proctor density test
普氏系数 Protodyakonov's number
普氏压实法 Proctor method
普氏压实效力 Proctor compactive effort
普氏岩石坚固性系数 Protodyakonov's coefficient of rock strength
普氏硬度 Proctor hardness
普氏硬度试验 Proctor hardness test
普氏硬度系数 Protodiakonov's hardness coefficient
普氏针测含水率试验 Proctor needle moisture test
普氏最大干密度 Proctor maximum dry density
普世城 <指按某种理想规划的城市> ecumenopolis
普式最佳含水量 Proctor optimum moisture
普适常数 universal constant
普适气体常量 universal gas constant
普适曲线 universal curve
普特 <俄罗斯的重量单位,=16.38076 千克或 36.1128 磅,波兰 16.329 千克或 36 磅> pood
普通 commonality
普通按钮 regular button
普通板 simple plate
普通绑扎 common seizing
普通包裹 general parcel
普通保险条款 general condition
普通暴风雨 <在美国是 5 年或 10 年一遇的暴风雨> ordinary storm
普通暴雨 ordinary storm
普通比例尺 natural scale
普通变量 common variable

普通病室 general ward
普通波特兰水泥 normal Portland cement; ordinary Portland cement; regular Portland cement
普通玻璃 ordinary glass; ordinary glazing; simple glass; soft glass
普通玻璃窗 ordinary glazing
普通玻璃门 sash door
普通玻璃纸袋 common glassine bag
普通玻璃质量 ordinary glazing quality
普通材料 common material
普通采购 ordinary purchasing
普通采样钻 common auger
普通仓库 free warehouse
普通侧铣刀 plain sidemilling cutter
普通测地学 elementary geodesy
普通测量学 common survey(ing); elementary geodesy; elementary survey(ing)
普通测量仪器 common surveying instrument
普通层理 regular bedding
普通插入式针头 simple insert needle
普通铲斗 general purpose bucket
普通长石 common feldspar
普通潮汐 ordinary tide
普通车床 center[centre] lathe; common lathe; engine lathe; plain lathe
普通车刀 regular turning tool
普通车道 general lane
普通沉积 plain sedimentation
普通成层琢石砌体 regular coursed ashlar masonry
普通成层琢石圬工 regular coursed ashlar masonry
普通成员 rank and filer
普通程序设计语言 common programming language
普通稠度 normal consistency
普通出生率 crude birth rate
普通厨房 common kitchen
普通储藏 common storage
普通传递装置 plain pick-up transfer
普通传动后轴 plain live rear axle
普通椽(木) common rafter
普通椽屋顶 single roof
普通窗玻璃 drawn glass; sheet glass
普通床身式车床 plain bed lathe; straight bed lathe
普通瓷料 normal porcelain
普通磁场 general magnetic field
普通搓 <钢绞线捻向与钢丝绳捻向相反> ordinary lay
普通搓捻法 <钢丝绳> ordinary lay
普通搭接 common lap
普通大潮 ordinary spring tide
普通大缆 hawser laid rope; ordinary laid rope
普通大气条件 average atmospheric(al) condition
普通带 normal tape
普通贷款 conventional loan; simple loan
普通单铧犁 regular single-mo(u)ld board plow
普通单开(侧向)道岔 common single turnout; ordinary simple lateral turnout
普通单开转辙器 simple points
普通单一水淬法 simple time quenching
普通担保合同 general warranty deed
普通蛋白石 semiopal
普通刀架 plain cutter holder; plain slide rest; plain tool rest
普通刀具 universal cutter
普通导爆索 ordinary detonating cord
普通导火线(一种缓燃导火线) common fuse
普通导数 general derivative
普通导线 plain conductor
普通导向钻头 plain pilot
普通道钉 <即钩头道钉> spike
普通的 common; exoteric; general purpose; general service; general utility; mediocre; vulgar; workaday
普通的成本核算方法 normal cost accounting method
普通灯光 common lighting
普通灯泡 clear bulb; standard lamp
普通等级 common grade; general schedule; ordinary grade; ordinary quality
普通等斜屋顶 span roof
普通低合金钢 common low alloy steel; ordinary low-alloy steel
普通低合金结构钢 common low alloy structural steel
普通低水位 ordinary low-water level
普通迪克型辐射计 simple Dicke-type radiometer
普通底架 normal undercarriage
普通底沙输移 general bed load transport
普通抵押 common law mortgage
普通抵押债券 general mortgage bond
普通地理学 general geography
普通地貌图 common geomorphologic(al) map
普通地图 general(purpose) map; custom-made map <指用普通方法制作的>
普通地图集 general atlas
普通地图制图学 general cartography
普通地形图 general(unit) topographic(al) map
普通地震学 general seismology
普通地质学 general geology; physical geology
普通点观测精度 observation accuracy of ordinary station
普通电话 normal telephone; ordinary call
普通电极系 general device
普通电极系电阻率测井 general device resistivity log
普通电极系电阻率测井曲线 general device resistivity log curve
普通电量滴定法 simple coulometric method
普通电子管 general vacuum tube
普通钉 common nail
普通顶尖 regular center[centre]
普通定期班轮码头 conventional liner terminal
普通定影液 plain hypo
普通动力观测船 conventionally powered ship
普通动力学 gross dynamics
普通动轴 plain live axle
普通毒物 ordinary in toxicity
普通毒物测定 ordinary toxicity test
普通独立舱 common independent tank
普通渡线 ordinary crossover
普通短螺撑 common short stay bolt
普通对绞多心型 ordinary twin type
普通对开闸门 regular-type biparting gate
普通对数 common logarithm
普通多圆锥投影 ordinary polyconic(al) projection
普通多支电路 general purpose branch circuit
普通二进制 normal binary; ordinary binary; pure binary; regular binary
普通二进制代码 natural binary code
普通发动机 convectional engine
普通发票 plain invoice
普通阀 block valve; section valve
普通法 common law
普通法留置权 common law lien
普通法人 ordinary corporation
普通法上的兑换 common law exchange
普通法系 common law system
普通反捻法 ordinary lay
普通反射复制法 plain reflectography
普通方钉 common cut nail
普通方法 common method; conventional method
普通方格纸 arithmetic(al) plotting paper
普通方块防波堤 concrete normal block gravity wall
普通方块码头 concrete normal block quay wall
普通方石 common ashlar
普通放电器 plain spark-gap
普通非成本制度 general noncost system
普通非金属矿产 ordinary non-metallic commodities
普通肥料 common fertilizer
普通费 flat rate
普通费用 general expenses; ordinary charges; overhead charges
普通分度 plain indexing
普通分度头 plain dividing head
普通分度装置 plain dividing apparatus
普通分类账 general ledger
普通分粒器 simple classifier
普通分数 common fraction; vulgar fraction
普通分析 regular analysis
普通粉刷 common plaster; common stucco
普通符号 ordinary symbol
普通辅瓦 plain floor tile
普通复利法 ordinary compound interest method
普通钙校正 correction of common calcium
普通干货 general dry cargo
普通干货船 general dry cargo ship
普通干燥设备 common equipment of crop drying
普通杆式规 plain bar type ga(u)ge
普通杆式量规 plain bar type ga(u)ge
普通感光膜 ordinary coating
普通刚玉 common corundum
普通钢 common steel; ordinary steel; plain steel; simple steel
普通钢材 common iron
普通钢筋 conventional steel; regular reinforcement; untensioned bar reinforcement; untensioned steel (reinforcement)
普通钢筋混凝土 ordinary reinforced concrete
普通钢丝钉 common wire nail
普通钢线钉 common steel wire brad
普通钢珠 plain ball
普通杠杆 common lever
普通搁梁 common joist
普通格栅 bridging; common grate; common joist
普通工地的提前开挖工程 preexcavation for common sites
普通工工时 common labo(u)rer hour
普通工具 general tool
普通工(人) general labo(u)r; unqualified man; unskilled labo(u)r; common labo(u)r; odd-job worker
普通工作队 bull gang
普通公差 commercial tolerance
普通公差等级 normal tolerance class
普通公害内容 ordinary-hazard content
普通公路 average highway
普通公司 ordinary partnership
普通供电系统 general distribution system
普通拱 common arch
普通拱架 common centering
普通钩 ordinarily hook
普通狗头道钉 common dog spike; cut spike
普通股 common(capital) stock; common shares; community stock; ordinary shares
普通股本 capital common stock
普通股比率 common stock ratio
普通股产权 common equity
普通股成本 cost of equity capital
普通股东 active partner; ordinary partner; ordinary shareholder
普通股利 common stock dividend
普通股能分到的收益 earned income for ordinary stock
普通股票 common stock; common [shares]; equity stock; general stock; ordinary stock
普通股票基金 common stock fund
普通股票投资资金 common stock fund
普通股收益 earned for ordinary stock
普通股收益报酬率 rate of return on common stock equity
普通股与优先股收益差额 yield gap
普通股折价账 discount on common stock account
普通骨料 common aggregate; conventional aggregate; normal aggregate
普通固定式间隙规 plain fixed gap ga(u)ge
普通固定资产 general fixed assets
普通管理费用 general administrative expenses
普通光 common light; simple light
普通龟裂(纹) common crack; common fissure
普通规则 blanket rule
普通硅酸盐水泥 general purpose Portland cement; geographic(al) silicate cement; normal Portland cement; ordinary Portland cement; Portland cement; standard Portland cement
普通轨枕 regular sleeper; regular tie
普通滚筒装置 plain drum winch
普通滚削 conventional hobbing
普通滚珠轴承 plain ball bearing
普通滚柱轴承 straight roller bearing
普通锅炉 plain cylindric(al) furnace
普通锅炉管 plain boiler tube
普通过磷酸钙 normal superphosphate; ordinary superphosphate
普通海图 general chart
普通海洋等深线图 general bathymetric(al) chart of the oceans
普通海洋学 general oceanography
普通海员 ordinary sailor; ordinary sea(s)man; sail before the mast
普通函数 general function
普通焊丝 solid wire
普通合伙 general partnership
普通合伙人 active partner
普通合接 ordinary splice grafting
普通河流灌溉计划 ordinary river irrigation scheme
普通荷载 ordinary load(ing); usual load
普通桁架 common truss
普通横木 plain rail
普通红心木 yard heart common
普通洪水 ordinary flood; frequency flood; generalized flood
普通呼叫 general call to all stations; ordinary call
普通虎钳 plain vice[vise]
普通滑道 common slipway
普通化学 general chemistry

普通划线 general crossing
普通划线机 conventional layout machine
普通划线盘 regular surface ga(u)ge
普通划线支票 cheque crossed generally
普通环衬 ordinary ring
普通环规 plain ring ga(u)ge
普通灰浆 ordinary lime mortar
普通辉石 augite;common augite
普通辉石橄榄岩 augite peridotite
普通辉石苦橄岩 augite picrite
普通辉石苏长辉长岩 augite noritegabbro
普通辉石岩 augite pyroxenite
普通回填 ordinary backfill
普通汇票 clean bill of exchange;clean draft
普通会计 general accounting
普通会计制度 general accounting system
普通混合肥 common mixed fertilizer
普通混凝土 <与轻或重混凝土有所区别> ordinary concrete; average concrete; common concrete; conventional concrete;dense aggregate concrete;medium concrete;normal concrete;normal weight concrete; plain concrete
普通混凝土过梁 normal concrete lintel
普通混凝土建筑构件 normal concrete build element
普通混凝土空心块 normal concrete hollow block
普通混凝土楼板 normal concrete floor
普通混凝土砌块 normal concrete block
普通混凝土墙 normal concrete wall
普通混凝土预制件 normal cast
普通混凝土柱 hooped column
普通混凝系统 convectional coagulation system
普通活力 common seed viability
普通活塞 pot-type piston
普通活性污泥法 common activated sludge method;conventional activated sludge process
普通火山碎屑岩 common pyroclastic rock
普通火灾隐患 ordinary-hazard contents
普通货车 general wagon
普通货船 conventional vessel
普通货轮 general cargo carrier;general cargo ship
普通货(物) general cargo; ordinary goods;general freight
普通货物列车 ordinary goods train; slow freight train;slow goods train
普通货物托运 slow goods consignment
普通货物运输 slow goods traffic
普通货物站 general freight station
普通机床 general machine tool;general purpose machine tool;paver lath
普通机件 general parts of machine
普通机械 standard machinery
普通机械锯 custom-built saw
普通机型 common type
普通机砖 building brick
普通基床 ordinary bedding
普通基金 general fund
普通及协定关税 general and conventioned tariff
普通级的 ordinary level[O-level]
普通级精度 plain grade
普通极性 general polarity
普通集料 common aggregate;conventional aggregate;normal aggregate
普通计算 routine calculation
普通计算机 general computer
普通记录 general record;ordinary record
普通加工室 general machining cell
普通加工中心 general purpose machining center[centre]
普通夹具 plain clamp
普通价目的一半 half ordinary rate
普通间壁 common partition
普通间接人工费用 general indirect labo(u)r cost
普通监督 general supervision
普通减水剂 water-reducing admixture
普通剪钳 plain-cut nippers
普通检修 trip service
普通碱性钢 plain basic steel
普通建筑 ordinary construction
普通建筑风格 everyday architecture
普通建筑水泥 ordinary builder's cement
普通建筑用纸 general-use paper
普通建筑砖 common brick
普通鉴定表 common score card
普通交叉 normal crossing
普通交叉口 regular intersection
普通交捻法 ordinary lay
普通浇注法 regular cast method
普通胶合板 all-veneer plywood;raw plywood
普通角闪辉石岩 hornblende pyroxenite
普通角闪苦橄岩 hornblende picrite
普通角闪石 breadalbaneite;common hornblende; gemeine hornblende; hornblende
普通角闪石橄榄辉长岩 hornblende olivine gabbro
普通角闪石橄榄岩 cortlandite
普通角闪石粒玄岩 hornblende dolerite
普通角闪石闪长岩 hornblende diorite
普通角闪石云母橄榄岩 hornblende mica peridotite
普通角闪石云母苦橄岩 hornblende mica picrite
普通角闪石云霞正长岩 hornblende miascite
普通角闪岩 hornblendite
普通角闪云母辉石岩 hornblende mica pyroxenite
普通角型梯级 regular angle-type step
普通绞链式舱盖 ordinary hinged hatch cover
普通桷木 common rafter
普通脚手架 common scaffolding
普通铰链 ready-made hinge
普通教堂 convectional church
普通接杆 solid jaw
普通街道 ordinary street
普通结构 general structure;ordinary construction; ordinary structure; simple structure
普通结构钢 general structural steel
普通结构混凝土 general structural concrete; ordinary structural concrete
普通解【数】 general solution
普通金融机构 ordinary financial institution
普通金属 plain metal;common metal
普通紧固件 common fastener
普通劲度 ordinary stiffness
普通进给 plain feed
普通经纬仪 general theodolite;ordinary theodolite; ordinary transit; plain theodolite;plain transit
普通颈轴承 plain journal bearing
普通鸠尾榫 common dovetail
普通卷边接缝 common lock seam
普通决算 general closing
普通决算表 general purpose statement
普通掘进法 common excavating method
普通掘进掏槽法 ordinary cut
普通卡车 on-highway truck;ordinary truck
普通开沟器体 plain shank
普通开挖 common excavation;general excavation
普通抗硫酸盐水泥 geographic(al) resistant sulfate cement
普通克立格法 ordinary Kriging method
普通客车 way train
普通客轮 tourist class
普通客票 ordinary ticket
普通空气清洁器 normal-duty air cleaner
普通空箱 empty container
普通空心块 normal hollow block
普通空心铣刀 plain hollow mill
普通空中交通 general air traffic
普通扣板 common clip
普通矿产 ordinary commodities
普通矿石 normal ore
普通矿物润滑油 plain mineral oil
普通矿物油 plain mineral oil
普通矿渣骨料 ordinary cinder aggregate
普通矿渣集料 ordinary cinder aggregate
普通矿质肥 common mineral fertilizer
普通喇叭管 plain horn
普通雷管 blasting cap;plain detonator
普通肋骨 ordinary frame
普通立窑 ordinary shaft kiln;Schneider kiln
普通利润率 ordinary rate of profit
普通利息 ordinary interest;usual interest
普通利息率 usual interest rate
普通沥青 common pitch;plain asphalt
普通砾石 ordinary gravel
普通连杆 plain connecting rod
普通联锁 normal interlocking
普通联轴器 regular coupling
普通链环 common link
普通梁 common beam
普通列车 local train;ordinary train; slow train
普通磷 ordinary phosphorus; white phosphorus
普通流量 common discharge
普通龙头 common hydrant
普通楼梯 access stair(case)
普通炉渣骨料 ordinary cinder aggregate
普通炉渣集料 ordinary cinder aggregate
普通驴队 <美国常用双驴同套> average mule team
普通旅客列车 accommodation train; ordinary slow passenger train;slow passenger train
普通铝及烤漆饰面 natural alumin(i)um and baked enamel finish
普通卵石 ordinary pebble
普通轮胎 common tire[tyre];ordinary tire[tyre];tire[tyre] of conventional construction
普通螺钉 plain screw
普通螺母 plain nut
普通螺栓 common bolt;plain bolt
普通螺纹 common thread; parallel thread;regular screw thread
普通螺旋铣刀 plain helical milling cutter
普通螺旋线 circular helix
普通墁涂 common stucco
普通漫射光照明装置 general diffuse luminaire(fixture)
普通毛石砌体 ordinary rubble masonry
普通毛石圬工 ordinary rubble masonry
普通每日雨量计 ordinary daily rain ga(u)ge
普通门 conventional door
普通门窗侧板 ordinary jamb lining
普通门窗框 ordinary casing
普通密封 common seal
普通密封件 common seal
普通密实混凝土 ordinary dense concrete
普通密纹木材 common dense timber
普通描述 common description
普通瞄准具 iron sight
普通名词 appellative
普通模板 ordinary shuttering
普通模壳 ordinary formwork;ordinary shuttering
普通模铸 orthodox casting
普通磨床 plain grinding machine
普通磨削 plain grinding
普通木板 common board
普通木材 open-grained wood
普通木节 standard knot
普通木枕 regular tie
普通木砖 common ground
普通目录 general catalog(ue)
普通内圆磨床 plain grinding machine internal
普通耐火浇注料 conventional refractory castable
普通耐火砖 common chamotte brick
普通耐热混凝土 ordinary refractory concrete
普通能力倾向成套测验 general aptitude test battery
普通泥浆 conventional mud
普通年金 ordinary annuity
普通年金现值 present value of ordinary annuity
普通年金终值 amount of ordinary annuity
普通黏[粘]合 regular bond
普通黏[粘]土 medium clay;ordinary clay;adamic earth <特指红黏[粘]土>
普通黏[粘]土砖 common clay brick; fired clay brick;ordinary brick;ordinary clay brick
普通黏[粘]土锥状屋瓦 tapered plain clay roof(ing) tile
普通捻 <钢绞线捻向与钢丝绳捻向相反> regular lay;ordinary lay
普通捻钢丝绳 non-spinning wire rope; ordinary lay cable; ordinary lay rope;regular lay rope
普通捻钢索 ordinary lay cable;ordinary lay rope
普通捻向 plain-laid
普通扭 <钢绞线捻向与钢丝绳捻向相反> ordinary lay
普通扭绞 ordinary lay
普通扭纹钢丝索 common strand wire rope;laid wire rope
普通排灌站 common pumping station
普通盘形砂轮 plain straight wheel
普通喷嘴 plain nozzle
普通膨胀水泥 expansive cement general;general expansive cement
普通品类 regular grade
普通品种汽油 house-brand gasoline
普通平板玻璃 flat-drawn sheet glass
普通平车 ordinary flat wagon
普通平底钢刨 regular smooth bottom steel bench plane

普通平面规 regular surface ga(u)ge
普通平刷 pound brush
普通坡度 common pitch
普通铺地砖 plain floor tile
普通曝气 conventional aeration
普通起落机构 plain lift
普通气象学 general meteorology
普通汽车间 common garage
普通汽油 regular gas
普通砌合 common bond; regular bond
普通砌体 ordinary masonry;plain masonry
普通砌筑式 common bond
普通砌砖法 American bond;common bond
普通砌砖工程 common brickwork
普通砌砖式 American bond;common bond
普通铅 common lead
普通铅测年法 the common-lead dating method
普通铅法 common lead method
普通铅校正 correction of common lead
普通嵌接 plain scarf
普通强度 regular strength;regular tenacity;single strength
普通强度玻璃 single strength glass
普通强度钢丝 common strength steel wire
普通倾斜仪 plain clinometer
普通情况 general case
普通球粒陨石 ordinary chondrite
普通曲柄钻 ordinary brace
普通全封闭式机壳 plain total enclosure
普通燃烧室 plain combustion chamber
普通燃油超级车 regular fuel super car
普通热处理 conventional heat treatment
普通热镀锌 conventional galvanizing
普通人造石 plain artificial stone
普通刃磨 convectional sharpening
普通日记账 general journal; proper journal
普通容重 common volume-weight
普通入级检验 general classification test
普通润滑油 plain oil
普通三通阀 plain triple valve
普通伞齿轮 plain bevel gear
普通砂浆 normal mortar; ordinary mortar
普通砂轮 plain grinding wheel
普通栅极 common grate
普通山地森林土 common mountain-forest
普通闪石安山岩 hornblende andesite
普通商品 general goods
普通商务语言 common business oriented language
普通设备 conventional equipment
普通射线 ordinary ray
普通审计 general audit
普通升降台式钻床 plain knee-and-column type drilling machine
普通生态学 general ecology
普通生铁 common iron
普通生物滤池 general biofilter; low-rate biologic(al) filter
普通施用 general application
普通石安山岩 augite andesite
普通石膏 common gypsum
普通石膏墙板 regular gypsum wallboard
普通石灰 common lime
普通石灰砂浆 ordinary lime mortar
普通石油沥青 wax containing asphalt
普通时号 common time signal
普通实验室方法 general laboratory method
普通食盐泉 common salt spring
普通式 plain type;general expression【数】
普通式收音机 people's receiver
普通式样 ordinary form
普通事故保险 ordinary accident insurance
普通视度 general visibility
普通视距 ordinary sight distance
普通试验 conventional test
普通饰面 dead finish
普通饰面用熟料石灰 normal finishing hydrated lime
普通适用的规划 blanket rule
普通手工具 common hand tool
普通手摇钻 ordinary brace
普通输入许可证 open general license [licence]
普通树脂 general-use resin
普通水 light water;normal water;ordinary water
普通水道 common watercourse
普通水工工程理事会 <英> Directorate of General Water Engineering
普通水井 bored well; common well; open well
普通水泥 general purpose cement; gray cement; mass cement; normal cement; normal Portland cement; ordinary cement; ordinary Portland cement; Portland cement; pozzolan(a) cement; regular cement; standard cement; common cement
普通水泥混凝土 Portland cement concrete
普通水泥浇注料 medium cement castable
普通水泥路面 Portland cement pavement
普通水手 foremastman
普通水准标石 ordinary benchmark
普通水准测量 ordinary level(l)ing
普通水准尺 ordinary level(l)ing rod
普通水准仪 general level
普通税率 common tariff; general rate;general tariff
普通税则 general tariff
普通私务电话 ordinary private calls
普通锶 common strontium
普通死亡率 crude death rate; crude mortality rate
普通松节油 common turpentine
普通松香 common rosin
普通损坏 general damage
普通梭口 plain shed
普通所得 ordinary income
普通锁 convectional lock
普通塌陷 common collapse
普通碳钢 ordinary carbon steel; plain carbon steel; straight carbon steel
普通碳钢管 common carbon steel pipe
普通碳素钢 common (straight) carbon steel
普通碳素结构钢 ordinary carbon structural steel
普通陶瓷 ordinary ceramics
普通讨价还价 arm's-length bargaining
普通套管接头 plain coupler
普通剃齿 conventional shaving
普通天文学 general astronomy
普通条件 usual terms
普通条款 general clause
普通铁轨 stock rail
普通铁条 common bar iron
普通投标 ordinary bid; ordinary tender
普通涂料 convectional paint
普通涂抹 common stucco
普通土壤 common earth
普通退火玻璃 general annealing glass
普通脱水辊 plain extractor roll
普通挖方 <指不含岩石的挖方> common excavation
普通瓦 common tile; double lap tile; normal tile
普通外径规 plain snap ga(u)ge
普通往来款项 open credit
普通位置 general position
普通温度计 dry bulb thermometer
普通文字 plain language
普通卧倒门 standard falling gate
普通卧室 ordinary bedroom; ordinary seat sleeping car <列车上的>
普通握力计 grip dynamometer
普通圬工 ordinary masonry
普通屋面 common roofing
普通屋面坡度 <37.5°>【建】ordinary pitch
普通无筋混凝土 plain concrete
普通物理参数 common physical parameter
普通吸扬式挖泥船 plain suction dredge(r)
普通习惯 ordinary practice
普通洗衣间 common laundry
普通铣床 plain milling machine
普通下水道 common sewer
普通显示器 conventional display
普通显微镜看不出的 submicroscopic
普通险 general insurance
普通现金 general cash
普通现象 common place
普通线划图 conventional linework
普通线路交叉 <只有行车钢轨的交叉> common rail crossing
普通项 general term
普通消费者系统 common consumer system
普通消化 general digestion
普通写法 longhand
普通信号 general signal
普通信号继电器 conventional signalling relay
普通信托基金 general trust fund
普通信托投资基金 common stock fund
普通信息 general information
普通信用证 ordinary credit
普通行政费用 general government expenses
普通形式 ordinary form
普通型 plain type
普通型钢 conventional steel section
普通型式 common form
普通型液力偶合器 general type coupling
普通性原则 general law
普通修理 general repair
普通修理用工具 common repair tool
普通需求函数 ordinary demand function
普通许可证 common license[licence]
普通许可证合同 non-exclusive license[licence] contract; simple license[licence] contract
普通许可协议 simple licensing agreement
普通蓄水库 common reservoir
普通玄武岩 parabasalt
普通悬架 conventional suspension
普通旋压 conventional spinning
普通选择器 regular selector
普通学校 school of general instruction
普通循环 normal round
普通压地滚 common land roller
普通压力管 normal pressure pipe
普通压缩环 plain compression ring
普通压条法 common layerage
普通压榨 plain press
普通烟囱 domestic chimney
普通延发雷管 regular delay cap
普通验收标准 general acceptability criterion[复 criteria]; general acceptance criterion[复 criteria]
普通燕尾榫 common dovetail
普通样板 ordinary formwork
普通摇臂钻床 plain radial drilling machine
普通业务 general purpose; general service;general utility
普通液施法 common liquid application
普通萤石 ordinary fluorite
普通硬砖 hard stock brick
普通涌出 common gas outflow
普通用户 domestic consumer
普通用户电度表 house-service meter
普通优惠关税 general preferential tariff;general preferential duties
普通优惠制 general system of preference
普通优先度 routine priority
普通邮件 surface mail
普通油井水泥 common oil well cement;ordinary oil-well cement
普通油漆 commercial paint; common paint; convectional paint; general service paint;ordinary paint
普通有杆首锚 common bower
普通鱼尾钻头 plain fishtail bit
普通语言 plain language
普通预浇混凝土 normal cast concrete
普通预算 ordinary budget
普通预制空心砖瓦 normal cast hollow tile
普通预制块 normal cast block
普通元素 common element
普通圆钢 smooth bar
普通圆螺母 plain round nut
普通圆盘 plain disk
普通圆柱塞规 plain cylindric(al) plug ga(u)ge
普通圆柱式门锁 plain cylindric(al) lock;Yale lock
普通圆柱形 plain cylindric(al) form
普通圆锥投影 simple conic(al) projection
普通云母 common mica
普通运价 general tariff;normal rate
普通运价表 common tariff; general tariff
普通运价率 general rate
普通杂货船 conventional break bulk ship; conventional cargo ship; conventional ship
普通杂货码头 conventional cargo terminal
普通载驳货船 lighter aboard ship
普通载体 common carrier
普通载物台 plain stage
普通凿井法 normal sinking
普通凿井施工测量 construction survey for conventional shaft sinking method
普通凿岩机 <相对于潜孔凿岩机的> ordinary rock drill; top hammer (drill)
普通责任的债务 general obligation bond
普通炸药 conventional explosive; orthodox explosive;low explosive
普通债权人 general creditor; ordinary creditor;simple contract creditor
普通债券 straight bond
普通债券本息账类 general bonded-debt and interest group of accounts
普通债务 general debt
普通站台 general goods platform

普通账户 general account
普通照明 common illumination
普通照明灯具 ordinary luminaire
普通折旧 ordinary depreciation
普通折扣 usual discount
普通辙叉 common crossing; ordinary crossing
普通枕木 common wooden sleeper
普通支票 open check; open cheque; ordinary check
普通纸复印机 plain paper copier
普通指数 general index number
普通制图 common mapping
普通制造方法 general fabrication method
普通中心式磨床 plain center[centre]-type grinding machine
普通种属 common genera
普通重混凝土 normal heavy concrete
普通重力波 ordinary gravity wave
普通重量骨料 normal weight aggregate
普通重量混凝土 normal weight concrete
普通重量集料 normal weight aggregate
普通重量耐火混凝土 normal weight refractory concrete
普通重箱【港】laden container
普通轴 plain shaft
普通轴衬 plain bush
普通轴承 journal bearing; parallel bearing; plain bearing
普通轴承轴台 plain bearing pillow block
普通轴瓦 plain bushing
普通住户 ordinary household
普通柱锉 regular pillar file
普通铸钢 normal steel casting
普通铸铁 common cast iron; plain cast iron
普通铸铁犁铧 plain cast iron share
普通专利 common monopoly
普通砖 London stock; normal brick; ordinary quality brick; soft brick; stock brick
普通砖定型面积 carrelage
普通砖工程 stock-brick work
普通砖块 normal block
普通砖石砌合 normal block bond
普通转印法 ordinary transfer
普通装配方法 general fabrication method
普通锥形附件 plain taper attachment
普通锥形埋头键 plain taper sunk key
普通琢石 common ashlar
普通资本来源 ordinary capital resource
普通资费 flat rate
普通自航式挖泥船 convectional hopper dredge(r)
普通自卸汽车 conventional dump truck
普通钻头 common bit
普通坐标 arithmetic(al) coordinate
普通座席车 ordinary seat coach
普瓦里埃氏橙 Poirriers' orange
普瓦里埃氏蓝 Poirriers' blue
普伟布洛砂岩 <一种产于美国科罗拉多州的浅灰色砂岩> Pueblo sandstone
普型砖 normal shape brick; normal type brick
普选 general election
普用仪 universal instrument

谱斑 <太阳光球层上的白斑> plage

谱斑辐射 plage radiation
谱斑面积 plage area
谱斑区耀斑 plage flare
谱斑走廊 plage corridor
谱半径 spectral radius
谱包络 spectral envelope
谱表示 spectral representation
谱参数 spectral parameter
谱测度 spectral measure
谱差 spectral difference
谱场 spectral field
谱成分 spectral composition
谱带 band; broadband; spectral band
谱带半宽度 half bandwidth
谱带包络 band envelope
谱带测量 band measurement
谱带常数 band constant
谱带点样管 band pipet
谱带分裂 band splitting
谱带基线 band origin
谱带及峰 band and peak
谱带级 band level
谱带宽度 effective bandwidth; spectral bandwidth
谱带扩展 band spread(ing)
谱带轮廓 band profile
谱带起始线 band origin
谱带伸长 leading peak
谱带伸前 band leading peak
谱带式声码器 channel vocoder
谱带头 band head
谱带拖尾 band tailing
谱带尾 tail
谱带尾色谱技术 tail chromatographic technique; tail assay
谱带吸收 band absorption
谱带系 band series
谱带压缩 band compression
谱带杂质 band impurity
谱带组 set of bands
谱的正则点 regular point of spectrum
谱定理 spectral theorem
谱段 spectral coverage
谱范数 spectral norm
谱放大系数 spectral amplification factor
谱分布函数 spectral distribution function
谱分解 spectral decomposition; spectral factoring; spectral resolution
谱分析 spectral analysis; spectrum analysis
谱分析法 spectrum analysis method
谱分析仪 spectral analyser[analyzer]
谱峰伸前 leading peak
谱幅 spectrum amplitude
谱幅度比 spectral amplitude ratio
谱辐照度曲线 spectral irradiation curve
谱功率 spectral power
谱功率谱 spectral power spectrum
谱估计 spectral estimate
谱函数 spectral function; spectrum function
谱荷载 spectral loading
谱级 spectrum level
谱加速度 spectral acceleration
谱加速度比 spectral acceleration ratio
谱检验 <数值解析> spectral test
谱角 spectral corner
谱矩 spectral moment
谱均衡 spectrum equalization
谱坑 spectrum dip
谱宽度 spectrum width
谱烈度 spectrum intensity
谱灵敏探测片 spectrum-sensitive foil
谱密度 spectral density; spectrum density
谱密度函数 spectral density function
谱密度级 spectral density level; spectrum density level
谱密度模型 spectral density model
谱漂移修正值 correction value of spectral drift
谱强度 spectral intensity
谱强度分布 distribution of spectral intensity
谱曲线 spectral curve
谱软化 spectrum softening
谱色 spectral colo(u)r; spectrum colo(u)r
谱色调 spectral hue
谱色与非谱色 spectral and non-spectral colo(u)rs
谱生成代数【数】spectrum generating algebra
谱失真 spectral distortion
谱识别 spectrum discrimination
谱衰减 spectral decay
谱速度 spectral velocity
谱台 music rack
谱特征 spectral character; spectrum signature
谱条件 spectral condition
谱外色 extra-spectrum colo(u)r; non-spectral colo(u)r
谱尾 tailing
谱位移 spectral displacement
谱系 family tree; genealogy; hierarchy; hierarchy system; pedigree
谱系采样 hierarchic(al) sampling
谱系带 lineage zone
谱系树 genealogical tree
谱系树枝图 hierarchic(al) dendrogram
谱系图 hierarchic(al) diagram
谱系学 genealogy
谱系枝带 lineage segment zone
谱线 spectral line; spectrum
谱线半高宽 half width of spectrum line
谱线比较式图像识别器 spectral comparative pattern recognizer
谱线变换 line reversal
谱线变宽 line broadening
谱线变窄 spectral line narrowing
谱线不对称性 line asymmetry
谱线测量 line measurement
谱线发射 spectral line emission
谱线发射源 spectral line source
谱线发射云 line emission cloud
谱线反转 line reversal
谱线分离技术 spectrum stripping
谱线分裂 line splitting
谱线覆盖 line blanketing; line blocking
谱线覆盖指数 line blanketing index
谱线干扰 spectral line interference
谱线轨道 spectrum locus
谱线轨迹 spectral locus; spectrum locus
谱线黑度 density of spectral line
谱线加宽 spectral line broadening
谱线间距 separation of spectra
谱线接收机 line receiver; spectral line receiver
谱线宽度 spectral [spectrum] line width; width of spectral lines; line breadth; line width
谱线宽度法 spectral line width method
谱线亮度 spectral luminance
谱线轮廓 line contour; line profile
谱线密度 spectral density
谱线频移 shift of spectral line
谱线强度 intensity of spectral lines; line intensity; line strength
谱线数目 spectrum line number
谱线尾波法 spectral line wake wave method
谱线位移 line displacement; line shift
谱线位置 spectrum line position
谱线稳定器 spectrum line stabilizer
谱线系 series of lines; spectral series
谱线相关 spectral line correlation
谱线形成 line formation
谱线形状 spectral line shape
谱线增宽 spectral line broadening
谱线证认 line identification
谱线中心 line center[centre]
谱线重叠 overlap of spectral lines
谱线自然宽度 spectrum line natural width
谱线自然形状 natural line shape
谱形状 spectral shape
谱型图 spectral type curve
谱压级 spectrum pressure level
谱仪 spectrograph; spectrometer
谱仪型号 spectrometer type
谱移 spectral shift
谱移堆 spectral shift reactor
谱移控制 spectral shift control
谱硬度 spectral hardness; spectrum hardness
谱硬化 spectral hardening
谱展宽 spectral widening; spectrum widening
谱指数 spectral index; spectrum index
谱置信区间系数 spectral confidence internal factor
谱坐标 spectral coordinates

镨矿 praseodymium ores

蹼板 <水车的> float board

瀑布 aerated nappe; aerated sheet of water; bold water; catadupe; chute; fall; linn; waterfall

瀑布别墅 <1936 年赖特设计的著名美国宾夕法尼亚洲的> falling water
瀑布冰川 cascading glacier
瀑布冲成的池潭 linn
瀑布带 fall line
瀑布的理论高度 theoretic(al) height of fall
瀑布湖 waterfall lake; plunge lake
瀑布假设 cascade hypothesis
瀑布模型 waterfall model
瀑布侵蚀 waterfall erosion
瀑布式 cascade
瀑布式滴滤池 plunging trickling filter
瀑布式喷嘴暖气池 jet aerator; plunging water
瀑布式梯级跌流 cascading flow
瀑布潭 plunge basin; waterfall lake
瀑布线 fall line
瀑布学说 waterfall sequence theory; waterfall theory
瀑洞 pot-hole
瀑流 cataract; chute
瀑落角 cataracting angle; cataracting point
瀑水钙华 cascade calc-sinter
瀑水钙华厚度 thickness of cascade calcareous-sinter precipitation
瀑泻 cataracting

曝辐量 radiant exposure

曝露 expose; exposure
曝露程度 <混凝土的> exposure degree
曝露龟裂 exposure cracking
曝露距离 exposure distance
曝露模式 pattern of exposure
曝露试验 exposure test

曝露试验台 exposure test fence
曝露于日光下 exposure to sunlight
曝露于污染物 exposure to pollutant
曝露在大气中 exposed to weather
曝露在空气中 exposed to air
曝气 aerate;aeration
曝气槽 aerated channel;aeration tank
曝气槽需气要求 aerated channel air requirements
曝气层 zone of aeration
曝气厂 aeration plant
曝气沉淀池 aeration-sedimentation tank;aeroaccelerator combined
曝气沉砂池 aerated grit chamber;aerating grit chamber;aeration grit settling tank
曝气池 <污水处理用> diffused air tank;aerated lagoon;aeration basin;aeration tank;aerator;aerotank;oxytank
曝气池池宽 width of aeration tank
曝气池池深 depth of aeration tank
曝气池尺寸 dimensions of aeration tank
曝气池出水 aerated pond effluent
曝气池分流管 aeration tank splitter box
曝气池负荷 aeration basin loading
曝气池谷脊式鼓风曝气 ridge and furrow air diffusion of aeration tank
曝气池横断面 aeration tank cross-section
曝气池桨板曝气 Sheffield paddles of aeration tank
曝气池宽深比 width and depth ratio in aeration tank
曝气池扩散器 aeration tank diffuser
曝气池排空管 drains in aeration tank
曝气池泡沫 foam in aeration tank
曝气池生化需氧负荷 aeration tank biological oxygen demand loading
曝气池停留时间 aeration tank detention time
曝气池外貌 aerating tank appearance;aeration tank appearance
曝气池污泥负荷 volume loading of aeration tank
曝气池污泥负荷率 volume loading rate in aeration tank
曝气池运转 operation of aeration tank
曝气池再曝气 reaeration of aeration tank
曝气池中挡板 baffles in aeration tank
曝气除砂 aerated grit removal;aerating grit removal
曝气处理 aerating treatment;aeration treatment
曝气处理泉水 aerated spring
曝气带入渗条件变差 deterioration of infiltration condition in aeration zone
曝气单元 aeration unit
曝气滴滤池 aerated trickling filter
曝气动力学 kinetics of aeration
曝气度 degree of aeration
曝气法 aerating process;aeration method;aeration process
曝气反应器 aeration reactor
曝气(范围)的水 suspended water
曝气废水 aerating waste(water)
曝气浮选(法) dispersed air flo(a)tation
曝气复氧 aerated reaeration
曝气沟 aeration ditch
曝气鼓风机 aeration blower
曝气固体接触 aerated solid contact
曝气管(道) aerated conduit;aerated pipe;aeration pipe;aerator pipe;aeration conduit
曝气灌溉 aeration irrigation
曝气湖 aerated lagoon
曝气活性炭滤池 aerated activated carbon filter
曝气机 aeration machine;aerator
曝气阶段 aeration phase
曝气接触床 aerated contact bed
曝气浸没 aerating submergence;aeration submergence
曝气冷却 aeration cooling
曝气量 aeration amount;aeration volume
曝气流槽 aerated launder;aerating launder
曝气漏斗 aeration cone;aeration funnel
曝气滤池 aerated filter;aerating filter
曝气排水 aeration drainage
曝气盘 aeration disc[disk]
曝气喷灌 aeration irrigation
曝气喷嘴 aerator nozzle
曝气撇油池 aerated skimming tank
曝气气浮法 aeration flo(a)tation
曝气器 aerator;air sparger;sparger
曝气器性能试验 aerator performance testing
曝气强度 aeration intensity;aeration strength
曝气区 aerating zone
曝气渠 aerated channel
曝气上浮法 aeration flo(a)tation
曝气设备 aeration equipment;aeration plant;aeration unit;aerator (blower);aerator fitting
曝气射流 aerating jet;aeration jet
曝气生物处理 aerated biological treatment
曝气生物过滤 aerated biofiltration
曝气生物滤池 biologic(al) aerated filter
曝气生物滤池出水 biologic(al) aerated filter effluent
曝气生物滤池后处理工艺 biologic(al) aerated filter post-treatment process
曝气生物滤池填料 biologic(al) aerated filter filling
曝气时间 aerating time;aeration period;aeration time;time of aeration
曝气实验 aeration experiment
曝气试验 aerating test;aeration test
曝气试验方法 aerating test method;aeration test method
曝气刷 <氧化塘处理污水用的> aeration brush
曝气刷系统 brush aeration system
曝气水流 aerated flow
曝气水流区 aerated flow region
曝气水舌 aerated nappe
曝气水体 aerated water body
曝气水域 aerated water body
曝气送风机 aerator blower;aerator fitting
曝气速率 aerating[aeration] rate;rating of aeration
曝气塔 aerating tower;aeration tower
曝气塘 aerated[aerating/ aeration] lagoon;aerated[aeration] pond;aerated pool;oxidation pond
曝气塘动力学 aerated lagoon kinetics
曝气调节 aerating regulation
曝气稳定池 aerated[aerating] stabilization basin
曝气污水池 aerated sewage lake
曝气污水(氧化)塘 aerated[aerating] sewage lagoon
曝气系数 aeration coefficient
曝气系统 aerating system;aeration system
曝气箱 aeration box
曝气效率 aeration efficiency
曝气需氧量 aeration-basin oxygen demand
曝气旋流沉砂槽 aerated[aerating] spiral flow grit channel
曝气循环型滤池 aeration circulation pattern filter cell
曝气氧化 aerating oxidation
曝气氧化塘 aerated lagoon
曝气因素 aeration factor
曝气预控制 anticipatory control of aeration
曝气站 aeration plant
曝气周期 <活性污泥法污水处理> aeration period;period of aeration
曝气柱 aerated column
曝气转子 aerating rotor;aeration rotor;rotor of aeration
曝气装置 aerated[aerating] apparatus;aerating[aeration] device;aeration facility;aerator
曝热起火 exposure to fire
曝晒 insolate;insolation;solarisation
曝晒场 exposure site
曝晒架 exposure rack;test-fence
曝晒架曝晒 test fence exposure
曝晒设备 exposure facility
曝晒设施 exposure facility
曝晒试验 exposure test
曝晒试验架 exposure test frame
曝晒试验样板 exposure panel
曝晒作用 solarisation[solarization]
曝射剂量 exposure dose
曝射量 exposure
曝声时间 exposure time
曝噪 noise exposure
曝噪预报 noise exposure forecast
曝置场 exposure site

P

Q

七倍的 septuple;sevenfold

七边的 septilateral
七边形 heptagon;septangle;septilateral
七侧向测井 laterolog seven[7]
七侧向测井曲线 laterolog seven[7] curve
七层门<芯板两面为夹芯板的> seven-ply door
七单元码 seven-unit code
七的 septimal
七点二次曲线平滑 quadratic smoothing with seven point
七点铅 minion
七点移动平均 seven points moving average
七分反射板<高空测风用> septenary foil
七分头 king closer;three-quarter bat;three-quarter brick
七个的 septenary
七股钢绞索 seven-wire strand
七股钢丝 seven-wire
七股钢丝钢绞线 seven-wire steel strand
七股钢丝绳 strand seven wire
七股绞合线 seven-strand wire
七股十九丝钢丝绳 seven by nineteen cable
七划编码磁性符号 seven-stroke coded magnetic character
七级风 high wind;moderate gale;near gale;scale-seven wind;wind of Beaufort force seven
七级浪 force-seven wave
七级能见度 visibility good
七级涌 heavy swell
七极管 pentagrid
七甲基壬烷 heptamethylnonane
七价 septavalency
七价的 septavalent
七架梁 seven-purlin(e) beam
七角的 septangular
七角棱镜 heptagonal prism
七角形 heptagon;septangle
七进数 septenary number
七进制 septenary system
七进制的 septenary
七进制数 septinary number
七铝酸十二钙 twelve to seven calcium aluminate
七氯 heptachlor
七氯呋喃 heptachlorofuran
七氯化二苯并呋喃 heptachlorinated dibenzofuran
七氯环氧化物 heptachlor epoxide
七氯联苯 heptachlorobiphenyl
七面体 heptahedron
七年生桐油 Abrasin oil
七十烷 heptatriacontane
七水胆矾 boothite
七水合硫酸铁 green vitriol
七水合硫酸锌 white vitriol;zinc vitriol
七水合硫酸亚铁 copperas;green vitriol;iron vitriol
七水硫酸镁 magnesium sulphate
七水硼砂 ezcurrite
七水铁矾 tauriscite
七水亚硫酸钠 sodium sulphide
七丝钢绞线 seven-wire strand
七丝钢绞线索 seven-wire strand cable
七台阶方锥体 seven-stepped pyramid
七台阶金字塔 seven-stepped pyramid
七天 seven-day;hebdomad
七天强度 seven-day strength
七通路型冷凝器 seven pass condenser
七弯矩方程 seven-moment equation
七维的 septuple
七维空间 septuple space
七位对数 seven place logarithms
七位字节 septet
七小塔花饰山墙(式样) seven wreaths of short turret
七氧化二氯 perchloric acid anhydride
七叶灵 aesculin
七叶树 buck-eye;horse chestnut
七叶树吉贝 pochote ceiba
七叶树属<拉> Aesculus
七叶形饰物 septfoil
七圆顶朝圣教堂 seven-domed pilgrimage church
七种公害 seven public nuisances
七重 sevenfold
七重的 septuple;sevenfold
七柱式 heptastyle
七柱式建筑 heptastyle building;heptastylos
七子花属 heptacodium

栖白蚁冢动物 termitocol

栖海面的 pelagophoulus
栖海岩的 actophilus
栖湖沼的 limnicolous
栖荒漠的 eremophilous
栖件柱 island
栖居的动物 inhabitant
栖居环境 inhabited environment
栖菌动物 mycetocole
栖留地下河 perched subsurface stream
栖留地下水 perched groundwater
栖留地下水面 perched water table
栖留地下水位 perched water table
栖留含水层 perched aquifer
栖留河<地下水的> perched stream;perched river
栖留潜水面 perched water table
栖留泉 perched spring
栖留水 perched groundwater;perched water
栖留水水面 false ground water table
栖留水水位 false ground water table
栖沙的 ammocolous
栖身处 shelter
栖树的 dendrophilous
栖所指示 habitat indicator
栖息 perch
栖息场所 habitat
栖息处 haunt
栖息地 habitat
栖息地倒金字塔 inverse pyramid of habitat
栖息地管理 habitat management
栖息湖底的生物 bathylimnetic organism
栖息流群落 lotic
栖息密度 population density
栖息盆地 perch basin
栖息区 home range
栖息习性 habitation
栖霞灰岩<早二叠世>【地】Chihsia limestone
栖霞山 Chisia mountain;Qixia mountain
栖岩坡 perched boulder
栖于湿地的 mesic
栖滞地下水 perched groundwater

桤木 alder

桤木属<拉> Alnus
桤叶鼠李 alder buckthorn

期初 beginning of period

期初差额 beginning balance;initial balance
期初存货 beginning inventory;opening inventory;opening stock
期初贷款 front-end finance
期初的 opening
期初订货 initial order
期初费用 front-end fee;initial charge
期初加重收费 front-end loading
期初亏欠 deficit at the beginning period
期初年金 annuity in advance
期初平均法 beginning average method
期初投资 initial investment
期初应付年金 annuity due
期初盈余 initial surplus;surplus at opening(of the)period
期初余额 balance at the beginning of the period;beginning balance;initial balance;opening balance
期初资本 beginning capital
期待 await;bargain for;count on;count upon;expectance[expectancy];expectation;look forward to
期待结果 expected outcome
期待净收益 net expected gain
期待数 expected number
期待物权合同 catching bargain
期待系数 coefficient of expectation
期待已久的 overdue
期待原则 principle of expectancy
期待值 desired value;expected value;required value
期付款项 bill and account payable
期号 issue;period number
期后 after date
期后收缩 hysterosystole
期汇汇率 forward exchange rate
期汇净价 outright rate
期汇契约 forward contract
期汇业务 forward exchange transaction
期货 dealing future;forward;future goods;futures
期货保证金 cover cost
期货保值 hedging
期货报价(单)forward quotation;future quotation
期货变化率 futures delta
期货的远期合同 option forward contract
期货购买 contract purchasing
期货购入 future purchase
期货合同 arrival contract;forward contract;futures contract
期货合同有效期 life of contract
期货合约的期权交易 options on futures contract
期货汇兑合同 forward exchange contract;futures foreign exchange contract
期货汇率 forward rate
期货价(格)forward price;forward rate;future price
期货价值 position value
期货交割 future delivery
期货交易 bargain on term;clearing contract;dealing in future;forward business;forward trade;forward trading;forward transaction;future goods transaction;futures sale;futures business;futures deal;futures trading;futures transaction;position trade;position trading;sale for account;time bargain;commodity exchange<农产品等的>
期货交易到期日 forward maturities
期货交易合同 futures contract
期货交易所 futures exchange
期货金额及交付日期通知 prompt note
期货经纪商店 futures brokerage house
期货买卖 option market
期货买卖的期权 straddle
期货买卖者 position trader
期货卖方抢购 short squeeze
期货贸易 futures trading
期货抛出和买进选择权 put and call option
期货抛出和买进自营商 put and call dealer
期货期权 European option;forward option
期货契约 forward contract
期货取消 forward cancelled
期货商 technical trader
期货升水 contango
期货市场 contract market;forward market;futures market;terminal market
期货式期权 futures-style option
期货通知单 prompt note
期货外汇 forward exchange;future exchange
期货委托商店 futures commission merchant
期货销售 future sale
期货溢价 contango
期货装运期 position
期货总额 future sum
期价交易 call transaction
期间 duration;spell;term
期间成本 period cost
期间成本分配 periodic(al)allocation of cost
期间费用 period charges;period expense
期间费用审计 period expenses audit
期间分析 period analysis
期间患病率 period prevalence
期间会计核算 periodic(al)accounting
期间计划 period planning;period project
期间审核 periodic(al)audit
期间损益表 interim income statement
期间延长 extension of period
期交订货 order for future delivery
期刊 periodic;periodic(al)publications
期刊目录室 periodic(al)index room;periodic(al)index space
期刊书架 periodic(al)stack
期刊阅览室 periodic(al)reading room;periodic(al)room
期满 become due;come to an end;expiration;expire;fall due;run-out;termination
期满保险费 earned premium
期满兑付 payment in due course
期满付款汇票 bill on maturity
期满失效 lapse
期满通知(书)expiration notice
期满投资 maturing investment
期末 term end
期末报表 end-of-period statements
期末存货 closing inventory;closing stock;ending inventory;final inventory;inventory final

期末存货估计 closing inventory valuation; ending inventory valuation
期末截账 end-of-period cutoff
期末净损益部分 final net profit and loss section
期末库存量 closing inventory
期末审计 final audit
期末收益 yield to maturity
期末调整分录 end-of-period adjusting entry
期末余额 closing balance; ending balance; final balance
期末账单 final account
期内账单 interim account; intermediate account
期票 accommodation bill; after-date bill; bill (undue); note of hand; promissory note; time bill
期票簿 notebook
期票承兑银行 accepting house; acceptance house
期票到期日期 date of maturity
期票抵押贷款 factoring
期票附件 additional part of a bill
期票股利 scrip dividend
期票买价 buying rate for time bill
期票贴现 discount on a promissory note
期票支付时间 usance
期票支付所 domicile
期权交易 option dealing
期权买卖市场 option market
期收款项 bill and account receivable
期外收缩 extra systole
期望 expectance [expectancy]; expectation; hope; in prospect
期望报酬 expected return
期望边际利润 expected marginal profit
期望边际损失 expected marginal loss
期望变量 expecting variable
期望车速 desired speed
期望出行(次数)【交】desired trip
期望到达时间 expected approach time
期望的形式 desired form
期望电流 prospective current
期望反应 expected response
期望方差 expected variance
期望费用 expect cost
期望风险 expected risk
期望服务时间 expected service time
期望覆盖选位模型 expected covering location model
期望概率值 expected probit
期望故障数 expected number of failure
期望后悔值 expected regret value
期望环境浓度 expected environmental concentration
期望货币表 expected monetary table
期望货币值 expected monetary value
期望获得的资本 acquisitive capital
期望几率损失 expected opportunity
期望假说 expectancy hypothesis
期望间隔时间 <统筹方法中,某个任务估计需要的时间> expected elapsed time
期望简单变量 expecting simple variable
期望角 expected angle
期望解树 potential solution tree
期望金属量 expected quantity of metal reserves
期望矿石量 expected quantity of reserves
期望理论 expectancy theory
期望利润 anticipated profit
期望路线图 desire line chart
期望年度脱销费用 expected annual stockout cost
期望排队长度 expected length of the waiting line
期望破损数 expected breaks
期望强度 target strength
期望强酸阴离子浓度 strong acid anion concentration expected
期望曲线 expectation curve
期望生产费用 expected cost of production
期望时间 expected time
期望收益 expected revenue
期望寿命 expectation of life; expected life span; life expectancy
期望输出值 desired output
期望数据 expected data
期望水平 level of aspiration
期望税收 expected revenue
期望完成时间 expected performance time
期望线 desire line; expected line
期望消费 expected consumption
期望消逝时间 expected elapsed time
期望效益 expected utility
期望效用 expected utility
期望效用假设 expected utility hypothesis
期望效用值 expected utility value
期望信号与不期望信号之比 desired to undesired signal ratio
期望行为 expected behavio(u)r
期望延迟 expected delay
期望延误 expected delay
期望影响 desired impact
期望原理 expected principle
期望噪声级 expect noise level
期望正常价格 expected normal price
期望正态频数 expected normal frequency
期望支付 expectation payment; expected payoff
期望值 desired value; expectation value; expected value; value of expectation
期望值标准 expected value criterion
期望值方差 expectation variance
期望值模型 model of expected value
期望值内部收益率 expected internal rate of return
期望值准则 expectation criterion
期望质量水准 expected quality level
期望重对数 expected loglog
期望总报酬 expected total reward
期望最小半径 desirable minimum radius
期限 limited period; period of limitation; term; time limit
期限差距风险 maturity gap exposure
期限结构 term structure
期限条款 duration clause
期限未满的 unexpired
期限延长 renewal
期限终止 time-expired
期中报表 interim estimate; interim statement
期中财务报表 interim financial statement
期中筹措资金 interim financing
期中的 interim
期中付款 interim payment
期中付款凭证 interim payment certificate
期中付款申请 interim payment application
期中付款证书 interim payment certificate
期中工作报表 interim work sheet
期中股息 interim dividend
期中集资 interim financing
期中检验 intermediate survey
期中结账 interim closing
期中借贷 interim borrowing
期中评估 interim valuation
期中设计 intermediate design
期中审计 interim audit
期中损益计算书 interim income and loss statement
期中验收 interim acceptance
期中盈利报表 interim-earnings statement
期中预算 interim budget
期终水头损失 terminal head loss
期终余额 closing balance
期租 charter by time
期租人 time charterer
期租人的股息 time charterer's interest

欺

欺骗 cheat; defraud; double cross
欺诈 cheat; deception; fraud; fraudulent; jockey
欺诈行为 deceit
欺诈性投标 collusive tender
欺诈性误述 <船租> fraudulent misrepresentation

漆

漆斑 gum bloom
漆包的 enamel covering; enamelized; enamel(l)ed
漆包电缆 enamel-insulated cable; enamel(l)ed cable
漆包绝缘线 enamel(l)ed cable
漆包铝线 enamel-insulated alumin(i)um wire
漆包锰铜线 enamel manganin
漆包皮 enamel covering
漆包软管 flexible varnished tubing
漆包铜线 enamel(l)ed copper wire
漆包线 enamel-covered wire; enamel-insulated wire; enamel(led) wire; glazed wire; lacquer cable; varnished wire
漆包线电缆 enamel(l)ed wire cable
漆包线漆 wire coating enamel
漆包线漆涂层 wire coating
漆包线漆涂装 wire coating
漆包线绕电阻 enamel(l)ed wire wound resistor
漆不均匀的表面 cloudy
漆布 coated cloth; dermateen; leather cloth; lino(leum); oil cloth; oilcoat; tack rag; varnished cambric; varnished cloth; varnished fabric; coated fabric
漆布带 varnished cloth tape; varnished tape
漆布地面 lino(leum) flooring
漆布绝缘 varnished cloth insulation
漆布绝缘电缆 varnished-cambric covered cable; varnished-cambric insulated cable
漆布绝缘管 spaghetti
漆布片 sheet lino(leum)
漆布铺地 lino(leum) flooring
漆布油 lino(leum) oil
漆草 pearl plant
漆层 enamel(l)ed coating; paintcoat; paint layer
漆层烘烤 paint baking
漆层轻微收缩 cissing
漆铲 paint scraper; scraper
漆沉积 lacquer deposit; lacquering
漆(成)木纹 graining
漆成木纹的方法 graining
漆冲淡剂 lacquer thinner
漆疵病 defect of lacquer
漆带 varnish-treated tape
漆的烘干 baking of varnish
漆底 varnish base
漆雕 lacquer carving
漆房 japanning room; painting room
漆仿云石 graining
漆酚 urushiol
漆酚醛塑料 cashew nut aldehyde plastic
漆酚树脂 cashew resin
漆酚树脂涂料 urushoil resin coating
漆粉 paint power
漆封 stop-off lacquer
漆封剂 lacquer sealer
漆覆盖层 lacquer sheathing
漆干剂 paint drier[dryer]
漆干片 shellac(k)
漆干燥剂 paste drier[dryer]
漆膏 paste; paste paint(in oil)
漆革 enamel leather; patent leather; varnished leather
漆革第一层涂料 daub coat
漆工 japanner; lacquerer; lacquer man; painter
漆工刀 painter's knife
漆工工作服 painter's overalls
漆工火炬 painter's torch
漆工喷灯 painter's torch
漆工油膏 painter's ca(u)lk
漆工油灰 painter's putty
漆工棕 iron oxide brown; japanners' brown
漆管班组 dope gang
漆管机 dope machine
漆罐 paint bucket; paint can
漆滚筒 paint roller
漆锅 japanning kettle
漆过的 japanned
漆黑 inkiness; pitch black; pitch-dark; thick darkness
漆黑的 coal black; pitchy; jet black
漆画 lacquer painting
漆画木纹 combining; graining
漆绘木纹法 graining
漆绘木纹用具 gramer
漆基 coating base; paint base
漆基需要量 binder demand
漆浆 mill base; pigment grind; pigment paste
漆胶布 empire cloth
漆椒树 pepper tree
漆酵素 gummase
漆浸细麻布 varnished cambric
漆绝缘线 enamel-insulated wire
漆开式洗脸盆 knee operated faucet
漆孔 enamel eye
漆蜡 japan wax; lacquer wax; urushi tallow; urushi wax
漆料 coating vehicle; medium paint media; paint vehicle; vehicle
漆料混合试验 vehicle mixing test
漆料色浆 colo(u)ring varnish
漆料稀释剂 lacquer diluent; paint thinner; lacquer thinner
漆酶 laccase
漆面干燥过快 surface drying
漆面结膜微粒 spewing
漆面流坠 curtaining; sagging of paint
漆面硬质纤维板 enamel(l)ed hardboard
漆膜 painted film; paint film; varnish film; greasiness <缺乏亲和性的>
漆膜病态 film defect
漆膜剥离强度试验 film stripping test
漆膜测厚仪 film thickness ga(u)ge
漆膜穿孔 film punching; film puncture
漆膜垂落 film curtaining
漆膜的不连续性 film discontinuity
漆膜的搭接覆盖 bridge; bridging;

Q

lap;overlap
漆膜的易洁性 film cleanability
漆膜发雾 film milkiness
漆膜防霉剂 film preservatives
漆膜丰满度 film fullness
漆膜附着力 adhesion of film
漆膜附着力测定仪 adherometer;adohero ga(u)ge
漆膜干燥计 film drying meter
漆膜固化剂 film-hardening agent
漆膜刮痕试验 scratch test of paint film
漆膜刮卷试验 film knife-curl test
漆膜光雾值 film haze value
漆膜烘烤时起烟雾 fuming
漆膜厚度 film thickness
漆膜厚度测定仪 film thickness ga(u)ge
漆膜厚度检验 film thickness test
漆膜划伤试验 film scratching test
漆膜混浊度测量仪 hazemeter
漆膜击穿试验 film breakdown test
漆膜检测仪 film-detector;filming-inspector
漆膜抗张强度测定器 filmometer
漆膜可搬运干 dry-to-handle of film
漆膜连续性 filming integrity
漆膜裂纹 film cracking
漆膜疲软 cheesy
漆膜起泡 cratering
漆膜起雾 film hazing
漆膜缺陷 film defect
漆膜热修补 film burning-in
漆膜韧性 film toughness
漆膜试验 film test
漆膜试验计 film tester
漆膜损坏 paint coat failure
漆膜涂布器 applicator;doctor blade;film applicator(blade);film caster
漆膜完全干燥 dry to handle
漆膜下的丝状锈蚀 filiform corrosion under film
漆膜形成 film formation;varnish formation
漆膜研磨材料 paint film polishing media
漆膜颜色随角变化 colo(u)r travel
漆膜硬度 film hardness
漆膜整体性 film integrity
漆膜走丝 film silking
漆木纹的工人 grainer
漆黏[粘]合剂 paint binder
漆皮 coat of paint;enamel(led) covering;enamel(led) leather;film of paint;hornskin;japanned leather;paint skin;patent leather
漆皮布 enamel(l)ed cloth
漆皮线 varnished wire
漆片 paint flake
漆片雕 lacquer relief
漆起皱 lacquer lifting
漆器 japan;lacquered ware;lacquerware;lacquer work;lacker
漆器材 plain wood of lacquerware
漆器的胎 substrate
漆器用材 plain wood
漆墙工 wall painter
漆球壳属 <拉> Zignoella
漆溶剂 white spirit
漆溶液 lacquer solution
漆生成 lacquer formation
漆石涂装法 paint harling
漆树 cashew;Japanese varnish tree;lacquer plant;lacquer tree;Rhus verniciflua;Rhus verniciflua stokes;varnish-tree
漆树黄酮 fisetin
漆树科 Anacardiaceae;toxicodendron
漆树漆 Cheshu lacquer;Rhus lacquer
漆树酸 anacardic acid
漆树属 Sumac(h)
漆树子油 sumac(h) seed oil
漆刷 lacquer brush;paint(ing) brush;varnish brush
漆刷喷灯 painter's torch
漆刷清洗剂 brush cleaner
漆刷刷毛黏[粘]合剂 brush binder
漆刷油灰 painter's putty
漆刷蘸漆量 pick-up of paint
漆刷子 paint brush
漆丝 urushi silk
漆素 urushin
漆酸 urushic acid
漆套管 varnished sleeve
漆条纹 streak;stripe
漆条纹工 striper
漆条纹机 striper
漆桶 can of paint;paint-pot
漆桶中的硬沉淀物 bottom of paint pot
漆头 patent drier[dryer]
漆涂层 lacquer coat;lacquering
漆污 paint stain
漆雾 paint mist
漆烯 urusene
漆细工 lacquer work
漆线 <路上用漆划的标线> paint line
漆咬底 lacquer lifting
漆用腻子 painter's putty
漆用汽油 lacquer petroleum
漆用溶剂 lacquer solvent
漆用溶剂油 varnish maker's and painter's naphtha
漆用软化剂 lacquer softener
漆用石脑油 paint and varnish naphtha;varnish maker's and painter's naphtha
漆用树胶 kikekunemalo
漆用树脂 paint resin
漆用硝基纤维素 nitrocellulose for lacquer;pyroxylin(e)
漆油 urushoil
漆折带 kink band
漆脂 haze tallow
漆纸 varnished paper
漆纸板 varnished cardboard;varnished pressboard
漆状双晶【地】 geniculate twin
漆状褶皱【地】 knee fold

齐

齐岸 <水与岸平> bankfull
齐岸的 bankfull
齐岸宽度 bankfull width
齐岸流 bankfull flow
齐岸流量 bankfull discharge
齐岸水位 bankfull stage
齐奥普斯金字塔 <埃及> great pyramid of Cheops
齐柏林合金 Zeppelin alloy
齐柏林天线 <一端馈电的双馈线水平半波天线> Zeppelin antenna
齐爆 <数个炮眼同时爆炸> multiple firing;simultaneous blast(ing)
齐爆性能 overall firing performance
齐边钢板 universal mill plate;universal plate;universal steel plate
齐边机 edger
齐边拉门 flush door
齐边涂刷 cutting-in
齐边压力机 edge squeezer;slab squeezer
齐边轧制 edging
齐边柱厚板 universal plate
齐波夫定律 Zipf's Law
齐波夫律法 Zipf's law method
齐勃拉管 zebra tube
齐草地的配水龙头 flush lawn hydrant
齐次边界条件 homogeneous boundary condition
齐次变换 homogeneous transformation
齐次波动方程 homogeneous wave equation
齐次不等式 homogeneous invariant
齐次的 homogeneous
齐次的生产函数 homogeneous production function
齐次度 degree of homogeneity
齐次多项式 homogeneous polynomial;quantic
齐次方程 homogeneous equation
齐次方程组 homogeneous system of equations
齐次非平稳时间序列 homogeneous nonstationary time series
齐次非稳定过程 homogeneous nonstationary process
齐次规划 homogeneous program(ming)
齐次过程 homogeneous process
齐次函数 homogeneous function
齐次积分方程 homogeneous integral equation
齐次解 homogeneous solution
齐次微分方程 homogeneous differential equation
齐次系 homogeneous system
齐次系数 homogeneous coefficient
齐次线性方程 homogeneous linear equation
齐次线性方程组 homogeneous linear equations
齐次线性约束 homogeneous linear restriction
齐次效用函数 homogeneous utility function
齐次形式 homogeneous form
齐次型网络 homogeneous network
齐次一次方程 linear homogeneous equation
齐次增大 homogeneous accretion
齐次组 homogeneous system
齐次坐标 homogeneous coordinates
齐次坐标系统 homogeneous coordinates system
齐的 uniform
齐动图 simo chart
齐端 butt end
齐端堵缝 ca(u)lking butt
齐墩果 common olive
齐墩果属 olive
齐恩(锡基轴承)合金 Zinn
齐尔可帕克斯 <锌、锡、杂质乳浊剂的商品名> Zircopax
齐尔染色剂 Ziehl's stain
齐尔氏石碳酸[苯酚]品红染剂 Ziehl's carbol-fuchsin stain
齐发 simultaneous firing
齐发爆破 mass shooting;simultaneous blast(ing);volley
齐放 multiple shooting
齐分子量聚合物 oligomer
齐根器 butter adjuster;butter deflector
齐肩高的 shoulder-high
齐肩高炮眼 shoulder hole
齐焦 parfocalization
齐焦距离 parfocal distance
齐焦调节套筒 parfocality adjustment sleeve
齐焦调整 parfocality adjustment
齐焦物镜 parfocal objective
齐焦性 parfocality
齐口窗台 slip sill
齐勒测定法 <测定钢中非金属夹杂物> Zieler process
齐肋斯牌油毛毡 <一种加强的屋面油毛毡> Zylex
齐洛伊锻造锌基合金 zilloy
齐马尔锌基合金 zimal
齐马铝镁锌合金 zimalium
齐苗 full stand
齐明点【物】 aplanatic foci
齐明镜 aplanat
齐默尔曼法 Zimmerman process
齐默尔曼辊轴平衡桥 Zimmerman roller bascule bridge
齐默尔曼辊转跳开桥 Zimmerman roller bascule bridge
齐默尔曼滚柱竖旋桥 Zimmerman roller bascule bridge
齐默尔曼过程 Zimmerman process
齐姆公式 Theim formula
齐纳电流 Zener current
齐纳电压 Zener voltage
齐纳二极管 Zener diode
齐纳防爆栅 Zener barrier
齐纳击穿 Zener breakdown;Zener effect
齐纳模型 Zener model
齐纳区 Zener region
齐纳隧道效应 Zener tunnel(l)ing
齐纳效应 Zener effect;Zener efficiency
齐纳阻挡层 Zener barrier
齐纳阻抗 Zener impedance
齐平 <指勾缝> flush
齐平安装的 flush mounted
齐平凹拉手 flush-cup pull
齐平橱柜门闩 flush cupboard catch
齐平焊 flush weld
齐平接缝 flush(ed) joint;flushing up
齐平接榫 flush fixing
齐平接头 flush fixing
齐平连接的 flush-coupled
齐平路缘 flush kerb
齐平式 flush type
齐平式板把手 flush type slab handle
齐平式电路 flush type circuit
齐平式进水口 flush inlet
齐平式连接 flushing coupling
齐平镶板 <与框架齐平> flush panel
齐平圆线脚 quick bead
齐射 salvo
齐射地 broadside
齐射目标 salvo point
齐射散布 salvo dispersion
齐射散布界 salvo pattern
齐射扇面宽度 salvo breadth
齐射信号 salvo signal
齐式【数】 binary form
齐水面的 awash
齐(同)性 homogeneity
齐头 jump butt
齐头锉刀 blunt pointed file
齐头锯 trimmer
齐头六边形锉 lock file
齐头排版 flush
齐头平锉 blunt file
齐头三角锉 banking file
齐头圆锉 blunt round file
齐投 salvo(release)
齐投弃 salvo jettison
齐瓦牌钢材 <一种店面橱窗用的不锈钢> Zilva
齐心斗 center block
齐心协力 work hard together
齐行 justification;justified line;justify;line centering
齐行楔 justifying space;space band
齐性空间 homogeneous space
齐胸(部)炮眼 breast hole
齐胸高的 breast dup;breast-high
齐胸高的女儿墙 breast wall;face wall
齐腰高的 waist-high

齐腰深的 waist-deep
齐周期的 homoperiodic

芪【化】stilbene

其次最佳法规 next best rule

其他人员工资 other labo(u)r expenses
其他标志 other marks
其他不动产 other real estate
其他部件 miscellaneous parts
其他部门支出 other sector expenses
其他财务报表 other financial report
其他操作 miscellaneous operation
其他产品年产值 annual values of other products
其他产业保险 extended coverage insurance
其他产值 other values
其他沉积构造 other sedimentary structure
其他沉积结构 other sedimentary texture
其他沉积岩 other sedimentary rocks
其他储量术语 other reserve term
其他电气装置 additional electric(al) service
其他方案 alternative scheme
其他方法 other methods
其他费用 other charges;other cost;other expenses
其他负债 other liabilities
其他国家或地区 other nations or districts
其他荷载 miscellaneous load
其他回填 miscellaneous backfill
其他货币资金审计 other money capital audit
其他基金欠款 due from other fund
其他间接标志 other indirect indicators
其他间接费 other indirect expenses
其他建设 other building
其他晶体 other crystal
其他开支 miscellaneous expenses
其他扣款 other deduction
其他款项 miscellaneous
其他类型重力仪 other type gravimeter
其他利率 other interest rate
其他气候标志 other climate index
其他权利不受损害【拉】salvo jure
其他设备 other equipment
其他生产设备折旧费 other industrial equipment depreciation expenses
其他事故 other accident
其他事故数据 data of other troubles
其他收入 miscellaneous revenue;other revenue
其他收入预算 other income budget
其他收益 other income;other revenue;other yields
其他税 other tax
其他税率 other tax rate
其他损失 unmeasured loss
其他特征 other features
其他投资 other investment
其他投资审计 other investment audit
其他微量元素 other minor element
其他污染 other pollution
其他物料 unclassified store
其他形态 other forms
其他形状 other shape
其他岩石学概念 other petrologic(al) conception
其他遥感器 other sensor
其他业务收入 other business income
其他一切风险 any other perils
其他一切危险 any other perils
其他因素 other factors
其他应付款 other payables
其他应付款项 accounts payable others
其他应付款项审计 other accounts payable audit
其他应收款项 accounts receivable others
其他应收款项审计 other accounts receivable audit
其他影纹 other texture
其他有关参数 other related parameter
其他有关事项 miscellaneous provisions;other matters concerned
其他预算 other budget
其他杂项收入 other miscellaneous receipts
其他支出 other expenses
其他支出预算 other expense budget
其他直接费 other direct expenses
其他制造费用 other manufacturing overhead
其他中间色 other mesial colo(u)r
其他专题图 other thematic maps
其他资产 other assets

奇怪的 bizarre;strange

奇幻思维 magical thinking
奇货可居 board as a rare commodity;rare commodity worth boarding to corner the market
奇迹米 miracle rice
奇积分 singular integral
奇加性集函数 singular additive set function
奇克山式动臂油管塔 Chiksan loading arm
奇克山式铰接装油臂 Chiksan marine loading arm
奇利风 <突尼斯的一种干热南风> Chili
奇量子数 odd quantum number
奇零价格 odd price
奇论 paradox
奇螺纹 odd thread
奇米尔公式 <计算洪水强度的> Chamier's formula
奇妙的 bizarre
奇妙装置 contraption
奇诺胶 gum kino
奇诺树脂 kinoin
奇趣建筑 fantastic architecture
奇圈 odd cycle
奇缺 critical shortage of
奇缺的 critical
奇摄动 singular perturbation
奇石园 alpine garden
奇兽画或雕刻 <中世纪教堂的> bestiary
奇斯克石 chesserite
奇态 odd state; singular state; ungerade
奇泰纳 <一种盘式集装箱> Geetainer
奇特的 singular
奇特环境 exotic environment
奇特性 curiosity
奇形的 oddly shaped
奇形怪状的 Barock;baroque
奇形怪状的凸嵌 <一种嵌砖缝的方法> half tuck
奇形装饰 fanciful ornamentation;grotesque;grotesque ornament
奇异 oddness;strange
奇异边界条件 singular boundary condition
奇异边缘循环 singular boundary cycle
奇异变换 singular transformation
奇异部分 singular part
奇异测度 singular measure
奇异超平面 singular hyperplane
奇异初值问题 singular initial value problem
奇异单元 singular element
奇异的 arabesque;paradoxical;supernatural;unaccustomed
奇异的建筑 phantastic architecture
奇异点 distinguished vertex; singular point;singularity
奇异点法 singular point method
奇异顶点 singular vertex
奇异对应 singular correspondence
奇异二次超曲面 singular quadric hypersurface
奇异二次曲面 singular quadric
奇异二次曲线 singular conic
奇异方程 singular equation
奇异分布 singular distribution
奇异分量 singular component
奇异分母 singular divisor
奇异覆盖 singular cover
奇异覆盖表 singular cover list
奇异刚度矩阵 singular stiffness matrix
奇异轨道 singular orbit
奇异棍形藻 Bacillariaparadoxa
奇异函数 singular(ity) function
奇异核 singular kernel
奇异弧线 singular arc
奇异积分 improper integral; singular integral
奇异积分方程 singular integral equation
奇异积分流形 singular integral manifold
奇异积分算子 singular integral operator
奇异级数 singular series
奇异集 singular set
奇异解 singular solution
奇异矩阵 singular(ity) matrix
奇异类 singular class
奇异立方 singular cube
奇异链 singular chain
奇异络 singular set
奇异面 singular interface
奇异命题 singular proposition
奇异内函数 singular inner function
奇异平面 singular plane
奇异谱 singular spectrum
奇异切面 trope
奇异切线 singular tangent
奇异曲面 singular surface
奇异曲面的分类 classification of singular surfaces
奇异柔度矩阵 singular flexibility matrix
奇异上同调群 singular cohomology group
奇异摄动理论 singular perturbation theory
奇异摄动系统 singular perturbation system
奇异石 <叶蜡石的俗称> wonderstone
奇异数 strangeness number
奇异算子 singular operator
奇异态 singular state
奇异吸引子 strange absorber
奇异系 singular set
奇异下同调类 singular homology class
奇异下同调论 singular homology theory
奇异线 line of singularity
奇异线丛 singular complex
奇异线性顺序机【电】singular linear sequential machine
奇异行列式 singular determinant
奇异性 irregularity;singularity
奇异性守恒 strangeness conservation
奇异支集 singular support
奇异直积 singular direct product
奇异直射变换 singular collineation
奇异直线 singular line
奇异值 singular value
奇异值分解 singular value decomposition
奇异自同构 singular automorphism
奇异组 singular set

歧点 bifurcation point;cusp;cuspidal point;point of cusp

歧管 branch pipe; divided manifold; forked tube; manifold (branch); pipe branch
歧管壁浸湿 manifold wall wetting
歧管抽空 manifold depression
歧管点火试验 manifold ignition test
歧管垫密片 manifold gasket
歧管阀 manifold valve
歧管干燥器 manifold drying apparatus
歧管加热控制 manifold heat control
歧管加热器片 manifold heater plate
歧管夹 manifold clamp
歧管空气氧化 manifold air oxidation
歧管冷凝器 manifold condenser
歧管排气 manifold exhaust
歧管塞 manifold plug
歧管外罩 manifold hood
歧管线 lateral line
歧管压盖 manifold gland
歧管压力 manifold pressure
歧管压力调节 manifold pressure control
歧管真空 manifold vacuum
歧管装置 manifolding
歧化反应 disproportionation reaction
歧化高聚物 branched high polymer
歧化作用 branching action; disproportionate; disproportionation; dismutation
歧离 divergence;jitter;straggling
歧路 branched road;forked road
歧视 discrimination
歧视待遇 discrimination treatment
歧视条件 discriminative condition
歧视性的限制政策 restrictive policy of a discriminative nature
歧视性法律 discriminatory law
歧视性解雇 discriminatory discharge
歧视性配额 discriminatory quota
歧视性条款 discriminative clause
歧义表示式 ambiguous expression
歧义性误差 error ambiguity
歧异 divergence

祈唱堂 cha(u)ntry

祈祷壁龛 <指向麦加的> mehrab
祈祷地毯 <穆斯林教徒祈祷用> prayer rug
祈祷雕像 kneeling figure
祈祷空间 <伊斯兰教寺院中围隔的> maksoorah
祈祷室 large praying chamber;oratory;praying chamber
祈祷堂 <清真寺的> maqsura
祈祷用壁龛 prayer niche

耆那庙宇 Jain temple

耆那式建筑＜耆那教是起于印度的宗教＞ Jain architecture

脐带系索＜潜水员的＞ umbilic(al) cord

脐点【数】 umbilic(al) point
脐点曲面 umbilic(al) surface
脐腹小蠹 Seolytus schevyrewi

崎岖(不平) ruggedness

崎岖不平道路 bumpy road
崎岖不平地面 unmade ground
崎岖不平地区 broken country
崎岖不齐 ruggedness
崎岖道路 bumpy road;unmade road
崎岖的 bumpy;cragged;craggy;rugged
崎岖地 rugged area;scabland;badland
崎岖地带 broken ground;broken terrain;rough country;rough terrain;rugged country;rugged terrain
崎岖地区 badland;broken ground;broken terrain;rough area;rough country;rough ground;rough terrain;rugged terrain
崎岖地群落 tirium
崎岖地形 accidented relief;jagged terrain;rough terrain;rugged relief;rugged topography;scabland topography
崎岖地植物群落 hydrotribium
崎岖海滨 rugged shore
崎岖荒凉地 badland
崎岖浪蚀崖 jagged ware-eroded coast
崎岖路面 rugged path
崎岖山区 rough area
崎岖盐丘 salt hill
崎岖之地 scabland

畦 border(check);contour ditch;contour furrow;furrow;ridge;stitch

畦播 sowing in furrows;trench sowing
畦池曝气系统 ridge and furrow aeration system
畦床 ridge-up bed
畦埂 border;border dike[dyke];check levee;ribbing
畦沟 field drain(age);supply ditch
畦沟灌溉 ridge furrow irrigation
畦沟移植 trench transplanting
畦灌 basin check irrigation;block irrigation;border check irrigation;border flooding;border irrigation;check-flooding irrigation;check irrigation
畦灌法 basin check method;check irrigation method
畦块 border strip
畦田 border check;ridged field
畦田灌溉 basin irrigation
畦田漫灌 basin check irrigation;basin flooding irrigation;border flooding irrigation
畦条 border strip
畦条灌溉 strip irrigation
畦头未耕地 head land
畦植 drill planting;saddle planting;seeding
畦植法 bed system
畦状曝气池 ridge and furrow tank

骑背钉 piggyback

骑车率 rate of bicycle riding
骑缝钉 rivet
骑缝号 tally mark
骑缝线滚轮 dot wheel
骑缝证书 chirograph
骑楼 porch
骑楼人行道 arcade sidewalk
骑楼式人行道 arcaded sidewalk;canopy sidewalk;covered sidewalk
骑楼下人行道 sotto portico
骑马 promenade;riding
骑马的人 rider
骑马的小径 riding trail
骑马钉 dog anchor;iron ridging;scaffold nail;staple;timber dog
骑马钉板 plate staple
骑马钉小册子 pamphlet
骑马旅游 equestrain tourism
骑马螺钉 stirrup
骑马螺栓 bulldog grip;U-bolt
骑马螺丝 spring buckle
骑马牛仔 rancher
骑马针 dropper
骑马桩【测】 striding stake
骑码标尺 rider bar
骑师 rider
骑士塑像 equestrian statue
骑式水准器 stride level
骑式俘汽阀 riding cut-off valve
骑术学校 riding academy;riding school
骑田岭石 qitianlingite
骑像饰檐壁 horseman frieze
骑自行车 bicycle;bike
骑自行车的人 wheel man
骑自行车式 cycle
骑自行车者 cyclist;pedal cyclist
骑自行车者交通量 pedal cyclist traffic

棋布式消力墩 baffle block;energy dispersion baffle;energy dispersion block

棋格板 chequer plate
棋格划痕试验＜试验涂膜黏[粘]结性的＞ cross-cut adhesion test
棋盘 chessboard
棋盘(测试)信号 checkerboard pattern test signal
棋盘法 checkerboarding
棋盘方格 checkboard square
棋盘格 checker;gridiron pattern;pane
棋盘格测试图 cross hatch pattern
棋盘格的切头桩 chequered lost head nail
棋盘格花的 checked
棋盘格嵌石 pinning
棋盘格扫描 cheque board scan
棋盘格式 chess-board type
棋盘格式构造 chess-board structure;lineament structure
棋盘格纹饰 block check
棋盘格状的 tessellated
棋盘环线加对角线形线网 gridiron, ring and diagonal line type road network
棋盘加环形线网 gridiron and ring road network
棋盘门 checkerboard door
棋盘频率 checkerboard frequency
棋盘式 grid pattern;grid type
棋盘式城市规划体系 rectangular system of city planning
棋盘式城市建设 town built in blocks
棋盘式道路的城镇 gridiron town
棋盘式道路网 gridiron road system
棋盘式道路系统 gridiron road system
棋盘式对照表 articulation statement;spread sheet
棋盘式荷载布置 checkerboard loading
棋盘式街道 rectangular street
棋盘式街道布置 gridiron type street layout;grid street layout;rectangular system of street layout
棋盘式街道体系 checkerboard street system;rectangular street system
棋盘式街道网 rectangular layout of streets;rectangular street system;rectangular system of street layout
棋盘式街道系统 checkerboard street system;rectangular street system
棋盘式街区 check bound street system
棋盘式排列 check disposition;chequered order
棋盘式排列的 staggered
棋盘式砌合 basket weave bond;diaper bond;diaper work
棋盘式设计 chessboard design
棋盘式图案 checkerboard pattern;chequered pattern
棋盘式系统 gridiron system
棋盘式栽植 alternate planting
棋盘式账目分析表 articulation statement;spread sheet
棋盘式种植 check row
棋盘式装载 checkboard load
棋盘式钻进 checkerboard drilling
棋盘图像 checkerboard image
棋盘形 grid pattern
棋盘形布置 tessellation
棋盘形街道布置 gridiron type street layout
棋盘形铺嵌 tessellation
棋盘形铺设 chequer work
棋盘形嵌石铺面 tessellated pavement
棋盘形调制盘 checkerboard reticule
棋盘形细(木)工 checkerwork;chequer work
棋盘形线网 gridiron type road network
棋盘阵 checkerboard array
棋盘状斑 tessellated macula
棋盘状图案 checkerboard pattern
棋心盘屋顶 grey lime core roof

蛴螬 grub

旗 flag;ensign;banner＜行政区域名＞

旗布 flag cloth
旗地 banner land
旗杆 flag pole;flag post;flagstaff
旗杆灯 bow light
旗杆顶球 mast ball;ox ball
旗杆帽 truck
旗杆式天线 flagpole antenna
旗工 flagman
旗钩 flag hooks
旗冠型 flag crown form
旗柜 flag locker
旗号 banner;flag indicator;semaphore
旗号通信[讯] flag signal(l)ing
旗舰 flag-ship
旗宽 depth of a flag
旗缆铜扣 flag clip
旗令停车 flag halt;flag stop
旗流形 flag manifold
旗纱 bunting
旗绳 flag line;halliard[halyard];signal halyard
旗绳钩 snap hook;spring hook
旗绳滑车 bunting block
旗绳木扣 flag toggle
旗式表示器 flag type indicator
旗式道口自动信号机【铁】 autoflag;automatic flagman
旗手 bellboy;signalman
旗塔 flag tower
旗艇 flagboat
旗尾 fly
旗箱 flag chest;flag locker;signal chest;signal locker
旗信号 flag signal;waft
旗信号员 flagman
旗形表示器 banner type indicator
旗形地块 flag lot
旗形开关 flag switch
旗形装置 flag arrangement
旗型 flag form
旗型方向标志 flag type directional sign
旗鱼 billfish
旗语 flag signal;semaphore
旗语信号 semaphore code
旗语信号指示器 flag indicator
旗站 flag halt;flag stop;halt;flag station＜未规定列车停车时刻,只有在旗站显示停车信号时才停车＞
旗站停车信号机【铁】 flag stop signal
旗志工 flagger
旗帜 banner;colo(u)r;flag
旗帜的滥用 abuse of flag
旗帜底座 banner bracket
旗帜托座 banner bracket
旗帜线 banner line
旗帜箱 flag kit
旗状云 banner cloud;cloud banner
旗子 flag

鳍基条 basalia

鳍龙骨 bilge keel;skid fins
鳍片管 extended surface tube;finned tube;gilled tube
鳍片管省煤器 fin tube economizer
鳍片式节煤器 finned tubular economizer
鳍形导杆 flipper guide
鳍形墙 fin wall
鳍状标牌 fin sign
鳍状管 fin tube
鳍状加热器 fin heater
鳍状龙骨 fin keel
鳍状物 fin
鳍状整流器 flipper

企划部 planning department

企口 bezel;dowel(l)ed joint;groove;grooving;quirt;tongue and groove;weather check
企口板 dressed and matched boards;grooved slab;interlocking board;match(ed) board;matching;matchlining;tongue and groove board;tongue and groove plank
企口板壁 novelty siding
企口板顶棚 matched ceiling
企口板接合 table fishplate joint
企口板平顶 matched ceiling
企口板条 rebated boarding
企口板桩 grooved sheet pile;tongue(d) and groove(d) sheet pile
企口壁板 novelty siding
企口边 tongue and groove edge
企口侧边 tongued and grooved edge
企口(侧)墙板 tongue and groove sid-

ing
企口窗框 rabbeted jamb
企口挡条 rebate ledge
企口的 end matched; tongue-and-grooved; tongue and groove matched
企口的板墙 slab wall tongued and grooved
企口底板 checked ground
企口地板 grooved and tongued floor(ing); joint floor(ing); matched floor; rebated floor; T and G sub-floor; tongue and groove floor(ing); tongue floor
企口地板的底层 matched sub-floor
企口地(面)砖 cross-grooved floor(ing)(finish)
企口顶棚 tongued and grooved ceiling
企口顶棚板条 grooved match ceiling boarding
企口对接 rabbeted butt joint
企口风雨板 rebated weather-boarding
企口缝 flat-lock seam; groove seam; hinged joint; match joint(ing); rabbet; ribbet; tongue(d) and groove(d) joint
企口钢板桩 lock sheet piling
企口构造缝 keyed construction joint
企口护墙板 matched siding
企口护墙板装饰线条 bolection mo(u)lding
企口加工 matching and grooving
企口胶合板 grooved plywood; tongued and grooved plywood
企口角桩 rabbeted corner pile
企口接 tongue and groove joint
企口接材 matched lumber
企口接缝 groove and tongue joint; joint tongue and groove; matched joint; tongue and groove joint; keyed joint
企口接合 fillet and groove joint; fillistered joint; groove connection; grooving and tonguing; joining by rabbets; matched joint; matching joint; match joint(ing); ploughed-and-tongued joint; plow and tongue joint; rabbet(ed) joint; rebated grooved and tongued joint(ing); rebated joint; tonguing
企口接合的 grooved and tongued
企口接合木铺板 tongued and grooved deal
企口接合墙板 tongue and groove siding
企口接口 tongued and grooved joint; tongue joint
企口接头 rabbet joint; rebated jointing
企口块地板 tongued and grooved block flooring; tongue-groove flooring
企口块模板 tongued and grooved form
企口块木板桩 tongued and grooved wood sheet pile
企口块体 tongued and grooved block; tongue-groove block
企口连接 cornerlock joint; dado joint; groove and tongue connection; groove connection; grooving <混凝土路面的>
企口楼板 tongue and groove floor(ing)
企口模板 tongue and groove form
企口木板 matched boards
企口木地板 strip flooring
企口木料 tongued and grooved wood; matched lumber
企口刨 matching plane; rabbet plane; rebate(d) plane; tongued and grooved plane; tonguing and grooving plane
企口配合木材 matched lumber
企口拼合 dressed and matched
企口拼接 tongue and groove
企口铺板 tongue floor
企口铺路石板 cross-grooved flag(stone)
企口砌块 groove(d) and tongue(d) block
企口嵌缝 interlocking joint
企口墙板 grooved and tongued panel; interlocking building panel; tongued and grooved siding
企口墙面板 tongued and grooved shingle
企口石块 joggled stone
企口式模板 grooved and tongued form
企口榫 feather; straight tongue
企口榫接 groove and tonguing; grooving and tonguing
企口天花板 tongued and grooved ceiling
企口屋顶板 tongued and grooved shingle
企口屋面板 matched roof board
企口镶板机 match boarding machine
企口斜角缝 tongue miter[mitre]
企口仪器玻璃框 bezel
企口凿 feather cutter; tonguing chisel
企口砖 lug brick; tongued and grooved brick
企事业 enterprises and establishments
企图 contemplate; make an attempt; meditate
企业 corporate agency; corporation; undertaking
企业搬迁费 enterprise relocation expenses
企业背景 business background
企业标准 company standard; enterprise standard; manufacturer's standard; mill certified
企业标准的 non-certified
企业标准化情报工作 standard information work in enterprise
企业财产 business property
企业财务 business finance
企业财务公开 opening of business finance
企业财务效益 enterprise financial benefit
企业策略 business strategy
企业偿付能力 business solvency
企业偿债能力 business solvency
企业偿债能力分析 enterprise repayable ability analysis
企业场所 business location
企业成本 business cost; cost in business; enterprise cost
企业成长 business growth
企业城(市) company town
企业储备基金 enterprise reserve fund
企业倒闭 business failure; enterprise failure
企业倒闭率 business mortality
企业的合作 consortium
企业的全部费用 outright cost of an undertaking
企业调查 census of business; establishment survey
企业发展基金 enterprise developing fund; enterprise expansion fund; venture expansion fund
企业法 business law; enterprise law
企业法规 business law and regulation
企业法人 business entity
企业分工 division of enterprise
企业分配指标 allocate income ratio for enterprise
企业风险和保险 business risk and insurance
企业负责人 pilot firm
企业改制 restructuring of enterprises
企业管理 business administration; business management; enterprise management; factory administration; factory management; shop administration
企业管理费用 administrative overhead; enterprise administration expenses
企业管理费账户 administration expense account
企业管理合同 business management contract
企业管理体系 business management system; enterprise management system
企业管理学 business engineering
企业管理因素 entrepreneurial factor
企业管理制度 career plan; managing system of enterprise
企业管理咨询 business management consulting
企业管理自动化 automation of enterprise management
企业规程 business regulation
企业规模 size of business
企业规章 operating provision
企业国有化 nationalization of enterprise
企业合并 business combination; business merger
企业合并法 merger law
企业和事业收入 revenue from enterprise and undertaking
企业环境 business environment; condition of business
企业环境管理 business environmental management; environmental management for enterprise
企业会计 enterprise accounting
企业活动 business activity
企业机构 business institution; establishment
企业机能 business function
企业积蓄金 business savings
企业基金 business fund; enterprise fund
企业基金银行存款 bank deposit on enterprise fund
企业基金制 enterprise funds system
企业集团 business conglomerate; enterprise conglomerate; enterprise group
企业集团合并 conglomerate merger
企业计划 business planning
企业家 business structure; enterpriser
企业家决策 entrepreneurial decision
企业间工资结构 external wage structure
企业间信贷 inter-enterprise credit
企业教育 business education
企业结构 business structure; pattern of enterprise
企业结业清算 enterprise termination and liquidation
企业界 business circle; business interests; business world; enterprise circle
企业经济 business economy
企业经济责任制 system of economic responsibility of enterprise
企业(经营)管理 management sponsor
企业经营管理制度 system of enterprise operation and management
企业经营合同 management contract
企业经营者 undertaker
企业精神 bossmanship; entrepreneurship
企业可用收入 disposable business income
企业库存的变动 change in business inventory
企业劳动分工 divide the work
企业类型 type of business
企业利润的提成 enterprise profit drawing
企业利润分析 analysis of business profit
企业利润留成 enterprise profit partly reserved; portion of profits retained by enterprise
企业利润提取 enterprise profit drawing
企业联合体 konzern
企业联合组织 syndicate; undertaking syndicate
企业联营协议 amalgamation process
企业领导 business leader
企业留成 profits retained by enterprise
企业民主管理 democratic management of enterprise
企业名称 fictitious name
企业模拟 business simulation
企业目的 enterprise objective
企业内部筹款 self-finance
企业内部存款 deposits within the company
企业内部训练 training within industry
企业内联网 intranet
企业破产 business failure; business insolvency; enterprise bankruptcy
企业区 enterprise zone
企业人力 business manpower
企业任务 enterprise objective
企业软件工程师 enterprise software engineer
企业设备 business equipment
企业社会责任 corporate social responsibility
企业生产能力 plant capacity
企业实体 business entity
企业收入 receipts from enterprise
企业首脑人物 top management
企业寿命保险信托 business life insurance trust
企业税 tax on enterprise
企业所得税 corporate income tax; enterprise income tax
企业所属机构 corporate structure
企业统计 business statistics
企业投资 enterprise investment
企业投资者 business investor
企业外部网【计】extranet
企业网 intranet
企业往来账户 business account
企业物资流通体系 business logistics
企业效绩单项评价 monomial evaluation of enterprise performance
企业效绩综合评价 comprehensive evaluation of enterprise performance
企业信托 business trust
企业行为科学 behavio(u)r theories of the firm
企业研究 business research
企业业务 business events
企业一般管理费 overhead
企业盈余 business earnings
企业拥有的港口 privately owned port
企业运用的净资本 net capital employed
企业运用资金 funds at the disposal of enterprise
企业债务 enterprise debt

企业占用的土地 land on which the enterprise are situated
企业战略计划 business strategy plan
企业站 enterprise station
企业招待费 business entertainment
企业诊断 business diagnosis
企业证券 business vouchers
企业(职工)住宅 firm's dwelling unit
企业主 business owner;undertaker
企业资本积累 business capital formation
企业资产管理 enterprise asset management
企业资产管理软件系统 enterprise asset management software system
企业资金的流动性 business liquidity
企业资信情况 business standing
企业资源 business resources
企业自备车 private car
企业自备车及租用车 self-provided or leased wagon of enterprise
企业自备电厂 industrial power plant
企业自备集装箱 shipper-owned container
企业自筹资金能力 enterprise's ability of self-financing
企业自有资金 corporate's ownership funds; enterprise's ownership resources
企业自主权 decision-making power of enterprise
企业总效率 overall plant efficiency
企业组建合同 contract of establishment
企业组织 business organization

启

启闭 make-and-break

启闭方式 start-stop system
启闭杆 lift(ing) ram; start and stop lever
启闭机 actuator;headstock gear
启闭设备 hoisting device; hoisting equipment;lifting device
启闭设备的标准改正力 normal margin of start-stop apparatus
启闭失真度 degree of start-stop distortion
启闭式双流通路 start-stop double-current channel
启闭台 hoist platform
启闭压力 trip pressure
启闭闸门的液压机 gate cylinder
启闭转换器 start-stop switch
启闭装置 start-stop system
启程 departure;outgoing;setting out
启程前检查 pre-trip inspection
启程时间 departure time
启船港 port of sailing
启动 actuating; actuation; commissioning;firing;initiating;pulse on;putting into operation; start-up; trigger;cranking;priming
启动按钮 activate button;initiate button;starter button
启动板 start-up plate
启动棒 start-up rod
启动泵 prime(r) pump; start pump; start-up pump
启动操作 start-up function; start-up operation
启动程序 start-up procedure
启动触发器 initiate trigger
启动磁力机 booster magneto
启动磁铁 driving magnet
启动地址 enabling address
启动点 breakaway point
启动电流 pick-up current
启动电路 start-up circuit
启动电平 triggering level
启动电位 initiate potential
启动电压 trigger voltage
启动端 enable end
启动发动机 start the engine; to fire an engine
启动发动机操作机构 starter engine control
启动方式 start-up mode
启动费 mobilization fee
启动分配器 start distributor
启动负荷 pick-up load
启动杆 actuating lever;projecting bar
启动感应器 priming inductor
启动给水泵 start-up feed pump
启动功率 initial power
启动功能 start-up function
启动管 start-up main
启动管路 primer line;start-up conduit
启动管系 start-up pipe
启动管线 start-up pipeline
启动过程 start-up procedure
启动虹吸(管) priming siphon
启动黄灯号 starting amber
启动簧 starting spring
启动机控制 starter control
启动机钥匙 starter key
启动计数 enabling counting
启动计数存储器 enabling count memory
启动寄存器 enable register
启动加热用的轻油燃烧器 warming-up light oil burner
启动尖头信号 trigger pip
启动键 activate key;start key;initiate key
启动经验 start-up experience
启动开关 enable switch
启动空气分配器 starting air distributor
启动力矩 starting moment
启动逻辑 enable logic
启动脉冲 initiating pulse
启动脉冲波形 trigger waveform
启动脉油 trigger pip
启动门 enabling gate
启动磨损 start-up wear
启动能力 start-up capability
启动平台 deck initiation
启动期 starting phase; start-up period;start-up stage
启动气泵 booster air pump
启动汽缸 setting cylinder
启动器 actuator;starter;trigger
启动前须知 prestarting instructions
启动前指南 prestarting instructions
启动请求 activation request; initiate log-on request
启动区 promoter
启动曲线 start-up curve
启动燃料 priming fuel;starting fluid
启动任务 initiating task
启动任选项 start option
启动溶液 priming solution
启动乳化剂 priming emulsion
启动乳化液 priming emulsion
启动设备 start-up component
启动绳 <一种钻机用的> jerk line
启动时间 acceleration time; attack time;pick-up time;start-up time
启动市场 enlivening the markets
启动事故 running-up accident; start-up accident
启动试验 starting characteristics test
启动输入 enable input
启动输入输出指令 start input/output instruction
启动数据 log-on data
启动数据传送和处理的程序 start over data transfer and processing program(me)
启动数据通信[讯]指示器 start-data-traffic indicator
启动水平 trigger level
启动速度 start-up velocity; toggle speed
启动算符 enabled operator
启动损失时间 start-up lost time
启动特性曲线 acceleration curve
启动条件 entry condition; start-up condition
启动调节阀 intercepting valve
启动调整工 start-up man
启动位 start bit
启动位置 enable position
启动纹 dimple
启动系统 start-up system; starting system
启动线 enable line
启动向量 start vector
启动信号 actuating signal; enabling signal;initial-of-message signal;initiating signal;signal-enabling;starting signal
启动信息 log-on message
启动性能曲线 starting characteristic curve
启动延迟 start delay
启动液 primer fluid
启动阴极 starter cathode
启动用自耦变压器 starting autotransformer
启动有准备 ready for operation
启动预告信号 signal of prestart
启动源 start-up source
启动指令 enabled instruction
启动终止 start-stop
启动终止程序 initiator-terminator
启动注水 initial charge
启动注油泵 fuel pressure injection
启动转矩 pull-in torque
启动装置 actuating apparatus; actuating device;starting device
启动子 promotor
启动子部位 promoter site
启动作业流 start-up job stream
启发 Elicitation;enlightenment
启发程序 heuristic program(me)
启发的 developmental
启发函数 heuristic function
启发模式 heuristic model
启发能力 heuristic power
启发式查找技术 heuristic search technique
启发式程序 heuristic procedure
启发式程序设计 heuristic programming
启发式的 heuristic
启发式的搜索 heuristic search
启发式规划 heuristic program(me)
启发式规则 heuristic rule
启发式(论据) heuristics
启发式判读样式 associative key
启发式算法 heuristic arithmetic
启发式探讨 heuristic approach
启发式推理 heuristic inference
启发式样片 associative key
启发式优化技术 heuristic optimization technique
启发性方法 heuristic method
启发性搜索 heuristic search
启发性知识 heuristic knowledge
启封 unseal
启航 haul-away; set sail; undock; weigh anchor
启航港 port of departure;port of embarkation;port of sailing
启航密度 sailing intervals
启航命令 sailing order
启航日(期) date of departure;sailing date
启键 activate key;start key
启开压力 cracking pressure
启亮时差 <联动信号灯> difference of offset
启门机 gate hoist;gate lifting device;hoister;lifter
启门机构 lifting gear
启门机架 hoist(ing) frame
启门机离合器 hoist clutch
启门机室 gate house; hoist(ing) chamber
启门力 lifting power
启门链 hoist chain
启门梁 lifting beam
启门设备 lifting device
启门索 hoist cable
启门塔 gate tower
启蒙 enlightenment
启蒙老师 abecedarian
启明星 morning star;phosphorus;Venus
启莫里阶 <晚侏罗世>【地】 Kimmeridgian stage
启莫里奇煤 Kimeridge coal
启莫里奇黏[粘]土 Kimeridge clay
启瓶器 bottle opener
启跳压力 opening pressure
启停风阀 on-off damper
启通的 unblanking
启通脉冲 unblanking pulse
启用 invocation; phase in; put into service
启用费资本化 capitalizing the front-end fee
启用旧号 reused number
启用前检验 <新机器安装后的> readiness review
启用日期 commissioning date
启用申请 commissioning application
启用申请表格 commissioning application forms
启用试验 commissioning test
启用线 enable line
启用状态 initiate mode
启运 start shipment
启运地船边交货价 free alongside ship
启运地点 place of shipping
启运地法 law of the place of dispatch
启运点交货价 free on board shipping point
启运港 port of departure;port of embarkation; port of commencement of carriage
启运港船边交货(价格) free on steamer
启运港码头交货 free on board quay
启运国 country of departure;country of origin;country of shipment
启运机场 airport of departure
启运日期 date of shipment
启运通知书 <保险> declaration
启运重量 shipping weight; loading weight
启止同步 start-stop synchronization

杞

杞柳 bitter willow;purple osier;purple willow;red osier;red willow

起

起岸 debark;discharge;disembark;landed;landing

起岸成本 landed cost;start-up cost
起岸费 landing charges; landing expenses;landing hire
起岸费用条款 landed terms
起岸后品质条件 after landing quality terms
起岸价格 landing price

起岸品质条件 landed quality term
起岸清单 landing book
起岸数量条件 landed quantity term
起岸税捐付讫条件 landed duty paid term
起岸证明书 landing certificate
起岸重量 landed weight
起凹槽 fluted roller
起凹痕 pitting
起拔道机 lifting and lining track scarifier; track lifting and lining machine
起拔套管 pull(-up the)casing
起斑点的褪色 mottled discolo(u)ration
起保费 original premium
起爆 denotation; detonate; detonating; explosive initiation; firing; fuse action; fuze action; go-off; ignition; initiation; priming
起爆材料 initiation material
起爆材料贮存室 primer house
起爆层 arousing blast layer; bomb breaking layer; layer induce blast
起爆成分 detonating composition
起爆传播 propagation of detonation
起爆导火线 detonation train
起爆点 fire point; initiation point; priming point
起爆电池 shot-firing battery
起爆电路 blasting circuit; detonation chain
起爆电源 firing current
起爆反应 detonation reaction
起爆方法 method of initiation; priming method
起爆方式 initiation pattern
起爆粉 priming powder
起爆感度 sensitivity to initiation
起爆管 blasting cap; starter cartridge
起爆火药 detonating powder; priming powder
起爆机 blasting(battery)machine
起爆机构 fuse mechanism
起爆剂 blowing agent; detonating agent; detonating composition; detonator; ignition powder; initial detonating agent; initial explosive; initiator; primer composition; priming composition
起爆继电器 detonating relay
起爆接头 priming adapter
起爆孔 leading hole of blast
起爆雷管 allways fuse; blaster cap; blasting cap; primer cap; primer detonator unit; detonate tube; detonating cap; detonator
起爆灵敏度 priming sensibility
起爆灭弧室 explosion pot
起爆能 energy of initiation
起爆能力 detonating capacity; initiating ability
起爆钮 knocker
起爆器 blaster; blasting machine; blasting unit; detonating primer; detonator; exploder; firing device; first fire; initiator; powder monkey; primer; priming apparatus
起爆器材 initiating equipment; priming material
起爆器材储存室 primer house
起爆枪 exploding gun
起爆软线 detonating cord
起爆设备 detonating equipment
起爆时间 break time; bursting time
起爆时间离散 ignition time scattering
起爆时滞 breaking lag
起爆室 blast(ing)chamber
起爆顺序 delay pattern; shot-firing program(me)
起爆速度 detonation speed; velocity of detonation
起爆速率 detonation rate
起爆索 igniter cord
起爆态 detonation state
起爆体 <包括起爆药线与雷管的> primadet
起爆筒 blasting cap
起爆网路 priming circuit
起爆温度 detonation temperature; flash point
起爆系统 initiation system; priming system
起爆线 detonating fuse
起爆信管 allways fuse
起爆信号 detonating signal
起爆性 detonation property
起爆药 booster; burster; detonating composition; detonating compound; detonating powder; ignition powder; initial explosive; initiating agent; initiating composition; initiating explosive; initiating charge; primary high explosive; primer charge; priming explosive
起爆药包 detonating charge; detonating primer; igniting primer; primer; primer cartridge; priming charge; priming compound
起爆药包放置在炸药包底的爆破法 bottom priming
起爆(药)装雷管 priming
起爆引线 blaster fuse; blasting fuse; detonation fuse; primacord
起爆引信 allways fuse; detonating fuse
起爆炸筒 primer cartridge
起爆炸药 detonated dynamite; primary explosive; priming powder; detonating explosive; primed charge
起爆炸药卷 primer cartridge
起爆炸药筒 blasting cartridge; detonator dynamite; primer cartridge
起爆装药 detonating charge
起爆装置 detonating device; detonator; fuse mechanism; initiation device; priming device
起标点 punch mark
起柄三通 drop tee
起柄弯头 drop elbow
起波区 generating area
起波水面 wave-covered water
起波纹 crisp
起波纹的 corrugated
起搏点 pacemaker; pacer
起步 breakaway; first time step; initial starting; initial step; starting and acceleration performance; take-off
起步板 riser
起步段 start-up section
起步反应 initiative reaction
起步级【建】 curtail step
起步开关 start button
起步力 breakaway force
起步平台 <金属楼梯的> subplatform
起步数 starting number
起步速度 starting speed
起步踏步 headstep
起步梯级 bottom step; curtail step <指一端或两端为半圆形或旋涡形者>
起步同步 start synchronism
起步阻力 breakaway force; breakaway resistance
起槽 riffling
起槽缝 groove seam
起草 drafting; draw up; make out a draught of
起草计划 draw up a plan
起草皮铲 sod lifter
起草人 drafter; draftsman
起草委员会 drafting committee
起草者 draftsman; draughtsman
起测基点【测】 starting datum mark
起层 layering
起潮力 tide-generating force; tide-producing force
起潮时刻 time of rise
起尘 <路面等磨损后造成的> dusting; escape of dust; dust emission
起尘的 dusty
起尘性 dustiness
起程 depart(ure); set out; start on journey
起程点 departure station
起程经度 longitude from; longitude left
起程坡度 starting grade
起程纬度 latitude from; latitude left
起承荷载【船】 sue load
起尺寸钎头 over-sized bit
起冲点 center mark
起出 lifting
起出井管 stripping the well
起出输送管 lifting of pipe lines
起磁 magnetization; magnetize; magnetizing
起磁场 magnetizing field
起磁电流 magnetizing current
起磁力 magnetizing force
起促进作用的 favo(u)rable
起大浪 surge
起刀箱 feeding box
起导间隙 starter gap
起到更有效的作用 pay a more effective role
起道 raise; raising of track; shovel packing; track lifting; track raiser
起道拨道机 track lifting and lining machine
起道钉器 spike extractor; spike puller
起道工作 track lifting work
起道机 rail jack; rail winch; track jack; track lifting jack; track raiser; track raising device
起道修理 lifting repair
起道抓轨器 rail clamping device
起道装置 track lifting device
起道作业 track lifting work; track raising operation
起点 origin; originating point; origination; point of origin; point of beginning; initial point; jumping-off point; launching pad; orifice zone; starting place; starting point; starting post; zero point
起点泵站 source pump station
起点定向 initial alignment
起点读数 reference mark; zero reading
起点发生区带 origin-generation zone
起点级 zero level
起点距 distance form initial point
起点距测量仪 productimeter
起点流域 originating basin
起点-目的地调查 origin and destination study; origin and destination survey
起点区 orifice zone; origin zone
起点调整 initial alignment
起点突变 zero point mutation
起点位置 initial position; starting position
起点线 zero line
起点站 stub-end station
起点/终点 origin/destination
起点子午线 zero meridian
起电 electrification; electrify; electrization; electrize
起电的 electromotive
起电机 induction machine
起电盘 electrizer; electrophorus
起吊 hoisting; lift; slinging
起吊把手 lifting lug
起吊扁担 hanger
起吊槽 slinging groove
起吊长度 handling length
起吊窗 lifting window
起吊窗设备 lifting window fitting
起吊窗五金 lifting window fitting
起吊打捞法 mechanical-hoisting salvage
起吊点 pick-up(and drop)point
起吊方法 lifting method
起吊方式 pick-up arrangement
起吊废料 hoisting spoil
起吊费 slinging
起吊高度 height of lift(ing)
起吊工具 means of slinging
起吊工作 crane lifting service
起吊滑门附件 lifting sliding door hardware
起吊滑门设备 lifting sliding door furniture
起吊环 hoisting ring
起吊机插锁 lift latch
起吊机房 lift machine room
起吊机构 hoist mechanism
起吊机理 hoist mechanism
起吊机械 hoist machinery
起吊卡车车身的起重机 truck hoist
起吊开关 lifting key
起吊孔 lifting hole
起吊块石螺栓 stone lifting bolt
起吊块石用的螺栓 stone lifting bolt
起吊垃圾 hoisting muck; hoisting spoil
起吊力 lifting force
起吊梁 lifting beam
起吊能力 lifting capacity
起吊强度 lifting strength
起吊设备 lifting gear; lifting rig; lift installation
起吊绳 pick-up line
起吊索 <附在结构上的> hand line
起吊污物 hoisting muck
起吊销子 <埋塞在大石块内的> lewis pin
起吊应力 <打桩工程> handling stress
起吊用预埋件 lifting inserts
起吊与降落重物 <起重机> up-down
起吊轴 suspension shaft
起吊轴颈 lifting trunnion
起吊柱 lift mast
起吊转角能力 luffing capacity
起吊装置 sling
起吊作业 crane lifting service
起钉锤 claw hammer
起钉斧 claw hatchet; shingling hatchet
起钉杆 wrenching bar
起钉棍 nail bar
起钉机 extractor
起钉器 claw; dog; nail extractor; nail puller; pinch bar
起钉钳 nail extractor; nail nippers; nail puller
起钉撬棍 crowfoot bar; wrecking bar
起钉錾 box chisel
起钉凿 box chisel
起锭器 claw; hanger
起动 starting; start-up; actuate; commissioning; cranking; fusing; initial operation; initiation of motion; priming; setting-in motion; set to work; triggering
起动按钮 launching button; start(ing) button; starter button; starter push; starting push-button; trigger button
起动把 starting handle
起动板 starting panel

起动保护接触器 starting protection contactor
起动保护系统 opposed-voltage protective system
起动泵 prime(r) pump; priming pump; start(ing) pump
起动泵的气室 starting air vessel
起动泵阀 priming valve
起动泵时的注油旋阀 priming cock
起动变加速度 starting-up of speed
起动变扭器 starting torque converter
起动变压器 starting transformer
起动变阻器 starter rheostat; starting rheostat
起动柄 starting bar
起动并加速 starting-up to speed
起动波 starting wave
起动补偿器 starting compensator
起动不良 poor starting
起动部分 actuating section
起动部件 starting unit
起动操纵 starting control
起动操纵盘 starting panel
起动操作 start operating
起动车 launching troll(e)y
起动沉没深度 starting submergence
起动程序 initiator; starting sequence; start-up procedure
起动持续时间 duration of starting
起动齿轮 starter gear; starter toothed wheel; starting gear
起动齿轮拨叉 starting gear shifting bolt
起动齿轮圈 gear wheel for starting
起动齿圈 starter gear rim
起动充电发电机 starterdynamo
起动冲量 inrush; starting impulse
起动抽气机 starting ejector
起动抽气器 starting ejector
起动触发 initiating
起动触发器 start trigger
起动传感器 start sensor
起动吹管 starting torch
起动磁电机 booster; magneto; starter magneto; starting magneto
起动次数 number of starts
起动单元 starting unit
起动弹射器 air ejector for starting
起动导轨 shoe
起动灯 start light
起动点 breakaway point; firing point; starting point
起动点火系统 start-ignition system
起动点火线圈 booster coil
起动点火照明用电池 starting-lighting-ignition cell
起动电池组 starting battery
起动电动机 actuating motor; starter armature; starter motor; starting motor
起动电动机固定带 starter motor strap
起动电动机控制 starter motor control
起动电动机整流子 starter motor commutator
起动电弧 starting arc
起动电机 cranking motor; electric(al) starter; electric(al) starting motor; starter motor; starting dynamo
起动电极 starting electrode
起动电空阀 starting electro-pneumatic valve
起动电阀 starting electrovalve
起动电缆 firing cable; starter cable
起动电流 initial firing current; pick-up current; starting current; triggering current
起动电路 firing circuit; initiation circuit; starting circuit
起动电码 start code
起动电气路 activation circuit
起动电容器 starting condenser
起动电位 starting potential
起动电线 starting cable
起动电压 initial voltage; starting voltage; trigger voltage; cranking voltage <起动机转动曲轴所需的>
起动电阻 starting resistance
起动电阻器 starting resistor; start relay box
起动定时器 starting timer
起动动力 starting power
起动扼流圈 starting reactor
起动发动机 ato unit; pilot engine; rev up the engine; starter engine; starting engine
起动发动机传动装置 starting engine transmission
起动发动机的操纵连杆 starting engine operating linkage
起动阀 go-valve; priming valve; start(ing) valve; stop-and-go valve; kick-off valve <空气升液器的>
起动范围 starting range
起动方法 priming method; starting method; starting technique
起动方式 start mode
起动分配器 starting distributor
起动风阀 starting damper
起动风扇 starting fan
起动风速 minimum speed of wind for carrying sand
起动辅助器 starting aid
起动辅助设备 starting aid; start-pilot device
起动辅助装置 starting aid
起动负荷 starting duty; starting load
起动负载 starting duty
起动附加阻力 additional resistance for starting
起动概率 starting probability
起动杆 actuating lever; launching beam; priming lever; starting lever; starting rod; throw-in lever; tripping lever
起动杆换向联锁装置伺服马达 starting lever reversing interlock servomotor
起动杆簧 starting lever spring
起动杆联锁 starting lever interlock
起动杆液压联锁 starting lever hydraulic interlock
起动给油装置 priming device
起动工况 state of starting operating
起动工作 starting work
起动功率 cranking horsepower; initial power; inrush; launching power; starting capability; starting capacity; starting duty; starting power
起动管路 primer line
起动惯性 starting inertia
起动过程 starting process
起动过载系数 starting overload ratio
起动和负荷限制装置 starting and load-limiting device
起动和停止按钮 start-stop push button
起动虹吸 priming siphon[syphon]
起动虹吸阀 priming siphon valve
起动虹吸管 priming siphon[syphon] pipe
起动后检查 post-firing check
起动环节 starting link
起动缓坡【铁】flat gradient for starting
起动挥发性 starting volatility
起动混合气 starting mixture
起动火舌 pilot flame; starting torch
起动机 hauler; motor starter; prime mover; starter; starting engine; starting machine; starting motor
起动机按钮护套 starter button guard
起动机变速杆 starter shift lever
起动机操纵杆 starting lever
起动机齿轮 starter gear
起动机齿轮拨叉 starter gear shifting fork
起动机齿轮传动壳体 starter gear housing
起动机传动 starter drive
起动机传动臂 starter motor shift lever
起动机传动轴棘爪 starter dog
起动机磁场线圈 starter field coil
起动机磁场线圈电刷 starter brush of field coil
起动机磁场线圈接头 starter field coil connection
起动机磁场线圈中间接头 starter field coil equalizer
起动机磁极瓦 starter pole shoe
起动机地线电刷 starter ground brush
起动机电磁继电器盖 starter solenoid relay cover
起动机电磁线圈 starter solenoid; starter solenoid coil
起动机电缆 starter cable
起动机电容器 starter condenser
起动机电枢 starter armature
起动机电刷 starter brush; starting motor brush
起动机电刷簧 starter brush spring
起动机电转子轴隔片 starter armature shaft spacer
起动机盖 starter motor cover
起动机构 starter mechanism; starting mechanism
起动机互锁电路 starter interlock circuit
起动机互锁机构 starter interlock mechanism
起动机环齿轮 starter ring gear
起动机继电器开关 starter relay switch
起动机架 starter mounting
起动机架穿过长螺栓 starter frame through bolt
起动机接地片 starter ground strap
起动机接合杆 starter shifting lever
起动机接线端子 starter terminal post
起动机接线盒盖 starter terminal cover
起动机开关 starter switch
起动机开关接触片 starter switch contact
起动机开关接线端子 starting switch terminal
起动机开关推杆 starter switch push rod
起动机壳 starter housing
起动机控制 starter control
起动机控制杆 starter control lever
起动机离合弹簧 starter clutch spring
起动机离合器 starter clutch
起动机连接盘 starting motor flange
起动机扭矩 starting motor torque
起动机钮 starting button
起动机喷嘴 starter nozzle
起动机前盖 starter front cover
起动机驱动齿轮 starter driving gear; starting motor pinion
起动机驱动弹簧 starter drive spring; starting drive spring
起动机弹簧 starter spring
起动机小齿轮 starter pinion
起动机小齿轮及离合器轭 starter pinion and clutch yoke
起动机移动叉弹簧 starter shift spring
起动机钥匙 starter key
起动机罩箍带 starter cover band
起动机整流器端盖 starter commutator end cover
起动机整流器护罩 starter commutator shield
起动机中心轴承板 starter center[centre] bearing plate
起动机中心轴承衬套 starter center [centre] bearing bushing
起动机轴 starter shaft
起动机轴壳衬套 starter drive housing bushing
起动机肘节开关 starter toggle switch
起动机主轴齿轮 starter main shaft gear
起动机主轴轴承 starter main shaft bearing
起动机转机 starter motor
起动机转子 starter armature
起动机租赁公司 crane hire company
起动机座 starter frame block
起动及换向手柄 throttle and reverse handle
起动及燃油控制手柄 starting and fuel control handle
起动棘爪 starter dog
起动技术 starting technique
起动继电器 accelerating relay; initiating relay; starter relay; starting relay
起动继电器箱 start relay box
起动加速度 starting acceleration
起动加油器捏手柄 primer knob
起动加注剂 priming composition
起动加注燃油 priming charge
起动键 activate key; start key
起动降压站 start-up reducing station
起动降压装置 start-up reducing equipment
起动交流声 starting hum
起动接触器 starting contactor; starting switch
起动距离 starting course; starting distance
起动开闭器 starter
起动开关 firing key; firing switch; fire switch; starter switch; starting switch; switch starter
起动开关触片 starter switch contact
起动空气 starting air
起动空气的辅助继电器 auxiliary relay for starting air
起动空气阀机构 starting air valve mechanism
起动空气分配阀 starting air timing valve
起动空气分配器 air starting distributor
起动空气分配器传动装置 starting air distributor drive
起动空气隔离阀 starting air isolation valve
起动空气管路 starting air line
起动空气管系 starting air piping
起动空气罐 starting receiver
起动空气耗量 starting air consumption
起动空气滑阀 starting air slide valve
起动空气截止阀 starting air stop valve
起动空气控制阀 starting air pilot valve
起动空气瓶 starting air bottle; starting air container; starting air cylinder; starting air tank; starting air vessel
起动空气系统 starting air system
起动空气消声器 starting air silencer
起动空气压缩机 starting air compressor
起动空气指示器 starting air indicator
起动空气主操纵手轮 starting air master handwheel

起动空气主阀 starting air master valve
起动控制 starting control
起动控制滑阀 starting control slide valve
起动控制盘 cranking panel; starting panel
起动控制器 starter controller
起动控制设备 starting and control equipment
起动困难 starting difficulty
起动拉索夹 starting wire clip
起动力 breakaway force; speeding-up force; starting effort; starting force; starting force effort; starting power
起动力矩 starting moment; starting torque
起动力矩变化系数 coefficient of starting torque conversion
起动联锁继电器 start interlocking relay
起动链 starting chain
起动流量 dynamic(al) threshold discharge
起动流速 critical tractive velocity; incipient velocity; starting velocity; threshold velocity; pick-up velocity; competent velocity <泥沙的>
起动流速比例 competent velocity scale
起动路线 starting circuit
起动率 starting rate
起动马达 cranking motor; electric(al) starter; electric(al) starting motor; starter motor; start(ing) motor
起动马达传动簧 starter motor drive spring
起动马达导线 starter motor wire
起动马力 start horsepower
起动码元 start bit
起动脉冲 anti-paralyse pulse; drive pulse; driving pulse; enable pulse; enabling pulse; initiating pulse; starting impulse; start(ing) pulse; trigger pulse; tripping pulse
起动脉冲鉴频器 trigger discriminator
起动脉冲作用 starting-pulse action
起动模式 activation pattern
起动摩擦力 starting friction
起动摩擦损失 starting friction loss
起动摩擦系数 coefficient of starting friction
起动磨损 starting friction
起动能力 initiating ability; starting ability; starting capability
起动能源 starting energy
起动扭矩 starting torque
起动扭力 starting torque
起动钮 starter button
起动盘 starting board; starting panel
起动配气器阀 starting air distributor valve
起动喷口螺塞 priming hole plug
起动喷口螺塞孔 priming plug hole
起动喷嘴 pilot burner; starting nozzle
起动频率 starting frequency
起动瓶 <压缩空气> starting receiver
起动坡度 grade for starting; starting grade
起动期 starting period
起动气阀 starting air valve
起动汽化器 primary carburet(t)or [caruret(t)er]; priming carburet(t)or[caruret(t)er]
起动器 compensator starter; cranking motor switch; exciter; primer; starter; starting apparatus; starting vessel
起动器按钮 starter push button
起动器触点 starter contact
起动器电池 starter battery
起动器电磁线圈 starter solenoid
起动器电磁线圈开关 starter solenoid switch
起动器发电机 starter generator
起动器固定夹螺钉 screw for starter fixing clamp
起动器绝缘螺纹 stator screw insulator
起动器开关 starter switch
起动器控制杆 starter control lever
起动器皮带轮 starter pulley
起动器顺序 starter sequence
起动器踏板回动弹簧 starter pedal return spring
起动器箱盖密垫 starter casing cover packing
起动牵引力 starting traction effort; starting tractive effort; starting tractive power
起动前 prestart
起动前操作 prestart operation
起动前的 prestarting
起动前检查 prestart check; prestarting inspection; pre-start-up check
起动前检查单 prestart checklist
起动曲柄 cranking lever; hand crank; starting bar; starting crank
起动曲柄棘轮机构 starting crank ratchet
起动曲柄孔盖 starting crank socket cap
起动曲柄螺帽 starting crank nut
起动曲柄托架 starting crank bracket
起动曲柄销 starting crankpin
起动曲柄摇手 starting crank handle
起动曲柄轴 starting crankshaft
起动曲柄轴承 starting crank shaft bearing
起动曲柄轴弹簧 starting crankshaft spring
起动曲柄爪 starting crank jaw
起动曲线 starting curve
起动燃料 starting fuel
起动燃料点火 auxiliary fluid ignition
起动燃料供给 starting fuel supply
起动燃料供给控制 starting fuel control
起动燃料管 primer fuel tube
起动燃料箱 starting tank
起动燃烧室 starting chamber
起动绕组 starter winding; starting winding
起动润滑 initial lubrication; starting lubrication
起动设备 starting equipment; starting outfit; starting unit; trigger
起动深度 priming depth
起动时断相保护装置 starting open-phase protection
起动时反转 back-fire
起动时间 actuating time; rise time; run-up time; starting duration; starting period; starting time; time in starting; time of priming
起动时间抖动 starting time jitter
起动时间间隔 starting time interval
起动时燃料过量注入 overprime
起动时投料 primed charge
起动时限 starting time limit
起动时轴重向后轴转移 load transfer to rear axle during starting
起动试验 running-up test; starting test
起动室 priming chamber
起动手柄 crank arm; crank handle; operating crank; priming handle; starter grip; starter handle
起动手轮 starting hand-wheel
起动手摇曲柄 starting crank handle
起动输入 enable input
起动输入输出指令 start input/output instruction
起动水深 priming depth
起动水位 <虹吸管的> priming level
起动顺序 starting sequence
起动瞬变量 starting transient
起动伺服马达 starting servomotor
起动速度 cranking speed; cutting-in speed; priming speed; starting speed; starting velocity
起动算法 starting algorithm
起动损失 start(ing) loss
起动索 starting rope
起动锁簧 starter dog
起动踏板 starter paddle; starter pedal; starting paddle
起动台 firing platform
起动特性 starting characteristic; starting property
起动条件 starting condition
起动调节变阻器 regulating starting rheostat
起动调节电阻 starting and regulating resistance
起动调汽阀 <复涨式机车> intercepting valve
起动停机控制 start-stop
起动停止操纵装置 stop-go control unit
起动停止及转速控制 start-stop and speed control
起动停止失真度 degree of start-stop distortion
起动凸轮 starting cam
起动拖曳力 threshold drag
起动网路 trigger circuit
起动位 start bit
起动位点 initiation site
起动位间隔 start interval
起动位置 enable position; start(ing) position
起动温度开关 starting temperature switch
起动稳压器 starting manostat
起动涡流 starting vortex
起动系统 cranking system; launching system; starting system
起动缝隙 starter gap
起动先导阀 start pilot
起动线 starter cable; starting wire
起动线路 trigger; trigger circuit
起动相似准则 similarity criterion of initiation of motion
起动相位 starting phase
起动箱 starter box
起动小齿轮 starter pinion
起动卸载器 starting unloader
起动信号 activating signal; actuating signal; enabling signal; initiating signal; start(ing) signal
起动信号发送器 starting signal transmitter
起动信号线 enable line
起动性能 startability; starting ability; starting performance
起动旋塞 starting cock
起动压风机 starting compressor
起动压力 starting pressure
起动压力表 starting pressure ga(u)ge
起动压力计 starting pressure ga(u)ge
起动延迟 starting delay
起动延误 starting delay
起动延滞 start lag
起动阳极 <整流器> starting anode
起动摇把 hand crank; starting handle
起动摇把棘爪 starting ratchet
起动摇把爪销 starting crank jaw pin
起动摇柄 starter crank; starting crankshaft
起动摇手柄 starting crank handle; starting grip; starting handle
起动摇手柄支架 starting handle bracket
起动摇转超限 overcrank
起动液 starting fluid
起动用变阻器 rheostatic starter
起动用电阻箱 starting box
起动用继电器 starting guide relay
起动用空气压缩机 starting air compressor
起动用燃油 priming fuel
起动用凸块 starting dog
起动用涡轮 starting turbine
起动用蓄电池 starting battery
起动用压缩空气瓶 starting receiver
起动用压缩空气消耗量 starting air consumption
起动用自耦变压器 starting autotransformer
起动油 starting-up oil
起动预燃室 starting prechamber
起动预热灯喷嘴 starting preheater lamp nozzle
起动预热锅炉 starting preheater boiler
起动预热锅炉支架 starting preheater boiler bracket
起动预热火盆 starting preheater fire pan
起动预热器 starting preheater
起动预热器喷灯 starting preheater burner
起动元件 initiating element; initiating unit; start(ing) element
起动原油 starting crude
起动圆盘 actuator disc[disk]; starting ring
起动源 starting source
起动运行 start operation
起动运转 starting operation
起动闸门 start gate
起动栅压 starting grid voltage
起动真空泵 pump primer
起动震动 starting shock
起动值 pick-up value
起动指令 starting order
起动中止螺线管 starting cut-out solenoid
起动周期 starting cycle
起动轴 starting shaft
起动助推器 launching booster
起动注料泵 primer pump; priming pump
起动注水 priming; priming charge
起动注水泵 primary pump; priming pump
起动注水阀 priming valve
起动注水器 priming arrangement
起动注液 priming
起动注液泵 primer pump; priming pump
起动注油 priming
起动注油泵 fuel priming pump; primary pump; prime pump; priming pump
起动注油不足 underpriming
起动注油阀 primer valve
起动注油管 priming pipe
起动注油管路 priming connection
起动注油开关 priming cock
起动注油器 primer
起动注油器的保险器 primer lock
起动注油塞 priming plug
起动注油旋塞 priming cup
起动爪 starting dog; starting jaw
起动转矩 detent torque; pick-up velocity; starting moment; starting torque
起动转数 cutting-in speed
起动转速 starting speed
起动转速控制阀 starting speed con-

Q

trol valve
起动装置 launcher;starter gear;starting apparatus;starting device;starting equipment;starting gear
起动装置电压 starter voltage
起动状况 starting condition
起动状态 starting condition
起动状态区域 period level
起动准备 preparation for starting;starting preparation
起动自耦变压器 starting autotransformer;starting compensator
起动阻抗继电器 starting impedance-relay
起动阻力 starting resistance
起动阻尼绕组 starting amortisseur
起动钻机 starting drill
起动钻头 starting bit
起读线 reference line
起端 origin
起额征税 over-tax
起垩 chalking
起阀器 valve extractor;valve lifter;valve remover
起阀器导管 valve lifter guide
起阀器滚子 valve lifter roller
起阀弹簧 valve lifter spring
起阀座器 valve seat grab
起反应 reacting
起反应物 reacting substance
起反作用 react
起飞 take-off;fly-off;lift-off
起飞安全区 take-off safety zone
起飞场 lift-off site;take-off site
起飞地带 take-off area
起飞方向 direction of take off
起飞海拔高度 take-off altitude
起飞航迹 take-off flight path;take-off path
起飞航线 take-off pattern
起飞航向 take-off heading
起飞滑跑 starting run;take-off roll
起飞滑跑距离 take-off run available
起飞滑跑距离列线图 take-off ground run distance chart
起飞滑行 launching slide;take-off run
起飞滑行距离 take-off distance
起飞机场 original base;take-off point
起飞加速器 jato;thruster
起飞监控系统 take-off monitoring system
起飞降落甲板 take-off and landing deck
起飞降落跑道 landing runway
起飞阶段 take-off phase
起飞距离 take-off distance available
起飞空速 take-off airspeed
起飞控制仪 take-off control equipment
起飞离地距离 lift-off distance;take-off distance
起飞马力 take-off horsepower;take-off power
起飞跑道 airway;departure runway;flight strip;take-off runway;take-off strip
起飞跑道坡度 take-off runway gradient
起飞喷射装置 catapult launching
起飞偏航 take-off drift
起飞前的 preflight
起飞前滑行 <飞机> exit taxiing
起飞前试车区 run-up area
起飞时刻 departure time
起飞时间 departure time
起飞速度 take-off speed
起飞推力 take-off thrust
起飞位置 start position;take-off setting
起飞位置指示灯 take-off position indicator lamp
起飞线塔台车 runway portable tower
起飞性能 take off ability
起飞需用滑跑距离 required take-off run
起飞需用跑道长度 required take-off distance
起飞许可 take-off clearance
起飞仰角 lift-off attitude
起飞用跑道的基本长度 basic take-off runway length
起飞噪声 take-off noise of aircraft
起飞指示器 take-off monitor
起飞指示系统 finger-nail system
起飞重量 <飞机的> all-up weight;take-off weight
起飞准备的 preflight
起飞着陆循环 landing and take-off cycle
起分割作用的 dividing
起风浪 raise the wind
起峰时段 flood to peak interval
起伏 ebb and flow;fluctuation;heave and set;jitter;rise-and-fall;rolling;sag and swell;sinuosity;undation;undulate;ups and downs;wave;wriggle
起伏比 fluctuation ratio
起伏冰 pressure ice;rough ice
起伏补偿 compensation of undulation;heave compensation
起伏不定 <物价、温度等> see-saw
起伏(不定)坡度 undulating grade
起伏不平的 broken;up-and-down
起伏不平的地形 wavy terrain
起伏不平度 undulation
起伏参数 roughness parameter
起伏差 relief difference
起伏成分 fluctuating component
起伏成型 embossing;raising
起伏程度 degree of fluctuation
起伏大的地貌 accidented relief
起伏大的地区 high-relief area;strong relief area
起伏道路 undulating road
起伏的 kinked;undulated;undulating;up-and-down;wavy
起伏的土地 undulating land
起伏等温退火 rising and falling isothermal annealing
起伏等值功率 fluctuation equivalent power
起伏地 rolling land;uneven ground;uneven terrain
起伏地面 rolling ground;rolling surface;rough ground;rugged terrain;undulating ground
起伏地区 rolling region;rough area;rough terrain
起伏地形 accidented relief;hummock-and-hollow topography;rolling topography;rugged topography;undulating topography;wavy terrain
起伏电场 fluctuating electric(al) field
起伏电流 fluctuating current
起伏电压 fluctuation voltage
起伏度 prominence [prominency];waviness
起伏反射波 fluctuation echo
起伏范围 range of fluctuation
起伏负荷 fluctuating load
起伏干扰 scintillation jamming;elephant <俚语>
起伏干扰频谱 scintillation jamming spectrum
起伏海岸 rugged coast;undulating coast
起伏回波 fluctuating echo;fluctuation echo
起伏角 relief angle
起伏校正仪 heave compensator
起伏理论 fluctuation theory
起伏力 fluctuating force
起伏量 undulating quantity
起伏流动 fluctuating flow
起伏路线 wavy trace
起伏率 fluctuation rate
起伏模 forcer
起伏目标探测 fluctuating target detection
起伏频率 fluctuating frequency
起伏坡度 rolling grade;undulation grade;undulating grade
起伏丘陵 rolling hills
起伏散逸定理 fluctuation-dissipation theorem
起伏沙丘 sand undulation
起伏数据 fluctuating data;fluctuation data
起伏衰落 scintillation fading
起伏衰落时间 decay time of scintillation
起伏水舌 undulated nappe
起伏系数 hilliness coefficient
起伏现象 fluctuation
起伏线路 undulating route;undulating track
起伏小的地区 low-relief area
起伏小山 rolling hills
起伏信号 fluctuating signal
起伏运动 phugoid motion;undulation
起伏运动控制 phugoid control
起伏运动曲线 phugoid curve
起伏噪声 fluctuation noise;shot noise
起伏噪声电皿 fluctuation noise level
起伏噪声水平 fluctuation noise level
起伏振荡 heaving oscillation
起伏值 fluctuating value
起伏周期 cycle of fluctuation;period of undulation
起伏阻尼参数 fluctuation damping parameter
起浮 ascending;floating-up
起浮布置 arrangement for raising a wreck
起盖螺栓 cover parting bolt
起隔子 burnt-on sand
起拱 arch camber;bow wave;bridging;camber;rise of span;rising of arch;sprung;turning an arch
起拱板 <支待拱用小弯度的> camber piece;camber board
起拱边扰动 disturbance at the springing
起拱标高 impost level
起拱处线脚 impost mo(u)lding
起拱大梁 cambered girder
起拱道路 crowned road
起拱的 haunched
起拱点 arch springer;arch springing;edge bend;impost;point of springing;spring bend;springer;springing;springing of curve;springing point;spring point;sprung arch
起拱度 excess clearance;camber
起拱高 amount of camber
起拱高度 arch springing height;springing height;spring level
起拱桁架 raised-chord truss
起拱接点 impost joint
起拱接缝 impost joint
起拱力矩 hogging moment
起拱梁 haunched beam;springing line
起拱楼盖 cambered ceiling
起拱路面 crowned road
起拱路面横坡 transverse slope
起拱面 edge bend
起拱(砌)层 springing course
起拱砌块 springing block
起拱石 arch springer;footstone;impost springer;skewback;springer;springing;spring stone;tas-de-charge
起拱图 camber diagram
起拱线 arch springing line;springing (line)
起拱线标高 arch springing level
起拱效应 bulging effect
起拱中心 center[centre] of springing
起拱作用 arch action;arching
起沟 corrugate
起钩杆 uncoupler lever
起钩杆链 uncoupled lever chain
起管夹 pull yoke
起管卡 pull yoke
起管器 rod gun;rod puller
起轨拨道机 track lifting and lining machine
起轨杆 track lever
起轨机 track jack;track winch
起轨绞车 track lifting winch;track winch;rail winch
起轨器 rail lifter;rail winch;track jack;track lifter;track lifting jack
起轨千斤顶 track lifting jack
起航点 departure point;point of departure
起核 nuclei of origin;nuclei origin
起横棒 rapping
起红丝 filiform corrosion
起弧 arcing;arc starter;arc starting;scratch start;shoot-out;strike
起弧板 run-on tab
起弧电流 striking current
起弧电位 striking potential
起弧电压 arc voltage;sparking voltage;striking voltage
起弧极电压降 ignitor drop
起弧距离 striking distance
起弧稳定器 <焊接> arc booster
起弧线圈 striking winding
起化学反应 attachment attack;attack
起划分作用的 dividing
起灰 dusting
起灰机 ash hoist engine
起灰装置 ash hoist
起辉器 starter
起火 catching fire;fire breakout;firing;outbreak of fire;sitting on fire
起火单位 fire originating unit;fire originator
起火后火灾危险模型 post-ignition fire risk model
起火后模型 post-ignition model
起火剂 igniting composition
起火原因 fire cause
起火源 fire origin
起货 debark;unload
起货费 landing charges
起货港 port of discharge;port of unloading;port of discharging
起货钩 cargo hook;crowbar;load binder;pry;granny bar <俚语>
起货滑轮 cargo block
起货机 cargo winch;loading winch;winch
起货机台 winch platform;winch table
起货绞车 cargo winch
起货卷扬机 cargo hoist
起货码头 landing pier;landing stage
起货设备初次检验和试验 initial survey and test of cargo gear
起货设备动负荷试验 dynamic(al) test for cargo gear
起货设备检验和试验 inspection and test of cargo gear;survey and test of cargo gear
起货设备静负荷试验 static test for

cargo gear
起货设备临时检验 contingent survey of cargo gear;occasional survey of cargo gear
起货设备每四年的检验 quadrennial cargo gear survey
起货设备证书 certificate for cargo gear
起货升降机 goods lift
起货桅 derrick(crane)
起货桅杆 derrick mast
起货栈单 landing account
起货证 permit to land
起脊 ridging
起脊台地【地】ridge terrace
起加劲作用的墙 reinforcing wall
起夹子 scab
起碱 saltpetering
起降跑道 landing strip
起降区 landing area
起脚石状态 stepping-stone state
起晶温度 nucleation temperature
起井架 set-up derrick
起净化作用的 aseptic
起居凹室 zotheca
起居室 drawing room;living area;living room;lounge;parlo(u)r;sitting room;state room;serdab <近东地区住宅地下室中>;calefactory <寺院内的>;area of activity <在一幢公寓楼内的>
起居室-餐室-厨房 living-dining-kitchen
起居室兼餐室 living dining
起居室兼厨房 living kitchen
起居室门厅 living room-entrance hall
起居室外的壁龛 zotheca
起桔面 <轧制表面的> alligatoring
起决定作用的技术 pacing technology
起开口销器 split pin extractor
起开尾销器 split pin extractor
起颗粒 graining;seediness
起壳 addling;incrusting
起坑 pitting
起空间位隔作用的增量剂 spacing extender
起块 <粉刷> bittiness
起拉螺钉 draw peg;draw screw
起浪 sea gets up;swell;undulate;uprise;waviness
起浪点 surge point
起勒墙 plinth walling
起肋玻璃 ribbed glass
起肋钢板 ribbed steel
起肋镶轧缝 ribbed seam
起棱 ridging
起棱螺栓 ribbed bolt
起棱纹 corded
起棱纹理 raised grain
起楞羊毛毡 ribbed wool felt
起励 field flashing
起粒 graininess;graining;seediness;seeding;seedy
起梁钩 girder dogs
起裂 crack initiation
起裂纹 checking
起鳞 delamination;fish scaling;flaking;scaling;slabbing
起鳞程度 scaliness
起鳞现象 fish-scaling effect
起流相角 angle of ignition
起垄 ridge forming
起垄机 ridger
起垄犁 mo(u)ld board ridger;ridge plow
起垄犁体 ridging bottom
起垄器 ridge-former;tie ridger
起漏压力 start-to-leak pressure
起螺母机 nut driving machine
起螺丝器 screw extractor
起落 lift;rise-and-fall;up-and-down
起落摆动 <起重时的> luff
起落臂 lift arm
起落臂锁 lift arm lock
起落场地 pad
起落吊钩 drop hanger
起落飞行区 landing area
起落杆 landing bar
起落拱架 lifting arch
起落构造 elevator tectonics
起落航线 traffic pattern
起落横木 landing beam
起落滑车绳 casing line
起落环 lifting eye
起落活动区 aircraft manoeuvring area
起落机构 lifting gearing;raising machinery;raising gear
起落架 <飞机> alighting carriage;alighting gear;undercarriage;undercart;under-chassis;lander-carriage;landing chassis;landing gear;chassis frame;lifter;load gear;tripping device;undercarriage;gear <飞机着陆用>
起落架舱 bell;undercarriage bay
起落架舱门 undercarriage door
起落架车 landing gear dolly
起落架导向板 guide gib
起落架防扭臂 landing gear torque arm
起落架减震 undercarriage cushioning
起落架减震器 landing gear shock absorber
起落架拉条 undercarriage bracing
起落架拦阻装置 undercarriage arrester
起落架轮 castorite
起落架轮距 donut;undercarriage track
起落架轮胎 gear tire
起落架倾斜液压缸 laydown tilt cylinder
起落架刹车装置 undercarriage brake
起落架弹性装置 undercarriage springing
起落架支柱 landing gear post
起落架制动器 undercarriage brake
起落架中心 undercarriage center[centre]
起落架状态指示器 undercarriage indicator
起落器凸轮 lifer cam
起落千斤顶 rim lift jack
起落区 land strip
起落式牵引装置 lift-up drawbar
起落式闸门 guillotine type gate
起落弹簧 set-up spring
起落弯臂 lift crank
起落蓄能器 lift accumulator
起落液压缸 laydown cylinder
起落油缸 lift ram
起落油缸活塞 lift piston
起落装置 landing gear
起落自动器 lift clutch
起麻点 cratering
起码里程 minimum charged distance
起码利润 threshold return
起码票价 minimum fare
起码收费 minimum charges
起码收费额 minimum charge quantum
起码运费 minimum freight
起码知识 rudimental knowledge
起码重量 minimum weight
起码资金 <购置房地产的> front money
起毛 fluff(ing);fuzz;picking up
起毛纹理 fuzzy grain;woolly grain
起锚 cast off;disanchor;heave-ho;heave up anchor;loose;purchase an anchor;raise anchor;retrieving anchor;running the anchors;unmoor;weigh(anchor);heave away the anchor;lift anchor;up anchor
起锚车 anchor capstan
起锚传令钟 anchor telegraph;mooring telegraph
起锚船 anchor barge;anchor boat;anchor hoy
起锚吊车 anchor crane
起锚杆 cathead shaft
起锚工作 cathead job
起锚滑车 cat
起锚滑轮 cathead line sheave
起锚机 anchor windlass;capstan crab;capstan engine;winch capstan;windlass
起锚机舱 windlass room
起锚机挡板 cross timber
起锚机绞盘 anchor capstan
起锚机绞筒 wildcat
起锚机离合器 cathead clutch
起锚机推杆 windlass heaver
起锚机(系)柱 carrick bitts;range heads;windlass bitts;carrick heads
起锚检视 sighting anchor;sight the anchor
起锚绞车 anchor hoist;anchor winch
起锚绞盘 anchor capstan;anchor winch;capstan windlass;vertical windlass;windlass capstan
起锚起重机 anchor crane
起锚设备 anchor gear
起锚索 cathead line;catline
起锚艇 anchor boat
起锚柱 anchor davit;cat davit
起锚装置 anchor gear;cathead installation
起模 pattern draw(ing);rapid the pattern
起模板 bull block;draw(ing) plate;lifting plate;rapping plate;stripping plate
起模棒 draw stick;rapping bar;rapping iron
起模钉 draw nail;spike
起模顶杆 lift pin;stripper pin;stripping pin
起模杆 ejector pin;jemmy
起模缸 stripper cylinder
起模工位 stripper station
起模机 drawing machine;stripping machine
起模机构 pattern drawing mechanism
起模架 draw arm
起模框架 draw frame
起模螺钉 draw screw;lifting screw;pattern screw
起模木棒 draw peg
起模器 ejector
起模时间 stripping time
起模式震压造型机 jolt-squeeze pattern drawing machine
起模手柄 strip handle
起模松动量 rapping allowance
起模台 stripping table
起模同步机构 equalizer
起模斜度 pattern draft
起模行程 draw stroke;pattern draw travel
起模性 liftability
起模用毛笔 floor swab
起模胀砂 rappage
起模针 draw stick;lift pin;picker
起模装置 die lifter
起膜 skinning
起膜的油漆面 blooming
起沫 airing;frothing
起沫剂 frother;frothing agent
起炮器 shot-firing battery
起泡 barbotage;beading;blister;blowing;blub;boil;bubble;bubbling;burble;dobying;ebullience;ebullition;foaming;free-swelling;poulticing;spackling;blistering <油漆>
起泡包装 blister packaging
起泡捕收剂 frothing collector
起泡的 blistered;bubbly;effervescent
起泡的肥皂水 buck
起泡点 bubble point;bubbling point
起泡度 foaminess;frothability
起泡翻滚 popple
起泡分离法 foaming and separation method;foaming flo(a)tation method
起泡浮选机 bubble machine
起泡腐蚀 blister corrosion
起泡剂 blowing agent;foamer;foam-forming liquid;foaming adjutant;foaming agent;froth agent;frother;frothing(re) agent;gas-forming admixture
起泡加速试验箱 blister house
起泡金属 sparkle metal
起泡沫 effervesce;foam;froth(formation);frothily;popple;sponging;spume;effervescence;lather
起泡沫的 foamed;foaming;foamy;frothing;spumescent
起泡沫的黏[粘]土 effervescing clay
起泡沫(能)力 foaming ability;foaming capacity
起泡沫性 frothiness
起泡沫油 frothing oil
起泡能力 bubbling potential
起泡黏[粘]土 bubbly clay
起泡盘 bubble deck
起泡器 bubble former;bubbler;ebullator
起泡倾向 tendency to bubble
起泡式肥皂配出器 lather-type soap dispenser
起泡试验 foaming test
起泡试验箱 blister box
起泡收集器 frothing collector
起泡熟铁条 blister bar
起泡速度 lather quickness
起泡稳定性 bubbling stability
起泡物 foaming substance
起泡系数 foaming coefficient
起泡性 vesicular nature
起泡性开裂 blister cracking
起泡性能 frothing capacity
起泡性指数 frothability index
起泡皂水 soapsuds
起泡状态 bubblement
起泡作用 aeration;barbotage;foaming action;foaming process
起疱 pimpling
起皮 blister;peeling;scaling;shelling;skinning <油漆筒内表面>
起偏光镜 polarizer
起偏角 polarizing angle
起偏器 polarizer
起偏振单色器 polarizing monochromator
起偏振镜 polariscope;polariser[polarizer];polarizator
起偏振棱镜 polarizing prism
起偏振目镜 polarizing eyepiece
起偏振片 polarizing plate
起偏振器 polarizer
起偏振作用 optically active
起坡 track raising
起破碎作用的球 breaking ball
起期 attachment
起气泡 aeration
起讫表 origin-destination table
起讫点 origin-destination
起讫点调查 origin-destination study;origin-destination survey
起讫点交通表 origin-destination traf-

Q

fic table
起讫点交通量 origin-destination traffic volume
起讫点模型 origin(-and)-destination model
起讫点直接问询方法的调查 direct interview origin and destination survey
起讫点资料 origin(-and)-destination data
起讫港 terminal port
起讫联票 coupon origin-destination
起讫点 beginning-and-ending point
起桥绞车 ladder hoist winch
起桥塞 retrieving a bridge plug
起球 pilling
起区分作用的 dividing
起圈花边 looped lace
起圈花线并捻机 loop yarn twister
起圈绒头 loop pile
起圈线 springing line
起确认作用的 confirmatory; confirmative
起燃 combustion; combustion initiation; initiation
起燃点 firing point
起绒 napping; tease
起绒表面结构 raised surface structure
起绒厂 gig mill
起绒粗呢 <用于室内装饰和家具套> frieze
起绒地毯 frieze carpet
起绒机 pile and nap lifting machine; raising gig
起绒织物 pile weave; raised fabric
起熔块 starting block
起软木塞的起子 corkscrew
起软木塞钻 corkscrew
起润滑作用 lubricate
起砂 <又称起沙> dusting; sand streak(ing); seedy; sanding <混凝土表面的>
起砂风速 starting wind-speed
起砂混凝土板 dusting
起砂混凝土表面 sand streak
起砂器 sand hoist
起伸缩臂节 telescoping boom
起升 lifting
起升导向轮装置 lifting guiding wheel device
起升道轨 lifting guiding runways
起升范围【机】 lifting range
起升钢丝 hoist wire rope
起升钢丝绳 hoist rope
起升高度 lifting altitude; hoisting height; load-lift height
起升高度超越报警器 over hoist alarm device
起升高度曲线 lifting height curve
起升高度限位器 lifting height limiter
起升高速度 high load-lifting speed
起升荷载 lifted load
起升机构 hoisting mechanism
起升机构同步装置 synchronous device of hoisting mechanism
起升加速度 lifting acceleration
起升卷扬机 lifting winch
起升螺旋 jack screw
起升时间 lifting time; rise time
起升速度【机】 hoisting speed; load lifting speed
起升索 hoisting cable
起蚀速度 eroding velocity
起始 entrance; inception; incipience; initialize; initiation; onset; start; threshold
起始斑点 start spot
起始板 starting sheet
起始半数致死浓度 asymptotic(al) LC50; incipient LC50; ultimate median tolerance limit
起始饱和下限 threshold saturation
起始边 initial side
起始边相对中误差 relative mean square error of starting side
起始变形 original deformation; original variable
起始标记 head flag; initialization token; start mark
起始拨号信号 start dialing signal
起始步骤 initialized
起始层 starting course
起始常数 primary constant
起始超载压力 initial overburden pressure
起始沉陷 initial set
起始成本 initial cost
起始程序 initialize routine; initial order; initiator program(me)
起始充电 initial charge
起始冲量 initial impulse
起始出力试验 inertia test; initial output test
起始处理 threshold treatment
起始触点 home-position contact
起始触发器 initiating trigger
起始磁导率 initial magnetic permeability
起始磁道号 starting track number
起始磁化率 initial susceptibility
起始磁化曲线 initial magnetization curve
起始磁化特性 initial magnetization characteristic
起始存储桶 home bucket
起始大地点 initial geodetic point
起始大地数据 initial geodetic data
起始带 initial tape
起始带长度 starting length
起始单元 start element
起始单元数 start element number
起始导磁率 initial permeability
起始的楼梯踏板 starting step
起始(登记)卡 start card
起始地点 starting place
起始地址 initial address; origin
起始地址信息 initial address message
起始地址译码器 starting address decoder
起始点 initial line; initial point; initial station; point of departure; start(ing) point
起始点校正 start point correction
起始点控制 starting control
起始电离 initial ionization
起始电流 initial current
起始电容 primary capacitance
起始电势 onset potential
起始电压 pick-up voltage
起始电压响应 initial voltage response
起始电晕电压 corona-starting voltage
起始顶点 initial vertex
起始定位装置 start ga(u)ge
起始冻胀含水量 initial water content of frost heaving
起始读 initialize read; initread
起始读出 start readout
起始读数 initial reading; opening reading
起始短路电流 initial-short circuit current
起始段 initial section
起始反馈 reference feedback
起始反向电压 initial inverse voltage
起始方法 initial mode
起始方位角 initial azimuth
起始方位角中误差 mean square error of initial azimuth
起始方位线 <图解辐射三角测量> construction line
起始方向 base direction; initial direction; prime direction; zero direction
起始方向线 initial directed line
起始分解温度 kick-off temperature
起始符的概率分布 start symbol probability distribution
起始符号 start symbol
起始复合物 initiation complex
起始感应 Ferri's induction; starting induction
起始格式控制 start-of-format control
起始拱度 initial camber
起始故障 incipient failure
起始轨道 original orbit
起始过程 initiating process
起始航向 datum course; initial course; original heading
起始和(数) initial sum
起始滑行阶段 initial slip stage
起始回车 initial carriage return
起始回授 reference feedback
起始回填料 initial backfill
起始激励时间 initial actuation time
起始脊瓦 starting ridge tile
起始记录 home record; initial record
起始加热负荷 starting load of heating up
起始加速度 starting acceleration
起始栅极发射 primary grid emission
起始阶段 initial period
起始节点 start node
起始经线 primary meridian
起始静压力 initial static pressure
起始开裂力矩 initial cracking moment
起始空白 start margin
起始空化数 inception cavitation number
起始孔隙率 initial porosity
起始控制 initial control
起始控制器 start controller
起始控制装置 initiation control device
起始块 fillet; starting block
起始力矩 starting moment
起始励磁响应 initial excitation response
起始链 start-of-chain
起始裂纹 incipient crack
起始流速 initial velocity
起始馏分 starting fraction
起始(楼)板 starting slab
起始码元 start element
起始脉冲 initial(im)pulse; original impulse
起始密度 initial density
起始密码 initiator code
起始密码子 start codon
起始面 initial surface; original surface; reference area
起始命令 initialization command
起始模量 initial modulus
起始磨损 initial wear
起始黏[粘]度 initial viscosity
起始黏[粘]结力 initial bond
起始牛顿黏[粘]度 initial Newtonian viscosity
起始扭矩 initial torque; starting torque
起始浓度 fresh concentration; initial concentration; threshold concentration
起始泡沫高度 initial foam height
起始偏差 initial deviation
起始频率 initial frequency; starting frequency
起始坡顶线 original crest
起始坡降 initial gradient
起始破裂点 starting point of rupture
起始气蚀 incipient cavitation
起始器 initiator
起始戗脊瓦 starting hip tile
起始倾角 original inclination
起始区判别器 initiation area discriminator
起始区域 threshold zone
起始曲线 initial curvature
起始去除 initial removal
起始燃烧 incipient combustion
起始任务指标 initial task index
起始烧结 incipient sintering
起始时间 datum time; starting period; start time; time of origin
起始输入程序 initial input program(me); initial input routine
起始数据 initial(izing) data
起始数据地址 starting data address
起始数据误差 initial data error
起始水力梯度 initial hydraulic gradient
起始水平 base level
起始水位 initial stage
起始水准面 initial level
起始瞬间 zero time
起始瞬态的 subtransient
起始瞬态电抗 subtransient reactance
起始速度 inception velocity; initial speed; initial velocity; original velocity; starting velocity
起始速率 initial speed
起始酸度 initial acidity
起始梯度 start gradient
起始天文点 initial astronomical point
起始条件 initial condition; starting condition
起始条件电路 initial condition circuit
起始条件调节器 initial condition setter
起始条件调整 initial condition adjustment
起始条件码 initial condition code
起始调整 initial adjustment; initial setting
起始跳越区 primary skip zone
起始同步 start step synchronism
起始投资 initial investment
起始推力 initial thrust; starting thrust
起始弯度 initial camber
起始未知 initial ignorance
起始位 start bit
起始位置 home; initial position; null position; reference(position); starting location; starting position; zero position
起始温度 initial temperature; starting temperature
起始涡流 starting vortex
起始无载垂度 initial unloaded sag
起始物 initiator
起始物料 starting material
起始物质 initial substance
起始误差 initial error
起始稀释 initial dilution
起始线 initial line
起始相位 starting phase; start-up phase
起始相稳定区 starting bucket
起始写 initialize write; initwrite
起始信号 initial signal; initiating signal; initiation signal; start signal
起始信号单元 initial signal unit; start signal element
起始序列 homing sequence
起始压力 initial pressure
起始延迟 initial delay
起始岩芯管 starting barrel
起始页 home page
起始硬度 initial hardness
起始于 date from
起始运动 <泥沙的> incipient motion

起始运价率 < 其中以终点费用为最大组成部分 > initial rate
起始增益 initial gain
起始增益时间 start gain time
起始站 station of origin
起始振荡 initial oscillation
起始蒸发点 point of evapo(u)ration
起始值 initial value; original value; starting value; threshold value; yield value
起始指令 initial instruction;initial order
起始致死浓度 incipient lethal concentration
起始致死水平 incipient lethal level
起始致死温度 incipient lethal temperature
起始滞后 initial lag
起始置定 initial setting
起始置位器 initial setter
起始砖 starter brick
起始装置 initial setting
起始状态 initial condition;initial condition mode;initial state;start state
起始子午线 first meridian;Greenwich meridian; initial meridian; prime meridian
起始字 start of word
起始坐标 origin(al) coordinates
起首的 incipient
起霜 bloom(ing); bloom out; chalking; crystalline bloom; efflorescence;frosting;salt crystallization
起霜程度 degree of efflorescence
起霜的 frosted
起霜工艺 frost process
起霜花 frosting
起霜油 bloom oil
起霜作用 bloom
起诉 bring action;bring an action against somebody; bring a suit against somebody; bring suit; commencement of action; implead; indictment; lawsuit; legal action; plaint;proceed;prosecute;prosecution; take legal action; take proceeding against;taking civil action
起诉国 applicant state
起诉期限 prescription
起诉人 complainant;prosecutor
起诉申请书 application for instituting legal proceedings
起诉书 application instituting proceeding; indictment; statement of charge
起诉者 suitor
起诉状 plaint;statement of claim
起算边 initial side
起算点 reference point
起算利息日期 date of value
起算量 initial quantity
起算日 dies a quo; initial day; zero date
起算时间 time to count
起抬压力 inception pressure
起条 streak
起条纹次砖 brindled brick
起条纹劣砖 brindled brick
起跳板 take-off board
起跳压力与回复压力差 blowback
起停 on-off;start-stop
起停按钮 start-stop button
起停操纵杆 start-stop lever
起停阀 stop-and-go valve
起停管理程序 start-stop supervisor
起停控制 bang-bang control; on-off control;start-stop control
起停时间 start-stop time
起停式传输 start-stop transmission
起停式计数器 start-stop counter
起停位 start-stop bit
起停装置 starting and stopping mechanism
起停自动控制 bang-zero-bang control < 按最大值、零值或最小值的 >; bang-bang control < 按最大值或最小值的 >
起头 priming
起头横列 starting-up course
起头器 starting looper
起凸 blistering
起凸纹的 gilled
起凸线 corded
起土平路机 elevating grader
起托梁作用的多跨构架 multiple bay frame as spandrel beam
起拖港 port of commencement of towage
起弯钢筋 bent reinforcement bar; bent(-up) bar
起网船 net-hauling ship
起网机 net hauler
起纹 boarding; graining; cissing[sissing]
起纹板 armboard
起纹道面 textured paving
起纹金属板 textured metal sheet
起纹理 textured
起纹理的涂料 textured paint
起纹膜片 corrugated diaphragm
起纹漆 texture paint
起纹器 grainer
起纹涂料 texture paint
起卧室 bed sit(ter);bed-sitting room
起雾 becoming foggy;bloom;fogging;hazing
起息期 attachment of interest
起息日 date of value;initial day;value date
起息日期 date of value
起下套管滑车 casing pulley
起下套管游动滑车 casing block
起下钻和钻进时间负荷周期 hoisting and drilling load cycle
起下钻时间 time on trips;trip time
起线 initial line;starting line
起线刨 sash plane
起线刷 marking brush
起限制作用的因素 limiting factor
起箱 lift
起削弱作用的 infirmatory
起卸场所 stripping yard
起卸方法 hoisting device
起卸后货物质量条款 landed quality term
起卸口 < 货物 > hoist hole;hoistway
起卸口开关 hoistway access switch
起旋凹线 rifle
起(盐)霜 effloresce
起(盐)霜的 efflorescent
起氧化皮 scaling;weigh scaling
起音速扩散锥 shock-forming nose
起油管 pull out the tubing string
起于两导体间的 peristaltic
起缘机 flanger
起缘机械 flange machinery
起源 center of origin; filiation; genesis;head spring;origin(ation);primary source;source
起源地 source area;source region
起源国 country of origin
起源目的指令【计】 source-destination order
起源形式 manner of origin
起运 shipping;start shipment
起运地点 place of dispatch; place of origin; place of shipping; starting place for shipping
起运地点交货价格 free on board shipping point
起运地法 law of the place of dispatch
起运吨数 originated tonnage;originating tonnage;tons originated
起运港 port of shipment
起运港船上交货价格 free on board; free on board harbo(u)r; free on board in harbo(u)r
起运港与目的港 both ends
起运国 country of departure;country of origin
起运机 skyline crane
起运机场 airport of departure
起运局 originating administration
起运空气分配器 starting air distributor
起运路 originating road
起运码头船上交货价 free on board quay
起运通知 letter of advice
起运运输 originating traffic
起运载重 shipping load; shipping weight
起运站 initial station;loading station;originating station;point of origin
起运重量 shipping weight
起晕电压 corona-starting voltage
起脏 scumming;tinting
起涨 beginning of rise
起涨点 < 洪水的 > point of rise
起涨量 initial flow
起褶 puckering
起针刀片 lifting blade
起针孔 < 釉面的 > pin-holing
起针三角 needle lifter; needle lifting cam;needle raising cam;raise cam
起枕器 tie remover
起振 resonant rise;start(ing) oscillation
起振力 vibrating force; vibromotive force
起振力调节 vibrating regulator
起振力调节装置 vibrating regulator
起振器 exciter;vibration generator
起振前电流 preoscillation current
起振条件 starting condition for oscillation;start-oscillation condition
起征界限 tax threshold
起直达票 book through
起止操作 start-stop operation
起止传输 start-stop transmission
起止的 start-stop
起止多谐振荡器 start-stop multivibrator
起止机构 start-stop mechanism
起止畸变测试 start-stop distortion-measuring
起止畸变度 degree of start-stop distortion
起止扫描 start-stop scanning
起止失真 start-stop distortion
起止时间 start-stop time
起止式 start-stop type
起止式电报机 start-stop apparatus
起止式电传打字机 start-stop telephotography;start-stop teleprinter
起止式电传机系统 start-stop teletypewriter system
起止式解调 start-stop restitution
起止式启闭系统 start-stop system
起止式启闭装置 start-stop system
起止式通信[讯] start-stop correspondence
起止式网络 start-stop network
起止式印字电报机 start-stop printing telegraph
起止式自同步码 start-stop self-phasing code
起止调制 start-stop modulation
起止同步 start-stop synchronism
起止同步传递 start-stop transmission
起止同步传输 start-stop transmission
起止同源的 amphicentric
起止位 start-stop bit
起止线 start-finish line
起止信号 start-stop signal
起止信号发生器 start-stop signal generator
起止信号畸变测试器 start-stop signal distortion tester
起止装置 start-stop apparatus; start-stop system
起止装置同步边界 synchronous margin of start-stop apparatus
起终点表 origin-destination table
起终点调查 origin-destination study; origin-destination survey
起终点推算模型【交】 origin-destination estimation model
起重 hoisting;jack-up;lifting
起重按钮 load button
起重扒杆 derrick kingpost;gib pole; standing derrick; derrick mast; gin pole
起重拔桩机 jack pile puller
起重把杆 derrick kingpost; derrick mast;gin pole
起重半径 handling radius
起重抱杆 gin pole
起重臂 cargo boom;derrick boom;elevator boom;erector beam;gibbet; jib(boom); lifting boom; sheer legs;erector arm;loading boom
起重臂部件 jib component
起重臂长度 boom length;jib length
起重臂的附加桩锤 boom extension ram
起重臂的角度显示器 boom angle indicator
起重臂的卷盘起重 boom hoist drum
起重臂的下弦杆 bottom boom member
起重臂的限制起重量开关 boom hoist limit switch
起重臂底部的卷筒 boom foot spool
起重臂底部枢轴 boom pin
起重臂吊索 boom harness
起重臂顶点 boom point
起重臂顶尖滑轮 boom point sheave
起重臂顶尖及锚固点 boom point with fixed point
起重臂顶尖枢轴 boom point pin
起重臂端点 boom point
起重臂杆 drawbeam
起重臂高度 boom height
起重臂桁架 boom brace
起重臂桁架式结构 boom bracing; boom brake
起重臂角度指示器 jib angle indicator
起重臂脚鼓 jib foot spool
起重臂结构 crane boom structure
起重臂截面 boom section
起重臂靠边滑车轮 boom side sheave
起重臂拉索 boom dragline
起重臂缆 jib cable
起重臂缆索 boom line;boom rope
起重臂起落摆动 luff
起重臂墙叉口 < 安装副吊臂用 > boom tooth
起重臂倾角 boom inclination angle
起重臂上卷扬机 boom winch
起重臂上起重小车 boom with crab
起重臂伸长 boom extension
起重臂伸缩机构 boom telescoping device
起重臂升降机构 boom raising and lowering gear
起重臂式铲土机 boom type shovel
起重臂式挖沟机 boom type trenching machine
起重臂提升阀 boom lift valve

起重臂提升及下降 boom raising and lowering
起重臂提升及下降齿轮 boom raising and lowering gear
起重臂提升及下降电动机 boom raising and lowering motor
起重臂提升及下降缆索 boom raising and lowering cable
起重臂提升及下降速度 boom raising and lowering speed
起重臂提升缆 jib lift cable
起重臂提升缆索 boom lift cable
起重臂提升能力 blocked capacity
起重臂提升限制开关 boom lift switch
起重臂提升圆筒 boom lift drum
起重臂提升桩锤 boom lift ram
起重臂调幅 derricking
起重臂调节 boom control
起重臂头部 boom head
起重臂托座 jib bracket
起重臂位置 boom position
起重臂斜度 boom inclination
起重臂移动轴 boom shipper shaft
起重臂制动闸 jib brake
起重臂转动 boom swing
起重臂转动角度 boom swing angle
起重臂撞头 boom ram
起重驳船 derrick barge; crane barge
起重操作 derricking motion
起重叉车兼牵引车 forklift truck and tractor
起重车 carriage hoist; crane carrier; crane wagon; derrick car; lift truck; mobile crane; wagon crane
起重秤 crane weigher
起重船 barge crane; barge derrick; crane ship; derrick boat; floating derrick; pontoon; derrick crane barge; floating crane
起重船打捞法 lifting with floating crane
起重磁盘 crane magnet; lifting magnet
起重磁铁 crane magnet; lifting magnet
起重打桩机 crane pile driver
起重单滑轮 gin wheel
起重导杆 lifting guide pillar
起重电磁盘 lift magnetic disc
起重电磁铁 lifting electromagnet; lift magnet
起重电动机 bridge motor; elevator motor; hoist motor; lifting motor
起重吊车 crane trolley; lifting crane; jack lift
起重吊杆 boom derrick; jack boom; gin pole
起重吊杆柱 derrick kingpost
起重吊杆转向盘 bull wheel
起重吊钩 crampo(o)n; hoisting hook; lift hook; load hook
起重吊环 lifting lug
起重吊机塔 derrick tower
起重吊链 crane chain
起重吊索 hoisting sling; lifting sling
起重吊塔 crane tower
起重吊桶 crane bucket
起重顶杆 hoisting ram; jack leg
起重动滑轮 hoisting block
起重动作 derrick(ing) motion
起重斗 lifting bucket
起重发动机 hoist engine
起重费用 cranage
起重浮箱 camel
起重附加装置 lifting attachment
起重复滑车 purchase tackle
起重杆 cargo mast; cathead shaft; dead-load lever; derrick(boom); derrick pole; gibbet(tree); gib(crane); hoisting mast; jack boom; jib boom; lifting rod
起重杆的升降 luffing
起重杆钢索 luffing cable
起重杆卷筒 boom drum
起重杆支柱底座 mast bottom
起重钢绳 hoisting rope
起重钢丝绳 <装卸货物用> cargo-handling wire rope; carrier cable
起重杠杆 jack lever
起重工(人) lifting worker; derrick worker; crane operator; winchman; hooktender
起重工长 hooktender
起重工作 crane work; lifting work
起重工作的操纵 crane function control
起重工作的控制 crane function control
起重钩 crane hook; grab; grab hook; lifting hook
起重钩防滑配件 mousing
起重钩上的安全索 mousing
起重箍 stirrup for hoisting
起重鼓轮 hoist(ing) drum; load drum
起重桁架 longitudinal brace system
起重横梁 lifting beam
起重葫芦 chain block; chain grab; hoist crane; hoist(ing) block; hoist tackle; lifting tackle
起重滑车 derrick block; gin block; gin tackle; gun tackle; hoist(ing) block; hoist tackle; lifting(pulley) block; lift(ing) tackle; pull(ey) tackle; purchase block; purchase tackle; tackle-block; Fidler's gear <水下吊装大石块等的>
起重滑车绳索应变的松弛 slacking off
起重滑车组 hoisting tackle; winding tackle; lifting tackle
起重滑轮 cathead line sheave; hoisting block; hoisting pulley; hoisting sheave(jenny wheel); lift block
起重滑轮偏离导向架 out of lead
起重环 lifting ring
起重活塞 hoist piston; lifting piston
起重活塞杆 hoist piston rod
起重活塞夹杆 hoist piston rod tie bar
起重活塞系杆 hoist piston rod tie bar
起重货车 crane wagon
起重机 crane; crane-type machine; elevator machine; haulier; hoist; hoist-away; hoisting crane; hoisting engine; hoisting gear; jammer; lifting block; lifting device; lifting jack; lifting machine; material hoist; alligator
起重机安全负荷 crane safe working load
起重机安装 crane erection
起重机搬运的铁水包 crane ladle
起重机臂 arm of crane; cathead; crane arm; crane jib; fly jib; gib arm; gib arm of crane; jib; jib arm of boom; lever of crane; outreach; boom; crane boom; cathead
起重机臂的俯仰、旋转、变幅 luffing
起重机臂的转动与起落 luffing
起重机臂低于水平线的角度 negative boom angle
起重机臂工作半径 handling radius (of crane); reach of crane; span of crane; swinging radius
起重机臂架伸出长度 boom-out
起重机臂架伸出极限 boom-out
起重机臂架伸出极限长度 boom reach
起重机臂倾角变化范围 luffability
起重机臂倾斜度 pitch of boom
起重机臂伸出长度 outreach; outreach of crane; reach of crane
起重机臂伸出极限长度 boom-out
起重机臂伸距 boom reach; reach of crane; span of crane
起重机臂伸缩范围图 range diagram
起重机臂提升机 block-lifting machine
起重机臂转向 luffing
起重机臂最大伸距 boom-out; boom reach
起重机变幅附件 luffing attachment
起重机变幅滑轮组 pulley block luffing gear
起重机驳船 crane barge
起重机部件 crane piece
起重机操纵的 crane-rigged
起重机操纵驾驶室 crane operator cab
起重机操纵室 crane cab(in); crane control compartment; crane man's house; hoisting box
起重机操纵司机室 crane operator cab
起重机操作的振捣器 crane-operated vibrator
起重机操作工 tagman
起重机操作手册 crane rating manual
起重机侧臂 side boom
起重机产量 crane output
起重机厂 crane factory
起重机车 crane car; crane locomotive; jack cart; lift truck
起重机车道横梁 crane runway girder
起重机承载能力 crane capacity
起重机称(重器) crane weigher
起重机船 crane barge; floating crane; pontoon; shear hulk
起重机大梁 crane girder
起重机的俯仰 luffing
起重机的荷载计量设备 portable crane scale
起重机的机臂转角 angle of boom
起重机的卷索鼓轮 hoist drum
起重机的跑车 travel(l)ing crab
起重机的起重臂 gib arm of crane
起重机的起重高度 stroke of crane
起重机的移动小车 travel(l)ing crab
起重机的主滑车 head block
起重机的主要构架 carrier frame
起重机底脚 crane base
起重机底脚支座 crane base
起重机底盘 crane carrier
起重机底座 crane pedestal
起重机电磁铁 lifting magnet
起重机电动机 crane motor
起重机电缆卷筒 cable drum of crane
起重机电气起重磁盘 crane electric lifting magnet
起重机吊臂 boom of crane; crane jib
起重机吊臂延伸的副吊臂 jib boom
起重机吊杆 boom; derrick boom; transverse member
起重机/吊杆驳船 crane/derrick barge
起重机吊杆的斜角度 boom angle
起重机吊杆倾角范围 luffability
起重机吊钩 crane hook; hoisting plug
起重机吊钩工作 lifting crane hook work
起重机吊钩工作范围 crane hook coverage
起重机吊钩滑轮 crane hook block
起重机吊钩回行滑轮 crane hook return block
起重机吊架 crane boom
起重机吊索 crane cable; crane fall; crane rope
起重机吊塔 crane tower
起重机吊装 crane loading
起重机顶部滑车组 crown block
起重机定额 crane rating
起重机动臂调幅半径 crane radius
起重机额定起重量 crane rating
起重机额定起重能力 rated load
起重机翻斗车 crane tipping skip
起重机费 cranage
起重机扶梯 crane ladder
起重机附加设备 attachment for cranes
起重机附件 crane accessories; crane attachment
起重机复滑轮 gin tackle
起重机钢轨 crane rail
起重机钢丝绳 crane rope; hoist rope
起重机钢丝索 hoist rope
起重机钢索 derrick rope
起重机工 crane man
起重机工具 hoist machinery
起重机工时 craneage
起重机工作半径 crane radius
起重机构 hoisting mechanism; jack mechanism; lifting mechanism; loading lifting mechanism; lifter
起重机构台 ga(u)ntry
起重机构台腿 gantry legs
起重机构造 crane configuration
起重机固定塔架 guyed mast
起重机轨道 crab runway; crane runway; crane track; crane way; tilting track; crane rail
起重机轨道标高 crane track height
起重机轨道梁 crane(track) beam; crane beam; crane girder
起重机轨道系统 crane trackage system
起重机轨道线 crane trackage
起重机轨道与铁路线的交叉 crane track and railway crossing
起重机轨道中心线 centre line of crane rail
起重机轨距 crane ga(u)ge
起重机夯板 crane falling plate; crane tamping plate
起重机荷载 crane load
起重机后部旋转半径 tail radius of crane
起重机后轨道梁 rear crane rail beam
起重机滑车 crane pulley; crane trolley
起重机滑道 crane runway
起重机滑轮 fiddler's gear; hoist(ing) block
起重机滑轮组 gin tackle
起重机缓冲装置 crane buffer
起重机回旋装置 crane slewing gear
起重机回转臂 swinging boom
起重机回转机构制动器 slewing gear brake
起重机活塞 hoist piston
起重机活塞杆 hoist piston rod
起重机活塞杆块 hoist piston rod head block
起重机活塞环 hoist piston ring
起重机活塞柱夹杆 hoist piston rod tie bar
起重机机臂 jib arm
起重机机架 crane frame; gantry; hoisting frame
起重机加油工 crane oiler
起重机驾驶室 crane cab(in); crane control compartment; crane man's house; hoisting box
起重机驾驶台 crane bridge
起重机驾驶员 crane driver; crane man; crane operator; hoistman; hoister
起重机架 crab girder; derrick tower; hoist frame
起重机减摩轴承 anti-friction crane bearing
起重机减速器 hoist reduction gear
起重机绞车 crane crab; crane winch
起重机接触导线 crane trolley wire
起重机结构部件 crane structure com-

ponent
起重机卷索鼓轮 hoisting drum
起重机卷筒 hoisting drum;hoist roller
起重机开关 crane switch
起重机控制阀托架 hoist control valve bracket
起重机控制器 crane controller
起重机宽度 crane width
起重机缆风 derrick(ing)guy
起重机缆绳 hoisting line
起重机缆索 derrick(ing)guy;hoist line
起重机链 crane chain;jack chain
起重机梁 crane beam;crane girder
起重机料斗 crane bucket
起重机临时走行轨道 temporary crane track
起重机(埋入式)轨道槽 crane rail pocket
起重机锚碇装置 anchoring equipment of crane
起重机门架 crane gantry;portal gantry
起重机门形构架 crane gantry
起重机门座 crane portal
起重机能力 cranage
起重机平台 crane platform
起重机坡度指示器 crane slope indicator
起重机破坏锤 wrecking ball
起重机破坏球 wrecking ball
起重机起吊 crane loading
起重机起吊半径 crane radius
起重机起吊荷载 crane snatch load
起重机起吊能力图 capacity chart
起重机起重臂 crane boom
起重机起重高度 stride of crane
起重机起重量 crane load;hoisting power of crane;lifting capacity of crane
起重机起重能力 crane lifting capacity;crane output;hoisting power of crane;load capacity of crane;lifting capacity
起重机桥(架) crane bridge;loading bridge
起重机驱动侧桥架 driving side bridge
起重机全幅 overall width of crane
起重机全高 overall height of crane
起重机润滑剂 crane grease
起重机三脚架 sheer legs
起重机上部结构 crane superstructure;crane upper structure
起重机上墩下水设备 crane plant
起重机设备 hoist machinery
起重机伸臂 jib of crane
起重机伸臂长度 boom reach
起重机伸臂活动半径 crane radius
起重机伸距 handling radius;outreach of crane
起重机生产率 crane output
起重机绳索 hoisting cable;crane cable;fall
起重机使用费 cranage
起重机式动臂 crane-type boom
起重机式拉索铲挖掘机 cable crane dragline
起重机式起重机 crane-type loader
起重机式主臂 crane-type boom
起重机式装载机 crane-type loader
起重机手 crane man;crane operator
起重机手操纵室 crane driver's cabin
起重机受电器侧的桥架 collector side bridge
起重机水平转盘 bull wheel
起重机司机 crane operator;hoistman;hoister
起重机司机室 crane cab
起重机索 crane cable;fall
起重机塔 crane tower;derrick mast;derrick tower
起重机台架 gauntree;ga(u)ntry
起重机台面 decking
起重机提升能力 crane lifting capacity
起重机提升踏板 jumping shoe
起重机挺杆 jib of crane;sheer legs;jib boom
起重机腿 crane legs
起重机外伸支腿的千斤顶 outrigger jack
起重机外伸支腿的水平梁 outrigger beam
起重机万向节 hoist universal joint
起重机桅杆 crane mast;derrick kingpost;derrick mast
起重机尾部旋转半径 rear end radius;tails wing
起重机稳定性 crane stability
起重机小车 crane crab;monkey;crane trolley
起重机小车滑轮组 troll(e)y block
起重机小车上提升用的钢丝绳 carriage rope
起重机斜撑 gib arm of crane
起重机械 crane machinery;elevator machinery;hoisting machinery;hoist mechanism;lifting equipment;lifting machinery;lifting plant
起重机械安全规程 safety rules for lifting appliances
起重机械能力 hoisting power
起重机械装置 lifting mechanism
起重机行车 crane trolley;travel(l)ing crane
起重机行车大梁 crane girder;crane runway girder
起重机行程限位器 crane travel(l)ing limiter
起重机行走机构 crane travel(l)ing gear;crane bogie;undercarriage
起重机悬臂 arm of crane;gib arm of crane
起重机悬臂长度 boom length
起重机悬臂伸距 crane radius
起重机悬臂升降机构 jib adjusting gear
起重机悬臂提升索 boom lift cable
起重机悬吊的打桩导向架 crane-suspended lead
起重机悬挂的夯实机 crane suspended compactor
起重机旋臂 jib boom
起重机旋臂半径和安全起重量指示器 radius-and-safe-load indicator
起重机旋臂制动器 jib stopper
起重机旋转臂 swinging boom
起重机液压泵 crane pump
起重机移动 crab traversing
起重机移动架 gantry travel(l)er
起重机移动同步装置 synchronous apparatus for crane travel(l)ing
起重机移距 crane travel distance
起重机用钢绳 crane rope
起重机用索 elevator rope
起重机用液压组合件 hydraulic crane cart
起重机油泵 hoist pump
起重机油泵操纵阀 hoist pump control valve
起重机油泵操作杆 hoist pump operating lever
起重机油泵传动轴衬套 hoist pump drive shaft bushing
起重机油泵盖 hoist pump cover
起重机油泵盖垫密片 hoist pump cover gasket
起重机油泵控制杆 hoist pump control rod
起重机油泵歧管 hoist pump manifold
起重机油泵体 hoist pump body
起重机油泵体衬套 hoist pump body bushing
起重机油泵托架 hoist pump bracket
起重机油泵止回阀 hoist pump check valve
起重机油泵止回阀体 hoist pump check valve body
起重机油泵中间轮 hoist pump intermediate gear
起重机油泵中间轴 hoist pump intermediate shaft
起重机油泵主动齿轮 hoist pump drive gear
起重机油泵主动轴 hoist pump drive shaft
起重机油缸 hoist cylinder;hoist oil cylinder;jack cylinder
起重机油缸垫密片 hoist cylinder gasket
起重机油缸盖 hoist cylinder head
起重机油缸活塞簧 hoist cylinder piston spring
起重机油缸填线圈 hoist cylinder packing
起重机油缸行程 jack cylinder range
起重机油缸油压安全阀 hoist cylinder oil relief valve
起重机油缸注入塞 hoist cylinder filler plug
起重机有效伸距 effective reach of crane
起重机运行机构 crane travel mechanism
起重机运转 crane movement
起重机载运车 crane carrier
起重机载重量 crane rating
起重机闸 hoist brake
起重机栈桥 crane trestle
起重机支架 sheer legs
起重机支腿 outrigger
起重机支柱 crane pillar;crane post;crane shaft;crane stalk
起重机支座 crane base
起重机直角杠杆 hoist bell crank
起重机指挥者 crane man;crane rigger
起重机制动器 crane brake
起重机制动闸 crane brake
起重机中心 crane center[centre]
起重机重型钢轨 heavy duty crane rail
起重机轴 crane shaft
起重机主柱 crane post;crane shaft;gin pole
起重机柱 crane post;crane stake
起重机抓斗 crane grab;grab(bing) crane bucket
起重机转盘 derrick bullwheel
起重机转轴钢架 hanging leader
起重机转柱 crane pillar
起重机装配工 crane fitter
起重机装卸 cargo-handling by deck crane;lift-on/lift-off
起重机装运 crane loading
起重机装载半径 load radius
起重机装载方式 crane system
起重机装置 hoist machinery
起重机纵梁 longitudinal crane girder
起重机走道 crane runway
起重机走道集电器 runway collector
起重机走动 crane running
起重机最优生产率 optimized crane productivity
起重夹具 crampo(o)n
起重夹钳 grab tongs
起重架 crane trestle
起重绞车 crab;crab machine;crab winch;crane trolley;hoisting crab;lifter winch;loading trolley;loading winch;winch hoist;winch lift
起重绞车架 crab derrick
起重绞盘 crab capstan
起重井架 derrick
起重卷筒 hoisting drum;load hoist drum
起重卷扬机 derrick winch
起重卡车 lifting truck
起重开挖两用机 crane-excavator
起重控制器 crane controller;hoist(ing)controller
起重框架 lifting stirrup
起重力 elevating force;elevating power;lifting power;portative force
起重力矩 hoist moment;load moment
起重链 hoisting chain;jack chain;lift(ing)chain
起重链条 lifter chain
起重梁 drawbeam;hoisting beam;lifting beam
起重量 carrying power;hoisting capacity;hoisting duty;hoisting power;lifting power;load-carrying ability;loading capacity;load of lifting;luffing capacity;lifting capacity
起重量限制器 load lifting limiter
起重量指示器 load indicator
起重辘轳 hoist tackle;lifting tackle
起重螺杆 jack bolt;lift(ing)bolt
起重螺母 jack nut
起重螺栓 jack(ing)bolt;lifting bolt
起重螺丝 jack screw
起重螺旋 jack screw;lift screw
起重马达 hoist motor
起重木(杆) shear log pole
起重木托座架 jib bracket
起重能力 crane capability;crane output;elevating capacity;hoisting capacity;hoisting power;load-carrying capacity;loading capacity;weight-lifting ability;weight-lifting capability;weight supporting capacity <钻塔>
起重能力表 capacity chart
起重平衡臂 jack arm
起重铺料机 crane spreader
起重汽车 autohoist;ladder truck;loader truck;lorry-mounted crane;mechanized lorry;truck crane
起重器 forcing screw;ground jack;hoist;hoisting jack;jack;lifting jack
起重器垫(块) jack(ing)pad
起重器具 lifting device
起重器支承点 jack lift point
起重器支垫 jacking pad
起重千斤顶 hoisting jack
起重钳 hoisting tongs;lifting tongs
起重桥 crane bridge
起重桥台 footpath platform
起重(容)量 lifting capacity
起重容量表 capacity chart
起重三角架 gin;jack;sheers
起重三脚架 sheer legs;sheers
起重设备 crane equipment;erecting equipment;hauling-up device;hoisting apparatus;hoisting device;hoisting equipment;hoisting installation;hoisting unit;lifting appliance;lifting device;lifting equipment;lifting rig;weight handling equipment
起重升降装置 lifting gear
起重绳 hoisting line;hoisting rope;holding line;load line
起重手柄 lifting handle
起重竖井 hoisting shaft
起重丝杠 jack screw;lifting spindle
起重速度 hoisting speed;lifting speed;load lifting speed
起重索 cathead line;catline;derrick rope;elevator rope;fall line;hoist-

Q

(ing) cable; hoist(ing) rope; lifting cable; lifting fall; load hoist; main cable; operating rope; winding rope; hoist(ing) line; holding line < 抓斗用 >; grapple line
起重索超速保护装置 hoist overspeed device
起重索起重绳 hoisting line
起重索套 bridle sling
起重塔 lift tower; light tower hoist
起重塔架 derrick tower; hoist(ing) tower
起重台架构造 gantry construction
起重台架建筑(物) gantry construction
起重特性曲线 lifting performance curve
起重天秤 lifting beam
起重调节器 overhung governor
起重推杆 hoisting ram
起重托架 jack lift
起重挖掘机械协会 < 美 > Power Crane & Shovel Association
起重网 lifting net
起重桅(杆) derrick kingpost; derrick mast; derrick pole; gin pole; standing derrick; steeve
起重稳定索 tag line
起重吸盘 lifting magnet
起重系索 hoisting sling; lifting sling; load hoist system
起重箱 skip
起重小车 crab; crane carriage; crane crab; hoist trolley; saddle
起重小车集电器 trolley collector
起重小滑车 monkey block
起重楔 lewis bar
起重信号工 bank(s)man
起重行车 hoisting truck
起重行车桥面 crane deck
起重悬臂 cantilever boom
起重液压泵 hoist pump
起重液压缸 jack cylinder
起重用部件 lifting fitting
起重用绳子和滑轮 whip-and-derry
起重油缸 lift cylinder; lifting oil cylinder
起重有效荷载 overweight payload
起重运输机械 elevating and conveying machinery
起重运输机械荷载 crane and transporter load; crane and vehicle load
起重(运输)设备 handling equipment; handling facility
起重振捣器 crane-operated vibrator
起重蒸汽机 hoisting steam engine
起重支撑 derrick brace
起重支架 derrick leg; lifting bracket
起重直升机 flying crane
起重制动(器) lifting brake
起重制动闸 lifting brake
起重轴 jackshaft
起重柱 crane stake; sam(p)son post; sheers
起重爪 lewis(anchor); lewis of crane; lewisson; lifting pin; lifting dog
起重转臂 craning boom; lifting boom
起重装车斜坡台 lifting loading ramp
起重装卸拖车 lift trailer
起重装置 elevating gear; gin; hauling-up device; hoist(ing) device; hoist(ing) gear; hoist(ing) plant; lifting apparatus; lifting appliance; lifting device; lifting gear; lifting set; lifting tackle; purchase; tackle gear; whim
起重装置制动器 hoisting gear brake
起重作业 cranage
起皱 buckle; cockling; corrugate; corrugation; crease; crisp; crumpling; gof(f)er; pucker(ing); purse; riveling; shrinkage; wrinkle; wrinkling; blub < 指刷灰墙时有气泡 >
起皱的 corrugate; corrugated
起皱机 jacking-machine
起皱加工 cockle finish
起皱纹 crinkle; drape; furrow; crinkling
起皱纹的次砖 brindled brick
起皱(纹)油漆 crawling paint
起转 run-up
起子 driver
起子槽 driving slot
起租检验 survey on hire
起租期间 rent-up period
起钻 pull out
起钻护箍 lifting cap
起钻前钻杆泵入一段重泥浆 < 使卸钻杆时钻杆无泥浆 > slug the pipe
起作用 functionate; make a difference
起作用的 operative
起作用约束 active constraint

气

气疤 body scab
气扳机 air wrench; pneumatic impact wrench
气斑 gas mark
气扳机扳轴 anvil
气扳机方头 square drive
气拌池 pneumatic cell
气拌库 aerated blending silo
气瓣 air flap
气瓣活塞 valve bucket
气包 air holder; air pocket; air receiver
气包底座 dome base
气包盖 dome cap; dome cover
气包盖座 dome covering
气包焊条 gas shielded electrode
气包加强板 dome liner
气包上盖板 dome head
气包梯 dome ladder
气包体 dome
气包压力 dome pressure
气包罩 dome casing
气胞 air cell
气保护 gas shield(ing)
气保护金属极电弧焊 gas metal-arc welding; metal-arc gas-shielded welding
气保护碳弧焊 gas carbon-arc welding; shielded carbon arc welding
气保护柱钉焊接 gas-shielded stud welding
气爆 airing; gas explosion
气爆搅动 gaseous burst
气爆雷管 gas exploder
气爆引信 gas exploder
气爆震源 dinoseis; gas exploder
气爆震源测深系统 gas gun system
气泵 air(-driven) pump; gas pump; heat pump; pneumatic pump; steam pump
气泵表 gas pump meter
气泵阀 air pump valve; valve for air pump
气泵法兰 flange for air pump
气泵活塞杆 piston rod for air pump
气泵喷雾器 pneumatic sprayer
气泵皮碗 leather cup for air pump
气泵弹簧 spring for air pump
气泵调节器 air pump governor
气泵旋塞 plug for air pump
气泵循环 gas pump cycle
气笔 air brush
气闭段 gas-tight section
气闭封接 hermetic seal
气闭式 gas force closed type
气闭套管 gas-tight sleeve
气闭头 gas-tight block
气壁 air wall
气表 gas flow meter
气表革 gas meter leather
气波 air wave; gas wave
气波增压器 comprex
气布比 air-to-cloth ratio; gas-cloth ration; specific gas flow rate
气舱 air chamber
气藏 gas pool
气藏平面图 planimetric(al) map of gas pool
气藏剖面图 sectional drawing of gas pool
气槽 gas channel; pneumatic transport
气测井仪 gas logger
气测录井间距 gas log interval
气测轻烃测井 gas log light hydrocarbon
气测全烃测井 gas log all hydrocarbon
气测重烃测井 gas log heavy hydrocarbon
气层 gas-bearing bed; gas bed; gas sand
气层深度 gas bearing depth
气层有效厚度 effective thickness of gas bed
气铲 air chipping hammer; chipping hammer; pneumatic chipping hammer
气尘比 gas-to-dust ratio
气尘复合体 gas-dust complex
气尘云 gas-dust cloud
气沉积涂层法 vapo(u)r deposited coating
气成 pneumatolysis
气成包体 pneumatogenic enclosure
气成的 pneumatogenic; pneumatolytic
气成矿床 pneumatolytic deposit
气成矿物 pneumatogenic mineral; pneumatolytic mineral
气成热液 pneumatolytic solution
气成-热液交代变质岩 pneumatolito-hydrothermal metasomatite
气成岩 atmolith; atmospheric rock
气成异常 gasogenic anomaly
气承建筑 air-inflated building; air-supported building; pneumatic structure
气承结构 air-supported structure
气承式液柱黏[粘]度计 suspended level visco(si)meter
气充空间 gas filling of spaces
气冲剪 pneumatic polisher; pneumatic puncher
气冲注射器 gas flush syringe
气出口 escape orifice
气储 gas reservoir
气储集层 gas reservoir
气传颗粒物 air-borne particulate
气传微粒 air-borne particle
气传污染物 air-borne contaminant; air-borne pollutant
气传噪声 air-borne noise
气囱罩 gas shaft hood
气窗 air end; dormer; fanlight; fanlight transom(e) window; gas shaft; louver[louvre] window; outlet ventilator; pocket; scuttle; sublight; sunburst light; transom(e); transom light; transom window; ventilating window; ventilator; venting window; ventlight; vent louver [louvre]; window ventilator; dream hole < 仓库等的 >; transom < 美 >; fairlight < 英 >
气窗插销 transom catch
气窗窗框 upper sash
气窗垂直启闭器 transom lifter
气窗扉 ventilation casement
气窗拱 port roof
气窗开关联动装置 window gearing
气窗联动开关 window gearing
气窗链 transom chain
气窗启闭链条 transom chain
气窗上的冷凝水集水槽 condensation sinking
气窗梃销 transom lifter
气窗托架 transom bracket
气窗转动轴 surface sash center[centre]
气床 air-bed
气吹 air blast
气吹断路器 gas-blast circuit breaker
气吹弧开关 air blast switch
气吹磨碎机 aerofall mill
气吹油 blown oil
气锤 air chipper; air(drop) hammer; air impact hammer; air ram(mer); atmospheric stamp; chipping hammer; jack hammer; pneumatic breaker; pneumatic chipper; pneumatic gun; pneumatic hammer; pneumatic rammer; spader
气锤病 pneumatic hammer's disease
气锤打桩 pneumatic hammer pile driving
气锤打桩机 pneumatic hammer pile driver
气锤捣打成型法 air ramming
气锤捣打成型 air ramming
气锤捣打机 air rammer press
气锤活塞 hammer piston
气淬 air cooling; air quenching
气淬溶渣 air quenching clinker
气带 air belt; gas zone; unsaturated zone
气袋 air bag; air pocket; air sac; gas bag; gaseous envelope; gas pocket
气袋压实 pressure bag compacting
气弹模型 aeroelastic model
气刀 air knife
气刀刮涂法 air knife coating
气刀刮涂机 air knife coater
气刀辊涂机 air knife coater
气刀涂布机 Kohler coater
气刀涂层 air knife coating
气导 air-conduction
气导式抗声传声器 air-conduction anti-noise microphone
气道 aerial port; air channel; air draft; air drain; air flue; air passage; air port; air space; airway; gas channel; gas duct
气道堵塞 airway obstruction
气道管 airway tube
气道滤网 air inlet screen
气道阻力 airway resistance
气的净浮力 net buoyancy of gas
气灯 air-turbo lamp; gas lamp; hydrogen lamp; vapo(u)r lamp
气灯灯船 gas boat
气灯浮标 gas buoy
气灯火焰 gas jet
气笛 air horn; air siren; air whistle; compressed-air horn; exhaust whistle; hooter; siren
气笛阀 whistle valve
气笛信号 air whistle signal
气电 pneumoelectric
气电电动车 gas electric(al) automobile
气电动力学 electrogas dynamics
气电焊 electrogas welding; gas arc welding
气电焊焊嘴 gas arc welding gun
气电立焊 electrogas enclosed welding
气电联合焊接 gas electric(al) welding
气电量规 pneumo-electric(al) ga(u)ge
气电式示功器 gas electric(al) type indicator

气电信号转换器 pneumatic-electric-(al)signal converter
气电转换器 pneumatic-electric(al) convertor; pneumatic-to-current converter; pneumo-electric (al) convertor
气电自动车 gas electric(al) automobile
气垫 air bearing; air buffer; air cushion; air mat; air pillow; air spring; air suspension; air-tight seal; cushion; cushion of air; gas bearing; gas cushion; inflatable cushion; pneumatic cushion
气垫爆破 air shooting; cushion blasting; cushion shooting
气垫避震器 pneumatic shock absorber
气垫驳船 air cushion barge; GEM lighter; hoverbarge
气垫车 aerocar; air cushion automobile; air cushion car; cushioncraft; ground cushion vehicle; ground effect machine; hovercar; hovercraft; skimmer; surfaced-effective vehicle; terraplane; air cushion vehicle
气垫车软围裙 flexible skirt
气垫车外周喷嘴 peripheral nozzle
气垫船 air cushion boat; air cushion(ing) craft; air cushion ship; air cushion vehicle; air cushion vessel; air-supported vessel; cushioncraft; hovercraft; hovering craft; hovermarine; hovership; hydroskimmer; surface effect ship <美>
气垫船非排水状态 air cushion vessel in non-displacement mode
气垫船港口 hoverport
气垫船海上行驶能力 hovercraft seakeeping capability
气垫船技术作业 hovercraft technology
气垫船码头 hoverport
气垫床 air-bed
气垫淬火 gas cushioned tempering
气垫淬火系统 gas hearth system
气垫的 air-cushioned
气垫登陆船 air cushion vehicle
气垫底板 <英> hoverpad
气垫电缆 gas cushion cable
气垫吊架 air spring suspension
气垫渡船 hoverferry
气垫方驳 air cushion pontoon
气垫飞行器 hovercraft; hover pallet
气垫飞行器展览 hovershow
气垫缝 <混凝土路面的> air cushion joint
气垫缸 cushion cylinder
气垫钢化 air-cushioned tempering; air-supported tempering; gas cushioned tempering
气垫挂车 air cushion trailer
气垫滑轮 hover pulley
气垫缓冲器 pneumatic shock absorber
气垫火车 aerotrain; air cushion train; hovertrain
气垫机 air cushioning machine
气垫减振器 pneumatic cushion shock absorber
气垫减震器 air cushion shock absorber; pneumatic cushion shock absorber
气垫交通工具 air cushion vehicle
气垫(胶带)输送机 air cushion belt conveyer[conveyor]
气垫举升高度 hover height
气垫块 air suspension block
气垫列车 aerotrain; air cushion train; air cushion vehicle; cushion car; hovertrain
气垫列车试验线路 air cushion train test track
气垫轮渡 hoverferry
气垫排水渠 air cushion culvert
气垫平台 hover platform
气垫破冰平台 air cushion ice breaking platform
气垫起重移位器 hover pallet
气垫气行器 hover pallet
气垫汽车 aeromobile; air car; air cushioning machine
气垫上压力顶杆 pressure pin
气垫式的 gas cushioned; air-cushioned
气垫式调压室 air cushion surge chamber
气垫式飞行器 air cushion vehicle
气垫式滤池 air cushion filter
气垫式喷雾器 air cushion sprayer
气垫式喷嘴热风拉幅机 floating jet stenter
气垫式输送机 air film conveyer[conveyor]
气垫式水上飞机 air cushion seaplane
气垫式挖泥船 air cushion dredge(r); hover dredge(r)
气垫式液压破碎机 gas cushioned hydraulic breaker
气垫式液压破岩机 gas cushioned hydraulic breaker
气垫式运输 air cushion high speed ground transportation
气垫式运输机 hovercraft
气垫式运载车 aeromobile; air cushion vehicle
气垫式运载工具 aeromobile; air cushion vehicle
气垫输送机 air cushion conveyer [conveyor]
气垫输送装置 air cushion pad; air cushion transporter
气垫弹簧 air spring cushion
气垫(腾空)能力 hovering power
气垫调平阀 air spring level(l)ing valve
气垫铁路 air cushion railway
气垫艇 air cushion boat; air cushion craft; hovercraft; surface effect vehicle
气垫托板 air(cushion) pallet
气垫托盘 air(cushion) pallet
气垫托盘运输 air pallet transport
气垫挖泥船 hover dredger
气垫微震 air ram jolt
气垫围板 hover skirts
气垫围裙 hover skirts
气垫系统 lift system
气垫压力 air cushion pressure
气垫窑 air cushion kiln; cushion kiln; gaseous cushion kiln
气垫应急救助船 air cushion crash rescue vehicle
气垫有轨(公交)系统 track system supported by air cushion
气垫圆盘 air-supported puck
气垫约束系统 aircushion restraint system
气垫运送机 air cushion transporter
气垫运载工具 air cushion pallet
气垫载重车 hovertruck
气垫增压火箭 ullage rocket
气垫帐篷 air tent
气垫支撑 air cushion support
气垫轴承 air cushion bearing
气垫着陆系统 air cushion landing system
气垫座(椅) pneumatic seat; air-cushioned seat
气顶 gas cap; gas dome; pneumatic booster; holder-on <压气铆钉的>
气顶动态 gas cap behavio(u)r
气顶高度 height of gas cap
气顶面积 area of gas cap
气顶膨胀 gas-cap expansion
气顶驱动 gas-cap drive
气顶收缩 gas-cap shrinkage
气顶油藏 gas-cap reservoir
气顶油减震缸筒 shock absorber air/oil cylinder
气顶油式 pneumato-hydraulic
气顶油式制动器 air hydraulic brake; air over hydraulic brake; air over oil brake
气顶油液压制动总泵 air over hydraulic brake
气顶油增压器 air-oil booster
气顶油制动总泵 air-powered master cylinder
气动 air starting
气动扒煤机 air slusher
气动拔火耙 air-rack-raking device
气动靶 pneumatic target
气动摆斗式装载机 pneumatic flipover bucket loader
气动摆式铲斗装载机 pneumatic overshot loader
气动扳手 air-operated impact wrench; air wrench; impact wrench; pneumatic impact spanner; pneumatic impact wrench; pneumatic wrench
气动报警给定器 pneumatic alarm set station
气动泵 air-driven pump; air-operated pump; air-powered pump; gas lift pump; mammoth pump; pneumatic pump
气动泵挖泥船 pneumatic pump dredge(r)
气动臂 pneumatic arm
气动变送 pneumatic transmission
气动变送及控制 pneumatic transmission and control
气动变送元件 pneumatic transducing element
气动变向器 pneudyne(positioner)
气动表面振捣器 air external vibrator
气动拨叉 air fork
气动波阻 aerodynamic(al) wave drag
气动剥岩铲 air stripping blade
气动薄膜调节阀 pneumatic diaphragm control valve
气动薄膜阀 pneumatic diaphragm valve
气动薄膜式恒温器 pneumatic membrane type thermostat
气动薄膜式温度调节器 pneumatic membrane type thermostat
气动采取 gas lift recovery
气动操舵装置 pneumatic steering gear
气动操纵 air-operated control
气动操纵导阀 pneumatically released pilot valve
气动操纵换向阀 pneumatically operated direction valve
气动操纵设备 gas operated device
气动操纵式草籽播种机 pneumatically operated grass seed drill
气动操纵装置 aerodynamic(al) controlling device; gas operated device
气动操作的 pneumatically operated
气动操作蝶阀 pneumatically operated butterfly-valve
气动操作机构 pneumatic operator
气动测车器 pneumatic detector
气动测量 pneumatic ga(u)ging
气动测量仪表 pneumatic ga(u)ge
气动测头 ga(u)ging head; plug jet; snap jet
气动测微仪 air ga(u)ge; air micrometer; pneumatic micrometer
气动差动变送器 pneumatic differential transmitter
气动差温式探测管 pneumatic rate-of-rise tubing
气动拆毁镐 <拆除房屋等用> pneumatic demolition pick(hammer)
气动铲 pneumatic digger; pneumatic spade(hammer)
气动铲土机 pneumatic spader
气动车门标准电气设备【铁】 electric-(al) standard equipment for pneumatic train/door control
气动车门控制【铁】 pneumatic train/door control
气动程序给定器 pneumatic program-(me) set station
气动秤 pneumatic scale
气动齿轮钻机 pneumatic gear rotary drilling machine
气动充填 pneumatic stowing
气动充填的管子 pneumatic stowing pipe; pneumatic stowing tube
气动充填机 pneumatic stowing machine
气动充填设备 pneumatic stowing plant
气动冲床 pneumatic punching machine
气动冲击锤 pneumatic impact hammer
气动冲击工具 pneumatic percussion tool
气动冲击破碎机 pneumatic impact breaker
气动冲击器 air hammer
气动冲击式凿岩机 air hammer drill; pneumatic hammer drill; pneumatic percussion drill
气动冲击式钻机 pneumatic hammer drill
气动冲击式钻眼 pneumatic percussive drilling
气动冲砂器 air ram(mer); sand rammer
气动抽水泵 air lift pump
气动稠化器 pneumatic thickener
气动除冰 pneumatic decking; pneumatic deicing
气动除冰设备 pneumatic de-icer
气动除冰装置 pneumatic de-icer
气动除尘装置 pneumatic dust removal system
气动除锈气锤 pneumatic scaling hammer
气动锄 compressed-air spade
气动储罐切换开关 pneumatic tank switcher
气动穿孔器 pneumatic perforator
气动传动 pneumatic actuator; pneumatic drive
气动传杆器 pneumatic transmitter
气动传感器 pneumatic transmitter
气动传输系统 pneumatic transmission system
气动传送管 pneumatic transfer tube
气动传送机 air slide; pneumatic slide
气动传送器 pneumatic rabbit
气动传送系统 pneumatic transfer system
气动传送旋转流量计 pneumatic transmitting rotameter
气动吹除装置 pneumatic scavenging gear
气动吹扫装置 pneumatic scavenging gear
气动吹送装置 pneumatic conveying device; pneumatic conveying plant
气动锤 compressed-air hammer; pneumatic rammer; air hammer
气动锉刀 pneumatic file
气动打夯机 air-earth hammer; pneumatic earth hammer; pneumatic

Q

rammer; pneumatic tamper
气动打磨机 air sander; pneumatic sander
气动打桩锤 air pile hammer; pneumatic pile hammer
气动打桩机 pneumatic hammer pile driver; pneumatic pile driver
气动带传送 pneumatic tape transport
气动单元组合仪表 pneumatic unit combination instrument
气动单针记录器 pneumatic single-pen recorder
气动导管 pneumatic tube
气动导轨式凿岩机 air drifter
气动捣棒 pneumatic dolly
气动捣固棒 pneumatic pole tamper
气动捣固机 air ram(mer); pneumatic rammer
气动道岔 pneumatic-operated switch
气动的 aerodynamic(al); air-actuated; air-driver; air-operated; air-powered; compressed-air-operated; gas-dynamic(al); pneumatic; pneumatically driven; pneumostatic; pressure-operated
气动低压控制 pneumatic low-pressure control
气动笛 pressure operated siren
气动递送 pneumatic dispatch
气动电动机 compressed-air motor; pneumatic motor
气动电键 pneumatic key
气动电子的 pneutronic
气动电子式损坏安全性系统 pneumatic electronic fail-safe system
气动雕刻刀 pneumatic engraving tool
气动吊车 air hoist; pneumatic crane; pneumatic hoist
气动蝶阀 butterfly valve with pneumatic actuator
气动钉钉机 air tacker
气动钉机 pneumatic stapler; pneumatic tacker
气动顶把 pneumatic hold-on
气动顶杆 air ram(mer)
气动动力传送带 air-activated gravity conveyer[conveyor]
气动动力输送机 air-activated gravity conveyer[conveyor]
气动堵缝 pneumatic ca(u)lking
气动堆料机 air cylinder pusher
气动对管器 air line-up clamp; alignment air clamp
气动盾构掘进 compressed-air shield driving
气动多轴工具 pneumatic multi-spindle tool
气动舵机 pneumatic servo
气动发电灯 pneumatic lamp
气动发电机 gas-driven generator
气动发动机 air motor
气动发送器 pneumacator
气动阀 air-operated valve; air-thrown valve; compressed-air operated valve; pneumatic-operated valve; pneumatic valve
气动阀门定位器 pneumatic positioner valve
气动阀门研磨机 air-operated valve grinder
气动阀位控制器 pneumatic valve positioner
气动法 gas lift method
气动翻斗车 air-dump car; pneumatic tip(ping) wagon
气动翻路机 pneumatic scarifier
气动翻卸车 air-dump car
气动防波堤 air breakwater
气动仿真器 pneumatic simulator
气动飞剪 air-operated flying shears
气动分离 pneumatic separation
气动分离器 pneumatic separator
气动分配器 pneumatic distributor
气动分选机 air-slide classifier
气动风阀操纵器 pneumatic damper operator
气动符号 pneumatic symbol
气动辅助支架 pneumatic auxiliary console
气动干燥装置 pneumatic drying system
气动杆 air leg
气动缸 air cylinder; pneumatic cylinder; pneumatic linear actuator
气动高速冲击磨 micronizer[microniser]
气动高温计 pneumatic pyrometer
气动高压控制 pneumatic high-pressure control
气动镐 pneumatic picker; pneumatic pick hammer
气动隔板 pneumatic barrier
气动给定元件 pneumatic setting element
气动给料 pneumatic feeding
气动工程 pneumatic engineering
气动工具 air power tool; air tool; pneumatic hand tool; pneumatic tool
气动工具厂 pneumatic tool plant
气动弓形钢锯 pneumatic hack saw
气动功率 pneumatic power
气动功率放大器 pneumatic power amplifier
气动攻丝工具 pneumatic tapping tool
气动攻丝机 pneumatic tapping tool
气动刮板卷扬机 air slusher
气动挂钩 gas lift hook-up
气动管道工程/安装 pneumatic pipework/installation
气动管道输送机 pneumatic conveyer[conveyor]
气动管道输送器 pneumatic conveyer[conveyor]
气动管道装卸系统 pneumatic conveyer[conveyor] system
气动管钳 spinning wrench
气动灌浆泵 air-driven grout pump
气动夯 compressed-air mechanical tamper; pneumatic compactor; pneumatic tamper
气动夯锤 air ram(mer); pneumatic rammer
气动夯具 compressed-air tamper
气动夯实 compressed-air ram; pneumatic ram
气动夯实机 pneumatic rammer; pneumatic tamping machine
气动夯土机 air soil hammer; compressed-air soil rammer; pneumatic earth hammer; pneumatic soil rammer
气动恒温器 air-operated thermostat; pneumatic thermostat
气动后卸式装载机 pneumatic overhead loader
气动葫芦 air hoist
气动虎钳 air-operated vice
气动滑板 air slide
气动滑槽 air slide
气动滑阀 pneumatic slide valve
气动滑行的 air slide
气动缓冲器 pneumatic buffer
气动换挡 pneumatic power shift
气动换气装置 pneumatic scavenging gear
气动换向阀 air-operated reversing valve; pneumatic reversing valve
气动黄油枪 air grease unit
气动灰浆喷射机 pneumatic mortar sprayer; pneumatic plaster-throwing machine
气动回路 pneumatic circuit
气动回转式凿岩机 pneumatic rotary drill
气动回转式钻机 compressed air rotary drill
气动混合 pneumatic mixing
气动混合激光器 gas dynamic(al) mixing laser
气动混合搅拌 pneumatic mixing and stirring
气动混料机 pneumatic mixer
气动混凝土机 pneumatic concrete placer
气动混凝土浇注机 pneumatic concrete placer
气动混凝土浇筑机 pneumatic concrete placer
气动混凝土破碎机 pneumatic concrete breaker
气动活顶漏斗车 air-slide covered hopper car
气动活塞 air ram(mer); air slide; air sliding; pneumatic piston
气动火警感察系统 pneumatic fire detection system
气动货油阀 pneumatically operated cargo valve
气动机 air engine; pneumatic motor
气动机构 pneumatic operating gear
气动机器人 pneumatic robot
气动机械 pneumatic machine; wind power machine
气动机械卸料库 pneumech silo
气动机油枪 pneumatic oil gun
气动激励器 air activator
气动激振 aerodynamic(al) excitation
气动挤压器 air squeezer
气动记录器 pneumatic recorder
气动技术 pneumatic engineering; pneumatics; pneumatics technique
气动继电器 air relay; gas-actuated relay; pneumatic relay
气动加减器 pneumatic adder-subtractor
气动加力制动器 air master
气动加料斗 pneumatic hopper loader
气动加料机 pneumatic feeder
气动加热 aerodynamic(al) heating
气动加压 pneumatic pressurizing
气动夹紧 pneuma-lock
气动夹紧器 pneumatic gripping
气动夹紧装置 air clamper
气动夹具 air-actuated clamp; air-actuated jaw; air jig; air-operated clamp; pneumatic clamp
气动夹钳 pneumatic gripping
气动夹头 pneumatic chuck
气动减速器 pneumatic retarder
气动减震器 pneumatic shock absorber
气动剪刀 air shears; pneumatic shears
气动浇灌 <混凝土> air placing
气动浇筑 pneumatic placement
气动绞车 air winch; capstan winch; compressed-air hoist; pneumatic hoist; pneumatic winch
气动搅拌器 pneumatic agitator
气动接触力 aerodynamic(al) contact force
气动进料马达 pneumatic feed motor
气动进料设备 pneumatic fill assembly
气动净化 compressed-air cleaning; pneumatic cleaning
气动净化装置 air scavenging gear
气动卷扬机 air(motor) hoist; air winch; compressed-air hoist; pneumatic hoist; pneumatic winch
气动掘岩机 pneumatic excavation
气动卡规 air snap
气动卡紧 pneumatic chucking
气动卡具 air-operated fixture
气动卡盘 compressed-air chuck; pneumatic chuck
气动卡爪 air-actuated jaw
气动开动阀 air-operated valve
气动开关 air-operated switch; air-operating switch; air-pressure switch; jettron; pneumatically operated switch; pneumatic switch
气动开关板 pneumatic switchboard
气动开门器 pneumatic door opener
气动控制 aerodynamic(al) control; air control; pneumatic control
气动控制单元 pneumatic control unit
气动控制阀 pneumatic control valve
气动控制换向阀 air-actuated direction valve
气动控制进样漏孔 pneumatically controlled sample leak
气动控制开关 pneumatic control switch
气动控制屏 pneumatic control manifold
气动控制器 air-operated controller; pneumatic controller
气动控制系统 air-actuated control system; pneumatic control system
气动控制仪表 pneumatic control instrument
气动快门 pneumatic shutter
气动扩胎机 air-operated spreader
气动拉铆机 pneumatic rivet puller
气动栏木 pneumatic gate
气动离合器 pneumatic clutch
气动离合式刹车 pneumatic clutch brake
气动离心式(沥青)抽提仪泵 air-powered centrifuge extractor
气动力 aerodynamic(al) force; air force
气动力补偿 aerodynamic(al) balance; aerodynamic(al) compensation
气动力不稳定性 aerodynamic(al) instability
气动力操纵 atmospheric control
气动力导数 aerodynamic(al) derivative
气动力方法 aerodynamic(al) tool
气动力负荷 air-load
气动力干扰 aerodynamic(al) interference
气动力荷载分布 aerodynamic(al) load distribution
气动力加热 hot blast
气动力焦点 constant pitching moment point
气动力喷射 air atomization
气动力起动机 air impingement starter
气动力试验 aerodynamic(al) test
气动力试验模型 aerodynamic(al) test model
气动力弹性 aeroelastic
气动力弹性干扰 aeroelastic interaction
气动力弹性力学 aeroelasticity
气动力天平 aerodynamic(al) balance
气动力调整片 aerodynamic(al) trimmer
气动力系数 coefficient of aerodynamic(al) force
气动力下沉效应 effect of aerodynamic(al) downwash
气动力学 pneumatics
气动力影响 aerodynamic(al) influence
气动力制动装置 air brake
气动力装置 jet power unit

气动力准则 aerodynamic(al) criterion[复 criteria]
气动力作用 aerodynamic(al) action
气动沥青喷浆法 asphalt gunite process
气动连接法 pneumatic connection
气动联锁 pneumatic interlocking
气动量测仪表 air ga(u)ge;air instrument
气动量规 snap jet
气动量塞 plug-jet
气动量水装置 <混凝土搅拌输送车的> pneumatic water measuring equipment
气动量仪 air ga(u)ge;pneumatic micrometer
气动溜槽 air slide;pneumatic chute
气动溜槽系统 pneumatic chute system
气动路面破碎机 air pavement breaker;pneumatic pavement breaker
气动路碾 pneumatic roller
气动履带式钻车 air-track drill
气动履带式钻机 air-track drill
气动罗盘 pneumatic compass
气动逻辑部件 pneumatic logic member
气动逻辑元件 pneumatic logic element
气动螺刀 air screwdriver
气动螺丝刀 pneumatic screwdriver
气动螺旋压砖机 pneumatic screw press
气动落锤 ceco-drop hammer
气动落砂机 pneumatic knockout
气动落砂器 pneumatic vibratory knockout
气动马达 air motor;compressed-air engine;pneumatic motor
气动马达驱动的 air motor powered
气动马达拖动 air motor drive
气动脉冲 air pulsing
气动铆锤 pneumatic hand riveter;pneumatic riveting hammer
气动铆钉锤 compressed-air riveting hammer;pneumatic riveting hammer;riveting machine hammer;windy hammer
气动铆钉顶 pneumatic dolly
气动铆钉顶座 pneumatic holder-on;pneumatic holder-up
气动铆钉机 air riveting hammer;pneumatic nailing machine;pneumatic riveting machine;pneumatic rivet(t)er
气动铆钉器 pneumatic riveter
气动铆钉枪 pneumatic compression riveter;pneumatic riveting gun
气动铆钉托 pneumatic dolly
气动铆机 air riveter;jam rivet(t)er
气动铆接法 compressed-air riveting;pneumatic riveting
气动铆接机 pneumatic riveter
气动铆枪 air riveter;jam rivet(t)er;pneumatic riveter
气动煤镐 pneumatic coal picker
气动模 gas dynamic(al) mode
气动模拟 pneumatic analog(ue)
气动模拟计算机 pneumatic analog(ue) computer
气动磨床 air grinder
气动磨光机 pneumatic sander
气动磨腻子机 orbital sander
气动抹灰工具 air float;air trowel
气动抹子 pneumatic float;pneumatic trowel
气动内部振动器 pneumatic internal vibrator
气动泥泵 pneumatic dredge pump
气动泥刀 pneumatic float;pneumatic trowel
气动逆向阀 pneumatic reversing valve
气动黏[粘]土挖掘机 pneumatic clay digger
气动排放调节器 pneumatic discharge regulator
气动排气装置 pneumatic scavenging gear
气动排钟式播种机 pneumatic drill
气动喷粉器 air duster
气动喷浆机 pneumatic mortar sprayer
气动喷漆 pneumatic spray painting
气动喷漆机 air-painter
气动喷枪 pneumatic gun
气动喷砂机 air peener;pneumatic peener
气动喷射泵 airlift ejector;pneumatic ejector
气动喷射器 airlift ejector;pneumatic ejector
气动喷头 pneumatic sprayhead
气动喷丸机 air peener;pneumatic shot blasting machine
气动喷雾器 pneumatic nebulizer
气动破碎机 pneumatic breaker
气动铺路夯(实机) pneumatic paving rammer
气动曝气 pneumatic aeration
气动起动机 gate lift
气动起动器 air-starter;pneumatic starter
气动起动装置 pneumatic starter
气动起重机 air(motor) hoist;compressed-air hoist;pneumatic crane;pneumatic hoist
气动起重器 air jack;air lift
气动千斤顶 pneumatic jack
气动牵引车 pneumatic tractor
气动嵌缝枪 air-operated ca(u)lking gun
气动桥式湿度表 pneumatic bridge hygrometer
气动切割嘴 pneumatic cutter
气动倾卸车 air-dump car
气动清管器 pneumatic pipeline pig
气动清洁 pneumatic cleaning
气动驱动 pneumatic drive
气动驱动器 air impeller
气动取样器 compressed-air sampler
气动热传递 aerodynamic(al) heat transfer
气动热弹性 aerothermoelasticity
气动热化学 aerothermochemistry
气动热力学 aerothermodynamics
气动热力压缩器 aerothermopressor
气动入口压力 inlet pressure
气动塞规 air plug ga(u)ge;ga(u)ging plug;ga(u)ging spindle;plug jet
气动塞瓶机 pneumatic capper corker
气动塞药棒 pneumatic pole tamper
气动三通电磁阀 pneumatic three-way solenoid valve
气动三叶片 pneumatic alloy spider
气动三针记录器 pneumatic tri-pen recorder
气动色带指示器 pneumatic ribbon type receiver indicator
气动砂冲子 sand rammer
气动砂带机 pneumatic sander
气动砂轮 air-operated grinder
气动砂轮机 air grinder
气动上翻式装载机 pneumatic flipover bucket loader
气动上料器 pneumatically controlled loader
气动烧蚀 aerodynamic(al) ablation
气动设备 air-powered equipment;pneumatic equipment
气动设备安装 pneumatic installation
气动设备的噪声 pneumatic equipment exhaust noise
气动射钉枪 pneumatic nail gun
气动升船机 pneumatic lift
气动升降机 compressed-air elevator;pneulift;pneumatic elevator
气动升降机构 air lifter
气动升力 aerodynamic(al) lift
气动升液泵 air lift pump
气动失稳性 aerodynamic(al) instability
气动施工 pneumatically applied
气动湿度控制器 pneumatic humidistat
气动式测压计 pneumatic piezometer
气动式沉降盒 pneumatic settlement cell
气动式传感器 pneumatic transducer
气动式电池 pneumatic cell
气动式舵机 pneumatic steering gear
气动式干扰物投放器 pneumatic chaff dispenser
气动式夯板 pneumatic compacting and finishing screed
气动式横动导丝控制 pneumatic traversing guide control
气动式灰浆泵 pneumatic mortar pump;pneumatic type mortar pump
气动式混凝土浇灌机 pneumatic concrete placer
气动式夹桩器 pneumatic chuck
气动式减震器 pneumatic shock absorber
气动式交通量计数器 pneumatic traffic counter
气动式孔压测头 pneumatic piezometer tip
气动式脉动器 pneumatic pulsator
气动式碾压机 pneumatic roller
气动式撒肥机 pneumatic fertilizer distributor
气动式升降卸载机 airlift unloader
气动式污水排水器 pneumatic sewage ejector
气动式自动驾驶仪 pneumatic autopilot
气动式自动转换阀 air relay valve
气动手持凿岩机 pneumatic hand hammer rock drill
气动输冰系统 pneumatic ice delivery system
气动输带装置 pneumatic pulldown
气动输气管 pneumatic tube
气动输送 pneumatic conveying;pneumatic transport;pneumotransport
气动输送泵 pneumatic pump
气动输送分配器 pneumatic distributor
气动输送管 <用于传送文件资料> lamson tube;pneumatic tube
气动输送管路 air conveying pipeline
气动输送管系统 pneumatic tube system
气动输送机 air-activated conveyer[conveyor];air activator;air conveyer[conveyor];air pad conveyer[conveyor];air slide;pneumatic conveyer[conveyor]
气动输送设备 pneumatic conveying equipment
气动输送系统 pneumatic conveyer[conveyor] system;pneumatic transfer system
气动数据 aerodynamic(al) data
气动刷子 bromak
气动双针记录器 pneumatic double-pen recorder
气动水泵 pneumatic water pump
气动水泥装卸 air cement handling
气动水窝泵 pneumatic sump pump
气动伺服操作 pneumatic servooperation
气动伺服机构 pneumatic servomechanism
气动伺服系统 pneumatic servosystem
气动伺服元件 pneumatic servoelement
气动伺服制动器 air master
气动伺服装置 pneumatic servo
气动送料装置 pneumatic discharge apparatus
气动速送器 pneumatic rabbit
气动速送器管道 pneumatic rabbit channel
气动速送容器 pneumatic shuttle
气动碎石锤 air chipper
气动锁闭 air actuated lockout
气动锁紧 pneuma-lock
气动踏板 gas pedal
气动弹射器 gas-powered catapult
气动弹性力学 aeroelastics
气动弹性学 aeroelasticity
气动弹性振动 aeroelastic vibration
气动探测器 pneumatic detector
气动提升泵 air lift pump
气动提升机 air hoist;air lift;airlift machine;pneumatic hoist;pneumatic lift
气动提升搅拌机 airlift agitator
气动提升器 seal leg
气动提水泵 air lift pump
气动替续器 air relay;pneumatic relay
气动填充机 air stuffer
气动调节 pneumatic control
气动调节阀 pneumatic control valve;pneumatic regulating valve;pneumatic valve
气动调节器 air-o-line;air-operated controller;pneumatic controller;pneumatic regulator
气动调节设备 pneumatic regulation unit
气动调位继电器 pneumatic positioning relay
气动调压器 pneumatic pressure regulator
气动投料 batch pneumatically
气动透平 aerodynamic(al) turbine
气动涂料喷射机 pneumatic paint sprayer
气动湍流 aerodynamic(al) turbulence
气动推进的架式钻机 air-feed drifter
气动推进凿岩机支架 pneumatic feed column
气动拖式卷扬机 air tugger(hoist)
气动脱模 air ejection
气动陀螺仪 air-driven gyroscope
气动挖掘机 pneumatic excavation;pneumatic pick;pneumatic shovel
气动挖泥船 pneumatic dredge(r)
气动挖土机 pneumatic pick
气动外部振动器 pneumatic external vibrator
气动外径量规 air snap ga(u)ge
气动喂料 pneumatic feed
气动温度调节器 pneumatic thermostat
气动稳定性 aerodynamic(al) stability
气动涡轮机 air turbine
气动涡轮牙钻 air turbine dental engine
气动雾化喷嘴 pneumatic nozzle;two-fluid nozzle
气动吸尘器 pneumatic cleaner
气动吸粮机 blower
气动洗涤器 pneumatic scrubber
气动铣刀 pneumatic mill
气动系数 aerodynamic(al) coefficient
气动系统 air system;pneumatic system
气动线锯 pneumatic sabre saw

气动小方石铺路夯实机 pneumatic sett paving rammer
气动卸货车 pneumatic discharge vehicle
气动卸料 pneumatic discharge; pneumatic emptying
气动卸料斗 air-operated bucket
气动卸料机 pneumatic discharger
气动卸料装置 pneumatic discharge apparatus
气动卸载式载重汽车 pneumatic discharge vehicle
气动型砂捣碎机 pneumatic sand rammer
气动修磨机 air-operated grinder
气动修整器 pneumatic finisher
气动修枝剪 pneumatic pruning shears
气动絮凝 pneumatic flocculation
气动压板 pneumatic clamp
气动压差变送器 pneumatic differential pressure transducer
气动压床 air press
气动压紧装置 air clamper
气动压力 pneumatic pressure
气动压力传感器 pneumatic pressure transmitter
气动压力机 pneumatic press
气动压力调节器 gas pressure regulator
气动压缩 compressed-pneumatic
气动牙钻 air dental drill
气动延时继电器 pneumatic time delay relay
气动岩石破碎机 pneumatic rock breaker
气动研磨机 pneumatic grinder
气动研磨器 pneumatic lapper
气动扬声器 electropneumatic loudspeaker
气动扬水法 air pumping method
气动摇臂式装载铲 pneumatic rocker type shovel(loader)
气动摇臂式装载机 pneumatic rocker type shovel(loader)
气动遥示器 pneumatic remote indicator
气动液动的 pneumohydraulic
气动液位控制 pneumatic level control
气动液压的 airdraulic; pneudraulic; pneumatic-hydraulic; pneumo-hydraulic
气动液压技术 airdraulics
气动液压控制器 pneumatic-hydraulic controller
气动液压控制系统 pneumatic-hydraulic control system
气动液压千斤顶 air hydraulic jack
气动液压式 pneumato-hydraulic
气动液压试验控制台 pneumatic-hydraulic test console
气动液压自动驾驶仪 pneudraulic autopilot; pneumatic-hydraulic autopilot
气动仪表 pneumatic meter
气动移动装置 pneumatic transfer device
气动印刷 pneumatic printing
气动油压开关 pneumatic oil switch; pneumo-oil switch
气动油脂枪 air grease unit; air-operated grease unit
气动元件 pneumatic components; pneumatic element; pneumatic equipment; pneumatic unit
气动远程发送器 pneumatic teletransmitter
气动远距离控制系统 pneumatic remote control system
气动远距离指示器 pneumatic remote indicator
气动运料 material conveyed by air
气动运输 pneumatic conveyer[conveyor]; pneumatic transportation; wind transport(ation)
气动运输机 air conveyer[conveyor]
气动运算放大器 pneumatic operational amplifier
气动錾 pneumatic chipping hammer
气动凿井 air sinking
气动凿井钻台 pneumatic shaft jumbo
气动凿毛机 bush hammer
气动凿岩机 air drill; air feed rock drill; compressed-air drill; pneumatic drill hammer; pneumatic rock drill
气动凿岩机腿架 pneumatic drill leg
气动凿岩钻车 pneumatic jumbo
气动噪声 aerodynamic(al) noise
气动噪声试验 aerodynamic(al) noise test
气动闸 pneumatic brake
气动闸门 pneumatic lock
气动张紧装置 air-operated take-up
气动胀开心轴 air-operated expanding mandrel
气动折门 air-operated folding door
气动折射计 pneumatic refractometer
气动针锤 needle gun
气动振打器 air shaker
气动振捣器 air vibrator; platform vibrator; pneumatic vibrator
气动振动器 air vibrator; pneumatic vibrator
气动振动匀面板 pneumatic-tyred vibrated finishing screed
气动振动整平板 pneumatic-tyred vibrated finishing screed
气动振动装置 pneumatic vibrator unit
气动震源地震剖面记录仪 gas source seismic profiler
气动正时系统 pneumatic timing system
气动正弦波发生器 pneumatic sine wave generator
气动支架风钻 air-driven post drill
气动执行机构 pneumatic actuator; pneumatic motor; pneumatic operator
气动执行机器 pneumatic actuator
气动执行机制 pneumatic actuator
气动纸带程序控制器 pneumatic tape controller
气动指示器 pneumatic indicator
气动制动 aerodynamic(al) braking
气动制动机构 pneumatic brake
气动制动执行机构 pneumatic brake actuator
气动重力输送机 air-activated gravity conveyer[conveyor]
气动重力运输机 air-activated gravity conveyer[conveyor]
气动轴承 pneumatic bearing
气动肘杆 pneumatic toggle link
气动肘杆压力机 pneumatic toggle press
气动主动轮 pneumatic capstan
气动助力的 air assisted
气动柱塞震击 air ram jolt
气动转换 air relay
气动转盘送料 pneumatically operated dial feed
气动转向助力装置 air-powered steering unit
气动转辙器 air-operated power switch
气动转子流量计 pneumatic rotameter
气动桩锤 pneumatic pile hammer
气动装订 pneumatic stapler
气动装卸货车 pneumatic discharge vehicle
气动装修升降平台 pneumatic lifting platform
气动装岩机 air loader
气动装载 pneumatic stowing; stow pneumatically
气动装载机 air loader; pneumatic stowing machine
气动装载机械 pneumatic loader
气动装载设备 pneumatic stowing plant
气动装置 air drive; air-moving device; pneumatic plant; pneumatics
气动琢毛机 bush hammer
气动自动的 pneumatic automatic
气动自动控制 pneumatic automatic control
气动自动控制系统 pneumatic automatic control system
气动自卸车 air discharge wagon; air-dump car
气动阻力 aerodynamic(al) drag; aerodynamic(al) resistance
气动阻尼 aerodynamic(al) damping
气动阻尼器 aerodynamic(al) damper; air damper
气动钻 pneumatic drill
气动钻法 pneumatic drilling method
气动钻杆提升器 pneumatic rod extractor
气动钻机 pneumatic drilling machine
气动钻架 feedleg feeding
气动钻进 air drilling; compressed-air drilling; pneumatic drilling
气动钻进马达 pneumatic feed motor
气动(钻进)腿架 pneumatic feedleg
气堵(塞) air binding
气段 air section
气阀 air cock; air valve; gas check valve
气阀瓣 air flap
气阀传动装置 air valve operating device
气阀导管 air valve guide
气阀盖 air valve cap
气阀盖座 air valve cage
气阀管 air valve box
气阀机构 air valve actuating gear
气阀间隙 air valve clearance
气阀井 air valve chamber
气阀控制杆 air valve control lever
气阀滤池 air valve filter
气阀螺钉 gas valve screw
气阀室 air valve cage
气阀体 air valve body
气阀旋转机构 rotocap
气阀摇臂 air valve rocker; air valve rocker arm
气阀摇杆 air valve rocker
气阀摇杆装置 air valve rocker gear
气阀圆锥面 air valve conic(al) face
气阀罩 air valve cap
气阀制销 air valve cotter; air valve key; air valve pin
气阀重叠 air valve overlap
气放炮 blowing
气肥煤 gas fat coal
气分机 air separator
气氛 ambience; atmosphere
气氛加压烧结法 gas pressure sintering
气氛控制器 atmosphere controller; oxy stop
气氛烧结 atmosphere sintering
气氛窑 controlled atmosphere kiln
气氛转变温度 atmosphere transfer temperature
气封 aerospace seal; air lock; air seal; air-tight; exclusion of draught; gas lock; gas seal(ing); gas-tight seal; glands; gland seal; seal leakage; vapo(u)r lock(ing); vapo(u)r seal; atmoseal
气封高低齿 stage tooth
气封口 air trap
气封漏气 packing leakage
气封圈 air seal ring
气封套筒 seal sleeve
气封系统 gas sealing system
气封蒸汽 seal steam
气封蒸汽加热器 gland steam heater
气封蒸汽冷却器 gland steam desuperheater
气封钟形盖 gas seal bell
气封装置 sealing gland
气缝 gas slot
气浮 air levitation; air flo(a)tation
气浮池 flo(a)tation basin; flo(a)tation pond; flo(a)tation tank
气浮池浮渣 <水处理> flo(a)tation crust
气浮池刮渣机 skimming machine of flo(a)tation pond; skimming machine of flo(a)tation tank
气浮除油池 aerated skimming tank
气浮处理技术 flo(a)tation treatment technique
气浮垫轴承 air bearing
气浮法 flo(a)tation process
气浮法的颗粒选择 particle selection by air flo(a)tation
气浮法分离器 air flo(a)tation classifier
气浮工艺 air flo(a)tation process
气浮挂车 air cushion trailer
气浮过程 air flo(a)tation process
气浮回流 flo(a)tation reflux
气浮接触室 flo(a)tation contact chamber
气浮浓缩 flo(a)tation thickening
气浮浓缩池 flo(a)tation thickener
气浮溶气罐 dissolved air vessel
气浮设备 flo(a)ration unit
气浮式往返循环列车 hovering shuttle
气浮-水解-接触氧化工艺 air flo(a)-tation-hydrolysis-contact oxidation process
气浮陀螺仪 air-supported gyroscope
气浮物化处理 air flo(a)tation physical-chemical treatment
气浮悬挂装置 air cushion suspension; air cushion suspension unit
气浮-厌氧-需氧工艺 air flo(a)tation-anaerobic-aerobic process
气浮约束系统 aircushion restraint system
气浮运输机 air cushion transporter
气浮桌 air floating table
气缚 air binding
气盖 gas cap
气干 air dry; air seasoning; natural seasoning
气干比重 specific gravity in dry air
气干变色 yard stain
气干材 air-dried wood; air dry wood; seasoned wood
气干的 air seasoned
气干法 air drying
气干基 air-dried basis
气干砾石 air-gravel
气干砾石混凝土 air-gravel concrete
气干木材 air-dried lumber; air dry wood
气干清漆 air drying varnish
气干扰 spherics
气干(色)漆 air drying paint
气干时间 air drying time
气干收缩 air shrinkage
气干状态 air-dried condition
气杆菌 aerobacter

气缸抽测 measurement of cylinder at random
气缸盖 cylinder head
气缸盖垫片 head gasket
气缸工作容积 displacement volume; stroke volume; swept volume
气缸工作容量 displacement
气缸架 cylinder frame
气缸口 cylinder passage
气缸排量 displacement volume; stroke volume; swept volume
气缸上盖 upper end plate
气缸套 cylinder liner
气缸注油器 cylinder lubricator
气镐 air hammer; air pick(er); air ram(mer); hammer pick; pneumatic digger; pneumatic pick; pneumatic pick hammer
气镐的钎头冲击 pick blow
气镐气缸 pick cylinder
气镐钎子 pick point
气割 air cut; autogenous cutting; autogenous fusing; burning; burning torch; flame cutting; oxyacetylene cutting; torch cutting; welding torch cutting; gas cut(ting) <氧乙炔切割>
气割槽 gas gouging
气割吹管 gas cutting torch
气割吹管嘴 gas cutting torch nozzle
气割的 autogenic; autogenous
气割法 oxyacetylene cutting
气割钢筋 burning reinforcement
气割工 burner
气割机 gas cutter; burner equipment; burning machine; flame cutter; gas cutting machine
气割炬 cutting blowpipe; oxyacetylene cutting torch; oxygen cutter
气割开坡口 flame grooving
气割器 gas cutting machine
气割枪 cutter; gas cutting torch; metal-cutting torch
气割清理 scarfing
气割下料 oxygen-acetylene cutting
气割嘴 cutting tip
气隔层 <平屋顶的> air casing
气隔开槽 gas gouging
气根 aerial root
气根毒藤 rhus orientalis schneid
气功锤 beche
气鼓泡 gas bell
气固比 air-solid ratio; air/solid ratio
气固层析 gas-solid chromatography
气固反应 gas-solid reaction
气固分配色谱法 gas-solid partition chromatography
气固接触 gas-particle contact; gas-solid contact
气固接触设备 gas-solid contactor
气固接触条件 gas-solid contact condition
气固结合力 air-to-solids bond
气固界面 gas-solid interface
气固流态化 gas-solid fluidization
气固膜 gas-solid film
气固黏[粘]附 gas-solid adhesion
气固平衡 gas-solid equilibrium
气固色谱法 gas-solid chromatography
气固吸附平衡 gas-solid adsorption equilibrium
气刮刀涂布 air doctor coating
气管 air pipe; air tube; gas hose; suction pipe; throttle
气管电磁阀 gas solenoid
气管剪钳 combination gas pliers
气管接头 air connection; gas type fitting
气管联结螺母 air union nut
气管滤器 air pipe strainer
气管路 air route
气管螺纹 gas pipe thread
气管内径 air hose inner diameter
气管钳 gas pliers
气管塞 gas shaft hood
气管式检数器 road tube detector
气管弯头 air connection tube
气管网 gas pipeline network
气管线 gas pipeline
气罐 air accumulator; air holder; air tank; gas storage holder; gas tank
气罐车 pressure car
气罐锥形顶 gas-holder bell
气柜 aerator tank; air accumulator; air collector; gas-holder tank; gasometer; gas receiver; gas tank; gas tube
气柜操作 gas-holder operation
气柜浮动顶盖 expensive roof of tank
气柜基础 gas-holder foundation
气柜润滑脂 gas-holder grease
气柜烧剥机 deseamer
气柜油 gas-holder oil
气柜钟罩 gas(-holder) bell
气棍 air wand
气锅炉上给水箱 feed head
气海界面 air-sea interface
气焊 acetylene weld(ing); air acetylene welding; electric (al) soldering; flame welding; gas welding; hot-gas welding; oxyacetylene welding; oxygen-acetylene welding; oxygen weld(ing); torch soldering; torch welding
气焊把 blow(er) pipe
气焊吹管带嘴 gas welding torch pipe with nozzle
气焊吹管嘴 gas welding torch nozzle
气焊的 <将金属熔化而接合> autogenous; autogenic
气焊的钢管 autogenous-welded steel pipe
气焊法 autogenous soldering; autogenous weld(ing)
气焊割炬嘴子 blowpipe head
气焊工 burner; gas welder
气焊工艺 gas welding technique
气焊工作 gas welding work
气焊管 gas-welded pipe
气焊焊缝 torch weld
气焊焊剂 gas flux
气焊焊炬 gas torch
气焊焊枪 gas torch
气焊焊丝 gas welding rod
气焊火焰 gas torch
气焊机 gas welder; gas welding machine; oxyacetylene welder; oxyacetylene welding outfit
气焊机组 gas welding outfit
气焊接头 gas welded joint
气焊进气硬管 gas welding torch butt
气焊气体 flare gas
气焊枪 blow pipe; brazing torch; gas jet; gas welding torch; heating torch; torch; welding torch
气焊设备 acetylene apparatus; gas welding device; gas welding equipment; gas welding outfit
气焊条 filler; gas welding rod; steel welding wire
气焊橡胶管 gas welding rubber hose
气焊氧乙炔焰焊接 oxyacetylene welding
气焊用护目镜玻璃 autogenous welding shield glass
气焊用氧气瓶颈阀 weld neck valve
气焊装置 gas welder
气焊嘴 torch head
气夯 air ram(mer); pneumatic beetle; pneumatic compactor; pneumatic ram; pneumatic tamper
气夯锤 pneumatic ram
气核 gas nucleus; vapo(u)r nucleus
气核浓度 concentration of gas nuclei
气黑 gas black
气喉型呼吸器具 airline breathing apparatus
气候 climate; clime; weather
气候保护 weather protection
气候变动 climate change; climate fluctuation; climate variation; climatic fluctuation; climatic oscillation; climatic variation
气候变化 climate change; climate fluctuation; climate variation; climatic change; climatic fluctuation; climatic variation
气候变化的限度 limitation of climate change
气候变化地球物理监测 geophysical monitoring for climate change
气候变化地球物理监测实验室 geophysical monitoring for climate change laboratory
气候变化和海洋委员会 Committee for Climate Changes and the Ocean
气候变化检测计划 climate change detection project
气候变化率 rate of climate change
气候变化全球监测方案 global monitoring for climate change program(me)
气候变化速度 rate of climate change
气候变化预测管 storm glass
气候变化韵律 rhythm due to climate change
气候变化周期 climate change cycle; climate cycle
气候变量 climatic variable
气候变暖 climate warming
气候变迁 climate change; climate fluctuation; climate variation; climatic fluctuation; climatic variation
气候标志 climatic indicator
气候表 climatological table
气候病 climatic pathology; meteoropathy
气候波动 climate fluctuation; climatic fluctuation; climatic variation
气候不良混线 weather contact
气候不稳定性 climatic instability
气候参数 climate parameter; climatological parameter
气候层 climatolith
气候长期趋势 secular trend in climate
气候长期研究制图计划 climate long range investigation mapping prediction study
气候长石砂岩 climatic arkose
气候成因 climatic factor
气候处理 seasoning
气候带 climate belt; climate zone; climatic belt; climatic zone; climatic region
气候带和气候类型 climatic zone and climatic type
气候的半球热力模式 hemi-spheric(al) thermodynamic (al) model of climate
气候的垂直分带性 vertical division of climatic
气候的可变性 climatic variability
气候的热力学模式 thermodynamic(al) model of climate
气候的湿润性分区 climatic division of humidification
气候的综合特点 complex characteristics of climate
气候等值线 isoclimate line; isoclimatic line
气候抵抗能力指标 resistance to weather
气候地层单位 climate stratigraphic(al) unit
气候地带性 climate zonation
气候地貌学 climatic geomorphology
气候地区 climatological region
气候顶极植被 climatic climax vegetation
气候动力学 weather dynamics
气候冻土层间融冻层 climafrost
气候对人类影响研究计划的行动计划 action plan for the program(me) of studies on weather impacting human
气候多年平均值 climatological normals
气候恶化 climatic degeneration
气候发生地貌学 climato-genetic geomorphology
气候反常 climate anomaly; climatic anomaly
气候防护电动机 weatherproofing motor
气候分带 climatic zoning
气候分界 climate divide; climatic divide
气候分类 classification of climate; climate classification
气候分区 climate area; climatic divide
气候分析 climate analysis
气候锋 climatic front
气候符号 climatic symbol
气候付区 climatic subdivision
气候改变 climatic shift
气候改良 climatic melioration
气候改善 climatic amelioration; climatic modification
气候干扰 weather interference
气候感应性 climatic sensitivity
气候观测 climatic observation; climatological observation
气候观测站 climatic observation station
气候过程 climate process
气候和人类问题专家会议 Conference of Experts on Climate and Mankind
气候和生物间关系的 bioclimatic
气候和天气 climate and weather
气候环境 climatic environment
气候环境资料 climatic environmental data
气候机构 climatic agencies
气候激烈变化 climate upheaval
气候极值 climatic extreme
气候计算技术和方法 climate computing technology and method
气候记录 climate record; climatic data; climatic record
气候记录站 climatological station
气候监测 climate monitoring
气候监测系统 weather monitoring system
气候监视 climate watch
气候阶地 climatic terrace
气候界限 climate divide
气候警报 climate alert
气候距平 climate anomaly; climatic anomaly
气候科学 climate science
气候控制 climatic control; climatic modification
气候老化 weather ag(e)ing; weathering
气候疗法 climatic treatment
气候模拟 climate simulation; climatic simulation; simulation of climate
气候模拟室 climate weathering cabinet
气候模型 climate model

气候耐受性 tolerance to climate
气候年 climatic year
气候偶变 climate accident
气候情况 climatic condition;climatological condition
气候情况测绘板 climatic situation board
气候情况条款 climatic conditions clause
气候区 climatic province;climatic region;weather zone
气候区分 climatic division
气候区域 climate region; climatic province; climate zone; climatic zone
气候(曲线)图 climatograph
气候趋势 climate trend; climatic trend
气候日 climatological day
气候(上)的 climatic
气候设施 climatic facility
气候生产指数 meteorologic(al) index of production
气候生活型 climatic life form
气候实验室 climatizer
气候事件 climatic event
气候试验 climatic test(ing)
气候试验室 climate cell;climatic laboratory
气候适宜(期) climatic optimum
气候适应 acclima(ta)tion;acclimatization;climatic adaptation
气候适应过程 climatization
气候适应性 climatization
气候适应性试验 climatic test(ing)
气候舒适 climatic comfort
气候数据 climate data;climatic data;climatological data
气候数据库 climate database
气候衰减 weather attenuation
气候所造成的误差 weather-related error
气候探测卫星 weather satellite
气候天气改变 climate-weather modification
气候条件 atmospheric condition;climatic condition;climatological condition;weather condition
气候条件和土壤条件 climate and soil conditions
气候条件模拟箱 climate cabinet
气候条件系数 weather condition factor
气候条件许可日数 weather working days
气候条件因素 weather condition factor
气候条件影响 effect of climatic conditions
气候统计学 climatological statistics
气候图 climatic chart;climatic map;climatological chart; synoptic (al) weather chart;climatic diagram
气候图表 climatic graph;climatological diagram; climograph; climatogram;climatograph
气候图册 climatological atlas
气候图集 climatological atlas
气候图解 climagram;climatic diagram
气候土壤形成 climatic soil formation
气候危害 climate hazard
气候卫生 climatic hygiene
气候温和 having a moderate climate
气候温和的 balmy
气候温暖 clemency
气候温暖的 clement
气候稳定性 resistance to weather
气候问题 climate issue
气候系列 climosequence
气候系数 coefficient of climate
气候系统 climate system; weather system
气候系统监测 climate system monitoring
气候显示仪 climatic indicator
气候现象 climatic phenomenon
气候限度 weather limitation
气候相同地带 isoclimatic zone
气候向性 meteorotropism
气候效应 climatic effect;climatological effect
气候型 climatic type
气候性成土作用 climatic soil formation
气候性土类 climatic soil type
气候性土壤 climatogenic soil
气候序列 climosequence
气候学 climatology
气候学和气象学 climatology and meteorology
气候学技术委员会 Technical Commission for Climatology
气候学家 climatologist
气候学委员会 Commission for Climatology
气候雪线 climate snow line;climatic snow line
气候驯化 acclimatement; acclimation;acclimatization
气候循环 climate cycle;climatic cycle
气候延误 weather delay
气候研究 climate study;research on climate
气候研究站站址选定 station identification for climate study
气候演替顶极 climatic succession climax
气候要素 climatic element
气候宜人 hospitality
气候异常 climate anomaly;climatic anomaly
气候因素 climatic element; climatic factor
气候因子 climatic factor
气候应变 climatic strain
气候应力 climatic stress; stress of weather
气候应用 climate application
气候应用检索系统 climate application referral system
气候影响 climate effect
气候影响性反应 meteorotropism
气候预报 climate forecast; climate prediction; climatic prediction; climatological prediction
气候预测 climate forecast; climate prediction;climatological prediction
气候预测管 storm glass
气候韵律 climatic rhythm
气候灾害 climatic scourge
气候(造成的)损坏 weather damage
气候展望 blind prognosis; climatic forecast
气候振动 climatic fluctuation
气候正常值 climatological normals; climatological normal value
气候植物区系 climate plant formation
气候指示物 climate indicator
气候志 climatography
气候治疗学 climatotherapeutics;climatotherapy
气候状况的变化 swing of climatic regime
气候资料 climate information;climatic data;climatic information;climatological data
气候资料管理 climatic data management
气候资料管理问题工作组 working group on climatic data management
气候资料中心 climatic data center [centre]
气候资源 climatic resources
气候最适度 climatic optimum
气候最适条件 climatic optimum
气候最适宜期 <距今 2000 年至 9000 年间> hypsithermal interval
气候最优期 climatic optimum
气候作用力 climate forcing
气弧保护焊 inert gas arc welding
气葫芦(俗称) air chamber
气护金属电弧焊 shielded metal arc
气滑式分级机 air-slide classifier
气滑式给料器 air-slide feeder
气滑式输送机 air-slide conveyer[conveyor]
气化 aerify;air slacking;gasify;transpiration;vapo(u)rize
气化变质 pneumatolytic metamorphism
气化不利 disturbance in gas transformation
气化带 gasification zone
气化的 pneumatolytic
气化点 point of evapo(u)ration
气化镀锌 vapo(u)r galvanizing
气化法 evaporating method
气化方式 gasification form
气化废水 gasification wastewater
气化分异作用 gaseous transfer differentiation;pneumatolytic differentiation
气化工艺 gasification technique
气化工艺指标 gasification technical parameter
气化固体 vapo(u)r solid
气化管 generating tube
气化管头 vapo(u)rizing tube nipple
气化光谱分析 air spectrography
气化过程 pneumatolytic process
气化灰水 gasification ash water
气化计 atomometer; evaporimeter [evaporometer]
气化剂 gas agent;gasifying agent
气化交代 pneumatolytic replacement
气化介质 gasifying medium
气化矿 pneumatolytic ore
气化矿床 pneumatolytic deposit
气化矿物 pneumatolytic mineral
气化-冷凝循环 vapo(u)rization-condensation cycle
气化炉 gasifier
气化率 vapo(u)rization rate
气化煤气 gasification gas
气化模 gasifiable pattern;gasified pattern;volatile pattern
气化模型 gasified pattern
气化能 vapo(u)rization energy
气化黏[粘]土 pneumatic clay;pneumatolytic clay
气化凝粒 <指金属气化凝结的颗粒> fumes
气化皮 blister
气化破拱装置 aeration device
气化期 pneumatolytic stage
气化器 gas generator;gasifier;grass fire;vapo(u)rizer;vapo(u)rator
气化潜热 gasification latent heat;latent heat of evapo(u)rization
气化强度 gasification strength
气化区(域) vapo(u)rization zone;gasification zone
气化燃料 vapo(u)rising fuel;vapo(u)rizing fuel
气化燃烧器 vapo(u)rizing burner
气化热(液) heat of gasification;heat of vapo(u)rization
气化热液变质作用 pneumatolytic hydrothermal metamorphism
气化热液矿床 pneumato-hydrothermal deposit; pneumatolytic hydrothermal deposit
气化热液作用地球化学 geochemistry of pneumato-hydrothermal processes
气化设备 regasification plant
气化渗镀 vapo(u)r plating
气化石灰 air hardening lime; air slaked lime;lime slaked in the air
气化式燃烧室 vapo(u)rizing combustion chamber; vapo(u)rizing combustor
气化式引燃器 carburet(t)ing pilot
气化速率 vapo(u)rization rate;vapo(u)rizing rate
气化损失 vapo(u)rization loss
气化特性 vapo(u)rizing property
气化调节 vapo(u)rization adjustment
气化涂膜反光镜 vapo(u)r-coated mirror
气化脱硫 gasificating desulfurization
气化外潜热 external latent heat
气化温度 vapo(u)rizing temperature
气化物 vapo(u)r
气化效率 gasification efficiency;rate of gasification;vapo(u)rization efficiency
气化型 pneumatic type
气化性 vapo(u)rability;vapo(u)rizability
气化性防锈剂 dichan
气化冶金 vapometallurgy
气化冶金方法 vapometallurgical process
气化液体 vapo(u)r
气化液体灭火器 vapo(u)rizing liquid extinguisher
气化液体燃料发动机 vapo(u)r engine
气化用煤 gasification coal
气化油 vapo(u)rizing oil
气化蒸发 steam raising
气化周期 gas making period
气化装置 vapo(u)rizing unit
气化状态 vapo(u)r state
气化作用 gasification;pneumatolysis;vapo(u)rization
气环 actuator piston ring; compression ring;gas ring;piston compression ring;pressure ring
气缓冲器 die cushion
气辉 air flow;airglow;night sky light
气辉现象 airglow phenomenon
气火山 air-volcano
气急 dyspnea
气胶溶体 gasoloid
气焦 gas coke
气结 air binding
气结构 pneumatic structure
气界 aerosphere; air sphere; atmospheric geology;gas sphere
气进水出概念 air in/water out concept
气进水出设计 air in/water out design
气浸电缆 gas impregnated cable
气井 gas input well;gasser;gas well
气井气 gas well gas
气井钻台 gas platform
气阱 air pocket;air trap
气举 air lift;gas lift
气举泵 gas bubble pump
气举采油间歇调节器 gas lift intermitter
气举出水井 gas lift flowing well
气举阀下入深度 depth of gas lift valve
气举阀形式 type of gas lift valve
气举反循环钻进 airlift reverse circulation drilling

Q

气举管 eduction column;eduction pipe;eduction tube;eductor
气举活塞 free piston
气举活塞泵 free piston pump
气举活塞泵送 free piston pumping
气举级数 stage of gas lift
气举间隙诱导流筒 gas lift intermitter
气举式泵 air lift pump
气举天然气 gas lift gas
气举压力 gas lift pressure
气举柱 eduction column
气举装置 gas unit
气炬 blow torch;gas torch;torch
气炬焊(接) gas torch welding;torch soldering;torch welding
气炬焊头 torch head
气炬钎焊 torch brazing
气炬切割 deseaming;flame cut;gas torch cutting;torch cutting
气炬烧剥 deseam(ing)
气炬烧割 torch cutting
气炬烧焊 torching
气炬硬焊 torch brazing
气锯 air saw;pneumatic saw
气开关 air switch
气坑 air pit;pocket of air
气空试验 <一种用于测试沥青胶结料和集料表面黏[粘]附力的方法> blister test
气孔 abscess;air cavity;air hole;air port;air vent;air void;blacking hole;bleeder hole;blister;blowhole;brogue hole;fumarole;gas cavity;gas pocket;gas pore;gas vent;open bubble;pore;vent(age);vesicle <矿物或岩石中的>
气孔测定 void test
气孔测定仪 porosimeter;porosity apparatus
气孔测头 air-probe
气孔池 gas cavity cell
气孔尺寸 pore size
气孔尺寸分布 pore size distribution
气孔带 stomatal band;stomatic band
气孔的 spiracular
气孔底式浮选机 blanket-type cell
气孔度测验器 porosity tester
气孔分布 pore size distribution
气孔缝 stomatic cleft
气孔构造 vesicular structure
气孔含量 <加气混凝土的> air void content
气孔焊工 broguer
气孔后的 postspiracular
气孔混凝土 aerocrete
气孔计 porometer
气孔间距 pore space
气孔间距系数 <加气混凝土的,以毫米计> distance factor between air voids
气孔间距因数 spacing factor
气孔检验 porosity test
气孔结构 aerated structure;pore structure
气孔炉箅 air grating;air-hole grate
气孔率 <以百分率表示> air void content;porosity;void content
气孔容积 void volume
气孔塞 air plug;Welch plug
气孔散发 stomatal transpiration
气孔疏松(缺陷) gassiness
气孔水 pore water
气孔图 porosity chart
气孔形成剂 <混凝土的> inflatable void former
气孔运动 stomatal movement;stomatic movement
气孔蒸腾 stomatal transpiration
气孔直径 pore diameter
气孔指数 stomatal index
气孔铸件 blown casting
气孔砖 cellular brick
气孔状构造 fumarolic structure
气孔阻力 stomatic resistance
气控 pneumatic control
气控阀门 pneumatically controlled gate valve
气控仪表 pneumatic control instrument
气控装料设备 pneumatic loading device
气控钻架 air-controlled rig
气口 riser;ventilation opening
气口布置 porting arrangement
气块 air parcel;gas in mass
气喇叭 air horn
气廊 air lane
气浪 air wave;avalanche blast;blast;puffing
气浪模拟 gas-wave analogy
气冷 air cooling;gas cooling
气冷壁 air-cooled wall
气冷变压器 dry-type transformer
气冷淬火 gas quenching
气冷的 blown
气冷电机 gas-cooled electrical machine
气冷电阻器 blown resistor
气冷发电机 gas-cooled generator
气冷阀 air cooling valve
气冷反应堆 gas-cooled reactor
气冷钢 air-cooled steel
气冷高炉矿渣 air-cooled blast-furnace slag
气冷高炉炉渣 air-cooled blast-furnace slag
气冷高炉熔渣 air-cooled blast-furnace slag
气冷管 air-cooled lamp;radiation-cooled tube
气冷(管道)系统 air cooling system
气冷核反应堆 gas-cooled nuclear reactor
气冷货(物) air-cooled cargo
气冷快中子增殖堆 gas-cooled fast breeder reactor
气冷矿渣 air-cooled slag
气冷炉渣 air-cooled slag
气冷磨 air-cooled mill
气冷凝器 gas condenser
气冷盘管 air coil
气冷器 air cooler
气冷(却)的 gas-cooled
气冷蛇形管 air coil;gas cooling coil
气冷式变压器 air-cooled transformer
气冷式的 air-cooled
气冷式电(动)机 air-cooled motor;gas-cooled machine
气冷式发动机 air cooling engine;air-cooled engine
气冷式(空气)冷凝器 air-cooled condenser
气冷式空调机 air-cooled conditioner
气冷式空调器 air-cooled air-conditioner
气冷式冷却机组 air-cooled chiller unit
气冷式铆钉锤 air-cooled riveting hammer
气冷式气缸 air-cooled cylinder
气冷式汽轮发电机 air-cooled turbogenerator
气冷式热交换器 air-to-air heat exchanger
气冷式水银整流器 air-cooled mercury rectifier
气冷式套筒 air-cooled jacket;cooling air jacket
气冷式天花板 air-cooled ceiling
气冷式涡轮机 air-cooled turbine
气冷式压缩机 air-cooled compressor
气冷式叶片 air-cooled blade
气冷式引燃管 air-cooled ignitron
气冷式制冷剂冷凝器 air-cooled refrigerant condenser
气冷系统 air-cooled system;air cooling system
气冷叶栅 air-cooled cascade blade
气冷硬化 air hardening
气冷硬化钢 air hardening steel;self-hardening steel
气冷转子 air-cooled rotor
气力 pneumatic power
气力泵 air-powered pump
气力操纵 pneumatic control
气力操纵的配料设备 pneumatically operated batching plant
气力操作 pneumatic handling;pneumatic operation
气力称重系统 pneumatic weighing system
气力抽吸机 pneumatic sucker
气力除灰系统 pneumatic ash removal system
气力传动 pneumatic power drive
气力传动提升机 pneumatic power drive lift
气力传送 pneumatic dispatch
气力吹开 blowing
气力垂直输送机 pneumatic vertical conveyer[conveyor]
气力锤 pneumatic hammer
气力抖动器 air knocker;pneumatic shaker
气力发动机 pneumatic engine
气力分级器 pneumatic classifier
气力分级装置 air classifier
气力分离器 air classifier
气力分选 pneumatic separation
气力辅助的 pneumatically assisted
气力给料 air-feed(ing)
气力谷粒运送装置 pneumatic grain handling unit
气力谷粒装卸装置 pneumatic grain handling unit
气力管道系统 pneumatic duct system
气力辊涂机 micro-jet roll coater
气力夯 pneumatic ram(mer)
气力缓冲器 inflatable diaphragm
气力回动装置 pneumatic reverse gear
气力混凝土浇筑 pneumatic concrete placing
气力混凝土浇筑机 pneumatic concrete placing machine
气力混凝土输送 pneumatic concrete conveying
气力混凝土运输机 pneumatic concrete handling machine
气力机械 pneumatic machine;wind power machine
气力集草车 wind stack
气力减震器 pneumatic shock absorber
气力搅拌机 compressed-air mixer
气力锯 pneumatic saw
气力掘凿机 pneumatic excavator
气力均化 pneumatic homogenization;pneumatic homogenizing
气力控制的 pneumatically controlled
气力控制的骨料料仓闸门 pneumatically controlled aggregate bin gate
气力控制系统 air-actuated control system
气力垃圾输送机 pneumatic waste conveyer[conveyor]
气力粒料运送装置 pneumatic grain handling unit
气力粒料装卸装置 pneumatic grain handling unit
气力螺旋输送机 pneumatic screw conveyer[conveyor]
气力铆钉机 pneumatic riveter;pneumatic riveting machine
气力弥雾 pneumatic atomization
气力弥雾器 air atomizer
气力密封 pneumatic seal
气力模垫 pneumatic die cushion
气力耙式搅拌机 air rake agitator
气力喷乳化沥青砂浆 pneumatically applied asphalt emulsion mortar
气力喷砂机 air peener;air sander;pneumatic sander
气力喷射 pneumatic injection
气力喷雾 air-atomizing
气力喷雾机 pneumatic sprayer
气力喷注机 air placer
气力起动 pneumatic cranking
气力起重机 air hoist;air lift;gas lift;pneumatic jack
气力起重器 air jack;compressed-air jack;pneumatic jack
气力千斤顶 air jack;compressed-air jack;pneumatic jack
气力牵引 pneumatic traction
气力清洁机 pneumatic separator
气力清洗 pneumatic cleaning
气力驱动 pneumatic drive
气力升降机 air lift;airlift system;pneumatic lift
气力升降装置 pneumatic elevating gear
气力升送机 pneumatic elevating conveyer[conveyor];pneumatic lifting device
气力式分离机 pneumatic separator
气力式分离器 pneumatic separator
气力式快速喷粉机 pneumatic high-speed duster
气力式喷雾机 pneumatic spraying machine
气力式输送机 wind conveyer[conveyor]
气力式调速器 pneumatic governor
气力式选粒机 pneumatic classifier
气力输冰系统 ice pneumatic conveying system
气力输送 air delivery;air slide;pneumatic conveying;pneumatic despatch[dispatch];pneumatic handling;pneumatic transmission;pneumatic transport
气力输送泵 pneumatic pump;pneumatic transport pump
气力输送法 pneumatic despatch[dispatch]
气力输送管 pneumatic carrier
气力输送管道 pneumatic piping
气力输送管路 air conveying line
气力输送管装置 pneumatic tube installation
气力输送烘干机 flash drier[dryer]
气力输送机 air conveyer[conveyor];air-pressure conveyer[conveyor];air-slide conveyer[conveyor];pneumatic conveyer[conveyor]
气力输送垃圾机 pneumatic waste conveyor
气力输送系统 pneumatic conveyer system;pneumatic transport system
气力输送装置 pneumatic transfer device;pneumatic transport device;pneumatic transporting equipment
气力水泥运送 pneumatic cement handling
气力水泥装卸 pneumatic cement handling
气力送料器 pneumatic feeder
气力送砂装置 pneumatic sand system
气力锁风阀 pneumatic air lock valve
气力弹射器 air catapult
气力调节器 pneumatic governor
气力提升 air lift

气力提升泵 air lift pump; pneumatic lifting pump
气力提升袋 pneumatic lifting bag
气力提升机 airlift machine; pneumatic elevating conveyer [conveyor]; pneumatic elevator; pneumatic lifting device; pneumatic vertical conveyer[conveyor]
气力提升装置 pneumatic elevating gear
气力推动 propulsion
气力吸粮机 pneumatic grain sucker
气力吸鱼泵 airlift pump for fish
气力系统压力表 compressed-air ga(u)ge
气力卸船机 pneumatic shipunloader
气力卸载机 vacuum unloader
气力修整器 pneumatic finisher
气力蓄能器 pneumatic accumulator
气力压送机 air placer; pneumatic placer
气力运输 pneumatic conveying
气力运输泵 pneumatic transport pump
气力增力器 air-booster
气力闸 pneumatic brake
气力制动器 aerodynamic(al) brake; air brake; pneumatic brake
气力制动器操纵阀 air brake control valve
气力制动器控制 air brake controls
气力致动的 air-actuated
气力助推的 pneumatically boosted
气力柱塞 pneumatic ram
气力装料机 pneumatic loading machine
气力装卸 pneumatic handling; pneumatic operation
气力装卸谷物设备 pneumatic grain handling equipment; pneumatic grain handling plant; pneumatic grain handling unit
气力装卸设备 pneumatic handling plant
气力自动装置 pneumatic automatic
气力作用的 air-actuated
气力座装置 airdraulic seat system
气粒 air particle
气帘 air curtain; air screen; draught excluder; gas-proof curtain
气帘防波堤 air-bubble breakwater
气帘式加热炉 curtain-type furnace
气炼熔融法 flame melting method
气量表 gas meter; gasometer; gas volumeter
气量测定 gasometric determination
气量滴定 gasometric titration
气量分析 gasometric analysis
气量计 aerometer; air meter[metre]; air(o)meter; gas ga(u)ge; gas meter; gasometer; drum gas meter < 拌和机干燥筒的 >
气量瓶 gasometer flask; gas volumeter
气裂 gas checking; gas crazing; throwing out
气流 aerial current; air blast; air current; air draft; air draught; air motion; air movement; air stream; blast of air; convection current; current of air; draft air; efflux; flow of gas; gas current; gas flow; steam flow; stream current; wind current; wind flow
气流泵 pneumatic pump
气流变幻不定 bumpiness
气流变形 flow distortion
气流表 air meter[metre]; anemometer
气流不均匀性 flow non-uniformity
气流参数 flow parameter; flow quality
气流测定 pneumatic measurement
气流层 air layer
气流差压 airflow differential
气流抄针器 pneumatic stripper
气流成网机 random web-laying equipment
气流冲击 gas shock
气流冲击荷载 air blast loading
气流冲击压力 impact air pressure
气流除尘器 pneumatic cleaner
气流除尘系统 pneumatic cleaning system
气流除杂机 super-jet cleaner
气流传感器 pneumatic sensor
气流传声器 pneumatic loudspeaker
气流窗花格 flowing tracery
气流床气化 airflow bed gasification; entrained particle gasification
气流吹洗管 air blowpipe
气流垂直运动干式静电吸尘器 vertical flow dry electrostatic precipitator
气流挡板 blast fence
气流导向叶片 < 管道转弯处设的 > airflow vane; air vane guide
气流倒灌 down draught
气流道 flow passage
气流的东西分量 easterly component of an air current
气流的非直接调节 indirect flow control
气流地形举升 orographic(al) lifting
气流电池 pneumatic cell
气流动力系数 dynamic(al) air flow factor
气流动力学 aerodynamics
气流堵塞 flow choking
气流煅烧 flash calcining
气流阀 baffler; draught damper
气流分布 airflow distribution
气流分层 flow separation
气流分级器 air classifier
气流分离 air blast separation; airflow breakaway; breakdown; burbling; flow separation; gas separation; separation bubble; separation of flow
气流分离点 burbling point; stalling point
气流分离器 air sifter; flow separator; stall spot; stream splitter
气流分离区 separation area
气流分离指示器 stallometer
气流分裂 burble
气流分裂区域 breakaway
气流分配 airflow distribution
气流分选 air sifting
气流分选器 air-separating mill
气流粉尘冷却反应堆 dust-cooled reactor
气流粉磨机 fluid energy mill
气流粉碎法 comminution by gas stream
气流粉碎机 aeropulverizer
气流负压仪 minimeter
气流干燥 flash drying; pneumatic conveying dry; pneumatic drying
气流干燥机 pneumatic conveying drier[dryer]; pneumatic drier[dryer]
气流干燥器 flash drier[dryer]; pneumatic conveyer drier[dryer]; pneumatic drier[dryer]
气流干燥设备 pneumatic conveying drier[dryer]
气流格栅 air grill(e)
气流隔断 draft excluder; draught excluder; draught lobby; draught preventer
气流管道 airflow line; ventilation stack
气流管式输送机 pneumatic tube conveyer[conveyor]
气流横断面 cross-stream dimension
气流环流 convective current
气流换向器 flow switch
气流混合器 fluidized bed mixer
气流畸变 flow distortion
气流激波 gas shock
气流计 air flow meter; air meter[metre]; air(o)meter; gas flow meter; wind instrument
气流计算 gas flow calculation
气流加压 pneumatic weighting
气流交黏[粘]法 air-felting
气流浇铸 flowing gas casting
气流角 flow angle
气流节制器 air damper
气流结构 flow pattern
气流截面积 flow area
气流介质 moving gaseous medium
气流均匀性 flow uniformity
气流控制 air current control
气流控制的 current-controlled
气流控制阀 control damper
气流控制器 draft regulator; gas flow controller
气流扩散 airflow diffusion; flow divergence
气流冷却指数 wind chill index
气流离体 breakaway; breakdown; flow separation
气流离体区 bubble
气流量 air flow; gas feed rate
气流流量表 gas flow meter
气流流量计 air flow meter
气流螺旋扩散 spiral air-flow diffusion
气流脉动 gas pulsation
气流密度 current density; flux density
气流敏感性 air draft sensitivity
气流模拟 flow simulation
气流模式 flow pattern
气流模型 airflow model; air-stream pattern; flow model
气流模型试验 air model study; air model test
气流磨 air current mill; air jet mill; jet milling
气流幕 draft curtain
气流能量 flow energy
气流排出 jet exit
气流喷吹工艺 blast attenuating process
气流喷雾机 airflow sprayer
气流喷雾器 gas stream atomizer
气流喷雾式干燥器 pneumatic type spray drier[dryer]
气流膨胀 flow expansion
气流偏导装置 air-stream deflector
气流偏转 flow deflection
气流品质 flow quality
气流平衡 airflow balancing
气流平面 flow plane
气流平行干燥机 parallel current drier[dryer]
气流铺装 air-felting
气流牵拉 pneumatic take-down
气流牵拉装置 pneumatic draw-off device
气流强度 current rate
气流侵蚀 air intrusion
气流清洁器 pneumatic cleaner
气流曲线 flow curve
气流筛分 air sifting
气流上升的 anabatic
气流射程 air throw
气流式不匀率试验仪 pneumatic irregularity tester
气流式打桩机 air blast gin
气流式干草输送器 pneumatic hay conveyer[conveyor]
气流式干束输送器 pneumatic sheaf conveyer[conveyor]
气流式干燥机 airflow drier[dryer]; pneumatic drier[dryer]
气流式干燥器 airflow drier[dryer]; flash dryer; pneumatic drier[dryer]
气流式谷物输送器 pneumatic grain conveyer[conveyor]
气流式锯齿型打桩机 air-blast saw gin
气流式喷射泵 ejector air pump
气流式清选 pneumatic cleaning
气流式燃烧 airflow combustion
气流式升运器 pneumatic elevator
气流式输送机 airveyor
气流式输送器 blast conveyer[conveyor]
气流式纤维细度仪 pneumatic fiber-fineness indicator
气流式轧花机 air blast gin
气流室 draft chamber
气流输送的 air conveying
气流输送分级机 pneumatic elutriator
气流输送粉末 fluidize flour
气流输送干燥机 pneumatic conveyer drier[dryer]
气流输送鼓风机 pneumatic conveyer blower
气流输送机 compressed-air conveyer [conveyor]
气流输送器 pneumatic conveyer[conveyor]
气流速度 air speed; air velocity; flow gas velocity; precipitator gas velocity
气流速度表 aerodromometer
气流速度计 air mileage unit; air speedometer; air velometer
气流速率 airflow rate
气流损失 draft loss; draught loss; windage loss
气流探测器 air-stream detector
气流特性 flow behavio(u)r
气流特性曲线 flow characteristic curve
气流提升 gas lift
气流调节板 gas flow adjusting flange
气流调节阀 barometric(al) damper
气流调节门 draft check damper
气流调节器 air damper; air regulator; damper; draught regulator; flow variator
气流调节系统 gas flow control system
气流通路 current path
气流图 flux map
气流弯曲应力 gas bending stress
气流温度 gas flow temperature; stream temperature
气流紊流度 flow turbulence; stream turbulence
气流稳定器 draught stabilizer
气流涡流 stream swirl
气流污染 air-stream contamination; air-stream pollution
气流下降 falling current of air
气流下流外延层 downstream film
气流下洗 down wash
气流线 flow line
气流限制 pneumatic restriction
气流向上排出的通风井 up-shaft
气流形 aerodynamic(al) form
气流形状 configuration of air flow
气流型式 airflow pattern
气流选粉机 airflow classifier
气流扬声器 pneumatic loudspeaker
气流仪 pneumatic tester
气流壅塞 flow choking
气流噪声 airflow noise; air-stream noise

气流增压 current boost
气流闸 draft damper;draught damper
气流张力装置 pneumatic tension device
气流罩 air bell
气流振动供料机 pneumatic vibrating part feeder
气流指示器 gas flow indicator
气流中的死区 dead air
气流转向器 draft diverter
气流转移 by-passing of gas
气流状态 current regime
气流阻挡阀 draft excluding threshold;draught excluding threshold
气流阻挡设施 draft excluder
气流阻力 airflow resistance; resistance to flow of steam
气流阻尼器 air damper
气流组织 air distribution; airflow makeup;airflow organization;space air diffusion
气馏 steam distillation
气楼 clear stor(e)y;clerestor(e)y; monitor top
气楼侧壁 dormer cheek
气滤 air filter
气路 air channel;air passage;gas circuit;gas path;steam way
气路操作阀 control valve of air passage
气路软管 pneumatic hose
气轮 air wheel
气轮电力机车【铁】 gas turbo-electric (al) locomotive
气轮式推土机 pneumatic-tyred bulldozer
气轮叶片 vapo(u)r vane
气螺刀 air screwdriver
气落磨 aerofall mill;autogenous mill; cascade mill
气落式磨矿机 aerofall mill
气马达驱动钻进 air motor drilling
气锚 gas anchor
气锚封隔器 gas anchor packer
气铆锤 pneumatic riveting hammer
气铆钉枪 airgun
气铆工具 pneumatic riveting appliance
气帽 air capping;gas cap;ventilating capping
气煤 bottle coal;gas coal;light coal
气煤钻 pneumatic auger
气门 aeriductus; air door; air drain; air gate; air lock; air port; air valve; breather hole; overflow gate;shrinkage head;spiracle;valve
气门安装工具 valve installing tool
气门摆杆【机】 valve rocker
气门扳手 valve key;valve spanner
气门板 air plate
气门保修用机具 valve maintenance tool
气门布置 valve arrangement
气门操作 air door operation
气门拆卸工具 valve lifter;valve remover
气门拆卸器 valve extractor
气门颤动 valve surge
气门颤动点 valve surging point
气门迟关 valve lag
气门传动机构 <发动机> valve actuating gear;valve gear
气门挡板 valve shield
气门刀具 valve cutter
气门导管 valve guide; valve shaft guide
气门导管测量表 valve guide ga(u)ge
气门导管拆卸工具 valve guide puller
气门导管拆卸器 valve guide remover
气门导管密封装置 valve guide seal
气门导管清扫器 valve guide cleaner
气门导管凸肩 valve guide shoulder
气门的 spiracular
气门的布置 valve location
气门蒂片 valve stem
气门吊装工具 valve fishing tool
气门定时角 valve timing angle
气门定时提前 valve lead
气门分配 valve control
气门分配相位的调节 valve timing
气门杆 valve lever;valve shaft;valve stem
气门杆系 valve linkage
气门工作面角度 valve face angle
气门环 valve ring
气门机构 valve mechanism
气门间隙 valve clearance;valve lash; valve play
气门间隙扳手 valve wrench
气门间隙调整 valve clearance adjustment;valve-lash control
气门间隙调整螺钉 valve adjusting screw
气门间隙调整器 valve gapper
气门进(出)气量 valve capacity
气门净空 valve clearance
气门控制进气口 shutter-controlled air inlet
气门口 valve port
气门密封件 valve seal
气门能力 valve capacity
气门排气装置 valve air relief
气门盘 valve cap
气门塞降压 chock release
气门塞释放阀 chock release valve
气门砂 grinding powder
气门室 valve box;valve chamber
气门室盖垫片 valve cover gasket
气门锁片 valve collet
气门弹簧 relief valve spring; valve spring; valve spring damper; valve tension spring
气门弹簧拆卸工具 valve spring remover
气门弹簧拆卸器 valve spring lifter
气门弹簧拆装工具 valve spring tool
气门弹簧颤动 valve spring surge
气门弹簧垫圈 valve spring washer
气门弹簧盖 valve spring cover
气门弹簧减振器 valve surge damper
气门弹簧键 valve spring key
气门弹簧力 valve spring force
气门弹簧起弹器 valve spring lifter
气门弹簧腔 valve spring chamber
气门弹簧试验器 valve spring tester
气门弹簧锁销 valve spring retainer lock
气门弹簧托盘 valve spring cap
气门弹簧托盘锁扣 valve retainer lock
气门弹簧托盘锁销 valve retainer cotter pin
气门弹簧销 valve spring cotter
气门弹簧压缩工具 valve spring compressor
气门弹簧罩 valve spring housing
气门弹簧座 cotter seat; valve spring cup; valve spring retainer; valve spring seat
气门弹簧座卡瓣 valve retainer lock
气门弹簧座圈 valve spring holder; valve spring retaining collar
气门弹簧座锁键 valve spring seat key
气门弹簧座锁片 valve spring retainer clamp
气门弹簧座销 valve spring seat pin
气门体 air valve body
气门调节 valve adjusting; valve setting
气门调节圆头螺钉 valve adjusting ball stud
气门挺杆 valve lifter
气门挺杆导管【机】 valve lifter guide
气门挺杆间隙调整机构 valve tappet clearance adjuster
气门头 valve disc[disk];valve head
气门推杆 valve follower; valve push rod
气门斜面 valve face
气门芯 valve core
气门芯套 valve core housing
气门行程 valve stroke
气门旋转装置 valve rotator
气门制销 valve lock
气门轴 air valve shaft
气门柱 valve lifter
气门嘴孔 valve hole
气门座孔内径 valve clear diameter
气门座圈 valve seat insert
气密 aerospace seal; air seal; air-tight; gas-tight seal; obturation; pressurization
气密薄膜 vapo(u)r-proof membrane;vapo(u)r seal membrane
气密部件 air sealing parts
气密材料 air locking material; vapo(u)r-tight material
气密舱 air-tight cabin
气密舱壁 air-tight bulkhead
气密操作 air-tight work
气密测试计 gas-tight lysimeter
气密层 air impervious liner;air resisting liner; air retaining wall; air-tight seal;gas seal
气密车辆 pressure-sealed car
气密挡板 gas tight damper
气密的 air proof;air-tight;gas-proof; gas-tight; vapo(u)r-proof; vapo(u)r-tight; hermetic; hermetically sealed; impermeable to air; impermeable to gas;leak tight;odo(u)r-tight;pressure-tight;sealed;staunch
气密电缆 pressurizing cable
气密度 air-tightness; tightness degree;vapo(u)r density
气密端盖 gas-tight end shield
气密封 hermetic seal
气密封闭 air-tight seal;hermetic seal
气密封罐 sealed container
气密封机 gas sealing mechanism
气密封剂 gas sealing agent
气密封接头 gas-tight joint;heermatic-(al) seal
气密封口 hermetically sealed
气密封炉 hermetic(al) seal furnace
气密封压力 seal-air pressure
气密封装 hermetic(al) package
气密盖 air-tight cover
气密干燥空腔 hermetically sealed dehydrated captive air space
气密隔板 air-tight bulkhead;pressure bulkhead
气密隔墙 air-tight partition
气密焊 pressure-tight weld
气密盒 pressure-tight box
气密混凝土 air-tight concrete
气密机座 gas-tight casing; gas-tight housing
气密加料机构 gas-tight feeding mechanism
气密件寿命 gas seal durability
气密胶 vacuum cement
气密接缝 air-tight joint;gas-tight joint
气密接合 air-tight joint;gas-tight joint
气密接头 air-tight joint;gas-tight joint
气密结构 air-tight construction
气密进出口 <电子微电镜的> air lock
气密进口 gas-tight entry
气密壳 gas-tight shell
气密口 pressurizing window
气密连接 air-tight joint; vapo(u)r-proof connection
气密炉 air-tight stove
气密螺纹 dry seal thread; gas-tight thread
气密门 air-tight access door
气密密封 air-tight packing;gas-proof sealing;gas-tight seal
气密幕 vapo(u)r-proof curtain
气密屏蔽包层 gas-tight shielded enclosure
气密青贮塔 gas-tight silo;sealed upright silo
气密容器 air-tight container;air-tight vessel;gas-tight container
气密式电机 vapo(u)r-proof machine
气密式分格卸料装置 air sealed rotary discharging device
气密式封罐机 air-tight separator
气密式回转卸料装置 air sealed rotary discharging device
气密式卸料阀门 air sealed discharging valve
气密式卸料装置 air sealed discharging device
气密试验 air-proof test;air test;air-tightness test;pneumatic test(ing)
气密室 air-tight chamber; hermetic cabin;sealed chamber
气密速闭门 air-tight quick acting door
气密填料 air packing
气密通道式连接 air-tight gangway connection
气密外壳 gas-tight casing
气密系统 leak-tight system
气密型机器 air-tight machine; gas-proof machine
气密性 air impermeability; air-tightness; gas impermeability; gas-tightness; impermeability; leakproofness; pressure-tightness; tightness to gas
气密性材料 air-tight material; hermetic material;gas-tight material
气密性阀门 air lock gate
气密性密封 air-tight seal
气密性能试验 air leak(age) test
气密性实验 air-tight experiment; gas impermeability experiment; gas-tight experiment; leakage experiment;soundness experiment
气密性试验 air-tight test; gas-tight test; tightness test; gas impermeability test;leak(age) test;soundness test
气密性试验压力 air-tight test pressure
气密性喂料机 air lock feeder
气密压盖 gas-tight gland
气密液度计 gas-tight lysimeter
气密轴封 air-tight shaft seal
气密装置 air locking;obturator
气密座舱 pressure cabin; pressurized cabin
气面摇杆装置 valve rocker gear
气苗 gas seepage;gas showings
气敏半导体 gas sensory semiconductor
气敏查漏仪 gas-sensitive leak detector;gas sensitive leak locator
气敏传感器 gas sensor
气敏电极 gas sensing electrode
气敏电极法 gas sensing electrode method
气敏电阻器 gas sensitive resistor
气敏金属 gas sensitive metal
气敏陶瓷 gas sensitive ceramics
气敏效应 gas sensitive effect
气敏元件 gas sensor

气膜 air film;gas blanket;gaseous envelope;gas film
气膜传质系数 gas film mass transfer coefficient
气膜垫 air film bearing
气膜控制 gas-film control(ling)
气膜冷却 film cooling; gaseous film cooling
气膜系数 gas-film coefficient
气膜阻力 gas film resistance
气沫橡胶 air foam rubber
气母岩对比 gas and source rock correlation
气幕 air curtain;air wall;draught excluder;gas curtain;gas-proof curtain
气幕法 air curtain method
气幕防波堤 pneumatic breakwater
气幕式炉 curtain-type furnace
气囊 air cell;air pocket;air sac;bellows;blader;bubble;envelope;gas bag; gas cell; gas envelope; gas pocket; pneumathode; pneumatic bag;tracheal sac;vapo(u)r pocket;void pocket
气囊窗 encapsulated window
气囊导管 balloon catheter
气囊防波堤 bubble breakwater
气囊式隔板 inflatable bulkhead
气囊式缓冲器 pneumatic cushion
气囊式货物保护装置 pneumatic lading protection device
气囊织物 gas cell fabric
气囊止水 water sealing with gasbag
气凝 air set
气凝的 air setting
气凝剂 aerofloc
气凝胶 aerogel
气凝胶体 aerocolloid
气凝聚层 gas condensate reservoir
气凝性 air setting
气凝性耐火胶结材料 air-setting refractory cement
气凝性水泥 air-setting cement
气暖 warm-air heating
气暖散热器 air-heating radiator
气刨 gas gouging
气刨枪 governing torch
气泡 abscess;air bladder;air blister;air bubble;air cavity;air hole;air pocket;air void;bladder;bleb;blister;blow-hole;bubble;bulb;cavity pocket;froth;gas bubble;gas pocket;gas pore;gas vacuole;pin-hole;popout; voids content; debiteuse bubble <玻璃熔制过程中槽子砖产生的>;air cell <从水底向上冒的>;air bell <显影时乳剂层表面上产生的>;entrained air <有意在混凝土或砂浆中产生的>
气泡斑 froth pit
气泡半径 bubble radius
气泡爆裂 bubble burst(ing)
气泡泵 air lift pump;gas bubble pump
气泡表面 bubble surface
气泡病 <鱼类> gas bubble disease
气泡玻璃 bubble glass
气泡参数 air-bubble parameter
气泡测斜仪 bubble(in)clinometer
气泡层 bubble layer
气泡迟滞 laziness of the bubble
气泡尺寸 air-bubble size
气泡串 bubble train
气泡大小 bubble size
气泡的形成 formation of gas bubble
气泡地平 bubble horizon
气泡点蚀 air-bubble pitting
气泡对流 bubble convection
气泡法 bubble method; gas bubble method; air-bubble technique <评定表面活性剂效用的>
气泡法检漏 bubble method leak detection
气泡法探漏 bubble method leak detection
气泡防波堤 air-bubble breakwater
气泡防波帘 pneumatic breakwater curtain
气泡防波设施 air breakwater
气泡防冻系统 air-bubbler system
气泡分布 bubble distribution;bubble spectrum
气泡分度规 bubble protractor
气泡分离器 bubble eliminator
气泡分散剂 bubble dispersant
气泡缝 <钢的> piping
气泡浮力 bubble floating force
气泡浮选槽 bubble cell
气泡概念 bubble concept
气泡感受元件 gas-bulb sensor
气泡管 air bleeder;bubble tube;level tube
气泡痕 bubble impression
气泡环带结构 pneumozonal texture
气泡回跃 bubble rebound
气泡计 bubble ga(u)ge
气泡计数 bubble counting
气泡间隔系数 void spacing factor
气泡间沟道 interbubble channel
气泡间距 air-bubble spacing;bubble spacing
气泡间隙 air-bubble spacing;bubble spacing
气泡检查仪 bubblemeter
气泡搅拌槽 bubble agitation tank
气泡搅拌池 bubble agitation tank
气泡校正【测】level correction;bubble correction
气泡结构 bubble structure
气泡径向运动方程 radial motion equation of gas bubble
气泡居中 centering[centring] of bubble
气泡聚合 bubbling polymerization
气泡卷流 plume of bubble
气泡孔眼 <混凝土> bug hole
气泡溃灭 cavitation bubble collapse
气泡帘 <用于消波、防淤等> pneumatic barrier; air-bubble curtain; bubble curtain
气泡帘幕 air-bubble screen
气泡灵敏度 sensitivity of the level
气泡流 bubble flow
气泡六分仪 bubble sextant
气泡脉冲 bubble pulse
气泡脉动 gas bubble pulsation
气泡密度 bubble density;bubble frequency
气泡幕 air-bubbler system
气泡内部能量 bubble internal energy
气泡黏[粘]度计 air-bubble visco(si)meter;bubble visco(si)meter
气泡黏[粘]附 bubble attachment
气泡扭变 bubble distortion
气泡排除装置 <铸件中的> air-bubble eliminator; air pocket eliminator
气泡喷出器 sparger
气泡喷射法 air-bubble injection
气泡喷射系统 air-bubble system;air bubbling system
气泡偏移 deviation of level bubble
气泡频率 bubble frequency
气泡屏 air-bubble screen; bubble screen
气泡屏蔽 air barrier
气泡破灭 bubble collapse
气泡谱 bubble spectrum
气泡曝气 bubble aeration
气泡器 bubbler
气泡腔 air chamber
气泡倾斜仪 bubble inclinometer;bubble type inclination
气泡群 swarm of bubbles
气泡砂 bubble sand
气泡生成 ebullition
气泡生成外加剂 gas-forming admixture
气泡式潮位计 bubble tide ga(u)ge
气泡式分度规 bubble protractor
气泡式浮选机 bubble flo(a)tation machine
气泡式汽化器 bubbling carburet(t)or
气泡式水位仪 bubble ga(u)ge
气泡式水准器 air level;bubble cell;spirit level
气泡式水准器视准系统 bubble collimating system
气泡室 bubble chamber
气泡室径迹 bubble chamber track
气泡室摄影 bubble chamber photography
气泡束 air-bubble plume
气泡栓塞 aeremia;aeroembolism;air embolism;bubble blockage;gas embolism
气泡水幕 bubble barrier
气泡水平器 spirit level
气泡水平仪 air-bubble level;bubble tube;sight level bubble
气泡水位计 bubble ga(u)ge;bubble stage recorder
气泡水准测量 air-bubble level(l)ing;air level(l)ing;bubble level(l)ing;spirit level(l)ing
气泡水准管 <如木工水平尺> chambered level tube;level tube
气泡水准器 air-bubble level; block level;bubble level;line level;spirit level
气泡水准仪 air-bubble level;air level;block level;boxed air level;bubble level;builder's level;spirit level;spirit-leveling instrument
气泡图 bubble diagram
气泡推斥式气枪 bubble push against airgun
气泡帷幕 air-bubble curtain
气泡稳定剂 bubble stabilizer
气泡吸收管 gas washing bottle
气泡析出浮选 gas precipitation flo(a)tation
气泡系统 bubbler system
气泡线 air line;hairline
气泡相合 coincidence of bubble
气泡消除器 air pocket eliminator
气泡消失 bubble collapse
气泡消失的时间 seed-free time
气泡形成 bubble formation
气泡型采样器 bubble type sampler
气泡型流量计数器 bubble type flow counter
气泡压力 bubble pressure;pressure in bubbles
气泡压力法 bubble pressure method
气泡验准器 bubble trier
气泡样的 physaliform
气泡影响 aeration
气泡(影像)重合 <水准仪的> coincidence of bubble
气泡噪声 bubble noise
气泡振荡 bubble oscillation
气泡直径 bubble size
气泡指示器 bubble ga(u)ge
气泡指数 bubble index;outgassing index
气泡滞后 hysteresis of the bubble
气泡置中 centering of bubble;centre bubble
气泡中心 bubble center[centre]
气泡柱色谱法 bubble-column chromatography
气泡状的 alveolar
气泡状结构 bubble structure
气泡状流动 bubble flow
气泡状汽化 nucleate boiling
气泡阻挡层 air barrier
气喷 air jet;gas blowout;gas gush;gas outburst;outburst
气喷的 air spray
气喷净法 air blast
气喷聚集 gas spurt
气喷漆枪 air-painting sprayer
气喷式筛分机 air jet screen
气喷制剂 aerosol
气膨胀避雷器 air expansion lightning arrester
气屏 air curtain
气屏蔽 gas shield
气瓶 air collector;balloon;bulb;cylinder;gas bottle;gas container;gas cylinder
气瓶标志 marking of cylinder
气瓶储存显示板 indication board of cylinder storage
气瓶阀 cylinder valve
气瓶集装架 cylinder rack
气瓶室 gas cylinder room
气瓶组 cylinder battery
气瀑 air cataract
气气对比 gas and gas correlation
气气空调设备 air and air-conditioning equipment
气气溶胶混合物 gas-aerosol mixture
气汽的 gas-vapo(u)r
气-汽-汽热交换器 triffux
气迁标志化合物 air-borne typochemical compound
气迁标志元素 air-borne typochemical element
气枪 airgun;chip-blower;gas exploder;gas gun
气枪波形 airgun waveform
气枪激发延迟时间 airgun fire delay time
气腔 air cavity
气锹 pneumatic spade(hammer)
气切 autogenous cutting
气切割 lancing
气侵泥浆 gas cut mud
气侵事故 gas cutting troubles
气侵液 gas cut fluid
气侵钻泥 gas cut mud
气丘 gas spurt
气球 air balloon
气球操纵 ballooning
气球测量 balloon survey
气球臭氧探测 balloon ozone sounding
气球传感器 balloon-borne sensor
气球船 airship
气球吊篮 aerial car;gondola
气球动力学 balloon dynamics
气球高度仪 balloon altimeter
气球观测 balloon observation
气球火箭探测 rockoon sounding
气球技术 balloon technique
气球驾驶员 aerostat
气球假目标雷达反射器 gulls
气球降落伞 ballute
气球库 balloon hangar
气球母船 balloon vessel
气球闪光三角测量 balloon-borne photoflash triangulation
气球摄影 balloon photography
气球摄影机 balloon camera
气球式卫星 balloon satellite
气球探测 balloon sounding
气球天文学 balloon(-based) astronomy
气球望远镜 balloon telescope

气球卫星 balloon satellite;satelloon
气球形(编组)线群 balloon tracks
气球形的 balloon
气球样变性 ballooning degeneration
气球仪 balloon equipment
气球用经纬仪 balloon theodolite
气球运载仪器 balloon-borne instrument
气球载红外望远镜 balloon-borne infrared telescope
气球照相术 balloon photography
气驱 gas drive
气驱动 <油面天然气驱动使油入井> gas cap drive
气驱油藏 gas-driven reservoir
气驱油田 gas controlled field
气圈 aerosphere; air sphere; gas sphere gas loop
气圈地球化学异常 atmogeochemical anomaly
气圈气 atmosphere gas
气泉 air fountain
气泉气 spring gas
气扰混沙分析仪 air siltometer
气热加固法 thermal stabilization method
气热声学 aerothermoacoustics
气溶粉尘 air-floated powder
气溶过滤效率 aerosol filtration efficiency
气溶胶 aerated solid;aerocolloid;aerogel; aerosol[airosol]; gas dispersoid;gasoloid
气溶胶包装 pressurized package
气溶胶包装食品 pressurized food
气溶胶采样 aerosol sampling; sampling of aerosol
气溶胶传播 aerosol transmission
气溶胶的化学组成 chemical composition of aerosol
气溶胶发生剂 aerosol propellant
气溶胶分光仪 aerosol spectrometer
气溶胶分析器 aerosol analyser[analyzer]
气溶胶罐 aerosolcan
气溶胶过滤器 aerosol filter
气溶胶含量 aerosol load
气溶胶化学 aerosol chemistry
气溶胶监测 aerosol monitoring
气溶胶监测器 aerosol monitor
气溶胶检测器 aerosol detector;aerosoloscope
气溶胶粒度谱 aerosol size spectrum
气溶胶粒径测定 measurement of size aerosol
气溶胶粒子 aerosol particle
气溶胶醚二甲苯 aerosolether-xylene
气溶胶浓度 aerosol concentration; aerosol load
气溶胶喷罐 aerosol dispenser;aerosol spray-can
气溶胶喷射剂 aerosol propellant
气溶胶喷射制剂 aerosol spray
气溶胶谱仪 aerosol spectrometer
气溶胶漆 aerosol lacquer; aerosol paint
气溶胶取样器 aerosol sampling device
气溶胶式喷涂 aerosol spraying
气溶胶态流 aerosol flow
气溶胶态污染物 aerosol pollutant
气溶胶涂料 aerosol coating; aerosol paint
气溶胶推进剂 aerosol propellant
气溶胶脱模剂 aerosol release agent
气溶胶微粒 aerosol particle
气溶胶吸入扫描 aerosol inspiration scanning
气溶胶系统【物】aerocolloidal system
气溶胶学 aerosology
气溶体 gaseous solution
气溶微粒 air-floated powder
气(溶)液界面 gas-solution interface
气褥 air-bed
气塞 aeroplug; air binding; air lock; air plug; gas lock; vapo(u)r lock-(ing)
气塞的 air-bound
气塞杆 gas lever
气塞入口 air lock entry
气塞系统 air lock system
气散时间 flash-off time
气扫式膜蒸馏 sweeping gas membrane distillation
气刹车 air brake;pneumatic brake
气刹车开关 air brake switch
气砂轮 pneumatic grinder
气升泵 air lift pump
气升萃取 gas lift recovery
气升挂钩 gas lift hook-up
气升回收 gas lift recovery
气升举 air lift
气升流动 gas lift flow
气升器 gas lift
气升式浮选机 aeration flotater[flotator];airlift machine
气升式搅拌器 airlift type agitator
气升式挖泥船 airlift dredge(r)
气升运输机 gas lift conveyer[conveyor]
气升柱浮选 bubble-column flo(a)tation
气升装置 gas lift unit
气生 aerial growth
气生的 aerial;aerophilic
气生根 aerial root
气生菌丝 aerial hypha
气生岩 atmogenic rock
气生藻类 aerophilic algae
气生植物 aero(plankto)phyte
气湿 air humidity
气湿订正值 corrected value of air humidity
气石 airstone;gas rock
气蚀 cavitation; cavitation erosion; corrosion by gases;gas etching
气蚀保证 cavitation guarantee
气蚀剥损 cavitation pitting
气蚀参数 cavitation parameter
气蚀传感器 cavitation sensor
气蚀等级 cavitation scale
气蚀度 cavitation degree
气蚀腐蚀 cavitation corrosion
气蚀痕迹 cavitation pitting
气蚀坑 cavitation pitting
气蚀麻面 cavitation damage;cavitation pitting surfaces
气蚀破坏 cavitation damage
气蚀气泡 cavitation bubble
气蚀强度 cavitation intensity
气蚀区域 cavitation range
气蚀试验 cavitation test
气蚀数 cavitation number
气蚀水流 cavitation flow
气蚀水箱 cavitation bank
气蚀碎裂现象 <混凝土> crepitation
气蚀损坏 cavitation damage
气蚀损伤 cavitation damage
气蚀特性 cavitation characteristic
气蚀系数 cavitation coefficient;cavitation sigma
气蚀现象 cavitation; cavitation phenomenon[复 phenomena]
气蚀限度 cavitation limit
气蚀效应 cavitation effect
气蚀性能 cavitation-free performance
气蚀穴 cavitation pocket
气蚀运行 cavitation-free operation
气蚀噪声 cavitation noise
气蚀噪音 cavitation noise
气蚀指数 cavitation index
气蚀滞后 cavitation hysteresis
气蚀中心 cavitation core
气势 gas potential
气势梯度 gas potential gradient
气室 air box; air compartment; air dome;air space;air vessel;gas cell;gas chamber;gas dome;haze dome;surge tank;air chamber <水准器>
气室泵 air chamber pump
气室柴油机 air cell diesel engine
气室门 air chamber door
气室式调压塔 air chamber type of surge tank
气室水准器 chambered level tube
气室型地板效应艇 plenum chambered craft
气输出 air-out
气刷 air brush
气刷涂布 air brush coating
气刷涂布机 air brush coater
气栓 air cock;gas lock
气栓症 air embolism
气水 aerated water
气水比 air-water ratio;gas water ratio;water-gas ratio
气水成的 hydatopneumatic; pneumato-hydatogenetic
气水冲洗 air-water washing
气水反冲洗 air-water backwashing
气水分离器 air trap; air-water separator;gas water separator;moisture separator;moisture trap;vapo(u)r-liquid separator;water separator
气水分离箱 air separation tank
气水过渡带厚度 thickness of gas water transition zone
气水化合物 gas hydrate
气水汇流 cocurrent air-water-flow
气水混掺 air-water mixture
气水混合喷射器 air-water jet
气水混合射流 air-water jet
气水混合式换热器 direct contact heat exchanger
气水混合体 air-water mixture
气水混合物 air-water mixture
气水交界面 gas water surface
气水交替作用 air-water interaction
气水接触(面) gas water contact
气水结合面 air-water surface; gas water surface
气水界面 air-water interface;gas water contact;gas water interface;gas water surface
气水界面张力 boundary tension between gas water contact
气水热泵 air-water heat pump
气水乳状液 air-water emulsion
气水砂清理 vapo(u)r blasting
气水生成的 hydratopneumatic
气水同层 gas water bed
气水系统 air-water system
气水相互作用 air-water interaction
气水压力容器 air-water pressure vessel
气送防污输送系统 air deliverable anti-pollution transfer system
气碎岩 atmoclastic rock
气锁 air lock;vapo(u)r lock(ing); air trap <燃油管路的>
气胎 air tire[tyre]; pneumatic tire[tyre]
气胎辊 pneumatic roller
气胎离合器 air clutch
气胎路碾 pneumatic roller;pneumatic-tired [tyred] roller; pneumatic wheel roller;rubber-tired compactor;rubber-tired roller
气胎轮辋 pneumatic tire[tyre] rim
气胎轮缘 pneumatic tire[tyre] rim
气胎碾 pneumatic-tired[tyred] roller; rubber-tired [tyred] roller; tired [tyred] roller
气胎式建筑 air-inflated building;air-supported building
气胎式压路机 pneumatic-tired[tyred] roller
气胎手推车 pneumatic(ally) tired [tyred] buggy
气胎压路机 air-tired[tyred] roller; pneumatic-tired[tyred] roller; tire [tyre] roller
气态 gaseity; gaseousness; gaseous state;gas phase;vapo(u)r phase
气态氨 gaseous ammonia
气态丙烯 propylene steam
气态材料 gaseous material
气态的 aeriform; gaseous; gasiform; gassy;vapo(u)rous
气态等离子体 gaseous plasma
气态电子检测器 gaseous electronic detector
气态发动机燃料 gas for motor fuel
气态发酵 gassy fermentation
气态反应 vapo(u)r reaction
气态方程 gas equation of state
气态放射性废物衰变槽 gas decay tank
气态废物 gaseous waste;off-gas
气态分散 gaseous dispersion
气态分散体 gas dispersoid
气态辐射废物 gaseous radiation waste
气态供热燃料 gaseous heating fuel
气态和液态石油产品混合物 gas-oil mixture
气态混合物 gaseous mixture
气态介质 gaseous medium
气态金属 gaseous metal
气态绝缘剂 gaseous insulant
气态矿 gaseous ore
气态扩散 gas diffusion
气态冷却剂 gaseous coolant
气态流化作用 gaseous fluidization
气态流体 gas(eous) fluid
气态氯 gas(eous) chlorine
气态氯化处理 gaseous chlorination
气态脉塞 gas maser
气态媒质 gaseous medium
气态膜 gaseous film; gaseous membrane
气态漂白 gaseous bleaching
气态平衡 mass balance
气态燃料 gaseous fuel
气态燃烧产物温度 product gas temperature
气态衰变槽 gas decay tank
气态水 pneumatolytic water; vapo(u)rous water
气态碳 gaseous carbon
气态烃类 gaseous hydrocarbon
气态污染 gaseous pollution
气态污染物 gaseous contaminant; gaseous pollutant
气态污染物控制 control of gaseous contaminant; control of gaseous pollutant
气态吸附 gaseous absorbent;gaseous adsorption
气态吸收 gaseous adsorption
气态氧 gaseous oxygen;gox
气态元素 elementary gas
气态杂质 gaseous impurity
气态蒸汽 gaseous steam
气态制冷剂 refrigerant gas
气态组分的循环比 recycle ratio of gaseous constituents
气潭 air pocket
气碳 gas carbon
气套 air casing;air jacket;lining【机】

气提 air lift
气提法 vapo(u)ring extract process
气提阶段 air stripping stage
气提升泵 gas lift pump
气提升流区 gas lift flow area
气提式反应器 airlift reactor
气提式膜生物反应器 airlift membrane bioreactor
气提式生物膜悬挂反应器 biofilm airlift suspension reactor
气提式循环反应器 airlift type circulating reactor
气提式氧化沟 airlift oxidation ditch
气提塔 stripping column; stripping tower
气体安全阀 gas relief valve
气体靶 gas target
气体摆动 oscillation of gas
气体搬运 gaseous transfer
气体包层 gas envelope
气体包裹体 gas inclusion
气体饱和(率) gas saturation
气体保持本领 gas retaining property
气体保护 gas(pressure) protection; Buchholz protection
气体保护层 protective gaseous envelope
气体保护电弧焊 electrogas welding; gas arc welding; shielded arc welding; gas shielded arc welding
气体保护电栓焊 shielded stud welding
气体保护焊 gas shield
气体保护焊方法 electrogas process
气体保护弧焊 gas shield welding; shielded arc welding
气体保护弧焊机 gas shielded arc welding machine
气体保护金属弧焊 gas shielded metal arc welding
气体保护金属极电弧焊 gas metal-arc welding; shielded metal arc welding
气体保护钨极电弧焊 gas tungsten-arc welding
气体保护自动埋弧焊 gas shield automatic submerged-arc welding
气体保留年龄 gas retention ages
气体爆击 fuel knock; gas knock
气体爆燃管 gas explosion tube
气体爆炸 gas burst; gaseous detonation
气体爆震 gas knock
气体倍增 gas multiplication
气体倍增系数 gas multiplication factor
气体比 gas ratio
气体比例自动调节 automatic gas ratio control
气体比率控制 gas ratio control
气体比重 gas specific weight
气体比重表 aerometer
气体比重测定法 aerometry
气体比重计 air(o)meter; aerometer
气体比重精密测定法 chancel flask method
气体比重瓶 gas balloon
气体变换 gas exchange
气体表面硬化法 nicarbing
气体波腹 gas loop
气体剥离 gas stripping
气体剥离器 gas stripper
气体剥离通道 gas-stripping canal
气体捕获表面 gas trapping surface
气体捕集器 gas trap
气体捕集器法 techniques of catch device on air
气体不饱和的 undersaturated with gas
气体采样 gas sampling
气体采样管 gas sampler; gas sample tube; gas sampling tube
气体采样器 atmosphere sampler; gas sampler
气体采样设备 gas sampler; gas sampling device
气体参数 gas parameter; gas standard
气体测定法 aerometry; eudiometry
气体测井 gas logging
气体测量 gasmetry; gas survey
气体测量的方法 methods of gas survey
气体测量的样品 samples of gas survey
气体测量和排放报告系统 aerometric and emission reporting system
气体测量学 aerometry
气体测温法 gas thermometry
气体层 gas blanket
气体查验 gas examination
气体掺杂 gas doping
气体掺杂技术 gas doping technique
气体产率 gas yield
气体产物 gaseous product
气体常量 gas constant
气体常数 gas constant; gas factor
气体尘埃云 gas-dust cloud
气体沉淀置换法 cementation by gases
气体沉降程度测定仪 micromerigraph
气体成分 composition of gaseous phase; gas component; gas composition
气体成分测定仪 dasymeter
气体成分分析 gas composition analysis
气体池频率标准 gas cell frequency standard
气体冲刷 air scour
气体冲洗 gas bleed
气体冲洗阀 gas bleed valve
气体冲洗法兰 gas bleed flange; gas scavenging flange
气体冲洗凸缘 gas bleed flange; gas scavenging flange
气体抽取 gas extraction
气体臭探器 air sniffer
气体臭味鉴定器 gas odo(u)rizer
气体出口 gas outlet; gas vent; pneumatic outlet
气体出口孔 gas outlet hole
气体处理 gas purification
气体处理法 gas treating process
气体处理系统 gas handling system; gas treating system
气体传导 gaseous conductance
气体传导整流器 gaseous conduction rectifier
气体传递动力学 kinetics of gas transfer
气体传输速率 gas transfer rate; rates of gas transfer
气体传输系统 gas transfer system
气体吹洗 gas purging
气体吹洗器 gas purger
气体纯度测定器 eudiometer
气体纯化 gas purification
气体纯化装置 gas purifying installation
气体纯化组列 gas purification train
气体纯物质 gas pure material; pure gaseous material
气体催化还原器 gas catalytic reduction equipment
气体催化氧化器 gas catalytic oxidation equipment
气体存储器 gas reservoir
气体存储元件 gaseous memory cell
气体打火机 gas lighter
气体大量散发点 boil point; open bubble point
气体导电 gaseous conduction
气体导电分析器 gaseous conduction analyser[analyzer]
气体导电体 gas(eous) conductor
气体导管 gas conduit; gas duct
气体导体 gaseous conductor
气体的 aeriform; gaseous; gasiform; gassy; pneumatic
气体的等压变化 isobaric change of gas
气体的对流作用 convection of gas
气体的辐射特性 radiative properties of gas
气体的回流 recirculation of gas
气体的解吸 desorption of gas
气体的抗电击穿强度 electric(al) breakdown strength of gas
气体的排出 gas outburst
气体的热转化 thermal conversion of gases
气体的形成 formation of gas
气体的逸出 liberalization of gas; liberation of gases
气体的运移 migration of gas
气体滴定 gas titration
气体滴定管 gas burette
气体地球化学测量 geochemical gas survey
气体点火 ignition of gas
气体点火器 gas ignitor
气体电池 gas cell
气体电池组 gas battery
气体电导率 gas conductivity
气体电极 gas electrode
气体电介质 gaseous dielectric
气体电离 gas exchange; gas ionization
气体电离常数 gas ion constant
气体电离电池 gas ionization battery
气体电离放大 gas ionization amplification
气体电离过程 electromerism
气体电离计数器 gas(eous) ionization counter
气体电离检定法 electroscopy
气体电离室 gas ionization chamber
气体电离探测器 gas ionization detector
气体电离显示 gas ionized display
气体电量计 gas coulomb-meter; gas voltameter
气体电流 gas current
气体电压计 gas voltameter
气体电子衍射法 gas electron diffraction
气体电子照射净化装置 gas purification equipment by electron beam irradiation process
气体定量的 gasometric
气体定量法 gasometry
气体定量分析法 gasometric method
气体定量器 gasometer
气体定律 gas law
气体动力摆动 aerodynamic-(al) oscillation
气体动力激光器 gas dynamic(al) laser
气体动力理论 kinetic theory of gases
气体动力平衡 aerodynamic(al) balance
气体动力平滑度 aerodynamic(al) smoothness
气体动力润滑 aerodynamic(al) lubrication
气体动力设备 gas dynamic(al) facility
气体动力特性 gas dynamic(al) behavio(u)r
气体动力稳定性 aerodynamic(al) stability
气体动力系数 aerodynamic(al) coefficient
气体动力学 aerodynamics; dynamics of compressible fluids; dynamics of gas flow; gas dynamics; kinetics of gases
气体动力(学的)波动 aerodynamic(al) oscillation
气体动力学碰撞 gas kinetic collision
气体动力学直径 gas-kinetic diameter
气体动力压力 dynamic(al) pressure
气体动力轴承 gas dynamic(al) bearing
气体动态流动 gas dynamic(al) flow
气体动压强 gas-kinetic pressure
气体动压轴承 aerodynamic(al) bearing
气体冻干捕集器 freeze-out gas collector
气体二次通过的炉算加热机 double gas passage travel(l)ing grate
气体二次通过的炉算加热炉 double pass grate
气体发电机 gas-driven generator
气体发动机发电机驱动 gas electric-(al) drive
气体发动机润滑油 gas engine oil
气体发光管 luminous discharge tube
气体发泡法 gas bubble method
气体发射 gas emission
气体发生 gas development; gas generation; gas making
气体发生法 gas evolution method
气体发生量管 gas evolution burette
气体发生炉 carbide-feed generator
气体发生炉燃料 generative fuel
气体发生瓶 gas-generating bottle
气体发生器 gas generator; gasifier; gas producer; hot-gas generator; producer gas generator
气体发生器活门的控制阀门开关 gas generator valve pilot valve switch
气体发生吸收法 gas generation absorption method
气体阀 gas trap
气体反流 gas reflux
气体反应 gas reaction
气体反应台 gas reaction stage
气体返流 gas back streaming
气体方程式 gas equation
气体放出 gas evolution
气体放出分析 gas evolution analysis
气体放大 gas multiplication
气体放大灯 vapo(u)r enlarge lamp
气体放大系数 gas amplification factor
气体放电 aerial discharge; gas(eous)-discharge
气体放电壁显示板 gas-discharge wallboard display
气体放电灯 gas(eous)-discharge lamp
气体放电电池 gas-discharge cell
气体放电电流 gas-discharge current
气体放电分析 gas-discharge analysis
气体放电管 connectron; gas(eous)-discharge lamp; gas (eous)-discharge tube; gravitron; luminous discharge tube
气体放电管噪声发生器 gas-discharge tube noise generator
气体放电光学微波激射 gas-discharge optic(al) maser
气体放电光源 gas-discharge source; light source of gas-discharge lamp
气体放电盒 gas-discharge cell
气体放电激光检测器 gas-discharge laser detector
气体放电继电器 flashtron
气体放电检测器 gas-discharge detector
气体放电离子源 gas-discharge ion source
气体放电理论 gas-discharge theory
气体放电器 gas-discharger; gas-dis-

charge device
气体放电区 gas-discharge zone
气体放电色彩检验法 gas-discharge colo(u)r method
气体放电设备 gas-discharge device
气体放电蚀刻 gas-discharge etching
气体放电式计数管 gas-discharge counter
气体放电特性曲线 gas-discharge characteristic curve
气体放电天线双工器 gas-discharge duplexer
气体放电显示 gas-discharge display
气体放电显示板 gas-discharge display panel
气体放电显示屏 gas-discharge panel
气体放电显示器 gas-discharge display
气体放电阴极 gas-discharge cathode
气体放电噪声 gas-discharge noise
气体放电噪声发生器 gas-discharge noise generator
气体放电噪声源 gas-discharge noise source
气体放电真空计 gas-discharge ga(u)ge
气体放射性测量计 gas radioactivity meter
气体放射性计 gas activity meter
气体分布器 gas distributor
气体分层 stratification of hot-gas
气体分出 gas bleeding
气体分级 gas classification
气体分离 breakaway;gas separating;gas separation;separation of gases
气体分离膜 gas separation membrane
气体分离器 gas separator;gas trap;gun barrel;knockout box;gas buster <钻井泥浆>
气体分离箱 knockout box
气体分离装置 fractionator;gas separation plant;gas separation unit
气体分流定则 principle of parallel vertical gaseous flow
气体分馏塔 gas fractionator
气体分馏装置 gas fractionation unit;gas plant
气体分配管 gas service pipes
气体分配器 gas distributor
气体分配系统 gas-distributing system
气体分配系统流量 gas distribution system capacity
气体分配装置 gas distribution installation
气体分配自动控制 gas distribution automatic control
气体分散器 gas disperser
气体分析 gas analysis;gasometry;gas test
气体分析法 eudiometry
气体分析记录器 gas alloying apparatus
气体分析记录仪 gas analysis recorder
气体分析器 catharometer;gas analyser[analyzer];gas analysis apparatus;gas analysis meter
气体分析探头 gas probe
气体分析样品 sample for gas analysis
气体分析仪 catharometer;gas analyser[analyzer];gas analysis indicator;gas analysis meter
气体分析装置 gas analytical apparatus
气体分压 partial pressure
气体分压定律 law of partial pressure
气体分子 gas molecule
气体分子流 molecular flow of gas
气体分子吸收分光光度计 gas-molecule-absorption spectrophotometer
气体分子运动论 kinetic theory of gases
气体焚烧器 gas incinerater[incinerator]
气体封隔器 gas packer
气体伏安计 gas voltameter
气体浮聚 flatulence
气体浮力 gas lift
气体辐射 gas(eous) radiation
气体辐射废水 gaseous radiation wastewater
气体辐射废物 gaseous radiation waste
气体辐射计数器 gas(eous) radiation counter
气体辐射系数 gaseous emissivity
气体腐蚀 corrosion by gases;gas attack;gas corrosion
气体复合 gas recombination
气体复合器 gas recombiner
气体干燥 gas drying
气体干燥厂 gas drying plant
气体干燥剂 gas drier[dryer];gas drying agent
气体干燥滤阱 gas dry filter trap
气体干燥瓶 gas drying bottle
气体干燥器 gas drying apparatus
气体干燥设备 gas drying apparatus;gas drying plant
气体干燥塔 gas drying tower
气体钢瓶 gas bomb
气体高度净化 fine gas cleaning
气体隔断 gas occlusion
气体工作压力 aero operating pressure
气体供应阀 gas grid
气体供应管路 gas grid line
气体骨料 <指加气混凝土中的气泡> gaseous aggregate
气体鼓泡搅拌器 gas sparger
气体固化涂料 vapo(u)r curable coating
气体固体界面 gas-solid interface
气体固体相互作用 gas-solid interaction
气体锢囚作用 occlusion
气体刮墨刀 air doctor blade
气体管 flue;gas pipe;gas tube
气体管道 gas conduit
气体管道脱水器 gas line dehydrator
气体管路 gas pipeline;gas piping
气体管线 gas line
气体光电管 gas phototube
气体光谱 gaseous spectrum
气体光谱分析 spectrochemical analysis for gases
气体光谱分析仪 gas spectrometer
气体光学脉泽 gas-discharge optic(al) maser
气体过饱和 gas supersaturation
气体过滤 gas filtration
气体过滤袋 gas filter bag
气体过滤管 gas filter tube
气体过滤器 gas filter;gas purifier
气体过滤效率 gas removing efficiency
气体含量 gas content
气体焊剂 gas flux;vapo(u)r flux
气体合金化处理 gas alloying
气体核子的形成 gas nucleation
气体后泄 blowback
气体化 aerification
气体化学 aerochemistry
气体化学分析 chemical analysis of gas
气体还原净化装置 gas reduction purification equipment
气体缓冲技术 gaseous buffering techniques
气体缓冲器 gas cushion
气体缓冲箱 gas buffer tank
气体回流 gas back streaming;gas reflux
气体回流装置 gas reflux apparatus
气体回路 gas return path
气体回收 gas recovery
气体回收洗涤器 process gas scrubber
气体回收系统 gas recovery system
气体混合高负荷污泥消化池 gas mixed high rate sludge digestion tank
气体混合设备 gas mixing device
气体混合物 gas(eous) mixture
气体活化 gas activation
气体活门 gas interlock
气体活性区反应堆 gas core reactor
气体火山 gaseous volcano
气体击穿 gas breakdown
气体积聚继电器 gas-accumulator relay
气体激光 gaseous laser
气体激光干涉仪 gas laser interferometer
气体激光高度计 gas laser altimeter
气体激光器 gas(eous) laser;laser tube cavity
气体激光系统 gas laser system
气体激活材料 gas active material
气体激活电池 gas-activated battery
气体激活探测器 gas activation detector
气体及损失量 gas and its loss
气体极化 gas polarization
气体集合管 gas manifold
气体集料 <指加气混凝土中的气泡> gaseous aggregate
气体计 air meter[metre]
气体计量 gas dosing
气体计量泄漏 gas dosing leak
气体计数 gas counting
气体计数管 gaseous counter
气体计数器 gas counter;gasometer
气体记录仪 air register
气体继电器 gas-actuated relay;gas detector relay;gas pressure relay;gas relay
气体加工厂 gas processing plant
气体加工装置 gas processing plant
气体加热回收量 recovery capacity on gas heater
气体加压系统 gas pressed system;gas pressurization system
气体加药器 chemical gas feeder
气体加药箱 chemical gas feeder
气体夹杂物 air inclusion;gaseous inclusion
气体间歇(喷)泉 gas geyser
气体监测 gas monitoring
气体监测器 gas monitor
气体检测 gas detection
气体检测锤 gas detection hammer
气体检测管 gas detecting tube;gas detector tube
气体检测继电器 gas detector relay
气体检测器 gas detector;gas locator;pneumatic detector
气体检测试剂 gas detect reagent
气体检定管 gas detector tube
气体检漏器 gas detector
气体检验光度计 gas referee's photometer
气体检验器 gas detector;gasoscope
气体检验证明书 gas certificate
气体鉴定器 gas tester
气体交换 gas(eous) exchange;gaseous interchange
气体交换扩散器 exchanger-diffuser
气体交换面积 gas exchange area
气体交换速度 air exchange rate
气体交换损失 gas exchange loss
气体交换系数 gas exchange quotient
气体交换障碍 gas interchange disturbance
气体交换作用 gas exchange action
气体节流式制冷器 throttling cooler
气体结晶作用 crystallization form vapo(u)r
气体介质压力 pneumatic pressure
气体警报器 gas alarm
气体净化 gas purge;gas purification;gas purifying;gas sweetening;gas treatment
气体净化法 gas treating process
气体净化管道 gas cleaning pipe
气体净化器 gas cleaner;gas cleaning device;gas purifier;gas purifying apparatus;gas washer
气体净化设备 air cleaning equipment;gas purifier;gas purifying equipment
气体净化系统 gas purge system;gas treating system
气体净化旋风器 gas-cleaning cyclone
气体净化装置 gas cleaning device;gas cleaning equipment;gas cleaning plant;gas purification equipment;gas purifying installation;gas treating system
气体静力润滑 aerostatic lubrication
气体静力学 aerostatics
气体静压轴承 aerostatic bearing
气体聚集 gas accumulation;gas focusing
气体聚焦示波器 gas-focused oscillograph
气体绝缘 gas(eous) insulation
气体绝缘变压器 gas-insulated transformer
气体绝缘电缆 gas-insulated cable
气体绝缘开关 gas-insulated switchgear
气体绝缘开关设备 gas-insulated switch
气体绝缘体 gaseous insulator
气体可爆性测定仪 explosimeter
气体空间 gas space
气体控制仪表 gas control instrument
气体矿产 gaseous commodity
气体扩散 gas dispersion;gas(eous) diffusion
气体扩散槽 gaseous diffusion cell
气体扩散电极 gaseous diffusion electrode
气体扩散定律 law of gas diffusion
气体扩散法 gaseous diffusion method;gaseous diffusion process
气体扩散分离器 gaseous diffusion separator
气体扩散环 gas diffusion ring
气体扩散级联 gaseous diffusion cascade
气体扩散计 diffusiometer;effusiometer;effusion meter
气体扩散率 diffusibility of gases
气体扩散设备 gaseous diffusion plant
气体扩散系数 coefficient of diffusion for gas
气体扩散源 gaseous diffusion source
气体扩散柱 gas diffusion column
气体冷冻 gas refrigeration
气体冷凝 condensation of vapo(u)r;gas condensation
气体冷凝净化装置 gas condensation purification equipment
气体冷凝器 gas condenser
气体冷却 cooling of gas
气体冷却淬火 gas quenching
气体冷却反应堆 gas-cooled reactor
气体冷却反应器 gas-cooled reactor
气体冷却剂 gas(eous) coolant
气体冷却控制器 gas-cooled monitor
气体冷却器 gas cooler
气体冷却装置 gas quench system
气体离解 gaseous dissociation
气体离心 gas centrifugation
气体离心法 gas centrifuge process

气体离心机 gas centrifuge
气体离子 gas(eous) ion
气体离子的复合 recombination of gaseous ions
气体离子化 gas ionization
气体力学 aeromechanics; pneumatics; pneumatology
气体力学的 pneumatic
气体帘幕热处理炉 gas curtain furnace
气体联合作用 joint action of gases
气体量 amount of gas
气体量管 gas burette
气体量热法 gas calorimetry
气体量热计 gas calorimeter
气体流 gas stream
气体流出物 gaseous effluent
气体流导 gas conduction
气体流动 gas flow
气体流化床 gas fluidized bed
气体流量测定仪 gas flow meter
气体流量测定站 gas metering station
气体流量方程 gas flow equation
气体流量计 gas flow counter; gas flow meter; gas measuring apparatus; gas measuring flowmeter; gas meter
气体流量指示器 gas flow indicator
气体流路 gas flow path
气体流速 gas flow rate
气体流速流量计 gas flow meter
气体流星余迹 gaseous train
气体流型 gas flow pattern
气体漏泄 gas leakage
气体录井 gas logging
气体滤清器 gas cleaner; gas filter
气体弥散器 gas disperser
气体密度 density of gases; gas(eous) density
气体密度测定法 manoscopy
气体密度测定术 manoscopy
气体密度测定仪 dasymeter; manoscope
气体密度测量器 gas density meter
气体密度计 elaterometer [elatrometre]; gas densitometer; gas density balance; gas density ga(u)ge; gas density meter
气体密度记录器 gas density recorder
气体密度记录仪 gas density recorder
气体密度天平 gas density balance
气体密度天平检测器 gas density balance detector
气体密封腔与压力比 enclosure and pressure proportion
气体密封装置 air-tight equipment; gas lock system
气体灭火 extinction using gas; gas-fire extinguishing
气体灭火装置 fire-smothering gear
气体灭菌剂 gaseous sterilant
气体模型 gas model
气体逆流装置 gas reflux apparatus
气体黏[粘]度 vapo(u)r viscosity
气体凝结器 gas condenser
气体浓度 concentration of gas content; gas concentration; gas strength
气体浓度透镜 gas lens
气体排除口 gas outlet
气体排放器 gas purger
气体排放物 gaseous effluent
气体旁路 gas by-pass
气体喷出 gas blowout
气体喷发 gas eruption
气体喷流 gas jet
气体喷洒器 gas sparger
气体喷射 gas injection; gas jet
气体喷射泵 gas jet pump
气体喷射法 gas jet method
气体喷射技术 gas jet technique
气体喷射器 gas ejector
气体喷射压缩机 gas jet compressor
气体喷头 gas tip
气体喷雾 gas atomization
气体喷洗室 spray chamber
气体喷嘴 gas nipple; gas nozzle
气体膨胀 gas amplification; gas expansion
气体膨胀的压力变化曲线 gas expansion line with variation of pressure
气体膨胀机 gas expander
气体膨胀式涡轮 gas expansion turbine
气体膨胀温度计 gas expansion thermometer
气体瓶车 gas cylinder cart
气体-气体汽化器 gas-gas caruret(t)er
气体-气体相互作用 gas-gas interaction
气体欠饱和的 undersaturated with gas
气体切割 flame cutter
气体清除 gas clean-up; scavenging with gas
气体清洗 blast purge; gas purge
气体氰化法 cyanide process; gaseous cyaniding
气体取样 gas sample; gas sampling
气体取样分接头 gas sampling tap
气体取样分析 gas sample analysis
气体取样管 gas sample tube
气体取样器 gas sample collector; gas sampler
气体取样设备 gas sampling device; gas sampling equipment
气体取样装置 gas sampling system
气体全分析仪 complete gas analyser [analyzer]
气体燃化计 eudiometer
气体燃料 fuel gas; gas(eous) fuel
气体燃料的比重 gaseous fuel specific gravity
气体燃料的发热值 gaseous fuel calorific value
气体燃料的露点 gaseous fuel dew point
气体燃料的水分含量 gaseous fuel water content
气体燃料电池 gas-fed fuel cell
气体燃料发动机 gaseous propellant engine
气体燃料矿产 gaseous fuel commodity
气体燃料炉 furnace for gaseous fuel
气体燃料喷嘴 gas fuel nozzle
气体燃料汽车 gaseous fuel automobile
气体燃料压燃机 gas-diesel engine
气体燃烧 combustion of gas and vapo(u)r; gas(eous) combustion
气体燃烧辐射器 gas-fire radiant
气体燃烧器 fume burner
气体热动力学 aerothermodynamics
气体热量计 gas calorimeter
气体热容量 gas heat capacity
气体热值 heating value of gas
气体容积辐射 cavity emissivity
气体容量 gas capacity
气体容量测定仪 gas-volumetric(al) measuring apparatus
气体容量法 gas volumetry
气体容量分析 gas-volumetric(al) analysis; manoscopy
气体容量分析法 gas-volumetric(al) method
气体容量计 gas volumeter
气体容器 gas container; gas vessel
气体容器的处置 disposal of gas container
气体溶度 gas solubility
气体溶剂反应 gas-solvent reaction
气体溶解 gas dissolving
气体溶(解)度系数 gas-solubility coefficient
气体溶解度因数 gas-solubility factor
气体溶解性 gas solubility
气体乳浊剂 gas opacifier
气体入口 gas access
气体软氮化 gas soft nitriding; Nitemper(ing)
气体软氮化处理 Nitemper(ing)
气体软氮化法 nicarbing
气体润滑 gas lubrication
气体润滑的 gas-lubricated
气体润滑轴承 gas-lubricated bearing
气体润滑转子 gas-lubricated rotor
气体散射 gas scattering
气体散射室 gas scattering chamber
气体散射损失 gas scattering loss
气体色层(分离)法 gas chromatography
气体色层分离仪 gas chromatograph
气体色层分析 gas chromatographic(al) analysis
气体色谱 gas chromatograph
气体色谱法 elution gas chromatography; gas chromatography
气体色谱分析仪 gas chromatograph
气体色谱质谱联用 gas chromatography combined with gas spectrometry
气体色散晕 gaseous dispersion halo
气体闪烁计数器 gaseous scintillation counter
气体闪烁探测器 gas scintillator detector
气体闪烁体 gas scintillator
气体射流 gas jet
气体渗出 gas seepage
气体渗氮 gas nitriding
气体渗氮处理 Nitemper(ing)
气体渗碳 gas carburization; gas carburizing; gas cementation
气体渗碳法 gas carbonizing method
气体渗碳剂 gas carburizer; gaseous cement
气体渗碳炉 gas carburizing furnace
气体渗碳装置 gas carburizing system
气体渗透率 gas permeability
气体渗透膜 gas-permeable membrane
气体升压机 gas booster
气体升液器 gas lift
气体升液装置 gas lift unit
气体石脑油 gas(oline) naphtha
气体释放 popping of gas
气体收集 collection of gases; gas collection
气体收集器 gas collector
气体输送 gas delivery; pneumatic transport
气体输送管 air shooter
气体输送机 gas transportation machine
气体输送机械 air-moving device
气体输送压缩设备 gas booster
气体输运 gas transport
气体水合过程 gas hydrate process
气体水合物 gas hydrate
气体速度 gas velocity
气体损害 gas damage
气体损伤 gas injury
气体弹性 elasticity of gases
气体弹性力学 aeroelastics
气体探测管 pneumatic cell
气体探测器 gas detector
气体探测系统 gas detection system
气体探测装置 gas detector system
气体特性 gas characteristic
气体提纯器 air purifier
气体提升泵 gas lift pump
气体体积 gas volume
气体体积定律 gaseous law of volumes
气体体积分析法 gas-volumetric(al) method
气体体积计 volumescope
气体体积色谱法 gas-volumetric(al) chromatography
气体体积修正系数 gas-volumetric(al) correction factor
气体天平 gas balance
气体调节 fume conditioning; gas control
气体调节阀 gas control valve
气体调节器 gas conditioner; gas controller; gas(eous) regulator; gas governor; gas modulator
气体调节塔 gas conditioning tower
气体停止 gas occlusion
气体通道 gas passage
气体透出 gas evolution
气体透过 gas permeation
气体透镜 gas lens
气体透平电力传动机车 gas turbine electric(al) locomotive
气体透平机 air turbine
气体褪色 gas fading
气体脱硫 gas sweetening
气体脱硫装置 gas sweetening unit
气体脱水 gas dehydration; gas dewatering
气体脱水器 gas dehydrator
气体危险浓度 hazardous concentration of gas
气体围井 gas trunking
气体温度 gas temperature
气体温度计 gas thermometer
气体温度探头 gas temperature probe
气体涡轮机 gas turbine
气体污染 gas contamination; gas pollution
气体污染物 gas contaminant; gas pollutant
气体污浊 gas mark
气体吸附 gas adsorbing; gas adsorption; adsorption of gas and vapo(u)r
气体吸附参数 gas adsorption parameter
气体吸附测定法 gas adsorption method
气体吸附阱 gas adsorption trap
气体吸附率 gas adsorption rate
气体吸附器 gas absorber
气体吸附装置 gas adsorption device
气体吸溜 gas occlusion
气体吸收 absorption of gas and vapo(u)r; gas absorption
气体吸收操作 gas absorption operation
气体吸收床 gas absorbent bed
气体吸收反应器 gas clean-up reactor
气体吸收剂 gas absorbent; gas-absorbing agent
气体吸收量 gas absorbed amount
气体吸收率 gas adsorption rate
气体吸收率计 absorptiometer
气体吸收器 gas absorber
气体吸收系数 absorption coefficient of gas; gas absorbed coefficient
气体吸收性 gaseous absorptivity
气体吸收油 gas absorber-oil
气体吸收元件 gas absorption cell
气体吸收装置 gas absorption equipment
气体吸住 gas occlusion
气体吸着 gas adsorption
气体洗涤 gas scrubbing; gas washing; scrub
气体洗涤鼓风机 gas washing blower
气体洗涤机 gas disintegrator
气体洗涤瓶 wash-bottle for gases
气体洗涤器 gas scrubber; gas washer
气体洗涤设备 gas scrubbing system;

Q

gas washing plant; installation of gas washing
气体洗涤塔 gas scrubber; gas scrubbing tower; scrubbing tower
气体洗涤系统 gas scrubbing system
气体洗涤油 gas scrubbing oil
气体洗涤装置 gas cleaning unit; gas washing installation
气体洗瓶 gas washing bottle
气体显示 gas display
气体显示板 gas panel
气体显影 gaseous development
气体线速 linear gas velocity
气体消除 gas freeing
气体消除证书 gas free certificate
气体行程 gas travel
气体型灭火系统 gas type fire extinguishing system
气体性的 aeriform
气体性污染物 gaseous contaminant; gaseous pollutant
气体性质 gas property
气体蓄压器供给系统 gas pressure feed system
气体悬浮预热器 gas suspension preheater
气体循环 gas circulation; gas cycle; gas recirculation
气体循环泵 gas circulator; gas recycle pump
气体循环堆 gas-cycle reactor
气体循环过程 gas recycle process
气体循环器 gas circulator
气体循环系统 gas-circulating system
气体循环装置 gas-circulating plant; gas-recycling plant
气体压差表 gas manometer
气体压力 gas(eous) pressure
气体压力保护 gas pressure protection
气体压力计 gas ga(u)ge; gas manometer; gas pressure meter; tensimeter; vacuum ga(u)ge
气体压力连接 gas pressure bonding
气体压力烧结 gas pressure sintering
气体压力调节器 gas pressure regulator
气体压力头 gas pressure head
气体压力温度计 gas-filled thermometer
气体压力指示器 gas pressure indicator
气体压入口 gas pressure inlet
气体压升出流区 < 地下水的 > gas lift flow area
气体压缩 gas compression
气体压缩机 gas compressor
气体压缩机房 gas reduction building
气体压缩循环 gas compression cycle
气体压头 gas head
气体压载泵 < 排压舱水用 > gas ballast pump
气体氧化净化装置 gas reduction oxidation equipment
气体样品 gas sample
气体样品进样 gas sample introduction
气体液化 gas liquefaction; liquefaction of gases
气体液化循环 gas liquefaction cycle
气体液体固体法 vapo(u)r-liquid-solid process
气体液体燃料量热器 calorimeter for gaseous and liquid fuels
气体液体相 gas-liquid phase
气体液体相互作用 gas-liquid interaction
气体液压式弹簧 gas hydraulic type spring
气体异常 gaseous anomaly
气体逸出 gas escape; gas evolution; gas spill
气体油 gas oil; gasol
气体油料泵 gas-oil pump
气体油料螺管阀 gas-oil solenoid valve
气体油料压力计 gas-oil pressure ga(u)ge
气体预热器 gas preheater
气体云膨胀和扩散 expansion and dispersion of gas clouds
气体运输船 gas carrier
气体运输规则 gas carrier code
气体运载法规 gas carrier code
气体杂质 gas impurity
气体载体 carrier gas
气体增硬处理 gas-hardening treatment
气体张力 gas tension
气体张力测量法 aerotonometry
气体张力计 aeroto(no)meter
气体直接燃烧器 gas direct combustion equipment
气体直接氧化器 gas direct oxidation equipment
气体指标 gas analyzed
气体制冷机 gas refrigerating machine
气体制冷循环 gas refrigeration cycle
气体治疗学 pneumatology
气体质谱计 solid-source mass spectrometer
气体滞留量 gas hold-up
气体滞留体积 gas hold-up
气体置换 gas displacement
气体置换试验 gas replacement testing
气体中和 neutralization with gas
气体中灰尘聚集 loading
气体中间冷却器 gas intercooler
气体重差计 gas gravimeter
气体重力分异 gas gravity differentiation
气体重力仪 gas gravimeter
气体重量单位 < 克瑞 > crith
气体轴承【机】gas bearing
气体轴承陀螺仪 air discharge gyro(scope); air-bearing gyro(scope)
气体注入 gas injection
气体转化 gas reforming
气体转化过程 gas conversion process
气体转化器 gas converter[convertor]
气体转移 gas transfer
气体转移速度 gas transfer velocity
气体转移系数 coefficient of gas transfer
气体状态 gaseous state
气体状态方程 equation of gas state; gas equation of state
气体阻尼 pneumatic damping
气体组成 gas composition
气体组分 gas component
气体钻井喷射式钻头 aerojet bit
气体钻头 < 钻井用 > aerobit
气田 gas field; gas reservoir
气调储藏 controlled atmosphere(gas) storage; gas storage
气调储运 transport and storage in controlled atmosphere
气调顶棚长槽形灯 air-handling troffer
气艇 air boat
气筒 carboy; inflator
气筒阀 air cylinder valve
气筒阀杆 air cylinder valve stem
气筒阀帽罩 air cylinder valve bonnet
气筒阀压盖 air cylinder valve gland
气筒旋塞 pneumatic cylinder cock
气团 air mass
气团变性 air-mass modification; air-mass transformation; modification of airmass
气团存在期 age of air-mass
气团的水平移动 advection
气团发源地 air-mass source
气团分级 air-mass classification
气团分类 air-mass classification
气团分析 air-mass analysis
气团轨迹 air-mass trajectory
气团号 air-mass number
气团及锋面分析 air-mass and front analysis
气团鉴别 air-mass identification
气团交换量 austausch
气团雷暴 air-mass thunderstorm
气团类型图解 air-mass type diagram
气团气候学 air-mass climatology
气团输送 air-mass transport
气团属性 air-mass property
气团特性 air-mass characteristic
气团停滞状态 stagnant atmospheric condition
气团图 air-mass chart
气团温度 air-mass temperature
气团雾 air-mass fog
气团性降水 air-mass precipitation
气团性质 property of air-mass
气团雨 air-mass rain
气团源地 air-mass-source region
气团源区 air-mass-source region
气团云雾 air-mass cloud fog
气团阵雨 air-mass shower
气团状流动 slug flow
气推油制动总汽缸 air-powered master cylinder
气腿 feed leg; feedleg feeding; pneumatic feedleg; pneumatic pusher; pusher; jack leg; air leg < 凿岩机的 >; air bar
气腿顶尖 air leg support
气腿快缩退回扳机 release rod
气腿连接体 air leg attachment
气腿区凿岩机 pusher leg
气腿式凿岩 air leg drilling
气腿(式)凿岩机 air feed leg drill; air-jackleg drill; air leg rock drill; air leg type rock drill; air-rider jack hammer; feed leg rock drill; jack hammer drill; jackleg drill; pusher-feed drill; pusher-feed rock drill; pusher leg drill; rock drill on air-feed leg; air leg mounted (rock) drill
气腿式钻机 air feed leg drill
气腿式钻架 air leg drill mounting
气腿支撑 air leg support
气腿支爪 air leg support
气腿子 pneumatic feedleg
气挖隧洞 < 用压缩空气开挖与出渣的 > air tunnel(ling)
气丸 air pills
气味 flavor; fume; odo(u)r; scent; smell; taint; tincture
气味不对 off odo(u)r
气味测定法 odo(u)rimetry; olfactometry
气味测定计 olfactometer
气味测定器 olfactometer
气味测定学 olfactronics
气味测量法 odo(u)rimetry
气味测量计 odo(u)rimeter
气味存留 odo(u)r retention
气味单位 olfact
气味的 aural
气味的感官评定 odo(u)r panel evaluation
气味度(指数) odo(u)r intensity index
气味分级 nasal rating
气味干扰 odo(u)r nuisance
气味公害 odo(u)r nuisance
气味和味道特征 characteristics of odo(u)r and taste
气味货 odo(u)rous cargo
气味计 scentometer
气味监测站 odo(u)r monitoring station
气味检查 odo(u)r detection
气味检索 detection of odo(u)rs
气味绝缘 odo(u)r barrier
气味控制 odo(u)r control
气味廓线分析 flavor profile analysis
气味密封 odo(u)r-tight
气味浓度 odo(u)r concentration; odo(u)rousness
气味屏障 odo(u)r barrier
气味强度 odo(u)r intensity; odo(u)r strength
气味强度法 odo(u)r intensity method
气味散发 odo(u)r emission
气味试验 odo(u)r test
气味试验袋 odo(u)r test bag
气味外泄 smelling out
气味物质 odo(u)r generation
气味吸收 odo(u)r absorption
气味限度 odo(u)r threshold
气味抑制 odo(u)r suppression
气味影响 effect of odo(u)rs
气味阈值 threshold odo(u)r
气味指数 odo(u)r index
气味组分 odo(u)r-producing component
气温 air temperature; atmospheric temperature
气温表 air thermometer
气温差 temperature difference
气温垂直递减率 adiabatic lapse rate of air temperature; vertical lapse rate of air temperature
气温倒布 atmospheric temperature inversion
气温递减率 temperature lapse rate
气温递减时期 temperature lapse period
气温改正 temperature correction
气温高度图 temperature altitude chart
气温计 air thermometer; free air temperature ga(u)ge
气温检测 air temperature measurement
气温距平 temperature departure
气温控制 climatic control
气温老化 weather ag(e)ing
气温逆增 air temperature inversion; atmospheric temperature inversion; temperature inversion
气温逆转 atmospheric temperature inversion
气温年变幅 annual range of temperature
气温年变化 annual variation of temperature
气温年较差 annual temperature range
气温前沿坡度 front slope of air temperature
气温日较差 daily temperature range
气温日循环 daily air temperature cycle
气温升高 climate warming
气温梯度 air temperature gradient
气温调节舱 climatized cabin
气温调节装置 weather adjustment device
气温突变 break-in temperature
气温突降 sudden drop in temperature
气温限制调节机构 air temperature limit governing mechanism
气温月平均等值线 isomenal
气温直减率 lapse rate of air temperature
气温周期 air temperature cycle
气文学 aerography
气涡 gas eddy

Q

气涡轮机 gas turbine
气涡轮机船 gas turbine ship
气涡轮机循环 Brayton cycle
气涡现象 cavitation
气涡旋 gas eddy
气窝 air lock; air pocket; gas pocket; pocket
气窝现象 cavitation
气污 gas checking
气雾 aerial fog; gas curtain
气雾捕集器 impingement
气雾剂 aerosol
气雾剂的挥发剂 aerosol propellant
气雾剂农药 aerosol pesticide
气雾喷漆 aerosol; aerosol lacquer
气雾(喷涂)涂料 aerosol coating
气雾器 reductionizer
气雾室 aerosol chamber; cloud chamber
气雾推进剂 aerosol propellant
气吸 aspiration
气吸的 suction pneumatic
气吸管道 aspirator
气吸机 pneumatic elevator; pneumatic suction; pneumatic unloader
气吸式风力清选机 aspirator cleaner
气吸式谷物输送器 aspirated-air grain conveyer[conveyor]
气吸式捡拾机 suction machine
气吸式捡拾器 suction pickup
气吸式输送器 suction pneumatic conveyer[conveyor]
气吸提升机 pneumatic elevator
气析微粒分析 micromerograph analysis
气息 breath; odo(u)r; smelling; tang; tinge
气息分析小组 <巡查酒醉驾驶员小组或小队> breath analysis squad
气洗 air purge; air wash; gas washing
气洗涤塔 gas wash tower
气洗供气管 air-scour supply pipe
气系统 gas distributed system
气隙 air-borne gap; air gap; air interstice; air space
气隙安匝 air-gap ampere turns
气隙比 air-space ratio; air-void ratio
气隙避雷器 air-gap lightning arrestor
气隙变压器 air-gap transformer
气隙不对称保护(装置) air-gap asymmetry protection
气隙测量装置 air-gap measuring device
气隙长度 gas length
气隙场 gap field
气隙磁场 air-gap field
气隙磁场方程 airfield equation
气隙磁导率 air-gap permeance
气隙磁动势 air-gap emf
气隙磁感应 air-gap induction
气隙磁化线 air-gap line
气隙磁密度 air-gap induction
气隙磁通 air-gap(magnetic) flux; gap flux; mutual flux
气隙磁通量密度 air-gap flux density
气隙磁阻 air-gap reluctance; gap reactance
气隙磁阻电动机 air-gap reluctance motor
气隙的磁感应密度 air-gap magnetic induction density
气隙电感线圈 air-gap inductance
气隙电压 air-gap voltage
气隙法 air-gap method
气隙放电器 gap arrester
气隙规 air-gap ga(u)ge
气隙击穿 air-gap breakdown
气隙间距 air-gap separation
气隙截面 cross-section of air-gap
气隙空间 inter-air space
气隙雷管 gas exploder
气隙量规 air-gap ga(u)ge
气隙漏磁 air-gap leakage
气隙漏磁通 air-gap leakage flux
气隙扭力计 air-gap torsion meter
气隙耦合器 kidney joint
气隙铁心线圈电感 aero-ferric inductance
气隙通量 air-gap flux; gap flux
气隙系数 Carter's coefficient
气隙圆筒 air-gap cylinder
气隙整流子 air-gap commutator
气隙直径 air-gap diameter
气显示 natural gas show
气相 gas(eous) phase; vapo(u)r phase
气相包体压力测定 measurement of gas phase inclusion pressure
气相比率 the percentage of gas phase
气相操作 vapo(u)r phase operation
气相测压法 gas content barometry
气相层析 gas chromatography
气相层析电子俘获检测法 gas chromatography with electron capture detection
气相层析火焰电离检测法 gas chromatography flame ionization detection
气相层析火焰光度检测法 gas chromatography flame photometric detection
气相层析火焰热离子检测器 gas chromatography flame thermionic detector
气相层析检测器 gas chromatographic(al) detector
气相层析谱图 gas chromatogram
气相掺杂技术 gas doping technique
气相沉淀 vapo(u)r deposition
气相沉积法 vapo(u)r deposition process
气相沉降 gas-phase sedimentation
气相处理 vapo(u)r phase treatment
气相传质 mass transfer in gas phase
气相纯化 vapo(u)r phase purification
气相催化氧化法 gas-phase catalytic oxidation process
气相等温线 gas-phase isotherms
气相滴定 gas-phase titration
气相缔合 vapo(u)r phase association
气相电量分析法 gas-phase coulometry
气相电子吸附 gas-phase electron adsorption
气相淀积 vapo(u)r phase deposition
气相淀积法 vapo(u)r deposition
气相法白炭黑 cabosil; fumed silica
气相反应 gas-phase reaction; vapo(u)r phase reaction
气相反应器 vapo(u)r phase reactor
气相防锈性 volatile inhibition
气相放射性分析 gas-phase radioassay
气相分离 gas-phase separation
气相分离计 vapo(u)r fractometer
气相分配层析 gas partition chromatography
气相分配色谱法 gas partition chromatography
气相辐射色谱仪 gaseous phase radiochromatograph
气相腐蚀 gaseous corrosion; vapo(u)r corrosiveness
气相干扰 vapo(u)r phase interference
气相固相相互作用 gas-solid interaction
气相过程 vapo(u)r phase process
气相合成金刚石 vapo(u)r phase synthetic(al) diamond
气相缓蚀剂 vapo(u)r phase inhibitor
气相回流管 vapo(u)r return
气相混合物 vapo(u)r phase mixture
气相活度 gas-phase activity
气相活度系数 gas-phase activity coefficient
气相胶 vapo(u)r phase gum
气相精炼 vapo(u)r phase refining
气相聚合 vapo(u)r phase polymerization
气相聚合法 gaseous polymerization
气相颗粒活性炭 vapo(u)r phase granular activated carbon
气相空间 <储罐液面上面的> ullage space
气相扩散 vapo(u)r phase diffusion
气相扩散渗镀 gas phase diffusion plating
气相扩散运移模式 migration model of gas phase diffusion
气相冷却 ebullition cooling
气相裂化 gas cracking
气相裂化汽油 vapo(u)r phase cracked gasoline
气相氯化 gas-phase chlorination
气相面积 the area of liquid phase
气相(膜)防锈剂 vapo(u)r phase inhibitor
气相摩尔分数 gas-phase mole fraction
气相浓度 gas-phase concentration
气相漂白 gas bleaching
气相频谱 vapo(u)r phase spectrum
气相氢化物原子系数分光光度法 gas hydride atomic absorption spectrophotometry
气相区 gas-phase region
气相燃烧 gas-phase combustion
气相染料激光器 vapo(u)r phase dye laser
气相染色 vapo(u)r phase dy(e)ing
气相热量滴定 gas-phase thermometric titration
气相溶液运移模式 migration model of gas phase solution
气相润滑 vapo(u)r phase lubrication
气相色层(分离)法 gas chromatography
气相色谱 gas-phase chromatograph
气相色谱层析法 gas chromatographic(al) method
气相色谱纯度 gas chromatographic(al) purity
气相色谱电子俘获检测法 gas chromatography with electron capture detection
气相色谱法 gas chromatography
气相色谱法负化学电离质谱法联用 gas chromatography-negative chemical ionization mass spectrometry
气相色谱法傅立叶红外光谱联用 coupling gas chromatography to Fourier transform infrared spectroscopy
气相色谱法质谱法单离子检测器联用 gas chromatography-mass spectrometry-singly ion detector
气相色谱法质谱法离子选择监测检测器联用 gas chromatography-mass spectrometry-selected ion monitor detector
气相色谱法质谱法联用 coupling gas chromatography to mass spectroscopy; gas chromatography-mass spectrometry
气相色谱法紫外光谱联用 gas chromatography-ultraviolet spectrometry
气相色谱分离法 vapo(u)r phase chromatography
气相色谱分析法 gas chromatographic(al) analysis; gas chromatography
气相色谱傅立叶红外光谱联机 gas chromatograph coupled with Fourier transform infrared spectrometer
气相色谱红外技术 gas chromatography-infrared technique
气相色谱化学电离质谱仪 gas chromatographic(al) chemical ionization detection
气相色谱火焰电离检测法 gas chromatographic(al) flame ionization detection
气相色谱火焰光度检测法 gas chromatography flame photometric detection
气相色谱火焰热离子检测法 gas chromatography with a flame thermionic detector
气相色谱技术 gas chromatographic(al) technique
气相色谱检测器 gas chromatographic(al) detector
气相色谱四极质谱仪 gas chromatograph-quad pole mass spectrometer
气相色谱探测器 gas chromatographic(al) detector
气相色谱图 gas chromatogram
气相色谱仪 gas chromatograph
气相色谱仪傅立叶变换红外光谱仪联用 gas chromatograph-Fourier transform infrared spectrometer
气相色谱仪质谱图数据系统 gas chromatograph-mass spectrogram data system
气相色谱仪质谱仪单离子检测器联用 gas chromatograph-mass spectrometer-singly ion detector
气相色谱仪质谱仪计算机联用 gas chromatograph-mass spectrometer-computer
气相色谱仪质谱仪联用 gas chromatograph-mass spectrometer
气相色谱质谱(分析)法 gas chromatography-mass spectrometry
气相色谱质谱联用仪 gas chromatograph-mass spectrometer
气相色谱质谱仪装置 gas chromatograph-mass spectrometer system
气相色谱柱 gas chromatographic(al) column
气相烧结 vapo(u)r phase sintering
气相渗透 gas-phase permeation
气相体积 the volume of gas phase
气相温度滴定法 gas-phase thermometric titration
气相吸附 gas-phase adsorption
气相吸附层析 gas adsorption chromatography
气相吸附色谱法 gas adsorption chromatography
气相系统 gaseous system; vapo(u)r phase system
气相硝化 vapo(u)r phase nitration
气相悬浮法 gas-phase suspension process
气相悬浮过程 gas-phase suspension process
气相循环冷却 vapo(u)r cycle cooling
气相循环制冷 vapo(u)r cycle cooling
气相压降 gas-phase pressure drop
气相氧化 gas-phase oxidation; vapo(u)r phase oxidation
气相氧化法 vapo(u)r phase oxidation process
气相抑制剂 vapo(u)r phase inhibitor
气相抑制纸 vapo(u)r phase inhibitor paper
气相易逸性 vapo(u)r phase fugacity
气相质量浓度 gas-phase mass concentration
气相重排 vapo(u)r phase rearrange-

ment
气相转移模型 gas transfer model
气相阻化剂 vapo(u)r phase inhibitor
气相组分 gas-phase concentration
气箱 air box
气箱式水击吸收器 air chamber water shock absorber
气箱式调压塔 air chamber type of surge tank
气象 meteorologic(al) phenomenon; weather
气象报告 meteorologic(al) report; meteorologic(al) summary; weather report; weather summary
气象变化 change of weather
气象标图接收机 weather plotter receiver
气象标准视距 meteorologic(al) standard range
气象表 meteorologic(al) table
气象参数 meteorologic(al) parameter
气象操纵员 weather operator
气象测量 meteorologic(al) survey
气象测量仪表 meteorologic(al) instrumentation
气象潮 <气压变化所引起的潮升变化> meteorologic(al) tide; surge; weather(-going) tide; wind set-up; wind tide
气象潮汐 meteorologic(al) tide
气象潮汐表 meteorologic(al) tide sheet; meteorologic(al) tide table
气象赤道 meteorologic(al) equator
气象穿越 weather penetration
气象传送 meteorologic(al) transmission
气象传真 weather facsimile
气象传真记录器 weather facsimile recorder
气象传真通信[讯]网 weather facsimile network
气象船 meteorologic(al) ship
气象代表性错误 meteorologic(al) representative error
气象的 meteorologic(al)
气象的极端情况 meteorologic(al) extreme
气象低压区 meteorologic(al) depression
气象地图 weather map
气象电报 meteorologic(al) message; meteorologic(al) telegraph; weather message
气象电传打字通信[讯]网 meteorologic(al) teleprinter network
气象电传通信[讯]网 meteorologic(al) telecommunication network
气象电传与传真系统 weather teletype and facsimile system
气象电码 meteorologic(al) code
气象电视系统 weather television system
气象电文 weather message
气象动力学 meteorologic(al) dynamics
气象队 weather service
气象方法预报地震 meteorologic(al) earthquake prediction
气象飞行 weather flight
气象分潮 meteorologic(al) constituent
气象分类 weather classification
气象分析 meteorologic(al) analysis
气象分析中心 weather analysis center[centre]
气象分析作业图 meteorologic(al) analysis working chart
气象风暴 meteorologic(al) storm
气象风洞 meteorologic(al) wind tunnel
气象风球 meteorologic(al) ball
气象服务 meteorologic(al) service; weather service
气象服务站 meteorologic(al) service station; weather service station
气象浮标 meteorologic(al) buoy; weather buoy
气象符号 meteorologic(al) sign; meteorologic(al) symbol; weather sign; weather symbol
气象辅助业务 meteorologic(al) aids service
气象改正量 meteorologic(al) correction
气象改正数 meteorologic(al) correction
气象工作者 meteorologist
气象公报 meteorologic(al) bulletin
气象观测 meteorologic(al) observation; weather observation
气象观测船 meteorologic(al) observation ship; weather observation ship; weather ship
气象观测飞机 meteorologic(al) aircraft
气象观测飞行 weather observation flight
气象观测台 meteorologic(al) observation station; meteorologic(al) observatory
气象观测员 meteorologic(al) observer; weather observer
气象观测站 meteorologic(al) station
气象观测站网 meteorologic(al) observation network
气象观察 weather observation
气象光学 meteorologic(al) optics
气象光学距离 meteorologic(al) optic(al) range
气象光学视距 meteorologic(al) optic(al) range
气象广播 meteorologic(al) broadcast
气象航线 meteorologic(al) route
气象环境背景调查 background survey on meteorologic(al) environment
气象回避 weather avoidance
气象回避雷达 weather avoidance radar
气象火箭 meteorologic(al) rocket
气象火箭网 meteorologic(al) rocket network
气象基本要素 meteorologic(al) element
气象基准平面 meteorologic(al) datum plane
气象及海洋考察船 aerologic(al) and oceanographic(al) research ship
气象极小值 meteorologic(al) minimum
气象计 meteorograph; weatherometer
气象计算 meteorologic(al) calculus
气象计算机 weather computer
气象记录 meteorologic(al) record
气象记录簿 weather log
气象记录器 meteorograph
气象记录曲线 meteorogram
气象记录图 meteorogram
气象记录仪 aerograph; meteorograph
气象监测网 meteorologic(al) network
气象监视台 meteorologic(al) watch office
气象检测器 weather detector
气象简语 weather briefing
气象景观 meteorologic(al) diversity scenery
气象警报 meteorologic(al) warning; weathering warning
气象警告 meteorologic(al) warning; weathering warning
气象局 Bureau of Weather Reports; Meteorologic(al) Bureau; Weather Bureau
气象控制 weather control
气象控制电报 weather controlled message
气象雷达 meteorologic(al) radar; weather radar
气象雷达示波器 parscope
气象雷达天线 weather radar scanner
气象雷达图 weather radar map
气象雷达系统 weather radar system
气象雷达显示 weather radar display
气象雷达站 meteorologic(al) radar station; thunderhead
气象类型 weather category
气象模拟 meteorologic(al) simulation
气象模型 meteorologic(al) model
气象能见度 meteorologic(al) visibility
气象年鉴 meteorologic(al) yearbook
气象旗号 meteorologic(al) flag; weather flag
气象气球 radiosonde
气象气压计 weather bar
气象情报 meteorologic(al) information; weather information; weather intelligence
气象情报接收网 meteorologic(al) information network and display; weather information network and display
气象区(域) meteorologic(al) district; meteorologic(al) region
气象日记簿 meteorologic(al) log book
气象色散 meteorologic(al) dispersion
气象哨 meteorologic(al) post
气象声学 meteorologic(al) acoustics
气象湿度 meteorologic(al) humidity
气象视程 meteorologic(al) visual range
气象视程表 meteorologic(al) visual range table
气象视距 meteorologic(al) visual range
气象试验 meteorologic(al) experiment
气象数据 air data; meteorologic(al) data; weather data
气象数据库 weather database
气象水 meteoric water; meteorologic(al) water
气象水文数据 meteorology and hydrologic(al) data
气象塔 meteorologic(al) tower; observing tower
气象塔层 meteorologic(al) tower layer
气象台 meteorologic(al) observatory; observation station; observatory; observing station; weather observatory; weather station
气象台站网 meteorologic(al) network
气象探测 meteorologic(al) sounding
气象探测火箭 meteorologic(al) rocket; rocket sonde
气象探测雷达 weather search radar
气象探测气球 meteorologic(al) balloon
气象特点 meteorologic(al) characteristic
气象条件 meteorologic(al) condition; weather condition
气象通报 meteorologic(al) message; weather advisory; weather bulletin
气象通信[讯]中心 weather communication center[centre]
气象图 aerographic(al) chart; aerography; climograph; meteorologic(al) chart; synoptic(al) weather chart; weathering map
气象图表 aerography
气象图传真机 weather chart facsimile apparatus; weather map facsimile equipment
气象图解 meteorogram
气象网 network of stations
气象卫星 meteorologic(al) satellite; meteosat; weather orbiter; weather satellite
气象卫星联合咨询委员会 Joint Meteorologic(al) Satellite Advisory Committee
气象卫星数据 weather satellite data
气象卫星双信道辐射计 meteosat dual-channel radiometer
气象卫星图像 meteorologic(al) satellite image
气象无线电报 meteorologic(al) radio telegram
气象无线电传播联合分委员会 Joint Meteorologic(al) Radio Propagation Subcommittee
气象系统 weather system
气象现象 meteorologic(al) phenomenon[复 phenomena]
气象效应 meteorologic(al) effect
气象信号 meteorologic(al) signal; weather signal
气象信息 meteorologic(al) information
气象信息服务 weather information service
气象形势 meteorologic(al) situation
气象选线 meteorologic(al) routing; weather routing
气象学 climatology; meteorology; synoptic(al) meteorology
气象学和气候专门应用委员会 Commission for Special Applications of meteorology and Climatology
气象学会 meteorologic(al) society
气象学家 aerologist; meteorologist
气象学应用计划 applications of meteorology program(me)
气象延迟系数 weather delay factor
气象谚语 meteorologic(al) proverbs
气象遥测仪 telemeteorograph
气象要素 meteorologic(al) element
气象要素变化过程曲线 curve of meteorologic(al) element change
气象要素图 meteorologic(al) chart; meteorologic(al) elements chart; meteorologic(al) map; weather chart
气象要素预报 prediction of various weather constituents
气象要素直方图 histogram of meteorologic(al) element
气象业务 meteorologic(al) service; weather service
气象仪(器) meteorologic(al) instrument; weather instrument
气象仪器设备 meteorologic(al) instrumentation
气象异常 meteorologic(al) anomaly
气象因素 meteorologic(al) factor
气象因子 meteorologic(al) factor
气象影响 meteorologic(al) effect; meteorologic(al) influence
气象预报 weather forecast(ing); weather prognosis
气象预报广播 weather service
气象预报间隔时间 forecast period
气象预报信号 weather flag; weather signal
气象预报站 weather forecasting center[centre]
气象预测 weather prognosis

气象员 aerographer; weatherman
气象灾害 meteorologic(al) calamity; meteorologic(al) disaster
气象噪声 meteorologic(al) noise
气象站 climatological station; meteorologic(al) office; meteorologic(al) station; weather facility; weather service; weather station
气象站高程 climatologic(al) station elevation
气象侦察 weather reconnaissance
气象侦察报告格式 weather reconnaissance code
气象侦察船 weather patrol ship
气象侦察飞行 meteorologic(al) reconnaissance flight; weather patrol; weather track
气象侦察飞行任务 weather-scouting mission
气象侦察机 weather reconnaissance aircraft
气象侦察任务 weather reconnaissance mission
气象侦察设备 weather reconnaissance system
气象指示器 weather indicator
气象中心 forecasting center[centre]; weather center[centre]
气象重现循环时间 climatic cycle
气象状况 meteorological condition; weather condition
气象状态 meteorologic(al) condition; meteorology; weather condition
气象资料 meteorologic(al) data; meteorologic(al) information; weather data
气象资料系列 meteorologic(al) data series
气象资料中心 meteorologic(al) data center[centre]
气象资料自动编辑中继系统 automatic data editing and switching system
气象自动报告站 meteorologic(al) automatic reporting station
气象自记仪 meteorograph
气象总局 Main Meteorologic(al) Office
气象组织本底空气污染监测网 background air pollution monitoring network
气象组织臭氧计划 meteorologic(al) organization ozone project
气消化石灰 air slaked lime
气心变压器 air transformer
气芯 gaseous core
气芯堆 gas core reactor
气胸 pneumothorax
气悬监视器 aerosol monitor
气旋 air whirl; cyclone; gas eddy
气旋暴风雨 cyclonic storm
气旋波 cyclonic wave; cyclotron wave; cyclone wave
气旋除沉器 cyclone dust separator
气旋的 cyclonic
气旋发生 cyclogenesis
气旋分离器 cyclone separator
气旋风 cyclonic wind
气旋风暴降水 cyclonic precipitation
气旋过滤器 cyclone filter
气旋集沉器 cyclone collector
气旋降水 cyclonic precipitation
气旋雷暴 cyclonic thunderstorm
气旋炉 cyclone furnace
气旋路径 cyclone path; cyclone track; track of a cyclone
气旋模式 cyclonic mode
气旋模型 cyclonic model
气旋前锋 cyclone front; cyclonic front
气旋曲度 cyclonic curvature
气旋塞 air faucet; air tap
气旋生成作用 cyclogenesis
气旋式环流 cyclonic circulation
气旋式集尘器 cyclonic collector
气旋通路 cyclonic path
气旋式运动 cyclonic motion
气旋涡 air pocket
气旋涡度 cyclonic vorticity
气旋涡流 cyclonic swirl
气旋消除 cyclolysis
气旋消失 cyclolysis
气旋形式 cyclone type
气旋性暴雨 cyclonic storm
气旋性风暴 cyclonic storm
气旋性风暴潮 cyclonic storm surge
气旋性环流 cyclonic swirl
气旋性降雨 cyclone precipitation; cyclone rain
气旋性涡度 cyclonic vorticity
气旋性雨 cyclonic rain
气旋眼 eye of cyclone
气旋移动路径方向 cyclone course
气旋中心 center[centre] of cyclone; cyclonic center[centre]
气旋中心指示盘 cyclonocope; cyclonograph
气旋周环 pericyclone; pericyclonic ring
气旋族 cyclonic
气旋作用 cyclogenesis
气穴 air cavity; air pocket; air sink; air void; cavitation; pocket of air
气穴曝气池 cavitator
气穴侵蚀 cavitation attack
气穴损失 cavitation(al) loss
气穴系统 system of air voids
气穴现象 cavitation
气穴噪声 cavitation noise
气压 air-pressure; atmospheric pressure; barometric(al) pressure; line pressure; pneumatic pressure; vapo(u)r tension
气压扳手 impact spanner
气压保持器 holding chamber
气压报警计 alarm manometer
气压泵 air lifter; air lift pump; pneumatic pump
气压变动 barometric(al) fluctuation
气压变化 air-pressure change; barometric(al) change; barometric(al) variation
气压变化图 pressure change chart; pressure tendency chart
气压变量表 pressure variometer
气压变量计 pressure variograph
气压表 air ga(u)ge; air-pressure ga(u)ge; dial barometer; manometer; pressure ga(u)ge; rain glass; vapo(u)rimeter; weather glass; storm glass; barometer; barometer table <指表格>; weather ga(u)ge <测大气用>; gas pressure ga(u)ge; gas pressure meter
气压表测高 level(l)ing by means of barometer
气压表的 barometric(al)
气压表的振荡 pumping of the barometer
气压表高度 height of barometer
气压表管 barometric(al) leg
气压表真高度 true height of the barometer
气压病 aeroembolism; air bends; bends; dysbarism
气压波 barographic wave; barometric(al) wave; pressure wave; surge
气压波线 surge line
气压补偿 barometric(al) compensation
气压补偿器 barostat
气压步级定位器 pneumatic step positioner
气压步级位置控制器 pneumatic step positioner
气压参数 pneumatic parameter
气压舱沉井 pneumatic shaft sinking
气压舱检查法 pressure cabin examination
气压操纵 air-operated control
气压操纵系统 air control system
气压槽 pressure trough
气压测定 barometric(al) determination
气压测定法 barometry
气压测定器 baroscope
气压测高 barometric(al) determination of altitude; barometric(al) determination; barometric(al) level(l)ing
气压测高表 pressure altimeter
气压测高常数 barometer constant
气压测高导线 barometric(al) traverse
气压测高点 barometric(al) level(l)ing point
气压测高法 barometric(al) altimetry; barometric(al) hypsometry; barometric(al) method; barometric(al) height measurement
气压测高公式 barometric(al)(heighting) formula
气压测高计 altigraph; aneroid altimeter; aneroid barometer; barometric(al) altimeter; pressure altimeter
气压测高仪 altimeter; barometer; barometric(al) altimeter
气压测高站 barometric(al) station
气压测孔法 bubble pressure method
气压测深器 air micrometer; baroscope; pneumacator
气压测深仪 pneumatic sounder; teledepth
气压测温计 barothermograph
气压测验器 baroscope
气压层 barosphere
气压差 draught head; pressure difference
气压常数 vapo(u)r pressure constant
气压场 field of pressure
气压潮 atmospheric pressure tide
气压沉箱 air caisson; air lock caisson; compressed-air caisson; pneumatic caisson
气压沉箱病 caisson disease; compressed-air disease
气压沉箱法 pneumatic caisson method; pneumatic caisson process
气压沉箱工程 pneumatic work
气压沉箱基础 compressed-air caisson foundation; compressed-air foundation; pneumatic caisson foundation
气压沉箱密封顶板 air deck
气压沉箱气闸 airlock in caisson
气压沉箱桥墩 pneumatic caisson pier
气压沉箱施工 foundation work under compressed air
气压沉箱式岸壁 pneumatic caisson quay wall
气压沉箱下沉 air sinking; pneumatic sinking
气压沉箱作业室 air working chamber; working chamber of pneumatic caisson
气压成型 gas pressure compacting; pneumatic forming
气压冲洗水箱 pneumatic flush tank
气压抽水泵 pulsometer pump
气压抽水机 air lift; pulsometer
气压抽水挖掘机 pulsometer dredger
气压除尘 pneumatic dedusting
气压穿孔机 pneumatic perforator
气压传动 air drive; pneumatic transmission
气压传动的 air driven
气压传动装置 pneumatic actuator; shuttle
气压传感的 pneumatically sensed
气压传感器 barocepter[baroceptor]; barometric(al) pressure sensor
气压传送机 barocepter[baroceptor]; pneumatic conveyer[conveyor]
气压传送器 air-slide conveyer[conveyor]
气压传送装置 pneumatic shuttle; shuttle
气压锤 pneumatic hammer
气压打孔机 pneumatic perforator
气压打桩 compressed-air sinking; pneumatic piling
气压带 zone of pressure
气压单位 barometric(al) unit; pressure unit; barometric(al) millimeter [millimetre] of mercury <以毫米水银柱计>
气压导布装置 pneumatic cloth feeder
气压导纳 barometric admittance
气压的 barometric(al); pneumatic
气压的日际变化 interdiurnal pressure variation
气压等变线 allobar
气压等高线 pressure contour
气压等压图 isobaric chart
气压电动装置 electropneumatic actuator
气压电缆 gas pressure cable
气压电阻器 baroresistor
气压顶铆器 air dolly
气压顶升机 air jack
气压顶升器 air jack; pneumatic jack
气压订正 barometric(al) reduction; pressure reduction
气压读数 barometer reading; barometric(al) reading
气压堵缝 pneumatic calking
气压盾构 air-pressure shield; compressed-air shield; pneumatic shield; pneumatic shield with compressed air; shield with air pressure
气压盾构的跑气事故 blow in air compressed shield
气压盾构法掘进 air shield driving
气压盾构法开挖 air shield driving
气压盾构施工法 compressed air shield method
气压盾构推进 pneumatic shield driving
气压发动机 air motor; air-pressure engine; compressed-air engine; air compressing engine
气压阀 air-pressure valve; pneumatic valve; pressure lock; pressure-operated valve
气压法 <隧道施工> compressed-air method; pneumatic process; pneumatic sealing method
气压方程 barometric(al) equation
气压防波堤 pneumatic barrier
气压非周期变动 barometric(al) nonperiodic(al) changes
气压风暴表【气】 barocyclo(no)meter
气压风暴计 barocyclonometer
气压负荷 air-load
气压感受器 air-speed head
气压感应器 baroreceptor
气压缸 pneumatic cylinder; servocylinder
气压高程 barometric(al) altitude;

barometric(al) elevation
气压高程表 sensitive altimeter
气压高程测量 atmospheric-pressure level;barometric (al) height level-(l) ing; barometric (al) height measurement;barometric(al) level-(l)ing
气压高程计 barometric(al) altimeter; barometric(al) level
气压高程仪 orometer
气压高度 barometer height; barometric (al) height; pressure altitude; pressure height
气压高度变差 pressure-altitude variation
气压高度表 barometric(al) altimeter; barometric (al) level; pressure-altitude indicator
气压高度传感器 pressure-altitude sensor
气压高度公式 barometric (al) height formula
气压高度计 barometric(al) altimeter; barometric(al) level
气压高度曲线 pressure-height curve
气压给水 pneumatic water supply
气压给水设备 pneumatic water supply equipment; pneumatic water supply installation
气压给水系统 air-pressure water supply system; pneumatic water supply system
气压给水箱 pneumatic water supply tank
气压给水装置 pneumatic water supply installation; pressurized type water supply unit
气压工程 pneumatic work
气压工具 pneumatic tool
气压供给系统 gas pressure feed system
气压供水系统 pneumatic water supply system
气压供油 gas-pressurized
气压谷 col
气压固结 gas pressure consolidation
气压管车辆检测器 pneumatic tube vehicle detector
气压管道 air-pressure duct
气压管路 pneumatic circuit
气压管式火灾报警系统 pneumatic tube fire alarm system
气压灌浆泵 boogie pump
气压罐 barometric (al) tank; pneumatic pressure tank; pneumatic tank
气压归正 barometric (al) reduction; pressure reduction
气压轨枕夯具 pneumatic (tie-) tamper
气压辊 air-loaded roller
气压过渡舱 air lock
气压焊机 gas pressure welding machine
气压焊(接) gas press (ure) weld-(ing); pressure gas welding; gas shielded welding
气压焊接机 gas pressure welding machine
气压夯锤 pneumatic ram
气压毫巴 <气压单位> baromil
气压回转仪 barogyroscope
气压混凝土浇灌机 pneumatic concrete placer
气压活塞 pneumatic piston
气压活塞传动装置 pneumatic piston actuator
气压机 aerostatic press; air compressor; air placer; air placing machine
气压唧筒 pulsometer
气压基础下沉工作 pneumatic sinking
气压计 air ga(u)ge; altimeter; atmosphere ga(u)ge; barograph; barometer; barometrograph; barostat; pressure ga(u)ge; pressure meter; vapo-(u)rimeter; weather ga(u)ge; weather glass
气压计杯 barometer cistern
气压计标度 barometric(al) scale
气压计标度因素 barometric(al) scale factor
气压计测高程 barometric (al) level-(l)ing
气压计的 baric; barometric(al)
气压计读数 pressure reading
气压计读数改正 correction for aneroid readings
气压计飞行高度 flying altitude of barometer
气压计高差仪 barometer statoscope; barometric(al) statoscope
气压计高程 barometric(al) height
气压计高程控制 aneroid height control
气压计高程控制点 aneroid height control point; controlled barometer height
气压计高度 barometer altitude; barometric(al) height
气压计观测 barometer observation; barometric(al) observation
气压计管 barometric(al) tube
气压计盒 aneroid chamber
气压计继电器 barometric(al) relay
气压计刻度 barometer scale
气压计膜盒 aneroid chamber
气压计水银槽 barometer cistern
气压计误差订正滑尺 gold slide (scale)
气压计液压指示器 manometer liquid level
气压计用管 barometer tube
气压记录 barometric(al) recording
气压记录器 barograph; barometric-(al) recorder; barometrograph; pressure gradient; pressure-graph; stormograph
气压记录图 barogram
气压继电器 air relay; barometric(al) pressure switch; pneumatic relay
气压加载系统 pneumatic loading system
气压夹具 air clamp; air jig
气压夹头 compressed-air chuck
气压监警设备 gas pressure supervision alarm system
气压减压器 air reducer
气压减振 pneumatic cushioning
气压减振器 pneumatic bumper
气压减震器 pneumatic shock absorber
气压减震柱 pneumatic shock absorber strut
气压浇灌 pneumatic placing
气压浇灌混凝土 air placed concrete
气压浇灌机 compressed-air placer
气压浇灌砂浆 pneumatically placed mortar
气压浇搅拌池 pneumatically placed mixing cell
气压浇注 pressure pouring
气压浇注混凝土 air placed concrete
气压浇注炉 press(ure) pouring furnace
气压浇注砂浆 pneumatically placed mortar
气压浇筑 pneumatic placing
气压浇筑混凝土 air placed concrete; pneumatic placed concrete
气压浇筑砂浆 pneumatically placed mortar
气压胶结 gas pressure bonding
气压卷扬机 pneumatic hoist
气压开关 baroswitch; pneumatic switch
气压控制 air control; barometric(al) control; pneumatic control
气压控制继电器 pneumatic control relay
气压控制器 gas pressure regulator
气压控制系统 pneumatic control system
气压快门开关 air-pressure release
气压扩孔机 air reamer
气压扩张机 air reamer
气压力输送 pneumatic conveying
气压联动装置 pneumatic linkage
气压量仪 air ga(u)ge
气压流 baric flow
气压滤池 pneumatic filter
气压轮胎 pneumatic tire[tyre]
气压脉冲 pulse of air
气压铆锤 air riveter
气压铆钉 pneumatic riveting
气压铆钉锤 pneumatic rivet (ing) hammer
气压铆钉机 air riveter; air riveting hammer; pneumatic rivet (er) machine; air jam
气压铆钉器 pneumatic riveter
气压铆钉枪 air riveter; pneumatic riveting gun
气压铆接 pneumatic riveting
气压密封 pressurized seal
气压描记器 baro(metro) graph
气压敏感元件 barocepter[baroceptor]
气压黏[粘]合 gas pressure bonding
气压排(除)污水法 pneumatic system of sewerage
气压排(除)污水系统 pneumatic system of sewerage
气压排水系统 air-pressure drainage system; pneumatic sewage system
气压排液管 barometric(al) leg; barometric(al) pipe; leg
气压喷浆 pneumatically applied mortar; pneumatic mortar
气压喷漆器 air brush; pneumatic paint brush
气压喷漆刷 pneumatic paint brush
气压喷枪 air lance
气压喷砂浆 pneumatic mortar
气压喷砂器 air sand blower
气压喷射 air blast; pneumatic fed <混凝土>
气压喷射泵 pneumatic ejector
气压喷射口冷凝器 barometric(al) jet condenser
气压喷射器 compressed-air ejector; pneumatic ejector
气压喷涂 air-pressure spray
气压喷雾 air blast atomizing
气压喷雾器 air blast atomizer
气压喷刷色器 air brush
气压喷油 air injection
气压坡度 <气压等压线之间的垂直距离> pressure gradient
气压曝气 pneumatic aeration
气压起重机 compressed-air crane
气压起重器 air jack
气压气温自记计 barothermograph
气压千斤顶 air jack; pneumatic jack
气压嵌缝 pneumatic calking
气压强 vapo(u)r pressure
气压强度及气密试验 air-pressure test for strength and tightness
气压倾向 barometric (al) tendency; pressure tendency
气压情况 barometric (al) information; barometric(al) condition
气压区 baric area; barometric(al) area
气压驱动 air-pressure drive; gas pressure drive
气压趋势 barometric (al) tendency; pressure tendency
气压圈 climbing curve
气压燃料调节器 barometric(al) fuel regulator
气压扰动 barometric (al) disturbance; barometric(al) fluctuation
气压日变化 barometric (al) day change
气压日差 barometric (al) diurnal range
气压润滑法 gas pressure lubrication feed
气压润滑器 gas pressure lubricator
气压伤 barotrauma
气压上升 increase of pressure; pressure rise
气压上升区 anallobar
气压设备 pneumatic installation
气压升降 fluctuation in atmospheric-(al) pressure
气压升降机 air lift; pneumatic elevator
气压升降率 barometric(al) rate
气压生波器 <波浪水工模型用> pneumatic wave generator
气压式 vapo(u)r pressure type
气压式测压计 pneumatic piezometer
气压式沉降仪 pneumatic settlement cell
气压式储水箱 air-water storage tank
气压式防波堤 <海底埋设排气管,排出压缩空气来破浪> pneumatic breakwater
气压式化学电离质谱分光光度计 atmospheric-pressure chemical ionization mass spectrophotometer
气压式缓冲器 pneumatic cushion
气压式混凝土泵 pneumatic concrete pump
气压式检测器 pneumatic detector
气压式近炸引信 pressure proximity fuze
气压式孔隙水压力计 pneumatic piezometer; pneumatic type piezometer
气压式孔隙水压力仪 pneumatic piezometer
气压式控制装置 pneumatic controller
气压式冷凝器 barometer condenser; barometric(al) condenser
气压式起动器 gas pressure type (self) starter
气压式取土器 soil sampler of air pressure
气压式探测器 pneumatic detector
气压式调节计 pneumatic controller
气压式调速器 pneumatic governor
气压式调压室 pneumatic surge chamber
气压式通风控制 barometric(al) draft control
气压式涂料泵 pneumatic paint pump
气压式系统 pressure-type system
气压式岩芯取样器 pneumatic core sampler
气压式引信 pressure fuze
气压式制动操纵装置 pneumatic brake operation device
气压式自动控制 pneumatic automatic control
气压式自动控制系统 pneumatic automatic control system
气压试验 air-pressure test; air test; atmospheric-pressure test; gas pressure test; pneumatic test(ing); vapo(u)r pressure test
气压试验室 altichamber; altitude

Q

chamber
气压试验台 pneumatic test stand
气压试验箱 pressure box
气压室 air chamber; compression chamber
气压释放阀 pneumatic release valve
气压输冰系统 pneumatic ice delivery system
气压输泥 pneumo-slurry transport
气压输送 delivery by compressed air; pneumatic transport
气压输送管 pneumatic tube
气压输送机 pneumatic conveyer[conveyor]
气压输送浇筑机 pneumatic transport placer
气压输送器 pneumatic carrier
气压输送文件管 pneumatic document conveyer
气压输送系统 gas pressure feed system; gas-pressurized system
气压输送滞后 pneumatic transmission lag
气压输送装置 pneumatic conveyer [conveyor]
气压水罐 autopneumatic cylinder
气压水泥喷枪 air cement gun
气压水枪 aero-hydraulic gun
气压水箱 hydropneumatic storage tank; pneumatic pressure tank; pneumatic(storage) tank
气压水箱冲洗池 pneumatic tank water closet
气压水准测量 barometric(al) level(l)ing; barometric(al) surveying
气压伺服电动机 pneumatic servomotor
气压伺服系统 pneumatic servo
气压伺服装置 air-powered servo; pneumatic servo
气压隧道 compressed-air tunnel
气压隧道开凿法 compressed-air method of tunnel
气压隧道施工 compressed-air tunnel(l)ing
气压弹簧 gas spring; pneumatic spring
气压梯度 baric gradient; barometric(al) gradient; pressure gradient
气压梯度力 barometric(al) gradient force; pressure gradient force
气压提升泵 air lift pump
气压提升法 airlift method
气压提升机 air hoist; air lift; pneumatic hoist
气压提升绞车 air hoist
气压提升起重机 air hoist
气压替续器 air relay
气压填缝 pneumatic calking
气压调节阀 air-pressure regulating valve; draught regulator
气压调节杆 pneumatic nozzle link
气压调节器 gas pressure regulating governor; air governor; air-pressure regulator; barometric(al) pressure controller; barostat; gas pressure regulator
气压调节治疗室 gas pressure regulating medical chamber
气压调解门 barometric(al) damper
气压通风 ventilation by forced draft
气压痛 dysbarism
气压图 barogram; pressure chart
气压脱水机 pneumatic thickener
气压陀螺仪 barogyroscope
气压微扰动 microbarm
气压维护通信[讯]设备 gas pressure supervision alarm system
气压温度计 barothermograph
气压温度记录器 barothermograph
气压温度记录仪 barothermograph
气压温度曲线 barothermogram
气压温度湿度表 barothermohygrometer
气压温度湿度计 barothermohydrograph
气压温度湿度记录器 barothermohydrograph
气压温度湿度曲线 barothermohygrogram
气压温度图 barothermogram; emagram
气压温度自动记录仪 barothermograph
气压雾化 air atomization
气压雾化器 pneumatic atomizer
气压吸振器 pneumatic shock absorber
气压系数 barometric(al) coefficient; barometric factor
气压系统 baric system; pneumatic circuit; pressure system
气压系统操作压力 line pressure
气压系统检查 pressure check
气压匣 compression chamber
气压下沉 compressed-air sinking
气压下降 atmospheric depression; barometric(al) depression; drop in atmospheric(al) pressure
气压箱 pneumatic pressure tank; pneumatic tank
气压效率 barometric(al) efficiency
气压效应 barometric(al) efficiency
气压形势 baric topography
气压型避振器 air type shock absorber
气压型鼓膜液面控制器 diaphragm box level controller
气压型造波机 air-pressure type wave generator
气压性鼻窦炎 barosinusitis
气压性耳炎 barotraumatic otitis
气压性中耳炎 aerootitis
气压蓄压器 pneumatic accumulator
气压扬水法 airlift method
气压扬水机 aquathruster; pulsometer
气压异常 pressure anomaly
气压影响 atmosphere influence
气压壅水 barometric set-up
气压涌升 pressure jump
气压涌升线 pressure jump line
气压运输管 pneumatic tube; pressure tube
气压凿 pneumatic chisel
气压凿孔机 pneumatic perforator; pneumatic pick
气压增稠器 pneumatic thickener
气压闸 draft regulator
气压闸室 air lock
气压张胎器 pneumatic-tired [tyred] expander
气压振动器 platform vibrator; pneumatic vibrator
气压支架 compressed-air leg
气压指示计 dial barometer
气压指示器 gas pressure indicator
气压制动 pneumatic braking
气压制动器 air brake; air-pressure brake; pneumatic brake
气压中心 baric center[centre]
气压钟 pneumatic clock
气压重力仪 barometric(al) gravimeter; barometric(al) gravity meter; gas pressure gravimeter
气压柱 barometric(al) leg
气压柱冷凝器 barometric(al) leg condenser
气压柱下降 falling glass
气压转换开关 baroswitch
气压桩 pneumatic pile
气压桩锤 air pile hammer
气压装卸 pneumatic handling
气压自动记录器 barometrograph
气压自动记录仪 baro(metro)graph
气压自记器 barograph; gas register
气压自记曲线 barogram; barograph trace
气压自停卷布机 pneumatic cut-off winder
气压最低值 barometric(al) minimum
气压最高值 barometric(al) maximum
气烟焊接 gas flame welding
气烟末 gas black
气眼 air bleeder; air cavity; air drain; air gate; air hole; air pocket; air vent; gas hole; gas orifice
气眼针 vent wire
气焰煤 gas flame coal
气焰钎焊 gas flame brazing
气焰热电偶 gas flame thermopile
气焰铜焊 gas brazing
气焰硬钎焊 gas brazing
气样 air sample
气样的 aeriform
气样洗涤器 gas sample scrubber
气液包裹体 gas fluid inclusion
气液包裹体成矿温度测定法 method of ore-forming temperature determination for gas-liquid inclusion
气液比 vapo(u)r-liquid ratio
气液层析 gas-liquid chromatograph
气液成的 pneumato-hydatogenetic
气液成核 vapo(u)r-liquid nucleation
气液抽提 vapo(u)r-liquid extraction
气液法 gas-liquid system
气液反应 gas-liquid reaction
气液分离 gas-liquid separation
气液分离器 flash chamber; gas-liquid separator; surge drum; surge header; vapo(u)r-liquid separator
气液分离器内压力 trap pressure
气液分配 gas-liquid partition
气液分配层析 gas-liquid partition chromatography
气液分配色谱法 gas-liquid partition chromatography
气液分配系数 gas-liquid partition coefficient
气液固层析 gas-liquid-solid chromatography
气液固工艺 vapo(u)r-liquid-solid technique
气液固色谱法 gas-liquid-solid chromatography
气液固生长 vapo(u)r-liquid-solid growth
气液混合流 concurrent gas-liquid flow
气液混合物 liquid vapo(u)r mixture
气液混合液 gas(-and)-liquid mixture
气液交换法 gas-liquid exchange process
气液交换过程 vapo(u)r-liquid exchange process
气液交换器 gas-liquid heat exchange
气液界面 air-liquid interface; gas-liquid interface; vapo(u)r-liquid interface
气液界面面积 area of liquid-gas interface
气液面积比 gas-liquid area ratio
气液喷雾接触器 gas-liquid spray contactor
气液平衡 gas-liquid equilibrium
气液千斤顶 air hydraulic jack
气液色层分离法 gas-liquid chromatography
气液色谱法【化】 gas-liquid chromatography
气液色谱仪 gas-liquid chromatograph
气液弹性力学 aero-hydro-elasticity; gas-liquid elastomechanics
气液体积比 gas-liquid volume ratio
气液相层析 gas-liquid chromatography
气液相间反应器 gas-liquid interphase reactor
气液相色层分析 gas-liquid chromatography
气液相色谱法 gas-liquid chromatography
气液相色谱仪-质谱仪联用 gas-liquid chromatograph-mass spectrometer
气液蓄能器 gas loaded accumulator
气液压力转换缸 airdraulic actuator; air-oil actuator
气液压型联动弹簧 gas hydraulic type spring
气液增力器 air over hydraulic booster; air over hydraulic intensifier
气液制动器 air over hydraulic brake
气液制动器 air over hydraulic brake; air-oil actuator
气液柱 gas-liquid column
气衣 pressurized suit; supplied-air suit
气以吸气 sniff
气翼 aerofoil; air foil
气翼船 aerofoil boat; winged vehicle
气硬 self-hardening
气硬的 air hardened; air setting; hardened in air
气硬钢 self-hardening steel
气硬水泥 air-hardening cement
气硬性 air hardened quality; air hardening; air set(ting)
气硬性的 air hardened; air hardening; air-hardenable
气硬性胶泥 air setting
气硬性胶凝材料 air-hardening binding material
气硬性耐火胶结材料 air-hardening refractory cement
气硬性耐火泥 air-setting refractory mortar
气硬性耐火砂浆 air-setting refractory mortar
气硬性耐火水泥 air-hardening refractory cement; air-setting refractory cement
气硬性耐火水泥砂浆 air-setting refractory cement mortar
气硬性黏[粘]接材料 air-setting jointing material
气硬性砂浆 air-setting mortar
气硬性石灰 air hardening lime
气硬性水泥 air-hardening cement; air-setting cement
气硬性质 air-hardening quality
气油比 gas-oil ratio
气油藏 gas-oil pool
气油接触面 gas-oil surface
气油界面 gas-oil contact line; oil-gas boundary
气油渗透率 gas-oil permeability
气油悬浮体 air-oil suspension
气油增回路 air-oil booster circuit
气浴 aeration; gas bath
气浴法 vapo(u)r method
气浴设备 air-bath equipment
气浴装置 air-bath device; air-bath equipment; air-bath unit
气域 air space
气源 air supply
气源故障 air failure
气源扩散 gas source diffusion
气源岩 gas source rock
气源总管 air supply manifold
气晕法 gas halo method
气晕图 gas halo diagram
气载尘埃 air-borne dust
气载放射性 air-borne radioactivity

气载废物 air-borne waste
气载浮游细菌 air-borne bacteria
气载颗粒污染 air-borne particulate contamination;air-borne particulate pollution
气载颗粒污染物 air-borne particulate contaminant; air-borne particulate pollutant
气载污染 air-borne contamination
气载污染物 air-borne contaminant; air-borne pollutant
气载污染物浓度 air-borne contaminant concentration;air-borne pollutant concentration
气载噪声 air-borne noise
气錾 chipping hammer;pneumatic chisel
气錾锤 chipping hammer
气凿 air chipper;compressed-air brake; compressed-air chiseling hammer; pneumatic chipper;pneumatic chisel
气闸 air(cushion)brake;air lock;air-pressure brake; compressed-air brake;damper plate;gas lock;man lock; pneumatic brake; slay key; damp(en)er
气闸操作员 lock tender
气闸阀 air lock valve
气闸进料器 air lock feeder
气闸门 air lock door; pneumatic brake door
气闸潜水钟 air locking diving bell
气闸软管 air brake hose
气闸软管螺纹接套 air brake hose nipple
气闸润滑脂 air brake grease
气闸设备 lock equipment
气闸式调节风门 damper regulator
气闸式装置 damper gear
气闸室 pneumatic brake tank;air lock tank;air lock chamber <气压沉箱的>
气闸室闸门 air lock gate
气闸调节器 damper regulator
气闸喂料器 air lock feeder
气闸系统 air lock system
气闸用马达 air motor
气闸闸门 air lock gate
气闸钟 air lock bell;pneumatic brake bell
气胀 air bulking
气胀棒坯 puffed bar
气胀的 flatulent
气胀孔 gas expansion hole
气胀轮胎 inflated tire
气胀式救生带 inflatable life-belt
气胀式救生筏 air inflatable raft;air inflation raft
气胀式救生服务站 service station of inflatable life rafts
气胀式救生艇 inflatable life-boat
气胀式救生衣 air inflatable jacket;air inflation jacket
气胀式橡皮工作艇 inflatable rubber dinghy
气胀式橡皮艇 inflatable rubber boat
气胀术 ballonetment;ballooning
气胀现象 ballooning
气胀压力 inflation pressure
气胀压力强度 intensity of inflation pressure;intensity of inflation pressure
气胀压坯 puffed compact
气障 air barrier;air-bond;vapo(u)r lock(ing)<发动机的>
气罩 air bell;air feed mask;gas hood;gas shield
气针 blower tube;air tube
气针孔 gas pin
气枕 air cushion;air pillow;pneumatic cushion
气振钻探器 vibro-corer
气镇 gas ballasting
气镇泵 gas ballast pump
气镇粗抽泵 gas ballast roughing-holding pump
气镇阀 gas ballast valve
气镇空气 gas ballast air
气镇控制 gas ballast control
气镇口 gas ballast port
气镇流 gas ballast flow
气镇器件 gas ballast device
气镇入口 gas ballast inlet
气镇原理 gas ballast principle
气蒸汽水混合流体 gas-steam-water fluid
气制动分泵 brake chamber
气制动控制装置 pneumatic brake controller
气制动(器) air brake; air-pressure brake;pneumatic brake
气质 nature
气质论 craseology[crasiology]
气质模式 air quality model
气致皱纹 <饰面时> gas checking
气帚 air broom
气柱 air column;gas column
气柱高度 height of gas column
气转 cyclostrophic
气状 gaseity;gassy
气锥 gas coning
气阻 air binding; air lock; air resistance;choke;vapo(u)r lock(ing)
气阻的 air-bound
气阻倾向 vapo(u)r locking tendency
气阻式转数计 air drag tach(e)ometer
气钻 air drill; borer; hammer drill; pneumatic drill
气钻进机 gas drilling rig
气钻气缸 drill cylinder
气钻转动套 drill chuck bushing
气嘴 air cock; air tap; cycle valve part;fume cock;gas tap

讫点/吸引区带 destination/attraction zone

弃采原因 cause of discarded mining

弃船 abandon a ship;abandonment of a ship
弃船部署 abandon ship station
弃船人 ship abandoner
弃船声明 notice of abandonment
弃船信号 abandon ship signal
弃船演习 abandon ship drill
弃负荷 load rejection; load thrown off
弃耕地 derelict land;old-field
弃耕地演替 abandoned field succession
弃荷 rejection of load
弃荷涌浪 rejection surge
弃货 <船舶遇险时> jettison
弃链 slip a cable
弃链器 cable releaser
弃料 quarry rubbish
弃料槽 reject chute
弃料地 derelict land
弃锚 abandon an anchor; slipping an anchor
弃锚出港 slip from an anchor
弃锚开航 cut-and-run
弃锚器拉索 anchor trip(ping)line
弃泥槽 spoil chute
弃泥地 dumping ground;spoil ground
弃权 abandon;abstain;abstention;disclaim; nonuser; release; renunciation;waive
弃权的 nonvoting
弃权声明信 letter of renunciation
弃权书 waiver
弃权条款 waiver clause
弃权证 waiver
弃水 surplus water
弃水坝 waste dam
弃水槽 waste channel
弃水出口 waste outlet
弃水道 wasteway;water escape
弃水陡槽 waste chute
弃水堆 waste pile
弃水门 waste gate
弃水渠 waste canal; waste channel; wasteway
弃水式压力调节器 water-wasting pressure regulator
弃水输送器 spoil conveyer[conveyor]
弃水竖管 standing waste
弃水堰 waste weir
弃水闸 waste sluice
弃水闸门 waste gate
弃土 dredged material;dredged matter; dredged spoil; muck; spoil-(ing); spoil material; spoil waste; surplus earth; surplus soil; surplus spoil;waste
弃土场 dump area; spoil area; spoil dump;spoil ground;waste area
弃土场围堤 surrounding embankment of a spoil ground
弃土车 waste truck
弃土处理 disposal of spoil;waste disposal
弃土岛 spoil island
弃土堤 waste bank
弃土堆 banket(te);banquette;dump; earth deposit;side casting pile;side piling;spoil bank;spoil heap;spoil pile; waste bank; waste deposit; waste pile;waste yard
弃土费 cost of spoil; cost of waste disposal;cost of wasting
弃土分离 locking out of the spoil
弃土卮斗 spoil hopper
弃土还耕 waste bank refarming
弃土量 spoil volume;volume of spoil
弃土溜槽 spoil chute
弃土漏斗 spoil hopper
弃土区 deposit(e)area; deposit(e) ground;disposal area;disposal site; dumping place; dumping site; dumping space; spoil ground; spoil site;waste area;spoil deposit area
弃土输送器 spoil conveyer[conveyor]
弃位【船】dead space
弃物 mud
弃渣 debris;excavation waste;muck; spoil;waste;waste rock
弃渣场 excavation waste dump;muck disposal;spoil bank;waste disposal area
弃渣场地 muck bank
弃渣车 muck car
弃渣处理 loading;mucking
弃渣堆 muck pile
弃渣坑 dump pit
弃渣面线 surface of spoil bank
弃渣坡脚 spoil toe
弃渣位置 location of spoil
弃渣装运系统 muck handling system
弃渣装载机 muck loader
弃宅文书 abandonment of homestead
弃置 elimination
弃置车站 <英> ghost station
弃置废物或污染液体的井 disposal well
弃置废物区 disposal site
弃置废物区的研究 disposal site designation study
弃置河(流)defeated stream;defeated river
弃置混凝土 waste concrete
弃置浚挖物质 spoil disposal
弃置设备或设施 retire
弃置危害性废物(封闭)设施 disposal facility

汽包 barrel;dome(barrel);drum; separator; steam drum; steam header;steam manifold

汽包挡板 drum baffle
汽包封头 drum end;drum head
汽包观察孔 drum manhole
汽包管接头 drum stub
汽包锅炉 drum boiler
汽包锅筒 dome course
汽包内部汽水隔板 internal steam/water baffle
汽包内部装置 drum internals
汽包式锅炉 drum-type boiler
汽包式蒸汽锅炉 drum steam generator
汽包水位许用限制值 limit of drum level
汽包筒身 drum shell
汽包压力 drum pressure
汽包蒸汽联箱 drum steam header
汽泵 heat pump;steam pump
汽铲 steam digger; steam navvy; steam shovel
汽车 autocar;automobile;mobile <美>; automotive vehicle; locomobile; motorcar; motor truck; motor vehicle; motor wagon;wheels;wide body
汽车安全带 shoulder-belt;shoulder harness <驾驶员的>;lap belt <美>
汽车安全试验 safety vehicle test
汽车安全性 auto safety
汽车百吨公里油耗 detecting-fuel consumption per hundred tonkilometers
汽车百公里油耗 detecting-fuel consumption per hundred kilometers
汽车摆头 wig-wag motion
汽车搬运车道 trucking lane
汽车搬运成本 trucking cost
汽车搬运桥 trucking bridge
汽车半悬板簧 cantilever spring
汽车半悬钢板 cantilever spring
汽车保险 automobile insurance; car insurance;motorcar insurance
汽车保险杠 bumper
汽车保修 truck warranty
汽车保养维修 servicing of cars
汽车保用 truck warranty
汽车备件 automotive supplies
汽车备用轮 <英> stepney
汽车背后的座位 rumble seat
汽车比赛场 motor-drome
汽车变速杆 strick shift
汽车变速器头挡、二挡、三挡或倒车变速叉 transmission first,second,third or reverse shift fork
汽车玻璃 autoglass;automotive glass
汽车餐馆 drive-in restaurant
汽车操纵轻便性 vehicle control easiness
汽车操纵失灵 understeer
汽车操纵稳定性 vehicle handling stability
汽车槽车 tank truck;tank wagon
汽车侧滑试验台 side-skid tester
汽车侧倾敏感度 vehicle roll suscepti-

bility
汽车侧向滑移 sidewise skidding
汽车差速齿轮 automobile differential gear
汽车柴油发动机 automobile diesel engine
汽车铲土机 truck shovel
汽车铲运机 wagon-scraper
汽车长列 motorcade
汽车厂工人 autoworker
汽车厂名牌 name plate
汽车场 motor depot; motor-drome; truck farm
汽车车顶天线 over-car antenna
汽车车架 automobile frame
汽车车间 automobile shop
汽车车流 vehicle flow
汽车车轮定位仪 wheel alignment meter
汽车车轮转矩计 road wheel torque meter
汽车车身 coach
汽车车身保险 motor hull insurance
汽车车身的设计与制造 coachbuilding
汽车车身的设计、制造和装配 coach work
汽车车身容积 capacity of body
汽车车身制造厂 coachbuilder; coach maker
汽车车位 car stall
汽车车型 automotive type
汽车成组交错停车 group staggered parking
汽车城 <美国底特律的俚语> Motor City
汽车乘客险 motor vehicle passenger insurance
汽车秤 lorry scale
汽车尺寸 automobile dimension
汽车冲洗 car washing
汽车出售场所 auto sales lot
汽车出行 car trip
汽车出租处 carracing rental service
汽车传动系 automotive transmission
汽车传动油 transmission oil
汽车存放率 total car ratio
汽车大梁 automotive frame; frame
汽车大修 vehicle major repair
汽车代用燃料 alternative motor fuel
汽车单伽马能谱法 single gamma spectrum method with truck
汽车挡风玻璃 car windscreen
汽车挡风玻璃清洗设备 screen washer
汽车挡泥板 fender; wheel guard; wing
汽车倒下的碎石 lorry-dumped riprap
汽车道连接线 motorway link
汽车道(路) automobile highway; bypass motorway; driveway; motorway
汽车道用地 right-of-way-of-motorway
汽车道与铁路的立体交叉 bus-rail interchange
汽车的 automotive; mobile
汽车的长蛇阵 motorcade
汽车的发展趋向 motor trend
汽车的离手速率 <汽车在超高曲线上行驶时自己转向不需驾驶人在方向盘上加力所需的一定速率> hands-off speed
汽车的磨合 running-in of automobile
汽车的爬坡能力 climbing capacity
汽车的所有人 car owner
汽车的行驶平顺性 ride smoothness of vehicle
汽车的性能规格 specifications of a car
汽车灯玻璃 lens
汽车灯泡 autobulb
汽车登记 motor vehicle registration
汽车底板 floor board
汽车底架天线 running board antenna
汽车底架运输平车 saddle-back
汽车底盘 automobile chassis; motor vehicle chassis
汽车底盘框架 carrier frame
汽车底盘式挖掘机 truck-mounted shovel
汽车底盘下的天线 under-car antenna
汽车底卸自动清洗机 automatic underbody washer
汽车地板脚坑 foot well
汽车地磅 platform truck scale; truck scale; truck weighing scale
汽车地秤 platform truck scale
汽车地理信息系统 mobile geographic information system
汽车第三者责任险 motor third party liability insurance
汽车颠簸 tramp
汽车点火(开关)接线头 ignition terminal
汽车点火系统 automotive ignition system
汽车点火整步器 synchroscope
汽车电磁波干扰 automobile electromagnetic interfere
汽车电动喇叭 motor-operated horn
汽车电气配线 harness
汽车电视接收机 mobile TV receiver
汽车电影院 drive-in theater[theatre]
汽车吊 autocrane; autohoist; autolift; crane truck; crane mobile; lorry crane; lorry-mounted crane; mobile crane; truck crane; truck-mounted crane; wheel crane
汽车吊车 automobile crane
汽车吊架 car sling
汽车吊具 automobile sling
汽车调度场 car pool
汽车调度员 car boss
汽车动力定额 ability rating
汽车动力特性曲线 driving force characteristic curve
汽车动力性(能) automobile dynamic(al) quality; vehicle dynamic(al) quality
汽车动力学 automobile dynamics
汽车动力因素 driving force factor
汽车渡船 automobile ferry; car ferry crossing; vehicle ferry
汽车渡船型 form of car ferry; shape of car ferry; type of car ferry
汽车渡口 car ferry crossing; car ferry-(ing)
汽车渡轮 autoferry; automobile ferry; car ferry boat
汽车渡站 car ferry terminal
汽车队 autofleet; motor squadron
汽车队列 train of cars
汽车队伍 motocade
汽车(队)运行 fleet operation
汽车对方向盘反应过敏的 oversteer-(ing)
汽车发动机 automobile engine; automotive engine; travel(l)ing engine
汽车发动机干扰 motorcar interference
汽车发动机加快 revving
汽车发动机燃烧过程 automotive engine combustion process
汽车发动机润滑油 mobile oil; mobiloil
汽车发动机在车身中部的 mid-engined
汽车发散物 automotive emission
汽车翻车保护杆 roll bar
汽车反应时间 vehicle response time
汽车饭店 drive-in restaurant; motor hotel; motor inn
汽车方向指示器 trafficator
汽车防滑链 skid chain
汽车房 garage
汽车房设备 garage equipment
汽车放气孔 louver[louvre]
汽车废气 automobile effluent; automobile exhaust gas; automotive emission; vehicle exhaust
汽车废气催化净化器 automobile exhaust catalytic cleaning cartridge
汽车废气净化 purification of automobile exhaust gas
汽车废气净化器 automobile exhaust gas cleaner; automobile exhaust gas purifier
汽车废气控制 <防止大气污染> vehicle emission control
汽车废气排放 exhaust emission; automobile emission
汽车废气排放标准 exhaust emissions standard
汽车服务 motor service; servicing of cars
汽车服务站 automobile service station; automotive service; car service station; service center[centre]; service station; servicing center [centre]; servicing depot
汽车浮筏 car float
汽车负前束 to-out
汽车副变速器变速杆 range selector
汽车改进 automobile improvement
汽车干道 motor-trunk road; primary route
汽车钢板 autobody sheet; automobile body sheet; automobile leaf-spring
汽车钢板弹簧 leaf spring
汽车跟踪模式 car-following model
汽车工厂 automobile machine shop; autoplant
汽车工程 automotive engineering
汽车工程师 <英国期刊名> Automobile Engineer (Design, Production, Materials)
汽车工程学 automobile engineering
汽车工业 auto(mobile) industry; automotive industry; car industry
汽车工业联合会 <美> Motor Vehicles Manufacture's Association
汽车公路 auto-highway
汽车公司提单 trucking company bill of lading
汽车公用组织 carpool
汽车挂车 automobile trailer; motor carrier trailer
汽车管理员 <美国私营公共汽车的车主、负责人> conductor
汽车号牌 auto number plate; number plate
汽车合乘 carpool
汽车合用 car pooling
汽车荷载 carload; traffic load
汽车荷载模拟器 mobile load simulator
汽车横越轨道平交跑道 truck runway at grade across the track
汽车衡 loadometer
汽车后部备用的折叠小椅 dick(e)y
汽车后部行李箱 boot
汽车后灯 rear lamp; rear light
汽车后视镜 driving mirror
汽车后束 to-out
汽车后轴档 rear axle bumper
汽车后座壁板小窗 opera window
汽车滑溜 skidding
汽车化 motorization; motorize
汽车化比率 motorization rate
汽车化载运 truck hauling
汽车换挡 change of speed
汽车回气管 muffler
汽车混凝土泵 mobilcrete pump
汽车货柜 auto-container
汽车货运服务区 trucking area
汽车货运公司 trucking company
汽车货运线交通 truck-line traffic
汽车货运业 road haulage industry
汽车货运业务 trucking operation; truck-line service
汽车货运价率 motor-truck rate
汽车货运终点站 truckhead
汽车货运装卸站 motor transport handling station
汽车机动车路网 motorway network
汽车极限技术状况 limiting technical condition of vehicle
汽车急转弯 quick turning
汽车集中调度场 motor pool; physical motor pool; physical pool
汽车集装箱 auto(mobile) container; car container; vehicle rack container
汽车技术 automotive engineering
汽车技术员 truckmaster
汽车加速度 pick-up
汽车加油修理服务站 service station; servicing center[centre]
汽车加油站 automobile filling station; filling station; gas filling station; gasoline service pump; gas station; motor vehicle service station
汽车驾驶 automobilism
汽车驾驶动力学 ride dynamics
汽车驾驶人 motorist
汽车驾驶人导向通信[讯]系统 car driver guidance information system
汽车驾驶试验环路 ride and handling loop
汽车驾驶室 chauffeurs' room
汽车驾驶室前壁 dash board
汽车驾驶台仪表板 dash unit
汽车驾驶员 autoist; automobilist; motorist; motor vehicle driver; wheel man; car driver; chauffeur <小型的>
汽车驾驶员紧急呼援系统 emergency motorist call system
汽车间 carport; garage; motor garage; stall
汽车间门配件 garage door furniture
汽车减低速度 change down
汽车检测站 vehicle detecting test station
汽车检修工 grease monkey
汽车检修坑 engine pit
汽车检修台 inspection rack
汽车交通 automobile transportation; interurban traffic; mechanical transport; motor service; motor (vehicle) traffic; road service; vehicle traffic; vehicular traffic
汽车交通控制【交】automobile traffic control
汽车交通控制系统【交】automobile traffic control system
汽车交通量 automobile traffic
汽车交通隧道 motor traffic tunnel
汽车交通综合控制系统 <日本> comprehensive automobile traffic control system
汽车绞盘 motor cable winch
汽车搅拌混凝土 truck-mixed concrete
汽车搅拌机 mixer-lorry; truck agitator
汽车阶级 autocrat
汽车接收机 mobile receiver
汽车接运 trucking connection
汽车节能 vehicle energy saving
汽车节油技术 vehicle fuel saving technique
汽车节油装置 fuel saving device

汽车经济速度 vehicle economical speed
汽车经济性 vehicle economy
汽车竞技运动会 autocross
汽车竞赛路 brick;racing track
汽车酒店 motor hotel
汽车剧场 drive-in theater[theatre]
汽车空档 neutral position
汽车空气调节 automotive air conditioning
汽车空气污染 automotive air pollution
汽车空箱里程 vehicle empty container kilometer
汽车库 automotive depot;garage;motor depot; motor garage; motor pool; motor vehicle storage; stock of cars
汽车库供暖 garage heating
汽车库门 garage door
汽车狂 auto-crazy
汽车喇叭 automobile horn
汽车喇叭声 honk
汽车蜡 car wax
汽车蜡克漆 motorcar lacquer
汽车离合器 automobile clutch
汽车犁 motor plough
汽车力学 automobile mechanics
汽车连续一昼夜循环运行 round-the-clock bus shuttle service
汽车列车 auto-train;combination vehicle;road train;track tractor;tractor trailer combination;trailer truck; vehicle serial;motorail <承办汽车托运的列车>
汽车列车拖挂条件 artic condition
汽车零件 automobile hardware;auto parts
汽车零件库 mobile parts store
汽车零件商店 mobile parts store
汽车漏气回流管 exhaust gas recirculation pipe
汽车路 autoroad; coach road; motor road;motorway
汽车路况巡查 motor patrol
汽车路桥 motorway bridge
汽车旅馆 auto court;cabin court;hotel-motel-motor hotel;motel motor-hotel; motor court; motor hotel; motel <专为汽车旅客开设的路边旅馆>
汽车旅游旅馆 motor inn
汽车旅游营地 automobile camp
汽车轮渡 autoferry;automobile ferry-steamer;car ferry crossing;car ferry(ing); car ferry-steamer; drive on/drive off ship; motorcar ferry; truck-ferry
汽车轮渡码头 car ferry terminal
汽车轮渡跳板 tail-gate ramp
汽车轮渡栈桥 ferry bridge of automobile;transfer bridge
汽车轮胎 automobile tire[tyre];motor tire[tyre]; dount <俚语>; doughnut <美国俗称>
汽车轮胎噪声 automobile tire[tyre] noise
汽车码头标志 vehicular ferry symbol
汽车门窗边 reveal lining
汽车门式自动清洗机 gate type automatic car washer
汽车面漆 automotive top coat
汽车名词 automotive nomenclature
汽车内搁脚空间 foot well
汽车内胎 rubber tire[tyre]
汽车耐久性 vehicle durability
汽车能量利用率 vehicle energy utilization factor
汽车能量平衡 vehicle energy balance
汽车排长队 autocade;motorcade
汽车排放的一氧化碳 carbon monoxide[CO] from motor vehicles
汽车排放分析仪 vehicle exhaust analyser[analyzer]
汽车排放物 vehicle emission
汽车排放转换器 automobile exhaust gas converter
汽车排气 autoemission;automobile emission;auto(mobile) exhaust; automotive emission; exhaust gas from car
汽车排气采样 sampling of auto exhaust
汽车排气分析 automobile exhaust analysis
汽车排气分析仪 auto exhaust analyser[analyzer]; exhaust gas analyser [analyzer]
汽车排气监测 monitoring of automobile exhaust emission
汽车排气取样 sampling of auto exhaust
汽车排气污染 vehicle pollution
汽车排气氧化催化剂 auto exhaust oxidation catalyst
汽车排气噪音 auto exhaust noise
汽车排气转化器 automobile exhaust converter
汽车牌号 mark of car
汽车牌号对照法 license[licence]-number-matching method
汽车牌照 automobile plate;licence[license] plate
汽车牌照灯 license[icence] light; license[licence] lamp
汽车牌照法 motor license[licence] plate method
汽车牌照(号码)对照法 <交通观测的> license[licence]-number-matching method
汽车抛光蜡 car polish
汽车跑道 carracing track
汽车配件 motor vehicle accessories
汽车棚 carport
汽车平均技术速度 vehicle average technical speed
汽车坡道 car ramp
汽车坡路防滑装置 norol
汽车坡路停车防滑机构 norol; hill holder
汽车破碎机 car shredder
汽车漆 automotive coating
汽车(企业)协会 Automobile Association
汽车起动摇柄 starting crane
汽车起重机 autocrane;crane car;motorcrane; tire crane; truck crane; truck hoist; truck-mounted crane; truck-mounted tower crane; tyre crane; vehicle crane; wheel crane; wheel-type truck crane
汽车汽化器 automobile carburet(t)or
汽车汽油 mobilgas
汽车器材 automotive supplies
汽车千斤顶 car jack;motorcar jack
汽车牵引的 motor drawn; motor-driven
汽车前灯 automobile headlight
汽车前灯减光器 head lamp dimmer
汽车前轮定位 front-wheel alignment
汽车前桥 front axle
汽车前轴挡 front axle bumper
汽车前轴梁 axle beam
汽车窃贼 carnapper
汽车倾卸机 car dumper;dumper
汽车清洗处 car wash
汽车清洗机 car washer
汽车清洗间 auto(mobile) laundry; automotive laundry
汽车驱动力 driving force of motor vehicle
汽车燃料利用率 vehicle fuel utilization factor
汽车燃料税 motor fuel tax
汽车燃料消耗定额 vehicle fuel consumption rating
汽车润滑站 lubritorium;lubritory
汽车闪光涂料 metallic automotive coating
汽车上的反光镜 cheat
汽车设计 mobile design
汽车社会 autopia
汽车升降机 automobile elevator;lorry-mounted elevator
汽车时代城市 motor-age city
汽车识别系统 motor vehicle identification system
汽车实际行驶时间 travel(l)ing time
汽车使用 automobilism; motor vehicle use
汽车使用调查 <亦称道路使用调查> motor vehicle use study
汽车使用税 tax on automobiles
汽车使用者 automobilist
汽车示廓灯 perimeter lighting
汽车式拌和机 truck agitator; truck mixer
汽车式铲土机 truck-mounted shovel
汽车式触探器 car-jack type sounding apparatus
汽车式的转向机构 automotive type steering
汽车式吊车 wheel-mounted crane
汽车式反向铲 truck back hoe
汽车式风动钻机 air truck drill
汽车式钢丝操纵的挖掘机 mobile cable-operated excavator
汽车式混凝土拌和机 motor truck concrete mixer;mixer-lorry
汽车式混凝土泵 truck-mounted concrete pump
汽车式混凝土搅拌机 truck concrete mixer
汽车式搅拌车 transit truck
汽车式搅拌机 lorry mixer; motor-mounted mixer; motor truck concrete mixer; truck mixer plant; truck-mounted mixer; truck-transit mixer
汽车式搅拌机给料的折叠式料斗 folding truckmixer feeding funnel
汽车式沥青喷布机 truck-mounted distributor
汽车式(轮胎式)自动回填机 machine-mounted automatic backfiller
汽车式平地机 <平地机刀片安装在卡车上> truck grader
汽车式起重机 autocrane; caterpillar truck;crawler truck;mobile crane; pneumatic-tyred mobile crane; rubber tyred truck crane; truck (-mounted) crane;truck with lift; autohoist; autolift; automobile crane; crane truck; derrick car; lorry crane;lorry-mounted crane
汽车式塔式起重机 truck-mounted tower crane
汽车式挖掘机 shovel truck;traxcavator;truck-mounted shovel
汽车式挖土机 truck-mounted shovel
汽车式住宅 <用汽车牵引的居住用车厢> motor caravan
汽车式桩架 tuck-mounted pile frame
汽车式装载机 lorry loader
汽车式钻堡 truck-mounted jumbo
汽车式钻车 pneumatic-tired[tyred] wagon drill
汽车式钻机 lorry-mounted crane; truck-mounted drill(ing rig); wagon drill
汽车式钻机架 main drill frame
汽车事故 automobile accident;motor vehicle accident
汽车事故死亡率 automobile death rate;motor vehicle fatality
汽车试车场 autodrome;motor-drome
汽车试验场 automobile test ground; motor-drome
汽车试验里程 test mile
汽车试验跑道 automobile test track; automotive test track
汽车试验专用轮 fifth wheel
汽车试用期 zero device-miles
汽车手(按)喇叭 hand horn
汽车手刹车 parking brake
汽车输送 truck move
汽车数 number of(motor)vehicles
汽车数字电话网 digital mobile telephone network
汽车双驱动桥 bogie[bog(e)y]
汽车税 motor vehicle tax
汽车司机 automobilist; chauffeur of motor-car;speed-merchant <美>
汽车司机室信号 cab signal
汽车四挡速率 fourth speed
汽车速度表 autometer
汽车速度表试验台 speedometer tester
汽车速度计 autometer;metrograph
汽车速度监视站 speed trap
汽车损耗 vehicle wear-out
汽车所有(权) auto ownership
汽车塔吊 truck-mounted tower crane
汽车踏板挡泥板 running board shield
汽车踏脚板 run board
汽车台钻 truck-mounted jumbo
汽车弹簧 motor spring
汽车弹簧钢板螺杆 perch bolt
汽车提升机 automotive lift
汽车天线杆弯曲 car rod aerial bent
汽车停车场 automobile parking space;motor park;parking space
汽车停车场标高 stop level
汽车停车车位 car stall
汽车停车防滑机构 norol
汽车停车间 motor vehicle hangar
汽车停车库 motor vehicle hangar
汽车停车棚 open parking structure
汽车停放 car parking
汽车停放场 automobile parking space
汽车停放场地 parking provision
汽车停放场所 parking stall
汽车停放计时器 parking meter
汽车停放建筑物 automobile parking structure
汽车停放设备 parking provision
汽车停放收费计 parking meter
汽车停放提升机 automobile parking lift
汽车停放展销市场 auto park or market
汽车头挡速率 first speed
汽车头灯 <汽车正面的大灯> dazzle lamp;dazzle light;autobulb
汽车头灯的炫目灯光 dazzle lighting
汽车头灯下的小光 headlight lower beam
汽车头灯小光灯泡 head lamp dimming light bulb;head lamp parking bulb
汽车涂料 automotive coating
汽车拖车 automobile trailer
汽车拖带的住房 trailer dwelling; trailer house
汽车拖带的活动住房 manufactured home;mobile home
汽车拖的活动房子 dependent trailer coach
汽车拖动的 motor-borne
汽车拖挂的活动房屋 automobile

trailer
汽车拖挂住房聚居地 caravan city
汽车拖拉机 truck tractor
汽车拖拉机工业 motor-tractor industry
汽车拖运的 motor-borne
汽车拖着的活动住房<美> trailer
汽车驮运系统 piggyback system
汽车挖掘机 shovel truck
汽车挖土机 truck excavator
汽车外表装饰 exterior automotive trim
汽车外胎 auto-tire[tyre] casing; casing
汽车微迹法 minute trace survey with truck
汽车维护规范 vehicle maintenance norms
汽车维护级别 vehicle maintenance grade
汽车维护区<运输企业中划分出来专供汽车维护用的区域> service island
汽车维修 motor vehicle care
汽车维修工程 vehicle maintenance engineering
汽车维修站 motor vehicle service station
汽车尾部挡板式石屑撒布器 tail-gate spreader of truck
汽车尾窗 hatchback
汽车尾灯 stop light; taillight
汽车尾气排放 motor vehicle exhaust
汽车尾气污染控制 control of automobile exhaust gases
汽车污染 auto(mobile) pollution
汽车污染物 automotive effluent
汽车污染源控制 auto-pollution source control
汽车无线电机 mobile radio
汽车无线电设备 motorcar set
汽车无线通信[讯] communication between fixed points and mobiles
汽车洗刷废水 wastewater from car washer
汽车舷梯 car ladder
汽车厢灯<车门打开即自动亮灯> courtesy light
汽车厢式车厢 van body
汽车小五金 automobile hardware
汽车小修 vehicle current repair
汽车卸货车位 unloading dock
汽车卸料斗 truck unloading hopper
汽车卸载 truck unloading
汽车行车道 vehicular traffic lane
汽车行程 vehicle trip
汽车行程时间<包括停车及受阻时间> travel time
汽车行李箱<英> boot
汽车行列<美> motorcade
汽车行驶的必要条件 requirement of motion of motor vehicle
汽车行驶平顺性 vehicle smooth running
汽车行驶区间 automobile restricted zone
汽车行驶线 auto route
汽车行驶阻力 resistance to motion of motor vehicle
汽车形式 mobile version
汽车型式 type of automobile
汽车型踏板 automotive pedal
汽车性能 characteristics of motor vehicle
汽车性能检验场 proving ground
汽车修补漆 automobile repair coating; automotive refinish paint
汽车修补(涂漆) autorefinish
汽车修理 servicing of cars
汽车修理厂 garage; motor garage; motor repair shop; service park
汽车修理车间 automobile (work)-shop; auto repair (work) shop; garage; motor repair; repair garage; motor vehicle repair shop
汽车修理店 automobile(work) shop
汽车修理工 automobile mechanic; garage mechanic
汽车修理工厂 motor vehicle repair shop
汽车修理工场 carburet(t)er overhaul shop
汽车修理基地 vehicle repair depot
汽车修理库 garage
汽车修理站 automobile repair station
汽车修配厂 garage; motor repair shop; service park
汽车修配工 motor car fitter
汽车修配工具 grease monkey
汽车悬架 automotive suspension
汽车羊角臂 knuckle arm
汽车叶子板 mud apron; mud guard
汽车仪表 motor meter
汽车仪表板<英> fascia[复 fa(s)-ciae/fa(s)cias]
汽车椅背 head rest
汽车银行 drive-in bank
汽车银行出纳窗 drive-in teller window
汽车引擎的水箱 radiator
汽车营地 automobile camp
汽车营运费 cost of motor operation
汽车营运速度 service speed
汽车拥有率 car ownership ratio
汽车用电梯 car elevator
汽车用(喷)漆 automobile lacquer
汽车用坡道 car ramp
汽车用闪光灯 winker
汽车用收音机 automobile receiver; autoradio
汽车用无线电机 motorcar set
汽车游客旅馆 automobile court; motel-tourist cabins; motor court; motor hotel; motor inn; motor lodge; tourist court
汽车越野赛 autocross
汽车运单 truck receipt
汽车运费 trucking charges
汽车运价管理体制 motor transport rate management system
汽车运价率 motor transport tariff rate
汽车运价体系 motor transport tariff system
汽车运输 automobile transportation; auto traffic; car transportation; motoring; motor service; motor transport(ation); motor vehicle transportation; road transport; trucking
汽车运输保养 motor transport maintenance
汽车运输边际成本 motor transport marginal cost
汽车运输车辆大修折旧 motor transport vehicle overhaul depreciation
汽车运输车辆费用 motor transport vehicle cost
汽车运输车型成本 motor transport cost by vehicle type
汽车运输成本范围 motor transport cost scope
汽车运输成本分析 motor transport cost analysis
汽车运输成本计算单位 motor transport cost accounting unit
汽车运输成本计算对象 motor transport cost accounting object
汽车运输成本降低额 motor transport cost reduction amount
汽车运输成本降低率 motor transport cost reduction rate
汽车运输成本结构 motor transport cost structure
汽车运输成本控制 motor transport cost control
汽车运输成本项目 motor transport cost item
汽车运输船 autocarrier; car carrier; vehicle carrier
汽车运输单车成本 motor transport unit vehicle cost
汽车运输的 motor-borne
汽车运输队 truck organization
汽车运输辅助生产成本 motor transport auxiliary production cost
汽车运输工程 automotive transport exploitation; automotive transport exploitation engineering
汽车运输公司 motor carrier; truck firm; truck line; van line; fleet operation firm
汽车运输换算和计算表 motor transport conversion and computation chart
汽车运输机会成本 motor transport opportunity cost
汽车运输计划成本 motor transport planned cost
汽车运输里程时间换算图表 motor transport distance and time conversion chart
汽车运输连 truck transport company
汽车运输列车 truck train
汽车运输排 truck platoon
汽车运输企业成本 motor transport enterprise cost
汽车运输器材库 motor transport depot
汽车运输设备 motor transport facility
汽车运输税 vehicle tax
汽车运输投入成本 motor transport sinking cost
汽车运输往返 truck turn-around
汽车运输线 motor carrier
汽车运输斜坡道 truck ramp
汽车运输业 trucking industry
汽车运输业条例<美> Motor Carrier Act
汽车运输业务 road transportation commitment
汽车运输业务成本 motor transport operation cost
汽车运输业者 trucker; truckman
汽车运输营 motor transport battalion
汽车运输用器材 trucking facility
汽车运输预测成本 motor transport forecast cost
汽车运输中心 trucking center[centre]
汽车运输终点站 trucking terminal
汽车运输装卸成本 motor transport handling cost
汽车运输综合成本 motor transport comprehensive cost
汽车运输总站 trucking terminal
汽车运送拌和机 truck-transit mixer
汽车运送的 motor-borne
汽车运行 automobilism
汽车运载 trucking
汽车运载车 wagon carrier truck
汽车运载船 car-carrier
汽车运载工具<一种装运汽车、木材、管道等货物的特殊敞式长大集装箱> automobile carrier
汽车匝道 car ramp
汽车载运 truck hauling; trucking
汽车(载运的)混凝土泵 truck-mounted concrete pump
汽车载运活动范围 trucking activities
汽车造成的大气污染 air pollution by automobile
汽车噪声 automobile noise
汽车责任保险 motorcar liability insurance
汽车展览 motorama
汽车展销站 automobile sales lot
汽车占用率 car occupancy
汽车站 automotive depot; motor depot
汽车罩面漆 automobile finish
汽车折叠式车顶 folding top
汽车真空式风挡刮水器 vacuum wiper
汽车诊断站 vehicle diagnostic station
汽车振动式喇叭 vibrator horn
汽车振动 thump
汽车执照 auto(mobile) license[licence]
汽车止动能力 stopping ability
汽车制动器 automotive brake
汽车制动自稳效应 self-stabilizing effect in braking
汽车制造 car building
汽车制造厂 automobile factory; motor vehicle plant; motor works
汽车制造厂商 motormaker
汽车制造废水 car manufacturing waste water
汽车制造商 automaker; automan
汽车制造商协会 Automobile Manufacturers Association; Car Manufacturers Institute
汽车制造业 automobile engineering; automotive industry; motorcar industry
汽车制造者 automan; carmaker
汽车中后轴平衡悬架 bogie unit
汽车中修 vehicle medium repair
汽车终点站 auto-terminal
汽车重量 vehicle weight
汽车重量功率比 weight horsepower ratio(of motor vehicle)
汽车重量马力比 weight horsepower ratio(of motor vehicle)
汽车重量税 automobile weight tax
汽车轴间差速器 interaxle differential
汽车轴重限度 axle weight limit
汽车主传动 main drive; main gear
汽车住房 mobile home
汽车住宅 motor home
汽车贮藏库 garage storage
汽车专用道 automobile road
汽车专用公路 motorway
汽车专用货车 car-carrier wagon
汽车专用路 auto route; autostrada
汽车专用区 autopia
汽车专用运输船 pure car carrier
汽车转矩管传动 torque-tube drive
汽车转向 motor steering; motor turning
汽车转向节 spindle
汽车转向系 automotive steering
汽车转运的 carborne
汽车装备 motor vehicle equipment
汽车装车机 vehicular loader
汽车装配 automotive mount
汽车装卸费 car loading charges
汽车装卸设备 automobile handling facility
汽车装卸台 truck platform
汽车装卸台高度 truck platform height
汽车装运车 car-carrier
汽车装载 car load(ing); car stowage
汽车装载机 truck loader
汽车状况监控 vehicle condition monitoring
汽车自动传动装置 drive master
汽车(自动)升降机 autolift
汽车总数 stock of cars
汽车总重 gross vehicle weight
汽车阻塞 auto rush

汽车组会志 <英期刊名> Proceedings of the Automobile Division
汽车最小回转圆 <通常以直径表示> minimum turning circle of car
汽车最小转弯半径 minimum turning radius
汽车最小转弯直径 minimum turning circle of car
汽车左转驶出行驶 left-off movement
汽车座前空当 knee room
汽锄 steam plough
汽船 motor boat; motor dory; motor ship; steam boat; steamer; steam ship; steam vessel
汽船交货价格条件 free on steamer
汽船轮渡 motor ferry
汽船声 motorboating; popping
汽锤 block hammer; gun; steam hammer
汽锤打桩机 ram steam pile driver; steam hammer piling machine; steam pile driver; Nasmyth pile; Nasmyth pile-driver
汽锤废气 exhaust steam from steam hammer
汽锤活塞 hammer piston
汽锤基础结构 hammer foundation structure
汽锤汽缸 hammer cylinder
汽锤砧 steam hammer anvil
汽醇混合物 petrol/alcohol mixture
汽灯 hurricane lamp; oil burner; pressure vapo(u)rizer lamp; storm lamp; storm lantern
汽灯罩 gas mantle
汽笛 buzz(er); hailer; hooter; siren [syren]; steam siren; steam whistle; whistle
汽笛按钮 whistle control
汽笛长声 long blast; prolonged blast
汽笛浮标 whistle buoy; whistling buoy; howling buoy
汽笛符号 whistle code
汽笛杆 whistle lever
汽笛杆轴臂 whistle lever shaft arm
汽笛杆轴臂键 whistle lever shaft arm key
汽笛杆轴键 whistle lever shaft key
汽笛杆轴托 whistle lever shaft bracket
汽笛开关阀 whistle valve
汽笛拉把 whistle lever
汽笛拉把杆 whistle lever rod
汽笛拉绳 whistle cord; horn cord
汽笛拉索 whistle pull
汽笛声 blast
汽笛响声 hoot
汽笛信号 whistle signal
汽垫 steam cushion
汽垫车 air cushion car; ground effect machine; terraplane
汽垫工程车 steam cushion construction vehicle
汽动泵 steam-driven pump
汽动锤 steam hammer; steam ram
汽动锤体 steam ram
汽动的 steam-driven; steam-propelled
汽动吊车 steam hoist
汽动辅助泵 donkey pump
汽动给水泵 steam feed pump
汽动鼓风机 steam-driven blower
汽动夯具 air tamp
汽动铆钉机 steam-driven riveting machine
汽动推钢机 steam ram
汽动挖掘机 steam excavator; steam navvy
汽动摇尺机 steam setwork
汽动凿毛机 steam bush hammer
汽动撞头 steam ram
汽阀 steam brake; steam valve
汽阀 T 形环 valve tee ring
汽阀垫圈 steam valve packing
汽阀杆 steam valve stem
汽阀控制器 steam valve controller
汽阀帽 steam valve bonnet
汽阀盘防松螺母 steam valve disk lock nut
汽阀青铜 steam valve bronze
汽阀体 steam valve body
汽阀填密 steam valve packing
汽阀填密压盖 steam valve packing gland
汽阀无余面的位置 line and line
汽阀压盖 steam valve gland
汽阀中心件 steam valve center piece
汽阀轴 steam valve spindle
汽封 steam binding; steam lock(ing); steam seal gland; steam seal(ing); vapo(u)r lock(ing); vapo(u)r seal
汽封抽汽器 gland steam exhauster
汽封的 vapo(u)r-bound
汽封供汽 gland sealing steam supply
汽封供汽系统 gland sealing steam supply system
汽封环 labyrinth ring
汽封间隙 labyrinth clearance
汽封冷凝器 gland steam condenser
汽封漏汽 gland leak-off; gland-packing leakage; packing leakage
汽封排汽 gland sealing steam exhaust
汽封片 labyrinth strip; sealing strip
汽封设备 vapo(u)r lock device
汽封室 gland body; gland steam pocket
汽封损失 labyrinth gland loss
汽封弹簧 gland spring
汽封套 gland sleeve
汽封套筒 seal sleeve
汽封体 gland casing; gland housing
汽封蒸汽 gland steam
汽封蒸汽集汽管 gland sealing steam collecting pipe
汽封蒸汽冷却器 sealing steam desuperheater
汽封蒸汽连接管 gland steam connexion
汽封蒸汽凝汽器 gland steam condenser
汽封蒸汽排汽管 gland steam exhauster
汽封蒸汽室 sealing steam box
汽封蒸汽调节阀 gland sealing steam control valve
汽封蒸汽调节器 gland steam regulator
汽封蒸汽调节系统 gland steam control regulator system
汽封装置 steam seal gland
汽干机 steam drier[dryer]
汽缸 air cylinder; casing; cylinder; jug; shell; steam cylinder
汽缸鞍座 cylinder saddle
汽缸保温套 cylinder lagging; lag; lagging jacket
汽缸壁 casing wall
汽缸壁厚度 cylinder wall thickness
汽缸壁温度 cylinder wall temperature; wall temperature
汽缸布置 alignment; cylinder arrangement
汽缸测径器 cylinder ga(u)ge
汽缸车床 cylinder lathe
汽缸衬 cylinder bush
汽缸衬垫 cylinder liner
汽缸衬垫扩孔 counter for cylinder liner
汽缸衬套 cylinder liner; cylinder lining
汽缸衬套内径 cylinder liner diameter
汽缸衬筒 cylinder bushing; cylinder liner
汽缸成对铸造式缸体 pair-cast cylinder
汽缸充满程度 admission
汽缸冲程 cylinder stroke
汽缸床 cylinder head gasket
汽缸单元 cylinder unit
汽缸导流片 cylinder baffle
汽缸的工作容积 swept of cylinder
汽缸的配置 cylinder layout
汽缸底 cylinder back cover; cylinder bottom; cylinder end
汽缸底盖 cylinder bottom head
汽缸底汽管 cylinder bottom steam pipe
汽缸底箱 cylinder under casing
汽缸底座 cylinder base
汽缸垫 cylinder gasket; cylinder ring
汽缸垫密片 cylinder gasket
汽缸垫片 cylinder head gasket
汽缸动力 cylinder power
汽缸端盖 cylinder end cap
汽缸对套 cylinder lagging
汽缸对置发动机 boxer engine
汽缸对置式发动机 opposed-cylinder (type) engine
汽缸阀 cylinder valve
汽缸放水口 cylinder drainage
汽缸放水旋塞 cylinder drain cock
汽缸附件 cylinder fitting
汽缸盖 cylinder cap; cylinder cover; cylinder head; cylinder lid; junk head
汽缸盖凹顶 cylinder head pocket
汽缸盖衬垫 cylinder head gasket
汽缸盖垫密片 cylinder cap gasket
汽缸盖垫(片) cylinder head gasket
汽缸盖端部 cylinder head end
汽缸盖固定螺钉 cylinder head anchor pin
汽缸盖固定螺栓 cylinder head fixed stud
汽缸盖紧固件 cylinder head fastener
汽缸盖进油口 cylinder head oil inlet
汽缸盖螺栓 cylinder head bolt
汽缸盖喷水嘴 cylinder head water nozzle
汽缸盖起卸螺母 cylinder head jack nut
汽缸盖塞 cylinder head plug
汽缸盖水套塞 cylinder head water jacket plug
汽缸盖套 cylinder head casing
汽缸盖罩 cylinder head casing
汽缸盖柱螺栓 cylinder head stud
汽缸隔热层格 lag
汽缸隔热(外)套 steam-cylinder lagging; cylinder lagging
汽缸工作面划伤 scuffing of cylinder bore
汽缸工作容积 actual displacement; cylinder capacity; displacement chamber; piston displacement; piston swept volume; stroke capacity; swept volume of cylinder; cylinder displacement
汽缸工作容量 cylinder capacity
汽缸工作室盖 displacement head
汽缸固定点 casing anchor point
汽缸规 cylinder ga(u)ge
汽缸珩磨机 cylinder hone grinder
汽缸珩磨头 cylinder hone
汽缸换气量 piston swept volume
汽缸活塞 piston; steam-cylinder piston
汽缸活塞杆冲程 cylinder stroke
汽缸积炭刮除器 cylinder carbon scraper
汽缸加油法 cylinder oiling
汽缸夹 cylinder head
汽缸夹套 cylinder jacket
汽缸间隙 cylinder clearance
汽缸减压旋塞 cylinder relief cock
汽缸绞刀 cylinder reamer
汽缸脚猫爪 cylinder foot
汽缸进气阀 cylinder suction valve
汽缸进水接头 cylinder water inlet connection
汽缸径规 cylinder ga(u)ge; cylindric(al) ga(u)ge
汽缸镜面擦伤 scuffing of cylinder bore
汽缸口 cylinder port
汽缸联通管 cross-over pipe
汽缸裂缝 cylinder block cracking
汽缸螺栓 cylinder bolt
汽缸螺栓凸缘 cylinder bolting flange
汽缸马力 cylinder horsepower
汽缸猫爪 casing lug
汽缸密封垫圈 cylinder seal
汽缸密封装置 cylinder seal
汽缸磨床 cylinder grinder; cylinder grinding machine
汽缸磨石 cylinder hone
汽缸磨损 cylinder wear
汽缸(内)壁 cylinder wall
汽缸内活塞环切口间隙 ring gap in bore
汽缸内径 cylinder bore; cylinder diameter
汽缸内径规 cylinder bore ga(u)ge
汽缸内径量规 cylinder ga(u)ge
汽缸内腔 cylinder chamber
汽缸排 bank of cylinders; cylinder bank; cylinder block; inblock cylinder
汽缸排量容积 cylinder displacement volume
汽缸排列 cylinder arrangement
汽缸排气 cylinder displacement
汽缸排气阀 cylinder discharge valve
汽缸排(气)量 piston swept volume; cylinder displacement
汽缸排水操纵阀 cylinder cock valve
汽缸排水阀 port cock
汽缸排水阀风缸 pneumatic cylinder
汽缸排水塞门 cylinder cock
汽缸排水旋塞把 cylinder cock lever
汽缸盘锥根盒 cylinder stuffing box
汽缸刨床 cylinder shaping machine
汽缸配件 cylinder attachment
汽缸膨胀 cylinder expansion
汽缸膨胀指示器 shell expansion indicator
汽缸偏置 cylinder offset
汽缸气孔密封垫 cylinder gasket
汽缸气力起动器 cylinder pneumatic actuator
汽缸牵引力 cylinder tractive force
汽缸前盖 front cylinder head
汽缸全套 cylinder complete; cylinder complete unit
汽缸容积 cylinder volume; swept volume of cylinder
汽缸容积比 cylinder ratio
汽缸容量控制 cylinder capacity control
汽缸容量调节 cylinder capacity modulation
汽缸润滑 steam-cylinder lubrication
汽缸润滑法 cylinder lubrication
汽缸润滑器 steam-cylinder lubricator
汽缸润滑旋塞 cylinder lubrication cock
汽缸润滑油 cylinder oil
汽缸塞 cylinder plug
汽缸散热片 cylinder cooling fin; cylinder fin
汽缸上部润滑油 upper cylinder lubricant
汽缸上装气门的发动机 valve-in-

block engine
汽缸升压 boost
汽缸实际排量 actual displacement
汽缸数 number of cylinders
汽缸水套 cylinder water jacket
汽缸镗床 cylinder (re) boring machine
汽缸镗杆 cylinder boring bar
汽缸镗磨床 cylinder boring and honing machine
汽缸套 cylinder bush; cylinder casing; cylinder gasket; cylinder jacket; cylinder lagging; steam-cylinder jacket; steam jacket
汽缸套拆卸器 cylinder-sleeve puller
汽缸套带 steam-cylinder jacket band
汽缸套导向表面 liner lead
汽缸套的上定位环带 liner neck
汽缸套拉出器 cylinder liner puller
汽缸套冷却水 jacket(-cooling) water
汽缸套冷却水泵 jacket water pump
汽缸套冷却水柜 jacket water tank
汽缸套冷却水冷却器 jacket water cooler
汽缸套冷却系统 jacket-cooling system
汽缸套毛坯 liner blank
汽缸套密封环 cylinder packing ring
汽缸套密封件 cylinder liner seal
汽缸套筒 liner of cylinder; cylinder sleeve
汽缸套凸缘 liner flange
汽缸套泄水阀 jacket drain valve
汽缸套修理工具 cylinder liner service tool
汽缸套压力表 < 冷却水 > jacket ga(u)ge
汽缸体 block; cylinder block; cylinder body; cylinder casing; engine block
汽缸体衬套 cylinder block liner
汽缸体盖 cylinder block cover; cylinder head
汽缸体上部 cylinder deck
汽缸填密函 steam stuffing box
汽缸填密片 steam-cylinder gasket
汽缸通道 cylinder port
汽缸通道口 cylinder passage
汽缸通汽旋塞 cylinder cock
汽缸筒 bore; cylinder barrel
汽缸头 cylinder end; cylinder head; head of cylinder; junk head; steam-cylinder head
汽缸头出水接头 cylinder head outlet connection
汽缸头垫 cylinder head gasket
汽缸头盖 cylinder head cover
汽缸头盖罩 cylinder head cover
汽缸头密封 cylinder end seal
汽缸头小盖 cylinder head cover
汽缸头压板 cylinder head clamp
汽缸头罩 cowl hood
汽缸托架 cylinder bracket
汽缸外盖 cylinder head cover
汽缸外壳 cleating of cylinder; cylinder baffle
汽缸外套 cylinder casing; steam-cylinder lagging
汽缸外罩 case shell; cylinder clothing
汽缸温度指示器 cylinder temperature indicator
汽缸泄水管 cylinder feed pipe
汽缸旋塞 cylinder cock
汽缸压衬圈 cylinder liner packing ring
汽缸压力 cylinder pressure
汽缸压力比 pressure ratio in cylinder
汽缸压力表 compression ga(u)ge; cylinder compression ga(u)ge; cylinder pressure ga(u)ge
汽缸压力计 cylinder compression ga(u)ge
汽缸压缩比 compressive ratio in cylinder
汽缸压缩表 cylinder compression ga(u)ge
汽缸压缩试验器 cylinder compression tester
汽缸研磨机 cylinder lapping machine
汽缸油 cylinder oil; locomotive oil; mineral turpentine
汽缸油管 cylinder oil pipe
汽缸油料 steam-cylinder stock
汽缸油原料 cylinder oil stock
汽缸有水套部分 jacketed portion
汽缸余隙 noxious space of cylinder
汽缸余隙空间 cylinder clearance space
汽缸罩 cleating of cylinder; cylinder cowl
汽缸直径 bore diameter; cylinder diameter
汽缸中分面支持 centring support
汽缸中心距 bore spacing
汽缸装置 cylinder arrangement; cylinder unit
汽缸自动注油器 impermeator
汽缸总成 cylinder assembly
汽缸组 bank of cylinders; cylinder (in-) block
汽缸组机油孔 cylinder block oil hole
汽缸作用 cylinder action
汽缸座 cylinder base; cylinder block; cylinder frame; cylinder seat
汽缸座架配合 cylinder frame fit
汽缸座凸缘 cylinder base flange
汽鼓 steam drum
汽管 auxiliary steam pipe
汽管丁字管节 steam pipe tee
汽管堵 steam pipe choke
汽管隔离保安阀 steam pipe isolating valve
汽管环 steam pipe ring
汽管架 steam pipe support
汽管接头 steam connection
汽管节垫密片 steam union gasket
汽管节垫圈 steam union gasket
汽管连管接 steam pipe union
汽管联管节 steam union
汽管塞 steam pipe plug
汽管隧道 steam tunnel
汽管缩径丁字管节 steam reducing tee
汽管套 steam pipe sleeve
汽管系 steam piping
汽管旋转接头 steam pipe swivel
汽管穴 steam pocket
汽管支架 steam pipe support
汽管柱螺栓 steam pipe stud
汽管装配 steam fitting
汽管装配工 steam fitter
汽柜 steam box
汽锅 boiler; steam boiler; steamer
汽锅防爆装置 hydrostat
汽锅间 stokehold
汽锅炉联管箱 boiler header
汽锅身 shell
汽锅室 stokehold
汽锅用煤 navigation-coal
汽焊工 burner
汽耗 consumption of steam; steam consumption
汽耗率 specific steam consumption; steam rate
汽化 boil(ing)-off; carburation < 内燃机中 >; carburat(t)ing; carburetion; carburate < 使气体与碳氢化合物混合 >; evaporate; evaporation; steam generation; steaming
汽化的 vapo(u)r-bound
汽化点 vapo(u)rizing point
汽化度 evaporation discharge
汽化度测定 evaporation test
汽化极限 vapo(u)r-bound
汽化计 atm(id)ometer
汽化降温法 evapor method; vapo(u)r method
汽化空气 carburet(t)ed air
汽化冷却 evaporation cooling
汽化冷却系统 evaporated cooling system
汽化率 evaporation rate; rate of evapo(u)ration
汽化平面 plane of vapo(u)rization
汽化器 carb; carburet(t)or[carburet(t)er] < 内燃机 >; evaporator; evaporizer; steam evaporator
汽化器打油泵 carburet(t)or tickler; tickler
汽化器的防冰器 carburet(t)or anti-icer
汽化器的扩散器 carburator barrel
汽化器的雾化装置 diffuser
汽化器风门控制 carburet(t)or throttle control
汽化器回火制止器 carburet(t)or backfire arrester
汽化器混合器 carburet(t)or mixing chamber
汽化器接头 carburet(t)or[carburet(t)er] fitting
汽化器节流板 carburet(t)or throttling plate
汽化器节气门 carburet(t)or throttle
汽化器进口 carburet(t)or intake
汽化器进气阀杆 carburet(t)or air lever
汽化器进气口 carburet(t)or air scoop
汽化器空气喷嘴 carburator air funnel
汽化器扩散管 carburet(t)or muff
汽化器滤网 carburet(t)or strainer
汽化器喷(油)口 carburet(t)er[carburet(t)or] jet
汽化器喷嘴 carburet(t)or jet; carburet(t)or nozzle
汽化器歧管 carburet(t)or manifold
汽化器起动注油器 carburet(t)or primer
汽化器气温 carburet(t)or air temperature
汽化器去冰器 carburet(t)or deicer
汽化器式发动机 carbureted engine; carburet(t)or engine; non-diesel engine
汽化器试验台 carburet(t)or flow bench
汽化器调整针 carburet(t)or adjusting needle
汽化器吸入管 carburet(t)or intake adapter
汽化器主射口 carburet(t)or main jet
汽化器阻风门 carburet(t)or choke
汽化前缘 evaporization front
汽化热 enthalpy of vaporization; evaporation heat; heat of evapo(u)ration; heat of vapo(u)rization; vapo(u)rization heat
汽化熵 entropy of evapo(u)ration
汽化升压 pressure build-up vapo(u)rized
汽化升压阀 pressure build-up vapo(u)rized valve
汽化室 evaporizer
汽化水 boiling water; evaporous water
汽化速率 rate of evapo(u)ration
汽化损耗 boil off
汽化温度 evaporating temperature; steaming temperature; vapo(u)rization temperature; volatilization temperature
汽化型发动机 carburetor(-type) engine
汽化性 evaporability
汽化压力 evaporating pressure; saturation steam pressure
汽化油 vapoil
汽化值 evaporation number
汽化作用 carburation
汽阱 steam trap
汽口 steam port
汽冷壁 steam-cooled wall
汽冷涡轮机 bleeder condensing turbine
汽犁 steam plough
汽力 steam power
汽力铲 steam shovel
汽力厂 steam plant
汽力打桩机 steam pile driver
汽力锯 steam saw
汽力掘凿机 steam digger; steam navvy
汽力炉箅摇动机 steam grate shaker
汽力起重机 steam crane
汽铃 compressed ring
汽流分路过热器 superheater with division of the steam current
汽流喷雾器 steam-jet atomizer
汽馏的 steam distilled
汽馏松节油 light wood oil
汽炉片 steam radiator
汽炉围栏 enclosure of radiators
汽路 steam passage
汽路排水孔 steam passage drain
汽轮泵 steam-jet air pump; steam-jet pump; turbine pump; turbopump
汽轮发电机 steam-turbine-driven alternator; steam-turbine generator; steam turbo-alternator; steam turbo-generator; turbine generator; turbo-generator; turbonator
汽轮发电机组 steam turbo-generator set; steam turboset; turbo-generator set; turboset; turbo-unit
汽轮鼓风机 turbo-blower
汽轮机 air turbine; steam turbine; turbine
汽轮机舱 steam-turbine room
汽轮机抽气 bleeder steam
汽轮机出力 output of turbine
汽轮机船 turbine ship; turbine steamer
汽轮机电气驱动 steam-turbine-electric(al) drive
汽轮机乏汽参数 exhaust condition
汽轮机房 turbine hall
汽轮机功率 output of turbine
汽轮机机车 steam-turbine locomotive
汽轮机继动器 turbine servomotor
汽轮机金属温度 turbine metal temperature
汽轮机进汽量 throttle flow
汽轮机控制 turbotrol
汽轮机零件 steam-turbine parts
汽轮机轮船 turbine steamer
汽轮机启动盘 turbine start-up panel
汽轮机驱动起锚机 steam-turbine-driven windlass
汽轮机-燃气轮机联合循环 combined steam and gas turbine cycle
汽轮机润滑 steam-turbine lubrication
汽轮机润滑系统 steam-turbine lubricating system
汽轮机通流部分间隙 flow passage clearance
汽轮机叶片 turbine blade
汽轮机油 steam-turbine oil; turbine oil
汽轮机蒸汽管道 steam-turbine lead
汽轮机主凝结水 full flow condensate
汽轮机转子 turbine rotor
汽轮机装置 steam-turbine plant
汽轮机自动控制 turbomat
汽轮机自动遥控系统 steam-turbine

automatic remote control system
汽轮机组 turboset
汽轮交流发电厂 turbo-alternator plant
汽轮交流发电机组 steam turbo-alternator;turbo-alternator set
汽轮式混合器 turbo-mixer
汽轮式搅拌器 turbine-type agitator
汽轮压缩机 steam turbo-compressor;turbo-compressor
汽轮增压器 turbocharger
汽轮整流机 turbo-converter
汽门 port(hole);steam port;throttling port;valve
汽门顶杆 valve plug
汽门定时 port timing
汽门杆 valve link
汽门开度 throttle lift
汽门磨床 valve grinder
汽门挺杆导孔 valve-tappet guide
汽门行程 valve stroke
汽门研磨膏 valve grinding compound
汽门座磨光机 valve seat grinder
汽门座镶圈 valve seat insert
汽密 steam packing
汽密的 steam tight
汽密灯罩 steam tight globe
汽密封 steam packing
汽密接合 steam tight joint
汽密性 steam tightness
汽密性试验 steam tight test
汽囊 steam pocket
汽泡 boiler drum;steam bubble;steam void;vapo(u)r bubble
汽泡崩裂 bubble crack
汽泡式浮选机 bubble flo a tation machine
汽泡式量角器 bubble protractor
汽泡式座舱罩 bubble canopy
汽泡水准测角仪 bubble quadrant
汽膨胀 expansion of vapo(u)r
汽气混合物 steam-gaseous mixture
汽汽热交换器 steam-to-steam heat exchanger
汽热机 steam heater
汽热器 steam heater
汽塞 steam binding;steam lock(ing);vapo(u)r lock(ing)
汽射器 steam injector
汽蚀 cavitation;cavitation pitting;steam corrosion
汽蚀保证 pitting guarantee
汽蚀剥蚀 pitting due to cavitation
汽蚀程度 level of cavitation;degree of cavitation
汽蚀等级 level of cavitation
汽蚀工况 cavitation condition
汽蚀机理 mechanism of cavitation
汽蚀浸蚀 cavitation erosion
汽蚀坑 pitting due to cavitation
汽蚀坑体积 pitting volume
汽蚀类型 kind of cavitation
汽蚀模型试验 model test of cavitation
汽蚀破坏作用 cavitation attack
汽蚀失重 pitting volume
汽蚀试验 net positive suction head test
汽蚀特性曲线 cavitation characteristic curve
汽蚀系数 pitting coefficient
汽蚀仪 cavitation meter
汽蚀余量 net positive suction head
汽室 steam chamber;steam chest;steam dome;steam pocket
汽室衬套 steam chest bushing
汽室垫圈 dome base
汽室阀座 steam chest valve seat
汽室盖 steam chest cover
汽室盖衬垫 steam chest cover gasket
汽室平衡板 steam chest balance plate
汽室头 steam chest head
汽室凸缘 dome flange
汽室油堵 steam chest oil plug
汽室罩 steam chest casing
汽室蒸汽养护室 steam chamber
汽室座 steam chest seat
汽水 aerated water
汽水冲击 steam-water shock
汽水阀 steam trap;trap
汽水分界面 disengagement surface
汽水分离 steam-water separation
汽水分离器 catchwater;drying screen;dry separator;steam screen;steam separator;steam-water separator;automatic return trap <蒸汽采暖>
汽水分离器/再热器 moisture separator/reheater
汽水分离装置 steam-water separator
汽水隔离管 anti-priming pipe
汽水共腾 priming
汽水换热器 steam-water(type) heat exchanger
汽水混合式换热器 steam-water(type) mixed heat exchanger
汽水混合物 steam-water mixture
汽水机 aerated water machine
汽水加热器 steam-to-water heating element;steam water heater
汽水瓶 pressurized bottle
汽水填密胀圈 steam-and-water gland
汽水系统 boiler circuit
汽水制造机 gasogene
汽态 steam state
汽态冷却 vapo(u)r phase cooler
汽套 steam casing;steam jacket
汽套加热 jacket heating
汽提 steam stripping;stripping
汽提车间 stripper plant
汽提除油器 fuel oil stripper
汽提段 stripper zone
汽提法 stripping method
汽提法除氨 ammonia removal by air stripping
汽提法除氮 nitrogen removal by air stripping
汽提法处理 stripping method
汽提废水 stripping wastewater
汽提裂化 stripping cracking
汽提溶极谱分析 stripping dissolving polarographic analysis
汽提闪蒸罐 stripper flash drum
汽提试验 stripping test
汽提塔 stripper(tower);stripping column;stripping drum;stripping tower
汽提塔效率 stripper efficiency
汽提塔再沸器 stripper reboiler
汽提脱附 desorption by stripping
汽提蒸馏器 stripping still
汽提装置 stripper plant
汽艇 autoboat;gas boat;gasoliner;launch;motor boat;motor launch;power-boat;steam boat;steam launch;steam yacht
汽艇旅馆 boatel
汽艇声 motorboating
汽艇型工作船 launch-type work boat
汽艇游客旅馆 boatel
汽相分解 vapo(u)r decomposition
汽相罐 vapo(u)r phase can
汽相刻蚀 gas etching
汽穴 steam void
汽穴衰退 vapo(u)r-cavity decay
汽穴消减 vapo(u)r-cavity decay
汽压泵 pulsometer pump
汽压表 steam ga(u)ge;steam pressure ga(u)ge
汽压表安装座 steam gage stand
汽压表配件 steam gage fitting
汽压表塞门 steam gage cock
汽压计 steam ga(u)ge;tonometer
汽压计旋塞 steam ga(u)ge cock
汽压计座 steam ga(u)ge plate;steam ga(u)ge stand
汽压力 steam pressure
汽压试验 steam pressure test
汽压水泵 pulsometer pump
汽眼 steam seep
汽液分离喷嘴 steam-and-water separation nozzle
汽液界线 steam-limit curve
汽液旁承 hydropneumatic side bearing
汽液平衡常数 vapo(u)r-liquid equilibrium constant
汽油 benzin(e);benzoline;gas oil;gasoline(oil);liquid power;mobile oil;motor spirit;petroleum spirit;petrol spirit;petrol(eum) <英>
汽油爆击性试验 knock test of gasoline
汽油爆震 knocking
汽油爆震性 gasoline knocking
汽油泵 gasoline feed pump;gasoline(lift) pump;gasoline pump;gasoline supply pump;gas pump;petrol pump
汽油泵沉淀杯 gasoline pump settling bowl
汽油泵沉淀玻璃杯 gasoline pump settling glass bowl
汽油泵壳 gasoline pump case;petrol pump case
汽油泵膜片保护垫圈 gasoline pump diaphragm protecting washer
汽油泵油杯U形钢丝夹 gasoline pump glass-bowl U-clamp
汽油泵总成 gasoline pump assembly
汽油变质 gasoline deterioration
汽油表 gasoline content ga(u)ge;gasoline ga(u)ge;gas tank ga(u)ge;petrol-depth ga(u)ge;petrol ga(u)ge;petrol-level ga(u)ge
汽油表开关 gas ga(u)ge switch
汽油表刻度盘 petrol-ga(u)ge dial
汽油表盘支架 gasoline ga(u)ge dial bracket
汽油表指示器 gasoline ga(u)ge dash unit
汽油驳船 gasoline barge;petrol barge
汽油不溶物 naphtha insolubles
汽油不足 gasoline hungry;gasoline shortage
汽油仓库 gas dump
汽油槽 gasoline tank
汽油槽车 gasoline carrier
汽油槽油面指示器 gasoline tank ga(u)ge
汽油掺四乙基铅 tetraethyl lead blending of gasoline
汽油掺添加剂 doping of gasoline;gasoline doping
汽油产率 gasoline yield
汽油铲掘机 gasoline shovel
汽油铲土机 gasoline shovel
汽油车 gasoline car
汽油车辆 gasoline vehicle
汽油沉淀试验 gasoline precipitation test
汽油沉降器 petrol sediment bulb
汽油秤 gasoline scale;petrol balance
汽油冲击式凿岩机 <手动的> gas hand hammer(rock) drill
汽油储备 gasoline stocks
汽油储存量的稳定性 petrol storage stability
汽油储存桶 gasol(ine) storage can
汽油储存稳定性 gas(oline) storage stability
汽油储存箱 gasoline storage tank
汽油储油器 gasoline reservoir
汽油船 gasoline tanker
汽油吹管 gasoline blow pipe torch
汽油代用品 gasoline substitutes
汽油单耗英里数 <汽车每耗1加仑汽油所能行驶的平均英里数;1英制加仑≈4.546升,1美制加仑≈3.785升;1英里≈1.609千米> gasoline mileage
汽油的供应 petrol delivery
汽油的加速馏分 gasoline pick-up fraction
汽油的精制 refining of gasoline
汽油的抗爆剂 tetraethyl-lead
汽油的抗爆性 anti-knock property of gasoline
汽油的凝结剂 gasoline coalescer
汽油的稳定化 stabilization of gasoline
汽油的质量 quality of gasoline
汽油的中间馏分 gasoline intermediate fraction
汽油灯 gasoline burner
汽油等级 gasoline grade
汽油滴 gas(oline) drippings
汽油滴漏 gasoline leak
汽油电车 petrol-electric(al) car
汽油电动车辆 gasoline-electric(al) car;gasoline-electric(al) vehicle
汽油电动的 gasoline-electric(al)
汽油电动公共汽车 gasoline-electric(al) bus
汽油电动卡车 gas-electric(al) truck
汽油电力车 gasoline-electric(al) vehicle
汽油电力动车 gas electric(al) rail-car
汽油电力牵引 petrol-electric(al) traction
汽油电力牵引车 gasoline electric(al) tractor
汽油动力机 gasoline-engine driven machine
汽油堆场 gasoline dump
汽油堆栈 gas dump;gasoline dump
汽油发电机 gasoline generator;petrol generator
汽油发动的挖掘机 gasoline shovel
汽油发动机 carbureted engine;carburet(t)or engine;gas(oline) engine;gasoline motor;petrol(eum) engine;petrol motor
汽油发动机废气排放 petrol engine exhaust emission
汽油发动机驱动的 petrol-driven
汽油发动机驱动的路碾 petrol-driven roller
汽油发动机驱动的压路机 petrol-driven roller
汽油发动机驱动的载重汽车 petrol-driven truck
汽油发动机载货汽车 gasoline powered truck
汽油发动机组 gasoline generator set
汽油发动性 kick
汽油阀(门) gasoline valve;petrol cock
汽油防爆剂 dopes for gasoline
汽油防冻混合物 gasoline anti-freeze mixture
汽油防冻添加剂 gasoline anti-icing additive
汽油废水 petrol sewage
汽油分离阱 gasoline trap
汽油分离器 gasoline separator;gasoline trap;petrol separator;petrol trap
汽油分馏塔 gasoline splitter
汽油分配站 gasoline distributing point

汽油分配装置 gasoline dispensing equipment
汽油分散剂 gasoline dispersant
汽油分送设备 gasoline dispensing facility
汽油浮标表 gasoline ga(u)ge with float
汽油浮力系统 gasoline buoyancy system
汽油供给管 gasoline feed pipe
汽油供给量调节阀 gasoline adjustment valve
汽油供应员 gasoline officer
汽油管 gasoline pipe;petrol pipe
汽油管路 gasoline(pipe)line
汽油罐 gasoline can;gasoline cistern;gasoline container;gasoline tank;jerry can;petrol container
汽油罐车 gas tank truck
汽油柜 gasolene tank;petrol tank
汽油过滤器 gasoline filter;petrol filter
汽油含硫试验 doctor test
汽油壶 petrol can
汽油挥发性试验 gasoline volatility test
汽油混凝土拌和机 gasoline concrete mixer
汽油混凝土搅拌机 gasoline concrete mixer
汽油机 carburet(t)ed engine;carburet(t)or engine;petrol(eum)engine;petrol motor
汽油机叉车 petrol engine fork truck
汽油机车 gasoline locomotive;gasoline powered truck;petrol motor car
汽油机船 gasoline-engine boat;petrol engine boat
汽油机电机发动的牵引车 gasoline electric(al)tractor
汽油机电机公共汽车 petrol-electric(al)bus
汽油机电机驱动 petrol-electric(al)drive
汽油机动力铲 gasoline shovel
汽油机发电机(机)组 petrol-electric(al)generating set;gasoline generator set;generating set with petrol engine
汽油机发动的起重机 gasoline crane
汽油机发动的拖拉机 gasoline tractor;petrol-engined tractor
汽油机公共汽车 gasoline bus
汽油机加油系统 fuel system of gasoline engine
汽油机排气 gasoline-engine exhaust
汽油机起动 gasoline starting
汽油机汽车 gasoline automobile;petrol motor car
汽油机牵引车 gasoline tractor
汽油机驱动 gasoline-engine driven
汽油机驱动泵 petrol(-driven)pump
汽油机驱动操作打桩锤 petrol operated pile driving hammer
汽油机驱动插入式振捣器 petrol poker vibrator
汽油机驱动插入式振动器 petrol poker vibrator
汽油机驱动的插入式混凝土振动器 petrol spud vibrator for concrete
汽油机驱动的绞车 petrol-powered winch
汽油机驱动的自动装卸车 petrol-powered lift truck
汽油机驱动内部振捣器 petrol poker vibrator
汽油机驱动内部振动器 petrol poker vibrator
汽油机驱动手扶冲击式钻机 petrol hand hammer rock drill
汽油机式夯板 petrol compacting and finishing screed
汽油机式压路机 gasoline roller;pedestrian-controlled roller
汽油机式振动平板夯 gasoline vibro plate
汽油机式振动压路机 gasoline vibratory roller
汽油机拖拉机 petrol tractor
汽油积聚 piling up
汽油集材机 gasoline skidder
汽油计量泵 gasoline meter pump
汽油计量表 gasoline meter
汽油加热器 gasoline heater
汽油加油车 gasoline bowser;gasoline dispenser;petrol bowser;refueller
汽油加油站 gasoline filling station;petrol filling station;petrol station
汽油加注口盖 petrol filler lid
汽油加注量桶 gasoline measure
汽油胶 gasoline gum
汽油胶管 petrol hose
汽油胶管接头 gasoline rubber hose coupling
汽油胶质测定杯 gasoline gumming test cup
汽油节油器 gasoline economizer
汽油截留设施 petrol interceptor
汽油精制 gasoline refining
汽油净化器 gasoline separator
汽油卷扬机 gas hoist
汽油抗爆 anti-detonation
汽油抗爆剂 anti-knock;knock suppressor
汽油抗震性 gasoline knocking
汽油空气混合气 petrol-air mixture
汽油空气混合物 gasoline mixture
汽油库 gasoline storage;petrol depot;petrol dump;petrol storage
汽油库模型 gas reservoir model
汽油冷凝器 gasoline condenser
汽油离析器 gasoline separator
汽油里程 <每加仑汽油行驶的英里数;1 英制加仑≈4.546 升,1 美制加仑≈3.785 升;1 英里≈1.609 千米> gasoline mileage
汽油漏油 gas drippings
汽油滤杯 gasoline filter bowl
汽油滤(清)器 gas(oline)filter;gasoline interceptor;gasoline separator;gasoline trap;petrol filter;petrol strainer;petrol trap
汽油路碾 gasoline roller;petrol motor roller;petrol roller
汽油内燃机 gasoline engine
汽油凝气阱 gasoline trap
汽油配给 gasoline rationing;petrol rationing
汽油喷灯 benzine blow lamp;gasoline blast burner;gasoline torch;gasoline torch lamp;petrol burner
汽油喷射 gasoline injection
汽油喷射泵 gasoline-injection pump
汽油喷射式发动机 gasoline-injection engine
汽油启动机 gas starting engine;petrol starting engine
汽油起动性质 gasoline startability
汽油起重机 gasoline crane
汽油汽 petrol trap
汽油汽车 gasoline motor car
汽油驱动 gasoline-driven
汽油驱动插入式振捣器 petrol immersion vibrator;gasoline internal vibrator
汽油驱动插入式振动器 petrol immersion vibrator;gasoline internal vibrator
汽油驱动的 gas-driven
汽油驱动的打桩锤 gasoline operated(pile driving)hammer
汽油驱动的打桩机 gas operated(pile driving)hammer
汽油驱动的卷扬机 gasoline-powered winch
汽油驱动的内部振动器 petrol internal vibrator
汽油驱动的起动机 gasoline starting engine
汽油驱动的提升机 petrol hoist
汽油驱动的自动装卸车 gasoline-powered lift truck
汽油驱动杆状振动器 petrol needle vibrator
汽油驱动卷扬机 gas-powered winch;gasoline hoist
汽油驱动卡车 gas(oline)-driven truck
汽油驱动提升机 gasoline hoist
汽油驱动压路机 gas(oline)-driven roller
汽油容量 gasoline capacity
汽油容器 gas vessel
汽油深度计 gasoline depth ga(u)ge;petrol-depth ga(u)ge
汽油生产设备 petrol generating set
汽油生产装置 petrol generating set
汽油试蚀杯 gasoline corrosion cup
汽油试验器 gasoline tester
汽油试验装置 gasoline testing outfit
汽油室 gasoline chamber
汽油输送 gasoline carrying
汽油税 gasoline tax
汽油损失控制 gasoline loss control
汽油特性 gasoline performance
汽油提升机 gas hoist
汽油添加剂 additive for gasoline;gasoline additive
汽油添加剂污染 pollution by gasoline additives
汽油调和料 gasoline blending stock
汽油调和组分总和 gasoline pool
汽油桶 gas can;gas(oline)drum;petrol can;petrol drum
汽油拖拉机 gasoline tractor
汽油脱硫设备 gasoline sweetener
汽油污水 gas(oline)sewage
汽油箱 gasolene compartment;gasoline tank;petrol tank
汽油箱放泄阀 gasoline tank dump valve
汽油箱浮子 gasoline tank float
汽油箱盖 gas(oline)(tank)cap
汽油箱盖吊链 gasoline tank cap chain
汽油箱加油口管 gasoline filler neck
汽油箱加油器 gasoline tank filler
汽油箱架 gasoline tank bracket
汽油箱滤油器 gasoline tank settler
汽油箱气门 gasoline tank valve
汽油箱容积 gasoline tank capacity
汽油箱油位表 gasoline tank ga(u)ge
汽油箱支架 gasoline tank bracket
汽油橡胶管 gasoline rubber hose
汽油消耗 consumption of gasoline
汽油消耗定额 gasoline ration
汽油消耗量 gallonage;gasoline consumption;petrol consumption
汽油消耗量测定装置 gasoline mileage tester
汽油消耗率 gasoline consumption rate;specific gasoline consumption
汽油辛烷值 gasoline octane number;gasoline octane rating;octane number;octane rating
汽油旋塞 gasoline cock;petrol cock
汽油压力计 gasoline pressure ga(u)ge;petrol-pressure ga(u)ge
汽油压力手泵 gasoline pressure hand pump
汽油压路机 gasoline roller;petrol motor roller;petrol roller
汽油压缩机 gasoline plant
汽油液面指示器 gasoline ga(u)ge;petrol level indicator
汽油引擎 gasoline engine;petrol(eum)engine
汽油油槽车 petrol tanker
汽油油滴 petrol dripping
汽油油罐车 gasoline tank car;gasoline tank truck;gas tanker;gas truck
汽油油量表传感线 gasoline ga(u)ge line
汽油油量表开关 gasoline ga(u)ge switch
汽油油面指示器 gasoline level indicator
汽油油位 gasoline level
汽油油位表 gasoline depth ga(u)ge;gasoline level ga(u)ge
汽油油位表传感器 gasoline ga(u)ge take unit
汽油原油 petroleum benzine
汽油运输车 gasoline carrying vehicle
汽油运输船 gasoline tanker
汽油站 bulk gasoline plant;gas(oline)station;petrol dump;petrol station
汽油振捣器 gasoline-proof;gasoline vibrator
汽油振动器 gasoline vibrator
汽油蒸气 petrol fume;petrol trap
汽油正规品种 house-brand gasoline
汽油中的显胶 present gum in gasoline
汽油中毒 gasoline poisoning
汽油中胶质测定 green test
汽油中硫醇的吸收塔 gasoline mercaptan absorber
汽油中硫含量的测定 gasoline sulfur test
汽油中铅 lead in petrol
汽油中四乙铅含量的测定 gasoline tetraethyl lead test
汽闸 steam brake
汽蒸 steaming
汽蒸法 steam evapo(u)ration;steaming process
汽蒸机 decatizer
汽蒸器 steamer
汽蒸室 steaming chamber
汽蒸松节油 gum turpentine oil
汽蒸圆筒 steam tumble
汽转球 aeolipile
汽桩锤 steam pile hammer
汽阻 vapo(u)r lock(ing)
汽阻损耗 steam friction loss

契比舍夫多项式 Chebyshev polynomial

契比舍夫分布 Chebyshev distribution
契比雪夫平差 <改正数之和为最小的平差> Chebyshev adjustment
契合法 method of agreement
契据 bond;contract;deed;escrow;title deed;title deed for land
契据本文 body of deed
契据登记簿 cartulary;deed books
契据登记费 recording fee
契据登记员 registrar of deeds
契据副本 antigraphy
契口键 border
契连柯夫计数管 Cerenkov's counter
契税 contract tax;deed tax
契约 agreement;bargain(ing);charter;compact;concordat;contract;deed;indenture;muniment;obliga-

tion;pact(ion)
契约标的 subject matter of contract
契约丙方 party C
契约持有人 escrow holder
契约担保<房地产> deed warranty
契约当事人 contracting party;contraction party
契约的 contractual
契约法 contractual law;conventional law;escrow contract
契约废案证书 acquittance
契约费用 escrow cost
契约副本 subcontract
契约规定 contractual specification
契约规定储备金 contractual reserve
契约规定的义务 contractual duty;contractual obligation
契约规定股利 contractual dividend
契约规定航线 contractual route
契约规定佣金 stipulated commission
契约规定准备金 contractual reserve
契约货币 currency of contract
契约货品 contract goods
契约甲方 party A
契约价格 contract of price
契约见证人 attesting witness
契约解除 dissolution
契约劳工 contract labo(u)r;indenture labo(u)r
契约履行期限 contract date;term of contract
契约落空 frustration;frustration of contract
契约期满 termination of agreement
契约期满日 terminal of an agreement
契约曲线 contract curve
契约融资 contract financing
契约商 contractor
契约上的罚金 contractual fines
契约上的价值 contractual value
契约上的认可 contractual acknowledgement
契约上的收益 contractual revenue
契约生效 execution of contract
契约式合营 contractual joint undertaking
契约式合资经营 contractual joint(ad)venture
契约式合资企业 contractual joint(ad)venture
契约收入 contract revenue
契约收益 benefit of the bargain
契约手续费 escrow fee
契约受托人 indenture trustee
契约受益方 covenantee
契约书 contract note
契约谈判 contract bargaining
契约条件 conditions of contract
契约条款 contract term;covenant;stipulation
契约条款释意信 letter of intent(ion)
契约文字 contractual language
契约无效 frustration of contract
契约限定 deed restriction
契约箱 deed box
契约效力 validity of contract
契约协议 agreement of deed;contractual agreement
契约形式的合同 contract by deed
契约性的初步协定 preliminary contractual agreement
契约性投入 contractual input
契约性投资 contractual investment
契约学徒 articled clerk
契约要素 essence of the contract
契约乙方 party B
契约有效期间 duration of contract
契约运费制度 contract rate system
契约债务 contract debt
契约中的废止条款 defeasance clause
契约终止 termination of a contract;termination of an agreement
契约终止(日)期 terminal date of an agreement
契约自由 freedom of contract
契约总价协定 stipulated sum agreement

砌

半石基底 half-bed
砌壁方法 walling system
砌壁凸出<玻璃制品缺陷> bulged side
砌壁砖 lining brick
砌扁石工 ragwork
砌层单元 unit of bond
砌层段 unit of bond
砌衬 lining-up
砌成拱顶 wedging of arch top
砌成拱形 spring an arch
砌成墙角的砖石 coin
砌出一些 set out
砌窗台砖 window sill brick
砌道 causeway;causey
砌底砖 field tile
砌端梁 beam with both ends built-in;built-in-end beam
砌墩块材 pier block
砌多角石圬工 trapezoidal masonry
砌垛工 pillar man
砌方块防波堤 block masonry breakwater
砌方块墙 masonry blockwork wall
砌方石 ashlaring
砌缝 masonry joint;paving seam;face joint;mason's joint
砌拱垫块 camber block;camber piece;camber slip
砌拱洞 spring an arch
砌拱弧形模板 turning piece
砌拱脚手架 arch center[centre]
砌拱模架 camber slip;soffit scaffolding
砌拱圈 spring an arch
砌拱屋顶 sprung roof
砌拱楔块 voussoir
砌拱用的楔形块材 ring stone
砌拱用的楔形砖 ring stone;end arch
砌拱用石 arch stone
砌拱用砖 arch brick
砌拱支架 soffit scaffold(ing)
砌拱砖 arch brick;skewback;springer
砌沟石 gutter sett
砌合 bonding;set in bond
砌合层<砖或石的> bond course;bond layer
砌合传递 bond transfer
砌合单元 unit of bond
砌合的 bonded
砌合丁头层 bonding header course
砌合丁(头)砖 bonding header
砌合法 bonding method
砌合方块 bonded block;fitted block
砌合拱 bonded arch
砌合夹钳 chain bond
砌合结构 bonded construction;bonded structure
砌合空斗墙 masonry bonded hollow wall
砌合块体 fitted blocks
砌合类型 bond type
砌合路面 bonded surface
砌合锚 bonding anchor
砌合面层 bonded surface
砌合黏[粘]土 bond clay
砌合强度 bond strength
砌合砂浆 pointing mortar
砌合石 bond stone
砌合饰面 bonding finish
砌合头(石)咬合砌(石)法 bond header
砌合屋顶 bonded roof
砌合形式<砖的> type of bond
砌护 pitch(ing)
砌护坡 slope paving
砌花墙 pattern bond
砌混水墙的空心砖 backup tile
砌结面 bonding plane
砌井壁 hanging of lining
砌井砖 well brick
砌空隙垫层 shell bedding
砌口 column offset
砌块 block;building block;masonry block;masonry mass;masonry unit
砌块岸壁 block wall
砌块驳岸 block-type quay
砌块层 block-layer
砌块厂<混凝土> building block factory;building block plant;building tile factory;building tile plant
砌块衬砌 block lining
砌块成型机 block-making machine
砌块尺寸 block format
砌块大小 block format
砌块堤岸 block-type quay
砌块钉 masonry nail
砌块堵填 ca(u)lking of segment
砌块隔断 block wall;blockwork partition
砌块隔墙 block partition(wall);block wall;blockwork partition
砌块工厂<混凝土> building block factory;building block plant
砌块工场 block yard
砌块工程 block masonry(work);blockwork;building block masonry(work);building tile masonry(work)
砌块工事 blockwork
砌块拱 block arch;voussoir arch
砌块构造 block construction
砌块构造(方)法 block construction method
砌块规格 block format
砌块过梁 block lintel;building block lintel;building tile lintel
砌块间 block house
砌块接缝宽度 bed-and-joint width
砌块举重器 segment erector
砌块抗压强度 block compression strength
砌块路面 block pavement
砌块码头 block-type quay;block-type wharf
砌块门扉 block leaf
砌块面层 block pavement
砌块模数 building block module
砌块模数单位 modular masonry unit
砌块模数法 building block module method
砌块模压制 block press
砌块黏[粘]结剂 adhesive for blocks
砌块拼装机械 segment erection machinery
砌块铺砌路面 block pavement
砌块砌合 block bonding
砌块砌墙 block laying
砌块砌体 block masonry(work)
砌块砌筑 block laying
砌块强度 block strength
砌块墙 building block wall
砌块墙工 block-walling
砌块墙体 blockwork;block masonry wall
砌块切削机 block splitter
砌块提升车 block-lifting machine
砌块圬工 block-walling
砌块圬工墙 block masonry wall;building block masonry wall;building tile(masonry)wall
砌块屋 block house
砌块坞墙 block-type dock wall
砌块烟囱 block chimney;block stack
砌块仰拱 segmental invert
砌块用量测定法 mason's measure
砌块预留孔 block joint
砌块振动 block vibration
砌块制造 block making
砌块制造机 block machine
砌炉材料 fireplace material
砌炉圬工工作 fireplace masonry work
砌路机 road paver
砌路用石块 paving stone block
砌乱毛石墙 rag rubble walling
砌毛石工 rubble masonry
砌面 encasing;facing(bond);finish on a wall;revet(ment)
砌面河岸 paving bank
砌面明沟 lined canal
砌面墙 veneered wall
砌面石 cover stone
砌面用灰浆 tiling plaster
砌面砖 cutter
砌面砖用灰浆 tiling plaster
砌木砖 battening;nogging
砌片石墙 coffer works
砌平面 tiling the plane
砌坡 slope paving
砌墙 carrying up;walling up
砌墙标杆 keeping ga(u)ge;keeping the ga(u)ge
砌墙到规定标高 carrying up to regulated elevation
砌墙的泥刀 narrow London
砌墙封住 close with brick wall
砌墙高度规准尺 keeping the ga(u)ge
砌墙工人 fixer
砌墙块石 bonder;walking block
砌墙留的齿形施工缝 toothing
砌墙泥 pise
砌墙石 bonder;bonding header
砌墙石层 course of bonders
砌墙石工 walling mason
砌墙石料 wall stone
砌墙时留齿缝待接 racking back
砌墙时留出阶梯形接缝 racking back
砌墙圬工 wailing masonry
砌墙用空心砌块 wall pot
砌墙用空心砖 pot for walls
砌墙用纤维织物 bonder fibre fabrics
砌墙砖 wall building tile
砌墙砖层 course of bonders
砌清水墙 dry wall(ing)
砌入 bedding into;boxing-in;built-in
砌入的 built-in
砌入端 tailing down;tailing in
砌入梁 fully restrained beam
砌入路面中央分道线 built-in center[centre] line
砌入墙内的托梁垫板 wall plate
砌入墙体中的 embedded in masonry
砌入石块或混凝土中 bedding into stone or concrete
砌入式窗框 built-in frame
砌入式肥皂粉配出器 recessed powdered soap dispenser
砌入式肥皂配出器 recessed soap dispenser
砌入式生活设备<尤指与房屋结构成为一体的全套卫生设备> built-in comfort
砌入式饮水器 recessed drinking fountain
砌入式纸巾配出器 recessed paper towel dispenser
砌入照明 built-in lighting
砌入柱墩的墙 wall bonded to piers
砌塞墙洞 immuration
砌石 laying block;placed rockfill;stone

Q

masonry; stone pitching; pitched stone
砌石岸坡 beached bank
砌石坝 bonder dam; masonry dam; stone masonry dam
砌石板圬工 ragwork
砌石边坡 pitched slope
砌石挡土墙 stone masonry retaining wall
砌石导流堤 placed stone jetty
砌石法 stone bond
砌石工 mason; rubble mason; stone setter
砌石工程 assize; rubble works; stonework; stone masonry
砌石工人 stone mason; stone setter
砌石工用活动悬吊脚手架 stonesetter's adjustable multiple-point suspension scaffold
砌石拱 stone arch
砌石拱圬工 stone arching
砌石勾缝工具 sword
砌石涵洞 masonry-stone culvert
砌石护岸 beaching; stone revetment
砌石护岸壁 masonry quay wall
砌石护码头 masonry quay
砌石护面 pitching; placed stone facing; protective pitching; stone facing; stone-pitched facing; stone pitching; stone paving
砌石护面的 stone-faced
砌石护面堤 mound breakwater with laying stones armo(u)r
砌石护坡 beaching; masonry pitching of slope; pitched face; pitched slope; pitched works; pitching; stone pitching
砌石护坡分水岛 pitched island
砌石护坡基床 bed of pitching
砌石护坦 stone apron
砌石灰浆 stuc mixture; stuc stuff
砌石接缝垫层 joint bedding
砌石接缝宽度 bed-and-joint width
砌石节制闸 masonry check
砌石结构 masonry structure; stonework
砌石块工人 blockman
砌石码头 masonry quay wall
砌石面 revetment
砌石面层 stone paving; stone surfacing
砌石排水沟 stone drain
砌石配水闸 masonry check
砌石墙 stone wall(ing)
砌石墙工人 dyker
砌石突出层 oversailing course
砌石土 terracing with stones
砌石圬工 cut-stone masonry; masonry-stone; stone masonry
砌石整形 wasting
砌石支护 masonry support
砌石重力坝 stone masonry gravity dam
砌体 bricking-up; brickwork; masonry (envelope)
砌体岸壁 masonry wall
砌体半分接角法 mason's miter[mitre]
砌体表面假缝 false joint
砌体表面清洁整齐 fair faced
砌体层收分 set in
砌体垂直缝 cross joint; straight joint
砌体(大块)结构 blockwork structure
砌体带状勾缝 ribbon pointing
砌体单位 unit masonry
砌体单元 unit masonry
砌体的名义尺寸 nominal dimension of masonry
砌体底部砂浆找平层 bedding course
砌体垫层砂浆 bedding mortar
砌体端头竖缝 head joint
砌体断线 hacking
砌体封顶层 blocking course
砌体附件 masonry fixing
砌体钢筋 masonry reinforcement
砌体工程 masonry
砌体工程单元 masonry unit
砌体工程的可调式多点悬挂脚手架 mason's adjustable multiple-point suspension scaffold
砌体工程计量方法 mason's measure
砌体工程拉杆 masonry tie
砌体工程墙 masonry wall
砌体工程墙顶 masonry capping
砌体工程墙角护条 masonry work angle bead
砌体工程砂浆勾缝 masonry pointing
砌体工程石凿 masonry bit
砌体工程饰面墙板 masonry veneer
砌体工程饰面砖 masonry veneer
砌体工程用锤 mason's hammer
砌体工程用钉 masonry nail
砌体工程用石灰 mason's lime
砌体工程用水泥 masonry cement
砌体工程用细琢石锤 mason's ax
砌体工程用准线 mason's line
砌体工程油灰 mason's putty
砌体工程预留齿形接合面 masonry toothing
砌体勾缝 masonry pointing; tooled joint
砌体勾缝面以 45 度向下倾斜 weathered pointing
砌体灰缝 masonry joint
砌体基础 foundation of masonry
砌体基底 masonry base
砌体结构 masonry structure
砌体结构房屋 masonry structure
砌体结合 binding of stones
砌体截面净面积 net cross-sectional area of masonry
砌体金属系板 masonry metal tie
砌体绝缘材料 masonry insulation
砌体拉杆 masonry tie
砌体力 body force
砌体连系杆 masonry tie
砌体漏泄 setting leak
砌体马牙槎 masonry toothing
砌体马牙接头 masonry toothing
砌体锚碇件 masonry anchor
砌体面层块体 facing unit
砌体膨胀 expansion of brickwork
砌体砌缝 mason's joint
砌体强度 masonry strength; resistance of masonry
砌体墙 masonry wall
砌体墙的隔热隔音层 masonry insulation
砌体墙改变方向点的异形砌块或石料 kneeler
砌体墙盖顶 masonry capping
砌体墙面板 masonry wallboard
砌体墙内填塞的块石 moellon
砌体墙上留出放置脚手架横向支承的凹处 putlog hole
砌体墙上预埋钩子 wall hook
砌体墙转角钢拉杆 angle bond
砌体砂浆 masonry mortar
砌体砂浆用砂 masonry aggregate
砌体伸缩 movement in masonry
砌体竖缝 head joint
砌体水平缝 bed joint
砌体填充材料 backing-filling material
砌体铁拉件 hoop-iron bond
砌体凸出墙面的带状层 projecting belt course
砌体斜接缝 mason's miter[mitre]; mason's stop
砌体斜接面 mason's miter[mitre]
砌体泄水缝 weatherboard joint
砌体泻水缝 <勾缝面以 45 度向下倾斜> weather struck joint
砌体用钉 masonry nail
砌体用木条连接 timber bond
砌体用碎瓷瓦贴面 opus testaceum
砌体胀缩变形 movement in masonry
砌体中的暗销 bed dowel
砌体中的锚梁 beam anchorage in masonry
砌体中间的垂直空洞中浇筑钢筋混凝土形成能抗水平力的墙 Quetta bond wall
砌体装饰面层 veneer
砌体装饰面层的拉杆 veneer tie
砌外墙用大理石 exterior cubic(al) marble
砌小拱的弧形板 turning piece
砌斜坡 slope paving
砌窨井扇形砖 arch brick for manholes
砌有彩色面砖的内墙 veenered wall
砌造 lay
砌遮阳光花墙的瓷砖 solar screen tile
砌遮阳光花墙的(琉璃)瓦 solar screen tile
砌筑 laying; masonry
砌筑坝 masonry dam
砌筑车间 block plant
砌筑单位 masonry unit
砌筑单元 unit masonry work
砌筑堤道 causeway
砌筑地沟 brick-laying channel
砌筑法 <其中之一种> keying mix
砌筑方法 laying technique
砌筑方块工程 bonded blockwork
砌筑防浪墙的黏[粘]土砖 parapet brick
砌筑工 masonry
砌筑工厂 block plant
砌筑工程 masonry work
砌筑工作 laying work
砌筑拱 masonry arch
砌筑拱桥 masonry arch(ed) bridge
砌筑管道的空心黏[粘]土块 conduit hollow clay(building) block
砌筑和填筑混合坝 combination-type dam
砌筑护坡 pitching
砌筑灰膏 mason's putty
砌筑技术 laying technique
砌筑胶结剂 bonding medium for laying
砌筑胶泥 masonry grout(ing)
砌筑脚手架 mason's scaffold
砌筑脚手架跳板 mason's runway
砌筑井壁 walling of a shaft
砌筑壳体模板 jack lagging
砌筑壳体用的模板 jack lagging
砌筑空心墙 masonry of hollow units
砌筑面层 masonry veneer
砌筑面积 laying area
砌筑内衬工作 lining work
砌筑女儿墙的黏[粘]土砖 parapet brick
砌筑砌体围地 masonry wall enclosure
砌筑砌体围圈 masonry wall enclosure
砌筑墙(体) masonry panel; masonry wall
砌筑穹隆 masonry vault
砌筑人行道 causeway
砌筑砂浆 bonding mortar; brick mortar; masonry grout(ing); masonry mortar; mason's mortar
砌筑砂浆用砂 building sand
砌筑石灰 masonry lime; mason's lime
砌筑水泥 brick cement; masonry cement
砌筑圬工 marshalling masonry
砌筑形式 <砖墙或圬工> pattern bond
砌筑烟囱 masonry chimney
砌筑用工具 masonry tool
砌筑用骨料 masonry aggregate
砌筑用集料 masonry aggregate
砌筑用石灰 masonry lime
砌筑用石灰石 masonry limestone
砌筑用熟石灰 mason's hydrated lime
砌砖 brick bonding; bricking; brick lay(er)ing; brickwork; installing brick; lay(ing) brick; laying tile; neat work
砌砖层 bond course
砌砖长尺 joint(ing) rule
砌砖的 bricky
砌砖吊盘 bricking scaffold
砌砖队 brick-laying crew
砌砖法 brick bond
砌砖辅助工 tupper
砌砖辅助拱 brick relieving arch
砌砖格 checkwork
砌砖工 bricklayer; brick mason; mason layer
砌砖工程 brickwork
砌砖工的锤 comb hammer
砌砖工的工具 bricklayer's tool
砌砖工的工作 bricklayer's work
砌砖工的脚手架 bricklayer scaffold(ing)
砌砖工的拉线 bricklayer line
砌砖工的砂浆板 fat board
砌砖工的梯子 bricklayer ladder
砌砖工脚手架 bricklayer's square scaffold
砌砖工具 bricklayer's tool; brick-laying tool; brickmason's tool; brick tool
砌砖工人的助手 hodman
砌砖工用锤 brick hammer
砌砖工助手 <搬运砖瓦、灰泥> hod carrier; hodman
砌砖工作 brickwork
砌砖工作队 brick-laying crew
砌砖挂线 line and pin
砌砖灌浆 grouted brick
砌砖规尺 jointing rule
砌砖灰浆 brick mortar
砌砖火泥 brick mortar
砌砖机 brick layer
砌砖挤浆 shove
砌砖挤浆缝 shove joint in brickwork
砌砖浆 mortar
砌砖拉结件 <用于砌空心墙> brick clamp
砌砖拉线 mason's lead
砌砖镘刀 brick trowel
砌砖面层 brick veneer
砌砖内衬 brick lining
砌砖泥刀 laying trowel
砌砖铺灰板 mortar tray
砌砖墙 line with bricks; parapet brick
砌砖墙用的铅直线及水平线 pins and line
砌砖砂浆 brick mortar
砌砖石的底层 grade course
砌砖式码堆 brick pattern
砌砖竖缝 header joint
砌砖体 brick bond; brick masonry
砌砖体 brick setting
砌砖图 drawing of brick works
砌砖用的手锤 club hammer
砌砖用线垫 tingle
砌砖杂工 bricklayer's labo(u)rer
砌砖铸造 loam casting
砌琢石(墙) ashlaring[ashlering]
砌琢石墙面 ashlar walling

碛【地】hammada

碛坝 shingle bank; shingle flat

槭单宁 aceritannin

槭鹅耳枥 maple hornbeam
槭木 maple
槭树 maple
槭属 <拉> Acer

器壁摩擦 wall friction

器壁凸出 bulged ride
器壁效应 container wall effect; wall effect
器材 equipment and materials; fittings; stock
器材调度员 expediter[expeditor]
器材堆场 plant depot
器材耗存报表 stores balance report
器材库 storage facility
器材损坏 property damage
器材预定单 advance material order
器材站 equipment station
器差订正表 calibration table
器官 organ
器官的 organic
器官系数 acropetal coefficient
器件 device; organ
器件布局 device placement
器件号 parts number
器件名称 device name
器件寿命 device lifetime
器具 apparatus; appliance; furnishing; furniture; implement; tackle; tool; utensil; ware
器具库 utensil storage
器具通气管 fixture vent
器乐 instrumental music
器量 caliber[calibre]
器皿 utensil; ware
器皿玻璃 ware glass
器物的侧面 breast
器物的颈状部 neck
器物上的提梁 bali
器械 apparatus; device; equipment; hickey; implement; instrument; tackle; tool
器械操作 instrumentation
器械设备 appliance
器械室 apparatus
器质性疾病 organic disease

憩潮 slack tide; slack water

憩潮期 slack-water time
憩流 slack tide; slack water; stand of tide
憩流段 area of slack water
憩流区 area of slack water
憩流水位 slack-water level

掐入 <指函数的曲线等> pinching-in

恰当布置 aptitudal station

恰当的例证 a case in point
恰当的数据 pertinent data
恰当点 exact point
恰当方程 exact equation
恰当(燃料)混合气比 chemically correct fuel-air ratio
恰当溶液条件 exact solution condition
恰当时刻 in season; proper time
恰当微分 exact differential
恰当微分方程 exact differential equation
恰当值 proper value
恰尔爱公式 Chele formula
恰可识别的色差 just noticeable difference
恰脱阶 <晚渐新世>【地】Chattian

洽商 make arrangement with; takeover with

洽商服务 services for business negotiations; services for trade talks
洽谈业务 business negotiation; negotiate business
洽询出票人 refer to drawer

千安(培) kiloampere

千奥 <磁场强度单位> kilooersted
千巴 <压力单位> kbar; kilobar
千靶(恩) kilobarn
千板尺量度 thousand-foot board measure
千板英尺 <木材计量单位> thousand feet board measure
千板英尺量材法 thousand-foot board measure
千板状构造 phyllitic structure
千棒计划 kilorod program(me)
千棒设施 kilorod facility
千磅 <力或重量的单位> kip; kilopound
千磅/平方英寸 kips per square inch
千贝克勒尔 <放射性活度单位> kilobecquorel
千比特 kilobit
千波德 <信息传输单位> kilobaud
千磁力线 kiloline
千次指令运算每秒 kips kilo-instruction per second
千达因 <力的单位> kilodyne
千单位 kilounit
千岛海沟 Kuril trench
千岛海流 Kuril current
千岛海盆 Kuril basin
千电子伏特 kiloelectron volt
千度 megawatt hour
千吨 kiloton
千吨以下 subkiloton
千二进制位 kilobit
千乏表 kilovar instrument
千乏时 kilovar-hour
千乏时计 kilovar-hour meter
千乏小时 quadergy
千分 millesimal
千分比 permill(age)
千分比较仪 minimeter
千分表 dial(type) indicator; amesdial; clock ga(u)ge; dial ga(u)ge; micrometer dial; micrometer ga(u)ge; micrometer with dial indicator; micron micrometer
千分表架 dial ga(u)ge stand; dial holder
千分表卡规 dial snap ga(u)ge
千分表平板 comparator plate
千分表切削深度调整 micrometer cutting depth adjustment
千分测径规 micrometer cal(l)ipers
千分尺 beam micrometer; feeler ga(u)ge; microcal(l)ipers; micrometer[micrometre]; micrometer cal(l)ipers; micron micrometer; milscale; scale micrometer
千分尺测量 miking
千分尺定位器 micrometer stop
千分尺架 micrometer stand
千分尺可换测砧 anvil piece
千分尺能伸缩的隔套 micrometer space
千分尺套圈 micrometer collar
千分尺制动把 clamp pin
千分尺轴 microspindle
千分尺座 micrometer stand
千分垫 clearance ga(u)ge; feeler ga(u)ge; feeler stock; thickness feeler
千分定位器 micrometer stop
千分度盘 milscale
千分伏特计 voltascope
千分规 ga(u)ge microsize
千分角 mil
千分卡(尺) microcal(l)ipers; micrometer cal(l)ipers
千分卡规 micrometer cal(l)ipers; micrometer ga(u)ge
千分卡头 micrometer head
千分卡微调机构 micrometer-driven tuning mechanism
千分刻度盘 micrometer dial
千分率 permill(age)
千分螺杆 microspindle
千分秒表 chronoscope
千分内径规 inside micrometer ga(u)ge
千分深测规 micrometer depth ga(u)ge
千分丝杠 micrometer screw
千分误差 mills error
千分压力表 micro-pressure-ga(u)ge
千分压力计 micro-pressure-ga(u)ge
千分之几 parts per thousand
千分之一 millesimal; parts per thousand; per mille
千分之一伏 millivolt
千分之一秒 millisecond
千分之一瓦 milliwatt
千分之一英寸 mil-inch
千分之一英寸水头 mil-inch head of water
千分之一英亩 <英美制面积单位,43.56平方英尺,1平方英尺=0.0929平方米> milacre
千峰翠色 thousand peaks in emerald colo(u)r
千伏安 kilovolt-ampere
千伏安额定容量 kilovolt-ampere rating
千伏安时计 kilovolt-ampere-hour meter
千伏安(小)时 kilovolt-ampere-hour
千伏电压 kilovoltage
千伏(电压)表 kilovolt meter
千伏峰值 kilovolt peak; kilovolt peak value
千伏计 kilovolt meter
千伏(特) kilovolt
千高斯 <磁感应或磁场单位> kilogauss
千格冷 <重量单位> kilograin
千公乘 <千立方米> kilostere
千公斤 metric(al) ton
千公里 megameter
千赫 kilocycle; kilocycles per second; kilohertz; kilohertz thousand cycles per second
千花玻璃 millfore glass
千焦耳 kilojoule
千斤椽 trimming rafter
千斤调整索 topper; topping rope
千斤顶 block-lifter; block screw; capsular jack; erector beam; hoisting jack; jack; jack and pinion rack; jacking apparatus; lifting jack; rising feeds; screw block; screw jack; screw prop; shifting cylinder; track jack
千斤顶齿柱 jack rack
千斤顶垫板 jacking plate; jack pad
千斤顶垫块 jack block; jacking disc [disk]
千斤顶垫木 jack bar
千斤顶垫座 abutment for jacks; jack(ing) pad
千斤顶顶进速度 jack propelling speed
千斤顶顶头 jacking header
千斤顶顶桩 jacking of pile
千斤顶法 <岩体中应力量测用> jack method
千斤顶法封顶 closure by jacking and sealing-off crown
千斤顶防漏油阀 scavenger valve
千斤顶杆 jack rod
千斤顶荷载 jack load
千斤顶后座 abutment for jacks; pad shoe
千斤顶加力杆 jack iron; jack lever
千斤顶加载法 jack-loading method
千斤顶举升力 jacking force
千斤顶螺杆 jack screw; lifting screw
千斤顶螺母 jack nut
千斤顶螺栓 jack(ing) bolt
千斤顶螺丝 jack screw
千斤顶帽 jack bit
千斤顶木垫块 growler board; jacking block
千斤顶入桩 jacked pile
千斤顶设备 jack(ing) equipment
千斤顶试验 jack(ing) test
千斤顶手柄 jack handle
千斤顶损失 jack loss
千斤顶提升杆 jack ram; jack rod
千斤顶托架 jack-carrying frame
千斤顶小车 jack lift
千斤顶压板 jacking platen
千斤顶压杆 jacking rod
千斤顶压力 jacking pressure
千斤顶压力盒 jack load cell
千斤顶应力 jacking stress
千斤顶油缸 jack cylinder
千斤顶油缸行程 jack cylinder range
千斤顶载架 jack-carrying frame
千斤顶张拉 jacking; jack tensioning
千斤顶张拉(钢筋)垫板 jacking plate
千斤顶张拉(钢筋)垫块 jacking block
千斤顶枕木 jack board
千斤顶支架 jack-carrying frame; jacking yoke
千斤顶支柱 jack leg; roof jack
千斤顶支座 jacking bracket
千斤顶作业垫板 jacking plate
千斤顶座 jack pad assembly
千斤顶座孔 <安放千斤顶进行起顶的预留孔> jacking pocket
千斤对顶 jack pair
千斤格栅 trimming joist; trimmer <美>
千斤链 span chain; topping lift chain
千斤绳 jack line
千斤索 toplift; topping lift; topping line
千斤索定位绞车 topping lift winch
千斤索滑车 topping lift block
千斤索绞车 topping winch
千斤索绞辘 topping lift purchase
千斤索具 span rigging
千斤索拉端 topping lift fall
千斤索升降绳 topping lift bull rope
千斤索眼板座 topping lift eye bracket
千斤顶调整 jack adjustment
千居里 <旧的放射性活度单位> kilocurie
千居里源 kilocurie source
千钧一发的 critical
千卡 <热量单位> grand calorie; great calorie; kilocalorie; large calorie; major calorie
千卡/秒 <热功率单位> kcal/second
千卡/时 <冷冻率单位> frigorie
千卡/小时 kilocalorie per hour

Q

千克 kilogram
千克米 <功的单位> kilogram-meter [metre]
千克质量 kilogram-mass
千克重 kilogram-weight
千克当量 kilogram-equivalent; kilogram-equivalent weight
千克卡 kilogram-calorie
千克力 kilogram force; kilopond
千克力米 kilogram force meter
千克/立方米 kilograms per cubic(al) meter
千克力每平方米 <非法定单位> kilogram-force per meter squared
千库(仑) kilocoulomb
千立方米 kilostere
千立方英尺小时 kilo cubic feet per hour
千粒重 thousand-grain weight
千量材尺 mille; thousand board measurement feet
千流明 <光强单位> kilolumen
千流明时 kilolumen-hour
千伦琴 <放射性物质的照射量单位> kiloroentgen
千枚板岩 phyllite slate
千枚糜棱岩 phyllite mylonite; phyllonite
千枚石英岩 phyllite quartzite
千枚岩 phyllite
千枚岩骨料 phyllite aggregate
千枚岩集料 phyllite aggregate
千枚岩与石英岩互层 phyllitic quartzite and quartzitic phyllite band
千枚岩状构造 phyllitic structure
千枚岩状花岗岩 phyllitic granite
千枚状板岩 phyllitic slate
千枚状变质粉砂岩 phyllite metasiltstone
千枚状变质火山岩 phyllite metavolcanitic rock
千糜岩 pyllonite
千米 kilometer[kilometre]
千米吨 kilometer-ton
千米波 kilometer wave; kilometric wave
千米的 kilometric(al)
千秒 kilosecond
千秒差距 kiloparsec
千摩尔 kilomol(e)
千尼尔沉积【地】Chenier deposit
千年 millennium
千年矮 common box tree
千年虫 millennium bug
千年期 millennium
千年桐 Aleurites montana
千年一遇的洪水 a thousand-year flood
千牛(顿) kilonewton
千牛(顿)·米 kilonewton meter
千欧(姆) kilo-ohm
千帕(斯卡) <压力单位,1 千帕(斯卡) = 103 牛/平方米> kilopascal [kPa]; pieze
千千伏安 megavolt ampere
千千瓦 megawatt
千千万 zillion
千涉波 interfering wave
千升 kiloliter[kilolitre]
千瓦 kilowatt
千瓦电能 kilowatt of electric energy
千瓦计 kilowatt-meter
千瓦年 kilowatt-year
千瓦日 kilowatt-day
千瓦时计 kilowatt-hour meter
千瓦损耗 kilowatt loss
千瓦(小)时 <功率单位> kilowatt-hour
千瓦小时表 kilowatt-hour meter
千瓦小时电度表 kilowatt-hour meter [KMH-meter]
千英厘 kilograin
千元美钞 big one
千张石板瓦 thousand slates
千兆 giga; giga billion; kilomega
千兆电子伏特 billion electron-volts
千兆分之一 billi
千兆赫 gigacycles per second; gigahertz; kilomega hertz
千兆赫计算机 gigahertz computer
千兆欧(姆) begohm; gigaohm
千兆瓦 gigawatt
千兆瓦日/吨 <燃耗单位> gigawatt-day/tonne
千兆位 billibit; kilomegabit
千兆位速率 gigabit rate
千兆周 billicycle; gigacycle; kilomega cycle
千兆周计算机 gigacycle computer
千兆周/秒 gigacycles per second; gigahertz
千兆字节 gbyte; gigabyte
千周 giga; kilocycle; kilohertz
千周年纪念 millenary; millennium
千字【计】k-word
千字节 <1024 字节> kilobyte
千自乘五次的数 <法国、美国> quadrillion

迁

迁出 dispossess; emigration; outmigration; vacation

迁出国 country of emigration
迁出洄游 emigration migration
迁出通知 notice to quit
迁地保护 ex-situ conservation
迁动速度 migration velocity
迁飞高峰期 peak of flight
迁建标准 relocation criterion
迁建规划 relocation plan(ning)
迁建规模 magnitude of relocation
迁建准则 relocation criterion[复 criteria]
迁居 change of residence; migration
迁离农村 rural exodus
迁入 in-migration
迁入国 country of immigration
迁徙 transmigration
迁徙河流 undefined course of river
迁徙湖 migration lake
迁徙鸟 migration bird
迁徙水流 migration flow
迁徙行为 migration behavio(u)r
迁徙性洲滩 shifting bar
迁延地址 deferred address
迁延剂量 protraction dose
迁延型 persisting type
迁延性比例失调 chronic imbalances
迁延移行 delayed migration
迁延照射 prolonged exposure; protracted irradiation
迁移 migrate; migrating; migration; removal; remove; take-off; transference; transplant; wandering
迁移本能 migratory instinct
迁移波 migrating wave
迁移波痕纹层 ripple laminae in-drift
迁移补偿费 compensation for removal
迁移部分 transport piece
迁移长度 migration distance; migration length
迁移常数 migration constant; mobility constant; transfer constant
迁移沉积层 displaced sediment
迁移成本 moving cost
迁移传递 transfer and transport
迁移的 migratory
迁移电池 transference cell
迁移电流 migration current
迁移动力 migration agent
迁移度 mobilance
迁移法 transfer method
迁移范围 migration circle
迁移方程 transport equation
迁移方向 direction of migration; migratory direction
迁移费补贴 moving expense allowance
迁移费(用) compensation for removal; moving expense
迁移概率 migration probability
迁移管 migration tube
迁移函数 transfer function
迁移河 migrating river
迁移河曲 migrating meander
迁移活动 migratory movement
迁移机构 transport mechanism; travel(l)ing mechanism
迁移机理 migration mechanism
迁移机械和设备 machinery and equipment moving
迁移基准线 line of movement data
迁移交错层理构造 drift cross-bedding structure
迁移居民 resettlement of the residents
迁移矩阵法 transfer matrix method
迁移距离 migration distance
迁移卷 migration volume
迁移量 volume of migration
迁移路线 migratory route
迁移率 drift mobility; migration rate; mobility; mobility ratio; transport factor
迁移模式 transfer mode
迁移模型 transfer model
迁移能 migration energy
迁移农业 ladang
迁移赔偿 removing indemnity
迁移赔偿费 compensation for removal
迁移起点 migration origin
迁移强度 migration intensity
迁移倾向 migration aptitude
迁移区 migration area; zone of migration
迁移取样法 transfer method
迁移取样技术 transfer technique
迁移群 migratory group
迁移群聚 symporia
迁移群落 migratory community
迁移热 transfer heat
迁移人口 migrant; population relocated
迁移人口总数 volume of migration
迁移试验 migration test
迁移数 transference number; transport number
迁移速度 drift velocity; migration rate; migration velocity
迁移速度公式 migration rate equation
迁移速率 migration rate
迁移速率常数 migration rate constant
迁移速率理论 migration rate theory
迁移损失 migration loss
迁移通告 announcing removal
迁移途径 migratory highway
迁移系数 migration coefficient; mobility ratio
迁移系统 migratory system
迁移现象 transport phenomenon[复 phenomena]; transshipment phenomenon[复 phenomena]
迁移相对系数 relative coefficient of migration
迁移效率 transport efficiency
迁移行为 migratory behavio(u)r; transfer behavio(u)r
迁移形式 mode of removal
迁移型 migration form
迁移性 migratory aptitude
迁移性的生活 colonial life
迁移性高气压 migratory anticyclone
迁移抑制因数 migration inhibition factor
迁移影响函数 transport kernel
迁移元素 migration elements
迁移增强外延 migration enhancement epitaxy
迁移植物 migrant
迁移质量通量 migrational mass flux
迁原设备 restorer

钎 poker bar

钎柄 bit shank; drill shank
钎缝 brazing seam; soldering seam
钎缝间隙 joint clearance; joint gap
钎缝金属 brazed metal
钎杆 chisel bar; drill; drill guide; drilling stem; drill shank; jack rod; rock drill steel; rod; shaft bar; steel stem; drill steel
钎杆导管 drill steel guide
钎杆架 steel rack
钎杆进尺 rod-feet; rod footage
钎杆进尺数 rod meter
钎杆进米数 rod meter
钎杆连接器 drill adapter
钎杆寿命 drill steel life; rod-length
钎杆套座 chuck driver
钎杆消耗量 drill steel consumption
钎杆英尺 rod footage
钎杆直径 steel diameter; steel ga(u)ge
钎杆组 drilling string
钎杆座 sleeve bushing
钎杆座套 rotary sleeve; rotation sleeve
钎钢 drill(ing) steel
钎钢断面 steel section
钎钢加热炉 drill steel furnace
钎焊 braze; braze welding; brazing; electric(al) soldering; solder; sweating; torch brazing
钎焊板 brazing sheet
钎焊笔 soldering pencil
钎焊表面间隙 clearance between surfaces to be brazed
钎焊表面清理 cleaning of brazed surfaces
钎焊挡板 brazing flap
钎焊的 brazed
钎焊点 soldering point
钎焊粉 brazing powder
钎焊缝 brazing seam; solder joint
钎焊膏 soldering paste
钎焊工 solderer
钎焊焊剂 soldering paste
钎焊焊接器 soldering hammer
钎焊焊炬 brazing torch; soldering torch
钎焊合金 brazing alloy; brazing metal; solder alloy
钎焊机 solderer; soldering machine
钎焊剂 brazing flux; soldering flux; soler
钎焊间隙 joint clearance of brazing
钎焊接 brazed welding
钎焊接头 brazed joint; soldered connection; soldered fitting; soldered joint; sweated joint
钎焊金属 brazing metal
钎焊金属料 brazing filler metal
钎焊炬 soldering torch
钎焊烙铁 solderer; soldering copper; soldering iron
钎焊连接 brazed joint; solder connection
钎焊连接强度 strength of brazed

joint
钎焊料 hard solder
钎焊炉 soldering oven
钎焊面 faying face
钎焊喷灯 brazing torch
钎焊片 soldering lug
钎焊片状板 brazing flap
钎焊枪 brazing torch;soldering gun
钎焊青铜焊料 Sifbronze
钎焊容器 brazed vessel
钎焊容器限制 limitations of brazed vessels
钎焊时间 holding time
钎焊丝 cored solder wire
钎焊填充金属 solder
钎焊铜焊 braze
钎焊温度 brazing temperature
钎焊性 braze ability;solderability
钎焊用具 solder set
钎焊用喷灯 brazing torch; soldering lamp
钎焊用酸 soldering acid
钎焊用铜 soldering copper
钎焊油 soldering paste
钎剂 brazing flux;flux;soldering acid
钎夹 bit holder;rod holder
钎尖 lance point
钎肩 collar;drill collar;rod shoulder; shank collar;shoulder;steel collar
钎肩式钎尾 bolster-type shank;collar shank
钎肩式钻头 <拧在钻杆上> shoulder-type bit
钎角 fillet
钎接 lead joint
钎接焊 braze welding
钎炬 brazing torch
钎距 drill change
钎卡 drill steel retainer; lug chuck; steel holder;steel retainer;retaining latch <凿岩机>
钎卡住 steel seizure
钎料 brazing alloy;brazing filler metal;solder;spelter
钎料棒自动铸造性 castomatic method
钎料脆性 solder embrittlement
钎料丝 solder wire
钎柠座套 chuck sleeve
钎刃 bit edge
钎刃的刃角 bit cutting angle;cutting edge angle
钎刃的修磨角 bit cutting angle
钎刃负载 edge load(ing)
钎刃厚度 wing thickness
钎刃角 bit taper; edge angle; insert angle
钎刃硬度 tip hardness
钎刃锥度 winged ga(u)ge taper
钎台 collar;drill collar
钎探 rod sounding
钎探检验 pin exploration
钎套 chuck bushing;chuck liner;drill chuck; rotation chuck bushing; sleeve bushing
钎铜焊 braze welding
钎头 bit (head); bore bit; boring head;drill(ing)bit;drill(ing)head; rock drill steel
钎头端部 <凿岩机> bit end
钎头更换 drill change
钎头横刃部 chisel point
钎头后角 clearance angle
钎头几何形状 bit geometry
钎头接头 bit adapter
钎头卡住 bit seizure
钎头量规 drill ga(u)ge
钎头磨刃角 bit taper
钎头磨损 ga(u)ge loss
钎头平锤 bit fuller
钎头切削刀 bit of the drill head
钎头刃面 bit face angle
钎头寿命 bit life;drilling bit life
钎头头部 bit head
钎头外形 bit geometry
钎头外周球齿 ga(u)ge row
钎头修理工 bit setter
钎头修磨 bit resharpening
钎头修整 bit reconditioning
钎头直径 bit diameter;size bit
钎维石膏矿石 fibrous gypsum ore
钎尾 bit shank; chuck adapter; drill shank; drill steel shank; shank (end);steel shank
钎尾尺寸 shank size
钎尾对准器 shank alidade
钎尾对准仪 shank aligner
钎尾杆 adapter rod
钎尾规格 dimensions of drill shank
钎尾连接器 pulling-tool adapter shank adapter
钎尾密封套 shank packing
钎尾钎肩、钻头挡 shank collar
钎尾套 rotation chuck bushing
钎尾套筒 chuck bushing;chuck nut
钎尾凿 adapter
钎尾锥度 shank taper
钎翼 bit wing angle
钎着率 ratio of brazing area
钎子 bot pick;drilling rod;drill steel; gad picker; hole digger; jumper; jumper bit; knockout bar; tap-hole rod
钎子冲击 pick blow
钎子淬火油 steel hardening oil
钎子挡环 retainer collar
钎子的尾端 bit shank
钎子镦粗 setting bit
钎子钢 drill steel; hollow drill shank steel;rock drill steel
钎子架 steel rack
钎子头刃角 bit cutting angle
钎子尾 steel shank
钎子折断 steel breakage
钎子组 set of drills
钎子组每根长度差 length increment
钎子组直径的公差 ga(u)ge change
钎字杆 chisel-tipped steel

牵

牵车 traction vehicle

牵车机 <代替调车机车> car puller
牵出功率 pull-out power
牵出列车 <准备推峰或解体> set out train
牵出扭矩 pull-out torque
牵出试验 pull-out test
牵出同步 pulling out of step;pulling out of synchronism
牵出线【铁】 dragline track;draw-out track;pull-out lead;shunting neck; switching lead; turnout track; pull-out lead track;lead track <调车场>
牵出线调车 shunting on shunting neck
牵出线改编能力 marshalling capacity of switching lead;resorting capacity of lead track
牵船柱 warped bollard;warping bollard
牵道 tow path
牵动 dragging
牵斗索 <挖掘机> dragline
牵帆 staysail
牵杆 drawhead
牵钢 pull-iron
牵挂式运粮车 tractor trailer
牵管 stay tube
牵航式拖船 tracting tug
牵弧技术 drag technique
牵环 stay collar
牵簧 draw spring
牵紧螺母 right-and-left nut;turnbuckle nut
牵开器 retracter
牵开器压紧器 impacter for retractor
牵开器支撑器 retractor support
牵拉船舶上岸 <英> slipway
牵拉杆 stayed pole
牵拉绞车 pull-rope winch
牵拉结构 tensioned structure
牵拉结构织物 tension structure fabric
牵拉卷扬机 pull-rope winch
牵拉设备 puller device;towing equipment
牵拉绳索的滚筒 reeving drum
牵(拉)桩 stay pile
牵累 encumbrance;encumber
牵力 pulling force; pull-up; tension; towing force
牵力对 tension pair
牵力计 tensometer
牵力顺联 traction bracing
牵连 entanglement; implication; involvement
牵连加速度 acceleration of following
牵连速度 velocity of following
牵链 guy chain
牵牛花 convolvulus althaeoides; morning glory
牵牛花式柱 morning-glory column
牵牛花形(竖井式)溢洪道 morning-glory shaft spillway
牵牛花形堰顶 morning glory sill
牵切机 stretch breaking machine
牵入并联运行 pulling-in parallel operation
牵入试验 pull-in test
牵入同步 lock-in synchronism; pulling-in step; pulling into synchronism
牵入转矩 pull-in torque
牵涉各方面的 sophisticated
牵涉面较广的 sophisticated
牵伸 drafting;drawing;extension
牵伸比 drawdown ratio;extension ratio;ratio of drawing
牵伸辊 draft roller
牵伸簧 extension spring
牵伸机构 drafter
牵伸黏[粘]度 tractive viscosity
牵伸气流 attenuating blast
牵伸器 extension apparatus;orthoterion
牵伸箱 draw box
牵绳 guy line
牵式除荆机 brush plough
牵收缆索 <缆索开挖机的> inhaul cable;inhaul line
牵收索 inhaul
牵手式安全装置 pullback; pull-out guard
牵索 back guy;backstay cable;exterior tieback;grab line;guy rope;guy wire; hauler; launching cable; pull rope; snivvey; stay guy; staying wire; stay rope; steel wire stay; suspending wire; suspension wire; tieback; towing cable; trailing cable;wire back-tie;lanyard <系于安全带上的绳索>
牵索测量 tagline survey
牵索铲斗挖掘机 dragline bucket dredge(r)
牵索的 guying
牵索调位 guying
牵索钢烟囱 guyed steel stack
牵索轨道 cog railway
牵索环 ring groove
牵索绞车 towing winch
牵索结构 guyed structure
牵索起重机 guyed derrick
牵索桥 <原为悬桥的一种形式，现代在大跨度预应力钢筋混凝土桥中采用> bridle chord bridge
牵索人字起重机 guyed derrick crane
牵索式桅杆起重机 guyed derrick crane
牵索塔 guyed tower
牵索套管 guy thimble
牵索铁烟囱 guyed iron chimney; guyed steel stack
牵索挖斗 dragline scoop
牵索桅杆式旋臂起重机 guy derrick
牵索转臂起重机 guyed derrick crane
牵胎钩 embryulcus
牵条 cleck rod;distance bar;stringer; tension brace;wind brace
牵条管 stay tube
牵条螺栓 cotter bolt; distance sink bolt;stay bolt
牵头承保人 lead underwriter
牵头公司 leading company
牵头经理 lead manager
牵头银行 lead bank
牵线法 tagline method
牵型引花纹 traction tread
牵曳 drafting;draw
牵曳式绞车 warping winch
牵引 draft (ing); drag; draw (ing); haul(ing);pull(ing);tote;towage; tow (ing); traction (drive); trailing;tug
牵引安全杆 drawbar safety bar
牵引搬运 traction transport
牵引板 draw plate;linkage drawbar; pull blade
牵引绷带 traction bandage
牵引臂 towed arm;draft arm
牵引变电所 railway substation; traction substation;tractive substation
牵引变电站 railway substation; traction substation
牵引变压器 traction transformer
牵引变压所 traction substation
牵引不足 hypotraction
牵引部件 traction member
牵引材料 pulling material
牵引侧向阻力 side draft
牵引叉宽度 fork width
牵引铲 traction shovel
牵引车 draft vehicle; driving truck; front truck; haulage motor; hauling engine; hauling truck; hauling unit; industrial tractor; motive unit; motor tractor; mover; prime power; pull tractor; self-propelled engine; tow car; towing machine; towing troll(e)y; towing vehicle; tow tractor; tracking truck; traction engine; traction machine (ry); tractor-truck; tractor; trailer-hauling tractor
牵引车半拖车机组 tractor semitrailer
牵引车挂车 tractor trailer
牵引车和底盘车系统 tractor and chassis system
牵引车荷重 weight on driver
牵引车后挂车 truck trailer
牵引车后悬 rear axle overhang of a towing vehicle
牵引车机 primary mover
牵引车及半挂车 tractor-semitrailer
牵引车及挂车 truck-trailer combination
牵引车架 draft frame
牵引车铰接的拖车 articulated lorry; tractor-trailer
牵引车辆现有数 <包括机车、动车等> tractive stock

牵引车辆用轮胎 lug type tire
牵引车轮距 tractor tread
牵引车配半拖斗的机组 tractor-semi-trailer combination
牵引车起动 priming
牵引车手扶拖拉机 tractor-truck
牵引车与挂车的连接 tractor-trailer engagement
牵引车与拖车之间的安全铰接装置 jackknife
牵引出术 extraction
牵引穿模机 pulling-in machine
牵引传动(设备) haulage gear
牵引传送皮带 traction-transmitting belt
牵引船 drag boat;tractor tug
牵引床 traction table
牵引磁导计 traction permeameter
牵引带 traction strip
牵引单元 traction unit;tractive unit
牵引道 sledge-way;towing pass;tow way
牵引的 dragging;drawn;tractive
牵引的飞机 towing plane
牵引的拖车 tow trailer
牵引的总运量 gross traffic hauled
牵引的总重吨公里 <除机车外> gross ton-kilometers [kilometres] hauled
牵引点 hitch point;towing point
牵引电磁铁 tractive electromagnet; tractive magnet
牵引电动机 electric(al) vehicle motor;traction motor
牵引电动机磁场分路接触器 traction motor field shunting contactor
牵引电动机的机械性能 mechanical performance of traction motor
牵引电动机发热 traction motor heating
牵引电动机励磁机方法 traction motor exciter method
牵引电动机联结开关 coupling switch
牵引电动机通风机 traction motor blower
牵引电机 pulling motor
牵引电机电流 traction motor current
牵引电机电压 traction motor voltage
牵引电机风扇 traction motor fan
牵引电机架承式悬挂 lateral suspension motor
牵引电机解体组装设备 mounting and dismounting device for traction motor
牵引电机试验台 traction motor test stand
牵引电机速度控制 pulling motor speed control
牵引电机悬挂 traction motor suspension
牵引电流 propulsion current;pull-in current;traction current
牵引电流导接线 propulsion bond; traction rail bond
牵引电流系统 traction current system
牵引电路 pulling circuit;traction circuit
牵引电压 traction voltage
牵引电压变换 traction voltage changeover
牵引吊索 traction suspension rope
牵引定额 tonnage rating(of traction)
牵引定数 tonnage rating(of traction)
牵引动车 tractive motor truck
牵引动力 traction power
牵引动力电池 traction battery
牵引动力照明变电所 traction power lighting substation
牵引动力装置效率 drawbar efficiency
牵引斗式装载机 tractor-bucket machine
牵引渡船 cable ferry
牵引段 puller section
牵引吨数 haul tonnage; tonnage of traction;tonnage rating(of traction)
牵引吨数表 haulage chart
牵引吨英里 trailing ton-mile
牵引发电机 traction generator
牵引发动机 traction engine
牵引法 <流速仪标定的> towing method
牵引范围 hold range;lock-in range; pull-in range
牵引方式 form of traction; mode of traction
牵引方向 lead
牵引费 towage
牵引幅度 amplitude of drag
牵引负荷 traction load
牵引复滑车 pull-and-out block
牵引复位 traction reduction
牵引杆 coupling bar; draft bar; draft connection; draft pole; drag bar; drag link; drag rod; draught bar; draw(ing) bar; hitch pole; linkage drawbar; pull rod; stay rod; tow bail;towbar;tow(ing) rod;traction pole;traction rod;trail
牵引杆摆角 <推土机> swing angle
牵引杆板 drawbar plate
牵引杆弹簧 drawbar spring
牵引杆导板 drawbar guide
牵引杆高度可调式挂车 trailer with adjustable drawbar height
牵引杆接头 hitch head
牵引杆拉力 drawbar pull
牵引杆拉力重量比 drawbar pull-weight ratio
牵引杆联结销 hitch pin
牵引杆螺旋连接套管 drawbar spiral coupling sleeve
牵引杆螺旋套管 drawbar spiral sleeve
牵引杆马力 drawbar(horse)power
牵引杆式挂车 rigid vehicle
牵引杆受拉能力 drawbar capacity
牵引杆托架 drawbar cradle
牵引杆销 drawbar pin
牵引杆销键 draw pin key
牵引杆销孔 hitch pin hole
牵引杆座 drawhead
牵引感 tugging
牵引钢缆 carrying cable
牵引钢丝绳 drag rope; load cable; traction wire-rope; haulage rope; hauling rope;pulling rope
牵引钢丝索 traction steel cable
牵引钢索 drag cable;pull line;traction steel cable
牵引工 haulier
牵引工程 propulsion engineering
牵引工具 draft implement; trailing implement drawn
牵引工况 traction condition
牵引工作 traction duty
牵引工作圆筒 hoisting draw works drum
牵引弓 traction bow
牵引功率 drawbar(horse)power; hauling power;power at the drawbar;traction power;tractive power
牵引功率比油耗测定 specific drawbar power fuel consumption measurement
牵引供电 traction feed
牵引供电臂 feeding section
牵引供电系统 tractive power supply system
牵引拱架 gooseneck
牵引钩 coupling hook; drag hook; draw-hook; hook; pintle hook; pull hook;towing hook;towing shackle; tow loop
牵引钩钉 dragging pin
牵引钩环 tow(ing)shackle
牵引钩联结销 clevis pin
牵引钩销 draw-hook pin
牵引钩装置 draw-hook bar
牵引构件 hauled element;traction element;tractive element
牵引刮土机 hauling scraper
牵引挂车的能力 trailer towability
牵引管 traction tube
牵引辊 carry-over pinch roll;traction roller
牵引滚筒 haulage drum;hauling drum
牵引荷载 hauled load; towed load; traction load; tractive load; trailing load
牵引荷重 tractional load
牵引横板 hitch crossbar
牵引横梁 <铲运机> draft tube
牵引横柱 draft tube
牵引弧 drag arc
牵引滑轮 drawer roller;straining pulley
牵引环 bail; clevis drawbar; draw ring; eye hitch; ring drawbar; towing eye; towing ring; towing shackle;tow lug
牵引环孔 lunette eye
牵引缓冲弹簧 traction spring
牵引缓冲装置 draw and draft gear
牵引簧 extend working spring;extension spring
牵引回流 traction return current
牵引回流电缆 traction return cable
牵引回流电路 traction return current circuit
牵引回流轨 traction return rail
牵引机 dragger;draw machine;haulage machine; hauler; hauling machine;prime mover;propelling machine;pulling machine;traction engine;traction machine;tractor
牵引机车 hauling engine; loco(motive);traction locomotive
牵引机构 drag gear;haulage gear
牵引机架 draft frame
牵引机牵引的半挂车 tractor semi-trailer
牵引机械 hauling machine;propelling machinery;traction machinery
牵引机型 tractor type
牵引机组 tractive unit
牵引机组合模块 puller unit module
牵引计 traction meter
牵引计算 tractive tonnage calculation
牵引继电器 traction relay
牵引夹 haulage clip
牵引夹头 tractor dog
牵引架 tongue; towing bracket; towing hitch
牵引架球头关节 drawbar ball joint
牵引架球头(关)套节 drawbar ball joint socket
牵引降压混合变电所 mixed traction and step-down substation
牵引角 angle of drag;angle of draw; angle of pull; angle of traction; haulage angle <皮带运输机上的>
牵引绞车 hauling winch; pull-rope winch; rack hoist; traction winch; barney car <翻斗车的>
牵引绞盘 towing capstan
牵引接地系统 traction earthing system
牵引净重 net weight hauled
牵引卷筒 draw drum;hauling drum; hitch wheel;rack drum
牵引卷扬机 pull-rope winch; rack winch;pulling winch
牵引均衡横木 evener
牵引卡车 <拖运违章车的车辆> tow truck
牵引孔 towing eye(let)
牵引控制单元 propulsion control unit
牵引控制系统 traction control system
牵引控制线 traction control line
牵引拉杆 anchor arm; anchor bar; drag link;draw tongue
牵引拉力试验 towing dynamometer test
牵引拉削 pull-broaching
牵引缆 <拉铲的> pull cable
牵引缆索 tow;traction cable <空运索道的>
牵引缆索的尾部锚碇 tail anchor
牵引犁 drag plough;drag plow
牵引理论 theory of traction
牵引力 draft force; draft load; drag force; drag(ging) power; drawbar pull; drawing force; effective tractive effort;force of traction;hatching power; haulage capacity; hauling force; hauling power; intensity of draught;net tractive effort;power of traction; pulling; pulling effort; pulling force; pulling power; pull rating; towing force; traction; traction force; traction modulus; traction power; tractive ability; tractive effort; tractive force; tractive power; tractive pull; towing traction
牵引力变动 tractive force variation
牵引力侧向分力 side draft
牵引力测力计 towing dynamometer; traction dynamometer
牵引力-车速-爬坡变关系表 rimpull-speed-gradeability chart
牵引力传感 draft sensing
牵引力传感器 draft responsible member; draft-sensing device; drawbar load sensing mechanism; traction sensor;tractive force sensor
牵引力电流特性曲线 tractive force-current characteristic curve
牵引力调节器 draft regulator;draught regulator
牵引力调节装置 traction control unit
牵引力范围 traction limitation
牵引力公式 formula of tractive effort
牵引力极限 traction limitation; tractive force limit
牵引力计 pull tension ga(u)ge;towing dynamometer;traction indicator
牵引力计算 calculation of tractive effort; traction calculation; tractive effort calculation
牵引力控制电路 throttle circuit
牵引力平衡 tractive force balance
牵引力平衡图 tractive force diagram
牵引力启动 tractor power take-off
牵引力试验跑道 drawbar test course
牵引力-速度曲线 speed-tractive effort curve; tractive effort-speed curve
牵引力损失 loss of traction
牵引力图 tractive force chart
牵引力图解法 tractive force graphic method
牵引力系数 coefficient of traction
牵引力自动调节 automatic draft control
牵引连接 tractive connection
牵引链 drag(line)chain;haulage chain; hauling chain;pull(ing)chain
牵引链钳 pulling-in dogs

牵引梁 draft beam;draft sill
牵引梁连接板 draft sill tie plate
牵引梁内侧磨耗板 draft gear pocket liner
牵引量 haulage
牵引料车 ground-hog
牵引溜放调车 flying switch
牵引流 tractive current
牵引流的 CM 图像 CM pattern of traction current
牵引轮 draft wheel;pulling-wheel
牵引轮胎 traction tread
牵引螺栓 draw bolt
牵引马达 traction motor
牵引马力 traction horsepower;drawbar horsepower
牵引锚 kedge
牵引模数 traction modulus
牵引能力 drive power;driving capacity;driving power;haulage capacity;lugging ability;pulling ability;pulling capacity;traction capacity;tractive ability;tractive capacity;tractive power;hauling capacity
牵引能量 haulage capacity;hauling capacity
牵引能需要量 demand for traction energy
牵引逆变器 propulsion inverter[inverter]
牵引逆变器试验装置 propulsion inverter test equipment
牵引耙 drag harrow
牵引盘 capstan
牵引片 leader film
牵引平地机 pull-grader
牵引起动 tow starting
牵引起重机 tractor crane
牵引起重机柱顶板 spider plate
牵引汽车和挂车成套运输装置 < 指用八轴特殊平板车运送 > lorry-and-trailer rig
牵引器 extension apparatus;puller;tractor
牵引千斤顶 pull-in jack
牵引钳 traction tongs
牵引强度 intensity of drag;intensity of draught
牵引区段 traction district
牵引驱动旋转扫路机 traction-driven rotary sweeper
牵引任务 traction duty
牵引设备 hauling equipment;towing equipment;traction equipment;tractor-allied equipment;tractor-based equipment
牵引生产力 traction capacity
牵引绳 carriage rope;chest rope;drag cable;dragline;drag rope;flying rope;gift rope;guest rope;guest warp;handling line;hauling cable;pull-cord;pull line;pull rope;pull wire;rack line;snake line;straining cord;towline;traction rope
牵引绳夹紧器 pulling rope clamp
牵引绳索 drag rope;inhaul line
牵引绳子 bull rope
牵引时脱开制动器的联动装置 surge brake tow bar package
牵引式 pull-behind;pull(ing) type;trail-type
牵引式半拖车 tractor-semitrailer
牵引式叉车装载机 tractor forklift
牵引式铲斗 drag scraper bucket
牵引式铲土和装载机 drag scraper and loader
牵引式铲土机 traction shovel
牵引式铲运机 drag road-scraper;drag scraper;hauling scraper;pull-type scraper;towed scraper
牵引式铲运机械 drag scraper machinery
牵引式铲运机装置 drag scraper installation
牵引式铲运起重机 drag scraper hoist
牵引式的 trail-behind
牵引式吊锚架 draw works cat head
牵引式反铲挖土机 tractor shovel backactor
牵引式反向铲 tractor backhoe
牵引式飞机 tractor aeroplane
牵引式割草机 traction mower;trailer mower
牵引式公共汽车 trailer bus
牵引式刮路机 drag road-scraper
牵引式刮土机 carryall
牵引式刮土平地机 pull-grader
牵引式刮运装置 hauling scraper
牵引式挂车运输 tractor-trailer trucking
牵引式滑坡 drawing landslide;retrogressive(land) slide
牵引式货车 dragon wagon
牵引式机具 semi-self-propelled machine
牵引式机械 traction machine;trailer-type machine
牵引式犁 trailed plough;trailed plow
牵引式沥青混凝土和骨料摊铺机 pull bituminous concrete and aggregate spreader
牵引式搂草机 pullrake
牵引式轮 traction wheel
牵引式轮胎 traction-type tyre
牵引式螺旋桨 tractor propeller
牵引式碾压机 articulated roller
牵引式碾子 articulated roller
牵引式农机具 rear-hitch tool;trailed implement
牵引式农具 trailing implement
牵引式喷灌机 towed sprinkler
牵引式喷雾机 trailer sprayer
牵引式皮带运输机 drawing belt conveyer[conveyor]
牵引式平板车 trailer platform
牵引式平地机 drawn grader;towed grader;traction grader;tractor grader
牵引式平路机 tractor grader
牵引式平土机 drawn grader;traction grader
牵引式破碎机 traction crusher
牵引式撒播机 trailing broadcaster
牵引式升降机 traction-type elevator
牵引式输送机 tow conveyer[conveyor]
牵引式饲料装运车 feed loader
牵引式松土机 trailed ripper
牵引式碎石机 traction crusher
牵引式碎石撒布机 drag spreader
牵引式碎石撒布机箱 drag spreader box
牵引式提升机 tractor elevator
牵引式挖沟机 traction ditcher;tractor ditcher;drag trencher
牵引式挖掘装载机 tractor loader
牵引式挖土机 traction shovel
牵引式压力机 tractor-compressor
牵引式压路机 tow behind roller;tractor roller
牵引式羊足碾 towed-type sheep's-foot roller;tractor-pulled sheep's-foot < 拖拉机式 >
牵引式圆盘犁 trailed disk plow
牵引式圆盘耙 drag disk harrow
牵引式振动羊足碾 towed-type vibrating sheep's foot roller
牵引式装载机铲斗 drag loader shovel
牵引式装载挖土联合机 tractor-loader-digger unit
牵引式钻车 traction jumbo;tractor jumbo
牵引式钻机 traction drill(ing machine)
牵引试验 draft test;drawbar test;pull test;towing test;traction test
牵引试验台 traction tested
牵引手册 index manual
牵引输送机 tow conveyer[conveyor]
牵引栓 pulling plug
牵引双联压路机 tractor-drawn two unit articulated roller
牵引顺序 indexed sequential
牵引死点 dead point of traction;dead spot of traction
牵引速度 haulage speed;hauling speed;hauling velocity;pulling speed;trailing speed
牵引索 carriage rope;chest rope;flying rope;gift rope;guest rope;guest warp;handling line;haulage cable;haulage rope;hauling cable;hauling rope;hauling wire;operating rope;pull line;pull rope;pull wire;rack line;snake line;straining cord;towing cable;towing line;towline;traction line;trailing cable;travel cable;travel rope;drag rope;traction rope;pull cable < 拉铲的 >
牵引索导轮 haulage rope guide wheel
牵引索滚轮 haulage cable roller
牵引索滚子组 haulage rope roller battery
牵引索滑车 main-line block
牵引索卷盘 haulage rope winding disc
牵引索套 traction rope sleeve
牵引弹簧 stiff spring
牵引弹簧杠杆 lever draft gear
牵引特性 drawbar performance;tractive characteristic;traction characteristic
牵引特性曲线 curve of traction characteristic;pull curve;tractive performance diagram
牵引条 draught strip
牵引调节板 hake
牵引推移力 traction power;tractive power
牵引拖车 track tractor;tractor trailer
牵引拖杆 draw tongue;towbar
牵引挖沟机 drag trencher
牵引挖掘机 tractor shovel;tractor shovel
牵引网 electric(al) traction network;traction network
牵引系数 coefficient of traction;drag coefficient;traction coefficient;tractive coefficient
牵引系统 propulsion system;traction system;hauling system
牵引系统动力学 traction system dynamics
牵引现象 traction phenomenon
牵引线 cord;draft line;pull wire
牵引箱 drag box
牵引销 towing pin
牵引小车 pushing carriage;trailer wagon;hauling-in-trolley;lead-in trolley
牵引效力 tractive effort
牵引效率 haulage efficiency;tractive efficiency
牵引效应 coupling hysteresis effect;restraining effect
牵引楔 drag wedge
牵引型 trailed model
牵引型电池 motive power battery;traction-type cell
牵引型轮胎 traction-type tire
牵引型液力耦合器 traction-type hydraulic coupling
牵引性能 drawbar performance;towing performance
牵引循环 inhaul cycle
牵引应力 traction stress
牵引运行 running under power
牵引张力 pulling tension
牵引褶皱 drag fold
牵引指示灯 traction lamp
牵引制动开关 traction-braking switch
牵引质量 hauled load;load hauled;traction mass;trailing load
牵引种类【铁】 category of traction;form of traction;kind of traction;type of traction
牵引重量 load hauled;trailing load;trailing weight;weight hauled
牵引重物时前端翘起 rearing
牵引主销 king pin
牵引助力机 traction booster
牵引柱 traction pole
牵引砖 draw bar
牵引转矩 traction torque
牵引转向架 < 车辆的 > trailing bogie
牵引装载机 tractor loader
牵引装置 draft device;draft hitch;drag gear;draught attachment;drawbar coupling;draw gear;haulage plant;pulling device;tension gear;towing machinery;towing unit;traction equipment;traction gear;draft gear < 美 >
牵引装置部件 drawbar assembly
牵引装置长度 drawgear length
牵引装置下挂结点 upper hitch point
牵引装置液压操纵 hydraulic drawbar control
牵引总载重 gross trailing load
牵引总重 gross load hauled;gross weight hauled;load hauled;trailing load
牵引阻力 draft resistance;drag resistance;resistance to traction;towing resistance;traction resistance;tractive resistance
牵引阻力系数 coefficient of tractive resistance
牵引作业 towage
牵引作用 tractive effort
牵运机 hauling engine;traction engine
牵运力 hauling capacity;hauling power
牵张反射 myotatic reflex;stretch reflex
牵张感受器 stretch receptor
牵制 containment;diversion;divert;pin down
牵制式半拖车 tractor-semitrailer
牵制式继电器 restraint relay
牵制效应 drag effect;restraining effect
牵制作用 holding action
牵纵拐弹簧 escapement crank spring
牵纵拐肘 escapement;escapement crank

铅

铅阿尔发粒子年代测定法 lead-alpha age method

铅按钮式冲水箱 lead bell type flush(ing) tank
铅钯矿 plumbopalladinite
铅靶管 sensicon;vistacon
铅白 basic lead white;flake white;lead white;white lead
铅白纯度试验 white lead purity test
铅白粉 < 调油漆用 > ground white lead(in oil)
铅白片 white lead flakes
铅白色试验 white lead test

铅白碳酸 ceruse
铅白油灰 white lead putty; whiting putty
铅白油涂料 flake lead; sharp coat
铅白与亚麻籽油 <作白油漆用> white lead and linseed oil
铅白云石 plumbodolomite
铅板 grid plate; lead plate; lead sheet; milled lead; stereotype; grid <电池的>
铅板防水条 <利用铅作为密封材料> lead weathering
铅板合金 grid metal
铅板间接口 welt
铅板屏 lead shielding
铅板铺面 lead paving
铅板试验 lead plate test; testing of grids <蓄电池>
铅板瓦 lead slate
铅板炸孔 explosion hole on lead plate
铅版 stereotype
铅版合金 stereotype alloy
铅版印刷 stereotype
铅版(印刷)的 stereotype
铅版整修机 plate finishing machine
铅版制造 stereotype
铅棒 lead rod
铅包玻璃窗格条 lead-clothed glazing bar; lead-covered glazing bar
铅包的 lead-covered; sheath
铅包电缆 lead cable; lead-covered cable; lead-sheathed cable
铅包电缆穿孔器 lead-covered cable borer
铅包钢带 lead-sheathed steel-taped
铅包加强的电缆 lead-covered armo(u)red cable
铅包绝缘电缆 lead-covered insulated cable
铅包皮 lead coat; lead covering; lead sheath(ing)
铅包皮的 lead-sheathed
铅包三芯电力电缆 lead-sheathed triple core cable; SL-cable
铅包铜板 lead-coated copper sheet
铅包外壳的 lead encased
铅包线 lead-covered wire; lead-sheathed wire
铅包纸绝缘电缆 paper-lead cable
铅包装 lead casing
铅保险丝 lead fuse(wire)
铅堡 lead castle
铅贝塔石 plumbobetafite
铅背滨鹬 lead back dunlin
铅钡玻璃 lead barium glass
铅褙条板 lead back lath
铅比 leading ratio
铅笔 pencil; plumbago
铅笔柏 eastern red cedar; pencil cedar; red cedar
铅笔草图 pencil sketch
铅笔淡彩 pencil-outlined wash drawing
铅笔刀 pencil sharpener
铅笔对中 centering[centring] of pencil
铅笔高度控制器 height controller of pencil
铅笔刮痕硬度试验 lead pencil scratch test
铅笔盒 pencil-case
铅笔画 pencil drawing; pencil(l)ing; pencil sketch
铅笔绘的 pencil drawn
铅笔桧 eastern red cedar; pencil cedar; red cedar
铅笔级滑石 pencil-grade talc
铅笔雷 booby trap
铅笔刨花 vandyke pieces
铅笔坯料 pencil stock
铅笔劈理 pencil cleavage
铅笔漆 pencil lacquer
铅笔式滑动 pencil glide
铅笔式话筒 pencil microphone
铅笔式三极管 pencil-type triode
铅笔束天线 pencil-beam antenna
铅笔松 pencil cedar
铅笔图画 pencil-work
铅笔线花饰 pencil(l)ing
铅笔线花样 pencil(l)ing
铅笔橡皮 pencil eraser
铅笔形波 pencil beam
铅笔形浇口 pencil gate
铅笔硬度 pencil hardness
铅笔硬度计 pencil tester
铅笔硬度试验 pencil hardness test
铅笔硬度试验机 pencil scratching tester
铅笔原图 pencil draft
铅笔圆规 lead compasses
铅笔窄板条 pencil slit
铅笔状射束 pencil beam
铅冰铜 lead copper matte
铅玻璃 barium flint glass; English crystal; flint glass; lead glass
铅玻璃纤维 lead glass fiber
铅箔 lead foil
铅箔衬里的隔热抹灰板 insulating plasterboard with foil back
铅箔增感屏 lead foil screen
铅薄板 flat lead; milled lead
铅槽 lead trough
铅槽雨水沟 lead box rainwater gutter
铅测锤 lead
铅测深锤 lead sinker
铅层 lead layer
铅插入层 lead insert(ion)
铅插入物 lead insert(ion)
铅尘 lead dust
铅沉积 lead deposit; lead fouling
铅衬 lead lining; lead works
铅衬槽 lead tank
铅衬的 lead-lined
铅衬垫 lead filler; lead washer
铅衬管 lead-lined pipe
铅衬混凝土电解槽 concrete lead-lined cell
铅衬里 lead lining
铅衬门 lead-lined door
铅衬门框 lead-lined frame
铅衬轴瓦 lead-lined journal bearing
铅橙 orange lead; orange mineral
铅冲水管 lead flushing pipe
铅窗 lead window
铅垂 vertical
铅垂变化 vertical variation
铅垂的 plumbed
铅垂方向 vertical direction
铅垂敷设 vertical laying
铅垂基准 plumb datum
铅垂流形 plumbed manifolds
铅垂面 vertical plane
铅垂面防波堤 vertical face breakwater
铅垂偏差 deflection of vertical; station error
铅垂球【测】 lead plummet; plummet
铅垂水准 plummet level
铅垂水准器 plumb level
铅垂弹性 vertical elasticity; vertical resilience
铅垂位置 plumb position
铅垂涡量 vertical vorticity
铅垂误差改正量 correction deflection of vertical
铅垂线 frontal-profile line; geographic(al) vertical; lead line; lead plumb; plumb(ing)(bob) line; plummet; vertical line
铅垂线校正 plumbing; plumb perpendicular
铅垂线偏差 plumb line deviation
铅垂线水平器 plumb level
铅垂向统计振动级 vertical statistical vibration degree
铅垂准线 bobwire
铅锤 bob(weight); dummy; gravity weight; lead-hammer; lead plumb; lead plummet; lead sinker; lead weight; plumb (bob); plummet (body)
铅锤测量 plumb(ing)
铅锤测深 sounding by lead
铅锤垂线 plumb bob vertical
铅锤点 plumb bob point
铅锤点三角测量 plumb point triangulation
铅锤对准器 plumb aligner
铅锤球 plumb bob
铅锤绳 lead line
铅锤索 lead line; sounding wire
铅锤索测深法 leadline method
铅锤线 lead line; plumb bob cord; plumb line
铅锤线偏差 plumb line deviation
铅锤线用高强钢丝 music wire
铅锤噪声 plumbing noise
铅锤重量 plummet weight
铅催干剂 lead drier
铅淬钢丝 lead patented wire; patented wire
铅淬火 lead bath; patenting; patentizing
铅淬火槽 lead bath
铅淬火钢丝 patented steel wire; patented wire
铅淬火高强度钢丝 best plough steel wire; plow steel wire
铅淬火高强度钢丝绳 plow steel wire rope
铅淬火炉 patenting furnace; patentizing furnace
铅淬硬化 hardening in lead bath
铅存水弯 lead trap
铅锉 lead file; shave hook
铅丹 lead paint; mineral orange; minium; red lead; saturnine red
铅丹打底的椽子 red lead-based spar
铅丹底房屋胶黏[粘]料 red lead-based building mastic
铅丹底涂料 red lead-based paint
铅丹底氧化物 red lead-based oxide
铅丹底油漆 red lead-based paint
铅丹防锈漆 red lead anti-corrosive paint
铅丹铬酸锌 red leaded zinc
铅丹铬酸锌防锈漆 red leaded zinc chromate anti-corrosive paint
铅丹铬酸锌涂料 red leaded zinc chromate anti-corrosive paint
铅丹炉 colo(u)r oven
铅丹(油)漆 minium paint; red lead paint
铅刀 lead knife
铅的 plumbic; plumbous
铅的叠氮化物 azide of lead
铅的树脂酸盐 resinate of lead
铅的氧化物 lead oxide
铅底板 lead safe
铅底漆 lead primer
铅垫 lead pad; lead safe
铅垫板 lead filler; lead insert(ion)
铅垫块平头钉 lead tack
铅垫片 lead insert(ion); lead washer
铅垫圈 <防水表壳内用> lead gasket; lead washer
铅钉 scupper nail
铅顶盖 lead coping
铅锭 lead bullion; lead ingot; lead pig; pig lead
铅锭标准重量 fodder
铅毒 lead poisoning
铅堵 lead plug
铅断面冲压机 lead section press
铅法地质年代测定法 lead method of age determination
铅法年龄 lead ages
铅矾 anglesite; lead vitriol
铅泛水 <管道穿过屋顶处的> lead slate
铅泛水片 lead flashing(piece)
铅泛水条 lead flashing(piece)
铅防护围裙 lead apron
铅废料 scrap lead
铅粉 glass putty; graphite; lead powder; lead white; powdered lead; whiting
铅粉袋 mo(u)lder's blacking bag
铅粉底漆 lead powder primer paint
铅粉漆 metallic lead
铅封 lead ca(u)lking; lead gasket; lead seal(ing); leadwork; seal lead
铅封按钮 sealed button
铅封环 sealing ring
铅封检查 checking of seals
铅封接 leaded joint
铅封钳 lead sealing pliers; sealing tongs
铅封圈 lead seal band
铅封手钮 sealed knob
铅封丝 seal wire
铅封铁丝 detective wire
铅封完整 seal on
铅封旋钮 sealed knob
铅封压铸机 lead seal casting machine
铅封已破 seal off
铅氟硅酸盐 fluosilicate of lead; lead fluorosilicate
铅腐蚀 lead corrosion
铅腐蚀试验 lead corrosion test
铅负荷 lead load
铅覆盖层 lead coating
铅覆盖的穹顶 lead-covered cupola
铅覆盖的穹隆 lead-covered dome
铅干料 lead drier
铅膏 lead plaster
铅隔汽具 lead trap
铅铬 <黄橙色颜料> lead chrome
铅铬橙 chrome orange
铅铬红 chrome red
铅铬黄 Brunswick green; lead chromate; Leipzig yellow; post yellow
铅铬绿 chrome green; lead chrome green
铅工 plumber
铅工技艺 plumbing
铅构件压制 lead unit press
铅管 lead pipe[piping]; lead tube[tubing]
铅管导爆索点火器 lead spitter fuse lighter
铅管工 lead pipe work; plumber; lead burner
铅管工厂 plumbery
铅管工场 plumbery; workshop for lead pipes
铅管工程 plumbing
铅管焊接 lead pipe soldering
铅管矫直木棒 drift plug
铅管接口抹刀 grozing iron
铅管接头 lead pipe joint; wiped joint
铅管开口塞 turning pin
铅管调直器 bending iron; bending pin
铅管弯头 lead elbow; lead pipe elbow
铅管系统 plumbing
铅管压力(试验)机 lead pipe press
铅管样强直 lead pipe rigidity
铅管业 plumbery
铅管支架 lead pipe support
铅管制造 plumbing

铅管装设 plumbing
铅管装置设备 plumbing fixture
铅罐 pig
铅害 air pollution with lead particles
铅焊 burning of lead; lead burning; lead welding
铅焊接 lead soldering
铅焊接头 lead burned joint
铅焊料 coarse solder; kupper solder; lead solder(ing)
铅焊条 burning bar of lead
铅合金 lead(ed) alloy
铅合金包皮 lead-alloy sheath(ing)
铅合金包皮电缆 lead-alloy-sheathed cable
铅合金电缆包皮 lead-alloy cable sheath
铅合金管 lead-alloy pipe; lead-alloy tube
铅护裙 lead apron
铅护套 lead sheath; lead shield
铅华 flake lead
铅化合物 lead compound
铅环 lead collar; lead ring
铅环氧树脂 lead-filled epoxy resin
铅黄 litharge; massicot; yellow lead (oxide)
铅黄甘油胶合剂 litharge-glycerin cement
铅黄铜 lead(ed) brass; ledrite; Manil(l)a gold; rule brass
铅灰 lead ash; lead dust
铅灰色 lead gray[grey]
铅辉石 alamosite
铅基巴氏合金 Isoda metal; lead-base babbitt; white metal bearing alloy
铅基白合金 Dandelion metal
铅基白合金衬层 lead-base white metal linings
铅基白色轴承合金 lead base
铅基底漆 lead-base(priming) paint
铅基钙钡轴承合金 Lurgi alloy; Lurgi metal alloy
铅基合金 lead-base alloy
铅基碱土金属轴承合金 union metal
铅基抗爆震添加剂 lead-antiknock additive
铅基抗摩擦合金 lead-base anti-friction alloy
铅基漆 lead-base(priming) paint
铅基润滑剂 lead-soap lubricant
铅基锑锡轴承合金 Dandelion metal
铅基涂层 lead-based coatings
铅基油漆 lead-base paint
铅基轴承合金 Gliever bearing alloy; graphite metal; Jacama metal; lead-base babbit metal; palid; Tego; termite
铅极板 Plante-type plate
铅极电量计 lead coulombmeter
铅极酸性蓄电池 lead-lead acid cell
铅加厚液 lead intensifier
铅碱玻璃 lead alkali glass
铅碱土金属轴承合金 Frary's metal
铅浇注模 lead pourer's mo(u)ld
铅铰链 lead hinge
铅接口 lead joint
铅接口管线 lead jointed pipe line
铅晶质玻璃 lead crystal(glass)
铅绝缘层 lead insulation
铅壳 lead sheath
铅块 lead block; lead pig; lead regulus; pig lead; quadrat; regulus lead
铅块平头钉 lead tack
铅矿 lead(ing) ore; lead mine; lead mineral
铅矿脉 lead vein
铅矿熔炼(工)厂 lead works
铅矿石 lead ore
铅矿渣 lead smelting slag
铅矿渣混凝土 lead slag concrete
铅蓝矾 caledonite
铅累积 lead accumulation
铅离子 lead ion
铅粒 lead shot
铅粒填封 <卸扣的> lead pellet
铅棂条 came; fret(ted) lead; lead came
铅硫同位素相关性 lead-sulfur isotope correlation
铅锍 lead matte
铅绿矾 caledonite
铅毛 lead wool; lead yarn
铅毛堵缝 lead wool ca(u)lking
铅冒口 lead riser
铅帽 lead hat; chapeau de plumb
铅帽钉 lead-capped nail
铅帽盖 lead capping; lead coping
铅锰催干剂 concentrated drier[dryer]
铅锰干燥剂 lead manganese drier
铅锰钴催干剂 zinc manganic cobaltic drier
铅锰钛铁矿 senaite
铅锰钛铁石 senaite
铅密封的 lead-tight
铅棉堵缝 lead wool ca(u)lking
铅模打印 fishing die(s)
铅磨削器 lead sharpening machine
铅末 lead powder
铅母 saturnic red
铅铌铁矿 plumboniobite
铅排水沟 lead gutter
铅排水管 lead discharge pipe; lead draining pipe
铅盘 lead pan; lead safe; lead tray
铅盘管 lead coil
铅硼玻璃 nonex
铅硼釉 lead borate glaze
铅披水片 lead flashing(piece)
铅铍闪石 joesmithite
铅皮 flat lead; lead-coating; leading; lead sheet; milled lead; sheet lead; strip
铅皮电缆 inarmo(u)red cable; lead-coated cable; lead-covered cable
铅皮电线 electric(al) lead cover wire
铅皮钉 lead nail
铅皮泛水 lead flashing(piece)
铅皮防潮层 lead damp course
铅皮防水层 lead damp course
铅皮分包电缆 separately leaded cable
铅皮管脚泛水 lead slate; lead sleeve
铅皮立墙泛水 lead soaker
铅皮门 lead core door
铅皮平屋顶 lead flat; lead-flat roof
铅皮胎箱 tin-lined case
铅皮套管 <落水管的一段排水套管> lead spitter
铅皮天沟 lead gutter; metal valley
铅皮铜板 lead-coated copper sheet
铅皮外层 <玻璃窗格条的> lead sheath
铅皮屋顶 lead(-sheet) roof
铅皮屋顶天沟 lead roof gutter
铅皮屋顶沿边滴水槽 bottle-nose curb
铅皮屋脊 lead ridge
铅皮屋面 lead(-sheet) roof(ing)
铅皮线 lead-coated wire; lead-covered wire; tinned wire
铅片 lead flake; lead mill; mill lead
铅平头钉 bale tack
铅屏 lead shield(ing)
铅屏蔽 lead foil screen; lead shielding
铅屏蔽舱 lead shield compartment
铅屏蔽墙 lead shielding wall
铅屏蔽室 lead shield compartment
铅屏厚度 lead screen thickness
铅漆 lead paint; white paste
铅汽缸 lead cylinder
铅器 plumbery
铅铅等时线 Pb-Pb isochron
铅墙 lead curtain of chamber
铅鞘 lead sheath
铅青铜 lead bronze
铅绒 lead wool; spongy lead
铅熔焊接 burned lead joint
铅熔块 lead frit; reguline of lead
铅熔炼 lead smelting
铅熔丝 lead fuse(wire)
铅塞 lead plug
铅塞承口 socket for lead joint
铅塞子 lead plug
铅色 lividity
铅色的 leaden; livid
铅砂 lead ash
铅砂管 aloxite tube
铅烧绿石 plumbopyrochlore
铅砷钯矿 borishanskiite
铅砷磷灰石 hedyphane
铅生物防护屏蔽 lead biologic(al) shied
铅生物防护墙 lead biologic(al) shielding wall
铅绳 lead rope
铅石墨和油的合金 ledaloyl
铅室 lead chamber
铅室底盘 lead chamber pan
铅室法 lead chamber process
铅室法硫酸 lead chamber process sulfuric acid
铅室结晶 lead chamber crystals
铅室晶体 lead chamber crystal
铅室气 lead chamber gases
铅室容积 lead chamber space
铅室酸 chamber acid
铅手 leadsman
铅手套 lead gloves
铅树 lead tree
铅栓 lead dot
铅水嘴 lead spitter
铅丝 galvanized iron wire; galvanized wire; galvanizer; lead wire; tinned wire
铅丝玻璃 toughened glass; wired glass
铅丝捆扎的包件 wirebound package
铅丝笼 wire box; wire cage
铅丝纱 wire screen
铅丝筛 wire-mesh screen
铅丝绳股 galvanized strand
铅丝石笼结构 <护岸用的> stone-mesh construction
铅丝网石笼坝 sausage dam
铅丝网石笼工程 sausage engineer
铅丝网石笼建筑 sausage construction
铅酸 <二价铅> plumbous acid
铅酸电池 lead-acid battery; lead-acid cell
铅酸钙 calcium plumbite
铅酸钴蓄电池 lead-acid cobalt battery
铅酸钠 plumbite of soda; sodium plumbite
铅酸钠精制过程 plumbite process
铅酸钠溶液的再生 regeneration of doctor solution
铅酸铅 lead plumbate
铅酸蓄电池 lead-acid battery; lead-acid cell; lead acid storage battery
铅酸盐 plumbite
铅酸盐处理 plumbite treatment
铅榫 lead dowel
铅钛矿 macedonite
铅糖 sugar of lead
铅套 lead shield
铅套管 lead sleeve
铅套筒 lead sleeve
铅套筒连接 lead sleeve joint
铅锑黄 antimony yellow
铅锑硫矿 playairite
铅锑烧绿石 mauzeliite; plumboan romeite
铅锑锡轴承合金 magnolia metal
铅天沟 lead trough
铅添加剂 lead additive
铅填充物 lead filler
铅填缝 lead ca(u)lking
铅条 bonding strip; clump; lead bar; leading; lead strip; pig lead; cames <用以嵌合窗上玻璃用>【建】
铅条玻璃 leaded glass
铅条玻璃窗 leaded glass window
铅条玻璃嵌镶 mosaic of leaded glass
铅条镶嵌玻璃 lead glazing
铅条镶嵌玻璃窗 leaded light
铅贴面 lead(sur)facing
铅铁矾 plumbojarosite
铅铁矿 plumboferrite
铅铁锗矿 bartelkeite
铅同位素单阶段模式 lead isotope single-stage model
铅同位素年代 leading-isotope age
铅铜合金 cuprolead
铅铜减摩合金 Lubrimetal
铅铜矿石 Pb-bearing copper ore
铅铜锍 lead copper matte
铅桶 lead cask
铅头钉 lead-capped nail; lead(head) nail
铅涂面层 lead facing
铅托架 lead safe
铅弯头 lead elbow
铅污染 lead pollution; pollution by lead
铅污染物 lead pollutant
铅污水管 lead soil pipe; lead waste (pipe)
铅屋脊 lead hip
铅屋面钉 lead nail
铅屋檐沟 lead roof gutter; lead trough
铅吸收体 <船尾的> lead absorber
铅锡碲探测器 lead-tin-telluride detector
铅锡铋易熔合金 Rose's metal
铅锡铋易融合金 onions alloy
铅锡碲化物 lead-tin-telluride
铅锡碲化物探测器阵列 <热成像用的> lead-tin telluride detector array
铅锡碲晶体 lead-tin-telluride crystal
铅锡镀层 terne plating
铅锡镉合金 ternary alloy
铅锡各半软钎料 lead-tin half-and-half solder
铅锡焊料 lead-tin solder; plumber's solder; tinman's solder
铅锡焊料合金 tinman's solder alloy
铅锡合金 <1.6:64> lead-tin alloy; terne (alloy); terne metal; terne plate; tinsel tin-plate
铅锡合金管 compo pipe
铅锡合金焊剂 plumber's solder
铅锡合金焊料 wiping solder
铅锡青铜 <其中的一种> Redford's alloy
铅锡软焊料 Slicker solder
铅锡锑合金 pewter
铅锡锑青铜 <其中的一种> Retz
铅锡涂复层 lead-tin overlay
铅锡轴承合金 ley; Morgoil
铅细晶石 plumbomicrolite
铅下水管 lead flushing pipe
铅纤维 lead wool
铅线 lead line; lead wire
铅限量(色)漆 lead-restricted paint
铅镶窗的玻璃圆片 rondelle
铅橡胶 lead rubber
铅橡胶布 lead-rubber sheet
铅橡胶手套 leader-rubber gloves
铅橡胶围裙 lead-rubber apron
铅橡胶支座 lead-rubber bearing
铅橡皮手套 protective gloves
铅橡皮围裙 protective apron

铅销 lead button;lead dot;lead rivet
铅销钉 lead dowel
铅楔 bat;lead bat;lead wedge
铅斜沟槽 lead valley(gutter)
铅斜屋脊 lead hip
铅屑板 lead shielding
铅芯架 lead holder
铅锌白 leaded zinc oxide
铅锌底漆 lead zinc primer
铅锌合金 hard lead
铅锌矿石 Pb-Zn ore
铅修饰物的压制 lead trim press
铅蓄电池 lead accumulator;lead-lead acid cell;lead storage battery
铅旋管 lead coil
铅循环 lead cycle
铅压(封)顶 lead capping
铅压接管 lead sleeve
铅压载 lead ballast
铅亚油酸盐 linoleate of lead
铅盐 lead salt;salt of lead
铅颜料 lead pigment
铅檐沟 lead(rainwater)gutter
铅液螺旋搅拌器 lead rotor
铅翼 lead wing
铅银矿石 Pb-Ag ore
铅银砷镍矿 animikite
铅引起的 plumbic
铅印 letter press; letter press printing; letter printing; typographic printing
铅印油墨 letter press ink
铅硬锰矿 coronadite
铅油 lead oil;lead paint;red lead putty;white paste
铅铀比 leading uranium ratio
铅铀年代测定法 leading uranium age method;Pb-U dating
铅铀云母 przhevalskite
铅黝帘石 hancockite
铅釉(条镶嵌玻璃)lead glaze(d finish)
铅鱼 elliptic(al)type weight; fish lead; streamlined sinker; streamlined weight;torpedo-type weight; torpedo sinker < 流速仪用 >;Columbus-type weight < 用于测深和流速仪上的 >
铅雨水斗 lead cesspool
铅雨水管 lead rainwater gutter
铅浴处理的钢丝 lead bathed steel wire
铅浴淬火 lead hardening;lead patenting
铅浴回火 lead tempering
铅浴索氏化处理 lead patenting
铅圆筒 lead cylinder
铅皂 lead soap
铅皂低黏[粘]度润滑脂 mobile grease
铅皂润滑 mobile grease
铅皂润滑脂 lead-base grease; lead grease
铅渣 lead slag
铅爪钉 lead plug
铅罩 lead shield
铅直板 lamina perpendicular
铅直的 plumb;vertical
铅直的环境问题 environmental implication
铅直地层离距 vertical stratigraphic(al)separation
铅直断面 vertical section
铅直对流 vertical convection
铅直风速表 anemoclinometer;vertical anemometer
铅直风速计 anemoclinograph
铅直风速仪 vertical anemoscope
铅直荷载 vertical load(ing)
铅直厚度 vertical thickness
铅直环流 vertical circulation
铅直截面 vertical section
铅直式砌合 plumb bond
铅直输送系数 vertical transport coefficient
铅直速度梯度 vertical velocity gradient
铅直梯度 vertical gradient
铅直线 plumbing line
铅直准线 vertical alignment
铅制的 leaden;leady
铅制垫片 lead button
铅制垫圈 leaden washer
铅制定位物 lead button
铅制间隔物 lead button
铅制排水沟 < 墙洞上边防止木材腐蚀的 > head guard
铅制品 leading;lead works
铅制屋檐水槽 eaves lead
铅制线脚型板 lesbian rule
铅制檐沟床 gutter bed
铅质底层漆 lead priming paint
铅质垫片 lead button
铅质气溶胶 lead aerosol
铅质热辐射罩具 lead radiation shield
铅质压力管 lead pressure pipe
铅质造型压床 lead profile press
铅致空气污染 air pollution by lead
铅中毒 lead poisoning;plumbism;saturnism
铅珠 lead shot
铅柱 lead block
铅柱膨胀试验 lead block expansion test
铅柱压缩试验 lead block compression test
铅铸窗的施工大样 cut line
铅铸法 lead cast
铅铸销 < 固定屋瓦用 > lead dot
铅铸型法 lead mo(u)lding
铅铸造厂 lead foundry
铅坠 lead weight
铅准直器 lead collimator
铅字 font(type);letter
铅字合金 type metal
铅字面 typeface
铅字盘 typecase
铅字印样 typeface
铅座 lead safe

签

签标器 endorser

签到簿 attendance book
签到工资 reporting pay
签订 enter into;given
签订合同 award a contract;awarding contract; award of contract; contract award; contract letting; make a contract;placing of contract;sign a contract;signing(of contract)
签订合同的时间 dotted-line time
签合同后的阶段 post-contract stage
签订合同阶段 contract definition phase
签订合同前的准备 precontract preparation
签订合同前计划 precontract planning
签订合同前阶段 precontract stage
签订合同权 contract authorization
签订合同通知书 notice[notification] of award of the contract
签订协议 conclude an agreement; make an agreement
签定 judgement
签定合同 awarding contract; award of contract;contract letting;make a contract;sign a contract
签定生效的副本 executed copy
签发 draw on;value on
签发处 issuing service
签发地 issue at
签发港口 port issued
签发汇票 issue of a bill of exchange
签发货单 issuing invoice
签发票据 issuance of a note
签发凭单 issue a warrant
签发清单 issuing invoice
签发日期 issuing date;date issue
签发账单 issuing invoice
签发证书 grant a certificate; issue a certificate
签号 lot number
签合同 sign up
签核人 approved officer
签护照 endorse a passport;visa passport
签名 affix one's signature;sign;signature;subscribe;superscript
签名单据 documents signed
签名盖章 signed and sealed; under one's hand and seal
签名密封 signed and sealed
签名人 the undersigned
签名使契据生效 execute a deed
签名使契约生效 execute a deed
签名于下 undersign
签票房 < 换车、船、飞机的 > ticket-transfer office
签收 receipt;sign in
签收据 receipt
签收章 acknowledgement stamp
签署 affix; autograph; endorsement; per pro; signature; subscribe; subscription
签署发票 signed invoice
签署合同 execute contract;sign a contract;sign contract
签署票据的人 endorser
签署权 authority of sign
签署声明 signed declaration
签署样本 specimen of authorized signature
签署意见 comments on the contracts
签署者 subscriber;signatory < 协议、条约等的 >
签署证书 sign a certificate
签条 docket;label;tag
签约 sign a contract
签约承运者 contract hauler
签约各方 contracting parties
签约雇用 sign on
签约国 signatory
签约后承包商依法占有工地以便施工 possession of site
签约前观察 pre-contract view
签约前谈判 precontract negotiation
签约人 parties to a contract
签约日期 contract date;date of contraction
签约自由 freedom of contract
签账卡 debit card
签证 attestation;endorse of ticket;legalize;visa
签证本 attested copy
签证发票 certification of invoice;certified invoice;legalized invoice
签证费 certificate fee
签证机关 signatory authorities; visa-granting office
签证卡 visa card
签证签发地 city where visa was issued
签证人 attestor
签证手续费 consulage;consular fee
签证条款 attestation clause
签证银行 appointed licensing bank;licensing bank
签证支票 certified check
签证种类 visa type class
签注 special endorsement
签字 per pro;signature;sign(ing)
签字费 signature payment
签字蜡封合同 signed and sealed contract
签字蜡封式的要约 offer under seal
签字权 power to sign;signing authority
签字人 signatory
签字式样证明书 letter of identification;letter of indication
签字条款 sign articles
签字文件 sign articles
签字样本 specimen signature
签字样卡 signature card
签字仪式 signing ceremony
签字作废 cancel a signature

前

前阿巴拉契亚海洋 pre-Appalachia ocean

前鞍 pommel
前岸 foreland;foreshore
前凹的 procoelous
前凹后凸的 platycelian;platycelous
前奥陶纪 Pre-Ordovician
前八字方向 broad on the bow
前坝 counter dam
前扳机 fore trigger
前班累计 previous total
前板 fore-plate
前板桩带卸荷平台码头 relieving platform with external bulkhead wharf
前板功能 front panel function
前板微程序 front panel microroutine
前板桩带卸荷平台岸壁 relieving platform with front sheet pile wall
前板桩减压平台岸壁 relieving platform with front sheet pile wall
前板桩式高桩码头 front sheet-piling platform
前半升高甲板 raised foredeck
前保护板 front fender
前保险杆 front bumper
前保险杆臂 front bumper arm
前保险杆高度 front bumper height
前保险杆减振器 front bumper shock absorber
前保险杆套环 front bumper arm grommet
前保险杆支架 front bumper bracket
前保险杠 front guard
前保险构架 front guard frame
前北方期【地】 preboreal epoch
前贝壳灰岩的 precray
前背部 antedorsal
前辈 predecessor;senior
前泵 front pump
前辟导洞 pilot tunnel
前壁 anterior wall; antetheca; front wall;leading wall
前壁导洞 tunnel pilot
前壁光电池 front wall cell
前壁隧道工程 tunnel pilot
前臂 forearm
前臂长 forearm length
前边浮箱 foreside pontoon
前边锚 forward side anchor
前变址 preindexing
前标志灯 front identification lamp; side lamp
前表面 front surface
前表面镀膜镜 front-coated mirror
前表面镀银射镜 front-silvered mirror
前表面反光镜 first-surface mirror; front surface mirror
前表面反射 front surface reflection
前表面反射镜 front surface mirror
前表象 prerepresentation
前滨 < 高潮线和低潮线之间的滨海地

Q

带 > foreshore;beach face
前滨沉积 foreshore deposit
前滨出水口的流水槽 rill way
前滨带 perezone
前冰期沉积物 preglacial deposit
前波 head wave
前玻板 face glass
前玻璃挡板加热器 windshield heater
前剥离器 prestripper
前部 anterior part; fore; forehead; forepart; forequarter; front; head; antechurch <古教堂的>
前部安装的 forward-mounted
前部安装凿岩机和有单独压风机的自行式凿岩设备 front-mounted self-contained unit
前部船舱 fore cabin;fore hold
前部船体 forebody
前部船体吃水 fore draft; forward draft
前部的 anterior
前部垫块 head block
前部钉法 head nailing
前部端梁 <车身底框的> front-end sill
前部房屋 front building
前部功率输出端 front power take-off
前部横向输送器 front cross conveyer [conveyor]
前部火焰的消退 retrogression of flame front
前部件 forepiece
前部静重量 front-end static weight
前部楼座 loge
前部帽衬 front cap gasket
前部取暖器 fore warmer
前部上层建筑 forward superstructure
前部死荷载 front-end dead weight
前部死重量 front-end dead weight
前部脱险舱 forward refuge compartment
前部外观 frontal appearance
前部悬挂电机 nose suspended motor
前部压缩 frontal compression
前部支撑 <隧道木支撑> horsehead-ing;forepoling
前部重量 front-end weight
前部装备 front-end work attachment;front equipment
前部阻力 foredrag
前舱 forebay; fore cabin; fore hold; nose cabin; nose compartment; nose nacelle; outer lock of recompression chamber <加压舱的>
前舱壁 front bulkhead; scuttle bulkhead
前舱口 forehatch
前舱面 foredeck
前操作期 preoperation period
前槽 foredeep
前侧包厢 proscenium box;stage box
前测 fore measurement
前测边 preceding line
前差速器 front differential
前觇标 front target
前场(舞台)泛光照明 apron flood lighting
前超梁 forestope
前车 leader;leader vehicle
前车道板 front vehicular slab
前车架 fore carriage
前车桥外壳 front axle housing
前撑柱 foreshore
前成的 paragenetic
前成岩作用 prediagenesis
前吃 <超前成圈的俗称> reverse timing;advanced timing;early timing
前吃水 fore draft;forward draft;forward draught
前池 forebay; forward spring; head bay; head pond; headrace (bay); stilling pond; upper pool; upper lock approach
前池护坦 forebay apron
前池区 forebay area
前池渠道 forebay channel
前池水库 forebay reservoir
前池水位 forebay level
前尺 <测量拉尺前面的测工> leader
前尺手 forward tapeman;head tapeman
前冲 forge ahead; kick ahead; preshoot
前储水空间 front water space
前处理 fore treatment;pretreatment
前处理程序 preprocessor
前处理底漆 pretreatment primer
前处理与后处理 pre and post-processing
前传动 front drive
前囟点 bregma
前窗玻璃刮水器 windshield wiper
前床 furnace vessel; forehearth <鼓风炉等的>
前吹 fore-blow
前唇砖 front lip tile
前次扫描结果 last look
前存河谷 preexisting valley
前带 prozone
前挡 front stop
前挡风玻璃洗刷器 front windshield washer
前刀 leading edge
前导边缘 preceding limb
前导标 front leading mark
前导部 <拉刀的> front pilot
前导部分 leading portion
前导车 leading vehicle
前导承 top guide
前导灯 front leading lamp;front leading light
前导灯标 forward light
前导轨 former rail
前导坑 <开凿隧道时的先掘部分> advancing gallery
前导零 leading zero
前导轮 front jockey wheel;ga(u)ge wheel
前导轮架 fore carriage
前导轮架式单铧犁 roll gallows
前导轮架式双向犁 reversible fore-carriage plow
前导轮缘 leading flange
前导轮轴 front roller shaft
前导轮组 leading wheel sets
前导码 lead code
前导判定 leading decision
前导数据 lead data
前导图形字符 leading graphic
前导网络 precedence network
前导网络图 precedence diagram
前导问题 introductory problem
前导系统 vanguard system
前导向灯 front range light
前导向轮 front idler; front jockey wheel
前导轴 leading axle
前导子 conductor
前导字符 leading character
前导钻头 pilot bit
前倒缆 bow spring;fore back spring; fore head spring; fore spring (line);forward spring;head spring
前灯 front lamp; front light; heading light;head lamp;headlight <汽车>; paragenetic light
前灯变光器 headlight dimmer
前灯玻璃 headlight lens
前灯玻璃固定器 headlight lens locator
前灯导线 front lamp wire
前灯导线管 headlight wiring conduit
前灯反射罩 head lamp reflector
前灯盖 headlight door
前灯护罩 headlight hood
前灯减光开关 headlight dimmer switch
前灯减光器 headlight dimmer
前灯接头 head lamp connector;headlight adapter
前灯卡 headlight clip
前灯壳 head lamp body
前灯汽轮机 headlight turbine
前灯栅罩 head lamp brush guard
前灯闪光 headlight glare
前灯上光线 head lamp upper beam
前灯调光器 headlight dimmer
前灯调节螺钉 head lamp adjusting screw
前灯托架 front lamp bracket
前灯远光 high beam of head lamp
前灯照明 headlighting
前灯照明视距 head lamp bright viewing distance
前灯照射距离 range of headlamp
前灯遮罩 head lamp mask
前灯支架 head lamp bracket
前灯桩 front light tower
前堤 front embankment
前底板 nose plate
前地 foreland
前地槽构造层【地】pregeosynclinal structural tectonic layer
前地槽阶段 pregeosynclinal stage
前地台阶段 preplatform stage
前第四纪的 antediluvial
前点 front pole
前点测工 head chainman
前电势 prepotential
前电位 prepotential
前垫架 fore poppet
前殿 <寺庙的> antetemple
前顶焦距 front vertex distance;front vertex focal distance
前顶角 front top rake
前定变量 predetermined variable; predetermined variate
前定点 prefixed point
前定论 predetermination
前定时装置 front timing gear
前定位 prelocalization
前定位垫 front locating pad
前动臂 front boom
前洞 head race
前端 fore-end; front end; front side; leading end
前端板 fore stand plate; front-end panel
前端板侧 fore stand side
前端板吊钩 fore stand-hanging hook
前端板支腿 fore stand leg
前端壁 front-end wall
前端部分 fore-end
前端舱壁 front bulkhead
前端侧方联合装载 combined front-and-side loading
前端铲斗 forward shovel bucket
前端铲运配件 forward shovel attachment
前端铲运器件 forward shovel fitting
前端沉笼丁坝 <防冲沉笼裹头,新西兰> sinker groin
前端处理 front-end processing
前端处理机 front-end processor
前端处理器 front-end processor
前端挡板 front guard
前端刀板 <平地机> front blade
前端导向轮护罩 front idler shield
前端点火 headend ignition
前端电路 front-end circuit
前端定位装置 front lay
前端动力输出装置 front power take-off
前端发火 headend ignition
前端翻卸 front dumping
前端反应 front reaction
前端费 front-end cost
前端附加装置 front-end attachment
前端盖 front-end housing
前端工作装置 frontage
前端护 front guard
前端集中器 front-end concentrator
前端计算机 front-end computer
前端加速度 leading-end acceleration
前端夹板 leading-end clamp
前端夹具 front jig
前端角 angle of throat;nose angle
前端角铁 front-end angle ring
前端壳体 front head
前端可辅助驱动装置 front accessory drive
前端框架 front-end frame
前端离地净高 <汽车车身> front clearance
前端面 front face(side)
前端内倾 toe-in
前端黏[粘]接型锚杆 point bonded bolt;wedge-type rock bolt
前端排气孔 <凿岩机钻头> front head air release
前端配重块 front weight
前端驱动 front drive
前端驱动装置 front power unit
前端燃烧 front-end combustion
前端设备 front(-end) rig;headend equipment;front-end equipment
前端式设备 front-end equipment
前端式装载机 front loader; tower loader;front-end loader <装载斗装在前端>
前端调谐器 front-end tuner
前端通信[讯]处理机 front-end communication processor; front-end processor
前端推土板 <平地机> front blade
前端弯钩 leading hook
前端位置 front position
前端斜角 <螺丝钢板> throat angle
前端泄气口 front-head release port
前端信贷 front-end finance
前端信号电线束 front-signal harness
前端型拖拉装载车 headend type tractor-loader
前端悬挂铲斗 front-mounted bucket
前端压缩 front compression
前端压载 front weight
前端圆木装载装置 log loading front
前端圆头 forward dome
前端窄 inswept
前端窄缩式车架 inswept frame
前端整流罩 muzzle door
前端支持 front-end support
前端直径 <螺纹的> point diameter
前端中心千斤顶 front center jack
前端轴套 front pressing ring
前端转变装置 front-end conversion unit
前端转换装置 front-end conversion unit
前端装备 front-end work attachment;front equipment
前端装车机 front-end lift
前端装卸式铲斗 front-end bucket
前端装载 front loading
前端装载机 front-end loader; front-end shovel;front-end type loader
前端装载装置 front-end loader attachment
前端自卸车 front-end dumper
前段 forepart;leading portion

前段动臂 foreboom
前惰轮 front idler
前额 forehead
前额支撑垫距离 forehead pad distance
前二班时 dog watch first
前发动机 front engine
前发脉冲 preliminary impulse
前伐 advance cutting; advanced felling; prelogging
前翻斗手推车 end tipper; front tipper
前翻式机罩 front-hinged bonnet
前翻式矿车 front tipper
前方 forward; frontage
前方(闭塞)区间 section block in advance
前方闭塞信号(机)【铁】 advance block signal
前方仓库 pier shed; quay shed; transit shed
前方场地 forefield
前方超吊 front lifting
前方车站 station in advance
前方承台 front platform
前方出口 front exit
前方调度 front despatching[dispatching]
前方堆场 apron; yard; marshalling yard
前方堆场计划 marshalling plan
前方飞机场 advanced landing ground
前方观察员激光测距机 forward observer laser rangefinder
前方货场 pier shed; quay shed; transit shed
前方降落场 advanced landing ground
前方交会【测】 forward intersection; foresight reading
前方交会法 forward intersection; method of forward intersection; method of intersection
前方库场 quayside shed and yard
前方能见度 forward visibility
前方倾卸轨道 front-dump track
前方区段 advance section
前方施工标志 men working sign; work ahead sign
前方式 premode; previous mode
前方视距 forward visibility
前方视野 field of front vision
前方位角【测】 forward angle; forward azimuth
前方线路所 box in advance; forward box
前方线瞄准器 primary flight line viewfinder
前方信道 forward channel
前方信号(机)【铁】 advance signal
前方信号楼 box in advance; forward box
前方医院 clearing hospital
前方桩基承台 front piled platform
前方作业标志 work ahead sign
前防护器 front guard; front protection
前房角 angle of anterior chamber
前风挡玻璃的安装角 inclination angle of a windscreen
前锋绞刀 end reamer
前扶垛 but buttress; buttress
前扶墙 buttressed wall
前复理石硅泥质组合【地】 preflysch silica muddy association
前盖 <制动缸的> non-pressure head
前盖布 front tarpaulin
前概型 prescheme
前冈 entrance hall
前缸盖 front cylinder cover
前钢板 front spring
前钢板弹簧第四片卡扣 front spring fourth leaf clamp
前钢板弹簧卷耳孔衬套 front spring eye bushing
前钢板弹簧橡胶块 front spring pad
前钢缆绞车 front cable winch
前钢桩【疏】 forward spud
前港 <闭式港池或坞前直接闸门的外港> avant port
前港池 <有闸港池口门外的水域> vestibule basin; vestibule dock
前哥特式的【建】 pre-Gothic
前更新伐 pre-regeneration cleaning
前工业城市 preindustrial city
前工作辊道 front mill table
前工作孔径 front operating aperture; operating front aperture
前功能主义(者设计的)建筑 pre-functional building
前拱 face arch; fore-arch; coking arch <层燃炉膛>
前拱脚 front abutment
前拱形 T 波 plateau T wave
前拱座压力 front abutment pressure
前构件 leading frame
前古生界【地】 primary group
前鼓轮 head pulley
前挂式 front-mounted
前管板 front tube plate; front tube sheet
前管板拉条 front tube sheet brace
前光 <观众厅的灯光总称> front spot
前轨 waterside rail
前辊 front roll; preliminary roller
前滚动率 front roll rate
前滚动中心 front roll center[centre]
前滚轮 front roll
前滚式潜水法 forward roll entering water
前磙 front drum
前过梁 <承托空心墙外层的过梁> front lintel
前海角 front range
前海滩 <高潮线和低潮线之间的地带> foreshore
前寒武纪【地】 Pre-Cambrian era; Pre-Cambrian period
前寒武纪地盾 Pre-Cambrian shield
前寒武纪地质年代表 Pre-Cambrian geochronologic(al) scale
前寒武纪基底 Pre-Cambrian basement
前寒武纪结晶基底 crystalline basement of Pre-Cambrian
前寒武纪片麻岩 Precambrian gneiss
前寒武统【地】 Pre-Cambrian series
前寒武系【地】 Azoic; Pre-Cambrian system
前寒武系的 Cryptozoic; Pre-Cambrian
前横档 front cross member
前横缆 bow breast; fore breast(line)
前横梁 front lintel
前洪积世的 <前第四纪> antediluvial
前后 fore-and-after
前后保持安全距离陆续通过 <车、船等> under continuous headway
前后比 front-to-rear ratio
前后不符 inconsequence
前后不一致的影响 front-to-back effect
前后参照 cross-index; cross reference
前后车轴架 rear and front axle stand
前后重叠 end(over) lap; fore-and-aft overlap; forward(over) lap; front-to-back lap; longitudinal overlap
前后传动 tandem drive
前后传动链 tandem drive chain
前后传动轴套 tandem drive housing
前后从板座 front-and-back stop
前后挡 front-and-back stop
前后地 back and forth
前后颠簸 nosing; pitching; plunge
前后颠倒的 preposterous
前后端板接管 fore stand and end plate nozzle
前后端下沉的平衡(车辆) trimming
前后对比调查 before and after study
前后对比法 before and after contrast approach
前后对比分析 before and after analysis; before and after study
前后对比研究法 <交通实效改善> before and after study
前后对照 cross reference
前后法 before and after method
前后方向 fore-and-aft direction
前后方向均可使用的司机座和操纵台 two-way-and control seat
前后方向牵引力 fore-and-aft traction
前后方向调整 fore-and-aft adjuster
前后杆 forward-reverse lever
前后鼓轮 terminal pulley
前后关系 context
前后弧焊 tandem arc weld
前后换置式装载机 alternative front or rear loader
前后间距 <工作部件> fore-and-after clearance
前后交接班 front-and-back shift
前后接点 dependent contact; front-back dependent contact
前后结合式悬架装置 combined suspension
前后紧接 lockstep
前后进导架 <凿岩机的> guide cell
前后径 anteroposterior diameter
前后距离 longitudinal separation
前后来回 back and forth
前后两车之间的时距 headway; time headway
前后两级变速箱 two-stage forward and reversing transmission
前后两用刀架 tool-post for front and rear slide
前后列柱式 amphiprostyle; amphiprostylos
前后轮负荷反应 front-and-rear wheel reaction
前后轮(结合)转向 front-and-rear wheel steer(ing)
前后轮距 wheel-base
前后轮协同转向的平地机 front-and-rear wheel steer grader
前后轮协同转向的装载机 front-and-rear wheel steer loader
前后轮制动压差警报阀 pressure differential warning valve
前后轮转向式平地机 front-wheel and rear-wheel steer grader
前后论证 argument back and forth
前后矛盾 antilogy
前后徘徊 backward and forward
前后排柱建筑 amphiprostyle
前后排柱廊神庙 amphiprostyle temple
前后排座位错开的布置形式 staggered seating
前后平衡调整 fore-and-aft trim
前后坡度 front-to-back slope
前后桥差动器 front-and-rear axle differential
前后倾斜计 fore-and-aft inclinometer
前后扫描方式 back-and-forth mode
前后视差距 difference between fore and back sighting distances
前后视距累积差 accumulated sum of differences between fore and back sighting distances
前后(视)水准测量 fore and back level(l)ing
前后四柱廊神庙 amphiprostyle tetrastyle temple
前后送料可倾式冲床 open back inclinable press
前后送料可倾式压床 open back inclinable press
前后索连推土机 <俚语> yo-yo
前后台处理 foreground-background processing
前后通风 cross ventilation
前后推导 forward and backward deduction
前后位 anteroposterior position
前后文无关的 context-free
前后向的 sagittal
前后向曲轴 cross crank
前后效果对比 contrast before and after
前后檐墙檐沟 boundary wall gutter
前后演变序列 sequential relationship
前后移动 fore-and-aft shift; shuttle <列车、飞机的>
前后移位 fore-and-aft shift
前后影响 front-to-back effect
前后有排柱而两边无柱的建筑 <两排柱式建筑> amphiprostylos
前后运动 see-saw
前后振动 porpoise
前后支索 fore-and-aft stay
前后支销 fore-and-aft pivot pins
前后轴重量分配 front-rear weight distribution
前后转动 reciprocate; reciprocating
前后纵列排列 tandem arrangement
前弧 frontal arc
前护板 apron plate; front fender; front wearing plate
前护堤 waterside banquette
前护墙 apron wall
前护墙衬砌 apron wall lining
前护墙封板 apron wall lining
前护墙镶板 apron wall panel
前护墙装修 apron wall facing
前护栅 front guard; front protection
前花园 front garden
前滑 forward slip
前滑角 advance angle; angle of advance
前环境角色 former surrounding case
前回声 pre-echo
前回填 initial backfill
前回转半径【机】 front fitting radius
前活塞杆导架 piston rod guide
前货舱 fore hold
前机舱 forward engine room
前机罩 front housing
前积层 foreset bed
前积层沉积 foreset deposit
前积层理 foreset bedding
前积坡 foreset slope
前积纹层 foreset laminae
前基座时间 front porch interval
前级 backing stage; forestage; preceding stage
前级泵 backing pump; prepump
前级抽气 forepumping
前级除尘器 pre-collector
前级放大器 prime amplifier
前级工质透平 top-fluid turbine
前级管道 backing line; foreline
前级管道阀 foreline valve
前级管道法兰 foreline flange
前级管道罐 foreline tank
前级管道阱 foreline trap
前级管道软管 foreline hose
前级互调 front-end intermodulation
前级激励 predrive
前级空间技术 backing space technique
前级冷井 backing side trap
前级冷凝器 backing condenser

前级耐压 forepressure tolerance
前级视频放大器 head amplifier
前级透平 top turbine
前级系统 backing system
前级压力 forepressure
前级压强 back(ing) pressure; forepressure
前级压强测量 forepressure measurement
前级压强端 forepressure side
前级压强破坏 forepressure breakdown
前级压强特性 forepressure characteristic
前级真空 backing vacuum; fore-vacuum
前级真空泵 fore-vacuum pump; forepump
前级真空表 fore-vacuum ga(u)ge
前级真空抽气系统 fore-vacuum pump system
前级真空的 preevacuating
前级真空阀 fore-vacuum valve
前级真空管道 fore-vacuum line; fore-vacuum pipe
前级真空管线 foreline
前级真空规 forepressure ga(u)ge; fore-vacuum ga(u)ge
前级真空空间 fore-vacuum space
前级真空冷阱 fore-vacuum trap
前级真空冷凝器 fore-vacuum condenser
前级真空冷却器 fore-vacuum cooler
前级真空连接 fore-vacuum connection
前级真空连接口 fore-vacuum port
前级真空腔 fore-vacuum cylinder
前级真空容器 fore-vacuum vessel
前级真空软管 fore-vacuum tubing
前级真空系统 backing system; fore-vacuum system
前级真空压强 fore-vacuum pressure
前级真空转子 fore-vacuum rotor
前级真空组件 fore-vacuum subassembly
前极尖 <电机磁极的> leading pole tip
前集电弓 leading bow
前加里东海洋【地】Pre-Caledonian ocean
前加里东期【地】Pre-Caledonian period
前甲板 fore-board; forecastle; foredeck
前架 side arm
前尖舱 forepeak tank
前尖嘴 fore beak
前焦点 front focus
前焦距 front focal distance; front focal length
前焦距计 front focometer
前焦平面 front focal plane
前角 front angle; hook angle; rake angle; top rake; angle of cutting edge
前绞车 lead winch
前脚 <船首柱底部> forefoot
前脚板 forefoot plate
前脚肘 forefoot knee
前接点 front contact; make contact; making contact; top contact
前接点闭合 front contact closing
前节 lead-unit
前节点 forward nodal point; front nodal point; nodal point of incidence
前节距 front pitch; front span
前结圈 ash ring; discharge ring; dust ring
前进 advance; debouch; fetch headway; forge ahead; go ahead; go forward; headway; make headway; make progress; make way; progression
前进半轴 <拖拉机转弯时> leading half axle
前进变质作用【地】progressive metamorphism
前进表面波 progressive surface wave
前进冰川 advancing glacier
前进冰碛 moraine of advance; push moraine
前进并联位置 forward parallel position
前进波 advanced wave; advancing wave; forward wave; progressive wave; travel(l)ing wave
前进波动 motion progressive wave
前进波说 progressive wave theory
前进波运动 motion of progressive wave
前进超覆 progressive overlap
前进潮波 progressive tidal wave
前进车速 forward speed
前进成岩作用 anadiagensis
前进齿轮 forward gear
前进齿轮(传动)系 forward gear train
前进冲程 forward stroke
前进串联位置 forward series position
前进大门 fore gate
前进挡 forward gear
前进挡变速比 forward gear ratio
前进挡传动 <机动车> straight shift transmission
前进挡的传动装置 forward drive
前进挡功率输出装置 forward output gear
前进挡离合器 forward clutch
前进挡速度 forward speed
前进挡位置 forward position
前进导坑 heading for advancing
前进倒车换挡杆 transmission forward-reverse lever
前进倒退式传动 forward-reverse transmission
前进倒退式主要传动 forward-reverse primary transmission
前进的 progressive
前进灯 go light
前进第二挡 second forward gear
前进第一挡 first forward gear
前进舵效试验 ahead steering test
前进二【航海】ahead two; half ahead
前进法 move-up; progressive method
前进方向 direction of advance; direction of travel; heading; working direction
前进分期施工 progressive stage construction
前进分析 forward analysis
前进跟踪 track up
前进海岸 anagenesis advancing coast
前进焊 forehand welding; forward welding; progressive welding
前进航向 course of advance
前进和停止 <交通指挥灯> go-stop
前进后退传动轴 input shaft forward reverse
前进后轴【机】forward rear axle
前进回动杆 forward and reverse lever
前进机场 forward airfield; jump-up base
前进激波前 advancing shock front
前进棘爪 pawl feed(ing)
前进继电器 forward relay
前进加速度 forward acceleration
前进桨叶 advancing blade
前进角 angle of advance; angle of approach; approach angle
前进接触角 advancing contact angle
前进进化 progressive evolution
前进控制 progression control
前进口舱门 front intake door
前进拉力 ahead pull
前进离合器气管 ahead clutch line
前进力 ahead power
前进流 forward flow
前进路 forward path
前进履带 leading truck
前进抛双锚 running moor
前进平面波 progressive plane wave
前进气活门栅 <脉动式发动机的> frontal grill(e)
前进气口 nose air intake
前进汽轮机 ahead turbine
前进牵引力 forward drawbar pull
前进切碎机 go-devil
前进绕组 progressive winding
前进撒砂电空阀 forward sanding electropneumatic valve
前进三【船】full ahead; ahead three
前进色 advancing colo(u)r
前进沙丘 fore dune
前进砂波 progressive sand wave
前进式长壁采煤法 advancing longwall mining
前进式长壁开采法 advancing longwall mining
前进式潮汐波动 progressive type of tidal oscillation
前进式分岔 progressive splitting
前进式回采 mining advancing
前进式回采工作面 longwall advancing
前进式开采 advance mining
前进式开采法 advancing
前进式开采面 advance workings
前进式开采系统 following mining system
前进势 advanced potential
前进速度 forward speed; forward velocity; headway; rate of advance; speed of advance; velocity of progress
前进速度调节 ground speed control
前进速度调节杆 traction speed lever
前进速度调节器 ground speed regulator
前进速度调速 ground speed control
前进铜镍合金 advance copper-nickel alloy metal
前进凸轮 ahead cam
前进系数 advance coefficient; advance constant
前进相位 travel(l)ing phase
前进向规划 forward planning
前进项目 advance item
前进信号【铁】proceed signal
前进行程 forward stroke
前进性演化 anangenesis; progressive evolution
前进压力 forward pressure
前进演化 anagenesis; progressive evolution
前进演替 progressive successive
前进一【船舶】ahead one; slow ahead
前进运动 advancement; forward motion; headway; progressive motion; forward travel
前进中的困难 growing pains
前进重力波 progressive gravity wave
前进爪 forward pawl
前进装置 forward gear
前进阻力 resistance to forward motion
前景 foreground(detail); outlook; prospect; vista
前景标定点 foreground collimating mark
前景点 foreground point
前景调查 anticipation survey
前景放映法 front-screen projection
前景框标 foreground collimating point
前景摄像机 foreground camera
前景图 foreground picture
前景细节 foreground detail
前景信号 foreground signal
前景星 foreground star
前景旋转 foreground rotation
前镜系统 auxiliary lens system
前酒海纪【地】prenectarian
前据者 predecessor
前卷扬机 <起重机> lead winch
前开式 front open type
前开式的 open-fronted
前客舱 fore cabin
前馈 feed forward
前馈控制 feed forward control
前馈控制系统 feed forward control system
前馈系统 feed forward system
前馈原理 feed forward principle
前馈最优化 feed forward optimalizing
前拉策略 pull strategy
前拉杆 fore stay; front tension bar
前拉钩 front pull hook
前拉紧装置 front bridle chains
前拉索 fore stay; front cable; head guy
前缆口 bow chock
前廊 foregallery; antechurch <古教堂的>; narthex <教堂的>; vestibule <美>
前肋骨 fore frame
前冷却器 fore cooler
前例 precedent
前连接杆 front connecting rod
前脸墙 backwall; front wall
前链员【测】chain-leader; forward chainman; forward tapeman; head tapeman
前梁 front beam
前列叶片 ahead blading
前裂谷期岛弧【地】prerift arc
前馏段 forerun
前馏分 fore running; overhead
前楼梯 front stair(case)
前露放大器 prime amplifier
前炉 forehearth
前炉床 I-column
前炉连接砖 forehearth connection block
前炉前床 H-shaped forehearth
前炉燃烧 front wall firing
前陆 foreland
前陆粗砂岩 foreland grit
前陆盆地 foreland basin
前陆褶皱作用 foreland folding
前路耙 front rake
前麓地 foreland
前掠 sweepforward
前轮 fore wheel; front-wheel; nose wheel
前轮摆动 front-wheel wobble; shimmy of front wheels
前轮摆动测定器 shimmy detector
前轮摆动阻尼器 shimmy damping device
前轮保修作业 front-end job
前轮舱 nose-gear nacelle
前轮叉 front fork
前轮传动 front-wheel drive
前轮传动式搂草机 front-drive rake
前轮挡油毡垫 front-wheel oil seal felt washer
前轮定位 front-wheel alignment
前轮定位工具 front-wheel aligner
前轮定位试验器 front-wheel angle tester
前轮定位仪 front-wheel angle tester

Q

前轮对准 front-wheel alignment
前轮防尘垫圈 front-wheel dust washer
前轮负荷束 toe-out
前轮毂盖 front-wheel hub cap
前轮毂轴承螺帽锁紧垫圈 front-wheel bearing nut lock washer
前轮轨距 front ga(u)ge
前轮荷载 front-wheel weight
前轮横向偏离角 slip angle
前轮护脂圈 front-wheel grease retainer
前轮滑移角 front slip angle
前轮机构 front-wheel mechanism
前轮及后桥组转向式平地机 front-wheel and rear bogie steer grader
前轮及铰接转向的平地机 front-wheel and articulated frame steer grader
前轮架 fore carriage
前轮结合后双桥转向 front-wheel and rear bogie steer
前轮结合铰接车架转向 front-wheel and articulated steer
前轮距 front ga(u)ge
前轮可辅助驱动装置 front accessory drive
前轮轮毂 front-wheel hub
前轮轮距 front tire[tyre] tread;front track;front-wheel ga(u)ge;front-wheel tread
前轮轮胎 front tire[tyre]
前轮内轴承 front-wheel inner bearing
前轮内转角 inside lock
前轮配重块 front-wheel ballast weight
前轮漂移角 creep angle
前轮起动汽车 front-drive automobile
前轮前束测定器 toe-in ga(u)ge
前轮前束量尺 toe-in ga(u)ge
前轮倾斜 front-wheel lean
前轮倾斜角度 front-wheel lean angle
前轮倾斜控制箱 front-wheel lean control housing
前轮倾斜连杆 front-wheel lean tie bar
前轮倾斜锁闭阀 front-wheel lean lock valve
前轮倾斜弯臂 front-wheel lean lifter adjusting
前轮倾斜系杆 front-wheel lean tie bar
前轮倾斜液压缸 front-wheel lean cylinder;wheel lean cylinder
前轮倾斜运行状态 <平地机> crab position
前轮倾斜支架 front-wheel lean rack
前轮倾斜轴 front-wheel lean shaft
前轮倾斜转向的平地机 leaning wheel grader
前轮倾斜装置 <平地机> front-wheel lean device
前轮驱动 front axle drive; front-wheel drive
前轮驱动斗式装载机 front-wheel drive loading shovel
前轮驱动汽车 front-drive vehicle
前轮驱动式 front drive; front-wheel drive type
前轮驱动式车辆 front-wheel drive vehicle
前轮驱动式拖拉机 front-wheel drive tractor
前轮驱动式装载机 front-wheel drive loader
前轮驱动系 front-wheel drive system
前轮驱动小客车 front-wheel drive car
前轮驱动载货汽车 front-drive truck
前轮刹车 front-wheel brake
前轮锁止机构 front-wheel locking mechanism
前轮跳动行程 front-wheel bump travel
前轮外倾 camber
前轮外倾度 front-wheel pitch
前轮外倾角 front-wheel camber
前轮外倾角变化率 rate of camber change
前轮外缘转弯圆周 curb clearance circle
前轮限止器 front-wheel retainer
前轮(心轴)内轴承 front-wheel spindle inner bearing
前轮(心轴)外轴承 front-wheel spindle outer bearing
前轮油缸 front-wheel cylinder
前轮缘 front-wheel felloe
前轮载荷 front-wheel weight
前轮闸 front-wheel brake
前轮闸鼓护油罩 front-wheel brake drum grease guard
前轮罩板总成 front-wheel house complete panel
前轮支枢销轴承 front-wheel pivot pin bearing
前轮制动分泵缸 <液压制动系> front brake wheel cylinder
前轮制动鼓带轮毂 front brake drum with hub
前轮制动管路三通接头 front brake tube T-union
前轮制动控制器 front brake limiter switch;front-wheel brake control
前轮中心距 front track
前轮轴 knuckle spindle;front axle
前轮轴承 front-wheel bearing
前轮轴承隔片 front bearing spacer
前轮轴大支销 front-wheel king pin
前轮轴回转架 front axle linked point
前轮转动底盘 front-drive chassis
前轮转向 front-wheel steer(ing)
前轮转向到头 hard over
前轮转向式侧卸卡车 front-wheel steer side-dump
前轮转向式铲运机 front-wheel steer scraper
前轮转向式底卸卡车 front-wheel steer bottom-dump
前轮转向式反铲挖掘装载机 front-wheel steer backhoe-loader
前轮转向式后卸卡车 front-wheel steer rear-dump
前轮转向式平地机 front-wheel steer grader
前轮转向式装载机 front-wheel steer loader
前逻辑的 prelogical
前螺杆 front screw
前螺距 front pitch
前锚 bower anchor
前锚灯 stem light
前门 front door;front entrance;front gate
前门廊 antenave; anticum; front porch
前门台阶 front steps
前面 anterior surface;elevation;face;front (age); frontal area; front side;lip surface
前面板 front panel
前面板命令 front panel command
前面玻璃 front glass
前面测量标志 survey party sign
前面敞开式坟墓 open-fronted tomb
前面齿 <牙轮的> front flank
前面道路施工标志 road work ahead sign
前面的 anterior;fore;foregoing;frontal
前面顶圈 front ring
前面观 anterior aspect
前面接线 <家庭用电表等> front connection
前面接线开关 front connected switch
前面接线式 front connection type
前面(临时)停止标志 stop ahead sign
前面(临时)停止通行标志 stop ahead sign
前面路肩施工标志 shoulder work ahead sign
前面让路标志 yield ahead sign
前面绕圈 foreturn
前面施工 construction ahead
前面双向交通 two way traffic ahead
前面瓦 front tile
前面效应 front surface effect
前面信号联结 forward linking
前面修路 road work ahead
前面有暗礁 breakers ahead
前面有信号标志 signal ahead sign
前瞄 lead sight
前瞄准器 foresight
前模 cover mo(u)ld(ing);front mo(u)ld
前耙 front rake
前排球齿 nose row
前排席位 stall
前排席位末满 <教堂或戏院> stall end
前喷口 bow chock
前片 front section
前偏移 prestack migration
前平面 frontal plane
前坡 foreslope;front slope
前期 earlier stage;early days;preliminary phase; previous period; prior period; prophase; upper stage; protophase【生】
前期拔款 planning advance
前期处理 preliminary treatment
前期放大器 preamplifier
前期服务成本 prior service cost
前期付款 planning advance
前期工程 antecedent engineering; avant project;preliminary works
前期工作时间 lead time
前期公平 prior equity
前期股利 preceding dividend
前期固结 preconsolidation
前期固结比 preconsolidation ratio
前期固结土 preconsolidated soil;preconsolidation soil
前期固结压力 preconsolidation pressure
前期滚存损益 losses and gains brought forward
前期滚结账(目) brought forward account
前期含水量 antecedent moisture;antecedent moisture content;antecedent wetness
前期加载填土 preloading fill
前期降水量 antecedent precipitation; past precipitation; precipitation in early days
前期降水指标 antecedent precipitation index
前期降雨 antecedent rainfall
前期降雨径流 carry-over precipitation flow
前期降雨指数 antecedent precipitation index
前期结算 preceding settlement
前期结转利润 profit brought forward
前期结转损失 loss carried forward from the last term
前期结转余额 balance brought over from the last account
前期径流 antecedent flow;antecedent runoff
前期决算 preceding settlement
前期可行性分析 prefeasibility study
前期可行性研究 prefeasibility study
前期累结账目 brought forward account
前期流量 antecedent discharge;antecedent flow;carry-over flow
前期绿灯信号显示 leading green signal phase
前期排水 antecedent drainage
前期生长 early growth
前期湿度 antecedent moisture; antecedent wetness
前期疏伐 advanced thinning
前期水分 antecedent moisture
前期损益 profit and loss for preceding term; profit and loss for the previous period
前期损益调整 prior-period adjustment
前期条件 antecedent condition
前期调整项目 prior-period adjustment
前期土壤水分 antecedent soil moisture
前期温度 antecedent temperature
前期温度指数 antecedent temperature index
前期研究 pre-implementation study
前期应力场 former stress field
前期余额法 previous balance method
前期雨量 antecedent precipitation
前期渣 early slag;pre-slag
前期转来金额 amount brought forward
前期准备工作 advance preparation; first phase preparation; make first-phase preparation
前起落架支柱左右转动角 castor angle
前起落架左右转动角 nose-gear steering angle
前碛【地】 front(al) moraine
前前后后 up-and-down
前墙 breast wall;gable wall
前墙砖垛 door jamb
前桥 front axle shaft
前桥半径杆 front axle radius rod
前桥保修车间 front-end shop
前桥侧偏角 front axle slip angle
前桥差速器 front axle differential
前桥差速器壳及盖 front axle differential case with cap
前桥撑杆 front axle stay rod
前桥传动 front axle drive
前桥传动离合叉 front-drive axle clutch fork
前桥传动箱 transfer case
前桥传动轴 front axle propeller shaft
前桥工字梁 axle I beam
前桥横拉杆 front axle tie rod
前桥横拉杆末端 front axle tie rod end
前桥结合 front axle engagement
前桥壳带半轴壳 front axle housing with axle shaft housing
前桥壳盖 front axle housing cover
前桥梁 axle beam;front axle beam
前桥楼 fore bridge
前桥配重决 front axle weight
前桥驱动 front axle drive
前桥驱动式 front drive
前桥驱动式铲运机 front axle drive scraper
前桥驱动式底卸卡车 front axle drive bottom-dump
前桥驱动式后卸卡车 front-rear dump axle drive rear-dump
前桥驱动装置 four-wheel drive
前桥润滑油量 front axle oil capacity
前桥托架 front axle carrier
前桥万向节球 front axle universal

joint ball
前桥万向节中心球 front axle universal joint center[centre] ball
前桥悬挂 front axle suspension
前桥压载 front axle weight
前桥摇摆梁总成 front axle radius beam assembly
前桥右支挡 right-hand axle stop
前桥支枢 front axle pivot
前桥重量分配大的汽车 front-heavy car
前桥轴 front axle
前桥轴负荷 front axle load
前桥转向节 front axle steering knuckle
前切口 fore-edge
前切削面 front face(side)
前侵 encroachment
前倾 anteversion;down by the head;fore rake;forward rake;nose down
前倾边 overhang
前倾铲斗 tilting front end bucket
前倾的 anteverted;forward tipping
前倾的料斗 forward-tip(ping) bucket
前倾度 head trim;positive rake
前倾翻斗车 front-dump lorry;front-dump truck;front tipper;front-tipping wagon
前倾焊 angle forward welding;forward welding
前倾角 front rake;top rake
前倾面 front rake
前倾式铲斗 front-tipping bucket
前倾式驾驶室 tilt cab(in)
前倾双冲构造 foreland-dipping duplex
前倾斜纵帆 dipping lug
前倾叶片组 forward blader
前丘 frontal dune
前区 prezone
前曲叶片 forward curve blade
前驱波 precursor
前驱动 front drive
前驱动轮 front-drive wheel
前驱山脉 Montes harbinger
前驱水射 pilot jetting
前驱物 precursor
前驱涌 forerunner
前渠 channel-type head race;headwater channel
前圈 front coil
前群 pregroup
前群发信 pre-group transmitting
前群解调器 pre-group demodulator
前群收信 pre-group receiving
前群调制 pre-group modulation
前人 forefather;predecessor
前任 predecessor
前任的 former
前任主席 preceding president
前入口 forward entrance
前三点(起落架式)飞机 aircraft with nose wheel
前三角洲 prodelta
前三角洲沉积 prodeltaic deposit
前三角洲相 prodelta facies
前刹车 front brake
前山甲 front range
前山间盆地 front intermontane basin
前熵 forward entropy
前上标 presupercript
前上端梁<集装箱> front top end rail
前哨 outpost;picket;piquet
前哨地区 outpost
前哨工事 detached work
前哨基地 outpost
前哨线 picket line
前摄的 proactive
前摄抑制 proactive inhibition
前伸 protract;protrusive occlusion
前伸臂<码头前沿装卸桥> above-water cantilever
前伸臂俯仰装置 boom hoisting mechanism
前伸角 negative sweep
前伸距 front outreach
前伸平衡 protrusive balance
前伸式叉车 reach truck
前伸吸管式挖泥船 forward stretching suction dredger
前伸移动 protrusive movement
前身 precursor
前生树 advance(d)growth
前十年 previous decade
前示灯 side lamp
前视【测】 foresight;forward sight;forward looking;forward vision;front shot;fore observation;minus sight
前视读数【测】 foresight reading;forward reading
前视红外系统 forward looking infrared system
前视红外线 forward looking infrared
前视激光传感器 forward looking laser sensor
前视雷达 forward looking radar
前视轮廓图 front outline
前视剖面图 front sectional elevation
前视图 elevational drawing;front elevation;front elevation drawing;front elevation view;front outline;front view;head-on view;longitudinal plan;view front
前室 antechamber;anteroom;forebay;fore-chamber;forehearth;lobby(area);prechamber;pronaos <古建筑>;antecabinet<通往密室的>
前手背书人 prior endorser
前束合取范式 prenex-conjunctive normal form
前束角 angle of toe-in
前束析取范式 prenex-disjunctive normal form
前述 antecedent
前述事项 foreground
前双桥组 tandem front axle
前司尺员【测】 head staffman;head tapeman;chain-leader
前苏联(触探试验)套筒锥头 USSR[Union of Soviet Socialists Republics] mantle cone
前所未闻的 unheard-of
前台 fore apron;forward ground;foreground
前台部分 foreground partition
前台操作方式 foreground mode
前台程序【计】 foreground program(me)
前台处理 foreground processing
前台调度程序 foreground scheduler
前台方式 foreground mode
前台后台程序 foreground-background program(me)
前台划分 foreground partition
前台监控程序 foreground monitor
前台例行程序 foreground routine
前台启动 foreground initiation
前台启动程序 foreground initiator
前台启动的后台作业 foreground-initiated background job
前台区 foreground partition;foreground region
前台任务 foreground task
前台设置 foregrounding
前台信息处理程序 foreground message processing program(me)
前台作业 foreground job
前台座 front stand
前坍垒 protalus
前滩 beach face;forebeach;fore shore;nearshore;foreland
前弹簧 front spring
前坦 fore apron
前探的 overhanging
前探梁 forepole;forestope;foretop
前探炮眼 trim hole
前探平硐 pilot tunnel
前探支架 cantilever timbering
前探钻孔 province(bore)hole
前堂 foreroom
前套环 front ring
前梯格林寒冷期 Pretiglian cold epoch
前提 precondition;prerequisite;presupposition;lemma[复 lemmata/lemmas]【数】
前提 proposition
前提否定 denial of antecedent
前体 forebody;precursor
前体碳氢化合物 parent hydrocarbon
前体污染物 precursor pollutant
前条件作用 preconditioning
前厅 antechamber;antehall;anteroom;entrance lobby;foyer;lobby(area);narthex;prostas;vestibule;pronaos <古建筑>;antetemple<寺庙>;zaguan<西班牙建筑的>
前厅的 vesticular
前厅休息室 anteroom
前亭<建筑物前部突出体> avant-corps
前庭【建】 antecourt;forecourt;mandapam<印度庙前的>;atrium[复 atria/triums];front yard
前庭刺激 vestibular stimulation
前庭的 vesticular
前同步(信号)preamble
前头部 front head;drill front head <凿岩机>
前头减光灯 head lamp dimmer
前透镜 front glass;front lens
前透镜调焦 front-cell focusing
前透镜框 front element mount;front lens mount
前推动器叶片 front pusher blade
前腿<三脚架的> front leg
前拖钩 front tow hook
前挖斗式装载机 front bucket loader
前瓦形支块 front shoe
前弯叶轮 forward-curved impeller
前弯叶片 forward-swept vane
前弯叶片风机 forward-curved blade fan
前往目的地国 go to destination country
前桅 foremast;head mast;head post
前桅的中段 foretopmast
前桅灯 forward masthead light;headlight;masthead light
前桅顶帆 fore-royal
前桅帆 foresail
前桅帆桁吊索 forelift
前桅帆脚索 foresheet
前桅冠 foretruck
前桅航行灯 fore masthead light;fore steaming light
前桅卷帆索 fore-gear
前桅瞭望员 barrel man
前桅楼 foretop
前桅上的横桁 head yard
前桅哨 foretopman
前桅索具 fore rigging;head-rigging
前桅天帆 fore-skysail
前桅下帆桁 foreyard
前桅与烟筒间支索 monkey stay
前桅张帆杆 foreboom
前桅支索 fore stay
前桅中桅帆 foretopsail
前桅主帆 fore-course
前稳索 fore stay
前乌拉尔海洋 pre-Ural ocean
前下标 presubcript
前下端梁<集装箱> front bottom end rail
前下缘<进气道的> lower lip
前线 front
前线堡垒 front fort;presidio
前线机场 front line airfield
前线通信[讯] lateral communication
前限动装置 front bridle chains
前向 forward direction
前向波 forward wave
前向波放大器 forward wave amplifier
前向叉车 front loading forklift
前向差分 forward difference
前向差分法 forward difference method
前向差分算子 forward difference operator
前向差公式 forward formula
前向拆线信号 clear forward signal
前向串扰 forward crosstalk
前向导数 forward derivative
前向电导 forward conductance
前向电流 forward current
前向动态规划 forward dynamic(al) programming
前向反射 forward reflection
前向反射照相术 forward reflection radiography
前向方程 forward equation
前向分差 forward difference
前向辐射 forward radiation
前向观察员激光测距仪 forward observer laser rangefinder
前向监视信号 forward supervision signal
前向解方法 forward solution procedure
前向纠错 forward error correcting;forward error correction
前向纠错方式 forward error correction mode
前向纠错设备 forward error correction equipment
前向纠错译码器 forward error correction decoder
前向纠错制 forward error correcting system
前向联结<线控制信号的> forward linking
前向链 forward chaining
前向流速曲线 forward velocity tracing
前向内插 forward interpolation
前向散射 forescatter(ing);forward scatter
前向散射反馈 forward-scattering feedback
前向散射角 forescattering angle;forward-scattering angle
前向散射碰撞 forward-scattering collision
前向散射器 forescatter(ing)
前向射线锥 forward ray cone
前向式循环 forward type cycle
前向特征线 forward characteristic line
前向替换 forward substitution
前向通路 through path
前向微分方程 forward differential equation
前向误差分析 forward error analysis
前向信道 forward channel
前向选择 forward selection
前向元件 forward element
前向障碍检测 frontal obstacle detection
前向振铃信号 ring-forward signal

Q

前向逐步法 forward stepwise method
前向逐步回归 forward stepwire regression
前向阻抗 forward impedance
前项 antecedent
前象限角【测】forward bearing
前斜电篱笆 forward electric (al) fence
前斜杠 forward slash
前斜夹条 front beveled gib
前斜角 front bevel angle
前斜缆 fore stay
前斜面 front bevel
前斜桩 front raking pile
前斜桩力 front raking pile force
前斜桩式板桩码头 sheet pile quaywall with batter piles in front
前卸 front dump
前卸式铲斗 front-dump bucket
前卸式斗手推车 front tipper
前卸式矿车 drop-end car
前卸式料车 forward dump
前卸式耙斗 front-dump scraper
前卸式装载机 front-end loader
前卸推料车 front-dump lorry; front-dump truck; front lorry
前行 precession
前行式柴油压路机 advance diesel (road) roller
前行星齿轮架 front planetary gear carrier
前行掩蔽 forward masking
前悬 <汽车车身> front overhang
前悬挂 front suspension
前悬挂式 front-mounted
前悬挂式的 push type
前悬挂装置 front-end equipment
前悬式振动器 front-mounted exciter
前悬装土机 front-mounted loader
前悬装载机 front-end loader
前压力 forepressure
前压(碎)板 front crushing plate
前言 foreword; preamble; preliminary marks; prolegomenon
前沿 advancing edge; advancing front; entering edge; forefront; foreside; frontage; front edge; front porch; leading edge; rising edge; quay surface <码头的>
前沿波带的复合记录剖面 complex record section of wavefront
前沿波带复合记录 front-waveband synthesized record
前沿长度 front overhang
前沿道路 <沿临街房屋前面的地方道路或辅助道路> frontage road (way); service road (way)
前沿地 head land; foreland <码头的>
前沿地层序 foreland sequence
前沿地带 frontal zone
前沿地相 foreland facies
前沿陡度 width of transition steepness
前沿放置 front lay
前沿分析 frontal analysis
前沿分析法 frontal analysis method
前沿峰 leading peak
前沿跟踪 leading edge tracking
前沿轨道 frontal orbital
前沿过冲 leading edge overshoot
前沿横护梁 fascia beam
前沿科学 frontier science
前沿码头 apron of dock
前沿脉冲 leading edge pulse
前沿坡度 slope of front
前沿切割板 leading edge septum
前沿切换 front porch switching
前沿区轨道 apron track
前沿时间 leading edge time
前沿瞬变 leading transient
前沿下凹的码头 pier with concave side
前沿线 <码头的> face-line
前沿要塞 presidio
前沿因子 front factor
前沿阵地 advance position; forward position
前沿支撑 advance timbering
前沿作业地带 apron
前檐 <盾构的> shield-hood
前演 foreshore; lower beach; lower shore
前演化 anagenesis
前演要素 foreshore feature
前阳台 front balcony
前样板 <抹灰的> front screed
前一闭塞区段 advance block section; ahead block section; section in advance
前一闭塞区段预报 advance section information
前一区段 advance section
前一区段消息 advance section information
前一天晚上的 overnight
前一页 preceding page
前伊斯兰建筑 pre-Islamic architecture
前移 advance; forward lead; forward shift; moving-up
前移补偿 forward motion compensation
前移动透镜 front moving lens
前移绞车 forward hauling winch
前移距 forward lead distance; forward moving distance; forward shift distance; length of advancement; length of forward movement
前移距指示仪 forward shifting distance indicator
前移量 amount of forward displacement; amount of forward shift; leading
前移式叉车 reach truck
前移式单向压实机 forward travel compactor
前移速度 rate of advancement; rate of forward shift
前移延迟系数 advancing delay factor; rate of delay shift
前移桩 <挖泥船> stepping spud
前意识 fore-consciousness; preconscious
前翼 forelimb
前翼板上的后视镜 wing mirror
前因子 prefactor
前雨海纪【地】preimbrian
前渊 foredeep
前园 front garden
前圆形便器座 closed front seat
前缘 advancing edge; anterior border; costal margin; entering edge; front edge; leading edge; rising edge
前缘边坡 foreslope
前缘波 leading wave
前缘地 foreland
前缘地脚螺栓 foundation bolt with nose
前缘分析法 frontal analysis
前缘轨迹 leading edge locus
前缘肋 nose rib
前缘木 <楞堆底层前面的第一根木材> head log
前缘剖面 nose section
前缘嵌条 nose fillet
前缘驱动 frontal drive
前缘散热器 leading radiator
前缘推进动态 frontal-advance performance
前缘吸力式压力分布 frontal suction pressure distribution
前缘斜坡 foreslope
前缘异常 front anomaly
前缘翼缝 eye brow
前缘元素 front element
前缘晕 front halo
前缘走峭波 steep front wave
前院 entrance court; forecourt; foreyard; front (courty) yard
前院空地 front court space
前月 ultimo
前闸 front brake; lock front
前闸杆 front brake lever
前闸轨 front snap guard
前闸门 early gate
前瞻性研究 prospective study
前张索 forward guy
前沼地 foreland
前兆 forerunner; forewarning; harbinger; omen; precursor; premonition; presage; prognostication
前兆断层 precursor fault
前兆滑动 premonitoring slip
前兆活动 precursor activity
前兆价值 predictive value
前兆脉冲 precursor pulse
前兆模式 precursor model
前兆现象 precursor(y) phenomenon; premonitoring phenomenon
前兆效应 precursor effect
前照灯 head lamp; headlight
前照灯明视距 head lamp bright viewing distance
前照准器 front sight
前者 the former
前振动器 front vibrator
前震 earthquake foreshock; forerunner earthquake; foreshock; pre-earthquake
前震波 foreshock wave
前震旦纪【地】pre-Sinian period
前震旦纪基底【地】pre-Sinian basement
前震旦系【地】pre-Sinian system
前震活动 activity of fore-shock
前震活动性 foreshock activity
前震余震型 foreshock-after-shock pattern
前震主震型 foreshock-main-shock type
前震主震型地震 fore-main shock earthquake
前支泵 holding pump
前支撑帆 jumbo
前支杆 front support rod
前支护板 forepoling board
前支护桩 forepole
前支架 fore-stock; front frame; front stand; front support
前支架连接 forearm connection
前支架索 head stay
前支轮 front stand wheel
前支索 fore-guy
前支索的三角帆 stay foresail
前支腿 front outrigger
前支柱 front standing pillar
前趾板 toe slab
前制动销 front stop
前质 precursor
前致癌物 pre-carcinogen
前置 advanced; lead; preposition; topping
前置白噪声化滤波器 prewhitening filter
前置泵 booster pump; forepump; roughing pump
前置标志 advance(d) sign
前置波 prewave
前置补偿器 predistorter
前置抽气 forepumping
前置触发器 pretrigger
前置触发数据 pretrigger data
前置代码 prefix code
前置单色器 premonochromator
前置导孔 advanced feed-hole
前置导孔带 advanced feed
前置导孔纸带 advanced feed tape
前置的开拓道路 advance heading
前置的开挖道路 advance heading
前置的开挖平巷 advance heading
前置灯泡式水轮机 upstream-bulb turbine
前置电路 front-end circuit
前置发动机汽车 front engine vehicle
前置发动机前轮驱动式车辆 front engine front drive vehicle
前置发动机位置 front engine location
前置发动机小客车 front engined car
前置发动机支承横梁 front engine support cross member
前置法自导引 predicted-point homing
前置反硝化池 pre-denitrification pool
前置防热板 forward heat shield
前置放大 preamplification
前置放大级 preamplifier stage
前置放大器 head amplifier; preamplifier; preliminary amplifier; prime amplifier
前置放大器输入 preamplifier input
前置放大器阻塞 preamplifier disable
前置放大作用 preamplification
前置飞机场 advanced landing ground
前置分光器 predisperser
前置割草装置 front mower attachment
前置功率 topping power
前置过滤器 pre-air filter; prefilter
前置火花塞 leading plug
前置机 front-end processor
前置机翼 forewing
前置机组 superposed plant; topping plant
前置级 prestage
前置寄存器 prefix register
前置尖头信号 preshoot
前置键 advance key
前置降落场 advanced landing ground
前置交通标志 advance traffic sign
前置角 advance angle; angle of advance; angle of lead; lead angle
前置警告标志 advance warning sign
前置镜 auxiliary lens; supplementary lens
前置距离 <标志牌等的> advance distance
前置聚焦透镜 prefocus lens
前置均衡器 pre-equalizer
前置控制 predictive control; predictor control
前置类目 preferred category
前置棱镜 prism attachment
前置冷却器 precooler
前置立式除茎叶器 vertical front haulm stripper
前置量 prediction
前置量计算机 deflection computer
前置流量限制阀 preset flow limit valve
前置炉 cell; furnace cell; precell
前置炉膛 expansion furnace; extended furnace; furnace extension
前置滤波 prefiltering
前置滤波器 prefilter
前置路名标志 advance street-name sign
前置脉冲 prepulse
前置平衡柱 pre-equilibration column
前置屏蔽 front shield
前置氢技术 pre-hydrogen technique
前置区 prefix area
前置驱动电路 predrive circuit

前置燃烧室 extended furnace; furnace cell
前置散热器 front-mounted radiator
前置升运器 front elevator
前置声频放大器 audio preamplifier
前置时间 lead time
前置式拔取机 push-type puller
前置式动力油缸 front-located ram
前置式装载机 front(end) loader
前置式装载机的提升臂 lift arm
前置式装载垃圾车 front loading refuse truck
前置调节 predictor control
前置调节器 preregulator
前置涡轮机 top turbine
前置消磁器 head eraser
前置信号放大器 signal preamplifier
前置信号机【铁】 advance signal
前置选择器 preselector
前置液 preflush
前置液名称 type of preflush
前置液用量 consumption of preflush
前置运算符 prefix operator
前置指路标志 advance direction(al) sign
前置中频放大器 intermediate frequency preamplifier
前置柱 pre-column
前置装置计数器 preset decimal counter
前中桥驱动式铲运机 front and center[centre] axle drive scraper
前中桥驱动式后卸卡车 front and center[centre] axle drive rear-dump
前中亚蒙古海洋 precentral Asia Mongolia ocean
前重后轻的投标价 front loading
前轴 fore axle; front axis; leading axle
前轴叉 front axle fork
前轴承 fore bearing; front bearing; large metal
前轴传动 front axle drive
前轴定位器 aligner
前轴对准工具 front axle aligning tool
前轴方形转向架 steering axle quadrangle
前轴工字梁 axle I beam
前轴号码 front axle number
前轴架 front axle bracket
前轴控制杆 front axle control lever; front axle control rod
前轴梁 axle beam
前轴面 front pinacoid
前轴驱动的 front drive
前轴枢销 front axle pivot pin
前轴悬挂 front axle suspension
前轴之前 ahead of front axle
前轴转向桥 steering axle
前主梁架 front main frame
前柱 front column; head mast; head post
前柱廊式建筑物的柱 prostyle column
前柱廊式庙宇 prostyle temple
前柱式 prostyle
前柱式的 prostylar
前柱式建筑 prostylos
前转角架 leading truck
前转向灯 front directional
前转向架 front truck; leading bogie; leading truck
前转向信号灯 front-turn signal lamp
前桩【疏】 forward spud
前装铲车 front-end shovel
前装反铲挖掘机 loader backhoe
前装机 front(-end) loader
前装甲 front armo(u)r
前装履带式铲土机 front-end crawler shovel
前装牵引装载机 front-end tractor loader
前装式铲车 front-end bucket
前装式铲斗 front-end bucket
前装式飞轮 front-mounted flywheel
前装式装载机 forward loading machine; front end loader <在船舱内用>
前装式自备动力钻机 front-mounted self-contained unit
前装载机 front loader
前装载机构 front loading mechanism
前锥体 nose cone
前缀标记法 prefix notation
前阻尼器 front damper
前组 front element
前作 fore crop
前作业期 preoperation period
前座 front seat
前座驾驶公共汽车 <驾驶员坐于引擎旁> forward drive bus
前座架配重块 front saddle weight
前座架压载 front saddle weight

钱

钱包 purse
钱币 coin; king's picture
钱布雷布 chambray
钱布雷绸 chambray
钱袋 purse
钱柜 cashbox; money box; money locker; till
钱柜后 till
钱货两讫 collect on delivery
钱夹 bill fold
钱粮 land tax; revenue
钱羟硅铝钙石 chantalite
钱箱 cashbox; vault
钱庄 exchange shop; money house; native bank

钳

钳板座架 nipper frame
钳堡(筑城) tenail(le)
钳臂 tong arm
钳床工人 bench worker
钳点焊 poke weld(ing); push weld(ing)
钳颚 tongs jaw
钳杆 claw beam; nippling lever; peel
钳杆锁固长度 locking length of bolt
钳工 fitter (fitment); mechanician; pitman; smith
钳工操作 benchworking
钳工车间 metal working shop
钳工锤 engineer's hammer; fitter's hammer; peen hammer
钳工锉 machinists file
钳工工段 fitter's shop
钳工工具 bench tool; fitter's tools; small tool
钳工工人 bench worker
钳工工作 bench work; fitter's work
钳工工作台 bench; file bench; fitter's bench; vice bench
钳工刮刀 machinist's scraper
钳工加工 benching
钳工间 benchwork department
钳工手锤 bench hammer
钳工台 bench
钳工台安装 bench assembly
钳工台长柄剪 lopping shears
钳工箱 vice table cage; vice table caisson
钳工小锤 bench hammer
钳工錾 bench chisel; cold cutter; top chisel
钳工作业 bench work
钳夹 binding clip; forceps holder
钳夹式长大货车 schnobel car
钳夹头 bar hold
钳夹制动器 clamshell brake
钳夹抓手 <机械手的> vise tool
钳角 angle of nip; jaw angle
钳接 cramp joint
钳具 stirrup
钳卷材 seam roll
钳口 chops; jaw; mouth of tongs; vise jaw
钳口凹入部分 sprue recess
钳口板 claw wedge
钳口带槽持针钳 grooved needle holder
钳口开度 span of jaw
钳拉电平转换 level switch
钳牢 clip
钳连接 clipped joint
钳镊缸 forceps jar
钳砌石层 chain course
钳取机构 grip
钳取器 protractor
钳入角 angle of nip
钳入式轮胎 clincher; clincher tyre
钳入式轮辋 clincher rim
钳式安培计 clamp-on ammeter; snap-on ammeter; tong-test ammeter
钳式带卷吊具 lifting grab
钳式点焊头 pliers spot welding head
钳式电流表 clamp tester; tong-type ammeter
钳式电流计 tang-type ammeter
钳式吊车 crab crane; dogging crane; tong crane
钳式风动缓行器 jaw-type pneumatic rail brake
钳式轨道缓行器 jaw-type rail brake
钳式轨道减速器 jaw-type rail brake
钳式轨道制动器 jaw-type rail brake
钳式起重机 claw crane
钳式索结 choker hitch
钳式外部振捣器 vice clamp vibrator
钳式压铆机 C-yoke squeezer
钳式液压缓行器 jaw-type hydraulic rail brake
钳式移动绞车 crab travel(l)ing winch
钳式闸瓦 shoe brake
钳式抓斗 pincers-like grab
钳头 binding clip
钳位 clamp; clamp-on; direct current restoration
钳位电流 clamp current
钳位电路 clamper; clamper circuit; clamping circuit
钳位电平 clamp(ing) level
钳位电平调整 level control
钳位放大器 clamped amplifier
钳位管 clamper tube; clamping tube
钳位脉冲 clamp pulse
钳位偏压 clamp bias
钳位偏置电平 clamp bias potential
钳位偏左 clamp bias to left
钳位期间 clamping interval
钳位器 clamper
钳位时间 clamping interval
钳位输出电压 clamp output voltage
钳位信号电平 block level
钳位油膏 clamping paste
钳位周期 clamping interval
钳位装置 clamp device
钳位阻抗 clamped impedance
钳位作用 clamping action
钳形的 forcipate(d); hook on
钳形电流表 clamp-on amperemeter; clamp tester; clip-on ammeter
钳形伏安表 clip-on voltammeter
钳形杆 caliper lever
钳压电路 clamper circuit; clamping circuit; voltage clamping circuit
钳压法 forcipression; forcipressure
钳压管 clamper tube; crimped connector
钳压密封 clipper seal
钳制 rein
钳制接头 clip joint
钳制梁 restrained beam
钳制砌合 clipped bond
钳制桩 restrained pile
钳住 bite; clench[clinch]; holdfast
钳爪 chela
钳状水系模式 pincerlike mode
钳状物 pincers
钳桌 vice bench
钳子 chain dog; choker; clippers; combination pliers; cutting pliers; jaw; jaw vice[vise]; Mexican speed wrench; nippers; pair of nippers; pincers; pliers; forceps
钳子扳手 plier wrench
钳子叉头 wrench jaw
钳子扣钉 cramp iron
钳子牙板 tongs jaw; wrench jaw
钳子牙口 tongs jaw; wrench jaw

掮

掮客 broker; business tout; commission broker; corn-factor; go-between; middleman; running broker; scrivener

Q

箝

箝闭 incarceration
箝紧 clip
箝入角 angle of nip
箝入式轮胎 clincher tire[tyre]
箝套 nest
箝位 clamp; level control
箝位电路 clamping circuit
箝位电平开关 level switch
箝位放大器 clamper amplifier
箝位管 clamper tube
箝位脉冲 clamping pulse
箝位脉冲持续时间 duration of the clamping pulse
箝位脉冲发生器 clamp-pulse generator
箝位器 clamper; clipper
箝位油膏 clamping paste
箝位作用 clamping action
箝位作用强度 strength of clamping action
箝压二极管 catching diode; clamping diode
箝制 tie-down
箝制砌合 clip bond
箝住 bracketing

潜

潜坝 barrier; bottom sill[cill]; controlled weir; drowned dam; low dam; low embankment; low jetty; sill [loss]; submerged dam; submerged dike [dyke]; submerged groyne; underground dam; underwater dike[dyke]
潜坝深度 sill depth
潜坝型沙洲 barrier bar
潜变形 virtual deformation
潜冰 anchor ice; frazil; frazil ice; submerged ice; subsurface ice
潜藏水 concealed water
潜沉法 Berg's diver method; diver method
潜程长 potential path length
潜出 sneak-off
潜出车道 sneak-off lane
潜催化剂 latent catalyst
潜堤 low jetty; submerged breakwater; submerged dike [dyke]; sub-

merged embankment; submerged mole; undersea jetty; underwater dike[dyke]; submerged wall
潜底性动物 infauna
潜电荷 latent electric charge
潜丁坝 ground sill rocked in the bank; submerged dike[dyke]; submerged groin; submerged pier; submerged spur; weir jetty
潜动 creep; shunt running
潜对舰 underwater-to-ship
潜反应性 potentially reactivity
潜防波堤 submerged breakwater; submerged dike[dyke]
潜缝 locked-seam
潜伏 burrow; dormancy; incubation; slumber
潜伏本征根 latent root
潜伏冲突点 point of potential collision
潜伏出流 submerged efflux
潜伏的 dormant; latent
潜伏地震 hiding-in earthquake
潜伏毒性 delayed toxicity
潜伏发育 latent development
潜伏辐射反应 latent radiation effect
潜伏构造 buried structure
潜伏化作用 latentiation
潜伏活动断裂 potential active fault
潜伏火灾 dormant fire
潜伏胶 potential gum
潜伏阶地 plunging terrace
潜伏阶段 latency stage
潜伏力 potential energy
潜伏流行过程 latent epidemiogenesis
潜伏隆起 buried uplift
潜伏隆起带 zone of potential heave
潜伏隆起油气田 burial dome oil-gas field
潜伏面 surface of potential
潜伏期 cincubative stage; dormant period; incubation period; latency period; latent period; latent phase; period of incubation; time of occurrence
潜伏期的 prepatent
潜伏期延长 prolongation of latency
潜伏侵染 latent infection
潜伏区 incubation zone
潜伏生长期 latent growth period
潜伏时间 latent time
潜伏水硬性胶合剂 latent hydraulic binding agent
潜伏水硬性胶结剂 latent hydraulic binding agent
潜伏水硬性物质 latent hydraulic substance
潜伏凸起 buried convex area
潜伏突变 concealed mutation
潜伏污染物 potential pollutant
潜伏型 resting form
潜伏性固化剂 latent curing agent
潜伏造山幕 eoorogenic phase
潜伏状态 delitescence; latency; latent condition
潜埂 submerged ridge; underwater ridge
潜埂型急滩 rapids of submerged ridge pattern; submerged-ridged-caused rapids
潜谷 submerged valley
潜固化剂 latent curing agent
潜管 drowning pipe; submerged tube
潜管锅炉 submerged tube boiler
潜管冷凝器 submerged tube condenser
潜管蒸发器 submerged tube evapo(u)rator
潜光 latent light
潜含量 latent content
潜涵病 caisson disease; air illness; bends
潜航 submerge; submerged navigation
潜航深度 submerged depth
潜航速度 submerged speed
潜航员 aquanaut
潜河 disappearing stream
潜弧 buried arc; sunken arc
潜弧焊(接) hidden arc welding; submerged arc welding
潜弧自动焊 quasi-arc welding
潜环礁 drowned atoll
潜火山 subvolcano
潜火山的 cryptovolcanic; subvolcanic
潜火山地震 cryptovolcanic earthquake
潜火山构造 cryptovolcanic structure
潜火山作用 cryptovolcanism
潜极性 latent polarity
潜脊圈闭 buried crestal trap
潜槛 subaqueous sill; submerged sill; underwater sill
潜浆式曝气 submerged paddle aeration
潜浆式曝气器 submerged paddle aerator
潜浆式曝气设备 submerged paddle aerator
潜礁式防波堤 breakwater of submerged reef type
潜结构 latent structure
潜晶磷酸铝石 zepharovichite
潜晶质 cryptocrystalline
潜科学 latent science; potential science
潜孔 down-the-hole
潜孔冲击式凿岩机 down-the-hole hammer drill
潜孔冲击钻机 downhole hammer; down-the-hole hammer
潜孔锤 downhole hammer; down-the-hole hammer
潜孔锤式钻机 downhole hammer drill
潜孔锤钻进 downhole hammer drilling
潜孔锤钻头 downhole hammer bit
潜孔电钻钻机 submerged drilling rig
潜孔口 submerged orifice
潜孔扩孔器 downhole reamer
潜孔气动冲击器 down-the-hole air hammer
潜孔气动凿岩机 down-the-hole air hammer
潜孔式 <凿岩机> in-the-hole
潜孔式凿岩 down-the-hole drilling
潜孔式凿岩机 down-the hole drill; down-the-hole rock drill
潜孔式钻机 down-the hole drill
潜孔岩石钻机 down-the-hole rock drill
潜孔凿岩机 downhole drill
潜孔振动钻机 downhole vibration drill
潜孔钻 downhole drill
潜孔钻机 downhole driller; downhole drilling machine; in hole perforator; in-the-hole drill; submersible boring rig
潜孔钻架 down-the-hole drill rig
潜孔钻进 downhole drilling
潜孔钻具 bottom-hole drilling tools
潜孔钻设备 down-the-hole percussive unit
潜孔钻头 down-the-hole bit
潜孔钻岩机 down-the-hole type rock drill
潜矿物 potential mineral
潜亏 latent loss
潜力 hiding power; latent capacity; latent force; potentiality; potentialness; reserve capacity
潜力大的 high potential
潜力晶核 potential nucleus
潜力均等性 equipotentiality
潜力面分析 potential surface analysis
潜力上均等的 equipotential
潜流 bottom current; density flow; drowned flow; groundwater flow; potential flow; sneak current; submerged current; subsurface drainage; subsurface flow; subterranean drainage; subterranean river; subterranean stream; underaround river; undercurrent; underflow; underground stream; undermining flow; underrun; underwater current
潜流的 cryptor(h)eic
潜流放电 creepage discharge
潜流孔口 submerged orifice
潜流量 submerged discharge
潜流人工湿地 subsurface flow constructed wetland
潜流人工湿地处理系统 subsurface flow constructed wetland treatment system
潜流湿地 subsurface flow wetland
潜流式出水口 submerged outlet
潜流式分水建筑 submerged flow diversion works
潜流式分水岭 effluent flow diversion work
潜流式分水设施 submerged flow diversion works
潜流水 buried stream; underflow water
潜流水道 <在河底以下> underflow conduit; underground water course
潜流水流 underflow flow
潜流系统 subsurface flow system
潜埋井 buried well of dewatering
潜埋水雷 ground mine
潜没 subduction
潜没带 subduction zone
潜没浮标 submerged buoy; submerged float; subsurface float
潜没涵洞 submerged culvert
潜没加热法 immersion heating
潜没冷凝器 submerged condenser
潜没滤水网框格 submerged crib
潜没深度 immersion depth
潜没式出水口 submerged outlet
潜没式导流板 submerged type bottom vane; submerged vane
潜没式进水口 submerged intake
潜没式挑流鼻坎 submerged bucket
潜没体 immersed body; submerged body
潜密度 immersed density
潜能 latent energy; potency; potentiality; proficiency; potential energy
潜能均等性 equipotentiality
潜盘水淬法 shallow plate water extraction method
潜丘 buried hill
潜燃火 hangover fire
潜热 critical heat; heat-latent; hidden heat; internal latent heat; latent heat; potential heat
潜热负荷 latent heat load; moisture ton
潜热荷载 latent load
潜热量 latent heat quantity
潜热量器 latent heat calorimeter
潜热流 latent heat current
潜热释放 latent heat release
潜热损失 latent heat loss
潜热通量 latent heat flux
潜热值 potential heat value
潜容重 immersed density; submerged unit weight
潜溶剂 cosolvent; latent solvent
潜溶性 cosolvency
潜熔 latent fusion
潜入 sneak-on
潜入车道 sneak on lane
潜入点 <入库泥沙> plunge point
潜入河 lost river; lost stream
潜入式驳船 submersible barge
潜入式电动机 submersible electric(al) motor
潜入式电机 submersible machine
潜入式发动机 submersible motor
潜入式喷嘴 immersed nozzle
潜入式水力穿孔器 submersed hydraulic punch
潜入式振捣器 submersed vibrator
潜三角洲 undersea delta
潜色 colo(u)r potential
潜山【地】 burial hill; buried hill
潜山差异压实背斜圈闭 anticlinal trap by differential compaction over buried hill
潜山带 buried-hill zone
潜山倾向坡圈闭 buried dip-slope trap
潜山圈闭 buried hill trap
潜山油(气)藏 burial hill pool
潜山油气田 burial hill oil-gas field
潜射流 submerged jet
潜伸 creep
潜蚀 concealed erosion; internal scour; latent corrosion; subsurface erosion; suffusion; underground erosion; water creep
潜蚀漏斗 suffusion funnel
潜蚀破坏 failure by subsurface erosion
潜蚀作用 suberosion
潜式丁坝 submerged groin
潜式防波堤 submerged breakwater
潜式龙头 discharge cock
潜式洗舱机 submerged washing machine
潜式堰 controlled spillway
潜势 latent influence; potentiality
潜势晶核 potential nucleus
潜水 dive; diving <指水下工作等>; groundwater; phreatic groundwater; phreatic water; underground water; free ground water; perched water; phreatic wave; submerge; subsoil water; subsurface water; subterranean water <指地下水>
潜水班组 diving crew
潜水泵 bottom-hole pump; deep-well pump; diving pump; drowned pump; immersed motor pump; immersible pump; sink(ing) pump; subaqueous pump; submerged motor pump; submerged pump; submergible [submersible] pump; underwater pump; sump pump <排除油池积水用>
潜水闭锁(装置) diving lock
潜水标高 groundwater elevation
潜水表 submersion watch
潜水病 bends; diver's disease; diver's paralysis; submarine sickness
潜水波 phreatic wave
潜水波动 phreatic fluctuation
潜水波动带 belt of phreatic fluctuation; zone of phreatic fluctuation
潜水补给 seepage flow
潜水补给的水体 effluent-impounded body
潜水补给河 effluent river; effluent stream
潜水舱 diving chamber; diving compartment
潜水长 diving team leader; head of diving team
潜水衬衣 diver's underwear

潜水成壤作用 soil formation of phreatic water
潜水成壤作用过程 process of soil formation by phreatic water
潜水承压井 gravity-artesian well
潜水承压水 unconfined-confined water
潜水出流 phreatic water discharge
潜水储存量的变化量 storage change of phreatic water
潜水储量 groundwater capacity; groundwater storage
潜水储气罐 diver's air bottle
潜水船 underwater craft; underwater ship
潜水蝽 water creeper
潜水打捞 salvage dive
潜水带 phreatic zone
潜水的 phreatic; subaqueous
潜水的分带性 zoning of phreatic water
潜水灯 submarine light(ing)
潜水等高线图 water-table map
潜水等水位线图 water-level contour map
潜水堤 submerged breakwater
潜水电泵 electric(al) submersible pump; submersible electric(al) pump
潜水电动机 diving motor; immersible motor; wet motor
潜水电话 diving telephone
潜水电机 <地下连续墙> submersible electric(al) motor
潜水电机钻头 submersible motor drill
潜水吊笼 diving cage; diving skip
潜水动态 phreatic fluctuation; regime(n) of phreatic water
潜水动态曲线峰点 phreatic high
潜水动态曲线谷点 phreatic low
潜水对讲机 diving intercommunicator
潜水墩 submerged pier
潜水舵 diving plane; diving rudder; horizontal rudder; hydroflap; hydroplane
潜水防波堤 submerged breakwater
潜水非稳定流不完整井公式 partial penetrating well formula of unsteady phreatic flow
潜水肺 aqualung
潜水分界面 groundwater divide; phreatic divide
潜水分界线 groundwater divide; phreatic divide
潜水分水界 phreatic divide
潜水分水岭 groundwater divide; phreatic divide
潜水服 diver's suit; diving dress; diving suit; immersion suit; submarine armo(u)r
潜水服装 diving suit
潜水浮标 diving buoy; submerged buoy
潜水浮桥 underwater buoyant bridge
潜水高水位 phreatic high
潜水工 diver
潜水工程钻机 diving engineering drill
潜水工作 diver's work; submarine works; submerged work
潜水工作船 diver's service boat; diving boat; diving ship
潜水工作艇 diver's service boat; diving boat; diving ship
潜水供应 submerged water supply; underground water supply
潜水管线 submerged pipeline
潜水管线末端 end of submerged pipeline; terminal of submerged pipeline
潜水灌溉 subbing; subirrigation
潜水含水层 groundwater compartment; phreatic aquifer; unconfined aquifer; water-table aquifer
潜水呼吸纯度标准 diving breathing gas purity standard
潜水呼吸管 schnorkel
潜水呼吸器 aqualung; underwater breathing apparatus
潜水护目镜 lunettes
潜水化学作用分带 chemical processes zoning of phreatic water
潜水换能器 submersible transducer
潜水灰壤 groundwater podzol
潜水灰壤作用 groundwater podzolization
潜水货船 submerged cargo vessel
潜水机动泵 submersible motor pump
潜水基础 submerged foundation
潜水极限深度 diving depth limit
潜水集水井 tapping of groundwater
潜水急升失血病 caisson disease; diver's paralysis
潜水计划 diving plan; diving program(me)
潜水技师 artificer diver
潜水技术员 diving technician
潜水监督 diving supervisor
潜水减压 decompression
潜水减压病 diving decompression sickness
潜水减压舱 diving decompression chamber; submersible decompression chamber
潜水减压速度 diving decompression speed
潜水检查 diving inspection
潜水结构 submerged structure
潜水浸润曲线 depression curve of phreatic water
潜水浸润曲线方程 saturation curve equation of phreatic water
潜水井 phreatic water well; water-table well
潜水径流 seepage flow; subsurface flow; subsurface runoff; underground flow
潜水具 diving tool
潜水均衡方程 balance equation of phreatic water
潜水开关装置 immersible switchgear
潜水客轮 passenger submarine
潜水空气泵 diving air pump
潜水空气软管 diving air hose
潜水空气压缩机 dive air compressor
潜水盔 diving armor; diving helmet; subaquatic helmet; subaqueous helmet
潜水来水量 inflow of phreatic water
潜水离心泵 submerged centrifugal pump
潜水临界深度 critical depth of phreatic water
潜水流 diving current; underground flow; flow with water table; undermining flow; water-table stream
潜水流量 phreatic(water) discharge
潜水埋藏深度图 buried depth map of phreatic water
潜水埋深 <指地下水> depth of phreatic water
潜水埋深图 depth to water-table map
潜水帽 diver's cap; diving helmet; hard hat
潜水面 <指地下水> free surface; groundwater level; groundwater surface; groundwater table; level of saturation; level of subsoil water; line of seepage; phreatic surface; saturated surface; underground water level; water plane; water-table; groundwater plane; offshoot; phreatic water surface; underground water table
潜水面波 water-table wave
潜水面波动 water-table fluctuation
潜水面波动带 belt of water table fluctuation
潜水面季节变化 season recovery
潜水面降落锥 water-table depression cone
潜水面露头 water-table outcrop
潜水面坡度 gradient of water table
潜水面上升 water-table rise
潜水面升降 fluctuation of water table; phreatic fluctuation
潜水面升降变化 phreatic fluctuation
潜水面梯度 gradient of water table; water-table gradient
潜水面下降 decline of water table; water-table decline
潜水母船 diving depot ship
潜水内衣 under-wear
潜水能 latency energy
潜水泥泵 submerged dredge pump
潜水凝结补给量 recharge quantity of water table condensation
潜水排出 phreatic discharge
潜水排污泵 immersible sewage pump
潜水盆地 unconfined water basin
潜水皮管 diving hose
潜水平台 submersible platform
潜水坡度 water-table slope
潜水剖面 water-table profile
潜水蹼板 flipper
潜水器 diving apparatus; diving instrument; underwater vehicle
潜水器观测 submersible observation
潜水器具 diving apparatus; scaphander
潜水器械 diving appliance
潜水丘 groundwater hill; groundwater mound
潜水球 bathysphere
潜水区【地】 phreatic zone
潜水泉 water-table spring
潜水缺氧症 diving anoxia
潜水燃烧蒸发 submerged-combustion evapo(u)ration
潜水燃烧蒸发器 submerged-combustion evapo(u)rator
潜水人 ducker
潜水日志 diver's log book
潜水上升 <指地下水> phreatic rise
潜水设备 diving apparatus; diving equipment; diving outfit; diving plant
潜水深度 diving depth
潜水深度极限 diving depth limit
潜水深埋带 phreatic deep buried zone
潜水生理学 diving physiology
潜水施工 diver's work
潜水湿生植物 phreatophyte
潜水式电动机 submersible motor
潜水式机器 submergible machine
潜水式离心鱼泵 diving centrifugal fish pump
潜水式排水泵 submergible drainage pump; submersible drainage pump
潜水式拖管法 submerged tow method
潜水事故 diving accident
潜水视察员 diving inspector
潜水手表 diver's wristwatch
潜水手套 diving mittens
潜水守护 diving stand-by and support
潜水水界 phreatic divide
潜水水面 groundwater level; level of ground water
潜水水热喷发 phreatic hydrothermal eruption
潜水水头 potential water head
潜水水位 level of groundwater; phreatic water level
潜水水位变动带深度 depth of phreatic water level change zone
潜水水位等高线 contour of groundwater table
潜水送气管 snorkel
潜水台 diving stage
潜水探测器 bathyscaph(e)
潜水挑水坝 <稳定河床的> submerged groin-dam
潜水条例 regulations for diving
潜水艇 submarine; submerged craft; submersible; U-boat; undersea boat; diver <俗称>
潜水艇沉浮箱 ballast tank
潜水艇供养船 submarine tender
潜水艇救生船 submarine rescue vessel
潜水艇瞭望台 conning tower
潜水艇逃生舱 escape trunk
潜水艇演习区 submarine sanctuary
潜水艇用热中子反应堆 submarine thermal reactor
潜水头 potential head
潜水头盔 diver's helmet; diving helmet
潜水土 groundwater soil
潜水拖艇 underwater tow vehicle
潜水挖泥泵 submerged dredge pump
潜水挖泥船 submersible dredge(r)
潜水位 groundwater elevation; groundwater level; groundwater line; groundwater table; phreatic water level; water-table
潜水位变化带【地】 zone of variable phreatic level
潜水位变化周期 phreatic cycle
潜水位波动 fluctuation of water table
潜水位等高线 water-table contour
潜水位等深线 water-table isobath
潜水位等值线 isobath of water table
潜水位降落 <抽水影响所引起的>锥 water-table depression cone
潜水位降落漏斗 water-table depression cone
潜水位漏斗 water-table depression cone
潜水位起伏 wave of the table
潜水位上升 phreatic rise; raising of ground water; rise of water-table
潜水位势 diving trim
潜水位下降 groundwater regression; phreatic decline
潜水位线 groundwater line
潜水涡轮 submerged turbine
潜水涡轮泵 submersible turbine pump
潜水系统 diving system
潜水下沉带 phreatic sinking zone
潜水下降 phreatic decline; water-table decline
潜水下潜 diving descent
潜水线 <指地下水> phreatic line
潜水箱 coffer(dam); diving box; pontoon
潜水小刀 dive knife
潜水鞋 diver's shoes; diving shoes
潜水心理学 diving psychology
潜水信绳员 diver helper; diver tender
潜水旋转泵 submerged rotor pump
潜水靴 diver('s) boots
潜水循环 phreatic cycle
潜水压铅 diving weight
潜水压铅带 laden belt
潜水研究船 diving research ship
潜水堰 submerged weir
潜水衣 armo(u)r; diving armor; diving dress; diving suit; subaqueous suit
潜水医务保证 diving medical security; diving medical support
潜水医学 diving medicine
潜水溢出带 area of groundwater discharge; phreatic overflowing zone

Q

潜水隐污染物 latent pollutant
潜水营救艇 undersea rescue vehicle
潜水用具 diving appliance; diving gear; diving goods
潜水用空气供应系统 hookah
潜水油船 submarine oiler; tanker-submarine
潜水油轮 oiler submarine
潜水鱼泵 submersible fish pump
潜水员 aquanaut; artificer diver; diver; frogman; helmet diver; submariner
潜水员病 bends; caisson disease; compressed-air illness; decompression sickness; diving disease; screws
潜水员刀 diver's knife
潜水员的升降索 life line
潜水员耳炎 diver's barotitis
潜水员呼吸供气管 diver's breathing supply hose
潜水员护目镜 lunette
潜水员减压 decompression of diver
潜水员紧急上浮 free ascent
潜水员水下居住舱 underwater habitat
潜水员瘫痪 diver's palsy
潜水员梯 diver's ladder
潜水员通信[讯] diver communications
潜水员应急转运系统 diver's emergency transfer system
潜水罩 diving bell
潜水钟 diving bell
潜水员助手 diver helper; diver tender; lift line man; diver's linesman <潜水员线路指示通信[讯]员>
潜水员装备 submarine armo(u)r
潜水装具 diving apparatus
潜水源程序 source program(me) of phreatic water
潜水源水河 groundwater run
潜水运动 underwater sport
潜水噪声记录仪 sonodivers
潜水闸 submerged sluice
潜水罩 schnorkel
潜水者 diver
潜水振荡器 submarine oscillator
潜水蒸发 phreatic evapo(u)ration
潜水蒸发量 evaporation quantity of phreatic water
潜水支援船 diving support vessel
潜水植物 phreatophyte
潜水钟 caisson; diving bell; pontoon; submarine bell
潜水钟式基础 diving bell foundation
潜水钟信号 submarine bell signal
潜水钟作用 diving bell operation; submarine bell operation
潜水种类 kind of diving
潜水周期<指地下水> phreatic cycle
潜水主管 in-charge of diving
潜水注入河 influent stream
潜水装备 diving equipment; diving gear
潜水装具 diving apparatus; diving equipment; underwater kit
潜水装置 undersea device
潜水坠落 diving fall
潜水滋育泥炭 basin peat
潜水组 diving group
潜水钻机 underwater drill; under water rig
潜水钻孔机 diving boring machine
潜水钻探船 submersible drilling barge
潜水作业 aquanaut work; diver's work; diving (operation); water diving work
潜水作业班 diving gang; diving party; diving team
潜水作业保障船 diving depot ship
潜速度 potential velocity
潜塑 hidden plastic
潜塑态 latent-plastic state
潜态【化】abeyance
潜态电位测量计 potentiostat
潜滩加高 heightening of submerged bar
潜弹性形变 latent elastic deformation
潜掏 undermining
潜体 immersed body; submerged body
潜体表面的压力 pressure force on submerged surface
潜体积变化<土体积变化的潜能> potential volume change
潜听哨 listening post
潜艇 submarine; underwater ship; oscar<俗语>
潜艇安全航线 submarine safety lanes
潜艇安全监控系统 submarine safety monitoring system
潜艇安全深度探测器 submarine secure depth sounder
潜艇被动火力控制声呐 submarine passive fire control sonar
潜艇被动声呐 submarine passive sonar
潜艇被动探测与跟踪设备 submarine passive detection and tracking set
潜艇部队 silent service
潜艇测高雷达 submarine height finding radar
潜艇测距计 submarine range finder
潜艇打捞浮筒 submarine lifting pontoon
潜艇浮力 trim
潜艇光电桅杆 submarine optoelectronic mast
潜艇活动锚泊地 moving submarine haven
潜艇接收装置 submarine receiving set
潜艇紧急上浮系统 submarine emergency buoyancy system
潜艇救难舰 submarine rescue ship
潜艇救生舱 submarine rescue chamber
潜艇雷达 dete
潜艇锚 submarine anchor
潜艇母舰 submarine depot boat; submarine tender
潜艇勤务支援船 submarine tender
潜艇搜索声呐 submarine search sonar
潜艇探测 submarine detection
潜艇探测器 detectoscope; submarine detecting set; submarine detector
潜艇探测器室 asdic dome
潜艇探测卫星 submarine detection satellite
潜艇探测系统 submarine detecting system
潜艇探测装置 submarine detection gear
潜艇逃生与沉水救援设备 submarine escape and immersion equipment
潜艇天线驱动系统 submarine antenna drive system
潜艇通气管 schnorkel
潜艇脱险设备 submarine escape equipment
潜艇演习区 submarine exercise area
潜艇噪声 submarine noise
潜通路 sneak path
潜突堤 submerged jetty; submerged mole; submerged pier
潜挖 undercut; undermine; undermining
潜挖爆破 undermining blast(ing)
潜挖限制 limit of undermining
潜洼地 crypto-depression
潜望测距仪 periscopic range-finder
潜望电视 periscopic television
潜望高度 periscope height
潜望镜 hydroscope; hyperscope; kleptoscope; periscope; periscopic sight
潜望镜导流罩 periscope fairing
潜望镜的 hydroscopic
潜望镜方位圈 periscope azimuth circle
潜望镜跟踪 periscope tracking
潜望镜观察 periscopic vision
潜望镜光学系统 periscope optic(al) system
潜望镜航迹 periscope wake
潜望镜护板 periscope guard
潜望镜检查钻孔 periscopic inspection of drill holes
潜望镜孔 periscope hole
潜望镜浪花 periscope feather
潜望镜瞄准线稳定 stability in periscopic aiming line
潜望镜平台 periscope stand
潜望镜深度 periscope depth
潜望镜升距 periscope rising distance
潜望镜使用航速 periscope service speed
潜望镜式六分仪 periscopic sextant
潜望镜式偏流计 periscopic drift sight
潜望镜式双筒望远镜 periscope binocular
潜望镜视角 periscope angle
潜望镜天线 periscopic antenna
潜望镜图像稳定 periscope image stabilization
潜望镜弯曲补偿 periscope bending compensation
潜望镜引起的微波 feather
潜望镜罩 periscope housing
潜望镜中心 periscope center[centre]
潜望镜自动卫星导航仪 periscope automatic stellite navigator
潜望雷达 periscope radar
潜望瞄准镜 periscopic sight
潜望深度 periscope depth
潜望式测微器 periscopic sight
潜望式风速观测器 periscopic wind ga(u)ge sight
潜望式风速计 periscopic wind ga(u)ge
潜望式偏航测角器 periscopic drift angle
潜望式偏航角观测器 periscopic drift-angle sight
潜望透镜 periscopic lens
潜望状态 periscope condition
潜望最大深度 periscopic depth
潜线<画法几何> hidden line
潜线干扰 spectral line interference
潜箱<修理船底用> caisson
潜像 latent picture
潜像处理 latensification
潜像存储器 latent image memory
潜像稳定性 latent image stability
潜像增强 latensification
潜行 sneak
潜行电流 sneak current; sneak out current
潜穴 burrow
潜岩浆 latent magma
潜岩螨 Petrobia lateens
潜堰 barrier; controlled weir; drowned weir; groundsel; ground sill; subaqueous dike[dyke]; submerged weir; barrette
潜移能力 creep capacity
潜移特性 creep behavio(u)r; creep characteristic
潜移作用 creeping
潜异重流 density underflow
潜溢洪道 submerged spillway
潜隐污染物 latent pollutant
潜隐陷落 crypto-depression
潜应力 hidden stress; latent stress
潜影 latent image
潜影强化 intensification
潜育 incubation
潜育层<土壤的> glei[gley](horizon)
潜育过程 gleyed process
潜育化 gleying; gluing
潜育化草甸土 gley meadow soil
潜育化土(壤) gleyed soil
潜育化作用 gleization
潜育期 incubation period
潜育生草灰化土 gley sod-pozolic soil
潜育水稻土 groundwater rice soil
潜育土<排水不良条件下形成的> glei[gley] soil; groundwater soil
潜育土类 gleysolic soil
潜育状灰壤 gley podzol
潜育棕色冲积土 glei alluvial brown soil
潜育作用 gleying
潜在坝址 potential dam site
潜在变量 latent variable
潜在不稳定性 latent instability
潜在财产 latent property
潜在财富 potential wealth
潜在产量 potential output
潜在沉降 potential subsidence
潜在成本 implicit cost
潜在储量 potential ore
潜在氮肥污染指数 nitrogen fertilizer pollution potential index
潜在的贷款方 potential lender
潜在的肥力 potential fertility
潜在的核能 latent nuclear energy
潜在的滑移系统 latent glide system
潜在的化合物组成 potential compound composition
潜在的借款方 potential borrower
潜在的进入 potential entry
潜在的社会生产率 potential social productivity
潜在的渗透速率 potential infiltration rate
潜在的生产资本 latent productive capital
潜在的资源 potential resources
潜在敌手 potential adversary
潜在地热区 latent geothermal area
潜在地下水污染 possible groundwater pollution
潜在地震区 potential earthquake zone
潜在电解质 potential electrolyte
潜在电能需要 latent power demand
潜在顶极群落 potential climax
潜在毒性 potential toxicity
潜在断裂 potential break
潜在反应 potential reaction
潜在非点污染 non-point pollution potential
潜在非点源污染 non-point source pollution potential
潜在风险 potential risk
潜在付税人 potential taxpayer
潜在工程地址 potential site
潜在工作人口 potential working population
潜在功率 potential power; virtual power
潜在供给 potential supply
潜在购买力 potential purchasing power
潜在股票 potential stock
潜在故障 incipient fault; latent defect
潜在故障预防措施 potential trouble measure
潜在顾客 potential customer
潜在国民生产总值 potential gross national product

潜在过剩人口 latent overpopulation; potential overpopulation
潜在含糊性 latent ambiguity
潜在后果 latent consequence
潜在户(口) concealed household
潜在滑动面 potential slip surface; potential surface of sliding
潜在滑溜性 <可能产生的最大滑溜程度> potential slipperiness
潜在滑坡 potential slide
潜在滑坡区 potential slide area
潜在环境影响指数 environmental impact potential index
潜在环境致癌物质 potential environmental carcinogen
潜在环境作用 potential environmental impact
潜在活动断层 potentially active fault
潜在活化部位 latent activation site
潜在活性 latent reactivity
潜在活性磷 potential mobile phosphorus
潜在火险 potential fire hazard
潜在货币资本 latent money capital
潜在获益评价法 potential acquisition valuation method
潜在价格 shadow price
潜在价值 intrinsic(al) value; potential value
潜在碱反应性 potential alkali reactivity
潜在健康危害物 potential health hazard
潜在交通量 <可能达到的最大交通量> potential traffic volume
潜在交通流量 <可能达到的最大交通强度> potential traffic flow
潜在交通需求 latent traffic demand
潜在晶核 potential nucleus
潜在竞争 potential competition
潜在竞争者 potential competitor
潜在拒付 latent dishonour
潜在卷曲 latent crimp
潜在客货流 potential traffic flow
潜在空气 entrapped air
潜在矿量 potential ore
潜在矿物组成 potential mineral composition
潜在矿物组成分析 potential mineral composition analysis
潜在来源 potential sources
潜在力 potential force
潜在利益 potential income
潜在磷负荷 potential phosphorus loading
潜在磷负荷源 source of potential phosphorus loading
潜在磷回收 potential phosphorus recovery
潜在能力 latent capacity; potential capacity
潜在能源不足 potential energy shortage
潜在农药污染 pesticide contamination potential
潜在排放 potential emission
潜在膨胀能力 shrink-swell potential
潜在膨胀势 shrink-swell potential
潜在碰撞点 potential collision point
潜在破坏面 potential failure surface
潜在破裂面 potential failure surface
潜在骑自行者 potential cyclist
潜在强度 potential strength
潜在亲气性 <指浮选粒子> latent aerophilic quality
潜在倾向 undertone
潜在缺陷 hidden defect; latent defect
潜在热效应 potential thermal effect
潜在生产量 potential production
潜在生态风险评价 potential ecological risk assessment
潜在生态风险指数 potential ecological risk index
潜在失业 latent unemployment; potential unemployment
潜在时间 latency time
潜在市场 potential market
潜在市场需求 potential demand of the market
潜在事故 latent defect; potential hazard
潜在事故指数 damage potential index
潜在势力 undercurrent
潜在收入 potential income
潜在收缩 potential shrinkage
潜在疏水性 <指浮选粒子> latent hydrophobic quality
潜在熟料矿物组成 potential clinker composition
潜在竖胀量 potential vertical rise
潜在水力能量 potential hydroenergy
潜在水泥成分分析 potential cement analysis
潜在水污染 potential contamination of water
潜在水污染指示物 indicator of potential contamination of water
潜在水硬性 latent hydraulicity
潜在水硬性胶结剂 latent hydraulic binding agent
潜在塑性铰区 potentially plastic hinge zone
潜在酸度 potential acidity
潜在通货膨胀 hidden inflation; latent inflation
潜在通行能力 potential traffic capacity
潜在危害 potential hazard
潜在危害能力 hazard potential
潜在危险 potential hazard; potential risk
潜在危险地点 potentially hazardous situation
潜在威胁 latent threat
潜在污染 pollution potential
潜在污染特性 potential staining characteristics
潜在污染危害 hazard of potential pollution
潜在污染物 latent pollutant; potential pollutant
潜在污染物源 latent pollutant source; potential pollutant source; potential source of pollutant
潜在污染效应 potential pollution effect
潜在污染源 potential contamination source; potential pollution source; potential source; potential source of contamination
潜在污染源位置 potential contaminant source location
潜在污染指数 pollution potential index; potential pollution index
潜在细菌再生长 potential regrowth of bacteria
潜在瑕疵 latent defect
潜在瑕疵条款 latent defect clause
潜在像 latent image
潜在销售量 potential sales
潜在效益 potential benefit
潜在效应 potential effect
潜在效用 potential utility
潜在型中毒 subclinical intoxication
潜在性 potentiality
潜在性污染物 passive pollutant
潜在需求量 potential demand
潜在盐土 hidden solonchak
潜在意识的 subconscious
潜在因素 latency
潜在应力 <固体在无外力情况下潜在的应力> latent stress; potential stress
潜在应力场 latent stress field
潜在影响 potential impact
潜在有毒化学品 potentially toxic chemicals
潜在有毒化学品国际登记中心 International Registry of Potentially Toxic Chemicals
潜在有毒挥发性有机化合物 potentially toxic volatile organic compound
潜在有害物质 potential hazardous materials
潜在有效物质 potentially harmful substance
潜在有效硝酸盐 potentially available nitrate
潜在渔业资源 potential fishery resources
潜在原因 potential source
潜在源岩 potential source rock
潜在蒸发 potential evapo(u)ration
潜在蒸发蒸腾总量 <指土壤水分> potential evapotranspiration
潜在致癌物质 potential carcinogen
潜在资本 potential capital
潜在资金需求 latent demand for funds
潜在资源 dormant resources; potential resources
潜在资源价值 potential value of mineral resources
潜在资源量 potential resource amount
潜在自然植被 potential natural vegetation
潜蒸发 latent evapo(u)ration
潜值 potential value
潜周期性 hidden periodicity
潜洲 submerged bar; submerged shoal; sand bar

浅

浅暗冰 light nilas
浅暗火操作 semi-invisible flame operation
浅凹 dimple
浅凹槽 <楼面上的> sinkage
浅凹形(地面) shallow bowl
浅坝港埠 barge port
浅白色 lunar white
浅白色的 albescent
浅杯凸试验 <检查钢板表面组织的> bulge test
浅边缘海 shallow marginal sea
浅变质带 epizone
浅变质带的 epizonal
浅变质作用 epimetamorphism; epizonal metamorphism
浅表的 superficial
浅表损伤 superficial injury
浅剥离层 thin overburden
浅钵 shallow bowl
浅钵形便池 shallow bowl toilet
浅播 upper seeding
浅薄 shallow; thin
浅薄的 amateurish; trivial
浅部 superficial part
浅部地下水 shallow groundwater
浅部构造 shallow structure
浅部混合岩化方式 epimigmatization way
浅部开采工作 shallow work
浅部露天开采 shallow open-cut surface mining
浅部砂矿床 shallow gravel
浅部水流系统 shallow flow system
浅部挖掘船 shallow dredge
浅部巷道 shallow workings
浅仓 bunker
浅槽 shallow ridge slot
浅槽段【给】 crossing
浅槽分级机 shallow-pocket classifier
浅槽漏板 shallow bushing
浅槽式 <梯田的一种型式> shallow-channel type
浅槽式(罐式)集装箱 shallow tank container
浅草黄的 pale straw yellow
浅层 shallow; shallow layer; superficial layer
浅层爆破 chip blasting
浅层爆炸 shallow explosion
浅层采样 near surface sampling
浅层沉淀 shallow depth sedimentation
浅层冲断层 shallow thrust belt
浅层次 shallow level
浅层地道 high-level subway
浅层地下爆炸 shallow underground burst
浅层地下水 shallow(-bed) groundwater
浅层地下水的 phreatic
浅层地下水位起伏 phreatic wave; wave of the table
浅层地下铁道 underground line in shallow
浅层地震 shallow earthquake
浅层地震法 shallow seismic exploration method
浅层地震反射技术 shallow seismic reflection technique
浅层地震勘探 shallow seismic prospecting
浅层地震折射 shallow seismic refraction
浅层冻结 superstructure freezing
浅层冻土 shallow freezing; superstructure freezing soil
浅层反射 superficial reflex
浅层构造【地】 epigenetic structure; epi-tectonic; superstructure; suprastructure
浅层构造图 suprastructure map
浅层刮泡 shallow scraping
浅层管井 shallow tube well
浅层灌浆 shallow grouting
浅层滑动 shallow slide
浅层滑坡 shallow (layer) landslide; sheet landslide
浅层环流 shallow circulation
浅层喀斯特 shallow karst
浅层埋藏 shallow underground burial
浅层排水井系统 shallow well system
浅层平板载荷试验 shallow plate loading test
浅层剖面 subbottom profile
浅层剖面勘探 subbottom profile exploration
浅层剖面探测 subbottom profile exploration
浅层剖面仪 subbottom profiler
浅层曝气 Inka aeration
浅层曝气活性污泥法 shallow aeration activated sludge process
浅层气 shallow gas
浅层热储 shallow reservoir
浅层渗透 shallow percolation
浅层生态学 surface ecology
浅层时窗长度 length of time window of shallow layer
浅层事件 shallow event
浅层土 shallow soil
浅层温度场 shallow temperature field
浅层修补 shallow patching
浅层压实 shallow compaction
浅层岩溶 shallow karst
浅层岩石爆破 chip blasting

浅层展开基础 shallow spread footing
浅层褶皱带 shallow fold belt
浅潮地带 littoral belt
浅沉砂池 shallow grit chamber
浅成带【地】epizone;epibelt
浅成的 hypabyssal; hypergence; supergene
浅成低温热液矿床 epithermal deposit
浅成动力变质作用 epizonal dynamometamorphism
浅成高温热液的 xenothermal
浅成高温热液矿床 xenothermal deposit
浅成花岗岩 epigranite
浅成脉状矿床 shallow vein zone deposit
浅成侵入体 hypabyssal intrusive body
浅成热沉积 epithermal deposit
浅成热的 epithermal
浅成热液作用 epithermal process
浅成砂矿 shallow placer
浅成型 shallow recessing
浅成岩 epizonal rock; hypabyssal rock;supergene rock
浅橙黄色 orange clear
浅吃水的 midship draught
浅吃水船 shallow-draft craft; shallow-draft ship;shallow-draft vessel
浅吃水肥大型船舶 wide beam ship
浅吃水开底泥驳 shallow-draft bottom-dumping barge; shallow-draft split barge
浅池 shallow pan;shallow tank
浅池泵 sump pump
浅穿透高功率换能器 <一种海上地球物理探测方法的高频低穿透震源> pinger
浅床 shallow bed;surface bed
浅床法 shallow-bed technique
浅脆沥青 epi-impsonite
浅淬硬钢 shallow-hardening steel
浅大陆架 shallow continental shelf
浅带变质岩 epirock
浅带变质作用 epizonal metamorphism
浅带片麻岩 epigneiss
浅袋型自由沉降分级机 shallow-pocket free settling classifier
浅淡影像 weak image
浅的 shallow
浅的检查人孔 shallow manhole
浅的教堂后半圆室 shallow apse
浅的人孔 shallow manhole
浅低压 shallow-depression; shallow low
浅底沉积物 shallow bottom sediment
浅地层处置 shallow underground disposal;shallow understand disposal
浅地层剖面法 shallow layer profile method
浅地层剖面仪 shallow layer profile hydrophone
浅点 shallow spot;shoal point
浅雕 low relief
浅碟 saucer
浅碟形的 pateriform
浅碟形物 saucer
浅斗货车 low-sided wagon
浅斗裙式输送机 bucket plate apron conveyer[conveyor]
浅断层 shallow fault
浅断层作用 shallow faulting
浅粉红色 almond pink
浅粉色 hermosa pink
浅肤色皮肤 lightly pigmented skin
浅浮槽 basse relief;basse-relievo;low relief carving;shallow bowl
浅浮雕 bas(se)-taille; basse-relief; basse-relievo; demi-relief; low relief;stiacciato
浅浮雕带 bas-relief frieze
浅浮雕珐琅 basse-taille enamel
浅浮雕品 basso-relief; basso-relievo; basse-relief
浅浮雕檐壁 low relief frieze
浅浮雕油饰 grisaille
浅覆盖层 thin overburden
浅橄榄绿 sap green
浅镉橙颜料 cadmium light orange
浅铬黄 chrome yellow
浅根的 shallow-rooted
浅根植物 shallow-rooted plant
浅耕 light tillage;shallow ploughing;shallow tillage;shallow work
浅耕粗作 shallow ploughing and careless cultivature
浅拱 shallow arch;skene arch
浅拱顶棚 ceiling saucer dome
浅沟 bunker; shallow ditch; shallow ploughing;swale;thank-you-madam <道路上的>
浅沟法灌溉 corrugation-method irrigation
浅沟灌溉 corrugation(furrow) irrigation
浅沟漏斗 ribbed funnel
浅沟斜栽苗木 heeling in
浅谷 flat valley;slack
浅管井 shallow tube well
浅贯入 light penetration
浅灌 light irrigation
浅硅铝层 epiderm
浅海 epeiric sea; epicontinental sea; shelf sea
浅海爆炸声 airy phase
浅海边钻井 pier drilling
浅海冰层 shelf ice
浅海波 conoidal wave;ellipsoidal trochoidal wave; elliptic(al) trochoidal wave; shallow sea; shallow sea wave
浅海簸脊圈闭 shallow-winnowed-crestal trap
浅海潮汐 shallow water tide
浅海沉积土 shallow sea sedimentary soil
浅海沉积(物) neritic deposit; neritic sediment;shallow sea deposit;shallow water sediment;surface sea deposit
浅海沉积作用 neritic deposition;epicontinental sedimentation
浅海沉箱 shallow caisson
浅海带 ellitoral zone; neritic zone; shallow zone of sea; shoal water zone;sublittoral zone;subtidal zone
浅海的 <通常指小于 200 米水深> neritic; epicontinental; infraneritic; shallow;subtidal
浅海底 neritic bottom
浅海底的 epibenthile
浅海底环境 epibenthic environment
浅海底栖生物 epibenthic organism; epibenthos
浅海地带 sublittoral zone
浅海地区 neritic area
浅海电缆 shallow water cable;shore-end cable
浅海动物区 marine littoral faunal region
浅海动物区系 neritic fauna
浅海浮游动物 neritic zooplankton
浅海浮游生物 neritic plankton
浅海伽马能谱测量 gamma-spectrum survey in shallow sea
浅海海底采矿用履带式织车 bottom crawler
浅海海流 current in shallow water
浅海环境 infraneritic environment; neritic environment; shallow sea environment
浅海环流 neritic circulation
浅海建造 neritic formation
浅海陆架 neritic shelf
浅海陆棚相 neritic shelf facies
浅海盆地 neritic marine basin
浅海区 neritic province; neritic region; neritic zone; on soundings; shallow sea
浅海群落 littoral community
浅海上层的 epactile
浅海水层群落 neritic community; neritopelagic community
浅海水区 shallow sea water
浅海水文学 shallow water hydrography
浅海水下水热喷发 shallow submarine hydrothermal eruption
浅海碎屑沉积 shallow marine clastic deposit
浅海碎屑沉积模式 shallow sea clastic sedimentation model
浅海碳酸盐沉积模式 sedimentation model of shallow-sea carbonate
浅海体系 neritic system
浅海铁锰结核 neritic Fe/Mn nodule
浅海外环境 infraneritic environment
浅海湾 shallow bay
浅海相 neritic facies;shallow sea facies
浅海型 shallow marine type
浅海岩相 undathem facies
浅海植物 littoral vegetation
浅海浊积岩圈闭 shallow-turbidite trap
浅海钻井设备 offshore rig
浅含气带 shallow gas zone
浅含油带 shallow oil zone
浅焊 shallow weld
浅河槽 shoaling channel
浅褐黑色 light brown-black
浅褐红色 light brown-red
浅褐黄色 light brown-yellow
浅褐灰色 light brown-grey
浅褐蓝色 light brown-blue
浅褐绿色 light brown-green
浅褐色 Beige;biscuit;ecru;light brown
浅褐色的 spadiceous
浅褐色黑松 <新西兰产> matai
浅褐色橡木 <澳大利亚产> mountain ash
浅褐紫色 light brown-violet
浅黑 somber
浅黑的 darkish
浅黑色碳土 white rendziness
浅红 pale red
浅红宝石 balas(ruby)
浅红褐色 light red-drown
浅红黄色 light red-yellow
浅红灰色 light red-grey
浅红晶石 balas(ruby)
浅红栗子色 light reddish chestnut
浅红色 light red
浅红色的 ruddy
浅红色花岗岩 light red granite;Imperial Mahogany <产于美国明尼苏达州>
浅红紫色 light red-violet
浅湖 mere;shallow lake
浅湖沉积 shallow lake deposit
浅湖相 shallow lake facies
浅花纹 shallow pattern
浅化 shoaling;aufhallung;clarification
浅化波浪 shoaling wave
浅化系数 shallow coefficient;shallow factor; shoal coefficient; shoaling factor
浅黄褐色 light yellow-brown
浅黄红色 light yellow-red
浅黄灰色 light yellow-grey
浅黄绿色 light yellow-green;pistachio
浅黄绿色的 chartreuse
浅黄三彩 tricolo(u)r with tender yellow
浅黄色 buff;light yellow;pale yellow
浅黄色标准砖 buff standard brick
浅黄色的 lurid;yellowish
浅黄色海相灰岩 Caen-stone
浅黄色滤光镜 light yellow filter
浅黄色石灰石 slightly yellow limestone
浅黄色细纹木材 <做家具用> yacca
浅黄釉 cane glaze;pale yellow glaze
浅黄砖 buff brick
浅灰 light gray[grey]
浅灰白色 light grey-white
浅灰褐色 light grey-brown
浅灰黑色 light grey-black
浅灰红色 light grey-red
浅灰黄色 light grey-yellow
浅灰蓝色 light grey-blue
浅灰绿色 light grey-green
浅灰青 pale grey-blue
浅灰色 dove; French grey; light grey [gray]; oyster grey [gray]; slightly grey
浅灰色的 griseous
浅灰色橡木(面) limed oak
浅灰细纹花岗岩 Vinal Haven granite
浅灰紫色 light grey-violet
浅混溶作用 epimigmatization
浅基槽 shallow trench
浅基(础) flat foundation; shallow footing;shallow foundation
浅基础结构 shallow foundation structure
浅基脚 shallow footing
浅检修井 shallow manhole
浅胶 white factice
浅礁 shallow reef; shoal reef; shoal rock
浅角焊缝 light fillet
浅脚刨 ogee plane(iron)
浅结 shallow junction
浅结扩散 shallow diffusion
浅近突破 shallow penetration
浅浸式水翼 shallow submerged hydrofoil
浅浸水翼艇 surface-piercing(hydro) foil craft
浅浸效应 free surface effect
浅井 bore pit; incomplete well; posthole well;shallow well <英>
浅井泵 shallow well pump
浅井采样 pit sampling
浅井地质编录 geologic(al) documentation of shallow shaft
浅井工程系统 exploring shallow system
浅井灌注 shallow well injection
浅井勘探 gophering;randing
浅井排水系统 shallow well system
浅井喷射式水泵 shallow well jet pump
浅井喷射水泵 shallow jet pump
浅井手摇泵 pitcher pump
浅井数 number of shallow wells
浅井素描图 sketch map of shallow mine;sketch of pit
浅井提升 shallow hoisting
浅井提升机 shallow shaft hoist
浅井提升机提升 hoisting by shallow shaft elevator
浅井油井水泥 shallow well oil well cement
浅井展开图 extending map of shallow mine
浅井注入 shallow well injection
浅井总深度 total depth of shallow wells
浅景深 shallow depth of field

浅橘黄色二氧化钛颜料 buff titanium dioxide
浅开挖 shallow cut; shallow excavation
浅刻 light engraving; scotch
浅刻痕 scotch
浅刻技术 line cutting
浅坑 posthole well; shallow excavation; shallow hole; shallow pit; chock hole <自行式钻机车轮下面的>
浅空心板 shallow cellular deck
浅孔 short hole
浅孔爆破 chip blasting; shallow blasting; shallow borehole explosion; short-hole blasting
浅孔爆破法 short-hole blasting method; short-hole method
浅孔测氡法 determining radon method in shallow hole
浅孔伽马测量 shallow hole gamma ray survey
浅孔灌浆 short-hole grouting
浅孔留矿法 surface hole shrinkage method
浅孔能谱测量 shallow hole spectrum survey
浅孔式进水口 high-level intake
浅孔温度 shallow hole temperature
浅孔注水泥 short-hole grouting
浅孔注水泥法 short-hole method
浅孔钻机 short-hole drill
浅孔钻进 <深度 30 米以内> shallow hole drilling; short-hole drilling; short-hole work
浅孔钻探 shallow exploratory boring
浅扩基础 shallow spread foundation
浅蓝褐色 light blue-drown
浅蓝黑色 light blue-black
浅蓝红色的 bluish red
浅蓝灰色 light blue-grey
浅蓝灰色砂岩 <产于美国俄亥俄州的> Gray Canyon
浅蓝绿色 light blue-green
浅蓝(色) baby blue; light blue; cornflower blue; powder blue; Russian blue; saxe blue; watery blue
浅蓝色玻璃 light blue-glass
浅蓝色的 bluish; light blue
浅蓝印样 light blue pull
浅蓝釉 light blue glaze
浅蓝紫色 light blue-violet
浅浪 surf
浅肋板 shallow floor
浅肋骨 shallow frame
浅肋形板 thin ribbed plate
浅犁 shallow ploughing
浅栗色 light maroon
浅粒岩 leuco granoblastite
浅梁 shallow beam
浅裂纹 checking
浅流化床层 teetered bed
浅陆缘海 shallow marginal sea
浅滤池 shallow filter
浅路堑 shallow cut
浅绿 pale green
浅绿褐色 light green-brown
浅绿黑色 light green-black
浅绿黄色 light green-yellow
浅绿灰色 light green-grey
浅绿瓶 pale green bottle
浅绿色 light green
浅绿色玻璃 light green-glass
浅绿色的 greenish
浅绿釉 light green glaze; pale green glaze
浅螺帽 shallow nut
浅埋 shallow depth; shallow embedment
浅埋藏的 shallow-lying
浅埋地下铁道 underground railway in shallow subway
浅埋结构 buried structure
浅埋隧道 shallow-buried tunnel; shallow depth tunnel; shallow tunnel; tunnels at shallow depth
浅埋作用 shallow burialism
浅密度跃层 shallow pycnocline
浅面 superficial surface
浅皿 flatware
浅木桶 keeler
浅内陆海 shallow inland sea
浅能级 shallow energy level; shallow level
浅啮合 shallow depth
浅柠檬黄 lemon yellow pale
浅牛皮色 cream buff
浅盘 celery; plate; shallow pan; tray
浅盘沉淀池 shallow tray settling basin
浅盘过滤器 shallow pan filter
浅盘式便池 shallow pan closet
浅盘式便桶 shallow pan closet
浅盘式池 shallow tray basin
浅盘式除气器 tray-type deaerator
浅盘式通气器 tray aerator
浅盘型焦炭曝气器 coke tray aerator
浅炮眼 block hole; shallow hole; short hole
浅盆 shallow pan
浅盆湖 saucer lake
浅盆形集水坑 receptor
浅片 lobe
浅撇泡沫 shallow scraping of foam
浅平基(础) open cut foundation; shallow spread foundation; spread footing; spread foundation
浅平器皿 flatware
浅器皿 tagliere
浅潜流洞 shallow phreatic cave
浅潜水带 zone of vadose water
浅桥面 shallow floor
浅切 light cutting
浅切锯 dimension saw
浅切口 shallow cut
浅切削 light cut; shallow cut
浅倾 low oblique
浅倾航片 low oblique aerial photograph
浅倾摄影 low oblique photography
浅倾摄影相片 low oblique photograph
浅穹隆顶棚 ceiling light cupola
浅球壳 shallow shell
浅区 shallow zone; shoal area
浅区岩 hypabyssal rock
浅染 tinge; understain
浅人孔 shallow manhole
浅溶蚀洼地【地】park
浅熔池 shallow bath; shallow pool
浅色 pale colo(u)r; pale; tint; tint colo(u)r; undertint; undertone
浅色安山岩 leucoandesite
浅色斑点 scumming
浅色层 bleached bed
浅色长苏长岩 leucotroctolitic norite
浅色处理面 highlighting
浅色大理石 light-colo(u)red marble; cipol(l)in(o) <罗马产含云母层和硅质的>
浅色的 light-colo(u)red
浅色调 pale tone; tint-tone
浅色橄长岩 leucotroctrolite
浅色橄榄辉长岩 leucooivine gabbro
浅色橄榄苏长岩 leucoolivine norite
浅色高黏[粘]度润滑油配料 bright stock
浅色花岗岩 light-colo(u)red granite
浅色辉长苏长岩 leucogabbro; leucogabbro norite
浅色辉长岩 labradorite
浅色角闪辉长岩 leucohornblende gabbro
浅色聚合油 pale-boiled oil
浅色矿石 <长石、石英石等> felsic
浅色矿物 light-colo(u)red mineral
浅色沥青路面 light bituminous carriageway
浅色路面 light surface
浅色毛 bright wool
浅色泥炭 light peat
浅色漆 light tint paint; pale paint
浅色侵入岩 light-colo(u)red intrusive rock
浅色清油 pale-boiled oil
浅色乳状液 light cream
浅色润滑油 pale oil
浅色闪长岩 leucodiorite
浅色(石油)沥青 albinobitumen
浅色熟油 pale-boiled oil
浅色水藓泥炭 light sphagnum peat
浅色松香 pale rosin
浅色苏长辉长岩 leuconorite gabbro
浅色苏长岩 leuconorite
浅色素皮肤 lightly pigmented skin
浅色土 light-colo(u)red earth; light-colo(u)red soil
浅色团 hypsochrome
浅色团作用 hypsochromy
浅色微晶闪长岩 leucomicrodiorite
浅色位移 hypsochromic shift
浅色相 light shade; pastel shade
浅色效应 hypsochromic effect
浅色玄武岩 leucobasalt
浅色岩 leucocratic rock; light-colo(u)red rock
浅色移动 hypochromic shift
浅色正长岩 leucosyenite
浅色脂环饱和烃树脂 <抗热及绝缘的> Arkon
浅色铸石 light cast stone
浅砂层 shallow sands
浅沙滩 sand shoal
浅闪石 edenite
浅深度饱和潜水 shallow saturation diving
浅深度空气饱和潜水 shallow air saturation diving
浅深度潜水 shallow diving
浅生的 supergene
浅石色 light stone
浅石滩 ledge; reef flat
浅式设备 shallow fixture
浅试坑 shallow trial pit
浅饰 gloss
浅竖管 shallow shaft
浅竖井 shallow shaft
浅水 fleet water; shallow water; shoaling water; shoal sounding
浅水爆炸 shallow underwater burst
浅水波 <以波深小于半波长前进的重力波> shallow water wave; conoidal wave; shoaling wave; water in intermediate water
浅水波浪区 seaway
浅水测深 shallow cast; shallow sounding
浅水测深杆 shallow sounding rod; wading rod
浅水测深设备 shallow sounding apparatus
浅水测深仪 shallow sounder
浅水长波 shallow water long wave
浅水长期淤积 shallow water stands for a long time
浅水长涌 ground swell
浅水潮汐 shallow water tide
浅水沉积(物) shallow deposit; shallow sediment; shallow water deposit; shallow water sediment
浅水沉积作用 shallow sedimentation
浅水池 shallow basin; shallow pond; shallow pool; wading pool <供儿童玩乐的>
浅水处 shallows
浅水船 light draft craft; shallow-draught boat; shallow water vessel
浅水船闸 shallow-draft lock
浅水带 shallow water zone; shoal water zone
浅水的 midship draught
浅水底的 epibenthic
浅水底电缆 shallow water cable
浅水底栖动物区系 epibenthic fauna
浅水底栖生物 epibenthos
浅水底栖生物种群 epibenthic population
浅水订正 shallow water correction
浅水动物区系 shallow water fauna
浅水防波堤和海堤 breakwater and sea-wall in shallow water
浅水分潮 shallow water component; shallow water constituent
浅水港 dry harbo(u)r; shallow(-draft) harbo(u)r; stranding harbo(u)r; shallow-draft port
浅水海湾 etang
浅水航道 shallow-draft waterway; shallow-draught waterway
浅水航行 shallow-draft navigation
浅水河床 shallow riverbed
浅水河道 shallow-draught waterway; shallow-draft waterway
浅水河段 shallow reach
浅水河口 lagoon mouth; shallow estuary
浅水河口港 lagoon harbo(u)r; lagoon port
浅水黑视 shallow water blackout
浅水湖(泊) shallow lake
浅水湖带 shallow zone of lake
浅水回声测深仪 shallow sounding apparatus
浅水机动 shallow water maneuvering
浅水校正 shallow water correction
浅水井 shallow well
浅水理论 shallow water theory
浅水轮船 light draught steamer
浅水码头 shallow-draft quay; shallow-draft wharf; shallow water quay; shallow water wharf
浅水锚地 high bottom
浅水泥浆泵 shallow water mud pump
浅水破碎波 shallow breaking waves; shoaling breaking wave
浅水潜水 shallow water dive
浅水区 shallow water zone; shoal area; shoal(ing) waters
浅水区测量 shallow water survey
浅水区涌浪 ground swell
浅水三角洲 shallow water delta
浅水散射层 shallow scattering layer; shallow water scattering layer
浅水珊瑚 shoal water coral
浅水涉渡 shallow fording
浅水疏浚 shallow dredging
浅水滩 wade
浅水体 shallow body of water
浅水湾 wash
浅水系数 <波浪> shoaling coefficient; shoaling factor
浅水效应 shallow water effect
浅水型冷却池 shallow cooling pond
浅水修正系数 shallow water correction factor
浅水影响 shallow water effect
浅水域 shallow waters
浅水运河 boat canal; shallow canal; shoal canal
浅水种 shallow water species

Q

浅水驻波 shallow water clapotis
浅水浊流沉积 shallow water turbidity deposit
浅水作用 shoaling
浅司型重介质锥形分选机的搅拌器 Chance cone agitator
浅燧石 chert
浅榫眼 stump mortise;stub mortise <不穿透木材>;corpsing <材料表面的>
浅台 platform
浅台名称 name of platform
浅滩 bendway;bottom bank;cross-over;flats;ford;high bed;ledge;minor pool;riffle;shallow ground;shallow patch;shallows;shallow shoal;shelving beach;shoal
浅滩标志 shoal indicator;shoal mark
浅滩冰 stranded ice
浅滩残余砂体 shoal retreat massif
浅滩的冲淤变化 scour and fill changes of shoal
浅滩地 shoaly land
浅滩动荡 surf beat
浅滩段 shallow reach;shoal reach
浅滩浮标 bar buoy
浅滩海底 high bottom
浅滩海湾 bayou(lake)
浅滩和沙洲 sand shallows and holms
浅滩河槽的年变化 annual variation of shoal channel;yearly variation of shoal channel
浅滩河段 shoal reach;shoal section
浅滩后退砂体 shoal retreat massif
浅滩环礁 bank atoll(on)
浅滩角砾岩 shoal breccia
浅滩开挖 shallow dredging
浅滩名称 name of shoal
浅滩年际变化 interannual shoal variation;interannual variation of shoal
浅滩年内变化 annual variation of shoal
浅滩区 shoal area
浅滩日际变化 interdiurnal shoal variation;interdiurnal variation of shoal
浅滩珊瑚礁 <海边珊瑚缘带> reef bank
浅滩疏浚 shallow dredging
浅滩水深 bar draft;depth of shoal;water depth of shallow shoal
浅滩挖除 shallow dredging
浅滩效应 shoaling effect
浅滩演变 evolution of shoal;shoal evolution;shoal variation
浅滩演变分析 analysis of shoal evolution;analysis of shoal process
浅滩演变规律 regularity of shoal process
浅滩演变过程 shoal process
浅滩影响 shoaling effect
浅滩整治 drift regulation;regulation of shoal;shoal regulation
浅滩整治水位 shoal regulation stage
浅滩指示浮标 bar buoy
浅滩治理 regulation of shoal
浅滩周期性变化 periodic(al) variation of shoal
浅滩组成 shoal materials
浅探井 trial pit
浅探坑 shallow pit
浅桃红 light pink
浅填 shallow fill
浅填焊接 light welding
浅填角焊缝 light filler;light filler block
浅调制磁迹 low-modulation track
浅涂油漆层 piss coat
浅土黄 <指颜料> gold ocher
浅挖 light cutting;shallow cut(ting);shallow digging
浅挖方 shallow cut
浅挖机 shallow dredge
浅挖基础 shallow spread foundation
浅挖掘机 scimmer scoop
浅挖土 shallow cut;shallow dredging
浅洼地 pod
浅弯 shallow bending
浅湾贝壳沉积 tangue
浅湾钙质泥 tangue
浅位侵入体 shallow plutonic body
浅坞 shallow dock
浅雾 land fog;shallow fog
浅箱 flat;pan;shallow box;tray
浅型充气浮选槽 pneumatic cell without pump body
浅锈黄 rust yellow
浅岩基带 epibatholite zone
浅岩浆的 epimagmatic
浅盐湖 shott
浅眼爆破 blast-hole blasting
浅眼凿进 short-hole drilling
浅眼凿岩机 popholing drill
浅眼钻凿 short-hole drilling
浅腰船 shallow waisted
浅游泥层 shallow muck
浅浴盆 hip bath
浅源 shallow source
浅源地震 normal earthquake;shallow focus earthquake
浅源岩浆 pyromagma
浅源远震 shallow teleseism
浅源震 shallow source earthquake
浅栈 short stack
浅沼地 swale
浅沼泽 banados;pocosin
浅照明装置 shallow light fitting;shallow luminaire(fixture)
浅折 shallow folding
浅震 normal earthquake;shallow earthquake;shallow focus earthquake;shallow shock
浅震反射 reflection of shallow earthquakes
浅状异常 linear anomaly
浅紫光泽彩 lilac colo(u)r luster[lustre]
浅紫褐色 light violet-brown
浅紫黑色 light violet-black
浅紫红色 light violet-red;peony red
浅紫灰色 light violet-grey
浅紫蓝色 light violet-blue
浅紫色 light violet;lilac colo(u)r
浅棕色 hazel;light brown
浅棕色巴斯石灰石 <萨默塞特产> stoke ground
浅棕色石灰岩 light brown limestone;chilmark <英国威尔特郡产>
浅棕色硬砂岩 light brown graywacke;bramley fall <产于英国约克郡>
浅棕色有光泽的硬木 light brown lustered hardwood;chickrasey <印度及缅甸产>
浅钻孔 shallow borehole;shallow drill hole
浅嘴 shoal point;shoal spit

遣

遣返 repatriate;repatriation

遣返费 cost of repatriation
遣返回国 repatriate
遣散 demobilization;demobilize;disband;dismissal;dismission
遣散费 compensation for removal;pay a ransom;release pay;severance pay
遣散薪金 separation pay
遣送费 compensation for removal
遣送回国 repatriate

谴

谴责 accusation;condemn(ation);denounce;denunciation;jump(up) on

谴责行为 act of denunciation

欠

欠安全的设计 underdesign

欠饱和的 undersaturated
欠饱和励磁机 undersaturated exciter
欠饱和流体 undersaturated fluid
欠饱和(现象) undersaturation
欠饱和液体 undersaturated liquid
欠曝光 under-exposure
欠曝光的 under-exposed
欠拨销售量 back order sales
欠补偿 undercompensation
欠补偿的 undercompensated
欠补偿的积分控制 undercompensated integral control
欠财务代理人款 due to fiscal agent
欠测值 missing value
欠产 shortfall in output
欠超高 cant deficiency;deficient superelevation;unbalanced superelevation;under superelevation
欠程 undershoot
欠处理 undercure;undercuring
欠吹 young blow
欠代理店款 due to agencies
欠单 accommodation bill;accommodation kite;accommodation note;accommodation paper;debit instrument
欠单交易 accommodation trading
欠电流保护 undercurrent protection
欠电流断路器 undercurrent circuit breaker
欠电流继电器 undercurrent relay
欠电流释放 undercurrent release
欠电流自动断流器 minimum cutout
欠电压 undertension;under-voltage
欠电压保护 low-voltage protection
欠电压解扣 under voltage trip
欠定的 underdetermined
欠定方程组 underdetermined system of equations
欠定值 underdetermined value
欠定组 underdetermined system
欠发达国家 less developed countries
欠发达状态 underdevelopment
欠费用户 defaulting subscriber
欠幅失真 underthrow distortion
欠付工资 back pay;back salary
欠复激 undercompound
欠复激电动机 underexcited compound motor
欠复激发电机 underexcited compound generator
欠复激绕组 undercompound winding
欠复励 undercompound excitation
欠复励磁 uncle compound excitation
欠复励的 undercompound
欠复励发电机 undercompound generator
欠复绕 undercompound
欠固化 undercure
欠固结黏[粘]土 underconsolidated clay
欠固结土 underconsolidated soil
欠固结土层 underconsolidated soil deposit layer
欠固结土沉积 underconsolidated soil deposit
欠光彩 ganosis
欠国外分支机构款 due to foreign branch
欠烘的 underbaked
欠户 debtor
欠换向 under-commutation
欠火 under-burning;under-firing
欠火的 unburnt;under-burned;underburnt;underfired
欠火低质砖 salmon brick
欠火建筑砖 grizzle
欠火劣质砖 peckings
欠火砖 baked brick;chuff(brick);cob;grizzle brick;place brick;soft burnt brick;unburnt brick;unburnt tile;under-burnt brick;underfired brick
欠激 underexcitation;unexcited
欠激励保护(装置) underexcitation protection
欠激励磁发电机 underexcited generator
欠激励发电机 underexcitation generator
欠激励状态 underexcitation state
欠浆的 undersized
欠交 non-delivery
欠交订货 back order
欠交订货通知单 back order memo
欠交数量 quantity due
欠交税款 back duty
欠胶的 undersized
欠胶接头 starved joint
欠胶劣质接缝 starved joint
欠缴或滞纳税款利息 interest on tax underpaid or postponed
欠缴上级运输进款 transport income due to senior administrative agency
欠精炼 underrefining
欠据 accommodation kite;accommodation paper;debit instrument
欠(聚)焦 under focus
欠款 amount due from somebody;arrearage;arrears;balance due;debt;outstanding amount
欠款报单 debit advice
欠款的延缓偿付期 moratorium
欠款金额 amount in arrear
欠款金额证明书 certificate of amount owing
欠款利息 interest on arrears
欠款通知 debit note
欠款余额证书 reduction of mortgage certificate
欠励磁 underexcitation;underexcite
欠连接 underlap
欠联属及联营企业款 due to affiliated and associated companies
欠硫化 undercure
欠密的 underdense
欠耦合 deficient coupling;undercoupling
欠膨胀喷管 underexpanded nozzle
欠平衡的 underbalance
欠缺材料 critical material
欠缺断面 unfilled section
欠缺流量 deficient flow
欠热 subcool;under-firing
欠人 balance due to
欠烧 under-burning;under-firing
欠烧材料 under-burned material
欠烧的 unburnt;under-burned;underburnt;undercalcined;underfired;unfired;unsound
欠烧软砖 malm grizzled brick
欠烧石灰 under-burnt lime
欠烧熟料 under-burned clinker
欠烧透砖 light-burnt brick
欠烧物料 unfired material
欠烧砖 sandal brick;soft brick;under-burnt brick
欠时效 under-ag(e)ing
欠熟 undercuring
欠熟材料 under-burned material

欠税 tax credit
欠税土地 tax-delinquent land
欠塑化 undercure
欠酸洗 underpickling
欠调量 undershoot
欠调制 undermodulation
欠调制的 undermodulated
欠挖 under-excavation;underbreak
欠往来账户款 due to correspondent's account
欠稳定的 understable
欠息 debit interest;debt interest
欠削榫舌 undercut tenon
欠薪 back pay;back salary
欠压 under-pressure;under-roll
欠压保护 under-voltage protection
欠压电路器 under-voltage circuit breaker
欠压动作 low-voltage
欠压断路器 under-voltage circuit breaker;under-voltage release
欠压继电器 under-voltage relay
欠压检测器 under-voltage detector
欠压密 under-compacted
欠压密黏[粘]土 underconsolidated clay
欠压实带 underconpaction zone
欠压实相 underconpaction facies
欠压释放 under-voltage release
欠压脱扣器 under-voltage release
欠压无闭合断路器 under-voltage no-close release
欠压状态 under-voltage condition
欠养护 undercuring
欠养护混凝土 undercured concrete
欠养生 undercure
欠氧 oxygen deficit
欠硬化 under-ag(e)ing
欠釉 short glaze;starved glaze
欠载 underload(ing)
欠载的 underloaded
欠载断路器 underload breaker
欠载继电器 underload relay
欠载开关 underload switch
欠载特性曲线 characteristic underload
欠载系数 undercapacity factor
欠载运行 underrun
欠载运行出错 underrun error
欠债 arrearage;owe
欠债期限 tenor
欠债人 debtor
欠债(违约)的铁路 defaulting railroad
欠债者 obliger[obligor]
欠账 bills due;bought for account;credit account;outstanding account
欠折射 sun reflection
欠重 short weight;underweight
欠重次数 times of lacking in weight
欠重叠 underlap
欠重吨数 tonnage of lacking in weight
欠重列车 train of lacking in weight
欠轴列车 underloaded train
欠注射 short shot
欠资 postage due
欠资的 short paid
欠租 back rent
欠阻尼 periodic(al) damping;under-damping
欠阻尼体系 underdamped system
欠阻尼响应 underdamped response

茜草红色淀 madder lake

茜草玫瑰红色 casino pink
茜草(染料)madder
茜草色喷漆 madder lacquer
茜草色清漆 garancine lacquer;madder lacquer
茜草色硝基清漆 garancine zapon
茜草植物 madder plant
茜草棕色淀 brown madder lake
茜红色 madder
茜丽绸 grosgrain
茜素 alizarin(e)
茜素沉淀色料 alizarin(e) lake
茜素橙 alizarin(e) orange
茜素橙色 Chinese orange
茜素黑 alizarin(e) black
茜素红 alizarin(e) red
茜素红色淀 alizarin(e) madder lake
茜素黄 alizarin(e) yellow
茜素蓝 alizarin(e) blue
茜素玫瑰红 rose madder;rose lake
茜素染料 alizarin(e) colo(u)r;alizarin(e) dye(stuff)
茜素色淀 alizarin(e) lake;thraquinone lake
茜素试剂 alizarin(e) complexone
茜素鲜红 alizarin(e) astrol
茜素颜料 alizarin(e) pigment
茜素棕 alizarin(e) brown

堑 moat

堑槽运河 canal in a cut(ting)
堑道滑坡 sliding in cut
堑顶 top of cutting slope
堑沟 fault trough
堑沟护板 trench shuttering
堑壕 entrenchment;trench
堑壕边缘 trench edge
堑壕底 trench bottom
堑壕法 trench method
堑壕后崖 trench back
堑壕回填 refilling of trench
堑壕喷火器 trench flame thrower
堑壕前崖 front slope
堑壕潜望镜 trench periscope
堑壕体系 trench system
堑壕线 trench line
堑路 sunken road
堑式道路 depressed road(way)
堑式高速公路 depressed freeway
堑式公路 depressed highway
堑土 chiseling
堑形的 trenched
堑形断层 trench fault
堑凿加工 bush-hammering

嵌 enchase

嵌板 panel(board);rocker panel;shingle panel
嵌板工作 panel work
嵌板框架 panel(l)ed framing
嵌板平顶 panel(l)ed ceiling
嵌板散裂试验 panel spalling test
嵌板式屋面系统 panelized roof system
嵌板细工(分段法)panel(l)ing
嵌板细木工 boiserie
嵌板子 panel
嵌报警线发光安全玻璃 alarm-wire laminated safety glass
嵌壁开关 flush switch
嵌壁式火警装置 wall-mounted fire warning device
嵌边卷筒形饰边 roll-and-fillet mo(u)lding
嵌玻璃 pane
嵌玻璃槽 glazing groove
嵌玻璃的槽口尺寸 rabbet size
嵌玻璃的沟边 bezel
嵌玻璃沟缘 bezel
嵌玻璃皮条 wash leather glazing
嵌玻璃条 glazing bead
嵌玻璃屋顶 glazed roof
嵌玻璃用油灰 back putty;bed putty
嵌玻璃油灰 glass putty;glazing putty
嵌玻璃针 glazing sprig
嵌玻璃砖混凝土构造 glass concrete construction
嵌补 inserting
嵌补石膏裂缝的墙粉 alabastine
嵌彩色玻璃窗的铅条骨架 fret lead
嵌槽 ca(u)lking pocket
嵌槽拱 sconcheon arch
嵌槽接头 dado;dado joint;dapped joint;housed joint;housing joint
嵌槽口<木梁的> sconcheon
嵌槽线圈 formed coil
嵌插杆 insert pin
嵌插型连接机构 fix-fix coupler type
嵌长靠椅 wall settle
嵌衬 inlay
嵌成花纹 tessellate
嵌成花纹路面 tessellated pavement
嵌齿 cog;pin tooth
嵌齿的<一般指圆锯> tipped
嵌齿锯 inserted tooth saw
嵌齿扣合 cocking
嵌齿扣合屋檐 bellcast eaves
嵌齿连接 coggea
嵌齿轮 cog gear;cog(ged) wheel;mortise wheel
嵌齿轮润滑脂 cog-wheel grease
嵌齿轮套 cog-wheel casing
嵌齿效应 cogging effect
嵌窗玻璃的有槽铅条 came
嵌窗玻璃铅条 came
嵌窗铅条 fretted lead
嵌带 band
嵌档靠码头 going alongside between berthed ships
嵌雕 inlay carving
嵌叠 nesting
嵌钉 scarf tack;stud
嵌段共聚物【化】block copolymer;periodic(al) copolymer
嵌段共聚作用 block copolymerization
嵌段聚合物 block polymer
嵌方石块的混凝土墙 reticulatum opus
嵌封垫料 back bedding
嵌缝 ca(u)lking(joint);fillet;fullering;joint pointing;key(ed) jointing;pointing joint;sealing;striking off
嵌缝板 filler board;filling board;matching
嵌缝泵插接头 ca(u)lking socket joint
嵌缝边 ca(u)lked edge
嵌缝材料 ca(u)lking compound;ca(u)lking material;glazing compound;joint material;sealant compound;sealer;sealing material
嵌缝槽 ca(u)lking groove
嵌缝承插接头 ca(u)lked socket joint
嵌缝锤 ca(u)lking hammer
嵌缝钉 ca(u)lking nail
嵌缝法 filleting;joint filling
嵌缝防水构造图 structural drawing of construction joint waterproofing
嵌缝粉 filling powder
嵌缝膏 ca(u)lking compound;ca(u)lking mastic;paste filler;sealant
嵌缝膏施工法 application of building mastics;application of ca(u)lking compound
嵌缝工 ca(u)lker
嵌缝工具 ca(u)lking filter;ca(u)lking iron;ca(u)lking tool
嵌缝勾缝 keyed pointing
嵌缝骨料 choker aggregate;key aggregate
嵌缝化合物 ca(u)lking compound
嵌缝灰膏 putty in plastering
嵌缝混合料 joint filling compound
嵌缝混凝土 sealed concrete
嵌缝集料 choker aggregate;key aggregate
嵌缝胶 ca(u)lking compound
嵌缝(胶)枪 ca(u)lking gun
嵌缝胶条 joint strip
嵌缝胶注射器 ca(u)lking gun
嵌缝连接 staking
嵌缝料 ca(u)lk;joint filler;mastic;performed joint filler;sealer;spackling compound;filler material
嵌缝料成分 filling composition
嵌缝麻絮 canker's oakum
嵌缝铆钉 ca(u)lked rivet
嵌缝木 wooden filler
嵌缝木条 fixing fillet;wood slip
嵌缝配筋 joint reinforcement
嵌缝片 sealing plate;slip feather;spline
嵌缝枪加料器 ca(u)lking gun loader
嵌缝砂石 blinding sandstone
嵌缝绳 sealing rope
嵌缝石 blinding stone;choke stone;filler stone;keystone;key block
嵌缝树脂清漆<高干燥剂的> gold size
嵌缝碎石 key aggregate
嵌缝填料 key aggregate;ca(u)lking compound
嵌缝填料筒 ca(u)lking cartridge
嵌缝条 ca(u)lking strip;joint filler tape;joint fillet;jointing strip
嵌缝线 striking-off lines
嵌缝压条 chamfer strip;cant strip;plane strip
嵌缝油膏 ca(u)lking compound
嵌缝凿(刀)ca(u)lking chisel;plugging chisel
嵌缝凿子 fillet chisel
嵌缝纸带 joint binding tape
嵌工用石 abaciscus
嵌工用瓦 abaciscus
嵌固 build(ing) in;constraining
嵌固板 fixed-edge-slab
嵌固边缘 built-in edge
嵌固玻璃橡皮条 zipper strip
嵌固长度 length of restraint
嵌固窗扇 built-in sash
嵌固的 built-in
嵌固点 point of fixity
嵌固端 built-in end;end restraint;tailing down;tailing in
嵌固拱 built-in arch
嵌固基础 embedded footing
嵌固件 first fixings;fixings
嵌固接头 stop joint
嵌固梁 built-in beam
嵌固深度 fixity depth
嵌固凿 ca(u)lking chisel
嵌固支点 fixed support
嵌固桩 fixed pile
嵌固作用 wedge action
嵌含的组构 poikilotopic fabric
嵌含晶 poikilotope
嵌焊 scarf weld(ing)
嵌合 jog;tabling;tumble in
嵌合灯 clearance lamp
嵌合锻接 tongue weld
嵌合接头 female joint
嵌合肋骨 jogged frame
嵌合轮胎 beaded tire[tyre];clincher tyre
嵌合石 bond stone
嵌合式发育 mosaic development
嵌合体 chimera
嵌合突起 mosaic process
嵌合显性 mosaic dominance
嵌合现象 chimerism
嵌合型 mosaic type

嵌合性 chimerism
嵌和反射镜 tessellated mirror
嵌花 applique;enchase;intarsia
嵌花板条 inlaid-strip
嵌花的 mosaic
嵌花地板 overlay flooring
嵌花地砖 faience mosaic
嵌花横机 intarsia flat knitting machine
嵌花花纹 intarsia pattern; panel design
嵌花铺面 Roman mosaic; tessellated pavement;tessellated paving
嵌花式地面 mosaic pavement
嵌花式块石路面 Kleinpflaster
嵌花式路面 mosaic pavement
嵌花式砌块 Kleinpflaster
嵌花式砌石 Kleinpflaster
嵌花式石块 Kleinpflaster
嵌花式小方石块 <立方形,3～4英寸见方,1英寸≈0.0254米> durax stone block
嵌花式小方石路面 durax pavement; durax paving
嵌花式小方石铺砌 durax paving
嵌花式小方石(铺砌)路面 duraxcube pavement
嵌花镶板 inlaid panel;parquetry panel
嵌花镶板线脚 inlaid panel mo(u)lding
嵌花油地毯 inlaid linoleum
嵌花装饰 enchased decoration
嵌环 ferrule;porcelain thimble;sleeve piece;thimble
嵌灰缝碎石片 galleting
嵌灰圬工 point(ing) masonry
嵌挤层 packing course
嵌挤式面层 interlocking surface
嵌件 insert
嵌件定位针 insert pin
嵌键勾缝 keyed pointing
嵌键斜角缝 keyed miter joint;keyed mortise and tenon
嵌接 foliated joint;halving joint;rabbet(joint);rebate;scarf(connection); scarfed joint; scarfing (joint); scarf together; scarph; socketing;table(d) joint
嵌接窗框 rabbeted jamb
嵌接的 scarfed
嵌接方块 bond(ed) block
嵌接缝 scarf-joint
嵌接焊 scarf weld(ing)
嵌接合 scarf together
嵌接机 scarfing machine
嵌接件一端 scarf
嵌接接头 scarf joint(ing)
嵌接片 scarfer
嵌接砌石 joggle work
嵌接深度 rabbet depth
嵌接式套管 inserted type of screw casing
嵌接榫 joint scarfed
嵌接套管 insert(ed) joint casing
嵌接头 scarfer
嵌接线 bearding line; rabbet line; stepping point
嵌接鱼尾板 tabled fish plate
嵌金刚石的圆锯 diamond impregnated circular saw; diamond saw (splitter)
嵌金银丝饰 filigree
嵌金属丝网的波纹玻璃 corrugated wire(d) glass
嵌金属条的梁 filigree girder
嵌金属条的楼面 filigree floor
嵌紧的缝 ca(u)lked joint
嵌进 keyed into;nest;telescope;work its way into
嵌进砾石 gravel(l)ing
嵌进木材 tumble in timber;tumbling-in timber
嵌进镶板 sunk panel
嵌晶 poikilocrysal
嵌晶构造 lineage structure;poikilitic structure
嵌晶胶结物 poikilitic cementation
嵌晶胶结物结构 poikilitic cement texture
嵌晶结构 lineage structure;poikilitic texture
嵌晶结构的 poikilitic
嵌晶状的【地】poecilitic[poikilitic]
嵌晶状结构【地】poikilitic texture
嵌锯 fret saw;pad saw
嵌口接合板 tongue and groove plank
嵌块 insert(ion) piece;plugged impression;quad(rat)
嵌料缝 reglet
嵌漏嘴 tip insertion
嵌轮芯套 bush
嵌埋式砌石护坡 embedded stone pitching
嵌密缝 ca(u)lked joint
嵌木地板 parquet
嵌木缝泥脂 filling;wood filler
嵌木细工 intarsia; marqueterie; marquetry;tarsia
嵌木制品 intarsia;marquetry;tarsia
嵌腻子 sanding sealer
嵌片 dummy slider; filling piece; insert
嵌片层 <大石块基层上的> packing course
嵌片基层 choke stone
嵌铺法 lay-fit method
嵌砌方块 bonded block
嵌砌块体 bonded blockwork
嵌砌块体工程 interlocking blockwork
嵌砌砖层 tumbled-in brick course; tumbling-in brick course
嵌铅环 keying ring
嵌墙长椅 wall settee
嵌墙存衣柜 recessed locker
嵌墙灯 built-in lamp; wall-mounted light(ing)
嵌墙墩 engaged pier
嵌墙风扇 partition fan
嵌墙炉灶 wall-mounted cooker
嵌墙木砖 built-in nailing block
嵌墙散热器 built-in radiator;recessed radiator
嵌墙式家具 built-in furniture
嵌墙式煤气炉 built-in gas fire
嵌墙式旗杆 wall-type flagpole
嵌墙樘子 <部分墙体小于洞口> keyed-in frame
嵌墙照明设施 built-in lighting
嵌墙柱 attached column;column engaged to the wall;demi-column;engaged column;inserted column
嵌圈 union thimble
嵌入 bedding into; build in; built-in; embed(ding);encase;encasing;entrench;housed;imbed(ment);implant; incuneation; infix; inserting; intercalation; interpose; jog down; plug-in
嵌入棒 setting-in;setting-in brick;setting-in stick
嵌入壁炉采暖炉 fireplace insert;inset fire
嵌入边(缘) built-in edge
嵌入不整合 inlaid unconformity
嵌入部分 embedded parts;tailing <砖在墙内的>
嵌入长度 embedded length; embedment length;insert length
嵌入齿轮轮箱 build-in hear box
嵌入冲头 insert punch
嵌入挡板 let-in flap
嵌入导体 embedded conductor
嵌入的 built-in; bulkheaded; intrenched;nested;set in
嵌入的磁铁 inserted magnet
嵌入的大梁 built-in girder
嵌入的铺面混合料 laid paving mixture
嵌入的铺面混合物 laid paving mixture
嵌入的石屑 imbedded chips
嵌入灯 built-in lamp;flush light;inset light
嵌入垫片 inserted shim
嵌入定理 embedding theorem
嵌入端 built-in end
嵌入端固定支座 built-in support
嵌入短管接头 pup joint
嵌入对象 embedded object
嵌入阀座 inserted valve seat
嵌入法 embedded method;embedding technique;imbedding method;inlay
嵌入法安放 insert process method
嵌入分量 embedded component
嵌入盖 insert cover
嵌入格栅 inserted grill(e)
嵌入构造 telescope structure
嵌入箍 backup strip
嵌入谷 entrenched valley
嵌入关系 intrenched relationship
嵌入柜 built-in cupboard
嵌入过程 telescopying
嵌入河 entrenched river; entrenched stream;intrenched river;intrenched stream
嵌入河道 in-and-out channel
嵌入河谷 intrenched valley
嵌入河流 intrenched stream
嵌入河曲 entrenched meander; incised meander
嵌入滑轮 built-in sheave
嵌入或镶进 trim in
嵌入集团 insertion group
嵌入剂 intercalating agent
嵌入加强块 inserted reinforcing piece
嵌入件 embedded parts;insert;liner; embedment
嵌入键 embedded key;sunk key
嵌入阶地 entrenched terrace; intrenched terrace
嵌入接头 housed joint
嵌入结构 <洪积扇的> telescope structure
嵌入介质 embedding medium
嵌入矩阵 embedded matrix
嵌入卷边 tuck-in
嵌入孔 insert hole
嵌入控制通路 embedded control channel
嵌入宽度 insert width
嵌入连接 housed joint
嵌入链 embedded chain
嵌入梁 built-in beam;encased beam; encastre;stemple
嵌入梁或板 built-in beam or slab
嵌入楼梯斜梁 housed stringer
嵌入滤器 insert strainer
嵌入路面标志 insert pavement marker
嵌入螺母 inserted num
嵌入螺栓 stud bolt
嵌入马尔可夫链【数】embedded Markov chain
嵌入码 embedded[imbedded] code
嵌入面层的透明或半透明预制板 pavement light;vault light
嵌入命令【计】embedded command
嵌入模内的 mo(u)lded-in
嵌入能力 embedability
嵌入平顶 inserted ceiling
嵌入平拉门边框的把手 edge pull
嵌入砌体的受钉木块 nailing block
嵌入强度 embedment strength
嵌入墙内 tailing in
嵌入墙内固定悬臂构件的型钢 tailing iron
嵌入墙内的柱 inserted column
嵌入墙体的木砖 wood nog
嵌入曲流 entrenched meander;inherited meander;intrenched meander
嵌入圈 backup ring
嵌入绕组 imbedded winding
嵌入舌片 inserted tongue
嵌入深度 insert depth
嵌入式 embedded type; flush-bonding; flush type; inset type; resettype
嵌入式安装 recessed fitting
嵌入式板式灶 built-in hotplate
嵌入式标志 insert mark
嵌入式测温计 embedded temperature detector
嵌入式插头 flush plug
嵌入式插座 flush plug; flush plug consent;flush plug socket
嵌入式撑杆 ribbon board
嵌入式齿轮箱 built-in gearbox
嵌入式橱柜 built-in kitchen cabinet
嵌入式的 flush mounted; flush recessed
嵌入式的冲水水箱 recessed flush cistern
嵌入式的磨刃装置 built-in sharpener
嵌入式电路 flush type circuit
嵌入式顶棚 lay in ceiling
嵌入式多值相关性 embedded multivalue dependency
嵌入式发券 built-in arch
嵌入式阀 <能用特殊设备在管道运行时嵌入的> insert valve
嵌入式防溅板 built-in antisplash
嵌入式肥皂盆 recessed soap dish
嵌入式格栅 inset grate
嵌入式格子窗 inserted grill(e)
嵌入式工具 inserted tool
嵌入式供暖 panel heating
嵌入式供暖器 panel heater
嵌入式供热和供冷 panel heating and cooling
嵌入式拱 built-in arch
嵌入式固订单元 in-built unit
嵌入式柜橱 in-built cabinet
嵌入式洪积扇 entrenched fan
嵌入式灰缸 in-built ashtray
嵌入式灰盆 built-in ashtray
嵌入式基脚 embedded footing
嵌入式计算机 embedded computer
嵌入式加热元件 embedded heating element
嵌入式减速齿轮箱 interleaved reduction gear
嵌入式检温计 embedded temperature detector
嵌入式铰链 insert hinge
嵌入式结构 flush type construction
嵌入式开关 flush switch
嵌入式空白 embedded blank
嵌入式控制器 embedded controller
嵌入式宽边浴盆 recessed wide-ledge (bath) tub
嵌入式冷柜 built-in refrigerator
嵌入式冷却系统 panel cooling
嵌入式帘格 inset grate
嵌入式淋浴间 in-built shower stall
嵌入式楼梯斜梁 housed string
嵌入式路肩 inserted spandrel
嵌入式轮胎 clincher tire[tyre]
嵌入式螺帽 inserted nut
嵌入式煤气灶 built-in appliance
嵌入式门环 flush handle
嵌入式门套 built-in casing

嵌入式磨石装置 built-in sharpener
嵌入式球形恒温器 insert thermo-bulb type thermostat
嵌入式软件 embedded software
嵌入式散热器 insert radiator; panel radiator
嵌入式生活设备 built-in comfort
嵌入式生活设施 built-in comfort
嵌入式手纸架 recessed toilet paper holder
嵌入式榫尖 slip nose
嵌入式锁 mortise-type lock
嵌入式条状双异质结构 embedded stripe double-heterostructure
嵌入式碗橱 in-built cupboard
嵌入式温度计 embedded temperature detector
嵌入式洗涤盆 inset sink
嵌入式洗面器 countertop lavatory
嵌入式系统 built-in system; embedded application; embedded system
嵌入式小淋浴间 built-in shower stall
嵌入式斜撑 let-in brace
嵌入式烟灰缸 in-built ashtray
嵌入式阳台 inset balcony
嵌入式衣橱 built-in wardrobe
嵌入式仪表 flush type instrument; flush type meter
嵌入式异质结构 embedded hetero-structure
嵌入式饮水器 recessed drinking fountain
嵌入式浴缸 in-built tub; recessed bathtub
嵌入式浴盆 built-in bathtub; recessed tub
嵌入式圆形凹模固定板 button die retainer
嵌入式照明 built-in lighting
嵌入式照明设备 recessed lightings
嵌入式照明设施 recessed lightings
嵌入式钻头 insert bit
嵌入素除子 embedded prime divisor
嵌入素理想 imbedded prime ideal
嵌入榫 false tenon; loose tongue; spline
嵌入榫头 inserted tenon
嵌入台地 intrenched terrace
嵌入图案 inlay
嵌入温度传感器 embedded temperature detector
嵌入温度探测器 embedded temperature detector
嵌入物 inlay
嵌入楔块 inserted wedge
嵌入楔块层 block-in-course
嵌入楔块层砌合 block-in-course bond
嵌入碹 inserted arch
嵌入指示符 embedded pointer
嵌入轴颈 built-in journal
嵌入装置 flush mounting
嵌入子波 embedded wavelet
嵌入作用 embedding action; imbedding action
嵌塞 ca(u)lk; impaction
嵌塞的 impacted
嵌塞解除法 disimpaction
嵌塞碎石片 pinning-in
嵌塞小木块 glut
嵌上 key-in
嵌石片 choke stone
嵌石铺面 tessellated pavement; tessellated paving; tessera[复 tesserae]
嵌石装饰 tessellation
嵌实的 ca(u)lked
嵌实缝 ca(u)lked joint; ca(u)lked seam
嵌饰 abaciscus; abaculus
嵌丝玻璃 filigree glass; pattern(ed) wire(d) glass; wire(d) glass
嵌丝玻璃板 flat wired glass; wired glass pan; wired sheet glass
嵌丝玻璃采光穹隆 wired glass light cupola
嵌丝玻璃采光圆顶 wired glass light cupola
嵌丝玻璃格子顶棚 wired glass caisson; wired glass cassette; wired glass coffer; wired glass waffle
嵌丝玻璃穹隆形天窗 wired glass domed rooflight; wired glass dome light; wired glass saucer dome
嵌丝玻璃天窗 wired glass roof light
嵌丝玻璃用钢丝网 wire glass mesh
嵌丝玻璃藻井 wired glass caisson
嵌丝的 filigree
嵌丝叠层玻璃 wired laminated glass
嵌丝抛光平板玻璃 wired polished plate glass
嵌丝平板玻璃 wired plate glass
嵌丝乳白玻璃 wired-opaque white glad
嵌丝图案玻璃 wired figured glass; wire(d) pattern(ed) glass
嵌丝装饰玻璃 wired decorative glass; wired ornamental glass
嵌榫接合 feather joint; ploughed-and-tongued joint
嵌榫拼接 feather joint; ploughed-and-tongued joint
嵌锁 bridge; interlock(ing)
嵌锁力 interlocking resistance
嵌锁木块 <门上> lock block
嵌锁式板 interlocking board
嵌锁式混凝土管 interlocking concrete pipe
嵌锁式混凝土(块)路面 interlocking concrete block pavement
嵌锁块面层 interlocking block pavement
嵌锁式路面 interlocking surface; macadam aggregate type
嵌锁式砌块 interlocking block
嵌锁型混凝土块体 interlocking concrete block
嵌锁型预制面层混凝土块 interlocking paving stone
嵌套 nestification; nest(ing)
嵌套层 nesting level; level of nesting
嵌套程序 nested procedure
嵌套程序结构 nested procedure structure
嵌套存储 nesting store
嵌套单元 <拉合尔花边机梳栉的> nesting unit
嵌套的 nested
嵌套递归式 nested recursion
嵌套分程序 nested block
嵌套过程 nested procedure
嵌套宏调用 nested macro call
嵌套宏定义 nested macro definition
嵌套级 nesting level
嵌套监督程序 nested monitor
嵌套监督程序调用 nested monitor call
嵌套结构 nested structure
嵌套块 nested block
嵌套临界区 nested critical section
嵌套文件传送格式 nested file transfer format
嵌套循环 monitoring loop; nested loop; nesting loop
嵌套循环方式 nested loop mode
嵌套映射 embedding mapping
嵌套映象 embedding mapping
嵌套栈自动机 nested stack automa-(tiza)tion
嵌套子程序 nested subroutine; nesting subroutine
嵌套子程序链 nested subroutine linkage
嵌套作用域 nested scope
嵌体 inlay
嵌体雕刻刀 inlay carver
嵌体蜡 inlay wax
嵌填 pointing
嵌填层 <涂料> filler coat
嵌填膏泥 <用于木石孔隙中> badigeon; badijum
嵌填勾缝模板 pointing template
嵌填灰缝 pointing
嵌填灰膏 badigeon; badijum
嵌填假缝用地沥青砂胶板 asphalt mastic board dummy joint
嵌填金属丝工程 filler wire construction
嵌填金属丝缆索 filler wire rope
嵌填料 stopper; stopping
嵌填塞 stopping
嵌填物 insertion
嵌填用韧性刮铲 flexible filling blade
嵌条 casing bead; filler rod; filler strip; fillet; insert ring; moss fringe; mo(u)lding; panel strap; panel strip; rubber filling; shim; counter lathing <常规灰板条间的>
嵌条材料 mo(u)lding material
嵌条单件 fillet unit
嵌条的带肋楼面 filigree rib(bed) floor
嵌条断面 fillet section
嵌条接缝 welted seam
嵌条接合 filleted joint
嵌条刨 mo(u)lding plane
嵌条切割器 mo(u)lding cutter
嵌条通路 slug pathway
嵌条外形 fillet shape
嵌条线脚 gradetto
嵌条镶边 fillet trim
嵌条砖 pallet brick
嵌凸缝 tuck and pat pointing
嵌凸缝工具 tuck point joint filler tool
嵌钨钢合金片钻头 tungsten-carbide insert bit
嵌物焊 slugging
嵌线 beading; fillet; inserted line; inserting winding; wind cork <车身接缝>
嵌线珐琅 enclosed style enamel
嵌线工具 inserting tool
嵌线机 coil assembling apparatus; coil inserting apparatus; slot inserter
嵌线卷边机 wiring press
嵌线器 slot inserter
嵌线条 dimpling
嵌镶靶 mosaic target
嵌镶板 mosaic plate
嵌镶表面 mosaic surface
嵌镶玻璃 mosaic glass
嵌镶成像 mosaic imaging
嵌镶瓷砖 mosaic tile
嵌镶的 mosaic
嵌镶光电阴极 mosaic photocathode; photomosaic
嵌镶夹 glazing clip
嵌镶结构 mosaic structure
嵌镶金合金 mosaic gold alloy
嵌镶晶体 mosaic crystal
嵌镶块 abaculus; mosaic block; tessera[复 tesserae]
嵌镶切削工 mosaic cutter
嵌镶设计师 mosaicist
嵌镶式 mosaic
嵌镶式红外系统 mosaic infrared system
嵌镶式滤色器 checkerboard of colo(u)r filters
嵌镶式仪表 panel meter
嵌镶式照相底片 mosaic film; mosaic photographic(al) film
嵌镶图摄影学 mosaic photography
嵌镶细工 mosaic
嵌镶阴极 mosaic cathode
嵌镶荧光屏 phosphor dot array
嵌镶钻头 bit setting; insert bit; set bit
嵌销 loose tongue
嵌销接合 tenon dowel joint
嵌小方块马赛克 opus sectile
嵌芯狭辫带 cord gimp
嵌芯窄辫带 gimp
嵌岩灌注桩 bored pile embedded rock
嵌岩桩 rock-socketed pile; socketed pile
嵌样法 <铺砌装修地板法> inset
嵌油灰 filling with putty; puttying
嵌油灰刀 stopping knife
嵌油灰于 slush
嵌有铁丝网的玻璃 armo(u)red glass
嵌云石块细工 opus tesselatum
嵌在里面 built-in
嵌在墙上 built-in wall
嵌值 crest
嵌制砌合的人字屋顶 clipped gable roof
嵌制细粒金刚石钻头 impregnated bit
嵌置 nestion
嵌置的黏[粘]土层 interspersed clay layer
嵌置的铺面混合料 laid paving mixture
嵌砖混凝土墙 reticulated work; opus latericium
嵌装 setting-in
嵌装的 impregnated
嵌装滑轮 built-in sheave
嵌装件 insertion
嵌装螺栓暗保险锁 flush-encased dead bolt block
嵌装式 flush type
嵌装式石膏装饰板 decorative gypsum board for lay-in installation
嵌装式无弹簧门锁 flush-encased dead bolt block
嵌装式仪表 flush type instrument
嵌装式照明设备 recessed luminaire
嵌装锁 flush-encased block
嵌装碗橱的墙 cupboard wall
嵌装物 insertion
嵌装小柜的墙 cupboard wall
嵌装在墙上的旗杆 wall-mounted flagpole

歉

歉年 partial mast
歉收 bad crop; bad harvest; crop failure; poor crop; poor harvest
歉收年 fail year; fall year; lean year; bad year
歉准区 anti-pleion

羌

羌颗粒荧光屏 grainless phosphor screen

羌族建筑 architecture of the Qiang nationality

戗

戗道 bench; berm(e)

戗道护面 trench fill revetment
戗道排水沟 berm(e) ditch; berm(e) spillway
戗道修整 berm(e) trimming
戗堤 back levee; postdam
戗脊 angle ridge; diagonal ridge gable and hip roof; hip
戗脊(椽)角 hip bevel
戗脊椽木 hip rafter; pined rafter
戗脊端饰 hip knob

Q

戗脊泛水片 soaker;soaker at hip
戗脊盖条 hip capping
戗脊盖瓦 angle tile;arris tile;bonnet;bonnet tile;hip capping;hip tile
戗脊挂瓦钩 hip hook;hip iron
戗脊脊瓦 angle hip tile
戗脊上斜石板 hip bevel stone
戗脊石 hip stone
戗脊瓦 angular hip tile
戗脊线脚 hip mo(u)ld(ing)
戗脊应力 hip stress
戗脊锥形筒瓦 cone hip tile;cone tile
戗架拉杆 hip vertical
戗金 incising and filling with gold dust
戗兽【建】hip-mounted animal ornament
戗台 banquette;berm(e) <马道>;dike ledge
戗台式堤脚护坦 berm(e) toe apron
戗台式建筑 berm(e) type structure

枪

枪柄式 pistol grip
枪柄式气钻 pistol-grip handle drill
枪点 lance point
枪管 gun barrel
枪尖 lance point
枪晶石 cuspidite
枪孔钻 gun drill;woodruff drill
枪孔钻床 gun drilling machine
枪矛 gavelock
枪铆铆钉 gun-driven rivet
枪木 lancewood
枪前托 gun fore-end
枪桥 gun bridge
枪射铆接 gun-driven riveting
枪式吹灰机 gun-type soot blower
枪式吹灰器 gun-type soot blower
枪式焊接器 pistol
枪式焊接区域 pistol range
枪式喷燃器 gun burner
枪式取土器 gun-type coring machine
枪式丝锥 gun tap;spiral pointed tap
枪式土样器 gun-type coring machine
枪式小梁 dragon beam
枪式油燃炉 gun-type oil burner
枪式油燃器 gun-type oil burner
枪刷 bore brush
枪栓 bolt
枪筒 cartridge;gun barrel
枪筒钻 boring drill;gun drill
枪托状门窗边梃 gun stock;gunstock stile
枪型电烙铁 soldering gun
枪眼 barbican;crenel(le);embrasure;kernel;loophole;port;oylet <采光的、通气的>;machicolation <古堡女儿墙中的>
枪眼形 machicolated form
枪用密封剂 gun-grade sealant
枪注 gun-injection
枪注法 <木材防腐> gun-injection process;Cobra process
枪状物 lance

腔

腔背 hollow
腔肠动物 Coelenterata
腔倒空 cavity dumping
腔倒空器 cavity dumper
腔的 cavitary;luminal
腔间的 intercavernous
腔结构 cavity configuration
腔进式污泥泵 progressing-cavity pump
腔控振荡器 cavity controlled oscillator
腔口 muzzle
腔内饱和吸收 intracavity saturated absorption
腔内破碎 breaking chamber
腔式反应堆 cavity reactor
腔压 <旁压仪的> cell pressure
腔中倍频 intracavity frequency doubling
腔中熔化压力 in-cavity melt pressure

蜣

蜣螂线 scarabee

强

强白光 lime light
强暴的 violent
强爆小碎石 burn cut
强爆炸 high explosion
强北风 norther
强变畸变特性 strength-deformation characteristic
强表面耦合的核模型 strong surface-coupling nuclear model
强波 high amplitude wave
强不可达序数 strongly inaccessible ordinal number
强侧风 extreme crosswind
强层 competent bed
强插信号 offer signal
强拆 forced releasing
强拆电键 cancelling key
强拆电路 interception circuit
强拆继电器 break-in relay;forced release relay
强拆忙音继电器 peremptory interrupter relay
强颤振 hard flutter
强场 intense field;strong field
强场场致电子放射弧 high field electron emission arc
强场畴发光 high field domain luminescence
强场电致发光 high field electroluminescence
强场法 strong-field method
强场式升压机 high field type booster
强场效应 high electric(al) field effect
强超辐射 strong superradiation
强超声波 intense ultrasonic wave
强潮 force tide;strong tide
强潮河口 strong tide river mouth
强潮流 tidal race;tide race
强潮汐环境 strong tidal environment
强沉淀水合氧化铁 freshly precipitated hydrous iron oxide
强沉淀水合氧化物 freshly precipitated hydrous oxide
强充电 soaking
强冲击 thump
强冲洗 flushing washing;vigorous flush(ing);vigorous wash
强抽运 heavy pumping
强初级射线 heavy primary ray
强磁半导体 strong magnetic semiconductor
强磁场 intense magnetic field
强磁化 hard magnetization
强磁金属棒 high magnetic metal rod
强磁限制器 excitation forcing limiter
强磁性 ferro-magnetic
强磁性材料 ferro-magnetic material;strong magnetic material
强磁性矿物 strong magnetic mineral
强磁性铁镍合金 resist permalloy
强磁性物料 strongly magnetic material
强磁性物质 ferro-magnetic substance
强磁致伸缩材料 giant magnetostrictive material
强大的 atomic;powerful
强大功率 ample power;great power
强大拉力 high tension
强大数定律 strong law of large numbers
强大压力 strong pressure;crunch
强大压力的 arm-twisting
强单边带 strong sideband
强单调 strongly monotone
强单调函数 strongly monotonic function
强单调性 strong monotonic property
强导电层 dense conducting layer
强的 competent
强低压 intense depression
强低压系统 strong low-pressure system
强地面震动 strong ground motion
强地震 macroseism
强地震动 strong earthquake motion
强地震区 strong earthquake zone
强地震图 strong seismogram
强地震学 strong seismology
强地震运动 strong earthquake motion
强点光源 junior
强电场 high field
强电磁性矿物 strong electromagnetic mineral
强电干扰 high-voltage interference
强电工程 heavy current engineering
强电弧 forceful arc;hard arc
强电解质 strong electrolyte
强电流 heavy current;intense current;power current;strong current
强电流电缆 heavy current cable;power supply cable
强电流绝缘子 heavy current insulator
强电流馈送 heavy current feedthrough
强电流线路 heavy current line
强电流引入 heavy current lead-in
强电室 strong electricity room
强电视信号区 class A signal area
强调 accentuation;emphasize;highlight;lay emphasis;make a point that;place emphasis;stress;underscore
强调独特风格的建筑 Mannerist architecture
强调符 bullet
强调顾客利益推销法 benefit method of selling
强调实用的主张 functionalism
强调线 underline
强调谐点 strong tuning point
强调形式 emphatic form
强调指出 accentuate
强动地震仪 strong motion seismograph
强冻胀 strong frost heaving
强度 dependability;intensity;intensity grade;robustness;strength
强度安全系数 safety factor of strength;strength assurance coefficient;strength safety coefficient
强度包(络)线 intrinsic(al) shear strength curve;Mohr's envelope;strength envelope
强度包线函数 intensity envelope function
强度保留率 strength retention rate
强度比 <通常为模量比> strength ratio;intensity ratio
强度比率 volumetric(al) efficiency
强度变化范围 range of strength
强度变换 intensity transformation
强度变形特性 strength-deformation characteristic
强度辨差阈差异试验 difference limen difference test
强度辨差阈试验 difference limen test
强度标度 intensity scale
强度标法 intensity-scale calibration method
强度标号 strength grade
强度表 scale of intensity
强度不够的 under-strength
强度不足 undercapacity
强度参数 intensive parameter;parameter of strength;strength parameter
强度测定仪 intensity measuring device
强度测量计 intensity meter
强度测量仪 ratemeter
强度层 stress bearing layer
强度差 intensity difference
强度常数 strength constant
强度成熟系数 strength-maturity factor
强度吃水 strength draft
强度持续因素 duration-strength factor
强度储备 degree of safety;margin of safety;reserve strength;strength margin;strength reserves
强度储蓄 margin of safety
强度传递函数 intensity transfer function
强度的分散 dispersion in strength
强度的形成 development of strength
强度等级 intensity level;intensity scale;scale of intensity;strength grade
强度低的岩石 low-strength rock
强度递减 decrease of strength;strength retrogression
强度递增 strength progress
强度电压关系 intensity-voltage dependence
强度定额 strength rating
强度对比度 intensity contrast
强度对重量比 strength-weight ratio
强度发展过程 strength development process
强度范围 strength range
强度方式 strength-wise
强度放大 intensity amplification
强度放牧 hard grazing
强度分布 intensity distribution
强度分布曲线 strength distribution curve
强度分级 strength class
强度分级材 stress graded lumber
强度分类 strength classification;strength group
强度分路比率 intensity bridge ratio
强度分析 intensive analysis;strength analysis
强度峰值 peak strength
强度干涉测量法 intensity interferometry
强度干涉仪 intensity interferometer
强度干舷 strength freeboard
强度高的岩石 high-strength rock
强度更迭 intensity alternation
强度功能 strength power
强度函数 intensity function
强度和气密性水压试验 water pressure test of strength and tightness
强度和酸度试验 strength and sedimentation test
强度核心 central core of strength
强度恢复 strength recovery;strength regain
强度级 intensity level;stage of strength;strength level;volume level
强度极低的岩石 very low-strength rock
强度极高的岩石 extremely high-strength rock
强度极限 breaking point;strength ceiling;strength limit;strength margin;ultimate strength

强度极限面 strength limit surface
强度极限状态 strength limit state
强度计 intensitometer
强度计算 calculation of strength; intensity calculation; strength calculation
强度计算压力 strength pressure
强度继电器 intensity relay
强度减半厚度 half-value thickness
强度减弱 strength decrease
强度减弱的板 weakened plate
强度减弱截面 attenuation cross section
强度减小 strength decrease
强度检验 strength check(ing)
强度渐近线 strength asymptote
强度鉴别 intensity discrimination
强度鉴定 intensity determination
强度降低 strength reduction
强度降低系数 strength reduction factor
强度降落 intensity dip
强度交变 alternation
强度校核 strength check(ing)
强度校准装置 intensity calibrating device
强度卷积积分 intensity convolution integral
强度坑 intensity dip
强度控制 intensity control
强度控制电极 intensity-controlling electrode
强度控制闸 intensity-controlling gate
强度理论 strength theory; theory of strength
强度历时 intensity-duration
强度历时公式 intensity-duration formula
强度历时曲线 intensity-duration curve
强度量 intensive quantity
强度淋溶 strongly leached
强度龄期关系 strength-age relationship
强度铆接 strength riveting
强度-密度比 strength-density ratio
强度模数 strength modulus
强度评定 intensity assessment
强度谱 intensity spectrum
强度谱分布 distribution of spectral intensity
强度侵蚀 deep erosion
强度侵蚀度 deepening erosion ratio
强度曲线 intensity curve
强度曲线上的下陷部分 intensity dip
强度热导比 strength conductivity ratio
强度色度章度 intensity-hue-saturation
强度上升 rising strength
强度上限值 <轻集料/骨料混凝土的> strength ceiling
强度设计 strength design
强度设计法 strength design method
强度设计准则 strength design criterion
强度射场 high radiation field
强度时间公式 intensity-duration formula
强度时间曲线 intensity-duration curve; intensity vs time curve; strength-duration curve
强度试验 pressure test(ing); reliability trial; strength test
强度试验机 strength tester; strength-testing machine
强度试验器 strength tester
强度试验试样 tensile specimen
强度试验仪 strength tester
强度数值 strength value
强度衰减 strength in decay; strength reduction
强度衰退 strength retrogression
强度水平 strength level
强度随时间增加 time-dependent strength
强度损失 loss of strength; strength loss
强度特性 strength character(istic); strength property
强度特征值 strength characteristic
强度提高率 strength enhancement ratio
强度条件 strength condition
强度调整 intensity adjustment; stiffness control
强度调整继电器 intensity regulation relay
强度调制 intensity modulation
强度调制电极 intensity modulating electrode
强度调制电流 intensity modulated current
强度调制电路 intensity modulation circuit
强度调制光束 intensity modulated beam
强度调制器 intensity modulator
强度调制指示器 intensity-modulated indicator
强度同向 strength isotropy
强度退化 retrogression of strength; strength retrogression
强度问题 strength problem
强度系数 coefficient of strength; quality coefficient; severity factor; strength factor
强度线 intensity line; line of intensity; failure envelope <土的>
强度限 intensity limitation
强度相等 equal in strength
强度相对数 intensity relative quantity
强度削减 strength reduction
强度削减处理程序 strength reduction processor
强度削减扫描 strength reduction pass
强度性能 strength capability; strength character
强度性质 strength property; strength quality
强度压缩器 volume compressor
强度研究 strength investigation
强度盐渍化 strong salinization
强度要求 strength requirement
强度异常 strength anomaly
强度异向 strength anisotropy
强度因数 strength factor
强度因素 intensity factor; strength factor
强度因子 strength factor; intensity factor
强度与重量比(率) ratio of strength to weight
强度与重量之比 strength-to-weight ratio
强度雨量关系曲线 intensity-rainfall curve
强度预测 strength forecast; strength prediction
强度预计 strength prediction
强度增长 development of strength; gain in strength; strength gain; strength increase; strength-increasing
强度增长率 rate of development of strength; rate of strength development
强度增长时间 strength-gaining time; strength-increasing time
强度增长值 values of gaining strength
强度增加速率 rate of strength gain
强度增进剂 strength improver
强度增益 strength gain
强度折减系数 strength reduction factor
强度征用 compulsory acquisition
强度值 amount of strength; strength value
强度指标法 intensity target method
强度指数 intensity index
强度终值 ultimate strength value
强度重量 strength(-to)-weight
强度重量比 ratio of strength to weight; strength-weight ratio
强度重现期 intensity return period
强度轴比例尺 scale of intensity axis
强度柱状图 intensity histogram
强度组分 strength component
强煅烧的 high fired
强对比度景物 strong object
强对比度图像 hard image; hard picture
强对偶空间 strong dual space
强恶臭 strong odo(u)r
强反差的 strong contrast
强反差滤光镜 contrast filter
强反差图像 contrast image
强反差显影剂 contrast developer
强方法 strong method
强方向射束发射机 link transmitter
强方向性天线 highly directional antenna
强方向性系统 highly directional system
强放管 power amplifier tube; power tube
强放级 power amplifier stage; power stage
强放射性 high-level radioactivity; hot
强放射性材料 hot material
强放射性测量仪器 high-level instrumentation
强放射性点 high-spot
强放射性废物 high-level radioactive waste; high-level waste; hot waste
强放射性废液 high activity liquor; high-level radioactive liquid waste
强放射性辐照器 high-level irradiator
强放射性核废料 high-level nuclear waste
强放射性粒子 hot particle
强放射性区 region of high activity
强放射性热室 high-level cell
强放射性设备 hot facility
强放射性试验 strong radioactive test
强放射性物质(研究)实验室 hot lab(oratory)
强非线性 strong non-linearity
强非线性类型 strong nonlinear patterns
强非线性微分方程 strongly non-linear differential equation
强非线性系统 strongly non-linear system
强沸腾金属液 wild metal
强分量 strong component
强分散 strong dispersion
强分支 strong branch; strong component
强风 <蒲福风级表六级风,风速 10.8 ~ 13.8 米/秒> strong breeze; air blowing; extreme wind; high wind; fresh gale; line blow; stiff wind; strong wind
强风暴 heavy storm; intense storm
强风暴观测 severe-storm observation
强风飑 thick squall
强风潮 surge
强风化层 badly weathered layer; badly weathered stratum; highly weathered stratum; strongly weathered layer; strongly weathered stratum
强风化带 strong weathered zone; strong weathering zone
强风化的 badly weathered; highly weathered; strongly weathered
强风化花岗岩 badly weathered granite; strongly weathered granite
强风化泥岩 badly weathered mudstone; strongly weathered mudstone
强风化黏[粘]磐土【地】 nitosol
强风化砂岩 badly weathered sandstone; strongly weathered sandstone
强风化土 badly weathered soil; strongly weathered soil
强风化岩石 highly weathered rock
强风化页岩 badly weathered shale; strongly weathered shale
强风警报旗 information signal
强风图 <记录每秒 10 英里以上风向、风力、风速,1 英里≈1 609.34 米> high wind diagram
强风下沉型烟羽 strong wind fumigation plume
强风向 direction of gales; direction of strong wind
强辐射 intense radiation
强辐射带 severe radiation belt
强辐射气体 highly radiating gas
强辐射夜 radiation night
强腐蚀 deep-etching
强腐蚀性液体 strongly corrosion liquid
强富水溢出带 abundant watery overflow zone
强伽马辐射 intense gamma radiation
强干扰 strong interference; strong jamming
强干扰阻塞 strong interference blocking
强感觉 annoying
强功率 high power; super-power
强功率管 super-power tube
强功率无线电台 high-power radio station
强功率整流器 heavy-duty rectifier
强功能操作系统 powerful operating system
强功能的 powerful
强功能结构 powerful architecture
强功能外部设备 powerful peripheral
强功能外围 powerful peripheral
强功能指令系统 powerful instruction set
强构件 strong component
强固的 fortifying; solid; strong
强固焊缝 structural weld
强固焊接 strength weld
强固铰 reinforced hinge
强固紧密焊缝 tight-strong seam
强固联结 rigid joint
强固铺板 solid apron
强固销 reinforced hinge
强光 bright light; dazzle light; glare; hard light; highlight; strong light
强光曝光 highlight exposure
强光灯 flood; flood lamp; flood light; major light; spotlight; accent light
强光灯标 strong light
强光电弧 high-intensity arc
强光反射 emergency light reflex
强光反射镜 intensive reflector
强光激光器 highlight laser
强光力望远镜 fast telescope; high-speed telescope
强光力物镜 high-power object lens
强光亮度 highlight brightness
强光脉冲 intense light pulse

强光前灯 dazzle lamp
强光强度 highlight intensity
强光区 highlight
强光视觉 daylight vision;photopia vision
强光试验 highlight test
强光通量 highlight flux
强光投射灯 high-intensity projector lamp;high-power projector lamp
强光透镜 fast lens;high aperture lens
强光物镜 high-power objective
强光隐灭试验 disappearing highlight test
强光荧光灯 hot cathode
强光源 intense light source;powerful light source
强光泽 high luster[lustre]
强光泽瓷漆 full gloss enamel
强光泽的釉瓷饰面 specular enamel finish
强光泽乳胶瓷漆 full gloss latex enamel
强光泽涂料 high gloss paint
强光照明 high key light(ing)
强海流 strong current
强含水层 good aquifer
强函数 majorant(e);strong function
强夯 dropping weight; dynamic (al) compaction; heavy compaction; heavy tamping
强夯法 dynamic (al) consolidation; heavy drop-hammer compaction; heavy tamping; dynamic compaction
强夯法压实地基 dynamic(al) consolidation ground
强夯加固法 dynamic (al) consolidation method
强夯实法 dynamic (al) compaction method;dynamic (al) consolidation process
强夯置换法 dynamic(al) replacement
强耗热量 wall loss
强厚滑车 clump block
强弧(光)光灯 intensive arc lamp; kliegshine
强滑坡 bad slip
强化 consolidation;forcing;hardening constituent; intensification; intensify(ing); reinforcement; rigidization; ruggedization; ruggedize; seeding; strengthening
强化保险 reinforcement bumper
强化玻璃 hardened glass; strengthened glass; tempered glass; tempered plate glass;toughened glass
强化玻璃钢 glass-reinforced plastics
强化玻璃器皿 tempered glassware; toughened glass ware
强化采液泵 enhanced recovery pump
强化操作窑 hard driven kiln
强化层 strengthening layer
强化层积木材 densified laminated wood;compressed laminated wood
强化常规水处理工艺 enhanced conventional water treatment process
强化沉淀设备 compelled sedimentation equipment
强化除氨氮 enhanced ammonia nitrogen removal
强化除磷 enhanced phosphorus removal
强化磁通变动率继电器 field-forcing relay
强化刺激 reinforcement stimulus
强化的试验 intensified test
强化的硬质纤维板 tempered hardboard
强化地表水处理法规 enhanced surface water treatment rule
强化电场金属层 field-intensifying metal
强化丁苯胶乳 fortified styrene-butadiene[SB] latex
强化定理 strengthenization theorem
强化煅烧 forced kiln operation;forcing of kiln operation
强化煅烧窑 hard burning driven kiln; hard driven kiln
强化堆积结构 crescumulate texture
强化堆芯 seed core;spiked core
强化堆芯反应堆 seed core reactor
强化发动机 jazz the motor; uprated engine
强化发展计划(工作) re-enforcing development effort
强化反应堆 spiked reactor
强化分子臭氧反应法 enhanced molecular ozone reaction method
强化工艺 strengthening process
强化管 enhanced tube
强化规律 hardening rule
强化规模律 law of reinforcement size
强化过程系列 contingency of reinforcement
强化过滤 enhanced filtration; intensive filtration
强化过滤效果 enhanced filtration efficiency
强化还原降解 enhanced reductive degradation
强化环境管理政策 policy of tightening up environmental management
强化混凝 enhanced coagulation
强化混凝生物铁法 enhanced coagulation-bio-ferric process
强化活性污泥法 enhanced activated sludge process; enhanced activated sludge system; intensified activated sludge process; strengthening activated sludge process
强化活性污泥系统 intensified activated sludge system
强化机理 strengthening mechanism
强化剂 hardening agent;reinforcer
强化交通流 intensification of traffic movement
强化胶 fortified glue
强化胶合板 metallized plywood;reinforced plywood
强化阶段 strain-hardening range
强化绝缘 heavy insulation
强化开采 enhanced recovery; excessive production of; forced production;intensified mining
强化快速重力渗滤 enhanced rapid gravity filtration
强化冷却设备 intense cooling plant
强化冷却装置 intense cooling plant
强化理论 reinforcement theory
强化梁 consolidating beam
强化炉子的操作 push the heater
强化面包 fortified bread
强化农业 intensive farming
强化皮管 reinforced hose
强化企业管理 strengthening business management
强化曲流 enforced meander
强化群呼 enhanced group call
强化燃烧 overheavy firing; supplemental firing
强化燃烧室喷嘴 afterburner nozzle
强化烧结 intensified sintering
强化审计 tighten auditing
强化生物除磷 enhanced biological dephosphorization; enhanced biological phosphorus removal
强化生物除磷生物量 enhanced biological phosphorus removal biomass
强化生物处理 enhanced bio-treatment
强化生物降解 enhanced biodegradation
强化试验 acceptable test; intensity test;strenuous test
强化试验寿命 accelerated life
强化树脂 fortified resin
强化数据变换网络 augmented data manipulator
强化水旋澄清池 enhanced swirling clarifier
强化松香 fortified rosin
强化松香胶料 fortified rosin size
强化塑料 reinforced plastics
强化塑性砂浆 reinforced plastic mortar
强化酸 fortified acid
强化微电解法 strengthened microelectrolysis
强化温室 forcing house
强化稳定性 strengthening stability
强化物 fortifier
强化纤维 reinforcing fiber[fibre]
强化线芯 reinforced core
强化效应 strengthening effect
强化絮凝沉淀 enhanced flocculation deposition
强化悬浮预热器 reinforced suspension preheater
强化学风化层 loipon
强化学氧化剂 strong chemical oxidizing agent
强化窑操作 forced kiln operation
强化窑煅烧 force a kiln
强化窑工作 push a kiln
强化因子 intensifier
强化营养去除高级生物工艺 advanced biological enhanced nutrient removal process
强化硬木板 tempered board
强化与增韧 strengthening and toughening
强化重整 powerforming
强化桩 < 用以加固地基的短桩 > consolidating pile
强化作用 invigorating effect
强还原环境 strong reducing environment
强还原剂 strong reducing agent; strong reductant
强混合 strong mixing
强混合假设 strong mixing hypothesis
强混合性 strongly mixing;strong mixing property
强混合性自同构 strongly mixing automorphism
强火花 fat spark; good spark; hot spark
强火山作用 ultra-vulcanian
强积分 majorant for integral
强基 strong basis
强激波 intense shock wave; sharp shock;strong shock
强激光辐射 intense laser radiation
强激励 soaking
强加 obtrude
强加的变形 imposed deformation
强加函数 forcing function
强加荷载 imposed load
强加显示 forced display
强间断 strong discontinuity
强间断面 strong discontinuity surface
强剪弱弯 strong shear capacity and weak bending capacity
强碱 strong base
强碱的 alkaline
强碱度 strong basicity
强碱剂 alkaline agent
强碱水 strong liquor;strong lye
强碱瓮 strong vat
强碱型离子交换树脂 strong-base type ion exchanger
强碱性 strong basicity
强碱性离子交换树脂 strong basic ion-exchange resin
强碱性水 strong alkaline water
强碱性土 strongly alkaline soil
强碱性阴离子交换树脂 strong-base [basic] anion-exchange resin; lewatit M2
强碱液 strong liquor;strong lye
强碱阴离子 strong-base anion
强碱再生过程 Marx regeneration process
强碱中毒 corrosive alkaline poisoning
强碱中和值 strong-base number
强健的 vigorous
强键 excellent bond
强胶结 strong cementation
强胶结的 strongly cemented
强结合水 strongly bound water
强结合水连接 strong bound water bond
强介电性微晶玻璃 strong dielectric(al) glass-ceramics
强金属络合剂 strong metal-complexing agent
强劲 hardiness
强劲甲板 strength deck
强径流带 strong runoff zone
强径流带地质类型 geologic(al) type of strong runoff zone
强竞争吸附质 strongly competing adsorbate
强飓风 hurricane-force wind
强聚光 kicker light
强聚合油 long oil
强聚焦 strong focusing
强聚焦分光仪 high order focusing spectrometer
强聚焦轨道 strong focusing orbit
强聚焦回旋加速器 strong focusing cyclotron
强聚焦加速器 strong focusing accelerator
强聚焦同步加速器 strong focusing synchrotron
强聚焦透镜 strong focusing lens
强聚焦原理 strong focusing principle
强聚焦注流 well-focused beam
强聚束离子源 intensely bunched ion source
强均衡徐缓型 strong balanced slow type
强均衡迅速型 strong balanced rapid type
强抗力导线 cadmium copper wire
强抗张力导线 cadmium copper wire
强可测函数 strongly measurable function
强可测性 strong measurability
强可调红外源 intense tunable infrared source
强可微性 strong differentiability
强控制【数】 robust control
强(控制)函数 dominant function
强矿化水 strong mineralized water
强拉的 dead-drawn
强拉钢丝 dead-drawn wire
强拉力 high tension
强浪 < 海浪五级 > rough sea
强劳动力 able-bodied labo(u)rer
强离子束 intense ion beam
强力 brute force;double strength
强力爆破 heavy blasting
强力背板 stiffback;strong back
强力背材 strong back
强力玻璃 double-strength glass
强力场 strong force field

强力冲洗阶段 <下金刚石钻头时的> conditioning period
强力除草剂 diquat <一种接触性除草剂>
强力磁电机 heavy-duty magneto
强力粗切削 heavy roughing cut
强力醋酯纤维 strong acetate
强力刀柄 full back arbor
强力的臭氧消耗物质 potent ozone depletion substance
强力电机 powerful machine
强力镀层墙布 rexine
强力发动机 hopped-up engine
强力泛光灯 <剧院用> olivette
强力非塑化掺和机 intensive non-fluxing blender
强力干燥 forced drying
强力钢 strong steel
强力高频振动器 heavy high-frequency vibrator
强力构件 strength member
强力鼓风机 ensilage dump blower
强力焊缝 strength weld
强力夯实法 dynamic(al) consolidation
强力回程凸轮 main and return cam
强力货币 high-powered money
强力夹具 strong holding device
强力甲板 strength deck
强力胶合 unbreakable bond
强力搅拌 intense agitation;strong agitation
强力搅匀 <稀释时> breaking up
强力聚焦 brute force focussing
强力扩声器 stentorphone
强力雷达 brute force radar
强力类凝集素 proagglutinoid
强力利用率 strength efficiency
强力流 high flux
强力铆钉 power rivet;strength rivet
强力霉素废水 doxycycline wastewater
强力磨削 abrasive machining
强力拧紧(螺丝) power-tight
强力起爆 heavy initiation
强力起重机 goliath(crane)
强力起重设备 heavy lift equipment
强力千斤顶 beaver-tail
强力切削 heavy cut
强力切削工具 full back cutter
强力伸长曲线图 strength elongation diagram
强力伺服机构 power servomechanism
强力锁 heavy-duty lock
强力探照灯 floodlight projector
强力通风 forced draught;sharp draft
强力通风冷却 forced air cooling
强力吸气容积 forced inspiratory volume
强力铣床 rigid mill
强力纤维 strong fiber[fibre]
强力修枝剪 heavy-duty shears
强力旋压 power roll forming;power spinning;shear spinning
强力扬声器 loud-hailer
强力油泵 heavy duty pump oiler
强力噪声 forced noise
强力照明 floodlighting;prestige lighting
强力整流管 heavy-duty rectifier
强力制冷器 sharp freezer
强力自动记录计 autodynamograph
强连接 strong connection
强连接成分 strongly connected component
强连接区域 strongly connected region
强连接时序机 strongly connected sequential machine
强连接系统 strongly connected system
强连接自动机 strongly connected automaton
强连通的图 strongly connected graph
强连通复形 strongly connected complex
强连通性 strong connectedness
强连通有向图 strongly connected digraph;strongly connected direction graph
强连续半群 strongly continuous semigroup
强连续表示 strongly continuous representation
强连续随机过程 strongly continuous stochastic process
强连续映射 strongly continuous mapping
强梁 strength beam; strong beam; web beam
强梁弱撑 strong girder-weak brace
强烈 high intensity;violence
强烈暴雨 extraordinary storm; intense storm
强烈爆震 severe knock
强烈崩坍 climax avalanche; severe avalanche
强烈不可制止型 strong unrestrained type
强烈颤动 jerking motion
强烈沉降阶段 intense sinking stage
强烈程度 severity
强烈冲击 sharp pounding
强烈冲刷 extraordinary scour
强烈的 intensive;strong
强烈的转向运动 heavy turning movements
强烈地震 intense earthquake;moving ground excitation; severe earthquake;severe knock;strong earthquake; strong motion earthquake; very strong earthquake; violent earthquake
强烈地震区 intense earthquake zone
强烈地震事件 extreme earthquake event
强烈地震运动 extreme earthquake motion
强烈对照 strong relief
强烈反对 counterblast
强烈泛光灯 <剧院用> olivette
强烈沸腾钢 wild steel
强烈沸腾熔炼 wild heat
强烈风 storm gale
强烈风暴 severe storm;tearer;war of element
强烈风化的 much-weathered
强烈辐射 acute irradiation
强烈感染 storming
强烈滑坡 bad slip
强烈混合 intensive mixing
强烈活动阶段 strong activity stage
强烈搅拌 vigorous agitation;violent stirring
强烈阶段 intensive stage
强烈抗议 counterblast
强烈喷出的喷汽孔 strong fumarole
强烈起伏 highly relief
强烈侵蚀 active erosion;deep erosion
强烈燃烧 high-intensity combustion; strong deflagration
强烈上升区 intensive uplifting region
强烈湿陷 highly collapsible
强烈使用 hard usage
强烈收敛的 strongly convergent
强烈西风 <指南纬 40°~50°一带的强烈西向风> fresh zephyr
强烈西南风 sou's wester;southwester
强烈下降区 intensive depressing region
强烈性 intensive property
强烈油流 heavy oil flow
强烈运动 intensive movement
强烈造成污染的货物 pollution-intensive goods
强烈噪声 intensive noise
强烈照明 dazzle lighting
强烈照射 acute irradiation
强烈振动 judder;severe jolt
强烈震波图 strong motion seismogram
强烈震动 sharp pounding;strong motion earthquake;violent earthquake
强裂隙带 dense fissure zone
强裂隙通道 passage of heavy fracture
强淋溶土 leached brown earth
强流 copious current; high current; strong current
强流电子光学 high-density electron beam optics
强流接头 heavy current connector
强流微波加速器 high current microtron
强龙卷 intense tornado
强络合剂 strong compexing agent
强脉冲 flash;high-power pulse
强密度放射性废水 high-level radioactive waste
强密连接 tight joint
强摩擦性岩层 extremely abrasive ground
强逆定理 strong converse theorem
强黏[粘]胶泥 fat mortar
强黏[粘]结性煤 strong caking coal
强黏[粘]力砂 strong sand
强黏[粘]砂浆 fat mortar;rich mortar
强黏[粘]土 gumbo;strong clay
强黏[粘]性 strong viscidity
强黏[粘]性土 tenacious clay
强捻 heavy twist
强耦合 close-coupled;close coupling; close linkage;tight coupling
强耦合电路 overcoupled circuit
强耦合近似法 strong-coupling approximation
强耦合理论 strong-coupling theory
强耦合模型 strong-coupling model
强耦合系统 strongly coupled system
强耦合转动模型 strong-coupling rotational model
强膨胀熔融黏[粘]结 strong expansive fusion caking
强膨胀水泥 high-expansion cement
强平稳的 strongly stationary
强平稳过程 strictly stationary process
强平稳随机过程 strictly stationary random process; strong stationary stochastic process
强迫 compel;compulsion;forcing
强迫摆动 constrained oscillation; forced vibration
强迫变形 compelled deformation
强迫波动 forced oscillation
强迫掺气 forced aeration
强迫抽气通风 ventilation by forced draft
强迫磁致伸缩 forced magnetostriction
强迫的 coercionary; coercive; compulsory;forced;forcible
强迫动作 obsessive action
强迫对流 forced convection
强迫对流沸腾 forced convection boiling
强迫对流换热 forced convection heat transfer
强迫对流空气冷却器 forced convection cooler
强迫对流冷却 forced convection cooling
强迫反应 compulsive reaction;forced response
强迫方法改正桩位 forcible correction to pile
强迫风冷管 forced air cooling tube
强迫付款 forced payment
强迫复位 compulsory restoration
强迫干燥 force dry
强迫雇佣劳动 labo(u)r feather bedding
强迫雇用 feather bedding
强迫观念 compulsive idea
强迫换向 forced reversing
强迫降落 forced landing
强迫交变电流 forced alternating current
强迫缴税 lay under contribution
强迫接触 enforced contact
强迫解锁 forced release
强迫借贷 forced loan
强迫进风 force-in air
强迫就位阀 backseated valve
强迫空间激振 forced spatial excitation
强迫空气冷却 forced air cooling
强迫控制 forcing control
强迫馈电 constrained feed
强迫劳动 forced labo(u)r
强迫冷却 forced cooling
强迫力 coercive force
强迫立位 forced standing position
强迫励磁 automatic field forcing;excitation forcing;field-forcing
强迫纳税 compulsory assessment
强迫扭曲 impressed distortion
强迫扭转振动 forced torsional vibration
强迫配合 forcing fit
强迫频率 forced frequency; forcing frequency
强迫气冷管 forced air-cooled tube
强迫气流 forced draft
强迫迁移 forced migration
强迫清算 forced liquidation
强迫让道指令 forced-off command
强迫润滑拉模组合装置 forced lubricating die unit
强迫蛇曲 enforced meander
强迫式加热温度调节 temperature-modulated forced-air heating
强迫释放 forced release
强迫收敛法 forced convergence method
强迫手段 compulsory measure
强迫衰减振动 forced damped vibration
强迫水冷 forced water cooling
强迫水流 induced flow
强迫顺序多频通信[讯] compelled sequence multi-frequency code signal(l)ing
强迫送料的拉拔机 push pointer bench
强迫锁模 forced mode locking
强迫体位 forced position;forced posture
强迫调整 positive governing
强迫跳伞 forced jump
强迫停车 compulsory stop
强迫停机率 forced outage rate
强迫停止 involuntary stop
强迫通风 forced air (circulation); forced draught;forced ventilation
强迫通风炉箅 grate with forced draught
强迫同步系统 general locking
强迫退休 mandatory retirement
强迫歪曲 constrained distortion;impressed distortion
强迫弯曲波 forced flexural wave

强迫蜿曲 enforced meander
强迫蜿蜒 forced meandering
强迫涡流运动 forced vortex motion
强迫下沉（沉井）enforced settlement
强迫线圈 forcing coil
强迫响应 forced response
强迫谐振 forced harmonic motion; forced harmonic vibration
强迫卸货 forced discharge
强迫性计数 compulsive counting
强迫性检查 compulsive checking
强迫性人格 obsessive-compulsive personality
强迫循环 forced circulation
强迫循环系统 forced circulation system
强迫要求 gavel
强迫一致 forced congruence
强迫涌浪 forced surge
强迫油冷 forced-oil-cooling
强迫油冷变压器 forced-oil-cooling transformer
强迫油冷强迫风冷式冷却 forced-oil forced-air cool
强迫油冷式套管 forced-oil-cooled bushing
强迫油循环 forced-oil circulation
强迫运动 forced motion; forced movement
强迫债券 forced bond
强迫振 constrained vibration
强迫振荡 constrained oscillation; forced oscillation
强迫振动 forced vibration
强迫振动测试 forced vibration test
强迫振动频率 forced vibration frequency
强迫震荡 forced oscillation
强迫整流 forced commutation
强迫阻尼振动 forced damped vibration
强破裂区 highly fractured region
强气流 air blast; big wind
强气流布料器 air blast distributor
强气流防护 blast protection
强气流分布器 air blast distributor
强亲水的 strong hydrophilicity
强曲流 competent meander
强取 extort
强取的款项 exaction
强热 heat flash
强热带风暴 severe tropic(al) storm
强热失量 loss on ignition
强韧的（金属）tenacious
强韧细绳 cut line
强韧橡胶外皮电缆 tough rubber sheathed cable
强溶剂 good solvent; strong solvent
强熔结火山角砾岩 strong welded volcanic breccia
强熔结集块结构 strong welded agglomeratic texture
强熔结集块岩 strong welded agglomerate
强熔结角砾结构 strong welded breccia texture
强熔结凝灰结构 strong welded tuffaceous texture
强熔结凝灰岩 strong welded tuff
强融沉 strong melt-settlement
强入 trunk offering
强弱变光 undulating light
强弱连续 hemi-continuous
强弱振幅反射模式 high and low amplitude reflection mode
强弱指标 relative strength index
强弱作用 high low action
强色 intense colo(u)r; rich shade
强色调 strong shade
强上升流 strong upwelling; uprush

强蛇曲 competent meander
强伸度 strength and elongation
强渗期 carburizing period
强生物活性 intense bioactivity
强声学 macrosonics
强使 impose
强势货币 hard money
强收敛 strong convergence
强衰减的 heavily damped
强衰落 deep fade
强双氧水 perhydrol
强霜冻侵蚀 cryoplanation
强水流 heavy flow of water
强水硬性石灰 eminently hydraulic lime; strong hydraulic lime
强水跃 strong hydraulic jump; strong jump
强送风 muzzler
强塑性黏[粘]土 strong clay
强酸 strong acid; spirit of acids
强酸度 strong acidity
强酸浸出 strong acid leaching
强酸量 amount of strong acid
强酸-钠离子交换 strong acid-sodium ion exchange
强酸熟化浸出 strong acid cure leaching
强酸酸度 strong acid acidity
强酸塔 strong acid tower; strong tower
强酸型离子交换剂 strong acid type ion exchanger
强酸型阳离子交换树脂 strongly acidic cation-exchange
强酸性 strong acidity
强酸性的 highly acidic
强酸性离子交换树脂 strong acid ion exchange resin
强酸性溶液 strongly acidic solution
强酸性树脂 strong acid resin
强酸性水 strongly acidic water
强酸性阳离子交换树脂 strong acid cation exchange resin
强酸性阳离子交换柱 strong acidic cation exchange column
强酸性氧化土 acrox
强酸阳离子 strong acid cation
强酸阴离子 strong acid anion
强酸阴离子浓度 strong acid anion concentration
强酸中毒 corrosive acid poisoning
强酸中和值 strong acid number
强台风 <最大风力在 12 级以上> strong typhoon
强弹簧 strong spring
强天电干扰 heavy sferics disturbance
强透水层 strongly permeable stratum
强拓扑 strong topology
强纬向环流 high zonal recirculation
强紊流火焰 high-intensity turbulent flame
强稳定性 stiff stability
强污水 strong sewage
强西风环流 high zonal recirculation
强吸附 strong adsorption
强吸附剂 fresh adsorbent
强吸面 hungry
强吸收 strong absorption
强吸收介质 strongly absorbing medium
强吸收区 strongly absorbing region
强洗刷坡 strong-wash slope
强咸度 supersalinity
强咸水 super-saline water
强显示偏好公理 strong axiom of revealed preference
强相干背景 strong coherent background
强相关 strong correlation
强相互作用 strong interaction
强相互作用动力学 strong interaction dynamics
强相互作用理论 strong interaction theory
强相互作用模型 strong interaction model
强相互作用效应 strong interaction effect
强相容的 strongly consistent
强消毒产物生成势 high disinfection byproduct formation potential
强消毒（剂）vigorous disinfection
强挟放电 intense pinch discharge
强泻气流 down rush
强信号 high signal; large signal; strong signal
强信号检波 power detection
强信号检波器 high amplitude detector; power detector
强信号理论 strong-signal theory
强信号栅极检波 power grid detection
强信号噪扰 blanketing
强信号阻塞 strong signal blocking
强行的 forced
强行断电 forced interruption
强行换相信号 force off-phase signal
强行加入 barge in
强行进入 muscle
强行靠泊 forced approach and run alongside
强行控制 forcing control
强行励磁 automatic field forcing; high-speed excitation; quick-response excitation
强行励磁机 impact exciter; shock exciter
强行励磁装置 compelled excitation increasing device
强行留置权 involuntary lien
强行实现 bull
强行收回 actual eviction
强行送电 forced power transmission
强行推销 hard sell
强行推销的 high-pressure
强行推销性的广告宣传 admass
强行下管 snubbing piping; snubbing tubing
强行下管固定辅助钢绳支柱 snubbing post
强行下钻具 snubbing equipment
强行显示 forced display
强行征用 impress(ment)
强行置码 force
强型 strong pattern
强性变分 strong variation
强性极值 strong extremum
强性切力极限测 pachimeter Epokimito pace
强性相对极小 strong relative minima
强性褶皱 competent fold
强旋涡 strong eddy
强压给水式锅炉 uniflow boiler
强压锚具 ramming-fixture
强压式 pusher-type
强压送风机 forced draught fan
强压缩 strong compression
强压通风机 pusher-type fan
强岩层 competent bed; competent rock bed; competent rock stratum [复 strata]
强岩（石）competent rock
强研磨性地层 strong abrasive formation
强盐渍土 strongly salified soil
强阳离子交换 strong cation-exchange
强氧化剂 powerful oxidant
强氧化性 strong oxidizing property
强氧化性渣 strongly oxidizing slag; strong oxidation slag
强氧化自由基 strong oxidative free radical

强摇 rack
强异常 strong anomaly
强阴离子交换剂 strong anion exchanger
强荧光水域 highly fluorescing waters
强硬的 hard boiled
强涌 high swell; rough swell
强有力 heavy-duty
强有力的 powerful
强有力的支持层 strong supporting layer
强有向图 strong digraph
强余震 strong after shock
强闸函数 strong barrier function
强占 deforce; dispossession; usurp
强胀缩性土 strong swelling-shrinkage soil
强褶皱 competent folding
强褶皱构造 highly folded structure
强阵雨 heavy passing shower
强振荡 hard oscillation
强振荡形态 highly oscillatory mode
强振动 strong vibration
强振动声 judder
强振幅 black deflection
强振幅连续反射波模式 high amplitude continuous reflection mode
强振观测 strong vibration observation
强震 magaseism; strong (motion) earthquake; strong shock; violent earthquake; violent shock
强震波 strong earthquake wave
强震场 macroseismic field
强震持续时间 duration of strong shaking
强震传感器 strong earthquake sensor
强震带 strong earthquake belt; strong earthquake zone; pleistoseismic zone
强震地震学 macroseism seismology
强震反应谱 strong earthquake response spectrum
强震观测 macroseismic observation; strong earthquake measurement; strong motion earthquake instrumentation; strong motion measurement; strong motion observation
强震观测台网 strong motion observation network
强震观测台阵 ground motion array; strong motion observation array
强震观测委员会 <日本> Strong-Motion Earthquake Observation Council
强震宏观调查 macroseismic investigation of strong earthquake
强震活动性 macroseismicity
强震计 strong motion seismograph
强震记录 strong earthquake record
强震加速度计 strong motion accelerograph
强震加速度图 strong motion accelerogram
强震加速度仪 strong motion accelerograph
强震加速度仪台网 network of strong motion accelerograph
强震前兆 forerunner of strong earthquake
强震区 highly seismic region; macroseismic area; macroseismic zone; meizoseismal area
强震位移 strong displacement; strong motion displacement
强震效应 macroseismic effect
强震信号检索系统 strong-motion information retrieval system
强震学 strong motion seismology
强震遥测台网 strong earthquake tele-

metric network
强震仪 macroseismograph; strong motion seismograph
强震仪台阵 strong earthquake instrument array
强震震中 macroseismic epicenter[epicentre]
强震资料 macroseismic data
强征 exaction
强枝 strong shoot
强直性痉挛 tetanus
强指向传声器 rifle mike
强制 coerce; compelling; compulsion; constraint; enforce(ment); forcing; gavel; impel; pressure
强制 O 形环密封 forced O-ring seal
强制摆动 forced oscillation
强制保险 compulsory insurance; forced insurance; obligatory insurance
强制保养制度 compulsory maintenance system
强制崩顶 blowdown
强制崩落 forced caving; positive caving
强制比较 coercive comparison
强制闭锁装置 forced locking device
强制编码 forced coding
强制编码程序 forced coding program(me)
强制变异 impressed variation
强制表面波 forced surface wave
强制波 forced wave
强制剥离试验 forcible separation method
强制裁决法 forced decision
强制铲土的浅铲斗 positive-action low-bowl
强制潮汐波 forced tidal wave
强制(车辆从匝道)驶 enforced diversion
强制车流 enforce traffic flow; forced flow
强制成型焊接 enclosed welding
强制出口管制 mandatory restrictions on exports
强制出力 must run output
强制储蓄 compulsory saving; forced saving
强制储蓄金融机构 contract thrift institution
强制处理要求 mandatory treatment requirement
强制传动 positive drive
强制磁化 constrained magnetization; forced magnetization
强制磁化条件 constrained magnetization condition
强制磁致伸缩 forced magnetostriction
强制存款 forced deposit
强制打开 positive opening
强制怠速工况 forced idling mode
强制贷款 forced loan
强制的 coercive; compulsory; forced; forcible; mandatory
强制点火 forced ignition; positive ignition; spark ignition
强制点火发动机 positive ignition engine
强制电流 impressed current
强制电能 must run energy
强制定向排列 active orientation
强制对流 forced convection
强制对流传热 forced heat convention
强制对流沸腾 forced convection boiling
强制对流换热 forced convection heat transfer
强制对流加热 forced convection heating
强制对流加热器 forced convection heater
强制对流空气冷却器 forced convection cooler
强制对流冷却 forced convection cooling
强制对流汽化 forced convection vapo(u)rization
强制对流式烤箱 forced convection oven
强制对流退火窑 forced convection lehr
强制对流装置 forced convection unit
强制对应 coercive correspondence
强制对中 forced centering[centring]; positive centering[centring]
强制对中法 method of three-tripods
强制对中器 constrained centering [centring] device
强制对中装置 forced centering[centring] device
强制发信号 compelled signalling
强制法 brute force approach
强制反转 power reversing
强制风 forced air
强制风冷 air blast cooling
强制风冷式发动机 forcedly air-cooled engine
强制风冷油浸变压器 forced-air-cooled oil immersed transformer
强制符合 connection under constraint
强制附合 constrained annexation
强制附合条件 conditions for constrained annexation
强制干燥 forced dry(ing)
强制干燥挥发性漆 forced drying lacquer
强制干燥温度 forced drying temperature
强制给进 force feed
强制给进机构 <转盘钻机> pull-down equipment
强制耕作 compulsory cultivation
强制工作 forced working
强制供油润滑器 forced feed lubricator
强制供油系统 forced feed oil system
强制购买 compulsory purchase
强制鼓风 forced air blast; forced-in air
强制鼓热风炉 forced warm air furnace
强制固定中心点 positive centering
强制关闭 positive closing
强制管理数据系统 force administration data system
强制灌浆 grout intrusion
强制过程 forced process
强制过滤 forced filtration; mandatory filtration
强制过滤速度 forced filtration velocity
强制函数 coercive function; forcing function
强制和解 compulsory conciliation
强制和谐振动 forced harmonic vibration
强制横摇 forced rolling
强制换能器 forced transducer
强制换气 forced air change
强制换向 forced commutation; positive reversing
强制换向逻辑 forced commutation logging
强制火焰钻进 forced-flame drilling
强制减振 forced damped vibration
强制(桨叶)式搅拌 non-lift mixing action
强制接受业务 assigned risk
强制节约 forced frugality
强制结晶 forced crystallization
强制解散清理 compulsory winding-up
强制借款 compulsory loan
强制进给 forced feed
强制进料 forced feed; forced vortex; positive feed
强制进料系统 forced feed system
强制进气排气通风系统 pressurizing and ventilating-fan system
强制竞争 enforced competition
强制空气 forced air
强制空气加热 forced air heating
强制空气冷却 air blast cooling; forced air cooling
强制空气冷却的 fan-cooled
强制空气通风 forced air ventilation
强制空气循环 forced air circulation
强制控制调节 rationing
强制扩散 forced diffusion
强制劳动 forced labo(u)r
强制冷风降温 forced air cooling
强制冷却 artificial cooling; controlled cooling; forced cooling; pump cooling
强制冷却系统 pressure cooling system
强制力 coercive force; compelling force
强制连接 connection under constraint
强制联合形式 forcible form of union
强制联锁 positive locking
强制联锁式差速器 positive-locking differential
强制领港权 compulsory pilotage
强制流(动) compulsory flow; forced flow
强制流动加热器 induced-flow heater
强制流动液体色谱法 forced-flow liquid chromatography
强制流通 compulsory circulation; forced circulation
强制滤率 compulsory filtration rate; forcing filter rate
强制滤速 compulsory filtration rate; forcing filter rate
强制内接 compulsory inscribing
强制啮合 positive catch
强制啮合齿轮 forced gear engagement
强制啮合机构 positive catch
强制凝结沉降装置 compelled coagulation sedimentation equipment
强制拍卖 execution sale; forced auction; forced sale
强制排量 positive displacement
强制排料斗式提升机 positive-discharge bucket elevator
强制排流 compelled current drainage; forced current drainage; forced drainage
强制排流器 forced electric(al) drainage
强制排水 forced drainage
强制排水系统 forced drainage system
强制排烟 forced draught
强制偏心摇晃筛分机 positive throw eccentric type screen
强制破产 involuntary insolvency
强制气干 <木材的> forced air drying; fan-drying
强制气冷 forced air cooling
强制弃权 obligatory abstention
强制清算 enforced liquidation; force liquidation
强制驱动 positive drive
强制驱动电梯 positive drive lift
强制权 right of coercion
强制让道指令 force-off command
强制热风供暖 forced air heating; forced warm air heating
强制热散射 forced thermal scattering
强制热水供暖 forced hot water heating
强制热水供应 forced hot water heating
强制润滑(法) forced lubrication; forced feed lubrication; forced feed oiling; mechanical lubrication; pressure-feed lubrication; pump lubrication
强制刹车 positive stopping
强制审计 compulsory audit
强制驶出 enforced diversion
强制式拌和法 positive mixing
强制式拌和机 compulsory mixer; forced action mixer; forced mixing type mixer; stirring-type mixer; turbine-type mixer
强制式单管供暖 one-pipe forced heating
强制式风冷 forced air cooling
强制式供气 forced air supply
强制式混凝土拌和机 forced concrete mixer
强制式混凝土搅拌机 compulsion type concrete mixer; forced concrete mixer; pug concrete mixer
强制式混凝土搅拌设备 forced concrete mixing plant
强制式加燃料 forced fuel feed
强制式减速器 dynamic(al) retarder
强制式搅拌机 compulsory mixer; forced action mixer; forced mixing type mixer; pan mixer; stirring-type mixer; turbine-type mixer
强制式离合器 positive clutch
强制式起吊和降落 <起重机> power up/down
强制式通风 forced air supply; forced ventilation
强制式稳定土厂拌设备 forced stabilized soil mixing plant
强制式循环气冷系统 forced circulation air cooler
强制式循环系统 forced circulation system
强制式液压支腿 power hydraulic outrigger
强制水冷却 forced water cooling
强制水循环 forced water circulation
强制顺次多频信号 compelled sequence multi-frequency code signal(l)ing
强制司法解决 obligatory judicial settlement
强制送电 forced line charger
强制送风 forced air supply
强制送风采暖 forced air heating
强制锁闭 positive locking
强制摊派的公债 forced loan
强制剃齿 brake shaving
强制条件 constraining condition; constraint condition; non-geometric(al) condition
强制条款 compulsory clause
强制调谐 pulling into tune
强制停车 compulsory stop
强制停机 positive shutdown
强制通风 blast air; forced air change; forced draft; force dry; forced ventilating; induced draught; mechanical ventilation; plenum[复 plenums/plana]; positive draft; positive ventilating; positive ventilation; sharp draft; ventilated by forced draught
强制通风的冻结装置 air blast freezer
强制通风电动机 forced ventilation motor; force-ventilated motor
强制通风电源转换器 inverter for pow-

Q

er supply for emergency ventilation
强制通风冻结 air blast freezing
强制通风风扇 induced-draft fan
强制通风管道 forced draft duct
强制通风锅炉 forced draft boiler
强制通风机 forced draft fan
强制通风空气冷却器 forced draught air cooler
强制通风冷却 forced draft cooling; forced draught cooling
强制通风冷却器 forced air cooler
强制通风冷却塔 forced draft(type) cooling tower; forced draught cooling tower
强制通风冷却系统 force-draft cooling system
强制通风冷却装置 forced draught cooling arrangement
强制通风凉水塔 forced draft cooling tower
强制通风炉口 forced draft front
强制通风燃烧法 down-draft combustion process
强制通风燃烧器 forced draft burner; induced draught burner
强制通风式 forced draught type
强制通风式电动机 forced-ventilated motor
强制通风式干燥机 forced air dehydrator
强制通风式冷凝器 forced draught condenser
强制通风式热风炉 forced air furnace
强制通风双流式冷却塔 double flow induced-draft cooling tower
强制通风水冷却塔 forced draft water-cooling tower
强制通风脱气塔 forced draft degasifier
强制通风系统 induced draught system
强制通风制冷 forced draught cooling
强制同步 forced synchronization; forced synchronizing; genlocking
强制同步电路 lock-in circuit; lock-on circuit
强制同步系统 general locking
强制同步自保持电路 lock-on circuit
强制同步自动跟踪电路 lock-on circuit
强制退解 forced unwinding
强制脱钩 forced release
强制脱扣 forced release
强制喂料 positive feed
强制喂料机 brute force feeder
强制涡流 forced vortex
强制下降 positive lowering
强制显示 forced display
强制销售 compulsory sale
强制卸货 compulsory discharge; compulsory unloading
强制卸料 positive discharge
强制卸土 positive discharge
强制卸载 positive discharge
强制行动 coercive action; enforcement action
强制型混凝土搅拌机 compulsory type concrete mixer
强制型曲流 unfree meander
强制性 coerciveness; coercivity; enforceability
强制性保留 peremptory reservation
强制性标准 mandatory standard
强制性操作 imperative operation
强制性处理过程 mandatory treatment process
强制性措施 compulsory measure
强制性的 compulsory; imperative; mandatory; obligatory
强制性对流 forced convection
强制性工会会员资格 compulsory union membership
强制性规定 coercive regulation; mandatory provision
强制性规范 peremptory norm
强制性规费 compulsory fee
强制性规则 mandatory rule; obligatory rule
强制性交通流 forced traffic flow
强制性开支 mandatory expenditures
强制性空气循环加热取暖系统 forced warm-air heating system
强制性控制 coercive control; mandatory control
强制性流通 coercive circulation
强制性拍卖 compulsory sales by auction
强制性破产清理 involuntary bankruptcy
强制性破坏 forcing failure
强制性迁移 compulsory relocation
强制性牵引 positive traction
强制性认证 compulsory certificate
强制性守则 mandatory rule
强制性条款 obligatory term
强制性通风 forced ventilation
强制性挖掘 power digging
强制性项目 mandatory particular
强制性许可 compulsory licensing
强制性要求 mandatory requirement
强制性要求居民将垃圾分类以便回收 mandatory recycling
强制性褶皱变形 forced folding
强制性振动 forced vibration
强制性支出 mandatory expenditures
强制性支付 compulsory payment; enforce payment; mandatory expenditures
强制性执行 compel enforcement; forcible execution; compulsory execution
强制性指示标记 mandatory instruction sign
强制性制裁 compulsory arbitration; compulsory sanction; mandatory arbitration; mandatory sanction
强制性转让 involuntary transfer
强制性准备金 compulsory reserve
强制旋摆式筛分机 positive circle throw type screen
强制旋摆运动 positive circle throw gyratory movement
强制旋回运动 positive circle throw gyratory movement
强制漩涡 forced vortex
强制循环 forced assisted circulation; controlled circulation; induced circulation; positive circulation; pump circulation
强制循环的热处理炉 forced circulation furnace
强制循环干燥窑 forced circulation kiln; forced draft kiln; forced draught kiln
强制循环锅炉 assisted circulation boiler; forced circulation boiler
强制循环空气加热盘管 forced circulation air heating coil
强制循环空气冷却器 force circulation air cooler
强制循环冷却 forced circulation cooling
强制循环汽化冷却 pressure-type circulation evapo(u)rative cooling
强制循环热水供暖 forced circulation hot water heating
强制循环润滑 self-contained circulation lubrication
强制循环润滑系统 pressure circulation lubricating system
强制循环式干燥窑 forced recirculating dry kiln
强制循环式蒸发器 forced circulation type evapo(u)rator
强制循环水加热系统 forced circulation water heating system
强制循环调风器 forced circulation register
强制循环系统 forced circulation system
强制循环再沸器 forced circulating reboiler
强制循环蒸发 forced circulation evapo(u)ration
强制循环蒸发器 forced circulation evapo(u)rator
强制循环蒸汽发生器 forced circulation steam generator
强制压尖 push pointing
强制要点 compulsory points
强制引风机 forced draft fan
强制引航 compulsory pilot
强制引水 compulsory pilot
强制油冷却 forced-oil-cooled
强制油循环 forced-oil-air circulation; forced-oil circulation
强制油循环吹风冷却 forced-oil-air cooling
强制油循环式变压器 forced-oil transformer
强制语句 imperative statement
强制运动 constrained motion; forced movement; positive motion
强制运行 forced service
强制再循环 forced recirculation
强制噪声 forced noise
强制增压 positive charge
强制振荡 forced oscillation; forced vibration
强制振动 constrained vibration; forced oscillation; forced vibration
强制振动捣实法 forced vibratory compaction
强制振动法 forced vibration method
强制振动扭摆 forced vibration torsion pendulum
强制振动器 forced vibrator
强制整流 artificial commutation
强制制动 positive stopping
强制仲裁 compulsory arbitration; mandatory arbitration; obligatory arbitration
强制仲裁解决 compulsory arbitration settlement
强置性组构 imposed fabric
强重比 ratio of strength to weight; strength-weight ratio
强重叠共振 strongly overlapping resonance
强重正化效应 strong renormalization effect
强轴 major axis; strong axis
强柱 strong column
强柱弱梁 strong column and weak beam
强子【物】 hadron
强阻尼 heavy damping
强阻尼的 heavily damped
强阻尼电路 heavily damped circuit

墙 wall(ing)

墙鞍 wall saddle
墙凹槽 wall recess
墙凹口 wall pocket
墙凹台 scarcement
墙凹凸部 scarcement
墙板 building board; gerwood; panel(l)ing; panel plate; plate for walls; sheathing; siding; trab; wall(ing) board; wall(ing) panel; wall(ing) plank; wall(ing) slab; spandrel <相邻柱或窗之间的>; wall plate
墙板安装 panel erection; wallboard installation
墙板锤 wallboard hammer
墙板的可互换性 panel interchangeability
墙板垫木 panel furring
墙板钉 wallboard nail
墙板分隔线条 applied mo(u)lding
墙板缝上盖条 splat
墙板构架 panelled framing
墙板护板 filling-in panel
墙板架 panel frame
墙板架设 panel erection
墙板角护条 wallboard corner bead
墙板接缝黏[粘]带 sparked tape
墙板接合处的覆盖层 scrim; scrimp
墙板结构 slab-wall structure
墙板结构系统 panel structural system
墙板立面 panel facade
墙板联结件 panel-to-panel connector
墙板面层 face ply
墙板模 panel mo(u)ld
墙板模板的固定夹具 panel clamp
墙板设计 pan design
墙板施工 panel construction
墙板施工方法 panel construction method
墙板施工系统 panel construction system
墙板体系 panel wall system
墙板外貌 panel facade
墙板圬工墙 panel masonry wall
墙保温衬 wall insulation lining
墙背 back of a wall
墙背衬 backing; backing of wall
墙背后排水 back drain
墙背面竖砌砖的墙 rowlock arch
墙背摩擦角 angle of wall friction
墙背摩擦力 wall friction
墙背排水(管) back drain(age)
墙背排水设施 back drain(age); back drain of wall
墙背土压力 back pressure
墙背有顶撑的钢板桩围堰 internally braced sheet pile cofferdam
墙箅子 wall grill(e)
墙壁 mural; wall
墙壁背景 mural background
墙壁变形 wall deformation
墙壁插头 wall outlet
墙壁插座 wall socket
墙壁潮湿度 wall moisture
墙壁衬板 wall lining board
墙壁衬里材料 wall lining material
墙壁穿通 wall breakthrough
墙壁传热 heat-transfer through walls
墙壁(打)底漆 wall primer
墙壁大理石 wall marble
墙壁的 mural
墙壁的凸出部 set-off
墙壁电缆 block cable
墙壁墩子 pilaster
墙壁反射 wall reflection
墙壁防潮层 wall damp-proof course; wall vapor barrier
墙壁防湿层 wall vapor barrier
墙壁防蒸汽层 wall vapor barrier
墙壁扶垛 erisma
墙壁浮雕 wall relief
墙壁辐射采暖 wall-panel heating
墙壁辐射供暖 wall-panel heating
墙壁辐射管 wall radiant tube
墙壁高炉 hearth furnace
墙壁隔声 sound insulation of walls
墙壁厚度 wall thickness
墙壁护板 wall cladding panel

墙壁花园 drywall garden
墙壁灰泥混合物 wall plaster mix(ture)
墙壁混合灰泥 wall mixed plaster
墙壁尖刺 cheval-de-frise
墙壁间(距)的 intermural
墙壁建筑材料 wall material
墙壁建筑构件 wall building component; wall building member; wall building unit
墙壁接齿缝 racking back
墙壁开裂 cracking of walls
墙壁拉毛粉刷 wall stucco
墙壁连接 wall junction
墙壁龙头 wall tap
墙壁绿化 wall greening
墙壁马赛克 wall mosaic
墙壁面层 wall facing
墙壁面积 wall area;wall space
墙壁内灯 built-in lamp
墙壁内电线管 wall wiring conduit
墙壁内装设电线 wall wiring
墙壁平柱 piedroit
墙壁漆 wall coating;wall paint
墙壁砌块 building block
墙壁砌块制造机 block machine
墙壁强度 wall strength
墙壁热绝缘 wall heat insulation
墙壁热损失 heat loss through walls; wall loss
墙壁溶蚀 wall corrosion
墙壁上的 mural
墙壁声导纳 wall admittance
墙壁声阻抗 wall impedance
墙壁湿度 wall humidity
墙壁石板 wall slate
墙壁式 wall type
墙壁式厚度 wall-type thickness
墙壁式消防栓 wall-type hydrant
墙壁式消火栓 wall-type hydrant
墙壁水泥抹面 hardwall plaster
墙壁填料 wall filler
墙壁挑檐 wall cornice
墙壁凸出部分 set-off
墙壁突出底座 canting strip
墙壁图案 wall pattern
墙壁涂层 wall coat
墙壁涂料 wall coating; wall paint; Permoglaze <一种透气防水的>
墙壁温度 wall temperature
墙壁稳定性 wall stability
墙壁屋顶粉面 false panel
墙壁下部 <装饰与上部不同的> wainscot
墙壁纤维盖板 wall fiberboard sheathing
墙壁斜面 splay of a wall
墙壁修理工 waller
墙壁悬挂式挂毯 wall hanging type tapestry
墙壁用金属箔 wall foil
墙壁油画 mural painting
墙壁油漆工 wall painter
墙壁油毡 wall lino(leum)
墙壁凿缝器 wall channeler
墙壁照明 wall lighting
墙壁照明表 wall lighting table
墙壁照明灯具 wall light fitting
墙壁照明装置 wall light fitting
墙壁支柱 perpeyn wall
墙壁柱 mid-wall column
墙壁装饰 hang; wall decoration; wall enrichment
墙壁装饰板 wall cladding panel
墙壁装饰件 wall decorative fixture
墙壁装饰术 wall ornamentation
墙边保护勒脚 wall-protection kerb
墙边插座 power point
墙边卧材 girt strip
墙边斜隅石 squint quoin of wall
墙表层 veneer of wall
墙表面 wall surface
墙表面处理 wall surfacing
墙表面加工 wall surfacing
墙部件 corner element
墙材 walling
墙槽 wall chase
墙槽沟 wall channel
墙草 wall pellitory
墙插座 flush receptacle;wall plug
墙衬 wall lining
墙衬纤维板 wall fiberboard sheathing
墙撑 springing wall;flying shore <墙间的,用于房屋修理等>
墙撑杆 wall clamp
墙撑脚手架 wall jack scaffold
墙撑柱撑垫 footing piece;sole piece
墙承重 wall-bearing
墙承重建筑 wall-bearing construction
墙承重结构 wall-bearing construction;wall-bearing structure
墙窗断面 wall and window section
墙粗糙度 wall roughness
墙带 veneer tie
墙挡板 head guard;set flashing
墙的凹凸面 jog
墙的保护措施 wall guard
墙的垂直断面 wythe
墙的钝形转角石块 obtuse quoin of (masonry) wall
墙的带形基础 wall footing
墙的钢拉杆 metal wall tie
墙的构架 wall framing
墙的后面 back of wall
墙的基础板 sole plate
墙的建造 wall erection
墙的浇筑 wall pouring
墙的结构 wall composition
墙的截面积模量 section modulus of wall
墙的坡度 wall slope
墙的伸缩缝 slip joint
墙的施工型式 wall construction type
墙的收缩缝 slip joint
墙的受压面 back of wall
墙的水泥抹面 hard plaster
墙的顺砖层 stretcher course
墙的填充(物) wall stopping
墙的通道 <墙侧的一个临时通道> wall run
墙的凸出部分 attachment to a wall; set-off
墙的凸出物 attachment to a wall;set-off
墙的凸面压顶 convex coping of a wall
墙的外板 wall sheathing
墙的斜率 slope of a wall
墙的有效高度 effective wall height
墙的有效厚度 effective thickness of wall
墙的圆形压顶 bahut
墙的造型 wall-forming
墙的总厚度 total wall thickness
墙的组成 wall composition
墙灯 wall bracket lamp
墙底部排水口 scupper
墙底脚 footing of wall
墙顶 wall crest;wall top;top of wall; wall coping
墙顶半圆形转角 lunette
墙顶藏灯凹槽 coving
墙顶承梁板 wall plate
墙顶窗 eyebrow window
墙顶挡板 wall coping plate
墙顶的椽木垫板 rafter plate
墙顶垫板 wall plate;top plate
墙顶垫木 raising piece
墙顶盖瓦 weather tile
墙顶花池 wall garden
墙顶混凝土块体 concrete coping block
墙顶尖刺 cheval-de-frise
墙顶角饰 sprung mo(u)lding
墙顶路 vamure
墙顶内灯 built-in lamp
墙顶砌块 cap block
墙顶石 cornice;cap stone
墙顶石盖板 <防天气侵蚀的> weather slating
墙顶挑出的凸带 lorymer
墙顶以上有荷载的挡土墙 surcharged wall
墙顶预留格栅支托 wall hanger
墙洞 loophole; wall aperture; wall cavity;wall hollow
墙洞口 wall opening
墙洞上的肘钉 reveal pin
墙洞栅 wall grill(e)
墙堵拱 surface arcade
墙端 <在壁端柱终止的> parastas
墙端短梁 tailpiece
墙端短托梁 tail trimmer
墙端上的装饰 end ornament
墙墩 antal;perpeyn wall;pier;speroni;wall column;wall pier
墙墩柱头 wall pier capital
墙垛 attached pier; bridge pier; buttress;pier;return wall
墙粉 calcimine;distemper;plaster
墙缝 fullering;wall joint
墙缝白漆线条 pencil(l)ing
墙缝盖板条 cover-up
墙附着力 wall adhesion
墙盖顶 wall capping
墙干燥后浮垢 drier scum
墙干燥器 wall dryer[drier]
墙高 height of wall;wall height
墙搁板 wall shelf
墙隔板 wall separating board
墙工槌 club hammer;lump hammer
墙工锤 walling hammer
墙工用锤 waller's hammer
墙拱 arc formeret; scoinson arch; wall arch
墙拱斜肋 diagonal rib of wall-arch
墙勾缝机 wall-pointing machine
墙钩 crampet; wall anchor; thumb screw <使铅板贴到墙面上>;wall hook
墙构件 wall element;wall unit
墙构筑法 wall-panel system
墙箍 tie iron;wall tie
墙箍扎 wall tying
墙刮刀 wall scraper
墙挂壁橱 wall-hung closet;wall-hung toilet
墙挂冲水箱 wall-hung water closet
墙挂绞车 overhung wall winch
墙挂梁座 wall hanger
墙挂石板瓦 weather slating
墙挂式 wall-hung shape; wall-hung type
墙挂式冲落式小便器 wall-hung wash down urinal
墙挂式锅炉 wall-hung boiler
墙挂式交换机 wall-type switchboard
墙挂式莲蓬头 wall shower
墙挂式淋浴器 wall shower
墙挂式器具 wall-hung fixtures
墙挂式散热器 wall radiator
墙挂式设备 wall-hung fixtures
墙挂式卫生间设备 wall-hung lavatory
墙挂式洗脸盆 wall-hung lavatory
墙挂式洗脸器 wall-hung lavatory
墙挂式小便器 wall-hung urinal
墙挂毯 wall carpet
墙挂碗橱 wall-hung cupboard
墙管 wall coil
墙轨 wall rail
墙合 brick seat
墙和最近的梁(或柱列)之间的空间 tail bay
墙横断面 wall cross-section
墙后回填式岸壁 backfilled bulkhead
墙厚度 wall thickness
墙厚度计算 calculation of wall thicknesses
墙护板 stop foot
墙花格 wall tracery
墙环梁 <支承圆顶的> wall ring
墙灰泥装饰工程 pargeting
墙混凝土浇筑 wall concreting
墙机 wall(-type) telephone set
墙基 base course;base of wall;bench table;wall base;wall bed
墙基础 foundation of wall;wall footing;wall foundation
墙基(础)层 plinth course
墙基处理 footing dressing
墙基夯实机 trench compactor
墙基厚度 base thickness of wall
墙基脚 lodgment table; footing of wall;wall footing
墙基深度 base depth of wall
墙基石 head stone;plinth course
墙基座 surbase
墙脊竖砌砖墙 rowlock-back
墙加固板 straight jacket
墙架 ledge
墙架立柱 pendant post;pendent post
墙间的 intermural
墙间空调室 air-conditioner room
墙间柱 wall column
墙交叉 crossing of walls;wall crossing
墙角 corner; dado base; lead; wall corner
墙角板 corner board
墙角半砖 angle closer
墙角(包角)护板 corner bead
墙角壁炉 angle fire place
墙角椽子 corner rafter
墙角顶砖 corner header
墙角墩 anta;antae
墙角墩前列柱神庙 amphi-antis temple
墙角方线条 square staff
墙角防护短石柱 hurter
墙角防护桩石 hurter
墙角敷带工具 tape corner tool
墙角扶壁 corner buttress
墙角刮刀 corner slicker
墙角盥洗盆 corner lavatory basin
墙角柜 corner cabinet;encoignure
墙角护板 wall corner guard; corner board
墙角护条 angle bead; corner bead(er); corner guard; plaster head; plaster staff
墙角护条紧固凸缘 beaded edge
墙角护圆扁条 beaded flat
墙角基脚砌块 skirting block
墙角基石 head stone;corner stone
墙角加强件 corner reinforcement
墙角家具 corner furniture
墙角金属线脚 metal corner bead
墙角砍砖 angle closer
墙角拉筋 angle bond
墙角勒角砖 corner plinth brick
墙角连接板 corner link plate
墙角连接件 angle bond
墙角门框砖 corner door frame brick
墙角面板 corner facing slab
墙角面砖 corner facing tile
墙角抹灰工具 corner tool
墙角木椽 angle staff
墙角砌法 angle bond;quoin bonding
墙角砌合 corner bond;quoin bonding
墙角砌合石 corner bondstone
墙角砌合砖 corner bonder

墙角散热器 corner radiator
墙角石 coign(e);quoin;quoin stone;corner stone
墙角式散热器 corner radiator
墙角饰条 base shoe corner
墙角挑檐 dragon piece
墙角突头砖 bull header
墙角瓦 corner tile
墙角洗涤盆 corner sink
墙角洗脸盆 corner wash basin
墙角线脚夹 corner bead clip
墙角线条 corner bead
墙角用空心砌块 corner(return) block
墙角用空心砖 <一端和两侧均为实心外面> corner block
墙角浴盆 corner bath
墙角圆饰条 bull-nose
墙角圆线条 corner beader
墙角支墩 corner buttress
墙角柱 corner column
墙角砖 coign brick;edge brick;quoin brick;corner block
墙脚 feet;patten;wall footing;walls below grade
墙脚板 baseboard;sole or sole plate;wash board
墙脚板电(加)热器 electric(al) baseboard heater
墙脚板供暖装置 base heating installation
墙脚处理 footing dressing
墙脚散热器 baseboard radiator;skirting radiator
墙脚水准标志 benchmark on wall
墙接 marrying
墙接合砖 wall abutment tile
墙结合木 bonded timber;bond timber
墙界线 wall line
墙筋 wall stud
墙筋板条 battening
墙筋交错布置的隔墙 staggered partition
墙筋锚固件 stud anchor
墙筋抹泥 stud and mud
墙劲性 wall stiffness
墙镜 wall mirror
墙卡子 wall clamp
墙孔 porthole;wall hole
墙孔上部的线脚 head mo(u)ld(ing)
墙拉杆 wall clamp;tie iron;wall tie
墙栏杆 wall rail
墙勒脚外角砌合 angle bond
墙肋 arc formeret <拱结构中的>;wall rib
墙肋梁 wall rib
墙连接 crossing of walls
墙梁 summer;wall beam;wall girder;wall string(er)
墙两边的阶梯 stile
墙裂缝 wall crack
墙裂开 cracking of walls
墙留齿缝 racking back
墙楼梯斜梁 wall string(er)
墙螺栓 wall screw
墙螺丝 wall screw
墙锚 joist anchor;wall anchor
墙锚固钢板 joist anchor steel plate
墙锚件 wall hanger
墙锚栓 wall anchor;wall screw
墙帽 capping stone;cope(stone);coping(stone);crest table;starling coping;tabling;wall coping
墙帽轮廓 profiled coping
墙帽斜线脚 skew fillet
墙面 wall face;wall space
墙面 V 形勾缝 mason's V-joint pointing
墙面凹凸不齐的砌砖方式 scintled brickwork
墙面板 cladding slab;clapboard;faceplate;panel board;shingle;shingle nail;shingle panel;wall shingle;weather-board(ing);wall board;siding shake <手工劈成的>
墙面板的固定板条 gripfase
墙面板卡规 clapboard ga(u)ge
墙面板扭曲 tin-canning
墙面板起拱 tin-canning
墙面保护 wall guard
墙面玻璃 structural glass
墙面不平 bulge
墙面材料 cladding material;wall covering;wall-facing material;wall surfacing material
墙面瓷砖 furring tile
墙面瓷砖制造厂 wall-facing tile factory
墙面粗琢饰面机 wall scabbler
墙面打毛 hacking off
墙面挡板 splash board
墙面吊瓦 tile hung wall
墙面定斜杆 batter stick
墙面泛光灯具 wall-wash luminaire
墙面泛光照明 wall-washing
墙面泛水 wall flashing
墙面粉饰 wall plaster
墙面粉刷 wall plaster
墙面粉刷用的熟石灰物 finishing hydrate
墙面风化起霜 whiskering
墙面风口 wall register
墙面辐射供暖 radiant wall heating
墙面覆盖 vertical cladding;wall cover
墙面覆盖板 wall-facing board
墙面覆盖材料 cladding element;wall covering material
墙面覆盖层 wall cladding
墙面覆盖层黏[粘]结剂 wall cladding adhesive;wall cladding cementing agent
墙面覆料 wall covering
墙面缸砖 wall-facing quarry tile
墙面格间 wall panel
墙面挂瓦 tile hanging;weather tiling
墙面广告 wall sign
墙面柜 wall cabinet;wall-hung cabinet
墙面花圃 wall garden
墙面划痕 plumb scratch
墙面灰浆缝 wall joint
墙面混合材料 wall surfacing compound
墙面积指数 wall area index
墙面及天棚快速吸声材料 Akoustolith tile
墙面加撒豆石工 pebble dashing
墙面假联拱 surface arcade
墙面交接线 neat line;neatline of wall
墙面接搓突石 tusses
墙面金属连接件 metal wall tie
墙面开关 surface switch
墙面刻痕 clouring;picking
墙面冷水 wall flashing
墙面亮度 wall luminance
墙面龙骨卡 wall furring base clip
墙面门碰头 wall-type door stop
墙面摩擦 wall friction
墙面摩擦角 angle of wall friction;wall friction angle
墙面摩擦力 wall friction
墙面摩阻角 angle of wall friction
墙面磨光机 wall grinder
墙面抹灰 wall plastering
墙面抹灰整平板 wall scraper blade
墙面抹面 finishing of wall
墙面木板 weather shingling
墙面黏[粘]贴料 wallcovering
墙面黏[粘]贴料支持层 wallcovering support
墙面黏[粘]土砖块 wall tile clay body
墙面喷水器 <消防用> sidewall sampler
墙面平镶连接 wall mounting flush
墙面铺板 wall surfacing board
墙面铺石板 slate hanging
墙面齐平门 gib door;jib door
墙面取暖器 wall heater
墙面扇轮 spheric(al) sector
墙面上被腐蚀层 incrustation
墙面上吊挂的石板 hung slating
墙面上吊挂的瓦片 hung tiling
墙面上外面的电线套管 surface raceway
墙面石膏盖板 wall gypsum sheathing
墙面刷洗 wall-washing
墙面顺砖砌合 outbond;outbond of wall
墙面塑料粉刷 plastic wall plaster
墙面挑出铁件 tailing iron
墙面贴砖 wall tiling[tile]
墙面凸出部分 ressaut
墙面凸起连接 wall mounting-projected
墙面突出部分 ressaut
墙面涂层 wall covering
墙面涂料 wall coating;wall covering
墙面涂釉层 glazed wall coat(ing)
墙面位置线 wall surface line
墙面吸声系数 wall sound absorption coefficient
墙面吸收系数 wall absorption coefficient
墙面镶嵌装饰 opus musivum
墙面小块石板 <装饰用> small ashlar
墙面小柱 wall shaft
墙面斜凹槽 skew notch on wall
墙面斜交砖 skintled
墙面修饰 wall surface finish
墙面牙石 tusks;tusses
墙面颜料 distemper
墙面饮水器 wall drinking fountain
墙面用印刷品装饰的房间 print room
墙面油漆前糊布 cloth hanging
墙面有色抹灰粉饰 terra nova wall plaster
墙面照明装置 wall-wash luminaire
墙面支承穹肋的小柱 wall shaft
墙面纸 garnet paper
墙面终饰 wall finish
墙面重饰 wall finish
墙面抓毛 plumb scratch
墙面砖 facing tile;finish tile;wall-facing tile
墙面砖胶黏[粘]剂 wall tile cementing agent
墙面砖块 wall tile body
墙面砖黏[粘]结剂 wall tile adhesive;wall tile bonding adhesive
墙面砖镶铺工作 wall tilework
墙面装饰 ornamentation on wall;placage;wall decoration
墙面装饰釉砖 wall tile
墙面装修 finishing of wall;wall finish
墙面准线 neat line
墙面最后粉饰 wall surface finish
墙面最后涂刷 wall surface finish
墙模板 wall form(work);wall shuttering
墙模定位拉结杆 wall spacer
墙模工程 wall formwork
墙模壳 wall shuttering
墙模型 wall sample
墙摩擦系数 coefficient of wall friction
墙木栓 wall dowel
墙幕 dossal
墙内暗拱 back arch
墙内暗线 wall wiring
墙内承梁板 raising plate
墙内灯 built-in lamp
墙内电器盒 beam box;wall box;wall frame
墙内丁砖 blind header
墙内风道 wall duct
墙内风门 ventilating damper in wall
墙内管线槽 wall chase
墙内厚纸板隔层 paper sheathing
墙内加热器 wall heater
墙内空气扩散管 wall air diffuser
墙内空气扩散器 wall air diffuser
墙内梁座 <墙上安装梁的凹穴> pocket in wall;wall pocket
墙内锚碇装置 wall anchor
墙内锚固 wall anchorage
墙内木砖 anchor block;inside stop <固定门窗的>
墙内木桩 inside stop
墙内暖气设备 recessed wall heater
墙内取暖器 wall heater
墙内取暖设备 recessed wall heater
墙内散热器 built-in radiator
墙内竖管 wall stack
墙内(水电)管道 wall conduit
墙内通风道空心砖 air brick
墙内柱 wall column
墙内座梁 pocket in wall;wall pocket
墙黏[粘]着力 wall adhesion
墙排管 wall coil;wall grid
墙盘管 wall coil;wall grid
墙皮线 wall line
墙剖面 wall section
墙漆 wall paint
墙砌到规定高度 carry up to regulated height
墙砌块 wall(building) block
墙砌体 walling
墙砌型钢 tailing iron
墙嵌梁座 wall hanger
墙腔板条 cavity batten
墙腔木板条 cavity batten
墙倾倒残渣盆 wall slope bowl
墙裙 dado;main wall panel;tabulatum;wainscot;wall-protection kerb
墙裙板 subbase;wainscot(t)ing
墙裙板框 dado framing
墙裙缝 dado joint
墙裙帽 dado cap(ping)
墙裙木条 dado rail
墙裙下部 dado base
墙裙线脚 breast mo(u)lding;dado mo(u)lding
墙裙压顶线 dado cap(ping);surbase
墙裙座 foot block
墙上安置 wall mounting
墙上安装 wall mounting
墙上安装梁的凹穴 wall box
墙上暗榫 wall dowel
墙上凹槽 raglet
墙上凹处 scarcement
墙上凹进式烟灰盒 recessed wall urn ash tray
墙上凹口 raglet
墙上凹面 ingoing
墙上板条 wall furring
墙上标志 wall sign
墙上玻璃窗 wall glazing
墙上槽口 raggle;reglet
墙上插口 wall socket
墙上插头 wall plug;Plug-it <商品名>
墙上插座 switch plug;wall outlet;wall rosette;wall socket
墙上撑头木 wall piece
墙上承梁板 wall plate
墙上承梁件 wall hanger
墙上承梁箱 wall box
墙上的 wall;wall type
墙上的悬臂旗杆 outrigger wall set

flange pole
墙上灯座 wall lamp socket;wall plug;wall socket
墙上电气插座 wall outlet;wall socket
墙上电气开关盒 wall box
墙上电源插座 wall outlet; power point <英>
墙上垫木 crown piece
墙上吊架 wall shelve
墙上钉板条的板 lathing board
墙上洞口的斜面 cant of a wall
墙上方的水平装饰带 frieze
墙上方挂镜线条 picture mo(u)lding
墙上固定件 wall fittings
墙上固定铅泛水的铅楔块 lead bat
墙上挂灯 wall-mounted light(ing)
墙上挂毯 wall fabric;wall hanging
墙上管道出口 stack head
墙上横板 horizontal panel
墙上横木 platt
墙上火警装置 wall fire warning device
墙上加热器 wall heater
墙上碱霜 wall white
墙上绞车 bracket winch
墙上脚手架(横木)孔 putlog hole(of wall);putlog holing
墙上进线导管 wall tube
墙上卷扬机 wall crab
墙上绝缘子 wall insulator
墙上开关 wall(board)switch
墙上孔口 wall opening
墙上梁托 wall hanger
墙上梁穴 beam box
墙上梁支座 beam box
墙上梁座 wall box
墙上裂缝 cranny
墙上裂口 cranny
墙上楣构 entablature
墙上煤气热水锅炉 wall gas geyser
墙上门后夹 wall door catch
墙上木质摇头 wooden wall plug
墙上牛腿 wall bracket
墙上暖热器 wall heater
墙上排气(风)扇 wall fan
墙上喷水器 wall sprinkler
墙上披水槽 raggle;raglet
墙上平架 scarcement
墙上起重机 wall crane
墙上取暖器 wall heater
墙上伸缩缝 wall expansion
墙上施工通道洞 <临时性的> wall run
墙上竖缝 wall joint
墙上竖梯 cat ladder
墙上水准点 benchmark built-in wall
墙上通风机 wall ventilator
墙上通风孔 wall ventilator
墙上凸出小塔楼 bartizan
墙上凸(肚)窗 wall oriel
墙上涂料 wall paint
墙上托架 wall bracket
墙上托架支承的脚手架和模板系统 bracket form scaffold
墙上污水贮槽 wall-mounted waste receptacle
墙上洗手盆 wall wash basin
墙上线脚 wall mo(u)lding
墙上线脚转延侧面 reprise
墙上消防栓 wall hydrant
墙上悬吊物 mural hanging
墙上旋臂起重机 wall slewing crane
墙上饮水器 wall drinking fountain
墙上预留插铁 wall dowel
墙上预埋木砖 wall dowel
墙上圆线条 beaded mo(u)lding
墙上照明设备 wall illumination
墙上支点 <构件的> wall hold
墙上支托 wall beam
墙上抓条 wall grab bar
墙上转臂起重机 wall-derrick
墙上装的止门器 wall door stop
墙上装饰品 wall ornament
墙上装饰品支架 plate shelf
墙身 body of wall
墙身凹鼓 buckling of wall
墙身保温 insulation for wall;wall insulation
墙身沉陷 settlement of walling;wall settlement
墙身沉陷裂缝 cracks owing to wall settlement; settlement cracks of walling;wall settlement fissure
墙身尺寸 dimension of wall
墙身顶撑 shoring to walls
墙身断面 cross-section of wall
墙身防潮 dampproofing of wall;wall damp-proofing
墙身防潮层 wall damp-proof course
墙身防水 wall waterproofing
墙身钢筋 stellar bar;stem bar
墙身横撑 needling to wall
墙身空腔 wall cavity
墙身饰面 wall facing
墙身收分 wall batter;batter of wall
墙身支座 wall bearer
墙施工方法 wall construction method
墙石板瓦 weather slating tile
墙石膏踢脚板 wall gypsum baseboard
墙式便桶 corbel closet
墙式插座 wall socket
墙式的 wall type
墙式地下连续墙 wall-type underground diaphragm wall
墙式电话机 wall(-type)telephone set
墙式防波堤 wall-breaker;wall-breakwater
墙式护栏 guard wall;wall fence
墙式基础 wall-type foundation
墙式交换机 wall-pattern switchboard
墙式梁 wall beam
墙式末端装置 wall-mounted terminal
墙式配电盘 wall-type switchboard
墙式设备 wall-mounted equipment
墙式通道 wall passage
墙式消火栓 hose station
墙试验样品 wall sample
墙饰面层锚固件 veneer tie; veneer wall tie
墙树 espalier
墙树间距 spacing between espaliers; trees
墙塔 wall tower
墙台 bench table
墙太阳方位 wall solar azimuth
墙套管 wall sleeve
墙体材料 walling;walling material
墙体层高 course depth
墙体长细比 wall slenderness ratio
墙体长细比极限 wall slenderness limit
墙体衬板 wall sheathing
墙体防雨 rain exclusion of walling
墙体封底漆 wall sealer
墙体工程 walling;walling work
墙体构造 wall configuration
墙体加固 wall tying
墙体建筑单元 walling component
墙体结构 wall construction
墙体开裂 wall cracking;wall crazing
墙体门框 bucks
墙体模板 wall form
墙体抹灰 wall plastering
墙体配筋 wall reinforcement
墙体砌块 walling unit
墙体强度 strength of wall
墙体渗透性 penetration through wall
墙体式样 walling pattern
墙体饰面 finishing of wall
墙体贴面砖 vertical tiling
墙体图案 walling pattern
墙体系 wall system
墙体系安装 wall system installation
墙体系统 walling system
墙铁件 wall iron
墙通风孔 <管道、管沟、房顶等的> wall vent
墙头饰 bratticing
墙头挑出瓦 tile creasing
墙头挑檐 wall cornice
墙头瓦 tile hanging
墙头压顶 <哥特式卷筒形装饰的> roll and weathered
墙托【建】backing of wall
墙托钩 wall hook
墙托架 beam box;wall box
墙托架轴承 longitudinal wall hanger bearing;post bearing
墙托梁 wall beam
墙外的 extramural
墙外吊楼 bartizan
墙外隔热层 outsulation
墙外角护角线条 staff bead
墙外升降机 wall lift
墙外狭道 scarcement
墙网 fence net
墙围工事 immuration
墙帷 dossal;wall hanging
墙席 wall mats
墙下单独基础 single foundation in wall
墙线 walling thread
墙象限仪 mural quadrant
墙橡胶 wall rubber
墙效应 wall effect
墙斜撑 wall brace
墙斜拉索 wall brace
墙心顶块接砌 heart bond
墙心圬工 backing masonry(work)
墙心砖 backer brick; backing block; backing brick; backup brick; building brick
墙压顶 coping; plinth; wall coping; wall crest;wall crown
墙压顶托架 coping bracket
墙压顶瓦 cope tile;copping tile
墙压顶斜面 splayed coping
墙压顶斜线脚 skew fillet
墙牙岔 tusking
墙烟囱 wall stack
墙岩 dikite
墙檐 barge course
墙腰板线条 <上部装饰> dado mo(u)lding
墙腰箍 wall cramp
墙仪 mural circle
墙用混凝土强度 wall concrete strength
墙用空心混凝土砌块 cavity concrete block for walls
墙用空心混凝土砖 cavity concrete tile for walls
墙用空心砌块 cavity block for walls
墙用软木板 cork slab for walls
墙与墙之间(机械或车辆的)最小转弯直径 wall-to-wall turning diameter
墙园 wall garden
墙垣干砌 dry wall(ing)
墙垣无浆砌 dry wall(ing)
墙障 wall hitch
墙罩面 hearting
墙支撑 wall bracing;wall support
墙支承 wall support
墙支墩 wall pillar
墙支柱 wall pillar
墙支座 wall support
墙纸 paperhanging; papering; surface-coated wall paper;wall paper
墙纸裁切 wall-paper sheet(ing)
墙纸的底子 background wallpaper
墙纸覆盖线条 wall-paper cover mo(u)lding
墙纸浆糊 paste for paper hanging
墙纸卷 wall-paper roll
墙纸黏[粘]结剂 paperhanging adhesive
墙纸黏[粘]贴剂 paste for wall paperhanging
墙纸切边刀 casing knife
墙纸图案错位 missetting
墙纸修剪工 wall-paper trimmer
墙纸修剪器 wall-paper trimmer
墙纸用晶纹清漆 crystal paper varnish
墙纸制造厂 wall-paper factory
墙趾 toe;toe of wall;wall toe
墙中凹梁座 columbarium
墙中凹座 columbarium
墙中承梁短板 template;templet
墙中空气通道 wall duct
墙中梁端垫块 torsel
墙中木栓 wooden wall plug
墙中木楔 wooden wall plug
墙中心线 centre line of wall
墙钟 wall clock
墙踵 wall heel
墙重 wall weight
墙柱 antal;applied column;wall post
墙柱脚 wall column footing
墙柱砌口 column offset
墙柱上的凸出线条缘饰 plateband
墙柱预留安装预制梁的缺口 beam pocket
墙砖 wall brick;wall tile;majolica mosaic wall tile <涂有不透明釉的>
墙砖角砌 skintled brickwork
墙转角踏步 corner step
墙桩 <采用地下连续墙施工方法制成的桩> Barrette
墙装插头 wall plug
墙装插座 wall socket
墙装冲落坐便器 wall-mounted washdown W.C. pan
墙装吊架 wall bracket crane
墙装动臂式起重机 pole derrick
墙(装)绞车 wall winch
墙装卷扬机 wall crab
墙装起重机 wall crane
墙装散热器 wall radiator
墙装式马桶 wall toilet
墙装式倾倒残渣盆 wall-mounted slop bowl
墙装式坐便器 wall-mounted washdown W.C. pan
墙装饰 wall ornament
墙装台案 console table
墙装镗床 wall boring machine
墙装托架 wall bracket
墙装摇臂钻床 wall radial drilling machine
墙装转臂起重机 wall-derrick
墙装钻床 wall drilling machine
墙座 wall base; wall socle; wall support

蔷

蔷薇 Japanese rosa;rose;wild rose

蔷薇苯胺 fuchsin(e)
蔷薇彩 rose colo(u)r
蔷薇花饰 rosace;rosette
蔷薇花形 rosaceous
蔷薇花型的 rose flowered
蔷薇黄锡矿 rhodostannite
蔷薇辉石 hermannite; hydropite; manganese spar;manganolite;paisbergite;rhodonite
蔷薇辉石矿床 rhodonite deposit

蔷薇辉石岩 rhodonite rock
蔷薇榴石 landerite;xalostocite
蔷薇木 amboina rosewood;Andaman rosewood; Burmacoast padauk; Burmese rosewood;rose wood
蔷薇色酸 aurin;rosolic acid
蔷薇石英 Bohemian ruby;rose quartz;rosy quartz
蔷薇水晶 rose quartz
蔷薇油 attar of roses;rose oil
蔷薇园 rose garden
蔷薇属 rose
蔷薇状断口 rosette fracture
蔷薇状共晶组织 rosette
蔷薇状石墨 rosette graphite
蔷薇紫 rose purple

抢

抢低潮施工 tidal work(ing)
抢定额者 ratebuster
抢夺者 grabber
抢风 tack
抢风航法 windward sailing
抢风航驶 sail on a bowline
抢风行驶 luff;tack
抢购 quick-take; scare buying; shopping rush
抢购风潮 panic purchase;stampede
抢购物资 panic buying of goods;rush for goods
抢救 emergency treatment; rescue; salvage;salve
抢救班 rescue party
抢救包 emergency set
抢救车 crash truck; crash wagon; emergency service vehicle
抢救车起重机 retriever's hoist
抢救措施 emergency measure; first-aid measures
抢救队 emergency team;rescue crew; rescue party; salvage crew; wrecking screw
抢救工作 rescue work
抢救古物工程 salvage archaeology
抢救快艇 crash boat
抢救人员 rescue crew;rescue personnel;rescuer;salvage crew
抢救设备 salvage appliance
抢救推车 emergency cart
抢救用具 salvage appliance
抢救站 rescue station
抢救装备 rescue outfit
抢救组 rescue party
抢去 snatch
抢时间作业 work against time
抢收 rush-harvest
抢水 water-pirating
抢滩 beaching;voluntary stranding
抢先 anticipate;forestall;outfoot;preemption
抢先成交 beating the gun
抢先的 preemptive
抢先调度 preemptive scheduling
抢先调度策略 preemptive scheduling strategy
抢先输入 anticipated input
抢先算法 preemptive algorithm
抢先优先权 preemptive priority
抢险 rush work for emergency
抢险车 emergency car; emergency service vehicle;emergency vehicle; hurry-up wagon; retriever; wrecking car
抢险吊车 emergency crane
抢险队 crash crew;emergency squad
抢险工程 emergency works
抢险工程车 breakdown lorry
抢险机车 emergency locomotive
抢险救援车 disaster unit
抢险救援消防车 emergency tender; emergency truck
抢险救灾 rush to deal with an emergency and provide disaster relief
抢险列车 emergency train;wrecking train
抢险起重机 breakdown lorry
抢修 emergency repair;first aid;first-aid repair; recovery service; rush repair
抢修车 breakdown gang;breakdown van; emergency repair car; emergency repair truck;hurry-up wagon
抢修队 breakdown gang; wrecking crew
抢修工程 <对损坏工程的补救或加固> rush-repair work;salvaging
抢修工程车 breakdown lorry;breakdown vehicle
抢修公路和桥梁 make urgent repair on highway and bridge
抢修列车 breakdown train
抢修起重机 breakdown crane
抢修站(车辆)breakdown service
抢占寄存器 register contention

羟

羟胺 azanol;hydroxylamine;oxammoniam
羟丙基甲基纤维素 hydroxypropyl methyl cellulose
羟丙基纤维素 hydropropyl cellulose
羟胆矾 brochantite
羟碲铜矿 cesbronite
羟碲铜锌石 quetzalcoatlite
羟碘铜矿 salesite
羟丁氨酸 threonine
羟钒石 duttonite
羟钒铁铅矿 mounanaite
羟钒铜矿 turanite
羟钒锌铅石 descloizite
羟氟硅灰石 bultfonteinite
羟氟磷钙镁石 panasqueiraite
羟钙石 portlandite
羟钙钛矿 kassite
羟铬矿 bracewellite
羟硅钡石 muirite
羟硅铋铁矿 bismutoferrite
羟硅钙钪石 cascandite
羟硅钙钠石 kvanefjeldite
羟硅钙铅矿 ganomalite
羟硅钙石 dellaite
羟硅磷灰石 hydroxylellestadite
羟硅铝钙石 vuagnatite
羟硅铝锰石 akatoreite
羟硅铝钇石 vyuntspakchkite
羟硅锰镁石 gaugeite
羟硅锰石 jerrygibbsite
羟硅锰铁矿 deerite
羟硅钠钙石 jennite
羟硅硼钙石 howlite
羟硅铍钙石 jeffreyite
羟硅铍石 bertrandite
羟硅铍石铍矿石 bertrandite ore
羟硅铅石 plumbotsumite
羟硅铈矿 toernebohmite
羟硅锑铁矿 chapmanite
羟硅铁锰石 balangerite
羟硅铁石 howieite;macaulayite
羟硅铜矿 shattuckite
羟硅钇石 iimoriite
羟黑锰矿 janggunite
羟基 hydroxy;hydroxyl group
羟基胺 alkylol amine
羟基苯胺 anrinophenol; hydroxyaniline
羟基苄磺酸甲醛树脂 hydroxybenzylsutfonic acid formaldehyde resin
羟基丙酮 acetone alcohol
羟基当量 hydroxyl equivalent
羟基丁二酸 malic acid
羟基脯氨酸 hydoxyl proline
羟基化合物 hydroxylate
羟基化聚合物 hydroxylated polymer
羟基化油 hydroxylated oil
羟基己酸内酯 caprolactone
羟基磷灰石 hydroxylapatite
羟基醚 alcohol ether
羟基木质树脂 hydroxylated wood resin
羟基溶剂 hydroxyl solvent
羟基数 hydroxyl value
羟基酸 oxy(gen)acid
羟基污染物 hydroxylated contaminant
羟基乙醛 glycolaldehyde
羟基乙酸酯 glycollate
羟基游离基 hydroxyl radical
羟基有机酸 hydroxyl organic acid
羟基振动 hydroxyl group vibration
羟基脂肪酸 hydroxylated fatty acid
羟基值 hydroxyl value
羟镓石 soehngite
羟甲基 hydroxymethyl;methylol
羟甲基丁二酸 itamalic acid
羟甲基化 hydroxymethylation
羟甲基化过氧氢 hydroxymethyl hydroperoxide
羟甲基膦酸二聚氧乙烯酯 di(polyoxyethylene) hydroxymethyl phosphonate
羟甲基脲 methylol urea
羟甲基纤维素 hydroxymethyl cellulose
羟碱铌钽矿 rankamaite
羟离子 hydroxide ion;hydroxidion
羟磷钙铍石 glucine
羟磷灰石 hydroxyapatite
羟磷锂铁石 tavorite
羟磷铝钡石 jagowerite
羟磷铝钙石 foggite
羟磷铝锂钠石 tancoite
羟磷铝锂石 hydroxylamblygonite
羟磷铝钠石 natromontebrasite
羟磷铝石 bolivarite;trolleite
羟磷铝锶石 goedkenite
羟磷铝铁钙石 samuelsonite
羟磷镁石 althausite
羟磷锰石 triploidite
羟磷钼锰石 ernstite
羟磷铍钙石 hydroherderite; hydroxyl-herderite
羟磷铅铀矿 dumontite
羟磷铁锰石 kryzhanovskite
羟磷铁钠石 kidwellite
羟磷铁石 wolfeite
羟磷铜铁矿 andrewsite
羟磷硝铜矿 likasite
羟磷锌铜石 kipushite
羟硫氯铜石 connellite
羟铝矾 basaluminite
羟铝钙镁石 wermlandite
羟铝黄长石 bicchulite
羟铝锑矿 bahianite
羟铝铜钙石 papagoite
羟铝铜铅矾 osarizawaite
羟氯碘铅石 schwartzembergite
羟氯铅矿 laurionite
羟氯铜铅矿 diaboleite
羟镁硫铁矿 tochilinite
羟镁铝石 meixnerite
羟锰矿 pyrochroite;pyrohroite
羟锰铅矿 quenselite
羟萘甲酸 oxynaphthoic acid
羟镍矿 hydronicite
羟配聚合物 olation polymer
羟硼锰石 wiserite
羟硼锶石 olshanskyite
羟铍石 behoite
羟蔷薇辉石 santaclaraite
羟桥聚合作用 olation
羟砷铋石 arsenobismite
羟砷钙铍石 bergslagite
羟砷钙锌石 prosperite
羟砷镧锰石 retzian-(La)
羟砷锰矿 eveite
羟砷锰石 arsenoclasite
羟砷钕锰石 retzian-(Nd)
羟砷铁铅矿 gabrielsonite
羟砷铜矿 cornubite
羟砷铜石 cornubite
羟砷锌钙石 gaitite
羟砷锌铅石 arsendescloizite
羟砷锌石 adamite
羟砷钇锰矿 retzian
羟水钒钼矿 alranite
羟水铁矾 hohmannite
羟钛钒矿 tivanite
羟钽铝石 simpsonite
羟碳铝矿 scarbroite
羟碳锰镁石 desautelsite
羟碳镍石 otwayite
羟碳铈矿 hydroxylbastnaesite
羟碳铁镁锌矾 hauckite
羟碳锌石 hydrozincite
羟碳锌铜矾 schulenbergite
羟碳钇铀石 bijvoetite
羟碳钴镍石 comblainite
羟铁矿 amakinite
羟钍石 thorogummite
羟硒铜铅矿 schmiederite
羟锡钙石 burtite
羟锡矿 hydroromarchite
羟锡镁石 schoenfliesite
羟锡锰石 wickmanite
羟硝锌矿 salmoite;tarbuttite
羟锌蒙脱石 zincsilite
羟锌锰矾 lawsonbalumite
羟锌锰镁矾 torreyite
羟氧钴矿 heterogenite
羟乙磺酸酯 isethionate
羟乙(基)磺酸 isethionic acid
羟乙基纤维素 hydroxyethyl cellulose
羟乙醛 glycoladehyde
羟铟石 dzhalindite
羟鱼眼石 hydroxyapophllite
羟锗铅矾 itoite
羟锗铁铝石 carboirite
羟锗铁石 stottite
羟种铁矾 bukovkyite

悄

悄平隆起圈闭 subdued uplift trap

跷

跷高 stilting
跷拱 rampant arch;rising arch
跷起拇指要求搭车 <美> thumb
跷跷板 see-saw;teeter-totter

敲

敲棒 loosening bar
敲扁铆钉 flattened rivet
敲铲船底 bottom cleaning
敲成 knock
敲成碎片 reduce to scrap
敲出 beat out;tap-out
敲打 beat; chap; drub; hammer at; knock;crimping tool
敲打检查 hammering test
敲点涂法 stippling
敲掉 striking
敲掉下来 knock away
敲掉楔子 striking wedges
敲钉穿孔器 nail punch
敲钉工人 nailer
敲钉器 nail gun
敲钉用的 nailing
敲顶 roof tapping

敲顶锤 rapper
敲顶棍 tapping bar
敲定价格 strike price
敲缸 engine knock; fuel knock; knocking; piston knock (ing); piston slap
敲花细工 repousse; repousse work
敲击 beat; hammering; knock (ing); slap; tap
敲击棒 stamp
敲击波 knock wave
敲击法 < 土工试验 > tapping method
敲击管钳 rap wrench
敲击火石 poll
敲击记数 number of blows
敲击器 knocker; shell knocker
敲击声 slap
敲击式起子 impact screwdriver
敲击试验 acoustic (al) test; knocking test
敲击信号 knocking; strike signal
敲接 struck joint
敲进 drive-in
敲进铆钉 laying-up a rivet
敲进去 drive home
敲牢 knock home
敲棱角 skiffling
敲裂试验 clip test
敲落 abate; knock off; knockout
敲门砖 stepped stone; stepping stone
敲模棒 rapping bar; rapping iron
敲模垫板 rapping plate
敲模孔 rapping hole
敲拍声 rap
敲平 beat out; clench
敲平锤 planishing hammer; smoothing hammer
敲平钉子 clench nailing
敲平铆钉尾部的金属板 rove or rove
敲去 knock off
敲入 knock-in
敲入螺丝钉 drive screw
敲水垢锤 chipping hammer; scaling hammer
敲碎 batter; crack; knap; shiver
敲碎的燧石 knapped flint
敲脱刀 crack-off iron
敲脱工具 cracking tool
敲弯 clenching; clench nailing
敲弯的钉 clench[clinch]
敲弯钉 clench nail
敲弯钉头 clench
敲下 knock off
敲响 peal
敲锈棒 scaling bar
敲锈锤 chipping hammer; scaling hammer
敲锈眼镜 chipping goggles
敲渣锤 chipping hammer; slag hammer
敲钟报时 strike a bell
敲钟拉绳 bell pull
敲钟索 bell rope

锹把 shovel shaft

锹钉 thumb pin
锹夯(实)混凝土 spaded concrete
锹尖端 spade tip
锹式采样器 spade sampler

橇棍长凿 jumper

橇式刮路器 sled drag
橇挖 broaching
橇形船首 cutaway boat
橇形船头 < 驳船 > swim end
橇形减震器 skid shock absorber
橇装加气站 skid-mounted LNG filling station
橇装引擎 engine skid

乔柏属 western red cedar

乔里斯基分解法 Cholesky decomposition
乔林混农作业 combined cropping system
乔麦 erogonum ovalioium
乔姆斯基层次结构 Chomsky hierarchy
乔木 < 树高 5 ~ 8 米以上；建筑用材胸径 20 厘米以上 > crop of formed trees; arbor; macrophanerophytes; tree
乔木的 arboreal; arboreous
乔木林 high forest; timber forest
乔木树种 tree species
乔木状 arborescence
乔木状的 arboreous
乔其纱 georgette
乔松 bhotan; Himalayan pine
乔治赵石 georgechaoite
乔伊连续采煤机 Joy continuous miner
乔伊型连续挖掘机 Joy continuous miner
乔治大理石 Georgian marble
乔治敦 < 圭亚那首都 > Georgetown
乔治时代建筑 Georgian architecture
乔治时代建筑风格 Georgian style
乔治式拱 Georgian arch
乔治式(夹丝)玻璃 Georgian glass
乔治统【地】Georgian series
乔治王朝建筑 Georgian architecture
乔治亚斗车 Georgia buggy
乔治殖民式建筑 Georgian colonial architecture

侨汇 immigrant remittance; remittance by migrants

侨居国 country of residence; host country
侨居国外的 oversea(s)
侨居种 alien species
侨民 immigrant; resident
侨民税 alien duty
侨团 immigrant community
侨资企业 overseas Chinese venture

桥 T 形网络 bridged-T network

桥坝 bridge dam
桥板 bridge board; bridge plate; catwalk
桥板最低处的排水口 scupper
桥臂【电】bridge arm
桥臂单元 bridge arm unit
桥臂元件 bridge arm unit
桥臂支路 arm path
桥变 bridging
桥测 bridge measurement
桥拆装梁 bridge dismountable beam
桥长 length of bridge
桥秤 weigh(ing) bridge
桥簇 family of bridges
桥大梁 bridge girder
桥挡护木 chafing plate at well side
桥档 ladder well
桥档龙门 ladder-well gantry
桥的岸座 bank seat
桥的防水 bridge seal(ing)
桥的护墩 fender pier
桥的活动支座 expansion bearing
桥的绝热 bridge insulation
桥的跨度 cut bay
桥的框架 frame of bridge
桥的模板工作 bridge form(work)
桥的模型 bridge model
桥的排水 bridge drainage
桥的坡度 bridge gradient
桥的人行道悬臂 footway cantilever bracket
桥的提升 bridge raising
桥的斜度 skew of bridge
桥的悬臂人行道 cantilever foot path
桥的振动 bridge vibration
桥的中间跨 intermediate bay
桥的主梁 bridge beam
桥的座梁 seat beam
桥吊 bridge crane; bridge type gantry crane; crane bridge; travel(l)ing bridge crane
桥吊的小车行驶 troll(e)y travel
桥吊的小车行驶大梁 troll(e)y girder
桥钉 bridge pin
桥洞 aperture of bridge; arch; bridge opening
桥洞净空 portal clearance
桥渡 bridge crossing; transporter bridge < 吊在桥上的车渡设旋 >
桥渡冲刷 scour at bridge crossing
桥渡河段 bridging crossing reach; river reach of bridge crossing
桥渡设计 design of bridge crossing
桥渡水文学 hydrology of bridge crossing
桥对河流的斜角 angle of skew
桥墩 bank pier; bridge abutment; bridge column; bridge pier; foundation pier; land abutment; pier; pier abutment; substructure of bridge
桥墩鼻尖 pier nose
桥墩边角护板 rigidfix
桥墩沉箱 bridge-pier caisson
桥墩冲刷 pier scour
桥墩处滞水 backwater at bridge pier
桥墩处滞水作用 backwater effect of bridge pier
桥墩的破冰构造 ice apron; ice guard
桥墩底座 pier footing
桥墩顶 bridge cap
桥墩顶板 bridge cap
桥墩顶部的破冰体 ice apron
桥墩顶帽 coping
桥墩顶台 pier table
桥墩顶凸出处 crest
桥墩顶座 bridge cap; pier cap
桥墩定位 location of bridge abutment; location of bridge pier; pier location
桥墩墩尖 pier nose
桥墩墩身 body of bridge pier; stem of bridge pier
桥墩墩身斜度 batter of pier body
桥墩墩头 pier head
桥墩方位 pier orientation
桥墩分水鼻 pier nose cutwater
桥墩分水尖 breadthwise water; breakwater; ice apron; pier breakwater; upstream nose
桥墩分水头 upstream nose
桥墩钢筋混凝土底板 reinforced concrete footing course of bridge pier
桥墩钢筋混凝土顶板 reinforced concrete coping course of bridge pier
桥墩护栏 pier fender; starling
桥墩护桩 sterling
桥墩混凝土 pier concrete
桥墩基础 pier base; pier footing; pier foundation
桥墩基础模箱 pier footing form
桥墩基座 pier base
桥墩尖端 nose; nose of pier; pier nose; starling
桥墩尖端分水桩 starling cutwater pile
桥墩尖头 breakwater glacis
桥墩间距 pier spacing
桥墩局部冲刷 local scour around pier; local scour near pier
桥墩棱体 nose of pier
桥墩帽 bridge cap; bridge support cap; pier coping
桥墩排水 pier drainage
桥墩排水设施 pier drainage facility
桥墩破冰结构 icebreaker of bridge pier nose
桥墩破冰体 ice apron of pier
桥墩上游端 upstream pier nosing
桥墩上游分水尖 upstream pier nosing
桥墩上游破冰设施 ice breaker
桥墩式栈桥码头 bridge type wharf
桥墩损失 pier loss
桥墩(台)顶安放支座处 bridge seat
桥墩台顶帽 cap of pier
桥墩台托盘 pier tray
桥墩体 pier body
桥墩跳升滑模板 pier-shaft jump formwork; pier-shaft lift formwork
桥墩圬工 butment masonry
桥墩下的冲刷 scour under a pier
桥墩小型基础围堰 Chicago caisson
桥墩形式 pier type
桥墩沿围 starling cope
桥墩摇座 rocking pier
桥墩翼墙 water ring of bridge pier; water wing; wingwall of bridge pier
桥墩迎水面 pier head(cutwater)
桥缝焊接 bridge seam weld
桥工铰刀 bridge reamer
桥工铆钉绞刀 bridge reamer
桥工铆钉扩孔钻 bridge reamer
桥拱 bridge arch; camber
桥拱下沉 sagging of arch
桥管 cross-over main
桥管清扫器 gooseneck cleaner
桥规 bridge ga(u)ge
桥涵按跨径分类 classification of bridge and culvert according to span length
桥涵标 bridge and culvert sign; bridge mark; bridge opening mark
桥涵洞测量 bridge-culvert survey
桥涵计算跨径 computed span
桥涵净跨(径) clear span
桥涵孔径 span of bridge and culvert
桥号标志 bridge number sign
桥桁架 bridge truss
桥桁架限界 bridge truss clearance
桥环 bridge ring
桥环大环内酯 ansamacrolide
桥环的 endocyclio
桥环化合物 bridge compound; endocyclic compound
桥机板 < 混凝土 > bridge slab
桥基 abut(ment); bridge foundation
桥基沉降观测 bridge foundation settlement observation
桥基冲刷 scouring at bridge foundation
桥基开挖 bridge excavation
桥基稳定性评定 bridge foundation stability evaluation
桥架 < 剧院舞台上绘制布景用的 > paint bridge
桥架大梁 < 起重机的 > bridge beam
桥架导线 < 沿起重机的 > bridge conductor
桥架耳轴 ladder trunnion
桥架绞车速度控制装置 ladder winch speed controller
桥架起重机 bridge type crane
桥架设用的工具 bridge assembly equipment
桥架枢轴 ladder pivot

桥架行程 <起重机的> bridge travel
桥间差速器 center differential
桥建筑高度 construction height of bridge
桥建筑限界 clearance above bridge floor
桥键 bridge bond;bridging
桥脚 bridge pile
桥接 bridge connection;bridge joint;bridge over;bridging;in bridge
桥接T形抽头 bridged-T tap
桥接T形电路 bridged-T circuit
桥接T形电桥 bridged-T bridge
桥接T形滤波器 bridged-T filter
桥接T形水龙头 bridged-T tap
桥接T形网络 bridged-T network
桥接T形陷波器 bridged-T trap
桥接保险丝 bridge fuse
桥接岔路 hybrid
桥接岔路线圈 hybrid coil
桥接岔路型 hybrid-type
桥接抽头 bridge tap
桥接触点 bridge contact
桥接抵消法 bridge neutralizing
桥接点 bridge contact;close-before-open contact;extended contact;make-before-break contact
桥接电缆 jumper cable
桥接电流 bridge current
桥接电路 bridge circuit
桥接法 bridge grafting;bridge method
桥接放大器 bridging amplifier
桥接故障 bridge fault
桥接管 bridge pipe
桥接焊 bridge welding
桥接焊缝 weld bridge seam
桥接弧刷 bridging wiper
桥接计算机 bridge machine
桥接件 bridgeware
桥接接点 double break contact
桥接开关 gapping switch
桥接旁路 bridge branch
桥接片 bridge piece;bridging piece
桥接器 bridge
桥接塞孔 bridge jack;bridging jack
桥接石墨 bridged graphite
桥接式触点 bridge contact
桥接式双工 bridge duplex
桥接双工器 bridge diplexer
桥接双工系统 bridge duplex system
桥接双工制 bridge polar duplex system;bridge duplex system
桥接四端网络 bridge X-network
桥接损耗 bridging loss
桥接条 bridge connector
桥接网络 bridge network
桥接线 bridge;bridge connector;bridging line;connection strap
桥接线路 bridge connection【电】;bridge circuit <探头的>
桥接线圈 bridging coil
桥接装置 bridge set
桥壳 axle case;axle housing
桥孔 aperture of bridge;bridge hole;bridge opening
桥孔净长 clear length of bridge opening
桥孔压缩 bridge opening contraction
桥跨 <以墩分隔的> bridge span;span of bridge;balk;bridge opening;bridge space
桥跨结构 bridge superstructure
桥跨结构拱度 rise of span
桥跨净空 clearance of space;clearance of span
桥跨上部结构 bridge superstructure
桥跨限界 clearance of span
桥跨堰 weir with overhead bridge
桥框架 bridge frame
桥栏杆 bridge handrail;bridge railing;parapet of bridge
桥栏杆柱 spindle
桥缆地锚 ground anchor of bridge cable
桥缆(索) bridge rope;bridge strand;bridge cable
桥连 bridging
桥连离子 bridging ion
桥联理论 bridging theory
桥联裂纹 bridging crack
桥/联络开关 bridge/coupler switch
桥联式电路【物】 bridge circuit
桥梁 bridge(girder)
桥梁安装的倒退分析 retrogressive analysis for bridge erection
桥梁安装的前进分析 progressive analysis for bridge erection
桥梁安装监测 bridge erection monitoring
桥梁安装容许误差 bridge erection tolerance
桥梁摆振 bridge-oscillation
桥梁摆轴支座 pendulum bearing for bridge
桥梁板梁 bridge girder
桥梁板式橡胶支座 laminated rubber bearing for bridge
桥梁保护体系 bridge protective system
桥梁边沟 bridge gutter
桥梁边索倾角 angle of backstay
桥梁编号 bridge numbering
桥梁标 bridge post
桥梁标志 bridge marker;bridge sign
桥梁标准活荷载 standard live load for bridge
桥梁标准跨径 standard span of bridge
桥梁标准设计 standard bridge design;standard design of bridge
桥梁病害 bridge defect
桥梁病害诊断 bridge defect diagnosis
桥梁病害整治 bridge fault repairing
桥梁布置 bridge layout;layout of bridge
桥梁部(构)件 bridge parts
桥梁材 wood used for bridges
桥梁测试车 bridge testing laboratory vehicle
桥梁插座 rocker
桥梁长度 bridge length
桥梁厂 bridge plant;bridgework
桥梁车间 bridge shoe
桥梁车行道 bridge carriageway;decking;traffic decking
桥梁车行道沥青铺面层 asphalt bridge carriageway pavement
桥梁承台 bearing platform
桥梁承载能力 load(ing) capacity of bridge
桥梁承载能力极限状态 ultimate limit state of bridge carrying capacity
桥梁承重等级标志 bridge class sign
桥梁冲毁 bridge washout
桥梁冲刷 bridge washout
桥梁出水净高 clearance of bridge opening;headroom under a bridge
桥梁道砟槽 ballast trough
桥梁的拆除 removable of bridge;removal of bridge
桥梁的横断截面 bridge cross-section
桥梁的极限状态设计 limit state design of bridge
桥梁的美观设计 aesthetic design of bridges
桥梁的桥面系统 <包括桥面及其支承部件> floor way
桥梁的上部结构 superstructure of bridge
桥梁的中央跨 centre span
桥梁的重建计划 bridge rebuilding program(me)
桥梁底脚 bridge footing
桥梁雕像 bridge statue
桥梁顶推施工法 incremental launching
桥梁动力响应试验 bridge response to forced vibration
桥梁动载试验 bridge dynamic loading test
桥梁断裂 bridge fracture
桥梁墩顶梁段 <俗称零号块> hammerhead section
桥梁墩台定位 location of pier and abutment
桥梁墩台防撞 collision prevention of pier and abutment
桥梁墩座 pier base
桥梁翻新 bridge retrofitting
桥梁方案设计 bridge conceptual design
桥梁防护信号 bridge protection signal
桥梁防护信号机【铁】 bridge protection signal
桥梁分级 rating of bridge
桥梁分孔 proportioning of the bridge spans
桥梁分跨 proportioning of the bridge spans
桥梁分类 classification of bridges
桥梁副竖杆 subvertical member
桥梁改建 bridge reconstruction;reconstruction of bridge
桥梁概率极限状态设计法 probabilistic limit state design method of bridge
桥梁杆件 bridge member
桥梁钢 bridge steel
桥梁钢板 bridge plate;bridge steel plate
桥梁钢面系 battle deck(bridge) floor
桥梁高跨比 rise-span ratio
桥梁工长 bridge foreman
桥梁工程 bridge construction;bridge project;bridge works
桥梁工程队 bridge-gang
桥梁工程局 Bureau of Bridge Construction
桥梁工程师 bridge engineer
桥梁工程学 bridge engineering
桥梁工地 bridge construction site;worksite of bridge construction
桥梁工区 bridge working area
桥梁工事 bridge works
桥梁拱势 bridge camber
桥梁构架 bridge framework
桥梁构件 bridge member
桥梁固定支座 fixed bearing for bridge
桥梁顾问工程师 bridge consulting engineer
桥梁观测 bridge observation
桥梁管理人员 bridgemaster;bridge tender
桥梁管理系统 bridge management system
桥梁规划 bridge planning
桥梁轨枕 bridge tie
桥梁辊轴支座 multiple roller bearing for bridge
桥梁过水孔径 bridge waterway
桥梁合成橡胶支座 elastomeric bridge bearing
桥梁合龙 closure of bridge structure
桥梁河段 bridge waterway
桥梁荷载 bridge load;load on bridge
桥梁荷载检定 bridge rating
桥梁荷载谱 bridge load spectrum
桥梁荷载系数设计法 load factor design method of bridge
桥梁盒式橡胶支座 pot rubber bearing for bridge
桥梁桁架 bridge truss
桥梁恒荷载 dead load on bridge
桥梁横断面 cross-section of bridge
桥梁横向刚度 lateral rigidity of bridge
桥梁弧形支座 rocker bearing for bridge
桥梁护轨 bridge guard rail;guard rail of bridge
桥梁护栏 bridge rail(ing);bridge barrier
桥梁护木 felloe plank;guard timber;guard timber of bridge
桥梁护墙 bridge parapet wall
桥梁护墙破碎器 bridge parapet remover
桥梁滑曳架设 launching erection
桥梁环行线 bridge loop
桥梁活动支座 expansion bearing for bridge
桥梁活荷载 live load on bridge
桥梁基础 bridge foundation
桥梁及辅助结构外观形状咨询委员会 <英> Advisory Committee on the Appearance of Bridges and Associated Structures
桥梁技术档案 bridge technical file
桥梁寄主 bridge host
桥梁加固 bridge improvement;bridge reinforcement;bridge reinforcing;bridge strengthening
桥梁加宽工程 broadening of bridge
桥梁夹板 bridge fishplate
桥梁架空高度 <高水位时> bridge clearance
桥梁架设 bridge erection;erection of bridge
桥梁监测系统 bridge monitoring system
桥梁检测车 bridge inspection vehicle
桥梁检查 bridge inspection
桥梁检查规则 bridge inspection regulation
桥梁检查类别 bridge inspection category
桥梁检查员 bridge inspector
桥梁检查周期 bridge inspection cycle
桥梁检定 bridge rating;rating of bridge
桥梁检定承载系数 rated load-bearing coefficient for bridge
桥梁检定试验 bridge rating test
桥梁检修车 bridge inspection car
桥梁简化分析法 bridge analysis simplified
桥梁简化计算 bridge analysis simplified
桥梁建设项目 bridge project
桥梁建造 bridge building;bridge construction
桥梁建筑 bridge construction
桥梁建筑高度 <桥面系顶面到结构底面间的距离> construction height of bridge;height of bridge superstructure
桥梁建筑工地 bridge construction site
桥梁建筑限界 bridge construction clearance;bridge truss clearance;delimitation of bridge superstructure
桥梁建筑艺术 aesthetics of bridges
桥梁鉴定 bridge checkup;bridge rating;rating of bridge
桥梁交通容量 capacity of bridge
桥梁脚手架 bridge falsework;bridge shoring
桥梁铰接摇座 articulated rocker
桥梁节点 bridge joint
桥梁结构 bridge construction;bridge structure

桥梁结构安装控制 bridge structure erection control
桥梁结构设计 bridge structure design;structural design of bridge
桥梁结构试验 bridge structural test
桥梁捷径 bridge cut-off
桥梁经济断面 economic section of bridge
桥梁经济跨径 economic span of bridge
桥梁净空 bridge clearance;clear opening of bridge
桥梁净空高度 clearance height of bridge
桥梁净空界限 clearance limit of bridge
桥梁净孔(径) clear opening of bridge
桥梁净跨 clear span of bridge
桥梁净跨径 clear span of bridge
桥梁静载试验 bridge static loading test
桥梁开挖 bridge excavation
桥梁勘测 bridge survey
桥梁看守室 bridge house
桥梁抗震加固 bridge aseismatic strengthening
桥梁抗震设计 aseismic stability of bridge
桥梁抗震稳定性设计 aseismatic stability of bridge
桥梁空间结构 space structure for bridge
桥梁孔径不足 insufficient span of bridge
桥梁跨度 bridge space; bridge spalling;bridge span;span of bridge
桥梁宽跨比 aspect ratio of bridge
桥梁框架 frame of bridge
桥梁类别 rating of bridge
桥梁立体模型 spatial model of bridge
桥梁立柱 tower column
桥梁临时顶撑 bridge shoring
桥梁领工员 bridge foreman
桥梁绿色 bridge green
桥梁螺旋线 bridge spiral
桥梁脉动测量 bridge pulsation measurement
桥梁锚碇 bridge anchor
桥梁铆钉钻孔机 bridge rivet reamer
桥梁美学 aesthetics of bridges;bridge aesthetics;bridge architecture
桥梁面步行板 foot plank
桥梁模板工作 bridge shuttering
桥梁模型风洞试验 bridge model wind tunnel test
桥梁模型试验 bridge model test; model test of bridge
桥梁耐久性 bridge durability
桥梁挠度 deflection of bridge span
桥梁挠度曲线 bridge deflection curve
桥梁排水 bridge drainage
桥梁疲劳剩余寿命 fatigue residual life of bridge
桥梁拼装 bridge fabrication
桥梁评级 rating of bridge
桥梁评价系统 bridge evaluation system
桥梁破坏 bridge collapse;bridge failure
桥梁漆 bridge coating
桥梁浅基保护 bridge shallow foundation protection
桥梁浅基(础) shallow foundation of bridge
桥梁浅基防护 unsafe depth foundation protection
桥梁全长 overall length of bridge;total length of bridge;whole length of bridge
桥梁全宽 overall width of bridge
桥梁群体 family of bridges
桥梁人行道构架 footway framing of bridge
桥梁人行道悬臂 footway cantilever bracket of bridge
桥梁上部结构 bridge superstructure
桥梁上承行车道 roadway above
桥梁上拱度 bridge camber;camber of bridge span
桥梁设计 bridge design
桥梁设计工程师 bridge design engineer
桥梁设计规范 bridge design specification
桥梁设计洪水频率 design flood frequency for bridge
桥梁设计流量 design discharge of bridge
桥梁设计人 bridge designer
桥梁伸缩缝 expansion joint for bridges
桥梁伸缩接缝 movable joint for bridges
桥梁伸缩接头 movable joint for bridges
桥梁施工 bridge construction
桥梁施工概算 general estimate of bridge construction
桥梁施工工程师 bridge field engineer
桥梁施工技术 bridge building practice;bridge construction practice
桥梁施工流程图 flow diagram of bridge construction
桥梁施工三角(网)测量 triangulation survey for bridge construction
桥梁施工图 construction drawing of bridge
桥梁施工员 bridge foreman
桥梁实验车 bridge test car
桥梁使用能力极限状态 serviceability limit state of bridge
桥梁使用年限 operational life of bridge
桥梁试验 bridge test(ing)
桥梁试验列车 bridge testing train
桥梁试运行荷载 test run loading for bridge
桥梁枢轴支承 pivot(al) bearing of bridge
桥梁数据 bridge data
桥梁数据库 bridge data bank;bridge database
桥梁水毁 bridge disaster by flood
桥梁隧道公路 bridge-tunnel highway
桥梁损坏 bridge failure
桥梁坍塌 bridge collapse
桥梁弹性橡胶支座 elastomeric bridge bearing
桥梁套线 bridge loop
桥梁梯 bridge stair(case)
桥梁通过能力 trafficability of bridge
桥梁图 bridge drawing
桥梁图式 bridge diagram
桥梁外部支承 exterior bridge support
桥梁维护 bridge maintenance
桥梁维修 bridge maintenance;bridge repair
桥梁维修顶升设备 lifting equipment for bridge maintenance
桥梁维修喷漆设备 paint spraying device for bridge maintenance
桥梁细部设计 bridge detail design
桥梁下部结构 bridge substructure; substructure of bridge
桥梁下承行车道 roadway below
桥梁下净空 bridge clearance
桥梁限载 bridge load limit
桥梁橡胶支座 rubber bridge bearing
桥梁信息管理系统 bridge information management system
桥梁行车道沥青铺(面)层 asphaltic-bitumen deck surfacing
桥梁修复 rehabilitation of bridge
桥梁验收荷载试验 bridge acceptance loading test
桥梁验算荷载 check loading of bridge
桥梁养护 bridge maintenance
桥梁摇轴支座 rocker bearing;tilting bearing
桥梁摇座 rocker;rocker bearing
桥梁艺术 bridge art
桥梁翼墙终端圆柱墩 newel
桥梁引道 bridge approach; road approach
桥梁引道路堤 approach bank
桥梁引路 bridge approach; road approach
桥梁膺架 bridge falsework
桥梁膺架支撑垫块 bridge camber block
桥梁壅水 bridge backwater
桥梁用钢 bridge steel
桥梁(迂)回线 bridge loop
桥梁预测构件 bridge predicting unit
桥梁预拱度 bridge precamber
桥梁预示信号 bridge annunciating signal
桥梁载重等级标志 structure-classification symbol
桥梁载重量 loading of bridge
桥梁载重能力 capacity of bridge
桥梁载重试验 load test of bridge
桥梁胀缝 expansion joint for bridges
桥梁照查实施要点 bridge formula
桥梁遮断信号 bridge obstruction signal
桥梁枕木 bridge sleeper;bridge tie
桥梁振动 bridge vibration
桥梁振型分析 bridge vibration mode analysis
桥梁支承 bearing of bridge; bridge bearing
桥梁支承靴 bridge bearing shoe
桥梁支座 bearing of bridge; bridge bearing;bridge pad;bridge support
桥梁支座衬垫 bridge bearing pad
桥梁支座垫板 bridge bearing plate
桥梁制造 bridge manufacture
桥梁中跨 middle span of bridge
桥梁中线 center[centre] line of bridge
桥梁中心线 bridge axis
桥梁中央跨 center span
桥梁重建 bridge replacement
桥梁轴承 bridge bearing
桥梁轴线测定 bridge axis location
桥梁桩基 piled foundation of bridge
桥梁自动化设计系统 design automation system of bridges
桥梁自振频率 self-excited vibration frequency of bridge span
桥梁自振频率测量 bridge natural frequency measurement
桥梁自振周期 natural vibration period of bridge
桥梁总体规划 bridge overall planning
桥梁纵梁 way beam
桥梁纵坡 longitudinal grade of bridge
桥梁最大横向振幅 maximum lateral amplitude of bridge
桥梁作用 bridge action;bridging action
桥楼【船】 bridge; bridge house; bridge space;fore bridge
桥楼挡雨幔 bridge dodger
桥楼后舱壁 bridge after bulkhead
桥楼甲板 bridge deck(ing)
桥楼甲板上层建筑 bridge deck superstructure
桥楼建筑安装 bridge erection
桥楼前舱壁 bridge front bulkhead
桥楼上层建筑 bridge superstructure
桥楼伸展平台 bridge extension
桥楼无线电室 bridge receiving room
桥路 bridge circuit
桥路电流 bridge current
桥路法 bridge method
桥路检波器 bridge detector
桥路中电键 bridge key
桥门 portal
桥门撑杆 portal bracing
桥门撑架 portal bracing
桥门架 bridge portal;frame of bridge; portal frame;portal frame of bridge
桥门架效应 portal effect;portal frame effect
桥门净空 portal clearance
桥门联结系 portal bracing
桥门入口 entrance portal
桥门限界 portal clearance
桥门斜撑 portal brace
桥门斜支架 portal strut
桥门支撑系 portal bracing
桥面 bridge carriageway pavement; bridge carriageway surfacing; bridge floor; bridge road; floor; floor deck;floor platform;roadway
桥面安装 bridge flooring
桥面板 bridge deck slab;bridge floor plank; bridge plank; bridge slab; decking;deck slab;floor board(of bridge); flooring; floor plank (of bridge); floor slab; bridge deck(ing)
桥面板大梁 floor girder
桥面板混凝土 bridge deck concrete
桥面板下轨梁 deck bottom rail
桥面板主梁 floor girder
桥面标高 deck elevation;elevation of bridge deck
桥面步行板 foot plank of bridge floor
桥面层 bridge floor
桥面单点加载装置 single point loading device on deck
桥面底层 blind floor;subfloor
桥面底线 sodium line;soffit line
桥面吊杆 floor suspender
桥面吊机 on deck crane
桥面防水层 layer of insulation of bridge deck
桥面分车带 parting strip of bridge deck
桥面更换 bridge deck renew(al)
桥面拱腹 deck soffit
桥面构架 deck framing
桥面构造 deck construction; floor construction; floor construction of bridge
桥面夯实机器 bridge tamper
桥面横梁 floor beam; floor beam of bridge;floor beam of bridge deck; floor joist
桥面横坡 deck transverse slope;lateral grade of bridge deck;transverse slope of bridge deck
桥面横向排水 transverse drainage for bridge surface
桥面活动铺板 free floor of bridge
桥面结构 deck structure
桥面净空 clearance above bridge deck;clearance above bridge floor; deck clearance;horizontal and vertical clearance above bridge deck
桥面净宽 clear width of deck;width of bridge carriage-way
桥面抗滑性能 skid-resistant characteristics of bridge floor
桥面跨度 deck spalling; deck span; floor span
桥面跨径 deck span;floor span
桥面宽度 bridge deck width

Q

桥面沥青铺面层 asphalt deck pavement
桥面连续措施 decking bridge continuous measure
桥面梁格系 grillage system of bridge floor
桥面路 road of bridge
桥面落水管 downspouts for bridge surface water
桥面模板 bridge deck formwork
桥面磨耗层 wearing coat of bridge deck
桥面磨损层 wearing floor
桥面排水 bridge deck drainage; floor drainage; floor drainage of bridge
桥面排水系统 deck drainage system
桥面平整度测定仪 travel(l)ing-beam testing device for deck irregularity
桥面铺板 bridge deck pavement; chess
桥面铺砌小方石 bridge paving sett
桥面铺装 bridge deck pavement; bridge flooring; deck pavement; flooring; pavement
桥面铺装层 covering of deck
桥面人行板 foot plank
桥面人行道 sidewalk of bridge floor
桥面上净空 overhead clearance; upper clearance
桥面伸缩缝 deck expansion joint; expansion joint of bridge deck
桥面伸缩装置 bridge floor expansion and contraction installation; bridge floor with telescopic device; deck expansion installation
桥面系 bridge deck(ing); bridge deck(ing) system; bridge floor system; floor construction; floor grid
桥面系大梁 floor girder of bridge
桥面系高度 floor depth of bridge
桥面系跨径 floor span of bridge
桥面系统 bridge floor system
桥面系纵横梁 beam and stringer in floor system of bridge
桥面下横梁 beam under bridge; needle
桥面限界 horizontal and vertical clearance above bridge deck
桥面行人板 foot board
桥面引车道下的纵梁 track stringer
桥面纵梁 deck stringer; floor-stringer of bridge; floor-stringer of bridge deck
桥面纵坡 deck profile grade; longitudinal grade of bridge deck
桥面纵向排水 longitudinal drainage for bridge surface
桥面组件 deck module
桥名 bridge name
桥名牌 bridge name plate
桥铺板搁条 cross bearer
桥前壅水高度 backwater height in front of bridge
桥墙 bridge wall; fire bridge
桥墙盖板 bridge wall cover
桥区航道 bridge approach channel; bridge crossing channel
桥区水道 bridge crossing; bridge waterway
桥渠摇座 rocker support
桥塞 bridge plug
桥塞下入深度 depth of the bridge plug
桥上避车台 refuge; refuge bay
桥上薄板 bridge slab
桥上步道 bridge sidewalk
桥上步行板 bridge board
桥上车道 bridge road
桥上道路 bridgeway
桥上护栏 guard rail on bridge
桥上交通线路 bridged traffic route
桥上绞合缆索 bridge strand
桥上净空 clearance of bridge
桥上沥青路面 bitumen deck surfacing
桥上楼梯 bridge stair(case)
桥上路面受热 bridge pavement heating
桥上铺面 deck surfacing
桥上人行道 banket(te); banquette; bridge sidewalk; bridge walkway; catwalk
桥上铁格子 bridge grating
桥上无缝线路 continuous welded rail track on bridge
桥上行车道路面 bridge carriageway pavement
桥上行车道铺面 bridge carriageway surfacing
桥上用小方石铺路面 bridge set paving
桥设计人 bridge designer
桥式 bridge type
桥式坝 bridge dam
桥式爆炸发生器 exploding bridge wire initiator
桥式布置 layout of bridge style
桥式测高阻电表 bridge megger
桥式差接变压器 bridge transformer
桥式承窝 bridge socket
桥式垂直升船机 bridge type crane for boat lift
桥式锤 bridge type hammer
桥式的 bridged; bridge type
桥式电动起重机 electric(al) overhead crane
桥式电动抓斗起重机 electric(al) overhead travel(l)ing grab crane
桥式电流 bridge current
桥式电路 bridge circuit
桥式电路测定 bridge measurement
桥式电线机车 overhead wire locomotive
桥式吊车 bridge crane; bridge tramway; gantry crane; movable crane; over travel(l)ing crane; traversing crane; conveyor bridge; movable crane; overhead crane; overhead traveling crane; shop traveler; traverse table
桥式断路 bridge cut-off
桥式断路继电器 bridge cut-off relay
桥式断路器 bridge breaker
桥式堆料系统 bridge stacking system
桥式多层立体交叉 multiple-bridge intersection
桥式反馈 bridge feedback
桥式防护棚 bridge protection canopy
桥式放大器 bridge amplifier; bridging amplifier
桥式浮桥 bridge type pontoon
桥式复接配置 bridged-multiple arrangement
桥式钢轨接头 bridge rail joint
桥式钢丝耙 bridge cable rake
桥式高阻表 bridge megger
桥式格栅 bridging joist
桥式刮板机 travel(l)ing scraper
桥式刮板取料机 bridge type scraping reclaimer
桥式刮泥机 bridge type sludge scraper
桥式光电放大器 bridge type photoelectric(al) amplifier
桥式硅整流器电路 bridge silicon rectifier circuit
桥式环形立体交叉 bridged rotary intersection; bridged-on intersection; bridged rotary
桥式换接过程 bridge transition
桥式集装箱起重机 container bridge crane
桥式计算电路 bridge calculating circuit
桥式检测器 bridge type detector
桥式接点 bridge-type contact
桥式接法 double way connection
桥式接合 bridge-adapter
桥式接头 bridged joint
桥式接线 bridge connection
桥式可移动的混凝土搅拌机 bridge removable concrete mixer
桥式孔形挤压模 bridge die
桥式矿砂堆集机 ore stocking bridge
桥式矿砂装卸机 ore handling bridge
桥式立体镜 bridge type stereoscope
桥式连接 bridged joint
桥式连接器 connecting bridge
桥式联结 bridge linkage
桥式联结器 bridge connector; bridge fittings
桥式梁 bridge beam
桥式量规 bridge ga(u)ge
桥式滤波器 lattice-type filter
桥式路由器【计】 brouter
桥式轮斗取料机 the rotary bucket wheel reclaimer
桥式迈格计 bridge megger
桥式平衡直流放大器 bridge-balanced direct current amplifier
桥式坡道 bridge ramp
桥式铺路机 bridge paver
桥式起重机 bridge crane; gantry crane; loading bridge; overhead (traveling) crane; overhead travel(l)er; overhead travel(l)ing crane; top-running crane; transporter bridge; transporter crane; travel(l)er; travel(l)ing bridge crane
桥式起重机大梁 overhead crane girder
桥式起重机导线 bridge crane conductor
桥式起重机的跨度部分结构 span structure of crane bridge
桥式起重机钢轨 bridge rail; crane bridge rail
桥式起重机轨道 bridge rail; crane bridge rail
桥式起重小车 crane carriage
桥式驱动器 bridge driver
桥式取料机 bridge reclaimer
桥式全波整流器 bridge full-wave rectifier
桥式双工系统 bridge duplex system
桥式体视镜 bridge type stereoscope
桥式稳频振荡器 bridge-stabilized oscillator
桥式限幅器 bridge limiter
桥式卸船机 unloading bridge
桥式行车 bridge crane
桥式堰 bridge weir
桥式移动装船机 bridge-type travelling shiploader
桥式运送机 bridge transporter
桥式振荡器 bridge oscillator
桥式整流器 bridge rectification
桥式支架 bridge support
桥式抓斗吊车 bridge grab crane; overhead travel(l)ing grab crane
桥式抓斗起重机 bridge grab crane; bridge type grab crane; transported grab crane
桥式抓斗卸煤机 bridge type coal grab-unloader; mantrolley type coal unloader
桥式转换触点 double break double-make contact
桥式装船机 ship loading bridge
桥式装料机 loading bridge
桥式装卸机 handling bridge
桥式装载机吊车 loading bridge
桥式自钻旁压仪 Camkometer
桥式座架照相机 camera with overhead bed
桥枢 bridge pin
桥隧领工区 subdivision for bridge and tunnel
桥隧维修工队 bridge and tunnel maintenance team
桥隧引道开挖 approach cutting
桥索 bridge cable; bridge stay
桥索垂跨比 sag ratio of bridge cable
桥塔 bridge head; bridge pylon; pylon bridge tower; pylon tower; tower; bridge tower < 悬索桥的 >
桥塔顶端 pylon head
桥塔立柱 tower column
桥台 abut; abutment; abutment piece; abuttal; bank of pier; bank pier; bridge abutment; butment (wall); foundation pier; land abutment; land(ing) pier; pylon
桥台板 < 铁路桥梁的 > open deck
桥台背墙 abutment backwall
桥台壁 abutment wall; butment wall
桥台边跨 abutment span
桥台边墙 abutment wall
桥台变形 abutment deformation
桥台挡土墙 abutment retaining wall
桥台的背墙 backwall of abutment
桥台的顶墙 backwall of abutment
桥台底板 footing of bridge abutment
桥台定位 abutment location
桥台耳墙 abutment seat back wall; cantilevered return wall
桥台拱 abutment arch
桥台横跨 abutment bay
桥台横向节间 abutment bay
桥台后回填 back filling behind abutment
桥台后压力 back abutment pressure
桥台接缝 abutment joint
桥台跨度 < 用于分离式桥台 > abutment space; abutment span; abutment spalling
桥台廊道 abutment gallery
桥台锚固 abutment anchorage
桥台锚固栓钉 abutment anchor bar
桥台排水设施 abutment drainage
桥台前墙 front wall; front wall of abutment
桥台前趾墙 abutment toe wall
桥台墙 abutment wall
桥台石 abutment stone
桥台四分之一锥体填坡 quarter cone filling
桥台台帽 abutment cap(ping)
桥台台帽前缘的矮墙 dwarf wall
桥台台帽上填土与上部构造间隔墙 ballast wall
桥台台身 abutment body
桥台填堆体 abutment quarter cone filling
桥台填土 abutment fill
桥台位移 displacement of abutment
桥台斜翼墙 aisle wall
桥台胸墙 front wall; front wall of abutment
桥台压力 abutment pressure
桥台翼墙 abutment wall
桥台主身 main body of bridge abutment
桥台锥坡 quadrant of truncated cone
桥台锥体 bridge abutment cone
桥体 bridge body
桥头 bridge head; land abutment
桥头堡 barbican; bridge head; bridge tower; tower portal of bridge
桥头搭板 approach slab; bridge end transition slab; bridging slab; transition slab; transition slab at bridge head
桥头定线 bridgehead alignment
桥头回填设计 bridge end backfilling design
桥头检修梯 inspection ladder of abut-

ment
桥头建筑 bridgehead construction
桥头接坡 road approach
桥头警卫室 bridge house
桥头路堤 approach embankment; bridge approach embankment
桥头绿化 bridgehead greening
桥头踏板 approach cushion
桥头填方 approach fill
桥头跳车 bump at bridge-head
桥头挖方 approach cutting
桥头线向 bridgehead alignment
桥头引道 bridge approach
桥头引道接线 approach alignment; bridge approach alignment
桥头引道路堤 bridge approach embankment
桥头(引)路 road approach
桥头引线 approach embankment; bridge approach
桥头引线定线 approach alignment
桥头锥体 approach cone
桥头锥体护坡 bridge gore
桥头锥形护坡 bridge gore; cone-shape pitched slope at bridge abutment
桥托 bridgework
桥弯硅藻 cymbella
桥弯藻属 cymbella
桥尾护木 fender log
桥位 bridge location; bridge position; bridge site
桥位测设 setting of bridge location
桥位地形 topography of bridge site
桥位地质查勘 geologic(al) exploration at bridge site
桥位勘测 bridge site survey; bridge survey
桥位水流方向 current direction at bridge site
桥位选择 bridge site selection; choice of bridge site; selection of bridge site
桥位桩 bridge site stake
桥下 underbridge
桥下部结构 bridge substructure
桥下道路 undercrossing
桥下防护架 cutwater
桥下建筑部件 knocked down building components
桥下净高 clear headway of bridge; clear height of bridge
桥下净空 bridge clearance; clearance of space; clearance of span; clearance under bridge (superstructure); clearance under space; clearance under span; clear headway; clear headway of bridge; headroom under a bridge; underbridge clearance; underclearance; underclearance of bridge
桥下净空高度 free height under the bridge
桥下孔道 bridge underpass
桥下通道 bridge underpass; underpass
桥下一般冲刷 general scour under bridge opening
桥线焊 bridge seam weld
桥箱总成 axle housing assembly
桥形钢轨 bridge rail
桥形接片 jumper
桥形结构 bridge configuration
桥形结线 bridge connection
桥形连接 bridge connection; double way connection
桥形通道<机器旁的> catwalk
桥形网络 bridge network; lattice network
桥形网络滤波器 lattice filter
桥形支架 straight beam bracket
桥形装卸机 handling bridge; material transporting bridge
桥型 bridge type
桥型连接件 bridge connector
桥型滤波器 lattice-type filter
桥型网络 lattice-type network
桥型装卸机 handling bridge
桥悬臂部分 cantilever arm
桥靴 bridge shoe
桥氧 bridging oxygen
桥枕 bridge sleeper; bridge tie
桥枕护木<明桥面的> spacing timber
桥支撑 bridge branch
桥支座 bridge bearing; bridge seat; bridge support
桥址 bridge location; bridge site
桥址测量 bridge site survey
桥址的水力研究 hydraulic study of bridge site
桥址地形测量 topographic(al) survey of bridge site
桥址勘察 investigation of bridge
桥址水文观测 hydrologic(al) observation at bridge site
桥址稳定性评定 bridge site stability evaluation
桥轴 bridge axis
桥轴大梁 axle beam
桥轴负荷 axle load(ing)
桥轴容许荷载 allowable axle loading
桥轴锁闭 axle lockout
桥轴线 bridge axis
桥轴线测量 bridge axis survey
桥轴箱盖 axle-box lid
桥轴支销 axle pivot pin
桥柱灯 bridge pier light
桥砖 bridge block; bridge brick; mantle block
桥桩 bridge pile
桥纵梁 sleeper
桥族 family of bridges
桥座 bearing of bridge; bridge seat; land abutment
桥座背 back of bridge seat
桥座大梁 bridge seating girder; bridge setting girder

巧

巧克力方盘吞型构造 chocolate-tape boudinage
巧克力红色瓦片<法国北方常用的> Beauvais tile
巧克力色含铅颜料 chocolate lead
巧克力色饰油 chocolate varnish
巧妙的 skil(l)ful
巧妙设计的 tactical
巧手工 handy man

壳

壳斑藻阶段 conchocelis stage
壳板 shell plate
壳板加工车间 shell plate shop
壳板装配 shell erection
壳板装配工 ceiler; plater
壳瓣 valve
壳孢 spore
壳壁 shell wall
壳扁片 bill
壳冰 shell ice
壳菜(属) Mytilus
壳层 lamella; sheathing; shell
壳层结构 shell(ly) structure
壳层模型 shell model
壳层序列 shell sequence
壳程 shell pass
壳的对壁结构 double-skin construction
壳顶 shell roof; umbo
壳斗 cupule
壳斗科<拉> Balanopsidales
壳端孔 pendant hole
壳断裂 crustal fracture
壳二孢属<拉> Ascochyta
壳箍 boss ring
壳冠 shell crown
壳管 package
壳管式 shell-and-tube
壳管式换热器 shell-and-tube exchanger; shell-and-tube heat exchanger; shell-tube heat exchanger
壳管式减温器 shell-type attemperator
壳管式冷凝器 shell-and-tube condenser; shell-and-tube type condenser
壳管式热交换器 shell-and-coil heat exchanger
壳管式热交器 shell-and-tube heat exchanger
壳管式水冷凝器 shell-and-tube water condenser
壳管式预热器 shell-and-tube preheater
壳管式蒸发器 shell-and-coil evapo(u)rator; shell-and-tube evapo(u)rator
壳环 girdle
壳灰岩 biostromal limestone; muschelkalk
壳灰岩阶【地】 Muschelkalk stage
壳灰岩统<三叠纪>【地】 Muschelkalk series
壳荚输送器 hull conveyer[conveyor]
壳尖 beak
壳间距离 shell-to-shell distance
壳脚圈梁 footing
壳接 shell joint
壳口 faucal; shell aperture
壳块 crustal cupola
壳矿 crust ore
壳粒 capsomere; virocapsomer
壳裂 shell shake
壳裂的 shelly
壳-幔混合花岗岩 crust-mantle mixed granite
壳模 case mo(u)ld
壳模法 shell mo(u)lding
壳模型 shell model
壳模铸造 shell moulding
壳膜 shell membrane
壳内构造运动 tectonism in crust
壳内熔融 intracrustal melting
壳内叶轮 box-shrouded impeller
壳囊孢属<拉> Cytospora
壳排泄管 case drain line
壳牌(公司)老化试验 Shell ag(e)ing test
壳牌滑板式测微黏[粘]度计 Shell sliding plate microviscometer
壳牌沥青质量控制九面图 QUALAGON
壳牌流体催化裂化 Shell fluid catalytic cracking
壳盘管式氨冷凝器 shell-and-coil ammonia condenser
壳(盘)管式冷凝器 shell-and-coil condenser
壳盘管式蒸发器 shell-and-coil evapo(u)rator
壳皮 hulk
壳圈 geosphere
壳熔法 skull melting method
壳蠕孢属<拉> Hendersonia
壳石灰岩 muschelkalk
壳式泵 shell pump
壳式变压器 shell-core transformer
壳式超压开关 casing overpressure switch
壳式建筑物 shell structure
壳式盘管冷凝器 shell-and-coil condenser
壳式强制油冷变压器 shell-type forced-oil-cooled transformer
壳式热电偶 shell thermocouple
壳式蛇管冷凝器 shell-and-coil condenser
壳式铁心 shell-type core
壳式抓斗 clamshell
壳糖 chitose
壳套 body sleeve
壳体 case; housing; body case; carcase; casing; hard surface; monoblock; shell
壳体摆动 shell oscillation
壳体半径 shell radius
壳体边界 shell boundary
壳体边界应力合成 shell-boundary stress resultant
壳体边界效应 edge effect of shell
壳体边缘 shell edge
壳体薄膜 shell membrane
壳体薄膜理论 membrane theory of shell
壳体材料 shell material
壳体衬板 casing liner
壳体承载体系 shell-bearing system
壳体承重体系 shell weight-carrying system
壳体城堡高楼 shell keep
壳体城堡主楼 shell dungeon
壳体城堡主塔 shell dungeon
壳体顶部 shell key; shell top
壳体顶点 shell crown; shell point; shell vertex
壳体断片 shell segment
壳体法兰 shell flange
壳体封头 shell cover; shell head
壳体钢筋 shell reinforcement
壳体工程 shell work
壳体拱度 shell crown; shell vertex
壳体荷载 shell load
壳体横断面 shell cross-section
壳体厚度 shell thickness
壳体基础 hull foundation; shell foundation
壳体加固 shell reinforcement
壳体建筑 shell construction
壳体接缝 body seam
壳体结构 shell(-type) structure; shell(-type) construction
壳体结构的模板 form of shell structure
壳体结构体系 shell structural system
壳体结构音乐台 band shell
壳体静力学 statics of shells
壳体静载 shell dead load
壳体宽度 house width; shell width
壳体理论 shell theory; theory of shells
壳体力 shell force
壳体面积 shell area
壳体前法兰盘 front barrel flange
壳体曲度 shell curvature
壳体圈带 shell belt
壳体人孔 shell manhole
壳体扇形面 shell sector
壳体施工 monocoque construction; shell work
壳体式结构 monocoque construction
壳体体系 shell system
壳体外形 shell configuration
壳体弯曲理论 shell bending theory
壳体问题 shell problem
壳体屋顶 shell structure roof
壳体效应 shell effect
壳体斜度 shell slope
壳体泄油管 case drain line
壳体形状 shell form; shell shape
壳体压曲 shell buckling
壳体研究 shell research
壳体噪声 case noise

壳体振荡 shell oscillation
壳体振动 shell vibration
壳体支承体系 shell supporting system
壳体轴线 shell axis
壳体作业 shell work
壳筒 sleeve piece;thimble
壳筒滤器 thimble filter
壳微体 capsomere
壳纹 stria[复 striae]
壳下流 subcrustal current
壳下型重力构造 subcrustal type of gravitative tectonic
壳线间隙【地】stria[复 striae]
壳相 shelly facies
壳箱 housing box
壳屑体 liptodetrinite
壳芯 shell core
壳芯吹制机 shell core blower
壳芯电热炉 wrap-around oven
壳芯法 shell core process
壳形 shell like
壳形联轴器 clamp coupling
壳形砂 shell sand
壳形筒联轴节 split muff coupling
壳形演奏台 band shell;orchestra shell
壳形音质反射体 <室内舞台或露天剧场的> orchestra shell
壳形造型 shell mo(u)lding
壳形造型机 shell mo(u)lding machine
壳形铸件 shell-mo(u)lded casting
壳形铸造的 shell-mo(u)lded
壳形铸造法 crab process
壳形铸造用模型 shell mo(u)lding pattern
壳形钻 shell bit;shell drill
壳锈菌属 <拉> Physopella
壳源 shell source
壳源花岗岩 crust-derived granite
壳源模型 shell source model
壳源岩浆 crust-density magma
壳罩结构 geodesic construction
壳针孢属 <拉> Septoria
壳质 chitin
壳质层 chitinous layer
壳质的 chitinous;crustaceous
壳质化 chitinization
壳质煤化沥青 exinonigritite
壳质煤素质 exinite;liptinite
壳质体(焦炭)exinite
壳质微亮煤 clarite E
壳质组 exinite;exinoid group
壳柱 columella
壳桩 encased sheet pile
壳装带螺旋机构的阀 non-rising stem valve
壳状 shelly
壳状的 crustose
壳状悬墙 sheath-type curtain wall
壳椎亚纲 Lepospondyli
壳子板 shuttering
壳钻 <又称清孔钻> shell auger

峭岸 bold shore

峭壁 beetling cliff;beetling wall;cliff;crag;escarpment;mural precipice;overhanging bank;precipice;promontory;rock cliff;sheer cliff;steep;vertical cliff;bold cliff;precipice
峭壁顶部 crest of bluff
峭壁谷 steep-walled valley
峭壁海岸 cliff coast
峭壁居所 cliff dwelling(settlement)
峭壁泉 cliff spring
峭壁峡谷 steep-walled canyon
峭度 kurtosis
峭峰 brink
峭河岸 brink
峭缘 brink

翘 flower arm;petal

翘板 wane
翘边 edge lift
翘变缝 warping joint
翘变接缝 warping joint
翘成杯形 cupping
翘动 tilting
翘动管嘴 tilting nozzle
翘飞椽 cantilever eaves rafter
翘角 rake angle
翘面 warping
翘棚 cantilever roof
翘起 buckling;tilting;upwarp(ing);warpage
翘起地块【地】tilted block
翘起断块 tilt block;tilted fault block
翘起构造 upwarped structure
翘起荷载 tilting load(ing)
翘起结构 tilt-up construction
翘起力矩 tilting moment
翘起装配 tilt-up
翘曲 bowing;buckling;cambering;curl(ing);hogging;warp(age);warping;sprung;cupping;bow crook <木料的>
翘曲变形 buckling deformation
翘曲波长 buckle wave length
翘曲导缆器 warping chock
翘曲的窗玻璃 warped sheet
翘曲地震 warping earthquake
翘曲度 angularity
翘曲分析 warpage analysis
翘曲缝 warping joint
翘曲刚度 warping rigidity
翘曲过渡进口(段)warped transition approach
翘曲和压碎试验 buckling and crushing tests
翘曲湖 lake due to warping
翘曲畸变 distortion by warping
翘曲渐变进口(段)warped transition approach
翘曲力矩 warping moment
翘曲裂缝 warping crack
翘曲面 warp(ed)surface
翘曲木材 warped timber
翘曲木地板 sprung floor
翘曲扭力 warping torsion
翘曲盆地 warped basin
翘曲强度 buckling strength
翘曲试验 camber test;warping test
翘曲图式 buckle pattern;buckling pattern
翘曲图形 buckle pattern;buckling pattern
翘曲瓦 warped roof tile;warped tile
翘曲系数 <混凝土路面> warping coefficient
翘曲线 warp line
翘曲协调方程 equation of compatibility of warping
翘曲应力 buckling stress;curling stress;warping stress
翘曲约束 warping constraint
翘曲运动 warping movement
翘曲阻力 buckling resistance
翘曲作用 warping effect
翘头 nose;rearing <车辆>
翘尾护板 raised rear fender
翘檐板 cocking piece
翘砖 chuffy brick

撬 jumbos;prise[prize]

撬棒 crown bar;hand spike
撬出 pry-out;picking up
撬动 pry
撬浮石 trimming
撬杆 bodger;chisel bar lever;crowbar;pike pole;pinch bar;pry bar
撬杆保护 crowbar protection
撬杆开关 crowbar switch
撬杠 crow;crowbar;iron bar;prize;pry bar;ripping bar
撬杠拔正管材连接 crowbar corrected connection
撬棍 caster;chisel bar;claw bar;claw hammer;crowbar;crowfoot bar;dwang;gavelock;granny bar;hand spike;iron claw;knish;nail bar;nail puller;pinch bar;prise[prize];pry bar;ripper;ripping bar;track liner;wrecking bar
撬货千斤顶 cargo jack
撬具 pry;prying tool
撬开 break open;lever;prize
撬开作用 prying action
撬落 barring down
撬落矿石 bar down ore
撬落松煤 ragging off coal
撬落危石 scaling-down loose rock
撬木 sleight
撬起 pry
撬起作用 prying action
撬石工 barman
撬式刮路器 sledge drag
撬式摊铺装置 averaging ski assembly
撬松作用 prying action
撬锁工具 picklock
撬胎棒 tire[tyre] spoon
撬卸扣栓的棍 shackle crow
撬岩工作 rock-removal chore
撬整工作面 bar down

鞘翅目 <昆虫> Coleoptera

鞘蛾 case bearer
鞘锈菌属 <拉> Coleosporium
鞘褶皱 sheath fold
鞘柱 sheath column
鞘状结构 sheath-like structure

切岸 cutbank

切凹口 notch cutting
切凹口压力机 notching press
切板机 block cutter;lath cutter
切棒机 clipping machine
切绷带机 cloth slitting machine
切比雪夫逼近法 Chebyshev's approximation
切比雪夫不等式 Chebyshev's inequality
切比雪夫多项式 Chebyshev's polynomial
切比雪夫级数展开 Chebyshev's series representation
切比雪夫集合 Chebyshev's set
切比雪夫均衡器 Chebyshev's equalizer
切比雪夫滤波器 Chebyshev's filter
切比雪夫模 Chebyshev's norm
切比雪夫平滑法 Chebyshev's smoothing
切比雪夫微分方程 Chebyshev's differential equation
切边 bead cut;cut-off the edge;cutting edge;edge cut;edge-notched;edge trimming;nipping;paring;prune;rasping;scrap edge;side cut(ting);side scrap;slit selvage;trimming cut
切边板材 sheared plate
切边成型模 wipe-down trim die
切边冲模 trim die
切边冲头 trimming punch
切边刀 trimming knife
切边的中厚板 shear plate
切边定位法 trim stop
切边钢板 sheared steel plate;trimming steel plate
切边工 trimmer
切边机 bead cutter;chipping edger;debeader;edge trimmer;mitring machine;trim cutter;trimmer;trimming press;print trimmer <照片>
切边剪 end trimming shears
切边剪机 trimming machine
切边卷取 balling up
切边卷取机 baller;balling machine;scrap baller
切边料 edge trims
切边裂纹 parch crack
切边模 comb die;cropping die;cut-off die;shaving die
切边器 verge cutter
切边压力机 clipping press
切边圆锯 edge circular saw
切边凿 fettling chisel
切边整割机 trimming sliver
切边中厚板轧机 sheared plate mill
切边砖 side cut brick
切边装置 bulb edge severing device
切变波 rotational wave;shear(ing) wave
切变不稳定性 shearing instability
切变场 shear field
切变磁滞 sheared hysteresis
切变断裂 shearing failure
切变干涉仪 shearing interferometer
切变降解 shear degradation
切变角 angle of shear
切变结构 shear structure
切变拉伸 shear stretch
切变理论 <材料破坏的> shear theory
切变力 shear force
切变裂缝 shear crack
切变裂痕 shear lip
切变流(动) shear(ing) flow;shearing current
切变流均化库 shearing stream homogenizing silo
切变流理论 shear flow theory
切变流紊动 shear flow turbulence
切变率 rate of shear
切变面 shear plane
切变模量 modulus of shear deformation;shear(ing) modulus
切变黏[粘]滞系数 shear-viscosity coefficient
切变强度 shear strength
切变区 shear zone;slipped region
切变时间 switching time
切变矢量 shear vector
切变速度 shear velocity
切变速率 rate of shear;shear rate
切变弹性 shear elasticity
切变推力 shear thrust
切变系数 coefficient of shear;shear coefficient
切变线 shear line
切变相变 shear transformation
切变向量 shear vector
切变形 shear deformation
切变性 sheariness;sheary
切变絮凝 shear-flocculation
切变应力 shear stress
切变原理 shearing principle
切变运动 shearing motion;shearing movement
切变重力波 shear-gravity wave
切标 trimscript

切冰机 ice cutter
切冰刃 ice cutter
切玻璃刀 glazing knife
切薄片机 microtome
切补修复 excision repair
切布刀 rag knife
切裁 cut(ting)
切裁玻璃用金刚石刀 cutting diamond
切裁尺寸 cut size
切裁刀 cutter blade;cutting stone
切裁工 capper;cut-off man
切裁规格 cut size
切裁后的压延玻璃 rough glass
切裁间 cutter's bay
切裁率 percent of shear cut;rate of cutting and sizing
切裁损耗 cutting loss
切裁装置 cutting device
切槽 fluting; grooving; initial cut; necking-groove; notch cutting; recess;shotted;slot
切槽车刀 necking bit;recessing bit
切槽刀 gashing tool;grooving cutter; grooving tool; slotting cutter; stick bit
切槽的 notched
切槽工具 grooving tool;necking tool
切槽机 grooving machine; recessing machine;slot-cutting machine
切槽间距 kerf spacing
切槽角 angle of rifling
切槽锯 drunken saw
切槽锯床 drunken sawing machine
切槽口刀 groove cutter
切槽螺母 slit nut
切槽弯曲试验 nick bend test
切槽压力机 notching press
切槽装置 groover
切草簇 turf cutting
切草机 chaff cutter; silage cutter; straw chopper
切草皮器 sod cutter
切层边坡 slope of strata dip cross excavation
切层滑动 insequent slide
切层滑坡 cut bedding plane landslide;insequent landslide
切成薄层 skive
切成薄片 laminate;shave;shive;skive
切成层板 knife-cut veneer;sliced veneer
切成长条 sliver
切成平行六面体的石材 parallelepipedal cut stone
切成扇形 scallop(ing)
切成碎片 chipping;cut to pieces
切成小方块 dice;dicing
切齿 cutting teeth;gear cutting;incisive tooth;tooth cutting
切齿安装距 cutting distance
切齿刀 gear cutter
切齿方法 gear cutting consideration
切齿机 gear cutter;gear cutting machine;tooth cutting machine;wheel cutting machine
切齿链节 cutter link
切齿轮 gear cutting machine
切齿起始孔 starting hole of gear cutting
切齿铣刀 nicked teeth milling cutter
切齿型 secodont
切齿中心钻头 center[centre] bit of claw type
切出线脚 stuck mo(u)lding
切除 ablation;abscission;cutting (out);excide;exscind
切除机 buster
切除坯体余泥 scraping
切除术 resection
切触 tangency
切穿逆断层 break thrust
切锉刀具 stinger bit
切锉刀片 stinger bit
切大理石厂 marble sawing plant
切刀 chipping knife;cutter;cutter and rubber; cutting bit; cutting blade; cutting knife;cutting-off blade;diffusion knife;gad;parting tool;slitter;trimmer
切刀边 cutting blade
切刀法 cutting ring method
切刀开关 chopper switch
切刀头 cutter head
切刀钻 cutter drill
切导 incisal guidance;incisal guide
切到基准线 cut-off to grade
切的 tangent
切底作用 undercutting
切点【数】point of tangency;tangency point;tangent point;point of contact
切点弦 chord of contact
切电缆铅皮用刀 cable lead cover cutter
切掉 cutaway
切丁机 cuber
切钉 cut nail
切短 shorten cut
切短的侧板 cutback side plate
切断 cut(ting)-off;cut(ting) out;discoupling; elimination; key-off; key out; lockout; make dead; out-of-work; overcut; parting; scission; segmentation; sever; shearing-off; shut-off;slicing;sunder;switch off; throw off; throw-out; trip-out; truncation; turn-off; turnout; disconnect <切断联络、电路等>
切断按钮 cancel(ling) key;split key
切断摆锯 swing cut-off saw
切断边 distributing edge
切断并列运行 parallel off
切断操作 rupturing operation
切断插刀 parting slotting tool;parting tool
切断长度 shearing length
切断车床 cutting-off lathe
切断车刀 parting-off tool
切断车面机 cutting-off and facing machine
切断传动 release the drive
切断锤 cutting hammer
切断刀 cut-off tool;parting-off tool
切断刀片 chopping blade; chopping knife
切断的 killed;out-of-gear
切断低压【气】cut-off low
切断点 point cut-off;cutoff point
切断电磁铁 tripping magnet
切断电磁线圈 cut-off solenoid
切断电弧 break arc;breaking arc
切断电键 cut-off key; cut-out key; split key
切断电力额定值 cut-off power rating
切断电流 de-energise[de-energize]; killing;power-off;rupturing current
切断电流的 deenergized;without current
切断电流时的火花 break spark
切断电流装置 shut-down feature
切断电路 clear the circuit; lockout circuit;put out of circuit
切断电源 cut-off the electricity supply; dump; memory dump; power cut;power cut-off;power dump
切断电源的 power-off
切断电源开关 power disconnect switch
切断电源试验 input rejecting test
切断电源状态 power-off condition
切断电阻组合开关 resistance switch group
切断定心机 cutting-off and centering machine
切断短路 cutting-off of short-circuit
切断阀 dump valve;isolating valve;isolator valve
切断法 process of chopping
切断分路 prune branches
切断负载 dropping of load
切断高压【气】cut-off high
切断功率 rupturing duty
切断功能 turnout function
切断供给 sever supply
切断沟 kerf[复 kerve]
切断过程 cutting-off process
切断厚度 shearing thickness
切断环 cutting ring
切断机 cutter;cut(ting-off) machine; guillotine
切断机床 cutting-off machine
切断机构 shut-down mechanism
切断继电器 cut-off relay;lockout relay;shut-down relay;trip(ping) relay
切断加长 jumboisation[jumboization]
切断检验 cutting test
切断件 blank
切断键 cut-off key
切断交通 break communication
切断接长 jumboisation[jumboization]
切断金属线或钢丝绳的工具 wire axe
切断井下钢丝绳的割刀 hook rope knife
切断锯 amputating saw;cut-off saw
切断开关 cut-off cock;cut-off switch; cut-out cock;cut-out switch;disconnecting switch; disconnector; discontinuity switch;isolating cock;isolating link;isolating switch;isolation switch;shut-off cock
切断空气 exclusion of air
切断拉线 releasing the pull
切断螺管线圈 cut-off solenoid
切断螺线管 shut-off solenoid
切断脉冲 break impulse; chopped pulse
切断铆钉 cutting out of rivets
切断面 cutting plane;section
切断模 cut-off die;knife edge die
切断模具 parting-off grinder
切断膜膛 cutter
切断(磨)轮 cut(ting)-off wheel
切断能力 breaking capacity
切断刨刀 cut-off shaping tool;parting planing tool
切断平面 secant plane
切断器 breaker;cutter;guillotine;microtome
切断钳 excising forceps
切断区 key-out region
切断燃料 fuel shut-off
切断若干线路的电流供应 <电源过载时> load-shedding
切断塞孔 break jack
切断砂轮 cutting-off wheel
切断时间 break time; cut-off time; disconnecting time
切断术 amputation
切断瞬间 breaking moment
切断速率 chopping speed
切断网络管 shut-off solenoid
切断位置 off position
切断铣刀 disc cutter
切断纤维 staple fiber[fibre]
切断线路 killing line;line disconnection
切断销 cutting pin
切断信号 disconnected signal;disconnecting signal; shut-off signal; clipping signal
切断序列 shut-off sequence
切断序数 cut number
切断旋塞 corporation cock;curb cock; cut-off cock; cut-out cock; isolating cock;shut-off cock
切断压力机 cutting out press
切断研磨机 parting-off grinder
切断用金属丝 cutting out wire
切断圆板 cut-off wheel;cutting wheel
切断周边 boundary cut-off
切断装置 shut-off device;shut-off unit
切断作用 chopping-off action
切废料装置 scrap cutter
切废纸机 paper shredder
切分 slitting
切分机 slicing machine
切分剪 dividing shears
切分孔型 slitting pass
切缝 curf; kerf[复 kerve]; lancing; open slot;joint-cutting;joint-outing <整体大面积的>
切缝法 joint-cutting
切缝痕 drag of kerf
切缝机 joint cutter;jointer;joint sawing machine;transverse joint cutter; groover; grooving machine; joint-cutting equipment; joint-cutting machine <用于切割混凝土路面>
切缝模 lancing die;slitting die
切缝模片 slotted templet
切缝铣刀 metal-slitting saw
切缝栽植 slit planting
切钢轨锯 railroad saw
切钢丝绳刀 rope knife
切割 chipping;cut(ting);dicing;dissection;incise;incision;knifing;slicing cutting
切割板型 septum-type
切割板型分离器 septum-type separator
切割边(缘) cut(ting) edge;steel cutting curb;steel drum curb <钢井筒或套管底部的>
切割标记 cut mark
切割波痕 dragline
切割不齐 cutting flare
切割部分 cutting part
切割长度 cut length
切割成较小的用料 cutstuff
切割成小块 dice;dicing
切割尺寸 length of cut
切割吹管 cutting blowpipe;flame cutter;cutting torch
切割错误 miscut
切割刀 parting knife;parting tool
切割刀具 parting tool
切割刀轮 cutting wheel; scoring wheel
切割刀盘 cutter disc[disk];cutter head
切割刀片磨石 sickle grinder
切割的 dissected
切割的材料 material to be cut
切割的地形 dissected topography
切割的构造层 cutting tectonic
切割低平原 dissected low plain
切割地 broken ground
切割地面 rough country
切割地区 cross country; intersected country; intersectional country; rough area
切割地形 dissected topography
切割点 cutting point
切割电缆 cut cable
切割电流 cutting current
切割陡坡地 rough broken land
切割方法 cutting method
切割费用 cutter cost

切割缝 kerf[复 kerve]
切割杆件机 bar cutting shears
切割钢筋 cutting of reinforcement; burning reinforcement <用氧气乙炔切割炬>
切割高度调节杆 cutting height control
切割高原 dissected plateau
切割工程 cutting engineering
切割工具 cutter
切割工作面 face preparation
切割公差 cutting tolerance
切割谷 dissecting valley
切割管 cutting pipe
切割规 cutting ga(u)ge
切割滚刀 kerf cutter
切割滚筒 cutting cylinder
切割火焰 cutting flame
切割机 chopper; cutter; cutting machine; scoring machine; knife cutter <制砖用的>; guillotine
切割机构 cutting mechanism
切割机头 cutter head
切割加工车间 cutting and forming shop
切割角 angle of cutting; cutting angle
切割晶体 cut crystal; sliced crystal
切割镜片 cut-off lens
切割炬 cutting torch
切割锯 slitting saw
切割锯齿 gulleting
切割力 cutting force
切割沥青 asphalt cutting
切割轮 cut-off wheel; cutter wheel
切割螺纹 cut thread
切割螺旋 cutting auger
切割毛边 edge as cut
切割密度 density of dissection
切割面 cutting plane; discontinuity; dissected plane
切割抛物面天线 cut-parabolic(al) antenna
切割配筋 cutting of reinforcement
切割喷枪 cutting burner
切割喷嘴 cutting tip
切割盆地 dissected basin; dissecting valley
切割平原 dissected plain
切割坡度 slope of cutting
切割气流 cutting steam
切割器 apparatus cutting; colter; cutter(bar); cutting knife; cutting machine; knife bar
切割器小镰刀 sickle
切割器传动离合器 cutterbar clutch
切割器刀杆 knife bar
切割器的传动装置 cutterbar drive
切割器的定刀片 knife plate
切割器的动刀片 sickle section; sickle knife
切割器对心 knife register
切割器护刃器 sickle guard
切割器起落操纵杆 cutterbar control lever
切割器前斜伸量 cutterbar lead
切割器式割草机 cutterbar mower
切割器压刀板 knife clip
切割器支撑滑脚 knife support
切割器总成 cutterbar assembly
切割墙板 cutting wallboard
切割容度 cut capacity; cut value
切割沙片 cutting send disc[disk]
切割砂轮 abrasive cut-off wheel; cut-off wheel; stripping wheel; cutting wheel
切割筛 cutting screen; cutting sieve
切割设备 cutting apparatus; cutting equipment; scoring equipment
切割射流 <水力开挖的> cutting jet
切割射线流 <探测器的> flux traverse
切割深度 cut penetration; depth of cut; depth of dissection
切割石材的锯 stone cutting saw
切割石工 stone cutter
切割石(块) cut stone
切割式井壁取芯器 core slicer
切割式掘进机 cutting-type tunneling machine
切割术 stereotomy
切割束 chopped beam
切割速度 cutting speed; scoring speed
切割损失 cutting loss
切割台 cutting table
切割碳棒 carving carbon
切割套管 casing severing
切割头 cutting head
切割头部 cutting head
切割图 cutting drawing
切割挖掘机 cutter grab
切割屋瓦与斜沟斜面接合 cut-and-mitered[mitred] valley; cut-and-valley
切割屋瓦与斜脊斜面接合 cut-and-mitered[mitred] hip
切割线 cut line; line of cut
切割线飞行 cross flying
切割飞行和切割线控制网飞行 cross flying and control-network flying
切割线间距 distance between cross line
切割线偏差 cutline deviation
切割镶面板 loose cut veneer
切割镶嵌 cutting mosaic
切割修补 excision repair
切割焰 cutting flame
切割焰吹管 cutting flame torch
切割氧 cutting oxygen
切割硬度 cutting hardness
切割用电极 cutting electrode
切割用刮板 cutting screed
切割用砂轮 parting wheel
切割圆盘 slitting disc[disk]; slitting wheel
切割圆形玻璃设备 trammel
切割针 cutting needle
切割制榫 joint-cutting; joint-outing
切割砖的钢丝 brick cutting wire
切割转盘 chopping wheel
切割桩头长度 pile cutoff
切割桩头高程 cut-off level
切割装载机 cutter loader
切割装置 cutting equipment
切割准平原 dissected peneplain
切割桌 cutter's table; cutting frame; cutting table
切割桌上刻度尺 cutter's table ruler
切割桌用尺　cutter's lath; cutter's straight edge
切割作用 cutting action; dissection
切隔磁铁 septum magnet
切根伐树机 uprooter
切根机 root cutter
切沟 kerfing
切沟治理 reclamation of gull(e)y
切管刀 casing knife; pipe cutter
切管和车螺纹机 pipe cutting and threading machine
切管和车丝机 pipe cutting and threading machine
切管机 inside cutter; pipe cutter; pipe cutting machine; tube cutter; tube cutting(-off) machine
切管器 casing knife; inside cutter; pipe cutter; pipe cutting machine; tube cutter
切管钳 pipe cutting pliers
切光器 chopper
切轨机 rail cutting machine
切轨锯 railroad saw; rail saw
切轨圆 tread circle
切过平面山 mendip
切合测距仪 cut-image range finder
切合实际的计划 sensible plan
切合实际的设计 realistic design
切痕 notch
切横缝机 transverse joint cutter
切花 cut flower
切花园 cut flower garden
切环槽 fraising
切环检验 ring section examination
切换 cut-over; switching over; switchover; throwback; throw over; transfer
切换程序 switching program(me)
切换触点 double-throw contact; transfer contact; two-way contact
切换电弧 switching arc
切换电路 commutation circuit
切换阀 reversing valve; transfer valve
切换方程 switching equation
切换方式 switching mode
切换杆 changer lever
切换功率 switching power
切换构件 switching member
切换机构 shifter; switching mechanism
切换计划 switching schedule
切换继电器 change-over relay; drive control relay; switching relay; transfer relay
切换阶段 switch step
切换接点 two-way contact
切换矩阵 switching matrix
切换开关 changer; commutator; commutator changeover switch; switch
切换控制键 toggle control key
切换录纸刀 chart cutter
切换脉冲发生器 switching pulse generator
切换母线 transfer bus
切换片 transfer strip
切换屏幕 flipping screens
切换期 switching period
切换时间 switching period; switching time; transit time
切换时间效应 transit time effect
切换顺序检测器 switching sequence detector
切换台 cutbank
切换通道 switching channel
切换位置 switching position
切换旋钮 switch knob
切换循环 switching cycle
切换原理 switching principle
切换装置 reversing service; shifter
切击式水轮机 tangential turbine
切击式转轮 tangential wheel
切机 cutting machine; dimension saw
切架 cutting frame
切肩刨刀 shoulder plane
切肩用具 setting shoulder tool
切趾滤波 apodization filtering
切趾器 apodizer
切角 canted corner; chamfer; clipped corner; corner cut; cut-off corner; cutting corner; dubbed corner; run-off corner; tangential angle
切角边 feathered edge
切角的 corner cutting
切角(对接)接头 chamfered joint
切角法 cut-off method
切角机 angle cutter
切角面终止石 stop-chamfer
切角平搭瓦 shouldering
切角曲线 isoptic curve
切角调节 cutting angle adjustment
切角镶边砖 mitred closer
切角柱 canted column
切接 cut grafting; veneer crown grafting
切接刀 grafting knife
切金属片机 snips
切进刀 cutting-in tool
切经线 tangent meridian
切晶片 pellet
切径平面 diametral plane
切距 length of tangent(ial)(line); tangent distance
切距的 orthoptic(al)
切距曲线 orthoptic(al) curve
切距圆 director circle; orthoptic(al) circle
切绢 strick
切开 ablation; chop off; cleave; cut-and-carve; cut(ting)-off; cutting out; excide; incise; incision; kerf[复 kerve]; shear-out; slice; slicing; slit
切开边 slit edge
切开采区 blockout
切开触点 cut-off contact
切开的格式纸 cut form
切开的混凝土 severing concrete
切开的砌块 split block
切开断面 broken-out section
切开格式纸方式 cut-forms mode
切开挂线法 incision and thread drawing therapy
切开剪 incision scissors
切开砌块 splitting block
切开试验 cut-up test
切砍 chopping
切刻 incise
切空间 tangent space
切口 cant bay; cut(-out); dap; groove; incision; kerf[复 kerve]; lancing; notch(ing); outcut; pitching-in; rebate; recess; scarf(ing); scotch; slit orifice; slot
切口板 slotted plate
切口棒冲击试验 notch-bar impact test
切口棒试验 notch bar test
切口比率 notching ratio
切口边 cutting shoulder
切口编码法 direct coding
切口槽 groove of the notch
切口冲击黏[粘]度 notch impact toughness
切口冲击强度 notched impact strength
切口冲击强度值　notch impact strength value
切口冲击试验 notch shock test; nick-break test; notched bar test
切口脆性 notch brittleness
切口脆性试验 notch sensitivity test
切口搭接焊 slotted lap weld
切口刀 recessing bit; recessing tool
切口的 notched
切口断裂试验 nick-break test
切口杆件 notched bar
切口焊 slot weld
切口焊缝 slot welding
切口环 <盾构的> cutting edge; shield-hood
切口交变挠曲试验 alternating notch bending test; alternating notch flexure test
切口宽度 kerf[复 kerve]
切口拉力强度 notched bar strength
切口拉伸强度 notched tensile strength
切口拉伸试验 notched tensile test
切口拉伸试样 notched tensile specimen
切口梁 <以备联结其他构件> coped beam
切口(量水)堰 notched weir
切口(量水)堰板 notch plate
切口裂开 disruption of wound
切口灵敏度 notch sensitivity
切口面积 notched area

Q

切口敏感度 notch sensitivity
切口敏感性 notch sensitiveness
切口敏感性试验 notch sensitivity test
切口模具 louvered die
切口挠曲试验 notch bending test
切口疲劳强度 notch fatigue strength
切口疲劳试验 notch fatigue test
切口强度 notch strength
切口曲线 notching curve
切口韧性 notch ductility
切口韧性试验 notch toughness test
切口锐度 notch acuity
切口上缘 cutting shoulder
切口深度 notch depth
切口式导管 resected duct
切口试棒 notched bar
切口试验 notch test
切口试样 notched specimen
切口试样冲击试验 notched bar test
切口收缩 notch contraction
切口天线 notch antenna
切口弯曲试验 notched bar test
切口铣刀 slotting cutter
切口系数 notch factor
切口线对 cut-out
切口效应 notched effect
切口溢流坝 notch spillway dam
切口应力 notch stress
切口应力集中效应 notch stress-concentration effect
切口影响 notch effect
切口增长阻抗 cut growth resistance
切口震动试验 notch shock test
切口锥形套筒 split conic(al) bushing
切块产量 cut-out yield
切块法 slab method
切块机 dicer; divider; slabber; slabbing machine
切拉树脂 churrah(gum)
切蜡器 paraffin(e) cutter
切力 shear(ing) force
切力系数 shear force factor
切粒机 dicer
切料 blanking
切料冲头 shearing punch
切料机 stock cutter
切料头 cropping
切料头机 cropping machine; cropping shears
切料头推出机 crop end pusher
切流旋塞 shut-off cock
切路缝机 road joint cutter
切伦科夫玻璃 Cerenkov's glass
切伦科夫辐射 Cerenkov's radiation
切伦科夫探测器 Cerenkov's detector
切伦科夫效应 Cerenkov's effect
切螺栓头的大凿 slogging chisel
切螺纹 chase; chasing; screw cutting
切螺纹角 angle of rifling
切螺旋槽刀杆 rifling bar
切绵 strick
切面 broken-out section; chamfer; profile; section; sectional plane; shearing-off; tangent(ial) plane; cutting plane
切面导数 tangential derivative
切面法 tangent system; cutting plane method
切面方程 tangential equation
切面尖拱 tangent ogive
切面角 chamfer
切面面积 cross-section area
切面曲率 tangential curvature
切面铁芯 cut core
切面投影 tangent plane projection; tangent projection
切面图 sectional elevation
切面线 section line
切模机 mo(u)lding machine
切泥饼设备 disc-cutting unit
切泥弓 bowl-type cutter; harp
切泥坯机 cut-off table
切泥台 cutting-off table
切诺韦思桩 Chenorveth pile
切坯台 cutter station
切片 cut film; cutlet; cut sheet; cutting; microtome section; paring; pellet; piece; sectioning; slab(bing); slice; sliced veneer; slicing; shiving <分层进行挖掘>
切片标本 section preparation
切片带 section ribbon
切片刀 microtome; slicer; slicer knife; slicing knife
切片刀刀架 microtome knife holder; microtome knife rack
切片刀痕 relief lines
切片刀自动研磨器 automatic microtome knife sharpener
切片法 section-cutting; sectioning; slicing; microtome method; microtomy <测表面浓度>
切片分析 slice analysis
切片光面 tight side
切片机 chipper; chipping machine; section cutter; sheeter; slicer; waferer
切片机刀片 microtome knife
切片锯 halving saw; slicing saw
切片器 slicer
切片染色 section staining
切片术 microtomy
切片台 sheet cutting table
切片质量 chipping qualities
切平的 truncated
切平面 tangent plane
切平山嘴 blunted spur
切齐寸 trimmed size
切气管 cutting tube
切铅刀 lead cutting knife
切钳 nippler pliers
切乔特公式 Chechotta formula
切球面 tangent sphere
切曲面 tangent surface
切取馏分 cut fraction
切去 mutilate; slicing
切去边缘的板 trimming plate
切去部分的 cutaway
切去棱角 cant
切去棱角的 canted
切去下部 undercut
切去一部分 slice
切去一部分的 cutaway
切去隅角的 cutaway
切热金属凿刀 hot chisel
切入 cut-over; cut(ting)-in; pitching-in; plunge; undercut
切入河曲 incised meander
切入或挑出 undercutting
切入角 digging angle
切入力 shearing force
切入深度 depth height
切入式破坏 break-in failure
切入水道沉积 cut-in-channel deposit
切入靴 <厚壁管前端的> cutting shoe
切梢小蠹属 <拉> Blastophagus
切身利益 vital interest
切深 <切深孔型中的> cutting-in; knifing
切石 stone cutting
切石刀 lithotome
切石的面 panel
切石法 stereotomy; stone cutting method
切石工作 rock cutting
切石机 rock cutter; split-face machine; stone cutter; stone cutting machine
切石锯 rock saw
切石锯条 stone saw blade
切石术 lithotomy
切石圆锯 slitting disc[disk]
切蚀基准面 base level of corrosion
切熟铁条 cabbling
切水 <油水分界面> oil-water interface
切丝板牙 thread die
切四边 square cut
切速 cutting speed
切速稠化 shear-rate thickening
切速触变性 shear-rate thixotropy
切速软化 shear-rate softening
切速稀化 shear-rate thinning
切速硬化 shear-rate hardening
切碎 crumble; cut to pieces; disintegrate; disperse; shred(ding); slash
切碎储藏 cut-and-stored for later use
切碎吹送机 chopper-blower
切碎刀 shredder bar
切碎的谷草 chaff
切碎滚筒 chopping drum; knife cylinder; knife drum; shredding cylinder; shredding drum
切碎混合搅拌装置 cutter-mixer agitator
切碎机 breaking machine; chopper; comminutor; devil; knife mill; macerater [macerator]; mincing machine; shredder; shredding machine
切碎机定刀 shredder bar
切碎螺旋 cutting auger
切碎木块 chopping block
切碎碾磨机 chopper mill
切碎抛送机 cutter thrower
切碎器 chopper; cutter head; reducer; slasher
切碎器刀片 chopper knife
切碎台 cutting bench
切碎脱粒机 chopper thresher
切碎脱粒装置 chop-thresher plant
切碎卸载机 cutter unloader
切碎植物撒布机 mulch spreader
切碎装载机 cutter loader
切碎装置 cutter head; lacerating machine; shredding mechanism
切碎装置刀片 chopper knife
切碎装置导向架 knife guide
切碎装置外壳 cutter body
切碎装置轴 cutter shaft
切榫刀 cope cutter
切损 chipped
切胎缘机 debeader
切滩 cut across the bar
切滩裁弯 chute cut-off
切题文献 pertinent literature
切条 slit(ting); sliver
切条刀 slitting knife
切条机 bar cutter; slitter; slitting shear machine
切条件 tangent condition
切头 butt; crop(end); cutter head; discard; front-end crop; lost head; top crop; mill scrap <轧材的>
切头机 cropper
切头砖 clip; cove header
切透 cut-through
切土部件 ground penetrating blade
切土铲刀 stinger
切土刮刀 cutter blade
切土管头 cutting shoe
切土环刀 circular soil cutter
切土机 clay cutter
切土角距离 cutting edge distance
切土角调整 <推土机> pitch adjuster
切土犁 cutting plow
切土深度 depth of cut
切土深度调整 <推土机> tip adjuster
切土式挖泥船 clay-cutter dredge(r)
切土式挖泥机 clay-cutter dredge(r)
切土式吸泥船 clay-cutter(suction) dredge(r)
切土筒 clay cutter
切土头 cutter head
切土圆盘刀 ground-working disk blade
切土嘴 cutting shoe
切挖射流 cutting jet
切尾 back end crop; crop; crop end
切吸式挖泥机 cutter-suction dredge(r)
切下的小片 snippet
切下来 chip-off
切线 intersecting line; line of contact; tangent; tangent(ial) line; tangent ray
切线棒机构 tangential bar-mechanism
切线逼近法 tangential approximation (method)
切线变形模量 tangent(ial) modulus of deformation
切线波路径 tangential wave path
切线长度 length of tangent(ial)(line); tangent(ial) length
切线超高缓和段 tangent(ial) superelevation run-off
切线超高延伸段 tangent(ial) superelevation run-out
切线车刀 tangential bit; tangential cutter
切线垂距 tangent(ial) offset
切线丛 tangential complex
切线道路 tangential road
切线的 tangent(ial)
切线的斜率 tangential slope
切线点 point of tangency [tangent]; tangential point
切线冻胀力 tangential frost heaving force
切线法 tangential method
切线法标定曲线 pegging out a curve from the tangent
切线法则 tangential rule
切线反力 tangential reaction
切线方程 tangential equation
切线方位角 bearing of tangent
切线方向 tangential direction
切线方向角 bearing of tangent
切线方向进刀剃齿法 right-angle traverse shaving
切线分割规 tangential division scale
切线分量 tangential component
切线辐轮 tangential spoke wheel
切线刚度 tangential stiffness
切线公式 tangential formula
切线规则 rule of tangential line
切线缓和段 tangential run-off
切线绘角法 plotting of angle by tangents
切线极坐标 tangential polar coordinate
切线加速度 tangential acceleration
切线剪 cropping shears
切线剪应力 tangential shearing stress
切线交叉口 intersection of tangents
切线交点 intersection of tangents; point of intersection of tangents
切线角 angle of tangent; grazing angle; angle of contingence
切线截距 tangential intercept
切线进场法 tangential approach principle
切线进给 tangential feed
切线进口 tangential inlet
切线进料 tangential feed
切线进入式旋流集尘器 tangential-entry type cyclone
切线距离 tangent(ial) distance
切线锯开 tangential saw
切线锯木法 tangential sawing; tangent-sawn
切线孔材 tangential porous wood
切线孔旋涡喷雾器 tangential hole swirl atomizer

切线力 circumferential force; tangential force
切线连接式环形交叉 tangential roundabout
切线联结式环形交叉 tangential roundabout
切线轮辐 tangential spoke
切线螺旋 tangent(ial) screw
切线面 tangential plane
切线模量 tangent(ial) modulus
切线模数 tangent(ial) modulus
切线摩擦力 tangential friction force
切线内浇口 tangential gate
切线挠度 tangential deflection
切线挠距 tangent(ial) deflection
切线偏移 tangent deflection; tangential offset
切线起点 point of tangency
切线曲率 tangential curvature
切线矢量 tangential vector
切线式燃烧器 tangential burner
切线视准 tangent line collimation
切线收缩 <木材的> tangential shrinkage
切线水流 circumferential flow
切线弹性模量 tangent(ial) modulus of elasticity
切线剃齿 tangential shaving
切线斜率 <速率分布曲线的> skewness index
切线型薄壁组织 tangential parenchyma
切线性空间 tangential linear space
切线性质 tangential property
切线压力 circumferential pressure; peripheral pressure; tangential pressure
切线延伸段 tangential run-off
切线样板精车刀 skiving tool
切线叶片式流量计 tangential vane wheel type flowmeter
切线应力 tangential stress
切线影 tangential shadow
切线运动 tangential motion
切线增压泵 tangential pump
切线折弯机 tangential bender
切线支距 tangential distance; tangent(ial) offset
切线支距法 method of tangent(ial) offsets; tangent(ial) offset method
切线支座 tangential support
切线铸造 tangential casting
切线锥近似法 tangent-cone approximation
切线锥面 tangential cone
切线纵距 tangential ordinate
切线阻尼 tangential damping
切线坐标 tangential coordinates
切向闭合光线跟踪 tangential close ray tracing
切向壁 tangential wall
切向变形 tangential deformation
切向波 tangential wave
切向槽 tangential slot
切向场 tangential field
切向车刀 tangential bit; tangential cutter; tangential tool
切向成型磨削 tangential form grinding
切向储库 tangential storage silo
切向传动带 tangential belt
切向刀夹 tangential tool-holder
切向导数 tangential derivative
切向的 circumferential; tangential
切向点 tangential point
切向点荷载 tangential point load
切向冻胀力试验 tangential frost heaving force test
切向断层 tangential fault
切向法 tangential method
切向反光镜 tangential mirror
切向反力 tangential reaction
切向非整周进水式水轮机 tangential partial turbine
切向分辨率 tangential resolution; tangential resolving power
切向分力 tangential component
切向分量 tangential component
切向分裂 tangential division
切向分速度 tangential component velocity; tangential speed (component); whirl(ing) speed; whirl(ing) velocity
切向风应力 tangential stress of the wind; tangential wind stress
切向负荷 circumferential load; tangential load(ing)
切向刚度系数 tangential stiffness coefficient
切向供给 tangential admission
切向光束 tangential pencil of rays
切向光线 tangential ray
切向光线聚点 tangential ray focus
切向滚刀 tangential hob
切向滚削 tangential hobbing
切向过滤 tangential filtration
切向荷载 circumferential load; tangential load(ing)
切向滑距 tangential slip
切向机动 tangential maneuver
切向畸变 tangential distortion
切向加速度 tangential acceleration
切向剪力 tangential shear
切向剪应力 tangential shearing stress
切向键 tangential key
切向交错层理 shear distortion; tangential cross-bedding; tangential distortion
切向焦线 tangential focal line
切向接管 tangential nozzle
切向接合 tangential joint
切向截面 tangential section
切向进刀机构 tangential feed mechanism
切向进刀架 tangential cutter head
切向进刀架丝杠 screw of tangential cutter head
切向进给 picker feed; tangential admission
切向进给滚刀 taper hob(bing-cutter)
切向进给机构 tangential feed mechanism
切向进给磨床 tangential grinder
切向进给磨法 tangential feed method grinding
切向进给蜗轮滚刀 tangential feed worm hob
切向进给装置 tangential feed attachment
切向进口 tangential inlet
切向进汽 tangential admission
切向进入 tangential admission
切向锯 tangent-saw
切向孔 tangential hole
切向孔道 tangential channel
切向控制器 tangential controller
切向拉伸强度 tangential tensile-strength
切向力 circumferential force; tangential force
切向量 tangential vector
切向量丛 tangential vector bundle
切向量空间 tangential vector space
切向流 slip stream
切向流动 tangential flow
切向流动式汽轮机 tangential flow turbine
切向流动式通风机 tangential flow fan
切向流动式透平 tangential flow turbine
切向流动式涡轮机 tangential flow turbine
切向螺带式搅拌机 tangential spiral agitator
切向螺纹 tangential screw
切向螺纹梳刀 Landis chaser; tangential chaser
切向螺旋 tangential screw
切向模量 tangential modulus
切向摩擦力 tangential friction force
切向逆断层 lateral thrust
切向黏[粘]性阻力 tangential viscous resistance
切向耦合 pick-off coupling; tangential coupling
切向配置电刷 tangential brush
切向喷管 spin nozzle; tangential jet; tangential nozzle
切向喷射 peripheral jet
切向偏移 tangential deflection
切向汽轮机 tangential turbine
切向牵引力 tangential tractive force
切向切环 tangential cut ring
切向切片 tangential section
切向侵蚀 tangential erosion
切向曲率 tangential curvature
切向燃烧 tangential combustion; tangential firing
切向燃烧器 tangential burner
切向入射 grazing incidence; tangential incidence
切向射线 tangent(ial) ray
切向收缩 tangential shrinkage
切向束 tangential beam
切向水流 tangential flow
切向水流空心纤维超滤 tangential flow hollow fibre ultrafiltration
切向水轮 tangential water-wheel
切向水轮机 tangential turbine
切向速度 tangential velocity
切向调节器 tangential actuator
切向同性 tangential isotropy
切向透平 tangential turbine
切向推力 tangential thrust
切向弯曲 tangential bending
切向位移 tangential displacement
切向涡流式喷头 turbulent sprayhead
切向涡轮机 tangential turbine; tangential wheel
切向误差 tangential error
切向线荷载 tangential line load
切向像场弯曲 tangential field curvature
切向楔 tangential wedge
切向卸料 tangential discharge
切向压力 tangential pressure
切向压应力 circumferential compressive force
切向叶片 tangential vane
切向叶片间距 tangential blade spacing
切向移动 tangential movement
切向引进 tangential admission
切向引力 tangential traction
切向应变 shear strain; tangential strain
切向应力 circumferential stress; shear stress; stress of tangential direction; tangential stress
切向运动 tangential motion; tangential movement
切向运动条件 tangency condition; tangential motion condition
切向振荡 tangential oscillation
切向支承 tangential bearing
切向阻力 tangential resistance
切像 cut-image
切像测距仪 cut-image range finder
切削 cut(ting); pulling down; remotion; shive; stock removing
切削半径 <挖土机的> radius of clean-up
切削比压 specific cutting pressure
切削边沟坡面 back sloping
切削表面 cutting surface
切削波 cutting liquid
切削部分 cutting tip
切削材料 cutting agent
切削操作 cutting operation
切削槽 cutter slit
切削层 cutting layer; lay of cutting
切削长度 cutting length; length of cut
切削常数 machining constant
切削成型 shaping by stock removal
切削尺寸 cutting size
切削齿 cogging; cutting tooth
切削齿轮 machine cut gear
切削挡板 chip guard
切削刀 bite; cutting edge; cutting lip
切削刀痕 cutter mark
切削刀具 cutter; cutting tool; rebating cutter
切削刀具磨床 cutter grinder
切削刀片 cutting blade; cutting tip
切削刀纹 cutter mark
切削导刃 tracer edge
切削点 cutting point
切削动作 stock removal action
切削端 <钻头的> cutting nose
切削方向 cutting direction
切削工具 cutter; cutting tool
切削工具钢 cutting-tool steel
切削工具和砂轮用研磨料 abrasive for cutting and grinding wheels
切削工具冷却油 cutting-tool coolant oil
切削工作 cutwork
切削功率 cutting power
切削沟纹 chipped grain
切削规 cutting ga(u)ge
切削规范 cutting specification
切削轨线 cutting trajectory
切削过程角 working angle
切削过的砖 cutting brick
切削合力 resultant tool force
切削荷载 chip-load
切削痕(迹) revolution mark; tooth mark; cutter mark
切削后角 back angle
切削厚度 cutting depth; cutting thickness; thickness of cutting
切削厚度比 cutting ratio
切削弧 cutting arc
切削护罩 chip guard
切削机 cutting machine; cutting milling; milling machine; block splitting machine <石块加工用>
切削机床 cutting machine tool; stock-removing machine
切削机构 cutting mechanism
切削极限 cutting limit
切削加工 cutting work; machine work; machining
切削加工单 cutting list
切削加工的 machine cut
切削加工性 workability for cutting
切削角 cutting angle; digging angle; gouge angle
切削金刚石 cutting diamond
切削金属冷却油 cutting metal cooling oil
切削进给刀架 tool slide
切削矩 cutting torch
切削具的后缘 back edge of tool
切削具切痕 tool mark
切削具寿命 cutter life
切削抗力 cutting resistance; resistance to cutting
切削颗粒 cutting grain
切削宽度 cutting width

切削冷却润滑液 cutting cooling lubricant
切削冷却液 cutting coolant
切削力 cutting effort; cutting force; tool thrust
切削力测定仪 mecalix
切削量 cutting output; stock removal
切削率 cutting rate; metal removal rate
切削轮 cutting wheel
切削螺纹 screw cutting; thread bolt; thread cutting
切削面 cutting face; cutting profile
切削面积 area of cut
切削能力 chip capacity; cutting ability; cutting action; cutting power; cutting property; cutting quality; cutting value; pulling force
切削能量 cutting energy
切削扭矩 cutting moment of torque
切削盘 cutting disc [disk]; cutting wheel
切削片 cutting blade
切削平面 cutting plane
切削器 cutter
切削强度 cutting hardness; cutting strength
切削曲面 tangent surface
切削去边缘的钢板 trimming plate
切削刃 cutting chin; cutting edge; cutting end; cutting lip; cutting profile; tool edge; wearing blade; chisel edge < 钻头 >
切削刃定向 edge orientation
切削刃钝化 cutting edge rounding
切削刃钝圆半径 rounded cutting edge radius
切削刃高度 lip height
切削容量 work capacity
切削润滑液 cutting compound; cutting fluid; cutting lubricant
切削润滑油 cutting oil
切削深度 cutting depth; depth of cut(ting)
切削时间 cutting time; machine time
切削试验 cutting test
切削寿命 working durability
切削数据 milling data
切削速度 cutting rate; cutting speed; rate of cutting; speed of cutting
切削速度指示计 cut meter
切削损伤 cutting damage
切削条件 cutting condition; machining condition
切削头 cutter head; cutting head
切削头刃 cutter tip
切削头转筒 cutter drum
切削凸轮 cutting cam
切削挖土机 cutter dredge(r); cutter excavator
切削位置 cutting position
切削温度 cutting temperature
切削效率 cutting efficiency; stock-removing efficiency
切削斜度 cutting slope
切削行程 cutting stroke
切削型掘进机 breasting wheel machine
切削型取芯钻头 cutting-type coring bit
切削型钻头 bit drag; bit with wings; drag bit
切削性 machinability; machining quality; tooling quality
切削性能 cutting property; cutting quality; cutting value; turning ability
切削性试验 machinability test
切削雄榫器 cope cutter
切削压力 cutting pressure
切削牙齿 cutting prongs
切削叶片 cutting blade
切削液 cooling liquid; cutting oil; cutting solution; cutting fluid
切削液挡板 splash guard
切削液添加剂 cutting fluid additive
切削应力 cutting stress
切削硬度 cutting hardness
切削用量标准 quantity standard used in cutting
切削用乳液 fatty cutting oil
切削用润滑冷却剂 cutting paste
切削油 cutting compound; cutting lubricant
切削油废水 cutting oil waste
切削油分离器 cutting oil separator
切削油添加剂 cutting oil additive
切削余量 machine-finish allowance; machine-machining allowance; over-size for machining
切削元件 cutting element
切削原理 cutting principle
切削圆 cutting circle
切削值 cutting value
切削制品 cut product
切削轴 cutter shaft
切削装置 topping mechanism
切削阻力 cutting resistance
切削钻 cutting drilling
切削钻头 cutter bit; drill pointer
切削作用 cutting action
切楔法 tangent-wedge method
切斜角机 mitre machine
切屑 chips; cuttings; cutting scrap; paring; shavings; swarf; trimming
切屑变形 chip deformation
切屑槽 chip tray; chip trough; cuttings shoot
切屑的裂程 path of shear
切屑堆 scrap baling; scrap briquetting
切屑防护 chip guard; chip protection
切屑防护器 chip guard
切屑分离 chip separation
切屑分离器 shavings separator
切屑钩 hook for cleaning chip(ping)s
切屑荷载 chip-load
切屑厚度 depth of cut; depth of impression
切屑厚度比 cutting depth ratio
切屑间隙 chip clearance
切屑接收器 cuttings chute; cuttings pit
切屑卷边 built-up edge
切屑抗力 chip-load
切屑坑 cuttings shoot
切屑篮 chip basket
切屑量 scrap rate
切屑流 chip flow
切屑瘤 built-up edge
切屑率 rate of metal removal
切屑螺旋 chip curl
切屑排出 swarf extraction
切屑排出通路 passage of chip
切屑排除 swarf removal
切屑盘 chip panel; chip tray; cutting tray; swarf tray
切屑收缩比 chip compression ratio
切屑输送器 cuttings conveyer [conveyor]
切屑填塞 chip packing
切屑通道 chip area
切屑桶 chip bucket
切屑箱 swarf box
切屑形成 chip formation
切屑压块机 scrap briquetting press
切屑余隙 swarf clearance
切样器 core cutter; core lifter
切叶机 foliage cutter
切应变 shear(ing) strain
切应变速率 shear rate
切应力 shear(ing) stress
切应力差法 shear difference method
切应面 shear plane
切余管 off-cut pipe
切余异常 truncated anomaly
切元素 element of contact
切圆 circle of contact; tangential circle
切圆角机 round cornering machine
切圆燃烧 tangential firing
切圆图形 image of tangential circle
切圆柱投影 tangent cylindrical projection
切圆锥投影 tangent conic(al) projection
切缘 cutting edge
切錾 parting tool
切展线 involute
切张量 tangent tensor
切枕木机 tie cutter
切纸刀 paper knife; scudding knife
切纸机 cutting machine; guillotine; paper cutter; paper cutting machine; slitter
切纸器 paper cutter
切至基准线 cut-off to grade
切制的齿轮 cut gear
切制的镶面层板 cut veneer
切制钉 cut nail
切制螺丝装置 chasing bar
切制螺纹 positive thread
切制螺纹最大孔径 tapping capacity
切制油沟装置 oil groove attachment
切柱面 tangent cylinder
切砖 clip; cut brick
切砖机 brick cutter; cut-off table
切砖锯 brick saw
切砖台 cutter's station
切锥 cone of tangents
切锥面 tangent cone
切嘴工程 spit cut-off works; spit cutting works

茄苳【植】katang

茄苳树白蚁 < 拉 > Glyptotermes fuscus
茄勒木 jarrah
茄类蔬菜 solanaceous vegetable
茄皮紫 aubergine purple; Jun purple
茄替胶 ghatti gum; Indian gum
茄子 mad apple; eggplant
茄紫色 dull purplish black

切实地 down-to-earth

切实可行的 workable
切实可行的方案 a realistic plan
切实可行的计划 workable plan
切勿倒置 keep upright
切勿接近饲料和食品 keep away from feed or food
切勿平放 not to be laid flat
切勿压挤 do not crush
切勿坠落 do not drop; no dropping

窃蠹属 Xestobium

窃取情报 wiretap
窃水 pirating
窃听装置 listening-in device; wiretap
窃用周期 stealing cycle

亲本材料 parent material

亲笔的 autographic(al); genuine; holographic(al)
亲笔签名 autograph; subscription
亲笔签名的 onomastic
亲笔签字 hand-written signature
亲笔文件 holograph
亲笔信 autographed letter
亲笔证书 holograph
亲笔字据 chirograph
亲潮 Kurile (cold) current; Oyashio current < 日语 >
亲等 < 亲属亲疏分等 > degree of kinship
亲电性 electrophilicity
亲电子的 electrophilic
亲电子反应 electrophilic reaction
亲电子试剂 electrophile; electrophilic reagent
亲硅酸盐性 lithophylic property
亲硅元素 lithophile element
亲海元素 thalassophile element
亲和 affiliation
亲和标记 affinity labelling
亲和层析法 affinity chromatography
亲和常量 affinity constant
亲和的 compatible
亲和力 affinity; avidity; chemical attraction; liking; mutual affinity; appetence【化】
亲和力强度 avidity strength
亲和力轴 axis of affinity
亲和曲线 affinity curve
亲和色谱法 affinity chromatography
亲和势 affinity
亲和数 amicable number; friendly numbers
亲和图法 affinity diagram
亲和物 affinant
亲和系数 affinity coefficient
亲和性 affinity; compatibility; hydrophilicity; liking
亲和性转移 affinity transgression
亲和液 affinitizing solution
亲和指数 index of affinity
亲核的 nucleophilic
亲核试剂 nucleophile
亲核性 nucleophilicity
亲花岗岩元素 granitophile element
亲硫的 thiophil
亲硫构造地球化学场 thiophilic tectono-geochemical field
亲硫元素 sulphophile element; thiophile element
亲卤素的 halophile
亲气元素 atmophile element
亲切 good-wilt
亲热需氧消化量 thermophilic aerobic digestion
亲生物元素 biophile element
亲石的 lithophile
亲石元素 lithophile element
亲水 hydrophile; water loving; water-like
亲水表面 water receptive
亲水材料 hydrophilic material
亲水的 hydrophil; hydrophilous; water-attracting; water-sensitive
亲水端基 terminal hydrophilic group
亲水法 hydrophilic method
亲水非极性芯胶束 hydrophilic oleomicelle
亲水分数 hydrophilic fraction
亲水改性 hydrophilic modification
亲水骨料 hydrophilic aggregate
亲水管 hydrophilic channel
亲水化 hydrophiling
亲水化法 hydrophilization process
亲水基(团) hydrophilic group(ing); lyophilic radical
亲水极 hydrophilic pole
亲水集料 hydrophilic aggregate
亲水胶 hydrophilic gel

亲水胶体 hydrophile; hydrophilic colloid
亲水胶质 hydrophilic solute
亲水链 hydrophilic chain
亲水面 water avid surface
亲水气体电极 hydrophilic gas electrode
亲水亲油平衡 hydrophile-lipophile balance
亲水亲油平衡值 <选择乳化剂的一种指标> hydrophile-lipophile balance value
亲水亲脂特性 lypohydrophilic character
亲水溶胶 hydrophilic sol;lyophilic sol
亲水疏水平衡值 hydrophilic-lyophobic balance
亲水填料 hydrophilic filler
亲水微胶粒 hydrophilic micelle
亲水物 hydrophile; water-attracting substance
亲水物亲脂物平衡 hydrophile-lipophile balance
亲水物质 hydrophilic substance
亲水物种 hydrophilic species
亲水吸附剂 hydrophilic adsorbent
亲水吸附质 hydrophilic adsorbate
亲水系数 hydrophilic coefficient
亲水系统 hydrophilic system
亲水性 affinity for water; hydrophilicity;hydrophilic nature;water affinity
亲水性的 hydrophilic
亲水性多孔载体 hydrophilic porous carrier
亲水性粉尘 hydrophilic dust;lyophilic dust
亲水性粉末 hydrophilic powder
亲水性胶体 lyophilic colloid
亲水性接触镜 hydrophilic contact lens
亲水性颗粒 hydrophilic particle
亲水性软膏 hydrophilic ointment
亲水性有机化合物 hydrophilic organic compound
亲水性质 hydrophilous nature
亲水有机胶体 hydrophilic organogel
亲水植物 hydrophilic plant; water-loving plant
亲水状态 hydrophily
亲水组分 hydrophilic fraction
亲酸的 acidophil(e)
亲体 parent
亲铁的 siderophile
亲铁元素 siderophile element
亲铜元素 chalcophile element
亲压体【化】barophile
亲岩的 lithophile
亲岩元素 lithophile element
亲氧构造地球化学场 oxyphile tectono-geochemical field
亲液保护胶体 lyophile protective colloid
亲液补体 lyophile complement
亲液材料 lyophilic material
亲液的 lyophilic
亲液胶体 lyophile(colloid);lyophilic colloid
亲液凝胶 lyophilic gel
亲液溶胶 lyophilic sol
亲液物【化】lyophile
亲液性聚合体 lyophilic polymer
亲油的 oleophilic;oleophylic
亲油肥料 oleophylic fertilizer
亲油分散基 lipophilic dispersant
亲油基 oleophilic group
亲油极性芯胶束 lipophilic hydromicelle
亲油胶凝剂 oleophylic gellant
亲油胶体 oleophylic colloid
亲油墨的 ink-receptive
亲油石墨 oleophylic graphite
亲油树脂 oleophilic resin
亲油性 lipophilicity;lipophilic nature
亲油性保护层 oleophilic coating
亲油液体 lipophile liquid
亲有机质的 organophilic
亲鱼 anadromous fish; brood fish; parent fish
亲鱼群体 parent fish population
亲缘关系 affiliation;genetic(al) relationship;natural affection
亲缘转换 affine transformation
亲脂的 oleophilic
亲脂性 lipophilicity
亲脂性的 lipophilic
亲质子溶剂 protophilic solvent
亲质子物 protophile
亲属 relation(ship)
亲自勘察 autopsy
亲族关系 consanguinity

侵

侵彻度 biting angle;penetration
侵出岩 extruded rock
侵犯 encroachment;impinge(ment); inroad;invade;trespass;violate;violation
侵犯版权诉讼 copyright infringement suit
侵犯财产罪 crime against property
侵犯地图版权 infringement of map copyright
侵犯环保规定 violate a law of environmental protection
侵犯权利 injustice
侵犯商标(专用)权 infringement of trademark;trademark infringement
侵犯所有权 infringement of title
侵犯他人财产 piracy
侵犯他人权利 interlope
侵犯污染控制法 encroach on standards of pollution control
侵犯者 trespasser
侵犯专利 patent infringement
侵犯专利权 infringement of patent rights;infringe upon patent rights; patent infringement;piracy
侵害 encroachment;infestation;inroad
侵害农作物 make inroads on crop
侵害生态平衡 encroach on ecologic(al) balance
侵害现象 infestation
侵害行为 injurious act;wrongful act
侵害征兆 disturbance indicator
侵界 encroachment
侵界点 invading point
侵进水 aggressive water
侵进岩浆 aggressive magma
侵陆潮水 tide water
侵陆海 sea of transgression; transgression sea
侵略 aggression
侵略的 aggressive
侵权 infringement of patent
侵权非法行为 tortious delinquency
侵权行为 act of tort;delict;infringement;tort
侵权行为的赔偿责任 tort liability
侵权行为诉讼 action of tort
侵权行为性质的损害 delictual damage
侵权行为引起的债权 delictual claim
侵染 infection;infestation
侵染的【地】disseminated
侵染现象 infection
侵染循环 cycle of infection
侵染状构造 dissemination structure
侵染作用 dissemination
侵扰 infestation
侵入 encroach(ment); inbreak; ingression;inroad;inrush;raid;trespass;transgress <侵入边界等>
侵入冰 intrusive ice
侵入波 send wave
侵入带 intrusive zone;invaded zone
侵入带校正 invaded zone correction
侵入带直径 diameter of invaded zone
侵入地带 invaded zone
侵入动力变质 kinetic metamorphism
侵入方式 mode of entry
侵入构造 intrusion structure; intrusion tectonics
侵入灌浆 intrusion grouting
侵入过渡层 ingress transition
侵入过滤 intrusive filtration
侵入海 invading sea
侵入海岸线 invading sea; invading shoreline
侵入害虫 exotic insect
侵入后构造 post-intrusion structure
侵入花岗岩 intrusive granite
侵入火成岩 intrusive igneous rock
侵入火山活动 intrusive vulcanicity
侵入基坑的水 inrushing water
侵入角砾岩 intrusive breccia
侵入接触【地】intrusive contact;faulted contact
侵入接触圈闭 intrusion contact trap
侵入空气 invading air
侵入浪 swash
侵入雷电波过电压 ingressed lightning wave overvoltage
侵入力 invasiveness
侵入脉 intrusive vein
侵入门户 portal of entry
侵入面积 encroachment area
侵入片 intrusive sheet
侵入期 stage of invasion
侵入泉 intrusion spring
侵入山 intrusive mountain
侵入深度 <钻进液渗入地层深度> invasion depth;depth of intrusion
侵入生长 interposition growth;intrusive growth
侵入生长时间 intrusive growing season
侵入式浇灌 intrusion grouting
侵入试验 invasive test(ing)
侵入水 aggressive water;inrush of water;intruding water;invading water
侵入体【地】intrusion; intrusive body;intrusive mass;pluton
侵入体剥蚀深度 erosional depth of intrusive body
侵入体产状 occurrence of intrusive body
侵入体分异分类 classes of differentiated intrusive
侵入体构造 structure of intrusive body
侵入体贯入 intrusive injection
侵入体矿化特征 mineralization feature of intrusive body
侵入体年龄 age of intrusion
侵入体侵入期次 emplacement stage of intrusion
侵入体形成深度 formation depth of intrusive body
侵入体岩相分带 facies zone within intrusive body
侵入体与围岩接触关系 contact relation between intrusive body and country rocks
侵入同化混染作用 intrusive assimilation and contamination
侵入偷猎 poach
侵入途径 routes of entry;routes of invasion
侵入位移 intrusion displacement
侵入物 invader
侵入线 encroachment line
侵入限界 fouling of clearance gauge; infringement
侵入限界绝缘 insulated joints located within the clearance
侵入相【地】intrusive facies
侵入岩【地】intruded rock;intrusion rock;intrusive rock;irruptive rock
侵入岩层 intrusive sheet
侵入岩产状 intrusive rock occurrence
侵入岩床 intrusive sheet;intrusive sill
侵入岩脉 injected dike[dyke]; intrusive dike[dyke]; intrusive vein; irruption vein
侵入岩墙 intrusive dike[dyke]
侵入岩热储 intrusive igneous reservoir
侵入岩体 intrusive body; intrusive rock
侵入岩体环形体 circular features of intrusive bodies
侵入岩体相带异常 lithofacies gone anomaly of intrusive body
侵入岩席 intrusive sill
侵入与喷出 intrusion and extrusion
侵入杂岩 intrusive complex
侵入种 adventicous species
侵入作用【地】incursion; intrusion; syntectonic intrusion
侵蚀 aggressiveness; attack; bite; corrode; corrosion; corrosive attack; destruction; eat(ing) away; eat into; eat out; encroach; etch; fret; scour; weathering; denude; gain (up) on【地】;aggression
侵蚀比 erosion(al) ratio
侵蚀边界 erosional boundary
侵蚀表面 denudation level
侵蚀剥离面 erosional level
侵蚀剥削面 erosional level
侵蚀不整合 erosional unconformity
侵蚀残留的孤立岩石 klippe[pl. klippen]
侵蚀残余 remnant
侵蚀层面接触 erosional bedding contact
侵蚀程度指数 erodibility index
侵蚀程序 erosivity process
侵蚀冲断层 erosion(al) thrust
侵蚀处 wasout
侵蚀带 zone of erosion
侵蚀的 eroded; erodent; erosional; erosive;rodent
侵蚀的山 mountain of erosion
侵蚀的原动力 agent of erosion
侵蚀的支柱 erosion pillar
侵蚀等级 erosion(al) class; staining class
侵蚀地 <海水的> encroachment
侵蚀地带 belt of erosion
侵蚀地改良 reclamation of eroded area
侵蚀地坪 denudation terrace
侵蚀地形 autogenetic topography; erosional landform;erosion landfill
侵蚀点 corrosion pitting
侵蚀调查 erosional investigation;erosional survey
侵蚀动因 agent of erosion
侵蚀度 degree of aggressiveness;erodibility; erosion(al) index; erosiveness;erosivity
侵蚀堆积物 washed drift
侵蚀墩 erosional pillar
侵蚀反应 etching reaction
侵蚀防护 erosion(al) protection
侵蚀防治 erosional control
侵蚀改造水流 eroding-reworking current

侵蚀高原 erosion(al) plateau; peneplain; plateau of erosion
侵蚀沟(槽) erosional groove; erosion(al) gull(e)y
侵蚀沟类型 type of erosion(al) groove
侵蚀沟形态要素 morphologic(al) elements of erosion(al) groove
侵蚀构造【地】 etching structure
侵蚀古陆块 old land
侵蚀谷 destructional valley; eroded valley; erosion(al) valley
侵蚀过程 erosional process
侵蚀海岸 erosional coast; erosional shore
侵蚀海进 erosional transgression
侵蚀海湾 abrasion embayment; corrasion embayment
侵蚀河岸 eroding bank
侵蚀河槽 eroding channel; erosional channel
侵蚀河流 corrading stream; corroding stream
侵蚀河漫滩 erosional flood plain
侵蚀河湾 abrasion embayment; corrasion embayment
侵蚀痕 erosional mark; erosion(al) scar
侵蚀后退 erosional retreat
侵蚀湖 erosional lake; interior lake; lake due to erosion; scooped lake
侵蚀化 eat up
侵蚀火山 eroded volcano
侵蚀火山口 erosional crater
侵蚀基面 base level of erosion; erosional base; erosional basis; erosion basin
侵蚀基准面 base level; base level of corrosion; base level of erosion
侵蚀基准面以上 above the erosion(al) basis
侵蚀基准面以下 under the erosion(al) basis
侵蚀剂 aggressive agent; erodent; etchant; etching agent; etching reagent
侵蚀间断 erosional gap; erosional hiatus
侵蚀交叉点 <河床的> erosional nick-point; knick point
侵蚀礁坡 erosional ramp
侵蚀阶地 destructional bench; destructional terrace; erosional bench; erosion(al) terrace; erosive terrace
侵蚀接触 erosional contact
侵蚀介质 aggressive medium; corrosion medium
侵蚀景观 destructional landscape
侵蚀壳 erosional valley
侵蚀坑 eroded crater; erosional crater; erosion(al) pit
侵蚀孔 tafoni
侵蚀控制 erosion(al) control
侵蚀力 agents of erosion; eroding agent; eroding force; erosional force; erosive agent; erosive force
侵蚀流失 erosion(al) loss
侵蚀流速 erosive velocity
侵蚀率 erosional index; erosion(al) ratio; penetration rate
侵蚀轮回 cycle of erosion; erosion cycle
侵蚀面 erosion(al) surface; planation surface
侵蚀面冲断层 erosion thrust
侵蚀模量 erosional modulus; erosive modulus; modulus of erosion
侵蚀模数 erosional modulus; erosive modulus; modulus of erosion
侵蚀木材的小甲虫 death watch beetle
侵蚀幕 episodic erosion
侵蚀能力 aggressive power; erosive capacity; erosiveness; erosive power; scouring capability
侵蚀盆地 erosion(al) basin
侵蚀平原 destruction(al) plain; erosion(al) plain; plain of erosion; plan of erosion; scouring plain
侵蚀坡度 erosion slope
侵蚀破坏地区 erosion(al) ravaged area
侵蚀破火口 erosional caldera
侵蚀强度 erosion(al) intensity
侵蚀泉 erosional spring; valley spring
侵蚀三角洲 destructional delta; erosional delta
侵蚀山隘 eroded saddle
侵蚀山口 erosion(al) gap
侵蚀深度 depth of erosion
侵蚀生长物 encroaching growth
侵蚀试验 erosion test
侵蚀寿命 erosional life
侵蚀水 eroding water; aggressive water
侵蚀速度 eroding velocity; erosion(al) velocity; erosive velocity; rate of attack
侵蚀速率 erosional rate; etching rate; rate of erosion
侵蚀台地 destructional terrace; erosional platform
侵蚀特性 erosional characteristic
侵蚀特征 erosional feature
侵蚀涂层 etching resist
侵蚀土地 eroded field
侵蚀土(壤) eroded soil; erosional soil; wash-off soil
侵蚀危险 risk of scour(ing)
侵蚀物 aggressive agent; erodent
侵蚀系数 erosivity coefficient
侵蚀峡谷 eroded canyon
侵蚀线 encroachment line
侵蚀陷斑 etched dimple; etch pit
侵蚀像 etched figure
侵蚀小区 erosion plot
侵蚀斜坡 erosional slope
侵蚀形成的地层 eroded formation
侵蚀形成的湖 erosion lake
侵蚀形式 erosional form
侵蚀形态 erosional pattern; erosion form; etching pattern
侵蚀型三角洲 destructive delta
侵蚀性 aggressive action; aggressivity; corrodibility; corrosivity; credibility; erodibility; erosiveness; erosivity
侵蚀性边坡 eroded slope
侵蚀性大气状况 aggressive atmospheric(al) condition
侵蚀性的 aggressive; corrosive
侵蚀性二氧化碳 aggressive carbon dioxide; corrosive carbon dioxide; erosional carbon dioxide
侵蚀性海岸 erosion beach
侵蚀性碰撞 erosive collision
侵蚀性评价等级 erosiveness evaluation grade
侵蚀性气氛 severe atmosphere
侵蚀性气候 aggressive climate
侵蚀性溶液 aggressive solution
侵蚀性熔渣 scouring cinder
侵蚀性水 active water; aggressivity [aggressive] water; corrosive water
侵蚀性碳酸 aggressive carbonic acid
侵蚀性特性 aggressive characteristic
侵蚀性物质 aggressive matter; aggressive substance; rodents
侵蚀性盐 aggressive salt
侵蚀性液体 aggressive liquid
侵蚀旋回 cycle of denudation; cycle of erosion; erosional cycle; geographic(al) cycle
侵蚀旋回幼年期 infancy period of erosion(al) cycle
侵蚀循环 cycle of denudation; cycle of erosion; erosion(al) cycle
侵蚀崖 erosion scarp
侵蚀严重的荒(芜土)地 <美国南达科他州及内布拉斯加州部分地区> badland
侵蚀岩屑积层 erosion(al) pavement
侵蚀野溪 undermining-torrent
侵蚀因素 agent of erosion; erosion(al) agent; erosion(al) factor; factor of erosion
侵蚀因子 erosional factor
侵蚀营力 erosional agent
侵蚀原动力 eroding agent; erosional agent
侵蚀指数 erosion(al) index
侵蚀治理 erosional control
侵蚀终期 senility
侵蚀柱 hoodoo
侵蚀铸瘤 erosion scab
侵蚀总量 total amount of erosion
侵蚀作用 agents of erosion; aggressive action; corrosiveness; eating; eroding agent; erosional action; erosional effect; erosional process; erosive action; work of erosion; erosion
侵填体 thylose; tylose
侵吞 embezzlement; peculate
侵吞公款 defalcate
侵位 emplacement; invasion
侵位机制 emplacement mechanism
侵位年龄 emplacement age
侵吸层 blotter coat
侵袭 invade; irruption
侵隐色 anticryptic colo(u)r
侵油层 blotter coat
侵越 transgress(ion)
侵占 encroach; entrainment; trespass; usurp
侵占邻车道 NOT-yield-right-of way
侵占邻地 encroachment

钦

钦诺克风 <美国落基山东坡的> Chinook wind

钦诺克风拱状云 Chinook arch
钦式控制制动联合阀 <地下铁道用> Cineston controller brake valve

琴

琴钢丝 piano wire

琴键开关 piano-key switch
琴键式按钮 piano-key push
琴键式(开关) piano type
琴式铰链 piano hinge
琴弦 music wire
琴弦钢丝 music wire

禽

禽兽养护地 <禁猎区> game preserve

禽畜废物收集 poultry and livestock waste collection
禽畜排泄物 poultry and livestock manure
禽畜养殖废物 poultry and livestock farm waste

勤

勤务兵 orderly

勤务楼梯 service stair(case)
勤务信道 service channel
勤杂工 dusty-butt; floor boy; handy man; odd-job man; orderly

擒

擒料辊 gripping roll

擒上钻 escape wheel cock jewel
擒下钻 escape wheel foot jewel
擒纵叉 escapement lever
擒纵叉左叉瓦 left pallet stone
擒纵齿轮 escape pinion
擒纵机 escapement
擒纵机锉 escapement file
擒纵机挡料器 escapement stop
擒纵机构 detent; escapement(mechanism); lever movement
擒纵机构冲击 escapement impulse
擒纵机构夹板 escapement plate
擒纵机构误差 escapement fault
擒纵机构效应 effect of the escapement
擒纵机构噪声 escapement noise
擒纵机构转动 trip
擒纵机轮 escapement wheel
擒纵角 escape angle
擒纵轮 escape wheel(and pinion)
擒纵轮齿 escape wheel teeth
擒纵轮齿轴 escape wheel pinion
擒纵轮护板 escape wheel guard
擒纵轮夹板 escape wheel bridge
擒纵轮磨床 escape wheel grinding machine
擒纵轮相位角 phase angle of the escapement
擒纵器 dog catch; escapement motion; holding latch; release catch
擒纵曲柄 escapement crank
擒纵式挡料器 escapement stop
擒纵调速组件 distributing and regulating module
擒纵误差 escapement error
擒纵系统 escapement system
擒纵装置 detent; escapement device; release

寝

寝居甲板 mess deck

寝具 bedding
寝具消毒整理中心 bed disinfection and clean up center[centre]
寝具整理工作 bed-making
寝室 bedroom; dormitory
寝室层 chamber stor(e)y

揿

揿按钮 activate button

揿板 push plate
揿刺叭表示 honk
揿钉 drawing pin; tack; thumb tack
揿钮 key button; snap fastener; snapper
揿钮报警器 push-bottom-operated alarm
揿钮操纵 push-bottom control
揿钮键 knob
揿钮接头 snap terminal
揿钮开关 push-bottom switch
揿下手钮 press knob; push knob
揿下旋钮 press knob; push knob
揿压器 depressor
揿针 thumb-tack needle for subcutaneous embedding

青

青白瓷 bluish white porcelain; greenish-white porcelain; shadow blue glaze porcelain

青白花瓷器 Qinghai mottled porcelain
青白黏[粘]土 camstone

青白(色) bluish white
青白石 greenish-white marble
青白釉广寒宫枕 <瓷器名> Qinghai glaze moon palace pillow
青白釉褐斑 Qinghai glaze with brown mottles
青变 blu(e)ing;blue stain
青变材 blue lumber
青变菌 blue sapstain fungus
青材料 green material
青草 grass
青草覆盖的土地 grass covered land
青草饲料 grass fodder
青草味的 gramineous
青春 youth
青春的 vernal
青春后期的 postpubertal
青春期 <指河流> neanic stage
青春期河谷 adolescent valley;neanic valley;youthful valley
青春前期 preadolescence;prepuberty
青瓷 <一种中国著名的瓷器> celadon(ware)
青瓷扁壶 celadon flak
青瓷单柄壶 celadon ewer with single handle
青瓷莲花形盅 celadon lotus shaped cap
青瓷色 China blue
青瓷痰盂 celadon spittoon
青瓷唾壶 celadon spittoon
青瓷釉 celadon glaze
青葱的 lush
青翠的 viridian
青蛋白石 girasol
青岛港 Port of Qingdao
青岛岩 tsingtauite
青的颜料 blue pigment
青地白龙纹尊 white dragon design on blue ground vessel
青点 bluish-green spotted
青豆色的 pea-green
青矾 green alum;iron vitriol
青腐 <木材的> green rot;blue rot
青冈(木) ring-cupped oak
青刚柳 osier willow;Salix viminalis
青光铁蓝 Prussian blue
青函隧道 <日本> Seikan tunnel
青河石 qingheiite
青褐色硬木 <产于巴西> nargusta
青黑粉红色 livid pink
青黑色 lividity
青黑色的 livid
青黑檀 green ebony
青黑紫红色 livid purple
青黑紫蓝色 livid violet
青黑棕色 livid brown
青红色 green lake
青花 blue and white;underglaze blue
青花瓷器 blue and white porcelain;nankeen;nankin
青花大盘 large dish with underglaze blue
青花斗彩 blue and white with overglaze colo(u)rs
青花梵文罐 underglaze blue jar with Sanskrit design
青花玲珑 blue and white with pierced decoration; blue and white with rice pattern
青花三星 three Gods in white-and-blue
青花塔式盖瓶 underglaze blue vase with pagoda shaped cover
青花碗 underglaze blue bowl
青花五彩莲池鸳鸯图碗 blue and white with copper red colo(u)rs
青花釉里红 blue and white with copper red colo(u)rs
青花釉里红盖罐 underglaze blue and underglaze red covered jar
青花站济公 a standing funny monk in blue-and-white
青花折枝花果纹瓶 underglaze blue vase with fruit and floral sprays
青灰 grey lime
青灰背平屋顶 grey lime roof deck
青灰背屋顶 grey lime roofing
青灰色 iron blue;lividity;steel gray;steel grey
青灰色的 caesious;livid;lurid
青灰色灰浆 grey plaster
青灰色细质花岗岩 <产于美国佐治亚州的> Lithenia granite
青辉石 anthochroite;violan(e)
青荚叶 Japan helwingia
青金石 azure stone;lapis(lazuli);lasurite
青金石器皿 lapis lazuli ware
青壳纸 fish paper
青空法 blue-sky law
青蓝 blueness
青蓝色 smalt blue
青篱竹 tonkin(cane)
青篱竹属 cane;Arundinaria <拉>
青莲色 heliotrope;pale purple
青龙灰岩 <早三迭世> Chinlung limestone
青龙木 amboina rosewood; Amboyna; Andaman rosewood; bubinga; Burmacoast padauk; Burmese rosewood
青铝闪石 crossite
青绿 verdure
青绿彩画 blue and green decorative painting
青绿色 turquoise;turquoise blue
青绿色的 virid
青绿石 turquoise
青绿饲料 green feed;green forage
青马大桥 <位于香港特别行政区,主跨 1377 米的悬索桥, 1996 年建成> Tsing Ma Bridge
青霉 blue mo(u)ld
青面纤维层 cambium layer
青苗 green schools of(food)grains;young crops
青苗补偿 crop loss compensation
青苗补偿费 young crops compensating fee
青苗法 Young Crops Law
青苗牧地 seedling pasture
青苗赔偿费 compensation for crops
青木 Japanese aukuba
青泥 blue mud
青泥灰岩 blue marl
青泥石 aerinite
青年地形 topographic(al)adolescence
青年河 adolescent river; adolescent stream; youthful river; youthful stream
青年旅店 youth hotel
青年期 adolescence;young period
青年期的 adolescent;neanic
青年期河谷 adolescent valley;neanic valley;youthful valley
青年期溪谷 adolescent valley
青年商业社 Junior Achievement
青年晚期 <河流的> late youth
青年早期 <地表侵蚀的> early youth
青年招待所 youth hostel
青年中心 youth center[centre]
青黏[粘]土 blue clay;blue earth
青磐岩 propylite
青磐岩化 propylitization
青皮负属 <拉> Aithaloderma
青铅矿 linarite
青铅皮 lead sheet
青钱枫柏 diskfruit wingnut
青钱柳 diskfruit wingnut
青钳矾 linarite
青色 cyan;green
青色层 cyan layer
青色的 azury
青色明度 cyan brightness
青色磨石 blue grindstone
青色纤维层 cambium layer
青砂石 green sandstone
青砂岩 subgraywacke
青砂岩石 green sandstone
青少年教养院 detention home
青少年拘留所 remand home
青石 blue stone
青石棉 blue asbestos;cape asbestos;cape blue asbestos;crocidolite asbestos; griqualandite; krokidolite;crocidolite <钠闪石的石棉状变种>
青水泥 grey cement
青饲料 feeder green;green holder;silage;soiling food
青饲料发酵 silage fermentation
青饲料作物 ensilage crop;green crop
青饲作物 silage crop;soiling crop
青锁龙属 <拉> Crassula[复 Crassulae]
青苔 Chlorophyceae;lichen;moss <绿藻类>
青苔色 mossy green
青钛闪石 heikolite
青檀 wingceltis
青天蓝 cerulean blue
青铜 bronze;gun metal;metal bronze <铜锡合金的>
青铜摆 bronze pendulum
青铜板 bronze sheet
青铜包裹 bronze surround
青铜保持架 bronze cage;gunmetal cage
青铜背轴承 bronze-backed bearing
青铜病 bronzed disease
青铜箔 bronze foil
青铜薄板 bronze sheet panel
青铜薄片 bronze flake
青铜衬 bronze liner
青铜衬套 bronze bush(ing)
青铜衬套的 bronze-bushed
青铜窗 bronze window
青铜导板 bronze guide
青铜导承 bronze guide
青铜导杆 bronze guide
青铜导轨 bronze guide
青铜导线 bronze connector
青铜垫片 bronze washer
青铜雕饰门 relief bronze door
青铜锭 bronze ingot
青铜粉 bronze powder;copper-bronze powder;powdered bronze
青铜浮雕 bronze relief
青铜工作 bronze-working
青铜构件 bronze unit
青铜管 bronze tube
青铜光泽彩 bronze lustre
青铜焊 bronze welding
青铜焊料 spelter bronze
青铜合金 bell metal;bronze alloy
青铜护墙 bronze curtain wall;bronze sheet panel
青铜花格 bronze grill(e)
青铜花格窗格栅 bronze grill(e)
青铜环 bronze ring
青铜辉石 bronzite
青铜基滑动轴承 bronze-backed metal bearing
青铜建筑设备 bronze builders furniture
青铜匠 bronzer
青铜胶 bronze paste
青铜铰链 bronze hinge
青铜金属护墙板 bronze sheet panel
青铜蓝 bronze blue
青铜连接件 bronze connector
青铜连接器 bronze connector
青铜螺栓 bronze bolt
青铜螺旋 bronze screw
青铜锚(固) bronze anchor
青铜门 bronze door
青铜模 bronze mo(u)ld
青铜摩天楼 bronze skyscraper
青铜幕墙 bronze curtain wall
青铜配件 bronze fitting
青铜坯块 bronze compact
青铜披挡板 bronze sheathed
青铜旗杆 bronze flagpole
青铜器 bronze vessels;bronze ware
青铜(器)时代【地】 Bronze Age
青铜(器)时代宫殿 bronze-age palace
青铜(器)时代建筑 bronze-age building
青铜(器)时代陶器 bronze age pottery
青铜色 bronze(colo(u)r)
青铜色玻璃 bronze glass
青铜色的 aereous;bronzy
青铜色的铝粉涂料 alumin(i)um bronze paint
青铜色腊克 bronzing lacquer
青铜色喷漆 bronzing lacquer
青铜色漆 bronze lacquer
青铜色陶器 bronzed pottery
青铜色颜料 bronze pigment
青铜色氧化 bronzing
青铜色泽 bronze sheen
青铜石墨电接触器材 bronze-graphite contact material
青铜蚀刻装饰 incised decoration in bronze
青铜双式弹簧门 bronze swing door
青铜推力轴环 bronze thrust collar
青铜瓦 bronze roof tile
青铜外形 bronze profile;bronze section;bronze shape;bronze unit
青铜屋面瓦 bronze roof tile
青铜系材 bronze tie
青铜线 bronze wire
青铜线脚 bronze mo(u)ld
青铜镶板 bronze sheet panel
青铜镶边 bronze trim
青铜镶框 bronze trim
青铜镶嵌 bronze surround
青铜像 bronze statue
青铜小五金 bronze hardware
青铜楔子 bronze wedge
青铜旋转双耳 bronze swivel with eye;bronze swivel with two trunnions
青铜烟灰色 bronze smoke
青铜硬件 bronze hardware
青铜圆箍线 <枯环饰> bronze annulet
青铜造型 bronze mo(u)ld
青铜止推环 bronze thrust collar
青铜制品 bronze
青铜轴承 bronze bearing
青铜轴承合金 Volvit
青铜柱环饰 bronze shaft-ring
青铜柱环形线脚 bronze shaft-ring
青铜铸件 bronze casting
青铜铸造 bronze casting
青铜转门 bronze swing door
青铜装饰 bronze trim
青土 blue clay;blue earth
青瓦 black tile;gray[grey] tile
青羊 <野羊> goral
青杨 cathay poplar
青釉瓷器 blue glazed porcelain
青釉褐斑 celadon with dark mottles
青鱼 black carp;herring
青鱼油 herring oil
青榆 Manchurian elm
青玉(色) sapphire

青玉色的 sapphirine
青肿 bruise
青贮(护苗)ensilage
青贮窖 horizontal silo
青贮饲料 silage
青贮饲料挂车 silage trailer
青贮饲料技术 ensilage technique
青贮料 ensilage
青贮作物 ensilage crop;silage crop
青砖 black brick;blue brick;flashed brick;gray[grey] brick;sewer brick
青砖铺面 blue brick paving
青紫 cyanosis
青紫色颜料 violet blue

氢

氢铵矾 letovicite
氢爆皮 hydrogen blistering
氢泵 hydrogen pump
氢标度 hydrogen scale
氢标准电极 hydrogen electrode
氢波 hydrogen wave
氢铂电极系统 hydrogen platinum electrode system
氢超电压 hydrogen overvoltage
氢传递 hydrogen transfer
氢脆 hydrogen attack;pickle brittleness;hydrogen embrittlement <钢的>
氢脆破坏 hydrogen embrittlement failure
氢脆性 hydrogen brittleness;hydrogen embrittlement
氢存储束管 hydrogen storage beam tube
氢氮环氧化物<治疗酒精中毒的化合物> chlordiazepoxide
氢当量 hydrogen equivalent
氢的 hydric
氢的氧化物 hydrogen oxides;oxides of hydrogen
氢的优先氧化 preferential oxidation of hydrogen
氢的制造 hydrogen production
氢灯 hydrogen discharge lamp
氢等离子弧过程 hydrogen plasma arc process
氢等离子流 hydrogen plasma jet
氢等离子体 hydrogen plasma
氢地冕 hydrogen geocorona
氢碘酸 hydriodic acid
氢碘酸分解 decomposition with HI
氢碘酸盐 hydriodate
氢电极 hydrogen electrode
氢对流层 hydrogen convection layer
氢发动机 hydrogen engine
氢发射区 hydrogen emission region
氢放电灯 hydrogen discharge lamp
氢分布 hydrogen distribution
氢氟硅酸盐 hydrofluosilicate
氢氟化钾 potassium hydro-fluoride
氢氟化钠 sodium acid fluoride;sodium bifluoride;sodium hydro-fluoride
氢氟化器 hydrofluorinator
氢氟化物 hydrofluoride
氢氟酸 fluorhydric acid;hydrofluoric acid;hydrogen fluoride acid
氢氟酸测斜玻璃试管 glass tube
氢氟酸测斜试管 angle test tube
氢氟酸测斜仪 hydrofluoric acid bottle inclinometer
氢氟酸分解 decomposition with hydrofluoric acid
氢氟酸腐蚀液 hydrofluoric acid corrosive liquid;Kroll corrosive liquid
氢氟酸和硫黄注入工艺 hydrofluoric acid and sulphur impregnation
氢氟酸蚀刻 hydrofluoric acid etch
氢氟酸铜 fluorous salt of copper
氢氟酸洗涤 hydrofluoric acid wash
氢氟酸盐 hydrofluoride
氢氟酸灼伤 hydrofluoric acid burn
氢氟碳化合物 fluorohydrocarbon;hydrofluorocarbon
氢腐蚀 hydrogen attack;hydrogen corrosion
氢负离子转移 hydride ion-transfer
氢复合线 hydrogen recombination line
氢钙铀云母 hydrogen autunite
氢高铁氰酸盐 hydroferricyanate
氢光谱 hydrogen spectrum
氢过氧化枯烯 cumene hydroperoxide
氢过氧化钠 sodium hydroperoxide
氢过氧化物 hydroperoxide
氢过氧化作用 hydroperoxidation
氢过氧基游离基 hydroperoxyl radical
氢含量 hydrogen content
氢焊 hydrogen soldering;hydrogen-welding
氢焊条 hydrogen electrode
氢弧灯 hydrogen arc lamp
氢化 hydriding;hydrogenate;hydrogenising;hydrogenization;hydrogenize
氢化钡 barium hydride
氢化蓖麻油 hydrogenated castor oil
氢化丙烯四聚物 hydrogenated propylene tetramer
氢化处理 hydrotreat(ing)
氢化处理法 hydride process
氢化催化剂 hydrogenation catalyst
氢化脆性 acid brittleness
氢化的 hydrogenant
氢化芳香系 hydroaromatic series
氢化芳香族化合物 hydroaromatic compound
氢化非晶碳 hydrogenated amorphous carbon
氢化钙 calcium hydride;hydrolith
氢化镉 cadmium hydride
氢化共叠合物 hydrocodimer
氢化硅氟酸 hydrosilicofluoric acid
氢化硅酸盐 hydrosilicate
氢化硅烷化 hydrosilylation
氢化剂 hydrogenant agent;hydrogenating agent
氢化钾 potassium hydride
氢化间苯二酚 hydroresorcinol
氢化聚丁烯 hydrogenated polybutene
氢化聚合物 hydropolymer
氢化聚合作用 hydropolymerization
氢化锂 lithium hydride
氢化裂解 hydrocrack(ing)
氢化铝锂还原 lithium aluminium hydride reduction;reduction by lithium aluminium hydride
氢化铝钠 sodium alumin(i)um hydride
氢化煤 hydrogenous coal
氢化镁 magnesium hydride
氢化钠 sodium hydride
氢化萘 hydrogenated naphthalene
氢化偶氮苯 hydrazobenzene
氢化硼 boron hydride
氢化硼钾 potassium borohydride
氢化硼钠 sodium borohydride
氢化破损 hydriding failure
氢化镨 praseodymium hydride
氢化漆酚 hydrourushiol
氢化器 hydrogenator
氢化燃料 hydrogenated fuel
氢化热 heat of hydrogenation
氢化溶剂 hydrogenation solvent
氢化润滑油 hydrolube
氢化三联苯 hydrogenated terphenyl
氢化石脑油 hydrogenated naphtha
氢化松节油 hydroterpins
氢化松香 hydrogenated rosin
氢化松脂醇 hydroabietyl alcohol
氢化塑料取芯器 hydroplastic corer
氢化钛 titanium hydride
氢化钛烧焊 titanium hydride brazing
氢化锑 antimony hydride
氢化提纯 hydrofine
氢化萜二醇 hydroterpine
氢化萜品 hydroterpine
氢化脱硫作用 hydrodesulphurisation [hydrodesulphurization]
氢化物 hydride
氢化物发生 hydride generation
氢化物分离-原子吸收法 hydride separation-atomic absorption method
氢化橡胶 hydrogenated rubber
氢化异构现象 hydroisomerisation
氢化植物油 hydrogenated vegetable oil
氢化装置 hydrogenation apparatus
氢化作用 hydrogenation
氢还原 hydrogen reduction
氢还原的铁粉 H-iron
氢还原粉末 hydrogen-reduced powder
氢还原高压釜 hydrogen reduction autoclave
氢还原铁粉 H-iron;hydrogen-reduced iron
氢回收过程 hydrogen recovery process
氢火焰电离检测器 hydrogen flame ionization detector
氢火焰检测器 hydrogen flame detector
氢火焰离子化 hydrogen flame ionization
氢火焰离子检测器 hydrogen flame ionization detector
氢火焰温度检测器 hydrogen flame temperature detector
氢激光器 hydrogen laser
氢极化 hydrogen polarization
氢集管 hydrogen header
氢加工【化】 hydroprocessing;hydrogen processing
氢加工工艺 hydroprocessing technology
氢键 hydrogen bond
氢键键合 hydrogen bonding
氢键力 hydrogen bonding force
氢解作用 hydrogenolysis;hydrogenesis
氢醌 hydroquinone;quinol
氢醌电极 quinhydrone electrode
氢冷却 hydrogen cooling
氢冷却剂 hydrogen coolant
氢冷却器 hydrogen cooler
氢冷却装置 hydrogen cooling system
氢冷式电机 hydrogen-cooled machine
氢冷式汽轮发电机 hydrogen-cooled turbo-alternator
氢冷式同步调相机 hydrogen-cooled synchronous condenser
氢冷式涡轮发电机 hydrogen-cooled turbine-generator
氢冷式旋转变流机 hydrogen-cooled rotary converter
氢冷系统 hydrogen gas cooling system
氢冷转子 hydrogen-cooled rotor
氢离子 hydrion;hydrogen ion
氢离子比色计 ionocolo(u)rimeter
氢离子测定器 hydrogen ion determination apparatus
氢离子的 hydrionic
氢离子电位计 hydrogen ion potentiometer
氢离子回渗 hydrogen ion back diffusion
氢离子活度 hydrogen ion activity
氢离子活度常数 activity coefficient of hydrogen ion
氢离子计 pH meter
氢离子碱度 hydrogen ion concentration
氢离子交换器 hydrogen ion exchanger
氢离子交换容量 hydrogen cation exchange capacity
氢离子跨膜梯度 transmembrane gradient of hydrogen ion
氢离子浓度<pH 值> hydro(gen)-ion concentration
氢离子浓度测定法 method for hydrogen ion concentration
氢离子浓度负对数值<表示酸碱度> pH value
氢离子浓度计 hydrogen ion concentration meter;ionometer;pH acidimeter;pH meter
氢离子浓度记录仪 pH-recorder
氢离子浓度值 hydrogen ion concentration value
氢离子浓度指数<表示酸碱度> pH value;hydrogen ion exponent
氢离子浓度指数比色标准序列 standard colo(u)rimetric series for pH measurement
氢离子浓度指数调节 pH control
氢离子浓度自动控制 automatic pH control
氢离子浓度自动调节 automatic pH control
氢离子浓度自动调节仪 pH automatic controller
氢离子浓度自记仪 hydrogen ion concentration recorder
氢离子移转 hydrogen ion transfer
氢离子指示剂 hydrogen ion indicator
氢离子指数 hydrogen ion index
氢离子指数试纸 pH test paper
氢硫化铵 hydrosulfide of ammonia
氢硫化钠 sodium hydrosulfide;sodium sulphhydrate
氢硫化物 hydrosulfide;sulfhydrate
氢硫酸 hydrosulfuric acid
氢氯化反应 hydrochlorination
氢氯化(合)物 hydrochloride
氢氯化橡胶 pliofilm
氢氯酸 chlorhydric acid;hydrochloric acid;muriatic acid
氢氯酸的 hydrochloric
氢煤法 hydrogenating
氢密封 hydrogen-tight
氢密封机座 hydrogen-tight house
氢模型 hydrogenic model
氢内冷 hydrogen inner cooling;internal hydrogen cooling
氢内冷发电机 hydrogen inner-cooling generator
氢内冷系统 hydrogen inner-cooling system
氢能 hydrogen energy
氢黏[粘]土 hydrogen clay[H-clay]
氢硼化锂 lithium borohydride
氢硼化钠 sodium borohydride
氢硼化物 borohydride
氢平衡 hydrogen balance
氢气 hydrogen;hydrogen gas
氢气保护热处理 hydrozing
氢气保护热镀铝法 Fink process
氢气焙烧 hydrogen firing
氢气处理 hydrogen gas processing
氢气纯度 hydrogen purity
氢气纯度调整 hydrogen gas purity control
氢气动力大客车 hydrogen-powered bus
氢气发动机 hydrogen engine
氢气发生和储配系统 hydrogen production and distribution system
氢气发生器 hydrogen gas generator;hydrogen generator;steam iron generator
氢气放电管 hydrogen discharge lamp

Q

氢气放空系统 hydrogen gas automatic blow down system
氢气氛 hydrogen atmosphere
氢气复合器 hydrogen recombiner
氢气钢瓶 hydrogen gas cylinder
氢气过压 hydrogen excess pressure
氢气缓冲罐 hydrogen gas buffer
氢气冷却发电机 hydrogen-cooled generator
氢气冷却塔 hydrogen gas cooling tower
氢气离子 hydrated ion
氢气流 hydrogen stream
氢气炉 hydrogen furnace
氢气密封 hydrogen seal
氢气泡法 hydrogen bubble method; hydrogen bubble technique
氢气泡技术 hydrogen bubble technique
氢气泡室 hydrogen bubble chamber
氢气瓶 hydrogen cylinder
氢气钎焊炉 hydrogen brazing furnace
氢气球 hydrogen balloon
氢气热处理 hydrozing; hydryzing
氢气烧结 hydrogen sintering
氢气深冷提纯过程 hydrogen cryogenic upgrading process
氢气体辐射值 radiation value of radon daughter
氢气筒 hydrogen cylinder
氢气退火 hydrogen annealing
氢气温度计 hydrogen thermometer
氢气系统 hydrogen gas system
氢气压缩机 hydrogen gas compressor
氢气阻火器 hydrogen gas fire arrester
氢弃场 dumping site
氢钎焊 hydrogen brazing
氢桥 hydrogen bridge
氢侵蚀 hydrogen attack
氢氰化作用 hydrocyanation
氢氰酸 hydrocyanic acid; prussic acid
氢氰酸中毒 hydrogen cyanide poisoning
氢燃料 hydrogen fuel
氢燃料电池 hydrogen fuel cell
氢燃料汽车 hydrogen-powered vehicle
氢燃料系统 hydrogen-fueled system
氢燃烧 hydrogen burning
氢燃烧器 hydrogen burner
氢蚀致脆 hydrogen embrittlement
氢受体 hydrogen acceptor
氢水化学 hydrogen water chemistry
氢酸 hydracid
氢损伤 hydrogen damage
氢损失 hydrogen loss
氢碳石 kratochvilite
氢碳原子比 atomic ratio of hydrogen to carbon; hydrogen-carbon atomic ratio
氢碳重量比 weight ratio of hydrogen to carbon
氢同位素地温计 hydrogen isotope geothermometer
氢外冷 hydrogen outer cooling
氢外冷汽轮发电机 hydrogen outer cooling turbogenerator
氢微波激射器 hydrogen maser
氢微闪检测器 hydrogen microflare detector
氢温标 hydrogen scale
氢稳定同位素标准 standard of stable hydrogen isotope
氢细菌 hydrogen bacteria
氢效率 hydrogen efficiency
氢溴的 hydrobromic
氢溴化物 hydrobromide
氢溴酸 hydrobromic acid; hydrobromie
氢溴酸分解 decomposition with HBr
氢溴酸浸渍试验 hydrobromic acid immersion test
氢溴酸盐 hydrobromate
氢溴酸中和试验 hydrobromic acid neutralization test
氢循环 hydrogen cycle
氢压力 hydrogen pressure
氢压力调整曲线 hydrogen pressure regulating curve
氢压下裂化 hydrocracking
氢焰曝光表 hydrogen flame photometer
氢焰光度计 hydrogen flame photometer
氢焰离子化检测器 hydrogen flame ionization detector
氢焰温度检测器 hydrogen microflare detector
氢阳离子交换器 hydrogen cation exchanger
氢氧饱和潜水 hydrogen oxygen saturation diving; oxyhydrogen saturation dive[diving]
氢氧爆炸气 oxyhydrogen gas
氢氧吹管 oxyhydrogen blowpipe
氢氧的 oxyhydrogen
氢氧电池 hydrogen oxygen cell
氢氧电量计 oxyhydrogen voltameter
氢氧钙石 portlandite
氢氧根离子吸收 hydroxyl ion absorption
氢氧焊 oxygen-hydrogen weld; oxyhydrogen weld
氢氧焊接 oxyhydrogen welding
氢氧焊炬 oxyhydrogen blowpipe
氢氧焊枪 oxygen-hydrogen; oxygen-hydrogen blowpipe; oxyhydrogen blowpipe; oxyhydrogen torch; torch
氢氧化铵 ammonium hydroxide; aqueous ammonia
氢氧化钯 palladium dydroxide
氢氧化钡 barium hydrate; barium hydroxide; caustic baryta
氢氧化苯汞 phenylmercuric hydroxide
氢氧化铋 bismuth hydrate; bismuth hydroxide
氢氧化铂 platinic hydroxide
氢氧化二氨合银 silver diamminohydroxide
氢氧化钙 calcium hydroxide; caustic lime; drowned lime; dry hydrated lime; hydrated lime; slaked lime; white lime
氢氧化钙糊剂 calcium hydroxide paste
氢氧化钙晶体 calcium hydroxide crystal
氢氧化高钴 cobaltic hydroxide
氢氧化高镍 nickelic hydroxide
氢氧化高铈 ceric hydroxide
氢氧化隔 cadmium hydroxide
氢氧化铬 chromic hydrate; chromic hydroxide; chromium hydrate; chromium hydroxide
氢氧化铬绿 chromium hydrate green
氢氧化汞 mercuric hydroxide
氢氧化钴 cobalt hydrate; cobalt hydroxide; cobaltous hydroxide; molybdenum hydroxide; molybdenum trihydroxide; molybdic hydroxide
氢氧化合物 hydroxide; oxyhydroxide
氢氧化钾 caustic potash; potassium hydrate; potassium hydroxide
氢氧化钾酒精溶液 alcoholic potash
氢氧化钾熔融 fusion with KOH
氢氧化钾值 potassium hydroxide number
氢氧化金 gold hydroxide
氢氧化铑 rhodium hydroxide
氢氧化铝 alucol; alugel; alumigel; alumina cream; alumin(i)um hydroxide; hydrate of alumin(i)um; Riopone
氢氧化铝八面体层 gibbsite layer
氢氧化铝凝胶 gel alumin(i)um hydroxide
氢氧化铝洗液 whitewash
氢氧化镁 caustic magnesia; caustic magnesite; magnesium hydrate; magnesium hydroxide
氢氧化镁浮悬液 magnesium magma
氢氧化镁合剂 mistura magnesia hydroxid
氢氧化镁铝 magaldrate
氢氧化镁乳液 magnesium oxide slurry
氢氧化钠 caustic soda; sodium droxide; sodium hydrate; sodium hydroxide
氢氧化钠处理 sodium hydroxide treatment
氢氧化钠法 sodium hydroxide method
氢氧化钠检验 <对混凝土骨料进行的> Abram's test
氢氧化钠溶液 soda solution; soda-lye
氢氧化钠熔融 fusion with NaOH
氢氧化镍 nickel(ic) hydroxide; nickelous hydroxide
氢氧化钕 neodymium hydroxide
氢氧化硼 boron hydroxide
氢氧化铍 beryllium hydroxide
氢氧化铅 lead hydrate; lead hydroxide
氢氧化铷 rubidium hydroxide
氢氧化三环己锡 cyhexatin
氢氧化三烃基硫 sulfonium hydroxide
氢氧化铯 cesium hydrate; cesium hydroxide
氢氧化钐 samaric hydroxide
氢氧化铈 cerium hydroxide; cerous hydroxide
氢氧化锶 strontium hydroxide
氢氧化四乙铵 tetrathylammnonium hydroxide
氢氧化铊 thallium hydroxide; thallous hydroxide
氢氧化钛 titanic hydroxide; titanium hydroxide
氢氧化铁 ferric hydroxide; ironic hydroxide
氢氧化铁法脱硫 ironic hydroxide desulfurizing
氢氧化铜 copper blue; copper hydroxide; cupric hydroxide
氢氧化物 caustic; hydrated oxide; hydroxyl group; oxyhydrate
氢氧化物沉淀 precipitation of hydroxide
氢氧化物沉淀法 hydroxide sedimentation method
氢氧化物固体 hydroxide solid
氢氧化物和含水氢氧化物 hydroxides and hydrousoxides
氢氧化物碱度 hydrate alkalinity; hydroxide alkalinity
氢氧化锡 stannic hydroxide
氢氧化锌 zinc hydroxide
氢氧化锌沉淀物 zinc hydroxide precipitation
氢氧化亚铂 platinous hydroxide
氢氧化亚铁 ferrous hydroxide
氢氧化亚铜 cuprous hydroxide
氢氧化亚锡 stannous hydroxide
氢氧化银 silver hydroxide
氢氧化脂肪酸 hydroxylated fatty acid
氢氧化重氮苯 diazobenzene hydroxide
氢氧基 hydroxy; hydroxyl group
氢氧离子 hydrated ion; hydroxide ion
氢氧离子活度 hydroxide ion activity
氢氧离子碱度 hydroxide alkalinity
氢氧镁石 brucite
氢氧气 oxyhydrogen
氢氧切割 oxyhydrogen cutting
氢氧同位素相关性 hydrogen oxygen isotope correlation
氢氧铜盐矿 diaboleite
氢氧焰 oxyhydrogen flame
氢氧焰焊接 oxyhydrogen flame welding; oxyhydrogen welding
氢氧焰切割 oxyhydrogen cutting
氢游离的 hydro(gen)-ionized
氢原子电弧焊接 atomic hydrogen arc welding
氢原子焊接 atomic H welding; atomic hydrogen welding
氢原子微波激射器 atomic hydrogen maser
氢原子振荡器 hydrogen atomic oscillator
氢原子质量 mass of hydrogen atom
氢原子钟 hydrogen atomic clock
氢云 hydrogen cloud
氢晕 hydrogen halo
氢载体 hydrogen carrier
氢闸流管 hydrogen thyratron
氢值 hydrogen value
氢指数 hydrogen exponent; hydrogen index
氢质土 hydrogen soil
氢致裂纹 hydrogen induced cracking
氢致损伤 <金属> hydrogen degradation
氢转移 hydrogen migration; hydrogen transfer(ence)
氢转移反应 hydrogen transfer reaction
氢转移聚合作用 hydrogen shift polymerization
氢自养反硝化 hydrogenotrophic denitrification

轻

轻埃洛石 termierite

轻百叶窗 light shutter
轻斑边材 light-stained sapwood
轻斑材料 light-stained sapwood
轻磅镀锡薄钢板 light tinned plate; lights
轻磅皮 lightweight hide
轻磅纸 featherweight paper; lightweight paper
轻包货物运输费用 cubic(al) freight transport cost
轻钡火石玻璃 light barium flint glass
轻钡冕玻璃 light barium crown glass
轻比重集料 lightweight aggregate
轻便 ease; handiness; lightness; lightweight; simplicity
轻便 X 射线荧光分光计 portable X-ray fluorescence spectrometer
轻便扒斗 portable scraper
轻便摆式抗滑测验仪 <英> pendulum-type portable skid resistance tester
轻便搬运机 portable conveyer[conveyor]
轻便泵 portable pump
轻便泵车 portable pump carriage
轻便臂式输送机 portable boom conveyer[conveyor]
轻便标尺 portable staff
轻便采样器 portable sampler
轻便槽 portable bath
轻便测波仪 portable oceanographic(al) recorder
轻便测汞仪 portable mercury detector
轻便测氦仪 portable helium detector
轻便插床 portable slotting machine
轻便茶桌 <十八世纪的> ambulante
轻便敞篷汽车 voiture
轻便车 cart

轻便车床 portable lathe
轻便冲击钻机 spudding drill
轻便冲击钻岩机 jackhammer drill
轻便抽水泵 handy billy; portable pump
轻便除草机 portable brush cutter; portable clearing saw
轻便触探试验 light sounding test; portable sounding test
轻便触探试验锤击数 blow count of light sounding test; light sounding test blow count
轻便磁性测厚仪 portable magnetic thickness ga(u)ge
轻便打孔器 bear
轻便带式运输机 portabelt
轻便带式运送机 portable belt conveyer[conveyor]
轻便单马车 carryall
轻便道轨 portable track
轻便的 handy; light; man-portable; mobile; portable; removable; runabout
轻便的艇首缆 harbo(u)r painter; lazy painter
轻便的转向机构 handy steering mechanism
轻便灯 portable lighter; service lamp
轻便灯具 portable lighter
轻便地磁台 portable magnetic observatory
轻便地震仪 portable seismograph
轻便点焊机 portable spot welder
轻便电锤 portable electric(al) hammer
轻便电动爆炸装置 magneto exploder
轻便电动工具 portable electric(al) tool
轻便电动磨轨机 portable motor driven rail grinder
轻便电动钻牙机 portable electric(al) dental engine
轻便电话机 buzzerphone
轻便电机 compacting machine
轻便电锯 bayonet saw; portable electric(al) saw
轻便电离室 portable ionization chamber
轻便电力工具 portable electric(al) tool
轻便电桥 portable bridge
轻便电台 hand transmitter receiver; portable radio station
轻便电钻机 portable electric(al) drilling machine
轻便吊车 portable yard crane
轻便定标器 portable scaler
轻便动力设备 lightweight power unit
轻便锻炉 field forge; travel(l)ing forge
轻便顿钻机 spudder
轻便发电机组 portable generator set
轻便发动机 donkey engine; portable engine
轻便放射性密度指示器 portable radioactive density indicator
轻便分析仪器 portable analytic(al) instrument
轻便风速表 hand anemometer; pocket anemometer
轻便风速计 hand anemometer; pocket anemometer
轻便风速仪 hand anemometer; pocket anemometer
轻便风钻 light weight pneumatic drill
轻便服务车 portable server
轻便高温计 portable pyrometer
轻便工程车 pick-up engineering truck
轻便工具 light instrument; lightweight tool; portable tool
轻便工具机 portable machine tool
轻便鼓风机 fire fan; portable blower
轻便鼓风器 portable blower
轻便刮铲 light stopper
轻便灌溉放水栓 portable irrigation hydrant
轻便轨 light rail
轻便轨车 light railway van
轻便轨道 apron track; light railway; portable track; sectional track
轻便轨道车 <窄轨的> decauville truck; light railway truck; light railway van; portable track truck
轻便轨道三轮车 velocipede
轻便轨面研磨机 portable rail surface grinder
轻便锅炉 portable boiler
轻便化学罐 portable chemical cylinder
轻便化仪器 portable instrument
轻便回照器 porte-lumiere
轻便活动桥 portable bridge
轻便货车 light truck; runabout
轻便货物 low-density cargo
轻便机车 dink(e)y locomotive; field loco(motive); light locomotive
轻便机车驾驶员 <俚语> dinky skinner
轻便机器 portable machine
轻便机械 portable machine
轻便基本轨研磨机 portable stock rail grinder
轻便计数管 portable counter
轻便加热器 portable heater
轻便监测仪 portable monitor
轻便剪切仪 portable shear apparatus
轻便检测器 portable detector
轻便建筑 portable structure
轻便胶带输送机 portable belt conveyer[conveyor]
轻便绞车 portable winch; tugger hoist
轻便绞辘 handy tackle; jigger tackle; watch tackle
轻便脚手架 horse scaffold; portable scaffold
轻便铰 pocket butt
轻便铰车 portable winch
轻便搅拌机 portable mixer
轻便接收机 mobile receiver; portable receiver
轻便结构 lightweight construction
轻便金刚石钻机 <可背运的> pack-back diamond drill
轻便经纬仪 portable theodolite; portable transit
轻便井架 portable derrick; portable mast; open-front derrick <前边敞开的>
轻便井架加固杆 shear pole
轻便净水器 portable water purifier
轻便锯 portable saw; portasaw
轻便锯机 portable saw
轻便卡车秤 portable truck scale
轻便客车 light bus; pick-up coach
轻便空气压缩机 portable air compressor
轻便快速测试装置 free acceleration test
轻便快艇 gig
轻便缆 lazy line; portable cable
轻便缆索钻探 portable cable drilling
轻便冷却器 portable cooling unit
轻便沥青拌和设备 transportable black-top plant
轻便漏斗车 light hopper car; light hopper wagon; portable hopper car
轻便旅行包 flight bag
轻便旅行车 estate car
轻便罗盘 portable compass
轻便落地吊车 canton crane; portable floor crane
轻便马车 buggy; voiture
轻便锚 portable anchor
轻便铆钉炉 portable riveting forge
轻便铆(接)机 portable riveter; portable riveting machine
轻便门 <非水密的> joiner door
轻便灭火器 hand portable extinguisher; portable extinguisher; portable fire-fighting appliance
轻便灭火设备 portable extinguishing equipment
轻便摩托车 scooter; trail bike
轻便磨床 portable grinder
轻便挠性轴磨轨机 portable flexible-shaft grinder
轻便抛绳枪 line throwing gun; shoulder gun
轻便刨 pocket plane
轻便刨床 portable planer
轻便平板卡车 pick-up and flat-bed truck
轻便骑 <美> roadster
轻便(起)吊具 pull lift
轻便起动机 portable starter
轻便起重机 pick-up crane; portable crane; runabout crane
轻便起重架 pick-up frame
轻便气焊装置 portable gas welding unit
轻便气压计 portable barometer
轻便器械 portable appliance
轻便桥 portable bridge
轻便切片机 handy microtome
轻便倾斜仪 portable tiltmeter
轻便取暖设备 recessed wall heater
轻便取土器 portable soil sampler
轻便取样器 portable sampler
轻便人字井架加固杆 shear pole
轻便三脚架 portable tripod
轻便三轮马车 <爱尔兰的> rid car; side-car
轻便筛分成套设备 portable screening plant
轻便设备 light equipment; portable set
轻便摄像机 field camera
轻便伸臂起重机 portable telescopic(al) crane
轻便伸缩臂式起重机 portable telescopic(al) crane
轻便升降装置 handy lift unit
轻便示波器 pocketscope
轻便式 portable type
轻便式标尺 portable rod; portable staff
轻便式采暖用具 portable heating appliance
轻便式沉落微尘探测器 transportable fallout detector
轻便式的 portable
轻便式电钻 portable electric(al) drill
轻便式读表系统 portable meter reading system
轻便式工程机械 portable equipment
轻便式构架桥 portable frame bridge
轻便式构造 package-type construction; potable type construction
轻便式海洋测流系统 portable oceanographic(al) survey system
轻便式降雨模拟器 portable rainfall simulator
轻便式空气压缩机 portable air compressor
轻便式螺旋输送器 portable auger conveyer[conveyor]
轻便式喷粉器 portable duster
轻便式棚屋 portable cabana
轻便式起重机 portable crane
轻便式气动工具 portable air tool
轻便式气体压力计 portable gas pressure ga(u)ge
轻便式扇风机 portable blower
轻便式生命保障设备 portable life support system
轻便式湿度滴定器 portable moisture titrator
轻便式手剪 portable hand-shear
轻便式输送机 portable conveyer[conveyor]
轻便式水准尺 portable rod; portable staff
轻便式通风设备 portable ventilating equipment
轻便式万能刨削机 portable universal shaping machine
轻便式运输机 portable conveyer[conveyor]
轻便式熨斗 lightweight iron
轻便式转臂吊车 portable derrick crane
轻便式装载机 portable loader
轻便式自记潮位计 portable automatic tide ga(u)ge
轻便式自记验潮仪 portable automatic tide ga(u)ge
轻便手动绞盘 dwarf capstan
轻便手力铆接机 hand portable riveter
轻便手推车 floor truck; portable truck
轻便手用器具 portable hand appliance
轻便手用器械 portable hand appliance
轻便输送机的可调车架 adjustable under-carriage construction for portable conveyors
轻便双程材料试验机 dual range portable tester
轻便双轮马车 gig
轻便水尺 portable staff
轻便水箱 dam
轻便水质监测装置 portable water quality monitoring system
轻便四轮马车 chariot
轻便酸碱度及电位计 portable pH-eH meter
轻便索道 simple aerial cableway
轻便台 knapsack station
轻便台地整平机 light terracing grader
轻便台架 portable rig
轻便台式圆锯机 portable rack circular saw
轻便探坑取样器 portable pit sampler
轻便梯 catwalk; portable ladder; rung ladder
轻便梯凳 footstool
轻便提升设备 pull lift
轻便天幕 portable awning
轻便铁道 decauville railway; light railroad; light railway; narrow-ga(u)ge railroad[railway]; portable railway
轻便铁道桥梁 light railway bridge
轻便铁轨 light railway track
轻便铁路 decauville; decauville railway; light-ga(u)ge railway; light railroad; light railway; narrow-ga(u)ge railroad[railway]; portable railway
轻便铁路电动牵引机 decauville motor tractor
轻便铁路系统 field railway system
轻便铁路线 field railway track
轻便铁索 simple aerial cableway
轻便铁索路 wire tramway
轻便通风干湿计 pocket aspiration psychrometer
轻便通风干湿仪 pocket aspiration

psychrometer
轻便图板 portable map board
轻便拖车式钻机 portable trailer-mounted drill; portable trailer-mounted drill rig
轻便弯轨机 portable rail bending machine
轻便万能钻床 portable universal drilling machine
轻便温盐测量计 portable salinity-temperature meter
轻便温盐测量仪 portable salinity-temperature meter
轻便吸尘器 portable dust collector; portable dust suctor; portable dust arrestor
轻便消防泵 handy billy
轻便小飞机 runabout
轻便小货车 pick-up truck;runabout
轻便小锚 portable anchor; small anchor
轻便小汽车 runabout
轻便小汽艇 runabout
轻便小铁道 field railway
轻便小型钻机 pocket drill
轻便小支架 tabo(u)ret
轻便卸货吊车 rowl
轻便信号灯 portable signal lamp;Aidis lamp
轻便型布告牌 portable tackboard
轻便型钢丝绳冲击式钻机 portable churn drill
轻便型回转起重机 portable slewing crane
轻便型机车整备车 <为调车机车服务> portable servicing truck
轻便型激光器 lightweight laser
轻便型控制仪表板 portable console
轻便型离子透入器 portable iontophoreser
轻便型落地吊车 portable floor crane
轻便型伸臂 portable boom
轻便型斜坡台 portable ramp
轻便型照相机 lightweight camera
轻便型蒸汽起重机 portable steam crane
轻便型紫外线辐射器 portable ultraviolet radiator
轻便性 portability;portableness
轻便性布告板 portable tackboard
轻便修边机 walking edger
轻便悬臂起重机 portable cantilever floor crane
轻便旋臂起重机 portable slewing crane
轻便遥测水温计 portable remote hydrotemp
轻便液压器 portable hydraulic power unit
轻便液压中轴打卷驱动机 portable hydraulic center[centre]-shaft rolling machine
轻便仪表 easy device; pocket instrument;portable instrument
轻便婴儿车 go-cart
轻便硬度计 duroscope
轻便油灰刀 light stopper
轻便有口门水管 <用于田间灌水的> portable gated pipe
轻便越野汽车 bantam car
轻便运输带 tripper belt conveyer
轻便运送机 portable conveyer[conveyor]
轻便凿 short shank chisel
轻便轧石厂 portable crushing plant
轻便栈桥 catwalk;portable catwalk
轻便帐篷 tabernacle
轻便折凳 campstool
轻便折椅 camp chair
轻便蒸汽起重机 portable steam crane
轻便转杯风速表 portable cup anemometer
轻便转臂吊机 portable derrick crane
轻便转臂起重机 portable derrick crane
轻便装药 portable charge
轻便装载机 portable loader
轻便装置 easy device;portable unit
轻便钻 portable drill
轻便钻机 bobtail rig; portable drill; portable drilling rig;portable rig
轻便钻架 portable rig
轻便钻孔变位计 portable borehole deflectometer
轻便钻孔攻丝机 portable drilling and tapping machine
轻便钻孔挠度计 portable borehole deflectometer
轻便钻塔 portable derrick
轻便钻探设备 portable rig
轻便钻岩机 lightweight rock drill
轻便座椅 glideover seat; pushover seat;slideover seat;walkover seat
轻薄防水布雨衣 mack
轻薄织物 light fabric
轻擦 scumble
轻材料铲斗 light material bucket
轻柴油 gas oil; light diesel fuel; light diesel oil
轻产品 float product
轻车快速交通 light fast traffic
轻沉排 light mattress work
轻触电子电路式编码 light touch electronic encoding
轻触刮涂 doctor kiss coating
轻触刮涂器 doctor kiss coater
轻触辊涂 kiss roll coating
轻触式薄膜键盘 membrane foil pressure sensitive keyboard
轻触涂装 kiss coating
轻触涂装机 kiss coater
轻船水线 light(water)line
轻锤击实试验 standard Proctor(compaction)test
轻打 wring
轻打配合 light-drive fit; light-press fit;wringing fit
轻淡色调 pastel shade
轻淡运量 light road traffic
轻的 light duty;weightless
轻电缆 lightweight cable
轻定额 loose standard
轻动接点 fly contact
轻动配合 free running fit; light running fit
轻冻 light freeze
轻度 light
轻度不正常的 subclinical
轻度的 low grade
轻度冻土 low-ice soil
轻度放牧 under grazing
轻度风化的 slightly weathered
轻度腐蚀的 mildly corrosive
轻度覆盖的 lightly covered
轻度干旱地区 mildly arid region
轻度钢化玻璃 lightly toughened glass; semi-tempered glass
轻度龟裂 cissing[sissing]
轻度灰化土 slightly podzolized soil
轻度碱性土壤 mildly alkaline soil
轻度开发的群体资源量 abundance of lightly exploited stock
轻度刻磨 grey cutting
轻度冷拔 light drawn
轻度冷藏舱 chilled hold
轻度冷藏货 chilled cargo
轻度利用 light use
轻度裂化 mild cracking
轻度淋溶 weakly leached; weakly leaching
轻度淋溶棕壤 weakly leached brown soil
轻度敏感的 slightly sensitive
轻度扭伤 wrick
轻度破坏 slightly damaged
轻度强化玻璃 semi-tempered glass
轻度失水 mild dehydration
轻度失效 minor failure
轻度失真 low distortion
轻度收敛 hypostypsis
轻度收敛的 hypostyptic
轻度凸面轧辊 full roll;rull roll
轻度弯曲 hettocyrtosis
轻度污染 light pollution; mild pollution
轻度污染带 zone of mild pollution
轻度污染河流 semi-healthy stream
轻度污染区 zone of mild pollution
轻度洗涤 light washing
轻度修剪 light pruning
轻度锈蚀 lightly rusted
轻度盐碱土 mildly alkaline soil
轻度盐渍的 brackish
轻度氧化 mild oxidation
轻度氧化沥青 semi-blown asphalt
轻度影响 slight effect
轻度沼泽化低地 everglade
轻度中毒 calomel poisoning; mild intoxication
轻度中暑 mild heat stroke
轻度着色的 light-colo(u)red
轻度阻尼结构 lightly damped structure
轻而薄的防水织物 mack
轻筏木 <比软木轻、制救生圈用的> balsa wood
轻帆 kite
轻帆布 duck
轻帆船 frigate
轻防腐蚀涂料 light-duty coating
轻非关节式流线型列车 non-articulated streamline train
轻粉壤土 light silty loam
轻粉质亚黏[粘]土 light silty loam
轻风 <蒲福风级表二级风> gentle breeze; light air; light wind; slight breeze; light breeze; dog vane <桅顶>
轻敷 dab
轻敷灰浆 mortar dab
轻浮货物 light and bulk freight; light and bulky goods
轻负荷 light burden; light duty; light-(ing)load(ing)
轻负荷工作 lighter-duty
轻负荷区 light loaded district; light loading district
轻负荷时 light hours; off-peak hours; slack hour
轻负荷试验 light-duty test
轻负荷小时 light hours;slack hour
轻负荷运行 light-duty service
轻负载 light duty; light(er)loading; lighting load(ing)
轻负载补偿 light load compensation
轻负载的 underloaded
轻负载断续试验 light intermittent test
轻负载试验 light-duty test
轻负载调节 light load adjustment
轻负载网络 light loaded network
轻负载循环 light-duty cycle
轻负载运行 light-duty service
轻钢薄壁结构 light-ga(u)ge steel construction
轻钢大梁 light steel girder
轻钢单元 light steel unit; lightweight steel unit
轻钢搁架 bar joist
轻钢构件 light-ga(u)ge steel member;light steel unit
轻钢货架 light steel pallet
轻钢集装箱 light steel container
轻钢结构 light steel structure; lightweight structure
轻钢龙骨 galvanized steel stud; lightage steel joist;metal frame
轻钢龙骨石膏板间隔墙 light steel keel gypsum board partition
轻隔墙 light partition; light partition wall
轻工业 light industry
轻工业(产)品 light industrial products;light industry products
轻工业机械厂 light industrial machinery plant
轻工业区 light industrial district
轻工作循环 light-duty cycle
轻工作制 light duty
轻汞膏 arquerite
轻拱 blank arch
轻垢刮除器 light scale tube scraper
轻谷物 light grain
轻骨构造 balloon framing
轻骨料厂 lightweight aggregate plant
轻骨料混凝土 all-light(weight)-aggregate concrete; light aggregate concrete;pumiceous concrete
轻骨料混凝土建筑砌块 lightweight aggregate concrete building block
轻骨料混凝土砌块 lightweight aggregate concrete block
轻骨料混凝土试验方法 test method for light(weight)aggregate
轻骨料预制块 lightweight aggregate block
轻刮 light blading
轻灌 light irrigation
轻轨 light rail
轻轨车 light rail vehicle
轻轨车辆 light rail vehicle
轻轨道 tramway
轻轨地铁 light metro
轻轨斗车 decauville truck; decauville tub;decauville wagon
轻轨高速运输系统 <一般在城市内及郊区> light rail rapid transit system
轻轨交通 light rail traffic; light rail transit
轻轨交通量道路 light traffic road
轻轨交通桥 rapid transit bridge
轻轨交通系统 light rail transit system
轻轨快速交通 light rail rapid transit
轻轨料车 decauville truck; decauville tub; decauville wagon; industrial railcar;industrial skip;jubilee truck
轻轨螺栓扳手 light rail track bolting wrench
轻轨铁道 light-ga(u)ge railroad; light-ga(u)ge railway; light-ga(u)ge track;light railroad
轻轨铁路 light-ga(u)ge railway; light railway
轻轨铁路器材 light railway material
轻轨铁路系统 light railroad system; light railway system
轻轨铁路线 light rail line
轻轨系统 light rail system
轻轨线路 light rail line
轻轨型垫式打磨机 orbital pad sander
轻轨运输 light rail transport
轻旱境 mildly arid region
轻焊缝 light weld
轻焊接 light welding
轻合金 light(-metal)alloy
轻合金板 light alloy metal plate
轻合金缸体 light alloy housing
轻合金活塞 light alloy piston
轻合金结构 light alloy structure

Q

轻合金梁 light alloy girder
轻合金梁桥 light alloy girder bridge
轻合金冶金学 light alloy metallurgy
轻合金铸造 light alloy casting
轻荷的 underloaded
轻荷耐火砖 low-duty fireclay brick
轻荷载 light duty;lighting load(ing);light load;lightweight loading;under load
轻荷载打桩机 light-duty pile driver
轻荷载地板 light-duty floor(ing)
轻(荷)载试验 light-duty test
轻滑配合 easy slide fit
轻划伤 fine scratch
轻混凝土 light concrete
轻混凝土空心块体 lightweight concrete hollow block
轻混凝土砌块 light concrete block
轻混凝土墙板 light concrete wall panel
轻火石玻璃 light flint
轻火石光学玻璃 light optic(al)flint
轻货 light freight;smalls;measurement cargo
轻货运汽车 light truck
轻霍乱 cholerine
轻击 tat
轻击锤 tapper
轻机车 light locomotive
轻机油 white oil
轻集料厂 lightweight aggregate plant
轻集料混凝土 light aggregate concrete;pumiceous concrete
轻集料混凝土砌块 lightweight aggregate concrete block
轻集料混凝土试验法 test methods for light(weight)aggregate concrete
轻集料预制块 lightweight aggregate block
轻加载 light(er)loading
轻甲板 spar deck;working deck
轻甲板船 hurricane-deck vessel;spar decker;spar deck ship
轻甲板梁 spar deck beam
轻间隙负载 light intermittent duty
轻浆 light sizing;light starching
轻浆整理 light finishing
轻交通道路 lightly trafficked roads
轻交通量 light traffic volume
轻交通量道路 light traffic road
轻交通路线 less trafficked route
轻焦油 coal-tar light;tar light oil
轻结构船 light scantling vessel
轻捷构架 balloon frame;balloon framing
轻捷构架房屋 balloon frame house
轻捷骨架【建】balloon framing
轻捷骨架构造【建】balloon framed construction
轻捷结构【建】balloon construction
轻捷木骨架 balloon framing
轻捷木骨架构造 balloon framed construction
轻捷型构造 balloon construction
轻金属 light metal
轻金属百叶板 light-metal louvers [louvres]
轻金属百叶板窗 light-metal blind slat
轻金属百叶窗 light-metal jalousie
轻金属百叶窗条板 lightweight metal blind slat
轻金属百叶门 light-metal shutter door
轻金属板 light metal plate
轻金属板条百叶窗 light-metal slated blind
轻金属玻璃窗 light-metal glazing
轻金属薄板 light sheet metal
轻金属窗 light-metal window
轻金属窗台 light-metal window sill
轻金属带腹板的梁 light-metal plain webbed beam
轻金属杠杆式(门)拉手 light-metal lever handle
轻金属格构桁架 light-metal lattice-(d)girder
轻金属隔墙 light-metal partition (wall)
轻金属关门装置 light-metal door closer
轻金属管材 light-metal tube
轻金属管脚手架 light-metal tubular scaffold(ing)
轻金属合金 light-metal alloy
轻金属桁架大梁 light-metal trussed girder
轻金属活塞 lightweight piston
轻金属家具 light-metal furniture
轻金属建筑 light-metal construction
轻金属建筑工人用的家具 light-metal builder's furniture
轻金属建筑工人用的五金 light-metal builder's hardware
轻金属建筑结构型材 light-metal structural section
轻金属结构 construction of light metal;light-metal construction
轻金属结构部件 light-metal construction section
轻金属结构单元 light-metal structural unit
轻金属结构工程 light-metal structural engineering
轻金属卷帘式百叶窗 light-metal slated roller blind
轻金属客车 lightweight metal coach
轻金属空间承重结构 light-metal space load bearing structure
轻金属矿产 light-metal commodities
轻金属矿石 light-metal ore
轻金属框架 light-metal frame
轻金属立面 light-metal facade
轻金属连接构件 light-metal connecting element
轻金属轮辋 light-metal rim
轻金属帽盖 light-metal cap
轻金属模板 light-ga(u)ge metal mo(u)ld
轻金属模型 light-metal shape
轻金属配件 light-metal fittings
轻金属皮带提升输送机 light-metal elevating belt conveyer[conveyor]
轻金属平腹板大梁 light-metal plain webbed girder
轻金属实心腹板大梁 light-metal solid webbed girder
轻金属实心腹板梁 light-metal solid webbed beam
轻金属涂料 coating for light metal
轻金属网格梁 light-metal lattice(d) girder
轻金属屋架 light-metal roof truss
轻金属屋面表层 light-metal roof cladding
轻金属五金制品 light-metal hardware
轻金属小五金 light-metal fittings
轻金属冶金学 metallurgy of light (weight)metals
轻金属装饰 light-metal trim
轻精构造 balloon construction
轻举妄动 blindness
轻聚油 <汽油轻馏分的聚合物> exanol
轻叩 chuck
轻快 lightness;volatility
轻快的立面 light facade
轻快列车 light rapid train
轻快列车运行 light rapid train running
轻快外形 airy appearance
轻快小汽车(或艇)runabout
轻快小艇 gig
轻快雪犁 speed snow plough
轻快钻进 easy drilling
轻矿物 light mineral
轻框架 lightweight frame
轻浪 <海浪二级> slight sea
轻离子 small ion
轻沥青 alyphite
轻粒子 lepton
轻连续焊接 beading weld(ing);light closing welding;light continuous welding
轻量 featherweight;lightweight
轻量吨位 lightweight tonnage
轻量谷类 lightweight grain
轻量关节流线型列车 articulated streamline train of light weight
轻量管子 lightweight pipe
轻量化 lightweight
轻量化制造法 lightweight fabrication
轻量混凝土 lightweight concrete;Cheecol <其中之一种>
轻量货品 light cargo
轻量货物 light goods
轻量货运输 lightweight shipment
轻量交通 light traffic
轻量绝热混凝土 lightweight insulating concrete
轻量木片 lightweight chip
轻量瓶 lightweight bottle
轻量型钢 light-ga(u)ge steel
轻料 <浮选中的> float material
轻料铲斗 wood chip bucket
轻料料堆 light stockpile
轻料用推土板 light material blade
轻硫砷银矿 trechmannite
轻馏分 light cuts;light fraction
轻馏分油 benzin(e)
轻滤料 light filtering medium;light filter material
轻率尝试 to attempt…rashly
轻率的计划 a wild schematization;a wild scheme
轻率的线性化 blind linearization
轻率的行为 reckless conduct
轻率的意见 crude opinion
轻煤 light coal
轻煤渣骨料 light cinder aggregate
轻煤渣集料 light cinder aggregate
轻冕(光学)玻璃 light crown glass
轻木 balsa wood;corkwood;fat wood;light wood
轻木焦油 soluble tar
轻木油 light wood oil
轻木属 balsa wood
轻泥浆 lightweight mud
轻黏[粘]壤土 light clay(ey)loam
轻黏[粘]土 light clay
轻碾裂缝 light roller check
轻拍 clap;dab;pat
轻排工 light mattress work
轻刨 light blading
轻泡货 light bulky cargo;light cube cargo;light cargo;measurement cargo
轻配电箱 light distribution box
轻瓶 lightweight container
轻气体油 light gas oil
轻汽油 benzine-ligroin;light petrol;naphtha;petroleum naphtha;wash oil <清洗油舱用的>
轻汽油溶剂 solvent naphtha
轻潜水 lightweight diving
轻潜水设备 lightweight diving apparatus;lightweight diving equipment;lightweight diving gear;lightweight diving outfit
轻潜水员安全所 lightweight dive haven
轻敲 clap;dab;tapping
轻敲锤 <取样用的> rapper
轻敲机构 lapping mechanism
轻敲声 rap
轻敲以减少残余应力 peening to reduce residual stresses
轻切 light cutting
轻侵蚀 slight attack
轻氢 light hydrogen
轻轻加压 <模制砖时> nudging
轻轻敲掉 tip-off
轻轻摇晃 jiggle
轻燃料 low-gravity fuel
轻燃氧化镁 light-burned magnesia
轻染 tinge
轻如羽毛之物 feather
轻入流 hypopycnal inflow
轻褥工 light mattress work
轻软的悬挂织物 drapeable fabric
轻润滑油 light lubricant
轻三合土 lightweight lime concrete
轻砂 lightweight sand
轻砂黏[粘]土 light sandy clay
轻砂壤土 light sandy loam
轻砂质壤土 light sandy loam
轻烧 light-burning;soft burning;soft flame
轻烧白云石 light-burned dolomite;light calcined dolomite;soft-burned dolomite
轻烧的 light-burned;soft-burned
轻烧矾土熟料 light-burned bauxite
轻烧镁砂 activated magnesia;calcined magnesia;caustic magnesia;caustic magnesite;light-burned magnesia;soft-burned magnesia
轻烧耐火制品 light-burned refractory ware
轻烧器皿 soft-fired ware
轻烧石灰 light-burnt lime;soft-burned lime
轻烧熟料 light-burned clinker;soft-burned clinker
轻烧氧化镁 light magnesia;quicklime
轻烧油浸砖 light-burned impregnated brick
轻烧砖 light fire brick;soft burnt brick
轻石 float stone;pumicate;pumice;pumice stone
轻石粉 powdered pumice(stone)
轻石灰三合土 light lime concrete
轻石混凝土 pumeconcrete;pumecrete;pumice concrete
轻石砂 pumice sand
轻石土 pumice soil
轻石油 ligroin(e);petroleum leve
轻视 contempt
轻水 <即普通水> light water;ordinary water
轻水反应堆 light-water reactor
轻水慢化堆 light-water moderated reactor
轻水慢化剂 light-water moderator
轻水泥制品 <分隔用> bubblestone
轻水增殖堆 light-water breeder reactor
轻松混凝土块体 lightweight concrete block
轻松使用信道 lightly utilized channel
轻松土 light soil
轻燧石玻璃 light flint glass
轻碳酸镁 magnesium carbonate,light
轻体直馏矿物油 straight mineral oil of light body
轻烃成岩作用阶段 light hydrocarbon diagenesis stage
轻同位素 light isotope
轻土料用的通用型推土机 light material universal bulldozer
轻推 jog

轻推配合 easy push fit
轻拖车 small tractor
轻挖 spuddle
轻瓦 light tile
轻微 lightness
轻微白腐材 spunk
轻微爆震 light knock
轻微变化型 minimal change group
轻微剥落 light scaling
轻微掺杂 light dope
轻微的 slight;trivial
轻微的表面涂布不均 cissing[sissing]
轻微的色度 tinge
轻微的涂膜疙瘩 cissing[sissing]
轻微地震声 brontide(s)
轻微发送冲突【计】 low distribution friction
轻微固结黏[粘]土 lightly over-consolidated clay
轻微故障 minor failure
轻微环裂 slight shake
轻微积雪 light accumulation of snow
轻微畸变 light distortion
轻微搅动 mild agitation
轻微结合的 softly cemented
轻微井喷 gas kick;well kick
轻微裂隙化 slightly fissured
轻微漏光 light bias
轻微脉动电动势 ripple electromotive force
轻微囊状腐朽 light white pocket
轻微排土桩 low displacement pile; nondisplacement pile
轻微偏态 slight skewness
轻微破坏 minor damage; slight damage
轻微倾侧 a slight heel
轻微缺点 incident defect
轻微缺陷 minor defect
轻微扰动 slight disturbed
轻微伤害的道路事故 slight-injury road accident
轻微湿陷 light collapsible
轻微事故 minor accident
轻微受损 slightly damaged
轻微损坏 slight damage
轻微违约 minor breach of contract
轻微瑕疵 blemish
轻微萧条周期 mild depression cycle
轻微压力 light pressure; slight pressure
轻微压下 shallow draft
轻微胀缩地基 very slightly swelling-shrinkage foundation
轻微振动板 light vibrating plate
轻微振动整平板 light vibrating screed
轻污染 mild pollution
轻污染级 be polluted to some degree
轻污染水域 oligosaprobic waters
轻物料用的 U 形推土板气动力 light material U-dozer
轻物料用的 U 形推土机气动力 light material U-dozer
轻雾(霭) <能见度 1000 ~ 2000 米> thin fog;light fog;light frost;mist
轻雾凇 soft rime
轻稀土 light rare earth
轻稀土矿 light rare-earth ores
轻稀土亏损 light rare earth element deficiency
轻心轴油 light spindle oil
轻芯材 light core
轻型 light-ga(u)ge; mild form; mild type
轻型 C 字麻袋 light C bag
轻型 E 字麻袋 light E bag
轻型板 light basis weight plate; low-mass panel
轻型板桩墙 camp sheathing column
轻型半自动氧气切割机 portable semi-automatic flame cutting machine
轻型表面活性剂 soft type surfactant
轻型驳船 foist
轻型舱壁 screen bulkhead
轻型槽钢 junior channel; light channel steel; thin-walled channel
轻型草地耙 grass comb
轻型插板 sheeter
轻型拆除工作 light demolition work
轻型柴油机 lightweight diesel; lightweight diesel engine
轻型车床 light-duty lathe;light lathe
轻型车辆 light vehicle; lightweight vehicle
轻型车辆的运输 <公路上> light-vehicular traffic
轻型车辆交通 light vehicle traffic
轻型车辆结构 lightweight vehicle construction
轻型车轮 light type wheel
轻型持针钳 light pattern needle holder
轻型冲击锤 light(hammer) drill
轻型冲击式凿岩机 light(hammer) drill
轻型穿孔机 buzzer
轻型打桩锤 light pile hammer
轻型打桩机 light-duty pile driver
轻型打桩架 light type pile frame
轻型单塔式起重机 light mono tower crane
轻型单翼机 grasshopper
轻型单座旋翼机 rotorcycle
轻型挡块 lightweight stopper
轻型导向电车线路 guided light tramway
轻型导向交通系统 guided light transit; light guideway transit system
轻型的 light duty; light-sized; lightweight; low-duty
轻型的水下勘探装置 aqualung
轻型地板 light-duty floor(ing)
轻型电动机械手 light-duty power manipulator
轻型电动压路机 light-duty power roller
轻型电钻 light-duty electric drill
轻型吊杆装置 light-duty derrick
轻型吊杆座 boom step bracket; gooseneck band
轻型动力触探 light dynamic(al) sounding
轻型二次破碎机 lightweight secondary crusher
轻型发电机组 portable electric(al) set
轻型发动机 light-duty engine
轻型帆布篷车 canvased light van
轻型反射镜 lightweight mirror
轻型方格 light panel
轻型飞机 light airplane
轻型飞机简易跑道 light aircraft strip
轻型风动式钻机 light pneumatic drilling machine
轻型风钻 light-duty pneumatic drill; sinker
轻型覆土耙 bush harrow
轻型钢 light ga(u)ge section steel
轻型钢材 lightweight section(al) steel; lightweight steel shape
轻型钢拱架 light steel arch
轻型钢构架 light-ga(u)ge steel framing
轻型钢构件 lightweight steel member; steel lightweight member
轻型钢结构 light-ga(u)ge steel construction; light-ga(u)ge steel framing; light-ga(u)ge steel structure; light section steel structure; lightweight steel structure; steel light structure; steel lightweight structure
轻型钢筋弯曲 light bending
轻型钢冷轧 cold-rolling
轻型钢梁 steel lightweight girder
轻型钢绳冲击钻架 light cable percussion rig
轻型工字钢 lightweight I-beam
轻型拱 light vault
轻型拱顶 light vault
轻型构造 balloon construction; light construction
轻型骨架 balloon framing; light framing
轻型挂车 light-duty trailer
轻型惯性导航系统 lightweight inertial navigation system
轻型光图像 light pattern
轻型轨道车 light-duty motor troll(e)y; light rail vehicle; light section car
轻型轨道内燃牵引车 light rail motor tractor
轻型焊 concave weld
轻型合页 shutter butt
轻型护卫舰 corvette
轻型混凝土建筑 light concrete building
轻型混凝土结构 light concrete structure
轻型混凝土梁 light concrete beam
轻型混凝土楼板 waffle floor
轻型混凝土砌块 light concrete building block
轻型混凝土烟囱筒体 light concrete chimney pot
轻型混凝土柱 light concrete column
轻型活动修理车 light mobile machine shop
轻型货车 light truck; light type truck; pick-up truck; van
轻型击实试验 Proctor compaction test
轻型机车 light locomotive; light rail motor tractor
轻型机械 light-duty machine
轻型机械化桥 mobile treadway bridge; rolling bridge
轻型架桥梁辎重队 light bridge train
轻型剪切机作业线 light shear line
轻型剪式屋架 light scissors truss
轻型检查车 light-inspection car
轻型建筑 light construction
轻型建筑板 light building board
轻型建筑板材 light building board
轻型建筑方法 light construction method
轻型建筑构件 light component
轻型建筑砌块 light building block
轻型建筑物 lightweight structure
轻型胶轮电车线路 tried[tyred] light tramway
轻型角钢 lightweight angle steel
轻型绞车 light type winch
轻型脚手架 ladder jack scaffold; light-duty scaffold; light scaffolding
轻型轿车道 light vehicle lane
轻型接触导线 light overhead contact line
轻型结构 balloon construction; balloon structure; light structure; light (weight) construction; light (weight) structure
轻型结构方法 light construction method
轻型金属客车 lightweight metal coach
轻型经济结构 light economic structure
轻型井点 light well point; vacuum well point
轻型井点排水 well point drainage
轻型警戒雷达站 lightweight warning radar station
轻型卷扬机 light type winch
轻型掘岩机 buzzer
轻型卡车 light-duty truck; light truck; pick-up truck
轻型可变深度拖曳式声呐 lightweight variable-depth towed sonar
轻型客货两用车 pick-up truck
轻型快速交通 light fast traffic; light rapid transit
轻型快速交通系统 light rapid transit system
轻型框架 horsehead
轻型框架木料 light framing lumber
轻型框架式打桩机 light type frame pile driving plant
轻型拉毛粉刷 light stucco
轻型立柱磨床 pedestal grinder
轻型连续焊 light continuance welder
轻型梁 junior beam
轻型料斗 light material bucket
轻型临时支架 false set(ting)
轻型灵活旋转门 light-duty flexible swing door
轻型楼面 light-duty floor(ing)
轻型漏斗车 light hopper wagon
轻型轮(胎)式拖拉机 light wheel tractor
轻型马 light horse
轻型面防御 light area defense
轻型面罩 lightweight mask
轻型模板 lightweight form
轻型摩托车 lightweight motorcycle
轻型摩托车胎 moped
轻型木构架 balloon frame[framing]
轻型木构架结构 balloon frame construction
轻型木骨架 balloon framework
轻型木碎料板 low-density wood chipboard
轻型耐火结构 light noncombustible construction
轻型黏[粘]土砖 light clay brick
轻型平板车 platform jubilee wagon
轻型平地机 high lift blade grader; lightweight grader
轻型平碾 light smooth wheel roller
轻型平台脚手架 needle beam scaffold
轻型平屋架 flat light truss
轻型破碎机 light-duty crusher
轻型普查钻机 <钻 15 ~ 20 米深> reconnaissance drill
轻型起重滑车 light hoisting tackle
轻型起重机 light-duty crane; light hoisting gear
轻型起重机架 pick-up frame
轻型气动风挡帚 light type air-operated screen wiper
轻型气动风屏帚 light type air-operated screen wiper
轻型汽车 light car
轻型汽车交通 light-vehicular traffic
轻型牵引机 mule;mule cart
轻型潜水服 light diving suit
轻型桥墩 light type pier
轻型桥台 light abutment; light type abutment; supported type abutment; thin-wall abutment
轻型轻油油轮 handy-size clean carrier
轻型穹顶 light-cupola
轻型取芯管 lightweight corer
轻型燃气轮机 lightweight gas turbine
轻型散装货轮 handy-size bulk carrier
轻型设备 light equipment; light facility; light fixture
轻型伸缩式凿岩机 lightweight stop-

per
轻型实验车 light experimental vehicle
轻型手持凿岩机 small hand drill
轻型受拉杆系结构 geodetic construction
轻型水泵 handy billy
轻型水接头 lightweight type water swivel
轻型水龙头 lightweight type water swivel
轻型送货车 light delivery truck
轻型塔吊 light tower hoist
轻型塔式起重机 light tower crane; light tower hoist
轻型提升机 light hoist
轻型填角焊 light fillet weld
轻型弹簧垫圈 light type spring washer
轻型铁路 light railway
轻型同轴电缆 lightweight coaxial cable
轻型推进器 light feed
轻型托架 light bracket
轻型围网 semi-purse net
轻型桅杆式转臂起重机 light mast crane
轻型圬工结构 light block masonry
轻型圬工墙板 <不承重> masonry wall panel
轻型屋顶 light roof
轻型屋架 light roof truss
轻型系列 light-duty series
轻型细直径钢筋弯曲 light bending for fine diameter bars
轻型纤维板 insulating board; softboard
轻型厢式送货车 light van
轻型向上式凿岩机 lightweight stopper
轻型小货车 pickup truck
轻型小客车 light car; lightweight car
轻型型钢 light-ga(u)ge section steel; light section steel; lightweight section steel
轻型旋转塔式起重机 light revolving tower crane; light rotating tower crane
轻型巡洋舰 light cruiser
轻型移动式输送器 light portable conveyer[conveyor]
轻型移动式钻机 drill scout
轻型油浴式空气滤清器 light-duty oil bath air cleaner
轻型有轨系统【交】light rail system
轻型预制产品 light prefabrication product
轻型预制构件 light cast member
轻型圆锥动力触探 portable cone sounding test
轻型运货车 light cargo carrier
轻型运货汽车 light goods-vehicle; light type truck
轻型运货升降机 dumb waiter
轻型运转状态 mild-duty operating condition
轻型载货汽车 light-duty vehicle; light goods-vehicle; pick-up truck
轻型载重货车 light vehicle
轻型载重卡车 light-duty truck
轻型载重汽车 light-duty lorry; light-duty truck; light truck; light-duty vehicle
轻型凿岩机 buzzer; light (hammer) drill; light rock boring machine; light rock drill; portable rock drill
轻型凿岩机操作工 rockman
轻型凿岩机凿岩 light drilling
轻型凿岩设备 light drilling equipment
轻型侦察机 jeep
轻型振动平板夯 lightweight vibro plant
轻型整体盒子结构 lightweight monolithic box
轻型整治 light-duty regulation
轻型整治工程 light-duty regulation works
轻型支臂 light boom
轻型支腿方脚板 square lightweight float
轻型直升机 light helicopter
轻型织物加强的小直径软管 clothe-inserted tubing
轻型中心增强电缆 lightweight center strength cable
轻型重力式艇架 free-standing gravity davit
轻型重油油轮 handy-size dirty carrier
轻型轴 light axle; plain spindle
轻型柱 light column
轻型铸件 light casting
轻型铸铁井盖 light-duty cast-iron well cover seating
轻型抓斗 light grab
轻型砖 light brick
轻型桩架 light type pile frame
轻型装备 light equipment
轻型装备抢救组 light recovery section
轻型装甲车 combat-car
轻型装甲汽车 light armored car
轻型装配构造 package-type construction
轻型装饰 light trim
轻型装载机 low loader
轻型自动氧气切割机 portable automatic flame cutting machine
轻型钻臂 light drilling boom
轻型钻车 light wagon drill
轻型钻杆安全夹持器 light-duty safety clamps
轻型钻机 light drill; jack hammer
轻型钻机钻进 light drilling
轻型钻机钻眼 light drilling
轻型钻架 jack leg
轻型钻探机 drill scout; X-ray drill
轻型钻探设备 light rig
轻修 light pruning
轻巡洋舰 light cruiser
轻循环油 light cycle oil
轻压盖 light cap
轻压力洛氏硬度试验机 superficial Rockwell hardness tester
轻压黏[粘]着带 pressure-sensitive adhesive tape
轻压配合 extra-light drive fit; finger press fit; light-drive fit; light force fit; light press(ure) fit
轻压涂装 kiss coating
轻压印版 kiss plate
轻亚黏[粘]土 clayey silt; light clay(ey) loam; sandy loam(soil)
轻亚砂土 light mild sand; light sandy loam
轻岩粉 light solids
轻摇 jiggle; jog(ging); joggle <卡片等的>
轻咬 nibble
轻药焊条 lightly coated electrode
轻页岩 light shale
轻液 light liquid
轻液层 layer of light liquid
轻液分布管 light liquid dispersion pipe
轻液入出口 light liquid inlet(outlet)
轻液状石蜡 light liquid paraffin
轻盈的 evanescent
轻油 cleaning oil; gas oil; light oil; light spirit; thin oil; white products
轻油车 <铁路用> light rail car
轻油罐车 petrol tank car
轻油柜 clean tank; light oil tank
轻油机 light oil engine
轻油加热器 light oil heater
轻油裂化 light oil cracking
轻油馏分 light oil distillate
轻油起泡剂 light frother
轻油强化法 Lomax method
轻油油轮 clean tanker
轻油组分 light oil constituent
轻油最大限度提取法 <洛马克斯法> light oil maximizing method
轻元素 light element
轻运货车 pick-up truck
轻载 fractional load; light load; underloaded
轻载补偿装置 light load compensating device
轻载吃水 light draught
轻载断路器 underload circuit-breaker
轻载公路 light-duty road
轻载荷 light load
轻载继电器 underload relay
轻载排水量吨位 light displacement tonnage
轻载期间 light load period
轻载启动负载 light starting duty
轻载气 light carrier gas
轻载驱动桥 light-duty axle
轻载试验 light load test
轻载条件 light condition
轻载调整 light load adjustment
轻载运行 light running
轻载运转 light running
轻载轴承 spigot bearing
轻载状态 light condition
轻凿 soft peening
轻轧 saddening
轻震 minor shock; weak shock
轻蒸汽雾 steam mist
轻制 <沥青的> cutback
轻制柏油 cutback tar
轻制产品 cutback product
轻制溚 cutback tar
轻制地沥青 asphalt cutback; cutback asphalt
轻制地沥青混合料 cutback pitch
轻制焦油沥青 cutback tar
轻制焦油沥青脂 cutback pitch
轻制沥青 cutback asphaltic bitumen; cutback bitumen; liquid asphalt
轻制沥青法 cutback asphalt principle
轻制沥青混合料 cutback asphalt binder; cutback group
轻制沥青混合物 cutback asphalt group
轻制煤焦油沥青 cutback coal tar
轻制煤焦油沥脂 cutback coal tar pitch
轻制(石油)沥青黏[粘]合料 cutback asphalt binder
轻制硬煤沥青 cutback coal tar pitch
轻质柏油 light tar oil; tar light oil
轻质板 light board; light plate; light-weight panel; lightweight sheet; softboard; Thistle board
轻质板材 light board; light plate
轻质板摇溶的 Thistle board
轻质板桩 lightweight sheet piling
轻质保温砖 lightweight insulating brick
轻质表壳 lightweight case
轻质材料 lightweight material
轻质材料的铲斗 light material bucket
轻质材料的料斗 light material bucket
轻质柴油燃料 light diesel fuel
轻质产品 light-end products
轻质衬底制 lightweight substrate system
轻质成品油 clean petroleum product
轻质纯碱 light soda ash
轻质瓷器 bone china
轻质粗骨料 light(weight) coarse aggregate
轻质粗骨料及砂的混凝土 sand-lightweight concrete
轻质粗集料 light(weight) coarse aggregate
轻质大梁 lightweight girder
轻质单位 lightweight unit
轻质的 lightweight
轻质的构成体 light component
轻质地板 lightweight fines
轻质地板梁 lightweight floor girder
轻质地板填充花砖 lightweight floor filler tile
轻质地土 light-textured soil
轻质煅烧氧化镁 light-burned magnesia
轻质多孔混凝土 cellular concrete
轻质多孔黏[粘]土砖 light porous clay brick; lightweight porous clay brick
轻质发动机油 light engine oil; light filtered cylinder oil
轻质粉刷 lightweight plaster
轻质粉状水泥 light powdered cement
轻质风干砖坯 light adobe brick
轻质覆盖层 siding
轻质刚玉砖 lightweight corundum brick
轻质钢材断面 lightweight steel section
轻质钢地板 lightweight steel floor
轻质钢构件 lightweight steel component
轻质钢结构 lightweight steel construction
轻质钢筋混凝土 lightweight reinforced concrete
轻质钢框架 lightweight steel frame
轻质钢梁 lightweight steel beam; lightweight steel girder
轻质钢梁地板 lightweight steel girder floor
轻质钢面板 lightweight deck(ing)
轻质钢桥面板 lightweight bridge deck(ing)
轻质高岭土 light kaolin
轻质高铝砖 alumina bubble brick; lightweight high alumina brick
轻质高强材料 light(weight) high strength material
轻质隔墙 light(weight) material partition; light(weight) partition wall
轻质隔热混凝土 lightweight insulating concrete
轻质拱顶 lightweight vault
轻质构件 lightweight member
轻质构造 lightweight construction
轻质构造材料 lightweight constructional material
轻质构造方法 lightweight constructional method
轻质骨料 bloated swollen aggregate; light(weight) aggregate; aglite <用膨胀性黏[粘]土制成>
轻质骨料的预浸 presoaking of light (weight) aggregate
轻质骨料混凝土 lightweight aggregate concrete
轻质骨料浇筑混凝土 lightweight aggregate poured concrete
轻质刮板 light screed; lightweight screed
轻质管子 lightweight pipe
轻质罐 lightweight pot
轻质硅酸钙砖 lightweight calcium silicate brick
轻质硅酸盐混凝土 light silicate concrete
轻质硅酸盐砖 Rhenish brick
轻质硅砖 lightweight silica brick
轻质辊 light top roller

轻质合金 light alloy;lightweight metal alloy
轻质核心 lightweight core
轻质桁架大梁 lightweight trussed girder
轻质黄砂石灰砖 lightweight sand-lime brick
轻质灰浆 lightweight plaster
轻质灰泥 lightweight plaster
轻质灰砂砖 light lime-sand brick; light sand-lime brick
轻质混合煅烧施工建筑灰泥 lightweight mixed calcined gypsum building plaster
轻质混凝土 aerated concrete;floated concrete; ytong; lightweight concrete
轻质混凝土板 lightweight concrete plank
轻质混凝土产品 lightweight concrete product
轻质混凝土车库 lightweight concrete garage
轻质混凝土承重框架 lightweight concrete load bearing frame; lightweight concrete weight carrying frame
轻质混凝土大梁 lightweight concrete girder
轻质混凝土的气泡 air cells of lightweight concrete
轻质混凝土顶板 lightweight concrete topping slab
轻质混凝土骨料 light concrete aggregate
轻质混凝土骨料工程 lightweight concrete aggregate works
轻质混凝土罐 lightweight concrete pot
轻质混凝土过梁 lightweight concrete lintel
轻质混凝土核心 lightweight concrete core
轻质混凝土集料 light concrete aggregate
轻质混凝土建筑 light concrete building
轻质混凝土建筑构件 lightweight concrete building member
轻质混凝土浇制单元 lightweight cast concrete unit
轻质混凝土浇制复合构件 lightweight cast concrete component unit
轻质混凝土浇制构件 lightweight cast concrete member; lightweight concrete building component
轻质混凝土浇筑构件 lightweight cast concrete component
轻质混凝土浇筑 lightweight concrete cast(ing)
轻质混凝土浇筑制品 lightweight cast concrete ware
轻质混凝土搅拌机 lightweight concrete mixer
轻质混凝土结构 light concrete structure;lightweight concrete structure
轻质混凝土结构框架 lightweight concrete supporting frame
轻质混凝土空心砌块 lightweight concrete cavity block; lightweight concrete hollow block
轻质混凝土空心填充块 lightweight concrete hollow filler
轻质混凝土空心砖 lightweight concrete cavity tile; lightweight concrete hollow tile
轻质混凝土块 lightweight concrete block
轻质混凝土框架 lightweight concrete frame
轻质混凝土立面 lightweight concrete facade
轻质混凝土梁 light concrete beam; lightweight concrete beam
轻质混凝土楼板 lightweight concrete floor slab;lightweight concrete slab
轻质混凝土楼板填充块 lightweight concrete floor filler
轻质混凝土砌块 light concrete building block; lightweight concrete block
轻质混凝土嵌板 lightweight concrete panel
轻质混凝土墙 lightweight concrete wall
轻质混凝土墙板 lightweight concrete wall slab
轻质混凝土桥 lightweight concrete bridge
轻质混凝土实心块 lightweight concrete solid block
轻质混凝土实心楼板 lightweight concrete solid floor
轻质混凝土填充板 lightweight concrete filler slab
轻质混凝土填充砖 lightweight concrete filler tile
轻质混凝土屋面板 lightweight concrete roof(ing) slab
轻质混凝土烟囱 lightweight concrete chimney
轻质混凝土烟囱套 lightweight concrete chimney pot
轻质混凝土烟囱筒体 light concrete chimney pot
轻质混凝土预制件 lightweight concrete cast member
轻质混凝土找平材料 lightweight concrete screed material
轻质混凝土找平层 lightweight concrete screed(topping)
轻质混凝土柱 light concrete column; lightweight concrete column
轻质混凝土砖 lightweight concrete block
轻质火泥 lightweight castable
轻质货物荷载 light density load
轻质货物列车编组 light train composition
轻质机械油 light machine oil
轻质基 <石油> lighter body
轻质基座 lightweight base
轻质基座体系 lightweight base system
轻质集料 aglite;bloated swollen aggregate;light(weight) aggregate
轻质集料的预浸 presoaking of light (weight) aggregate
轻质集料混凝土 lightweight aggregate concrete
轻质挤压型材 lightweight extrusion
轻质建筑板材 lightweight building board;lightweight building slab
轻质建筑薄板 lightweight building sheet
轻质建筑材料 lightweight construction material
轻质建筑单元 lightweight building unit
轻质建筑构件 lightweight building member
轻质建筑绝缘板材 lightweight building insulation(grade) board
轻质建筑块材 lightweight building block
轻质建筑立面 lightweight facade
轻质建筑砌块 light building block
轻质建筑油毡 lightweight building felt
轻质建筑预制板 lightweight building panel
轻质浇制单元 lightweight cast unit
轻质浇制构件 lightweight cast member
轻质浇制墙板 lightweight cast wall slab
轻质胶合板 lightweight plywood
轻质焦油 light tar
轻质焦油沥青 light tar
轻质脚手架 lightweight scaffold(ing)
轻质结构 lightweight construction; lightweight structure
轻质结构混凝土 light(weight) structural concrete; structural light (weight) concrete
轻质结构用黏[粘]土产品 lightweight structural clay product
轻质金属 lightweight metal
轻质金属百叶窗 lightweight metal jalousie; lightweight metal louvers [louvres]
轻质金属板 lightweight metal plank
轻质金属断面 lightweight metal profile
轻质金属钢框玻璃窗 lightweight metal glazing
轻质金属杠杆手柄 lightweight metal lever handle
轻质金属格构大梁结构 lightweight metal lattice(d) girder
轻质金属隔墙 lightweight metal partition(wall)
轻质金属家具 lightweight metal furniture
轻质金属结构 lightweight metal construction
轻质金属结构单元 lightweight metal construction unit
轻质金属结构断面 lightweight metal construction section
轻质金属结构装饰 lightweight metal construction trim
轻质金属宽板大梁 lightweight metal plain webbed girder
轻质金属宽板横梁 lightweight metal plain web(bed) beam
轻质金属框架 lightweight metal frame
轻质金属立面 lightweight metal facade
轻质金属门闭合装置 lightweight metal door closer
轻质金属器具 lightweight metal hardware
轻质金属上卷门 lightweight metal roll-up overhead door
轻质金属设备零件 lightweight metal fittings
轻质金属网格桁架 lightweight metal roof lattice truss
轻质金属屋面覆盖板 lightweight metal roof cladding
轻质金属屋面覆盖层 lightweight metal roof sheathing
轻质金属屋面盖板 lightweight metal roof covering
轻质金属五金 lightweight metal hardware
轻质金属型材 lightweight metal section
轻质绝缘层 lightweight insulating screed
轻质绝缘层材料 lightweight insulating screed material
轻质绝缘分度板 lightweight insulation(grade) board
轻质绝缘混凝土 lightweight insulating concrete
轻质卡路尼 <一种坚硬地板材料> light Karuni
轻质空心混凝土砌块 lightweight hollow concrete block
轻质空心楼板单元 lightweight core-out floor unit
轻质空心砌块 lightweight hollow block
轻质空心塑料楼板单元 lightweight core-out plastic floor unit
轻质空心填充块 lightweight hollow filler
轻质空心砖 lightweight cast cavity tile
轻质块材隔墙 lightweight block partition(wall)
轻质块材圬工 lightweight block masonry(work)
轻质矿物骨料 light(weight) mineral aggregate
轻质矿物集料 light(weight) mineral aggregate
轻质矿渣粉末 light slag powder
轻质矿渣水泥 lightweight slag cement
轻质拉毛粉刷 lightweight stucco
轻质立面构件 light facade element
轻质梁 lightweight beam
轻质楼板体系 lightweight deck(ing) system
轻质垆坶 light loam
轻质炉渣 light slag
轻质炉渣混凝土 light slag concrete
轻质路油 light road oil
轻质煤焦油 light tar oil
轻质煤渣混凝土 breeze concrete
轻质煤渣混凝土大砖 breeze block
轻质煤渣混凝土块 breeze block
轻质镁氧 light magnesia
轻质面层骨料 coated type aggregate
轻质面层集料 coated type aggregate
轻质木材 balsa;light wood
轻质木材纤维板 beaver board; light wood fiber[fibre] board
轻质耐火材料 lightweight refractory (material)
轻质耐火产品 lightweight refractory product
轻质耐火黏[粘]土保温砖 light fireclay insulating brick
轻质耐火黏[粘]土砖 light fire-clay insulating brick
轻质耐火砖 light chamot(te) brick; lightweight fire brick; lightweight refractory brick
轻质泥沙 lightweight sediment
轻质泥土砖 light mud brick; lightweight adobe brick
轻质黏[粘]土 light clay
轻质黏[粘]土墙 lightweight loam wall
轻质黏[粘]土砖 light adobe brick; light clay brick; lightweight chamotte brick;lightweight clay brick; lightweight fireclay brick; lightweight mud brick
轻质泡沫黏[粘]土制品 light foamed dayware
轻质泡沫塑料 lightweight foamed plastics
轻质膨胀混凝土骨料 lightweight expanded concrete aggregate
轻质膨胀混凝土找平层 lightweight cast cellular expanded concrete screed
轻质膨胀黏[粘]土骨料 light expanded clay aggregate
轻质膨胀黏[粘]土集料 aglite
轻质膨胀塑料 lightweight expanded plastics
轻质(铺)路油 light road oil
轻质汽油 benzin(e);benzoline;light gasoline
轻质砌块 lightweight block

轻质砌块隔墙上的管托 pipe bracket for lightweight block partition
轻质砌块假平顶 false ceiling of lightweight blocks;false ceiling of lightweight slabs
轻质墙 light wall;lightweight wall
轻质墙板 lightweight wall slab
轻质清水混凝土 lightweight fair-faced concrete
轻质清水混凝土砌块 lightweight fair-faced concrete block
轻质清水混凝土砖 lightweight fair-faced concrete tile
轻质燃料 light fuel
轻质燃油 light fuel oil
轻质壤土 light loam
轻质熔渣 lightweight slag
轻质熔渣粉末 lightweight slag powder
轻质熔渣骨料 lightweight cinder aggregate
轻质熔渣混凝土 lightweight slag concrete
轻质熔渣混凝土砌块 lightweight slag concrete block
轻质熔渣混凝土砖 lightweight slag concrete tile
轻质熔渣建筑砌块 lightweight slag building block
轻质熔渣块 lightweight slag block
轻质熔渣砖 lightweight slag tile
轻质润滑油 light lubricating oil
轻质砂 lightweight sand
轻质砂混凝土 sand-lightweight concrete
轻质砂土 light soil
轻质烧结集料 lightweight sintered aggregate
轻质石膏粉刷 lightweight gypsum plaster
轻质石膏隔墙板 lightweight gypsum partition tile
轻质石膏灰泥 lightweight gypsum plaster
轻质石灰混凝土 light lime concrete
轻质石灰混凝土砌块 lightweight lime concrete block
轻质石灰混凝土砖 lightweight lime concrete tile
轻质石灰三合土 lightweight lime concrete
轻质石油 light oil;clean oil
轻质石油产品 clean oils;oils
轻质石油沥青 asphalt cutback
轻质实心黏[粘]土砖 solid light (weight) clay brick
轻质实心砌块 lightweight solid block
轻质实心砖 light solid tile; lightweight solid tile; solid light (weight) brick
轻质竖向空心黏[粘]土砖 lightweight vertical coring(clay) brick
轻质苏打 light ash
轻质塑料 lightweight plastics
轻质塑料泡沫 lightweight plastic foam
轻质塑料球 lightweight plastic ball
轻质碳酸钙 precipitated calcium carbonate
轻质碳酸镁 light magnesium carbonate
轻质陶粒 lightweight sintered aggregate
轻质陶粒集料 light expanded clay aggregate
轻质填充板材 lightweight filler slab
轻质填充块材 lightweight filler block
轻质填充料 lightweight filler
轻质填充楼板 lightweight filler floor
轻质填方 light weight fill
轻质填料 floated filler;light filler
轻质烃 light hydrocarbon
轻质烃混合物 petroleum spirit
轻质土(壤) light soil; light-textured soil
轻质瓦斯油 light gas oil
轻质外粉刷 lightweight external rendering
轻质外粉刷墙 lightweight external plaster
轻质外露混凝土片砖 lightweight exposed concrete tile
轻质外露混凝土砌块 lightweight exposed concrete block
轻质外露混凝土瓦 lightweight exposed concrete tile
轻质烷基化物 light alkylate
轻质圬工 lightweight masonry(work)
轻质屋顶 lightweight roof(ing)
轻质屋顶窗 lightweight roofing window
轻质屋顶断面 lightweight roofing section
轻质屋顶形式 lightweight roofing shape
轻质屋面 lightweight roof(ing)
轻质屋面安装管架 lightweight roofing tube
轻质屋面安装管子脚手架 lightweight roofing tubular scaffold
轻质屋面窗台 lightweight roofing window sill
轻质屋面单元 lightweight roofing unit
轻质屋面桁架大梁 lightweight roofing trussed girder
轻质屋面结构工程 lightweight roofing structural engineering
轻质屋面空间承重结构 lightweight roofing spatial loading bearing structure
轻质屋面三向荷载承重 lightweight roofing three-dimensional load bearing
轻质屋面实心腹板大梁 lightweight roofing solid web girder
轻质屋面瓦 lightweight roofing tile
轻质屋面装饰 lightweight roofing trim
轻质无定形材料 lightweight castable material
轻质细粒 lightweight fine grain
轻质细粒砂 lightweight fine grained sand
轻质下部结构体系 lightweight substructure system
轻质斜坡屋顶 lightweight sloping roof
轻质型材 lightweight profile; lightweight section;lightweight shape
轻质型钢 light section(al)steel;lightweight section(al)steel
轻质循环进料 light recycle stock
轻质亚黏[粘]土 light(weight)loam
轻质亚砂土板 lightweight loam slab
轻质氧化镁 light magnesium oxide
轻质液状石蜡 light liquid petrolatum
轻质硬导线管 light rigid conduit
轻质油 low oil
轻质油船 clean ship
轻质油回收 recovering of vapo(u)rized hydrocarbons
轻质油回收装置 condensate extractor
轻质油蒸馏锅 reducing still
轻质预浇混凝土构件 lightweight precast concrete component
轻质预应力混凝土单元 lightweight prestressed concrete unit
轻质预应力混凝土构件 lightweight prestressed concrete member
轻质预制单元 lightweight precast unit
轻质预制混凝土产品 lightweight precast concrete product
轻质预制混凝土复合单元 lightweight prefabricated concrete compound unit
轻质预制混凝土复合构件 lightweight precast concrete compound unit
轻质预制混凝土构件 lightweight precast concrete member
轻质预制混凝土制品 lightweight precast concrete ware
轻质预制件 light cast member
轻质预制墙板 lightweight precast wall slab
轻质原油 light crude
轻质渣 foam(ing)slag
轻质找平层 lightweight screed
轻质珍珠岩 popped pearlite
轻质珍珠岩幕墙 lightweight pe(a)rlite curtain wall
轻质织物 lightweight fabric
轻质铸造零件 lightweight cast component
轻质砖 floating brick; light(weight) brick
轻质砖隔墙 lightweight tile partition (wall)
轻质砖瓦 lightweight tile
轻质砖瓦圬工 lightweight tile masonry(work)
轻质装修 lightweight trim
轻重缓急次序 order of priority
轻重货物的配装 match loading of light and heavy goods
轻重量车身 lightweight body
轻重量高能推进剂 high energy light (weight) propellant
轻轴 light axle
轻砖 light(weight) brick
轻砖瓦 light tile
轻转配合 easy running fit; free running fit; light running fit; slack-running fit
轻装甲 light armo(u)ring
轻装甲防护 light armo(u)red protection
轻装料 light burden
轻装潜水 skin diving
轻装潜水鞋 lightweight shoes
轻装潜水员 frogman;skin diver
轻装头盔 lightweight helmet
轻撞 jog
轻琢石面 stugging
轻子【物】 lepton
轻组分 more volatile component
轻钻井管 lightweight drill tubing
轻作业 light operation;light work

倾

倾杯式雨量计 tipping-bucket rain (fall) ga(u)ge; tipping cup rain ga(u)ge
倾槽拌和机 swing bucket mixer
倾槽式拌和机 tipping trough mixer
倾槽式搅拌机 tipping trough mixer
倾槽式输送机 tipping tray conveyer; tipping trough conveyer[conveyor]
倾槽式运输机 tipping tray conveyer; tipping trough conveyer[conveyor]
倾侧 heel;list;roll displacement; swag;tilt
倾侧边界 tilt boundary
倾侧的 lopsided
倾侧的钢轨 canted rail
倾侧度 inclination
倾侧机构 tilting mechanism
倾侧计 clinometer
倾侧检修 careen
倾侧角 angle of bank; heeling angle; angle of heel <船的>
倾侧角读数起点 roll datum
倾侧角杆 roll angle
倾侧角基准点 roll datum
倾侧角加速度 rolling acceleration
倾侧力矩 heeling moment; inclining moment; moment in roll; rolling couple;upsetting moment
倾侧力矩的力臂 heeling lever
倾侧水线面【船】 heeled water plane
倾侧调整 heeling compensation
倾侧线 heel line
倾侧运动 rolling motion
倾侧运动阻尼 lateral damping
倾侧闸门 tilting gate
倾侧状态 heeling condition
倾侧阻尼 lateral damping
倾侧阻尼器 roll damper
倾差 heeling deviation;heeling error
倾差校正磁铁 heeling corrector;heeling magnet
倾差校正仪 dipping needle instrument;heeling adjuster
倾差系数 heeling coefficient
倾差仪 <磁罗经> heeling error instrument;vertical force instrument
倾出 effuse
倾船场 careen site
倾吹碱性转炉钢 side blow basic converter steel
倾刀 tilted blade
倾倒 dump(ing); knocked down; overturn; pour-out; random dumping; tilting motion; tip; up-end; bleeding <将袋装谷类倾入散装舱>
倾倒槽 dump tank
倾倒场 dump area;dumping ground
倾倒车 dump cart;tip car
倾倒车身转轴 tip beam;tip shaft
倾倒的 off-balance
倾倒点 pour(ing) point
倾倒堆垛车 lorry dump piler
倾倒(方)法 method of dumping;pouring process
倾倒放射性废物场所 radioactive cemetery
倾倒废物 dumping of wastes
倾倒垃圾 sanitary pitting; sanitary tipping
倾倒垃圾装置 dumping gear
倾倒平板培养基 pour plate medium [复 media]
倾倒破坏 tilting failure
倾倒器 dumper;inclinator
倾倒千斤顶 tilting jack
倾倒区(域) dump(ing) area; dump(ing) ground; spoil area; spoil ground
倾倒曲流 inclined meander
倾倒设备 tilter
倾倒式吊罐 tipping bucket
倾倒式反射炉 tilting reverberatory furnace
倾倒式加料桶 dump bucket
倾倒式浇注桶 tipping ladle
倾倒式搅拌机 turn-batch mixing drum
倾倒式开底车 car dumper
倾倒式模板 tilt up formwork
倾倒式泄水闸门 tilting sluice gate
倾倒式溢流堰闸门 tilting spillway gate
倾倒式雨量计 tilting bucket rain ga(u)ge
倾倒式闸门 flap gate;tilting gate
倾倒试验 pour test

倾倒台拖车 tilt deck trailer
倾倒填土 dumped fill;tipped fill
倾倒卸货车 car dumper
倾倒卸物 discharge by pumping
倾倒用液压缸 tilt cylinder
倾倒渣车 dump cinder car
倾倒闸门 oscillating flashboard;tumble gate
倾倒者 midnight dumper
倾倒装置 tipper;tipping gear
倾点 pour point
倾点回升 pour point reversion
倾点试验 pour point test
倾点温度 flow temperature; pour point temperature
倾点下降剂 pour point depressant additive;pour point depressor
倾点下限 minimum pour point
倾点抑制剂 pour point depressant
倾动地块【地】tilted block
倾动流铁槽 tilting spout
倾动式保温炉 kettle furnace
倾动式坩埚炉 tilting crucible furnace
倾动式浇包 labiate ladle;tilting ladle;top-pour ladle
倾动式前炉 ladle forehearth; tilting forehearth;tilting ladle
倾动试验 flow test
倾动铸型 tilt mo(u)ld
倾动装置 tilting arrangement; tilting equipment;tilting machinery
倾斗车 trough tip wagon
倾斗式秤 weighing machine with tipping hopper
倾斗式雨量计 tipping-bucket rain-(fall) ga(u)ge
倾斗式雨量器 tipping-bucket rain-(fall) ga(u)ge
倾斜线 tilt line
倾度 angularity;batter;inclination;incline grade; incline gradient; obliqueness;obliquity
倾度变位法 method of slope deformation
倾度风 gradient wind
倾度计 dip needle
倾度(计算)法 < 坝工 > inclination method
倾度角 angle of elevation
倾度线 isostatics
倾度仪 dip(ping) compass
倾翻 dump; overturning; roll-over; tilting
倾翻铲斗 tilting dipper bucket
倾翻车 trough tipping wagon
倾翻车厢 dump body
倾翻出料式灰浆搅拌机 tilting-type mortar mixer
倾翻挡 tip stop
倾翻点 tipping point
倾翻顶杆 tilting ram
倾翻斗 tipping bin
倾翻惰轮 tilting idler
倾翻负荷 tipping capacity
倾翻荷载 overturning load
倾翻机构 dumping gear;ingot tipper; tilter; tilting gear; tilting mechanism; tipper; tripping mechanism; tumbler
倾翻角度 angle of overturn;angle of tip;tipping angle
倾翻力偶 roll couple
倾翻料斗 tilting skip
倾翻平台 tilting bed
倾翻器 dumper;trip
倾翻式拌和机 tilting mixer
倾翻式铲斗 roll-over bucket
倾翻式车厢 tilt cab(in)
倾翻式底卷 tipping chassis
倾翻式方筒搅拌机 mixer cube
倾翻式滚筒混凝土搅拌机 tilting drum concrete mixer
倾翻式混凝土搅拌机 tilting concrete mixer
倾翻式机罩 tilt hood
倾翻式搅拌机 tilting mixer
倾翻式升运器 turnover elevator
倾翻式推土机 tilting dozer
倾翻式熄焦车 tilt-bed quenching car
倾翻试验 overturning test
倾翻速度 tipping speed
倾翻桶 tilting tank
倾翻稳定性试验 overturning stability test
倾翻系统 tilting system;tipping system
倾翻线 tipping line
倾翻卸料式灰浆搅拌机 tilting mortar mixer
倾翻轴 tipping pitman
倾翻装置 tipping device;tipping unit
倾翻作用 roll-over action
倾废区 dumping zone
倾覆 cap size;jumping up;overset;overthrow;overturn(ing);tipping; topple; tumble; turnover; upset-(ting);pitch;plunge;rake;bottom up【船】
倾覆安全系数 safety factor against overturning
倾覆背斜 plunging anticline
倾覆槽 tipping trough
倾覆船扳正扒杆 sheer legs for righting a capsized vessel
倾覆倒转性模式 plunge-reversal mode
倾覆的 overturned
倾覆地层 overturned bed
倾覆点 upsetting point
倾覆动作 tilting motion
倾覆端 plunging end
倾覆端位置 position of plunging crown
倾覆断层 plunging fault
倾覆荷载 tilting load(ing); tipping load
倾覆及滑走 stability against
倾覆角 angle of overturn; capsizing angle;upsetting angle;pitch;plunge angle;plunging angle
倾覆力 overturning force; upsetting force
倾覆力臂 upsetting arm;upsetting lever
倾覆力矩 capsizing moment; disturbing moment; lifting moment; load tipping moment; moment of overturning;overturning moment;tilting moment; turning moment; turning movement; upending moment; upsetting moment
倾覆力偶 overturning couple;turning couple;upsetting couple
倾覆临界荷载 < 起重机的 > critical tipping load
倾覆临界条件 critical tipping condition
倾覆扭矩 overturning torsional moment
倾覆破坏 toppling failure
倾覆区 plunge area
倾覆式闸板 oscillating flashboard
倾覆式闸门 oscillating flashboard
倾覆推力 overturning thrust
倾覆稳定系数 overturning factor of safety
倾覆稳定性 overturning stability;tilt-(ing) stability; stability against overturning
倾覆向斜 pitching syncline;plunging syncline
倾覆崖 plunging cliff
倾覆褶额 plunging crown
倾覆褶皱【地】pitching fold;plunging fold
倾覆轴 topple axis;tumble axis
倾覆阻力 overturning resistance
倾覆作用 overturning effect
倾航 careenage
倾虹吸 inverted siphon[syphon]
倾滑断层面【地】dipping fault plane
倾滑断裂 dip slip fault
倾毁 dilapidation
倾计 incline level
倾架 tilter
倾筒式混凝土拌和机 tilting drum concrete mixer
倾浆 draining
倾角 angle of declination;angle of depression;angle of dip;angle of obliquity; depression angle; descending vertical angle; incidence; inclination angle; inclined angle; rake angle;tilt(ing) angle;tip(ping) angle;tip-tilt;dip angle【地】;amount of inclination【地】; angle of lean < 井架 >
倾角表 incidence meter
倾角测定 determination of tilt; dip test
倾角测量仪 dipmeter[dipmetre]
倾角读数 dip reading
倾角法 tilt angle method
倾角法测量剖面平面图 profile-plan figure of results of dip-angle method
倾角法测量剖面图 profile figure of results of dip-angle method
倾角改正 adjustment for tilt; correction for angle of tilt;tilt adjustment
倾角计 banking inclinometer;banking indicator; clinometer; drift indicator;incidence meter;inclinometer; tilt(o)meter
倾角计算法 inclination method
倾角减小 decrease in dip
倾角校正 adjustment for tilt;tilt adjustment
倾角校正图 dip-corrected map
倾角近似法 approximation for tilt
倾角可变电感器 incline variometer
倾角控制系统 tilt angle control
倾角罗盘 inclination compass
倾角挠度法 slope-deflection method
倾角剖面曲线 profile curve of dip angle
倾角圈 tilt circle
倾角式混凝土搅拌机 tilting drum concrete mixer
倾角误差 error of tilt;inclination error
倾角系数 < 筛子以与水平面成 15°角为 1 > S-factor[factor of screen incline]
倾角线 incidence wire;stagger wire
倾角相干加强 slant coherent enhancement
倾角向量 dip vector
倾角修正 incidence correction
倾角仪 clinometer; dip compass; dip needle; inclinator; inclinometer; wire angle indicator < 吊索式 >
倾角仪记录 dipmeter log
倾角圆 tilt circle
倾角章动 nutation in obliquity;nutation of inclination
倾角指示器 angle incidence indicator; incidence indicator; tilting angle indicator
倾角指针 inclination needle
倾角走向符号 dip-strike symbol
倾角钻井记录 dip log
倾截面 oblique section
倾靠椅 < 剧场用 > tip-up seat
倾力推销 high-pressure selling
倾料场 dump position
倾料车 dump hopper
倾料斗 tipping hopper
倾滤 decantation
倾落式 < 煤炭装船设备的一种形式 > drop system
倾盘式过滤机 tilting disc filter
倾盘式配水器 tipping tray distributor
倾盆大雨 cloudburst; deluge; downpour; drench; driving rain; heavy downpour; pelting rain; pouring rain;rain in torrents;rattler;sluicy rain; spate; torrential rain; torrent of rain
倾盆大雨的湿度 pelting rain humidity
倾盆状流水 water sheet
倾泼的水 slop
倾弃垃圾填地 landfill dumping;landfilling
倾弃于海洋 sea disposal
倾弃于环境 environmental release
倾腔 cavity dumping
倾腔激光器 cavity-dumped laser
倾翘洒水盘 tipping tray
倾翘式溢水堰闸门 tilting spillway gate
倾全力于 buckle to
倾群【生】cline
倾入 pitch;plunge;rake
倾入槽 pan tank;receiving tank;rundown tank(storage)
倾摄照片 oblique photograph
倾竖褶皱【地】vertical fold
倾榫【建】oblique tenon
倾坍 dilapidation
倾填 dumped fill
倾填方材料 deposit(e) fill material
倾听 listening-in
倾筒式分批拌和机 tilting drum batch mixer
倾筒式分批搅拌机 tilting drum batch mixer
倾筒式混凝土搅拌机 tilting drum concrete mixer
倾筒式机械 tilting drum mechanism
倾筒式搅拌机 tilting drum mixer;tilting mixer;tilting-type mixer;tipping mixer
倾头现象【船】dipping
倾土堆 tipped fill
倾析 decant;outwell
倾析阀 decanting valve
倾析法 elutriation;decantation
倾析瓶 decanting glass
倾析器 decanter
倾析溶液 decanted solution
倾析试验 decantation test
倾箱机 box dumper
倾箱式取样机 tilting-box sampler
倾向 angle of dip;bias;dip direction; disposition; leaning; lurching; propensity;swing;tendency;trend
倾向爆破 dip blasting;dip shooting
倾向出口 side outlet
倾向倒转 roll-over
倾向地质剖面图 dip geologic(al) profile
倾向断层 dip fault
倾向断距 dip separation
倾向方程 tendency equation
倾向风险 risk proneness
倾向隔距【地】dip separation
倾向谷 dip valley
倾向河 dip river;dip stream
倾向河流上游的丁坝 groyne pointing upstream
倾向河流上游的防波堤 groyne point-

ing upstream
倾向河流下游的丁坝 groyne pointing downstream
倾向河流下游的防波堤 groyne pointing downstream
倾向滑动【地】 dip slip
倾向滑动拖曳 dip slip drag
倾向滑断层【地】 dip slip fault;hade-slip fault
倾向滑距 dip slip; normal displacement
倾向滑逆断层 dip slip reverse fault
倾向滑移 preferential slip
倾向滑震源机制 dip slip source mechanism
倾向检验 trend test
倾向节理 dip joint
倾向离距 dip separation
倾向落差 dip throw
倾向玫瑰图 dip rose diagram
倾向面山足泉 dip-foot spring
倾向坡 back slope; outface; dip slope【地】
倾向剖面 dip section
倾向/倾角 <表示岩层、断层或裂隙产状的> dip angle
倾向上游 dipping toward upstream
倾向上游的丁坝 groin pointing upstream
倾向上游的防波堤 groin pointing upstream
倾向下游 dipping toward down stream
倾向下游的丁坝 groin pointing downstream
倾向下游的防波堤 groin pointing downstream
倾向线 line of dip;line of trend
倾向线理 dip lineation
倾向相反 contrary dipping direction
倾向性 tendentiousness
倾向性方案 preferred scheme
倾向一方 lopside
倾向一方的 lopsided
倾向一致 same dipping direction
倾向移距 dip shift
倾向移位【地】 dip shift
倾向于稀的 <混合气> leaning out
倾向自卸车 side-tip dumper
倾销 dumping;memory dump;unload
倾销差额 margin of dumping
倾销电力 dump power
倾销功率 dump power
倾销价格 dumping price
倾销商品 dumped goods
倾销市场 dumping ground
倾销税 dumping exchange duty
倾销政策 dumping policy
倾斜 batter(ing); bevel(l)ing; cant; declension; declination; declivity; dip(ping); divagation; downhill; droop; hanging; inclination; lean-(ing); luff; lurching; lye-bye; nose in; offsetting; out-of-level; pre-shoot; rake; ramp; shelving; sloping; sway(ing); tilt; heeling【船】; fall <地面等的>; underlay <矿脉等的>
倾斜岸滩 shelving shore
倾斜暗场照明 oblique dark ground illumination
倾斜坝 tilting dam
倾斜百分率 percentage inclination; percent slope
倾斜板 hang plate; rake board; sloping panel
倾斜报警器 tilt alarm
倾斜暴 tilt storm
倾斜背斜 inclined anticline
倾斜比 rake ratio
倾斜壁 tilting wall
倾斜臂 tilt arm
倾斜边 lopside
倾斜边的 lopsided
倾斜边界 tilt boundary
倾斜边坡 battered slope
倾斜变换点 point of change of gradient; point of vertical curvature; point of vertical curve
倾斜变换线 break line; inclination change line
倾斜变形 tilt distortion
倾斜标尺 inclined ga(u)ge; sloping ga(u)ge
倾斜标度盘 bank scale
倾斜表 declinometer;nauropemeter
倾斜冰山 tilted iceberg
倾斜玻璃窗 sloping glazing
倾斜玻璃切台 tilting glass cutting table
倾斜补偿地震计 tilt-compensation seismometer
倾斜补偿器 heel compensator; inclination compensator
倾斜补偿装置 slope compensating device
倾斜不整合 angular unconformity
倾斜不足 cant deficiency
倾斜舱壁 sloping bulkhead
倾斜操纵 bank control
倾斜草稿底图 oblique sketch master
倾斜测定器 gradienter; tilt finder; gradiometer
倾斜测定仪 tilt finder
倾斜测高仪 oblique height finder
倾斜测量 declivity survey; fall measurement; tilt survey
倾斜测流槽 tilted flume
倾斜测图仪 oblique plotter; oblique plotting instrument; oblique plotting machine
倾斜测压管 tilting ga(u)ge
倾斜测压计 inclined tube manometer
倾斜层 dipping layer; inclined stratum; sloping course; sloping layer; tilted stratum
倾斜层理 incline(d) bedding; inclined stratification
倾斜差 tilting error
倾斜产业政策 priority guidelines of industrial policy
倾斜场 inclined field; tilt field
倾斜车厢 oblique compartment
倾斜承影面 tilting easel
倾斜冲击试验 inclined impact test
倾斜处 brae
倾斜传播 oblique propagation
倾斜传感器 <地下连续墙> pick-up for deflection indicator; tilt sensor
倾斜船 listing ship
倾斜船台 inclined building berth; slope berth
倾斜磁化 inclined magnetization
倾斜磁迹 skewed track
倾斜粗洗淘金槽 tom
倾斜催化剂管道 sloping catalyst line
倾斜打入 driven on the rake
倾斜打桩 driving on the rake
倾斜挡板 sloping baffle
倾斜挡风玻璃 oblique windshield
倾斜导板 inclined guiding board
倾斜导洞 inclined pilot tunnel
倾斜导轨 inclined guide
倾斜导坑 inclined drift
倾斜的 acclive; acclivitous; acclivous; angular; atilt; banked; battered; declivous; Fastigiated; hading; lopsided; oblique; out-of-square; pronate; prone; ramped; shelving; slant-(ing); slantwise; sloped; sloping; tilted; downhill
倾斜的壁炉铁栏 inclined grate
倾斜的窗 rising window
倾斜的导轨 sloping rack
倾斜的鼓式浸出器 inclined drum leacher
倾斜的黑色面层铺料机 slope black top paver
倾斜的建筑基坑 sloped building pit
倾斜的坑道入口 sloping adit
倾斜的沥青路面铺路机 slope black top paver
倾斜的排水(拦污)栅 inclined sewer screen
倾斜的气水界面 dip gas-water interface
倾斜的墙 tilt-up wall
倾斜的墙帽 raking coping
倾斜的桥 <由于墩台沉陷所致> tilted bridge
倾斜的散热管 canted radiator tube
倾斜的散热器护栅 sloping radiator grille
倾斜的上游面 battered upstream
倾斜的水平旋翼 tilting rotor
倾斜的线脚 raked mo(u)lding
倾斜的泄水涵洞 canal rapids
倾斜的檐板 raking cornice
倾斜的油水界面 dip oil-water interface
倾斜的正面大楼梯通道 inclined moving stair(case)
倾斜的支架 sloping rack
倾斜底板 inclined floor
倾斜底层 inclined floor
倾斜地 aslant; slantways; aslope
倾斜地板 ramped floor; slopping floor; tilting floor
倾斜地层 dipping stratum[复 strata]; edge seam; inclined stratum [复 strata]; tilted stratum [复 strata]; tilting of strata; cline stratum [复 strata]
倾斜地段 sloping site
倾斜地面 inclined floor; sloping floor; sloping ground
倾斜地下水面 sloping water table
倾斜地形 depth shape; oblique landform; sloping topography
倾斜地震带 inclined earthquake zone; inclined seismic zone
倾斜电刷 angular brush
倾斜电压 tilt voltage
倾斜电子透镜 tilted electron lens
倾斜垫 tilting pad
倾斜垫板 canted tie plate
倾斜垫圈 grade washer
倾斜吊杆 <起重机的> inclined jib; inclined boom
倾斜吊装 tilt-up
倾斜叠加 slant stack
倾斜斗 tipping bucket
倾斜斗式提升机 inclined bucket elevator
倾斜度 amount of inclination; angularity; degree of dip; degree of inclination; degree of pitch; degree of slope; degree of tilt; grade incline; gradient; gradient of slope; gradient pitch; inclination pitch; rake; shelving; obliquity; slant; batter <墙壁等的>
倾斜度测定器 gradienter
倾斜度测量仪 tilting ga(u)ge
倾斜度大的 high dipping
倾斜度改正 grade correction; inclination correction
倾斜度调节器 tilt adjuster
倾斜度调整 adjustment of trim/heel; tilt control
倾斜度调整的液压管道 <推土机> tilt line
倾斜度误差 bank error
倾斜度指示器 tilt indicator
倾斜端头 ramp
倾斜段落 downhill section
倾斜断层【地】 dip fault; inclined fault
倾斜断层错动 dislocating inclined fault
倾斜堆石 dumped rockfill; tipped rockfill
倾斜法 method of careening; tilted-up
倾斜反射镜 inclined mirror
倾斜反射率 inclined reflectivity
倾斜范围 tilting range
倾斜方位角传感器 slant angle bearing sensor
倾斜方向 dip direction; direction of dip; direction of tilt; sloping direction; trend of the tilt
倾斜飞行 heeling
倾斜分层 incline cut and fill stoping
倾斜分层崩落回采法 inclined slicing
倾斜分层充填 inclined fill; inclined cut-and-fill
倾斜分层充填采场 inclined cut-and-fill stope
倾斜分层充填法 inclined cut-and-fill method
倾斜分层回采 rill cutting
倾斜分划板 tilt graticule
倾斜分级 slope rank
倾斜风挡 slanting windshield
倾斜腐蚀 bevel etch
倾斜附件 tipping attachment
倾斜改正 grade correction; correction for grade; correction for slope; reduction to the horizon(tal); slope correction; tilt adjustment
倾斜改正因数 inclination correction factor
倾斜杆 tilt arm; tilting lever
倾斜杆状引信 tilt rod fuze
倾斜杠杆 tilt lever
倾斜高度计 inclined ga(u)ge
倾斜割理 dip cleat
倾斜格筛 inclined grizzly
倾斜隔水底板 inclined impervious bottom bed
倾斜工作面 diagonal face; inclined cut
倾斜工作台 tilting table
倾斜拱 inclined arch
倾斜构件 sloping member
倾斜古平原 inclined fossil plain; sloping paleoplain
倾斜鼓筒 <搅拌机的> non-tilting drum; tilting drum
倾斜刮土板推土机 tilt dozer
倾斜关节 inclination joint
倾斜观测 declivity observation; inclination observation; incline(d) sight; tilt observation
倾斜观察 oblique observation; oblique view(ing)
倾斜管 dip pipe
倾斜管式压力计 inclined tube manometer
倾斜轨道 inclined orbit; inclined track
倾斜轨迹 inclined trajectory
倾斜柜 heeling tank
倾斜辊道 gravity conveyer[conveyor]
倾斜过度 overbank
倾斜海滩 sloping beach
倾斜含水层 sloping aquifer; tilted aquifer
倾斜函数 ramp function
倾斜函数响应 ramp-forced response
倾斜焊 inclined position welding
倾斜航迹起飞 oblique take-off
倾斜航空摄影 oblique aerial photography
倾斜航空摄影测量 oblique air sur-

vey;perspective air survey
倾斜航空照片 oblique air photograph
倾斜航空照片带 oblique air photograph strip
倾斜航空照相 oblique photography
倾斜航片 tilt aerial photo;tilting photo print
倾斜航摄测量 oblique air survey;perspective air survey
倾斜航摄像片 oblique aerial photograph;oblique air photograph
倾斜航摄照片 oblique(aerial)print
倾斜航行 receding log
倾斜荷载 dip load;downhill loading;oblique load(ing);tilting load(ing)
倾斜桁架 pitched truss
倾斜后加速度 post deflection acceleration
倾斜后角 tilt-back angle
倾斜护岸工程 battered bank system
倾斜戽斗 tip bucket
倾斜滑槽 tip chute
倾斜滑尺 tilt slide rule
倾斜滑动 dip slip
倾斜滑移断层 oblique-slip fault
倾斜绘图机 oblique plotting machine
倾斜混波器 tilt mixer
倾斜混凝土镶板 tilt-up concrete panel
倾斜活门 oblique valve
倾斜机构 leaning device
倾斜基床 sloping bed;sloping floor
倾斜计 angle ga(u)ge;angle ga(u)ge block;angle meter;drift indicator;gradienter;inclination ga(u)ge;inclined ga(u)ge;incline level;inclinometer;tiltmeter
倾斜计测井记录 dipmeter log
倾斜计算器 tilt calculator
倾斜计算装置 tilt calculator
倾斜记录仪 clinograph
倾斜继动器 tilt servomotor
倾斜建筑物 slanting construction
倾斜交叉 tilted intersection
倾斜浇注 inclined casting;tilt casting
倾斜浇(注)法 inclined casting;tilt casting;tilt cast method
倾斜角 angle of bank;angle of declination;angle of dip;angle of gradient;angle of heel;angle of inclination;angle of oblique;angle of obliquity;angle of rake;angular pitch;bank angle;bevel angle;elevation angle;pitch angle;skewed angle;slope;tilt angle;tipping angle;vertical drift;angle of tilt(ing);angle of tip;slant angle <垂直孔指顶角,水平孔指倾角>
倾斜角传感器 slant angle sensor
倾斜角的人工调节 <推土机> manual tilt adjuster
倾斜角度 angle of slope
倾斜角度控制 tilt control
倾斜角度试验机 overturn angle tester
倾斜角积分器 roll integrator
倾斜角伺服系统 slant angle slaved system
倾斜角增大 increase in dip
倾斜角指示器 rate of roll indicator
倾斜校正 correction for inclination;correction for slope;grade correction;inclination correction;tilt correction;correction for grade
倾斜校正器 tilt corrector
倾斜校正装置 correction device for tilt
倾斜阶跃波 ramp step
倾斜接触 inclined contact
倾斜接缝 inclined joint;sloping joint
倾斜节理【地】 cutter;dip joint
倾斜结构 slanting construction
倾斜截面 oblique section
倾斜界面 tilted interface
倾斜金箔静电计 tilted gold-leaf electrometer
倾斜进水口 inclined intake
倾斜茎 bent stem
倾斜晶界 tilt crystalline boundary
倾斜镜 tilting mirror
倾斜镜头纠正仪 tilting-lens rectifier
倾斜镜头拍摄 canted shot
倾斜镜装置 mirror-tilting mechanism
倾斜距离 inclined distance;slant range
倾斜锯台 canting saw table
倾斜锯子桌 canting saw table;tilting saw table
倾斜掘进 driving on the rake
倾斜看台 amphitheater[amphithcatre]
倾斜可见距离 oblique visibility;oblique visual range;slant visibility
倾斜坑道 inclined drift;inclined opening
倾斜空中照片 oblique aerial photograph
倾斜孔 downward sloping hole
倾斜控制逻辑 tilt command logic
倾斜控制器 tilt controls
倾斜块断式构造【地】 tilted block structure
倾斜矿层 inclined seam
倾斜矿柱 slope pillar
倾斜框架 tilting frame
倾斜力 inclined force;tilting force
倾斜力矩 heeling moment;inclining moment;moment of forces tending to capsize;overturning moment;tilting moment
倾斜立体测图仪 stereoblique plotter
倾斜连杆 tilt link
倾斜连接斗式提升机 inclined continuous bucket elevator
倾斜帘子 inclined lattice
倾斜量 amount of inclination;amount of precipitation;pitch dimension
倾斜料斗 tilting skip
倾斜灵敏度 slope sensitivity
倾斜流 slope current;slope flow
倾斜流液洞 taper throat
倾斜漏斗 tipping hopper
倾斜炉 uphill furnace
倾斜炉箅 inclined grate
倾斜炉箅煤气发生炉 inclined-grate-type gas producer
倾斜炉底 sloping bottom
倾斜炉栅 inclined grate
倾斜炉栅冷却器 inclined grate cooler
倾斜率 gradient of slope;percentage inclination;percentage of inclination
倾斜轮 angled wheel;leaning wheel
倾斜轮式侧向搂草机 oblique reel-side delivery rake
倾斜轮式平地机 leaning wheel grader
倾斜螺杆混合机 inclined screw mixer
倾斜螺纹 leaning thread
倾斜螺旋 <水准仪的> gradienter screw;level(l)ing(foot)screw;gradient screw;tilting screw
倾斜螺旋法 method of tilting screw
倾斜螺旋浓缩机 hydradenser
倾斜螺旋视距仪 tach(e)ometer with gradient screw;tangent-reading tachymeter
倾斜煤层 inclined coal seam
倾斜煤巷 slant coal road(way)
倾斜面 declivity;dip face;inclined plane;oblique plane;rake face;raking surface;slant plane;sloping portion
倾斜面板 sloping desk
倾斜面产状 attitude of inclined surface
倾斜面纠正 affine rectification
倾斜模型 tilted model
倾斜木纹 interlocked grain;short grain
倾斜能见度 approach visibility;oblique visibility
倾斜逆转 dip reversal;reversal of dip
倾斜凝汽阀 tilting steam trap
倾斜浓密箱 inclined lamellar thickener
倾斜排水层 inclined drain
倾斜排水沟 swept valley
倾斜排水铺盖 sloping blanket drain
倾斜盘防逆阀 tilted disc check valve
倾斜盘式干燥机 inclined tray drier
倾斜盘式离心机 inclined disk centrifuge
倾斜抛物线 tilted parabola
倾斜抛物线波形 tilted parabola waveform
倾斜炮眼 angling hole;grip hole
倾斜喷管 angled nozzle
倾斜喷砂 blasting with angular grit
倾斜盆地 deflection basin
倾斜皮带输送机 inclined belt conveyer[conveyor]
倾斜皮带提升机 belt slope lifter
倾斜片状结构 inclined fibrous texture
倾斜偏差 slope deviation
倾斜拼合模板 depressed panel form
倾斜平板法 gradient plate method
倾斜平面 dip plane
倾斜平原 clinoplain;dip plain
倾斜坡 dip slope;outface
倾斜破坏 failure by tilting
倾斜破碎岩层型 type of inclined bed and fractured rocks
倾斜旗杆 tilting flagpole
倾斜起重机 inclined hoist
倾斜气压计 inclined draft ga(u)ge
倾斜汽车螺旋形楼梯通道 inclined motor stair(case)shaft
倾斜汽缸销 tilt cylinder pin
倾斜器 inclinator;tilt(o)meter;trip
倾斜嵌入 oblique embedding
倾斜嵌条 cant strip
倾斜墙 batter-curved wall;battered wall;battering wall
倾斜墙壁 batten wall
倾斜切刃两端高度差 shear height
倾斜切削 inclined cutting
倾斜球磨机 inclined ball mill
倾斜容限 tilt tolerance
倾斜入口 inclined intake
倾斜入射 oblique incidence
倾斜入射光栅 grating with oblique incidence
倾斜扫描 oblique scan
倾斜色散 inclined dispersion
倾斜筛 inclined screen
倾斜筛分作用 slope sorting action
倾斜筛选 downhill screening
倾斜山 tilted mountain
倾斜栅网 inclined screen
倾斜上游面 inclined upstream face
倾斜射线 inclined ray
倾斜射线途径 inclined ray path
倾斜摄影 oblique photography
倾斜摄影测量 oblique photogrammetry
倾斜摄影航片 oblique air photograph;perspective air photograph;tilted photograph
倾斜摄影照片 wing photograph
倾斜摄影照片转绘仪 oblique sketch master
倾斜伸臂 inclined jib
倾斜升降回转控制器 bank-and-climb gyro control unit
倾斜升降机 inclined lift
倾斜升料机 inclined elevator
倾斜时差 oblique equation
倾斜矢量 tilt vector
倾斜式拌和机 tilting mixer
倾斜式差示压力计 slanting leg manometer
倾斜式窗框 hopper sash
倾斜式挡土墙 angular inclined retaining wall
倾斜式调压室 sloping surge chamber
倾斜式防波堤 mound-type breakwater
倾斜式浮标 tilting float
倾斜式鼓形搅拌机 tilting drum mixer
倾斜式滚道输送机 gravity roller conveyer[conveyor]
倾斜式混料器 tilter mixer
倾斜式活套槽 sloping loop channel
倾斜式驾驶室 tilting cab(in)
倾斜式搅拌机 tilted mixer;tilting(drum)mixer
倾斜式搅拌器 tilted mixer;tilted mixing device;tilting mixer
倾斜式卡车 dump truck
倾斜式缆道 luffing cableway
倾斜式缆索 lifting cableway
倾斜式冷床 rake-type cooling bank;rake-type cooling bed
倾斜式链斗挖沟机 inclined chain-and-bucket trench digger
倾斜式搂草机 oblique rake
倾斜式炉 tilting furnace
倾斜式炉箅子冷却机 inclined grate cooler
倾斜式炉箅子冷却器 inclined grate cooler
倾斜式滤网 inclined screen
倾斜式螺旋桨 tilting propeller
倾斜式平板拖车 tilting low-bed trailer
倾斜式球磨机 inclined ball mill
倾斜式燃烧器 tilting burner
倾斜式筛分机 rake classifier
倾斜式升降台 tilting table
倾斜式输送机 incline conveyer[conveyor]
倾斜式水尺 inclined ga(u)ge;slope ga(u)ge;tilting ga(u)ge
倾斜式炭化炉 inclined retort
倾斜式停车场 angle parking
倾斜式筒形混合机 tilted cylinder mixer
倾斜式推土机 tilt(ing)dozer
倾斜式微压计 tilting micromanometer
倾斜式压力计 inclined manometer;tilting manometer
倾斜式液压机 tilting hydraulic press
倾斜式圆筒筛网清灰机 inclined cylinder screen cleaner
倾斜式运输带分离机 cinder separator
倾斜式闸门 flap gate;tilting gate
倾斜试验 heeling experiment;inclining experiment;inclining test <浮体的>;tilting test
倾斜试验台 inclined test bed;stand
倾斜手车 tipping cart
倾斜手推车 tip cart
倾斜输送带式分离机 inclined conveyer separator
倾斜输送带式清选机 inclined conveyer separator
倾斜输送机 inclined conveyer[conveyor];slope conveyer[conveyor]
倾斜树 nodding tree
倾斜水标尺 inclined ga(u)ge
倾斜水槽 tilted flume
倾斜水尺 inclined staff ga(u)ge;sloping ga(u)ge;tilting ga(u)ge
倾斜水锤泵 tilt hydraulic hammer
倾斜水平仪 tilting level
倾斜水舌 inclined nappe

倾斜水压扬汲机 tilt hydraulic hammer
倾斜水准器 fore-and-aft level
倾斜伺服电动机 tilt servomotor
倾斜送料机 inclined elevator
倾斜索道 gravity cable(way);inclined cableway
倾斜索面 oblique cable plane
倾斜台 sloping desk;tiltboard;tilted platform
倾斜台触止 tilting table stop
倾斜台阶 raking riser
倾斜台式干燥机 sloping platform drier[dryer]
倾斜台试验 tilt table test
倾斜梯 inclined ladder
倾斜梯级 inclined step;oblique step
倾斜梯田 gradient terrace
倾斜提升机 inclined elevator
倾斜体 hanger;tilter
倾斜天窗 inclined clerestory
倾斜天线 inclined antenna
倾斜铁路 <建在山坡上运送材料等> incline railway
倾斜铁拖车 tipping trailer
倾斜通道 rampway
倾斜筒式谷物干燥机 inclined column grain drier
倾斜头 <三脚架上的> tilt head
倾斜投影 oblique projection
倾斜投影赤道 oblique projection equator
倾斜投影恒向线 oblique rhumb line
倾斜投影器 oblique projector
倾斜突变 tilt step
倾斜拖尾 tilting hangover
倾斜陀螺仪 roll gyroscope
倾斜瓦块式推力轴承 Kingsbury-type thrust bearing;Michell thrust bearing
倾斜桅 raked mast;raking mast
倾斜位移 tilt displacement
倾斜位置 inclined position;oblique position;tilted position;tilting position
倾斜稳定试验 inclination stability test
倾斜稳定陀螺仪 vertical flight gyroscope
倾斜圬工墙 battered masonry wall
倾斜屋顶 slanting roof
倾斜误差 error of tilt;oblique error;tilt error;heeling error
倾斜洗矿台 ragging frame
倾斜洗选槽 log washer
倾斜系数 inclination factor
倾斜下开式闸门 inclined undercut gate
倾斜下陷 tilt sag
倾斜掀断盆地 tilt-block basin
倾斜显示器 bank indicator
倾斜线 camber line;line of dip;oblique line;parallax
倾斜线脚 canted mo(u)lding
倾斜线圈 tilt coil
倾斜线圈型 inclined coil type
倾斜巷道 inclined drift
倾斜像片 oblique photograph;tilted photograph
倾斜像片坐标系统 tilted photo coordinate system
倾斜消槽 tip chute
倾斜效应 gap tilt effect;obliquity effect
倾斜楔体 inclined wedge
倾斜形状 tilted shape
倾斜型 apsacline;clinoform
倾斜性调整 slanting readjustment
倾斜修船码头 careening wharf
倾斜悬吊 inclined suspension
倾斜悬挂(车体)tilting suspension
倾斜压顶 sloped coping
倾斜压力计 inclined manometer;pressure inclination ga(u)ge;sloping ga(u)ge
倾斜压力角 angle of obliquity of action
倾斜岩层 inclined bed
倾斜岩面 inclined rock face
倾斜岩石巷道 inclined stone drift
倾斜眼掏槽 inclined hole cut
倾斜仰焊 upwardly inclined weld
倾斜叶片 inclined blade
倾斜液压缸管道 tilt cylinder line
倾斜液压油缸 tilt hydraulic ram
倾斜仪 batter finder;batter level;clinometer;clinoscope;dip compass;dipmeter[dipmetre];inclinator;inclinatorium;inclinometer;tilt ga(u)ge;tilting ga(u)ge;tilt(o)meter <测量地面倾斜度用>
倾斜仪测量 dipmeter survey
倾斜因数 inclination factor;obliquity factor
倾斜因子 inclination factor;obliquity factor
倾斜应力 inclined stress
倾斜油缸 <叉车> tilting oil cylinder
倾斜油路 tilt circuit
倾斜油田 dipping reservoir
倾斜于垂直线的 arake
倾斜余角 hade;underlay;underlie
倾斜与弯曲阴影补偿 tilt-and-bend shading
倾斜与转向指示器 bank-and-turn indicator
倾斜裕量 inclined allowance;lurching allowance
倾斜圆筒混合机 tilted cylinder mixer
倾斜缘条 tilting fillet
倾斜运输带 inclined belt conveyer[conveyor]
倾斜运输道 inclined haulageway
倾斜运输机 inclined conveyer[conveyor]
倾斜造型 slope mo(u)lding
倾斜闸板 rake damper
倾斜闸门 rake gate
倾斜照明 oblique illumination
倾斜照片 oblique photograph
倾斜照相机 oblique camera
倾斜折叠 canted folding;sloped folding
倾斜褶皱 inclined fold
倾斜褶皱的 sloping folding
倾斜振动控制系统 tilt vibration controlling system
倾斜正交桁架 inclined orthogonal truss
倾斜支承板 <采矿巷道的钢筋混凝土> inclined side slab
倾斜指示表 bank indicator
倾斜指示器 bank indicator;incidence indicator;indicator of inclination
倾斜轴 axis of tilt;inclined shaft;pitch axis;pitch brace;sloping shaft
倾斜轴承 inclined bearing
倾斜轴瓦轴承 tilting pad bearing
倾斜主应力轴 inclined principal axes
倾斜柱塞泵 inclined piston pump
倾斜铸造 sloping casting
倾斜转动架 tilted turret
倾斜转动式窗 tilt and turn window
倾斜转筒 tilting drum
倾斜转筒机构 tilting drum mechanism
倾斜转弯 banked turn
倾斜转弯仪 bank-and-turn indicator;bank indicator
倾斜装车台 inclined slide
倾斜装甲板 sloping plate
倾斜装卸式拖车 tilt-loading trailer
倾斜装置 tilting device;tilting gearing;tipping service
倾斜状态 heeling condition
倾斜自差 heeling deviation;heeling error
倾斜自差调整器 dipping needle instrument;heeling adjuster;heeling error instrument;vertical force instrument
倾斜自动同步机 tilt autosyn
倾斜走向 cant course
倾斜走向符号 <地图的> dip strike symbol
倾斜钻进 inclined drilling
倾斜钻孔 inclined hole
倾斜钻炮眼 angled drilling
倾斜钻探 drilling on the rake;inclined drilling
倾斜作用 tilting
倾斜坐标 oblique angle coordinates
倾泻角 angle of drain
倾泻水流 tumbling flow
倾泻闸沟 sluice
倾卸 discharge by pumping;dump(by tilting);dumping;memory dump;tilting;tip(ping);tilter <自卸卡车的>
倾卸板 dumping board
倾卸边缘 <堤、路基的> dump bank
倾卸驳船 dump barge
倾卸仓 tipping bin
倾卸槽 chute;tipping trough
倾卸车 dump cart;dumped truck;dumping car;dump lorry;dump wag(g)on;hopper wagon;jubilee wagon;side-tip wagon;tilter;tilting cart;tip car(t);tipler;tipper;tipping truck;tripper car
倾卸车斗 tipping cradle
倾卸车警告灯 dump body warning light
倾卸车辆 tipping cart
倾卸车身 dump(ing)body;tipping body
倾卸车身底板斜度 dump body bottom plate slope
倾卸车身加热 dump body heating
倾卸车身升起汽缸 dump body hoist cylinder
倾卸车身位置灯光 dump body position light
倾卸车身运货车 tipping lorry
倾卸车身支撑 dump body prop
倾卸车厢 tilter
倾卸船 dump scow;water-borne dumper
倾卸的 bevel
倾卸点 tipping point
倾卸斗 dumping bucket;skip bucket;tilt bucket;tipping bucket;tipping skip
倾卸斗挖泥机 hopper dredge(r)
倾卸堆石 dumped rockfill;tipped rockfill
倾卸堆石坝 dumped rockfill dam
倾卸高度 dump(ing)height
倾卸挂车 dumping trailer
倾卸管 tipping tube
倾卸荷载 tipping load
倾卸滑槽 tip chute
倾卸滑车 dump sheave
倾卸滑道 tip chute
倾卸滑轮 dump sheave
倾卸混凝土车车身 dumpcrete body
倾卸货车 car dumper;rear-dump wagon;tilting cart;tilting vehicle;tipper;tip(ping)wagon
倾卸机 discharger;dumper;tipping apparatus
倾卸机构 dumping mechanism;tilting mechanism;tipping mechanism;tippler;tripping mechanism
倾卸机械装置 dump(ing)mechanism;tipping mechanism
倾卸加料机 skip charger
倾卸甲板驳 side-dump scow;tipping barge
倾卸架 tilter
倾卸角(度)dump(ing)angle;angle of dump
倾卸卡车 jubilee truck;tipping truck
倾卸开关 tilt switch
倾卸矿车 box type tipping wagon
倾卸矿山车 side-dump mine car
倾卸垃圾车 dumping cart;dumping wagon
倾卸料仓 tipping bin
倾卸料斗 tipping bucket;tipping hopper
倾卸溜管 tipping tube
倾卸漏斗 tipping hopper
倾卸炉箅 clinker grate;drop grate;dump grate;tipping grate
倾卸炉箅燃烧器 dump grate stoker
倾卸乱石 tipped stone rubble
倾卸乱石堆 dumped stone riprap
倾卸码头 <突出式> tipping jetty
倾卸盘 tipping tray
倾卸抛石护岸 dumped stone riprap
倾卸平台 tilting platform
倾卸起重机 dump crane;dump hoist
倾卸汽车 dump car;dump truck;motortilter;tipper;tipping vehicle
倾卸设备 tipping device;tipping plant;tipping unit
倾卸式 dump-type
倾卸式半拖(挂)车 dump articulated trailer
倾卸式拌和机 tipping through mixer
倾卸式车厢 dump body
倾卸式挂车 tilt trailer
倾卸式滚筒混凝土搅拌机 tilting drum concrete mixer
倾卸式混合器 tilted mixer
倾卸式货车 dump truck
倾卸式货车皮 troll(e)y
倾卸式搅拌机 tipping through mixer
倾卸式阱 tilt trap
倾卸式料槽 tilting pan
倾卸式泥浆泵 dump bailer
倾卸式平板拖车 tilt deck trailer
倾卸式撒布机 <筑路用> tip spreader
倾卸式手推车 tilting cart
倾卸式拖车 flat trailer;tilt(ing platform)trailer
倾卸式挖土机 <转载用的> dumping grab(for rehandling)
倾卸式运料车 dumper
倾卸式运输车 tilting vehicle;tipper vehicle
倾卸式运输机 tripper type conveyer[conveyor]
倾卸式载重汽车 tilting vehicle;tipper vehicle
倾卸手推车 tip(ping)barrow;tipping cart
倾卸速度 rate of discharge;tipping movement
倾卸台 dump(ing)platform;jigging platform;tipping platform
倾卸提升加料机 skip charger
倾卸填料 dumped fill
倾卸填土 tipped fill
倾卸桶 dumping bucket;tipping hod
倾卸土斗车 tipping skip
倾卸拖车 dump trailer;tipping trailer
倾卸位置 dumping position;emptying position
倾卸限位器 dump kick-out
倾卸箱 tip box
倾卸箱车 box body dump car
倾卸小车 dump buggy;skip car;tilt-

Q

(ing) cart
倾卸斜斗 tipping bucket
倾卸许可证 licence [license] to discharge
倾卸液压缸 dump hydraulic ram; tip hydraulic cylinder
倾卸液压缸管道 tilt hydraulic cylinder line
倾卸液压油缸 tipping hydraulic ram
倾卸用液压设备 hydraulic equipment for tipping
倾卸油缸 dump(ing) ram
倾卸运货车 dump body track
倾卸运料车 throw-off carriage
倾卸者 dumper
倾卸轴 drop shaft; tipping pitman
倾卸轴小车 dumping shaft skip
倾卸装置 dump(ing) device; dumping gear; tipper; tipping device; tipping gear; tipping service; tippler; tripper; tripping device; tripping gear
倾修费 careenage
倾轧费率 discriminator rate
倾置场 dumping site; tipping site
倾置模型 tilted model
倾轴式柱塞泵缸体 inclined cylinder block
倾轴误差 error of axis of tilt
倾注 pour; decantation; transfuse; transfusion
倾注法 tilt-pour process
倾注炉 tilting furnace
倾注培养法 pour plate method
倾注平皿 pour plate
倾注桶 top-pour ladle
倾注洗涤 decantation washing
倾注(洗涤)器 decanter
倾注造球 shotting
倾转架 tilter
倾转浇注 tilt pouring
倾转式浇包 converter-type ladle; tilt ladle
倾转式炉子 tilting furnace
倾转台 swivel angle plate
倾转轴 tilting axis

清仓查库 check-up of warehouses; check warehouse stocks; make an inventory of warehouse; take stock
清仓关闭 flush closedown
清仓贱卖 cheap clearance sale
清仓廉售 cheap clearance sale
清仓拍卖 clearance sale; rummage sale
清仓损失 loss on clearing warehouse
清舱【船】 clean a ship; clean-up; clear a hold; hold stripping
清舱泵 cargo stripping pump; stripper pump
清舱浮定 cleaning compartments for wreck to keep afloat positively
清舱工作 clean up work on the bottom
清舱机 hold cleaning machine
清舱量 quantity of clean-up
清舱量比重 percentage of clean-up
清舱内油气 gas freeing
清舱设备 stripping facility
清舱抓斗 trim type bucket
清舱装置 clean-up device
清舱作业 digging out of barge; hold cleaning
清槽 channel cleanout
清查 check over; probe; sterilization
清查存货 stock taking
清查证 jerk note; jerque note
清拆作业 clearance operation
清产 liquidation
清产大减价 liquidating sale
清产核资 appraisal of property and funds; liquidation of property and appraisal of assets; reappraise the stocks and assets of enterprises
清铲 chipping; peeling off; snag(ging)
清铲工 cleanser
清铲凿 peening tool
清偿 acquit(tal); amortization; clear off; clear up; discharge; liquidate; payoff; satisfaction; settlement
清偿差额 liquidity balance
清偿成本 settling up cost
清偿费 amortization charges; amortization fund
清偿负债 satisfy the liabilities
清偿基础 liquidity basis
清偿基础总差额 overall balance on liquidity basis
清偿价格 redemption price
清偿力不足 liquidity shortage
清偿力约束 solvency constraint
清偿能力 clear-up ability; liquidity; repayment ability
清偿能力比率 liquidity ratio
清偿能力分析 repayment ability analysis
清偿能力高 high-level of liquidity
清偿期 amortization period
清偿期限 <债务的> payoff period; date of discharge
清偿契据 payoff escrow
清偿损失额 liquidated damages
清偿因子 amortization factor
清偿债务 liquidation of debts; acquittal; clear off debts; clear the debt; discharge (of) a debt; liquidation of the debt; meet engagement; payoff; settlement of debt; work off a debt
清偿债务备忘录 memorandum of satisfaction
清偿债务之要求 liquidated demand
清偿账款 settlement account
清偿中的困境 liquidity dilemma
清偿准备金 liquidation reserve
清场 site-clearing
清场地 clearing ground; clearing site
清车时间 <交叉口> clearance interval; clearance period; clearance time
清车损失时间 <交叉口信号相位间，任何方向车辆都不通车的时间> clearance loss time
清澈的 clean; clear
清澈度 clarity; limpidity
清澈如水的 waterclear
清澈时间 clearing time
清澈透明 clean/clear
清尘器 dust cleaner
清晨班 dawn shift
清齿刮铲 scotch cleaner
清齿环 stripper loop
清出 clearage
清出的垃圾 cleansings
清舱工作 clean up work on the bottom
清缝内砂浆或混凝土 raked out
清出口 clean-out door
清除 cleaning (up); clean-out; cleansing; clean-up (and move out); clear; clearage; clearance; clear(ing) away; clearing off; clear(ing) out; clear off; erase; evacuation; initialization; initialize; memory dump; obliterate; obliteration; purging; rain-out; removal; reset; scavenge; scavenging; stripping; unset
清除矮树丛 bush cleaning
清除按钮 reset button
清除被吸附的杂草 removal of adsorbed impurities
清除泵 clean-up pump; scavenger pump; scavenging pump; sink evacuator
清除标志 clear flag
清除表格 erase list
清除表目 clear entry
清除表土 removal of top soil
清除冰箱内余冰 deicing the bunkers
清除参考位 reset reference bit
清除残损物费用 debris clearing cost
清除残渣工 mucker
清除操作 sweep operation; clear operation【计】
清除草木 clearing
清除场地 clear the site
清除场地内不需要保留的植物 clearing the site
清除场地内多余的表土 clearing away
清除场内灌木、树枝、垃圾 brush out; swamp out
清除沉船障碍 wreck obstacles removal
清除程序 dump program(me); dump routine
清除冲程 scavenging stroke
清除储器 clear memory; clear storage
清除船底 graving
清除船底污垢 grave fouls at the bottom of ship
清除粗石 cleaned off rough stone
清除存储指令【计】 clear store
清除错误 erase error
清除打印 memory dump
清除待发 clear to send
清除道砟 ballast removal
清除的场地 cleared area
清除地段 clearance site
清除地面 clearing ground; land clearing
清除地面树木 land clearance
清除地面障碍物 clearing and grubbing
清除掉 dispose
清除毒性 toxicity elimination
清除读出器 clear reader
清除堵塞物 <井管壁或滤水管上的> declogging
清除堵头 cleaning plug
清除发送【计】 clear to send
清除发送线路 clear-to-send circuit
清除伐 refining
清除阀 purge valve
清除方式 reset mode
清除放射性废水 radioactive cleaning wastewater
清除放射性设备 decontamination device
清除放射性污染 decontamination
清除飞边 deburr(ing)
清除废气发动机 scavenging engine
清除废石 rudding
清除废物弯管 waste elbow
清除废物装置 <设在隐蔽处的> waste fitting
清除废墟(瓦砾) removal of ruins
清除服务 scavenging service
清除浮面 scaling
清除浮石 removal of float stone; rock removal; trimming of loose rock
清除符号 erase character
清除覆盖层 stripping overburden
清除干扰 interference elimination
清除根株 snag
清除工具 erasing tool
清除工作 clearing operation; clearing work
清除鼓风机 scavenging blower
清除管 clean-out tube; scavenge tube
清除灌木 brush cutting
清除灌木、树枝的工人 swamper
清除灌木机 tree-dozer
清除光泽 deluster(ing)
清除规程 clear procedure
清除过程 reset procedure; scavenging process
清除过载 stripping overburden
清除和洗涤系统 purge and wash system
清除黑鳞 descaling
清除灰尘 dusting
清除灰尘与噪声 dust and noise abatement
清除机 scavenger; scavenging machine
清除剂 remover; scavenger; scavenging agent
清除剂沉淀 scavenger precipitation
清除剂效应 scavenger effect
清除键 cancel key; clear(ing) key; reset key
清除结焦 clean-out
清除金属油渍 metal degreasing
清除荆棘 brush removal
清除静水污泥 hydrostatic sludge removal
清除旧沥青面层 asphalt pavement removal
清除旧漆面 clean-out old paint surface; strip
清除旧砂浆缝的松散部分 tuck pointing
清除拒爆的炮眼 relieving shot
清除开关 clear switch
清除开口管桩中的土 preexcavation soils in open pipe piles
清除坑 clean-out pit
清除空气 scavenging air
清除孔 cleaning eye; cleaning hole; clean-out eye
清除控制 clear control
清除控制逻辑 clear control logic
清除口 cleaning port; clean-out port <模板>; manhole; scavenge port; side pocket
清除口堵头 clean-out plug
清除口净距 clean-out clearance
清除宽度 clearing width
清除垃圾 refuse removal; rubbish removal
清除垃圾的人 dustman
清除垃圾的通道口 cleaning eye
清除垃圾工人 garbologist
清除垃圾弯管 waste elbow
清除垃圾装置 waste fitting
清除砾石装置 gravel separator
清除滤料垢壳 decrusting of a filter
清除率 clearance rate
清除乱石 rubble removal
清除轮 clearing wheel
清除螺钉 clean-out screw
清除脉冲 reset pulse
清除毛边 defrasing; frazing; reaming
清除毛刺 burr removal; deburring; removing burrs
清除毛刺用凿 burring chisel
清除毛石 cleaned off rubble
清除煤粉工 bugduster
清除门 eraser gate
清除密码键 clear cryptographic(al) key
清除面板键 clear panel key
清除面层 desurfacing
清除泥沙 disloading of sediment; dislodging of sediment
清除(排列)表 clear list
清除披缝 remove flash
清除气流 purge stream
清除气油 freeing tanks of gas

清除弃土 spoil removing
清除器 cleaner; clearer; remover; scavenger
清除切头 crop disposal
清除请求 clear request
清除区 clear area; override < 机场跑道两端的备用地区 >
清除曲线 clearance curve
清除缺陷 removal of defects; remove flaw
清除确认 clear confirmation
清除日期 purge date
清除熔渣通条 skimmer bar
清除润滑油 lubricant removal
清除塞 cleaning plug
清除森林 deforestation
清除沙槛 bar removal
清除上部沉积 stripping overburden
清除施工现场 clearing and grubbing
清除施工现场(的植物、树根、漂石等杂物) clear(ing) and grub(bing)
清除时间 checkout time; erasing time
清除试验 clearance test
清除输入键 clear entry key
清除树根 clearing and grubbing; grubbing; removal of tree stumps
清除树根与岩块的耙子 land-clearing and rock rake
清除树桩 grubbing; removal of tree stumps
清除刷花痕迹 bossing
清除水垢 scale-handling; scaling handling
清除水雷 mine clearance
清除水面油污的自动平衡刮集装置 self-level(l)ing unit for removing oil pollutants
清除水中残留物 remove debris from water
清除水中沉树、残根的抓斗船 snag boat
清除水中障碍船 snag boat
清除速率 washout rate
清除碎砖 rubble removal
清除条件 cleared condition
清除筒脚车 bobbin stripper
清除图像 recall image
清除涂料剂 paint remover
清除网络 annulling network
清除危石 clearing dangerous rock
清除位置 stripping position
清除污泥 de-sludging
清除污染 abatement of pollution; decontamination; depollute; depollution
清除物料 material removal
清除吸附气体 desorption
清除系统 scavenge(r) system
清除显示 clear display
清除现场 clearance site; site-clearing
清除现场建筑物 site cleared of buildings
清除限制因素 removing restraint
清除小树丛 brushwood clearing
清除效应 clean-up effect
清除信号 erasure signal; reset signal
清除信息污染 cleaning up information pollution
清除烟道内灰垢 coring out
清除岩粉 bugdusting; clean cuttings
清除氧化皮 removal of oxide scale; scale-handling; scale removal; scaling handling
清除溢油 oil removal
清除因素 clearing factor
清除因子 clearing factor
清除应力退火 clear stress annealing
清除油泥 de-sludge
清除油污 unclog
清除淤泥 de-sludge
清除杂质 removal of contamination
清除杂质输送器 trash mover
清除障碍 snagging
清除障碍物 unclog; snagging
清除障碍物的设备 snagging equipment
清除直接存取存储设备 direct access storage device erase
清除植物 removal plant
清除指令 clearance instruction; clearance order; clear instruction; clear order; erase command
清除指示符 clear indicator
清除中断 clear(ing) interrupt
清除中断请求 interrupt request clearing
清除装置 clean-out fitting; clearing device; removal device; remover; scavenge unit; purge unit < 不冷凝气体的 >
清除状况 cleared condition
清除状态 reset mode
清除字符 clear character
清除阻碍 unclog
清除钻孔石尘 removal of rock cuttings in borehole
清除钻孔石屑 removal of rock cuttings in borehole
清除作用 scavenging action
清吹封闭空间 purging enclosure
清脆的短音 blip
清袋 bag cleaning
清袋器 bag cleaner
清单 balance sheet; bill; clean bill; data book; detailed account; detailed list; inventory; inventory table; list; manifest; muster; repertoire; specification
清单材料账 list of material account
清单目录 check list
清单文件 inventory file
清单主目溢出 inventory master overflow
清淡期间 off-peak period
清淡时间 off-peak hours
清淡优美的色调 pastel shade
清淡运输 light traffic
清淡运输线(路) lightly used line; light traffic line; light traffic route; line carrying light traffic
清蛋白 albumin
清道车 street cleaning vehicle; sweeping car; sweeping vehicle
清道电报 line-clear telegram
清道电报闭塞法 blocking by line-clear telegrams
清道夫 street sweeper; street sweep(ing) machine
清道工(人) scavenger; street cleaner; street sweeper; dustman; street-orderly < 英 >
清道机 brush machine
清道机车 pilot engine
清道解锁 line-clear release
清道艇 < 挖泥船的 > snag boat
清道用锄 street-hoe
清道证 line clear permit
清底 bottoming; cleaning of bottom
清底半径 radius of clean-up at floor
清底销售 clearance sale
清底钻具 clean-out bit
清点 tally
清点材料 checking list; checking of materials; detailed account
清毒(杀菌)剂量 sterilization dose
清缝 joint cleaning; open joint; raked joint; rake-out a joint; raking-lout(of) joint; raking out of joint
清缝机 crack-cleaning machine; joint cleaner; joint cleaning machine
清缝耙 joint raker
清缝器 joint cleaner
清缝填封 rout and seal
清缝凿 hook(ing) iron
清稿 fair copy
清耕法 clean till; clean tillage
清沟 ditch cleaning
清沟班 crumb gang; crumbing crew; crumb party; crumb screw
清沟杓 ditch cleaning bucket
清沟铲 crumber
清沟铲斗 ditch cleaning bucket
清沟斗 ditch cleaning bucket
清沟工人 mud lark
清沟机 crowder; crumber; ditch cleaner; ditch cleaning machine; ditch dredge(r); ditch sweeper; gutter cleaner; sewer cleaner
清沟平土板 < 挖掘机 > crumber
清沟器 crumber
清沟器履板 < 挖掘机 > trench cleaner shoe
清沟桶 ditch cleaning bucket
清沟靴 ditch shoe
清沟用铲斗 trench-cleaning bucket
清垢工具 scaling tool
清垢频率 clean-out frequency
清垢器 stay-furring tool
清关 clearance; customs clearance
清关货场 clearance depot
清管 cleaning pipe; pigging operation
清管衬法 cleaning and lining process
清管发送器 shooter
清管法 pigging process
清管刮刀 go-devil
清管规定 cleaning guide
清管合格证 cleaning certificate
清管和衬管过程 cleaning and lining process
清管机 cleaning go-devil; cleaning pig; cleaning scraper; pipe cleaning machine
清管器 go-devil; pipe cleaning pig; pipe go-devil; pipeline cleaner; pipeline pig; sweeper; tube cleaner
清管器材料 pigging material
清管器发送 pig trap launch
清管器发送器 go-devil launcher; pig launcher
清管器接收 pig trap receive
清管器收发站 go-devil station
清管器位置信号器 pig alert; pig signal(l)er
清管球 ball go-devil; ball pig; ball scraper
清管刷 pig fitted with wire brushes; pipe brush
清管系统 pigging system
清管信号 pig signal
清管信号发生器 pig signal(l)er
清管信号装置 go-devil signal(l)er
清管用过滤器 strainer
清管用气体 buffer gas
清光漆 spar varnish
清轨器 runway cleaner
清航 clearance of navigation obstructions
清烘漆 baking varnish
清还借款 repayment of loan
清灰 ashing; ash removal; de-ashing
清灰工 ashman
清灰孔 cinder cleaning hole
清灰口 cleaning door
清灰门 soot door
清灰器窝形出口 socket outlet for cleaners
清灰装置 cleaning device
清绘 black overlay; delineation; fair draught; ink drafting; ink drawing; nib pen drafting; smooth drafting; fair draft; fine drawing; fair drawing
清绘图 finalized run
清绘原图 drawing original; fair sheet; final drawing; final manuscript; hand-drawn original; ink manuscript
清绘原图审校 drafting editing
清绘着墨 pen-and-ink drafting
清货 clear stock; stock taking
清货减价 clearing stock sale
清机 clear
清机键盘开关 clear key switch
清机数据输出 clear data output
清基 cleaning of construction site; foundation preparation; stripping; clearing and stripping < 水库 >
清基面 stripped surface
清基深度 excavating base depth
清基线 stripping line
清加 reservoir and add; reset and add
清减 reservoir and subtract; reset and subtract
清洁 brushing; despumate; sanitation
清洁白液体制品 clear liquid preparation
清洁标准 cleanliness standard
清洁车 ash car; emptier; garbage truck
清洁城市 healthy city
清洁处理 cleaning
清洁船舶 clean ship
清洁单据 clean documents
清洁的 sanitary
清洁度 cleanliness; cleanliness grade
清洁度水平 cleanliness level
清洁发展机制 clean development mechanism
清洁房间 clean room
清洁房舍 < 指牛棚、马厩 > neat house
清洁分散剂 detergent dispersant
清洁蜂 cleaning bee
清洁浮动 clean floating
清洁浮动汇率 clean floating exchange rate
清洁膏 clean(s)ing paste
清洁工具柜(橱) broom closet
清洁工(人) clean(s)er; dumper; dustman; sanitation worker; scavenger; street cleanser; wiper; maintenance cleaner; orderly; porter; swamper; sanitationman < 维护城市环境卫生的 >
清洁工业废水 non-polluted industrial wastewater
清洁公债 clean bond
清洁骨料 clean aggregate
清洁刮刀 cleaning doctor
清洁管道的通道口 cleaning eye
清洁辊 clearer; scraper roller
清洁河流 clean river; clean stream; healthy stream; stream health
清洁环 wiper ring
清洁活动 clean-up activity
清洁货物 clean cargo
清洁机 cleaner
清洁基 clear base
清洁集料 clean aggregate
清洁技术 cleaner technology
清洁剂 cleaning agent; cleaning solvent; cleanser; clearing agent; detergent; detersive
清洁间 broom closet
清洁金属用粉 permag
清洁井 clear well
清洁可转让提单 clean negotiable bill of lading
清洁空气 clean air; scavenging air
清洁空气系统 clean air system
清洁孔 clearing hole; lancing door

清洁孔盖板 clean-out cap
清洁孔盖闩 cleaning hole cover latch
清洁孔净距 clean-out clearance
清洁口 clean-out
清洁口堵头 clean-out plug
清洁能源 clean energy;pollution-free energy source
清洁票据 clean bill
清洁器 clean(s)er;devapo(u)rizer
清洁器杆 cleaner rod
清洁区 clear area;clear band
清洁燃料 clear fuel
清洁燃油舱 clean fuel oil tank
清洁人员 caretaker
清洁砂 clean sand
清洁生产促进法 cleaner production promotion law
清洁生产方案 cleaner production option
清洁生产审核 cleaner production audit
清洁生产与环保产业 cleaner production and environmental protection industry
清洁式通风 clean-type ventilation
清洁室 cleaning chamber;clean room
清洁刷 cleaning brush
清洁水 clear water;clear effluent
清洁水法令 Clean Water Act
清洁水恢复法案 Clean Water Restoration Act
清洁提单问题 problem of clean bill of lading
清洁提(货)单 clean bill of lading
清洁添加剂 detergent additive
清洁挖泥船 cleanup dredger
清洁卫生费 sanitation service fee
清洁物权 clean title
清洁系统 scavenger system
清洁箱 garbage bin
清洁信用 clean credit
清洁信用证 clean letter of credit
清洁休闲 full fallow
清洁巡逻 patrol cleaning
清洁压载舱 clean ballast tank
清洁压载水 clean ballast water
清洁液 cleaning solution
清洁用溶剂 clearer's solvent
清洁用石脑油 cleaner's naphtha
清洁运输单据 clean transport documents
清洁藻 clean water algae
清洁罩 cleaning cover
清洁装船单证 clean shipping documents
清洁装运提单 clean on board bill of lading
清洁装置 cleaning mechanism
清洁走廊 clean corridor
清洁租船合同 clean charter
清洁钻孔 clean (drill) hole;clean borehole
清洁作业 cleaning operation
清结 settle account
清结代理 clearing agent
清结账户 balance account;clearing account
清解锚链 clearing a hawse
清劲风 <蒲福风级表五级风,风速 8.0 ~ 10.7 米/秒> fresh breeze
清净机 branner
清净剂 abluent;abstergent;cleanser;detersive
清净空气 scavenging air
清烤漆 bakelite varnish;baking varnish
清孔 borehole cleaning;hole cleaning
清孔工作 clean-up job
清孔器 wimble;hole cleaning tool
清孔钻 clean-out auger;shell auger
清孔钻取土样法 shell-and-auger boring
清孔钻头 clean-out auger;clean-out bit;scouring bit
清库 <水库蓄水前> reservoir clearance
清库工作 stripping of reservoir
清缆 clearing of wire rope
清理 cleaning-off;cleaning up;clean-up and move out;clearage;clearance;clear up;desintering;disentangle;dressing-off;liquidate;liquidation;rake-out (of joint);refine;unscramble;untangle;wind-up;chipping
清理边沟 cleaning ditch;ditch cleaning out
清理变产表 liquidation account
清理部分剩余物 partial slash disposal
清理财产接管人 equity receiver
清理拆模 clearing and stripping
清理铲 cleaning blade
清理铲斗 cleaning bucket
清理场地 clearing (of site);disposal point;grub(bing);land clearance;land clearing;place cleaning;preparation of site;site-clearing;stripping;winding up
清理场地用铲斗 clean-up bucket
清理车间 dressing shop
清理成本 disposal cost
清理程序 liquidation procedure
清理存货 take stock
清理的现场 cleared site
清理地面 clearing ground;flogging
清理吊货绞辘重铁 overhauling weight
清理顶板 roof cleaning
清理伐 final clearing
清理废墟 demolition cleanup;removal of demolition waste
清理费(用) clearing expenses;clearance cost
清理风錾 dresser
清理杆 stripper bar
清理工 fettler
清理工段 conditioning department;conditioning site;conditioning yard
清理工作 clean-up work
清理工作面 place cleaning
清理沟渠 cleaning ditch;ditch cleaning
清理刮板 cleaning blade
清理滚筒 cleaning cage;clean-up barrel;finishing barrel;rumbler;tumbling barrel
清理过表面缺陷的坯料 conditioned billet
清理过的场地 cleared area
清理焊缝磨床 lap grinder
清理河床 clearing of a river bed
清理灰缝 opening joint;rake-out (of joint)
清理会计师 accountant in bankruptcy
清理火场 mopping up
清理火场时间 mopping up time
清理货运办法 <规定重车挂运和空车配送办法> clearance of goods traffic
清理机 cleaning eye;descaling machine;sweeper
清理价值 break-up value;liquidation value
清理假集体工作 work of screening pseudo-collectives
清理浇铸口錾 foundry chisel
清理绞辘 overhaul a tackle
清理孔 cleaning eye;clean-out
清理孔内岩石碎块的钻具 <活塞抽筒状> cavings filler
清理跨 scalping bay
清理漏油 clearing up oil spill
清理毛刺 deburr(ing)
清理毛口 deburr(ing)
清理毛口齿轮 deburring wheel
清理毛口工具 deburring tool
清理毛刷 remove burrs
清理门 cleaning door
清理炮眼 scrap out a hole
清理赔偿 claims settlement
清理赔偿费用 claim expenses
清理切边 crop handling
清理切头 crop handling
清理人 liquidator;official liquidator
清理散料的工作半径 <挖掘机> clean-up radius
清理散料的工作范围 <挖掘机> clean-up radius
清理设备 cleaner;cleaning equipment;clearing device
清理升运器 cleaner-elevator
清理施工现场 cleaning of construction site
清理时间 clearance period
清理受理员 official receiver
清理损益 liquidation gain or loss;liquidation profit and loss
清理台 cleaning table;dressing table;fettling bench;fettling table
清理坍方 mucking
清理土地 land clearance
清理完毕 clean-out
清理完毕信号灯 all-clear signal lamp
清理污水井管的木球 sewer pill
清理污物的海底阀 clearance of fouling of sea valve
清理现场 clean-up;clearing of site;site cleaning;site-clearing
清理现场多余石料 clouring
清理行人时间 pedestrian clearance time
清理烟道 <施工中> coring out
清理岩面 cleaning rock surface
清理岩石面 rock cleanup
清理沿线各站货运办法 <规定重车挂出和空车配送办法> clearance of roadside traffic
清理用铲斗 cleaning bucket
清理逾期票据 clean-up past due bill
清理与安装时间 stream-to-stream time
清理在建项目 checking up on the projects under construction
清理债权 settle claims
清理债务 clear debt;clear-up debt;consolidation of debts;liquidation of debts
清理账户 liquidation account
清理账目 adjust accounts;clearing account;square account
清理整顿公司 sorting out and consolidating companies
清理支出 outlay for liquidation
清理中固定资产 fixed assets in liquidation
清理铸件 fettle casting
清理装载机 cleaner-loader
清理装置 cleaning mechanism;cleaning plant
清理资产 liquidate assets
清理钻具 cleanout(jet) auger
清理钻孔 open a hole
清理作业 clean-up operation
清链 clear a hawse;open a hawse
清凉剂 refrigerant
清凉可口 nice taste
清凉饮料厂 soft-drink plant
清凉饮料厂废水 soft-drink industry wastewater
清粮机 dressing machine
清粮机刷 grain-brush
清粮口 clean-out door
清粮器传动连杆 shoe pitman
清粮筛 cleaning shoe
清粮设备 cleaning unit
清粮室 cleaning shoe;cleaning unit
清亮区 clear zone
清料升运器 cleaning elevator
清零 clear;minimum clearing;zero clearing
清零电路 clear circuit
清零开关 reset switch
清零区 clear area
清零条件转移 branch on condition clear
清零状态 cleared condition
清垄器 row cleaner
清炉 boiler cleaning;prepurging
清炉壁 chipping-out
清炉除灰 clean ash;clean fire
清炉渣块 barring;cobbing
清炉装置 cupola drag
清滤器 secondary cleaner
清滤液 clear filtrate
清路机车【铁】 pilot engine
清棉机 picker
清面 sand down
清面板键 clear panel key
清模剂 mo(u)ld cleaning compound
清泥机 mud cleaning machine
清耙机 raking equipment
清盘 winding up
清喷漆 clear lacquer
清漆 lac varnish;varnish;varnish lacquer;varnish-type paint;vernix
清漆薄膜 lacquer film
清漆催干剂 varnish drier[dryer]
清漆打磨光面工艺 French polish
清漆的油度 oil length of varnish
清漆底子 varnish base
清漆干燥 varnish-drying
清漆干燥剂 varnish drier[dryer]
清漆工人 varnisher
清漆固化 varnish cure
清漆罐 varnish kettle;varnish pot
清漆锅 varnish kettle;varnish pot
清漆红颜料 dragon's blood
清漆介质 varnish medium
清漆类别 varnish system
清漆毛刷 varnish brush
清漆媒介剂 varnish vehicle
清漆媒介液 varnish vehicle
清漆磨光打底 flat varnish
清漆磨光作业 flatting varnish
清漆磨退 rubbing-down of varnish;varnish rubbing
清漆漆基 varnish base
清漆漆面延长剂试验 varnish renovator test
清漆漆膜发白 milkiness
清漆轻度发浑 hazing
清漆缺陷 <指苍白色或不透明性缺陷> milkiness
清漆色 varnish colo(u)r
清漆试验 varnish test
清漆熟化 varnish cure
清漆树 varnish-tree
清漆树脂 varnish gum
清漆刷 lacquer brush;varnishing brush
清漆涂层 clear coat;varnish coating;varnish cover;varnish finish
清漆涂料 varnish colo(u)r;varnish paint
清漆型着色剂 varnish stain
清漆亚麻仁油 varnish linseed oil
清漆用石脑油 varnish maker's naphtha
清漆用树脂 varnish resin
清漆用炭黑 varnish black
清漆用油 varnish oil

清漆与油漆用石脑油 varnish maker's and painter's naphtha
清漆原材料 varnish raw material
清漆罩面 clear finish
清漆制造 varnish manufacture
清漆制造厂 varnish maker
清漆组成 varnish formulation
清讫 account balanced; account settled
清欠收据 acquittance
清扫 clean down; clean (ing) up; clean-out; clearing (up); refine; scavenge; sweeping; sweep out
清扫把 clearing rake
清扫板 cleaning flap
清扫便道 <俚语> chunk out
清扫并运出工地 clean-up and move out site
清扫车 cleaning truck; motor sweeper; sweeper
清扫锉刀的钢丝刷 file card
清扫费 cleaning expenses
清扫风机 purging air fan
清扫盖 access plate
清扫工人 swabber
清扫刮板 scraper cleaner
清扫管线 flushing circuit; flushing conduit
清扫和洗刷设备 cleaning and washing plant
清扫机 scavenging machine; sweeper; sweeping machine
清扫机械 sweeping mechanism
清扫级喷砂 brush off blast
清扫级喷射除锈 sweep blast
清扫间 access chamber
清扫街道 street cleaning
清扫井 clean-out chamber
清扫孔 access pit; bottom door; cleaning hole; clean-out opening
清扫口 access eye; cleaning eye; clean-out cover; clean-out hole; clean-out opening; clear out door; handhole; rodding eye; clean-out hatch【船】
清扫口堵头 access screw
清扫口盖板 clean-out plate
清扫口丝堵 access screw
清扫宽度 sweeping width
清扫垃圾者 scavenger
清扫毛刷 brush sweeper
清扫门 cleaning door; clean-out door
清扫面积 sweep-out pattern
清扫跑道 runway clearance; runway sweeping
清扫平盘 cleaning berm
清扫器 sweeper
清扫设备 clearing device
清扫室 access chamber
清扫刷 brush cleaner
清扫水深 clearing depth
清扫图形 flood-out pattern
清扫效率 sweep efficiency
清扫性放炮 brushing shot
清扫用具箱 broom box
清扫装置 cleaning device; sweep device
清色 clear colo(u)r
清砂 sand cleaning; fettling
清砂工具 sand cleaning tool
清砂滚筒 cleaning drum; rumbling mill
清筛道床 ballast cleaning
清筛道砟 ballast cleaning
清筛辊 screen roll
清筛机 cleaner; screen scarifier
清筛机构 sieve cleaning mechanism
清筛器 cleaner; screen cleaner; screen rack; sieve cleaner
清筛刷 screen brush
清筛用撞击器 screen bumper
清石工 lasher
清石机 derocker
清刷段 brush segment
清刷机 brusher
清刷洗涤机 scrubbing machine
清爽锚地 clean anchorage; clear anchorage
清水 dear water; drinking water; fresh water; sweet water; water alone
清水泵 clarified water pump; clear water pump; treated water pump
清水池 clean water basin; clean-water reservoir; clean water tank; clearing water basin; clearing water reservoir; clearing water tank; clear water basin; clear water reservoir; clear water tank; distribution reservoir; filtered water basin; filtered water reservoir; finished-water reservoir; purified water tank; reservoir of clean water; settling basin; treated water tank
清水冲刷 clear water scour
清水带 clear water zone; katharobic zone
清水导管 treated water conduit
清水定床模型 clear water fixed-bed model
清水对缝 keeping perpend; keeping the perpends
清水阀 filtered water valve
清水隔墙 drywall partition
清水(勾缝)砖砌体 fair-faced brickwork
清水河(流) clean river; clean stream; clear river; clear stream; sediment-free river; sediment-free stream
清水湖 clean-water lake; clear water lake
清水混凝土 exposed concrete
清水混凝土面 as-cast finish
清水井 clear well
清水库 clean-water reservoir; clear-(ing) reservoir; dear-water reservoir
清水面 self-finish
清水面混凝土 as-cast-finish concrete; fair-faced concrete
清水灭火器 plain water extinguisher
清水模型 clear water model
清水泥 plain cement
清水泥浆 freshwater mud; water mud
清水泥涂料 Portland cement paint
清水泥屋面瓦 plain cement roofing tile
清水砌体建筑 exposed masonry(work)
清水墙 drystone wall; drywall(ing); exposed masonry (work); fair-faced wall; plain wall; pointing masonry; unsheathed wall
清水墙壁结构 drywall construction
清水墙承包人 drywall contractor
清水墙承包商 drywall contractor
清水墙勾缝圬工 neat work
清水生物 catarobia; katharobe; katharobiont
清水水流 sediment free flow
清水碎石路 <无结合料的碎石路> plain macadam
清水隧洞 clean water gallery
清水污染 freshwater pollution
清水洗井 plain water flush; water flush(ing)
清水洗井法 water flush system
清水洗井钻进 water flush boring; water flush drilling
清水洗孔 plain water flush; water flush(ing)
清水洗孔法 water flush system
清水洗孔钻进 water flush boring; water flush drilling
清水系统 freshwater system
清水箱 clean water tank
清水性能 clean water performance
清水循环时间 time for fresh water circulation
清水压裂处理 riverfrac treatment
清水用量 amount of fresh water
清水藻 clean water algae
清水砖工 brick and brick; ga(u)ged work; ga(u)ged brickwork
清水砖拱 ga(u)ged arch
清水砖砌体 fair-faced brickwork
清水砖墙 brick wall without plastering; neat work; plain brick wall
清水砖墙勾缝圬工 neat work
清水砖墙水泥勾缝 plain brick wall pointed with cement mortar
清水钻进 clear water drilling
清水钻进水龙头 clear water type swivel
清算 account settled; clearance; clearing; clear off; liquidate; liquidation; square account; wind-up
清算安排 arrangement for settlement
清算变量 settlement variation
清算变现表 liquidation account
清算标准 clearing standard
清算拨款 liquidate appropriation
清算代理人 clearance agent; clearing agency; clearing agent
清算单价 liquidation unit price
清算单价法 liquidated unit price method
清算方式选择权 settlement option
清算分摊额 liquidation dividends
清算风险 liquidity risk; settlement risk
清算工程量 squaring up
清算公司 clearing corporation
清算机构 clearing house
清算基金 clearing fund
清算价格 clearing price; settlement price
清算价值 abandonment value; disposal value; liquidation value; liquidation price <破产企业资产的>
清算交易 clearing contract
清算金额 amount of settlement
清算流动资金 clearing current capital
清算美元 clearing dollar
清算拍卖价值 forced sale value; forced value
清算票据 clearing of bills
清算期间 settlement period
清算人 liquidating partner; liquidator; receiver
清算日 make-up day; settlement date; settlement day; settling day
清算商行 clearing firm
清算式资产负债表 statement of affairs
清算事务 liquidation affairs
清算收入 clearing revenue; liquidation income
清算收入子系统 subsystem of liquidated revenue
清算书 statement of settlement
清算损失账 deficiency account; deficiency statement
清算损益 liquidation gain or loss; liquidation profit and loss
清算所成员 clearing member
清算所得 income at liquidation
清算所基金 clearing house funds
清算条件 settlement terms; terms of settlement
清算同盟 clearing union
清算完结 closure of liquidation
清算项目 clearing item
清算协定 clearing agreement
清算业务 clearing operation
清算银行 clearing bank; reimburse bank
清算盈余 surplus at liquidation
清算余额 balance of clearing
清算债务 settle a claim
清算债务数 clearing debits
清算账户 clearing account; liquidating account
清算账户协定 open account agreement
清算账目 accounting
清算支票 settlement check
清算指标 clearing index; clearing target
清算制度 clearing system
清算资产负债表 condition of affairs; liquidation balance sheet
清算资金 settlement fund
清通井 clean-out chamber
清土板 clean-up scraper
清污分流系统 effluent segregation system
清污杆 sewer rod
清污耙 trash rack rake
清污耙起重机 trash-rake hoist
清污耙起重机平台 rake hoist platform
清污耙起重台架 rake gantry
清污泥水 filter sludge water
清污设备 decontamination facility; trash-removal device
清晰背景 plain background; plain ground
清晰边沿 sharp edge
清晰的 clear; distinct
清晰的白线标志 <道路上> plastic white line markings
清晰的白线合成物 <道路上> plastic white line composition
清晰的(道)路(划)线 plastic roadline
清晰的扫迹 sharp trace
清晰的条带 plastic strip
清晰的条纹 plastic strip
清晰度 acuity; clearness; clearness number; clearness of articulation; definition power; distinctness; intelligibility; legibility; limpidity; readability; resolution; sharp definition; sharpness; sharpness of definition; articulation; asperity <声音的>
清晰度百分数 articulation score
清晰度标准 resolution standard
清晰度参考当量 articulation reference
清晰度测试 resolving power test
清晰度测试卡 definition chart; identification resolution chart
清晰度测试条 resolution bars
清晰度测试图 resolution pattern
清晰度测试楔形束 definition wedge
清晰度差 bad sharpness
清晰度定焦法 definition determining focal length method
清晰度好 good sharpness
清晰度降低 articulation reduction; decrease of definition; degradation of resolution; detailloss; loss in definition; loss of resolution
清晰度降低系数 reduction factor of articulation
清晰度欠佳 lack of resolution
清晰度锐度 sharpness definition
清晰度试验 articulation test(ing)
清晰度损失 detailloss
清晰度调整 adjustment for definition
清晰度系数 sharpness factor
清晰度下降 decrease of definition;

sharpness fall(ing) off
清晰度响应 resolution response
清晰度楔 resolution block; resolution wedge
清晰度增强器 detail enhancer
清晰度指数 articulation index; resolving index
清晰度中等 middle sharpness
清晰范围 definition range
清晰分离 sharp separation
清晰回声 resolved echo
清晰景象 clear through vision
清晰区 circle of good definition
清晰区域 clear area; zone of sharpness
清晰视界 clear view
清晰视距 clear sight distance
清晰视野 clear view
清晰图 circle of good definition
清晰图像 brilliant image; distinct image; full-resolution picture; picture rich in detail; sharp image; sharply focused image
清晰显示 clear display
清晰显示雷达 clear scan radar
清晰像 sharply defined image
清晰效率 articulation efficiency
清晰性 legibility
清晰阴影 hard shadow
清晰影像 bright image; brilliant image; clear display; sharp image
清晰指数 articulation index
清洗 ablution; cleaning; cleanse; flush out; launder; purge; purgation; purge; purging; purification; purify(ing); rinse; scavenging; washing-up
清洗板 cleaning plate
清洗倍率 rinsing ratio
清洗泵 scavenger pump; scavenging pump
清洗残余 erase residual
清洗操作 cleaning operation
清洗槽 rinse tank
清洗厂废物 cleaning plant reject
清洗场 cleaning area
清洗车 rinse truck
清洗程度 degree of cleaning
清洗池 rinsing tank
清洗窗的平台 window cleaning balcony
清洗窗墙吊架 cleaning cradle
清洗道砟 track ballast cleaning
清洗的 abluent
清洗阀 purge valve; scavenger valve
清洗法 ablution
清洗方法 cleaning process
清洗分级机 washer-grader
清洗干燥机 washer-drier[dryer]
清洗干管 rinsing main
清洗工场设备 cleaning plant equipment
清洗工具 cleaning means
清洗工具橱 cleaning equipment closet
清洗工具室 cleaning equipment room
清洗工序 matting
清洗工作 clean-up work
清洗管 cleaning tube; purging pipe [piping]; purging tube [tubing]; rinsing pipe; rinsing tube; scavenge pipe
清洗管道工具 pipe cleaning device
清洗管路 detergent line; wash line
清洗管路堵塞 clearance of pipeline blockage
清洗管子 chase the pipe
清洗罐 cleaning tank
清洗滚筒 cleaning cage; washing barrel
清洗过程 rinsing process
清洗过的 washed drown
清洗海绵 cleaning sponge
清洗回路 scavenger circuit
清洗活塞 cleaning piston
清洗机 cleaning machine
清洗机油的洗涤剂 oil detergent
清洗(积垢)空气 scavenging air
清洗及处理程序 cleaning and disposal procedure
清洗剂 abluent; cleaner; cleaning agent; washing agent
清洗剂使用 detergent application
清洗加热器 cleaning heater
清洗间 washing room
清洗阶段 rinse stage
清洗接合器 adapter for cleaning
清洗金属 wash metal
清洗卷筒 cleaning web
清洗卡车 washing truck
清洗坑 cleaning pit
清洗孔 cleaning opening; clean-out port; washout hole
清洗孔板 demister (of vehicle window)
清洗孔机具 clean-out tools
清洗列车线 train washing siding
清洗溜槽 clean-up chute
清洗炉 wash heating furnace
清洗螺钉 cleaning screw
清洗螺纹 chase the threads
清洗盘 clean-out disc[disk]
清洗喷射泵 scavenging ejector
清洗喷嘴 washing nozzle
清洗气 purgative gas
清洗器 badger; cleaner; eraser; purger; purifier; rip
清洗枪 <压缩空气> cleaning gun
清洗球 cleaning ball
清洗溶剂 cleaning solvent; clean-up solvent
清洗熔炼 wash heat
清洗乳液 cleaning emulsion; cleansing emulsion
清洗筛 cleaning screen
清洗设备 cleaning and washing plant; cleaning equipment; purger
清洗时期 rinse time
清洗室 purge chamber; purge tank
清洗水 rinse water; rinsing water
清洗水定额 rinsing water norm
清洗水封 scavenge water seal
清洗水库 clearing of reservoir
清洗丝网 cleaning web
清洗(丝)网指示器 cleaning web indicator
清洗速度 rinsing velocity
清洗速率 rinsing rate
清洗碎石机 washer-stoner
清洗台 clean bench; rinsing table
清洗通道 washing tunnel
清洗系统 scavenger system
清洗旋塞 delivery cock
清洗旋塞及环 <过滤器> wash plugs and rings
清洗液 cleaning fluid; cleaning liquid; cleaning solution; cleansing solution; washing liquid
清洗液接收器 cleaning catcher
清洗液入口 filter washing water inlet
清洗用定额 rinsing water norm
清洗用软管 flushing hose
清洗用水 cleaning water
清洗油漆剂 paint remover
清洗周期 cleaning frequency; rinsing cycle; rinsing period
清洗装置 cleaning device; purger; purge unit; purifier; rinser; rinsing device; rinsing equipment; rinsing unit; washing installation
清洗作业线 cleaning line
清洗作用 cleaning action
清箱 emptying container
清屑杆 cleaner bar
清(新微)风 fresh breeze
清选 riddle
清选分级机 cleaner-grader
清选分离机 cleaner-separator
清选滚筒 cleaning cylinder
清选机 cleaning plant; dressing shoe; mill; scalping machine
清选筛 sorting screen
清选设备 cleaning equipment
清选升运器 cleaner-elevator
清选输送器链 cleaner chain
清选筒 <脱粒机的> scourclean
清选脱粒机 finishing thresher
清选脱粒机备件 spare parts for finishing thresher
清选锥体 cleaning cone
清雪设备 snow clearing equipment
清烟道器 flue cleaner
清岩机 ballast loader
清岩耙 rock rake
清样 clean proof; fair copy; press proof; repro; reproduction proof; slick paper proof
清液 clear liquid; clear solution; liquor
清液层 supernatant layer; supernatant liquid
清液出口 purified liquor outlet
清阴沟的工人 gutter man
清油 boiled oil; bunghole oil; kettle-boiled oil; paint oil
清油漆 oil varnish
清淤 cleaning of sediment; clean-out; clear up sludge; desilt(ing); dredging; removal of sediment; sediment removal; silt clearance; desilting
清淤河底 dredging bottom
清淤机械 sweeping mechanism
清淤孔 inspection eye
清淤口 access eye
清淤口塞子 cleanout plug
清淤水泵 dredging pump
清早 bright and early
清渣 mucking; brushing; detritus clearing; scarfing cinder; slag remove; spoil removing
清渣车 cinder bogie
清渣阀 washout valve
清渣工 mucker
清渣机 ballast cleaning machine
清渣口 breast hole; slag pocket
清渣口异形砖 skimmer block
清渣门 cleaning door
清渣器 dross extractor; ballast cleaner
清账 clear account; close account; general ledger; square up
清障 <水下等处的> snagging
清障船 snag boat
清障吊杆 <疏浚船吸口> snagging boom
清真食堂 Mostem restaurant
清真食堂穆斯林 Muslims' canteen
清真寺 masjid; mosque; musjid
清真寺壁龛 <面向麦加墙内的> mehrab
清真寺拱 mosque arch
清真寺光塔 minaret
清真寺建筑 mosque architecture
清真寺讲经坛 mimbar; Muslim pulpit
清真寺穹顶 Muslim dome
清真寺食堂 Muslims' canteen
清真寺中的妇女楼座 tecassir
清整 trim(ming)
清整动作 grooming behavio(u)r
清整工 fettler
清整工段 dressing yard
清整转筒 cleanser mill

鲭天 <鱼鳞天,一种自然风景> mackerel sky

情报安全性 information security

情报处理中心 information-handling center[centre]
情报处理装置 information processing equipment
情报分析中心 <美> Information Analysis Center
情报交换中心 information exchange center[centre]
情报交易所 clearing house
情报局 information bureau; intelligence bureau
情报设施 information facility
情报室 information office; information section
情报资源中心 information resource center[centre]
情景 case; condition; situation; state; status
情境角色 situation role
情况 case; circumstance; condition; situation; status
情况报告 circumstantial report; information report; situation report; status report <指工程方面>
情况变量 situation variable
情况调查 condition survey
情况调查表 survey questionnaire
情况调查工作 casework
情况恶劣的 ill-conditioned
情况反馈 state feedback
情况分析 regime(n) analysis; scenario analysis; situation analysis
情况管理原则 situation approach to management
情况介绍 brief briefing
情况良好 in good order
情况良好的 well conditioned
情况码 condition code
情况名 condition name
情况明细表 bordereau
情况剖析 case study
情况室 situation room
情况说明书 fact sheet
情况显示 situation display
情况显示管 situation display tube
情况显示计算机 situation display computer
情况显示台 situation display console
情况选择开关 condition switch
情况研究 case study
情况正常 good order; in sound condition; under control
情况证据 circumstantial evidence
情况周报 weekly condition report
情况综合报告 situation summary
情绪创伤 emotional trauma

晴空 azure; blue sky; clear air; clear sky; cloudless sky

晴空日 day of clear sky
晴空天蓝色 sky-blue
晴空湍流 clear-air turbulence
晴空雨 serein
晴朗 clearness; cloudless
晴朗的 clear; shiny
晴朗时期 spell of fine weather
晴朗无变化的 set fair
晴日 clear days
晴天 clear days; clear sky; clear weather; fine day; fine weather; sky clear; fair weather
晴天的 fair weather

晴天工作日 fair weather working days
晴天积云 fair cumulus
晴天径流 fair weather runoff
晴天卷云 fair weather cirrus
晴天流量 dry-weather flow
晴天霹雳 thunderclap
晴天天数 number of clear days
晴天停泊时间 weather lay days
晴天通车道路 < 雨天封锁交通 > fair weather road
晴天污水 dry weather sewage
晴天污水量 dry-weather flow
晴天效应 fine weather effect
晴天装卸作业天数 weather handling working days
晴通雨阻路 seasonally run road
晴夜空 firmament; starry sky
晴雨表 barometer; rain glass; storm glass; weather glass
晴雨表式价格领导 barometric (al) price leadership
晴雨计 barometer; rain glass; storm glass; weather glass
晴雨通车 all-weather service
晴雨通车道 all-weather road; road usable all year round
晴雨通车路 all-weather road; road usable all year round
晴雨通车路线 all-weather route
晴雨指示箱 weather box; weather house

氰 cyanogen; dicyan

氰氨法 cyanamide process
氰氨化钙 calcium cyanamide; nitrolime
氰氨化铅 lead cyanamide
氰氨化铅防锈漆 lead cyanamide antirust paint; lead cyanamide rustproof paint
氰胺 cyanamide
氰白 cyamelide
氰丙基 cyanopropyl
氰丙基硅酮 cyanopropyl silicone
氰醇 cyanalcohol; cyanhydrin
氰带 cyanogen band
氰的 cyanic
氰定 cyanidin
氰毒气 blue gas
氰仿 cyanoform
氰钙粉 cyanogas
氰高钴酸钾 potassium cobalticyanide
氰高钴酸盐试纸 cobalticyanide paper
氰高铁酸盐 hydroferricyanate
氰根离子 cyanide ion
氰钴胺 cyanocobalamine
氰胍 param
氰化 carbonitriding; cyanide carburizing; cyaniding
氰化氨 cyanamide
氰化铵 ammonium cyanide
氰化钡 barium cyanide
氰化槽 cyanidation vat
氰化碘 cyanogen iodide; iodine cyanide
氰化法 cyanidation; cyaniding process
氰化钙 black cyanide; calcium cyanide
氰化钢 carbonitrided steel; cyanided steel
氰化高钴试纸 cobalticyanide paper
氰化汞 mercuric cyanide
氰化合物 cyanogen compound
氰化甲苯基汞 cresyl mercury cyanide
氰化甲汞 agrosol; methylmercury cyanide
氰化甲烷试验 acetonitril test
氰化钾 cyanide potassium
氰化钾中毒 potassium cyanide poisoning
氰化金 gold cyanide
氰化金属 metallocyanide
氰化钠 sodium cyanide
氰化钠金 sodium gold cyanide
氰化钠中毒 sodium cyanide poisoning
氰化镍 nickel(ous) cyanide
氰化镍钾 nickel potassium cyanide
氰化铅 lead cyanide; plumbous cyanide
氰化氢 hydrocyanic acid; hydrogen cyanide
氰化热处理 cyanide carburizating
氰化溶液槽 cyanide bath
氰化纱布 cyanide gauze
氰化锶 strontium cyanide
氰化铁 ferric cyanide
氰化烃 hydrocarbyl cyanide
氰化铜 copper cyanide; cupric cyanide
氰化物 cyanide; cyanogen compound; prussiate
氰化物表面硬化剂 cyanide hardener
氰化物的臭氧分解 cyanide decomposition by ozone
氰化物的细菌分解 cyanide attack bacteria
氰化物电镀废水 cyanide plating wastewater
氰化物电镀铜 cyanide copper
氰化物电解铜 cyanide copper
氰化物淀积 cyanide deposit
氰化物法 cyanide process
氰化物废水处理 cyanide effluent treatment
氰化物分解细菌 cyanide attack bacteria
氰化物分解一次反应 first reaction of cyanide decomposition
氰化物监测仪 cyanide monitor
氰化物矿泥 cyanide pulp
氰化物离子 cyanide ion
氰化物络合物 cyanide complex
氰化物浓度 cyanide concentration
氰化物脱除法 cyanide elimination method
氰化物微粒 cyanide slime
氰化物污染 cyanide pollution
氰化物污染物 cyanide pollutant
氰化物污水处理 cyanide effluent treatment
氰化物消除法 cyanide elimination method
氰化物氧化 cyanide oxidation
氰化物硬化 cyanide hardening
氰化物中毒 cyanide poisoning
氰化物总量 total cyanide
氰化锌 zinc cyanide
氰化锌汞 Lister's anti-septic; mercury-zinc cyanide
氰化锌钾 zinc potassium cyanide
氰化亚铂 platinous cyanide
氰化亚金钾 gold potassium cyanide; potassium gold cyanide
氰化亚铁 ferrous cyanide
氰化亚铁的铁制件 ferrocyanide ferrous article
氰化亚铁金属 ferrocyanide metal
氰化亚铜 cuprous cyanide
氰化银 silver cyanide
氰化银钾 silver potassium cyanide
氰化浴中快速渗碳法 Shimer process
氰化正亚铁 ferriferous cyanide
氰化作用 cyanation; cyanogenation
氰基 cyan(o)-group
氰基丙烯胶泥 < 一种应变片的黏[粘]贴剂 > cyanoacrylate cement
氰基丙烯树脂黏[粘]结剂 cyanoacrylate adhesive
氰(基)丙烯酸酯 cyanoacrylate
氰基传感器 cyanosensor
氰基的 cyanophoric
氰基胍 dicyandiamide
氰基类树脂灌浆 cyanaloc grout
氰基树脂 < 防水剂 > cyanaloc
氰基烃 cyanocarbon
氰基乙酸 cyanoacetic acid; malonic mononitrile
氰基乙酸甲酯 malonic methyl ester nitrile; methyl cyanoacetate
氰基乙烯 acrylonitrile
氰基乙烯、丁二烯和苯乙烯共聚物 < 一种热塑料管原料 > ABS[acrylonitrile-butadiene-styrene] copolymer
氰基乙酰胺 cyanoacetamide; malonamide nitrile
氰离子分析法 analysis of cyanide ion
氰硫基 thiocyanato
氰络合物 complex cyanide
氰脲蓝 cyanurin
氰脲(三)酰胺 cyanuric triamide
氰脲酸 cyanuric acid
氰脲酸酯 cyanurate
氰脲酰胺 ammelide; cyanuramide; cyanuric amide
氰脲酰氯 cyanuric chloride
氰凝堵漏剂 low polymerized polyurethane leakproofing agent
氰凝防渗剂 low polymerized polyurethane leakproofing agent
氰凝灌浆 polyurethan(e) grouting
氰配合物 cyanocomplex
氰酸 cyanic acid
氰酸铵 ammonium cyanate
氰酸化物 cyanate
氰酸钾 potassium cyanate
氰酸钠 sodium cyanate
氰酸盐 cyanate
氰酸银 silver cyanate
氰酸酯 cyanate
氰铁酸盐 ferricyanide; prussic salt
氰亚铁氯化物 ferrocyanide chloride
氰亚铁酸蓝 ferrocyanide blue
氰亚铁酸锌 zinc ferrocyanide
氰乙基纤维素 cyanoethyl cellulose
氰乙酸 cyanoethanoic acid
氰乙酸肼 cyanazine
氰印照相法 cyanotype
氰硬化 cyanide hardening
氰酯 hydrocyanic ester

擎盖 < 汽车的 > bonnet

擎缆索 stopper
擎链器 controller; riding chock
擎逆轮 ratchet wheel
擎檐柱 eaves-supporting post

请拨车辆数 wagon requisitioned

请拨单 call slip
请查问出票人 refer to drawer
请大家上车 all aboard
请大家上船 all aboard
请戴口罩 respiratory protection
请戴手套 hand protection must be worn
请付款 please pay; request for payment
请购单 buying acquisition; invoice requisition; purchase requisition; purchasing requisition
请购单号 number of purchase requisition
请购书 purchase requisition
请汇支票付款 kindly remit by check
请即答复 return on post
请即付款 kindly remit
请即回示 by return of post
请即刻答复 please reply at once
请假 ask for leave
请交换 please exchange
请款单 requisition for money
请款书 application of funds statement; cash requirement
请料单 requisition; requisition for materials
请领车票单 requisition for tickets
请你注意 for your attention
请求包 request packet
请求保护专利范围 claim in a patent application
请求保释 application for bail
请求报头 request header
请求参数表 request parameter list
请求参数表出口例行程序 request parameter list exit routine
请求参数表串 request parameter list string
请求操作 solicit operation
请求操作信息 action message
请求偿还 apply for reimbursement
请求处理 demand processing; request stacking; time-sharing process(ing)
请求传输数据 request data transfer
请求待处理灯 request pending light
请求单号 requisition number
请求单元 request unit
请求单元部分 request unit portion
请求单元链 request unit chain
请求担保 application for bail
请求的 solicited
请求堆积 request stacking
请求队列 request queue
请求发送 request to send
请求发送电路 send-request circuit
请求发送线路 request-send circuit
请求分程序 request block
请求分段 demand staging
请求分配开销 request allocate overhead
请求服务系统 on-demand system
请求付款 application for payment
请求赋能 request enable
请求干预消息 action message
请求更改合同 contract change request
请求更正 asking for correction
请求共用 share request
请求管理程序 supervisor call
请求呼叫 request call
请求级 request level
请求即付资金 callable capital
请求加工 request stacking
请求价格引证 request for price quotation
请求监控器 request monitor
请求减免(税) abatement claim
请求接通线路 request on-line
请求控制块 request control block
请求块 request block
请求扩充 query enhancement
请求雷达导航 request a radar vector
请求联机【计】 log in
请求零页面 demand zero page
请求排队 request queue
请求判断 request judgement
请求评议 request for comments
请求人 applicant; claimant; petitioner
请求日期 date of application
请求时付款 pay on application
请求式喷墨头 on-demand ink gun
请求式系统 on-demand system
请求书 requisition
请求输入键 request entry key
请求损害赔偿的诉讼 action for damages

请求调页 demand paging
请求调职 request for transfer
请求停机 request stop
请求图形 demand graph
请求退还多收费用 application for repayment of an overcharge
请求信号 request signal
请求援助 request for assistance
请求者 applicant;demander;petitioner
请求支付 present for payment
请求指示径路 request for route
请求准许列车前进 request permission for a train to proceed
请确认 please confirm
请提建议 request for proposal
请帖 letter of notices
请勿动手 hands-off
请勿用钩 use no hooks
请勿张贴 stick no bills
请询背书人 refer to endorser
请询问承兑人 refer to acceptor
请与出票人洽询 refer to drawer
请与我联系 reach me
请证 confirmation request
请指导 please advise

庆

庆祝会 celebration

Q

磬

磬折形 gnomon

穷

穷国 poor country

穷竭法 method of exhaustion
穷竭检查 exhaustive test
穷竭搜索 exhaustive search
穷举 exhaustivity
穷举测试 exhaustive testing
穷举法 exhaust algorithm;method of exhaustion
穷举法文法推断 grammatical inference by enumeration
穷举归组 exhaustive grouping
穷举搜索 exhaustive search
穷举索引 exhaustive index
穷举调试 exhaustive testing
穷举证法 proof by exhaustion
穷苦 distress
穷山恶水 poor mountains and torrential rivers
穷乡僻壤 byplace;hinterland

穹

穹半筋 < 土木建筑 > formeret

穹半肋 formeret;wall rib
穹苍 vault
穹的露出肋 surface rib
穹地 dome
穹顶 cupola;dome;kupola;pericline;sky-dome;vault;doming < 落料缺陷 >
穹顶仓库 storage dome
穹顶的分隔间 < 哥特式建筑 > severy bay
穹顶的鼓座 dome-drum;drum of dome
穹顶的肋 springer
穹顶的楔形构件 gore
穹顶的座圈 drum
穹顶灯 dome lamp
穹顶底面 soffit of vault(ing);vault soffit
穹顶地窖 crypt;vaulted basement
穹顶泛水采光 coved lighting
穹顶分隔间 < 哥特式建筑穹隆顶棚的开间 > civery
穹顶分块 severy
穹顶工程 vaulted work;vaulting
穹顶构造 periclinal structure
穹顶厚度 depth of vault
穹顶弧形肋 pendentive cradling
穹顶混凝土料仓 dome silo
穹顶建筑 dome construction;domed building
穹顶脚圈梁 footring
穹顶孔眼 eye of dome
穹顶跨度 span of vault
穹顶肋 vault rib
穹顶连拱坝 multiple dome dam
穹顶门廊 < 伊斯兰教清真寺的 > iwan
穹顶模板 dome form
穹顶深坑 aven
穹顶十字架 < 教堂 > lantern cross
穹顶矢高 rise of vault
穹顶天窗孔 dome light
穹顶圬工 vaulting masonry
穹顶线 crown line(of vault)
穹顶形拱坝 domed arch dam
穹顶形式 dome form
穹顶支承结构 tholobate
穹顶枝肋 lierne rib
穹顶中的水平应力 belt stress
穹顶中两支承券间的三角形部分 panache
穹顶状的 dome like
穹顶状底脚 bottom dome
穹顶状结构 dome-like structure
穹顶坐圈 drum of dome
穹断 arched fault
穹拱 arch of vault
穹脊在同一水平上的两个穹 level ridge vaults
穹肩高低的穹顶 rampant vault
穹角形托架 pendentive bracketing
穹壳 vaulted shell
穹棱 groin(e);groyne
穹棱顶棚 groined ceiling
穹棱工作 groining
穹棱拱 groined arch
穹棱拱顶 groined vault
穹棱间曲面 sectroid
穹棱交点 groined point
穹棱肋 groin rib
穹棱天花板 groin ceiling
穹棱屋顶 groined arch;groined roof
穹隆【建】 arch;cove;coving
穹隆坝 domed(arch)dam
穹隆背斜群 domal anticline zone
穹隆薄壳 dome shell
穹隆布景 kuppel horizon
穹隆的 domical;domy
穹隆的边 dome edge
穹隆的顶窗 domed roof-light
穹隆的拱脚横带 vaulting course
穹隆的拱脚石层 vaulting course
穹隆的一角的形状 trompe
穹隆底部 basal parts of the dome
穹隆地板 dome-shaped floor
穹隆顶 cupola;dome key;interdome
穹隆顶窗 eye;eye of dome
穹隆顶大厅教堂 domed hall church
穹隆顶的 vaulted
穹隆顶底座 tholobate
穹隆顶点 dome apex
穹隆顶孔 eye of a dome
穹隆顶棚 vault(ed)ceiling
穹隆顶棚中央的浮雕(饰)scutcheon
穹隆分部 dome segment
穹隆风格建筑 domed style
穹隆副肋 lierne
穹隆副肋作业 lierne vaulting
穹隆杆 dome bar
穹隆拱肋 vault rib
穹隆构造 dome structure;vaulting
穹隆构造地貌 landform of domal structure
穹隆构造环形体 circular features of dome structures
穹隆和构造盆地 dome and basin structure
穹隆横肋 cross springer
穹隆环 dome ring
穹隆建筑 vault architecture
穹隆开间 vault bay;vaulting cell
穹隆孔眼 eye of a dome
穹隆理论 vault theory
穹隆连合体 fornicommissure
穹隆门饰 archivolt
穹隆圈 dome ring
穹隆圈闭 dome trap
穹隆山 dome mountain
穹隆式基础 dome foundation
穹隆天花板 coved ceiling;groined ceiling
穹隆屋顶 domed roof;groined roof;vault(ed)roof
穹隆形 archivolt
穹隆形坝 cupola dam
穹隆形板 domical slab
穹隆形的 fornicate
穹隆形底 dome-shaped bottom
穹隆形顶塔楼 lantern tower
穹隆形分布 dome type distribution
穹隆形墓穴 burial vault
穹隆形屋顶 dished roof
穹隆用空心砖 vaulting tile
穹隆油气田群 oil-gas fields group of dome
穹隆照明 vault light
穹隆中央高地 central domal highland
穹隆柱 fornicolumn
穹隆柱顶 vaulting capital
穹隆柱墩 dome impost
穹隆状 bosslike;domed form
穹隆状板 domed(-shaped)slab
穹隆状薄壳 dome-shaped shell
穹隆状采光顶楼 < 屋顶上 > cimborio
穹隆状的 domal;fornicate;vaulted
穹隆状地下室 under-croft
穹隆状结构 dome-like structure
穹隆状热储 dome reservoir
穹隆状屋顶 dome-shaped roof
穹隆状形式 dome-shaped form
穹隆纵肋 longitudinal rib of vault
穹隆作顶的建筑物 domed structure
穹隆作顶的体育场 domed stadium
穹隆作顶的小教堂 domed chapel
穹隆作用 doming
穹面灯 dome lamp;dome light
穹面反射器 dome reflector
穹内凹顶 concha
穹盆构造 dome-basin structure
穹起 arch
穹起断陷式构造 arch-subsidence structure
穹起褶皱 arched up folds
穹起轴 axis of arch
穹倾斜 quaquaversal dip
穹丘 dome
穹式 dome type
穹式结构 dome format
穹式屋顶 dome roof
穹束 fornix
穹屋顶 < 由中柱旋转成的 > radial arch roof
穹屋顶的坐圈 drum of dome
穹形坝 dome-shaped dam
穹形齿 castellation
穹形的【地】 quaquaversal
穹形地下室 under-croft
穹形顶板 kettle back
穹形反射器 dome reflector
穹形拱坝 domed(arch)dam
穹形拱顶 domed vault
穹形构造【地】 quaquaversal;quaquaversal structure
穹形火山 domed volcano
穹形建筑 dome construction
穹形鳞片 fornix
穹形墓穴 burial vault
穹形山 dome mountain
穹形饰 ornamental barrel vault
穹形物 pavilion
穹形圆山 ballon
穹形褶皱【地】 quaquaversal fold
穹形砖 dome brick
穹形状结构 dome-like structure
穹翼 arch limb
穹隅 < 圆屋顶过渡到支柱之间的渐变曲面 > pendentive
穹褶 arched fold
穹状 bosslike
穹状的 dome like;periclinal;quaquaversal
穹状断块山系 dome-block mountain system
穹状构造【地】 domal structure;dome structure;periclinal structure;quaquaversal structure
穹状火山颈 tholoid
穹状隆起 domal uplift;dome-like upheaval
穹状沙丘 dome dune
穹状圆顶 quaquaversal

琼

琼胶 agar

琼楠属 tawa;Beilsehmiedia < 拉 >
琼斯-莫特反应 Jones-Mote reaction
琼斯波罗花岗岩 < 一种产于美国缅因州的浅红色花岗岩 > Jonesboro
琼斯-帛特兰斯打浆机 Jones-Bertrams beater
琼斯大锥度精磨机 Jones Berkshire refiner
琼斯等效噪声功率 Jones noise equivalent power
琼斯高低速双转盘水力碎浆机 Jones pulper
琼斯格槽缩样器 Jones riffle
琼斯计算法 Jones calculus
琼斯矩阵 Jones matrix
琼斯炉 Jones furnace
琼斯区 Jones zone
琼斯矢量 Jones vector
琼斯试剂 Jones reagent
琼斯缩分器 Jones splitter
琼斯温差电偶 Jones thermocouple
琼泰风 Junta
琼脂 agar(-agar);Bengal gelatine;black test;gelose;japan ager
琼脂包埋剂 agar embedding material
琼脂杯法 agar cup method
琼脂电泳 agar electrophoresis
琼脂管 agar tube
琼脂光电光度计 agar photoelectric(al)photometer
琼脂划线法 agar streak method
琼脂胶 agaropectin
琼脂胶体强度计 agar jelly strength tester
琼脂空斑技术 agar plaque technic
琼脂块法 agar block method
琼脂块技术 agar block technique
琼脂扩散法 agar diffusion method
琼脂扩散试验 agar gel diffusion test
琼脂凝胶 agar gel
琼脂凝胶电泳法 agar gel electrophoresis
琼脂凝胶反应 agar gel reaction
琼脂培养 agar culture
琼脂培养基 agar medium
琼脂片法 agar disc[disk] method
琼脂平板 agar plate
琼脂平面培养 agar plate culture
琼脂平面培养法 agar plate method
琼脂平皿 agar plate
琼脂强度 agar strength

琼脂色谱法 agar chromatography
琼脂深层培育 agar deep culture
琼脂双扩散测定法 agar double diffusion technique
琼脂水块试验法 agar block test
琼脂素 gelasin
琼脂酸 agaric acid agaricin
琼脂糖凝胶 agarose coagulation
琼脂斜面 agar slant
琼脂斜面培养 agar slant culture
琼脂斜面培养基 agar slant culture-medium
琼脂悬块 agar hanging block
琼脂针刺培养 agar stab; agar stab culture

丘巴斯科雷暴 <中美西海岸伴有雷雨的强阵风> Chubasco

丘侧砾岩 plaster conglomerate
丘齿 bunodont
丘的顶 knap
丘碲铅铜石 choloalite
丘顶 hill top; summit; knap
丘吉尔河 Churchill River
丘勒勃计程仪转子 Cherub rotor
丘陵 hill; hump; mound; ridging ground; kopje <南非>
丘陵草地 hilly grassland
丘陵草原 downs
丘陵侧面 hillside
丘陵城堡 motte castle
丘陵的 hilly; undulating
丘陵地 broken country; broken ground; broken terrain; down land; hilly land; rugged terrain; wold; downs <英>
丘陵地带 closed country; difficult country; foot hill; hilly terrain; knob; rolling terrain; rough country; undulating terrain; hilly country; rolling ground; undulating ground
丘陵地带的 hillocky
丘陵地段 hilly terrain; undulating terrain
丘陵地区 difficult country; difficult ground; difficult terrain; hilly area; hilly country; intersected country; moderate hills; undulating terrain; hilly ground; rolling country
丘陵地区道路 hilly road
丘陵地区河流 hill stream
丘陵地区选线 hilly land location; location of line on hilly land
丘陵地形 rolling land; rolling topography; undulating topography
丘陵沟渠 motte ditch
丘陵海岸 hilly coast
丘陵或石山 <受冰河包围的> nunatak(k)
丘陵间的低地 <尤指沿河适宜耕作的地方> intervale; interval land
丘陵间洼地 glen
丘陵景观 hilly landscape
丘陵路段的加速设备 accelerating device for humps
丘陵木纹 quilted figure
丘陵泥炭 hill peat
丘陵平原岩溶 hilly and plain karst
丘陵坡 hillside
丘陵起伏 a chain of undulating hills
丘陵起伏的 downy
丘陵起伏线 undulating line
丘陵区 brae; hilly area; intersected country; low-relief terrain; rolling country; rolling ground; rolling terrain; undulating area
丘陵区道路 hilly road
丘陵区河流 hill river
丘陵森林地带 rough wooded country
丘陵山脉 range of hills
丘陵沼泽 everglade
丘陵沼泽地 fell
丘奇法 <一种测定航摄仪倾角的方法> Church method
丘泉 knoll spring
丘群 cumulus
丘砂 down sand
丘洼地形 knob-and-kettle topography
丘形地面 hummocky surface
丘翼圈闭 fill-flank trap
丘状的 buninoid
丘状浮冰 hummocked ice
丘状交错层理构造 hummocky cross bedding structure
丘状盆地 hummock-and-hollow topography; knob-and-basin topography
丘状盆地地形【地】 knob-and-basin topography
丘状栽植 mound planting

邱园 <英国皇家植物园> Kew garden

秋材 autumn timber; autumn wood

秋材率 autumn wood ratio; late wood ratio
秋发芽 germinating in autumn
秋分潮 autumn(al) equinox tide
秋分大潮 equinoctial tide
秋分点 autumnal equinox; September equinox; first point of Libra
秋分期 autumnal equinoctial period
秋耕 autumn ploughing; autumn plow; fall plowing
秋耕日 fall plowing time
秋耕闲地 black fallow
秋耕休闲 dead fallowing
秋灌 fall irrigation
秋海棠 begonia
秋海棠红色 carajura
秋洪 autumn flood; fall flood
秋后 after autumn
秋湖 <月球> lacus autumn
秋华柳 Salix variegata
秋积尺 computing scale
秋吉尔造山旋回【地】 Akiyoshi orogeny cycle
秋吉尔造山运动【地】 Akiyoshi orogeny
秋季 autumn; fall
秋(季采伐的木)材 late wood
秋季等温线 isometropal
秋季对流 autumn(al) autumn overturn; fall overturn
秋季翻转 autumnal turnover; fall overturn; fall turnover
秋季灌溉 fall irrigation
秋季洪水 autumn flood; fall flood
秋季环流 autumnal circulation
秋季降雨量 fall precipitation
秋季垃圾 autumn refuse
秋季品种 autumn variety
秋季热 autumnal fever
秋季施工 autumn construction
秋季相 autumn aspect
秋季循环 autumn overturn; fall overturn
秋季循环期 autumnal circulation period
秋季作物 autumn crops
秋剪 fall pruning
秋津造陆运动【地】 Akitsu epeirogeny
秋景园 autumn garden
秋葵 <美> gumbo
秋兰姆 thiuram
秋老虎 <秋季中天晴干燥的一个热期> old wives' summer; Indian summer; after summer; after heat(ing)
秋涝 water-logging from autumnal rains
秋千 swing
秋色 autumn colo(u)or; fall colo(u)r
秋沙鸭 merganser
秋收 autumn harvest
秋天 autumn; fall
秋汛 autumn flood; fall flood
秋植 autumn plumage
秋庄稼 autumn crops

蚯蚓 earthworm; rainworm

蚯蚓生物滤池 earthworm biofilter; vermi biofilter

囚犯 prisoner

求包(工程) bid for

求并运算【数】 cup
求部分和 subtotaling
求参模型 evaluating parameter model
求长法 rectification
求偿汇票 reimbursement draft
求偿权 right of claim; right of recourse
求偿银行 claiming bank
求导(数) derivation; differentiate
求定值 required value
求读数 plotting
求二次曲线的面积 quadrature of a conic
求反操作 complementary operation
求反程序 negate routine
求反触发器 complementing flip-flop
求反器 complementer
求反运算 complementary operation
求反运算符 complementary operator
求反指令 negate instruction
求反子波 inverse wavelet estimation
求方补助 demand-side subsidy
求分力 resolution of forces
求负 negate
求高基金 go-go fund
求根 extraction; rooting; extract a root
求根法【数】 extraction of root
求根解算器 root solver
求根器 root solver
求根仪 isograph
求公式 derivation of equation
求过于供 demand exceeds supply
求过于供的市场情况 oligopoly
求和 summation; summing
求和测量法 summation metering
求和存储计数器 adding storage register
求和存储器 sigma memory; sigma storage
求和点 summing junction; summing point
求和电路 summing circuit
求和法 method of summation; sum formula; summation method; summation process
求和法则 sum rule
求和放大器 summing amplifier
求和公式 summation formula
求和机构 summing gear
求和积分法 summation method of integration
求和积分器 summing integrator
求和检查 sum check; summation check
求和检验 summation check
求和校验 sum check; summation check
求和器 adder; adding machine; summer
求和求差网络 sum-and-difference net
求和求积分法 integration by summation method
求和缩写法 summation convention
求和网络 summation network; summing network
求和仪 totalizing instrument
求和元件 summation element; summator; summing element
求和约定 summation convention
求和指令 summarizing instruction; summation instruction
求和指数 summation index
求和装置 adder; totaliser[totalisator]
求积法 mensuration; planimetry; stereometry
求积分 quadrature
求积分法 method of quadrature
求积(分)器 integrator
求积公式 quadrature formula; summation formula
求积计数器 totalling meter
求积理论 quadrature theory
求积器 multiplicator
求积图 planimetric(al) map
求积仪 integrating instrument; integrator; planimeter[platometer]; productimeter
求积仪侧轮 planimeter drum
求积仪常数 planimeter constant
求积仪导轮 integrating wheel
求积仪极点 anchor point of planimeter
求积仪描迹臂 planimetric(al) arm
求极大值 maximizing
求极限过程 limit process
求极小值 minimization
求极小值法 minimization process
求几何面积法 quadrature
求解 compute
求解过程 solution procedure; solving process
求解器 equation solver
求解仪 resolver
求解仪转子 resolver rotor
求解运算 derivation
求救灯号 distress light
求救呼号 distress call
求救旗号 flag of distress
求救信号 distress signal; waft; weft
求救信号波 distress wave
求救信号方向 direction of weft
求距角 <三角形中的> distance angle
求绝对值函数 absolute function
求均值 averaging
求勒型压力浮选机 Juell pressure flo(a)tation cell
求面积 quadrature; squaring
求面积法 area method; mensuration; method of quadrature
求面积器 squaring device
求面积仪 planimeter
求逆 inversion
求偶 courtship
求频带比 band rationing
求平方根算法 square-rooting algorithm
求平均数 averaging
求平均数算子 averaging operator
求平均值 averaging
求平均值法 averaging method
求曲线长 rectifying
求曲线长度 rectify
求容积法 cubage; cubature
求商图 quograph

求时间平均 time averaging
求树法 tree approach
求体积 cubing
求体积法 cubage; cubature; mensuration
求体积公式 cubature formula
求同趋势 tendency to identify
求同思维 convergent thinking
求微分 differentiate; differentiation
求像法 image construction
求心规 center[centre] head
求业者 employment
求圆的面积 quadrature of a circle
求援 recourse
求援信标 emergency beacon; rescue beacon
求援信号装置 distress signal warning device
求值 evaluate; evaluation
求值程序 evaluation program(me)
求值的运算顺序 operational order of evaluation
求值卡 evaluation card
求值设备 valuator device
求职 job hunting; job wanted
求职成本 search cost
求职广告 positions wanted
求职人员 job applicant; job hunter
求职申请 application for the position; application of job; bidding
求职申请表 application for employment
求职申请书 application for employment; application form
求职途径 job search channel
求职者 job hunter
求中的 centripetal
求助 bespeak; recourse; resort
求助菜单 help menu
求助程序【计】help program(me)
求助的对象 recourse
求助复制 help copy
求助功能 help function
求助过程 help procedure
求助键 help key
求助卷 help screen
求助码【计】help code
求助屏幕 help screen
求助设备 help facility
求助系统 help system
求助字条 help script
求最大值 maximizing
求最佳参数 optimization
求最小值 minimization
求最小参数值 minimizing
求最小树算法 min-tree finding algorithm
求最小值的方法 minimization process

泅色 colo(u)r bleeding; crocking; feathering

球凹 < 硬度试验的 > ball impression

球凹面积 area of ball imprint
球拗门闩 knob bolt
球拗门栓 knob bolt
球拗门锁 knob lock
球拗碰锁 knob latch
球把插锁 knob latch
球把垫圈 knob rose
球把门锁 knob lock
球摆 spheric(al) pendulum
球摆试验 ball pendulum test
球半径 radius of sphere
球半径系数 spheric(al) radius factor
球瓣 ball clack
球瓣共振腔 spheric(al) sector resonator
球棒模型 ball-and-stick model
球棒形手把 bat handle
球包界 sphere-packing bound
球贝塞尔函数 spheric(al) Bessel function
球鼻冰川 bulb glacier
球鼻船首 bulbous bow; bulbous stem
球鼻(首船体) bulbous hull
球鼻首隔板 diaphragm in bulbous bow
球鼻首肋骨 bulbous bow frame
球鼻首前端板 bulb front plating
球鼻艏 bulbous bow; bulbous stem
球鼻艏船 bulb-bowed ship
球鼻形尾柱底部 aftfoot bulb
球壁硬度 rigidity of ocular wall
球边 bead
球边角材肋骨 angle bulb frame
球扁钢 bulb steel; flat bulb steel
球扁铁 flat bulb iron
球波函数 spheric(al) wave function
球测头 ball contact tip
球测硬度 bail hardness
球差系数 spheric(al) aberration coefficient
球场 ball park; court; ball-game ground < 球场总称 >
球超柱 spheroidal hypercylinder
球掣 ball catch
球承 ball joint
球承式机械手 ball joint manipulator
球承窝 ball socket
球齿钎头 button bit
球齿形钎头 button bit
球齿钻头 button bit
球冲击 pellet impact
球雏晶【地】globulite
球带 spheric(al) zone; zone of sphere
球带函数 zonal spheric(al) function
球带调和函数 zonal spheric(al) harmonics
球带调和数 zonal harmonic
球胆 bladder
球胆破坏 bladder wrack
球的 spheric(al)
球的撞击 sphere impact
球滴 globule
球滴定管 chamber burette
球底冲头 round bottomed punch
球底垫圈 spheric(al) washer
球底料仓 spheric(al) bottom bin
球底料斗 spheric(al) bottom bin
球底面模 spheric(al) punch
球点吸水量 ball point water demand
球垫 friction(al) ball
球顶补缩内浇口 ball gate
球顶储罐 globe-roof tank
球顶浮标 staff and globe surmounted buoy
球顶(杆)状浮标 globe buoy
球顶式活塞 spheric(al) head piston
球顶形扬声器 done loudspeaker
球度 < 表示集料等表面积对体积的关系 > sphericity; roundness
球度极限值 sphericity limit
球端 ball; pommel[pummel]; ball end【机】
球端扁杆 flat pummel bar
球端定位螺钉 ball point setscrew
球端杆 joystick lever
球端心轴 centre pin
球对称 spheric(al) symmetry
球对称透射法测量 sphere transmission method measurement
球对称加积 spheric(al) accretion
球二孢属 < 拉 > Botryodiplodia
球阀 ball-and-socket valve; ball check; ball check valve; ball cock; ball plug; ball tap; ball valve; bullet valve; globe cock; relief ball; sphere valve; spheric(al) valve; straight-flow rotary valve
球阀垫圈 ball seat gasket
球阀式取土器 ball valve sampler; soil sampler of spheric(al) value
球阀铜浮子 ball valve copper float
球阀止逆器件 ball stop
球阀装置 ball-cock assembly; ball-cock device
球阀组件 ball-cock assemble
球阀座 ball retainer; ball seat(ing); valve with ball seat
球法 marble process
球法拉丝法 marble melt process
球法拉丝工艺 marble making process
球反射镜 ball mirror
球反射面 spheric(al) reflector
球房虫 globigerina[复 globigernae]
球分布 spheric(al) distribution
球分规值 ball subga(u)ge
球辐射 spheric(al) radiation
球辐射器 spheric(al) radiator
球杆 ball arm
球隔流阀 ball lever
球根 napiform root
球根瓶 bulbous vessel
球根栽植器 bulb planter
球根桩 bulb pile
球关节【机】globe joint
球冠圆板式橡胶支座 ball-topped circular plate-type rubber bearing
球管 buke; bulb tube; pipet(te)
球管磨机 ball tube mill
球管平衡 glomerulotubular balance
球管气压计 bulb barometer; vessel barometer
球管温度计 bulb thermometer
球管嘴 ball nozzle
球罐 spheric(al) tank
球规 ga(u)ge ball
球硅钙石 radiophyllite
球滚光 ball-burnishing
球棍状的 club-shaped
球果植物 coniferophyte
球函数 spheric(al) function; spheric(al) harmonics
球焊 ball bonding; ball welding; nail-headed bond(ing)
球焊接头 ball bond
球耗 ball wear
球和座 < 深井泵的 > ball-and-seat
球护圈 ball retainer
球滑车 jewel block
球化 balling; globuling; nodularization; nodulizing; pelletization[pelletisation]
球化不良 under-nodularizing
球化处理 ductile treatment; spheroidisation [spheroidization]; spheroidizing
球化处理钢 spheroidized steel
球化机 nodulizer
球化剂 nodularizer; nodulizing agent; spheroiditic agent; spheroidizing medium
球化颗粒 spheroidized particle
球化率快速测定试棒 microlug
球化渗碳体 spheroidized cementite
球化衰退 degradated spheroidization
球化水冷退火 water annealing; water softening
球化碳化物 spheroidized carbide
球化退火 carbide annealing; sphericizing annealing; spheroidal annealing; spheroidized annealing; globurizing
球化作用 spheroidization; spheroidizing
球环 ball ring
球环带外形 equatorial profile
球环法 < 沥青软化点试验的 > ball-and-ring method
球环磨 ball-and-race mill
球环软化点测定法 ball-and-ring method
球环软化点测定器 ball-and-ring apparatus
球环软化点测试 ball-and-ring test
球基桩 bulb pile
球极平面射影 stereographic(al) projection
球极平面投影 stereographic(al) projection
球极平面投影尺 stereographic(al) projection ruler
球极平面投影网 stereographic(al) net; stereographic(al) projection grid; stereographic(al) projection net
球极投影 stereographic(al) projection
球极投影格网 stereographic(al) grid
球极投影图 stereogram
球极投影网 stereographic(al) net
球极映射 stereographic(al) mapping
球极坐标 polar spheric(al) coordinates; spheric(al) polar coordinates
球极坐标系 spheric(al) polar coordinate system
球剂 globule; globulus
球夹 ball chuck
球夹式提引器 ball type holding dog; ball type pulling dog
球间区 interglobular areas
球间隙 sphere gap
球铰 ball pivot; cup-and-ball joint; globe joint
球铰承座 ball-and-socket bearing
球铰接 swivel coupling
球铰接合 articulation by ball-and-socket; ball-and-socket joint
球铰接屋架杆件 ball joint roof bar
球铰接支承 ball joint prop
球铰节 ball bonding
球铰链 spheric(al) hinge
球铰碗 ball cup
球铰轴承 ball socket bearing
球铰转节球头 fulcrum ball
球铰座 spheric(al) seating
球校准 ball sizing
球接 ball couplet
球接点网架 ball joint net frame
球接活接头 swing union
球接头 ball joint; spheric(al) joint
球节 ball coupling; ball pivot
球节点 ball joint; globe joint; roller joint; sphere node; spheric(al) joint
球节盖 ball joint cover
球节盖闩 ball joint cover latch
球节机械手 ball joint manipulator
球节夹 ball joint clamp
球节裂 split pattern
球截角锥体 frustum of sphere
球截体 segment of a sphere
球截形 spheric(al) segment
球近似 spheric(al) approximation
球茎 bulb; corm
球晶 spheric(al) crystal; sphero-crystal
球晶辐向生长速度 spherulite radial growth rate
球晶生长 spherulitic growth
球晶生长速率 spherulite growth rate
球颈轴承 ball journal bearing
球径测量术 spherometry
球径度 sphericity
球径规 globe cal(l)ipers; spherometer cal(l)ipers

球径计 spherometer
球径量规 ball cal(l)ipers
球径率 sphericity
球径平面 diametral plane of a sphere
球径仪 spherometer
球菌 cocci;coccus
球菌抱球虫软泥 coccolithglobigerina ooze
球颗 variole
球颗化作用【地】variolitization
球颗结构 variolitic texture
球颗玄武岩 variolite
球壳 spheric(al)shell
球壳式水轮机 globe cased turbine
球壳式涡轮机 globe cased turbine
球壳属 <拉> Sphaerella
球控道岔自动转换装置 ball-controlled point setting apparatus
球类运动厅 ball-games hall
球粒 pellet; spherulite <火成岩中呈球状的矿物集合体>【地】
球粒斑岩 spherophyre
球粒尺寸 nodule size
球粒构造【地】spherulitic structure
球粒花岗岩 pudding granite
球粒化 pelletization[pelletisation]
球粒灰泥 pellet-lime mud
球粒灰岩 pelleted limestone
球粒剂 globular powder
球粒结构【地】spherulitic texture; pellet texture
球粒亮晶灰岩 pelsparite
球粒亮晶砾屑灰岩 pelsparrudite
球粒磷块岩 pellet phosphoraite
球粒铝质岩 pellet aluminous rock
球粒泥晶灰岩 pelmicrite
球粒泥质结构 spherulitic pelitic texture
球粒泥状铝质岩 pellet pelitomorphic aluminous rock
球粒熔结凝灰岩 sperolitic welded tuff
球粒体 spheroplast
球粒微晶灰岩 pelletal-micritic limestone;pelmicrite
球粒微亮晶灰岩 pelmicsparite
球粒玄武岩的 variolitic
球粒岩 spherulite rock
球粒陨石 chondrite;chondritic meteorite
球粒陨石标准化丰度 chondrite-normalized abundance
球粒状 spherulitic
球粒状磷块岩 pelletoidal phosphoraite
球料比 ratio of grinding media to material
球磷钙铁矿 egueite
球磷铝石 sphaerite
球菱钴矿 sphaerocobaltite
球菱铁矿 iron spar
球笼万向节 Rzeppa constant velocity joint
球帽座 spheric(al)nut seat
球煤 coal ball
球门 goal
球门柱 goal post
球门自动阀 disk hinged valve; flap valve;paddle valve
球密计 dasymeter
球面 sphere;spheric(al)face;spheric(al)plane;spheric(al)surface
球面凹凸透镜 spheric(al)meniscus
球面坝 spheric(al)dam
球面摆 spheric(al)pendulum
球面半径 spheric(al)radius
球面变换公式 spheric(al)transformation formula
球面表示 spheric(al)representation
球面波 spheric(al)wave
球面波倒易校准 spheric(al)wave reciprocity calibration
球面波函数 spheric(al)function
球面波近场效应 spheric(al)wave proximity effect
球面波前 spheric(al)wave front
球面波形喇叭 spheric(al)wave horn
球面波照明 spheric(al)wave illumination
球面不符值 spheric(al)discrepancy
球面擦准法 spheric(al)lapping
球面参考共振腔 spheric(al)reference cavity
球面参数 spheric(al)parameter
球面车床 globe lathe; spheric(al) lathe;spheric(al)turning lathe
球面齿轮 Hindley worm gear
球面窗式光电倍增管 spheric(al) window photomultiplier
球面磁罗经 spheric(al)compass
球面带 spheric(al)strip
球面带光通量 zonal light flux
球面带谐函数 surface zonal harmonics;zonal spheric(al)harmonics
球面挡水墙 spheric(al)dam
球面刀 ball cutter;spheric(al)cutter
球面的 spheric(al);steradian
球面的平均烛光 mean spheric(al) candle power
球面底 dome head;rounded end
球面地 spheric(al)earth
球面电极头 radius tip
球面垫圈 ball faced washer;spheric(al)washer
球面碟形盖板 spherically dished cover plate
球面度 <立体角度单位> steradian
球面对称分布 spherically symmetric(al)distribution
球面对称三角形 symmetric(al)spheric(al)triangles
球面多边形 spheric(al)polygon
球面二次曲线 sphero-conic
球面二角形 lune;lune of a sphere;spheric(al)lune
球面发散校正 spheric(al)divergence correction
球面发射强度 steradiancy
球面反射 spheric(al)reflection
球面反射光 spheric(al)reflector light
球面反射镜 spheric(al)mirror;spheric(al) reflecting mirror; spheric(al)reflector
球面反射镜腔 curved-mirror cavity
球面反射率 spheric(al)reflectivity
球面反射漆膜 spherically reflective paint film
球面反射器 spheric(al)mirror;spheric(al) reflecting mirror; spheric(al)reflector
球面反射器天线 spheric(al)reflector antenna
球面反射式物镜 spheric(al)mirror objective
球面反射系统 spheric(al)reflecting system
球面反向天线阵 spheric(al)retrodirective array
球面反照率 spheric(al)albedo
球面方差函数 spheric(al)variance function
球面方位角 spheric(al) azimuth; spheric(al)bearing
球面飞行 spheric(al)flying
球面分布 spheric(al)distribution
球面辐射测温计 globe thermometer
球面辐照强度计 spheric(al)irradiance meter
球面副 spheric(al)pair
球面概率误差 spheric(al)error probability
球面干涉仪 sphericity interferometer
球面弓形 lune of a sphere; spheric(al)segment
球面拱顶 spheric(al)calotte
球面光度计 sphere photometer
球面光强 spheric(al)luminous intensity
球面光学 spheric(al)optics
球面光栅 spheric(al)grating
球面光照度 spheric(al)luminosity of light
球面滚柱轴承 spheric(al)roller bearing
球面滚子 spheric(al)roller
球面滚子轴承 spherangular roller bearing;spheric(al)roller bearing
球面函数 spheric(al)function
球面航迹计算法 spheric(al)sailing
球面换算系数 spheric(al)reduction factor
球面会聚 spheric(al)convergence
球面活管接旋转接头 spheric(al)union swivel
球面活接头 spheric(al)union
球面积分 surface integral
球面激波 spheric(al)shock
球面极 spheric(al)polar
球面极坐标 polar spheric(al)coordinates;spheric(al)polar coordinates
球面几何学 geometry of sphere; spheric(al)geometry;spherics
球面计 spherometer
球面交叉滑块机构 conic(al)crossed slider chain
球面浇射损耗 spheric(al)diffraction loss
球面角 spheric(al)angle
球面角超 spheric(al)excess
球面角度 <立体角单位> sterad;steradian
球面角盈 spheric(al)excess;spheroidal excess
球面铰刀 ball reamer
球面接地 spheric(al)earth
球面镜 spheric(al)mirror
球面镜干涉仪 spheric(al)mirror interferometer
球面镜共振腔 spheric(al) mirror cavity
球面镜口径 aperture
球面镜片 spheric(al)glass
球面镜头 spheric(al)lens
球面镜谐振腔 spheric(al)mirror resonator
球面距离 spheric(al)distance
球面距离改化 reduction of spatial distances
球面聚光镜型分光仪 spheric(al) condenser-type spectrometer
球面壳体 spheric(al)shell
球面扩展 spheric(al)spreading
球面立体角 spheric(al)solid angle
球面六角螺母 ball faced hexagonal nut
球面螺杆传动 globoid worm gear
球面螺帽 spheric(al)nut
球面螺母 ball nut
球面螺旋线 spheric(al)helix
球面镘刀 bacca box smoother;button sleeker
球面磨损 surface ablation
球面穹面 spheric(al)dome
球面曲柄链系 spheric(al)crank chain
球面曲率 spheric(al)curvature
球面曲率计 spherometer
球面曲线 spheric(al)curve
球面三角法【数】spheric(al)trigonometry
球面三角形 astronomic(al)triangle; spheric(al)triangle
球面三角形测量 spheric(al)triangulation
球面三角学 spheric(al)trigonometry;spherics
球面三角余弦公式 cosine formulas of spheric(al)triangle
球面三角正弦公式 sine formulas of spheric(al)triangle
球面散射表示量 spheric(al)indicatrix of scattering
球面散射指示量 spheric(al)indicatrix of scattering
球面色像差 spherochromatic aberration
球面枢轴 spheric(al)pivot
球面数据 sphere data
球面衰减校正值 correction values of spheric(al)attenuation
球面双曲线 spheric(al)hyperbola
球面四边形 spheric(al)quadrangle; spheric(al)quadrilateral
球面四方形 spheric(al)quadrangle
球面天文学 positional astronomy; spheric(al)astronomy
球面天线 spheric(al)antenna
球面调和函数 spheric(al)harmonics
球面投影 azimuthal orthomorphic(al) projection; globular projection; stereographic(al)projection; stereoprojection
球面投影地图 globular chart;stereographic (al) chart; stereospheric(al)chart
球面透镜 spheric(al)lens
球面透镜精加工设备 spheric(al)lens smoothing machinery
球面透镜制造设备 spheric(al)lens generating machinery
球面透视 spheric(al)perspective
球面推力轴承 spherically mounted thrust bearing
球面弯接头 ball elbow
球面弯头活管接 spheric(al)bend union; spheric(al) elbow union; spheric(al)ell union
球面蜗杆 cone worm;globoidal worm; Hindolet worm;hour-glass worm
球面蜗杆传动 globoid worm gear
球面蜗杆滚轮式转向器 Marles steering gear
球面蜗杆啮合 globoidal worm toothing;Hindley worm toothing
球面蜗杠 ball face worm
球面蜗轮 globoid worm gear; hourglass worm wheel
球面蜗轮传动装置 globoid worm gearing
球面蜗轮减速机 cone-worm unit
球面蜗轮减速箱 enveloping wormgear reducer
球面蜗轮蜗杆 Hindley worm gear
球面五联公式 five parts formula
球面铣刀 rose cutter
球面象限三角形 quadrant spheric(al)triangle
球面像差 spheric(al) aberration; spheric(al)defect
球面像放大镜 spheric(al)image amplifier
球面谐函数 surface spheric(al)harmonics
球面心射投影 central projection of sphere
球面悬链线 spheric(al)catenary
球面旋涡 spheric(al)vortex
球面研磨点 spheric(al)grinding point
球面仪 spherometer

球面应力区 bulbous zone of stress
球面影像 spheric(al) image
球面预解式 spheric(al) resolvent
球面元素 spheric(al) element
球面圆 circle of sphere
球面圆点曲线 sphero cyclic
球面圆盘 spheric(al) disk
球面圆柱头螺钉 cheese head screw; oval fillister head screw
球面运动 spheric(al) motion; spheric(al) movement
球面运动机构 spheric(al) mechanism
球面运动链 spheric(al) chain
球面照度 spheric(al) illumination
球面折射 refraction at spheric(al) surface
球面折算因数 spheric(al) reduction factor
球面阵 spheric(al) array
球面正余弦公式 five parts formula
球面支柱式推力轴承 Kingsbury spheric(al) thrust bearing
球面支柱式止推轴承 Kingsbury spheric(al) thrust bearing
球面支座 spheric(al) bearing
球面直角三角形 right-angle spheric(al) triangle; right spheric(al) triangle; quadrant spheric(al) triangle
球面直角坐标 rectangular spheric(al) coordinates; spheric(al) rectangular coordinates
球面直径 spheric(al) diameter
球面中心支枢 pillow pivot
球面轴承 globe bearing; self-aligning bearing; spheric(al) bearing
球面轴颈 spheric(al) journal
球面烛光 spheric(al) candlepower
球面锥 spheric(al) cone
球面坐标 spheric(al) coordinates
球面坐标基面 fundamental plane of spheric(al) coordinates
球面坐标系 spheric(al) coordinate system
球面座 ball seat(ing)
球模式构造 geodetic construction
球磨床 ball grinder(machine); ball grinding mill
球磨的填充系数 percentage loading of mill
球磨粉碎机 ball mill pulverizer; ball crusher
球磨机 ball grinder(machine); ball grinding machine; ball grinding mill; ball mill; ball tube pulverizer; bowl mill; globe mill; grinding cylinder; grinding mill; pebble mill; sphere-grinding mill; preliminator < 水泥 >
球磨机衬层 ball mill lining
球磨机滚筒 balling drum
球磨机加工 ball milling
球磨机球磨作用 ball action
球磨机球配量 ball rationing
球磨机添加物料 ball mill addition
球磨机研磨 ball milling
球磨(机用)石球 crushing boulders
球磨机转速 ball rotation speed; drum's speed of rotation
球磨机装球量 ball charge
球磨精研机 ball mill refiner
球磨精制机 ball mill refiner
球磨碎石机 ball crusher
球磨炭黑 attrited black
球磨作用 < 球磨机的 > ball action
球墨 spheric(al) graphite
球墨生铁 nodular cast iron
球墨铁 ductile iron
球墨轧辊 spheroidal graphite roll
球墨铸铁 ductile cast iron; ductile iron; graphite cast iron; nodular graphite cast-iron; spheric(al) graphite cast iron; spheroidal graphite cast iron; spheroidal iron
球墨铸铁弓形支撑 < 盾构的 > ductile segment
球墨铸铁管 ductile iron pipe
球墨铸铁管片 ductile cast iron segment; ductile iron segment
球墨铸铁环 spheroidal graphite iron ring
球墨铸铁轧辊 spheroidal graphite roll
球墨铸铁铸造厂 malleable foundry
球墨组织 nodular graphitic structure
球黏[粘]土 ball clay
球盘磨 ball-race mill
球旁器 juxtaglomerular apparatus
球泡霏细岩 pyromeride
球泡酸钴矿 spherocobaltite
球坯 glass gob for marble making
球钳 globe pliers
球腔菌属 < 拉 > Mycosphaerella
球驱动旋转喷洒头 ball drive rotary head
球塞 ball plug; ball plunger; ball tap
球塞泵 ball piston pump
球塞槽 ball plunger groove
球塞规 ball plug ga(u)ge
球塞式 ball-and-spigot
球塞式电动机 ball piston motor
球塞式液压马达 ball piston type hydraulic motor
球栅分析器 spheric(al) grid analyser [analyzer]
球上荷载 ball load
球砷锰石 akrochordite
球石 ball stone
球石软泥 coccolith ooze
球石藻类 Coccolithophorida
球式工作台 ball table
球式光度计 globe photometer
球式接合器 pellet bonder
球式倾斜计 ball inclinometer
球式倾斜指示器 ball bank inclinometer
球式示倾器 ball bank indicator
球式首柱 bulb stem
球式输送机 ball conveyer[conveyor]
球式硬度试验机 ball hardness tester
球式转运台 ball transfer table
球势阱 spheric(al) well
球饰 balloon; pommel
球枢 ball pivot
球枢轴 ball stud
球碳镁石 dypingite
球套 ball sleeve
球套管节 ball sleeve tubing union
球体 globe; globe body; sphere; spheric(al) body; sphericity; spheroid; spherome
球体半径 radius of sphericity
球体表面两点之间最短的线 geodetic line
球体波 spheroidal wave
球体波函数 spheroidal wave function
球体导磁率 permeability of sphere
球体导电率 conductivity of sphere
球体的 spheroidal
球体的综合参数 synthetic(al) parameter of sphere
球体耳轴 ball gudgeon
球体贯入试验 < 测定新鲜混凝土稠度或已硬化混凝土表面强度用 > ball penetration test
球体函数 spheroidal harmonic
球体解算器 ball resolver
球体孔 spheric(al) pore
球体平均光度角 Russell's angle
球体式振荡 spheroidal oscillation
球体调和函数 spheric(al) harmonic function; spheroidal harmonic function
球体投影图 planisphere
球体谐函数 solid spheric(al)-harmonics
球体形成 spheroiding
球体旋转轴 rotational axis of the sphere
球体质量 spheroid mass
球体状态 spheroidal state
球体坐标 spheric(al) coordinates; spheroidal coordinates
球调节器 ball governor
球头 ball(head); bulb; pommel; round head
球头 T 钢 bulb rail steel
球头扁钢 bulb plate; flat bulb iron
球头撑杆 ball head stay
球头锤 ball hammer; ball-pane hammer
球头丁字钢 bulb rail steel
球头工字钢梁 bulb steel I-beam
球头工字梁 bulb I-beam
球头挂板 ball-clevis
球头挂钩 ball-hook
球头挂环 ball-eye
球头轨 bulb rail
球头滚针 needle roller with ball end
球头角钢 angle bulb iron; bulb angle; bulb angle iron; bulb angle steel; bulb iron
球头节 ball joint; spheric(al) connection
球头联节 coupling socket
球头螺栓 ball-bolt; ball head bolt; ball stud
球头式水准点 round-head bench mark
球头式振捣棒 bullet-bossed vibrating bar; bullet-nosed vibrator
球头式振捣器 bullet-nosed vibrator
球头枢轴 ball end
球头铣刀 rose cutter
球头形绝缘子 ball head insulator
球头圆铁钉 ball wire nail
球头状绝缘子 ball head insulator
球头锥形螺栓 ball headed conic(al) bolt
球头钻 rose drill
球投影 spheric(al) projection
球透镜 globe lens
球凸轮 ball cam
球土 ball clay
球团 agglomerate; agglomeration; pelletizating
球团化 pelletization[pelletisation]
球团化骨料 pelletized type aggregate
球团化集料 pelletized type aggregate
球团矿 pellet; pellet ore; spheric(al) agglomeration
球团矿砂 ore pellet
球窝 ball-and-socket; ball socket
球窝承座 ball-and-socket bearing
球窝关节 articulation by ball-and-socket; ball-and-socket joint; socket joint; cup-and-ball joint; ball joint
球窝活节 ball-and-socket joint; cup-and-ball joint
球窝活节接合 articulation by ball-and-socket
球窝活节球头操纵杆 ball arm
球窝机械手 ball manipulator
球窝基座 ball socket base
球窝夹头 spheric(al) grips
球窝铰链 universal joint
球窝接合 articulation by ball-and-socket; ball joint; cup-and-ball joint; globe joint; socket joint; ball-and-socket joint
球窝接口 cup-and-ball joint; globe joint
球窝接头 ball-and-socket head; ball-and-socket joint; ball joint; cup-and-ball joint; globe joint
球窝节 ball-and-socket jointing; universal and socket joint
球窝节理 ball-and-socket joint
球窝节头 ball-and-socket head
球窝结合 socket joint
球窝连接 globe joint
球窝联结 ball-and-socket attachment; ball-and-socket joint
球窝联结器 ball-and-socket coupling
球窝式 ball-and-socket type
球窝式变速杆 ball-and-socket gear shifting
球窝式调档 ball-and-socket gear shifting
球窝式悬式绝缘子 ball-and-socket type suspension insulator
球窝万向节 self-contained cardan joint
球窝芯轴 mandrel socket
球窝支座 ball-and-socket bearing; socket bearing
球窝轴承 ball-and-socket bearing; socket bearing
球窝座 ball-and-socket base
球吸入阀 ball suction valve
球戏场 spheric game place
球限制堵 ball choke
球霰石 vaterite
球销式 ball-and-spigot
球效应 spheric(al) effect
球谐方程 spheric(al) harmonic equation
球谐函数 spheric(al) harmonic function; spheric(al) harmonics
球谐函数法 spheric(al) harmonic method
球谐函数级数 spheric(al) harmonics series
球谐函数近似法 spheric(al) harmonics approximation
球谐函数系数 spheric(al) harmonic coefficient
球谐函数展开 spheric(al) harmonic expansion
球谐计算 spheric(al) harmonic calculation
球谐位扰动函数 spheric(al) harmonic potential disturbing function
球谐展开 spheric(al) harmonic expansion
球心 center[centre] of sphere
球心阀 globe valve
球心花饰 ballflower
球心角 angle at spheric(al) center [centre]
球心角体 spheric(al) sector
球心埋深 depth of sphere center[centre]
球心投影 gnomonic projection
球心投影地图 gnomonic map
球心投影海图 gnomonic chart; great-circle chart
球心像 image of spheric(al) center [centre]
球心轴承 spheric(al) center[centre] bearing
球形 bulbous; globosity; globular shape; roundness; sphere; spheric(al) shape; spheroidicity
球形安全阀 ball relief valve; ball safety valve; spheric(al) safety valve
球形把手 ball knob; knob(bling); turn knob
球形把手柄 knob shank
球形把手止定螺栓 side knob screw
球形白炽灯 incandescent globe
球形包裹体 spheric(al) inclusion body

球形包面 spheric(al) envelope
球形爆破法 spheric(al) explosion
球形爆破桩 bulbous pile
球形爆炸波 spheric(al) blast wave
球形泵 ball pump
球形避雷器 spheric(al) arrester
球形表面 spheric(al) surface
球形柄 knob
球形波 spheric(al) wave
球形玻璃缸 bubble bowl
球形玻璃容器 balloon
球形薄壳 spheric(al) shell
球形补偿器 ball type compensator; expansion ball joint
球形测针 spheric(al) head probe
球形岔管 spheric(al) bifurcated pipe
球形车轮车 spheric(al) wheel vehicle
球形衬 globe lining; spheric(al) bush
球形衬里 tubbing
球形承口 ball socket
球形承窝 spheric(al) socket
球形冲头端 ball point
球形储罐 spheric(al) holder; spheric(al) storage vessel; spheric(al) tank
球形储罐瓣片 the plate pieces of spheric(al) storage tank
球形储罐组群 group of spheric(al) tanks
球形储能器 spheric(al) accumulator
球形储气罐 spheric(al) gas-holder
球形穿孔机 bullet perforator
球形传声器 spheric(al) microphone
球形船首 bulb(ous) bow
球形船头 bulb(ous) bow; bulbous stem
球形船尾 bulb stern
球形锤头 ball peen hammer
球形催化剂 spheric(al) catalyst; spheroidal catalyst
球形存水弯 ball trap
球形打印机 spheric(al) typewriter
球形大玻璃瓶 balloon
球形单向阀 ball check; ball valve
球形导体 spheric(al) conductor
球形的 bulbous; conglobate; globate; globose; globular; spheric(al); spheroidal; globoid
球形灯 globe lamp; globular light; spheric(al) lamp; globe lantern
球形灯泡 globular bulb
球形灯罩 globe holder
球形灯座 globe holder
球形等离子粒团 spheric(al) plasmoid
球形地面辐射表 spheric(al) pyrgeometer
球形地面衰减 spheric(al) earth attenuation
球形地面因数 spheric(al) earth factor
球形电动机 spheric(al) motor
球形电弧法 globule method of arcing
球形电极 spheric(al) electrode
球形电离室 spheric(al) ionization chamber
球形电枢 spheric(al) armature
球形垫块 ball washer
球形雕饰 knob
球形吊钩 hook ball
球形丁字管节 globe tee
球形顶 globe-roof
球形顶盖 spheric(al) cap
球形顶式活塞 spheric(al) head piston
球形顶饰 pomelo; pommel
球形度 degree of sphericity; sphericity
球形端 spheric(al) end; knob top <把手的>
球形堆 spheric(al) reactor
球形对称 spheric(al) symmetry
球形发光 globe type luminescence
球形发射体 spheric(al) emitter
球形阀(门) ball cock; ball valve; globe cock; globe valve; rotary valve; ball check; ball type valve; spheric(al) valve; ball clack; globe body valve
球形阀闸 ball valve sluice
球形阀座 ball seat(ing)
球形反应器 global reactor
球形房屋 ball house; spheric(al) house
球形放电 globular discharge
球形分离舱 spheric(al) capsule
球形分离器 spheric(al) separator
球形粉 spheric(al) powder
球形风化 spheric(al) weathering
球形风头 spheric(al) head
球形封头 dome head; egg end; hemispheric(al) head
球形浮标【港】 spheric(al) buoy; ball float; floating ball; globe buoy; pivoted float
球形浮子液面计 ball-float(liquid-) level meter
球形辐射计装置 spheric(al) radiometer package
球形钙化层状小体 globular calcified lamellar bodies; Schaumann's bodies
球形盖 bubble cap
球形坩埚 spheric(al) crucible
球形坩埚技术 spheric(al) crucible technique
球形隔电子 ball insulator
球形跟踪器 tracker ball; tracking ball
球形共振腔 spheric(al) resonator
球形构造的 orbicular
球形刮管器 ball-and-chain crawler
球形挂环 ball with eye
球形关闭件 ball closure member
球形关节 sphere joint
球形管 globe pipe; globe tube
球形管板 spheric(al) tubesheet
球形管接头 pipe connection with ball joint
球形管嘴 ball nozzle
球形罐 hoctonspheres; hocton spheroid; plain hocton spheroid; spheric(al) tank
球形光度计 globe photometer; spheric(al) photometer
球形柜 spheric(al) tank
球形海流计 spheric(al) current meter
球形和片状铰链 ball and sheet hinge
球形核 spheric(al) nucleus
球形盒式计数器 pill-box counter
球形弧 globular arc
球形花饰 ballflower
球形化晶胞 sphericized lattice cell
球形化学气相沉积金刚石晶体 ball-shaped chemical vapo(u)r deposit diamond crystal
球形话筒 spheric(al) microphone
球形环状 spheric(al) annulus
球形缓冲器 radial buffer
球形混凝土稠度试验 ball test for concrete consistence[consistency]
球形活络把手 drop ring
球形活塞泵 ball piston pump
球形火花放电器 sphere spark-gap
球形火花隙 ball spark gap
球形机壳 globe case
球形基态 spheric(al) ground state
球形加热器 ball heater
球形加压容器 spheric(al) pressure container
球形减压舱 decompression sphere
球形减压室 decompression sphere
球形建筑物 sphere building; Perisphere <1939 年纽约世界博览会建筑>
球形胶体 sphere colloid
球形铰 ball hinge; spheric(al) hinge
球形铰接支座 spherical knuckle bearing
球形铰链 ball couplet
球形铰链换挡杆 ball change
球形校正锤 cambered flatter
球形接合 ball bonding; ball joint; pillow joint; spheric(al) joint
球形接头 ball-and-socket joint; ball bonding; ball joint; hitch ball; pillow joint; socket and spigot joint(ing); spheric(al) joint
球形结构 orbicular structure; pellet formation; spheric(al) structure
球形截流阀 globe stop valve
球形截水器 ball type interceptor
球形截止阀 lock valve globe
球形截止止回阀 globe stop check valve
球形解算器 ball resolver; spheric(al) resolver
球形界面 globular interface
球形进料阀 spheric(al) feed valve
球形进料开关 spheric(al) feed valve
球形聚光器型能谱仪 spheric(al) condenser-type spectrometer
球形聚合物 globular polymer
球形聚焦测井 spheric(al) focused log
球形聚焦测井曲线 spheric(al) focused log curve
球形绝缘器 globe insulator
球形绝缘子 ball insulator; globe insulator
球形卡座 globe holder
球形颗粒 spheric(al) granule; spheric(al) particle; spheroidal particle
球形壳体 domed shell; spheric(al) shell
球形可变电感器 ball variometer
球形刻度盘 spheric(al) scale
球形空腔谐振器 spheric(al) cavity resonator
球形(空)穴 spheric(al) cavity
球形控制台打字机 spheric(al) console typewriter
球形扣环 globe retaining ring
球形快门 spheric(al) shutter
球形矿渣 pelleted slag
球形扩脚柱 bulk column
球形扩脚桩 pedestal pile; bulb pile
球形榔头 ball hammer
球形冷凝器 ball condenser
球形连接 ball joint; pillow-jointing
球形连接器 ball adaptor
球形联轴器 spheric(al) coupling
球形量规 ball ga(u)ge
球形邻域 spheric(al) neighbo(u)rhood
球形零件 bulb
球形流量计 ball flow meter
球形轮廓 spheric(al) profile
球形螺帽 spheric(al) nut
球形脉冲电离室 spheroidal pulse ionization chamber
球形镘刀 bacca box smoother
球形铆钉头 globe rivet head
球形冒口 spheric(al) feeder; spheric(al) riser
球形帽罩 spheric(al) calotte
球形煤气表 spheric(al) gasometer
球形煤气罐 spheric(al) tank
球形门把手 ball door knob; door knob
球形门柄 door nob
球形门拉手 door knob
球形门拉手零件 door knob furniture
球形模锻机 ball header
球形磨 pot mill
球形磨口玻璃接头 spheric(al) ground glass joint
球形磨砂玻璃灯泡 spheric(al) depolished glass globe
球形磨砂玻璃灯罩 spheric(al) depolished glass globe
球形磨头 globe grinding head
球形耐拉绝缘子 globe strain insulator
球形逆止阀 ball check; ball check valve
球形捏手 hand knob; knobboss
球形偶核 spheric(al) even nuclei
球形排出阀 ball delivery valve
球形泡 spheric(al) bulb
球形泡沫色谱法 spheric(al) foam chromatography
球形胚 globular embryo
球形配流阀表面 spheric(al) valving surface
球形配水室 spheric(al) distribution chamber
球形喷嘴 ball nozzle
球形碰头 dome door stop
球形偏转电子束 spherically deflected electronic beam
球形平面轴承 spheric(al) plain bearing
球形屏蔽 spheric(al) shield
球形瓶 balloon
球形剖面 bulbous section
球形期 globular stage
球形气量表 spheric(al) gasometer
球形气瓶 storage sphere
球形气球 spheric(al) balloon
球形汽缸 spheric(al) casing
球形潜水器 bathysphere
球形腔 spheric(al) cavity
球形倾斜仪 ball inclinometer; ball bank indicator <利用离心力原理测定弯道上车速的一种仪器>
球形穹顶 bulbous dome; spheric(al) cupola; spheric(al) dome; spheric(al) vault
球形区域 spheric(al) region
球形曲柄链系 spheric(al) crank chain
球形全辐射表 spheric(al) pyradiometer
球形全向传声器 eight ball
球形燃料反应堆 pebble-bed reactor
球形燃料气冷反应堆 pebble-bed gas cooled reactor
球形燃料元件 spheric(al) fuel element
球形燃烧室 spheric(al) combustion chamber; spheric(al) segment combustion chamber
球形容器 globe; spheric(al) container; spheric(al) vessel
球形容器裙式支座 skirt support of spheric-(al) vessel
球形容器支座 supports of spheric(al) vessel
球形容水器 watersphere
球形塞 ball plug
球形塞门 globe cock
球形三通 globe tee[T]
球形栅壁计数管 spheric(al) grid-wall counter
球形伞齿轮 spheric(al) bevel gear
球形筛底 crown bottom
球形闪电 globular discharge
球形烧瓶 balloon flask
球形烧燃室 spheric(al) combustion chamber
球形射气源 spheric(al) emanation source
球形渗透计 ball permeameter
球形十二面体 spheric(al) icosahedron

球形石墨铸件 spheroidal graphite casting
球形示倾器 ball bank indicator
球形收集器 spheric(al) collector
球形手柄 balanced handle;ball handle
球形艏 bulbous bow
球形枢轴 spheric(al) pivot
球形枢轴颈 pivot joint
球形树 globe-shaped tree
球形刷 knot brush
球形水龙头 globe tap
球形水听器 ball hydrophone
球形水箱 spheric(al) tank
球形水准器 box bubble
球形送风口 ball diffuser
球形送话器 spheric(al) microphone
球形穗饰 ball-fringe
球形探头 spheric(al) probe
球形探针法 globe probing method
球形体 globoid; sphericity; spheroid; toroid
球形天线 spheric(al) antenna
球形填充物 ball packing
球形填料 ball packing
球形挑料机 ball gatherer
球形铁粒 spheric(al) iron particle
球形铜环 spheric(al) brass cup
球形头 ball attachment;ballhead;spheric(al) head;knob top <把手的>
球形头快速管接头 quick coupling with spherical heads
球形头皮托管 spheric(al) Pitot probe
球形投影 globular projection
球形凸管 bulb tube
球形凸轮 globe cam; globoid cam; spheric(al) cam
球形土压力分布曲线 bulb earth pressure distribution curve
球形推力轴承 ball-thrust bearing; spheric(al) thrust bearing
球形陀螺分子 spheric(al) top molecule
球形外壳 globe housing
球形弯月面 spheric(al) meniscus
球形万向接头 ball-and-socket coupling; ball-and-socket joint; universal ball joint
球形万向节 ball-and-socket coupling; ball-and-socket joint; universal ball joint
球形万向节夹角 ball joint inclination
球形万向联轴节 ball-and-socket coupling;ball-and-socket joint
球形网架 geodesic dome
球形微壳靶 spheric(al) microshell target
球形艉 bulbous stem
球形温度计 globe thermometer
球形蜗杆 globoid worm
球形蜗轮蜗杆 globoid worm gear
球形握把 jaw stick
球形屋顶 bulbous dome; spheric(al) calotte; spheric(al) dome; spheric(al) roof
球形物 ball; bowl; buke; bulb; globe; glomeration; orb; sphere
球形铣刀 ball cutter
球形系船浮筒 spherical mooring buoy
球形系留气球 spheric(al) captive balloon
球形系数 sphericity factor
球形氙灯 spheric(al) xenon lamp
球形消防栓 ball hydrant
球形消音防逆阀 globe type silent check valve
球形谐振器 spheric(al) resonator
球形信号 <旧式进站信号> ball signal
球形蓄能器 spheric(al) accumulator
球形蓄压器 spheric(al) accumulator
球形旋塞 ball tap;globe cock
球形旋塞阀 globe cock
球形旋涡 spheric(al) vortex
球形旋转漂白器 globe rotary bleacher
球形压力罐 spheric(al) pressure tank
球形压头 spheric(al) indenter
球形阳极 ball anode
球形样板 proof sphere
球形摇座 spherical rocker bearing
球形液舱型液化天然气运输船 moss-type LNG carrier
球形椅 globe chair
球形硬度计 ball durometer
球形油罐 sphere can; spheric(al) oil tank
球形游丝 spheric(al) balance spring
球形圆顶 spheric(al) cupola
球形造粒 marumerizer;pelletizer
球形造粒机 marumerizer
球形闸门 ball lock
球形窄颈瓶 carafe
球形照明体 sphere illumination
球形折射面 spheric(al) refracting surface
球形蒸煮器 rotary spheric(al) digester;spheric(al) boiler
球形支承螺帽 spheric(al) seating nut
球形支座 ball-frame carriage; ball socket; beaded support; free bearing; globular bearing; knuckle bearing; spheric(al) bearing
球形止回阀 ball check valve
球形止回喷嘴 ball check nozzle
球形止逆阀 ball stop; ball valve
球形轴承 spheric(al) bearing
球形轴颈 globe journal
球形烛光 spheric(al) candlepower
球形转子 spheric(al) spinner
球形状柱 bulbous pile
球形锥 spheric(al) cone
球形字锤 type ball
球形总日射表 spheric(al) pyranometer
球形座 spheric(al) seat
球型瓣膜 ball type valve
球型打印机 ball type printer; type ball printer
球型单向阀 ball type check valve
球型防逆阀 ball type check valve
球型高斯轨道 spheric(al) Gaussian orbital
球型连接盘式绝缘子 cap-and-pin insulator
球型头 spheric(al) type head
球型总阀 ball type corporation valve
球锈菌属 <拉> Sphaerophragmium
球旋塞 ball cock;globe cock
球旋型粗碎机 coarse gyrasphere crusher
球压焊 ball bond
球压滤法 pressure-filter-bulb method
球压式硬度试验 ball(pressure) hardness test;ball tester
球压试验 ball (indentation) test; Brinell's impact test
球压硬度 ball indentational hardness
球压载 spheric(al) ballast
球压轴承 ball-thrust bearing
球研磨 ball grinding
球样的 globoid
球样体 globoid body
球窑 marble furnace
球衣菌膨胀 Sohaerotilus bulking
球印法 ball indentation method
球印贯入度 <测路面硬度用> sphere penetration
球印器 indenter
球印试验 ball pressure test;ball test
球印硬度 ball hardness
球印硬度试验 ball hardness test; ball (indentation) test; dynamic (al) indentation test; indentation test; static indentation test; ball pressure test
球印硬度试验法 ball hardness testing method
球印硬度试验机 ball hardness testing machine
球印硬度值 ball hardness number
球应力 isotropic (al) stress; spheric(al) stress
球硬度试验 <压痕法> static indentation test
球缘 bulb
球缘 T 形材 bulb tee;T-bulb-bar
球缘扁钢 bulb bar
球缘扁铁 bulb bar
球缘钢板 bulb plate
球缘角钢 angle bulb;bulb angle(-bar)
球缘首柱 bulb stem
球缘铁 bulb iron
球载红外望远镜 balloon-borne infrared telescope
球载平台 balloon platform
球载望远镜 balloon-borne telescope
球载仪器 balloon-borne instrument
球闸门 ball valve
球张量 spheric(al) tensor
球罩灯照明 globe lighting
球针壳属 <拉> Phyllactinia
球枕构造 ball-and-pillow structure
球振荡 spheroidal vibration
球支承 ball support
球支承的桌子 ball table
球支枢 ball pivot
球支座 ball bearing;tumbler bearing
球止回阀 ball check valve
球轴 ball spindle
球轴承 ball bearing; spot contact bearing
球轴承保持架 ball-bearing retainer; ball holder
球轴承噪声 ball-bearing noise
球轴承罩 ball-bearing cage
球轴颈 ball journal; ball pin; ball pivot; spheric(al) journal
球轴颈铰链支承 ball jointed rocker bearing
球轴套 ball bushing
球轴座 ball-and-socket base
球珠导轨 spheric(al) guide
球柱床 <四角短柱各饰大圆球的> cannonball bed
球柱面冰凸透镜 sphero-cylindrical lenticular
球柱面透镜 sphero cylindrical lens
球柱体 spherocylinder
球铸机 ball-casting machine
球状 globosity; globularity; sphericity; varihedroid
球状把手 ball knob
球状爆炸前峰 spheric(al) detonation front
球状崩解作用 granular disintegration
球状避雷器 sphere gap
球状冰 ball ice
球状波 spheric(al) wave
球状剥离 ball structure parting
球状部件 pellet part
球状沉淀 nodular precipitation
球状成长相 nodular growth phase
球状次系 spheric(al) subsystem
球状带 glomerular zone
球状的 bulbous; bulb-shaped; globoid; globular; knoblike; orbed; orbicular; orbiculate; pelletized; spheric(al); spheroidal
球状地球 spheroidal earth
球状电极 sphere pole
球状电闪 ball lightning; spheric(al) lightning
球状叠层石 oncolithes
球状碟形封头 spherically dished head
球状端点 ball terminal
球状堆积 sphere packing
球状方解石 vaterite
球状放电 globular discharge
球状放电器 sphere gap
球状粉末 globular powder; spheric(al) powder
球状风化 concentric(al) weathering; onion weathering; spheric (al) weathering; spheroidal weathering
球状风化体 spheroidal weathering body
球状浮标 globe buoy
球状浮体 ball float
球状浮子 ball float; spheric(al) float
球状辐射器 spheric(al) radiator
球状复原 spheroidal recovery
球状盖 bubble cap
球状感觉 spheresthesia
球状刚果金刚石 Congo rounds
球状共晶晶粒 spheric (al) eutectic grain
球状构造 ball structure; globular structure; orbicular structure; spheric(al) structure; spheroidal structure
球状光度计 sphere photometer; spheric(al) photometer
球状锅炉 spheric(al) boiler
球状核 globose nucleus; nucleus globosus
球状花 globular flower
球状化 spheroidizing
球状灰岩 globulitic limestone
球状辉长岩 napoleonite; orbicular gabbro
球状活性炭 spheric(al) active carbon
球状集合体 spheric(al) aggregate
球状碱安岩 bulgarite
球状胶束 globular micelle; spheric(al) micelle
球状铰接机械手 ball manipulator; ball type handler
球状铰链机械手 ball type handler
球状节理【地】 spheric(al) jointing; spheroidal structure; ball structure parting; globular joint(ing); spheroidal jointing
球状结构 globular texture
球状结合 ball coupling
球状结核 spheric(al) nodule
球状结晶 <指钢铁> spheroidal cementite
球状晶粒 spheric(al) grain
球状晶体 spherulite
球状矩阵排列 ball grid array
球状颗粒 spheric(al) particle
球状粒 peloid
球状粒组构 spheric(al) fabric
球状连接器 ball coupling
球状裂纹 spheric(al) crevasses
球状螺母 ball nut
球状冒口 ball feeder; sphere riser
球状煤 ball coal; pebble coal
球状模式 spheroidal mode
球状模型 spheric(al) model
球状囊 ball pocket
球状黏[粘]土 ball clay
球状捏手 ball knob
球状气体储罐 ball gas tank; Horton sphere
球状曲柄 ball crank
球状热敏电阻 thermistor bead
球状容器 spheroid
球状塞门体 spheric(al) cock-shell
球状闪长岩 corsite; miagite; napoleonite
球状闪电 ball lightning; globe light-

ning;spheric(al)lightning
球状渗碳体 modulous cementite;spheroidite
球状石墨 globular graphite;graphite in sphere form;nodular graphite;spheric(al)graphite;spheroidal graphite;spherulitic graphite
球状石墨铸钢 centra steel;spheroidal graphite cast iron
球状石墨铸铁 nodular cast iron
球状手柄 ball grip
球状苏格兰胶 Scotch glue in pearl form
球状碳化物 globular carbide;spheroidal carbide
球状体 globule;sphere
球状体的 globoid
球状填充 sphere packing
球状投影 globular projection
球状凸模 spheric(al)punch
球状蜗杆啮合 Hindley worm toothing
球状物 ball;bulb;globe;pill
球状物表面 spheric(al)surface
球状物料 pelletized material
球状细菌 sphere bacteria
球状小粒 coccode
球状压机 ball header
球状氧化物夹杂 globular oxide inclusion
球状药包 spheric(al)charge
球状引导物 ball guide
球状硬沥青 pelleted pitch
球状油罐 spheroid tank
球状圆顶 round dome
球状炸药 blasting pellet
球状振荡 spheric(al)oscillation
球状支点 ball socket
球状珠光体 beaded pearlite;spheroidal pearlite
球状装药 spheric(al)charge
球状准则 global criterion
球状子系 spheric(al)component
球状组织 spheroidal structure
球锥滚柱轴承 spherangular roller bearing
球锥函数 sphero-conic(al)harmonics
球锥剖面 spheric(al)conic(al)section
球锥坐标 sphero-conic(al)coordinate
球组构 spheric(al)fabric
球钻 round bur
球坐标分解器 ball resolver;spheric(al)resolver
球坐标系统 global coordinate system
球座 ball seat(ing);ball socket
球座衬垫 ball seat sealing ring
球座阀 valve with ball seat
球座螺母 spheric(al)seat nut
球座式 spherically seated
球座弹簧 ball seat spring
球座罩 ball shell
球座支承 ball bearing

琉基苯并噻唑 mercaptobenzothiazole

裘布意承压水井公式 Dupuit confined well formula

裘布意方程 Dupuit equation
裘布意-弗舍伊默假定 <地下水力学> Dupuit-Forchheimer assumption
裘布意公式 Dupuit's equation
裘布意关系式 Dupuit relation
裘布意假设 Dupuit assumption
裘布意潜水井公式 Dupuit phreatic water well formula
裘布意潜水平面流公式 Dupuit planar flow formula of phreatic water
裘布意微分方程 Dupuit differential equation
裘皮 furskin

区保留 block reservation

区变量 area variable
区标 block mark;trim
区标格式 zoned format
区标志 distinctive emblem
区标准时 zone standard time
区别 contradistinction;difference;differentiate;differentiation;discriminate;discrimination
区别标记 distinctive mark
区别不同土质 selective digging
区别出 pick out
区别对待 differential treatment
区别对待的 discriminatory
区别对待的价格 price discrimination
区别对待性关税 discriminating duty
区别反应 distinguishing reaction
区别机 selector-repeater
区别记号 diacritical sign
区别铃声 distinctive ringing
区别旗 distinguishing flag
区别时序 distinguishing sequence
区别试验 distinguishing test
区别树 distinguishing tree
区别吸收 differential absorption
区别显著的颜色 distinctive colo(u)r;distinguishing colo(u)rs
区别信号 distinguishing signal
区别性 distinctiveness
区别性的 distinctive
区别性强化 differential reinforcement
区别性特征 distinctive feature
区别种类 identifying species
区别状态 distinguishing state
区层取样器 zone sampler
区长途电话局 zone center[centre] for long distance call
区城合作税收制度 regional cooperation tax system
区城稳定条件 stability conditions of region
区抽水站 district pump station
区代码 area code
区带电泳 zone electrophoresis
区带离心法 zonal centrifugation
区道 district road
区的 regional
区的移动 zone movement
区地址 regional address
区调度所 district control office
区调图幅【地】sheet of regional geologic(al)reconnaissance
区段 block;link of levels;portion;section of level(l)ing;sector;segmentation;district;zone;part of deposit <指矿段>
区段保护 segment protection
区段边界线 <楼群> lot line
区段标记 sector mark;segment mark
区段标识符 sector marker
区段标志 sector marker
区段长度 block length
区段长度违章标志 segment length violation flag
区段出清 clearing of section
区段穿孔 zone punch
区段道班 section gang
区段地址方式 sector address system
区段电话装置 division communication system
区段调度员 district traffic controller;section controller
区段断路器 section circuit breaker
区段二 section two;district 2
区段方式 sector mode;segmented mode
区段格式 sector format;zone format
区段工长 section(fore)man
区段工程师 district engineer;division engineer;section engineer
区段管理策略 segment management policy
区段管内车辆输送计划 local cars dispatching plan in district
区段灌注 impregnation of zone
区段号 segment number
区段荷载 load on section
区段或部分的移交 taking-over of sections or parts
区段货物列车 depot-to-depot through goods train
区段基地址 segment base address
区段寄存器 sector register
区段间运输 zone service
区段监督盘 section supervising board
区段监视盘 section supervising board
区段校核继电器 section check relay
区段界限 segment limit
区段空闲 section unoccupied
区段里程折半计算法 method of compacting wagon kilometres at halt of the length of district
区段链接 sector chaining
区段列车 district train;local train;sectional train
区段名 field name
区段内甩挂作业安排 arrangements for attaching and detaching of wagon within the district
区段排队 sector queuing
区段偏移量地址 segment offset address
区段票 district passenger ticket
区段平均货物运输密度 average cargo transportation density in a section
区段平均速率 space mean speed
区段平面图 district plan
区段入口 segment entry
区段十进制 zoned decimal
区段式处理机 segmented processor
区段式计算机 segmented computer
区段数 sector number;zone digit
区段顺序 sector sequence
区段锁闭 section locking
区段锁闭防护法 section locking protection
区段特征 segment attribute
区段特征字段 segment attribute field
区段通信[讯] district communication
区段位 section bit;zone bit
区段线路工长 division road master
区段小运转列车 district transfer train
区段信号 block signal
区段序列 sector sequence
区段遥控 sectional traffic remote control
区段遥信 remote surveillance for section;sectional remote signal(l)ing
区段一 district 1;section one
区段允许速度 permissive speed in district
区段运价率 tariff rate of railway section
区段摘挂车辆 <美> short load
区段摘挂列车 district local train
区段占用表示 section occupancy indication
区段站 depot station;sectional station;locomotive terminal【铁】;district station <机务段所在站>
区段直通列车 district through train
区段中间的公共汽车站 mid-block bus stop
区段中心 sectional center[centre]
区段柱 section pillar
区房管局 district housing management bureau
区分 decollate;delimitation;differentiate;differentiation;discrimination;division;jig;parting;partition(ing);plot;repartition;secern;separation;specification;zoning
区分标记 separator
区分部分 specification part
区分的 dividing;divisional
区分地下径流 <在过程线上的> separation of groundwater flow
区分电平 pedestal level
区分符 specificator;specifier
区分结构 specification configuration
区分类 region class
区分能力 separating capacity
区分频率 cross-over frequency
区分嵌合体 sectorial chimera
区分系统 compartment system
区分效应 differentiating effect
区分性溶剂 differentiating solvent
区分用户区域 partition user
区分语句 specification statement
区分阈值 distinctive threshold value
区格点 panel point
区格式照明 panel lighting
区隔市场 market segment;segment the market
区公路 district highway
区公所 ward office
区公园 district park;regional park
区供热厂 district heating plant
区供热管道 district heating duct;district heating line
区号 zone description
区号标志管理员 block post keeper;tower man
区划 block plan;compartment;delimitation;demarcation;regionalism;regionalization;zoning
区划单位 enumeration unit
区划灯 division lamp
区划地块的界线 front line of a zone lot
区划地图 regionalization map
区划法 block system
区划法规 zoning ordinance
区划分析 compartment analysis
区划分线 <交通调查用> cordon
区划副线 minor ride
区划化 regionalization
区划平面图 zoning plan
区划圈 zone circle
区划申诉委员会 Board of Zoning Appeals
区划市区 zoning district
区划授权 zoning by right
区划图 block diagram;block plan;plot plan;regional plan;zonation map
区划土地 plot
区划外 overzoning
区划网 network of accessory frame
区划线 layout line;section line
区划行政官 zoning administrator
区划修正 zoning amendment
区划许可证 zoning permit
区划掩蔽所 zone pen
区划中的相互制约关系 locational interdependence
区划住房法规 zoned housing code
区划准则 zonation criterion
区际交通 inter-zonal traffic
区际交通流量 inter-zonal flow
区际贸易 inter-regional trade
区间 interval;siding-to-siding block;space interval;section【铁】

区间闭塞 end of siding to end of siding block; section blocked; station-to-station block; track block
区间闭塞信号法 track block signal(l)ing
区间变异 between-state variance
区间标尺 intervaled scale
区间岔线 intermediate siding; lay by(e); outlying siding; siding within a block section; siding with section
区间岔线凭证锁闭器 token siding lock
区间长度 section length; siding-to-siding block length
区间超时 interval time-out
区间车 interzonal vehicle; shuttle bus
区间车速 overall speed; overall travel speed; sectional travel(l)ing speed; shuttle-bus speed; shuttle train speed
区间车行时间 overall travel time
区间尺度 interval scale
区间抽样理论 interval sampling theory
区间出清 clearing of section; section clear
区间出行 inter-regional trip; interzonal trip; interzone trip
区间道岔 turnout in a block section
区间道岔防护信号机【铁】 distant switch signal
区间的 interzonal
区间的长 length of an interval
区间的末端 extremity of an interval
区间电话 track-side telephone
区间定时器 interval timer
区间定时询问计时器 interval polling timer
区间方式 interval mode
区间分半法 interval halving
区间分半检索 half-interval search
区间分割技术 interval dividing technique
区间分析 interval analysis
区间封锁 section cleared up; section closed up
区间服务值 interval service value
区间公共汽车 shuttle; shuttle bus
区间沟 quarter drain
区间估计 estimate by an interval; estimation of interval; interval estimate; interval estimation
区间估计量 interval estimate
区间估值 interval estimate
区间轨道 section track
区间函数 interval function
区间荷载 load on section
区间积分 interval integral
区间集散 feeder
区间集散港 feeder port
区间集散运输 feeder service
区间计量法 interval measurement
区间计时器 interval timer
区间间距 interzone spacing
区间监督继电器 block supervisory relay
区间交叉(口) midpoint crossing
区间交通 inter-zonal traffic; interzone traffic; zone-to-zone traffic; zone-to-zone travel
区间交通流量 inter-zonal flow
区间解 interval solutions
区间金属线槽安装 tunnel lighting metal pipe
区间紧急照明灯 tunnel emergency light
区间径流 local runoff
区间距离 zone distance
区间开通 section clear
区间客流负荷 linkload; passenger link load
区间空闲 section cleared; section unoccupied
区间馈电线 sectional feeder
区间馈线柜 sectional feeder cabinet
区间来水 intermediate inflow; intervening area inflow; local inflow
区间联络开关屏 section coupler panel
区间联系电路 connecting circuit with block signal(l)ing
区间列车 local train; shuttle train
区间流入 local inflow
区间绿灯定时推进式信号联动系统 transit system
区间轮询计时器 interval polling timer
区间面积 interbasin area; intermediate basin; intervening area; local area
区间排水沟 quarter drain
区间迁移 interval migration
区间清算资金 interdistrict settlement fund
区间入流 inflow into reach; intermediate inflow; intervening (area) inflow; local inflow
区间施工作业安排 construction plan in section
区间收缩 interval contraction
区间疏远业务 feeder service
区间数 interval number
区间水量 local inflow
区间四则运算 interval arithmetic
区间速度 overall speed
区间隧道 running tunnel
区间隧道通风系统 ventilation system for sectional tunnel
区间隧道维修配电箱 tunnel maintenance sub-distribution box
区间锁闭 detector locking; section locking
区间套 nested intervals; nested sequence of intervals
区间套序列 nested sequence of intervals
区间调焦 zone focusing
区间通道 crosshead
区间通过能力 carrying capacity of a section; carrying capacity of the block section
区间通话 junction call
区间图 interval graph; zoning plan
区间推定 interval estimation
区间拓扑 interval topology
区间线性规划 interval linear programming
区间信号点 wayside signal location
区间信号 track-side signal; section signal; wayside signalling
区间行程 interzone trip
区间性网络 regional computer network
区间引航员 branch pilot
区间映射 interval mapping
区间预测 interval forecast; interval prediction
区间运输路线 feeder route
区间运算 interval arithmetic
区间运行时间 running time in the section
区间占用 block occupancy; section occupied
区间占用表示 block occupancy indication; section occupancy indication
区间照查闭塞【铁】 sectional check block system
区间照明灯 section light
区间照明放线 section lighting wiring
区间照明配电箱 section lighting distribution box
区间照明配管 section lighting piping
区间折半 interval halving
区间支架 <架空索道的> intermediate mast
区间值 interval value
区间值班员 section controller
区间值函数 interval valued function
区间值扩展 interval valued extension
区间值向量函数 interval vector valued function
区间制动器 spacing rail brake
区间装卸车计划 program(me) of loading and unloading operation in section
区间阻截信号【交】 block signal
区间阻塞 blocked section; section congestion
区间最大通过能力 maximum carrying capacity in the section in the pairs of trains or number of trains
区交换 area exchange
区接线器 district connector
区结束 end of extent
区结束符 end of extent
区截机 block apparatus
区截运行 block movement
区截指示器 block indicator
区截装置 blocking device
区界 county boundary; zone boundary
区界调查 contour survey
区界交通出入量调查 traffic cordon count
区局局长 <美国陆军工程师团> division engineer
区距 offset
区力 field forces
区立学校 district school
区路 district road
区名 realm name
区牧师聚会厅 synodal hall
区内拆箱 terminal devanning
区内出线 outgoing rural line
区内出行 intrazonal trip; intra-zone trip
区内电力网 regional network
区内调运 <集装箱的> terminal transit
区内辅助性调节 area supplementary control
区内辅助性控制 area supplementary control
区内装箱 <集装箱的> terminal vanning
区内交通 intra-zone traffic
区内选呼电话 local line selective call telephone
区内自动电话局 community automatic exchange
区气象站 regional weather office
区熔 zone melt(ing)
区熔法装置 zone melting apparatus
区熔技术 zone melting technique
区熔空段法 zone-void process
区熔模拟计算机 zone-melting analog computer
区熔色谱法 zone melting chromatography
区熔生长法 zone melting growth
区熔提纯 zone purification; zone-refine
区熔提纯法 floating zone refining method
区熔液体 zone fluid
区时 zone time
区时 12 点 standard noon
区时制 zone time system
区识别 zone identification
区锁国 zone-locked state
区头向量 display
区外交换服务 foreign exchange service
区位 location; zone bit
区位分析 locational analysis
区位格式 zoned format
区位科学 regional science
区位理论 location(al) theory
区位三角形 locational triangle
区位商(数) locational quotient
区位系数 location quotient
区位需求 locational requirement
区位一体化 locational integration
区位因素 factor of location; locational factor
区系 fauna[复 faunas/faunae]; flora
区系动物地理学 faunal zoogeography
区县学校 district school
区域 region; district; domain; realm; territory; zone
区域半宽度 peak width at half-height
区域保护 locality protection
区域报告 regional forecast
区域报警器 local fire alarm control panel
区域背景 background of region; regional background
区域背斜 regional anticline
区域泵房 local pumping station
区域泵站 local pumping station; regional pumping station
区域比例系数 local scaling factor
区域避雷器 sectional lightning arrestor
区域边界 zone boundary
区域边界角 regional corner; section corner
区域编码 regional code; regional coding
区域变电所 areal power substation
区域变量 area variable
区域变质岩 regional metamorphic rock
区域变质岩类 regional metamorphic rocks
区域变质作用【地】 regional metamorphism; normal metamorphism
区域标识符 area identification
区域标识码 regional identify code
区域标志器 zone marker
区域标准 regional standard
区域表 region list
区域表征 area attribute
区域波束 zone beam
区域波束反射器 zone reflector
区域波束覆盖范围 zone coverage
区域波束接收机 zone receiver
区域剥蚀 areal degradation
区域补偿 regional compensation
区域不整合 regional unconformity
区域材积表 regional volume table
区域采暖 district heating; zone heat
区域参数 region parameter
区域操作员 domain operator
区域操作装配区 region job pack area
区域测量 area survey; regional survey; town-site survey
区域层型 limitotype
区域差距 areal differentiation
区域差异 area(l) differentiation; spatial variation
区域产量值 area yield value
区域长度 zone length
区域超覆 regional overstep
区域超周期 regional supercycle
区域沉淀法 zone precipitation
区域沉积构造【地】 zone sedimentary tectonics
区域沉降 zone settlement; zone settling
区域沉降量 area precipitation
区域沉降速度 zone settling velocity
区域沉降速率 zone settlement rate;

zone settling rate
区域成矿学 regional metallogeny
区域成矿预测图 map showing regional mineralization prediction
区域成矿远景图 map showing regional mineralization expectation
区域成煤预测图 map showing regional coal-forming prediction
区域成煤远景图 map showing regional coal-forming expectation
区域成型 drape forming
区域城市规划 regional town planning
区域乘数 regional multiplier
区域冲积海岸 regional alluvial coast
区域重新发展规划 area redevelopment program(me)
区域抽取 region extraction
区域抽样 block sample
区域抽样法 area sampling;territorial sampling;zonal sampling
区域传输法 zone transport method
区域纯化 zone refining
区域磁异常 regional magnetic anomaly
区域磁异常图 regional magnetic anomaly chart
区域代表站 regional representative station
区域代码 area code
区域单位 area unit;enumeration unit;regional unit
区域导航 area navigation
区域导航计算机 area navigation computer
区域导航站 area navigation station
区域道路网 regional road network
区域的 sectional;zonary;zoned
区域地层 regional stratum
区域地层学 regional stratigraphy
区域地壳稳定问题 problem of stability of regional earth's crust
区域地壳稳定性 stability of regional crust
区域地壳稳定性分级 grade of stability of regional crust
区域地壳稳定性评价 evaluation of stability of regional crust
区域地块分界线 boundary line of zone lot
区域地理学 regional geography
区域地貌学 physiography;regional geomorphology
区域地球化学 regional geochemistry
区域地球化学测量 regional geochemical survey
区域地球化学调查 regional geochemical survey
区域地球化学调查性质 character of regional geochemistry survey
区域地球化学异常 regional anomaly of geochemistry
区域地球化学作用 regional geochemical process
区域地球物理调查 regional geophysical survey
区域地球物理调查方法 method of regional geophysical survey
区域地球物理调查类型 type of regional geophysical survey
区域地球物理勘探 regional geophysical prospecting
区域地球物理异常 regional geophysical anomaly
区域地球物理异常评价 assessment of regional geophysical anomaly
区域地图 regional map;chorography
区域地图集 regional atlas
区域地下热流 regional subsurface heat flow
区域地下水 local groundwater;local ground watering
区域地下水调查 local ground water survey
区域地下水流系统 regional ground water system
区域地震 regional earthquake
区域地震地层学研究 study of regional seismic stratigraphy
区域地震地质 regional-seismic geology
区域地震活动性 regional seismicity
区域地震台网 regional seismic network
区域地震危险性 regional seismic risk
区域地震系数 regional seismic coefficient
区域地震学 regional seismology
区域地震震级 local magnitude
区域地址 regional address
区域地质 regional geology;geologic(al) province
区域地质调查 regional geologic(al) survey(ing)
区域地质调查报告 report of regional geologic(al) survey
区域地质调查报告附本 appendixes of regional geologic(al) survey report
区域地质调查比例尺 scale of regional geologic(al) survey
区域地质调查成果 results of regional geologic(al) survey
区域地质调查方法 methods of regional geologic(al) survey
区域地质调查项目 project of regional geologic(al) survey
区域地质调查性质 feature of regional geologic(al) survey
区域地质构造 areal geologic(al) structure;tectonics
区域地质构造环境 regional tectonic setting
区域地质填图 regional geologic(al) mapping
区域地质图 areal geologic(al) map;regional geologic(al) map;regional map
区域地质学 area(l) geology;regional geology
区域电厂 supercentral station
区域电话局 area exchange
区域电话制 district telephone system
区域电话中心局 regional center[centre]
区域电信枢纽 regional telecommunication hub
区域电站热力系统 central-station cycle
区域调查 area research;regional survey
区域定位 zone location
区域动力变质作用 regional dynamometamorphism
区域动热变质作用 regional dynamothermal metamorphism
区域断陷线 regional lineament
区域对比 regional correlation
区域发展 regional development
区域法 area method;field method
区域反射 regional reflex;segmental reflex
区域范围 area coverage
区域方式覆盖 local-mode coverage
区域防空 area defense[defence]
区域分布 area distribution
区域分割 region-dependent segmentation
区域分隔带 area separator
区域分配图 area allocation diagram
区域分析 regional analysis
区域分析信息 area analysis information
区域分组 territorial classification
区域赋值 area assignment
区域覆盖波束 zone coverage beam
区域覆盖天线 spot beam antenna
区域改正 chart amendment patch;chartlet;block correction <指海图修正贴纸>
区域概查阶段 regional gross survey stage
区域概化 regional generalization
区域干路 regional distributor
区域钢化玻璃 partially tempered glass;zoned tempered glass
区域港 regional harbo(u)r
区域工程地质图 regional engineering geologic(al) map
区域工程地质图上岩层分级 rock formation grade in engineering geology map
区域工程地质学 regional engineering geology
区域工程地质因素 factors of regional engineering geology
区域工程地质因素分带性 zonality of the factors of regional engineering geology
区域公路 district highway
区域公用事业 regional utility
区域供电 block supply
区域供电临时改进措施 area assist action
区域供冷 district cooling
区域供冷和供暖 district cooling and heating
区域供暖 area heating;district heating;zone heat
区域供暖锅炉房 district heating plant
区域供暖系统 district heating system
区域供热 district heat supply;district thermal heating
区域供热供冷系统 district heating and cooling system
区域供热锅炉房 regional heating plant
区域供热系统 district heating system
区域供水 regional water supply
区域供水系统 regional water supply system
区域构造【地】 areal structure;formation of country;zonal structure
区域构造单元名称 name of regional tectonic unit
区域构造地质学 regional tectonics
区域构造纲要图 outline map of regional structure
区域构造特征 feature of regional tectonics
区域构造图 regional structure map
区域构造线方向 lineational orientation
区域构造研究 study of regional structure
区域构造应力场 regional tectonic stress field
区域构造运动 regional tectonic movement
区域估计 region estimation
区域管辖 territorial jurisdiction
区域规定 zoning
区域规划 region(al) plan(ning);territory plan(ning)
区域规划方案 regional planning program(me)
区域规划阶段 regional planning stage
区域规划控制 zoning control
区域规划委员会 regional planning commission
区域规划执照 zoning permit
区域规模 regional scale
区域锅炉房 district boiler room;regional boiler plant
区域锅炉房供热系统 heat-supply system based upon heating plant
区域海洋 regional sea
区域海洋学 areal oceanography;regional oceanography
区域航空测量 area aerial survey
区域航空图 <比例尺 1:50 万> sectional aeronautical chart;local aeronautical chart;local chart;regional air chart
区域合作开发组织 Regional Cooperation Organization for Development
区域和堆场排水 area and yard drains
区域河道管理局 regional water authority
区域洪水预报 regional flood forecast(ing)
区域洪水预测 regional flood forecast(ing)
区域洪水预估 regional flood forecast(ing)
区域互连 regional interconnection
区域化 regionalization
区域化变量 regionalized variable
区域化变量理论 theory of regionalized variables
区域划定 regional assignment
区域划分 regionalism;regionalization;zoning
区域划分图 zoning plan
区域环境 regional environment;regional setting
区域环境规划 regional environmental planning
区域环境基线值 baseline value of regional environment
区域环境监测 regional environmental monitoring
区域环境评价 regional environmental assessment
区域环境噪声 regional environmental noise
区域混合岩化作用 regional metamorphism
区域火力发电站 local thermal power plant
区域火山地质 regional volcanic geology
区域火山活动概况 summary of volcanic activities
区域火山组合 regional volcanic complex
区域火灾报警箱 extension fire alarm box
区域基本天气观测站网 regional basic synoptic network
区域基础设施 regional infrastructure
区域级 region class
区域集 range set
区域集中供热 district heating;regional heating
区域给水 regional water supply
区域给水和废水项目 regional water supply and waste project
区域给水系统 regional water supply system
区域剂量测定法 area dosimetry
区域价格政策 spatial pricing policy
区域假定 regional hypothesis
区域间的 inter-regional
区域间公路 inter-regional highway
区域间合作 inter-regional cooperation
区域间呼叫 interzone call
区域监测 area monitoring
区域监察器 area monitor
区域检查 range check(ing)
区域检定所 regional verification office
区域检索 area retrieval;area search

Q

区域简化 regional generalization
区域降水量 area(l) precipitation; regional precipitation
区域降雨量 area(l) precipitation; regional precipitation
区域交代作用 regional metasomatism
区域交换中心 zone switch(ing) center[centre]
区域交通观测 area-wide count
区域交通管理系统 area traffic management system
区域交通控制 <电脑分区指挥城市交通> area traffic control
区域交通控制系统 area traffic control system
区域校核 range check
区域校正 block correction; regional correction
区域结构 regional construction
区域界限 area boundary; district board; district border
区域进近管制中心 area approach center[centre]
区域经济结构 regional economic structure
区域经济协调发展 coordinated development of regional economy
区域经济学 regional economics
区域经济影响 regional economic impact
区域精炼 zone purification; zone-refine; zoning
区域精炼法 zone melt(ing); zone refining
区域精炼金属 zone melted metal
区域精炼炉 zone refiner
区域精制 zone purification; zone refining
区域净化 zone purification
区域径流系数 regional runoff coefficient
区域境界线 area delimiting line; cordon
区域静校正 region static correction
区域聚焦 zone focusing
区域均衡 regional isostasy
区域均衡异常 regional isostatic anomaly
区域均化 zone level(l)ing
区域均化单晶 zone leveled single crystal
区域开发 regional development
区域开发计划 regional development plan
区域开发银行 regional development bank
区域开关 section switch
区域勘测 areal study; area reconnaissance; regional reconnaissance; regional survey
区域勘察 areal study; area reconnaissance; regional reconnaissance; regional survey
区域勘探 areal study; area reconnaissance; regional reconnaissance; regional survey
区域科学 regional science
区域空调器 zone air conditioner
区域空段法 zone-void method
区域空段提纯器 zone-void refiner
区域控制 area control; areawide control; regional control; zone control
区域控制机 zone control unit
区域控制任务 region control task
区域控制员 area controller
区域控制站 district control station
区域控制装置 zone control unit
区域宽度 peak width
区域宽化 zone broadening
区域矿产远景评价 prospect evaluation of regional mineral resource
区域扩展 zone spreading
区域拦阻射击 sector barrage
区域类别 area classification
区域烈度 regional seismicity
区域领航 regional air navigation
区域滤光器 zonal filter
区域码 region code
区域煤田预测图 map showing regional coalfield prediction
区域描述体 realm description entry
区域名 area-name
区域目标 area target
区域内的 intradistrict
区域凝固 zone freezing
区域排水系统 regional sewerage system
区域配电网 regional distribution network
区域配额制 regional quota
区域平衡方程 free body balance equation
区域平滑技术 <一种回归分析方法> local smoothing technique
区域平均法 zone level(l)ing
区域平均温度 average regional temperature
区域平均压力 areal average pressure
区域平面示意图 pictorial directory
区域平面图 regional plan
区域平整法 zone level(l)ing
区域评估 mass appraising
区域评价 area assessment; regional assessment
区域坡度 regional slope
区域坡面沉积 regional slope deposit
区域普查阶段 regional survey stage
区域气候学 regional climatology
区域气象监视 area meteorological watch
区域气象站 regional weather office
区域气象中心 regional meteorologic(al) center[centre]; regional meteorologic(al) office
区域迁移 zone migration
区域强化 local tempering
区域倾斜 normal dip; regional dip
区域群落 local community; regional community
区域燃烧 zonal combustion
区域扰动 regional disturbance; zone perturbation
区域热力站 regional thermal substation; branch line thermal substation
区域热流量 regional heat flow
区域热水供应 district heat water supply
区域热源 regional heat source
区域熔化法 zone melt(ing)
区域熔化技术 zone melting technique
区域熔炼 zone refining
区域熔炼器 zone refiner
区域熔融 zone fusion; zone melt(ing); zone pass
区域熔融法成长 zone melting growth
区域熔融结晶 zone melting crystallization
区域熔融提纯 zone melting purification
区域扫掠 sector scan(ning)
区域扫描 sector scan(ning); zone scanning
区域烧结 zone sintering
区域设施 regional facility
区域设置 zone-setting
区域射线 zonal ray
区域深部构造背景 regional deep tectonic background
区域深层含水层系统 deep regional aquifer system
区域剩余校正值 regional-residual correction value
区域时间 zone time
区域识别码 area identifying number
区域市场 spot market
区域式十进数格式 zoned decimal number format
区域试验 regional trial
区域试验方案 regional testing program(me)
区域输运法 zone transport method
区域输运提纯器 zone transport refiner
区域属性 area attribute
区域数据 area data
区域水 regional water
区域水化学异常 regional hydrochemical abnormality
区域水量平衡 regional water balance; regional water budget
区域水平 regional level
区域水文地质调查 regional hydrogeologic(al) investigation
区域水文地质化学 regional hydrogeochemistry
区域水文地质历史法 regional hydrogeology history method
区域水文地质剖面图 regional hydrogeologic(al) profile
区域水文地质图 regional hydrogeologic(al) map
区域水文地质学 regional hydrogeology
区域水文学 regional hydrology
区域水文循环 regional hydrologic(al) cycle
区域水文预报 basic hydrologic(al) forecast(ing)
区域水系 regional water system
区域搜索 zonal scan
区域速度 zone velocity
区域台阵 regional array
区域探查 regional search
区域探索 area search
区域套 nested domain; nested region
区域特征 provincial characteristic
区域梯度 regional gradient
区域提纯 zone purification
区域提纯材料 zone-refined material
区域提纯法 zone refining
区域提纯金属 zone-purified metal
区域提纯器 zone refiner
区域提纯设备和装置 zone refining unit and equipment
区域提纯装置 zone purification device
区域填冲法 area method of landfill
区域填图单位划分 division of units in regional mapping
区域条件 area condition
区域调焦 zone focus
区域调节器 zone actuator
区域通风 zonal ventilation
区域通过 zone passage
区域同位素平衡 regional isotope equilibrium
区域统计图法 dasymetric technique
区域图 areal map; regional map; zone diagram
区域图案 zone map
区域土 zonal soil
区域土质学 regional geotechnique
区域网【测】 block
区域网布点 control point distribution for block aerotriangulation; point layout for aerial triangulation blocks
区域网法 block triangulation
区域网航空三角测量 block aero triangulation
区域网解析空中三角测量 block analytic(al) aerial triangulation
区域网空中三角测量 block aerial triangulation
区域网络 local area network
区域网络系统 local area network system
区域网平差 area adjustment; block adjustment
区域位 zone bit
区域位置指示器 zone position indicator
区域温差 space temperature variation
区域稳定性程度评价 evaluating for regional steady degree
区域污染 area(l) pollution
区域污染模式 regional pollution model
区域污染源 area emission sources; regional pollution sources; area sources
区域污水处理 regional sewage disposal; regional sewage treatment
区域污水处理设施 regional sewage treatment facility
区域污水处置 regional sewage disposal
区域污水系统 regional sewage system
区域无线电信标 Consolan; regional wireless beacon
区域物化探资料 data of regional geochemistry and geophysics
区域系数 coefficient of region; zonal coefficient; zoning factor
区域系统 district system
区域现象 zone phenomenon
区域线 regional wire
区域线性构造 regional lineament structure
区域相对优势 comparative advantage of a region
区域像片 area picture; regional picture
区域消毒 locality disinfection
区域消去 region elimination
区域消元法 block elimination
区域效应 area effect
区域形式 area format
区域型分布 regional type distribution
区域型组织 organization by location
区域性安排 regional arrangement
区域性编组站【铁】 regional marshalling station
区域性变化 areal change; areal variation; regional change; regional variation
区域性波道 regional channel
区域性补偿计划 regional compensation scheme
区域性产品 regional products
区域性沉降 regional settlement
区域性冲断层接触 regional thrust contact
区域性储存中心 regional distribution center[centre]
区域性大断裂 geofracture
区域性道路 district road; regional road
区域性的 regional; zonal
区域性的经济合作 regional economic cooperation
区域性地面沉降 general area settlement of ground surface
区域性地面倾斜 regional slope
区域性地球化学异常 areal geochemical anomaly; regional geochemical anomaly
区域性地下水位下降 descend of regional groundwater level; descent of regional ground water table; regional fall of underground water level

区域性地震一览表 earthquake s-regional catalogue
区域性地质 areal geology
区域性地质-水文地质调查 regional geology-hydrogeologic(al) survey
区域性地质图 areal geologic al map
区域性调查 regional research
区域性定价 zone pricing
区域性发电厂 regional power station
区域性发展战略 regional development strategy
区域性防油污中心 The Regional Oil Combating Center[Centre]
区域性分布 regional distribution
区域性分流道路 district distributor; regional distributor
区域性分流街道 regional distributor
区域性风向 local wind direction
区域性港口 regional harbo(u)r; regional port
区域性隔水层 regional aquifuge
区域性公路 provincial highway; regional highway
区域性公约 regional convention
区域性供热站 district-heating station
区域性购物中心 regional shopping center[centre]
区域性广播 regional broadcast
区域性海湾水污染 regional water pollution in the Gulf
区域性航图 regional aeronautic(al) chart
区域性合作 regional cooperation
区域性洪水 regional flood
区域性呼叫 geographic(al) area call
区域性环境监测传感器 place environment(al) monitoring sensor
区域性环境异常 regional environmental anomaly
区域性混合岩化 regional migmatization
区域性价格 zone price
区域性检验 regional test
区域性经济一体化 regional economic integration
区域性开发战略 regional development strategy
区域性扩充设施 extended area service
区域性喷发 areal eruption
区域性劈理 regional cleavage
区域性品牌 local brand
区域性剖面 regional profile
区域性气候 regional climate
区域性迁移 regional migration
区域性人口特征 regional demographic characteristic
区域性商业中心 regional shopping center[centre]
区域性生物群 regional biota
区域性市场 regional market
区域性试验 regional test
区域性输送(分)线 regional distributor; district distributor
区域性水文地质调查 regional hydrogeologic(al) survey
区域性水准测量 regional level(l)ing survey
区域性通道 regional channel
区域性图件 regional maps
区域性土(壤) regional soil; zonal soil; zone soil
区域性污染 regional pollution
区域性信息 regional information
区域性修正系数 regional correction factor
区域性岩层倾斜 regional slope
区域性异常 regional anomaly
区域性因素 regional factor
区域性有轨交通 regional rail transit
区域性雨量变化 areal variation in rainfall
区域性运动 regional movement
区域性运输 local transportation
区域性证券交易所 regional securities exchange
区域性专业 regional specialization
区域性走向 regional trend
区域选择 sector selection
区域选择器 discriminating selector; district selector; zone selector
区域选择性 region selectivity
区域选择性保护系统 discriminating protective system
区域选择性反应 region selective reaction
区域岩石学 regional petrology
区域岩土力学 regional geotechnique
区域研究 areal study
区域研究程度图 map showing regional research level
区域样本 area sample
区域遥感图像及遥感地质资料索引 index of regional remotely sensed images and geologic(al) results
区域遥控 remote control of an area
区域夷平 zone level(l)ing
区域异常下限 regional threshold
区域银行 district bank
区域应变 area strain; region strain
区域应力 regional stress
区域应力方位 orientation of regional stress
区域应力性质 character of regional stress
区域影响分析 regional impact analysis
区域硬化合金 zone-hardened alloy
区域预报 area(l) forecast(ing); regional forecast
区域预报系统 area forecast system
区域预报中心 area forecast center [centre]
区域预测 regional prediction
区域元 zone bit
区域约束 range constraint
区域匀化 zone-leveled
区域匀化器 zone leveler
区域运输规划 regional transportation planning
区域运输计划 regional transportation planning
区域运行程序 domain operator
区域再结晶法 zone recrystallization method
区域站 regional station
区域照明 area illumination
区域照片 area picture
区域遮没 zone blanking; zone obstruction
区域蒸发 regional evapo(u)ration
区域蒸汽供应 district steam supply
区域正常化 area normalization
区域植被 zonal vegetation
区域指点标 zone marker beacon
区域制 zone system; zoning
区域致匀 zone level(l)ing
区域中心 regional center[centre]
区域中心城市 regional hub city
区域重力 regional gravity
区域重力测量 regional gravity survey
区域重力异常 regional gravity anomaly
区域重力异常图 regional gravity anomaly map
区域重心 zone centroid
区域周期 regional cycle
区域主干断裂及强震带分布图 distribution of the regional major faults and pleistoseismic zones
区域专业气象中心 regional specialty meteorological center[centre]
区域转移 areal transfer
区域状况 area condition
区域准周期 regional paracycle
区域资源调查 regional resource inventory; regional resource investigation
区域自然公园 regional natural park
区域综合 regional generalization; regional synthesis
区域总平面图 general location plan
区域走向 regional strike
区域阻滞 regional block
区域作图机 area composition machine
区长 rector; warden; chief executive, district government
区镇居民 town dweller
区镇所在地 <美> townsite
区政府 district government
区中心 district center[centre]
区转换中心 zone switch(ing) center [centre]
区字号 zone letter
区组 block
区组化 blocking
区组间分析 interblock analysis
区组内信息 within block information
区组设计 block design
区组数 block number
区组预选器 group line switch

曲

曲孔 polyconcave pore

曲板 bent plate; curved slab; curve plate
曲板检验器 curve tester
曲板屋顶 curved plank roof
曲棒绞盘 crank capstan
曲背手锯 skewback saw
曲壁 curved wall
曲壁效应 curved wall effect
曲臂 crank arm; crank radius; crank web; radius bar
曲臂夹紧螺栓 maneton
曲臂开关器 <窗的> crank arm operator
曲臂压砖机 toggle press
曲边机 collaring machine
曲边三角形 curved triangle
曲边四边形 curved quadrilateral
曲别针 clip
曲柄 angle crank; crank; crank lever; hooked bar; winch
曲柄把手 crank handle
曲柄扳手 brace wrench; cranked wrench
曲柄半径 crank radius; crank throw; radius of crank; throw of crankshaft
曲柄泵 crank pump
曲柄臂 cheek of crank; crank cheek; crank web; crank arm
曲柄布置 crank arrangement
曲柄操纵的压床 crank-operated press
曲柄插床 crank slotter
曲柄车床 crank shank lathe
曲柄传动 crank drive
曲柄传动的 crank-driven
曲柄传动装置 crank(ed) gear
曲柄锤 crank-operated hammer
曲柄带动的 crank-driven
曲柄导槽 crank guidance
曲柄导向槽 crank-guide way
曲柄导向装置 crank guide
曲柄动力机构 crank type power unit
曲柄动作 toggle action
曲柄端换挡 crank side shift
曲柄飞轮泵 crank-and-flywheel pump
曲柄杆 cranked rod
曲柄杠杆 angle lever
曲柄杠杆机构 toggle; toggle mechanism
曲柄杠杆压机 toggle lever press
曲柄杠杆装置 toggle joint; toggle-joint unit; toggle mechanism
曲柄毂 crank boss
曲柄弧口凿 bent gouge
曲柄滑块 crank block
曲柄滑块机构 slide-crank mechanism
曲柄回转 crank up
曲柄回转力(矩) crank effort
曲柄回转速度 <起动时> cranking speed
曲柄机构 crank mechanism
曲柄颊板 crank cheek
曲柄角 angle at which the cranks are fastened
曲柄角度的调整定位 crank-fastening angle
曲柄角间距 crank spacing
曲柄接合板 cranked fish-plate
曲柄颈 crank web
曲柄颈轴承合金 crank bearing metal
曲柄颈轴承铜衬 crank bearing brass
曲柄连杆 crank connecting link; crank link; toggle lever; toggle link
曲柄连杆传动 slot-and-crank drive
曲柄连杆机构 crank connecting rod mechanism
曲柄连杆式启闭机械 operating machinery with crank arm and connecting rod
曲柄连杆式压力机 knuckle press
曲柄连杆支承 crankpin bearing
曲柄连接 coupling crank
曲柄连接板 crank web; web of crank; crankshaft web
曲柄链系 crank chain
曲柄龙门刨床 crankplaner
曲柄轮 crank wheel
曲柄螺旋钻 crank auger
曲柄牧草棍 crook
曲柄牛头刨床 crank shaper
曲柄刨床 crank planing machine
曲柄平衡 crank counter balance
曲柄平衡器 crankshaft balancer
曲柄起动杆 cranking lever
曲柄起动器 crank starter
曲柄起子 offset screwdriver
曲柄驱动的 crankshaft drive
曲柄式冲床 crank type press
曲柄式剪切机 crank shears
曲柄式气动马达 air cranking motor
曲柄式压床 crankshaft press
曲柄式压力机 crankshaft press
曲柄手摇钻 brog
曲柄手钻 breast drill
曲柄顺序 sequence of crank
曲柄缩绒机 crank fulling mill
曲柄镗床 crank boring machine
曲柄头扳手 brace wrench
曲柄头架 crank head
曲柄弯程 crank throw
曲柄箱 crankcase; crank chamber; sump
曲柄箱导门 crank case door
曲柄箱防爆门 explosion-proof crank case
曲柄箱“放炮” crankcase explosion
曲柄箱换气法 crankcase scavenging
曲柄箱加热器 crankcase heater
曲柄箱通气管 crankcase breather
曲柄箱稀薄化 crankcase dilution
曲柄箱用油 crank case oil
曲柄箱用油的稀释 crankcase oil dilution
曲柄箱油槽 engine pit

曲柄销 crank pin;wrist pin
曲柄销车床 crankpin lathe
曲柄销衬套 crankpin bush(ing)
曲柄销衬套合金 crankpin metal
曲柄销垫圈 crankpin collar;crankpin washer
曲柄销滚柱 crankpin roller
曲柄销滑脂 crankpin grease
曲柄销离心润滑 banjo lubrication
曲柄销(轮)座 crankpin wheel seat
曲柄销螺钉 crankpin stud
曲柄销螺母 crankpin nut
曲柄销螺栓 crankpin bolt
曲柄销猫头 crankpin spool;wrist-pin spool
曲柄销磨床 crankpin grinder
曲柄销配置角度 crankpin angle
曲柄销修整工具 crankpin returning tool
曲柄销压机 crankpin press
曲柄销支承 crankpin bearing
曲柄销轴承 crankpin bearing
曲柄销轴瓦 crankpin step
曲柄销座 crankpin seat
曲柄行程 crank throw;throw
曲柄形扒钉 cranked dog
曲柄形的 cranked
曲柄旋削机 crankpin turning machine
曲柄压床 crank press;eccentric crank press;eccentric shaft press
曲柄压机 toggle-joint press;toggle lever press;toggle press
曲柄压力机能力 capacity of crank press
曲柄摇把 crank handle
曲柄摇杆 crank rocker
曲柄摇杆机构 crank and rocker mechanism
曲柄摇块机构 crank-swing block mechanism
曲柄摇手 crank handle
曲柄移位装置 barring gear
曲柄圆 crank circle;crankpin path
曲柄圆盘 crank disc;crank web
曲柄月牙板 crank guide
曲柄运动 crank-motion
曲柄凿 bent chisel;bent gouge
曲柄支托 knee grip
曲柄执手 crank handle
曲柄执行机构 crank type power unit
曲柄制动器 knee brake
曲柄中心 crank center[centre]
曲柄轴 crank axle;crankshaft;pitman shaft;jackshaft < 电力机车 >
曲柄轴承 bearing of the crankpin;crank(shaft) bearing
曲柄轴承合金 crankshaft bearing metal
曲柄轴承护油圈 crank bearing oil seal
曲柄轴承瓦 crank bearing liner
曲柄轴承罩 crank bearing casing
曲柄轴导框联结杆 jackshaft pedestal binder
曲柄轴对准器 crankshaft alignment
曲柄轴横梁 jackshaft cross-tie
曲柄轴护油圈 crankshaft oil seal retainer
曲柄轴环 crankshaft collar
曲柄轴棘爪 crankshaft ratchet
曲柄轴颈 crank journal;throw bearing
曲柄轴平衡块 jackshaft counterweight
曲柄轴前端密封(装置) crankshaft front seal
曲柄轴瓦 crank bearing liner
曲柄轴线 crank axis
曲柄轴谐和平衡器 crankshaft harmonic balancer
曲柄轴轴承 jackshaft bearing
曲柄轴轴颈 crank journal neck;crankpin
曲柄轴钻孔机 quartering machine
曲柄转动角 crank angle
曲柄转角 crank angle
曲柄装置 crank arrangement;crank bearing
曲柄状楼梯梁 cranked string
曲柄钻 belly brace;bit brace;brace drill;breast borer;breast drill;click bore;crank brace
曲柄钻孔器 brace bit
曲柄钻丝锥 crank tap
曲柄钻头 pin tin
曲柄钻钻头 nose bit
曲材 knee
曲槽 cranked slot
曲槽刨 compass rebate plane
曲差 stuffy nose
曲撑杆 butterfly
曲池 hypertension;paralysis of upper extremities
曲尺 bevel ga(u)ge;carpenter's square;jointed rule;square;surveyor's rod;trisquare;try(ing) square;zigzag rule
曲尺楼梯 angle stair(case);L stair(case);quarter-turn stair(case)
曲尺楼梯平台 quarter pace;quarter pace landing;quarter-space landing
曲尺形栈桥 L-headed jetty
曲齿联结轴 < 燃气轮机车 > curved tooth coupling shaft
曲齿耙 curve-lined harrow
曲船首 clipper bow
曲椽 crook rafter;knee piece;knee rafter
曲床 koji bed
曲导程 < 自道岔曲线切点起至辙叉理论尖端止的距离 > curved lead
曲导轨 curved approaching rail;curved closure rail;curved lead rail
曲的 flexuose[flexuous]
曲钉 hooked nail
曲度 camber;curvature;degree of curvature;degree of curve
曲度法 curvature method
曲度规 radius ga(u)ge
曲度计 curtometer;cyrtometer[kyrtometer]
曲度检验 curve test
曲度角 angle of curvature
曲度限制 limitation of curvature
曲度远视 curvature hyperopia
曲断层 curved fault
曲断层面 curved fault surface
曲多面体 curved polyhedron
曲辐 curved spoke
曲辐带轮 curved armed pulley
曲干材 compass timber
曲杆 bent element;bent lever;bent member;bent rod;curved bar;curved rod;curve member;deflecting bar;deflecting rod;hook(ed) bar;knee lever
曲杆臂 toggle-lever arm
曲杆杠 bell crank
曲杆结构 curved member structure
曲杆式擒纵机构 crank lever escapement
曲杆温度计 angle-stem thermometer
曲根材 sabre butt
曲拱的 arched
曲拱石桥 arched stone bridge
曲拐 bell crank;crank;side leg;throws
曲拐扳手柄 speed handle
曲拐板 cranked slab
曲拐臂 radius bar
曲拐机构 reciprocating mechanism
曲拐式搅拌机 sigma mixer;Z-blade mixer
曲拐水阀 bell crank water tap
曲拐弯头 end anchorage of bars
曲拐箱 crank box;crankcase
曲拐销子 < 水阀 > bell crank pin
曲拐形叶片混合机 sigma blade mixer
曲管 angle pipe;bend tube;deflecting pipe;knee pipe
曲管卷 lap of coil
曲管水平仪 bent tubular lever
曲管送料器 elbow feeder
曲管温度表 bent stem thermometer
曲管温度计 curved thermometer
曲规 curved ruler
曲轨滑油 curve grease
曲棍球场 hockey area;hockey ground
曲合拢轨 curved closure rail;curved lead rail
曲滑 bedding plane slip;flexural slip
曲滑褶皱 flexural slip fold
曲化平面 curved surface reduced to horizontal plane
曲簧联轴器 Bibby coupling
曲架 curved frame
曲角墙 canted wall
曲角线 return bead
曲解 distort;misconstruction;misinterpret;perversion;torture;twist
曲晶石 cyrtolite
曲颈管 trap
曲颈夹 retort clamp
曲颈瓶 flask;retort
曲颈甑架 retort holder
曲颈蒸馏器 retort
曲径 caracol(e);labyrinth;maze;meander(ing);tortuous path;winding path;zigzag course
曲径环 labyrinth ring
曲径迷宫 labyrinth
曲径密封 labyrinth seal
曲径瓶 retort
曲径气封压盖 labyrinth gland
曲径汽封 labyrinth seal
曲径汽封片 labyrinth fin
曲径式密封 labyrinth seal
曲径式密封套 labyrinth gland
曲径式密封箱 labyrinth box
曲径式轴封 leak-off-type shaft seal
曲径堰顶 labyrinth crest;labyrinth sill
曲径轴封 labyrinth packing
曲壳硅藻属 Achnanthes
曲块 knee piece;knee rafter
曲拉斯温泉 Thermae of Trajan
曲廊 winding corridor;zigzag corridor;zigzag vernada
曲肋 curved rib
曲肋桁架 curved-rib truss
曲连接杆 bent joint bar
曲联结杆 looped coupling link
曲梁 bending beam;bridge span in curved plan;curved beam;curved girder
曲梁桥 curved beam bridge
曲梁式 curved beam type
曲裂石板 yorky
曲流 meander(curve);meandering;meander line
曲流摆幅 meander amplitude
曲流半径 meander radius
曲流比 meander ratio
曲流壁龛 meander niche
曲流边滩 meander bar
曲流波长 meander wavelength
曲流裁弯 meander cut-off;neck cut-off
曲流裁直 neck cut-off
曲流侧移 swinging of meander
曲流层 current sheet
曲流冲蚀岸坡形成的洞穴 meander niche
曲流冲蚀山崖或河谷形成的标志 meander scar
曲流带长度 meander length
曲流带宽度 meander belt width
曲流的 meandering
曲流的环状河道 meander loop
曲流地带 meander belt;meander zone
曲流陡岸 meander bluff
曲流段 meandering reach
曲流朵体 tongue
曲流幅度 amplitude of meander
曲流河 meandering river;meandering stream;snaking river;snaking stream
曲流河层序 meandering stream sequence
曲流河沉积 meandering stream deposit
曲流河沉积模式 meandering river sedimentation model
曲流河段 meander belt
曲流河谷 meandering valley
曲流痕 meander scar
曲流湖 meander lake;meander scroll
曲流环绕岛 cut-off meander spur;meander core
曲流阶地 meander terrace
曲流颈 meander neck
曲流颈桥 natural bridge
曲流龛 meander niche
曲流宽度 meander width
曲流类型 meander types
曲流率 meander ratio
曲流率半径 radius at bend
曲流内侧坝 meander scroll
曲流内侧沉积 accretionary meander point
曲流内侧沉积埂 meander scroll
曲流内侧泛滥带平原 scroll
曲流内侧尖形沙洲沉积 point bar deposit
曲流内侧沙洲 meander bar;point bar
曲流内侧一连串新月形【地】scroll
曲流平原 meander plain
曲流期 period of meander
曲流迁移 meander migration;migration of meander
曲流沙坝 meander bar;point bar
曲流沙坝层序 point bar sequence
曲流山嘴 meander spur
曲流舌 lobe
曲流舌尖 tail land
曲流深槽 meandering thalweg channel
曲流系数 curvature factor
曲流系统 twisting meander system
曲流下移 swinging of meander
曲流现象 meandering phenomenon
曲流线 meander line
曲流型河床 meander river-bed
曲流要素 meander element
曲流游荡 meander migration;migration of meander
曲流迂回扇 flood plain scroll;meander scroll
曲流运动特征 motion characteristic of turbidity current
曲流指数 index of meandering
曲流作用 meandering
曲路 meander(ing)
曲率 curvature;curvity;degree of curvature;degree of curves;measure of curvature;rate of curvature;rate of curves
曲率半径 bending radius;curvature radius;curve radius;radius at bend;radius of bend;radius of curvature;radius of curvature for turn-off < 飞机场出租汽车车道出

口处的>
曲率半径法 radius of curvature method
曲率比 ratio of curvature
曲率变化 change of curvature;curvature change
曲率标量 curvature scalar
曲率表 radius indicator
曲率不变量 curvature invariant
曲率不大的曲线 sweeping curve
曲率不稳定性 buckling instability
曲率测度 measure of curvature
曲率测量 curvature measurement
曲率测量器 rotameter
曲率差 curvature difference; difference of curvature
曲率常数 buckling constant
曲率场 field of curvature
曲率超高综合设计 curvature-superelevation design
曲率的急剧变化 rapid rate of curvature
曲率辐射 curvature radiation
曲率改正 curvature correction
曲率公式 flection formula
曲率桁架 curved chord truss;curved-rib truss
曲率计 curvemeter; flexometer; flexure meter
曲率角 angle of curvature;curvature angle
曲率校正 curvature correction
曲率量测 curvature measure
曲率面积法 curvature-area method
曲率容许量 allowance for curvature; curvature coefficient;curvature factor
曲率设计 curvature design
曲率矢量 buckling vector; curvature vector
曲率损失 curvature loss
曲率图 flexography
曲率系数<土的级配曲线的> coefficient of curvature; coefficient of curvity;curvature factor
曲率弦 chord of curvature
曲率线 lines of curvature
曲率向量 curvature vector
曲率延性 curvature ductility
曲率延性比<钢筋的> curvature ductility ratio
曲率仪 circumferentor[circumferenter]; curvature meter; curvometer; flexometer;flexure meter
曲率圆 circle of curvature
曲率张量 contravariant tensor;curvature tensor
曲率指示器 radius indicator
曲率指示线 curvature indicatrix
曲率中心 center[centre] of curvature
曲率重心 barycenter[barycentre] of curvature
曲率轴 axis of curvature
曲率阻力系数 coefficient of curvature resistance
曲轮轴 camshaft
曲麻莱构造结 Qumarleb tectonic knot
曲霉<拉> Aspergillus
曲霉酸 aspergillic acid
曲面 bent surface; camber; curved face;curved plane;curved surface; hood face;toroidal
曲面板 curved plate;outside plank
曲面板压机 buckle plate press
曲面逼近 surface approximation
曲面表示法 representation of surface
曲面玻璃 bend glass;bent glass;curved glass
曲面薄板 curved corrugated sheet
曲面测量法 cyrtometry
曲面测量计 cyrtometer[kyrtometer]
曲面超高缓和段 warped transition
曲面成型 curve generating
曲面挡板 curved baffle
曲面挡风玻璃 wrap-around windscreen
曲面导轨 curved guide
曲面的 cambered
曲面的包络面 envelope of surfaces
曲面的第二基本形式 second fundamental form of a surface
曲面的法线 normal to a surface
曲面的法线矢量 normal vector of a surface
曲面的拐度 flexion of surface
曲面的基本型 fundamental forms of a surface
曲面的内蕴几何 intrinsic(al) geometry of a surface
曲面的内蕴性质 intrinsic(al) property of a surface
曲面的抛物点 parabolic(al) point of a surface
曲面的平点 planar point of a surface
曲面的奇点 singular point of a surface
曲面的切平面 tangent plane to a surface
曲面的球面表示 spheric(al) representation of a surface
曲面的贴合 application of a surface
曲面的叶 sheet of curved surface; sheet of a surface
曲面的主曲率 principal curvature of a surface
曲面法线 normal to a curve; normal to surface;surface normal
曲面法线向量 normal vector of a curve
曲面反射镜干涉仪 curved-mirror interferometer
曲面反射式全息光学元件 curved reflection holographic(al) optic(al) element
曲面回转 surface rotation
曲面混凝土屋顶 curved concrete roof
曲面积分 surface integral
曲面基板 curved substrate
曲面加工装置 curvature generator
曲面件 curved work
曲面桨叶 cambered blade
曲面胶合板 curved plywood; mo(u)lded plywood
曲面镜 curved mirror
曲面镜共振腔 curved-mirror resonator
曲面控制 camber control
曲面裂缝沉箱式防波堤 curved slit caisson breakwater
曲面轮传动 globoid gearing
曲面面积 area of a curved surface; surface area
曲面摩擦 curvature friction
曲面抹子 circle trowel
曲面拟合 surface fitting
曲面抛光机 bending planer
曲面抛光器 curved polisher
曲面刨 circular plane;compass plane; spokeshave
曲面刨包铁 spokeshave
曲面刨床 radius planer
曲面刨光机 bending planer
曲面坯块 curved compact
曲面片生成 surface patch generation
曲面剖面图程序 surface section program-(me)
曲面千斤顶 curved jack
曲面千斤顶法 curved jack technique
曲面铅版浇铸机 casting box for curved stereos
曲面墙<一侧直墙,一侧曲面> curved batter
曲面墙栅<断面呈规定曲线型,能使斜冲车辆归入原车道> new Jersey barrier
曲面求积法 complanation
曲面全图息 wrap-around hologram
曲面全息光学元件 curved holographic(al) optic(al) element
曲面上的挠点线【数】 torsal line of a surface
曲面上的双曲点 hyperbolic point on a surface
曲面上的虚圆点 circular point on a surface
曲面说 theory of surfaces
曲面随动件 mushroom follower
曲面陶瓦 clay curved roof(ing) tile
曲面陶瓦屋顶 clay curved tile roof
曲面体 curved surface body
曲面体层摄影 pantomography
曲面同方向的 synclastic
曲面投影 curved surface projection
曲面透镜 toroidal lens
曲面透平式搅拌器 curved blade turbine type agitator
曲面图 surface chart;surface diagram
曲面图程序 surface chart program(me)
曲面屋顶 curved roof(ing)
曲面型多层胶合木 curved laminated wood
曲面修整 curved surface trimming
曲面修整工具 camber correction tool
曲面移门 curved sliding door
曲面阴极 curved cathode
曲面银幕 curved screen
曲面印刷机 flexographic(al) press
曲面印刷(术) flexographic printing
曲面应力 quadric stress
曲面元素 surface element
曲面运动 non-plane motion
曲面造型 surface modeling
曲面中心 center[centre] of surface
曲面砖 curved brick
曲面坐标 surface coordinate
曲木<弯曲成型而非加工成型的> bent wood
曲木屋架 cruck
曲挠性模数 modulus of flexure
曲盘 koji tray
曲片 knee piece
曲桥 winding bridge;zigzag bridge
曲切削面 curved cutting face
曲屈因数 buckling factor
曲刃剪 hawkbill snips;Hawk snips
曲刃刨刀 curved spoke shaver
曲扇状流 curved fan
曲勺 angle scoop
曲射 rigging-angle fire
曲射线路径 curved path
曲式横撑 curved batter
曲试验器 flex tester
曲手把 offset hand lever
曲弹性模量 modulus in elasticity in bending
曲条 knee
曲条跳甲 cabbage flea beetle
曲调 curve gain regulation; curvilinear regulation
曲调电路 curvilinear regulation circuit
曲贴 crimp
曲贴角钢 crimped angle
曲铁 knee iron
曲铁桩砧 hatchet stake
曲头钉 brad; brad nail; round lost head nail;wire brad
曲头钉打钉器 brad pusher
曲瓦 bent tile
曲瓦管 curved earthenware pipe
曲尾车床鸡心压头 bent-tail lathe dog
曲纹贴面板 butt veneer
曲涡流层 curved vortex sheet
曲涡线 curved vortex line
曲系 bowed pastern
曲弦 curved chord
曲弦桁架 broken chord truss;curved chord truss; curved-rib truss; Parker('s) truss
曲线 bandy;bight;curved line;curviline;wave rule
曲线板 curve board;circular-arc ruler; curved drawing instrument; curve(d) ruler; curve template; drawing curve; flexible curve; French curve; gabarite; irregular curve;irregular French curve;plotting board;profile ga(u)ge;spline
曲线板头<楼梯底级踏步的> curtail
曲线半径 curve radius; radius of curve
曲线笔<画等高线用的> contour pen;contour instrument;swivel pen
曲线笔绘图 swivel pen drafting
曲线边界 curved boundary;curvilinear boundary
曲线变换 curvilinear transformation
曲线标 curve post
曲线标尺 sloping ga(u)ge
曲线标高 curve elevation
曲线标绘 curve plotting
曲线标石 curve stone
曲线标志 curve sign
曲线表 curve table
曲线并集 curve union
曲线补偿 curve compensation
曲线裁弯取直 curved cut-off
曲线参数 parameter of curve
曲线槽 cam slot
曲线测长计 purgemeter;rotameter
曲线测定 curve test
曲线测定器 curve tester
曲线测量 curve survey;setting-out of curve
曲线测量设备 arrangement of curve
曲线测设 arrangement of curve;curve ranging; curve setting; keying off curve; laying off curves; layout of curve; ranging of curve; setting-out of curve
曲线测设用表 table for laying off curves
曲线插值法 curve interpolation; curvilinear interpolation
曲线产生器 curve generator
曲线长度 curve length; length of curves
曲线长度仪<图上量长度用> opisometer;stadi(o)meter;curvometer
曲线超高 banked crown on curve; banking of curve; cant; cant of curve;curve super-elevation;superelevation on curve; superelevation of curve
曲线超高段单向横断面 straight-topped section on the curve
曲线超限 curve super-elevation
曲线车行道 light-ga(u)ge carriageway
曲线尺 curves ruler;spline
曲线齿锥齿轮 conic(al) gear with curved teeth
曲线出闸 leaving lock in a curvilinear way
曲线船首 clipper stem
曲线窗格 curvilineal tracery;curvilinear tracery

曲线窗花格 curvilinear tracery; flowing tracery; tracery; undulating tracery
曲线垂距 offset
曲线丛 complex of curves
曲线簇 curve family; family curves; series of curves
曲线大梁 curved girder
曲线单位阻力 specific curve resistance
曲线挡板 sloping baffle
曲线道岔 curved points; curved switch; turnout from curved track
曲线的 curved; curvilineal; curvilinear
曲线的包络 envelope of curves
曲线的长度 length of a curve
曲线的顶点 hump
曲线的法线 normal to a curve
曲线的反弯点 point of reflection on the curve
曲线的反演 inversion of curves
曲线的方向 direction of a curve
曲线的分支 branch of a curve
曲线的峰值 peak of a curve
曲线的共切点联结 tangential connection of curves connection
曲线的拐点 point of inflexion on a curve
曲线的极线 line polar or a curve
曲线的阶 order of a curve
曲线的结点 node of a curve
曲线的亏格 genus of a curve
曲线的邻域 neighbo(u)rhood of a curve
曲线的内蕴方程 intrinsic(al) equations of a curve
曲线的挠率 torsion of a curve
曲线的平稳点 stationary point on a curve
曲线的平稳段 plateau of curve
曲线的平直部分 plateau[复 plateaus/ plateaux]
曲线的平直段 plateau[复 plateaus/ plateaux]
曲线的奇性 singularity of a curve
曲线的起点 springing of curve
曲线的曲率 curvature of a curve
曲线的上凹形 concave upward shape of curve
曲线的射影理论 projective theory of curves
曲线的弯曲处 knee
曲线的弯折 kink of a curve
曲线的协变式 covariant of a curve
曲线的寻常点 ordinary point of a curve
曲线的直线段 tangential path
曲线的转向点 turning point on a curve
曲线的转折点 curve break
曲线点 point of curvature; point of curve
曲线顶点 apex of bend; vertex of curve
曲线定规 curve ruler
曲线定位装置 pull-off
曲线定线 curve ranging
曲线陡度 curve steepness; steepness of a curve
曲线读出器 curve reader
曲线度 angularity
曲线度数 curve degree; degree of curve
曲线端(点) end of curve
曲线段 curved section; curve segment; section of curve
曲线段加宽 widening of curve
曲线对 pair of curves
曲线对比(方)法 curve correlation method of dip log; curve-matching method
曲线发生器 curve generator
曲线法 curve method
曲线方程 curvilinear equation
曲线方程次数 degree of curvature; degree of curve
曲线方程的类型 type of curve equation
曲线方程中的指数 degree of curvature; degree of curve
曲线放大 curve enlargement
曲线放宽 curve enlargement
曲线放线 <路线测量的> ranging a curve; staking out a curve; curve ranging
曲线放样 <线路测量的> staking out a curve; arrangement of curve; ranging a curve
曲线分析 tracing analysis
曲线分析器 curve analyser[analyzer]
曲线峰 peak of a curve
曲线缝制 zigzag stitching
曲线附加阻力 additional resistance for curve
曲线附加阻力换算坡度 equivalent gradient of additional resistance on curve
曲线复示器 curve follower
曲线副法线球面指标 spheric(al) indicatrix of binormal to a curve
曲线改造 curve improvement
曲线高度 height of curve
曲线高架桥 curved viaduct
曲线格子梁桥 curved grillage girder bridge
曲线跟随器 curve follower
曲线跟踪 curve following; curve tracing
曲线跟踪器 curve follower
曲线共切点连接 tangential connection of curves
曲线共切线段 tangential path
曲线股道 curved track
曲线拐点 break-in a curve; curve break; knee of a curve
曲线拐肘 curve crank
曲线关系的假设检验 testing hypothesis of curvilinear relationship
曲线管 curved pipe
曲线光滑 line smoothing
曲线光滑函数 smoothing function of a curve
曲线光顺 fairing of curves
曲线规 curved ruler; curve ga(u)ge; French curve; spline; banjo frame <放样用>
曲线轨道 curved track
曲线轨迹 curvilinear path
曲线(轨距)加宽 ga(u)ge widening on curve
曲线函数 curvilinear function
曲线焊缝 curve welding seam
曲线航迹 curved path
曲线航行 traverse
曲线航行法 traverse sailing
曲线护轨 curve guard rail
曲线画法 curve plotting; curve tracing
曲线缓变弯折处 soft knee
曲线缓和 curved easement; curve reduction; easing
曲线缓和长度 length of runoff
曲线换算坡度 converted gradient of curve
曲线回归 curvilineal regression; curvilinear regression
曲线回路输送机 curved conveyer[conveyor]
曲线汇 congruence of curves
曲线绘图机 curve plotter
曲线绘图仪 curve plotter
曲线绘制器 plotting device
曲线绘制仪 plotting device
曲线机锯 jig saw
曲线积分 curvilinear integral; line integral
曲线几何偏移 throw on curve
曲线计 <测量地图等曲线距离用> opisometer; curvimeter; curvometer
曲线记录式电流表 curve drawing ammeter
曲线记录仪 chart-recording instrument
曲线记录装置 tracer
曲线加宽 curve widening; widening at curve track; widening on curve
曲线加宽缓和段 transition zone of curve widening
曲线加强 curve strengthening
曲线加速率 acceleration in curves
曲线尖端 cusp of curve
曲线尖轨 curved diamond rail; curved switch point; curved switch rail; curved tongue
曲线尖轨转辙器 curved split switch
曲线间的夹直线 straight between two curves
曲线剪床 scroll shear
曲线检验 curve test
曲线渐近 curvilinear asymptote
曲线交叉 curve(d) crossing; curve(d) intersection
曲线角 angularity; curvilinear angle
曲线校正 curvature correction
曲线结构 light-ga(u)ge structure
曲线截锯 lock saw
曲线近似压缩法 curve fitting compaction
曲线进闸 entering lock in a curvilinear way
曲线锯 compass saw; coping saw; ribbon saw; scroll saw; sweep saw; turning saw
曲线靠山一边 high side of curve
曲线可调连接杆 offset adjustable link
曲线控制点 curve controlling point
曲线宽口接杆 offset wide jaw
曲线类型的鉴别 identification of curve type
曲线理论 curve theory
曲线连接杆 offset link
曲线梁 camber beam
曲线梁桥 curved beam bridge
曲线菱形交叉 curved diamond
曲线流(动) curved flow; curvilinear flow; quasi-flow; flow of curves
曲线轮廓 curved profile
曲线论 theory of curves
曲线美 line of beauty
曲线描绘 curve tracing
曲线描绘器 curve plotter; curve tracer
曲线描绘台 plotting table
曲线描绘仪 curve tracer
曲线描绘针 curve following stylus
曲线描迹法 curve tracing
曲线描记器 kymograph
曲线模板控制器 cam inversor
曲线模型 curve model
曲线磨床 profile grinder
曲线磨削 camber grinding
曲线木作 curved carpentry; shaped work
曲线内侧 inside of curve
曲线内插法 curvilinear interpolation
曲线内轨 low rail
曲线内角 subtended angle
曲线内接 inscribe to curves
曲线拟合 fitting curve; logistic curve fitting; fitting of a curve; curve fit(ting)
曲线拟合程序 curve fitting program(me); curve fitting routine
曲线拟合法 curve fitting method; curve-matching method; method of curve fitting
曲线拟合技术 curve fitting technique
曲线拟合精简数据法 curve fitting compaction
曲线扭折 kink of curve
曲线牌 curve board
曲线配合 curve fit(ting)
曲线喷流 curved jet
曲线劈裂的石板 yorky
曲线平坦段 flat region
曲线平移 curvilinear translation
曲线平整 curve flattening
曲线坡度折减 <曲线上的纵坡折减> compensation of curve; compensation of gradient on curve section; curve compensation
曲线坡度折减率 reduction(ratio) of gradient on curve
曲线起点 beginning of curve; orifice of curve; origin of curve; point of curvature; point of curve; spring of curve; tangent to curve; curve point
曲线起点标 beginning of curve sign; curve begin post
曲线起点标志 beginning of curve sign
曲线起讫点 starting, ending points of curve-section; tangent points of curves; tangent points of the curve
曲线桥 bridge of circular(plan) form; curved bridge
曲线切点 tangent points of a curve
曲线切剖面 curve cutting face
曲线切线的球面指标 spheric(al) indicatrix of tangent to a curve
曲线求长 rectification of a curve
曲线求积法 volume-curve method
曲线求律法 curve fit(ting)
曲线曲率 curvature of curve; curve curvature
曲线曲值 knee of curve
曲线趋势 curvilineal trend; curvilinear trend
曲线锐度 sharpness of curve
曲线上的钢轨磨耗 wear of rail on curves
曲线上的极大点 maximum point on a curve
曲线上的极小点 minimum point on a curve
曲线上的尖叫声 <有轨车> squealing on curve
曲线上的列车速度 speed of trains on curves
曲线上的偏移量 distance deflected on curve
曲线上的视距 visibility on curves
曲线上的纵坡折减 curve compensation
曲线上点的散布 spread of points
曲线上高度 curve elevation
曲线上轨距加宽 ga(u)ge widening on curve
曲线上轨距加宽量 easement; ga(u)ge easement
曲线上极小值 trough
曲线上升斜率 rate of rise
曲线上向内倾覆 inward overturning
曲线上向外倾覆 outward overturning
曲线上纵坡折减 grade compensation for curve
曲线设定 curve setting
曲线设计 curve design
曲线绳整法 realignment by string line and versine offset method; string lining of curve
曲线始或终点标 curve “begin” or “end” post

曲线式 curvilinear style
曲线式空气分级机 curvilinear air classifier
曲线式样精简数据法 curve-pattern compaction
曲线式振动 flexural vibration
曲线饰 curvilinear ornament
曲线收敛 curve convergence
曲线输出机 curve follower
曲线束 pencil of curves
曲线数据 curve data
曲线水道 curved waterway
曲线水道的最小半径 minimum radius of curved waterway
曲线四边形 curvilinear quadrilateral
曲线隧道 curved tunnel; curvilinear tunnel
曲线所包围的区域 region enclosed by a curve
曲线特性 curve characteristic
曲线调整 realignment of curve
曲线通过 curve negotiation
曲线通过速度 curve negotiation speed; curve passing speed
曲线图 curve chart; curve diagram; curve graph; diagram curve; graph; graphic (al) chart; plot; plotted curve; trace
曲线图表 curve
曲线图程序 curvilinear figure program(me)
曲线图示 diagram of curves
曲线图形 curvilinear figure
曲线图形压缩 curve-pattern compaction
曲线图状态点 condition point
曲线涂油器 curve lubricator; rail lubricator
曲线推销 metamarketing
曲线瓦 curvilinear tile
曲线外部区域 exterior to curve
曲线外轨【铁】 high rail
曲线外轨超高 super-elevation of outer rail on curve
曲线外轨超高度 super-elevation
曲线外轨一侧【铁】 high rail side of curve
曲线弯曲点 knee point
曲线弯曲度 <自记仪器的> embroidery
曲线弯折部分 kink of curvature
曲线网(格) curved mesh; curvilinear net(work); net of curves
曲线网文件 boundary network file
曲线维修 curve maintenance
曲线屋顶 curved roof(ing)
曲线屋面排水沟 swept valley
曲线物件 rover
曲线系 system of curves
曲线系列 series of curves
曲线细部 curvilinear detail
曲线细部点测设 setting-out detail points of curve
曲线下的面积 area under a curve
曲线显示 graphic(al) display
曲线线长 curve length
曲线线脚 rover
曲线线路 curved track; curvature line
曲线相关 curvilinear correlation
曲线箱形钢梁桥 curved steel box girder bridge
曲线箱形梁桥 curved box girder bridge
曲线协调的皮带运输机 curve-negotiating belt conveyer[conveyor]
曲线协调的转向架 curve-negotiating bogies
曲线斜率 rate of curves; slope of a curve
曲线斜坡 curved batter
曲线行程 curvilinear travel
曲线行走转向架 <吊车的> curve going bogies
曲线形 <又称曲线型> shaped form
曲线形凹衬板 curved concave head
曲线形薄壳 curved shell
曲线形船首 fiddle bow
曲线形的 archy; arrondi
曲线形房屋 curved building
曲线形固定辙叉 curved rigid-type frog
曲线形滑槽 curved chute
曲线形畸变 curvilinear distortion
曲线形简单回归 curvilinear simple regression
曲线形建筑 curved block
曲线形金属托 knee
曲线形楼梯外斜梁 wreathed handrail(ing); wreathed string
曲线形模型 curvilinear model
曲线形喷管 curved nozzle
曲线形上弦杆 curved top chord
曲线形式 curve profile
曲线形式系数 curve form factor
曲线形输送机 curved conveyer
曲线形态 tracing pattern
曲线形屋盖 compass roof
曲线形下弦杆 curved bottom chord
曲线形堰 curved weir
曲线形预应力钢索 curved tendon
曲线形障碍 curvilinear obstacle
曲线形纸盆 curvilinear cone
曲线形轴线 curved axis
曲线形钻架 camber of truss
曲线性 curvilinearity
曲线修匀 graduation of curve
曲线修整 graduation of curve
曲线选配 curve fit(ting)
曲线(鸭嘴)笔 curve pen
曲线样板 sweep ga(u)ge
曲线要素 curve element; essential of curve; key element of curve
曲线仪 contour tracing apparatus; curvimeter; map measurer
曲线应变计 curvilinear strain meter
曲线应力 stress due to curving
曲线阅读器 curve follower; graph follower
曲线运动 curve motion; curvilinear motion; movement in a curved line; snaking motion
曲线运动型有轨巷道堆垛机 curve-negotiating street/rail machine
曲线折点 break point; knee of a curve
曲线折断 break-in curve
曲线折减 compensation on grade curve
曲线折减率 curve compensation ratio
曲线折减坡度 grade compensation for curve
曲线辙叉 curved frog
曲线振动均数 mean value of tracing oscillation
曲线整正 curve adjusting; realignment of curve
曲线整正法 method of curve realignment
曲线整正计算器 curve lining calculator
曲线正交坐标 curvilinear orthogonal coordinates
曲线正矢 curve versine; middle ordinate
曲线直线图 curve-line graph
曲线指示牌 curve board
曲线指示器 curve pointer
曲线中点 middle of curve; midpoint of curve; point of secant
曲线中心 center[centre] of curve; center[centre] of curvature
曲线中心角 curve center[centre] angle
曲线终点 curve to tangent; end of curve; point of tangency; point of tangent
曲线终点标 curve end post; end of curve sign
曲线轴 curvilinear axis
曲线主点 main point of curve; principal point of curvature
曲线转向 curve hand
曲线转折点 break-in a curve
曲线转辙器 <折叠门导轨> curved switch
曲线桩 curve stake
曲线状 curvilinear
曲线准则 criterion of curve
曲线族 family curves; family of curves; series of curves; set of curves
曲线族的矩 moment of a family of curves
曲线族的模 module of a family of curves
曲线阻力 curvature resistance; curve resistance
曲线组 family curves; family of curves; series of curves; set of curves
曲线组合 assemblage of curves; curve composition
曲线作图 curve tracing
曲线坐标 curve coordinates; curvilineal coordinates of curve; curvilinear coordinates
曲斜撑 curved batter
曲斜齿轮 skew bevel gear
曲斜面墙 curved batter
曲形部件 knee piece
曲形槽 shaped groove
曲形草捆滑槽 curved bale chute
曲形封闭板 curved closure plate
曲形光墁刀 <修光滑的> sleek
曲形胶合板 curved panel
曲形耙 zigzag harrow
曲型层积木材 curved laminated wood
曲型胶合板 curved laminated wood
曲型线条 falling mo(u)ld
曲性 convexity
曲翼面 aerocurve
曲隅部 bend
曲折 anfractuosity; anfractuous; detour; flex; meandering; serpentine turnings; sinuosity; tortuosity; twists and turns
曲折岸 indentation
曲折边沿 ragged edge
曲折波线 zigzag path of wave
曲折舱壁 zigzag bulkhead
曲折长度 meander length
曲折成型 <装饰性的> zigzag mo(u)lding
曲折穿行 twist
曲折带宽度 meander breadth; meander width
曲折挡板室 zigzag baffle chamber
曲折道路 road diversion
曲折的 meandering; tortuous; zigzag
曲折的回纹形线条 zigzag mo(u)lding
曲折的回纹形装饰 zigzag mo(u)lding
曲折的线条 zigzag
曲折地层 twisted stratum
曲折地流 meander
曲折雕带 zigzag frieze
曲折定线 crooked align(e)ment
曲折度 sinuosity
曲折断层 zigzag fault
曲折反射 zigzag reflection
曲折缝合 zigzag closing
曲折缝金属屋面 chev(e)ron design metal roof; chev(e)ron seam metal roof
曲折隔断 zigzag partition
曲折隔墙 zigzag partition
曲折拱 zigzag arch
曲折构型 zigzag configuration
曲折海岸 broken coast; indented coast
曲折海岸线 broken coastline; indented coastline
曲折航线 zigzag coast; zigzag course
曲折航行 zigzagging
曲折航行计算法 traverse sailing
曲折河 meandering stream; snaking stream; meander
曲折河道 sinuous channel; tortuous channel
曲折回纹(饰) key pattern; labyrinth fret; meander
曲折槛 zigzag sill
曲折接法 interconnected star connection; Z-connection; zigzag connection
曲折接法滤波器 zigzag filter
曲折接线 zigzag connection
曲折接线电力变压器 zigzag power transformer
曲折经纬仪 broken transit
曲折宽度 meander width
曲折连接 zigzag connection; zigzag coupling
曲折连接绕组 interconnected star winding
曲折连续掺和机 zigzag continuous blender
曲折链 zigzag chain
曲折裂缝 meandering crack; zigzag crack
曲折流 tortuous flow
曲折流线 zigzag path
曲折漏磁 zigzag leakage
曲折漏磁通 zigzag leakage flux
曲折漏抗 zigzag reactance
曲折路 diversion road
曲折路线 zigzag path
曲折率 meander ratio
曲折密封 labyrinth seal
曲折密封圈 labyrinth ring
曲折平地海岸 flat inclined coast; flat indented coast
曲折气隙泄漏 zigzag air gap leakage
曲折砌合 rake bond
曲折前室 maze
曲折绕组 interstar winding
曲折三角形结线 zigzag-delta connection
曲折式 dioptric type
曲折式过滤器 zigzag paper filter
曲折式密封 labyrinth packing
曲折试验 bending and unbending test
曲折试验器 <涂膜的> bending tester
曲折饰 dancette
曲折水道 zigzag lead; tortuous channel
曲折水流 sinuous flow; winding stream
曲折送料 zigzag feed
曲折天线 bent aerial; bent antenna; zigzag antenna
曲折通路 tortuous passage
曲折图案 dancette; zigzag; zigzag pattern
曲折图形 zigzag pattern
曲折位错 zigzag dislocation
曲折系数 buckling factor; zigzag coefficient
曲折下降 irregular lower
曲折线脚 chev(e)ron mo(u)lding; dancette; dancette mo(u)lding; zig-

zag mo(u)lding
曲折线(路) zigzag line; meander line
曲折线行波管 meander-line travel(l)-ing wave tube
曲折巷道 zigzag roadway
曲折星形连接 zigzag-star connection
曲折行车 hunting
曲折行进 crankle
曲折形 indentation
曲折形拉杆 pigtail tie
曲折形锚杆 zigzag anchor bar
曲折形铆钉 reeled riveting
曲折形褶皱 zigzag fold
曲折型 meander configuration
曲折型河流 sinuous stream
曲折窑 zigzag kiln
曲折移动 meandering movement
曲折因数 buckling factor
曲折应力 buckling stress; crippling stress
曲折运动 sinuous movement
曲折栅栏 worm fence
曲折折光计 elbow refractor
曲折褶皱 zigzag fold
曲折轴垫 labyrinth packing
曲折柱槽 zigzag fluting
曲折状电阻 zigzag resistance
曲折状电阻器 zigzag resistance unit
曲折状排水系统 zigzag drainage
曲折阻漏 labyrinth packing
曲枝压条 bowed-branch layering
曲直线轨道间菱形交叉 diamond crossing with one curved track
曲中点 point of secant
曲周 anterio-superior to the ear; curvature of hairline
曲轴 bent axle; bent shaft; cambered axle; crank; crankshaft(axle)
曲轴板 crankshaft cheek
曲轴泵 bent-axis pump
曲轴臂 crankshaft arm
曲轴变位传动装置 barring gear
曲轴车床 crankshaft lathe
曲轴齿轮 crankshaft gear; crankshaft toothed wheel
曲轴冲力平衡器 crankshaft impulse neutralizer
曲轴传动(装置) crankshaft drive; crankshaft gear
曲轴的曲拐 throw of crankshaft
曲轴的自由端 free end
曲轴电子枪 bent gun
曲轴电阻弧花压焊 crankshaft flash butt welding
曲轴定时齿轮 crankshaft timing gear
曲轴端部推力 crankshaft end thrust
曲轴法兰盘 crankshaft flange
曲轴飞轮泵 crank-and-flywheel pump
曲轴管的压力控制阀 pressure-control of crankcase
曲轴横放压力机 cross shaft press
曲轴回转 foaming shaft up
曲轴回转力矩 crankshaft rotational moment
曲轴减振器 crankshaft vibration damper
曲轴键 crankshaft key
曲轴铰光用刀 crankshaft returning tool cutter
曲轴颈 crank journal; crankshaft pin; maneton
曲轴连杆机构 crankshaft connecting rod system
曲轴连杆马达 crankshaft connecting rod motor
曲轴链轮 crankshaft sprocket
曲轴链系 crankshaft chain
曲轴磨床 crankshaft grinder
曲轴扭转加工机床 crankshaft twisting machine
曲轴刨床 crankplaner
曲轴皮带轮 belt pulley of crank shaft; crankshaft pulley
曲轴平衡机 crankshaft balancer
曲轴平衡块 crankshaft balancer
曲轴平衡器 crankshaft balancer
曲轴平衡重 crankshaft balance weight
曲轴起动爪 crankshaft starting claw; crankshaft starting jaw
曲轴前轴承盖 crankshaft bearing cap
曲轴前轴承油封 crankshaft front bearing seal
曲轴驱动筛 crankshaft drive screen
曲轴甩油环 crankshaft oil slinger
曲轴调整器 crankshaft governor
曲轴推力轴承 crankshaft thrust bearing
曲轴弯程 crankshaft throw
曲轴弯头 throw of crankshaft
曲轴维修止推板 crankshaft thrust-plate
曲轴铣床 crankshaft milling machine
曲轴箱 crankcase; crankshaft case [casing]; crankshaft housing
曲轴箱舱 crankcase section
曲轴箱冲洗用油 crankcase conditioning oil
曲轴箱的下面部分 lower part of crankcase
曲轴箱分隔室 crankcase compartment
曲轴箱盖板 crankcase cover plate
曲轴箱护罩 crankcase guard
曲轴箱加热器 crankcase heater
曲轴箱检查盖 crankcase inspection cover
曲轴箱排放物 crankcase emission
曲轴箱强制通风 positive crankcase ventilation
曲轴箱强制通风装置 positive crankcase ventilator
曲轴箱润滑系统 crankcase lube system
曲轴箱润滑油 crankcase lubrication
曲轴箱润滑油压 crankcase lube pressure
曲轴箱扫气 crankcase scavenging
曲轴箱扫气二冲程发动机 crankcase scavenged engine
曲轴箱双头螺栓 crankcase stud
曲轴箱通风 crankcase ventilation
曲轴箱通气管 crankcase breather
曲轴箱通气装置 crankcase breather; crankshaft breather
曲轴箱压力 crankcase pressure
曲轴箱压力传感器 crankcase pressure sensor
曲轴箱压力敏感元件 crankcase pressure sensor
曲轴箱压缩 crankcase compression
曲轴箱用油起沫试验 crankcase oil foaming test
曲轴箱(油)槽 crankcase sump
曲轴箱轴承座 crankcase bearing seat
曲轴销 crankshaft pin
曲轴压力机 crankshaft press
曲轴压直机 crankshaft straightening press
曲轴(引)导轴承 crankshaft pilot bearing
曲轴油孔 crankshaft oil hole
曲轴油路 crankshaft oilway
曲轴轴承 crankcase bearing; cranked bearing; crankshaft bearing
曲轴轴承合金 crankshaft bearing metal
曲轴轴颈 crankshaft journal
曲轴轴颈测量器 crankshaft ga(u)ge
曲轴主轴颈 main bearing journal
曲轴转角度 crankshaft degree
曲轴转速 speed of crankshaft
曲肘 toggle
曲肘形木材 knee timber
曲肘形木梁 knee timber
曲柱式防撞桩 buckling column fender
曲桩一销 crank and pin

驱

驱傲雪 blowing snow
驱冰航行 slewing
驱尘 dust-repelling
驱虫的 insect proof
驱虫剂 insectifuge; insect repellant
驱虫药 worm-tablet
驱除的 repellent
驱除(动物和昆虫)药 repellent
驱除剂 expellent
驱动 activation; drive; propulsion; run
驱动斑纹 drive pattern
驱动板 driving plate
驱动臂 actuating arm
驱动表 table drive
驱动部分 drive part
驱动部件 drive part; driver element
驱动侧 drive side; driving side
驱动侧桥架 driving side bridge
驱动车轴 drive axle
驱动程序【计】 driver
驱动齿轮 drive gear; driving gear; output gear
驱动齿轮轴 driving gear shaft
驱动齿条的小齿轮轴 <挖掘机> shipper shaft
驱动齿条的小齿轮轴座 shipper shaft saddle
驱动磁铁 drive magnet; driving magnet
驱动带 rotating band
驱动单元 driver element
驱动导杆 driver guide arm
驱动的 driving; operated
驱动点 drive point
驱动点导纳 driving point admittance; driving point mobility
驱动点函数 driving point function
驱动点阻抗函数 driving point impedance function
驱动电动机 actuating motor; actuator motor; driving motor; drive motor
驱动电动势 <阴极防蚀中的> driving emf
驱动电流 drive current
驱动电路 blocking-oscillator driver; drive circuit; driver; driving circuit; propulsion circuit
驱动电平 drive level
驱动电汽路 drive circuit
驱动电梯的动力机械 driving machine for elevator
驱动电压 slaving voltage
驱动端 drive end; driving end
驱动短轴 live stub axle
驱动方法 drive system; means of driving
驱动方式 driving method
驱动方向 direction of compaction; driving direction
驱动放大器 driving amplifier
驱动风扇涡轮 fan-driven turbine
驱动负荷 driven load
驱动杆 actuating arm; drive rod; drive shaft; drive spindle; hound rod
驱动隔离 drive isolation
驱动隔离开关 drive isolator
驱动工具 means of driving
驱动功率 driving power; motive power; propulsion power
驱动构件 driving member
驱动鼓轮 <缆索铁路及架空缆道> driving drum
驱动鼓筒的平衡索 drum counterweight rope
驱动管路 drive line
驱动辊 drive rod; driving drum; driving roll
驱动滚轮 drive roll; driving roll
驱动滚筒 drive pulley; driving drum; driving roller; head roll
驱动和传动装置 driving and transmission system
驱动滑轮 <缆索铁路和架空缆道> driving pulley
驱动环节设计 tow ring layout
驱动机 driving machine
驱动机构 actuating mechanism; drive gear; drive mechanism; driving mechanism
驱动机件 <深井泵的> power head
驱动机件快门 driving mechanism of shutter
驱动机理 drive mechanism; expulsion mechanism
驱动机器辅助设备的附加荷载 parasitic load for auxiliary equipment of driven machine
驱动机械 propelling machinery
驱动机械装置 drive mechanism
驱动机组 actuating unit; drive unit; driving unit
驱动极 driver; driving stage
驱动介质 driving medium
驱动进程 driver process
驱动进料三角皮带轮 feed pulley
驱动开启窗 multiple-operated window
驱动控制 propulsion control
驱动块 driving block; driving dog
驱动力 drive force; drive power; driving energy; driving force; driving influence; motion-promoting force; propulsive effort
驱动力矩 driving moment; driving torque
驱动力指数 driving force index
驱动链 chain drive; drive link; messenger chain
驱动链节 driver link
驱动链轮 drive chain sprocket
驱动轮 bull wheel; drive tumbler; drive wheel; driving wheel; leading wheel; sprocket; traction wheel; tumbler; wheel drive; compression roll <压路机的>
驱动轮齿节 sprocket segment
驱动轮传动轴 traction drive shaft
驱动轮荷重 weight on driving wheel
驱动轮护罩 sprocket guard
驱动轮加载的调节 traction control
驱动轮加载系统 traction-booster system
驱动轮加载装置 traction booster; traction device
驱动轮式推土机 high drive bulldozer
驱动轮胎 drive tire
驱动轮中心距 drive wheel track; driving track
驱动轮轴 live axle; sprocket axis; sprocket shaft
驱动轮轴箱 driving box
驱动轮总成 complete sprocket
驱动轮组合 complete sprocket
驱动螺杆 drive screw
驱动马达 drive motor; driving motor; power motor
驱动脉冲 drive impulse; drive pulse; driving pulse
驱动脉冲发生器 drive-pulse generator

驱动能力 driving power
驱动能量 propelling energy
驱动碾轮 drive drum;drive roll
驱动扭矩 driven torque
驱动盘 driving flange
驱动皮带 driving belt
驱动皮带轮 drive pulley
驱动皮带轮侧功率 power-on driving pulley side
驱动频率 driving frequency
驱动汽轮机 driving steam turbine
驱动器 actuating apparatus; driving mechanism; passenger train operator;driver【计】
驱动器编号 drive number
驱动器磁头回零系统调用 home drive system call
驱动器代号 drive letters
驱动器电平 driver level
驱动器号 drive letters
驱动器检测 drive sense
驱动器时差 driver skew
驱动器速率 driver speed
驱动器选择 driver selection
驱动器组件 actuator assembly
驱动千斤顶 driving jack
驱动桥 driving axle;driving unit;axle
驱动桥定位臂 axle arm
驱动桥扭矩管 axle torque tube
驱动桥套管 axle tube
驱动桥系统 drive axle system
驱动桥轴 drive axle;driving axle;live axle; moving axle; transaxle; powered axle
驱动桥轴毂 drive axle hub
驱动桥轴座架 drive axle mount
驱动绕组 drive winding
驱动三极管 driver triode
驱动扫描 driven sweep
驱动设备 drive unit
驱动式滚柱输送机 driver roller conveyer[conveyor]
驱动式转子输送机 driven roller conveyer[conveyor]
驱动手段 means of driving
驱动输入阻抗 driving point impedance
驱动速度 actuating speed
驱动锁簧 driving dog
驱动踏板 treadle
驱动弹簧 driving spring
驱动特性 drive characteristic
驱动烃 substituted hydrocarbon
驱动同步机 driving synchro
驱动头 driving head
驱动凸轮 driving cam
驱动拖车 driving trailer
驱动(误差)图形 drive pattern
驱动系统 actuating system;drive system;driving system
驱动线 drive wire
驱动线路 driver line
驱动线圈 drive coil
驱动销 driving pin
驱动小齿轮 drive pinion;driving pinion
驱动斜盘 <装在轴上的> swash cam
驱动信号 drive signal;driving signal
驱动星轮 driving star
驱动行走轮 ground engaging wheel
驱动循环 driving cycle
驱动压差 driving pressure differential
驱动压力 actuating pressure; driving pressure
驱动压路机 drive drum;drive roll
驱动压气机的涡轮机 compressor turbine
驱动压气机涡轮 compressor turbine
驱动扬程 <水泵的> driving head
驱动用凸铁 energizing lug
驱动元件 drive element; motor element
驱动(原)点阻抗 driving point impedance
驱动圆盘 driver plate;driving plate
驱动指令 driving instruction
驱动钟 driving clock
驱动轴 actuating shaft; drive shaft; driving axle; driving shaft; head shaft;live axle;power axle;propel;propeller shaft; prop shaft; transmission output shaft
驱动轴齿轮 axle shaft gear
驱动主控制器 driving master controller
驱动转矩 driving torque
驱动转向机构 drive/steer configuration
驱动转向结构 drive/steer configuration
驱动转向两用桥轴 drive/steer axle
驱动装置 actuate;actuating device;actuator;drive set;drive unit;driving agent;driving gear;driving machine; driving unit; propulsion plant;propulsion unit
驱动装置形式 drive type
驱动装置罩 driving unit enclosure
驱动子 driver element
驱动自动电梯的动力机械 driving machine
驱动阻力 driving resistance
驱力刺激 drive stimulus
驱力降低作用 drive-reduction
驱力引导的行为 drive-oriented behavio(u)r
驱气俘获毛细管柱气相色谱法 purge and trap capillary column gas chromatography
驱铅剂 deleading reagent
驱潜快艇 corvette
驱潜艇 mosquito boat;mosquito-craft
驱入步骤 drive-in procedure
驱散 drive off
驱射气 de-emanation
驱绳轮 driving rope wheel
驱使 actuate;impel
驱水剂 water repellent
驱替效率 efficiency of displacement
驱雾器 defogger
驱油气 gas-oil displacement
驱油试剂 repressuring medium
驱轴 propeller shaft
驱逐 destroy;dispossession;expel;expulsion;turnout
驱逐出境 deport;deportation
驱逐舰 destroyer
驱逐舰船尾型 destroyer stern
驱逐舰队 flotilla
驱逐舰母舰 destroyer depot ship
驱逐领舰 frigate

屈

屈变力 yield strength

屈从 compliance;resignation
屈从的 servile
屈地关系 geotropic(al)relation
屈度计 vertometer
屈服 yield;climb;crumple;give way;stoop;submission;submit;submittal
屈服安全系数 yield factor of safety
屈服比 yield ratio
屈服变形 yield(ing)deformation
屈服程度系数 yielding level coefficient
屈服带 yield band
屈服单位 yield unit
屈服的 yielding
屈服点 breakdown point;creep point;ductility limit; ductility point; flow limit;limit of yielding;proof stress;time-yield; yield(ing)point; yield point strength; yields point; yield value;point of fluidity <金属的>
屈服点变形 yield point strain
屈服点荷载 yield point load
屈服点力 yield point force
屈服点流限 yield limit
屈服点强度 yield point strength
屈服点伸长 yield point elongation
屈服点升高 rise of yield point
屈服点下限 lower yield point
屈服点延伸 yield point elongation
屈服点应变 yield point strain;yield sign strain
屈服点应力 yield point stress
屈服点与推力强度之比 ratio of yield point to tensile strength
屈服点值 yield point value
屈服阀 yield valve
屈服概率 probability of yielding
屈服刚度比 yielding stiffness ratio
屈服拱 <隧道> yielding arch
屈服轨迹 yield locus
屈服过程 offset procedure
屈服过程线图 yield-time diagram
屈服函数 yield function
屈服荷载 yield load(ing)
屈服后断裂力学 post-yielding fracture mechanics
屈服后刚度 post-yielding stiffness
屈服后效应 post-yielding effect
屈服后性能 post-yielding behavio(u)r
屈服后性状 post-yielding behavio(u)r
屈服后延性 post-yielding ductility
屈服机构 yielding mechanism
屈服极限 limit of yielding;offset limit;yields point;yield strength;yield value;yield limit
屈服极限安全系数 yield safety factor
屈服加速度 yielding acceleration
屈服铰 yield hinge
屈服结构 yielding structure
屈服界线 yield boundary
屈服抗力 resistance to yield
屈服扩散 yielding spreading
屈服类型 yielding pattern
屈服力矩 yielding moment
屈服流动 yielding flow
屈服率 yield rate
屈服面 limiting surface
屈服模量 yield modulus
屈服挠度 deflection at yield;yielding deflection
屈服平台 yield plateau;yield point elongation;yield point jog
屈服破坏 yield failure
屈服破坏荷载 failure load by buckling
屈服强度 yield(ing)intensity; yield(ing)strength
屈服强度比 yielding strength ratio
屈服强度下限 lower limit of yield strength
屈服区 creep zone;yield region;yield zone;plastic zone
屈服曲率 yielding curvature
屈服曲面 yield surface
屈服曲线 yield curve
屈服上限 maximum yield;upper yield point
屈服伸长 yield elongation in tension
屈服伸长点 yielding point elongation
屈服伸长度 elongation at yield
屈服伸长率 elongation at yield
屈服试验 high-level testing;yield test
屈服塑性阶段的平均应力 yield stress level
屈服台阶 yield plateau;plastic plateau
屈服条件 <试验材料的弹性弯度等> yield condition
屈服弯矩 yield moment
屈服位置 yielding locus
屈服温度 yield temperature
屈服稳定性 buckling stability
屈服线 yield line
屈服线法 <按板面辐向负弯矩屈服线设计水泥混凝土路面的方法,即瑞典 Losberg 法,属于塑性理论的范畴> yield line method
屈服线理论 yield line theory
屈服效应 effect of yielding
屈服性能 yield behavio(u)r
屈服压力 yield pressure
屈服压力强度 yield compression strength
屈服应变 yield strain
屈服应力 flow stress;yield stress
屈服应力比 yield-stress ratio
屈服应力面 limiting surface
屈服应力模型 yield stress model
屈服应力黏[粘]结 <在高于屈服点应力的压力作用下进> yield stress controlled bonding
屈服应力值 yield value of stress
屈服支承 yield support
屈服值 yield point value;yield value
屈服状况 yield condition
屈服状态 yield condition
屈服准则 yield criterion[复 criteria]
屈光的 dioptric
屈光度 diopter[dioptre]; dioptric strength
屈光度测定器 dioptometer
屈光度计 dioptometer
屈光率单位 diopter[dioptre]
屈光器 dioptric apparatus
屈光透镜 dioptric lens
屈光系统 dioptric system
屈光学 dioptrics
屈光学的 dioptric
屈光组 dioptric system
屈后强度 postbuckling strength
屈挠不能 acampsia
屈挠角度 angle of flexure
屈挠试验机 bending machine
屈强比 <物体屈服力与拉力的比率> yield ratio;yield/tensile ratio
屈曲 buckling;curvature;flex;flexion;flexuosity;twisting buckling
屈曲变形 curvature distortion
屈曲断裂 rupture in buckling
屈曲反射 flexion reflex
屈曲过度 hyperflexion;super-flexion
屈曲荷载 buckling load
屈曲畸形 flexion deformity
屈曲节 flexure
屈曲力矩 moment of deflection;moment of flexure
屈曲强度 buckling strength
屈曲试验 buckling test
屈曲图式 buckle pattern
屈曲危险 risk of buckling
屈曲向量 buckling vector
屈曲型式 buckling mode
屈曲性 flexibility
屈曲应力 buckling stress;yield stress
屈曲域 buckling domain
屈伸起重机 lazy jack
屈伸式起重机 lazy jack
屈伸运动 flexion and extension
屈氏体(金相)troostite
屈氏珠光体 troostitic pearlite
屈铁锤【机】dumper
屈压常数 buckling constant
屈压荷载 collapsing load
屈折 refraction;swerve
屈折的光线 broken ray
屈折性 refrangibility
屈折角 angle of bend

屈折力 refracting power
屈折率 index of refraction
屈折螺旋灯丝 wreath filament
屈折性 refrangibility

蛆室 maggot chamber

躯干 carcase;torso;trunk

趋避冲突 approach-avoidance conflict

趋表分布 epistrophe
趋触性 stereotaxis; telotaxis; thigmotaxis
趋地性 geotaxis
趋电性 electropism; electrotaxis; galvanotaxis; galvanotropism
趋风性 anemotaxis
趋肤扩散 skin diffusion
趋肤深度 skin depth
趋肤效应 conductor skin effect; Kelvin(skin) effect; skin effect
趋肤效应衰减 skin effect attenuation
趋肤效应损耗 skin effect loss; skin loss
趋肤效应校正 skin effect correction
趋肤指数比 index ratio
趋光的 lucipetal; photokinetic; phototatic
趋光反应 phototactic reaction
趋光节律 phototactic rhythm
趋光群聚 symphotia
趋光性 photocinesis [photokinesis]; phototaxis; phototropism
趋光性的 phototactic
趋光运动 phototactic movement
趋合 closure
趋近 approach
趋近法 approach method
趋近角 approach angle
趋近速度 velocity of approach
趋离板块界线 divergent plate boundary
趋流感受器 rheoreceptor
趋流性 rheotaxis
趋气性 pneumotaxis
趋热性 thermotaxis
趋日性 heliotaxis
趋渗的 osmophilic
趋渗性 osmotaxis
趋湿性 hygrotaxis
趋实体性 stereotaxia
趋势 course of things; tend; tendency; tide; trend
趋势百分率 trend percentage
趋势比率 trend ratio
趋势比率法 trend ratio method
趋势变动 trend variation
趋势变量 trend variable
趋势成分 trend component
趋势法 tendency method; trend method
趋势方程 tendency equation
趋势分析 trend analysis
趋势分析法 trend analysis method
趋势分析模型 trend analysis model
趋势更正法 trend correction method
趋势计算 trend computation
趋势记录 trend record
趋势检验 test of trend
趋势廓线 tendency profile
趋势面 trend surface
趋势面次数 degree of trend surface
趋势面方程 equation of trend surface
趋势面分析 tendency plane analysis; trend surface analysis
趋势面分析法 tendentious face analysis method
趋势面拟合法 fitting of trend surfaces
趋势模型 trend model
趋势平均比率法 method of average ratios to trend
趋势曲线 trend curve
趋势时距 tendency interval
趋势调整 adjustment for trend
趋势调整指数 trend adjusted indexes
趋势图 tendency chart; trend map
趋势外延法 trench extrapolation
趋势线 line of trend
趋势线外推法 trend extrapolation
趋势效果 trend effect
趋势循环 trend cycle
趋势循环比率法 trend-cycle ratio method
趋势预报 persistence forecasting
趋势预测 trend forecast(ing)
趋势预测法 trend-based forecasting
趋势直线 trend line
趋势值 trend value
趋水性 hydrotaxis
趋同 convergence
趋同进化 convergent evolution
趋同理论 convergence thesis; convergency theory
趋同适应 adaptive convergence; convergent adaptation
趋同特性 convergent character; convergent characteristic
趋同同型 convergent homeomorphy; convergent homomorphism
趋同性 homoplasy
趋同作用 convergence
趋完满律 law of pregnancy
趋完形律 law of pregnancy
趋温的 thermotactic
趋温性 thermotaxis
趋稳蠕变曲线 transient creep curve
趋稳性 sticking
趋向 tendency
趋向监测 tendency monitoring
趋向平衡 tend to be balanced
趋向平缓 to level out
趋向试验 trend test
趋向衰退 downhill
趋向于纵向颠簸 prone to pitch
趋阳性 heliotaxis
趋氧性 aerotaxis
趋药性 chemotaxis
趋异 divergence
趋于成熟阶段 drive to maturity stage
趋于零 vanish
趋于零地 vanishingly
趋于平缓 tabling off
趋于增加的 multiplicative
趋张力性 tonotaxis
趋中性 centrotaxis

渠 dike[dyke]; ditch; trench; fossa(e)

渠岸 channel bank
渠岸出水高度 canal freeboard
渠岸护坡 canal slope protection
渠岸修整机 canal slope trimming machine
渠岸余幅 canal freeboard
渠槽 canal basin
渠槽底坡 bottom slope of canal
渠槽底坡(管道内底)高程 invert elevation
渠槽粉面 channel lining
渠槽渐变段 channel transition
渠槽戗道 berm(e) for channel
渠槽修整机 canal trimmer
渠床(底) canal bottom
渠床杂草 channel weed
渠道 canal; artificial watercourse; channel; conduit; duct; flow passage; irrigation ditch; killesse; leat; ravine; sike; split pipe; water course
渠道安全流速 safe velocity of canal
渠道岸坡 canal slope
渠道倍增器 channeltron
渠道比降 canal gradient; channel grade
渠道边坡 canal side slope; side slopes of canal
渠道边坡修整机 canal slope trimmer; canal slope trimming machine
渠道边墙 canal wall
渠道布置 alignment of canal
渠道糙率 channel roughness
渠道测量 canal survey
渠道长度 canal length
渠道沉积 channel deposit
渠道衬砌 canal lining
渠道冲刷 channel cutting; channel erosion; channel scour
渠道出口 canal exit
渠道出水口 canal off-let
渠道的隔断探井 disconnecting manhole
渠道底部 bottom of bed
渠道地质测绘 geology mapping of canal
渠道跌水 canal drop
渠道定线 alignment of canal; channel alignment; canal alignment
渠道斗门 canal off-let
渠道堵塞 obstructed channel
渠道渡槽 canal crossing
渠道断面 canal cross-section; canal section
渠道放水口 canal off-let
渠道放水门 canal off-let
渠道废水 canal wastage
渠道分叉口 turnout
渠道分流 canal diversion
渠道分支 channel segment
渠道附属建筑物 canal appurtenance
渠道附属设施 canal appurtenance
渠道改道 canal diversion
渠道改线 channel realignment; channel relocation
渠道割切 channel cutting
渠道工程 canal engineering; canal works
渠道工程地质勘察 engineering geologic(al) investigation of canal
渠道过水能力 canal capacity; carrying capacity of canal; channel capacity
渠道横截面 channel cross-section; cross-section of channel; canal cross section
渠道护岸 channel revetment
渠道护面 channel revetment; lining of canal
渠道护坡 canal slope protection
渠道滑动模板 canal slipform
渠道化 canalization; channelization
渠道汇合点 junction of channels
渠道混凝土浇筑机 canal concrete paver
渠道集水井 drop pit
渠道几何条件 channel geometry
渠道加宽 canal widening
渠道建筑 canal construction
渠道渐变段 canalized transition; canal transition
渠道建筑物 canal construction
渠道交叉 canal crossing
渠道交叉建筑物 canal crossing structure
渠道节制阀 canal check; channel check
渠道节制闸 canal check; channel check
渠道截面 canal cross-section; canal section; channel section
渠道进水口 canal inlet; canal intake; channel intake
渠道进水量 channel inflow
渠道进水闸门 head gate
渠道径流 channel runoff
渠道开挖 canal construction; channel digging
渠道开挖机 trencher
渠道抗冲 channel stabilization
渠道抗冲措施 channel stabilization measure
渠道控制 channel control
渠道口工程 canal headwork
渠道口流量 canal capacity at the headwork
渠道宽度 channel span
渠道扩大(段) channel expansion; expansion in channel
渠道沥青衬砌 asphalt canal lining
渠道流量 canal capacity
渠道流量公式 capacity formula
渠道流量(逐渐)增长 <渗流引起的> channel flow accretion
渠道路线 channel way
渠道轮灌 canal rotation
渠道排灌能力 canal capacity; carrying capacity of canal; channel capacity
渠道排沙设施 ejector
渠道排水 gutter drainage
渠道旁边的路 berm for channel
渠道配件 channel fittings
渠道配水系统 canal distribution system
渠道坡度 channel slope
渠道剖面 canal cross-section; canal section
渠道砌筑机 canal lining machine
渠道前池 canal pond
渠道桥 canal bridge
渠道清淤 canal desilting; canal silt clearance; channel cleanout
渠道容量 canal capacity
渠道容许冲刷流速 canal permissible scouring velocity
渠道容许流速 permissible canal velocity
渠道入口 canal entrance
渠道深度 channel depth
渠道渗流损失 canal seepage loss; channel loss; channel seepage loss
渠道渗漏 canal seepage; channel seepage
渠道渗漏损失 canal seepage loss; channel seepage loss
渠道施工 canal construction
渠道式电子倍增器 channel electron multiplier
渠道疏浚 canal cleaning; training of channel
渠道输水量 delivery of canal
渠道输水能力 canal capacity; carrying capacity of canal; channel capacity
渠道输水损失 channel loss
渠道刷深 channel degradation
渠道水 canal water; channel water
渠道水力特性 canal hydraulics
渠道水量损失 canal wastage
渠道水流 channel flow
渠道水深 depth of water in channel
渠道隧洞 canal tunnel
渠道损失 canalized wastage
渠道弯曲 canal curve; channel curve
渠道网 canal network
渠道稳定化 channel stabilization
渠道稳定性 channel stability
渠道系统 canalization; channel system; canal system
渠道泄水闸 canal scouring sluice
渠道修整机 canal trimmer

渠道畜水量 conduit storage;channel storage
渠道养护常数 constant of channel maintenance
渠道养鱼 channel fish culture
渠道引水式电站 power canal development
渠道淤积 canal sedimentation;canal silting;channel accretion;channel deposit;channel sedimentation;channel silting
渠道约束条件 channel constraint condition
渠道展宽 channel widening
渠道整治 regulation of channel
渠道轴线 canal axis
渠道自动闸门 automatic gate for channel
渠道纵断面 profile of canal
渠道纵坡线 canal grade line;grade line of canal
渠堤超高 canal freeboard
渠底 canal bed;canal bottom;channel bed;channel bottom
渠底宽(度) canal bottom width
渠底渗透计测定法 lysimeter beneath channel bed
渠段 canal pond;canal pool;canal reach;channel pond;channel reach;reach of channel
渠盖 drain cover
渠工学 canal engineering
渠灌 canal irrigation;irrigation with ditch
渠灌井排 irrigation with ditch and drainage with well
渠化 canalize;canalization;channelization;channelize
渠化Y形交叉 channel(l)ized Y intersection
渠化坝 canalization dam
渠化标记 channelizing marking
渠化标线 channelizing marking;channel(l)ized mark(ing)
渠化布置 channel(l)ized layout
渠化车道 channel(l)ized lane
渠化程度 degree of canalization;degree of channelization
渠化工程 canalization project;canalization works;canalized engineering
渠化航道 canalized channel;canalized waterway
渠化河道 canalization river;canalized channel;canalized waterway
渠化河段 canalized river section;canalized river stretch;channel(l)ized river section;channel(l)ized stream section;channel(l)ized stretch
渠化河流 canalized river;channel(l)ized river;channel(l)ized stream;locked river;locked stream
渠化交叉口 channelize the intersection;channel(l)ized intersection
渠化交通 channelizing traffic;channel(l)ized traffic
渠化交通岛 channelizing island;directional island
渠化交通道路 canal road
渠化交通荷载 channel(l)ized load
渠化交通交叉口 channelizing intersection
渠化路带 channel(l)ized[channelizing] strip
渠化剖面 canalization section
渠化设计 channel(l)ized layout
渠化水道 canalized waterway
渠化水流 channel(l)ized flow
渠化梯级水头 water head canalized step
渠化线 channelizing line
渠化状态 pool stage
渠浇地 river-irrigated land
渠口 outfall
渠口船闸 basin lock
渠宽 width of channel
渠流隧洞 free surface flow tunnel;non-pressure tunnel
渠排 drainage with ditch
渠旁盐渍区 salinization range along ditch
渠桥 aqueduct bridge
渠式道路 canal road
渠式干船坞 canal dry dock
渠首 canal head;canal intake;head bay;head of canal;head race
渠首部分 head reach
渠首部设施 canal headwork
渠首处渠道过水能力 canal capacity at the headwork
渠首分流结构 diversion structure
渠首工程 canal headwork;diversion works;headworks;intake head
渠首工程结构 headworks structure
渠首构筑物 headworks
渠首过水能力 canal capacity at the headwork
渠首建筑物 canal headwork
渠首进水闸 head regulator;intake sluice
渠首毛灌水率 head-gate duty of water
渠首排沙设施 excluder
渠首前池 canal pond
渠首调节建筑物 head regulator
渠首堰 diversion weir
渠首引水率 gross duty(of water);head-gate duty
渠首用水率 gross duty(of water)
渠首闸 inlet sluice;water intake sluice
渠首闸门 canal head gate;crown gate;head gate
渠首闸门灌水率 head-gate duty of water
渠头控制水量建筑物 headworks
渠网水质监测 water quality canal network monitoring
渠系 canal system
渠系建筑物 canal structure
渠系(用水)效率 project efficiency
渠相径流 channel-phase runoff
渠闸 canal lock
渠状断层 moatlike fault

曲

曲调 tune
曲艺场 variety hall

取

取保候审 recognizance
取保护 fetch protection
取钓分析 ladle analysis
取补装置 complementer
取操作 extract operation;load operation
取操作数 fetch operand
取操作数功能块 operand fetch module
取操作数周期 operand fetch cycle
取长补短 cross fertilisation;trade off
取车时间 collection time
取程序 program(me)fetch
取出 dislodge;dislodging;draw-off;extraction;fetch;fish out;take-off;take-out;taking-out
取出不用 take-out of service
取出岔线货车 setting-out of wagons at sidings
取出机 knockout machine
取出机构 unloading device
取出馏分 take-off the-fraction
取(出)码 code fetch
取出盘 take-off reel
取出燃料 <指放射性燃料> defueling
取出时间 access time
取出试样 draw trial
取出位 fetch bit
取出序列 fetch sequence
取出岩芯 drill out;removal of core
取出指令【计】 extraction instruction
取出周期 fetch cycle
取出爪 extracting jaw
取出装入跟踪 fetch-load trace
取出装置 withdrawal device
取代 replacement;replacing;subrogate;supersede;supplant
取代苯 substituted benzene
取代材料 substitute material
取代产物 substitution product
取代的 substitutive
取代度 degree of substitution
取代反应 reaction of substitution;substitution reaction
取代分子 substituent molecule
取代酚 substituted phenol
取代固溶体 substractional solid solution
取代厚度 displacement thickness
取代基常数 substitution constant
取代技术 replacement technique
取代扩展法 displacement development
取代鞣剂 replacement tanning agent
取代酸 replacing acid
取代碳氢化合物 substituted hydrocarbon
取代物 substitute
取代系数 substitution index
取代线路 override circuit
取代效应 substituting effect
取代衍生物 substitutive derivative
取代债权人 subrogation
取代展开法 displacement development
取代者 substituent
取代值 substitution value
取代中心 substitutional center[centre]
取代字符 substituted character
取代作用 metalepsis;metalepsy;substitution【化】
取导数 differentiation
取得版权 acquisition of copyright
取得保险单 effect a policy
取得车辆 wagon collected
取得成本 acquisition cost
取得存款 collection of deposits
取得的权利 acquired right
取得递盘 secure a bid
取得额调整数 acquisition adjustment
取得过程 get procedure
取得技术 acquisition of technology
取得交换价值 acquisition of exchange value
取得控制权 acquisition of control
取得利润 extraction of the profit;securing of the profit;take profit
取得联络 make a junction
取得垄断 monopolize
取得赔偿 recover damage
取得平衡 level off
取得时效 acquisitive prescription
取得收入的工作 revenue effort
取得现代技术 access to modern technology
取得协议 conclude a bargain
取得循环 fetch cycle
取得语句【计】 GET statement
取得指令 get instruction
取得专利 patent
取得专利权者 patentee
取得资格 qualify
取得资料的方法 access method
取灯泡器 lamp extractor
取缔 ban;lid
取缔暴利法令 anti-profiteering ordinance
取缔超速汽车的警察 <美俚> speed cop
取缔垄断 trust busting
取电器 troll(e)y
取钉锤 nail puller
取锭机构 withdrawal mechanism
取锭器 ingot adapter;ingot retractor
取读数 <仪表上> make readings;take readings
取费标准 service fee norm;standard for collecting fee
取分离器气样 separator gas sampling
取分离器水样 separator water sampling
取分离器油样 separator oil sampling
取粉管 calyx;mud bucket;mud pipe;mud tube;sediment pipe;sediment tube
取粉管接头 sludge barrel head
取杆器 <用以取出钻井时断掉的钻杆> beche;bell jar
取杆器舌门 <钻机> clapper
取管器 bell jar
取过试样的混凝土梁 cored beam
取过钻芯试样的混凝土梁 cored beam
取海水设备 seawater intake facility
取换工具 replacer
取回 drawing out;recall;recapture;resumption;subduce
取回被扣押财物 replevy
取回股份 reacquired stock
取回绳 retrieving line
取回文据 reconveyance deed
取回销售收入 recouping sales income
取回债券 reacquired bond
取火盒 tinderbox
取火镜 burning glass
取货单 carrier's note
取货收据 receipt for freight
取货通知 carrier's note
取货证 carrier's note
取货作业 unpack;unstuffing
取极小值 minimalization
取件装置 pick-off unit
取胶辊 pick-up roll
取截面 cross-sectioning
取尽 exhausting
取井底样 bottom-hole sampling
取井口样 well-head sampling
取景 peep-sight
取景窗口 viewfinder window
取景地点 view-endowed site
取景镜片 finder lens
取景镜头 landscape lens;viewing lens
取景孔 viewing aperture
取景框 camera aperture
取景器 finder;optic(al) finder;picture finder;viewer;viewfinder
取景器光圈 viewfinder indication of stop in use
取景器光线 finder light
取景器目镜 viewfinder eyepiece
取景器相框 viewfinder mask
取景释放钮 viewfinder release button
取景调节器 frame hold
取景物镜 viewing lens
取决于市价订单 resting order
取决于土壤情况 depending on the condition of the soil
取决于土壤温度 dependent on soil temperature
取款 collection;withdrawal
取款单 bill of credit;withdrawal slip
取款费 collection charges

取款凭单 bill of credit
取款凭证 bill of credit
取款人 remittee
取矿机 reclaimer
取离散值 quantization;quantize
取料地坑 reclaiming tunnel
取料端 flowing end
取料辊 pick-up roll
取料机 reclaimer
取料机械设备 reclaiming unit
取料胶带输送机 reclaiming belt conveyer[conveyor]
取料坑 borrow
取料量 reclaiming rate
取料耙 scraping rake
取料器 bunk;dispenser
取料试验 spoon-proof
取料输送机 reclaiming conveyer[conveyor]
取料喂料机 withdrawal feeder
取料系统 reclaim system
取料叶桨 reclaiming paddle
取料运输机 retriever conveyer[conveyor]
取流 current pickup
取码周期 code fetch cycle
取幂 exponentiate;exponentiation
取幂符号 exponentiation sign
取幂运算 exponentiation operator
取模 delivery;impression taking;taking a squeeze
取模函数 mod
取模斜度 pattern taper
取模针 picker
取某一固定的方向 canalize
取木钻 increment borer
取泥浆样筒 slurry sampler
取泥样 mud sampling
取黏[粘]土岩芯管 clay coring barrel
取暖 heating
取暖插塞 heater plug
取暖带 heating strip
取暖电炉 electric(al) space heater;radiator
取暖费补贴 subsidies for heating expenses
取暖锅炉 heating boiler
取暖火炉 latrobe
取暖机组 unit heater
取暖煤 fire coal
取暖耦合器 heater coupler
取暖器 heater;heating apparatus;radiator;warmer
取暖器管道 heated line
取暖热水管 heating water pipe
取暖设备 heating installation
取暖室 <寺院内的> calefactory
取暖系统 heat(ing) system
取暖用具 heating appliance
取暖用煤 fire coal
取暖用热水 heating water
取暖油炉 oil-filled radiator
取暖与通风 heating and ventilating
取暖照明及动力成本 heat, light and power cost
取暖装置 heater
取排水设备 facility for water intake and drainage
取平均数 take the mean
取平均值 averaging
取去 remove
取权 weighting
取权因子 weighting factor
取热器 heat extractor
取砂袋 sediment bulb
取砂坑 sandy borrow
取砂区 sand borrow area
取砂(样)器 <又称取沙(样)器> sand sampler;silt-sampler
取舍 alternative;trade off;volition
取舍标价 alternate bid
取舍线 <计算不规形面积的> give-and-take lines
取数 access;fetch;load;peek;taking-off
取数臂 access arm
取数方法 access method
取数方式 access mode
取数据 data fetch;memory fetch
取数据时间【计】data age
取数据信号 data taken signal
取数孔 access hole
取数扫描【计】access scan
取数时间 access time;data age;latency time
取数违法 fetch violation
取数位 access bit
取数指令 load instruction
取水坝 intake dam
取水泵房 intake pump house
取水层数 number of pumping strata
取水池 intake basin
取水道 water intake gallery
取水(道)首部 intake heading
取水段长度 length of pumping sections
取水段间距离 interval between pumping sections
取水段深度 depth of pumping sections
取水段数 number of pumping sections
取水阀门 intake gate
取水阀塔 valve tower
取水干管 intake main
取水工程 intake works;water-diversion project
取水工程布置图 arrangement map of water withdrawal engineering
取水构筑物 intake structure;intake works;water intake structure;water intake works
取水管 hydrant;intake pipe
取水涵洞 intake culvert
取水戽斗 intake of well
取水建筑物 intake structure;intake works;water intake structure
取水井 intake well
取水井中级配良好的沙砾滤层 filter pack
取水口 <取水头部> intake;intake header;water catchment;water intake;water inlet
取水口格栅 intake screen
取水量 quantity of water intake;rate of draft;water intake;water withdrawal
取水量线 draft(ing)line
取水率 draft rate;intake rate
取水渗井 soak well
取水隧洞 intake tunnel
取水塔 draw-off tower;intake tower;take-off tower;tower intake
取水头(部)intake
取水许可制度 licence system of using water;license system of using water
取水样 water sampling
取水样瓶 water sample bottle
取水样器 water sampler
取水用螺旋钻 Archimedean drill
取水闸门 head gate
取顺时针方向为正的 positive clockwise
取送车费及其他 cars collection and delivery charges
取送调车 moving in and out cars;taking-out and placing-in of cars
取锁法 interlocking
取土 borrow;borrow soil;sampling soil
取土杓 clay scoop
取土场 borrow area
取土场地 borrow site
取土地点 borrow site
取土方法 sampling method
取土方数 borrow yardage
取土费用 cost of borrow
取土分样器 sample splitter
取土沟 earth supply ditch
取土管 soil sampling tube
取土开挖 borrow cut;borrow excavation
取土坑 borrow area;borrow pit;earth pit;pit;side borrow
取土孔 soil sampling borehole
取土量 quantity of borrow
取土螺(旋)钻 earth auger;soil auger
取土面积 borrow area
取土器 geotome;sampler;sampling tube;soil sampler
取土器衬套 sampler liner;sampler tube
取土器的基本参数 parameters of soil sampler
取土器吊绳 swinging core line
取土器端部 sampler head
取土器面积比 area ratio of the sampler
取土器容纳管 sampler tube
取土器升降绳 swinging core line
取土器种类 type of soil sampler
取土区 borrow area
取土筒 soil sample barrel
取土样 sampling;soil sampling
取土样的麻花钻 earth screw;twisted auger
取土样孔 tube sample boring
取土样器 core cutter;core lifter
取土样设备 soil sampling equipment
取土样钻 soil auger
取土样钻孔 tube sample boring
取土钻 soil auger
取土钻头部 sampling tip
取挖方回填 reclaiming
取伪 type B error
取未扰动砂样 undisturbed sand sampling
取位 fetch bit
取物滑轮匣 <起重机吊具> lower load block
取物及堆垛的附属器具 article grappling and stacking attachment
取物器 catcher
取下 dismount
取下螺栓 unbolt
取下外壳 desheathing
取下修复再装上 remove fair and refit
取下悬挂物 unhang
取现 encashment
取向 orientation;steer
取向标记 description point;orientation mark
取向衬度 orientation contrast
取向错误 misorientation
取向度 degree of orientation
取向法则 orientation rule
取向范围 orientation range
取向附生 epitaxy
取向附生的 <晶体> epitaxial
取向关系 orientation relationship
取向核 oriented nuclei
取向极化 orientation polarization
取向极化率 orientation polarizability
取向角 angle of orientation
取向接长的 <晶体> epitaxial
取向矩阵 orientation matrix
取向聚合物 orientated polymer
取向均一性 orientation uniformity
取向力 dipole-dipole force;Keesom force;orientational force
取向律 orientation law
取向无序 orientation disorder
取向吸附 orientation adsorption
取向效应 orientation effect
取向性硅钢片 grain-oriented electrical steel
取向因素 orientation factor
取向运动 orientation movement
取像透镜 taking lens
取消 abolish(ment);abolition;abrogate;annul(ment);back-out(of);blank out;bust;cancel(lation);countermand;cross-out;delete;disannul;liquidate;nullification;nullify;quash;recall;repeal;rescind;rescission;retract(ion);revoke;undo;vacate;voidance;withdrawal
取消……的资格 disqualify;dishabilitate
取消按钮 cancellation button;cancelling button
取消报价 withdraw an offer
取消闭塞 block cancel(l)ed;block cleared;block released;cancel a block;clearing of section;giving "line clear";open block;unblock
取消闭塞电流 unblocking current
取消拨款 strike-off an appropriation from the budget
取消出口 cancelling
取消传输记号 cancel transmission
取消传送 cancel transmission
取消传送记号 cancel transmission
取消船级 class withdrawal;raze a vessel
取消存储 storage cancellation
取消的叶 cancelled leaf
取消登记 cancellation of registration
取消抵押品赎回权 foreclose;foreclosure
取消抵押品赎回权的法律手续 judicial foreclosure
取消地租 abolition of ground-rent
取消订单 countermand of an order
取消订货 cancel an order;cancellation of the order;countermand;kill order
取消订货合同 cancellation of purchase contract
取消定额分配 deration
取消对……的管(限)制 unfreeze
取消飞行 <因设备故障> abort
取消封港 taking-off the embargo
取消符 erase character;ignore character
取消符号 cancel character;suppression symbol
取消符号开关 cancel mark switch
取消股利 rescission of dividends
取消国际标准 withdrawal of international standard
取消合同 annulment of contract;cancel a contract;cancellation of contract;contract cancellation;discharge of contract;rescind a contract;rescission of contract
取消交织 cancellation interlace
取消进路 cancel a route
取消列车 cancellation of train
取消临时工 decasualization
取消旅行计划 bump
取消判决 reversal of judgment
取消歧视待遇 elimination of discrimination treatment
取消千斤顶 retract jack
取消前令 countermand
取消前有效 good-till-cancelled;good until cancelled
取消商品经济 abolish commodity economy
取消手柄 cancelling handle
取消赎取权 foreclose
取消特别提款权 cancellation of special drawing right
取消条款 cancellation clause

取消通话 cancellation of a call
取消图(边)界 map dissolve
取消下潜状态 secure the diving stations
取消限速 derestrict
取消限制 derestrict;ease
取消项目 cancellation of project
取消协议 abrogate the agreement
取消信号 cancellation signal;cancelling signal;withdrawal of signal
取消银行贷款 take-out
取消营造执照 cancellation of building licence[license]
取消债务 cancellation of indebtedness;debt cancellation
取消者 nullifier
取消征用 derequisition
取消支票 cancel a cheque
取消注册的档案 cross-out file
取消装置 cancellation device
取消资格 disqualification;disqualify
取消字符 cancel character
取芯 core lifting;take a core
取芯棒 cored driver
取芯次数 coring time
取芯风钻 core pneumatic drill
取芯工具 coring tool
取芯管 core barrel
取芯回次 core run
取芯夹具 core catcher
取芯进尺 cored footage;coring footage
取芯井 core well
取芯井段 coring interval
取芯孔 cored hole
取芯内管 <三层岩芯管> core shell
取芯器 corer;core taker;coring apparatus;coring device
取芯设备 core equipment
取芯式涡轮钻具 coring type turbodrill
取芯双管钻具 double tube core barrel
取芯筒 core barrel;core taker
取芯楔 core-lifter wedge
取芯牙轮钻头 coring roller bit;roller-cone core bit
取芯样 coring
取芯凿岩机 hammer drill;hammer plugger
取芯凿岩钻 core hammer bit
取芯总次数 total times of coring
取芯钻 corduroy cutter;corduroy drill;core tool;coring tool;drifter drill;core cutting machine
取芯钻机 coring machine
取芯钻夹具 core catcher
取芯钻进 core drilling
取芯钻进速度 coring rate
取芯钻井 core drilling
取芯钻具 core bit;coring tool;core drill
取芯钻孔 cored borehole
取芯钻探 core boring;core-drilling exploration
取芯钻探法 corduroy-drilling;corduroy drill method;core drilling;core drill method
取芯钻头 core bit;coring bit;coring crown;crown bit
取芯钻头压力 coring bit pressure
取芯作业 coring
取雪管 snow sampler;snow tube
取雪样 snow sampling
取压点 <测压管的> tapping point
取压分接管 pressure tap(ping) hole
取压孔 pressure port
取岩粉样(品) sludge sampling
取岩石 <用杓斗> bite
取岩屑 cuttings sampling
取岩芯 core extraction;coring;recovery of core;running coring
取岩芯冲浆钻探 core wash boring
取岩芯定向定位仪 oriented-core barrel
取岩芯法 core method;core removal
取岩芯管 core barrel
取岩芯设备 coring apparatus;coring device
取岩芯爪压盖 core catcher gland
取岩芯装置 coring device
取岩芯钻 core cutter;core lifter
取岩芯钻头 boring core bit;core-barrel bit;cored drill bit;rock core bit
取岩样 boring sample
取样 essaying;exampling;outgoing material;purposive sampling;sample collection;sample drawing;sample taking;sampling;sampling taking;take sample;taking of samples;thief
取样靶 sampling target
取样保持 sample-hold
取样保持电路 sample-and-hold circuit
取样保持技术 sample-and-hold technique
取样保存 sample reservation
取样报告 sampling report
取样杯 sampling cup
取样泵 sampling pump
取样比 sample ratio;sampling rate;sampling ratio
取样边 sample edge
取样编录 sampling documentation
取样变换器 sampling switch
取样标绘器 sampling plotter
取样步骤 sampling procedure
取样部分 sample segment
取样参数计算 sampling parametric(al) computation
取样槽 testing tray
取样层位 sample horizon;sampling horizon
取样长度 recovery length
取样程序 sample program(me);sampling procedure;sampling program(me)
取样出口 sampling outlet
取样传感器 sampling detector
取样垂线 sampling vertical(line)
取样袋 sample bag;sampling bag
取样单位 sampling unit
取样单元 sampling unit
取样地 sample plot(ting);sample section
取样地点 location of sample
取样点 sample dot;sample point;sampling location;sampling point;sampling site
取样点类型 sampling point type of water
取样点密度 density of sampling points
取样电极 sampling electrode
取样电路 sample circuit;sampling circuit
取样电气法 electric(al) method of sampling
取样电压 sampling voltage
取样定理 sampling theorem
取样定向流速仪 sampling directional current meter
取样动作 sampling action
取样斗 sampling spoon
取样断面 sample section
取样阀(门) sampling valve;sampling tap
取样阀值 sampled threshold
取样法 exploration method by sampling;method of taking samples;induct method <在风道取样长度内试验或测量方法>
取样方案 sampling plan
取样方法 method of sampling;sampling;sampling method
取样方法和工具 sampling method and tool
取样方法与设备 method of obtaining sample and equipment(collecting method of sample)
取样分布 sampling distribution
取样分析程序 sampling and analysis program(me)
取样分析仪 sampler-analyzer
取样伏特计 sampling voltmeter
取样盖 thief hatch
取样杆 sample rod
取样高程 altitude of sample
取样工 sample taker
取样工具 bleeding iron;sample tool;sampling tool
取样观测 sampling observation
取样管 coupon;pipe sampler;probe tube;sample pipe;sampler conduit;sampler pipe;sampler tube;sampling nozzle;sampling tube;stopple coupon;sampling pipe
取样管道 sampling line
取样管路 sampling line
取样棍 bleeding iron
取样过程 sampling process
取样函数 sampling function
取样环刀 sampling ring cutter
取样环勺 sampling ring spoon
取样火花室 sampling spark chamber
取样机 sampler;sampling rig;mechanical sampler
取样机构 sampling mechanism
取样畸变 sampling distortion
取样计划 sample program(me)
取样计数器复位 sampling counter clear
取样记录 sample log
取样技术 sampling technique
取样间隔 sampling spacing;sample interval
取样检测器 sampling detector
取样检查 check by sampling;sampling inspection;spot-check
取样检验 apart check;inspection by sampling;probe inspection;sampling inspection;take a sample for examination and test
取样交换机 sampling switch
取样井 sampling well
取样距离 sample distance;sample spacing
取样开关 drip cock
取样空间 sampler space;sample space
取样孔 sample hole;sampling aperture;sampling borehole;thief hatch;thief hole
取样控制 sampling control
取样口 sample connection;sample port;thief hatch;thief hole
取样类型 sample type
取样冷却器 sampling cooler
取样理论 theory of sampling
取样连接管 sampling connect;sampling connection
取样连接器 sampling connector
取样龙头 sample cock
取样路径 sample path
取样率 rate of draft;sample rate;sampling fraction
取样脉冲 burst flag;sampling pulse
取样脉冲发生器 sampling pulse generator
取样脉冲宽度 sampling pulse width
取样脉冲相位 sample pulse phase
取样模 sampling mo(u)ld
取样模型 sampling pattern
取样喷嘴 sample nozzle;sampling nozzle
取样片 coupons
取样偏差 sampling bias
取样频率 frequency of sampling;sampling frequency;sampling rate
取样平均 sample averaging
取样平均迭代法 sample-averaging iteration method;sample mean iteration method
取样平均值 sample mean
取样瓶 sample bottle;sample bulb;sampling bottle;bottle sampler
取样瓶取样 bottle sampling
取样瓶水样 <美国泥沙测验用语,直接从河中取样> dip sample
取样剖面 sample profile
取样剖面记录 sample log
取样器 chief;core cutter;sample collector;sample grabber;sampler;sample splitter;sample taker;sample thief;sampling appliance;sampling device;sampling instrument;sampling spoon;sniffer;thief;trier;verifier
取样器长度 length of sampler
取样器衬套 sampling liner
取样器面积比 area ratio of the sampler
取样器直径 diameter of sampler
取样枪 sampling gun
取样全套工具 sampling kit
取样扰动 sample disturbance;sampling disturbance
取样扰动土样 taking disturbance soil sample
取样扰动样 retrieving disturbance sample;taking disturbance sample
取样日期 date of sampling;sampled date;sampling date
取样勺 sample spoon;say ladle;sampling spoon
取样设备 sampling appliance;sampling device;sampling equipment;sampling installation
取样设计 sampling design
取样摄像机 sampling camera
取样深度 sample(d)[sampling] depth
取样深度标高 depth of elevation
取样时间 sampling time
取样时间表 sampling schedule
取样时间间隔 sample period;sample interval
取样示波器 sampling oscilloscope;sampling scope;samploscope
取样试验 pick test;sampling test
取样室 sampling room
取样数据 sampled data
取样数据控制系统 sampled-data control system
取样数据滤波器 sampled-data filter
取样数据系统 sampled-data system
取样伺服机构 sampling servomechanism
取样速率 sampling rate
取样损失 sampling loss
取样锁相环 sampled phase-locked loop
取样套管 drive pipe
取样条件 sampling condition
取样同步脉冲 sampling synchronization pulse
取样筒 core taker;sampler barrel;sample spoon;sampling barrel;sampling spoon
取样头 sampling head;sampling probe
取样图案 sampling tessellation
取样图像 sampling image
取样维持电路 sample-and-hold circuit
取样位置 sampling location;sampling

point;sampling position
取样位置图 map of sample site
取样误差 sample error;sampling error
取样系统 sampling system
取样系统小型化 sample system miniaturization
取样线路 sampling line
取样限数计数器 sampling counter maximum
取样信号 sampled signal
取样信号网络 sampling signal network
取样信息的期望值 expected value of sample information
取样型式 sampling pattern
取样旋塞 sampling cock
取样旋转器 sample spinner
取样选通电路 sampling gate
取样雪样器 snow sampler
取样验收 acceptance sampling
取样仪器 sampling instrument
取样与同步放大器 sample-and-hold amplifier
取样员 sample grabber;sampler(jerker)
取样原理 sampling theorem;theory of sampling
取样原则 sample theorem
取样站 sample station;sampling station
取样站网 network of sampling stations
取样针 sampling probe
取样正态分布 sampling normal distribution
取样值 sample value
取样指示计 sampling valve indicator
取样质量 sampling quality
取样钟 sampling clock
取样舟 sampling boat
取样周期 period of sampling;sample period;sampling period
取样柱 sample stand
取样铸件 test casting
取样爪 sample cutter
取样转换器 sampling switch
取样装置 sampling device;sampling unit
取样装置列 sampling train
取样组件 sampling assembly
取样钻机 sampler drill;sampling drill
取样钻进 sample boring;sample drilling
取样钻探 sample boring;sample drilling;sampling drilling
取样钻头 chisel bit
取样钻眼 sample drilling
取样作用 sampling action
取用土料 borrow material
取邮包器 mail catcher;pouch catcher
取邮包器架 mail catcher bracket
取油样器 oil thief
取油样旋塞 oil sampling cock
取原状土样 original soil sampling;sampling undisturbed soil;dry sampling
取整 round off
取整的 rounded-off
取整数 rounded figure;round number;round-off number
取整误差 rounding error;round-off error
取之不尽的能源 nondepletable energy source
取直 cut-off
取值范围 sampling range;span
取值过程 sampling process
取值速率 sampling rate
取指令 instruction fetch;load instruction
取指令部件 instruction fetch unit
取指令操作 instruction fetch operation
取指令程序 instruction fetch routine
取指令阶段 instruction fetch phase
取指令时间 instruction fetch time
取指令微操作 instruction fetch microoperation
取指令周期 instruction fetch phase
取主转移 pivot transformation
取砖器 off-bearer
取装跟踪 fetch-load trace
取字时间 word time
取自然界的装饰主题 ornamental motif taken from nature
取自自然界的装饰主题 decorative motif taken from nature
取走 off bear;take-off
取组合水样 composite water sampling
取钻器 <从钻孔里钩出断钻杆的工具> drill extractor

去

去凹器 dent remover
去白云石化作用 dedolomitisation [dedolomitization]
去饱和作用 desaturation
去爆噪声 elimination of burst noise
去表面层 scalping
去冰 deicing
去冰器 deicer
去冰系统 deicing system
去冰盐 deicing salt
去冰用水 defrost water
去玻璃处理 de-glassing
去玻璃化 devitrification;devitrify
去玻璃化作用 devitrification
去补体 de-complementize
去不掉的 indelible
去残渣工具 sludger
去草剂 terbutryn(e)
去草净 terbutryn(e)
去层理作用 de-stratification
去颤 defibrillation
去颤器 defibrillator
去潮 dehumidification
去潮气器 moisture trap
去潮器 moisture eliminator
去尘 dustproofing
去尘通风 dedusting ventilation
去程电路 go circuit
去程通路 forward path
去程运费 out freight
去齿的螺栓 skinned bolt
去齿轮毛边机 gear burr machine
去翅机 burring machine
去臭 deodo(u)rizing
去臭剂 deodo(u)rizing material
去臭味剂 deodo(u)rant
去除 removal;remove
去除百分率 percentage removal
去除百分数 percentage reduction
去除表面瑕疵 tease
去除表土 stripping
去除擦痕 scratch removal
去除擦伤 scratch removal
去除方法 removal method
去除飞边 trim
去除覆盖层 de-coat(ing)
去除痕量有机污染物 removing trace organism contaminant
去除痕量有机物 removing trace organism
去除颗粒 particle removal
去除粒面 de-grains
去除率 clearance;removal rate
去除毛刺 de-flashing;flash removal;deburr(ing)
去除毛结 burling
去除冒口 feeder removal;remove feeder;remove riser
去除披缝 remove flash
去除器 remover
去除取样法 removal sampling
去除热原法 de-pyrogenation
去除水分 moisture-removal
去除水垢 scaling off
去除凸纹 remove the burr
去除为增加混凝土和易性而超用的水 excess water removal
去除污染 decontaminate
去除污染物 pollutant removal;removal of pollutants
去除锡层 de-tinning
去除系数 removal coefficient
去除效率 removal efficiency;retention efficiency
去除氧化皮 scale removal
去除抑止 de-inhibition
去除油灰 removal of putty
去除油污 degrease
去除有机污染物 removing organism contaminant
去除杂质改进材料的化学物理性质 beneficiation
去磁 de-gauss(ing);de-magnetism;de-magnetization;de-magnetize;depolarize
去磁安匝 de-magnetizing ampere turns
去磁场 degaussing field
去磁磁场 demagnetizing field
去磁电缆 degaussing cable
去磁电流 demagnetizing current
去磁电路 degausser
去磁电阻 de-magnetizing resistance
去磁扼流圈 degausser
去磁机 degausser
去磁力 de-magnetization force
去磁能 de-magnetizing energy
去磁器 degausser;de-magnetizer;magnetic eraser
去磁曲线 de-magnetization curve
去磁绕组 bucking winding
去磁损失 de-magnetization loss
去磁系数 de-magnetization factor
去磁线圈 bucking coil;degaussing coil;demagnetizing coil;octagonal coil
去磁线匝 de-magnetizing band
去磁效应 de-magnetizing effect
去磁因数 de-magnetizing factor
去磁匝数 demagnetizing turns
去磁装置 de-magnetizer
去磁状态 de-magnetized state
去磁作用 back induction
去刺 burring
去氮法 denitrification;denitrogenation
去氮作用 denitrification
去道钉头【铁】 necking
去电 de-electrifying
去电离 de-ionize
去电离电势 extinction potential
去电离电位 de-ionizing potential
去电离熔丝 de-ion fuse
去电子作用 de-electronation
去掉边材的木料正面 heart face
去掉废石的原矿 run-of-mill ore
去掉夹杂铁 tramp-iron removal
去掉水分的 unwatered
去掉水分以便于储藏 remove the moisture for easy storage
去掉盐分 freshen
去掉引信 defuse
去掉障碍物 unplug
去扼流作用 unchoking effect
去防水胶合板 perfectly water-proofing plywood
去防锈剂 anti-tarnishing agent
去飞边 de-flashing;edging;trimming
去肥 denudation
去分化 dedifferentiation
去分配 deallocate
去封闭因子 de-blocking factor
去氟作用 defluorination
去负载 unloading
去铬 dechromisation[dechromization]
去梗机 de-stemmer;stemming machine
去沟道效应 de-channel(l)ing
去垢 anti-sludge;denudation;detergency[detergence];skim
去垢本领 detergent power
去垢的 abstergent;abstersive;antisludging;pellant
去垢粉 Dutch cleanser
去垢工具 scaling tool
去垢机 scaling machine
去垢剂 abstergent;anti-lithic;degreaser;deincrustant;detergent (agent);scaler
去垢能力 descaling capability;detergency[detergence]
去垢添加剂 detergent addition;detergent additive
去垢性 detergency[detergence]
去垢油 detergent oil
去垢作用 detergency[detergence];detergent action;detergent effect
去箍压力机 press for removing tyres
去骨 bone;boning
去管口毛刺绞刀 pipe burring reamer
去光 <旧漆面> sand down
去光剂 matting agent
去光泽 deluster(ing);frosting;matting;mottling
去光泽的 tarnishing
去光泽法 felting down
去光泽面 frosting
去光泽油 tarnishing oil
去硅 desilicification;desiliconization
去硅反应 desilication
去锅炉水锈用锤 boiler scaling hammer
去焊缝 deseaming
去焊枪 de-solder-ring gun
去焊(药)剂 de-flux
去核机 corer
去荷 pressure relieve;unload
去黑废水 deinking waste(water)
去痕石 Scotch stone
去厚机械刨 <木工> thicknessing machine
去花岗岩化方式 de-granitization way
去话长途电路 originating toll circuit;outgoing trunk circuit
去话电路 outgoing circuit
去话呼叫 outgoing call
去话话务员 outgoing operator;outward operator
去话台 A-position;outgoing position;outward position;A station
去环作用 decyclization
去灰分 de-ash;de-ashing
去灰雾 defogging
去挥发分作用 devolatilization
去活化 deactivate
去活化作用 deactivation
去激电路 deenergizing circuit
去激(发)de-excitation
去激发光子 de-excitation photon
去激发截面 de-excitation cross- section
去激发效应 de-excitation effect
去激化 deactivate
去激励 deactivation;de-energise[de-energize];deenergizing;deenergization
去激励的 currentless;deenergized

去激作用 deactivation
去极化电极 impolarizable electrode
去极化分量电平 level of the de-polarized component
去极化剂 depolarizer[depolorizer]; depolarizing agent
去极化器 depolarizer[depolorizer]
去极化系数 depolarization factor
去极化因子 depolarizing factor
去极化装置 depolarizer[depolorizer]
去极化作用 depolarization; unpolarizing
去极剂 battery de-polarizer;depolariser
去加重 deaccentuation;de-emphasis
去加重电路 de-emphasis circuit
去加重器件 de-emphasis parts of an apparatus
去加重网络 de-emphasis network
去假频滤波 anti-alias filtering
去碱 alkali elimination;lixiviation
去胶 de-gumming;striping
去胶结作用 de-cementation
去胶试验 stripping test
去焦 defocussing
去角机 muley saw
去角锯 muley saw
去角(纵剖)砖 king closer
去矫 de-emphasis
去结合 debond(ing)
去禁溜线信号 shunting signal for prohibitive humping line
去静电器 destaticizer
去聚焦 defocus
去聚器 de-buncher
去聚束 de-bunched beam
去聚束脉冲 de-bunched(beam) pulse
去壳 dejacket
去壳器 scaler
去空气法 <制砖等用> deaering
去矿化 demineralize
去矿化装置 demineralizer
去矿化作用 demineralization
去矿质 demineralization
去蓝滤光片 minus-blue filter
去离子器 de-ionizater;de-ionizer
去离子水 de-ionised[de-ionized] water;demineralized water
去离子柱 de-ionizing column
去离子装置 de-ionizer
去离子作用 de-ionization
去力机齿轮 power takeoff gear
去励磁电流 drop-out current
去励磁电压 drop-out voltage
去劣去杂 roguing the weak; sickly and off-type plants
去磷作用 dephosphorylation
去鳞 descaling
去流段 after run
去流角 angle of run
去硫 desulfidation;desulfurization[desulphurization]
去硫的 off-sulphur
去硫化 devulcanization
去硫铸铁 off-sulphur iron
去硫作用 desulfuration [desulphidation]
去路通畅 all-clear
去氯 anti-chlorination;dechlorinate
去氯剂 anti-chlor
去螺钉器 screw remover
去毛边 burring;deburr(ing)
去毛刺 flash;hone out;ragging;trimming;burring
去毛刺刀片 deburring blade
去毛刺工序 deburring operation
去毛刺滚齿 burring toothing
去毛刺机 burr removing machine;deburring machine;dressing machine;flash trimmer;radiusing machine
去毛刺手铰刀 deburring reamer
去毛刺整孔钻 burring reamer
去毛和净面刀 unhairing and scudding knife
去毛机刀轴 unhairing cylinder
去毛口 burring
去毛刷 hone out brush
去毛用刮皮板 unhairing beam
去镁 demagging
去锰 demanganization;demanganizing
去敏化 desensibilization
去敏作用 desensitisation[desensitization]
去模糊 de-blurring
去模糊法 de-blurring method
去模糊滤光器 de-blurring filter
去模混凝土桩 peerless pile
去膜 strip
去沫 defoam;skim
去沫剂 defoamer(agent)
去墨剂 deinking
去墨水液 eradicator
去木质素作用 delignification
去能 de-energise[de-energize]; deenergization
去泥剂 de-sludging agent
去年全年总损益 total profit and loss for the previous year
去年同比 comparable month a year earlier
去黏[粘]剂 viscosity remover
去镍 denickeling
去凝作用 deflocculating
去浓器 deconcentrator
去偶 depair
去耦 uncouple
去耦电路 decoupling circuit
去耦电容器 decoupling capacitor;decoupling condenser
去耦电阻 decoupling resistance
去耦发生器 padded generator
去耦合 decouple;decoupling
去耦滤波器 decoupling filter
去耦器 decoupler;isolator
去泡工 debubblizer
去泡剂 defoamer agent;defoaming agent;de-frothing agent
去泡沫 defoam
去泡沫的 defoaming
去泡沫剂 defoamant;defoaming agent
去皮 hull; peel(ing); pele; shelling; skinning;debarking <木材的>
去皮材 disbarked wood
去皮刀 peeling knife;slabbing cutter
去皮的 peeled;skinless
去皮工 peeler
去皮工具 skinner
去皮机 huller; seed huller; sheller; skin eliminator; skinning machine; skin-remover
去皮筛 scalping screen
去皮圆木 peeler log
去皮直径 <木材> core diameter;diameter under bark;inside diameter
去皮质 decorticate
去偏斜 de-skewing
去偏振 depolarize
去偏振效应 depolarization effect
去偏振作用 unpolarizing
去平行性 decollimation
去屏蔽 de-shielding;unmask
去漆工具 paint remover
去漆剂 paint remover
去漆器 paint remover
去漆水 paint remover
去启动 deactivation
去气 air elimination;deaerate;deaeration;de-airing;degas(sing);outgas
去气法 degasification
去气混凝土 <即用真空法抽去气泡的混凝土> deaerated concrete; deaired concrete;de concrete
去气挤压机 de-airing extruder
去气剂 air-detraining admixture; airdetraining compound
去气黏[粘]土 de-aired clay
去气器 air eliminator; deaerator; degasser
去气枪 degassing gun
去气室 deaerating chamber; deaeration chamber
去气桶 deaerator
去气砖 de-aired brick
去气作用 degasification
去铅 de-lead(ing)
去氢电焊条 dehydrogenized welding rod
去热 remove heat
去韧皮机 decorticator
去溶剂化 de-solvation
去溶剂作用 de-solvation
去乳化 de-emulsification
去锐边 back-off;blunt-edged
去色 colo(u)ration removal; decolo(u)r;scumbling
去色作用【化】decolo(u)ration
去砂 desanding;shakeout
去砂器 desander
去梢 topping
去射气 de-emanate
去升华作用 desublimation
去湿 dehumidification; dehumidify; dewetting; exsiccation; release of humidity
去湿槽 moisture-catcher
去湿的 desiccant
去湿法 <木材的> boulton process
去湿机 desiccating machine
去湿剂 dehumidifier;dehumidizer
去湿气的 desiccant
去湿器 moisture separator
去湿效果 dehumidifying effect
去湿型空气调节器 dry-type room air conditioner
去湿装置 dehumidifier;moisture-catcher;moisture-removal device
去势器 emasculator
去霜器 defroster
去水 dehydrate; expulsion of water; water removal
去水分的 desiccant
去水浮游生物 dry plankton
去水垢 disincrustant
去水剂 dehydrolyzing agent
去水煤沥青 dehydrated tar
去水物 dehydrate
去水锈工具 boiler scaling tool
去水作用 dehydration;dehydrolysis
去弹性钢丝 <经机械处理> killed wire
去碳 carbon elimination
去碳作用 decarbonation
去条带 de-striping
去调幅 demodulation
去调幅器 demodulator
去铁 de-iron;iron removal
去烃作用 de-alkylation
去同步 de-synchronizing
去同步化作用 de-synchronization
去铜 decopper(ing)
去头 topping
去涂料工具 paint remover
去涂料剂 paint remover
去尾 crop end
去味 taste suppression
去味器 destinker
去稳定化作用 destabilization
去稳效应 destabilizing effect
去污 soil removal
去污斑化合物 anti-dust compound
去污斑胶合剂 anti-dust binder
去污沉积物 decrustation
去污的 detergent;detersive
去污点 stain removal
去污废物 decontamination waste
去污粉 abstergent; cleaner; cleanser powder; clean(s)ing powder; household cleanser; putty powder; scouring powder
去污机理 detergency mechanism
去污剂 abstergent;decontaminant;decontaminating agent; detergent; detersive
去污净化剂 decontaminant
去污力 detergency[detergence]
去污面积 decontamination area
去污能力 soil removability
去污染 decontamination;depollution
去污染试验 detergency test
去污染柱 decontaminating column
去污设备 decontaminating device;decontaminating equipment; decontaminating plant; decontaminating unit
去污系数 decontamination factor
去污效果 detergent effect
去污效率 soil removal efficiency
去污性 detergency[detergence]
去污仪 decontaminating apparatus
去污因数 decontamination factor
去污因子 decontamination factor
去污指数 decontamination index
去污装置 decontaminating apparatus
去污作用 decontamination; depollution;soil-removing action
去雾 demist
去雾导管 deforging duct
去雾滤光片 haze filter
去雾器 demister(of vehicle window)
去锡 de-tin
去锡作用 de-tinning
去相干 de-coherence
去相关 decorrelation
去相关器 decorrelator
去向不明的船舶 missing ship
去向增音机 outgoing repeater
去像散的 anastigmatic
去像散偏转系统 anastigmatic de-flection system
去像散透镜【物】anastigmat;anastigmatic lens
去像散透镜组 anastigmat
去斜处理 de-skew processing
去谐 detuning
去谐滤波器 harmonic filter
去屑槽 chip room
去心板材 center[centre] board
去心厚板 center[centre] plank
去心邻域 deleted neighbourhood
去芯锯法 <木材的> boxed heart
去芯器 corer
去锌 dezincify;de-zinkify
去锈 derusting;scaling;stain removal
去锈的 derusted
去锈机 scaling machine
去锈药水 rust remover
去压机构 decompression gear
去烟橱 exhaust fume hood; fume hood
去烟罩 fume hood
去氧 deoxygenation
去氧钢 dead setting steel;deoxidized steel
去氧化皮 descaling
去氧剂 deoxidizer;oxygen scavenger
去叶 defoliation
去应力 de-stress(ing)[distressing]; stress-relieving
去应力带 de-stressed zone;distressed zone
去应力退火 relief annealing

Q

去荧光 de-blooming
去油 deoil(ing);get rid of oil;oil removal[removing];withdrawal of oil
去油粉 pounce
去油垢洗手皂 < 含熔岩粉 > lava soap;pumice soap
去油垢皂 < 含浮石粉 > lava soap;pumice soap
去油剂 degreaser;grease remover
去油墨剂 type wash
去油器 degreaser
去油溶液 degreasing solution
去油污 degrease;degreasing
去油污工人 degreaser
去油污剂 degreaser
去油污渍 spotting
去油脂 degrease;degreasing
去油装置 degreaser
去杂质作用 decontamination
去噪电容器 anti-hum capacitor
去噪开关 silent switch
去噪声电容器 anti-hum condenser
去沾染 detoxify
去振荡线圈 anti-shunt field
去脂 grease removal
去职补偿费 golden handshake
去中继 out-trunk
去中继线 outgoing trunk

趣味 taste

圈把剪刀 handle shears

圈板 girth sheets
圈闭 closure; encirclement; entrap; entrapment;oil trap;trap
圈闭闭合度 closure of trap
圈闭长度 length of trap
圈闭成因分类 genetic(al) classification of trap
圈闭的封闭因素 sealed factor of trap
圈闭的角隅 confined quarters
圈闭顶埋深 buried depth of trap top
圈闭海盆 entrapped basin
圈闭合油气性 oil-gas bearing condition of trap
圈闭宽度 width of trap
圈闭流粉磨 closed circuit grinding
圈闭评价 evaluation of trap
圈闭其他分类 other classification of trap
圈闭容积 volume of trap
圈闭相 entrapped phase
圈闭形成时间 tine of formation trap
圈闭形态分类 morphologic(al) classification of trap
圈闭型储油结构 structural trap
圈闭油 < 石油 > enclosure of oil
圈出 iris out
圈存资金 earmark
圈地 rodeo
圈地法 inclosure act
圈定 delineation
圈定边界 contouring
圈定成矿远景分级 defining minerogenetic prospect grade
圈定成煤远景分级 defining coal-forming prospect grade
圈定法 contouring
圈定含油气远景分级 defining oil-gas accumulation prospect grade
圈定路线 inclosing route
圈定炮眼 peripheral hole
圈肥 barnyard manure
圈粪 barnyard manure;yard manure
圈缝 girth seam
圈拱座柱 arch abutment
圈划 iris wipe
圈簧支柱 coil spring standard
圈间绝缘 turn-to-turn insulation
圈栏 girth rail
圈梁 belt course; collar beam; collar tie;collar tie beam;girt(h);ledger board; perimeter beam; peripheral beam;periphery beam;ring beam;ring girder;ring-shaped beam;skirt beam;dropped girt < 楼板格栅下的 >
圈梁荷构造体系 ring cell system
圈梁结合 circular beam connection
圈鳞 cycloid scale
圈流 closed circuit
圈流粉磨 feedback grinding;grinding in closed circuit
圈流粉磨系统 closed circuit grinding system
圈流湿法粉磨系统 closed circuit wet grinding system
圈流湿法原料粉磨 closed circuit wet raw grinding
圈拟阵 cycle matroid
圈盘旋梯 circular geometrical stair(case)
圈绕 coil winding
圈绒地毯 loop-pile carpet; round wire carpet
圈绒面 loop pile
圈入 iris in
圈纱 loop yarn
圈烧 edge firing
圈式信息交流网络 circle communication network
圈数 cyclomatic number; cylinder number;number of turns < 绕组 >
圈套 pitfall
圈套钩住 hook
圈套器 snare
圈套绳结 running knot
圈条齿轮 coiler wheel
圈条器 can coiler
圈筒群结构 multicell-framed tube structure
圈网 catch net;hoop-net
圈围地 fold
圈线 loop line
圈向量 cycle vector
圈形成 formation of rings
圈形物 wreath
圈形物件抓具 coil grab
圈形烟缕 looping plume
圈状的 cycloid;round
圈状物 wreath
圈足 ring foot
圈座 ring support

全 U 形铲刀 full U blade

全安装 full install
全凹式车门把手 fully recessed door handle
全靶场点 whole range point
全白口 complete chill
全白土 carclazyte
全白土催化剂 carclazyte catalyst
全白岩 hololeucocrate
全百叶门 full-louvered door
全摆幅 full swing
全斑结构 euporphyric texture
全板接合 full splice joint
全板门 flush door
全板门构造 solidcor;Solite
全半宽 total half-width
全包的楼梯梁 closed string(er)
全包号表示法 fully parenthesized notation
全包合同 all-in contract
全包价格 all-round price
全包建投标 turnkey bid
全包楼梯梁 curb string
全包原则 principle of all inclusiveness
全包运费 freight in full
全饱和绿灯期 fully saturated green period
全保的 fully insured
全保条件提单 full terms bill of lading
全保险 full insurance;fully insured
全备的 well-appointed;well-found
全背式 whole back
全闭 full cut-off
全闭额定压力差 close-off rating
全闭合 complete closure
全闭合位置 full application position
全闭塞区段重叠 full block overlap
全闭式舱 totally enclosed cabin
全闭式司机室 totally enclosed cab
全闭式叶轮 completely shrouded impeller
全闭式照准仪 sight alidade
全闭式座舱 totally inclosed cockpit
全闭压头 shut-off head
全闭状态 full-shut position; nodischarge state
全便携农用喷洒设备 fully portable agricultural sprinkling installation
全变差 total variation
全变差递减 total variation diminishing
全变差递减格式 total variation diminishing scheme
全变差非增 total variation bound
全变差非增格式 total variation bound scheme
全变换 total transform
全变速范围的变速器 full-range transmission
全变质作用 paramorphism
全标称速度 full nominal speed
全标度 full-scale
全标记 all mark
全表面金刚石钻头钻孔 full face diamond bit drilling
全表面侵蚀 sheet erosion
全并行存储器 fully parallel memory
全并行模/数转换器 all parallel analog(ue)/digital converter
全并行相联处理机 fully parallel associative processor
全拨号 < 自动电话 > complete selection
全波 all wave;double wave;full wave
全波倍压器 full-wave voltage doubler
全波补偿器 full-wave compensator
全波带接收机 all-wave receiver
全波(带)振荡器 all-wave oscillator
全波的 biphase
全波电压电击 < 避雷器的 > full-wave voltage impulse
全波电源 full-wave power supply
全波段 all band; all wave; all-wave band
全波段的 all wave
全波段辐射计 panradiometer
全波段干扰 barrage jamming
全波段接收机 all-wave receiver;full-waveband receiver
全波段调谐器 full-range tuner
全波段无线电台 comprehensive radio
全波放大器 full-wave amplifier
全波辐射测量仪 panradiometer
全波辐射器 total radiator
全波汞(弧)整流器 full-wave mercury rectifier
全波计数管 long counter
全波检波 double-wave detection
全波可控硅电源 full-wave thyristor power supply
全波控制 full-wave control
全波偶极子 full-wave dipole; full-wave doublet
全波片 full-wave plate
全波频率 full-wave frequency
全波平方律检波器 full-wave square-law detector
全波平衡放大器 full-wave balanced amplifier
全波曲线 curve of total wave
全波收音机 multiwave receiver
全波天线 all-wave antenna
全波无线电台 comprehensive radio
全波相位控制 full-wave phase control
全波照明控制 full-wave lighting control
全波振荡 full-wave oscillation
全波振动器 full-wave vibrator
全波振子 full-wave dipole
全波整流 all-wave rectification; biphase rectification;full-wave rectification
全波整流电路 full-wave rectifying circuit
全波整流电桥 full-wave rectifier bridge
全波整流管 full-wave rectifier tube; Raysistor;Raytheon tube
全波整流器 biphase rectifier;diametric(al) rectifier;full-wave rectifier
全波整流式 X 射线机 full-wave rectifying type X-ray apparatus
全波周期 time of one complete oscillation
全玻璃的 all-glass
全玻璃顶棚采光 all-glass ceiling luminaire;all-glass light fitting
全玻璃顶棚装配 all-glass light fitting
全玻璃建筑构件 all-glass building unit
全玻璃结构 all-glass construction
全玻璃立面 < 建筑物的 > all-glass facade
全玻璃门 all-glass door; full glass door;solid glass door
全玻璃摩天大楼 all-glass skyscraper
全玻璃幕墙 full glass curtain wall
全玻璃热绝缘 all-glass thermal insulation
全玻璃推拉窗 all-glass sliding sash
全玻璃推拉扇 all-glass sliding sash
全玻璃纤维纸 all-glass paper
全玻璃显像管 all-glass kinescope
全玻璃悬墙 sheath-type curtain wall
全玻璃制品 all-glass work
全玻璃质的 holohyaline
全玻纤纸 complete glass fiber[fibre] paper
全玻质的 holohyaline
全剥开采法 mountain top removal
全补偿接触网 all autotensioned messenger wire and contact wire with balance
全补偿链形悬挂装置 all autotensioned catenary equipment
全补偿运算放大器 fully compensated operational amplifier
全不变列 fully invariant series
全不变子群 fully invariant subgroup
全不变子群列 fully invariant series of subgroups
全不连通的 totally disconnected
全不连通图 totally disconnected graph
全布面装订的 full cloth bound
全步行相位 all walk phase
全部 alpha and omega;bodily;entirety;every bit;full;omnium
全部安装玻璃的 fully glazed
全部包建的工厂 turnkey factory
全部包建的工程承包方式 turnkey

全部包装 full packed
全部保险 full insurance
全部保险费 in full premium
全部曝光 burn-out
全部爆破 clean blast
全部崩溃 overall collapse
全部编制 full commission
全部不动产的信托抵押借据 package trust deed(mortgage)
全部布线 through-wiring
全部财产的移交 general assignment
全部财产信托契据 all-inclusive trust deed
全部财力 all financial resources
全部财力概念 all financial resources concept
全部参与人 full participant
全部差额 overall balance
全部拆卸检修 complete overhaul
全部产品和劳务 all goods and services
全部产品试验 commercial test(ing)
全部产品综合定价 pricing entire product package
全部偿清的抵押贷款 fully amortized mortgage
全部敞开的 full opening
全部车辆 rolling stock
全部车轮均可转向 all-wheel steerable
全部车站停车的往返循环列车 all-station shuttle
全部成本 absorption cost;complete cost;overall cost;full cost
全部成本法 full absorption method
全部成本计算法 absorption costing;full costing
全部成本减折旧 full cost less depreciation
全部成本原则 full cost principle
全部成本制 full cost system
全部承包合同 all-in contract;turnkey contract
全部承重墙系统 system in which all the wall carry loading
全部持续时间 full duration
全部尺寸<长宽高> overall dimension
全部充填 whole fill
全部出力 full output
全部处理的 fully processed
全部串联位置 full-series position
全部床沙质 total bed sediment load
全部存货价值 value of all stocks
全部抵押 blanket mortgage
全部电话用户停止通话 total failure
全部电气化 all-electric(al)
全部电气联锁装置 all-electric(al) interlocking
全部调查 complete investigation
全部调度集中制<在通常调度集中制外,加装各种自动化装置> total traffic control system
全部丁砖层 full header(course)
全部动产 goods and chattels
全部反应 total overall reaction
全部返回<冲洗液> full return
全部防直射式天花板照明 louver all ceiling lighting
全部费用 all expenses;charged in full;outright cost;outright expenses;overall charges;total cost
全部费用付讫 all charges paid
全部费用价值 value of total cost
全部费用已付 all charges paid
全部费用在内发票 franco invoice
全部费用在内价格 franco
全部付款 full payment
全部付讫 fully paid;payment in full
全部付清 fully paid-up;pay up
全部赋税 fully tax-exempt treasury securities
全部负债 full liability
全部附件保持供给 full complement of accessories
全部工程 all-work
全部工程方案预算 budget of project
全部工程费用 total project cost
全部工程量清单的汇总金额 main summary
全部工作时间 full time
全部工作小时统计 statistics of total hours worked
全部公差 all-in allowance
全部功能失效 flop
全部购进 buy up
全部购买 portfolio acquisition
全部股本 total equity
全部固体的 holosteric
全部归还 integral restitution
全部国外损失冲减 recapture of overall foreign loss
全部过程 whole process
全部过硬 mass hardness
全部焊接船体 all-weld hull
全部焊接和淬火的辙叉 completely welded and flame hardened crossing
全部航行设备已经试验完毕 all-gear tested
全部合并 total incorporation;total merger
全部合成 total synthesis
全部呼叫电键 simultaneous calling key
全部糊精化淀粉 fully dextrinised starch
全部换气 complete air exchange
全部或大部分的债券或欠款的到期日 balloon maturity
全部机加工的 machined all over
全部机内联用程序和信息控制系统 total online program(me) and information control system
全部机械化 full mechanization
全部机组发电 full power
全部基金综合平衡表 all funds-combined balance sheet
全部级配骨料 fully graded aggregate
全部级配集料 fully graded aggregate
全部计时法 overall timing method
全部技能 repertoire
全部继电联锁 all-relay interlocking
全部加热的 warm to the tread
全部加温的 warm to the tread
全部加压润滑的 full pressure lubricated
全部检查 complete inspection
全部检修 complete overhaul
全部检验 one-hundred percent inspection
全部建成地区 completely built-up area
全部交换 complex exchange
全部缴清的股份 full-paid stock
全部校验 complete verification
全部接收 blanket
全部结清 all-squared
全部截面 gross section
全部解掉 let go everything
全部解法 total solution
全部进气汽轮机 full admission turbine
全部进气透平 full admission turbine
全部进气涡轮机 full admission turbine
全部浸没腐蚀试验 total immersion corrosion test
全部浸入浸渍法 full dip infiltration
全部旧桶 all second hand drums
全部矩阵代数 totally matrix algebra
全部竣工 complete in place;finish all over
全部竣工项目 completed project
全部开动的<如机械等> fully actuated
全部开工 at full capacity
全部开支 total expenditures
全部控制进入 full control of access
全部劳动消耗 overall labo(u)r
全部累进率 slab scale
全部利率 all-in rate
全部林木<一地区或一国的> silva
全部流动资金周转率 turnover rate of whole circulating funds
全部流量 full discharge
全部馏分蒸出时的温度 full-boiling point
全部履行 full performance
全部履行义务 full discharge of liability
全部落锻 all drop forging
全部买进 buying outright
全部门面积<包括门框> door area
全部面积 entire area;gross area;whole area
全部盘存 complete inventory
全部刨光的 planed all round
全部破坏 overall collapse
全部铺瓷砖的 fully tiled
全部铺有护面石防波堤 all armo(u)r rock mound breakwater
全部弃荷 total load rejection
全部砌体墙只从一侧搭脚手架进行砌筑 overhand work
全部契约的条款和全部授权证书 full covenant and warranty deed
全部轻骨料混凝土 all-lightweight concrete
全部清偿 pay up
全部清除 full-scale clearance
全部清除存储信息 full-scale clearance
全部清单 repertory
全部清理 complete liquidation
全部权利要求 blanket claim
全部热量 net quantity of heat
全部认可 blanket approval
全部容差 all-in allowance
全部溶化 off-bottom
全部尚未归还的放款 all outstanding advance
全部设备接线图 general connection diagram
全部审查 complete verification
全部审计 complete audit
全部生锈 rust through
全部施工时间 overall construction time
全部时间备用音频链路 full-time reserved voice-frequency link
全部时间(工作)的 full time
全部时间双工连接 full-time duplex connection
全部时间信息交换 full-time message switching
全部实际资本 real capital stock;stock of real capital
全部使用费 overall operational cost
全部使用期的修理费 life repair cost
全部使用寿命<固定资产> whole service life
全部市价会计 all current accounting
全部试样 total sample
全部输入信号 full-scale input signal
全部数据集授权 full data set authority
全部甩负荷 total load rejection
全部说明的有限状态机 fully specified finite-state machine
全部损失 actual total loss;total loss
全部调节完毕 adjustment free
全部调整完毕 adjustment free
全部通过筛号的骨料等级 total-passing gradation
全部通过筛号的集料等级 total-passing gradation
全部同步 all synchro
全部同意 blanket approval
全部投产 full operation;fully put into production
全部投入电阻 all-in resistance
全部投入使用 full operation
全部投资 all investment;total investment
全部脱叶 complete defoliation
全部完成 finish all over
全部完工 finish all over;final completion
全部完工毛利计算法 completed contract method
全部挽牢 all(made) fast
全部蜗壳 full scroll(case)
全部下套管井 cased through well
全部线路占线 no-lines
全部限额 overall limits
全部详细审核 complete detailed audit
全部消费支出 total consumption expenditures
全部效率 overall efficiency
全部效用 total utility
全部泄流能力 full outflow capacity
全部卸载 complete discharge;full discharge
全部薪酬 overall remuneration
全部性能 overall property
全部延时序列<洪水频率计算用> complete duration series
全部验收 completion acceptance
全部样品 total sample
全部溢价计划 one-hundred percent premium plan
全部硬化 through-hardening
全部佣金 total commission
全部用丁头砖石砌的墙 perpend wall
全部用金属制成的 all-metal
全部油口打开 all-port open
全部油液回流 full return flow
全部预期成本 all expected cost
全部预期代价 all expected cost
全部预期效益 all expected benefit
全部预支信用证 clean payment credit;clean payment letter of credit
全部载货容量 full reach and burden
全部载重量 all told
全部在球体部分 entirely in spheric(al) portion
全部责任 full liability
全部债券 global bond
全部展开 full development
全部展开的 fully developed
全部占线 all busy;no-lines;no trunks;all trunks busy
全部占用时间 total holding time
全部折旧完的资产 fully depreciated assets
全部支撑 forepoling
全部制动管路缓解 train line release
全部制造成本 complete manufacturing cost
全部重置成本 full replacement cost
全部转储 total dump
全部准确度 overall accuracy
全部着色 integral colo(u)ring
全部仔细检修 complete overhaul
全部资金利税率 rate of both profits and taxes on entire funds
全部自动的 complete automatic;purely automatic;fully automatic
全部租用 full charter
全部最高荷载<包括施工荷载> maximum rated load

全部作业评级法 whole job ranking
全财产继承人 universal legatee; universal successor
全采光系数 total daylight factor
全采样法 total sampling method
全彩【计】true colo(u)r
全彩色 full colo(u)r
全彩色影像 full-colo(u)r image
全彩条信号发生器 full-bar generator
全参数 population parameter
全参与优先股 fully participating preferred stock
全舱口船 all hatch vessel
全操作结构 for-all structure
全操作数 full operand
全槽格式集装箱船 full cellularized container ship
全槽焊 complete-penetration groove weld
全侧向压力 omnilateral pressure
全侧缘活塞 full skirted piston
全测 borehole survey(ing)
全测地超曲面 totally geodesic hypersurface
全测地的 totally geodesic
全测度 total measure
全测回法 set method
全测站 total station
全层 holostrome
全层公寓 floor-through
全层(混凝土)建筑工程 full-course construction work
全层混凝土路面 full-course concrete pavement
全层雪崩 ground avalanche
全差分方程 total difference equation
全差示分光光度法 whole differential spectrophotometry
全拆散式输出 complete knockdown export
全拆卸铲杆 rod uncoupling
全拆卸钻杆 rod uncoupling
全产量 full production capacity
全长 aggregate length; completed length; entire length; full distance; length overall; overall length; total length
全长淬火轨 full-length heat-treated rail
全长的 <从一端到另一端> fore and aft; full-length
全长度 out-to-out
全长接触 full-length contact
全长镜子 full-length mirror
全长锚固 all-length anchorage
全长上的接触 full-length contact
全长式缸套 full-length liner
全长式汽缸水套 full-length cylinder water jacket
全长式汽缸套 full-length cylinder liner
全长途占线 all trunks busy
全厂定额 plant standard
全厂工资汇总表 summary of factory wages
全厂工资税汇总表 summary of factory payroll taxes
全厂间接费用率 plant-wide overhead rate
全厂热耗率 plant heat rate
全厂热效率 plant thermal efficiency
全厂性奖励计划 plant-wide bonus plan
全场分析法 whole-field analysis
全超车视距 full overtaking sight distance
全潮 full tide
全潮码头 <任何潮汐都可靠泊> tidal quay
全车道补坑 full-lane patch
全车道流量 full carriageway traffic volume
全车动式信号 full traffic-actuated signal
全车动式信号控制器 full traffic-actuated controller
全衬砌隧道 fully lined tunnel
全称 legal name
全称闭包 universal closure
全称规定规则 rule of universal specification
全称量词 universal quantifier
全称推广规则 rule of universal generalization
全称永真公式 universally valid formula
全称永真前提 universally valid premise
全成本定价法 full cost pricing
全成型领圈 fully shaped collar
全承包工程 turnkey job; turnkey project
全承包合同 turnkey contract
全承包设计 turnkey project
全承式 Y 形分叉管 all bell wye branch
全承载车身 unitary body
全程 omnidistance; omnirange; round trip
全程变量 global variable
全程变量符 global variable symbol
全程变量引用 global variable reference
全程出行 <从起点到终点的一次出行> linked trip
全程传输衰耗等效值 overall transmission equivalent
全程单据 through document
全程单元 global location
全程导航 all-way guidance; all-way navigation
全程定位衰耗【铁】overall location loss
全程多次反射 <地震勘探> long path multiple reflection
全程多种方式联运 through multimode transport
全程工作衰耗【铁】overall location operation loss
全程轨迹 track history
全程航行历时 <灌水沟的> advance time
全程净衰耗测量 overall net loss measurement
全程距离 whole range distance
全程均衡器 mop-up equalizer
全程螺杆 full-flighted screw
全程票 through ticket
全程频率响应 overall frequency response
全程区头向量地址 global display address
全程润滑的 lifetime lubricated
全程升降窗 full drop window
全程时间 <公路工程> through journey time
全程试验 full distance test
全程水道运价 all-water rate
全程水运 all-water transportation
全程搜索 global search
全程调速器 all-speed governor
全程调整 overall system adjustment
全程铁路运价 all-rail rate
全程未使用的客票 wholly unused ticket
全程向量表 global vector table
全程旋转 full rotation
全程引用 global reference
全程运费 through freight
全程运费率 through rate
全程运输 through movement; through shipment; through traffic; through transit
全程振鸣 end-to-end singing
全程直通运输 <美> through carriage
全程值 global value
全程制导 all-way guidance
全程追踪 whole course tracing
全程自养脱氮 full autotrophic denitrification
全尺寸 boxed dimension; full-scale; full size; overall dimension
全尺寸部件 full-size component
全尺寸测量 full-scale measurement
全尺寸的 full-sized
全尺寸读数 full-scale reading
全尺寸堆 full-scale reactor
全尺寸发动机试验 full-scale engine test
全尺寸风洞 full-scale tunnel
全尺寸焊缝 full-sized weld
全尺寸建筑构造 full-size construction; full-scale structure
全尺寸建筑结构 full-size construction; full-scale structure
全尺寸流 full-scale flow
全尺寸脉冲 full-signal pulse
全尺寸模拟 full-scale simulation
全尺寸模型 full-scale mock-up; full-size model
全尺寸碰撞试验 full-scale crash test
全尺寸扫描 full-size scanning
全尺寸实物模型 full-size physical model; mock-up
全尺寸试验 full-scale experiment; full-scale test; full-size test
全尺寸条件 full-scale condition
全尺寸条件下的研究 full-scale investigation
全尺寸图 full-size draft
全尺寸造型模型 full-scale styling representation mock-up
全尺度 full size
全齿顶高小齿轮 all-addendum gear
全齿高 total depth; whole depth
全齿高齿 full-depth tooth
全齿高齿轮 full-depth gear
全齿高加强板齿轮 full-depth strengthening shroud
全齿高渐开线制 full-depth involute system
全齿高系数 whole depth coefficient
全齿根高齿轮 all-dedendum gear; recess-action gear
全齿轮车床 geared-head lathe
全齿轮传动 all-gear drive; all-gear system
全齿轮床头箱 all-geared headstock
全齿轮立式钻床 all-geared upright drill
全齿轮主轴箱 all-geared headstock
全齿啮合 full engagement
全齿式抓斗 whole tine[tyne] grab
全充电电池溶液比重 full charge specific gravity
全充电蓄电池 fully charged battery
全充气位 full charging position
全充填 perfect filling
全冲程安全阀 full-stroke safety valve
全穿孔 all mark; lace punch
全穿孔卡片 laced card
全穿孔(纸)带 fully perforated tape
全穿透井 fully penetrating well
全船船员 all told
全船广播系统 general announcing system
全船通风 hull ventilation
全船下沉 hull sunk
全船油漆 all painting
全船战备部署 general quarters
全串行模/数转换器 all serial analog/digital converter
全纯 holomorphism
全纯函数【数】holomorphic function
全纯函数环 ring of holomorphic functions
全磁场 full field
全磁道 fluxoid
全磁化 holomagnetization
全磁迹 full track
全磁控制器 full magnetic controller
全磁射线管 all-magnetic tube
全磁通 fluxoid
全淬硬 full hardening
全淬硬钢 full-hardened steel
全存储 store through
全搭接接合 full lap
全打滑状态 fully sliding condition
全单板胶合板 all-veneer plywood
全单板结构 all-veneer construction
全单流式发动机 full uniflow engine
全单位模矩阵 totally unimodular matrix
全导电性 complete conductivity
全导数 total derivative
全等 congruence[congruency]
全等变换 congruent transformation
全等的 identically equal
全等公理 congruence axiom
全等角 congruent angle
全等三角形 congruent triangles
全等时线 complete synchrone
全等式 identical relation
全等图形 congruent figures
全等效伏安 total equivalent volt-amperes
全等直射变换 congruent collineation
全等转变 congruent transformation
全地铁 full metro
全地形 all terrain
全地形车 all-terrain vehicle
全地形车辆 all-terrain vehicle
全地形交通工具 all-terrain vehicle
全地址 full address
全地址转移 full address jump
全碘化碳 periodo-carbon
全碘乙烷 periodo-ethane
全碘乙烯 periodo-ether
全电动的 all-electric(al)
全电动舵机 all-electric(al) steering gear
全电动控制 all-electric(al) control
全电加热炉 all-electric(al) furnace
全电加热熔融 all-electric(al) melting
全电控制 all-electric(al) control
全电离 full-ionization
全电力驱动的 all-electric(al) drive
全电流 full current
全电流定律 law of total current
全电路 complete circuit
全电路定律 all circuit law
全电气号志系统 all-electric(al) signal(l)ing system
全电气化操作 all-electric(al) operation
全电气化厨房 all-electric(al) kitchen
全电气化厨房用具 all-electric(al) kitchenware
全电气化的 all-electric(al)
全电气化系统 all-electric(al) system
全电气化运转 all-electric(al) operation
全电气联锁机 all-electric(al) interlocking frame
全电气设备 all-electric(al) plant
全电气制 all-electric(al) system
全电容 plenary capacitance
全电熔 all-electric(al) melting
全电熔窑 all-electric(al) furnace
全电势 full potential
全电视信号 composite video signal
全电位 full potential
全电信号装置 all-electric(al) signal-

(l)ing
全电压 full voltage
全电压绝缘绕组 full insulated winding
全电压起动电动机 full voltage starting motor
全电压起动法 method of full voltage starting
全电压起动器 line starter
全电压效应 full voltage effect
全电子的 all electronic;full electronic
全电子电话交换机系统 full electronic switching system
全电子化的 all electronic
全电子式电话交换机 full electronic telephone exchange
全电子交换机 full electronic switching system
全电子交换系统 full electronic switching
全电子式交换 full electronic switching
全电子式交换制 full electronic switching system
全电子系统 all-electronic system
全电子制 all-electronic system
全电阻 impedance;total resistance
全吊挂炉顶 fully suspended roof
全丁砖皮层 header course
全丁砖砌层 full header
全丁砖砌合法 header[heading] bond
全定额 global quota
全定向(道路)立体交叉 all-directional interchange
全定向互通式立交 fully directional interchange
全定向交叉(口) all-directional interchange;all-directional intersection
全定向立体交叉 all-directional intersection
全定向三叉互通式立交 fully directional three-leg intersection
全定向型互通立交 fully directional interchange
全定性分析 total qualitative analysis
全动平尾 stabilator
全动射电望远镜 fully steerable radio telescope
全动式水平尾翼 all-moving tail
全动视力 vision with driver and object
全动物型营养 holozoic nutrition
全动轴 all axles motored
全冻结 all frozen
全短路 dead-short circuit
全断 full cut-off
全断裂 through crack
全断面 full cross-section
全断面爆破 full face blast(ing);full section blast(ing)
全断面衬砌 full face lining;full-round placing system
全断面的 full face;full section
全断面的构件 full cross-sectional element
全断面盾构 full shield
全断面法 full face method;full section method
全断面分块开挖法 Austrian method;bottom heading-over-head bench
全断面工程 <隧洞开挖> full face work
全断面工作 <隧道开挖> full face work
全断面混合法 sectional mixing method
全断面掘进 drive-in full section;full advance;full-bottom advance;full face drilling;full face driving;full face attack <隧洞的>
全断面掘进采矿法 mole mining
全断面掘进法 full face tunnel(l)ing method;full face driving method
全断面掘进机 full face machine;tunnel borer
全断面掘进炮眼组 full face round
全断面开挖 excavation of full section;full face bore;full face cutting;full face excavation;full face tunnel(l)ing;full section excavation
全断面开挖法 full face excavation method;full face tunnel(l)ing method
全断面开挖隧道 full face drilling
全断面联合掘进机 full face machine
全断面炮孔组 full face round
全断面取水 full cross-sectional pumping
全断面隧道掘进法 full face tunnel(l)ing
全断面隧道掘进机 full face tunnel(l)er;full face tunnel(l)ing machine;tunnel boring machine for full section
全断面隧道开挖 full face excavation tunnel;full face method;full face tunnel(l)ing
全断面隧道开挖法 full face tunnel(l)ing method
全断面隧道施工 full face tunnel(l)ing
全断面隧洞开挖 full circle mining of tunnel heading
全断面隧洞挖掘机 fullface tunnel(l)ing machine
全断面图 complete sectional view
全断面无岩心钻头 full hole bit;full hole rock bit
全断面一次爆破法 one-blast full-face method
全断面一次开挖法 full face excavation method
全断面钻爆法 full face drill and blast method
全断面钻进 full face boring;full face drilling
全断面钻探 full face boring;full face drilling
全对称 hologrammetry;holosymmetry;total symmetry
全对称晶形的 holohedral
全对称现象 pantomorphism
全对称形态 holomorphism
全对称性 holohedrism;pantomorphism
全对称轴 axis of total symmetry
全对数尺度 full logarithmic scale
全多孔微珠载体 porous microbeads support
全多孔型填充剂 totally porous packing
全多余 total float
全额 full protection policy
全额保险 full insurance
全额保证金 full margin
全额成本 absorption cost
全额成本法 full absorption costing
全额成本计算 absorption costing
全额担保 full coverage
全额定载货容量 full-rated capacity
全额分期付款 instalment in full
全额付清 pay in full
全额负担保险单 full protection policy
全额共同保险 full coinsurance
全额股票 full stock
全额及差额试算表 trial balance of totals and balances
全额累进税制 taxation system based on progressive rates
全额利润分成 sharing the entire profit;sharing the total profit
全额赔偿 first dollar coverage
全额社会所得税 entire social income tax;full social income tax
全额试算表 trial balance of totals
全额息票债券 full coupon bond
全额信用证 full-value letter of credit
全额佣金 full commission
全额运费 full freight
全额支付 full installment
全额转让 total transfer
全额转让通知书 advice of total transfer of credit
全额准备制 total reserve system
全额租赁 full pay out lease
全发射 total emission
全发生 hologony
全伐 clearance;clear cutting;clear felling
全伐法 clear cutting
全法兰 Y 形支管 all-flanged Y-branch
全法兰三通 all-flanged tee
全法兰四通 all-flanged cross
全帆行驶 stretch
全反馈 unity feedback
全反射 perfect reflection;total reflection
全反射比 total reflectance
全反射测量计 perflectometer
全反射层 total reflection layer
全反射层的 totally reflected
全反射层堆 fully reflected reactor
全反射层反应堆 fully reflected reactor
全反射长光程毛细管分光光度法 total-reflection long capillary cell in absorption spectrophotometry
全反射带 whole reflecting zone
全反射法 total-reflection method
全反射共振腔 total reflecting resonator
全反射计 total reflectometer
全反射角 angle of total reflection
全反射镜光学系统 all-mirror optics
全反射聚光镜 holophote
全反射聚光体系 holophotal system
全反射棱镜 total reflection prism
全反射临界角 critical angle of total reflection
全反射率 total reflectivity
全反射面 fully reflecting surface
全反射损耗 loss at total reflection
全反射条件 total reflection condition
全反射系数 total reflection factor
全反射仪 total reflectometer
全反射折射计 total reflection refractometer
全反应 total reaction
全反转 total inversion
全范围 full range;gamut
全范围量程 full-scale measuring range
全范围内的 nationwide
全范围调谐装置 full-range tuner
全方位 all-direction;omnibearing
全方位辐射器 isotropic(al) radiator;omnidirectional radiator;spheric(al) radiator
全方位航向指示器 omnirange course indicator
全方位距离导航 omnibearing distance navigation
全方位距离导航系统 omnibearing distance navigation(al) system
全方位距离电台 omnibearing distance station
全方位开放 open to the outside world in an all-round way
全方位潜水器 fully directional submersible vehicle
全方位线 omnibearing line
全方位信标 omnibearing beacon
全方位旋转式起重机 full slewing crane
全方位选择器 omnibearing selector;radial selector
全方位指示器 omnibearing indicator
全方向 omnirange
全方向普照灯 omnidirectional light
全方向太阳电池 omnidirectional solar cell
全方向无线电导航标 omnidirectional radio beacon
全方向信标(灯) non-directional beacon light
全方向信号灯 non-directional beacon light
全防爆型电动机 fully flameproof motor
全防护型的 fully guarded
全防护型电机 fully guarded machine
全放射性 gross radioactivity
全放完电蓄电池 fully discharged battery
全放下襟的翼 full-flap
全肥 <包含氮磷或钾> complete fertilizer;NPK nutrition
全沸点标准燃油 <试验用> full-boiling range reference fuel
全费 in full
全费率 full rate
全分辨度的黑白图像 full-resolution monochrome picture
全分辨能力的样板 full-resolution template
全分节驳船 fully integrated barge
全分节顶推驳船队 fully integrated tow
全分离流冰 very open ice
全分离信号方式 fully dissociated signaling
全分流系统 entire split system
全分配制会议电话 telephone conference of full distribution system
全分散式交通控制系统 totally distributed traffic control system
全分析 bulk analysis;full analysis;overall analysis;total analysis
全分析水样 water sample of comprehensive analysis
全分析样品 sample for full analysis
全风化冰脊 totally enclosed
全风化层 full weathered layer
全风化带 halo-weathering zone
全风化岩 badly weathered rock
全风险应收账款 all risks account receivable
全风压 full wind pressure;total air pressure
全封闭 full access control;full closure;full control of access
全封闭不通风的 totally enclosed non-ventilated
全封闭处理设施 totally enclosure treatment facility
全封闭垂直提升机 totally enclosed vertical elevator
全封闭的 canned;hermetically sealed;totally enclosed
全封闭电池 leakproof battery
全封闭防喷器 blind rams blow(ing) out preventer
全封闭防水 fully closed waterproofing
全封闭风扇冷却 totally enclosed fan-cooled
全封闭风扇冷却鼠笼式电动机 totally enclosed fan-cooled squirrel cage motor
全封闭风扇式电机 totally enclosed fan-cooled type machine
全封闭管子通风式 totally enclosed

pipe-ventilated
全封闭气体绝缘开关 full enclosed gas insulated switchgear
全封闭扇冷式防爆电动机 totally enclosed fan cooling motor
全封闭式 totally enclosed type; totally enclosured type
全封闭式电(动)机 totally enclosed machine; totally enclosed motor; totally enclosured motor
全封闭式分相母线 enclosed isolated phase bus
全封闭式感应电动机 totally enclosed type induction motor
全封闭式机动救助艇 totally enclosured motor propelled survival craft
全封闭式机器 totally enclosed machine
全封闭式救生艇 totally enclosured motor lifeboat
全封闭式空气绝缘开关 full enclosed gas insulated switchgear
全封闭式制动器 full wrap-around brake
全封闭式制冷剂压缩机 full hermetic refrigerant compressor
全封闭输送机 complete enclosure of conveyer
全封闭水冷却式 totally enclosed water cooled
全封闭索股结构钢索 full-locked coil construction rope
全封闭通风式电动机 ventilated totally enclosed motor
全封闭外壳 totally enclosed frame
全封闭线路 controlled access route; fully isolated route; limited access route
全封闭型曲轴箱通风系 full closed crankcase ventilating system
全封闭闸板 <防喷器的> blind rams
全封密的 totally enclosed
全封密风冷式电动机 fan-cooled motor; motor totally enclosed
全蜂窝式货柜船 fully cellular container ship
全蜂窝式集装箱船 fully cellular container ship
全氟丙二烯 perfluoroallene
全氟丙烯二氟乙烯共聚物 perfluoropropylene vinylidene fluoride copolymer
全氟代烃 perfluoro-hydrocarbon
全氟二甲基环丁烷 perfluorodimethyl cyclobutane
全氟化碳示踪剂 perfluorocarbon tracer
全氟化物 perfluoro-compound
全氟化作用 perfluorination
全氟环丁烷 perfluorocyclobutane
全氟环醚 perfluorocyclicether
全氟硫酸膜 perfluorosulfonic acid membrane
全氟煤油 perfluorokerosense; perfluorokerosine
全氟碳化物 perfluorocarbon
全氟烃 perfluorocarbon
全氟烷基化作用 perfluoroalkylation
全氟烷氧基聚合 perfluoroalkoxy polymer
全氟烷氧基树脂 perfluoroalkoxy resin
全氟烷氧基烷烃共聚物塑料衬黑色金属管 perfluoro alkoxyalkane copolymer plastic lined metal pipe
全氟乙丙烯 fluorinated ethylene propylene
全氟乙醚 perfluoroether
全氟乙烷 hexafluoroethane; perfluoroethane
全氟乙烯 perfluoroethylene
全氟乙烯-丙烯共聚物塑料衬黑色金属管 perfluoroethylene-propylene copolymer plastic lined metal pipe
全氟异丁烯 perfluorpospbutylene
全氟有机金属化合物 perfluoroorganometallic compound
全浮充制 full-floating system
全浮动 full floating
全浮动机械填料 full-floating mechanical packing
全浮动式翻斗车 fully-floating tip lorry
全浮动式桥轴 full-floating axle shaft
全浮动尾翼 all-flying tail
全浮后(轮)轴 full-floating rear axle; fully floating rear axle shaft
全浮轮轴 full-floating axle shaft
全浮式半轴 full-floating semi-axle
全浮式传动轴 full-floating-drive axle
全浮式活塞销 full-floating gudgeon pin; full-floating piston pin; full-floating wrist pin
全浮式机 full floating machine
全浮式联轴节 full-floating coupling
全浮式喂料器 full-floating feeder
全浮选 bulk flo(a)tation
全浮选法 all-flo(a)tation process
全浮游生物 holophankton
全浮轴承 full-floating bearing
全幅度 peak-to-trough amplitude
全幅图 full sheet
全幅应变 peak-to-peak strain
全幅值 <正负峰间的> double amplitude; peak-to-peak value
全辐射 total radiation
全辐射辐射计 total radiation radiometer
全辐射高温计 rayotube pyrometer; total heat radiation pyrometer; total radiation pyrometer
全辐射功率 total radiant power
全辐射器 total radiator
全辐射体 full radiator
全辐射温度 total radiation temperature
全付 full payment; payment in full
全付条款 full term
全负荷 full-connected load; full loading; gross load
全负荷的 full load
全负荷电流 full-load current
全负荷生产定额 full load production standard
全负荷试验 full load test; full-scale load(ing) test
全负荷速度特性 full load velocity characteristic
全负荷运行 high-power operation
全负载 full(-connected) load; gross load
全附件发动机 fully equipped engine
全复接 full multiple
全傅立叶区间 full Fourier interval
全富位置 full-rich position
全覆盖 complete veneer
全盖大石堆石堤 all-armo(u)r rock mound
全概率【数】total probability
全概率定理 total probability theorem
全概率定律 breakdown law
全概率公式 total probability formula
全干 all dry; bone dry; dry-through
全干比重 absolute dry specific gravity
全干法(工艺)水泥厂 all-dry cement mill; all-dry cement plant
全干集材 whole-stem logging
全干浸渍 impregnation on dried basis
全干密度 oven-dry density
全干木材 absolutely dry wood; bone-dry wood
全干硬度 full hardness
全干燥 white drying
全感应(交通信号) fully actuated
全感应式控制器 <在接近交叉口的所有进口道上,全部装车辆检测器> full traffic-actuated controller
全感应式信号 <装置在交叉路口的> full actuated signal
全感应式信号控制 fully actuated signal control
全感应信号 <指交通> full traffic-actuated signal
全感应(信号)控制机 full actuated controller
全刚性结构 fully rigid framing
全刚性框架结构 fully rigid framed structure
全钢 all-steel
全钢表 full stainless steel watch
全钢薄壳车体 all-steel sheu coach body
全钢车身 all-steel body
全钢导杆 all-steel leader
全钢导管 all-steel leader
全钢底架 all-steel under frame
全钢货车 all-steel wagon; sheeted wagon
全钢建筑物 all-steel construction
全钢结构 all-steel construction
全钢客车 all-steel carriage
全钢链条 all-steel chain
全钢楼梯 all-steel stair(case)
全钢幕墙 all-steel curtain wall
全钢棚车 all-steel box car; all-steel covered wagon
全钢倾卸车 all-steel tipper
全钢悬墙 all-steel curtain wall
全钢制的 all-steel
全钢自动倾卸车 all-steel tipper
全高 height overall; total height
全高齿 full-depth tooth
全高齿渐开线制 full-depth involute system
全高度 overall height
全高度舱室 full head room
全高度的窗 full-height window
全高开采 full dimension mining
全高铅质玻璃 high lead crystal glass
全高螺母 full nut
全格舱式集装箱船 fully cellular container vessel
全隔声司机室 fully insulated cab
全工程统包 project-packaging; turn-key
全工序油漆法 bodying in
全工业经济计量模型 industry-wide econometric model
全工种 all-work
全工作范围 full operating range
全工作面 full face
全工作面一次爆破 full face blast(ing)
全工作面引爆法 full face firing
全工作日的 full time
全公司范围内(计算机)网络 company-wide network
全功电度表 apparent energy meter
全功率倒车 full power astern
全功率的 full power
全功率电动机 all-watt motor
全功率反应 full power response
全功率飞行 full power flight
全功率工作时间 full flow endurance
全功率换挡 full power shift
全功率时间 full power turn
全功率试车 full power trial
全功率试验 full power trial
全功率输出 full power output
全功率响应 full power response
全功率液压转向 full power hydraulic(al) steering
全功率正车 full power ahead
全功率转弯 live-power turn
全功率转向 full power steering
全功能 global function
全功能齿轮系 full-functional gear train
全功能传动系 full-functional gear train
全功能方式 full capability mode
全功能网络 fully functional network
全功能相关性 full-functional dependence[dependency]
全供水位 full supply level; normal pool level
全拱 full arch
全鼓风 full blast
全固定的 fully restrained; fully stable
全固定端 fully fixed end
全固定构件 fully fixed member
全固定喷灌系统 fully permanent sprinkler system
全挂车 complete trailer; independent trailer; rigid vehicle
全挂式拖车 full trailer
全关 complete shut-down
全冠 complete crown; full crown
全贯穿对焊 complete penetration butt weld
全贯流式水轮机 rim-generator(tubular) turbine; straflo turbine
全贯入 complete penetration; complex penetration
全惯性导航系统 all-inertial guidance system; all-inertial navigation system
全惯性制导 all-inertial guidance
全光 full gloss
全光瓷漆 full gloss enamel
全光反射的 holophotal
全光反射法 holophotal system
全光反射设备 holophotal apparatus
全光反射系统 holophotal system
全光反射装置 holophote
全光胶乳瓷漆 full gloss latex enamel
全光漆 full gloss paint; high gloss paint
全光栅 full raster
全光饰面 full gloss finish
全光通信[讯] all optic(al) communication
全光系计算机 all optic(al) computer
全光泽 full gloss
全光泽精修 full gloss finish
全光泽涂料 full gloss paint
全光泽油漆 full gloss paint
全光照 full exposure
全规模生产 full-scale production
全规速聚合物 holotactic polymer
全规整 holotactic
全轨 full track
全轨距 full ga(u)ge
全国 throughout the country
全国安全理事会 <美> National Safety Council
全国安全委员会 <美> National Safety Council
全国标准化技术委员会 National Technical Committee for Standardization
全国波道 national channel
全国财产估价者协会 National Association of Assessing Officers
全国产出量 national output
全国长途电话网 national telephone system
全国成本会计师协会 National Association of Cost Accountants
全国城市经济发展委员会 <美> National Council for Urban Economic

Development
全国城市联盟 National League of Cities
全国储蓄及贷款联盟 National Savings and Loan League
全国大气监视网 National Air Surveillance Network
全国道路建设程序 national road-building program(me)
全国道路建设大纲 national road-building program(me)
全国道路建设方案 national road-building program(me)
全国道路建设计划 national road-building program(me)
全国道路网 national road network
全国道路研究协作计划 <美> National Cooperative Highway Research Program(me)
全国电话总局 national center[centre]
全国电视系统委员会制彩色电视 National Television System Committee colo(u)r television
全国反对住房歧视委员会 National Committee against Discrimination in Housing
全国范围的 countrywide;statewide
全国范围的调查 countrywide survey
全国范围内的长途拨号 nationwide toll dial(l)ing
全国防火协会 <美> National Fire Protection Association
全国防蚀工程师协会 <美> National Association of Corrosion Engineers
全国房地产独立收费评估员协会 National Association of Independent Fee Appraisers
全国房地产经纪人协会 National Association of Regional Estate Brokers
全国房地产经理联合协会 National Association of Cooperate Real Estate Executives
全国房地产评估员学会 National Society of Real Estate Appraisers
全国房地产特许律师协会 National Association of Real Estate License Law Officials
全国房地产投资信托协会 National Association of Real Estate Investment Trusts
全国房地产委员会 National Realty Committee
全国房屋建造商协会 National Association of Home Builders
全国各县联络协会 National Association of Counties
全国工业化房屋建造业者协会 National Association of Home Manufacturers
全国工业污染控制委员会 National Industrial Pollution Control Council
全国工业运输联盟 <美> National Industrial Traffic League
全国公共交通研究开发合作计划 <美> Cooperative Transit Research and Development Program
全国公共交通研究与发展协作计划 <美> National Cooperative Transit Research and Development program(me)
全国公路交通安全管理机构 <美> National Highway Traffic Safety Administration
全国公路交通安全管理局 <美> National Highway Traffic Safety Administration
全国公路研究协作计划 <美> National Cooperative Highway Research Program(me)
全国公寓协会 National Apartment Association
全国管道螺纹协会 Association of National Pipe Thread
全国规模 national scale
全国国防运输协会 <美> National Defense Transportation Association
全国海洋资料中心 National Oceanographic(al) Data Center[Centre]
全国互助储蓄银行协会 National Association of Mutual Savings Banks
全国环境监测条例 Regulations on Administration of National Environmental Monitoring
全国环境空气质量标准 national ambient air quality standard
全国环境卫星中心 National Environmental Satellite Center[Centre]
全国黄铜铸造协会 National Brass foundry Association
全国货物运输计划 national plan for goods traffic
全国货物运输联合企业 <英> National Freight Consortium
全国货运公司 <英国,对铁路和公路货运时间负责> National Freight Corporation
全国技术情报服务处 <美> National Technical Information Service
全国建筑工人联合会 National Federation of Building Trades Operatives
全国建筑金属生产者协会 National Association Architectural Metal Manufacturers
全国居者有其屋基金会 <美> National Home Ownership Foundation
全国科技情报系统 nation-wide scientific information system
全国科学基金会 National Science Foundation
全国科学院全国研究协会 <美> National Research Council
全国空气取样网 National Air Sampling network
全国空气有害污染物排放标准 national emission standards for hazardous air pollutants
全国联播 national dissemination
全国煤炭协会 <美> National Coal Association
全国每年空气污染物排放量 national annual emissions of air pollutants
全国农村住房联盟 National Rural Housing Coalition
全国排放数据系统 national emission data system
全国普选 national poll
全国气象中心 National Meteorologic(al) Center[Centre]
全国情报网 national information network
全国人口普查 national population census
全国人民代表大会 National People's Congress
全国砂石协会 <美> National Sand and Gravel Association
全国商品混凝土协会 <美> National Ready Mixed Concrete Association
全国社区发展协会 National Association for Community Development
全国石灰石协会 <美> National Limestone Association
全国石灰协会 <美> National Lime Association
全国市场 national market
全国熟练泥瓦工协会 <英> National Association of Master Masons
全国水井协会 National Water Well Association
全国水平 national level
全国陶土管研究所 National Clay Pipe Institute
全国铁道科学研究院 <苏联,设在莫斯科> All-Union Railway Scientific Research Institution
全国铁路不同运输方式联合运输协会 <美> National Railroad Intermodal Association
全国铁路旅客协会 <美> National Association of Railroad Passengers
全国铁路统计资料汇编 collection of national railway statistic data
全国通信[讯]光缆骨干网 a key national network of communications optic(al) cables
全国通行的车辆 cross-country vehicle
全国统一地图投影 national (map) projection
全国统一法律 <美> uniform law
全国统一价格 national uniform price
全国统一商业准则 <美> Uniform Commercial Code
全国卫生基金会 National Sanitation Foundation
全国行政区际公路委员会 <美> National Interregional Highway Committee
全国性报表 state report
全国性抽样调查 nation-wide sample survey
全国性的 countrywide;nationwide
全国性的传染病 pandemic
全国性工业公司 national industrial corporations
全国性公路 national highway
全国性观测网 national observation net
全国性样本 cross-country sample
全国研究理事会 National Research Council
全国养老基金协会 National Association of Pension Funds
全国银行电汇电脑中心 <日本> Data Telecommunication Center
全国银行账户 all-bank account
全国优秀产品 national quality product
全国预制混凝土协会 <美> National Precast Concrete Association
全国轧石协会 <美> National Crushed Stone Association
全国展览中心 <英> National Exhibition Centre
全国证券交易所 <美> National Securities Exchange
全国证券市场体系 Nation Market System
全国制造商协会 <美> National Association of Manufacturers
全国智能交通运输系统通讯规程 <美> National Transportation Communication/ITS Protocol
全国主要物资合理运输 rational transport flow diagram for major products in the country
全国住房更新协会 National Housing Rehabilitation Association
全国住房及经济发展法律服务中心 National Housing and Economic Development Law Project
全国住房及开发商协会 National Association of Housing and Redevelopment Officials
全国住房信用社协会 National Association of Housing Cooperatives
全国专利权委员会 <美> National Patent Commission
全国装饰品协会 National Decorating Products Association
全国资源计划委员会 National Resources Planning Board
全国自动道路系统协会 <美国 1994 年为开发自动道路系统成立的机构> Automated Highway System Consortium
全国综合开发计划 comprehensive national development plan
全国综合平衡 national comprehensive equilibrium
全国综合指数 national composite indices
全国总产量 gross national product
全国总产量增长率 gross national product growth rate
全国总动员 national mobilization
全国租赁住房协会 National Leased Housing Association
全过程 overall process
全过程招标 inviting bids of whole procedure;whole stage tendering
全海洋综合监视 integrated total ocean surveillance
全焊钢桥 all-welded steel bridge
全焊桁架桥 all-welded truss bridge
全焊加劲钢箱梁 all welded steel stiffened box girder
全焊接储料仓 all-welded storage
全焊接船舶 all-welded ship
全焊接船体 all-welded hull
全焊接的 all-welded
全焊接钢结构 all-welded steel structure
全焊接钢梁或桁架 all-welded steel girder or truss
全焊接锅炉 all-welded boiler
全焊接结构 all-welded structure;all-welded construction
全焊接式浮船坞 all-welded floating dock
全焊金属的焊缝拉力试验 all-weld-metal tension test
全焊梁 all-welded beam
全焊桥 all welded bridge
全焊透 complete penetration and fusion in welding
全焊无缝门 fully welded seamless door
全航程 round voyage
全河道渠化 continuous canalisation [canalization]; continuous channelization
全河剖面 long profile of river
全荷特性 full load characteristic
全荷载时最高速度 maximum full load speed
全黑的 black-filled
全黑生料 complete black meal
全黑信号 <电视图像> black signal
全黑岩 holomelanocrate
全痕 <遗迹化石> full relief
全横向式系统 fully transverse system
全横向通风 full transverse ventilation;transverse ventilation
全红灯 <对所有交通显示红灯> all red
全红信号【交】all-red signal
全红信号时间 all-red period
全厚回收 full depth reclamation
全厚开采 full-seam extraction
全厚膜化集成电路 all-thick-film integrated circuit
全厚式地沥青混凝土路面 total asphalt pavement
全厚式沥青补坑 full-depth bituminous patch
全厚式沥青混凝土铺面 full-depth asphalt concrete pavement
全厚式沥青路面 <从基层到面层,整

个路面厚度都用沥青混合料铺成 > full-depth asphalt pavement
全厚式沥青铺面 full-depth asphalt pavement;total asphalt pavement
全厚式铺筑 full-depth paving
全厚式水泥混凝土路面 full depth concrete pavement
全厚修补 full depth patching
全厚再生 full depth reclamation
全呼 general call; all ship calls < 对船 >
全呼按钮 all-call push-button
全呼键 general calling key
全呼叫 complete call
全互连网络 fully connected network
全护面堆石堤 all-armo(u)r rock mound
全护面堆石防波堤 all armo(u)r rock mound breakwater
全护面防波堤 all-armo(u)r rock mound
全画闪光曝气 full frame flash exposure
全环境 total environment
全环食 total-annular eclipse
全缓冲通道 fully buffered channel
全缓存储器 global buffer
全缓解 direct release;full release
全缓解时间 entire release time; full release time
全缓解位 full-release position
全缓解与阶段缓解定位盖 direct and graduated release cap
全黄铜阀 all-brass valve
全黄铜泄水口 all-bronze drain
全灰度等级 full grey scale
全挥发水处理 volatile treatment;zero solids treatment
全辉照度计 holophane lumeter
全回波抑制器 full echo suppressor
全回风系统 close cycle system
全回火 deep drawing
全回流 infinite reflux;total reflux
全回流操作 total-reflux operation
全回转 full circle swinging; full revolving;full swing < 挖土机的 >
全回转(浮式)起重机 full circle slewing floating crane
全回转螺旋桨 duck propeller; shuttle propeller;Z-axis propeller
全回转式反铲挖掘机 full-revolving back hoe excavation
全回转式浮吊 full circle slewing floating crane
全回转式起重机 full circle crane
全回转式挖掘机 swing excavator
全回转式挖土机 swing excavator
全回转式转塔 full-revolving turret
全回转通用单斗挖掘机 universal shovel
全回转型拖轮 fully turntable tug
全会 plenary assembly
全毁 burnt down;total collapse
全混 perfect mixing
全混合作用 holomixis
全混凝土的 all-concrete
全混凝土路面 all-concrete pavement
全活动性组分 perfectly mobile component
全活叶螺旋桨 full feathering propeller
全货柜船 full container ship
全机构的 organization wide
全机械化设备 fully mechanized equipment
全级试样 < 取自各个水平位置 > all-level sample
全集 collected edition;collected works; complete set; complete works; universal set
全集成薄膜电路 completely integrated thin-film circuit
全集成电路 completely integrated circuit
全集成数字网 fully integrated digital network
全集中控制方式 all common control system
全集中控制系统 fully centralized control system; fully integrated control system
全集装箱船 all container ship; full container;full container ship
全集装箱货轮 full container ship
全计件工资 entire pay calculated by piecework
全记录方式 full recording mode
全记录监测 full race monitoring
全记录频谱 whole record spectra
全季放牧 continuous grazing
全继电器法 all-relay method
全继电器式的 all-relay
全继电器式选择器 all-relay selector
全继电器制 < 自动电话 > relay system
全继电系统 all-relay system
全继电制 all-relay system
全寄生植物 holoparasite
全加法电路 full adder circuit
全加法器 three-input adder; full adder
全加工轧辊 necked-and-rough turned roll
全加浓铀 full enriched uranium
全加器 one-position adder; three-input adder
全家用处理装置 whole-house treatment unit
全家用设备 whole-house equipment
全家用装置 whole-house unit
全价 full rate
全价电报 full rate message
全价票 full fare ticket
全价票旅客 regular-fare passenger
全减法器 full subtracter; three-input subtracter
全减振座位 fully sprung seat
全碱性的 all-basic
全浆橡胶 whole latex rubber
全交变疲劳极限 completely reversed fatigue limit
全胶合板构造 all-veneer construction
全胶合板结构 all-veneer construction
全胶结式锚杆 completely grouted rockbolt
全胶乳 whole latex
全胶橡胶 whole latex rubber
全胶制品 all-rubber article
全角焊 full fillet weld
全绞车锚泊系统 all-winch mooring system
全轿驱动式后卸卡车 all axle drive rear-dump
全阶段交易税 all-phase transaction tax
全接地 full ground contact
全接受角 total acceptance angle
全接型 holostylic;holostyly
全节矩绕组 diametral winding; full-pitch winding
全节距 diametric(al)pitch;full pitch
全节(距)线圈 whole coil
全节流 full throttle;throttle full open
全节流发动机特性 gross performance engine characteristic
全节吸量管 whole pipet(te)
全结构检索 specific search
全结晶的 holocrystalline
全截面 bulk cross-section;total cross-section
全截面掘进 full advance
全截面拼接 < 构成截面各部分在一处拼接 > full splice
全截止 full cut-off
全解 complex solution;complete solution
全解析空中三角测量程序 fully analytic(al) aerotriangulation program(me)
全介质层滤光片 all-dielectric(al) filter
全介质对称干涉滤光片 symmetric-(al) all-dielectric(al) interference filter
全介质多层高反射膜 all-dielectric-(al) multi-layer high reflection film
全介质干涉滤光片 all-dielectric(al) interference filter
全金属波导 all-metal waveguide
全金属车辆 all-metal car
全金属车身 all-metal body
全金属车厢 all-metal coach
全金属的 all-metal
全金属反射镜 all-metal mirror
全金属高容量烧嘴 all-metal high capacity burner
全金属加热器 all-metal heater
全金属建筑物 all-metal building; all metallic construction
全金属焦面快门 all-metal focal-plane shutter
全金属结构 all-metal construction
全金属客车 all-metal coach
全金属立面 all-metal facade
全金属耐烘烤阀 Alpert bakable valve
全金属设备 all-metal plant
全金属镶板 all-metal panel
全金属闸衬片 all-metal brake lining
全进程 full process
全进料的 fully charged
全进位 complete carry
全浸法 full dip process
全浸没式染色机 fully flooded dye machine
全浸式卷染机 immersion jig
全浸式水翼装置 fully submerged hydrofoil system
全浸式翼艇 submerged hydrofoil craft
全浸式蒸发器 flooded evapo(u)rator
全浸试验 total immersion test
全浸温度计 total immersion thermometer
全浸渍绝缘 fully impregnated insulation
全晶 holohedral crystal
全晶斑状 holocrystalline-porphyritic
全晶玻璃 full crystal;real crystal
全晶的 holocrystalline
全晶粒岩 granomerite
全晶体 holocrystalline
全晶体管 all-transistor
全晶体管点火系 full-transistor ignition system
全晶体管电路 all-transistor circuitry
全晶体管化 fully transistorized
全晶体管计算机 all-transistor computer
全晶体管照相机 all-transistor camera
全晶体结构【地】 wholly crystalline texture
全晶形 holohedry
全晶质 pleocrystalline
全晶质玻璃 full crystal
全晶质的 holocrystalline
全晶质霏细斑岩 red rock
全晶质结构 eucrystalline texture; holocrystalline texture; pleocrystalline texture
全晶质岩 holocrystalline rock
全井径地层测试 full hole testing
全井孔 full hole
全井套管 full hole casing
全井注水泥 full hole cementing
全井总进尺 total footage
全井总台时 total drilling hours
全景 full scene;full view;group shot; overall look;overall perspective;overall view; panoramic view; total view;whole scene
全景比较 panoramic comparison
全景玻窗 panoramic window
全景玻璃窗 panoramic glass window
全景窗 panoramic window
全景的 panoramic
全景电视 panoramic television
全景电影 cinerama;panorama
全景电影院 panorama theatre; panoramic cinema
全景分析器 sweeter
全景分析仪 panalyzer[panalyzer]
全景概要 general view
全景航空摄影 panoramic aerial photography
全景航摄仪 panoramic aerial camera
全景畸变 panoramic distortion
全景技术 panoramic technique
全景驾驶室 full vision cab(in)
全景监视器 panoramic monitor
全景接收机 ether scanner;panoramic receiver
全景镜头 following shot;follow shot; full shot; panoramic lens; panoram(a)
全景宽银幕 panoramic screen
全景宽银幕电影 cinepanoramic
全景宽银幕电影院 cinepanoramic house
全景廓线图 panoramic profile map
全景雷达 all-round looking radar; panoramic radar; scent spray; stinky
全景瞄准镜 panoramic sight
全景拍摄 full shot;panning;pan-shot
全景全息图 full-view hologram
全景乳剂 panoramic emulsion
全景扫调接收机 panoramic receiver
全景摄像 panoramic photography; pan-shot
全景摄影 panoramic photography; pan-shot;photographic(al) panorama;scan(ning)
全景摄影机 panorama camera;panoramic camera;pantoscopic camera
全景摄影畸变 panoramic photographic(al) distortion
全景视差立体图 panoramic parallax stereogram
全景适配器 panoramic adapter
全景衰减器 panoramic attenuator
全景搜索接收机 < 带有阴极射线管的 > ether scanner
全景探测计划实验室 panoramic detector program(me) laboratory
全景调整 pan control
全景透视 panoramic perspective
全景透视图 panoramic perspective drawing
全景图 panoramagram; panorama (sketch); panoramic sketch; panoramic view
全景图片 panorama
全景图像 panoramic picture
全景望远镜 rotary telescope
全景显示 panoramic videomapping
全景像片 panoramic photograph; panoramic picture; photographic (al) panorama
全景效应 panoramic effect
全景照片 distant view photograph; panoramic photograph; panoramic

picture
全景照片图 panoramic-photograph map
全景照相机 cyclograph;panoramic camera;pantoscopic camera
全景照相器 pantoscope
全景装置 panoram(a)
全径 full ga(u)ge;full hole size;overall diameter
全径偏斜钻头 full ga(u)ge deflecting bit
全径钎头 full face bit
全径切的 fully quarter-sawn
全径造斜钻头 full ga(u)ge deflecting bit
全径钻孔 full ga(u)ge drill hole
全径钻头 full ga(u)ge bit;full-sized bit
全静电光导摄像管 all-electrostatic vidicon
全静电射线管 all-electrostatic tube
全静压管入口 Pitot tube mouth
全静压联合式探头 combined pitot-static probe
全局编址标头 globally addressed header
全局变化 global change
全局变换 global transformation
全局变量 global variable
全局变量符(号) global variable symbol
全局变量引用 global variable reference
全局标准化矩 standardized moments of the range
全局不稳定性 global instability
全局步进乘法 global step multiplication
全局参数 global parameter
全局参数缓冲器 global parameter buffer
全局测试 global test
全局查找 global search
全局乘法运算 global multiply operation
全局程序控制 global program(me) control
全局处理程序 global processor
全局处理机 global processor
全局传送命令 global transfer command
全局存储(器) global storage
全局代码 global code
全局的 global;in the large
全局地址 global address
全局地址向量 global address vector
全局段 global section
全局段数据基 global section data base
全局方式 global mode
全局服务 global service
全局符号 global symbol
全局符号表 global symbol table
全局复写操作 global copy operation
全局共享资源 global shared resource
全局共用子表达式 global common subexpression
全局关系 holotopy
全局观点 global view
全局环境 global environment
全局极大值 global maximum
全局极小点 global minimum point
全局极小值 global minimum
全局寄存器优化 global register optimization
全局加法 global addition
全局减法 global subtraction
全局减法命令 global subtraction command
全局检索 global search
全局渐近稳定 asymptotically stable in the large
全局渐近稳定性 global asymptotic stability
全局结合 global binding
全局控制 global control
全局控制段 global control section
全局控制线 global control line
全局控制总线 global control bus
全局利益 general interest;interests of the whole
全局灵敏度 global sensitivity
全局流分析 global flow analysis
全局路径选择表 global routing table
全局名称 global name
全局模型 comprehensive model
全局目标 global object
全局目录 global dictionary
全局区域 global area
全局时钟 global clock
全局收敛 global convergence
全局数据 global data
全局数据库 global data base
全局数据流分析 global data flow analysis
全局数据模块 global data module
全局搜索 global search
全局锁 global lock
全局通信[讯]业务 global communication service
全局网络 global net(work)
全局稳定性 stability in the large;global stability
全局误差 global error
全局限制 global restriction
全局行为 global behavio(u)r
全局性 globality
全局性准则 global criterion
全局虚拟协议 global virtual protocol
全局寻址方案 global addressing scheme
全局引用 global reference
全局优化 global optimization
全局约束 global restriction
全局栈顶 global stack top
全局栈顶单元 global stack top location
全局整体理论 global theory
全局正规化 global regularization
全局知识 global knowledge
全局指令 global command
全局主过程 global main process
全局最大(值) global maximum
全局最小(值) global minimum
全局最优化理论 global optimization theory
全局最优解 globally optimal solution
全局最优(值) global optimum
全局坐标 world coordinates
全矩阵 complex matrix;full matrix
全矩阵环 complete matrix ring
全矩中心 center[centre] of range
全距 full pitch;overall spread;range;whole range
全距离 full distance;total distance
全距系数 coefficient of range
全距线圈 full-pitched coil
全绝缘 insulation integrity
全绝缘的 all-insulated
全绝缘电线 all-insulated wiring
全绝缘开关 all-insulated switch
全绝缘铠装电缆 all-insulated sheathed cable
全绝缘铠装电线 all-insulated sheathed wiring
全绝缘绕组 fully insulated winding
全开 complete opening;full opening
全开的 full gate;full-open;wide open
全开度 complete opening;full gate;total opening
全开度襟翼 full depression wing flap
全开度运行 full gate operation
全开额定压力 relief set pressure
全开阀 clearway valve
全开风门 full throttle
全开风门运转 run at full throttle
全开钩 full-open hook
全开接点 full-open contact
全开节气门 full open throttle
全开进气 full admission
全开口角接头 full-opened corner joint
全开口式桶 full aperture drum
全开门棚车 all door box car
全开-全关控制 on-off control
全开-全关控制系统 on-off control system
全开图纸 double double folio;double double crown;quad crown <10 英寸×40 英寸,1 英寸=0.0254 米>;double double cap <28 英寸×34 英寸,1 英寸=0.0254 米>
全开吸气 full admission
全开油门(运转) full throttle
全开状态 full gear;full open state
全抗磁性 complete diamagnetism
全颗石 holococcolith
全壳口 entire aperture
全可变多址 fully variable multiple access
全可变多址连接方式 full-variable multiple access
全可凝集的 panagglutinable
全可约矩阵代数 fully reducible matrix algebra
全刻度 full-scale
全刻度范围 full-scale range
全刻度偏转(仪表) full-scale deflection
全坑法 gross pit sampling
全空斗墙 all-rowlock wall
全空号 all-space
全空间 total space
全空间延拓 all space continuation
全空泡螺旋桨 fully cavitating propeller
全空气方式 all air system
全空气高压控制设备 all air high-pressure control equipment
全空气热回收系统 all air heat recovery system
全空气系统 all air system
全空气诱导箱 all air induction box
全空调的 fully air-conditioned
全空心的 through-voided
全空心的盖 through-double cap
全空心式绝缘器 full-cored insulator
全孔采取率 percentage recovery of full hole
全孔法 gross hole sampling
全孔回填封闭 back-stuffing and sealing in whole well
全孔径 full aperture
全孔径钻铤 full hole drill collars
全孔流动 full hole flow
全孔树脂锚杆 full column resin grouted bolt
全孔套管 full hole casing
全孔隙度 total porosity
全孔隙压力比 full pore-pressure ratio
全孔钻进 full hole drilling
全控桥 fully controlled bridge
全口径 foil aperture;full aperture
全扣打捞丝锥 tap bottoming
全跨度走廊板 full-span corridor panel
全宽 overwide;extreme breadth,registered breadth
全宽度 breadth extreme;breadth maximum;extreme breadth;full duration;full width;overall width
全宽度护车板 full-width cab-protection plate
全宽度轮毂制动器 full-width hub brake
全宽度铺路机 full-width paver
全宽建筑 <在路面全宽度上同时施工> full-width construction
全宽桥 full-width bridge
全宽施工 full-width construction
全宽式混凝土路面板 full-width concrete slab
全宽式混凝土铺筑 full width concrete paving
全宽梯台 half landing
全宽(同时)压实 full-width compaction
全宽型 full-width type
全宽堰 full-width weir
全矿层回采 full-seam mining
全矿泥化 all sliming
全扩散 perfect diffusion
全扩散单块集成电路 all-diffused monolithic integrated circuit
全扩散面 perfectly diffusing plane
全扩散体 perfect diffuser
全扩展的流 full developed flow
全拉伸变形丝 fully drawn texturing yarn
全拉伸矫直 pure stretch level(l)ing
全拉伸丝 fully drawn yarn
全缆定深精度 all streamer fathom accuracy
全劳动力 able-bodied farm worker;full labo(u)r power;full-time labo(u)r
全类 universal class
全累积 total cumulation
全力以赴 with might and main
全力以赴地 hammer and tongs
全立方点群 full cubic point group
全利用度 full availability
全励磁转速 full field speed
全砾磨机 all-pebble mill
全粒 whole grain
全连单用主卫星系统 fully connected single primary satellite system
全连接网络 completely connected network;fully connected network
全连续线性变换 completely continuous linear transformation
全连续自动压力机 perfect transfer press
全连主卫星系统 fully connected primary satellite system
全连铸 sequence casting
全链式 full chained
全量 full dose;population parameter
全量程 full range;full-scale(range);gamut
全量分析 complete analysis;gross analysis;total analysis
全量化学分析 complete chemical analysis
全量浸注法 <木材的> full-cell process
全量理论 total strain theory
全量水分析 complete water analysis
全量组成 total composition
全料 complete feed
全裂 holoblastic cleavage
全林渐伐作业法 uniform system
全林伞伐作业 shelter-wood uniform system
全磷 total phosphorus
全磷模型 total phosphorus model
全灵敏度对比度 full sensitivity contrast
全零信号 all-zero signal
全零状态 all-zero state
全龄的 all-aged
全流 integral current;whole current

全流阀 full flow valve
全流方法 whole current method
全流过滤燃料油 full flow filtering fuel
全流理论 whole current theory
全流量 full flow
全流量高性能离子净化器 full flow polishing demineralizer
全流量过滤 full flow filtration
全流量过滤器 full flow filter
全流式 full flow type
全流式分级机 whole current classifier; whole current settler
全流式过滤 full flow filtration
全流式机油滤清器 full flow oil filter
全流式滤清器 full flow filter
全流式热交换器 full flow heat exchanger
全流液压油滤清器 full flow filter; full flow hydraulic(al) filter
全流域的 basin-wide
全流域性洪水 valley-wide flood
全流域性水情 basin-wide hydrological regime
全硫 total sulfur
全硫含量分级 total sulfur content graduation
全硫化氢天然气 sour gas
全硫砷酸钠 sodium thioarseniate
全硫碳酸钠 sodium thiocarbonate
全硫锑酸钠 sodium thioantimonate
全楼层 full floor; full stor(e)y
全漏失 lost circulation; total loss
全卤化的 fully halogenated; perhalogenated
全卤化碳氢化合物 perhalogenated fluorocarbon
全卤化烷烃 fully halogenated alkane; perhalogenated alkane
全卤化物 perhalide
全路的 system-wide
全路列车运行图和时刻表 system-wide graphical timetable
全路旅客发送量 volume of sending passengers by the railway
全路面照明灯 full road lighting
全路通用会计核算及管理信息系统 railway general accounting and management information system
全路网每线路公里平均货物吨公里数 freight ton-kilometers per route kilometer (average for entire network)
全路运输进款 transport income all the railway
全路运输信息系统 system-wide information for traffic
全铝导体 all alumin(i)um conductor
全铝建筑物 all-alumin(i)um building
全铝饮料罐 all-alumin(i)um beverage can
全履带车 full track
全履带车辆 full-track laying vehicle
全履带车轮 full-track vehicle
全履带底盘 full-track chassis
全履带牵引车 fully track vehicle
全履带式车辆 full-track vehicle
全履带式拖拉机 full-track tractor
全履带运输工具 full-track carrier
全氯 total chlorine
全氯代聚氯乙烯 perchlorinated polyvinyl chloride
全氯代烃 perchloro-hydrocarbon
全氯丁二烯 perchloro-butadiene
全氯化的 perchlorinated
全氯化石蜡 perchloroparaffin
全氯化作用 perchlorination
全氯甲硫醇 perchlormethyl mercaptan
全氯甲烷 perchloromethane
全氯萘 perchloronaphthalene; perna
全氯碳化物 perchlorinated hydrocarbon; perchlorocarbon
全氯乙醚 perchlorether
全氯乙烷 perchlorethane
全氯乙烯 perchlorethylene
全轮操纵 all-wheel steer(ing)
全轮传动 all-wheel drive
全轮传动式车辆 all-wheel drive vehicle
全轮驾驶 all-wheel steer(ing)
全轮廓的 full-sized
全轮气压刹车 all-wheel compressed-air brake; all-wheel pneumatic brake
全轮气压制动 all-wheel compressed-air brake; all-wheel pneumatic brake
全轮牵引车 all-wheel tractor
全轮驱动 all-wheel drive
全轮驱动搬运车 all-wheel drive wagon
全轮驱动车辆 all-wheel drive vehicle
全轮驱动的推土机 all-wheel(bull) dozer
全轮驱动汽车 all-wheel drive vehicle
全轮驱动式运输车 all-wheel drive wagon
全轮驱动挖土机 all-wheel drive excavator
全轮驱动运货车 all-wheel drive truck
全轮胎再生胶 whole-tire[tyre] reclaim
全轮推土机 all-wheel(bull) dozer
全轮拖拉机 all-wheel tractor
全轮行星齿轮驱动 all-wheel planetary wheel drive
全轮转向 all axle steering; all-wheel crab steer; all-wheel steer(ing)
全轮转向式平地机 all-wheel steer grader
全轮自卸车 all-wheel dumper
全轮自卸汽车 all-wheel dump car; all-wheel tipping lorry
全螺距螺旋 full-pitch auger
全螺纹 full thread; perfect thread
全螺纹(螺)栓 full screw bolt; full thread bolt
全螺旋法 full spiral cut
全落时期 period of full draw-down
全买 buying up
全脉冲 overall pulse
全脉冲转发 full transponder circuit
全满 clean full
全漫反射 total diffuse reflection
全漫反射比 total diffuse reflectance
全漫反射体 perfectly diffuse reflector
全漫辐射体 perfectly diffuse radiator
全漫射 perfect diffusion
全漫射光 complete diffusion
全漫射面 perfect diffuser
全漫射球 perfectly diffusing sphere
全漫射体 perfect diffuser
全忙 all busy; all-paths-busy
全忙电路 all busy circuit
全忙呼叫 overflow call
全忙计数器 overflow meter
全忙接点 overflow contact
全忙信号 overflow signal(l)ing
全忙装置 busying arrangement
全毛垫 all-hair pad
全锚式 fully anchored
全貌 complete picture; full view; general view; integrated picture; overall look; overall perspective; overall view
全貌窥视窗 overall viewer
全貌图 close up view
全美洲的 Pan-American
全门棚车 full door box car
全迷向的 totally isotropic
全迷向子空间 totally isotropic subspace
全密闭座舱 totally enclosed cockpit
全密封 omniseal
全密封泵 canned pump
全密封工作服 fully impervious clothing
全密封式救生艇 totally enclosed lifeboat
全密封室 hermetically sealed chamber
全面保付代理 full factoring
全面保证契据 general warranty deed
全面财务计划 overall financial plan
全面采矿法 breast stopping
全面策略 overall strategy
全面拆修 capital repair
全面长壁开采法 full longwall mining
全面成本管理 overall cost control
全面抽出式机械通风 general exhaust mechanical ventilation
全面抽样 overall sampling
全面处理 complete treatment
全面淬火钢轨 fully heat-treated rail
全面的 across-the-board; all-embracing; all-out; all round(all); all sided; comprehensive; full-scale; overall
全面的分析 rounded analysis
全面的强制性制裁 comprehensive mandatory sanction
全面的条件 all-embracing term
全面点火 area ignition
全面电气化 full electrification
全面调查 all-round investigation; complete enumeration; complete investigation; complete survey; overall investigation; overall statistical survey
全面调查团 general survey mission
全面冻结 freeze over; ice up
全面对称 holohedral symmetry; holohedry
全面发展 all-round development; comprehensive development; integrated development; overall development
全面发展的个人 fully developed individual
全面发展计划 comprehensive plan(ning)
全面方案 overall project
全面费用管理 overall cost control
全面分析 comprehensive analysis; multianalysis; overall analysis; through analysis
全面腐蚀 general corrosion
全面负责制 comprehensive services
全面复查 comprehensive review
全面观点 comprehensive view
全面管理 general management; total management
全面管理系统 total management system
全面规划 comprehensive plan(ning); overall plan(ning)
全面规划方案 comprehensive planning program(me); overall project
全面规划、合理布局原则 principle of overall planning and rational layout
全面滚切 plunge cutting
全面好转 general improvement
全面后退式 full-retreat type
全面后退式开采 full retreat mining
全面后退式开采法 full retreat mining system
全面后退式盘区 full retreat panel
全面化 generalization
全面换气次数 number of complete air changes
全面回报率 overall rate of return
全面或综合的预算 overall or comprehensive budget
全面机械通风 total mechanical ventilation
全面积 gross area
全面积负荷 load over the entire area
全面基脚 mat footing
全面计划(方案) overall project
全面计划管理 overall planning management
全面加荷 all-around loading
全面价格 all-round price
全面减税 across-the-board tax cut
全面检查 complete inspection; general inspection; general overhaul; overall check; overall inspection; run down; total inspection
全面检修 general overall; general overhaul; major overhaul; major tune-up; through overhaul; top overhaul
全面检验 complete survey; overall inspection
全面建筑观 total scope of architecture
全面鉴定 complete verification
全面降低关税 across-the-board tariff reduction
全面降价 all-round decline
全面浇水 overall watering
全面接触 full face contact
全面接近式架空地板 full access raised floor
全面结冰 ice up
全面截弯 wholesale cut-off
全面解决办法 overall solution
全面戒备 <机场> all-out alert
全面近似值 global approach
全面进给法 plunge cut
全面进磨法 plunge grinding
全面进行生产 full-scale operation
全面经济效果 overall economic effect
全面经营管理 total operational management
全面晶形 holohedral form
全面警戒 full-scale alert
全面剧院 <建筑大师格罗派斯 1927 年设计的> Total Theater[Theatre]
全面掘进法 breasting method
全面均衡 general equilibrium
全面均衡分析 general equilibrium analysis
全面开采法 breasting method; full longwall
全面开发计划 all-out development plan; all-out development program(me); comprehensive development plan; comprehensive development program(me)
全面开工 at full capacity; full-scale production
全面开拓 full development
全面开挖 <隧道> full face digging
全面开挖法 breasting method
全面开展 in full swing
全面勘察法 overall investigation method
全面考虑 overall consideration
全面扩建 general extension
全面论规划 holistic plan(ning)
全面锚碇 all-round anchorage
全面黏[粘]合 full bonding
全面黏[粘]接型锚杆 fully bonded bolt
全面耙松 overall discing[disking]
全面排风 general exhaust ventilation
全面盘存 complete inventory; wall-to-wall inventory
全面平衡分析法 general equilibrium

analysis
全面平衡体系 general equilibrium system
全面铺筑的 all-paved
全面普查 general census
全面起道 out-of-face raise
全面起道捣固 out-of-face surfacing
全面汽车化 general motorization
全面前进式 full-advance type
全面前进式开采法 full-advance method;full-advance system
全面前进式盘区 full-advance panel
全面清仓 clear out all the holding
全面屈服 general yield
全面趋势 overall climate
全面审查 comprehensive review
全面审计 complete audit;final audit;full audit
全面生产 full-scale production
全面生产率 overall productivity
全面生产性维修 total productive maintenance
全面施工 full-scale construction
全面施用 overall application
全面试验 complete test;comprehension test;full-scale test
全面损失控制 total loss control
全面体 holohedron
全面条款 all-embracing term
全面调节 full regulation
全面调整关税 across-the-board tariff changes
全面通风 entirely ventilation;general air change;general ventilation
全面通货膨胀 generalized inflation
全面投资 full invested
全面涂漆于…… overpaint
全面网 continuous net
全面维修 out-of-face maintenance
全面紊流 fully developed turbulence
全面误差 global error
全面限额 overall limitation
全面像 holohedry
全面削减 across-the-board cut
全面萧条 full-blown depression
全面协定 comprehensive agreement
全面行车安全 overall driving safety
全面行政管理 full-scale civil management
全面形 holomorphism
全面形晶体 holohedral crystal
全面型的 holohedral;holosymmetric;holosystemic
全面性 all-sidedness
全面性能 all-round property
全面性能试验 comprehensive test;exhaustive test;full-scale experiment
全面修测 complete revision
全面修检 general overhaul
全面修整机 scalper
全面延伸式运输机系统 full dimension extensible conveyor system
全面异极性 holohedral hemimorphism
全面余额 overall balance
全面预算 comprehensive budget;overall budget
全面预算活动 overall budgetary operation
全面预算控制 complete budgetary control
全面运输业务 <指铁路兼营公路业务等> full-service transportation
全面运转 full-plant operation;full-scale operation
全面战略 overall strategy
全面涨价 all-round advance
全面掌管业务 comprehensive grasp of business
全面照明 general illumination
全面罩 full face mask;full face piece mask;total mask
全面罩潜水头盔 full face diving helmet
全面罩式 full face mask type
全面指示速干道 through briefing
全面制裁 comprehensive sanction
全面质量保证 total quality assurance program(me)
全面质量管理 total quality control;total quality management
全面质量管理圈 total quality control circle
全面质量管理制 total quality control system
全面质量控制 total quality control
全面中央信息问题 overall central information problem
全面专业责任 overall professional responsibility
全面咨询服务 full consultancy service
全面钻进 full hole drilling
全面钻进钻头 full hole bit
全面钻头 full face bit
全苗 full stand
全民产权 right to property owned by the whole people
全民福利 national well-being
全民环境教育 environmental education for the whole people
全民所有制 ownership hy the whole people
全民所有制经济 economic sector owned by the whole people
全民(族)的 nation-wide
全模标本 cotype;syntype
全模式 syntype
全木的 all-wood;timber positive
全目 general view
全苜蓿叶式 full clover-leaf
全苜蓿叶式立体交叉 full clover-leaf crossing
全内反射 total internal reflection
全内反射导光纤维 total internal reflection light fibre
全内反射共振器 total internal reflection resonator
全内反射角 alinternal reflection angle
全内反射腔 total internal reflection cavity
全能 all-function;omnipotence
全能保险 all risks insurance;insurance against all-risks
全能曝光控制器 universal intervalometer
全能比长仪 universal comparator
全能测量仪 universal plotting instrument
全能测图仪 universal plotting machine
全能磁力仪 universal magnetometer
全能淬火炉 all case furnace
全能代理人 universal agent
全能的 all-around;all-purpose;all round(all);almighty;omnipotent
全能地面摄影机 universal terrestrial camera
全能电站 total energy plant
全能法单模型布点 universal method for point layout in single model
全能法摄影测量 universal method of photogrammetry
全能法(摄影)测图 universal method of photogrammetric mapping;universal method of photogrammetry;universal photo
全能法双模型布点 universal method for point layout in double model
全能飞机 general purpose plane
全能服务银行 full-service bank
全能耗 total energy consumption
全能绘图机 parallel motion protractor;universal drafting machine;universal plotting machine
全能绘图仪 parallel motion protractor;universal drafting machine;universal plotting machine
全能加速器 omnitron
全能胶片摄影机 universal film camera
全能经纬仪 universal theodolite;universal transit;universal instrument
全能力生产 full production
全能立体测图仪 universal stereoplotter
全能量法 total energy method
全能量系统 total energy concept;total energy system
全能龙门起重机 full gantry crane
全能模拟摄影纠正系统 universal analog photographic rectification system
全能起落 pantobase
全能渗碳炉 all case furnace
全能生产 full capacity
全能试验机 universal test(ing) machine
全能通信[讯]系统 all-purpose communication system
全能像片转绘仪 universal sketch master
全能仪器 multipurpose instrument;universal apparatus
全能照片转绘仪 universal sketch master
全能资产负债表 balance sheet for all purposes
全能自动测图仪 universal autograph;universal automatic map compilation equipment
全能钻床 pantodrill
全泥浆化 all sliming
全逆转装置 all-round reversing gear
全年 all the year round;annual;yearly
全年的 year-(a) round
全年度日数 annual degree days
全年放牧地 year-long range
全年费用 annual charges
全年负载曲线 annual load curve
全年负载因数 annual load factor
全年耕作 full-time farming
全年降水天数 days of precipitation all over the year
全年降雨分布 distribution of rainfall throughout the year
全年浇筑量 <混凝土> year-round pours
全年空气调节 all-year air conditioning
全年空调 year-round air conditioning
全年空调器 year-round air conditioner
全年平均气温 annual mean temperature
全年平均效率 mean annual efficiency
全年气温变化幅度 annual range of temperature
全年日程表 calendar
全年所得额 annual income
全年通车的路面 year-round surface
全年通航 all-year navigation
全年通用式 year-round type
全年维修 yearly maintenance
全年夏季工作日交通量 summer annual weekday daily traffic
全年修理量 annual repair
全年需水量 continuous demand
全年需修量 annual repair
全年阴影 perpetual shadow
全年渔业 year-round fishery
全年增长 annual growth;increase during the year
全年支出额 annual expenditures
全年住宅供给 all-the-year-round housing
全年总费用 total annual charges;total annual cost;total annual expenses
全年总收入 gross annual income
全年总雨量 total precipitation
全黏[粘]屋面 fully adhered roofing
全黏[粘]着 full adhesion
全黏[粘]着机车 locomotive providing total adhesion
全啮合离合器 all-geared clutch
全凝集 panagglutination
全凝集素 panagglutinin
全凝器 complete condenser;total condenser
全欧道路 European road
全欧列车控制系统 European train control system
全耦合线圈 unity-coupled coil
全排列解码 permutation decoding
全排列译码 permutation decoding
全排水系统 full drainage(system)
全排泄型泉 complete drainage spring
全盘财产信托契据 all-inclusive deed of trust
全盘电气化 full electrification
全盘电气化的 fully electrified
全盘否定 complete denial
全盘管理系统 total management system
全盘机械化 all-round mechanization;comprehensive mechanization;system mechanization
全盘考虑 overall consideration
全盘情况 overall perspective
全盘商业情况预测 overall business forecast
全盘转臂吊机 full-rotating derrick
全盘自动化 all-round automa(tiza)tion;complete automa(tiza)tion;full automation;integrated automation;integrated automatization
全抛物面反射器 full-parabolic(al) reflector
全赔保险 whole coverage
全喷砂 full blast
全膨胀波 complete rarefaction wave
全皮纤维 only-coat fiber[fibre]
全片纠正 full image rectification
全片可变匀光 full-range variable dodging
全偏轨箱形梁 full bias rail box girder
全偏激磁 full-biased excitation
全偏心 full eccentric type
全偏心传动轴 full eccentric drive shaft
全票费 full fare
全票价 full fare
全贫混合料 full lean mixture
全贫混合物 full lean mixture
全频程记录范围 full-frequency recording range
全频带扬声器 full range speaker
全频道调谐 all-channel tuning
全频道天线 all-channel antenna
全频范围记录 full-frequency range recording
全频偏信道电平 full deviation channel level
全平衡舵 full balanced rudder
全平均 population average
全屏板 full-screen panel
全屏幕 full screen
全屏幕形式 full-screen form
全剖面 complete section;full section
全剖视图 full section(ed) view
全铺 T 形交叉 all-paved T cross
全铺 Y 形交叉 all-paved Y cross

全铺的 all-paved
全铺路面宽度 full-surfaced width
全铺式交叉(口) all-paved crossing; all-paved intersection
全铺装(的道路) paved full
全期放牧 season long grazing
全期获利率 maturity yield; yield to maturity
全期记录 full-time record
全期收益率 yield to maturity
全期油封轴承 sealed-for-life bearing
全气动制动器 full air brake
全气候板条 all-weather strip
全气候带材 all-weather strip
全气候胶合板 all-weather plywood
全气化煤气 completed gasification gas
全气压盾构 all-round pressurized shield; compressed-air shield; shield with compressed air
全汽化过程 gas integral process
全牵引车 full trailer tractor
全铅晶质玻璃 full lead crystal; full lead crystal glass
全潜式驳船平台 submersible barge platform
全潜式平台 completely submerged platform
全潜式岩芯钻机 submersible core drill
全潜式钻探驳船 submersible drilling barge
全潜式钻探平台 submersible drilling platform
全堑式快速(干)道 <净空全部位于地面以下> fully depressed expressway
全嵌铰链 full mortise hinge
全强度 full strength
全强度道面 full-strength pavement
全强度焊接 full-strength welding
全橇式起落架 all-skid landing gear
全桥驱动 all-axle drive; all drive
全切断按钮 stop-all button
全切开 slitting-up
全切开法 slitting up method
全青铜泵 all-bronze pump
全氢化蒽 perhydroanthracene
全氢化松香 perhydrogenated rosin
全轻骨料混凝土 all-light(weight)-aggregate concrete
全轻质混凝土 all-lightweight concrete
全清零 all-clear
全球 worldwide
全球安全 global security
全球保护 global protection
全球变化 global change
全球变化及生物多样性环境保护 global change and biodiversity conservation
全球变暖 global warning
全球波束 global beam
全球波束发射天线 global beam transmit antenna
全球波束覆盖范围 global beam coverage
全球波束接收天线 global beam receive antenna
全球波束天线 global beam antenna
全球测绘保障方案 world-wide topographic(al) readiness concept
全球长期趋势 long-term global trend
全球城市 ecumenopolis
全球臭氧层 global ozone layer
全球臭氧分布 global ozone distribution
全球臭氧观测系统 global ozone observation system
全球大地控制网 world-wide geodetic net
全球大地坐标系统 world geodetic system
全球大陆地带 land area of the globe
全球大气 global atmosphere
全球大气监视网 global atmosphere watch
全球大气研究计划 global atmospheric(al) research program(me)
全球代理 global representation
全球代理人 universal agent
全球带宽 global bandwidth
全球导航系统 global navigation system
全球导航与计划图 global navigation and planning chart
全球的 global; world-wide
全球的残留鼓胀 global fossil bulge
全球地表活动性 global seismic activity
全球地层学 geostratigraphy; global stratigraphy
全球地理(坐标)参考系统 world geographic(al) reference system
全球地震活动图 global seismicity map
全球地震活动性 global seismic activity
全球地震监测 global seismic monitoring
全球地震监视 global seismic monitoring
全球地质构造 global tectonics
全球电信系统 global telecommunications system
全球定位卫星 global positioning satellite
全球定位系统 global positioning system[GPS]
全球定位系统测绘 GPS mapping
全球定位系统差分修正业务 GPS differential correction service
全球定位系统接收机 GPS receiver
全球定位系统空中三角测量 GPS aero-triangulation
全球定位系统控制网 GPS control network
全球定位系统实时差分接收机 GPS real-time differential receiver
全球动力系 world-wide dynamic(al) system
全球多圆锥投影坐标网 world polyconic(al) grid
全球范围收入 world-wide income
全球方式覆盖 global-mode coverage
全球放射性沉降 world-wide radiation fallout
全球辐射 global radiation
全球覆盖天线 global beam antenna
全球干线通信[讯]质量 global trunk quality
全球跟踪网 global tracking network
全球跟踪站 global surveillance station
全球共有物 global common
全球构造尺度 global tectonic scale
全球构造图 world-wide structural pattern
全球构造学 global tectonics
全球构造域 global tectonic regime
全球观测系统 global observation system; global observing system
全球观察站 global surveillance station
全球广播 global broadcast(ing)
全球规划 global planning
全球规模 global scale
全球轨道导航卫星系统 global orbiting navigation satellite system
全球海浸期 global transgression period
全球海面温度计划 global sea surface temperature project
全球海面相对变化 global relative change of sea level
全球海平面观测系统 global sea level observing system
全球海上遇险安全系统 global maritime distress and safety system
全球海退期 global regression period
全球海洋环境污染调查 global investigation of pollution in the marine environment
全球航空导航图 global navigation chart
全球航线图 global navigation chart
全球航行警告业务 world-wide navigation warning service
全球合作 global cooperation
全球化 globalization
全球话务 global telephony traffic
全球环境 global environment
全球环境监测计划 earth watch program(me)
全球环境监测系统 global environmental monitoring system
全球环境质量影响评估 global environmental quality assessment
全球环流模式 global circulation model
全球监测 global monitoring
全球监测系统 global monitoring system
全球监视系统 global surveillance system
全球降水气候学计划 global precipitation climatology project
全球降水气候学中心 global precipitation climatology center[centre]
全球救助警报网 global rescue alarm net
全球均衡 global equilibrium
全球科学技术情报系统 universal system for information in science and technology
全球裂谷系 global rifting system
全球流通手段 global liquidity
全球硫收支 global sulfur budget; global sulphur budget
全球罗兰导航图 global Loran navigation chart
全球贸易 global trade
全球能量和水循环试验 global energy and water cycle experiment
全球农药污染 global pesticide contamination
全球排放量 global emissions
全球配额 all-allocated quota; global quota
全球平均地幔热流 global mean mantle flow
全球平均热流 global mean heat flow
全球平均温度 average global temperature
全球气候变化 global climate change; global climatologic change
全球气候的起源和模型 origins and patterns of global climate
全球气候改变 global climatologic change
全球汽车通信[讯]系统 global system for mobile communication
全球情报系统 global information system
全球扰动 world-wide disturbance
全球热储系 global reservoir system
全球人造卫星(通信[讯])网 world-wide network of satellites
全球社区 world community
全球社团 world community
全球升温潜能值 global warming potential
全球生态学 global ecology
全球时间同步 global time synchronization
全球时间同步系统 global time synchronization system
全球市场 global market
全球水环境监测系统 global environmental monitoring system for water
全球水伙伴 <期刊> Global Water Partnership
全球水量平衡 global water budget
全球税制 global system
全球台网 world network
全球谈判 global negotiations
全球天气 global weather
全球天气勘察 global weather reconnaissance
全球天气试验 global weather experiment; global weather test
全球天气搜索 global weather reconnaissance
全球天气侦察 global weather reconnaissance
全球通信[讯]范围 global communication coverage
全球通信[讯]卫星系统 global communication satellite system
全球通信[讯]系统 globecom
全球通信[讯]业务 global traffic
全球同步 world-wide synchronization
全球投影 globular projection
全球图 global map
全球网 global net(work)
全球微粒散落 world-wide fallout
全球卫星定位系统 global positioning system[GPS]
全球卫星定位系统接收机 global positioning system receiver
全球卫星通信[讯]系统 global communication system[Globecom]
全球卫星通信[讯]业务 global satellite traffic
全球卫星系统 global satellite system
全球系统 global system
全球信息网 world-wide web[www]
全球形变模型 global deformation pattern
全球性安排 global arrangements
全球性变化问题特别委员会 Special Committee on Global Change
全球性大气污染 global air pollution
全球性断层网 global grid of fault
全球性分保合同 world-wide treaty
全球性分布 world distribution
全球性干涉政策 globalism
全球性公约 global convention
全球性海面变化 global eustatic movement
全球性海面升降旋回 global eustatic cycle
全球性海面相对上升 global relative rise of sea level
全球性海面相对稳定 global relative stillstand of sea level
全球性海面相对下降 global relative fall of sea level
全球性海平面变化 eustatic change of sea level
全球性海水进退 global scale of transgressions and regressions
全球性环境问题 global environmental problem
全球性环境异常 global environmental anomaly
全球性环流 global circulation
全球性混合组织 globe wide mixed organization
全球性剂量分担 world-wide dose commitment
全球性经济衰退 global recession
全球性巨成矿带 global metallogenetic belt

全球性空气污染监测 global monitoring of air pollution
全球性扩散 global dispersion
全球性气候 global climate
全球性散落 global fall-out
全球性危机 global crisis
全球性温度 global temperature
全球性污染 global contamination; global pollution
全球性污染物 global contaminant
全球性影响 global effect
全球性运动 whole earth movement
全球循环 global cycle
全球研究站网 global network of research station
全球移动电话系统 global system mobile
全球移动通信[讯]系统 global system for mobile communication
全球雨水线 global isohyetal line
全球预报 global prediction
全球云区 global cloud fields
全球运送 global transport
全球招标 global tendering
全球植物地理区系 global floral realm
全球制移动电话 global system mobile
全球质量 global quality
全球中心式(跨国公司)geocentrism
全球重力基本点网 world-wide gravity base station network
全球重力异常 global gravity anomaly
全球资料处理系统 global data-processing system
全球资源信息数据库 global resource information database
全球自然灾害报警系统 world-wide natural disaster warning system
全球最大(值)global maximum
全球最小(值)global minimum
全区穿孔 lace
全区划式集装箱船 fully cellularised container ship
全区间防护法 full block protection
全区间隔离法 full block protection
全区性截流 regional water intercepting
全区域谈判 areawide bargaining
全区照明 general lighting
全曲率 all curvature;total curvature
全曲面 complementary surface
全驱动脉冲 full drive pulse
全圈旋转式起重机 full circle slewing crane
全权 full power;full authority;general power <委托人授予代理人的>
全权处理 sole discretion
全权大使 plenipotentiary
全权代表 plenipotentiary;universal agent
全权代理人 authorized agent;universal agent
全权的 omnicompetent
全权高级人员 authorized senior member
全权公使 envoy;minister plenipotentiary
全权合伙人 managing partner
全权托管 discretionary trust
全权委任 carte blanche
全权委托书 general power of attorney
全权信托 discretionary trust
全权拥有 entire tenancy
全权证书 credential;general warranty deed;warranty deed
全权租船契约 bare boat charter
全裙活塞 full skirted piston
全燃气 full gas
全燃气轮机 all gas turbine
全燃气轮机推进装置 all gas turbine propulsion
全让线 passing track
全热 total heat
全热带的 tropicopolitan
全热换热器 air-to-air total heat exchanger
全热量 full heat quantity
全日本航空公司 All Nipon Airways
全日波 diurnal wave
全日潮 daily type of tide;diurnal(type of)tide;single day tide
全日潮流 regular diurnal tidal current;diurnal current
全日单向最高断面流量 daily directional maximum sectional passenger volume
全日分潮 diurnal component; diurnal constituent
全日服务 full period service
全日工改为两个半日工 job-splitting
全日工作 full-time service
全日工作职位 full-time jobs
全日(光)照 broad daylight;full sun
全日客运量 daily passenger volume
全日停车 all-day parking
全日效率 all-day efficiency
全日循环 day-night cycle
全日制 full time
全日制工作 full-time employment
全日制和非全日制教育 full-time and part-time education
全日制学生 full-time student
全日驻车 all-day parking
全容罐 full containment tank
全容积 all capacity;total volume
全容量 full capacity
全容量操作 capacity operation
全容量抽头 full capacity tap
全容吸移管 whole pipet(te)
全溶液 total solution
全熔合焊 full-fusion welding
全熔化 running down
全熔(炼)fine melt
全熔融年龄 total fusion age
全熔树脂 fine melt resin
全熔透 complete penetration; full penetration
全熔岩浆 hypersolvus
全熔质试样 all-weld-metal test specimen
全三维偏移 full three-dimensional migration
全散粒噪声<在统计的独立时间内通过一表面的电子流的起伏> full shot noise
全散射 total scattering
全散射系数 total scattering coefficient
全散(装)件 complete knockdown
全扫描电视摄像机 full-scan television camera
全扫描电视摄影机 full-scan television camera
全色 all colo(u)r;full colo(u)r
全色玻璃 pot-metal glass
全色薄膜 panchromatic film
全色差 total colo(u)r difference
全色的 orthopan; ortho-panchromatic;panchromate;panchromatic
全色底片 panchromatic plate
全色电视摄像机 panchromatic television camera
全色调颜料 full-strength colo(u)r
全色调颜色 full-strength colo(u)r
全色调原版 full-tone original
全色负片 panchromatic negative
全色复制本 full-tone copy
全色干片 panchromatic plate
全色感光板 panchromatic plate
全色感光膜 panchromatic coating
全色红外片 panchromatic infrared film
全色激光显示 full-colo(u)r laser display
全色减蓝摄影 panchromatic minus blue photography
全色胶卷 orthopan film;panchromatic film
全色胶片 orthopan film;panchromatic film
全色滤光镜 panchromatic vision filter
全色盲 achromasia; achromatopsia; achromatopsy; monochromasia; monochromasy;monochromatism
全色盲的 monochord;monochromatic
全色盲者 monochromat
全色片 panchromatic photographic(al)material
全色全息(摄影)术 full-colo(u)r holography
全色染剂 panchrome stain
全色乳剂 panchromatic emulsion
全色软片 panchromatic film
全色散 total dispersion
全色素 holochrome
全色图 colo(u)r gamut;full-colo(u)r picture
全色图像 full-colo(u)r image
全色响应 panchromatic response
全色像片 panchromatic photograph
全色信号 composite colo(u)r signal
全色性 all-chromatism; panchromatism
全色性乳剂 panchromatic emulsion
全色印样 composite print
全色原版 continuous tone original
全色原件 full-tone original
全色再现 panchromatic rendition
全色泽 full tone
全色照片 panchromatic photograph
全色重显 panchromatic rendition
全沙 total loading;total sediment load
全沙模型 model with all sediment;overall sediment model;river model with all sediment
全沙模型律 law of overall sediment model;law of total sediment transported model
全沙模型试验 overall sediment model test
全烧 clean burn
全烧无烟煤锅炉机组 all-anthracite unit
全烧烟煤锅炉机组 all-bituminous unit
全社会工资增长的幅度 range of society-wide wage increases
全社会劳动生产率法 whole-society-productivity method
全射程角 whole range angle
全射镜 holophote
全射通风方式 portal-to-portal ventilation system
全涉及子图 spanning subgraph
全摄 group shot
全伸出长度 fully extended length
全身的 systemic
全身辐射 total body radiation;whole-body radiation
全身辐射剂量计 whole-body radiation meter;whole-body radiometer
全身辐照量 whole-body exposure; whole-body irradiation
全身挤压伤 body crush injury; body squeeze
全身计数器 whole-body counter
全身剂量 whole-body dose
全身监测仪 whole-body monitor
全身污染 systemic contamination
全身显微镜 holographic(al) microscope
全身锈蚀 rust through
全身照射剂量 whole-body radiation dose
全身振动 general vibration
全深处理的路面结构 fully treated pavement structure
全深度插入式混凝土路面振捣器 full-depth internal concrete pavement vibrator
全深度施工 full-depth construction
全深灌浆 full-depth grouting
全深集总采样 depth-integration sampling
全深沥青 full-depth asphalt
全深沥青补坑 full-depth bituminous patch
全深沥青混凝土 full hole asphalt concrete
全深沥青基层 full-depth asphalt base
全深沥青路面 full-depth asphalt pavement
全深铺砌 full-depth paving
全深一次注浆 grouting all the depth once
全深(用结合料)处理的路面结构 fully treated pavement structure
全渗滤 block percolation
全渗透 block percolation
全升(程)阀 full lift valve
全升举位置 fully lift position
全升力 total lift
全生料熔制玻璃 glass melted from batch only
全生命周期测试 full life cycle test
全生态系统 entire ecologic(al) system;total ecosystem
全生物带 holontozone
全声波偏移 full acoustic(al)migration
全省 all province
全省范围的 statewide
全省范围内的 province-wide
全盛期 noon; noon tide; noontime; spring tide;spring time
全盛时期 culmination;heyday;meridian;palmy days
全湿的 panhygrous
全湿法水泥厂 all-wet cement plant
全湿工艺水泥厂 all-wet cement plant
全石料防波堤 all-rock breakwater
全石墨反射层 all-graphite reflector
全实船 full scantling vessel
全蚀带 belt of totally;eclipse
全食 total eclipse
全食的时间 totality
全矢量插入 full vector offering
全使用期费 whole-life cost
全使用期内修理费 life repair cost
全世界 worldwise
全世界范围内的 UNIX 系统网 UUNET UNI-to-UNIX net
全世界各大洋 seven seas
全世界航行警告业务 the world-wide navigational warning service
全世界商船总数 world merchant fleet
全世界自然灾害警报系统 world-wide natural disaster warning system
全式提单 long form bill of lading
全事件 whole event
全视差 overall parallax;total parallax
全视差角 overall parallax angle
全视度盘 full vision dial
全视界 all-round coverage;all-round view;all-round vision
全视界座舱 all-round view cab(in)
全视面罩 whole vision mask
全视全息片 full-view hologram
全视图 full view;general view
全视野 all-round coverage; all-round view

全视野舱 <车厢> circle vision cab(in)
全视野侧窗 full-view side window
全视野叉车门架 full vision fork portal
全视野的车身前部 full-view front
全视野风挡玻璃 all-round view windscreen
全视野机舱 space-view cab(in)
全视野景象 all-round vision
全视野穹隆 full vision dome
全视野座舱 all-round cab(in); all-round view cab(in)
全视野座室 all-round cab(in)
全室地毯 fitted carpet; wall-to-wall carpeting
全室换气次数 complete air changes
全室空调 all room air conditioning
全收缩高平顶型 holosystolic high plateau type
全收缩期的 holosystolic
全寿命成本 whole-life cost
全寿命费用计算 life cycle cost(ing)
全寿命费用计算法 life cycle cost method
全寿命恢复 full life restoration
全寿命经济分析 life cycle economic analysis
全寿命期费用 life cycle cost(ing)
全舒张期的 holodiastolic
全熟料配合料 raw cullet
全数 in full; whole number
全数必中界【数】 hundred-percent rectangle; rectangle of dispersion
全数调查 compete enumeration
全数好质量 as found; tale quale
全数检验 one-hundred percent inspection
全数据集特许权 full data set authority
全数码呼叫 all-number calling
全数收清 received in full; receipt in full
全数字仿算 all-digital simulation
全数字(号码)呼叫 all-number calling
全数字化摄影测量 full digital photogrammetry; soft copy photogrammetry
全数字化摄影测图 full digital mapping
全数字化自动测图系统 full digital automatic mapping system
全数字盘 full numeral dial
全数字显示 all-digital display
全数字相关 all-digital correlation
全栓架 perfect frame(work)
全双层车站 full double-level design; station of full two level design
全双工 full-duplex
全双工操作 full-duplex operation
全双工传输 full-duplex transmission
全双工电传 full-duplex teletype
全双工电路 full-duplex circuit
全双工服务 full-duplex service
全双工工作方式 full-duplex mode
全双工链路 full-duplex link
全双工通道 full-duplex channel
全双工线路 full-duplex line
全双工线路转接器 full-duplex line adapter
全双工信道 full-duplex channel
全双工性能 full-duplex performance
全双工终端 full-duplex terminal
全双工主站 full-duplex primary station
全双曲型 totally hyperbolic
全双曲型微分方程 totally hyperbolic differential equation
全双向的 full-duplex
全水层采样器 integrating water sampler
全水方式 all-water system
全水分 total moisture
全水分煤样 total moisture sample
全水冷 full water cooling
全水冷壁炉膛 fully water-cooled furnace
全水冷发电机 full water cooling generator
全水冷炉膛锅炉 integral-furnace boiler
全水冷炉膛水管锅炉 integral-furnace water-tube boiler
全水冷汽轮发电机 fully water-cooled turbogenerator
全水流过滤 full flow filtering
全水流通量陶瓷膜 full flow flux ceramic membrane
全水头 full head
全水系统 all-water system
全水压机 all-hydraulic press
全顺砖砌合 all-stretcher bond
全顺砖砌体 plumb bond
全丝织物 all-silk goods
全松弛处理后尺寸 fully relaxed dimension
全松位置 full-release position
全苏道路科学研究院(路面设计)公式 <苏联> SOJUSDORNII equation
全速 flat-out; full(operating) speed; maximum speed; top speed
全速冲刺列车 <美国的一种不同运输方式联运货物列车> sprint train
全速倒车急转 crash-a-head maneuver
全速电报电路 full speed telegraph circuit
全速工作 full speed operation
全速航行 under full steam
全速后退 full reversing; full speed astern
全速排放 full discharge
全速前进 full speed ahead; highball
全速前进信号 <火车的> highball
全速试航 full speed trial
全速行进信号 highball
全速运行 full speed running
全速运转 full speed operation; race
全速正车/急转全速正车 crash-astern maneuver
全塑承载结构 all-plastic bearing structure
全塑多层板 all-plastic sandwich panel
全塑夹层板 all-plastic sandwich panel
全塑结构 all-plastic structure
全塑冷釉墙面涂料 all-plastic cold-glazed wall coating
全塑路水车 all-plastic road tanker
全塑路油车 all-plastic road tanker
全酸洗钢板 full pickled sheet
全碎屑岩 holoclastic; holoclastic rock
全损 total average loss; total loss
全损保险 insurance against total loss only
全损赔偿 free from all average; free of all average
全损时补缴全部保险费 full premium if lost
全损险 free from all average; total loss cover; total loss only; total risk
全损险及共同海损 total loss only and general average
全锁 full lock
全锁钢丝索 fully locked cablewire
全台剧场 total theatre[theater]
全太阳 full sun
全碳电刷 all-carbon brush
全碳酸 total carbonic acid
全陶粒混凝土 all-haydite concrete
全套 complete set; complex; full set; major combination
全套搬运线 complete handling line
全套板牙架 die stock set
全套备件 complete spare parts
全套部件 integral unit
全套测量设备 instrumentation complex
全套程序组件 complete program(me) package
全套出租 furnished house for rent; renting with furniture <住宅和家具设备>
全套穿孔 gang punch
全套炊具 kitchen
全套单据 full set of documents
全套的住房 overall housing
全套多倍仪 multiplex(plotting) set
全套发射设备 launching complex
全套辅助工具 slave kit
全套耕 complete tillage
全套工具 implement; outfit; set of tools; tool set
全套工具装备 complete tool equipment
全套固定式悬设备 full stationary boom
全套管成孔机 full casing tube boring machine
全套管钻机 all casing(earth) drill
全套管钻孔机 Benoto method boring machine
全套绘图仪(器) drafting set
全套机组 complete unit
全套家具 fitted furniture
全套建筑单一风格 monostyle
全套搅拌设备 all mixing plant
全套雷达设备 radar complex
全套冷却设备 complete coolant equipment
全套罗盘仪 set of compasses
全套螺母 full nut
全套螺丝攻和螺丝板牙 tap and die set
全套螺丝绞板 die stock set
全套配件 complete spare parts
全套器械 instrumentarium
全套切缝模片 slotted templet set
全套设备 complete equipment; complete plants and equipment; complete unit; outfit
全套设备部件零件表 equipment component list
全套设备操作人员 complex facility operator
全套设备试验 complex coordination test
全套设计图 complete plan
全套套筒扳手 socket wrench set
全套提单 complete set of bill of lading; full set bill of lading
全套投标书 bid package; tender package
全套文件 complete documentation
全套用品 outfit
全套轧辊 set of rolls
全套照明设备 complete lighting equipment
全套照明装置 luminaire
全套装置 complete unit; complex; whole set
全特性 total external characteristic
全特性图 four quadrant characteristic diagram
全特征线理论 full characteristic theory
全提花 full-jacquard
全提前点火位置 full-advance position
全体 alpha and omega; body; complement; entire; overall; population; totality; whole
全体层照相术 pantomography
全体乘车者 ridership
全体乘客 ridership
全体乘务员 crew
全体船员 all hands; crew; officers and crews
全体大会 general assembly
全体故障 total failure
全体雇员的工时数 exposure hours
全体国民 commonwealth
全体合伙人 general partners; partnership
全体呼叫电键 general calling key; simultaneous calling key
全体会议 general assembly; general meeting; plenary assembly; plenary meeting; plenary session; plenum[复 plenums/plana]
全体会员 membership
全体技术员 technical staff
全体叫通 general calling
全体劳力 labo(u)r in the aggregate
全体利益 corporate profit
全体纳税人 all taxable individuals
全体人员 personnel; faculty <从事某一专门职业的>
全体所有制 corporate ownership
全体线性群 full linear group
全体效率 general efficiency
全体选民 constituency
全体一致 unanimity
全体一致的 unanimous
全体职工大会 general meetings of the staff and workers
全体职员 personnel
全天对策控制 <按全天交通流量不同，而分成几个程序> time of day strategic control
全天工作 full day's operation
全天恒星图 star chart of whole sky
全天候百叶窗 all-weather slatted blind
全天候保护层 all-weather coat
全天候保护机栅 fall weather protection
全天候标准司机室 standard all-weather cab
全天候材料 <一种土石坝填筑材料> all weather material
全天候测计 all-weather ga(u)ging device
全天候测距仪 all-weather terrestrial rangefinder
全天候测量能力 all-weather measuring capacity
全天候测器 all-weather ga(u)ging device
全天候导航 all-weather navigation
全天候道路 all-weather highway; all-weather road
全天候道路基层 all-weather subbase
全天候的 all-weather; weatherproof; resistance to weather
全天候电话机 all-weather telephone set; weatherproof telephone set
全天候飞机 all-weather aircraft; all-weather plane
全天候(飞)机场 all-weather airfield
全天候飞行 all-weather operation
全天候服务 all-weather service
全天候港口 all-weather port
全天候工作 all-weather operation
全天候工作能力 all-weather capability
全天候公路 all-weather highway
全天候海上目标搜索系统 all-weather sea target acquisition system
全天候航空港 all-weather port
全天候机场跑道 all-weather runway
全天候基层 all-weather subbase
全天候降陆 all-weather landing
全天候胶合板 all-weather plywood;

weatherproof plywood
全天候救生艇 all-weather lifeboat
全天候可用性 all-weather availability
全天候空运 all-weather air service
全天候良好视度 all-weather visibility
全天候路基 all-weather base
全天候路线 all-weather route
全天候码头 all-weather quay; all-weather wharf
全天候能见度 all-weather visibility
全天候跑道 all-weather runway
全天候设备 all-weather equipment; all-weather facility
全天候司机室 all-weather environmental cab
全天候通车路面 year-round surface
全天候稳定性 all-weather stability
全天候性 weatherproofness
全天候巡逻艇 all-weather patrol boat
全天候压路机 all-weather compactor
全天候压实机 all-weather compactor
全天候运输 all-weather transport
全天候制导系统 all-weather guidance system
全天候装备 all-weather equipment
全天候着陆系统 all-weather landing system
全天候作业 all-weather operation
全天呼叫次数 all day calls
全天开行计划 daily running plan
全天空摄影机 whole-sky camera
全天空照相机 all-sky camera
全添加法 full additive method
全填充 full packing
全填充电缆 jelly filled cable
全填充式光缆 fully filled optic(al) fibre cable
全调测 full line up
全调幅 complete modulation
全调节式座椅 full adjustable seat
全调谐 complete modulation
全调整 complete modulation; full tune-up
全调制 complete modulation; full modulation
全调制射频信号 fully modulated radio frequency signal
全跳动 total run-out
全铁 total iron
全铁泵 all-iron pump
全停 full cut-off
全停信号时间 all-red period
全通 fully on
全通传感器 all-pass transducer
全通船楼 complete superstructure
全通道的 all-channel
全通道式阀 full-way valve
全通道译码器 all-channel decoder
全通的 all-pass
全通电的 fully energized
全通格子网络 all-pass lattice
全通换能器 all-pass transducer
全通甲板船 flush deck vessel; full deck vessel
全通梁 through beam
全通滤波器 all-pass filter
全通路离心泵 full-way centrifugal pump
全通四端网络 <所有频率都能通过的四端网络> all-pass lattice
全通网络【计】 all-pass network
全通位置 full application position
全通岩芯管 full flow core barrel
全通元件 all-pass element
全通轴 through-going shaft
全通转换器 all-pass transducer
全同步 full synchronizing
全同步变速箱 fully synchronized transmission
全同部件 identity unit
全同操作 identity operation
全同单元 identity unit
全同规整度 isotacticity
全同聚丙烯纤维 isotactic polypropylene fiber[fibre]
全同立构 isotaxy
全同立构的 isotactic
全同立构规整度 isotacticity
全同立构聚合物 isotactic polymer
全同门 identity gate
全同态映射 equimorphism
全同形的 panidiomorphic
全同选择 full-sib selection
全同元件 identity element
全同运算 identity operation
全筒式活塞 full skirt piston
全透 full impregnated
全透镜内窥镜 all-lens endoscope
全透射 full radiograph
全透射因数 total transmission factor
全透视 full perspective
全透水井 fall penetrating well; fully penetrating well
全图 complete image; full figure; general drawing; general view; overall view; total graph【数】
全图示控制面板 full graphic(al) panel
全图示面板 full graphic(al) panel
全图像记录方式 image recording model
全退火 dead soft; full annealing; true annealing
全退火薄钢片 true annealed sheet
全退火机 dead annealing
全拖车 full trailer
全拖车方式 full trailer system
全拖车连接车 full trailer combination
全拖车牵引车 full trailer tractor
全拖式钢索集材 ground lead cable logging
全拖式集材 ground skidding
全拖鞋式裙部活塞 full-slipper piston
全脱硫焙烧 sweet roasting
全脱位 complete dislocation
全脱氧的 fully killed
全脱氧钢 dead-melted steel; dead steel; fully deoxidized steel; fully finished steel; fully killed steel; killed steel; perfectly killed steel; quiet steel
全脱氧合金 fully killed alloy; killed alloy
全挖式断面 full cut section
全瓦特电动机 all-watt motor
全外延光电晶体管 all-epitaxial phototransistor
全弯曲强度 total deviation intensity
全微分法 total differential; complete differential; perfect differential
全微分方程式 total differential equation
全微粒化 all sliming
全微商【数】 total derivative
全围式滑动轴承 full-journal bearing
全伪序 total pseudo-ordering
全位错 perfect dislocation; whole dislocation
全位能最小法 minimization of total potential energy
全位置焊接 all-position welding
全位置焊条 all-position electrode
全温度区间 interval of overall temperature
全文 full text
全文集 corpus[复 corpora]
全文记录 verbatim record
全文检索 full text retrieval
全稳定导体 full stabilized conductor
全稳定的 fully restrained; fully stable
全稳定卫星 fully stabilized satellite
全吸收 complex absorption; total absorption
全吸收法 <防腐> full-cell process; full-cell treatment
全吸收分光计 total absorption spectrometer
全吸收光电管 black body photocell
全息凹面反射光栅 holographic(al) concave reflection-type grating
全息编码 holographic(al) encoding
全息编码板 holographic(al) coding plate
全息变形 holographic(al) deformation
全息波纹法 holographic(al) Moire technique
全息彩色储存 holographic(al) colo(u)r storage
全息参数 holographic(al) parameter
全息测图系统 holographic-(al) mapping system
全息成像 hologram imaging; holographic(al) imaging
全息处理 holograph processing
全息传输 hologram transmission
全息存储 holographic(al) storage
全息存储系统 hologram memory system
全息存储与存取系统 holographic(al) storage and access system
全息存储装置 holographic(al) memory device
全息带帧 holotape frame
全息的 holographic(al)
全息等高线 holographic(al) contouring
全息底片 holo film
全息地图 anaglyph map
全息地图显示 holographic(al) map display
全息地震 holoseismic; seismic holography
全息地震法 holoseismic method
全息地震仪 holographic(al) seismograph
全息电视 holographic(al) television
全息电影 holographic(al) movies
全息电影摄像术 holographic(al) cinematography
全息电影照相术 cineholographic(al) method
全息多重像 holographic(al) multiplication
全息反射镜 holographic(al) mirror
全息分层照相术 holographic(al) tomography
全息干扰测量法 holographic(al) interferometry
全息干扰计量 holographic(al) interference measuring
全息干扰仪 holographic(al) interferometer
全息干涉测量法 holographic(al) interferometry
全息干涉度量学 holographic(al) interferometry
全息干涉法 holographic(al) interferometry
全息干涉估价法 holographic(al) interferometric evaluation method
全息干涉图 holographic(al) interference pattern
全息干涉仪 hologram interferometer
全息共轭关系 holographic(al) conjugate relation
全息观察仪 holographic(al) viewer; holoviewer
全息光谱学 holographic(al) spectroscopy
全息光栅 hologram grating; holographic(al) grating
全息光栅扫描器 holographic(al) grating scanner
全息光线追迹方程式 holographic-(al) ray-tracing equation
全息光学 holographic(al) optics
全息光学元件 holographic(al) optic-(al) elements
全息光弹(方)法 holographic(al) photoelasticity; photo-holoelasticity
全息光弹试验 holographic(al) photo-elastic test
全息光弹性法 holo-photoelasticity
全息(光)照像干涉测量术 holometry
全息互补色立体照片 holographic-(al) anaglyph picture
全息畸变 holographic(al) deformation
全息激光护目镜 holographic(al) laser visor
全息激光器 hololaser
全息激光束扫描器 holographic(al) laser beam scanner
全息记录参数 holographic(al) recording parameter
全息减去法 holographic(al) subtraction
全息胶合件 holographic(al) doublet
全息胶片 holo film
全息结构干涉 constructive interference
全息解卷积 holographic(al) deconvolution
全息雷达 hologram radar
全息立体互补图 holographic(al) anaglyph
全息立体模型 holographic(al) stereo-model
全息立体显示 holographic(al) stereo display; holographic(al) three dimensional display
全息录像带 holotape
全息录像盘 holographic(al) video disk
全息录音机 holophone
全息滤波 holographic(al) filtering
全息滤波器 holographic(al) filter
全息滤光片 holographic(al) filter
全息轮廓法 holographic(al) contouring
全息模拟 holographic(al) simulation
全息莫尔法 holographic(al) Moire technique; holographic(al) Moire technology
全息耦合器 holographic(al) coupler
全息匹配滤波器 holographic(al) matched filter
全息片透射比 hologram transmittance
全息平视显示器 holographic(al) head up display
全息去褶积 pholographic(al) deconvolution
全息全景立体照片 holographic(al) panoramic stereogram
全息扫描 holographic(al) scan; holoscan
全息扫描法 holographic(al) scanning method
全息扫描像差校正 holographic(al) scan aberration correction
全息摄影 homography
全息摄影测量 hologrammetry; holographic(al) mensuration
全息摄影存储器 holographic(al) storage
全息摄影的 holographic(al)
全息摄影法 holographic(al) method
全息摄影光学 hologram optics
全息摄影机 holocamera
全息摄影记录 holographic(al) recording
全息摄影记录装置 holographic(al)

Q

recording device
全息摄影技术 hologram technique; holographic(al) technique
全息摄影术 hologram photography; holography
全息摄影系统 holographic(al) system
全息摄影员 holographer
全息摄影原理 principle of holography
全息摄影栅 hologram grating
全息摄影坐标系统 holographic(al) coordinate system
全息术无损检验 holographic(al) nondestructive testing
全息数据存储器 holdor; holographic(al) data storage
全息天线 holographic(al) antenna
全息透镜 holographic(al) lens; hololens
全息透镜系统 holographic(al) lens system
全息图 hologram; hologram interferometer; hologram view
全息图的衍射效率 diffraction efficiency of hologram
全息图的组合性质 associative property of hologram
全息图检查系统 holocheck system
全息图面 hologram page
全息图模式 hologram pattern
全息图像 hologram picture; holographic(al) image
全息图像再现 holographic(al) image restoration
全息图再现 hologram reconstruction
全息系统 holographic(al)-base system
全息纤维 holographic(al) fibre
全息显示 holographic(al) display
全息显示系统 holographic(al) display system
全息显微镜 holographic(al) microscope
全息显微照相术 holomicrography
全息相关滤波 holographic(al) correlation filtering
全息相片 hologram; holograph
全息像差 hologram aberration; holographic(al) aberration
全息信息存储 holographic(al) information accumulation[storage]
全息学 holography
全息学原理 principle of holography
全息衍射光栅 holographic(al) diffraction grating
全息掩模技术 holographic(al) mask technology
全息元件 holographic(al) element
全息原版片 hologram master
全息照明波 hologram illuminating wave
全息照片 hologram; holograph
全息照片矩阵雷达 hologram matrix radar
全息照片乳胶收缩 hologram emulsion shrinkage
全息照相 hologram
全息照相材料 hologram material
全息照相存储器 holographic(al) storage; holographic(al) memory
全息照相的 holographic(al)
全息照相底片 hologram photoplate
全息(照相)电视 hologram television
全息照相干扰测量术 holometry
全息照相横向剪切干涉仪 holographic(al) lateral shear interferometer
全息照相换能器 holographic(al) transducer
全息照相换能器分析仪 holographic(al) transducer analyser[analyzer]
全息照相机 holocamera; holographic(al) camera; holography camera; holoscope
全息(照相)记录材料 hologram recording material
全息照相栅 holographic(al) grating
全息(照相)术 holography
全息照相图像 hologram image
全息照相系统 holographic(al) system
全息(照相)显微术 holographic(al) microscopy
全息照相装置 holographic(al) device
全息照像 holograph
全息照像存储器 holographic(al) memory
全息照像电视 hologram television
全息照像机 holoscope
全息照像术 holography
全息照像图像 hologram image
全息阵列 hologram array
全息振动分析 holo-vibration analysis
全息帧 holoframe
全铣 omni mill
全系 complete set
全系数 overall coefficient
全系统 total system
全系统的 system-wide
全系统范围的控制 system-wide control
全系统费用 total system cost
全系统振动试验 global vibration test
全先行进位 full look ahead carry
全纤维素 holocellulose
全显示 push-through presentation
全险 against all risks; all risks(insurance)
全险保(险)单 all risks covers; open policy
全线 full line
全线函数 holomorphic function
全线开工 all-work
全线宽 whole-line width
全线例行测试 overall circuit routine test
全线旅游客票 <无一定方向> all line rover ticket
全线圈绕组 whole coiled winding
全限定名 fully qualified name
全相关 full correlation; total correlation
全相关分析 full correlation analysis
全相关因子 total correlation factor
全相关振荡 fully correlated oscillation
全相联 total association
全相联(高速)缓冲存储器 fully associative buffer storage
全相联高速缓冲存储器 fully-associative cache
全镶板构造 all-veneer construction
全响应 total regression
全向 all-round; omnibearing
全向测距 omnidirectional ranging
全向测距导航设备 omnibearing distance facility
全向测距仪 omnidirectional range set
全向测向电台 omnidirectional range station
全向传声器 omnidirectional microphone
全向导航 omnidirectional range; omnirange
全向导航台 omnidirectional beacon; omnidirectional range station; omnirange
全向导航系统 omnibearing distance navigation; rho-theta navigation; R-theta navigation
全向的 all-directional; omnidirectional
全向灯光 omnidirectional light
全向灯光设备 omnidirectional lighting
全向方位变换器 omnibearing converter
全向方位距离导航设备 omnibearing distance facility
全向方位指示器 omnibearing indicator
全向方向性可控制的天线 full steerable antenna
全向辐射 spheric(al) radiation
全向辐射槽缝式天线 omniguide antenna
全向辐射器 isotropic(al) radiator; spheric(al) radiator
全向辐射天线 omnidirectional antenna
全向辐射仪 omnidirectional radiometer
全向光学扫描器 omnidirectional optic(al) scanner
全向号筒式扬声器 omnidirectional horn loudspeaker
全向换能器 omnidirectional transducer
全向经纬仪 omnimeter
全向可控天线 full coverage steerable antenna; full steerability antenna
全向力 omniforce
全向流量 omnidirectional flux
全向六角转台 all-round turret
全向麦克风 omnidirectional microphone
全向起重机 universal crane
全向前位置 full-forward position
全向扫掠设备 omnidirectional scanner
全向声源 omnidirectional sound source
全向式雷达预测 omnidirectional radar prediction
全向式无线电信标 omnirange
全向式无线电指向标 omnidirectional beacon
全向视程无线电设备 omnivisual range radio
全向收集器 omnidirectional collector
全向受拉状态 state of all-round tension
全向数字雷达 omnirange digital radar
全向太阳电池 omnidirectional solar cell
全向天线 non-directional antenna; omni aerial; omniantenna; omnidirectional antenna; omni directive antenna
全向图天线 circular diagram aerial
全向推进器 all direction propeller
全向无线电导航航道指示器 omnirange course indicator
全向无线电距离 omnidirectional radio range
全向无线电信标 circular radio beacon; non-directional radio beacon; omnidirectional radio beacon
全向无线电信标分解器 omnirange resolver
全向无线电信标伺服系统 omnirange servo
全向无线电信标天线 omnirange antenna
全向无线电指向标 non-directional radio beacon; omnidirectional radio beacon
全向无线电指向标台 circular radio beacon station
全向相控阵 omnidirectional phased array
全向响应 omnidirectional response
全向效应 omnidirectional effect
全向信标 non-directional beacon; non-directional radio beacon; omnidirectional beacon; omnidirectional radio beacon; omniranger
全向信标无线电设备 omnirange radio set
全向信号 omnidirectional signal
全向旋转无线电信标 omnidirectional rotating radio beacon
全向选择器 omnibearing selector
全向压力 omnidirectional pressure
全向扬声器 omnidirectional loudspeaker
全向有效辐射功率 effective isotropically radiated power
全向指向标 omni beacon
全向制 omnidirectional system
全向转动炮塔 all-round turret
全向转换器 omnibearing converter
全巷法 bulk method; whole tunnel method
全项目标定 full calibrating; full proving
全像 complete image; full figure
全像的 panoramic
全像偏离 total image run-out
全消光 total extinction
全消耗型燃烧器 total consumption burner
全消耗性喷灯 total consumption burner
全消落期 <水库> period of full draw-down
全消色差的 pantachromatic
全楔形环 full keystone ring
全谐振 complete resonance
全心材的 all heart
全心木材 all heart; clear-all heart
全心全意 heart and soul
全芯纤维 full core fibre
全新的 brand-new; fire-new
全新风 fresh air
全新风机控制箱 control box for fresh air fan
全新世【地】Holocene; Holocene epoch
全新世沉积 Holocene sedimentation
全新世地壳运动 Holocenic crust movement
全新世活动断裂 Holocene epoch active fault
全新统【地】Holocene series
全新鲜机油润滑 full-fresh oiling
全薪 full pay
全信道译码器 all-channel decoder
全信号 <时间分割制线路上的信号> aggregate signal
全信息对策 game with complete information; game with perfect information
全行程 full stroke; total distance; total excursion
全行程的干草压捆机 full-stroke hay press
全行程进给 full stroke admission
全行程时间 overall travel time
全行缓冲器 full line buffer
全行(显示)方式 full line mode
全形的 euhedral
全形性 pantomorphism
全形褶皱 holomorphic folds
全型 holomorph; holotype
全休 complete rest
全溴丙酮 perbromo-acetone
全溴乙醚 perbromo-ether
全溴乙烷 perbromo-ethane
全溴乙烯 perbromo-ethylene
全虚域 totally imaginary field
全需求量运转 full demand operation
全序 linear order; simple order; total order
全序的【数】simply ordered; totally

ordered
全序关系 total ordering relation
全序集 simply ordered set;totally ordered set
全悬挂 integral
全悬挂式的 fully mounted
全悬挂式犁 fully mounted plough
全旋转 full swivel
全旋转的 full-swinging
全旋转起重船 fully rotating crane barge
全旋转式的 full rotating
全旋转式接点 full-rotary contact
全旋转式起重机 full circle crane
全旋转式挖土机 swing excavator
全旋转式装料机 full(circle) revolving loader
全旋转桅杆式起重船 fully revolving crane barge
全选电流 full selected current
全选(读)脉冲 full-read pulse
全循环 complete alternation; complete cycle;full circle;full cycle
全循环边周点谱 fully cyclic peripheral point spectrum
全循环操作 extinction recycle operation
全循环湖 holomictic lake
全循环时间 complete cycle time
全压 total pressure
全压比(率)full pressure ratio
全压测针 impact pressure probe
全压服 full pressure suit
全压管 impact tube;ram-air pipe;total pressure probe
全压和静压测深管 combined pitot-static probe
全压接收管 ram intake
全压静压差 Pitot-static difference
全压静压管系 Pitot-static system
全压静压探测管 combined pitotstatic
全压力 full pressure;ram pressure
全压力进给 full force feed
全压力润滑 full force feed lubrication
全压力润滑系统 full pressure lubrication system
全压力梯度 total pressure gradient
全压力增减率 total pressure gradient
全压起动 full voltage starting
全压起动电动机 across-the-line motor
全压曲线 Pitot curve
全压润滑 full pressure lubrication
全压润滑系统 full pressure lubricating system
全压式塑模 positive mo(u)ld
全压试验 full-scale test
全压受感器 ram intake
全压损失 Pitot loss
全压缩高 solid height
全压头 total differential head; total head pressure
全压头管 total head tube
全压头值 total head value
全压位置 full cock position
全压效率 total pressure efficiency
全压循环润滑系统 full pressure circulating lubrication system
全压与静压差 pilot static difference
全压运转 full pressure operation
全淹没 oversubmergence
全延迟点火 full retard
全延迟位置 full retard position
全岩 bulk rock;whole rock
全岩等时线 whole rock isochron
全岩分析 bulk rock analysis
全岩年龄 whole rock ages
全岩溶 holokarst
全盐量 total salinity;total salt
全盐量测定 total salt determination
全盐组【地】Salina formation;Salinian formation
全扬程 total pump(ing) head; total lift <装卸桥的>
全氧助燃窑 oxy-fuel fired furnace
全样 bulk sample
全要素板 combination plate;composite plate
全要素定位 composite position
全要素效率 total factor productivity
全要素样图 composite proof
全要素原图 composite drawing
全野外布点 full field control point distribution;allocating point of all-field
全页的 full-page
全页图 full sheet
全页显示 full-page display
全液冷 full liquid cooling
全液冷发电机 full liquid-cooling generator
全液力转向 full fluid steering
全液膜润滑区 full fluid film region
全液压操作 full hydraulic operation
全液压传动 full hydraulic transmission
全液压传动静力触探 all-hydraulic driven static cone penetration test
全液压的 all-hydraulic;fully hydraulic
全液压电梯 full hydraulic lift
全液压反铲挖沟成型机 all-hydraulic trench-forming shovel
全液压反铲挖沟机 all-hydraulic trench-hoe;all-hydraulic ditching shovel
全液压反铲挖掘成型机 all-hydraulic trench-forming shovel
全液压反铲挖掘机 all-hydraulic backacter;all-hydraulic back digger;all-hydraulic(trenching)hoe
全液压反铲挖土机 all-hydraulic backacter
全液压混凝土拌和机 all-hydraulic concrete mixer
全液压混凝土泵 all-hydraulic concrete pump
全液压混凝土搅拌机 all-hydraulic concrete mixer
全液压举升器 full hydraulic lift
全液压拉铲挖掘机 all-hydraulic dragshovel;all-hydraulic pullshovel
全液压拉铲挖土机 all-hydraulic dragshovel;all-hydraulic pullshovel
全液压履带式挖土机 all-hydraulic crawler-mounted excavator
全液压驱动装置 fully hydraulic drive
全液压式 all-hydraulic
全液压式起重机 all-hydraulic crane
全液压式挖掘机 all-hydraulic excavation
全液压式凿岩机 all-hydraulic drill
全液压式装载机 skid steer loader
全液压式钻机 all-hydraulic drill
全液压索铲挖掘机 all-hydraulic pullshovel
全液压拖铲挖掘机 all-hydraulic dragshovel;all-hydraulic pullshovel
全液压拖铲挖土机 all-hydraulic dragshovel;all-hydraulic pullshovel
全液压挖沟机 all-hydraulic ditcher; all-hydraulic trencher
全液压岩芯钻机 all-hydraulic core drilling rig
全液压移动式挖掘机 all-hydraulic mobile excavator
全液压移动式挖土机 all-hydraulic mobile excavator
全液压增压器 oil-to-oil booster;oil-to-oil intensifier
全液压转向 full hydraulic steering
全液压装载机 all-hydraulic loader; skid steer loader
全液压钻机 hydraulic driven driller
全移动喷灌系统 fully portable sprinkler-system
全移位能力 full shift capability
全译本 cover to cover translation
全译码地址 full decode address
全翼弦层流 full chord laminar flow
全翼展 extreme span
全翼展襟翼 full-span flap
全翼展开缝襟翼 full-span slotted flap
全音阶 gamut
全音域 gamut
全殷钢空腔谐振器 all-invar cavity
全隐含图 universal implication graph
全印数 print run;production run
全印刷(电路)all print
全影 umbra[复 umbrae/umbras]
全影调 full tone
全影技术 panoramic technique
全影立体照片 panoramagram
全影区 complete shadow
全硬淬透钢 fully hardening steel
全硬化 full hardening;pansclerosis
全硬式设施 fully hardened installation
全用途电子计算机 all-purpose computer
全用橡胶制成 all-rubber made
全优工程 meeting the quality standards projects; projects meeting all quality standards
全优先文法 total precedence grammar
全由构架组合的结构 frame-covered structure
全油道旋塞 plug cock
全油分析 complete oil analysis
全油浮选 bulk-oil flo(a)tation;oil flo(a)tation
全油门加速 full throttle acceleration
全油膜润滑 full fluid film lubrication
全有或全无的限制 all-or-none embargo
全有或全无分配法【交】all-or-nothing(assignment)method
全有或全无继电器 all-or-nothing relay
全有机化合化学物 all-organic compound chemicals
全有全无 with full and without full
全有效集 universal effective set
全余能 total complementary energy
全预应力 full prestressing
全预应力混凝土 full prestressing concrete;fully prestressed concrete
全预应力设计 full prestressing design
全预张力的 full-scale pretensioned
全域 macrocosm;universe;universe set
全域关系 universal relation
全域规则 across rule
全域极小 global minimum
全域理论 population theory
全域值 universe value
全域总体 universe population
全员安全管理 whole staff safety management
全员动力准备程度 installed power per employee
全员动力准备水平 installed power per employee
全员劳动合同制 system of all employees signing labo(u)r contracts
全员劳动生产率 all personnel labo(u)r productivity
全员劳动效率 total numbers labo(u)r efficiency
全员培训 training all workers and staff;training of entire staff
全员效率管理法 efficient management method for all personnel
全圆闭合(差)horizon closure;close of horizon
全圆测回法 full-angle set method
全圆读数的【测】circle swing
全圆方向观测法 close of horizon; method of direction observation in rounds
全圆分度器 whole circle protractor
全圆观测法 method of round
全圆回转的 circle swing
全圆角 full angle
全圆刻度垂直(度盘)full vertical circle
全圆喷洒器 circle circular sprinkler
全圆喷头 circle circular sprinkler
全圆形液压伸缩式模板 full circle expanding hydraulic form
全圆(周)方位角 whole-circle bearing
全圆周回转 all-around traverse
全圆周喷头 circular sprinkler
全圆周(上的)接触 full circle contact
全约束 complete contraction; perfect restraint
全约证书 deed of contact
全运价 full freight
全运行工况 full operating condition
全载波 full carrier
全载波值 full carrier value
全载的 full-laden
全载特性 full load characteristic
全责联运提单 full liability through bill of lading
全栅栏 full barrier
全占用时间 busy period
全站型电子速测仪 electronic tach(e)ometer;electronic tachymeter; total station optic (al) electronic tach(e)ometer;total station optic(al)electronic tachymeter
全站型光电速测经纬仪 total station optic (al) electronic tachometric theodolite;total station optic(al)electronic tachymetric theodolite
全站仪 electronic tacheometer; total station instrument
全张 whole paper
全张翻版印 work-and-turn
全张翻转印 work-and-tumble
全张图 full sheet
全张像片清晰度 edge-to-edge sharpness
全罩式 all cover
全罩翼阀 fully shrouded trickle valve
全遮式雨篷 covered canopy
全折射 total refraction
全折射波 complete refraction wave
全阵环 full matrix ring
全振动地板 fully sprung floor
全振峰 double amplitude
全振幅 peak-to-peak; peak-to-trough amplitude; total amplification; total amplitude;full amplitude
全振高 full wave
全镇静钢 dead steel;fully deoxidized steel;fully finished steel;fully killed steel;quiet steel
全蒸发(过程)pervapo(u)rization
全蒸发膜 pervapo(u)rization membrane
全整色 panchromate
全整数规划 all-integer programming
全整数问题 all integer problem
全帧电视录像 frame television video recording
全支承 full bearing;full support
全支座 full support
全直径水路 full diameter waterway
全直桩 all vertical pile
全值 total head
全值保险 full value insurance

全值变量 full-value variable
全值抵押 < 房地产的 > mortgaging out
全值绘图机控制 absolute plotter control
全值计算机 absolute value machine; absolute value computer
全植物式营养 holophytic nutrition
全指数 thorough index
全制动 full application of brake; full brake application; full braking
全制动时间 entire brake time
全制动试验 full brake trial
全制动位 maximum reduction position
全制动作用 full application braking
全制调速器 isochronous governor
全致密的 fully dense
全终端回波抑制器 full terminal echo suppressor
全重叠法 full overlap method; fully eclipsed form
全重挂车 full trailer
全重力出矿矿块 full gravity block
全重量 all-up weight; full weight; gross weight; in the substance; total weight
全州范围的 statewide
全州公路设备 < 美 > state-wide highway plant
全周 complete cycle
全周倒车杆装置 all-round reversing gear
全周焊接 weld-all-over
全周进汽 < 汽轮机的 > full admission arc admission
全周进汽启动装置 full arc admission starting device
全周进汽汽轮机 uniform-admission turbine
全周进汽涡轮机 full admission turbine
全周期 complete cycle; complete period
全周式 full circular
全周转动起重机 full-rotating crane; full circle crane; full-rotating derrick
全洲的 continent-wide
全洲联络(公路)网 continent-wide connected system
全洲联络系统 continent-wide connected system
全轴的 holoaxial
全轴距 total wheel base
全轴驱动 all(-axle) drive
全主元法 complete pivot(ing)
全柱冷阱法 whole column cryotrapping method
全铸的 all-cast
全砖的 all-brick
全砖建筑(物) all-brick building
全转 complete turn
全转换 full conversion
全转驱动式 all-wheel drive type
全转式扒杆起重机 full-rotating derrick
全转式电铲 full-revolving electric(al) shovel
全转式翻车机 full-revolution dumper
全转式反铲挖掘机 full-revolving back hoe excavator
全转式反铲挖土机 full-revolving back hoe excavator
全转式机铲 full circle shovel
全转式起重机 full circle slewing crane; full-rotating crane; transit crane
全转弯状态 full turn position
全转移 total transfer
全装备车 all-equipped car
全装备复式塞孔盘 complete multiple
全装备重量 fully equipped weight
全装料的 full charge
全装配式建筑 total-prefabricated construction
全装配式结构 total-prefabricated structure
全装载的最大限度 full load maximum
全状态 total state
全锥形座面的轮辋 full tapered bead seat
全着色力 full tinting strength
全资子公司 wholly owned subsidiary
全字界 full-word boundary
全自动 full automaticity
全自动拌和厂 perfect automatic batcher plant
全自动编目技术 fully automated cataloguing technique; perfect automated cataloguing technique
全自动编译技术 fully automatic compiling technique
全自动变速器 fully automatic gearbox
全自动表 fully automatic watch
全自动拨号 full automatic selection; fully automatic selection
全自动操纵 fully automatic control
全自动操作 fully automatic working
全自动长途电话交换机 automatic toll switching system; fully automatic toll switching system
全自动车 fully automatic vehicle
全自动车床 fully automatic lathe
全自动处理 fully automatic processing
全自动触发器 full auto trigger
全自动穿扣机 fully automatic drawing-in machine
全自动的 all-automatic; fully automatic; perfectly automatic
全自动的滑动门装置 fully automatic sliding door installation
全自动滴定管 full-automatic buret
全自动点火方法 fully automatic firing process
全自动电传打字机通信[讯]系统 fully automatic teletypewriter communication system
全自动电镀 fully automatic plating
全自动电弧焊 fully automatic arc welding
全自动复凿孔机交换制 fully automatic reperforator switching
全自动高质量翻译机 full automatic high-quality translation(machine)
全自动工作 fully automatic working
全自动国际用户电报业务 fully automatic international telex service
全自动焊 fully automatic welding
全自动焊接装置 full auto-bonding system
全自动化 complete automa(tiza)-tion; full automation
全自动化编译技术 fully automatic compiling technique
全自动化传动 supermatic drive
全自动化的 completely automatic; fully automatic; purely automatic
全自动化电子判定器 fully automatic electronic judging device
全自动化工作过程 unattended operation
全自动化锅炉 fully automatic boiler
全自动化垃圾车 fully automated refuse truck
全自动化生产过程 fully automatic processing
全自动化弯管机 fully automatic pipe bender
全自动化系统 complete automatic system
全自动化组件 full automation module
全自动混凝土搅拌机 full automatic concrete mixer; perfect automatic batcher
全自动机床 fully automatic machine
全自动计算机程序 fully automated computer program(me)
全自动记录 full recording
全自动技术 fully automatic technique
全自动夹钳 fully automatic tongs
全自动交换 fully automatic switching
全自动交换网 fully automatic switching network
全自动交换系统 fully automatic switching system
全自动搅拌厂 perfect automatic batcher plant
全自动接合 full auto-bonding; fully automatic bonding
全自动接合装置 full auto-bonding system; fully automatic bonding system
全自动(可变)光阑 fully automatic diaphragm
全自动控制 completely automatic control; fully automatic control
全自动犁 fully automatic plough
全自动立体测图仪 < 加拿大 > Gestalt photo mapper
全自动流水作业技术 fully automatic flow technique
全自动螺丝机床 fully automatic screw machine
全自动门 fully automatic door
全自动膜盒 fully automatic diaphragm
全自动磨床 fully automatic grinder
全自动浓度比色计 fully automatic colo(u)r densitometer
全自动频率 full automatic frequency
全自动平交道口 fully automated grade crossing; fully automatic grade crossing
全自动汽车检测系统 computerized vehicle inspection system
全自动曲柄磨床 crank-o-matic grinding machine
全自动升降装卸车 full free lift
全自动式 full rotation type
全自动提升 fully automatic winding
全自动同步机 all synchro
全自动凸轮磨床 camomatic grinder
全自动系统 fully automatic system
全自动信息检索系统 fully automatic information retrieval system
全自动压机 fully automatic press
全自动压接机器 fully automatic compression machine
全自动遥控 fully automatic telecontrol
全自动闸门 fully automatic gate
全自动展绘格网 fully automatic grid plotting
全自动照相机 fully automatic camera
全自动纸条转接设备 fully automatic tape relay set
全自动制 fully automatic system
全自动转塔式车床 fully automatic turret screw machine
全自动转塔铣床 fully automatic turret milling machine
全自动转台阶梯 fully automatic turntable ladder
全自动装配线 fully automated assembly line
全自动装置 complete automatic device
全自动阻风门控制系统 fully automatic choke system
全自感 complete inductance
全自检查电路 totally self-checking circuit
全自形的 panidiomorphic
全自形粒状 panautomorphic granular; panidiomorphic granular
全自形粒状结构 panidiomorphic granular texture
全自形岩石 panidiomorphic rock
全阻力 overall drag
全阻尼 full damping
全组 complete set; full set
全组构 total fabric
全组合 all possible combination
全组合测角法 angle measurement method in all combination; method of angle observation in all combinations
全组合键 full compound key
全组合角度测量法 measurement of angles in all combination
全组合角度平差 adjustment of angles in all combination
全组合式驳船队 fully integrated tow
全坐标 full coordinates

权 weight

权编码器 weight encoder
权变惩罚律 law of contingent punishment
权变的 contingent approach
权变关系 contingency relationship
权变管理方法 contingency management approach
权变理论 contingency theory
权变强化律 law of contingent reinforcement
权变战略 contingency strategies
权差 weighted error
权单位 weight unit
权倒数 weight reciprocal
权度 measures and weights; weights and measures
权度法 weights and measures law
权方程 weight equation
权杆调节器 weight-lever regulator
权函数 weighting function
权函数法 weighting function method
权衡 balance; trade-off; weigh(ing)
权衡的 weighed
权衡法 trade-off
权衡过程 trade-off process
权衡利弊 trade-off; weigh the advantages and disadvantages; weigh the pros and cons
权衡轻重 weigh
权衡误差 weigh error
权衡因素 weighing factor
权衡值 trade-off value
权衡资源 trading-off resources
权级 power level
权界 right of way
权矩阵 weighing matrix; weight matrix
权力 authority; capability; power
权力变数 power variable
权力导向 authority oriented
权力动态 dynamics of power
权力管理程序 management program(me)
权力和利益的分散 decentralization of power and interest
权力划分 division of power
权力计算系统 capability computing system
权力结构 power structure
权力模式 authority pattern

Q

权力委托 authority to delegate
权力下放 delegation of authority;devolution
权力向量机 capability vector machine
权力主义管理理论 authoritarian theory
权利 right;droit;entitlement
权利安排 arrangement of rights
权利保留 reservation of right
权利财产 choses in action
权利初始界定 initial definition of rights
权利船 privileged vessel;right-of-way vessel;stand on vessel
权利代位 subrogation
权利代位书 subrogation form
权利担保 warranty of title
权利的剥夺 abridg(e)ment
权利的消灭时效 prescription of rights
权利的中断 interruption of a right
权利法 law of interest
权利法案 bill of rights
权利范围 interest field
权利汇合 merger of title
权利继承人 successor in title
权利金 option money;premium;royalty;royalty payment
权利客体 object of right
权利能力 capacity for rights;legal capacity
权利凭证 document of title
权利人 obligee
权利申请书 petition of right
权利要求 claim of right
权利与义务 rights and obligations
权利与义务的平衡 balance of rights and obligations
权利暂时停止 suspense of rights
权利债券 bonds with warrants
权利证明要约书 abstract of title
权利证书托存 deposit(e) of title-deeds
权利终止期 cesser
权利主体 subject of rights
权利转让 subrogate;subrogation
权利转让书 subrogation form
权利转移 conversion privilege;subrogation;subrogation of rights
权码 weighted code
权能 capability;competence
权逆 weight reciprocal
权偏差 weight bias error
权势等级 pecking order
权数 weight;weight number
权数抽样方案 demerit sampling plan
权数单位 unit of weight
权图 weighted graph
权威 dominance
权威案例引证表 list of authorities
权威代表团 delegation of authority
权威法 <影响他人改变行为> authoritarian method
权威工程师 authority engineer
权威检验机构 authoritative inspection organization
权威结构 authority structure
权威人格 authoritarian personality
权威性 authenticity
权威性的 authentic;definitive
权威著作 standard works
权系数 weighting coefficient
权系数矩阵 weight coefficient matrix
权限 competence[competency];extent of authority;extent of power;jurisdiction; limits of authority; purview;terms of reference
权限范围 bailiwick;scope of authority
权限内的 competent
权限凭证 authority credential
权限值 authority credential
权相关偏倚 weight correlation bias
权向量 weight vector
权序列长度 length of weight sequence
权宜办法 half measure
权宜从众 <当团体压力解除就私下说出真正意见> expedient conformity
权宜措施 expedience[expediency];expedient measure; makeshift; stopgap measure
权宜的改善措施 interim improvement
权宜之计 expedience[expediency];expedient; expedient measure; makeshift(device); matter of expediency;modus vivendi;stopgap
权宜之计的 jerry
权益保留 reservation
权益保留条款 <授权人、出租人的> reddendum
权益保障金 equity margin
权益比率 equity ratio
权益成本 equity cost
权益顶让书 letter of subrogation
权益购买人 equity purchaser
权益交易 equity transaction
权益联营 pooling of interests
权益取得 acquisition of right
权益融资 equity financing
权益投资者 equity investor
权益委托书 letter of subrogation
权益增加 equity build-up
权益值 equity value
权益转让书 letter of subrogation
权益资本 equity capital
权余法 method of weighted residuals
权与责的划分 division of authority and responsibility
权责不明 ambiguity of power and duties
权责层次 hierarchy of authority
权责发生额 accrual
权责发生概念 accrual concept
权责发生会计制 account on accrual basis
权责发生制 accounting on the accrual basis; accrual accounting; accrual basis;accrual system;accrued basis
权责法 law of responsibility and authority
权责集中 centralization of authority and responsibility
权责利 responsibilities and interests of an enterprise;rights
权责流动过程 flow process of authority and responsibility
权责平衡 parity of authority and responsibility
权责体制 accrual system
权责相称 authority commensurate with responsibility
权责相当原则 accrual parity principle
权责已发生的 accrued
权责应计制 accrual basis
权责应计制惯例 accrual convention
权责应计制会计 accrual basis accounting
权重 weight
权重的 weighted;weighting
权重叠加 weighted superposition
权重法 method of weighting
权重函数 weighting function
权重计数子 weight enumerator
权重矩阵 weighing matrix
权重曲线 weighting curve
权重误差 weighted error
权重因数 weight coefficient; weight(ed)factor;weighting factor
权重因子 weight(ed)factor
权重噪声 weighted noise
权状 warrant

泉

泉出露的地形条件 topographic(al)condition of spring emerging
泉的补给含水层 recharge aquifer for spring
泉的出露高程 altitude of spring emerging
泉的丰水期流量 spring discharge in raising season
泉的枯水期流量 spring discharge in drought season
泉的流量 spring discharge
泉的流量变化率 variation rate of spring discharge
泉的稳定性系数 steady coefficient of spring
泉的溢出量 quantity of spring overflow
泉的涌水量 yield of spring
泉的最大流量 maximum discharge of spring
泉的最小流量 minimum discharge of spring
泉点 spring site
泉华 <矿泉中沉淀的结晶岩石>【地】sinter;mineral deposit
泉华波痕 sinter ripple mark
泉华沉淀物 sintered deposit
泉华沉积 sinter deposit
泉华唇 sinter lips
泉华蛋 sinter eggs
泉华的成因 origin of sinter
泉华的分类 composition classify of sinter
泉华的形态分类 morphology classify of sinter
泉华堤 sinter barrier
泉华洞穴 sinter cavity
泉华盾 sinter shield
泉华管 sinter duct
泉华厚度 thickness of sinter
泉华化 crenitic
泉华阶地 sinter terrace
泉华阶地高度 height of sinter terrace
泉华块体 sinter block
泉华坡 sinter slope
泉华峭壁 sinter cliff
泉华丘 sinter dome
泉华裙 sinter apron
泉华扇 sinter fan
泉华台 sinter flat
泉华席 sinter sheet
泉华形成时间分类 forming time classify of sinter
泉华垣 sinter rim
泉华冢 sinter mound
泉华锥 sinter cone
泉集河 confluence of springs
泉间地 interfluve
泉间分水区 interfluve
泉井 artesian well; blowing well; flowing well;well hole
泉径流 spring runoff
泉坑 catch pot;spring pit
泉口 spring vent
泉口个数 number of spring mouth
泉口漏斗 spring funnels
泉口排列方式 arrangement of spring mouth
泉流量衰减 attenuation of spring discharge
泉流量衰减方程法 method of spring attenuation equation
泉流量下降期 period of spring discharge decreasing
泉流量增加期 period of spring discharge increasing
泉群 group of springs
泉群分布方向 direction of spring distribution
泉群分布线 spring line
泉群分析 line of springs
泉群最小流量比拟法 spring group minimum flow analogy method
泉上冷藏所 <建在冷泉上的> springhouse
泉石华 plombierite
泉蚀凹壁 alcove
泉室 <取泉水的> spring chamber;well chamber
泉水 aqua fontana;brook;font;fountain;fountain water;natural water;piping water; spring bend; spring water
泉水编号 number of spring
泉水补给 spring feed
泉水补给的间歇河 spring-fed intermittent stream
泉水补给河 spring fed stream
泉水采样 sampling of spring water
泉水沉淀物名称 name of spring precipitation
泉水沉积 bateque
泉水沉积物 spring sediment
泉水出口 outlet of spring
泉水出水量 output of spring
泉水调查 spring investigation
泉水调查点 observation point of spring
泉水调查记录 investigation record of spring
泉水动态 spring water regime(n)
泉水动态分析 spring regime(n)analysis
泉水动态分析法 spring regime(n)analysis method
泉水动态观测 behavio(u)r observation of spring
泉水冻结时间 freezing time of spring
泉水冻结状态 freezing state of spring
泉水堆积土 spring accumulation soil
泉水多的 springy
泉水干枯年份 year of dried spring
泉水干枯月份 month of dried spring
泉水观测点 observation point of spring water
泉水华 plombierite
泉水汇集 catchment of spring
泉水活动 spring activity
泉水检漏 soaped for leakage
泉水枯竭 spring exhaustion
泉水类型 type of spring
泉水类型及流量 type and discharge of springs
泉水利用状况 condition of spring utilization
泉水流出通道 flow passage of spring
泉水流出状态 flow out state of spring
泉水流量 artesian discharge;yield of spring
泉水流量的变率 change rate of spring discharge
泉水流量的递减速度 rate of progressively decrease of spring discharge
泉水流量的年变幅 annual change of spring discharge
泉水流量过程曲线 curve of spring flow
泉水流量减小值 decrease of discharge from spring
泉水流量总和法 spring discharge summation
泉水面 spring level
泉水名称 name of spring
泉水排泄 spring drainage
泉水气体逸出状况 condition of gas escape from spring
泉水区 artesian province

泉水室 spring chamber
泉水卫生 sanitation of spring water
泉水位置 location of spring
泉水温度 temperature of spring
泉水物理性质 physical property of spring
泉水系统 spring system
泉水线 spring line
泉水消耗系数 coefficient of spring depletion
泉水消耗系数法 spring decay coefficient method
泉水样 spring water sample
泉水涌出地面高度 height of spring gushed above the surface
泉水注入的池槽 spring-fed pond
泉水最小流量比拟法 analogy method of minimum spring discharge
泉塘 holding pond;spring bowl
泉溪(流)spring run
泉线 spring line
泉眼 mouth of spring; orifice of spring;outlet of spring;spring eye;spring opening
泉眼密集点 closest discharge point
泉眼钻车 spring-hole drill rig
泉涌 gush
泉源 font;fountain head;spring head;well spring
泉源间歇河 spring-fed intermittent stream
泉源侵蚀 spring-sapping
泉渣 adarce;spring residue
泉最大流量出现时间 emerging period of maximum discharge of spring
泉最小流量出现时间 emerging period of minimum discharge of spring

拳击场 boxing arena;boxing field;boxing ring;prize ring

拳石 boulder;sett <铺路用>
拳石基层 boulder base
拳石块 blockage
拳石铺面 bouldering
拳头产品 fist products

蜷片板 wafer;wafer board

蜷线【物】spiral

醛醇 aldehyde alcohol

醛基 aldehyde group
醛缩醇 ethvlidene ether
醛缩醇类 acetal
醛亚胺 aldimine
醛氧化作用 aldehyde oxidation

犬齿接 dogtooth joint

犬齿式砌合层 dogtooth course
犬齿饰 dogtooth;tooth ornament
犬齿砖饰 dog's-tooth course; dogtooth course
犬牙键节 dog link
犬牙交错 rough dentation
犬牙交错缝 indented joint
犬牙离合器 dog clutch
犬牙联杆 dog link
犬牙联轴器 dog coupling
犬牙式接合 dog clutch
犬牙式砌合【建】dogtooth bond
犬牙式饰 dogtooth ornament
犬牙饰 dogtooth
犬牙形 dogtooth
犬牙形篱笆 Virginian fence
犬牙状线脚 dogtooth mo(u)lding

劝导疗法 persuasive therapy

劝告车速 advisory speed
劝告车速标示 advisory speed indication
劝告车速标志 advisory speed sign
劝告信号 advisory signal
劝诱投资 induced investment

券涵 arch culvert

券门【建】arch door
券模 arch centering[centring]

炔丙醇 propargyl alcohol

炔烃 alkyne
炔属醇 acetylenic alcohol
炔属酸 acetylenic acid
炔属烃 acetylene series
炔属酮 acetylenic ketone

缺本目录 desiderata list

缺变元 missing argument; missing variable
缺测降水资料 missing precipitation data
缺测年份 unobserved year
缺层 non-sequence;stratigraphic(al) break;stratigraphic(al) hiatus
缺除 decount
缺带 low tape
缺氮 nitrogen deficiency
缺点 blemish;defect;demerit;dereliction; disadvantage; drawback; failing;flaw;glitch;mar;objectionable feature;shortcoming;speck;tache;vice;weakness
缺点检验 spot-check
缺点预防 defects prevention
缺点允许率 lot-tolerance percent defective
缺碘 iodine deficiency
缺电 electric(al) power deficiency;power shortage
缺顶函数 lacunary function
缺顶级数 lacunary series
缺顶结构 lacunary structure
缺顶空间 lacunary space
缺段中断 missing segment interrupt
缺额 shortage;vacancy;wantage
缺额材积 underrun
缺额罚款 shortage penalty
缺额条款 deficiency clause
缺乏 dearth;default;deficiency;drop short of; shortage; stop short of stop;stringency;void
缺乏材料 lack of materials
缺乏存货 absence of stock; lack of materials
缺乏的东西 desideratum[复 desiderata]
缺乏对价<指合同> absence of consideration
缺乏法律上的行为能力 under a legal incapacity
缺乏方向性 devoid of directionality
缺乏供应品 lack of supplies
缺乏经验 inexperience
缺乏逻辑的 illogic(al)
缺乏耐性 lack of tolerance
缺乏平衡 lack of balance
缺乏人力 lack of manpower
缺乏人手 short-handed
缺乏润滑 lack of lubrication
缺乏色彩的 colo(u)rless
缺乏生物学上成熟的树木 lack of biologically mature trees
缺乏实际根据 to depart from actual realities
缺乏弹性 inelasticity
缺乏卫生设施系统 deflective plumbing
缺乏卫生用水带来的疾病 water-washed disease
缺乏现金 short of cash
缺乏训练 indiscipline
缺乏一致性 lack of conformity
缺乏饮用水 shortage of drinking water
缺乏植物滋养<湖泊> oligotrophic
缺乏助熔质砖泥 foul clay
缺乏资金 deficiency in funds; out of funds; out-of-pocket; shortage of funds
缺乏资源 be deficient in resources
缺光地区 aphotic zone
缺号 lacking number
缺货 out of stock;scarcity
缺货成本 shortage cost;stockout cost
缺货单 want slips
缺货费 shortage penalty cost
缺货价格 scarcity price;scarcity value
缺货价值 scarcity value
缺货率 out of stock rate; rate of shorts
缺货市场 scarcity price
缺货行市 famine price
缺货造成的高价 famine price
缺胶层合板 dry laminate
缺胶接层 starved joint
缺胶接头 dry joint;starved joint
缺角 broken corner;unfilled corner
缺角的 wan(e)y
缺角的金刚石 broken-in diamond
缺角方木边 waney edge
缺角四边形的 quarter octagonal
缺角镶边方木 wane-edged wood
缺斤 weight shortage
缺径 broken end;end out;missing end
缺刻 indent;inden(ta)tion;indenture
缺刻状的 eroded;erose;gnawed
缺刻状年轮 indented ring
缺口 breach;cant bay;dap;gap;nick;notch;rabbet
缺口坝 notch dam
缺口板 notch plate
缺口半径 root radius
缺口表尺 open sight
缺口冲击强度 notch impact strength
缺口冲击试验 nick-break test;notch impact teat
缺口冲击试样 notched impact specimen
缺口冲击值 notched bar value
缺口脆性 notch brittleness;notch embrittlement
缺口脆性试验 notch brittleness test;notch sensitivity test
缺口导流 dam-gap diversion
缺口跌水 notch drop
缺口断裂试验 notch breaking test
缺口墩 notch pier
缺口分析 gap analysis
缺口拱 broken arch
缺口焊接宽板试验 notched and welded wide plate test
缺口铧 notch-edged share
缺口降落 notch fall
缺口卡片 scored card
缺口抗拉强度 notched tensile strength
缺口抗张试验 notched bar tensile test
缺口拉力试验 notched bar pull test
缺口拉伸试验 notch tension test
缺口拉伸试样 notched tensile specimen
缺口量水坝 notch dam
缺口量水堰 notched weir
缺口(量水)堰板 notch plate
缺口裂度 notch acuity
缺口落差 notch fall
缺口脉冲 serrated pulse
缺口面积 notched area
缺口敏感的 notch sensitive
缺口敏感性 notch sensitivity
缺口敏感性试验 notch sensitivity test
缺口敏感性系数 notch sensitivity factor
缺口耙片 notched disk[disc]
缺口疲劳因素 fatigue-notch factor
缺口平头喷雾器 notched flat headed atomizer
缺口强度 notch strength
缺口圈 broken ring
缺口韧性试验 notch ductility test
缺口三角楣饰 broken pediment
缺口式流量计 notch ga(u)ge
缺口试棒 notched bar
缺口试样 notched specimen
缺口试样试验 test with notched test piece
缺口缩颈 lateral contraction
缺口弯曲试验 nick bend test
缺口销 notch pin
缺口效应 notch effect
缺口形状 notch geometry
缺口修复 gap repair
缺口延性 notch ductility
缺口堰 notched weir
缺口样板 female ga(u)ge
缺口应力 notch stress
缺口应力集中效应 notch stress-concentration effect
缺口有效应力集中系数 notch stress-concentration coefficient
缺口圆犁刀 notched colter
缺口圆盘 cutaway disk;serrated disk
缺款 to be out of cash
缺蓝 minus blue
缺棱 edge defect;waney edge
缺棱材 waney lumber
缺棱的 waney;wany
缺量 shortage; shortage in weight;short weight;ullage
缺裂锥 breached cone
缺林地 opening
缺磷指数 phosphate deficiency index
缺零件表 parts short list
缺码 code absence
缺煤地区 coal scarce area
缺面的 merohedral; merosymmetric(al)
缺面体 merohedrism
缺面形 merohedral form
缺面性 merohedrism
缺能源的工业先进国 energy-hungry advanced country
缺气沉积 anaerobic sediment
缺气的 anaerobic
缺钱 short of money
缺勤 absenteeism
缺勤工人 absentee operator
缺勤工日 man days of absence from work
缺勤工日数 man days in absence
缺勤率 absence rate;absenteeism rate;percentage of absentees to total staff in service
缺勤人员 absenteeism operator
缺勤少时期 bull weeks
缺勤时间 time absent
缺勤者 absentee

缺区 missing plot
缺燃料油信号灯 lack of fuel oil signal lamp
缺色性 hypochromism
缺少 absence; disappearance; hard up; shortage; shortfall
缺少包装 lack of packing
缺少报盘 lack of offer
缺少臭氧的空气 ozone poor air
缺少订货 lack of order
缺少空间 lack of space
缺少空气的 airless
缺少量 wantage
缺少水分 moisture density
缺少现款 short of cash
缺少营养的湖泊 oligotrophic lake
缺少资金 be in low water; lack of capital
缺省【计】default
缺省说明文件 default specification file
缺省图注 default legend
缺省网关 default gateway
缺省选项 default option
缺省选择设备数据 default choice device data
缺省值 default value
缺省值等级 default class
缺省字符串设备数据 default string device data
缺失 deficiency; deletion; hiatus; lacuna[复 lacunae]; omission < 地层 >
缺失的地层 missing stratigraphy; hidden layer; phantom
缺失样品 missing sample
缺水 anhyetism; deficiency of water; lack of water; shortage of water; water deficiency; water deprivation; water scarcity; water-short; water shortage
缺水报警器 low-water alarm
缺水的 hydropenic; unwatered
缺水地 thirst-land; water deficient area
缺水地区 water deficient area; water deficit region; water-scarce area; water-short area
缺水量 deficient draft
缺水期 dry spell
缺水热 exsiccation fever; thirst fever
缺水性脱水 hypertonic dehydration; water deficient dehydration
缺酸的 acid-deficient
缺损 defect; defection; wane
缺损边缘 waney edge
缺损的 defective
缺损概率分布 defective probability distribution
缺纬 mispick
缺位 omission; vacancy
缺位固溶体 defect solid solution; omission solid solution; vacancy solid solution
缺位浓度 vacancy concentration
缺席 absence; absentee; absenteeism
缺席部门号 default department number
缺席裁决 a judgment by default
缺席或不能担任工作 absence or incapacity
缺席判决 default judgement; judgement by default
缺席判罪 condemnation par contumace
缺席情况下作出的仲裁裁决 award rendered by default
缺席人 defaulter
缺席任务 default task
缺席说明 default declaration
缺席说明文件 default specification file
缺席文件 absent file
缺席文件属性 default file attribute
缺席系统控制区 default system control area
缺席选项 default option
缺席选择 default optimum
缺席者 absentee; defaulter
缺席者所有权 absentee ownership
缺席值 default value
缺席仲裁 arbitrate by default; award rendered by default
缺席属性 default attribute
缺席组 default group
缺陷 blemish; blot; defect; deficiency; deformity; drawback; fault; flaw; pitfall; shortcoming; bug < 机器等的 >; kink < 设计或施工中的 >
缺陷百分数 defective per hundred unit
缺陷半导体 defect semiconductor
缺陷辨认 defect recognition
缺陷标记员 defect marker
缺陷表面 blemish surface
缺陷不连续性 discontinuity of defect
缺陷产生 defect production
缺陷丛 defect cluster
缺陷带 imperfect tape
缺陷单元 defective unit
缺陷导电 defect conduction
缺陷的清除 removal of defects
缺陷的准化学平衡 quasi-chemical equilibrium of defect
缺陷点阵 defect lattice
缺陷定位 flaw location
缺陷分布 failure distribution; flaw distribution
缺陷分极法 defect system
缺陷分类 classification of defects
缺陷改正竣工证书 completion of defects certificate
缺陷构造 defect structure
缺陷焊补 reweld(ing)
缺陷化学 defect chemistry
缺陷回波 flaw echo
缺陷火山口 breached crater
缺陷监测数据 flaw detection data
缺陷检测 error detection
缺陷检测仪 defectometer
缺陷检查 defect detecting test; detection of defects
缺陷检查仪 deflectoscope
缺陷检验 defect detecting test; fault detect(ion)
缺陷角 reentrant angle
缺陷矫治 correction of defect
缺陷矫治率 correction rate of defect
缺陷结构 defect structure
缺陷晶格 defect lattice
缺陷扣计额 defect deduction
缺陷量测(仪) defectoscopy
缺陷灵敏度 flaw sensitivity
缺陷率 defects per unit; rate of defect
缺陷密度 defect concentration
缺陷敏感区 defect susceptible area
缺陷木材 defective wood
缺陷评定 defects assessment
缺陷切除剪 nibbler shears
缺陷清理工 chipper
缺陷区 < 结构方面的 > kinked region
缺陷烧结 defect sintering
缺陷树 fault tree
缺陷数 defective number; number of defects
缺陷探测 fault detection
缺陷探测光电装置 aniseikon
缺陷跳跃 defect skip
缺陷通告 preliminary warning notice
缺陷通知 notice of defects
缺陷吸收 defect absorption
缺陷项目 defective item
缺陷小孔检查器 pin-hole detector
缺陷信号 flaw indication
缺陷修补 defect repair
缺陷修复完工证书 completion of defects certificate
缺陷修理 repair of defects
缺陷修整工 chipper
缺陷羊毛 defective wool
缺陷预测 failure prediction
缺陷运动 defect motion
缺陷责任 defects liability
缺陷责任期 defects liability period
缺陷责任期的延长 extension of defects liability period
缺陷责任完成证书 defects liability certificate
缺陷责任证书 defects liability certificate
缺陷责任终止证书 defects liability release certificate
缺陷值 defective value
缺陷指示 discontinuity indication of defect; flaw indication
缺陷中心 defect center[centre]
缺项 lacuna[复 lacunae]; lacunarity; missing data
缺项的 lacunary
缺项估计 estimation of missing data; estimation of missing value
缺项函数 lacunary function
缺项汇票 inchoate bill
缺项级数 lacunary series
缺项空间 lacunary space
缺序 absent order; missing order
缺养分湖 oligotrophic lake
缺氧 oxygen deficiency; oxygen depletion; oxygen lack; oxygen deficit
缺氧保藏法 anaerobious preservation
缺氧层 layer of oxygen deficient; oxygen-poor layer
缺氧沉积 anoxic deposit
缺氧池 anoxic pond
缺氧代谢 anaerobic metabolism
缺氧的 anoxic; anoxybiotic; oxygen deficient
缺氧底泥 anoxic sediment
缺氧地带 anoxic zone; deoxygenated area; deoxygenated zone
缺氧地区 anoxic zone; deoxygenated area; deoxygenated zone
缺氧反硝化 anoxic denitrification
缺氧反应器 anoxic reactor
缺氧废水 anaerobic wastewater
缺氧腐蚀 oxygen starvation
缺氧工艺 anoxic process
缺氧过程 anoxic process
缺氧/好氧工艺 anoxic/oxic process
缺氧/好氧/好氧工艺 anoxic/oxic/oxic process
缺氧好氧(污水)处理法 anoxic-oxic treatment process
缺氧/好氧系统 anoxic/oxic system
缺氧环境 anoxic environment; anoxygenous environment; oxygen deficient environment
缺氧活性污泥 anoxic activated sludge
缺氧降解 anoxic degradation
缺氧量 oxygen deficiency; quantity of oxygen deficit; quantity of oxygen lack; oxygen deficit
缺氧流化床反应器 anoxic fluid bed reactor
缺氧灭活污泥 anoxic inactivated sludge
缺氧期 anoxic length
缺氧情况 anaerobic condition; oxygen deficient condition
缺氧区 anoxic area; anoxic basin; anoxic zone
缺氧燃烧 anoxycausis
缺氧生境 anaerobic habitat
缺氧生物降解 anoxic biodegradation
缺氧时间 anoxic time
缺氧适应 acclimatization to anoxia
缺氧水 anaerobic water; anoxic water
缺氧水体 anoxic waters
缺氧水域 anoxic waters
缺氧条件 anoxia condition; oxygen deficient condition
缺氧污泥 anoxic sludge
缺氧吸磷 anoxic phosphorus uptake
缺氧厌氧卡鲁塞尔式氧化沟 anoxic-anaerobic-Carousel oxidation ditch
缺氧折流板反应器 anoxic baffled reactor
缺氧症 anoxia
缺氧指示器 oxygen deficient indicator
缺氧状态 anaerobic condition; anoxic condition
缺叶 afoliate
缺页中断 missing page interrupt; missing page interruption
缺一边角的木材 waney edged wood
缺因子 missing factor
缺营养的 oligotrophic
缺营养湖泊 oligotrophic lake
缺油 oil starvation
缺油的 oil-free; oilless
缺油而磨损(停车) starve
缺油墨指示 low toner indicator
缺油止挡 lack oil stopper
缺釉 exposed body
缺釉痕 pluck
缺雨 lack of rainfall
缺雨区 anhyetism
缺雨性 anhyetism
缺圆挡板 segmental baffle
缺圆拱 scheme arch; segmental arch
缺圆孔板 segmental orifice plate
缺月形 wane
缺月形枕木 wane tie
缺账户 short-landed account
缺纸带 low tape
缺纸检测 low paper detection
缺纸指示灯 low paper indicator
缺中径颗粒骨料 gap-graded aggregate
缺中子同位素 neutron-deficient isotope; neutron poor isotope
缺重量 shortage in weight; short weight
缺柱面 cylinder fault

雀

雀斑 fleck
雀麦 oat
雀木 < 中国古典木建筑中,梁柱交接处的托座 > column bracket[que-ti]
雀替【建】column bracket; decorated bracket; sparrow brace
雀啄纹 sparrow peck

确

确保安全 perfect safety
确保安全的 foolproof
确保大堤安全性 safety of guaranteed main dike[dyke]
确保的上界 guarantee upper bounds
确保正常运转的 fail safe
确报 train list information after departure
确报数 final figure
确报站 train list information reporting station
确定 ascertain; confirmation; define; determination; determine; fix; fixedness; settlement; verify; vouch
确定……数量 quantify

确定保险单 definite policy; named policy; specific policy; valued policy
确定保险费率 insurance rate-making
确定报废 final rejection
确定报价 firm offer; firm quotation; firm quote
确定备件表 definited spare parts list
确定备件一览表 definite spare parts list
确定变异 definitive variation
确定标高 take the altitude to
确定标准 standard setting
确定波浪的参数 wave parameterisation
确定采矿方法 definition mining method
确定采矿技术 definition mining technique
确定测量仪器(视准线)高度的方法 collimation method
确定偿还期 payout period determination
确定偿债能力 solvency determination
确定成本对售价的比率 determining ratio cost to selling price
确定成色 assay
确定承诺 definite undertaking
确定承诺出价 firm-commitment offering
确定乘积区 fixed-product area
确定尺寸 size discrimination
确定尺寸大小 sizing establishment
确定出价 firm bid
确定储量 measured reserves
确定传动 positive driving
确定当量 certainty equivalence
确定的 clear-cut; definite; definitive; deterministic; unquestionable
确定的表达式 deterministic expression
确定的储量 established reserves
确定的堆栈自动机 deterministic stack automaton
确定的二项式系数 definite binomial coefficient
确定的非抹除堆栈自动机 deterministic non-erasing stack automaton
确定的负荷率 establishment burden rate
确定的惯例 established customs
确定的函数 deterministic function
确定的价格 firm price
确定的买卖 firm bargain
确定的抹除堆栈自动机 deterministic erasing stack automaton
确定的赔偿金 liquidated damages
确定的契约 firm contract
确定的权利 established right
确定的时序机 definite sequential machine
确定的事件 definite event
确定的手段 determination method
确定的数据 established data
确定的投资量 established investment
确定的下推自动机 deterministic push-down automat
确定的线性界限自动机 deterministic linear bounded automaton
确定的销售合同 firm sale contract
确定的要约 firm offer
确定的依据 deterministic base
确定的有穷自动机 deterministic finite automaton
确定的重量 ascertained weight
确定等值 certainty equivalent
确定地层岩性的波形处理 waveform processing of determine stratigraphic(al) lithology
确定电位离子 potential-determining ions
确定订货 firm order
确定断层深度手段 methods of determining depth
确定法 deterministic approach
确定方向 orienting
确定费率 fix rates
确定浮土厚度和基岩性质的钻探 bedrock test
确定高程 take the altitude
确定工程建设任务 project definition
确定工程数量方法 method of determination project number
确定古典柱型比例 modulation
确定固定年限的年金 certain annuity
确定故障点 localization of faults
确定关系曲线 determination of relation curve
确定焊缝尺寸用样板 welding ga(u)ge
确定航向 vector
确定核 definite kernel
确定荷载标准 rule for determining load
确定荷载规则 rule for determining load
确定横向(剪)力 transverse force determination
确定回答 definite response
确定混凝土配合比的立方体试验 preliminary cube test
确定火灾模型 deterministic fire model
确定机构 caging mechanism
确定技术经济阶段 definition technical economic phase
确定结构条件的岩芯钻进 structural core drilling
确定解 determinate solution
确定金额 <税款、罚款的> assessed amount
确定(井内)被卡点 locating stuck point
确定距离时间 position learning time
确定开拓方案 definition development program(me)
确定空间坐标 dimensional orientation
确定孔深 bottom up; determinate depth of borehole
确定论 determinism
确定罗盘偏差 adjustment the compass
确定脉冲 acknowledging(im) pulse
确定年金 certain annuity; fixed annuity
确定权重 definitive weight
确定任务 determination of the mission; project definition
确定任务阶段 definition mission phase
确定申报 final return
确定申告 final declaration
确定时间 definitive time
确定试验 confirmed test
确定适合曲线 curve fit(ting)
确定适宜曲线 curve fit(ting)
确定收入单证 fixed letter of revenue
确定授标 confirmation of award
确定输入 specified input
确定数据 specified data
确定数量 quantification
确定水底地形 bottom definition
确定税款平等 equalization of assessments
确定税率 tax rating
确定损失 ascertain loss; ascertainment of damage; ascertainment of loss
确定体系 determinate system
确定条件下的单准则 single criterion under certainty
确定条件下的决策 decision-making under certainty
确定(土地)分级的因素 class-determining factor
确定误差 definite error
确定系统 definite system; determinate system
确定相对位移的依据 basis for determining relative displacement of fault
确定项目 project identification
确定像主点 collimate
确定销售 firm sale
确定销售契约 firm sale contract
确定信道 deterministic channel
确定信号 deterministic signal
确定型 condition of certainty
确定型等价(事件) certainty equivalent
确定型递归模型 recursive deterministic model
确定型经验模型 deterministic empirical model
确定型决策 decision under certainty
确定型决策过程 deterministic decision process
确定性 certainty; definiteness; determinacy; reliability
确定性部分 deterministic component
确定性成分 deterministic component
确定性的 conclusive; deterministic
确定性等价 certainty equivalence
确定性等价特性 certainty equivalence characteristic
确定性地球化学模型 deterministic geochemical model
确定性调度 deterministic schedule
确定性动态规划 deterministic dynamic(al) programming
确定性动态系统 deterministic dynamic(al) system
确定性仿真 deterministic simulation
确定性非线性规划 deterministic non-linear programming
确定性分布 deterministic distribution
确定性分量 deterministic component
确定性分析 deterministic analysis; deterministic parsing
确定性分析(方)法 deterministic analysis method
确定性副本 determinacy counterpart
确定性概率统计污染物输移模型 deterministic-probabilistic contaminant transport model
确定性估算 definitive estimate
确定性故障 determinate fault
确定性关联 deterministic interconnection
确定性过程 deterministic process
确定性函数 deterministic function
确定性技术 deterministic technique
确定性检索 deterministic retrieval
确定性决策 certainty decision; decision-making under certainty
确定性库存模型 deterministic inventory model
确定性路径选择 deterministic routing
确定性模拟 deterministic simulation
确定性模拟模型 deterministic simulation model
确定性模式 deterministic mode
确定性模型 determinacy model; determinate model; deterministic model
确定性农业非点源污染评估 deterministic assessment of agricultural non-point source pollution
确定性排队系统 deterministic queuing system
确定性情况 determinacy case
确定性算法 deterministic algorithm
确定性系统 deterministic system
确定性需求 deterministic demand
确定性与值态的独立性 certainty and valuewise independence
确定性语言 deterministic language
确定性预报 cause-and-effect forecast(ing); deterministic forecast(ing)
确定性振动 determination vibration; deterministic vibration
确定性状态图 deterministic diagram
确定性自顶向下文法 deterministic topdown grammar
确定性自动机 deterministic automation
确定选矿方法 definition dressing method
确定选矿技术 definition dressing technique
确定选矿流程 definition flow-sheet of mineral dressing
确定询价 definite inquiry
确定冶炼方法 definition metallurgical method
确定冶炼技术 definition metallurgical technique
确定冶炼流程 definition flow-sheet of metallurgy
确定依据 basis of determination
确定因素 determining factor
确定应答 definite response
确定源 deterministic source
确定站立点 determination of standing point
确定障碍点 fault localization; localization of a failure; trouble localization
确定折扣的总价法 gross price method of discount recognition
确定震中 determination of epicentre [epicenter]
确定政策 formulated policy
确定中断 precise interruption
确定准确位置 spotting
确定资本 vested proprietorship
确定资源 defined resources
确动 positive motion
确动盘形凸轮 positive motion disk cam
确动平板凸轮 positive motion plate cam; positive plate cam
确动式增压器 positive driven type supercharger
确动凸轮 positive motion cam
确立 build; elaboration; establishing; establishment
确立机构 caging mechanism
确立可靠生 established reliability
确立连接 established connection
确切的证据 tangible proof
确切法 exact method
确切分离 clean-cut separation
确切通知 definite advice
确切凸轮 positive motion cam
确切性 determinacy; exactness; trueness
确切应答 definite response
确切值 explicit value
确切中断 precise interruption
确认包 ack(nowledge) packet
确认保证 affirmative warranty
确认报告 validation report
确认本身的能力 sense of identity
确认闭塞 acknowledgement of blocking
确认标志 confirmatory sign; identification mark <检验机构认可的>
确认部件 acknowledging unit
确认传输前的等待 wait before transmit positive acknowledgement

确认传输延迟 wait before transmit positive acknowledgement
确认传信 confirmation signalling
确认的 acknowledged
确认的延迟 round-trip delay
确认电路 acknowledging circuit
确认订货 confirm order;book confirmation
确认发送 delivery confirmation
确认符号 acknowledge character
确认函 letter of acknowledgment
确认极限 realizable limit
确认继电器 acknowledging relay
确认接触器 acknowledging contactor
确认开关 acknowledger switch; acknowledging switch
确认可以调整的条款 <承包合同的> escape clause
确认控制器 acknowledging controller
确认列车到达 acknowledgement of train arrived
确认霉菌酸 recognizes mycolic acid
确认你方的报盘 confirm your offer
确认汽笛 acknowledging whistle
确认契据 deed of confirmation
确认器 acknowledger
确认前须经验证无风险 subject to approval no risk
确认时间 acknowledging time
确认式信息传递服务 acknowledged information transfer service
确认试验 confined test; confirmed test;convinced test
确认书 confirmation deed; confirmed order; confirming order; letter of confirmation
确认输出 acknowledge output
确认输入 acknowledge input
确认水深 verified depth
确认损失 ascertainment of damage
确认通知书 confirmation sheet
确认图纸 certified drawing
确认信号 acknowledgement signal; acknowledging signal
确认信用证 confirmed letter of credit
确认序号 acknowledged sequence number
确认样本 confirmatory sample; confirmed sample
确认样品 confirmation product; validating product
确认用户身份的合法性 validation of a user's identity
确认有效条款 attestation clause
确认允许 acknowledge enable
确认运行标志 acknowledged run flag
确认责任 acknowledgement of liability
确认债务 acknowledgement of debt
确认致癌物 proved carcinogen; recognized carcinogen
确认致癌原 proved carcinogen; recognized carcinogen
确认重量 ascertained weight
确认装置 acknowledging device
确认字段 affirmative field
确认字符 acknowledge character
确实储量 positive reserves
确实可靠的情报资料 reliable information
确实可靠资源 reasonably assured resources
确实完成日期 physical completion date
确实性 authenticity; credibility; reliability;tangibility;validity
确实性检查 validity check
确实原则 sure-thing principle
确数 exact quantity
确限 fidelity
确限种 exclusive species
确信数据接收部件 data assurance unit
确凿证据 valid evidence
确证试验 confirmatory test
确证条款 conclusive evidence clause

阙【建】 watch tower

阙楼 side tower

裙板 apron(board); apron plate; breast lining;dado;shirt

裙板式供料器 feed apron
裙板式进料机 feed apron
裙板式输送机 apron conveyer[conveyor]
裙板式输送机托辊 apron roll
裙板式喂料机 apron feeder
裙板式喂料机的漏料 apron feeder spillage
裙板线脚 apron lining;apron mo(u)lding
裙板消能工 shirt dissipator
裙臂 <轮动式铲运机> apron arm
裙部 skirt(section)
裙部凹槽 skirt relief
裙部间隙 skirt clearance
裙部开槽活塞 split skirt piston
裙衬 shirt lining
裙带关系 nepotism
裙房 podium[复 podiums/podia]
裙管 <烟囱的> petticoat pipe; draft pipe
裙礁 fringing reef;shoe reef
裙梁 pitching piece
裙料 skirting
裙墙 apron wall;sill wall
裙墙玻璃 spandrel glass
裙圈 apron ring
裙式的 apron-type
裙式隔电子 petticoat insulator
裙式给料器 apron feeder
裙式进料器 apron feeder
裙式绝缘子 bell-shaped insulator; petticoat insulator
裙式输送机 apron conveyer[conveyor]
裙式运输机 apron conveyer[conveyor]
裙式支承压力容器 shirt-supported pressure vessel
裙式支座 skirt support
裙索 <轮动式铲运机> apron cable
裙形垫底 trapezoid pedestal
裙形阀 petticoat valve
裙形花边的 scolloped
裙形夹板 skirt-type joint bar
裙桩 <外海平台护裙结构的> skirt pile
裙桩套管 sleeve for skirt pile
裙状绝缘子 petticoat; petticoat insulator
裙状屋顶 <墙上挑出的假屋顶> skirt-roof
裙状物 petticoat
裙座 skirt
裙座长度 length of skirt
裙座的设计 design of vertical vessels skirt
裙座计算 calculation of skirt support
裙座支承的立式容器 vertical vessel supported by skirt

群变换设备 group translating equipment

群变频器 group frequency converter
群表示 group representation; representation of a group
群拨号 group dialing
群波 group wave
群波速度 group(wave)velocity
群不变性【数】 group invariance
群常数 group constant
群重复周期 group repetition interval
群穿孔机 gang punch
群传播时间 group propagation time
群丛 association;ecologic(al)association;population
群代数 group algebra
群带通滤波器 group bandpass filter
群导频 group pilot frequency
群导频报警器 group pilot alarm
群导频显示器 group pilot indicator
群导频振荡器 group pilot oscillator
群岛 archipelago; chain of islands; group of islands; island group; pulau-pulau
群岛的 archipelagian;archipelagic
群岛国 archipelagic state
群岛海区 archipelago;archo
群岛基线 archipelagic baseline
群岛名称 name of archipelago
群岛水域 archipelagic waters
群到时 group arrival time
群的表示法 representation of a group
群的不可约表示 irreducible representation of a group
群的共轭元 conjugate elements of a group
群的阶 order of a group
群的实现 realization of a group
群的秩 rank of a group
群的作用【数】 action of a group
群地址 group addresses
群地址信息 group address message
群迭代法 group iteration;group iterative method
群钉 gang nail
群动 group motion
群动力学 group dynamics
群短闪光 group short flashing light
群墩 bundle piers; clustered piers; group piers
群多路分解器 group demultiplexer
群多路复用器 group multiplexer
群发送 packet transmission
群方程 group equation
群方向特性 group-directional characteristic
群放大器 group amplifier
群分类 heap sort
群分离符 group separator character
群合闪光灯 group-flashing light
群合转换器 group converter
群核 group germ
群呼(叫) group call(ing)
群呼叫电键 group calling key
群环境温度 group ambient temperature
群黄色 colonial yellow
群回波 group echo
群火山群 cluster of cones
群集 afflux(ion);constellation;crowd; throng
群集本能 herd instinct
群集场所 concourse
群集城市 combined city
群集抽样 mass sample
群集带 cenozone
群集度 stacking
群集过程 clustering procedure
群集技术 clustering technique
群集节 cluster knot
群集控制器 cluster control unit;cluster controller
群集控制器节点 cluster controller node
群集控制装置 cluster control unit
群集密度 aggregation density;packing density
群集模型 cluster model
群集器 cluster
群集筛选 mass screening
群集栅格 cluster lattice
群集式平面布置 cluster plan
群集死亡 mass mortality
群集系数 cluster factor;percentage of men,women and children
群集现象 social phenomenon
群集效应 crowding effect; group effect
群集终端 cluster terminal
群继电器 group relay
群检测器 group detector
群简并波 group degenerate modes
群件 group wares
群礁 reef cluster
群节 group knot
群结构 group structure
群截面 group cross-section
群解调 group demodulation
群解调滤波器 group demodulator filter
群解调器 group demodulator
群解调器滤波器 group demodulator filter
群井 group of wells
群井平台 multiwell platform
群井系统 multiple well system
群居本能 gregariousness; herd instinct
群居的 communal
群居动物 social animal
群居寄生(现象) social parasitism
群居物种 colonizing species
群居性 gregarous
群矩阵 group-matrix
群锯切削 gang-saw cutting
群聚 aggregation;clustering
群聚参数 bunching parameter
群聚电荷 bunched charge
群聚器 buncher
群聚生态学 synecology
群聚束 buncher beam
群聚态 cybotactic state
群聚团 bunch
群聚团密度 bunch density
群聚性 cybotaxis
群决策 group decision-making
群可加效用函数 group additive utility function
群孔抽水 group grilling pumping
群孔抽水试验 well group pumping test
群孔干扰抽水试验 interfering wells pumping test
群控变换 group control change
群控节点 cluster controller node
群控器 cluster control unit
群控(制) group control; cluster control;direct numerical control
群控制继电器 group control relay
群控制屏 group control panel
群扩散法 group diffusion method
群扩张 group extension
群连接器 group connector
群联动穿孔 gang punch
群量分布学 synchrology
群列 series
群滤波器 group filter
群路带通滤波器 group bandpass filter
群路混合线圈 group hybrid coil
群路调制解调设备架 group modem equipment bay
群论 theory of group;group theory
群落 coen;coenosium;community
群落成分 community component; community composition

群落代谢 community metabolism
群落带 coenozone
群落的类型 coenosium type;community type
群落的水平结构 horizontal structure of community
群落的稳定性 community stability
群落的周期性 periodicity of community
群落地 locality
群落动态 community dynamics
群落多样性 community diversity
群落发生 syngenesis
群落分布学 synchorology
群落分类 community classification
群落分区 aggregated subarea;aggregated subregion
群落复合体 community complex;phytocoenose complex
群落过渡区 ecotone
群落和环境梯度 gradient of communities and environments
群落环 circle of vegetation
群落渐变群 coenocline
群落交错 alterne
群落交错区 ecotone
群落结构 community structure
群落结构调整 community structure regulation
群落结构图 phytograph
群落结构与功能 community structure and function
群落净生产力 net community productivity
群落类型 coenotype;phytocoenosium type
群落类型转变顺序过程 orderly process of community change
群落滤水率 community filtration rate
群落迁移 community migration
群落区 community zone
群落生活 coenobiosis
群落生境 biotope;community habitat
群落生境和生态系统 biotopes and ecosystem
群落生态 community ecology
群落生态群 coenocline
群落生态系统 community ecosystem
群落生态学 synecology
群落生物量 community biomass
群落生物学 coenobiology
群落特征 assemblage characteristic
群落同化度 community assimilation number
群落外貌 physiognomy of community
群落污染量 community pollution value
群落系数 coefficient of community
群落系统发育 phylocoenogenesis
群落消退指数 community degradation index
群落形成单位 colony-forming unit
群落型 coenotype
群落选择 selection of community
群落学 coenology
群落演替 community succession;syngenesis
群落与生态系统模型 community and ecosystem model
群落指数 aggregation index
群落周期性 community periodicity
群落最小面积 minimal area
群码 group code
群码记录 group code record
群忙 group busy
群忙灯 group-busy lamp
群忙继电器 group-busy relay
群忙设备 group busying facility
群忙时呼叫 group busy(ing) hour call
群忙音 group busy tone
群锚 group anchorage
群锚锚具 multianchor anchorage
群锚千斤顶 multianchor jack
群锚效应 group effect of anchors
群模成型 gang-die forming
群能 group energy
群能间隔 group energy interval
群凝集 group agglutination
群配线架 group distribution frame
群频(率) group frequency;train frequency
群频转译 group translation
群桥 multiple-bridge
群桥交叉 <由几个立体交叉所组成,一般用于多层交通,各交叉道路直接相互连接> braided intersection; multiple-bridge intersection; braided interchange
群青 blue ultramarine
群青的 ultramarine
群青黄 ultramarine yellow
群青蓝 ultramarine blue
群青绿 ultramarine green
群青色 colonial blue;ultramarine;ultramarine blue
群青色的灰 ultramarine ash
群青紫 ultramarine violet
群青棕 ultramarine brown
群上平均 group averaging
群摄 group shot
群石海岸 boulder coast
群时 group time
群时延 envelope delay;group delay
群时延变化 group delay variation
群时延测试仪 group delay measuring set
群时延校正 group delay correction
群时延校正器 group delay corrector
群时延均衡 group delay equalization
群时延均衡器 group delay equalizer
群时延频率 group delay frequency
群时延频率特性 group delay-frequency characteristic
群时延失真 envelope delay distortion;envelope distortion;group delay distortion
群时延失真校正 group delay distortion correction
群时延时间 group delay time
群时延数据 group delay data
群时延响应 group delay response
群时延展宽 group delay spread
群束 cluster
群松弛 group relaxation
群速度 envelope velocity;group velocity
群速度法 group velocity method
群速度频散值 group velocity dispersion value
群特征 group character
群特征符号 group character symbol
群体 cohort;colony;population
群体暴增 population explosion
群体变异 group modification
群体查点法 method of population enumeration
群体沉降 group settling
群体沉陷 group settlement
群体抽样 cluster sampling
群体的 colonial
群体的自由组合 free massing of group
群体调查 mass survey
群体动态 population dynamics
群体多线性效用函数 group multilinear utility function
群体反应 group response
群体改良 population improvement
群体高速运输 group rapid transit
群体共生 group symbiosis
群体合理性 group rationality
群体计数法 method of population enumeration
群体检查 mass examination
群体建筑 annexed building
群体交叉 group interaction
群体结构【物】 group structure
群体结核 colonial nodule
群体离散 population dispersion
群体力学 population dynamics
群体连续体 continuum
群体每年平均增长百分率 mean annual growth rate percent
群体密度 population density
群体模型 group model
群体平均数 population mean
群体普查 general survey
群体散布 population dispersal
群体珊瑚 colonial coral; compound coral
群体生态学 synecology
群体生物 colonial organism
群体生物学 population biology
群体试验 population experiment
群体特性 group property
群体温度 bulk temperature
群体限额 group allowance
群体效应 group effect
群体效用函数 group utility function
群体行为 group behavio(u)r;population behavio(u)r
群体幸存 cohort survival
群体选择 group selection
群体压力 population pressure
群体因子 group factor
群体隐蔽照明 group occulting light
群体值函数 group value function
群体住房建设 group housing development
群天线 group antenna
群调制 group modulation
群调制反调制设备 group modem equipment
群调制器 group modulator
群调制器滤波器 group modulator filter
群同步 group synchronization
群同构 group isomorphism
群屋 insula[复 insulae]
群系 formation
群系统 group system
群现象 group phenomenon
群像 imagery
群信号继电器 group marking relay
群行列式 group determinant
群性因子 groupiness factor
群选(择) group selection;mass selection
群选择器 group selector
群芽 group germ
群延迟 envelope delay;group delay
群延迟测量设备 group delay measuring equipment
群延迟均衡器 group delay equalizer
群延迟频率 group delay frequency
群延迟失真 group delay distortion
群延迟时间 group delay time
群延迟特性 envelope delay characteristic
群延迟特性测量装置 group delay characteristic measuring equipment
群抑制 group suppression
群遇险报警 group distress alerting
群载频 group carrier frequency
群载频供给架 group carrier supply bay
群占线信号 group-busy signal
群占线音 group busy tone
群折射率 group index;group refractive index
群震 clustering of earthquake;clustering of seismic events
群植 group planting;mass planting
群指数法 group index method
群终端机架 group terminal bay
群终端设备 group terminal equipment
群终接 group connection
群终接器 group connector
群柱 bundle of columns; bundle of pillars;clustered columns;clustered piers; grouped columns; pile cluster;pile columns;reinforced pier
群柱排架 cluster bent
群柱形墙墩 bundle pier
群转接滤波器 group through-connection filter
群桩 clump of piles;cluster of piles;group(ing) of piles; group piles;multi(ple)piles;pile group
群桩承载力折减系数 reduction factor of bearing capacity for piles in group
群桩的整体破坏强度 block failure strength(of pile group)
群桩的整体强度 block failure strength (of pile group)
群桩的桩尖下土的压力泡 bulb of pressure
群桩荷载力 bearing capacity of group piles
群桩基础 multicolumn piers foundation; group piles foundation; clustered piles foundation
群桩基础式建筑 pilotis building
群桩排列法 columniation
群桩效率 group efficiency
群桩效率系数 efficiency factor of pile group
群桩效应 group pile effect
群桩折减系数 reduction factor of pile group
群桩柱 pile columns
群桩作用 action of pile group;group action of piles
群状采伐 group felling method
群状采伐方式作业 group felling system
群状母树作业法 group parent tree cutting
群状伞伐作业 shelter-wood group felling system
群状栽植 block planting
群状择伐 group-selection cutting; group-selection felling; group shelterwood cutting;group shelterwood felling
群状择法作业法 group-selection system
群字母 group alphabet
群组抽样 cluster sampling
群组调查研究 group study
群组分析 cluster analysis; cohort analysis
群组观察 cohort observation
群组技术 group technology
群组取样 cluster sampling
群组死亡率 cohort mortality
群组系统 group system
群组研究 cohort study
群钻 gang drill

R

燃丙烷气的灯浮标 propane gas lighted buoy

燃材 fire wood;fuel wood
燃除 burning off
燃灯含硫量试验 lamp sulfur test
燃灯油 signal oil
燃点 burning point;flammability point;flare point;focal point;ignition point;inflammation point;inflammation temperature;kindling temperature;point of ignition
燃点测定 determination of ignition point
燃点确定 flash point determination
燃点试验器 ignition point tester
燃点温度 ignition temperature
燃点线圈 ignition coil
燃耗 burn-up
燃耗比 burn-up fraction
燃耗分析 burn-up analysis
燃耗份额 burn-up fraction
燃弧角 angle of ignition
燃弧时间 arcing time
燃弧时间测定装置 arc timer
燃尽 burning through;combustion
燃尽了 all-burnt
燃尽率 burn-off rate
燃具 gas appliance;gas-fired equipment
燃具饱和率 appliance saturation
燃具改装 change-over of appliance
燃具试验 gas appliance test
燃坑 flare pit
燃料 combustible charge;combustible material;firing;fuel(material);heating agency
燃料板 fuel plate;fuel slab
燃料棒 briquet(te);fuel rod
燃料棒定位板 fuel rod locating plate
燃料棒剪切机 fuel rod shearer
燃料棒束 fuel bundle
燃料棒组件 fuel strainer assembly
燃料包壳 fuel can;fuel sheath
燃料包壳结合层 fuel cladding bond
燃料包壳温度计 fuel cladding temperature meter
燃料保存 fuel conservation
燃料爆震 fuel knock
燃料杯 fuel bowl
燃料倍增时间 fuel doubling time
燃料泵 fuel actuator;fuel charger;fuel pump;petrolift
燃料泵操纵杆 fuel pump lever
燃料泵操纵杆夹 fuel pump lever clamp
燃料泵槽 fuel pump sump;fuel pump tank
燃料泵沉淀杯 fuel pump sediment bowl
燃料泵传动链轮 fuel pump drive sprocket
燃料泵调节机构 fuel pump metering mechanism
燃料泵给油杆 fuel pump priming lever
燃料泵固定带 fuel pump strap
燃料泵滤清装置 fuel pump filter
燃料泵偏心轮 fuel pump eccentric
燃料泵心轴溅油环 fuel pump spindle oil thrower
燃料泵心轴联轴节 fuel pump spindle coupling
燃料泵心轴弹性接头 fuel pump spindle resilient member
燃料泵主壳体 fuel pump main housing
燃料比 fuel ratio
燃料比冲量 fuel specific impulse
燃料比功率 fuel rating;specific power
燃料比耗 specific fuel consumption
燃料比(率)控制 fuel ratio control
燃料比热 specific heat of fuel
燃料比热容 enthalpy of fuel
燃料变化率 rate of fuel change
燃料表 fuel ga(u)ge;fuel quantity ga(u)ge
燃料驳 fuel barge
燃料驳船 fuel barge
燃料补充 fuel make-up;fuel refreshment
燃料补充剂 fuel extender
燃料补给 fuel make-up
燃料补给港 fuel(l)ing port
燃料补给距离 fuel distance
燃料不渗透的 fuel tight
燃料不足 fuel shortage;fuel starvation;short of bunker
燃料部分 <释热元件的> fuel meat;meat
燃料仓 fuel bin;fuel bunker;fuel magazine
燃料仓斗 fuel storage hopper
燃料仓库 fuel depot
燃料舱 bunker;bunker hold;fuel bunker;fuel compartment
燃料舱平舱 trimmed in bunkers
燃料舱容量 bunker capacity
燃料层 bed of fuel;fuel bed;fuel body
燃料层厚度 fuel bed depth
燃料层厚度调节 fuel bed control
燃料层着火燃烧 fuel bed combustion
燃料产汽率 fuel evapo(u)ration rate
燃料长期储存稳定性 fuel storage stability
燃料场 fuel yard
燃料车 fuel tank truck
燃料成本 fuel cost
燃料成分 fuel constituent;propellant composition
燃料成分比 fuel-air ratio
燃料稠化剂 fuel thickener
燃料储备 fuel reserve
燃料储备量 propellant capacity
燃料储备能力 fuel endurance
燃料储备箱 fuel reserve tank
燃料储备行程 fuel endurance
燃料储藏 fuel storage
燃料储存 fuel conservation;fuel storage
燃料储存表 bunkering schedule
燃料储存池 fuel storage pool
燃料储存室 fuel storage cell
燃料储存箱 fuel storage tank
燃料储运 fuel handling
燃料处理 fuel handling;fuel processing
燃料处理热室 fuel process cell
燃料处理装置 fuel treating equipment
燃料传输容器 fuel transfer cask
燃料床 fuel bed
燃料粗滤器 preliminary filter
燃料存储 fuel conservation
燃料代用电能 fuel replacement energy
燃料单元 fuel unit
燃料当量 fuel equivalent
燃料的爆炸特性 explosive characteristics of fuel
燃料的掺和 blending of fuel
燃料的反应性能 worth of the fuel
燃料的估算价值 imputed value of fuel
燃料的含硫量 fuel sulphur content
燃料的评价 rating of fuel
燃料的雾化 fuel atomizing
燃料的熄灭 extinction of fuel
燃料的预热 preheating of fuel
燃料的杂质 fuel impurity
燃料的质量 quality of fuel
燃料的主要部分 main bulk of fuel
燃料等级 fuel grade
燃料地窖 fuel cellar
燃料地下室 fuel cellar
燃料地质学 fuel geology
燃料点火性质 fuel ignition quality
燃料电池 fuel battery;fuel cell
燃料电池车 fuel cell vehicle
燃料电池催化剂 fuel cell catalyst
燃料电池的燃料 fuel cell fuel
燃料电池底 fuel cell bottom
燃料电池电解质 fuel cell electrolyte
燃料电池动力装置 fuel cell power plant
燃料电池功率容积比 fuel battery power-to-volume ratio
燃料电池检测器 fuel cell detector
燃料电池汽车 fuel cell car;fuel cell powered vehicle
燃料电池陶瓷 fuel cell ceramics
燃料电阀 fuel electrovalve
燃料电极 fuel electrode
燃料叠堆 fuel stack
燃料动力 fuels and energy
燃料动力工业 fuel and power industries
燃料斗 fuel hopper
燃料段 fuel section
燃料堆 fuel assembly;fuel bank
燃料对喷 contrainjection
燃料发火性 fuel ignition quality
燃料发热值 calorific value of fuel
燃料阀 fuel valve
燃料阀填料 fuel cock packing
燃料反馈控制系统 fuel feedback control system
燃料防爆剂 fuel dope
燃料放出 fuel jettison
燃料放出阀 fuel discharge valve
燃料放出塞 fuel draining plug
燃料放泄喷嘴 fuel jettisoning nozzle
燃料废油 fuel oil waste
燃料废渣 fuel slag;waste fuel
燃料费(用) fuel cost;bunkerage;cost of fuels;cost of power;fuel expanses
燃料费账单 fuel bill
燃料分布 fuel distribution
燃料分解作用 fuel decomposition
燃料分类 fuel grade
燃料分离 fuel segregation
燃料分配 fuel distribution
燃料分配开关 fuel distributing cock
燃料分配器 fuel distributor;oil distributor
燃料分区 fuel zoning
燃料分送设备 fuel dispensing facility
燃料分析 fire assay;fuel analysis
燃料粉化 fuel pulverization
燃料粉化系统 pulverised[pulverized] fuel system
燃料封装机 fuel encapsulating machine
燃料辐照度 fuel irradiation level
燃料腐蚀 fuel corrosion
燃料附加费 bunker surcharge
燃料改进 fuel modification
燃料改善 fuel modification
燃料更新 fuel rejuvenation
燃料工程 fuel engineering
燃料工业 fuel industry
燃料工艺学 fuel engineering
燃料功率 fuel horse power
燃料供给 fuel feed(ing);fuel supply
燃料供给泵 fuel feed pump
燃料供给操纵杆 fuel control linkage
燃料供给的滞后 fuel lag
燃料供给管路 fuel feed line
燃料供给过量 overfueling
燃料供给控制 fuel feed control
燃料供给控制膜 fuel control diaphragm
燃料供给系统【机】 fuel feed system
燃料供给中断 failure of fuel;fuel failure
燃料供给装置 fuel supply system
燃料供给自动调节 automatic fuel metering control
燃料供送 fuel charge
燃料供应 bunkering;fuel supply;supply of fuel
燃料供应驳 fuelling barge
燃料供应泊位 fuel berth
燃料供应部门 fuel supply service
燃料供应车 fuel servicing truck
燃料供应船 bunkering boat;fuel ship
燃料供应点 fuel supply point
燃料供应公司 fuel supplier
燃料供应量 fuel duty
燃料供应设施 bunkering facility
燃料供应系统图 fuel system diagram
燃料供应站 fuel(l)ing station
燃料供应者 fuel supplier
燃料关闭阀操纵杆 fuel shut-off valve rod
燃料管 fuel hose;fuel line;fuel pipe;fuel tube
燃料管道 fuel canal;fuel channel;fuel (pipe) line
燃料管理 fuel handling;fuel management
燃料管理系统 fuel management system
燃料管路 fuel(pipe)line
燃料管线 fuel(pipe)line
燃料惯性 fuel inertia
燃料罐 fuel storage tank
燃料罐储存能力 fill capacity
燃料罐能力 fuel tank capacity
燃料规格 fuel specification
燃料过滤器 fuel filter;fuel strainer
燃料含硫量 sulfur content in fuel
燃料耗尽 run-out of gas
燃料耗尽的发射 fuel depletion flight
燃料耗尽信号器 fuel runout warning device
燃料耗量 fuel consumption;rate of firing
燃料耗量表 rate of fuel flow indicator
燃料耗量调节器 fuel rate controller
燃料耗率 fuel rate
燃料合金 fuel alloy
燃料和空气混合物 fuel and air mixture
燃料和慢化剂的分界面 fuel moderator interface
燃料和润滑剂 fuel and lubricant
燃料和润滑剂供应卡车 fuel and oil servicing truck
燃料和润滑剂卡车 fuel and lubricant truck
燃料核算 fuel accounting
燃料荷载 fire load
燃料盒 fuel cassette
燃料褐煤 fuel brown coal
燃料后处理 fuel reprocessing
燃料后处理厂 fuel reprocessing plant;fuel reprocessor
燃料后处理废物 fuel reprocessing waste
燃料后处理生产线 fuel reprocessing loop
燃料后处理室 fuel reprocessing cell
燃料后处理装置 fuel reprocessing facility

燃料化学 fuel chemistry
燃料环 fuel ring
燃料灰 fuel ash
燃料灰沉积 fuel ash deposition
燃料灰的自硬性 self-hardening characteristics
燃料挥发度调节 fuel volatility adjustment
燃料挥发性 fuel volatility
燃料挥发性调整 fuel volatility adjustment
燃料回路 fuel circuit
燃料回收 forward recovery; fuel recovery
燃料回收热室 fuel recovery cell
燃料混合 fuel combination
燃料混合比 fuel mixture ratio
燃料混合剂调节 fuel mixture control; mixture control
燃料混合气密度 combination density; combined density
燃料混合气瞬间倒流 fuel blowback
燃料混合器 fuel mixer
燃料混合物 fuel mixture
燃料混烧率 mixed fuel burning ratio
燃料基体 fuel matrix
燃料及制造工艺技术标准 materials and workmanship specification
燃料激荡试验 slosh test
燃料集中分配器 fuel centralizer
燃料计 fuel meter
燃料计量 fuel metering
燃料计量泵 fuel metering pump
燃料计量的闭环控制 closed loop control of fuel metering
燃料加工废物 fuel processing waste
燃料加料器 fuel charging mean; fuel feeder
燃料加热器 fuel oil heater
燃料加热装置 fuel(l)izer
燃料加入(量) fuel input
燃料加注场 fueling area
燃料加注车 fuel servicer; propellant-servicing vehicle; servicer
燃料加注艇 refueling boat
燃料价格附加费 surcharge on fuel prices
燃料价值 worth of the fuel
燃料间 fuel room
燃料检验 fuel inspection
燃料焦炭 fuel coke; fuel tar
燃料焦油 fuel tar
燃料接收储存站 fuel receiving and storage station
燃料接收装置 fuel(l)ing gear
燃料节省 saving of fuel
燃料节省器 fuel economizer
燃料节省装置 fuel saving device
燃料节省自动调整器 automatic fuel saving device
燃料节约 fuel conservation; fuel economy; fuel saving
燃料节约器 fuel economizer
燃料紧急自动切断器 emergency fuel trip
燃料近似分析 proximate fuel analysis
燃料进给 fuel feed
燃料进口 fuel inlet
燃料进口钟 fuel feeding bell
燃料进入阀 fuel admission valve
燃料经济性 fuel economy
燃料经济性的改善 fuel economy improvement
燃料经济性的缺陷 fuel economy penalty; fuel penalty
燃料净化 fuel detergenting; fuel purification
燃料净化设备 fuel purification unit
燃料净化性 fuel detergency
燃料净化装置 fuel purging system; fuel purification unit
燃料酒精 fuel alcohol
燃料开杯闭杯闪点测定 open/closed test of flash-point of liquids
燃料抗爆性 fuel anti-knock quality
燃料颗粒 fuel particles
燃料坑 fuel pit
燃料空气爆炸物 fuel-air explosive
燃料空气比控制 fuel-air ratio control
燃料空气比指示器 fuel-air ratio indicator
燃料空气混合比 fuel-air mixing ratio
燃料空气混合物 fuel-air mixture
燃料空气混合物分析器 fuel-air mixture analyser[analyzer]
燃料空气循环 fuel-air cycle
燃料空气重量比 fuel-air weight ratio
燃料空腔比 fuel to cavity
燃料控制 fuel control
燃料控制阀 fuel control valve
燃料控制杆 fuel control lever
燃料控制器 flue control assembly
燃料控制装置 fuel control device; fuel control unit
燃料口 fuel port
燃料库 fuel bank; fuel depot; fuel reservoir; fuel storage; fuel store; stock yard
燃料库存价值 value of fuel stocks
燃料块 fuel brick; fuel clump; fuel slug; slug
燃料快速压燃 rapid compression ignition of fuel
燃料矿产 fuel commodities; fuels ore
燃料冷却 fuel-cooling
燃料冷却喷管 fuel-cooled nozzle
燃料利用(率) fuel availability; fuel utilization
燃料利用系数 factor of fuel utilization; fuel efficiency; fuel utilization factor
燃料利用效率 efficiency of fuel utilization
燃料沥青型操作 fuel pitch operation
燃料量 fuel quantity
燃料量计 fuel ga(u)ge
燃料量热计 fuel calorimeter
燃料量指示器 fuel ga(u)ge indicator
燃料流动性 fuel slippage
燃料流量 flow of fuel
燃料流量表 fuel metering device
燃料流量调节 fuel flow control
燃料流量调节器 fuel governor
燃料流量计 fuel flow meter; fuel metering device; fuel oil flow meter
燃料流量率 fuel flow rate
燃料流量指示器 fuel flow indicator
燃料流速计 fuel oil flow meter
燃料漏泄 fuel leak
燃料滤网 fuel filter strainer; fuel sieve
燃料马力 fuel horse power
燃料煤 bunker coal; fuel coal
燃料门 fuel door
燃料密实现象 fuel densification phenomenon
燃料敏感性 fuel sensitivity
燃料木柴 fire wood
燃料能力 fuel capacity
燃料黏[粘]合剂 fuel binder
燃料配给政策 fuel allocation policy
燃料配量 fuel metering
燃料配量装置 fuel metering unit
燃料喷气系统 fuel injection system
燃料喷射 expulsion of fuel; fuel injection
燃料喷射泵传动轴 fuel injection pump driving spindle
燃料喷射泵调节器 fuel injection pump governor
燃料喷射泵筒 fuel injection pump barrel
燃料喷射泵凸轮轴 fuel injection pump camshaft
燃料喷射定时机构 fuel injection timing mechanism
燃料喷射阀 fuel injection valve
燃料喷射阀阀身 fuel injection valve body
燃料喷射阀阀体 fuel injection valve body
燃料喷射阀喷嘴 fuel injection valve nozzle
燃料喷射孔 fuel orifice
燃料喷射器 fuel injector
燃料喷射式发动机 fuel injection engine
燃料喷射束 fuel spray beam
燃料喷射图 fuel(ling) injection pattern
燃料喷射系统 fuel injection system
燃料喷射压力 fueling injection pressure
燃料喷射引燃点火器 fuel injector igniter
燃料喷雾 fuel spray; injected fuel spray
燃料喷雾的均匀性 uniformity of fuel spray
燃料喷雾结构 fuel spray formation
燃料喷雾器 fuel atomizer; propellant injector
燃料喷注 propellant spray
燃料喷嘴 fuel injection nozzle; fuel injector; fuel nozzle; fuel orifice; fuel spray nozzle; propellant injector; propellant orifice
燃料喷嘴孔径 fuel hole diameter
燃料片 fuel sheet
燃料贫化 fuel depletion
燃料品质 fuel quality
燃料平衡表 balance of fuel; fuel balance
燃料平面有效密度 fuel planar smear density
燃料气 fuel gas; cylinder gas
燃料气的精制 fuel gas treatment
燃料气化器 propellant gasifier
燃料气系统 fuel gas system
燃料气中硫氧化物和二氧化硫测定法 method for determination of total sulfur oxides and sulphur dioxide in fuel gas
燃料器 refueller
燃料清理装置 fuel purging system
燃料球 briquet(te); fuel sphere
燃料球芯块 fuel pellet
燃料取样器 fuel sampler
燃料缺乏 lack of fuel
燃料燃烧层 active fuel bed
燃料燃烧过程 fuel process combustion
燃料燃烧量 fuel-fired power
燃料燃烧率 fuel firing rate
燃料燃烧器 fuel combustor
燃料燃烧设备 fuel-burning equipment
燃料燃烧室 fuel combustor
燃料燃烧所需实际空气量 actual air consumption for fuel combustion
燃料燃烧系统设备 combustion equipment
燃料燃烧效率 fuel economy
燃料热含量 enthalpy of fuel
燃料热效率 fuel thermal efficiency
燃料热值 calorific power of fuel; fuel efficiency; heat(ing) value of fuel; fuel value
燃料热值测定器 fuel calorimeter
燃料日报单 daily fuel report
燃料容积 fuel capacity; fuel space
燃料容量 propellant capacity
燃料溶体 fuel solution
燃料溶液 fuel solution
燃料熔盐泵 fuel salt pump
燃料乳化剂 fuel emulsifier
燃料软管 fuel hose; fuel lead
燃料润滑油消耗比 fuel oil consumption ratio
燃料深度 fuel depth
燃料生热 heat generated by fuel
燃料剩存量 fuel inventory
燃料石油 oil fuel
燃料试验 fuel test(ing)
燃料试验(反应)堆 fuel assay reactor
燃料试验器 fuel tester
燃料收集管 fuel collecting pipe
燃料寿命 fuel lifetime
燃料输入孔 fuel port
燃料输送 feed of fuel; fuel haul
燃料输送调节器 flow proportioner
燃料输送挂车 fuel trailer
燃料输送管 fuel supply pipe; fuel supply tube; fuel tube
燃料输送管道 fuel delivery pipeline
燃料输送管路 fuel pump deliver line
燃料输送管系 fuel piping
燃料输送卡车 fuel hauling truck; fuel truck
燃料输送器 fuel conveyer[conveyor]
燃料输送设备 fuel hauling equipment
燃料输送系统 expulsion system of fuel; fuel delivery system; propellant feed system
燃料数据 fuel data
燃料税 fuel tax
燃料损伤 fuel damage
燃料塔 fuel filling column
燃料替换 fuel substitution
燃料替换燃烧炉 conversion burner
燃料添加剂 fuel doping; fuel additive
燃料添加器 fuel feeder
燃料条款 bunker clause
燃料调节 fuel adjustment; fuel metering; metering; meting
燃料调节盒 fuel control package
燃料调节器 flue control assembly; fuel governor; fuel regulator; fuel trimmer
燃料调节装置 fuel control assembly
燃料调整因素 bunker adjustment factor
燃料停供 fuel shut-off
燃料停供开关 fuel cutoff switch
燃料停止输送 fuel cut-off
燃料桶 propellant bottle
燃料筒 propellant bottle
燃料涂层 fuel coating
燃料脱硫作用 desulfurization[desulphurization] of fuel; fuel desulfurization[desulphurization]
燃料温度系数 fuel temperature coefficient
燃料雾 fuel fog
燃料雾化 fuel atomization; fuel atomizing; fuel pulverization
燃料雾化喷燃器 fuel atomizing burner
燃料雾化速度 atomized fuel velocity
燃料吸进帽密垫 fuel suction hood packing
燃料稀释试验 fuel dilution test
燃料系数 fuel coefficient; fuel factor; fuel system
燃料系统测试装置 fuel system test rig
燃料系统的故障 fuel system trouble
燃料系统试验 fuel system run
燃料限制器 fuel limiter
燃料箱 fuel cell; fuel container; fuel tank
燃料箱的容量 propellant capacity
燃料箱放油口盖 fuel tank drain cover

燃料箱盖密垫 fuel tank cap packing
燃料箱加注孔盖 fuel tank filler cap
燃料箱平衡管 fuel tank balance tube
燃料箱容量 fuel capacity
燃料箱油封液 fuel tank preserving fluid
燃料箱总容量 total tankage
燃料消耗 fuel depletion;fuel investment
燃料消耗测定仪 electric(al) fuel ga(u)ge
燃料消耗定额 fuel consumption quota; fuel consumption rating; norm of fuel consumption
燃料消耗计 fuel consumption meter
燃料消耗里程测试计 fuel consumption mileage tester
燃料消耗量 consumption of fuel;fuel consumption; fuel discharge; fuel exhaustion
燃料消耗量曲线 fuel consumption curve
燃料消耗率 consumption rate; fuel consumption rate;fuel rate;propellant flow rate; rate of fuel consumption;specific fuel consumption
燃料消耗率试验 economy test
燃料消耗试验 fuel consumption trial
燃料消耗系数 fuel coefficient
燃料消耗指数 fuel consumption index
燃料效率 fuel efficiency
燃料泄出装置 fuel draining provision
燃料泄漏 fuel leak
燃料卸车滑坡台 fuel ramp
燃料卸料机 fuel unloading machine
燃料芯核 fuel kernel
燃料芯块 fuel pellet
燃料芯块柱 fuel stack
燃料型减压塔 fuel type vacuum tower
燃料性能 fuel performance
燃料需要 fuel need;fuel requirement
燃料需要量 fuel requirement
燃料学会 <英> Institute of Fuel
燃料循环 fuel cycle
燃料循环泵 fuel circulating pump
燃料循环成本 fuel cycle cost
燃料循环工艺学 fuel cycle technology
燃料循环经济学 fuel cycle economics
燃料循环系统 fuel circulating system
燃料压块 fuel compact
燃料压力 fuel pressure
燃料压力表 fuel pressure ga(u)ge; fuel pressure indicator
燃料压力降低解脱器 fuel pressure low trip device
燃料压头 fuel head
燃料研磨机 fuel pulverizing mill
燃料盐 fuel salt
燃料样品辐照盒 fuel specimen capsule
燃料液面指示器 fuel indicator
燃料液位指示计 fuel indicator
燃料用量 firing rate
燃料用量测定仪 fuel metering unit
燃料油 boiler oil;bunker oil;furnace oil;furol;heating oil;fuel oil
燃料油泵 fuel oil pump;fuel oil service pump
燃料油泵(送)站 fuel oil pumping station;heating oil pumping station
燃料油驳 fuel oil barge
燃料油补偿系统 fuel oil compensating system
燃料油残渣 fuel oil residue
燃料油仓库 heating oil store
燃料油产率 fuel oil yield
燃料油沉淀箱 fuel oil settling tank
燃料油成品 fuel oil product
燃料油初(级)滤(清)器 fuel oil primary filter
燃料油储罐 fuel oil(service)tank
燃料油当量 fuel oil equivalent
燃料油的储藏 fuel oil storage
燃料油的具体要求 detailed requirement for fuel oil
燃料油滴 fuel oil particle
燃料油堆栈 heating oil store
燃料油二级滤清器 fuel oil secondary filter
燃料油分析 fuel oil analysis
燃料油供给 fuel oil supply
燃料油供应系统 fuel oil supply system;heating oil supply system
燃料油管 fuel oil pipe;fuel(l)ing oil hose
燃料油管接头 fuel tank joint
燃料油管线 fuel oil pipe
燃料油罐 fuel oil drum; heating oil (storage)tank
燃料油罐警笛 fuel oil tank whistle indicator
燃料油过滤器 fuel oil filter
燃料油回路 fuel oil return
燃料油及铺路油 Furol[fuel and road oil]
燃料油集中供热 fuel oil-fired central heating
燃料油截流管 fuel oil interceptor
燃料油截流器 fuel oil interceptor
燃料油解析塔 fuel oil stripper
燃料油库 fuel oil store; heating oil storage
燃料油库房 heating oil store
燃料油粒 fuel oil particle
燃料油量控制 fuel quantity control
燃料油馏出物 fuel oil distillate
燃料油炉渣 fuel oil slag
燃料油面指示器 fuel level indicator
燃料油喷射泵 fuel injection pump
燃料油气焦油 fuel oil-gas tar
燃料油汽提塔 fuel oil stripper
燃料油取样 fuel oil sampling
燃料油燃烧器 fuel oil burner
燃料油闪蒸塔 fuel oil flash tower
燃料油商店 heating oil store
燃料油添加剂 fuel oil additive
燃料油脱硫 desulfurization from fuel oil;desulfurization of fuel oil
燃料油位 fuel level;fuel oil level
燃料油位尺 fuel level plunger
燃料油位杆 fuel level plunger
燃料油位指示器 fuel level indicator
燃料油稳定剂 fuel oil stabilizer
燃料油洗提器 heating oil stripper
燃料油箱 oil storage tank
燃料油箱接头 fuel tank joint
燃料油消耗量 fuel oil consumption
燃料油泄塞杯 fuel drain cook bowl
燃料油卸料器 heating oil stripper
燃料油压力 fuel flow pressure
燃料油油罐车 fuel dispenser
燃料油渣 fuel oil slag
燃料油装船港口 fuel oil bunkering port
燃料油装料系统 fuel oil filling system
燃料油阻挡装置 fuel oil barrier
燃料与空气的混合气 fuel-air mixture
燃料与物料 fuel and stores
燃料与照明用电指数 fuel and light indices
燃料元件 fuel cartridge;fuel element
燃料元件处理 fuel element processing
燃料元件导向装置 fuel element guide
燃料元件端孔架 fuel element nozzle
燃料元件端帽 fuel element cap
燃料元件盒 fuel element case
燃料元件架 fuel rack
燃料元件结合 fuel element bond
燃料元件孔道 fuel element hole
燃料元件破裂检测器 burst-can detector;leak detector
燃料元件破损探测器 fuel element rupture detector
燃料元件清洗 fuel element cleaning
燃料元件容器 fuel box;fuel element flask
燃料元件生产工厂 fuel fabrication plant
燃料元件输送管 fuel element transfer tube
燃料元件损伤仪 fuel element failure instrumentation
燃料元件外壳 fuel container;fuel element jacket;fuel jacket
燃料元件细棒 fuel pencil
燃料元件芯体 fuel element core
燃料元件悬挂棒 fuel element hanger rod
燃料元件转换器 fuel element converter
燃料元件装盒 fuel channel(l)ing
燃料元件装卸口 fuel handling port
燃料原料 breeder material
燃料源物质 fertile material
燃料运输器 fuel conveyer[conveyor]
燃料运输系统 haul of supply
燃料蕴藏量 fuel reserve
燃料杂质 fuel impurity
燃料再处理 fuel reprocessing
燃料再分布效应 fuel relocation effect
燃料再合成 fuel resynthesis
燃料再生 fuel reproduction
燃料再生因子 fuel reproduction factor
燃料增浓 fuel densification
燃料增压泵 fuel booster pump
燃料增殖 fuel breeding
燃料渣 fuel waste;waste of fuel
燃料站 fuel station
燃料真空供给 fuel vacuum feed
燃料真空进给 fuel vacuum feed
燃料蒸发 fuel vapo(u)rization
燃料蒸发量比 fuel evapo(u)ration rate
燃料蒸发器 fuel evapo(u)rator;fuel vapo(u)rizer
燃料蒸发物 fuel evaporative emission
燃料蒸馏罩 fuel distillation bell
燃料蒸气 fuel vapo(u)r
燃料蒸气泡 fuel vapo(u)r pocket
燃料制备 fuel preparation
燃料质点 fuel particles
燃料质量 fuel quality
燃料质量标准 fuel quality standard
燃料钟 fuel bell
燃料重力供油箱 fuel gravity tank
燃料重量与发动机重量比 charge-weight ratio
燃料注入工具 fuel charging mean
燃料注射泵柱塞 fuel injection pump plunger
燃料贮藏站 bunker station
燃料贮槽 fuel bunker
燃料柱 fuel column
燃料砖 briquet(te);fuel briquette
燃料转换率 fuel conversion factor
燃料转换系数 fuel conversion efficiency;fuel conversion factor
燃料转换因数 conversion fuel factor
燃料转运泵 fuel transfer pump
燃料转运池 fuel transfer pond
燃料转注 fuel(l)ing
燃料转注干线 fuel(l)ing main
燃料装料管道 fuel port tube
燃料装料机 fuel(l)ing machine
燃料装载 fuel inventory;fuel loading
燃料自动调节器 automatic fuel regulator
燃料自动控制 automatic fuel control
燃料总管 fuel manifold
燃料总投入量 fuel inventory
燃料总指示器 fuel totalizer ga(u)ge
燃料阻塞 fuel clogging
燃料组成 fuel composition
燃料组成分析 proximate fuel analysis
燃料组分 fuel component
燃料组元 fuel component
燃煤 fire coal
燃煤舱 coal bunker
燃煤的 coal-fired
燃煤的燃气轮机 coal-burning gas turbine;coal-fired gas turbine
燃煤发电厂 coal-burning power station;coal-fired power station[plant]
燃煤发电机 coal-fired power station
燃煤供暖 coal heating
燃煤锅炉 coal-burning boiler; coalfield boiler;coal-fired boiler
燃煤锅炉机组 coal-fired unit
燃煤火电工程 coal-fired project
燃煤火电站 coal-fired power station
燃煤火炉 coal-fired stove
燃煤机车 coal-fired locomotive
燃煤(加热)炉 coal-fired furnace
燃煤炉灶 coal-fired range
燃煤率 coal consumption rate
燃煤气的火力发电站 gas fired power plant
燃煤气的炉子 gas-fired stove
燃煤气锅炉烟道 gas-fired boiler flue
燃煤热风炉 coal firing hot-air generator
燃煤室容积 combustion-chamber space
燃煤装卸台 coaling stage
燃煤装置 coal-burning installation
燃木 fire-brand
燃泥煤发电厂 peat burning power station
燃泥煤发电站 peat burning power station
燃泥炭发电站 peat-fired power station
燃气 burned gas; combustible gas; gas;heating gas
燃气板 hot plate
燃气表 flow meter; gas counter; gas meter
燃气冰箱 gas refrigerator
燃气采暖锅炉 gas-fired heating boiler
燃气采暖设备 gas-fired unit heater
燃气产量 gas yield
燃气产率 gas yield
燃气车辆 gas-fueled vehicle
燃气储存 gas storage
燃气处置垃圾 refuse for combustion
燃气的 gas-fired
燃气灯光浮标 gas(light)buoy
燃气点火棒 gas poker
燃气点燃 gas ignition
燃气短管 discharge nozzle
燃气对流器 gas converter[convertor]
燃气对液体换热器 gas-to-liquid heat exchanger
燃气对液体热交换器 gas-to-liquid heat exchanger
燃气舵 gas vane;internal controller
燃气额定压力 gas rated pressure
燃气发电厂 gas power plant; gas power station
燃气发电站 gas power station
燃气发动机 gas motor;gas(power) engine
燃气发动机船 gas engine ship
燃气发生 gas generation
燃气发生炉透平机车 turbo-diesel locomotive
燃气发生器 combustion pot;gas generator;gasifier; gasifier nozzle diaphragm;gas producer
燃气发生器点火 ignition of gas generator

R

燃气发生器涡轮机 gasifier turbine
燃气发生器转速 gasifier speed
燃气发生器装置 gas generator unit
燃气发生站 gas-generating station
燃气发生装置 gasifier section
燃气阀 gas valve
燃气放热器 gas radiator
燃气分配 distribution of gas
燃气分析仪 gas alloying apparatus
燃气辐射管 gas-fired radiant tubes
燃气负荷 gas demand;gas load
燃气工业炉 industrial gas furnace
燃气供暖法 gas firing;gas heating
燃气供暖机组 gas-fired unit heater
燃气供暖器 gas space heater
燃气供应 gas supply
燃气管 gas pipe;gas line
燃气管道 gas pipeline;gas piping
燃气管路 gas pipeline
燃气管网系统 gas distribution system;gas network
燃气管线 burning line
燃气罐 gas tank
燃气锅炉 gas boiler;gas-fired boiler
燃气烘炉 gas stove
燃气烘箱 gas fired oven
燃气红外线供暖系统 gas type infrared heating system
燃气互换性 interchangeability of gas
燃气混合风机 gas mixing blower
燃气火电工程 gas-fired project
燃气计量 gas metering
燃气加热器 combustion heater; gas heater
燃气加热设备 gas-fired heating unit
燃气空气混合物 gas-air mixture
燃气冷柜 gas refrigerator
燃气连接阀 gas linking valve
燃气连续流动量热器 gas continuous-flow calorimeter
燃气量测定 gas quantity measurement
燃气流 combustion gas stream;gas jet
燃气流量 gas consumption; gas flow rate;gas weight flow
燃气漏泄 gas escape
燃气炉 gas-fired stove;gas furnace
燃气炉安全(送气)装置 pilot safety device
燃气轮柴油机车 gas turbine diesel locomotive
燃气轮单元列车 gas turbine train unit
燃气轮的 gas turbine
燃气轮电力传动机车 gas turbine electric(al) locomotive; gas turbine locomotive with electric(al) transmission
燃气轮电力机车 gas turbo-electric(al) locomotive
燃气轮动车组 gas turbine railcar
燃气轮发电机 gas turbine driven generator;gas turbine generator
燃气轮发动机 gas turbine engine
燃气轮发动机的履带车辆 gas turbine tracked vehicle
燃气轮机 combustion turbine; gas turbine; internal combustion turbine;gas turbine engine
燃气轮机泵组 gas turbine pump system
燃气轮机舱 gas turbine room
燃气轮机船 gas turbine ship
燃气轮机电力驱动 gas turbo-electric(al) drive
燃气轮机电气机车 gas turbine electric(al) locomotive
燃气轮机(动)车 gas turbine railcar
燃气轮机动力装置 gas turbine power plant;gas turbine power unit
燃气轮机发电厂 fuel-fired gas-turbine plant; gas turbine (electric(al)) power plant; gas turbine power plant;gas turbine power station
燃气轮机发电机 gas turbine powered generator
燃气轮机发电站 gas turbine power plant
燃气轮机鼓风机组 gas turbo-blower
燃气轮机机车 gas turbine locomotive
燃气轮机加力推进装置 gas turbine booster propulsion set
燃气轮机控制盘 gas turbine control board
燃气轮机喷气发动机 gas turbine jet
燃气轮机起动机 gas turbine starter
燃气轮机汽车 gas turbine automobile; turbine-powered automobile; turbo-car
燃气轮机牵引 gas turbine propulsion
燃气轮机燃料 gas turbine fuel
燃气轮机润滑剂 gas turbine engine lubricant
燃气轮机小客车 gas turbine powered car
燃气轮机循环 gas turbine cycle
燃气轮机压气机组 gas turbo-compressor
燃气轮机压缩机 gas turbine compressor
燃气轮机叶轮 gas turbine wheel
燃气轮机液体燃料燃烧器 gas turbine liquid-fuel burner
燃气轮机原理 gas turbine principle
燃气轮机载货汽车 gas turbine truck
燃气轮机增压锅炉 gas turbine supercharged boiler
燃气轮机装置 gas turbine installation
燃气轮机组 compressor-turbine unit
燃气轮机组装件 gas turbine module
燃气轮列车 turbotrain
燃气轮列车组 gas turbine train-set
燃气轮喷嘴 gas turbine nozzle
燃气轮牵引装置 gas turbine traction unit
燃气轮式干燥机 gas turbine drier [dryer]
燃气轮循环 gas turbine cycle
燃气面积 gas area
燃气摩尔质量 number of moles of gas
燃气内燃机 gas(-diesel) engine
燃气排送机 exhauster
燃气喷流 gaseous jet
燃气喷嘴 gas burner
燃气启动机 gas starting engine
燃气-燃气热交换器 gas-to-gas heat exchanger
燃气热电厂 gas-fired thermal plant
燃气热电站 gas-fired thermal plant
燃气热风机组 gas-fired unit heater
燃气热辐射器 gas radiant tube heater
燃气热水器 gas-fired water heater; gas geyser;gas water heater
燃气散热器 gas radiator
燃气烧锅炉 boiler gas burner
燃气设备 gas appliance;gas fittings; gas installation
燃气生产 gas production
燃气试样 gas sample
燃气室的比热强度 specific calorific intensity of combustion chamber
燃气室透平 combustion chamber turbine
燃气室涡轮机 combustion chamber turbine
燃气收集器 combustion header
燃气输送管道 gas transmission pipeline
燃气输送管线 gas main;gas transmission line
燃气水泥窑 gas-fired kiln
燃气炭黑 impingement black
燃气调节器 gas pressure regulator
燃气调压器 governor
燃气透平 combustion gas turbine;gas turbine
燃气温度调节器 gas temperature controller
燃气涡轮 combustion turbine
燃气涡轮发电机 gas turbo-generator
燃气涡轮发电站 gas turbine power station
燃气涡轮发动机 gas turbine engine; gas turbine unit
燃气涡轮鼓风机 gas turbo-blower
燃气涡轮机 combustion gas turbine; gas turbine
燃气涡轮机设备 gas turbine plant
燃气涡轮交流发电机 gas turbo-alternator
燃气涡轮离心式压缩机 gas turbine centrifugal compressor
燃气(涡)轮列车 gas turbine train
燃气涡轮喷气发动机 gas turbine jet
燃气涡轮起动装置 combustion turbine starter
燃气涡轮压气机 gas turbine compressor
燃气涡轮叶片 gas turbine blade
燃气涡轮转子 gas turbine rotor;reaction wheel
燃气系统 fuel gas system
燃气需用量 gas demand;gas load
燃气旋风炉 gas-fired turbofurnace unit
燃气循环 gas cycle
燃气压力 gas pressure
燃气压力表 gas ga(u)ge
燃气压力计 gas ga(u)ge
燃气压力调节 gas pressure regulation
燃气压力调节阀 gas pressure regulator
燃气压力调节器 gas pressure regulator
燃气压缩机 gas compressor
燃气窑 gas-fired furnace; gas-fired kiln
燃气引燃器 gas pilot
燃气用具 gas appliance
燃气与氧混合产生的 oxy-house-gas; oxy-paraffin
燃气杂质 gaseous impurity
燃气灶 gas cooker;gas kitchener;gas oven;gas range
燃气振捣器 gas vibrator
燃气振动器 gas vibrator
燃气质量 gas quality
燃气重量流量 gas weight flow
燃气装置 gas fixture
燃烧 ablaze;blaze out;burn(a);combustion;inflame;inflammation;kindle;kindling;scorch;take fire;deflagration <黑色炸药爆炸引起的>
燃烧安全控制器 combustion safety controller
燃烧安全装置 flame safeguard
燃烧爆击 combustion knock
燃烧比 ratio of combustion
燃烧比耗 specific fuel consumption
燃烧比率 combustion ratio;fuel ratio
燃烧表面 burning surface; combustion surface
燃烧波 combustion wave
燃烧不稳定性 combustion instability
燃烧材料 incendiary material
燃烧参数 combustion parameter
燃烧残留 combustion residue
燃烧残余(物) residue of combustion; combustion residue
燃烧残渣 combustion residue;ignition residue
燃烧层 burning zone;fire layer;zone of combustion
燃烧产热量 combustion heat
燃烧产生的核 combustion nuclei
燃烧产物 combustion products;products of combustion
燃烧常数 burning constant
燃烧车间 combustion plant
燃烧尘埃 combustion dust
燃烧程度 burning degree; depth of burning
燃烧池 combustion cell
燃烧持续时间 burning duration;duration of combustion;fire duration
燃烧持续试验 fire-duration test
燃烧冲程 firing stroke; combustion stroke
燃烧除臭处理 odo(u)r treatment by fire
燃烧除漆 paint burning
燃烧除氧 oxygen removal by combustion
燃烧处置垃圾 refuse for combustion
燃烧穿透测试 fire penetration test
燃烧粗暴性 combustion roughness
燃烧带 burning zone; combustion zone;firing zone;zone of combustion
燃烧带长度 length of burning zone
燃烧导火线 combustion train
燃烧道 fire trough
燃烧的 burned; comburant; flaming; flammable;incendiary;burning
燃烧的化学计量燃料(与)空气混合物 combusted stoichiometric fuel-air mixture
燃烧的理想配比燃料(与)空气混合物 combusted stoichiometric fuel-air mixture
燃烧的木头 fire-brand
燃烧的颜色 burning colo(u)r
燃烧的自动控制 combustion automatic control
燃烧等离子体 combustion plasm(a)
燃烧点 burning point; firing point; flammability point;flare point;ignition point; point of ignition; fire point;kindling point
燃烧动力学 combustion kinetics; kinetics of combustion
燃烧度 degree of burning
燃烧段 burning zone
燃烧堆 burner reactor
燃烧发动机 combustion engine
燃烧反应 combustion reaction; reaction of combustion
燃烧范围 burning range;limit of flammability
燃烧方程式 equation of combustion
燃烧方法 burning process; combustion method
燃烧废料发电厂 refuse-burning plant; waste recovery power plant
燃烧废气 burned gas;burnt gas;combustion emission;combustion gas
燃烧废气的烟囱 flambeau[复 flambeaus/flambeaux]
燃烧废油坑 burn pit
燃烧分解法 decomposition by combustion
燃烧分析 combustion analysis
燃烧分析器 combustion analyser[analyzer]
燃烧封层法 <一种路面养护措施> burn and seal
燃烧负荷率 specific combustion intensity
燃烧工程 combustion engineering
燃烧工艺 burning process

燃烧拱 combustion arch
燃烧管 combustion pipe; combustion tube
燃烧管锅炉 flue boiler
燃烧管冷却 combustion tube cooling
燃烧管炉 combustion tube furnace
燃烧罐 explosion chamber
燃烧过程 burning process; combustion process; flame mechanism; mechanism of combustion
燃烧过程的摄影研究 photographic(al) study of combustion
燃烧过程控制 combustion processing control
燃烧过的气体 burned gas; burnt gas
燃烧过的燃料 spent fuel
燃烧过度的 hard-burned
燃烧合成 combustion synthesis
燃烧核 combustion nucleus
燃烧后的单位重 <耐火混凝土> fired unit weight
燃烧后气体 combustion product gas
燃烧环 fire ring
燃烧混合物 ignition mixture
燃烧混合物的润湿 moistening of mixture
燃烧火箭弹 incendiary rocket
燃烧火焰 combustion flame
燃烧机理 combustion mechanism; mechanism of combustion
燃烧极度 ignition limits
燃烧极限 flammability limit
燃烧计算 combustion calculation
燃烧记录法 burning recording
燃烧记录器 combustion recorder
燃烧剂 incendiary agent
燃烧加热系统 firing system
燃烧阶段 combustion stage; stage of combustion
燃烧结束 completeness of combustion
燃烧进程机构 flame mechanism
燃烧进风管 combustion tube
燃烧开裂 fire crack
燃烧开裂痕迹 fire crack mark
燃烧空间 combustion space
燃烧空气 air of combustion; combustion air
燃烧空窝 raceway
燃烧孔 burner port
燃烧控制 control of combustion; fire control
燃烧控制器 combustion controller; fuel control unit
燃烧控制系统 combustion control system
燃烧控制仪 combustion control instrument
燃烧口 burner port; port
燃烧口砖 burner ring; burner tile
燃烧垃圾发电厂 refuse-burning plant
燃烧冷凝器 burner-condenser
燃烧力强度 intensity of burning power
燃烧量 burning capacity; quantity combusted
燃烧料 comburant
燃烧裂缝 burning crack
燃烧炉 burner; burning furnace; burning kiln; combustion furnace
燃烧炉接口 burner port
燃烧炉容量 burner capacity
燃烧炉试验 furnace test
燃烧铝矾土 burnt bauxite
燃烧率 burning rate; burning speed; rate of burning
燃烧煤气的对流加热器 gas-fired convector
燃烧煤气的室内采暖 gas-fired space heating
燃烧煤油 burning kerosene
燃烧面 combustion front; fire face; firing level
燃烧面递减性 combustion front regressivity
燃烧面积 burning area
燃烧模式 combustion mode
燃烧能力 burning capacity
燃烧排气测定器 combustion flue gas apparatus; combustion fuel gas apparatus
燃烧盘 fire tray
燃烧喷嘴 atomizer burner; burner nozzle; fuel injector
燃烧膨胀 burning expansion
燃烧膨胀比 combustion expansion ratio
燃烧评定试验 fire evaluation test
燃烧气的湿式净化 wet purification of combusting gas
燃烧气溶胶 combustion aerosol
燃烧气体 burning gas
燃烧气体理论容量 theoretic(al) combustion gas volume
燃烧汽化油的炉子 primus stove
燃烧器 burner; combustor; inflamer; oil burner
燃烧器本体 burner body
燃烧器布置 burner arrangement
燃烧器衬砌 combustion liner; combustor liner
燃烧器的鼓风机 burner blower
燃烧器的减声装置 burner muffle block
燃烧器风箱 burning mechanism
燃烧器火道 burner tunnel
燃烧器火焰 burner flame
燃烧器降燃因数 burner turndown factor
燃烧器截面 burner section
燃烧器具 burner system
燃烧器耐火砖 burner tile
燃烧器能力 burner capacity
燃烧器喷尖 burner tip
燃烧器喷口 burner throat
燃烧器喷嘴 burner head; burner jet; burner opening; burner tip; firing nozzle
燃烧器燃料管 burner manifold
燃烧器容量 burner capacity
燃烧器软管 burner hose
燃烧器套管 jacket of burner
燃烧器熄灭保护 burner protection
燃烧器效率 burner efficiency
燃烧器装置 burner unit
燃烧器自动控制 automatic burner control
燃烧器自动控制装置 automatic burner control system
燃烧器组 burner bank
燃烧器最大热负荷 burner capacity
燃烧强度 combustion intensity; firing rate; intensity of combustion
燃烧区 burning area; burning zone; combustion area; combustion zone; firing zone; flame zone
燃烧区温度 combustion zone temperature
燃烧曲线 burning diagram; combustion curve
燃烧曲线图 combustion diagram
燃烧全进程 complex mechanism of combustion
燃烧热 combustion heat; heat of combustion; heat of content; heat output; ignition heat
燃烧容积 combustion volume
燃烧溶滴 flame drip
燃烧三要素 fire triangle
燃烧设备 burning appliance; burning plant; combustion device; combustion equipment; firing device; fuel-burning equipment; furnace arrangement
燃烧设备室 burning appliance room
燃烧设施 combustion plant
燃烧(生成的)气体 combustion gas
燃烧时的喷射 ejection during combustion
燃烧时放出的热量 quantity of combustion
燃烧时间 combustion time; combustion distance; duration of combustion; firing time; flame duration; hours of combustion; time of firing
燃烧式燃气浓度测定器 burning method gas analyser
燃烧事故 pyrophoricity accident
燃烧势 combustion potential
燃烧试验 burning test; combustion test
燃烧试验暴露于火的强度 fire test exposure-severity
燃烧试验程序 fire test procedure
燃烧室 blast chamber; burning chamber; burning compartment; chamber; combustion box; combustion chamber; combustion chamber volume; combustion space; combustion tube; combustor; compression chamber; fire box; fire cell; fire chamber; fire pot; firing box; firing chamber; fuel chamber; furnace; furnace cavity; furnace chamber; furnace room; hearth; heater; heating chamber; hotbulb; ignition chamber
燃烧室壁 chamber wall
燃烧室壁负荷 chamber-wall load
燃烧室壁温度 chamber-wall temperature
燃烧室壁应力 chamber-wall stress
燃烧室侧墙 chamber jamb
燃烧室衬砌 combustion liner; combustor liner
燃烧室衬套 burner inner liner; combustion liner
燃烧室的表面积 combustion chamber surface area
燃烧室的换算长度 characteristic chamber length
燃烧室额定压力 rated chamber pressure
燃烧室负压 combustion chamber draft
燃烧室工况 combustion chamber performance
燃烧室拱顶 ignition arch
燃烧室锅筒 combustion chamber
燃烧室过热器 combustion chamber superheater
燃烧室横截面积 chamber cross section
燃烧室积灰 combustion deposit
燃烧室积炭 combustion chamber deposit
燃烧室集气管 combustion header
燃烧室几何形状 chamber geometry
燃烧室计算条件 rated chamber condition
燃烧室颈 combustor cover
燃烧室空间 bags
燃烧室空间热力强度 strength of thermal force within the combustion chamber
燃烧室口 burner port
燃烧室拉条 combustion chamber brace
燃烧室冷却 chamber cooling
燃烧室冷却系统 chamber coolant system
燃烧室内表面积 chamber surface area
燃烧室耐火层 combustion liner
燃烧室气体温度 chamber gas temperature
燃烧室砌体 burner setting
燃烧室前部 nose cap
燃烧室前端 chamber front end
燃烧室墙 heating wall
燃烧室热力强度 thermal intensity of combustion chamber
燃烧室热容强度 specific combustion intensity
燃烧室容积 chamber capacity; chamber volume; combustion volume; volume of combustion chamber
燃烧室收缩段 constrictor
燃烧室特性【机】combustion chamber characteristic
燃烧室特征长度 characteristic chamber length
燃烧室条件 chamber condition
燃烧室通道 chamber passage
燃烧室头(部) chamber head
燃烧室外壳 outer chamber shell
燃烧室外形 chamber configuration
燃烧室温度 chamber temperature; combustion chamber temperature
燃烧室消能 combustion chamber performance
燃烧室效率 burner efficiency
燃烧室形状 combustion chamber shape
燃烧室性能 chamber performance
燃烧室压力 chamber pressure; pressure in combustion chamber
燃烧室压力记录 chamber pressure record
燃烧室压力-时间曲线 chamber pressure-duration curve
燃烧室压力损失 combustion chamber pressure loss
燃烧室蒸汽压 chamber steam pressure
燃烧收缩 burning shrinkage
燃烧四面体 fire tetrahedron
燃烧速度 burning speed; burning velocity; combustion velocity; rate of combustion; rate of firing; speed of burning; speed of combustion; velocity of combustion
燃烧速度试验 firing time test; rate-of-burning test
燃烧速率 burning rate; rate of burning; combustion rate; firing rate; rate of combustion
燃烧速率系数 burning-rate constant
燃烧损伤 burning defect
燃烧损失 burning loss; combustion loss; loss of ignition; loss on ignition
燃烧所需的空气量 air need for combustion
燃烧所需的实际空气量 active amount of air for combustion
燃烧所需空气 combustion air
燃烧特性 burning behavio(u)r; burning characteristic; combustion behavio(u)r; combustion characteristic
燃烧体 combustible component
燃烧天然气 flared gas
燃烧调节 combustion control
燃烧头 combustion head
燃烧图【机】combustion chart
燃烧脱臭治理 odo(u)r treatment by fire
燃烧脱蜡炉 investment burn out furnace
燃烧完全 complete combustion
燃烧完全度 completeness of combustion
燃烧碗 burner cup

R

燃烧温度 burning point;burning temperature;combustion temperature;ignition temperature;inflammation temperature;temperature of combustion
燃烧稳定性 combustion stability
燃烧物 blazer;comburant;inflamer
燃烧物质的灰 inherent ash
燃烧系统 burning system;combustion system;firing system
燃烧现象 after burning
燃烧箱 firing chamber
燃烧消耗量 propellant flow
燃烧效果 burning effect;firing effect
燃烧效率 combustion efficiency
燃烧效应 effect of combustion
燃烧芯 burner insert
燃烧行为 firing behavio(u)r
燃烧性 flammability
燃烧性按时间分级 combustibility grading period
燃烧性能 combustibility;combustion performance
燃烧性试验 combustibility test
燃烧性质 ignition quality
燃烧修饰法 fiery finish
燃烧需空气量 air for combustion;air of combustion
燃烧需要的空气 air of combustion
燃烧旋管 combustion spiral
燃烧压力 combustion pressure;firing pressure
燃烧窑 burning kiln
燃烧用第二次空气 secondary air for combustion
燃烧用送风机 combustion fan
燃烧油 burner oil
燃烧元件芯体 fuel core
燃烧元素分析 ultimate fuel analysis
燃烧原理 mechanism of combustion
燃烧源 combustion source
燃烧噪声 combustion noise;flame noise
燃烧沼气的热水器 gas-fired water heater
燃烧值 calorific power of fuel;calorific value of fuel;combustion value;fuel value;quantity of combustion
燃烧指示器 combustion indicator
燃烧指数 fire burning index
燃烧质 combustible matter
燃烧质量指数 burning-quality index
燃烧舟 combustion boat
燃烧助剂 combustion adjuvant
燃烧装置 burning installation;burning plant;combustion device;firing device;firing unit
燃烧准备期 delay period
燃烧着 one fire
燃烧着的 conflagrant
燃烧着的煤块 live coals
燃烧自动控制 automatic combustion control
燃烧总热损失 total loss on ignition
燃烧嘴 burner
燃烧最低点 point of lowest burning
燃烧作用 burning action;burning effect;incendiary;incendiary effect
燃素 phlogiston
燃完 after-combustion
燃屑 ember
燃用煤粉的 dust fired
燃用煤气 lighting gas
燃用树脂 preformed gum
燃油 burning oil;fuel;oil fuel
燃油泵 fuel charger;fuel lift pump;fuel priming pump;fuel pump
燃油泵杯 fuel pump bowl
燃油泵齿轮 gear for fuel oil pump
燃油泵阀支片 fuel pump valve retainer
燃油泵分析器 fuel pump analyser
燃油泵继电器 fuel oil pump relay
燃油泵校准机 calibrating machine for fuel pump
燃油泵接触器 fuel oil pump contactor
燃油泵接头 fuel pump adapter
燃油泵进油阀 fuel pump suction valve
燃油泵开关 fuel oil pump switch
燃油泵壳 fuel pump case
燃油泵壳体 fuel pump housing
燃油泵空气室 fuel pump air dome
燃油泵联轴节 fuel pump coupling
燃油泵量油杆 fuel pump dip stick
燃油泵滤网 fuel pump screen
燃油泵膜片 fuel pump diaphragm
燃油泵膜片弹簧 fuel pump diaphragm spring
燃油泵曲臂 fuel pumpcam
燃油泵体 fuel pump body;oil pump housing
燃油泵体完 fuel pump cover
燃油泵调节螺钉 fuel pump adjusting screw
燃油泵调节器 fuel oil pump governor
燃油泵挺杆 fuel pump tappet
燃油泵摇臂联杆 fuel pump rocker arm connecting arm;fuel pump rocker arm link
燃油泵罩垫片 fuel pump cover gasket
燃油泵支架 fuel pump bracket
燃油泵主动芯轴 fuel pump drive spindle
燃油泵主动轴 fuel pump drive shaft
燃油泵组 oil fuel unit pump
燃油泵座垫片 fuel pump base gasket
燃油比 fuel ratio
燃油比控制系统 fuel ratio control
燃油比重度数 American Petroleum Institute gravity
燃油变换指示器 fuel oil change over indicator
燃油补给船 fuel oil tanker
燃油采暖 oil-burning heating
燃油采暖器 oil-fired heater
燃油残滴 fuel dribbling;fuel oil residue
燃油残渣 fuel oil residue
燃油舱 fuel compartment;fuel tank;oil bunker;oil fuel tank
燃油舱总容量 gross oil fuel capacity
燃油沉淀柜 oil fuel settling tank
燃油池 fuel oil tank
燃油齿条指示器 fuel oil rack indicator
燃油初滤器 primary fuel filter
燃油初滤芯 primary fuel element
燃油储存柜 fuel oil storage tank;heating oil storage tank
燃油储存量 capacity of fuel oil
燃油储罐 fuel oil storage tank;heating oil storage tank
燃油储量指示灯 fuel level indicator lamp
燃油船 bunkering tanker;fuel tanker
燃油粗滤器 fuel preliminary filter;primary fuel filter
燃油存量表 fuel content ga(u)ge
燃油存量传感器 fuel-quantity transducer;fuel sensor
燃油单元采暖器 oil-fired unit heater
燃油导管 fuel duct
燃油的 oil-burning;oil-fired
燃油的雾化 fuel oil atomization
燃油灯 fuel-burning lamp
燃油电动机泵组 fuel oil motor-pump group
燃油电加热器 electric(al)oil heater
燃油动力机具 oil fueled engines
燃油二级滤清器 secondary fuel filter
燃油发电机 oil-fired power plant
燃油发电站 oil-fired power station
燃油阀 fuel cock;fuel valve
燃油阀体 fuel cock body
燃油阀座 fuel cook seat
燃油反射炉 oil-fired air-furnace
燃油防爆剂 fuel dope
燃油防腐剂 fuel dope
燃油费用 fuel cost
燃油附加费 bunker surcharge
燃油工具 oil-burning appliance
燃油供给 fuel delivery
燃油供给泵 fuel supply pump
燃油供给特性 fuel delivery characteristics
燃油供暖 oil-burning heating
燃油供暖系统 oil-burning heating system;oil-fired heating system
燃油供热系统 oil-burning heating system;oil-fired heating system
燃油供热烟囱 oil heating chimney
燃油关闭(阀)fuel shut-off
燃油管(道)fuel pipe;fuel(oil)line;fuel passage
燃油管接头 fuel oil pipe connector
燃油管联管节 fuel pipe union
燃油管系 fuel oil service and transfer system
燃油罐 fuel canister
燃油柜 fuel tank;oil fuel bunker
燃油锅炉 oil-burning boiler;oil-fired boiler
燃油锅炉房 oil boiler house;oil-burning boiler house
燃油过耗警告系统 fuel pacer system
燃油过滤器 fuel(oil)filter;fuel strainer
燃油过滤装置 fuel strainer
燃油过热器 oil-fired superheater
燃油和空气节制器 fuel and air restrictor
燃油挥发性 fuel evapo(u)rability
燃油回流管 fuel return pipe
燃油火电工程 oil-fired project
燃油火焰 oil-fired fire
燃油机 oil engine
燃油机车 oil-burning locomotive
燃油机发动 oil engine drive
燃油集管 fuel oil header
燃油集中供暖法 oil-fired central heating
燃油集中供热 heating oil-fired central heating
燃油加热器 fuel-fired heating equipment;fuel(oil)heater;oil heater
燃油加热蒸煮器 oil-heated steamer
燃油加温器 fuel heater
燃油加注表 fuel management chart
燃油进口 fuel inlet
燃油进口接头 fuel inlet fitting
燃油经济性的改善 fuel economy improvement
燃油精滤器 secondary fuel oil filter
燃油净化器 fuel oil purifier
燃油开关 fuel cock
燃油开口尺寸 fuel filler opening
燃油抗爆值 fuel anti-knock value
燃油空气混合比 fuel-air ratio
燃油空气混合气分析仪 fuel-air mixture analyser[analyzer]
燃油控制系统 fuel control linkage system
燃油口开度 fuel filler opening
燃油库 heating oil store
燃油垃圾焚化炉 oil refuse incinerator
燃油累积流量计 fuel flow totalizer
燃油里程 fuel mileage
燃油量表 fuel ga(u)ge;fuel quantity indicator
燃油量计 fuel oil level ga(u)ge
燃油量孔阀针 fuel pin
燃油量控制齿条 fuel rack
燃油量控制套筒 fuel oil quantity control sleeve
燃油量针 fuel metering needle
燃油流 fuel flow
燃油流量传感器 fuel flow transducer
燃油流量调整 fuel flow trim
燃油流量计 fuel flow meter
燃油流量控制阀 fuel flow control valve
燃油流失量 fuel spillage
燃油漏失量 fuel spillage
燃油炉 oil burner;oil-burning furnace;oil-burning stove;oil-fired furnace;oil-heated furnace;oil oven;rotary oil burner
燃油炉的烟囱 oil stack
燃油滤清杯 fuel filter bowl
燃油滤(清)器 fuel filter
燃油滤清器加热器 fuel filter heater
燃油滤清器滤芯 fuel filter element
燃油滤网 fuel screen;fuel strainer
燃油滤芯 element of fuel filter
燃油马达 oil motor
燃油铆钉炉 oil burning rivet heater
燃油没烧透 imperfect combustion
燃油密封 heating oil-tight
燃油膜 fuel film
燃油囊 fuel bag
燃油黏[粘]度 fuel viscosity variation
燃油暖气供热装置 oil warm air heater
燃油排放管道 fuel drain line
燃油旁通调节器 fuel by-pass regulator
燃油配给 gasoline rationing;petrol rationing
燃油喷入与排出系统 fuel injection and exhaust system
燃油喷射 fuel injection;fuel jet;fuel oil injection
燃油喷射泵 fuel injection pump;oil fuel injecting pump
燃油喷射泵凸轮轴 fuel injection camshaft
燃油喷射阀 fuel injection valve
燃油喷射式二行程发动机 fuel injected two cycle engine
燃油喷射特性 fuel spray characteristic
燃油喷射系统 fuel oil injection system
燃油喷射压力 fuel injection pressure
燃油喷射装置 fuel injection equipment
燃油喷雾形状 fuel injection pattern
燃油喷注管道 fuel injection line
燃油喷注器 fuel injector
燃油喷嘴 fuel injection nozzle;fuel jet;fuel nozzle;nozzle burner;oil burner
燃油喷嘴调节 fuel injector modulation
燃油歧管 fuel manifold
燃油气焦油沥青 fuel oil-gas tar
燃油器 oil appliance;oil burner
燃油器吹扫塞门 oil burner blow-out cock
燃油器进油塞门 oil burner cock
燃油切断装置 fuel cut device
燃油燃烧泵 fuel oil burning pump;oil fuel transfer pump
燃油燃烧器 oil fuel burner
燃油燃烧特性 fuel consumption characteristic
燃油热电厂 oil-fired thermal plant
燃油热风加热器 oil-burning warm air heater;oil-fired warm air heater
燃油热风器 hot-air furnace;oil-fired unit heater
燃油热水器 oil-water heater
燃油润滑油比 fuel to oil ratio
燃油烧透 complete combustion
燃油设备 oil-burning appliance;oil-

burning equipment
燃油设施 liquid fuel burning appliance
燃油射束 fuel jet
燃油深舱 fuel oil deep tank
燃油升压泵 fuel booster pump
燃油式路面热管器 oil-fired road burner
燃油式路面热炙器 oil-fired road burner
燃油手泵 fuel hand pump
燃油输送 fuel transfer
燃油输送泵 fuel oil shift pump;fuel oil transfer pump; fuel supply pump;fuel transfer pump;transfer-pump for fuel
燃油输送管 fuel feed pipe
燃油输送管卡子 fuel line clip
燃油输送管路 fuel delivery line;fuel line
燃油输送软管 flexible fuel line
燃油双重过滤器 duplex fuel filter
燃油水分分离器 fuel oil purifier
燃油税暂行条例 The Interim Regulations on the Fuel Oil Tax
燃油添加剂 fuel oil additive
燃油调节阀 fuel regulator valve
燃油调节器 fuel knob;fuel regulator
燃油调节系数 fuel adjustment factor
燃油调节旋钮 fuel knob
燃油通道 fuel gallery
燃油桶 fuel drum
燃油凸轮 fuel cam
燃油位调节 fuel level adjustment
燃油雾化 fuel spreading
燃油雾化喷射 fuel spraying
燃油雾化器 fuel oil atomizer
燃油雾化蒸汽 fuel oil atomizing steam
燃油吸入 fuel induction
燃油稀释 fuel dilution
燃油稀释机油 fuel dilution of oil
燃油系【机】fuel system
燃油系故障 failure of fuel;fuel failure
燃油系统 fuel oil system;oil-burning assembly
燃油系统充油驱气 prime the fuel system;priming the fuel system
燃油箱 fuel oil tank;fuel tank
燃油箱抽油泵 fuel tank evacuating pump
燃油箱分配阀 fuel tank selector valve
燃油箱浮子 fuel tank float
燃油箱盖 fuel tank lock
燃油箱后支架 fuel tank rear bracket
燃油箱加油盖 fuel tank filler cap
燃油箱加油口 fuel tank filler
燃油箱卡箍 fuel tank strap
燃油箱卡箍减声器 fuel tank strap anti-squeak
燃油箱口接头 fuel adapter
燃油箱滤网 fuel tank strainer
燃油箱容量 fuel tank capacity
燃油箱通排放阀管道 fuel tank to drain valve tube
燃油箱限压阀 fuel tank pressure relief valve
燃油箱支杆 fuel tank support bar
燃油箱支架 fuel tank support
燃油箱作用底座 fuel tank base
燃油消耗计量 fuel consumption metering
燃油消耗量 fuel oil consumption
燃油消耗率测定 fuel consumption test
燃油效率 fuel efficiency
燃油辛烷值试验机试验 CFR[cooperative fuel research committee] engine test
燃油续航力 fuel range
燃油循环 fuel cycle
燃油压力 fuel pressure
燃油压力泵 fuel pressure pump
燃油压力表 fuel pressure ga(u)ge
燃油压力计 fuel oil pressure ga(u)ge
燃油压力开关 fuel pressure switch
燃油压送泵 fuel oil delivery pump
燃油压头 head of oil fuel
燃油液面传感器 fuel level transmitter
燃油液面高度 fuel level
燃油溢流阀 fuel oil overflow valve
燃油引入 fuel entry
燃油油泵滤杯固定卡 fuel pump bowl clamp
燃油油量计 fuel quantity meter
燃油油位指示器 fuel indicator
燃油油压计 fuel oil pressure ga(u)ge;fuel pressure ga(u)ge
燃油预过滤器 fuel preliminary filter
燃油预热器 fuel oil preheater;oil heater
燃油预热器滴管 oil heater drip pipe
燃油再处理 fuel oil reprocessing
燃油再加注 refill of fuel
燃油增压泵 fuel oil booster pump;oil fuel supercharge pump
燃油闸门 fuel lock
燃油站 fuel oil station
燃油针阀 fuel pin
燃油真空供给 vacuum feed
燃油真空进给 vacuum feed
燃油蒸发的有害排出物 fuel evapo(u)rative emission
燃油蒸发管 vapo(u)rizer tube
燃油直接喷射系统 direct injection fuel system
燃油重力自供系统 fuel gravity system
燃油贮存柜 oil fuel storage tank
燃油贮罐 oil fuel storage tank
燃油转运泵 fuel transfer pump
燃油装置 oil-burning assembly;oil-fired equipment
燃重油锅炉 heavy oil fired boiler
燃轴 hot box;hot jewel
燃着的 alight

冉

冉赛条纹 Ramsey fringes
冉斯登环 Ramsden's circle
冉斯登目镜 Ramsden's eyepieces;Ramsden's ocular
冉斯登圈 Ramsden's circle;Ramsden's disc

染 dye

染白料 Spanish white
染病 contracted a disease;get a disease
染厂 dye-works
染尘度 dustiness
染尘法 dust exposure method
染成木纹色 engrain
染成深红色 crimson
染成棕色 brown
染毒 toxicant exposure
染毒柜 toxicant exposure cabinet
染毒室 toxicant exposure chamber
染法 staining
染坊 dye-house;dye-works
染缸 dye vat
染工 stainer
染红 dy(e)ing flower;ruddle
染迹试验 stain test
染剂 stain
染架 staining stand
染菌水 contaminated water
染料 colo(u)rant;colo(u)ring agent;colo(u)ring matter;dye(stuff);dyeware;tinct;tint' dye
染料厂 dyestuff works
染料池 dyestuff lake
染料传递法 imbibition
染料单色的 monogenetic
染料的磷光 phosphorescence of dyes
染料废水 dye waste(water);wastewater from dyestuff industry
染料废水处理 dye wastewater treatment
染料工业 dyestuff industry
染料工业废水 dye manufacturing wastewater
染料盒 dye cell
染料糊 colo(u)r paste
染料激光器 dye laser
染料碱 colo(u)r base
染料媒介剂 mordant
染料敏化 dye sensitization
染料木树 fustics
染料木素 genistein
染料耦合 dye coupling
染料排泄功能试验 dye exclusion test
染料破坏法 dye destruction process
染料企业管理 enterprise management of dye-stuffs industry
染料溶液 solution of dyestuff
染料溶液法 colo(u)r dilution method
染料桑树 fustics
染料渗透 dye penetration
染料渗透试验法 <测定陶瓷气孔率> dye absorption;dye penetration
染料生产废水 dye manufacturing wastewater
染料生产废水处理 wastewater treatment of dyestuff industry
染料示踪剂 dye tracer
染料试验计 fugitometer
染料索引 colo(u)r index
染料索引号 colo(u)r index number
染料索引名 colo(u)r index name
染料调色法 dye toning
染料透入 dye penetration
染料脱色 dye decolo(u)rization
染料稀释曲线测定 dye dilution curve determination
染料线条 dyeline
染料性颜料 dye pigment
染料悬浮跟踪 dye dispersion and float tracking
染料悬浮体 dyestuff suspension
染料隐色基 leuco dye
染料印流 bleeding of colo(u)r
染料中间体 dyestuff intermediate
染料中间体废水 dyestuff intermediate wastewater
染料着色试验 dye stain test
染料作物 dye crop
染漂法 dye bleaching process
染浅色的 hypochromatic
染色 bedye;colo(u)ration;colo(u)ring;colo(u)r staining;dye;stain;tinction;tincture;tintage
染色半体 chromatid
染色本领 staining power
染色标本 stained preparation
染色标记 dye marker
染色玻璃 stained glass
染色玻璃纤维纱 dyed glass yarn
染色槽 staining bath;staining trough
染色测速法 colo(u)r method of measuring velocity
染色层 dye layer
染色掺和 colo(u)r admixture
染色掺料 colo(u)r admixture
染色衬纸 ingrain lining paper
染色的 painted;tinctorial
染色的木材 stained wood
染色碟 staining dish
染色法 decoration method;dy(e)ing method;stained method;staining method;staining technique;staining
染色反应 staining reaction
染色废水 dy(e)ing wastewater
染色分涂法 dy(e)ing retouching process
染色(辅)助剂 dy(e)ing assistant
染色腐蚀 stain etch
染色观察 dye penetrant inspection
染色机 dyer
染色技术 staining technique
染色剂 colo(u)ring agent;stainer;staining agent;staining material;tinter;dye tracking
染色架 staining rack;staining stand
染色间 dye-house
染色减速剂 dye-retarding agent
染色检查 chromoscopy;dye test
染色胶片 ink film
染色卷流 dye plume
染色刻图膜 dyed scribe-coating
染色扩散 dye diffusion
染色离散研究 dye dispersion study
染色力 staining power
染色栎 gall oak
染色粒 chromomere
染色皿 staining plate
染色明胶滤色片 dye-gelatin filter
染色能力 colo(u)ring power
染色品 dy(e)ings
染色器 stainer;tinter
染色墙纸 ingrain wallpaper
染色溶液 staining solution
染色砂(沙) colo(u)red-sand;dyed sand
染色砂(沙)实验 colo(u)red-sand experiment
染色烧杯 dye beaker
染色渗透剂 dye penetrant
染色渗透检查 dye penetrant inspection
染色渗透试验 dye penetrant test;dye penetration test
染色渗透液 dye penetrant;visible dye
染色师 colo(u)rman
染色示踪 dye tracing
染色示踪剂 dye tracer
染色试验 dye test;staining test
染色水体 dyed volume
染色体【生】chromosome
染色性 chromaticity
染色液体 dyed fluid
染色值 colo(u)r number
染色质 chromatin
染色助熔剂 colo(u)red flux
染色组织 colo(u)rant
染深色的 hyperchromatic
染深性能 pile-on property
染渗法 <检验电焊质量的染料渗入法> dye penetrant
染水标志 dye marker
染苏丹的 sudanophilic;sudanophilous
染透 imbue
染污 discolo(u)ration;fouling;stained
染污薄板 soiled plate
染污剂 <测去污力用> soiling agent
染污险 risk of contamination with other cargo
染液 dye;dyestuff solution;staining solution
染液瓶 colo(u)r solution bottle
染印法假彩色合成 dye print false colo(u)r composite
染印有木纹的衬里纸 engrain lining paper
染印有木纹的墙纸 engrain wallpaper
染整 dy(e)ing and finishing
染纸颜料 paper colo(u)r

壤良 soil luck

壤黏[粘]土 loam clay
壤土 doras;loam
壤土的 loamy
壤土的夯实工程 rammed loam construction
壤土的夯实施工 rammed loam construction
壤土荒漠地 loam desert
壤土荒野地 loam heath
壤土路 loam road
壤土密封 loam seal(ing)
壤土墙 loamy bottom
壤土砂石 loamy sandstone
壤土楔(块)loam wedge
壤土芯 loam core
壤土芯墙 loam core wall
壤土质的 loamy
壤土质底 loam ground
壤土质灰泥 loamy marl
壤土质结构 loamy texture
壤土质砾石 loamy gravel
壤土质砂 loam(y)sand
壤土(筑)墙 loam wall(ing)
壤土砖 loam brick
壤性土 latosol
壤质 loamy texture
壤质粗砂 loamy coarse sand
壤质化 loamification
壤质结构 loamy texture
壤质黏[粘]土 clay-loam soil; loamy clay
壤质砂土 loam(y)sand
壤质土 loamy soil
壤质细砂 loamy fine sand
壤质细土 loamy fine soil
壤中流 interflow; prompt subsurface runoff; storm seepage; subsurface runoff; subsurface (storm) flow; through-flow
壤中气 soil gas
壤中气测量 soil gas survey
壤中水 subsurface water

让步比 odds ratio

让车标(志)线 give-way line
让车侧线【铁】passing siding
让车道 <道路加宽部分> passing zone;refuge manhole;passing bay; lay-by track;passing place;passing tracking;passing lane
让车回线 passing loop
让车线 passing track
让出 abalienate;resign
让出所有权 yield possession
让出条款 relinquishment clause
让出位置 give way
让出有限用地权产权一方 servient estate
让船处 lay by(e)
让船道 passing place;passing lane
让船水域 passing basin
让刀 cutter back-off;cutter relieving
让刀量 relieving amount
让渡 abalienate; abalienation; alienation;assignment;dedition;expropriation; pass-through; release; remise;transference;turnover;yield
让渡财产 execute;execute an estate
让渡人 <权利或财产> alienator;releasor;transferor
让渡日 name day
让渡手续费 negotiation charges
让渡所有权 alien
让渡条款 assignment clause
让给 setover
让购 negotiate;negotiation
让购(汇票)银行 negotiating bank
让购信用证 assign letter of credit;negotiation credit
让股 transfer of shares
让股人 transferor
让管道悬挂 leave the pipe hanging
让价 amicable allowance; better the price;concessional rate;price concession
让开 keep clear;keep out of the way
让开航路 get out of the way; give way;keep out of the way
让利 interest concessions;transfer of profits
让利(接受)定货 accepting at sacrifice
让路 give way;make way
让路标志 <从支路或匝道进入干道时让干道交通先行> yield sign;giveway marking
让路船 burdened vessel;give-way vessel;obliged vessel;oblique vessel
让路系数 give-way coefficient
让路信号 signal of giving way
让清 past and clear
让受方 acquirer;acquiring party
让受人 assignee;grantee;releasee
让头 discount
让位墙 demising wall
让压可缩性钢支柱 yielding steel prop
让压矿柱法 yield-pillar system
让压性拱形支架 sliding roadway arch;yielding arch
让压性金属支架 yielding steel arch
让压支架 compressible support;cushion support;pliable support
让压支柱 yield timbering
让与 alienate; concede; concession; demise;let;remise;transfer;transference
让与权取得 acquisition of concessions
让与人 conveyer[conveyor];granter[grantor]; surrenderor; transferror[transferrer]
让与所有权 yield possession
让与条款 alienation clause; assignment clause;release clause
让与者 assignor;granter[grantor]
让与证据 conveyance
让与证书 grant certificate; transfer certificate
让与证书制作业 conveyancing
让租租约 bare boat charter party

饶舌 chatter

饶头重量 tret

扰动 bump;commotion;destabilization;disturbance;mixing;perturbation;pulsative oscillation;turbulence;turbulent motion

扰动比 disturbance ratio
扰动变量 disturbance variable
扰动变量的补偿 disturbance-variable compensation
扰动冰 disturbed ice
扰动波 disturbance wave;disturbed wave; disturbing wave; perturbation wave
扰动波痕 interference ripple mark
扰动参数 excitation parameter
扰动层理 disturbed bedding
扰动程度 degree of disturbance
扰动传播 propagation of disturbance
扰动大气 rough atmosphere
扰动的 disturbed
扰动的生成 generation of disturbances
扰动的逐渐减弱 decay of a disturbance
扰动地层 disturbed stratum[复 strata]
扰动度 degree of disturbance; degree of remo(u)lding; disturbance degree
扰动法 method of perturbation; perturbation method
扰动范围 range of disturbance
扰动方程 perturbation equation
扰动分布 perturbed distribution
扰动负荷 fringe load
扰动高层大气 disturbed-upper atmosphere
扰动共鸣 disturbing resonance
扰动过轨道 disturbed track
扰动函数 disturbance function
扰动加速度 disturbance acceleration
扰动减弱 remo(u)lding loss
扰动扩散 eddy diffusion
扰动理论 perturbation theory
扰动力 disturbed force; perturbing force; disturbing force; exciting force【船】
扰动力矩 disturbing moment
扰动量 disturbance quantity
扰动流 erratic flow
扰动脉冲 annoying pulse
扰动面 disturbed(sur)face;Mach stem
扰动泥晶灰岩 dismicrite
扰动年龄 disturbed age
扰动黏[粘]土 remo(u)lded clay
扰动频率 disturbance frequency; forcing frequency
扰动破碎带 disturbed fractured zone
扰动剖面 disturbed profile
扰动气流 rough air
扰动器 perturbator
扰动情况 disturbance case
扰动区 disturbance zone; disturbed area
扰动砂 disturbed sand;disturbance sand
扰动深度 depth of disturbance
扰动势函数 disturbance potential function
扰动试件 remo(u)lded sample
扰动试件强度 remo(u)lded strength
扰动试样 disturbed sample;remo(u)lding sample
扰动疏浚 agitation dredging; disturbing dredging
扰动速度 disturbance velocity
扰动损失 turbulence loss; remo(u)lding loss <土壤强度的>
扰动太阳 disturbed sun
扰动体 disturbing body
扰动图形 disturbance pattern
扰动土(壤)disturbed soil; remo(u)lded soil
扰动土样 disturbed sample;disturbed sample of soil; disturbed soil sample;disturbing sample of soil
扰动微晶灰岩 dismicrite
扰动位 disturbing potential; perturbing potential; potential of disturbing masses; potential of random masses
扰动位函数 disturbing potential function
扰动稳定性 stability for disturbance
扰动问题 perturbed problem
扰动系数 coefficient of disturbance; perturbation coefficient
扰动现象 disturbing phenomenon
扰动线 disturbance line;wavelet
扰动向量 disturbance vector
扰动项 disturbance term
扰动协方差 disturbance covariance
扰动谐振 disturbing resonance
扰动信号 disturbing signal
扰动型三合镜 perturbed triplet
扰动岩层 disturbed stratum[复 strata]
扰动样品 disturbed sample
扰动抑制 disturbance rejection
扰动因素 disturbance factor
扰动影响 disturbing influence
扰动运动特性 transient response data
扰动运动振幅 response excursion
扰动振幅 response excursion
扰动指数 disturbance index;disturbed index;remo(u)lding index
扰动中心 center[centre] of disturbance
扰动周期性模型 disturbed-periodicity model
扰动轴 stirring shaft
扰动状态 state of disturbance
扰动锥 Mach cone
扰断 disruption
扰流 burble;disturbed flow;overfall
扰流板 intercepting plate;spoiler
扰流颤振 <扭转振动为周期性的,挠曲振动为不规则的> buffeting flutter
扰流空化 burbling cavitation
扰流空蚀 burbling cavitation
扰流器 baffle; vortex generator; spoiler <装在汽车上防止回形滑行>
扰流器噪声 spoiler noise
扰流子 turbolator
扰乱 derangement; disturb; perturb(ation);undo;violation
扰乱反射体 confusing reflector;radar confusion reflector
扰乱共鸣 disturbing resonance
扰乱共振 disturbing resonance
扰乱计划 disconcert
扰乱经济秩序 disrupt economic order
扰乱生境 habitat dislocation
扰乱市场 destabilize the market; disrupt the market;raid the market
扰乱水系 deranged drainage;deranged drainage pattern
扰乱行车 disturbance to operation
扰乱治安 disturbing the peace
扰码器 scrambler
扰磨损力 resistance to abrasion
扰频的 scramble
扰频器 frequency scrambler; scrambler

绕 X 轴旋转 X-rotation

绕坝渗漏量 leakage around dam abutment
绕板式圆筒 spirally coil-layered cylinder
绕半个轮轴的绳索传动 half wrap drive
绕避 passing round
绕侧履带转向 pivot turn
绕缠的草木 voluble herb
绕缠的灌木 voluble shrub
绕成 coiled;reel up
绕成螺旋 coiling
绕城高速公路 around-the-city expressway
绕城公路 around-the-city road; city ring road
绕城环形路 <英> orbital road
绕程 back haul
绕赤道轴转动惯量 moment of inertia about equatorial axis
绕出 reel off

绕垂直轴的转动 yawing rotation
绕锤辊 winding weight drum
绕带 strip winding
绕带机 strip winding machine
绕带器 tape winder
绕带式容器 strip winding vessel
绕带式液压机 wire-wound hydraulic press
绕道 detour; diverting; git; rat-run; run-around way
绕道的 roundabout
绕道的渠 by-pass canal
绕道附加费 canal surcharge
绕道可走的道路 by-pass street
绕道输沙 sand bypass
绕道通过 pass-by
绕道行驶 detour
绕道行驶的机动车道 by-pass motorway
绕道运河 bypass canal;diversion canal
绕道指示标 by-pass marker
绕道指向 detour direction
绕地(球)earth-orbit
绕地球的 circumterrestrial
绕地(球)轨道会合 earth-orbital rendezvous
绕地(球)轨道运行 earth-orbital operation
绕法 winding
绕杆 turnstile
绕杆式馈源 turnstile feed
绕杆式天线 doughnut antenna; turnstile antenna
绕杆天线 super-turnstile antenna
绕钢丝机 wire-winding machine
绕管 wound tube
绕过 by-pass(ing);pass-by;round
绕过拐角处的填角缝焊 end return
绕过滑车的绳索 running line
绕过回车滑轮的缆索 tail rope
绕过仪表的水管 meter bypass
绕过式弯管 passover bend
绕管式换热器 spiral heat exchanger; coil-wound heat exchanger
绕焊 boxing; contour welding; end turning
绕航 deviation(sailing)
绕航报告 deviation report; report of deviation
绕航权 diversion privilege
绕航条款 deviation clause; diversion clause
绕好绞盘杆 swifter a capstan
绕好绞盘杆围绳 swift a capstan
绕花饰群柱 annulated column
绕簧机 torsion machine
绕回 wrap-around
绕机翼的环流 motion round wing
绕极海流 circumpolar current
绕极水 circumpolar water
绕极星 circumpolar star
绕极旋涡 circumpolar vortex
绕极轴转动惯量 moment of inertia about polar axis
绕接 solderless wrapped connection; wire wrap connection; wire wrapping
绕接板 wire wrapping board
绕接插孔板 wire wrapped socket board
绕接法 wrapped connection
绕接方式 wire wrapping system
绕接工具 wire wrapped tool
绕接机 wire wrapper; wire wrapping machine
绕接模件 wire wrappable module
绕接模件板 wire wrappable module board
绕接装置 wire wrappable assembly
绕结敷层 sintered coating
绕结环 sintered ring
绕结裂纹 sintering crack
绕结性能 sintering character
绕茎卷须线脚饰 twining stem mo(u)lding
绕锯 turning saw
绕卷轴工人 spooler
绕缆索机 cable wrapping machine
绕裂 flex crack(ing)
绕流 circumfluence; circumfluential motion; flow around a body; flow over a body; peripheral system
绕流的 circumfluent(ial); circumfluous
绕流管 by-pass
绕路 side-track(ing)
绕片式翅支管 tension wrapped fin tube
绕起 reel up
绕起绞辘绳索 render a tackle
绕球机 ball winder
绕曲刚度 flexural rigidity
绕圈 winding
绕圈比 ratio of winding
绕圈轨道 roundabout trajectory
绕纱角度 winding angle
绕纱圈层 winding coil
绕射 diffract
绕射 P 波 diffracted P wave
绕射波 diffracted wave; diffraction wave
绕射测微计 eriometer
绕射初波 < 地震纵波 > diffracted P wave
绕射传播 diffraction propagation
绕射电阻器 wire-wound resistor
绕射光 diffracted light
绕射光谱 diffraction spectrum; diffuse spectrum
绕射光栅 diffraction grating
绕射花样【物】diffraction pattern
绕射积分偏移 diffracted integrated migration
绕射级 order of diffraction
绕射计 diffractometer
绕射角 angle of diffraction; diffraction angle
绕射截面积 diffraction cross-section area
绕射理论 diffraction theory
绕射面积 diffraction area; diffraction cross-section
绕射频谱 diffraction spectrum
绕射散射 diffraction scattering
绕射扫描 diffraction scan
绕射射束 diffracted beam
绕射栅 diffraction grating
绕射深度偏移 diffraction depth migration
绕射束 diffracted beam
绕射损耗 diffraction loss
绕射条纹 diffraction fringe
绕射图 diffraction diagram
绕射图像 diffraction pattern
绕射系数 diffraction coefficient
绕射现象 diffraction phenomenon
绕射线 diffracted ray
绕射效应 diffraction effect
绕射作用 diffraction
绕渗 by-pass seepage
绕绳的碰垫 keckle
绕绳方法 reeve
绕绳防滑 snub
绕绳滚筒 whim
绕绳索滚筒 rope runner
绕绳索柱 roping pin
绕绳筒 rope drum
绕数 winding number
绕丝 wire winding
绕丝等静压机 wire-wound isostatic press
绕丝定位件 wire wrap spacer
绕丝高压容器 wire-wound high pressure vessel
绕丝焊条 wrapped electrode
绕丝机 filament winder; reeling machine; spiral filament forming machine; spooling machine; stranding machine; winding machine; wire-coiling machine
绕丝架 lacing stand
绕丝胶管 wire wrapped screen
绕丝框架 wire-wound frame
绕丝炉 wire-wound furnace
绕丝式容器 wire-wound vessel
绕丝筒 forming tube; collector < 玻璃纤维 >
绕丝头 feeding head
绕丝油箱 filament winding tank
绕索鼓轮 load drum
绕索绞盘的制动器 rope drum brake
绕索轮 spinning wheel
绕索小车 spinning carriage
绕贴标签 wrap-around label
绕线 coiling; reeving; winding; wire wrapping
绕线棒刮涂器 wirebar applicator; wire-rod applicator; wire-wound doctor; wire-wound rod
绕线操作 winding operation
绕线车 cable winding cart; reel; supply reel; swift
绕线磁极式电动机 wound field motor
绕线导架 winding guide
绕线的 wire-wound
绕线电阻器 wire-wound resistor
绕线定位机 winding positioning machine
绕线法 wire wrapping method
绕线工具 wire wrap tool
绕线管 bobbin; reel; winding tube
绕线管高度 bobbin height
绕线管进给器 bobbin feeder
绕线滚筒驱动机械 winding drum machine
绕线过滤器 wire filter
绕线滑轮 winding pulley
绕线机 cable winder; coiling machine; coil winder; coil-winding machine; reeling machine; spooling machine; winder; winding apparatus; winding machine; wire-coiling machine; wire wrap(ping) machine; wrapping machine; merry-go-round equipment < 圆形容器,预应力张拉用 >
绕线机构 winding mechanism
绕线机解线工具 unwrapping tool wire wrap stand
绕线架 drum(spreader); reel; winding former
绕线架进给器 bobbin feeder
绕线间 winding department
绕线节距 winding pitch
绕线连接 wire wrap connection; wrapping wiring
绕线模 winding former
绕线盘 wire reel; wire spool
绕线盘传动装置 reel drive
绕线器 winder
绕线圈 spool(ing)
绕线圈机 coil-winding machine
绕线设备 spooling equipment
绕线式电枢 wound armature
绕线式冷凝器 wire and tube condenser
绕线式滤油器 wire-wound filter; wire-wound strainer
绕线式脉冲传感器 peak strip
绕线式转子电动机 wound-rotor type motor
绕线式转子感应电动机 wound-rotor induction motor
绕线筒 winding drum; winding reel; winding roll
绕线筒进给器 bobbin feeder
绕线图 winding diagram
绕线位移 winding displacement
绕线系数 winding factor
绕线预应力 prestressing by winding
绕线纸板 wire-wound card
绕线轴 cop; spool
绕线轴式控制阀 spool-type control valve
绕线柱 wrapping post
绕线转筒 winding drum
绕线转子 wound rotor
绕行 by-passing; pass-by; passing round; rerouting; rounding; turning movement
绕行标志 detour sign
绕行道 < 环形公路 > by-pass; detour
绕行道路 by-pass road; road diversion
绕行的 circuitous; diverted
绕行地段 detouring section; round section
绕行电路 detour circuit; winding-around circuit
绕行方位群 round of bearings
绕行公路 by-pass highway
绕行管 jump over; return offset
绕行控制 detour control
绕行路 by-pass(route); by-path; detour road
绕行路计划 detour plan
绕行路平面图 < 施工期的 > detour plan
绕行水渠 bye channel
绕行线 alternative trunking; circuitous line; detour
绕行线路 diversion line
绕行小道 byway
绕行左转 by-pass left turn
绕型 < 玻璃钢 > geometric(al) pattern
绕一道 round one turn
绕一定点左右摆动 hunting
绕一圈 round one turn
绕越 passing avoiding; pass round
绕越道路 loop road
绕匝索环 Flemish eye; made eye; selvage eye
绕匝索套 salvage strap; selvage strop
绕在线管上 spool
绕扎 riding seizing; round seizing
绕扎电缆 wrapped cable
绕支点处力矩 moment about point support
绕植株采样 sampling around the plant
绕制线圈 coiling
绕轴旋转 axial rotation; pivoting; revolve around an axis
绕轴转动 circuition
绕住 cling; enwind
绕柱身的线脚条 cymbia
绕转 circumrotation; revolution; revolving; spin
绕转角频率 angular revolution frequency
绕转频率 revolution frequency
绕转周期 revolution period; rotation period
绕转轴 axis of revolution
绕子 < 捆扎用的枝条 > withe
绕组 coil; pirn winding; winding
绕组保护 winding protection
绕组比 ratio of winding
绕组布置 winding arrangement; winding layout

绕组长度 winding length
绕组常数 winding constant
绕组抽头 tapping
绕组出线端 winding terminal
绕组磁漆 coil enamel
绕组单元 winding element; winding section
绕组的匝数 turn of a winding
绕组电流 winding current
绕组电路 winding circuit
绕组电压 winding voltage
绕组端部 winding head; winding over-hang
绕组端部漏抗 end connection reactance
绕组端部区域 end winding region
绕组端部支撑 winding overhang support
绕组对 pair of windings
绕组对地电容 winding-to-earth capacitor
绕组分布 winding distribution
绕组分布系数 winding distribution ratio
绕组功率 winding power
绕组护罩 winding shield
绕组间的 interwinding
绕组间电容 interwinding capacity
绕组间隙 winding space
绕组节距 winding pitch
绕组结构 winding construction
绕组紧密度 closeness of winding
绕组浸渍漆 coil impregnating varnish
绕组静电绝缘 electrostatic winding insulation
绕组绝缘 winding insulation
绕组空间 winding space
绕组利用率 winding availability; winding utilization
绕组模 winding former
绕组清漆 coil varnish
绕组试验器 winding tester
绕组数 number of windings
绕组数据 winding data
绕组损耗 winding loss
绕组套筒 coil sleeve
绕组体积 winding volume
绕组图 winding diagram
绕组温度跳闸 winding temperature trip
绕组温度指示器 winding temperature indicator
绕组涡流损失 winding eddy current loss
绕组系数 breadth coefficient; differential factor; winding coefficient
绕组线 winding wire
绕组因数 winding coefficient; winding factor
绕组用导线 winding material
绕组元件 element of winding; winding element
绕组占空系数 stacking factor; winding space factor
绕组支撑 winding support
绕组支路断线保护 divided-conductor protection

惹丁那阶 < 早泥盆世 >【地】Gedinnian(stage)

惹芙耳溶液 eau-de-Javelle
惹烯 retene
惹卓碱 retrorsine

热爱汽车交通的 auto-crazy

热拔的 hot drawing; hot-drawn
热拔无缝 seamless hot
热白光 warm white light
热柏油炉 tar heater
热柏油涂刷 mopping of hot asphalt
热斑 hottest spot; warm spot; hot spot < 钢锭缺陷 >
热斑模型 hot spot model
热板 heating platen; hot plate
热板控制 hot plate control
热板设备 hot plate apparatus
热板式崩裂性试验 hot plate spalling test
热板式干燥机 hot plate drier
热板压泥机 hot plate press
热半径 thermal radius
热拌 hot mix(ing)
热拌地沥青砂混合料 hot-mixed sand-asphalt mixture
热拌法 hot-mix(ing) method
热拌封层 carpet course; hot-mix seal coat
热拌(工)厂 < 沥青混凝土 > hot-mix plant
热拌和 mixed hot; hot mix(ture)
热拌和厂 hot plant
热拌和料 hot mixture
热拌和铺面材料 hot mix(ture) paving material
热拌混合料 hot-laid mix(ture); hot mix(ture)
热拌混合料铺路 hot-mix construction of road; hot-mixed construction of roads
热拌混合料修复路面法 hot-mix recap method
热拌混凝土 heated concrete; heating concrete; hot-mixed concrete
热拌焦油沥青混凝土 hot tar concrete
热拌冷补的 hot patch; hot repair
热拌冷铺的 hot-mix-cold-laid
热拌冷铺式 < 沥青混合料 > hot-mixed cold-lay type
热拌沥青 hot-mix asphalt
热拌沥青衬砌 hot-mix asphalt lining
热拌沥青底层 hot-mix asphalt base course
热拌沥青骨料厂 hot-mixed asphalt plant
热拌沥青混合料 hot-mixed asphalt mixture
热拌沥青混合料工厂 hot-mix plant
热拌沥青混合料路面 hot-mix surfacing
热拌沥青混合料面层 hot-mix surface [surfacing]; hot-mix topping
热拌沥青混凝土 hot asphalt concrete; hot-mixed asphalt(ic) concrete; hot-mixed bituminous concrete
热拌沥青基层 hot-mix asphalt base course
热拌沥青集料厂 hot asphalt plant
热拌沥青砾石 hot asphalt-coated gravel; hot bitumen-coated gravel
热拌沥青面层 hot-mix asphalt surface course
热拌沥青铺面 hot-mix bituminous pavement
热拌沥青砂 hot asphalt-coated sand; hot bitumen-coated sand
热拌沥青碎石 hot-mixed bituminous macadam
热拌路面 hot-mixed surface
热拌面层 hot-mixed topping
热拌热铺 hot-mixed construction
热拌设备 hot-mixed plant
热拌碎石 hot-mixed macadam
热拌修复路面法 hot-mix(ed) recap method
热拌再生利用 hot-mixed recycling
热棒 hot pin; thermostick
热棒测定法 hot bar method
热棒效应 thermal rod effect
热包工艺 hot foiling
热包裹法 hot pack(ing)
热保持 heat retaining
热保护层 heat-protection layer; thermal protective coating
热保护电动机 thermal guard motor
热保护器 thermal protector
热保护装置 thermel protection device
热保护自动开关 thermosnap
热保温集装箱 thermal container
热保险装置 thermocut-off; thermocut-out
热爆 thermal explosion
热爆浸取法 heat-decrepitation extraction method
热爆裂 thermal spalling
热爆墙 hot wall
热爆炸 thermal explosion
热备用 hot bank; hot standby; standby heat
热备用锅炉 banded fire; banked boiler
热备用系统 hot standby system
热备用制 hot standby system
热倍加器 thermomultiplicator
热泵机组 heat pump unit
热泵式干燥机 heat pump drier[dryer]
热泵式供暖 heat pump heating
热泵式空调 heat pump air conditioning
热泵式空(气)调(节)器 heat pump air conditioner; packaged heat pump
热泵式热水器 heat pump water heater
热泵系统 heat pump system
热泵性能系数 coefficient of performance of heat pump
热泵蒸发器 evaporator with heat pump
热比容偏差 thermosteric anomaly
热比色计 thermocolorimeter
热比重计 thermohydrometer
热壁反应器 hot wall reactor
热壁管式炉 hot wall tube furnace
热壁外延 hot epitaxy; thermal wall epitaxy
热边界层 thermal boundary layer
热变 pyrometamorphism; thermal change
热变电阻 thermal resistance; thermistor
热变电阻流量计 thermistor flowmeter
热变电阻器 heat variable resistor; temperature-dependent resistor; thermal resistor; thermister [thermistor]
热变定 hot set; thermoset
热变定法 thermoset
热变法 pyromorphism
热变幅 bolometric amplitude
热变化 thermal change
热变换 thermal conversion
热变换器 thermal converter [convertor]; thermocouple converter; thermoelectric(al) generator; thermoelectric(al) power generator; thermoelement
热变换器制 thermal converter system
热变漆 chameleon paint
热变色 thermocolo(u)r
热变色现象 thermochromism
热变色(油)漆 heat-sensitive paint
热变涂料 thermoindicator paint
热变位 thermal movement
热变相 pyrometamorphism
热变形 heat collapse; heat deformation; heat distortion; thermal deformation; thermal distortion
热变形点 heat distortion point
热变形模具钢 hot die steel
热变形温度 heat distortion temperature; thermal distortion temperature
热变形学 thermomechanics
热变性 heat denaturation; thermotropy
热变性试验 heat denaturation test
热变性作用 thermodenaturation
热变颜料 chameleon paint; heat-sensitive paint
热变指数 thermal alteration index
热变质成矿作用 thermally metamorphosed metallization
热变质低挥发分蒸汽煤 heat-altered low volatile steam coal
热变质废水 thermally altered seawater
热变质中挥发分煤 heat-altered medium-volatile coal
热变质作用 thermal metamorphism
热标记 heat label
热表 hot list
热表面 hot surface
热表皮探测 heat skin detection
热波 heat wave; thermal wave
热波测流计 < 测污水流量用 > thermal wave flowmeter
热波成像技术 thermal wave imaging technique
热波动 thermal shock
热波段 heat band
热波流量计 < 测污水流量用 > thermal wave flowmeter
热剥 hot soarfing
热剥落 thermal spalling
热补 burning-on; hot patch(ing); hot repair; hot vulcanizing
热补偿 compensation of thermal expansion; heat compensation; thermal compensation
热补偿合金 thermal compensation alloy
热补偿夹套 heat(ing) compensating jacket
热补偿器 bellows joint; thermal compensator
热补给 heat-supply
热补给量 heat influx
热补机 vulcanizer
热补胶水 thermosetting glue
热补轮胎 vulcanization
热补设备 hot patch outfit
热补胎胶 tire [tyre] hot patch; tire [tyre] vulcanizing cement
热补养护 hot patch maintenance
热补用胶水 hot-setting glue
热不利因子 thermal disadvantage factor
热不平衡 thermal unbalance
热不稳定的 heat-labile
热不稳定试验 heat-labile test; heat unstability test; thermolability test
热不稳定物质 thermally unstable substance
热不稳定性 heat instability; thermal instability
热部位 hot spot
热材料 hot material
热材料输送机 conveyor for hot materials
热参数 thermal parameter
热残渣 hot residue
热仓库 heat storage capacity
热藏 hot storage
热槽 heat channel
热槽处理 boulton process
热槽镀锌 hot-dip galvanize
热槽形低压系统 thermal through system

热测法 thermal method
热测绘技术 thermal mapping technique
热测井 thermal log
热测井曲线 curve of heat logging
热测力计 heat dynamometer
热层 hot deck;thermal layer
热插入 hot plugging
热插头 <马达上的> heat plug
热差点滴加油器 thermal drop feed oiler
热差电流 thermocurrent
热差分析 differential thermal analysis
热差值 thermal difference
热产率 heat productivity
热产生 pyretogenesis;thermogenesis
热产生的 pyretogenetic;thermogenetic
热场 thermal field
热超负荷 thermal overload
热超荷载 warm-overloading
热超限 thermal overrun
热潮红 hot flush
热车间 hot workshop
热车熄火 hot stall
热沉 heat sink
热沉层 heat-sink shell
热沉淀 thermoprecipitation
热沉淀法取样 sampling by thermal precipitation
热沉淀器 thermal precipitator
热沉淀硬化 warm hardening
热沉淀原 thermoprecipitinogen
热沉淀载片 <尘末显微检查> thermal precipitation slide
热沉积涂层 pyrolytic coating; pyrolytic plating
热沉降装置 thermal precipitation device
热沉系统 heat-sink system
热沉系统设计 heat-sink(system) design
热沉罩 heat-sink shield
热陈化 thermal ag(e)ing
热成层 thermal stratification;thermosphere
热成层顶 thermopause
热成风 thermal wind
热成风方程 thermal wind equation
热成积云 heat cumulus
热成畸变 thermal distortion
热成急流 thermal jet
热成雷暴雨【气】 warm thunderstorm
热成土 thermogenic soil
热成涡度平流 thermal vorticity advection
热成像 heat imaging; thermal imagery;thermal imaging
热成像器件 thermoimage device
热成像扫描仪 thermal mapping scanner
热成像摄像管 thermal imaging camera tube
热成像系统 thermal imaging system
热成像响应 thermal imaging response
热成像装置 thermal imaging device
热成型 hot forming;hot working; thermoforming
热成型的 hot shaped
热成型的型材 hot-finished shape
热成型(方)法 heat forming process; hot forming process; thermoforming process
热成型工艺 heat forming process;hot forming process; thermoforming process
热成型过程 heat forming process;hot forming process; thermoforming process
热成型机 hot shaper; hot former; thermoforming machine
热成型机械 thermoforming machinery
热成型塑料板 thermoforming plastic sheet
热成型性 hot-forming property
热成因气 thermogenic gas
热成引导 thermal steering
热成云 heat cloud
热成装置 heat imaging device
热弛时间 thermal relaxation time
热弛豫 thermal relaxation
热迟延 thermal retardation
热赤道 heat equator;thermal equator
热冲击 heat shock; hot impact; thermal impact;thermal shock
热冲击裂缝 thermal shock crack(ing)
热冲击屏蔽 thermal shock shield
热冲击破断 thermal shock fracture
热冲击试验 thermal shock test
热冲击试验设备 thermal shock rig
热冲击性 thermal shock
热冲击影响 thermal shock effect
热冲洗 heat flush
热冲压 drop stamping;hot stamping
热冲压箔 hot stamping foils
热重整 thermal reforming
热重整装置 thermal reforming plant
热抽除法 thermal substractive process
热抽提 heat extraction
热稠化 <干性油> heat body
热稠化油 heat-bodied oil
热臭石 pyrosmalite
热除盐法 removal of salt by heating
热储备 hot reserve
热储补给区 feeding area of geothermal reservoir
热储存 <材料的> hot storage
热储的水位高度 high of reservoir level
热储动力模型 dynamic(al) reservoir model
热储工况 reservoir performance
热储机制 reservoir mechanics
热储积 heat accumulation
热储积能力 heat accumulation capacity;heat accumulation power
热储积性能 heat accumulation property
热储建造温度 reservoir formation temperature
热储界面 reservoir interface
热储静水位 reservoir rest level
热储开发过度 reservoir overdraft
热储类型 reservoir type
热储量 heat reserve;heat storage capacity
热储量计算方法 calculated method of heat reserve
热储料斗 hot storage bin
热储流体 reservoir fluid
热储流体补给 reservoir fluid recharge
热储渗透性 reservoir permeability
热储围岩 reservoir country rock
热储温度 reservoir temperature;temperature of reservoir
热储系统 steam reservoir system
热储系统岩石 rock of reservoir system
热储形状 reservoir configuration
热储岩石 reservoir rock
热储贮存能力 retaining capacity of reservoir
热处理 carburizing; case-hardened; heat-strengthened; heat-toughened; heat-treating; slack quench(ing); thermal process(ing); thermal treatment;Kogel process <路面防滑的>
热处理保护涂层 heat-treatment protective coating
热处理保护涂料 heat-treatment protective coating
热处理玻璃 heat-treated glass; tempered glass
热处理层 heat-treatment layer
热处理车间 heat-treating department; heat-treatment department; heat-treatment shop
热处理尺寸变化率 size-changing rate in heat-treatment
热处理的 heat-strengthened
热处理的合金钢 heat-treated alloy steel
热处理钉 heat-treated nail
热处理法 heat-treatment method
热处理废物 heat-treating waste
热处理辅助硬质纤维板 tempered service hardboard
热处理钢 heat-treatable steel; heat-treated steel;case steel
热处理钢轨 heat-treated rail
热处理钢筋 heat-treated steel bar
热处理钢履带 heat-treated steel track
热处理工 temperer
热处理工艺曲线 heat-treatment cycle curve
热处理规范 heat-treatment regime(n);specification of heat-treatment
热处理过程 heat-treatment process
热处理过的 heat-treated
热处理过的表面 heat-treated surface
热处理过的高碳钢 high-carbon steel heat treated
热处理过烧 burning
热处理合金钢 heat treated alloy
热处理和冷却费 heating and cooling cost
热处理机 heat-treating machine
热处理技术 heat-treatment technics
热处理建筑合金钢 heat-treated construction alloy steel
热处理金属 heat-treated metal
热处理金属箔 heat-treated metal foil
热处理净化 purification by heat-treatment
热处理绝热炉 insulated furnace
热处理裂纹 heat-treatment crack
热处理鳞状结构 mill scale
热处理炉 furnace for heat-treatment; heat-treating furnace; heat-treatment furnace;treating oven
热处理铝合金 heat-treatable alumin(i)um alloy
热处理履带瓦 heat-treated track shoe
热处理履带销 heat-treated track pin
热处理率 heated treatment rate
热处理铆钉 burnt rivet
热处理膜 heating treating film
热处理平板玻璃 heat-treated flat glass
热处理区 heat-treatment zone
热处理设备 heat-treating facility; heat-treatment fixture
热处理施行 heat-treatment practice
热处理时间 heat treatment duration
热处理实验室 thermal treatment laboratory
热处理试验室 heat treatment laboratory
热处理条件 heat-treating condition; heat-treatment condition
热处理温度 heat-treatment temperature
热处理消除应力 stress-relieving by beat treatment
热处理效果 thermal effectiveness
热处理性 heat-treatability
热处理验证试验 heat-treatment verification test
热处理氧化膜 heat-treating oxidation film
热处理液 heat-treatment liquor
热处理硬质纤维板 tempered hardboard
热处理用油 heat-treating oil; heat-treatment oil
热处理鱼尾板 heat-treated joint bar
热处理辙叉 heat-treated frog
热处理制度 heat-treating regime
热处理状态 condition of heat-treatment; heat-treated condition; heat-treatment condition
热处理资料 information of heat-treatment
热处理组织 heat-treated structure
热处理作业线 heat-treatment line
热处治 hot cure
热触点 thermal contact
热穿孔 hot piercing
热穿透 thermal break-through
热穿透窗断面 window section with thermal break
热传播 propagation of heat
热传导 calorific conduction; conduction of heat; egress of heat; heat conduction; heat transfer; heat-transference; heat transmission; thermal conductance; thermal conduction; thermal transmission; heat conductance
热传导比测器 thermal comparator
热传导表面系数 surface coefficient of heat transmission
热传导测定器 thermal conductivity detector
热传导的 heat-conducting
热传导阀计算 computing by heat transfer
热传导方程 equation of heat conduction;heat conduction equation;heat conductivity equation;heat-transfer equation
热传导金属 heat-transfer metal
热传导警报器 conductivity alarm
热传导理论 heat-transfer theory
热传导量 capacity of heat transmission
热传导流 heat-conducting flow
热传导率 coefficient of heat transmission;thermal conductivity;thermoconductivity
热传导模拟 heat conduction analogy
热传导能力 capacity of heat transmission
热传导式探测器 thermal conductivity detector
热传导试验 thermal transmittance test
热传导体 heat conductor
热传导问题 heat conduction problem
热传导系数 coefficient of heat conductivity;coefficient of heat transfer; coefficient of heat transmission;coefficient of thermal conductivity;coefficient of thermal transmission;thermal conductivity coefficient
热传导性 heat conductivity; thermal conductivity
热传导压力表 heat conductivity pressure ga(u)ge
热传导盐 heat-transfer salt
热传导用钢管 steel tubes for heat transfer
热传导与表面摩擦比 ratio of heat transfer to skin friction
热传导真空计 thermal conductivity vacuum ga(u)ge
热传导值 thermal transmittance
热传递 heat passage; heat transfer; heat transmittance; thermal transfer;thermal transformation;thermal transmission; thermal transmittance;transmission of heat

R

热传递公式 thermal transport formula
热传递计算机 heat-transfer computer
热传递率 heat transfer
热传递器 heat transmitter
热传递速率 rate of heat transfer
热传递系数 thermal transmissivity
热传递性质 thermal transport property
热传递装置 heat-transfer arrangement
热传感器 heat sensor; thermal sensor
热传入 heat penetration
热传输 heat transfer
热传输流体 heat transport(ing) fluid
热传输系数 coefficient of heat transfer; heat-transfer factor
热传输系统 heat transport system
热传送 heat convection; heat transfer
热传元件 heat-transfer element
热窗 <戴克里先公共浴场的> thermal window
热床 cooling table; hot bed
热吹 hot blow
热吹风干手器 blown warm air hand drier[dryer]
热锤击 hot peening
热锤击尺寸整形 thermosizing
热锤钻孔 heat hammer drill
热纯化点 thermal inactivation point
热纯化温度 thermal inactivation temperature
热唇分子束炉 hot lip K-cell
热磁 pyromagnetic; thermomagnet
热磁材料 thermal magnetizing material
热磁测定法 thermomagnetometry
热磁处理 thermomagnetic treatment
热磁的 thermomagnetic
热磁电动机 thermomagnetic motor
热磁发电机 pyromagnetic generator; thermomagnetic generator
热磁合金 calmalloy; thermomagnetic alloy
热磁化 thermomagnetization
热磁换能器 thermomagnetic converter; thermomagnetic transducer
热磁开关 thermal magnetic breaker
热磁控超高真空计 hot cathode type magnetron ga(u)ge
热磁流体 hydrothermomagnetic
热磁流体波 hydrothermomagnetic wave
热磁探测 pyromagnetic detection
热磁探测器 pyromagnetic detector
热磁物质 pyromagnetic substance
热磁现象 thermomagnetic phenomenon; thermomagnetism
热磁效应 pyromagnetic effect; thermomagnetic effect
热磁效应的 thermomagnetic
热磁性 pyromagnetism; thermomagnetic property; thermomagnetism
热磁选矿法 thermomagnetic preparation
热磁学 thermomagnetism
热磁学法 thermomagnetometry
热磁氧分析仪 thermomagnetic oxygen analyser[analyzer]
热磁振子 thermal magnon
热磁转换 thermomagnetic conversion
热刺激电导率 thermally stimulated conductivity
热刺激电流 thermally stimulated current
热催化 thermocatalysis
热催化活性镍合金片 Raney nickel film
热催化检测器 thermocatalytic detector
热催化净化作用 thermocatalytic purification
热催化聚合 thermocatalytic polymerization
热催化排气装置 thermocatalytic exhaust device
热催化作用 heat catalysis
热脆 hot brittlement; hot quenching; hot short
热脆材料 hot brittle material; hot short material
热脆的 hot brittle; nesh; short red brittle; hot short; red short; short-brittle
热脆钢 hot brittle iron; red-short steel
热脆化 heat embrittlement
热脆铁 hot short iron; red-short iron
热脆性 heat shortness; hot brittleness; red brittleness; red shortness; short-brittleness; hot embrittlement; hot shortness
热脆性的 hot short
热淬(火) hot quenching; thermal quenching
热打铁 hot peening
热打箱 hot shake-out
热大气层 thermosphere
热大气层顶部 thermopause
热带 burning zone; burnt zone; hot zone; intemperate zone; torrid zone; tropic(al) belt; tropic(al) zone
热带包装 tropic(al) pack
热带包装的 tropically packed
热带暴露试验 tropic(al) exposure testing
热带闭合环流 tropic(al) closed circulation
热带飑 breather
热带病 tropic(al) disease
热带材 south sea timber; tropic(al) timber; tropic(al) wood
热带草生灌丛 thorny scrub
热带草原 savanna(h); tropic(al) grassland
热带草原气候 savanna(h) climate; tropic(al) grasslands climate; tropic(al) savanna(h) climate; tropic(al) wet and dry climate
热带(产)木材 tropic(al) timber
热带潮 tropic(al) tide
热带潮汐 tropic(al) tide
热带大草原 savanna(h)
热带大草原气候 tropic(al) savanna(h) climate
热带大陆空气 tropic(al) continental air
热带大陆气团 continental tropic(al) air; tropic(al) continental air; tropic(al) continental air-mass
热带大西洋过渡气团 transitional tropic(al) Atlantic air-mass
热带大西洋气团 tropic(al) Atlantic air-mass
热带大雨林 ceja
热带淡水满载吃水线 tropic(al) freshwater load line
热带淡水满载吃水线标志 tropic(al) freshwater load line mark
热带淡水载重线 tropic(al) freshwater load line
热带淡水载重线标志 tropic(al) freshwater load line mark
热带的 tropic(al)
热带低(气)压 tropic(al) depression
热带地方菌 tropicopolitan
热带地区 torrid area; tropic(al) region; tropic(al) zone; tropics
热带地区多潮湿 usually humid in tropic(al) area
热带东风带 subtropic(al) easterlies; tropic(al) easterlies
热带动物 tropic(al) animal
热带对流层顶 tropic(al) tropopause
热带多雨气候 tropic(al) rain(y) climate
热带伐木工舍 bangalow
热带反气旋 tropic(al) anti-cyclone
热带非洲紫檀树脂 kinoin
热带风暴 revolving storm; tropic(al) revolving storm; tropic(al) storm
热带风暴中心 storm vortex
热带风观测系统 tropic(al) wind observation system
热带风景 tropic(al) cyclone
热带锋 equatorial front; intertropic(al) front; tropic(al) front
热带服装 tropic(al) wear
热带辐合带 international convergence; intertropic(al) convergence zone; tropic(al) convergence
热带辐合区 equatorial convergence zone; international convergence; intertropic(al) convergence zone
热带腐殖黑黏[粘]土 grum(m) usol
热带腹泻 tropic(al) diarrhea
热带钙层土 tropic(al) pedocalic soil
热带干舷 tropic(al) freeboard
热带高草草原 high grass savanna
热带高地气候 climate of tropic(al) table-land
热带高空气团 tropic(al) superior air-mass
热带高密草原 tropic(al) grass
热带高山冻原 tropic(al) alpine tundra
热带高山灌丛 tropic(al) alpine shrub
热带高山植物 oreithalion tropicum
热带高压 tropic(al) anti-cyclone
热带果树栽培 tropic(al) fruit culture
热带过渡高空气团 transitional tropic-(al) superior air-mass
热带海区 tropic(al) seas; tropic(al) zone
热带海洋 tropic(al) ocean
热带海洋过渡气团 transitional tropic-(al) maritime air-mass
热带海洋空气 tropic(al) marine air
热带海洋气候 tropic(al) marine climate; tropic(al) ocean climate
热带海洋气团 maritime tropic(al) air(mass); tropic(al) maritime air(-mass)
热带海域 tropic(al) seas
热带旱地水稻土 tropic(al) upland rice soils
热带旱生灌木丛 thorny scroll
热带旱生林 thorn forest
热带旱树林 thorn forest
热带和温带的干旱地区 arid regions in tropic(al) and temperate zones
热带和亚热带辐合区 inter-and-subtropic(al) convergence zone
热带河流 tropic(al) river
热带黑黏[粘]土 tropic(al) black clay
热带黑色土 tropic(al) black soil
热带黑土 tropic(al) black earth
热带红壤土 tropic(al) red loam
热带红土 tropic(al) red earth
热带湖泊 tropic(al) lake
热带化的 tropicalized
热带环境 torrid environment; tropic(al) environment
热带环流 tropic(al) cell
热带荒漠 tropic(al) desert
热带汇流区 intertropic(al) confluence zone
热带棘林 thorn forest
热带季风 tropic(al) monsoon
热带季风林 tropic(al) monsoon forest
热带季风气候 tropic(al) monsoon climate
热带飓风 tropic(al) hurricane
热带聚集区 intertropic(al) convergence zone
热带卷运输机 hot coil conveyer[conveyor]
热带绝缘 tropic(al) insulation
热带空气 tropic(al) air
热带冷却器 tropic(al) radiator
热带林带 tropic(al) forest zone
热带林地 tropic(al) woodland
热带满载吃水线 tropic(al) load line
热带密灌丛 brush; bush; fourre; mallee; thicket; tropic(al) scrub
热带墨西哥湾过渡气团 transitional tropic(al) Gulf air-mass
热带墨西哥湾气团 tropic(al) Gulf air-mass
热带木材 tropic(al) wood
热带漂白粉 tropic(al) bleach
热带漂白土 tropic(al) bleached soil
热带曝晒试验 tropic(al) exposure testing
热带漆 tropic(al) finish
热带气候 tropic(al) climate; tropic(al) weather
热带气候处理 tropicalize
热带气候试验 hot-weather trial
热带气候适应性试验 tropicalization test
热带气候学 tropic(al) climatology
热带气团 tropic(al) air-mass
热带气团雾 tropic(al) air-mass fog
热带气象学 tropic(al) meteorology
热带气旋 tropic(al) cyclone; tropic(al) hurricane
热带气旋登陆地 tropic(al) cyclone landfall
热带气旋计划 tropic(al) cyclone program(me)
热带气旋中心 boiling pot; center[centre] of tropic(al) cyclone
热带浅水 tropic(al) shallow water
热带区带 tropic(al) area; tropic(al) zone
热带扰动 tropic(al) disturbance
热带热病 calenture
热带容许吃水 tropic(al) allowance draught
热带伞伐作业 tropic(al) shelter wood system
热带森林 tropic(al) forest
热带森林生态系(统) tropic(al) forest ecosystem
热带森林土 tropic(al) forest soil
热带森林行动计划 tropic(al) forestry action plan; tropic(al) forestry action program(me)
热带沙漠气候 tropic(al) desert climate
热带山区 mountainous regions of tropics
热带山岳林地 mountain forest terrain of torrid zone
热带生态系统 tropic(al) ecosystem
热带生态学 tropic(al) ecology
热带生物带 tropic(al) life zone
热带使用 tropic(al) use
热带始成土 tropept
热带试验 tropicalization test; tropic(al) trial
热带适应性 tropicalization
热带疏林 park land; savanna(h) woodland; tropic(al) savanna(h); tropic(al) woodland
热带树木 tropic(al) tree
热带水 tropic(al) water
热带水果 tropic(al) fruit(tree)
热带水域 tropic(al) waters
热带水族馆 tropic(al) aquarium

热带太平洋过渡气团 transitional tropic(al) Pacific air-mass
热带太平洋气团 tropic(al) Pacific air-mass
热带铁铝土 tropic(al) ferriallitic soil
热带土 tropic(al) earth; tropic(al) soil
热带外地区 extra-tropic(al) belt
热带温度层结 tropic(al) temperature lamination
热带涡漩 tropic(al) vortex
热带无风带【气】 tropic(al) calm zone
热带稀树草原 savanna(h)
热带稀树草原气候 tropic(al) savanna(h) climate
热带稀树干森林群落 tropodrymium
热带型驾驶舱 tropic(al)-type driver's cab(in)
热带性 tropicality
热带性气候 tropic(al) climate
热带旋风 tropic(al) cyclone
热带旋转风暴 tropic(al) revolving storm
热带亚高山森林 tropic(al) subalpine forest
热带亚热带亚带 tropic(al) to subtropic(al) subzone
热带岩溶 tropic(al) karst
热带研究组织 Organization for Tropic(al) Studies
热带医院 tropic(al) hospital
热带以外的 extra-tropic(al)
热带硬木 bitch wood; red mercanti; tropic(al) hardwood
热带用开关 tropic(al) switch
热带用途 tropic(al) use
热带鱼 tropic(al) fish
热带鱼类 tropic(al) fishes
热带雨 tropic(al) rain
热带雨林 equatorial rain forest; euhylacion; hileia; hylaea; hylea; rain forest; selva; tropic(al) rain forest
热带雨林带 megathermal zone
热带雨林气候 megathermal climate; tropic(al) rain forest climate; tropic(al) wet climate
热带雨气候 tropic(al) rain climate
热带云团 tropic(al) cloud cluster
热带云雾林 tropic(al) cloud forest
热带载重线 tropic(al) load line; tropic(al) water line
热带载重线标志 tropic(al) load line mark; load line in tropical zones
热带藻 orchella
热带植丛 jungle
热带植物 megatherm; tropic(al) plant; tropic(al) vegetation
热带植物地理区系 tropic(al) floral realm
热带植物群落 macrothermophytia
热带植物温室 palm house
热带植物亚区 tropic(al) floristic subregion
热带植物园 tropic(al) plants garden
热带装备 tropic(al) outfit
热带自动气象站 tropic(al) automatic weather station
热带棕色土 tropic(al) brown soil
热带作物 tropic(al) crop
热单板再干机 hot plate veneer redrier
热单对流混合湖 warm monomictic lake
热单位 heat(ing) unit; hot cell; thermal unit
热氮氧化物 thermal nitrogen oxides; thermal NOx
热当量 caloric value; calorific value; calorimetric value; combustion value; equivalent heat; equivalent of heat; heat equivalent; heat(ing) value; heat of combustion; thermal equivalent; thermal value; thermic equivalent
热当量光谱仪 thermal equivalent spectrum level
热当量值 thermal equivalent value
热导 heat-transfer conductance; thermal conductance; thermal conductivity
热导测定 thermal rating
热导池 thermal conductivity cell
热导池检测器 thermal conductivity cell detector
热导的 diathermal
热导法 thermal conductivity method
热导分析器 thermal conductivity analyser[analyzer]
热导管 thermal pipe; duct heater; heat pipe <一端靠蒸发液体吸收热量,另一端由冷凝蒸汽发出热量>
热导计 catharometer; conductometer; katharometer
热导检测器 katharometer; thermal conductivity detector
热导流 thermally induced flow
热导率 coefficient of thermal conductivity; conductivity; heat conductivity; heat diffusivity; temperature conductivity; thermal conductance
热导率测量方法 method of heat conductivity measurement
热导率测量术 katharometry
热导率桥 thermal conductivity bridge
热导模件 thermal conduction module
热导纳 thermal admittance
热导式气体分析器 thermal conductivity gas analyser[analyzer]
热导式气体分析仪 thermal conductivity gas analyser[analyzer]
热导式真空规 heat conductivity vacuum ga(u)ge
热导体 thermal conductor
热导系数 heat conductivity factor; thermometric conductivity; thermal conductivity
热导现象 heat conduction
热导性系数 coefficient of thermal conductivity
热导引头 heat seeker
热导真空规 thermal conductivity ga(u)ge
热导真空计 thermal conductivity ga(u)ge
热导值 conductivity
热岛 heat island; thermal island
热岛现象 <城市的> heat island
热岛效应 heat island effect; hot-island effect; thermal island effect
热捣法 heat vibration method
热道 heat passage
热的 austral; calorific; hot; thermal; thermic
热的补充 provision of heat
热的产生 development of heat; heat production
热的厂拌 <沥青材料> hot-plant mixing
热的厂拌法 hot-plant mixing method
热的储存 <材料的> heat storage capacity
热的传导 conduction of heat
热的传递 heat transmission
热的地带 hot belt
热的对流 convection of heat
热的发射 giving up the heat
热的反常 thermal anomaly
热的放出 evolution of heat
热的辐射 radiation of heat
热的感受性 susceptibility to heat
热的供给 provision of heat
热的刮平 hot screed
热的回收 recovery of heat
热的回收利用 pick-up the heat
热的机械当量 conversion constant; mechanical equivalent of heat
热的交换 heat transform
热的绝缘 heat insulation; insulation against heat; thermal insulation
热的可透度 permeability to heat
热的控制 thermal control
热的利用 utilization of heat
热的抹平 hot screed
热的耐受性 susceptibility to heat
热的排出 heat abstraction
热的排除 heat abstraction
热的强度 intensity of heat
热的散失 heat abstraction
热的适应 thermal acclimation
热的消散 dissipation of heat
热的有效性 heat availability
热的轴承油脂 hot bearing grease
热的转移 heat transfer
热灯 thermolamp
热灯浆 hot breakdown
热灯丝 glow wire; hot filament
热灯丝电离真空计 hot filament ionization ga(u)ge
热等静压 high-temperature insostatic pressing; hot isopressing; hot isostatic compaction; hot isostatic pressure
热等静压处理 hot isostatic pressing treatment
热等静压机 hot isostatic press(ing)
热等静压黏[粘]结 hot isostatic bonding
热等静压设备 hot isostatic apparatus
热等静压制法 hot-isostatic-bonding and pressing process
热等梯度 thermoisogradient
热等压成型 hot isostatic compaction
热低压【气】 heat low; thermal low
热堤 heat dam
热滴定法 pyrotitration; thermometric titration
热滴定分析 pyrotitration analysis
热滴胶质试验 hot drip gum test
热滴滤器 heat-trickling filter
热地 thermal ground; thermal land
热地板干燥器 hot floor drier[dryer]
热地沥青灌注 hot asphalt bitumen impregnation; hot asphalt impregnation
热点 hot point; hot spot(ting)
热点火塞 hot plug
热点火系统 thermal ignition system
热点式进气歧管 hot spot manifold
热点温度 hot spot temperature
热电 pyroelectricity; thermal electricity; thermal power
热电比测器 thermoelectric(al) comparator
热电比较器 thermoelectric(al) comparator
热电变换 thermoelectric(al) inversion
热电变换器 thermoelectric(al) converter
热电表 electrothermal meter
热电材料 pyroelectric(al) material; thermoelectric(al) material
热电测量 thermoelectric(al) measurement
热电常数 pyroelectric(al) constant
热电厂 cogeneration power plant; heating and power plant; steam power plant; steam power station; steam supply and power generation; steam supply and power generating plant; thermal generating station; thermal generation station; thermal power plant; thermoelectric(al) plant; thermoelectric(al) power plant; thermoelectric(al) power station
热电厂供热系统 heat-supply system based upon heating power cogeneration plant
热电成像系统 pyroelectric(al) imaging system
热电池 thermocell
热电池组 thermobattery
热电充电 pyroelectric(al) charging
热电传感器 thermal electric(al) transducer; thermoelectric(al) sensor
热电传感元件 thermoelectric(al) sensing element
热电磁泵 thermoelectromagnetic pump
热电磁的 thermoelectromagnetic
热电次序 thermoelectric(al) series
热电当量 electric(al) equivalent of calorie; thermoelectric(al) equivalent
热电挡板 thermoelectric(al) baffle
热电导 pyroconductivity; thermal conductance
热电导线 thermoelectric(al) wire
热电导性 pyroconductivity
热电的 pyroelectric(al); thermoelectric(al)
热电电极 thermoelectrode
热电电流计 thermoelectric(al) galvanometer
热电动的 thermoelectromotive
热电动力 thermal electromotive force
热电动力系统 thermoelectric(al) power system
热电动势 thermal electromotive potential; thermoelectromotive force; thermoelectromotive potential
热电动势法钢材材质检定仪 identometer
热电镀 hot galvanizing
热电堆 thermoelectric(al) pile; thermopile
热电堆发电机 thermopile generator
热电发电机 thermoelectric(al) generator
热电发热器 thermoelectric(al) heating device
热电发射 hot-electron emission; thermoelectronic emission
热电放射 thermionic emission
热电放射效应 Edison effect; thermionic emission effect
热电非均匀效应 thermoelectrically heterogeneous effect
热电腐蚀 thermogalvanic corrosion
热电高温表 thermoelectric(al) pyrometer
热电高温计 thermocouple pyrometer; thermoelectric(al) pyrometer
热电功率 thermoelectric(al) power
热电供暖 thermoelectric(al) heating
热电光度计 thermoelectric(al) photometer
热电化学 thermoelectrochemistry
热电机车 thermal electric(al) locomotive
热电极 hot electrode; thermode
热电计 thermoelectrometer
热电继电器 pyroelectric(al) relay
热电加热 thermoelectric(al) heating
热电检波器 thermodetector
热电交换器 thermal current converter; thermoelectric(al) converter
热电晶体 pyroelectric(al) crystal; thermoelectric(al) crystal

热电警报器 thermoelectric(al) alarm
热电空调器 thermoelectric(al) air conditioner
热电冷却器 thermoelectric(al) cooler
热电离 thermoionization
热电离层 <地球上空约 80 千米以上，一直伸展至外太空的大气层> thermosphere
热电离能 thermal ionization energy
热电离源 thermal ionization sources
热电离质谱计 thermal ionization mass spectrometer
热电离质谱仪 thermal ionization mass spectrometer
热电联产 cogeneration; heat and power cogeneration
热电联供 combined heat and power
热电联合生产 pyroelectric(al) production
热电联合中心 heating and power producing centre
热电量计 electrocalorimeter
热电流 heating current; thermocurrent; thermoelectric(al) current
热电流表 thermoammeter
热电流计 thermoammeter
热电能 thermoelectric(al) power
热电偶 electric (al) thermocouple; flame couple; heat electric(al) couple; pyrometer couple; temperature plug; thermal converter; thermal couple; thermal electric(al) couple; thermal element; thermocouple; thermoelectric(al) couple; thermoelement; thermopair; thermopile; pyod; couple
热电偶安培计 thermoammeter; electrothermal ammeter
热电偶安装 thermocouple attachment
热电偶保护管 protecting tube; thermocouple protection tube
热电偶比较器 thermocouple comparator
热电偶材料 thermocouple material
热电偶参考表 reference table for thermocouples
热电偶测量 thermocouple measurement
热电偶测温法 thermocouple thermometry
热电偶传感器 thermocouple sensor
热电偶传声器 thermocouple microphone
热电偶导线保护 thermocouple wire protection
热电偶的冷端 cold junction
热电偶的热端 hot junction
热电偶电表 thermocouple meter
热电偶电池组 thermojunction battery
热电偶电流计 thermogalvanometer
热电偶定温火灾探测器 fixed temperature detector using thermocouple
热电偶发电器 thermocouple generator
热电偶辐射计 thermoelectric(al) radiometer
热电偶高温计 thermocouple pyrometer; thermoelectric(al) pyrometer
热电偶功率计 thermocouple wattmeter
热电偶管 thermocouple tube; thermocouple well
热电偶海流计 thermocouple current meter
热电偶合金 thermocouple metal
热电偶护管 secondary protection tube; thermocouple protection tube
热电偶恢复器 thermocouple restorer
热电偶计 thermocouple needle; heat meter
热电偶继电器 thermocouple relay; thermoswitch
热电偶检测器 thermocouple detector
热电偶检流计 thermogalvanometer
热电偶接点 thermal cross; thermocouple junction; thermojunction; thermal crossing
热电偶接合 thermal crossing; thermal junction
热电偶接头 thermocouple junction; thermoelectric (al) junction; thermojunction
热电偶结 thermocouple junction
热电偶孔道 thermocouple well
热电偶冷端 cold end
热电偶冷端补偿 cold junction compensation
热电偶冷端温度补偿 cold junction compensation
热电偶冷接点 cold end
热电偶联结珠 bead thermister[thermistor]
热电偶流速仪 thermocouple current meter; thermocurrent meter
热电偶模拟器 thermocouple simulator
热电偶气压计 thermocouple ga(u)ge
热电偶热端 pyrometer fire-end
热电偶热接点 thermojunction
热电偶塞 thermocouple plug
热电偶湿度计 thermistor psychrometer
热电偶式 thermojunction type
热电偶式安培计 thermocouple ammeter; thermojunction ammeter
热电偶式测量计 thermojunction type meter
热电偶式测试仪器 thermocouple instrument
热电偶式电表 thermocouple meter
热电偶式真空计 thermocouple(type) vacuum ga(u)ge
热电偶水冷器 thermoelectric(al) water cooler
热电偶探头 thermocouple probe
热电偶套管 thermocouple sheath; thermocouple tube; thermowell
热电偶瓦特计 thermocouple wattmeter
热电偶温度计 electropyrometer; electrothermometer; thermocouple thermometer
热电偶线 thermocouple wire
热电偶仪表 thermocouple ga(u)ge; thermocouple instrument; thermocouple meter
热电偶元件 thermocouple element
热电偶圆筒加热器 thermocouple cartridge heater
热电偶针 thermocouple needle
热电偶砖 pyrometer block; thermocouple block
热电破碎 thermoelectric(al) comminution
热电器件 thermoelectric(al) device
热电牵引 thermoelectric(al) traction
热电桥 <测无线电干扰用> thermal bridge
热电日射表 thermoelectric(al) actinography; thermoelectric (al) actinometer
热电容 thermal capacitance
热电容量 thermal capacitivity
热电色度计 thermoelectric(al) colo(u)rimeter
热电生产 thermoelectricity generation
热电拾取器 thermoelectric(al) pick-up
热电式 thermoelectric(al) type
热电式传感器 thermoelectric(al) sensor; thermoelectric(al) transducer
热电式发动机 thermoelectric (al) generator
热电式感温元件 thermoelectric(al) cell
热电式光导摄像管 pyroelectric(al) vidicon
热电式空(气)调(节)系统 thermoelectric(al) air conditioning system
热电式空(气)调(节)装置 thermoelectric(al) air conditioning plant; thermoelectric(al) air conditioner
热电式示波器 heat writing oscillograph
热电式仪表 thermal meter; thermal type meter; thermoelectric (al) instrument
热电式制冷系统 thermoelectric(al) refrigeration system
热电势 thermoelectric(al) potential
热电势法 thermoelectric(al) method
热电太阳能电池 thermoelectric(al) solar cell
热电探测器 thermal detector
热电陶瓷 thermoelectric(al) ceramics
热电特性表 thermoelectric(al) series
热电体 pyroelectrics
热电铁氧化体 pyroferrite
热电图 thermoelectric(al) diagram
热电微型组件 thermomodule
热电温度计 potentiometer pyrometer; thermoelectric (al) thermometer; thermel <装有热电偶的>
热电物质 pyroelectricity; thermoelectricity
热电现象 pyroelectricity; thermoelectricity
热电效应 pyroelectric (al) effect; thermoelectric(al) effect
热电效应仪表 thermal instrument
热电性 pyroelectricity; thermoelectricity
热电性能 thermoelectric(al) property
热电序 thermoelectric(al) series
热电学 pyroelectricity; thermoelectricity
热电学法 thermoelectrometry
热电循环 thermoelectric(al) cycle
热电循环冷却水 thermoelectric(al) circulating cooling water
热电压 thermal voltage
热电压计 thermovoltmeter
热电阴极 thermal cathode; thermionic cathode
热电元件 thermoelectric (al) cell; thermoelectric (al) element; thermoelement; Le Chatelier thermocouple
热电站 heat and power station; heating and power center[centre]; heat(ing) power plant; heat power station; steam electric (al) generating station; steam supply and power generation plant; steam supply and power generating plant; thermal electric(al) (power) plant; thermal electric(al) station; thermal power plant; thermal power station; thermal project; thermoelectric (al) power station; thermoelectric (al) station
热电站废水 wastewater of heating and power stations
热电站清洗 steam electric (al) plant cleaning
热电制冷 thermoelectric(al) cooling; thermoelectric(al) refrigeration
热电制冷器 thermoelectric (al) cooler; thermoelectric(al) cooling module
热电转换 thermoelectric (al) conversion
热电转换器 thermoelectric (al) converter
热电装置 thermoelectric(al) device
热电子 thermoelectron
热电子发射 thermal electron emission; thermoionic emission
热电子发射体 hot electron emitter
热电子管 thermionic tube(valve)
热电子阴极 thermionic cathode
热电子蒸发速率测量器件 heated electron evapo(u)ration rate sensing device
热电阻 hot resistance; resistance to heat
热电阻焊 electric(al) resistance weld(ing)
热电阻丝 hot wire
热电阻丝型分析仪 hot-wire analyser [analyzer]
热电阻温度计 resistance temperature detector
热电阻线 hot resistance wire
热电阻线风速仪 hot-wire anemometer
热电作用 thermoelectric(al) action
热垫层 thermocushion(ing)
热雕全息图 thermally engraved hologram
热雕全息照片 thermally engraved hologram
热钉 thermal spike
热顶 hot top
热顶棚 heated ceiling
热定标 thermostaking
热定额 thermal rating
热定径 hot size
热定形 heat set(ting); hot sate; thermohardening
热定形机 heat setter; heat-setting machine
热定形挤塑法 thermoset extrusion
热定形黏[粘]合 heat-setting bonding
热定形纱 thermosetting yarn
热定形性 heat settability
热锭炉 ingot heat furnace
热动的 heat-actuated; thermodynamic(al)
热动浮漂式疏水器 float thermostatic trap
热动关闭器 thermocut-off
热动机 pyromotor
热动继电器 temperature relay; thermal relay
热动接点 heat-operated contact; thermal contact
热动开关 heat-actuated switch; thermal switch; thermostat
热动力 heat power; thermal power
热动力变质 dynamothermal metamorphism; thermodynamic(al) metamorphism
热动力的 thermokinetic; thermomotive
热动力关系 thermodynamic(al) relation
热动力平衡 thermodynamic(al) equilibrium
热动力式疏水器 steam trap, thermodynamic type; thermodynamic (al) (type steam) trap
热动力势 thermodynamic(al) potential
热动力疏水器 thermodynamic (al) trap
热动力梯度 dynamothermal gradient
热动力条件 thermodynamic(al) condition
热动力稳定性条件 thermodynamic(al) stability condition
热动力系数 kinetic thermal coeffi-

cient
热动力学 thermodynamics;thermokinetics
热动力学分析 thermokinetic analysis
热动力学模型 thermodynamic(al) model
热动力循环 thermodynamic(al) cycle;heat cycle
热动力循环效率 thermodynamic(al) cycle efficiency
热动力装置循环 heat cycle
热动式 thermal type
热动式电流计 thermal type galvanometer
热动式调节阀 thermal valve
热动式仪表 thermal type meter
热动态 heat regime
热动装置 thermal device
热洞 hot cave
热斗 hot aggregate conveyer[conveyor]
热陡度 thermal gradient
热度 degree of heat;heat;heat degree;hotness
热度计 heat counter
热度探测器 heat detector
热镀 heat plating;hot-dip(coating)
热镀铝 calorize
热镀铝法 Aldip process;alumin(i)um coating
热镀铝钢板 aludip
热镀模具钢 hot work tool steel
热镀锡 hot-dip tinning;hot tinning
热镀锡薄钢板 hot-tinned strip
热镀锡车间 hot-dip tinning plant
热镀锡装置 hot-dip tinning stack
热镀锌 hot-galvanize;hot galvanizing
热镀锌法 hot galvanizing
热镀锌钢丝 hot-dip galvanized steel wire
热镀锌混凝土铺路板 footpath concrete paving flag
热端 fire end;hot end;hot junction;hot side
热端操作 hot end operation
热端镀膜 hot end coating
热端灰尘 hot end dust
热端灰泡 hot end dust
热端扩大型窑 kiln with enlarged discharge end
热端周缘密封 hot end circumferential seals
热短路 thermal short circuit
热短时额定电流 thermal shock time current rating
热段 hot arc
热断锤 hot chisel
热断开 thermal cutoff
热断开关 hot override switch
热断流器 thermal cutout;thermocutout
热断路 thermal trip
热断路装置 thermal trip
热锻 forge hot;heating training;hot forging
热锻分级淬火 hot peening marquenching
热锻件图 heat forging drawing
热锻模钢 hot die steel;hot-working die steel
热锻模具钢 hot work tool steel
热锻钎头锻钎机 jackmill for hot-milling jackbits
热锻压 hot shaping
热锻压机 hot-press forge
热锻用具 hot set
热堆 thermal reactor
热堆积 hot stack
热堆快中子转换器 doughnut;dount
热对称设计 thermally symmetrical design
热对称性 thermal symmetry
热对流 free convection;gravitational convection;heat convection;thermal convection;thermal conversion;thermoconvection
热对流暴(风)雨 thermal convection rainstorm
热对流风暴 thermal convection storm
热对流海流 thermal convective current
热对流湖流 thermal convective current
热对流气流 thermal convective current
热对流系数 coefficient of heat convection;heat convection coefficient
热对流循环法 thermosiphon circulation
热对流原理 thermosiphon theorem
热对流值 heat mixed value
热镦头 hot heading
热钝化 thermal inactivation
热剁下料 hot cropping
热惰性 thermal inertia
热惰性指标 index of thermal inertia;inertial index of heat;thermal inertia factor
热额定电流 thermal current rating
热而潮湿的 sultry
热发动机 hot engine;thermomotor
热发光 thermoluminescence
热发光法 krypton-bromine method
热发光方法 thermoluminescence method
热发光剂量计 thermoluminescent dosimeter
热发光年龄 thermoluminescence ages
热发磷光 thermophosphorescence
热发泡塑料 thermally foamed plastics
热发散 heat emission
热发射安全玻璃 safety reflective glass
热发射反应堆 thermoemission reactor
热发射功率 thermal emissive power
热发射离子源 thermal emission ion source
热发射率 heat emissivity;thermal emissivity
热发射涂层 heat-reflective coating
热发射系数 heat emissivity coefficient
热发生 heat generation;pyretogenesis
热发生的 pyretogenetic;pyrogenetic
热发生器 heat generating appliance;heat(ing) generator
热发生设备 heat generating device
热发生速率 rate of heat generation
热发声法 thermosonimetry
热发声器 <测定水体不同深度处的温度用> thermophone
热阀 thermal valve
热法 hot process
热法表面处理 hot surface treatment
热法表面处治 hot surface treatment
热法补坑 hot patch(ing)
热法除气器 thermal deaerator
热法浸提器 calorizator
热法精制 hot refining
热法净化器 hot process purifier
热法拉 thermal farad
热法磷酸 phosphoric acid by furnace process
热法路面处理 hot surface treatment
热法磨木浆 hot ground pulp
热法漂白 hot bleaching
热法汽油 thermal gasoline
热法软化 hot process softening
热法脱气器 thermal deaerator
热法芯合 hot box
热法熏制 hot smoking
热法再生 thermal reactivation;thermal reclamation
热法钻孔 thermic borehole;thermic boring
热法钻探 thermic boring
热反弹 thermal bounce
热反馈 temperature feedback;thermal feedback
热反滤光器 heat reflection filter
热反气旋 thermic anti-cyclone
热反射玻璃 heat-reflecting glass;heat-reflective glass
热反射材料 heat-reflecting material
热反射的 heat-reflecting
热反射镜 hot mirror
热反射率 heat reflectance;heat reflectivity
热反射能力 heat reflection
热反射漆 heat-reflecting paint
热反射涂层 heat-reflective coating
热反射性 heat reflectivity
热反射性质 heat-reflecting property
热反应 hot reaction;thermal reaction
热反应模拟 thermal response simulation
热反应器 thermal reactor
热反应系数 thermal response
热反应性 heat reactivity
热反应性酚醛树脂 heat reactive phenolic resin
热反应性树脂 heat reactive resin
热防护 thermal protection;thermal shield(ing)
热防护玻璃 heat-filtering glass
热防护材料 thermoprotective material
热防护层 heat shield(ing);thermal protection shield;thermal protective coating
热防护服 heat-protective clothing
热防护面 heat-protective surface
热防护屏 heat shield(ing)
热防护系统 thermal protection system
热放出 heat release
热放光参数 heat luminescence parameter
热放散 thermolysis
热放散的 thermolytic
热放射系数 heat emissivity coefficient
热放射效率 thermionic activity
热废水 heated effluent;heated waste;thermal waste;thermal wastewater
热废水综合利用 multiple utilization of thermal wastewater
热废物 hot waste
热废液 heated effluent
热分辨率 thermal resolution
热分布 distribution of heat;thermal distribution
热分层 thermal stratification
热分解 heat decomposition;pyrolysis;pyrolytic breakdown;pyrolytic decomposition;thermal breakdown;thermal decomposition
热分解法 pyrolytic process;thermal decomposition method
热分解炉 thermal decomposition furnace
热分解器 thermal decomposer
热分解区 pyrolysis zone
热分解物 pyrolysate
热分解作用 pyrolysis;thermolysis
热分离色谱 thermofractograph
热分量 thermal component
热分裂 thermal crack(ing);thermofission
热分流器 heat shunt;thermal shunt
热分配 heat distribution
热分散 heat dispersion
热分析 thermoanalysis
热分析图 thermogram
热分析仪 thermal analyser[analyzer]
热分析仪器 thermal analysis instrument
热分子压力 thermomolecular pressure
热分子真空计 thermomolecular ga(u)ge;thermomolecular vacuum ga(u)ge
热焚烧 thermal incineration
热焚烧炉 thermal incinerator
热风 blast heating;heated air;heating air;hot-air(blast);hot blast(air);hot wind;thermal wind;warm(ed) air
热风摆动张布机 hot-air jig tenter
热风采暖 air heating;hot-air heating;hot-blast heating;plenum heating;warm-air heating
热风采暖炉 air furnace
热风采暖设备 forced air heating;forced warm air heating plant
热风采暖系统 forced warm air heating system;hot-air heating system;plenum heating system;warm-air heating system
热风采暖装置 space heater;warm-air installation
热风层控制器 hot deck controller
热风冲天炉 blast-heating cupola;hot-blast cupola
热风出口 hot-blast outlet
热风吹送机 heat blower
热风吹送器 heat blower
热风道 hot-air duct
热风地板采暖 air floor heating
热风洞气候试验 hot tunnel climatic test
热风二次通过的烘干机 double pass drier[dryer]
热风发生器 heat blower
热风阀 hot-blast valve
热风放射器 air-heating radiator
热风分布器 hot-air distributor
热风分配 warm-air distribution
热风分配通道 warm-air distribution duct
热风分向干式纺丝机 split-draft metier;split-flow metier
热风干燥法 <木材的> hot-air seasoning;heated air drying;hot-air drying
热风干燥器 heat gun;hot-air drier[dryer]
热风干燥室 hot-air chamber;mansard
热风高压(暖气)系统 hot blast fan high pressure system
热风供暖 air heating;blast heating;hot-air heating;hot-blast heating;warm-air heating
热风供暖炉 hot blast heater
热风供暖器 air-heating radiator;hot-air heater
热风供暖系统 air-heating system;fan furnace system;hot-air heating system;hot-air system;plenum heating system;warm-air fan system;warm-air heating system
热风供暖烟道 hot-air heating flue
热风供暖站 forced warm air heating plant
热风供暖装置 hot-air heater;hot-air heating system;warm-air heating plant
热风鼓风机 hot-air blower;warm-air blower
热风鼓风口 hot-air blast hole
热风管(道) hot-air duct;warm-air duct(ing);warm-air pipe;hot-blast main
热风辊 hot-air roll
热风焊接 hot-gas welding;hot jet

R

welding
热风(和)热水混合供暖器 combined stoves and hot-water heater
热风烘干 hot-air seasoning
热风烘箱 hot-air oven
热风烘燥机 airflow drier[dryer]; hot flue drier[dryer]
热风烘燥拉幅机 hot-air stenter
热风化铁炉 hot-blast cupola
热风机 air heater; calorifier; hot-air blower; hot-air fan; hot-air generator
热风及热水混合供热器 combined stoves and hot-water heater
热风集中供暖 central fan heating
热风集中加热 warm-air central heating
热风技术 hot-blast technique
热风加热套 hot-air jacket
热风加脂转鼓 hot-air stuffing mill
热风炼铁 hot blown iron
热风炉 air heater; air-heating furnace; air-heating stove; air stove; blast heater; calorifier; cowper; heat generator; hot-air-blown oven; hot-air furnace; hot-air oven; hot-air stove; hot-air unit; hot-blast furnace; hot blast heater; hot-blast stove; warm-air furnace; warm-air stove
热风炉算停止送风 shutting out of hot-blast furnace
热风炉工 stovetender
热风炉供暖 warm-air furnace heating
热风炉燃烧器 hot-blast stove gas burner
热风炉烧嘴 hot stove burner
热风幕 heated air screen; warm-air curtain
热风暖气 hot-blast heating
热风暖气系统 hot-blast system
热风排出口 warm-air outlet
热风喷射烘燥机 hot-air jet drier[dryer]; nozzle drier
热风喷射式农产品干燥 jet crop drier
热风气垫式烘燥机 lay-on-air drier [dryer]
热风器 air blast heater; air heater; air-heating radiator; blast heater; heat gun; hot-air generator; hot-air heat
热风取暖 hot-air heating
热风散热器 air-heating radiator
热风扇供暖系统 warm-air fan heating system
热风上升管 warm-air rising duct
热风设备 air-heating installation; warm-air heating installation
热风生铁 warm air pig iron
热风式地面供暖系统 hot-air floor heating system
热风式干燥机 heated air drier
热风式浆纱机 hot-air sizing machine; hot-air slasher
热风收缩 hot-air shrinkage
热风输出量 heated air output
热风输送式干燥机 hot-air conveying type dryer
热风速计 anemotherm
热风速仪 heated thermometer anemometer; thermal anemometer
热风调节器 warm-air register
热风通道 warm-air duct(ing)
热风系统 hot-air system; hot-blast system
热风循环 heated air circulation; hot-air circulation; warm-air circulation
热风循环干燥机 hot-air circular drying machine
热风循环烘干炉 recirculation drying stove
热风循环炉 recirculation furnace
热风循环器 warm-air circulator
热风循环系统 hot-air circulation system
热风养护 hot-air curing
热风制冷系统 heat-operated refrigerating system
热风转换炉 changing-over stove
热风装置 hot-air apparatus; hot-air plant; warm-air circulator
热风总管 hot-blast main
热封 heat-seal
热封包装机 hot-seal packing machine
热封闭作用 heat-blocking action
热封缝材料 hot sealing compound
热封焊 thermal heat sealing
热封合 heat seal(ing)
热封机 heat seal(ing) machine
热封缄器 heat-sealer
热封接 heat seal(ing)
热封蜡 hot sealing wax
热封式高压釜 hot-seal pressure vessel
热封塑膜包装 heat-wrapping
热封涂层 heat seal(ing) coating
热封性能 <塑料薄膜袋> heat sealability
热锋(面)【气】 thermal front; warm front
热峰 thermal spike
热敷 fomentation; hot application; hot compress
热敷布 stupe
热敷层 heat-applied coating; hot-applied coating
热敷(路)面处理 hot surface treatment
热伏打 thermovoltaic
热伏探测 thermovoltaic detection
热浮选 hot flo(a)tation
热浮选回路 hot flo(a)tation circuit
热辐射 bolometric radiation; caloric radiation; calorific radiation; emission of heat; heat emission; heat emissivity; heat radiation; radiation of heat; temperature radiation; thermal emission; thermal radiation
热辐射板 radiant heating panel
热辐射波 heat radiating wave; heat wave
热辐射玻璃 radial glass; radiant glass
热辐射材料 thermal radiating material
热辐射测量计 bolometer
热辐射测量图 bologram; bolograph
热辐射成像 heat radiating imaging; thermal imaging
热辐射成像仪 radiometric mapper
热辐射传感设备 heat radiation sensing device
热辐射导引系统 heat-seeking guidance system
热辐射的 thermal radiating
热辐射防护屏 thermal radiation shield
热辐射高温计 heat radiation pyrometer
热辐射烘干 radiant heating
热辐射计 heat radiometer; kampometer; thermal radiometer
热辐射计电桥 bolometer bridge
热辐射距离 heat distance
热辐射量 thermal exposure
热辐射敏感元件 heat radiation sensing device; heat-sensitive eye
热辐射破坏距离 thermal radiation destruction distance
热辐射谱 thermal spectrum
热辐射器 heat(ing) radiator; thermal radiator
热辐射铅质遮热墙板 lead radiation shielding wall
热辐射强度 caloradiance; heat radiation intensity
热辐射区域 heat-banking area
热辐射伤害 thermal radiation injury
热辐射设备 heat radiating equipment
热辐射探测器 heat-sensitive eye; thermal radiation detector
热辐射体 heat radiator; heat unit; thermal radiator
热辐射透过率 thermal transmittance
热辐射系数 emissivity; heat emissivity coefficient; thermal emissivity
热辐射现象 thermal radiation phenomenon
热辐射线 caloradiance
热辐射寻的制导 heat homing guidance
热辐射仪 thermal radiometer
热辐射源 infrared source
热辐射侦察 thermal reconnaissance
热辐射指示器 heat cell
热辐射装置 radiant heating installation
热辐射灼伤 thermal radiation burn
热辐射自导引 heat-seeking
热辐射自导引头检测器 heat-seeking head detector
热辐摄影 thermal photograph
热辐探测器 thermal type radiation detector
热腐蚀 erosion of thermal; heat erosion; thermal etching
热负荷 firing rate; heat burden; heat duty; heat flux; heat(ing) load(ing); heat input; heat-transfer rate; refrigeration duty; thermal load
热负荷测试炉 hot load testing furnace
热负荷计划 thermal load scheme
热负荷可调节的燃具 hot load range-rated appliance
热负荷能力 thermal load capacity
热负荷试验 hot load test
热负荷图 heating load diagram
热负荷延续时间图 heating load duration graph
热负极 antilogous pole
热负载 heating load; thermal load
热负载额定值 thermal burden rating
热赋能 thermal forming
热覆膜砂 hot precoated sand
热干的 heat dried
热干化 heat drying
热干碾压法 hot and dry rolling
热干漆 heat set ink
热干清漆 stoving finish
热干扰 heat nuisance
热干岩 hot dry rock
热干油墨 heat set ink
热干燥机 thermal dryer[drier]
热干燥漆 baking varnish
热感 sense of hotness; thermal inductance
热感测定仪 heat sensimeter; thermosensitive meter
热感觉 thermal sensation
热感觉平均标度预测值 predicted mean vote
热感受器 Ruffini end organ; thermoreceptor
热感探测仪 thermosonde; thermosounder
热感探空仪 thermosonde
热感系数 heat perception coefficient
热感应 thermal induction
热感应变 thermally induced strain
热感应光学畸变 thermal induced optic(al) distortion
热感应塞 heat ga(u)ge
热感应线圈 work coil
热感应折射指数梯度 thermally induced refractive-index gradient
热感照相机 thermal camera
热钢板回火 hot plate tempering
热钢材冷却台架 hot rack
热钢锭剥皮机床 hot ingot peeling machine
热钢锭车 hot slab car
热高压 thermal high
热膏体沥青 hot bitumen cement
热格 heating grid
热隔层 thermofin
热隔层热敏元件 thermofin
热隔绝 heating isolation
热各向异性 thermal anisotropy
热跟踪 thermal tracking
热跟踪头 heat seeker
热工 pyroprocessing
热工参数 thermal process parameter
热工厂 heat power plant
热工车间 hot(work) shop
热工的 thermotechnic(al)
热工过程 thermic process
热工计算 heat engineering calculation; thermal calculation
热工检查 thermal examination
热工介质 thermodynamic(al) medium
热工设备 pyroprocessing plant
热工设计 thermal performance design
热工摄影术 thermography
热工水力学 thermohydraulics
热工特性 thermal characteristic
热工性能 thermodynamic(al) property
热工学 heat(power) engineering; pyrology; thermal engineering
热工仪表 thermal meter; thermodynamic(al) instrument
热工艺 thermal technique
热工用钢 hot working steel
热工指数 thermodynamic(al) index
热工制度 thermal condition; thermal regulation
热功当量 heated equivalent work; heat equivalent of work; mechanical equivalence of heat; mechanical equivalent of heat; thermal equivalent of work
热功率 calorific power; heat capacity; heat output; heat power; heat rating; thermal capacity; thermal output; thermal power; thermal rating
热功转换 heat-to-work conversion
热功转换常数 heat-to-work conversion constant
热供应 heat application
热沟 heat channel
热箍器 tyre heater
热箍式钢管 hot-banded steel pipe
热箍缩效应 thermal pinch effect
热箍应力 thermal hoop-stress
热骨料 hot aggregate
热骨料(储)仓 hot aggregate storage bin
热骨料提升斗 hot aggregate elevator
热骨料提升机 hot aggregate elevator
热鼓风 heat blast; hot blast; hot blow
热鼓风机 hot-air blower
热鼓风炮眼爆破 hot-hole blasting
热固成型粉 thermosetting mo(u)lding powder
热固的 thermosetting
热固粉末涂层 thermosetting powder coating
热固复合材料 thermosetting composite
热固复合塑料 thermosetting compos-

ite
热固化 hot set;hot cure(curing);hot sate;thermocuring;thermofixation;thermosetting
热固化胶粘剂 hot-setting adhesive
热固化黏[粘]合剂 warm-setting adhesive
热固化系统 heat curing system;thermal curable system
热固化性 thermoset
热固胶(合剂)thermosetting adhesive
热固胶粘剂 thermal setting adhesive
热固结 hot consolidation; thermal consolidation
热固聚合物 thermosetting polymer
热固绝缘 thermosetting insulation
热固式结合料 thermohardening binder
热固树胶 hot-setting glue
热固树脂 thermoset
热固塑料 thermoset(ting) plastics
热固塑料板 thermosetting plate
热固型塑料 thermosetting material
热固性 heat-convertibility;thermoset(ting property)
热固性丙烯酸树脂 thermosetting acrylic resin
热固性玻璃纤维增强塑料 glass fiber reinforced thermoset plastics
热固性材料 thermoset(ting material)
热固性的 heat-cured;heat set(ting);thermoset
热固性酚醛树脂黏[粘]结剂 hot-setting phenol resin adhesive
热固性合成树脂 thermosetting synthetic resin
热固性火泥 heat-setting mortar
热固性胶 thermosetting cement
热固性胶粘剂 hot-setting adhesive;thermosetting adhesive
热固性聚合物 thermosetting polymer
热固性聚酯树脂 thermosetting polyester resin
热固性绝缘 heat-cured insulation
热固性沥青-环氧树脂结合料 thermosetting bitumen/epoxy binder
热固性模塑化合物 thermosetting mo(u)lding compound
热固性模塑料 thermosetting mo(u)lding compound
热固性耐火材料 heat-setting refractory
热固性黏[粘]合剂 resinoid;thermosetting adhesive
热固性黏[粘]合胶 thermosetting cement
热固性黏[粘]结剂 hot-setting adhesive;thermosetting adhesive
热固性清漆 thermoset varnish
热固性树脂 heat-convertible resin;resinoid;thermosetting resin;heat-hardenable resin; heat-hardening resin;hot-setting resin;thermohardening resin
热固性树脂粉末 thermosetting resin powder
热固性树脂复合材料 thermosetting resin composite
热固性树脂基复合材料 thermosetting resin matrix composite
热固性树脂胶合材料 resinoid bond
热固性树脂黏[粘]结剂 thermosetting resin adhesive
热固性塑料 thermosetting plastics
热固性塑料板 thermoset plate
热固性塑料去飞边 thermoset deflashing
热固性涂料 thermohardening lacquer;thermosetting coating
热固性物质 thermosetting material
热固性纤维 thermosetting fibre[fiber]
热固性油墨 thermosetting ink
热固性油漆 thermosetting varnish
热固性增塑剂 thermosetting plasticizer
热固着 heat set;thermal fixation
热固着油墨 heat set ink
热刮泡沫分配器 thermal shaving foam dispenser
热观察仪 thermal viewer
热管 heating tube;heat pipe;thermotube
热管材整径机 hot reeling machine
热管承放式干燥器 pipe-rack drier[dryer]
热管道输送 hot piping
热管换热器 heat-exchange of heat pipe
热管冷却 heat pipe cooling
热管冷却电动机 heat pipe motor
热管理 heat management
热管式空气-空气热交换器 heat pipe air-to-air heat exchanger
热管式空气预热器 heat pipe preheater
热管式热交换器 heat pipe heat exchanger
热管系数 hot channel factor
热贯 hot penetration
热贯沥青碎石 hot penetration bituminous macadam
热贯流 overall heat transmission
热贯(入)法 <沥青的> hot penetration method
热贯填缝料 molten filler
热贯性 thermal inertia
热惯量 thermal inertia
热惯量成像仪 thermal inertia mapper
热惯量图 thermal inertia map
热惯性 thermal lag
热灌 hot penetration
热灌柏油 hot tarring
热灌的 hot-poured
热灌法 hot penetration method
热灌缝 heat seal(ing)
热灌焦油沥青 hot tarring
热灌接口 hot-poured joint
热灌沥青 hot penetration bitumen
热灌沥青的 hot penetration bituminous
热灌沥青路面施工 hot penetration construction
热灌沥青碎石路 hot penetration bituminous macadam
热灌式 hot-poured type
热灌填缝料 hot-poured crack filler;molten filler;hot-poured joint sealant
热灌筑 heat application;hot application
热灌装 hot filling
热光 calorescence;hot light;thermal light
热光不稳定性 thermooptic(al) instability
热光常量 thermooptic(al) constant
热光常数 thermooptic(al) constant
热光电元件 thermophotovoltaic cell
热光电转换 thermophotovoltaic conversion
热光度 bolometric luminosity
热光伏变换 thermophotovoltaic conversion
热光伏元件 thermophotovoltaic cell
热光化学强化芬顿反应 thermal and photochemically enhanced Fenton reaction
热光畸变 thermal optic(al) distortion
热光谱分析 thermal spectrum analysis
热光生电的 thermophotovoltaic
热光弹性 photothermoelasticity
热光稳定性 thermooptic(al) stability
热光系数 thermooptic(al) coefficient
热光像差 thermooptic(al) aberration
热光效应 thermooptic(al) effect
热光性能 thermooptic(al) property
热光性质 thermooptic(al) property
热光学 thermooptics
热光学法 thermoptometry
热光源 incandescence
热光制 hot finished
热龟裂 thermal checking
热规 thermal ga(u)ge
热轧扁钢 <厚 5 毫米以上,宽 75 ~ 150 毫米> flat hot-rolled bar
热辊成型 thermoroll forming
热辊锻 hot forge rolling
热辊压焊接 roll welding
热辊压榨的 hot-rolled
热辊轧制 hot-groove rolling
热滚麻花钻 hot-rolled drill
热滚筒定形 hot cylinder setting
热滚压印机 roll hot stamping machine
热裹法 hot pack(ing)
热过程 hot process
热过剩 excess of heat
热过载 thermal overload
热过载保护 thermal overload protection
热过载继电器 thermal overload relay
热过载容量 thermal overload capacity
热害 thermal harm
热害区级别 rank of geothermal hazard area
热害源类型 type of heat hazard source
热含 total heat
热含量 enthalpy; heat content; heat enthalpy;heat(ing) capacity;thermal capacity;thermal content
热函 enthalpy;heat content;heat enthalpy;total heat;thermal content
热函电位 enthalpy potential
热函浓度图 enthalpy-concentration diagram
热函数 heat function
热函增加 enthalpy gain
热焊 heatweld;thermit(e) welding
热焊封袋 hot weld encapsulation
热焊钢轨接头 thermit(e) welded rail joint
热焊工人 thermit(e) welder
热焊技术 thermo-welding technique
热焊剂 heat flux
热焊接 heat seal(ing);thermal welding;thermoweld; wiped joint;hot welding;themite welding
热焊接头 thermit(e) welded joint
热焊接性 thermal weldability
热耗 heating consumption;heat(ing) rate;heating sink;lost heat;rate of heat loss
热耗量 heat consumption
热耗率 heat consumption rate; heat rate;specific heat consumption
热耗散 burn-off; heat dissipation; thermal exhaustion;thermal run-away
热耗试验 heat consumption test
热耗损 thermosteresis
热合 thermal seal(ing)
热合成 thermosynthesis
热合法 heat seal method
热合金 thermalloy
热合镶嵌 applique
热核爆炸 thermonuclear explosion
热核变化 thermonuclear transformation
热核弹 fusion bomb
热核的 nuclear thermal;thermonuclear
热核电站 thermonuclear power plant
热核动力 thermonuclear power
热核堆 thermonuclear reactor
热核发电厂 fusion power plant
热核反应 fusion reaction;nuclear fusion;thermonuclear fusion;thermonuclear reaction
热核反应堆 fusion reactor;thermonuclear reactor
热核反应堆再生区 thermonuclear reactor blanket
热核反应速度 thermonuclear reaction rate
热核化学工程 thermonuclear chemical engineering
热核火力 thermonuclear fire
热核技术 thermonucleonics
热核聚变 thermonuclear fusion
热核聚变反应 thermonuclear fusion reaction
热核能 fusion energy
热核燃料 fusionable material;thermonuclear fuel
热核设备 thermonuclear apparatus
热核条件 thermonuclear condition
热核温度 thermonuclear temperature
热核武器 fusion weapon;thermonuclear weapon
热核炸弹 thermonuclear bomb
热核中子源 thermonuclear neutron source
热核转化 thermonuclear transformation
热核装料 thermonuclear charge
热核装置 thermonuclear device;thermonuclear machine
热核子 thermonuclear
热核子学 thermonucleonics
热核钻进 nuclear drilling
热荷载 thermal force;thermal load(ing)
热荷载试验 hot load test
热盒制芯法 hotbox process
热壑 heat sink
热轰击 thermal bombardment
热烘干 thermal drying
热红外波长 thermal infrared wave length
热红外波段 thermal infrared band
热红外成像系统 <包括红外和微波探测器> thermal infrared mapping system
热红外成像仪 thermal infrared mapper
热红外窗口 thermal infrared window
热红外多谱段扫描仪 thermal infrared multispectral scanner
热红外分布图 heat map;thermal infrared distribution map
热红外扫描 thermal infrared scanning
热红外扫描数据 thermal infrared scanning data
热红外扫描仪 thermal infrared scanner
热红外探测器阵列 thermal infrared detection array
热红外探测元件 thermal detection element
热红外通道 <波段为 3 ~ 5 微米,8 ~ 14 微米> thermal infrared channel
热红外图 thermograph
热红外图像 thermal infrared image(ry);thermography
热红外线 thermal infrared
热红外影像 thermal infrared image(ry)
热虹吸 thermal siphon;thermosyphon
热虹吸管 thermosiphon [thermosyphon]
热虹吸管冷却 thermosiphon cooling

R

热虹吸管再煮锅 thermosiphon reboiler
热虹吸冷却系统 thermosiphon cooling system
热虹吸器 thermosyphon
热虹吸水冷却系统 gravity-system water cooling
热虹吸蒸发器 thermosiphon[thermosyphon] evapo(u)rator
热后成型 post forming
热后效应 thermal after-effect
热湖 <水温在40℃以上> warm lake
热互变 thermotropy
热互换 heat interchange
热互换器 interchanger of heat
热互作用 thermal interaction
热滑 hot-running
热化 calorization; thermalization [thermolization]; thermalizing
热化反应堆 heat reactor
热化区 thermalization range
热化时间 thermalization time; thermalizing time
热化时间常数 thermalization time constant
热化位置 thermalized position
热化系数 thermalization coefficient
热化学 thermal chemistry; thermochemistry
热化学处理 heat chemical treatment; thermochemical treatment
热化学的 thermochemical
热化学动力学 thermochemical kinetics
热化学发动机 thermochemical engine
热化学反应 thermal reaction; thermochemical reaction
热化学反应式 thermochemical equation
热化学方程 thermochemical equation
热化学方法 thermochemical method
热化学卡 < =4.184 焦耳 > thermochemical calorie
热化学理论 thermochemical theory
热化学平衡 thermochemical equilibrium
热化学气体火焰 hot chemical gas flame
热化学剩余磁化 thermochemical remanent magnetization
热化学同位素效应 thermochemical isotope effect
热化学性能 thermochemical property
热化学循环 thermochemical cycle
热化中子 thermalized neutron
热化柱 thermalization column
热还原 thermal recovery
热还原反应 thermal reduction reaction
热环 hot ring
热环化反应 thermal cyclization reaction
热环境 thermal environment
热环境评价 evaluation of thermal environment
热环流 thermal circulation; thermal cycle; thermodynamic(al) circulation; thermosiphon <水体内垂直的>
热环流器 thermocirculator
热环流热水器 thermosiphoning water heater
热缓冲器 thermal buffer
热换能器 thermal transducer
热换算因数 heat conversion factor
热荒漠 warm desert
热挥发分析 thermal volatilization analysis
热挥发作用 thermal volatilization
热辉光曲线 thermal glow curve
热辉光特性 thermal glow characteristic
热回放 heat soak-back
热回放温度 heat soak-back temperature
热回流 hot reflux
热回流管 hot reflux condenser
热回路 hot loop
热回收 heat recovery; heat recuperating
热回收风机 recuperation fan
热回收冷凝器 heat recovery condenser
热回收轮 heat-recovery wheel
热回收箱 heat recovery box
热回收装置 heat radiation device; heat reclamation system; heat reclaiming device; heat recovery device; heat-recovery unit; heat recuperator
热回授 thermal regeneration
热汇 heat sink
热绘图器 thermal plotter
热混合料储仓 hot-mix silo
热混合料贮仓 hot-mix silo
热混乱度 thermal randomness
热混凝土 hot concrete; warm concrete
热混凝土升降机 hot elevator
热混凝土提升机 hot cement elevator; hot elevator
热活化 thermal activation
热活化化学气相沉积 thermally activated chemical vapo(u)r deposition
热活化胶黏[粘]剂 heat-activated adhesive
热活化黏[粘]结剂 heat-activated adhesive
热击穿 thermal breakdown; thermal run-away
热击穿保护电路 thermal shut-down circuit
热机 combustion engine; heat machine; thermal engine; thermal machine; heat engine
热机处理钢筋 thermomechanically treated steel
热机的 mechanocaloric; thermomechanical
热机发电机 thermomechanical generator
热机马达法 hot motoring method
热机生产流程 thermomechanical manufacturing process
热机陶瓷 heat engine ceramics
热机械处理 thermomechanical treatment
热机械方法 thermomechanical method
热机械分析 thermomechanical analysis
热机械分析法 thermomechanical analysis method
热机械分析仪 thermomechanical analyser[analyzer]
热机械加工 thermal mechanical working
热机械效应 thermomechanical effect
热机械性能 thermomechanical property
热机械性能极限 thermomechanical restriction
热机械性质 thermomechanical property
热机械学 thermomechanics
热机械循环效率 efficiency of heat engine cycle
热机循环 engine cycle; heat engine cycle
热迹 heated spot; oxidized spot
热积聚 heat accumulation; heat build-up; heat localization
热积云 thermic cumulus
热畸变 heat distortion
热畸变温度 heat distortion temperature
热激波 thermal barrier; thermal layer; thermal shock
热激电池 thermal cell
热激电致发光 thermoelectrolumines; thermoelectroluminescence
热激发 thermal activation; thermal agitation; thermal excitation
热激发光 thermal stimulated luminescence; thermoluminescence
热激发能量 thermal excitation energy
热激活 heat activation; thermal activation
热激活电池 heat-activated battery
热激活能 thermal activation energy
热激活蠕变 thermally activated creep
热激活性鞣剂 heat-activable tanning agent
热激励 thermal excitation
热激噪声 thermal agitation noise
热及热电发电机 thermionic and thermoelectric(al) generator
热极磁扰动器 hot plate magnetic stirrer
热极交流声 heater hum
热集料 hot aggregate
热集料(储)仓 hot aggregate storage bin
热集料输送机 hot aggregate conveyer [conveyor]
热挤 hot extrude; hot extrusion
热挤出成型 hot extrusion forming
热挤加工 hot extrusion
热挤模塑 hot extrusion mo(u)lding
热挤塑法 thermoset extrusion
热挤压 hot extrusion
热挤压制模法 hot hobbing
热脊 <锅炉水箱的> thermic syphon tube
热计量 metering of heat supply system
热计算 heat calculation
热计算元件 thermal computing element
热记录笔 hot stylus
热际 thermal shock
热剂 thermit(e); thermit(e) mixture
热剂反应 thermit(e) reaction
热剂焊 thermit(e) welding
热剂焊用坩埚 thermit(e) crucible
热剂模 thermit(e) mo(u)ld
热剂熔焊 fusion thermit welding
热剂铸焊 full-fusion thermit welding; non-pressure thermit welding; thermit(e) fusion welding
热季 hot season
热继电器 overheat switch; thermorelay
热继电器断路器 thermal circuit-breaker
热寂 heat death; thermal death
热寂时间 thermal death time
热加工 heat process; hot forming; hot machining; thermal treatment; hot work(ing)
热加工处理 hot-working treatment
热加工的 hot worked
热加工范围 working range
热加工工序 hot procedure
热加工管 hot-finished tubing
热加工过程 hot procedure
热加工铣床 hot miller
热加工性 hot work ability
热加工性能 hot-working character
热加固(法) thermal stabilization
热加料 hot feed
热加曲线 heating curve
热加速泵 <采暖系统> accelerating pump
热加油器 thermal lubricator; thermal oiler
热加脂 hot stuffing
热夹具焊接技术 hot fixture soldering technique
热价 caloric value; calorific value; heat price
热剪(切) hot shear(ing)
热检波器 heat detector; thermodetector
热检测器 heat detector; thermal detector; thermodetector
热检验 hot inspection
热碱槽 hot caustic bath
热渐退 febrile lysis
热键 hot key
热浆处理 hot stock treatment
热浆浮选 hot pulp flo(a)tation
热降(低) heat drop
热降解 thermal degradation; thermal degrading
热降解作用带 thermal degradation zone
热交换 exchange of heat; heat(er) exchange; heating transfer; heating transmission; heat interchange; heat transmission; interchange of heat; thermal exchange
热交换篦子机 heat-exchanger grate
热交换表面 heat-transfer surface
热交换不稳定性 heat-exchange instability
热交换测定 heat-exchange measurement
热交换池 heat-exchange tank
热交换动力学 heat-exchange dynamics
热交换法 heat-exchange method
热交换工作程序 heat-exchanging process
热交换管 heat-exchange tube; recuperator tube
热交换管束 heat-exchange tube bundle
热交换回路 heat-exchange circuit
热交换理论 heat-transfer theory
热交换流体 heat-exchange fluid
热交换率 heat exchange ratio; rate of heat exchange
热交换面 heat-exchange surface
热交换面积 area of heat transfer
热交换器 exchanger; heat-exchange equipment; heat-exchange facility; heat-exchanger; heat-exchanging apparatus; heat interchanger; heat transfer; thermal converter
热交换器管组装 heat-exchanger pipe assembly
热交换器回热度 heat-exchanger effectiveness; heat-exchanger thermal ratio
热交换器盘管 heat-exchanger coil
热交换器容量 capacity of heat exchanger
热交换器蛇形管 heat-exchanger coil
热交换器探漏器 heat-exchanger leak detection probe
热交换器线圈 heat-exchanger coil
热交换器效率 thermal ratio
热交换器延迟 <反应堆的> heat-exchanger lag
热交换器有效度 heat-exchanger effectiveness
热交换器有效性 heat-exchanger effectiveness
热交换器原理 heat-exchanger principle
热交换器滞后 heat-exchanger lag

热交换蛇形管 heat-exchange coil
热交换式低温冷却器 exchanger-type subcooler
热交换塔 heat-exchanger tower
热交换系数 heat-exchange coefficient
热交换系统 heat-exchange system
热交换效率系数 coefficient of heat exchange efficiency
热交换循环 heat-exchange cycle
热交换窑炉 recuperative furnace
热交换有机液体 perolene
热交换站 heat-exchange station
热交换装置 heat-exchanger installation;heat-exchanging system
热浇 hot application;hot-pour
热浇柏油 hot tarring
热浇道模 hot-runner mo(u)ld
热浇的 hot-poured
热浇灌沥青 hot bitumen grout
热浇混合料 hot-poured compound; hot pouring compound
热浇焦油 hot tarring
热浇沥青 hot tar
热浇注密封膏 hot-poured sealant
热胶 warm glue
热胶合 heat bonding;hot gluing
热胶黏[粘]剂 thermal adhesive
热胶土 melt-bonding soil
热焦斑 thermal focal spot
热焦油 hot tar
热矫正 heat straightening
热矫直 hot straightening; thermal straightening
热角保护 thermal corner protection
热搅拌 hot mix;mixed hot
热搅拌机 heat pugmill(mixer)
热搅动 thermal agitation
热校正压力机 hot-sizing press
热校正系数 thermal correction factor
热校准 thermal calibration
热窖 hotpit
热阶度 temperature gradient;thermal gradient
热接 hot joining
热接触 thermal contact; thermocontact
热接触变质晕 contact aureole;thermal aureole
热接触电阻 thermal contact resistance
热接触阈限值 heat exposure threshold limit value
热接点 fire end; hot end; hot junction; measuring junction; thermal contact;thermojunction
热接点温度 hot junction temperature
热接缝 hot joint
热接合 heat joining
热接片机 hot splicer
热接受器 thermal acceptor
热节 hot spot;thermal center[centre]
热结 heat spot;hot junction;thermal junction
热结层 temperature stratification
热结合 <元件的> thermal bond
热结合层 thermal bond
热结节 hot nodule
热截面 hot cross-section
热解 pyrolytic decomposition; pyrolyze; thermolysis; thermolytic dissociation
热解产物 pyrolysis product; pyrolytic product; pyrolyzate; thermal decomposition product
热解沉积法 pyrolytic deposition process
热解沉积技术 pyrolytic coating technique
热解粗汽油 pyronaphtha
热解单元 pyrolysis unit
热解氮化硼 pyrolytic boron nitride
热解定向石墨 pyrographite
热解法 thermal decomposition method
热解法二氧化硅 pyrogenic silica
热解反应 pyrolysis reaction;pyrolytic reaction
热解反应器 pyrolysis reactor
热解反应性 thermolytic reactivity
热解分析 pyrographic analysis;pyrolysis analysis
热解光谱 pyrolysis spectrum; pyrolytic spectrum;pyrolyzate spectrum
热解过程 pyrolytic process
热解技术 pyrolytic technique
热解聚合物 pyrolytic polymer;pyrolyzed-polymer
热解聚合作用 thermal depolymerization
热解扣 thermal trip
热解离 pyrolysis;thermodissociation
热解炉 pyrolysis oven;pyrolyzing furnace;thermal decomposition furnace
热解气相色谱法 pyrolysis gas chromatography; pyrolytic gas chromatography; thermal cracking gas chromatography
热解气相色谱联机 pyrolysis gas chromatography combination
热解汽油 pyrolysis gasoline
热解汽油馏分 pyrolysis naphtha
热解器 pyrolysis apparatus; pyrolyzer;pyrolyzing apparatus
热解曲线 pyrolysis curve; thermogram;thermolysis curve
热解取样带 pyro-probe-ribbon
热解色层 pyrography
热解色层法 pyrographic method
热解色层分析 pyrographic analysis; pyrolytic chromatography
热解色谱 pyrograph
热解色谱法 pyrography; pyrolytic chromatography
热解色谱气测井 pyrolysis-chromatogram gas logging
热解色谱图 pyrogram
热解渗滤 pyrolytic infiltration
热解石墨 pyrolytic black; pyrolytic carbon;pyrolytic graphite
热解石墨合金 pyrographalloy
热解时间 pyrolysis time
热解室 pyrolysis chamber
热解水 water of hot decomposition
热解水收率 thermolysis water yield
热解速度 pyrolysis rate
热解碳化硅 pyrolytic silicon carbide
热解碳膜电阻器 pyrolytic carbon film resistor
热解碳涂敷颗粒 pyrolytic carbon-coated particle
热解图 pyrogram;pyrolysis diagram; pyrolytic spectrum
热解温度 pyrolysis temperature
热解物 pyrolyzed substance
热解吸 thermal desorption;thermodesorption
热解吸法 thermal desorption method
热解吸收作用 thermal desorption
热解吸物 thermal desorption substance
热解消除 pyrolytic elimination
热解旋管 pyrolysis coils
热解液体产物 pyrolysis liquids
热解蒸镀 coating by vapo(u)r decomposition
热解质谱测量 pyrolysis-mass spectrometry
热解重量分析 thermal gravimetric analysis;thermogravimetric analysis
热解重量分析术 thermogravimetry
热解重量仪 thermal gravimeter;thermogravimeter
热解组件 pyrolysis unit
热解作用 pyrogenetic decomposition; pyrolysis
热介质 heat medium;thermal medium
热介质旋流分离器 heavy-medium cyclone separator
热金属辊道 hot run table
热金属锯机 hot-metal sawing machine
热金属丝切割法 hot-wire cutting
热进料 hot feed
热浸电镀钢 hot-dip galvanized steel
热浸(镀) hot dip(ping)
热浸镀保护层 hot-dip coating
热浸镀层法 hot-dip coating process
热浸镀金属法 metallic coating by hot-dipping process
热浸镀铝 aludip;hot-dip aluminizing
热浸镀铝层 hot-dip alumin(i)um coating
热浸镀铝法 hot-dip alumin(i)um process
热浸镀铝钢板 aludip
热浸镀铅锡合金层 hot-dipped terne
热浸镀锡 hot-dip tinning
热浸镀锡薄板 hot-dipped tinplate
热浸镀锌 galvanizing by dipping; hot-dip(ped) galvanizing; hot-dip zincing;hot galvanizing;flame scaling
热浸镀锌法 hot-dip process
热浸镀锌钢 hot-dip galvanization steel
热浸镀锌钢丝 hot-dip galvanized wire
热浸法 hot dipping method;hot-maceration method
热浸合金过程 hot-dip alloying
热浸环氧树脂保护层 hot-dip epoxy coating
热浸金属涂层 hot dip coating
热浸金属涂层钢板 hot-dip metallic coated sheet steel
热浸金属涂层制品 hot-dip metallic coated product
热浸净化 pyroclean
热浸沥青施工 hot penetration construction
热浸渗法 hot penetration method
热浸施工法 hot penetration construction method
热浸蚀 heat etching;thermal etching
热浸蚀试验 hot etching test
热浸涂 hot-dip coating
热浸涂镀 hot-dip plating
热浸涂法 hot-dip coating process
热浸涂可剥涂层 hot-melt coating
热浸脱脂 pyroclean
热浸渍镀锌 hot-dip galvanizing
热浸渍镀锌的 hot-dip galvanized
热浸渍镀锌钢筋 hot-dip galvanized reinforced bar
热浸渍脱模剂 hot-dip striping compound
热浸渍型防锈剂 hot-dip compound; hot-dip type rust preventive
热经济性 heat economy
热经济指标 heat economy figure
热精炼 thermal refining
热精馏 thermal rectification
热精压 hot-coining
热精整 hot trimming
热精制浆粕 hot-refined pulp
热井 <贮存冷凝水的装置> hot well; heat well
热井泵 hot well pump
热警报(信号) heat alarm
热净反应堆 hot clean reactor
热净临界 hot clean criticality
热净装载 hot clean loading
热静力学 thermostatics
热锯 hot drop saw; hot saw(ing); warm saw
热锯机 hot-metal sawing machine; hot sawing machine
热聚合 thermal polymerization
热聚合干燥 hot polymerization drying
热聚合橡胶 hot rubber
热聚合油 heat-bodied oil;heat polymerized oil
热聚合作用 heat polymerization;hot polymerization; thermopolymerization
热聚焦光谱学 thermal blooming spectroscopy; thermal lensing spectroscopy
热聚结 thermal coalescence
热聚物 heat polymer; hot polymer; pyrolytic copolymer
热聚橡胶 heat polymerization rubber
热卷 hot rolling
热卷边 hot crimping
热卷流 thermal plume
热卷取机 hot coiler;hot reeling machine
热绝缘 heat(ing) insulation
热绝缘板 board insulation
热绝缘板材 slab insulant
热绝缘板和毡 paxboard and paxfelt
热绝缘材料 heat insulating material; heat insulator; thermal insulation material;thermic insulant
热绝缘的 heat-insulating
热绝缘底槽板 bottom grid
热绝缘及隔声材料 Akousticos
热绝缘泡沫聚苯乙烯板 foamed polystyrene board for thermal insulation purposes
热绝缘体 heat insulator; thermal; thermal insulant;thermal insulator
热绝缘涂层 heat barrier coating;heat insulating coat(ing)
热绝缘系数 coefficient of thermal insulation
热绝缘子 heat insulator
热喀斯特 thermal karst;thermokarst
热喀斯特地形 thermokarst topography
热喀斯特坑 thermokarst pit
热卡 calorie
热卡强度 calorific intensity
热卡值 calorific power
热开关 heat switch;thermoswitch
热坑 thermal pit
热空气 hot-air;warm air
热空气出口管 hot-air outlet duct
热空气处理 hot-air treatment
热空气导管 hot-air duct
热空气定形 hot-air setting
热空气发动机 hot-air engine
热空气干湿表 heated air psychrometer
热空气干燥法 hot-air seasoning;hot-air drying
热空气干燥烘茧机 hot-air cocoon drying machine
热空气干燥机 hot air dryer
热空气干燥炉 air oven
热空气供给 warm-air feed
热空气供暖器 hot-air heater
热空气供暖系统 hot-air(heating) system
热空气供热系统 hot-air heating system;warm air heating system
热空气管(道) warm-air pipe
热空气焊接 hot-gas welding
热空气焊接法 hot-air welding method
热空气焊接机 hot-air welding machine
热空气烘干 hot-air seasoning
热空气环流系统 warm-air circulating

system
热空气环流制 warm-air circulating system
热空气加热设备 hot-air type heater
热空气拉软机 hot-air staking machine
热空气喇叭 hot-air horn
热空气老化 air oven ag(e)ing; heat air ag(e)ing
热空气老化试验 air oven ag(e)ing test
热空气流 thermal air current
热空气硫化 dry hot vulcanization; hot-air cure; hot-air vulcanization
热空气炉 hot-air furnace
热空气灭菌法 hot-air sterilization
热空气喷流 hot blast
热空气喷射干燥 drying by hot-air jet
热空气气球 montgolfier
热空气钎焊 hot-air soldering
热空气取暖设备 hot-air heater; hot-air type heater
热空气射流 hot blast
热空气式供暖 hot-air heating
热空气式加热 hot-air heating
热空气式取暖 hot-air heating
热空气收缩 hot-air retraction
热空气熟化 hot-air cure
热空气调节阀 hot-air control valve
热空气脱漆器 hot-air stripper
热空气系统 hot-air system
热空气箱 hot-air box
热空气消毒 hot-air sterilization
热空气循环对流加热器 convector
热空气养护 hot air cure
热空气制 hot-air system
热空穴 hot hole
热空转 hot idle
热孔 hot hole
热控测辐射计 thermal detection radiometer
热控多层滤光器 heat control multilayer filter
热控放气门 thermal relief
热控放泄门 thermal relief gate
热控管 thermister[thermistor]
热控开关点燃器 thermal-switch starter
热控涂层 thermal control coating
热控制阀 heat control valve
热控(制)开关 heat control switch; thermal switch; thermoswitch
热控制器 heat controller; thermal controller
热库 heat reservoir; thermal reservoir
热库仑 thermal coulomb
热矿泉 hot mineral spring
热矿水 thermomineral water
热框格玻璃构件 <门窗上的> thermopane unit
热亏损 heat debt
热馈遥控泵 hot-feed remote controlled pump
热扩管机 hot tube expanding machine
热扩撒 temperature scattering; thermal diffusivity
热扩散泵 hot diffusion pump
热扩散比 thermal diffusion ratio
热扩散常数 thermal diffusion constant
热扩散电流 thermal diffusion current
热扩散法 thermal diffusion method
热扩散分级 thermal diffusion fractionation; thermodiffusion fractionation
热扩散分离 thermal diffusion fractionation
热扩散分离管 thermal diffusion column
热扩散工厂 thermal diffusion plant
热扩散管 thermal diffusion tube
热扩散交换柱 thermal diffusion exchange column; Clusius column
热扩散距离 thermal diffusion length
热扩散流 thermal diffusion flow
热扩散率 heat diffusivity
热扩散区 thermal diffusion zone
热扩散深度 thermal diffusion depth
热扩散势 thermal diffusion potential
热扩散系数 thermal diffusion coefficient; thermodiffusion coefficient; thermodiffusion factor; thermal diffusivity
热扩散系统 thermal diffusion system
热扩散性 thermal diffusivity; thermodiffusivity
热扩散循环 thermodiffusion cycle
热扩散因数 thermal diffusion factor
热扩散因子 thermal diffusion factor
热扩散柱 Clusius-Dickel column; thermal diffusion column; thermogravitational column
热扩散作用 diffusion of heat; heat diffusion; heat dissipation; thermal diffusion; thermal diffusivity; thermodiffusion
热拉 heat stretch; hot draw(ing)
热拉的 hot-drawn
热拉幅定形 hot stenter setting
热拉钢丝 hot drawing wire; hot-drawn steel wire
热拉管 hot-drawn tube
热拉伸 hot stretch; thermal stretch(ing)
热拉伸合成帘子线 hot-stretched cord
热拉伸机 heat stretching machine
热拉伸强度 hot tensile strength
热拉伸纤维 heat-stretched fibre
热拉无缝钢管 seamless hot-drawn steel pipe
热拉直法 hot stretching
热蜡(彩)画法 encaustic
热蜡彩画釉烧成装饰处理(的砖、瓷砖、玻璃、陶器等) encaustic decoration
热缆 heat cable
热浪 hot spell; warm wave; heat wave; hot wave
热老化 heat deterioration; heat(ing) ag(e)ing
热老化变色 heat-aged discolo(u)ration
热老化试验 heat ag(e)ing test
热烙除法 thermocauterectomy
热烙气管切开术 thermotracheotomy
热烙器 thermocautery
热烙术 thermocauterization; thermocautery
热烙铁焊接 heated-tool welding
热雷暴 heat thunderstorm; thermal thunderstorm
热垒 thermal barrier
热冷槽处理 hot and cold open-tank treatment
热冷槽(浸注)法 <木材防腐处理> hot and cold bath(e)
热冷槽浸渍(处理) <木材防腐处理> hot and cold steeping; hot and cold soak process
热冷负载法 hot and cold load technique
热冷加工 hot-cold work(ing)
热冷浸泡法 <木材防腐处理> hot and cold soak process
热冷开口槽 hot and cold open tank
热冷溶解 hot-cold lysis
热离解法 thermal deposition
热离解作用 thermal dissociation
热离子 thermion
热离子变换 thermionic conversion
热离子变换器 thermionic converter
热离子传导 thermionic conductance; thermionic conduction
热离子的 thermionic
热离子电池 thermionic cell
热离子电流 thermionic current
热离子发电 thermionic generation
热离子发射 thermionic emission
热离子发射电流 thermionic emission current
热离子发射检测器 thermionic emission detector
热离子发射体 thermionic emitter
热离子反应堆 thermionic reactor
热离子放电 thermionic discharge
热离子功函数 thermionic work function
热离子管 thermionic tube; thermionic valve
热离子管电压表 thermionic voltmeter
热离子管检波器 thermionic detector
热离子换能器 thermionic converter
热离子活度 thermionic activity
热离子激活性 thermionic activity
热离子及热电偶发电机 thermal ionic and thermoelectric(al) generator
热离子继电器 thermionic relay
热离子检测 thermionic detection
热离子检测器 thermionic detector
热离子交换器 thermionic converter
热离子器件 thermionic device
热离子燃料元件 thermionic fuel element
热离子室 thermoionic cell
热离子调节器 thermionic controller
热离子效应 thermionic effect
热离子学 thermionics
热离子仪器 thermionic instrument
热离子阴极 thermionic cathode; hot cathode
热离子元件 thermoionic element
热离子源 thermionic source
热离子真空管 thermionic vacuum tube
热离子真空计 thermionic vacuum ga(u)ge
热离子整流器 thermionic rectifier
热离子转换 thermionic conversion
热力 heating power; thermal force; warmth
热力保护 thermal protection
热力泵 heat pump; thermal pump
热力变质煤 heat-altered coal
热力变质岩 thermal metamorphic rock
热力变质作用【地】 thermal metamorphism; pyrometamorphism; thermometamorphism
热力剥落 heat spall
热力参数 thermal parameter
热力层次 thermodynamic(al) stratification
热力场 thermal field
热力潮 thermal tide
热力沉降 thermal precipitation
热力除尘器 thermal collector
热力除气 thermal deaeration
热力除气器 steam deaerator
热力除氧气 steam deaeration; heating deaeration
热力储存 accumulation of heat
热力穿孔 fusion piercing; jet-piercing; thermal piercing
热力穿孔机 jet-pierce machine; jet-piercing machine
热力穿孔器 thermal penetrator
热力穿孔设备 fusion piercing equipment
热力的 thermodynamic(al)
热力低压 thermal depression
热力点 heating substation
热力电动牵引 thermoelectric(al) traction
热力电力机车 thermal electric(al) locomotive
热力电位 thermal electric(al) potential
热力动车【铁】 thermoelectric(al) rail car
热力对流 free convection; thermal convection
热力发电 thermoelectric(al) power generation
热力发电机 thermoelectric(al) power generator
热力发电站 thermal power station; thermoelectric(al) generation station
热力发动机 heat engine; thermal engine
热力阀 thermal valve
热力法穿孔钻机 fusion piercing method drill
热力法钻进 fusion piercing
热力方程 heat equation; thermodynamic(al) relation
热力风化 thermal weathering
热力负荷 thermal loading
热力干燥的 heat dry
热力干燥机 thermal dryer[drier]
热力工程 heat engineering; heat power engineering; thermal engineering
热力工况 thermodynamic(al) condition
热力供热 thermodynamic(al) heating
热力供热系统 thermodynamic(al) heating system
热力供应 heat power supply
热力构造 thermal structure
热力管 heating pipe
热力管沟 heating trench
热力过程 thermal process
热力红斑 erythema caloricum
热力化学致死曲线 thermochemical destruction curve
热力环流 thermodynamic(al) circulation
热力活性 thermodynamic(al) activity
热力机 heat engine
热力机械的 mechanocaloric
热力计算 heat(ing) calculation
热力继电器 bimetallic strip relay; thermal relay
热力加成作用 thermal addition
热力浆化法 thermomechanical puling
热力接点 heat-operated contact; thermal contact
热力结构 thermal structure
热力开挖 thermal excavation
热力老化 thermal ag(e)ing
热力类岩溶 thermal karst
热力离解 thermal dissociation
热力量 thermodynamic(al) quantity
热力灭菌法 heat sterilization
热力模拟 thermodynamic(al) analogy
热力模型 thermodynamic(al) model
热力浓度 thermodynamic(al) concentration
热力偶塞 thermal plug
热力喷气发动机 thermal jet engine; thermojet
热力膨胀阀 thermostatic expansion valve
热力平衡 thermal equilibrium
热力屏障 thermal barrier
热力破坏曲线 thermal destruction curve
热力气旋生成 thermocyclogenesis
热力牵引发电机组 heat engine trac-

tion unit
热力强度 strength of thermal force; thermal intensity; thermal strength
热力情况 thermal regime
热力曲线 thermal curve
热力驱动力 thermodynamic(al) driving force
热力燃烧 flame combustion; thermal combustion
热力燃烧法净化有机废气 control of organic waste gas by thermal combustion
热力扰动 thermodynamic(al) disturbance
热力蠕变 thermal creep; thermal sweep
热力入口 building heating entry; consumer heat inlet
热力设计准则 thermal design criterion
热力失调 thermal misadjustment
热力式定时解锁器 thermal time release
热力式疏水器 steam trap, thermostatic type
热力式限时解锁器 thermal time release
热力势 free energy; thermal potential; thermodynamic(al) potential
热力试验 thermal test(ing); thermodynamic(al) investigation
热力抬升 plume rise by buoyancy force
热力探矿 thermal prospecting
热力梯度 thermal gradient
热力图 energy diagram; thermodynamic(al) chart
热力图解 thermodynamic(al) diagram
热力推进系统 thermodynamic(al) propulsion system
热力完善度 thermodynamic(al) perfect degree
热力网 heat-supply network
热力位 thermodynamic(al) potential
热力系数 thermodynamic(al) coefficient
热力系统 heat system; cycle arrangement <热力设备或电站的>
热力响应 thermal response
热力消毒 thermal disinfection
热力效率 thermodynamic(al) efficiency
热力性 tenacity
热力学 thermal mechanics; thermodynamics; thermomechanics
热力学变量 thermodynamic(al) variable
热力学表 thermodynamic(al) table
热力学冰球温度 thermodynamic(al) ice-bulb temperature
热力学参数 thermodynamic(al) parameter
热力学的 thermodynamic(al)
热力学的熵 thermodynamic(al) entropy
热力学第二定律 second law of thermodynamics; second principal of thermodynamics
热力学第三定律 third law of thermodynamics
热力学第一定律 equivalence principle; first law of thermodynamics
热力学定律 law of thermodynamics; thermodynamic(al) law
热力学定义 thermodynamic(al) definition
热力学法 thermodynamic(al) method
热力学分配系数 thermodynamic(al) partition coefficient
热力学分析 thermodynamic(al) analysis
热力学概率 thermodynamic(al) probability
热力学关系 thermodynamic(al) relation
热力学管 thermodynamic(al) duct
热力学过程 thermodynamic(al) process
热力学函数 thermodynamic(al) function
热力学活性 thermodynamic(al) activity
热力学极限 thermodynamic(al) limit
热力学家 thermodynamicist
热力学理论 thermodynamic(al) argument; thermodynamic(al) theory
热力学露点温度 thermodynamic(al) dew-point temperature
热力学能 thermodynamic(al) energy
热力学能量方程 thermodynamic(al) energy equation
热力学平衡 thermodynamic(al) equilibrium
热力学平衡常数 thermodynamic(al) equilibrium constant
热力学平衡方程 thermodynamic(al) equilibrium equation
热力学平衡状态 thermodynamic(al) equilibrium state
热力学侵蚀 thermal etching; thermodynamics erosion
热力学容许过程 thermodynamically admissible process
热力学溶液理论 thermodynamic(al) solution theory; thermodynamic(al) solution theorem
热力学上的退化 thermodynamic(al) degeneration
热力学湿球温度 thermodynamic(al) wet-bulb temperature
热力学势(能) thermodynamic(al) potential
热力学试验 thermodynamic(al) test
热力学数据 thermodynamic(al) data
热力学酸度 thermodynamic(al) acidity
热力学态函数 state parameter; state variable; thermodynamic(al) function of state; thermodynamic(al) variable
热力学体系 thermodynamic(al) system
热力学通量 thermodynamic(al) flux
热力学同位素效应 thermodynamics isotope effect
热力学图 thermodynamic(al) diagram
热力学图表 thermodynamic(al) chart
热力学推论 <计算吸附作用> thermodynamic(al) reasoning
热力学温标 thermodynamic(al) scale; thermodynamic(al) scale of temperature; thermodynamic(al) temperature scale
热力学温度 <绝对温度,符号 T> thermodynamic(al) temperature; absolute temperature
热力学温度单位 Kelvin degree
热力学稳定相 thermodynamically stable phase
热力学稳定性 thermodynamic(al) stability
热力学物态方程 thermodynamic(al) equation of state
热力学系统 thermodynamic(al) system
热力学相似 thermodynamic(al) similarity
热力学效率 thermodynamic(al) efficiency
热力学性能 thermodynamic(al) property
热力学性质 macroscopic property; thermodynamic(al) property
热力学循环 heat cycle; thermodynamic(al) cycle
热力学压力 thermodynamic(al) pressure
热力学研究 thermodynamic(al) study
热力学原理 thermodynamic(al) principle
热力学约束 thermodynamic(al) restriction
热力学转化 thermodynamic(al) change
热力学状态 thermodynamic(al) state
热力学准则 thermodynamic(al) criterion
热力学坐标 thermodynamic(al) coordinates
热力循环 thermodynamic(al) circulation; thermodynamic(al) cycle
热力压缩机 thermal compressor
热力延时继电器 bimetal time delay relay; time element thermal relay
热力研究 thermodynamic(al) investigation
热力有效浓度 thermodynamic(al) concentration
热力再流动法 thermal reflowing process
热力增强器 heat booster
热力站 substation of heat supply network; thermal substation
热力指示器 heat indicator
热力制冷 heat-operated refrigeration
热力制冷系统 heat-operated refrigerating system
热力装置 thermal device
热力自动调节 automatic heat regulation
热力钻机 fusion piercing drill; jet pierce drill; jetting drill; piercing drill; thermodrill; jet piercing drill
热力钻进 thermal boring; thermal drill(ing)
热力钻进设备 fusion piercing equipment
热力钻进速度 piercing rate
热力钻进效率 piercing efficiency
热力钻孔 thermal boring; thermic boring; thermodrill
热力钻探 thermal boring; thermal drilling
热力钻眼 jet drilling; jet piercer drilling; jet-piercing drilling; thermal jet piercing drilling
热历史 heat history; thermal history
热利用率 fuel heat utilization; heat availability; heat utilization efficiency; thermal utilization factor
热利用系数 available heat factor; coefficient of heat efficiency
热利用效率 efficiency of heat utilization
热沥青 heated asphalt; hot asphalt; hot asphaltic bitumen; hot bitumen; hot stuff
热沥青灌浆法 heat bitumen grouting method
热沥青灌注 hot bitumen impregnation
热沥青混合物 hot-mix asphalt
热沥青混凝土 hot asphaltic concrete; hot bitumen concrete
热沥青浇灌 hot bitumen grout
热沥青胶泥 hot asphalt cement; hot bitumen cement
热沥青铺路 hot tarring paving
热沥青砂浆 hot asphaltic mortar
热沥青涂刷 mopping of hot asphalt
热粒料 hot aggregate
热炼 cooking; heat bodying
热炼规程 cooking schedule
热炼锅 cooking kettle
热炼聚合速度 bodying velocity
热炼温度 bodying temperature; cooking temperature
热链接 hot link
热量 amount of heat; heat; heat quantity; quantity of heat
热量保存 heat conservation
热量保存启闭器 heat conservation shutter
热量变化 heat vibration
热量变化率 heat drop; heat gradient
热量补给 heat input
热量不足 shortage of heat
热量测定 calorimetric measurement; calorimetry
热量测量 calorimetric measurement
热量产生 heat production
热量储存 storage of heat
热量传递机制 heat-transfer mechanism
热量传动 heat movement
热量单位 caloric unit; calorie; heat unit; thermal unit; calorific unit; unit of heat
热量的 caloric; calorific; calorimetric; thermal
热量的净损失 net loss of heat
热量滴定法 thermometric titration
热量对照表 heat balance
热量发射 emission of heat; heat emission
热量法 calorimetry
热量分布 distribution of heat; heat distribution
热量分析 calorimetric analysis
热量概算 heat budget
热量功率 caloric power; calorific power
热量耗散率 heat dissipation rate
热量换算系数 conversion coefficient of heat
热量回收 heat reclaim; heat recovery; heat regenerator; heat return; regeneration of heat
热量回收的 heat recovering
热量回收塔 heat recovering tower
热量回收系数 reheat factor
热量回收系统 heat reclamation system; heat recovery system
热量获得和损失 heat gain and loss
热量积聚带 heat-collecting zone
热量计 caloric meter; calorimeter; calorimeter assembly; calorimeter vessel; heat ga(u)ge; heat meter; hydropyrometer; thermal flowmeter
热量计算 heat account; heat budget; heat calculation
热量交换 heat exchange; interchange of heat
热量交换计算 calculating of heat exchange
热量节省 heat economising[economizing]
热量节约 heat conservation; heat economising[economizing]
热量控制 heat control
热量扩散率 heat diffusivity
热量浪费 heat waste
热量利用 heat utilization efficiency; utilization of heat
热量流失 heat loss
热量密度 heat density
热量排除 heat removal
热量排放 heat discharge
热量排放率 heat emission rate
热量膨胀系数 thermal coefficient of

R

expansion
热量平衡 caloric balance; caloric of heat; calorimetric balance; heat budget; heat balance; thermal balance
热量平衡定律 law of thermal equilibrium
热量平衡法 <蒸发计算的> heat budget method
热量平衡试验 heat balance test
热量平衡数据 heat balance data
热量器 calorimeter; caloriscope
热量强度 calorific intensity
热量强度分布线 heat-transfer rate distribution
热量取出 heat extraction
热量商(数) caloric quotient
热量摄取 calorie intake
热量生产 heat generation
热量生成速率 rate of heat generation; rate of heat production
热量试验 calorimetric test
热量释放 liberation of heat; release of heat
热量收支 heat budget
热量守恒 heat conservation
热量输出 heat output
热量输入 heat input
热量输送 heat transfer; heat transport
热量损耗 thermal loss
热量损失 heat loss; loss of heat
热量探测器 calorimetric detector; heat detector
热量梯度 heat gradient
热量梯度计 thermal gradiometer
热量调节 heat regulation
热量调节器 heat regulator; thermoregulator
热量调节装置 heat regulation device
热量外流通道 heat bridge
热量温度 heat temperature
热量吸取 heat abstraction
热量吸收 heat abstraction
热量系统 heat system
热量限制自动开关 thermal limit switch
热量消除法 heat elimination
热量消耗 consumption of heat; heat consumption; heat exhaustion
热量信号 thermal signal
热量需要 caloric requirement
热量蓄积 accumulation of heat
热量移动 heat movement
热量引入区 thermal intake length
热量与物料平衡 heat and material balance
热量预测器 heat anticipator
热量再生 heat regeneration
热量再生装置 thermal regenerator
热量值 calorie value
热量指示器 heat indicator
热量指数 heat index
热量转换系数 calorie conversion factor
热量状况 heat condition; heat regime
热量总和 heat summation; sigma heat
热量总和恒等原理 principle of constant-heat summation
热量总损失 overall heat loss
热料仓 hot aggregate bin
热料工 hot stuff man
热料筛 hot screen
热料筛析 hot screen separation
热料升送机 hot elevator
热料提升机 hot aggregate elevator; hot elevator
热料振动筛 hot aggregate vibration screen
热烈 violence
热烈的 strenuous; violent
热烈的活动 feverish activity
热裂 fire crack(ing); heat check(ing); hot crack; auto crack; decrepitate; pull crack; shrinkage crack(ing); thermal crack(ing); thermal fragment(ation); thermal stress cracking
热裂变钚 thermally fissile Pu
热裂地沥青 cracked asphalt
热裂法 cracking process
热裂化 thermal crack(ing)
热裂化粗汽油 pressure naphtha
热裂化过程 thermal cracking process
热裂化馏出物 pressure-still distillate
热裂化汽油 pressure gasoline
热裂化石脑油 thermal naphtha
热裂化装置 thermal cracker; thermal cracking unit
热裂解 pyrolytic cracking; thermal splitting; thermo-fracture; thermal crack(ing)
热裂解气 thermally cracked gas
热裂解气相色谱法 thermal cracking gas chromatography
热裂谱法 thermofractography
热裂(倾向性)试验 <金属焊接中的> hot-crack(ing) test
热裂设备 cracking unit
热裂炭黑 thermal black; thermatomic black
热裂纹 fire check; heat crack; hot check; thermal crack(ing)
热裂(纹)试验 hot-crack(ing) test
热裂性 hot crackability
热裂作用 pyrolytic cracking
热临界带 thermal critical range
热临界点 thermal critical point
热临界堆 hot critical reactor
热磷光体 thermophosphor
热灵敏 heat-sensing
热灵敏度 heat sensitivity
热灵敏器件 heat sensing device
热灵敏性 thermal sensibility
热灵敏装置 heat sensing device
热流 flow of heat; flux of heat; heat(ing) current; heat stream; heat transmission; hot flow; rate of heat transfer; thermal flow; thermal flux; warm flow
热流变的 thermorheologic(al)
热流变性能 thermorheologic(al) property
热流变学 thermorheology
热流槽 heated launder
热流测量 heat flow measurement; heat flow survey
热流传感器 heat flow sensor; heat flow transducer; heat flux transducer
热流道 hot runner; insulated runner
热流道模 hot-runner mo(u)ld
热流动 heat flow; heat flux
热流方程 heat equation; heat flow equation
热流分布 distribution of heat flow; distribution of heat flux; heat flux distribution
热流概念 concept of heat flow
热流节热器 heat regenerator
热流量 heat flow rate; heat flux; heat-transfer rate; rate of heat flow; thermal discharge; thermal flux; thermoflux
热流量变化 heat flow variation
热流量单位 heat flow unit
热流量等值线图 map of heat flow contours
热流量计 heat flow meter; thermal flowmeter
热流量密度 density of heat flow rate
热流率 rate of heat flow
热流密度 heat flux density
热流平衡 heat flow balance
热流强度 heat flow rate per unit area
热流闪烁 heat shimmer
热流式差示扫描量热法 heat flux differential scanning calorimetry
热流速率 heat flow rate
热流速率密度 density of heat flow rate
热流特性 heat flow property
热流体 hot fluid; thermal fluid
热流体动力不稳定性 thermohydrodynamic(al) instability
热流体动力的 thermohydrodynamic(al)
热流体动力方程 thermohydrodynamic(al) equation
热流体系统 thermofluid system
热流体压损 pressure loss of heat-transporting fluid
热流通道 heat passage
热流通量 heat flux
热流图 heat flow diagram
热流温度 heat flow temperature
热流问题 heat flow problem
热流线路 heat flow path
热流向量 heat flow vector
热流循环 thermal convection cell
热流异常 heat flow anomaly
热流异常带 heat flow anomaly zone
热流逸 thermal transpiration
热流远距离测量 heat flow remote measurement
热流值 heat flow value
热流值研究 thermal flow study
热流状况 condition of heat flow
热流阻力 heat flow resistance
热硫化 heat cure; heat curing; heat vulcanization; hot curing; hot vulcanization
热硫化促进剂 heat curing accelerator
热硫化法 heat curing system
热硫化胶 heat vulcanizate
热漏失 heat leak(age)
热漏泄 heat leak(age)
热漏泄系数 thermal leakage coefficient
热炉次 hot heat
热炉顶 hot top
热卤化作用 thermal halogenation
热卤水 hot brine
热卤水沉积 hot brine deposit
热卤水沉积作用 hot brine deposition
热卤水成矿说 hot brine ore-forming theory
热卤水成矿作用 mineralization of thermal brine
热卤水矿调查 hot brine mineral survey
热卤水田 hot brine field
热铝焊接 aluminothermic welding
热氯化 thermal chlorination
热轮 thermal wheel
热落差 heat drop
热马力 thermal horsepower; truemotor load
热霾 heat haze
热脉冲 thermal pulse
热脉冲测定法 heat impulsive method
热脉冲法 thermal pulse method
热脉冲法导热系数测定仪 thermal conductivity measuring apparatus by heat pulse
热铆 heat riveting; hot driving; rivet hot
热铆的铆钉 hot driven rivet
热铆钉接送工 rivet catcher
热铆固 hot-riveted
热铆合 hot riveting
热铆铆钉 hot driven rivet; pop rivet
热冒口 hot riser; live riser
热帽发热剂 feedex; lunkerite
热帽钢锭 hot topped ingot
热媒 heat carrier; heat carrying agent; heating element; heat(ing) medium; medium of heat transmission
热媒参数 heating medium parameter
热媒平均温度 mean temperature of heat medium
热煤焦油覆盖 hot-applied coal tar coating; hot coal-tar coating
热煤气 heating gas; hot coal gas
热煤油干燥 hot kerosene drying
热门股票 blue chip; fancy stocks; hot issue
热门货(品) fast-selling products; goods in great demand; goods which sell well; hot items; hot number; hots; popular ware; products become hot on the market
热门商号 blue chip firm
热门证券 blue chip
热密度波动 thermal density fluctuation
热面 back surface; hot face; hot side
热面层 hot face zone
热面点火机 surface ignition engine
热面器 surface heater
热面切割系统 hot face cutting system
热面预热 hot spot application
热面造粒系统 hot face pelletizing system
热灭活点 thermal death inactivation point
热灭活法 thermoinactivation
热灭菌 heat sterilization
热敏 sensitive to heat
热敏材料 sensitive to heat; thermosensitive material
热敏参数 thermal sensitive parameter
热敏成像法 thermography
热敏传感器 heat pick-up; heat-sensitive sensor; heat-sensitive transducer; thermosensitive transducer; heat sensing device
热敏式打印机 thermal printer
热敏的 heat-sensitive; heat variable; temperature sensing; temperature sensitive; thermally sensitive; thermosensitive
热敏电缆 heat-sensitive cable
热敏电阻 critesistor; heat-variable resistance; heat variable resistor; sensistor; temperature-dependent resistor; thermally sensitive resistance; thermistance; thermistor; thermal resistance
热敏电阻玻璃温度计 glass thermistor thermometer
热敏电阻补偿 sensistor compensation; thermistor compensation
热敏电阻材料 thermistor composition; thermistor material
热敏电阻测辐射热计 thermistor bolometer
热敏电阻测辐射热仪 thermistor bolometer detector
热敏电阻测热器 thermistor heat detector cell
热敏电阻传感器 thermistor probe; thermistor temperature sensor
热敏电阻电桥 thermistor bridge
热敏电阻定温火灾探测器 fixed temperature detector with heat sensitive resistance
热敏电阻风速计 hot-wire anemometer
热敏电阻辐射热测量器 thermistor detector
热敏电阻功率计 thermistor power

meter
热敏电阻合金 thermistor alloy
热敏电阻检测器 thermistor bolometer; thermistor controller; thermistor detector
热敏电阻列线图 thermistor nomogram
热敏电阻器 heat variable resistor; thermosensitive resistor; thermal resistor; thermister[thermistor]
热敏电阻桥形电路 thermistor bridge circuit
热敏电阻式测量仪 thermistor ga(u)ge
热敏电阻式辐射热测量计 thermistor bolometer
热敏电阻式压力计 thermal ga(u)ge; thermistor ga(u)ge
热敏电阻衰老特性 ag(e)ing characteristics of thermistor
热敏电阻温度表 thermistor thermometer
热敏电阻温度计 thermistor thermometer
热敏电阻型电动机保护装置 thermistor type motor protection
热敏电阻延迟继电器 thermistor time delay relay
热敏电阻真空计 thermistor vacuum ga-(u)ge
热敏电阻振荡器 thermistor oscillator
热敏电阻座 thermistor mount
热敏度 thermal sensitivity
热敏阀 thermovalve
热敏复制术 thermic copying
热敏感器 temperature sensitization apparatus
热敏化剂 heat sensitizing agent
热敏化装置 temperature sensitization apparatus
热敏化作用 heat sensitization
热敏混炼 heat sensitized mixing
热敏剂 thermal sensitizer
热敏继电保护装置 thermal relay
热敏继电器 thermorelay
热敏金属 thermometal
热敏开关 thermal-responsive switch; thermoswitch
热敏控制剂 thermocontroller
热敏蜡笔 tempil stick
热敏黏[粘]合剂 heat-sensitive adhesive agent
热敏器件 heat-sensitive device
热敏区 sensitive volume
热敏摄像管 thermicon
热敏式风速计 thermistor anemometer
热敏塑料 heat-sensitive plastics
热敏探测器 heat responsive detector
热敏探示器 thermistor probe
热敏陶瓷 thermal sensitive ceramic; thermosensitive ceramics
热敏铁氧体 heat-sensitive ferrite
热敏涂料 thermosensitive paint
热敏效应 thermal sensitive effect; thermosensitive effect
热敏型恒温器 thermistor type thermostat
热敏性 heat sensitivity; susceptibility to heat; thermal sensitivity
热敏性材料 heat-sensitive material
热敏性胶乳 heat-sensitive latex
热敏油漆 heat-sensitive paint; heat-sensitive varnish; temperature sensitive paint; thermocolo(u)r; thermopaint; thermopress paint; thermosensitive paint
热敏元件 heat responsive element; heat-sensitive element; heat-sensitive eye; heat-sensitive sensor; heat sensor; temperature bulb; temperature detector; temperature sensing element; temperature sensitive element; temperature sensor; thermal element; thermal sensitive element; thermal sensor; thermic element; thermoelement; thermosensor
热敏原料 sensitive stock; thermal sensitive stock
热敏真空计 thermistor vacuum ga(u)ge
热敏纸 heat-sensitive paper
热敏转印纸 heat-sensitive transfer sheet
热敏装置 temperature sensing device
热模 heated mo(u)ld
热模吹制 hot(iron)mo(u)ld blowing
热模锻 die forging; hot (closed-) die forging; hot forming; hot mill; hot-press forge
热模锻造 warm-die forging
热模横切割 hot die-face cutting
热模精压 hot size; pants press
热模具钢 hot working steel
热模拟 thermal simulation
热模拟法 thermal analogy method
热模拟计算机 thermal analog computer
热模拟实验 thermal simulation experiment
热模塑 hot mo(u)ld
热模烫印 die stamping
热模型 hot mo(u)ld; thermal model
热模压 hot forming; hot mo(u)lding; pressure forging
热模压法 heated-die pressing process
热模压机 extrusion press
热模压加工 extrusion pressing
热模压制表面镶片 mo(u)lded facing
热模压制(离合器)表面镶片 mo(u)ld facing
热模养护 heating form maintenance
热模制 hot mo(u)lding
热模制品 hot iron mo(u)ld ware
热膜风速表 hot film anemometer
热膜风速计 hot film anemometer
热膜流速仪 hot film current meter
热磨合 hot breaking-in
热磨水泥 hot ground cement
热目标 hot target
热内边界层 thermal internal boundary layer
热那亚港 Port Genoa
热耐久性 thermal endurance
热挠曲试验 heat flexibility test
热挠曲温度 heat deflection temperature
热能 caloric power; heated energy; heat(ing) energy; steam energy; temperature energy; thermal energy; thermal power; thermopower
热能变成电能过程的有效作用系数 heat-to-power efficiency
热能变换器 thermal energy converter
热能产量 heat production
热能储存率 storage rate of thermal energy
热能当量 thermal energy yield
热能对流传递量 convective transport
热能发散 emission of heat
热能方程 thermal equation of energy
热能分析 thermal energy analysis
热能辐射 thermal radiation
热能辐射衰减云 thermal radiation attenuating cloud
热能感受器 caloreceptor
热能供应 heat-supply
热能供应网 heat-supply network
热能红外线扫描器 thermal infrared scanner
热能化 thermalization; thermalize
热能回收 heat recovery
热能级 thermal level
热能价格 heat price
热能交换器 thermal energy converter
热能聚合作用 thermal polymerization
热能勘测任务 heat capacity mapping mission
热能可逆性 thermal reversibility
热能流程图 heat flow diagram
热能流动 heat conduction
热能脉冲 heat pulse
热能密度 heat density; thermal energy density
热能平衡 thermal energy balance
热能区 thermal energy range; thermal energy region
热能生产 heat generation; heat production
热能生产量 heat production rate
热能输入 thermal energy input
热能输送成本 heat transmission cost
热能损失 heat energy loss; loss on heat
热能头 thermal head
热能隙 thermal energy gap
热能消耗 thermal energy consumption
热能需求量 caloric requirement; heat requirement
热能液流系统 transport system
热能中子 thermal neutron; thermal velocity neutron
热能转换 conversion of thermal energy; heat conversion
热泥 hot mud; warm sludge
热泥流 hot lahar; hot mud flow
热泥塘 swamp of hot mud
热黏[粘]度 thermal viscosity; thermoviscosity
热黏[粘]度计 thermoviscosimeter
热黏[粘]附 thermal adhesion
热黏[粘]合 heat bonding; thermal adhesion; thermal bonding
热黏[粘]合剂 hot binder; thermal adhesive
热黏[粘]合料 hot binder; hot bonding composition; hot bonding compound; hot cement(ing) composition
热黏[粘]合纤维 heat bondable fiber [fibre]
热黏[粘]合织物 heat bonded fabric
热黏[粘]结 heat bonding
热黏[粘]结材料 hot adhesive composition
热黏[粘]弹性 thermoviscoelasticity
热黏[粘]纤维 heat bonded fiber
热黏[粘]性 hot tack; thermal viscosity
热黏[粘]着性 heat-blocking
热碾 hot rolling
热碾的 hot-rolled
热碾钢板 hot rolled sheet
热碾压 hot rolling
热碾压沥青基层 hot-rolled asphalt base layer
热捏和机 heat pugmill(mixer)
热凝材料 thermoset material
热凝的 thermosetting
热凝固 heat set(ting); hot set
热凝固剂 heat coagulant
热凝固术 thermocoagulation
热凝结 heat set(ting); thermal coagulation
热凝结胶 hot-setting glue
热凝聚 thermal coagulation
热凝黏[粘]合剂 hot-setting adhesive
热凝黏[粘]结剂 hot-setting adhesive
热凝砂浆 hot-setting mortar
热凝物 thermoset
热凝性 thermosetting property
热扭变 distortion under heat; heat distortion
热扭变点 heat distortion point
热扭变温度 heat distortion temperature
热浓碱 hot concentrated alkali
热偶 thermal electric(al) couple; thermoelectric(al) couple
热偶电池 thermoelectric(al) battery; thermogenerator
热偶堆供电的无线电接收机 thermopile radio
热偶发电器 thermogenerator
热偶伏特计 thermovoltmeter
热偶偶极子 thermocouple dipole
热偶式土壤温度计 thermosoil-hygrometer
热偶真空计 thermocouple ga(u)ge
热耦合系数 heat couple coefficient
热爬距 creeping length
热排放 heat effluent; hot driving; thermal discharge; heat discharge
热排放物 thermal effluent
热排聚形 thermotaxy
热排性 thermotaxy
热盘管 heating coil; hot coil
热盘管表 heat balance sheet
热刨机 hot planing machine
热炮管套 thermal tube jacket
热泡 thermal
热泡沫 thermal foam
热配套管 shrunk-on sleeve
热喷镀 thermal spray
热喷淋机 hot spraying machine
热喷铝 alumin(i)um metal spray
热喷漆 hot lacquering; hot spray paint
热喷射 thermojet
热喷射噪声 heat injection noise
热喷涂 heating spraying; hot gunning; thermal spraying; thermojet
热喷涂道路划线漆 hot spray traffic paint; melt coating traffic paint
热喷涂法 hot spray process
热喷涂料 hot spray(ing) material
热喷涂溶解型船底漆 hot venetian type ship bottom paint
热喷涂涂料 hot spray coating
热喷涂装置 hot spray apparatus
热喷丸 hot peening
热喷雾 thermal spray; warm spraying
热喷雾机 hot spraying machine
热喷雾枪 hot spraying pistol
热喷雾液相色谱法质谱法联用 thermospray liquid chromatography-mass spectrometry
热喷型路标漆 hot spray traffic paint
热膨辐射探测 thermal expansion detection
热膨胀 heat expansion; temperature expansion; thermal dilatation; thermal expansion
热膨胀补偿器 thermal expansion compensator
热膨胀测定 thermal expansion measurement; thermal expansion test
热膨胀测量仪 expansion indicator
热膨胀产生的裂缝 thermal fracture
热膨胀的适应性 thermal expansion compatibility
热膨胀阀 thermal expansion valve
热膨胀法 thermodilatometry
热膨胀反应性系数 thermal expansion reactivity coefficient
热膨胀分析 thermal expansion analysis; thermodilatometric analysis
热膨胀缝 thermal expansion joint
热膨胀及热收缩 thermal expansion and contraction
热膨胀继电器 thermal expansion relay

热膨胀率 thermal expansivity
热膨胀模塑成型 thermal expansion mo(u)lding
热膨胀曲线 thermal expansion curve
热膨胀试验 thermal expansion test
热膨胀特性 thermal expansion character
热膨胀系数 coefficient of expansion by heat; coefficient of heat expansion; coefficient of thermal expansion; factor of expansion; heat expansion coefficient; heat-stretch factor; thermal coefficient of expansion; thermal expansion factor; thermal expansivity; thermal expansion coefficient
热膨胀线性系数 linear coefficient of thermal expansion
热膨胀性 heat expansibility; thermal expansibility; thermal expansivity
热膨胀仪 thermal dilatometer; thermal expansion instrument
热膨胀因数 factor of expansion; thermal expansion factor
热碰撞 thermalizing collision
热疲劳 heat fatigue; thermal fatigue
热疲劳断裂 thermal fatigue fracture
热疲劳试验 thermal fatigue test
热片安培计 hot-strip ammeter
热偏转 thermal deflection
热漂移 heat drift; thermal drift
热平衡 heat balance [balancing]; balance of heat; calorific balance; heat equilibrium; thermal budget; thermal equilibration; thermal equilibrium
热平衡表 heat balance table
热平衡法 method of heat budget
热平衡恢复 recovery of thermal equilibrium
热平衡计算 computation of heat balance; heat balance computation; heat account
热平衡流量计 heat balance flowmeter
热平衡器 heat balancer; heat compensator
热平衡时间 thermal balance time
热平衡条件 thermal equilibrium condition
热平衡图 heat balance diagram
热平整机 hot level(l)ing machine
热屏 shielding heat; thermal shield
热屏蔽 heat barrier; heat shielder; heat shield(ing); thermal shield(ing); thermoscreen; thermoshield
热屏蔽层 thermal shield(ing)
热屏蔽阴极 heat-shielded cathode
热坡度 thermal slope
热泼 hot application
热泼法 heat pour method
热破坏 heat collapse; heat damage; heat erosion; thermal break; thermal collapse; thermodestruction
热破坏性试验 heat-rupture test
热破裂 thermal fracture
热剖面 thermal profile
热铺 hot laying
热铺表面 warm-laid surfacing
热铺粗粒焦油混凝土 hot-laid coarse tar concrete
热铺的 hot-laid; laid hot
热铺地沥青混凝土路面 hot asphaltic concrete pavement
热铺法 <建筑沥青路面用的> hot process; hot-laid method
热铺法施工 hot-laid process of construction
热铺沥青 hot-applied bitumen; hot-laid mixture
热铺沥青混合料 hot-laid asphalt mixture
热铺沥青混合物 hot-laid asphalt mixture
热铺沥青混凝土 hot asphalt concrete; hot-laid asphaltic concrete; hot-laid bituminous concrete
热铺沥青砂浆 hot bituminous mortar
热铺沥青碎石路 hot-laid bituminous macadam
热铺路面 hot-laid pavement
热铺密封剂 hot-poured sealing compound
热铺面层 hot-laid surfacing
热铺碾压沥青 hot-laid rolled asphalt
热铺细粒焦油混凝土 hot-laid fine tar concrete
热谱 pyrograph; thermal spectrum
热谱法 pyrography; thermography
热漆 hot varnish
热漆划线机 hot paint road marking machine
热奇异点 thermal singularity
热启动 warm boost; warm start
热启动灯 hot start lamp
热启动排放物 warn start emissions
热启动装置 heat-actuated device; heat-actuated means
热起动灯 hot start lamp
热起伏 thermal fluctuation
热气 hot gas; reek
热气标准循环 hot-air standard cycle
热气剥漆器 hot-air stripper
热气测量 thermal gas lens measurement
热气出口 heat outlet
热气传递 transference of heat
热气传送量 capacity of heat transmission
热气动控制系统 thermopneumatic control system
热气动弹性力学 thermoaeroelasticity
热气发生设备 hot-gas plant
热气防冰设备 hot-air deicer
热气防冻设备 hot-air deicer
热气放射 heat emission
热气干燥 desiccation
热气干燥法 hot-air seasoning
热气干燥窑 hot-air drying kiln
热气供暖器 hot-air heater
热气供暖系统 perimeter warm air heating system
热气管 caliduct
热气管线 hot-gas line
热气焊接 hot-gas welding
热气烘干 <木材> hot-air drying; hot-air seasoning
热气烘箱 hot-air cabinet
热气化器 thermal vapo(u)riser[vapo(u)rizer]
热气机 heat engine; hot ah engine
热气集合器 hot-air collector; hot-air couplection
热气集合体 hot-air couplection
热气加热器 hot-air heater
热气进口 hot-air intake
热气净化 hot-gas purification
热气烤干 flue curing
热气流 hot current; thermal current
热气流分量 hot-gas fraction
热气流焊接 hot-gas reflow soldering; hot-gas soldering
热气流喷射变形工艺 hot fluid jet texturing process
热气漏斗 hot-air funnel
热气滤池 thermal aerobic filter
热气灭菌器 hot-air sterilizer
热气排除 heat elimination
热气旁通管 hot-gas by-pass
热气泡 thermal bubble
热气喷枪 hot-air gun
热气驱动发电机 hot-gas driven generator
热气融霜 hot-gas defrosting
热气融霜管路 hot-gas line for defrosting
热气式轻便弥雾机 hot pneumatic type portable sprayer
热气试验 hot-air test
热气室 laconicum
热气伺服机构 hot-gas servo
热气伺服系统 hot-gas servosystem
热气速干 flash drying with hot air
热气体除霜 hot-gas defrosting
热气体多级增压机 hot-gas multi-stage booster
热气体分层 stratification of hot-gas
热气体管 hot-gas line
热气体火焰 hot-gas flame
热气体旁路 hot-gas by-pass
热气体透镜 thermal gas lens
热气团云 pelean cloud
热气稳定系统 hot-gas stabilizing system
热气相色谱法 thermogas chromatography
热气消耗 heat consumption
热气循环法 hot-gas recycle process
热气循环过程 hot-gas recycle process
热气压成型 hot-gas pressure compaction
热气压结 hot-gas squeeze
热气养护 hot-air treatment
热气蒸汽硫化 hot-air steam cure
热气轴承 hot-gas bearing
热气装置 hot-gas system
热气总管 hot gas main
热汽缸 heat cylinder; hot cylinder
热汽灭菌法 hot-air sterilization
热汽浴 hot-air bath
热汽浴室 sudatorium[复 subdatoria/subdatory]
热迁移 thermal migration; thermophoresis; transfer heat
热前置 thermal lead
热嵌配合 shrunk fit
热嵌石屑法 hot-pressed chippings
热强度 heat intensity; het flux; high-temperature strength; intensity of heat; thermal intensity
热强度指标 heat stress index
热强度指数 heat stress index; hot strength index
热强钢 heat-resisting steel; refractory steel
热强化 thermal tempering
热强化玻璃 heat-strengthened glass
热强性 heat resistance
热桥 heat bridge; heat channel; heat leak(age)
热切刀 hot knife
热切的 earnest
热切法 hot cut method
热切割 hot cutting; thermal cutting
热切扩口机 hot cut flare machine
热切毛边 hot trimming
热切削 hot cutting
热侵入 hot intrusive
热侵入体 hot intrusion
热侵蚀 heat erosion; thermal erosion
热清除 heat cleaned
热清洁处理 thermal cleaning
热清洁法 heat cleaning
热清洗 heat cleaning; thermal cleaning; warm wash
热情 warmth
热球 hot-bulb
热球点火 hot-bulb ignition
热球(式)柴油机 hot-bulb engine; hot surface ignition engine
热球式电风速计 heat-bulb electric(al) velometer; hot-bulb anemometer; thermoglobe anemometer
热球式发动机 semi-diesel engine; surface ignition engine
热球式风速表 hot ball type anemometer
热球式温度计 thermal bulb-type thermometer
热球装置 hot-bulb arrangement
热区 hot space; hot zone; thermal range; thermal region
热驱动 thermal drive
热驱动的 thermally driven
热驱动防火门 heat-actuated fire door
热驱动装置 thermal driven device
热去磁 thermal demagnetization
热去极化 thermally depoled
热圈 gasket
热全分析 total thermal analysis
热泉 hot spring; thermal mineral spring; thermal spring; thermcale
热泉气 hot spring gas
热泉气雾 hot spring vapo(u)r
热泉泉华 hot spring sinter
热泉水 thermal water
热泉沼泽 hot spring marsh
热缺陷 thermal defect
热染色 heat tinting
热扰动 thermal agitation; thermal disturbance
热人 <耐燃试验用人体模型> thermal man; thermoman
热韧化(处理) heat-toughened
热日 thermal day
热容 thermal capacitance
热容比 heat capacity ratio; rate of specific heat; ratio of specific heat
热容成像 heat capacity mapping
热容绝缘 capacity insulation
热容量 caloric receptivity; calorific capacity; calorific receptivity; capacity for heat; heat absorption capacity; heat carrying capacity; heat enthalpy; heating capacity; thermal capacity; caloricity; capacity of heat; heat(storage) capacity
热容量比 ratio of heat capacities
热容量成像辐射计 heat capacity mapping radiometer
热容量成像卫星 heat capacity mapping mission
热容量的量子理论 quantum theory of heat capacity
热容量定额 thermal capacity rating
热容量额定值 thermal capacity rating
热容量分析 thermovolumetric analysis
热容量系数 heat content coefficient
热容量值 thermal capacity value
热容量制图卫星 heat capacity mapping mission
热容量滞后 heat capacity lag
热容流率 capacitance rate
热容率 heat capacitivity
热容强度 volumetric(al) heat release rate
热容式冷却 capacitive cooling
热溶剂箱清洗 hot solvent tank cleaning
热溶剂浴 hot solvent bath
热溶胶 thermosol
热熔保险器 thermal cutout
热熔穿透计 penetrometer
热熔断路器 thermal cutout
热熔断器 thermal cutoff
热熔法 torch-applied method
热熔分离制牙边法 hot-wire scalloping
热熔管 temperature bulb
热熔焊(接) fusion welding; sweat(ing) soldering; thermofusion weld-

ing;exothermic welding
热熔滑动 thaw frozen landslide
热熔滑坡 thaw frozen landslide
热熔滑坍 thaw frozen landslide
热熔化 heat melting-down;hot melt
热熔挤出机 hot-melt extruder
热熔胶 hot-melt adhesive;hot glue
热熔胶合剂 heat-activated adhesive
热熔接缝 heat refused joint
热熔料 hot melt
热熔密封膏 hot-melt sealant
热熔黏[粘]合环氧涂层 heat-fusion-bonded epoxy coating
热熔黏[粘]合剂 hot-melt adhesive
热熔黏[粘]接剂 hot-melt adhesive
热熔黏[粘]结法 hot-melt method
热熔黏[粘]结剂 hot-melt adhesive
热熔喷镀法 flame spraying
热熔强度 hot-melt strength
热熔融 hot melt
热熔融涂装法 hot-melt coating
热熔融性胶黏[粘]剂 melted cement
热熔施工 hot-melt application
热熔丝 heating fuse;thermofuse
热熔塑料条 hot-melt plastic stripe
热熔塑性 thermoplasticity
热熔体 hot melt
热熔体涂布机 hot-melt coater
热熔体涂装器 hot-melt applicator
热熔铜焊 sweating brazing
热熔涂布 heat seeling coating
热熔涂装 hot-melt painting
热熔线圈 heat(ing) coil
热熔线圈断路器 thermal circuit-breaker
热熔型道路划线 melt coating traffic paint
热熔型路标漆 hot-melt road marking paint
热熔型温度计 fuse-type temperature meter
热熔性 hot melt
热熔性胶黏[粘]剂 heat seal(ing) adhesive; heat-welding adhesive; hot adhesive; melted cement; hot-melt adhesive
热熔性喷枪 hot-melt gun
热熔性涂布 hot-melt coating
热熔印刷 thermography
热熔油墨 plastisol ink
热熔油毡 torch-applied asphalt felt; torching roofing felt
热融滑塌 heat-induced collapse
热融滑坍 solifluction [solifluxion]; thaw frozen landslide
热柔韧性试验 heat flexibility test
热软化 thermal softening
热软化点 heat softening point
热软化挤出装置 hot-melt extruder
热软化剂 hot melt agent
热软化树脂 heat softened resin
热润滑器 thermal lubricator
热塞柴油机 hot plug engine
热三氯乙酸 hot trichloroacetic acid
热散布 thermal dispersion
热散发 thermal run-away
热散焦 thermal defocussing
热散射 heat scatter(ing);temperature scattering;thermal scatter(ing)
热散射器 heating radiator
热散逸 heat diffusion; heat dissipation; thermal dispersion; thermal run-away
热骚动 thermal agitation;thermal excitation
热骚动电压 thermal agitation voltage
热扫描仪 thermal scanner
热色 hot colo(u)r
热色棒 tempil stick
热色谱法 chromatothermography;hot chromatography
热色显示 thermochromic display
热色现象 thermochromism
热色效应 thermochromatic effect
热色性 thermochromatism; thermochromy
热色性能 thermochromatic property
热杀 heat kill
热杀的 heat killed
热杀菌时间 thermal reduction time
热沙漠 hot desert
热砂浆 heated mortar
热砂试验 hot desert test
热闪 heat lightning
热闪蒸法 thermoflash process
热熵 thermal entropy
热上升气流 thermal updraft
热烧结 thermal sintering
热设计 thermal design
热射病 heat stroke;thermoplegia
热射电辐射 thermal radio waves
热射流 thermojet
热射式风洞 hotshot wind tunnel
热射线 heat ray
热射线摄影术 thermoradiography
热摄影法 thermographic(al) process
热摄影术 thermography
热伸长 thermal elongation; thermal expansion; thermal stretcher; thermal stretch(ing)
热渗 thermoosmosis
热渗透 heat channel;heat leak(age)
热渗透性 heat permeability; thermoosmosis
热渗透作用 thermoosmosis
热渗析 thermodialysis
热渗系数 coefficient of thermoosmotic transmission
热渗眼 thermal seepage
热生应力 thermally induced stress
热声法 heat-sound method
热声筛选机 heat sonic screening machine
热声学法 thermoacoustimetry
热声阵 thermoacoustic array
热声自动记录仪 heat-sound automatic recorder;thermoacoustic(al) recording meter
热剩磁 thermoremanent magnetism
热剩余磁化强度 thermoremanent magnetization
热剩余磁性 thermoremanent magnetism
热剩余磁性的自反转 self-reversal of thermoremanent magnetism
热剩余再磁化 thermoremanent remagnetization
热失控 thermal breakdown; thermal run-away;thermorunaway
热失稳 thermal buckling
热失重 thermal loss in weight
热失重法 thermogravimetry
热失重分析 thermogravimetric analysis
热失重曲线 thermogravimetric curve
热施工沥青 hot-applied bitumen
热施工密封膏 hot-applied sealant
热施工涂料 hot-applied coating
热施型防锈剂 hot application type rust preventive
热湿比 angle scale;heat and humidity ratio;heat moisture ratio
热湿传导 thermohydro conduction
热湿度表 thermal hygrometer
热湿交换 heat and moisture transfer
热湿气候 hot humid climate
热湿润 hot wetting
热石膏 plaster
热石灰软化法 warm lime softening
热时标 thermal time scale
热时间常数 thermal time constant
热时间继电器 thermal time relay
热时效 heat ag(e)ing
热实验室 <强放射性物质实验室> hot lab(oratory)
热蚀 thermal abrasion; thermal erosion
热蚀法 thermal etching
热史 thermal history
热式尘埃计 thermal dust precipitator
热式风速表 thermal type anemometer
热式复制术 raised-letter printing
热式聚尘器 thermal dust collector
热式流量计 thermal flowmeter
热势 thermal potential
热势差 thermal potential difference
热视 thermovision
热试车 heat run;hot firing
热试法 heat test(ing)
热试验 heat test(ing)
热试验法 heat test method
热试验间 firing bay
热试验容器 hot chamber
热试转 heat run
热室 hot cave;hot cell;hot chamber; hot laboratory cave;hot room
热室压铸法 hot chamber die casting
热室压铸机 hot chamber die casting machine;hot chamber machine
热适应 acclimation to heat; heat acclimatization
热适应系数 thermal accommodation coefficient
热适应作用 heat adaptation
热释测辐射热器 pyroelectric(al) bolometer
热释传感器 pyroelectric(al) sensor
热释电 pyroelectricity
热释电材料 pyroelectric(al) material
热释电晶体 pyroelectric(al) crystal
热释电式光传感器 pyroelectric(al) optic(al) sensor; pyroelectric(al) optic(al) transducer
热释电式温度传感器 pyroelectric(al) temperature sensor; pyroelectric(al) temperature transducer
热释电陶瓷 pyroelectric(al) ceramics
热释电体 pyroelectrics
热释电效应 pyroelectric(al) effect
热释电性 pyroelectricity
热释电学 pyroelectricity
热释发光法 heat release luminescent method
热释放 heat release; thermal discharge
热释放动力学 heat release kinetics
热释放率 rate of heat release
热释放器 thermal releaser
热释分析 thermal release analysis
热释辐射探测器 pyroelectric(al) radiation detector
热释复制法 pyroelectric(al) copying process
热释光 thermal glow;thermoluminescence
热释光材料 material of heat releasing light
热释光测量效率 effectiveness of heat releasing light method
热释光测年法 thermoluminescence dating method
热释光的 thermoluminescent
热释光等值图 contour map of heat releasing light
热释光峰 thermal glow peak
热释光剂量计 thermoluminescent dosimeter
热释光剂量学 thermoluminescent dosimetry
热释光判断年代法 thermoluminescence dating
热释光探测器 thermoluminescence detector; thermoluminescent detector
热释光总光量 total light amount of heat releasing light
热释技术 pyroelectric(al) technology
热释晶体 pyroelectric(al) crystal
热释能量计 pyroelectric(al) energy meter
热释热探测器 pyroelectric(al) thermal detector
热释视像管 pyroelectric(al) vidicon
热释探测器 pyroelectric(al) detector
热释显示 pyroelectric(al) display
热释效应 pyroelectric(al) effect
热释信号 pyroelectric(al) signal
热释装置 pyroelectric(al) device
热收集器 heat trap;thermal trap
热收缩 thermal contraction; thermal pinch;thermal shrinkage
热收缩包装 thermal shrink packaging
热收缩材料 heat-shrinkage material
热收缩差异 thermal shrinkage differential
热收缩纤维 heat-shrinkable fiber[fibre]
热收缩效应 thermal pinch
热收缩应力 thermal shrinkage stress
热寿命 thermal endurance; thermal life
热受体 thermal acceptor
热受主 thermal acceptor
热疏水 thermal drain
热舒适 thermal comfort
热舒适通风 thermal comfort ventilation
热舒适指标 thermal comfort index
热输出 heat egress
热输出功率 heating output
热输出力 heating output
热输出量 heating output
热输入 firing rate
热输入量 heat input
热输送 heat transmission
热输运 heat transport
热刷 hot spraying
热刷层 mopping coat
热衰减 heat fade
热衰竭 heat exhaustion;heat prostration
热衰退 heat fading
热水 aqua fervens;hot-water;thermal water; warmed water; hydrothermal water
热水泵 heat-exchanger pump;hot-water pump
热水泵缸衬 hot-water pump cylinder liner
热水泵缸盖 hot-water pump cylinder cover
热水泵活塞 hot-water pump piston
热水泵活塞填密盖 hot-water pump piston follower
热水表面 hot-water surface
热水采暖系统 hot-water heating system
热水采暖装置 hot-water heater
热水槽 hot-water storage; hot-water tank
热水沉积 hot-water deposit; hydrothermal deposit
热水成本 hot-water cost
热水池 hot pool;hot-water basin;hot-water tank;warm water lagoon
热水池泵 hot well pump
热水冲洗 hot water injection
热水冲洗方案 hot-water wash scheme
热水抽吸泵 hot-water extraction

R

pump
热水出口 outlets of thermal water
热水储备设备 hot-water storage plant
热水储槽 hot-water storage tank
热水储存系统 hot-water storage system
热水储水箱 hot-water storage tank
热水储蓄设备 hot well
热水处理(法) hot-water cure; hot-water treatment; hydrothermal treatment
热水袋 hot-water bag
热水的 hydrothermal; hydrothermic
热水电加热器 electric(al) hot water
热水对流器 hot-water convector
热水发生器 hot-water generator
热水放热器 hot-water heat emitter
热水浮选 hot-water flo(a)tation
热水腐蚀 corrosion by thermal water
热水腐蚀性侵蚀 corrosion attack by thermal water
热水干管 hot-water supply main
热水干燥槽 hot-water drying tank
热水供给 hot-water service; hot-water supply
热水供给干线 hot-water supply main
热水供给管线 hot-water supply line
热水供暖 hot heating; hydronic heating; water heat(ing)
热水供暖泵 hot-water heating pump
热水供暖负荷 hot-water heating load
热水供暖管 hot-water heating pipe
热水供暖立管 hot-water heating riser
热水供暖器 hot-water calorifier
热水供暖散热器 hot-water heating radiator
热水供暖设备 hot-water heating equipment
热水供暖设计 hot-water heating design
热水供暖时间表 hot-water heating schedule
热水供暖温度 hot-water heating temperature
热水供暖系统 hot-water heating system; hydronic heating system
热水供暖运行 hot-water heating operation
热水供暖装置 hot-water heating installation
热水供热方法 warm water method
热水供热系统 steam heat-supply system; hot water heating system
热水供水管 hot-water supply
热水供应 hot-water service; hot-water supply
热水供应负荷 hot-water supply load
热水供应热负荷 hot-water heating load
热水供应设备 hot-water apparatus
热水供应室 hot-water service room; hot-water supply room; kettle room
热水供应温度 temperature of hot water supplying
热水供应系统 hot-water supply system
热水供应站 hot-water supply station
热水沟 hot creek
热水管 hot-water pipe; hot-water tube
热水管静止端节 dead leg
热水管网 hot-water network
热水管线 hot-water(pipe) line
热水罐 hot-water cylinder; expansion tank <能承受膨胀的>
热水柜 hot-water tank
热水锅炉 hot-water boiler; hot-water heater; water boiler; water heater
热水河 thermal river
热水湖 thermal lake
热水回路 hot-water return
热水回水 hot-water return
热水回水分流管系统 separate flow and return system
热水回水管 hot-water return pipe
热水加热 hot-water heating
热水加热炉 hot-water heater; hot-water stove
热水加热盘管 hot-water heating coil
热水加热器 hot-water heater; hot-water stove; water-heated calorifier; water heating appliance
热水加热装置 hot-water heating system
热水解 pyrohydrolysis; thermal hydrolysis
热水解脱水技术 thermal hydrolytic dewatering technology
热水浸解 warm water wetting
热水井 hot(-water) well
热水聚合 polymerization under hot-water
热水聚集器 hot-water accumulator
热水空气加热器 hot-water air heater
热水口 hot-water inlet
热水库 hydrothermal reservoir
热水矿床调查 hydrothermal mineral survey
热水扩容系统 hot-water flashing system
热水冷却渠道系统 cooling canal system for hot-water
热水力学效应 thermohydraulic effect
热水淋洗 hot-water rinsing bath
热水淋浴 hot shower; hot-water rinsing bath
热水硫化 water cure; water curing; water vulcanization
热水龙头 hot tap
热水漏斗 hot-water funnel
热水炉 hot-water generator; hot-water preparer; back boiler <安装在明火或火炉后面的>
热水幕 hot-water curtain
热水泥 hot cement
热水暖气系统 wet heating system
热水排放 heated water discharge; thermal outfall
热水喷射器 hot-water injector
热水瓶 thermobottle; thermoflask; thermos bottle; thermos flask; vacuum bottle; vacuum flask; hot-water bottle
热水瀑布 hot-waterfall
热水器 calorifier; geyser; hot-water apparatus; hot-water calorifier; hot-water generator; hot-water preparer; water heater; injector hot-water lifter <能自动送水至高处>; service water heater
热水器放水管 heater discharge pipe
热水侵蚀 erosional by thermal water
热水侵蚀腐蚀联合作用 erosion-corrosion by thermal water
热水清洗 hot-water rinse
热水清洗装置 hot cleaner
热水取暖 hot-water heating
热水取暖炉 hot-water heater
热水热量 hot-water heat
热水热网 hot-water heating network; hot-water heat-supply network
热水容器 hot-water vessel
热水溶解的浸膏 hot soluble extract
热水溶蚀 hot-water corrosion
热水溶液 hydrothermal solution
热水散热器 hot-water radiator
热水设备 hot-water apparatus
热水蚀变异常 anomaly of hot-water alteration
热水式 hot-water type
热水式干燥炉 water-oven
热水式供暖 hot-water heating
热水试验 hot-water test
热水熟化 water cure
热水塘 hot pool
热水通道 hot-water channel
热水筒 hot-water cylinder
热水筒上盖 hot-water upper cover
热水洼 hot-water cups
热水温度 hot water temperature
热水污染 heat water pollution; thermal water pollution
热水吸管 hot-water suction pipe
热水溪 jot stream
热水洗 hot-water washing
热水洗涤 hot wash
热水系统 hot-water network; hot-water system
热水箱 heating water tank; hot-water tank; hot-water vessel; hot well (tank)
热水箱的管组 <炉灶上> water front
热水消耗量顶点 point of hot water consumption
热水型 type of hot-water
热水循环 hot-water circulating; hot-water circulation
热水循环泵 hot-water circulating pump
热水循环流 hot-water circulating flow
热水循环流量 hot-water circulating rate
热水循环器 hot-water circulator
热水循环取暖系统 hot water circulating heating system
热水循环系统 hot-water circulation system; circulating hot water system
热水养护 hot-water curing; warm water curing
热水浴 hot bath
热水再用中的冷却渠道系统 cooling canal system for hot-water reuse
热水蒸汽散热片 hot-water steam radiator
热水支管 hot-water branch
热水止回阀 hot-water check valve
热水重力循环 gravity circulation of hot-water
热水重力循环系统 hot-water gravity system
热水贮槽 boiler
热水贮存箱 hot-water storage tank
热水柱 hydrothermal plumes
热水装置 hot-water apparatus
热水综合利用 multipurpose utilization of thermal water
热水总管 hot-water main
热税 heat tax
热顺序 heated succession
热瞬变过程 thermal transient
热瞬态 thermal transient
热丝 heater; hot wire
热丝变压器 heater transformer
热丝变压器绕组 heater transformer windings
热丝池 hot-wire cell
热丝电离规 hot-wire ionization ga(u)ge
热丝电流 heater current
热丝电流表 electrothermal meter
热丝电路 heater chain; heater circuit
热丝电压 heater volt; heater voltage
热丝电源 heater supply
热丝电源电路 heater power circuit
热丝电阻 heater resistance
热丝风速计 hot-wire anemometer
热丝功率 heater power
热丝焊 hot-wire welding
热丝化学气相沉积法 hot filament chemical vapo(u)r deposition
热丝极发射 heater emission
热丝检测器 hot-wire detector
热丝流量计 hot-wire flowmeter
热丝热解器 filament pyrolyzer
热丝压强计 hot-wire pressure ga(u)ge
热撕裂 hot tear(ing)
热死点 thermal death point
热死率 thermal death rate
热死温度 thermal death point
热速度 thermal velocity
热塑 hot mo(u)lding
热塑变形点 chamot(te) point
热塑薄膜 thermoplastic film
热塑薄型板 thermoplastic sheet
热塑材料 thermoplastic material
热塑车身汽车 thermoplastics bodied vehicle
热塑成型 thermoforming
热塑存储系统 thermoplastic memory system
热塑带 thermotape
热塑的 thermoplastic
热塑地板花砖抛光 polishing of thermoplastic floor tile
热塑地板砖 thermoplastic floor(ing) tile
热塑防护带 thermoplastic protective tape
热塑改性结合料 thermoplastic modified hinder
热塑管 thermoplastic pipe
热塑光电导体 thermoplastic photoelectric conductor
热塑光阀 thermoplastic light valve
热塑划线机的 thermoplastic road marking machine
热塑胶片 thermoplastic film
热塑腈聚氯乙烯 thermoplastic nitrile-polyvinyl chloride
热塑绝缘(胶)带【电】thermoplastic insulating tape
热塑炼 heat plasticization; heat softening; hot mastication; thermal plasticization
热塑料薄层 thermoplastic layer
热塑料光阀 thermoplastic light valve
热塑料树脂黏[粘]合剂 ger-bond
热塑流 thermoplastic flow
热塑螺纹盖 press-twist cap
热塑膜 thermoplastic film
热塑(木)材 heat-stabilized wood
热塑片 <填封料> thermoplastic sheet
热塑全息照片 thermoplastic hologram
热塑热熔黏[粘]合剂 thermoplastic hot-melt adhesive
热塑收缩包装 shrink wrapping
热塑树脂 Formvar
热塑塑料 thermoplastic material
热塑塑料族 thermoplastic family
热塑梯度 thermal gradient
热塑体 thermoplastics
热塑显示 thermoplastic display
热塑性 pyroplasticity; thermal plasticity; thermoplast; thermoplasticity; thermoplastic nature; thermosoftening; hot ductility <混凝土>
热塑性板 thermoplastic sheet
热塑性变形 pyroplastic deformation; thermoplastic deformation
热塑性丙烯酸树脂 thermoplastic acrylic resin
热塑性材料 thermoplast; thermoplastic; thermoplastic material
热塑性层压 thermoplastic laminate

热塑性衬里 thermoplastic backer
热塑性船底防污漆 hot plastic antifouling paint
热塑性瓷釉 thermofluid enamel; thermoplastic enamel
热塑性的 thermoplastic
热塑性地面(饰面)砖 thermoplastic tile
热塑性酚醛树脂 novolac; novolac resin; thermoplastic phenolic resin
热塑性粉末 thermoplastic powder
热塑性粉末敷层 thermoplastic powder coating
热塑性氟塑料 fluorinated thermoplastics
热塑性隔条 thermal plastic spacer; thermoplastic spacer
热塑性光导体 thermoplastic photoconductor
热塑性焊条 thermoplastic welding strip
热塑性合成橡胶材料 thermoplastic synthetic(al) rubber material
热塑性胶 thermoplastic cement
热塑性聚氨酯 thermoplastic urethane
热塑性聚氨酯织物 thermoplastic urethane fabric
热塑性聚丙烯系纤维 thermoplastic acrylic
热塑性聚合混合物 thermoplastic polymer mixture
热塑性聚酰胺 thermoplastic polyamide
热塑性聚酯 thermoplastic polyester
热塑性绝缘电缆 thermoplastic insulated cable
热塑性沥青制品 thermoplastic bitumen product
热塑性硫化体 thermoplastic vulcanizer
热塑性路标 thermoplastic marking
热塑性路标材料 thermoplastic traffic marking material
热塑性模塑料 thermoplastic mo(u)lding compound
热塑性内包头 thermoplastic box toe; thermoplastic puff
热塑性黏[粘]合剂 thermoplastic adhesive
热塑性铺地板 thermoplastic floor(ing) tile; thermoplastic tile
热塑性试验 hot-ductility test
热塑性树脂 thermoplastic resin
热塑性树脂复合材料 thermoplastic resin matrix composite
热塑性树脂胶 thermoplastic resin adhesive
热塑性树脂黏[粘]合剂 thermoplastic resin adhesive
热塑性塑料 thermoplastic plastics; thermoplast(ics)
热塑性塑料板 thermoplastic plate
热塑性塑料薄膜 thermoplastic film
热塑性塑料丁字模头挤出 crosshead extrusion of thermoplastics
热塑性塑料隔热 thermoplastic insulation
热塑性塑料胶 thermoplastic glue
热塑性塑料轮毂帽 thermoplastic hubcap
热塑性塑料片 thermoplastic sheet
热塑性塑料提浓物 thermoplastic concentrate
热塑性塑料微球体 thermoplastic microsphere
热塑性弹性材料 elasto-plastics; thermoplastic elastomer
热塑性弹性体 thermal plastic elastomer; thermoplastic elastomer
热塑性涂层 thermoplastic coating
热塑性屋面料 thermoplastic tile
热塑性烯烃 thermoplastic olefin
热塑性纤维 thermoplastic fibre[fiber]
热塑性纤维表面毡 thermoplastic veil
热塑性纤维结合力 thermoplastic fiber bonding
热塑性橡胶 thermoplastic rubber
热塑性乙烯树脂 hot vinyl
热塑性油灰 thermoplastic putty
热塑性增强树脂 reinforced thermoplastic resin
热塑性装订 thermoplastic binding
热塑乙烯丙烯二烯单体 thermoplastic ethylene-propylene diene monomer
热塑乙烯饰面砖 thermoplastic vinyl tile
热塑云母 hot-mo(u)lding mica
热塑(制)板 thermoplastic tiling
热塑装饰法 thermoplastic decoration
热酸 hot acid
热酸处理 hot acid treatment
热酸回收系统 hot acid recovery system
热酸聚合过程 hot acid polymerization process
热酸蚀试验 hot etching test
热损害 pyrolytic damage
热损耗 heat loss; heat rejection; loss of heat; heat waste; thermal loss
热损耗估算 estimate of heat loss
热损耗计算 heat loss computation
热损耗率 rate of heat loss
热损坏 thermal deterioration; thermal distress
热损墙面 hot-damaged wall
热损伤 heat injury; thermal damage
热损失 egress of heat; heat dissipation; heat leak(age); heat rejection; thermal loss; thermosteresis; waste of heat; heat waste
热损失测定 measurement of heat loss
热损失估测仪 dummy man; eupatheoscope
热损失估计 heat loss estimate
热损失函数 heat loss function
热损失机理 mechanism of heat loss
热损失计算 calculation of heat losses; heat loss calculation
热损失量测仪 eupatheoscope
热损失率 rate of heat loss
热损失设计 design heat loss
热损失速率 rate of head loss
热损失系数 U-factor
热损仪 eupatheoscope
热缩 firing shrinkage; hot shrinkage
热缩的 heat-shrinkable
热缩锻 thermal upset
热缩拉力 hot shrinkage-tension
热缩裂隙 thermal contraction crack
热缩塑料包 shrink wrap
热缩塑料薄膜包装 shrink wrap
热缩塑性防护层 heat-shrink plastic shield
热缩塑性套管接头 heat-shrink plastic tubing joint
热缩性管材 heat-shrinkable tubing
热缩作用 pyrocondensation
热塔 thermal tower
热台显微镜 hot stage microscope
热态 hot condition; thermal state
热态备用机组 hot reserve
热态变形 hot deformation; hot shaping
热态常数 hot constant
热态反应堆 hot reactor
热态检修 hot repair
热态锯切 hot saw(ing)
热态模型试验 hot model test
热态起动 hot start
热态起动时间 hot-start-up time
热态汽轮机 heat-soaked turbine
热态强度 hot strength
热态时可浇的 hot pourable
热态试验 hot test(ing)
热态有效性曲线 hot curve
热弹性 thermal elasticity; thermoaeroelasticity
热弹性变形 thermoelastic distortion
热弹性带 thermoplastic tape
热弹性后效应 thermoelastic aftereffect
热弹性理论 thermoelastic theory
热弹性力学 thermoelasticity
热弹性逆转 thermoelastic inversion
热弹性逆转点 thermoelastic inversion point
热弹性效应 thermoelastic effect
热弹性应变 thermoelastic strain
热弹性应力 thermoelastic stress
热弹性应力-应变规律 thermoelastic stress-strain law
热弹转换 thermoelastic inversion
热态运行 run hot
热碳层 hotbed of carbon
热炭黑 thermal black
热探测辐射计 thermal detection radiometer
热探测设备 heat detecting equipment
热探测仪 thermosounder
热探头 infrared detection unit; thermal probe; thermoprobe
热探针 thermal probe
热探针法 hot probe method
热碳酸钾过程 hot potassium carbonate process
热碳酸盐过程 hot carbonate process
热碳酸盐水浴 Nauheim bath
热汤渗透法 hot-water permeation method
热搪瓷涂层 hot enamel coating
热套 hot jacket; shrinkage fit(ting); shrink fit
热套表面压力 shrink fit pressure
热套构件 shrink member
热套管接头 shrunk-on-pipe joint
热套环 shrink ring
热套配合 shrink(age) fit(ting); shrunk fit
热套配合压力 shrink fit pressure
热套圈 shrink ring
热套式多层圆筒 shrunk-on multilayered cylinder
热套式圆筒 shrunk-on cylinder
热套轴 shrunk-in shaft
热套装 skrink on
热套装挡圈 shrunken collar
热套装的 shrunk-on
热特性 thermal behavio(u)r; thermal character; thermal performance
热特性测量 thermal characteristic measurement
热梯度风 thermal slope wind
热梯度探测仪 thermal gradient measuring probe
热梯度探头 thermal gradient measuring probe
热体 heating element; hot body
热体积变化 thermal volume change
热体制 thermal regime
热天浇灌混凝土 hot application of concrete; hot-weather concreting
热天浇注混凝土 hot application of concrete; hot-weather concreting
热天浇筑混凝土 hot application of concrete; hot-weather concreting
热天平 thermal balance; thermobalance
热天施工 hot-weather construction
热田的净质量输出 net mass withdrawal from field
热田地质剖面图 geothermal geologic(al) section
热田地质图 geothermal field geologic(al) map
热田开发所诱发的地面沉降 land subsidence induced by geothermal field exploitation
热田开发图 map of geothermal field exploitation
热田排放系统 field drainage system
热田寿命 field longevity
热田性状 field behavio(u)r
热条痕 thermal striae
热条件 heat condition; thermal condition
热条式电表 hot band meter
热调管 thermally tuned valve
热调节 heat conditioning
热调节阀 thermo-regulating valve
热调节杆 heat regulating lever
热调聚反应 thermal telomerization
热调谐 thermal tuning
热调谐常数 thermal tuning constant
热调谐时间常数 thermal tuning time constant
热通风装置 warm register
热通量 flux of heat; thermal flux; thermoflux; heat flux; thermal conductance
热通量方程 heat flux equation
热通量向量 heat flux vector
热同素异形 thermometamorphism
热头点火 hot-head ignition
热头发动机 hot-head engine
热头引擎 hot-head engine
热透补偿 thermal lensing compensation
热透镜 thermal lens
热透明点 hot clear end-point
热透气性 hot permeability
热透(入) heat penetration
热透射 heat transmitting; transmittance[transmittancy]
热透射率 heat transmittance
热透效应 thermal lens effect; thermolens effect
热图 thermal map
热图像 <红外线暗视器摄取的图像> heat picture; thermal picture; thermograph
热涂 hot-applied
热涂层 hot coating
热涂煤焦油 hot-applied coal tar
热涂塑性涂料 hot plastic paint
热涂装 hot coating
热土 hot soil
"热土豆"式路由选择 hot potato routing
热湍流 thermal turbulence
热湍流度 thermal turbulence
热团矿 hot briquetting
热退磁 thermal demagnetization
热退磁仪 thermal demagnetizer
热退期 defervescence; defervescent stage
热褪色 thermal discoloration
热脱附 thermal desorption
热脱附谱 thermal desorption spectrum
热脱附谱术 thermal desorption spectroscopy
热脱浆 heat desizing
热脱落 thermal spalling
热脱硝 thermal denitration
热椭圆体 thermal ellipsoid
热弯的弯头 hot bent bend
热弯曲 hot bending; thermal flexure
热弯曲试验 hot bending test; warm bending test
热顽磁 thermoremanence
热网 heat distributing network

热网补给水泵 feed pump for heating network
热网水压图 hydraulic diagram for heat supply
热网运行管理 operation and maintenance of heating network
热网站 heat-supply distribution and control station
热网中间泵站 relay pump station of heating network
热微粒分析法 thermoparticulate analysis method
热尾流 thermal wake
热位 thermal potential
热位差 temperature head; thermal head
热位能 thermodynamic(al) potential
热位移 thermal movement; thermal walking
热温差电偶 thermoelement
热温除冰设备 thermal deicer
热温激活设备 heat actuating device
热温控装置 heat-actuated device
热温式风速计 thermoanemometer
热紊流 thermal turbulence
热稳定 thermostabilization
热稳定的 heat-proof; heat stable; temperature-resistant; thermally stable; thermostable
热稳定电流 thermal current
热稳定防老剂 heat stable anti-oxidant
热稳定钢 oxidation-resistant steel
热稳定剂 heat stabilizer; heat stable inhibitor; thermal stabilizer
热稳定漆 thermostat varnish
热稳定性 heat endurance; heat fastness; heat stability; heat steadiness; hot stability; resistance to thermal shocks; stability to heat; temperature stability; thermal endurance; thermal shock resistance; thermal stability; thermostability
热稳定性测定法 thermal stability measuration
热稳定性辅助指标 thermal stability help out parameter
热稳定性极差煤 least thermal stability coal
热稳定性较差煤 less thermal stability coal
热稳定性良好煤 good thermal stability coal
热稳定性曲线 thermal stability curve
热稳定性试验 heat stabilization test; heat-resistance test; heat stability test; reheat test; thermal stability test
热稳定性指标 thermal stability parameter
热稳定性中等煤 mid thermal stability coal
热稳定轴承 heat-stabilized bearing
热稳定状态 thermal steady state
热稳性 high-temperature strength; therm stability
热涡度 thermal vorticity
热污染 calefaction; heat pollution; thermal pollution
热污染的影响 effect of thermal pollution
热污染监测 heat pollution monitoring
热污染控制 thermal pollution control
热污染效应 heat pollution effect
热污染源 heat pollution source; thermal pollution source
热无序的 thermally disordered
热物理常数 thermal physical constant
热物理的 thermophysical
热物理性能 thermophysical property
热物理性试验 thermal physical property test
热物理学 thermophysics
热吸附剂 heat-adsorbent
热吸附作用 thermoadsorption
热吸率 heat efficiency
热吸收 heat absorption; thermal absorption
热吸收管 heated absorption tube
热吸收剂 heat absorbent
热吸收率 heat absorptance; thermal absorptivity
热吸收器 heat absorber
热析 ex(s)udation; sweat-back; sweat(ing); sweat out
热析出 heat evolution
热析浮渣 sweating dross
热析沥青体 exsudatinite
热析炉 sweat furnace
热析珠 sweating out beads
热稀释法 thermodilution
热洗净 warm wash
热洗炉 chipping-out furance
热徙动 thermomigration
热系数 thermal coefficient
热系统 thermal system
热系统空气动力学 aerodynamics of heat system
热隙透 thermal effusion
热隙透效应 thermal feedback effect
热显示区 thermal display area; thermal display zone
热显微镜 heatable stage microscope
热显微镜学 thermomicroscopy
热显微照片 thermomicrograph
热显影 heat development; thermal development; thermic development
热线 heat ray; hot-line; hot ray; on-line <与信息网络互联>【计】; receptionist <接待>
热线安培计 hot-wire ammeter
热线测量 hot-wire measurement
热线测试技术 hot-wire technique
热线测速仪 hot-wire anemometer
热线测针 hot-wire probe
热线传声器 hot-wire microphone; thermal microphone
热线点火 hot-wire ignition
热线点火棒 hot-wire lighter
热线点火器 hot-wire fuse lighter
热线电话机 hot-wire telephone; thermal telephone
热线电阻地震计 hot-wire resistance seismometer
热线法 hot-wire method
热线风速表 hot-wire anemometer
热线风速计 hot-wire anemometer
热线风速仪 heated wire's air speed meter; heated wire type anemometer; hot-wire probe
热线服务 hot-line service
热线检流计 hot-wire galvanometer
热线检流器 barretter
热线灵敏度 hot-wire sensitivity
热线流量计 hot-wire flowmeter
热线流速仪 hot-wire anemometer; hot-wire current meter
热线能力 direct access capability
热线膨胀系数 linear coefficient of thermal expansion
热线圈 hot coil; hot-wire coil; sneak-current arrestor
热线式传声 hot-wire microphone
热线式传声器 thermal type microphone
热线式电流表 hot-wire ammeter; hot-wire galvanometer
热线式电流计 hot-wire ammeter; hot-wire galvanometer
热线式电桥 hot-wire bridge
热线式风速表 hot-wire type anemometer
热线式风速仪 hot-wire type anemometer
热线式伏特计 thermovoltmeter
热线式话筒 hot-wire microphone; thermal microphone
热线式继电器 hot-wire relay
热线式检波器 hot-wire detector
热线式示波器 hot-wire oscillograph
热线式受话器 thermophone
热线式瓦特计 hot-wire wattmeter
热线式扬声器 hot wire loudspeaker
热线式仪表 hot-wire instrument; hot-wire meter
热线式真空开关 hot-wire vacuum switch
热线受话器 hot-wire telephone; thermal telephone
热线送话器 hot-wire microphone; thermal microphone
热线通信[讯]业务 hot-line service
热线图 steam flow diagram
热线性热敏电阻 thermilinear thermistor
热线压力表 hot-wire manometer
热线压力计 hot-wire ga(u)ge; hot-wire manometer; hot-wire pressure ga(u)ge; hot-wire pressure meter
热线引火塞 glow plug
热线噪声发生器 hot-wire noise generator
热线胀系数 thermal coefficient of linear expansion
热线真空计 resistance manometer
热线支持 hot-line support
热线作业 hot-line job
热限 thermal boundary
热相似 heat similarity
热箱(试验用) hot box
热箱压铸 hot chamber die casting
热响应时间 heat responsive time; thermal adjustment time
热响应元件 heat responsive element
热像 thermal image
热像差 thermal aberration
热像管 thermal imaging tube
热像检测 thermographic(al) inspection
热像通用组件 thermal imager common module
热像图 thermogram
热像图仪 thermoviewer
热像仪 thermal imager; thermal imaging system
热像照相机 thermocamera; thermographacamera
热消除 thermal elimination
热消毒 heat sterilization
热消毒器 thermal death equipment
热消毒时间 thermal death time
热消费率 heat rate
热消耗 heat consumption; heat dissipation; heat expenses; heat reflection
热消耗量 thermal loss; heat loss; heat consumption
热消散 heat dissipation
热消散能力 heat dissipating ability
热消散系数 heat dissipation factor
热消退 pyretolysis
热销房产 hot listing
热效换能器 thermal transducer
热效率 calorific efficiency; efficiency of cycle; fuel efficiency; heat efficiency; thermal efficiency; thermischer; thermoefficiency
热效率比 thermal efficiency ratio
热效率系数 coefficient of thermal efficiency; thermal efficiency coefficient
热效率指数 thermal efficiency index
热效闪光器 thermal flasher
热效式电流计 thermal galvanometer
热效系数 heating coefficient
热效性 heat availability
热效应 calorific effect; fuel factor; heat(ed) effect; heat efficiency; heating effect; influence of heat; thermal effect; thermal result
热效应断路器 thermal trip
热效应仪器 thermal instrument
热楔 thermal wedge
热歇期 apyrexia
热胁变 thermal strain
热写记录仪 heat writing recorder
热泄出物 thermal discharge; thermal drift
热芯盒 heated core box; hot(core) box
热芯盒造型法 hotbox process
热信号 thermal signal
热星 hot star
热星等 bolometric magnitude
热星等变幅 bolometric amplitude
热星等改正 bolometric correction
热形变 thermal deformation
热型 <用于薄膜焊接> heatweld
热型锻 hot swaging
热型火花塞 hot plug; hot spark plug; hot-type plug
热型气候 thermal climate
热性的 pyrogenic; pyrogenous
热性肥料 warm manure
热性肥料处理 warm fertilizer treatment
热性灰浆 warm mortar
热性能 hot property; thermal behavio(u)r
热性胀缩 thermal expansion and contraction
热性针叶林 hot needle-leaf forest
热修 heat work
热修边模 hot trimming die
热修补 hot repair
热修整 hot trimming
热需要量 heat requirement
热蓄积 heat storage capacity
热旋锻 hot-swage
热旋压 hot spinning
热旋压温度图 hot spinning temperature chart
热选法 heat separation
热学 calorifics; thermology; thermotics
热学测试仪器 thermal measuring instrument; thermal testing instrument
热学常数 thermal constant
热学定律 heat law
热学分析 thermal analysis
热学特性 thermal characteristic
热学性能 thermal property
热学性质 heat character; thermal property
热循环 circulation of heat; heat circulation; thermal cycle(cycling); hot cycle
热循环泵 hot recycle pump
热循环试验 thermal cycling test
热循环稳定性 thermal cycling stability
热循环效率 efficiency of cycle; thermal cycle efficiency
热循环效应 heat cycle effect
热循环装置 thermocirculator
热循环阻力 resistance to thermal cycling
热压 heat pressure; hot briquetting; hot compression; hot-press(ing); hot-pressure; stack effect pressure; thermal pressure; thermocompression

热压板 hot pressboard
热压波纹机 hot dimpling machine
热压箔 hot-press printing; hot stamping
热压薄 hot ironing
热压车间 hot stamping shop
热压成型 hot forming; hot mill; hot-press(ing); thermoforming
热压成型胶合板 post-formed plywood
热压冲杆 hot-press ram
热压处理的石棉水泥硅酸钙板 autoclaved asbestos cement calcium silicate board
热压氮化硅 hot-pressed silicon nitride
热压氮化铝 hot-pressed alumin(i)um nitride
热压的 hot-rolled
热压地沥青混凝土 hot-rolled asphalt
热压碲化镉陶瓷 hot-pressed cadmium telluride ceramics
热压垫块 dummy block
热压定形机 boarding press
热压锻造 hot-press forging
热压法 heat-die pressing process; heat pressure method; hot-press approach; hot-pressed process; hot-press method; pressure sintering
热压法连接 thermocompression
热压氟化镧陶瓷 hot-pressed lanthanum fluoride ceramics
热压氟化锶陶瓷 hot-pressed strontium fluoride ceramics
热压工具 hot-pressing tool
热压锅 press heater
热压过程 hot-pressed process
热压焊(接) hot-pressure welding; hot-press welding; thermocompression bond(ing); thermocompression welding
热压焊接合 thermocompression bonding
热压焊结 pressure-welded junction
热压合金 hot-pressed alloy
热压合金粉坯 hot-pressed alloy powders
热压红外元件 hot-pressing infrared element
热压黄铜 hot-pressed brass
热压黄铜合金 hot stamping brass
热压机 hot-platen press; hot-press(ing); thermocompressor
热压机操作工 hot-presser
热压机装板时间 hot-press loading time
热压机座 hot-press bed
热压加气混凝土 autoclaved aerated concrete
热压胶板机 hot process plywood press
热压胶合 hot-press gluing; thermocompression bond
热压焦炭 form-coke
热压接合器 thermocompression bonder
热压结 sinter(ing)
热压结的 sintered
热压金 hot-pressed gold
热压金属陶瓷 hot-press cermet
热压块坯 hot-pressed compact
热压力 thermal pressure
热压力计 heat ga(u)ge
热压力加工 hot press working
热压沥青砂毡层 hot-rolled sand carpet
热压零件 hot-pressed parts
热压硫化锅 press heater
热压轮圈 shrunk ring
热压灭菌法 autoclaving
热压灭菌器 autoclave sterilizer
热压模 hot die; hot-pressing mo(u)ld
热压模塑 hot-press mo(u)lding
热压模制 hot-press mo(u)lding
热压黏[粘]结 hot gluing
热压配合 shrink(age) fit(ting); shrinking on; shrunk fit
热压喷漆 thermopress paint spraying
热压膨胀(率) autoclave expansion
热压汽器 thermal compressor
热压器 autoclave
热压青铜 hot-pressed bronze
热压青铜块坯 hot-pressed bronze compact
热压轻混凝土 autoclaved lightweight concrete
热压曲 thermal buckling
热压曲面机 hot dimpling machine
热压杀菌器 pressure sterilizer
热压烧结 hot-pressed sintering
热压烧结法 sintering process with hot pressing
热压烧结制品 hot-pressed and sintered product
热压试验 <耐火材料的> hot load test
热压树脂 hot-press(ure) resin
热压树脂塑料胶板 laminated plastic board
热压缩 hot reducing
热压缩机 thermocompressor
热压缩蒸发器 thermocompression evapo(u)rator
热压陶瓷 hot-pressed ceramics
热压套环 shrink ring
热压铁坯块 hot-pressed iron compact
热压头 thermal head
热压头压烫机 hot-head press
热压凸印刷 thermography
热压温度 hot-pressing temperature
热压纤维板 pressed wood
热压型毛坯 hot mo(u)lded blank
热压压力 hot-pressing pressure
热压印 hot padding
热压油法 hot oil expression
热压云母 hot-pressed mica
热压蒸发 thermal compression evapo(u)ration
热压蒸馏法 thermocompression distill
热压蒸馏器 thermocompression distiller; digester
热压制品 hot-pressed product
热压铸成型 hot injection mo(u)lding; hot-pressure casting; injection mo(u)lding
热压铸机 hot injection mo(u)lding machine
热压装置 hot-press arrangement
热压作业 hot pressing
热烟道 heating flue; hot flue
热烟道气体 hot flue gas
热烟流 thermal plume
热烟气 hot gas
热烟气管 hot-gas pipe
热烟云 hot plume
热延迟 heat lag
热延迟时间 thermal time lag
热延伸区 <直接成条机的> heat stretch zone
热延时继电器 thermal time delay relay
热延时器 thermal delay
热延性 hot ductility
热岩溶 thermal karst; thermokarst
热岩水库 hot-rock reservoir
热岩土工程学 thermal geotechnics; thermo-geotechnology
热研磨 hot grinding
热盐的 thermohaline
热盐对流 thermohaline convection
热盐环流 thermohaline circulation
热盐水 hot brine
热盐水溶解试验 hot saline solubility test
热演化 thermal evolution
热焰喷射法 jet flame process
热焰喷射钻孔 jet-piercing
热焰喷射钻孔法 <开石方用> jet-piercing method
热焰射穿钻孔 thermal jet piercing drilling
热焰射穿钻孔法 thermal jet piercing method
热焰钻孔装置 jetting drill
热养护 heat ag(e)ing; heat cure[curing]; hot cure[curing]; thermal protection
热养护设备 heat curing instrument
热氧化 thermal oxidization; thermooxidizing
热氧化锆 hot zirconia
热氧化降解 thermal oxidative degradation
热氧化稳定性 thermooxidative stability
热氧化作用 thermal oxidation
热冶金实验室 hot metallurgy laboratory
热冶金学 pyrometallurgy
热液 hydrotherm
热液变质(作用)【地】 katogenic metamorphism; catogenic metamorphism; hydrothermal metamorphism
热液成矿的 hydatogenous
热液成矿作用 hydatogenesis
热液充填 hydrothermal filling
热液处理 hydrothermal treatment
热液的 hydrothermal
热液合成(法) hydrothermal synthesis
热液活性 hydrothermal activity
热液交代变质 pyrometasomatism
热液交代的 pyrometasomatic
热液交代矿床 pyrometasomatic deposit
热液交代作用 hydrothermal metasomatism
热液浸渍 hot dip
热液晶体生长 hydrothermal crystal growth
热液矿藏 hydrothermal deposit
热液矿床 hydrothermal deposit; hydrothermal ore deposit
热液矿田构造 hydrothermal orefield structure
热液矿物 hydrothermal mineral
热液脉状锑矿床 hydrothermal veined antimony deposit
热液锰结核 hydrothermal manganese [Mn] nodule
热液期 hydrothermal stage
热液溶解作用 hydrothermal solution
热液石英 keatite
热液蚀变【地】 hydrothermal alteration
热液水 hydrothermal water
热液同位素成分 hydrothermal isotope composition
热液系统温度变化 temperature change of hydrothermal system
热液系统压力变化 pressure change of hydrothermal system
热液消光 hydrothermal delustring
热液循环 hydrothermal circulation
热液压成型 thermohydroforming
热液压系统 hot hydraulic system
热液作用 hydrothermal action; hydrothermalism; hydrothermal process
热液作用地球化学 geochemistry of hydrothermal processes
热移定性试验 heating warpage test
热移动 <构件因温度变化而膨胀或收缩> thermal movement
热移动系数 coefficient of thermal movement
热异常 thermal anomaly
热异常带 thermal belt
热异常区 thermal locality zone
热异常图像 thermal picture
热异常状态 state of thermal anomaly
热异构化 thermal isomerization
热逸散 thermal breakdown; thermal run-away
热阴磁控管真空计 hot cathode magnetron ga(u)ge
热阴极 heated cathode; incandescent cathode; thermionic cathode
热阴极 X 射线管 hot cathode X-ray tube
热阴极充气二极管 hot cathode gas-filled diode; phanotron
热阴极充气管整流器 hot cathode gas-filled rectifier
热阴极充气整流管 hot cathode gas-filled rectifier tube; hot cathode gas rectifier tube
热阴极充气整流器 hot cathode gaseous rectifier
热阴极磁控电离规 hot cathode magnetron ionization ga(u)ge
热阴极灯 hot cathode lamp
热阴极电离规 hot cathode ionization ga(u)ge
热阴极电离计 hot cathode ionization ga(u)ge; Lafferty ga(u)ge
热阴极电离真空计 hot filament ionization ga(u)ge
热阴极电子倍增管 thermionic multiplier tube
热阴极电子管 hot cathode lamp; thermionic valve
热阴极电子学 thermionics
热阴极电子仪器 thermionic instrument
热阴极放电 hot cathode discharge
热阴极放电管 hot cathode discharge tube
热阴极汞弧整流器 hot cathode mercury-arc rectifier
热阴极汞汽整流管 hot cathode mercury vapo(u)r rectifier tube
热阴极汞汽整流器 hot cathode mercury vapo(u)r rectifier
热阴极管 hot-cathode tube
热阴极辉光放电管 hot cathode glow tube
热阴极快速启动灯 hot cathode rapid-start lamp
热阴极离子源 hot cathode ion source
热阴极氙 hot cathode lamp
热阴极氖灯 neon arc lamp
热阴极水银放电灯 hot cathode mercury discharge lamp
热阴极四极管 quadratron
热阴极钨丝荧光灯 hot cathode
热阴极预热灯 hot cathode preheat lamp
热阴极整流器 hot cathode rectifier
热阴影 thermal shadow
热引导 thermal steering
热引发反应 thermal booster reaction
热引发剂 thermal booster
热引发聚合 thermal initiated polymerization
热引发作用 thermal initiation
热应变 hot straining; temperature strain; thermal strain
热应变脆性 hot straining embrittlement
热应变计 thermal strain ga(u)ge; thermal strain meter

热应变消除法 elimination of thermal strain
热应激 heat stress
热应力 heat stress; temperature stress; thermal load; thermal stress
热应力抵抗因子 heat stress resistance factor
热应力分布 thermal stress distribution
热应力分析 thermal stress analysis
热应力功能 thermal stress duty
热应力缓和 thermal stress reduction
热应力计 thermal stress ga(u)ge; thermal stress meter
热应力开裂 thermal stress cracking
热应力量测法 determination of thermal stress
热应力裂隙 thermal crack(ing)
热应力疲劳 thermal stress fatigue
热应力引起开裂 thermal stress cracking
热应力指标 heat stress index
热应力指数 heat stress index
热影响 heat action; heat affecting; heat influence; influence of heat
热影响区 heat-affected zone
热影响区裂缝 heat-affected zone crack(ing)
热影响区裂纹 heat action zone crack
热影像 thermal imagery
热硬度 hot hardness; red hardness; thermal hardness
热硬钢 red-hard steel
热硬化 heat embrittlement; hot set; thermohardening; thermosetting
热硬化性 thermosetting property
热硬塑料 thermosetting plastics
热硬塑料板 thermosetting plate
热硬性 hot hardness; red hardness
热硬性黏[粘]接材料 heat-setting jointing material
热硬性水泥砂浆 heat-setting mortar
热硬性物 thermosetting plastics
热硬性质 hot hardness property
热泳现象 thermophoresis
热用 hot application
热用保险丝 heating fuse
热用的 hot-applied
热用户 heat consumer
热用户连接方式 connecting method of counter with heat-supply network
热用胶 cooked glue
热用焦油(沥青) hot application tar
热用接蜡 warm mastic wax
热用沥青 hot bitumen grout
热油 hot oil
热油泵 hot oil pump
热油淬火 hot oil quenching
热油导管 hot oil duct
热油干材料 oil seasoning
热油干(木)材法 oil seasoning
热油干燥法 <木材> oil drying
热油机 hot oil machine
热油离心泵 hot oil centrifugal pump
热油器 oil heater
热油洗井 hot oiling
热油轧染 hot oil dy(e)ing process
热油真空干燥法 <木材> oil vacuum seasoning
热油蒸馏 hot oil distillation
热油装置 <冬季注油用> hot oil plant
热诱导 thermal induction
热浴淬火 hot-bath quenching; martemper(ing); thermoquenching
热浴淬火时效 hot-bath quench aging
热浴镀层 hot-dip coating
热浴镀锌薄板钢材 hot-dipped galvanized sheet steel
热浴炉 heat bath furnace
热浴时效处理 direct quench aging
热浴室 caldarium; heat bath; stew
热预混 warm premixing
热预算 heat budget
热预算方法 heat budget method
热元件 fuel cell; temperature element; thermal element; thermister [thermistor]; thermoelement
热原料 hot charge
热原质 pyrogen
热原子反应 hot atom reaction
热原子炭黑 thermatomic carbon black
热原子退火 hot atom annealing
热源 heater; heat power supply; heat producer; heat reservoir; heat source; heat-supply; origin of heat; radiant; source of heat; thermal resource; thermal source
热源靶机 heat source drone target
热源成因 origin of heat source
热源检查法 pyrogen test
热源的 pyretogenic; thermogenic
热源屏障 heater screen
热源探测器 heat source detector
热源消防栓 heater plug
热源岩体 heat source rock
热源移动 heat source movement
热源种类 kinds of heat source
热约束 thermal confinement
热跃迁 thermal transition
热云 hot cloud
热运动 heat motion; thermal moon; thermal motion
热运动能 energy of thermal motion
热运转 heat run
热晕 thermal blooming
热晕补偿 thermal blooming compensation
热晕法 thermal halo method
热晕轮 thermal aureole
热晕图 thermal halo diagram
热熨平 hot ironing
热杂波 thermal noise
热载体 heat carrier; heat carrying agent; thermal medium
热载体出口 heating medium outlet
热载体出入口 heat-transfer medium inlet and outlet
热载体储罐 thermal medium storage tank
热载体锅炉 thermal medium boiler
热载体回流口 thermal medium return nozzle
热载体加热炉 thermal medium boiler; thermal medium heater
热载体加热器 thermal fluid heater
热载体入口 heating medium inlet
热载体入口总管 header of thermal medium inlet
热载体油 heat medium oil
热载体蒸发器 thermal medium vapo(u)rizer
热载体中间贮槽 intermediate thermal medium tank
热载体贮罐 thermal medium storage tank
热再流平涂层 heating reflow coating
热再生电池 thermal regenerative cell
热錾 hot chisel
热凿 hot chisel
热灶风 foehn
热噪电压 Johnson noise voltage
热噪声 agitated noise; Johnson noise; Johnson-Nyquist noise; Nyquist noise; resistance noise; temperature noise; thermal agitation; thermal noise
热噪声测试器 thermal noise tester
热噪声测温术 Johnson noise thermometry
热噪声电压 Johnson noise voltage; thermal noise voltage
热噪声发生器 thermal noise generator
热噪声功率 thermal noise power
热噪声温度计 thermal noise thermometer
热噪限制工作 thermal noise limited operation
热噪音 thermal noise
热燥风 Chinook wind; foehn
热增长速度 rate of heat development
热增耗 heat increment
热增耗率 heat increment rate
热增加 heat gain
热增宽效应 thermal broadening effect
热增量 heat gain
热增强作用 thermal enhancement
热增殖 thermal breeding
热增殖堆 thermal breeder
热轧 hot reducing; hot reduction; hot rolling; hot shaping
热轧板 hot-rolled plate
热轧半成品 re-roiling quality
热轧扁钢 hot-rolled flat product
热轧变形中碳钢筋 deformed hot rolled mild steel bar
热轧变形钢筋 hot-rolled deformed (steel) bar
热轧变形高强钢筋 deformed hot rolled high yield bar
热轧变形圆钢 hot-rolled deformed (steel) bar
热轧表面加工 hot-rolled finish
热轧薄(钢)板 black plain steel sheet; hot-rolled sheet(steel)
热轧材料 hot-finished material; hot rolled material
热轧槽钢 hot-rolled channel steel; rolled steel channel
热轧厂 hot mill; hot-rolling mill
热轧厂废水 hot-rolling wastewater
热轧成品薄板 hot-rolled finished sheet
热轧带材 hot-rolled band; hot-strip
热轧带钢 black strip; hot-strip
热轧带钢机 hot-strip mill
热轧带钢卷取机 hot-strip reels
热轧带卷 hot-rolled coil
热轧的 hot-rolled; hot shaped
热轧电动机 hot mill motor
热轧钢板 hot-rolled sheet(steel)
热轧钢材 hot-rolled steel; hot-strip
热轧钢窗框 hot-rolled window sash
热轧钢带 hot-rolled band
热轧钢锭的表面氧化皮 mill scale
热轧钢废水 hot-rolling wastewater; wastewater from hot rolling
热轧钢管 hot-finished steel pipe; hot-finished steel tube; hot-finished tubing
热轧钢筋 hot-rolled bar; hot-rolled reinforced bar; hot-rolled steel bar
热轧钢经过热处理后形成的氧化皮 mill scale
热轧钢龙骨 hot-rolled joist; hot-rolled steel joist
热轧钢切削 hot saw(ing)
热轧钢丝 hot-rolled wire
热轧钢条 hot-rolled strip
热轧高强异形钢筋 deformed hot-rolled high-yield bar
热轧工字钢梁 rolled steel joist (beam)
热轧管 hot-rolled pipe; hot-rolled tube
热轧光 hot calendering
热轧光洁度 hot-rolled finish
热轧光面钢丝 plain hot rolled wire
热轧硅钢板 ferrosil; hot-rolled silicon steel sheet
热轧辊 hot roll
热轧辊颈润滑脂 hot roll neck grease
热轧(后)退火 hot-roll annealing
热轧机 hot-rolling mill; hot mill
热轧件的火焰清理 hot scarfing
热轧件火焰清理机 hot scarfer
热轧结构型钢 hot-rolled structural section
热轧金属板 hot-rolled sheet metal
热轧金属薄板 <0.45~0.55毫米> latten
热轧精轧机 hot finisher
热轧精轧机座 hot finishing mill
热轧盘条 green rod; hot-finished rod
热轧软钢 hot-rolled mild steel
热轧设备 hot mill
热轧特薄板 <厚小于0.45毫米> extra latten
热轧条材 hot-rolled bar
热轧铁鳞 mill scale
热轧铜 hot-rolled copper
热轧铜板用锭坯 copper wedge cake
热轧铜门 hot-rolled bronze door
热轧涂层扁钢制品 coated hot rolled flat product
热轧涂层钢板 coated hot rolled flat product
热轧无镀扁钢 hot-rolled uncoated flat product
热轧线材 hot-rolled long product; hot-rolled wire
热轧型钢 hot-rolled (steel) section; rolled steel section
热轧氧化皮 mill scale
热轧异形钢筋 hot-rolled deformed (steel) bar
热轧异形高强钢筋 deformed hot rolled high yield bar
热轧异形中碳钢筋 deformed hot rolled mild steel bar
热轧用轧辊 hot-rolling roll
热轧圆钢 hot-rolled bar; hot-rolled round steel
热轧缘边 mill edge
热轧造船钢板 hull plate
热轧轧辊 roll for hot-rolling
热轧窄带钢 mill coil
热轧窄钢条 hot-rolled narrow strip
热轧中厚板 hot-rolled plate
热轧中碳异形钢筋 deformed hot-rolled mild steel bar
热轧装置 hot-rolling arrangement
热榨 hot mo(u)lding
热榨油 hot-pressed oil
热沾染 thermal pollution
热张力 hot tensile strength
热涨落 thermal fluctuation
热胀 thermal expansion
热胀补偿向心球轴承 thermoexpansion compensation angular ball bearing
热胀断路器 thermal circuit-breaker
热胀锡 hot swelling
热胀冷缩 expand with heat and contract with cold; expansion caused by heat and contraction caused by cold
热胀裂 thermal crack(ing)
热胀漆 <一种防火漆> intumescent paint
热胀装配 expansion fit
热障 heat barrier; heat dam; temperature barrier; thermal barrier; thermal boundary; thermal break; thermal curtain; thermal dam; thermal thicket; thermodynamic(al) barrier
热障涂层 thermal barrier coating
热照相术 thermophotography

热遮板 heat shroud
热折断 < 挤压缺陷 > thermal break-off
热真空 hot vacuum; thermal vacuum; thermovacuum
热真空成型 thermovacuum forming
热真空法 heat-vacuum method
热真空容器 thermal vacuum chamber
热真空蒸发器 thermovac evapo(u)rator
热振波动 temperature vibration
热振荡 heat-driven oscillation; thermal oscillation
热振荡频率 thermal oscillation frequency
热振动 temperature vibration; thermal vibration
热振幅衰减倍数 thermal amplitude decrement
热震参数 thermal shock parameter
热震(荡) heat shock; thermal shock; temperature shock
热震断裂 thermal shock fatigue
热震试验 thermal shock test
热震碎 thermal spalling
热蒸发 thermal evapo(u)ration
热蒸发薄膜 thermal evapo(u)rated thin film
热蒸发分离 thermal assisted evaporative separation
热蒸发器 hot vapo(u)rizer
热蒸馏 thermal distillation
热蒸馏釜 thermal still
热蒸气线 hot-vapo(u)r line
热蒸汽 heating steam
热蒸汽磷化 steam phosphating
热蒸汽灭菌器 autoclave sterilizer
热正极 analogous pole
热支出 heat expenses
热值 caloric value; calorie value; caloric content; caloricity; caloric power; calorific capacity; calorific efficiency; calorific power; calorific value; calorimetric value; combustion value; heated value; heat generating capacity; heat(ing) value; heat number; heat of combustion; heat output; thermal value
热值测定 heating value determination
热值测定器 heating effect indicator
热值测定仪 calorimeter
热值范围 heat range
热值公式 heat value formula
热值试验 calorimeter test
热值仪 heat value meter
热指标 heating index
热指数 heat number
热制备 hot preparation
热制动 thermal arrest
热质 caloric; heat calorie
热质量 thermal mass
热质流量计 thermal mass flowmeter
热质学说 caloric theory of heat
热致变色 thermochromism
热致磁性 pyromagnetism
热致的 thermal; thermic; thermotropic
热致电离 thermal ionization
热致发光 thermoluminescence
热致发光的 thermoluminescent
热致发光剂量玻璃 radio thermoluminescent dose glass
热致发光判断年代法 thermoluminescence dating
热致发声器 thermophone
热致击穿 thermorunaway
热致剂量计 thermoluminescence dosimeter
热致聚合物 heat polymer
热致宽 thermal broadening
热致粒子数减少 thermal depopulation
热致裂缝 heat crack
热致密化 hot consolidation; hot densification
热致内磁性 pyromagnetism
热致扭曲 heat distortion
热致破坏 thermorunaway
热致曲线 thermoluminescence curve
热致蠕动 hot creep
热致收缩 thermostriction
热致死 thermal death
热致死点 thermal death point
热致死率 thermal lethal rates
热致死时间 thermal death time
热致弯曲 thermal flexure
热致相分离 thermal induced phase separation
热致液晶聚合物 thermotropic(al) liquid crystal polymer
热致振动 heat-driven oscillation
热滞 heat stagnation; thermal lag < 在热敏电阻中的电流 >
热滞后 thermal hysteresis; thermo-lag
热滞后现象 thermal hysteresis
热滞弹性 thermoanelasticity
热中和定律 law of thermoneutrality
热中和性 thermoneutrality
热中心 thermal center[centre]
热中心反应堆 thermal center reactor
热中子 slow neutron; thermal energy neutron
热中子薄层照相法 thermal neutron laminagraphy
热中子成像照相探测器 photographic(al) thermal neutron image detector
热中子堆 thermal neutron reactor
热中子反应 thermal neutron reaction
热中子放射性 thermal activity
热中子俘获 thermal capture
热中子俘获截面 thermal neutron capture cross section
热中子激活 thermal neutron activation
热中子校准 thermal neutron calibration
热中子截止 thermal cutoff
热中子利用因子 thermal utilization factor
热中子裂变 thermal fission; thermal neutron fission
热中子裂变产额 thermal fission yield
热中子裂变物质 thermal fisser
热中子裂变因子 thermal fission factor
热中子漏逸率 thermal neutron leakage factor
热中子谱 thermal neutron spectrum
热中子区 thermal neutron range; thermal range
热中子试验反应堆 thermal test reactor
热中子寿期 thermal neutron lifetime
热中子束 thermal beam; thermal neutron beam
热中子衰变时间测井记录 thermal decay time log
热中子探测器 thermal neutron detector; thermal detector
热中子通量 thermal neutron flux
热中子通量标准 thermal neutron flux standard
热中子吸收 thermal neutron absorption
热中子泄漏 thermal neutron leakage
热中子循环 thermal neutron cycle
热中子源 thermal source
热中子增殖(反应)堆 thermal breeder reactor
热中子转换反应堆 thermal neutron converter reactor
热中子作用下可裂变性 thermal fissionability
热衷庞大计划者 projectite
热衷于 high on
热重法 thermogravimetry
热重量变化曲线 thermal weight change curve
热重量分析 thermogravimetric analysis; thermogravimetry
热重量分析法 thermogravimetry
热重曲线 thermogravimetric curve
热重仪 thermogravimeter
热周环反应 thermal pericyclic reaction
热周期 heat cycle
热轴 heating of axle box; hot(axle) box; hot jewel
热轴承检测器 hot bearing detector
热轴定位器 hotbox locator
热轴定位器盘 hotbox locator panel
热轴检测器 < 在驼峰顶或到达站入口处 >【铁】hotbox detector
热轴检测扫描器 detector scanner
热轴颈用润滑油 hot-neck grease
热轴颈指示灯 hot journal light
热轴颈指示器 hot journal indicator
热轴识别电路 bearing identification circuit
热轴箱 hot axle box
热皱缩 thermal shrinkage
热皱折 thermal buckling
热骤变 heat shock
热骤退 febrile crisis
热注封缝材料 hot-applied joint pouring compound
热贮藏 heat storage capacity
热贮存 hot storage
热贮料斗 hot storage bin
热驻波比 heat standing wave ratio; hot standing wave ratio
热驻极体 thermal electret
热柱 thermal column; thermalizing column; thermal plume
热柱中子束 thermal column beam
热铸 hot-cast
热转化 thermal inversion; thermal transition
热转化过程 thermal conversion process
热转化性 heat-convertibility
热转换 thermal conversion
热转换面积 area of heat transfer
热转换能 conversion of heat into power
热转换器 heat inverter unit; thermal converter
热转换系统 heat-transfer system
热转配合 thermorunning fit
热转送 transference of heat
热转移 thermal pumping
热转移操作 hot-transfer operation
热转移介质 heat-transfer material
热转移系数 coefficient of heat transfer; heat-transfer coefficient
热转移印花 heat-transfer printing; thermal transfer printing
热转印印刷 thermal printing; thermal transfer printing
热装 shrinkage fit(ting)
热装罐 heat pack
热装罐法 hot-pack method
热装环座 shrunk-on ring carrier
热装接头 shrunk joint
热装料 hot charging; molten charge
热装配合 shrinking on
热装置 heat facility
热状况 thermal regime(n)
热状态 heat condition; thermal condition
热状态方程 thermal equation of state
热锥比值 pyrometric cone equivalent value
热灼剂 thermocaustica
热灼伤 thermal burn
热自动补偿活塞 autothermic piston
热自动瞄准头 heat-homer
热自聚焦 thermal self-focusing
热自流系统 thermoartesian system
热自作用效应 thermal self-action effect
热渍电镀 hot-dip plating
热总量 heat budget
热阻 heat dam; heat(-transfer) resistance; thermoresistance; thermal resistance
热阻分离法 heat-resistance-separating method
热阻抗 thermal impedance
热阻力系数 coefficient of thermal resistance; heat-resisting coefficient; temperature resistance coefficient
热阻率 heat-resistivity; thermal resistivity; thermal resistance
热阻系数 thermal resistivity
热阻现象 thermal chocking
热阻性 thermal resistivity
热阻值 resistance value; R-value; thermal resistance
热钻法 thermic boring; heat drilling
热钻孔 heat drilling; thermal drilling; thermic boring
热作 hot working
热作锤 hot set hammer
热作的 hot worked
热作模具钢 hot die steel
热作模具合金钢 alloy hot-die steel
热作业 hot working
热作业用模具钢 hot work mo(u)ld steel
热作用 heat action
热作用具 hot cutter

人班 < 工时计算单位 > man-shift

人本心理学 humanistic psychology
人本主义 humanism
人播的 anthrochrous
人才 manpower; qualified personnel; talent; talented person
人才储备表 inventory of manpower
人才存留率 rate of brain retention
人才的成长 human development; trained personnel can make professional progress
人才管理 human resources management
人才集中地 talent highland
人才交流服务中心 personnel exchange service center[centre]
人才开发 developing talents; talent development
人才流动 flow of personnel
人才流动的评价 valuation of skilled resource flow
人才流动法规 personnel mobility law
人才培训 manpower training
人才培养 manpower training
人才外流 brain drain
人才优势 make good use of talent
人操纵特性 human-control characteristics
人差【测】human error; individual error; personal bias; personal difference; personal equation; personal error; respective error; subjective error
人差改正 personal correction
人差计 idiometer
人车分隔设计法 Radburn method

人车分离设计 <首先在美国拉德柏恩使用> Radburn
人车分流(通行) pedestrian-vehicular segregation; pedestrian segregation
人车分流设计 Radburn
人车合用信号显示 combined pedestrian-vehicle phase
人车路三项式 user vehicle-road trinomial
人车通行信号显示 combined pedestrian-vehicle phase
人车站 major stop
人次 man-time; person-time
人道主义管理 humanitarian management; human-management
人的财产 personal property
人的差异管理 management of human difference
人的担保 personal guarantee
人的感觉级别 human sensation level
人的感觉能力 man's sensory capability
人的工作能力 human performance
人的工作效率 human performance
人的供给 supply of people
人的环境 human environment
人的记忆 human-memory
人的价值 human value
人的监督管理 management by personal supervision
人的减速度死亡极限 human survival limit
人的决策 human decision-making
人的劳动力 human labo(u)r power
人的老化 human aging
人的模拟 human simulation
人的模型 model of man
人的耐久力 human endurance
人的能力发展 personal development
人的潜力 human potential
人的全面发展研究 research in man's development in an all-round way; research on man's all-round way development
人的生产率 man-productivity
人的视觉 human vision
人的舒适性 human comfort
人的推理 human reasoning
人的效率 human efficiency
人的信息处理系统 human information processing system
人的信息输出通道 human information output channel
人的信息输入通道 human information input channel
人的行动数学化模拟理论【计】 game theory
人的因素 human element; human factor; personal factor
人的因素工程 human factors engineering
人的因素工程学 human engineering
人的因素学会 human factors society
人的振动敏感性 human sensitivity of vibration
人的质量 man qualities
人的智能 human intelligence
人的准动产 chattels personal
人等量 man equivalent
人地比率 man-land ratio
人地关系 man-land relationship
人地系统 man-earth system
人地相称 harmony of population and land
人地相关原理 theory of man-land relationship
人丁税 poll tax
人定概率 personal probability
人动视力 vision with driver moving
人洞 manhole
人渡 passenger ferry
人对地震反应 human reaction of earthquake
人对信息的处理系统 human information processing system; information processing system
人对噪声的容忍度 human tolerance to noise
人对振动的敏感性 human sensitivity to vibration
人耳响应 response of the ear
人防 civil air defense [defence]; people's air defense[defence]
人防办公室 office of civil air defense [defence]
人防措施 air defence measure
人防地道 air-raid underground shelter
人防地下室 air-raid shelter; air-raid sheltering basement; bombproof basement
人防工程 civil air defence works
人防建筑 civil defense construction
人防建筑物 civil defense shelter
人防连通道 people's air defense subway
人防隐蔽所 civil defense shelter
人粪尿 human excrement; human waste; night soil
人粪尿处理 night soil treatment
人粪尿处理厂 night soil treatment plant
人粪尿污泥 night soil sludge
人格表面特性 surface trait
人格层 hierarchy of personality
人格化 humanisation; personification
人格化资本 personified capital
人格品质 personality trait
人格完整 personal integrity
人格主义心理学 personalistic psychology
人工 labo(u)r; manpower; manual work; work done by hand
人工矮化 nanization
人工安装 hand set(ting)
人工安装队 hand laying gang
人工岸 artificial coast
人工按长臂节 manual section
人工扒动 hand raking
人工扳道【铁】 manual setting of points
人工搬运 hand haulage
人工拌和法 hand mixing; hand mix procedure
人工拌和砂浆耙 mortar hoe
人工拌制的混凝土 hand-mixed concrete
人工包扎 hand packing
人工包扎的 hand-packed; artificial-packed
人工包装 hand filling
人工爆炸 secondary reflection; shoot
人工备件 manual backup unit
人工比较装置 manual collator
人工闭塞 manual block; manual-controlled block; non-interlocked block
人工闭塞区域 manual block territory
人工闭塞系统 handworked block system; manual block system
人工闭塞信号系统 manual block signal system
人工闭塞行车规则 manual block signaling rule
人工闭塞制 handworked block system; manual block system
人工闭塞装置 manual block apparatus
人工闭锁 manual block
人工边界 artificial boundary
人工边坡 artificial slope
人工编制程序 manual programming
人工变量 artificial variable
人工变量法 artificial variable method
人工变速传动 manually shifted transmission
人工变性 experimental sex several
人工标石 artificial monument
人工标图 hand plotting
人工标志(点) artificially marked point; artificial point; artificial target; panel point; signalized point
人工标准 labo(u)r standard
人工标准成本 labo(u)r standard cost
人工表 labo(u)r statement
人工冰核 artificial ice nucleus
人工波动 artificial fluctuation
人工剥膜 manual stripping of membranes
人工播撒机 hand distributor
人工播云 cloud seeding
人工补充的地下水 artificially recharged groundwater; artificially recharged underground water
人工补给 artificial recharge; induced recharge; artificial renourishment <地下水的>
人工补给层厚度 thickness of artificial recharge stratum
人工补给层时代 age of artificial recharge stratum
人工补给层岩石 rock of artificial recharge layer
人工补给地下水 artificial groundwater recharge; artificial recharge; artificial recharged groundwater
人工补给方法 measures of artificial recharge
人工补给工程 artificial recharge project; engineering of artificial recharge
人工补给量 artificial recharge
人工补给目的 purpose of artificial recharge
人工补炉 hand fettle
人工补沙方法稳定岸线 shoreline stabilization by artificial nourishment
人工补水 artificial recharge
人工材料费保单 labo(u)r and material payment bond
人工裁切 manual cut(ting)
人工裁弯 artificial cutoff
人工采光 artificial daylight; artificial lighting
人工采掘面 hand-mined face
人工采试样 hand sampling
人工操纵 hand control; human operation; manual control; manual handling; manually control; manual maneuvering; manual operation
人工操纵臂板信号 manually operated semaphore; manually operated semaphore signal
人工操纵的 manually guided; manually operated
人工操纵的气动夯具 manually operated pneumatic tamper
人工操纵减速器 manually operated retarder
人工操纵交通信号 manually operated signal
人工操纵推进 man operated propulsion
人工操作 hand-operating; hand operation; manhandle; manipulation; manual drive; manual handling; manual manipulation; manual operation
人工操作焙烧炉 handworked roaster
人工操作程序 manual procedure
人工操作的 hand-operated; manually operated; manual acting
人工操作电路 manually operated circuit
人工操作阀 manually operated valve
人工操作反射炉 hand reverberatory furnace
人工操作粉刷 manual plastering
人工操作夯 manually operated rammer
人工操作机器 hand-operated machinery; manually operated machinery
人工操作抹灰 manual plastering
人工操作配料(拌和)机 manual batcher
人工操作时间 handing time
人工操作手柄 manual lever
人工操作线路 manually operated circuit
人工操作信息处理中心 manual processing station
人工操作要求 manual operational requirement
人工操作振捣器 hand-manipulated vibrator
人工糙率 artificial roughness
人工草地 artificially sown pasture; artificial pasture; sown pasture; tame pasture
人工草皮 artificial grass
人工草原 seeded pasture
人工测定 labo(u)r measurement
人工测制 mapping by free hand
人工差异 labo(u)r variance
人工产卵场 <鱼类> artificial spawning ground
人工产品 man-made products; artificial products
人工产物 artefact; artifact
人工产物的 artifactitious
人工铲削清理 hand chipping
人工长途记录 manual toll recording
人工长途交换台 manual toll board
人工场 artificial field
人工场法 artificial field method
人工车辆统计 hand traffic count
人工陈化 artificial ag(e)ing; artificial seasoning; preag(e)ing
人工成本 cost of labo(u)r; labo(u)r cost
人工成本百分比法 labo(u)r cost percentage method
人工成本比率 labo(u)r cost ratio
人工成本单 labo(u)r cost sheet
人工成本汇总表 summary of labo(u)r cost
人工成本节约计划 labo(u)r cost saving plan
人工成核 artificial nucleus
人工成核作用 artificial nucleation
人工成熟 artificial maturation
人工成型 artificial forming; chair work
人工程序 manual program(me)
人工池(塘) artificial pond; cistern; man-made pond; man-made pool; manual pond; ponding
人工充电 artificial recharging
人工充水通道 passage of artificial water filling
人工充填 artificial renourishment; hand gobbing; hand packing
人工充填带 hand-packing strip
人工冲刷 man-made erosion
人工冲洗 artificial flush(ing)
人工冲淤 artificial flush(ing)
人工重调 hand reset; manual reset
人工抽样 artificial sample; artificial sampling; hand sampling
人工出渣 hand mucking
人工除霜 man-made defrosting; man-

ual defrosting
人工储存量 artificial storage
人工处理 manual handling
人工触发器 manual trigger
人工穿孔卡 hand punched card
人工传递边界 artificial transmitting boundary
人工传染 artificial infection
人工船道 artificial shipway
人工床层 artificial bed
人工吹管法 blow cylinder process
人工吹筒摊平法 <平板玻璃> hand cylinder method; hand cylinder process
人工吹制 artificial blowing; hand blowing; manual blowing; mouth blowing
人工吹制玻璃 hand blown glass
人工吹制法 hand blown process
人工磁化法 artificial magnetization method
人工促进降水 precipitation enhancement
人工催化降水 artificial inducement(al) of rainfall; artificial precipitation stimulation; precipitation stimulation; rain stimulation by artificial means; stimulation of precipitation
人工催雨 artificial inducement(al) of rainfall; artificial precipitation stimulation; rain stimulation by artificial means; stimulation of precipitation
人工存储器 manual memory
人工搭配差异 labo(u)r mix variance
人工打包 hand press-packing
人工打电话支局 attended telephone substation; manual telephone substation
人工打夯机 hand rammer
人工打键 manual manipulation
人工打捆 manual banding
人工打眼 hand holing
人工打眼法 hand-holding
人工打桩 hand spike
人工打桩机 bell rope hand pile driver; common pile driver; hand pile driver; manually operated pile driver; manual pile driver; ringing engine; ringing pile engine
人工大气老化 artificial weathering
人工代用装置 override
人工岛法 artificial island method
人工岛港 port island; artificial island harbor
人工岛(屿) artificial island; man-made island
人工捣固(夯实) hand tamping
人工捣实 hand compaction; hand puddling; hand punning; hand tamping
人工捣实的 hand-compacted
人工捣实混凝土 hand-compacted concrete
人工道路 made-up road
人工的 artificial; factitial; factitious; false; made-up; man-made; non-natural; human-made
人工的鲜艳色彩 technicolo(u)r
人工低温测定 freezing test
人工堤 man-made dike[dyke]; barrier
人工堤防 embankment
人工地表 man-made surface
人工地基 artificial foundation; artificial ground; artificial subgrade; artificial subsoil
人工地基密实 artificial consolidation
人工地基上的梁和板 slab-end-beam on artificial foundation
人工地面运动 artificial ground motion
人工地面噪声 artificial ground noise
人工地平 artificial horizon
人工地物 anthropogenetic form; anthropogenic form; artificial feature of terrain; cultural detail; cultural feature; human element; man-made feature
人工地物版 culture board
人工地下灌溉 artificial sub-irrigation
人工地下水 artificial groundwater
人工地下水补给 artificial groundwater recharge; artificial recharge of ground water
人工地下水回灌 artificial groundwater recharge; artificial recharge of ground water
人工地震 artificial earthquake; man-made earthquake; miniature earthquake
人工地震动 artificial ground motion
人工点火燃烧器 manual burner
人工点缀石 ornamental artificial stone
人工电报机 manual telegraph set
人工电话 manual telephone
人工电话机 manual telephone set
人工电话交换机 manual exchange
人工电话局 manual central office; manual exchange
人工电话区 manual area
人工电话所 manual telephone office
人工电话系统 manual telephone system
人工电话支局 attended substation; manual substation
人工电话制 manual system; manual telephone system
人工电话中心 manual telephone center[centre]
人工电路 manual circuit
人工电压调节器 manual voltage regulator
人工淀积方解石 artificially precipitated calcite
人工顶板 artificial roof
人工冻结地基 artificial freezing of ground
人工冻土 artificial frozen soil
人工洞室 artificial cavern; grotto
人工读入【计】 manual read
人工短路 artificial short circuit; man-made short circuit
人工短路跳闸 fault throwing
人工堆肥 artificial compost
人工堆积型 fill type
人工堆砌 hand packing
人工堆砌巷道 <采空区内的> brattice road(way)
人工堆石 hand-packed rockfill; hand-packed stone
人工堆筑的抛石护坡 hand-placed riprap
人工钝化处理 artificially blunted
人工发报速度 hand speed
人工发热 artificial fever
人工发热器 hypertherm
人工繁殖 artificial propagation; culture
人工方法 hand labo(u)r method
人工方块 modified cube
人工防冲铺盖 armo(u)r
人工防护 artificial defence; manual shield
人工防火线 fire guard
人工防治 artificial control
人工放顶 artificial caving
人工放入 manual insertion
人工放射性 artificial radioactivity; induced activity
人工放射性产物 artificial radioactive product; artificial radioactivity product
人工放射性核素 artificial radionuclide
人工放射性示踪剂 artificial radioactive tracer
人工放射性同位素 artificial radiation isotope
人工放射性元素 artificial radioactive element; artificial radio element
人工放射源 man-made source
人工费(用) artificial labo(u)r cost; cost of labo(u)r used; labo(u)r cost; manpower cost
人工分洪 deliberate division
人工分解 artificial disintegration
人工分局 manual substation
人工分类法 artificial system
人工分路 man-made shunt; manual shunt
人工分配 labo(u)r distribution
人工分配单 labo(u)r distribution sheet
人工分群 artificial swarm
人工分选 hand sorting separation; manual separation
人工分(钟) man-minute
人工粉碎泥炭 hand-cut peat
人工风化 artificial weathering
人工风蚀 artificial weathering
人工敷设 hand laying; manual laying
人工浮船道 artificial shipway; ringing shipway
人工浮床 man-made floating island
人工辐射 artificial radiation
人工辐射带 artificial radiation belt
人工辐射剂量 artificial radiation dose
人工辐射能 artificial radioactive energy
人工辐射源 artificial radioactive source
人工负荷 dummy load
人工复式交换台 manual multiple type switchboard
人工复位 manual reset
人工复位继电器 manual-reset relay
人工复原 hand reservoir; manual restoration; manual restore
人工复原式 hand reset system
人工覆盖层 artificial mulch
人工覆盖物 artificial cover
人工改变天气 weather modification
人工改变系统结构 manual reconfiguration
人工改道 artificial diversion of river; artificial realignment; artificial rerouting
人工改小骨料尺寸 coning and quartering
人工改造气候法 weather modification process
人工干材法 artificial seasoning
人工干预 manual intervention
人工干燥 artificial drying out; exsiccation
人工干燥的 artificially dried
人工干燥法 artificial drying; artificial seasoning
人工干燥炉 artificial drying oven
人工干燥施工法 artificial drying
人工干燥室 artificial drying oven
人工干燥装置 hand drying apparatus
人工钢筋切断机 hand bar cutter
人工港 artificial harbo(u)r; artificial port; Mulberry harbo(u)r; port island
人工港布置 layout for artificial harbo(u)r
人工港池 artificial basin
人工港湾 artificial harbo(u)r
人工高地-湿地污水处理系统 constructed upland-wetland wastewater treatment system
人工高架单轨系统 hand suspended monorail system
人工给料 hand feed
人工跟踪 hand tracing; hand tracking; manual following; manual tracking
人工跟踪转速计 manual-tracking tach(e)ometer
人工更新 artificial regeneration
人工工时标准 labo(u)r quantity standard
人工工时差异 labo(u)r quantity variance
人工工资率标准 labo(u)r rate standard
人工工资率差异 labo(u)r rate variance
人工工作 manual working
人工工作时间 handing time
人工构筑物 artificial structure; artificial works
人工骨料 artificial aggregate; artificially crushed aggregate
人工固定化工程菌 artificial immobilized engineering bacteria
人工固结 manual consolidation
人工管理 labo(u)r control; labo(u)r management
人工灌溉 artificial irrigation; artificial watering
人工灌浆 manual grouting
人工灌浆胶结 artificial cementation; artificial cementing
人工灌注 artificial recharge; artificial recharging; manual priming
人工灌装 hand filling
人工光 artificial light
人工光源 artificial light source; artificial source of illumination
人工光源乳剂 emulsion for artificial light source; photoflood emulsion
人工光照 artificial lighting
人工国际交换台 manual international exchange
人工过船道 artificial shipway
人工海岛 artificial island
人工海滩 artificial beach
人工海滩养护 artificial nourishment
人工焊机 hand welder
人工焊接 manual welding
人工夯 hand ram
人工夯具 set ram
人工夯实 artificial compaction; artificial tamping; hand compaction; hand packing; hand tamping
人工夯实的 hand-packed
人工夯实的基层 hand-packed bottoming
人工夯实的基础 hand-packed bottoming
人工夯实的石填料 hand-packed hardcore
人工夯实的泰福式基层 hand-packed Telford type subbase
人工航道 artificial channel; artificial course; artificial navigable waterway; artificial navigation canal; artificial waterway
人工合成 artificially synthesized; artificial synthesis
人工合成工业 synthetic(al) industry
人工合成化合物环境标记法 environmental labeled method of artificial compound
人工合成胶粘剂 artificial adhesive
人工合成树脂胶 artificial glue
人工合成弹性体 synthetic(al) elastomer

R

人工合成土 synthetic(al)soil
人工河槽 artificial channel
人工河道 artificial channel; artificial stream; leat; sloot
人工河流 artificial stream
人工河网 artificial drainage
人工核 artificial nucleus
人工核法 artificial nuclear survey method
人工核反应 artificial nuclear reaction
人工横膈 diaphragmatic prosthesis
人工烘干 artificial drying
人工后备 manual backup
人工后援 manual backup
人工呼叫 manual call(ing)
人工呼吸 artificial respiration
人工呼吸器 biomotor; inhalator; spirophore
人工湖 artificial lake; barrier lake; dam lake; impoundment; man-made lake; storage lake
人工护坡 artificial guard a slope
人工滑道 artificial slipway
人工环境 artificial environment; man-made environment; manual environment
人工环境动植物园 vivarium
人工环境生态学 hemeroecology
人工环境室 climatic chamber; environmental chamber
人工缓解 manual release
人工换挡 manual shift
人工回灌 artificial recharge; induced recharge
人工回灌冲淡已污染水质 diluting polluted water by artificial recharge
人工回灌的水源 water resources for artificial recharge
人工回灌工程 <地下水的> artificial recharge project
人工回灌井 artificial recharge well
人工回填 artificial fill; hand filling
人工汇接制 manual tandem system
人工混合 hand mixing
人工混凝土拌和 hand concrete mixing
人工混凝土骨料 <如煤渣,矿渣,陶粒> artificial aggregate
人工混凝土块体 artificial concrete block
人工混响 artificial reverberation; synthetic(al) reverberation
人工混响器 artificial echo unit
人工火警系统 manual fire alarm system
人工击碎 cobbing
人工机械夯 hand-held mechanical tamper
人工积肥 artificial manure
人工基础 artificial footing; artificial foundation
人工基底 artificial basis
人工基地 artificial base; artificial ground
人工激振 man-excited vibration
人工及材料担保书 <由保险公司出具的> mechanic's lien surety bond
人工及材料费用留置权 mechanic's and material man's lien
人工级别 job grade; labo(u)r grade
人工级配骨料 artificially graded aggregate
人工级配集料 artificially graded aggregate
人工极化 artificial polarization
人工集料 <包括副产品集料如煤渣、矿渣等以及人造集料如陶粒> artificial aggregate; artificially crushed aggregate
人工集水(面积) artificial catchment
人工集水区 artificial catchment
人工计量 artificial ga(u)ging; manual ga(u)ging
人工计数 artificial counting; manual counting
人工计算 artificial calculation; artificial computation
人工计算和解释 manual calculation and interpretation
人工记录 manual record
人工记录表 man record chart
人工加糙 artificial roughness
人工加固 artificial consolidation; artificial strengthening
人工加固土 artificially improved soil
人工加料 hand feed(ing)
人工加煤的 hand-fired; hand fixed
人工加煤锅炉 hand-fired boiler
人工加煤机 manual stoker
人工加煤炉 hand-fired furnace
人工加热电路 manual heating circuit
人工加速度 artificially generated acceleration
人工(加速)时效 accelerated ag(e)-ing
人工架设的脚手架 hand erected scaffold
人工假顶 artificial roof
人工尖晶石 Hope sapphire
人工监控装置 manual supervisory control
人工检查 human check; human inspection
人工检索 <系统> manual retrieval
人工建筑 <海图标志> culture
人工建筑叠标 artificial range
人工建筑物 artificial structure; civil works; works of man
人工建筑物符号 culture symbol
人工键控 manual manipulation
人工将地下水引到地面 developed water
人工降低地下水位 artificial dewatering; predraining
人工降水 artificial precipitation
人工降雨 artificial inducement(al) of rainfall; artificial precipitation; artificial rain(fall); cloud seeding; design rain; overhead damping; precipitation enhancement; rain-making; simulated rainfall
人工降雨法 artificial rain method; nucleation
人工降雨工作人员 rainmaker
人工降雨灌溉 overhead irrigation; sprinkler irrigation
人工降雨罐 overhead tank
人工降雨机 raining; sprinkling machine
人工降雨喷管装置 sprinkler pipe system
人工降雨器 rainer; rainfall simulator; rain gun; rainulator; sprinkler
人工降雨设备 overhead irrigating machine; rainmaker; rainulator
人工降雨试验法 method of artificial rainfall experiment
人工降雨试验小区 sprinkled plot
人工降雨装置 artificial rain device; artificial simulator; rainer; rainfall simulator; rainmaker; rain-making machine; rain simulator; sprinkler; sprinkler rig; water sprinkler
人工交换 manual operation
人工交换机 manual exchanger; manual switchboard
人工交换台 attendant's switchboard; manual desk; manual switchboard
人工交通量计数 manual traffic counting
人工浇水 artificial watering
人工浇筑 manual placement
人工浇筑混凝土 hand-placed concrete
人工浇铸 hand teem(ing)
人工礁石 artificial reef
人工胶凝材料 artificial cementing agent
人工搅拌 hand mixing; hand rabbling; hand stirring; manually operated mixing
人工搅拌炉 hand-rabbled furnace
人工搅拌水泥浆 hand-mixed cement grout
人工搅拌台 banker
人工校准 manual adjustment
人工阶地 artificial terrace
人工接地 artificial earth(ing)
人工接地体 artificial earthed body
人工接替 manual backup
人工接线员 manual operator
人工接续 manual closing operation; manual operation
人工节流阀 hand throttle
人工节约法 labor-saving device
人工结构(物) artificial structure
人工截断 manual intercept
人工截弯 artificial cutoff
人工解释 manual interpretation
人工解锁 hand release; manual release
人工解锁按钮 manual release push button
人工解锁表示 manual release indication
人工介质 artificial media
人工金刚砂(磨料) artificial abradant; artificial abrasive
人工金属电弧焊接 manual metal arc welding
人工进料砂光机 hand block sander
人工晶体 artificial crystal; artificial lens; synthetic(al) crystal
人工精选上等石灰 best hand-picked lime
人工景观 artificial landscape
人工径流场 artificial runoff field
人工厩肥 artificial manure
人工局 manual office
人工举升 artificial lift
人工锯管机 hand pipe cutter
人工绝缘和包缠 hand doping; hand laying up
人工掘进 advance by hand; hand sinking
人工掘井 hand pit
人工掘土 hand mucking
人工卡片式缓行器控制 card-manual type of retarder control
人工开辟场地 made ground
人工开采 quarrying by hand
人工开动 hand drive
人工开发的地下水 developed water
人工开挖 hand excavation; hand mucking; manual excavation
人工开挖沉箱 hand-dug caisson; hand-dug shaft
人工开挖的港口 artificially excavated harbo(u)r; artificially excavated port
人工开挖的水道 artificial waterway; rhine; rine
人工开挖盾构 artificial excavation shield; manual shield
人工开挖沟 excavated artificial trench
人工开挖河道 canalized river; canalized stream
人工开挖井孔 hand-excavation shaft
人工开挖竖井 hand-excavated shaft
人工开挖钻机 hand drill(ing machine)
人工可航水道 artificial navigable waterway
人工空气 artificial atmosphere
人工控制 artificial control; hand control; manual control; manual manipulation
人工控制板 manual control panel
人工控制的 hand-guided
人工控制地面沉降方法 ground settlement method of artificial control
人工控制机构 hand control unit
人工控制交通信号 manually operated signal; manually controlled traffic signal
人工控制开环 manual control open loop
人工控制盘 manual control panel
人工控制配料 manual mixture control
人工控制器 manual controller
人工控制设备 manually controlled plant
人工控制天气 control of weather
人工控制系统 hand controls; hand control system
人工控制信号机【铁】 manual-controlled signal
人工口盖 obturator
人工会计 labo(u)r accounting
人工快速渗滤系统 constructed rapid infiltration system
人工馈送 hand feed
人工馈送穿孔机 hand feed punch
人工拉长 artificial elongation
人工拉绳式打桩机 bell rope hand pile driver
人工拉索式打桩机 bell rope hand pile driver
人工拉索式打桩器 bell rope hand pile driver
人工拦截沿岸流的沙埂 man made littoral barrier
人工拦沙埂 man made littoral barrier
人工劳动 hand labo(u)r
人工老化 accelerated ag(e)ing; accelerated weathering; artificial weathering; preag(e)ing; tempering; artificial ag(e)ing <美>
人工老化机 weatherometer
人工老化机试验 weatherometer test
人工老化铝 artificially aged alumin(i)um
人工雷达干扰 window
人工冷却 artificial cold; artificial cooling; attemperation; positive cooling
人工立时接通制 demand service
人工砾料 artificial gravel material
人工连接灰泥板 plasterboard manual jointing
人工联锁 manual interlocking
人工量程选择器 manual range selector
人工裂解 artificial lysis
人工裂隙 artificial fissure
人工林 artificial afforestation; artificial forest; man-made forest; plantation
人工淋雨试验 rain test
人工溜冰场 artificial ice rink; artificial skating rink
人工流场 artificial flow field
人工露头 artificial exposure; artificial outcrop
人工露头观测点 observation point of artificial outcrop
人工芦苇湿地 constructed reed wetland
人工炉 hand furnace; hand-raked fur-

nace
人工滤料 artificial filter media
人工滤膜 artificial filtrable membrane
人工路基 artificial subgrade
人工绿柱石 igmerald
人工码垛 hand stowage
人工镘平 hand floating
人工镘涂 hand floating
人工毛石铺底 hand-packed bottoming
人工煤化试验 artificial coalification experiment
人工煤气 manufactured gas
人工密实法 artificial consolidation
人工模拟 manual simulation
人工模拟产率曲线 hydrocarbon productivity curve of artificial simulate
人工膜 artificial membrane
人工磨床锯 bucking saw
人工磨矿 bucking
人工磨圆金刚石 shaped stone
人工抹面的混凝土 hand-finish concrete
人工木材干燥法 artificial seasoning
人工目标 artificial target; man-made target
人工牧场 artificial pasture; seeded pasture
人工耐火建筑材料 artificially refractory construction material
人工脑 <即计算机> mechanical brain; artificial brain
人工能源 man-made energy source
人工黏[粘]接剂 manufactured adhesive
人工黏[粘]性 pseudo-viscosity
人工黏[粘]性方法 pseudo-viscosity method
人工黏[粘]性压力 pseudo-viscous pressure
人工排水 artificial drainage
人工排水沟 artificial drain
人工排水系统 artificial drainage system
人工排泄 artificial discharge
人工抛光盘 hand polisher
人工抛石 hand-packed rockfill; hand-packed rubble; hand-placed riprap; man-packed rockfill
人工培养 artificial culture; artificial nourishment; culture
人工培养基 artificial culture medium; synthetic(al) medium
人工培养珍珠 cultured pearl
人工培植 nutrient culture
人工培植植物 cultigen
人工配合 hand fit
人工配料拌和机 manual batcher
人工配料器 manual batcher
人工配料箱 manual batcher
人工配砂 synthetic(al) sand
人工喷泉系统 deluge system
人工喷雾器 hand shower
人工偏斜补取岩芯 side-tracking coring
人工频率测深异常曲线 anomaly curve of frequency sounding with a source
人工平路机 hand grader
人工坡 anthropogenic slope
人工坡降 artificial grade
人工破坏程度 artificial disturbance
人工破碎 cobbing; hand breaking
人工铺砌(斜坡或护岸的块石或混凝土块体) pitching
人工铺砌的 hand-placed
人工铺砌的基础 hand-pitched foundation
人工铺砌基础 hand-laid foundation
人工铺砌块石 hand-packed rubble
人工铺砌碎石基层 hand-pitched stone subbase
人工铺撒石屑 hand chipping
人工铺砂机 hand gritter
人工铺设 manual laying
人工铺设的 hand-placed
人工铺设的防冲乱石 hand-placed stone riprap
人工铺置乱石 hand-placed rubble
人工铺置石块 hand-placed stone
人工铺筑 hand set(ting)
人工铺筑填石沥青 hand-laid stone-filled asphalt
人工瀑布 artificial fall; artificial waterfall
人工曝气 artificial aeration
人工曝气池 artificial aeration basin
人工曝气滴滤池 artificially aerated trickling filter
人工曝晒机 weatherometer
人工起重小车 hand crab
人工气候 artificial atmosphere; artificial climate
人工气候控制室 weather master
人工气候试验 artificial weathering test
人工气候试验室 climate cell
人工气候室 artificial climate chamber; climate chamber; climatic laboratory; climatron; environmental control chamber; man-made climate room; phytotron(e); psychrometric room; sychometric room
人工气流 air steam
人工气象 artificial meteorology
人工砌石 hand-packed rockfill; hand-packed rubble; hand-placed rock
人工砌石面层 hand rock facing
人工砌筑的岩石面层 hand-packed rock facing
人工敲碎 cobbing
人工敲碎的 hand-pitched
人工敲碎的石块 hand-broken metal
人工切割 manual cut(ting)
人工切换 manual switching
人工倾斜式金属桶 hand-tipping type ladle
人工清除 hand cleaning; manual clear
人工清除式格栅 hand-cleaned rack
人工清除装置 manual cleaning device
人工清理 hand cleaning; hand mucking
人工清扫 hand sweeping
人工清扫设备 manual cleaning device
人工清筛 hand screening
人工清洗滤网 hand-cleaned screen
人工驱动的 hand-operated
人工渠槽 artificial channel
人工渠道 artificial canal; artificial channel; hand channel
人工取代安全装置 overriding a safety device
人工取代控制 overriding control
人工取消 manual cancellation
人工取样 hand sampling; manual sampling
人工取样器 hand sampler; manual sampler
人工全息图 artificial hologram
人工全息照片 artificial hologram
人工泉 bored spring
人工缺陷 artificial defect
人工群落 artificial community; culture community
人工燃气 manufactured gas
人工燃烧 hand-firing
人工热储流体 artificial reservoir fluid
人工热源照射法 ignisation
人工日 man-day
人工日报单 daily labo(u)r report
人工日光 alpine light; artificial light
人工溶解 induced lysis
人工融雪 artificial snowmelt
人工柔光 candle light(ing); soft artificial light
人工揉练泥料 wedging
人工润滑 manual application of lubricant; manual lubrication
人工润滑法 hand lubrication
人工润滑器 manual lubricator
人工散布 manual spreading
人工扫除 hand brooming
人工砂 <人工粉碎的细骨料的总称> artificial sand; crushed sand; manufacture(d) sand
人工砂岛 artificial sand island
人工沙滩 artificial beach
人工筛分的骨料 artificially graded aggregate
人工闪电 artificial lighting
人工伤害事故保险 insurance against accident to workmen
人工商用电话局 manual commercial office
人工烧火 hand-firing
人工设置进路 manual route setting
人工生产日 man-day
人工生成加速度图 artificially generated accelerogram
人工生火 hand-firing
人工生火的 hand-fired
人工生态系统 artificial ecological system; artificial ecosystem
人工生物法 artificial biological method
人工湿地 constructed wetland
人工湿地处理系统 constructed wetland treatment system
人工湿地垃圾渗滤液 constructed wetland treating landfill leachate
人工石板 artificial flagstone
人工石墨 electrographite
人工石屑撒布机 hand gritter
人工时间差异 labo(u)r time variance
人工时间法 man-hour method
人工时效(处理) artificial ag(e)ing; artificial seasoning
人工时效过度 overag(e)ing
人工时效硬化 artificial aging
人工识别 artificial cognition; artificial identification; artificial perception
人工式 manual mode
人工示踪剂 artificial tracer
人工试验筛 hand testing sieve
人工试样 hand specimen
人工收集<废物的> piggyback collection
人工收缩臂节<起重机> manual retracted section
人工手调复位 manual-reset adjustment
人工输入 manual entry; manual input
人工输入寄存器 manual input register
人工输入键 manual input key; manual load key
人工输入设备 manual input generator; manual input unit; manual number generator
人工输入装置 manual input device; manual input unit; manual word generator
人工输入坐标 manual feeding of coordinate
人工输沙 sand by-passing
人工树脂胶交换器 artificial resin exchanger
人工竖井回灌 artificial recharge through well
人工数<特定工程各项需用的> labo(u)r constant
人工数据 artificial data
人工数据收集系统 manual data collection system
人工数据输入 manual data input
人工数据输入设备 manual data input equipment
人工数据中继中心 manual data relay center[centre]
人工数列 artificial series
人工双折射 artificial birefringence; synthetic(al) birefringence
人工水槽 artificial water tank
人工水草 artificial aquatic mat
人工水道 artificial watercourse; artificial waterway; leat; man-made waterway; water gang
人工水沟 leat
人工水晶 synthetic(al) quartz
人工水库 artificial reservoir; man-made lake; man-made reservoir
人工水利资源 artificially created water resources
人工水流控制 artificial flow control
人工水凝灰浆 artificial hydraulic mortar
人工水驱 flooding; water-flooding
人工水系 artificial drainage
人工水下培养器 artificial substrate
人工水闸渠道 sluiceway
人工水中给养基 artificial substrate
人工饲养 artificial feeding
人工送料 hand feed
人工搜索 manual search
人工速度辅助跟踪 manual rate-aided tracking
人工速度调节 manual speed adjustment
人工碎石 hand-broken stone
人工碎石集料 artificially crushed aggregate
人工碎石块铺底层 hand-pitched base
人工台 manual board
人工太阳光 artificial sun light
人工太阳能岛 artificial solar energy island
人工滩淤 beach rehabilitation
人工碳化处理 artificial carbonation
人工掏槽法 hand-holding
人工淘汰 artificial culling
人工梯田 artificial terrace
人工提高金属疲劳强度法 coaxing
人工提升 artificial lift; hand hoisting
人工提升的 hand pulled
人工天井掘进 hand raising
人工天空 artificial sky
人工添加剂 man-made additive
人工填充 artificial replenishment; hand filling
人工填地 made land; made-up ground
人工填方 artificial fill
人工填砾井 artificial gravel-pack well
人工填料 hand-stuff
人工填土 artificial earth fill; artificial made land; artificial refilling; artificial soil; fill; made ground
人工填筑 manual placement
人工填筑地 made land
人工填筑垫层 structural fill
人工填筑土层 make-up ground
人工挑选 hand picking
人工条件反射 conditioned reflex
人工调节 hand governing; hand regulation; manual adjuster; manual adjustment; manual regulation
人工调节均衡器 manual equalizer
人工调控 artificial regulation
人工调谐 manual tuning
人工调谐线圈检测器 manual-tuned loop detector
人工调整 hand regulating; hand regulation; hand reservoir; manual ad-

justment; manual regulation
人工跳汰(选) hand jigging
人工通风 artificial draft; artificial draught; artificial ventilation; artificial venting; assisted draught; forced air supply; forced draft; inducted draft; positive draft
人工通风风扇 induced-draft fan
人工通航水道 artificial navigable waterway
人工通炉 hand poking
人工同步 manual synchronizing
人工同位素 artificial isotope; man-made isotope
人工投煤 hand shovel(l)ing of coal
人工投入线性平衡法 line balancing
人工透射边界 artificial transmitting boundary
人工图形 artificial graphics
人工涂敷 manual spreading
人工土方工 hand mucker
人工土方工程 man-made earthwork
人工土壤 artificial soil
人工土壤测定 test of artificial soil
人工团聚体 artificial aggregate
人工拖动 hand drive
人工挖除淤泥 manual mucking(out)
人工挖掘 hand digging; hand mucking; manual excavation
人工挖掘井 stuff dug shaft
人工挖掘式盾构 hand-mined shield
人工挖孔灌注桩 manual boring grouted pile
人工挖孔桩 hand-dug caisson; hand-dug shaft
人工挖泥 artificial dredging; bag and spoon dredging
人工挖泥船 bag and spoon dredge(r)
人工挖泥耙 hand dredge(r)
人工挖深 hand sinking
人工挖土石方 hand excavation
人工挖运软土 hand mucking
人工弯轨机 jim crow
人工弯筋机 stuff bending machine
人工弯曲木 bent wood
人工围堤填土 artificial nourishment
人工喂料 hand feed
人工喂料辊 hand-drawn roller; stuff drawn roller
人工温度控制 manual temperature control
人工温度应力 artificial temperature stress
人工紊流水槽 artificial turbulence flume
人工稳定 servo-stabilization
人工稳定同位素 artificial stable isotope
人工稳定性指数 labo(u)r stability index
人工污泥 artificial sludge
人工污染 artifact pollution; artificial pollution
人工污染试验 artificial pollution test
人工污染物 artificial contaminant
人工无水石膏 artificial anhydrite
人工洗砂 sand cleaning by hand
人工系统 man-made system; manual system
人工下水道 artificial drain
人工纤维 artificial fiber[fibre]
人工现实法【计】virtual reality method
人工线(路) artificial line; bootstrap
人工相关成本 labo(u)r-related cost
人工橡胶填充剂 adhesive putty
人工削坡 artificial pare a slope
人工消磁 manual degaussing
人工消耗 artificial drainage
人工消化 artificial digestion
人工小交换机 intercommunication plug switchboard
人工小时 man-hour
人工小时法 labo(u)r hour method; man-hour base; man-hour method; worker-hour method
人工小时率 labo(u)r hour rate; man-hour rate; work(er)-hour rate
人工小太阳 xenon vapo(u)r lamp
人工效率 labo(u)r efficiency; man efficiency
人工效率报告 labo(u)r efficiency report
人工效率差异 labo(u)r efficiency variance; labo(u)r time variance
人工效应物 artifact
人工卸货 artificial discharging
人工信号 manual signal
人工信号控制机 manual controller
人工型流程程序图 man-type flow process chart
人工修补 manual patching
人工修改 manual completion
人工修管工具 hand pipe tool
人工修配 hand fitting
人工修坯 hand fettle
人工修整 hand chipping; hand finishing; manual amendment
人工畜道 cattle walkway
人工蓄水 artificial storage
人工蓄水池 artificial lake; dew pond
人工旋压法 conventional spinning; manual spinning
人工选择 artificial selecting; artificial selection; manual selection
人工选择性调整 manual selectivity control
人工压浆处理 artificial cementing; artificial solidification
人工压浆处理法 artificial cementing method
人工压紧的岩石面层 hand-packed rock facing
人工压实土 human-made compacted embankment
人工延长 artificial elongation
人工盐 artificial salt
人工养蚝场 stew
人工养滩 artificial beach nourishment
人工养殖 artificial culture
人工养殖项链 cultured pearl necklace
人工样本 artificial sample
人工业务 manual service
人工抑制 artificial suppression
人工音量控制 hand volume control
人工引变 artificial induction of hereditary changes
人工引出水量 quantity of artificial exported water
人工引入水量 quantity of artificial inputted water
人工引用的河水 developed water
人工引种植物 hemerophyte
人工印花 block printing; hand blocking
人工应答 manual answering
人工应急能力 manual intervention capacity
人工影响降水 precipitation stimulation; rain stimulation by artificial means
人工影响气候 climate modification
人工影响天气 man-induced weather modification; weather modification
人工影响天气方案 weather modification program(me)
人工硬化 artificial ag(e)ing; warm hardening
人工用棒捣实 hand rodding
人工用量差异 labo(u)r usage variance
人工诱发 man-induced
人工诱发地震 man-induced earthquake
人工诱发裂变径迹密度 induced fission track density
人工淤滩 artificial nourishment; artificial renourishment
人工淤填 artificial deposition; artificial silting
人工语言 artificial language; synthetic(al) language
人工育种 biologic(al) engineering
人工预算 labo(u)r budget
人工源遥感 active remote sensing
人工月 man-month
人工云 artificial cloud
人工运河 artificial canal
人工运量计数器 hand traffic counter
人工运输 hand haulage
人工运送 hand haulage
人工栽培的 tame
人工凿井 hand sinking
人工凿掘 hand picking
人工凿平 hand chipping
人工造成活套 hand looping; manual looping
人工造肥 artificial fertilizer
人工造林法 artificial reforestation; reforestation
人工造石 stuc
人工造斜 whipstocking
人工造雪 artificial snow-making
人工造雪机 snow maker
人工造雨 rain-making
人工造雨法 artificial rainmaking process; nucleation
人工造雨器 rainmaker
人工造云 cloud seeding
人工增甜剂 artificial sweetener
人工增压 pressure trapping
人工增雨 artificial precipitation enhancement
人工障碍物 artificial barrier
人工招募计划 personnel recruitment program(me)
人工照明 artificial illumination; artificial light(ing)
人工照明的 artificially lighting
人工照明过渡 artificial lighting transition
人工照明要求 requirements of artificial illumination
人工振动 artificial vibration
人工振铃 manual ringing
人工振铃电话 manual ringing telephone
人工振铃器 manual ringer
人工振铃制(电话) manual ring down basis
人工振源 artificial vibration source
人工震波 artificial earthquake wave
人工震源 artificial seismic source
人工整平的河床 artificially flattened bed
人工整平的基床 artificially flattened bed
人工整修 hand finish
人工整治 artificial regulation
人工支出 labo(u)r expenditures
人工支座 artificial abutment
人工知觉 artificial perception
人工知识 artificial knowledge
人工制 manual system
人工制版 manual plate making
人工制干草 hay drying
人工制冷 artificial cold
人工制品 artefact[artifact]; manufactured article
人工制砂 sand fabrication
人工制造的 hand-fabricated; made-up
人工制造的玻璃 hand-made glass
人工制作的金属格栅 handicraft-type metal grille
人工致裂 artificial fracturing
人工致雨 artificial inducement(al) of rainfall; artificial precipitation stimulation; rain stimulation by artificial means; stimulation of precipitation
人工智能 artificial intelligence
人工智能的应用 application of artificial intelligence
人工智能法 artificial intelligence approach
人工智能模拟 human simulation
人工智能系统 artificial intelligence system
人工智能修剪 alpha cut-off
人工智能语言 artificial intelligence language
人工智能专家系统软件 expert system
人工重砂采样 artificial placer sampling
人工重砂样品 artificial placer sample; sample for person heavy mineral
人工周转成本 cost of labo(u)r turnover
人工周转率 labo(u)r turnover
人工注水 artificial water flooding; artificial water injection
人工注油枪 hand grease gun
人工筑岛 <沉井施工方法之一> artificial island
人工筑岛法 <深基础施工时下沉井，沉箱用> artificial island method
人工筑岛进行于施工 buildup land
人工转换程序 manual changeover procedure
人工转换信号 manual changeover signal
人工转接 human relay working
人工转向力 manual steering force
人工转运 man-handling of materials
人工装罐 hand caging
人工装货 artificial loading
人工装料 hand charging; manual loading
人工装入 hand insertion; hand introduction
人工装填石块 hand-packed hardcore; hand-packed stone
人工装卸 manual handling
人工装岩 loading by hand
人工装载 hand-fill; manual loading
人工装载法 artificial loading
人工装载工作面 hand-filled face
人工着色 artificial colo(u)ring
人工琢石 hand-pitched stone
人工自动的 manual-automatic
人工自动开关 attent-unattent switch
人工自动转换继电器 manual-automatic relay
人工阻尼 artificial damping; synthetic(al) damping
人工组成砂 synthetic(al) sand
人工钻探 manual boring; manual drilling
人工作日 man-day
人工作业 manual working
人工座席 manual position
人公里 passenger-kilometer[kilometre]
人公里运价 tariff per passenger-kilometer; tariff per person-kilometre
人过流面积 circulation area
人和生物圈 man and biosphere
人化的 humanized
人环节特性 human component characteristic
人货两用电梯 service elevator
人祸 <如战争、暴动等> act of public

enemy
人机 man-machine
人机程序图 man-machine process chart
人机调查图 census map
人机对话 interface communication; man-computer dialog(ue); man-machine conversation; man-machine dialogue; man-machine communication; man-machine interaction
人机对话编辑 interactive editing
人机对话程序 interactive program(me)
人机对话交通调试系统 interactive debugging system
人机对话模拟器 interacting simulator
人机对话式查询 interactive query
人机对话数字图像处理系统 interactive digital image manipulation system
人机对话图 interactive graphics
人机对话图形处理 interactive graphic processing
人机对话制图编辑 interactive cartographic(al) editing
人机分布 distribution of population
人机分布图 distribution map
人机工程学 human engineering; man-machine engineering; ergonomics; ergonomy
人机共存 man-machine symbiosis
人机关系 man-machine rapport; man-machine relationship
人机环境系统【交】 man-machine-environment system
人机交互 interact
人机交互处理 interactive processing
人机交互动作 interactive action
人机交互方式【计】 interactive mode
人机交互界面 man-machine interface
人机交互系统 man-machine interactive system
人机交互作用 man-machine interaction
人机接口 man-machine interface
人机接口动作站 man-machine interface workstation
人机界面【计】 human-computer interface
人机可靠性 man-machine reliability
人机控制 ergonomics; ergonomy
人机控制系统 man-machine control system
人机联合控制 man-machine control
人机联系 interface communication; man-computer interaction; man-machine commission; man-machine interface; man-machine communication; man-machine interaction
人机联系编辑系统 interactive editing system
人机联系操作 interactive operation
人机联系处理机 interactive handler
人机联系方式 interactive way
人机联系光标 interactive pointer
人机联系绘图系统 interactive graphics system; interactive mapping system
人机联系计算机制图系统 interactive computer graphic(al) system; interactive computer mapping system
人机联系解释 interactive interpretation
人机联系控制器 interactive controller
人机联系模型法 interactive modelling
人机联系数据编辑系统 interactive data edit system
人机联系系统 interactive man-machine system; man-machine system
人机联系系统编辑 interactive display and edit facility
人机联系显示控制 interactive display control
人机联系显示控制器 interactive display controller
人机联系修改 interactive modification
人机联系要素抽出系统 interactive feature extraction system
人机联系制图 interactive graphics; interactive graphy
人机联系装置 man-machine interface
人机联系自动制图系统 automated interactive drafting system; interactive graphy
人机模拟 man-machine analogy; man-machine simulation
人机配合 human machine adaption
人机配合设计 man-computer interaction design
人机匹配 man-machine matching
人机软件包 man machine package
人机数字系统 man-machine digital system
人机调节系统 man-machine processing system
人机通信[讯] man-computer communication; man-machine communication
人机通信[讯]系统 man-machine communication system; man-machine system
人机通信[讯]语言 man-machine language
人机问答 man-machine interrogation
人机相容 man-machine compatibility
人机协调 man-machine harmony
人机直接检索 on-line search
人机组合 man-machine complex
人迹稀少的 lone
人激振动试验 man-excitation test
人际关系 human relation; interpersonal relationship
人件 liveware
人居 human habitat
人居环境 human settlements
人居环境科学 science of human settlements
人聚环境 human settlements
人均 per capita
人均产量 per capita output; product per capita
人均产值 per capita product
人均出行率 average person trip rate
人均道路面积 road area per capita; road area per citizen <城市>
人均道路占有率 road area per capita
人均观念 conception of per capita
人均国民生产总值 gross national product per capita; per capita gross national product
人均国民收入 national income per capita; per capita national income
人均国内生产总值 per capita gross domestic product
人均居住面积 average living floor area per capita; living space per capita; per capita living space
人均可支配实际收入 per capita disposable real income
人均劳动生产率 productivity per worker
人均利润率 profit margin per person
人均量 quantity per capita
人均量的多项比较 multilateral per capita quantity comparisons
人均林地面积 forest area per capita
人均留利 per capita retained profit; retained profit per capita
人均每夜住宿收费 average receipts per person-night
人均能源消费量 per capita energy consumption
人均年收入 annual per capital income
人均实际产量 per capita real output
人均实际国民收入 real national income per capita
人均实际可支配收入 real disposable per capita income
人均实际收入 real income per capita
人均使用面积 floor space per person
人均收入 income per capita; income per head; per capita income
人均税率 per capita rate
人均物量指数 per capita quantity index
人均消费量 consumption per head
人均消费支出 per capita consumption expenditures
人均消耗量 consumption per head; per capita consumption
人均需求法 per capita demand method
人均需求量 per capita demand
人均用电量 per capita household electricity consumption
人均占地 land holdings per capita
人均支出 per capita expenditures
人均指标 per capita index
人均周转额 turnover per capita
人均住房面积数 occupancy factor
人均总产量 per capita gross product
人可接触的安全电压 contact voltage
人可进出的烟筒 man-sized chimney
人孔 access hatch; access hole; access opening; clean-out hatch; clean-out opening; colluviarium; handhole; hatch; inlet manhole; inspection chamber; man head; manhole; manhole entrance; manway; manway nozzle; sight hole; scuttle hole <通向屋顶的>
人孔壁 manhole wall
人孔壁及仰拱 manhole walls and invert
人孔撑架 crossbar; dog back; manhole dog; strong back
人孔尺寸 manhole size
人孔的砖石工程 manhole masonry (work)
人孔底部 floor of manhole
人孔顶部 manway head
人孔复式封盖 double seal manhole cover
人孔覆盖 manway covering
人孔盖 access door; chute gate; clean-out cover; companion; inspection plate; manhole cover(ing); manhole head; manhole plate; manlid; manway cover
人孔盖板 clean-out plate; cover plate for manhole; manhole cover plate; manhole dog
人孔盖合页 manhole hinge
人孔盖架 manhole frame and cover
人孔盖口 companion
人孔盖压板 manhole dog
人孔盖座 manhole(cover) ring; manway(cover) ring
人孔格栅 hatch grating; manhole grid; manhole grill(ag)e
人孔横杆 manhole cross bar
人孔环 manhole ring
人孔加强圈 manhole ring
人孔口 access port; hatch; manhole opening
人孔扣夹 manhole dog
人孔框架 manhole frame; manway frame
人孔栏栅 manhole guard
人孔类型 type of manhole
人孔里盖 catch pan
人孔门 manhole door
人孔内部 invert of manhole
人孔内底 invert of manhole
人孔墙 manhole wall
人孔圈 manhole ring
人孔室 chamber of manhole
人孔踏步 manhole step
人孔台 manhole step
人孔(探视)室 manhole chamber
人孔套管 manhole junction box
人孔梯(子) manhole ladder; manhole step; ladder to manhole
人孔铁口 eruption; manhole frame
人孔铁踏步 manhole iron step
人孔围板 manhole coaming
人孔围壁 manhole coaming
人孔下水道砖 manhole sewer brick
人孔(楔形)拱砖 arch brick
人孔形拱砖 arch brick for manholes
人孔闸 man lock
人孔中的备份电缆 slack in the manhole
人孔最小尺寸 minimum size of manhole
人孔座 manhole collar
人控部分 personal sub-system
人控的姿态控制 manned attitude control
人控功能 override facility
人控机动小车 man trolley
人控机器 man-controlled machine
人控机器人 man-controlled mobile robot
人控撒布机 hand sprayer
人控式起重小车 man trolley
人控试验设备 manual test equipment
人控系统 man-manageable system; manual system
人控相位 override phase
人控制观测气球 kytoon
人口 population
人口百岁图 age pyramid
人口爆炸 demographic(al) explosion; population explosion
人口变动 change of population; population change
人口变动统计 statistics of population change; statistics of population movement
人口变化 demographic(al) transition
人口变迁 demographic(al) change
人口参数 population parameter
人口操纵渡线 hand throw crossover
人口城市规划 urban planning of population
人口城市化 population urbanization
人口城市集中 urbanization of population
人口稠密城市 over-crowding city
人口稠密的 densely populated; over-peopled; populous; thickly inhabited
人口稠密的中心 pressure core
人口稠密地带 highly populated corridor
人口稠密地区 congested area; congregated district; densely inhabited district; densely populated area; district thickly inhabited; region of dense population
人口稠密区 developed area
人口稠密市区 densely populated urban area
人口出生率 birth rate; fertility
人口垂直分布 perpendicular distribution of population
人口当量<工厂污水量换算为人口污水量> population equivalence;

R

population equivalent
人口的地理分布 geographic(al) distribution of population
人口的对数增长 logarithmic increase in population
人口的管理 human care
人口的社会增长 social increase of population
人口登记册 population register
人口登记站 census registration station
人口地理分布学 anthropography
人口地理学 population geography
人口地区分布 spatial distribution of population
人口地区构成 region structure of population
人口地图 population map
人口调查 demographic(al) census; demographic(al) survey
人口调查表 census paper
人口调查分析 census analysis
人口调查计算机 census computer
人口调查数字 intercensal figures
人口调查统计法 census statistic method
人口调查研究 intercensal study
人口动力学 population dynamics
人口动态 dynamics of population; population behavio(u)r; population dynamics
人口动态统计 dynamic(al) statistics of population
人口动态预测模型 model of population dynamic(al) projection
人口都市化 urbanization of population
人口多产分配 prolificacy distribution; prolificness distribution
人口发展 population development
人口分布 population distribution
人口分布结构 demographic(al) structure
人口分布统计图表 <如性别、年龄等> population pyramid
人口分析 demographic(al) analysis; population analysis
人口分组 grouping of population
人口负增长 negative population growth
人口概况 population profile
人口高度集中的大城市 mega(lo) polis
人口高密度区 densely inhabited district
人口构成 composition of population; demographic(al) composition; population composition; population structure
人口估计 population estimate
人口估算 population estimate
人口惯性 inertia of population
人口规划 population planning; program(me) for planning population growth
人口规律 population law
人口过程 population process
人口过多 overpopulation
人口过多的 overpeopled
人口过密集城市 over-crowding city
人口过少 underpopulation
人口过剩 excess of population; overpopulation; overspill; surplus population
人口过剩区 over-populated zone
人口过稀城市 over sparse city
人口过稀区 underpopulation area; underpopulation zone
人口和工业过度集中 excessive concentration of population and industry
人口-环境-发展系统 population-environment-development system
人口环境容量 environmental capacity of population
人口环境政策 population-environmental policy
人口机械迁移 mechanical migration of population
人口激增 demographic(al) explosion
人口集合体 population aggregate
人口集中 concentration of population; population concentration
人口集中(稠密)地区 densely inhabited district
人口集中点 population center
人口减少 decrease in population; depopulate; depopulation
人口减少的 unpeople(d)
人口健康序列 population-health sequence
人口结构 population structure
人口金字塔 population pyramid
人口净密度 net population density
人口静态 state of population
人口就业序列 population-employment sequence
人口居住点 agglomeration of population
人口聚集势 population potential
人口聚居点 agglomeration of population
人口靠拢势 <人/公里> population potential
人口控制 population control
人口控制目标 goal of population control
人口劳动构成 working structure of population
人口老(龄)化 ag(e)ing of population
人口类型 type of population
人口理论 demographic(al) theory; population theory
人口历史地理 demographic(al) historic geography
人口零增长 zero population growth
人口流动 mobility; movement of population
人口流动率 mobility rate
人口流动性 population mobility
人口毛密度 gross population density
人口密度 density of population; population density
人口密度比较指标 comparative density index
人口密度等值线 isarithm
人口密度分布 population density distribution
人口密度分区地图 dasymetric map
人口密度理论 theory of population density
人口密度曲线 partial population curve
人口密度图 isarithmic map
人口密集 concentration of population
人口密集地区 densely inhabited district; densely settled area
人口模式 population pattern
人口内挤 population implosion
人口内向暴增 population implosion
人口年龄构成 population composition by age
人口膨胀 population expansion
人口平衡 population equilibrium
人口平均值 mean of population
人口普查 census; head count; national census; population census
人口普查表 census paper
人口普查街区 census block
人口普查区段 census tract
人口普查日 census day
人口迁移 migration of population; population migration; population movement
人口迁移流动率 rate of migration flow
人口趋势 population trend
人口趋势指标 indicators of demographic(al) trend
人口群 population cluster
人口生态学 human ecology; population ecology
人口生物学 human biology
人口事故率 population accident rate
人口收入序列 population-income sequence
人口疏散 population dispersal; population evacuation
人口税 capitation; capitation tax
人口死亡率 human mortality
人口素质 population quality
人口特征 population characteristic
人口统计 demographic(al) statistics; demographics; population statistics; statistics of population; vital statistics
人口统计参数 demographic(al) parameter
人口统计的 demographic(al)
人口统计及社会趋势 demographic(al) and social trend
人口统计年鉴 demographic(al) yearbook
人口统计培训 demographic(al) training
人口统计区 census enumeration district
人口统计趋势 demographic(al) trend
人口统计数字 population census
人口统计特性曲线 demographic(al) characteristic
人口统计特征 demographic(al) characteristic
人口统计图 census map; demogram
人口统计学 demography
人口统计学家 demographer
人口统计学研究 demographic(al) study
人口统计转化期 period of demographic transition
人口统计资料 demographic(al) data
人口投资 population investment
人口图 population map
人口推算 estimation of population
人口外溢 overspill
人口稀少的 thinly inhabited; underpeopled; unpeople(d)
人口稀少地区 sparsely populated area; underpopulation area; underpopulation zone
人口稀少国家 scarcely populated country
人口稀少区 underpopulation
人口稀少区边缘 underpopulation front
人口稀疏地区 low population area
人口现象 demographic(al) behavio(u)r
人口向城市集中 urban concentration of population
人口学 demography
人口亚群 population subgroup
人口移居 population migration
人口影响气候 man's impact on climate
人口预报 population forecast
人口预测 demographic(al) projection; population forecast; population prediction; population projection
人口预测法 population projection method
人口预测图解法 graphic(al) method of population projection
人口预计 population forecast
人口再生产 population reproduction; reproduce the population
人口增长 demographic(al) increase; growth of population; population growth
人口增长率 accretion of population; growth rate of population; rate of population growth; ratio of population in crease
人口增加率 accretion of population
人口增殖 propagation of population
人口增殖力的下降 fertility decline
人口政策 demographic(al) policy
人口指标 target population
人口众多发达地区 heavily developed populated area
人口骤增 population boom; population explosion
人口状况 demographic(al) situation
人口资料 population key
人口自然迁移 natural migration of population
人口自然增长 natural increase of population; natural population growth
人口自然增长率 natural increase rate; rate of natural increase of population; rate of natural population growth
人口总增长率 overall population growth rate
人口组成 composition of population; demographic(al) composition; population composition
人类 human being; humanity
人类安全工程学 human safety engineering
人类产业权利 industrial rights of man
人类尺度 human scale
人类地理学 anthropogeography
人类对气候的影响 man's impact on climate
人类发生的 anthropogenetic; anthropogenic
人类废物 human waste
人类分布学 anthropography
人类粪便的利用 human excrement use
人类粪便无害化处理 decontamination of human excreta
人类干预 human intervention
人类工程活动 engineering activities of human
人类工程计量学 ergonometrics
人类工程师 human engineer
人类工程手册 human factors handbook
人类工程学 ergonomics; ergonomy; human(factor) engineering
人类工程学调查 ergonomic survey
人类工效学 ergonomics; ergonomy
人类共同感 sense of community among men
人类环境 human environment
人类环境宣言 Declaration on the Human Environment
人类活动 anthropic activity; cultural activity; human activity; man's activity
人类活动的 anthropogenic
人类活动污染 pollution by human activity
人类活动学 anthropokinetics
人类活动因素 action factor of human
人类活动影响 effect by human activi-

ty;human impact;human influence
人类疾病 human disease
人类纪 Anthropogene
人类健康 human health
人类进入宇宙时代 man-in-space project
人类经常居住区 ecumene
人类居住类型 human settlement pattern
人类居住模式 human settlement pattern
人类居住区 human settlement
人类居住学 ekistics
人类历史 human history
人类利益 benefit of mankind;human interest
人类排泄物 <如汗、粪便、尿等> human excrement;human excreta
人类气候学 anthopoclimatology
人类躯体学 somatology
人类圈 anthroposphere;noosphere
人类群居学 ekistics
人类群落 human community
人类社会经济系统 human socioeconomic system
人类社会生物学 human society
人类社会文化生态学 human sociocultural ecology
人类生存空间 human living space
人类生活环境 human habitant;human habitat
人类生活环境模式 human habitat model
人类生活空间 human living space
人类生境 human habitat
人类生境模型 human habitat model
人类生态系统 human ecological system
人类生态学 anthroecology;anthropecology;human ecology
人类生物地理学 human biogeography
人类生物学 anthropobiology;human biology
人类声学 human acoustics
人类声学测量 human acoustic measurement
人类时代 Anthropogenic age
人类时期 human period
人类寿命 human longevity
人类文化学 ethnology
人类洗涤污水 flush human sewage
人类行为 human behavio(u)r
人类行为学 praxeology
人类需要层次 hierarchy of human need
人类学 anthropology
人类学博物馆 Museum of Anthropology
人类因素 human factor
人类引起的环境退化 human induced degradation
人类营造力 anthropogenetic force
人类影响 anthropogenic influence
人类政治地理学 geoanthropolitics
人类住区的环境方面 environmental aspect of human settlement
人类住区的社会经济方面 socioeconomic aspects of human settlements
人类住区管理 human settlements management
人类住区展览 exposition on human settlement
人类资源保护 conservation of human resources
人类自我保护 human conservation
人力 human force;human power;labo(u)r power;manual effort;unit of power;manpower
人力安排规划 manpower planning
人力安排计划 job cover plan
人力拌制的灰浆 hand applied(mixed) plaster
人力补充 manpower replenishment
人力不可抗拒事故 force majeure event
人力不足 manpower deficit;shortage of manpower
人力财富 human wealth
人力操纵的 manually controlled;manually operated
人力操纵活塞泵 hand piston pump
人力操作 manual operation
人力操作挡水板 handstop
人力操作的 hand-operated
人力操作悬挂式脚手架 manual suspended scaffold
人力车 a two-wheel vehicle drawn by man;jinrikisha;rickshaw
人力冲击钻进 hand percussion boring;percussion hand boring;percussion hand drilling
人力冲击钻探 percussion hand boring
人力铡草器 hand straw cutter
人力储备 stock of manpower
人力船 manpowered boat
人力打夯机 ramming engine
人力打炮眼 hand auger work
人力打桩机 ringing engine
人力的 man-drawn
人力调配 manpower scheduling
人力短缺 manpower shortage;shortage of manpower
人力方面的比率 manpower ratio
人力费用 manpower cost
人力分配 man-assignment
人力分配法 manpower allocation;manpower allocation procedure
人力分析 manpower analysis
人力负荷拉平 manpower leveling
人力负荷图 personnel loading chart
人力干预 manual intervention
人力割捆机 manual reaper binder
人力工程学 ergonomics;ergonomy;human engineering
人力工作 manual work
人力供给来源 source of labo(u)r supply
人力供应 supply of manpower
人力供应来源 source of labo(u)r supply
人力固结 manual consolidation
人力管理 human resources management;manpower control;manpower management
人力规划 manpower planning
人力过剩 manpower surplus
人力过时 human obsolescence
人力夯 hand ram(mer);hand tamp
人力夯具 bishop
人力夯样板 hand-operated screed
人力号笛 manual whistle
人力混凝土夯实板 concrete tamper
人力激发 human-motivation
人力棘轮钻 ratchet drill
人力集材 hand skidding
人力加载的 hand(ing)loaded
人力绞磨 bar capstan;hand capstan;jack roll;man-capstan
人力绞盘 bar capstan;hand capstan;jack roll;man-capstan
人力绞滩 hand rapids-warping;manpower rapids-heaving;rapids warping by manpower
人力节约 labo(u)r saving
人力紧压捆包 hand press-packed bale
人力救生艇属具 equipment of manual lifeboat
人力卷扬机 hand hoist
人力开动 manhandle
人力开发 manpower development
人力开挖 simple hand excavation
人力开挖深井式基础 hand excavation of deep shaft foundation
人力控制 hand control;manual control
人力控制带马达的小车 power barrow
人力控制器 manual controller
人力拉紧装置 manual takeup
人力来源 source of personnel;sources of manpower
人力利用 manpower utilization;utilization of manpower
人力利用调查 manpower utilization survey
人力链条传动(装置) hand chain drive
人力路碾 hand(-guided) roller
人力螺旋钻 hand auger
人力螺旋钻钻孔 hand auger boring
人力密集的 manpower-intensive
人力培训 manpower training
人力配备表 Manning table
人力配置 man-assignment
人力喷雾器 hand sprayer
人力铺砌块石 hand-placed stone;pitching stone
人力启动 manual starting
人力启动器 manual starter
人力起动 hand starting
人力起动曲柄 manual starting crank
人力起动摇把 manual starting crank
人力起锚机 hand windlass
人力切草机 clipper;hand straw cutter
人力轻便铆钉器 hand portable riveter
人力倾覆装置 hand tilting device
人力清单 manpower inventory
人力驱动升降机 hand powered lift
人力缺乏 manpower shortage
人力扫除 hand brooming
人力升降机 hand power elevator
人力收割机 hand reaper;manual reaper
人力输入输出分析 manpower input-output analysis
人力淘汰 hand jigging
人力提升 dead lift
人力提升机 hand elevator;hand-powered lift
人力提升悬挂式脚手架 manual scaffold
人力投入产出分析 manpower input-output analysis
人力投资 human capital (investment);human investment;investment in human resources
人力土钻 hand earth auger
人力推车 hand putting;hand tramming;manual haulage
人力推车运输 hand tramming
人力推动 manhandle
人力推动的 manhandled
人力推进 manual feed
人力推钻 manhandle boring;manpowered boring
人力推钻钻孔 manhandle boring hole
人力拖铲 hand scraper
人力脱轨器 derailable by hand
人力脱粒 hand threshed
人力挖井 hand-dug well
人力挖泥机 hand dredge(r)
人力物力的消耗 drain on manpower and material resources
人力物力资源 human and material resources
人力系统 manpower system
人力效率 manpower effectiveness
人力需求 manpower demand
人力旋紧管接头 hand tight
人力训练 manpower training
人力压路机 hand(-guided) roller;manpower roller
人力压榨机 hand squeezer
人力压装的包 hand press-packed bale
人力移动脚手架 manually propelled mobile scaffold
人力引导的 manually guided
人力预算 manpower budget(ing)
人力运输 hand haulage;hand tramming
人力震害 man-made seismic hazard
人力政策 manpower policy
人力支付 manpower payoff
人力转向 manual steering
人力装料的 hand-loaded
人力装卸 manual handling
人力资本 human capital;manpower capital
人力资本成型 human capital formation
人力资本的价值 value of human capital
人力资本法 human capital approach
人力资本赋予量 endowment of human capital
人力资本经济学 economics of human capital
人力资本流动 human capital flow
人力资本投资 investment in human capital
人力资产会计 human asset accounting
人力资产会计师 human asset accounter
人力资源 labo(u)r resources;manpower resources;resources of manpower;staff resources;human resources;human capital
人力资源部 human resources department
人力资源成本 cost of human resources;human resources cost
人力资源的衡量 measuring of human resources
人力资源的潜力 human potential
人力资源的确定 identifying of human resources
人力资源管理 human resources management
人力资源价值 human resources value
人力资源开发 human resources development
人力资源开发管理 human resources development and management
人力资源会计 accounting for human resources;human resources accounting;manpower resources accounting
人力资源模式 human resources model
人力资源平衡表 balance of manpower resources
人力资源情报 human resources information
人力资源资本计划 human resources capital plan
人力钻 hand brace
人力钻机 manual auger
人力钻探 manual boring;manual drilling
人流 flow of pedestrians;passage of crowd;pedestrian flow;stream of people
人流理论 theory of crowd passage
人流历时 duration of inflow
人流密度 density of passenger flow
人流疏散道路 emergency exit
人流速度 crowd walking speed
人流速率 people flow rate
人面狮身像 androsphinx;sphinx
人民币 people's currency[RMB]
人民代表 people's representative
人民防空 civil defense[defence] against air raids
人民防空工程 civil defense[defence]

construction
人民防卫 civil defense
人民英雄纪念碑 Monument to the People's Heroes; People's Heroes Monument; the Monument to the People's Heroes
人名单 roster
人名导卡 name guide card
人名地名目录 name catalog
人名地名研究 onomastics
人名分类 personal ledger
人名录 directory
人名牌 name plate
人名索引 index of persons;name index
人名账户 personal account
人名总账 personal ledger
人命安全电报 safety of life telegram
人命损失 loss of human life
人脑模拟 brain simulation
人年 <劳动量单位,一个人在一年内完成的工作量> man-year;person-year
人年工作量 man-year
人年劳动当量 man-year equivalent
人排泄物 human excrement
人气味 aura[复 aurae/auras]
人千米收入 revenue passenger-kilometer
人欠 balance due from…
人群 cohort; concourse; confluence; crowd;throng
人群调查 pedestrian group size study
人群关系 human relation
人群关系研究所 institute of human relations
人群归因危险度 population attributable risk
人群荷载 load by human crowd;load form crowd;pedestrian load(ing)
人群汇合处 confluence
人群极度拥挤的负载 crush load
人群靠(街)左行标志 walk on left sign
人群通行理论 theory of crowd passage
人日 person day
人日单价 man-day rate
人身安全 personal safety;physical security
人身安全防护用品 personal safety supplies;safety supplies
人身安全容许电流 let go current
人身安全设备 protective equipment
人身安全装置 personal safety equipment
人身保护 personnel protection
人身保护权 <拉> habeas corpus
人身保护状 habeas corpus writ
人身保险 bodily injury insurance; personal insurance
人身保险合同 contract of life insurance;life insurance contract
人身保险退保现值 cash surrender value of life insurance
人身不可侵犯 personal inviolability
人身产热 heat generated by people
人身防护 personal protection
人身关系 personal relations
人身权利 rights of the person
人身伤害 bodily injury
人身伤害责任 <交通事故> bodily injury liability
人身伤亡 personal injury
人身伤亡事故 personal injury accident
人身事故 accident to person(s);accident to workmen; fatal accident; fatal crash; health hazard; personal damage;personal injury by accident
人身事故责任 liability for injury
人身受伤 personal injury
人身受伤事故 personal injury accident
人身死亡险 life insurance
人身损害 damage to persons;personal injury
人身意外保险 personal accident insurance
人身意外伤害 personal accident
人身意外伤害保险 personal accident insurance
人身意外死亡的双重赔偿条款 double indemnity clause
人身因素 <卫生、健康、安全等> human factor
人身与财产损害 damage to person and property
人身自由 autonomy
人生地理(学) anthropogeography; human geography
人生环境 human environment
人时 man-hour
人时产量 output per man-hour; production per man-hour
人时成本 man-hour cost
人时单价 man-hour rate
人时分配任务报告 man-hour distribution task report
人时评价法 man-hour evaluation
人事处 personnel department; staff department
人视差 <眼睛移动引起的视差> personal parallax
人手不够的 short-handed; undermanned;understaffed
人手不足的 underhanded; undermanned
人手电容 <人手的电容影响> hand capacitance;hand capacity
人手缺乏的 short-handed
人首马身像 centaurus
人寿保险 insurance for life;insurance till death;life insurance;life assurance
人寿保险单 life insurance policy;life policy
人寿保险单下的保险 insurance under a life policy
人寿保险的风险额 amount at risks
人寿保险费 life insurance premiums; life insurance with dividend; life rate
人寿保险服务费用估计数 imputed service charges for life insurance
人寿保险公司 life insurance corporation
人寿保险公司的死亡率统计表 experience table
人寿保险精算准备金 actuarial reserves in respect of life insurance
人寿保险赔偿的第一受益人 prima beneficiary
人寿保险赔偿方式选择 optional modes of settlement
人寿保险死亡比率 mortality ratio
人寿保险退保解约金值 cash surrender value
人寿保险信托 life insurance trust
人寿保险业 life office
人寿定期保险 endowment insurance [assurance]
人寿险额外保险费 loading
人兽雕像装饰 <古典建筑中半凸出的> protome
人兽饰盘壁 zoophilous
人兽图像装饰的 historiated
人数 poll;population
人丝树皮皱片 bark crepe
人四轮游览车 <美> surrey
人踏步 manhole step
人体变态性反应 allergic reaction
人体参数 human parameters
人体测量数据 anthropometric data
人体测量学 anthropometry
人体尺度 dimensions of human figure;human dimensions
人体电容 body capacitance;body capacity
人体防护 physical protection
人体辐射剂量检查 personnel monitoring
人体负荷 body burden
人体感应噪声 man-made radio noise
人体工程学 human engineering
人体功率学 ergonomics;ergonomy
人体功率学的 ergonomical
人体计数器 human counter
人体健康 human health
人体健康风险评价 human health risk assessment
人体接触污染物 human exposure to pollutants
人体冷冻技术 cryonics
人体伦琴当量 anthropomorphous phantom;manikin
人体模型 phantom; phantom line; man(n)ikin <试验用的>
人体模型冲击试验 dummy test; human dummy impact test
人体耐受量 <指对放射性等的> human tolerance
人体屈曲度 body flex
人体热损失 body heat loss
人体容许曝露程度 permissible level of human exposure
人体容许曝露水平 permissible level of human exposure
人体散热量 heat gain from occupant
人体散湿量 moisture gain from occupant
人体生理反应 human physiological reaction
人体生物气候学 human bioclimatology
人体特征 physical feature
人体体型椅 contour chair
人体污染 human pollution
人体污染负荷 human pollution burden
人体系统 human system
人体效应 body effect
人体组织 human organism
人天 man-day
人头 poll
人头飞牛雕像 <美索不达米亚宫殿庙宇门口> Lamassu
人头模型试验 head-form test
人头人身鱼尾的海神美人鱼雕像 <希腊神话> Triton
人头数 caput
人头税 capitation;capitation tax;head money;head tax;poll tax
人为报废 artificial obsolescence
人为贝塔射线 man-made beta radiation
人为背景 human setting
人为边界线 artificial boundary
人为变化 man-induced change; man-induced variation
人为变量 artificial variable
人为表层 anthropic epipedon
人为波动 man-induced fluctuation
人为财富 artificial wealth
人为操作 pseudo-operation
人为场 artificial field
人为冲蚀 human erosion; man-induced erosion;man-made erosion
人为冲刷 human erosion; man-induced erosion;man-made erosion
人为传播 anthropochory
人为传播植物 anthropochore
人为粗糙率 artificial roughness
人为错误 mistake;human error
人为错误故障 human error failure
人为大气污染物 man-made atmospheric contaminant
人为的 anthropic;anthropogenic;artificial;man-induced;man-made
人为的地震 man-made earthquake
人为的工作 made work
人为的供应紧张 man-made shortage of goods
人为的河槽变化 man-induced channel change
人为的溜滑 <调车时> false skid
人为的物质 substance of anthropogenic origin
人为的稀缺性 artificial scarcity;contrived scarcity
人为的直接破坏 direct harm done by purely man-made factors
人为低价 artificially low price
人为地表 man-made surface
人为地面沉降 man-made land subsidence
人为地面形态 hemeroecology
人为地平 artificial horizon
人为地物 man-made feature
人为地质灾害 man-made geologic(al)hazard
人为电离 artificial ionization
人为电子干扰 electronic jamming
人为顶极 anthropogenic climax
人为定价 arbitrary pricing
人为防治 artificial control
人为废弃物 anthropogenic wastes; man-made wastes
人为分布 brotochore
人为分类法 artificial classification
人为分配资金 artificial allocation of funds
人为风险 human risks
人为负载 artificial load
人为富营养化 anthropogenic eutrophication
人为干扰 active jamming; deliberate interference; jamming; man-made interference
人为干扰台 jammer
人为故意破坏 vandalism
人为故障 man-made fault; personal committed failure
人为故障发生率 error rate for human incurring fault
人为归化植物 artificial naturalized plant
人为洪水 artificial flood water
人为洪水波 artificial flood wave
人为滑坡 artificial landslide
人为环境 human setting; man-made environment
人为环境水文地质问题 hydrogeologic(al)problem of artificial environment
人为环境问题 man-in environment problems
人为环境污染物 man-made environment(al) contaminant; man-made environment(al)pollutant
人为环境异常 artificial environmental anomaly
人为环境影响 anthropogenic environmental effect
人为活动 anthropic activity; human activity;man's activity
人为激发运动 technogenous movement
人为价格 imputed price
人为接地 artificial earth; artificial ground
人为静电干扰 man-made statics

人为空气污染 man-made air pollution
人为空气污染源 man-made source of air pollution
人为来源 man-made source
人为灭绝 extermination by man
人为泥石流 artificial mudflow
人为排放 anthropogenic discharge; artificial discharge
人为偏差 personal bias
人为偏倚 human bias
人为破坏 man-made sabotage; vandalism
人为气候变化 man-induced weather modification
人为气候室 psychrometric room
人为侵蚀 accelerated erosion; artificial shift erosion; human erosion; man-induced erosion; man-made erosion
人为区 anthropic zone
人为扰动 artificial disturbance; man-made disturbance
人为散布 anthropochory
人为生态系统 man-made ecosystem
人为实物单位 unit of produced kinds
人为事故 human element accident; human failure
人为水平 artificial horizon
人为酸化 anthropogenic acidification
人为损坏 vandalism damage
人为淘汰 artificial selection
人为特性 artificial characteristic
人为条件 artificial condition
人为调节 human adjustment
人为通信[讯]扰乱 communication jamming
人为蜕变 artificial disintegration
人为稳定 artificial stability; artificial stabilization
人为污染 artefact pollution; artificial pollution; human contamination; man-made contamination; man-made pollution
人为污染(来)源 anthropogenic pollution sources; anthropogenic sources of pollution; man-made pollution sources; sources of man-made pollutant
人为污染物 anthropogenic contaminant; anthropogenic pollutant; artificial contaminant; man-made contaminant; man-made pollutant
人为污染物浓度 anthropogenic contaminant concentration
人为无线电干扰 radio countermodulation
人为误差 human-caused error; human error; human incurring error; personal error
人为误差信号 artificial error signal
人为现象 artefact[artifact]
人为限制 artificial constraint; institutional constraint
人为效应 man-made effect
人为(形成的)沙漠 man-made desert
人为修剪 applied pruning
人为修饰 contrivance
人为演替 anthropogenic succession
人为因素 anthropic factor; anthropogenic factor; anthropolfactor; human element; human equation; human factor; personal factor
人为因素事故 human element accident
人为影响 anthropogenic influence; man-made influence; man's activity
人为影响臭氧 ozone modification
人为原因 man-induced cause
人为约束 man-made constraint
人为灾害 disaster of human origin
人为灾难 human-made disaster; man-made disaster
人为噪声 man-made noise
人为障碍 artificial obstruction; artificial adverse; artificial barrier; man-made obstruction
人为振动 man-made vibration
人为值 artificial value
人为植物群丛 anthropogenic association
人为指令 pseudo-instruction
人为资本 artificial capital
人为资源 man-made resources
人卫 artificial satellite
人卫测地 satellite geodesy
人卫测高 satellite altimetry
人卫多普勒导航系统 satellite Doppler navigation system
人卫跟踪 satellite tracking
人卫跟踪照相机 satellite tracking camera
人卫激光测距 satellite laser ranging
人卫监视 moon watch
人卫近站点 closest approach point
人卫日冕照相术 satellite coronagraphy
人卫三角测量 satellite triangulation
人卫天线 satellite antenna
人卫像片 extraterrestrial photograph
人卫照相机 satellite camera
人文 human culture
人文地理学 human geography
人文动物学 ethnozoology
人文工程 human engineering
人文工程学 <研究机械对其操作者影响的科学> ergonomics; ergonomy
人文景观 human-culture landscape; human landscape; scenery of humanities
人文要素 cultural feature
人文因素 institutional factor
人文因素工程 human factors engineering
人文主义 humanism
人文主义心理学 humanistic psychology
人文资源 human resources
人物造型 figurine
人像或装修座 acroter
人像柱 figure post; telamon[复 telamones]
人心果【植】Sapodilla
人行板 board walk
人行避难通道 passenger refuge way
人行边道 sidewalk
人行便道 catwalk; access side
人行岛 pedestrian island
人行道 bridle; bridle path; bridleway; access side; crosswalk; foot walk; Irishman's sidewalk; manway; parapet; pavement; pedestrian path; pedestrian walk; pedestrian way; side path; side pavement; side view; sidewalk; walk; walking way; walk path; sideway <道路用地范围内的>; banquette; causeway <高于路面的>; pavement <英>
人行道安全线 carriageway marking
人行道板 pedestrian slab; sidewalk slab
人行道边缘 footpath edging; sidewalk edging
人行道标牌 curb mark(ing)
人行道玻璃砖砌块 sidewalk glass
人行道步板 foot plank
人行道侧行人(升降)出入口 sidewalk manhole
人行道尺寸 manway size
人行道的缺口 curb break
人行道灯 sidewalk light
人行道地面水阀门箱 curb box
人行道地下室采光棱镜 pavement prism
人行道电梯 sidewalk elevator
人行道顶 crown of sidewalk
人行道范围 sidewalk space
人行道格栅 sidewalk joist
人行道公共电话亭 sidewalk kiosk
人行道公共书报亭 sidewalk kiosk
人行道构架 footway framing
人行道规则 rule of footway
人行道滚压机 footpath roller
人行道荷载 sidewalk loading
人行道横坡度 transverse sidewalk grade
人行道混凝土铺面板 footpath concrete flag(stone)
人行道混凝土石板 sidewalk concrete flag(stone)
人行道记数员 sidewalk teller
人行道(检查孔)的砖石结构 manway masonry(work)
人行道(检查孔)环 manway ring
人行道(检查孔)套管 manway junction box
人行道(检查孔)阴沟用砖 manway sewer brick
人行道交通 pedestrian traffic
人行道交通管制 pedestrian control
人行道交通控制 pedestrian control
人行道咖啡座 sidewalk café
人行道宽度 width of sidewalk
人行道栏杆 pedestrian guardrail
人行道栏木 sidewalk gate; sidewalk (gate) arm
人行道路面 footpath paving; side pavement
人行道路面板 pedestrian deck
人行道路碾 sidewalk roller
人行道路网 pedestrian network
人行道绿化 sidewalk greening
人行道门 sidewalk door
人行道碾压机 sidewalk roller
人行道棚 sidewalk shed
人行道铺板 walk plank
人行道铺路板 footpath flag
人行道铺路工 sidewalk paver
人行道铺路机 sidewalk paver
人行道铺路石板 footpath flagstone
人行道铺面 footpath paving; sidewalk pavement; sidewalk paving
人行道铺面用天然石板 natural stone sidewalk paving flag
人行道铺装 footpath paving; sidewalk pavement; sidewalk paving
人行道人孔 sidewalk manhole
人行道伸缩缝 sidewalk expansion joint
人行道升降平台 sidewalk elevator
人行道石板 footpath flagstone; sidewalk flag(stone)
人行道树种植 street planting
人行道水阀门箱 curb-valve box
人行道托架 sidewalk bracket; walkway bracket
人行道托梁 sidewalk bracket; walkway bracket
人行道温度缝 sidewalk expansion joint
人行道斜坡 pedestrian ramp
人行道悬臂 footway cantilever
人行道悬臂牛腿 sidewalk cantilever bracket; footway cantilever bracket
人行道悬臂托架 sidewalk cantilever bracket; footway cantilever bracket
人行道有效宽度 effective walking width
人行道雨水沟 walking way rainway gutter
人行道指数 footway index
人行道转角通行面积 circulation area
人行道纵梁 sidewalk stringer
人行的 pedestrian
人行地道 passenger subway; pedestrian subway; pedestrian tunnel; pedestrian underpass; underpass; walk-through
人行地下道 foot subway; foot tunnel
人行分段平巷 manway subdrift
人行浮动码头 floating passenger landing stage
人行浮动平台 floating passenger landing stage
人行浮式码头 floating passenger landing stage
人行格 manway compartment
人行格梯子平台 manway landing
人行格子尺寸 manway compartment size
人行拱桥 arch foot bridge
人行过道 pedestrian crossing; pedestrian pass
人行过街道 foot crossing; pedestrian crossing; pedestrian crosswalk
人行过街道信号 pedestrian crossing beacon
人行过路设施 pedestrian crossing facility
人行旱桥 skywalk
人行巷道 manway
人行荷载 lightest load; load by human crowd; load from crowd
人行横道 crosswalk; foot crossing; footway crossing; pedestrian crossing; pedestrian crosswalk; pedestrian passenger; xing pedestrian; zebra crossing
人行横道标志 crosswalk sign
人行横道线 crosswalk line; pedestrian crossing line; pedestrian passenger
人行交叉道信号 pedestrian crossing beacon
人行交通安全 pedestrian safety
人行交通道路 pedestrian traffic way
人行交通门道 pedestrian traffic door
人行交通入口 pedestrian traffic door
人行交通设施 pedestrian facility
人行井 manway shaft
人行跨线桥 passenger foot-bridge
人行廊道体系 gallery system
人行立交 overpass for pedestrians
人行立体交叉 grade-separation of pedestrian
人行林荫路 pedestrian mall
人行楼梯间 pedestrian stair(case)
人行路带 walking strip
人行路线 pedestrian path
人行平台 pedestrian deck
人行坡道 pedestrian ramp
人行桥 footpath bridge; walkover; pedestrian bridge <横贯车站>; foot bridge <只供行人通过的小桥>
人行桥钢构件 foot bridge steelwork
人行桥护网 pedestrian overcrossing screen
人行时期 <道路上只许人行的时期> pedestrian period
人行手推叉车 pedestrian fork lift truck
人行隧道 foot tunnel; pedestrian subway; pedestrian tube; pedestrian tunnel; subwalk
人行隧洞 pedestrian tube
人行索道 walk-through cableway
人行梯 access stairway
人行梯子 ladder way
人行天井尺寸 manway shaft size
人行天桥 elevated pedestrian crossing; foot bridge; overbridge; overcrossing; overpass for pedestrians; passenger foot-bridge; passenger overpass(bridge); pedestrian overbridge; pedestrian overcrossing; pedestrian overpass; pedestrian sky-

R

way;skywalk
人行天桥护网 pedestrian overcrossing screen
人行挑梁 pedestrian overhanging beam
人行跳板 gangway
人行通道 pedestrian access;pedestrian pass; pedestrian passageway; footway;walkway
人行小道 footway;foot path
人行小路 foot walk;pedestrian lane; pedestrian path
人行小门 man door
人行小桥 footbridge
人行匝道 pedestrian ramp
人行闸 man lock
人行栈道 catwalk
人行栈桥 walkaround
人行走道 <坑道里的> manway
人形步行机 man-shaped walking machine
人形壶 ewer in human form
人形机 humanoid
人形机器人 anthropomorphic robot
人形石 ningyoite
人型机器人 humanoid robot
人性化设计 customer friendly design
人畜粪便 animal and human excreta
人畜排泄物 animal and human excreta
人烟稀少的 thinly inhabited
人眼分辨能力 acuity of eye;resolution of eye
人眼调节 accommodation of human eye
人仪差 personal and instrumental equation; personal and instrumental error
人因分析 human factor analysis
人因工程学 ergonomics; ergonomy; human engineering; human factor engineering
人英里 seat mile
人与机器作业图 man-machine chart
人与机械控制 ergonomics;ergonomy
人与生物圈计划 man and biosphere program(me)
人员 personnel
人员安排 personnel placement
人员安全防护 personnel safety guard
人员编制 personnel management
人员变动率 turnover rate of personnel
人员不足的 andermaned; undermanned;understaffed
人员测位系统 personnel location system
人员超编制 over-strength
人员重新配置 redeployment of personnel
人员出入闸 man lock
人员的更换 changes in personnel
人员的提供 supply of personnel
人员调配 cannibalize [cannibalise]; personnel assignment
人员定额 personnel authorization
人员定位(信)标 personal locator beacon
人员费用 personal expenditures;personal expenses;staff cost
人员服务 personnel service
人员负荷图 personnel loading chart
人员更换成本 personnel replacement cost
人员管理 man-management; personnel management
人员管理分析 man-management analysis
人员规范 personnel specification
人员或物料不足 deficiency of men or stores
人员间偏倚 between analyst bias
人员紧急撤离飞机的路线 rescue path
人员紧急运载工具 personal carrier
人员进出用气闸 <气压沉箱或隧道施工的> man lock
人员聚集场所 place of assembly
人员力量 people power
人员流动 personnel mobility; undertake transfer of personnel
人员流动渠道 personal channel
人员落水 man overboard
人员落水信号灯 man overboard light
人员培训 personnel training
人员配备 Manning; manning level; staffing
人员配备计划 personnel scheduling
人员漂浮设备 personal flo(a)tation device
人员评价选择中心 assessment center [centre]
人员去污废物 personnel decontaminated waste
人员伤亡率 <行车事故> casualty rate
人员申请 personnel requisition
人员升降机 man-machine
人员识别 personal identification
人员事故保险 insurance against accident to workmen
人员事故报告 personnel occurrence report
人员疏散 personnel evacuation
人员输送车 personnel appliance; personnel carrier
人员宿舍 crew hut
人员提升 man winding; promotion of personnel
人员挑选 short-list
人员通道 personnel aisle
人员推销 personal selling
人员退休 retire
人员携带者 personnel carrier
人员行动 personal mobility
人员选择 personnel selection
人员选择咨询师 personnel selection consultant
人员训练 personnel training
人员运输车 passenger car
人员运送 man-riding haulage; man trip
人员运载能力 passenger capacity; passenger carrying capacity
人员运载装置 personnel carrier
人员载运舱 personnel transfer capsule
人员载重 man-load
人员直接费用 direct personnel expenses
人员直销 direct personal selling
人员指挥 staffing
人员组成 Manning
人员组织机构图 staffing organogram
人员组织图表 staff organization chart
人月单价 man-month rate
人造坝座 abutment block
人造白榴火山灰 man-made pozz(u)olana
人造柏油 artificial bitumen
人造板 building board;hard board
人造宝石 artificial gem; hard mass; imitation jewel; paste jewel; synthetic(al)gem
人造背景 artificial background
人造标石 artificial monument
人造冰 artificial ice;water ice
人造冰场 artificial ice rink
人造冰晶石 artificial cryolite;cryolith
人造冰洲石 artificial Iceland spar
人造材料 artificial material;synthetic(al)material
人造草皮 artificial turf
人造稠黄油 whipped toppings
人造磁场 artificial magnet field
人造磁铁 artificial magnet
人造粗骨料 artificial coarse aggregate
人造措施 artificial measure
人造大理石 artificial marble;artificial travertin(e);art marble;biancola; imitation marble; man-made marble; manu-marble; marble resin; marezzo;marezzo marble; scagliola;scagliola marble
人造大气 artificial atmosphere
人造的 artificial; fabricated; man-made; allotriomorphic granular; non-natural;synthetic(al)
人造的结构材料 man-made structural material
人造的结构产品 man-made structural product
人造地 man-made land
人造地表 man-made ground
人造地沥青 <即焦油沥青> artificial asphalt; artificial bitumen <美国,即焦油沥青>
人造地沥青骨料混合料 artificial asphalt-aggregate mix(ture)
人造地面砖 artificial stone tile floor cover(ing)
人造地平 artificial horizon
人造地球卫星 artificial earth satellite; circumterrestrial satellite; earth-circling vehicle;earth satellite vehicle; manmade earth satellite; sputnik
人造地球卫星地球物理观测站 orbiting geophysical observatory of artificial earth satellite
人造地球卫星天文学 satellite astronomy
人造地球卫星望远镜 moonscope
人造地球卫星运动理论 theory of artificial earth satellite motion
人造地热流体储层 artificial geothermal fluid reservoir
人造地热系统 man-made geothermal system
人造地物 man-made culture
人造地震 artificial earthquake
人造地震时程 artificial time history
人造地砖 artificial stone tile
人造电离层 artificial ionization
人造短纤维 staple fiber[fibre]
人造短纤维切断机 staple fiber[fibre] cutting machine
人造短纤维清洗机 staple fiber[fibre] washing machine
人造短纤维条 staple fiber[fibre] top
人造堆肥 artificial compost
人造多肢块体 <一种防波堤护面块体> artificial multilegged block
人造凡士林 artificial vaseline
人造方块堤 artificial block dike[dyke]; human-made block dike[dyke]
人造放射性同位素 artificial radioactive isotope;induced radioisotope
人造飞船 artificial spacecraft
人造肥料 artificial fertilizer; artificial manure;synthetic(al)fertilizer
人造肥料撒播机 artificial fertilizer distributor
人造沸石 artificial zeolite; permufit; permutite
人造辐射带 artificial radiation belt
人造腐殖质 synthetic(al)humus
人造钙华 manufactured travertine
人造干燥 artificial drying out
人造刚玉 aloxite; alundum; artificial corundum; boule; carbonrundum; Corubin;synthetic(al)corundum;alundum <用黏[粘]土、木炭和铁屑在电炉中烧成的>
人造刚玉砂轮 alundum wheel
人造刚玉石 artificial corundum
人造港口 man-made harbo(u)r
人造港湾 artificial harbo(u)r; man-made harbo(u)r
人造革 artificial leather; coated fabric; dermateen; imitation leather; leather cloth; leatheret(te); leatheroid; pegamoid; synthetic (al) leather;support vinyl <衬垫用织物的乙烯制品>
人造革挂幔 artificial leather hanging
人造革挂幕 artificial leather hanging
人造革挂毯 artificial leather hanging
人造古迹 artificial monument
人造骨料 artificial aggregate; manufactured aggregate; synthetic (al) aggregate
人造冠 artificial crown
人造光 artificial light; candle light(ing)
人造光源 artificial light source
人造海水 artificial seawater;synthetic-(al)seawater
人造海滩 artificial beach
人造汗液 synthetic(al)perspiration
人造航路 artificial navigable waterway
人造黑沥青 artificial gilsonite
人造黑素 melanoid
人造红宝石 artificial ruby;synthetic-(al)ruby
人造洪水波 artificial flood wave
人造琥珀 amberoid;ambrain[ambroin]
人造琥珀胶 ambrain cement;ambrain cement
人造花岗石 granitoid; granolithic concrete;granolithic face
人造花岗石面 granitic plaster
人造花岗岩石面 allotriomorphic granular
人造花岗石饰面 granolithic finish
人造花砖地面装修 artificial stone tile floor cover(ing)
人造滑冰场 artificial ice rink; artificial skating rink
人造环境 artificial environment;man-made environment
人造环境室 environmental cabinet
人造环形山 artificial crater
人造黄油 margarine; oleomargarine; oleo oil;vegetable butter
人造回波 artificial echo
人造回声器 artificial echo unit
人造混凝土粗骨料 artificial concrete coarse aggregate; manmade coarse concrete aggregate
人造混凝土骨料 artificial concrete aggregate; manmade concrete aggregate
人造混凝土集料 artificial concrete aggregate; man-made concrete aggregate
人造火山灰 artificial pozzolana; permutite pozzolana
人造火山灰(质)材料 artificial pozzolanic material; man-made pozzolanic material
人造机器 man-made machine
人造极光 man-made aurora
人造集料 artificial aggregate; manufactured aggregate; synthetic (al) aggregate;manufacture aggregate <如陶粒>
人造集料混凝土 synthetic(al)aggregate concrete
人造加速度图 synthetic(al)accelerogram
人造岬角防波堤 headland breakwa-

ter;headland breakwater
人造岬角形成的海湾 subbay bay
人造假山石 artificial stone
人造假石 French stuc
人造假石面 French stucco
人造尖晶石 emerada;synthetic(al) cast spinel;synthetic(al)spinel
人造建筑材料 man-made building material;man-made construction material;manufactured construction materials
人造建筑制品 man-made building product; man-made construction product;man-made structural product
人造胶结材料 man-made cementing
人造胶结剂 man-made cementing agent
人造胶结料 artificial bonding adhesive;artificial bonding agent
人造胶乳 artificial latex
人造胶水 man-made glue
人造胶体 artificial colloid
人造胶粘剂 man-made additive
人造胶质 artificial gum
人造焦油沥青 artificial bitumen
人造礁石 artificial reef
人造角闪石 artificial amphibole
人造结构 artificial formation
人造介质 artificial dielectrics
人造金 imitation gold
人造金刚砂 artificial carborundum; artificial corundum;carborundum
人造金刚砂粉 alundum powder
人造金刚石 artificial diamond;man-made diamond;synthetic(al) diamond
人造金刚石粒 synthetic(al)grit
人造金刚石钻头 synthetic(al) diamond bit
人造金红石 synthetic(al)ruffle
人造晶体 artificial crystal
人造厩肥 synthetic(al)manure
人造空间飞行器 man-made spacecraft
人造空心小球形轻骨料 cenosphere
人造块体 artificial block
人造矿块 agglomerate
人造矿物 artificial mineral;manufactured mineral
人造矿物玻璃纤维 man-made mineral vitreous fibre
人造矿物绒 man-made mineral wool
人造蜡 synthetic(al)wax
人造蓝宝石 artificial sapphire;synthetic(al)sapphire
人造立体观测 artificial stereoscopy
人造立体镜 artificial stereoscope
人造沥青 synthetic(al)asphalt
人造裂缝 man-made fracture
人造林 artificial forest
人造流星 artificial meteor
人造楼面砖 artificial stone tile floor cover(ing)
人造陆地 artificial land;made land
人造落差 artificial fall
人造煤气 artificial gas;manufactured gas
人造棉 artificial cotton;staple rayon
人造棉纱 spun rayon
人造棉细布 staple fiber muslin
人造模型 manikin;phantom
人造磨(刀)石 artificial grind stone
人造磨料 artificial abrasive;manufactured abrasive
人造磨石 artificial grit;Norton pulpstone
人造木材 man-made wood;imitation wood
人造奶油厂废水 margarine factory wastewater
人造奶油废水 margarine wastes
人造耐油橡胶 ameripol
人造黏[粘]合剂 man-made bonding medium;manufactured adhesive
人造黏[粘]结料 artificial bonding adhesive;artificial bonding agent
人造黏[粘]土砾石 artificial clayed gravel
人造黏[粘]土碎砖 manufactured clay brick sand
人造凝灰石 artificial travertin(e); man-made travertin(e)
人造泡沫 artificial foam
人造膨润土 hand-made bentonite
人造膨润土粒 hand-made bentonite ball
人造皮革 dermatine
人造偏光板 polaroid
人造偏振箔 polaroid foil
人造偏振片 polaroid;polaroid film
人造平原 man-made plain
人造铺地(面)石 granolith
人造铺地石基底 granolithic base
人造起偏振镜 polaroid polarizer
人造气候 artificial climate
人造气溶胶示踪技术 artificial aerosol tracer technique
人造气体 artificial gas;manufactured gas
人造轻骨料 artificial light(weight) aggregate; synthetic (al) light (weight) aggregate
人造轻骨料混凝土 synthetic(al) aggregate light(weight) concrete
人造轻集料 artificial light(weight) aggregate; synthetic (al) light (weight) aggregate
人造轻集料混凝土 synthetic(al) aggregate light(weight) concrete
人造清漆 synthetic(al) varnish
人造群青(蓝) factitious ultramarine
人造燃料 synthetic(al) fuel;artificial fuel
人造燃气 manufactured fuel gas
人造热水循环系统 artificial hot-water circulation system
人造日光 artificial daylight
人造润滑油 artificial lubricating oil
人造沙漠 man-made desert
人造砂 artificial sand;crusher screenings; manufactured sand; stamp sand
人造砂石 artificial sandstone
人造闪电发生器 artificial lighting generator;lightning generator
人造湿地 <一种污水处理方法> constructed wetland;artificial wetland
人造石 artificial stone;imitation stone; man-made stone; masoned cast stone;patent stone;protean stone; reconstituted stone; reconstructed stone;synthetic(al)rock;synthetic (al) stone;Maycoustic <隔声用>
人造石板 artificial flagstone;artificial slate
人造石表皮层 patent stone skin
人造石材 artificial stone block;precast stone;cast stone
人造石材分裂机 splitter for reconstructed stone
人造石地板面层 artificial stone flooring) finish
人造石地面 granolithic concrete surface; granolithic finish; granolithic finish floor;granolithic floor(ing)
人造石地面砖 patent stone tile floor cover(ing)
人造石防水层 artificial stone waterproofer
人造石膏石 protean stone
人造石工场 artificial stone shop;patent stone shop
人造石工艺 artificial stone work
人造石工作 patent stone work
人造石构件 artificial stone work
人造石花砖地面装修 artificial stone floor(ing) finish
人造石花砖饰面 artificial stone floor cover(ing)
人造石灰华 imitation travertine;man-made travertin(e); manufactured travertine
人造石灰石 artificial limestone
人造石混凝土地面 granolithic concrete screed
人造石混凝土楼面砖 granolithic concrete flooring tiling
人造石混凝土(铺)地砖 granolithic concrete tile[tiling]
人造石混凝土铺路 granolithic concrete paving
人造石混凝土铺面层 granolithic concrete course; granolithic concrete layer;granolithic concrete topping
人造石混凝土梯级踏步 granolithic concrete tread
人造石块 artificial stone block;cast stone
人造石楼板面层 artificial stone floor(ing) finish
人造石楼面 granolithic floor(ing)
人造石楼面修整 artificial stone floor dressing
人造石楼梯 artificial stone stair (case);patent stone stair(case)
人造石棉 artificial asbestos;man-made asbestos
人造石面层 artificial stone skin
人造石墨 artificial graphite; Delanium; delanium graphite; manufactured graphite;synthetic(al) graphite
人造石墨电刷 electrographite brush
人造石铺面 artificial stone pavement;granolithic finish;granolithic pavement; granolith; granolithic paving
人造石铺面层 granolithic layer
人造石铺面的 granolithic
人造石铺面找平尺 granolithic screed
人造石铺面整平板 granolithic screed
人造石铺砌层 artificial stone pavement
人造石饰地面 granolithic concrete
人造石饰面 cast stone finish;granolithic finish
人造石踏步 granolithic tread
人造石英 man-made quartz;synthetic(al) quartz
人造石英玻璃 synthetic (al) silica glass
人造石英晶体 artificial quartz crystal;synthetic(al) quartz crystal
人造石油 alternate fuel;artificial petroleum;synthetic(al) petroleum
人造石油沥青 artificial asphalt
人造石砖(瓦) patent stone tile;artificial stone tile
人造食物 artificial nourishment
人造世界 man-made world
人造饰面板 artificial decorative board
人造手 magic hand
人造树脂 artificial resin; man-made resin;manufactured resin;synthetic(al) resin
人造树脂基胶合剂 manufactured resin-based cementing agent
人造树脂基胶粘剂 man-made resin-based adhesive
人造树脂基黏[粘]合剂 manufactured resin-based bonding medium
人造树脂类黏[粘]结剂 man-made resin-based adhesive
人造树脂砂轮 resinoid wheel
人造数据 generated data
人造双折射 synthetic(al) birefringence
人造水晶 synthetic(al) crystal; synthetic(al) quartz
人造水泥 artificial cement; artificial cementing agent
人造水泥大理石 cement man-made marble
人造丝 artificial silk;fiber[fibre] silk; rayon;rayon filament
人造丝厂 rayon mill; rayon silk works;rayon works
人造丝厂废水 rayon mill waste
人造丝工厂废水 wastewater from rayon mill
人造丝浆 rayon pulp
人造丝浆筛渣 rayon reject
人造丝轮胎线 rayon tire[tyre] yard
人造丝络筒油 rayon coning oil
人造丝绳 rayon cord
人造丝纤维 rayon fiber[fibre]
人造丝芯运输带 rayon belt
人造丝织物 artificial silk fabrics;rayon fabric
人造丝制造厂 rayon manufacturer
人造塑料 man-made plastic
人造塑料材料 man-made plastic material
人造太阳灯 artificial sunlight lamp
人造太阳行星 solar space vehicle
人造碳 artificial carbon
人造碳化硅 <研磨用> cystolon
人造碳化硅磨料 crystolon
人造碳氢化合物 artificial hydrocarbon
人造天空 artificial sky
人造天体 artificial object
人造天体天文代号 astronomic(al) number of artificial heavenly bodies
人造铁氧体 artificial ferrite
人造庭园 man-made ground
人造同位素 artificial isotope; transmutation product
人造图形 artificial graphics
人造土地 artificial ground;man-made land
人造土壤 anthropic soil
人造卫星 artificial satellite; man-made satellite;orbital vehicle;satellite vehicle;space craft;sputnik
人造卫星测高法 satellite altimetry
人造卫星大地测量学 artificial satellite geodesy
人造卫星的载运火箭 satellite-launching vehicle
人造卫星地面站 satellite ground station
人造卫星固定跟踪站 fixed tracking station
人造卫星加注燃料站 refueling satellite station
人造卫星壳体涂层 satellite coating
人造卫星试验场 satellite test range
人造卫星探测器 satellite probe
人造卫星位置显示屏 spascore
人造卫星用电池 satellite battery
人造卫星站 satellite station
人造卫星整流罩 satellite fairing
人造卫星自动跟踪天线 satellite automatic tracking antenna
人造圬工体 man-made masonry unit

R

人造圬工用砖石 artificial masonry unit
人造无机粉尘 artificial inorganic dust
人造无水石膏 artificial anhydrite
人造无烟煤 artificial anthracite
人造系统 man-made system
人造细骨料 artificial fine aggregate; artificial fine grain
人造细骨料混凝土 man-made fine concrete aggregate
人造细集料 artificial fine aggregate; artificial fine grain
人造细颗粒 artificial fine aggregate; artificial fine grain
人造纤维 artificial fiber[fibre]; man-made fiber[fibre]; nylon; rayon; lanital <酪素纤维制造的>
人造纤维板 beaver board
人造纤维材料 man-made fiber material
人造纤维厂 rayon factory
人造纤维单丝 rayon monofil
人造纤维废水 rayon wastewater
人造纤维(缆)绳 synthetic(al) fiber[fibre] rope; man-made fiber[fibre] rope
人造象牙 artificial ivory
人造橡胶 artificial gum; artificial rubber; duprene; elastomer; man-made rubber; masticated rubber; synthetic(al) chloroprene rubber; synthetic(al) rubber
人造橡(胶)浆 neoprene latex
人造橡胶块 synthetic(al) rubber
人造橡胶密封 neoprene seal
人造橡胶密封垫片 neoprene sealing gasket
人造橡胶嵌缝料 synthetic(al) rubber filler
人造橡胶乳液 synthetic(al) rubber latex
人造橡胶套 neoprene sleeve
人造橡皮 ameripol; manufactured rubber
人造小行星 artificial asteroid
人造行星 artificial planet; man-made sun satellite
人造杏仁油 artificial almond oil
人造玄武岩 haplobasalt
人造雪崩 artificial snow slide
人造岩芯 synthetic(al) core
人造研磨料 artificial abradant
人造羊毛 cellulose wool; lanital
人造氧化铝 Borolon; synthetic(al) alumina
人造液体燃料 alternative fuel
人造银朱 vermil(l)ionette
人造荧光树脂 lucite
人造硬沥青 artificial gilsonite
人造硬石膏 artificial anhydrite; by-product anhydrite
人造有机骨料 synthetic(al) organic aggregate
人造有机集料 synthetic(al) organic aggregate
人造有机物 anthropogenic organics
人造宇宙飞船 man-made space-craft
人造雨 artificial precipitation; artificial rain(fall); artificial watering; man-made rain
人造语言 <如世界语、计算机语言等,与习用的自然语言相对> artificial language
人造月球卫星 lunar satellite; moon satellite
人造云 artificial cloud; cloudier
人造云母 artificial mica; build-up mica; micanite; mo(u)lded mica
人造云石 artificial travertin(e)
人造运动场地 athletic surfacing
人造障碍物 man-made obstacle
人造珍珠 olivet(te)
人造珍珠石 synopal
人造支座 artificial abutment
人造织品 man-made fiber fabric
人造纸 synthetic(al) paper
人造制品 artifact
人造重力 artificial gravity; quasi-gravity
人造砖石 artificial masonry unit
人造砖石单元 man-made masonry unit
人造装饰石 decorative artificial stone
人造资本 artificial capital
人造资源 man-made resources
人造鬃毛漆刷 synthetic(al) bristle brush
人造祖母绿 synthetic(al) emerald
人造钻石 man-made diamond
人闸 personnel lock
人证 human testimony; testimony of witness
人种地理学 geography of race
人字扒杆 sheer legs
人字把杆 A-frame; sheer-leg frame
人字把杆吊车 A-frame crane
人字把杆起重机 breast derrick
人字坝 rafter dam
人字布置排水管 herringbone drain
人字槽 chev(e)ron notch
人字撑 herringbone bracing; herringbone bridging; herringbone strut(ting)
人字齿 double helical gear tooth; herringbone gear tooth
人字齿轮 angular gear; double helical gear; herringbone wheel; herringbone gear
人字齿轮泵 herringbone gear pump
人字齿轮减速器 herringbone reducer
人字齿轮刨床 double helical gear planer; herringbone gear planer
人字齿轮切齿机 herringbone gear cutting machine
人字齿轮铣床 herringbone gear milling machine
人字齿轮座的齿轮轴 mill pinion
人字齿伞齿轮 double helical bevel gear
人字椽屋顶 rafter roof
人字窗槛 mitred sill
人字点 lambda
人字吊臂起重机 shear-leg crane; shear-leg derrick
人字吊杆 A-framed derrick; A-framing derrick
人字吊杆起重船 floating sheer legs; floating sheers
人字吊杆起重机 shear-leg crane
人字动臂起重机 jinniwink
人字缝 herringbone joint; lambdoidal suture
人字缝尖 lambda
人字拱 mitre arch; pediment arch
人字构架 A-frame
人字滚轴运输机 herringbone roller conveyer[conveyor]
人字焊纹 herringbone pattern
人字桁架 roof truss
人字花形 herringbone effect
人字脊 pediment apex
人字架 A-bracket; principal rafter; propeller shaft bracket; propeller shaft stay; propeller shaft strut
人字架起重设备 shear legs
人字架中柱 broach post; crown post
人字架转臂起重机 A-framed derrick
人字尖 vee crossing; V-piece
人字槛 mitre sill
人字结构堰 Thomas weir
人字门 leaf gate; miter[mitre] gate
人字门处于关闭位置 mitered[mitred] position
人字门槛 clapping sill; clap sill; lock sill; mitered[mitred] sill
人字门门槛 <船闸> miter sill
人字门门叶 miter[mitre] gate leaf
人字门推拉杆 miter[mitre] gate strut
人字门液压启闭机 hydraulic machinery of miter gate
人字密封环 herringbone seal ring
人字密封圈 chev(e)ron ring
人字木 principal rafter; truss principal
人字木屋顶 <椽子承重的> rafter roof
人字木屋架系梁杆节点 collar joint
人字木之间的系梁 principal rafter beam
人字坡 double spur grade; gable slope
人字坡墙 hip and gable roof
人字铺面工程 herringbone work
人字铺砌工 herringbone work
人字起重机 A-framed derrick; derrick (crane); derrick tower gantry; sheer legs; sheers
人字起重机柱 derrick post
人字起重架 sheer legs; sheers; sheers derrick
人字起重架绑绳 shear head lashing
人字起重架船 shear bulk
人字砌合 raking bond; zigzag bond
人字墙 gable wall; miter[mitre] wall; pediment
人字山墙与门头间的三角部分 tympanum enclosed by pediment
人字山头 gable
人字山头砌筑墙 gable masonry wall
人字式 herringbone fashion; herringbone pattern; mitering type
人字式船闸闸门 miter[mitre] lock gate
人字式活动坝 bear-trap dam
人字式矿房布置 herringbone room arrangement
人字式路面 herringbone pavement
人字式码头布置 herringbone wharf layout
人字式铺地 herringbone masonry; herringbone pavement
人字式铺砌 spicatum opus
人字式铺砌法 herringbone paving
人字式铺砌路面 herringbone pavement
人字式砌合 herringbone bond
人字式闸门 miter[mitre] gate
人字式转臂起重机 derricking jib crane
人字榫 herringbone joint
人字梯 double ladder; folding ladder; roof ladder; standing ladder; step ladder; stack ladder <俗称高凳>
人字条纹 chev(e)ron bar tread
人字头 gavel
人字凸环 herringbone convex ring
人字桅 bipod mast; sheer mast
人字尾轴架 two-leg propeller strut
人字文饰 zigzag mo(u)lding
人字纹修琢 <石面> herringbone dressing
人字纹装饰 zigzag ornament
人字稳定体 <一种防波堤异形块体> stabit
人字坞门 miter[mitre](dry-) dock gate
人字坞门的斜接柱 miter[mitre] post
人字屋顶 collar roof; comb roof; double pitch roof; ridge(d) roof; ridged wall
人字屋顶窗 gabled roof dormer
人字屋顶房屋 gabled house
人字屋顶上小塔 gable roof ridge turret
人字屋脊上小塔 gable roof ridge turret
人字屋架上弦的紧密联结 close couple
人字纤维板 beaver board
人字线 <在高坡地区,列车进站后调头转至另一方向时所用的一种设计> switchback line
人字斜纹 arrowhead twills; chev(e)ron type
人字星物检定器 artificial star-field calibrator
人字形 chev(e)ron; chev(e)ron type; herringbone; herringbone fashion; lambda-type
人字形编织 twilled herringbone weave
人字形布置 herringbone pattern
人字形采矿法 herringbone stoping; herringboning
人字形超重机柱 derrick post
人字形车站 switchback station
人字形撑 herringbone bracing
人字形齿轮 double helical gearing; herring(bone) gear; double helical spur gear
人字形挡板 chev(e)ron baffle
人字形的 chev(e)ron; lambdoidal; pedimented
人字形的多跨构架 multiple span gabled frame
人字形地下排水系统 mitered[mitred] drainage
人字形地下渗水系统 mitered[mitred] drainage
人字形顶撑 herringbone strut
人字形断口 chev(e)ron
人字形堆料法 chev(e)ron method
人字形缝 herringbone stitch
人字形格栅撑 double bring
人字形构造 Lambda-type structure
人字形管式沉降计 chev(e)ron tube settler
人字形管式沉降器 chev(e)ron tube settler
人字形灌溉系统 herringbone irrigation system
人字形过滤器 herringbone filter
人字形桁架 A-truss
人字形桁架屋顶 trussed rafter roof
人字形花样 chev(e)ron pattern
人字形混凝土块体 <防波堤护面的> stabit
人字形激波 shock wave
人字形脊瓦 angle hip tile; angle tile; arris hip tile
人字形建筑部分 gable
人字形交错层理 chev(e)ron cross bedding; herringbone cross bedding
人字形交错层理构造 chev(e)ron cross-bedding structure
人字形接合 oblique bond; saddle joint; mitered[mitred] joint
人字形结构 chevron; herringbone structure; herringbone texture
人字形金属板条 chev(e)ron slat
人字形矿房采矿法 herringbone method
人字形老虎窗 gable(d) dormer(window)
人字形裂缝 herringbone crack
人字形码头布置 herringbone wharf layout
人字形密椽屋顶 coupled roof
人字形密封 chev(e)ron seal
人字形面砖排列 diagonal bond; herringbone bond
人字形木板条 chev(e)ron slat
人字形木屋架 rafter set
人字形扭曲变形 herringbone distor-

tion
人字形排架 gabled bent;raking trestle
人字形排列的板 herringbone planking
人字形排水 herringbone drainage
人字形排水法 herringbone drain(age) system
人字形排水沟 herringbone drain; chev(e)ron drain
人字形排水管 herringbone drain; chev(e)ron drain
人字形排水管系 chev(e)ron drainage;herringbone system
人字形排水系统 chevron drain; herringbone (drainage) system; herringbone system of drains
人字形拼花 herringbone matching
人字形拼花地板 herringbone parquet floor
人字形拼接 herringbone matching
人字形拼(木)地板 herringbone parquetry
人字形铺面 herringbone work
人字形铺砌路面 herringbone pavement;herringbone paving
人字形铺砌砌体 herringbone masonry
人字形铺砌圬工 herringbone masonry
人字形铺砖 opus spicatum
人字形企口 herringbone matching
人字形起重架 sheer legs
人字形砌工 herringbone work
人字形砌合 diagonal bond; herringbone(masonry)bond
人字形砌砖 herringbone brickwork
人字形嵌砖细工 opus spicatum
人字形绕组 herringbone winding
人字形散热片 chev(e)ron fin
人字形饰 chev(e)ron
人字形水门 miter gate
人字形塔 gable tower
人字形梯线 V-ladders
人字形图案 herringbone pattern
人字形瓦管排水系统 herringbone tile drainage system
人字形网孔 herringbone mesh opening
人字形网眼 herringbone mesh
人字形纹(样)chev(e)ron
人字形屋顶 gable roof; tent-shaped roof;gable roof type <畜舍>
人字形屋顶窗 gable dormer
人字形屋架 gable roof truss
人字形线脚 zigzag mo(u)lding
人字形斜撑 herringbone strut(ting)
人字形斜纹 herringbone twill; serpentine twill
人字形斜纹组织 twilled herringbone weave
人字形压实齿 chevron pattern
人字形掩护支架 L-type shield
人字形油槽轴承 herringbone bearing
人字形闸门 miter[mitre] gate
人字形障板 chev(e)ron baffle
人字形褶皱 chevron fold;kink fold
人字形支撑 herringbone strut(ting)
人字形支承 miter[mitre] bearing
人字形支护法 herringbone timbering
人字形支架 cockering; herringbone timbering
人字形砖层 herringbone masonry course
人字形砖砌体 herringbone brickwork
人字形琢石面墙 herringbone ashlar
人字形琢石墙面 herringbone ashlar
人字形组织 herringbone weave
人字缘 lambdoid margin
人字闸槛 miter(ing) sill
人字闸门 miter(ed)[mitre(d)](type) gate;mitering[mitring] gate; miter(ed)[mitre(d)] sill;roof gate
人字闸门挡柱 mitre[mitre] post
人字闸门门槛 mitre[mitre] sill
人字闸门斜接柱 miter[mitre] post; meeting post
人字辙叉 vee crossing
人字支架 A-frame gantry; herringbone timbering <隧道>

仁

仁川港 <朝鲜> Inchun Port

壬

壬醇 <通常指壬醇-1> nonylalcohol;nonanol

壬二酸 anchoic acid; azelaic acid; nonanedioic acid
壬二酸二辛酯 dioctyl azelate
壬二酸二异辛酯 diisooctyl azelate
壬二酸甘油酯 azelain
壬二酸酯 azelate
壬基 nonyl
壬炔 nonyne
壬烷 nonane

忍

忍冬【植】honeysuckle

忍冬饰 anthemion(mo(u)lding);honeysuckle ornament
忍耐范围 tolerance range
忍受响应曲线 tolerance response curve

荏

荏油 perilla oil

刃

刃边扩散函数 edge-spread function

刃边磨损 ga(u)ge wear
刃叉式连杆 blade and fork rod
刃长 wing length
刃锉 edge file;joint file
刃带 land
刃刀低举式平地机 low-lift blade grader
刃刀高举式平地机 high-lift blade grader
刃冬饰 honeysuckle ornament
刃沸石 cowlesite
刃锋 knife edge
刃钢 shear steel
刃沟 flute
刃后角 edge clearance angle
刃脊 knife-edge crest;razor back
刃尖 cutting nose
刃角 angle of throat; basil; edge angle;angle of cutting edge <刀具>; bezel <平刨或其他切削工具的>; taper angle <凿岩机>
刃角和刃半径 edge angle and edge radius
刃脚 curb shoe; cutting edge; cutting shoe;cutting shore;knife edge
刃脚踏面 <沉井的> knife-edge tread
刃具 cutter;cutting tool
刃具钢 cutlery steel;cutting-tool steel
刃口 cutting edge <沉井等的>; cutting point;jackbit insert;knife edge
刃口冲裁模 finish blanking
刃口负荷 edge load(ing)
刃口角度 blade angle; cutting edge angle
刃口磨损 wear across the edge
刃口平尺 toolmaker's straight edge
刃口式磁力仪 knife-edge magnetometer
刃口寿命 bit life
刃口(线)荷载 knife-edge load(ing)
刃口样板 honing ga(u)ge
刃口圆弧半径 rounded cutting edge radius
刃口支承 knife-edge support
刃宽 tread
刃棱面 land
刃岭 arete
刃面 active face
刃磨 sharpen(ing)
刃磨导柱 sharpening guide
刃磨机床 cutter grinding machine
刃磨间 grinding booth
刃磨角 tool angle
刃磨器 cutter sharpener;sharpener
刃切割力 edge-cutting force
刃式刮路机 blade drag;blade grader
刃式继电器 knife-edge relay
刃式平路机 blade grader
刃天青凝乳试验 resazurin-rennet test
刃形避雷器 knife-shaped lightning arrester
刃形触点 knife-edge contact
刃形的 sharp-edged
刃形继电器触点 knife-edge relay contact
刃形枢轴 knife-edge pivot
刃形位错 edge dislocation
刃形堰 sharp-edged weir
刃形折射 knife-edge refraction
刃形支承 bearer blade; blade bearing; knife-edge bearing; knife-edge support
刃形支承边 knife edge
刃形指针 knife-edge pointer; knife-shaped needle;knife-shaped pointer
刃靴 <沉井、沉箱等的> curb shoe
刃缘 cutting edge;knife edge
刃状的 bladed
刃状构造 bladed structure
刃状位错 edge dislocation
刃状习性 bladed habit

认

认保单 covering note

认定 firmly believe; hold; identification
认定份额 subscription quota
认定股利 consent dividend
认付 <支票> accept; acceptance; marking
认付范围 acceptability limit
认付费 acceptance fee
认付汇票 bill for acceptance
认付汇票人 accepter
认付时交还抵押单的汇票 document acceptance bill
认付时交货单据 document against acceptance
认付支票 certified check
认购证券 subscription certificate
认购资本 subscribed capital
认股 subscription
认缴资本 capital contribution
认捐(额)subscription
认可差别 recognition differential
认可抽样检验 lot-acceptance sampling
认可的标志 approved marking
认可的等效方法 approved equivalent method
认可的工程要求 accepted engineering requirements
认可的股份 authorized shares
认可的锅炉压力 authorized boiler pressure
认可的急救包 approved first-aid outfit
认可的集装箱条件 acceptable container condition
认可的价值 acceptable value
认可的交易商 recognized dealer
认可的买主 authorized buyer
认可的(投)标书 accepted bid
认可的印鉴 authorized signature
认可的争执 recognition dispute
认可服务处 recognition service
认可机构 accreditation body
认可价格 subscription price
认可扩建 authorized extension
认可某种牌子的产品为标准设备 brand standardization
认可期限 subscription period
认可试验 approval test;warranty test
认可书 award
认可术语 permitted term
认可体系 accreditation system
认可同等替换 approved equal
认可图(纸)acceptance drawing; approval drawing
认可样品 approval of sample
认可银行 recognized bank
认可元件索引 recognized component index
认可债券 indorsed bond
认可账额 account stated
认可者 licensor;sustainer
认可值 certificate value;certified value
认可准则 accreditation criterion[复 criteria]
认可资本 authorized capital
认可资产 admissible assets
认可资料 approved data
认赔书 back letter;counter letter;letter of indemnity
认人支票 check to order
认赎 tender
认识 acquaintance with;awareness
认识标记 identification marker;identification marking
认识差异产品 differential products
认识功能 recognizing ability
认识过程 cognitive process
认识货物的标志 leading mark
认识科学 cognitive science
认识论 <与"本体论"相对> epistemology;theory of knowledge
认识模型 cognitive model
认识上的制约 cognitive limits
认识时间差滞 recognition lag
认识系统 recognition system
认识性好奇(心)epistemic curiosity
认收继电器 acknowledging relay
认收接触器 acknowledging contactor
认收开关 acknowledging switch
认同的转折点 identity crisis
认同作用 identification
认债 admission of liability
认真测试 close control
认真地尽责 duty of care
认真管理 close supervision
认真交货的 delivery-conscious
认证 attestation;authentication
认证代理人 authentication agent
认证费 certification fee
认证活动 certification activity
认证机构 certification body
认证计划 certification scheme
认证人员 auditor
认证体系 certification system
认证体系的成员 member of certification system
认证体系的利用 access to certification system
认证行为 act of authentication
认证要求 certification requirement
认知定向 cognitive orientation
认知制图 cognitive mapping

任

任何一种建筑物 <英> biggin(g)

任何与空气湿度有关的气象 hydro-

meteor
任命 appoint; appointment; commission; designate; instate; nominate; nomination; post(ing)
任命者 installant; nominator
任命助理 appointment of assistants
任期 incumbency; tenure; term of office; term of service
任期满后接任的人 hold-over
任期目标管理制 management by objectives during term of service
任期制 fixed term appointment system
任其自己运行 let it work itself
任取 option
任取点 arbitrary assignment point
任停 optional stop
任务 assignment; darg; duty; job; mandate; mission; role; task; work load
任务备忘录 task memorandum
任务变量 task variable
任务表 active task list
任务程序 task program(me)
任务大纲 terms of reference
任务单 job order
任务调度 task scheduling
任务调度程序 task dispatcher; task scheduler
任务调度排队 task scheduler queue
任务调用 task call
任务定向群体 task-oriented group
任务定义表 task definition table
任务队列【计】task queue
任务分解结构 work breakdown structure; work breaking structure
任务分配表 work-distribution chart
任务分析 assignment analysis; mission analysis; task analysis
任务挂起 task suspension
任务观念 task idea
任务管理 task management
任务管理程序 task management program(me); task supervisor
任务管理系统 task management system
任务宏指令 task macro
任务计时器 task timer
任务简要讲解图 brief chart
任务建立 task creation
任务结束 task termination
任务解释表 task dictionary table
任务津贴 assignment allowance
任务控制程序 task control program(me)
任务控制程序段 task control block
任务控制卡片 task control card
任务控制块 task control block
任务控制中心 mission control center [centre]
任务流程图 mission flow diagram
任务描述符 task descriptor
任务名字 task name
任务命令单 work sheet
任务内容 task definition
任务排队【计】task queue
任务期限 mandate; mission duration
任务强制转移 task force
任务确定 task determination
任务时间 mission time; task time
任务书 assignment; job order; mission statement; prospectus; work sheet
任务输入输出表 task input/output table
任务数据 mission data; task data
任务数据卡 mission data card
任务说明书 task description
任务速成 project crashing
任务条【计】taskbar
任务停止 task quit
任务同步 task synchronization
任务完成日期 task completion date
任务委托书 mission assignment
任务显示器 role indicator
任务选择 task option; task selection
任务异常终止 abnormal end of task
任务优先级 task priority
任务优先(权) priority of task
任务优先数 priority value of task
任务再执行程序 task retry routine
任务再指定 task reassignment
任务暂停 task suspension
任务照明 task illumination
任务执行存储器 task execution memory
任务指示器 role indicator
任务中断控制 task interrupt control
任务终止 task termination; termination of task
任务属性 task attribute
任务转换 task switch
任务转接 context switching
任务转接最小时间 minimum context switching time
任务状态 task status
任务状态索引 task status index
任务族 task family
任性的 arbitrary; selfish
任选 option
任选板 optional board
任选半指令 optional half instruction
任选标号 optional label
任选并行仪器总线 optional parallel instrument bus
任选成分 optional member
任选成员 optional member
任选传动箱 optional transmission
任选串 optional string
任选的 optional
任选多级中断 optional multilevel interruption
任选附件 optional attachment; optional extras
任选附则 optional annex
任选港 optional port
任选港的装船 optional shipment
任选港附加费 optional charges; optional fee
任选(港)货物 optional cargo
任选港交货 optional delivery
任选港提单 optional bill of lading
任选工作装置 optional attachment
任选功能 optional feature; optional function
任选功能板 optional function board
任选股利发付方法 optional dividend
任选机构 optional feature
任选记号常数 optionally signed constant
任选件 optional parts
任选接口 optional interface
任选局部外围设备 optional local peripheral
任选空白 optional blank
任选零件 optional parts
任选目的港 optional destination; optional port
任选配套装置 optional implement
任选特点 optional feature
任选条款 optional clause
任选停止 optimal stop
任选停止指令 optional stop instruction
任选物 option
任选项 option
任选样品 sample taken at random
任选要求 optional requirement
任选优先 optional priority
任选中断 optional interrupt
任选属性 optional attribute
任选装置 optional equipment
任选字 optional word
任一保险 anyone risks
任一船舶 anyone vessel
任一事件 anyone event
任一损失 anyone loss
任一向通信[讯] either-way communication
任意安放 free-standing
任意保险 voluntary insurance
任意背书 facultative endorsement
任意比例尺 indefinite scale
任意变化 arbitrary variation
任意变形 arbitrary deformation
任意标度 arbitrary scale
任意拨款 voluntary appropriation
任意波形 random waveform
任意布置 random pattern; willful arrangement
任意采用效率试验 <在运行中> free use of efficiency tests
任意参数 arbitrary parameter
任意长度材 random length
任意常数 arbitrary constant
任意场地起落飞机 all-terrain aircraft; pantobase aircraft; pantosurface aircraft
任意抽查 arbitrary inspection; random check; random inspection
任意抽取件 random sampled parts
任意抽选 haphazard selection
任意抽样 convenience sampling; free sampling; optional sampling; random sample
任意抽样法 method of random sampling
任意处理 randomization; randomize
任意处理的订单 discretionary order
任意穿孔的 random perforated
任意传动齿轮 optional transmission gear
任意传动装置 optional transmission gear
任意次序 arbitrary sequence; random order
任意存取 arbitrary access
任意大小网点目板 random dot grain screen
任意带 arbitrary zone
任意单位 arbitrary unit
任意的目标航线 random target course
任意地面点 arbitrary ground point; random ground point
任意点 arbitrary point
任意点法 arbitrary point method
任意堆积填料塔 random-packed column
任意堆石料 random rockfill
任意多边形 arbitrary polygon
任意法兰 optional flange
任意方向机动 omnidirectional maneuver
任意分布 random distribution
任意分段文件 arbitrarily sectioned file
任意分数 arbitrary fraction
任意港 optional port
任意格排梁桥 arbitrary grillage girder bridge
任意格网 arbitrary grid
任意个随机变量和 sum of a random number of random variables
任意购买力 discretionary purchasing power
任意管理 rule-of-thumb
任意规格 optional specification
任意轨道 wild trajectory
任意函数【数】arbitrary function
任意函数发生器 arbitrary function generator
任意航向计算机 arbitrary course computer
任意合同 contract at discretion
任意荷载 arbitrary load
任意横向荷载 arbitrary lateral load
任意环流分布 arbitrary circulation distribution
任意基面 arbitrary datum
任意基准线 arbitrary datum line
任意基准值 arbitrary reference value
任意激励 arbitrary excitation
任意几何形状 random geometry
任意几何状态 random geometry
任意加税 slap additional taxes
任意价格 capricious value
任意假定 arbitrary assumption
任意假设 arbitrary assumption
任意角 arbitrary angle
任意角域 random angle domain
任意刻度 arbitrary scale
任意宽度 random width
任意拉格朗日-欧拉方法 arbitrary Lagrangian-Euleuan method
任意滥发贷款 abuse in the granting of loans
任意连接 random connection
任意联测导线 arbitrary traverse; random traverse
任意量 arbitrary quantity
任意料 random material
任意料填土 random fill
任意零值 arbitrary zero
任意流 arbitrary flow
任意面 arbitrary surface
任意浓度 arbitrary concentration
任意排列 random disposition; random pattern; willful arrangement
任意偏离中点与面元中心之间的距离 distance between centers [centres] of area element and arbitrary offset middle point
任意平均数 arbitrary average
任意平面 oblique plane
任意铺砌的马赛克 block random mosaic
任意牵伸 random draft
任意球面三角形 oblique spheric(al) triangle; scalene spheric(al) triangle
任意取样 chance sample
任意取样法 random sampling
任意权数 arbitrary weight; haphazard weight
任意权数滞后型式 arbitrary-weight lag scheme
任意三角形 oblique triangle
任意三色浓度 arbitrary three-colo(u)r density
任意色 random colo(u)r
任意色调 arbitrary hue
任意失配 random mismatch
任意石料堆筑 random rockfill
任意石料抛筑 random rockfill
任意石料填筑 random rockfill
任意时间 random time
任意试件 random sample
任意试验 random test
任意手工方法 random manual mode
任意输入 arbitrary input
任意数 arbitrary number
任意数据点 arbitrary data point
任意数序列【计】random number series
任意数字转化器 random digitizer
任意水位 arbitrary level
任意顺序 arbitrary sequence; random order
任意顺序计算机 arbitrary sequence computer
任意填筑 random fill

任意停机指令 optional stop order
任意投影 aphylactic map projection; arbitrary projection; free projection
任意弯矩图 arbitrary moment diagram
任意伪圆柱投影 pseudo-cylindric(al) arbitrary projection
任意位置 arbitrary position; optional position
任意相角功率继电器 arbitrary phase-angle power relay
任意相位 arbitrary phase
任意相位关系 arbitrary phasing
任意卸货港货物 optional cargo
任意卸货港交货 optional delivery
任意形状 arbitrary shape
任意型号 disposable type
任意型键入 unsolicited keying
任意型信息 unsolicited message
任意性 haphazard
任意选择 any selection; arbitrary selection; optional selection
任意选择条款 optional clause
任意选址 random access discrete address
任意选址同步卫星 random access stationary satellite
任意选址增量调制 random access delta modulation; random access discrete modulation
任意样本 arbitrary sample
任意异分子聚合物 random copolymer
任意因素 arbitrary factor
任意引航制 free pilotage
任意盈余 free surplus
任意盈余指拨 discretionary appropriation
任意游览客票 travel-at-will ticket
任意游走 drift
任意原点 arbitrary origin; arbitrary point of origin
任意账户 discretionary account
任意账户交易 discretionary account transaction
任意折旧法 arbitrary depreciation method; depreciation arbitrary method; voluntary depreciation method
任意折旧计算 free depreciation caculation
任意征税 arbitrary duty; arbitrary taxation
任意政策 discretionary policy
任意支配所得 discretionary income
任意值 arbitrary value
任意指拨 voluntary appropriation
任意中央子午线 arbitrary central meridian
任意仲裁 voluntary arbitration
任意周期 cycle-free
任意轴子午线 arbitrary axis meridian
任意装载条款 optional stowage clause
任意状态 free position(retarder)
任意准备 voluntary reserves
任意资金 discretionary funds
任择强制管辖 optional compulsory jurisdiction
任择条款 optional clause
任择议定书 optional protocol
任职 incumbency
任职期间 <董事、理事> directorship
任职首次薪金 starting pay

韧

韧测单位燃料消耗量 brake specific fuel consumption
韧测燃料消耗率 brake specific fuel consumption
韧脆过渡 tough-brittle transition
韧脆转变 ductile-brittle transition
韧钉耳属 <拉> Ditiola
韧度 temper; tenacity; toughness; tough value
韧度回火 temper
韧度率 toughness factor
韧度模量 <在受力破坏时物体每单位体积所吸收的应变能量,常以应力-应变曲线所包括的面积来表示> modulus of toughness
韧度试验 rupture test; toughness test
韧度系数 toughness coefficient; toughness factor
韧度因素 toughness factor
韧度因子 toughness factor
韧度指数 toughness index
韧革菌属 <拉> Stereum
韧合金钢 tough alloy steel
韧化 tempering
韧化玻璃 tempered glass; toughened glass
韧化处理 toughening
韧化淬火处理 negative hardening
韧化钢 annealed steel
韧化聚苯乙烯 toughened polystyrene
韧化铜 annealed copper
韧化铜线 annealed copper wire
韧化氧化锆陶瓷 toughened zirconia ceramics
韧化作用 annealing; malleableize; malleablization; toughen(ing)
韧力 strength
韧沥青 wurtzilite; tabbyite
韧炼铸铁 annealed cast-iron
韧木 tough wood
韧皮 bast
韧皮薄壁组织 bast parenchyma; phloem parenchyma
韧皮部【植】phloem; bast
韧皮(部)射线 phloem ray
韧皮工人 tanner
韧皮射线 bast ray
韧皮纤维 bast-fiber[fibre]; phloem fiber[fibre]
韧皮纤维扫帚 bass-broom
韧皮坐垫 bast mat
韧铅 tough lead
韧青铜 tough bronze
韧塑黏[粘]土 tough plastic clay
韧塑性砖 fibroplastic tile
韧铁 tough iron
韧铁退火 malleablising
韧铜 flat set copper; tough cake; tough copper cake; tough pitch; tough pitch copper
韧铜线 annealed copper wire
韧弯试验 temper bending test
韧性 ductility; malleability; stiffness; tenacity; toughness
韧性变形缘 ductile bead
韧性材料 ductile material; tough material
韧性脆性转变 toughness-brittleness transition
韧性的 ductile; tough
韧性低 low ductility
韧性低碳钢 malleable mild steel
韧性断层 ductile fault
韧性断口 gliding fracture; tough fracture
韧性断裂 ductile rupture; gliding fracture; tough fracture
韧性锻铁 malleable wrought iron
韧性钢 ductile steel; malleable steel; notch ductile steel; tough steel
韧性高 highly ductility
韧性合金钢 tough alloy steel
韧性剪切带 ductile shear belt; ductile shear zone
韧性结持度 tenacious consistency; tough consistency
韧性结构 ductile structure
韧性金属 ductile metal; tough metal
韧性率 toughness factor
韧性模量 modulus of toughness
韧性黏[粘]土 tenacious clay
韧性破坏 ductile failure
韧性破裂 ductile fracture
韧性软钢 malleable mild steel
韧性试验 ductile test; ductilimeter test; pliability test; toughness test
韧性树脂基体 toughened resin matrix
韧性塑料 toughness plastics
韧性铁 <可锻铸铁> ductile iron; tough iron; malleable iron
韧性铁管 malleable iron pipe
韧性退火 malleablizing
韧性锡青铜轴承合金 Tourun Leonard metal
韧性系数 toughness coefficient; toughness factor
韧性橡胶 toughening rubber
韧性橡胶护皮 sheathing
韧性橡胶护套绝缘电缆 tough rubber sheathed rubber insulated cable
韧性橡皮护皮 tough rubber sheath
韧性橡皮绝缘电缆 tough rubber sheathed cable
韧性要求 toughness requirement
韧性因数 toughness factor
韧性硬度 toughness hardness
韧性值 toughness value
韧性指数 toughness index; toughness number
韧性中等 middle ductility
韧性铸件 malleable casting
韧性铸铁 annealed cast-iron; malleable cast-iron; malleable(hard) iron; malleable pig
韧性转变 ductility transition
韧硬度 tough hardness
韧硬木 dog wood
韧致辐射 bremsstrahlung
韧致辐射效应 bremsstrahlung effect

扔

扔出 outthrow

仍

仍未完成 remain unfulfilled
仍未执行 remain unperformed
仍未装船 remain unshipped

日

日班 day pair; day shift; daytime shift; day turn
日班工 days man
日班工人 day man; day pair
日班工作 run day
日班计划 daily and shift plan; daily and shift traffic working plan
日班交接 day's work joint
日班制 day work system
日斑 sunspot
日斑周期 sunspot cycle; sunspot periodicity
日保费 daily premium
日报 daily paper; daily return
日报表 daily report; daily sheet; daily statement; daily drilling report <钻探>
日报单 daily report
日本 JLPA 标准 Japan standard JLPA
日本扁柏 hinoki cedar; Japanese cypress; obtuse ground cypress
日本标准时 Japanese standard time
日本标准锥 Japanese standard cone
日本不动产银行 Nippon Fudosan Bank
日本长期信贷银行 Long-Term Credit Bank of Japan
日本城市规划协会 City Planning Association of Japan
日本赤松 Japanese red pine
日本船舶用品检定协会 Nippon Hakuyohin Kentei Kyokai
日本船级社 Nippon Kaiji
日本瓷器 japan; Japanese porcelain
日本大漆 A jap-A-lac
日本大漆催干剂 japan drier[dryer]
日本地槽 Japan geosyncline
日本地物理学特别委员会 Special Committee on Solar-Terrestrial Physics
日本地下水协会 Japanese Association of Groundwater Hydrology
日本地震分级 class of earthquake of Japan
日本地震工程促进协会 Japan Society of Earthquake Engineering Promotion
日本地震烈度表 Japanese intensity scale
日本地震学会 Seismologic(al) Society of Japan
日本电工委员会标准 Japanese Electric(al) Committee Standard
日本电离层探测卫星 Japan Ionosphere Sounding Satellite
日本电气公司 Nippon Electric(al) Company
日本电气学会 Japanese Electrotechnical Committee
日本东方租赁公司 Japan Oriental Leasing Company
日本一东非航线 Japan/East Africa route
日本东海道铁路新干线 <标准轨距高速铁路> Tokaido Shinkans(h)en
日本东亚双壳类地理亚区 Japanese-east Asia bivalve subprovince
日本短期货币市场 Gen-saki
日本椴 Japanese linden
日本法律 Japanese law
日本防腐协会 Japan Association of Corrosion Control
日本榧树 Japanese torreya
日本佛教庙宇的主屋 kondo
日本冈谷快速硬化水泥 o'kaycrete
日本钢管公司 Nippon Kokan
日本钢管株式会社 Nippon Kokan Kabushiki kaisha
日本港湾协会 Japanese Port and Harbour Association
日本工程标准 Japanese Engineering Standard
日本工会总评议会 General Council of Trade Unions of Japan[SOHYO]
日本工商会 Chamber of Commerce and Industry; Japan Chamber of Commerce and Industry
日本工业标准 Japanese Engineering Standard; Japanese Industrial Standard[JIS]
日本工业标准托盘 Japanese Industrial Standard flat pallet
日本工业规格 Japanese Industrial Standard[JIS]
日本工业银行 Industrial Bank of Japan
日本公认会计师协会 Japanese Institute of Certified Public Accountants
日本古典建筑 <木结构> Nipponese classical architecture(timber structure)
日本股票指数期货 Nikkei stock aver-

age futures
日本雇主协会联合会 Japan Federation of Employers' Association
日本管理学会 Japan Management Association
日本国际协力事业团 Japan International Cooperation Agency
日本国(家)铁(道)标准 Japan National Railways Standards
日本国家铁路公司 Japanese National Railways
日本国石油公团 Japan National Oil Corp
日本国有铁路 Japanese National Railways
日本海沟 Japan trench
日本海暖流 Japan current
日本海盆 Japan basin
日本海上保安厅 Japan Maritime Safety Agency
日本海上灾害防止中心 Japan Maritime Disaster Prevention Center[Centre]
日本海事检定协会 Nippon Kaiji Kentei Kyokai
日本海事协会 Japanese Marine Corporation
日本海水线 Japan Sea Cable
日本海棠 Japanese quince
日本海外经济协力基金 Japanese Overseas Economic Cooperation Funds
日本海外铁道技术协力会 Japan Railway Technical Service
日本海外运输咨询业者协会 Japan Transport Consultants Association
日本海峡地区航线 Japan/Strait Area route
日本海型地槽 Japan sea type geosynclines
日本海洋工程学会 Marine Engineering Society in Japan
日本海洋资料中心 Japanese oceanographic(al) Data Center[Centre]
日本海运集会所 Japan Shipping Exchange Inc
日本海重工业株式会社 Nippoki Heavy Industries
日本航海学会 Nautical Society of Japan
日本航空货物通关资料处理制度 Nippon Air Cargo Clearance System
日本航空协会 Japan Air Society
日本核燃料公司 Japan Atomic Fuel Corp.
日本红 <含硅氧的氢氧化铁> Japanese red
日本红漆 red japan lacquer
日本红松 akamatsu
日本红土 Juraku zuchi
日本湖沼学会 Japanese Society of Limnology
日本花柏 sawara cedar
日本槐木 tamo
日本环境法 Environmental Law of Japan
日本环境管理体制 Japan system of environmental management
日本环境厅 Japan Environmental Agency
日本环境协会 Japan Environmental Association
日本黄杉 Japanese Douglas fir
日本回纹饰 Japanese fret
日本会计研究学会 Japan Accounting Association
日本货币单位 Japanese yen
日本机械工程师学会 Japanese Society of Mechanical Engineers
日本机械工程学会 Japan Society of Mechanical Engineering
日本检定协会 Nippon Kaiji Kyokai
日本建设咨询业者协会 Japan Construction Consultants Association
日本建筑 Japanese architecture
日本建筑学会 Architectural Institute of Japan
日本建筑咨询业者协会 Japan Construction Consultants Association
日本金属粉末 Japanese metal powder
日本进出口银行 Export-Import Bank of Japan
日本经营者团体联盟 Japanese Federation of Employers' Associations
日本开发银行 Japan Development Bank
日本蜡 haze tallow;japan tallow;japan wax;sumac wax
日本冷杉 Japanese(silver)fir;strong white fir
日本栗 Japanese chestnut
日本烈度表 intensity scale of Japan
日本柳杉【植】cryptomeria
日本柳杉脂 sugi gum
日本罗汉柏 hatchet-leaved arbor-vitae;Japanese thuja
日本罗汉柏叶油 hiba oil
日本落叶松 Japanese larch;larch-tree
日本贸易振兴会 Japan External Trade Organization
日本农业标准 Japanese Agricultural Standard
日本女贞【植】Japanese privet
日本暖海流 Japan warm current;Kuroshio(current)
日本泡桐 karri-tree;kiri;Princess tree
日本七叶树 Japanese horse-chestnut
日本漆 japan
日本漆浆 paste in japan
日本漆树 Rhus verniciflua stokes
日本气象厅 Japan(ese)Meteorologic(al) Agency
日本山茶 japonica
日本山毛榉 Japanese beech
日本商社 Japanese Trading Company
日本商事仲裁协会 Japan Commercial Arbitration Association
日本神殿 Japanese temple
日本生产本部 Japan Productivity Center[Centre]
日本生漆 japan varnish
日本绳纹文化建筑式样 <一种传统的日本建筑风格> Jomon style
日本石柯 Japanese tanoak
日本石油技术协会 Japanese Association of Petroleum Technologists
日本石油学会 Japan Petroleum Institute
日本式 Japanese style
日本式隔扇 fusuma
日本式建筑 Japanese architecture;Japonica style
日本式精细工艺品 japan
日本式盆景 bonsai
日本式褥垫 <榻榻米> Japanese mat
日本式屋架 Japanese roof truss
日本水污染研究学会 Japan Society on Water Pollution Research
日本松 matsu
日本松节油酸 japopinic acid
日本酸 japanic acid
日本陶石 toseki
日本天皇住所 dairi
日本天然漆 Japanese lacquer;japan lacquer
日本天然漆色浆 japan colo(u)r
日本贴梗海棠 dwarf Japanese quince;Japanese quince
日本铁道工程学会 Japanese Railway Engineering Association
日本铁路工程协会 Japanese Railway Engineering Association
日本铁路技术公司 Japanese Railway Technology Corporation
日本铁路新干线客运行车制度(系统)Shinkansen system
日本铁路新高速干线 <新建的标准轨距高速干线> Shinkansen
日本桐油 Japanese tung oil;Japanese wood oil
日本涂料工业协会 Japanese Paint Industrial Association
日本涂装(工业协会)标准 Japanese Painting Standard
日本污水工程局 Japan Sewage Works Agency
日本五须松 Japanese strobus pine
日本新干线 Japanese High-Speed Train Networks;Shinkansen Lines
日本新闻中心 Nippon Press Center
日本新潟水俣病 niigataminamata disease
日本型企业组织 Japanese organizations
日本血吸虫病 Katayama disease
日本一般能源研究委员会 General Energy Research Council of Japan
日本一揽子原油进口平均价格 Japan crude cocktail
日本银行 Bank of Japan
日本樱花 Japanese cherry;Japanese flowering cherry;someriyoshine
日本油墨者协会色谱 Japan Printing Ink Maker's Association Colo(u)r
日本原子能保险组织 Japan Atomic Energy Insurance Pool
日本原子能委员会 Japan Atomic Energy Committee
日本原子能研究所 Japan Atomic Energy Research Institute
日本藻类学会 Japanese Society of Phycology
日本债券信用银行 Nipoon Credit Bank
日本罩面黑漆 japanned finish
日本罩面漆 japan finish
日本—(中国)台湾航线 Japan/Taiwan route
日本中央标准时 Japan central standard time
日本钟楼 Japanese bell tower
日本重量单位 kin;monme
日本专利中心 Japan Patent Centre
日本自然保护协会 Nature Conservation Society of Japan
日本租赁国际公司 Japanese Lease International Corporation
日本租税研究协会 Japan Tax Association
日变程 daily course;daily fluctuation;daily variation;diurnal amplitude;diurnal variation
日变动态分析【交】day-to-day dynamic(al)
日变动系数 daily variation coefficient
日变短 diurnal variation
日变法 diurnal variation method
日变风 diurnal wind
日变幅 amplitude of the variation;daily amplitude;daily range;diurnal amplitude;diurnal fluctuation;diurnal range
日变改正 diurnal correction
日变改正值 diurnal correction value
日变化 daily change;daily fluctuation;daily variation;diurnal change;diurnal variation
日变化起伏 diurnal fluctuation
日变化系数 daily variation coefficient;daily variation factor
日变键控单元 daily keying element
日变量 variation per day
日波 diurnal wave
日波动 daily fluctuation;diurnal fluctuation
日补偿 daily compensation
日不均系数 daily unbalance factor
日参差功率 daily diversity power
日差程 daily range
日差率 daily rate;rate of chronometer;watch rate
日差率测定 rating of chronometer
日拆 day call;day loan;day-to-day loan
日拆资金 overnight at call
日缠 solar equation
日产机械材料公司 Nissan Kizai Co., Ltd.
日产量 current yield;daily capacity;daily flow;daily output;daily production;daily turnover;daily working capacity;daily yield;day output;output per day;single day production
日产上班范围 daily commuting sphere
日产桶数 <石油> barrels per day
日长 length of the day
日长度 day length
日长石 aventurin(e) feldspar;heliolite;sunstone
日常安全 routine safety
日常保养 current maintenance;daily maintenance;daily service;day-to-day maintenance;minor maintenance;operational maintenance;running maintenance
日常必需品 various daily necessaries
日常波动 current fluctuation
日常操作 routine practice
日常产量 current output
日常处理报告 daily transaction reporting
日常贷款 day-to-day accommodation
日常当地供应 daily local supply
日常调查 day-to-day investigation
日常订货服务 standing order service
日常费用 cost of upkeep;current expenditures;current expenses;ordinary everyday expenses;running cost;running expenses
日常分录 current entry
日常分析 daily analysis;routine analysis
日常工艺管理 daily process management
日常工作 daily routine;daily work;day-to-day work;everyday routine;regular work;routine work
日常工作惯例 routine
日常供给 daily supply
日常供应 daily supply
日常供应罐 daily supply tank;day tank
日常故障检查 diagnostic check
日常管理 day-to-day management
日常过财 day-to-day posting
日常活动 day-to-day activity
日常记录 current entry
日常检查 current check;current control;routine check;routine inspection;routine operating inspection;running check;running control;daily inspection
日常检修 daily inspection;daily operating repair;maintenance repair;routine maintenance;routine overhaul;running repair
日常检验 daily inspection;day-to-day test;routine check;running repair
日常经营 day-to-day operations
日常会计职能 routine accounting function

日常垃圾 consumer waste
日常例行工作 daily round
日常流量 natural discharge
日(常)平均温度 mean daily temperature
日常勤务 running service
日常(燃料)损耗控制 day-to-day loss control
日常润滑 daily lubrication
日常审计 current audit;daily audit
日常生产 current production;day-to-day production
日常生活 daily life;everyday life
日常生活必需品系数 coefficient of daily necessities
日常生活范围 daily living sphere
日常生活圈 daily living sphere
日常使用的 informal
日常使用费用 running cost
日常事务 daily pursuits;daily round
日常试验 routine test
日常收入 daily receipts
日常疏浚 routine dredging
日常水位 natural water level
日常调整 routine adjustment
日常通信[讯] day-to-day communications
日常维护 constant maintenance;current maintenance;daily maintenance work;day-to-day maintenance;maintenance overhaul;routine attention;routine maintenance
日常维护检修 operating repair
日常维修 current repair;daily maintenance;day-to-day repair;maintenance overhaul;maintenance repair;operating repair;ordinary maintenance;permanent repair;regular maintenance;routine maintenance;routine repair;running maintenance;running repair;running service;scheduled maintenance
日常维修工作 maintenance work
日常文书工作 paper work
日常消耗 current consumption
日常修理 current repair;maintenance overhaul;operating maintenance;operating repair;running repair
日常需求的基础结构 infrastructure for daily needs
日常需求量 domestic demand
日常需要 daily needs;daily wants
日常养护 day-to-day maintenance
日常养护费 cost of upkeep
日常样品 routine class sample
日常要求 routine request
日常业务 current undertaking;daily routine;day-to-day business
日常业务管理 day-to-day management
日常营业费 current operation expenditures
日常运行 day-to-day working
日常支助性服务 routine supporting services
日潮 day wave;diurnal tide;solar tide
日潮波 solar tidal wave
日潮不等 daily inequality of tides;diurnal inequality of tides
日潮不等潮龄 age of diurnal inequality;age of diurnal tide;diurnal age
日潮不等性 daily inequality;diurnal inequality
日潮差 daily inequality;diurnal inequality;diurnal range;low-water inequality
日潮港 diurnal tide harbo(u)r
日潮高潮不等 mean diurnal high-water inequality
日潮力 diurnal force
日潮龄 diurnal age of tide
日潮流 diurnal tidal current
日潮流(速)差 diurnal inequality
日潮时滞后 lagging;daily retardation of tide
日潮推迟 lagging;daily retardation of tide
日潮位差 diurnal inequality
日潮型 daily type of tide;diurnal type of tide
日成风 heliotropic(al) wind
日承【天】 circumhorizontal arc
日程 program(me)
日程安排 scheduling program(me)
日程安排型式 scheduling pattern
日程表 schedule;calendar;calendar progress chart;schedule list;schedule sheet;time schedule;timing schedule
日程法 calender method
日程计划 schedule planning;schedule programming
日程控制 time schedule control
日程控制图表 schedule control chart
日程图 itinerary map
日池窑 day tank
日持续曲线 daily duration curve
日出风 solarie;souledras
日出力 daily output
日出料量 daily output
日出没方位角 eastern amplitude of the sun
日出效应 sunrise effect
日出与日落过渡期 sunrise and sunset transition
日储藏所 daily storage
日储量 daily storage
日处理量 daily capacity;day output
日磁层 helio magnetosphere
日大气潮 solar atmospheric(al) tide
日当量轮数 daily number of equivalent wheel
日低潮不等性 diurnal low water inequality
日地关系 solar-terrestrial relationship;sun-earth relationship
日地观测与气候学卫星 sun-earth observation and climatology satellite
日地环境 solar-terrestrial environment
日地距离 solar distance
日地物理学 solar-terrestrial physics
日定量 ration
日度 day-degree
日吨数 tons per day
日耳曼式褶皱 German type folds
日耳曼型造山运动 German type orogeny
日耳曼银 <锌镍铜合金> German silver
日珥 lowitz;prominence[prominency];solar prominence
日发的 quotidian
日发电量 day current
日发煤量 daily coal shipment
日罚款额 <承包合约中延误工期每日定额罚款> liquidated damages
日范围 daily range
日放逸量 <某种空气污染质的> emission inventory
日风 solar wind;sun wind
日峰荷 daily peak load
日峰荷曲线 peak-day-load curve
日峰荷调节 daily peaking operation
日辐射强度 intensity of solar radiation
日辐射强度计 pyranometer
日付款 daily payment
日负荷 daily load;diurnal load
日负荷变动 daily load fluctuation
日负荷波动 daily load fluctuation
日负荷率 daily load factor;diurnal load factor
日负荷曲线 daily load curve
日负荷图 diurnal load diagram
日负荷系数 daily load factor
日负荷预测 daily load prediction
日负载 daily load
日负载率 daily load factor
日负载曲线 daily load curve
日负载系数 daily load factor
日高潮不等性 diurnal high water inequality
日高潮位 daily high tide
日高峰 daily peak
日高构造带 Hidaka tectonic zone
日高造山运动 Hidaka orogeny
日工 <做日班的工人> day labo(u)r;day wage work;daywork;peon;time work;day pair
日工制 day labo(u)r system;day work system
日工资 daily earning;daily wage;wages per day
日工资率 daily rate of pay;daily wage rate;day rate
日工作缝 <一日工作结尾时与次日工作衔接时所作的接缝> day work joint
日工作量 daily work load
日工作末的超载 end-of-day surcharge
日工作时间登记卡 daily time card
日供应量 day of supply
日光 daylight(ing);solar light;sunbeam;sun light;sunshine
日光白炽灯 daylight incandescent lamp
日光保护操作 sun control work
日光保护装置 sun control device
日光曝光 daylight exposure
日光泵 sun-pump
日光泵激光器 sun-pumped laser
日光泵 solar pumping;sun-pump(ing)
日光泵能 sunlight pumping energy
日光变化 daylight changing
日光病 heliopathia
日光玻璃(暖)房 solar house
日光采暖 solar gain
日光测定 measurement of daylight
日光层【气】 heliosphere
日光抽运 solar pumping
日光的 solar
日光灯 artificial daylight;daylight lamp;fluorescent lamp;fluorescent light;heliolamp;sun lamp
日光灯管 fluorescent tube
日光灯管座 daylight base
日光灯启动器 daylight starter;glow starter
日光灯照明 fluorescent lighting
日光灯整流器 rectifier for daylight lamp
日光定位法 position by sun sights
日光发电厂 helioelectric power plant
日光发电机 solar generator
日光阀 sun valve
日光反射 sun reflection
日光反射比 daylight reflectance;solar reflectance
日光反射信号 heliogram
日光反射信号法 heliography
日光反射信号机【铁】 heliograph
日光反射信号器 heliograph;heliotrope
日光反射信号仪 heliograph
日光反射养护膜 solar reflective curing membrane
日光反照通信[讯]镜 daylight signalling mirror
日光放大印像机 daylight enlarger printer
日光放射区 sunburst area
日光分光光谱图 spectroheliogram
日光辐射 solar light radiation;solar radiation;sun radiation
日光辐射(率)计 actinometer
日光辐射强度计 pyrheliometer
日光辐射强度仪 pyrheliograph;pyrheliometer
日光辐射试验 sunlight radiation test
日光负荷 solar load
日光干燥 solar drying
日光感光度 daylight speed
日光格子窗 solar grille
日光管 solar cell <检验弯沉仪的>;Sunray <一种专利灯管>
日光光度学 daylight photometry
日光光谱镜 spectrohelioscope
日光光束 sun path
日光光隙 daylight slits
日光过滤玻璃 daylight filtering glass
日光回照器 heliotrope
日光计 heliograph
日光甲板 sports deck;sun deck
日光胶版 heliograph
日光胶版术 heliography
日光节约 daylight-saving
日光开关 sun switch;sun valve;twilight switch <航标灯的>
日光控制 daylight control
日光控制玻璃 solar control glass
日光控制器 daylight controller
日光蓝颜料 daylight blue
日光亮度 brightness of daylight
日光榴石 helvine
日光滤光片 daylight filter
日光滤光器 daylight filter
日光滤色镜 daylight filter
日光路径 sun path
日光罗盘 sun compass
日光敏感性 heliosensitivity
日光模拟仪 <研究建筑日照用> solarscope
日光能 energy of the sun
日光能技术 heliotechnics
日光能量测定器 actinometer
日光能量测定仪 actinograph
日光屏 sunscreen
日光谱线消色差 actinic achromatism
日光强度 intensity of sunlight
日光强度自动记录器 actinograph
日光曲线 daylighting curve
日光热 sun heat
日光入射 incoming solar irradiation
日光晒印 heliographic(al) print;sunlight print;sun printing
日光摄谱计 spectroheliograph
日光摄谱仪 spectroheliograph
日光摄影法 heliography
日光摄影机 heliograph
日光摄影术 heliography
日光石 sunstone
日光石英玻璃 solar silica glass
日光视觉 photopic vision
日光室 sun parlo(u)r;sun porch
日光束 sunbeam
日光调节器 sun control
日光投影仪 sun projector
日光温度表 heliothermometer
日光温室 heliogreenhouse;sunlight greenhouse
日光系数 daylight factor;daylight ratio
日光线 sunbeam
日光效应 daylight effect;sun effect
日光型 daylight type
日光型彩色胶片 daylight film;daylight type colo(u)r film
日光型彩色片 daylight colo(u)r film
日光型碳弧 <老化试验机> sunshine carbon arc weathering tester

R

日光型荧光涂料 daylight fluorescent paint
日光型荧光颜料 daylight fluorescent pigment
日光耀日 sunlight glare
日光仪 heliograph
日光因数 daylight factor
日光浴 insolation;siriasis;sun bath
日光浴板 sun deck
日光浴场 sunbathing area;sunbathing patio;sunbathing terrace;sun patio
日光浴廊 sun-patio(u)r;sun porch
日光浴内院 sunbathing patio
日光浴平台 sun deck
日光浴室 solarium[复 solaria/solariums];sun-room;sunbathing room;sun parlo(u)r;sun porch
日光浴室用弯玻璃 solarium curved glass
日光浴台 sun deck
日光浴屋面 sun deck
日光圆顶 daylight dome
日光照明 daylight illumination
日光照明的 daylighted
日光照射 downpour
日光照射的 daylighted
日光遮蔽 solar screening
日光治疗室 solarium[复 solaria/solariums]
日光柱 sun pillar
日光自动开关 astronomic(al) time clock
日光作用 daylight effect
日圭(标杆) gnomon
日圭原理 gnomonics
日圭指针 gnomon
日规 azimuth dial;sun dial
日规壳属 <拉> Gnomonia
日晷 analemma;horologe;sun dial
日晷投影 gnomonic projection
日晷仪 mean daily;sun dial
日晷仪指时针 gnomon
日过闸耗水量 daily locking water volume
日旱流量 daily dry weather flow
日旱流污水 daily dry weather flow
日耗水高峰 peak consumptive use
日耗水量 daily water consumption
日荷载率 daily load factor;diurnal load factor
日洪峰 daily peak
日洪峰流量 daily flood peak
日华 corona;solar corona
日环境负荷 daily environmental load
日环食 annular eclipse
日辉 dayglow;solar flare
日极值 daily extremes
日给工资 daily pay
日给养 ration cycle
日给制 day rate plan;day wage system
日计表 daily account;daily trial balance
日计划 daily plan;schedule of the day
日计划兑换率 the rate of fulfilment for daily passenger transport plan
日记簿 day book;journal account;journal book;journal
日记簿分录 journal entry
日记法 diary method
日记分录过账 posting journal entries
日记账 daily accounting;day book;journal book
日记账传票 journal voucher
日记账电传打印机 journal teleprinter
日记账分录 journal entry;journalizing
日记账检查 proof of journal
日记账调整分录 adjusting journal entry
日记账页数 journal folio
日记总账 combined journal and ledger;journal ledger
日记总账核算形式 journal-ledger form of accounting
日技术服务费 daily technical service fee
日际变化率 interdiurnal variability
日剂量 daily dose
日价限幅 daily price limit
日间 daytime
日间残渣 day residue
日间传输 day transmission
日间负荷 day(time) load
日间交通 daytime traffic
日间交易 day trading
日间亮度 brightness of daylight
日间列车 day train
日间频道 day channel
日间频率 day frequency
日间人口 daytime population
日间视度 daylight visibility
日间停车场 day parking
日间投送电报 day letter(telegraph)
日间透支 daylight overdraft
日间托儿所 day nursery
日间雾 daytime fog
日间信号 daytime signal
日间信号灯 daylight signalling light;day signal light
日间型 diurnal pattern
日间医院 day hospital
日间治疗中心 daytime treatment center[centre]
日降角 solar depression angle
日降水量 daily precipitation;daily rainfall
日降雨量 daily precipitation;daily rainfall
日交通量 daily traffic;daily traffic volume;daily volume
日交通量变化图 daily traffic pattern;diagram of daily traffic variation
日交通量变化型 <显示 24 小时的小时交通量> daily traffic pattern
日较差 daily amplitude;daily range;diurnal amplitude;diurnal range
日结存 daily balance
日结算余额 daily balance
日界线 calendar line;date line;international date line
日界线通过日 meridian day
日津贴 daily allowance
日进尺 <钻探> daily footage;daily advance;daily drilling progress
日进度 daily progress
日进度表 calendar progress chart
日进度率 daily progression rate
日进展 daily advance
日径流量 daily runoff;daily flow
日径流深 drainage modulus
日掘进量 daily advance
日掘土方量 daily yardage
日均差 diurnal inequality
日均收付差额 average daily balance
日均已收票据差额 average daily collected balance
日均值 daily mean value
日均装车数 average daily car loadings;daily average number of car unloading
日刊 diurnal
日客运总量 daily total passenger volume
日控制 daily control
日劳动价值 value of a day's labo(u)r
日历 almanac;calendar
日历变动 calendar variation
日历标准 calendar standard
日历差幅 calendar spread
日历工日数 calendar man-days;calendar workdays;number of calendar worker-days
日历计划 calendar plan
日历计算人日数 calendar man-days
日历年度 calendar year
日历日 calendar day
日历日数 number of calendar days
日历时间 calendar hours;calendar time
日历手表 calendar watch
日历台日数 calendar machine-time;machine days in accordance with calendar days
日历诱导变动 calendar induced variations
日历月 calendar month
日历指示器 date indicator
日历指示器护板螺钉 screw for indicator guard
日历钟 time-of-day clock
日历周 calendar week
日历装车安排 calendar loading arrangement
日历装车计划 calendar loading plan
日历总时数 total calendar hours
日立公司 <日本> Hitachi
日立网络体系结构 Hitachi network architecture
日立造船工程公司 Hitachi Shipbuilding and Engineering Co. Ltd.
日立制作所 <日本> Hitachi Limited
日立租赁公司 <日本> Hitachi Lease Corporation
日量程 daily range
日量 daily amount
日裂 sun crack
日流量 daily discharge;daily rate of flow;daily water flow;day flow
日流量过程线 daily hydrograph
日率制 per diem
日轮 solar disc[disk];sun's disc[disk]
日轮望远镜 disk telescope
日落 sun set
日落黄 sunset yellow
日落辉 crepuscular ray
日落角 solar depression angle
日落效应 sunset effect
日落预算法 <系零基预算法的别称> sunset budgeting
日冕 corona;solar corona;solar crown
日冕的 coronal
日冕低层 lower corona
日冕洞 solar coronal hole
日冕高层 high corona
日冕观测镜 coronascope
日冕光 coronal light
日冕光电光度计 coronal electrophotometer
日冕光学偏振 coronal optic(al) polarization
日冕红线 coronal red line;red coronal line
日冕技术 coronagraphic(al) technique
日冕加热 coronal heating
日冕禁线 coronal forbidden line
日冕连续光谱 coronal continuum
日冕亮点 coronal bright point
日冕绿线 coronal green line
日冕模型 corona model
日冕凝聚区 coronal condensation
日冕凝聚物 coronal condensation
日冕平衡 coronal equilibrium
日冕谱线 coronal line
日冕圈 solar coronal loop
日冕射线 coronal ray;streamer;streamline
日冕瞬变 coronal transient
日冕增强区 coronal enhancement
日面 solar disc[disk]
日面北极 north heliographic(al) pole
日面边缘 solar limb
日面分布 heliographic(al) distribution
日面极 heliographic(al) pole
日面经度 heliographic(al) longitude;heliolongitude;solar longitude
日面图 heliographic(al) chart
日面纬度 heliographic(al) latitude;heliolatitude
日面坐标 heliographic(al) coordinates
日没 sun set
日没时的幅角 western amplitude of the sun
日暮 dark
日内变动 daily variation
日内精度 day average precision
日内调峰运行 <水电站> daily peaking operation
日内瓦玻璃 Geneva crystal
日内瓦公约 Geneva Convention
日内瓦回合 Geneva Round
日内瓦贸易会议 Geneva Trade Conference
日内瓦命名法 Geneva nomenclature
日内瓦石 genveite
日内瓦式发条停止上条装置 Geneva stop work
日内瓦式发条限上装置 Geneva stop
日内瓦式机芯 Geneva movement
日内瓦天文台 Geneva observatory
日内瓦通电 Geneva Circular Telegram
日内瓦通函 Geneva Circular
日内瓦通知书 Geneva Notification
日内瓦运动 geneva motion
日内瓦总秘书处 <国际电信联盟> Geneva Bureau
日内限额 intra-day limit
日排出量 daily output
日排水量 daily drainage;drainage coefficient
日排污量限制 daily average effluent limitation
日牌价 price of day
日盘 solar disc[disk]
日喷 spray
日偏食 partial eclipse of the Sun;partial solar eclipse;solar partial eclipse
日平均 daily mean;diurnal mean;mean daily
日平均海平面 daily mean sea level
日平均降水量 daily average precipitation
日平均降雨量 daily average rainfall
日平均交通量 daily average traffic
日平均交通密度 daily traffic density
日平均进尺 average footage per day
日平均精度 day average precision
日平均流量 daily mean discharge;daily mean flow;mean daily discharge;median influent flow per day
日平均气温 daily mean air temperature
日平均容许浓度 average day permissible concentration;daily average permissible concentration
日平均声级 day-night average sound level
日平均湿度 mean daily humidity
日平均疏浚量 average daily dredging quantity;mean daily dredging quantity
日平均数 daily average
日平均水位 mean daily stage
日平均温度 average diurnal temperature;daily mean temperature
日平均运量 average daily traffic;mean daily traffic

日平均值 daily average value;diurnal mean value
日平均综合温度 average daily solar temperature
日平均最高浓度 daily average maximum concentration
日平均最高容许浓度 daily average maximum allowable concentration
日平均最高允许浓度 daily average maximum permissible concentration
日曝 sun exposure
日期 date
日期戳 date mark;dater;date stamp
日期戳记 date mark
日期戳子 dater
日期和时间 date and time
日期时间组 date time group
日期提前票据 antedated bill
日期提前提单 antedated bills of lading
日期提前支票 antedated check
日期填迟 dating backward
日期填迟支票 post-dated check
日期填早 antedating; dating ahead; dating forward;prochronism
日期条件 date term
日气温变幅 difference of daily temperature
日气温变化 diurnal temperature change
日勤货运调度员 day shift goods dispatcher[despatcher]
日勤制 day's shift system
日全食 solar total eclipse; total eclipse of the Sun;total solar eclipse
日燃烧周期 daily burning cycle
日容许摄入量 acceptable daily intake;allowable daily intake; allowance daily intake
日晒 insolation;solarization
日晒不褪色的 sunlight-fast
日晒的 sunbaked
日晒法 solarization method
日晒风化 insolation weathering
日晒风化砾石 insolilith
日晒干草 sun-cured hay
日晒干燥的 sun cured
日晒红斑 erythema solare;sunburn
日晒裂 sun crack
日晒试验 sun test
日晒性皮炎 dermatitis solaris
日晒夜露 weather exposure
日晒遮风的花园 sun trap
日晒遮风的露台 sun trap
日晒蒸发 solar evapo(u)ration
日晒蒸馏 solar distillation
日晒作用 solarization
日射 insolation; solar radiation; sun radiation
日射表 actinometer;solarimeter
日射病 heliosis; insolation; siriasis; sunstroke
日射测量法 actinometry
日射观测 solar radiation observation
日射计 actinograph
日射率 insolation
日射能量 insolation
日射强度计 pyrheliometer;solarimeter
日射曲线图 actinogram
日射温度 solar temperature
日射温量法 actinometry
日射仪 actinograph
日射阴影面积 sun shadow area
日射总量表 solarimeter
日射总量计 solarigraph
日射总量曲线 solarigram
日摄入容许量 acceptable daily intake
日升效应 sunrise effect
日生产量 daily capacity;day capacity
日生产能力 daily working capacity
日生潮力 diurnal force
日施工缝 day work joint;end-of-day joint
日时钟 time-of-day clock
日蚀 eclipse of the sun;solar eclipse
日食 eclipse of the sun;solar eclipse
日食观测队 solar eclipse expedition
日食天气 eclipse weather
日食限 solar eclipse limit
日式瓦 Japanese tile
日式屋架 Japanese roof truss
日视检查 sight control
日输出量 daily output
日输入量 daily input
日输送量 daily capacity;daily throughput
日水流量 daily discharge;daily flow
日速 daily rate
日损失 daily loss
日调节 daily regulation
日调节池 daily regulating pond
日调节电站 daily storage plant
日调节库容 balance storage;regulator storage;daily pondage
日调节容量 adequate pondage
日调节水池 storage regulator
日调节水库 daily regulating reservoir;daily storage
日调节蓄水量 balance storage;regulator storage
日推迟时间 daily retardation
日吞吐量不平衡系数 daily unbalance coefficient of cargo handled at the port
日托机构 day care facility
日托所 day care home
日托托儿所 child day care home; nursery day school
日托中心 day-care center[centre]
日挖方量 <以立方码计>【疏】daily (dredged) yardage
日温(测量)计 <测太阳温度的> heliothermometer
日温差 daily temperature range
日温度 day temperature
日温计 pyrheliometer
日温计太阳热量计 pyrheliometer
日纹理 diurnal lamination
日乌头碱【化】japaconitine
日污水量 daily dry weather flow
日污水流量 daily discharge
日息 daily interest;interest per diem
日息率 per diem rate
日下点 subsolar point
日下点高度 subsolar point altitude
日下点射电温度 radio subsolar temperature
日下点温度 subsolar point temperature
日下位置 subsolar position
日消耗 day's expenditures
日(消)耗量 daily consumption
日销售额 daily sales
日效率 daily efficiency
日校 day school
日卸车数 number of cars unloaded per day
日心的 heliocentric
日心轨道 heliocentric orbit
日心角 heliocentric angle
日心经度 heliocentric longitude
日心径 central meridian path
日心距(离) central meridian distance;heliocentric distance
日心历表 heliocentric ephemeris
日心视差 annual parallax;heliocentric parallax;stellar parallax
日心视向速度 heliocentric radial velocity
日心说 heliocentric theory;heliocentricism
日心速度 heliocentric velocity
日心体系 heliocentric system
日心天体位置 heliocentric place
日心天象 heliocentric phenomenon
日心纬度 heliocentric latitude
日心位置 heliocentric position
日心系统 heliocentric system
日心引力常数 heliocentric gravitational constant
日心子午线 central meridian
日心坐标 heliocentric coordinates
日薪 datal
日薪工 daywork;wage work
日需量 daily demand; daily requirement
日需水量 daily water demand
日需要量 <水电> daily demand
日蓄能 daily storage
日蓄水量 adequate pondage; daily storage
日循环 diurnal cycle
日要车计划 daily car requisition plan;daily wagon requisition plan
日耀 sunlit
日耀表面 sunlit surface
日耀极光 sunlit aurora
日野 Hino
日叶石 heliophyllite
日夜班交接 night header joint
日夜办公处 permanent-service office
日夜不停的 round-the-clock
日夜成像 day-and-night image
日夜分收费率 <电话白日与夜间费率不同> two-rate tariff
日夜服务 continuous service
日夜工作的可靠性 day-in and day-out dependability
日夜勘测 day-and-night (-time) reconnaissance
日夜连续试验 day-to-day test
日夜平均声级 day-night average sound level
日夜施工 around-the-clock job;round-the-clock job
日夜效应 day-night effect
日夜营业电报(话)局 permanent-service office
日夜转换键 day-and-night transfer key
日益增多的工作量 expanding workload
日引力潮 semi-diurnal tide-producing force
日英里 <汽车每日行驶的英里数> daily mileage
日营业额 daily sales
日影环绕区 periscian region
日影曲线 sun shadow curve
日影双向区 amphiscian region
日影仪 heliodon
日影异向区 heteroscian region
日映云辉 anthelion
日用泵 daily service;general service
日用必需品 daily necessaries[necessities];household necessities
日用玻璃 domestic glass
日用玻璃器皿 domestic glassware;utility glassware
日用玻璃制品工业 tableware and domestic glassware industry
日用瓷 household china; household porcelain
日用瓷器 daily porcelain ware; domestic porcelain
日用电器 utilization equipment
日用发电机 service generator
日用工业品 industrial consumer goods;manufactured articles of daily use;manufactured goods for daily use
日用锅炉 domestic boiler
日用化工废水 cosmetic wastewater
日用精陶 domestic fine pottery
日用开支 general expenses
日用量油箱 day fuel tank; day oil tank;day tank
日用流水账 journal day book;journal folio
日用煤堆 live storage pile
日用漂白织物 white goods
日用品 articles for daily use;artificial of everyday use; convenience goods; day-to-day goods; everyday object
日用品供应舱 ready-issue room
日用品价格 consumer price
日用品价格指数 consumer price index
日用器皿 utility ware
日用燃油泵 fuel oil service pump
日用燃油输油泵 fuel oil service supply pump
日用软件 utility software
日用商品 commodity; general merchandise
日用炻器 domestic stoneware
日用水泵 service pump
日用水舱 daily service tank; service tank
日用水柜 daily service tank; daily supply tank;top tank
日用水量 daily water consumption; water consumption per day
日用陶瓷 ceramics for daily use;daily porcelain ware;domestic ceramics
日用陶器 domestic earthenware
日用消费品 goods for everyday consumption
日用油柜 free-standing fuel tank;service oil tank;service tank;top tank
日用油箱 free-standing fuel tank;service oil tank;service tank;top tank
日用杂货 various household supplies
日预报 daily forecast
日元 Japanese yen
日元基准 Yen base
日元债券 Yen bond
日原料量 daily crude capacity
日月变化 lunisolar variation
日月潮(汐) lunisolar tide; lunisolar diurnal tide
日月合成日周期分潮 lunar solar-diurnal tide
日月每日变化 lunisolar daily variation
日月摄动 lunar perturbation;lunisolar perturbation
日月升落潮汐仪 volvele
日月食的数学理论 sciametry
日月食理论 sciametry
日月岁差 lunisolar precession
日月效应 lunisolar effect
日月谐波分量 lunisolar harmonic component
日月引力 gravitational attraction of moon and sun;lunisolar attraction
日月引力能 attraction energy
日月引力摄动 lunisolar gravitational perturbation
日月影响 lunisolar influence
日月运动 lunisolar motion
日月章动 lunisolar nutation
日晕 aureola[aureole];halo;solar halo
日增 circumzenithal arc
日增长率 daily progression rate
日增加量 daily increase
日增重 daily gain
日增重率 daily gain rate
日涨落 daily fluctuation

R

日照 incoming radiation;incoming solar radiation; insolation; solar insolation;sun exposure;sunshine
日照百分率 relative sunshine duration
日照变形测量 sunshine deformation survey
日照标准 sunlight ordinance
日照病 sunstroke
日照长度 duration of day;duration of sunshine;length of day
日照长度适宜 proper duration
日照持续时间 insolation duration;duration of sunshine
日照单位 sunshine unit
日照得热量 <通过窗户和结构材料传入室内的太阳能> solar heat gain; sun effect
日照地球背景 sun-illuminated earth background
日照法 <美国有关劳资纠纷的法律> sunshine law
日照计 actinometer; heliograph; sunshine recorder
日照记录 sunshine record
日照记录器 sunshine recorder
日照加热 solar heating
日照间距 minimum distance for sunlight;sunshine spacing
日照监测 monitor skylight
日照角 solar angle;sun angle
日照控制 solar control;sunshine control
日照宽度 daylight width
日照累积器 sunshine integrator
日照历时 duration of sunshine
日照量 solar irradiation
日照率 percentage of bright sunshine; percentage of possible sunshine;percentage of sunshine
日照面积 insolation area;sunshine area
日照目标 sun-illuminated target
日照期 sunshine period
日照器 helioscope
日照强度计 pyranometer
日照穹顶 daylight dome
日照射 incoming solar irradiation
日照时间 day length;insolation duration; sunshine duration; sunshine time
日照时间变化 daylight change
日照时间充足 enough sunshine duration
日照时数 duration of sunshine;hours of daylight;hours of sunshine;sunshine duration;sunshine hours
日照水平 sunshine standard
日照条件 sunshine condition
日照调节 sunshine control
日照图 sun chart
日照温度 solar temperature
日照吸收剂 solar absorber
日照系数 daylight factor;sky factor
日照小时 hours of daylight;hours of sunshine;sunshine hours
日照仪 pyrheliometer; sunshine recorder
日照因数 daylight factor
日照因素 daylight factor
日照预测 daylight prediction
日照云层背景 sun-illuminated cloud background
日照云层亮度 sun-illuminated cloud brightness
日照纸 heliogramma
日照总量 solarimeter
日振幅 daily amplitude;diurnal amplitude
日震学 helioseismology
日蒸发量 daily evapo(u)ration discharge
日志 daily record;day book;journal
日志电传打印机 journal teleprinter
日志缓冲器 journal buffer
日志任务 log task
日志输出 journal output
日志文件 journal file
日致气压变化 solar barometric(al) variation
日中 midday;midnoon
日钟 day clock
日周潮 diurnal tide;lunisolar diurnal tide;single day tide
日周潮潮龄 diurnal age;age of diurnal tide
日周期 diurnal period
日周期节律 circadian rhythm;diurnal rhythm
日周期性 diurnal periodicity
日周期循环风 diurnal winds
日周转量 daily turnover
日柱 light pillar;sun pillar
日转方向 sunwise
日转风 heliotropic(al) wind
日灼病 sun scald
日字(环节)链 stud link chair;stud type chain
日字形链节 stud link
日字形七划图形 seven-bar pattern
日总计水流量 daily totalized flow volume
日租率 daily rental rate
日最大暴雨量 maximum daily storm rainfall
日最大量 daily maximum; diurnal maximum
日最大流量 maximum daily discharge
日最大污染物负荷总量 total maximum daily load
日最大值【气】 diurnal maximum; daily maximum
日最低气温 daily minimum air temperature
日最高 diurnal maximum
日最高潮 diurnal maximum tide
日最高气温 daily maximum air temperature
日最高温度 maximum daily temperature
日最小量 daily minimum; diurnal minimum
日最小流量 minimum daily discharge
日最小值 daily minimum; diurnal minimum

绒

绒布 flannelet(te);velveteen
绒布擦坯 flannelling
绒布磨轮 branners
绒布选种机 draper cleaner
绒衬洗矿槽 blanket strake
绒垫 waste pad
绒垫润滑法 waste pad lubrication
绒盖多孔菌 velvet top fungus
绒高 wire height;pile height <从衬垫至地毯面高度>
绒辊 clearer roller;felted fabric roller
绒辊弹簧 clearer spring
绒聚 flocculate
绒聚的 flocculent
绒聚剂 flocculating agent;flocculent; floc-forming chemical reagent
绒聚试验 flocculation test
绒聚体 flocculating constituent
绒聚物 floccules
绒聚系数 flocculation factor
绒毛 fluff;fuzz;villus
绒毛白蜡树 pumpkin ash
绒毛净化器 clearing machine
绒毛状沉淀物 floccules
绒毛状的 downy;flocculent;fluffy
绒毛状组织 flocculating constituent
绒面处理 velour finishing
绒面革 <棕色> suede
绒面硅太阳能电池 textured silicon solar cell
绒面呢 broadcloth
绒面涂料 suede coating
绒面织物 velour
绒球 <用于装饰> pompon
绒铜矾 cyanotrichite
绒头 tufted
绒头地毯 pile carpet;tuffed rug
绒头抗拔强力 tuft pull strength
绒头抗压性能 pile crush
绒头纱线 pile yarn
绒头压陷 <地毯> pile crushing
绒头织物 pile weave
绒纤维 linter
绒屑 flock
绒状的 velvet
绒缀条 moss fringe

茸

茸瑚菌属 <拉> Lachnocladium
茸毛 fuzz

荣

荣誉部长 minister without portfolio
荣誉大臣 minister without portfolio
荣誉承兑 acceptance for honour;acceptance for supra-protest; act of honour
荣誉承兑人 accepter for honour
荣誉付款 payment of honour
荣誉付款人 payer for honour
荣誉感 sense of honor
荣誉花环 laurel-leaf swag
荣誉座 <古罗马> bisellium

容

容标正比电容器 straight-line capacity condenser
容差 admittance;allowance;tolerance; tolerance limit
容差分析 tolerance analysis
容差极限 tolerance limit
容差试验 allowance test
容尘量 clogging capacity;dust capacity; dust holding; dust holding capacity
容错 fault-tolerance;fault toleration
容错法 fault tolerant approach
容错计算 fault tolerant computing
容错计算机【计】 fault tolerant computer
容错技术 fault tolerant technique
容错匹配 permissive matching
容错软件 fault-tolerant software
容错系统 fault tolerant system
容错性能 fault freedom
容电抗 condensance
容电器 condenser; electric(al) condenser
容电式验电器 condensing type electroscope
容度 bulk unit weight;specific volume
容度函数 content function
容根层 root room
容荷 capacitive load(ing)
容积 bulk; bulk volume; capacity; combustion chamber volume; containment; cubage; cubic (al) content;cubic measure;holding capacity;measure volume;solid measure; volume;volume capacity;volumetric(al) capacity; volumetric (al) content;voluminal;internal volume <集装箱>
容积百分比 percent by volume; percent volume to volume; volume percent(age)
容积百分率 volume percent(age)
容积百分数 volume percent(age)
容积泵 air-displacement pump; positive-displacement pump
容积比冲 density specific impulse
容积比(率) rate by volume; volume ratio; volumetric (al) proportion; volumetric(al) ratio
容积比热 specific heat by volume; volume specific heat
容积比推力 volume specific thrust
容积比重 volume weight
容积变更 volumetric(al) change
容积变化 volumetric(al) change
容积变化测定管 eudiometer
容积表 strapping table
容积不变性 constancy of volume
容积残余变形率 rate of residual volumetric(al) deformation
容积残余变形值 value of residual volumetric(al) deformation
容积测定仪 volumescope
容积测量法 volumetric(al) measurement;volume measurement
容积测流法 volumetric(al) method
容积尺码 cubic(al) dimension
容积大的 voluminous
容积单位 volume unit
容积导电 volume conduction
容积导体 volume conductor
容积的 volumetric(al);voluminous
容积的倒数 inverse volume
容积的压力测量 manometry
容积电量计 volume voltameter
容积度 voluminosity
容积吨 capacity ton; spare ton; tonnage capacity; volume ton; volumetric(al) ton; measurement ton(nage) <船只装载单位,一个容积吨相当于 40 立方英尺,或 1.13 立方米>
容积吨法 measurement ton method
容积吨位 cubic (al) measurement; volumetric(al) tonnage
容积吨运费率 measurement rate
容积发射 volume emission
容积法 <测量混凝土空气量的> cubature;volumetric(al) method
容积法测流 gravimeter measurement;volumetric(al) measurement of discharge; volumetric(al) method of measuring discharge
容积放热率 volumetric(al) heat release rate
容积分配 proportioning by volume
容积分区 bulk zoning
容积分数 fraction by volume;volume fraction
容积分析 analysis by measure;volumetric(al) analysis
容积辐射率 cavity emissivity
容积负荷 volume loading;volumetric-(al) loading
容积负荷率 volume loading rate;volumetric(al) loading rate
容积含水量 volumetric(al) water content
容积含水率 <土壤的> moisture volume percentage
容积核算 mass accounting
容积恒定性 constancy of volume
容积换算 cubic(al) conversion
容积回收率 volume recovery
容积混合 volume mix
容积货(物) measurement cargo

容积积分 volume integral
容积基位 volume basis
容积激发 volume excitation
容积极限 volume limit
容积计 bulkmeter; volume ga(u)ge; volume meter; volumenometer; volumeter
容积计量供料装置 volume dosing unit
容积计量喂料 volumetric(al) feed
容积计量喂料机 volumetric(al) feeder
容积计量装置 volumetric(al) measuring device
容积减小 reduction in bulk; volume reduction
容积校正 capacity correction
容积校准 volumetric(al) proving
容积控制器 volumetric(al) governor; volumetric(al) regulator
容积量 cubic(al) capacity; measurement capacity; volumetric(al) capacity
容积量测法 measuring by volume
容积量斗 < 测定流量用的 > volume pit
容积流量 volume discharge; volume flow; volumetric(al) flow
容积流量计 volume-displacement meter; volume flowmeter
容积流量调节器 vlume-and-flow controller
容积流率 volume flow rate; volume rate of flow; volume rating of flow; volumetric(al) flow rate
容积率 floor area ratio; floor space index; plot ratio; volumetric(al) rate
容积密度 bulk density; volume density
容积免检 volume exemption
容积描记法 plethysmography
容积描记器 plethysmograph; recorder of volume
容积描记图换能器 plethysmogram transducer
容积模量 bulk modulus; volumetric-(al) modulus
容积模数 volumetric(al) modulus
容积摩尔浓度 molar concentration
容积浓度百分率 percentage by volume; volumetric(al) concentration percentage
容积配比 proportioning by volume
容积配合法 proportioning by volume
容积配料 proportioning by volume
容积配料斗 volumetric(al) batcher
容积配料法 batching by volume; volume mix method
容积膨胀系数 coefficient of volume expansion
容积频率曲线 volume frequency curve
容积平衡式伸缩节 volume balanced slip
容积区域 volumetric(al) district
容积曲线 volume curve
容积热容量 volumetric(al) heat capacity
容积-深度分布曲线 volume-depth distribution curve
容积式泵 displacement pump; positive-displacement pump
容积式地震计 capacity seismometer
容积式电加热器 electric(al) storage heater
容积式发动机 positive-displacement engine
容积式防护水表 displacement type meter
容积式风机 positive-displacement fan
容积式鼓风机 displacement blower
容积式计量阀 positive-displacement metering valve
容积式加热器 storage-type calorifier
容积式流量表 volumetric(al) meter [metre]
容积式流量计 displacement flowmeter; positive-displacement (flow) meter; volumetric (al) flowmeter; volumetric(al) meter[metre]
容积式煤气表 positive-displacement meter; positive meter; volumetric-(al) meter[metre]
容积式膨胀机 positive-displacement expansion engine
容积式热交换器 storage-type water heater
容积式热水锅炉 cubic(al) type hot water boiler
容积式水泵 volumetric(al) water pump
容积式水表 displacement meter; positive-displacement type water meter; volumetric(al) meter[metre]; volumetric(al) water ga(u)ge; volumetric(al) water meter
容积式压气机 displacement compressor; positive-displacement compressor
容积式压缩机 displacement compressor; positive-displacement compressor
容积式液压电动机 positive-displacement motor
容积式液压马达 positive-displacement hydraulic motor; positive-displacement motor
容积式增压器 displacement blower
容积式制冷机 positive-displacement refrigerator
容积室 dummy volume chamber
容积室换热器 volumetric(al) heat exchanger
容积收缩 volume shrinkage
容积松胀系数 < 堆石料开采的 > bulking factor
容积损失 volume loss; volumetric-(al) loss
容积弹性 elasticity of bulk; elasticity of volume; volume elasticity; volumetric(al) elasticity
容积弹性模量 bulk modulus of elasticity; modulus of elasticity of bulk; modulus of volumetric(al) elasticity; volume elastic modulus
容积弹性模数 modulus of elasticity of bulk; modulus of volumetric(al) elasticity; volume elastic modulus
容积调节器 volume governor; volumetric(al) regulator
容积调速 volumetric(al) control speed
容积图 capacity plan
容积稳定的 volumetrically stable
容积稳定度 bulk stability
容积系数 bulk coefficient; coefficient of volume; volume coefficient; volumetric(al) coefficient; volumetric-(al) factor
容积限制 space limit
容积效率 volume efficiency; volume effort; volumetric(al) efficiency
容积性质 volumetric(al) behavio(u)r
容积絮凝 volume flocculation
容积压缩因素 bulk factor
容积因素 bulk factor
容积应变 cubic(al) strain; volumetric-(al) strain
容积应力 volumetric(al) stress
容积余隙 clearance volume; volume clearance
容积与载重量 reach and burden
容积与载重之比 < 立方米/吨 > volume-to-load ratio
容积运费 freight by measurement
容积增大 volume gain
容积增大曲线 bulking curve
容积增大系数 bulking factor
容积增加 volume gain
容积增量 bulk increase
容积直径 volumetric(al) diameter
容积指示器 volume indicator
容积指数 volume index
容积质量关系 volume-mass relationship
容积重 volumetric(al) weight
容积重力仪 capacity gravimeter
容积重量 volume weight; weight by volume
容积重量百分数 weight by volume percent
容积重量表 measurement and weight list
容积贮存量 volumetric(al) storage
容积组成 volumetric(al) composition
容计频率 tolerance frequency
容解气体 dissolved gases
容抗 capacitance; capacitive impedance; capacitive reactance; capacity reactance; captance; condensance; condenser component; condenser reactance; condensive reactance
容克式发动机 Junker's engine
容克式量热器 Junker's calorimeter
容克式喷油泵 Junker's pump
容克式气体量热计 Junker's gas calorimeter
容克式水冷锭模 Junker's mo(u)ld
容克式水力测功器 Junker's water dynamometer
容克式水流量热器 Junker's water flow calorimeter
容克斯川式汽轮机 Ljungstrom steam turbine
容矿构造类型 type of ore-containing structures
容矿岩 host rock
容量 bulking figure; capability; capacity; capacity value; containment; content; cubic(al) capacity; cubic (al) measure; cubic(al) yardage; holding capacity; measure of capacity; volume capacity; volumetric (al) capacity; tankage < 一槽、一箱、一罐等的 > ; cubic(al) content
容量百分比 volume percent(age)
容量包装 capacity packing
容量比 capacitance ratio; capacity ratio
容量边 capacitive edge
容量变化 capacity variation
容量变化范围 capacity variable range
容量标准 volumetric(al) standard
容量表 volumeter
容量玻璃器皿 volumetric(al) glass (ware)
容量不等的群体 unequal clusters; unequal sized clusters
容量不足 off-capacity; undercapacity
容量不足的样本 undersized sample
容量测定 capacity measure; solid measure; volumetric(al) determination; volumetry
容量(测流)法 volumetric(al) method
容量测试 volume test
容量差异 capacity volume variance
容量常数 capacity constant
容量沉淀法 volumetric(al) precipitation method
容量储备 capacity reserve
容量大小 volume level
容量单位 unit of capacity
容量的 volumetric(al)
容量滴定法 volumetric(al) titrimetry
容量度 voluminosity
容量度量 measure of capacity
容量法 cubature; titration method; volumetric(al) procedure; volumetry
容量反应速度系数 volumetric(al) rate of reaction coefficient
容量范围 range of capacity
容量分析测定法 volumetric(al) determination
容量分析法 measure analysis; titrimetric analysis; volumetric(al) analysis; volumetry
容量分析试验 volumetric(al) analysis test
容量分析仪器 volumetric(al) analysis apparatus
容量风缸 capacity reservoir; volume reservoir
容量负荷 volumetric(al) loading
容量负载 bulk load
容量负载和调度程序系统 capacity loading and scheduling system
容量感受器 volume receptor
容量估计 capacity estimate
容量或重量检定费 metage
容量极限 capacity limitation
容量计 volume meter; volumeter; volumometer
容量计划系统 capacity loading and scheduling system
容量计算 capacity rating
容量计重 checking weight by volume
容量记录 capacity record
容量交换 capacity conversion
容量校准因素 volumetric(al) correction factor
容量-进水比 capacity-intake ratio
容量克分子浓度 volumetric(al) molar concentration
容量刻度 containing mark
容量利用 utilization of capacity
容量利用率 < 设备的 > rate of capacity
容量利用系数 < 公交车辆实际乘客量与所提供的容量之比 > capacity utilization factor
容量摩尔浓度 volumetric(al) molar concentration
容量膨胀 capacity expansion
容量膨胀系数 coefficient of cubic-(al) expansion
容量平衡模型 volume balance model
容量瓶 measuring flask; volumetric-(al) flask
容量器皿 volumetric(al) apparatus
容量曲线 capacity curve; storage curve; volume curve
容量试验 capacity test
容量水平 volume level
容量送料器 volumetric(al) feeder
容量损失 capacity loss
容量探示器 capacity probe
容量调节 capacity regulation; volume control
容量调节阀 capacity regulating valve
容量调节器 capacity regulator
容量调节装置 capacity regulating device
容量调整 displacement setting
容量维数 capacity dimension
容量吸移管 volumetric(al) pipet(te)
容量系数 capacity coefficient; capacity factor; volume efficiency; volume effort; volumetric(al) efficiency
容量限点 capacity point
容量限制 capacity limitation; capacity restraint
容量效率 volumetric(al) efficiency

容量型隔膜泵 volumetric(al) type diaphragm pump
容量修整系数 capacity correction factor
容量选择器 volume selector
容量压差曲线 capacity head curve
容量压头曲线 capacity head curve
容量因数 capacity factor;volumetric(al) factor
容量因子 capacity factor
容量有效荷载 capacity payload
容量与流速控制器 volume and flow controller
容量证明 capacity certification
容量置换计 volumetric(al) displacement meter
容量重量证明书 certificate and list of measurement and/or weight
容量注入机 volumetric(al) filler
容量装填法 volume filling
容量装填机 volumetric(al) filler;volumetric(al) filling machine
容量装填装置 volumetric(al) filling device
容量装置 volumetric(al) apparatus
容量纵剖面 capacity profile
容留所 shelter facility
容模的 molar
容纳 acceptance;contain;hold;recipience[recipiency]
容纳更多的劳动力 absorb more labo(u)r power
容纳量 carrying capacity;holdup;intake capacity
容纳流 carrying current
容纳埋入涵管的槽 planter
容纳能力 accommodation;hold capacity
容纳淤积能力 capacity of holding deposit
容耐试验 tolerance test
容气管 cross-over pipe
容气量 air capacity
容气器 gas container
容汽器 steam receiver
容器 vessel;container;containment;holder;outer casing;receiver(element);receptacle;recipient;repository;reservoir;store holder
容器本体 vessel reality
容器壁 vessel wall;wall of container
容器壁厚 wall thickness
容器壁孔上人孔 shell manway
容器壁同接管的装配 assembly of nozzle necks to vessel walls
容器变形 distortion of vessel
容器表面的曲率 curvature of vessel surface
容器表面缺陷 irregularities in vessel surface
容器玻璃 container glass
容器材料 container material;vessel material
容器侧 vessel side
容器测土壤的容积 container soil volume
容器衬垫 container liner
容器衬里 container lining
容器成本 container cost
容器充满部分 shell innage
容器的超压极限 overpressure limit for vessel
容器的界限点 termination point of a vessel
容器的球形部分 spheric(al) section of vessel
容器的弯曲 bend of vessel
容器的展开 unrolling of container
容器镀层 container coating
容器吨 vessel ton
容器法兰 vessel flange
容器浮造法 jar flo(a)tation
容器盖 vessel cover;mud ring <俚语>
容器隔间 conveyance compartment
容器罐拖车 tank trailer
容器环 container ring
容器混合机 container mixer
容器静态试验 chamber test
容器开口 vessel port
容器类别 category of vessel;vessel class
容器密封 seal(ing) of vessel
容器内壁保护层 container lining
容器内径 inside diameter of vessel
容器球形盖 domed tank roof
容器容量 container capacity;volume of vessel
容器设备 tankage
容器式量水计 displacement meter
容器试板 vessel test plate
容器试板的复试 retest of vessel test plate
容器填充高度指示器 cut-off level ga(u)ge
容器筒节 shell ring
容器外壳 container case;shell of tank
容器未充满部分 shell outage
容器压瘪 crushing
容器压力计 container manometer
容器压实 container compaction
容器摇架 container cradle
容器液位计 tank ga(u)ge
容器育苗苗圃 container nursery
容器栽植 container planting
容器在役时的安装 installation on vessels in service
容器罩 jar
容器支承 support of vessel
容器制造 manufacture of vessel
容器制造厂的出厂合格证明 certification by vessel manufacturer
容器中层的蒸汽 middle tank air
容器中状态 condition in container
容器重(量) tare(weight)
容器组件 container assembly
容让性材料 cushioning material
容忍次品比率 lot-tolerance percentage defective
容忍度 tolerance
容忍风险 risk-bearing
容忍剂量 tolerant dose
容水度 moisture capacity;specific moisture capacity;water capacity <单位水头变化引起的含水量变化>
容水量 entrance capacity;water-bearing capacity;water capacity
容酸性 acid acceptance
容体积度量法 cubic measure
容筒 volumetric(al) cylinder
容膝空档书桌 knee-hole desk
容膝孔 knee hole
容隙 allowance;tolerance
容限 allowance;margin;tolerable limit;tolerance(limit)
容限电压 margin voltage
容限频率 tolerance frequency
容箱 tank
容泄水流最底溶氧度 critical reach
容屑槽 chip flute
容性传感器 capacitive pick-up
容性电路 capacitive circuit
容性耦合 capacitive coupling
容性天线 condenser antenna
容许 admissible;permission;permit;sufferance;tolerance(allowance)
容许安全系数 allowable factor of safety
容许暴露极限 permissible exposure limit
容许曝光 allowable exposure
容许爆破 permitted explosive
容许爆炸 permitted explosive
容许背景水平 allowable background level
容许本底水平 allowable background level
容许闭合误差 closure tolerance
容许闭塞 permissive block
容许闭塞系统 permissive(block) system
容许闭塞信号法 permissive block signal(l)ing
容许闭塞制 permissive block;permissive(block) system
容许变分 admissible variation
容许变化 permissible variation;admissible variation <指尺寸>
容许变量 permissible tolerance;permissible variation
容许变位(量) allowable displacement
容许变形 allowable deformation;permit deformation
容许变形量 fairness limit
容许变异 admissible variation;allowable variation
容许标准 compliance criterion;permissible criterion;permissible level
容许表示灯 permissive light
容许不冲刷流速 allowable non-scouring velocity
容许不清晰度 permissible unsharpness
容许不淤流速 allowable speed of non-silt
容许采伐量 allowable cut
容许参数 admissible parameter;admittance parameter;allowable parameter
容许参数空间 admissible parameter space
容许残留量 allowable residue limit
容许残留值 tolerance value
容许策略 admissible strategy
容许层间变形 allowable storey deformation
容许差 allowable difference
容许差度 permissible deviation
容许差异沉降量 allowable differential settlement
容许差异额 tolerance
容许颤动 allowable flutter
容许长度 admissible length
容许超高 allowance over-height
容许超高欠量 permissible deficiency of superelevation
容许超宽 allowance over-width;horizontal tolerance
容许超深 allowable over-depth;vertical tolerance
容许超载 admissible overload;overload allowance;permissible overload
容许车辆尺寸 permissible vehicle dimensions
容许沉降 acceptable setting
容许沉降量 allowable settlement;allowance for settlement;permissible settlement;tolerable settlement
容许沉降量要求 requirement of tolerable settlement
容许沉落量 permissible settlement
容许沉陷 acceptable settlement
容许成本 allowed cost
容许承压力 allowable bearing pressure
容许承压应力 allowable bearing stress
容许承载 safe bearing load
容许承载力 allowable bearing capacity;allowable bearing power;allowable bearing pressure;allowable bearing value;permissible soil pressure;safe bearing capacity;safe bearing power
容许承载量 allowable bearing capacity
容许承载值 allowable bearing value
容许程度 allowable degree;tolerance degree
容许吃水 permissible draft;admissible draught
容许持续电流 allowable constant current
容许尺寸上限 maximum limit of size;upper limit of size
容许尺寸误差 acceptance off-size;dimensional tolerance;permissible dimensional error
容许尺寸下限 lower limit of size;minimum limit of size
容许齿隙游移 tolerable backlash
容许充装量 rated filling weight
容许冲击负荷 load impact allowable;load impact allowance
容许冲击负荷值 impact allowance
容许冲击荷载 allowable impact load;impact allowance load;load impact allowance
容许冲击量 impact allowance
容许冲剪 allowable punching shear
容许传输带宽 allowable transmissible bandwidth;transmissible bandwidth
容许疵病 allowed defect;tolerable defect
容许单位接触应力 allowable contact unit stress
容许单位结合应力 allowable bond unit stress
容许单位扭曲应力 allowable unit stress for buckling
容许单位扭应力 allowable unit stress for torsion
容许单位扭转应力 allowable twisting unit stress
容许单位土压力 allowable unit soil pressure
容许单位弯曲应力 allowable flexural unit stress;allowable unit stress for bending
容许单位压力 allowable unit pressure
容许单位压碎应力 allowable crushing unit stress
容许单位压应力 allowable compressive unit stress;allowable unit stress for compression
容许单位应力 allowable unit stress
容许单位支承应力 allowable bearing unit stress
容许单桩荷载 allowable pile bearing load
容许的 acceptable;allowable;facultative;permissible;permissive
容许的捕捉量 allowable catch
容许的反向峰值电压 permissible peak inverse voltage
容许的钢材类型 permissible for types of steels
容许的极限偏差 acceptable tolerance
容许的通话 permissible call
容许的误差 tolerance
容许的作用力 admissible action
容许灯光 permissive light
容许地貌变化 admissible relief variation
容许点负荷 permissible point-load
容许点荷载 permissible point-load
容许电流 allowable current;permissible current;safe current
容许电压 allowable voltage
容许电压试验 withstand test
容许定时制 permissive timing system

容许断裂应力 admissible stress to rupture
容许对准误差 permissible misalignment
容许恶化率 acceptable degradation rate;acceptable malfunction rate
容许范围 acceptability limit;allowable range;allowance limit;allowance range;permissible range;tolerable limit;tolerance range
容许放射性物质水平 acceptable level of radioactive material
容许分布 tolerance distribution
容许符号 admissible mark
容许幅度标准 safe haven
容许辐照限度 permissible exposure limit
容许负荷 accepted load;admissible load;allowable load(ing);permissible load;safe load
容许负荷量 carrying capacity;safe carrying capacity
容许负荷率 possible loading rate
容许负载 admissible load;allowable load(ing);permissible load;safe load
容许负载电流 carrying current
容许负载量 safe carrying capacity
容许高度 permissible height
容许工作荷载 permissible working load;safe working load(ing)
容许工作压力 allowable working pressure;safe working pressure;maximum working pressure
容许工作应力 permissible working stress
容许公差 accepted tolerance;allowable tolerance;permissible tolerance;tolerance limit
容许功率 allowable power
容许供给 permissible feed
容许估计 admissible estimate
容许估计量 admissible estimation
容许固体沉积物 possible precipitation of solid
容许固着力 allowable pull out capacity
容许故障等级 acceptable malfunction level
容许故障率 acceptable failure rate;acceptable malfunction rate
容许故障水平 acceptable malfunction level
容许挂接速度 permissive coupling speed
容许管内最大工作压力 <管线的> maximum allowable operating pressure
容许光谱 allowed spectrum
容许轨道 allowed orbit;permissible orbit
容许轨道不平整度 allowance of track irregularity
容许过载 permissible overload
容许含量 allowable concentration;permissible level;permissible tolerance
容许含铅量 lead tolerance
容许函数 admissible function;admittance function
容许耗损 permissive waste
容许合闸电流 allowable switching current
容许荷载 accepted load;admissible load;allowable load;charge of surety;permissible load;permitted load;safe bearing load;safe(ty) load(ing);safety weight;safe weight
容许恒定电流 allowable constant current
容许后继者 allowed successor
容许环境 acceptable environment
容许环境保护水平 acceptable level of environmental protection
容许环境极限 acceptable environment(al) limit
容许环境限度 acceptable environment(al) limit
容许环境影响 acceptable environment(al) impact
容许灰尘浓度 permissive limit of dust concentration
容许回弹弯沉 allowable rebound deflection
容许回弹弯沉值 allowable rebound deflection value
容许混合区 permited mixing area
容许活载 acceptable live load
容许基准 permissible criterion[复 criteria]
容许畸变 tolerable distortion
容许极限 acceptable limit;acceptance limit;allowable limit;allowance limit;margin;permissible limit;tolerance limit
容许极限定律 <环境因素或条件的> law of tolerance
容许集 admissible set
容许集中度 acceptable concentration
容许集中荷载 permissible point-load
容许剂量 acceptable dose;permissible dose;tolerance dose
容许加工偏差 permissible manufacturing deviation
容许加速度(值) allowable acceleration
容许假设 admissible hypothesis
容许间隙 admissible clearance;allow clearance;safety clearance;tolerance
容许剪力 allowable shear;permissible shear
容许剪应力 allowable shearing stress
容许检验 admissible test
容许降度 permissible drop
容许降落深度 permissible drawdown
容许降(落)水位 permissible drawdown
容许接触应力 allowable contact stress
容许接点磨损 contact wear allowance
容许接近速度 permission approach speed;permissive approach speed
容许结构 admissible structure
容许结合应力 allowable bond stress
容许解法 admissible solution;feasible solution;permissible solution
容许界限 allowable limit;tolerance limit;tolerance range
容许紧急剂量 acceptable emergency dose
容许近似值 allowable approximation
容许进入 admission
容许进水率 permissible rates of water entry
容许浸水深度 allowable flooding
容许经济限度 permissible economic limit;permissive economic limit
容许精度 acceptable precision;permissible precision
容许净承压力 allowable net bearing pressure
容许净承载力 allowable net bearing pressure
容许开采量 permissible yield
容许颗粒尺寸 acceptable particle size
容许可靠性程度 acceptable reliability level
容许可靠性水平 acceptable reliability level
容许客舱容量 allowable cabin load
容许控制 admissible control
容许控制系统 admissible control system
容许扣除额 allowable deduction
容许宽度 allowable width
容许拉力 allowable tension
容许拉应力 allowable tensile stress
容许缆索拉力 permissible line pull
容许离差 admissible deviation;tolerance deviation
容许离线偏差 permissible lateral deviation
容许利用度 allowable use
容许连挂速度 <溜放车辆> Acceptable coupling;acceptable coupling speed
容许连接速度 permissive coupling speed
容许量 allowance;limit of tolerance;permissible dose;throughput;tolerance
容许量标准 tolerance standard
容许量范围 tolerance domain;tolerance range
容许裂缝宽度 allowable crack width;permissible crack width
容许邻居使用的私人道路 occupation road
容许磷负荷 acceptable phosphorus loading
容许流量 permissible discharge
容许流速 acceptable flow;acceptable velocity;allowable velocity;allowance velocity;permissible velocity;safe(ty) velocity
容许硫含量 permissible sulfur
容许漏失量 allowable leakage
容许路签 permissive staff
容许路签附加装置 permissive attachment
容许轮胎载荷 allowable tire load
容许慢行标志 slow-speed permissive sign
容许毛重 allowable gross weight
容许磨耗 tear-and-wear allowance;wear allowance
容许磨损 wear allowance
容许磨损量 allowable wear;permissible wear
容许内插约束 admissible interpolation constraint
容许耐力 permissible tolerance
容许挠度 allowable(rebound) deflection;allowance deflection;permissible deflection
容许挠性 permissible flexibility
容许能带 allowed energy band
容许能级 allowed energy level
容许能力 throughput
容许能谱形状 allowed spectrum shape
容许黏[粘]度 tolerable viscosity
容许黏[粘]结力 allowable bond
容许黏[粘]结应力 allowable bond stress
容许黏[粘]着应力 allowable bond stress
容许扭矩 allowable torque
容许扭曲应力 allowable buckling stress
容许扭应力 allowable torsional stress
容许扭转负荷 torsional capacity
容许扭转应力 allowable twisting stress
容许浓度 acceptable concentration;admissible concentration;allowable concentration;permissible concentration
容许浓度水平 allowance level of concentration
容许浓度限度 acceptable concentration limit;allowable concentration limit;permissible concentration limit
容许浓度指数 allowable concentration index;allowance concentration index
容许排放 allowable emission;permissible discharge
容许排放流量 allowable emission rate
容许排放率 allowable emission rate
容许排污率 allowable emission rate
容许判决函数 admissible decision function
容许偏差 allowable deviation;allowable variation;allowance deviation;permissible deviation;permissible deviation;tolerable deviation;tolerance deviation;tolerance limit;tolerant deviation
容许偏小误差 negative allowance
容许偏心距 acceptable eccentricity
容许漂移 tolerable fluctuation
容许频率变化 tolerable frequency variation
容许平均最大压力 permissible mean maximum pressure
容许凭证闭塞机 permissive token block instrument;permissive token instrument
容许凭证机 permissive token instrument
容许坡度误差 slope tolerance
容许破坏概率 allowable failure probability;permissible probability of failure
容许破坏机理 admissible failure mechanism
容许谱线 permitted line
容许曝露极限 permissible exposure limit
容许曝噪时间 allowable noise exposure time
容许汽车连接总重量 permissible gross combination weight
容许前置符 allowed predecessor
容许欠超高 permissible deficiency of superelevation
容许强度 allowable strength;allowance strength
容许切削速度 permissible cutting speed
容许倾斜度 allowable heeling
容许区间 tolerance interval
容许区域 admissible region;tolerance region
容许曲面 admissible surface
容许曲线 admissible curve
容许缺陷 allowable defect
容许缺陷标准 acceptable defect level
容许缺陷数目 tolerance number of defects
容许热限度 allowable heat limit
容许日摄入量 acceptable daily intake
容许容量 allowable capacity
容许蠕变 allowable creep strain
容许蠕变应变 permissible creep strain
容许色差 acceptability of colo(u)r match;colo(u)r tolerance
容许上限 high limit of tolerance
容许设计 allowable design;permissible design
容许伸长度 allowable elongation
容许深度 allowable depth
容许渗漏 allowable leakage
容许渗漏准则 allowable seepage criterion[复 criteria]
容许渗水量 allowable water leakage
容许时间 allowed time

R

容许使用荷载 allowable working load; permissible working load
容许示像 permissive aspect
容许事故等级 acceptable malfunction level
容许试样量 allowable sample size
容许适应率 permissible response rate
容许收缩量 allowable for shrinkage; shrinkage allowance
容许输出水平 permissive output level
容许输入 acceptable input
容许输水流量 transferable discharge permit
容许数 admissible number
容许竖向误差 vertical tolerance
容许水平 acceptable level
容许水平误差 horizontal tolerance
容许水位降落 permissible drawdown
容许水温 acceptable water temperature
容许速度 acceptable velocity; allowable speed; allowable velocity; allowance velocity; permissible speed; permissible velocity
容许速率 permissible speed
容许损耗 permissible loss
容许讨论的问题 open question
容许调节 permissible control
容许调整工资制 permissive-wage adjustment
容许停车 permissive stop
容许通量 tolerance flux
容许通行车间隔 allowable space between running cars
容许通行证 permissive card
容许统计假设 admissible statistical hypothesis
容许透射比 <隧道> admissible transmittance
容许土(壤)承载力 bearing capacity of permissible soil; permissible soil bearing capacity; permissible soil bearing pressure; permissible soil pressure
容许土壤流失量 soil loss tolerance
容许土压力 allowable earth pressure; allowable soil pressure; permissible soil pressure
容许弯沉 allowable deflection; permissible deflection
容许弯曲单位应力 allowable flexural unit stress
容许弯曲曲率半径 allowable bending radius
容许弯曲应力 allowable flexural stress
容许危险性 accepted risk; tolerable risk
容许位移 allowable displacement; allowable drift
容许温差 allowable temperature differential
容许温度 allowable temperature; permissible temperature
容许温升 rated temperature-rise
容许温升电流 rated temperature-rise current
容许握裹应力 allowable bond stress
容许污染 permissible contamination
容许污染极限 permissible limit of pollution
容许污染水平 allowable contamination level
容许污染物浓度 admissible pollutant concentration
容许污染物排放量 allowable quantity of pollutant discharged
容许污染限度 permissible limit of pollution
容许误差 acceptance deviation; admissible error; allowable error; allowance error; leeway; margin of tolerance; permissible error; tolerance
容许误差范围 limit of allowable error; range of allowable error
容许误差级判据【数】acceptable defect level measure
容许误差极限 limit of admissible error; limit of allowable error; limit of permissible error
容许误差限度 acceptable error limit
容许误差修正量 allowance
容许吸出高度 permissible draft head
容许吸出水头 permissible draft head
容许吸附质 possible adsorbate
容许吸光系数 allowable extinction coefficient
容许吸入量 breathing tolerance
容许系数 tolerance factor
容许显示 permissive aspect
容许限度 acceptable limit; allowable limit; permissible limit; tolerable limit; tolerable range; tolerance limit; tolerance range
容许限量 permissible error; permissible limit
容许相对变形 allowably relative deformation
容许消光系数 extinction coefficient of allowable smoke concentration
容许消落深度 permissible drawdown
容许泄降深度 permissible drawdown
容许泄漏量 allowable leakage
容许泄水量 permissible discharge
容许卸货时间 time of discharge allowed
容许信号 permissive signal; tonnage signal
容许信号机【铁】permissive signal
容许信噪比 limit of signal to noise ratio
容许行动 admissible action
容许形状 admissible set
容许修建柱基(础)的土地 column lots
容许锈蚀(损耗) allowance for corrosion
容许压力 admission pressure; allowable pressure; authorised[authorized] pressure; licenced pressure; permissible pressure; safe pressure; set pressure; allowable compression
容许压力差 allowable pressure difference; permissible pressure difference
容许压强 admission pressure; allowable pressure; authorized pressure; permissible pressure; safe pressure
容许压碎应力 allowable crushing stress
容许压缩比 allowable compression ratio; allowance compression ratio
容许压应力 allowable compressive stress
容许烟雾浓度 allowable smoke concentration
容许延迟 delay allowance
容许延期付款 indulge
容许摇摆量 <桥梁设计用> lurching allowance
容许应变 permissible strain; strain tolerance
容许应力 admissible stress; allowable stress; permissible stress; permitted stress; proof stress; safe(ty) stress; safe working stress
容许应力法 allowable stress method
容许应力设计 allowable stress design; permissible stress design; working stress design
容许应力设计法 allowable stress design method; permissible stress design method; permissible allowable stresses method; working stress method
容许应力提高系数 increasing coefficient of allowable stress
容许营养负荷 allowable nutrient loading
容许硬度 tolerance on hardness
容许渔获量 permissible yield
容许淤积量 allowance for siltation
容许预测函数 admissible prediction function
容许域 tolerance domain
容许跃进 permitted transition
容许跃迁 allowable transition; allowed transition; permitted transition
容许运输期限 permissive period of transport
容许运行 permissive run
容许运行地震 operating basis earthquake; operating level earthquake
容许运行温度 safe operating temperature
容许载货量 allowable cargo load
容许载货重量 permitted payload
容许载重 admissible load; allowable load; permissible load
容许载重量 legal payload
容许噪声 allowable noise
容许噪声暴露时间 allowable noise exposure time
容许噪声标准 acceptable noise level; permissible noise level
容许噪声级 acceptable noise level; acceptable sound level; permissible noise level
容许噪声接触时间 allowable noise exposure time
容许胀隙 allowed space for expansion
容许照射 allowable exposure; permissible exposure
容许照射剂量 radiation tolerance
容许真空 marginal vacuum
容许振动极限 permissible limit of vibration
容许振动加速度 allowable vibrating acceleration; allowable vibration acceleration
容许振幅 acceptable amplitude; allowable amplitude
容许正偏差 accommodate positive tolerance
容许支承力 allowable bearing capacity
容许支承压力 allowable bearing pressure; allowable pressure
容许支承应力 allowable bearing stress
容许直径 admissible diameter
容许值 acceptable value; admissible value; allowable value; allowed value; permissible value
容许质量标准 acceptable quality level
容许质量负荷 allowable mass load
容许质量级 acceptable quality level
容许轴荷载 admissible axle-load
容许转变 allowed transition
容许转矩 allowable torque
容许转数 permissible revolutions
容许转速偏差 speed tolerance
容许装载时间 loading time allowed
容许状态 competent condition
容许资产 admitted assets; estate at sufferance
容许子群 allowable subgroup
容许自动闭塞系统 permissive automatic block system
容许自动闭塞制 permissive automatic block system
容许总重 allowable gross weight
容许阻力 allowable resistance
容许最大沉降量 allowable maximum settlement
容许最大成本 maximum allowable cost
容许最大工作压力 maximum allowable working pressure
容许最大流速 maximum allowable velocity
容许最大浓度 maximum allowable concentration
容许最大坡度 maximum allowable slope
容许最大水泥含量 maximum cement content
容许最大限额 allowable maximum amount
容许最大运行压力 maximum allowable operating pressure
容许最低误差 lower tolerance
容许最高点误差 upper tolerance
容许最高速度 permissible maximum speed
容许最高污染水平 maximum contamination level
容许最稀站网 minimum network
容许最优控制 admissible optimal control
容许坐标 allowable coordinates
容易变成粉的 powdery
容易辨认的 well-identifiable
容易操作的 idiot proof
容易拆卸的 easy off
容易产生冻胀的 frost susceptible
容易产生误差的 error-prone
容易达到目的的途径 high road
容易堵孔 <筛子> proneness to clogging
容易堵塞 <筛子> proneness to clogging
容易发生事故 exposed to accident possibilities
容易防治的杂草 easily controlled grasses
容易风化的 eugeogenous
容易感觉到 easily noticeable
容易感染的 liable to infection
容易加工的玻璃 easily workable glass
容易加工的岩石 freestone
容易检修 easy servicing
容易胶黏[粘] inclined to puddle
容易接触的 exposed to contact
容易接近 easy access
容易进入的 readily accessible
容易控制程度 ease of control
容易抛光的 polish-susceptible
容易受虫害的 liable to insect injury
容易受灾地区 disaster-prone area
容易受灾害区域 disaster-prone region
容易刷的涂料 slip under the brush
容易塌落的 free caving
容易脱模模板 collapsible formwork
容易挖掘 easy digging
容易维护 easy maintenance
容易洗去的织物油墨 washout ink
容易运输 easy transport
容易种植 easy to raise
容易自然塌落的 free caving
容油量 oil capacity
容载限点 capacity point
容胀间隙 gap allowed for expansion
容重 apparent density; box weight; bulk density; bulking figure; bulk specific gravity; gravity of volume;

R

load-carrying ability; unit weight; volume density; volumetric (al) weight; volume weight; voluminal mass; weight per unit volume
容重测定法 volumenometer method
容重法测流 gravimeter measurement
容重控制 density control
容重试验 unit weight test
容阻时间常数 capacity resistance time constant

溶残席 <洞穴中方解石溶解残余物> palette; shield

溶槽 fluid bowl; karst trench; solution groove
溶出 digest; dissolve out; stripping
溶出残渣 digestion residue; digestion slime
溶出分析 stripping analysis
溶出伏安法 stripping voltammetry
溶出酸性水 stripped sour water
溶出物 dissolving-out substance
溶出系统 digestion series
溶道 solution channel
溶滴 fused drop
溶靛素红紫 indigosol red-violet
溶靛素金黄 indigosol golden yellow
溶靛素印染黑 indigosol printing black
溶靛素印染蓝 indigosol printing blue
溶冻土壤 tabet soil
溶洞 karst cave; solution cave; solution cavity; solution channel; solution crevice; solution opening
溶洞充水矿床 mineral deposit of karst grotto
溶洞的位置 abra position
溶洞调查 investigation of karst cave
溶洞顶板厚度 strata thickness above karst
溶洞横剖面图 transverse section of grotto
溶洞连通试验 karst cave connection test
溶洞平面图 grotto plan
溶洞泉 exsurgence; solution channel spring
溶洞水 cavern flow; cavern water; solution channel water
溶洞稳定性分级 grade of cave stability
溶洞岩石 cavern rock
溶洞纵剖面图 longitudinal section of grotto
溶斗 dolina [doline]; sink hole; solution funnel
溶度 solubility
溶度参数 solubility parameter
溶度等温线 solubility
溶度分析 solubility analysis
溶度焊配件 solvent weld fitting
溶度积 solubility product
溶度积常数 solubility product constant
溶度积规则 rule of solubility product
溶度积原理 solubility product principle
溶度计 lysimeter
溶度剂水分 solvent water
溶度间隔 solubility gap
溶度曲线 solubility curve
溶度商 solubility quotient
溶度系数 coefficient of solubility; solubility coefficient
溶度效应 solubility effect
溶度指数 solubility index
溶方解石 luo-calcite
溶峰 dissolved peak
溶敷金属 deposited metal
溶敷系数 deposition coefficient
溶沟 karren; karst ditch; solution channel; solution groove
溶垢剂 scale solvent
溶谷 karst valley; solution valley
溶管 pipe
溶焊缝 solvent-welded joint
溶焊助焊板 runoff plate
溶合 colliquefaction
溶合作用 solvation
溶痕 corrosion mark
溶化 break (ing)-up; defreeze; deliquesce; dissolution; dissolve; lysis; sleak(ing); solvation
溶化的 dissolved; lytic
溶化分数 leaching fraction
溶化锅 dissolution boiler
溶化热 heat of fusion
溶化烧蚀 flame ablation
溶化水 medium water
溶化土层 mollisol
溶化物 dissolved matter
溶化性 deliquescence
溶混能力 miscibility
溶混性 <气体与液体的可混合程度> miscibility
溶混性间隔 miscibility gap
溶剂 dissolvant; menstruum [复 menstruums/menstrual]; resolver; solution medium; solvate; solvent; thinner
溶剂包藏作用 solvent inclusion
溶剂饱和的 solvent saturated
溶剂泵 solvent pump
溶剂比 ratio of solvent
溶剂补充 solvent makeup
溶剂不平衡 solvent imbalance
溶剂擦洗法 solvent wipe-off method
溶剂参数 solvent parameter
溶剂槽 flux bath; solvent trough
溶剂层 solvent layer
溶剂成型 solvent mo(u)lding
溶剂程序变换器 solvent programmer
溶剂池 solvent cell
溶剂冲击 solvent shock
溶剂抽出物 solvent extract; solvent extractable matter
溶剂抽提 solvent extraction
溶剂储槽 solvent feed tank
溶剂纯化 solvent purification
溶剂萃取 extraction into solvent; solvent exhaustion; solvent extraction
溶剂萃取的 solvent extracted
溶剂萃取发生器 solvent extraction generator
溶剂萃取法 solvent extraction
溶剂萃取分离 solvent extraction separation
溶剂萃取过程 solvent extraction process
溶剂萃取监测器 solvent extraction monitor
溶剂萃取器 solvent extraction contactor
溶剂萃取塔 solvent extraction tower
溶剂萃取同位素发生器 solvent extraction generator
溶剂萃取物 solvent extract
溶剂萃取物分析 solvent-extract analysis
溶剂萃取油 solvent extracted oil
溶剂萃取原液 aqueous feed
溶剂萃取柱 solvent extraction column
溶剂萃取装置 solvent extraction plant
溶剂存留性 solvent retention
溶剂的精制设备 refining plant for solvents
溶剂的盐析 salting-out of solvent
溶剂点滴试验 solvent spot test
溶剂冬化法 solvent winterization
溶剂对油比 solvent oil ratio
溶剂二次处理抽出物 solvent re-treated extract
溶剂法 solvent method
溶剂法制塑料薄片 solvent laminating
溶剂反萃取 solvent stripping
溶剂防毡合整理法 solvent anti-felt finishing
溶剂分解 solvent cracking
溶剂分解作用 solvolysis
溶剂分离法 solvent segregation
溶剂分离器带水计量槽 solvent separator with calibrated tank for water
溶剂分馏 solvent fractionation
溶剂分配 solvent partitioning
溶剂分散 solvent dispersion
溶剂分析 solvent analysis
溶剂分子 solvent molecule
溶剂浮选 solvent flo(a)tation
溶剂辐解 solvent radiolysis
溶剂腐蚀法 solvent etching
溶剂负荷 solvent loading
溶剂干燥 solvent drying; solvent seasoning
溶剂过敏 solvent sensitivity
溶剂焊接 solvent weld(ing)
溶剂合物 solvate
溶剂化度 solvability
溶剂化反应 solvolytic reaction
溶剂化能 energy of solvation
溶剂化能力 solvability
溶剂化物 solvate; solvolyte
溶剂化显色(现象) solvatochromism
溶剂化学说 solvate theory
溶剂化皂 solventized soap
溶剂化作用 solvation
溶剂挥发量低的涂料 low-emission coating
溶剂挥发型丁基橡胶密封膏 solvent release butyl sealant
溶剂挥发造型法 solvent mo(u)lding
溶剂回收 solvent reclamation; solvent recovery; solvent recuperation
溶剂回收厂 solvent recovery plant
溶剂回收法 solvent recovery process
溶剂回收器 solvent reclaimer
溶剂回收塔 solvent recovery column
溶剂回收柱 solvent recovery column
溶剂回收装置 solvent recovery plant
溶剂混合料 solvent mix(ture)
溶剂混合物 solvent mix(ture)
溶剂混合油 miscella
溶剂混溶法 solvent mixing method
溶剂活化胶粘剂 solvent-activated adhesive
溶剂活性胶 solvent-activated adhesive
溶剂基 solvent base
溶剂基黏[粘]合剂 solvent-based adhesive
溶剂基涂料清除剂 solvent-based paint remover
溶剂激发 solvent excitation
溶剂己烷 solvent hexane
溶剂夹带物 solvent entrainment
溶剂监测 solvent monitoring
溶剂浆液萃取 solvent-in-pulp
溶剂胶 solvent adhesive
溶剂胶结材料 solvent cement
溶剂胶结料连接接头 solvent cemented joint
溶剂结晶法 solvent crystallization
溶剂解 lyolysis; solvolysis
溶剂浸提作用 solvent extraction
溶剂精炼煤法 solvent refining coal method
溶剂精制 solvent refining; solvent treating; solvent treatment
溶剂精制产物 solvent raffinate
溶剂精制厂 solvent treating plant
溶剂精制法 solvent process
溶剂精制剂 solvent refining agent
溶剂精制油 solvent refined oil
溶剂聚合法 solvent polymerization method
溶剂聚合膜 solvent polymeric membrane
溶剂可萃取有机物 solvent-extractable organics
溶剂可溶性 solvent-soluble
溶剂空白 solvent blank
溶剂矿浆萃取 solvent-in-pulp
溶剂阔幅连续处理设备 solvent wide-continuous plant
溶剂冷凝器 condenser for solvent; solvent condenser
溶剂离解作用 solvation; solvolytic dissociation
溶剂量 solvent amount
溶剂料斗 solvent hopper
溶剂料液 solvent feed
溶剂敏感的 solvent susceptible
溶剂敏感性胶粘剂 solvent sensitive adhesive
溶剂模塑法 solvent mo(u)lding
溶剂膜 solvent membrane
溶剂黏[粘]度 solvent viscosity
溶剂黏[粘]合 solvent bonding
溶剂黏[粘]接 solvent weld(ing)
溶剂黏[粘]结 solvent cementing
溶剂浓度 solvent strength
溶剂偶 solvent pairs
溶剂抛光 <漆面的> pulling over
溶剂漂白 solvent bleaching
溶剂平衡 solvent balance
溶剂漆 solvent paint
溶剂气割 flux
溶剂气焊 flux
溶剂气体焙固整理 solvent vapo(u)r cure finishing
溶剂汽油 petroleum naphtha; solvent naphtha
溶剂前沿 solvent front
溶剂强度 solvent strength
溶剂侵蚀 bite
溶剂清洗 solvent cleaning
溶剂染料 solvent dye
溶剂染色法 solvent dyeing process
溶剂容忍度 solvent tolerance
溶剂容许溶解度 solvent tolerance
溶剂溶化黏[粘]合剂 solvent-borne adhesive
溶剂溶化皂 solvent soluble soap
溶剂溶解力 solvent power
溶剂溶解染料 solvent soluble dye
溶剂溶胀 solvent swell
溶剂入口 solvent inlet
溶剂石脑油 solvent naphtha
溶剂释放性 solvent release
溶剂收集槽 solvent catch tank
溶剂输送系统 solvent delivery system
溶剂水分 solvent water
溶剂损失 solvent loss
溶剂塔 solvent column; solvent tower
溶剂提出物 solvent extract
溶剂提取 solvent extraction
溶剂提取法 solvent extraction method; solvent extraction process
溶剂提取塔 solvent extract tower
溶剂提取脱盐 desalination by solvent extraction
溶剂停流 solvent immobilization
溶剂同位素效应 solvent isotope effect
溶剂涂层 solvent coat(ing)
溶剂脱蜡 solvent dewaxing
溶剂脱蜡法 solvent dewaxing process
溶剂脱沥青 solvent deasphalting
溶剂脱树脂法 solvent deresining
溶剂脱碳 solvent decarbonizing

溶剂脱碳过程 solvent decarbonizing process
溶剂脱油 solvent deoiling
溶剂脱脂 solvent degreasing
溶剂稳定化算符 solvent stabilization operator
溶剂稳固的 solvent-fast
溶剂污染 solvent contamination; solvent pollution
溶剂吸附体 solvent absorber
溶剂吸附作用 lysorption
溶剂吸着 solvent uptake
溶剂稀释极限 solvent tolerance
溶剂稀释型防锈剂 solvent cut-back type rust preventive
溶剂稀释型涂料 solvent thinned coating
溶剂洗涤 solvent scouring; solvent washing
溶剂洗涤剂 solvent detergent
溶剂洗涤器 solvent washer
溶剂洗净剂 solvent cleaner
溶剂洗提 solvent washing
溶剂洗提效应 solvent washing effect
溶剂效率 solvent efficiency
溶剂效应 solvent effect
溶剂芯焊条 flux core-type electrode
溶剂型丙烯酸密封膏 solvent acrylic sealant
溶剂型防水涂料 solvent type waterproofing paint
溶剂型封口 solvent-based compound
溶剂型环氧树脂清漆 solvent epoxy varnish
溶剂型建筑涂料 solvent architectural coating
溶剂型胶黏[粘]料 solvent cement
溶剂型胶粘剂 solvent adhesive; solvent cement; solvent-activated adhesive
溶剂型黏[粘]合 solvent adhesive
溶剂型涂料 solvent-based coating; solvent paint; solvent type coating
溶剂型脱模剂 solvent release agent
溶剂型有机硅清漆 solvent silicone varnish
溶剂型增塑剂 solvent type plasticizer
溶剂性胶粘剂 solvent adhesive
溶剂性渗色 solvent bleeding
溶剂性涂布 solvent coating
溶剂性银纹 solvent induced crazing
溶剂选择法 selective solvent method
溶剂选择分离 solvent fractionation
溶剂选择分析 <用以测定焦油沥青中不溶成分> selective solvent analysis
溶剂选择性 solvent selectivity
溶剂冶金 lyometallurgy
溶剂移动距离 solvent migration-distance
溶剂用脉石英 vein quartz for flux
溶剂用石英岩 quartzite for flux
溶剂油 megilp; solvent naphtha; solvent oil
溶剂与液体比 solvent/liquid ratio
溶剂运输船 solvents carrier
溶剂载体比 solvent support ratio
溶剂载体的交换体系 solvent carrier exchange system
溶剂再生 solvent regeneration
溶剂再生法 solvent regeneration process
溶剂蒸发 solvent evapo(u)ration
溶剂蒸馏液 solvent distillate
溶剂蒸气 solvent vapo(u)r
溶剂蒸气处理过程 solvent vapo(u)r treatment process
溶剂滞留 solvent hold-up
溶剂滞留性 solvent retention
溶剂中掺和物 solvent blend
溶剂组成 solvent composition
溶剂组配沥青 extracted and group blended asphalt; solvent extracted and group blended asphalt
溶胶 colloidal sol; collosol; sol; suspensoid
溶胶化作用 solation
溶胶沥青 bitusol
溶胶粒子 sol particle
溶胶凝胶 sol-gel
溶胶凝胶变换 sol-gel transformation
溶胶凝胶法 sol-gel method
溶胶凝胶过程 sol-gel process
溶胶凝胶技术 sol-gel technique
溶胶凝胶转变 sol-gel transformation
溶胶溶液 sol solution
溶胶体橡胶 sol rubber
溶胶涂装法 sol coating(method)
溶胶原 procollagen
溶胶原的 collagenolytic
溶接 cementing
溶解 colliquation; decomposition; digestion; dissolve; dissolving; hold in solution; liquefy; solution; solvation
溶解本领 dissolving power; solution power; solvency
溶解崩塌 solution collapse
溶解变形的 lytomorphic
溶解波痕 solution ripple mark
溶解采矿法 solution mining
溶解槽 dissolving tank; solution tank
溶解产物 lysate
溶解沉淀机理 dissolution precipitation mechanism
溶解沉析传质机理 dissolution reprecipitation material transfer mechanism
溶解沉陷 solution subsidence
溶解成分 dissolved constituent
溶解池 dissolving tank; solution basin; solution tank
溶解传感器 dissolved sensor
溶解氮 dissolved nitrogen
溶解的 dissolved; dissolvent; lytic; solvent
溶解的空气 dissolved air
溶解的气体 dissolved gas
溶解的水 dissolved water
溶解的无机物 inorganic dissolved substance
溶解的无机组分 inorganic dissolved component
溶解的物理显影 solution physical development
溶解电解槽 dissolution cell
溶解动力学试验速度方程 dissolution kinetics test velocity equation
溶解动力学试验方法 dissolution kinetics test method
溶解度 dissolubility; dissolvability; solubility; solubleness
溶解度表 solubility table
溶解度参数 solubility parameter
溶解度测定 solubility test
溶解度定律 solubility law
溶解度法 solubility method
溶解度积 solubility product
溶解度极限 solubility limit
溶解度曲线 solubility curve; solvus
溶解度试验 dissolubility test; solubility test(ing)
溶解度温度曲线 solubility-temperature curve
溶解度温度系数 temperature coefficient of solubility
溶解度与 pH 函数关系图 diagram showing the solubility as a function of pH
溶解度指数 solubility exponent
溶解二价阳离子浓度 dissolved divalent cation concentration
溶解法 dissolution method
溶解反应 dissolution reaction
溶解方法 method of attack; solvent process
溶解辐射物种 dissolved cadmium species
溶解腐殖质 dissolved humic materials
溶解负载 solution load
溶解汞物种 dissolved mercury species
溶解固体 deliquescent solid; dissolved solid
溶解固体浓度 dissolved solid concentration
溶解硅 dissolved silicon
溶解过程 solvent process
溶解荷载 dissolved load
溶解湖 solution lake
溶解化学需氧量 chemical oxygen demand dissolved
溶解回收法 solution reclaiming method
溶解活性磷 dissolved reactive phosphorus
溶解活性磷酸盐 dissolved reactive phosphate
溶解机 dissolver
溶解集合 solvent polymerization
溶解剂 deliquescent material; dissolver; lytic agent
溶解加速剂 solutizer
溶解角砾岩 solution breccia
溶解节理 solution joint
溶解金属离子 dissolved metal ion
溶解金属浓度 dissolved metal concentration
溶解开采 solution mining
溶解空气浮选法 dissolved air flo(a)tation
溶解孔隙 solution porosity
溶解矿物质 dissolved mineral
溶解离子 dissolved ion
溶解理 solution cleavage
溶解理论 solution theory
溶解力 solution power; solvency power
溶解量【地】 dissolved load
溶解料 deliquescent material
溶解裂缝 solution crack
溶解裂隙 solution channel; solution fissure
溶解磷 dissolved phosphorus
溶解硫 dissolved sulfur[sulphur]
溶解硫化物 dissolved sulfide
溶解氯化钠 dissolved sodium chloride
溶解氯化物活性 dissolved chloride activity
溶解氯离子 dissolved chlorine ion
溶解面 solution plane
溶解木质素 dissolved lignin
溶解能力 dissolving capacity; solvability; solvency (power); solvent power; dissolving power
溶解浓度 dissolved concentration
溶解配位体 dissolved ligand
溶解盆地 solution basin
溶解期 breaking-in period; resolution stage; stage of dissolution
溶解气 dissolved gas; gas in solution; solute gas; solution gas
溶解气驱动 solution gas drive
溶解气驱油藏 solution gas drive pool
溶解气驱油田 solution gas drive pond
溶解气体 deliquescent gas; solute gas
溶解气体成分 dissolve gas component
溶解气体驱动 dissolved gas drive
溶解气油藏 solution gas reservoir
溶解器 dissolver
溶解亲水性有机物 dissolved hydrophilic organic substance
溶解氢 dissolved hydrogen
溶解去污剂 solvent removes
溶解圈 solusphere
溶解燃料电池 dissolved fuel cell
溶解热 dissolving heat; heat of dissolution; heat of solution; integral heat of solution; solution heat; total heat of solution
溶解热法 heat of solution method
溶解溶环氧涂层 fused epoxy coating
溶解溶质 dissolved solute
溶解生物需氧量 biologic(al) oxygen demand dissolved
溶解疏水性有机物 dissolved hydrophobic organic substance
溶解速度 dissolution rate; dissolution speed; solution rate
溶解速率 dissolution rate
溶解损失量 loss by solution
溶解塌陷 solution subsidence
溶解态 dissolved form; dissolved state
溶解态铝 soluble alumin(i)um
溶解态组分 aqueous species
溶解碳 dissolved carbon
溶解碳酸盐碳 carbon from dissolved carbonate
溶解陶瓷 solution ceramics
溶解微分热 partial heat of solution
溶解微量金属 dissolved trace metal
溶解微生物的 microbivorous
溶解温度 dissolution temperature; solution temperature
溶解瓮 dissolving vat
溶解污染物 dissolved contaminant
溶解污染物浓度 dissolved contaminant concentration
溶解污染物输移 dissolved contaminant transport
溶解无机氮 dissolved inorganic nitrogen
溶解无机固体 dissolved inorganic solids
溶解无机磷 dissolved inorganic phosphorus
溶解无机碳稀释系数 dilution coefficient of dissolved carbon
溶解无机物的去除 removal of dissolved inorganic substance
溶解无机物质 dissolved inorganic substance
溶解无机盐 dissolved inorganic salt
溶解无机质 dissolved mineral
溶解物 dissolved solids
溶解物质 deliquescent matter; dissolved matter; dissolved substance; solute
溶解物总量 total dissolved solid value
溶解吸附质 dissolved adsorbate
溶解吸附质浓度 dissolved adsorbate concentration
溶解系数 solubility factor
溶解系数值 dissolve coefficient value
溶解相 dissolved phase
溶解相污染物浓度 dissolved phase contaminant concentration
溶解箱 dissolving box; solution tank
溶解小丘 etch hillock
溶解效应 solubility effect
溶解锌离子 dissolved zinc ion
溶解型 lysotype
溶解性 dissolubility; dissolvability; solubility; solubleness
溶解性固体 soluble solid
溶解性固体量 dissolved solid content
溶解性灰分量 soluble ash content
溶解性铁 dissolved iron
溶解性蒸发残留物 soluble evapo(u)rated residue

溶解性蒸发残渣 soluble evapo(u)rated residue
溶解性总固体 total dissolved solid
溶解压 solution tension
溶解压力 solution pressure
溶解压强 solution pressure
溶解亚硫酸盐纸浆 dissolving sulphite pulp
溶解盐 deliquescent salt; dissolved salt
溶解盐的去除 removal of dissolved salt
溶解盐类 dissolved salts
溶解盐浓度 dissolved salt concentration
溶解阳离子 dissolved cation
溶解氧 dissolved oxygen
溶解氧饱和百分率 percentage of saturated(dissolved)oxygen
溶解氧饱和百分数 percentage of saturated(dissolved)oxygen
溶解氧饱和差 dissolved oxygen concentration deficit;dissolved oxygen deficit
溶解氧饱和差基值 dissolved oxygen deficit contribution
溶解氧饱和度 saturation of dissolved oxygen
溶解氧饱和量 dissolved oxygen saturation(capacity); oxygen saturation capacity
溶解氧饱和值 air saturation value
溶解氧标准 dissolved oxygen standard
溶解氧测定 dissolved oxygen determination
溶解氧测定仪 dissolved oxygen device;dissolved oxygen meter
溶解氧传感器 dissolved oxygen sensor
溶解氧的还原 reduction of dissolved oxygen
溶解氧滴定 titration of dissolved oxygen
溶解氧电极 dissolved oxygen electrode
溶解氧分布 dissolved oxygen distribution
溶解氧分布曲线 dissolved oxygen profile
溶解氧分布图 dissolved oxygen profile
溶解氧腐蚀 dissolved oxygen corrosion
溶解氧含量 dissolved oxygen content
溶解氧极谱分析仪 polarographic oxygen analyser[analyzer]
溶解氧计 dissolved oxygen meter
溶解氧检测仪 dissolved oxygen monitor
溶解氧挠度曲线 dissolved oxygen sag curve
溶解氧浓度 dissolved oxygen concentration
溶解氧气体分析仪 dissolved oxygen gas analyser[analyzer]
溶解氧摄取率 dissolved oxygen uptake rate
溶解氧水平 dissolved oxygen level
溶解氧下垂曲线 dissolved oxygen sag curve
溶解氧下跌曲线 oxygen-sag curve
溶解氧消耗量 dissolved oxygen consumption
溶解氧消耗(曲线) dissolved oxygen depletion
溶解氧消耗速率 dissolved oxygen consumption rate
溶解氧效应 dissolved oxygen effect
溶解氧需求量 dissolved oxygen demand
溶解氧演算模型 dissolved oxygen routing model
溶解氧昼夜变化 duration change of dissolved oxygen in day and night
溶解液 solution
溶解乙炔 dissolved acetylene
溶解营养盐类 dissolved nutrient salts
溶解优势物种 dominant dissolved species
溶解油乳化剂 soluble oil emulsifier
溶解有机氮 dissolved organic nitrogen
溶解有机分子 dissolved organic molecule
溶解有机化合物 dissolved organic compound
溶解有机化合物输移 transport of dissolved organic compound
溶解有机磷 dissolved organic phosphorus
溶解有机卤素 dissolved organic halide
溶解有机碳 dissolved organic carbon
溶解有机碳浓度 dissolved organic carbon concentration
溶解有机物 dissolved organic material; dissolved organic matter; dissolved organic substance
溶解有机物分布 dissolved organic distribution
溶解有机物含量 dissolved organic matter content
溶解有机物浓度 dissolved organic matter concentration
溶解原纤维的 fibrillolytic
溶解杂质 dissolved impurity
溶解张力 solution tension
溶解脂类 dissolved lipids
溶解值 solubility value
溶解纸浆厂废水 dissolving pulp mill wastewater
溶解质 chemical load;dissolved load; dissolved solids
溶解质定律 solubility law
溶解质径流 chemical flow
溶解质径流量 dissolved solid yield
溶解质总量 total dissolved solid
溶解中性物种 neutral dissolved species
溶解转移 solution transfer
溶解装置 dissolver
溶解状态 dissolving state
溶解自由电子浓度 free dissolved electrons concentration
溶解组分 solution composition
溶解作用 dissolution;solution effect; lysis;solubilization;solvent action
溶浸沉淀 leach-precipitation
溶浸法开采矿山 solution mine
溶浸工序 leaching process
溶浸化学 leach chemistry
溶浸剂 leachant;leaching agent
溶浸搅拌机 leaching agitator
溶浸结晶法 leaching-crystallization process
溶浸时间 drenched corrosion time
溶浸物 leach material
溶井 aven;ponor[复 ponore]
溶菌反应 bacteriolytic reaction
溶菌效率 lysis efficiency
溶菌作用 baeteriolysis
溶坑湖 sinkhole lake
溶孔 dissolution pore; solution opening
溶块 clinker
溶炼不合格 off-heat
溶菱镁矿 luo-magnesite
溶菱锰矿 luo-diallogite
溶菱铁矿 luo-chalybite
溶流热处理 solution heat treatment
溶流生长法 solution method of growth
溶流温度传感元件 solution temperature sensing element
溶滤 leach out;lixiviation
溶滤带 leached zone
溶滤面 leached surface
溶滤圈闭 leached trap
溶滤水 leaching water
溶滤作用 leaching;lixiviation
溶媒 dissolvant; dissolvent; menstruum[复 menstruums/ menstrual]; resolvent;resolver;solvent
溶媒回收 solvent recovery
溶媒黏[粘]结剂 solvent adhesive
溶媒漆 solvent paint
溶媒吸附作用 lyosorption
溶媒压 solvent pressure
溶敏性胶粘剂 solvent sensitive adhesive
溶凝灰岩 welded tuff
溶凝胶 sol-gel
溶盘岩溶 corrosional pan karst
溶漆剂 lacquer solvent
溶气浮选法 dissolved air flo(a)tation
溶气罐 dissolved air tank;dissolving-air tank
溶气计 <液体> absorptiometer
溶气驱动油藏 a bubble point reservoir;dissolved gas-drive reservoir
溶气释放阀 dissolved air releasing valve
溶气真空气浮法 dissolved air vacuum flo(a)tation
溶铅水 <一种能溶解管道及配件铅质的溶剂> plumbo
溶丘洼地 hill depression
溶失量 losing solution; loss by solution
溶失量测定 loss by solution test
溶石的 litholytic
溶石液灌注器 litholyte
溶石英 fused quartz
溶蚀 corrosion;denudation;erosion
溶蚀边 corrosion border;corrosion rim
溶蚀(残)谷 steep head
溶蚀残余层 katatectic layer
溶蚀槽 solution channel
溶蚀带 corrosion zone
溶蚀地貌 solution landform
溶蚀洞穴 solution cave;solution cavity
溶蚀方程计算参数 calculating parameter of corrosion equation
溶蚀沟 solution groove
溶蚀锅穴 solution potholes
溶蚀痕 solution trace
溶蚀湖 solution lake
溶蚀基准面 base level of corrosion; base of karstification
溶蚀角砾岩 ablation breccia
溶蚀坑 pocket; solution pit; weather pit
溶蚀孔隙 dissolution pore
溶蚀孔穴 dissolution pore and cave
溶蚀量 quantity of dissolution
溶蚀裂缝 corroded fissure; dissolution fissure;solution fissure
溶蚀裂隙 corroded fissure; dissolution fissure;solution fissure
溶蚀裂隙充水矿床 mineral deposit of karst-fissure inundation
溶蚀鳞片状剥落 solution scaly peeled off
溶蚀漏斗 corroded funnel; solution funnel;doline
溶蚀率 corrosion rate
溶蚀落水洞 solution doline
溶蚀面 corrosion surface
溶蚀盆地 dissolution basin; solution basin
溶蚀平原 corrosional plain; karst plain
溶蚀评价 corrosion evolution
溶蚀评价指标 indices of corrosion evaluation
溶蚀气罐 air saturator; dissolved air vessel
溶蚀器 denuder
溶蚀强度 corrosion intensity
溶蚀深槽 helk
溶蚀试验 corrosion test
溶蚀试验过程 test process of corrosion test
溶蚀试验试样规格 sample standards of corrosion test
溶蚀试验试样类别 sample type of corrosion test
溶蚀试验项目 item of corrosion test
溶蚀竖井 karst sinkhole;doline
溶蚀速度 corrosion velocity
溶蚀速度计算法 calculus of corrosion velocity
溶蚀潭 solution pool
溶蚀通道 dissolution gallery;karst corridor;karst street;solution opening
溶蚀洼地 corroded depression; karst depression;solution depression;uvala
溶蚀现象【地】 solution phenomena
溶蚀岩片埋放位置 place to bury rock lump
溶蚀岩柱 pillar
溶蚀液化学成分 chemical composition of corrosion solution
溶蚀液体积 volume of corrosion solution
溶蚀作用类型 corrosional type
溶水性有机酸 water dissolvable organic acid
溶酸水盘 drip tray
溶潭 karst pond
溶提层 eluvial horizon
溶体 solution
溶体处理 solution treatment
溶体化退火 solution annealing
溶铜水 <能溶解水管及配件铜质的溶剂> cupro-solvent
溶析渣 scoria[复 scoriae]
溶隙 karst fissure;solution crack
溶纤剂 cellosolve
溶线 solution line
溶陷湖 sinkhole lake
溶性玻璃 soluble glass
溶性残渣产物 soluble residual product
溶性氮 soluble nitrogen
溶性淀粉 starch soluble
溶性毒性金属 soluble toxic metal
溶性惰性化学需氧量 soluble inert chemical oxygen demand
溶性惰性磷 soluble unreactive phosphorus
溶性腐殖质 soluble humus
溶性干燥剂 soluble drier[dryer]
溶性硅酸盐 soluble silicate
溶性化合物 soluble compound
溶性化学需氧量 soluble chemical oxygen demand
溶性活性硅 soluble reactive silicon
溶性活性硅氧 soluble reactive silica
溶性活性磷 soluble reactive phosphorus
溶性活性有机阴离子 soluble reactive organic anion
溶性即时需氧量 soluble immediate oxygen demand
溶性金属总量 total amount of soluble metal
溶性克耶达定氮量 soluble Kjeldahl nitrogen
溶性蓝 soluble blue
溶性沥青 soluble bitumen

溶性磷 soluble phosphorus
溶性磷酸盐 soluble phosphate
溶性磷酸盐浓度 soluble phosphate concentration
溶性硫化物 soluble sulfide
溶性硫化物析出 soluble-sulfide precipitation
溶性络合物 soluble complex
溶性钠百分率 soluble sodium percentage
溶性羟络合物 soluble hydroxo complex
溶性生化需氧量 soluble biochemical oxygen demand
溶性生物需氧量 soluble biological oxygen demand
溶性微生物代谢产物 soluble microbial metabolic product
溶性污水生化需氧量 soluble effluent biochemical oxygen demand
溶性无机磷 soluble inorganic phosphorus
溶性无机物 soluble inorganic matter; soluble matter
溶性五日生化需氧量 soluble 5-day biochemical oxygen demand
溶性物质 soluble material; soluble substance
溶性物种活度 activity of soluble species
溶性物种浓度 soluble species concentration
溶性吸附质浓度 soluble adsorbate concentration
溶性相 soluble phase
溶性盐类 soluble salts
溶性阳离子 soluble cation
溶性油 soluble oil
溶性油质乳胶 soluble oil emulsion
溶性有机氮 soluble organic nitrogen
溶性有机剂 soluble organic substance
溶性有机碳 soluble organic carbon
溶性有机物 soluble organic matter
溶性有机物需氧量 soluble organic oxygen demand
溶性杂质 soluble impurity
溶性蒸发残渣 soluble evapo(u)rated residue
溶锈剂 rust solvent
溶穴 karstic cave
溶岩洞 sink-hole
溶岩孔洞 lava pore and cavern
溶盐池 salt dissolving tank
溶盐空穴 leached salt cavern
溶氧反应 oxygen dissolution reaction
溶氧率 oxygen dissolution rate
溶氧下垂线 oxygen sag curve
溶液 aqua; colloidal sol; collosol; solution; solvent fluid
溶液 pH 值 solution pH
溶液泵 solution pump
溶液变化过程 solution behavio(u)r
溶液标定 standardization of solution
溶液表面 solution surface
溶液补充 compensation of solution
溶液采矿 solution mining
溶液残渣 solution remains
溶液残渣法 solution residue technique
溶液残渣技术 solution residue technique
溶液槽 solution tank
溶液沉淀(物) solution deposit; solution precipitate
溶液成分 solution composition
溶液成分测量计 solution composition ga(u)ge
溶液池 solution tank
溶液充气 aeration
溶液抽提法 liquid solvent extraction
溶液抽吸槽 solution pump sump
溶液储槽 solution storage tank
溶液储槽泵 solution sump pump
溶液处理 solution treating; solution treatment
溶液萃取(法) solvent extraction
溶液萃取技术 solvent extraction technique
溶液导电性定碳分析 conductometric analysis
溶液的标定 standardizing of solution
溶液的标准浊度 standard turbidity of solution
溶液的冰点降低 solution depression
溶液的再生 revivification of solution
溶液电导法 electric(al) conductivity method of solution
溶液电导检测器 solution conductivity detector
溶液电解 electrolysis of solutions
溶液毒物 solution poison
溶液法 solution method
溶液法测流 dilution ga(u)ging
溶液分析 liquor analysis; solution assay
溶液管道 solution line
溶液灌浆 solution grout; solution injection
溶液光谱 solution spectrum
溶液过滤泵 solution filter pump
溶液过滤器 solution strainer
溶液含一定量的酸 solution contains a certain amount of acid
溶液和溶剂比 solution to solvent ratio
溶液厚度 sample path length
溶液化 solubilization; solubilize
溶液化剩余污泥 solubilized excess sludge
溶液化显色 solvatochromy
溶液化学 solution chemistry
溶液回吸法 solution reclaiming method
溶液混合槽 solution mixing tank
溶液混合池 solution mixing tank
溶液混合器 solution mixer
溶液技术 solution technique
溶液加料器 solution feeder
溶液加热器 solution heater
溶液加入器 solution feeder
溶液碱度 solution alkalinity
溶液胶粘剂 solution adhesive
溶液解吸 solution stripping
溶液介质 solution medium
溶液进料剂量 solution feed dosage
溶液进料器 solution feeder
溶液净化 solution purification
溶液聚合 solution polymerization
溶液控制阀 solution control valve
溶液冷凝液泵 aqua condensate pump
溶液冷却器 solution cooler
溶液量热计 solution calorimeter
溶液量热器 solution calorimeter
溶液流量 liquid inventory
溶液流速 solution flow rate
溶液络合物 solution complex
溶液密度检定器 solution density detector
溶液膜 solution film
溶液浓度 concentration of solution; solution concentration; solution strength; strength of solution
溶液浓度检定器 solution concentration detector
溶液排放孔 solution drain hole
溶液培养 hydroponic culture; hydroponics; solution culture; water culture
溶液培养学 hydroponics
溶液培植 hydroponics
溶液平衡 solution equilibrate
溶液平衡成分 solution equilibrium composition
溶液热处理 solution heat treatment
溶液热量计 solution calorimeter
溶液容纳量 holdup of solution
溶液/溶剂黏[粘]度比 solution/solvent viscosity ratio
溶液闪烁体 solution scintillator
溶液生长 solution growth
溶液生长法 solution growth method
溶液生长晶体 solution grown crystal
溶液势 <土壤水的> solute potential
溶液试验 solution test
溶液试样 liquor sample
溶液收集器 solution catcher
溶液输送泵 solution transfer pump
溶液缩聚 solution polycondensation
溶液体积 solution volume
溶液天平 solution balance
溶液条件 solution condition
溶液涂渍法 solution coating method
溶液稳定性 stability of solution
溶液吸附 adsorption from solution
溶液吸附法 solution adsorption method
溶液相 solution phase
溶液相关系 solution phase relationship
溶液相平衡 solution phase equilibrium
溶液箱 solution tank
溶液型(反应)堆 solution type reactor
溶液型(反应)堆罐 solution type reactor tank
溶液型陶瓷涂层 solution ceramic coating
溶液型涂料 solution coating
溶液性浆液 solute grout
溶液循环 solution circuit
溶液压力 solution pressure
溶液盐 solution salt
溶液样品 solution example
溶液荧光法 solution fluorimetry
溶液硬化 solution hardening
溶液栽培 hydroponics; water culture
溶液再沸器 solution reboiler
溶液再生长法 solution regrowth technique
溶液再生长过程 solution regrowth process
溶液再生长技术 solution regrowth technique
溶液再生塔 solution regenerator
溶液直接蒸发技术 direct vapo(u)rization technique for solution
溶液中的固体 solid in solution
溶液中的天然气 natural gas in solution
溶液中的物质 matter in solution
溶液组成 composition of solution; solution composition
溶液组分 composition of solution; solution composition
溶夷作用 panplanation
溶移 solution transfer
溶油浮选 dissolved matter; dissolved solids; solvend
溶油浮选法 dissolved oil flo(a)tation
溶于石油馏出物中的沥青 asphaltic cutback
溶于石油馏出油的沥青 asphalt cutback; bitumen cutback
溶于酸液的 acid-soluble
溶原反应 lysogenic response
溶原化状态 lysogenized state
溶原性细菌 lysogenic bacterium
溶原转变 lysogenic conversion
溶跃层 lysocline
溶跃面 lysocline
溶藻细菌 algicidal bacteria
溶渣 bottom ash; clinker; dross
溶渣地沥青混合料 clinker asphalt mixture
溶渣骨料 slag aggregate
溶渣膨胀器 expanding device of slag
溶渣砂 granulated blast furnace slag
溶胀 swelling
溶胀度 degree of swelling; swelling index
溶胀剂 sweller; swelling agent
溶胀(能)力 swelling power
溶胀性 swelling index
溶胀值 swelling value
溶真菌细菌 mycolytic bacteria
溶质 dissolved matter; solute; dissolved solids
溶质边界层 solute boundary layer
溶质捕获 solute trapping
溶质层 solute zone
溶质场 solute field
溶质沉淀(物) solution deposit
溶质传递 solute transfer
溶质传输 solute transport
溶质的吸附 adsorption of solute
溶质分布 solute distribution
溶质分布预测 prediction of solute distribution
溶质分凝 solute segregation
溶质分子 solute molecule
溶质衡算 solute balance
溶质挥发干扰 solute-volatilization interference
溶质激发 solute excitation
溶质降解速率 solute degradation rate
溶质径流 flow of dissolved matter; flow of dissolved substance; solute runoff
溶质扩散 solute diffusion; solute dispersion
溶质流 solute flux
溶质黏[粘]度因子 solute viscosity factor
溶质浓度 solute concentration
溶质浓度差 solute concentration difference
溶质浓度矢量 solute concentration vector
溶质偏析 solute segregation
溶质气团 solute atmosphere
溶质汽化干扰 solute-volatilization interference
溶质势 solute potential; solute pressure
溶质水相互作用 solute-water interaction
溶质吸附 solute sorption
溶质吸力 solute suction
溶质线条 solute striation
溶质相 solute phase
溶质烟羽 solute plume
溶质移动 solute movement; solute travel
溶质硬化 solute hardening
溶质与溶质相互作用 solute-solute interaction
溶质原子 solute atoms
溶质运移 solute transfer; solute transport
溶质运移方程 solute transport equation
溶质运移模型 solute transport model
溶质运移时间 solute travel time
溶质再分配 solute redistribution
溶致液晶 lyotropic liquid crystal
溶珠反应 bead reaction
溶柱 dissolved pillar; stone forest
溶浊交代型碳酸盐岩油气藏趋向带 leached-replacement pool trend of carbonate rock
溶组织的 histolytic

榕木 bunyan tree

榕属 <拉> Ficus
榕树 banian; banyan (tree); fig-tree; India laurel fig (tree); pagoda-tree
榕树蜡 gondang wax
榕树网络 banyan network

熔 fluxion

熔边 edge fusion; edge melting
熔边机 edge melting machine
熔边狭条 fused ribbon
熔冰电流 ice melting current
熔玻璃炉 calcar
熔玻璃窑 calcar
熔箔法 fusible foil method
熔补 burning-on; cast-on; tinkering
熔补法 burning-on method
熔槽 fused groove
熔层 crucible zone
熔车装置 <用火烤法处理废弃车辆的> carbecue
熔成结 fused junction
熔成磷肥 fused tricalcium phosphate
熔成率 batch changing into melt rate
熔池 crucible; fused bath; liquid bath; melted pool; melting bat; molten bath; molten pool; puddle; steel bath; weld crater; weld puddle
熔池测试 in-bath measurements
熔池导管 melting duct
熔池电弧焊 molten-bath arc welding
熔池分析 bath analysis
熔池焊 molten-bath arc welding
熔池取样 spoon sample
熔池液面 bath line; bath surface
熔次 heat
熔滴 droplet; fused bead; globule; molten drop; smelter drippings
熔滴过渡 globular transfer
熔点 fusing point; fusion temperature; melting temperature; point of fusion
熔点测点器 melting point apparatus
熔点测定 fusing point test; melting point test
熔点测定管 melting point tube
熔点测定计 meldometer
熔点测定器 melting point apparatus
熔点低的 low-melting
熔点降低 melting point depression
熔点块 melting point bar
熔点曲线 melting point curve
熔点图 melting point diagram
熔点指标 melt index
熔点指示器 melting point indicator
熔洞 concavity; dissolution cavity; fluxing hole; pit
熔度 flexibility; fusibility; fusibleness; meltability; meltable; meltableness
熔度标 scale of fusibility
熔度表 fusibility scale; fusing chart
熔度图 fusibility curve; fusibility diagram; melting diagram
熔断 blowout; burn-out; fusing
熔断报警器 fuse alarm
熔断点 striking point
熔断电流 blowing current; blowout current; fusing current
熔断距离 fuse breaking distance
熔断开关 fuse cutout; fused disconnect switch; safety cutout; switch fuse
熔断片 fuse link; strip fuse
熔断器 cartridge fuse; cut-out; fuse; fuse block; fuse box; fuse plug; fuse protector; fusible cutout; safety cutout; thermal element
熔断器板 fuse panel
熔断器断丝 fuse burn-out
熔断器断丝报警 alarm for burn-out of a fuse
熔断器盖 fuse cover
熔断器及导线组件 fuse and wire kit
熔断器夹爪 fuse grip jaw
熔断器盘 fuse board
熔断器配件 fuse mounting
熔断器匣 cut-out box
熔断器支座 fuse holder
熔断时间 fusing time
熔断式喷水龙头 fused sprinkler head
熔断丝 blown fuse; fusible link; safety fuse
熔断丝隔离开关 disconnector with fuse
熔断系数 fusing factor
熔断线 fuse wire
熔断指示器 blown fuse indicator
熔封 sealing by fusing
熔缝 flow line; weld mark
熔敷层 overlay
熔敷钢 deposit (e) steel
熔敷焊道 bead; weld bead
熔敷金属 deposit (e); deposited metal; metal deposit
熔敷金属试件 deposited metal test specimen
熔敷顺序 built-up sequence; deposition sequence
熔敷速度 deposition rate; rate of deposition
熔敷系数 deposition efficiency; melting efficiency
熔敷效率 deposition efficiency; melting efficiency
熔缸 crucible
熔割 fusion cutting
熔割盘 fusing disc[disk]
熔挂法 burning-in; melting-in
熔管 fusion tube
熔罐 fusion pot
熔罐腹 crucible belly
熔埚 fire pot; skillet
熔锅 melting pot; melting vessel
熔焊 autogenous weld (ing); burning-in; flame weld (ing); fusing solder (ing); fusion-welded; fusion welding; heat seal (ing); puddle welding); sweating; thermoweld; flow welding
熔焊层 weld bead
熔焊法 fusing process of intermediate
熔焊钢锻件 welded steel forging
熔焊技术 fusion welding technic
熔焊接 heat-seal; fusion welding
熔焊接合 sweated joint
熔焊接头 sweated joint
熔焊金属 fusion welding metal
熔焊料 flux solder; molten solder
熔焊炉 welding furnace
熔焊区 fusion zone
熔焊热 sweating heat
熔焊碳化钨粉末 interspersed carbide
熔焊锡浴槽 bath of molten solder
熔焊线 weld junction; weld line
熔焊液 welding fluid
熔号 melt number
熔合 alloying; fuse; fusing; fusion; merging; sew; alligation <金属的>
熔合板 rigid plate
熔合比 penetration ratio
熔合不良 lack of fusion
熔合部 fusion zone
熔合的 fusional
熔合法配制合金 fusion alloying
熔合反应 fusion reaction
熔合光电晶体管 fused phototransistor
熔合过程 alloying process
熔合接触 bond contact
熔合接点光电晶体管 fused contact phototransistor
熔合面 bond interface; fusion face
熔合区 bond; fusion area
熔合双焦点透镜 fused bifocals
熔合双焦镜片 fused bifocals
熔合物 rafting
熔合线 bond line; fusion-line; line of fusion; melt run; weld bond; weld junction
熔合杂质光电晶体管 fused impurity phototransistor
熔核 nugget
熔核直径 nugget size
熔化 burn-off; colliquation; flux; fusation; fusing (melting); fusion; liquate; liquification; melt-blown; melt (ing); smelting; swale
熔化比 melting ratio
熔化铋冷却 molten-bismuth cooling
熔化玻璃 molten glass
熔化玻璃试池 fused glass cell
熔化不良 lack of fusion
熔化不全 lack of fusion
熔化部 melting end; melting zone
熔化操作 melting operation
熔化槽 melt tank
熔化层 melting zone; smelting zone
熔化池 melting chamber; melting tank
熔化池壁外帮砖 melting tank backup lining
熔化池温度控制 melter temperature control
熔化穿孔 fusion piercing; piercing
熔化穿孔机 fusion piercing drill; jet (-piercing) drill; Linde drill
熔化淬火法 melt-quenching method
熔化淬火过程 melt-quench process
熔化带 fusion zone; melted zone; melting section; melting zone
熔化单元 melting unit
熔化的 fusional; liquefied; molten
熔化的冰碛土 melt-out till
熔化点 melting point
熔化电极 consumable electrode
熔化电解质 molten electrolyte
熔化电流 fusion current; melting current
熔化端部 melting end
熔化法 fusion method; melting process
熔化范围 fusion range; melting range
熔化分界面 melting interface
熔化分离 liquidate
熔化釜 melting kettle
熔化坩埚 melting cup
熔化坩埚装置 melting cup assembly
熔化高温计 fusion pyrometer
熔化工 melter; metal tender
熔化工具 melting tool
熔化罐 melting chamber; melting pot
熔化硅延迟线 fused silica delay line
熔化锅 melting kettle; melting pot
熔化锅炉 melting boiler
熔化过程 fusion process; melt-back process; melting process
熔化焓 fusion enthalpy
熔化混合物 fusion mixture; melting compound
熔化极电弧焊 consumable electrode welding
熔化极惰性气体保护焊 metal-inert-gas welding; welding
熔化极脉冲氩弧焊 pulsed metal argon-arc-welding
熔化极(气体)保护焊 gas metal-arc welding
熔化极自动保护电弧焊 shielded metal arc welding
熔化加热器 fusing heater
熔化阶段 melting stage
熔化金属 deposit (e) metal; melt down; molten metal
熔化颗粒沉积 molten particle deposition
熔化孔 melt-off pore
熔化离心机 melting and fusing centrifuge
熔化沥青 melted asphalt
熔化沥青的锅 bitumen kettle; bitumen melting kettle
熔化沥青的锅炉 bitumen melting boiler
熔化沥青箱 bitumen melting tank
熔化量 melting capacity; melting quantity; the amount melted
熔化料柱 melting stock column
熔化炉 flowing furnace; fritting furnace; fusion furnace; liquation furnace; melting furnace
熔化铝热焊 fusion thermit welding
熔化率 melting capacity; melting rate; melt-off rate; specific melting efficiency
熔化玛琋脂锅炉 mastic melting boiler
熔化面积 melting area
熔化能 fusing energy
熔化能力 melting capacity
熔化黏[粘]合料的锅 binder cooker; binder heater
熔化期 melting period; period of melting
熔化器 melter
熔化潜热 latent heat of fusion; latent heat of melting
熔化切割 fusion cutting
熔化清除(氧化皮) wash heating
熔化区 fusion zone; melting section; melting zone
熔化曲线 fusion curve; melting curve
熔化热 fusion heat; heat of fusion; heat of melting; melting heat; sweating heat
熔化溶剂 molten solvent
熔化熵 fusion entropy
熔化烧蚀 flame ablation; melting ablation
熔化设备 melting unit
熔化深度 depth of fusion
熔化生长 flux growth
熔化石英 fused quartz
熔化石英延迟线 fused quartz delay line
熔化石油沥青 molten bitumen
熔化时间 fusing time; melt-down time; melting time
熔化实验 fusion experiment
熔化试验 fusion test
熔化室 melting chamber; melting compartment
熔化水 melt water
熔化水晶 fused quartz
熔化水泥 melt cement
熔化速度 melting speed; speed of melting
熔化速率 melting rate; rate of melting
熔化损失 fusion loss; loss on melting; melting loss
熔化台 melting platform
熔化特性 melting characteristic
熔化体 melting solid
熔化条件 melting condition
熔化图 melting diagram
熔化温度 fusing temperature; fusion point; fusion temperature; melting point; melting temperature; squatting temperature
熔化温度传送器 melting temperature transducer
熔化温度点 fusing temperature point

R

熔化温度曲线 melting temperature curve;temperature curve of melting
熔化温度制度 temperature curve for melting; temperature regulation of melting
熔化物 fusant;fused ore;molten mass
熔化系数 fusing coefficient; melting coefficient
熔化效率 fusion efficiency; melting efficiency
熔化玄武岩 fused basalt
熔化压接 fusion pressure welding
熔化压力传送器 melt pressure transducer
熔化压力控制器 melt pressure controller
熔化岩浆 molten magma
熔化颜料 vitrifiable colo(u)r
熔化再凝结 melting and refreezing
熔化指标 melt index
熔化珠 molten drop
熔化状况 melting condition
熔化状态 melting condition; melting state;molten state
熔化自由能 free energy of melting
熔化钻机 fusion drill
熔化钻进法 fusion piercing
熔化钻眼法 fusion drilling
熔化嘴 fusion nozzle
熔灰燃烧 slag tap firing
熔灰岩台地 ignimbrite plateau
熔毁 melt-blown;melt down
熔积金属 deposited metal
熔剂 agent of fusion;flux;fluxing agent; flux material; furnace addition;fusing agent;fusion agent
熔剂处理 fluxing
熔剂法 flux growth method
熔剂粉 ground flux
熔剂覆盖层 flux cover
熔剂刚玉 phermitocorundum
熔剂辊 flux roll
熔剂焊条 fluxed electrode
熔剂尖晶石 phermitospinel
熔剂精炼 flux-refining
熔剂矿物 fluxing mineral
熔剂脉石英 flux vein quartz
熔剂枪 flux gun
熔剂石英砂 flux silica sand
熔剂石英砂岩 flux silicarenite
熔剂石英岩 flux quartzite
熔剂燧石岩 flux flint
熔剂芯焊条 flux core-type electrode
熔剂因子 flux factor
熔剂用灰岩 limestone for flux
熔剂用蛇纹岩 serpentinite for flux
熔剂原料矿产 flux raw material commodity
熔胶锅 glue boiler
熔接 autogenous brazing;burn(ing)-in; butt fusion; frit; fusing; fusion jointing;fusion splicing;sealing-in; thermoweld;weld
熔接棒 sealing bar
熔接尺寸 weld dimension
熔接的 welding
熔接的双焦眼镜 fused bifocals
熔接的纤维束 fused fibre bundles
熔接法 autogenous soldering;autogenous weld(ing);burning-on method;soldering welding
熔接缝 weld line
熔接焊 join by fusion
熔接机 heat seal(ing) machine;sealing machine
熔接金属 added metal; build-up the metal
熔接炬 welding torch
熔接密封 frit seal
熔接头 welded joint
熔接线 weld line
熔接整流器 welded-contact rectifier
熔结 alloying;clinkering;close burning;frit(ting);sintering
熔结玻璃 sintered glass
熔结带 clinkering zone
熔结的 clinkery
熔结点 clinkering point; smelting point
熔结多孔玻璃 fritted glass
熔结多孔玻璃过滤器 fritted glass filter
熔结工厂 sinter plant
熔结黑曜岩 welded obsidian
熔结灰 clinkering ash
熔结火山碎屑岩 welded proclastic rock
熔结集块岩 welded agglomerate
熔结剂 agglomerant
熔结角砾岩 welded breccia
熔结块层 clinker bed
熔结临界频率 critical frequency
熔结煤 clinkering coal
熔结纳克斯胶 <一种电木> pertinax
熔结凝灰结构 welded tuffaceous texture
熔结凝灰岩 flood tuff;ignimbrite;tuff lava;welded tuff
熔结松脂岩 welded pitchstone
熔结碎屑结构 welded clastic texture
熔结温度 clinkering point
熔结物 frit
熔结岩渣 welded scoria
熔结指数 sintering index
熔解 fluxing;fusion;liquefaction;liquidate; liquidation; liquification; melt(ing)
熔解参数 fusion parameter
熔解产品 fused product
熔解分析 liquation
熔解分析器 fusion analyser
熔解量 meltage
熔解黏[粘]性 melt viscosity
熔解黏[粘]滞度 melt viscosity
熔解频率 fusion frequency
熔解热 fusion heat;heat of fusion;latent heat of fusion;melting heat
熔解熵 entropy of fusion
熔解温度 fluxing temperature;fusion temperature; melting temperature; temperature of fusion
熔解物 liquefactant;meltage
熔解系数 flux factor
熔解线 <冶金学的平衡面线图解中表示金属材料熔解温度的面线> solidus[复 solidi]
熔界 pyrosphere
熔金属面 heel of metal
熔浸 melt dipping
熔开焊接处 unsolder
熔开温度 melting-out temperature
熔块 clinker;clinker clew;clinker clue;frit;fused block;fusion cake; regulus[复 reguli/reguluses]
熔块玻璃 fritted glass
熔块瓷 fritted china; fritted porcelain;glassy porcelain;pate frit
熔块淬冷 quenching of frit
熔块的 reguline
熔块法 ingot process
熔块骨料 clinker aggregate
熔块集料 clinker aggregate
熔块料 smelt
熔块炉 smelter
熔块窑 calcar;frit kiln
熔块釉 fritted glaze
熔块渣 clinker
熔宽 fused breadth
熔矿炉 crucible;ore furnace
熔矿炉底石 hearth block
熔蜡炉 wax melter
熔蜡器 paraffin(e)melter
熔棱装饰 fire-cord decoration
熔离 fuse off;liquation
熔离矿床 liquation deposit
熔离作用 liquid immiscibility
熔粒再结合耐火材料 rebonded fused-grain refractory
熔炼 dry metallurgy;fire metallurgy; founding; fusion metallurgy; melting;plaining;smelting
熔炼操作 melting practice
熔炼产物 smelted product
熔炼厂 smeltery;smelting works
熔炼车间 meltshop
熔炼成分不合格 off-melt; off-smelting
熔炼法 fusion process; method of smelting
熔炼废品 off-heat;off-melt
熔炼废渣 smelting waste
熔炼分析 heat analysis
熔炼坩埚 melting crucible; melting kettle;smelting pot
熔炼工 melter
熔炼焊剂 fused flux;smelting flux
熔炼炉 melter;melting furnace;smelter; smelter hearth; smelting furnace hearth
熔炼炉飞尘 smelter dust
熔炼炉号 hearth number
熔炼能力 smelting capacity
熔炼期 smelting period
熔炼区 smelting zone
熔炼设备 smelting unit
熔炼试验 smelting trial
熔炼室 working chamber
熔炼水晶 smelting crystal; smelting quartz
熔炼速度 speed of melting
熔炼损耗 melting loss
熔炼序次 heat
熔炼原料矿产 smelting raw material commodities
熔炼周期 heat cycle
熔炼总损耗 total melting loss
熔料 melting stock;molten material
熔料长石 <长石熔成的玻璃> maskelynite
熔料串炉 whirling of fusing charge
熔料分料器 melt extractor
熔料坩埚 melting pot
熔裂 melt fracture
熔流法 fusion flow method
熔炉 crucible; founding furnace; liquid bath furnace; melter; melting receptacle; melt kettle; smelter; smelting furnace
熔炉出铁口 tap hole
熔炉钢 furnace steel
熔炉隔墙 <桥墙> bridge wall
熔炉观测镜 periscope
熔炉过程 run of a furnace
熔炉渣 air-cooled slag
熔炉渣骨料 air-cooled slag aggregate
熔炉渣集料 air-cooled slag aggregate
熔滤潜水带 leaching zone of groundwater
熔媒剂 melt catalyst
熔面焊接 wash welding
熔模 fusible pattern;investment(mo(u)lding) pattern
熔模材料 expendable pattern material
熔模法 permanent mo(u)lding method
熔模精密铸造 precision-investment casting
熔模壳型 plycast
熔模壳型铸造法 invest shell casting method
熔模涂料 precoating
熔模铸过程 lost wax process
熔模铸型 invested mo(u)ld
熔模铸造 fusible pattern mo(u)lding;investing;investment mo(u)lding;lost wax casting;lost wax mo(u)lding;investment casting
熔模铸造法 investment casting method;lost wax process
熔模铸造涂料 investment compound
熔模铸造用蜡 investment casting wax
熔能线 <金属材料熔解的面线> solidus[复 solidi]
熔黏[粘]环氧树脂面层保护钢筋 fusion-bonded epoxy coated reinforcement
熔凝 fused seal;fused together;fusing together
熔凝硅石 <熔烧过的石英> fused silica
熔凝硅石丝 fused silica filament
熔凝硅石纤维 fused silica fiber[fibre]
熔凝灰岩 welded tuff
熔凝壳 fusion crust
熔凝柯巴脂 run copal
熔凝石英 fused quartz;vitreosil
熔凝石英玻璃 fused quartz glass
熔凝树脂 fused colophony;fused copal
熔凝水泥 fused cement
熔凝松香 fused colophony
熔凝松脂 fused copal
熔凝氧化硅 fused silica
熔凝氧化硅反射器 fused silica reflector
熔凝氧化硅共振腔 fused silica cavity
熔凝氧化硅基质 fused silica substrate
熔凝氧化铝磨料 fused alumina
熔凝作用 melting and refreezing
熔配合金 fusion alloying
熔喷法非织造织物 melt-blown fabric
熔片 fuse element;fuse piece
熔气焊 flame welding
熔铅 molten lead
熔铅淬火 molten lead quench
熔铅锅 lead melting kettle
熔铅结合 lead burning
熔铅炉 lead melting furnace
熔铅喷雾盖板 lead splash lap
熔铅润滑拉丝法 lead lubrication process
熔铅设备 lead melting equipment
熔球试验 button test;flow button test
熔球试验法 fusion-flow test
熔区 molten zone
熔区长度 zone length
熔区传输精炼炉 zone transport refiner
熔区回流 zonal reflux
熔区间间距 interzone spacing
熔区空段 zone-void
熔区空段法 zone-void method
熔区空段精炼炉 zone-void refiner
熔区偏析 zone segregation
熔区通过 zone pass
熔区形状 zone shape
熔区移动 travel of zone;zone movement
熔区移动机构 molten zone moving mechanism;zone travel mechanism
熔热 fusion heat;heat of heat
熔热力 melting heat
熔溶胶 pyrosol
熔融 fuse; fusing; fusion; liquate; liquation;melt(ing);melt-out
熔融包裹体 melt inclusion
熔融包裹体成岩温度测定法 method of rock-forming temperature determination for melt inclusion; petrogenic temperature determination method of melting inclusion
熔融冰铜 molten matte
熔融玻璃镶衬 melted-glass lining

熔融槽 fusion tank
熔融层压法 fusion lamination
熔融处理 sintering treatment
熔融床气化 fusion-bed gasification
熔融粗铅 molten bullion
熔融催干剂 fused drier[dryer]
熔融催化剂 fused catalyst
熔融淬火法 melt-pulling method; melt-quenching method
熔融带 fusion zone
熔融的 fused; igneous; igneous plutonic; molten
熔融的苛性碱 fused caustic
熔融点 fuse point; fusion point; melting temperature
熔融电解精炼 molten slag electrolysis refining
熔融电解铁 fused electrolytic iron
熔融电解质 fused electrolyte
熔融对接 butt fusion
熔融二氧化硅 fused silica
熔融法 fusion method
熔融矾土骨料 fused alumina aggregate
熔融矾土集料 fused alumina aggregate
熔融反应 frit reaction
熔融范围 fusion range
熔融分段屏蔽物 fused grading shield
熔融分解 decomposition by fusion
熔融分解法 melting decomposition method
熔融粉尘 molten dust
熔融氟化物燃料 molten fluoride fuel
熔融复合 melt compounding
熔融刚玉骨料 fused corundum aggregate
熔融刚玉集料 fused corundum aggregate
熔融高温计 fusion pyrometer
熔融光纤接头 fused fiber[fibre] splice
熔融光学纤维面板 fused fiber[fibre] optics plate
熔融硅酸盐 molten silicate
熔融硅砖 fused silica brick
熔融焊接金属 molten weld metal
熔融焊接药粉 ignition powder
熔融焊料 molten solder
熔融合金 molten alloy
熔融后拉断 fuse off
熔融混合物 molten mixture
熔融挤出 melt extrusion
熔融挤塑 melt extrusion
熔融挤压纺丝法 fused extruding spinning process
熔融减摩垫 molten pad
熔融浇注模 fusion casting mo(u)ld
熔融浇注耐火材料 molten-cast refractory
熔融胶合法 fusion-bonding process
熔融角砾岩 fusion-breccia
熔融结 fused junction
熔融金属 hot metal; liquid metal; molten metal
熔融金属加料 hot-metal charge
熔融金属接收器 hot-metal receiver
熔融金属浸涂 molten metal dip coating
熔融金属浸渍 molten metal dipping
熔融金属流量计 molten metal flowmeter
熔融金属去壳 molten metal decladding
熔融金属液位指示器 molten metal level indicator
熔融精制 melt refining
熔融精制过程 melting refining process
熔融聚合物 molten polymer
熔融聚合作用 above-the-melt polymerization
熔融颗粒耐火材料 fused grain refractory
熔融矿渣 melted gangue; melted slag; pottsco; molten slag
熔融扩散 melting diffusion
熔融料 melting charge
熔融磷酸盐电解槽 fused phosphate bath
熔融流 melt-flow
熔融流动性 <塑料等的> melt fluidity
熔融铝 molten alumin(i)um
熔融铝水过滤布 molten alumin(i)um filtration fabric
熔融莫来石 fused mullite
熔融耐火材料 fused refractory
熔融黏[粘]度 melting viscosity
熔融黏[粘]度计 fusion viscosimeter
熔融黏[粘]结 bond vitrified
熔融黏[粘]结工艺 melt-bonding process
熔融尿素泵 melt urea pump
熔融凝聚 <玻璃管> fused on
熔融喷吹法 melting and blowing process
熔融喷镀覆层 fused spray deposit
熔融硼砂 borate; borax glass
熔融膨胀 melting expansion
熔融平衡 melt equilibrium
熔融钎料 molten solder
熔融潜热 latent heat of fusion
熔融区 region of melting
熔融去壳 melting decladding
熔融热 heat of fusion
熔融烧碱 molten caustic
熔融烧注玄武岩 fusion-cast basalt
熔融生长 melt growth
熔融生长的 flux-grown
熔融生长红宝石 flux-grown ruby
熔融石 obsidianite; tectite[tektite]
熔融石英 fused quartz; fused silica; melting quartz; vitreous silica
熔融石英毛坯 fused silica blank
熔融石英砖 fused quartz block; fused silica brick
熔融时间 melting time
熔融试验 fusion test
熔融水泥 fused cement; melted cement; molten cement
熔融水泥料 fused cement
熔融速率 rate of melting
熔融损失 loss in smelting
熔融缩聚 melt polycondensation
熔融态玻璃 molten glass
熔融提纯法 fusion purification process
熔融体 fused mass; fuse link
熔融铁 molten iron; molten pig
熔融铜 molten copper
熔融涂装 melt coating
熔融外延法 epitaxial growth by melting
熔融温度 fusion temperature; temperature of melting
熔融物 fusant; fused material; liquid melt; molten mass
熔融物料刮窑皮 coat the lining with fused material
熔融物质 fused mass; melt mass
熔融消失 consume
熔融效率 melting efficiency
熔融锌 molten zinc
熔融性 fusibility; meltability
熔融玄武岩 cast basalt; fused basalt; molten basalt
熔融旋压法 melt spinning
熔融岩浆 igneous intrusion
熔融岩石 molten rock
熔融盐 fuse salt
熔融盐槽 molten salt bath
熔融盐电精制 fused salt electrorefining
熔融盐类 molten salts
熔融氧化硅反射器 fused silica reflector
熔融氧化铝 aloxite; fused alumina; molten alumina
熔融氧化镁 fused magnesium oxide
熔融液 melt water
熔融硬度 remelting hardness
熔融在熟料上的粉尘 dust fused on the clinker
熔融渣 liquid slag
熔融指数 melt index
熔融酯基转移作用 melt-transesterification
熔融转移印花 melt transfer printing
熔融装料 melt loading
熔融装药 fused charge
熔融状态 melting state; molten condition; molten state
熔融锥 fusion cone
熔入 fusing into
熔入法 frit-in method
熔塞 fused plug; fusible plug
熔栅 melt(ing) grid
熔栅盘管 melting grid spiral
熔烧机理 mechanism of roasting
熔烧料层 feed bed
熔烧炉膛 roaster hearth
熔勺 melting ladle
熔射 meltallizing
熔深 <指焊接> depth of fusion; penetration; fused depth; weld penetration
熔深不足 insufficient penetration
熔石的 saxifragant
熔蚀 burn-out; resorption
熔蚀斑晶 corroded crystal; corrosive phenocryst
熔蚀边 corrosion border; resorption border
熔蚀池 resorption border; resorption rim
熔蚀化石 corroded fossil
熔蚀晶 corroded crystal
熔蚀孔 corrosion cavity
熔蚀面 corrosion surface
熔蚀石英 vermicular quartz
熔释体系 melt release system
熔丝 fuse(element); fuse wire; thermal element; wire fuse
熔丝板 fuse panel
熔丝保险器 fuse cutout; fusible cutout
熔丝报警继电器 fuse alarm relay
熔丝报警器 fuse alarm
熔丝灯 cartridge lamp
熔丝断路器 fuse block; fuse cutout; fusible circuit breaker; fusible cutout; safety cutout
熔丝额定值 fuse rating
熔丝阀 fire valve; fusible link valve
熔丝隔断开关 fuse disconnecting switch; fusible disconnecting switch
熔丝更换器 fuse tongs
熔丝管 cartridge(-type) fuse; fuse cartridge; tube fuse
熔丝管插座 fuse receptacle
熔丝管座 fuse socket
熔丝盒 cartridge; fuse block; fuse box; fuse cabinet; fuse cartridge; fuse case; fuse chamber
熔丝及保安器组 fuse and protector block
熔丝架 fuse holder; fuse rack; hook guard
熔丝开关 fuse switch; fusible switch
熔丝开关动力箱 fusible switch power panel
熔丝链 fuse link
熔丝链接 fusible link
熔丝耐火材料 fuse refractory
熔丝盘 fuse board; fuse panel
熔丝配合 fuse coordination
熔丝熔断报警 fuse burn-out alarm
熔丝熔断指示灯 fuse break lamp
熔丝塞(子) fuse plug; plug fuse; safety plug
熔丝试验开关 fuse test switch
熔丝特性 fuse characteristic
熔丝信号 fuse signal
熔丝型温度计 fuse-type temperature meter
熔丝异离体 resorption schlieren
熔丝支持器 fuse holder
熔丝座 fuse base
熔态金属染色 molten metal dy(e)ing
熔态锌 molten zinc
熔潭 puddle
熔体 flux; fondant; fusant; fused mass; fusion; melt; melt mass; molten bath; molten mass
熔体表面 bath surface
熔体成分 bath composition
熔体处理法 melt treatment process
熔体动态分类 flow regime classification
熔体逗留时间 melt residence time
熔体纺丝 melt spinning
熔体覆盖物 cover for fusions
熔体滑动 melt slippage
熔体化学计量学 melt-stoichiometry
熔体混合物 melt blend
熔体挤出 melt extrusion
熔体挤塑 melt extrusion
熔体挤压纺丝法 melt spinning by extrusion method
熔体计量 melt-stoichiometry
熔体加入剂 melted addition
熔体结构 structure of melt
熔体结晶作用 crystallization from melt; melt crystallization
熔体浸渍 melt impregnation
熔体聚合作用 melt polymerization
熔体均匀性 melt uniformity
熔体拉伸法 melt-pulling method
熔体料斗 melt hopper
熔体流动不稳定性 melt flow instability
熔体流动指数 melt flow index
熔体黏[粘]度 melt viscosity
熔体黏[粘]度计 melt viscometer
熔体破坏 melt fracture
熔体生长 melt growth
熔体试样 bath sample; molten test sample
熔体弹性 melt elasticity
熔体潭 melt pool
熔体添加剂 melted addition
熔体通道 melt canal
熔体突然膨胀 Barus effect; melt swell
熔体涂覆 melt coating
熔体旋凝工艺 melt spinning process
熔体运动 bath movement
熔体增白剂 melt-additive brightener
熔体指数 melt index
熔体指数测定仪 extrusion type capillary visco(si)meter; melt indexer
熔体着色法 melt pigmentation
熔体着色技术 melt colo(u)ration technique
熔体组成 bath composition
熔体组分 bath component
熔铁厂 iron-smelting factory
熔铁炉 cupola; iron-melting furnace
熔铁炉底 bottom of blast furnace
熔铁炉渣 cupola slag
熔铜焊 fusing brazing

熔透 penetration
熔透焊道 penetration bead; uranami bead
熔透焊缝 melting through weld; melt thru weld; penetration weld
熔透型等离子弧焊 fusion type plasma arc welding
熔透型焊接法 fusion type welding
熔脱 melt-off
熔析 aliquation; liquate; liquidation; segregation
熔析槽 liquation bath
熔析出 liquate out
熔析锅 liguating pot; liquating kettle
熔析精炼 liquation refining; refining by liquation
熔析炉 liquation hearth
熔析铅 liquated lead; liquation lead
熔析铅锅 liquated-lead kettle
熔析设备 liquating apparatus
熔析铜 liquated copper
熔线 fuse strip; fuse wire; solvus
熔线板 fuse board
熔线保护 fuse protection
熔线报警 fuse alarm
熔线插入件 fuse wire insert
熔线额定值 fuse rating
熔线管钳 fuse tongs
熔线盒 cut-out case; fuse block; fuse box; fuse cartridge
熔线架 fuse rack
熔线开关 fuse switch
熔线排 strip of fuses
熔线片 fuse strip
熔线熔断指示器 blown fuse indicator
熔线塞 fuse plug; plug cartridge; plug fuse; safety plug
熔线塞座 safety plug socket
熔线时间电流试验 fuse time-current test
熔线式隔离开关 fusible disconnecting switch
熔线型热动继电器 fuse-type temperature relay
熔线元件 fuse element
熔陷 melt sinking
熔锌池 bath of molten zinc
熔锌炉 zinc furnace
熔性 flexibility; fusibility
熔岩 latite; lava; molten rock
熔岩暗道 lava tunnel
熔岩表壳构造 surface structures of lava
熔岩饼火山角砾岩 lava-cake breccia
熔岩饼集块岩 lava-cake agglomerate
熔岩玻璃 lava glass
熔岩层 lava bed
熔岩沉积物混合体 peperite
熔岩床 lava bed
熔岩的固化作用 solidification of lava
熔岩的凝固 solidification of lava
熔岩堤 lava levee
熔岩洞(穴) lava cave
熔岩堵成的湖 lake ponded up by lava
熔岩泛(滥) lava flood
熔岩放热 emission of lava
熔岩高原 lava plateau
熔岩鼓包 lava tumulus
熔岩管 <熔岩流空的管状孔道> lava tube
熔岩海 lava sea
熔岩湖 lava lake
熔岩湖活动 lava lake activity
熔岩荒野 lava desert; lava field
熔岩灰 lava ash
熔岩火山 lava volcano
熔岩火山湖 lava cauldron
熔岩急流 torrent
熔岩棘 lava spine
熔岩脊 pressure ridge
熔岩浆 molten magma
熔岩浆体 asthenolith
熔岩阶地 lava terrace
熔岩颈 lava neck
熔岩巨砾结构 lava boulder texture
熔岩坑 lava pit
熔岩口 bocca
熔岩块 block lava
熔岩块渣 clinker
熔岩砾状结构 lava psephitic texture
熔岩粒 <冲积层中的> buckshot
熔岩流【地】 lava(l) flow; coulee; lava stream; nappe
熔岩流类型 type of lava flow
熔岩脉 lava streak
熔岩囊 pocket of molten rock
熔岩凝灰结构 lava tuffaceous texture
熔岩泡 lava blister
熔岩泡沫玻璃 lava foam glass
熔岩喷发 lava discharge
熔岩喷溢 lava effusion
熔岩喷涌 fire fountain; lava fountain
熔岩平原 lava plain
熔岩瀑布 lava fall
熔岩气泡 lava blisters
熔岩穹丘 lava dome
熔岩丘 lava cone
熔岩区 malpais
熔岩泉 lava fountain; rise; rising
熔岩砂状结构 lava aleuritic texture
熔岩舌 tongue
熔岩石笋 lava stalagmite
熔岩隧道 lava tunnel
熔岩塔 spine
熔岩台地 lava plateau
熔岩土壤 lava soil
熔岩外喷 extravasate; extravasation
熔岩围堵湖 lave-dam lake
熔岩席 lava sheet
熔岩相 lava phase
熔岩楔 lava wedge
熔岩玄武岩 lava basalt
熔岩穴【地】 pit; lava hole
熔岩崖 lava scarp
熔岩岩颈 lava plug
熔岩原 lava field
熔岩渣 lava slag
熔岩趾 rock toe
熔岩钟 barrow
熔岩钟乳 lava(l) stalactite
熔岩肿瘤 tumulus[复 tumuli]
熔岩柱 lava column
熔岩砖 lava brick
熔岩锥 lava cone
熔盐 fused salt; molten salt
熔盐传热剂 molten heat-transfer salt
熔盐萃取 fused salt extraction; molten salt extraction
熔盐电化学 molten salt electrochemistry
熔盐电解 fused salt electrolysis; fusion electrolysis
熔盐电解槽 fused salt bath; molten salt container
熔盐电解法 fused salt electrolysis process
熔盐电解还原 reduction of fused salts
熔盐电解精炼 fused salt electrolytic refining
熔盐电解液电池 molten electrolyte cell
熔盐电迁移 molten salt electromigration
熔盐电切削 molten salt electrocutting
熔盐堆 fused salt reactor
熔盐法 flux-grown; molten salt method
熔盐反应堆 fused salt reactor; molten salt reactor
熔盐反应堆实验 molten salt reactor experiment
熔盐伏安法 molten salt voltammetry
熔盐化学 fused salt chemistry
熔盐极谱法 fused salt polarography
熔盐急冷 salt growth
熔盐计时电位滴定法 molten salt chronopotentiometry
熔盐加热器 molten salt heater
熔盐介质 fused salt medium
熔盐冷却剂 molten salt coolant
熔盐裂解 molten salt pyrolysis
熔盐裂解炉 molten salt cracker
熔盐炉裂解 molten salt cracking
熔盐取样器 molten salt sampler
熔盐燃料 molten salt fuel
熔盐热载体 molten heat-transfer salt
熔盐生长法 fused salt growth; molten salt growth
熔盐循环泵 molten salt circulating pump
熔盐液相 fused salt liquid phase
熔盐浴处理 molten salt bath treatment
熔盐增殖(反应)堆 molten salt breeder reactor
熔盐转换反应堆 molten salt convertor
熔窑 melter; melting furnace
熔窑放液侧 exhaust side
熔窑烤窑系统 furnace preheating system
熔窑流量 furnace pull
熔窑热负荷 heat load of furnace
熔窑热耗 heat load of furnace
熔窑热效率 heat efficiency of furnace
熔液表面 molten surface
熔液池 liquid bath
熔液试样 molten test sample
熔油罐 melting chamber
熔油箱 melting tank
熔浴法气化 molten-bath gasification
熔渣 breeze; clinker aggregate; dross; fused slag; lead dross; molten slag; scoria [复 scoriae]; sinter; slag(ging); sullage
熔渣包体 slag inclusion
熔渣保护型焊条 slag-covered electrode
熔渣材料 clinker material
熔渣成分 slag composition
熔渣成粒 slag granulation
熔渣处理法 slag handling process
熔渣处理体系 slag handling system
熔渣床 clinker bed
熔渣的 scoriaceous
熔渣的成粒水淬 granulating; granulation
熔渣地沥青(混合料) clinker asphalt
熔渣地面 clinker floor
熔渣堆 slag heap
熔渣分离 slag separation
熔渣分离器 slag separator
熔渣粉 slag powder
熔渣腐蚀 slag action
熔渣盖层 clinker coating
熔渣构造 scoriaceous structure
熔渣骨料 clinker aggregate
熔渣骨料混凝土 clinker aggregate concrete
熔渣固化 solidifying of slag
熔渣罐 slag receiver
熔渣硅酸盐水泥 clinker-bearing slag cement; metallurgic(al) cement
熔渣混凝土 clinker concrete; slag concrete
熔渣混凝土板 clinker slab
熔渣混凝土块 clinker concrete block
熔渣混凝土砖 clinker tile
熔渣集料 clinker aggregate
熔渣集料混凝土 clinker aggregate concrete
熔渣孔 clinker hole; slag hole
熔渣块体 clinker block
熔渣(块)圬工 clinker masonry
熔渣流动性 flow of the slag; fluidity of the slag; slag fluidity
熔渣路 clinker road; slag road
熔渣路面 clinker pavement; scoria pavement; scoria surface; slag surface
熔渣面层 scoria surface
熔渣砌块 clinker block
熔渣侵蚀作用 slag action
熔渣热度 cinder heat
熔渣砂 slag sand
熔渣石 slag stone
熔渣水淬 slag granulation
熔渣水淬法 water treatment of slag
熔渣水泥 clinker slag cement; metallurgic(al) cement; permetallurgical cement; slag cement
熔渣水泥砂浆 slag cement mortar
熔渣物理热 physical heat of slag
熔渣纤维 silicate cotton
熔渣形成 clinker formation
熔渣型高铝骨料 aluminous aggregate of cinder type
熔渣型高铝集料 aluminous aggregate of cinder type
熔渣溢出 slag overflow
熔渣骤冷装置 quencher
熔渣砖 clinker brick; iron brick; slag brick
熔渣状的 scoriform
熔渣状态 clinker phase
熔胀 intumescence
熔制(玻璃等) found
熔制车间 glasshouse; melting furnace shop
熔制锭 melted ingot
熔制工 founder; furnace-man; metal tender; teaser
熔质电池 fused electrolytic cell
熔珠反应 bead reaction
熔珠试验 bead test
熔珠下裂纹 underbead crack
熔注成型耐火材料 cast refractory
熔注耐火材料 fusion-cast refractory
熔注装药 cast loading; melt loading
熔铸 casting; found(ing); fusion cast(ing)
熔铸的 fusion cast; melt-casted
熔铸电解铁 fused electrolytic iron
熔铸法 fusion-cast process
熔铸锆刚玉砖 fused cast zirconia-alumina-silica brick
熔铸铬刚玉耐火材料 fused cast chrome-corundum refractory
熔铸工 melter; smelter
熔铸工业金属 fused industrial metal
熔铸合成云母 fusion-cast synthetic(al) mica
熔铸碱性耐火砖 fused basic brick; fused cast magnesite-chrome brick
熔铸块 fused cast block
熔铸镁铬砖 fused cast magnesite-chrome brick
熔铸模 fusion casting mo(u)ld
熔铸莫来石 fused mullite
熔铸莫来石砖 fused cast mullite brick
熔铸耐火材料 castable refractory; cast-fused refractory; fusion-cast refractory; molten-cast refractory
熔铸耐火材料成品 fusion-cast refractory product
熔铸耐火砖 fused cast block; fused cast refractory brick; molten-cast refractory
熔铸十字形格子砖 fused cast cruciform
熔铸玄武岩 fused cast basalt
熔铸氧化铝(耐火)砖 fused cast alu-

mina brick
熔铸砖 fused cast brick; molten-cast brick
熔锥 fusible cone; melting cone; pyrometric cone
熔锥比值 pyrometric cone equivalent
熔锥当量 pyrometric cone equivalent
熔着磨损 scuff wear
熔阻丝 fusible resistor
熔嘴 consumable nozzle
熔嘴电渣焊 electroslag welding with consumable nozzle

融冰 break up of ice; deicing; ice melt(ing); melting ice; thawing ice

融冰槽 dip tank
融冰剂 ice melting agent
融冰径流 ice runoff
融冰流 debacle; ice gang; ice motion; ice run
融冰能力 ice melting capacity
融冰器 defroster; deicer; thawing apparatus
融冰设备 deicing equipment
融冰洼地 asgruben
融冰装置 thawing apparatus
融沉 melt settlement; thaw collapse; thaw subsidence
融沉湖 thaw lake
融沉降 thaw settlement
融沉土 sagging soil
融沉系数 coefficient of melt-settlement; coefficient of thaw subsidence
融沉因数 thaw subsidence factor
融沉因素 subsidence factor thaw; thaw subsidence factor
融点 thaw point
融动作用 congeliturbation
融冻 frost-melting
融冻崩解岩块 congelifractate
融冻崩解作用 frost bursting; frost riving; frost shattering; frost splitting; frost weathering; frost wedging; gelifraction; gelivation; congelifraction【地】
融冻层 active layer; thawed layer; thawing layer
融冻层深度 depth of frozen-thaw layer
融冻沉陷漏斗 thaw settlement funnel
融冻沉陷盆地 thaw settlement basin
融冻沉陷洼地 thaw settlement depression
融冻堆积物 congeliturbate
融冻管 thawing pipe
融冻湖 cave-in lake; thaw lake
融冻泥流阶地 solifluction terrace
融冻泥流作用 congelifraction; congeliturbation; cryoturbation; frost churning; frost stirring; geliturbation; gelivation; solifluction[solifluxion]
融冻泥石 congeliturbate
融冻时期 frost-melting period
融冻试验 thawing and freezing test
融冻桶 thawing tank
融冻土层 tabetisol; talik
融冻洼地 thaw depression
融冻作用【地】 solifluxion
融洞 thaw cave
融沟 thawing grooves
融固结作用 thawed consolidation
融合 interfuse; interfusion; merge
融合的 confluent; fusional
融合地形景色的坡面修整 transitional grading with interfusing landscape
融合核 fusion nucleus
融合剂 fluxing agent; fusion agent
融合问题 blending problem
融合现象 fusion phenomenon
融合约束条件 blending constraint
融化 deliquescence; melting; thaw-(ing); thaw out
融化测量仪 ablatograph
融化层 thawed layer
融化沉陷 melt settlement
融化带 melting band
融化点 melting point; thaw point
融化冻土流 solifluction
融化高度 melting level
融化固结 thaw consolidation
融化固结试验 thaw-consolidation test
融化季(节) thawing season
融化间 thaw house
融化解冻管 thaw pipe
融化壳 milky crust
融化面 thawing surface
融化盘 thaw bowl
融化期 melting period; melting season; period of thaw; thaw period
融化深度 thaw depth
融化水 melt water; thaw(ing) water
融化水径流 melted water runoff
融化速率 melt rate
融化塌陷 thaw collapsibility
融化天气 thaw weather
融化土石法 <建筑隧道用> melting-rock-and-earth method
融化温度 melted temperature
融化稳定永冻土 thaw stable permafrost
融化系数 coefficient of thaw
融化下沉 thaw subsidence
融化下沉系数 thaw settlement coefficient
融化雪泥 snow-broth
融化压力系数 thaw compressibility coefficient
融化压缩 thaw compression
融化压缩试验 thaw compression test
融化压缩系数 coefficient of thaw compression
融化压缩因数 factor of thaw compression; thaw compression factor
融化压缩因子 factor of thaw compression; thaw compression factor
融化指数 <融化季中,累计度一日曲线上最低点和最高点之间的差,以度·日计> thawing index
融化状态 thawed condition
融季 thaw season
融解 deliquesce; melting; thaw(ing); blowout
融解洞 thawing hole
融解期 stage of melting
融解水 <冰雪的> melt water; thawing water
融解土层 tabet soil
融解温度 thaw temperature
融解压力计 manocryometer
融坑 thaw sink
融流动 thaw flowing
融硫黄混凝土 sulfur[sulphur] modified concrete
融硫混凝土 sulfur[sulphur] impregnated concrete
融硫砂浆 sulfur[sulphur] impregnated mortar
融黏[粘] fusion bonding
融泡 thaw bulb
融洽生境 harmonic habitat
融区 thawed zone; talik <多年冻土上的>
融区地下水 groundwater in area of thaw
融熔颗粒耐火材料 fused grain refractory
融熔深度 depth of fusion
融蚀斑晶 brotocrystal
融蚀冰碛 ablation drift; ablation moraine
融蚀层带 ablation zone
融蚀范围 ablation zone
融蚀面 ablation surface
融蚀区 ablation zone
融蚀速率 ablation velocity
融蚀形态 ablation form
融蚀锥 ablation cone
融霜 defrosting
融霜热气管线 hot-gas line for defrosting
融霜水盘 defrost(ing) pan
融霜系统 defrosting system
融水 thawing water
融水沉积 melt-water deposit
融水灌溉 thawing water irrigation
融水河流 meltwater stream
融水湖 melt water lake
融态电解质 molten electrolyte
融碳酸盐法 molten carbonate process
融通 refinancing
融通票据 accommodation check; finance bill <美>
融通资金安排 financing arrangement
融通资金总额 total financing
融土 thawed soil
融线 thaw line
融陷 thaw collapse
融陷湖 cave-in lake; thaw lake
融陷坑深度 depth of thaw-collapse pit
融陷坑数量 number of thaw-collapse pit
融陷土 sagging soil
融陷系数 coefficient of thaw subsidence
融陷性 thaw collapsibility
融陷因子 thaw subsidence factor
融雪 melted snow; snow-broth; snow melt; thaw of snow
融雪崩 warm avalanche
融雪补给河(流) snowmelt-fed river; snowmelt-fed stream; snowfed river; snowfed stream
融雪持续时间 length of melt period
融雪储量 snowmelt storage
融雪风 snow eater
融雪管 snowmelting pipe
融雪洪水 flood of melted snow; snow flood; snowmelt flood; snow water flood; spring flood
融雪后退分析 recession analysis for snowmelt
融雪机 snow melter
融雪径流 melted snow runoff; runoff from snow; runoff of melting snow; snowmelt runoff; snow runoff
融雪径流过程线 snowmelt hydrograph
融雪径流预报 forecast(ing) of melting snow runoff; forecast of snowmelt runoff
融雪库 thawing house
融雪率 snowmelting rate
融雪面积 area of snowmelt
融雪期 melt(ing) period; melting season; period of thaw; snowmelt period; snowmelt season
融雪器 snow melter
融雪强度 snowmelt intensity
融雪区 area of snowmelt
融雪热 heat of fusion of snow
融雪水 snow melt
融雪水补给 recharge of melted snow
融雪水补给河 spring fed stream
融雪水产流量 snowmelt excess
融雪水流 snowmelt release
融雪(速)率 melt rate
融雪雾 snow eater
融雪系数 coefficient of snowmelt
融雪系数法 degree-day method
融雪箱 snowmelting tank
融雪岩 ice melting salt
融雪油 snowmelting oil
融雪预报 snowmelt forecast
融雪终止 ending of snowmelt
融雪装置 snow melter; snowmelting plant
融资 financing; flo(a)tation of loan
融资成本 financing cost
融资方针 credit extending policy
融资费(用) financial charges; financial cost; commitment fee; financing charges
融资合同 financing agreement; financing contract
融资机构 fund raising institution
融资计划 financing plan
融资利率 financing interest rate
融资商 factor
融资商佣金 factorage
融资手段 financing means; means of financing
融资条件 financing term
融资协议 financing agreement
融资业务 provide financing services
融资账户 financing accounts
融资总额与资产价值之比 total financing-to-value ratio
融资租赁 capital lease[leasing]; equipment leasing by financing; finance lease; financial lease[leasing]; financing lease[leasing]

冗长乏味 tedium

冗杆 redundant bar
冗力法 method of redundant reaction
冗力矩 redundant moment
冗量 redundant quantity
冗饰 super-fluidity of ornamentation; superfluity of ornamentation
冗余 redundance
冗余编码 redundance[redundancy] encoding; redundant coding
冗余标引 redundant indexing
冗余并联 piggyback connection
冗余测试 redundance[redundancy] testing
冗余程序设计 redundant programming
冗余的 redundant; superfluous
冗余度 degree of redundance[redundancy]; redundance[redundancy]
冗余度压缩 redundance[redundancy] reduction
冗余反力 redundant reaction
冗余方程 redundant equation
冗余分析 redundant analysis
冗余符号 redundance[redundancy] character; redundant character; redundant symbol
冗余杆件 redundant member
冗余故障 redundant fault
冗余桁架 redundant truss; redundant frame
冗余环 redundance[redundancy] loop
冗余技术 redundance[redundancy] technique
冗余检查位 redundant check bit
冗余检验 redundance[redundancy] check; redundant check(ing)
冗余校验 redundance[redundancy] check; redundant check(ing)
冗余节点 redundant node
冗余力矩 redundant moment

冗余联系 redundant link
冗余率 redundance[redundancy] rate
冗余码 redundance [redundancy] code;redundant code
冗余人员解雇费 redundance[redundancy] payment
冗余设备 redundance[redundancy] unit
冗余设计 redundance [redundancy] design
冗余数 redundant number
冗余数据 redundant data
冗余通道 redundant channel
冗余未知力 redundant unknown force
冗余位 redundance[redundancy] bit; redundance[redundancy] digit; redundant bit;redundant digit
冗余位数校验 redundance[redundancy] check
冗余系杆 redundant tie
冗余系统 redundant system
冗余相位记录 redundant phase record
冗余项 superfluous term
冗余信息 redundant information
冗余性 redundance[redundancy]
冗余应力 redundant stress
冗余约束 reductant constraint
冗余运算消除 redundant operation elimination
冗余支点 redundant support
冗余支杆 redundant strut; redundant support
冗余支座 redundant support
冗余轴 redundant axis
冗余状态 redundant state
冗余字符 redundance [redundancy] character;redundant character
冗余自动机 redundance[redundancy] automaton
冗员 redundance[redundancy]
冗轴 redundant axis

柔布机 cloth breaking machine

柔层 soft layer
柔层摆动机构 soft storey sway mechanism
柔度 flexi(bi)lity;softness
柔度法 flexibility approach;flexibility method
柔度矩阵 flexibility matrix
柔度系数 flexibility factor; softness coefficient;softness factor
柔度影响系数 flexibility influence coefficient
柔杆电钻 electric(al) drill with flexible pipe
柔杆电钻钻进 flexible stem electro-drilling
柔杆钻进 bendable pipe drilling
柔光 lambency; soft light; subdued light
柔光镜头 soft-focus lens
柔光聚焦镜 soft-focus lens
柔光谱 soft-focus attachment
柔光器 lighting diffuser
柔光照明 soft lighting
柔和 mellowing;softness;tone-down
柔和的色调 pastel tone
柔和反差 soft contrast
柔和空气支承 compliant air bearing
柔和磨料 mild abrasive
柔和黏[粘]土 mild clay
柔和曲线 comfort curve
柔和色 soft colo(u)r
柔和色调 tender tone;undertint
柔滑的 sleeky
柔焦 soft focus
柔焦镜 soft-focus attachment;soft-focus filter
柔焦镜头 soft-focus lens
柔焦像 soft-focus image
柔金属软管 flexible metal hose
柔量 compliance[compliancy]
柔流 flowage
柔流变质 rheomorphism
柔流褶皱 flowage folding
柔毛山核桃 big-bud hickory; black hickory
柔膜菌属 <拉> Helotium
柔黏[粘]土 blunge
柔曲 flexure
柔曲计 fleximeter
柔曲性 flexibility
柔韧 pliability
柔韧的 flexible;pliable
柔韧度 degree of flexibility
柔韧剂 plasticiser[plasticizer]
柔韧铜辫 copper pigtail;pigtail
柔韧性 flexibility;litheness;pliability; pliableness;pliancy;softness
柔韧性试验 flexibility test
柔韧支承结构 flexible supporting structure
柔软 pliability
柔软处理 mellowing
柔软的 doughy; downy; ductile; Limp;lithe;lithesome;mild;plastic; pliable;soft;supple;velvety
柔软电阻 flexible resistor
柔软度 scratchability;softness
柔软度试验机 softness tester
柔软度与刚度测定仪 softness-stiffness tester
柔软分路器 flexible shunt
柔软感涂料 soft-feel coating
柔软剂 softener;softening agent
柔软结持度 soft consistence[consistency]
柔软经营 soft management
柔软绝缘 flexible insulation
柔软膜片 flexible diaphragm;limp-diaphragm
柔软态 limp state
柔软填料 soft packing
柔软性 compliance;ductility;flexibleness;limpness;pulpiness;softness
柔软性能测定 softness test
柔软羊皮手套 mocha
柔软云母板 flexible mica
柔软折叠 soft fold
柔软整理 <少用或不用浆料> soft finish
柔弱 softness;weakness
柔顺 compliance
柔顺常数 compliance constant;elastic constant
柔顺的 pliable;yielding;supple
柔顺性 flexibility;pliability
柔体 deformable body
柔弦 flexible chord
柔性 smooth feature
柔性岸壁 flexible bulkhead; flexible tie-back bulkhead
柔性坝 flexible dam
柔性摆(动式)门 flexible swing door
柔性板 flexible board;flexible sheet; flexible slab;flexiboard
柔性板荷载法 flexible plate loading method
柔性变幅机构 flexible derricking mechanism
柔性变形 dough deformation;ductile deformation;plastic deformation
柔性标志 <标志横立在路面上,车辆经过时弯倒,之后弹回> flexible sign
柔性表面 flexible surfacing
柔性表面设计 flexible surfacing design
柔性表面设计法 flexible surfacing design method
柔性波导管 interlocked type waveguide; squeezable waveguide; flexible waveguide
柔性薄膜衬垫 flexible membrane liner;flexible membrane lining
柔性薄膜衬里 flexible membrane liner
柔性材料 flexible material
柔性层 flexible layer
柔性插销及套管连接器 flexible plug and socket connector
柔性沉排 flexible mattress
柔性衬砌 flexible fining;flexible lining
柔性承台 flexible platform; flexible relieving platform
柔性(出)料槽 <可弯折的出料槽> flexible spout
柔性传动 wrapping connector driving;flexible drive
柔性传动系统 flexibility drive system
柔性传像束 flexible image guide
柔性搭板 soft patch
柔性挡板 flexible choker
柔性挡土墙 flexible retaining wall
柔性导管 flexible conduit
柔性导体 flexible conductor
柔性道路 flexible road
柔性道路工程公司 flexible road industry
柔性道路施工 flexible road construction
柔性的 flexible;non-rigid
柔性底层 flexible base course;flexible first storey;flexible substratum
柔性底层结构 flexible first storey construction
柔性地板 non-rigid floor
柔性地板材料 flexible flooring
柔性地板复面层 asbestos-vinyl floor cover(ing)
柔性地基 flexible base
柔性电缆 flexible cable
柔性电阻器 flexible resistor
柔性垫块 flexible block
柔性垫块浮体防波堤 flexible mat floating breakwater
柔性垫子 flexible mattress
柔性垫子或系缆浮子防波堤 flexible mat or tethered float breakwater
柔性墩 flexible dolphin;flexible pier
柔性墩的设计 design of flexible dolphin
柔性防潮层 flexible damp course
柔性防冲衬垫 flexible fender cushion
柔性防水 flexible water-proof
柔性防水套管 flexible water-tight sleeve
柔性防水屋面 built-up roof(ing)
柔性防撞垫 flexible fender cushion
柔性非金属管 flexible tubing
柔性风管 flexible duct
柔性缝 flexible joint
柔性复合面层 flexible composite pavement
柔性盖板 boot
柔性盖层 flexible cover;flexible overlay
柔性杆 deformable bar; deformable rod;flexible rod
柔性钢拱 non-rigid steel arch
柔性钢管靠船墩 flexible steel cylinder dolphin
柔性钢结构拱 flexible steel arch
柔性钢筋 flexible reinforcement
柔性钢筋混凝土基础 flexible reinforced(concrete)foundation beam
柔性钢钎 flexible drill steel
柔性钢索 flexible cable
柔性钢桩靠船墩 flexible steel pile dolphin
柔性隔热材料 flexible insulation
柔性隔热隔音板 insulation batt
柔性拱 flexible arch; yielding roadway arch
柔性拱刚性梁桁架 Langer girder; Langer truss
柔性拱桥 flexible arch(ed)bridge
柔性构件 flexible member; flexure member
柔性构造 flexible construction;flexible structure
柔性固体 flexible solid;plastic solid
柔性管 flexible pipe; flexible tube; hose;metal hose
柔性管充气式探测器 road tube detector
柔性管道 flex-duct;flexible conduit; flexible duct
柔性管缆 <疏通排水管和排气管用> plumbing snake
柔性管连接 flexible hose connection
柔性管套 flexible pipe section
柔性硅应变仪 flexible silicon strain ga(u)ge
柔性滚动轴承 flexible roller bearing
柔性过渡管接头 flexible transition coupling
柔性海漫 flexible apron
柔性涵洞 flexible culvert
柔性涵管 flexible conduit
柔性巷道拱构件 yielding roadway arch
柔性荷载 flexible load
柔性护岸 flexible armo(u)red revetment;flexible revetment
柔性护栏 flexible safety fence
柔性护面 flexible apron
柔性护面接缝 flexible facing joint
柔性护木系统 flexibility-fender system;flexible fender system
柔性护坡 flexible revetment
柔性护舷系统 flexibility-fender system;flexible fender system
柔性护罩 flexible shroud
柔性混凝土 flexible concrete
柔性活节 flexible joint
柔性基层 flexible base; flexible base course
柔性基层材料 flexible base material
柔性基层路面 flexible base pavement
柔性基础 flexible base; flexible footing;flexible foundation
柔性基底材料 flexible base material
柔性集装器 flexible container
柔性加工系统 flexible manufacturing system
柔性加荷面 flexible loading plane
柔性夹板 flexible spline
柔性减振器 flexible shock absorber
柔性胶合板 flexible plywood
柔性胶粘剂 flexible adhesive
柔性焦距透镜组 spectacle lenses
柔性铰 flexure hinge
柔性接头 flex connector;flexible connection; flexible coupling; flexible joint
柔性节点 flexible joint;flexible node
柔性结构 flexible construction;flexible structure
柔性金属 flexible metal
柔性金属板 flexible metal sheet
柔性金属板屋面 flexible metal roofing
柔性金属管 flexible metal hose;flexible metallic hose
柔性金属管道 flexible metal conduit
柔性金属软管 flexible metallic hose

柔性金属套管 flexible metallic conductor
柔性金属屋面 flexible metal roofing
柔性静流 stationary plastic flow
柔性矩阵 flexibility matrix; flexible matrix
柔性聚氨基甲酸乙酯 flexible polyurethane
柔性聚氯乙烯摆(动式)门 flexible polyvinyl chloride(swing)door
柔性聚氯乙烯薄膜 flexible polyvinyl chloride film
柔性聚氯乙烯树脂 flexible polyvinyl chloride resin
柔性聚氯乙烯弹簧门 flexible polyvinyl chloride(swing)door
柔性聚亚氨酯 flexible polyurethane
柔性聚乙烯 flexible poly(e)thene
柔性聚乙烯管 flexible poly(e)thene pipe
柔性开口肋 soft-open rib
柔性铠装电缆 flexible armo(u)red cable
柔性靠船墩 flexible dolphin
柔性靠船架 flexibility dolphin; flexible dolphin
柔性可调出料槽 flexible spout
柔性控制 flexible control
柔性框架 flexible frame
柔性缆索 flexible cable
柔性肋 soft rib
柔性连接 flexible connection
柔性连接管 flexible tubing
柔性连接器 flexible connector
柔性连接润滑器 flexible lubricator
柔性连接线 lamp cord
柔性联轴节 flexibility coupling; flexible connection; flexible coupling; flexible joint
柔性链段 soft segment
柔性溜料槽 flexible drop chute
柔性楼板整修 asbestos-vinyl floor(ing)finish
柔性楼层 flexible stor(e)y
柔性楼房 flexible stor(e)y
柔性楼盖 flexible floor
柔性楼面整修 asbestos-vinyl floor(ing)finish
柔性路标 flexible sign
柔性路面 flexible pavement; flexible surface; non-reinforced pavement
柔性路面地基 flexible pavement subgrade
柔性路面厚度设计法 thickness design methods for flexible pavement
柔性路面设计 flexible pavement design
柔性路面设计方法 flexible pavement design method
柔性路面弹性动态 elastic behavio(u)r of flexible pavement
柔性路面弹性作用 elastic behavio(u)r of flexible pavement
柔性路养护系统 <英国一种用于干线公路等的> Computerised Highway Assessment of Ratings and treatment
柔性履带 resilient track
柔性氯丁橡胶摆(动式)门 flexible neoprene(swing)door
柔性氯丁橡胶弹簧门 flexible neoprene(swing)door
柔性密封(条)wiping seal
柔性面板接缝 flexible facing joint
柔性面层 flexible carpet; flexible cladding; flexible pavement; flexible surface[surfacing]
柔性模量 compliance modulus
柔性膜 flexible membrane
柔性模(具)flexible die; flexible mo(u)ld
柔性模子 flexible mo(u)ld
柔性母线 string bus
柔性木桩靠船架 flexible timber pile dolphin; flexible wood pile dolphin
柔性黏[粘]合剂 flexible adhesive
柔性黏[粘]土 mild clay
柔性耦联 flexible coupling
柔性盘联轴节 disk coupling; flexible disc[disk] coupling
柔性泡沫 flexible foam
柔性泡沫材料 flexible foam
柔性泡沫塑料 flexible plastic foam
柔性碰垫 flexible fender
柔性铺面 flexible pavement
柔性铺面护岸 flexible pavement revetment
柔性起动器 soft starter
柔性墙 flexible wall; limp wall
柔性墙的被动面 passive face of flexible wall
柔性墙的主动面 active face of flexible wall
柔性曲线规 flexible curve rule; flexible spline
柔性驱动屏幕 flexible drive screen
柔性驱动筛 flexible drive screen
柔性全履带车 flexible full track
柔性入口 flexible throat
柔性软管 flexible hose
柔性塞缝片 flexibility spline; flexible spline
柔性砂浆 flexible mortar
柔性砂岩 flexible sandstone
柔性梢排 flexible mattress
柔性石墨 flexible graphite
柔性石墨垫片 flexible graphite gasket
柔性受弯建筑物 flexural building
柔性塑料 flexible plastics; flexiplast
柔性塑料摆动式门 flexible plastic door
柔性塑料排水管 flexible plastic drain
柔性塑料磁盘 diskette; flexible plastic disc; floppy disc[disk]
柔性隧道支护 flexible tunnel lining
柔性索缆 flexible cable
柔性弹簧门 flexible swing door
柔性弹性体泡沫绝热材料 flexible elastomeric cellular thermal insulation
柔性体积法 flexible volume method
柔性天线反射器 flexible reflector curtain
柔性填缝料 flexible joint filler
柔性调节 soft readjustment
柔性调节剂 flexibility modifier
柔性透空式桩基结构 flexible open-piled structure
柔性腿(柱)flexible leg
柔性弯曲 flexural bending
柔性弯头 flexible bend
柔性万向节 rubber universal joint
柔性无缝管(道)flexible seamless tubing
柔性系泊浮筒 flexible dolphin
柔性系船墩 flexible mooring dolphin
柔性系数 flexibility coefficient; flexible coefficient; flexible factor; ratio of slenderness; slenderness ratio
柔性下部结构 flexible understructure
柔性下料管 flexible drop chute
柔性纤维 curved fibre[fiber]; flexible fiber[fibre]
柔性弦 flexible chord
柔性线绝缘 flexible insulation
柔性橡胶摆(动式)门 flexible rubber(swing)door
柔性橡胶垫圈 flexible rubber gasket
柔性橡胶弹簧门 flexible rubber(swing)door
柔性卸料槽 flexible spout
柔性芯墙 flexible core wall
柔性型式 flexible type
柔性悬浮 flexible suspension
柔性悬挂 flexible suspension
柔性悬索 flexible suspension
柔性掩护支架 flexible type shield
柔性液囊 towed flexible barge
柔性液轴输送机 limberoller conveyer[conveyor]
柔性仪 plastometer
柔性翼 flex-wing
柔性影响系数 flexible influence factor
柔性预制混凝土桩靠船架 flexible precast concrete pile dolphin
柔性毡 flexible mat
柔性毡垫状保温层 bat insulation
柔性毡垫状隔音层 bat insulation
柔性张量 compliance tensor
柔性罩面 flexible overlay
柔性折棚装置 closure
柔性振动桩锤 flexible type vibratory pile hammer
柔性支撑 resilient mounting
柔性支承 flexible support; sinking support
柔性支承端 flexible supported end
柔性支承墙 flexible supported wall
柔性支承系统 flexible supported system
柔性支护 flexible support
柔性支架 flexible support
柔性支腿 flexible leg; sway leg
柔性止水 flexible seal
柔性止水缝 flexlock
柔性止水片 flexible plate
柔性制造系统【机】flexible manufacturing system
柔性轴 flexible axle; flexible shaft
柔性轴承 flexible bearing
柔性主轴 principal axis of compliance
柔性柱 flexible column
柔性转向盘轴管 collapsible steering column
柔性转子 flex rotor
柔性桩台 flexible platform on piles
柔性装配设备 flexibly mounted equipment
柔性自动化 flexible automation
柔性组件 flexible unit
柔影 soft shadow
柔枝 splint
柔质夹心板 flexible coreboard
柔质乳胶 cement rubber latex; fleximer

揉

揉拌作用 rolling and mixing action
揉布机 breaker
揉擦剂 anatriptic
揉成团(的油灰、陶土等)dough
揉搓式混砂机 kneading machine
揉搓式磨浆机 curlator
揉搓作用 kneading action
揉好的陶土 dough
揉和槽 kneading trough
揉和机 kneading compactor; malaxator
揉混 malaxation
揉混机 malaxator; masticator
揉麻泥工具 loam beater
揉麻梳麻机 breaker scutcher
揉面 duff
揉泥碾 chaser
揉泥台 wedging table
揉泥围堰 puddle cofferdam
揉捏 kneading
揉捏法 kneading; malaxation; petrissage
揉捏混合作用 pressure mixing
揉捏机 pug
揉捏了的 kneaded
揉曲性 flexibility
揉曲性黏[粘]合剂 flexible glue
揉压法 kneading method
揉压机 kneading compactor
揉轧涂装 fading
揉皱 corrugation
揉皱层理 corrugated bedding
揉皱带 crumpling zone
揉皱的 crumpled
揉皱结构 corrugation texture; crumpled texture
揉皱作用 crumpling
揉纵杆轴 control lever shaft

鞣

鞣革 tannage; tanning
鞣革厂 tannery
鞣革废物 bating process waste; tanning waste
鞣革工厂废物 beamhouse waste
鞣花单宁 ellagitannin
鞣花单宁酸 ellagitannic acid
鞣花酸 ellagic acid; gallogen acid
鞣化过程废水 bathing process wastewater
鞣化过程废物 bathing process waste
鞣剂 tanning agent
鞣料 tan; tanning material; tanning matter
鞣皮 curry
鞣皮厂 curriery
鞣皮业 curriery
鞣酸 gallic acid; gallotannic acid; tannic acid; tannin <处理混凝土模板用>
鞣酸皮渣 <可用来筑路> spent tan; tan
鞣酸侵蚀 tannin attack
鞣酸渣 tan
鞣液 tan(ning)liquor
鞣液比重计 barkometer
鞣制 tan
鞣质的 tannic
鞣质体 phlobaphinite

肉

肉吊架 troll(e)y meat rail
肉豆蔻属 nutmeg
肉豆蔻酸 myristic acid; tetrade canoic acid
肉豆蔻油 myristic oil
肉附座壳菌属 <拉> Sarcoxylon
肉桂【植】sweet bay; cassia-bark-tree; Chinese cinnamon
肉桂树 cassia bark
肉桂酸苄脂 benzyl cinnamate
肉桂烯 cinnamene
肉桂棕色 cinnamon brown
肉红色 incarnadine; yellowish pink
肉红玉髓 sard; sardachate; sardine
肉架 <冷藏车> beef rail; meat rail; hanging bar
肉类包装工业 meat packing industry
肉类船 meat ship
肉类罐头厂 meat cannery
肉类果品加工包装厂 pack-house
肉类加工 meat curing; meat packing; meat processing
肉类加工厂 meat plant; meat work
肉类加工厂废水 packing house waste(water)
肉类加工厂副产品 meat processing byproduct
肉类加工场 meat packing plant
肉类加工废水 abattoir wastewater; meat industry wastewater; meat processing wastewater

R

肉类加工废物 abattoir waste;meat industry wastes;packing house wastes
肉类加工副产品 meat by-product
肉类加工工业 meat product industry
肉类冷藏库 meat cooling plant
肉类冷藏所 packing house
肉类冷冻 meat freezing
肉类联合(加工)厂 meat packing plant;meat processing factory
肉类装运船 meat carrier;meat ship
肉盘菌属 <拉> Sarcosoma
肉色 carnation; carne(o)se; flesh; flesh colo(u)r
肉色正长石 paradoxite
肉色柱石 sarcolite
肉食包装厂 packing house
肉食包装废料 meat packing waste
肉食加工厂 meat ware factory;packing house;packing plant
肉丝白粉菌属 <拉> Leveillula
肉头 depth of finish
肉托果壳酚 bhilawanol
肉托果壳油 bhilawan nut shell oil
肉眼 naked eye; unadjusted eye; unaided eye
肉眼测量法 visual measurement
肉眼观测 naked eye observation
肉眼观察 by sight;naked viewing;perusal; unaided view(ing); visual observation
肉眼观察法 macrography method
肉眼检查 check by sight;eye assay;eyeball assay; gross examination; gross inspection;lens examination; macrograph; macrographic examination; macrography; macroscopy; naked eye examination; visual check;visual examination;visual inspection
肉眼鉴别 eye assay; visual appearance;visual examination;visual inspection
肉眼鉴定 eye assay; visual appearance;visual examination;visual inspection
肉眼鉴定盐渍度 salinity identification by the naked eye
肉眼镜煤 megascopic anthraxylon
肉眼勘察技术 visual exploration technique
肉眼看得到 <不用显微镜的> gross
肉眼可见的 apparent to the naked eye;macroscopic;megascopic;lucid【天】
肉眼可见裂缝 eye visible crack
肉眼孔隙 macroporosity
肉眼煤岩类型 macroscopic lithotypes of coal
肉眼判别法 megascopic method
肉眼识别 megascope;visual identification
肉眼识别的 macroscopic
肉眼识别法 macroscopic method
肉眼丝炭 megascopic fusain
肉眼图 macrograph
肉、鱼、家禽加工废物 meat, fish and poultry processing wastes
肉质刺灌丛 succulent thorn-scrub
肉质植物 succulent
肉棕色 flesh blond
肉座菌属 <拉> Hypocrea

如 飞檐(的字头) geisso

如镜的 specular
如镜面 glassy surface
如泡沫的 foamy
如期偿付 meet
如期付款 meet
如期交货 deliver the goods
如期完工 keep to program(me); keep to schedule;timely completion
如前所述 ditto
如球的 knoblike
如全景 panorama
如石的 stony
如数收讫 <支票的> value received
如丝绒的 velvet
如梯的 scalary
如系日周潮 lower low water
如意扳手 S(-type) spanner
如意钩 luck hook
如意牌 Ruyi brand
如圆丘的 knoblike
如针状的 needly
如砖大小的瓦 brick size tile

茹 拉夫斯基法 <俄国工程师,静定桁架的节点分析法> Jourawski's method

铷 87 的热产率 heat productivity of Rb-87

铷 87 含量 content of Rb 87
铷的热产率 heat productivity of Rb
铷矿床 rubidium deposit
铷矿(石) rubidium ore
铷频率标准 rubidium frequency standard
铷-锶测年法【地】 Rb-Sr [rubidium-strontium] dating method
铷-锶等时线 rubidium-strontium isochrone
铷-锶法 rubidium-strontium method
铷-锶模式 rubidium-strontium model
铷-锶年代测定法 rubidium-strontium age method
铷-锶年龄 rubidium-strontium age
铷蒸气磁力仪 rubidium vapo(u)r magnetometer

儒 可夫斯基变换 Joukowski transformation

儒略纪元 Julian epoch;Julian era
儒略历 Julian calender;Julius Caesar calendar;old style
儒略历书日期 Julian ephemeris date
儒略历书世纪 Julian ephemeris century
儒略历元 Julian epoch
儒略年 Julian year
儒略日 Julian day
儒略日历 Julian day calendar
儒略日期 Julian date
儒略日数 Julian day number; Julian number
儒略世纪 Julian century
儒略周期 Julian period

濡 湿力 wetting force;wetting power

濡损 sweat damage

蠕 变 creepage; creep(ing); plastic loss;after flow;time-yield;walk out

蠕变比 creep ratio
蠕变变形 creep deformation; creep strain
蠕变(变形)试验仪 creep deformation tester
蠕变泊松比 creep Poisson's ratio
蠕变(薄)层 creep lamella
蠕变差 creep value
蠕变常数 creep constant
蠕变沉降 creep settlement
蠕变成型 creep forming
蠕变持久试验 creep and stress rupture test
蠕变传感器 creep cell
蠕变的损失 loss due to creep
蠕变的温度影响 temperature dependence of creep
蠕变点 creep point
蠕变叠加 superposition of creep
蠕变度 specific creep
蠕变断裂 creep fracture; creeping crack;creep runway;creep rupture
蠕变断裂强度 creep breaking strength;creep rupture strength
蠕变断裂试验 creep rupture test
蠕变反挠度 creep camber
蠕变复原 creep recovery
蠕变观测 creep observation
蠕变过程 creep process
蠕变荷载 creep load
蠕变盒 creep cell
蠕变恒速区 secondary creep
蠕变滑动 creep(ing) slip(page)
蠕变恢复 creep recovery
蠕变回复 <卸载后蠕变随时间而减少> creep recovery
蠕变毁坏 creep collapse
蠕变机理 creep mechanism; mechanism of creep
蠕变机制 creep mechanism
蠕变极限 creep limit;creep strength; limited creep stress;limit of creep
蠕变极限应力 limited creep stress
蠕变极限值 creep limiting value
蠕变计 creep meter
蠕变阶段 creep stage
蠕变劲度 creep stiffness
蠕变距离 creeping distance
蠕变抗力 creep resistance
蠕变可逆性 creep reversibility
蠕变理论 creep theory
蠕变裂缝 creep crack
蠕变裂纹 creeping crack
蠕变流动度 creep fluidity
蠕变流动性 creep fluidity
蠕变率时间图 creep rate-time plot
蠕变模量 creep modulus;modulus of creep
蠕变模型 creep model
蠕变挠度 creep deflection
蠕变挠曲 creep deflection;creep deflexion
蠕变黏[粘]度 creep viscosity
蠕变疲劳 creep fatigue
蠕变破断寿命 creep rupture life
蠕变破坏 creep collapse;creep failure;creep fracture;creep rupture
蠕变破坏试验 creep rupture test
蠕变破坏因素 creep damage factor
蠕变破裂强度 creep rupture strength
蠕变强度 creep strength
蠕变曲线 creep(ing) curve;flow line
蠕变曲线类型 type of creep curve
蠕变屈曲 creep buckling
蠕变柔量 creep compliance
蠕变设计曲线 creep design curve
蠕变伸长 creep elongation
蠕变失稳 creep instability
蠕变失效准则 creep failure criterion
蠕变时间曲线 creep time curve
蠕变势(边坡) creep potential
蠕变试验 creep test; long duration static test;time loading test
蠕变试验机 creep tester;creep test(ing) machine
蠕变寿命 creep life
蠕变寿命消耗率 creep damage factor
蠕变松弛 creep relaxation
蠕变速度 creep speed; creep velocity;rate of creep;velocity of creep
蠕变速率 creep rate;rate of creep
蠕变特性 creep property
蠕变特性试验 creep behavio(u)r test
蠕变条件 creep condition
蠕变位移仪 creep displacement meter
蠕变温度 creep temperature
蠕变系数 creep coefficient;creep factor
蠕变纤维 creep fibre[fiber]
蠕变现象 creep phenomenon; flow phenomenon
蠕变线 line of creep
蠕变效应 creep effect
蠕变行为 creep behavio(u)r
蠕变性(能) creep property
蠕变性状试验 creep behavio(u)r test
蠕变压力 creep pressure
蠕变压曲 creep buckling
蠕变延伸 creep extension
蠕变岩体 creeping rock mass
蠕变仪 creep apparatus;creep meter
蠕变因数 creep factor
蠕变引起的预应力损失 prestressing loss due to creep
蠕变应变 creep(ing) strain
蠕变应变回复 creep strain recovery
蠕变应变-时间图 creep strain-time plot
蠕变应力 creep stress
蠕变影响 draw
蠕变预测 creep prediction
蠕变原理 theory of creep
蠕变再分布 redistribution of creep
蠕变再分配 redistribution of creep
蠕变增长 creep growth
蠕变值 creep value
蠕变指数 creep index
蠕变状态 creep state
蠕变阻力 creep resistance
蠕波 creeping motion;creeping wave
蠕虫 helminth;worm
蠕虫病 helminthic disease
蠕虫图 worm diagram
蠕虫形 worm
蠕虫铸型 eropoglyph
蠕虫状结构 myrmekitic texture;vermicular texture
蠕虫状黏[粘]土 netted texture clay; vermicular clay
蠕虫状影纹 vermicular texture
蠕动 crawl(ing); creep(age); creeping(motion); stable sliding; worm; wriggle;wring
蠕动泵 peristaltic pump
蠕动比 creep ratio
蠕动变形 creep deformation
蠕动波 peristaltic rushes; peristaltic wave;rushes
蠕动(薄)层 creep lamella
蠕动擦痕面 creeping slickenside
蠕动残渣物 creeping waste
蠕动迟缓 hypoperistalsis
蠕动的 enterokinetic;peristaltic
蠕动电荷耦合器件 peristaltic charges coupled device
蠕动钉住 crawling peg
蠕动断层 creep fault
蠕动方式 creep type
蠕动极限 creep limit
蠕动"咔嗒"声 click-slip chatter
蠕动空间 crawl space
蠕动量 cumulative amount of creep
蠕动流 <黏[粘]滞性很高的液体的> creeping flow
蠕动膜 creeping film

蠕动能力 creep capacity
蠕动坡积物 creeping slope wash
蠕动破裂 creep fracture
蠕动速率 creep rate;rate of creep
蠕动特性 creep behavio(u)r;creep characteristic
蠕动图形 crawling pattern
蠕动线 creep line;line of credit;line of creep
蠕动限制系统 crawling peg
蠕动效应 crawling effect;creep effect
蠕动压力 creeping pressure
蠕动应变 creep strain
蠕动皱纹 creep wrinkle
蠕滑 creep
蠕滑断裂 creeping slip fault
蠕滑阻力 creep resistance
蠕缓放电 creeping discharge
蠕流变 creeping flow
蠕流线 creep line;line of creep
蠕绿泥石 helminthes;lophaite;oncoite;prochlorite;ripidolite
蠕墨铸铁 vermicular cast iron
蠕蛇纹石 sungulite
蠕升 creeping;strain relaxation
蠕丝壳囱属<拉> Irene
蠕速弯沉 creep-speed deflection
蠕陶土 anauxite
蠕行波 earthworm-motion wave
蠕行速度 crawl speed
蠕行随机搜索 creeping random search
蠕移输沙量 silt discharge of creep-movement
蠕状的 myrmekitic
蠕状构造 myrmekitic structure
蠕状结构 myrmekitic texture
蠕状耐火黏[粘]土 burley clay
蠕状石 myrmekite

乳

乳胺 lactam
乳胺镉 phenylaminocadmium dilactate
乳吧间 milk bar
乳白 milky-white;white fraction
乳白玻璃 bone glass;cryolite glass;milk white glass;milk(y) glass;opacified glass;opalescent glass;opalesque glass;opaline;opaline glass;opaque glass;sponge glass
乳白玻璃灯(光) opal light
乳白玻璃纤维 opalescent glass fibre
乳白衬里琥珀玻璃板 plated amberian
乳白的 alabaster;opacified;opal
乳白灯泡 opal bulb;opal lamp;opal lamp bulb
乳白度 milk scale
乳白镀层 milky surface
乳白光 opalescence
乳白光的 opalescent
乳白化 creaming;opalescence;opalization
乳白(灰)玻璃 alabaster glass
乳白辉 arctic whiteout
乳白剂 opalizer
乳白刻度滴管 milk scale buret(te)
乳白蜡 opal wax
乳白磷铝石 taranakite
乳白面玻璃 milky surface of glass
乳白色 cream colo(u)r;cream white;ivory;ivory white;lacte;lunar white;milkiness;milk(y) white;opalescence
乳白色冰柱 rime
乳白色玻璃 bone glass;milk glass;opalescent glass;opal glass
乳白色的 milky
乳白色灯(泡) inside-frosted lamp;opal lamp
乳白色厚光纸 ivory paper
乳白色凸透镜 opal convex lens
乳白色油漆 broken white;ivory paint
乳白石英 milky quartz;opaline quartz
乳白水 milky ice
乳白燧石 opaline chert
乳白搪瓷 opaque enamel
乳白天空 milky weather;white-out
乳白天气 milky weather
乳白现象 white-out
乳白油 opal oil
乳白釉 egg white glaze;milky glaze;opal(ine) glaze
乳白纸 cream paper
乳比重计 galactometer
乳钵 mortar
乳产品 dairy product
乳蛋白石 milk opal
乳滴状结构 emulsion texture
乳儿室 suckling room
乳房构造【地】 mammillary structure
乳房状的【地】 mammillary
乳房状山【地】 mammillary hill
乳房状(云) mammatus
乳袱 beam tie
乳菇属<拉> Lactarius
乳光 opalescence
乳光玻璃 emulsion opal glasses;opaline glass
乳光玻璃板 marver
乳光的 opalescent;opaline
乳光度 degree of opalescence
乳化 emulsify(ing);emulsionize
乳化本领 emulsifying power
乳化玻璃 bone glass
乳化薄浆 slurry emulsion;slurry seal emulsion
乳化材料 emulsifiable material
乳化槽 emulsifying tank
乳化层 emulsion layer
乳化处理器 emulsion treater
乳化带 emulsion band
乳化的 emulsive
乳化的过程 emulsification
乳化地沥青 asphalt(ic) emulsion;emulsified asphalt;emulsified bitumen
乳化度 emulsifiability
乳化法 emulsion process
乳化反应 emulsion reaction
乳化方法 emulsifying process
乳化防腐油 creosote emulsion
乳化分解度 demulsibility
乳化分解性 demulsibility
乳化浮选 emulsion flo(a)tation
乳化改性剂 emulsified modifier
乳化膏 emulsifiable paste
乳化硅油 emulsified silicone oil
乳化过程 emulsifying process;emulsion process
乳化黑色油 black strap
乳化糊 emulsion thickener
乳化混合器 emulsifying mixer;emulsion mixer
乳化机 disperse mill;emulsifying machine;emulsion machine;homogenizer;mulser
乳化基质 emulsified base
乳化剂 dispersing agent;emulator;emulgator;emulphor;emulsifier;emulsifying agent;emulsion;emulsive agent;pentamul;wedding agent
乳化剂层 emulsifier layer
乳化剂灭火法<一种油面灭火法> emulsified system
乳化剂灭火系统 emulsified system
乳化剂黏[粘]度 emulsifier viscosity;emulsion viscosity
乳化剂使用 use of emulsifying agents
乳化浆 emulsion paste
乳化浆印花 emulsion printing
乳化胶 emulsion glue
乳化胶体<用于沥青涂层> emulcol;emulsifying colloid
乳化焦油沥青 tar emulsion
乳化搅拌器 emulsifying agitator
乳化校正曲线 emulsion calibration curve
乳化聚合作用 emulsification polymerization
乳化颗粒 emulsified particle
乳化蜡 emulsifying wax
乳化冷却剂 emulsified coolants
乳化冷却液 oil-in-water type coolant
乳化理论 emulsion theory
乳化力 emulsifying power
乳化沥青 asphaltic emulsion;bitumen emulsion;bituminous emulsion;cold bitumen emulsion;elestex;emulsion asphalt;inverted asphalt emulsion;tar emulsion
乳化沥青材料 emulsified bituminous materials
乳化沥青沉淀试验 settlement test for emulsified asphalt
乳化沥青防渗灌浆法 shell-perm process
乳化沥青混合料 emulsified asphalt mixture
乳化沥青浇灌法 bitumen emulsion injection
乳化沥青焦油 emulsion of tar
乳化沥青密封膏 asphalt emulsion slurry seal
乳化沥青喷剂 mulseal
乳化沥青漆 emulsified asphalt varnish
乳化沥青洒布机 emulsion sprayer
乳化沥青设备 asphalt emulsion system
乳化沥青(水泥)拌和试验 cement mixing test(for emulsified asphalt)
乳化沥青碎石 emulsified bitumen macadam
乳化沥青系统 asphalt emulsion system
乳化沥青蒸发仪 emulsified asphalt distillation apparatus
乳化沥青蒸馏试验后的残渣 residual from distillation
乳化灭火 extinguishment by emulsification
乳化磨 emulsifying mill
乳化能力 emulsifying ability;emulsifying capacity
乳化泥浆 emulsion mud
乳化黏[粘]合剂 emulsified binder
乳化漆 emulsifiable paint
乳化器 emlgator;emulgator;emulsifier;emulsifier injector;emulsor
乳化强度 emulsion strength
乳化切削油 emulsion cutting oil
乳化燃料 emulsion fuel
乳化燃烧器 emulsifying burner
乳化溶剂 emulsified solvent
乳化软膏 slurry emulsion;slurry seal emulsion
乳化软膏操作 slurry seal operation
乳化润滑油添加剂 emulsifying lubricants additive
乳化石蜡 emulsion wax
乳化试验 emulsifiability test;emulsification test;emulsion test
乳化试验器 emulsification testing apparatus
乳化室 emulsion chamber
乳化树脂 emulsified resin
乳化树脂黏[粘]结料 resin emulsion cement
乳化树脂涂料 resin-emulsion paint
乳化酸 emulsified acid
乳化塔 emulsifying tower;emulsion tower
乳化特性 emulsification property;emulsifying property
乳化添加剂 emulsifying additive
乳化烃 emulsifying hydrocarbon
乳化烃燃料 emulsified hydrocarbon fuel
乳化涂料 emulsified paint;emulsion coating
乳化脱脂 emulsion degreasing
乳化稳定剂 emulsion stabilizer
乳化物质 emulsifier
乳化橡胶 emulsified rubber
乳化硝酸铵 emulsified ammonium nitrate
乳化效率 emulsifying efficiency
乳化效应 emulsifying effect
乳化型金属加工液 emulsion type metalworking liquid
乳化性 emulsibility;emulsifiability;emulsifying ability;emulsifying property
乳化性渗透液 post-emulsification penetrant
乳化液 emulsion(fluid);oil-water emulsion;water-in-oil emulsion
乳化液泵 emulsion pump
乳化液层 emulsion membrane
乳化液充气 emulsion aeration
乳化液处理 treatment of emulsions
乳化液分离 separation of emulsions
乳化液概论 emulsion general
乳化液清洗 emulsion cleaning
乳化液染色 emulsion dy(e)ing
乳化(液)稳定性 emulsion stability
乳化抑制剂 emulsion inhibitor
乳化用盐 emulsifying salt
乳化油 cut oil;emulsible oil;emulsifiable oil;emulsified oil;emulsifying oil;roily oil;soluble oil
乳化油淬火 emulsified oil quenching
乳化油滴 emulsified oil droplet
乳化油废水 emulsified oil wastewater
乳化油漆 emulsion paint
乳化油液 oil emulsion
乳化淤渣 emulsion sludge
乳化原油 emulsified crude oil
乳化原油泥浆 oil emulsion mud
乳化原油钻井液 oil emulsion drilling mud
乳化再生剂 emulsified recycling agent
乳化炸药 emulsified explosive;emulsion explosive
乳化脂 emulsified grease
乳化珠光体 emulsified pearlite
乳化作用 emulsification;emulsifying effect
乳黄色 cream;milky-yellow
乳剂 aqueous emulsion;backing;emulsion;laitance
乳剂-玻璃板界面 emulsion-glass interface
乳剂层 emulsion coating;emulsion layer;photographic(al) layer
乳剂倒置 emulsion inversion
乳剂的 emulsive
乳剂分辨率 resolving power of emulsion
乳剂分层 emulsion creaming
乳剂分散性 emulsion dispersion
乳剂感光度 emulsion speed
乳剂搁置寿命 emulsion shelf life
乳剂工艺 emulsion technology
乳剂号 emulsion numbers
乳剂厚度 emulsion thickness
乳剂灰雾 base fog
乳剂基质 emulsion base

乳剂技术 emulsion technology
乳剂检验 emulsion testing
乳剂校正曲线 emulsion calibration curve
乳剂解像力 resolution of emulsion
乳剂聚合作用 emulsion polymerization
乳剂颗粒度 emulsion granularity; granularity of emulsion
乳剂颗粒性 grains of emulsion
乳剂沥青 asphaltic emulsion
乳剂流动性 emulsion flow property
乳剂面 emulsion side; emulsion surface; face
乳剂批号 emulsion batch number
乳剂漂移 emulsion shift
乳剂软膏基质 emulsion ointment base
乳剂湿润 emulsion dampening
乳剂收缩效应 effect of emulsion shrinkage
乳剂特性曲线 characteristic emulsion curve; emulsion calibration curve
乳剂位移误差 error due to emulsion displacement
乳剂稳定性 emulsion stability
乳剂颜料 emulsion paint
乳浆 emulsion slurry
乳胶 emulsion(glue); lactoprene; latex[复 latices/latexes]; rubber latex
乳胶安定性试验 emulsion stability test
乳胶保护剂 latex preservative
乳胶薄膜 emulsion membrane
乳胶层 emulsion layer
乳胶瓷漆 latex enamel
乳胶袋 latex rubber bag
乳胶地面涂层 latex floor(ing) finish
乳胶对比度 emulsion contrast
乳胶法 emulsion process
乳胶防腐剂 latex anti-septic; latex preservative
乳胶防凝剂 latex preservative
乳胶分解 emulsion break(ing); emulsion resolving
乳胶分解剂 emulsion breaker
乳胶分离 emulsion separation
乳胶分水 breakdown
乳胶封闭底漆 latex primer-sealer
乳胶改质混凝土 latex-modified concrete
乳胶干涉滤光片 emulsion interference filter
乳胶感光速率 emulsion speed
乳胶管 emulsion tube
乳胶光度计 emulsion photometer
乳胶光密度 emulsion density
乳胶硅酸盐水泥砂浆 latex-Portland cement mortar
乳胶化学 emulsion chemistry
乳胶灰雾 emulsion fog
乳胶混凝土 latex concrete
乳胶基 latex base
乳胶基底 emulsion base
乳胶激光存储器 emulsion-laser storage
乳胶计数 emulsion counting
乳胶技术 emulsion technique
乳胶胶凝剂 emulsion cement(ing agent)
乳胶介质 emulsion medium
乳胶径迹 emulsion track
乳胶聚合 emulsion polymerization
乳胶聚合物 emulsion polymer
乳胶聚合物胶粘剂 emulsion polymer adhesive
乳胶聚合作用 emulsion polymerizing
乳胶蜡 emulsion polish
乳胶沥青 latex bitumen
乳胶粒 emulsion particle
乳胶裂化 emulsion cracking
乳胶灵敏度 emulsion sensitivity
乳胶面 emulsion surface
乳胶黏[粘]合剂 emulsion binder; emulsion bonding adhesive; emulsion glue; latex mastic
乳胶黏[粘]合介质 emulsion bonding medium
乳胶黏[粘]合料 latex glue
乳胶黏[粘]结 emulsion adhesive; latex-bound
乳胶黏[粘]结剂 emulsion binder; emulsion binding medium; emulsion bonding agent; emulsion cement(ing agent); latex adhesive; emulsion adhesive
乳胶黏[粘]膜 emulsion tack coat
乳胶凝集反应 latex agglutination test
乳胶凝集抑制反应 latex agglutination inhibition reaction
乳胶配合工艺 latex compounding
乳胶配合物 latex composition; latex compound
乳胶片 photoplate; stripping film
乳胶漆 latex paint; latex water paint; rubber-emulsion paint; water emulsion paint
乳胶嵌缝剂 latex ca(u)lk
乳胶清漆 emulsion varnish
乳胶生产 emulsion product
乳胶试验 emulsion test
乳胶收集体 emulsion collector
乳胶收缩 emulsion shrinkage
乳胶树脂 latex resin
乳胶水 emulsion water
乳胶水泥浆 latex-modified cement paste
乳胶水泥密封复合料 latex-cement sealing compound
乳胶塑料 latex paint
乳胶体 emulsion colloid; pap; emulsoid
乳胶体分层 breaking of emulsion
乳胶体外相 external phase of an emulsion
乳胶涂层 <混凝土养护用> emulsion coating
乳胶涂覆 latex proofing
乳胶涂料 emulsion coating; latex coating; latex paint
乳胶(微)粒子 emulsion particle
乳胶位移 emulsion shift
乳胶稳定度 emulsion stability
乳胶稳定剂 emulsion stabilizing agent
乳胶稳定器 emulsion stabilizer
乳胶洗净剂 emulsion cleaner
乳胶响应 emulsion response
乳胶橡胶 latex rubber
乳胶型水性涂料 emulsion type water-carried paint
乳胶掩膜 emulsion mask
乳胶液 emulsoid; latex emulsion
乳胶油灰混合物 latex mastic compound
乳胶淤渣沉液 emulsion sludge
乳胶载体 emulsion carrier
乳胶增稠剂 latex thickener
乳胶炸药 emulsion explosive
乳胶找平层 latex screed
乳胶状的 emulsive
乳胶状漆 emulsion paint
乳胶阻化剂 emulsion inhibitor
乳井 milk well
乳酪 cheese
乳酪厂 creamery
乳酪焊剂 cream solder
乳酪色 cream
乳酪色坯体 cream body
乳酪色器皿 creamware
乳酪色陶瓷器皿 queen's ware
乳酪塑料 lactolite
乳酪涂饰剂 casein finish
乳酪状干燥 cheesiness
乳酪状膜 cheesiness
乳凝 coagulum
乳牛 milch cow; milch cattle
乳牛场 dairy farm
乳品厂 milk plant
乳品加工厂 dairy processing factory
乳品加工废水 milk product wastewater
乳品业 dairy husbandry; dairy industry
乳清白朊 serum albumin
乳清白朊胶 serum albumin glue
乳清废液 whey wastewater
乳色 opalescence
乳色斑 milk spots
乳色玻璃 alabaster glass; bone glass; obscured glass; opal
乳色的 lactescent; opal; opalescent; opalesque
乳色黑硅石 opaline chert
乳色石英 milky quartz
乳色釉面 opalescent glaze
乳砷铅铜石 bayldonite
乳石 galalith; mammary calculus; milk-stone
乳石英 greasy quartz; milky quartz
乳饰 pasties
乳酸 lactic acid
乳酸丁酯 butyl lactate
乳酸环己酯 cyclohexyl lactate
乳酸胶 casein glue
乳酸戊酯 amyl lactate
乳酸盐 lactate
乳酸乙酯 ethyl lactate
乳酸酯 lactate
乳糖 lactose
乳头状的 mam(m)illated
乳头状构造【地】 mammillary structure
乳头状隆起 nipple
乳头状泉华体【地】 mammillary sinter
乳头状突起 nipple
乳突 nipple
乳香 mastic gum; mastics
乳香黄连木 mastic shrub; mastic tree
乳香胶 gum mastic
乳香属 frankincense
乳香树脂 gum mastic
乳香油 frankincense oil
乳香棕色沥青漆 acratex
乳液 curdling; latex[复 latices/latexes]; milk
乳液表面处治 emulsion coating
乳液表面涂层 emulsion coating
乳液澄清器 demulsifier
乳液储存稳定性 emulsion storage stability
乳液处理 emulsion treatment
乳液的不安定性 <沥青的> lability of emulsion
乳液的不稳定性 <沥青的> lability of emulsion
乳液的分解速度 lability of emulsion
乳液分层 break-up of emulsion
乳液分裂 breaking of emulsion
乳液分裂速度 rate of break(for emulsion)
乳液共聚 emulsion copolymerization
乳液管 emulsion tube
乳液混合料 latex mix
乳液基料 emulsion binder
乳液聚合法 emulsion polymerization; emulsification polymerization
乳液聚合物 emulsion polymer
乳液聚合物胶粘剂 emulsion polymer adhesive
乳液聚合作用 emulsion polymerization
乳液颗粒尺寸 emulsion particle size
乳液离析 breaking of emulsion
乳液密封膏 emulsion sealant
乳液逆转 emulsion inverse
乳液腻子 emulsion putty
乳液黏[粘]度 emulsion viscosity
乳液黏[粘]附性 emulsion adhesivity
乳液黏[粘]结剂 emulsion binder
乳液黏[粘]结毡 emulsion bonded mat
乳液抛光剂 emulsion polish
乳液泡沫橡胶 latex foam rubber
乳液喷射器 emulsion sprayer
乳液喷雾橡胶 latex sprayed rubber
乳液喷嘴 emulsion spray nozzle
乳液破坏 demulsification
乳液破坏剂 demulsifying agent
乳液其他用途 emulsion miscellaneous uses
乳液清洁剂 emulsion cleaning agent
乳液清漆 emulsion varnish
乳液砂浆 emulsion slurry; latex mortar
乳液烧结纺丝法 emulsion sintering method of spinning
乳液涂料 emulsion paint; water emulsion paint
乳液脱模剂 mo(u)ld cream emulsion
乳液微粒粒径 emulsion particle size
乳液稳定性 emulsion stability; stability of emulsion
乳液稀砂浆 emulsion slurry
乳液型丙烯酸树脂漆 emulsion type acrylic resin
乳液型醇酸 emulsion type alkyd
乳液型胶合剂 emulsion type binder
乳液型洗净剂 emulsion type cleaner
乳液油 cream emulsion
乳液注入 emulsion injection
乳液转相 emulsion inverse; emulsion inversion
乳油 oil concentration; oil miscible concentrate
乳油分离器 skimming centrifuge
乳油化 creaming
乳油色硬木 <产于加纳> avodire
乳汁 latex[复 latices/latexes]
乳汁色 lactescence
乳脂 cream
乳制品废水 milk product wastewater
乳制品废物 milk product waste
乳状 emulsus; lactescence; milkiness; rufu
乳状斑 milk spots
乳状的 emulsive; lactescent; milky
乳状废油 emulsion waste oil
乳状浆液 cream
乳状胶体 emulsion colloid; liliquoid
乳状胶质流体 milky colloidal fluid
乳状蜡 emulsion wax
乳状沥青 emulsified asphalt
乳状流型 emulsion flow regime
乳状润滑油 emulsion type lubricant
乳状石蜡 wax emulsion
乳状水 emulsified water
乳状突 mamelon
乳状物 milk
乳状洗涤剂 emulsion cleaner
乳状悬浮液 creamlike suspension
乳状液 emulsion fluid; latices; milk sap
乳状液处理器 emulsion treater
乳状液封堵剂 emulsion blocking agent
乳状液类型 emulsion type
乳状液稳定度 emulsion stability
乳状液稳定剂 emulsion stabilizer
乳状液相转变温度 phase inversion temperature of emulsion
乳状汁 milk sap

乳浊 milkiness
乳浊玻璃 opacified glass
乳浊瓷釉 opaque porcelain glaze
乳浊的 emulsive
乳浊的陶瓷釉面砖 opaque ceramic glazed tile
乳浊灯泡 opal lamp bulb
乳浊度测定 opacity test
乳浊度计 opacity meter
乳浊灌浆 emulsion grouting
乳浊剂 emulsifying agent; opacifier; opacifying agent; opalizer
乳浊能力 opacifying power
乳浊搪瓷 opaque enamel
乳浊体 emulsoid
乳浊效应 opacifying effect
乳浊性 opacity
乳浊岩浆 emulsive magma
乳浊液 calpis; emulsoid
乳浊液破坏 breaking of emulsion
乳浊釉 opacified glaze; opalescent glaze; opaque glaze
乳浊釉瓦 opacified glaze tile

入

入波 incoming wave
入仓 warehousing
入仓配矿 ingredient ore of incoming deposit
入仓申请书 warehousing entry
入仓通知 housing note
入仓温度 placing temperature
入舱 bunker
入舱口 entry hatch
入藏号 accession number
入藏新书索引 accession index
入槽前径流 prechannel flow
入槽前水流 prechannel flow
入场费 admission; door money; entrance fee; gate money
入场权 entree
入场券 admission ticket; pass-check
入场税 admission tax
入超 advanced trade balance; adverse trade balance; excess of imports; import surplus; trade gap; unfavo(u)rable balance of trade
入超净额 net inflow
入车库的车道 garage drive
入出 input-output
入出库能力【机】productivity
入船坞(的) docking
入船坞盘木 docking keel block
入船坞设施 docking accommodation
入端 leading-in
入段线【铁】access; line for locomotive to shed; receiving line to depot
入风 inlet air; intake air
入风井 downcast
入风天井 fresh-air raise
入风网 inlet louver
入港 entering port; haven; sail in
入港报关单 bill of entry; loading list
入港避风 putting into port on account of bad weather
入港呈报表 bill of entry
入港呈报单 bill of entry
入港的禁止 restraint of princes
入港灯 harbo(u)r light
入港费 harbo(u)r dues; inward charges; inward port charges; keelage; port charges; port dues; port duties; port charges inward
入港航道 entrance channel
入港航行的传真 docking facsimile; docking television
入港河口 ocean outfall
入港货重通知单 dock weight note
入港及入城税 dock and town dues
入港及截载 gate-in operation
入港领港费 pilotage inward
入港申报表 bill of entry
入港申报(单) declaration of entrance; entry declaration
入港申报手续 entry formality
入港时燃料舱存煤量 bunker quantity on arrival
入港时燃料舱存油量 bunker quantity on arrival
入港时燃料存量 bunker on arrival
入港时剩余数量 boarding quantity on arrival
入港手续 entrance
入港税 harbo(u)r dues; port dues; port duties; keelage <英>
入港停泊 make harbo(u)r
入港停泊税 keelage
入港许可证 <检疫后发给的> certificate of pratique
入股 become partner; become shareholder; bring in; buy a share
入股申请人 applicant for shares
入馆证 admission card
入国籍 naturalization
入国境税 landing tax
入海冰川 tidal glacier; tide glacier; tide water glacier
入海的淡水(河)流 freshet
入海航道 seaward navigation canal; seaward navigation channel
入海河口 estuary
入海河口沉积(物) marine estuarine deposit
入海河流 land stream
入海口 entrance; ocean inlet; sea outfall; single outlet
入海泥沙 inflowing sediment
入海水系 external drainage
入海顺河段 sea search
入海通道 access to the sea
入河流量 channel inflow
入湖泥沙 inflowing sediment
入户导线 service feeder
入户电缆 service cable
入户管 service line; service pipe
入户管道 service pipe
入户管线 house branch
入户(给水)管线 service pipe line
入户配电设备 service equipment
入户配件 service fittings
入户线 service line; service wire
入户线接线盒 entrance switch
入户支管 service branch
入户总阀 <气、水> corporation cock
入会 affiliation; enrollment
入会费 admission; initiation fee
入会条件严格的工会 closed union
入伙 admission of partner
入伙许可证 occupation permit
入级 classification
入级船舶 class boat
入级符号 character of classification
入级检验 classification survey; class survey
入级修理 classification repair
入籍申请 petition for naturalization
入检门 access door
入井套管层数 casing program(me)
入境 enter the country; entrance; entry
入境办事处 office of entry
入境报关单 declaration inwards
入境车道 inbound lane
入境道路 access road
入境的 inbound; on bound
入境地点 points of entry
入境调查表 entry forms
入境河流 inflowing river
入境机场 aerodrome of entry; airport of entry
入境交通 import traffic; inbound traffic; incoming traffic; on bound traffic
入境坡道 on bound ramp
入境签证 entrance visa; entry visa
入境签证处 immigration office
入境申报单 customs declaration made at the time of entry
入境手续 entry procedures; immigration procedure
入局传号线 incoming order wire
入局电报 incoming message
入局电路 incoming circuit
入局多点设备 incoming multipoint unit
入局机架 incoming frame
入局接线器 incoming connector
入局链 incoming link
入局设备 incoming unit
入局通信[讯]量 incoming traffic
入局信道 incoming channel
入局选择器 incoming selector
入局选择器寻觅器 incoming selector hunter
入局选组器 incoming group selector
入局中继线 incoming trunk
入局转发器 incoming translator
入孔 inlet; notch
入孔冲击钻机 downhole drill
入口 adit; door opening; gateway; ingress; inlet aperture; inlet hole; inlet manhole; inlet opening; inlet orifice; inlet port; intake; port(al); propylon; roll-in; threshold; throat; way-in; entrance【港】; propylaeum[propylaea] <古希腊神殿的>; principal portal <尤指建筑物的正门>
入口半径 entry radius
入口边界 entrance boundary
入口边界点 entrance boundary point
入口变量 entry variable
入口标灯 <飞机场跑道的> threshold light
入口波 admittance wave; entrance wave
入口参数 suction parameter
入口操作温度 inlet operating temperature
入口侧 entrance side; inlet side
入口车门的布置 entrance door location
入口车行道 entrance driveway; entrance roadway
入口程序 entry program(me)
入口冲击 shock at entry
入口处 access point; conning tower; porch
入口处理 threshold treatment
入口处理机 gateway processor
入口处立面 entrance facade
入口处门阶 stoop
入口船闸 entrance lock; guard lock; tidal lock; tide lock
入口词表 entry vocabulary
入口大厅 entrance hall
入口单元 lay-in unit
入口导堤 entrance jetty
入口导向叶片 entrance blade
入口道路 entrance way; entry way
入口的正面 entrance front
入口低矮的 low-browed
入口地址 entry address
入口点 entry point
入口点存取 entry position access
入口点存取法 entry point access method
入口端 arrival end; entry end; incoming end; inlet end; intake terminal; receiving end
入口段 access zone; lead in section
入口段效应 entrance effect
入口断面 lay-in section; inflow face <岩芯筒>
入口阀 inlet valve
入口法兰 inlet flange
入口方向 upstream side
入口防漏环 suction leak-proof ring
入口分支 entrance branch
入口浮标 entrance buoy
入口符号 entry symbol
入口盖 inlet plate
入口干线 incoming trunk
入口港栈货价 ex importation dock price
入口供暖 entrance heating
入口拱 entrance arch
入口沟 geat
入口构造 inlet structure
入口管 entrance pipe; inlet pipe; inlet tube
入口管道通风 inlet duct ventilation
入口轨道 entrance track
入口过滤层 inlet filter
入口过滤器 suction strainer
入口过桥 access bridge
入口巷道 access gallery
入口横径 transverse diameter of pelvic inlet
入口花格大门 entrance lattice gate
入口环 throat ring
入口汇管 arrival manifold
入口集合管 inlet manifold
入口几何形状 entrance geometry
入口剂量 entrance dose
入口夹板 clamp type entry guide
入口检验比 entrance test ratio
入口建筑 portal building
入口渐变段 approach transition
入口角 angle of inlet; entrance angle; inlet angle
入口接管 inlet connection
入口接线器 access switch
入口节流 entrance orifice
入口节流圈 entrance orifice
入口节流式电路 meter-in circuit
入口节流式回路 meter-in circuit
入口结合 entry association
入口截面 mouth of entrance
入口界灯 <飞机场跑道的> threshold light
入口界石 access monument
入口井 inlet well
入口净空 portal clearance
入口坑道 access gallery
入口孔 access port; ingate
入口孔颈 entry neck
入口孔隙 entry pore
入口控制 inlet control; access control <高速公路>
入口块 entry block
入口宽度 entrance width; throat width; width of entrance
入口喇叭口 inlet barrel; inlet horn
入口廊 access balcony
入口廊道 access gallery
入口连接 inlet connection
入口联箱 induction manifold
入口铃形绝缘子 inlet bell
入口流量 entrance flow
入口滤器 inlet strainer
入口路 entry
入口毛细管压力 entrance capillary pressure
入口门 entrance door; entrance porch; entry door
入口面积 area of the inlet
入口名 entry name
入口浓度 threshold concentration
入口平面 entrance level; plane of inlet

R

入口平台 entrance terrace;terrace at entrance
入口屏蔽 entry mask
入口坡道<道路交叉处> access ramp;entrance ramp;in-ramp
入口坡道路面标示 entrance ramp marking
入口铺板 access board
入口气体 inlet gas
入口气温 inlet air temperature
入口壳体 suction casing
入口区 doorway zone;entrance region;entry zone;inlet region
入口区段 entrance zone;threshold zone
入口区域 entrance area
入口曲线 entrance curve
入口设计半径 entrance design radius
入口收费机 toll machine of entrance
入口属性【计】attribute of entry;entry attribute
入口数据 entry data
入口水头 entrance water head
入口水头损失 entrance loss
入口水柱 inlet water
入口顺序文件 entry sequenced file
入口速度 admission velocity;entrance velocity;entry speed;entry velocity;inlet velocity
入口速度压头 inlet velocity head
入口损耗 entrance loss;entry loss;inlet loss
入口损失 entrance loss;entry loss;inlet loss;intake loss
入口损失系数 intake loss coefficient
入口锁 entry lock
入口塔 access tower
入口台阶 entrance steps
入口态程序 entry mode program(me)
入口梯级 entrance steps
入口条件 entrance condition;entry condition
入口跳板 access board
入口铁爬梯 access hook
入口通道 access balcony;access channel;access road;entry
入口弯头 inlet bend
入口位置 entry position
入口温度 inlet temperature
入口涡流速度 entrance velocity of whirl
入口系数 inlet coefficient
入口狭缝 entrance slit
入口线 arrival line;entrance line
入口线脚砌筑层 intake belt course
入口箱 inlet box
入口消声器 intake silencer
入口蓄水池 inlet reservoir
入口选择器 access selector
入口压降 entrance pressure drop
入口压力 entrance pressure;inlet pressure
入口压力控制 inlet pressure control
入口压强 head pressure;inlet pressure
入口压头 inlet head
入口烟道 gas approach
入口引道 access way
入口雨篷 entrance canopy
入口语句 entry statement
入口缘 entry edge
入口匝道 access ramp;entrance ramp
入口匝道控制 entrance ramp control
入口匝道限流(控制) entrance ramp metering
入口闸门 entrance gate
入口张角 angle of throat
入口张紧装置 entry bridle
入口张开角 angle of throat
入口照明 entrance illumination;entrance lighting
入口正面 access front
入口直径 entry hole;eye diameter;inlet diameter
入口止回阀 inlet non-return valve
入口指令 entry instruction
入口终点 entrance terminal
入口装饰 lay-in trim
入口装置 inlet device
入口锥体 approach cone;inlet cone
入口锥形岛 entrance taper
入口走廊 entrance corridor
入口组级 access switch
入库保税品 warehouse bond
入库不平衡系数 unbalanced coefficient of reservoir inflow
入库(存货) be put in storage;entering warehouse;store;warehouse entry;warehousing
入库单 godown entry;warehousing entry
入库股票 reacquired stock
入库洪水 reservoir inflow flood
入库检验 incoming test;warehouse-in inspection
入库流量 reservoir inflow;reservoir inflow rate
入库流量过程线 reservoir inflow hydrograph
入库流量预报 forecast of reservoir inflow
入库泥沙 inflowing sediment;reservoir inflow sediment
入库凭单 stock debit notes
入库日期 date sent to library
入库水流 reservoir inflow
入库线 depot track
入库验收 warehouse entry inspection
入库原材料报告 materials received report
入款 income;receipts
入力取土螺钻 manual soil auger
入粒下限 bottom size
入链架 incoming link frame
入流 incoming flow;inflow;inflow current;water inflow
入流比 inflow ratio
入流储量出流曲线 inflow-storage-discharge curve
入流点 point of inflow
入流段 inflow(ing) reach
入流废水 incoming waste water
入流管 inlet pipe
入流过程线 inflow hydrograph
入流河 inflowing river;inflowing stream
入流河川 influent stream
入流集水管 influent header
入流角 inflow angle
入流联管箱 influent header
入流量 inflow rate;influent flow
入流流量 rate of inflow
入流率 intake rate;rate of infiltration
入流曲线 inflow curve;inflow discharge curve
入流三通 inflow tee-piece;inflow T-piece
入流污水水质 inflow wastewater quality
入流蓄水出流法<洪水追踪> inflow-storage-discharge method;inflow-storage-outflow method
入流蓄水出流曲线 inflow-storage-outflow curve
入流液分析 influent analysis
入流预报准则 inflow forecasting criterion
入炉焦比 ratio of putting coke into furnace
入霉 set in meiyu
入门 access;approach;doorway【建】;primer;guide;guidebook;introduction<指说明书等的入门知识>
入门费 initial payment
入门价格 threshold price
入门书 how-to book;primer
入门梯级 entrance steps
入门职工 threshold worker
入门指导书 guide manual
入磨粒度 mill feed size
入磨物料 mill feed material
入漆朱 Lithol red
入歧途 divagation
入气口 aspirating hole;aspirating mouth
入侵 encroachment
入侵超覆 onlap;transgressive overlap
入侵后的地下水型 groundwater type after intrusion
入侵后的矿化度 salinity after intrusion
入侵后氯离子含量 chlorine content after intrusion
入侵量 potential infiltration rate
入侵前的地下水型 groundwater type before intrusion
入侵前的矿化度 salinity before intrusion
入侵前氯离子含量 chlorine content before intrusion
入区交通 incoming traffic
入渠流量 channel inflow
入射 entrance;injection
入射半角 entrance half-angle
入射波 incident wave;incoming wave
入射波长 incident wavelength
入射波功率 incident power
入射波前 incident wavefront
入射场强 incident field intensity;incident field strength
入射潮流速度 incident current velocity
入射冲击波 incident blast wave
入射窗 entrance window
入射道 entrance channel
入射的 incident
入射地震 incident earthquake
入射点 entrance point;incidence point;incident point;point of incidence
入射电流 incident current
入射电子 incident electron
入射电子束 incident electron beam
入射电子束电流 incidence electron-beam current
入射法线 incident normal
入射反光镜 ingress reflector
入射方向 entrance direction;incidence direction;incident direction
入射辐射 incident radiation;incoming radiation
入射伽马光子能量 incident gamma photon energy
入射高度 height of incidence;incident height
入射功率 incident power
入射光 incident light;incoming ray
入射光测定读数 incident light reading
入射光路 input path
入射光强度 incident light intensity
入射光式曝光表 incident light meter
入射光束 entrance bundle of rays;entrance light beam;incidence beam;incident light beam;incident beam
入射光通量 incident flux
入射光(瞳)孔 entrance pupil
入射光线 incident ray
入射光线的入射角 incident angle of the entering ray
入射光线角 entrance ray angle
入射光锥 input ray pencil
入射光子 incident photon
入射核子 incident nucleon
入射弧 incident-in arc
入射激光束 incoming laser beam
入射激震前沿 incident shock front
入射剂量 incident dose
入射角【物】angle of arrival;angle of fall;angle of incidence;entrance angle;impingement angle;incidence angle;inlet angle;reentering angle;reference angle;striking angle;incidence;incident angle
入射节点 incident nodal point
入射介质 incident medium
入射孔径 input aperture
入射口 entrance port
入射棱镜 entrance prism
入射流 incident flow;incident flux
入射路径 incident path
入射率 incidence rate;rate of incidence
入射脉冲 incident pulse
入射面 entering surface;entrance face;incidence plane;incident face;plane of incidence
入射模式 incident mode
入射能量 incident energy;projectile energy
入射强度 incident intensity
入射(射)线 incident ray
入射声 incident sound
入射束 incoming beam
入射水流速度 incident current velocity
入射损失系数 incidence loss coefficient
入射太阳射线 incident solar ray
入射图像 incident image
入射狭缝 entrance slit
入射狭缝高度 entrance slit height
入射线 incident beam;incident line;line of incidence
入射信号 incident signal;incoming signal
入射压力 incident pressure
入射阳光 incident sunlight
入射影 incident image
入射余角<无线电波> grazing angle
入射运动 incident motion
入射噪声 incident noise
入射照明 incident illumination
入射照明器 incident light illuminator
入射重力波 incident gravity wave
入渗 influent seepage;ingoing;inleakage
入渗池 infiltration tank
入渗点 infiltration point
入渗过程 infiltration process
入渗河流 influent river;influent stream
入渗井 infiltration well
入渗理论 infiltration theory;theory of infiltration
入渗量 infiltration capacity
入渗流量 infiltration flow
入渗路径 infiltration path
入渗率 infiltration index
入渗面积 infiltration area;percolation area
入渗能力 infiltration capacity
入渗能力方程 infiltration capacity equation
入渗能力曲线 infiltration capacity curve
入渗强度 infiltration intensity
入渗渠 infiltration ditch
入渗试验小区 infitrometer plot
入渗水 infiltrated water;infiltration

water
入渗水量 infiltration volume
入渗水流 infiltrated flow
入渗速度 infiltration velocity
入渗速率 infiltration rate;rate of infiltration
入渗损失 infiltration loss
入渗推算 infiltration routing
入渗雨量 infiltrating rainfall
入渗指数 infiltration index
入渗作用 infiltration
入市交通 city-bound traffic
入市限价 position limit price
入水管 inhalant siphon[syphon]
入水角 immersion angle
入水绳 descending line;shot rope
入水时间 leave surface time;time of entering water
入睡 slumber
入睡前幻觉 hypnagogic hallucination
入土 be buried;be interred
入土长度 <指桩> embedded length;embedment length;length of embankment
入土深度 <指桩> buried depth;depth of embedment
入土深度限制器 depth band
入土桩 driven pile
入土桩承载试验 lateral pile load test
入挖池塘 dugout pond
入挖水池 dugout pond
入网信道 access channel
入围 acceptable range
入位衔铁 seated armature
入伍 enrollment
入坞吃水 docking draft;docking draught
入坞方法 method of docking
入坞费 dockage;dock charges;dock due
入坞检验 docking survey
入坞期 docking period
入坞设备 docking accommodation
入坞设施 docking facility
入坞时间 docking time
人字坞门 miter[mitre](dry-) dock gate
入坞修理通知书 docking order
入线 lead wire
入线复凿孔机 incoming line reperforator
入线复凿岩机 incoming line reperforator
入陷转移 trapping
入陷状态 trapping mode
入选的个体 intermate individuals
入选的投标者 selected bidders
入选矿石品位 milling ore grade
入选原矿分析 mill head assay
入学 enrollment
入窑料浆 finished slurry
入窑料浆泵 kiln-feed slurry pump
入窑料浆浓缩机 kiln-feed thickener
入窑料球 nodulized feed;pelletized feed
入窑生料 finished meal;kiln feed
入窑物料 kiln-feed material
入宅电源开关 service entrance switch
入站的 inbound
入站口 station entrance
入站线 up-line
入站线路 inbound line
入账 enter an item in an account;entering book;enter into the account book;enter up an accounts;keep account;pass entry;pass-through accounts;receipt on accounts
入账价格 entry price
入账日 up-to-date
入账受款人 account payee
入中继 in-trunk
入中继测试 incoming trunk test
入中继监视信号 backward supervision signal
入中继器 incoming trunk circuit
入中继台 incoming junction position
入中继套线继电器 incoming sleeve relay
入中继线 incoming junction line;incoming trunk line
入中继线继电器 incoming junction line relay
入住 residential occupancy
入座 chair

褥垫 mattress;pillow

褥垫排水 blanket drain
褥形供暖器 quilt heater
褥子 mattress

朊【化】protein

软白垩砖 malm rubber;marlm rubber;marl rubber;soft malm brick

软白胶 soft white gum
软白黏[粘]土 prian
软百叶窗 Venetian blind;Venetian shutter;boxing shutter <能藏在匣内的>
软百叶窗条板 slat of Venetian blind
软百叶帘 Venetian blind
软百叶片 slat of Venetian blind
软柏油脂 fatty pitch;soft pitch
软板 flexible slab;softboard
软包装 soft package
软雹(霰) graupel;snow pellets;soft hail
软保兑 soft confirmation
软保护套 sock
软币 less favorable currency;soft currency;soft money
软铋矿 sillenite
软铋铅钯矿 urvantsevite
软变孔径 soft aperture
软变速 soft-shift(transmission)
软标号 soft label
软冰 sludge ice;slush ice;soft ice
软波导(管) flexible waveguide;squeezable waveguide
软玻璃 soft glass
软补钉片 soft patch
软补贴 soft allowance;soft subsidy
软不锈钢 soft stainless steel
软布 mull;soft cloth
软布敷料 mull-dressing
软布面 limp cloth
软布团 <涂漆等用> tampon
软材 damp mass;softwood
软材板 softwood board
软材插条 softwood cutting;succulent wood cutting
软材干馏 softwood distillation
软材焦油沥青 softwood tar pitch
软材料 soft material
软材料坝 flexible dam
软材林 softwood forest
软材木炭 softwood charcoal
软材枕木 softwood sleeper
软彩 soft colo(u)r
软层 soft formation;soft layer;soft stratum;weak course
软层剥离 soft layer stripping
软车顶 cape top;soft top
软衬垫 cushioning
软成本 <建筑的> soft cost
软成分 soft component
软尺 flexible rule(r);tape
软冲积黏[粘]土 soft alluvium loam
软冲头法 <玻璃钢> flexible plunger method
软触控制 soft touch control
软传动 soft transmission
软锤 <锤头用软材料做的> soft hammer;dummy
软瓷 porcelain tendre;soft porcelain;tender porcelain
软瓷器 soft porcelain
软磁 soft magnetism
软磁材料 soft magnetic material
软磁材料总损耗 total loss of soft magnetic material
软磁钢 magnetically soft steel
软磁合金 magnetically soft alloy
软磁盘 flexible disc;floppy disc[disk]
软磁盘机 flexible disk cartridge;flexible disk drive;flexible disk unit;floppy disk drive
软磁盘文件目录 diskette file directory
软磁铁 soft magnet
软磁铁氧体 soft magnetic ferrite
软磁铁氧体单晶 soft magnetic ferrite single crystal
软磁性材料 magnetically soft material;soft magnetic material
软磁性材料总损耗 total loss of soft magnetic material
软磁性合金 non-retentive alloy
软磁性质 soft magnetic property
软搓 long lay;soft lay
软搓绳 long laid rope;soft laid rope;warp laid rope
软错误 soft error
软打桩 <在预先钻好的孔内打桩> soft driving
软带材 soft strip
软袋货柜 flexible bag container
软贷款 soft loan
软贷款窗口 <指对发展中国家的优惠贷款业务> soft-loan window
软氮化法 soft nitriding
软导管 duct hose;Nelaton's catheter;soft catheter
软导管母线 flexible conductor bus
软导线 flexible circuit conductor;flexible lead;pigtail
软道面跑道 soft surface runway
软道面用的起落架 soft-earth gear;soft-field gear
软的第三纪砂石 soft Tertiary sandstone
软的第三纪砂岩 soft Tertiary sandstone
软底 soft bottom
软底板 soft seat
软底板蜡 soft baseplate wax
软底蚀刻 soft ground etching
软底土 soft subsoil
软地 soft ground
软地层 soft formation;soft ground;soft subsoil
软地层隧道掘进机 soft ground tunnel(l)ing machine
软地层钻头 mud bit
软地道路 soft surface road
软地基 soft foundation;weak ground;weak subgrade
软地沥青 earth pitch;maltha;mineral tar;petrolene;soft asphalt;malthite;pissasphalt
软地幔 soft mantle
软地面 loose earth
软地拖拉机 soft-land tractor
软碲铜矿【地】vulcanite
软点 soft spot
软点火线 plastic igniter cord
软点子 soft dot
软电弧 soft arc
软电缆 cab-tyre cable;flexible cable;BX cable <安装用>
软电线 cord
软垫 bolster;cushion(ing);flexible disc;protecting piece
软垫病室 padded cell;padded room
软垫层 cushion course
软垫涂层 cushion coat
软垫座 cushion seat
软垫座椅 upholstered chair
软雕塑 soft sculpture
软吊线 flexible hanger
软调 soft;soft tone
软调底片 flat negative;soft negative
软调图像 low-key(ed) image;low-key(ed) picture
软调显影剂 low contrast developer
软顶板 tender roof
软顶尖 soft center[centre]
软顶式集装箱 soft top container
软定位器 curve steady arm
软锭剂 pastil(le)
软度 softness
软镀层 soft coating
软镀膜材料 soft coating material
软对接 soft docking
软而湿的 boggy
软而湿的小块 dab
软发抛光轮 soft hair polishing wheel
软反馈 elastic feedback;transient feedback
软反馈装置 compensation return
软方解石 mountain milk
软分段 soft sectoring
软酚醛树脂 soft phenolic resin
软粉刷砂浆 soft stuff
软风 <蒲福风级表一级风,风速 0.3 ~ 1.5 米/秒> light air
软风管 flex-duct;pneumatic hose
软风景 soft landscape
软封面 soft-cover
软封面书 limp
软辐射 soft radiation
软腐菌 soft-rot fungus
软腐朽 soft rot
软钙质砂岩 hassock
软感 meagre feeling
软钢 dead steel;low-carbon steel;malleable steel;mild steel;quiet steel;soft steel
软钢板 mild sheet steel;mild steel plate
软钢板火箱 mild steel fire box
软钢吊钩 mild steel hook
软钢钢筋 <含钢量 0.12% ~ 0.25%> mild steel reinforcement
软钢管 flexible metal(lic) hose;soft steel pipe
软钢家用锅炉 mild-steel domestic boiler
软钢皮 mild steel sheet
软钢卡爪 soft jaw
软钢卡爪卡盘 soft jaw chuck
软钢手柄 mild steel shank
软钢水套 soft steel water-jacket
软钢丝 annealed wire;stone dead wire;stone wire
软钢丝钩环 mild steel shackle
软钢丝绳 flexible steel wire rope
软钢索 soft wire rope
软钢铁锻件 blackwork
软钢纸 soft vulcanized fiber
软膏 inunction;light paste;ointment;paste;runny paste;slurry;soft grease;thin paste;unction;unguentum
软膏板 ointment slab
软膏刀 spattle;spatula;spatule
软膏罐 ointment jar
软膏盒 ointment box

软膏磨 ointment mill
软膏调刀 spatula
软膏锡管填充器 filling machine for collapsible tube
软膏状油垢 soft sludge
软革 soft leather
软格式 soft format
软垢水 soft filth water
软骨料 soft aggregate
软骨料颗粒 soft particle
软管 collapsible tube; couloir; flexible conduit; flexible duct; flexible hose; flexible pipe; flexible tube[tubing]; hog; hose; hose pipe; hose tube
软管扳手 hose spanner; hose wrench
软管包装 hose packing
软管操纵架 hose handling frame; hose handling gantry
软管操作设备 hose handling equipment
软管拆装机 mounting and dismounting device for flexible pipes
软管车 hose cart
软管冲洗 hose flushing
软管的折角阀 hose angle valve
软管电缆 cab-tyre cable
软管吊板 hose hanger
软管吊绳 hose hawser
软管吊柱 hose davit
软管帆布 hose duck
软管防护套 hose protector
软管防护罩 hose guard; hose protector
软管防护装置 hose protector
软管放油 hose down
软管附件 hose fitting
软管供气潜水 hose-fed diving
软管钩 hose cock
软管固定装置 hose holder
软管挂环 hose supporting clip
软管管道 hose(pipe)line
软管管路 hose(pipe)line
软管和传送带用帆布 hose and belts duck
软管虹吸 hose siphon
软管呼吸器 hose-apparatus
软管滑轮 hose whip
软管机 hose machine
软管加长节 hose extension
软管加油器 hose oiler
软管夹具 hose clamp
软管夹(子) hose clamp; hose clip; hose rack
软管架 <凿岩机> hose holder
软管架子 hose rack
软管绞车 hose reel
软管绞盘 hose drum
软管接合器 hose union
软管接头 connecting hose; coupling; flexible pipe coupling; hose adapter; hose assembly; hose connection; hose connector; hose coupler; hose coupling; hose-end fitting; hose fitting; hose joint; hose union
软管接头垫圈 hose coupling gasket
软管接头螺纹 hose thread
软管接嘴 hose bib
软管举升架 hose handling frame
软管卷盘 hose reel
软管卷盘系统 hose reel system
软管卷绕轮 hose reel
软管卷组 hose reel pack
软管连接 hose connection
软管连接阀 hose connection valve
软管连接器 flex connector; flexible connector; hose connector; hose coupling
软管连接器垫圈 packing ring
软管联结器防尘堵 dummy hose coupling
软管联管节 hose union; union-hose connector
软管联轴节 hose coupling
软管龙头 hose cock; sill cock
软管轮胎 hose pipe tire[tyre]
软管螺纹接套 hose nipple
软管螺纹接头 hose nipple
软管面罩 hose mask
软管末端压力 hose-end pressure
软管内部承受静水压力试验 hold test
软管内部机内卷绕 internal hose reeving
软管扭接 kinking of hose
软管配件 hose fittings
软管喷水工 hoseman
软管喷水嘴 outlet hose nozzle
软管喷嘴 hose director; hose nozzle
软管平衡器 hose balancer
软管卡箍 hose clamp; hose clip; hose collar
软管卡子 hose band
软管铅接合 lead encasing of hose
软管球形阀 hose globe valve
软管伸长节 hose extension
软管式泵 hose type pump
软管式机轮刹车 expander tube wheel brake
软管试验 hose test
软管试验槽 test device for flexible pipes
软管受拉后截面缩细 necking down
软管束 hose bundle
软管水准器 hose levelling instrument
软管套头 hose liner
软管涂料 collapsible tube paint
软管托架 hose retriever
软管弯头 hose elbow
软管线(路) hose line
软管小龙头 hose bib
软管修理工 hose mender
软管虚连接 dummy hose coupling
软管旋塞 hose cock; hose bib
软管旋转接头 hose swivel
软管印刷 collapsible tube printing
软管用白墨 white coat(ing)
软管用接头 coupling nipple; hose coupling nipple
软管运输小车 hose carrier
软管支架 hose clamp; hose support
软管中的钢丝 body wire
软管中间接头 union-hose connector
软管轴 hose reel; hose shaft
软管装配 hose assembly
软管组成 hose assembly
软管钻进 bendable pipe drilling
软管嘴 hose nipple
软罐头 soft can
软光电管 soft photo tube
软硅铜矿 bisbeeite
软海泥 soft marine mud
软含水层 bibbles
软焊 soft soldering; soldering; sweat soldering
软焊缝 soldered joint seam
软焊剂 soft solder flux
软焊接合金 soft soldering alloys
软焊接接头 wiped joint
软焊接料 soft solder
软焊料 fake; medium solder; molten solder; soft solder; tinman's solder; tin solder
软焊炉 sweat-soldering oven
软焊丝 soft wire
软焊条 flexible electrode; flexible welding rod; soft solder
软焊涂料 solder paint
软合金 mild alloy; slush metal
软横跨 flexible cross-span; head span; headspan suspension
软横跨承力索 headspan cross messenger wire
软横跨固定底座 headspan bracket
软横跨环型线夹 cross-span eye clamp
软横跨下定位索 lower span wire
软横跨张力补偿器 cross-span tensioning spring
软琥珀 gedanite
软滑砖 rubbed brick
软化 after-tack; bating; demineralize; ripening; pitting <耐火材料>; demineralization <水的>
软化材料 softening material
软化槽 softening tank
软化程度 softening degree
软化池 softening tank
软化处理 softening treatment
软化淬火处理 negative hardening
软化的 malactic; softening
软化地址 software address
软化点 melting point; sagging point; sintering point; softening point; softening temperature; yield point
软化点测定 softening point measurement
软化点测定管 softening point tube
软化点漂移 softening point drift
软化点试验 <沥青的> softening point test
软化点试验仪 <环球法> softening point tester[ring ball method]
软化点温度低的玻璃 soft glass
软化点温度计 softening point thermometer
软化度 softening degree; softness; softness number; softness value
软化法 softening method; softening process
软化范围 softening range
软化幅度 softening range
软化膏 flux paste
软化故障 <程序的> fail softly
软化光谱 degraded spectrum
软化锅 softening kettle
软化过程 softening process
软化过滤器 softening filter
软化回火 soft temper
软化极限曲面 limit surface of yielding
软化剂 cleansol; emollient; flexibilizer; macerating agent; mollifier; plasticizer; plastifier; soft agent; softener; softening agent; softer; water softener; soft flux <降低沥青稠度的>
软化剂的再生 regeneration of ion-change softener
软化价格 softening price
软化浆 flux paste
软化库巴树脂 run copal
软化炉 softener; softening furnace
软化滤水器 softening filter
软化率计 malakograph
软化能力 softening power
软化谱 soften spectrum
软化器 mollifier; softener; softening apparatus; demineralizer <用于水化学脱盐的>
软化铅 <脱除砷锑锡等杂质的铅> softened lead; softening lead
软化强度 yield strength
软化区 softened zone; softening range; softening register; soft zone
软化区间 softening interval
软化时期 <路基因冰冻融化而变软的时期> period of weakening
软化试验 softening test
软化水 demineralized water; softened water
软化水厂 softening plant; water softening plant
软化水处理 softening water treatment
软化水法 softening of water; water softening
软化退火 blue annealing; recrystallization annealing; softening annealing
软化温度 sintering temperature; softening point; softening temperature
软化温度范围 softening range
软化污泥 softening sludge
软化系数 coefficient of softening; coefficient of softness; softening coefficient; softening factor
软化现象 ruckbildung
软化橡胶 softened rubber
软化型 softening type
软化型双曲线系统 softening bilinear system
软化学技术 soft chemistry technique
软化岩石 softening rock
软化硬水 softening of water
软化站 softening plant
软化脂 plastic fat
软化脂膏 softener paste
软化值 softness value
软化装置 softener; softening apparatus; softening installation; softening plant; softening unit
软化作用 emollescence; ramollescence; soft action
软黄铜 low brass
软簧 reed spring; soft spring
软回火的薄钢板 quarter hard temper sheet
软回收 soft recovery; soft waste
软混凝土 quaking concrete
软火焰 soft flame
软货币 soft currency
软货物 soft goods
软击穿 soft break down
软基 soft foundation
软基地 soft base
软基加固 consolidation of ground; improvement of soft ground; soft soil treatment; stabilization of soft foundation
软级联 soft cascade
软极限 soft limiting
软极限积分器 soft limited integrator
软集料 soft aggregate
软给水 soft feed water
软脊 flexible
软脊无线装订机 flexibak
软技术 soft tech; soft technology
软加法器 soft adder
软钾镁矾 picromerite
软碱 soft base
软件 flexible disc; software; superpave software Superpave
软件安全性 software security
软件安装 software setup
软件包 software kit; software package; software packet
软件计划 software project
软件接口 software interface
软件支持系统 software support system
软件中断 software interruption
软件中心 software center[centre]
软件蛀船虫 molluscan borer; mollusk teredo
软件装置 software service
软件资料管理员 software librarian
软件资源 software resource; software source
软胶 flexible glue; soft gum
软胶辊 soft roll
软胶胶囊 soft gelatin capsule
软胶囊 soft capsule

软焦点 soft focus
软焦点镜头 soft-focus lens
软焦距透镜 soft-focus lens
软焦滤光片 soft-focus filter
软焦炭 soft coke
软焦物镜 soft-focus objective
软焦像 soft-focus image
软焦油沥青 soft pitch;soft tar pitch
软接触 soft contact;soft feeling
软接触控制 soft touch control
软接缝 <玻璃幕墙的> soft joint
软接管 connecting hose
软接头 flexible connector; flexible joint
软接线 flexible circuit conductor; flexible wire
软节臂 articulated arm
软节臂式磨石机 articulated arm-type stone grinder
软节臂式石头打磨机 articulated arm-type stone grinder
软结 soft junction
软结线控制 soft wired control
软金属 soft metal
软金属带天线 flexible metal tape antenna
软金属管 flexible metal hose
软金属钳口垫片 soft metal jaw
软金属屋面材 conic(al)roll
软金属屋面折缝 upstand
软金属线 soft annealed wire
软浸焊 dip soldering
软禁 house arrest
软晶体 soft crystal
软静止图像 soft still
软聚合物材料 soft polymetric material
软聚焦 soft focus(ing)
软聚氯乙烯 flexible PVC; plasticized polyvinyl chloride
软绝缘管 flexible tubing;tubing
软开销 <指基建中的保险费、利税、设计费等> soft cost
软开业 soft opening
软拷贝 soft copy
软靠背 <座席> upholstered back
软靠枕 air buffer
软科学 <包括现代管理学、系统分析、科学学、预测研究、科学技术论等> soft science
软颗粒 soft particle
软颗粒含量试验 soft grain content test
软控制 soft control
软块 mush
软矿物 sift mineral
软拉 soft drawn
软拉钢丝 mild drawn wire
软拉铜丝 soft drawn copper wire
软拉线 soft drawn wire
软蜡 amorphous wax;slack wax;soft wax
软蜡槽 soft wax tank
软沥青 earth pitch; maltha; mineral tar;petrolene;soft asphalt;soft bitumen;soft pitch
软沥青混凝土砖 soft pitch concrete tile
软沥青质 <沥青组分,溶于轻汽油部分> petrolene;malthene
软沥青砖 soft pitch tile
软粒喷砂 non-erosive blasting; seed blasting
软连接 flexible connection; flexible coupling;soft wiring
软连数控 soft wired number control
软连索 soft commissure
软连续管 hose connector
软练 mushy consistence[consistency]; plastic mortar; wet consistence [consistency]
软练胶砂 plastic mortar
软练胶砂强度试验法 plastic mortar strength test
软练砂浆 plastic mortar;wet mortar
软练砂浆立方体 plastic mortar cube
软练试验 test by wet mortar
软溜槽 flexible drop chute
软流 plastic flow
软流变质作用 rheomorphism
软流层【地】asthenosphere
软流带 zone of flowage;zone of rock flowage
软流分层 rheologic(al)stratification
软流圈【地】 asthenosphere; rheosphere
软流圈凸起 asthenosphere bump
软硫铋铅铜矿 gladite
软路肩 soft shoulder
软铝 annealed alumin(i)um
软铝棉 alumin(i)um soft wool
软铝绒 alumin(i)um soft wool
软铝线 annealed alumin(i)um wire
软绿泥石 strigovite
软绿黏[粘]土 soft blue clay
软绿砂岩 ca(u)lk stone
软绿脱石 mullerite
软轮胎 soft tire[tyre]
软麻布 lint
软麻粗梳机 combined spreader and softener
软麻机 softener;softening machine
软麻延展机 combined spreader and softener
软毛 fur;wool
软毛宽刷 badger-hair softener;badger softener
软毛皮 mole-skin
软毛刷 <油漆工用的> badger softener;banister brush
软煤 easy coal;free coal;soft coal
软煤焦油脂 soft coal-tar pitch
软煤沥青 soft base-tar; soft coal asphalt;soft coal pitch
软锰矿 manganese black; manganese dioxide;pyrohisite;pyrolusite
软锰矿矿石 pyrolusite ore
软锰矿颜料 manganese black pigment
软密封剂 flexible sealant
软棉布 muslin
软棉线绳 soft twisted-cotton string
软面锤 soft face hammer
软面钢锤 soft surface steel hammer
软面书 soft-covered book
软面轧辊 sand roll
软模 flexible mo(u)ld
软模成型 flexible die forming
软模理论 soft mode theory
软模拟 soft simulation
软膜 soft coat(ing)
软磨石 soft grinding stone
软磨毡层 cork carpet
软木 cork; non-porous timber; periderm;softwood;suber
软木柏油 <混有木屑的> cork tar
软木板 compressed cork;cork board; cork slab;softwood plank
软木背衬 cork backing
软木薄板 cork block sheet; cork sheet
软木薄片 cork shaving
软木布告板 cork tackboard; tack board
软木材 softening wood
软木材楼地面 softwood flooring
软木材体积标准 petrograd standard
软木尘肺 cork pneumoconiosis
软木衬垫 core gasket;cork gasket
软木衬料 vegetable cork
软木的 corky
软木底层 cork base
软木地板 cork flooring
软木地板面 cork carpet
软木地板饰面 cork floor finish
软木地面 cork flooring
软木地毯 cork carpet;corticin(e)
软木地毡 cork floor cover(ing);corticin(e) <用树胶制成的>
软木地砖 cork floor cover(ing)
软木垫 cork carpet; cork mat; cork packing;cork pad
软木垫层 cork underlay
软木垫底 cork base
软木垫底的 cork based
软木垫片 cork spacer;cork strip
软木垫圈 cork gasket;cork washer
软木粉 cork flour;cork powder
软木浮标 cork buoy
软木浮子 cork float;dips(e)y
软木附聚加工 cork worked by agglomeration
软木覆盖 cork cover(ing)
软木隔热 cork insulation
软木隔热舱 cork-insulated hold
软木隔热垫 <一种专卖品> coresil
软木管段 cork pipe section
软木轨枕 softwood sleeper;softwood tie
软木黑颜料 cork black
软木护舷材 cork fender
软木化 suberification;suberization
软木环 cork ring
软木基底 cork base
软木救生带 cork life-belt
软木救生圈 cork life buoy
软木救生衣 cork jacket;cork life-belt
软木绝热 cork insulation
软木绝缘 cork insulation
软木颗粒 granulated cork
软木块 cork block;cork brick
软木块板 cork block board; cork block slab
软木块地板 cork tile floor(ing)
软木块绝热 block-cork insulation
软木块绝缘 block-cork insulation
软木粒 cork granule
软木粒嵌缝料 granulated cork filler
软木楼梯踏步 cork stair tread
软木轮 cork wheel
软木毛地板 cork subfloor
软木密(封)垫 cork gasket
软木木垛 softwood pack
软木抛光轮 cork polishing wheel
软木碰垫 cork fender
软木碰球 cork fender
软木片 cork sheet
软木铺地砖 cork flooring tile
软木铺路小方块 cork paving sett
软木铺面 cork flooring
软木砌块墙 cork block wall
软木墙 cork wall
软木墙砖 cork wall tile
软木绒纤维 corkwood
软木塞 cork plug; cork stopper; go-devil;natural cork;tampon <涂漆等用>
软木塞垫 cork gasket
软木塞法 cork method
软木色 cork tan
软木砂浆 cork mortar
软木树 corktree
软木酸 suberic acid
软木榫凿 sash mortise chisel
软木套 cork jacket
软木套管 cork pipe
软木填充 cork fill(ing)
软木填缝板 cork board
软木填缝料 cork joint-filler
软木填料 cork filler;cork fill(ing)
软木贴面 cork flooring;cork tile
软木瓦 cork tile
软木橡胶 <密封用> cork rubber
软木橡木 corkoak
软木橡皮 <混有粒状软木的> cork rubber
软木小方块 cork setting
软木屑 cork dust; cork granule; ground cork
软木屑混凝土 cork dust concrete
软木屑涂料 cork dust paint
软木油地砖 cork linoleum
软木元件 cork cell
软木遮阳帽 helmet
软木脂 suberin
软木纸 cork sheet
软木制品 cork article;cork product
软木质的 suberose
软木质素 vinsol
软木砖 cork brick;cork tile
软木砖地板 cork tile floor(ing)
软木组织 suberic tissue
软目标 soft target
软耐火黏[粘]土 daugh
软能路线 soft energy paths
软泥 batter; bungum; muck; muck soil; ooze; ooze muck; slime; sludge;slush;soft mud
软泥薄层 ooze film
软泥采样器 ooze sampler
软泥成型法 soft-mud process
软泥的 oozy
软泥底 muck bottom;ooze bottom
软泥地 quagmire
软泥法(制砖)soft-mud process
软泥工艺 soft-mud process
软泥基中板排桩 camp sheathing
软泥挤坯 soft extrusion
软泥涂抹层 slime coating
软泥移动 muck shifting
软泥藻类 silt algae
软泥造砖机 soft mud brick machine
软泥植物区(系)silt flora
软泥制坯法 slop mo(u)lding; soft-mud process
软泥制砖法 soft-mud process
软泥质黏[粘]土 soft muddy clay
软泥砖 soft mud brick; soft mud process brick
软拟胶骨架 soft gel-like matrix
软黏[粘]土 bury;mickle[myckle];soft clay
软黏[粘]土层 soft clay stratum
软黏[粘]土地 soft clay ground
软黏[粘]土片岩 blacks
软黏[粘]土质土壤 pounson
软镍矿 nickhydroxide
软凝块 soft clot
软盘 diskette; flexible disc; floppy disc[disk]
软盘格式化 diskette formatting
软盘菌属 <拉> Mollisia
软盘驱动器 floppy disk drive
软泡沫聚氨酯塑料 flexible polyurethane foam
软喷砂处理 soft blast
软膨胀性天然橡胶 soft expanded natural rubber
软碰垫 fender rope;fisherman's fender
软坯 soft paste
软皮 buff; soft bark; wash leather; chamois <羚羊、山羊、鹿等的>
软皮布抛光 buffing
软皮布抛光机 buffing machine
软皮布抛光轮 buff wheel
软皮带 wash leather strip
软皮件 limp leather goods
软皮轮 buff wheel
软皮条 wash leather strip
软皮镶玻璃 wash leather glazing
软片 film;film negative;roll film
软片暗盒 film cartridge;film magazine

软片包 film pack
软片盒 film carrier
软片升降装置 film elevating mechanism
软片岩【地】cash
软片轴 film spool
软片状岩 rashings
软票 soft paper ticket
软铺排 soft mattress laid
软起动 soft start
软起动型 soft start type
软钎焊 soft soldering;solder(ing)
软钎焊缝 soldered seam; soldering seam
软钎焊接头 soldered joint
软钎焊涂层 flow brightening
软钎焊药 soft soldering flux
软钎剂 soldering flux
软钎料 low-melting solder;soft solder;solder
软铅 refined lead;soft lead
软铅板 soft lead plate
软铅管 soft lead pipe
软铅皮 soft lead sheet
软铅丝 soft lead wire
软铅压力管 soft lead pressure pipe
软钳位 soft clamping
软青铜 soft bronze
软驱接口【计】floppy disk controller
软去污剂 soft detergent
软燃料箱 fabric fuel tank
软绕组 mushy winding
软人行道饰面 soft walked-on finish
软任务 soft mission
软韧的(木料) punky
软韧橡皮管 band tubing
软韧性材料 soft ductile material
软绒布 outing flannel
软溶 flashing off < 耐火材料的 > ;soft heat < 平炉的 >
软熔发亮处理 flash melting; flow brightening; reflowing; thermal reflowing
软熔机组 reflow unit
软弱 flabbiness;weakness
软弱层 soft layer; soft stratum; weak formation
软弱层厚度 thickness of soft layer
软弱层埋藏深度 buried depth of soft layer
软弱承载层 poor bearing stratum
软弱的 flabby;weak
软弱底土 poor subsoil;soft subsoil
软弱地层 difficult ground; soft rock; soft stratum;weak formation
软弱地层测试仪器 soft ground instrumentation
软弱地层隧道 soft ground tunnel
软弱地层物理性质试验 soft ground physical property test
软弱地层钻井测试 soft ground downhole testing
软弱地基 flimsy ground;poor ground; poor subgrade; poor subsoil; soft foundation; soft ground; soft soil foundation;weak ground;weak subgrade; weak foundation; incompetent ground
软弱构造 weak formation
软弱集料 soft and weak aggregate
软弱夹层【地】incompetent(-ly intercalated) bed; soft intercalated bed; soft interlayer; soft vein; weak interbed; weak intercalated layer; weak seam;weak intercalation
软弱夹层地基 foundation with interbedded argillous soil
软弱夹杂物 soft inclusions
软弱结构面【地】weak structural plane;soft structural plane
软弱结构面的剪切试验 shear test of discontinuities; shear test of weak plane
软弱界面钻孔 line drilling
软弱颗粒 soft particle
软弱路基 weak subgrade
软弱路基土 poor subgrade soil
软弱路肩的平整 level(l)ing of soft shoulders
软弱路肩的找平 level(l)ing of soft shoulders
软弱面 plane of weakness;surface of weakness
软弱黏[粘]土 soft clay
软弱土层 < 难处理的 > difficult ground
软弱围岩 soft ground;soft surrounding rock;weak surrounding rock
软弱下卧层 soft layer underneath; soft substratum; soft underlying stratum
软弱线 line of weakness
软弱岩层 incompetent bed;weak rock
软弱岩组滑坡 landslide of soft rock group
软扫帚 soft broom
软色调 soft tone
软砂 soft sand
软砂质黏[粘]土 soft sandy clay
软商品 soft goods;undurable goods
软烧器皿 soft-fired ware
软烧质陶器 soft earthenware
软烧砖 soft burnt brick
软舌螺绝灭【地】hyolithid extinction
软设备 soft device; software; software device
软设备组件 software package
软射线 soft ray
软砷铜矿 trippkeite
软渗碳 mild carburizing
软渗碳剂 mild carburizer
软生铁 soft cast iron;soft pig iron
软绳 extension cord; extension flex; soft line;soft rope
软湿冰雪 slush
软湿地 boggy ground;quashy ground; springy place
软湿似沼泽的 quashy
软石 freestone;soft rock;soft stone
软石板瓦 soft rags
软石翻松 ripping of soft rock
软石开挖 soft-rock excavation
软石蜡 paraffin(e) molle;soft paraffin(wax)
软石油沥青 soft asphaltic bitumen
软石脂 petrolatum;vaseline
软炻器 soft stoneware
软矢量 soft vector
软式飞艇 blimp; non-rigid air-ship; pressure airship
软式减振 soft cushioning
软式汽艇 < 可折叠的 > blimp
软式扫床 flexible bed-sweeping
软式设施 soft installation
软式天线 soft antenna
软式系结 flexible connection
软式小飞艇 blimp
软式阵地 soft position
软树脂 barras;paste resin;soft resin
软刷子 dabber
软水 low hardness water; rainwater; softened water;softening;soft water
软水槽 demineralized water tank
软水出口 soft water outlet
软水处理 soft water treatment;water treatment
软水处理设备 water softening plant
软水法 softening;water softening method
软水罐 water softener;water softening tank
软水化学需氧量 soft chemical oxygen demand
软水剂 softener; water softener; water softening agent
软水铝石 boehmite
软水器 water softener
软水砂 permutite
软水设备 water softening plant
软水设施 water softening facility
软水室 water softening room
软水手帽 flat cap;flat hat
软水所 water softening plant
软水系统 water softener system
软水箱 softened water tank
软水循环泵 demineralized water circulation pump
软水站 water softening plant
软水装置 demineralized water treatment device;water softener;water softening unit
软水作用 demineralization
软税收 soft taxation
软松木 soft pine
软塑的 incompetent plastic;soft plastic
软塑料 non-rigid plastics;soft plastics
软塑塞 < 填充轨道螺栓孔 > philplug
软塑性体 molliplast
软酸 soft acid
软胎式钻车 rubber-type drill jumbo
软胎体 soft matrix
软弹簧系统 spring softening system
软弹性体 mollielast
软炭质页岩 marken;rashings
软套管 teleflex
软套利 soft arbitrage
软梯 disappearing ladder;jack ladder; Jacob's ladder;rope ladder
软体动物地理区 mollusks faunal province
软体动物壳 mollusk shell
软体动物清除剂 molluscicide
软填料 soft packing;wadding
软填料密封 soft packing seal; soft stuffing-box seals
软填料的填料箱 soft-packed stuffing box
软跳(接)线 flexible jumper
软铁 mild iron;soft iron
软铁磁路 soft iron circuit
软铁分路器 soft iron shunt
软铁杆 soft iron bar
软铁极片 iron pole piece
软铁矿 pyrolusite
软铁球 quadrantal sphere; soft iron ball;soft iron sphere
软铁丝 annealed wire;soft iron wire
软铁芯 soft iron core
软铁芯电磁测量仪表 soft iron instrument
软铁氧体 soft ferrite
软通风管 flex-duct
软通货 soft currency
软铜 annealed copper;soft copper
软铜复绞线 soft copper compound strand wire
软铜管 soft copper tube
软铜排 flexible busbar
软铜丝 annealed copper wire
软铜线 annealed copper wire
软桶 fabric tank;rubber drum
软头锤 soft-headed hammer
软头靠 upholstered head-rest
软图像 soft picture
软土 mollisol; soft ground; soft soil; weak soil;yielding soil
软土侧向挤出 lateral squeezing-out of soft soil
软土层 auger ground; soft deposit; soft soil layer; strata of soft soil; weak ground
软土地基 soft foundation;soft ground
软土地面 soft ground
软土工程学 soft soil engineering
软土工程指标 engineering index of soft soil
软土机动性 soft soil mobility
软土挤出 squeezing out of soft soil
软土夹层 drift band
软土掘进机 slurry boring machine
软土路堤极限高度 critical height of roadbed on soft soils
软土路基 roadbed on soft soils; soft roadbed
软土壤处理 soft soil improvement
软土隧道 soft ground tunnel
软土隧道工程 soft ground tunnel(l)ing
软土通过性能 < 车辆 > soft soil performance
软土挖运机 mucker
软土楔形基础 wedge-type foundation of yielding soil
软土阻滞材料 soft ground arrestor
软拖曳电缆 flexible trailing cable
软弯头 flexible bend
软微丝煤 soft fusite
软卧车 carriage with cushioned berths
软污泥 soft sludge
软稀释剂 < 使沥青变软,降低稠度 > soft flux
软锡管 collapsible tube
软锡钎料 soft tin
软席 cushioned seat;upholstered seat
软席卧铺 cushioned berth
软洗涤剂 soft detergent
软细胞组织 parenchyma
软纤维内窥镜 flexible fiber scope
软纤维芯 < 钢丝绳的 > soft fiber core
软显示字处理机 soft display word processor
软线 extension cord; extension flex; flex cord; flexible conductor; flexible cord; soft wire; tinsel cord; patch cord < 配电盘的 >
软线吊线 cord pendant
软线分流器 flexible shunt
软线跟踪头 free cursor
软限幅器 soft limiter
软橡胶 soft argtum;soft rubber
软橡胶插入物 soft rubber insert
软橡胶衬垫 soft rubber insert
软橡胶垫圈 soft rubber washer
软橡皮 argtum eraser
软心(材) soft heart
软心钢 soft-center steel
软心钢犁壁 soft-center steel mo(u)ld board
软心钢犁铧 soft-center share
软心轮胎 cushion tire[tyre]
软芯势 soft core potential
软芯轴 soft core;soft core spindle
软信贷 soft credit
软型表面活性剂 soft type surfactant
软型去垢剂 soft type detergent
软型去污剂 soft detergent
软型洗涤剂 soft detergent
软性 softness
软性布线 flexible wiring
软性车行道 flexible carriageway
软性贷款保险计划 flexible loan insurance plan
软性导线管 flexible conduit
软性的 mild;soft
软性底片 soft negative
软性电缆 flexible cable

软性电子管 soft tube
软性反射幕 flexible reflector curtain
软性废气管 flexible exhaust pipe
软性辐射 low-energy radiation; soft radiation
软性覆面 soft floor covering
软性钢索 flexible wire cable
软性隔壁 flexible partition
软性隔离电缆 shielded flexible cable
软性刮平整修刀 flexible knife
软性光学玻璃 flint glass
软性基础板 flexible foundation slab
软性基础梁 flexible foundation beam
软性接头 flexible coupling; soft joint
软性绝缘 flexible insulation
软性开关 flex cock
软性联系 flexible connection
软性路面公路 flexible highway
软性磨耗层 flexible carpet
软性木材 free stuff; softwood
软性排水管 flexible exhaust pipe
软性泡沫聚氨酯甲酸乙酯 flexible foamed polyurethane
软性泡沫塑料 flexible expanded plastics; flexible foamed plastics
软性泡沫填料 soft foam filling
软性喷射机 flexible sprayer
软性屏蔽电缆 shielded flexible cable
软性驱动 flexible drive
软性射线 low-energy radiation; soft radiation; soft ray
软性石 freestone
软性刷墙粉 soft distemper
软性水浆涂料 soft distemper
软性塑料 non-rigid plastics
软性调制器 soft-switch modulator
软性铁 mild iron
软性烷基苯 <指易为生物降解的直链烷基苯> soft alkyl benzene
软性万向节 fabric joint; fabric universal joint; flexible universal joint
软性洗涤剂 <易生物降解者> soft detergent; soft type detergent
软性显影液 soft-working developer
软性像纸 soft paper
软性消毒剂 mild disinfectant
软性压缩空气管道 flexible compressed-air line
软性压缩空气管线 flexible compressed-air line
软性阴影 soft shadow
软性印刷布线电路 flexible print wiring circuit
软性砖 flexing brick; soft brick
软性作用 soft action
软雪 ripe snow
软烟煤 run coal
软岩层 soft formation; soft stratum
软岩层用钻头 <直径 76 毫米> rotacore
软岩(石) soft rock; weak rock; incompetent rock
软岩脱落 sluff
软岩钻进工具 soft ground boring tool
软页岩 bury; coaly rashings; soft shale
软椅 soft chair
软饮料 soft drink
软饮料厂 soft-drink bottling plant
软饮料厂废水 soft-drink bottling wastewater
软饮料厂废物 soft-drink bottling waste
软硬垢水 soft and hard filth water
软硬互层 admixture of hard and soft layers; alternately hard and soft strata
软硬互层岩组滑坡 landslide of rock group of hard-soft interlacing
软硬木 soft hardwood
软硬酸碱原理 soft-hard-acid base concept
软油 consistent lubricant
软油箱 bag tank; bag-type cell; flexible cell
软有槽铜条 soft copper came
软淤泥 soft mud
软玉 greenstone; kidney stone nephrite; nephrite
软预算约束 soft budgetary restraint
软皂 soft soap
软渣 sludge
软振荡 soft oscillation
软织物制动衬层 soft woven lining
软脂酸 cetylic acid; palmitic acid; palmitinic acid
软至中硬度黏[粘]土 soft-to-medium clay
软制地沥青 <掺重质油类制品> fluxed asphalt; fluxing asphalt
软制湖沥青 fluxed lake asphalt
软制剂 fluxing agent
软制沥青 fluxed bitumen
软制沥青材料 fluxed asphalt bituminous material; fluxed bituminous material
软制熔剂 fluxing medium
软制天然地沥青 fluxed native asphalt
软制油 flux(ing) oil
软质白瓷土 soft white clay
软质刨花板 insulating-type particle board
软质玻璃 soft glass
软质彩釉精陶 soft faience
软质瓷 artificial porcelain; soft paste porcelain
软质地面面层 soft floor finish
软质顶板 clod
软质高岭土矿床 soft kaolin deposit
软质骨料 soft aggregate
软质集料 soft aggregate
软质精陶 soft fine earthenware; soft fine pottery
软质颗粒 <金属表面喷射加工用> soft grit
软质面层 <建筑> soft finish
软质面粉 soft flour
软质木板 plank
软质木材 softwood
软质木材焦油 softwood tar
软质木材焦油沥青 softwood pitch
软质木料尺寸 softwood size
软质耐火黏[粘]土 soft fireclay
软质黏[粘]土 mild clay; soft clay; tender clay
软质黏[粘]土矿石 soft clay ore
软质绕丝筒 soft forming tube
软质石灰岩 <产于美国> marianna
软质塑料 flexible plastics
软质酸 palmitic acid
软质炭黑 lamp black; vine black
软质探针 flexible probe
软质碳黑 soft black
软质陶土 soft white clay
软质天然地沥青 mineral tar; miner tar
软质纤维 soft fiber[fibre]
软质纤维板 fibre insulation board; low-density fiberboard; softboard; tex
软质橡胶 soft rubber
软质岩(石) incompetent rock; soft rock; weak rock
软质颜料 soft-textured pigment
软质乙烯地板材料 flexible vinyl flooring
软质轧辊 soft bowl
软中碳钢 soft medium carbon steel
软中碳结构钢 soft medium structural steel
软轴 flexible axle; flexible shaft
软轴传动 flexible shaft drive; flexible shaft transmission
软轴混凝土研磨机 flexible shaft concrete grinding machine
软轴连接 flexible shaft coupling
软轴内插式振动器 flexible shaft internal vibrator
软轴驱动插入式振捣器 <混凝土捣实用> flexible shaft drive internal vibrator
软轴砂轮 flexible grinder
软轴砂轮机 flexible shaft grinder
软轴式振动器 flexible type vibrator
软轴套 flexible shaft protecting hose
软轴研磨机 flexible shaft grinding machine
软注坯 flabby cast
软铸铁 soft cast iron
软砖 <指可切削及雕刻的软砖> cutter; rubbers; soft brick
软状态 soft state
软着陆 soft impact; soft-land(ing)
软着陆仪器站 soft landed instrument station
软自激振荡 soft self-excited oscillation
软总线 flexible bus
软组织木材 soft-textured wood
软组织木地板 soft-textured wood floor boarding
软组织木纤维混凝土 soft-textured wood fiber[fibre] concrete
软组织木屑混凝土 soft-textured chipped wood concrete
软坐软垫制作 upholster
软座车 carriage with cushioned seats
软座圈 soft seated valve
软座席 upholstered seat

芮

芮木泪柏 Rimu(pine)

锐

锐边 arris; sharp edge
锐边的 sharp-edged
锐边角钢 angle with sharp corners
锐边角型材 angle with sharp corners
锐边界模型 sharp boundary model
锐边孔板 sharp-edged orifice plate
锐边小锉 pinion file
锐变层 spring layer
锐波 sharp wave
锐匙 sharp spoon
锐匙口牙刮匙 sharped dental curet
锐齿 sharp mouth
锐齿菌属 <拉> Oxydontia
锐刀 sharp tackle
锐顶 sharp crest
锐顶量水堰 sharp-crested measuring weir; sharp-edged measuring weir
锐顶小窗 miniature lancet(window)
锐顶堰 knife-edge weir; sharp-crested weir; sharp-edged weir
锐度 acuity; acuteness; sharpness; acutance
锐度匹配 acuity matching; acutance matching
锐度曲线 acutance curve
锐度试验 acuity test
锐端细胞组织 prosenchyma
锐耳用钩 sharp ear hook
锐方向性波束天线 pencil-beam antenna
锐方向性射束 pencil beam
锐方向性射束式雷达 pencil shape beam radar
锐方向性天线 high directional antenna
锐辐照 acute irradiation
锐钢钎 sharpened steel
锐高 sharp height
锐拱 acute arch
锐拱式建筑 lancet architecture
锐共振 sharp resonance
锐光束 sharp beam
锐化 sharpen(ing)
锐化电路 sharpener
锐化放大器 sharpener amplifier
锐化装置 sharpener
锐积 sharp product
锐尖窗 lancet
锐尖拱 lancet arch
锐减 slash; steep decline
锐角 acute angle; close bevel; closed angle; sharp angle; sharp corner; straight angle
锐角道岔 acute turnout
锐角等分线 acute bisectrix
锐角断块 acute angle block
锐角附件 acute angle attachment
锐角拐肘 acute angle crank
锐角尖 tapered point
锐角交叉 acute angle crossing; acute angle intersection; scissors joint; scissors junction
锐角交会 acute-intersection
锐角接头 joint at acute angle
锐角螺纹 sharp screw; sharp thread; sharp V thread
锐角墙角 acute corner; acute-squint corner
锐角曲柄 acute angle crank
锐角入口 sharp-cornered entrance
锐角三角形 acute-angled triangle; acute triangle; oxygon(e)
锐角三角形的 oxygonal
锐角三通(管) double sweep tee
锐角势井 sharp-cornered well
锐角弯道 acute-angled turn
锐角弯管 knuckle bend
锐角弯曲 sharp bend
锐角弦长的缩尺 scale of chord
锐角效应 corner effect
锐角叶片 acute angle blade
锐角辙叉 acute angle crossing; acute frog
锐角支管 branching at an acute angle
锐角转角 sharp corner
锐角转角隅石 acute squint
锐截断模型 sharp cut-off model
锐截止 sharp cut-off
锐截止管 sharp-cutoff tube
锐截止滤波器 sharp cut-off filter
锐截止滤光片 sharp cut-off filter
锐截止式滤波器 sharp-cut-off filter
锐截止特性 sharp cut-off characteristics
锐截止五极管 sharp cut-off pentode
锐聚焦 sharp focus(ing)
锐聚焦的 fine-focus(s)ed
锐聚焦面 plane of sharpest focus
锐聚焦区 zone of sharp focus
锐聚透镜 sharp-focus lens
锐聚图像 sharply focused image
锐孔 orifice
锐孔板 orifice plate
锐孔挡板 orifice baffle
锐孔菌属 <拉> Oxyporus
锐孔流量计 sharp-edged orifice
锐孔流速计 orifice flowmeter; orifice meter [metre]; sharp-edged orifice meter
锐孔气体洗涤器 orifice gas scrubber
锐孔调节阀 orifice control valve
锐孔型黏[粘]度计 orifice-type visco(si)meter
锐口 orifice
锐口堰 sharp-crested weir
锐棱的 sharp angle
锐利 acuity; acuteness; sharpness

R

锐利的 piercing; sharp; sharp-edged; spiky
锐利工具 sharp instrument; sharp tool
锐利角 angle of keenness
锐利考验 acuity test
锐利钻头 sharp bit
锐六角星 oxyhexaster
锐敏的 subtile; subtle
锐敏度 acuity; keenness
锐敏性 acuition
锐谱线 sharp line
锐器解剖法 sharp dissection
锐器损伤 sharp instrument injury
锐曲线 sharp curve; steep curve
锐石 sharpstone
锐蚀高地 fretted uplands
锐双齿牵开器 two-prong sharp retractor
锐钛矿 anatase; octahedrite
锐钛型 anatase configuration; anatase modification; anatase variety
锐钛型二氧化钛 anatase titanium dioxide
锐钛型钛白粉 anatase titanium dioxide
锐调谐 fine tuning; hairbreadth tuning; sharp tuning
锐弯(管) sharp bend
锐弯头 sharp bend
锐线光谱 sharp line spectrum
锐线光源 narrow line source
锐线系 sharp series
锐线系光谱 sharp series
锐楔形队 sharp wedge formation
锐斜角 closed bevel
锐叶木兰 cucumber tree
锐缘 knife edge
锐缘测堰 sharp-edged measuring weir
锐缘的 sharp-edged
锐缘进气口 sharp-edged intake; sharp-lip inlet
锐缘孔口 sharp-edged orifice
锐缘量水堰 sharp-edged ga(u)ging weir
锐缘喷孔 sharp-edged orifice
锐缘堰 knife-edged weir; sharp-crested weir; sharp-edged weir
锐缘堰顶 sharp crest; sharp-edged crest
锐缘堰顶的 sharp-crested
锐缘圆孔口 sharp-edged circular orifice
锐缘柱槽 sharp fluting
锐缘锥体 sharp-edged cones
锐折曲线 quicken
锐锥 acute cone
锐锥度 abrupt taper
锐锥形射束天线 pencil-beam antenna

瑞

瑞德方程式 Reid equation

瑞德蒸气压力 Reid vapo(u)r pressure
瑞典薄壁取土器 Swedish foil sampler
瑞典沉降盒 <现场观测用的> Swedish box
瑞典法 <计算土坡稳定性的> Swedish method
瑞典工会联合会 Confederation of Swedish Trade Unions; Swedish Trade Union Confederation
瑞典工业联合会 Federation of Swedish Industries
瑞典雇主联合会 Swedish Employers' Confederation
瑞典滚珠轴承公司 Aktiebolaget Svenska Kullager-fabriken
瑞典国际车辆注册字 Sweden's international vehicle registration letter
瑞典国际开发署 The Swedish International Development Authority
瑞典国家动力局 Swedish State Power Board
瑞典海绵铁粉 Swedish iron powder
瑞典滑弧法 <土体稳定分析的> Swedish slip-circle method; Swedish cylindrical surface method
瑞典滑弧计算法 <计算土坡稳定性的> Swedish slip-circle method
瑞典混凝土协会 Swedish Concrete Association
瑞典加重式贯入度仪 Swedish weight penetrometer
瑞典建筑 Swedish architecture
瑞典静力触探 Swedish cone penetration test
瑞典里 <相当于 10 千米> Swedish mile
瑞典落锥法 Swedish fall-cone method
瑞典脉冲燃油系统 Swedish impulse oil firing system
瑞典木炭生铁 Swedish iron
瑞典木瓦 takspan
瑞典全国道路管理局 Swedish National Road Administration
瑞典塞格纳斯铁粉 Swedish Hoganas powder
瑞典砂油灰 Swedish sand putty
瑞典商业银行 Svenska Handelsbanken
瑞典式触探 <探头为土钻形状,可用来测得黏[粘]土稠度> Swedish sounding
瑞典式窗 <中间装有活动百叶的双层玻璃窗> Swedish window
瑞典式防撞栏 Swedish safe stop barrier
瑞典松木屋顶 takspan
瑞典梯形钻孔法 Swedish
瑞典投资银行 Investment Bank of Sweden; Sveriges Investerings Bank
瑞典土坡稳定性分片计算法 Swedish method of slices
瑞典现代派风格 <指家具和陈设> Swedish modern
瑞典信贷银行 Sveriges Kreditbank
瑞典学派 Swedish school
瑞典岩土工程研究所 Swedish Geotechnical Institute
瑞典银行 Bank of Sweden; Sveriges Riksbank
瑞典油灰 Swedish putty
瑞典油灰密封 Swedish putty seal(ing)
瑞典油灰涂膜 Swedish putty coat
瑞典油灰弯曲试验仪 Swedish putty bending tester
瑞典圆弧法 Swedish circle method; Swedish cylindrical surface method
瑞典圆柱面法 Swedish cylindrical surface method
瑞典中央银行 Sveriges Riksbank
瑞典轴承制作所 Svenska Kullager Fabriken
瑞典自动轧管机 Sweden mill; Swedish mill
瑞耳 <声阻抗率单位> rayl
瑞格耳(轴承)合金 Regel metal
瑞海德煤气红外线加热器 Rayhead
瑞基图 <即振动器的功率和频率同负载关系曲线图> Rieke diagram
瑞金(钢板橡皮夹层)缓冲器 Raykin buffer
瑞金式缓冲器 Raykin(fender) buffer
瑞金式消能垫 Raykin(fender) buffer
瑞克接地 rake reception
瑞利 R <发光强度单位> Rayleigh
瑞利斑 Rayleigh's disk
瑞利比(率) Rayleigh's ratio
瑞利比值 Rayleigh's ratio
瑞利波 Rayleigh's wave
瑞利波单一传递函数 Rayleigh's wave singlet transfer function
瑞利波法 Rayleigh's wave method
瑞利波群速度 Rayleigh's group velocity
瑞利波弹性品质因子 Rayleigh's elastic quality factor
瑞利波相速度 Rayleigh's phase velocity
瑞利常数 Rayleigh's constant
瑞利大气 Rayleigh's atmosphere
瑞利倒易定理 Rayleigh's reciprocity theorem
瑞利电流天平 <安培秤> Rayleigh's current balance
瑞利定律 Rayleigh's law
瑞利法 Rayleigh's method
瑞利方程 Rayleigh's equation
瑞利分辨极限 Rayleigh's resolution limit
瑞利分辨率 Rayleigh's resolution
瑞利分辨率判据 Rayleigh's criterion for resolution
瑞利分布 Rayleigh's distribution
瑞利分光计 Rayleigh's spectrometer
瑞利分馏 Rayleigh's fractionation
瑞利峰值分布 Rayleigh's distribution of peaks
瑞利干涉计 Rayleigh's interferometer
瑞利公式 Rayleigh's equation
瑞利过程 Rayleigh's process
瑞利函数 Rayleigh's function
瑞利耗散函数 Rayleigh's dissipation function
瑞利互易定理 Rayleigh reciprocity theorem
瑞利回线 Rayleigh's loop
瑞利极限 Rayleigh's limit
瑞利-杰恩斯定律 Rayleigh-Jeans law
瑞利-杰恩斯公式 Rayleigh-Jeans formula
瑞利近似 Rayleigh's approximation
瑞利-兰姆频方程 Rayleigh-Lamb frequency equation
瑞利-兰姆频谱 Rayleigh-Lamb frequency spectrum
瑞利棱镜 Rayleigh's prism
瑞利-里茨法 <分析板体和其他复杂结构的一种方法> Rayleigh-Ritz method
瑞利流(动) Rayleigh's flow
瑞利密度函数 Rayleigh density function
瑞利能量方法 Rayleigh's energy method
瑞利欧(姆) Rayleigh ohm
瑞利盘 Rayleigh's disk
瑞利判据 Rayleigh's criterion[复 criteria]
瑞利散射 Rayleigh's scatter(ing)
瑞利散射定律 Rayleigh's law of scattering
瑞利散射法 <X 射线分析> Rayleigh's scattering method
瑞利散射光谱 Rayleigh's scattering spectrum
瑞利散射函数 Rayleigh's scattering function
瑞利散射截面 Rayleigh's cross-section
瑞利散射器 Rayleigh's scatterer
瑞利散逸函数 Rayleigh's dissipation function
瑞利商数 Rayleigh's quotient
瑞利声盘 <测声强用> Rayleigh's dish
瑞利声盘法 Rayleigh's disc method
瑞利数 <表示产生和缓解浮动对流作用的各力之间关系的指数> Rayleigh's number
瑞利衰落 Rayleigh's fading
瑞利速度 Rayleigh's speed
瑞利天平 Rayleigh's balance
瑞利条件 Rayleigh's condition
瑞利威利斯关系 Rayleigh-Willis relation
瑞利线 Rayleigh's line
瑞利限度 Rayleigh's limit
瑞利效应 Rayleigh's effect
瑞利循环 Rayleigh's cycle
瑞利翼状散射 Rayleigh-wing scattering
瑞利原理 Rayleigh's principle
瑞利折射计 Rayleigh's refractometer
瑞利蒸馏 Rayleigh's distillation
瑞利周期 Rayleigh's cycle
瑞利准则 Rayleigh's criterion[复 criteria]
瑞利阻尼矩阵 Reyleigh's damping matrix
瑞玛建筑体系 Reema
瑞米图版法 Ramey type curve matching
瑞奇型旋转钻机 Reichdrill
瑞士 <欧洲> Switzerland
瑞士爱伯特式山区齿轨铁路 Abt system railway
瑞士产合成桐油 trienol
瑞士锤 Swiss style hammer
瑞士的州银行 cantonal banks
瑞士法郎 <货币名称,单位符号 SF> Swiss Franc
瑞士工会联合会 Swiss Federation of Trade Unions
瑞士工商联合会 Schweizerischer Handels-und Industrie-Verein
瑞士公式 Swiss formula
瑞士国家银行 Banque Nationale Suisse; National Bank of Switzerland; Schweizerischer National Bank
瑞士航空公司 Swiss Air
瑞士建筑 Swiss architecture
瑞士金法郎 Swiss gold franc
瑞士蓝 Swiss blue
瑞士蓝宝石 Swiss lapis lazuli
瑞士联邦 Swiss Confederation
瑞士联合银行 Union Bank of Switzerland; Union de Banques Suisse
瑞士柳 Swiss willow
瑞士青金 Swiss lapis
瑞士山中倾斜屋顶的木屋 palafitte
瑞士山中倾斜屋顶的农舍 palafitte
瑞士式村舍 Swiss cottage
瑞士式锉刀 Swiss pattern file
瑞士式茅屋 Swiss cottage
瑞士式平行沟瓦 Swiss parallel gutter tile
瑞士式熊阱堰 roof weir
瑞士松 cembra(n)pine; Swiss pine
瑞士松油 Swiss pine oil
瑞士五针松 cembra; cembra(n)pine; Swiss stone pine
瑞士信贷银行 Crédit Suisse; Swiss Credit Bank
瑞士型锉 Swiss pattern file
瑞士型自动机床 Swiss-type automatic
瑞士银行 Swiss Bank Corp
瑞士银行公司 Schweizerischer Bankverein
瑞士债券 rembrandt bonds
瑞士中央技术学院水力研究和土壤力学试验室 The Laboratories for Hydraulic Research and Soil Mechanics, Swiss Central Technique College
瑞特勒型连续运输机 Redler converter
瑞提阶 <晚三叠世>【地】 Rhaetian
瑞香红色 daphne red

闰

闰的 <如闰年、闰月> intercalary; bissextile

闰秒 leap second

闰秒调整 leap second adjustment
闰年 bissextile (year); embolismic year; intercalary year; leap-year
闰盘 intercalated disc
闰期 intercalation
闰日 <即 2 月 29 日> intercalary day; leap-day
闰余 epact
闰月 intercalary month; leap month
闰周 intercalary cycle

润

润版 dampening
润版辊 dampening roller
润版药水 dampening solution; dampening water
润彩(涂装)法 blending
润滑 lube; lubricate; lubricating; oiling
润滑杯 greaser
润滑泵 lubricant[lubricating] pump
润滑表面 lubricated surface
润滑玻璃管 lubrication tube glass
润滑不当 faulty lubrication
润滑部位 lubricant housings; oil site
润滑部位图 lube chart
润滑材料 greasing substance; lube; lubricant; lubricating material; unguent
润滑层 lubricating layer; slip layer
润滑层效应 lubricating layer effect
润滑车 greasing truck
润滑车间 lubrication shed
润滑程序 lubricating order; lubrication procedure
润滑齿轮 oil gear
润滑冲洗液【岩】 lubricating drilling fluid
润滑的 lubricant
润滑点 lube point; lubricating point; lubrication point; oiling point; oil site; point of lubrication
润滑点说明图 lubrication chart; oiling chart
润滑点指示图 lubrication point chart
润滑垫 lubricating pad
润滑阀 lube valve
润滑法 lubricate; lubricating method; lubrication
润滑服务车 lubrication service truck
润滑工 grease monkey; greaser; lube man; lubricator; oiler
润滑工人 greaser
润滑功能 lubricating function
润滑故障 lubrication trouble
润滑管道 lubrication line; lubrication piping
润滑管路 lubricating circuit; lubrication circuit; lubrication line
润滑规程 lubrication order; lubrication program(me)
润滑滚针 lubrication quill
润滑过程 greasing process
润滑过的管接头 grease coupler
润滑合成剂 lubricating composition; lubricating compound
润滑和冷却油系统 lubricating and cooling oil system
润滑糊 lubricating paste
润滑环 drip ring; latern(ring); lubricating ring
润滑剂 lubricant(base); anti-friction material; consistent lubricant; emollient; greasing substance; lube; lubricating agent; lubricating medium; lubricator; sliding agent
润滑剂变质 lubricant deterioration
润滑剂不足 lubricant starvation
润滑剂稠化 lubricant thickening
润滑剂的基础组分 lubricant base
润滑剂的添加剂 lubricant additive
润滑剂的荧光 lubricant bloom
润滑剂的油性 oiliness of lubricant
润滑剂分离器 lubricant separator
润滑剂封挡 lubrication dam
润滑剂工作性能 lubricant performance
润滑剂供应车辆 lubricant supply vehicle
润滑剂供应断绝 lubricant starvation
润滑剂减磨值 wear-reducing value of lubricant
润滑剂检验仪 lubricant tester
润滑剂可混用性 lubricant compatibility
润滑剂流速 rate of lubricant flow
润滑剂毛细管现象 lubricant capillarity
润滑剂膜 lubricant film
润滑剂耐热耐压试验机 Falex tester
润滑剂容器 lubricant container
润滑剂施用器 lubricant applicator
润滑剂试验仪 lubricant tester
润滑剂添加料 lubricant additive
润滑剂载体 lubricant carrier
润滑剂振荡试验机 oscillating machine for testing of lubricants
润滑剂注射器 lubricating syringe
润滑加油嘴 lubricating nipple
润滑间隔期 lubrication interval
润滑胶冻 lubricant jelly
润滑接头 lubrication connection
润滑介质 lubricating medium
润滑界限 lubricating[lubrication] limit
润滑近似法 lubrication approximation
润滑坑 lubrication pit
润滑孔 grease tap; lubricating hole; lubrication hole; oil cavity
润滑蜡 lubricating wax; Albany grease
润滑冷却剂 <用以冷却和润滑正在操作中的切割工具或机械> coolant
润滑理论 theory of lubrication
润滑力 lubricating capacity
润滑滤清器 lubrication filter
润滑螺旋 lubricating screw
润滑脉冲 lubricating impulse
润滑毛毡 lubricating felt
润滑模式 lubrication mode
润滑膜 lubricant film; lubricating film; slippery film
润滑膜破裂 lubricating oil breakdown
润滑磨损 lubricated sliding wear
润滑能力 lubricating ability; lubricating power; lubricating value; lubricity
润滑牛油 lubricating grease; Stauffer grease
润滑喷油嘴 lube oil spray jet
润滑喷嘴 lubricating nozzle
润滑期 lubrication interval
润滑期限 lubricant life
润滑器 greasing apparatus; lubricating coupler; lubricator; oil-holder
润滑器给油阀 lubricator feed valve
润滑器加油器 lubricator transfer filler
润滑器接头 lubricator adapter
润滑器具 greaser
润滑器开关 lubricator cock
润滑器配件 lubricator fitting
润滑器气孔盖 lubricator air hole cap
润滑器汽阀 lubricator steam valve
润滑器驱动装置 lubricator drive
润滑器示油玻璃管 lubricator show glass
润滑器推轴 lubricator driving shaft
润滑器托 lubricator bracket
润滑器箱 lubricator box
润滑器项 lubricator shank
润滑器指示玻璃窗 lubricator show glass
润滑清净添加剂 lubricating detergent additive
润滑容器 lube container; lube oil container
润滑纱线 lubricated yarn
润滑砂浆 lubricating mortar
润滑设备 lube equipment; lubricating utensil; lubrication equipment; oiling system
润滑失效 lubrication failure
润滑石墨 lubricating graphite
润滑使用期限 lifetime lubricated
润滑手册 lubrication guide; lubrication manual
润滑说明书 lubrication instruction; lubrication manual
润滑说明图 lube chart
润滑添加剂 lubricating additive
润滑填料 grease packing; oil pad
润滑条件下的磨损试验 lubricated wear test
润滑图 lube chart; lubrication drawing; oiling chart
润滑图表 lube plan; lubrication chart; lubrication diagram
润滑涂层 lubricant coating
润滑物质 greasing substance
润滑系减压阀 oil-pressure relief valve; oil relief valve
润滑系统 lube(oil) system; lubricating system; lubrication; lubrication system
润滑系统容量 lubrication system capacity
润滑系统图 lube chart; lubrication chart; oiling chart
润滑效率 lubricating efficiency
润滑效应 lubricating effect
润滑性 greasiness; lubricating ability; lubricity; oiliness; unctuosity
润滑性能 greasy property; lubricating property; lubrication property
润滑性能的保持 retention of lubricity
润滑性水分 <降低土黏[粘]聚力的水分> lubricating moisture; lubricating water
润滑性提高剂 lubricity additive
润滑性添加剂 oiliness additive
润滑性载体 lubricity carrier
润滑性增塑剂 lubricant plasticizer
润滑性质 lubricating property; lubricating quality
润滑旋塞阀 lubricated plug valve
润滑压力调节 lubrication pressure regulation
润滑液 lubricating fluid; lubricating liquid
润滑液供给装置 lubricant application device
润滑用具 lubricating equipment
润滑用设备 greasing equipment
润滑用液压 lubrication pressure
润滑油 anti-friction material; engine oil; fluid oil; grease oil; lube(oil); luboil; lubricant[lubricating/ lubrication] oil; mobile oil; motor oil; nonfluid oil; sunned oils; unguent; key paste <由黑色糖浆与石墨组成的>
润滑油包装 lube oil packaging
润滑油保持环 oil retainer(thrower)
润滑油保持性能 oil-retaining property
润滑油杯 grease cup; lubricating cup; lubricator cap; lubricator cup; oil cup
润滑油杯盖 grease cover
润滑油泵 grease lubricating pump; lube(oil) pump; lubricant pump; lubricating oil pump; lubrication pump; oil lubricating pump
润滑油泵接触器 lubricating oil pump contactor
润滑油比耗 lubricating oil consumption
润滑油补充口 lubricating oil charging port
润滑油仓库 lube oil warehousing
润滑油槽 lube sump; lubricant groove; lubricating chute; lubrication groove; lubrication sump; oil groove; oil trough; oil-way; lubricating groove
润滑油车 lubrication lorry
润滑油澄清器 oil clarifier
润滑油冲压机 compressor gun
润滑油抽吸警告开关 lubricating oil suction alarm switch
润滑油抽吸警报开关 lubricating oil suction alarm switch
润滑油稠度等级 grease grades
润滑油储槽 oil sump
润滑油储存柜 lube oil storage tank
润滑油处理 lubricating oil treatment
润滑油传送泵 lubricating oil transfer pump
润滑油粗滤器 lubricating oil strainer
润滑油存储能力 oil storage capacity
润滑油的加压 forcing of oil
润滑油的抗沫剂 oil anti-foaming agent
润滑油的黏[粘]度 oil body
润滑油的生产 lubricating oil processing
润滑油的铜含量 copper in lubricating oil
润滑油的氧化 oil ag(e)ing
润滑油的有机酸值 lubricating oil organic acidity
润滑油等级 lubricant class
润滑油底质 body of oil
润滑油发动机 engine testing of lube oil
润滑油分类 lube classification
润滑油分类方法 lubricant classification
润滑油分离器 lubricating oil separator; oil separator
润滑油分配器 oil header
润滑油分配箱 oil distributing box
润滑油分水机 lubricating oil purifier
润滑油分析 lubricating oil analysis
润滑油分杂机 lubricating oil clarifier
润滑油腐蚀性测定仪 jetometer
润滑油膏 lubricating cream
润滑油更换 oil change
润滑油工厂 lube plant
润滑油供给装置 oil feeding device
润滑油供油联动装置 lubricating oil supply linkage
润滑油沟 oil cavity; oil-way
润滑油故障 lubricating oil failure
润滑油管 lubricating oil pipe
润滑油管道 lubricating pipe; supply line
润滑油管垫 lube oil pipe grommet
润滑油管系 oil piping
润滑油罐 lubricating can
润滑油规格 lubricating oil specification
润滑油柜 lube oil tank
润滑油过滤器 lube oil filter; lubricating oil filter
润滑油含水量试验 lubricating oil water test
润滑油耗量 consumption of lubricating oil
润滑油和润滑脂分配器 oil and grease distributor
润滑油呼吸管 oil breather
润滑油环 lubricating ring; lubrication ring; splash ring
润滑油环流 lubricating oil circulation

润滑油回收系统 oil recovery system
润滑油回油泵 lube scavenge pump
润滑油积炭试验 carbonization test
润滑油基础油 lube base oil
润滑油基油 lubricant base
润滑油及活塞冷却油泵 lubricating oil and piston cooling oil pump
润滑油及冷却油泵 lubricating oil and cooling oil pump
润滑油挤出 pounding out
润滑油计量泵 oil-metering pump
润滑油加工装置 lubricating oil processing unit
润滑油加氢补充精制 lube hydrofinishing
润滑油加氢补充精制过程 lube oil hydrofinishing process
润滑油加氢处理过程 lube oil hydrotreating process
润滑油加氢精制 lube hydrotreating
润滑油加热器 lube oil heater
润滑油加油车 mobile lubrication truck
润滑油加油口 lubrication filler
润滑油结构 grease structure
润滑油金属含量试验 lubricating oil metal test
润滑油井 lubricating well
润滑油净化系统 oil purification system
润滑油抗老化剂 anti-oxidant for lube oil
润滑油控制环 oil control ring
润滑油口塞 grease hole plug
润滑油类别 lubricant class
润滑油类型 lube form
润滑油冷却放泄箱 lubricating oil drain and cooler tank
润滑油冷却器 lube oil cooler;lubricating oil cooler
润滑油冷却系统 oil cooling system
润滑油量计 oil ga(u)ge
润滑油流出 lubricant discharge
润滑油滤清器 lubricating oil filter;lubricating oil purifier; lubricator oil strainer
润滑油滤网 lubricating oil strainer
润滑油路 lubricant passage way
润滑油路图 lubricating chart
润滑油路压力表 oil circulation ga(u)ge
润滑油膜 fluid film;lubricant film;lubricating oil film; lubrication film; oil-bound film
润滑油膜导电性 oil film conductivity
润滑油黏[粘]度 lubricating oil viscosity
润滑油牌号 oil number
润滑油盘 lubrication pan
润滑油喷嘴 lubricating [lubrication] nipple
润滑油起沫 oil whip
润滑油器 lubricator
润滑油枪 grease gun; grease squirt; lubricating nozzle; lubricating oil gun;oil syringe; pressure gun; syringe for lubrication;lubricating gun
润滑油圈 splash ring;thrower ring
润滑油容量 lubricating oil capacity; lubrication oil capacity
润滑油容器 lube oil container
润滑油溶剂 lubricant container
润滑油溶剂精制抽出油 lubex
润滑油乳化试验 lubricating oil emulsion test
润滑油乳胶试验 lubricating oil emulsion test
润滑油入口 lubricating oil inlet
润滑油软管 lube hose
润滑油绳 lubricator wick
润滑油试验 lubricating oil test
润滑油室 grease chamber;oil chamber
润滑油收集器 grease trap; lubricant trap
润滑油输送泵 lube oil transfer pump; oil transfer pump
润滑油损耗量 lubrication oil consumption
润滑油套 oil jacket
润滑油添加剂 lubricating oil additive
润滑油调和 lube oil blending
润滑油调和汽油 lubricated gasoline; lubricating gasoline
润滑油调节 oil control
润滑油调整旋塞 lubrication regulating cock
润滑油通道 oil gallery
润滑油通路 oil passage
润滑油通气管 oil breather
润滑油桶 grease bucket
润滑油涂料 oil dope
润滑油弯管 lubricating elbow
润滑油污染 lubricant bloom
润滑油污渍 lube contamination
润滑油无灰分散剂 lubricant ashless dispersant
润滑油稀释 lubricating oil dilution
润滑油稀释量测定 oil dilution test
润滑油稀释用汽油 oil-diluent gasoline
润滑油系统 lubricating oil system
润滑油细滤器 lubricating oil fine filter
润滑油箱 grease box;grease tank;lubricant reservoir; lubrication box; lubrication(oil)tank;sump
润滑油消耗 lubricant consumption
润滑油消耗量 lube oil consumption; oil consumption
润滑油消耗率 rate of lubrication oil consumption
润滑油楔 lubricating oil wedge;lubrication wedge
润滑油泄漏 oil leak(age)
润滑油芯 lubricating oil wick; oil wick
润滑油型减压塔 lube type vacuum tower
润滑油性质 lubricating oil property
润滑油循环 lubricating oil circulation
润滑油压 lube pressure
润滑油压差控制 differential oil pressure control
润滑油压力 lubricating oil pressure; lubrication(oil)pressure
润滑油压力表 lubricating oil pressure ga(u)ge
润滑油压入器 lubricating press
润滑油溢流阀 lube relief valve
润滑油用橡胶 rubber for lubricating oils
润滑油油泵 grease pump
润滑油油池 lubricating oil sump
润滑油油密性试验 lubricating oil test for tightness
润滑油油膜强度试验 oil film test
润滑油油盘 lube pan
润滑油油位控制 lubricating oil level control
润滑油淤渣 lube oil sludge
润滑油预热器 lube oil preheater
润滑油再生车间 lube oil reclamation plant
润滑油再循环系统 oil recirculating system
润滑油再蒸馏装置 oil rerun unit
润滑油增效剂 oiliness compound
润滑油增压机 compressor gun
润滑油渣 slum
润滑油(毡)芯 lubricating felt wick
润滑油真空泵 oil vacuum pump
润滑油蒸发损失 lubricating oil evapo(u)ration loss
润滑油蒸馏 lubricating oil distillation
润滑油蒸馏釜 lubricating oil still
润滑油支管 lubricating oil branches
润滑油脂 grease;lubricant grease;lubricating grease;Stauffer grease
润滑油脂孔 grease hole; lubricating hole
润滑油脂自动加注管卷筒 lubreel;lubrication reel
润滑油制造工艺 lube oil technology
润滑油中提出的混合物 lube extract blend
润滑油装置 lube plant
润滑油渍 lubricant residue
润滑油总管 lubricating oil main
润滑油族 lubricating oil family
润滑油嘴 grease fitting; lubricating nozzle
润滑与检验车间 lubrication and inspection department
润滑站 lubrication station;lubritory
润滑脂 consistent fat; consistent lubricant; grease lubricant; grease substance;lube grease; oleosol; solid lubricant
润滑脂杯 compression cup; grease cup
润滑脂泵 grease pump
润滑脂泵送性能 grease pumpability
润滑脂表观黏[粘]度 grease apparent viscosity
润滑脂不溶物 grease insolubles
润滑脂槽 grease groove
润滑脂槽润滑 grease-well lubrication
润滑脂充填口 grease gate
润滑脂抽出器 grease extractor
润滑脂稠度 consistency of grease; grease consistence[consistency]
润滑脂稠度号 grease consistency numbers
润滑脂稠度计 grease consistometer
润滑脂稠化 grease thickening
润滑脂稠化剂 grease thickener
润滑脂穿透度测定计 grease penetrometer
润滑脂弹筒式润滑器 grease cartridge lubricator
润滑脂的成沟现象 grease channeling
润滑脂的脱水收缩 syneresis
润滑脂的相特性 phase behavio(u)r of greases
润滑脂的硬度 hardness of grease
润滑脂滴点 grease dropping point
润滑脂反应试验 reaction test for grease
润滑脂非皂基稠化剂 grease non-soap thickener
润滑脂分类 grease classification
润滑脂分离器 grease separator;grease trap
润滑脂分配系统 grease-dispensing system
润滑脂分析 grease analysis
润滑脂分油 grease bleeding; grease separation;separating of grease
润滑脂分油试验 separation test for greases
润滑脂工厂 grease-making plant
润滑脂工作器 grease worker
润滑脂供送性能 grease feedability
润滑脂固化 grease solidification
润滑脂锅 grease kettle
润滑脂过滤器 grease filter
润滑脂盒 grease basin
润滑脂壶 grease can
润滑脂护脂圈 grease retainer
润滑脂基 grease base
润滑脂加热试验 English heating test
润滑脂碱 grease alkali
润滑脂胶凝 grease gelling
润滑脂搅和机 grease working machine
润滑脂均匀化 grease homogenization
润滑脂抗氧化剂 grease anti-oxidant
润滑脂抗硬化安定性 grease hardening resistance
润滑脂块 block grease
润滑脂流出 grease creeping
润滑脂流动试验 grease dispensing test;grease flow test
润滑脂排出管嘴 lubricating nipple
润滑脂喷射润滑 grease spray lubrication
润滑脂品级 grease grades
润滑脂铅笔 grease pencil
润滑脂枪 grease gun;grease pump;lever grease gun; lubricating gun;lubricating screw;oil gun
润滑脂枪接头 lubricating gun adapter
润滑脂枪喷嘴 grease gun fitting
润滑脂腔 grease pocket
润滑脂溶胀 grease swelling
润滑脂熔点 grease melting point
润滑脂蠕升 creeping of grease
润滑脂润滑 grease lubrication; greasing
润滑脂润滑接头 grease lubrication fitting
润滑脂石墨混合物 grease-graphite mo(u)ld
润滑脂试验 grease testing
润滑脂水分 water bleeding from grease
润滑脂弹性 grease elasticity
润滑脂弹性试验 elasticity test of grease
润滑脂填料 grease filler
润滑脂调制设备 grease compounder
润滑脂调制装置 grease compounding unit
润滑脂筒给油嘴 grease cylinder feeding nozzle
润滑脂脱除剂 grease remover
润滑脂稳定剂 grease stabilizer
润滑脂稳定性 grease stability
润滑脂污染的油料 grease-spoiled oil
润滑脂箱 grease box
润滑脂性质 grease characteristic
润滑脂旋塞 grease cock
润滑脂用搅和器 grease worker
润滑脂有机填料 grease organic filler
润滑脂增稠剂 lubrication grease thickener
润滑脂沾污 grease contamination
润滑脂针入度测定 grease penetration test
润滑脂针入度测定器 grease penetrometer
润滑脂震击试验 shock test of grease
润滑脂蒸发损失 grease evapo(u)ration loss
润滑脂注入架 grease rack
润滑脂注入器 greaser
润滑脂注入嘴 grease coupler
润滑脂注油器 greaser
润滑脂装置 grease plant
润滑脂组成 grease composition
润滑脂组分混合 grease compounding
润滑脂嘴 grease nipple
润滑值 lubricating value
润滑重油 black oil
润滑周期 lube interval;lubrication interval;lubrication schedule
润滑转向节轴油嘴 steering knuckle bearing grease nipple
润滑装备 lubricating fitting
润滑装置 lube equipment; lube fit-

ting;lubricating arrangement;lubricating device;lubricating gear;lubricating set;lubricating system;lubrication apparatus;lubrication assembly;lubrication device;lubricator;oil cellar
润滑嘴 grease tap
润滑作业地沟 lubricating pit
润滑作业规程 lubricating instruction
润滑作业间 lubrication bay
润滑作用 lubricating effect;lubrication(action);oiling action
润滑座 lubricator stand
润磨液 grinding fluid;grinding lubricant
润磨油 grinding oil
润切液 cutting fluid
润色 embellishment;retouching;touching up
润湿 dabble;dampening;humectation;moisten(ing);wet without rain
润湿部件 wetted parts
润湿槽 wetting-out deck
润湿的 humid;springy;wet
润湿的岩石 fresh rock
润湿度 degree of wetting;hygroscopicity
润湿辊 dampener;dampening roll;damping roll
润湿剂 humidification agent;wetter;wetting agent
润湿角 contact angle;wetting angle
润湿接触角 contact angle of wetting
润湿力 wetting force
润湿率 percent wet-out
润湿能力 wetting ability;wetting capacity;wetting power
润湿器 fogger;humectant;moistener
润湿器垫板 moistener pad
润湿热 heat of wetting;wetting heat
润湿热焓 humid enthalpy
润湿时间 wetting time
润湿收缩 hygral shrinkage
润湿水 <含有强润湿剂的水,灭火用> wet water
润湿特性 wetting characteristics
润湿天平 wetting balance
润湿通风 humidity ventilation
润湿误差 wetting error
润湿效率 wetting efficiency
润湿性 water affinity;wettability
润湿性能 wetting property
润湿液 damp(en)ing solution;damp(en)ing water
润湿液体 wetting liquid
润湿张力 wetting tension
润湿指数 wetting index
润湿周边 wetting perimeter
润湿装置 dampening system;wetting device
润湿作用 wetting action
润饰 finishing touch
润饰光 subsidiary light
润周 wetted perimeter
润砖刷 stock brush

若丹明 rhodamine

若丹明红 rhodaminered
若丹明色淀 rhodamine lake
若丹明颜料 rhodamine pigment
若杜林蓝 Rhoduline blue;setoglaucine
若杜林紫 Rhoduline heliotrope
若尔盖地块 Zoiga block
若干抵押贷款集合 mortgage prortofolio
若干房屋、土地同时征用 blanket condemnation
若干木兰属 cucumber tree
若干年内分头损失 loss spread over a number of years
若全域 similar region

弱 ε 近似法 weak ε approximation

弱埃尔米特纯量积 weakly Hermitian scalar product
弱暗电流 low dark current
弱币 less favorable currency
弱闭包 weak closure
弱闭集 weakly closed set
弱边界条件 weak boundary condition
弱变分 weak variation
弱变形收缩核 weak deformation retract
弱变质岩 weak metamorphic rock
弱遍历测度空间 weakly ergodic measure space
弱波 wavelet;weak wave
弱波峰脊 slight crest
弱波脊 slight crest
弱不可达序数 weakly inaccessible ordinal number
弱颤场回旋加速器 weak-flutter field cyclotron
弱颤场加速器 weak-flutter field accelerator
弱颤加速器 weak-flutter accelerator
弱场 feeble field;weak field
弱场法 weak-field method
弱场条件 weak-field condition
弱潮河口 weak tide river mouth
弱潮汐环境 weak tidal environment
弱乘法 weak multiplication
弱冲波 Mach region wave
弱磁场 low intensity field
弱磁场磁选机 low intensity(magnetic)separator
弱磁场接触器 weak-field contactor
弱磁化 weakly magnetization
弱磁性 feeble magnetism
弱磁性矿物 weak magnetic mineral
弱磁性体 weak magnetic substance
弱磁性物料 weakly magnetic;weakly magnetic material
弱磁选 low intensity magnetic separation
弱磁转筒 low intensity drum
弱大数定律 weak large number law;weak law of large numbers
弱代入 weak substitution
弱单调函数 weakly monotonic function
弱单调性 weak monotonic property
弱单位 weak unit
弱导光波导 weakly guiding optic(al)waveguide
弱导数 weak derivative
弱的 faint;unable;weak
弱等价变换 weakly equivalent transformations
弱等效原理 weak equivalence principle
弱低气压 weak low pressure
弱低压 weak depression;weak low pressure
弱地震 weak earthquake;weak shock
弱地震区 weak earthquake zone;weak shock zone
弱点 chink;failing;fail point;frailty;gall;soft spot;vulnerable spot;weakness;weak point;weak spot;zone of weakness
弱点法 weak point method
弱点分析 vulnerability analysis
弱电持续充电器 trickle charger
弱电磁相互作用 electroweak interaction
弱电磁性矿物 weak electromagnetic mineral
弱电工程师 millampere man
弱电工程学 light-current engineering;weak-current engineering
弱电解溶液 weak electrolyte
弱电解质 weak electrolyte
弱电绝缘子 weak-current insulator
弱电控制 weak-current control
弱电控制系统 light-current system
弱电离气体 weakly ionized gas
弱电流 feeble current;light current;weak current
弱电流电缆 weak-current cable
弱电流电路 dry circuit
弱电流放电 low-current discharge
弱电流继电器 weak-current relay
弱电流开关 weak-current switch
弱电流热丝 low-current heater
弱电流数字电路 low-current digital circuit
弱电流线路 weak-current line
弱电流纸绝缘铅包皮电缆 weak-current paper-insulated lead-covered cable
弱电流装置 weak-current installation
弱电室 weak electricity room
弱定位【航海】weak fix
弱定向的 weakly oriented
弱定向天线 poorly directive antenna
弱冻胀 weak frost heaving
弱度盐渍化 weak salinization
弱对比度景物 weak object
弱对角线优势 weak diagonal dominance
弱发射 low emission
弱发育剖面 weakened profile
弱反差 faint negative;thin negative
弱反差像 soft picture
弱反耦合 loose anti-coupling
弱方法 weak method
弱方向性天线 moderate directivity antenna
弱放射 low emission
弱非等熵流 weak non-isentropic(al)flow
弱非黏[粘]性土 weak non-cohesive soil
弱非线性 small nonlinearity;weak nonlinearity
弱非线性微分方程 weakly non-linear differential equation
弱非线性系统 weakly non-linear system
弱分量 weak component
弱分散 weak dispersion
弱风化 slightly weathering
弱风化层 weakly weathered layer
弱风化带 slightly weathered zone;weakly weathered zone;weak weathering zone
弱风化的 moderately weathered
弱风化岩石 moderately weathered rock;slightly weathered rock
弱峰 slight crest
弱锋【气】weak front
弱富水溢出带 weak watery overflow zone
弱干扰 weak interference;weak jamming
弱各向异性波导管 weakly anisotropic(al)waveguide
弱各向异性共振腔 weakly anisotropic(al)resonator
弱共轭碱【化】weak conjugate base
弱光 dim light;faint light
弱光层 disphotic zone
弱光带 disphotic zone
弱光的 disphotic;pale
弱光灯 passlamp
弱光灯标 weak light
弱光点 weak spot
弱光强度 low-light intensity
弱光区 disphotic zone
弱光通量 weak light flux
弱光性 dysphotic
弱光源 faint light source;low-level light source;low-power light source;weak light source
弱含水层 aquitard;low-water bearing formation
弱函数 minorante
弱互易 weak reciprocity
弱化 weakening
弱化函数 attenuation kernel
弱化核 attenuation kernel
弱还原环境 weak reducing environment
弱灰池 weak lime
弱灰化土 slightly podzolized soil;weakly podzolic soil
弱回波 weak echo wave
弱回波区 weak echo region
弱混合 weakly mixing
弱混合假设 weakly mixing hypothesis
弱混合性 weak mixing property
弱混合自同构 weakly mixing automorphism
弱火 low fire;slow fire
弱火花 weak spark
弱激波 weak shock
弱激磁继电器 field economizing relay
弱极小 weak minimum
弱极值 weak extremum
弱假设检验 weak hypothesis test
弱间断 weak discontinuity
弱间断面 weak discontinuity surface
弱剪切 shear-weak
弱剪正交各向异性板 shear-weak orthotropic plate
弱碱 weak base
弱碱浓度 weak base concentration
弱碱物种 weakly basic species
弱碱型离子交换剂 weak base type ion exchanger;weak basic type ion exchanger
弱碱性 weak basicity
弱碱性的 alkalescent;weakly alkaline;weakly basic
弱碱性介质 weak basic medium
弱碱性离子交换树脂 weak basic ion exchange resin
弱碱性树脂 weak base resin;weak basicity resin
弱碱性水 weak alkaline water
弱碱性土 weakly alkaline;weakly alkaline soil
弱碱性阴离子交换树脂 weak base anion-exchange resin
弱碱性阴离子交换纤维 weakly basic anion exchange fibre
弱碱液 weak lye
弱碱阴离子交换器 weak-base anion exchanger
弱碱阴离子树脂 weak base anion resin
弱键 weak linkage
弱胶合的 weakly cemented
弱胶结的 softly cemented;weakly cemented
弱胶结的砂岩 semi-consolidated sandstone
弱胶结岩层 incompetent rock stratum
弱焦点 under focus
弱焦性煤 weakly coking coal
弱结合 low binding
弱结合的 weakly cemented
弱结合的分子 weak-bonded molecules

弱结合水 loosely bound water; weakly bound water
弱结合水联结 weak bound water bond
弱解 weak solution
弱紧集 weakly compact set
弱紧线性映射 weakly compact linear mapping
弱紧性 weak compactness
弱径流带 weak runoff zone
弱聚焦 weak focusing
弱聚焦磁铁 weak-focusing magnet
弱聚焦存储环 weak-focusing storage ring
弱聚焦电子感应加速器 weak-focusing betatron
弱聚焦回旋加速器 weak-focusing cyclotron
弱聚焦加速器 weak-focusing accelerator
弱聚焦同步加速器 weak-focusing synchrotron
弱苛性碱溶液 weak caustic solution
弱可测函数 weakly measurable function
弱可积函数 weakly integrable function
弱可积算子 weakly integrable operator
弱可积性 weak integrability
弱可微性 weak differentiability
弱矿化水 weak mineralized water
弱框架 weak framework
弱类空曲面 weakly space-like surface
弱黎曼对称空间 weakly symmetric(al) Riemannian space
弱力货币 low-powered money
弱励磁继电器 field economizing relay
弱连通的 weakly connected
弱连通有向图 weakly connected digraph
弱连续表示 weakly continuous representation
弱连续随机过程 weakly continuous stochastic process
弱梁 weak beam
弱流靶 low-current target
弱流量 hard flow; low flow; stiff flow
弱流束 low-current beam; low-density beam
弱螺旋回旋加速器 weak spiral cyclotron
弱螺旋几何形状 weak spiral geometry
弱螺旋(扇)加速器 weak-spiral accelerator
弱络合剂 weak complexing agent
弱脉动电流 ripple current
弱面 plane of weakness; weak plane
弱面缝 plane joint of weakness; plane-of-weakness joint; weakened plane joint
弱面缩缝 <混凝土路面的> weakened plane contraction joint
弱能识性 weak identifiability
弱拟不变测度空间 weakly quasi-invariant measure space
弱逆定理 weak converse theorem
弱黏[粘]附油膜 weakly adhering oil film
弱黏[粘]合的 weakly cemented
弱黏[粘]结 weakly caking
弱黏[粘]结的 weakly cemented
弱黏[粘]结母岩 weak-cementing matrix
弱黏[粘]结性煤 weakly caking coal
弱黏[粘](聚)性土 feebly cohesive soil
弱黏[粘]性 weak viscidity
弱黏[粘]性母岩 weak-cementing matrix
弱耦合 loose coupling; weak coupling
弱耦合回路 loose coupler
弱耦合器 loose coupler
弱耦合系统 weakly coupled system
弱配筋梁 under-reinforced bonded beam
弱平稳 weakly stationary
弱平稳过程 weakly stationary process
弱平稳过程的秩 rank of a weakly stationary process
弱平稳随机分布 weakly stationary random distribution
弱平稳随机过程 weakly stationary random process; weakly stationary stochastic process
弱平稳序列 weakly stationary sequence
弱平稳移动平均数 weakly stationary moving average
弱起动 <泥沙的> weak movement
弱气压梯度 weak pressure gradient
弱强度 weak intensity
弱强铁路问题 weak-and-strong-road problem
弱强制映射 weakly coercive mapping
弱切削加工 cutting down
弱亲水性 weak hydrophilicity
弱溶剂 weak solvent
弱溶解物质 slowly soluble material
弱熔结火山角砾岩 weakly welded volcanic breccia
弱熔结集块结构 weakly welded agglomeratic texture
弱熔结集块岩 weakly welded agglomerate
弱熔结角砾结构 weakly welded breccia texture
弱熔结凝灰结构 weakly welded tuffaceous texture
弱熔结凝灰岩 weakly welded tuff
弱融沉 weakly melt-settlement
弱肉强食的原则 jungle law
弱散锋 diffuse front
弱色 pale shade
弱收敛 weak convergence
弱衰减 periodic(al) damping; underdamp(ing)
弱衰减的 weakly damped
弱双潜带 weak-double band
弱双曲线算子 weakly hyperbolic operator
弱双曲型方程 weakly hyperbolic equation
弱双曲型微分算子 weakly hyperbolic differential operator
弱双图 weak digraph
弱双折射 weak birefringence
弱水河 misfit river
弱水硬性石灰 feebly hydraulic lime
弱水跃 weak hydraulic jump; weak jump
弱塑性土体 weak plastic soil
弱酸 weak acid
弱酸的 acid-deficient
弱酸度 weak acidity
弱酸/碱 weak acid/base
弱酸/碱偶 weak acid/base pair
弱酸/碱系统 weak acid/base system
弱酸浓度 weak acid concentration
弱酸溶液 weak acid solution
弱酸型阳离子交换剂 weak acid cation exchanger
弱酸性树脂 weak acidic resin
弱酸性水 weak acidic water
弱酸阳离子交换树脂 weak acid cation exchange resin
弱台风 weak typhoon
弱弹簧 weak spring
弱透水边界 weakly pervious boundary
弱透水层 aquiclude; aquitard; less permeable layer; low-water bearing formation
弱透水层水量不释放 non-water release from weakly permeable layer
弱透水层水量释放 water release from weakly permeable layer
弱透水层影响系数 influence coefficient of weakly pervious layer
弱透水性岩体 aquitard
弱拓扑 weak topology
弱完备的 weak complete
弱纬向环流 low zonal circulation
弱紊动水流 low turbulence stream (flow)
弱稳定性 weak stability
弱稳物种 weakly stable species
弱西风环流 low zonal circulation
弱吸附 weak adsorption
弱吸目标 weakly absorbing object
弱吸收 weak absorption
弱吸收物质 weakly absorbing material
弱洗刷坡 weak-wash slope
弱显示偏好公理 weak axiom of revealed preference
弱线 line of weakness
弱相等 weak equality
弱相对极小 weak relative minimum
弱相关 low correlation; poor correlation
弱相互作用 beta interaction; weak interaction
弱相互作用实验 weak-interaction experiment
弱相互作用物理学 weak-interaction physics
弱削磁场继电器. weakening field relay
弱小企业 twilight industry
弱信号 weak signal
弱信号检测 weak signal detection
弱信号探测 weak signal detection
弱性水硬石灰 feebly hydraulic lime
弱性炸药 <由消焰剂和少量敏化剂合成的> bikarbit
弱絮凝作用 weak flocculation
弱旋涡 weak eddy
弱压缩 weak compression
弱(岩)层 incompetent bed
弱岩石 incompetent rock
弱研磨性地层 weak abrasive formation
弱盐渍土 slightly salified soil
弱焰 laze flame
弱氧化环境 weak oxidizing environment
弱异常 weak anomaly
弱音响信号检测仪 correlator
弱优先文法 weak precedence grammar
弱胀缩性土 weak swelling-shrinkage soil
弱褶皱 incompetent fold
弱震 weak earthquake; weak shock
弱制动弹簧 weak brake spring
弱智者学校 school for feeble minded
弱重力场 reduced gravity field
弱轴 minor axis; weak axis
弱状污染物 fanning pollutant
弱阻尼 low damping; underdamping
弱阻尼的 weakly damped
弱最优解 weak optimal solution

S

撒播 broadcast(ing);broadcast sodding

撒播板<撒播机的> scattering board
撒播机 broadcaster;broadcast sower;seed broadcaster;surface drill
撒播器 dispenser
撒播装置 dispersing device;spreading gear
撒布 distribution;intersperse;scatter(ing);spreading;sprinkle;sprinkling
撒布板 spreading board
撒布材料 dumping;interspersed matter
撒布车 spreader;spreader car
撒布辊 spreading roller
撒布厚度 application thickness;distribution thickness
撒布机 mechanical spreader;spreader;spreading machine;bulk spreader<水泥稳定土壤的>
撒布剂 dust powder
撒布宽度 spreading width
撒布料 mineral sprinkling material;surfacing
撒布料斗 distributing bucket;spreading hopper
撒布料斗容量 spreading hopper capacity
撒布料漏斗 surfacing hopper
撒布轮 scattering beater;spreading rotor
撒布螺旋布料器 spreading screw
撒布能力 distributive ability
撒布盘 spreading disk
撒布器 dispenser;distributer[distributor];spreader;trowel tool
撒布生产率 spray capacity
撒布石屑 chip sprinkling
撒布用材料 sprinkle material
撒布用骨料 sprinkle aggregate
撒布用集料 sprinkle aggregate
撒布用(的)石料 cover-up;cover stone
撒布装置 dispensing device;spread device;widespreader
撒道碴车 ballast spreader
撒丁运动 Sardinian orogeny
撒豆石油毡屋面 felt and gravel roof(ing)
撒尔佛散 Salvarsan
撒肥机 fertilizer distributor
撒粉 dust formation;dusting(on);pounce;powder(ing);powder spraying
撒粉的 mealy
撒粉法 dusting method
撒粉覆盖层 dusted-on coat
撒粉罐 pounce-pot
撒粉机 powder spreader
撒粉末 dusting
撒粉器 duster
撒粉器小车 powder applicator trolley
撒粉刷 dusting brush
撒哈拉沙漠 Sahara
撒化学药液 sprinkling liquid chemicals
撒灰法 dusting method
撒金刚石粉器 stone dust apparatus
撒厩肥附加装置 muck-spreading attachment
撒克逊玻璃板 Saxon slab
撒克逊(环裂)木瓦 Saxon shake
撒克逊建筑<古英国> Saxon
撒克逊建筑形式 Saxon architecture
撒克逊木板屋面 Saxon shake
撒克逊式 Saxon style
撒克逊式窗 Saxon window
撒克逊式建筑 Saxon architecture
撒克逊式(建筑)饰面 Saxon facade
撒克逊式塔楼 Saxon tower
撒克逊式砖石工程 Saxon masonry work
撒拉逊尖拱 pointed Saracenic arch
撒料板 dispersion plate;scattering flight
撒料法 powder blow in technique
撒料机 mineraliser[mineralizer]
撒料机粉碎机构 spreader disintegrating mechanism
撒料溜槽 distributing chute
撒料盘 disc distributor;dispersion plate;distributing disc[disk];distributing plate;distribution plate;distributor plate;spreading plate
撒料圈 distribution ring
撒料勺 scattering flight;scattering scoop
撒料式烘干机 dispersion drier[dryer]
撒料式选粉机 dispersion separator
撒料箱 spreading box
撒料锥体 dispersion cone
撒落 fallout
撒落作用 dispersion
撒煤砖 spreader block
撒面骨料 cover aggregate
撒面集料 cover aggregate
撒农药 dust crops with an insecticide
撒铺 spreading
撒铺黏[粘]结料 binder spreading
撒砂<又称撒沙> sand dressing;sanding;stuccoing<熔模铸造时>
撒砂驳船 spreader barge
撒砂车 sanding vehicle
撒砂的 sand cloth;sanded
撒砂的跑道 sanded runway
撒砂的压实积雪 sanded packed snow
撒砂电路 sand sprinkling circuit
撒砂阀 sand valve
撒砂法 sand spreading;sand spraying
撒砂防滑 frost-gritting
撒砂服务 sanding service
撒砂钢轨 sanded rail
撒砂管 sand(er)pipe
撒砂机 gritting machine;sand distributor;sand dressing machine;sand spraying machine;sand-spreader;sand spreading machine
撒砂沥青油毡 mineral-surfaced asphalt felt
撒砂器 sander;sand(er)sprayer
撒砂器拦砂阱 sand trap
撒砂设备 sand dressing machine;sand spraying device
撒砂时间继电器 sanding time relay
撒砂在面上 strew
撒砂(罩)面 sand facing
撒施 broadcast distribution
撒施处理 broadcast treatment
撒施法 broadcast application;broadcast method
撒施距离 broadcasted distribution space
撒石灰 liming
撒石灰附加装置 lime-spreading attachment
撒石机 stone spreader
撒石渣车 ballast spreader
撒水灭火箱 sprinkler tank
撒水泥 dusting with cement
撒水头 sprinkler head
撒水系统 sprinkler system
撒土 soiling
撒土厕所 earth closet
撒岩粉 rockdusting
撒盐<路面> salt application;salting
撒盐车 wagon for spraying salt
撒盐处理 salt treatment
撒盐地 salt ground
撒盐机 salt distributor;salt spreader
撒盐路面 salty overlay pavement
撒盐跑道 salt-treated runway;sanded runway
撒盐器 salt spreader
撒盐试验<一种加速锈蚀试验> salt spray test
撒药不匀度 unevenness of application
撒药机 applicator
撒种设备 drop plate

洒柏油柜车 tar spraying tank

洒布车 spraying car
洒布定额 spray rate
洒布管喷嘴开关 spraybar nozzle valve
洒布机 distributor;hand sprayer;spray bar;sprayer;spraying machine
洒布宽度测定 distributing width measurement
洒布器 trowel tool
洒布速度 spreading rate
洒布温度 application temperature
洒滴滤池 sprinkling filter;trickling filter
洒滴(滤)池法 trickling filter process
洒滴滤床 percolating filter;percolation filter
洒粉 powdering
洒粉器 duster
洒管 sprayer tube
洒浆 dash
洒金玻璃<含金色细粒的不透明褐色玻璃> aventurin(e)
洒金紫 Cassius gold purple
洒沥青卡车 boot truck
洒沥青养护 oiled application maintenance
洒了水的 watered
洒泼 slush
洒铺沥青机 road oiling machine
洒器 sprinkle
洒砂地沥青(面层) sand-rubbed asphalt
洒施器 applicator
洒湿 sparge
洒水 flushing;sparge water;sprinkle;sprinkling;water(ing);water spray;water sprinkling
洒水泵 sprinkler pump
洒水车 spraying car;sprinkler(wagon);hose cock;motor flusher;motor water car;road sprinkler;sprayer;sprinkling car;sprinkling truck;sprinkling wagon;street sprinkler;tank car;water(ing)barrow;water(ing)car(t);water(ing)truck;water(ing)wagon;water sprayer;water sprinkler(car);water tanker
洒水车尾部的洒水管 spray bar
洒水除霜系统 water spray defrost system
洒水防尘 water spray for dustproofing
洒水管 sprinkler pipe;sprinkling bar;sprinkling pipe
洒水灌溉 dribbling
洒水过筛 wet screen
洒水壶 watering can
洒水机 flushing machine;sprinkler;water sprayer;water sprinkler;water sprinkler tank
洒水降温 water spray cooling
洒水冷却 cooling spray
洒水龙头 flushing tap;spraying cock
洒水密度 density of sprinkling
洒水灭火器 sprinkler tank
洒水灭火系统 fire-sprinkling system;sprinkler system
洒水灭火装置 sprinkler fire extinguishing system
洒水喷头 sprinkler head
洒水喷头装置 sprinkler head system
洒水器 distributor;drencher;rugosa;sprayer;sprinkler;watering can;water sprayer;wet sprinkler
洒水器供水管 sprinkler lateral
洒水扫集机 sprinkler-sweeper collector
洒水设备 sprayer;sprinkler;water sprinkler;wetting system;water distributor
洒水湿煤器 coal sprinkler
洒水刷 splash brush
洒水栓 sill cock
洒水退火 annealing with water
洒水系统 water spray system;water sprinkler system;wetting system
洒水小车 water barrow;water cart
洒水压实 compaction by watering
洒水养护 curing by sprinkling;spraying curing
洒水养护混凝土 water cured concrete
洒水抑制尘埃系统 dust suppression spray system
洒水装置 spraying gun;sprinkler system;watering device;water sprinkler
洒药厕所 chemical closet
洒药治害虫 spray insecticide for control insect pests
洒油率 rate of application;rate of bituminous application
洒油嘴 spray(ing)nozzle

萨巴蒂尔效应 Sabattier effect

萨巴蒂循环 Sabathe cycle
萨拜因隔声灰泥 Sabinite
萨比丘木<古巴一种质地坚硬的贵重木材> sabicu
萨波蒂拉装饰用硬木<产于洪都拉斯> Sapodilla
萨波特克建筑<中美洲> Zapotec architecture
萨波脱各种花色硬木<产于中美洲> Sapote
萨布哈 Sabkha
萨布哈沉积 Sabkha deposit
萨布哈成矿模式 Sabkha metallogenic model
萨布哈硫酸盐 Sabkha sulphate
萨布哈相 Sabkha facies
萨布里炸药<一种极强烈的炸药,炸力约为普通炸药的三倍> sabulite
萨布罗琼脂 Sabouraud's agar
萨布罗纸碟法 Sabouraud's pastille
萨尔茨曼系数 Saltzman conversion factor
萨尔贡王宫<亚西利亚古建筑> Palace of Sargon
萨尔玛法 Sarma method
萨尔玛特阶<晚中新世>【地】Sarmatian(stage)
萨尔姆阶【地】Salmian(stage)
萨尔坦冰阶【地】Sartan stade
萨尔坦冰期【地】Sartan glacial stage
萨尔瓦多 EL Salvador
萨尔温江 Salween River
萨尔兹吉特逆循环法 Salzgitter re-

S

verse circulation method
萨尔兹吉特逆循环钻井机 Salzgitter reverse circulation drilling rig
萨伐特片 Savart plate
萨伐特偏光镜 Savart polariscope
萨夫造山运动【地】Savian orogeny; Savic orogeny
萨福克插销 Suffolk latch
萨哈方程式 Saha's equation
萨哈林岛 Sakhalin
萨哈罗娃矿 sakharovaite
萨赫残余应力测定法 Sach's residual stress determination method
萨赫尔风 Sahel
萨赫尔观象台 Sahel observatory
萨加塞毛织物 sagathy
萨金特钢板 <一种木底衬的钢板> Sargent
萨金特循环 Sargent cycle
萨克定理 Sack's theorem
萨克管状浇口 Saxophone gate
萨克拉门托盆地 Sacramento basin
萨克马尔阶 <早二叠世>【地】Sakmarian(stage)
萨克森阶【地】Saxonian(stage)
萨克森造山运动【地】Saxonian orogeny
萨克斯拜-方默(集中)联锁机 Saxby and Farmer's interlocking machine
萨克斯拜-方默(集中)联锁架 Saxby and Farmer's interlocking frame
萨克斯板 <一种绝缘墙板> Sx board
萨克斯欧尼亚合金 <一种锌合金> Saxonia metal
萨拉门特炉 Salamander
萨拉门特式石棉制防火板 Salamander
萨拉门特式石棉装饰板 Salamander
萨拉萨尔铝合金 Thalassal
萨拉森建筑式 <回教建筑式> Moorish
萨拉森式拱 Saracenic arch
萨拉森式建筑 <回教建筑式> Saracenic architecture
萨拉橡胶树 ceara rubber
萨拉伊尔构造旋回【地】Salair cycle
萨拉伊尔构造作用幕【地】Salair orogeny
萨勒冰期 Saalian glacial epoch; Saalian glacial stage
萨勒姆石灰岩 Salem limestone; Spergen limestone
萨勒盆地 Saalian basin
萨勒造山运动 Saalian orogeny; Saalic orogeny
萨雷尔迈防水剂 <用于砖、石、水泥墙> Szerelmey stone liquid
萨雷尔迈石液 <一种专利材料> Szerelmey Stonecoat Encaustic
萨雷尔迈石用液 Szerelmey stone liquid
萨列阿法则 Sanio's law
萨列阿横梁 Sanio's beam
萨罗【化】salol; phenyl salicylate
萨罗斯周期 <日食和月食出现周期> saros
萨洛普统 <英国中志留世>【地】Salopian(series)
萨马洛夫冰阶【地】Samarovo stade
萨马洛夫冰期 Samarovo glacial stage
萨蒙风 <一种干热焚风> samoon; samun
萨蒙红褐色硬木 <印度产> Salmon wood
萨蒙胶 Salmon gum
萨米迟缸体设计 Siamese bore design
萨米特轴承青铜 Sumet bronze
萨母纳天文定位法 Summer method
萨母纳位置线 Summer line
萨那 <也门首都> Sanaa
萨尼特脱硫法 Saniter process
萨尼亚克干涉仪 Sagnac's interferometer
萨尼亚克实验 Sagnac's experiment
萨怒海 Sanu sea
萨诺达油毛毡 Sanodar
萨佩莱木 sapele
萨桑(王朝)建筑 Sassanian architecture
萨砷氯铅矿 syhinite
萨特科铅合金 Satco alloy
萨特帕耶夫石 satpaevite
萨图恩神庙 Temple of Saturn
萨图尼树 arbor Saturni
萨瓦金海渊 Sawakin deep
萨维斯坦宫 <波斯古建筑> Palace of Sarvistan
萨沃纽斯风车 Savonius windmill
萨沃纽斯转子 Savonius rotor
萨沃纽斯转子海流计 Savonius rotor sea current meter
萨沃纽斯转子流速计 Savonius rotor current meter
萨乌尔运动 Sawuer orogeny
萨西尼铝合金 Susini
萨彦岭 Sayan Mountains
萨伊定律 Say's Law

塞 cork; plug; seal lock

塞拔 corkscrew
塞扳手 plug spanner
塞板 filler
塞棒 stopper
塞棒铁芯 stopper rod
塞贝克系数 Seebeck coefficient
塞贝克效应 <温差电动势效应> Seebeck effect; Seebeck thermoelectric(al) effect
塞波特黏[粘]度计 Saybolt's visco(si)meter
塞伯特斯廷空心梁地板 Seibert Stinnes hollow beam floor
塞舱 chock
塞查德风 sechard
塞尺 clearance ga(u)ge; feeler; feeler ga(u)ge; measuring wedge; plug ga(u)ge; searcher
塞尺片 feeler leaf
塞茨特曼电解池 Szechtman cell
塞得紧紧的 jam-packed
塞德耳变量 Seidel variable
塞德耳程函 Seidel eikonal
塞德耳迭代法 Seidel iteration; Seidel iteration method
塞德耳格拉泽屈光学 Seidel-Glaser dioptrics
塞德耳公式 Seidel's formula
塞德耳光学 Seidel optics
塞德耳三级像差理论 Seidel third order theory of aberration
塞德耳像差 Seidel aberration
塞德耳像差理论 Seidel aberration theory
塞德耳像差系数 Seidel coefficient of aberrations
塞德罗斯海沟 Cedros trench
塞垫 <一种推拉窗垫套> knag gasket
塞钉 channel pin; plug pin
塞钉导接线 pinned bond; pin type bond; plug-in bond; plug-type bond; punch-driven bond
塞钉式钢轨接续线 plug-type rail bond
塞洞口的东西 stopgap
塞尔白栅筛 Serpa harp screen
塞尔茨堡的大学教堂 University church at Salzburg
塞尔顿氏和西特氏边坡稳定分析法 Sultan and Sead method of slope stability
塞尔马白垩层 Selma chalk
塞尔彭阶 <晚白垩世>【地】Selbornian
塞尔斯风 Cers
塞尔维亚建筑 Serbian architecture
塞耳迈耶尔方程 Sellmeier equation
塞耳温颗粒度系数 Selwyn granularity coefficient
塞阀 plug cock; plug(-type) valve
塞法戴克斯 Sephadex
塞缝 jag
塞缝材料 ca(u)lking material
塞缝骨料 key aggregate
塞缝集料 key aggregate
塞缝片 feather; slip feather; spline
塞缝石 keystone
塞缝条 ca(u)lking strip
塞缝小石块 sneck
塞夫勒蓝 Sevres blue
塞盖 gag; plug cap
塞盖衬垫 bung gasket
塞盖垫圈 bung washer
塞杆 plug stick; stopper rod
塞杆式底注浇包 stopper rod ladle
塞杠 piston rod
塞格尔测温锥 Seger cone
塞格尔瓷 Seger porcelain
塞格尔公式 Seger formula
塞格尔规则 Seger's rule
塞格尔绿 <耐1050℃的陶瓷彩料> Seger's green
塞格涅特盐 Seignette salt
塞格锥 pyrocone
塞管接缝铅条 ribbonite
塞规 chock ga(u)ge; feeler ga(u)ge; ga(u)ge feeler; ga(u)ge head; ga(u)ge plug; hole ga(u)ge; internal ga(u)ge; male ga(u)ge; plug ga(u)ge; rod ga(u)ge; thickness ga(u)ge
塞规自动定尺寸内圆磨床 ga(u)gematic internal grinder
塞规自动控制尺寸法 ga(u)ge-matic method
塞焊 jam weld; rivet welding; plug weld
塞焊焊缝 ca(u)lk(ing) weld; plug weld(ing)
塞焊接 plug weld(ing)
塞环 ring of plug
塞环线 ring wire
塞环引出线 ring lead
塞环引线 ring wire
塞尖引出线 tip lead
塞尖引线 A-wire; tip wire
塞接口 ca(u)lking joint
塞紧 chocking(-up); pluggage
塞紧的软木塞 plutonic plug
塞进 fill in; tuck
塞进去 tuck
塞径规 cork ga(u)ge
塞卡风 seac
塞卡水泥 <一种耐火水泥> Secar cement
塞可尔大地测量卫星 Secor-geodetic satellite
塞克斯传声器 Sykes microphone
塞孔 chink; consent; jack; nest; plug hole; plug socket; receptacle; tap hole; tip jack
塞孔板 jack base; jack board; jack panel
塞孔标记 jack marking
塞孔补缺的木块 dutchman
塞孔补缺的木料 dutchman
塞孔衬套 bush of jack
塞孔的塞套引线 test wire
塞孔的套管 bush
塞孔符号 jack marking
塞孔环 socket of jack
塞孔簧片 jack strip; switch spring
塞孔接触弹簧 jack spring
塞孔接点 female contact
塞孔盘 female receptacle; jack field; jack panel
塞孔圈 jack ring
塞孔栓 tie plug
塞孔套 sleeve of jack
塞孔套管 socket of jack
塞孔砖 closure
塞孔座 nest
塞块 chock; setting block
塞块振动器 chock vibrator
塞拉杰拉尔极性带 Serrageral polarity zone
塞拉杰拉尔极性时 Serrageral polarity chron
塞拉杰拉尔极性时间带 Serrageral polarity chronzone
塞拉利昂柯巴脂 sierra leone copal
塞拉卢明合金 Ceralumin(alloy)
塞拉坦风 Selatan
塞莱布里昂斯克极性带 Serebriansk polarity zone
塞莱布里昂斯克极性时 Serebriansk polarity chron
塞莱布里昂斯克极性时间带 Serebriansk polarity chronzone
塞朗克银锑合金 Silanca
塞勒传动 Seller's drive
塞勒涅斯的阿波罗神庙 Temple of G. T. at Selenius
塞勒锥度 Seller's taper
塞勒(锥形)联轴节 Seller's coupling
塞利特电阻器 Thyrite resistor
塞利特压变电阻避雷器 thyrite arrester[arrestor]
塞料磨轮 loaded wheel
塞流 choked flow; plug flow
塞流空化数 choking cavitation number
塞流线圈 line trap
塞漏材料 leak stopper
塞露蒂课题 Cerruti problem
塞鲁士墓 <波斯国王> Cyrus's tomb
塞璐珞(硝化纤维) celluloid
塞罗贝斯铋铅合金 Cerrobase(alloy)
塞罗本德合金 Cerrobend alloy
塞罗铋基低熔合金 Cerro
塞罗马特里克斯合金 Cerromatris alloy
塞罗泽土壤 Sierozem soil
塞洛 <英制加速度单位, =1英尺/秒> celo
塞洛尔透镜系统 Celor lens system; Gauss lens system
塞洛夫接收机 Selove receiver
塞珞玢 cellophane
塞满 chock full; choke; crowd; jampack; lade
塞满淤泥的 silt-laden
塞曼分离 Zeeman splitting
塞曼分裂 Zeeman splitting
塞曼缝隙 Zeeman slit
塞曼光谱片 Zeeman spectrogram
塞曼能级 Zeeman level
塞曼能量 Zeeman energy
塞曼谱线加宽 Zeeman spectrum broadening
塞曼态 Zeeman state
塞曼调谐激光器 Zeeman-tuned laser
塞曼图谱 Zeeman spectrogram
塞曼位移 Zeeman displacement
塞曼现象 Zeeman phenomenon
塞曼相干性 Zeeman coherence
塞曼效应 Zeeman effect
塞曼效应分裂能级 Zeeman-splitted level
塞曼效应激光器 Zeeman laser
塞曼效应位移 Zeeman shift

塞曼效应稳频 Zeeman effect frequency stabilization
塞曼效应校正 Zeeman effect correction
塞曼子线 Zeeman component
塞门 bib(b);bibcock;isolating cock;stop cock
塞门杆 cock spindle
塞门套 cock key seat
塞门体 cock body
塞门心 cock key
塞门心座 cock key seat;key seat
塞摩福流动床 Thermofor
塞摩福流动床白土灼烧再生过程 Thermofor clay burning process
塞摩福流动床连续裂化车间 Thermofor continuous cracking plant
塞摩福流动床连续渗滤过程 Thermofor continuous percolation process
塞默杜尔钴铁簧片合金 Semendur
塞姆纳法【航海】Sumner method
塞姆纳线 Sumner line
塞木节孔 knotting
塞拿蒙补偿器 Senarmont compensator
塞拿蒙棱镜 Senarmont prism
塞拿蒙偏振计 Senarmont polarimeter
塞内加尔河 Senegal River
塞内加尔树胶 Senegal gum
塞纳河 Seine River
塞泥 sludging
塞佩克斯碱性耐火材料 serpex
塞片 patch
塞浦路斯 < 亚洲 > Cyprus
塞浦路斯人的 Cypriot
塞普克窑 Sepulchre kiln
塞普坦尼克斯数学板 Septanix
塞齐分类法 Secchi's classification
塞齐试验圆板 Secchi's disc
塞奇威克拉夫特法 Sedgwick-Rafter method
塞奇威克拉夫特滤器 Sedgwick-Rafter filter
塞铅条 leading
塞钳 plug pliers
塞嵌碎石片 pinning-in
塞入 crush;forcing
塞入部分 tuck-in
塞入基础托换的枕梁 needling
塞入率 stuffing rate
塞入式挡土板 tucking board
塞入式接头 plug-in coupling
塞入数字 stuffing digit
塞上瓶塞 corkage
塞舌尔(群岛)Seychelles
塞绳 cord; flex; flexible cord; jack cord;plug cord
塞绳保持器 cord retainer
塞绳测试塞孔 cord test jack
塞绳灯 cord lamp
塞绳电路 cord circuit
塞绳端 cord terminal
塞绳夹头 cord clip
塞绳架 cord rack
塞绳接线柱 cord fastener
塞绳接续器 cord connector
塞绳结头 cord grip
塞绳连接图 cording diagram
塞绳式交换机 plug box;plug-selector
塞绳式选择机 plug-selector
塞绳式中继台 cord system trunk board
塞绳调节器 cord adjuster
塞绳线对 cord pair
塞绳修理 cord repair
塞绳选择器 cord chooser
塞绳增音机 cord circuit repeater
塞绳增音机架 cord circuit repeater bay
塞绳中继器 cord circuit repeater
塞绳转接电键 splitting key
塞氏公式 < 用于设计桥台翼墙平均厚度 > Sejourne formula
塞式保险器 plug cut-out
塞式井底封隔器 bottom-hole plug packer
塞式流动 plug flow
塞式喷管 plug-type nozzle
塞式熔断器 plug cover fuse cutout
塞式熔丝保险器 plug cut-out
塞栓 gland
塞栓插口 spigot
塞斯奥王一世祠庙 Temple of King Sethos I
塞斯图建筑 < 指意大利 17 世纪建筑 > Seicento
塞榫 plug tenon
塞套 plug bush;plug sleeve;sleeve of plug;transit plug
塞套线 S-wire
塞套引出线 sleeve lead
塞套引线 < 第三根线,即 C 线 > C-wire;private wire;sleeve wire
塞特洛夫混凝土强度增量特征值 Sadgrove maturity figure
塞体 cock body
塞条 shim
塞铁 gag
塞头 bung; chock plug; choke plug; stop end
塞头栓 stopper pin
塞头套 sleeve of plug
塞头砖 domed-head stopper;stopper;stopper brick;stopper head(brick)
塞韦奇原则 Savage principle; regret criterion
塞文桥 < 英 > Severn Bridge
塞沃弗雷克斯 < 三硫化二砷透红外材料 > Servofrax
塞形比重计 plug-type pycnometer
塞形刀头 plug bit
塞形密度计 plug-type pycnometer
塞形钻孔器 plug cutter
塞眼片 < 塞在窗门缝中以防风雨 > weather strip
塞药棒 stemmer; tamping pole; tamping stick
塞药杆 tamping bar
塞缘 margo
塞支强干 closing branches for strengthening main channel
塞住 blankoff; block up; bung; choke up with;cork;jam;plug(up);stop-off
塞住的 stopping
塞住油罐出口 tank outlet plugging
塞柱 stick harness
塞砖装配 plug brick assembly
塞状管丝锥 plug pipe tap
塞状流 plug flow
塞子 bib(b);blank plug; bod; bung; bysma; chock; closer; connector pin;connector plug;gag;plug;plug adapter; plug core; spigot; stopple; tamp(i) on; tap; stopper; crutch key < 螺旋式水龙头顶部呈 T 形的 >
塞子架 plug shelf
塞子螺钉 plug screw
塞子套 sleeve of plug
塞子统计 plug count
塞子脱落 bungs off
塞子砖 plugging block

腮

腮片散热器 gilled radiator

腮状的 gilled

噻

噻吩【化】thiophene

噻吩的溶解度 solution thiophene
噻吩烷 thiophane
噻唑(结构)染料 thiazole dye
噻唑啉 thiazoline
噻唑烷 thiazolidine

赛

赛白金 platinite;platinoid

赛贝尔萃取器 Scheibel-York extractor;York-Scheibel column
赛宾 < 建筑声学的表面声吸收单位,相当于一平方英尺的全吸声面 > sabin(e); square-foot unit of absorption;open-window unit
赛宾防锈法 Sabin process
赛宾公式 Sabine formula
赛宾混响公式 Sabine reverberation formula
赛宾吸声量 Sabine absorption
赛宾吸声系数 Sabine absorption coefficient
赛宾系数 Sabine coefficient
赛波特比色计 < 又称赛氏比色计 > Saybolt colo(u)rimeter
赛波特-富洛重油黏[粘]度 Saybolt-Furol viscosity
赛波特-富洛重油黏[粘]度计 Saybolt-Furol visco(si)meter
赛波特秒数 Saybolt seconds
赛波特黏[粘]度 Saybolt viscosity
赛波特黏[粘]度单位 Saybolt centistoke
赛波特黏[粘]度级 Saybolt scale
赛波特黏[粘]度计 Saybolt visco(si)meter
赛波特黏[粘]度秒 Saybolt seconds
赛波特热黏[粘]度计 Saybolt thermoviscosimeter
赛波特色 Saybolt colo(u)r
赛波特通用黏[粘]度 Saybolt universal viscosity
赛波特通用黏[粘]度计 Saybolt universal visco(si)meter
赛波特通用黏[粘]度计秒数 Saybolt universal seconds; seconds Saybolt universal;Saybolt seconds universal
赛波特温度计 Saybolt thermometer
赛波特重油黏[粘]度秒数 seconds Saybolt Furol
赛波特重油黏[粘](滞)度 Saybolt-Furol viscosity
赛波特重油黏[粘](滞)性 Saybolt-Furol viscosity
赛伯空间 < 全球计算机空间 > Cyber Space
赛车 kart
赛车场 autodrome;velodrome
赛车车身 racing body
赛车道 carracing track; racecourse; race track
赛车发动机 racing(-car)engine
赛车路线 racing course
赛车轮胎 racing-car tyre[tire]
赛车跑道 autodrome;automobile race course;automobile race track;car-racing track;dirt track
赛车起点栏 carcer
赛车燃料 racing fuel
赛车型泡沫橡皮燃油室 racing-type foam-filled rubber fuel cell
赛车用跑道 motor racing track
赛车运动员 racing driver; sporting driver
赛船会 regatta
赛船用划道 boat course
赛达铝锌合金 Cetal
赛得莫尔目镜 Scidmore eyepiece
赛德耳方法 Seidel method
赛德耳氏暗点 Seidel's scotoma(sign)
赛德利茨粉 Seidlitz powder
赛多维阶 < 植物阶 > Seldovian(stage)
赛尔卡铝合金 Sylcum
赛夫沙丘 seif(dune)
赛弗特 X 射线管 Seifert tube
赛弗特整流机 Seyfert rectifier
赛狗场 dog track
赛黄晶 danburite
赛黄石 danburite
赛金刚青铜 diamond bronze
赛金刚石合金 diamondite
赛康姆型胶轮驱动运输系统 Seccam
赛克斯比重计 Sikes hydrometer
赛库安亚阶【地】Sequanian(stage)
赛快车 racing car
赛勒分类法 Seyler's classification
赛勒螺纹 Seller's screw thread; US screw thread
赛龙陶瓷 < 含钇的硅铝氧氮陶瓷 > Sialon ceramics
赛璐里斯 cellulith
赛璐璐模片法 < 像片三角测量 > celluloid-templet method
赛璐珞 xylonite
赛璐珞废料 celluloid scrap;scrap celluloid
赛璐珞模板 celluloid template
赛璐珞片 celluloid sheet
赛璐珞漆 celluloid lacquer
赛璐珞填缝料 < 其中的一种 > polyfiller
赛璐特克吸声板 acoustic(al)celotex board;acoustic(al)celotex(tile)
赛纶 < 氯乙烯、二氯乙烯共聚纤维 > Saran
赛纶树脂 Saran
赛纶陶瓷 < 含钇的硅铝氧氮陶瓷 > syalon
赛马 horse race
赛马场 racecourse;race track;turf
赛马场大看台 turf grand stand
赛马的椭圆形运动场 < 古希腊 > hippodrome
赛蒙顿水泥 < 一种防火水泥 > Semmentum
赛帕油 Scheiber oil
赛跑场 < 古罗马希腊长 607 英尺,周围有台阶式看台 > stade
赛跑道 racecourse
赛跑运动场 < 古希腊的 > stadion
赛奇板 Secchi's disc
赛什克盆地 Saishike basin
赛氏标准通用黏[粘]度计 Saybolt standard universal visco(si)meter
赛氏厚油黏[粘](滞)度计 Saybolt-Furol visco(si)meter
赛氏滤器 Seitz filter
赛氏黏[粘]度测定 Saybolt-viscosity test
赛氏黏[粘](滞)度计 Saybolt visco(si)meter
赛氏黏[粘](滞)度试验 Saybolt-Furol viscosity test;Saybolt-viscosity test
赛氏通用黏[粘]度计秒数 Seconds Saybolt Universal
赛氏通用黏[粘]度计油管 oil tube for Saybolt universal viscosimeter
赛氏通用黏[粘](滞)度 Saybolt universal viscosity
赛氏通用黏[粘](滞)度计 Saybolt universal visco(si)meter
赛氏重油黏[粘]度计 Saybolt
赛特阶 < 早三叠世 >【地】Scythian stage
赛艇 gig;racing yacht
赛艇会 yacht regatta
赛威尔红 < 含铁的石英质色料 > Thiviers red
赛西法 Siacci method
赛霞漆 clear lacquer
赛扬处理器 Celeron processor

S

赛扬总片【计】Celeron processor
赛银的 white
赛银朱 vermil(l)ionette
赛泽逊 <奥地利新艺术的变异> Sezession

三MO 设计方案 TRIMO [modern, mobilia, modello]

三 T 试验台 three T test rack
三 T 形板 tri-slab;tri-tee slab
三 T 形梁 tri-tee beam
三安培计方法 three-ammeter method
三八面层 trioctahedral
三八面体 triakisoctahedron;trisoctahedron
三八面体层 trisoctahedronal layer
三八面体的 <结晶结构> trioctahedral
三百二十英亩土地 <根据美国政府测量法,1 英亩 = 4046.86 平方米> half section
三百六十度角 perigon
三百六十度转动管嘴 three-hundred and sixty degree nozzle rotation
三班 triple shift
三班倒的 around-the-clock
三班工作制 working system in three shifts;day system;three-shift work(ing);triple shift;three shift day
三班轮换浇注 <混凝土的> three-shift pouring
三班轮值制 three-watch system
三班作业 three-shift operation;three-shift work(ing)
三半波滤光片 triple half-wave filter
三瓣的 trivalvular
三瓣戽斗 orange-peel bucket
三瓣玫瑰线 three-leaved rose curve
三瓣(式)戽斗挖土机 orange-peel excavator
三瓣式模筒 three-way split former
三瓣凸轮 tri-lobe cam
三瓣形花饰 trefoil
三包 sub-subcontract
三包单位 sub-subcontractor
三包堆材料单元 three-pack unit
三包商 sub-subcontractor
三保险横臂吊车 triple-safe boom hoist
三杯风速表 three-cup anemometer
三杯风速仪 three-cup anemograph;three-cup anemometer
三北防护林带 Three-North Shelterbelt
三北偏角图 <地图上表示真北、磁北和坐标北的关系图> declination diagram
三倍 triplicate;triplicity
三倍倍频器 frequency tripler;trebler
三倍长度 triple-length
三倍长工作 triple-length working
三倍的 three-fold;treble;trinal;trinary;trine;triple(x);triplicate
三倍电路 tripler circuit
三倍精度 triple precision
三倍精度数 triple precision number
三倍量 triple;triplication
三倍量公式 triplication formula
三倍赔偿 treble damage
三倍频 triple frequency
三倍频电路 trebling circuit
三倍(频)器 tripler
三倍频效应 frequency tripled effect
三倍数 triple
三倍数谐波 triple-frequency harmonic
三倍体 triploid
三倍性 triploidy
三倍压电路 voltage tripler
三倍压整流 voltage tripler rectifier
三倍于原物的损害赔偿 treble damage
三倍柱底径空距柱廊 diastyle
三倍柱径距双列柱门廊 diastyle
三苯胺 triphenylamine
三苯基 triphenyl
三苯甲基纤维素 trityl cellulose
三苯甲基溴 trityl bromide
三苯甲烷 triphenylmethane;tritan
三苯甲烷染料 triphenylmethane dye(stuff)
三苯锑 triphenyl antimony
三笔记录仪 three-pen recorder
三闭塞区段 three-block
三闭塞区段信号系统 three-block signal system
三闭塞区段信号制 three-block signal system
三闭塞四显示自动闭塞信号 three-block four-indication automatic block signal
三篦床复合冷却机 three-grate combination cooler
三臂分度器【测】three-arm protractor
三臂分度仪 station pointer;three-armed protractor
三臂管 three-limb tube
三臂坯托 stilt
三臂平巷凿岩台车 three-boom drift-jumbo
三臂系船设施 trot mooring
三臂卸卷机 three arm unloader
三臂星形窑 stilt
三臂钻车 triple-boom drill rig
三边测量 trilateration;trilateration survey
三边测量定位法 trilateration position
三边测量法 trilateration method
三边测量网 trilateration network
三边承托 three-edge bearing
三边承重试验 <测定管子承重能力用> three-edge bearing test
三边的 trilateral
三边贸易 trilateral trade
三边旁面三角台【数】trilateral prismatoid
三边为半圆的四方平面布置形式 triconch
三边协定 trilateral agreement
三边形 trilateral;trilateral figure
三边形突肚窗 cant(-bay)window
三边压模型板 three-sided mold-board
三边样板 three-sided mold-board
三边支承的板 slab supported on three sides
三边支承的腹板 plate-stalk supported on three sides
三边支承试验 three-edge bearing test
三边轴承 three-edge bearing
三苄胺 tribenzylamine
三变的 trivariant
三变量 ternary
三变量分析 three-variable analysis
三变量模型 three-variable model
三变量生产 three-variable production
三变量正态分布 trivariate normal distribution
三变数的【数】ternary
三变体系 triovariant system
三标高点横断面 three-level section
三标两角定位法 three point problem
三标两角法 three-point fix method
三标准试样法 method of three-standard sample
三表轨计时器 triple dial timer
三饼滑车 three-fold block;treble block;triple block;trispast
三饼滑车组 tackle burton
三波 tricrotism
三波的 tricrotous
三波混合 three-wave mixing
三波曲的 triundulate
三补色 three-complementary colo(u)r
三不饱和甘油酯 triunsaturated glyceride
三步骤工程研究法 <确定概略范围、进行试验、分析研究三个步骤> three-stage engineering approach
三步作用 three-level action
三部 triplex
三部的 triregional
三部法 <冶金的> triplexing
三部分的 triple(x)
三部分定色系统 tristimulus system of colo(u)r specification
三部分组成的 three-component;trinal;trinary;trine;tripartite
三部分组成的穹顶 tripartite vault
三部收费电度计 triple tariff meter
三部收费制 three-part rate schedule
三才升 small block
三彩(考古)three-colo(u)r
三彩器皿 three-colo(u)red ware
三彩釉 three-colo(u)r glaze;tricolo(u)r glaze
三彩釉上装饰 three colo(u)rs overglaze decoration
三参数法 three-parameter method
三参数模型 three-parameter model
三仓称量分批量斗 triple weighing batcher
三仓磨 three-chamber mill;three-compartment mill
三仓磨机 three-compartment compound mill
三舱不沉制 three-compartment subdivision
三舱制船 three-compartment ship
三槽 trislot
三槽板间的面石 metope
三槽板间分隔片 femur
三槽板间距 <古希腊陶立克建筑上> ditriglyph
三槽板间平面 meros;metope
三槽板(浅槽饰)triglyph
三槽板上的小槽 canaliculus
三槽板上的小沟 canaliculus
三槽板陶立克柱式 triglyph
三槽板下短条线脚 regula(e)
三槽法 three-pit system
三槽机用铰刀 three-groove chucking reamer
三槽浸灰法 three-pit liming system;three-pit system of liming
三槽陇板间空间 intertriglyph
三槽螺母 tri-slot nut
三槽铣刀 three flute bit
三槽钻头 three-fluted drill;three-grooved drill
三侧向测井 laterolog 3
三侧向测井曲线 laterolog 3 curve
三层 three-layer;triple
三层安全玻璃 three-layer sandwich glass
三层板 three-ply board;triplex board;triplywood
三层板铲柄 triple-leaf standard
三层版 tripack
三层包线 triple covered wire
三层编包线 triple braided
三层表面处理 triple(course)surface treatment
三层表面处治 triple(course)surface treatment
三层玻璃 <二层玻璃中夹一层塑料的安全玻璃> laminated glass;triplex glass
三层玻璃窗 triple-glazed units
三层不碎玻璃 <两层玻璃中夹一层塑料作黏[粘]合剂> triplex
三层彩色胶片 tripack colo(u)r film
三层彩色片 tripack
三层舱客机 three-decker(tunnel) jumbo
三层重叠高增益天线 triple sleeve antenna
三层船壳板铺板法 treble planking
三层窗 triple-window
三层床 three-high berth
三层带式热风干燥机 three stages hot air conveying type dryer
三层道路交叉 triple-decker
三层的 three-ply;triple-deck;three-stor(e)yed;tri-level;triple(x)
三层底 triple bottom
三层地下室 three-level basement
三层镀膜 triple-layer coating
三层多机钻车 three-level jumbo
三层复合的屋顶覆盖层 three-ply built-up roof cladding
三层干筛 triple-deck dry screen
三层干燥床 treble kiln floor
三层感光材料 tripack material
三层钢 compound steel
三层构造 three-layer structure
三层罐笼 three-deck cage
三层光栅多色仪 triple grating polychromator
三层烘缸 triple-deck dryer[drier]
三层灰泥 three-coat plaster
三层夹板 triple laminate
三层甲板船 three-decker;three-deck vessel
三层架货车 <铁路用> rack car
三层讲坛 three-decker
三层胶合安全玻璃 three-piece laminated safety sheet glass
三层胶合板 three-layered panel;three-plywood;triple laminate
三层胶合板正面 three-layered panel facade
三层胶合皮带 triple(-ply)belt
三层结构板 three-layer board
三层金属轴承合金 trimetal
三层壳 three-layered shell
三层跨云车 three-high straddle carrier
三层矿物 three-layer mineral
三层立交桥 triple-deck grade separation structure
三层立体交叉 tri-level grade separation;three-level grade separation
三层立体交叉结构 triple-deck grade separation structure
三层量板法 three layers template method
三层硫化床干燥器 three stages fluidized bed drier[dryer]
三层楼房(屋)three-storied house
三层楼公寓 triplex apartment
三层楼面的 three-floored
三层楼厅 balcony
三层滤池 trinal layer filter
三层滤料滤池 tri-media filter
三层滤料终端处理滤池 tri-media polishing filter
三层铆接 three-ply riveting
三层迷宫式挡风圈 triple labyrinth seal
三层膜 trilamellar membrane;tripack film
三层抹灰 render, float and set
三层木片筛 triple-deck chip screen
三层浓密机 three-tray thickener
三层刨花板 three-layer board
三层皮带 triple(-ply)belt
三层坡度减小筛 trislope screen
三层铺 three-high berth;triple bunk
三层墙板 three-layered wall panel

三层绕杆式天线 super-turnstile antenna
三层绕组 triple-layer winding
三层纱包的 triple cotton-covered
三层筛 triple-deck screen
三层实心层叠隔墙 triple solid laminated partition
三层式 triple-decker
三层式表面处治 triple course surface treatment
三层式道路 triple-decker
三层式防碎玻璃 triplex safety glass
三层式构筑物 three-level structure
三层式立体交叉 tri-level grade separation;triple-decker
三层式密封环 triple sealing ring
三层式洗涤分级机 three-deck washing classifier
三层式振动筛 triple-deck vibrating screen
三层式纸杯配出器 triple paper cup dispenser
三层树形结构 tree height of three
三层双面波纹纸板 tri-wall corrugated board
三层丝包线 triplex silk covered wire
三层碎料板 three-layer particle board
三层台车 three-level jumbo
三层弹簧垫圈 triple spring washer
三层弹性体系 three-layer elastic system
三层涂抹<打底、镘平、结硬>【建】 render-float-and-set
三层瓦<檐口处> triple course
三层网架 triple-layer grid
三层网流 drift trammel net
三层卧铺 three tiers of berths
三层屋顶板 three-layered roof(ing) slab
三层系列 three-coat system
三层系统 three-layered system
三层纤维玻璃 triple-ply fibreglass
三层消球差镜 triple aplanat
三层斜纹 triple twill
三层型 three layers type
三层型结构 three-layer type structure
三层烟囱<由内烟道管、砖砌层及外套构成的> three-layered chimney
三层摇床 triple-deck concentrating table
三层凿井 three-deck sinking stage
三层振动筛 three-deck vibrating screen
三层织法 triple weave
三层织物 treble cloths
三层周缘嵌合体 tripericlinal chimera
三层钻车 three-level jumbo
三叉孢貭属<拉> Tripospporiopsis
三叉的 trident
三叉管 Y-pipe
三叉戟 trident
三叉戟飞机 trident
三叉戟式测距仪 trident
三叉戟饰 trisul
三叉控制转移 three-forked controlled jump
三叉裂谷系 three-arm rift system
三叉锹 three-pronged hoe
三叉曲线 trident
三叉式船坞 trident-type dock
三叉式港池 trident-type dock
三叉式装卸机 three-fork truck
三叉树 ternary tree
三叉体 triaene
三叉形轨道 Y-track
三叉鱼叉 trident
三叉抓斗 three-tine grapple
三叉转移 three-forked jump
三叉钻 bow drill
三叉钻头 center[centre] bit
三岔道 wye
三岔管 tee tube
三岔轨道【铁】 wrought track; wye track
三岔交叉(口) crossing at triangle
三差校正码 triple-error correcting code
三岔路互通式立交 three-leg interchange
三岔路口 fork in the road;junction of three-roads;three-way intersection
三岔路交叉口 three-leg intersection
三岔式立体交叉 three-leg interchange
三差改正 three-corrections for horizontal direction
三差消除器 eliminator of three-recording errors
三差异间接费分析 three-variance overhead analysis
三差异制造费用分析 three-variance overhead analysis
三产品测试 triadic product test
三产品跳汰机 three-product jig
三长度记录 three-length recording
三常数 three-constant
三车道 three-lane
三车道道路 three-lane road
三车道断面 three-lane section
三车道公路 three-lane road
三车道交通运行 three-lane traffic handling
三车道路面 three-lane pavement; three-lane road surface; three-line pavement
三车道双行道 three-lane dual highway
三车道一次铺设 three-lane at a time paving
三车线道路 three-lane road
三车相互作用<交通流量理论> three-car interaction
三撑的 tripartite
三撑电杆 tripartite pole
三承丁字管接 three ends bell T
三承三通 three-bell tee
三承十字管 boss branch; sanitary cross;three-socket cross pipe;three-socket cross tube
三承四通 boss branch;sanitary cross; three-socket cross pipe;three-socket cross tube
三乘的 triplicate
三程 triple pass
三程泵 three-throw pump
三程二次观测 three-way double observation
三程循环观测 three-way loop observation
三齿防转式 trilock type
三齿履带板 triple-grouser shoe
三齿轮钻头 three-cone bit
三齿耙路机 three-tooth ripper
三齿松土机 three-tooth ripper
三冲程泵 three-throw pump
三冲程深井泵 triple-stroke deep-well pump
三冲程深水泵 triple-stroke deep-well pump
三重 tern;triplex;triplicity
三重凹面光栅 tripartite concave grating
三重比圆 triplicate ratio circle
三重变频接收机 triple-conversion receiver
三重标度 triple scale
三重标积 triple scalar product
三重表达 three-fold representation
三重玻璃 triple glazing
三重参差调谐 staggered triple tuning
三重操纵系统飞机 triplex-system (ed) aircraft
三重操纵系统自动进场着陆 triplex autolanding
三重插入奇偶检误 triple interleaved parity
三重存取【计】 triple access
三重错误校正码 triple-error correcting code
三重的 ternary;three-ply;trinal;trinary; trine; tripartite; triple (x); triplicate;three-fold;treble
三重等温共轭曲面系 triply isothermal-conjugate system of surfaces
三重地 trebly
三重点 triple junction;triple point
三重叠切割锯木 triple-cut quarter sawn
三重叠系统 triple-reset system
三重对数 trilogarithm
三重对位 triple counterpoint
三重反应 triple reaction; triple response
三重访问 triple access
三重分度头 triple indexing center [centre]
三重分集 triple diversity
三重峰 trip peak
三重符合 three-fold coincidence;triple coincidence
三重复视 triplopia
三重割曲线 trisecant curve
三重割线 trisecant
三重根【数】 triple root
三重共轭曲面系 triply conjugate system of surfaces
三重沟道 triple channel
三重管 triple pipe;triple tube
三重管化学搅拌法 triple-pipe chemical churning process
三重管旋喷法 triple-pipe chemical churning process
三重光谱 tertiary spectrum
三重喉管<化油器的> triple diffuser;triple Venturi
三重喉管式化油器 triple Venturi carburetor
三重后齿轮传动装置 treble back gear
三重环流生物流化床 three-recycle flow biological filter
三重回收区域精炼炉 triple withdrawal refiner
三重回收提纯器 triple withdrawal refiner
三重积 triple product
三重积对合 triple-product convolution
三重积分【数】 triple integral
三重积分法 triple integration
三重积卷积 triple-product convolution
三重激态 triple excited state
三重甲板运费制 three-decker rate system
三重间接 indirect triple
三重简并 three-fold degeneracy
三重接头 triple junction
三重聚焦 triple focusing
三重聚焦质谱仪 triple focusing mass spectrometer
三重壳 triple case
三重刻度 triple scale
三重控制 triple(x) control
三重控制器 triple(x) controller
三重棱镜摄谱仪 three-prism spectrograph
三重离子 ion triplet
三重立体组 triple stereoset
三重联合开关 triple-coincidence switch
三重联机车 triple header
三重链滑车 triplex chain block
三重链接 triply linked
三重链接的树 triply linked tree
三重梁 tertiary beam
三重临界点 tricritical point
三重螺旋轴 screw triad
三重密封 triple seal
三重密封活塞环 triple seal piston ring
三重模板 triple mould
三重模块冗余度 triple modular redundancy
三重内积 triple scalar product
三重能力 tricapability
三重碰撞 triple collision
三重偏转磁铁 triplet bending magnet
三重频带 triband
三重器 tripler
三重切面 tritangent plane
三重切线 triple tangent
三重驱动 triple drive
三重去氧电焊条 triple deoxidized wire
三重染色 triple dye
三重染色法 triple staining
三重绕组 triplex winding
三重冗余 triple redundancy
三重蠕变 tertiary creep
三重散射 triple scattering
三重扫描 triple scan(ning)
三重色 triplet charm
三重栅极 triple grid
三重伸缩式管子钻塔 triplex design tubular derrick
三重圣坛层 three-fold altarpiece
三重矢 triad
三重矢式 triadic
三重数组 number triple
三重税 treble tariff
三重态 triplet state
三重替代路由 alternate triples
三重调谐耦合电路 triple-tuned coupled circuit
三重调制信标点 multibeacon
三重退火 triple annealing
三重线 triplet
三重线单谱线 triplet-singlet
三重线间隔 triplet interval
三重线谱 triplet spectrum
三重线圈 triple coil
三重线态 triplet state
三重相关 triple correlation
三重相关器 triple correlator
三重相关系数 triple correlation coefficient
三重向量积 triple vector product
三重性 ternary
三重序 ordered triple
三重岩芯管 triple-tube core barrel
三重摇杆 triple rocker
三重意义 three-fold purpose
三重余度的 triple-redundant
三重造影术 triplography
三重正交曲面系 triply orthogonal system of surfaces
三重正交曲面族 triply orthogonal family of surfaces
三重正交系 triply orthogonal system
三重轴 triad axis
三重组 triad
三触点塞孔 three-point jack
三触针井径规 three-fingered calipers
三船作业 three-ship operating
三床房间 three-bed room
三床(位)的 three-bed
三床位房间 triple bed room
三垂面反射镜 triple mirror
三垂线井下定线法 three point problem
三磁场发电机 three-field generator
三磁场直流发电机 three field DC generator

S

三次 cubic
三次倍频器 triductor
三次变电所 tertiary substation
三次变换 cubic(al) transformation
三次变频超外差接收机 triple superheterodyne
三次插值 cubic(al) interpolation
三次差异制造费用分析 three-variance overhead analysis
三次产业 tertiary industry
三次超静定的 three-fold statically indeterminate
三次超静定结构 three-fold statically indeterminate structure
三次重复 triplicate
三次重复试验设计 triplication design
三次重复小区 triplicated plots
三次的 cubic(al)
三次电流 tertiary current
三次电压 tertiary voltage
三次对称轴 axis of trigonal symmetry; triad axis
三次对称轴线 three-fold symmetry axis
三次多项式 cubic(al) polynomial; third-degree polynomial
三次反射 three-hop; triple reflection
三次反射镜 triplex reflector
三次方 cube; third angle projection; third power
三次方程 cubic; third-order equation
三次方程式 cubic(al) equation
三次方程式问题 cubic(al) equation problem
三次仿样函数 cubic(al) spline function
三次分枝 tertiary branching
三次风 recoup air; tertiary air
三次风管 recoup duct; tertiary air duct
三次浮选给料 tertiary float feed
三次干涉仪 triple interferometer
三次根 cube root; cubic root
三次过荷继电器 three times overload relay
三次函数 cubic; cubic(al) function
三次荷载弯曲试验 bending test under three-point loading
三次缓和曲线 cubic(al) equation transition
三次畸变 cubic(al) distortion
三次甲基三硝基胺 cyclonite; cyclotrimethylene trinitramine; hexahydro-1, 3, 5-trinitro-symtriazine; sym-trimethylene trinitramine; trinitrotrimethylenetriamine
三次节点 tertiary node
三次精选 triple cleaning
三次空气 tertiary air
三次扩散工艺 triple diffusion process
三次扩散技术 triple diffusion technique
三次幂 cube; third power
三次磨矿 tertiary grinding
三次内插 cubic(al) interpolation
三次拟合 cubic(al) fit
三次抛物线 cubic(al) parabola
三次抛物线缓和曲线 cubic parabolic transition curve
三次抛物线曲线 cubic(al) parabola curve
三次膨胀 triple-expansion
三次平滑法 triple smoothing
三次破碎机 tertiary crusher
三次切割 thrice-cut
三次曲面 cubic(al) surface
三次曲线 cubic; cubic(al) curve
三次曲线的切线割点 tangential point of a cubic
三次曲线回归 cubic(al) curvilinear regression
三次取样器 tertiary sampler
三次燃烧的 three-burn
三次绕组 tertiary winding
三次色 tertiary colo(u)r(s)
三次烧成 third firing
三次渗碳体 tertiary cementite
三次式 cubic(al)
三次收敛 cubic(al) convergence
三次双曲线 cubic(al) hyperbola
三次双生的 tergeminate
三次调谐 third harmonic tuning
三次图 cubic(al) graph
三次弯曲式除鳞机 <连续酸洗线上的> triple processor
三次线圈 tertiary coil
三次项 cubic term
三次谐波 third harmonic; triple-frequency harmonic
三次谐波生成 third harmonic generation
三次谐波序列 triplen
三次谐波振荡 third harmonic generation
三次行程回波 triple transit echo
三次行程信号 triple travel signal
三次修匀 triple smoothing
三次徐变 tertiary creep
三次循环 ternary cycle
三次研磨 tertiary grinding
三次掩蔽的 trimask
三次掩蔽结构 trimask structure
三次样条 cubic(al) spline
三次样条函数 cubic(al) spline function
三次再结晶 tertiary recrystallization
三刺激色度计 tristimulus colo-(u) rimeter
三刺激实测法 three-stimulate measurement method
三刺激数值 tri-stimulus value
三刺小蠹 Seolytus esuriens
三醋精 triacetin
三醋酸材料膜过滤器 triacetate metrical membrane
三醋酸甘油酯 triacetyl glycerine
三醋酸纤维 triacetate
三醋酸纤维素 cellulose triacetate
三醋酸酯基胶片 triacetate-base film
三醋酯短纤维 triacetate staple fiber [fibre]
三醋酯人造丝 triacetate rayon
三大件货车转向架 three-piece freight car truck
三大块法 three-blocks method
三大平衡 <指财政、信贷和物资平衡> balance of finance credits an materials
三代的 triatomic; tribasic; triple-substituted
三代店 <即代购代销代营店> triple agency
三代砷酸盐 tertiary arsenate
三带的 trizonal
三单干形 triple cordon
三单元天线 three-element aerial
三单元调制 three-position modulation
三单元信号(机)【铁】three-unit signal
三单元信号机构 three-unit signal head
三单元制码 ternary code
三弹头 triplet
三氮化氢 azoimide
三氮烯 triazene
三氮烯基 triazenyl
三氮烯纸 triazene paper
三氮杂苯 triazanaphthalene
三氮杂蒽 naphthotriazines
三氮杂菲 naphthisotriazine
三氮杂萘 triazanaphthalene
三挡 third(-speed) gear
三挡齿轮 third gear; three-range transmission
三挡传动 third gear
三挡速度 third speed
三挡速率 third speed
三刀单投开关 three-pole single-throw switch
三刀单掷 triple-pole single throw
三刀割草机 triple mower
三刀开关 three-pole switch; triple-pole on-off switch
三刀双掷 triple-pole double throw
三刀双掷开关 three-pole double-throw switch; triple-pole double-throw switch
三导阶跃函数 unit triplet function
三导线系统 three-wire system
三导线制 three-wire system
三岛法 mishima
三岛式泊位 three-island berth
三岛型(轮)船 three-inland ship; three-islander
三道比色计 three-channel colo(u)-rimeter
三道粗纺机 roving frame
三道粗纱机 rover
三道多点记录器 three-channel multi-point recorder
三道翻板阀 triple flap grate
三道粉刷 lath, plaster, float and set
三道粉刷中的第二道 topping coat
三道抹灰 float and set; render; render-float-and-set; three-coat; three-coat plastering
三道抹灰工作 three-coat (plaster) work
三道抹灰作业 three-coat work
三道筛 tertiary screen
三道锁风阀 triple air lock
三道体区 trivium
三道涂刷工作 three-coat work
三道卸料闸门 triple air lock; triple discharging gate
三道闸门喂料机 triple gate feeder
三道紫外-可见光分光光度计 three-channel ultraviolet visible spectrophotometer
三灯丝方法 triple filament method
三等边灯心草 three-square
三等舱 cabin class; third-class room
三等导线网 third-order traverse
三等的 third class
三等点 tertiary point
三等分【数】trisect(ion)
三等分的中部一等分 middle third
三等分的中间一份 middle third
三等分点荷载 third point load(ing)
三等分法(则) middle third rule
三等分角线【数】trisectrix[复 trisectrice]
三等分器 trisector
三等精度【测】third-order accuracy
三等距离支点 three-equidistant supports
三等(品) third class
三等三角测量 third-class triangulation; third-order triangulation; tertiary-order triangulation
三等三角点 third-order triangulation point
三等水准测量 third-order level(l)ing
三等水准点 third-order benchmark
三等水准观测手簿 third-order leveling field book
三等战备状态 secure condition of readiness
三等证券 third-class paper
三地址 three-address
三点比较式气味袋法 triangle bag method for odo(u)r sensing measurement
三点测验 three-point test
三点充氧 three point service
三点传动 three-point transmission
三点电路 three-point circuit
三点定位 three-point fix
三点定位法 three-point fix method
三点断面 three-level section
三点法 three point problem; three-point method <平板仪测量的>
三点法交会 three-point intersection
三点法引导 three-point guidance
三点方案 three-pronged program(me)
三点估计 three-time estimate
三点固定门锁 three-point lock
三点挂结 three-point hitch
三点观测 triparted observation
三点后方交会 trilinear survey(ing)
三点后方交会法 three-point resection
三点画齿规 three-pointodontograph
三点加油 three point service; triple-point refueling
三点加油机 three-point refueling tanker
三点加载 third point load(ing)
三点检查【铁】released by three sections
三点检验 three-point assay
三点交会法【测】method by intersection of three-directions
三点交会分析径向三角测量 analytic (al) tree-point resection radial triangulation
三点开关 three-point switch; three-way switch
三点抗折试验 three-point bending testing
三点控制法 three-point tone controlling
三点连接(机构) three-point linkage
三点联结 three-point attachment; three-point bond
三点落地 <飞机> three-point landing
三点平均 triadic mean
三点起动器 three-point starter
三点起动箱 three-point starting box
三点铅字 minnikin
三点曲线图 circle sheet
三点试验 three-point test
三点试验法 triangle test
三点投影 three-point projection
三点透视 three-point perspective
三点弯曲 three-point bending
三点问题 <后方交会法>【测】three point problem
三点五英尺轨距 <3.5 英尺折合 1.067 米> cape ga(u)ge
三点线素坐标 trilinear line coordinates
三点线性平滑 linear smoothing with three-point
三点悬挂 three-point suspension
三点悬挂法 three-point hitch
三点悬挂装置 three-link hitch; three-point linkage mounting; three-point link hitch
三点悬置 three-point suspension; trifilar suspension
三点荧光组 triad
三点折线断面 three-level section
三点支承 three-point bearing; three-point support
三点钟 three
三点着陆 three-point landing
三点着陆姿态 three-point attitude
三碘化铬 chromium triiodide
三碘化合物 teriodide; triiodo-compound

三碘化砷 arsenic triiodide
三碘化铊 thallic iodide;thallium triiodide
三碘化物 teriodide;triiodide
三碘化铟 indium triiodide
三碘甲烷 iodoform
三电动机铲 three-motor all-electric shovel
三电荷 tricharged
三电机电铲 three-motor shovel
三电极侧向测井仪 laterolog 3
三电极的 three-electrode
三电极弧光灯 three-electrode arc lamp
三电极火花 three-electrode spark
三电极屏障电流法测井曲线 laterolog 3 curve
三电路 three-circuit
三电路塞孔 three-way jack; triple circuit jack
三电台机组 triplet
三电压表法 three-voltmeter method
三电子枪管 three-gun tube
三弹性轴 three-elastic axes
三叠纪【地】Triassic(period)
三叠纪砂石 Triassic sandstone
三叠纪砂岩 Triassic sandstone
三叠接桩 triple-lap pile
三叠系【地】Triassic(system)
三丁基 tributyl
三定子绕组自动同步机 three-stator winding synchro
三动泵 triple acting pump
三动道岔 triple-working switches
三动力组电铲 three-motor shovel
三动式压力机 triple action presses
三动压床 triple action press
三斗称重分批配料器 triple weighing batcher
三斗给料机 triple-scoop feeder
三肚板门 three-pack door
三度表示 three-dimensional representation
三度存储器 three-dimensional memory
三度带 three-degree zone
三度的 three-dimensional triaxial;tridimensional
三度定向 dimensional orientation
三度割线 trisecant
三度红斑 third-degree erythema
三度结构 three-dimensional structure
三度空间 three-dimensional space
三度空间的 three-dimensional
三度空间雷达 volumetric(al)radar
三度空间模拟器 three-axis simulator
三度空间模式 three-dimensional space pattern
三度空间数据 tridimensional data
三度空间图 solid diagram;three-space diagram
三度空间应力 three-dimensional stress
三度空间追踪 three-dimensional follow-up
三度空间组合形态 three-dimensional shape of a cave
三度烧伤 third-degree burn
三度投影 trimetric projection
三度异常化二度异常 three-dimensional anomaly transforming to two-dimensional anomaly
三度运动 three-dimensional motion
三度重力量板 three-dimensional gravity graticule
三端 three-terminal
三端 npnp 开关<可控硅整流> trinistor
三端交流开关 triac
三端晶体闸流管 triode-thyristor
三端开关器件 trigistor
三端可控硅 triode-thyristor
三端快速半导体开关 trisistor
三端启动电阻箱 three-point starting box
三端起动器 three-point starter
三端双向可控硅开关(元件)triac
三端网络 three-terminal network
三端线路 three-terminal line
三端子半导体开关元件 triode-thyristor
三端子接点 three-terminal contact
三短截线调谐器 three-stub tuner
三短线变量器 triple-stub transformer
三段电渗析装置 three-stage electrodialysis unit
三段法 three-stage process
三段混炼 three-stage mixing
三段距离继电保护装置 three-step distance relays
三段楼梯 three-flight stair(case)
三段论法 syllogism
三段密封式防喷器 blowout preventer of three stage packer type
三段逆流反应器 three-stage counter-current reactor
三段排架 three-section framing bent
三段破碎 three-stage crushing
三段区域精炼炉 three-stage refiner
三段蠕变 third stage of creep
三段烧结 three-stage sintering
三段设计 three-stage design
三段升压(制动)three stage build-up
三段生物处理除氮工艺 three-stage biologic(al) treatment process for nitrogen removal
三段式浮船坞 three-section floating dock
三段式炉 triple-fired furnace
三段碎矿 three-stage reduction
三段掏槽 three-section cut
三段最小平方法 three-stage least squares method
三段最小平方估计 three-stage least squares estimates
三堆法 three-bin system
三对 three-pair
三对的 trimerous
三对角线矩阵 tridiagonal matrix;triple diagonal matrix
三对数坐标反应谱 tripartite logarithmic response spectrum
三吨载货汽车 three-ton truck
三垛式台 three-pedestal base
三舵船 ripple-rudder ship
三发动机 trimotor
三发动机布局 triple-engine layout
三发动机的 three-engined
三发动机飞机 three-engined plane;tri-motored airplane;triple-engined airplane
三发动机驱动 three-motor drive
三发动机式 triple-engined type;triple engine reaction
三发动机式起重机 three-motor type crane
三发动机移动式起重机 three-motor travel(l)ing crane
三发射架的发射场综合设施 three-missile-to-a-site complex
三发射井的发射场综合设施 triple-silo complex
三阀的 three-valve
三番五次 over and over again
三反射面馈源系统 three-reflector feed system
三方 trigonal
三方钡解石 paralstonite
三方单锥 trigonal pyramid
三方当事人 three-parties
三方呼叫 three-way calling
三方交易 tripartite transaction
三方结晶 trigonal crystal
三方经营 tripartite arrangement
三方晶系【地】trigonal system
三方竞争市场 three-market
三方蓝辉铜矿 trigodgenite
三方硫砷银矿 trechmannite
三方硫碳铅石 susannite
三方硫锡矿 berndtite
三方氯铜矿 paratacamite
三方面之间的 tripartite
三方硼砂 tinzalconite
三方偏方面体 tetragonal trapezohedron
三方签署的票据 three-name paper
三方羟铬矿 grimaldiite
三方羟磷镁石 holtedahlite
三方羟磷铁石 satterlyite
三方闪锌矿 matraite
三方式排气净化系统 triple mode emission control system
三方式循环 three-way circulation
三方双面锥 trigonal bipyramid face
三方双锥 trigonal bipyramid
三方水硼镁石 mcallisterite
三方碳钾钙石 buetschliite
三方铜铬矿 macconnellite[mcconnellite]
三方硒铋矿 laitakarite
三方硒镍矿 makinenite
三方向法 method of three-direction
三方协定 tripartite agreement
三方协议 tripartite agreement
三方氧钒矿 karelianite
三方柱 trigonal prism
三芳基甲烷 triarylmethane
三芳基甲烷染料 triarylmethane colo(u)ring matters
三芳甲基 triaryl methyl
三防<海、陆、空国防> full weapon protection;three antis
三房客 sublessee
三废<指废水、废气、废渣> the three wastes[waste water, waste gas and waste residue]
三废治理 three wastes utilization and disposal;treatment of three types of wastes
三废综合利用 multipurpose use of "the three wastes"; multipurpose use of three types of wastes; utilization of the three wastes
三废综合利用工厂 salvage shop
三分 trisect
三分点 middle third point
三分点荷载 third point load(ing)
三分点集 ternary set
三分点挠曲试验 third point flexural test
三分点准则 middle third rule
三分度取样 three-sampling intervals per decade
三分段式浮船坞 three-sectional dock
三分法 three-way classification; trichotomy
三分方位 trigonal aspect
三分方位点 trigonal point
三分格窗 three-light window
三分隔(空间)转门 three-compartment revolving door
三分角 trisectrix[复 trisectrice]
三分量 three-component; tri-component
三分量磁测 three-component magnetic survey
三分量地震测线剖面 seismic line profile of three-component
三分量地震检波器台阵 three-component geophone array
三分量地震仪 three-component seismograph
三分量动圈式地震仪 three-component moving coil-type seismometer
三分量风速仪 three-component anemometer
三分量加速度计 motion accelerometer;three-component accelerometer
三分量加速度仪 motion accelerometer;three-component accelerometer
三分量检波器 three-component geophone
三分量模拟深层地震仪 analog(ue) deep seismograph in three-azimuth
三分量强烈加速度仪 three-component strong-motion accelerograph
三分量自记加速表 three-component accelerograph
三分绕组 triplex winding
三分式 triplasy
三分向地震检波器台阵 three-component geophone array
三分向地震仪 three-component seismograph
三分之二公式 two-thirds formula
三分之二规则 two-thirds rule
三分之七定则 seven-thirds rule
三分之一 triplicate
三分之一倍频程 one-third octave band
三分之一波长变换器<由同轴线至对称线的变换器> triple coaxial transformer
三分之一的 subtriple
三分之一的面积 one-third of the area
三分之一法则<指桥台翼墙一定截面上平均厚度,常以翼墙的相应高度的"三分之一"来设计的惯例> one-third rule
三分之一高跨比 one-third pitch
三分之一拱 tiers-point arch
三分之一股道软横跨 one-third track headspan;one/three track headspan
三分之一和三分之二原则 one-third two-thirds rule
三分之一跨度点荷载 third point load(ing)
三分之一砌合 one-third bond
三分之一日潮 terdiurnal tide
三分之一最大波高 one-third maximum wave; average highest one-third wave height
三分中一<三等分的中部一等分> middle third
三分钟定时器 three-minute glass
三分子机理 termolecular mechanism
三份 triplet;triplicate
三份的 triplicate
三风道喷煤管 three-channel coal burner
三峰曲线 triple humped curve
三缝波淋头 three-slot boiling head
三缝燃烧器 three-slot burner head
三伏特计法 three-voltmeter method
三伏天 canicular days;dog days
三氟碘化铀 uranium monoiodotrifluoride
三氟硅烷 silicofluoroform
三氟化氮 nitrogen trifluoride
三氟化钒 vanadium trifluoride;vanadous fluoride
三氟化铬 chromium trifluoride
三氟化钴 cobaltic fluoride;cobalt trifluoride
三氟化合物 trifluoro-compound;trifluoride
三氟化磷 phosphorus trifluoride
三氟化氯 chloride trifluoride
三氟化硼 boron trifluoride
三氟化铈 cerous fluoride
三氟化铊 thallic fluoride;thallium trifluoride
三氟化钨 tungsten trifluoride

S

三氟化溴 bromine trifluoride
三氟化溴法 BrF3 method
三氟化氧钒 vanadyl trifluoride
三氟化铟 indium trifluoride
三氟甲烷 trifluoromethane
三氟硫化磷 phosphorus sulfofluoride
三氟氯甲烷 trifluorochloromethane
三氟三溴甲烷 bromotrifluorochloromethane
三氟硝胺 trifluralin
三氟溴乙烷 halothane
三氟氧化钒 vanadium oxytrifluoride
三氟氧化磷 phosphorus oxyfluoride
三氟一溴化铀 uranium monobromotrifluoride
三浮筒式水上飞机 triple-float-type seaplane
三幅式道路 triple carriageway road
三幅式路 trisected carriageway road
三幅一联画或雕刻 triptych
三辐体 triradiata
三副【船】third mate;third officer
三副室【船】third mate's room
三腹板箱式断面 three-webbed box section
三伽马函数 trigamma function
三钙盐 tricalcium
三甘氨酸 nitrilotriacetic acid;triglycine
三杆保险杠 three-bar bumper
三杆比拟法 <剪力滞计算方法> three-bar simulation method
三杆吊车 triplex derrick
三杆定位法 <三标两角定位的海图作业> station-pointer fix
三杆定位仪 station pointer;three-arm protractor
三杆法 <分辨率检查> tribar method
三杆分度器 station pointer;three-arm protractor
三杆分度器偏差 error of eccentricity in three-arm protractor
三杆分度器弯曲差 error of bend in three-arm protractor
三杆分度器隙动差 error of back lash in three-arm protractor
三杆分度仪 station pointer;three-arm protractor
三杆分度仪偏差 error of eccentricity in three-arm protractor
三杆分度仪弯曲差 error of bend in three-arm protractor
三杆分度仪隙动差 error of back lash in three-arm protractor
三杆节点 three-member nodal point
三杆联结 three-member joint
三杆牵引装置 three-in-one drawbar
三杆曲线 three-bar curve
三杆应变计 triple rod extensometer
三杆闸机 tripod fate
三缸泵 three-cylinder reciprocating pump;three-throw pump;triplex pump
三缸单作用泵 triplex single action pump
三缸二行程发动机 three-cylinder two-stroke engine;three-port two stroke engine
三缸发动机 three-cylinder engine
三缸复式机车 three-cylinder compound locomotive
三缸高速泵 three-throw high-speed pump
三缸机车 three-cylinder locomotive
三缸式(水)泵 triplex pump
三缸往复式泵 triplex reciprocating pump
三缸星形发动机 Y engine
三钢丝法 three-thread wire method;three-wire method
三高度层系统 three-layer system
三割卧室居住单元 three-bedroomed dwelling unit
三格蒸汽干燥机 three-compartment steam drier
三个半圆形内室 <教堂的> tri-apsidal chevet
三个半圆形室 tri-apsidal
三个成一组的穹顶 triparted vault
三个的 ternary
三个精确一致的光栅 three-exactly coincident rasters
三个具限分明的土层 three-well-defined horizons
三个棱镜反射器 triple-prism reflector;triplex reflector
三个明显层 three-distinct layers
三个羟乙苯胺 triethanolamine
三个水分来源 three-sources of water
三个卧室的住房 three-bedroomed house
三个相联的雕刻 triptych
三个一套 tern;triplet
三个一套的 ternary
三个一种 triplet
三个一组 triad;trial;triplet
三个一组的 tripartite
三个一组的拱 tripartite arch
三个一组穹顶 tripartite vault
三个月 trimenon
三个正厅 <教堂中的> three-nave
三个中殿 <教堂中的> three-nave
三个着丝点的 tricen fiber point
三个组成的 ternary
三个组成的穹顶 tripartite vault
三铬酸 trichromic acid
三铬酸盐 trichromate
三根一股钢绞线 triple twisted wires
三根钻杆组成的立根 three-joint unit
三工器 triplexer
三工制 triplex system
三工质热交换器 triffux
三工作边凸轮 tri-lobe cam
三弓形折流板 triple segmental baffle
三攻丝锥 third tap
三拱坝 triple arch dam
三拱桥 three arch bridge
三拱式 triforium
三构面领导理论 three-dimension leadership model
三股编绳 three-strand sennit
三股的 three-ply
三股分配斜槽 three-way distributing chute
三股绞花 triple cable
三股绞结 French shroud knot
三股绞线 three-wire strand;triple conductor
三股螺旋 triple helix
三股棉线 three-cord-thread
三股软线 three-way cord
三股绳 three-stranded rope
三股绳索 three-part line
三股绳索滑轮组 three-part line tackle
三股线 three-folded yarn;trifilar wire
三股油麻绳 house line;stuff;tarred hemp rope;three-yarn nettle
三股油麻绳填塞料 three-yarn nettle stuff
三股右旋绳 round line
三骨料库 three-aggregate bin
三鼓法 three-drum system
三鼓卷纸机 three-drum reel
三鼓式滚子链 three-strand roller chain
三鼓式快速换挡绞车 three-drum rapid shifting hoist
三鼓式快速换卷扬机 three-drum rapid shifting hoist
三鼓式蒸汽绞车 three-drum steam hoist
三刮刀刃管鞋下扩孔器 three-cutter underreamer
三刮刀钻头 tri-blade-drag bit;tri-drag-blade bit
三拐曲柄 three-throw crank
三拐曲柄泵 three-throw crank pump
三拐曲轴 three-throw crankshaft
三管采暖系统 three-pipe heating system
三管彩色摄像机 tricolo(u)r camera
三管空调系统 three-pipe air conditioning system
三管路系 triple circuit system
三管轮【船】third engineer
三管轮室 third engineer's room
三管配置方式 three-piping system
三管式摄像机 three-tube camera
三管线 three-pipe line
三管岩芯筒 triple-tube core barrel
三管治水系统 three-pipe water system
三光导视像管摄影机 three-vidicon camera
三光灯泡 three-way bulb
三光点内摆线 deltoid(al)
三光气 triphosgene
三光子衰变 three-photon decay
三硅酸镁 magnesium trisilicate
三硅酸盐 trisilicate
三轨上部接触式 third rail top contact;three-rail top contact
三轨实验(导轨)装置 three-rail test track
三轨试验(导轨)装置 three-rail test track
三轨下部接触式 third rail below contact
三轨制 third rail system
三轨中部接触式 third rail middle contact
三辊穿孔机 three-roll piercer
三辊的 three-high
三辊钢板轧机 three-high plate mill
三辊矫直机 three-roll unbender
三辊校直机 three-roll unbender
三辊开坯机 three-high bloomer;three-high cogging mill
三辊(拉伸机) trio
三辊劳特式钢板轧机 three-high Lauth plate mill
三辊轮式机械 three-roll type machine
三辊磨 triple-roll(er) mill
三辊破碎机 triple roll crusher
三辊式初轧机 three-high blooming mill
三辊式滚轧机 three-roll mill
三辊式机座 three-high house
三辊式精轧机组 three-high finishing train
三辊式卷取机 three-roll-type coiler
三辊式开坯机 three-high cogging mill
三辊式磨料机 three-roll grinder
三辊式配置 three-high arrangement
三辊式型钢轧机 three-high jobbing mill;three-high shape mill
三辊式轧钢机组 three-high rolling train
三辊式轧钢机座 three-high rolling stand
三辊式轧机 three-high mill;trio mill
三辊式装置 three-high train
三辊筒轧机 three-roll mill
三辊弯板机 three roll bending machine
三辊万能式轧机 three-high universal mill
三辊压路机 three drum roller;three-roll roller
三辊轧机 three-high mill
三辊制球机 three-cylinder marble machine
三辊中速磨 three-roller mill
三滚轮钻头 three-roller bit
三滚筒滚压机 three-wheel roller
三滚筒耙矿绞车 triple drum scraper hoist
三滚筒提升机 three-drum hoist
三滚筒压路机 three-wheel roller;triple-roll roller
三滚筒压实机 triple drum tamper
三滚筒羊脚压路机 triplex sheepfoot roller
三滚轴破碎机 triple roll crusher
三滚轴压缩机 triple roll crusher
三滚柱式螺纹量规 triroll ga(u)ge
三锅系 three-pan system
三国间的三角贸易 triangular trade
三国文字的 trilingual
三国语的 trilingual
三过磷酸钙 triple superphosphate
三过氧铬酸 triperchromic acid
三合安全气制动系 triple-safe air brake system
三合板 glued wood;ternary plate;three-layer board;three-plywood;triplywood
三合薄透镜 triplet thin lens
三合齿轮 triple gear
三合促进剂 triangular acceleration
三合单元 triple-unit
三合地震台网 triparatite seismic network
三合夹板 triplywood;three-ply board
三合镜 triplet lens
三合滤波器 triplener
三合气 tridyne
三合树 three-component tree;three-part tree;three-piece tree
三合台网观测 tripartite net observation
三合台阵 tripartite array
三合透镜 triplet;triplet glass;triplet lens
三合土 concrete;lime concrete
三合土基础 tri-composition foundation
三合物镜 triplet objective
三合星 triple star
三合盐 triple salt
三合一 triad
三合一槽式滚轴 <输送带的> triple throughing roll(er)
三合一长老席位 tripartite presbytery
三合一措施 three-in-one process
三合一式调节器 three-unit regulator
三合一制动阀 three-in-one brake valve
三和弦 triad
三核都市 tricity
三桁架桥 three truss bridge
三厚窗玻璃 <5~6 毫米厚窗玻璃> thick sheet glass;heavy sheet;crystal sheet glass
三弧法 three-plug method
三弧外旋轮线缸体 three-lobe epitrochoidal bore
三弧旋转活塞 <转子发动机的> three-lobe rotor
三花色信号形成设备 colo(u)rplexer
三滑轮滑车组 three-fold block
三化的 trivoltine
三化性 trivoltinism
三化油器发动机 three-carburettor engine
三环槽式磨碎机 three-race mill
三环的 tricyclic;trinuclear;trinucleated
三环芳香化合物 tricyclic aromatics

三环芳香烃 triaromatics
三环核 tricyclic ring
三环花纹轮胎 triple rib tire
三环化合物 tricyclic compound
三环环烷 tricyclic naphthene
三环己基甲烷 tricyclohexylmethane
三环己基氢氧化锡 tin tricyclohexylhydroxide
三环己氧基铀 tricyclopentadienyl-cyclohexyliloxy-uranium
三环连接装饰物 <一种家具织物装饰> frog
三环烃 tricyclic hydrocarbon
三环烷烃 tricycloalkane
三环戊二烯化物 tricyclopentadienide
三环戊二烯基丁氧基铀 tricyclopentadienyl-butoxy uranium
三环烯 tricyclene
三环系统 three-loop system
三环岩兰烯 tricyclovetivene
三环酯 triclazate
三磺酸 trisulfonic acid
三磺酸盐 trisulfonate
三簧片塞孔 three-point jack
三回程锅炉 three-pass boiler
三回出叶的 triternate
三回路调谐器 three-circuit tuner
三回路经济锅炉 three-pass economic boiler
三回羽状的 tripinnate
三混煤气 dreigas
三火式灯头 three-lit base
三机牵引 three-bagger;triple locomotives
三机小队 three-ship element
三机掩体 triple pen
三机照相 tri-camera photography
三机组 <由电动机、发电机、永磁发电机组成的机组> three unit
三机座串列式轧机 three-stand tandem mill
三基 triguaiacyl
三基色 three primary colo(u)rs
三基色析像能力 resolution requirement in the primary image
三基色信号比 ratio of the three-colo(u)r primary signals
三基色信号形成设备 colo(u)rplexer
三基色荧光点组 point triad
三基线法 three-base method
三激波进气口 triple-shock intake
三激波进气扩散段 triple-shock intake
三激光过滤器 tristimulus light filter
三激光器陀螺 laser triad
三激励规格曲线 tristimulus specification curve
三激励值 tristimulus value
三激源色度计 tristimulus colo(u)rimeter
三级胺 tertiary amine
三级泵 triple-stage pump; triplex pump
三级编址 three-level addressing
三级变矩器 three-stage torque converter
三级变扭器 three-stage torque converter
三级变速箱 three-ratio gear; three-speed gear box;triple-change gear
三级变速装置 three-speed gear
三级玻璃水银扩散泵 three-stage glass pump
三级差动抽气真空系统 three-stage differentially pumped vacuum system
三级产业 <指服务业> tertiary production
三级成矿远景区 the third grade of minerogenetic prospect
三级成煤远景区 the third grade of coal-forming prospect
三级齿轮驱动 triple geared drive
三级出水 tertiary effluent
三级处理 tertiary treatment;polishing stage <指废水处理>
三级处理池 tertiary treatment pond
三级处理的废水 tertiary treated wastewater
三级处理法 tertiary treatment method
三级处理后的污水 tertiary treated wastewater
三级处理塘 tertiary treatment pond
三级串联捻股机 triple tandem strander
三级存储 third-level storage
三级道路 tertiary road
三级道路网 tertiary network
三级的 three-stage
三级地址 third-level address(ing); three-level address(ing)
三级反射 tertiary reflex
三级反射面天线 tertiary reflector antenna
三级反应 third-order reaction
三级反应器 third-order reactor
三级返回系统 three-level return system
三级防治网 three-level control network
三级废水 tertiary wastewater
三级废水处理 tertiary sewage treatment
三级废水处理过程 tertiary wastewater treating process
三级粉碎 tertiary grinding; tertiary reduction
三级风 force-three-wind;gentle breeze; wind of Beaufort force three; slight breeze
三级风浪 moderate sea
三级风流分支 tertiary split
三级公路 tertiary highway; tertiary road; third class highway; third-class road
三级共态预测 three-level costate prediction
三级构造 third grade structure
三级管网系统 three-stage system
三级辊式破碎机 tertiary crusher with rolls
三级过程 <污水处理> ABC process
三级过滤 tertiary filtration
三级过滤器 tertiary filter
三级滑车组 triple pulley block
三级环流 tertiary circulation
三级畸变 third-order distortion
三级记录 three-level record(ing)
三级记录系统 three-level return system
三级减速器 triple reduction gear (box)
三级减速装置 triple reduction gear (ing)
三级降落伞 three-stage parachute
三级降压防滑制动系统 triple action skid control brake system
三级阶地 third terrace
三级结构 tertiary structure
三级结构面 grade three-discontinuity
三级结构体 grade three-texture body
三级进给箱 three-step feed box
三级精度配合 plain fit
三级静止状态 quiet condition three
三级开关 three-step switch
三级开路破碎 three-stage open-circuit crushing
三级刻度 third level calibration
三级空气压缩机 three-stage air compressor
三级控制器的准连续动作 quasi-continuous action of a three level controller
三级浪 force-three-wave;slight sea
三级力矩 <在刚架内力分析中,由杆件轴向变形产生的力矩> tertiary moment
三级梁 tertiary beam
三级六场编组站【铁】 three-stage/6-yard marshalling station
三级滤池 tertiary filter
三级路 tertiary road
三级锚链 grade three-chain
三级磨碎 tertiary comminution
三级木材 third grade timber [3rd grade timber]
三级能见度 moderate fog
三级逆流抽提系统 three-stage counter current extraction system
三级膨胀发动机 triple-expansion engine
三级膨胀式蒸汽机 triple-expansion engine
三级膨胀蒸汽泵 triple-expansion steam pump
三级破碎 tertiary crushing;three-stage crushing
三级破碎机 tertiary breaker; tertiary crusher
三级起步时差 <交通> triple offset
三级气密封系统 three-stage gas sealing system
三级渠道 tertiary canal
三级绕组 tertiary winding
三级三场编组站【铁】 three-stage-three-yard marshalling station
三级三角测量 tertiary triangulation; third-order triangulation
三级生物处理 biologic(al) tertiary treatment
三级市场 <美国术语,指大宗股票的场外交易> third market
三级式燃气轮机 three-stage gas turbine
三级税率系统 three-rate tariff system
三级碎石机 tertiary breaker; tertiary crusher
三级塘 tertiary maturation pond; tertiary pond
三级瓦斯煤矿 third category gassy mine
三级稳定塘 tertiary stabilization ponds
三级涡轮机 triple turbine
三级污水处理 tertiary sewage treatment
三级像差 third-order aberration
三级像差理论 third-order aberration theory
三级像差系数 third-order aberration coefficients
三级泄压循环 three-step decompression cycle
三级寻址 third-level addressing; three-level addressing
三级压力式汽轮机 three-pressure stage turbine
三级压缩机 three-stage blower; three-stage compressor
三级压延 tertiary reduction
三级研磨 tertiary grinding
三级氧化塘 tertiary oxidation pond
三级一四级管 tri-tet
三级涌 moderate swell short
三级油气远景区 the third grade of oil-gas prospect
三级增压机 three-stage
三级增压器 three-stage supercharger
三级轧碎机 tertiary crusher
三级振荡管 oscillion
三级振荡器 three-level generator
三级蒸发系统 three-boiling system
三级蒸馏 three-stage distillation
三级蒸馏水 triple distilled water
三级蒸汽喷射器 three-stage steam ejector
三级至二级中心线路 tertiary to secondary center[centre] circuit
三级制信号 three-grade system signal
三级质谱计 three-stage mass spectrometer
三级转矩变换器 three-stage torque converter
三级子程序 three-level subroutine
三级子例行程序 three-level subroutine
三级作用 three-level action
三极测量法 trielectrode arrangement
三极测深 three-electrode sounding
三极测深曲线 curve of three-electrode sounding
三极的 triple-pole;tripolar
三极电磁铁 tripolar electro magnet
三极电子管 <旧称> radiotron
三极电子枪 triode electron gun
三极反向器 triple pole reverser
三极反演机构 triple pole reverser
三极管 aerotron; audion; three-electrode tube; three element tube; triode
三极管电离真空计 triode ion ga(u)ge
三极管发射机 triode transmitter
三极管反馈线路 urraudion
三极管放大器 triode amplifier; triple-grid amplifier
三极管回授式检波器 ultraudion
三极管混频器 triode mixer
三极管激励的五极管 triode-driven pentode
三极管激励器 triode driver
三极管检波 triode detection
三极管检波器 audion detector; triode detector
三极管晶体振荡器 triode crystal oscillator
三极管气体激光器 triode laser
三极管时间选择器 triode time selector
三极管削波器 triode clipper
三极管振荡器 triode generator; triode oscillator
三极检波电子管 audion
三极开关 three-pole switch; triple-pole on-off switch; triple-pole switch
三极熔丝 triple-pole fuse
三极式 three-pole
三极式电子枪 triode gun
三极式铅钴电池 tripolar lead-cobalt battery
三极氩检测器 triode argon detector
三极一六极管 triode-hexode
三极一七极管 triode-heptode
三极一四极管 triode-tetrode
三极一五极管 triode-pentode
三极闸刀开关 three-pole knife switch; triple-pole knife switch
三极闸流管 three-electrode thyratron
三极真空管 three-electrode vacuum tube; triode
三极振荡管 oscillion
三极坐标 tripolar coordinates
三集料库 three-aggregate bin
三加一地址 three-plus-one address
三夹板 three-plywood; triple laminate; triplywood
三夹板箱 plywood case
三家合住房屋 three-family dwelling
三甲胺 trimethylamine
三甲苯酚 mesitol
三甲基丁烷 <高抗爆燃料> trimethylbutane; triptane

S

三甲基铝 trimethylaluminum
三甲基噻吩 trimethylthiophene
三甲基四氢化萘 trimethyltetrahydronaphthalene
三价 tervalence [tervalency]; trivalence
三价的【化】trivalent
三价铬的 chromic
三价钴的 cobaltic
三价镓的 gallic
三价钼的 molybdenic;molybdic
三价铌的 niobous
三价镍的 nickelic
三价茎 triad
三价铈的 cereus
三价酸 triatomic acid
三价铊的 thallic
三价钛的 titanous
三价铁 ferric iron
三价铁的 ferric
三价元素 triad
三驾马车 three-in-hand
三驾双轮马车 triga[复 triage]
三架梁 three-purlin(e) beam
三尖瓣的 tricuspid
三尖点四次线 tricusidal quartic
三尖顶拱窗 triple lancet window
三尖端捞绳矛 three-prong rope grab
三尖拱 tricuspid
三尖头的 tricuspid
三尖形 trinacriform
三间地带性 three-dimensional zonation
三间隔形式 triptych
三剪力 triple-shear
三检制 tripartite inspection system
三碱价的 tribasic
三碱磷酸钠 tribasic sodium phosphate
三碱式硫酸铅 tribasic lead sulfate
三碱酸 tribasic acid
三件式刀刃 three-piece cutting edge
三件式护口管接头 nut-and-sleeve flare fitting;three-piece flare fitting
三件式切削刃 three-piece cutting edge
三件式肘板 three-piece toggle plate
三件套浴室 three-fixture bathroom
三件一套 triplet
三件一套的 three-piece
三交 three-way cross
三交点交会法分析径向三角测量 analytic(al) three-point resection radial triangulation
三焦距的 trifocal
三角 trigonal;trigone;trigonum
三角鞍形棒 saddle
三角岸标 triangular beacon
三角凹口 triangular notch
三角凹凸槽接头 triangular rabbet joint
三角八面体 triakisoctahedron
三角板 marquois scale; plane triangle;set square;sliding square;triangle;triangle square
三角板下短条线脚 regula(e)
三角保护(器)angle guard
三角比 trigonometric(al) ratio
三角比例尺 triangular scale
三角闭合 closure of triangle
三角闭合差 closure error of triangle; triangle closing error
三角波 chopping sea; choppy sea; pyramidal wave; triangle wave; triangular wave
三角波发生 triangle generation
三角波极谱法 triangular-wave polarography
三角波脉冲 triangular pulse
三角波脉冲发生器 triangle generator
三角波扫描 triangular voltage sweep
三角波信号 triangular signal
三角波形 triangle waveform; triangular waveform
三角波噪声 triangular noise
三角薄壳 triangular shell
三角补偿交易 triangular compensation trade
三角不等式 triangle inequality; triangular inequality
三角槽板 V-block
三角槽口 vee notch
三角槽口堰 vee-notch weir
三角槽舌接合 triangular rabbet joint
三角槽铁 vee block
三角槽形断面 V-section
三角测高法 trigonometric(al) level(l)ing
三角测高仪 hypsometer
三角测量 triangulation; triangulation operation; triangulation work; trigonometric(al) survey(ing); trigonomical survey
三角测量标点 triangulation station; trigonometric(al) station
三角测量标记 triangulation mark
三角测量标志 triangulation signal
三角测量补充网 supplementary scheme of triangulation
三角测量觇标 triangulation signal; triangulation target; triangulation tower; trigonometric(al) beacon
三角测量成果表 triglist
三角测量导航法 navigation by triangulation
三角测量的测站 trig station
三角测量等级 class(ification) of triangulation; order of triangulation
三角测量点 triangulation point; station of triangulation
三角测量定位法 triangulation location
三角测量法 triangulation method
三角测量高程 triangulation height
三角测量弧 triangulation arc
三角测量基线 triangulation base (line)
三角测量基准(点) triangulation datum; trigdatum[复 trigdata]
三角测量计算 trigonometric(al) calculation
三角测量架 <觇标的> triangulation tower
三角测量交会法 triangulation intersection
三角测量校正 triangulation adjustment
三角测量控制 trigonometric(al) control
三角测量雷达系统 triangulation radar system
三角测量平差 triangulation adjustment
三角测量气球 triangulation balloon
三角测量摄影机 triangulation camera
三角测量数据 triangulation data; trig data; trigdatum[复 trigata]; trigonometric(al) data
三角测量锁 chain of triangulation
三角测量锁环 triangulation loop
三角测量图 triangulation sheet
三角测量外业组 triangulation field squad
三角测量网 triangulation net(work); trilateration net(work)
三角测量选点 triangulation reconnaissance
三角测量者 trigonometer
三角测量资料 tria-data; trig data; trigdatum[复 trigata]
三角测量作业 triangulation operation
三角测量作业组 triangulation field squad
三角测站 triangulation station; trig station
三角测站标志 triangulation station mark
三角插值多项式 trigonometric(al) interpolation polynomial
三角插座 shuttered socket
三角岔道 triangular junction
三角铲 triangular spade
三角场 trigonal field
三角撑 triangle tie
三角尺 marquois scale; set square; three-edged rule(r); triangular ruler
三角齿 peg tooth; triangular tooth
三角齿锯 peg-tooth saw
三角齿锯片 peg-tooth saw blade
三角除锈刮刀 triangular rust scraper
三角窗 triangular window
三角锉(刀) angle file; parting tool; three-square file; triangle file; triangular file; angular file; three-angular file; three-cornered file
三角代换 trigonometric(al) substitution
三角带 triangular belt; vee belt
三角带角度 angle of the V-belt
三角导线混合网 triangulation-traverse network
三角岛 triangular island
三角的 cuspate; cusped; cuspidal; triangular; trigonal
三角等积投影 triangular equivalent projection
三角底穹隆 triangular dome
三角底座 lampstand
三角地带 gore; gore area; gore lot
三角点 station point; triangular point; triangulation point; triangulation station; trigonometric(al) point; trigonometric(al) station; trig point; trig station
三角点标石 triangulation pillar
三角点标志 triangulation mark
三角点成果表 control data card; summary of trigonometrical points; table of trigonometric(al) points; triangulation publication; trig card; triglist
三角点等级 class of triangulation point
三角点埋石 marking of a trigonometric point
三角点水平方向及成果表 trig horizontal direction and list
三角点说明 trigonometric(al) point description
三角点一览表 triangulation publication
三角点展绘 plotting of triangulation point
三角点之记 trigonometric(al) point description
三角垫 delta-ring; V-pad
三角垫密封 delta gasket closure
三角垫木 arris fillet; doubling fillet; tilting fillet; triangular fillet; wedge block
三角垫圈 triangular washer
三角吊架 sulky derrick
三角钉 crowfoot[复 crowfeet]
三角顶帆 raffle
三角定规 plain triangle; triangular rule
三角定积分 trigonometric(al) definite integral
三角定律 triangle law
三角断面电缆 SO-cable
三角断面梁 triangular section girder
三角堆垛法 grib stacking
三角对称线 triangular symmetric curve
三角多项式 trigonometric(al) polynomial
三角垛式支架 three-member cog
三角法 triangle method; triangular method; trigonometry
三角法的 trigonometric(al)
三角帆 staysail
三角帆顶角撑木 headsticks
三角帆滑动上桅 <小艇的> gunter yard
三角方程 three-angle equation; triangle equation; triangular equation; trigonometric(al) equation
三角分度规 station pointer
三角分解法 triangular decomposition; triangular factorization
三角分线杆 pole strutted in pyramidal form; Y-section
三角枫 trident maple
三角浮标 triangular buoy
三角盖板 angle fillet
三角杆 triangle pole
三角钢丝 triangular steel wire
三角钢丝网 triangle mesh wire fabric
三角港 firth[frith]; negative delta
三角高程 <由三角高程测量测定的高程> triangulated height; triangulation height
三角高程标志 vertical angle benchmark
三角高程测量点 triangulation benchmark; vertical angle benchmark
三角高程测量(法) triangulated level(l)ing; trigonometric(al) level(l)ing
三角高程导线测量 trigonometric(al) height traversing
三角高程网 trigonometric(al) level(l)ing network
三角高程网平差 adjustment of trigonometric(al) level(l)ing network
三角高度测量 trigonometric(al) level(l)ing
三角格构 triangular frame
三角格构桁架 triangulation
三角格网 triangular grid
三角格网模型 triangular grid model
三角格子饰线脚 triangular fret mo(u)lding
三角拱 miter[mitre] arch; pediment arch; scoinson arch; triangular arch
三角拱式桁架 three-hinged arch truss
三角拱饰 pediment arch
三角股钢丝绳 triangular strand wire rope
三角刮刀 cant scraper; plumber's scraper; three-cornered scraper; triangular scraper
三角关系 trigonometric(al) relation
三角冠 tricuspid cap
三角光线追迹 trigonometric(al) ray tracing
三角光线追迹法 trigonometric(al) ray-trace method
三角规 plane triangle; triangle; triangular compasses
三角轨 arris rail
三角函数 circular function; trig function; trigonometric(al) function
三角函数表 table of natural; table of trigonometric(al) functions; trigonometric(al) table
三角函数的真数 natural trigonometric(al) function
三角函数的正交性 orthogonality of trigonometric(al) function

三角函数对数表 logarithmic trigonometric(al) function
三角函数曲线 trigonometric(al) curve
三角函数预测法 trigonometric(al) function forecast method
三角函数装置 trigonometric(al) device
三角焊缝 angle fillet
三角焊接拱 triangulated welded arch
三角航线 triangular routes
三角航线飞行 triangular flight
三角和 trigonometric(al) sum
三角恒等式 trigonometric(al) identify
三角桁架 English truss; triangular truss; triangulated roof truss
三角桁架桥 triangular truss bridge
三角弧测量 arc triangulation
三角花键 serration
三角滑块 cam block
三角化形式 triangularized form; trigonometric(al) function
三角化直接法 direct method of triangularization
三角划分法 triangular division method
三角环 clew iron; lashing triangle; span shackle
三角簧 triangular spring
三角回车线【铁】 triangular track; wye track
三角汇兑 three-cornered exchange
三角或导线网平差 adjustment of network
三角或弧形拱饰 pediment arch
三角基 triangular basis
三角基座 triangular base
三角畸变 triangular distortion
三角级数 trigonometric(al) series
三角计 trigonometer
三角岬 cuspate foreland
三角岬海岸 cuspate foreland coast
三角岬洲 cuspate foreland bar
三角架 brandreth; shear legs; trevet; triangular frame; trigonal frame; tripod
三角架顶 plate of tripod
三角架起重机 angle crane
三角架头 head of tripod
三角架屋顶 couple-close roof
三角架中间腿 gin popper
三角尖锉 knife gin saw file
三角交叉 triangular junction
三角胶带 bead apex(core)
三角胶带传动 V-belt drive
三角胶带传动装置 angular belting
三角接 triangular grafting
三角接法电流 delta current
三角接线电压 delta voltage
三角接线法 delta connection; triangle connection
三角结合方案 triangular association scheme
三角截面导体 triangular segmental conductor
三角金属丝 triangular wire
三角金属线 triangular wire
三角金字塔型填料 triangular pyramid-type packing
三角近似法 trigonometric(al) approximation
三角晶的 rhombohedral
三角晶系 trigonal crystal system
三角精轧孔型 triangular finishing pass
三角鸠尾连锁式线脚 dovetail mo(u)lding
三角矩 triangle moment
三角矩阵【数】 triangular matrix
三角均值差 circular mean difference
三角均值离差 circular mean deviation
三角开挖法 triangle excavating process
三角刻度尺 trigonometric(al) scale
三角坑 twist; warp; track twist; twist of track【铁】
三角孔 delthyrium
三角孔钢丝网 triangle mesh wire fabric
三角孔格构桁架 triangular lattice truss
三角孔桁架 diagonal member truss; triangular truss
三角孔梁 triangulated girder; Warren girder; zigzag girder
三角孔网 triangular mesh
三角孔屋架 triangulated roof truss
三角控制 triangulation control; trigonometric(al) control
三角控制点 triangulation control point; trigonometric (al) control point; trigonometric(al) fixed point
三角控制网 triangulation control network; trigonometric (al) frame (work)
三角框架堤 triangular frame dike [dyke]
三角栏杆 arris rail
三角浪 chopping sea; choppy sea; cockling sea; cross wave; intersecting wave; lop; lumpy sea; pyramidal wave; short sea; sugar-loaf sea
三角棱镜 triangular prism; trihedral prism
三角连杆 triangular coupling rod
三角连接法线电压 delta voltage; ring voltage
三角联轨站 triangular junction
三角联结 triangular configuration
三角链 chain of triangles; triangular chain
三角梁 fronto(o)n
三角量水堰 triangular weir
三角履带板 triangle-section crawler shoe
三角螺纹 angular thread; triangular thread; vee[V] thread
三角贸易 delta trade; merchanting trade; three-cornered trade; triangle trade; triangular trade
三角楣窗 Saxon window
三角门楣 fronto(o)n
三角面 triangular facet
三角面壳 triangular shell
三角面壳体结构 triangular shell structure
三角面砖 triangle tile
三角木 skid; trig; sprag <止轮用>
三角木块 angle block
三角木条 arris rail
三角内插法 trigonometric(al) interpolation
三角皮带 cogged belt; texrope; texrope belt; trapezoidal strap; triangle belt; triangular belt; V-belt; wedge-shaped belt; cone belt
三角皮带传动 texrope drive; V-belt drive
三角皮带轮 grooved pulley; pulley conical disk; triangular leather belt wheel; V-belt pulley; vee-grooved pulley; vee pulley; V sheave
三角皮带轮传动装置 grooved gearing
三角皮带轮带 V-groove pulley
三角皮带驱动 V-belt drive; wedge-shaped belt drive
三角皮带输送器 V-belt conveyer [conveyor]
三角皮带张力不足 insufficient V-belt tension
三角皮带折断 V-belt rupture
三角片 triangular plate
三角平板 triangular slab
三角瓶 triangular flask
三角剖分 triangulation
三角剖分问题 triangulation problem
三角旗 pendent; pennant; triangular flag
三角旗杆 pennant staff
三角企口接合 triangular tongue and groove joint
三角起重机吊索 gin strap
三角起重架 gin; gyn; triangle hoisting gallows
三角嵌条 arris rail
三角墙 gable
三角墙尖顶 <装饰用的> cusp
三角穹圆顶 pendentive dome
三角区 gore; triangular space
三角区划法 triangular division method
三角曲线 trigonometric(al) curve
三角日晷 trigon
三角软帆 leg of mutton
三角三八面体 trisoctahedron
三角三边测量 triangulateration
三角三角形接法 delta-delta connection
三角三四面体 triakistetrahedron; tristetrahedron
三角沙槛 cuspate barrier
三角沙洲 cuspate bar
三角沙嘴 cuspate bar
三角筛法 conic(al) sieve method
三角山墙楣饰 aetoma
三角山墙檐饰 aetos
三角山嘴 facetted spur
三角烧瓶 Erlenmeyer flask
三角舌槽接合 triangular rabbet joint
三角设计 triangular design
三角十二面体 trigondodecahedron
三角石板 triangular slab
三角式 trigonometric(al) expression
三角式吊杆 three-leg crane
三角视差 trigonometric(al) parallax
三角试块 chill block
三角试片 test wedge
三角试样 wedge test piece
三角饰顶点 <门窗的> fastigium
三角室 trigonulum
三角术 trigonometry
三角双锥 trigonal bipyramid
三角水准测量 trigonometric(al) level(l)ing
三角榫槽接合 triangular rabbet joint
三角榫槽接头 triangular rabbet joint; triangular tongue and groove joint
三角榫接 triangular tongue and groove joint
三角锁【测】 chain of triangles; chain of triangulation; triangulation chain; trigonometric(al) chain
三角锁条 triangulated strip; triangulation strip
三角锁网 network of chains
三角锁网平差 adjustment of network of triangulation chains
三角锁网起始边 initial side of triangulation
三角套汇 indirect arbitrage; three-point arbitrage; three-point arbitrage in foreign exchange
三角套筒 cam sleeve
三角填石木笼丁坝 triangular crib groin(e); triangular crib groyn(e)
三角条 cant strip; sprig; glazier's point <镶玻璃用>
三角调节板 three-pin clevis link
三角调制 delta modulation
三角铁 angle bar; angle iron; knee plate
三角听音测距声呐 triangulation-listening-ranging sonar
三角凸凹榫接合 triangular rabbet joint
三角凸凹榫接头 triangular rabbet joint
三角凸轮 triangular cam
三角图 axonometric(al) projection; triangular chart; triangular diagram
三角图解 triangular diagram
三角土壤分类图 triangular soil classification chart
三角湾 estuary; mere; negative delta
三角网【测】 grid; net of triangulation; triangle mesh; triangular grid; triangular mesh; triangular network; triangulation(net); triangulation net(work)
三角网闭合差 triangle closure
三角网篦 triangular mesh
三角网布设略图 base map; basic map; triangulation diagram
三角网测量 area triangulation; triangulation survey
三角网测站 triangulation (survey) station; trigonometric(al) station
三角网解算程序 triangulation program(me)
三角网链 triangular net chain
三角网络模型 triangular network model
三角网平差 adjustment of triangulation; net adjustment; triangulation adjustment
三角网强度 strength of triangulation
三角网图形 triangulation figure
三角网系 chain of triangles; chain of triangulation; triangulation chain
三角屋顶 column clamp; comb roof; gable roof
三角屋架 collar roof
三角洗面器 corner lavatory
三角洗手盆 corner lavatory
三角系 triangulation system
三角系点阵 trigonal lattice
三角系晶体 trigonal crystal
三角系统 triangular system; trigonometric(al) system
三角匣形梁 triangular box section beam
三角线【铁】 triangle track; turnaround wye; Y-track
三角箱 cam box
三角箱形梁 triangular box section beam
三角橡胶带 rubber cone belt
三角小旗 pennant
三角斜帆 thimble-headed trysail
三角芯股芯丝 triangle core wire
三角芯光纤 triangular-cored optic(al) fiber
三角星形接线变换 delta-star transformation
三角星形接线法 delta-star connection; mesh-star connection
三角星形连接法 mesh-star connection
三角形 triangle; trigon; trilateral
三角形 D 盒 triangular dee
三角形坝体断面 triangular dam profile
三角形板 triangular panel
三角形爆破 <地震勘探> triangle shooting
三角形闭合 triangular misclosure; triangle closure
三角形闭合(误)差 closure error of triangle; error of closure of triangle; triangle closing error; triangle error of closure; triangular error
三角形边沟 triangular side ditch
三角形边饰 triangular trim

三角形表 triangle table
三角形波 triangular wave
三角形波发生器 triangle generator; triangular-wave oscillator
三角形波痕 cuspate ripple mark
三角形不规则网络 triangulated irregular network
三角形部件 triangular element
三角形槽 vee gutter
三角形槽接合 triangular rabbet joint
三角形槽口 triangular notch
三角形槽口量水堰 triangular notch weir; triangular weir; V-notch weir
三角形测量 wye-track survey
三角形测算坝 triangular measuring weir
三角形岔道 triangular slip road
三角形觇标 triangle target
三角形铲刀 V-shaped
三角形沉箱 triangle caisson; triangular caisson
三角形(齿)花键轴 serration shaft
三角形冲积平原 triangular alluvial plain
三角形冲积扇 fan delta
三角形橱 corner cupboard
三角形瓷砖 triangular tile
三角形磁场 delta field
三角形锉刀 angle file
三角形搭接 delta matching
三角形大梁 triangular girder
三角形单锁 single triangulation chain
三角形单元 triangular element
三角形导轨 triangular guide
三角形的 deltoid(al); lambdoidal; leg of mutton; triangular; triquetrous; vee; trigonal; V-type <如边沟等>
三角形的底 base of a triangle
三角形的顶点 vertex of a triangle
三角形的高度线【数】 altitude of a triangle
三角形的内切圆 inscribed circle of a triangle
三角形的旁切圆 escribed circle of a triangle
三角形的旁心 escenter of a triangle
三角形的外心 circumcenter of a triangle
三角形底层平面 triangular ground plan
三角形电流 delta current
三角形电路 delta network
三角形垫板 triangular plate
三角形吊孔 triangular lifting eye
三角形叠加法 method of triangular superposition
三角形顶点 vertex of triangle
三角形断面槽 chamfered groove
三角形断面的重力坝 gravity dam of triangular section
三角形断面料堆 coned tent pile
三角形堆垛层错 triangular stacking fault
三角形多束天线 multiwire-triatic antenna
三角形垛式支护 triangular pigsty
三角形法 triangular method; triangulation method
三角形反射器 triangular reflector
三角形方程组 triangular system
三角形方向操纵杆 triangular steering control arm
三角形飞机 delta-winged plane
三角形分布 triangular element distribution
三角形分布荷载 triangular distribution load(ing); triangular load
三角形分布模型 triangular distribution model
三角形分布全荷载 triangular distribution full loading
三角形分布噪声 triangle noise
三角形分道点标志 gore sign
三角形分割绘制法 triangular divisor
三角形分解 triangular decomposition
三角形分水闸 V-notch turnout
三角形风道 triangular duct
三角形风管 triangular duct
三角形风嘴 triangular edge fairing
三角形腹板 triangular web
三角形腹杆 triangular web
三角形盖板 angle fillet
三角形杆 triangular section bar
三角形钢丝网 triangle mesh wire fabric
三角形钢柱 triangular steel post
三角形格构 triangular framing; triangulated lattice
三角形格构建筑 triangular latticed construction
三角形公理 triangle axiom
三角形拱 miter arch
三角形拱窗 triangular arched window
三角形拱饰 pediment arch
三角形拱支承的挑出层 triangular arched corbel-table
三角形构架 spandrel frame[farming]; triangular frame[framing]
三角形构件 triangular unit
三角形股钢索 triangle strand construction rope
三角形固定百叶窗 <设于屋脊处的> gambrel vent
三角形光栅 triangular grating
三角形轨道 wrought track
三角形过程线 triangular hydrograph
三角形函数发生器 triangular function generator
三角形荷载分布 triangle load distribution; triangular load distribution
三角形桁架 triangular truss; triangulated truss
三角形桁架空腹桁架 Vierendeel truss
三角形桁架龙门起重机 triangular truss gantry crane
三角形桁架桥 triangular truss bridge
三角形化 triangularization
三角形灰板条 triangular wood lath(ing)
三角形基 triangular basis
三角形激光器 triangular laser
三角形(建筑)部分【建】 gable
三角形降落伞 triangular parachute
三角形交叉(口) crossing at triangle; triangle crossing
三角形礁 plug reef
三角形角超 excess of triangle
三角形角密封片 triangular corner seal
三角形角盈 excess of triangle
三角形铰接拱 mitre arch
三角形铰接结构 triangulated pin-jointed structure
三角形接端 delta connector
三角形接法 A-connection; delta connection; mesh connection
三角形接线 A-connection; delta connection
三角形结构 triangular structure
三角形结线 delta connection
三角形截面 triangular section; trilateral cross-section; triple-box-section
三角形截面踏步 spandrel step
三角形解法 solution by a triangle; triangle computations
三角形解算 solution of triangle
三角形斤斗 triangular loop
三角形勘探网 triangular exploration grid
三角形空腹梁 triangular hollow-web beam
三角形块 gore
三角形块段法 triangular oreblock method
三角形矿柱 triangular stump
三角形框架 triangular framing; spandrel frame
三角形拉撑 triangular bracing
三角形蜡孔 tricerores
三角形立体交叉 delta interchange
三角形例行程序 triangulation routine
三角形连接 delta connection
三角形连接的 delta connected
三角形连接电动机 delta connected motor
三角形连接电动势 delta electromotive force
三角形连接法 delta connection
三角形联结 triangle junction
三角形联络线 Y-connection
三角形链 triplet
三角形梁 triangular beam
三角形量水堰 triangular weir
三角形路拱 triangular road camber
三角形轮廓 triangular profile
三角形螺纹 angular thread
三角形螺旋 V-thread screw
三角形螺旋撑 dog stay
三角形脉冲 triangular pulse
三角形脉冲发生器 triangle impulse generator
三角形脉冲函数 delta impulse function
三角形楣饰 fronto(o)n; pediment
三角形门廊 triangular porch
三角形密封垫 delta ring sealing gasket
三角形密封圈 delta-ring
三角形面砖 triangular tile
三角形模型 triangular shape
三角形木架 abates
三角形木架透水坝 abat(t)is
三角形木笼丁坝 triangular crib groin(e); triangular crib groyn(e)
三角形内浇道 triangular gate
三角形捏合块 triangular kneading block
三角形排列 in-line triangular pitch
三角形旁切圆心 escenter of a triangle
三角形皮带无级变速传动装置 adjustable V-belt drive
三角形匹配天线 delta matching antenna
三角形平均闭合差 <三角网的> mean triangle misclosure
三角形剖分图 triangulated graph
三角形剖面堰 triangular-profile weir
三角形普拉特桁架 triangular Pratt truss
三角形普拉特(式)屋架 triangular Pratt truss
三角形企口接合 birdsmouth joint
三角形嵌条 angle fillet; angle stile; triangular fillet
三角形切口接合 triangular rabbet joint
三角形穹隆 triangular dome
三角形缺口堰 triangular notch weir; V-notch weir
三角形缺陷 triangle defect
三角形绕组 delta winding
三角形人字墙 triangular pediment
三角形溶蚀坑 triangle solution concavity
三角形入口处 triangular porch
三角形沙坝 cuspate bar
三角形砂粒 pennant grit
三角形筛眼 triangular mesh
三角形山墙 triangular pediment
三角形栅格 triangular lattice
三角形式 triangular form
三角形饰 triquetra
三角形受压区 triangular compression zone
三角形枢纽 triangle type junction terminal
三角形数 triangular number
三角形速度分布 triangular velocity distribution
三角形算子代数 triangular operator algebra
三角形塔柱 delta tower
三角形踏步 spandrel step; triangular section step; turret step
三角形台站 triangular array station
三角形台阵 triangular array
三角形弹性板 triangular resilient plate
三角形弹性楔 triangular elastic wedge
三角形掏槽 triangular cut
三角形体系 triangular system
三角形天窗 shed dormer
三角形天线 delta(-type) antenna; triangle antenna
三角形填石木笼丁坝 triangular crib groin(e)
三角形填石木笼护岸 hen-cooping
三角形填实部分 <楼梯外斜梁以下的> spandrel
三角形条件方程 triangle condition equation
三角形统计图表 triangular graph
三角形凸点密封技术 triangular asperities technique
三角形突堡 triangular bastion
三角形图表 triangular chart; triangular diagram
三角形图解 triangular diagram
三角形图形 cusped profile
三角形图形强度 triangle strength
三角形图样 triangular diagram
三角形外形 triangular profile; triangular shape
三角形网格 triangle grid; triangular mesh; triangular net(work)
三角形网格桁架 truss with triangular web opening
三角形网(络) delta network; trigonal network; triangular network
三角形网络钢丝网 triangle mesh fabric
三角形网眼 triangle mesh
三角形屋架 triangular(roof) truss
三角形物 delta
三角形系统 triangular system
三角形纤维长度分布 triangular fibre-length distribution
三角形相图 triangular phase diagram
三角形相位图 triangular phase plot
三角形镶嵌物 trigonum
三角形小窗 <窗花格中开的> angel light
三角形星形接法 delta-star connection
三角形檐饰 pointed pediment
三角形檐饰悬挑 geison
三角形堰 V-notch weir
三角形样板 <由板条组成的> gun template
三角形翼板 <酒杯状翼板> wine-glass flange
三角形应力分布 triangular stress distribution
三角形圆屋顶 triangular
三角形运动 triangular motion
三角形栽培 triangular planting
三角形栽植 triangular spacing
三角形闸门 triangular-type gate; delta gate
三角形阵列 triangular array
三角形支撑 triangular tie

三角形支撑体系 triangular support system
三角形支承棒 triangle bar
三角形支持法 triangulation supported method
三角形支架法 triangular support system
三角形植树 planting in triangle;triangle-planting;triangular planting
三角形中线 median of a triangle;medium line of triangle
三角形柱 triangular column
三角形铸铁端片 triangular cast-iron end piece
三角形装潢 triangular decoration
三角形装饰 triangular cupola;triangular trim
三角形状 triangular shape
三角形撞击坑 triangle impact concavities
三角形组成的 triangulate
三角型 triangular form
三角性质 triangle property
三角碹 angle-shaped arch
三角学 trigonometry
三角学的 trigonometric(al)
三角学家 trigonometer
三角岩板 triangular slab
三角岩芯钻头 tricone rock bit
三角眼板 eye plate;monkey face;shamrock plate;triangular eye plate
三角眼铁 shamrock plate;triangular eye plate;eye plate;monkey face
三角堰 triangular(notch) weir;V-notch weir
三角叶杨 cotton wood
三角仪 triangulator
三角异形 triangle profile
三角油槽 triangular oil groove
三角油石 three-square oil stone;triangular oil stone
三角余函数 trigonometric(al) cofunction
三角元 triangular element
三角凿 parting tool
三角凿刀 burr;V-tool
三角凿嵌缝 V-tooled joint
三角轧头 triple clamp
三角轧头螺钉 triple-clamp bolt
三角闸门 wedge gate
三角闸门液压启闭机械 hydraulic machinery of wedge roller gate
三角债 chain debt;cross default debt;debt chain;tripartite debts
三角折线图 trilinear chart
三角阵 triangular matrix
三角支撑物 tripod
三角支架 A-frame
三角值 tristimulus value
三角制动梁 brake triangle
三角置换求积分法 integration by trigonometric substitution
三角洲 branch island;cuspate delta;cuspate foreland;delta;delta river mouth;tidal delta <潮流形成的>
三角洲坝 delta bar;delta dam
三角洲边缘平原 delta-marginal plain
三角洲冰碛 delta moraine
三角洲草本相 deltaic herbaceous facies
三角洲测量 delta measurement;delta survey
三角洲层理 delta(ic) bedding
三角洲层序 delta sequence
三角洲汊河 delta arm
三角洲沉积模式 deltaic sedimentation model
三角洲沉积土 delta deposit(e)
三角洲沉积(物) delta(ic) deposit(e);delta(ic) sediment
三角洲沉积削面结构 profile of sedimentary structure of delta
三角洲冲积扇 delta(ic) fan;deltoidal fan
三角洲冲积锥 delta(ic) cone
三角洲的 deltaic;deltoid(al)
三角洲地槽 delta geosyncline
三角洲地貌单元 geomorphologic(al) unit of delta
三角洲地区 delta area;delta(ic) region;delta(ic) tract
三角洲顶部 delta top;intradelta
三角洲顶点位置 position of deltaic apex
三角洲段 deltaic tract
三角洲堆积土 delta accumulation soil
三角洲朵体 delta lobe
三角洲发育 delta development
三角洲分港汊 arm of delta
三角洲海岸 delta coast;river deposition coast
三角洲河汊 arm of delta;delta branch;delta channel;delta river branch;deltoid branch
三角洲河口 delta(ic) river mouth
三角洲河口整治 delta estuary regulation
三角洲湖 delta lake
三角洲环境 deltaic environment
三角洲间的 interdeltaic
三角洲间湾沉积 interdelta bay deposit
三角洲交错层 deltaic cross bedding
三角洲阶地 delta terrace
三角洲靠内陆部分 delta top
三角洲类型 type of delta
三角洲路堤 deltaic embankment
三角洲名称 name of delta
三角洲模式 delta(ic) model
三角洲平原 delta(ic) plain
三角洲平原沉积 deltaic plain deposit
三角洲平原相 delta plain facies
三角洲平原沼泽地 delta plain swamp
三角洲前沉积 prodelta
三角洲前坡 foreset
三角洲前坡河床 foreset bed
三角洲前缘 delta front;delta point
三角洲前缘沉积 delta front deposit
三角洲前缘水下沉积黏[粘]性土 prodelta clay
三角洲前缘席状砂 delta front sheet sand
三角洲前缘席状砂沉积 sheet sand deposit of delta front
三角洲前缘相 delta front facies
三角洲前缘斜坡 slope of front delta
三角洲前缘斜坡沉积 delta front shape deposit
三角洲前趾 digitate margin of delta
三角洲砂体圈闭 delta sand trap
三角洲上部平原 upper delta plain
三角洲水道 delta channel
三角洲水上部分面积 subaerial area of delta
三角洲水下部分面积 subaqueous area of delta
三角洲体系 delta regime
三角洲天然堤沉积 natural levee deposit of delta
三角洲填土 delta fill
三角洲洼地公寓 delta flat
三角洲洼地平地 delta flat
三角洲席状砂圈闭 deltaic-sheet trap
三角洲相 delta facies
三角洲相序 delta facies sequence
三角洲形成(过程) delta building;deltafication;delta formation;deltation
三角洲形成作用 deltafication
三角洲形态 shape of delta
三角洲型河口 deltaic estuary
三角洲型盆地 delta basin
三角洲序列 delta succession
三角洲旋回 delta cycle
三角洲学 deltalogy[deltology]
三角洲样的 deltoid(al)
三角洲淤长 alluviation of delta;delta growth;upbuilding of delta
三角洲淤积 alluviation of delta
三角洲淤积物 deltaic deposit
三角洲增长 delta progradation;upbuilding of delta
三角洲沼泽沉积 delta marsh deposit
三角洲整治计划 delta project
三角洲支汊 delta arm;delta branch;deltoid branch
三角洲主汊(河) main delta arm
三角洲主支股 main delta arm
三角柱架 tripod derrick
三角柱镜 triple spiegel
三角柱面体支管 T-Pee lateral
三角柱体 <混凝土块用于防波堤> tribar
三角转子 <转子发动机的> three-apexed rotor
三角转子发动机 rotary polygonal piston engine;Wankel engine
三角状 cuspate
三角状岬 cuspate foreland
三角锥 broach;pyrometric cone
三角锥等值 pyrometer cone equivalent;pyrometric cone equivalent
三角桌 knee table;angular table;angle table <工具机上的>
三角子程序【计】 trigonometric(al) subroutine
三角钻钩 triplex drilling hook
三角钻架 drill tripod
三角钻头 tricorn bit
三角坐标 trigonometric(al) coordinates;trilinear coordinates
三角坐标分类图 <土的> triangular classification chart; triangular chart;trilinear chart
三角坐标分类系统 triangle coordinate classification system
三角坐标图 triaxial chart;triaxial diagram;trilinear chart;Ferett triangle <土的分类>
三角坐标图法 triangular diagram
三角坐标土壤分类法 triangle classification of soils
三角坐标纸 triangular graph paper
三脚矮橡皮轮车 tripod dolly
三脚插塞 three-point plug
三脚插头 three-pin plug;three-prong plug;tripod
三脚的 three-pinned
三脚凳 tripod
三脚电杆 triple mast
三脚吊滑轮 tripod sheave wheel
三脚吊货架 gin;hoisting gin
三脚吊架 shear legs
三脚顶升机具 three-leg jack-up rig
三脚碇泊浮筒 three-arm mooring;three-leg mooring
三脚放大镜 tripod magnifier
三脚规 three-legged dividers;triangular compasses
三脚夯锤 three-legged rammer;three-legged tamper
三脚桁架拱 trussed arch with three-hinge
三脚回填夯锤 three-legged backfill tamper
三脚混凝土预制块 tribar
三脚机架结构 tripod leg construction
三脚基座 tribranch
三脚几 teapoy
三脚架 tripod(legs);tripod mast;tripod mount(ing);tripod rest;three-way support; tribranch; triple stand;trivet;crowfoot[复 crow-feet];lampstand
三脚架 U 形环 tripod clevis
三脚架安装的 tripod-mounted
三脚架拔桩机 tripod puller
三脚架插孔 tripod receptacle
三脚架穿钉 tripod bolt
三脚架的节点 joint of trigonal frame
三脚架的腿 tripod legs
三脚架底座 tripod socket
三脚架吊杆 <俚语> monkey pole
三脚架回转式钻机 tripod rotary drill
三脚架基座 tripod base;three-arm base;three-screw base;level(l)ing base
三脚架码垛机 tripod stack
三脚架摄影机 camera on tripod
三脚架式垛草机 tripod loader
三脚架式凿岩机 tri-point rock drill
三脚架式植树挂淤 tripod planting
三脚架式钻机 tripod drill
三脚架式钻岩机 tri-point rock drill
三脚架适配器 tripod adaptor
三脚架调节器 tripod regulator
三脚架铁尖 tripod shoe
三脚架头 heap of tripod;level(l)ing head;tripod head
三脚架(桅杆)起重机 tripod derrick
三脚架屋顶 close-couple roof
三脚架下沉 settlement of tripod
三脚架信号 tripod signal
三脚井架 three-pole derrick
三脚井塔 three-pole derrick
三脚螺栓 tripod bolt
三脚起落架 tripod undercarriage
三脚起重机 gin;tripod crane;tripod derrick;tripod sheer
三脚起重机吊索 gin sling;gin strap
三脚起重架 triangle gin
三脚起重架的杆 gin pole
三脚起重爪 three-legged lewis
三脚千斤顶 tripod jack
三脚式起重机 shear-leg crane;shear-leg derrick
三脚式植树挂淤法 tripod planting
三脚式重力擒纵机构 three-legged gravity escapement
三脚台 tribranch;tripod
三脚体 sheer legs
三脚天线脚 triple mast
三脚铁架 <搁在火上的> trivet
三脚桅杆 tripod derrick;tripod mast
三脚桅座 sheer legs;sheers
三脚压路机 three-legged roller
三脚照相架 camera tripod
三脚支撑 trishore
三脚支架 three-legged support
三脚支架吊车 tripod derrick
三脚支架式凿岩机 tripod drill
三脚支架式钻车 tripod drill;tri-point rock drill
三脚桌 tripod table;trivet table
三脚钻机 tripod drill;tri-point rock drill
三脚钻机架 tripod drill mounting
三脚钻架 drill tripod;three-legged derrick
三脚钻塔 drill tripod;three-legged derrick
三脚座 trivet
三铰半框架 three-hinged half-frame;three-pinned half-frame; triple-hinged half frame
三铰的 three-hinged
三铰腹板拱梁 plate-webbed arched girder with three hinges
三铰刚架 three-hinged(rigid) frame;three-pin rigid frame

S

三铰钢格构刚架 three-pin steel lattice rigid frame
三铰钢拱 three-hinged steel arch; three-pinned steel arch; triple-hinged steel arch
三铰拱 arch with three-articulations; three-element arch; three-hinged arch; three-lobed arch; three-pinned arch; triple articulation arch; triple-hinged arch
三铰拱坝 three-hinged arch dam
三铰拱大梁 three-hinged arched girder; three-pinned arched girder
三铰拱桁架 three-hinged arch truss
三铰拱肋穹隆 three-hinged arch-ribbed dome; triple-hinged arch-ribbed dome
三铰拱桥 three-hinged arch bridge; three-pinned arch bridge
三铰构架 three-pin(ned) frame
三铰构架式桁架 barn truss
三铰桁架拱大梁 three-hinged trussed arched girder
三铰桁架拱梁 triple-hinged trussed arched girder
三铰桁架构架 three-hinged trussed frame
三铰桁架支撑的框架 three-pinned truss(ed) frame
三铰加劲桁架 three-hinged stiffening truss
三铰接的 triple-hinged
三铰接拱梁 triple-hinged arched girder
三铰接框架组成的多层结构系统 multifloor system composed of three hinge(d) frame
三铰接链 three-hinged braced chain
三铰矩形构架 three-hinged rectangular frame; triple-hinged rectangular frame
三铰矩形框架 three-pinned rectangular frame; triple-hinged rectangular frame
三铰框架 frame with three hinges; three-pinned frame
三铰链 triple-hinged
三铰门架 triple-hinged portal frame
三铰体系 three-hinged system
三铰屋顶 three-hinged roof; three-pinned roof; triple-hinged roof
三阶段工艺 three-stage process
三阶段设计 three-phase design; three-stage design; three-step design
三阶段隧道盾构 three-stage jumbo
三阶段预应力制管工艺 wire-wound concrete core-pipe process
三阶光混频 third-order light mixing
三阶互调失真【无】third-order intermodulation distortion
三阶目的 third-order goal
三阶通路 three-level channel
三阶相关 third-order correlation
三阶原理 third-order theory
三节的 three-pinned; triarticular; triarticulate; triplex
三节点湖震 trinodal seiche
三节点静震 trinodal seiche
三节拱 triple articulation arch
三节环 three-membered ring
三节假潮 trinodal seiche
三节间桁架 three-panel truss
三节滤波器 triple-section filter
三节排架 three-section framing bent
三节砂箱 three-part box
三节伸缩吊杆 triple telescopic(al) boom
三节式复涨机车 triplex compound locomotive
三节式可延伸梯 triple extension ladder
三节式门架 triplex mast
三节托辊 tripartite idler
三节线圈 three-section coil
三结点四次线 trinodal quartic
三结合设计小组 "three in one" design group
三结节的 tritubercular
三截 trisect
三截窗 three-light window
三解形节距 triangular-pitch
三金属 trimetal
三金属催化剂 trimetallic catalyst
三金属带 trimetallic strip
三金属轴瓦 plain tri-metal bearing
三筋履带板 triple grouser
三筋式履带板 triple-grouser pad
三进路房柱式采矿法 triple-entry room-and-pillar mining
三进位数系 ternary number system
三进展开 ternary expansion
三进制 ternary notation
三进制标度 ternary scale
三进制表示法 ternary representation
三进制乘法 ternary multiplication
三进制程序 ternary sequence
三进制(存储)单元 ternary cell
三进制代码 ternary code
三进制的【计】ternary
三进制计数器 ternary counter
三进制记数法 ternary notation
三进制加法 ternary addition
三进制逻辑 ternary logic
三进制逻辑电路 ternary logic circuit
三进制逻辑元件 ternary logic element
三进制码 ternary code
三进制全加器 ternary full adder
三进制设备 ternary device
三进制树 ternary tree
三进制数 ternary number
三进制数位 trit
三进制数系 ternary number system
三进制运算 ternary arithmetic
三进制增量表示法 ternary incremental representation
三晶 tricrystal; trimorphism
三晶片矿物 three-sheet mineral
三晶体电子干涉仪 three-crystal electron interferometer
三晶形物 triamorph
三晶型金属 trimorphous metal
三晶衍射仪 triple-crystal diffractometer
三颈瓶 three-necked bottle
三颈烧瓶 three-necked flask
三净租契 <财产税、保险费和维修费均由租户承担> triple net lease
三镜航空照相机 trimetrogon camera
三镜航摄机 trimetrogon
三镜航摄仪摄影测图法 tri-camera method
三镜摄影测图法 trilateration method
三镜式三角形环形共振腔 triangular three-mirror ring cavity
三镜筒观测镜 trinitron
三镜头 three-lens
三镜头航空摄影 tri-met
三镜头航空摄影测量学 trimetrogon photogrammetry
三镜头航空摄影测图 trimetrogon mapping
三镜头航空摄影机 trimetrogon camera
三镜头航摄机摄影法 trimetrogon method
三镜头航摄机摄影术 trimetrogon photography
三镜头航摄仪 trimetrogon
三镜头航摄照片 trimetrogon photograph
三镜头航摄照片组 trimetrogon set
三镜头空中照相 trimetrogon photograph
三镜头摄影机 three(-lens) camera; triplex camera
三镜头摄影像片 trimetrogon photograph; trimetrogon set
三镜头摄影仪 trimetrogon
三镜头照相机 triplex camera
三镜系统 three-mirror system
三镜照相 trimetrogon photography
三镜照相制图 trimetrogon charting
三居室的楼房 triplex building
三局双工传输制 three-station duplex transmission system
三聚 trimerization
三聚苯 trimeric benzene
三聚丙烯 tripropylene
三聚磷酸钠 sodium tripolyphosphate; three-polyphosphoric sodium; trimeric sodium phosphate
三聚磷酸盐 tripolyphosphate
三聚硫氰酸 trithiocyanuric acid
三聚硫酮 trithioacetone
三聚氢酰胺甲醛树脂 melamine formaldehyde
三聚氢酰胺树脂面层的刨花板 melamine faced chipboard
三聚氰胺 melocol; tripolycyanamide
三聚氰胺表面处理装饰层压板 melamine-surfaced decorative laminate
三聚氰胺表面涂层 melamine-surfaced
三聚氰胺层压板 melamine laminate
三聚氰胺醇酸树脂胶粘剂 melamine alkyd resin adhesive
三聚氰胺醇酸树脂涂料 melamine alkyd resin paint
三聚氰胺甲醛 melamino-formaldehyde
三聚氰胺甲醛胶 melamine-formaldehyde glue
三聚氰胺甲醛树脂 melamine-formaldehyde resin; melamine resin; ultrapas
三聚氰胺甲醛塑料 melamine-formaldehyde plastics
三聚氰胺甲醛纸 melamine-formaldehyde paper
三聚氰胺泡沫 melamine foam
三聚氰胺清漆 melamine resin varnish
三聚氰胺清漆树脂 melamine varnish resin
三聚氰胺树脂 melamine resin
三聚氰胺树脂胶 melamine resin adhesive
三聚氰胺树脂胶粘剂 melamine glue
三聚氰胺树脂黏[粘]合剂 melamine resin adhesive
三聚氰胺树脂塑料 melamine resin plastics
三聚氰胺树脂塑性装潢层 melamine resin decorative plastic film
三聚氰胺塑料 melaminoplast
三聚氰胺塑料面板 melamine plastic board; melaminoplastic board
三聚氰胺贴面 melamine facing
三聚氰胺贴面的面板 melamine-faced panel
三聚氰胺贴面碎木胶合板 melamine-surfaced chipboard
三聚氰胺贴面碎木刨花板 melamine-surfaced chipboard
三聚氰胺装饰板 melamine decorative laminate
三聚氰酸 cyanuric acid; pyrolithic acid; triazine triol; tricyanic acid
三聚氰酸三乙酯 triethyl cyanurate
三聚氰酰胺 cyanuramide; melamine
三聚体 tripolymer
三聚物 trimer
三聚盐 triple salt
三聚乙硫醛 sulfoparaldehyde; trithioacetaldehyde
三聚乙醛 para-acetaldehyde; paraldehyde
三聚异丁烯 triisobutene [triisobutylene]
三聚作用 terpolymerization
三卷变压器 three-circuit transformer
三卷筒绞盘 triple-headed capstan
三卡瓦卡盘 three-jaw chuck
三开道岔 double turnout; double turnout junction; symmetric(al) three-throw turnout; three-throw (split) switch; three-throw turnout; three-way points; three-way switch; three-way turnout; double points <美>
三开的 three-way
三开对称道岔 double bilateral turnout
三开间的 three-bay
三开模 three-part pattern; three-piece pattern
三开日期 third spudding date
三开砂箱 three-part flask
三开箱铸型 three-cavity mould
三开芯盒 three-piece core box
三开转辙器 three-throw (split) switch; three-way switch
三空间变量体 triplet of three-spatial variable
三空位 trivacancy
三孔插座出线口 triplex receptacle outlet
三孔磁元件 three-hole element
三孔干涉仪 three-aperture interferometer
三孔管道 three-way duct
三孔光栏 three-hole aperture
三孔滑车 dead eye
三孔联板 three-pin clevis link; three-pin connection strap
三孔模型 triple-cavity mo(u)ld
三孔喷嘴 triple jet
三孔砌块 three-core block
三孔隙度组合法 tri-porosity method
三孔楔子 three-hole wedge
三孔烟道 battery of three-flues
三孔制 three-span system
三孔砖 three-core brick
三控制点 trigonometric(al) control point
三口扳手 triple end wrench
三跨度 three-span
三跨拱 triple articulation arch
三跨框架 three-span frame
三跨缆索吊桥 three-span cable supported bridge
三跨连续大梁 three-span continuous girder
三跨连续梁 three-span continuous beam
三跨连续桥 three-span continuous bridge
三跨梁 beam of three-spans; three-span beam
三跨桥梁 three-span bridge
三跨相连的拱 triple articulation arch
三块板道路 triple carriageway road
三块式风挡 three-piece windshield
三缆定位装置【疏】Christmas tree
三雷达台接收系统 ratran
三雷达站接收系统 ratran
三肋板 tri-slab
三肋单箱式 single triple box
三肋履带板 three-grouser track shoe
三肋双箱式 double triple box
三类目标 tertiary target
三类容器 third category vessel

三类压力容器 third-class pressure vessel
三类有机体 three-groups of organisms
三棱 trigone
三棱板 deltoid plate
三棱(比例)尺 triangular plotting scale
三棱草【植】bolbosehoenus maritimus palla
三棱尺 architect's scale; engineer's rule; engineer's scale; three-edged rule(r); three-square rule; three-square scale; triangle scale; triangular scale
三棱单刃锉 cadet
三棱刀 three-square tool
三棱的 deltoid(al); three-cornered; triangular
三棱刮刀 cant scraper; striking knife
三棱刮皮刀 striking pin
三棱镜 prism glass; triangular prism; triple prism
三棱镜盘 prismatic(al) disc[disk]
三棱镜摄谱仪 three-prism spectrograph
三棱螺丝 three-square screw
三棱石 triangular stone; dreikanter
三棱体 prism
三棱形 prism; prismatic(al)
三棱直尺 trihedral(tetrahedral) toolmaker's straight edge
三棱柱 triangular prism
三棱锥(体) triangular pyramid; triangle pyramid
三棱锥网架 triangular pyramid space grid
三离子物 triple ion
三力矩定理 Clapeyron's theorem; theorem of three moments
三力矩法 method of three-moments; three-moment method
三力矩方程式 three-moment equation
三粒级煤 trebles
三连齿轮 three-speed gear
三连窗 tripartite window; triple-casement window
三连岛沙洲 triple tombolo
三连都市(城市群) <由三个邻近市镇联合而成一个城市> tricity
三连拱 tripartite arch; triple arch
三连拱的 triple-arched
三连接头 joint among three-members
三连晶 threeling; trill(ing)
三连曲柄 three-throw crank
三连闪光灯 triple flashing light
三连十字架 triple cross
三连通曲线 triply connected curve
三联 trigeminy; trilogy
三联苯 terphenyl
三联苯甲基 tridiphenylmethyl
三联泵 three-through pump; treble pump; triple pump
三联城市 tricity
三联齿轮 triple gear
三联齿轮泵 three-section gear-type pump; triple gear pump
三联齿轮传动 triple geared drive
三联带锯 triple band-saw
三联单 bills in three-parts
三联的 triplex
三联地震台网 triparted seismic network
三联点 triple junction
三联发射装置 triple launcher
三联法 three-furnace process; triplex-process
三联分路传动箱 triple gear divider
三联管 thribble
三联管工作板 thribble board
三联合同 tripartite indenture
三联画 triptych
三联脚架法 <导线测量的> three-tripod traversing; method of three-tripods
三联铰 tri-hinges
三联结头 joint between three-members
三联开关 three-way switch
三联可变电容器 triple variable capacitor
三联立体像片 stereo-triplet
三联炼钢法 triplexing; triplex-process
三联(木材防腐)剂 Triolith
三联炮塔 three-gun turret; triple turret
三联起网机 triplex net winch
三联浅槽饰 triglyph
三联式泵 triplex pump
三联式活塞泵 triplet piston pump
三联式汽化器 triplex carburetor[carbureter]
三联式拖车方式 triples
三联式住宅 triplex house
三联胎 tridymus
三联台阵 triparted array
三联体 triplet
三联体假说 triplet hypothesis
三联体理论 triplet theory
三联体密码 triplet code
三联拖车 triple trailer
三联现象 trigeminy
三联(销)轴 triplex spindle
三联柱塞泵 three-throw plunger pump
三联装的 triple-mounted
三联装架 triple mounting
三敛子 bilimbing
三炼桐油 triple boiled tung oil
三链杆 three links
三链式输送机 triple chain conveyer[conveyor]
三梁天平 triple beam balance
三量程的 three-range
三量统计 three-quantities of reserves estimates
三料斗摊铺机 three-batcher paver
三列舱口 three-row hatch
三列车冲突 <两列相撞后,第三列撞入> double collision
三列的 triserial; tristichous
三列绕组 three-range winding; three-tier winding
三列套筒滚子链 triple strand roller chain
三列凸纹针织物 three-miss blister fabric
三列轴承 three-row bearing
三裂 ternary fission; trilobation
三裂的 three-cleft
三裂缝的 trilete
三裂痕 trilete scar
三裂口 trilete aperture
三裂片的 trilobated
三裂漆树 skunkbrush
三裂叶 trilobated leaf
三磷酸钠 sodium tripolyphosphate
三菱流态层分解炉悬浮预热器 Mitsubishi fluidized calcinator suspension preheater
三菱面体的 trirhombohedral
三菱形符号 <货物标志> three-diamond
三菱重工业株式会社 Mitsubishi Heavy Industries
三零一树脂胶 three-zero-one resinous adhesive
三流假说 three-drift hypothesis
三硫化二磷 phosphorus trifulfide
三硫化二砷 arsenic trifulfide; orpiment
三硫化二铊 thallic trifulfide
三硫化二锑 stibous trifulfide
三硫化物 trisulfide[trisulphide]
三硫磷 carbophenothion; trithion
三硫酸盐 trisulphate
三硫型水化硫铝酸钙 ettringite
三陇板 <陶立克柱式的特征之一> triglyph
三陇板间距 ditriglyph
三陇板檐壁 triglyph frieze
三楼 three-stor(e)y; third stor(e)y <美>
三楼后房 <英> second-pair back
三楼楼厅 amphitheater[amphithcatre]; upper circle
三楼前房 two-pair front; second-pair front <英>
三漏斗货车 triple hopper car
三漏斗漏斗车 triple hopper
三炉联炼法 three-furnace process
三卤化合物 terhalide
三滤光镜比色计 three-filter colo(u)rimetry
三滤料滤床 tri-medium bed
三路传动器 tee-drive unit
三路道岔 three-way switch
三路的 three-way
三路电键 three-way key
三路电缆地下管道 three-way cable subway
三路电色谱法 three-way electrochromatography
三路定位方法 three-way positioning method
三路阀 three-way valve
三路管 three-way connection(pipe); three-way piece; three-way pipe
三路管道 <敷设三根电缆> three-way duct
三路管道瓦筒 three-way conduit tile
三路交叉口 three-way intersection
三路接线盒 trifurcating box
三路开关 three-way switch
三路控制阀 three-way control valve
三路龙头 three-way tap
三路门拱 three-way arch
三路栓 three-way stop cock
三路通报制 multiplex triode system
三路同轴线变压器 triple coaxial transformer
三路相交的交叉 three-legs intersection
三路叶轮泵 three-channel impeller pump
三路载波 three channel carrier
三氯 trichloride
三氯苯 trichlorobenzene
三氯苯酚 trichlorophenol
三氯(代)乙烯 trichlorethylene
三氯氟甲烷 trichlorofluoromethane
三氯化铬 chromium trichloride
三氯化金 gold chloride
三氯化钛 titanium trichloride
三氯化锑 <堵塞岩石空隙用的化学药剂> antimony trichloride; antimony butter
三氯化物 trichloride
三氯甲硅烷 trichlorosilane
三氯甲烷 chloroform; trichloromethane
三氯氢硅 trichlorosilane
三氯硝基甲烷 chloropicrin
三氯乙醛 chloral; trichloroacetaldehyde
三氯乙酸废水 trichloroacetic acid wastewater
三氯乙烷【化】trichloroethane
三氯乙烯 trichloroethylene; triclene <商品名>
三轮 third wheel
三轮泵 three-wheel pump
三轮铲车 three-wheeled fork-lift truck
三轮车 cycle rickshaw; pedicab; three-wheeler; tricar; tricycle; trisha(w); eyclopousse <东南亚一带的脚踏或机动>
三轮车场 cycle stand
三轮车车道加边 cycle path edging
三轮车式钻机 three-wheeled wagon drill
三轮齿轴 third pinion
三轮出租汽车 cyclo; three-wheeled taxi
三轮的 three-wheeled
三轮多用滚压机 three-wheeled all-purpose roller
三轮钢滑车 three sheaves steel block
三轮钢筋弯曲机 three roll bending machine
三轮滑车 three-fold block; three-sheave block; trispast
三轮滑车组 three-sheave block
三轮货车 tri-truck
三轮机(车) tricycle
三轮机器脚踏车 tricar; tricycle
三轮脚踏车 trisha(w); tricycle
三轮截管器 three-wheel pipe cutter
三轮静力压路机 three-wheeled static roller
三轮卡车 auto-tricycle; three-wheeled truck; tri-truck; three-wheeler
三轮犁 three-wheel plow
三轮路碾 three-wheel roller
三轮摩托车 motor tricycle; tricar; tricycle
三轮起落架 three-wheel landing gear; tricycle gear; tripod landing gear
三轮汽车 three-wheeled automobile; three-wheeled motor vehicle; three-wheeled vehicle; tricar; tri-truck
三轮人力车 pedicab
三轮式(飞机)着陆架 tricycle landing gear
三轮式起落架 tricycle landing gear
三轮式装载机 three wheel loader
三轮双轴路碾 three-wheel two axle roller
三轮双轴压路机 three-wheel two axle roller
三轮送货车 three-wheeled delivery van
三轮拖拉机 single front wheel tractor; three-wheel tractor
三轮小车 three-wheeler
三轮小汽车 cyclecar
三轮型中耕拖拉机 tricycle-type row-crop tractor
三轮压路机 three drum roller; three-wheel roller
三轮研光机 three-bowl calendar
三轮蒸汽压路机 three-wheeled steam roller
三轮着陆 three-wheel landing
三罗拉式大牵伸装置 three-line high draft system
三螺杆式水泵 three-screw pump
三螺栓通风式潜水装置 three-bolt ventilation diving equipment
三螺旋基座【测】three-screw base
三螺旋桨 triple screw
三螺旋桨船 triple-screw ship
三马达纸带传运机构 three-motor tape-transport mechanism
三马战车雕饰 triga[复 triage]
三脉冲编码 triple pulse coding
三煤酚 tricresol
三门进入口正面 three-portal facade
三门峡水利枢纽工程 Sammenxia Key Water Control Projects
三门厢式送货车 three-way van
三米直尺 tree-meter ruler

S

三面刀盘 triplex cutter
三面导槽 prismatic(al) guide; V-shaped guide
三面顶点 trihedral vertex
三面镀铬型活塞环 triple chrome piston ring
三面反射镜 trihedral reflector
三面隔水边界 impervious boundary in three-sides of an aquifer
三面毂 three-way hub
三面观众的舞台 thrust stage
三面回转接头 triple plane swivel joint
三面角 trihedral angle;tripod angle
三面临观众的舞台 arena stage
三面嵌入式浴缸 recess basin
三面刃键槽铣刀 three-lip keyway cutter
三面刃铣刀 face-and-side(milling) cutter;side cutter
三面体 trihedral;trihedron
三面体棱镜 triangular prism; trihedral prism
三面投影 three-plane projection
三面投影图 three-plane projection drawing
三面图 three-dimensional map;three-view drawing
三面涂层 back primed
三面下锯法 three-faced sawing
三面信号 three-way signal
三面形(体) three-sider; trihedral; trihedron
三面眼板 triangle plate
三面正交反射器 trihedral corner reflector
三面正投法 axonometry
三面正投影 three-plane orthographic projection
三面直角棱镜 corner cube prism
三名 trinomen
三名汇票 three-party draft
三明治 sandwich
三明治锚 <即麦尼尔锚,比利时 Magnel 教授所创> sandwich-plate anchorage
三明治式教程 <课堂学习和工厂实习交替进行> sandwich course
三模跟踪 three-mode tracking
三模控制 three-mode control
三膜片元件 three-diaphragm element
三膜片制动分泵 three-diaphragm brake chamber
三母线 triple busbar
三母线系统 triple-busbar system
三钠 trisodium
三能级 three-level
三能级固态量子放大器 three level solid state quantum amplifier
三能级光发射体 three-level light emitter
三能级激发 three-level excitation
三能级激光 three-level laser
三能级激光材料 three-level laser material
三能级激光结构 three-level laser configuration
三能级激光器材料 three-level laser material
三能级系统 three-level system
三能级荧光固体 three-level fluorescent solid
三能级振荡器 three-level generator
三能量瞬变现象 three-energy transient phenomenon
三年周期 three-year cycle
三捻 bilimbing
三捻花釉 tri-twist flower glaze
三钮板 three-button(filter) plate
三偶氮化合物 trisazo compound
三偶氮染料 trisazo dye
三耙式分级机 triplex rake classifier
三耙组合 three-harrow unit
三排的 three-row
三排横列舱口 three-row hatch
三排汽口涡轮机 triple flow turbine
三盘丁字管 three-collar tee fitting
三盘丁字管接 three ends flanges T
三盘三通 three-collar tee fitting
三盘十字管 three-collar cross pipe
三盘四通 three-collar cross pipe
三判决问题 three-decision problem
三跑楼梯 double L stair(case);three-flight stair(case); three-quarter-turn stair(case); tripartite stair (case);triple-flight stair(case)
三跑式楼梯 staircase of three-quarter turn type
三配位 three-fold coordination
三喷嘴喷枪 three-spray gun
三膨胀式蒸汽泵 triple-expansion steam pump
三劈裂 triplet splitting
三片层结构 trilaminar structure
三片层矿物 three-sheet mineral
三片的合模 three-part mold
三片法 <空中三角测量> triplet method
三片离合器 three-plate clutch
三片螺纹 triple thread
三片模型 triplet model
三片三组镜头 triplet lens
三片式离合器 triple plate clutch
三片物镜 three-lens objective
三偏磷酸钠 sodium trimetaphosphate
三偏磷酸盐 trimetaphosphate
三频带滤波器 triple-band filter
三频航电仪 trifrequency airborne electromagnetic system
三平面法 method of three-planes
三平面构筑物 three-level structure
三平面建筑作业系统 three-level building operational system
三平面结构 three-level structure
三平面作业 three-level operation
三平巷 triple entry;triple heading
三平巷房柱采矿法 triple-heading room-and-pillar mining
三平巷掘进法 three-heading system
三平巷掘进系统 triple-entry system
三坡老虎窗 hipped
三剖面的 trisected
三铺碹 juxtaposed arch
三七灰土 three/seven lime earth[3:7 lime earth]
三七灰土夯实 rammed 3:7 lime earth; rammed three-to-seven lime earth; three/seven lime earth rammed
三七墙 brick-and-a-half wall; thirty-seven cm brick wall
三期灌浆孔 tertiary grout hole
三歧式 trichasium
三气缸式机车阀动装置 three-cylinder locomotive valve gear
三气监测站 three-air monitoring station
三气门发动机 three-valve engine
三汽包锅炉 three-drum boiler
三汽缸泵 triple piston type pump; triplex pump
三汽缸发动机 three-cylinder engine
三汽缸机车 three-cylinder locomotive
三千秒差距臂 three-kps arm
三枪管 three-gun tube
三枪显像管 triniscope
三腔放大管 three-cavity amplifier tube
三腔化油器 triple-throat carburetor
三腔模 triple-cavity mo(u)ld
三腔模成型法 triple cavity process
三腔式 three-cell
三腔式断面 triple box section
三腔速调管 three-cavity klystron
三羟的 trihydric
三羟钒石 haggite
三羟基的 triatomic
三桥式侧卸卡车 three axles side-dump
三桥式铲运机 three axles scraper
三桥式底卸卡车 three-axles bottom-dump
三桥式后卸卡车 three axles rear-dump
三桥轴共同驱动 three axles all drive
三桥轴式 triple axle reaction
三切削刃管鞋下扩孔器 three-cutter underreamer
三氰酸 tricyanic acid
三球定 trisphaeridine
三球枢轴式万向节 three-ball and trunnion universal joint
三球悬铃木 sycamore
三区制 three-field system
三曲柄的 three-throw
三曲柄曲轴 three-throw crankshaft
三曲线划分法 <室内照明计算的> three-curve(calculation) method
三曲线计算法 three-curve(calculation) method
三曲翼面 triple cambered aerofoil
三取二 select 2 in 3
三取二系统 two-out-of-three system
三取一 select 1 in 3
三裙碗式绝缘子 triple petticoat bell shaped insulator
三群反应堆 three-group reactor
三群理论 three-group theory
三燃料发动机 trifuel engine
三燃料管线站用发动机 trifuel pipeline engine
三燃烧管锅炉 triple flue boiler
三燃烧室发动机 triple-barrel motor
三绕组 three-winding
三绕组变压器 three-circuit transformer;three-winding transformer
三绕组的 three-winding
三绕组发电机 three-coil dynamo; three-coil generator
三绕组继电器 triple-winding relay
三绕组调谐变压器 triple-tuned transformer
三绕组自耦变压器 three-winding autotransformer
三人并肩座椅 three-abreast seats;triple seat
三人舱 three-man module
三人公寓 three-person flat
三人脚踏车 triplet
三人手持夯 triple hand tamper
三人组合 triad
三人组合的 triadic
三刃钉 Smith-Petersen nail; three-flanged nail
三刃刮路机 three-blade drag; three-drag blade drag
三刃钎头 three-point bit; three-wing bit
三刃铣刀 triple-fluted bit
三刃钻头 three-point bit; three-wing bit
三日强度 <混凝土的> three-day strength
三日热 ephemeral fever; three-day sickness
三三制 triangular organization
三三制发射场综合设施 three-by three-complex
三三制配置 three-by three-arrangement
三三制配置形式 three-by-three-configuration
三色暗房灯 tricolo(u)r darkroom lamp
三色版 trichromatic edition
三色版印刷 three-colo(u)red printing
三色标示 tristimulus designation
三色标志特性 trichromatic specification
三色表示镜 three-light spectacle
三色重现装置 three-colo(u)r reproducer
三色部件 three-colo(u)r unit
三色测光 three-colo(u)r photometry
三色单元 three-colo(u)r unit
三色的 trichroic;trichromatic;tricolo(u)r(ed);tristimulus
三色灯 tricolo(u)red lantern;tricolo(u)red light
三色底片 monopack
三色叠加法 three-colo(u)r additive method
三色分离 three-colo(u)r separation
三色分析 trichromatic analysis
三色复原 three-colo(u)r display
三色感光乳剂 tricolo(u)r emulsion
三色光度测量 three-colo(u)r photometry
三色光度学 three-colo(u)r photometry
三色光学分离器 trichromatic optic(al) separator
三色环氧树脂 three-pack epoxy resin
三色激励 tristimulus
三色激励比色计 tristimulus colo(u)rimeter
三色激励测色法 tristimulus colo(u)rimetry
三色激励滤色片 tristimulus filter
三色激励系数 tristimulus coefficient
三色激励值 tri-stimulus value
三色接收 tricolo(u)r reception
三色结节纱 three-colo(u)r knob yarn
三色堇(植物) garden pansy
三色镜 trichromoscope
三色觉异常 anomalous trichromatism
三色理论 trichromatic theory
三色磷光点组 phosphor dot trio
三色录像机 three-colo(u)r recorder
三色滤光片 three-colo(u)r filter;tricolo(u)r filter
三色滤光器 tricolo(u)r filter
三色旗 tricolo(u)r
三色器 trichromat
三色全息图 three-colo(u)r hologram
三色乳胶 trichrom-emulsion;tricolo(u)r emulsion
三色色标管 Lawrence tube;tricolo(u)r chromatron
三色色度学 three-colo(u)r colo(u)rimetry
三色声光调制器 three-colo(u)r acousto-optic modulator
三色视觉型 trichromatism
三色视像管 tricolo(u)r vidicon
三色说 trichromatic theory
三色图 chromaticity diagram;trichromatic diagram;tristimulus diagram
三色系数 three-colo(u)r coefficient
三色系统 three-colo(u)r system;trichromatic system
三色显像管 three-colo(u)r tube;tricolo(u)r kinescope
三色现象 trichroism;trichromatism
三色响应 trichromatic response
三色响应理论 trichromatic response theory
三色信号灯 coston light
三色性 trichroism;trichromatism
三色原理 trichromatism

三色源色度计 tristimulus colo(u)rimeter
三色照相术 trichromatic photography
三色指示 tristimulus designation
三色制 three-colo(u)r system; trichromatic system
三色装饰 three-colo(u)r decoration
三色组 triad
三色组合 colo(u)r triad
三筛砂 three-screen sand
三筛式分级装置 three-screen sizing unit
三栅极管放大器 triple-grid amplifier
三扇窗 three-light window; triple-casement window
三扇搭接门板 three-speed door panel
三扇平开和一扇上悬腰窗 three-side hung sections and top hung fanlight
三扇伸缩门板 three-speed door panel
三扇式上下启闭窗 triple-hung sash window
三扇式转门 three-wing revolving door
三扇折叠式百叶门 three-leaf folding shutter door
三上四下辊子牵伸 three-over four roller drafting
三上四下曲线牵伸 three-over four curvilinear draft
三上一下斜纹 crow twill;swansdown twill
三射的 triradiata
三射缝 triradiate slit
三射痕 trilete marking
三射脊 triradiate ridge
三射流实验 triple-jet experiment
三射石灰骨针 triradiate calcareous spicule
三射束 three-beam
三射束光斑图像干涉仪 three-beam speckle pattern interferometer
三射线干涉 three-ray interference
三深裂的 triparted;tripartite
三生螺纹 tertiary spiral
三声道 triple-track
三声道立体声 three-channel stereo
三声过程 three-phonon process
三声速的<亚声速、跨声速和超声速> trisonic
三声速空气动力学 trisonics
三声吸收 three-phonon absorption
三十二进制(记数法) duotricemary notation
三十二开 thirty-twomo
三十二烷 dotriacontane
三十分之一半径(柱下部) part
三十基 triacontyl
三十六烷 hexatriacontane
三十面体 triacontahedron
三十三基 tritriacontyl
三十三烷 tritriacontane
三十酸盐 melissate
三十碳烯 triacontylene
三十天期的汇票 draft at thirty days sight
三十烷 melissane;triacontane
三十烷醇 melissyl alcohol; triacontanol
三十烷二酸 triacontanedioic acid
三十烷基 melissyl
三十烷酸 melissic acid; triacontanoic acid
三十五毫米彩色胶片 thirty-five colo(u)r film
三石塔<两石柱上架石梁的纪念碑> trilith(on)
三时估计数 three-time estimate
三示像 three-aspect
三示像部件 three-aspect unit
三示像单元 three-aspect unit
三示像二闭塞区段系统 three-aspect two-block system
三示像二闭塞区段制 three-aspect two-block system
三示像三闭塞区段系统 three-aspect three-block system
三示像三闭塞区段制 three-aspect three-block system
三示像信号 three-aspect signal
三示像信号机 three-aspect signal
三示像信号系统 three-aspect signal(l)ing system
三示像信号制【铁】 three-aspect signal(l)ing system
三示像自动闭塞 three-aspect automatic block
三示像自动闭塞系统 three-aspect automatic block system
三示像自动闭塞制 three-aspect automatic block system
三视图 three-view diagram
三试件压力盒 three-in-one cell
三试件压力室 three-in-one cell
三室储气筒 three-chamber air reservoir
三室的 three-roomed;trilocular
三室公寓 three-roomed apartment
三室居住单元 three-roomed dwelling unit
三室溜槽 three-compartment chute
三室炉 three-cell furnace
三室磨碎机 three-compartment mill
三室生活单元 three-roomed living unit
三室套房 three-roomed flat
三室调压井 three-chamber surge tank
三室箱梁 three cell box girder
三室箱形截面 three-box section
三室箱形桥梁 three-box section girder bridge
三枢直流电动机 three-armature direct current motor
三输入加法器 three-input adder
三输入减法器 three-input subtracter
三输入开关 three-input switch
三鼠笼转子 triple-squirrel-cage rotor
三束管 three-beam tube
三束光源 three-light-beam source
三数列定理 three-series theorem
三数组 triad
三刷发电机 three-brush generator
三刷式灯头 three-lit base
三刷式直流发电机 third-brush generator
三刷洗瓶机 three-brush bottle washer
三水胆矾 bonattite
三水钒矿 navajoite
三水分子 trihydrol
三水合物 trihydrate
三水合氧化铝 hydrated alumina; hydrated alumin(i)um oxide
三水菱镁矿 nesquehonite
三水铝矿 gibbsite;hydrargillite
三水铝耐火材料 gibbsite refractory
三水铝石 gibbsite;hydrargillite
三水铝石红土 gibbsitic laterite
三水钠锆石 hilairite
三水砷铅铜石 goudeyite
三水碳钙石 trihydrocalcite
三水碳酸钙 trihydrocalcite
三水型铝土矿矿石 gibbsite ore
三水氧化铝 gibbsite
三顺一丁砌法 Flemish garden wall bond;Sussex garden-wall bond
三顺一丁砌合 boundary wall bond
三顺一丁砌墙法 English garden-wall bond;mixed garden-wall bond
三顺一丁砌体 Flemish garden bond
三瞬处理 three-instantaneous processing
三瞬态正弦连续波法 three-transient-sinusoidal successive wave method
三瞬心 three instantaneous centers
三丝水准测量 three-wire level(l)ing
三丝水准测量法 three-wire level(l)ing method
三丝铜网 triple wire
三、四挡传动同步化 third and fourth gear synchronization
三四等水准测量 third to fourth-order leveling
三四面体 tristetrahedron
三四五定直角法 three-four-five rule
三四五直角定向法 three-four-five rule
三速变速箱 three-speed transfer case
三速唱机 three-speed player
三速齿轮 three-speed gear
三速齿轮箱 three-speed gear box
三速电动机 three-speed motor
三速度测井 three-dimensional velocity log;three-D log
三速分动箱 three-speed transfer case
三速风洞 trisonic wind tunnel
三酸甘油酯 triglyceride
三酸价的【化】 triacid
三酸式盐 triacid salt;trihydric salt
三酸盐 trisalt
三羧酸循环 tricarboxylic acid cycle
三索面 three-plane cable
三索套 three-leg-sling
三索悬挂式抓斗 three-rope suspension grab(bing) bucket
三塔高的 three-towered
三态 tristate
三态变量 ternary variable
三态的 ternary
三态点 triple point
三态缓冲器 three-state buffer
三态控制 tristate control
三态控制信号 three-state control signal
三态逻辑 tristate logic
三态逻辑电路 three-state gate
三态门 three-state gate;triple gate
三态启动 three-state enable
三态驱动器 three-state driver
三态使能 three-state enable
三态输出 three-state output
三态数据输出 three-state data output
三态性 trimorphism
三态总线 tristate bus line
三态总线驱动器 tristate bus driver
三弹性轴 three-elastic axes
三探针 three-point probe;triple probe
三探针法 three-probe method
三碳酸盐 tricarbonate
三陶品 tritopine
三套索吊装装置 three-leg-sling
三套爪夹盘 three-jaw collet chuck
三梯段楼梯 tripartite stair(case);triple-flight stair(case)
三体 trimer;trisome
三体比率 trisomic ratio
三体船 trimaran; triple-hulled ship; triple-hulled vessel
三体复合 three-body recombination
三体碰撞 three-body collision; triple collision
三体问题 problem of three-bodies; three-body problem
三体性 trisomy
三体再化合 three-body recombination
三天假期许可证 three-day pass
三条通道的<剧院观众座位间> three-aisled
三条通道教堂<古罗马> three-aisled basilica
三条腿装饰台 teapoy
三调位的铲斗连杆 three-position bucket linkage
三调谐电路接收机 three-circuit receiver;three-way receiver
三调制信标 triple modulation beacon
三萜(烯) triterpene
三萜系化合物 triterpenoid
三通 bifurcated forked pipe; branch cell; branch (ing) pipe; branch piece;pipe tee;tee;tee conduit;tee piece;T-fitting;wye
三通玻璃栓 three-way glass stopcock
三通道地震鉴别器 three-channel seismic discriminator
三通道放大器 three-channel amplifier; three-path amplifier
三通道喷煤管 three-channel coal fired burner
三通道燃烧器 three-channel coal burner
三通道自动驾驶仪 three-dimensional autopilot
三通的 three-way
三通电磁阀 three-way magnetic valve
三通电缆分线盒 three-way coupling box
三通阀 cross(-over) valve; intercepting valve; tee valve; three-throw tap; three-way valve; triple valve; 3-way valve
三通阀安装座 triple valve bracket
三通阀安装座垫木 triple valve bracket filler
三通阀垫 triple valve gasket
三通阀盖 triple valve cap;triple valve cylinder cap
三通阀盖垫 triple valve cylinder cap gasket
三通阀活塞 triple valve piston
三通阀联箱 three-way valve manifold
三通阀排气短管 triple valve exhaust stub tube
三通阀试验台 triple valve test rack
三通阀体 triple body;triple valve body
三通阀下体 triple valve bottom body
三通阀下体垫 triple valve bottom body gasket
三通阀与辅助风缸联结短管 triple valve nipple
三通阀支管 triple valve branch pipe
三通管 Y-pipe; Y-shaped duct; Y-tube; wye pipe; bifurcated pipe; branching pipe; branch tee; forked tube; single-sweep tee; T-bend; three-way connection; three-way piece; three-way pipe; T-piece; tee [T]-pipe; tee [T] branch(pipe); tee [T] branch(tube)
三通管 T 形管 tee fitting
三通管件 three-way fitting
三通管接头 T-branch; tee pipe coupling; three-way; three-way union; T joint; triple joint; union tee; V-branch; Y-connection; Y-junction; Y-pipe; Y-section
三通管试验 test tee
三通焊接 tee welding
三通横长 horizontal length of tee
三通滑阀 three-port slide valve
三通换向旋塞 three-way reversing cock
三通活管接 three-way union;T union
三通活接头 union tee
三通活栓 three-way cock
三通货协定 triparties currency agreement
三通接头 branch joint; branch tee; duplex fitting; T-connection; tee coupling; tee fitting; tee joint; tee pipe; tee union; three-way connection;T-juncture;T-piece;T union

S

三通接线夹 three-way connector
三通开关 three-way cock; three-way switch; three-way tap; three-way valve; triple valve
三通控制阀 three-way control valve
三通连接 tee junction
三通联结管 union T
三通龙头 three-way tap
三通路循环器 three-terminal circulator
三通歧管 three pipe manifold; three way manifold
三通塞门 T cock; three-way cock
三通式热风炉 three-pass stove
三通受汽管 Y-pipe
三通陶管 Y-tile
三通天线转发开关 triplexer
三通筒 triple cylinder
三通筒垫密片 triple-cylinder gasket
三通瓦管 Y-tile
三通弯管接 three pipe manifold; three way manifold
三通弯矩 Y-bend
三通弯头 three-way elbow
三通旋塞 tee cock; three-throw tap; three-way cock
三通循环器 three-terminal circulator
三通液体流量计 tee fluid flow meter
三通一平 < 水、电、道路要通，施工场地要平 > three-connections and the one level(l)ing[assuring that a construction site is connected to water/power and roads/and that the land is levelled before a building project is begun]
三通闸阀 treble ported slide valve
三通辙叉 crotch frog
三通支管 Y-branch
三通止阀 three way stop valve
三同步发射台组 triad; triplet
三同立构 triactic
三同时 < 指完成工程的同时，完成废水、废气、废渣的处理设施 > three-simultaneities[completion of a project simultaneously with the completion of the facilities for controlling waste water/ waste gas and industrial wastes]
三同时原则 three-simultaneous rule
三筒汽化器 triple-barrel carburetor
三筒式锅炉 tridrum boiler
三筒式快速换挡绞车 three-drum rapid shifting hoist
三筒式快速换卷扬机 three-drum rapid shifting hoist
三筒洗机 three-cylinder washer
三头捣锤 triplex tamper
三头连接 triple joint
三头螺纹 three-start screw; triple thread; triple threaded screw
三头螺线螺旋 triple-flight auger
三头螺旋环填料 triple-spiral ring packing
三头蜗杆 triple start worm
三头斜榫 triple bevel scarve
三投 three-throw switch
三透镜 three-lens
三透镜校正系统 three-lens corrector
三透镜聚光器 three-lens condenser; triple condenser
三透镜物镜 three-lens objective
三透镜系统 three-lens system
三透镶物镜 three-lens objective
三凸角形凸轮 tri-lobe cam
三凸轮接头 tricam connecter[connector]
三突瓣式水泵 three-lobe pump
三突齿凸轮 tri-lobe cam
三涂三烘工艺 three-coat three-bake
三推进器 triple screw propeller
三推进器船 triple screw propeller ship
三腿的 three-legged
三腿吊杆 three-legged derrick
三腿吊楔 three legged lewis
三腿回填夯 triplex backfill tamper
三腿起重机 three-legged derrick
三腿梯子 three trees
三腿形饰 triskelion[复 triskelia]
三腿摇臂起重机 three-legged derrick
三腿桌 trivet table
三腿钻塔 three-legged derrick
三托架测试 three-bracket testing
三托梁 beam with central prop
三陀螺罗经 triple gyrocompass
三陀螺稳定平台 three-gyro platform; three-gyro-stabilized platform
三陀螺组 triad of the gyro
三瓦楞铁皮 triple corrugated sheet iron
三瓦特计法 three-wattmeter method
三弯矩定理 three-moment theorem; theorem of three moments
三弯矩法 three-moment method
三弯矩方程 equation of three-moments; three bending moment equation
三弯矩方程式 three-moment equation
三弯叶形截面 triskelion cross section
三烷基胺 trialkylamine
三烷基化的 trialkylated
三烷基铝 trialkylaluminium
三烷氧基烷烃 trialkoxyparaffin
三腕板 tertibrach
三桅 schooner
三桅船 three-masted vessel; three-master
三桅的 three-masted
三桅帆船 bark; barque; tern schooner
三维 three-dimensional frame(d) load bearing structure; tridimension
三维 X 方向 three-dimensional X direction
三维 Y 方向 three-dimensional Y direction
三维 Z 方向 three-dimensional Z direction
三维变换 three-dimensional transformation
三维变形 three-dimensional deformation
三维变形状态 three-dimensional state of deformation
三维表示法 three-dimensional presentation
三维波动方程偏移 three-dimensional wave equation migration
三维薄层色谱法 three-dimensional thin layer chromatography
三维部件 three-dimensional element
三维彩色显示 three-dimensional colo(u)r display
三维参数单元 three-dimensional parameter unit
三维草图 three-dimensional sketch
三维测井 three-dimensional log; three-dimensional velocity log
三维测量法 three-dimensional measurement
三维成像 three-dimensional imaging
三维成像器 three-dimensional imager
三维承重建筑 three-dimensional bearing structure
三维承重结构 three-dimensional load-carrying structure
三维处理 three-dimensional treatment
三维穿透度 three-dimensional inter penetration
三维存储器 three-dimensional storage; three-dimension memory
三维大地测量 spatial triangulation
三维大地测量法 three-dimensional method of geodesy
三维大地测量学 three-dimensional geodesy
三维单元 three-dimensional element; three-dimensional unit
三维导线测量 three-dimensional traversing
三维的 three-dimensional; triaxial; tridimensional
三维等参数单元 three-dimensional isoparametric element
三维地貌模型 three-dimensional terrain modeling
三维地面运动 three-dimensional ground motion
三维地图 three-dimensional map; tridimensional map
三维地形坐标 three-dimensional terrain coordinates
三维地震勘探 three-dimensional seismic survey
三维地震模型 three-dimensional seismic model
三维地震剖面网络图 fence diagram
三维地震射线跟踪 three-dimensional ray tracing; three-dimensional seismic ray system; three-dimensional seismic ray tracing
三维地震资料的数字电影显示 digital movies from three-dimensional seismic
三维点阵 three-dimensional lattice
三维电导率 three-dimensional conductivity
三维电极 three-dimensional electrode
三维电极法 three-dimensional electrode method
三维电极反应器 three-dimensional electrode reactor
三维电缆 three-dimensional streamer
三维电路 three-dimensional circuit
三维顶盖结构 three-dimensional area-covering structure
三维定量信息 quantitative three-dimensional information
三维动画片 three-dimensional animation
三维动态频谱分析仪 three-dimensional dynamic spectrum analyser
三维断层 three-dimensional fault
三维二次漂移 three-dimensional quadratic drift
三维反演 three-dimensional inversion
三维仿形加工 three-dimensional profiling
三维非弹性分析 three-dimensional inelastic analysis
三维非线性分析 three-dimensional non-linear analysis
三维分布 three-dimensional distribution
三维分类法 three-dimensional classification
三维分析 three-dimensional analysis
三维杆件 three-dimensional bar; three-dimensional rod
三维港口模型 three-dimensional harbo(u)r model
三维各向同性 three-dimensional isotropy
三维跟踪 three-dimensional tracking
三维构架 three-dimensional frame
三维构件 three-dimensional element
三维构造图 three-dimensional tectonic map
三维构造系统 three-dimensional structural system
三维固结 three-dimensional consolidation
三维固体 three-dimensional solid
三维观测系统 three-dimensional layout
三维管线 three-dimensional pipeline
三维光测弹性学 three-dimensional photoelasticity
三维光栅 three-dimensional grating
三维光弹模型 three-dimension-optical elastic model
三维光弹性学 three-dimensional photoelasticity
三维光学变换 three-dimensional optic(al) transform
三维光致弹性 three-dimensional photoelasticity
三维桁条 three-dimensional purlin(e)
三维互连 three-dimensional interconnection
三维汇点 three-dimensional sink
三维绘图 three-dimensional plot
三维绘制 three-dimensional rendering
三维机动 three-dimensional maneuver
三维基本形 three-dimensional fundamental form
三维极坐标曲线 three-dimensional polar curve
三维集成电路 three-dimensional integrated circuit
三维几何学 three-dimensional geometry
三维计算机辅助设计项目 three-dimensional CAD project
三维计算机模型 three-dimensional computer model
三维加权偏移 three-dimensional weighted migration
三维建筑 three-dimensional construction
三维交联网络 three-dimensional crosslinked network
三维结构 three-dimensional structure; space structure
三维结构描述 three-dimensional structural description
三维结构模型 three-dimensional architectural model
三维晶核 three-dimensional nucleus
三维径向时基 three-dimensional radial timebase
三维矩阵 three-dimensional matrix
三维卷曲 three-dimensional crimp
三维空间 three-dimensional space
三维空间的 three-dimensional
三维空间定位 three-dimensional fix
三维空间刚架体系 three-dimensional (space) frame system
三维空间攻击 three-dimensional attack
三维空间构架 three-dimensional frame (work)
三维空间径向塑性流动 three-dimensional radial plastic flow
三维空间框架 three-dimensional frame(work)
三维空间立体图 three-dimensional plot
三维空间联动装置 linkage in three dimensions
三维空间模型 three-dimensional space model
三维空间体系 system in three-dimensional space; three-dimensional system
三维空间图 three-dimensional graph
三维空速管 three-dimensional Pitot tube
三维控制 three-dimensional control
三维框架 three-dimensional frame; space frame

三维拉紧 three-dimensional tensioning
三维拉伸 three-dimensional stretching
三维雷达 three-dimensional radar
三维力分布 three-dimensional distribution of forces
三维立体模型 three-dimensional space model
三维连续体 three-dimensional continuum
三维连续统 three-dimensional continuum
三维连续性 three-dimensional continuity
三维量测 tridimensional measurement
三维裂隙比 three-dimensional extent of fissure
三维檩条 three-dimensional purlin(e)
三维流(动) flow in three dimensions; three-dimensional flow
三维乱向 three-dimensional randomly orientated
三维乱向分布 three-dimensional random distribution
三维轮廓图 line drawing
三维轮廓仪 contourgraph
三维面 three-dimensional surface
三维面元记录 three-dimensional bin record
三维面元显示系统 three-dimensional bin display system
三维模拟 three-dimensional simulation
三维模拟计算机 three-dimensional analog computer
三维模拟计算装置 tree-dimensional analog computer
三维模式 three-dimensional model
三维模数式住房 three-dimensional module house building
三维模型 three-dimensional model
三维模型系统 three-dimensional model(l)ing system
三维偏移 three-dimensional migration
三维偏移的时间切片 time slice of three-dimensional migration
三维偏移速度图 three-dimensional migration velocity map
三维偏移系列剖面 three-dimensional migration series section
三维平方和开方法 root-sum-square of three-dimensional method
三维平行系统 three-dimensional parallel system
三维墙纸 three-dimensional wallpaper
三维倾角模型显示 three-dimensional slope angle model display
三维倾斜叠加分析 three-dimensional inclination stack velocity analysis
三维倾卸搅拌机 three-cone tilting mixer
三维球面跟踪系统 three-dimensional-spheric(al) tracking system
三维球面上的统计 statistics on three-dimensional sphere
三维区域网平差 stereoblock adjustment
三维曲面图 block diagram
三维曲线 three-dimensional curve
三维趋势分析 three-dimensional trend analysis
三维全静压管 three-dimensional Pitot tube
三维全息光栅 three-dimensional hologram grating
三维全息摄影 three-dimensional holograph
三维全息术 three-dimensional holography
三维全息图 three-dimensional hologram
三维全息照相存储器 three-dimensional holographic memory
三维散射 three-dimensional scattering
三维扫描仪 three-dimensional scanner
三维闪烁照相法 three-dimensional scintigraphy
三维设计 three-dimensional design
三维摄影 three-dimensional photography
三维渗流 three-dimensional seepage (flow)
三维渗透 three-dimensional penetration
三维实体模型 three-dimensional solid modeling
三维实体图 three-dimensional solid chart
三维视图 three-dimensional view
三维数据 three-dimensional data
三维数据体的切片 slice of three-dimensional data block
三维数据图解分析 graphic(al) analysis of three-dimensional data
三维数学模型 three-dimensional mathematic(al) model
三维数字图像 three-dimensional digital image
三维数字转化器 digitizer
三维数组 three-dimensional array
三维双曲线跟踪系统 three-dimensional-hyperbolic(al) tracking system
三维双翼 three-dimensional biplane
三维水池 three-dimensional tank
三维水工模型 three-dimensional hydraulic model
三维水质模型 three-dimensional water quality model
三维速度 three-dimensional velocity
三维速度测井 three-dimensional velocity log
三维弹性分析 three-dimensional elasticity analysis
三维体系 three-dimensional system
三维条纹 three-dimensional fringe
三维透视图 three-dimensional perspective(view map)
三维图 graphic(al) model; three-dimensional figure
三维图像 three-dimensional plot
三维图形操作 three-dimensional graphic operation
三维图样 three-dimensional model
三维弯曲 three-dimensional bending
三维万能组合图形 three-dimensional composite multiple-purpose figure
三维网(络) three-dimensional network
三维网络结构 three-dimensional net structure
三维问题 three-dimensional problem
三维物体 three-dimensional body
三维系统 three-dimensional system
三维显示 stereo display; three-dimensional display; three-dimensional representation; three-dimension display
三维显示系统 three-dimensional display system
三维镶嵌风格 three-dimensional panelling pattern
三维镶嵌式样 three-dimensional panelling pattern
三维向量 three-dimensional vector; three-vector
三维像 three-dimensional image
三维像点坐标 three-dimensional image coordinates
三维效应 three-dimensional effect
三维形态 three-dimensional configuration
三维渲染图 three-dimensional rendered picture
三维悬垂 three-dimensional drape
三维压力试验仪 triaxial apparatus
三维衍射图 three-dimensional diffraction pattern
三维一次漂移 three-dimensional linear drift
三维应变 three-dimensional strain
三维应力 stress in three-dimensions; three-dimensional stress(ing); triaxial stress
三维应力体系 three-dimensional stress system
三维应力状态 general state of stress; three-dimensional stress state
三维影像 three-dimensional image
三维有限元法 three-dimensional finite element method
三维预加应力 three-dimensional prestressing
三维域 three-dimensional domain
三维运动 three-dimensional motion
三维载重结构 three-dimensional weight-carrying structure
三维造型 three-dimensional model(l)ing
三维张紧 three-dimensional tensioning
三维张开 three-dimensional stretching
三维阵列 three-dimensional array; cubic(al) array
三维支承建筑 three-dimensional supporting structure
三维支承结构 three-dimensional supporting structure
三维织物 three-dimensional woven fabric
三维直方图 stereogram
三维直角坐标系 rectangular three-dimensional coordinate system
三维制导 three-dimensional guidance
三维质点运动的直线度 rectilinearity of particle motion in three-dimension
三维重现 three-dimensional reconstruction
三维转弯 three-dimensional turn
三维状态应力 three-dimensional state of stress
三维资料立体显示 three-dimensional data stereo-display
三维资料显示 display of three-dimensional data
三维自导引 three-dimensional homing
三维组合 space composition; three-dimensional arrangement; three-dimensional composition
三维组装 three-dimensional package
三维作用力系统 three-dimensional system of forces
三维坐标 three-dimensional coordinates
三维坐标数据 space data; spatial data
三维坐标图 three-dimensional graph
三位 three-position
三位臂板式表示器 three-position semaphore type indicator
三位臂板式复示器 three-position semaphore type indicator
三位编码 tri-bit encoding
三位的 three-position; triply-discharging
三位度数罗经盘 three-figure compass card
三位阀 three-position valve
三位辅助继电器 centre zero relay
三位开关 three-position switch
三位控制 three-position control; three-step control
三位式按钮 three-position push button
三位式闭塞机 three-position block instrument
三位式继电器 three-position relay
三位式控制器 tristate controller
三位式上向臂板信号 three-position upper-quadrant semaphore
三位式上向臂板信号(机)【铁】 three-position upper-quadrant semaphore
三位式无极发码接点 three-position neutral code contact
三位式下向臂板机构 three-position lower-quadrant semaphore mechanism
三位式下向臂板信号(机)【铁】 three-position lower-quadrant semaphore
三位式信号复示器 three-position signal repeater
三位式信号机 three-position signal
三位数码组 three-numeral code group
三位四通阀 three-position four-way valve
三位一体 trinity
三位一体蛇绿岩 trinity ophiolite
三位置 triolocation
三位置转换开关 three-position switch
三位准确度 three-place accuracy
三位字节 triplet
三位组 triad
三位作用 three-step action
三纹 treble cut
三纹螺旋 triple screw
三稳定设备 three-stable state device
三稳定性 three-stability
三稳态 tristable
三稳态设备 three-stable state device
三物分离 three-product separation
三洗跳汰机 tertiary jig; tertiary washbox
三系剪刀 triple-shear
三系剪力 triple-shear
三系弹簧装置 triple series spring equipment
三系斜杆 triple cancellation
三隙缝 treble-slot
三峡大坝 The Gorges Dam
三狭缝干涉 three-slit interference
三下采煤 three unders in coal mining
三弦桁架 three-chord truss
三弦桥 three-chord bridge; triangular truss bridge
三显示 three-aspect; three-indication
三显示部件 three-aspect unit
三显示单元 three-aspect unit
三显示二闭塞区段系统 three-aspect two-block system
三显示二闭塞区段制 three-aspect two-block system
三显示机车信号系统 three-indication cab signal system
三显示机车信号制 three-indication cab signal system
三显示三闭塞区段系统 three-aspect three-block system
三显示三闭塞区段制 three-aspect three-block system
三显示(示像)色灯 three-aspect colo(u)r light
三显示信号(机)【铁】 three-aspect signal; three-position signal
三显示信号系统 three-aspect signal(l)ing system

S

三显示信号制 three-aspect signal(l)ing system
三显示自动闭塞系统【铁】three-aspect automatic block system
三显示自动闭塞制 three-aspect automatic block system
三线包缝机 three-thread overlock
三线保险丝塞 three-way fuse plug
三线保险丝塞孔 three-way fuse socket
三线变流机 three-wire rotary converter
三线补偿器 three-wire compensator
三线测螺纹法 three-thread wire method; three-wire measurement; three-wire method; three-wire system
三线插孔 three-way jack
三线插座 three-plug connector
三线觇标 three-line target
三线的 three-wire; trilinear
三线点素坐标 trilinear point coordinates
三线电导率 three-dimensional conductivity
三线电缆 triaxial cable; trilead cable
三线读数水平测量法 three-wire leveling method
三线发电机 three-wire generator
三线发动机 three-wire generator
三线法 trilinear method
三线交叉 three-cross
三线交点 trijunction
三线接法 three-wire connection
三线均压器 three-wire balancer
三线控制 three-wire control
三线螺纹 triple thread
三线捻织 triple warp weave
三线圈 three-winding
三线圈变压器 three-winding transformer
三线圈调整 three-coil regulation
三线塞孔 three-point jack; three-way jack; triple circuit jack
三线式棒材轧机 three-strand rod mill
三线式电度表 three-wire meter
三线式发电机 three-wire system generator
三线式绝对容许闭塞区域 three-wire APB [absolute permissible block] territory
三线式控制继电器 three-wire control relay
三线式型钢轧机 three-strand rod mill
三线式选择器 three-wire selector
三线式仪表 three-wire meter
三线式中继线 three-wire trunk line
三线双交换 three strands double crossing-over
三线水准测量 three-wire level(l)ing
三线条纹 triple stripe
三线图 trilinear chart
三线网 triple chain wire
三线系泊 three-wire mooring
三线系统 three-wire system
三线线路 three-wire line
三线信号桥 three-track signal bridge
三线型滞回曲线 trilinear hysteretic curve
三线性的 trilinear
三线性读数水平测量法 three-wire level(l)ing method
三线性理想化滞流环线 trilinear idealized hysteresis loop
三线性理想化滞流图 trilinear idealized hysteresis loop
三线性曲线 trilinear curve
三线性退化刚度 trilinear degrading stiffness
三线性形式 trilinear form
三线性硬化模型 trilinear hardening model
三线制 three-wire system
三线制变压器 three-wire transformer
三线制电量计 three-wire meter
三线制直流装置 three-wire direct current equipment
三线中继线 three-wire junction; three-wire trunk
三线装置 three-wire installation
三线坐标 trilinear coordinates
三相半波整流器 three-phase half-wave rectifier
三相饱和的 trisaturated
三相变压器 three-phase transformer
三相并激电动机 three-phase shunt motor
三相并励换向电动机 three-phase shunt commutator motor
三相补偿器 three-wire compensator
三相不平衡功率 three-phase unbalanced power; three-unbalanced power
三相插头 three-prong plug
三相沉积(物)【地】ternary sediment
三相串励电动机 three-phase series motor
三相串励换向电动机 three-phase series commutator motor
三相磁放大器 three-phase magnetic amplifier
三相的 three-phase; triphase; triple state
三相低频焊机 three-phase low frequency welder
三相点 triple point
三相点轨迹 triple point path
三相点降低曲线 triple point depression curve
三相电表 three-phase kilowatt-hour meter
三相电动机 three-phase motor; triple-phase motor
三相电动势 three-phase electromotive force
三相电度表 three-phase kilowatt-hour meter
三相电弧焊接 three-phase arc welding
三相电弧炉 three-phase (electric-) arc furnace
三相电机 three-phase machine
三相电缆 three-phase cable
三相(电力传动)柴油机车 three-phase diesel
三相电力传动装置 three-phase electric-drive
三相电力机车 three-phase locomotive
三相电力牵引铁路 three-phase railway
三相电力输送 three-phase power transmission
三相电力网 three-phase network
三相电流 three-phase current; triphase current
三相电流互感器 three-phase current transformer
三相电炉 three-phase furnace
三相电路 three-phase circuit
三相电路振荡器 three-phase current vibrator
三相电熔炉 three-phase smelting furnace
三相电势 triphasic potential
三相电枢 three-phase armature
三相电压 three-phase voltage
三相电压互感器 three-phase potential transformer
三相电源 three-phase supply
三相定子 three-phase stator
三相短路 three-phase short-circuit
三相短路曲线 three-phase short-circuit curve
三相断路时间 three-phase breaking time
三相二次自动重合闸装置 three-phase two shot reclosing device
三相发电机 three-phase generator; three-wire generator
三相反向变流机 three-phase inverter
三相分布 three-phase distribution
三相分离区污泥层 three-phase separation region sludge layer
三相负载 three-phase load
三相感应 three-phase induction
三相感应电动机 three-phase induction motor
三相感应调节器 three-phase induction regulator
三相功率计 three-phase wattmeter
三相供电 three-phase supply
三相供电网 three-phase power supply net
三相故障 three-phase fault
三相好氧生物流化床 three-phase aerobic biological fluidized bed
三相恒流充电机 three-phase Flotrol
三相滑环式电动机 three-phase slip-ring motor
三相滑环式感应电动机 three-phase slip-ring induction motor
三相换向电动机 three-phase commutator machine; three-phase commutator motor
三相汇流排 three-phase bus
三相机车 three-phase current locomotive
三相机车传动(装置) three-phase locomotive transmission
三相激磁器 three-phase exciter
三相激励 three-phase excitation
三相加热 three-phase heating
三相尖峰负荷 three-phase peak load
三相交流等离子发生器 three-phase alternating current plasma generator
三相交流电 three-phase alternating current
三相交流电动机 three-phase alternating current motor
三相交流电机车 three-phase alternating current locomotive
三相交流电牵引 three-phase alternating current traction
三相交流电源 three-phase alternating-current supply
三相交流发电机 three-phase alternator
三相交流感应电动机 three-phase alternating current induction motor
三相开关 three-phase switch
三相励磁机 three-phase exciter
三相连接 three-phase connection
三相流化床 three-phase fluidized bed
三相流化床反应器 three-phase fluidized bed reactor
三相流态化 three-phase fluidization
三相六线制 three-phase six-wire system
三相泡沫 three-phase froth
三相平衡 three-phase equilibrium
三相七线制 three-phase seven-wire system
三相起动器 three-phase starter
三相起重电动机 three-phase hoist motor
三相牵引 three-phase traction
三相桥 three-phase bridge
三相桥式线路整流器 double-way rectifier
三相区 three-phase region
三相曲线 triple curve
三相驱动 three-phase driving
三相全波接法 three-phase full wave connection
三相全波整流器 three-phase full-wave rectifier
三相绕线式感应电动机 three-phase; winding induction motor
三相绕组 three-phase winding
三相三次自动重合闸装置 three-phase three-shot reclosing device
三相三角形连接 delta connection
三相三绕组接线牵引变压器 traction transformer for three phase three winding connection
三相三线制【电】three-phase three-wire system
三相生物流化床 three-phase biological fluidized bed
三相输电线路 three-phase line
三相输入的单相焊接变压器 three-to-single-phase welding transformer
三相鼠笼式电动机 three-phase squirrel-cage motor
三相鼠笼式感应电动机 three-phase squirrel-cage induction motor
三相四线制 three-phase four-wire system; three-phase with neutral wire 4 hole socket-outlet
三相四线装置 three-phase four-wire installation
三相四芯电缆 three-phase four core cable
三相碳精灯 three-phase carbon arc lamp
三相条绕组 three-phase bar-winding
三相同步发电机 three-phase synchronous generator
三相图 block diagram; skeletal diagram; three-phase diagram; skeleton diagram <土的>
三相土 three-phase soil; triphase soil
三相瓦特计 three-phase wattmeter
三相网络 three-phase network
三相系 three-phase system
三相线路 three-phase line
三相限流熔断器 three-phase current limiting fuse
三相一次自动重合闸装置 three-phase one shot reclosing device
三相异步电动机 three-phase asynchronous motor
三相油浸式变压器 three-phase oil immersed transformer
三相整流焊机 three-phase rectifier welder
三相整流器 three-phase rectifier
三相整流子电机 three-phase commutator machine
三相制 three-phase system
三相转换开关 three-phase change-over switch
三相自耦变压器 three-phase autotransformer; three-wire compensator
三箱式断面 triple box section
三箱造型 three-cavity mo(u)lding
三镶板门 three-pack door
三向毕托管(嘴) three dimensional Pitot probe
三向补充阀 three-way compensating valve
三向布筋 three-way reinforcement
三向布筋体系 three-way system of reinforcement
三向布筋制 three-way system of reinforcement
三向测缝计 three-dimensional joint meter
三向测站 tripartite station
三向呈圆边的面砖 triple round tile

S

三向承重木结构 wooden three-dimensional load-bearing structure; wooden three-dimensional weight-carrying structure
三向的 three-dimensional; three-way; tridimensional
三向灯 three-way lamp
三向等长 three-dimensional equality
三向地带性 three-dimensional zonation
三向顶嵌合与双向嵌合的瓦片 triple head-lock and double side-lock tile
三向度视力 three-dimensional seeing
三向阀 three-way valve; 3-way valve
三向翻斗车 three-way dumper
三向方差分析 three-way analysis of variance
三向分层总合法 three-dimensional layer-built total-sun method
三向分析 three-dimensional analysis
三向辐射 triradiation
三向构架 three-dimensional frame
三向固结 three-dimensional consolidation
三向刮路机 three-way drag
三向光弹性 three-dimensional photoelasticity
三向桁构 three-dimensional truss
三向花格网架 three-way lattice(d) grid
三向加载 triaxial loading
三向接头 T-piece
三向结构 three-dimensional structure
三向结合 three-way connection
三向筋 space reinforcement
三向开关 T cock; three-way switch
三向靠模铣床 three-dimensional profiling machine
三向空间网架 three-way space grid
三向连接 space linkage; three-way connection
三向连接铁板 triple grip
三向联结板 three-way strap
三向联结构造 triple junction structure
三向链合 space linkage
三向量 trivector
三向裂隙比 three-dimensional extent of fissure
三向流 three-dimensional flow
三向龙头 three-way cock; three-way tap
三向内插 triple interpolation
三向配筋 three-way reinforcement
三向配筋板 three-way flat slab
三向配筋体系 three-way system of reinforcement
三向配筋制 three-way system of reinforcement
三向皮托管 three-dimensional Pitot tube
三向平板 three-way flat slab
三向嵌合的瓦片 triple interlocking tile
三向倾卸 three-way dump discharge; three-way dumping
三向倾卸车 three-way tipper
三向倾卸车身 three-way dump body
三向倾卸货车 three-way dump truck
三向倾卸机构 three-way tipping gear
三向倾卸式车厢 three-way dump body
三向倾卸式挂车 three-way tipping trailer
三向倾卸拖车 three-way dump trailer wagon
三向倾卸装置 three-way tipper
三向曲线 curve in space
三向三通接头 side outlet ell
三向三通弯接头 side outlet bend
三向三通弯头 side outlet elbow

三向色镀层 trichroic coating
三向色性 trichroism
三向渗透固结 three-dimensional consolidation
三向式信号机【交】three-way signal
三向四通管接 four-way T; side outlet
三向缩聚 three-dimensional polycondensation
三向弹性问题 three-dimensional elastic problem
三向体系 three-dimensional system
三向投影的 axonometric(al)
三向投影法 axonometric(al) projection
三向投影图 axonometric(al) projection
三向透视图 axonometric(al) perspective
三向图 axonometric(al) perspective; axonometry
三向图的 axiometric
三向推土机 three-side(d)(bull) dozer
三向弯管 three-way pipe
三向弯头 three-way elbow; three-way pipe
三向网格的 tri-grid
三向问题 three-dimensional problem
三向无梁楼板 three-way flat slab
三向系板 three-way strap
三向卸料装置 three-way tipping device
三向卸载车 three-way dump truck
三向旋转型滑坡 three-dimensional rotational slide
三向压力 triaxial pressure
三向压力传感器 triaxial cell
三向压缩强度 strength under peripheral pressure
三向压制 three-dimensional compaction
三向延伸矿体 three-dimensional extended orebody
三向应变计 rosette strain ga(u)ge
三向应变仪 rosette strain ga(u)ge
三向应变针 rosette strain ga(u)ge
三向应力 three-dimensional stress; triaxial stress
三向应力部位 triaxial stress area
三向应力状态 three-dimensional state of stress; three-dimensional stress state; triaxial state of stress
三向预应力 three-dimensional prestressing; triaxial prestress
三向振动 three-dimensional vibration
三向振动台 tri-axial shaking table
三向织物 three-dimensional weave
三向直角头 three-directional optic(al) square
三向制 three-way system
三向主应力三轴仪 true triaxial apparatus
三向驻波 diagonal standing wave
三向转辙器 three-throw switch
三向自卸汽车 three-way dump truck
三项的【数】trinomial
三项递归 three-term recurrence
三项方程 trinomial equation
三项分布 trinomial distribution
三项控制 three-term control
三项控制器 three-term controller
三项式【数】trinomial
三像片盘立体坐标仪 three-stage stereo-comparator
三硝胺 trinitramine
三硝基苯 trinitrobenzene
三硝基苯胺 trinitroaniline
三硝基化合物 trinitro-compound
三硝基甲胺 trinitro-methylamine
三硝基甲苯 tolite

三硝基甲苯硝胺 tetryl[trinitrophenylmethylnitramine]
三硝基甲苯炸药 TNT[trinitrotoluene]
三硝基甲苯中毒 trinitrotoluene poisoning
三硝基甲酚 cresolite; cresylite; trinitrocresol
三硝基萘 naphtite; trinitronaphthalene
三硝酸甘油 glycerol trinitrate
三硝酸甘油酯 glonoine; trinitrin; trinitroglycerin
三硝酸盐 trinitrate
三硝酸酯 trinitrate
三硝油 trinitrol
三销架式构架 triple-pinned truss(ed) frame
三销接的 triple-pinned
三小时段指数 three-hour range index
三小时行星地磁指数 three-hour geomagnetic planetary index
三效 triple effect
三效蒸发 triple-effect evapo(u)ration
三效蒸发浓缩 triple-effect evapo(u)ration concentration
三效蒸发器 triple-effect evapo(u)rator
三楔边 triple wedge
三斜 clinorhomboidal; triclinic
三斜半面体类 triclinic hemihedral class
三斜钡解石 alstonite
三斜长石 triclinic feldspar
三斜的 triclinic; anorthic <晶体>
三斜底面 triclinic base
三斜度 triclinicity
三斜对称 triclinic symmetry
三斜多面柱镜 triclinic holohedral prism
三斜晶胞 triclinic cell
三斜晶的 triclinic
三斜晶体 anorthic crystal; triclinic crystal
三斜晶系 anorthic system; clinorhoboidal system; triclinic system
三斜晶系的 triclinic
三斜蓝铁矿 metavivianite
三斜磷钙石 monetite; whitlockite
三斜磷钙铁矿 anapaite
三斜磷铅铀矿 parsonsite
三斜磷锌矿 tarbuttite
三斜氯羟硼钙石 coll-tyretskite
三斜镁铁磷灰石 collinsite
三斜锰辉石 pyroxmangite
三斜硼钙石 meyerhofferite
三斜硼钙石减水剂 meyerhofferite water reducer
三斜全面体类 triclinic holohedral class
三斜三水铝石 triclinogibbsite
三斜闪石 aenigmatite; enigmatite
三斜砷钙石 weilite
三斜砷钴钙石 roselite
三斜砷铅铀矿 hallimomdite
三斜石 trimerite
三斜水钒铁矿 schubnelite
三斜水磷铍锰石 roscherite triclinic
三斜水硼石 veatchite-A
三斜铁辉石 pyroxferroite
三斜系 anorthic system
三斜霞石 carnegieite
三斜硬绿泥石 triclinochloritoid
三斜柱 triclinic prism
三斜柱体 triclinic prism
三斜组构 triclinic fabric
三心 three-core
三心插塞 three-pin plug
三心单元 three-core cell
三心的 three-centered; three-pinned; three-way

三心电缆 three-core cable
三心电缆与三根单心电缆的接线盒 trifurcating joint
三心复曲线 three-centered compound curve
三心拱 anse de panier; basket arch; basket handle arch; hance arch; multicentred arch; three-centered arch; three-element arch; three-segment arch
三心拱坝 three-centered [centred] arch dam
三心拱形断面 three-centered [centred] circular arch section
三心拱支架 three-centered [centred] circular arch support
三心花瓣拱 trefoil arch
三心平圆拱【建】three-centered[centred] arch
三心曲线 three-centered curve
三心塞子 three-contact plug
三心椭圆 <仿椭圆> pseudo-ellipse
三心椭圆拱 three-centered [centred] arch
三心圆拱 three-centred[centred] arch
三心直线尖顶拱 Tudor arch
三心柱变压器 three-column transformer
三心柱铁芯 <变压器> three legs type core
三心装饰拱【建】false ellipse arch
三芯测井电缆 three-cone logging cable
三芯导线 three-core conductor; triple conductor
三芯电缆 three-conductor cable; three-core cable; triple(-core) cable; triplex cable
三芯电缆平衡保护装置 core-balance protective system
三芯塞绳 three-way cord
三芯塑料线 three-core plastic wire
三芯同轴电缆 triple-concentric cable
三芯线终端套管 trifurcating box
三芯型电缆 triple type cable
三信道跟踪接收机 three-channel tracking receiver
三星标 asterism
三星齿轮 tumbler gear
三星定位 three-star fix
三星定位法 three-star problem
三星体 triaster
三星凸轮 cloverleaf cam
三星牙(齿轮) tumbler gear
三行 triplex row
三行程柱塞泵 triple throw plunger pump
三行的 three-row; triple row
三行键盘 three-row keyboard
三行铆钉 triple-riveted joint
三行铆钉对接 treble riveted butt joint
三行铆钉接合 treble riveted joint
三行铆钉搭接 triple-riveted lap joint
三行铆接 triple-riveted joint
三行平铆接 triple-riveted butt joint
三行起垄机 three-row ridger
三行区 three-row zone
三行区穿孔 zone punch
三行区打孔 overpunch
三行式铆钉 three-plier; triple rivet(ing)
三行柱的 tripteral
三形 trimorphism
三形的 trimorphous
三形花 trimorphic flower
三型钾霞石 trikalsilite
三溴胺 tribromamine
三溴苯 tribromo-benzene
三溴苯胺 tribromaniline

S

三溴苯酚 tribromophenol
三溴苯酯 tribromophenyl
三溴二氯乙烷 tribromo-dichloroethane
三溴酚铋 tribromophenol bismuth
三溴化钒 vanadous bromide
三溴化合物 tribromide; tribromo-compound
三溴化钼 molybdic bromide
三溴化铊 thallic bromide; thallium tribromide
三溴化铟 indium tribromide
三溴甲烷 bromoform; methenyl tribromide; tribromomethane
三溴片 tribromide
三溴氧化钒 vanadium oxytribromide
三溴乙酸 tribromoacetic acid
三溴乙烷 tribromoethane
三穴模具 triple-cavity mo(u)ld
三学科 <中世纪学校的> trivium
三循环 ternary cycle
三压区沟纹压榨 tri-vent press
三压区压榨 tri-nip press
三牙轮扩孔器 three-point roller reamer
三牙轮岩石钻头 three-cone rock bit; tricone rock bit
三牙轮硬合金球齿钻头 tricone bit with tungsten carbide inserts
三牙轮凿岩钻头 tricone rock bit
三牙轮钻头 three-cone bit; three-roller bit; tricone bit
三烟囱的 three-stack(ed)
三眼滑轮 dead eye
三眼环 eye plate; monkey face
三眼辘轳 dead eye
三眼木饼 blind pulley; dead eye
三氧二化物 sesquioxide
三氧化碲 telluric acid anhydride; telluric oxide
三氧化二铋 bismuth trioxide; bismuth yellow
三氧化二钚 plutonium sesquioxide
三氧化二氮 nitrogen trioxide
三氧化二钒 vanadium sesquioxide; vanadium trioxide
三氧化二铬 chrome green; chromic oxide; chromium oxide; chromium pentoxide
三氧化二镓 gallic oxide; gallium sesquioxide
三氧化二金 gold trioxid
三氧化二铼 rhenium sesquioxide
三氧化二磷 phosphorus anhydride; phosphorus trioxide
三氧化二硫 sulfur sesquioxide
三氧化二铝 alumina; alumin(i)um oxide; alundum
三氧化二铝饱和度 Al_2O_3 saturability
三氧化二铝含量 content of alumina
三氧化二锰 manganese sesquioxide; manganic oxide
三氧化二钼 molybdenum sesquioxide
三氧化二镍 nickelic oxide; nickel sesquioxide
三氧化二钕 neodymium sesquioxide
三氧化二铅 plumbous plumbate
三氧化二砷 arsenic trioxide; rude arsenic; white arsenic
三氧化二酸 phosphorous anhydride
三氧化二铊 thallium trioxide
三氧化二锑 antimony trioxide
三氧化二铁 ferric oxide; iron sesquioxide; roude
三氧化二铁含量 content of ferric oxide
三氧化二铟 indium trioxide
三氧化铬 chromium trioxide
三氧化铼 rhenium trioxide
三氧化硫 sulfuric[sulphuric] acid anhydrite; sulfur[sulphur] trioxide
三氧化硫和硫酸雾中毒 sulfur trioxide and sulfuric acid fume poisoning
三氧化硫烟雾 sulfur trioxide mist; sulphur trioxide mist
三氧化钼 molybdenum trioxide
三氧化钛 titanium peroxide; titanium trioxide
三氧化锑 antimony trioxide
三氧化铁 Indian red
三氧化钨 tungsten trioxide
三氧化物 trioxide
三氧化铱 iridium black
三氧化铀 uranium trioxide
三氧硫钨酸盐 trioxysulfotungstate
三样本理论 three-sample theory
三摇臂式酸洗机 three-arm pickling machine
三叶 three-leaf
三叶草 clover
三叶草形 cloverleaf pattern
三叶草栽培地 clover sod
三叶虫灰岩 trilobite limestone
三叶窗花格 three-lobe tracery
三叶的 three-bladed
三叶钉 three-flanged nail
三叶拱 trefoil arch
三叶毂 three-blade hub
三叶花饰 trefoil flower; Tudor flower
三叶花样【建】trefoil
三叶胶 paracaoutchouc; para-rubber tree; seringa
三叶铰链 three-leaf pin hinge
三叶搅拌桨 three-blade mixing paddle
三叶连拱廊 trefoiled arcade
三叶轮 trilobed wheel
三叶螺旋桨 three-bladed propeller; three-blader
三叶密封片 triple seal-tab
三叶片的 three-vaned
三叶片离心泵 three-vane(d) centrifugal pump
三叶片式泵 three-lobe pump
三叶片式水泵 three-lobe pump
三叶期 trefoil stage
三叶式布置 cloverleaf layout
三叶式窗饰 trefoiled tracery
三叶式花格窗 trefoiled tracery
三叶式活动闸门坝 three leaved bear-trap dam
三叶饰 trefoil(ornament)
三叶推进桨式搅拌器 three-blade marine type propelled agitator
三叶卧式闸门熊阱坝 three leaved bear-trap dam
三叶线【数】trifolium
三叶橡胶树 paracaoutchouc; para-rubber tree
三叶形 trefoil; trilobal
三叶形窗花格 three-foiled tracery
三叶形拱 trefoil arch
三叶形拱点 trefoilapsis
三叶形截面 trifoil cross-section
三叶形平面图 three-foiled ground plan
三叶形气缸 three-lobe chamber
三叶形饰的连拱廊 three-foiled arcade
三叶形饰拱 three-foiled arch
三叶形陶瓷热交换器 trefolate ceramic heat exchanger
三叶形轴承 three-lobe bearing
三叶植物 trefoil
三叶轴承 three-lobe bearing
三叶转子泵 cloverotor pump
三叶状平面图 trefoiled ground plan
三叶状塔 trefoil-shaped tower (block)
三页经面斜纹 three-leaf warp twill
三页纬面斜纹 three-leaf filling twill
三页斜纹 three-harness twill; three-leaf twill
三一统【地】Trinity series
三乙铋 triethyl-bismuthine
三乙醇胺 <一种混凝土早强剂> triethanolamine
三乙醇胺复合早强剂 complex accelerator based on triethanolamine
三乙醇胺盐 triethanolamine salt
三乙基铋 triethyl-bismuth
三乙基甲硅烷 triethyl-silicane; triethyl-silicon
三乙基金属 triethide
三乙基铝 triethylaluminium
三乙基氯化锡 triethylchlorotin
三乙基锡 hexaethylditin; tin triethyl; treithyltin
三乙镓 triethyl-gallium
三乙酸铝 alumin(i)um triacetate
三乙酸盐 triacetate
三乙锑 triethylantimony
三乙烯 triethylene
三乙氧基铀 tricyclopentadienyl-ethoxyuranium
三异丙醇胺 tri-isopropanolamine
三异戊胺 triisoamylamine
三翼的 three-winged
三翼钉打拔器 impactor extractor for trifin nails
三翼飞机 triplane
三翼刮刀钻头 three-winged drag bit
三翼岩芯钻头 soft-formation cutter head
三翼钻头 three-point bit; three-wing drill bit
三因次的 three-dimensional
三因次分析 three-dimensional analysis
三因次理论 three-dimensional theory
三因次弹性问题 three-dimensional elastic problem
三因次振动 three-dimensional vibration
三因分类 trifactor classification
三因素方差分析 three-factor analysis of variance
三因子交互影响 triple interaction
三音平面位置指示器 three-tone plan position indicator
三音速 trisonic
三音速气动力学 trisonics
三音速试验 trisonic test
三音信号 three-tone signal
三引擎组合 three-engine hook up
三引线连接器 three-entry connector
三英尺安全线 <甲板舱口周围划的安全线,1 英尺 = 0.3048 米> three-feet line
三英尺半轨距 three-feet and six inches ga(u)ge
三英尺长折尺 three-foot rule
三英尺轨距 <折合 0.950 米> three-feet line ga(u)ge
三英尺六英寸轨距 <1 英寸 = 0.0254 米> three-feet and six inches ga(u)ge
三英尺一英寸十二分之一的轨距 three-feet line ga(u)ge
三英寸大钉 tenpenny; tenpenny nail
三硬脂酸甘油酯 glyceryl tristearate; stearin(e); tristearin
三硬酯酸盐 tristearate
三用的 triple-purpose
三用电表 avometer
三用电桥 <测量电阻、电容和电感> component bridge
三用阀 <调节、减速、止回用> flow control deceleration check valve
三用分析仪 three-way analyser[analyzer]
三用机 three-way set
三用接收机 three-way radio; three-way receiver
三用门式起重机 <吊钩-抓斗-磁铁> hook grab magnet gantry crane
三用品种 triple-purpose breed
三用起重机 three-operating crane
三用沙发 studio couch
三用蒸发器 triple-effect evapo(u)rator
三用钟 three-part clock
三用装置 three-in-one unit
三油精 olein; triolein
三油酸甘油酯 olein
三油楔轴承 three-wedge bearing
三铀 triuranium
三元 three-element; unit triplet
三元胺 tertiary amine; tri-amine
三元玻璃形成区 region of ternary glass formation
三元财务技术 triotechnology
三元醇 trihydric alcohols; trivalent alcohol
三元催化反应器 three-way catalytic reactor
三元催化废气净化系统 three-way system
三元催化净化 three-way emission control
三元催化净化器 three-unique catalytic converter
三元催化排气净化器 three-way catalytic converter
三元代数 ternary algebra
三元的 ternary; three-component; triaxial; tribasic【化】; trihydric; trinal; trinary
三元电解质 ternary electrolyte
三元二次形式 ternary quadratic form
三元分布常数 ternary distribution constant
三元分配系数 ternary distribution coefficient
三元酚 triatomic phenol
三元复合肥料 three-nutrient compound fertilizer
三元复合毡 triple mat
三元钢 ternary steel
三元工程技术 triotechnology
三元共晶 ternary eutectic
三元共聚 ternary polymerization
三元共聚物 terpolymer
三元共聚物整理 terpolymer finish
三元固体 tertiary solid
三元关系 ternary relation
三元管理技术 triotechnology
三元光谱分类 three-dimensional spectral classification
三元过磷酸钙肥料 triple superphosphate fertilizer
三元合金 ternary alloy; ternary composition; three-part alloy
三元合金钢 ternary alloy steel
三元合金探测器 trimetal detector
三元化合物 ternary compound
三元化合物半导体 ternary semiconductor
三元化合物结构 ternary compound structure
三元环 triatomic ring
三元混合耐火材料 miscellaneous ternary refractory
三元混合染料 ternary dye
三元混合色 ternary colo(u)r
三元混合物 tertiary mixture
三元基 triad
三元级 triplet
三元件滤波器 three-element filter
三元件伺服系统 three-element servo-system
三元件物体 triplex
三元结构层序 three-component sequence

三元聚合物 terpolymer
三元论 trialism
三元络合物 ternary complex
三元络合物分析法 ternary complexes analysis method
三元码 ternary code;three-unit code
三元脉冲 triplet unit impulse; unit triplet impulse
三元面 three-dimensional surface
三元配料 ternary mix
三元配置 three-way layout
三元喷管 three-dimensional nozzle
三元碰撞 ternary collision;triple collision
三元取代 triple substitution
三元溶液 ternary solution
三元三次型 ternary cubic(al)form
三元色白 trichromatic
三元双一次形式 ternary bilinear form
三元水流 three-dimensional flow
三元四次型 ternary quartic form
三元四极透镜 triplet quadrupole
三元酸 triacid; ternary acid; tribasic acid;triprotic acid
三元酸酯 tribasic ester
三元羧酸 tribasic carboxylic acid
三元碳化物 double carbide
三元体系 ternary system;triad
三元天线 three-element antenna
三元调节 three-element control
三元图 ternary diagram
三元图解 ternary diagram
三元推进剂 tripropellant
三元万能测长机 Trioptic
三元紊流 three-dimensional turbulence
三元系 ternary system; three-part system
三元系区域精炼炉 three-component refiner
三元系陶瓷 ternary system ceramics
三元系压电陶瓷 ternary system piezoelectric(al)ceramics
三元相 ternary phase
三元相图 ternary phase diagram
三元型 ternary form
三元液体系统 ternary liquid systems
三元乙丙胶 ethylene-propylene terpolymer;trihydric ethylene-propylene rubber
三元乙丙胶垫 ternary ethylenepropylene rubber packing
三元乙丙橡胶 ethylene-propylene rubber
三元乙丙橡胶密封垫 ethylene-propylene rubber gasket
三元蒸气液体平衡 ternary vapo(u)r-liquid equilibrium
三元组 triad;triple
三元组表 triple table
三元组群 triad group
三元组同伦集 triad homotopy set
三元组形式 triple form
三原色 set of colo(u)rs; three-colo(u)r
三原色比色计 trichromatic colo(u)rimeter;trichrometric colo(u)rimeter
三原色彩色显像管 tricolo(u)r kinescope
三原色单位 trichromatic unit
三原色的 trichromatic
三原色法 tricolor system
三原色滤光片 elementary colo(u)r filter
三原色滤色片 trisimulus filter
三原色色度计 trichromatic colo(u)rimeter
三原色图像 trichromatic image
三原色系数 trichromatic coefficient
三原色系统 trichromatic system
三原色性 trichroism
三原色印花 trichromatic printing
三原色坐标 trichromatic coordinates
三原型 triarch
三原岩 miharaite
三原子的 triatomic
三原组织 <平纹斜纹缎纹三个基本组织> three-foundation weave
三圆边 <釉面砖> triple-round edge
三圆测角仪 third circle goniometer
三圆的直交圆 orthotomic circle
三圆心蝠线 three-centered curve
三圆心曲线 three-center curve
三圆柱模滚轧螺纹法 three-cylindrical-die thread rolling
三圆锥室磨碎机 tricone compartment mill
三缘空心钻头 three-lip core drill
三缘麻花钻 three-lipped twist drill
三缘钻头 three-lipped drill
三月检库 three-months inspecting shed
三匝蹬筋 triple-looped stirrup
三渣 lime-flyash treated broken stone; lime-flyash concrete <三渣俗称>
三闸板防喷器 triple ram preventer
三宅岩 migakite
三毡两油平屋顶 macasfelt
三张相联的图画 triptych
三张像片组 single triplet
三张纸试验 three-paper test
三胀式蒸汽机 triple-expansion engine
三账户制 three-account system
三折布抛光轮 triple buff
三折合页 three-ply butt
三折铰链 three-ply butt
三折喇叭形反射器天线 triple-folded hornreflector
三折书牒 triptych
三折书牒的一块板 volet
三折隐梯 triple-fold disappearing stair(case)
三者 three
三支边试验陶管法 three-edge-bearing method of testing clay pipes
三支点梁 beam with central prop
三支电路接线盒 trifurcating box
三支交叉口 three-legs intersection
三支铰 tri-hinges
三支裂谷系 three-arm rift system
三支式 triadic type
三支腿起重机 tripod crane
三支形饰 triskelion[复 triskelia]
三枝构成的辐射状图形 triskelion[复 triskelia]
三直角球面三角形 trirectangular spheric(al)triangle
三直角三面形 trirectangular trihedral
三直角四边形 trirectangular quadrilateral
三直角锥反射镜 corner cube mirror
三直角锥反射器 corner cube reflector
三直角锥棱镜 corner cube prism
三直线基线 trilinear primary curve
三值逻辑 three-valued logic
三值模拟 three-value simulation
三指标符号 three-index symbol
三指定律 Fleming's rule;three-finger rule
三指定则 three-finger rule
三指内径规 <用于测量钻孔断面的> three-fingered calipers
三趾马属 Hipparion
三种成分的环氧树脂 three-component epoxy resin
三种成分的水力黏[粘]合剂 three-part hydraulic binder
三种成分的液压黏[粘]合剂 three-part hydraulic binder
三种成分重量配料计量器 three-weight batcher;triple weighing batcher
三种费率计算器 three-kind rate meter
三种轨道病害 three-rail fault
三种声速范围的空气动力学 trisonic aerodynamics
三种时间估计法 three-time estimate
三种土壤水分状况 three-soil-water regime
三种压力平衡的压力控制阀 three pressure equalizing pressure control valve
三种压力平衡系统(制动)three-pressure equalizing system
三种音速范围 trisonic range
三种原料配料 ternary mix
三种主要吸收途径 three-main routes of absorption
三种作物的轮作周期 tree-crop cycle
三州间地区 tristate
三周年纪念 triennial
三周期函数 triply periodic function
三轴饱和磁力仪 triaxial flux-gate magnetometer
三轴参考系 three-axis reference system
三轴参照系 triaxial reference system
三轴测压仪 triaxial cell
三轴承电机 three-bearing machine
三轴承曲轴 three-bearing crankshaft
三轴承式 three-bearing type
三轴承小齿轮 three-bearing pinion
三轴传动装置 three-axle gear
三轴串联滚压机 three-axle tandem roller
三轴串联式压路机 three-axle roller; three-axle tandem roller; three-roll tandem roller;triaxial tandem roller
三轴串列式滚压机 three-axle tandem roller
三轴串列式压路机 three-axle tandem roller
三轴磁强针 three-axial magnetometer
三轴的 triaxial
三轴地震计 triaxial seismometer
三轴地震仪 triaxial seismograph
三轴电缆 triaxial cable
三轴定位器 tri-axis locator
三轴定向 triaxial orientation
三轴多缸汽轮机 triple cross-compound turbine
三轴发动机 three-shaft engine
三轴翻斗卡车 three-axle rear-dump truck
三轴分布 triaxial distribution
三轴骨针 triaxon
三轴固结排水压缩试验 consolidated-drained triaxial compression test
三轴挂车 three-axle trailer
三轴光弹量测法 triaxial photoelastic measurement
三轴后卸卡车 three-axle rear-dump truck
三轴货车 three-axle truck;three-axle wagon
三轴基准 three-axis reference
三轴激光陀螺装置 three-axis laser gyro package
三轴加荷 triaxial loading
三轴加速计 triaxial accelerometer
三轴加载 triaxial loading
三轴剪力 triaxial shear
三轴剪力试验 triaxial shear test
三轴剪力试验仪 triaxial shear apparatus;triaxial shear equipment
三轴剪力仪 triaxial shear equipment
三轴剪切 triaxial shear
三轴剪切机 triaxial shears
三轴剪切试验 triaxial shear test
三轴剪切试验仪 triaxial shear equipment;triaxial shear apparatus
三轴剪切仪 triaxial shear equipment
三轴静载压力试验 triaxial static compression test
三轴卡车 three-axle truck
三轴抗压强度 strength under peripheral pressure; triaxial compression strength
三轴抗压试验 triaxial compression test
三轴控制系统 three-axis control system
三轴快剪试验 quick triaxial(shear) test
三轴拉伸试验 triaxial extension test; triaxial tensile test
三轴力平衡式加速度计 triaxial forced balance accelerator
三轴联动机车 six-coupled locomotive
三轴模拟器 three-axis simulator
三轴排水"涂抹"试验 drained triaxial "smear" test
三轴平衡式加速度计 triaxial forced balance accelerometer
三轴汽车 six-wheeler;six-wheel vehicle;three-axle vehicle
三轴强度 triaxial strength
三轴桥梁绞车架 three-axle girder body
三轴切变仪 triaxial shear equipment
三轴穹顶 tripartite vault
三轴燃气轮机 three-shaft gas turbine
三轴伸长试验 triaxial extension test
三轴式侧卸卡车 three axles side-dump
三轴式铲运机 three axles scraper
三轴式底卸卡车 three-axles bottom-dump
三轴式后卸卡车 three axles rear-dump
三轴式挤压制管机 triaxial extrude pipe machine;triaxial pipe extruder
三轴试验 triaxial test
三轴试验的三种类型 three types of triaxial test
三轴试验方法 triaxial system;triaxial test method
三轴试验容器 triaxial testing cell
三轴试验仪 triaxial testing apparatus
三轴试验装置 triaxial apparatus
三轴受压试验 triaxial test
三轴图 triaxial chart;triaxial diagram
三轴陀螺平台 three-axis gyroplatform
三轴椭球体 three-axial ellipsoid;triaxial ellipsoid
三轴椭圆体 triaxial ellipsoid
三轴稳定平台 three-axis stabilized platform
三轴稳定自动驾驶仪 three-axis gyropilot
三轴稳定作用 three-axis stabilization
三轴铣床 three-head milling machine
三轴限制压力 triaxial confining pressure
三轴向负载感传器 triaxial load cell
三轴向抗压强度 triaxial compressive strength
三轴向无捻粗纱布 triaxial roving fabric
三轴向压力试验 triaxial compression test
三轴向因数 triaxiality factor
三轴向应力 triaxial stress
三轴向张力 hydrostatic tension;triaxial tension
三轴向织物 triaxial fabric
三轴向织造 triaxial weaving

三轴形变 triaxial deformation
三轴型 three-axle model
三轴性 triaxiality
三轴压力 triaxial compression
三轴压力传感器 triaxial cell
三轴压力盒 triaxial cell
三轴压力试验法 triaxial system
三轴压力试验仪 triaxial apparatus; triaxial compression apparatus
三轴压力室 triaxial cell; triaxial chamber
三轴压路机 three-axle roller; three-tandem roller
三轴压实仪 triaxial compaction apparatus; triaxial compaction test machine
三轴压缩 triaxial compression
三轴压缩剪切试验 triaxial compression shear test
三轴压缩试验 triaxial compression test
三轴压缩试验仪 triaxial compaction apparatus; triaxial compression (test) apparatus
三轴压缩仪 triaxial compression apparatus; triaxial compression machine
三轴压应力 triaxial compressive stress
三轴仪 triaxial apparatus; triaxial equipment
三轴仪压力盒 triaxial cell
三轴仪压力室 triaxial cell
三轴引张试验 triaxial extension test
三轴应变 triaxial strain
三轴应变计 triaxial strain cell
三轴应力 three-dimensional stress; triaxial stress
三轴应力场 triaxial stress field
三轴应力-应变图 triaxial stress-strain plot
三轴应力状态 triaxial state of stress; triaxial stress state
三轴预应力 triaxial prestress
三轴运输机 three-axle carrier
三轴载货挂车 three-axle truck trailer
三轴转向架 six-wheel truck; three-axle bogie; three-axle truck
三轴装置 three-axis mounting; triaxial mount
三轴紫外染料激光器 triaxial ultraviolet dye laser
三轴自动驾驶仪 three-axis autopilot
三轴钻孔变形计 triaxial bore-hole deformation ga(u)ge
三轴钻孔形变计 triaxial bore-hole deformation ga(u)ge
三轴坐标控制 three-axis control
三住户楼房 triplex building
三注意 <驾驶员的> three-L's
三注意测深 lead; location and look-out
三注意定位及瞭望 lead; location and lookout
三柱管 three-beam tube
三柱径式 diastyle
三柱块体 <混凝土> tribar
三柱塞泵 three-ram pump; three-throw plunger pump; triplex plunger pump
三柱十字桩 three-post cruciform bollard
三柱式 three-column type
三柱式车库用举升器 triple-post lift
三柱式檩支屋顶 purlin(e) roof with three posts
三柱式桥墩 tri-column pier
三柱式升车机 triple-post lift
三柱体 <防波堤护面块体> tribar
三柱铁芯 three-limb core
三柱头的 tristigmatic
三柱系统 system of three posts
三柱型铁芯 three-column core
三柱凿孔机 hand perforator; punch perforator
三柱桩 <屋顶架> three posts
三爪 three-jaw
三爪安全提引钩 triple suspension safety hook
三爪定心夹盘 three-jaw concentric lathe chuck; three-jaw independent lathe chuck
三爪分动卡盘 three-jaw independent chuck
三爪杆动夹盘 three-jaw lever-operated chuck
三爪夹盘 three-jaw chuck; universal chuck
三爪卡盘 cam-ring chuck; chuck with three-jaws; ring wheel chuck
三爪锚 triple-fluked anchor; triple grip anchor
三爪木材抓斗 three-tine log grappler
三爪内拉簧式夹盘 three-jaw draw-in type chuck
三爪破碎机 three-jaw crusher
三爪气动卡盘 delta air clutch
三爪钳 trielcon
三爪手动夹盘 three-jaw hand operated chuck
三爪提引钩 triplex hook
三爪同心夹盘 three-jaw concentric chuck
三爪同心卸轮器 three-jaw equalizing drive
三爪万能卡盘 three-jaw universal chuck
三爪自动定心卡盘 scroll chuck; three-jaw chuck
三爪钻卡 three-jaw drill chuck
三转热风炉 three-pass stove
三转速电唱机 triple-speed gramophone
三转子 triple-spool; trispool
三转子泵 tri rotor pump
三幢以上并联住宅 dwelling townhouse; row house
三锥 bottoming tap; third hand tap; third tap
三锥齿 triconodont
三锥齿轮钻头 tricone bit
三锥滚轮钻头 tricone roller bit
三锥轮凿岩钻头 tricone rock bit
三锥式分室 tricone compartment mill
三锥式磨机 tricone mill
三锥式球磨机 tricone mill
三锥象虫科 Brentidae
三锥型钻头 tricone bit
三锥牙 triconodont
三锥牙轮钻头 three-cone bit; tricone bit
三浊点 triple cloud points
三字点 three-letter point
三字点磁罗经 false point; intermediate point
三字母信号 <国际信号规则的> three-letter signal
三字母组 trigram
三自由度 three-degree of freedom
三自由度建筑物 three-degree of freedom structure
三自由度陀螺练习器 three-axis-degree-of-motion trainer
三自由度陀螺仪 free gyroscope
三自由度陀螺指示器 three-axis gyro indicator
三棕榈精 palmitin
三棕榈酸甘油酯 palmitin
三足离心机 tripod pendulum type batch centrifugal
三足起重机 three-legged derrick
三足支架 cockspur
三组 trio
三组并排椭圆弹簧 triplet elliptic spring
三组法 three-group method
三组分测量系统 three-component measuring system
三组分瓷器 triaxial porcelain
三组分单元 three-component unit
三组分的 tripartite
三组分的液压黏[粘]结剂 three-component hydraulic binder
三组分混合料 ternary blends; three-constituent mixture <多用水泥,高炉渣与火山灰>
三组分混合物 ternary mixture
三组分混凝土 three-component concrete
三组分力 triparted force
三组分陶瓷 triaxial ceramics; triaxial pottery
三组分系统 three-component system
三组分显微煤岩类型 trimaceral microlithotype
三组交通信号 three-way signal
三组锚碇桩 triple grouser
三组锚碇桩座 triple-grouser shoe
三组汽缸 three-bank cylinders
三组式变焦距镜头 three-component zoom lens
三组圆点 dot trio
三组镇压器 triple roller
三嘴包装机 three-spout packing machine
三尊窗 Venetian window
三作 triple-cropping
三作用催化转换器 triple mode catalytic converter
三作用桨叶式搅拌机 triple action paddle mixer
三作用千斤顶 three-acting jack; tri-actional screw jack; triple acting jack
三坐标测量仪 three-coordinates measuring machine
三坐标带控制 three-dimensional tape controlled
三坐标电子扫描固定阵雷达 three-dimensional electronically scan(ning) fix array radar
三坐标反应谱 tripartite response spectrum
三坐标夹板式送料装置 three-dimensional holding plate feeder
三坐标目标跟踪和指示雷达 three-dimensional target tracking acquisition radar
三坐标搜索与测高雷达 three-dimensional search and height finding radar
三坐标探测雷达 three-dimensional acquisition radar
三坐标外形加工 three-axis contouring
三座车身 cloverleaf body

伞

伞板式换热器 bevel plate heat exchanger
伞兵运输飞机 troop carrier
伞齿 bevel gear
伞齿节面角 helix angle
伞齿轮 angular wheel; bevel (led) gear; bevel (pinion); bevel wheel; cone gear; conic (al) gear; miter gear
伞齿轮传动 bevel drive
伞齿轮传动机构 bevel gear transmission
伞齿轮底角 root angle
伞齿轮基锥 generating cone
伞齿轮节锥顶 cone center[centre]
伞齿轮螺栓 bevel gear bolt
伞齿轮面角 face angle
伞齿轮刨齿机 bevel gear generator
伞齿轮刨刀刀盘 Gleason cutter
伞齿轮起重器 bevel gear jack
伞齿轮式起重器 ratchet type jack
伞齿轮速比 bevel gear ratio
伞齿轮行星减速机 bevel planetary gear drive
伞齿轮研齿机 bevel gear burnishing machine
伞齿轮闸门启闭机 bevel gear gate lifting device
伞齿轮轴 bevel gear shaft
伞齿轮主传动(机构) bevel gear main drive
伞齿轮转动 bevel wheel
伞齿铣刀 bevel cutter
伞传动齿轮 bevel wheel
伞代更新 compartment system; regeneration under shelterwood system
伞伐 shelter-wood felling
伞房花桉 swamp gum
伞盖状的 umbraculiferous
伞菌 agaric; agaric fungus
伞菌科 <拉> Agaricaceae
伞菌类 tlymenomycetes
伞菌属 <拉> Agaricus
伞式发电机 below bearing type generator; umbrella-type generator
伞式水力发电机 umbrella-type hydrogenerator
伞式水轮发电机 umbrella-type (hydraulic) generator; umbrella-type hydrogenerator
伞头螺栓 tumbler bolt
伞投物资 parabundle
伞形 umbrella shape; umbrella type
伞形薄壳 umbrella shell
伞形齿轮 bevel gear; bevel pinion; bevel wheel; crown gear <差动器侧面的>
伞形齿轮变速装置 bevel wheel change gear
伞形齿轮操纵阀 bevel gear operated valve
伞形齿轮差动器 bevel gear differential
伞形齿轮传动 bevel drive; bevel gear drive
伞形齿轮转换器 bevel gear reverse
伞形粗纱架 umbrella creel
伞形的 mushroom; umbellate; umbelliform
伞形防鼠板 rat guard
伞形防雨罩 rain hood
伞形风帽 cowl; rectangular cowl; weather cap
伞形拱顶 umbrella vault
伞形滑轮 umbrella pulley
伞形货棚(有顶无墙) umbrella shed
伞形交流发电机 umbrella-type alternator
伞形结构 umbrella form
伞形进风口 mushroom-type inlet
伞形绝缘器 umbrella-type insulator
伞形绝缘子 umbrella insulator
伞形壳(体) mushroom shell
伞形矿物 agaric mineral
伞形锚 mushroom anchor
伞形帽 cone cup
伞形喷口 cap jet
伞形棚 umbrella frame
伞形散流器 mushroom diffuser
伞形树 umbrella tree
伞形水轮发电机 umbrella-type alternator
伞形天线 umbrella aerial; umbrella antenna
伞形通风帽 mushroom ventilator

伞形通风筒 umbrella vent
伞形屋顶 mushroom roof; station roof; umbrella(-shaped) roof
伞形屋面壳 umbrella shell
伞形吸气罩 canopy hood
伞形相思树 umbrella acacia
伞形效应 umbrella effect
伞形斜拉桥 bundle type cable-stayed bridge; converging type cable-stayed bridge
伞形修枝 umbrella system
伞形圆形屋顶 parachute vault
伞形晕轮状灌浆 aureole grouting
伞形帐篷 umbrella tent
伞形照明 veiling luminance
伞形罩 canopy hood; cowl; umbrella hood
伞形真菌 toadstool
伞形支架 umbrella stull
伞形支柱 umbrella prop
伞形植物 umbellifer
伞形柱顶 mushroom slab
伞形桩锚 umbrella pile anchor
伞型 umbellate form; umbrella type
伞绣菌属 < 拉 > Ravenelia
伞衣 canopy
伞椅缠绕 seat and chute entanglement; seat and chute involvement
伞状的 umbrella-shaped
伞状模具 mushroom mo(u)ld
伞状泥芯 umbrella core
伞状石斑木 yeddo raphiolepis
伞状收集器 umbrella collector
伞状树 umbrella tree
伞状天线 mushroom antenna; umbrella antenna
伞状物 mushroom
伞状型芯 umbrella core

散

散包 bale off
散比重 bulk density
散边 fag-end
散兵壕 rifle-pit
散兵坑 fox hole; slit trench
散材料 bulk
散仓水运 bulk freight
散舱货 bulk cargo
散存料 bulk stock
散弹噪声电压 shot noise voltage
散岛 strewn island
散点交会定位法 location by forward intersection
散点石 scattered stone
散点图 dot graph; scatter diagram
散斗 small block
散度 divergence[divergency] vergence [vergency]【物】
散度场 field of divergence[divergency]
散度定理 divergence theorem[divergency]
散端 fag-end
散断股 < 钢丝缆中的 > molly
散堆货物 piled bulk produce
散堆密度 bulk density
散堆填料 random packing
散堆装货物 bulk goods
散放牛舍 loose barn; loose housing system; pen-type barn
散放棚 loafing shed
散放饲养棚 loose housing shed
散飞石块 < 爆破引起的 > fly rock
散分定货 split order
散工 casual worker; day labo(u)r(er); day's man; day taller; day wage work; daywork; jobber; job(bing) work; journal-man; journal-work; journey-man; journey work; odd job; roustabout < 矿山等 >
散光 diffused light; astigmia【医】
散光表面 diffusing surface
散光玻璃 diffusing glass; diffusion glass
散光的 astigmatic
散光灯 flood light
散光顶棚 light-diffusing ceiling
散光光度计 light scattering photometer
散光光弹性仪 scattered light polariscope
散光计 astigm(at)ometer
散光镜 astigm(at)oscope
散光镜检查 astigmatoscopy
散光目镜 diverging ocular
散光偏光镜 scattered light polariscope
散光器 diffusing unit; light diffuser
散光嵌板 diffusing panel
散光墙体 diffusing wall
散光圈 blur circle; circle of confusion
散光闪石 imerina stone; imernite
散光束 spreading beam
散光塑料板 diffusing plastic sheet
散光塑料膜 diffusing plastic sheet
散光透镜 dispersing lens
散光系数 coefficient of light diffusion
散光镶板 diffusing panel
散光照明 diffused illumination; diffusing illumination; floodlighting; louver lighting; stray lighting
散光罩 diffuser
散合式连接链环 detachable link
散化肥防漏抓斗 leakproof grab for bulk fertilizer
散货 bulk freight; material in bulk; bulk goods; bulk cargo
散货舱 bulk cargo hold
散货储存设施 bulk storage facility
散货船 bulk boat; bulk cargo ship; bulk carrier; bulker; bulk freighter; bulk vessel; freighter; bulk carrier
散货船队 bulk carrier fleet
散货船货 cargo shipped in bulk; goods shipped in bulk
散货吊船起重机 bulk unloading crane
散货吊船设备 bulk unloading device
散货堆场 bulk cargo yard
散货港(口) bulk cargo port; bulk freight port; bulk goods port; bulk port; bulk-harbo(u)r; port bulk; bulk material port
散货港区 bulk terminal; bulk cargo terminal
散货和矿石两用船 bulk-cum-ore carrier
散货和石油两用船 bulk-cum-oil carrier
散货集装器 bulk container
散货集装箱 bulk container; solid bulk container
散货码头 bulk cargo berth; bulk(cargo) terminal; bulk cargo wharf
散货密度 < 积载系数 > bulk density
散货平舱机 bulk distributor
散货/汽车/矿石船 bulk/car/ore ship
散货/汽车运输船 bulk/car ship
散货燃油船 bulk fuel oil carrier
散货容积 bulk capacity
散货石油(运输)船 bulk-oil carrier
散货停留地 < 装货码头的 > loading berth
散货卸车机 bulk cargo car unloader
散货卸船机 bulk cargo ship unloader
散货卸船链斗垂直提升机 marine leg elevator
散货卸船起重机 bulk cargo unloading crane; bulk unloading crane
散货卸船设备 bulk cargo unloading device; bulk cargo unloading facility; bulk unloading device
散货卸货门 bulker discharge hatch; bulker roof hatch
散货卸载机 bulk cargo unloader; bulk freight unloader
散货运输 bulk cargo transport; bulk freight transport
散货运输船 bulk carrier
散货载重汽车 truck-type bulk transporter
散货转载机 bulk cargo transfer machine
散货转载设备 bulk cargo transfer equipment
散货装车机 bulk cargo loader
散货装卸 bulk handling; handling of bulk cargo; handling of bulk freight; handling of bulk goods; handling of materials
散货装卸机械 bulk handling machinery
散货装卸设备 bulk handling equipment; bulk handling plant; bulk handling unit
散货作业机械 bulk handling machinery
散剂 powder
散加蒙间冰期【地】Sangamon
散见的 sporadic
散见多年冻土 sporadic permafrost
散见永冻土 sporadic permafrost
散件 parts and components; spare parts
散件出口 knock down export
散件货 piece goods
散件货物 break-bulk cargo
散件输出 knock down export
散件组装 manufactured parts for assembly
散捆 bale off; bundle off
散蜡 slack wax
散粒 particulate
散粒储存箱 bulk bin
散粒的 chessom
散粒货物 grain cargo
散粒矿物 disseminated values
散粒料 loose material
散粒黏[粘]土 shattered clay
散粒砂 friable sand
散粒体力学 mechanics of particulates
散粒物车厢 bulk body
散粒物料装载机 bulk loader
散粒物容器 bulk container
散粒效应 shot effect
散粒斜槽安装钩 grain spout hook
散粒噪声 shot noise; Schottky noise; shot effect
散粒噪声电流 short-noise current; shot noise current
散粒噪声电压 shot noise voltage
散粒噪声降低因数 shot noise reduction factor
散粒贮存箱 bulk bin
散粮 bulk grain
散粮舱容 < 船舶 > grain capacity
散粮船 grain carrier
散粮防漏抓斗 leakproof grab for bulk grain
散粮隔板 bin board
散粮码头 grain terminal; grain wharf
散粮气力卸货系统 pneumatic grain unloading system
散粮输送机 grain conveyer[conveyor]
散粮提升机 grain elevator
散粮筒仓码头 grain elevator terminal
散料 bulk cargo; unformed material
散料搬运 bulk handling
散料储仓 bulk storage silo
散料混合设备 bulk mixing equipment
散料料堆 light stockpile
散料卸载机 bulk unloader
散料运输车 bulk carrier
散料转运车 bulk carrier; bulk freighter
散料转运船 bulk carrier
散料装卸机 bulk handling machine; bulk solid loader and unloader
散料装载机 bulk loader
散列 hash
散列编码 hash coding
散列表 hash table
散列表类 hash class
散列表项 hash table entry
散列表元 hash table bucket
散列法 hashing(method)
散列符号 discrete symbol
散列函数 hash function
散列码 hash code
散列式居民点 dispersed settlement
散列数据表法 hash table method
散列索引 hash index
散列图解 scatter diagram
散列型 hash type
散列值 hashed value
散列注记 spaced name
散列总和 hashing total
散乱 debunching
散乱崩滑物 slurry slump
散乱边纹 fringe
散乱边纹区 fringe area
散乱的 riddled
散乱的杂物 litter
散乱点 shotgun pattern
散乱反射衰减 flutter fading
散乱范围 range of scatter
散乱回波 ghost echo
散乱介质 random medium
散乱片石铺砌 crazy pavement; crazy paving
散乱浅滩 scattered shoal
散乱入射灵敏度 random-incidence sensitivity
散乱扫描 random sampling; random scan(ning)
散乱输入 random input
散乱图案 scattered motif
散漫 sloppiness; sprawl
散漫的 diffusive; sloppy; desultory
散棉 bulk wool; loose wool
散木流放槽 needle drift chute
散配件 loose fitting
散批货物 broken lot
散片紫胶 free shellac
散漂材 drive
散绕线圈 mushwound coil; random wound coil
散容重 bulk specific gravity
散沙 strewing sand
散纱洗涤机 hank scouring machine
散砂 free flowing sand; loose sand; strewing sand
散射 dispersion; diverging; scatter(ing)
散射靶 scattering target
散射板 scatter plate
散射报知通信[讯]系统 scatter warning system
散射曝光表 light scattering photometer
散射本领 scattering power
散射比表 scatterance meter
散射比(率) scatterance
散射波 scattered wave; scattering wave
散射波通信[讯] scattering wave communication
散射波自洽场法 self-consistent field scattered wave technique
散射玻璃 diffusion glass
散射参数 scattering parameter
散射测定计 scatterometer
散射测量 scattering measurement; scatterometry
散射测浊法 nephelometry
散射层 scattered sheaf; scattering layer
散射长度 scattering length; scattering power

散射场 fringe field;scattered field
散射池 scattering cell
散射传播 beyond-the-horizon communication; beyond-the-horizon propagation;scatter(ing) propagation
散射传输 beyond-the-horizon transmission
散射大气 scattering atmosphere
散射带 zone of diffuse(d) scattering
散射灯 broad light
散射电子 drifting electron; scattered electron
散射定律 scattering law
散射反差 scattering contrast
散射反应 scattering reaction
散射幅度 scattering amplitude
散射辐射 diffused radiation;scattered radiation
散射伽马测井 scattered gamma-ray log
散射概率 probability of scattering
散射干扰 clutter
散射干涉 scattering interference
散射干涉仪 scatter interferometer
散射格栅 scattered trap
散射公式 scattering formula
散射功率 scattered power
散射光 broad light; diffuse(d) light; scattered light; soft light; spread light;stray light
散射光波 scattered light wave
散射光度计 light scattering photometer
散射光线 scattered ray
散射函数 scattering function
散射恒参信道 constant parameter channel
散射(横)截面 scattering cross-section
散射回波 scattered echo
散射机 scattering machine
散射机理 scattering mechanism
散射极坐标图 scattering polar diagram
散射几何条件 scattering geometry condition
散射几率 scattering probability
散射计 scatterometer
散射剂量 scattered dose
散射角 angle of dispersion; angle of scattering;scattering angle
散射角分布 scattering angular distribution
散射介质 scattering medium
散射矩阵 collision matrix; scattering matrix[S matrix]
散射扩散张量 scattering diffusion tensor
散射离子 scattered ion
散射理论 scattering theory
散射量子 scattered quantum
散射率 scattered power
散射率因素 emissivity factor
散射密度 diffuse density
散射面 scattering surface
散射面积 scattering area
散射面积比 scattering area ratio
散射面积系数 scattering area coefficient
散射模型 scattering model
散射(能)力 scattering power
散射频带 scatter band
散射频带宽度 scattered band
散射频率 scattering frequency
散射平均自由程 scattering mean free path
散射屏 diffuser screen; diffusing screen;diffusing sheet
散射器 diffuser;scatterer
散射强度 scattering intensity;scattering strength
散射求逆方法 scattering inverse method
散射区域 scattering region
散射圈 diffusion disc[disk]
散射日照 diffused solar radiation
散射式压力计 scattering-type pressure ga(u)ge
散射势垒 scattering potential
散射束 scattered(-out) beam
散射束流强度 scattering beam intensity
散射束锥角 beam angle of scattering
散射衰减系数 scattering attenuation coefficient
散射算子 scattering operator
散射损耗 scattering loss
散射损失 scattering loss
散射所致模糊 scatter unsharpness
散射探测 scatter sounding
散射探测器 scattering detector
散射探测与测距 scatter detection and ranging
散射特性 scattering property
散射体 diffuser; scatterer; scattering body;scattering material
散射体回波 scatter echo
散射体积 scattering volume
散射体系 scattering system
散射条纹 scattered striation
散射条纹干涉仪 scatter-fringe interferometer
散射通信 scatter communication
散射通信[讯]发射机 scatter communication transmitter
散射通信接收机 scatter communication receiver
散射通信[讯]系统 scatter communication system
散射通信[讯]信道终端 scatter communication channel termination
散射通信[讯]信道终端机 scatter communication channel terminal
散射透镜 divergent lens;diverging lens
散射图 scattergraph;scatter(ing) diagram
散射图样 scattering pattern
散射微差 scattering differential
散射尾部 scattering tail
散射吸收 scattering absorption
散射吸收系数 scatter absorption coefficient
散射系数 coefficient of scattering;scattering coefficient
散射系数表 scattering coefficient meter
散射线 scattered ray
散射相 scattering phase
散射相移 scattering phase shift
散射效率 scattering efficiency
散射效应 scattering effect;shot effect
散射信道模拟器 scatter channel simulator
散射信号 scattered signal
散射行波 scattered outgoing wave
散射型灯罩 flat lampshade
散射型光电感烟探测器 photoelectric scattering smoke detector
散射性衰减 dissipative attenuation
散射性质 scattering nature
散射修正值 scattering corrected value
散射仪 scattering meter;scatterometer
散射因数 scattering factor
散射因子 dispersion factor
散射影响 scattering effect
散射阈 scattering threshold
散射圆盘 circle of confusion
散射源 diffusing source
散射杂光 scattered stray light
散射噪声 scattering noise;shot noise
散射障板 scatter baffles
散射照明 stray illumination
散射振幅 scattering amplitude
散射指示量 indicatrix of scattering
散射中心 scattering center[centre]
散射状态 scattering state
散射(浊)度(单位) nephelometric turbidity unit
散射浊度计 nephelometer
散射着色力 scattering tinting strength
散射作用 scattering process
散生的混交林 scattered mixed forest
散湿量 moisture gain; moisture release
散石 field stone;scattered rock
散石排水沟 French drain
散碎的 arenaceous
散碎石块 loose rock
散索鞍 <悬索桥部件> cable splay saddle;splayed saddle
散索铸件 splay casting
散滩 scattered shoal
散陶普尔 <一种降凝添加剂> Santopour
散体积 bulk volume
散体积比重 bulk specific gravity
散体结构 dispersion texture; loosen texture
散土 spreading process
散尾葵 Cuban royal palm
散现反射 sporadic reflection
散现永冻层 sporadic permafrost
散卸 shoot loose
散屑 <使检波器恢复常态> de-coherence;decohere
散屑器 anticoherer;antiwherer;decoherer
散絮 deflocculate
散烟点 point source
散盐船 bulk salt carrier
散养牛棚 pen-stabling
散叶印刷品 leaflet
散页方式 cut form
散页胶片包装盒 film pack magazine
散页列车时刻表 timetable sheet
散页片 sheet film
散页资料 in sheets
散液船 tanker vessel;tank ship
散液集装箱 liquid bulk container;liquid cargo container;tank container
散运 in bulk;transport in bulk
散晕 bloom
散杂影像 ghost image;spurious image
散在的 sporadic
散在误差 sporadic fault
散张胶片 cut film;flat film;sheet film
散植 loose planting;scattered planting
散置 intersperse
散置城镇 dispersed town
散置的黏[粘]土层 interspersed clay layer
散重 bulk weight
散装 bulk(load);bulk pack;cargo in bulk;load(ing) in bulk;multiple lift packing
散装白云石船 bulk dolomite carrier
散装搬运 bulk handling;bulk transfer
散装比重 bulk specific gravity; bulk specific weight
散装材料 bulk material;bulk product
散装材料处理 handling of bulk materials
散装材料起重运输机械 mechanics of bulk materials handling
散装材料撒布机 bulk spreader
散装材料装卸 handling of bulk materials
散装仓库 bulk storage; bulk storage building;bulk(-cargo) warehouse
散装舱容 bulk capacity;grain capacity;grain cubic;grain space
散装产品 loose product
散装车辆船 bulk vehicle carrier
散装储存 bulk storage
散装储存仓 bulk bin
散装储存箱 bulk bin
散装储油 bulk-oil storage
散装船 bulk boat;bulk ship
散装船货 bulk ship cargo
散装船许可证 bulk ship cargo certificate
散装袋卸料机 bulk-bag unloading station
散装的 in bulk; laden in bulk; loose packed;unpackaged
散装冻结 bulk freezing
散装发货 bulk delivery
散装发运 despatch in bulk;loading in bulk
散装矾土船 bulk bauxite carrier
散装干燥曲线 bulk drying curves
散装隔舱板 grain bulkhead
散装谷物 bulk grain
散装谷物灌补舱 bulk grain feeder
散装谷物活动隔舱壁 portable grain bulkhead
散装觳仓 grain silo
散装固体 bulk solid
散装挂车 bulk trailer
散装过滤 bulk filtration
散装荷载 load in bulk
散装化肥 bulk chemical fertilizer;bulk fertilizer
散装化学品船 bulk chemical tanker
散装化学品法规 bulk chemical code
散装化学品分委员会 Subcommittee on Bulk Chemicals
散装化学品规则 bulk chemical code
散装混合料 bulk compound
散装货 bulk freight; bulkload; goods in bulk;loose cargo;loose stock
散装货仓库 bulk cargo; bulk storage warehouse;cargo in bulk
散装货车 bulk carrier
散装货车厢 bulk body
散装货船 bulk cargo carrier; bulk cargo ship; bulk carrier; bulk ship; bulk freighter
散装货袋卸系统 bag-out;bulk-in; bulk-in bag-out
散装货集装箱 bulk cargo container
散装货码头 bulk cargo terminal; bulkhead wharf; bulk material terminal;bulk cargo wharf
散装货倾倒装舱装置 tip loading installation
散装货输送机 bulk cargo conveyer [conveyor]
散装货条款 bulk cargo clause
散装货拖车 bulk trailer
散装货物 bulk cargo; bulk goods; goods in bulk; loose goods; loose product;loose cargo
散装货物港 bulk cargo port
散装货物换装终点站 bulk transfer terminal
散装货物集装箱 bulk container
散装货物列车 bulk train
散装货物码头 bulk cargo wharf
散装货物清单 bulk items list
散装货物线 goods-in-bulk loading track
散装货物运输底盘 bulkload chassis
散装货物载重 bulk load
散装货物装载机 bulk loader
散装货箱 bulk container
散装货运输 bulk transport
散装货运输工具 bulk carrier
散装货贮藏设备 bulk storage facility
散装货抓斗 bulk cargo grab-bucket
散装货装货槽 loading chute
散装货装货筒 loading chute

散装货装卸机 transporter
散装集装箱 bulk container
散装件 demounted
散装交货 bulk delivery
散装界线 bulk line
散装库存 bulk storage
散装粮谷 bulk grain
散装料 bulk material
散装磷酸盐船 bulk phosphate carrier
散装硫黄船 bulk sulphur carrier
散装码头 loading berth
散装满载 laden in bulk
散装煤船 bulk coal carrier
散装密度 apparent density;loose density
散装棉 loose wool
散装木材船 bulk timber carrier
散装木浆/硫酸船 bulk wood-pulp/sulphuric acid carrier
散装木片船 bulk wood chip carrier
散装泥浆船 bulk slurry carrier
散装镍矿船 bulk nickel carrier
散装燃油运输船 bulk fuel oil carrier
散装容积 bulk volume;grain capacity;bulk capacity <船舶>
散装容器 bulk container
散装入船 bulk loading
散装砂 unpackaged sand
散装石膏 bulk gypsum
散装石膏船 bulk gypsum carrier
散装石灰石船 bulk limestone carrier
散装石油 bulk petroleum
散装试样 random sample
散装输送 bulk conveyer[conveyor];bulk delivery;bulk handling;bulk transfer;bulk transport
散装输运 bulk handling;bulk transfer
散装水泥 bulk(loading of)cement;cement in bulk;loose cement
散装水泥驳船 bulk cement barge
散装水泥车 bulk cement truck;cement container car;cement delivery truck
散装水泥船 cement carrier;cement tanker
散装水泥负压输送机设备 suction pressure transfer installation for bulk cement
散装水泥罐 cement container;cement vessel
散装水泥罐车 bulk tanker;cement tanker;container vehicle
散装水泥货车 bulk cement lorry
散装水泥卡车 bulk cement lorry;bulk cement truck
散装水泥库 bulk cement silo
散装水泥螺旋输送机 screw conveyer for bulk cement
散装水泥气力卸料车厢 pneumatic discharged car of bulk cement
散装水泥汽车 automatic vehicle for bulk cement
散装水泥输送机 bulk cement transporter
散装水泥输送装置 bulk cement carrier
散装水泥筒舱 cement silo
散装水泥拖车单元 bulk cement trailer unit
散装水泥用裸线输送机 screw conveyer for bulk cement
散装水泥运输车 bulk cement vehicle;bulker
散装水泥运输工具 cement bulk transporter
散装水泥运送机 bulk cement transporter
散装水泥站 bulk cement loading station
散装水泥中转站 bulk cement supply station;bulk cement terminal
散装水泥专用船 bulk cement boat
散装水运 bulk shipment
散装饲料 bulk feed
散装糖船 bulk sugar carrier
散装体的倾倒容器 bulk tipping container
散装桐油特款 special clauses for wood oil in bulk
散装土样 bulk soil sample;random soil sample
散装拖运 bulk hauling
散装物铲斗 utility bucket
散装物秤 bulk weigher
散装物冷却器 bulk cooler
散装物料 bulk cargo;laden in bulk
散装物料装载机 bulk loader
散装物输送机 bulk handling machine;bulk transporter
散装物卸载机 bulk unloader
散装物运输底盘 bulkload chassis
散装物载重汽车 bulk lorry
散装物装载管道出口 bulk loading point
散装消石灰 bulk hydrated lime
散装液体 bulk liquid;liquid bulk
散装液体船 bulk liquid carrier;tanker vessel
散装液体货物 bulk liquid cargo;liquid bulk cargo
散装硬件 unbundled hardware
散装油站 bulk station
散装鱼粉船 bulk fishmeal carrier
散装原料 bulk raw material
散装运货车 truck-type bulk transporter
散装运输 bulk freight;bulk handling;bulk shipment;bulk transport;transshipment in bulk
散装运输船 bulk carrier
散装运输水泥 bulk shipping of cement;cement transport in bulk transporters
散装运输装置 bulk carrier
散装杂货船 general bulk carrier
散装杂货运载 bulk and general cargo loading
散装载运汽车 bulk carrier vehicle
散装炸药 bulk powder
散装贮藏 bulk storage
散装装车 bulk loading
散装装料器 bulk loader
散装装料站 bulk loading station
散装装卸 bulk handling;bulk transfer
散装装运 shipment in bulk
散装租船 bulk chartering
散装钻机 part element drill
散状陶瓷纤维 bulk ceramic fiber
散状纤维 loose fiber[fibre]

散斑 speckle

散斑场 speckle field
散斑干扰 speckle noise
散斑干涉 speckle interference
散斑干涉测量(术) speckle interferometry
散斑干涉法 speckle interferometry
散斑计量术 speckle metrology
散斑剪切 speckle-shearing
散斑剪切干涉计量法 speckle-shearing interferometry
散斑全息摄影 speckle holography
散斑随机相移器 speckle random phase shifter
散斑调屏假彩色编码 speckle modulating screen pseudocolo(u)r encoding
散斑图 specklegram;speckle pattern
散斑效应 speckle effect
散斑噪声 speckle noise
散斑照相 speckle pattern photography
散波 diverging wave
散播 seeding;strew
散播机 broadcaster;broadcast seeder
散播力 diffusibility
散播农药 spread pesticide
散布 diffuse;disperse;disseminate;intersperse;interspersion;scatterance
散布编码 diffuse coding
散布存储 scatter storage
散布点图 scatter plot
散布范围 scattered band;scattering field
散布分析 scatter diagram
散布覆盖 scattered covering
散布金箔 tinsel
散布矩阵 scatter matrix
散布卷积码 diffuse convolutional code
散布控制 spread control
散布力 vagility
散布面积 zone of dispersion
散布潜力 dispersal potential
散布区 range of scatter;scatter band
散布区域 dispersion zone
散布曲线 scattergram
散布式加煤机 spreader stoker
散布式建筑 sporadic building
散布速度 spread velocity
散布梯尺 dispersion ladder
散布面调整 pattern harmonization
散布图 dispersion pattern;scatter diagram;scattergram
散布误差 dispersion error
散布系数 coefficient of diffusion;dispersion factor;scatter coefficient;diffusion factor
散布性的 diffusive
散布中心 center[centre] of dispersion
散布作用 dissemination
散步 promenade;walk
散步长廊 promenade gallery
散步场 promenade
散步场所 promenade;public walk
散步道 foot walk;footway;promenade;trail;walk
散步道路 alameda
散步的 meandering
散步地带 esplanade
散步广场 promenade
散步甲板 promenade deck
散步路 esplanade;promnard
散步器 promnard;walk path
散步区 pedestrian space
散步小路 promnard
散步"行走"<行人过街的交通信号> walk
散步游廊 promenade gallery
散出 effluvium[复 effluvia]
散躲屏 diffuser screen
散发 efflux;shed
散发比 water-use ratio
散发计 potometer
散发监视 emission monitoring
散发流 transpiration(al) steam
散发率 rate of transpiration;transpiration ratio
散发频率 shedding frequency
散发气味的 odo(u)riferous
散发器 heat sink
散发热量 exothermic heat
散发物计量系统 emission measurement system
散发物控制 emission control device
散发物污染系数 emission factor
散发物资料 emission data
散发性的 exhalant
散发烟雾工业 smoke-emanating industry
散发仪 photometer
散反射 scattered reflection
散反铁磁性 asperromagnetism
散苇 bloom
散伙 partnership dissolution
散极化 polarization diversity
散集钢块 tramp iron
散集铁块 tramp iron
散胶色料 dispersal colo(u)r
散焦 blooming;debunching;defocus(sing);focus-out;misfocusing;spot defocusing
散焦斑点 defocused spot
散焦斑点照相术 defocused speckle photography
散焦编码 defocused code
散焦的 out-of-focus
散焦点 caustic
散焦电子束 defocused beam
散焦技术 defocusing technique
散焦灵敏度 activity for defocus;acuity for defocus
散焦脉冲响应 defocused impulse response
散焦区 region of defocusing
散焦锐度 acuity for defocus
散焦射束 defocused beam
散焦透镜 defocused lens
散焦图像 out-of-focus image
散焦系数 defocusing coefficient
散焦系统 defocused system
散焦像 defocused image;misfocused image
散焦效应 defocusing effect
散聚脉冲 de-bunched(beam)pulse
散聚束 de-bunched beam
散聚效应 debunching effect
散开 defocussing;dispersal;dispersion;divergence;fan;fan out;unravel;untwine;untwining
散开的 divergent
散开的礁石 scattered rock
散开队形 loose formations
散开光束 cone of light
散孔 diffuse porous
散孔材<木材的> diffuse porous wood
散裂 nuclear spallation;rotting;spallation;spall(ing)
散裂产物 spallation product
散裂抵抗力 spalling resistance
散裂反应 spallation reaction
散裂截面 spallation cross section
散裂试验 spalling test
散流 air diffusing;air diffusion;sporadic flow;spread flow
散流安全阀 diffusing relief valve
散流的 divertive
散流器 air diffuser;air radiator;diffuser;multidirectional ceiling diffuser
散流器送风 diffuser air supply;diffuser feeding
散流器外罩 enclosure of radiators
散流器最小通过面积 minimum diffuser free area
散流设备 air diffusing equipment
散落 fallout;spillage
散落孢子 seiospore
散落砂 spill sand
散落物 fallout;particle fall
散落物测定 fallout measurement
散落物测量 fallout measurement
散落物剂量率 fallout dose rate
散落物料溜子 spill chute
散能块 energy dispersion block
散能量分光计 energy dispersion spectrometer
散凝 deflocculation
散凝剂 deflocculant;deflocculating agent;deflocculation agent
散凝浇筑料 deflocculated castable
散凝作用 deflocculating;deflocula-

S

tion
散诺阶【地】Sanoisian(stage)
散排 break-up of raft
散泡器 sparger
散气板 filter plate
散气浮选法 dispersed air flo(a)tation
散气面积比 diffuser area ratio
散气片 diffuser vane
散气曝气 diffused air aeration
散气曝气法 diffused aeration system
散气器 diffuser
散去 drop-off
散热 abstract heat; abstraction of heat; cooling; dissipating heat; egress of heat; elimination of heat; heat abstraction; heat dissipation; heat emission; heat emissivity; heat evolution; heat radiation; heat-transfer by radiation; rejection of heat; remove(d) heat; recalescence <指钢材由热到冷却时热的放射>
散热板 cool sheet; heating panel; louver board
散热玻璃 radial glass; radiant glass
散热翅片 fin; radiating fin; cooling fin
散热当量面积 equivalent direct radiation
散热的 radiating
散热阀 radiating valve
散热法 heat elimination
散热反应 exothermic reaction
散热风门 radiator flap
散热风扇 cooling-down fan
散热供暖系统 radiator heating system
散热管 cooling coil; cooling tube; radiating collar; radiating pipe; radiator pipe; after-cooler <压缩空气的>; discharge pipe <总风缸管的>
散热过程 exothermal process; exothermic process; radiating process
散热剂 coolant
散热筋片 radiation rib; radiating rib
散热筋型电机 ribbed surface machine
散热口 thermovent
散热肋 air cooling fin
散热肋冷却型电动机 rib-cooled motor
散热肋片 fin; radiated rib
散热冷却 heat-sink cooling
散热量 heat dissipating capacity; heat release
散热量有效系数 coefficient of effective heat emission
散热率 heat emission rate; heat loss rate; heat output; rate of heat dissipation; rate of heat release; thermal diffusivity
散热面 cooling surface; heat delivery surface; heat-dispersing surface; radiating surface <蒸汽采暖散热器的>
散热面积 area of dissipation; heating surface area; radiating area
散热能力 heat-sinking capability; radiant capacity; radiating capacity
散热盘管 panel coil; radiating panel coil
散热片 air cooling fin; carbon fin; cooling plate; cooling rib; cool sheet; fins; louver board; radiated flange; radiating fin; radiating flange; radiation fin; radiator fan; thermal sink; heat transfer foil
散热片翅板 fin of radiator
散热片冷却 gill cooling
散热片冷却汽缸 fin-cooled cylinder
散热片冷却器 finned cooler
散热片汽缸 flange cooled cylinder
散热片式供暖设备 radiator system heating apparatus
散热片式冷却器 plate-fin cooler
散热片吸热 heat sink
散热片效率 fin efficiency
散热片型电动机 ribbed motor
散热器 cooler; energy radiator; exchanger; heat abstractor; heat emitter; heat-exchanger; heating radiator; heat sink; radiation; radiator; sink
散热器安全阀 radiator relief valve
散热器百叶窗 radiator blind; radiator damper; radiator grill(e); radiator shutters
散热器百叶窗板条 radiator blind plank
散热器百叶窗操纵 radiator shutter control
散热器百叶窗操纵手柄 radiator flap handle
散热器百叶窗架 radiator louver frame
散热器百叶窗减声片 radiator louver antisqueak
散热器百叶窗控制 radiator shutter control
散热器百叶窗轴 radiator louver shaft
散热器(百叶)气窗 radiator shutters
散热器保护栅 radiator screen
散热器保温帘 radiator screen
散热器保温罩 radiator apron; radiator cover; radiator muff; radiator shield; winter front
散热器部件 radiator fixtures
散热器采暖系统 radiator heating system
散热器衬垫 radiator pad
散热器撑杆固定架 radiator stay rod bracket
散热器撑杆固定销 radiator stay rod pin
散热器翅片 fin of radiator
散热器出水管 radiator outlet pipe
散热器出水管配件 radiator outlet fitting
散热器出水软管 radiator outlet hose
散热器出水软管夹 radiator outlet hose clamp
散热器出液软管 radiator outlet hose
散热器粗滤器 radiator strainer
散热器单片 radiator element
散热器挡板 radiator baffle
散热器挡风板 radiator air deflector
散热器挡风帘 radiator blind; radiator screen
散热器导管 radiator duct
散热器到立管的横向连接口 run-out
散热器的恒温阀 thermostatic radiator valve
散热器的可调百叶窗 adjustable shutters for radiator
散热器底漆 radiator priming paint
散热器底罩 radiator bottom guard
散热器吊钩 radiator hanger
散热器堵漏油灰 radiator repair cement
散热器耳轴 radiator trunnion
散热器防冲杆 radiator bumper rod
散热器防冻溶液 radiator antifreeze solution
散热器防护装置 radiator protection
散热器防漏剂 radiator leak stop
散热器防杀剂 radiator antifreeze solution
散热器防砂罩 radiator sand shield
散热器防锈溶液 radiator rust resister
散热器防锈液 radiator rust preventive
散热器放气阀 air cock on the radiator
散热器放气孔 radiator louver[louvre]
散热器放热阀 radiator vent
散热器放水开关 radiator drainage valve
散热器放水口 radiator drain
散热器放水塞 radiator draw-off (cock); radiator draw-off plug
散热器放水塞盖 radiator drain plug cover
散热器放水弯头 radiator drain outlet elbow
散热器放水旋塞 radiator drain cock
散热器风门片 radiate shutter; radiator flap
散热器风扇 radiator fan
散热器风扇传动装置 radiator fan drive
散热器风扇罩 radiator fan shroud
散热器蜂窝管 radiator core
散热器盖 radiator filler cap
散热器盖温度计 radiator cap thermometer
散热器格栅 radiator grid
散热器格栅套 grill(e) cover
散热器供暖 radiator heating
散热器供暖用的铜管 copper pipe for radiator heating
散热器供热支管 feeding branch to radiator; return branch to radiator
散热器固定架 radiator mounting
散热器管 radiator tube
散热器恒温器 radiator thermostat
散热器横条 radiator cross member
散热器护栅 radiator brush guard; radiator fender; radiator grate; radiator guard; radiator screen
散热器护栅嵌条 radiator grill molding
散热器护罩 radiator guard; radiator shield; radiator shroud
散热器回水支管 return branch of radiator
散热器及芯子试验台 radiator and core testing stand
散热器加水盖 radiator filler cap; radiator locking cap; radiator screw-cap
散热器加水口 radiator filler (with cap)
散热器加水口密封垫 radiator filler cap gasket
散热器夹条 radiator clamping strip
散热器角阀 angle radiator valve
散热器脚 radiator pedestal
散热器进出水软管 radiator inlet and outlet hose
散热器进水管配件 radiator inlet fitting
散热器进水滤清器 radiator inlet strainer
散热器进水软管 radiator inlet hose
散热器进水软管夹 radiator inlet hose clamp
散热器壳 radiator shell
散热器壳衬带 radiator shell lacing
散热器壳支架 radiator shell support
散热器空气通道 radiator air passage
散热器框架 radiator frame
散热器拉杆 radiator brace rod; radiator tie rod
散热器肋片 radiator element; radiator fin
散热器冷却的 radiator cooled
散热器冷却法 radiator cooling
散热器冷却管 radiator cooling tube
散热器冷却片 radiator cooling fin
散热器冷却水加热器 radiator coolant heater
散热器连接管 radiator connection
散热器螺母 radiator closing nut
散热器螺纹连接短管 radiator nipple; radiator screw nipple
散热器密封垫 radiator gasket
散热器面饰 radiator facing
散热器内螺纹接口 radiator blind nipple
散热器暖气装置 radiator system heating apparatus
散热器盘管 radiator coil
散热器皮管 radiator hose
散热器片 radiator fin; radiator section; radiator shutters
散热器平衡水箱 radiator surge tank
散热器前护栅 false front; radiator false front; radiator stoneguard; radiator grill(e)
散热器墙距 radiator wall spacing
散热器清洁器 radiator cleaner
散热器清洗剂 radiator cleaning compound; radiator flush
散热器软管 radiator hose
散热器软管夹箍 radiator clip
散热器软管接头 radiator hose connection
散热器散热片 radiator fin
散热器上集管板 radiator upper header
散热器上控制阀 radiator control valve
散热器上水箱 radiator header; radiator top tank; radiator upper tank
散热器上涂层 radiator coat
散热器上支架 radiator top frame
散热器蛇形管 radiator coil
散热器式变压器 radiator-type transformer
散热器饰件 radiator ornament
散热器疏水阀 radiator trap
散热器疏水器 radiator trap
散热器水位 radiator water-level
散热器水箱 radiator tank
散热器水箱盖测试器 radiator cap tester
散热器丝堵 screwed plug of radiator
散热器速度 sink rate
散热器所安装的风扇 radiator mounted fan
散热器弹性支架 radiator bumper
散热器调节阀 radiator regulating valve
散热器填料 heat-sink compound
散热器通风管 radiator vent pipe
散热器通风透气管 radiator vent pipe
散热器通气管路 radiator vent line
散热器托架 radiator bracket; radiator support bracket
散热器外壳 radiator shell
散热器外罩 enclosure of radiators
散热器温度计 radiator thermometer
散热器温度降落 radiator temperature drop
散热器洗涤剂 radiator cleaner; radiator flush
散热器洗涤装置 reverse flushing tool
散热器下部托座 bottom radiator bracket
散热器下集管板 radiator bottom header
散热器下集水槽 radiator bottom tank
散热器下水箱 radiator bottom header; radiator lower tank
散热器箱 radiator case
散热器箱形框 radiator case frame
散热器效率 sink-efficiency
散热器芯部清洗枪 radiator core cleaner gun
散热器芯护栅 radiator core protection grid
散热器芯护罩 radiator core guard
散热器芯片 radiator core fin
散热器芯子 radiator center[centre]; radiator core
散热器芯子单元 radiator core section
散热器芯子接头 radiator core adapter
散热器芯子支架 radiator core support
散热器型变压器 radiator-type transformer

S

散热器型芯 radiation core
散热器溢流管 radiator overflow tube
散热器溢流箱 radiator overflow tank
散热器用漆 radiator paint
散热器用液 radiator liquid
散热器油漆 radiator paint
散热器元件 radiator element
散热器毡帘 < 防冻帘 > radiator blanket
散热器罩 radiator cover; radiator enclosure; radiator guard; radiator shell; radiator shield; radiator bonnet; radiator box; radiator cowling
散热器罩的挡板 radiator shell apron
散热器罩嵌条 radiator shell molding
散热器罩支架 radiator hood ledge
散热器支杆 radiator bar
散热器支管 radiator branch
散热器支架 radiator bearer; radiator bracket; radiator support; radiator support bracket; radiator stand
散热器支架垫 radiator support cushion
散热器支架角撑 radiator brace
散热器支柱 radiator strut
散热器支座 radiator pedestal
散热器置 radiator hood
散热器注液枪 radiator filler gun
散热器阻板 radiator baffle
散热器阻气板 radiator air baffle
散热器组 radiator block; radiator panel
散热器组合 radiator assembly
散热器座 foot of radiator; radiator base
散热器座弹簧 radiator mounting spring
散热器坐垫 radiator mounting pad
散热强度 specific heat load
散热区 region of dissipation
散热曲线 exothermic curve
散热扇带 radiator fan belt
散热设备 heat emission equipment
散热设计 thermal design
散热式冷却装置 radiator-type cooling unit
散热试验 dissipation test
散热水 circulating water; cooling water
散热水槽 sink
散热速率 cooling rate; rate of heat dissipation; rate of heat release; heat release rate
散热损失 dissipation heat loss
散热踢脚板 washboard radiator
散热体 heating body; heat-transfer material
散热条 heat-sink strip
散热调整 regulating heat extraction
散热凸缘 cooling flange; radiated flange; radiating flange
散热途径 pathway of heat loss
散热稳定性 radiation stability
散热系数 coefficient of heat emission; heat emission coefficient; thermal diffusivity
散热系统 cooling system
散热叶片 fin
散热液 cooling fluid; cooling liquid
散热源 source of heat release
散热罩 radiator guard
散热砖 heat dissipating tile
散热装置 heat abstractor; heat sink
散热装置蛇管 radiator coil
散热组件 heat-sink block
散热作用 heat rejection; thermolysis
散声透镜 dispersing lens
散失 wantage
散失热 abstracted heat; dissipated heat; stray heat
散式流化床 dispersion fluidized bed; particulately fluidized bed
散式流态化 particulate fluidization
散式相 particulate phase
散势器导管 radiator duct
散势器底板 radiator bottom plate
散束 debunching
散束校正 debunching correction
散束力 debunching force
散束器 de-buncher
散束作用 debunching action
散水 splash block; water-table; apron < 房屋外墙脚坡 >
散水暗沟 spray drain
散水滤床 trickling filter bed
散水滤床法 trickling filter process
散水条 metal splash flat bar
散丝 loose filament; stray fiber[fibre]
散雾剂 fog disperser
散下绕组 mush winding
散下式绕组 random winding
散下线圈 mush coil
散纤维染色 stock dyeing
散纤维染色纱 stock-dyed yarn
散逸层 dissipative sort(e)y; mesosphere
散逸电子 drifting electron
散逸函数 dissipation function
散逸稳定液 circulation loss of stabilizing fluid
散逸作用 dissipation
散音 diffused sound
散云 scattered clouds

桑

桑德公式 Sander's formula
桑德森干燥时间测量仪 Sanderson drying time meter
桑德森-米尔纳空间色差方程 Saunderson-Milner zeta space colo(u)r difference equation
桑德森修正 Saunderson correction
桑迪马斯计算槽 < 用于测量明渠中含大量砂砾的水流量 > San Dimas flume
桑恩袋式剥皮机 Thorne barker
桑弗罗公式 Sainflou's formula
桑弗罗计算驻波压力公式 Sainflow formula
桑弗罗计算驻波压力理论 Sainflow theory
桑福特-迪型底卸式矿车 Sanford-Day car
桑干片麻岩 < 太古界 > Sangtean gneiss
桑给巴尔树胶 Zanzibar gum
桑木 mulberry
桑拿浴 sauna bath
桑拿浴室 sauna
桑尼 < 相位控制的区域顺序旋转无线电指向标 > Sonne
桑尼航标 Sonne beacon
桑尼无线电信标 Sonne beacon
桑葚 mulberry
桑葚色 murrey
桑萨风 sansar
桑森弗兰斯蒂投影法 Sanson Flamsteed projection
桑森弗兰斯蒂正弦曲线投影 Sanson-Flamsteed sinusoidal
桑森投影法 Sanson projection
桑树 Morus alba; mulberry; weeping mulberry; white mulberry
桑思韦特土壤水分指数 Thronth Waite moisture index
桑斯威特公式 Thornthwaite's formula
桑斯威特降水效率指数 precipitation effectiveness index; Thornthwaite precipitation effectiveness index
桑斯威特气候分类法 Thornthwaite's classification of climates
桑斯威特湿润指数 Thornthwaite moisture index
桑斯威特指数 Thornthwaite index
桑塔·劳伦斯地槽 Santa Lawrence geosyncline
桑塔·罗沙风暴 Santa Rosa storm
桑塔·玛丽亚深条纹的红色硬木 < 产于中美洲 > Santa Maria
桑塔·乔治盆地 Santa George basin
桑塔阿那风 Santa Ana
桑天牛 < 拉 > Apriona germari
桑田 mulberry field
桑托阶 < 晚白垩世 >【地】Santonian
桑托斯常数 < 筛分析法中 > Santos constant
桑托斯港 < 巴西 > Port Santos
桑园 mulberry field; mulberry orchard
桑属 mulberry; Morus < 拉 >

丧

丧失饱满度 loss of turgidity
丧失的利益 benefit forgone
丧失的水权 forfeited water rights
丧失工作能力 disablement
丧失劳动力的 disabled
丧失劳动能力 disability
丧失劳动能力恤金 < 工伤事故 > disability pension
丧失能力的 incapacitated
丧失能力的股东 incapacitated shareholder
丧失权利 divesting of a right
丧失稳定 < 压杆、壳、板等 > buckling
丧失稳性 loll
丧失销售 lost sales
丧失信用 break faith; discredit
丧失支付能力 failing

骚

骚动 convulsion; kick-up; pother; tempest; tumult; turbulence
骚动的 turbulent
骚动点 burbling point
骚动误差 agitation error
骚乱 turbulence; turmoil
骚扰 annoyance; commotion; outbreak

缫

缫丝厂 filature
缫丝机 filature

扫

扫瓣干涉仪 lobe sweep(ning) interferometer; swept-lobe interferometer
扫壁蒸发器 wiped film evapo(u)rator
扫壁蒸馏器 wiped wall still
扫舱 hold sweep(ning); stripping; clearance of residual mud in a wreck < 救捞 >
扫舱泵 cargo stripping pump; stripping pump; tripper pump
扫舱费 sweeping hold charges
扫舱工作 sweeping
扫舱货 cargo sweep(ning); spillage; sweepings
扫测 scan survey; sweeping survey; wire drag
扫测幅 swath(e)
扫测系统 < 多通道回声测深 > swath(e) sounding system
扫测线 sweep
扫车 rotary broom
扫尘刷 jamb duster
扫出 scan out
扫出效应 sweep-out effect
扫除 broom; cleaning; clean-up; mop; scavenge; sweeping; turnout
扫除电极 sweeping electrode
扫除法 balayage method
扫除干净 clean-out; scavenge
扫除工地渣土 clean-up site's residues
扫除机 mechanical sweeper; sweeper
扫除机械 clearer; remover
扫除积雪 snow clearing
扫除障碍 clear the way
扫除作业 clearing; sweeping
扫处理 broom finish
扫船底 arching; hogging
扫床 < 指探测河床有无障碍物 > bed sweep(ing); dragging; riverbed sweep(ing); fine-wire sweep
扫床测量 sweep survey
扫床测深仪 sweeping sounder
扫床船 bed-sweep(ing) boat; drag boat; dragger; snag boat
扫地机 mechanical sweeper; sweeper; broom
扫地水泥 cement sweep(ning)s
扫动接点 wipe contact; wiping contact
扫动式搅拌机 sweep-type agitator
扫动式淘汰盘 sweeping table
扫动位错 sweeping dislocation
扫过 sweep-over
扫过容积 sweep volume
扫海 bed sweep(ing); dragging; drag the sea; mine sweep(ing); sweeping of sea; seabed sweep(ning)
扫海测量 sea-floor scan survey; sweep(ing) survey; wire-drag survey
扫海船 sweeper
扫海杆 sweep bar
扫海钢梁 sweep beam
扫海钢索 drag; wire drag
扫海具 hydrographic(al) trawl; surveying trawl; survey sweep; sweeper
扫海锚 sweeping anchor
扫海区 swept area
扫海深度 sweeping depth
扫海趟 sweeping trains
扫海趟有效宽度 effective width of sweep(ing) trains
扫海趟重叠带 overlapping zone of sweep(ning) trains
扫海用具 sweeping appliance
扫回电路 sweep circuit
扫迹 trace
扫迹道 < 扫测水深的 > swath(e)
扫积型探测器 Sprite(signal processing in the element) detector
扫及区 swept area
扫集物 sweepings
扫接触点 wiping contact
扫街车 scavenger's car(t)
扫街机 scavenger; street sweeper; street sweep(ing) machine; sweeping machine; sweeper
扫孔【岩】drill off
扫孔器 drag twist
扫库水泥 cement sweep(ning)s
扫雷 mine clearance; mine sweep(ing)
扫雷灯 mine sweep(ing) light
扫雷浮标 mine sweep(ing) float
扫雷工具 sweep gear
扫雷航道中心线 centre line of mine-swept route
扫雷舰 mine sweeper
扫雷具 mine sweep(ing) gear
扫雷器式除荆机 flail-type cutter
扫雷器式旋转除荆机 flail-type rotary cutter
扫雷区 mine sweep(ing) area
扫雷艇 mine hunter; mine sweeper; mine sweep(ing) boat

S

扫雷装置 flail;sweeping gear
扫路 brooming
扫路车 road broom; street cleaner; street sweeper; street sweep(ing) machine; sweeper; sweeping car; sweeping machine
扫路机 broom; brooming machine; mechanical sweeper; motor sweeper; power broom; road broom; road sweeper; road sweeping machine; street sweeper; sweeper; track broom
扫路集尘机 road sweeper-collector
扫路集尘器 road sweeper-collector
扫路垃圾车 collecting sweeper; sweeper collector
扫掠波段 swept band
扫掠持续时间 duration in scan(ning)
扫掠抽运 swept pumping
扫掠干扰机 sweep-through jammer
扫掠机构 scanner
扫掠频率 swept frequency
扫掠气 sweep gas
扫掠时间 trace time
扫掠式干扰 sweep-through jamming
扫掠输入 ramp input
扫掠速度 sweep speed
扫掠天线 scanning antenna
扫掠透镜 scanning lens
扫盲运动 literacy campaign
扫毛 broom finish(ing); broom(ing)
扫毛混凝土 broom-finish concrete
扫面 broom finish
扫面层 broom finish
扫面处理 broom finish
扫描 interlacing; scan; sweep(ing)
扫描 X 射线微型分析器 scanning X-ray micro-analyser[analyzer]
扫描残迹 stutter
扫描测量 scanning survey
扫描测试 sweep test
扫描测微密度计 scanning microdensimeter[microdensitometer]
扫描绘图机 scan(ning) plotter
扫描机 scanning machine
扫描摄影术 scanning photography; smear photography
扫描声呐 scanning sonar
扫描声呐系统 scanning sonar system
扫描声学显微镜 scanning acoustic(al) microscope
扫描时间 duration scanning; scanning time; sweep interval; trace time
扫描时间范围 sweep time range
扫描室 scanning room
扫描输出 scan out(put)
扫描输出放大器 sweep-output amplifier
扫描输出级 scanning output stage
扫描输出总线 scan out bus
扫描输入 scan-in; scanner input
扫描输入总线 scan in bus
扫描束电流 scanning-beam current
扫描数字化 scan-digitizing
扫描数字化器 scanning digitizer; traverse scan
扫描数字化仪 scanning digitizer
扫描顺序 scanning sequence
扫描搜索 scanning search
扫描速度 pick-up velocity; scanning rate; scanning speed; scanning velocity; sweep rate; sweep speed; sweep velocity; velocity of scanning
扫描头 scanning slide
扫描透镜 scanning lens
扫描透射电子像 scanning transmission electronic image
扫描透射式电子显微镜 scanning transmission electron microscope
扫描图像 scan image; scanned picture; scanning pattern image
扫描图形 scanning pattern
扫描完整性检验 scanning integrity check
扫描准确度 scanning accuracy
扫频标志发生器 marker sweep generator
扫频测试 sweep check
扫频电压 sweep voltage
扫频发生器 frequency sweep generator
扫频反射计 sweep-frequency reflectometer
扫频放大器 panoramic amplifier
扫频干扰 swept jamming
扫频干扰机 sweep jammer
扫频干涉仪 swept frequency interferometer
扫频跟踪干扰 sweep lock-on jamming
扫频技术 sweep-frequency technique; swept frequency technique
扫频检验 sweep check
扫频接收机 panoramic receiver; sweep-frequency receiver
扫频瞄准式干扰 sweep-spot jamming
扫频式干扰机 sweep-through
扫频衰减测试技术 sweep-frequency attenuation measurement technique
扫频线性 sweep linearity
扫频信号发生器 swept signal generator
扫频信号图像调制 swept frequency picture modulation
扫频仪 sweeper; sweep generator; sweep signal generator
扫频振荡器 sweep generator; swept frequency oscillator
扫气 scavenge; scavenging
扫气泵 scavenge pump; scavenger (pump); scavenging pump
扫气比 scavenging ratio
扫气冲程 scavenging stroke
扫气导管 scavenging duct
扫气道 scavenging air belt
扫气调节风门 scavenging air damper
扫气阀 scavenge valve; scavenging air valve; scavenging valve
扫气阀箱 scavenge valve chest
扫气缸 scavenge cylinder
扫气鼓风机 scavenge(r) blower
扫气管 scavenging air pipe; scavenging duct
扫气管道 scavenge line
扫气空气 scavenging air
扫气空气量与发动机排量之比 scavenging air ratio
扫气空气止回阀 scavenging air non-return valve
扫气孔面积 port area; scavenge area
扫气口 scavenge port; scavenging port
扫气冷却器 scavenge air cooler
扫气期 scavenging period
扫气气流 <发动机> scavenge flow
扫气腔 scavenge space
扫气梯度 scavenge gradient
扫气通道 scavenge trunk
扫气系统 scavenge(air) system
扫气箱 scavenge(manifold); scavenging air box; scavenging(air) receiver
扫气效率 scavenging efficiency
扫气压力 scavenging pressure
扫气重叠 overlap for scavenging
扫气装置 scavenging arrangement
扫气总管 scavenging air trunk
扫浅 spot dredging; sweeping
扫清 clean-out; sweeping; weed
扫清道路 clearway; clear the way
扫清时间 checkout time
扫清树枝 brush clearing
扫清作用 scouring action
扫入 scan-in
扫射 mow
扫视 scan
扫视程序 scanner
扫视途径 scan path
扫水泥塞时间 drilling out cementing plug time
扫速校准器 sweep speed calibrator
扫调 pan
扫调比较 panoramic comparison
扫调比较器 panoramic comparator
扫调范围 pan-range
扫调放大器 panoramic amplifier
扫调附加器 panadapter[panadaptor]; panoramic adapter
扫调监视器 panoramic monitor
扫调接收机 panoramic receiver
扫调屏 panoramic screen
扫弯水 bend-rushing flow
扫尾工程 outstanding works
扫尾工序 finishing trade
扫线 line purging
扫线管 scavenger pipe
扫线球 pig
扫相干涉仪 phase-swept interferometer
扫选 scavenging
扫选精矿 scavenger concentrate
扫选跳汰机 scavenger jig
扫选循环 scavenger flo(a)tation circuit
扫雪 snow removal
扫雪板 flanger
扫雪壁柜 broom closet
扫雪铲运机 snow scraper
扫雪车 snow plough carriage; snow removal truck; snow sweeper; sweeping car; track sweeper
扫雪机 mechanical snow plough; snow blower; snow breaker; snow cleaner; snow fighter; snow-fighting vehicle; snow grader; snow plough; snow plow; snow pusher; snow remover; snow shifter; snow sweeper
扫雪机刀片 snow blade
扫雪机底盘 snow plough carriage
扫雪犁板 snow plough blade
扫雪汽车 snow plowing automobile
扫雪器 snow flanger; snow sweeper
扫雪设备 snow removal equipment
扫雪刷 snow brush
扫雪栅栏 collecting snow fence
扫雪装车机 snow loader
扫烟囱工人癌 chimney sweeper's carcinoma
扫油板 sweep boom
扫油双体船 mop-cat; mop-catamaran
扫余均衡器 mop-up equalizer
扫帚 hand-broom
扫帚把 broomstick
扫帚痕纹 < 饰面抹灰 > trace of rake
扫帚留存条痕 broom finishing

埽

埽工 brush work; fascine work; kidding; sunken fascine layer; wicker works

埽工护岸 fascine revetment; kidding
埽料 fascine; willow
埽蓐 mattress
埽褥 brush(wood) mattress; mattress
埽褥盖面 mattress covering
埽褥护岸 mattress protection; mattress revetment
埽褥基础 mattress foundation

色

色氨酸 tryptophan(e)

色斑 colo(u)red speck; colo(u)r spot; colo(u)r stripe; discolo(u)ration; mottling; speck of color; tarnish
色板 colo(u)r disc[disk]; colo(u)r slab; tint plate
色饱和度 chroma(saturation); colo(u)r saturation
色饱和度降低 colo(u)r dilution
色饱和度信号 chroma signal
色保护 colo(u)r protection
色保护环 guard ring
色比 colo(u)r index; colo(u)r ratio
色编码光束 colo(u)r-coded light beam
色变暗 subduing
色变点 colo(u)r transition point
色变法 colo(u)r variation method
色变换 colo(u)r transformation; colo(u)r transition
色变调制器 chrominance modulator
色变指数 colo(u)r alteration index
色标 colo(u)r code; colo(u)r control scale; colo(u)r grade; colo(u)r index; colo(u)r note; colo(u)r scheme; colo(u)r standard; colo(u)r wedge; tonal scale
色标度 colo(u)r scale
色标管 chromatron
色标识器 colo(u)r marker
色标数 colo(u)r code number
色标值 colo(u)r value
色标志 colo(u)r marking
色表 colo(u)r chart
色表面 colo(u)r appearance
色表示 colo(u)r specification
色别法 colo(u)r scheme
色波筋 streaky metal
色波里特锅炉 Serpollet boiler
色玻璃 colo(u)red glass; stained glass
色玻璃框 colo(u)r roundel spectacle
色玻璃罩 colo(u)r glass shade
色补偿技术 colo(u)r compensation technique
色彩 colo(u)r(ing); hue; tinge
色彩暗淡 dingy
色彩饱和度 depth of colo(u)r saturation
色彩饱和度和强度系统 hue-saturation and intensity system
色彩保持 colo(u)r retention
色彩保真度 colo(u)r fidelity
色彩逼真度 chromatic fidelity; colo(u)r fidelity
色彩比值 colo(u)r ratio
色彩变化范围调整 hue range control
色彩补偿 colo(u)r compensation
色彩补偿滤色镜 colo(u)r compensating filter
色彩不调和的 tinty
色彩测量装 colo(u)r measuring system
色彩测试机 colo(u)r testing machine
色彩层次 gradation of colo(u)r
色彩层次变化 gradation change
色彩的 chromatic
色彩的不同层次 gradations of colo(u)r
色彩的三项属性 <指色相、明度、彩度> three-attributes of colo(u)r
色彩的重量感 colo(u)r evaluation of weight
色彩对比度 colo(u)r contrast
色彩对比强烈 strongly contrasting colo(u)rs
色彩分辨率 chromatic resolving

色彩复制 tone copy
色彩光泽 colo(u)r gloss
色彩和谐 lightness hue
色彩还原 colo(u)r rendering
色彩混(杂)colo(u)r contamination
色彩基调 colo(u)r motif
色彩计 colo(u)rimeter
色彩加浓 benday
色彩加深 benday
色彩渐变花纹 graduated patterns
色彩角 hue angle
色彩校正 <像片的> chromatic correction;colo(u)r correction
色彩校正计算机 colo(u)r masking computer
色彩校正器 colo(u)r corrector
色彩亮度 chroma luminance;colo(u)r intensity;colo(u)r brightness
色彩密度 colo(u)r density
色彩密度比 colo(u)r density ratio
色彩模拟 colo(u)r simulation
色彩浓淡法 colo(u)r dynamics
色彩浓度 chroma
色彩浓艳的 highly pigmented
色彩平衡 colo(u)r balance
色彩圈 circle of hues
色彩柔和 lightness hue
色彩闪变 play of colo(u)r;schiller
色彩设计 colo(u)r cast;colo(u)r conditioning;colo(u)r design;colo(u)r planning;colo(u)r scheme;colo(u)r way
色彩视觉 colo(u)r vision
色彩调合 colo(u)r combination
色彩调和 colo(u)r harmony
色彩调节 colo(u)r conditioning;transformation of colo(u)r
色彩调配 colo(u)r conditioning;colo(u)r dynamics
色彩调色理论 colo(u)r harmonic theory
色彩调制箱 paint box
色彩特性 colo(u)r property
色彩图表 colo(u)r chart
色彩图例 pache
色彩鲜艳 in gay colo(u)rs
色彩鲜艳的 technicolo(u)red
色彩显示 colo(u)r display
色彩协调 colo(u)r coordination;colo(u)r harmony
色彩信息 hue information
色彩学 chromatics;chromatology
色彩颜料 colo(u)r pigment
色彩样品 colo(u)r sample
色彩影响 influence of colo(u)r
色彩与植物 colo(u)r and plant
色彩再现 colo(u)r rendering;colo(u)r reproduction
色彩质量 chromaticity
色彩组成部分 chromatic component
色测高温计 colo(u)r pyrometer
色测温度 colo(u)r temperature;temperature colo(u)r
色层 colo(u)r(ed) layer;hypsometric(al) layer
色层叠合打样法 cromalin colo(u)r proofing system;light-sensitive layer lamination proof
色层法 chromatographic method
色层分隔 chromatographic separation
色层分离【化】chromatographic fractionation
色层分离的 stratographic(al);chromatographic
色层分离堆 chromatopile
色层分离法 chromatography;stratography
色层分离法分析 stratographic(al) analysis
色层分离谱 chromatogram;chromatograph
色层分离仪 chromatograph
色层分析 chromatographic analysis;stratographic(al) analysis
色层分析法 chromatography
色层分析谱 chromatograph
色层分析图 chromatogram
色层分析仪 chromatograph
色层技术 chromatographic technique
色层谱法【化】chromatography
色层谱仪 chromatograph
色层球 chromospheres
色层试剂 reagent for chromatography
色层吸收 chromatographic absorption
色层(油漆)彩色涂层 colo(u)r-coat(ing)
色层柱 chromatographic column
色差 chromatic defect;colo(u)r aberration;colo(u)r difference;colo(u)r error
色差方程 colo(u)r difference equation
色差放大器 colo(u)r difference amplifier
色差公式 colo(u)r difference formula
色差计 colo(u)r difference meter
色差计算机 colo(u)r difference computer
色差角 hue-difference angle
色差灵敏度 colo(u)r difference acuity;colo(u)r difference sensitivity
色差校正 <镜头的> colo(u)r correction;chromatic correction
色差敏感度 chromatic sensitivity
色差曲线 chromatic curve
色差容限 colo(u)r tolerance
色差视觉 colo(u)r difference acuity
色差视力 colo(u)r difference acuity
色差像 chromatic image
色差消除 achromatization
色差信号 colo(u)r difference signal
色差信号电压 colo(u)r difference voltage
色差仪 differential colo(u)rimeter
色差阈 colo(u)r threshold
色差值 value of chromatism
色场校正器 colo(u)r field corrector
色场同时传送制 simultaneous colo(u)r television system
色衬度 colo(u)r contrast
色成分 colo(u)r content;component colo(u)r
色冲淡 colo(u)r dilution
色串孢属 <拉> Torula
色纯度 colo(u)rimetric purity;colo(u)r purity;excitation purity
色纯度调节 purity control
色刺激 colo(u)r stimulus
色刺激函数 colo(u)r stimulus function
色刺激值 colo(u)r stimulus specification;psychophysical colo(u)r specification
色带 chromatape;chromatobar;chromatopencil;colo(u)r strip;copying ribbon;ink ribbon;ribbon;typer ribbon;gird <木杆防腐用>
色带比色计 colo(u)r-cord colo(u)rimeter
色带(测流)法 colo(u)r band method
色带导向 ribbon guide
色带反绕控制器 ribbon reverse control
色带盒 ribbon cartridge
色带换向 ribbon reverse
色带卷 ribbon cartridge;ribbon spool
色带可逆控制 ribbon reverse control
色带控制键 restrainable key
色带馈送机构 ribbon feed mechanism
色带盘 ribbon bobbin
色带色区选择器 ribbon zone selector
色带升降 ribbon oscillation
色带矢量 colo(u)r bar vector
色带使用程度指示器 carbon ribbon apply indicator
色带输送机械装置 ribbon feed mechanism
色带输送棘轮 ribbon feed ratchet
色带提升导向器 ribbon lift guide
色带提升机构 ribbon lift mechanism
色带图 colo(u)r bar pattern
色带位置指示器 ribbon position indicator
色带信号发生器 colo(u)r strip generator
色带左导向 ribbon left guide
色带左向机械装置 ribbon left mechanism
色当量 colo(u)r equivalent
色的分解 dispersion of colo(u)rs
色的过度校正 chromatic overcorrection
色的均等密度 chromatic equidensity
色的控制 colo(u)r control
色灯 colo(u)r(ed) light
色灯部件 colo(u)r light unit
色灯单元 colo(u)r light unit
色灯电锁器联锁【铁】interlocking by electric(al) locks with light signal
色灯附近的交通蠕动 <即阻滞> blight creep
色灯机构 colo(u)r light unit
色灯示像 colo(u)red light aspect
色灯顺序 <在周期内显示各色灯信号的顺序> colo(u)r light sequence
色灯系统 colo(u)r system
色灯显示 aspect
色灯信号 colo(u)r(ed) light signal;light beacon;light signal;daylight signal <交通>;colo(u)r system
色灯信号机【铁】colo(u)red light signal
色灯信号机构 colo(u)r light signal head
色灯周期长 <以秒计> cycle length
色灯周期时间变换【交】cycle length change
色灯自动闭塞 automatic colo(u)r light block
色灯组合 colo(u)r light combination
色等级 colo(u)r gradation
色底控制 colo(u)r base control
色点 colo(u)r dot
色淀 colo(u)r lake;lake
色淀 C 颜料红 red lake C pigment
色淀橙 lake orange
色淀红 lake red
色淀蓝 lake blue
色淀染料 lake colo(u)rs;lake dyes
色淀酸性黄 lake of acid yellow
色淀调色剂 lake toner
色淀性红 red for lake
色淀颜料 lake colo(u)rs;lake pigment
色淀枣红 lake bordeaux
色调 colo(u)r hue;colo(u)ring;colo(u)r melody;colo(u)r scheme;colo(u)r tone;hue;paint clay;shading of colo(u)r;tinct;tint;tonality;tonal value;tone(value)
色调板 tint plate;tone plate
色调饱和度 hue content
色调饱满度 fullness of shade
色调逼真度 tonal fidelity
色调变化 colo(u)r change;gradation change;shading
色调变化范围调整 hue range control
色调变化规则 regular tone variation
色调不均 tint unevenness
色调不鲜明图像 soft image
色调参数 colo(u)rimetric parameter
色调差 hue difference;tonal difference
色调的 tonal
色调等级 tonal gradation
色调电平 hue level
色调度 tonal range
色调对比 colo(u)r contrast;contrast of tone;picture contrast;tonal relationship
色调范围 colo(u)r content;gamut of colo(u)r;hue range;tonal range;tonal content <图像的>
色调范围调整 hue range control
色调粉 toner powder
色调丰满度 fullness of shade
色调改变 tone reversal
色调光楔组 tone wedge set
色调含量 tonal content
色调和灰阶 tone and gray scale
色调和谐 lightness hue
色调环 hue circle
色调计 tint(o)meter
色调角 hue angle
色调校正 tint correction
色调较淡的 tinge
色调界限 <航测地形的> tonal boundary
色调均匀 homogeneous tone
色调均匀程度 tone homogeneous degree
色调均匀性 colo(u)r uniformity
色调控制 hue control;tonal control
色调控制剂 pitch control additive
色调控制孔径 tone-control aperture
色调控制器 hue controller;tonal controller
色调亮度 tonal brightness
色调密度 tonal density
色调偏移 hue shift
色调强度比 colo(u)r density ratio
色调深度 depth of shade
色调深浅比 picture contrast
色调失真 tonal distortion;tone distortion
色调适中图像 middle-keyed picture
色调调节 hue control;phase control
色调调整 hue control;tonal response adjustment
色调调整电路 hue adjust circuit
色调特征 key value;tonal signature;tone signature
色调梯度 tonal range
色调梯级 tonal gradation
色调紊乱 confused tone
色调误差 hue error
色调鲜明图像 harsh image
色调线 tone-line
色调线法 tone-line technique
色调楔 tone wedge
色调选择 hue selection
色调样片晒印 <供制镶嵌图用> master print
色调异常 tonal anomaly
色调荧光屏 tint screen
色调再现 tone rendering;tone reproduction
色调值偏差 tonal value shift
色动力学 chromodynamics
色度 chroma;chromaticity;chrominance;colo(u)r grade;colo(u)rity;tinting strength
色度编码器 chromacoder
色度变码 chrominance transcoding
色度标志 colo(u)r specification
色度标准 colo(u)rimetric standard;colo(u)r scale
色度标准液 colo(u)r standard solution
色度表 chromaticity diagram

S

色度波动 colo(u)r fluctuation
色度测定 colo(u)rity determination
色度测定辐射探测器 colo(u)rimetric radiation detector
色度测定仪 colo(u)r grader
色度测定域 colo(u)rimetry measurement loci
色度测量 colo(u)rimetry
色度测量计 colo(u)r grader
色度差 chromaticity difference
色度差图 chromaticity difference diagram
色度常数 colo(u)r constant
色度传感器 chromaticity sensor; chromaticity transducer
色度纯度 colo(u)rimetric purity
色度代码转换 chrominance transcoding
色度单元 chrominance unit
色度倒相器 chroma inverter
色度对比 chromatic contrast
色度法 chromatometry; colo(u)rimetric method
色度放大器 chrominance amplifier
色度分级 colo(u)r grading
色度分离器 chroma separator
色度分析 colo(u)rimetric analysis
色度副载波 chrominance subcarrier
色度副载波边带 chromaticity subcarrier sideband
色度副载波再生器 chrominance-subcarrier regenerator
色度感 chromaticness
色度跟踪电路 chroma tracking circuit
色度恒定性 chromatic constancy
色度基色 chrominance primary
色度计 chrom(at)ometer; colo(u)rimeter
色度鉴定 colo(u)r check; colo(u)r test
色度键 chroma key
色度键控信号发生器 chroma-key generator
色度解调 chrominance demodulation
色度镜 chromoscope
色度控制 chroma control; tint control
色度亮度 chroma luminance
色度亮度干扰 crosstalk
色度明度 chroma luminance
色度频率 chrominance frequency
色度清晰度 chrominance resolution
色度去除率 hue removal rate; rate of colo(u)r removal
色度矢量 chrominance vector
色度视频信号 chrominance video signal
色度试验 chromaticity test
色度术 colo(u)rimetry
色度锁相 chrominance lock
色度锁相效应 effect of chrominance lock
色度特性 colo(u)rimetric property
色度体系 colo(u)rimetric system
色度调节 chrominance adjustment
色度调整 chroma control; colo(u)r (-saturation) control
色度通道 chrominance channel
色度图 chromatic(ity) diagram; colo(u)r mixture diagram
色度图表 chromatometer
色度位移 colo(u)rimetric shift
色度温度 colo(u)r temperature
色度系数 chromaticity coefficient
色度相位 chrominance phase
色度消隐 chrominance cancellation
色度信道 chrominance channel
色度信号 carrier-chrominance signal; chroma(ticity) signal; chrominance signal
色度信号带宽 chromaticity signal bandwidth
色度信号解调器 chroma demodulator
色度信号配准 scan registration
色度信号矢量图 chrominance signal vector diagram
色度信号下边带 lower chroma sideband
色度信号载波 chrominance signal carrier
色度信息 chromatic information
色度学 chromatometry; colo(u)rimetry
色度学家 colo(u)rimetrist
色度仪 colo(u)r measuring instrument
色度与亮度时延差 chrominance and luminance delay inequality
色度与亮度增益差 chrominance and luminance gain inequality; chrominance luminance gain inequality
色度杂波测量 chrominance-noise measurement
色度载波基准频率 chrominance-(sub) carrier reference; colo(u)r-carrier reference
色度增益调整 chrominance gain control
色度章度和透明度 hue-saturation and intensity
色度值 chromatic value; colo(u)rimetric value
色度轴 chrominance axis
色度坐标 chromaticity coordinates
色对比度 colo(u)r contrast
色对称 colo(u)r symmetry
色多项式 chromatic polynomial
色反差 colo(u)r contrast
色反映 colo(u)r reaction
色放大率误差 colo(u)r magnification error
色分辨本领 chromatic resolving power
色分辨力 acuity of colo(u)r
色分辨力锐度 acuity of colo(u)r definition
色分辨率 chromatic resolving power
色分辨率光栅 resolving chromatic power grating
色分辨率棱镜 chromatic resolving power prism
色分辨能力 chromatic resolving power
色分复用 colo(u)r-division multiplexing
色分解 colo(u)r breakup
色分离 colo(u)r splitting
色分离系统 colo(u)r separation system
色分析 colo(u)r analysis
色粉 toner
色复现 colo(u)r reproduction
色改正 colo(u)r correction
色感度 response colo(u)r relation
色感(觉) colo(u)r sensation
色感一致性 colo(u)r constancy
色感应 colo(u)r induction; colo(u)r response
色感应曲线 colo(u)r response curve
色箍 shroud
色光 colo(u)red light; tone
色光调 shade
色光改变 colo(u)r light modification
色光光度计 leucoscope
色光模拟 colo(u)r light analogue
色光信号 colo(u)r light signal
色光中心吸收 colo(u)r center[centre] absorption
色规格 colo(u)r stimulus specification
色过饱和 colo(u)r overload
色过渡 colo(u)r transition
色过量 colo(u)r overload
色过载 colo(u)r overload
色黑而薄的冰 dark nilas
色恒定性 colo(u)r constancy
色恒量 colo(u)r constant
色环 colo(u)red ring
色环检查 check for interference fringes; put down in colo(u)r
色幻觉 chromatism; chromesthesia; colo(u)r hearing
色灰度 shade
色辉 tint
色辉计 tint(o)meter
色辉透镜 tinted lens
色辉荧光屏 tint screen
色混合 mixing of primary pigment colo(u)rs; mixture of colo(u)rs
色混合规律 law of colo(u)r mixing
色迹管 colo(u)r trace tube
色基 chromogen; chromophore; colo(u)r base
色畸变 chromatic distortion; colo(u)r distortion
色几何学 colo(u)r geometry
色剂 toner
色剂标记 toner mark
色剂呈色 pigment colo(u)ration
色剂焊条 covered electrode
色剂浓缩 toner concentrate
色剂容器 toner container; tone reservoir
色剂载体 toner carrier
色价 colo(u)r number
色鉴别 colo(u)r discrimination
色浆 colo(u)rant slurry; colo(u)r paste; distemper; paste colo(u)r concentrate; pigment paste; washable distemper
色浆稠度 colo(u)r consistency
色浆刮刀 colo(u)r doctor; colo(u)r knife
色浆涂布 paste coating
色浆涂刷 distemper(ing)
色浆装饰法 barbotine decoration
色胶 distemper
色胶刷 distemper brush
色阶 colo(u)r graduation
色金 shell gold
色镜 colo(u)red shade; shade glass
色镜误差 shade error
色矩 colo(u)r moment
色觉 chromatic sensation; chromatic vision; colo(u)r sensation
色觉比较镜 anomaloscope
色觉不全 chromatelopsia
色觉恒常 colo(u)r constancy
色觉计 chromatometer
色觉检测计 ophthalmoleukoscope
色觉检查 chromatoptometry; chromoscopy; colo(u)rimetry; colo(u)r sense test
色觉检查表 colo(u)r test cards
色觉检查法 chromometry
色觉检查器 chromoscope
色觉亮度 brightness
色觉疲劳 colo(u)r fatigue
色觉锐度 acuity of colo(u)r
色觉说 colo(u)r theory
色觉学说 theory of colo(u)r vision
色觉仪 chromatometer
色觉异常 colo(u)r anomaly
色卡 colo(u)r chip(ping)s; colo(u)r chart; shade card
色卡图册 colo(u)r atlas
色克特拉式体系建筑 Sectra system
色空间 colo(u)r space; colo(u)r spacing
色宽容度 colo(u)r tolerance
色扩散 colo(u)r diffusion
色牢度 colo(u)r fastness
色牢度评级 colo(u)r fastness grading; colo(u)r fastness rating
色棱锥 colo(u)r pyramid
色立体 <用三度空间表示色彩之间关系> colo(u)r solid
色立体图 colo(u)r solid
色联觉 chromesthesia; colo(u)r hearing
色亮度测试仪 colo(u)r-brightness tester
色量 colo(u)r specification
色料 colo(u)rant; colo(u)ring material; colo(u)ring matter; colo(u)ring pigment; pigment; stainer; tinter
色料粉化 colo(u)r chalking
色料混合 colo(u)r mix
色料浆 dispersal colo(u)r
色料扩散(涂料) bleeding
色料粒子 colo(u)r particle
色料片 colo(u)r pellet
色料清洗 colo(u)r purge
色料研磨机 colo(u)r grinder; colo(u)r grinding machine; colo(u)r mill
色临界图 colo(u)r-critical graph
色六角 colo(u)r hexagon
色滤光片 colo(u)r filter
色率 colo(u)r index
色乱 colo(u)r breakup; colo(u)r splitting
色轮 colo(u)r cycle
色罗提隔音板 celotex board
色罗铜 Therlo
色码 colo(u)r code
色码电缆 colo(u)r-coded cable
色码数 colo(u)r code number
色满 chroman(e)
色盲 achromatopsia; achromatopsy; blindness; chromatelopsia; colo(u)r blindness; colo(u)r defective vision
色盲镜 anomaloscope
色盲片 blind film
色描绘 colo(u)r rendering
色名 colo(u)r name
色明暗度调整 colo(u)r shading control
色明度 colo(u)r brightness
色母料 colo(u)r batch
色母片 colo(u)r chip(ping)s
色姆科型长螺旋钻孔压浆灌注桩 Cemcore auger-injected pile
色浓度 depth of colo(u)r
色盘 colo(u)r disc[disk]
色盘降落信号指示器 drop annunciator
色泡 colo(u)r lamp
色配制 colo(u)r rendering
色坯 colo(u)red body
色匹配 colo(u)r match; colo(u)r matching
色偏 colo(u)r cast
色偏差 colo(u)r bias; colo(u)r deviation
色偏移 colo(u)r shift
色偏振 chromatic polarization
色票插(破损车) defect card holder
色票扣修车 colol(u)r label
色品 chroma(ticity); chrominance; colo(u)r stimulus specification; Munsell chroma
色品差 chromaticity difference
色品差图 chromaticity difference diagram
色品度 chromaticness; colo(u)rimetric quality
色品分量 chromaticity component
色品键控 chroma key
色品控制 chroma control
色品亮度 chroma luminance
色品闪烁 chromaticity flicker
色品调整 chroma control; chrominance adjustment

色品图 chromaticity diagram;colo(u)r triangle
色品系数 chromaticity coefficient
色品系统 chromaticity system
色品像差 chromaticity aberration
色品信号调制器 chromaticity modulator
色品延时线 chrominance delay line
色品值 chromatic value
色品坐标 chromaticity coordinates
色评定 colo(u)r evaluation
色谱 chromatograph;colo(u)r grade;colo(u)r spectrum;colo(u)r standard
色谱板 chromatoplate;chromatosheet
色谱棒 chromarod;chromatobar;chromatopencil;chromato-stick
色谱操作人员 chromatograph operator
色谱测定法 chromatographic detection
色谱的 chromatographic
色谱堆 chromatopile;chromatostack
色谱法 chromatographic method;chromatography
色谱法分级 chromatographic fractionation
色谱法分析气体 gas chromatograph
色谱方程 colo(u)r equation
色谱方法 chromatographic process
色谱分离 chromatograph;chromatographic fractionation
色谱分离室 chromatograph chamber
色谱分析 chromatographic analysis
色谱分析类型 chromatographic species
色谱分析器 chromatogram analyser [analyzer]
色谱分析质谱测定法 chromatography mass spectrography
色谱峰区域宽度 peak width
色谱工作人员 chromatographer
色谱功能 chromatographic function
色谱管 chromatographic tube;chromatotube
色谱过程 chromatographic process
色谱函数 chromatographic function
色谱红外光谱联机 gas chromatographic(al) infrared spectrometry combination
色谱级 chromatographic grade
色谱极谱 chromatopolarograph
色谱集 colo(u)r atlas
色谱技术 chromatographic technique;tail
色谱扩散 chromato-diffusion
色谱盘 chromatodisk
色谱气测井 chromatogram-gas logging
色谱圈 colo(u)r circle
色谱溶剂 chromatographic solvent
色谱溶液 chromatographic solution
色谱术 chromatography
色谱特性曲线 colo(u)r response curve
色谱提纯 purification by chromatography
色谱条 chromatostrip
色谱图 chromatogram;chromatomap
色谱图计算 chromatogram calculation
色谱图扫描 chromatogram scan(ning)
色谱吸附 chromatographic absorption
色谱吸附柱 chromatographic column
色谱效应 chromatographic effect
色谱学 chromatographia;chromatographic science
色谱学的 chromatographic
色谱仪检测器 chromatographic detector
色谱仪器 chromatographic apparatus
色谱圆筒 chromatoroll
色谱纸 chromatographic paper
色谱纸束 chromatopack
色谱质谱红外光谱联机 gas chromatographic(al) mass spectrometry infrared spectrometry combination
色谱质谱联机 gas chromatographic(al) mass spectrometry combination
色谱-质谱联用 chromatography mass spectrometry
色谱质谱联用法 combined gas chromatography mass spectrometry
色谱柱切换技术 switching column technique
色漆 colo(u)red paint;pigmented coating
色漆料浆 ground paste
色漆涂层 coat of paint
色起霜 colo(u)r bloom
色迁移 colo(u)r migration
色强度 intensity of colo(u)r
色球暗条 filament of chromosphere
色球爆发 chromospheric eruption;flare
色球层【气】 chromosphere
色球层的发射光谱 chromospheric emission spectrum
色球差 spherochromatic aberration;spherochromatism
色球低层 lower chromosphere
色球高层 upper chromosphere
色球光斑 chromospheric facula
色球光谱 chromospheric spectrum
色球结 chromospheric knot
色球精细结构 chromospheric fine structure
色球模型 model chromosphere
色球抛射 chromospheric ejection
色球泡 chromospheric bubble
色球谱斑 chromospheric flocculus
色球日芒 chromospheric mottling
色球日冕过渡区 chromosphere-corona transition region
色球石 dat(h)olite[datolith]
色球网络 chromospheric network
色球望远镜 chromosphere telescope
色球温度 chromospheric temperature
色球物质 chromospheric material
色球旋涡 chromospheric whirl
色球耀斑 chromospheric flare
色球针状物 chromospheric spicule;chromospheric spike
色球蒸发 chromospheric evapo(u)ration
色区 colo(u)r zone
色群 colo(u)r group
色容限单位 colo(u)r tolerance unit
色融合 colo(u)r fusion
色弱 anomalous trichromatism;colo(u)r weakness;tritanomalous vision
色三角 colo(u)r triangle;Maxwell's triangle
色散 chromatic aberration;disperse;dispersion
色散本领 dispersive capacity;dispersive power
色散波长 dispersion wavelength
色散测定 dispersion measurement
色散测谱学 dispersive spectrometry
色散常数 dispersion constant
色散超声延迟线 dispersive ultrasonic delay line
色散点 dispersion point
色散度 degree of dispersion;dispersion degree;dispersity;dispersiveness
色散法 dispersion method
色散反射器 dispersing reflector
色散方程 dispersion equation
色散分光法 disperse spectrometry
色散分光计 dispersive spectrometer
色散公式 dispersion equation;dispersion formula
色散共振腔 dispersing resonator
色散关系 dispersion relation
色散光度计 dispersion photometer
色散光谱 dispersion spectrum;dispersive spectrometry
色散光谱调制器 dispersing spectromodulator
色散光谱仪 dispersive spectrometer
色散光学系统 dispersive optical system
色散计 dispersimeter
色散角 angle of dispersion;dispersion angle
色散介质 dispersion medium
色散棱镜 dispersing prism;dispersion prism;dispersive prism
色散理论 dispersion theory;theory of dispersion
色散力 dispersion force
色散量 dispersion measure
色散律 dispersion law
色散率 dispersive power
色散率倒数 reciprocal dispersion;reciprocal dispersive power
色散媒质 dispersion medium
色散能力 dispersive power
色散频率 dispersion frequency
色散谱线 dispersion line
色散器 disperser[dispersor]
色散曲面 dispersion surface
色散曲线 dispersion curve
色散射 chromatic dispersion
色散树脂 dispersion resin
色散衰减 dispersion attenuation
色散衰减因数 dispersion attenuation factor
色散调制器 dispersive modulator
色散梯度 dispersion gradient
色散透镜 dispersing lens;dispersive lens
色散位移单模光纤 dispersion shift single mode fiber[fibre]
色散系数 coefficient of dispersion;dispersion coefficient;reciprocal dispersive power
色散现象 chromatic dispersion
色散限制 dispersion-limited
色散限制作用 dispersion-limited operation
色散效应 dispersion effect
色散型分光辐射计 dispersing-type spectroradiometer
色散性 dispersivity
色散延迟线 dispersive delay line
色散仪 dispersing instrument
色散元件 dispersing component;dispersion element
色散增数 constringence[constringency]
色散组元 dispersing element
色砂 vitreous sand
色栅 colo(u)r grid
色闪动现象 schillerization
色升高 colo(u)r boost
色失聚 colo(u)r misconvergence
色失真 chromatic distortion;colo(u)r distortion;cross colo(u)r
色石渣 colo(u)r marble chips
色视差 chromatic parallax
色视场 colo(u)r field
色视觉 colo(u)r sense;colo(u)r vision
色视觉测验 colo(u)r vision test
色视力计 chromo-optometer
色试验 colo(u)r test
色适应 chromatic adaptation
色适应变化 change of chromatic adaptation
色适应位移 adaptive colo(u)r shift
色适应状态 state of chromatic adaptation
色数 chromatic number;colo(u)r number
色水测流(速)法 colo(u)r method of measuring velocity;colo(u)r-velocity ga(u)ging
色素 colo(u)rant;colo(u)r element;colo(u)ring agent;pigment
色素胶水 pigmentary size
色素粒 pigment granule
色素母体 pigment precursor
色素强度 absorbance of the colo(u)r
色素示踪 <用于测流、测波等> dye tracing
色素图案 pigment figure
色素吸收率 absorbance of the colo(u)r
色素细胞 chromatophore
色素显影剂 dye developer
色素形成 chromogenesis
色素形成作用 pigmentation
色胎瓷 colo(u)red body china
色套色法 chromatography
色特塞式 Surtseyan
色条 colo(u)r stripe;stripe;band of colo(u)r <指漆于航标,测流标杆等上>
色条试验 streak test
色条涂层 stripcoat
色条纹 streak
色条信号发生器 colo(u)r bar generator
色条信号图 stripe pattern
色听 colo(u)red hearing
色同步分离 burst separator
色同步控制振荡器 burst controlled oscillator
色同步脉冲 burst pulse
色同步门脉冲 burst gating pulse
色同步信号 burst;reference burst;colo(u)r burst
色同步形成器 burst former
色同步再生效率 burst-energy-recovery efficiency
色头 tinge;tone
色透镜 chromatic lens
色图 colo(u)r diagram;colo(u)r figure;colo(u)r graph
色陀螺 colo(u)r top
色维 colo(u)r dimension
色位灯(光) colo(u)r position light
色位灯信号(机)【铁】 colo(u)r position light signal
色位移 colo(u)r displacement
色位组合 colo(u)r position combination
色温变换玻璃 colo(u)r temperature changing glass
色温变换滤光片 conversion filter for colo(u)r temperature
色温标 temperature colo(u)r scale
色温测定 colo(u)r temperature measurement
色温度 colo(u)r temperature
色温计 colo(u)r temperature meter
色温转换滤光器 conversion filter for colo(u)r temperature
色纹 pigment figure
色污斑点 flecked
色污染 colo(u)r staining
色吸收 absorption of dyes
色吸收体 colo(u)r absorber
色锡 fin-coat
色系 colo(u)r system
色隙 colo(u)r space
色线 colo(u)r streak;colo(u)r stripe;filament line;streak(ing);streak line;streak of colo(u)r
色线试验 streak test
色相 colo(u)r hue;colo(u)r shade;hue;hue of colo(u)r

S

色相纯度 purity of hue
色相对比 contrast of hue
色相干 chromatic coherent
色相关误差 colo(u)r dependent error
色相环 colo(u)r cycle; hue circle; hue circuit; colo(u)r circle
色相鲜明度 hue content
色相指数 colo(u)r index
色响应 colo(u)r response
色像 colo(u)r image
色像差 chromatism; colo(u)r aberration; chromatic aberration
色像亮度 brightness colo(u)r image
色像频率 colo(u)r picture rate
色效应 chromatic effect
色楔 tonal wedge
色心 colo(u)r center[centre]
色心激光器 colo(u)r center[centre] laser
色心晶体 colo(u)r center crystal
色心种类 type of colo(u)r center [centre]
色型 colo(u)r pattern
色修正 colo(u)r correction
色修正系数 colo(u)r correction factor
色序率 colo(u)r-sequence rate
色嗅 colo(u)red olfaction; colo(u)-red smelling
色选法 colo(u)r-selection technique
色样 colo(u)r sample; colo(u)r swatch
色样本 colo(u)r atlas
色液贯入试验 dye penetrant test
色移 colo(u)r migration; gamut
色移性 migrating property
色荫 shade
色诱导 colo(u)r induction
色釉 colo(u)red glaze
色余 colo(u)r excess
色域 colo(u)r gamut; colo(u)r range
色阈 chromatic threshold
色原 chromogen; toner
色原黄 toner yellow
色原学说 colo(u)r theory
色原棕 toner brown
色晕 chromatic halo
色再现 colo(u)r reproduction
色泽 colo(u)r and luster[lustre]; tinct(ure); tinge
色泽变化灰色分级卡 grey scale of colo(u)r change
色泽复原 recovery of colo(u)r
色泽均匀性 uniformity of colo(u)r
色泽优良的 colo(u)ry
色泽诊断 chromodiangnosis
色泽准确性 accuracy of colo(u)r value
色增感 colo(u)r sensitization
色增感剂 colo(u)r sensitizer
色折射 chromatic refraction
色帧 colo(u)r frame
色帧频 colo(u)r-frame frequency
色知觉 colo(u)r perception; perception of colour
色值 chromaticity value; colo(u)r number
色值校正 correction of colo(u)r values
色值失真 falsification of tone values
色纸 chromopaper; pigmented paper; stained paper
色指示器 colo(u)r indicator; colo(u)r marker
色指数 chromatic index; colo(u)r index
色质 chromaticness; colo(u)r quality; stimuli; stimulus
色质镜 chromascope
色轴 colo(u)r axis
色转换滤色镜 colo(u)r conversion filter
色锥 colo(u)r cone
色渍 colo(u)r patch; colo(u)r staining
色组 colo(u)r class
色坐标 colo(u)r coordinates
色坐标变换 colo(u)r coordinate transformation

涩

涩度 acerbity; asperity

涩黏[粘]土 astringent clay
涩味 acerbity; asperity; astringency (taste)

铯

铯137辐射 radiation of 137Cs

铯沸石 pollucite
铯矿 cesium ore
铯矿床 cesium deposit
铯榴石 pollucite; pollux
铯绿柱石 morganite
铯锰星叶石 cesium-kupletskite
铯锑钽矿 cestibtantitie
铯盐 cesium salt
铯源 cesium source
铯蒸汽磁力仪 cesium vapo(u)r magnetometer
铯钟 c(a)esium clock

瑟

瑟科玛琋脂 <一种衬垫金属的接合剂> Secomastic

瑟克雷坦 <一种铝青铜> secretan
瑟库勒斯连接 <一种铜管连接> Securex
瑟雷调节器 Thury regulator
瑟利奥拱窗 Serlian motif
瑟利奥拱门 Serlian motif
瑟玛琋脂 <一种砖形铺地材料> Semastic
瑟莫科斯热电偶 thermocoax
瑟莫里特 <一种耐蚀铜镍锌合金> Seymourite
瑟斯顿尺度 Thurston's scales
瑟斯顿高锌黄铜 Thurston's brass
瑟斯顿个性测验表 Thurston's temperament schedule
瑟斯顿锡基轴承合金 Thurston's metal
瑟斯顿铸造锌合金 Thurston's alloy
瑟索板层岩群【地】 Thurso flagstone group
瑟瓦定 <一种沙巴达碱> cevadine
瑟韦尔孔斜计 Surwell clinograph

森

森泊姆恒合金 Senperm

森达洛 <一种硬质合金> sendalloy
森方法 Senn process
森吉米尔镀锌法 Sendzimir coating process
森吉米尔式极薄钢板多辊轧机 Sendzimir mill
森吉米尔式轧机 Sendzimir mill
森吉米尔行星轧机 Sendzimir planetary mill
森科高真空回转油泵 Cenco pump
森林 sylva; timber; timber land; woods
森林版 woodland board; woodland plate
森林保护 forest conservation
森林保护区 forest conservancy; forest reserve
森林保护学 forest-protection
森林保留地 woodland reserves
森林保险 forest insurance
森林边界 forest boundary
森林边缘地 purlieu
森林补植 forest replantation
森林财产 forest property
森林财政 forest finance
森林采伐更新管理办法 Managing Measures of Cutting and Reforestation of Forest
森林采伐权 forest concession
森林草原 forest steppe; sylvosteppe
森林草原带 forest steppe belt
森林测量 forest survey(ing)
森林产品 forest product
森林产权判定 forest settlement
森林产业 forest estate
森林成煤过程 forest-to-coal process
森林虫害 damage by forest-insects
森林垂直分布线 forest line; timber line; tree line
森林粗腐殖质 mor
森林村 forest village
森林存量 forest inventory
森林存量增长 growing stock of forest
森林带 forest belt; forest land; forest zone
森林道路 forest track; forest(ry) road
森林的 forestall
森林的表示 representation of forests
森林地 woodland
森林地被物 forest floor; forest litter
森林地带 land covered with forest; weald; backwoods <边疆或远离城市的>
森林地面 forest floor
森林地图 forest cover map; forest map; map of stock forest
森林地位级 site quality of forest
森林地植物学 forest geobotany
森林调查 forest description; forest survey
森林顶极 forest climax
森林(顶)梢枯死 forest dieback
森林动物 forest animal
森林动物的存在 presence of animal life in a forest
森林动物学 forest zoology
森林冻原 forest tundra
森林对土壤的影响 influence of forest on soil
森林多的 sylvan
森林伐除后牧地 cut-over pasture
森林法 forest law
森林法院 forest court
森林防火 forest-fire prevention
森林防火条例 regulation on forest fire-fighting
森林分布线 forest line; timber line; wood line
森林分区署 divisional forest-office
森林风 forest wind
森林风景区 forest scenic spot
森林符号 tree sign
森林抚育 tending of wood
森林抚育采伐 tending and cutting of forest
森林抚育更新 care and regeneration of forest
森林抚育间伐 forest thinning
森林腐殖土 forest humus soil
森林腐殖质 forest humus
森林覆被率 percentage of forest cover
森林覆被面积 cover area of forest
森林覆被(物) forest floor
森林覆被型 forest cover type
森林覆盖 forest canopy; forest conservancy; forest cover
森林覆盖程度 degree of vegetation
森林覆盖地区 areas covered with forests
森林覆盖地图 forest cover map
森林覆盖度 areal coverage of forest; forestation coefficient
森林覆盖率 forest acreage; forest coverage(rate); forest covering rate; forest percent; land area covered with trees; percentage of forest cover; percentage of land-covered forests
森林覆盖面积 forest covered area
森林覆盖面积统计 forest cover statistics
森林改进伐 forest improvement
森林改良 forest reclamation
森林干线公路 arterial forest road
森林更新 forest regeneration; forest renewal; reafforest
森林工业 forest industry
森林工艺学 forest technology
森林公路 forest highway
森林公园 forest park; park forest
森林估价 forest valuation
森林管理 forest husbandry; forest management
森林管理区 forest district
森林管理人 forest manager
森林管理系统 forest management system
森林管理学 forest administration
森林管理与侵蚀 forest management erosion
森林管理员 ranger; wood ranger
森林规划 forest planning
森林轨道 forest-tramway
森林害虫 forest pest
森林航测 forest aerial photogrammetry
森林航空测量 forest aerial photography; forest aerosurveying
森林和草原间的灌木过渡带 jarales
森林和野生动物型自然保护区管理 Administration Measures of Nature Reserves for Forest and Wild Animals
森林护养 forest conservation
森林环境 forest environment
森林环境卫生 forest sanitation
森林环境遥感 remote-sensing for forest environment
森林荒废 forest devastation
森林火灾 forest fire
森林火灾保险 forest-fire insurance
森林火灾的种类 kinds of forest fires
森林火灾防救 forest-fire control
森林火灾气象学 forest-fire meteorology
森林火灾烧成的空地 burn(a)
森林火灾云烟 forest-fire cloud
森林火葬场 forest crematorium
森林极限温度 forest limit temperature
森林技师 forest examiner; forest experiments officer; forest-master
森林监督管理 forest supervision
森林监督局 Forest Inspection Bureau
森林阶段 forest stage
森林截流 interception by forest
森林界 edge of a wood
森林界线 stand board; timber line; tree line
森林界限 forest limit; forest line
森林经济 forest economy; forest economics
森林经济学 forest economics
森林经营 forest management
森林经营单位 forest management unit
森林经营法 mode of treatment
森林经营图 management map
森林经营学 forest management
森林景观 forest landscape
森林景色 weald

森林警察 forest policeman
森林净化 forest purification
森林居民 forester
森林砍伐 deforestation;forest cutting
森林砍伐后牧地 stump pasture
森林砍伐率 rate of deforestation
森林滥伐 forest denudation; timber mining
森林立地指示植物 forest site indicator
森林利用学 forest utilization
森林淋余土 marron soil
森林流域 forest watershed
森林绿 forest green
森林美学 forest esthetics
森林面积 areal coverage of forest;area of woods;forest area;wooded area
森林面积统计 forestry area statistics
森林面积系数 coefficient of forestation
森林苗圃 forest(-tree) nursery
森林墓地 forest cemetery
森林泥炭 forest peat
森林破坏 forest deterioration; forest devastation
森林期望价值 forest expectation value
森林气候 forest climate
森林气候观测所 forest weather station
森林气象学 forest meteorology
森林潜育土 forested gley soil
森林区 forest area; forested region; silva[sylva];forest district
森林区的 silvan
森林区划 forest-division
森林区域 woodland
森林群落 forestry community
森林摄影测量 forest photogrammetry
森林生产率 forest productivity
森林生产资本 forest-cost-capital
森林生境 forest habitat
森林生态 forest ecology
森林生态系统 forest ecologic(al) system;forest ecosystem
森林生态学 forest ecology;silvics
森林生物群 forest biomass
森林生物学 forest biology
森林施肥 forest fertilization
森林使用费 royalty
森林使用权 forest-right;forest-servitude;right to use in forestry
森林事业年度 forest year
森林试验站 forest experiment station
森林受害 hazard of forest;hazard to forest
森林树木 forest tree
森林树木园 forest garden
森林树种 forest species
森林水文学 forest hydrology
森林损害 forest damage
森林所得税 forest-income tax
森林所有权 forest property
森林苔藓泥炭 forest moss peat
森林调节径流 regulation of streamflow by forest
森林调整 regulation of forest
森林条例 forest regulation
森林铁道 forest-railroad;forest-railway
森林铁路 forest logging railway; forest-railway;logging railway
森林铁路机车 logging locomotive
森林统计学 forest statistics
森林图 forest map
森林土 forest soil;sylvestre soil;sylvogenic soil;woodland soil
森林土壤 forest soil;mor;timbered soil
森林土壤改良 forest soil reclamation
森林退化 forest decay
森林卫生 forest hygiene
森林卫生学 forest hygienics
森林线 tree line
森林小气候 forest microclimate
森林小区 forest plot
森林休耕作业法 bush fallow system
森林蓄积量 forest growing stock;forest reserve
森林蓄积清查 management volume inventory
森林学 forestry
森林学家 silviculturist
森林学校 forest-school
森林巡视 forest-warden
森林巡视员 wood ranger
森林巡视制度 forest-guard system
森林演替 forest succession
森林养护 forest conservancy
森林遥感 forest remote sensing
森林益虫 useful forest insect
森林影响 forest influence
森林郁闭度 canopy density; crown density
森林原图 woodland drawing
森林造林率 forest percent
森林沼泽 forest bog;forest swamp
森林沼泽沉积 forest swamp deposit
森林沼泽相 forest swamp facies
森林遮盖 forest canopy
森林政策 forest policy
森林知识 woodcraft
森林职员 forest staff
森林植被 forest cover
森林植被型 drymion
森林植物 forest plant
森林植物带 forest region
森林植物群 forest vegetation
森林植物学 forest botany
森林志 silva[sylva]
森林制图 forest mapping
森林治水 regulation of streamflow by forest
森林致死遮荫度 forest fatal shade
森林中的落叶 duff
森林中无木处 alcove
森林逐渐失去活力 forest decay
森林逐渐衰败 forest deterioration
森林主伐 final cutting;final felling
森林资本 forest capital
森林资源 forest reserve; forest resources
森林资源保护 forest resources conservation
森林资源档案 files of forest resources
森林资源调查 forest inventory
森林资源丰富的 well-wooded
森林资源管理 forest inventory control
森林资源管理机构 management organs of forest resources
森林资源利用 utilization of forest resources
森林资源评估 forest resource assessment
森林资源清查 forest assessment;forest inventory
森林资源消长目标责任制 responsibility system with the aim of balancing depleting and growing forest resources
森林综合利用 integrated forest utilization
森林作物 forest crop
森林作业法 silvicultural system
森乃特 <硫氰乙酸异茨酯> Thanite
森诺曼阶【地】Cenomanian
森诺统 <晚白垩世>【地】Senonian (series)

僧

僧寮 dorter[dortour]
僧侣 monk;lay brethren <修道院里做杂役的>
僧侣住房 monastery
僧帽装饰顶 miter cap
僧尼层【地】Senni beds
僧室 monastic cell
僧院回廊 tresaunte

杀

杀病毒剂 viricide
杀草丹 benthiocarb
杀草剂 herbicide
杀草快 <一种除莠剂> Diquat
杀虫粉 insect powder
杀虫剂 biocide; disinfectant; insecticide; insectifuge; insect repellant; pesticide;mipafox <其中的一种>
杀虫剂残留物 pesticide residue
杀虫剂车间 <石油炼厂内> insect farm
杀虫剂用煤油 insecticide kerosene
杀虫胶 insect-paste
杀虫菌剂 pesticide
杀虫灵 <一种农药> acephate
杀虫灭菌剂 insectofungicide
杀虫漆 insecticidal lacquer; insecticidal paint;insecticide paint
杀虫涂料 insecticidal paint; insecticide paint
杀虫药 insecticide;pesticide
杀虫用熏剂 cimex
杀虫油 insect(icidal) oil
杀杆菌剂 bacillicide
杀害微生物的药剂 biocide
杀害物药剂 pesticide
杀寄生虫剂 parasiticide
杀寄生物的 parasiticide
杀价 force down price
杀菌 sterilize
杀菌处理 germicidal treatment
杀菌的 antiseptic(al); bactericidal; germicidal
杀菌灯 bactericidal lamp; germicidal lamp
杀菌灯玻璃 germicidal lamp glass
杀菌法 disinfection
杀菌活动 fungicidal property
杀菌剂 antiseptics;bactericidal agent; bactericide; biocide; disinfectant; germicidal agent; germicide; germifuge; microbicide; mycotox; sanitizing compound; sterilizing agent; disinfect
杀菌剂的 germicidal
杀菌剂化学稳定性 chemical stability of fungicide
杀菌剂喷洒器 fungicide sprayer
杀菌篮 retort basket
杀菌漆 bactericidal paint
杀菌器 sterilizer
杀菌曲线 killing curve
杀菌室 sterilized room
杀菌涂料 disinfectant paint;fungicidal paint;germicidal paint;germ repellent paint
杀菌物 germicide
杀菌洗涤剂 fungicidal wash
杀菌系数 bactericidal coefficient
杀菌效果 germicidal efficiency
杀菌效力 germicidal effect
杀菌效率 bactericidal efficiency
杀菌效应 bactericidal effect; sterilizing effect
杀菌盐剂 mycocide salt
杀菌作用 bactericidal action;germicidal action; sterilizing effect; disinfection;sterilization
杀螨剂 acaricide;miticide
杀霉菌剂 fungicide; mildewcide; mycocide
杀灭 killing
杀灭剂 killer
杀伤半径 kill radius
杀伤炸弹 fragmentation bomb
杀生物剂 biocide
杀鼠剂 rodenticide
杀鼠酮 pindone
杀水草剂 aquatic herbicide
杀水桩 starling
杀微生物剂 microbicide
杀蚊剂 culicide
杀蚜虫剂 aphicide
杀幼虫剂 larvicide[lavacide]
杀藻剂 algaecide;algicide
杀真菌剂 fungicide

沙

沙岸 hurst;sand bank;strand
沙岸植物 ammochthad
沙巴尔纤维 palmetto fiber
沙巴织物 saba
沙坝 bar; dam; hirst [hurst]; sand bank; sand bar (rier); sand reef; sand ridge
沙坝岛 barrier island
沙坝海岸 barrier coast
沙坝类型 type of bar
沙坝名称 name of bar
沙坝盆地 barred basin
沙坝平原 bar plain
沙坝台地 bar platform
沙坝滩 bar beach
沙坝形态 shape of bar
沙坝型转移 bar by-passing
沙坝阻隔理论 bar theory
沙斑 sand patch
沙包 bagwork
沙抱榆 bigfruit elm
沙暴 desert storm; dust storm; sand devil;sand flood
沙泵 sheet pump
沙壁 wall of sand
沙波 bed ripple;bed wave;sand ridge; sand ripple; sediment ripple; sand wave
沙波背流面 lee surface
沙波的形成 sand wave formation
沙波陡度 steeping of sand wave
沙波法 <测推移质的> dune tracking
沙波高度 height of sand wave
沙波前坡滑面 slip face
沙波数 wave number
沙波系 sand wave system
沙波向流面 stoss surface
沙波形状阻力 form drag of sand wave; form resistance of bed wave
沙波运动 sand ripple motion; sand ripple movement;sand wave movement
沙波周期性变化特性 periodic(al) change characteristics of sandwave
沙布 emery cloth
沙布袋 gauze bag
沙仓 sand depot
沙槽 sand trap
沙草 sand-grass
沙草原 sandveld
沙层中的黏[粘]土瘤 clay gall
沙尘暴 dust storm;sand haze
沙沉法 <一种消除漂浮在海面上油污的方法> sand sink
沙池散白蚁 Reticulitermes arenincola
沙船 large junk
沙床河流 sand-bed river
沙茨基海隆 Shatsky rise
沙带 sand ribbon;sand strip
沙袋 sand(-filled)bag
沙袋堤围堰 sandbag embankment cofferdam
沙袋灌浆护岸 grout filled fabric bag revetment

S

沙袋护岸 sack revetment
沙袋掩体 sand-bag bunker
沙挡 sediment barrier
沙岛 sand island
沙岛列 barrier chain
沙岛群 barrier chain
沙的百分比含量 sand equivalent
沙的沉降 sand fallout
沙的临界孔隙比 critical void ratio of sand
沙堤 sand dike[dyke];whaleback dune;sand levee
沙底锚地 sand bottom anchorage
沙底河弯 sand-bedded river meander
沙底水井 blind well
沙地 sandy land
沙地的 sand cloth
沙地锚 dory anchor; sand anchor; trawl anchor
沙地试验路线 sand course
沙地游戏场 sandlot
沙垫层 sandfill cushion
沙丁鱼 sardine;short-bodied sardine
沙丁鱼油 pilchard oil
沙堆 down sand-dune;drifting;dune;sand drift
沙堆模型 sand ionization formula
沙尔皮特过程 Salpeter process
沙尔硬木 < 印度产 > sal
沙发 settee;sofa
沙发床 davenport; sofa bed; studio couch
沙发套 anti-macassar
沙发椅 couch; padded armchair; upholstered armchair
沙发椅子 squab
沙发用钢线钉 steel wire upholsterer's tack
沙伐尔脱 < 音程单位 > savart
沙防 sediment control
沙菲尔德采样瓶 Shafferd flask
沙菲尔德防喷器 Shafferd cellar control gate
沙菲莱特玻璃 Saferite
沙沸 boiling of sand
沙费里塞 < 一种防水材料 > Sal-ferricite
沙风暴【气】sand storm
沙岗 sand dune
沙埂 barrier (of sand) ; sand reef; sand ridge;barrier bar
沙埂湖 barrier lake
沙埂列 barrier chain
沙埂沙滩 barrier island
沙钩 hook;recurved spit
沙谷 trough
沙管 sand gall
沙果木 crabwood
沙埂海滩 barrier beach
沙害 sand calamity;sand hazard
沙害防护 debris protection
沙蒿 sand sagebrush
沙痕 sediment ripple
沙洪 sand flood
沙化 sandification
沙荒 sand dune area; sandy waste (land)
沙荒地 lande
沙皇宫廷 Tsar's palace
沙黄 safranine
沙基培养 sand culture
沙棘 sea backthern;sea buckthorn
沙脊 sand ridge;whaleback
沙加拉海渊 Shagara deep
沙岬 barrier spit
沙槛冲除 bar removal
沙槛顶 sill crest
沙槛高程 sill elevation
沙槛盆地 barred basin
沙槛上升 sill raising
沙槛水深 bar depth;bar draft
沙姜土 shachiang
沙礁 cay;sand bar;sandkey
沙晋尔斯超速离心机 Sharples type ultracentrifuge
沙井 wash bore
沙阱 desilter;sand trap
沙阱式冲沙闸(门) sand trap scour gate;silt trap scour gate
沙颈岬 land-tied island;tombolo(cape)
沙卷 sand pillar
沙卷风 desert devil
沙坎 scarp
沙坑 jumping pit;sand hole
沙拉 salad
沙拉多建造 Salado formation
沙拉拿风 solano
沙拦河口 barred mouth
沙浪 bed wave;sand wave
沙浪运动 movement of sand-wave
沙类土 sandy soil
沙梨 sand pear
沙砾 sandy gravel
沙砾保护层 gravel envelope
沙砾糙率 grain roughness
沙砾覆盖层 sand and gravel overlay-(er)
沙砾干河床 sandwash
沙砾混合滩 sand and gravel foreshore
沙砾石 gritstone
沙砾石变滩 sideflat of shingle and sand
沙砾碎石法令 < 1893 年美国国会为防止水力采矿的沙砾碎石进入通航河道而制定的法令 > Debris Law
沙砾滩 gravel beach
沙粒 grains of sand
沙粒糙率 grain roughness
沙粒粗(糙)度 grain roughness
沙粒焚化 grit incineration
沙粒含量 sand content
沙粒计 sand-grain meter
沙粒雷诺数 grain-size Reynolds number
沙粒磨损率 wear rate of sand grains
沙粒器 sand catcher
沙粒小体 psammoma body
沙涟 ripples
沙廉榴石 Syrian garnet
沙量平衡 sediment budget
沙林 sarin
沙林尼亚地体 Salinia terrane
沙岭 dune ridge
沙流 sand drift;sand flow;sand stream
沙龙 salo(o)n
沙龙卷 sand tornado
沙隆 sand swell
沙垄 dune;sand dune
沙垄河床 dune covered riverbed
沙漏 hand glass;hour-glass < 计时用 >
沙漏构造 < 中国古代计时器 > sand-watch structure
沙漏线【数】hour-glass curve
沙漏形活塞 hour-glass piston
沙漏形反射器 hour-glass reflector
沙漏状断口 hour-glass fracture
沙卢克风 Shaluk
沙卵石覆盖层 sand-pebble protection layer
沙卵石河床 sand-pebble riverbed
沙洛风 Solore
沙埋 bury by sand;sand burying;sand cover
沙门氏杆菌属 salmonella
沙面 sand pavement
沙面波纹 sand ripple
沙漠 desert; sand barren; sand (y) desert
沙漠草原 desert steppe
沙漠沉积 desert deposit
沙漠带 desert belt;desert zone
沙漠地区 desert area
沙漠(地区)住房 desert residence
沙漠地下水 groundwater in desert
沙漠动物 desert animal
沙漠风 desert wind
沙漠风暴 sand storm
沙漠高原 desert plateau
沙漠灌丛 desert scrub
沙漠湖泊沉积 desert lake deposit
沙漠湖泊相 desert lake facies
沙漠化 desertification;desertization
沙漠化防治 anti-desertification
沙漠化控制 desertification control
沙漠环境 desert environment
沙漠蝗虫 desert locusts
沙漠居住 desert residence
沙漠控制 sand control
沙漠扩展 desert spread;spread of the desert
沙漠砾石表层 desert pavement
沙漠砾石盖层 desert pavement
沙漠绿洲 oasis[复 oases]
沙漠卵石覆盖层 desert mosaic; desert pavement
沙漠蔓延 desert creep
沙漠木麻黄 desert oak
沙漠盆地 basin desert;bolson
沙漠平原 desert peneplain;desert plain
沙漠漆 desert paint;desert patina
沙漠气候 desert climate
沙漠侵蚀准平原 desert peneplain
沙漠区径迹测量 track survey on desert
沙漠商队 caravan
沙漠试验 desert test
沙漠铁路 desert railway
沙漠土(壤) desert soil
沙漠洼地 kevir
沙漠相 desert facies
沙漠性气候 desert climate
沙漠(岩)漆 desert varnish
沙漠岩石表面锈层 < 一般呈棕黑色 > desert varnish
沙漠研究 desert survey
沙漠盐壳 patina
沙漠植被 desert vegetation
沙漠植物 desert plant
沙漠治理 control of desert; desertification control
沙漠中的绿洲 oasis[复 oases]; wadi [wady]
沙漠中的水泉 water hole
沙漠中的盐斑地 shott
沙漠中蘑菇石 gour
沙漠中小洼地 daia
沙姆维造山旋回 Shamvaian cycle
沙囊 ball;ballast
沙喷注 sand spout
沙坪 sand flat
沙坡 bed ripple
沙栖生物 psammobont;psammon
沙情 sediment regime
沙丘 downs; down sand-dune; rig; sand bank; sand drift; sand dune; sand hill;sowback;tibba;touradon; xerophorbium; climbing dune < 山坡迎风面的 >
沙丘保护 dune protection
沙丘层理 dune bedding
沙丘长湖 bajir
沙丘冲刷 dune erosion
沙丘的吹积 drifting of sand dunes
沙丘的落沙坡 sand falling side of dune
沙丘的形成 formation of sand dune
沙丘地带 sand dune terrain
沙丘地区 sand-hill region
沙丘地形稳定系数 dune topographic stability factor
沙丘地形稳定性 dune topographic stability
沙丘顶 dune crest
沙丘堆 dune
沙丘覆区 dune complex
沙丘高程 dune elevation
沙丘高度 dune height
沙丘工程 dune works
沙丘沟【地】gassi
沙丘固定 dune fixation; sand dune fixation;sand dune stabilization
沙丘海岸 dune coast;sand dune coast
沙丘和沙脊 sand dune and sand ridge
沙丘和沙梁 sand dune and sand ridge
沙丘河床 dune covered riverbed
沙丘湖 dune lake
沙丘滑落面 slip face
沙丘滑面顶 brink
沙丘荒漠 nefud
沙丘基底 plinth
沙丘脊 dune ridge
沙丘间风槽 gassi
沙丘间距 space of dune
沙丘间通道 < 地面常为泥沙 > straate
沙丘间洼地 bajir
沙丘间小湖 < 南非 > mier
沙丘宽度 dune width
沙丘链 barchan chain;chain like dune
沙丘流动性 mobility of sand dune
沙丘流沙 dune drift sand
沙丘面 dune surface
沙丘平原 dune plain
沙丘屏障 barrier dune
沙丘汽车 dune buggy
沙丘潜水 phreatic water in sand dune
沙丘侵蚀 dune erosion
沙丘群 dune complex
沙丘群落 thinium
沙丘群落的 thinic
沙丘沙 < 又称沙丘砂 > dune sand
沙丘施工 dune construction
沙丘围护 sand hedge
沙丘稳定(方法) sand dune stabilization
沙丘线 sand ridge
沙丘相推移 dune-phase traction
沙丘形成 dune formation
沙丘型海岸 duned coast
沙丘野麦 lyme-grass;meager oat
沙丘移动 sand migration
沙丘移动方式 movement way of dune
沙丘移动方向 direction of dune movement
沙丘移动速度 speed of dune movement
沙丘迎风面 slip face
沙丘运动 dune motion; dune movement
沙丘植被 dune vegetation
沙丘植物 thinophyte
沙丘植物群落 amanthium
沙丘状积砂 goz(gozes)
沙裙 sand apron
沙壤土剖面 profile of sandy loam
沙色的 sand-colo(u)red;sandy
沙沙声 rustling
沙生演替系列 psammon series
沙生植物 psammophyte
沙石泵 gravel pump
沙石配备厂 aggregate preparation plant
沙闩水深 bar draft
沙水硅锰钠石 shafranovite
沙松 sand pine;Siberian spruce
沙滩 alluvion;dry beach;hirst[hurst]; natural bar;overslaugh;sand bank; sand bar;sand flat;sand foreshore; sea beach;shoal
沙滩变陡 beach steepening
沙滩变平 beach flattening
沙滩补给 beach nourishment
沙滩地 bench land

S

沙滩海岸 sand beach
沙洲平原 bar plain
沙滩(轻便)汽车 <轮胎很大的轻型敞篷车> dune buggy
沙滩群落 sand-beach community
沙滩上的冲沟 rill mark
沙洲滩地 barrier flat
沙滩泳屋 bathing shed
沙特阿拉伯公式 Saudi Arabia
沙体坍塌 sand avalanche
沙田 sandy land
沙土浮动状态 quick condition
沙土内灌注膨润土泥浆以便下桩套管 mudding-in
沙土植物 silicicole
沙纹 bed dune; bed ripple; current ripple(mark);sediment ripple
沙窝 <波浪冲击沙滩形成的> sand drip
沙雾 sand mist
沙隙生物 mesopsammon
沙下湖 undersand lake
沙楔 sand wedge
沙性土 non-cohesive soil
沙性土壤液化 liquefaction of sandy soil
沙性物质 raw material
沙眼 blister;trachoma
沙样 sediment sample;sediment specimen
沙样防落弹簧 core catcher
沙样瓶 sample bottle;sample container
沙影 sand shadow
沙涌 blowout;sand boil
沙原 sand plain
沙灾 sand disaster
沙枣 desert date;oleaster;Russian olive
沙账 sediment budget
沙障 barrier beach;sand obstacle
沙枕 sand sausage
沙质沉积物 sandy sediment
沙质海滩 sand beach;sea sand
沙质荒漠 erg desert
沙质灰岩 sand limestone
沙质基础 sand foundation
沙质胶结物 arenaceous cement
沙质砾岩相 arenaceous-conglomeratic facies
沙质垆坶 sandy loam
沙质黏[粘]土 dauk[dawk];lam;sandy clay
沙质浅滩 sandbank
沙质沙漠 erg;k(o)um
沙质填土地 sandy land fill
沙中生长的 arenaceous;arenarious
沙钟 hour-glass
沙钟构造 sandglass structure; sandwatch structure
沙钟结构 hour-glass texture
沙洲 agger arenal; alluvion; alluvium [复 alluvia/alluviums]; bench land; bottom bank; coastal bar; high bed; hirst[hurst] <有树的>; natural bar; sand bank; sand bar; sand ridge;sand spit;shelf;shoal;cay <海上低潮露出的>; sand cay <海上低潮露出的>
沙洲标尺 bar ga(u)ge
沙洲沉积学说 hypothesis of bar deposition
沙洲吃水 bank draft
沙洲岛 wadden island
沙洲的起源 origin of bar
沙洲洞 pass
沙洲港 port with bars
沙洲拦阻的河口 barred river mouth
沙洲砾石 bar gravel
沙洲链 barrier chain
沙洲浅水区 shoal area of sand
沙洲生成 bar formation
沙洲台地 bar platform
沙洲头 bar head
沙洲突端 horn of bank
沙洲形成 bar formation
沙洲渔坝 bar weir
沙洲整治 regulation of sand bar
沙柱 sand column;sand spout;twister
沙状石膏 gypsum sand
沙资源 sand resources
沙嘴 barrier spit; beach cusp; point bar; sand horn; sand spit; shoal head;spit(of land)
沙嘴岸线 salient shoreline
沙嘴的形成 salient formation
沙嘴地带 salient area
沙嘴扭点 fulcrum[复 fulcra/fulcrums]
沙嘴沙坝相 spit and sand bar facies
沙嘴台地 spit platform
沙嘴形状 point form
沙嘴增长 spit growth

纱包 covering

纱包层 cotton sleeving
纱包电缆 cotton-covered cable; cotton insulation cable
纱包电线 cotton-covered wire
纱包绝缘 cotton cover insulation
纱包绝缘线 cotton-insulated wire
纱包漆包线 cotton-enamel covered wire;enamel(l)ed and cotton covered wire
纱包铜线 cotton-covered copper wire
纱包线 cloth-reinforced wire;cotton-insulated wire;cotton yarn covered wire
纱布 carbarsus;etamine;gauze;gauze fabric;pledget
纱布过滤 wick filtration
纱布扇形天平 yarn and cloth quadrant
纱厂 cotton mill;spinnery
纱橱 screen cupboard
纱窗 insect screen;insect wire screening; mosquito screen; screen window;window screen
纱窗窗扇 screen sash
纱窗窗框 screen frame(unit)
纱窗门 insect screen
纱窗下开关器 lever-under-screen operator
纱窗下推杆 under screen push bar
纱窗线脚 screen mo(u)lding
纱窗压条 screen mo(u)lding
纱窗阳台 screen porch
纱窗用金属丝布 screen wire cloth
纱带 cotton tape
纱锭 spindle
纱管坯 spool bar
纱管用材 spool bar
纱框 reel
纱框架 reel stand
纱罗边 leno-fastening
纱罗布边 leno edge;leno-selvedge
纱罗花纹 gauze effect
纱罗筘 gauze reed;leno reed
纱罗织法 leno weave
纱罗织机 leno loom
纱罗织物 gauze cloth; gauze fabric; leno(cloth)
纱罗组织 leno weave
纱门 door screen;screen door
纱门(门)闩 screen-door latch
纱门锁止器 screen-door latch
纱染的 ingrained
纱束包合性 yarn bundle cohesion
纱头 thrum
纱网 gauze
纱线 yarn
纱线测速仪 yarn speed meter
纱线长度测定器 yarn meter
纱线冲击强力试验 yarn ballistic test
纱线打包机 yarn bundling machine
纱线及织物上的斑点 burl
纱线支数 count of yarn;yarn size
纱线支数秤 yarn count balance
纱罩 fiddle
纱绉 moss crepe

刹车 arrestment; brake; braking (action); deadman control; keeper; spoke; trig

刹车板 braking vane
刹车臂 brake lever
刹车操纵机构 brake control mechanism
刹车带 black tape;brake band;brake ribbon;brake strap;brake lining
刹车挡油圈 brake grease baffle
刹车灯 stop lamp
刹车灯和尾灯 stop-and-tail light
刹车灯开关 brake lamp switch
刹车垫 brake lining
刹车阀 brake valve
刹车杆 brake lever
刹车工作液 braking fluid
刹车鼓 brake drum; friction(al) sheave
刹车鼓缓速器 brake drum retarder
刹车鼓宽度 width of brake drum
刹车鼓直径 diameter of brake drum
刹车荷载 braking load
刹车滑动 slipping of brake
刹车滑行 forward skidding;skid
刹车滑移 braking slip
刹车继电器 brake relay
刹车距离 braking distance; braking path; stopping distance; stopping length
刹车控制 brake control
刹车来令(俗称) brake lining
刹车力 brake load;braking force
刹车联动装置 brake linkage
刹车联轴器 brake coupling
刹车链 drag chain
刹车轮 brake disc[disk];brake rim
刹车轮毂 brake hub
刹车马力 brake horse power
刹车面 brake lining
刹车摩擦 drag friction
刹车盘 brake rim
刹车片 brake band lining;brake-shoe
刹车片固定座 brake block holder
刹车片制 brake block
刹车偏心轮 brake eccentric
刹车平均有效压力 brake mean effective pressure
刹车平均有效压强 brake mean effective pressure
刹车器 compressed-air for brake control
刹车伞 drag chute
刹车实验 braking test
刹车试验 wheels-locked test(ing)
刹车手把 brake crank
刹车手柄 brake lever
刹车输气管 brake pipe
刹车踏板 brake paddle;brake pedal
刹车踏板轴衬 brake pedal bushing
刹车涂料 brake dressing
刹车瓦 brake block
刹车系统 brake[braking] system
刹车橡皮 brake rubber
刹车橡皮带 brake rubber plate
刹车效率 braking efficiency
刹车性能 stopping performance
刹车旋转圆筒 brake drum
刹车用砂 track sand
刹车油 brake fluid;brake oil;hydraulic brake fluid
刹车重块 brake weight
刹车轴 brake axle
刹车装置 brake apparatus; brake assembly; braking device; brake device
刹车状态 braking position
刹车自控闸 overrun controlled brake
刹尖 closure by wedging-in crown
刹尖砌块 key segment
刹尖石 key segment
刹刹把 brake crank
刹水装置 cutwater
刹瓦 brake-shoe
刹住 lock
刹住车轮 lock the wheels
刹住的车轮 <制动时> locked wheel
刹住锚链 hold on the cable

砂 sand;grait

砂疤 blister sand;sand sticking
砂坝 sand bank;sand bar;sand dam;sand reef;sand ridge
砂坝圈闭 bar trap
砂斑试验 sand patch test
砂(拌)黏[粘]土 sand clay
砂拌乳液 sand-emulsion
砂包 bagwork;earth bag;sand-bag
砂暴 sand devil;sand storm
砂崩 sand avalanche
砂泵 mud hog pump; sand pump; sand-sucker;shell pump
砂泵采掘船 sand pump dredge(r)
砂泵阀 sand pump valve
砂泵螺旋钻探法 shell-and-auger boring
砂泵取样器 sand pump sampler
砂泵容器 sand pump container
砂泵挖泥船 sand pump dredge(r)
砂泵岩样 sand pumpings
砂闭路 sanded siding
砂壁状建筑涂料 sand textured building coating
砂滨港 sandy harbo(u)r
砂波 sand ripple;sand wave;sastrugi
砂波痕 sand ripple
砂波纹 sand ripple
砂玻璃 sand glass
砂驳 sand barge
砂铂矿石 platinum placer ore
砂铂矿床 platinum placer
砂布 abrasive cloth;coated abrasive;emery cloth;emery fillet;glass cloth;sand(ing)cloth
砂布表层 abrasive coating
砂布带 abrasive belt;sand belt;emery tape
砂布带打光机 finisher belt grinder
砂布底布 backing for abrasives
砂布加工 coated abrasive machining
砂槽 riffle;sand launder;sand trap
砂槽模型 sand box model;sand trap model
砂草原 <南非洲的> sandveld
砂(侧)线 sanded siding
砂层 layer of sand; sand body; sand layer;sand seam;sand(y) stratum;stratum of sand; sand mat(tress) <混凝土养护的>
砂层夯实 sand packing
砂层厚度 sand thickness
砂层回填 sand backfill
砂层积储藏法 sand method
砂层膨胀 sand expansion
砂层线 sand line
砂层压力 sand pressure
砂层压力控制 sand pressure control

S

砂层总厚的确定 sand count
砂碴铺垫 sand ballast
砂掺黏[粘]土稳定的道路 road of clay-stabilized sand
砂铲 sand shovel
砂场 check dam;sand dump;sand plant
砂炒法 boiling-on-grain
砂尘暴 sand storm
砂尘试验 dust test
砂沉淀池 sand settling tank
砂沉淀分析仪 sand washer
砂沉法 sand sink
砂衬 sand lining
砂承试验 <测定管子承载能力的> sand-bearing test
砂承载力试验 sand-bearing test
砂承重试验 sand-bearing test
砂充法 <测定密实度用的> calibrated sand method;sand replacement method
砂冲 sand-blasting
砂冲蚀 sand erosion
砂冲刷 sand scouring
砂处理工段 sand shop
砂处理设备 sand control equipment;sand preparing machine
砂处理系统 sand-conditioning system;sand preparing system
砂处理装置 sand preparation plant
砂川 sand glacier
砂川闪石 sadanaguite
砂床 mo(u)ld bed;sand bed(ding);sand floor;shift bed
砂床过滤器 sand-bed filter
砂床河川 erodible bed stream;erodible stream;sand-bed stream
砂床铸铁 sow
砂翠 sand percentage
砂打机具 sanding apparatus
砂打桩船 sand piling barge
砂代 replacement with sand
砂带 abrasive band;emery fillet;sand belt
砂带打磨 belt sanding
砂带加工 coated abrasive machining
砂带磨床 abrasive belt grinding machine;belt sander;coated abrasive grinder
砂带磨光 belt grinding
砂带磨光机 belt grinder;belt sander;finishing machine;resinder
砂带磨削 belt grinding
砂带抛光 abrasive band polishing;belt polishing
砂带抛光机 belt sander
砂带式 abrasive belt
砂带式抛光机 sanding machine
砂袋 earth bag;sand-bag;sand pack;sand pocket <水泥混凝土中缺乏水泥的砂团>
砂袋挡墙 sand-bag wall(ing)
砂袋垫层 sand-bag mattress
砂袋封堵 sand-bag damming
砂袋隔墙 sand-bag stopping
砂袋护岸 sand-bag revetment
砂袋护坡 sand-bag revetment
砂袋护墙 sand-bag wall
砂袋墙设置 sand-bag walling
砂袋圈围 <筑于涌砂处,处理管涌用> ring of sandbags
砂袋填心坝 sand-bag-core dam
砂袋填筑堤 sand-bag embankment
砂袋围堰 sand-bag cofferdam;sand weir
砂袋修筑隔墙 sand-bag damming
砂袋掩体 sand-bag bunker
砂袋筑坝 sand-bag damming
砂袋筑堤法 sand-bag damming
砂袋筑墙 sand-bag walling
砂当量 sand equivalent
砂当量试验 sand equivalent test
砂当量振动器 sand equivalent shaker
砂当量振动装置 sand equivalent shaker
砂岛 barrier island
砂岛法 <沉箱施工的> sand island method
砂岛群 barrier chain
砂道 sand track
砂道砟 sand ballast
砂的百分比含量 sand equivalent
砂的百分含量 percentage of sand
砂的不均匀性 non-conformity of sand
砂的等厚图 isopach map of sand
砂的固化 petrification of sand
砂的含水量与体积变化关系曲线 bulking curve
砂的混入 inclusion of sand
砂的级配 sand grading
砂的挤实比 sand compaction ratio
砂的夹进 inclusion of sand
砂的均匀系数 uniformity coefficient of sand
砂的粒度分析 mechanical analysis of sand
砂的粒径分析 mechanical analysis of sand
砂的流动试验 sand flow test
砂的平均粒径 average grain size of sand
砂的平面变形剪力图 plane strain shear diagram
砂的铺筑 sand placement
砂的侵害 encroachment by sand
砂的侵蚀作用 erosive action of sand
砂的筛分 sand separation
砂的湿胀 bulking of sand;sand bulking
砂的湿胀系数 bulking factor
砂的湿胀性 bulking of sand
砂的湿胀因素 bulking factor
砂的体胀 bulking of sand;sand bulking
砂的吸湿 absorption of sand
砂的细度模数 fineness modulus of sand
砂的相对密度 relative density of sand
砂的压气式取样器 compressed air sampler for sand
砂的有效粒径 effective grain size of sand;effective size of sand
砂的有效容重 effective unit weight of sand
砂的有效直径 effluent size of sand
砂的再生装置 sand reclaimer
砂的真空试验方法 vacuum method of testing sand
砂的制备 sand manufacture
砂堤 sand bank;sand bar;sand embankment;sand reef
砂底 sand(y) bottom
砂底的 sand-bottomed
砂底动物 epipsammon
砂地 sand;sandy ground
砂地储水 sand storage of water
砂地穿胶底帆布鞋 sandshoe
砂地基 sand foundation;sandy ground
砂地试验 test on sand
砂地图 sand map
砂地植被 sand vegetation
砂点 sand depot;sand point
砂垫 sand block
砂垫层 bedding sand;sand bed(ding course);sand blanket;sand carpet;sand cushion(base);sand mat(tress);sand subbase;sand underlay;sandy cushion
砂垫床 blanketing layer
砂斗 mo(u)lding-sand hopper;sand hopper
砂堵 sand plug
砂堵钻孔 sand-up well
砂堆 drifting;face of sand;hirst[hurst];mass of sand;sand dump;sand mass;sand mound;sand pile;sand wash
砂堆积 sand wash
砂钝 glazing
砂阀杆 sand valve stem
砂阀接头 sand valve connection
砂方解石【地】sand-calcite
砂坊 checking dam;sand check dam
砂沸 sand boiling
砂分 sand fraction
砂分配器 sand distributor
砂分选机 sand classifier
砂粉砂 mixture of sand silt and clay
砂-粉砂-泥 sand-silt-mud
砂粉水泥 sand cement
砂封板 sand apron;sand seal plate
砂封层 sand seal
砂浮法 sand flo(a)tation
砂负载 sand load
砂复热机 sand reheater
砂盖层 sand blanket;sand cover
砂干表面 <指油漆面干到不黏[粘]沙> sand-drying surface
砂干化床 sand-drying bed
砂干阶段 <油漆的> sand dry
砂坩埚 hessian
砂岗 sand hill
砂钢板 sand plate
砂隔层 parting sand;sand parting
砂耕 sand culture
砂埂 barrier
砂汞矿石 Hg placer ore
砂钩 jagger;lifter;sand hook;strickle
砂垢 sand scale
砂骨比 sand aggregate ratio;sand and total aggregate ratio;sand-coarse aggregate ratio
砂骨架的 sandy-skeletal
砂骨料 sand aggregate
砂鼓磨床 drum sander
砂管 sand pipe
砂光板 sanded board
砂光薄片法 sand polished thin section method
砂光机 sander
砂轨道 sand track
砂锅 casserole;olla
砂过滤 sand filtration
砂过滤层 sand filter layer
砂过滤器 permutite filter;sand filter
砂河 sand river
砂壶滴漏 sand glass
砂滑塌 sand avalanche
砂环 ring of sand
砂黄土 sand loess
砂灰比 sand-cement ratio
砂灰砌块 sand limestone
砂灰质黏[粘]土 adobe clay
砂灰质土 adobe soil
砂灰砖 sand-lime brick
砂灰砖墙 backing sand-lime brick
砂回收设备 sand reclaiming plant;sand recovery plant
砂混黏[粘]土 clay(ey) sand;sand with clay
砂混凝土 sand concrete
砂混凝土乳液 sand-concrete emulsion
砂火山 sand volcano
砂积矿床 placer deposit
砂积矿开采 placer-mining
砂积矿砂 placer sand
砂积矿石 placer mineral
砂基 sand foundation
砂基堰 weir on sand
砂集聚 sand concentration
砂集料 sand aggregate
砂脊 sand reef;sand ridge;sand streak
砂夹杂物 <铸件的> sand inclusion
砂岬 sand horn;sand spit
砂假型 sand match
砂尖柱 sand pinnacle
砂姜土 sajong soil
砂浆 grout;mortar(mix);paste;sanded grout;sand grout;sand mortar;sand pulp;sand slurry
砂浆板 mortar slab;reinforced cement mortar board;fat board <瓦工用>
砂浆拌灌机 grout mixer and placer
砂浆拌和板 spot board
砂浆拌和槽 mortar box
砂浆拌和厂 mortar mill
砂浆拌和机 cement mixer;grout mixer;mortar mill;mortar mixer
砂浆拌和设备 mortar mixing plant
砂浆棒试验 mortar bar test
砂浆包盖系数 wrapping coefficient of mortar
砂浆泵 grout pump;mortar pump;sand pump;slurry pump;mortar pumping machine
砂浆标号 grade of mortar;mortar grade;strength grading of mortar
砂浆标号试验 mortar cube test
砂浆表皮 skin of mortar
砂浆饼块 dabbed mortar
砂浆拨开系数 superfluous coefficient of mortar
砂浆材料 mortar material
砂浆槽 mortar trough
砂浆层 bedding mortar;screed <作为地面整饰或花砖垫层用>
砂浆层掺和料 screed admix(ture)
砂浆层促凝剂 screed accelerating agent
砂浆层底 screed base
砂浆层外加剂 screed admix(ture);screed agent
砂浆掺和料 mortar admixture
砂浆衬里 back mortaring;mortar lining
砂浆衬里钢管 mortar lining steel pipe
砂浆衬里铸铁管 mortar lining cast iron pipe
砂浆衬砌 mortar lining
砂浆成分 mortar composition
砂浆稠度 consistency of mortar;grout consistency[consistence]
砂浆稠度测定仪 apparatus for determining mortar consistency
砂浆稠度计 mortar consistometer
砂浆稠度仪 grout consistency meter;mortar consistency tester;mortar penetration tester
砂浆打底 mortar bedding
砂浆的泵送 pumping of mortar
砂浆的骨料 mortar aggregate
砂浆的骨料砂 mortar sand
砂浆的凝结 setting of mortar
砂浆的喷射使用 spray application of mortar
砂浆的湿稳定稠度 wet stable consistency
砂浆底层 setting bed
砂浆底座 mortar base
砂浆垫层 bedding course mortar;bedding mortar;mortar bed;setting bed <水磨石下层面的>;stringing mortar
砂浆垫层的锚筋 bed dowel
砂浆垫层铺砖路面 brick pavement on mortar bed
砂浆防冻剂 mortar anti-freezing admixture
砂假防水层 mortar waterproofer
砂浆防水工 mortar waterproofer
砂浆分层度测定仪 apparatus for determining stratification of mortar;mortar stratification tester

砂浆粉面 rendering
砂浆缝 abre(a)uvior <砌块中的>; mortar joint
砂浆缝凹槽 raggle
砂浆缝压实及修饰 tooling
砂浆敷面 mortar layer
砂浆附加剂 <促凝或缓凝用的> dope
砂浆覆盖层 mortar covering
砂浆工厂 mortar plant; mortar works
砂浆工程 mortar works
砂浆工作 mortar works
砂浆勾缝 mortar-calked joint; slushed joint
砂浆骨料 mortar aggregate
砂浆骨料结合力 mortar-aggregate bond(strength)
砂浆骨料结合强度 mortar-aggregate bond strength
砂浆固定液 mortar fixative
砂浆灌缝 slushing
砂浆含砂量 sand content of drilling fluid
砂浆混合料 mortar(ad)mixture
砂浆混凝土 mortar mixture; sand-bentonite slurry
砂浆基础 mortar base
砂浆集料 mortar aggregate
砂浆集料结合力 mortar-aggregate bond(strength)
砂浆集料结合强度 mortar-aggregate bond strength
砂浆减水剂 mortar water-reducing agent
砂浆胶砌的(块料)路面 mortar paving
砂浆搅拌厂 mortar mill; mortar mixing plant
砂浆搅拌场 mortar mill
砂浆搅拌灌注器 grout mixer and placer
砂浆搅拌机 mortar mill; mortar mixer; mortar mixing machine; pugmill mortar mixer
砂浆搅拌喷射器 grout mixer and placer
砂浆搅拌站 mortar mill; mortar mixing plant
砂浆接缝 mortar(ed) joint
砂浆接缝压光 tooling
砂浆结合层 bonding layer
砂浆结合层外加剂 bonding agent
砂浆结碎石路面 mortar-bound surface; mortar-bound surfacing
砂浆结碎石面层 mortar-bound surface; mortar-bound surfacing
砂浆浸润剂 mortar wetting agent
砂浆可湿剂 mortar wetting agent
砂浆空隙 mortar void
砂浆空隙比 mortar-void ratio
砂浆空隙法 mortar-void method
砂浆空隙配料法 mortar-void method
砂浆空隙试验 mortar-void test
砂浆类 mortar class
砂浆棱体试验 mortar prism test
砂浆立方块(荷重)试验 mortar cube test
砂浆立方试块 mortar cube
砂浆立方体试验 mortar cube test
砂浆料粉碎机 mortar mill
砂浆流动度 fluidity of mortar
砂浆锚杆 mortar anchor
砂浆锚固 cement mortar anchor
砂浆泌水率 bleeding capacity
砂浆面层 dash(-bond) coat
砂浆面层涂环氧漆 epoxy paint on screen
砂浆模制试件 mortar briquet(te)s
砂浆抹面 mortar top
砂浆抹平 cement mortar plaster(ing)
砂浆黏[粘]结强度试验仪 mortar bond strength device
砂浆黏[粘]卵 pebble ballast
砂浆配合比 mortar mix(ing) ratio
砂浆配料 mortar material
砂浆喷器 air blowpipe
砂浆喷射 mortar injection
砂浆喷射机 mortar injecting machine
砂浆喷涂机 mortar jet
砂浆铺砌 mortar paving; paving with mortar
砂浆砌面 mortar paving
砂浆砌筑 mortar bricking
砂浆砌砖 brick laid with mortar
砂浆嵌缝接头 grouted scarf joint
砂浆强度 mortar strength; strength of mortar
砂浆全长黏[粘]结式锚杆 completely grouted rockbolt
砂浆润湿剂 mortar wetting agent
砂浆设计方法 mortar design method
砂浆石 mortar stone
砂浆石子堵塞 spackling
砂浆式锚杆 bolt of sand grout
砂浆试棒 mortar bar
砂浆试棒法 mortar bar method
砂浆试块试验 mortar cube test
砂浆试验立方体 mortar cube
砂浆刷面 mortar rendering
砂浆塑化剂 mortar plasticizer; mortar workability agent
砂浆碎石嵌缝 garneting; garreting
砂浆添加粉 powder additive for mortar
砂浆添加剂 mortar additive
砂浆添加料 mortar additive
砂浆填充门框 grouted frame
砂浆填缝 slushed(-up) joint; mortar joint
砂浆条痕 streaks of mortar
砂浆桶 hod; mortar hod
砂浆涂层 dash(-bond) coat; mortar coating; mortar covering; parge
砂浆涂层处理 parge coat
砂浆涂复机 mortar coating machine
砂浆涂抹 mortar dab; slurrying
砂浆托板 mortarboard
砂浆外加剂 mortar admixture
砂浆握固力 bond stress of mortar
砂浆污斑 mortar stain
砂浆污点 mortar stain
砂浆稀释剂 dispersant
砂浆相 <混凝土中相对于集料的> mortar phase
砂浆小毛石墙 incertum opus
砂浆型锚杆 mortar bolt
砂浆养生 curing of the mortar
砂浆样品 mortar specimen
砂浆硬化 mortar setting
砂浆用砂 mortar aggregate; mortar sand
砂浆增湿剂 mortar wetting agent
砂浆找平 mortar level(l)ing
砂浆找平层 mortar bed
砂浆找平层开裂 screed cracking
砂浆制备车间 pulp-preparation plant
砂浆制备设备 mortar fabrication installation
砂浆置换法 mortar replacement method
砂浆重新拌和 grout remixing; retempering of mortar
砂浆注射 mortar injection
砂浆转运工人 hod carrier
砂浆着色剂 mortar stain
砂胶 mastic
砂胶层 mastic bed
砂胶地面 mastic floor
砂胶垫层 mastic cushion
砂胶隔离层 mastic insulation
砂胶勾缝 mastic joint(ing)
砂胶灌缝施工法 sand and mastic method
砂胶灌浆料 mastic grout
砂胶加热锅 mastic cooker
砂胶接缝 mastic joint(ing)
砂胶绝缘层 mastic insulation
砂胶块 mastic block
砂胶密封 mastic seal(ing)
砂胶铺地面 mastic flooring
砂胶碎石路 mastic macadam
砂胶填缝料 mastic tiller
砂胶填料 mastic filler
砂胶性限值 <成为玛蹄脂的最多填充料用量> masticity limit
砂礁 sand reef
砂搅拌机 <油层水力压裂时> sand-oil blender
砂金 alluvial gold; gold dust; gulch gold; placer gold; stream gold
砂金矿 Au placer ore; gravelmine; placer mine
砂金矿床 gold placer
砂金矿砂 gold-placer sand
砂金石 aventurin(e); aventurin(e) glass
砂金釉 aventurin(e) glaze; goldstone glaze
砂晶 sand crystal
砂井 column of sand; desilter; drain(age) pile; pile drain; sand-filled drainage well; sand well; vertical sand drain
砂井基础 sand well foundation
砂井加固法 sand compaction pile method
砂井排水 drainage by sand piles; sand drain
砂井排水法 sand drain method
砂井排水预压法 prepressing process of sand filled drainage well
砂井平面布置形状 plane distribution shape of sand drain
砂井渗水暗沟 sand drain
砂井预压加固 stabilization by sand drain and preloading
砂井真空排水法 sand drain vacuum method
砂咀圈闭 spit trap
砂锯 carborundum saw
砂锯石面 sand-sawed finish
砂壳混凝土 <砂浆或混凝土中的集料用低水灰比水泥浆壳包裹了的混凝土> sand enveloped with cement
砂坑 sand hole; sand pit
砂孔 pit; sand hole
砂口 sandgate
砂库 sand storage
砂块 sand(y) lump
砂块破碎 sand lump breaking; sand lump crushing
砂块破碎机 sand crusher
砂矿 ore of sedimentation; placer; placer deposit
砂矿边沿 rimrock
砂矿采样 placer sampling
砂矿床 placer deposit; placer formation; sand deposit
砂矿床富集作用 placer concentration
砂矿底岩 reef
砂矿底岩中开凿的巷道 reef drive
砂矿地质图 geologic(al) map of placer deposit
砂矿工作面 gravel face
砂矿勘探钻机 placer prospecting drill
砂矿矿车 placer
砂矿淘采 placer-mining
砂矿渣混合料 sand-slag mix
砂矿钻机 banka drill; empire drill
砂浪 sand ripple
砂类土 sandy soil
砂棱状堆 prism of sand
砂冷 sand cooling
砂冷却器 sand cooler
砂冷却设备 sand cooling plant
砂沥青 sand bitumen
砂沥青护面 sand asphalt facing
砂砾 sand gravel; sand grit; gravel (grit); grit (gravel); land waste; underlay mineral <上浇沥青屋面的>; chad; coarse sand; grail
砂砾保护层 gravel envelope
砂砾保护井 gravel envelope
砂砾材料 gritting material
砂砾槽 grit chamber
砂砾层 gravel layer; gravel sand stratum; gravel stratum [复 strata]; gravel band; grit stratum; sandy gravel stratum
砂砾层接触净化 gravel contact purification
砂砾层天然砂颗分曲线图 gradation curve of natural sand in sand and gravel layer
砂砾厂 sand and gravel plant
砂砾充填段顶深 top depth of gravel pack
砂砾充填完井 gravel packing completion
砂砾冲洗机 sand and gravel washer
砂砾冲洗器 sand and gravel washer
砂砾冲洗设备 sand and gravel washing plant
砂砾床人工湿地 gravel bed constructed wetland
砂砾床水栽发人工湿地 gravel bed hydroponic constructed wetland
砂砾带 gravel band
砂砾道 gravel track
砂砾的 gritty
砂砾的相对密度 relative density of sands and gravel
砂砾地 sandy ground
砂砾地层 sandy gravel stratum
砂砾地带渗井漫灌 well flooding
砂砾堆 debris
砂砾覆盖层 sand and gravel overburden; sand and gravel overlay(er); sand and gravel overlaying layer
砂砾管 sand pipe
砂砾过滤机 sand-gravel filter
砂砾过滤器 gravel filter
砂砾海滨 shingle beach
砂砾滑坡 debris slide
砂砾混合料 gravel-sand mixture
砂砾混合喷嘴 mixed sand-and-shingle spit
砂砾混合滩 sand and gravel foreshore
砂砾混凝土 sand-gravel concrete
砂砾矿床 sand and gravel deposit
砂砾类土 sand-gravelly soil
砂砾滤池 gravel filter
砂砾滤床 gravel and sand filter bed
砂砾路 sand-gravel road
砂砾路面 sand-gravel surface
砂砾面层 sand-gravel surface
砂砾铺撒车 grit spreading vehicle; gritting lorry
砂砾乳液 sand-gravel emulsion
砂砾生物滤池 grit biological chamber
砂砾石 sandy gravel; gritstone; sand and gravel
砂砾石冲积层 sand and gravel wash
砂砾石垫层 gravel-sand cushion; sand and gravel bedding; sandy gravel cushion
砂砾石骨料 sand-gravel aggregate
砂砾石混凝土 sand and gravel concrete
砂砾石集料 sand-gravel aggregate
砂砾石浅滩 sand-gravel shoal

S

砂砾水槽 gravel basin
砂砾滩 sand bank
砂砾填充水井 gravel packed water well
砂砾土 gravel soil;tabby
砂砾围护层 <水井的> gravel envelope
砂砾围护井 gravel envelope well
砂砾屋面 gravel roofing
砂砾洗涤 sand and gravel wash
砂砾屑 gritty dust
砂砾屑混凝土 gritcrete
砂砾油毛毡屋面 building-up roofing
砂砾运输泵 gravel transporting pump
砂砾质土 gritty soil
砂砾桩 sand-gravel pile
砂砾锥 debris cone
砂粒 graining sand; grain of sand; grog;sand grain;sand particle
砂粒表层滚磨 scuffing grind
砂粒表面 face of sand
砂粒糙率 grain roughness
砂粒冲蚀探头 erosion sand probe
砂粒大小 sand size
砂粒的有效粒径 effective size of sand grains
砂粒分级 sand fraction;sand grading
砂粒分离器 sand separator
砂粒焚化 grit incineration
砂粒滚圆度 rounding of grains
砂粒含量 sand content
砂粒机械分析 mechanical analysis of sand
砂粒级泵 sand pump
砂粒级金刚石 sand dust
砂粒级配 sand fraction; sand gradation;sand grading
砂粒计 sand grain meter
砂粒粒径 sand(y)size
砂粒粒组 sand fraction
砂粒流动的河床 sandy mobile bed
砂粒磨损率 wear rate of sand grains
砂粒器 sand(grain)meter
砂粒撒布器 aggregate spreader
砂粒渗透性 sand permeability
砂粒式沥青混凝土 sand particle bituminous concrete
砂粒碎屑岩 arenite
砂粒体积 sand-grain volume
砂粒细度 grain fineness number
砂粒形状 sand particle size
砂粒岩 arenyte
砂粒直径 sand diameter
砂粒质的 gritty
砂粒状的 granular
砂粒组 sand fraction;sand grain grade
砂涟 sand ripple
砂梁 sand ridge
砂量平衡 sediment balance
砂料堆集场 sandwich arrangement; sand yard
砂料开采 sand production
砂料螺旋复洗机 screw rewasher sand tank
砂料生产 sand production
砂流 sand flow
砂流磨耗法 sand stream abrasion method
砂瘤 sand blister
砂龙卷 sand tornado
砂滤槽 sand filter trenches
砂滤层 sand blanket;sand filter;sand filter bed;sand filter blanket;sand filter layer
砂滤层管沟网 sand filter trenches
砂滤层上面落下的絮凝泥层 mud blanket
砂滤池 sand filter
砂滤池净化机理 purification mechanism of sand filter
砂滤床 sand filter;sand filter bed
砂滤多孔石 airstone
砂滤法 sand filtration
砂滤沟 sand filtration ditch
砂滤介质 screen filtration media
砂滤器 sand filter
砂滤塔 sand tower
砂路 sand filtering; sand filtration; sand leach;sand road
砂路堤 sand embankment
砂率 percentage of sand;percent fines; sand-coarse aggregate ratio;sand percentage; sand-total aggregate ratio; sand ratio <水泥混凝土的>
砂卵石 sand-free gravel
砂卵石料 <由河床采取未经筛分的> river-run material
砂轮 abrasion wheel; abrasive disc [disk];abrasive grinding wheel;abrasive wheel; carborundum grinding wheel; emery cutter; emery wheel;grinding disk[disc];grinding stone; grinding wheel; grindstone; knife grinder; oilslip; sand disc [disk]; sanded wheel; sander; sanding disc[disk];sand reel;sandstone disk[disc];stone wheel;tool sharpener;whetstone
砂轮表面直线速度 surface wheel speed
砂轮成型工具 wheel forming tool
砂轮成型修整 form truing
砂轮传动箱 grinding wheel driving box
砂轮挡板 wheel guard
砂轮刀 crusher
砂轮等级 grinding wheel grade
砂轮电机 grinding wheel drive motor
砂轮垫板 wheel washer
砂轮垫盘 grinding wheel flange
砂轮法兰盘 grinding wheel flange
砂轮分级 grading abrasive wheels
砂轮杆 wheel shaft
砂轮鼓 emery cylinder
砂轮横向进给 wheel traverse
砂轮横移进给量 cross feed
砂轮护目镜 grinding wheel spectacles
砂轮滑座 grinding wheel slide
砂轮缓冲垫 wheel blotter
砂轮回转强度 rotation strength of grinding wheel
砂轮机 grinder; grinding machine; grinding mill of the edge runner; rotary sander;sand grinder;sanding machine;sharpener
砂轮挤压修整器 crush dress
砂轮架 grinding carriage
砂轮结构 grinding wheel structure
砂轮截断机 cut-off grinder
砂轮锯 abrasive-disc cutter
砂轮锯断机 abrasive sawing machine
砂轮磨床 sand grinder
砂轮磨光机 disk sander;sander;plane sander
砂轮磨钎机 grinding wheel dresser
砂轮黏[粘]结材料 grinding wheel bonding material
砂轮黏[粘]结剂 wheel bond
砂轮切断 abrasive cutting-off
砂轮切断机 abrasive cut-off machine
砂轮切割机 abrasive sawing machine; abrasive wheel cutting-off machine
砂轮清理 clearage with grinding wheel
砂轮清理台 grinding bed
砂轮试验机 grinding wheel testing machine
砂轮套筒 wheel sleeve
砂轮头 wheel-head
砂轮形状 grinding wheel shape
砂轮修整 crushing; grinding wheel dressing;wheel dressing
砂轮修整工具 truer;wheel truer
砂轮修整机组 grinding wheel conditioning unit
砂轮修整器 abrasive dresser;grinder-wheel dress;wheel dresser
砂轮修正 wheel truing
砂轮研磨机 grinding wheel dresser
砂轮研磨切割器 abrasive-disc cutter
砂轮用玻璃纤维纤维网格布 glass scrim for grinding wheel
砂轮罩 wheel guard
砂轮整形 grinding wheel dressing; grinding wheel truing
砂轮整形工具 crush dresser
砂轮整形器 grinding wheel dresser
砂轮轴承 grinding wheel bearing
砂轮轴承护盖 grinding wheel bearing protecting cover
砂轮轴套 grinding wheel spindle sleeve
砂轮主轴 grinding wheel spindle
砂轮主轴伸出 grinding wheel spindle extension
砂轮转速 grinding wheel speed
砂轮装置 grinding wheel unit
砂轮座 grinding wheel head;grinding wheel stand
砂螺旋钻钻探 shell-and-auger bore
砂埋 bury by sand
砂脉 sand dike
砂门子 hydraulic stowing partition
砂面 sand finish
砂面的 sand-faced
砂面沥青油毡 sanded bitumen felt
砂面黏[粘]土砖 sand-faced clay brick
砂面排水管道 sand-surface drainpipe
砂面墙 sand coated wall
砂面油毡 mineral-surfaced felt; sanded bitumen felt; sand surface asphalt felt
砂面砖 facing slip;sand-faced brick; sand-finished brick
砂面装饰 aggregate finish
砂模 sand-lined mo(u)ld; sand mo(u)ld(ing)
砂模拟 sand analogy
砂模生铁 sand-cast pig;sand mold pig iron
砂模旋制管 sand-spun pipe;sand-spun tube
砂模制砖法 pallet-molding
砂模铸管 sand-cast tube
砂模铸件处理 sand cast finish
砂模铸铁管 sand-cast pipe
砂模铸造 sand casting
砂模砖 sand mo(u)lded brick
砂磨 sand cut;sand grind;sand rubbing
砂磨带 sanding belt
砂磨光现象 sand polishing
砂磨耗层 sand carpet
砂磨机 bead mill; sand grinder; sand mill;stirred ball mill
砂磨机研磨 sand grinding
砂磨面 sand rubbed
砂磨面层 sand-rubbed finish
砂磨石 sand grinding stone
砂磨饰面 sand-rubbed finish
砂磨筒 sand cylinder
砂磨修整 sand-rubbed finish
砂磨与抛光机 sanding and polishing machine
砂漠边缘平坦黏[粘]土表面 <龟裂黏[粘]土层> flat clay desert land
砂目半色调板 grained half-tone plate
砂目表层 abrasive coating
砂目铝板 mill run alumin(i)um plate
砂目平板 grain plate
砂目锌板 grained zinc press plate
砂内灌浆 sand grouting
砂内含水量测定仪 water-in-sand estimator
砂内冷却 in sand cooling
砂囊 <混凝土中缺水泥的砂团> sand pocket
砂泥分离器 sand slime separator
砂泥浆 sand slurry
砂泥路面 sand clay (road) surface [surfacing]
砂泥石灰角砾岩 brockram
砂泥土地基 sand clay base
砂泥淤积 sand silting
砂泥质结构 sand argillaceous texture
砂泥质相 arenaceous-petitic facies
砂黏[粘]土 hazel earth
砂黏[粘]土灰泥 loam mortar
砂黏[粘]土混合物 sand clay mix(ture)
砂黏[粘]土路 sand clay road
砂黏[粘]土路面 sand clay pavement
砂黏[粘]土填料 loam fill(ing)
砂黏[粘]土土壤 sand clay soil
砂耙 sand drag
砂盘 sand table;sand tray <放蓄电池的>
砂盘调度 sand control
砂盘作业 sand table exercise
砂炮泥 sand stem
砂泡 sand blister
砂培法 gravel culture;sandy culture
砂配重 sand load
砂喷 sand-blasting
砂喷射器 sand ejector
砂喷嘴 sand nozzle
砂膨胀比 sand expansion ratio
砂皮 abrasive cloth;sand skin
砂皮状路面 sandpaper surfacing
砂平均粒径 average size of sand
砂坪沉积 sand flat deposit
砂坡 sand slope
砂铺底 sand underlay
砂铺地基层 sand underlay
砂栖蠕虫 sand worm
砂栖石 arenicolite
砂桥 sand bridge
砂桥卡钻 bridge up
砂侵 formation entry;sand cutting
砂丘 sand dune
砂丘固定 sand dune stabilization
砂丘区 sand-hill region
砂丘砂体圈闭 dune sand trap
砂丘围护 sand hedge
砂丘线 sand ridge
砂球和砂枕构造 ball-and-pillow structure
砂区 sand zone
砂裙 sand apron
砂壤 clayey sand
砂壤土 clayey sand; sabulous loam; sandy loam(soil)
砂塞 bod
砂筛 sand riddle; sand screen; sand shaker;sand sieve;sand sifter
砂上升 boil of sand
砂生植物 silicicole
砂石 crushed concreting sand; dinas; freestone;grinding stone;gritstone; rubstone;sand rock;sandstone
砂石搬运 transport of debris
砂石泵 aggregate pump;gravel pump
砂石比 <砂与粗集料之比> sand and coarse aggregate ratio; sand aggregate ratio; sand-coarse aggregate ratio;sandstone ratio
砂石驳 aggregate barge
砂石场 sand yard
砂石车 gravel wagon
砂石垫层 sand(-and)-gravel bedding;sand and gravel cushion;sand-gravel cushion

砂石堆场 aggregate stock; aggregate yard
砂石防护网 stone guard
砂石分级设备 sand classifying equipment
砂石分选机 sand separator
砂石粉 powdered sandstone
砂石膏灰浆 sand-gypsum plaster
砂石混凝土 sandstone concrete; stone concrete
砂石精选 sand and gravel extraction
砂石开采 sand extraction
砂石坑料 pit-run
砂石块 sandstone block
砂石矿床 sand and gravel deposit
砂石料仓 grit bin
砂石料厂 aggregate plant; aggregate reclaiming plant
砂石料场 aggregate reclaiming plant
砂石料冲洗机 washing machine for gravel and sand
砂石料开采 sand and gravel exploitation; sand and gravel mining
砂石料生产系统 aggregate production system
砂石料(运输)车 ballast truck
砂石流 avalanche of sand and stone
砂石路面 sand-gravel surface
砂石码头 aggregate handling dock
砂石配备厂 aggregate preparation plant
砂石铺撒车 gritting truck
砂石铺撒作业 gritting service
砂石水泥配料机 sand, stone and cement proportioner
砂石田 rubble land
砂石挖掘船 sand and gravel dredge(r)
砂石卸货码头 aggregate unloading dock
砂石碴 sand ballast
砂石碴碾压机 sand ballast type roll
砂石碴压路机 sand ballast type roll
砂蚀 sand cutting
砂饰面 sand finish
砂室 sand dome
砂收集器 sand collector
砂水分离器 sand-water separator
砂水泥混合浆 sand-cement mixture
砂水泥砂浆 sand-cement grout
砂烁 granule
砂松 sand pine
砂碎屑岩 arenite[arenyte]
砂塔 sand track
砂胎 close-over
砂胎模 sand match
砂滩 sand waste; sandy beach
砂滩地 sand accretion
砂滩过滤 bank filtration
砂滩泳屋 bathing shed
砂汤 sandbree
砂糖 granulated sugar; sand sugar
砂糖粉红色 candy pink
砂糖石 saccharite
砂糖松 gigantic pine
砂糖状 saccharoidal; sucrosic; sugary
砂糖状大理岩 saccharoidal marble
砂糖状结构【地】 saccharoidal texture
砂陶粒混凝土 sand-haydite concrete
砂提取 sand extraction
砂体 ore mass
砂体沉积环境 depositional environment of sand body
砂体累计等厚图 isopach map of total sandstone body thickness
砂田 sandy land
砂填 sand feed; sand sealing
砂填层 chocked sand layer; choked layer of sand
砂填充 sand fill
砂填料 sand filling
砂填实的井 sand-tamped well
砂填实的钻孔 sand-tamped well
砂条 sand streak
砂条石 rectangular sandstone
砂铁岩 carstone
砂桶 sand bucket
砂筒 < 某些工程结构中卸落支架用的设备 > sand cylinder
砂透镜 lens of sand
砂透镜体 sand lens
砂土 light soil; sabulous clay; sabulous soil; sandy ground; sand(y) soil; stonebrash; sand(y) clay < 砂和黏[粘]土混合物 >
砂土包裹体 sand inclusion
砂土崩落 < 钻孔内的 > sloughing of sand
砂土的密实(程)度 compactness of sand; density of sand
砂土地层隧道 sandy-earth tunnel
砂土地基 sand soil foundation
砂土骨架 sand skeleton
砂(土)回填 sand backfill
砂土混合料路面 sand clay surface [surfacing]
砂土混合料面层 sand clay surface [surfacing]
砂土结合料 sand clay matrix
砂土砾石 sand clay gravel
砂土流失 sand running(-down)
砂土路 sand clay road
砂土路面 sand clay surface[surfacing]
砂土面层 sand clay surface[surfacing]
砂土潜在液化能力 liquefaction potential
砂土水泥 sand cement
砂土水泥浆 sand-cement slurry
砂土相对密度 relative density of sand
砂土压力与孔隙比曲线 pressure-void ratio curve for sand
砂土液化 liquefaction of sand(soil); sand clay liquefaction; sand liquefaction(of soil); liquefaction of saturated soil
砂土振动液化 vibration liquefaction of sand
砂土植物 silicicole
砂土植物群 sandy soil vegetation
砂团 mass of sand; sandy lump
砂托 sand-bearing
砂脱坯砖 sand-struck brick
砂脱水 dewatering of sand
砂脱水机 sand dehydrater
砂脱水机螺杆 sand dehydrator screw
砂脱水器 sand dewaterer
砂脱水器螺杆 sand dewatering screw
砂瓦 abrasive segment
砂纹 sand ripple; sand streak(ing); ripple mark
砂纹波长 ripple spacing
砂纹层理 ripple lamination
砂纹陡度 ripple steepness
砂纹峰 ripple crest
砂纹序列 ripple train
砂窝 sand(y) pocket
砂钨 W placer ore
砂锡 grain tin; Sn placer ore; stream tin
砂锡矿床 tin placer
砂席 blanket sand; sand sheet; sheet sand
砂洗 sand-blasting
砂洗箱 wash box
砂系洗涤器 sand system scrubber
砂细度 grain fineness number
砂线 sand streak
砂箱 sand bin; sand container; sand holder; sand storage case; sand box; flask; foundry flask
砂箱传送带 flask conveyer[conveyor]
砂箱底板 bottom board; flask bottom board
砂箱顶座 sand jack
砂箱定位套 socket
砂箱定位销 box pin; flask pin
砂箱墩 sand block; sand jack
砂箱耳 flask trunnion
砂箱法 < 试验管压力 > sand-bearing method
砂箱分界线 parting
砂箱盖 sand box cover
砂箱管 sand box pipe
砂箱夹 flask clamp; mo(u)ld clamps
砂箱脚蹬 sand box step
砂箱孔盖 sand box top
砂箱框 upset frame
砂箱拉杆 sand box arm; sand box rod
砂箱滤器 sand box strainer
砂箱模型 sand box model
砂箱内壁凸条 sand grip; sand ledge
砂箱千斤顶 sand jack
砂箱填料 flask filler
砂箱托架 sand box bracket
砂箱箱筋 flask bar
砂箱造型 box mo(u)lding; flask mo(u)lding
砂箱罩 sand box casing
砂箱轴 flask trunnion; mo(u)lding box trunnion; sand box shaft
砂箱铸造 flask casting
砂箱作用杆 sand box arm
砂箱座 sand box base
砂屑 sandy clast
砂屑白云岩 doloarenite
砂屑构造【地】 psammitic structure
砂屑灰岩 calcarenite
砂屑结构 psammitic texture; sandy clastic texture
砂屑泥质岩 psammonpelite
砂屑泥质岩的 psammonpelitic
砂屑凝灰岩 tuffstone
砂屑片麻岩 psammite-gneiss
砂屑片岩 psammitic schist
砂屑砂岩 psammite
砂屑岩 arenaceous deposit; arenite; arenyte; psammyte
砂楔 sand wedge
砂芯 bore; sand core
砂芯盒 core box
砂芯烘架 core shelf
砂芯胶合 pasting of core
砂芯黏[粘]合膏 core paste
砂芯破碎机 core breaker
砂芯气孔 core blow
砂芯填料 core filler
砂芯头 core print
砂芯下沉 core sag
砂芯样品 core sample of sand
砂芯硬度计 core hardness tester
砂芯铸空的水套通道 sand-cored water jacket passage
砂型 sand-lined mo(u)ld; sand mo(u)ld
砂型表面强度试验仪 mo(u)ld strength tester
砂型吊钩 gagger
砂型钉 mo(u)lder's brad
砂型法 sand mo(u)lding
砂型骨 inserted piece
砂型加固圈 inside band
砂型假箱 oddside
砂型抗裂试验仪 mo(u)ld fracture tester
砂型控制 sand control
砂型离心浇注 sand-spun casting
砂型离心铸造法 sand-spun process
砂型皮下强度测定器 impact penetration tester
砂型生铁 sand mo(u)ld pig iron
砂型碎裂 crush
砂型套框 crib
砂型套箱 mo(u)ld jacket
砂型透气性测定仪 mo(u)ld void tester
砂型涂料 mo(u)ld blacking; sand wash
砂型箱 casting box
砂型芯 sand core
砂型修补 patching
砂型硬度 sand hardness
砂型硬度计 mo(u)ld hardness tester
砂型铸件 sand(mo(u)ld) casting
砂型铸铁管 sand-cast tube
砂型铸造 sand cast(ing)
砂型铸造法 sand-casting process
砂型铸造铝合金 alumin(i)um sand-casting alloy
砂性 grittiness
砂性的 sandy
砂性土 light-textured soil; non-cohesive soil; sandy soil
砂性土河岸 non-cohesive bank
砂性雪 sand snow
砂悬浮体 sand suspension
砂压载 sand ballast
砂岩 arenaceous sediment; malmstone; sand rock; sandstone; stone sand; grit
砂岩百分比平面图 plan of sandstone percentage
砂岩板 sandstone slab
砂岩包裹体 sandstone inclusion
砂岩采石场 sandstone quarry
砂岩层 sand stratum
砂岩储集层 sandstone reservoir
砂岩大圆石 < 常见于白垩区 > bridestone
砂岩地沥青 < 一种天然岩沥青 > sandstone rock asphalt
砂岩地面板 sandstone slab floor cover(ing)
砂岩方块 sandstone cube
砂岩方琢石 sandstone ashlar
砂岩粉 sandstone meal
砂岩粉末 sandstone powder
砂岩覆盖层 sandstone facing
砂岩骨料 siliceous aggregate
砂岩怪石 sarsen
砂岩及各类储集体分布图 map of distribution of sand rocks and various deposits
砂岩夹层 intercalated beds of sandstone; sandstone band; sand streak(ing)
砂岩结构 sandstone texture
砂岩矿床 sandstone deposit
砂岩路边石 sandstone curb[kerb]
砂岩路缘石 sandstone curb[kerb]
砂岩乱石 sandstone rubble
砂岩乱石幕墙 sandstone rubble curtain wall
砂岩脉 sandstone dike
砂岩毛石 sandstone rubble
砂岩幕墙 sandstone curtain(wall)
砂岩漂砾 sarsen
砂岩飘砾 greyweathers
砂岩墙 sandstone dike[dyke]
砂岩桥门 sandstone portal
砂岩球 sandstone ball
砂岩石 < 加于毛石混凝土中用 > sandstone plum
砂岩碎石 sandstone plum
砂岩隧道门 sandstone portal
砂岩筒 sandstone pipe
砂岩透镜体差异压实背斜圈闭 anticlinal trap by differential compaction over sand lens
砂岩透镜体圈闭 sand lens trap
砂岩土壤 sandstone soil
砂岩相 sandstone facies
砂岩小方石 < 铺路用 > sandstone paving sett

砂岩型铀矿石 U ore of sandstone type
砂岩岩床 sandstone sill
砂岩(岩)墙 sandstone dike[dyke]
砂岩遗址 <史前建筑物遗址> sarsen
砂岩油苗 sand shows
砂岩中巨大的钙结核 cracker
砂盐撒布机 sand and spreader
砂盐撒布器 sand and spreader
砂眼 sand hole; air bell; air blister; air bubble; air hole; air pocket; bleb; blister; blowhole; crush; gaul; open bubble; pit hole; porosity; push-up; sand blister; sand explosion; slag blowhole; trachoma; abscess <金属中的>
砂眼及夹杂物 sand hole and inclusions
砂眼裂纹检验 pore and crack detection
砂眼排除装置 air pocket eliminator
砂眼针孔 slag pin hole
砂养护 <混凝土> sand curing
砂样采取器 sand pump sampler
砂样防落簧 core catcher
砂样(品) sand sample; sand specimen
砂样瓶 sample container
砂样制备的棒捣法 plunger method; plunging method
砂样制取器 core cutter; core lifter
砂页比 sand-shale ratio
砂页岩 hazle
砂页岩比(率) sand-shale ratio
砂页岩切割阶梯地形 alcove lands
砂移动 sand shifting
砂印 <轧材缺陷> sand marks
砂影堆积 sand shadow; sand shallow
砂涌 blowout
砂与粗骨料比 sand and coarse aggregate ratio
砂与粗集料比 sand and coarse aggregate ratio
砂与骨料比 sand and aggregate ratio
砂与骨料总量比 sand(-and)-total aggregate ratio
砂与集料总量比 sand(-and)-total aggregate ratio
砂与砾石推移质 sand and gravel bed load
砂与其他轻骨料配合的混凝土 sand-lightweight concrete
砂浴锅 sand-bath
砂浴回火 sand bath tempering
砂浴(器) sand-bath
砂原 sand(y) plain
砂源 sand source
砂载法 <排水瓦管埋入砂中作压碎强度试验> sand-bearing method
砂造型 sand mo(u)lding
砂渣水泥 slag based cement
砂栅栏 sand fence
砂毡 sand blanket
砂涨地 sand accretion
砂障 sand barrier
砂罩面 sandy blanket
砂纸 abrasive paper; abrasive sheet; coated abrasive; emery paper; garnet paper; glass paper; polishing paper; sand cloth; sand paper; silicon paper
砂纸打光 sand(paper)ing
砂纸打光机 sand(paper)ing machine
砂纸打磨 sand(paper)ing
砂纸打磨产生的碎屑 swarf
砂纸打磨法 sand(paper)ing method
砂纸打磨机 sand(paper)ing machine
砂纸辊打磨机 roller sander
砂纸加工 coated abrasive machining
砂纸磨擦 sandpapering
砂纸磨光机 sandpapering machine
砂纸磨痕 sandpaper marks
砂纸抛光机 emery papering machine
砂纸状饰面 sandpaper surface
砂制模型 sand pattern
砂制砖 sand cloth brick; sanded brick
砂质板岩 sandy slate
砂质辫状河沉积 sandy braided-stream deposit
砂质滨 sandy shore
砂质层系 sandy formation
砂质产品 sand product
砂质沉淀物 sandy deposit
砂质沉积物 arenaceous sediment; sandy sediment; sandy deposit
砂质沉积岩 arenaceous sedimentary rock
砂质充填 sandy filling
砂质垂直排水 sand chimney
砂质的 arenaceous[arenarious] arenose; arenous; sabulous; sandish; sandy; tophaceous
砂质底 sandy bottom
砂质地段 arenaceous region
砂质地基 sandy subgrade
砂质地沥青(混凝土) sheet asphalt
砂质地沥青混凝土面层 sheet asphalt surfacing
砂质地区 arenaceous region
砂质动床 sandy mobile bed
砂质多孔石灰岩 buhrstone
砂质方解石 sand-calcite
砂质粉土 sandy silt
砂质腐泥 sapropsammite
砂质高岭土矿床 sandy kaolin deposit
砂质骨料 sandy aggregate
砂质过滤层 sandy filter
砂质海岸 sandy coast; sandy shore
砂质海滨 sandy shore
砂质海滨地带 psammolittoral zone
砂质海滩 sandy beach; sea sand
砂质河床 sand(y) bed(ding)
砂质河床河流 stream with movable bed
砂质荒漠 erg; sand(y) desert
砂质黄土 sandy loess
砂质灰岩 arenaceous limestone; sandy limestone
砂质灰岩碎屑 arenaceous limestone chip(ping)s
砂质基础 sandy foundation
砂质集料 sandy aggregate
砂质胶结物 arenaceous cement
砂质角砾岩 sandy breccia
砂质结构 arenaceous texture
砂质矿石 sandy ore
砂质砾石 sandy gravel
砂质砾岩 sandy conglomerate
砂质砾岩相 arenaceous-conglomeratic facies
砂质垆坶 sandy loam(soil)
砂质路基 sandy subgrade
砂质漠壤土 sandy desert soil
砂质泥灰岩 crag; sandy marl
砂质泥炭土 sand-soil-peat matures
砂质泥土防渗墙 sandy clay impervious core
砂质黏[粘]壤土 sand(y) clay loam
砂质黏[粘]土 arenaceous clay; arene; dauk[dawk]; douke; mild clay; rubber clay; sabulous clay; sandy clay; lam
砂质黏[粘]土地基 sand clay base
砂质黏[粘]土防渗芯墙 impervious sandy clay core
砂质黏[粘]土垆坶 sandy clay loam
砂质排水反滤层 inverted sand drain filter
砂质排水竖井 sandy chimney
砂质炮泥 sand stemming
砂质片地沥青 <即砂质地沥青混凝土> sand sheet asphalt
砂质浅滩 sand(y) shoal
砂质浅滩爆破 sandy shoal blasting
砂质壤土 sand(y) loam(soil)
砂质三角洲 sand delta
砂质沙漠 erg; sandy desert
砂质石灰石 arenaceous limestone; sandy limestone
砂质石灰石屑 sandy limestone chip(ping)s
砂质石灰岩 arenaceous limestone; sandy limestone
砂质石灰岩碎屑 arenaceous limestone chip(ping)s
砂质(石油)沥青 sand asphalt
砂质熟料 sandy clinker
砂质水泥 <含石英砂细粉的硅酸盐水泥> sandy cement
砂质燧石 sandy chert
砂质填缝料 sand filler
砂质土路 sand clay road
砂质土(壤) light-textured soil; sabulous soil; sand(y) soil
砂质亚黏[粘]土 sandy clay loam
砂质烟囱式排水 sand chimney
砂质岩 arenaceous rock; arenite; psammite[psammyte]
砂质页岩 arenaceous shale; doab <印度>; fake; rock bind; sandy shale
砂质淤泥 sandy mud; sandy silt
砂质宙积物 sandy sediment
砂中有机杂质 organ impurities in sands
砂洲 sand reef; sand ridge; sand shoal; shoal
砂柱 dropmade column; sand column; sand pillar; sand pinnacle; vertical drain <沼泽填土中的,供排水用>
砂柱加固法 sand compaction method
砂铸 sand cast
砂铸生铁 sand-cast pig
砂砖 sanded brick; bath brick <用于磨刀或打磨金属面的>
砂砖粉 brick dust
砂桩 sand column; sand pile; sand piling; compaction pile <普通砂井>
砂桩船 sand piling barge
砂桩基础 sand pile foundation
砂桩挤密法 extrusion method of sand pile
砂桩加固法 consolidated method of sand drains; sand compaction pile method; sand pile method; soil strengthened by sand drains
砂状表面 dry spray
砂状结构 psammitic texture
砂状磷块岩 sandy phosphoraite
砂状涂层 dusty coat
砂锥 abrasive cone
砂锥法 <量挖出的土或路面中洞穴体积的一种方法,测现场密度用> sand cone method
砂子 sand
砂子部分 sand fraction
砂子采运 sand transport
砂子沉降槽 sand thickener
砂子稠化器 sand thickener
砂子存放期 bench life of a sand mix
砂子打毛 sanding
砂子分级器 sand classifier
砂子分级设备 sand classifying plant
砂子分选机 sand classifier
砂子管理 sand control
砂子回收 sand recovery
砂子回收机 sand reclaiming machine; sand recovery machine
砂子回收螺旋装置 sand reclaiming screw; sand recovery screw
砂子级配区 zones for sand grading
砂子加工车间 sand processing plant
砂子加工机 sand processing machine
砂子加工设备 sand processing plant
砂子颗粒组成图 sand grading chart
砂子冷却滚筒 louvered drum sand cooler
砂子沥青混合料 sand asphalt
砂子粒度分布 sand grain distribution
砂子粒度分级 sand grading
砂子粒度分析 screen analysis
砂子粒径 particle size
砂子裂解 sand cracking
砂子膨胀 sand expansion
砂子平均粒度 sand fineness
砂子韧性读数 sand toughness number
砂子湿度 sand moisture
砂子输送设备 sand handling equipment
砂子调节用水 temper water
砂子细度 sand fineness
砂子循环系统 sand recycle system
砂子与湿红黏[粘]土混合土壤 bull's liver
砂子运输 sand transport
砂子制备机 sand producing machine
砂嘴 barrier spit; sand spit

莎

莎草泥炭 carex peat
莎草沼泽 carex marsh
莎草属 galingal; Cyperus <拉>
莎纶缆索 saran rope
莎禾属 <拉> Coleanthus

鲨

鲨肝油 shark liver oil
鲨革 shagreen
鲨烯 squalane
鲨油 oil of dogfish
鲨油酸 selacholeic acid
鲨鱼楔形器件 angelfish device

傻

傻瓜式(建筑) noodle style

唼

唼头 injection cup; injector cup; spray cup

筛

筛 bolter; classifying screen; cullender; gridiron; screener; shaker; sieve
筛板 screen board; screen deck; screen(ing) plate; sieve plate; sieve tray
筛板表面水平偏差 difference in floor lever of sieve tray
筛板萃取塔 perforated plate extraction tower
筛板孔径 hole diameter of sieve(perforated) plate
筛板孔距 hole pitch of perforated tray
筛板孔数 number of sieve tray holes
筛板孔形系数 T deck factor T
筛板孔中心距 hole distance in sieve tray; hole pitch in sieve tray
筛板拉杆 screen set bar
筛板面水平差 level difference of sieve tray floor
筛板筛孔面积 perforated area of screen plate
筛板塔 perforated plate tower; sieve-plate column; sieve-plate(column) tower; sieve tray tower
筛板效率 efficiency of sieve plate
筛板柱 sieve-plate column
筛比 sieve ratio; sieve scale
筛篦 sieve grating
筛布 bolting cloth; screen(ing) fabric; screen(ing) cloth; sieve cloth

筛布面积 wire cloth area
筛部 sieve part;sieve-portion
筛侧空气跳汰机 Baum jig
筛层 screen deck
筛碴机 ballast screening machine
筛铲 sieve shovel
筛场 sieve field
筛出 screen(ing)out
筛出粗块 scalping
筛出的 screened
筛出废料 screened refuse; screen overflow;screen oversize;screen reject
筛出过大颗粒 scalping
筛出物 sieved material; underflow of classifier
筛出岩粉 stone screenings
筛除 screen off
筛除粗粒 scalping
筛除粗物料 scalping
筛除法 screening
筛除物 screenings;tailover
筛的带式装载机 screen belt loader
筛的有效面积 clear area of screen; effective area of screen;useful area of screen
筛的有效面积百分率 percentage open area of screen
筛底 screen deck; screen(ing)bottom;sieve bottom
筛底安装 <指水厂、污水厂> fitting of screen bottom
筛底部分 section of screen(ing)deck
筛底加热 screen bottom heating; screen deck heating
筛底料 undersize;undersize material
筛耳 <每英寸筛孔数> mesh
筛法 sieve method;sifting method
筛分 bolting; grading; jigging; pan fraction; screen classification; screen(ing)(out); screen separation; settling; sieve analysis; sieve classification; sieving; sift; size up; sizing; mechanical analysis
筛分比 sieve ratio
筛分步骤 screening procedure
筛分部分 screen fraction
筛分材料 screened material; sieved material
筛分长度系数 D deck factor D
筛分厂 grading plant; screening plant; sieve plant
筛分车间 screen building; screen house; screening plant
筛分成大粒级范围 screening into wide size ranges
筛分尺寸 step sizing
筛分出来的颗粒物料 particle material to be screened
筛分出料 screen discharge
筛分带 picking belt
筛分的精细度 accuracy of separation; precision of separation
筛分的条件系数 <以含水率小于 4% 的统货坑采材料为 1> condition factor
筛分的准确性 precision of screen separation
筛分定时器 sieving timer
筛分法 method of sieving; screen(ing) method; sieve analysis; sieve method; sorting
筛分范围 grading envelope; grading range
筛分分级 screen sizing; sieve classification; sieve sizing
筛分分级材料 screen sized material
筛分分离装置 sieving extractor
筛分分析 cascade analysis; sieve analysis; mesh analysis; screen analysis
筛分析校正 sieve correction
筛分分析图 screen analysis chart
筛分分析图表 sizing plot
筛分浮沉大样 size-float-and-sink sample
筛分浮沉试验报告表 sieve float-and-sink analysis report table
筛分格栅 sizing grill
筛分工 screener; screen man
筛分工场 screening plant; screening works
筛分骨料 screening aggregate
筛分规范 sieving specification
筛分过的玻璃粉 graded glass powder
筛分和沉淀综合试验 combined sieve and sedimentation test
筛分机 balter; bolter; bolting machine; bolting mill; classifier; grading plant; grate; scalper; screen classifier; screener; screen grader; screening machine; screening plant; sieving machine; sifter; sifting machine
筛分机给料 screen feeding
筛分机给料仓 screen-feed bin
筛分机械 sieving machinery
筛分机织物品质 quality of screen cloth
筛分机织物质量 quality of screen cloth
筛分级 sieving classification
筛分级别 mesh scale
筛分级配 screen size gradation
筛分集料 screening aggregate
筛分计时器 sieving timer
筛分颗粒度分析 analysis by sieving; sieve analysis
筛分颗粒组分重量百分比 percentage fraction weight
筛分颗粒组分重量百分数 percentage fraction weight
筛分粒度 screen size; sieve size
筛分粒度级 mesh fraction; sieve fraction
筛分粒度率 mesh fraction; sieve fraction
筛分粒级 sieve fraction
筛分粒径 sieve particle-diameter
筛分粒径分布情况 analysis sieved
筛分粒径组分重量百分比 percentage fraction weight
筛分粒径组分重量百分数 percentage fraction weight
筛分料斗 screening hopper
筛分率 accuracy of separation
筛分煤台 picking table
筛分面积 sieving area
筛分模数 screen modulus
筛分能力 screen(ing) capacity; sieving capacity
筛分能力系数 <包括基本能量系数和一系列有关系数> screen capacity factor
筛分能量 screening capacity
筛分耙 classifier rake
筛分器 sand screen; sand shaker; sand sifter
筛分前煤样总重量 coal sample total gravimetry before sieve
筛分球磨机 screen discharge ball mill
筛分曲线 grain-size distribution curve; screening curve; sizing plot
筛分取出装置 sieving extractor
筛分设备 screening device; screening equipment; screening installation; screening machinery; screening plant; sieving machine; sizing device
筛分石膏 sized gypsum
筛分试验 mesh test; screen analysis; screen(ing) test(ing); sieve analysis test; sieve test(ing); sieving test; size analysis; size test; sizing test
筛分试验报告表 sizing test report table
筛分试验编号 sizing test number
筛分试验负责人 leading cadre of sizing test
筛分试验盘 pan
筛分试验曲线 sieve test result diagram
筛分试验图 plot of screen test
筛分试验振荡器 testing sieve shaker
筛分试验总样化验结果 sizing test total sample laboratory test result
筛分室 sieve compartment
筛分输送机 separator conveyer[conveyor]
筛分数据 screening data; sieving data; size data
筛分损失 screen loss
筛分特性 screening characteristic; size characteristic
筛分特性曲线 sizing characteristic
筛分脱水法 screening dewatering
筛分析 grain-size analysis; screen analysis; screening test; sieve analysis; sizing analysis; sizing test; test sieving
筛分析校正 sieve correction
筛分析机 sieve machine
筛分析曲线 sieve(analysis)curve
筛分系数 sorting coefficient
筛分系统 screening system
筛分细度 screen fineness; sieve fineness
筛分箱 screening cell
筛分效率 classification efficiency; efficiency of screening; screen(ing) efficiency
筛分效率系数 E efficiency factor E
筛分效应 sieve effect; sizing effect
筛分样 size sample
筛分样品 sieved sample
筛分用筛 separating screen
筛分与冲洗设备 screening and washing plant
筛分运输机 separator conveyer[conveyor]
筛分运转 screen motion
筛分渣 screened slag
筛分质量 screening quality
筛分转筒 screening drum
筛分装载机 loader screen; loadscreen
筛分装置 grading plant; screening arrangement; screening plant; screen(ing) unit
筛分作用 screening action; sieving action
筛粉筛 dressing sieve; seed sieve
筛格 grate
筛格尺寸 mesh width
筛工 sifter
筛垢 sieve scale
筛鼓 screen cylinder; sizing drum
筛刮 screen rack
筛管 perforated tail pine; sieve tube; sieve vessel
筛管成分 sieve element
筛管分子 sieve(tube) element
筛管节 sieve tube member; sieve tube segment
筛管组织 sieve tissue
筛过的 screened; sifted
筛过的灰泥 sifted plaster
筛过的混合砂 graded sand mix(ture)
筛过的混凝土骨料 screened concrete aggregate
筛过的混凝土集料 screened concrete aggregate
筛过的煤 screened coal
筛过的碎石 hoggin
筛过的物料 screening
筛过的细料 hurry gum
筛号 basic mesh size; mesh ga(u)ge; mesh marks; mesh number; mesh per inch; screen number; screen size; sieve number; sieve size; size of mesh; test sieve size; screen mesh
筛号标准 sieve series standard
筛号分级 screen size gradation; sieve size gradation
筛号纲目 screen mesh
筛核效应 screening effect
筛机 scalper; screening device; screening machine; sieving machine; sifter
筛级 screen grading
筛集粗料 screening refuse
筛架 deck support; screen frame(unit); screen shoe; screen structure; screen unit; sieve pan; sieve shoe
筛拣 garble
筛检 screening
筛检灵敏度 sensitivity of screening
筛检特异度 speciality of screening
筛检真实性 screening validity
筛浆 pulp cleaner; pulp screening
筛浆机 pulp screen
筛浆料 run to putty
筛节 strum
筛径 screen aperture; sieve diameter
筛孔 aperture of screen; mesh; mesh aperture; mesh opening; passage of a screen; screen aperture; screen hole; screen mesh; screen mesh opening; screen opening; screen perforation; sieve mesh; sieve mesh opening; sieve opening
筛孔板支架 screen holder
筛孔测定 sieve test(ing)
筛孔尺寸 mesh size; screen mesh size; screen size; sieve hole size; sieve size; size of mesh; size of opening; opening size
筛孔尺度 sieve fineness
筛孔尺码 size of mesh
筛孔大小 sieve mesh size
筛孔大小序次 screen size gradation
筛孔堵塞 blinding of screen; screen blinding
筛孔堵塞的筛子 dumb screen
筛孔度 aperture size
筛孔缝隙系数 slot factor
筛孔管 anti-priming pipe
筛孔号 sieve mesh number; sieve number
筛孔极限尺寸 limiting mesh
筛孔净面积 clear area of screen
筛孔卡粒 pegging
筛孔孔径 sieve size
筛孔面积 area of mesh; area of opening; screening area
筛孔面积百分率 percentage of opening; percentage open area of screen
筛孔排出 discharge of a screen
筛孔数目 screen number
筛孔筒 perforated cylinder
筛孔系列 mesh series
筛孔效应 sieve action; sieve effect
筛孔有效面积 working of mesh
筛孔与矿粒总表面比 ratio of opening to total surface
筛孔直径 sieve diameter
筛孔状分布 sieve-like distribution
筛矿机 jigger
筛矿器 jigger
筛矿室 dozing chamber
筛矿者 jigger
筛框 deck base; screen surface support frame
筛框式过滤器 screen-frame filter

S

筛框线脚 screen mo(u)ld
筛量规 sieve ga(u)ge
筛料 screen feed
筛料口面积 discharge area of screen
筛留百分率 percentage retained;retained percentage
筛留百分数 percentage retained;retained percentage
筛滤板 screen filter
筛滤材料 contact material
筛滤法 screen filtration method
筛滤管 screen pipe
筛滤灰膏 plaster's putty
筛滤机 passing machine
筛滤截留法 screen filtering
筛滤器 screen filter
筛滤网 strainer
筛落 riddle
筛麻机 tow shaker
筛煤场 tipple
筛面 screen deck;screening plane;screen surface;sieve sifter
筛面工作面积 effective screening area;screen open area;screen working area
筛面荷载 screen surface load
筛面混合料 face mix
筛面料 plus material
筛面面积 screen deck area
筛面倾角 screening angle
筛面有效筛分面积 effective screening area;open area
筛面支承框 screen surface support frame
筛膜 sieve membrane
筛磨 screen mill
筛目 mesh;mesh opening;mesh per inch;mesh screen;screen mesh;screen size;sieve mesh
筛目范围 mesh range
筛目规 mesh ga(u)ge
筛泥网 shale shaker
筛泥压碎机 screen sludge crusher
筛盘 screen tray;sieve tray;sifting disk
筛盘分级 pan fraction
筛盘式萃取塔 sieve tray extraction tower
筛盘塔 sieve tray column;sieve tray tower
筛去 screen off;screen out;tramp
筛去细粒 de-slurrying by screens
筛去新拌混凝土中大于规定尺寸的骨料 wet screening
筛容量 screen capacity
筛砂 sand-sifting
筛砂机 mechanical sand screen;sand screening machine;sand shaker;sand sieving machine;sand sifter
筛砂器 sand screen;sand shaker;sand sieve;sand sifter
筛砂松砂机 screenerator
筛上 plus mesh;plus sieve;over-size
筛上百分率 sieve residue percentage
筛上产品 sieve oversize;sieve ragging
筛上产品率 over-size factor
筛上冲洗 dillying
筛上粗物料 coarse material retained on screen;coarse material retained on sieve
筛上粉粒 over-size particle
筛上粉末 over-size powder
筛上颗粒尺寸 plus size of a screen
筛上累计百分数 accumulative total on the sieve percent
筛上料 over-sized particle;over-size material;plus mesh;shorts
筛上剩余物料 retained material on the sieve;sieve residue
筛上物 over-size product;plus mesh;riddlings;screen overflow;screen oversize;screen tailings
筛上物累计分布曲线 cumulative oversize distribution curve
筛上物料 material retained on screen;material retained on sieve
筛上物料率 matter rate on the sieve
筛上物中的筛底料 undersize in the oversize
筛上溢流 screen overflow
筛身 screen box
筛剩残渣 residue of sieve;residue on sieve
筛剩余渣 residue on the sieve
筛石 gravel screening
筛石厂 screening plant
筛石机 bolter;scalper;screening machine;sieving machine
筛石设备 stone sizing plant
筛石屑 riddlings
筛式凹板 cylinder grate
筛式沉降器 screen clarifier;screen decanter
筛式干燥器 screen-type drier
筛式滚筒 screen drum
筛式给料机 screen feeder
筛式滤选机 screen cleaner
筛式捏合机 screen kneader
筛式清选机 screen cleaner;sieve cleaner
筛式清选器 screen-type separator
筛式输送机 screen conveyer[conveyor]
筛式听力计 screening audiometer
筛式挖掘机 sievedigger
筛式洗矿机 screening washer
筛式圆盘喂料机 circular screen feeder;screen circular feeder
筛式运输机 screen conveyer[conveyor]
筛室 screen chamber
筛刷 sieve brush
筛体 screen frame(unit)
筛条 diagrid;grate bar;grating;grizzly bar;jigger bar;screen of bars
筛条隔隙 gap of screen strip
筛条间距 bar spacing
筛筒 perforated cylinder;riddle drum;screen drum;sizing drum
筛筒式离心机 screen-type centrifuge
筛筒转速 screen drum speed
筛网 bar screen;mesh cloth;mesh net;mesh screen;network of meshes;screen cloth;screen mesh;screen stencil;sieve cloth;sieve sifter;strainer;wire cloth;woven filter medium
筛网布 sieve cloth
筛网成套 screen series
筛网成组 screen series
筛网尺度 screen scale
筛网粗粒 shorts
筛网过滤膜 granular membrane
筛网过滤器 granular membrane;mesh filter
筛网号 grit number
筛网架 screen bed base
筛网结冰 icing of screen
筛网(金属)丝 screen wire
筛网孔净宽 aperture width
筛网孔径 screen cloth opening
筛网离心机 screen centrifuge
筛网漏勺 wire screen ladle
筛网滤清器 screen cleaner
筛网面积 screen area
筛网目数 sieve mesh number
筛网排料磨机 screen-type mill
筛网排料式磨矿机 screen discharge mill
筛网铅黄铜 leaded screen wire brass
筛网倾斜度 screen rack
筛网倾斜调节轴 screen slope adjusting shaft
筛网式撕碎机 screen shredder
筛网式通气塞 air release screen
筛网网眼尺寸 screen slot size
筛网系列 screen series
筛网印花 screen printing
筛网印花框机 printing troll(e)y
筛网圆筒烘干机 sieve drum drier[dryer]
筛尾 tail end
筛问题 sieve problem
筛析 fractional analysis;mesh analysis;particle-size analysis;screen analysis;sieve analysis;sieve classification;sieving
筛析法 sieve analysis method
筛析法颗粒分析 sieving method particle analysis
筛析粒度范围 sieve size range
筛析品 sized material
筛析曲线 screen analysis curve
筛析曲线图 screen analysis chart
筛析色谱法 exclusion chromatography
筛析试验 sieve test(ing);size test;sizing test
筛析图表 sizing plot
筛析效应 sizing effect
筛洗分析实验 screen washing analysis
筛洗器 screening washer
筛系列 mesh scale
筛隙宽度 aperture size
筛下 minus sieve;undersize
筛下产品 material passing the screen;material passing the sieve;screen undersize;through product;minus mesh
筛下产品回收率 undersize recovery
筛下产品破碎 screenings crushing
筛下的 undersized
筛下底流 screen underflow
筛下范围 screen underflow
筛下级物料 material passing the screen
筛下级物料尺寸<即小于筛孔的物料尺寸> minus size of a screen
筛下料 minus mesh;sieve residue;undersieves
筛下料百分率 undersize percentage
筛下气室跳汰机 Batac jig;Tacub jig
筛下水 underscreen water
筛下物 screen underflow;siftage;spigot;through product
筛下物粉碎机 screenings disintegrator
筛下物累积分布曲线 cumulative undersize distribution curve
筛下物料 discharge of undersize
筛下物密度试验 spigot density test
筛下细料百分率曲线 percentage undersize curve
筛下中的过大者 over-size in the undersize
筛箱 screen body;screen box;straining box
筛箱的工作面积 sieve box area
筛效率 screen capacity
筛屑 chippings;fines;riddlings;scalpings;screenings;screen tailings;screen-throughs;siftings
筛屑处置 tailings disposal
筛屑坑 tailings pond
筛型磨碎机 screen-type mill
筛序 screen size gradation;sieve series
筛选 bolting;dressing by screening;garble;jigging;preparation by screening;screen(ing);selection by sifting;sieve analysis;sifting
筛选板 screen plate
筛选部分 screening portion
筛选材料 processed material
筛选参数 filtering parameter
筛选残渣 screening refuse
筛选操作 filtering operation;grading operation;screening operation
筛选厂 screening building;screening plant
筛选程序 filter;screening procedure;screening process
筛选出料 screen discharge
筛选的玻璃粉 screened glass powder
筛选法 screening;screening method;screening technique;sieve analysis method;sieve method
筛选方法 screening practice;screening process
筛选分类法 sifting sort
筛选分析法 screening analysis
筛选附件 grading attachment
筛选工 screener
筛选工具 grading tool
筛选工序 screening operation
筛选工艺 screening practice
筛选工作面积 screen open area
筛选规格 screening specification
筛选过程 screening process
筛选过的物质 screenings
筛选好的砾石 prepared gravel
筛选和储存区 screening and storage
筛选机 bolter;jig;screen(ing) grader;screening machine;screening plant;sieve machine;sieving machine;sifting machine
筛选级 screening level
筛选技术 screening technique
筛选架 screen carriage
筛选假设 screening hypothesis
筛选检查 screening inspection;screening test
筛选进口商品 screening of import commodities
筛选进料 screen feed
筛选矿石 screened ore
筛选累加 filtered accumulation
筛选累加器 filtered accumulator
筛选离心机 screen-bowl
筛选力 screening force
筛选砾石 prepared gravel
筛选率 screening rate
筛选煤 sized coal
筛选模拟法 screening simulation method
筛选能力 screening capacity
筛选泥浆 screen sludge
筛选盆 screen basin
筛选器 screening washer
筛选曲线 screening curve
筛选砂 riddled sand
筛选设备 screen(ing) device;screen(ing) installation;sizing plant
筛选设计 screening design
筛选生物检定 screening bioassay
筛选石灰 screened lump lime
筛选石料 processed rock
筛选实验 screening test
筛选试验 screening experiment;screening test
筛选水槽 screening flume
筛选算法 filter algorithm
筛选损失 sieve loss
筛选塔 screen tower;sieve-plate column
筛选特性 screening characteristic
筛选天然碎石 screened natural rock
筛选条件 screening condition
筛选效率 grading efficiency;screen(ing) efficiency
筛选性能 screenability

筛选性质 sifting property
筛选样品 screening sample
筛选指示剂 screened indicator
筛选制版 screening
筛选质量 grading quality; screening quality
筛选装置 screener; screening device; screening equipment; screening plant; sieving mechanism
筛选作业 screening
筛眼 handhole; screen aperture; screen hole; screen (ing) opening; screen mesh; sieve mesh; sieve opening
筛眼尺寸 aperture size; sieve size; size of mesh; mesh size
筛眼大小 size of mesh
筛眼堵塞 jam up
筛眼分析 mesh analysis
筛眼号数 size of mesh
筛眼孔径 aperture size; mesh size; screen size; sieve size; size of mesh
筛余 residue of sieve
筛余百分率 over-size percentage; pan fraction
筛余百分数 percent retained; sieve residue
筛余粗粒 coarse of a screen
筛余粗料 rejects; rejected material; retained material; riddings; screen overflow; screen oversize; screen reject
筛余粗料槽 reject chute
筛余废石料 scalp rock
筛余粒径 retaining size
筛余量 retained amount; sieve residue content
筛余量控制 tailings control
筛余料 screen residue; plus mesh; screen tailings; shorts; sieve tailings; tailings; triage
筛余率 percentage retained; retained percentage
筛余碾碎物 comminution of screenings
筛余试验 residue-on-sieving test
筛余污泥压碎机 screen sludge crusher
筛余物 over-size (d) product; residue on sieve; screen floating; screen residue; screen tailings; sieve residue; tailings; through; knocking
筛余物结构 screenings texture
筛余物碾碎 comminution of screenings
筛余物破碎机 screenings disintegrator
筛余物脱水 screenings dewatering
筛余物压滤机 screenings press
筛余渣 sieve residue
筛余折旧 remainder depreciation
筛余重 retained weight
筛余重量曲线 weight-retained curve
筛域(区) sieve region; sieve area
筛渣 overs; screen residue; screen sludge; screen tailings; sieve residue; tailover
筛渣槽 screenings gutter
筛渣处置 disposal of screenings
筛渣捣碎机 screenings triturator
筛渣焚化 screenings incineration
筛渣焚化炉 screenings incinerator
筛渣粉碎机 screenings grinder
筛渣磨浆机 reject refiner
筛渣破碎机 macerater [macerator]; screenings disintegrator
筛渣数量 quantity of screenings
筛渣撕碎机 screening shredder
筛渣填料压机 screening bale press
筛渣脱水 screening dewatering
筛渣压机 screenings press
筛渣再磨机 screenings refiner
筛者 sifter
筛制 sieve scale
筛滞留 sieve retention
筛状板 cribriform plate
筛状变晶结构 diablastic texture
筛状的 cribriform
筛状浇口 strainer gate
筛状结构 sieve texture
筛状结构的 diablastic
筛状纹孔式 sieve pitting
筛状吸附剂 sieve sorbent
筛状芯片 strainer core
筛子 sieve; boult; mesh screen; screen (temse); service screen; griddle; sifter; sizer; sizing screen; temse; riddle
筛子安装平面 screen floor
筛子打孔 punching of screen
筛子大小 size of mesh
筛子的倾角系数 factor of screen incline
筛子的阻塞 blinding of screen
筛子定位器 screen lock
筛子抖动器 screen shaker
筛子堵塞 blinding of screen; screen blinding
筛子加料率 screen loading rate
筛子开口面积系数 open area factor
筛子生产能力 screen capacity
筛子试验结果图 plot of sieve test results
筛子室 hut(ch)
筛子斜度 screen inclination
筛子有效面积 open space of screen; screen area; screening surface
筛子撞击器 screen shaker
筛子阻塞 screen blinding
筛组 screen banks; set of sieves

晒白 bleach

晒斑 sunburn
晒板台纸 flat
晒版 plate copying; printing down
晒版机 printing frame
晒版架 printing frame
晒不褪色的 lightfast
晒草机 hay maker
晒场 drying yard
晒干 caking; desiccation; insolation; sunbaked; sun-curing; sun drying
晒干的 sun cured; sun-dried
晒干的田地 parched field
晒干泥砖 sunbaked brick
晒干土 sunbaked soil
晒干衣服的阳台 drying balcony
晒干砖 baked brick; sunbaked brick
晒干砖齿轮【机】 sun-baked brick gear
晒干砖坯 sunbaked brick; sun-dried brick
晒谷场 grain-sunning ground
晒黑 bronze; brown; sunburn; tan
晒黑的皮肤 tan
晒夹 peg
晒焦 sun blister; sunburn; tanning
晒焦的 torrid
晒咖啡豆的场地 barbecue
晒蓝 blueprint
晒蓝图 blueline process; blueprinting; blue tone process; cyanotype
晒蓝图工人 blueprinter
晒蓝图机 blueprint apparatus; blueprinter
晒蓝图器 blueprint apparatus
晒蓝图设备 blueprint apparatus
晒裂 <泥块的> sun crack
晒泥场 drying mud yard; sludge drying bed
晒片机 printer
晒全株玉米用的木架 shocking house
晒热的 torrid
晒日光浴 sunbathe
晒溶液 blueprint solution
晒伤 sun scald
晒台 deck roof; drying stage; terrace
晒太阳的露台 sun trap
晒图 blueprinting; contact print; print(ing); transmission print
晒图灯 blueprint lamp; copying lamp
晒图机 blueprinter; blueprint(ing) apparatus; blueprint (ing) machine; contact printer; printer; printing machine
晒图架 copyholder; printing frame
晒图间 printing room
晒图框 copy board
晒图设备 blueprint apparatus; blueprint machine
晒图室 blueprinting room; printing room; reproduction room
晒图涂层 print overlay
晒图员 blueprinter
晒图纸 blueprint (ing) paper; heliographic(al) paper
晒相片 plating
晒像灯 printing lamp
晒像管 printing tube
晒像柜 printing box
晒像机 photocopier; photoprinter
晒像架 copyholder
晒像框 copying frame; printing frame
晒烟 sun-cured tobacco
晒衣架 clothes horse
晒衣绳 clothes line
晒衣绳杆 clothes line pole
晒衣绳柱 clothes pole
晒印 copying; light printing; printing
晒印干燥器 print drier[dryer]
晒印绘图机 print plotter
晒印计数器 printing counter
晒印图 photostat copy
晒印员 photoprinter
晒印照片 photographic (al) print; photoprint
晒印装置 lightprinting device
晒制干草 hay curing
晒制盐 solar salt
晒砖场木栅 hack cap
晒转台旋车盘 turntable
晒棕图 brownprint

山霭 Scotch mist

山艾树 sagebrush
山隘 col; corfe; critical road elevation
山隘符号【地】 pass symbol
山隘口 mountain saddle
山隘上台地 balaghat
山鞍 nek; saddle
山凹 cirque; corrie; cove; gap; recess
山坳 col; punch bowl
山白果 Chinese filbert; Chinese hazel
山白树 Henry wilsontree
山白松 mountain white pine
山傍地图 <美国一种显示土地价格的地图城区道路选线时用或显示城市市区房屋建筑物及其层数和建筑类型的地图> Sanborn map
山背 yamase
山崩 avalanche; billow; debacle; land fall; landslide; landslip; mountain creep; mountain slide
山崩防户墙 avalanche baffle wall
山崩废砾堆 avalanche debris
山崩风 landslip wind
山崩湖 lake due to landslide; landslide lake
山崩流 landslide flow
山崩泉 landslide spring
山崩支挡结构 avalanche brake structure
山边 sidehill; sidelong ground
山边断面 sidehill section
山边湖 border lake
山边建筑 hypogee; hypogeum
山边开挖 side cutting
山边丘陵地带 foot hill
山边渠道 sidehill canal
山边全挖断面 full bench section
山边全挖横断面 full bench cross section
山边斜地 sidelong ground
山扁豆属 senna
山侧 sidehill; versant
山侧喷发 flank eruption
山侧斜坡 sidelong ground
山茶花 camellia; common camellia
山茶木炭 camellia carbon
山茶油 camellia oil
山茶属 <拉> camellia
山潮比 ratio of mountain runoff to tidal volume
山城 hill-city; mountain city; mountain town
山村 mountain village
山达板 <一种木纤维墙板的牌号> Sundeala
山达木 alerce
山达树脂 sandarac(h) gum
山达斯特合金 Sendust
山达酸 sandaracolic acid
山达脂 gum Sandarac; sandarac(h)
山岛 mountain island
山道 alpine road
山的 mountainous; mountainy
山的高度测量 orometry
山地 high land; massif; mountain land; mountainous region; mountainous area; mountainous country; mountainous ground; mountainous terrain; upland; wold
山地背风涡旋 lee vortex behind hills
山地冰 highland ice
山地冰川 alpine glacier; mountain glacier
山地草场 patana
山地草甸草原土 mountain meadow-steppe soil
山地草甸地 mountain meadow soil
山地草原 mountain land; upland meadow
山地草原土 mountain steppe soil
山地测高分类 classification of mountain height
山地冲沟 mountain gulch
山地冲刷 upland erosion
山地的 upland
山地地貌 hill features; mountainous region landform
山地地貌单元 geomorphic(al) unit of mountain
山地冻原 mountain tundra
山地多层地形 multileveled mountain landform
山地工程测氡法 determining radon method by mountainous project
山地工程控制点 control point of engineering in mountainous region
山地构造分类 structural classification of mountain
山地灌丛 montane thicket
山地果园 hillside orchard
山地海岸 mountain coast
山地河段 mountain tract
山地红壤 mountain red earth
山地湖(泊) mountain lake
山地机场 mountain airfield

山地急流的防护工程 torrent works
山地假山毛榉 mountain beech
山地降雨量 mountain precipitation
山地纠正仪 rectifier for mountainous terrain
山地空盒气压计 mountain aneroid
山地空气 mountain air
山地阔叶常绿林 cloud forest
山地犁 upland plow
山地联合收获机 hillside combine
山地牧场 down land
山地泥炭 mountain peat
山地农业 mountain farming
山地农作 mountain farming
山地平原 upland plain
山地气候 mountain climate
山地气象站 mountain weather station
山地气压表 orometer
山地气压计 mountain barometer
山地侵蚀 upland erosion
山地区 upland area
山地森林 mountain forest
山地森林土 mountain forest soil
山地生态系统 mountain ecosystem
山地生物群系 orobiome
山地石灰 mountain lime
山地水文学 orohydrography
山地松 spruce pine
山地隧道施工法 rock tunnel(l)ing
山地苔藓林 montane mossy forest
山地泰加林土 mountain taiga soil
山地土壤 mountain soil;orogenic soil
山地土系列 orogenic sequence
山地稀树草原 mountain savana
山地下沉形成的海湾 embayed mountain
山地效应 mountain effect
山地形成学 orology
山地形态分类 morphologic(al) classification of mountain
山地形态要素 orographic(al) element
山地形态组合 association of mountain-shape
山地岩屑 mountain waste
山地衍射 mountainous diffraction
山地演化类型 evolution type of mountain
山地影响 mountain effect;orographic-(al) effect
山地栽培 growing on hill-side
山地沼泽 mountain bog
山地自行车 mountain bike[bicycle]
山巅 pike peak
山顶 hill crest; hill point; hill top; mountain top; brow of the hill; crest; pike peak; pinnacle; poke peak; summit; summit of a mountain
山顶白云 cape cloud
山顶高度 elevation;height of summit
山顶火山口 summit crater
山顶火山喷发 summit eruption
山顶路 ridge way
山顶喷发 summit eruption
山顶平齐 summit concordance
山顶平原 mesa plain
山顶群落 lophium
山顶水池 hilltop reservoir; summit pond
山顶水库 hilltop reservoir
山顶线 summit line
山东棕壤 non-calcic brown soil; Shantung brown soil
山洞 abri;cave;cavern of mountain
山洞爆炸 cave shooting
山洞工程 grotto work
山洞建筑 hypogeal
山洞式锚碇 anchor in rock gallery
山多利尼管 Santorini's duct
山多宁火山灰 Santorin
山矾属 sweetleaf
山风 canyon wind;gravity wind;katabatic wind; mountain breeze; mountain wind;berg wind <南非洲与海岸的一种热风>
山峰 alp;ben;hill top;mountain peak;pike(peak);summit;top of hump
山峰高程 height of summit
山峰纹 chev(e)ron mark
山腹凹地 corrie
山腹冰川 corrie glacier
山腹洼地 corrie
山冈 down;dune;hillock;hump
山格烙 <一种专利的灯式供暖电热器> Sunglow
山根 mountain root;root of mountain
山根带 root zone
山沟 mountain trench; upland channel;vale
山狗洞 <爆破的一种> coyote hole
山狗洞式 <美国一种爆炸法> coyote blast-hole type
山谷 clough; court dock; dale; intermont;mountain valley;valley
山谷边碛 valley train
山谷边坡 valley flank
山谷冰川 intermontane glacier;valley glacier
山谷道路 valley way
山谷(底)鼓胀 valley bulging
山谷地形影响 valley relief effect
山谷风 anabatic wind;mountain(-and)-valley breezes; mountain and valley winds
山谷和风 valley breeze
山谷(横)断面 valley cross-section
山谷黄土 valley loess
山谷或海底谷最底部联线 thalweg
山谷降水 valley precipitation
山谷(路)线 valley route
山谷凝雾 valley precipitation
山谷神庙 valley temple
山谷天线 valley-span antenna
山谷线 valley line;water course line
山谷主河流 axial stream
山谷主要河流 axial stream
山谷主要水道 axial stream
山海湾 mere
山合欢木 eastern Indian walnut
山核桃 Cathay hickory;hickory
山核桃木 hickory;pecan
山核桃属 hickory;carya <拉>
山洪 freshet; hill torrent; mountain torrent;torrent;torrential flood
山洪暴发 flash flood;torrential burst
山洪陡槽 overchute
山洪警报 flash flood warning
山洪控制工程 torrent control works
山洪来石量 quantity of incoming detritus from mountain torrent
山后区 backland
山弧 mountain arc
山胡椒 mountain pepper
山胡桃 Japan walnut
山花 acroteria;acroterion
山桦 cherry birch; mahogany birch; sweet birch
山槐 lebbek-tree
山货 <山区土产> mountain produce
山矶 river cliff
山基坡 plinth
山脊 chine; dorsum; hogback; mountain crest; mountain ridge; mountain uplift; rand; ridge; sowback; cuesta <一边陡峭面另一边坡度很小的>
山脊道路 ridge way
山脊高度 ridge height
山脊捷径 summit cut-off
山脊口 col
山脊拉沟 summit-cut
山脊路 ridge road;ridge route;ridge way
山脊路线 ridge route
山脊线 crest line; ridge line; ridge route;ridge way;summit line
山脊线与河谷线对比 ridge versus valley location
山脊垭口 break-in ridge
山尖 gable peak
山尖饰 acroter(ion)[akroter(ion)]
山尖形装饰 acroterium
山间隘道 gut
山间隘路 narrow pass
山间凹地 intermont;lap
山间坳地 intermontane depression
山间坳陷 interdeep; intermontane deep; intermontane depression; intermountain deep
山间冰川 intramontane glacier
山间槽地 intermontane trough
山间的 intermont; intermontane; intermountainous
山间地槽 idiogeosyncline
山间地带 intermontane space
山间地区 intermontane area
山间地震带 intermountain seismic belt
山间高原 intermontane plateau;intermountain plateau;puna
山间海槽 intramontane trough
山间含煤建造 intermontane coal-bearing formation
山间(河)谷 intermontane valley
山间盆地 intermontane basin; intermountain basin; intermountainous basin
山间盆地冰川 intermont glacier
山间盆地地下水 groundwater in intermontane basin
山间平原 intermontane plain; intermountainous plain; intermountain plain
山间碛原 mountain pediment
山间水湖 tarn
山间洼地 intermont;intermontane depression;intermontane hollow
山间狭道 arrow pass
山间小湖 tarn
山间小径 notch
山肩 mountain shoulder;replat;shoulder of mountain
山涧 beck;brook;defile;hill stream; hill torrent;mountain creek;mountain stream;ravine stream
山脚 foot of a hill;hillside;mountain foot; mountain root; submountain region
山脚冰 piedmont ice
山脚沟 hillside ditch
山脚线 piedmont angle
山荆子 Siberian crab
山峻 promontary
山口 clough; col; corfe; critical road elevation;cross-over;gap;gha(u)-t;hawse;kluf;lak;mountain pass
山口风 mountain-gap wind
山口公路 mountain pass highway
山口区域 mountain pass area
山块 mount slab
山梨 mountain ash
山梨醇 sorbierite
山梨聚糖 sorbitan
山梨酸 sorbic acid
山梨酸废水 sorbic acid wastewater
山梨酸钾 potassium sorbate
山梨酸钾废水 potassium sorbate wastewater
山梨酸酯 sorbate;sorbicester
山梨(糖)醇 sorbitol;sorbol
山梨(糖)醇醇酸树脂 sorbitolalkyd
山梨糖醇酯 sorbitol ester
山理学 <即山岳成因学>【地】orology
山栎 chestnut oak;densiflora
山砾石 pit gravel
山链 mountain chain
山梁 ridge
山林地 mountainous area with woods
山林地区 weald
山岭 chine; flat-top ridge; mountain ridge;mountain uplift;ridge
山岭道路 alpine road
山岭的 mountainous
山岭地区 mountainous country;mountainous region; mountainous terrain; very difficult terrain
山岭地形 mountainous topography
山岭区 mountain region; mountain terrain
山岭隧道 mountain(ous) tunnel
山岭线 mountain line;mountain route
山岭造林 ridge planting
山柳树 Hupeh wingnut
山柳属 <拉> Clethra
山路 alpine road;gha(u)t;mountain pass;mountain road;nek
山路蛇曲盘旋 sinuosity
山麓 foothill;foot of mountain;mountain foot; mountain root; mountain slide;submountain region
山麓冰 piedmont ice
山麓冰川 pediment glacier;piedmont glacier
山麓冰川舌瓣 piedmont bulb
山麓冰坡 ice piedmont
山麓剥蚀面 piedmont denudation surface
山麓剥蚀平原 piedmont denudation plain
山麓冲积层 piedmont alluvial deposit
山麓冲积平原 bahada[bajada] <北美西南部的>;mountain apron;piedmont alluvial plain; piedmont glacier
山麓冲积平原冰川【地】piedmont alluvial plain glacier
山麓冲积平原角砾岩 bajada breccia
山麓冲积扇 alluvial apron; bahada; bajada;compound alluvial fan
山麓冲积物 piedmont alluvial deposit
山麓冲积洲 alluvial bench
山麓冲刷扇 alluvial apron;bajada
山麓粗砾 piedmont gravel
山麓带 foothill belt
山麓的 piedmont;submontane;submountain
山麓地带 piedmont belt
山麓断层面 terminal face
山麓断层崖 piedmont scarp
山麓堆积 cliff debris; piedmont deposit
山麓堆积泉 talus spring
山麓堆积碎屑物质 talus clastics
山麓堆积体蠕动 talus creep
山麓堆积细粒物质 talus fine
山麓高草莽丛 terai
山麓高原 piedmont plateau
山麓公路 foothill freeway
山麓湖 border lake;piedmont lake
山麓环境 piedmont environment
山麓缓斜平原 pediment
山麓汇合冰川 intermontane glacier
山麓基岩侵蚀面 rock pediment;rock plane
山麓角 piedmont angle
山麓角砾岩 talus breccia
山麓阶地 piedmont benchland;piedmont flat
山麓接触泉 hillside contact spring
山麓聚落 foothill settlement

山麓砾积平原 Sai
山麓砾石 piedmont gravel
山麓平台 piedmont flat
山麓平原 piedmont plain
山麓坡地 piedmont slope
山麓侵蚀面平原 pediplane
山麓侵蚀面作用 pediplanation
山麓侵蚀平原 mountain pediment; pediplain
山麓侵蚀平原的侵蚀面与相临较高地面之间的角 knick
山麓侵蚀坡 rock floor
山麓丘陵 foot hill
山麓区 foothill region
山麓取水口 mountain intake
山麓泉 vauclusian spring
山麓深井 gemma[复 gemmae]
山麓石堆 run-of-hill stone
山麓水文学 hillslope hydrology
山麓碎石 hillside waste
山麓碎石堆 scree;talus
山麓碎屑冲积平原 waste plain
山麓碎岩 < 悬崖下的崩坏岩石堆 > talus
山麓台地 piedmont plateau
山麓梯级 piedmont bench; piedmont flat; piedmont stairway; piedmont steps; piedmont treppe
山麓相 piedmont facies
山麓(小)坝 hillside dam
山麓小(山)丘 foot hill
山麓夷平作用 pedimentation; pediplanation
山麓园林 hill-foot garden
山峦 chain of mountains; multipeaked mountain
山萝卜属 scabious
山脉 chain of mountains; mountain belt; mountain chain; mountain range; mountain stream; range of mountains; ridge; cordillera < 并行的 > ;sierra < 美 >
山脉残脊 nubbin
山脉成因类型 origin structure of mountains
山脉的丘陵地带 foot hill
山脉发育阶段 developmental stage of mountain chain
山脉平顶 cooktop
山脉无烟炸药 < 用硝棉、甘油、石油脂制成 > cordite
山脉线 skeleton line
山脉线略图 map of skeleton lines
山脉形成 mountain-building
山脉走向 trend of mountain chain
山毛榉 beech;sallow
山毛榉板条楼面 beech strip floor covering;beech strip floor(ing)
山毛榉坚果 beechnut
山毛榉坚果油 beechnut oil
山毛榉木材 beechwood
山毛榉木槌 beechwood mallet
山毛榉木焦油 beechwood tar
山毛榉木踏步板 beechwood tread
山毛榉木条地板 beechwood strip floor cover(ing)
山毛榉墙面板 beech shingle
山毛榉镶木地板 beech parquet
山毛榉枕木 beech sleeper
山毛榉属 beech;Fagus < 拉 >
山毛柳 sallow
山帽云【气】 cap cloud; cloud cap; crest cloud; standing cloud; stationary cloud
山梅花 mock-orange; pniladelpnus sp
山梅酸 ximenic acid
山门 temple gate
山门上大塔 < 金字塔形 > gopura(m)
山名 oronym
山农 < 信息论中一组互斥事件的判定量以 2 为底的对数量度单位,如 $\log_2 8=3$ 香农 > Shannon
山农采样定理 < 每个信息周期至少需要采两次样 > Shannon's sampling theorem
山农第一定理 Shannon's first theorem
山农定理 Shannon's theorem
山农方程 Shannon's equation
山农投影机 Shannon's projector
山农韦弗指数 Shannon-Weaver's diversity index
山坡 backfall; brae; hillside; ide; mountain side; mountain slope; sidehill; side(hill) slope; slope of mountain; versant; sidelong ground
山坡半挖半填 hillside cut and fill
山坡保护工程 hillside covering works
山坡泵站 hill pump station
山坡标志 hill sign
山坡不对称定律 law of unequal slope
山坡采料场 hillside borrow
山坡采石场 hillside quarry
山坡草地 patana
山坡场所 hillside location
山坡城镇 hillside town
山坡冲刷 hillslope erosion
山坡挡水坝 hillside dam
山坡(道)路 hillside road
山坡地表水 hillside surface water
山坡地拖拉机 hillside tractor
山坡地中耕机 upland cultivator
山坡定位 hillside location
山坡陡谷 kloof
山坡断面 hillside section; sidehill section
山坡(防护)工程 hillside work
山坡覆盖层 < 防止雨水冲刷,土壤移动 > hillside covering works
山坡沟蚀后退 gull(e)y gravure
山坡护坡旁的截水沟 bench flume
山坡滑坡 hillside creep
山坡建筑 hillside architecture
山坡截水沟 hillside cut-off trench
山坡径流 hillslope runoff
山坡开采区 hillside workings
山坡开阶 side-benching
山坡开挖 sidehill cut; sidehill excavation
山坡勘测 hillside location
山坡砾石 hill gravel
山坡露天采场 mountain surface mine
山坡露天采矿 opencast of mountain slope
山坡面 hillslope; uphill side
山坡排水沟 hillside ditch; slope drain
山坡坡道 hillside ramp
山坡潜移 hillside creep
山坡侵蚀 hillslope erosion
山坡区段 hillside section
山坡取土坑 hillside borrow
山坡蠕动 hill-creep
山坡蠕滑 hillside creep
山坡砂矿 hillside diggings
山坡渗流 hillside seepage; sidehill seepage
山坡隧道 hill side tunnel
山坡台阶 hillside ramp
山坡坍方 hillside creep
山坡填土 hillside fill; sidehill fill
山坡筒仓 hillside silo
山坡土方 hillside work
山坡土爬 hill-creep
山坡土石方 sidehill work
山坡物质缓慢下滑 hill-creep; hillside creep
山坡雾 upslope fog
山坡现场部位 hillside location
山坡线 mountain-slope line; sidehill line
山坡斜度 hillside slope
山坡斜坡 hillside ramp
山坡斜坡道 hillside bin
山坡型泥石流 slope debris flow
山坡徐变 hill creep
山坡腰线 hillside line
山槭 mountain maple
山漆树 Rhus delavayi
山前 mountain front
山前坳陷 foremountain depression
山前坳陷型沉积建造 foredeep type formation
山前拗陷 piedmont depression
山前冲洪积平原地下水 groundwater in alluvial-pluvial piedmont plain
山前冲积平原 bajada
山前带 footage
山前地带 alpine piedmont
山前地带的 piedmont
山前高原 piedmont plateau
山前含煤建造 foremountain coal-bearing formation
山前平原 conoplain; pediment; piedmont interstream flat
山前坡地 piedmont slope
山前侵蚀平原 conoplain; pediment; piedmont interstream flat; piedmont plain
山前区河段 waterway in front of mountain
山前梯地 piedmont steps
山前洼地 piedmont depression
山墙 breast wall; flank wall; gable; gavel; head wall; overdoor; pediment
山墙百叶窗 gable louver
山墙边瓦 marginal tile for gables
山墙搏风板 verge fillet
山墙呈尖斜坡的 jerkin head
山墙窗 gable window
山墙(窗)花格 gable tracery
山墙挡风板 barge board
山墙顶部侧砌的砖 barge course; verge course
山墙顶部坡度的几何曲线 curvilinear gable
山墙顶部压顶石 springer
山墙顶端 fronto(o)n
山墙(顶)封檐板 gable board
山墙顶盖瓦 gable tile
山墙顶石 apex stone; crown stone; saddle stone; termed saddle stone
山墙顶饰 acroterion; acroterium; gablet
山墙端 gable end
山墙端盖顶石 fractable
山墙端框架 gable frame
山墙端挑石 skew block; skew putt
山墙泛水 gable flashing; skew flashing
山墙封檐板 parge board
山墙封檐石 parge stone
山墙封檐瓦 parge tile
山墙辅助瓦 end band
山墙盖板 skew table
山墙拱底石 gable springer
山墙挂瓦条 verge fillet
山墙河段 summit level reach
山墙绘画 gable painting
山墙基础 pediment foot
山墙基石 gable springer
山墙基石线脚 gable shoulder
山墙尖 < 呈斜坡的 > jerkin head; shread head
山墙尖呈斜坡的两坡式屋顶 jerkin head roof
山墙尖端 gable peak
山墙尖屋顶 < 呈斜坡的 > jerkin head
山墙角石 kneeler; kneestone; skew
山墙进气通风 gable inlet ventilation
山墙口桁条 verge purlin(e)
山墙连檐木 verge fillet
山墙面 gable side
山墙排气孔 exhaust hole in gable
山墙起拱支座 gable springer
山墙墙板 gable slab
山墙墙身 gable wall
山墙三角形檐饰 miter[mitre] arch
山墙上踏步式压顶 corbel steps
山墙上缘砖砌成排齿饰型 muistanden
山墙石板 gable slate
山墙式构架 gable dormer; gable frame
山墙饰 < 壁龛成小墙洞顶上的 > gablet
山墙饰内三角面 tympan; tympanum [复 tympana/tympanums]
山墙饰物 acroter(ion)[复 acroteria], akroter(ion)
山墙饰物衬托 acroterium
山墙饰物底座 acroterium
山墙饰线 gable mo(u)lding
山墙塔楼 gabled tower
山墙挑檐 outlooker
山墙挑檐椽 barge couple
山墙挑檐的托石 gable springer
山墙头顶窗 gable dormer
山墙头桁条 gable pole
山墙头十字架 gable cross
山墙凸石 bare stone; barge stone
山墙突瓦 verge
山墙涂漆 gable painting
山墙托臂 gable springer
山墙托肩 gable shoulder
山墙下端支承压顶的石块 footstone
山墙线脚 gable mo(u)lding; rake(d) mo(u)lding
山墙(小)柱 < 山头顶点的短柱 > gable post
山墙斜坡底部斜高石 footstone
山墙形窗 gable-shaped window
山墙形过梁 gable-shaped lintel
山墙形老虎窗 gable dormer; gable type dormer(window)
山墙形门窗楣 gable-shaped lintel
山墙压顶 gabble cope; gable coping
山墙压顶线脚 gable shoulder
山墙压顶砖层 barge course
山墙檐边(突瓦)【建】 verge
山墙檐边突瓦下的线脚 verge mo(u)ld(ing)
山墙檐口 verge
山墙檐口椽 barge couple; barge rafter
山墙檐口嵌条 verge fillet
山墙檐口线脚 verge mo(u)ld(ing)
山墙檐瓦 barge course; verge course; verge tilting
山墙阳台 gable wall balcony
山墙遮雨板 skew flashing
山墙钟楼 gable wall belfry
山墙砖 gable brick
山墙砖压顶 barge course; brick coping on gable; verge course
山墙装饰物 acroter
山墙装饰线 raked mo(u)lding
山墙装饰线脚 gable mo(u)lding
山墙装饰线条 rake mo(u)lding
山丘 massif; rock massif
山丘的影线 hill hachures
山丘特征图 representation of hill features
山区 hilly country; mountain belt; mountain land; mountainous area; mountainous district; mountainous region; mountainous terrain
山区暴雨 mountain storm
山区标准时 mountain standard time
山区城市 hilly city
山区道路 mountain road
山区道路定线 mountain road location
山区的 montane
山区地下水动态 groundwater regime in mountain area

山区地形 mountainous topography
山区定线 mountain location
山区段 mountain tract
山区钢缆铁道 mountain lift
山区高草丛 kar herbage
山区高草地 kar herbage
山区公路 mountain highway
山区航道 mountain(ous) channel; mountain(ous) waterway
山区河床改造 improvement of river-beds in mountain areas
山区河道 mountain channel
山区河道整治 mountainous river regulation
山区河谷地区选线 location of line in mountain and valley region; mountain and valley region location
山区河谷定线 location of line in mountain and valley region; mountain and valley region location
山区河流 hill stream; hill torrent; mountainous river; mountainous stream; piedmont stream
山区洪水 mountain torrent
山区急流 hill torrent; mountain torrent
山区景观 mountain landscape
山区可能最大降水 orographic(al) probable maximum precipitation
山区缆索铁道 mountain cable railway; mountain lift
山区缆索铁路 mountain cable railway
山区溜索桥 sliding rope bridge in mountainous district
山区流域 mountain watershed
山区路线 mountain route
山区绵亘 range
山区气候 mountain climate
山区桥梁 mountain bridge
山区日光时间 mountain daylight time
山区试验 mountain test
山区试验路线 mountain test course
山区水库 mountain reservoir
山区水文地理学 orohydrography
山区水文学 orohydrology
山区铁路 mountain railway
山区铁路定线 mountain railway location
山区峡谷 mountain gulch
山区线路 mountain route
山区行车 mountain driving
山区窑 country kiln
山区用绳索卷扬机 mountain rope hoist
山区治理 harness mountain
山泉 mountain spring
山群 mountain group
山柔皮 mountain leather
山蠕动 mountain creep
山乳 rock milk
山软木 mountain cork
山砂 mountain sand
山上给水池 mountain water tank
山上阶地 altiplanation terrace
山上牧场 <瑞士的> alp
山上水池 tarn
山上小湖 mountain tarn; tarn
山势特性 orographic(al) character
山水城市 landscape city
山水地形模型 orohydrographic(al) model
山水地形样图 orographic(al) model
山水画 landscape painting; mountains and waters painting
山水立体地图 orohydrographic(al) model
山水立体模型 orohydrographic(al) model
山水盆景 potted landscape
山松 mountain pine
山桃 myrtle
山体 massif; mass of natural mountain; mountain mass(if); natural ground
山体厚度 mountain thickness
山体稳定问题 problem of stability of mountain bodies
山体压力 mountain-pressure; pressure of ground
山体应力 orogen stress
山田 hillside plot
山头【建】 frontal; fronto(o)n; pediment; pike
山头顶窗 gable dormer
山头封檐板 verge board
山头(铅皮)泛水 skew flashing
山头云 helmcloud
山尾陡崖 spur-end cliff
山文学 orography
山狶榁 redwood of the mountains
山溪 beck; hill stream; hill torrent; mountain creek; mountainous stream; mountain river; mountain torrent; nai; ravine stream; torrent; torrential river; torrential stream
山溪河道整治 mountainous regulation
山溪进水结构 water intake in main-tain river
山溪取水口 stream intake
山溪性河道整治 mountainous river regulation
山溪性河流 flashy river; flashy stream
山溪整治 correction of mountain stream; mountain stream regulation; mountain stream training; mountain torrent improvement; regulation of torrent
山溪整治设施 torrent works
山溪治理 correction of mountain torrent
山系 chain of mountains; cordillera; mountain chain; mountain range; mountain stream; mountain system
山系略图 representation of hill features
山峡 col; defile; donga; gap; gorge; narrows; notch <美>
山狭河槽 neck channel
山形 chev(e)ron type; hill features; hill shape
山形百叶窗 gambrel vent
山形表示法 representation of hill features
山形部 backfall
山形穿法 point draft; reversed draft
山形穿综 return point
山形的【地】 orographic(al)
山形断层 orographic(al) fault
山形多层焊 cascade welding
山形法 orographic(al) method
山形法绘图 trachographic(al) relief drawing
山形符号 chev(e)ron
山形盖顶纸盒 gable top carton
山形拱 pediment arch; gable arch
山形隆起 orographic(al) lifting; orographic(al) uplift
山形墙侧边梯形突出物 corbie step
山形塔 gable(d) tower
山形图 orogram
山形瓦 gable tile; mountain tile
山形屋顶 gable roof; ridge roof type
山形线 sketching contour
山形斜纹 feather twill; herringbone twill; reverse twill
山形仪 orograph
山型 mountain type
山型缺口 chev(e)ron notch
山垭口 mountain pass saddle back
山岩 jagged rocks
山岩湾 rocky bay
山岩压力 rock pressure
山羊粪石 goatstone
山羊(座)【天】 goat
山阳 adret(to)
山杨木 aspen wood
山腰 brae; breast; halfway up the mountain; hillside; mountain slide; mountain slope; slope of mountain
山腰料仓 hillside bin
山腰露岩 sca(u)r
山腰泉 hillside spring
山腰砂矿 hillside placer
山腰砂砾 hillside gravel
山腰上半挖半填 cut-and-fill; hillside cut and fill; mountain side
山腰小树林 shaw
山腰岩屑 hillside waste
山腰沼泽 hillside swamp
山阴 opaco; ubac【气】
山阴区 backland
山樱桃 manchu cherry
山雨 mountain rain
山原 plateau mountain
山岳暴雨 orographic(al) storm
山岳冰川 mountain glacier; pediment glacier
山岳病 soroche
山岳成因学 orology
山岳地带 mountain belt; mountainous country
山岳地貌调查 survey of mountain morphology
山岳地区 mountain(ous) area; mountain(ous) region; orographic(al) region
山岳地形 mountain topography
山岳分界线 orographic(al) divide
山岳高度计 orometer
山岳构造线 orotectonic line
山岳海岸 mountain coast
山岳降水 orographic precipitation
山岳景观 alpine landscape; mountain landscape
山岳名称 oronym
山岳模式 orographic(al) model
山岳气候 highland climate; mountain climate
山岳气压计 orometer
山岳水文学 orohydrography
山岳抬升 orographic(al) lifting; orographic(al) uplift
山岳协会 Sierra Club
山岳形成作用 mountain folding
山岳形态的 orographic(al)
山岳形态条件 orographic(al) condition
山岳形态学 orography
山岳形态因素 orographic(al) factor
山岳影响 orographic(al) influence
山岳雨 orographic(al) precipitation; orographic(al) rain(fall)
山岳植物 orophyte
山岳志 orography
山云 orographic(al) cloud
山楂 hawthorn; maybush; maythorn; maytree
山楂木 crabwood
山楂树林 woodland hawthorn
山脂土 oropion
山志的 orographic(al)
山志学 orography
山中避暑地 <印度等地> hill station
山中峡道 defile
山中小湖 tarn
山中小屋 chalet; louse cage <伐木工的>
山茱萸 dog wood
山茱萸木 <北美洲产> tupeloe
山柱 center column; gable column
山庄 mountain house; mountain villa
山状 montiform
山字钩 ramshorn hook
山字块体 tribar
山字形 <站台雨棚> E-shape
山字形构造【地】 epsilon-type structure
山字型构造成分 compose of epsilon-type tectonic system
山字型构造体系【地】 epsilon-type structural system; epsilon-type tectonic system
山嘴 hoo; mountain spur; nab; nose; pike peak; spur
山嘴阶地 spur terrace
山嘴梯田 spur terrace
山嘴崖 spur-end cliff

删除文件 deleted file

删改 blue-pencil; editing; pruning
删改记录 deletion record
删改者 expurgator
删节 abridge; abridg(e)ment; retrench
删节版 expurgated edition
删去的条款 murder clause
删去器 deleter
删去前零 leading zero suppress

杉板 deal; deal board

杉篙 fir pole
杉硅钠锰石 saneroite
杉科 Taxodiaceae
杉木 cedarwood; chinafir; Chinese fir
杉木材 yellow cedar
杉木镶板门 cedar panelled door
杉形根(叶片) fir-tree root
杉属 chinafir

钐147 含量 content of 147Sm

钐矿 samarium ore
钐钕等时线 samarium-neodymium isochron
钐钕法 samarium-neodymium method
钐钕模式 samarium-neodymium model
钐钕年龄 samarium-neodymium age

珊瑚暗礁 cliff in coral; coral reef; coral shoal

珊瑚杯 anthocyathus
珊瑚壁 theca
珊瑚沉积 coralgal; coral sediment
珊瑚沉渣 coralgal; coral sediment
珊瑚虫 coral; coral insect; coral polyp; polyp
珊瑚瓷 coralline ware
珊瑚单体 corallite
珊瑚岛 coral island; reef island
珊瑚岛常绿林 evergreen forest on coral island
珊瑚的 coralline
珊瑚底 coral bottom
珊瑚动物地理区 coral faunal province
珊瑚粉红色 coral blush; coral pink
珊瑚港 coral harbo(u)r
珊瑚个体 solitary coral
珊瑚骨料 coral aggregate
珊瑚海 coral sea
珊瑚海岸 coral coast

珊瑚和藻类沉积 coralgal
珊瑚红(色)coral red
珊瑚环礁 coral atoll
珊瑚灰岩 coral limestone; coralline crag
珊瑚混凝土骨料 coral concrete aggregate
珊瑚混凝土集料 coral concrete aggregate
珊瑚集料 coral aggregate
珊瑚岬 coral head; coral knoll
珊瑚尖峰 coral pinnacle
珊瑚礁 bank reef; bioherm; cay; coral reef; kay; klint <包围岩石的>
珊瑚礁岸 coast coral reef
珊瑚礁滨线 coral reef shoreline
珊瑚礁岛 coral-reef island
珊瑚礁地貌 geomorphology of coral reef
珊瑚礁分带 zonation of coral reef
珊瑚礁海岸 coral coast; coral reef coast
珊瑚礁海岸线 coral reef coastline; coral reef shoreline
珊瑚礁海区航法 coral reef sea navigation
珊瑚礁环礁湖 coastal lagoon; coral-reef lagoon
珊瑚礁类型 type of coral reef
珊瑚礁生活环境 living environment of coral reef
珊瑚礁台地 coral reef platform
珊瑚礁土 coral reef soil
珊瑚礁泻湖 coral reef lagoon
珊瑚礁崖 cliff in coral
珊瑚礁油藏 coral reef oil pool
珊瑚角 coral head
珊瑚茎 anthocaulus
珊瑚菌 club fungi
珊瑚壳 theca
珊瑚玫瑰色 coral rose
珊瑚泥 coral mud
珊瑚丘 coral knoll; coral head
珊瑚区 coral zone
珊瑚群 madrepore
珊瑚色 coral red
珊瑚色大理石 corallite
珊瑚色的 coralline
珊瑚色陶器 coralline ware
珊瑚砂 <又称珊瑚沙> coral sand
珊瑚砂滩 coral sand beach
珊瑚石 corallite
珊瑚石灰岩 coral limestone; coral rag
珊瑚塔 coral pinnacle
珊瑚滩 coral foreshore
珊瑚体 anthoblast; corallum
珊瑚头 coral head
珊瑚相【地】coralline facies
珊瑚型耙头 coral type draghead
珊瑚崖堆 coral talus
珊瑚岩 coral rock
珊瑚藻 coralline algae
珊瑚藻的 coralline
珊瑚藻(石)灰岩 coralgal
珊瑚质材料 coralline material
珊瑚洲 coral shoal
珊瑚状 coralloid
珊瑚状环 coral ring
珊瑚状泉华 coral-sinter
珊瑚状细石墨 coral graphite

栅

栅 blocked grid limiter; grill(e); palisade

栅板 baffle; grating sheet; grid plate; paling; sieve plate; grating; pale
栅板回路调谐的 tuned grid
栅板节流塞 grid plug
栅板空腔 grid plate cavity
栅板跨导 grid plate transconductance
栅板耦合 grid plate coupling
栅板塔盘 grid tray
栅板特性 grid plate characteristic
栅板箱 grid hopper
栅板振荡器 grid plate oscillator
栅瓣 grating lobe
栅厂 hangar
栅车 covered wagon
栅车门固定设备 box car door fixture
栅磁方位角改正 grid magnetic azimuth adjustment
栅挡机 grid
栅导纳 grid admittance
栅底安装 <指水厂、污水厂> fitting of screen bottom
栅地自激振荡器 self-excited grounded grid oscillator
栅电极 gate electrode
栅电流 gate current
栅电容 gate capacitance
栅电压 gate voltage
栅电阻 gate resistance
栅顶 grid ceiling
栅缝扫描 slit scan
栅腐蚀 gate etching
栅负电压 bias
栅负压供给 grid bias supply
栅覆盖层 gate overlap
栅格 grid; lattice; panel; raster
栅格比值 grid ratio
栅格测试图 grid pattern
栅格反应堆 lattice reactor
栅格分光计 grid spectrometer
栅格光栏 bucky diaphragm
栅格化 rasterization; tessellate
栅格绘图 raster plotting
栅格间距 grid distance; grid spacing; lattice spacing
栅格交点 grid intersection
栅格节流式调压室 grate type throttling surge tank
栅格结构 lattice structure
栅格结合 lattice binding
栅格进口 grate opening
栅格孔口 grate opening
栅格冷却器 grate cooler
栅格人孔 rack well
栅格十字线 grid intersection
栅格式凹板 grid-type concave; lattice-type concave
栅格式舱盖 grid hatch cover
栅格式红外线加热器 grid-type radiant burner
栅格式节流阀 grate type throttling
栅格式节流设施 grate type throttling
栅格式进水口 grating inlet
栅格式幕墙 grid-type curtain wall
栅格式气体洗涤器 hurdle scrubber; hurdle-type washer
栅格式瓦管排水系统 gridiron tile-drainage system
栅格竖井 rack well
栅格数据 raster data
栅格填料 grid packing
栅格信号发生器 grill(e) generator
栅格形指示器 grid-type indicator
栅格型绕组 lattice winding
栅格照明 lattice lighting
栅格状瓦管排水系统 gridiron tile-drainage system
栅格组件 grill(e) sub-assembly
栅箍 bali
栅管 bank tube
栅环 grating ring
栅基 grid base
栅极 baffle; mesh electrode; grid electrode
栅极闭锁 grid extinguishing; grid locking
栅极边条 grid side-rods
栅极变压器 grid transformer
栅极不启动电压 gate nontrigger voltage
栅极材料 grid material
栅极槽路 grid tank
栅极侧剩余电压 grid residual voltage
栅极出线端 grid terminal
栅极触发电流 gate trigger current
栅极导纳 grid admittance
栅极到阳极间 grid-to-anode capacity
栅极到阴极电压 grid-to-cathode voltage
栅极电池 grid battery
栅极电池组 C-battery
栅极电导 grid conductance
栅极电流 grid current; grid electricity
栅极电流截止 grid-current cut-off
栅极电流截止电压 grid-current cut-off voltage
栅极电路 grid circuit
栅极电路时间常数 grid time constant
栅极电路削波 grid-circuit clipping
栅极电路阻抗 grid-circuit impedance
栅极电容 grid capacitance
栅极电位 grid potential
栅极电压 grid voltage
栅极电压截止 grid-voltage cut-off
栅极电压相位控制 phase control of grid voltage
栅极电源 grid-power supply
栅极电阻 grid resistance; grid resistor
栅极短路棒 grid shortening bar
栅极对地电容 grid-to-ground capacitance
栅极发射 grid emission
栅极反馈绕组 grid-feedback winding
栅极反向电流 grid inverse current
栅极分压器 grid potentiometer
栅极感应 grid induction
栅极隔直流电容器 grid blocking capacitor
栅极汞槽整流管 grid pool tube
栅极耗散 grid dissipation
栅极辉光放电管 grid-glow tube
栅极辉光放电继电器 grid-glow relay
栅极激励 grid drive; grid excitation
栅极激励电容器 grid excitation capacitor
栅极激励功率 grid driving power
栅极激励绕组 grid-excitation winding
栅极激励特性 grid-drive characteristic
栅极寄生振荡抑制器 grid stopper
栅极尖峰电压 grid peak-voltage
栅极减速 grid retardation
栅极检波 cumulative grid detection; grid detection; grid(-leak) rectification
栅极检波电子管伏特计 grid detection voltmeter
栅极检波器 grid detector; grid rectification detector
栅极检波特性 grid detection characteristic
栅极检波系数 grid detection coefficient
栅极键控法 grid keying
栅极交流声 grid hum
栅极接地电路 ground grid circuit
栅极接地放大器 grounded grid amplifier
栅极接地三极管放大级 grounded-grid triode stage
栅极接线 grid connection
栅极截止电压 grid cut-off voltage
栅极抗流圈 grid choke
栅极可变电感器 grid variometer
栅极空腔调谐器 grid-cavity tuner
栅极控制 grid control
栅极控制场 grid field
栅极控制的 grid-controlled
栅极控制电路 grid control circuit
栅极控制电压 grid control voltage; grid drive
栅极控制角 grid control angle
栅极控制特性 grid control characteristics
栅极控制整流管 grid-controlled rectifier; grid control rectifier
栅极扩伸机 grid stretcher
栅极联箱 grid plenum
栅极邻近效应 grid proximity effect
栅极脉冲 grid impulse; grid pulsing
栅极脉冲电压 grid impulse voltage
栅极脉冲调制 grid pulse modulation
栅极耦合 grid coupling
栅极旁路 grid bypass
栅极偏压电池组 grid-bias battery
栅极屏蔽 grid cover; grid shielding
栅极屏蔽法 grid shadowing method
栅极屏蔽罩 grid shielding can
栅极破坏 grid failure
栅极钳位 grid clamping
栅极绕组 grid winding
栅极式电平检波器 grid-type level detector
栅极式静电喷涂装置 grid electrostatic paint spray apparatus
栅极输入电容 grid input capacitance
栅极输入阻抗 grid input impedance
栅极损耗 grid loss
栅极损坏 grid failure
栅极特性 grid characteristic
栅极调谐 grid-tuned
栅极调谐式振荡器 grid resonance type oscillator
栅极调整 grid alignment
栅极调制 grid control; grid modulation
栅极调制电路 grid modulation circuit
栅极调制放大器 grid-modulated amplifier
栅极网孔 grid mesh
栅极线圈 grid coil
栅极限幅 grid limiting
栅极限幅器 grid limiter
栅极谐振频率 grid resonance frequency
栅极信号 grid signal
栅极信号波 grid wave
栅极选通 grid gating
栅极选通脉冲 grid gate pulse
栅极阳极着火电位 grid-anode ignition potential
栅极抑制器 <消除寄生振荡用的> grid suppressor
栅极阴极电压 grid-cathode voltage
栅极阴极着火电位 grid-cathode ignition potential
栅极引出帽 grid cap
栅极引出头 grid cap
栅极引线 grid connection; grid lead (wire); grid return
栅极引线屏蔽 grid-lead shield
栅极圆盘 grid disc
栅极再生检波方式 grid regenerative detection system
栅极噪声 grid noise
栅极罩 grid cage; grid cover
栅极振荡回路 grid tank
栅极直流 direct grid current
栅极中和 grid neutralization; Rice neutralization
栅极中和电路 Rice circuit
栅极中和法 grid neutrodyne method
栅极注频 grid injection
栅极阻抗 grid impedance
栅极阻塞 grid blocking
栅夹断电压 gate pinch-off voltage

栅桨式搅拌机 finger blade agitator
栅桨式搅拌器 finger blade agitator
栅截止键控 blocked-grid keying
栅截止限幅器 blocked grid limiter
栅截止信号分离器 grid cut-off separator
栅距 cascade spacer; cascade spacing;pitch;raster unit
栅孔 grating opening
栅控彩色显像管 chromatron
栅控电离压力计 grid-controlled ionization ga(u)ge
栅控法 grid control method
栅控汞弧管 plomatron
栅控管 grid control tube
栅控继电器管 grid-controlled relay tube
栅控脉冲 grid pulse
栅控特性(曲线) gate control characteristic
栅控系数 grid control ratio
栅控整流管 grid-controlled rectifier tube
栅篱 hedgerow;open support
栅篱架设机 support stringer
栅铃 grid bell
栅流 gate current;grid current
栅流分布起伏噪声 grid-interception noise
栅流感生起伏噪声 induced grid noise
栅流检波 grid-current detection
栅流降落式波长计 grid-dip wavemeter
栅流截止 grid-current cut-off
栅流脉冲 grid-current impulse
栅流容量 grid-current capacity
栅流失真 grid-current distortion
栅流特性 grid-current characteristic
栅流调制 grid-current modulation
栅流陷落式测试表 grid-dip meter
栅流引起的失真 grid-current distortion
栅笼 cage
栅笼壁条 cage bar
栅漏 grid leak
栅漏电容器 grid-leak capacitor;grid leak condenser
栅漏电阻 grid leak; grid-leak resistance
栅漏检波 grid-leak rectification
栅漏检波器 grid-leak detector
栅漏偏压 grid-leak bias
栅漏偏压分离 grid-leak bias separation
栅漏调制 grid-leak modulation
栅漏整流 grid-leak rectification
栅露明楼板 open stor(e)y
栅路电感 grid inductance
栅路感应噪声 grid induced noise
栅螺纹镶边 extruded rib trim
栅帽 grid cap;grid cover
栅帽接线柱 grid clip
栅门 bar; barred door; barred gate; barrel door;fence gate;grid mesh
栅门臂 <在不准通行时可以自动放下> gate arm
栅门齿条 barrack
栅门式张力器 gate tensioner
栅耙 rack rake
栅片间距 cascade spacer
栅偏电阻 grid bias resistance
栅偏压 C bias; gate bias; grid bias; grid bias voltage
栅偏压电池 grid-bias battery
栅偏压电源 C-power supply;grid bias supply;grid-voltage supply
栅偏压继电器 grid bias relay
栅偏压检波器 bias detector;grid-bias detector
栅偏压控制 grid-bias control
栅偏压调制 grid-bias modulation
栅屏 grid screen
栅屏耦合 grid plate coupling
栅墙 hoarding
栅塞 zare(e)ba
栅筛 grizzly
栅式布袋过滤器 cage bag filter
栅式接受站 grid reception stations
栅式进料器 bar feeder
栅式进水口 grated inlet
栅式行李架 baggage grid;luggage grid
栅丝间距 grid-wire spacing
栅(丝)距 grid pitch
栅探针 gridded probe
栅条 grate;grate bar;grid;grid rack; grizzly bar;screen bar;stake;stave
栅条坝 needle dam
栅条彩色显像管 trinitron
栅条管 grid tube
栅条盘分馏柱 grid tray column
栅条筛 bar screen
栅条式滚筒 cage cylinder
栅条式拦污栅 bar grizzly; bar type trash rack
栅条式挖掘铲斗 crib bucket
栅条塔板 grid tray
栅条填充柱 grid-packed column
栅条填料 grid packing
栅条填料塔 grid-packed tower
栅条蓄电池 grid accumulator
栅条堰 needle weir
栅网 aperture plate; grid mesh; matting;mesh;reseau
栅网参考点 grid reference point
栅网电离室 grid ionization chamber
栅网电阻元件 grid element
栅网端盖 grid end covering
栅网格絮凝池 grill(e) flocculating tank
栅网加载 grid-loaded
栅网加载波导 grid-loaded waveguide
栅网架 grid holder
栅网聚焦 grid focusing
栅网聚焦系统 grid focusing system
栅网聚焦性能 grid focusing property
栅网聚焦运行 grid-focused operation
栅网聚焦直线加速器 grid-focused linac;grid-focused linear accelerator
栅网十字 reseau crossing
栅网形离子源 grid-type ion source
栅线间距 grating space;grating spacing
栅线隙距 grating aperture
栅陷式波长计 grid-dip wavemeter
栅陷振荡器 dipmeter[dipmetre];grid-dip meter;grid-dip oscillator
栅形编组场 gridiron yard
栅形补偿摆 gridiron pendulum
栅形场信号发生器 grating generator
栅形场振荡器 cross hatch signal generator;grating generator
栅形抽气阀 grid-type extraction valve
栅形调车场线路 gridiron siding
栅形阀 grid valve
栅形干扰 railing
栅形扩散器 diffuser grid
栅形冷铁 grid chill
栅形炉箅 burning gate
栅形炉箅砖 sole-flue port brick
栅形排列 grip spread
栅形梯线 gridiron ladder
栅锈菌属 <拉> Melampsora
栅压 grid voltage
栅压摆幅 grid sweep;grid swing
栅压电源 grid-voltage supply
栅压截止 grid-voltage cut-off
栅压漂移 grid swing
栅压滞后 grid-voltage lag
栅阴电阻 grid-cathode resistance
栅阴二极管 grid-cathode diode
栅阴回路 grid-cathode circuit
栅阴极电压 cathode-grid voltage
栅阴极间电容 grid-cathode capacitance
栅阴极馈电放大器 grid-cathode-feed amplifier
栅阴间电导 grid-cathode conductance
栅阴间空隙 grid-cathode gap
栅阴空腔 grid-cathode cavity
栅阵 grating array
栅状薄壁组织 palisade parenchyma
栅状测云器 grating nephoscope
栅状带 palisade zone
栅状反射器 grating reflector; openwork reflector
栅状构造 mullion structure
栅状空气挡板 bulkhead air grill(e)
栅状调制盘 picket-fence retic(u)le
栅状图 fence diagram
栅状图程序 grid graph program(me)
栅状信号 grid wave

舢

舢板 dinge;pinnace;rowboat;sampan;saxboard;junk <中国民船>

舢板侧舷缘板 inwale;inwall
舢板船 junk
舢板搭钩 sampan hitcher
舢板底板 footing
舢板钩杆 hook pole
舢板横座板 bench
舢板锚 boat anchor
舢板旁小碰垫 fender
舢板棚 boat shed
舢板天幕 canopy;spray hood
舢板脱钩 detaching hook
舢板尾柱护环 skeg band
舢板舷侧桨门板 poppet;rowlock bolster;rowlock cover
舢板舷围绳 swifter
舢板租金 boat fare

煽

煽动 inflame

闪

闪闭法 blinking method

闪铋矿 bismuth blende;eulytite
闪避 duck;fend;run-around
闪变 flickering
闪变法 flicker method
闪变分光光度计 flicker spectrophotometer
闪变光度计 flicker photometer
闪变光度术 flicker photometry
闪变光束 flicker-beam
闪变频率 flicker frequency
闪变试验 flicker test
闪变系统 flicker system
闪变相位 flicker phase
闪变效应 flicker effect
闪变效应噪声 flicker noise
闪变噪声 excess noise;flicker noise
闪长安山岩 diorite andesite rock
闪长暗拼岩 dioritic appinite
闪长斑岩 diorite porphyry
闪长玢岩 diorite porphyrite
闪长煌斑岩 vogesite
闪长辉长岩 dioritic gabbro
闪长伟晶岩 dioritic pegmatite
闪长细晶岩 dioritic aplite;dioritite
闪长岩 black granite;whinstone
闪长岩的 dioritic
闪长岩含铁铜建造 diorite iron and copper-bearing formation
闪长岩类 diorite group
闪齿横割锯 lightning tooth cross-cut saw
闪冲砂矿法 booming
闪磁化 flash magnetization
闪道 lightning channel
闪灯电键 flashing key
闪灯继电器 blinker relay;flasher relay;flashing relay;lamp flashing relay
闪灯连呼 flash-recall
闪灯式二次呼叫 flash-recall
闪灯信号 flashing light signal
闪点 flash point;point of flammability
闪点杯 flash cup
闪点测定 determination of flashing point;flash test
闪点测定器 flashing point tester
闪点测定仪 flashing point tester;flash point apparatus; naphthameter <测定石油产品>
闪点产率曲线 flash point yield curve
闪点的测定 flashing test
闪点确定 flash point determination
闪点试验 flash test
闪点试验器 flashing point tester
闪点试验仪 flashing point tester
闪点试验仪器 flash point testing apparatus
闪点温度 flash temperature
闪点仪 <开口杯式> flash point apparatus;flash point tester
闪电 bolt; flashing lightning; lightning (flash)
闪电般发光 fulgurate
闪电保护 lightning protection;protection against lightning
闪电暴雷 lightning storm
闪电测量仪 fulgurometer
闪电叉 lightning fork
闪电电流特性记录器 fulchronograph
闪电电压 sparking potential
闪电防护器 lighting protector
闪电放电 lightning; lightning discharge
闪电管石 fulgurite
闪电害 damage by lightning
闪电火球 lightning ball
闪电记录器 lightning recorder; fulchronograph
闪电接地 lightning grounding
闪电雷雨 electric(al) storm
闪电链 lightning chain
闪电轮 lightning ring
闪电脉冲 lightning impulse
闪电熔石 fulgurite
闪电式清除场地 blitzed site clearance
闪电预示器 keraunophone
闪电战 blitz
闪电阵雨 shower of sparks
闪电状裂纹 lightning shake
闪动阀 clack valve;flap valve
闪动控制 bang-bang control; flicker control
闪动平衡机 strobodynamic balancing machine
闪动探照灯 flickering searchlight
闪动作 snap action
闪顿页岩 Shineton shale
闪躲【船】 glance
闪发 flash up;flash vapo(u)rization; instantaneous vaporization
闪发膨胀 flashing expansion
闪发气体 flash gas
闪发式蒸发器 flash type evapo(u)rator
闪发式蒸馏器 flash type distiller
闪发为蒸汽 flash-off steam
闪(发)蒸(发)室 flash chamber
闪发蒸发室集水底壳 flash sump
闪发蒸汽 flash steam
闪纺技术 flash-spinning technique

闪干 flash dry
闪光 blanker; blaze; blink; flash(ing); flicker(ing); fulguration; glance; glare; gleam; glimpse; glint; intermittent light; lightening; light flare; light(ing) flash; outburst; scintillation; shimmer; schillerization【地】
闪光暗点 teichopsia
闪光斑 <漆病> stardust
闪光(保护)挡板 flash barrier
闪光报警器 flash alarm; flasher
闪光报时 flashing light alarm
闪光曝光 flash exposure
闪光曝光计 flash meter
闪光曝光器 flashmatic
闪光标 flasher
闪光标灯 blinker; flasher unit; flashing light beacon; flashing(light) unit
闪光标度盘 luminous dial
闪光标志灯 flashing light beacon
闪光冰 glimmer ice
闪光冰长石 schillerization aduralia
闪光玻璃 actinic glass; flash(ed) glass
闪光部件 flashing light unit
闪光操纵装置 flash gun
闪光测距 flash ranging
闪光测频法 stroboscopic method; stroboscopy
闪光测频器 stroboscope
闪光(测速)仪 stroboscope
闪光层 flash layer
闪光插头及插座 flash connections
闪光插座 flashing jack
闪光长度 flash length
闪光持续时间 flash duration
闪光触点绝缘值 contact insulation value
闪光单元 flashing light unit
闪光弹 flash(light) bomb
闪光道口信号 flashlight crossing-signal
闪光灯 blinker; flash bulb; flasher; flashing lamp; flash tube; flickering lamp; photoflash; pocket flashlight; quick flashing light; signal lamp flicker; flash(ing) light
闪光灯标 code beacon; winker beacon
闪光灯插座 hot shoe
闪光灯触点 contact for flash unit
闪光灯电池 photoflash battery
闪光灯法 flashlight method
闪光灯架 flashlight bracket
闪光灯具 winker
闪光灯泡 flasher bulb; flash lamp
闪光灯塔 splash beacon
闪光灯系统 flashing light system
闪光灯照片 photoflash
闪光灯装置 flash unit
闪光灯座 flash socket
闪光点 flashing point
闪光电弧 flash arc
闪光电弧焊 flash welding
闪光电流 flashing current
闪光电压 flare voltage
闪光电源 flashing light supply; flashing power source
闪光读数器 flash apparatus
闪光度盘 glare index
闪光对焊 flash butt weld(ing); flash welding
闪光对焊机 flash butt welder
闪光对焊流水作业线 flash butt weld(ing) line
闪光对焊头 resistance flash-butt welding
闪光对接焊 resistance flash-butt welding; flash butt welding
闪光法 flicker principle; method of flashing
闪光反应 light break reaction
闪光方向信号机【铁】 flashing direction signal
闪光放电管 flashing discharging tube
闪光粉 flashlight powder
闪光粉点燃器 flash gun
闪光浮标 short-flashing buoy
闪光复原 glare recovery
闪光感觉 flicker sensation
闪光高速摄影 flash radiography
闪光构造 schiller structure
闪光观测 flash shot; flash spotting
闪光管 blinker tube; flash tube; spark tube; speed flash; speed light; stroboscope tube
闪光管抽运 flash tube pumping
闪光管点火 flash tube ignition
闪光管激发源 flash tube excitation source
闪光管频闪仪 flash tube stroboscope
闪光管型频闪观测法 flash type stroboscopy
闪光管型频闪观测仪 flash type stroboscope
闪光光度计 flicker photometer
闪光光度学 flash-photometry
闪光光解 flash photolysis
闪光光谱 flash spectrum
闪光光谱学 flash spectroscopy
闪光焊 flash butt weld; flashover welding; flash weld(ing); weld machined flash
闪光焊覆层 flash coat
闪光焊机 flash welder; flash welding machine
闪光焊接接头 flash joint
闪光焊毛刺 flash
闪光号笛浮标 light and horn buoy
闪光合金 flash alloy
闪光红宝石玻璃 flash ruby
闪光化 schillerization
闪光环 zipper
闪光机 flashing machine
闪光机构 flasher mechanism; flashing mechanism
闪光激发 light flash excitation
闪光计 glarimeter
闪光记录卡 flash recording card
闪光技术 flash technique
闪光剂 glister[glistre]; glitter
闪光继电管 glim relay tube
闪光继电器 flasher relay; flash(ing) relay; flicker relay
闪光继电器接点 flasher relay contact
闪光接触持续时间 shutter contact duration
闪光接触效率 efficiency of the shutter contact
闪光解吸光谱术 flash desorption spectroscopy
闪光警报灯 flashing warning lamp
闪光警告信号灯 warning winker
闪光警戒标 blinker
闪光警戒继电器 blinker relay
闪光卡片索引 flash-card indexing
闪光控制管 burst gate tube
闪光矿 lamprophane; lamprophanite
闪光矿石 spar
闪光拉长石 labradorite
闪光立标灯 flashing beacon
闪光留量 flashing allowance; flash loss
闪光率 luminous time ratio
闪光盲 flash blindness
闪光氖灯 <回声仪标尺上的> flashing neon tube
闪光扭力仪 flashlight torsionmeter
闪光泡 flash bulb; flash lamp
闪光碰焊机 flash welder
闪光片 diamant(e)
闪光(频)率 flashing rate
闪光屏蔽 flash barrier
闪光屏(挡) glare shield
闪光谱 flash spectrum
闪光气体 flare gas
闪光器 flash apparatus; flash gun; flash(ing) unit; flash mechanism; flash pistol; winker
闪光枪 flash gun
闪光强度 flash intensity
闪光融合 flicker fusion
闪光融熔法 flash-melting process
闪光三角测量 flare triangulation; flash triangulation
闪光三角网 flare triangulation
闪光色 iridescent
闪光烧伤 flash burn
闪光设备 flasher; flash unit
闪光射线分析 flash radiography
闪光摄影 flash shot
闪光摄影 X 光管 flash X-ray tube
闪光摄影术 flash photography
闪光摄影印刷机 flash photo printer
闪光石 <又称猫眼石> chatoyant
闪光时间 flash duration; flash(ing) time
闪光示像 flashing aspect
闪光式测波仪 neon tube wave height meter
闪光饰面 solar reflecting surface
闪光水泥 lightning cement
闪光速度 flashing rate
闪光特性曲线 flash curve
闪光体 twinkler; flashing light
闪光通信[讯] flashing light signal(l)ing
闪光同步 flash synchronization
闪光同步机构 flash synchronizer; synchroflash mechanism
闪光同步快门 flash-synchronized shutter
闪光头 flashing head
闪光透镜 flash-o-lens
闪光涂层 flash coating
闪光卫星 flashing light satellite; flashing satellite
闪光下拍摄的照片 flash
闪光显示器 flashing light indicator
闪光现象 chatoyance[chatoyancy]; schillerization
闪光相 flash phase
闪光效应 changeant; flash effect
闪光信标 oscillating beacon; splash beacon
闪光信号 blink(ing) signal; flare signal; flasher; flasher light signal; flashing light beacon; flashing signal; flickering signal; intermittent signal; light signal; nict(it)ation; flare; flashing indication; flashing light
闪光信号标灯 flashing beacon
闪光信号灯 blinker light; flash light; flash-signal lamp; flicker signal
闪光信号放大器 burst amplifier
闪光信号机【铁】 blink signal; flasher light signal; intermittent signal; nict(it)ation; flashing light signal
闪光信号示像 flashing signal aspect
闪光信号装置 beacon flasher
闪光星 pulsating star; twinkling
闪光序列 flashing sequence
闪光蓄电池 photoflash battery
闪光延续时间 burst width
闪光颜料 flitter
闪光仪 flasher; flash meter; flashometer; strobo
闪光因数 flash factor
闪光萤石饰面 spar finish
闪光余量 flashing allowance
闪光源 flashing light source
闪光源的测量 flashlight source measurement
闪光噪声 flick noise
闪光照明 flashlight illumination
闪光照片 photoflash
闪光照相机 flash camera
闪光照相术 flash photography
闪光照相 flash photography
闪光遮蔽 flash defilade
闪光振荡器 flashing oscillator
闪光织物 changeant
闪光纸 flash paper
闪光指路(方向)信号 flashing direction signal
闪光指示 flashing index
闪光指示灯 flashing light indicator; flashing trafficator【交】
闪光指示箭头 flashing indicating arrow
闪光指示器 flashing indicator; flashing light indicator
闪光指数 guide number
闪光指数盘 glare index
闪光指向标 splash beacon
闪光指向器 flashing light indicator
闪光指向信号 flashing direction signal
闪光钟响浮标 light and bell buoy
闪光周期 flash period
闪光砖 flashed tile
闪光转速表 stroboscopic tach(e)ometer
闪光转向信号灯 flasher lamp
闪光装置 flare package; flashing apparatus; flashing feature; flashing light system
闪光组 group-flashing light
闪焊 flash welding
闪弧 flashing; snapover
闪弧继电器 flasher relay
闪煌岩【地】 camptonite
闪晃 glister[glistre]; glitter; shimmer
闪辉沸霞斜岩 lugarite
闪辉黄煌岩 farrisite
闪辉响岩 apachite
闪辉斜长片麻岩 hornblende pyroxene plagioclase gneiss
闪辉正煌岩 vogesite
闪回 flash back
闪火 flash fire
闪火焙烧 flash roasting
闪火点 <油质的一项指标> flash point
闪火点试验 flash point test
闪火倾向 flash fire propensity
闪击 strike
闪击电势 striking potential
闪击电压 striking voltage
闪击距离 striking distance
闪击密度 stroke density
闪击式高压发生器 flash generator
闪击位 striking potential
闪激磁场绕组 flashing the field winding
闪急通报 lightning call
闪降 flash down
闪金属光泽涂层 metallizing
闪亮 blink
闪亮显微镜 blink microscope
闪流 flashing flow
闪流剂 flitter
闪绿斑岩 diorite porphyrite
闪绿岩 diorite; greenstone; whinstone
闪络 flashover; overflash; spark over
闪络电位 flashover potential
闪络电压 flashover voltage; shorting voltage
闪络故障 arcing fault
闪络继电器 flashover relay

S

闪络接触 flashover contact
闪络接地电流 flashover ground current
闪络接地继电器 flashover ground relay
闪络距离 flashover distance
闪络强度 flashover strength
闪络时间 arcing time
闪络试验 flashover test
闪络信号 flashover signal
闪凝 flash set(ting)
闪片岩 amphibole schist
闪频转速表 stroboscopic tach(e)ometer
闪千枚岩 amphibole-phylite
闪燃 flash burn;flash fire
闪燃点 flash(ing)point
闪燃管 flash pipe
闪热 heat flash
闪热解 flash pyrolysis
闪色 colo(u)r flash
闪色光 flashing colo(u)r light
闪色光灯 flashing colo(u)r light
闪闪发亮 glister[glistre];glitter
闪烧脱落 flash dewax
闪石 amphibole
闪石磁铁矿石 amphibole magnetite
闪石的 amphibolic
闪石化 amphibolization
闪石片麻岩 amphibolic gneiss
闪石石棉 amphibole asbestos
闪石岩 amphibolide
闪示法 flicker method
闪示和出示灯光 flash and show a light
闪示一下信号灯 flash and show a light
闪视 blink
闪视比长仪 blink comparator
闪视比较镜 blink microscope
闪视法 blinking method;flicker principle
闪视镜 blink comparator
闪视器 blinking device
闪视显微镜 blink microscope
闪视现象 blinking
闪视装置 flickering device
闪烁 blink; flaring; flashing; flashover;flickering;fulgurate;glimmer;glisten;glister[glistre]; glitter; scintilla; shimmer; sparkle; twinkling;waver;wink
闪烁曝光 flash exposure
闪烁玻璃 scintillation glass
闪烁材料 scintillant;scintillating material
闪烁测井计数管 scintilogger
闪烁灯丝法 flash filament method;flash filament technique
闪烁电键 flashing key
闪烁断层扫描图 scinti-tomogram
闪烁断层照相 scintitomography
闪烁法 scintigraphy;scintillation method
闪烁分光计 scintillation spectrometer
闪烁干扰 flicker disturbance
闪烁管 scintillation vial
闪烁光 flare light;flick(er)ing light
闪烁光标 blinking cursor
闪烁光度计 flicker photometer
闪烁光分析 scintillation analysis
闪烁光谱 scintillation spectrum
闪烁光谱仪 scintillation spectrometer
闪烁光焰 flickering flame
闪烁光源 flasher
闪烁过程 scintillation process
闪烁计测量 scintillometer
闪烁计数 scinticounting;scintillation counting
闪烁计数管 scintillation counter
闪烁计数监测仪 scintillation counter monitor
闪烁计数晶体 scintillation-counting crystal
闪烁计数器 <测定放射性强度用>【物】scintillation counter; flicker counter; photomultiplier counter; scintillating counter; scintillation detector;scintillometer
闪烁计数器晶体 scintillation counter crystal
闪烁计数器探头 scintillation counter head
闪烁计数器望远镜 scintillation counter telescope
闪烁计数望远镜 scintillation telescope
闪烁计算器 scintillation unit
闪烁技术 scintillation technique
闪烁剂 scintillator
闪烁检测器 scintillation detector
闪烁介质 scintillating medium
闪烁晶体 scintillation crystal
闪烁景 laurence
闪烁镜 scintilloscope
闪烁开关 flasher
闪烁控制 flicker control
闪烁裂变计数器 scintillation fission counter
闪烁磷光体 scintillation phosphor
闪烁率 flicker rate
闪烁脉冲 scintillation pulse
闪烁脉冲测氡仪 scintillation pulse survey radon meter
闪烁门限 flicker threshold
闪烁目标 scintillating target
闪烁频率 flicker frequency
闪烁屏 scintillation screen
闪烁谱仪 scintillation spectroscope
闪烁器 flasher;scintillator
闪烁器材料 scintillator material
闪烁曲线 scintigram
闪烁热 flash heat
闪烁溶液 scintillating solution
闪烁融合临界频率 flicker-fusion frequency
闪烁扫描 scintiscanning
闪烁扫描的 scintigraphic
闪烁扫描器 scintigraph; scintillation scanner;scintiscanner
闪烁扫描术 scintigraphy
闪烁扫描图 scintigram;scintiscan diagram
闪烁室 scintillation chamber
闪烁室法 scintillation chamber method
闪烁室体积 volume of scintillation chamber
闪烁寿命试验 cycling life test
闪烁探测器 fluorescent probe;scintillation detector;scintillation probe;scintillator
闪烁探头 scintillation detector
闪烁探针 scintillation probe
闪烁特性 blinking characteristic
闪烁体 scintillant; scintillating phosphor;scintillator
闪烁体光电倍增管探头 scintillator-photomultiplier probe
闪烁体快中子通量计 scintillator fast neutron fluxmeter
闪烁体慢中子探测器 scintillator slow neutron detector
闪烁体能量转换效率 energy conversion efficiency of a scintillator
闪烁体转换效率 scintillator conversion efficiency
闪烁统计学 scintillation statistics
闪烁物 scintillating medium
闪烁物质 scintillating material
闪烁误差 shimmer error
闪烁系数 flicker factor
闪烁现象 scintillation;blinking;flicker
闪烁限度 flicker threshold
闪烁效应 flicker effect; scintillation effect
闪烁信标 oscillating beacon
闪烁信号灯 flashing sign
闪烁型射气仪 scintillation chamber type emanation apparatus
闪烁液 scintillation solution;scintillator liquid
闪烁液体 scintillating liquid
闪烁仪 scintillator;scintilloscope
闪烁与亮度性能 flicker-brightness performance
闪烁阈 flicker threshold
闪烁噪声 flicker noise; scintillation noise
闪烁照相 scintiphoto(graph)
闪烁照相法 scintillography;scintiphotography
闪烁照相机 scinticamera;scintillation camera
闪烁照相术 scintiphotography
闪烁罩面 glitter finish
闪烁指数 scintillation index
闪烁指向标 oscillating beacon
闪烁转换器 scintillation converter
闪烁转换效率 scintillation conversion efficiency
闪铄 lambency;wink
闪铄镜 <计算粒子数的> spinthariscope
闪速焙烧 shower roasting
闪速焙烧炉 flash roaster
闪速对焊 flash weld
闪速浮选 flash flo(a)tation
闪速干燥 flash drying
闪速熔炼 flash smelting
闪速熔炼法 flash smelting process
闪炭 vitrain;vitri-fusin
闪锑铁锰矿 lamprostibian
闪铜光 bronzing
闪突起 pseudo-absorption
闪图 flash figure
闪脱 flash desorption
闪霞粒玄岩 kulaite
闪现 flash
闪像 flash figure
闪斜煌岩 spessartite
闪锌矿 black jack;blende;false galena;jack;lead marcasite;mock lead;mock ore; pseudo-galena; rosin jack; sphalerite; steel jack; zinc blende
闪星 flash star
闪锈倾向 tendency for flash rusting
闪岩 amphibolite
闪岩化 amphibolization
闪岩相 amphibole facies;amphibolite facies
闪耀 ablaze;blaze;crackle;flare;glare;glint;glitter <海面对点光源的反射>
闪耀波长 blaze wavelength
闪耀角 blaze angle
闪耀全息图 blazed hologram
闪耀衍射光栅 blazed diffraction grating
闪叶石 lamprophyllite
闪叶异霞正长岩 lamprophyllite-lujavrite
闪跃型 flash type
闪云斑状花岗岩 invernite
闪云橄榄岩 scyelite
闪云灰玄岩 buchonite
闪照射 flash irradiation
闪蒸 flash evapo(u)ration;flashing;flash vapo(u)rization; instantaneous vapo(u)rization
闪蒸薄膜浓缩器 flash film concentrator
闪蒸薄膜蒸发器 flash film evapo(u)rator
闪蒸部分 flash section
闪蒸残渣 flash bottoms
闪蒸槽 flash drum
闪蒸出 flash-off
闪蒸段 flash section;flash zone
闪蒸阀 flash valve
闪蒸反应器 flashing reactor
闪蒸分离 flash separation
闪蒸干燥器 flash drier[dryer]
闪蒸工发器 flash evapo(u)rator
闪蒸管 flash pipe
闪蒸罐 flash drum;flash pot;flash tank
闪蒸锅炉 flash boiler
闪蒸过程 flash process
闪蒸计算 flash calculation
闪蒸加热器 flash heater
闪蒸冷凝器 flash condenser
闪蒸冷却器 flash cooler
闪蒸馏 flash distillation
闪蒸排出 flashout
闪蒸气(体)flash gas
闪蒸气体冷冻 flash gas refrigeration
闪蒸汽油 flash gasoline
闪蒸器 flash vapo(u)rizer;flash vessel
闪蒸曲线 flash vapo(u)rization curve;flash(yield)curve
闪蒸入口 flash vapo(u)rization inlet
闪蒸设备 flashing apparatus
闪蒸石脑油 flash naphtha
闪蒸时间 flash-off time
闪蒸式中间冷却器 flash intercooler
闪蒸室 flash box; flashing chamber; flash trap;flash vessel
闪蒸水蒸气发生器 flash-steam generator
闪蒸塔 flash distillation column;flashing column;flash tower
闪蒸温差 flash range
闪蒸温度 flash(ing) temperature;flash vapo(u)rization point
闪蒸系统 flash system
闪蒸箱 flash tank
闪蒸压力 flashing pressure
闪蒸液器 flashing liquid vessel
闪蒸油 flash oil
闪蒸蒸发器 flash evapo(u)rator
闪蒸蒸汽 flash-off steam
闪蒸至常压 flashing to atmosphere
闪蒸装置 flash distillation plant;flashing plant

陕砷矿 schneiderhoehnite

陕叶蛇纹石 ferroantigorite

苫布 drop cloth

苫盖篷布(在货车上)sheeting

扇板测程仪 common log;hand log

扇板测速仪 common log;hand log
扇贝 scallop
扇贝形点线 scalloped rows of dots
扇贝形图案 scollop
扇贝形柱头 scalloped capital
扇贝养殖 scallop culture
扇步 <楼梯的> angle type step
扇出 fan out
扇出变元 fan-out argument
扇出干线 fan-out stem
扇出极限 fan-out limitation
扇出节点 fan-out node
扇出率 fan ratio
扇出逻辑函数 fan-out logic function

扇出特点 fan-out feature
扇出调制解调器 fan-out modem
扇出网络 fan-out network
扇出转移 fan-out branch
扇顶区 fan head
扇动对流加热器 fan convector heater
扇段 sector
扇风 fanning
扇风机 fan blower; fan engine; fanner; rotary fan; ventilator
扇风机布置 fan layout
扇风机操作工 fanner; fan runner
扇风机出风道 fan-outlet
扇风机出风口 fan discharge
扇风机出风扩散道 evase
扇风机出风扩散螺道 fan-volute
扇风机出风筒 fan chimney; fan evase; fan evasion stack
扇风机额定能力 fan rating
扇风机反转 fan inversion; fan reversal
扇风机房 fan chamber; fan house; fan room
扇风机风轮 fan wheel
扇风机工作轮轮壳 fan hub
扇风机节风闸门 fan shutter
扇风机静压力 fan static pressure
扇风机静压头 fan static head
扇风机扩散器 fan diffuser
扇风机通风 fan ventilation
扇风机通风道 fan drift
扇风机星形轮 fan spider
扇风机叶轮 fan impeller
扇风机引风道 fan drift
扇风机轴 fan shaft
扇风设备 fan unit
扇风试验 fan test
扇风效应 fan effect
扇风者 fanner
扇菇属 < 拉 > Panellus
扇弧形轨道 scalloped orbit
扇积砾(岩) fanglomerate
扇间洼地 interfan depression
扇间洼地沉积 interfan depression deposit
扇阶地 < 冲积的 > fan terrace
扇近端沉积 proximal fan deposit
扇锯 segment saw
扇开闸门 radius gate
扇壳饰 scalloped surface
扇控制字 fan control word
扇块动子制动器 segmented rotor brake
扇冷电机 fan-cooled machine
扇砾岩 fanglomerate
扇面 sector
扇面板 horizontal partition wall
扇面背拱 rere arch
扇面长度 sector length
扇面地址 sector address
扇面缓冲器 sector buffer
扇面开关 sector switch
扇面亮子的弦向窗棱 cot bar
扇面墙 horizontal partition wall
扇面形 fan-like pattern
扇面形嵌接 splayed scarf
扇坡泉 alluvial slope spring
扇球谐函数 sectorial spheric(al) harmonic function
扇区 sector
扇区标记信息 sector mark-information
扇区孔 sector hole
扇区软划分 soft sectoring
扇区图 sector chart
扇入 fan-in
扇入变元 fan-in argument
扇入网络 fan-in network
扇三角洲层序 fan delta sequence
扇三角洲沉积 fan delta deposit
扇三角洲相 fan delta facies
扇式对流加热器 fan convector heater
扇式加热器对流器 fan heating convector
扇束扇形光束 fan beam
扇尾 fantail
扇尾形火焰喷燃器 fantail burner
扇尾鲻 fantail mullet
扇谐函数 sectorial harmonic function; sectorial harmonics
扇心电缆 sector cable
扇形 quadrant; sector (-type); segment(of a circle); scallop
扇形凹口 scallops
扇形坝 sector dam
扇形摆动齿轮 oscillating gear
扇形摆动式电压调整器 rocking-sector regulator
扇形板 needle segment; quadrant; sector plate; segmental plate
扇形板给料器 fan-shaped plate feeder
扇形板轴 sector shaft
扇形棒 segmented-rod
扇形爆炸 < 地震勘探法的 > fan shooting
扇形臂 sector arm
扇形边界 sector boundary
扇形编组 < 交换机电缆 > fanning
扇形标志 fan maker beacon; fan mark
扇形表面 scalloped surface
扇形冰川 fan glacier
扇形冰川尾 bulb glacier
扇形冰水沉积 fan-outwash
扇形波 sectorial wave
扇形波射束 fan beam
扇形波束天线 fanned-beam antenna; harp antenna
扇形波束信标 fan beam beacon
扇形布孔 radial drilling pattern
扇形布孔钻进 fan drilling
扇形布设 < 地球物理勘探中拾振器的 > fan shooting
扇形布置 fan pattern
扇形布置的炮眼 fan holes; fan-shaped round
扇形布置的钻孔 fan holes; fan-shaped round
扇形餐馆帘褶 scalloped best pleat
扇形测量角法 sector method
扇形测试物 fan test object
扇形场 sector field
扇形场方向聚焦 sector-field direction focusing
扇形场分光镜 sector-field spectroscope
扇形场回旋加速器 sector-field cyclotron
扇形场畸变 sector imperfection
扇形场加速器 sector-field accelerator
扇形场仪器 sector instrument
扇形潮 sectorial tide
扇形齿板 sector rack; toothed quadrant; toothed segment
扇形齿轮 gear sector; gear segment; quadrant (gear); rack circle; sector gear; segmental gear; segmental rack; toothed quadrant; tooth sector; sector
扇形齿轮变幅 segment gear luffing
扇形齿轮传动 quadrant drive
扇形齿轮的齿 sector teeth
扇形齿轮架 swinging bracket
扇形齿轮轴 sector shaft
扇形齿轮轴承 sector gear bearing
扇形齿轮轴导向端 sector shaft pilot
扇形齿轮轴轴向间隙 sector shaft end play
扇形齿条 sector rack
扇形齿压路机 segmented drum roller
扇形充气区 aerating quadrant
扇形冲积 alluvial fan
扇形冲积地带 fan-shaped alluvial tract
扇形冲积阶地 fan alluvial terrace
扇形冲片 segmental punching; segmental stamping
扇形冲沙闸门 segmental sluice gate
扇形触角 antenna flabellate
扇形船尾 fan(tail) stern
扇形船闸 sector gate; sector lock
扇形船闸闸门 sector-type lock gate
扇形窗 fanlight transom(e) window; fan-shaped window; fan window
扇形窗撑杆 quadrant stay
扇形窗窗钩 fanlight catch
扇形窗开闭传动装置 fanlight opening gear
扇形窗拉杆 shadbolt
扇形窗上的扇形档 fanlight catch; fanlight quadrant
扇形磁场 sectorial magnetic field
扇形磁铁 sector electromagnet; sector magnet
扇形弹头 quadrant warhead
扇形导线 sector-shaped conductor; segmental conductor
扇形的 fan-like; fanned; fan-shaped; quadrantal; sectorial; segmental; segmentary; flabellate; flabelliform
扇形底灰盘 < 燃油机车 > segmental arc draft pan
扇形地 fan; submarine bulge; submarine delta
扇形地背斜轴 salient
扇形迭片 segmental lamination
扇形叠片磁轭转子 sector-rim type rotor
扇形顶砖 radial header; radiating bonder
扇形定向天线 fanned-beam antenna
扇形度盘 fan dial
扇形端子板 fanning end strip
扇形断面下水道系统 fan sewer system
扇形堆积(物) fan drift
扇形舵柄 quadrant
扇形发动机 fan-type engine
扇形阀(门) fan valve; segment(al) valve; sector valve
扇形反射器 fan reflector
扇形放射 fan shooting
扇形分划板 range deflection fan
扇形分离瓣 discarding petal
扇形分区充气系统 quadrant system of aeration
扇形分区充气装置 quadrant aeration unit
扇形分区搅拌系统 quadrant blending system
扇形芬克桁架 fan Fink truss
扇形风 sector wind
扇形盖 segment cover
扇形钢丝锚固区 < 预应力混凝土的 > anchorage zone of fanned-out wires
扇形港池 fan-shaped basin; fan-shaped dock
扇形格架 fan tracery
扇形格式 radial format
扇形格式穹顶 fan tracery vault
扇形格体 diaphragm cell
扇形工程用砖 radiating engineering brick
扇形功率计 vane wattmeter
扇形拱 fantail arch; ribbed arch; scheme arch; segmental arch
扇形拱顶 ribbed vault; rib vaulting; fan vault(ing)
扇形拱架 fan-shaped centering
扇形拱支肋 branch
扇形共振 sector resonance
扇形构造 fan-like structure
扇形股道 radiating track
扇形观测系统 fan layout
扇形惯矩 sector moment of inertia
扇形光 sector(ed) light
扇形光度计 sector photometer
扇形光束 fan-shaped beam; sectored beam
扇形光栅 fan-shaped grating
扇形含水层 fan-shaped aquifer
扇形航空发动机 fan-type aeroengine
扇形桁架 fan truss
扇形横向挖泥法 fan-shaped transverse dredging method
扇形弧 fan-shaped arc
扇形花格架 fan tracery; fanwork
扇形花格交叉拱顶 fan roof
扇形花格交叉屋顶 fan roof
扇形花格穹顶 fan tracery vault
扇形花格装饰 fan-shaped tracery
扇形花纹 scallop
扇形回采工作面 ring stope
扇形混凝土衬砌块 sector-type concrete ring
扇形或球形编组线线束布置 < 驼峰编组场 > group fan or balloon layout
扇形机车库 round house
扇形基岩 rock fan
扇形激光束 fan-shaped laser beam
扇形棘轮板 ratchet plate
扇形计程器 log-ship
扇形继电器 sector-type relay; vane relay
扇形加料器 sector feeder
扇形加速器 sector accelerator
扇形浇口 fan gate
扇形胶合板制材法 sectorwood
扇形角 angle of sector; angular sector
扇形校正因数 sector correction factor
扇形节制闸门 sector regulator
扇形结构 fan-folding; fan-structure; sector configuration; sector structure
扇形镜 sector mirror
扇形聚焦等时性回旋加速器 sector-focused isochronous cyclotron
扇形掘进 fan cut
扇形均化 quadrant homogenizing
扇形刻度盘 fan dial
扇形空窗撑杆 quadrant stay
扇形孔 ring hole; scallop hole
扇形孔板 segmental orifice plate
扇形孔掏槽 ring cut
扇形孔凿岩 fan drilling
扇形孔凿岩机 fan-drill; radial drilling machine
扇形孔凿岩台车 fan drill rig; fan jumbo
扇形块 cupola block; radiating block
扇形快门 sector shutter; segmental shutter
扇形拉索 fan-like stay cables; fan-type stay cable
扇形拉土法 fan-shaped earth pulling process
扇形喇叭(筒) sectorial horn
扇形雷达波束 beaver-tail
扇形雷达束 beaver-tail
扇形肋拱 ribbed arch
扇形肋穹顶 lierne vault(ing); rib vault(ing)
扇形理论 sector theory
扇形砾石滩 fan-shaped shingle bank
扇形连接器 fan connector
扇形联结通道 < 在平炉沉渣室与蓄热室之间的 > fantail
扇形量板法 sector graticule method
扇形裂缝 fan-shaped crack
扇形溜槽 sector sluice
扇形流 radiant flow
扇形流域 sectorial watershed; spreading basin

S

扇形露头砖 radiating header
扇形滤波 fan filter;velocity filter
扇形滤光器 fan filter
扇形轮 sectional wheel;sector wheel
扇形轮齿 quadrant tooth
扇形轮缘 felloe;felly
扇形轮组件 sector shaft module
扇形落纱装置 quadrant doffing motion
扇形码头 fan-shaped dock
扇形锚碇 fan-type anchorage
扇形锚固 fan anchorage
扇形门 clamshell-type door;fan-shaped door;fan-shaped gate
扇形面 sector
扇形面板 radial surface shingle
扇形面积 sector(ial) area
扇形模型 sector model
扇形摩擦片 friction(al) plate sector
扇形磨盘 segmental grinding ring
扇形纳污水域 receiving water segment
扇形排架 radial grating
扇形排列摄影机组 fan cameras
扇形排列型式 fan pattern
扇形排水管系统 fan sewer system
扇形排水系统 fan drain system
扇形盘 sector disk
扇形炮孔 fan holes
扇形炮孔爆破 ring blasting
扇形炮孔爆炸 ring blasting
扇形炮孔间距 fan burden
扇形炮孔组 fan-pattern holes;fan(-shaped) round;ring-fanned holes
扇形炮眼 blast-hole ring
扇形炮眼爆破 fan hole shooting
扇形炮眼组 fan-pattern holes;fan round
扇形喷灌机 part circle sprinkler;sector sprinkler
扇形喷灌器 sector rainer
扇形喷流 fan jet
扇形喷雾器 fan atomizer
扇形喷嘴 fan(-type) nozzle
扇形劈理 fan cleavage
扇形片 fanning strip
扇形片磁轭转子 <水轮发电机> segmental-rim rotor
扇形片铁芯 segmental core disk
扇形偏振片 polaroid sector
扇形平面 fan-shaped plan
扇形平面礼拜堂 radiant chapel
扇形平面位置指示器 sector plan position indicator
扇形铺地砖 fan-shaped floor tile
扇形铺砌(路面) fan-shaped paving;fanwise paving
扇形气窗 sunburst light;fanlight
扇形砌块 segment block
扇形嵌合体 sectorial chimera
扇形桥 radiating bridge
扇形穹顶【建】fan groining;fan vault(ing);palm vault(ing);ribbed vault
扇形穹拱 fan vault(ing)
扇形球面调和函数 sectorial spheric(al) harmonic function
扇形区 fan section
扇形区域特性曲线 sector characteristic curve
扇形热电偶 segmented thermocouple
扇形人孔 fan-shaped manhole
扇形三角洲 fan(-shaped) delta
扇形扫测声呐 sector scan(ning) sonar
扇形扫掠 sector display
扇形扫描 sector display;sectoring;sector scan(ning)
扇形扫描声呐 sector scan(ning) sonar
扇形扫描天线 rocking horse antenna
扇形扫描显示器 sector scan indicator
扇形砂轮 segmental wheel
扇形烧透砖 radial hard brick;radial well-burnt brick;radiating well-burnt brick
扇形射束 fan(ning) beam;fan-shaped beam
扇形射束天线 fanned-beam antenna;harp antenna
扇形射线 fan ray
扇形失真 sector distortion
扇形石块路面 radial sett paving
扇形石块铺砌 radial sett paving
扇形实心(砌)块 radial solid block;radiating solid block
扇形实心砖 radial solid brick
扇形式(排水)系统 fan system
扇形饰 fantail
扇形树 fan-shaped tree
扇形水系 fan-shaped drainage
扇形顺砖 radial stretcher;radiating stretcher
扇形搜索 sector scan(ning)
扇形搜寻方式 sector search pattern
扇形速度 sector velocity
扇形速度点相对于给定中心的速度 sector velocity
扇形榫头 fan-shaped tenon
扇形索 fan cable
扇形索斜拉桥 fan-type cable-stayed bridge
扇形锁 <转辙器> sector lock
扇形塌堆 fan talus
扇形踏步 balanced steps;dancing steps;diminishing steps;quadrant steps;radial steps;turn tread;wheel(ing) steps;winder
扇形踏步宽度平衡法 <螺旋形楼梯的> balanced steps
扇形台阶 radial step
扇形弹簧圈 segmented spring ring
扇形掏槽 fan-shaped round;fan cut <隧道开挖时>
扇形梯线 fan ladders
扇形体 quad(rant);segment
扇形体系 fan-shaped system
扇形天平 quadrant balance;quadrant scale
扇形天线 fan antenna
扇形铁芯 segment core
扇形铜片 copper segment
扇形头喷嘴 fanhead nozzle
扇形图 fan chart;fan diagram;pie chart;sector chart;sector diagram
扇形图案 fan-shaped pattern;scollop
扇形图解 angle diagram
扇形推进 fan advance
扇形推力轴瓦 thrust bearing segment
扇形托架 quadrant bracket
扇形瓦 <轴承的> segmental pad
扇形往复式压缩机 semi-radial reciprocating compressor
扇形尾(部) fantail
扇形蜗杆 sector worm
扇形蜗轮 worm sector
扇形蜗轮蜗杆 worm-and-sector
扇形屋顶 fantail roof
扇形无线电指向标 Consol;radio fan marker beacon;sector radio marker
扇形物 sector
扇形雾锥喷嘴 flat-fan nozzle
扇形吸入罩 fantail suction hood
扇形系统 fan system
扇形显示 sector display
扇形显示器 sector-display indicator
扇形线脚 quarter round
扇形线圈 sector-shaped coil
扇形线束布置 group fan layout
扇形相交的隔水边界 impervious boundary of fan-shaped intersection
扇形相交的透水边界 pervious boundary of fan-shaped intersection
扇形肖氏硬度计 quadrant style Shore durometer
扇形斜拉桥 fan-type cable-stayed bridge
扇形谐和函数 sectorial harmonics
扇形谐振腔 vane-type resonator
扇形泄水阀 sector sluice valve
扇形泄水闸门 segmental sluice gate
扇形芯电缆 sector cable
扇形芯线 sector conductor
扇形旋转轮 rotating sector wheel
扇形旋转式闸门 sector pivot gate
扇形烟囱砖 radial chimney brick
扇形烟羽 fanning plume
扇形岩堆 fan talus
扇形堰 sector dam;sector weir
扇形窑口护板 nose ring segment
扇形叶片 scalloped leaf
扇形仪表 sector instrument;sector pattern instruments
扇形仪表装置 sector instrument
扇形阴影 blind sector
扇形引出线 sector conductor
扇形圆锥齿轮 sector bevel gear
扇形闸门 clamshell type gate;sector gate;fan(-shaped) gate;radial gate;sector regulator;segment(al) gate;segmental sluice gate
扇形闸门堰 fan-shaped weir;sector gate weir
扇形闸堰 sector dam
扇形展开 fanning;fan out
扇形照明 sector light
扇形褶皱 fan-fold(ing);fan-structure
扇形蒸汽机车库 round house
扇形整枝 fan-shaped training
扇形支撑 radial strut in tunnel support
扇形支架 fan-shaped falsework;fan-type support
扇形指点标 fan marker;fan marker beacon;radio fan marker beacon;vertical fan marker beacon
扇形轴限制器调整 sector shaft thrust adjustment
扇形皱褶拱 scalloped arch
扇形助航灯 sector light
扇形砖 cupola brick;fan-shaped brick;radial brick;radius brick;sector brick
扇形砖工业烟囱 radial brick industrial chimney
扇形砖砌烟囱 big brick chimney
扇形砖烟囱 radial brick chimney
扇形转向装置 sectoring device
扇形转子 segmental rotor
扇形装船机 quadrant-type shiploader;slewing bridge type shiploader
扇形装船棚 radial loader
扇形锥状穹顶 conoidal vaulting
扇形棕榈 fan palm
扇形钻架 fan rig
扇形钻孔 fan drilling
扇形钻眼 fan drilling
扇形坐标 sectorial coordinate
扇形座 quadrant blocks
扇型继电器 vane-type relay
扇性静力矩 static moment of sector
扇性面积 sector area
扇岩砾 fanglomerate
扇域 vannal region
扇域的 vannal
扇远端沉积 distal fan deposit
扇闸 moulinet
扇中段沉积 mid-fan deposit
扇状变晶结构 fan-shaped blastic texture
扇状冲积层 alluvial fan
扇状冲积砂 alluvial fan
扇状单干形 fan trained cordon
扇状的 fan-shaped
扇状分流 fan-like distributory;fan-like diversion
扇状复式褶皱 fan complex fold
扇状格栅 radially cut grating
扇状拱顶 fan vault(ing)
扇状构造 fan-shaped structure
扇状阶地 fan terrace
扇状流 fan flow
扇状流域 radial basin
扇状脉的 fan veined
扇状磨损 scallop wear
扇状劈理 cleavage fan;fan cleavage
扇状坡积物 fan talus
扇状三角洲 delta(ic) fan;fan delta
扇状水系 fan-like drainage
扇状尾 fan-shaped tail
扇状物 fan
扇状下水道系统 fan sewer system
扇状岩(屑)堆 fan talus
扇状展开 fan cut
扇状展开的拱顶构造 fan vault(ing)
扇状褶曲 fan fold
扇状褶皱 fan(-shaped) fold
扇状整枝树 fan trained tree
扇足压路机 segmented drum roller

善本保有权 copyhold

善本图书馆 rare book library
善后措施 rehabilitative measures
善后转储 postmortem dump
善意第三方 bona fide third party;innocent third party
善意合约 bona fide contract
善意买方 bona fide purchaser;buyer in good faith
善意占用 bona fide possession
善意执票人 bona fide holder
善于应用色彩的画家 colo(u)rist
善于应用色彩的设计师 colo(u)rist

缮写尺寸 scribing dimension

缮写室 scriptorium

嬗变 permutation;transmutation

擅离职守 unauthorized absence

擅自复制 unauthorized duplication
擅自占地(或空屋)者 squatter

膳窗 service hatch

膳食 diet;meal
膳食标准 dietary standard
膳食学 dietetics
膳宿 boarding
膳宿费 pension
膳宿工资 board wages
膳宿公寓 board-and-lodging apartment;boarding house;pension <欧洲大陆的>
膳宿供应 accommodation
膳宿津贴 accommodation allowance
膳宿税 drinking and lodging tax;eating
膳宿学校 <欧洲大陆的> pension
膳宿杂费 schooling
膳务室 butler's pantry;butlery
膳务员 steward

赡家汇款 family maintenance remittance

赡养费 alimony
赡养契约 support deed
赡养者 supporter

鳝鱼绳 <旗索用> sennit line

伤斑 macula[复 maculae]

伤病缺勤事例 spell of sick absence
伤病员收容所 collecting post; collecting station
伤残 disability
伤残保险 disability insurance; insurance for impaired lives; invalid insurance
伤残补助 disability benefit
伤残率 disability rate
伤残频率 disabling injury frequency
伤残条款 disability clause
伤残严重程度率 disabling injury severity rate
伤残者穹隆 <1680 年～1691 年建于巴黎的文艺复兴时期穹隆建筑> Dome of the Invalides
伤腐 wound rot
伤害 damage; detract; detriment; disserve; disservice; harm
伤害保险 accident insurance; casualty insurance
伤害的性质 nature of injury
伤害工人事故 accident to workmen
伤害剂 injurant
伤害事故 injury accident(al)
伤害事故率 injure-accident rate
伤害阈值 damage threshold
伤害诊断 diagnosis of damage
伤寒 typhoid fever
伤寒病 enteric fever
伤痕 bruise; injury; scar
伤痕材 slash-cut
伤后时间 post burn
伤胶 wound gum
伤口凝血 gore
伤流压 bleeding pressure
伤蚀剂 anti-corrosion agent
伤损寿命 damage ag(e)ing
伤亡保险 bodily injury insurance
伤亡人数 toll
伤亡人员 casualty person
伤亡事故 casualty; fatal accident
伤亡事故降低率 reductive rate of injuries and deaths
伤亡事故人数 numbers of injuries and deaths
伤亡事故损失 casualty loss
伤员急救 first-aid to the injured

商办工业 business industry

商半群 quotient semigroup
商标 brand; chop; emblem; idiograph; merchandise mark; name plate; trade mark
商标保护 trademark protection
商标产品 branded goods
商标登记法 trademark registration law
商标法 merchandise marks act; trademark act; trademark law
商标方法 trademark method
商标符号 brand mark
商标管理法规 trademark regulations
商标名(称) trade brand name; brand name
商标权 ownership of trade mark; right of trade mark; trademark privileges; trademark right
商标识别 brand identification
商标(市场)占有率 brand share
商标图式证(明)书 certificate trade mark scheme
商标系统 trademark system
商标选择 brand choice
商标纸 label paper
商标注册 trademark registration
商标注册费 trademark registration fee
商标注册管理 trademark registration administration
商标转换 brand switching
商标转让 trademark leasing
商表示 quotient representation
商埠 commercial port; trading port
商测度 quotient measure
商层 quotient sheaf
商长度操作数 quotient length operand
商场 baza(a)r; covered mall center [centre]; emporium[复 emporiums/emporia]; market; mart
商场临时隔断 demising partition
商场中通中廊和外部的商店 anchor store
商车 commercial vehicle
商车队 caravan
商船 commercial marine; commercial vessel; merchantman trading vessel; merchant ship; merchant steamer; merchant vessel; trader
商船船长 master
商船船员 merchant man; merchant seaman
商船队 argosy; mercantile fleet; mercantile marine; merchant fleet; merchant marine; merchant navy
商船法 Merchant Marine Act; Merchant Shipping Act
商船高级船员 merchant marine officer
商船海员 merchant seaman
商船航运 merchant shipping
商船航运通告 merchant shipping notes
商船局 Merchantile Marine Office; Merchant Marine Department
商船旗 merchant ship flag
商船水手 merchant seaman
商船搜索救助手册 Merchant Ship Search and Rescue Manual
商船条例 Merchant Shipping Act
商船用反应堆 merchant ship reactor
商船运输 merchant shipping
商船运输法规 Merchant shipping act
商船注册的容积单位 ton
商船自动报告系统 automatic merchant vessel report system
商船自动救助系统 automated merchant vessel rescue system
商串 quotient string
商丛 quotient bundle
商簇 quotient variety
商代数 quotient algebra
商地址操作数 quotient address operand
商店 business house; commercial concern; department store; mercantile firm; sales depot; shop; outfitter <出售旅行及野营用具的>
商店橱窗 display window; shop window; show case; show window
商店橱窗装饰 window dressing
商店的辅助仓库 back room
商店房屋 store building
商店辅助仓库的储存品 back room stock
商店柜台橱窗照明 stallboard light
商店货架 stallboard
商店(兼用)住宅 combined dwelling
商店街 arcade of shops
商店廊 arcade of shops
商店楼面 shop stor(e)y
商店区 shopping district; shopping precinct
商店群 shopping complex
商店容差 shop allowance
商店所雇打听行情的人 shopper
商店所雇代客选购商品人 shopper
商店沿街正面 storefront
商店营业时间 shopping hours
商店用空气调节器 store air conditioner
商店直接贮存费用 direct store expenses
商店住宅并用 shophouse
商定 come to an agreement
商定补偿 agreed compensation
商定的期限 agreed period
商定的条款 provisions already agreed upon
商定国境税 border tax arrangement
商定价格 bargain
商定价值 agreed valuation
商定金额 agreed sum
商定律 quotient law
商定稳定价格 agreed stable price
商队 caravan; trade caravan
商队路 <沙漠地带> caravan route
商对象 quotient object
商法 business law; commercial law
商法则 quotient rule
商范畴 quotient category
商贩摆摊处 pitch
商符号 quotient symbol
商港 commercial harbo(u)r; commercial port; trading port
商港习惯 custom of port
商格 quotient lattice
商函数 quotient function
商行 business firm; commercial firm; correspondent; sales office; trading company
商行在国外的代理处 foreign agency
商号 business firm; business title; business unit; corporate name; firm; shop; store; trade name
商环 quotient ring
商会 chamber of commerce; commercial club; trade association; Board of Trade <美>
商会证明书费 chamber of commerce noting fee
商货位倾倒式翻车机 car dumper
商集(合) quotient set
商计数器 quotient counter
商检人员 cargo surveyor
商检员 cargo surveyor
商界 business career; business circle; business community; tradesfolk; tradespeople
商空间 factor space; quotient space
商扩张 quotient expansion
商李代数 quotient Lie algebra
商李群 quotient Lie group
商链复形 quotient chain complex
商量 consultation; deliberate; deliberation
商量协商 conference
商邻近 quotient proximity
商流 commercial distribution
商流形 quotient manifold
商路 trade route
商名 trade term
商模 module of quotients; quotient module
商品 article; article of trade; commercial article; commercial item; commodity; merchandise; merchantable product; shop primed; ware
商品包 commodity bundle
商品包装 commodity packaging
商品包装保标志 protective mark
商品包装和标签规定 packing and labelling regulation
商品保管 commodity custody
商品本位制 commodity standard
商品比价 parity rate of commodities; price parity of commodities; price ratios between commodities; relative price of commodities
商品比较法 product comparison
商品必备目录 catalog(ue) of necessary commodity item
商品编码 commodity code
商品标签 squib
商品标志 certification mark; commercial designation; merchandise mark
商品标准化 commodity standardization
商品材 <加工用的> merchantable timber; timber
商品材材积 merchantable volume
商品材高度 commercial height; merchantable height
商品材规格 merchantability specification
商品材料 proprietary concentrate
商品采购成本 merchandise procurement cost
商品采购垄断 oligopsony
商品采购预算 merchandise buying budget
商品菜园 market garden
商品差别化 commodity differentiation
商品差额 merchandise balance
商品差价 price differentials between commodities; price differentials of commodities
商品产出 commodity output; output of commodities
商品产出矩阵 commodity output matrix
商品产地证明书 certificate issue voucher; certificate of origin
商品产量 commodity output
商品产品 commodity product
商品产值 commodities output value; value of merchandise production
商品陈列馆 commercial museum
商品陈列架 store shelf
商品陈列室 display room; storeroom; wareroom
商品陈列所 baza(a)r; emporium[复 emporiums/emporia]
商品成本 cost of goods; merchandise cost
商品尺寸 market size
商品虫胶 commercial shellac
商品出口许可证 licence for the export of commodities
商品出口证明书 certificate issue voucher; certificate of origin
商品储藏室 wareroom
商品储蓄 commodity supply
商品存货 merchandise inventory
商品大理石 commercial marble
商品代号 commercial designation
商品贷款 commodity loan
商品的国际价值 commodity's international value
商品的国内价值 commodity's domestic value
商品的主观价值 subjective value
商品的自然属性 natural property of commodity
商品等级 class of commodity; market grade; order of goods; commercial grade; merchantable <美国木材的>
商品等级钢 commercial grade steel

商品等价物 commodity-equivalent
商品调拨计划 commodity allocation plan
商品发票 merchandise invoice
商品繁荣 commodity boom
商品范围 commodity coverage
商品房屋出卖 selling of ready-built house
商品肥料 commercial fertilizer
商品费用 merchandise expenses
商品分类 classification of goods; commodity classification
商品分配 physical distribution
商品符合 monomark
商品附录 commodity appendixes
商品概览 commodity survey
商品干材 merchantable bole
商品干污泥 commercially dry sludge
商品钢 commercial steel; merchant steel
商品钢板 commercial sheet
商品钢材 commercial steel; merchant shape
商品钢管 commercial steel pipe
商品高度 saleable height
商品个性标志 trade character
商品供给 commodity supply
商品供求比率 ratio of commodities' supply and demand
商品供求平衡 balance between supply and demand of commodities
商品供应 supply of commodities
商品供应充足 abundance commodity supplies
商品供应垄断 oligopoly
商品供应时间 pipeline time
商品供应线 pipe line
商品供应线上的库存量 pipeline inventory
商品构成 commodity composition
商品估价 merchandise valuation
商品管材 commercial pipe; merchant pipe
商品管理 control of merchandise; merchandise control
商品管子 commercial pipe
商品规格 commercial specification
商品过剩 marketable surplus; overproduction of commodities; overstore
商品合金 commercial alloy
商品红黄铜 commercial red brass
商品花岗岩 commercial granite
商品滑石 commodity talc
商品化 commercialization
商品化的加工制剂 commercial formulation
商品化工业化房屋 commercial industrial building
商品化管理模式 model of commodities
商品化过程 commercialized process
商品化计划 merchandising plan
商品化住宅 commercialized residence
商品黄铜 market brass
商品回转率 merchandising turnover
商品混合胶 ready-mixed glue
商品混合饲料 commercial mixed feed
商品混凝土 commercial concrete; commodity concrete; ready mixed concrete; truck-mixed concrete
商品混凝土拌运车 ready-mix truck
商品混凝土摊铺机 ready-mix distribution facility
商品混凝土运拌车 transit-mixing truck
商品混凝土运输时间 time of haul
商品混凝土站 ready-mixed concrete plant
商品货币 commodity money; merchandise money
商品货币关系 commodity-money relationship
商品货物跌价 loss of market
商品机械 commercial machine
商品积累 accumulation of commodities
商品价格 commodity price
商品价值 value of commodity
商品间的相互依赖 interdependence among commodities
商品检验 commodity inspection; quality control
商品检验法规 commodity inspection law
商品检验局 commodity inspection bureau; commodity inspection and testing bureau
商品检验证书 certificate of inspection
商品交易 commodity transaction; merchandise trade
商品交易会 commodities fair; commodity exchange; fair; trade fair
商品交易所 commodity exchange; Board of Trade <美>
商品交易展览会 trade exhibition
商品胶合板 commercial plywood
商品结构 commodity composition
商品金属 commercial metal
商品进口税 duty on imported goods
商品经纪人 commodity broker; merchandise broker
商品经济 commodity economy
商品经销特权 dealership
商品经营者 distributor of commodity
商品经营资本 commodity-dealing capital
商品空间 commodity space
商品矿石 merchantable ore
商品冷藏 commercial cooling
商品冷却 commercial cooling
商品沥青拌和厂 commercial asphalt plant
商品粮 commodity grain; marketable grain
商品流动 movement of goods
商品流动分析 commodity flow analysis
商品流量 commodity flow
商品流通 circulation of commodity; commodity circulation; commodity turnover
商品流通仓库 distribution warehouse
商品流通中心 distribution center [centre]
商品炉子 trade furnace
商品煤样品 salable coal sample
商品门窗细(木)工 package trim
商品名 proprietary name; trade brand name
商品名称 name of commodity; trade name; commercial name
商品木材 commercial wood; merchantable timber; wood of commerce; yard lumber
商品目录 catalog(ue); commodity catalog(ue); inventory; qualified products list
商品目录价格 catalog(ue) price
商品农场 commercial farm; commercial holding
商品盘存控制问题 inventory control problem
商品票据 <指银行本票和跟票提单> commodity paper
商品品质 commercial sort
商品平板玻璃 commercial flat glass
商品奇缺 famine
商品启运许可证 authorization for shipment of goods
商品汽油 merchantable gasoline
商品铅 market lead
商品倾销 dumping
商品燃料 commercial fuel; merchantable fuel
商品燃烧用具 trade burning appliance
商品软木 commercial cork
商品润滑油 commercial lubricating oil
商品三层玻璃 commercial triplex
商品色粉涂料 commercial distemper
商品生产 commercial manufacture; commodity production; production of merchandise
商品生产者 producer of commodity
商品石灰 commercial lime
商品石蜡 paraffin(e) grade wax
商品石棉 commercial asbestos
商品市场 commodity market
商品市场均衡 commodity-market equilibrium
商品试样 commercial sample
商品输出 commodity export
商品蔬菜栽培基地 base for truck farming; market gardening base
商品蔬菜种植业 market gardening
商品蔬菜作物 <用卡车装运的> truck crop
商品熟铁管 commercial wrought iron pipe
商品水 commercial water
商品水泥 commercial cement
商品税 commodity tax; hallage
商品说明书 description of commodity
商品酸 commercial acid
商品损耗 commodity wastage
商品损耗量 wastage
商品特别运价率 special commodity rate
商品提单 commodity draft
商品提价 appreciation of goods
商品替代率 rate of commodity substitution
商品条钢 merchant bar iron; merchant steel
商品铁 commercial iron; merchant iron
商品铁条 merchant bar
商品铜 commercial copper; merchant copper
商品统一分类和编码办法 harmonized commodity description and coding system
商品涂料 commercial paint
商品推销员 roundsman
商品推销站 roundsman depot
商品屋面料 ready roofing
商品纤维 commercial fiber[fibre]
商品项目 merchandise account
商品销售 realization of goods
商品销售区 zone of merchandise
商品销售预算 merchandise selling budget
商品销售总差距 gross merchandise margin
商品协议 commodity agreement
商品锌 spelter
商品信赖度分析 brand loyalty analysis
商品信息 commodity information
商品信用公司 commodity credit corporation
商品形式 commercial form
商品型号 monomark; marque <尤指汽车型号>
商品性 marketability
商品性生产 commercial production
商品需求量 commodity demand volume
商品畜群 commercial herd
商品宣传员 spieler
商品养禽业 commercial poultry production
商品样本 sales literature
商品要求 commercial requirement
商品油漆 commercial paint
商品有效期 shelf life
商品渔业 commercial fishery
商品与税率的详细分类 specialization of commodities and duty rates
商品运价表 <按个别商品规定运价> commodity tariff
商品运输 merchandise traffic
商品展览馆 commercial museum
商品展览会 commodities (for) fair; trade fair
商品展览室 stock room
商品展销会 commodities (for) fair
商品折扣 merchandise discount
商品质量 commercial quality
商品质量的标签说明 grade labelling
商品滞销 holdup in the sale of commodities
商品中央陈列台 gondola
商品种类 commodity description
商品种子 commercial variety
商品周转 merchandise turnover; merchandising turnover; turnover tax
商品主权的转移 passing of title
商品住宅 commercial residential buildings
商品铸件 commercial casting
商品转让 alienation of commodities
商品转移 merchandise transfer
商品资本 commodity capital
商品自然损耗率 rate of natural loss of commodity
商品栽培者 commercial grower
商品综合方案 integrated for commodities; integrated program(me) for commodities
商品总目录 consolidated trade catalog(ue)
商品组合 grouping of commodities
商品作物 cash crop; commodity crop; money crop
商情报告 business report
商情报告书 market letter
商群 factor group; quotient group
商群序列 sequence of quotient group
商人 businessman; dealer; merchant; monger; trader; tradesfolk; tradesman; tradespeople
商人的 mercantile
商人惯常法 law merchant
商人介入 dealer tie-in
商人信用证 merchant's credit
商人银行 commercial bank; merchant bank
商事法 business law
商事法院 commercial court; staple court; tribunal of commerce
商事庭 commercial court
商数 <除法的得数> quotient
商数继电器 quotient relay
商数寄存器 quotient register
商谈 conference; palaver
商讨 negotiation
商图 quotient graph
商妥 come to terms
商位分配 assignment of space
商务 business
商务备选方案 alternative of financial nature
商务标准化 commercial standardization
商务部 Board of Trade
商务参赞 commercial attaché; commercial counsellor
商务参赞处 commercial counsellor's office
商务舱 commercial class
商务处 commercial department

商务处理 business process
商务代办 commercial agent
商务代表 commercial agent;commercial representative;trade representative
商务代表处 office of commercial representatives
商务代表团 commercial delegation
商务仿真 business simulation
商务封锁 commercial blockade
商务负责人 < 商船上的 > supercargo
商务规章 commercial rules
商务函件 commercial correspondence
商务会谈 business talks
商务活动 commercial activity
商务检查 business check-up
商务交换电报 telex
商务纠纷 commercial dispute
商务纠纷索赔 claim for trade dispute
商务旅行 commercial trip
商务旅行速度 commercial trip speed
商务秘书 commercial secretary
商务契约 commercial agreement
商务洽谈 business confabulation
商务数据处理 business data processing
商务条例 commercial code
商务往来 commercial intercourse
商务文件 commercial documents
商务系统 business system
商务协定 commercial agreement
商务选择性报价 alternative of financial nature
商务运价表 commercial tariff
商务占用 mercantile occupancy
商务账 commercial account
商务仲裁 commercial arbitration
商务注册 commercial register
商务专员 commercial attaché;commercial counsellor;office of commercial attaché;trade commissioner
商务自动化 business automation
商务综合组 business affairs panel
商务作业 commercial operation
商务作业过程 < 如招揽、承运、填发货票等 > marketing technology
商系 quotient system
商线性空间 quotient linear space
商信 goodwill
商学硕士 master of commercial science
商学学校 business school
商学院 business school
商业 commercial pursuit; commercium; mercantile pursuits; merchandise[merchandize];trade;trading
商业案件 commercial case
商业包装 commercial packaging
商业保险 commercial insurance
商业保险单 commercial policy
商业报表格式 business form
商业报单 trade circular
商业报价 commercial offer
商业备咨 business reference; house reference;trade reference
商业本票 commercial paper
商业编译程序 business compiler
商业标准 commercial criterion;commercial standard
商业标准胶合板 commercial standard plywood
商业波动 business fluctuation
商业不景气 business depression
商业步行街 commercial pedestrian street;shopping precinct
商业部 Board of Trade; Ministry of Commerce;Department of Commerce < 美 >
商业部门 commercial department
商业簿记 business accounting; commercial book-keeping
商业财产 commercial property
商业财务计划 commercial financial plan
商业采购 commercial procurement
商业参与者 trade party
商业测量部门 commercial survey organization
商业策略 business strategy
商业场所 commercial establishment
商业成本 commercial cost
商业承兑汇票 trade acceptance draft
商业承兑票据 trader('s) acceptance
商业诚实保证书 commercial blanket bond
商业城市 commercial city; commercial town;trade city;trading town
商业程序设计 business programming
商业尺寸 trade size
商业筹措 commercial financing
商业出行 business trip
商业厨房 commercial kitchen
商业纯的 commercially pure
商业存款 business deposit
商业磋商 business consultation
商业大楼 business building; mercantile building
商业大厅 shopping hall
商业大厦 commercial building
商业呆滞 slackening
商业代表 commercial agency
商业代表团 commercial mission
商业代理合同 commission contract
商业代理机构 mercantile agency
商业贷款 commercial loan
商业贷款理论 commercial loan theory
商业贷款利率 commercial loan rate
商业单位 business enterprise; commercial unit
商业单证 business vouchers;commercial documents
商业档位 shop stall
商业道德 business ethics;commercial character;commercial morality
商业的 commercial; mercantile; merchant
商业登记 business registration; commercial registration;trade register
商业登记证 business registration certificate
商业等级 commercial grade
商业地产 business property
商业地带 commercial zone
商业地点 place of business
商业地理 commercial geography
商业地面站 commercial earth station
商业地区 commercial area
商业地位 commercial standing
商业电报 commercial telegram
商业电力 commerce power;commercial power
商业电视 commercial television;sponsored television
商业电网供电 commercial power;grid power
商业调度 business arrangement
商业订单 commercial order
商业订货 commercial order
商业短期押汇票 documentary commercial bill drawn at short sight
商业发票 commercial invoice
商业发展 business development
商业发展区 business development zone
商业法(规) commercial law
商业法庭 commercial court
商业繁荣 business boom;trade booming
商业反应堆 commercial reactor
商业房地产 commercial property
商业房屋 business building;commercial building;utilitarian building
商业废料 commercial waste
商业废水 commercial wastewater
商业废物 commercial waste
商业费用 distribution charges
商业分类 commercial sort
商业分配 commercial time-sharing
商业分析 commercial analysis
商业分析员 business analyst
商业焚化炉 commercial incinerator
商业风险 commercial risk
商业服务系统 commercial service system
商业符号 commercial symbols
商业负荷 commercial load
商业负债 trade liability
商业盖缝 commercial canopy
商业干污泥 commercially dry sludge
商业高层建筑街坊 commercial high-rise block
商业高层建筑区段 commercial high-rise block
商业工程师 sales engineer
商业公会 commercial guild
商业公司 business corporation; commercial company
商业供气 commercial service
商业拱廊 shopping arcade
商业骨料厂 commercial aggregates plant
商业固体垃圾 commercial solid waste
商业关系 commercial relation
商业管理 business management;commercial management
商业管理学院 school of business administration
商业管制 commercial control
商业惯常做法 commercial practice
商业惯例 commercial practice; commercial usage; customary business practice; customary commercial practice
商业广场 business parade; shopping square
商业广告 commercial advertising;commercial message
商业规程 business regulations
商业规划 commercial plan
商业规模 commercial scale
商业规模的 commercial scale
商业航空业 commercial aviation
商业航行 commercial navigation
商业耗水量 commercial consumption of water
商业合股 commercial partnership
商业合伙人 trading partner
商业和工业噪声 commercial and industrial noise
商业和生活综合体 shopping and living complex
商业核算 commercial accounting
商业互惠 commercial reciprocity
商业化 businesslike; commercialization
商业化社会 commercialized society
商业环境 business environment
商业回扣 commercial rebate
商业回信用明信片 business reply cards
商业汇兑 commercial exchange
商业汇价 merchant rate
商业汇率 commercial exchange rate
商业汇票 bill of exchange; commercial bill of exchange; commercial draft;trade bill
商业汇票市场 commercial bill market
商业会计 business accountancy;business accounting; commercial accounting
商业活动 business proceedings;commercial activity;merchantilism
商业机构 business establishment;business organization; commercial organization
商业机密 trade secret
商业机器 business machinery
商业机械 commercial machine;commercial production machinery
商业基本建设 commercial capital construction
商业基础设施 commercial infrastructure
商业基地 commercial base;marketing station
商业及行政中心 commercial and administrative center[centre]
商业集中地 entrepot
商业集中区 grouped commercial district
商业计算 business calculation
商业家 commercialist
商业价值 commercial value
商业间接费用 commercial indirect expenses
商业间谍活动 corporate spying
商业减价 trade discount
商业简报 business in brief
商业建筑 business block; business building; business premises; shopping building;free-standing building < 经营单一业务的 >
商业建筑物 commercial building
商业交付周期 commercial lead time
商业交往 commercial intercourse
商业交易 business dealing; business transaction; commercial deal; commercial transaction
商业交易法 law of commercial transactions
商业交易所 Commercial Sale Rooms
商业教育 business education
商业街(道) business street;commercial street; shopping mall; shopping street;commercial strip
商业街坊 commercial block;shopping block
商业街面 business frontage;commercial frontage
商业街区 shopping block;street commercial district
商业界 business circle;business interests; commercial society; commercial circle;commercial world
商业金融公司 commercial finance company;commercial financial company;commercial financial corporation
商业金融机构 commercial financial institution
商业进销差价 difference between commercial purchasing and selling prices
商业禁运 commercial embargo
商业经济 business economy;commercial economy
商业经济学 businesses economics
商业经营 commercial business
商业景气 business boom
商业竞赛 business game; commercial competition
商业竞争 business rivalry; business rivalship;commercial competition
商业决策 business decision
商业考查 business survey
商业库存 commercial inventory
商业垃圾 commercial refuse; commercial(solid) waste
商业廊亭 covered mall building
商业冷藏室 commercial refrigerated cabinet

商业冷机 commercial cooler
商业理论 merchantilism
商业利润 business profit;commercial profit
商业利益 commercial interest
商业联系 commercial intercourse
商业量 volume of trade
商业流动资金 commercial circulating funds
商业流通 commercial distribution
商业垄断 business monopoly
商业楼宇 commercial
商业旅行者 commercial travel(l)er
商业铝 <纯度大于 99%> commercial aluminum
商业冒险 business venture;venture
商业贸易 commercial transaction
商业贸易中心 mercantile trading center[centre]
商业美术 commercial art
商业秘密 business secret;trade secret
商业密集区 grouped commercial district
商业免税区 commercial free zone
商业面积 commercial space
商业名称 business name;trade name
商业名片 business card
商业内部调拨价格 transfer prices among commercial enterprises
商业年度 commercial year
商业排放污水 trade waste effluent
商业排放液体 trade effluent
商业配色 commercial colo(u)r match
商业骗局 bubble
商业票据 commercial bill; commercial draft; commercial paper; mercantile paper
商业票据承销公司 commercial paper house
商业票据交易所 commercial paper exchange
商业票据市场 commercial paper market
商业平价 commercial par
商业企业 commercial enterprise;commercial undertakings; concern; merchandising enterprise
商业企业保险费用 insurance premium of commercial enterprises
商业企业的资产 business opportunity
商业企业管理 management of commercial enterprises
商业企业管理体制 managerial system in commercial enterprises
商业企业核算形式 calculation forms in commercial enterprises
商业企业货币资金 monetary funds of commercial enterprises
商业企业基本职能 basic functions of commercial enterprises
商业企业经济核算 economic accounting of commercial enterprises
商业企业经济效益 economic efficiency of commercial enterprises
商业企业市场调查 market survey made by commercial enterprises
商业企业市场预测 market forecast of commercial enterprises
商业企业种类 type of commercial enterprises
商业气候 business climate
商业潜水 commercial diving
商业情报系统 business information system
商业区 business area;business block; business district; business quarter; business zone; center of commerce;commercial center[centre]; commercial district; commercial zone; industrial district; shopping area;shopping center[centre];trade area;shopping mall <禁止车辆通行的>
商业区段 commercial block
商业区交通终点站 shopping terminal
商业区街道 business frontage; business local street; business parade; business street; downtown street; shopping street
商业区街面 <临街房屋为商业性建筑> business frontage
商业区林荫路 shopping mall
商业区路标 business route marker
商业区路线指示标 business route marker
商业区与住宅区之间的地区 midtown
商业区中的最佳位置 one-hundred percent location
商业燃(气用)具 commercial gas appliance
商业人员 commerce business people
商业软管供气潜水 commercial hose diving
商业软件 commercial ware
商业上的抵押 commercial pledge
商业上的调节手段 commercial leverage
商业上的应用 business application
商业石油库 depots for commercial oil
商业时代 commercial age
商业使用 business occupancy; commercial occupancy
商业事务 commercial affair
商业事务所 business office; business remises
商业试验 commercial test(ing)
商业试验室 commercial laboratory
商业书信 business correspondence
商业数据处理 business data processing
商业衰退 business recession
商业水平 shopping level
商业税 business tax
商业诉讼 commercial case
商业太阳能建筑物 commercial solar building
商业摊位 shop stall
商业谈判 commercial negotiation
商业特权 special commercial privilege
商业提货单 commercial bill of lading
商业条件 commercial terms
商业调节(手段) commercial leverage
商业调整 business adjustment
商业条约 commercial treaty
商业贴现 commercial discount
商业停车场 commercial parking facility
商业停车处 commercial parking space
商业停车库 commercial garage
商业通道 shopping passage(way)
商业通信[讯] business communication; business correspondence; commercial communication; commercial correspondence
商业统计 business statistics; commercial statistics
商业投标不动产研究所 Commercial Investment Real Estate Institute
商业投机 adventure; business speculation
商业投资 trade investment
商业投资公司 commercial investment trust company
商业投资信托公司 commercial investment trust company
商业图表 business graphics; chart of business
商业土地开发 commercial land development
商业推销员 roundsman
商业网 commercial distributive network;network of trading establishments
商业网点 commercial network
商业网点规划 commercial network planning
商业网密度 <英> density of shops
商业往来 commercial exchange;commercial intercourse
商业往来账户 commercial account
商业危机 business crisis; commercial crisis
商业文件 commercial credit; commercial message
商业习惯 business customs; commercial customs; merchantile usages; merchantilism
商业习惯法 law of merchant
商业系统 business system
商业系统分析员 business-system analyst
商业萧条 commercial depression
商业销售 commercial disappearance
商业销售价值 value of business sale
商业小弄 shopping lane
商业小巷 shopping lane
商业新闻 commercial article
商业信贷 business credit; commercial credit;commercial loan;trade credit
商业信托 business trust; common law trust
商业信托公司 commercial credit company
商业信用 business credit; commercial credit; commercial standing; mercantile credit
商业信用保险 commercial credit insurance
商业信用查询 trade reference
商业信用单据 commercial credit documents
商业信用调查 credit inquiry
商业信用凭证 instrument of credit
商业信用情况 business standing
商业信用证 commercial letter of credit
商业信用状况 business standing
商业信用咨询所 mercantile inquiry agency
商业信誉 commercial reputation; goodwill
商业兴旺 brisk commerce
商业行为 business dealing; commercial action; commercial operation
商业形式 commercial form
商业型窗 commercial window
商业性捕鲸 commercial whaling
商业性采伐 commercial clear cutting
商业性建筑物 business building
商业性交通量 business volume
商业性林地 commercial forest land
商业性林业 commercial forestry
商业性旅馆 commercial hotel
商业性牛棚 commercial stable
商业性疏伐 commercial thinning
商业性水道 commercial water course; commercial waterway
商业性水路 commercial waterway
商业性损失 trade loss
商业性投资 businessman's investment
商业性运输 commercial carrying
商业性轧石机 commercial stone-crushing plant
商业学校 business school; commercial college; commercial school
商业学院 business school
商业循环 business cycle; trade cycle
商业循环波动 business cycle fluctuation
商业循环分析 business cycle analysis
商业循环模型 business cycle model
商业循环序列 cyclic(al) sequence
商业循环指示数字 business cycle indicators
商业循环周期 business cycle
商业样品进口证 export commercial sample carnet
商业意义 trade drift
商业银行 bank of commerce; business bank; commercial bank
商业银行贷款 commercial bank loan
商业银行信贷 commercial bank credit
商业银行业务 commercial banking
商业应用 business application; commercial application; commercial use
商业盈利率 commercial profitability
商业影响 influence
商业佣金 commercial brokerage
商业用窗 commercial window
商(业)用建筑物 commercial architecture; mercantile occupancy
商业用地 commercial acre; trading estate
商业用高层建筑 commercial tall building
商业用户 mercantile occupancy; commercial user
商业用旅馆 business hotel
商业用气 commercial utilization
商业用语 commercial language; commercial terms
商业园 business park
商业运输 business traffic; commercial traffic; commercial transport
商业运输工具 commercial vehicle
商业噪声 commercial noise
商业账簿 business books; commercial account; trade books
商业遮篷 commercial awning
商业折扣 commercial discount; trade discount
商业争端 business dispute
商业争议法 trade dispute act
商业征信所 commercial credit agency; commercial credit bureau
商业证券 commercial portfolio; instrument of credit
商业证券经纪行 commercial paper house
商业政策 business policy
商业知识 business knowledge
商业执照 trading certificate
商业指数 business index; index of business
商业制度 mercantile system
商业制冷机组 commercial refrigeration unit
商业制冷装置 commercial refrigerating plant; commercial refrigerator
商业中心 block of groped shops; business center[centre]; center of commerce; centerpot; emporium[复 emporiums/emporia]; hub of commerce; market; mart; mercantile center[centre]; seat of commerce; shopping center[centre]; staple
商业中心地 market place; staple
商业中心区 center of business district; central business area; central business district; shopping center area; town; commercial center[centre]; downtown(area)
商业中心区消防队 high-value company
商业仲裁 commercial arbitration
商业重潜水服 commercial heavy weight diving dress; commercial heavy weight diving suit
商业周期 business cycle
商业周期的缓和 moderation of business cycles

S

商业周期低潮 trough
商业周期低潮线 trough line
商业周期转折点 turning point
商业周期转折期 turning zone
商业周转总额 gross trade turnover
商业注册 commercial registration
商业铸件 casting of commerce
商业装置 commercial installation
商业资本 business capital; capital for trading purpose; commercial capital; merchant capital
商业资金 commercial finance; commercial loan
商业资料 business material
商业资料处理 business data processing
商业自由港 commercial free port
商业综合体 business complex; shopping complex
商业总产出 gross output of trade
商业总产值 gross business product
商业总担保 commercial blanket bond
商业租契 commercial lease
商业组织 commercial organization
商议 consultation; consult(ing); counsel
商议者 consultant; negotiator
商映射 quotient mapping
商用报文 commercial message
商用编译程序 commercial compiler
商用标准 commercial standard
商用冰箱 commercial refrigerator
商用不锈钢 commercial rustless steel
商用材高 merchantable height
商用长度 commercial length
商用车辆 commercial vehicle
商用车身 commercial body
商用船坞 commercial dock; merchant's dock
商用单位数 <相当于 10 个×120 个> mille
商用电动机 commercial motor
商用电话 business telephone
商用电路 commercial circuit
商用电台 commercial station
商用丁烷 commercial butane
商用锻钢坯 open forging steel
商用飞机 business aircraft; commercial aircraft
商用符号 commercial character
商用规格 commercial size; commercial standard
商用回收处理 commercial reprocessing
商用货车现有数 stock of commercial wagons
商用机器 business machine
商用机器语言【计】common business oriented language
商用计算机 business computer; commercial computer
商用结构钢 commercial structural steel
商用客运飞机 commercial passenger airplane
商用快中子增殖反应堆 commercial fast breeder reactor
商用冷柜 commercial refrigerator
商用冷凝机组 commercial condensing unit
商用码头 commercial dock
商用频率 commercial frequency
商用品种 commercial breed
商用青铜合金 commercial bronze
商用术语 merchantilism
商用数据管理系统 commercial data management system
商用通话 commercial call
商用通信[讯]网络 business network
商用通信[讯]卫星 commercial communications satellite
商用卫星 commercial satellite
商用消息 commercial message
商用型铁 commercial iron
商用仪器 commercial instrument
商用硬脂酸 stearin(e)
商用预应力构件 commercial prestressing member
商用云母 mica of commerce
商用轧制型钢 commercial rolled section
商用指令集 commercial instruction set
商用指令系统 commercial instruction set
商用制冷系统 commercial refrigerating system
商用致冷装置 commercial refrigerating plant
商用重量 commercial weight
商用字符 commercial character
商域 quotient field
商誉好 goodwill
商誉坏 badwill
商约 commercial treaty
商展开 quotient expansion
商栈 khan; trade post
商住楼 business-living room

墒 moisture in the soil

墒沟 dead furrow
墒情 soil moisture content
墒情好 good conditions for soil moisture
墒情预报 forecasting of soil moisture

熵 <热力学函数> entropy

熵比图 entropy ratio map
熵编码 entropy coding
熵变 entropy change
熵捕获 entropy trapping
熵产 entropy generation
熵单位 entropy crisis
熵分布 entropy contribution
熵分析 entropy analysis
熵函数 entropy function
熵流 entropy flow
熵模型【物】entropy model
熵权 entropy weight
熵弹性 entropy elasticity
熵图 entropy chart; entropy diagram
熵温曲线 entropy-temperature curve
熵温图 tephigram
熵污染 entropy pollution
熵析 exergy analysis
熵原理 entropy principle
熵跃 entropy spring
熵最小化 entropy minimization

赏罚条款 penalty-and-bonus clause

赏银 gratuity

上岸 debark(ation); disembark(ation); disembarking; go ashore; go on shore; landing

上岸费 landing charges
上岸港 port of disembarkation; port of landing
上岸码头 landing stage; landing wharf
上岸设施 landing accommodation
上岸跳板 landing ramp; shore board
上岸许可 liberty
上岸证明书 <海关发给货物的> landing certificate
上岸指示 <登陆或降落> landing order
上凹的 concave-up(ward)
上凹底 push-up
上凹下凸的波状花边 cyma recta
上凹下凸的双弧形线脚 Doric cyma(tium)
上凹下凸反曲线 sima recta; sima reversa
上凹型 upper depression
上奥陶纪【地】Upper Ordovician period
上奥陶统【地】Upper Ordovician series
上拔 up-pull
上拔力 uplift
上白垩砂岩 Upper Cretaceous sandstone
上白垩统【地】Upper Cretaceous series
上白釉 white glazed coat(ing); white glazing
上摆 sole plate
上摆动颚板面 upper swing jaw face
上摆钻 top balance jewel
上班 report for duty
上班高峰时间 home-to-work peak hour
上班公共汽车 works bus
上班即计时惯例 show-up time rule
上班交通 journal to work; journey-to-work
上班旅程 journey-to-work
上班时间 office hours; on-duty time; service hours; work hours
上班行程时间 journal to work travel time; journey-to-work travel time
上板铺 upper bunk
上版 adjustment of printing plate; press adjustment
上半部 upper half; upper part
上半部对接而下半部斜接的木工接缝 butt and miter joint
上半部平均长度 upper half mean length
上半部桥墩 bust pier
上半部曲柄箱 upper half crank case
上半齿面 addendum flank
上半端罩 upper end-shield
上半断面 top heading; upper section
上半断面超前施工法 top heading method; upper half method
上半断面开挖法 top heading method
上半格 upper semi-lattice
上半截装配玻璃的门 half-glass door
上半连续的 upper semicontinuous
上半连续点 upper semicontinuity
上半连续分解 upper semicontinuous decomposition
上半连续函数 upper semicontinuous function
上半面 upper half plane
上半模 male building form
上半模格 upper semi-modular lattice
上半汽缸 casing top half; upper cylinder half
上半球光通量 upper hemispheric flux
上半完备的 sup-complete
上半型 mo(u)ld cope
上半轴瓦 upper half-bearing
上半子午圈 upper branch
上拌下贯式(沥青)路面 penetration macadam with coated chip(ping)s
上包络 coenvelope; upper envelope
上包络线 maximum envelope curve
上保险 lock; safing
上背 upper back
上壁板 wainscot
上臂 arm
上臂围 upper-arm circumference
上边 top margin; upper margin; up side
上边板 upper edge board
上边舱 wing tank
上边带 upper side band
上边带波 upper side band wave
上边带谱 upper sideband spectrum
上边观测高度 <天体> observed altitude of upper limb
上边光 upper rim ray
上边机 glazer
上边界 coboundary; upper boundary
上边频 upper side frequency
上边声道 upper side track
上边声迹 upper side track
上边滩 side bar upstream of crossing; upper side flat
上边缘 upper limb
上边缘复(合)形 coboundary complex
上边缘模 module of coboundaries
上边缘算子 coboundary operator
上变频 up-conversion
上变频器 up-converter
上标 raising of indices; superior figure; superscript
上标表示法 superscript representation
上标(数)字 <如指数> superior figure
上表层 upper epidermis
上表面 top surface; upper face
上冰风 on-ice wind
上玻璃片 upper glass plate
上薄褶皱 <同沉积褶皱> supratenuous fold
上部 head space; top(ping); upper (class); upper curtate; upper portion; up side
上部安玻璃的门 half-glass door
上部半断面超前台阶施工法 heading and benching
上部半圆型沉箱式防波堤 round crown breakwater
上部边缘区 upper marginal area
上部标志灯 upward identification light
上部采光的圆屋顶 light-cupola
上部采区 upper workings
上部叉形联杆 upper wishbone link
上部超前掘进法 top heading; top heading method
上部车体 <挖掘机、起重机> upper wagon
上部沉积(物) overburden
上部沉积影响带 zone of affected overburden
上部衬 upper gasket
上部撑板 upper stay plate
上部充填带 high-side(d) pack
上部出车台 top landing
上部穿孔 overpunching
上部穿孔区 upper curtate
上部窗扇 top leaf
上部吹扫口 blow-up mouth
上部打孔 overpunch
上部带安装环的偏斜楔 ring-type wedge
上部导轨 <门的> top door rail
上部导轮 upper tumbler
上部地壳 upper crust
上部钉法 head nailing
上部堆芯围筒 upper core support barrel
上部阀盖 upper valve cap
上部反射层 top reflector
上部防护面层 <防波堤的> capping cover layer
上部防护屏 top shield
上部分层 higher slice
上部分段装药 upper-deck charge
上部浮圈 upper buoyancy chamber

S

上部副翼 top aileron;upper aileron
上部盖 upper cap
上部钢结构 steel superstructure
上部钢筋 top reinforcement; upper reinforcement
上部隔板 upper baffle plate
上部隔水层 upper confined bed
上部构造层次 upper tectonic level
上部鼓风 top blast
上部轨 upper rail
上部横撑 top transversal strut
上部横向支撑 upper lateral bracing
上部(湖面)温水层 epilimnion(layer)
上部(滑动)刀架 upper slide rest
上部回采平巷 upper development road
上部回转机构 rotating upper machinery
上部回转结构 revolving upperstructure;upper revolving structure
上部活动平衡装置 <起重机> sky horse
上部给料锤式粉碎机 top-feed hammer crusher
上部给料破碎机 top-feed hammer crusher
上部记录曲线 <地震图的> top trace
上部记录线 top trace
上部加热的坩埚窑 top-flame furnace
上部间断 upper break
上部建筑 permanent way;superstruction;superstructure;top-out
上部建筑成本 superstructure cost
上部建筑费用 superstructure charges
上部建筑物 superstructure work
上部角钢 <钢梁上部设置的> top angle
上部脚手架 upper falsework
上部阶段 top bench
上部结构 overhead structure; superstruction; superstructure; upper structure;upper wagon
上部结构、基础与地基共同作用分析 structure-foundation-subsoil interaction analysis
上部结构工程 superstructure works
上部结构混凝土 superstructure concrete
上部结构机架 <起重机> superstructure frame
上部结构框架 frame superstructure
上部结构面向后端的状态 <起重机> upper facing rear
上部结构体积 volume of superstructure
上部结构向前 <起重机> upper facing front
上部结掏操纵台 head-cab
上部截槽 overcut;top cutting
上部截槽截煤机 over coal cutter
上部截煤机 over coal cutter
上部进路 upper intake passage
上部进气百叶窗 top air inlet louver
上部井筒凿井 foreshaft sinking
上部净空 upper clearance
上部绝缘密封垫 upper insulator gasket
上部刻度 high scale
上部空气分布罩 upper-air hood
上部孔段 top hole
上部矿层 upper leaf;upper seam
上部框架构件 top frame member
上部框架结构 frame superstructure
上部拉板 <路牌机> top slide
上部累赘船具 top hammer
上部离合器 upper clutch
上部连接结构 upper coupling structure
上部联结系 super-connecting system
上部临界冷却速度 upper critical cooling rate
上部临界深度 upper critical depth
上部陆源建造 upper terringenous formation
上部履带 upper run of track
上部轮筒 upper tumbler
上部煤层 superjacent
上部门扇 top leaf
上部模板 upper mo(u)ld
上部模型 cope
上部炮塔位置 upper turret station
上部炮眼 back hole;header
上部平底相纹理构造 upper planebed facies lamellar structure
上部平台 upper brace
上部平巷 upper entry;upper gangway
上部平行平巷 top parallel entry
上部屏蔽 top shielding
上部汽缸润滑 upper cylinder lubrication
上部汽缸套带 upper steam cylinder jacket band
上部汽路 upper steam passage
上部砌块 upper segment
上部驱动装置 upper drive mechanism
上部缺失 upper break
上部裙筒 upper skirt
上部三棱镜 top prism
上部设施 superstructure facility;superstructure work
上部试样 upper sample
上部水流动态 upper flow regime
上部水平 upper level
上部水平支撑 top lateral bracing;upper lateral bracing
上部司机室 upper cab
上部陶槽爆破 burn cut near top of working face
上部提升筒 upper lift drum
上部调速环 upper speed ring
上部停机闸 <起重机> upper parking brake
上部通风的立窑 overdraft kiln
上部通风(装置) overdraft; overdraught
上部通信[讯]天线 upper communication antenna
上部凸角 upper lobe
上部凸轮 top jaw
上部土层压力 overburden pressure
上部危险 overhead hazard
上部围岩 overlying wall rock
上部桅 upper mast
上部纤维 upper fiber[fibre]
上部小窗 high window
上部斜面沉箱堤 sloping top caisson breakwater
上部卸载机 <青贮塔的> top unloader
上部旋转构架 upper revolving frame
上部压板 top hold-down
上部压紧胶辊 nipple roll upper
上部摇动筛 upper shaker screen
上部液体 top fluid
上部引入的燃烧空气 overfire air
上部油舱 topside tank
上部支撑 upper strut
上部重的荷载 top-heavy load
上部轴瓦 top brass
上部转台 <起重机的> revolving superstructure
上部装罐水平 upper-decking level
上部装载带型运输机 top-loading belt
上部装置驾驶室 upperstructure cab
上部锥体 top cone
上部缀板 upper stay plate
上部钻杆 top rod
上财政年度费用支出 expense belonging to the preceding financial year
上财政年度结余 surplus in preceding financial year
上彩虹 iridizing
上彩色涂料 colo(u)r wash
上菜窗口 serving hatch
上舱 upper-deck cabin
上舱口 upper hatch(way)
上舱钮 top gudgeon
上操纵杆 top lever
上侧 upper side
上侧拉紧的皮带 belt driving over
上侧梁 cant rail;side plate;side top chord;top chord;upper side rail <集装箱>
上侧梁矫正 <集装箱> straighten top rail
上侧梁嵌补 <集装箱> inserting top rail
上侧门 topside door
上侧门铰链 topside door hinge plate
上侧门折页 topside door hinge plate
上侧门止销及链 topside door staple pin and chain
上侧面 upthrow side
上侧片 alifera
上测晶体分辨率 resolution of up measured crystal
上测晶体体积 up survey crystal volume
上测站 up-station
上层 higher slice;overstor(e)y【建】;supercrust;superjacent bed;superstratum[复 superstrata];top bed;top deck;topping;upper bed;upper coat;upper layer;upper stratum;upper tier;distegia <古希腊、罗马剧院布景房的>;upper deck【船】
上层白垩 top chalk
上层舱 upper-deck hold
上层车厢 top deck;upper deck
上层池壁砖 top course of tank block
上层床 upper bed
上层(粗)钢筋 top bar
上层大气 outer atmosphere;upper atmosphere
上层大气空间 upper atmosphere
上层道砟 top ballast
上层的 upstairs
上层地板 <牲畜车> upper floor
上层地板侧梁 deck plate
上层地下室 upper basement
上层地下水 groundwater of the upper zone
上层迭置冰川 superimposed glacier
上层风 upper wind
上层浮游生物 autopelagic plankton;epiplankton
上层覆盖 upper cover
上层钢板弹簧 <汽车> helper spring
上层钢筋网 top mat;top mesh
上层格栅楼板 upper joist floor
上层格栅楼面 upper joist floor
上层管理 top management
上层轨道下层站房的车站 over-track station
上层含水量 perched aquifer
上层环行(路)top-rounding
上层混凝土 top concrete layer;top course concrete;toplift
上层机构 top level mechanism
上层级 upstage
上层甲板 upper deck
上层驾驶台 upper bridge
上层建筑 dead works;erection;superstruction;top-out;upper strake;upper works
上层建筑顶 top of superstructure
上层建筑甲板 superstructure deck
上层建筑甲板覆盖层 covering of superstructure
上层建筑领域 realm of superstructure
上层建筑物 superstructure
上层脚手架 upper falsework; upper flying
上层接结 upper binder
上层结构 superstructure;upper works
上层列板 upper strake
上层领导报告 top management report
上层流 upper current
上层楼板 single upper floor
上层楼的壁炉地面 upper floor hearth
上层楼的墙壁 upper floor wall
上层楼面 upper floor
上层楼平面图 upper floor ground plan;upper stor(e)y ground plan
上层楼走廊 upper gallery
上层路 overlying roadway
上层路基 upper subbase
上层码垛 top piling
上层木 overstor(e)y;overwood
上层木格栅楼板 wooden joist upper floor
上层泥渣 supernatant sludge
上层破裂面 upper failure plane
上层气流 upper current
上层气体 top tank air
上层桥面 high deck of bridge;upper deck;upper-deck of bridge
上层轻甲板 awning deck;hurricane deck; hurricane roof; promenad deck
上层清液 clear supernatant;supernatant;supernatant liquid;supernatant liquor;supernate
上层清液排出管 supernatant drawoff pipe
上层(取的)试样 upper sample
上层燃烧室 penthouse combustion chamber
上层绕组铜条 upper winding-bar
上层乳剂 upper emulsion
上层筛面 top scalping deck
上层渗水带 vadose zone
上层受光区 euphotic zone
上层疏伐 thinning from above
上层水 headwater;upper water;water above oil reservoir <油藏上的>
上层水位 upper-level of water
上层水样 upper sample of water
上层水域 superjacent waters
上层探测器 topside sounder
上层土层 upper soil layer
上层土(壤) top soil; upper horizon soil;upper layer of soil
上层(土)样 upper sample
上层托盘 top deck tray
上层瓦 top course tile
上层网 upper gauze
上层屋面木瓦层 straight course
上层纤维 top fiber[fibre]
上层线棒 top bar;upper bar
上层线圈 upper coil
上层小库 upper floor silo
上层遗留河 superimposed river;superposed stream
上层游泳生物 supranekton
上层釉 glaze coat
上层鱼 epipelagic fish
上层鱼类 epipelagic fishes
上层预应力标高 upper prestressing level
上层云 upper layer cloud
上层滞留地下水 perched(ground) water
上层滞留地下水位 perched water table
上层滞留含水层 perched aquifer
上层滞水 perched(ground) water;shallow detained water; stagnant ground water;suspended water;upstream water;vadose water

上层滞水带 vadose belt;vadose zone
上层滞水含水层 perched aquifer
上层滞水面 apparent water table;perched water table
上层滞水泉 perched spring;vadose spring
上层滞水位 apparent water table;perched water table
上层种 pelagic(al)species
上层组织 epithelium
上差 high limit of tolerance
上铲式单斗挖土机 crane navvy
上长 top growth
上场商品<指在交易所的> listed brand
上超层理模式 onlap bedding mode
上车 aboard;board(ing);entrain;heavy barrow;on board;upper wagon;upper structure<起重机>
上车驾驶室 upperstructure cab
上车架 superstructure frame;upper-structure frame
上车客流量 boarding passenger volume
上车向前<起重机> upper facing front
上承板 deck(ing)slab
上承板梁 deck plate beam;deck plate girder
上承板梁式桥 deck plate girder bridge
上承拱 upper loading arch
上承拱腹 deck soffit
上承桁架 deck truss
上承结构 deck structure
上承梁 deck beam
上承梁式桥 deck-girder bridge
上承锚底座 upper anchor bracket
上承桥 top road bridge;deck bridge
上承桥面 roadway on bottom boom
上承式 deck type
上承式板梁 deck plate girder
上承式板梁桥 deck plate girder bridge
上承式大梁 deck girder
上承式吊车梁 top-running crane beam
上承式叠板弹簧 overhung laminate spring
上承式公路桥 deck highway bridge;top road bridge
上承式桁架 deck truss;top supporting truss
上承式桁架桥 deck truss bridge
上承式架 deck beam
上承式孔 deck space;deck span
上承式梁桥 deck-girder bridge
上承式桥 deck(type)bridge;top road bridge
上承式桥跨结构 deck space;deck span
上承式桥面 deck surface
上承式桥式起重机 overrunning bridge crane
上承式行车路 roadway above
上承式悬臂桥 deck cantilever bridge
上承式转车台<铁路车站调车场> deck turntable
上承式纵梁<桥梁车行道的> deck stringer
上承压层<地下水的> positive confining bed
上池 upper pool
上池水位 level of upper pond
上池最低水位 lowest upper pool elevation
上齿槽管 superior alveolar canals
上齿槽指数 maxillo-alveolar index
上冲 upthrow;upthrust
上冲板块 overriding plate
上冲程 upward stroke
上冲断层【地】overfault;overthrust;overthrust fault;reversed fault;upthrust;upthrust fault
上冲断层背斜 overthrust anticline
上冲断层崖 upthrust fault scarp
上冲断块 overthrust block
上冲杆 upper plunger;upper punch
上冲流 uprush
上冲片体 overthrust slice
上冲头 upper punch
上冲推覆体 overthrust slice
上冲系数 run-up factor
上冲褶皱【地】overthrust fold;overfold
上抽的 updraft;up-draught
上抽发生炉 up-draught producer
上抽(风的)updraft;up-draught
上抽风式间歇砖窑 up-draught intermittent brick kiln
上抽风式窑 up-draught kiln
上抽煤气发生炉 up-draught gas producer
上抽气管 upper suction tube
上抽式燃烧 updraft combustion
上抽烟道 up-draught flue
上传动 upper transmission
上传动压力机 top-drive press
上传送钢辊 top steel feed roll
上船 aboard;boarding;embark(ing);embarkment;going aboard;on board
上船工作的装卸工人 shipman
上船后即付 pay-as-you-enter
上船检查登记簿 boarding book
上船检验 boarding
上船检验人员 boarding inspector
上船任职 join a ship
上船日期 day of embarkation
上船入口 embarkation entrance
上船载货的(横)码头 embarkation quay
上窗盖 top flap
上床铺 upper bunk
上吹的 upwind
上吹效应 upwind effect
上锤头 upper ram
上瓷釉 porcelain enameling
上次检修日期 date of last repairs
上淬 up-quenching
上淬法 up-quenching
上大下小的 big-end up
上大下小钢锭模 big-end-up mould
上带 upper band
上挡墙 top retaining wall
上档<门窗的> top rail
上刀 upper slitter
上刀架 top rest;upper slide rest
上刀片 top blade;top knife;upper blade
上刀片滑块 top knife block
上导槽 upper rail
上导出数 upper derivate
上导洞 top heading
上导洞法 bar and sill method;top heading method<开挖隧道>;top drift method<隧洞施工>
上导杆 top guide bar
上导函数 upper derived function
上导坑 top drift;top heading
上导坑法 bar and sill method;top heading method<隧洞施工>
上导坑台阶(开挖)法 top heading and bench cut method
上导轮 top tumbler;upper block
上导轮中心高 center height of upper tumbler
上导轮轴承 shaft bearing of upper tumbler
上导数 upper derivative;upper differential coefficient
上导线 upper conductor
上导向器 upper guide
上导叶轮 upper guide vane ring
上导轴承 top guide bearing;upper guide bearing
上导轴承架 upper bearing bracket
上导轴瓦乌金 upper guide-metal
上道 upper track
上道车 cars off-line
上的映射 onto mapping
上等白色煤油 prime-white kerosine
上等板纸<绘图用> Bristol board
上等玻璃条 patent glazing bar
上等材<无缺陷> prime log;upper stock
上等产品 real extra;real thing
上等粗糖 good muscovado
上等大块煤 fancy lump coal
上等单板 fancy veneer
上等的 choice;first-class;first rate;high class;superior
上等高线截面积 area of upper contour
上等黑料 drop black
上等(厚)玻璃 crown glass
上等汇票 prime bill
上等货 prime quality;real stuff;super-fine quality
上等精矿 top-grade concentrate
上等(锯)材 clear lumber
上等淋浴器 de luxe shower
上等煤 fancy coal
上等面粉 patent flour
上等木材 high grade stock;upper stock
上等木屑 excelsior
上等皮革 crown leather
上等品质 first quality
上等填料 crown filler
上等细布 lawn
上等细麻布 lawn
上等锌玻璃 zinc crown
上等鱼 prime fish
上等脂 primer jus
上等紫胶 fine shellac
上低音大号 baryton
上底 precoat(ing)<油漆的>;upper bottom
上底漆 prime lacquer;prime painting;shop painting;undercoating
上底色 ground laying
上底釉 engobe coating
上地壳层 supracrust;upper crust
上地壳的<指覆盖基底上的岩石> super-crustal
上地壳侵蚀 super-crustal erosion
上地幔【地】asthenosphere;earth upper mantle;peridotite shell;upper mantle
上地幔结构 upper mantle structure
上第三系【地】Neogene system;Neozoic
上颠<船体的> scend
上电 power turn-on
上电离层 upper ionized layer
上电自检 power-on self test
上叠层 superimposed layer
上叠的 superimposed
上叠灯标 upper range light
上叠地槽 superimposed geosyncline
上叠拱 superimposed arches
上叠关系 superimposed relationship
上叠河漫滩 superimposed flood plain
上叠阶地 superimposed terrace
上叠盆地 superposed basin
上叠沙丘 climbing dune
上叠式洪积扇 overlapping pluvial fan
上叠式逆冲扩展 over-step thrust propagation
上顶力 uplift force;uplift
上顶压力 uplifting pressure
上动防松闸 fixing brake
上动固定闸 fixing brake
上动皮带轮 head pulley
上动态 dynamic(al)behavio(u)r
上冻期 freezing season
上端 topping;upper bed;upper end;upper extreme;upper extreme point;upper side;upper surface
上端板 upper head plate
上端沉淀精矿 heading;head tin
上端阀座 upper valve seat
上端防松螺母 top lock nut
上端盖 upper end cover
上端局 up-station
上端局设备 upset
上端梁 body end plate;end plate;top-end rail<集装箱>;top-end transverse member<集装箱>
上端内存 upper memory area
上端气缸碗 top cylinder cup
上端球面的壁龛 spheric(al)headed niche
上端印有文字的信笺<如姓名、地址、联系电话等> letterhead
上端缘<敞车的> end top chord;end top rail
上端缘角钢 end top angle
上端轴 upper shaft
上端轴承 head bearing
上段 up side
上段肋板 frame head
上段运输带 top belt
上堆芯板 upper core plate
上墩 putting on the stocks;docking【船】
上舵杆 rudder stock;upper stock
上舵轮柄 king spoke;midship spoke;up-and-down spoke
上舵钮 top gudgeon
上舵栓 top pintle
上惰轮 upper idler
上惰轮辊筒 upper idler roller
上惰轮辊筒胎 upper idler roller tire
上惰轮辊筒轴 upper idler roller shaft
上颚板<压碎机的> upper cheek plate
上耳轴 upper gudgeon
上二层舱【船】upper tweendeck
上二迭纪【地】Upper Permian period
上二迭统【地】Upper Permian series
上发条 wind-up
上阀杆密封 upper stem seal
上珐琅的板 enamel(l)ed slate
上帆布输送带 upper canvases
上帆缩帆眼索 bull earing
上翻 upturning
上翻边梁 upstand string
上翻车门 side hinged lift door
上翻梁 inverted beam;upstand beam
上翻门 overhead door;tilt-up door;tip-up door;trap door;up-and-over door
上翻式淋浴器 overhead shower
上翻式汽车库门 overhead garage door
上翻洋流 upwelling oceanic current
上反角 allotriomorphic;anhedral angle;dihedral
上方 upstream;upward side;barge measure<指船放泥量>;volume on board<指船放泥量>;bin measurement【疏】
上方加煤机 over-feed stoker
上方交叉飞过 cross-over
上方净空 upper clearance
上方喷浆成型器 top former
上方水域 superjacent waters
上方晕 superjacent halo
上方值 upper value

上枋 upper fillet and fascia
上防水胶的 waterproof-glued
上飞机 boarding;enplane[emplane]
上(飞)机道路 enplaning road
上肥 dung
上肥皂 soap
上分层 top leaf
上分岔 up-splitting
上分配 overhead distribution
上分式 overhead distribution
上分式干管 overhead main
上分式双管(热水供暖)系统 double pipe dropping system
上分系统 down-feed system
上分下给式系统 <暖通空调> down-feed overhead system
上分支 topset
上粉 powdering
上风 fetch to;windward
上风岸 weather shore; windward bank;windward shore
上风侧 weather side
上风差分公式 upwind difference formula
上风道式通风机 updraft ventilator
上风的 weather side;windward
上风舵 lee helm
上风方向 upwind
上风防波堤 windward breakwater
上风桁架 windward truss
上风井 upcast;uptake;uptake breeching
上风井口扇风机 upcast fan
上风联杆 upper wind girder
上风炉箅 updraft grate
上风满舵 hard aweather;hard up of rudder
上风锚 weather anchor
上风面 weather face;windward side
上风汽化器 updraft carburet(t)or [carburet(t)er]
上风式的 updraft
上风首舷 weather bow
上风弦 windward chord; weather side
上风向防波堤 weather breakwater
上风柱 windward post
上封层 capping layer
上伏辊 top couch(roll);upper couch(roll)
上浮 floating-up;float upward
上浮法 rising method
上浮分离法 flotation separation method
上浮计算 ascent calculation
上浮力 uplift force;heave force
上浮时间 ascent time
上浮水雷 rising mine
上浮水流 upward water flow
上浮水压力 uplift water pressure
上浮速度 ascent rate;floating speed
上浮位置 <挖土机前端铲斗> float position
上浮下沉处理 float and sink treatment
上浮下潜时间 excursion time
上浮下潜重量系统 ascent decent weighting system
上覆 onlap
上覆不可透水层 overlying impervious bed
上覆层 overburden layer; overlying bed; overlying seam; superimposed bed;superstratum[复 superstrata]
上覆沉积物 overlying deposit;overlying sediment
上覆大气 overlying atmosphere
上覆的 incumbent;overlying
上覆地层 overlying bed; overlying formation;overlying stratum[strata]
上覆地层厚度 thickness of overlying strata
上覆地层压力 pressure of overlying strata
上覆垫层 overlying base course
上覆非渗透地层 overlying impermeable strata
上覆隔水层 upper confining bed
上覆构造 suprastructure
上覆荷载 overburden;overcharge
上覆可透水层 overlying pervious bed
上覆面 upper sheathing
上覆水 overlying water
上覆水域 superjacent waters
上覆土 overlying soil; superimposed soil
上覆土层 overlying soil
上覆(土)荷载 overburden
上覆土重压力 overburden earth pressure
上覆压力梯度 overburden pressure gradient
上覆岩层 overlying rock
上覆岩层厚度 depth of cover
上覆岩层压力 overburden pressure
上盖 top head; upper shield; upper cover <电机的>
上盖板 top lap; upper cover plate; keeper plate <小型桥梁支座>
上盖垫片 upper cover gasket
上盖夹紧装置 clamp for upper cover
上盖式车库门 up-and-over garage door
上盖压板 cover clamp
上盖针 top needle
上盖轴承 top bearing
上干燥室 primary drying chamber
上杆 upper boom
上杆灯 foot board
上杆钉 foot board;pole step
上杆脚板 foot board
上杆脚扣 grapnel;pole climbers
上缸 upper cylinder half; upper half casing
上告者 appellant
上格板 upper grid plate
上格状板 upper latticed plate
上隔板 upper spacer
上隔板梁 upstand diaphragm beam
上隔膜板 top diaphragm plate
上隔水层 positive confining bed
上更新世【地】 Epipleistocene; plio Pleistocene
上工作辊 top working roll
上弓 upbow
上弓形接线 supermartingale
上供或下供式双管系统 upfeed or down feed two pipe system
上拱 upwarp
上拱板 top arch bar
上拱度 camber
上拱高 depth of camber
上拱梁 camber(ed) beam; hog-backed beam
上拱竖曲线 hump vertical curve
上共轭水深 upper alternate depth
上钩 coupler head upper pivot lug
上估计值 upper estimator
上古【地】 Pal(a)eoid
上古构造体系 Pal(a)eoid tectonic system
上古生代【地】 Upper Paleozoic era
上古生界【地】 Upper Paleozoic erathem
上鼓轮 head pulley;top tumbler;upper tumbler
上固定颚板面 upper fixed jaw face
上固定支柱 <贯流式水轮机> upper stay column
上挂式轿厢框架 overslung car frame
上挂式离心机 suspended centrifuge
上冠 runner crown;upper canopy
上冠长 upper crown-length
上冠密封 rubber crown seal
上管钳 <拧开下管钳固定接头用> lead tongs
上光 glaze; glazing; lustering; polishing
上光板 glazing plate;glazing sheet
上光产品 glazing product
上光法 glazing method
上光花砖 <常为蓝色> azulejo
上光机 glazer;glazing machine
上光剂 brush polish; glazing agent; polish;Farromastic <一种专卖的用于金属窗的>
上光蜡 wax polish
上光轮 glazer
上光棉 glazed cotton
上光漆 finish vanish;gloss varnish
上光乳剂 polish emulsion
上光色料 glazing
上光丝 glazed silk
上光涂料 glazing paint
上光涂饰剂 glazed finish
上光线 glazed thread
上光像纸 glazed print
上光研磨 glaze
上光油 oil polish
上光照片 glazed print
上规 upper circle
上辊 top roll;upper roll
上辊筒 hang roll(er)
上辊压力 pressure of top roll
上滚波 uprushing wave
上滚轮 <升运机用> head pulley
上海港 Port of Shanghai
上海绿石砂层 upper greensand
上海-香港地区集装箱班轮航线 container liner service to Hong Kong area by Shanghai COSCO
上函数 superior function
上寒带【地】 upper frigid zone
上寒武统【地】 Upper Cambrian series
上好可销品质 good merchantable quality
上合 superior conjunction
上合下 above and under
上和 upper sum
上核 cokernel
上黑涂料 blackening
上横撑 top lateral brace; top transverse strut
上横档 top rail
上横号 superbar
上横梁 super-cross beam; top-end rail;up cross beam;upper beam
上横气道 upper horizontal flue
上横支撑 upper lateral bracing
上虹膜 iridizing
上弧线 upper camber
上护槛 upper guard sill
上护针三角 upper guard cam
上滑 upglide
上滑车 head block;top pulley
上滑道 bilge log;bilge way;sliding way
上滑道集材 ball hooting
上滑道作业 drydocking
上滑的 anabatic
上滑锋【气】 anafront
上滑距 upslip
上滑面 upglide surface;upslide surface
上滑坡道 ramp for climbing
上滑云 upglide cloud
上滑运动 upglide motion; upslope motion
上划线 overbar;overline
上环 upper ring
上环节组织 suprasegmental structures
上回转平台式塔式起重机 tower crane with slewing upper platform
上回转塔式起重机 high-level slewing tower crane
上回转台 upper turntable
上回转台压榨机 tilting head press
上混合共振吸收 upper-hybrid-resonance absorption
上火管锅炉 overhead fire tube boiler
上货 pick-up
上击 uptilt
上击肥轮 top beater;upper beater
上击式的 overshot
上击式水轮机 overshot water wheel
上机程序 coded program(me)
上机架 upper generator bracket; upper spider
上机架支臂 upper bracket-arm
上机壳 upper cover
上迹 <遗迹化石的> epichmia
上积 cup product
上基面 top base
上基座 top base
上级 higher-up; senior; superior; upper
上级拨入投资 investment allocated by senior administrative agency
上级补助收入 revenue from senior administrative agency subsidy
上级的 superior
上级机构 parent body
上级机关 higher authority; higher body
上级主管机关 competent authorities at higher level
上极 upper pole
上极部 upper pole piece
上极面 upper pole face
上极限 ceiling margin;limit superior; lines superior;superior limit
上极限尺寸 upper limit
上极限事件 superior limit event
上极限限位器【机】 lifting height limiter
上棘轮 top ratchet wheel
上集管(箱) upper header
上集水槽 upper tank
上集油箱 upper oil-header
上给 upfeed
上给料式燃烧 over-feed burning; over-feed combustion
上给煤 overstockering
上给燃料 overstockering
上给式 upfeed method
上给式供暖系统 upfeed system of heating
上给式供水系统 upfeed system
上给式双管热水(供暖)系统 two-pipe hot water upfeed system;two-pipe hot water system
上给式系统 upfeed system
上给下分式 upfeed method
上给下分式的 upfeed
上给下分式供暖系统 upfeed system of heating
上给下分式压力水箱系统 upfeed head tank system
上挤 force up
上夹板 train wheel bridge;upper plate
上夹板托 train wheel bridge support
上夹板位钉 train wheel bridge foot
上夹钳 top clamp
上夹钳尖 point of top clamp
上夹送辊 top pinch roll
上夹头 upper grip
上甲板 above deck
上甲板船首附近厕所 round house
上甲板间 upper tweendeck
上甲板梁 upper-deck beam
上甲板下吨位 under upper deck tonnage
上架 top carriage

上架耳轴轴承 top carriage trunnion bearing
上架拱圈 overarch
上剪片 top shear blade
上渐进线 upper asymptote
上槛 head of frame; head rail; head sill <门或窗的框顶上的横梁>
上槛木 straining sill
上浆 dressing;size;sizing;starch(ing)
上浆材料 sizing agent
上浆成分 sizing ingredients
上浆辊 doctor roll;sizing roller
上浆机 starching machine
上浆剂 slashing agent
上浆率 percentage size pick-up;sizing percentage
上浆配方 sizing instruction
上浆清漆 sizing varnish
上浆整理 sizing finish
上交叉 crossing above
上交成果资料 results and material of handing
上交税费 turnover taxes to the state
上交税金 tax delivery
上交税利 taxes and profits to be delivered to the sate
上交通运输污染源 transportation source of water pollution
上交通运输效率化法案 <美> Intermodal Surface Transportation Efficiency Act
上浇 top pour(ing)
上(浇)铸 top casting
上胶 gluing; proofing; size; sizing; sizing treatment; surface sizing; top with gum
上胶的 rubberized
上胶机 adhesive spreading
上胶机器 glue applicator
上胶量 spread
上胶密闭接头 joint rub
上胶盘 blocking cement
上胶器 sizer;sizing machine
上胶塞 top plug
上胶试验 sizing test
上礁 elevated reef
上角标 superscript
上绞综 top doup
上铰缝 hang
上铰链 top hinge
上铰链推窗 top hung
上搅拌翼 upper impeller
上缴国家利润 profits turned over to the state
上缴利润 profit delivery
上缴上级管理费 pay chief administrative expenses
上缴上级支出 pay senior administrative agency expenses
上接点 top contact
上接某页 continued from…sheet
上接头 top connection;upper contact
上节 upper segment
上截盘 overcutting jib
上界 upper bound
上界表达式 upper bound expression
上界操作数 high operand
上界法 upper bounding method
上界符号 upper bound of symbol
上界概念 broader term
上界解法 upper bound solution
上界失效频率 upper ineffective frequency
上金枋 upper purlin(e) tiebeam
上金檩 upper principal purlin(e)
上紧 tighten(ing)
上井行程 upstroke
上颈 upper hind neck
上警戒限 upper warning limit
上静点 center[centre] to center[centre];top center[centre]【机】
上举 raise;uplift
上举荷载 lifting load
上举力 lift; lift(ing) force; lifting power;uplift force;heave force
上举位置 <挖土机前端铲斗> float position
上锯轮 upper wheel
上卷 scroll
上卷机 up-coiler
上卷门 up-and-over door
上卷盘 upper drum
上卷式 overwind
上卷筒 coiling
上绝缘釉 insulating glazing
上开口 upper shed
上开口销 pinning
上开门 swing-up door;upward-acting door
上开桥 balanced drawbridge;bascule bridge;counterpoised drawbridge
上开式窗 sash window
上开式大门 fish-belly gate;fish-belly hinged-leaf gate
上壳 epitheca
上壳瓣 epivalve
上壳翼【地】carapace
上客 pick-up;on-load(ing)
上客港 port of embarkation
上客时间 boarding time
上坑柱 upper prop
上空 midair
上空结构 overhead structure
上空腔 upper plenum
上空使用权 air rights
上孔法 uphole method
上孔节气门 top-hole choke
上孔上盖 hatch cover
上孔阻风门 top-hole choke
上控制界限 upper control limit
上口 back cut
上口方板 ledg(e)ment
上口护条样 top-band pattern
上口线 top curve
上口鱼 superior mouth
上库 upper pool
上跨道路 high-flying highway
上跨公路 overpass
上跨公路桥 highway-overpass
上跨交叉 flyover crossing;overcrossing; overhead crossing; overhead passing
上跨交叉的斜坡道 overpass ramp
上跨交叉的匝道 overpass ramp
上跨立交 flyover conjunction;flyover crossing
上跨立交道路 cross-over road
上跨立交桥道路 <跨越铁路或道路的道路> cross-over road
上跨立体交叉 overhead crossing; overpass
上跨零点 upward zero crossing;zero up-crossing
上跨路 highway-overpass;overpass
上跨桥 overbridge; overhead bridge; overpass bridge;overspan bridge
上跨桥面 flying deck
上跨式车库门 overhead-type garage door
上跨式立交 overpass
上跨式立体交叉 overpass grade separation
上跨式立体交叉道路 cross-over road
上跨铁路立体交叉 overpass grade separation
上跨通道 overhead passing;overpass
上框 head rail;upper ledge;top rail <门的>
上扩式调压井 upper expansion chamber surge tank
上拉窗 lifting window
上拉电钮 upward button
上拉电阻箱 pull-up resistor
上拉法 up-draw sheet process
上拉杆 top connecting rod; top connection; upper connecting rod; upper link
上拉杆传感 top link sensing
上拉杆铰链 top link joint
上拉杆联结点 upper hitch point
上拉杆调节 <悬挂机构的> top link adjustment
上拉式叠库门 turnover door
上拉条 top bracing
上蜡 beeswax;cere;waxing
上蜡的 waxen
上蜡防雨布 wax cloth
上蜡机 wax-coating machine; waxing machine
上蜡橡皮线 waxed rubber wire
上栏 top rail
上蓝 blu(e)ing;bluing
上蓝剂 blu(e)ing;bluing
上蓝色 blue
上类 upper class
上立柱 upper column
上连接板 upper junction plate
上联结链 top connecting chain
上连续 continuity from above
上联管 upper-header pipe
上联箱 upper header
上敛复背斜 abnormal anticlinorium
上链 cochain;upper chain
上链复形 cochain complex
上梁 floor boarding joist;top bar;top beam
上梁庆宴 topping-out ceremony
上梁试验 earth beam test
上亮子 fanlight;transom
上了螺帽的 nutted
上了漆的板 enamel(l)ed slate
上了油的 oiled
上了釉的 glazed
上料车 charging wagon;skip car
上料斗 elevating hopper; feed hopper;feeding funnel
上料格栅 charging grate
上料刮板 feed shoe
上料管 feed pipe
上料辊 pick-up roll
上料机 feeder
上料机构 feed mechanism
上料井 charging well
上料卷扬机 charging hoist; loader winch
上料料车 charging wagon
上料皮带 charging belt
上料设备 charging equipment
上料设施 charging installation
上料系统 hoisting system
上料装置 loading attachment
上列 upper row
上邻异常 superjacent anomaly
上临界的 super-critical;upper critical
上临界冷却速度 upper critical cooling velocity
上临界流速 upper critical velocity
上流 upflow;upstream
上流泵站 upstream pumping unit
上流侧 upstream side
上流澄清池 upflow clarifier
上流过滤器 upflow filter
上流护床 rear apron
上流接触澄清池 solids contact upflow clarifier;upflow contact clarifier
上流节流 upstream restriction
上流流体 upflow fluid-catalyst unit
上流滤器 upflow baffle
上流石灰(石)床 upflow limestone bed
上流式沉淀池 upflow tank
上流式池 upflow basin
上流式的 updraft
上流式连续溶解器 upflow continuous dissolver
上流式滤清器 upflow filter
上流式泥渣层澄清池 upward-flow floc-banket clarifier
上流式系统 upflow system
上流式厌氧污泥床 upflow anaerobic sludge blanket
上流梯度试验 upward gradient test
上流原理 upflow principle
上楼梯 step-upstair(case)
上炉 drawing up; making-up; mudding up
上炉腹线 upper bosh line
上炉钩子 hook for starting up
上陆类群 disembarked group
上陆期 disembarked period
上绿砂岩 <下侏罗纪里阿斯统>【地】Upper Greensand
上轮 superior whorl;upper whorl
上轮压机 calender
上螺钉 screwing
上螺帽 nutting
上螺母 top nut;upper cap nut;upper nut
上螺母器 nut-runner
上螺栓 bolting
上落窗 sash window
上码头阶梯 berth access ladder
上毛布 top felt
上毛毯 glazing felt;upper felt
上毛毯压榨 top felt press
上冒头 head rail;top rail(of door); upper rail;frieze panel
上帽 top hat
上楣的下部 cornice soffit
上楣托座 cornice bracket
上楣(柱)cornice
上煤 coaling;fuel(l)ing
上煤粉机 pulverized coal feeder
上煤股道 coal(ing)track
上煤机 coal feeder
上煤器 coal pocket
上煤设备 coaling handling facility; coaling installation
上煤台 coaling station
上煤线 coaling road;coaling track
上煤站 coaling station
上煤装置 coaling gear
上门板压铁 topside door washer plate
上门服务 door-to-door service;door-to-service
上门接取包裹 collecting parcels
上门接送货物 door-to-door delivery
上门联运集装箱运输 door-to-door container traffic
上门取货 collection from domicile; pick-up from patron
上门取货付款凭单 collection voucher
上门取货证 collection permit
上门取送货物 collection and delivery
上门取送价率 collection and delivery rate
上门取送业务 collection and delivery service;pick-up and delivery service
上门取送运价表 collection and delivery tariff
上门取送运输 door-to-door conveyance;door-to-door traffic
上门收货服务 pick-up service
上密封盖 top cover labyrinth
上面 face;topside;up(per)side
上面板 top panel
上面的 overhead;upper
上面高 upper face height

S

上面居住楼层 upper residential floor; upper residential stor(e)y
上面刻齿动刀片 <切割器的> top-serrated knife
上面跨越 <道路立体交叉> crossing above
上面楼层 upper stor(e)y
上面排料 upper discharge
上面装模 top level die-filling
上皿式天平 top-loading balance
上模 male mo(u)ld; patrix; pattern cope;plunger die;top force;top form
上模板 cope match-plate pattern;cope plate;upper die plate
上模巢 top cavity
上模座 upper bolster;upper shoe
上膜具 top die
上磨面 top bevel
上磨盘 runner stone;upper millstone
上末道清漆 final varnishing
上墨 inking up
上墨辊 form roller;fountain roller
上墨手柄 ink lever
上墨水线 ink in;inking
上墨图 ink(ed) drawing
上墨系统 inking system
上墨装置 inking mechanism
上木 overwood;upper-growth
上木塞 <注水泥> wiper plug
上挠 bow up
上挠度 camber
上能带 upper energy band
上能态 upper state
上能态弛豫 upper state relaxation
上泥盆统【地】Upper Devonian series
上泥釉 slip coating
上逆 superinverse
上年度 antecedent year;prior year
上年结转余额 balance brought forward from last year
上枭【建】recta;upper cyma
上(拧)螺母 nut
上凝固点曲线 upper freezing-point curve
上爬 run-up;swash;uprush
上爬波 uprushing wave
上爬界限 limit of uprush
上爬区 swash zone
上爬水流 swash current
上排椽子 <双折屋顶> curb rafter
上排风管 upper exhaust duct
上排孔型 <轧制用语> top pass
上排污 upper drain
上攀沙纹交错纹理构造 climbing cross lamellar structure
上盘 hanger;hanging wall;upper circle;upper plate;vernier circle;vernier plate; hanging side【地】; top wall【地】; upper wall【地】; upper side <断层的>
上盘的地层 strata of hanging wall
上盘定位夹具 blocking fixture
上盘断块 upfaulted block
上盘废石 hanging wall waste
上盘立柱 hanging wall leg
上盘平峒 hanging adit
上盘平巷 hanging wall drift
上盘上升 hanging wall upthrow
上盘水准器 plate level
上盘微动螺旋 upper plate slow motion screw
上盘位移方向 displacement direction of hanging wall
上盘下降 hanging wall downthrow
上盘岩层 hanging layer
上盘晕 upper wall halo
上盘制动螺旋 <经纬仪的> alidade clamp;upper clamp
上盘主动上投 hanging wall active upthrow
上盘转动 upper motion
上旁承 body side bearer; body side bearing
上旁承梁 body side bearing beam
上旁承梁角钢 body side bearing beam knee angle
上旁滚 upside roll
上抛的 upcast
上抛物 upcast
上抛卸式垛草机 overshot stacker
上炮 deluge gun
上喷 upwelling
上喷式燃烧器 upshot-type burner
上喷水器 <在被轧制带材的上面> top water sprayer
上喷嘴 top nozzle
上喷嘴组 top sprays
上皮薄壁组织 epithelial parenchyma
上皮栏 terminal ledge
上偏 upward bias
上偏差 upper deviation; upper variation of tolerance
上偏的升降舵 up-elevator
上偏光镜 analyser[analyzer]; upper nicol
上撇号 apostrophe
上频率 upper frequency
上平板 upper flat plate
上平联 top bracing
上平台 upper mounting plate
上平行 coparallel(ism)
上平纵联 top lateral bracing
上坡 adverse grade; ascending gradient; ascent; climb; foreslope; grade-up; hill-climbing; run-up; summit grading; upgrade; upheaval; uphill; uphill grade; upslope
上坡侧 uphill side
上坡车道 crawler lane
上坡车速 speed on grade
上坡淬火 uphill quenching
上坡道 ascending gradient;rising gradient;upgrade;up(hill)gradient
上坡的 acclivous;uphill
上坡低速挡 grade retarder
上坡地段 upgrade section
上坡度 ascending grade; plus grade; rising gradient; up (hill) gradient; ascending gradient
上坡方向【铁】up(train)direction
上坡防退器 hill holder
上坡风 anabatic wind;upslope wind
上坡公路 highway with rising gradient
上坡管道 uphill line
上坡焊 uphill welding; upward welding in the inclined position
上坡交通 ascending traffic
上坡角 ramp angle
上坡扩散 uphill diffusion
上坡流 uphill flow
上坡路 track raising;uphill road
上坡锚 anchor cast on upward
上坡能力 ability to climb gradients; hill climbing capacity
上坡跑道 upslope runway
上坡平巷 upraise drift
上坡坡度 uphill gradient
上坡坡段 upgrade section;uphill section
上坡气流 upslope flow
上坡牵引车 barney;bullfrog;donkey; ground-hog;larry;mule
上坡牵引力 ascending tractive force; climbing power
上坡区段 upgrade section
上坡升压泵站 upgrade pumping station
上坡时的高档行驶速度 top-gear speed on gradient
上坡速度 ascending velocity;speed on upgrade
上坡雾 upslope fog
上坡下坡 ups and downs
上坡信号标志 <在双线上准许货物列车限速通过> grade signal marker
上坡行车 upward run
上坡行程 uphill journey;upwards journey
上坡运输 ascending traffic
上坡运行 upgrade run
上坡阻力 up(grade)resistance
上铺 upper berth
上铺挡板 bunk apron
上铺扶梯 upper berth ladder
上铺路面层 overlay of pavement
上铺嵌口 upper berth pocket
上期成本 historic(al)cost
上期结余 old balance
上期结转 balance brought forward; carry forward
上期增刊 previous supplement
上期转来 carried forward
上漆 dope;japanning;lacquering;painting
上漆辊(子)applicator roll
上漆炉 japanning oven
上起重滑轮 load upper block
上气道 gas uptake;uptake
上气筒垫密片 upper-air cylinder gasket
上汽包 steam drum
上汽缸垫密片 upper steam cylinder gasket
上汽缸起吊 lifting turbine cover
上汽锅 upper boiler;steam drum
上砌层 upper course
上铅小圆规 bow pencil
上铅釉 lead glazing
上前舱 upfront
上钳板 top nipper
上钳板臂 top nipper arm
上嵌缝料 tooling
上墙板 cornice panel;cornice subfascia;frieze
上桥斜坡 bridge ramp
上翘 camber;cocking-up
上翘活动桥 bascule bridge
上翘屋檐 upturned eaves
上切面 upper edge
上切式 up cut
上切式剪切机 up-cut shears
上倾 up-dip
上倾附着物 up-dip attachment
上倾尖灭型油气藏趋向带【地】up-dip wedge-out type pool trend
上倾流 upwash
上倾式履带装载机 crawler overshot loader
上倾式装载机 overhead loader;overshot loader
上清流量 supernatant flow
上清排出 supernatant draw-off pipe
上清漆 varnishing;varnish lacquer
上清漆的 varnished
上清液 supernatant liquid
上顷尖灭圈闭【地】up-dip wedge-out trap
上区 upper curtate
上区段 upper curtate
上曲 upsweep
上曲柄 upper crank
上曲面 top camber;top surface camber;upper camber
上曲线 upper curve
上曲轴 upper crank shaft
上曲轴箱 upper crankcase
上屈服点 upper yield point
上屈服应力 upper yield stress
上确界 least upper bound;supremum【数】
上燃料栈桥码头 bunker jetty
上人平屋顶 <指屋顶的一部分> roof deck(ing);accessible flat roof
上柔下弯墩 super-flexible sub-rigid pier
上乳白色釉的英国瓷器 queen's ware
上润滑油的 lubricating
上塞 top plug;upper plug
上塞法注水泥 moving-plug method of cementing
上塞注水泥(固井)法 top packer method
上三叠纪【地】Upper Triassic period
上三叠统【地】Upper Triassic series
上三分点 two-third point
上三角形矩阵 upper triangular matrix;upper triangulation matrix
上三角洲平原沉积 upper delta-plain deposit
上三叶轮增压器 upper three-blade rotor
上扫描线 upper tracer
上色 colo(u)ring; painting; tintage; tinting;to stain
上色打底机 <陶瓷制品> ground laying machine
上色辊 fuser
上色滚筒 colo(u)r drum
上色率 degree of dyeing
上砂 abrasive grain dispensing
上砂轨 sand track
上砂机 sanding machine
上砂设备 sanding equipment;sanding plant
上砂线 sand track
上砂箱 cope(box);cope flask
上砂装置 sand feeder
上筛 top sieve;upper sieve
上筛层 <多层排列筛分装置的> outer screen
上筛的延长部分 upper sieve extension
上筛骨 suprethmoid
上筛架 scalping shoe
上山 board up;raise;uphill
上山的 uphill
上山掘进 raise advance
上山开石 raise opening
上山坡(度)uphill slope
上山箱式缆车 gondola
上山巷道 bord-up
上山运输机 up brow conveyer[conveyor];uphill conveyer[conveyor]
上扇 upper fan
上扇滑动式窗扉 casement with sliding upper-sash
上上下下的 up-and-down
上舍入 round up
上设拱圈 overarch
上射式的 overshot
上射式水轮机 overshot(water)wheel
上射式水枪 overpumping
上身安全带 upper torso restraint
上深槽 upper deep;upper pool;upper trench
上蜃景 superior mirage
上升 arise; ascend; build-up; climb; going up;heave;lifting;move-up;override; rise; run-up; upfold; upheaval; upheave; uphill; uplift; upthrust;upturn
上升岸 rising coast
上升岸线 emerged shoreline;negative shoreline;shoreline of emergence
上升壁 unlifted wall
上升篦板 rising grate
上升边 leading edge;advancing edge; entering edge;front edge;rising edge
上升滨岸平原 raised beach plain; raised beach platform
上升滨面阶地 elevated shore face terrace

上升滨前阶地 elevated shore face terrace
上升滨线 negative shoreline; shoreline of elevation <海底或湖底上升的>; shoreline of emergence <陆地上升的>
上升波 rising wave
上升部分 incremental portion; upflow circuit <回路中的>
上升侧 uplifted side
上升侧滚 upward roll
上升齿 lifting teeth; vertical teeth
上升冲程 upstroke
上升船机作业 drydocking
上升垂直位移 <逆断层> upthrow
上升磁铁挡 stop for lifting magnet
上升次序 ascending order
上升岛 upheaval island
上升到顶 top
上升道 riser
上升道闸板 uptake damper
上升的 angle of climb; ascending; ascensional; hypogene; upward
上升的岸线 emerged shoreline
上升的滨前沉积阶地 elevated shore face terrace
上升的冲断层前缘 emergent thrust front
上升的气泡 coursing bubble
上升地块【地】 uplifted block; elevated block
上升地垒 uplifted horst
上升点 point of rise; rising point
上升电磁铁 <选择器的> vertical magnet
上升动作 up maneuver
上升动作断续接点 vertical interrupter contact
上升动作接点 vertical contact
上升段 ascending branch; rising section; up-leg; rising segment <曲线的>
上升段轨道 upward trajectory
上升断层 upthrow fault
上升断块 elevated block; uplifted block
上升断块差异压实背斜圈闭 anticlinal trap by differential compaction over uplifted block
上升范围 lifting range
上升飞行 upward flight
上升分线管 lifting pipe
上升风 anabatic wind
上升伏安特性 rising volt-ampere characteristic
上升杆 elevating lever
上升高原 chapada
上升割面 ascending face
上升功率 ascending power
上升管 riser circuit; riser conduit; riser pipe; riser piping; riser tube; riser tubing; upflow circuit; uptake pipe; uptake piping; uptake tube; uptake tubing
上升管岔道 riser turnout
上升管出口 riser outlet
上升管连接件 rising connector
上升管连接器 rising connector
上升管联箱 upcast header
上升管渠 uptake conduit
上升管线 rising pipeline
上升管支管 riser leg
上升轨道 ascending trajectory
上升海岸 coast of elevation; elevated coast; emergent coast; rising coast; uplift coast; coast of emergence
上升海岸阶地 elevated shore face terrace
上升海岸线 elevated shoreline; emerged shoreline; shoreline of elevation; shoreline of emergence
上升海滨线 elevated shoreline
上升海底 rising(sea) bottom
上升海滩 elevated beach; raised beach; elevated shore line
上升函数 increasing function
上升焊接法 raised welding method
上升航迹 upward path
上升航向 upward course
上升花被卷迭式 ascending aestivation
上升花序 ascending inflorescence
上升回路 riser circuit
上升集合 upgrade set
上升计数器 up-counter
上升记录 uptrace
上升继电器 vertical movement relay
上升架 raised floor(ing)
上升礁 elevated reef; raised reef; uplifted reef
上升角 angle of ascent; angle of climb; angle of rise
上升铰链 skew hinge
上升阶地 elevated bench; raised bench; uplift bench
上升阶段 uplifting stage
上升阶梯信号 riser staircase signal
上升节点 ascending node
上升径流 ascending flow
上升卷门 overhead door
上升开采 rise working
上升空气 ascending air
上升浪蚀台 elevated wave-cut bench
上升离位触点 vertical off-normal contacts
上升力 ascending power; climbing power; uplift force; upward force
上升力矩 lifting moment
上升立管 upfeed riser; uprise pipe
上升联锁继电器 raise interlock relay
上升流 ascending flow; updraft; updraught; upward flow; upwelling current; upwelling water
上升流动 rising flow
上升流过滤层 upward flow filter
上升流区 upwelling area; upwelling region
上升流速 ascending velocity; ascensional velocity
上升陆架 uplifted(continental) shelf
上升率 escalating rate
上升率传感器 rate-of-climb sensor
上升率控制 rate-of-climb control
上升螺旋 elevating screw; upward spin
上升螺旋桨 lifting airscrew
上升脉外平巷 rising stone drift
上升面 raised floor(ing)
上升磨削法 climb cut
上升能力 ascending ability; ascending power; upward capability
上升能量 ceiling capacity
上升暖气流 ascending warm current; thermal
上升排风道 uptake shaft
上升盘 upthrown block; upthrust block
上升盘地层厚度 strata thickness of upthrow side
上升盆地 elbasin
上升平原 uplifted plain
上升坡 uprise
上升坡度 ascending grade; plus grade
上升坡度的道路 road with rising gradient
上升期 uplifting period; period of rise <水文过程线的>
上升气流 anaflow; ascending air; ascending air current; ascensional air current; upcurrent; updraft; updraught; upflow; upstream; upward current of air; upward draft; ascending current
上升气体 uprising gas
上升区 upheaved region
上升曲线 ascending curve; ascent curve; up curve
上升趋势 upward trend
上升渠道 ascending trajectory
上升泉 ascending spring; ascension spring; hypogene spring; non-gravity spring; rising water
上升热流带 upwelling hot flow zone
上升日珥 rising prominence
上升三角洲 elevated delta
上升扫帚 vertical wiper
上升生成的 hypogeal; hypogene; hypogenic
上升时间 rise time; rising period; time of rise; upslope time
上升时间校正 rise-time correction
上升时间开关 rise-time switching
上升时间失真 rise-time distortion
上升时间跳动 rise-time jitter
上升时间响应 rise-time response
上升式采暖系统 rising heating system
上升式流液洞 lifted throat; step-up throat
上升式通风 upward ventilation
上升式脱轨器 lifting derail; lift-type derail
上升竖管 uprise conduit; uprise pipe [piping]; uprise tube[tubing]
上升水 ascending water; upward water; upwell(ing) water; hypogene water【地】
上升水量 upward water flow
上升水流 ascending current; upward current; upwelling flow
上升水流分级机 hydrosizer; upward current classifier
上升水流分离器 rising-current separator
上升水流跳汰机 upward current jig
上升水面曲线 rising surface curve
上升水位 rising stage
上升水下阶地 elevated shore face terrace
上升顺序 ascending
上升瞬态 rising transient
上升说 ascension theory
上升速度 ascending velocity; ascensional velocity; raising speed; rising velocity; rate of climb; rate of rise; rising rate; upward velocity; velocity of ascension
上升速度计 vertimeter
上升速度指示器 vertical speed indicator
上升速率 raising rate; rate of rise; rise rating; rising rate; rise rate
上升算符 raising operator
上升索道 hoistway
上升特性 rising characteristic
上升梯度变化率 rate of upward gradient
上升梯子 up-ladder
上升通风 ascensional ventilation
上升途径 elevated path
上升脱离目标 upward break
上升位置 lifting position
上升涡流 ascending eddy
上升污泥 raised sludge
上升雾 lifting fog
上升系数 climbing number
上升系统 upward system
上升下降延迟 rise-fall delay
上升咸水锥体 rising salt brine cone
上升线 ascending branch
上升限度 ceiling height
上升斜流 oblique inflow
上升斜率 rising slope
上升旋涡 ascending eddy; rising whirl
上升旋转选择器 two-motion selector
上升选线器 vertical selector
上升压力 unlifting pressure
上升烟道 funnel uptake; uptake; uptake breeching; uptake flue
上升岩巷 rising stone drift
上升沿 positive-going edge; rising edge
上升洋流成矿模式 upwelling-current metallogenic model
上升余弦 rised cosine
上升运动 ascending motion; lifting motion; lifting movement; positive movement; rising motion; upward movement
上升沼地 climbing bog
上升阵风 upgust
上升轴 elevating shaft
上升爪 vertical pawl
上升转为下降区 converting area from uplifting into depression
上升桩 uplift pile
上升准平原 elevated peneplain; uplifted peneplain
上升总管 rising main
上升总时间 total ascent time
上十字架隔屏高台的阶梯 rood stair(case)
上石膏垫料 laying on plaster
上石炭纪【地】 Conemauch
上石炭统【地】 Upper Carboniferous (series)
上市 marketing
上市程序 listing program(me)
上市代理人 <房地产> listing agent
上市的 on-sale
上市股票 <指在交易所挂牌的> listed stock
上市股票的总市值 aggregate value of listed stock
上市价格 listed price
上市商品量 pitch
上市证券 listed securities
上视回声探测仪 upward-looking echo sounder
上视图 top view
上首门槛 upper port sill
上枢轴 top pivot; upper pintle
上鼠笼 upper cage
上鼠笼条 top bar; upper bar
上述规定 afore mentioned rules
上竖撑 top vertical brace
上衰落 fading rise
上甩套的围盘 breakout repeater
上水 feed(ing) water; upriver
上水泵 charging pump
上水驳船队 upbound tow
上水出闸时间 upbound exit time
上水船(舶) upbound boat; upbound steamer; upbound vessel; upstream vessel
上水道 boosted water supply; water line; water supply line
上水的 upbound
上水(顶推)驳船队 upbound tows
上水管 rising water pipe; water pipe; water supply pipe
上水(管)道 water supply(pipe) line
上水航程 upbound journey
上水航驶 proceed up the channel
上水航行 upstream voyage
上水(货)运量 upstream traffic
上水检验 examination of sanitary water
上水进闸时间 upbound entry time
上水口 filling nozzle; filling pipe end
上水库 head reservoir
上水立管 water supply riser
上水流 headwater
上水率 rate of feed; rating of feeding (water)
上水石 tufa; tuff

上水系统 water supply system
上水箱 feed tank; overhead water tank; roof tank; upper water box
上水行驶 proceed up the channel; upbound sailing; upstream sailing
上水运量 upbound commerce traffic; upbound traffic
上水运输 upstream traffic
上税单 duty memo
上死点 top center[centre]; top dead center[centre]; upper dead center [centre]
上死点标记 top-dead-center indicator
上死点指示器 top center indicator
上四分位长度 upper quartile length
上四分位数 upper quartile
上饲(燃料) over-feed
上饲式燃烧 over-feed firing
上诉法院 court of appeals
上诉庭 < 房地产税 > appeals board
上速程 upper speed course
上塑限 upper plastic limit
上溯鱼 upstream-bound fish
上梭口 top shed
上锁保管 lockup
上锁臂 upper lock arm
上锁人 locker
上锁销 < 车钩 > top lock lifter
上锁作动筒 uplock cylinder
上塔楼梯 tower stair(case)
上台阶 top bench; upper bench
上抬杆铰接环 < 悬挂装置的 > top line rocker
上态 upper state
上态符号 upper state symbol
上滩 ascending a rapid; beaching; stranding grounding
上搪瓷釉 porcelain enameling
上梯段式开采法 top-benching
上提式(车钩) top uncoupling type
上提综杆 top jack; top lever; upper treadle motion
上天极 elevated pole
上挑鼻坎 upturned bucket
上挑的 upcurved
上挑丁坝 upward-angled spur dike [dyke]; upward-pointing groin; upward-pointing spur dike[dyke]
上条大钢轮 ratchet winding wheel
上条大钢轮摆杆 ratchet winding wheel swing lever
上条大钢轮摆杆衬圈 ratchet winding wheel swing lever ring
上条大钢轮摆杆簧 ratchet winding wheel swing lever spring
上条大钢轮簧 ratchet winding wheel spring
上条方柄头 winding square
上调和的 superharmonic
上调和函数 superharmonic function
上调压井 upper surge tank
上铁轭 upper yoke
上同调【数】 cohomology
上同调理论 cohomology theory
上同伦【数】 cohomotopy
上统【地】 Upper series
上筒体 upper shell
上筒体法兰 flange for upper shell
上投 upcast; upthrow; upslide
上投侧 upcast side; upthrow side
上投断层 jump-up fault; upcast fault; upthrow fault
上投断层的 upfaulted
上投断块 upfaulted block
上投滑距 upcast slip
上投距 upthrow
上透光层 upper photic zone
上凸 < 层面的 > epirelief
上凸梁 upstand(ing) beam
上凸轮 overhead cam
上凸轮轴 overhead camshaft
上凸模具 hump mo(u)ld
上凸下凹的波状花边 cyma reversa
上凸下凹反曲线 cima reversal
上凸下凹双曲线脚 cima-inversal
上凸型 upper convex type
上凸缘 top flange; upper flange
上图廓 top border; top margin; upper border; upper margin
上涂 < 石灰膏涂层 > setting coat (plaster)
上涂层 upper coat
上涂料 dope; doping; enamel(l)ing; slur(ring)
上土层 A-horizon
上推 push-up; upthrust
上推表 first-in first-out; push-up list
上推操作 push-up operation
上推插销 upper cat bar
上推存储 push-up store
上推存储器 push-up storage
上推的门 push-up door
上推断层 jump-up
上推队列 push-up queue
上推分类法 bubble sort
上推力轴承 upthrust bearing
上推门 lift-up door
上推排序法 bubble sort
上推式存储器 push-up storage
上推页 push-up leaf
上托 pop-up
上托板 mounting plate
上托辊 top idler; upper supporting roller
上托力 uplift; uplift force
上托力的抗力 resistance to uplift
上托力计 uplift cell
上托力矩 lifting moment
上托力强度系数 uplift intensity factor
上托线圈 repulsion coil
上托指令 pop instruction
上托钻 upper side stone
上驮 overriding
上驮板块 overriding plate
上挖 high cut; raising; up digging
上挖斗式挖土机 bucket excavator for upward scraping
上挖式多斗挖掘机 bucket excavator for upward scraping
上挖式链斗挖掘机 bucket excavator for upward scraping
上挖式挖掘船 up-boom dredge
上瓦层 < 紧挨屋脊 > top course tile
上弯度 bending deflection
上弯钢筋 bent-up bar
上弯构架 upswept frame
上弯管 < 压力管道的 > upper bend
上弯桁架 camel-back truss
上弯梁 moonbeam
上弯(倾) kick-up
上弯式支架 kick-up frame
上弯线 hang
上网 top wire; upper wire
上网潮流 load flow; power flow
上网电价 electricity price to network
上网挂浆 up former
上桅 topgallant mast; topmast
上桅帆 topsail
上桅水手 canvas climber
上桅台 upper mast table
上桅桅肩 topmast hounding
上桅扬帆结 topsail halyard bend
上维界 upper dimension bound
上纬度 upper latitude
上卫板 hanging guard
上位参量 epistatic parameter
上位词自动登录 up-posting
上位方差 epistatic variance
上位花 epigynous flower
上位机 master
上位集 co-level
上位矩阵 precedence matrix
上位开孔 superior-bore
上位离差 epistatic deviation
上位内存【计】 upper memory area
上位内存块 upper memory blocks
上位式 epigyny
上位锁 uplock
上位相互作用 epistatic interaction
上喂入齿耙 upper feed rake
上稳定性 upward stability
上沃尔特 < 非洲 > upper Volta
上午 ante meridian
上午测天 morning sight
上午的中段时间 midmorning
上午高峰(小时) AM peak(hour)
上吸 up-draught
上吸式拔气罩 updraft hood
上吸式化油器 updraft carburet(t)or [carburet(t)er]
上吸式均流侧吸罩 updraft uniform-flow lateral exhaust hood
上吸式汽化器 updraft carburet(t)or [carburet(t)er]
上吸式圆形回转罩 updraft circular rotary hood
上洗 upwash
上洗流 upwash
上下摆动 luffing
上下班报到地点 on-off duty point
上下班乘车出行 work trip
上下班出行【交】 commuting trip
上下班公车 commuter bus
上下班公共交通 commuter transit
上下班公共汽车 commuter bus
上下班计时 clock in and clock out
上下班交通 commuter traffic; commuting traffic
上下班交通拥挤 rush-hour overcrowding
上下班客流 commuting movement
上下班列车 commuter railroad
上下班路线 commuter route
上下班圈 commuter sphere
上下班人口 commuting population
上下班时间 commuter time; rush hours
上下班时间服务 rush-hour service
上下班数据 start and end shift data
上下班铁路 commuter railroad
上下班行程 commuting journey to work
上下波动 surge
上下不对称 up-down asymmetry
上下层窗间玻璃 spandrel glass
上下层窗间墙 spandrel wall(ing)
上下层窗空间【建】 spandrel[spandril]
上下层窗空间盖板 spandrel panel
上下层窗空间墙 allege
上下层窗之间的空间 window spandrel
上下车 boarding and alighting
上下车旅客 on-off passenger
上下扯窗(扇) double hung sash(window); double hung (counterweight) window; vertical sash
上下窜动 uphill and downhill travel
上下窗空间悬臂踏步 spandrel cantilevered step
上下窗之间的空间 spandrel
上下导洞法 top and bottom pilot tunnel(l)ing method
上下导坑先墙后拱法 Austrian method(of timbering)
上下的 up-and-down
上下等径的井 well of one diameter
上下等速烧割法 drop cut
上下颠倒 up-down reversal
上下对开行车 up-and-down traffic
上下法 up-and-down method
上下分流廊道 < 船闸中部 > over-and-under culvert
上下浮动周期 period of dipping and heaving
上下滑动吊窗 drop slide window
上下滑动套管 reciprocate the casing
上下活动遇阻钻具 work the string up and down
上下活动钻具 spudding up and down; work the pipe up and down
上下级冲突 subordinate-supervision conflict
上下级往来账户 account current among senior administrative and subordinate unit
上下极限 high low limit
上下结构探测器 detector with up and down structure
上下结合方式 top-bottom integrated approach
上下界 bound
上下界防护 datum-limit protection
上下界寄存器 boundary register; limit register
上下进整 approximation up or down
上下锯机 top and bottom saw
上下可动反射镜 tilting reflector
上下可以分别开关的两截门 Dutch door
上下客站台 loading platform
上下控制 up and down control
上下块之接头 tabled joint
上下拉窗的空心框架 cased frame
上下联杆 up-and-down rod
上下流活性炭吸附 up-flow-down-flow carbon adsorption
上下楼作业 interfloor operation
上下轮换开挖法 up-and-down alternating excavating process
上下落料法 up-and-down blanking process
上下冒头 upper and lower rail; ledge batten < 门的 >
上下模间的间隙 spew relief
上下配合模 counter locked die
上下坡 climb and fall
上下坡踏步 terrace slope
上下坡作业 up-down operation on incline
上下起伏 heaving
上下深槽交错 staggering of upper and lower deeps; staggering of upper and lower pools
上下式箱形断面 < 有上、下层的地下隧道断面 > over-and-under box section
上下视差 vertical parallax; Y-parallax; longitudinal parallax
上下视差滑尺 parallax slide; vertical parallactic slide; vertical parallax slide
上下视差螺旋 vertical parallax screw
上下双动卡木钩 boss dog
上下双动水压机 Bussman-Simetag press
上下水道 water and sewage pipeline
上下水道工程 water supply and sewerage works
上下水道立管 riser-and-stack
上下水道设备 supply and disposal services
上下水道系统 water-wastewater system
上下水管道系连通 cross-connection in piping system
上下四分点 upper and lower quartile
上下索 parbuckle
上下提动被卡钻杆使之解卡 work the pipe to free
上下推动牵引钩 up-and-down pull on the drawbar

上下推拉窗(扇) vertically sliding sash; balanced sash; double hung(counter-weight) window; sash window
上下位锁 up-and-down locks
上下文 context(ure)
上下文有关的菜单 context-sensitive menu
上下误差 up-down error
上下弦 half moon; quadrature【天】
上下弦低潮 low-water quadrature
上下弦非平行桁架 non-parallel-chord truss
上下弦平均高潮间隙 high-water quadrature
上下弦平行桁架 parallel chord truss
上下限 bound; upper and lower limit
上下限报警机构 upper and lower limit alarm mechanism
上下限地址寄存器 boundary register
上下限幅放大器 window amplifier
上下限寄存器 boundary register
上下行分开 directional separation
上下行交通 up-and-down traffic
上下行列车 up-and-down trains
上下行色谱 ascending-descending chromatography
上下型机组 cope and drag set
上下型箱架 cope and drag mount
上下摇摆 teeter
上下液面警报器 level alarm high low
上下移窗 vertical sliding sash
上下移动 reciprocate; wag
上下游河区 river reach
上下游水的一致性 hydro-solidarity
上下游隙 vertical play
上下游闸门枢轴之间的总长 pintle-to-pintle length
上下运动 lift movement; see-saw; see-saw motion; up-and-down motion; up-and-down movement; vertical motion
上下闸门距离 miter[mitre] to miter [mitre]
上下轴承 metal upper/lower
上下左右全反转的 upside-down and laterally reversal
上下作用的垂直力 up-and-down pull
上弦 dichotomy; principal rafter; top chord; top flange; upper chord; upper string; first quarter of the moon <月亮的>; gibbous
上弦侧向支撑 top lateral bracing
上弦承重悬臂桥 cantilever-deck bridge
上弦杆 top boom; upper boom; top chord; top chord member; upper chord; upper chord bar
上弦杆板 top boom plate; top chord plate; upper boom plate
上弦杆构件 top boom member
上弦杆件 upper chord member
上弦杆力 top chord force
上弦杆连接板 upper boom junction plate; upper chord junction plate
上弦横撑 upper lateral bracing
上弦横向水平支撑 top lateral bracing; upper lateral bracing
上弦剪刀撑 cross bracing of the tope chord
上弦节 top chord panel
上弦连接板 top boom junction plate
上弦木 principal; principal rafter
上弦小潮 first quarter neap tide
上弦与斜端的杆接点<桁架> hip point
上弦与斜端的结点<桁架> hip joint; hip point
上弦月 waxing moon; first quarter
上弦支撑 upper chord bracing
上弦纵撑 upper longitudinal bracing
上弦纵向水平支撑 top longitudinal bracing; upper longitudinal bracing
上现蜃景【气】 loom; superior mirage; upper mirage
上线 in-line
上线拉力 top tension
上限 high limit; superior limit; top limit; upgrade; upper(limit); upper bound
上限尺寸 high limit of size
上限的 limited to the right
上限点 upper change point
上限定理 upper bound theorem
上限动块 up-stop
上限法 upper bound method
上限公差 high limit of tolerance
上限积分 upper integral
上限寄存器 high limit register
上限价格 ceiling price
上限接受值 upper acceptance value
上限截止频率 higher cut-off frequency; upper cut-off frequency
上限解 upper bound solution
上限界 upper boundary; upper control limit
上限累加器 upper accumulator
上限临界速度 upper critical speed; upper critical velocity
上限滤波器 upper limiting filter
上限浓度 ceiling concentration; concentration upper limit
上限频率 upper limiting frequency
上限起始致死温度 upper incipient lethal temperature
上限甚高频 high-very-high frequency
上限温度 ceiling temperature
上限信号器 high alarm
上限越界 off-normal upper
上限值 upper limit value; upper range value
上陷型模 top swage
上相 top phase
上箱 top box; top tank
上箱下沉 sag
上镶板<有五块或更多镶板门的> frieze panel
上向 upper quadrant
上向臂板 upper quadrant arm; upper quadrant(semaphore) blade
上向臂板信号机【铁】 upper quadrant (semaphore) signal
上向采矿 upward mining
上向抽力 upward pull
上向单臂板信号机【铁】 upper quadrant single-arm semaphore signal
上向分层采矿法 upward slicing
上向分力 lifting component
上向风流 rising air current
上向工作面 up-face; upwall
上向关闭式滑板闸门 upward-closing slide gate
上向掘进 raise driving; raise mining; upraising; upward advance
上向掘进的井筒 upraise shaft
上向孔 uphole; upward hole
上向孔爆破 uphole shooting
上向孔测定器 uphole detector
上向孔凿岩 uphole drilling
上向孔钻进 overhead drilling
上向矿房 upraise room
上向连接井壁法 underpinning
上向流 upward current; upward flow
上向流过滤 upflow filtration
上向流活性污泥法 upper flow activated sludge process
上向流接触器 upflow contactor
上向流倾斜管 upflow slanted tube
上向流色谱法 upward-flow chromatography
上向流砂滤池 upward-flow sand filter
上向流砂滤器 upflow sand filter
上向炮孔 uphole
上向炮眼作业 uphole work
上向坡度 uphill gradient; upward gradient
上向渗滤 upward percolation
上向速度 upward velocity
上向台阶式开采 overhand mining
上向掏槽 draw cut
上向梯段回采法 overhand stoping
上向通风 upward draft; upward ventilation
上向运动 upward movement
上向凿井<自下向上的> raise boring; raise drilling; up-over
上向凿井提升机 raise lift
上向凿岩 overhead drilling; upward drilling
上向钻进 uphole boring; uphole drilling; upward drilling
上向钻孔 rising borehole; up-over borehole; upper borehole
上向钻孔工作 uphole work
上向钻眼 overhead drilling; upward drilling
上象限 upper quadrant
上橡胶涂层的 rubber-coated
上小钢轮 top crown wheel; upper crown wheel
上小钢轮螺钉 top crown wheel screw
上小下大的 big-end down
上小下大钢锭模 narrow-end-up mo(u)ld
上斜 upsweep
上斜撑 top diagonal brace
上斜的 acclivitous
上斜铰链 skew hinge
上斜炮眼 dry hole
上斜式输送机 upwardly inclined conveyer[conveyor]
上斜式运送机 upwardly inclined conveyer[conveyor]
上斜斜坡 acclivity
上心拱 stilted arch; stilted vault
上心拱桥 stilted arch bridge
上心拱形穹顶 stilted vault
上心盘 body-centered[centred] plate; male center[centred] plate; top center[centred] plate
上心盘中心距 distance between pivots
上心盘座 body bolster filler; bolster center[centre] filler; filling spider
上心穹顶 stilted vault
上芯头 cope print
上新世【地】 Pliocene epoch
上新世后【地】 Post-Pliocene
上新世时期【地】 Pliocene period
上新统【地】 Pliocene series
上信标灯 upper beacon light
上行 uphill; up-run; upstream; up traffic; forward run
上行波反褶积剖面 deconvolution section of upwave
上行波各道叠加 stack of upgoing traces
上行波和下行波的叠加 stacking of upgoing and downgoing events
上行波和下行波分离并拉直 separating and flattening of upgoing and downgoing events
上行波剖面 on-going wave section
上行驳船 ascending barge
上行驳船队 upbound tow
上行车道车流 upstream lane flow
上行车道流量 upstream lane flow
上行程 upstroke
上行充电带 upward run
上行冲程 upstroke
上行出闸时间 upbound exit time
上行船(舶) ascending boat; ascending vessel; upbound boat; upbound vessel; upstream vessel
上行船队 upbound fleet; upbound tows
上行道 up-line
上行的 anadromous; upstream; upbound
上行段 ascending branch
上行吨位 tonnage upstream
上行多普勒 up-Doppler
上行法 ascending method
上行方面 up side
上行方向 up direction; up(train) direction【铁】
上行分支 ascending branch
上行风流 upcast air
上行沟通<员工向上级报告工作情况> upward communication
上行管 ascending pipe; ascending tube; ascension pipe
上行管道 uphill line
上行海岸 upcoast
上行回路 upflow circuit
上行货流 upbound cargo flow
上行进闸时间 upbound entry time
上行开采 ascending ming
上行链路 up-chain; up-link
上行列车【铁】 uptrain; upbound train
上行路线 turn-up wiring
上行捻线机 uptwister
上行坡 acclivity
上行气流 ascending current
上行热流量 hot ascending flow
上行色谱法 ascending chromatography
上行色谱图 ascending chromatogram
上行式 upstriker
上行式采暖系统 upfeed heating system
上行式打手 upstroke heater
上行式的管理 bottom-up management
上行式电葫芦 top-running hoist
上行式钉式打手 upstroke pin type beater
上行式平板机 upstroke press
上行式起重机 top-running hoist
上行式通风机 upcast fan
上行式悬臂起重机 cantilever overhead travelling crane
上行式压力机 upstroke press
上行式振动运输机 uphill shaker
上行输电带 ascending belt
上行竖井 rising shaft
上行停止装置 up travel stop
上行通风 ascensional ventilation
上行下给式 upfeed down type; upfeed system
上行下给式供暖系统 drop system
上行下给式供热 upfeed system of heating
上行下给式双管系统 down-feed overhead two pipe system
上行下给式系统 attic main system
上行下给式蒸汽供暖系统 steam-heating down-feed system
上行下给式蒸汽重力供暖系统 steam-heating down-feed gravity system
上行下给系统 down-feed system
上行线 up track; up-line
上行线路 unline; unlink; up-line; up-link
上行线路保护率 up-link protection ratio
上行线路下行线路干扰量 up-link and down-link contributions
上行线路信号 up-link signal

S

上行线月台 up side
上行效应 bottom-up effect
上行斜坡 foreslope
上行咽喉 up throat
上行液压压力机 upstroke hydraulic press
上行运动 upstroke
上行展开 ascending development
上行展开法 ascending development method
上行正线 up main line;up main track
上行制气 up-run
上行注入站 upload station
上型 mo(u)ld top half
上型模 top swage
上型箱 cope;top box;top case;top flask
上型箱孔 cope hole
上胸射式水轮机 high breast water wheel
上蓄水池 upper pool;upper reservoir
上悬窗 horizontal projected window;projecting top-hung window;top-hinged casement;top-hinged swinging window;top-hinged window;top-hung casement;top-hung sash;top-hung window;top-pivoted window
上悬挂臂 upper suspension arm
上悬活动挑门 <汽车及仓库用> canopy door
上悬内开窗 top-hinged in swinging window
上悬潜水面 perched water table
上悬式(窗) top hung;top suspension
上悬式离心机 suspended centrifuge;top suspension(basket) centrifuge
上悬挑门 canopy door
上悬外撑 <窗扇的> top-hinged out swinging
上悬外撑窗 overhang sash window
上悬外开窗 top-hinged out swinging window
上悬外推窗 hopper window;hospital window
上旋窗 awning sash;awning window;canopy window;horizontal projected window;top-hinged casement;top-hinged sash (window);top-hinged window;top-hung sash;top-hung window;top-pivoted window
上旋门 overhead door of the swing-up type
上旋钮 turned-up button
上选品质 selected quality
上学出行 school trip
上旬 beginning of a month
上压 force up;overdraught
上压板 top board;upward acting plate
上压刀 top sword
上压力 top pressure;underdraft;upward pressure
上压式螺旋桨 upthurst type propeller
上压式压力机 up-packing press
上压式液体肥皂配出器 push-up-type soap dispenser
上压式液压机 hydraulic inverted press
上压式造型机 top squeeze molding machine
上压条 ceiling mo(u)lding
上烟道 overtop flue;upper flue
上沿时间 rise time
上研磨盘 top lap
上盐固定式皮带运输机 fixed belt conveyer for feeding salt
上盐釉 glazing by salting
上演 staging;staging
上鞅【电】 supermartingale
上扬 lift-off
上仰 nose-up pitch
上腰板 frieze rail;wainscot
上腰带 <客车外部> upper belt rail
上摇摆轮 upper wig-wag
上叶 superior lobe
上曳气流 up-draught
上曳气流速度 updraft velocity
上曳气流向上排气 updraft
上液限 flocculation limit
上一层 upper stor(e)y
上一次定位 last fix
上一次停工与下次开工之间的时间 stream-to-stream time
上一阶段 upper stage
上一年 pass year
上一位 top digit
上一页 preceding page
上移 shift up;upper shift
上移位操作 shift-up operation
上移位方式 shift-up mode
上遗谷 epigenetic valley
上遗河 epigenetic river;epigenetic stream;superimposed river;superposed stream
上遗水系 epigenetic drainage
上溢 overflow
上溢式引水闸门 overpour head gate
上溢式闸门 overpour gate
上溢水 overpouring water
上溢中断 interrupt on overflow
上翼 top flange;upper limb;upper wing;top wing <堆石坝截水墙的>
上翼面 top airfoil;upper aerofoil;upper(sur)face
上翼面副翼 upper surface aileron
上翼面减速板 upper surface brake
上翼片 upper panel
上翼坡 upper slope
上翼缘 top flange;upper flange;upper flange of girder
上翼(缘)板 top flange plate;upper flange plate
上翼缘连接板 upper flange junction plate
上翼展 upper span
上引的 above cited
上涌 run-up;upsurging;scend;uprush;upwelling <船在波浪作用下的一种运动>
上涌地点 upwelling site
上涌地区 upwelling region
上涌段 <波浪的> rising portion
上涌高程 run-up elevation
上涌界限 limit of uprush
上涌浪 upsurge
上涌隆起 upsurging swell
上涌水 upwell water
上涌水源 upwelling source
上涌现象 upwelling phenomenon
上涌洋流 upwelling oceanic current
上用圆刃楔形锤 top fuller
上油 apply oil;fat liquoring;oiling;topping up
上油处理 oil finish
上油盘 upper oil-pan
上油漆 painting
上油增艳处理 oil brightening
上油嘴 lubricator fitting
上游 above water;head race;upstream
上游坝脚 upstream toe of dam
上游坝壳 <土石坝的> upstream shell
上游坝坡 upstream batter
上游坝墙 upstream wall
上游坝体 <土石坝的> upstream fill
上游坝趾 upstream toe of dam
上游保护区 headwater area of protection
上游泵站 upstream pumping station;upstream pumping unit
上游鼻端 upstream nose;upstream nosing
上游边 upstream side
上游侧 upstream side
上游测量断面 upstream measuring section
上游承推墙 upper thrust wall
上游池 head basin;head bay;head pond;head pool
上游池底板 head bay floor
上游船闸 upstream lock;upper lock
上游导航墙 upper(guide)wall
上游导向叶片 upstream guide vane
上游的河岸所有者 upstream riparian
上游堤岸 upper bank
上游底板 upstream floor
上游地区 headwater area;upstream area
上游电站 upper station;upstream plant
上游端 head end;upstream end;upstream extremity
上游段 head reach;headwater section;upper course;upper reach
上游断面线 upstream range
上游法 upstream method
上游方向 updrift side;upriver
上游防冲铺砌 upstream apron
上游防洪措施 headwater control
上游防水面 waterproofing upstream face
上游分水尖 upstream nosing
上游港 <在河的上游> closed port
上游高程 upstream elevation
上游管道 sewer upstream
上游管线 upstream line
上游含水层 upper aquifer
上游河槽 headwater channel;upstream channel
上游河道 upper river course
上游河道截流 interception of upper reaches of river
上游河段 forebay;head bay;headwater(reach);mountain tract;upper course;upper reach;upper stream;upstream reach;upper river;upper pond
上游河流 head river;head stream;head waters of river;upper reach
上游河湾 upstream bay
上游护槛 upper guard sill
上游护面 upstream facing
上游护坡 upstream slope protection
上游护坦 upstream apron;upstream floor
上游积水 upstream ponding
上游节制闸 head regulator
上游(进)水池 upper pool
上游进水管 upper pool pipe
上游进水渠 <电站> headrace conduit;headwater channel
上游控制 headwater control;upstream control
上游口门 upstream entrance
上游库水位 level of upper pond
上游(来)水 upland water
上游立视图 upstream view
上游流量 upstream flow
上游流域 upland catchment
上游锚 upstream anchor
上游面 upstream face;waterside face
上游面板 <坝的> upstream deck
上游面坡度 upstream batter
上游泥沙流失总量 total upstream erosion
上游坡 upstream batter
上游坡护坡 upstream slope protection
上游坡脚 upstream toe
上游坡面 upstream slope
上游铺盖 upstream apron;upstream blanket
上游铺砌 upstream apron;upstream floor
上游墙 head bay wall
上游倾斜面 battered upstream face
上游区 headwater region;upstream zone
上游区段 upstream section
上游渠道 head race;headrace conduit;upriver canal
上游人字墙 upper mitre wall
上游式 upper course-type
上游受益 upstream benefit
上游水 head waters of river;upper stream water;upper water;headwater(of stream);upstream water
上游水池 upper pond;upper pool
上游水池水位 upper pond level
上游水道 head race;headrace channel;headwater channel
上游水库 headwater storage reservoir;upper basin;upper pond;upper pool;upper reservoir
上(游水)库水位 level of upper pond
上游水面 headwater
上游水塘 upper pool;upper pond
上游水塘水位 upper pond level
上游水位 affluent level;headwater-level;upper pond(water) level;upstream table;upstream water-level;forebay elevation <船闸的>;upper pond level <船闸的>;upper pool level <船闸的>
上游水位标 upstream ga(u)ge
上游水位高程 upstream level;upstream water-level;headwater elevation
上游水位恒定的浮筒自动弧形闸门 float auto-arc gate of upstream stable water-level
上游水位计 headwater-level ga(u)ge
上游水位控制 headwater control
上游水位线 upstream water line
上游水压力 headwater pressure
上游水域 headwater region
上游水源 upstream water source
上游水源保护条例 Regulations on Protection of Upper Reaches of Water Head
上游隧洞 <电站的> headrace tunnel
上游填土体 upstream fill
上游填筑体 upstream fill
上游调压池 head surge basin;upstream surge basin
上游调压井 headrace surge tank;head surge basin;head surge chamber;upstream surge basin
上游调压室 head surge chamber
上游调压水槽 headrace surge tank
上游停泊区 <船闸的> upstream garage
上游通航河段 navigation pool
上游围堰 upper cofferdam;upper coffer-wall;upstream cofferdam
上游围堰工程地质剖面图 engineering geological profile upstream cofferdam
上游围堰轴线工程地质剖面图 engineering geological section along axis of upstream cofferdam
上游污染源 upstream sources of pollution
上游衔接段 upstream transitional section
上游限制 upstream limit
上游向闸门扉 upstream leaf
上游楔形体 <坝的> upstream fill
上游斜反滤层 upstream slanting filter
上游型 <河流> upper course-type

上游蓄洪水库 upstream retarding reservoir
上游蓄水 upstream impoundment
上游蓄水量 upstream ponding
上游压力 <液压元件> upstream pressure
上游一边 upstream side
上游翼墙 upper wing wall; upstream wing wall
上游引航道 upper(lock) approach
上游羽状水系 barbed drainage pattern
上游圆顶 <大头坝的> upper round head
上游闸门 upper gate; upstream gate; upper lock gate
上游闸门门槽 upper gate recess
上游闸门门槛 upper gate sill
上游闸室 headroom
上游闸首 <船闸的> upper lock head
上游沼泽 high bog; moss moor
上游整流器 upstream fairing
上游正常水深 upper normal depth
上游直管 straight upstream pipe
上游滞洪水库 upstream retarding reservoir
上游注入水系 barbed drainage
上游总冲刷量 total upstream erosion
上游总水头 upstream total head
上游最低库水位 lowest upper pool elevation
上游最低水位 lowest upper elevation; upstream minimum water-level
上游最高库水位 highest upper pool elevation
上游最高水位 maximum headwater; upstream maximum water-level
上有色釉 colo(u)red glazing
上有土坡的挡土墙 retaining wall with surcharge
上釉 enamel impregnation; enamel(l)ing; glaze; glazing; gloss; vitrification
上釉部件 glazing unit
上釉材料 glazing material
上釉彩色饰面 glazed finish
上釉产品 glazing product
上釉瓷 glazed porcelain
上釉瓷的钢材 vitreous enamelled steel
上釉瓷的建筑镶板 vitreous enamelled building panel
上釉瓷砖 glazed ceramic tile
上釉的 enamel(l)ed; glazed; vitreous; vitrified
上釉的材料 glazed material
上釉的檩条 glazing purlin(e)
上釉的砌块 faience unit
上釉的水管 ceramic glazed sewer pipe; ceramic glazed sewer tube
上釉的陶器 faience ware
上釉的涂层 glazed coat(ing)
上釉的下水管 ceramic glazed sewer pipe
上釉的型材 glazing profile
上釉底料 engobing
上釉地砖 glazed floor tile
上釉法 glazing method
上釉方法 glazing technique
上釉工人 glazer
上釉硅酸盐水泥面层 Portland cement glazed coat(ing)
上釉过厚 overglazing
上釉涵洞 vitrified-clay culvert
上釉(火)炉 enamel(l)ing stove
上釉火石器 glazed stoneware
上釉技术 glazing technique
上釉加筋混凝土 glazing reinforced concrete
上釉建筑构件 glazed building units
上釉(结构)构件 glazed structural units
上釉精加工 enamel(l)ed tile
上釉空心砖 glazed hollow block
上釉黏[粘]土 vitrified clay
上釉砌块 shepwood
上釉前补水 body wetting before glazing
上釉前着色的 underglaze
上釉烧的 encaustic
上釉陶瓷 glazed ceramics
上釉陶瓷器 glazed ware
上釉陶管 glazed stoneware pipe
上釉(陶)管涵洞 vitrified-clay pipe culvert
上釉陶器 Delft pottery; glazed clayware; glazed earthenware; glazed stoneware; glazing pottery
上釉陶土 vitrified clay
上釉陶土管 glazed earthenware pipe; vitrified-clay pipe
上釉陶土泄水管 vitrified-clay drain tile
上釉瓦 encaustic tile
上釉瓦管 vitrified pipe; vitrified tile
上釉温度 glazing temperature
上釉窑 glaze kiln; glazing kiln
上釉硬质陶土管 glazed stoneware
上釉用的夹子 glazing gripper
上釉制品 glost ware
上釉砖 shepwood; vitrified brick
上域 codomain
上阈限 upper threshold
上元古代【地】Epiproterozoic
上元古亚代【地】Upper Proterozoic subera
上元古亚界【地】Upper Proterozoic suberathem
上缘 upper limb
上缘板 top flange plate
上缘抹角墙脚板 spaly skirting
上缘抹角踢脚板 splayed skirting
上缘踏步形楼梯梁 mitred-and-cut string
上缘弯曲的梁 hog-backed girder
上缘纤维 upper fiber[fibre]
上源 <多用复数> headwater(of stream)
上月 ultimo
上越流层 upper leakage layer
上载 upload
上载式带式运输机 top-loading belt conveyer[conveyor]
上凿式凿岩机 stopper hammer drill
上轧槽 top pass
上轧辊 topping roll
上轧辊的平衡 top-roll balance
上轧辊机构 top-roll-balancing mechanism
上轧辊平衡装置 top-roll-balance arrangement
上轧辊弹簧式平衡装置 top-roll spring balance arrangement
上轧辊液压式平衡装置 top-roll hydraulic balance arrangement
上轧辊重锤式平衡装置 top-roll counterweight balance arrangement
上轧辊组合部件 top-roll assembly
上闸 application of brakes
上闸门 upward-acting door
上闸室 top chamber; upper chamber
上闸首 head bay; upper gate bay; upper head; upper lock gate
上闸首人字闸门 crown miter gate; upper miter[mitre] gate
上闸首闸门 crown gate
上闸闸门 head gate; upper gate
上涨 run-up
上涨潮 rising tide
上涨的趋势 upward tendency
上涨价格 escalating price
上涨率 rate of rise
上涨趋势 up-trend
上涨水位 rising level; rising stage
上涨速度 rising velocity
上罩 upper shield
上账 keep books
上震旦统【地】Upper Sinian series
上支撑 overhead bracing
上支承 upper support
上支承辊 top backing up roll
上支架 top hold-down; upper bracket; upper support
上止点 top center[centre]; top dead center[centre]; upper dead center [centre]
上止点标记 top-dead-center[centre] indicator
上止点附近的气缸壁磨损 upper cylinder bore wear
上止机构 up-stop
上止漏环 rubber crown seal
上志留纪【地】Upper Silurian period
上志留统【地】Upper Silurian series
上置定位臂 upper control arm
上置阀门 overhead valve
上置谷 superimposed valley
上置河流 superimposed stream
上置荷载 surcharge
上置式秤料斗 overhead bucket
上置式气门 upper valve
上置式盛料桶 overhead bucket
上置式油箱 top petrol tank
上中平衡 intermediate purlin(e)
上中天 superior passage; superior transit; upper meridian transit; upper transit; upper culmination
上中天潮 direct tide; superior tide
上中心 above center[centre]
上中心列 upper central series
上重下轻 top-heavy
上轴 top roll; upper spindle; top shaft <铲土机>; upper shaft <铲土机>
上轴承 bearing upper; head bearing; upper bearing
上轴承端盖 top bearing cover
上轴承盖 upper ball cover
上轴承螺母 top bearing shaft nut
上轴承锁紧螺母 top bearing lock nut
上轴承调节螺母 top bearing adjusting nut
上轴承箱 top bearing housing
上轴承座 top chock
上轴瓦 top bearing shell; upper bearing bush; upper half
上侏罗纪【地】Upper Jurassic period
上侏罗统【地】Malm; Upper Jurassic series
上主星序 upper main sequence
上主翼 top main plane
上注 cast from the top
上注法 top teeming process
上柱列 entablature
上柱列节理 upper colonnade
上铸 direct casting
上铸法 downhill casting; top pouring
上转 superduct; supraverge; supravergence
上转换 up-conversion
上转换器 up-converter
上装 upper garment
上装料点 overhead loading rack
上浊点 upper cloud point
上子实层 epihymenium
上子午线 upper branch of meridian
上纵标集 upper ordinate set
上纵撑 upper longitudinal strut
上作用水阀 lever faucet
上座圈 upper ball race; upper race ring
上座套 upper bushing

尚

尚待履行的合同 executory contract

尚待完成的工作 arrears
尚待完成的合同 outstanding contract
尚蒂利细花花边 chantilly lace
尚可服务年限 residual mine age
尚可优化循环 still optimizable loop
尚难利用的储量 unavailable reserves
尚未被利用 fallowness
尚未被市场吸收的 undigested
尚未标引的 indexless
尚未分配的代码 code not allocated
尚未付款的发票 unpaid invoice
尚未干透的 tacky
尚未就任(的主席、会长、委员长等) chairman designate
尚未抹灰泥的墙 naked wall
尚未盛开的 unblown
尚未完税 <在保税仓库中> in bond
尚未硬化的混凝土 green concrete
尚未用完的拨款 backlog
尚未知道的损失 not known loss
尚须改正 undercorrection
尚须考虑 <拉> ad referendum

梢

梢板法 <制漏板> tip-plate process

梢杯 tip cup
梢鞭 fascine whip
梢变 shoot mutation
梢部失速 tip-stall
梢材 top log
梢端 small end
梢端干围 top girth
梢端原木 top log
梢端直径 top diameter
梢工 fascine work
梢棍 fascine roll
梢护岸设施 brush wicker-work; brush work
梢间【建】end bay; intermediate bay
梢节点 tip node
梢径 top diameter
梢枯病 top dry
梢捆 anchored tree; brush wood; brushwood faggot; bundle of branch; fascine; fascine bundle; kid
梢捆坝 fascine dam
梢捆保护沙丘 kidding
梢捆挡沙坝 brushwood check dam
梢捆挡水坝 brushwood check dam
梢捆垫 brush mattress
梢捆丁坝 fascine groin; fascine groyne
梢捆工程 fascine project; fascine work
梢捆护岸 brush and cable bank protection; brushwood revetment
梢捆护舷 brushwood fender
梢捆夹石铺在河岸岸坡的水下部分 kidding
梢捆建筑物 fascine work
梢捆排水沟 bavin drainage
梢捆堰 brushwood weir
梢捆堰坝 weir of fascines
梢篱 fascine hurdle
梢料 brush(wood); fascine(wood)
梢料坝 brush barge; brush(wood) dam
梢料柴排 brushwood mattress
梢料船 brush barge; fascine barge
梢料堤 brush dike[dyke]
梢料垫层 brush matting
梢料谷坊 brush check dam
梢料护面 brush paving
梢料基础 fascine foundation
梢料锚索护岸 brush and cable bank protection

S

梢料木桩坝 brush and pile dam
梢料木桩堤 brush and pile dike[dyke]
梢料排 brush matters
梢料排水暗沟 brushwood drain
梢料束狭工程 brush contracting works
梢料填层 fascine layer
梢料围堰 brushwood cofferdam
梢料堰 brushwood weir
梢龙 long fascine
梢笼 fascine roll;fascine whip
梢木坝 beaver-type timber dam
梢排 mat of bush
梢排护岸 brush revetment
梢蓐 brushwood mattress;fascine mattress
梢蓐护岸 fascine revetment
梢蓐料沉褥 fascine mattress
梢褥 brush mattress
梢扫沉排 brush mattress
梢埽沉排 brushwood mattress
梢石坝 brushwood-and-stone dam
梢石料驳船 brush and stone barge
梢石料船 brush and stone barge
梢速 tip speed
梢头材 topwood
梢头直径 <木材> small-end diameter;small-top diameter
梢涡空化 tip vortex cavitation
梢眼放大端头 eyebar head
梢枕 gravel core fascine;gravel core roll;stone core fascine
梢枝坝 brush dam
梢枝护岸设施 brush wicker-work
梢枝块石导水堰 brush and loose rock diversion weir
梢枝捆 bundles of fascine
梢枝排 brush mattress;fascine mattress
梢枝排护岸 brush revetment
梢枝束 fascine pole;fascine whip
梢枝堰 brushwood weir

烧 baking

烧暗 <荧光屏发光效率降低> dark burn
烧白云石 calcined dolomite
烧爆 decrepitate
烧爆作用 decrepitation
烧杯 beaker;Bunsen beaker
烧杯式取样器 beaker sampler
烧杯试验 jar test
烧边 <玻璃的> fire polishing
烧边器 heel burner
烧变煤 burnt coal
烧变岩 burnt stone
烧剥 flame chipping; hot scarfing; scarfing
烧剥室 scarfing dock
烧钵 fireclay container
烧钵体 saggar clay
烧钵土 sagger clay
烧不掉的 unburnable
烧彩玻璃 stained glass
烧彩瓦 encaustic tile
烧彩砖(瓦) encaustic tile
烧彩装饰 encaustic decoration
烧草机 weed burner
烧柴 brush
烧柴油的取暖器 diesel-fuelled heater
烧衬 eating-through
烧成 burning;firing;maturing 成 <陶瓷>
烧成板油 burnt plate oil
烧成车间 calcination plant
烧成的 burnt
烧成骨料 clinker aggregate
烧成灰 incinerate;incineration
烧成集料 clinker aggregate
烧成焦渣 crozzle
烧成裂隙 firing crack
烧成木炭 char
烧成耐火材料 burnt refractory
烧成膨胀 firing expansion
烧成品 burnt ware
烧成平地 burnt down
烧成器皿 burnt ware
烧成缺陷 defect on firing;firing defect
烧成熔块 burning to(cement) clinker
烧成石灰 calcination
烧成时间 soaking time
烧成收缩 burning shrinkage; firing shrinkage
烧成收缩率 coefficient of firing shrinkage
烧成收缩系数 firing constriction coefficient
烧成熟料 clinkering; finish burned clinker;well-burned clinker
烧成水泥熟料 firing to cement clinker
烧成炭 coal;char(ring)
烧成炭的 charred
烧成陶瓷涂料 fired ceramic coating
烧成体积收缩率 coefficient of firing volume shrinkage
烧成温度 burnt temperature
烧成温度范围 <陶瓷的> maturing range
烧成物 calcining
烧成线性收缩系数 coefficient of firing linear shrinkage
烧成压痕 kiln mark
烧成油浸碱性砖 burned impregnated basic brick
烧成赭色 burned umber
烧炽残留【物】 residue on ignition
烧出釉的 dead-burned;dead-burnt
烧除 burning off;burning-out;removal by burning
烧除油漆 burn-off paint
烧除指标 burning index
烧穿 burn(ing) through; burn-off; burnt through;flame breakthrough
烧穿炉衬 breakout;eating-through
烧瓷 japanning
烧瓷土 molochite
烧的 burned
烧掉 removal by burning;clean burning
烧掉的垃圾 burned refuse
烧断 burn-out
烧断继电器 burn-out relay
烧珐部 baked enamel
烧珐琅质的炉 enamel stove
烧饭用具箱 canteen
烧粉 grog
烧附图像 burned-in image; sticking picture
烧干 boil down
烧干土 baked clay
烧割 burn-off;burnt-cut
烧固 baking
烧固了的 baked
烧固体燃料的家用锅炉 domestic solid fuel boiler
烧管道煤气的家庭用具 domestic appliance burning town gas
烧光 burn-out;burnt down;clean burning
烧过的 burnt
烧过的材料 burnt material;fired material
烧过的富铁黄土 burnt sienna
烧过的木材 burnt wood
烧过的燃料 spent fuel
烧过火 overburning
烧过头 burn-through
烧焊 burn-in;flame welding;freeze
烧焊前 preweld
烧焊前热处理 preweld heat treatment
烧焊枪 welding head
烧褐色赭土 burnt umber brown
烧褐铁矿 burnt umber
烧褐土 burned umber
烧红的铁渣 red slag
烧红铆钉铆固 hot-riveted
烧后呈红色的黏[粘]土 terra sigillata
烧后颜色 burnt colo(u)r;colo(u)r after firing
烧化 burn-off; flash(ing); swealing <钢锭皮>
烧化余量 flashing allowance;total flash-off
烧画笔 scorched pencil
烧坏 burn(ing)-out; deflagration; overburning
烧坏的钻头 green bit
烧坏接点 sparkwear contact
烧荒 field burning;moorburn
烧荒农业 fire agriculture
烧荒用火 debris burning fire
烧灰场 ashery
烧毁 burning;burn-out;burnt down; burn-through; burn-up; overburning;overfire;sparkwear
烧毁表示 burn-out indication
烧毁表示器 burnt-out indicator
烧毁长度 damaged length
烧毁程度 burned degree
烧毁的 burned
烧毁等离子体 burn-out plasma
烧毁电阻 burn-out resistance
烧毁金刚石 burned diamond
烧毁金刚石钻头 burnt bit
烧毁面积 damaged area
烧毁热通量 burn-out flux; burn-out heat flux
烧毁寿命 burn-out life
烧毁指示器 burn-out indicator
烧火 stoking
烧火工 burner man
烧火工具 stoking tool
烧火工人 firer
烧火间 firing sector;stokehold
烧碱 caustic soda;hydrate of sodium; sodium hydrate;sodium hydroxide
烧碱法 soda process
烧碱浓缩装置 caustic soda concentration unit
烧碱生产 caustic soda production
烧碱石灰 natroncalk
烧碱石棉剂 ascarite
烧碱盐泥 caustic mud
烧碱液 caustic lye of soda;white liquor
烧僵 dead burn(ing)
烧僵白云石 dead-burned dolomite
烧僵的 dead-burned
烧僵的耐火白云石 dead-burned refractory dolomite
烧僵的石膏 <抹灰用的无水石膏> dead-burned gypsum
烧焦 burn(a);char(ring);dead burn(ing);scorch;singe;swale
烧焦处理 charring treatment
烧焦似的 scorching
烧焦味 <英国方言> smeech
烧结 agglutinate;cake;caking;cementation; cementing; clinkering; clotting;dead burn(ing); fritting; furnace run;fusing-in;nodulize;sinter(ing)
烧结白云石 dead-burned dolomite; doloma; dolomite clinker; sintered dolomite
烧结板 sintered plate;sintered slab
烧结板过滤体集尘器 sintered lamella filter
烧结棒 sintered bar;sintered rod
烧结焙烧 agglomeration roast; sinter roasting
烧结变形 clinkering strain; sintering warpage
烧结饼 sinter cake
烧结玻璃 fritted glass;sintered glass
烧结玻璃板漏斗 sintered glass funnel
烧结玻璃过滤坩埚 sintered glass filtering crucible
烧结玻璃过滤器 fritted glass filter; sintered glass filter
烧结玻璃器皿 fritted glassware
烧结玻璃纤维 sintered glass fiber
烧结玻璃心柱 sintered glass stem
烧结玻璃珠 sintered glass bead
烧结不锈钢 sintered stainless-steel
烧结材料 sintered material; sintering product
烧结材料贴面的离合器 sintered faced clutch
烧结彩饰 vignetting
烧结操作 sintering operation
烧结层 burn-in; sinter bed; sintered layer
烧结厂 agglomerating plant;ore-sintering plant;sinter(ing) plant
烧结车间 sintering mill;sintering plant
烧结衬里 sintered liner
烧结衬面 sintered facing
烧结程度 degree of sintering
烧结齿轮 sintered gear
烧结赤铁矿 sintered hematite
烧结触点 sticking of contacts
烧结纯铁 Pomet
烧结磁铁 ceramic magnet; sintered magnet
烧结磁性材料 sintered magnetic material
烧结磁性合金 sintered magnetic alloy
烧结磁性零件 sintered magnetic parts
烧结催化剂 sintered catalyst
烧结带 sintering belt;sintering zone
烧结氮化硅 sintered silicon nitride
烧结的 agglomerant;clinkery;sintered
烧结的粉煤灰骨料 lytag
烧结的粉煤灰陶粒 lytoy
烧结的耐火白云石 dead-burned refractory dolomite
烧结的坯块 sintered compact
烧结的砂子 fused sand
烧结点 clinkering point;sintering point
烧结点测定仪 sinter meter
烧结电接触器材 sintered contact material; sintered electric(al) contact material
烧结电流 sintering current
烧结电气零件 sintered electric(al) parts
烧结锭 sintered ingot
烧结锭条 sintered bar
烧结动力学 sintering kinetics
烧结锻造法 sinter forging
烧结多晶体金刚钻 sintered polycrystalline diamond
烧结多孔镍滤杯 sintered porous nickel cup
烧结多元碳化物 cemented multiple carbide
烧结法 sintering method; sintering process; burning method <地基处理>
烧结矾土 sintered alumina
烧结范围 clinkering range; sintering range
烧结方法 sintering method
烧结分解 decomposition by sintering
烧结粉煤灰 sintered fly ash; sintered pulverised[pulverized]-fuel ash
烧结粉煤灰骨料 sintered fly-ash aggregate; sintered pulverised[pulverized-fuel] ash aggregate

烧结粉煤灰集料 sintered fly-ash aggregate
烧结粉煤灰轻集混凝土 sintered fly-ash concrete
烧结粉煤灰制造轻骨料 lightweight aggregate made by sintered fly-ash
烧结粉煤灰砖 fired fly ash brick;sintered fly-ash brick
烧结粉末 sintered powder
烧结粉末磁铁 sintered powder magnet
烧结粉末金属 sintered powder metal
烧结粉末胎体(金刚石)钻头 sintered bit;sintering diamond crown
烧结粉末冶金 sinter powder metal
烧结敷层 sintered coating
烧结敷氧化物阴极 sintered oxide-coated cathode
烧结坩埚 sintered crucible
烧结刚玉 sintered corundum
烧结刚玉骨料 alundum aggregate
烧结刚玉模具 sintered alumina abrasive tool
烧结(缸)砖 clinker brick
烧结钢 pseudo-steel;sintered steel
烧结锆刚玉 bonded alumina-zirconia-silica
烧结锆刚玉砖 ceramic bonded AZS[alumina Zirconia Silica] brick
烧结工 agglomerant
烧结工序 sintering circuit
烧结骨架 sintered skeleton
烧结骨料 burnt aggregate; sintered aggregate
烧结贵重金属 sintered precious metal
烧结锅 sinter pot
烧结过程 sintering process
烧结过滤器 sintered filter
烧结海绵金属 sintered sponge
烧结含油轴承 sintered metal bearing
烧结含油轴承合金 sintered metal powder oil impregnated alloy
烧结焊剂 baked flux
烧结合成树脂搪瓷 baked-on synthetic(al) resin enamel
烧结合金 sintered alloy
烧结痕 kiss mark
烧结环 sintered ring
烧结黄铜 sintered brass
烧结黄铜坯块 sintered brass compact
烧结黄铜制品 sintered brass product
烧结灰 clinkering ash
烧结灰渣 sintered ash
烧结混合料 sintering mix(ture)
烧结混合物 sintered mix(ture)
烧结混凝土 sinter concrete
烧结活塞环 sintered piston ring
烧结活性 sintering activity
烧结机 sinter(ing) machine
烧结机理 sintering mechanism
烧结集料 burnt aggregate; sintered aggregate
烧结剂 agglutinant
烧结加气混凝土骨料 sintered aerated concrete aggregate
烧结加气混凝土集料 sintered aerated concrete aggregate
烧结加气混凝土陶粒 sintered aerated concrete aggregate
烧结键 sinter bond
烧结结合剂 sintered bond
烧结结合料 sintered binder
烧结界面 sintered interface
烧结金 sintered gold
烧结金刚刀头 sintered diamond insert
烧结金刚砂 sinter(ed) corundum
烧结金刚石 polycrystalline diamond compact
烧结金刚石钻头 sintering diamond bit;sintering diamond crown
烧结金属 bond metal;cemented metal;fritted metal;sintered metal;sintering metal;sinter-metal;synthetic-(al) metal
烧结金属磁铁 sintered metallic magnet
烧结金属磁芯 sintered metallic core
烧结金属过滤器 sintered metallic filter
烧结金属滤油器 sintered metal filter; sintered metal powder filter
烧结金属摩擦材料 sintered metal friction material
烧结金属抛光 buffing of sinter metals
烧结金属丝网片 sintered woven wire sheet
烧结金属纤维 sintered metal fiber
烧结金属学 ceramal; ceramet; cerametallics
烧结矩阵 sintered matrix
烧结控制 sintering control
烧结块 agglomerated cake; cake; caked mass; clinker; lump sinter; sinter;sinter cake;sintering briquet(te);frit
烧结块破碎器 sinter breaker
烧结矿 agglomerate;sintered ore;sintering ore
烧结矿冷却机 sinter cooler
烧结矿破碎机 sinter breaker
烧结矿渣 sintered slag
烧结矿渣泡沫 agglomerate-foam
烧结矿渣泡沫混凝土 agglomerate-foam concrete
烧结砾石 sintered gravel
烧结料 clinker aggregate; sintered charge;sintered material
烧结料层 sinter bed
烧结裂缝 clinkering crack
烧结裂纹 sintering crack
烧结菱苦土 sintered magnesite
烧结菱镁矿 sintered magnesite
烧结零件 sintered parts
烧结炉 fritting furnace;sinter hearth; sintering apparatus; sintering furnace;welding furnace
烧结炉箅 sintering grate
烧结炉衬 sintered lining
烧结炉底 fused hearth bottom
烧结炉栅 sintering grate
烧结铝 sintered alumin(i)um
烧结铝粉 sintered alumin(i)um powder
烧结铝镍铁制品 sintered alnico product
烧结铝制品 sintered alumin(i)um product
烧结煤 sintering coal
烧结煤矸石骨料 sintered colliery waste aggregate
烧结煤矸石砖 fired brick of colliery waste;sintered brick of gangue
烧结煤渣 sintered cinder
烧结煤渣墙板 sintered cinder wall slab
烧结镁砂 magnesite clinker
烧结镁砖 fire-bonded magnesite brick
烧结密度 sintered density
烧结面积 grate area;sintering area
烧结模塑 sinter mo(u)lding
烧结模型 sintering model
烧结膜 sintered membrane
烧结摩擦材料 sintered friction material
烧结摩擦零件 sintered friction parts
烧结磨细燃料灰 sintered pulverised [pulverized]-fuel ash
烧结耐高温材料 sintered high-temperature material
烧结耐火材料 ceramic bonded refractory
烧结耐磨材料 sintered wear-resistant material
烧结能力 caking capacity power;caking power;sintering capacity
烧结尼龙 sintered nylon
烧结黏[粘]土 aglite; burned clay; burnt clay; sintered clay <轻质混凝土骨料>
烧结黏[粘]土道砟 burned ballast; burnt ballast;burnt clay ballast
烧结黏[粘]土盖瓦 burned clay ridge tile;burnt clay ridge tile
烧结黏[粘]土弧形瓦屋面 burnt clay curved tile roof
烧结黏[粘]土混凝土 sinter clay concrete
烧结黏[粘]土脊瓦 burned clay ridge tile;burnt clay ridge tile
烧结黏[粘]土路面 burned clay pavement;burnt clay pavement
烧结黏[粘]土泥料粉 burned dust; burned filler;burnt dust
烧结黏[粘]土轻骨料 burnt clay light (weight) aggregate
烧结黏[粘]土轻集料 burnt clay light (weight) aggregate
烧结黏[粘]土曲瓦屋面 burned clay curved tile roof;burnt clay curved tile roof
烧结黏[粘]土熟料 trass
烧结黏[粘]土水泥 brick cement; burnt clay cement;sintered clay cement
烧结黏[粘]土瓦 burnt clay tile;fired clay tile
烧结黏[粘]土圬工 <砖建筑> burnt clay masonry(work)
烧结黏[粘]土屋脊盖瓦 burned clay hip tile;burnt clay hip tile
烧结黏[粘]土屋面弧形瓦 burnt clay curved roof(ing) tile
烧结黏[粘]土屋面曲瓦 burnt clay curved roof(ing) tile
烧结黏[粘]土制品 burned clay product; burnt clay article; burnt clay product
烧结黏[粘]土制造轻骨料 lightweight aggregate made by sintered clay
烧结黏[粘]土砖 burned brick; burnt clay brick
烧结镍 sintered nickel
烧结镍过滤杯 sintered nickel cup
烧结镍制品 sintered nickel product
烧结扭曲 sintering warpage
烧结盘 sintering pan
烧结泡沫金属 sintered foamed metal
烧结膨胀 clinkering expansion;sintering grow
烧结膨胀矿渣 sintered expanded slag
烧结膨胀黏[粘]土 <由黏[粘]土和粉状焦炭混合起来,烧成多孔烧块,加以破碎> aglite;sintered expanded clay
烧结膨胀性矿渣 sintered bloating slag
烧结膨胀性黏[粘]土 sintered bloating clay
烧结皮 sinter skin
烧结品 sinter
烧结气氛 sintering atmosphere
烧结气体转化器 gas converter in sintering
烧结砌块 fired block
烧结青铜 sintered bronze
烧结青铜饰面 sintered bronze facing
烧结轻金属 sintered light metal
烧结轻质混凝土 sintered light(weight) concrete
烧结球团废水 sintering and palletizing wastewater
烧结区 clinkering zone;sintering zone
烧结熔剂 staflux
烧结熔渣墙板 sintered clinker wall slab
烧结乳剂 sintering emulsion
烧结软磁材料 ferroxcube
烧结软木 agglomerated cork
烧结软木板 rock cork
烧结软木砖 agglomerated cork brick; baked cork brick
烧结色谱板 sintered chromatographic plate
烧结砂 chamot(te) sand; sand burning
烧结砂块 burning sand
烧结设备 agglomerating installation; sintering apparatus;sintering plant; sintering unit
烧结石板瓦 burnt slate
烧结石膏 dead-burned gypsum
烧结石墨青铜 Durex
烧结石英 sintered quartz
烧结试验 agglutinating test; sintering test
烧结试样 sintered specimen
烧结收缩 burning shrinkage; clinkering contraction; sintering shrinkage;thermal shrinkage
烧结熟料 sintered clinker
烧结水泥熟料 burning to(cement) clinker
烧结速率 rate of sintering
烧结碳化钙 sintered carbide
烧结碳钢 sintered-carbon steel
烧结碳化钨 cemented tungsten carbide;diamondite
烧结碳化物 <常指硬质合金> cemented carbide [carbite]; sintered carbide
烧结搪瓷涂饰 baked enamel finish
烧结陶瓷 sintering ceramics
烧结陶瓷涂料 fired ceramic coating
烧结陶土片条 slip tile
烧结陶瓦 sintered stoneware tile
烧结体 agglomerate; cake(d) mass; clinkered body; sintered body; sintering body
烧结添加剂 agglomerating agent
烧结条件 sintering condition
烧结铁 cemented iron; clinkering iron;sintered iron
烧结铁零件 sintered iron parts
烧结铁素体 sintered ferrite
烧结铁制品 sintered ferrous product
烧结铜 sintered copper
烧结涂层 sinter coating
烧结(团)块 agglomerate
烧结网领 sintered metal ring
烧结温度 fusion temperature; sintering point;sintering temperature
烧结温度范围 sintering temperature range
烧结钨棒 sintered tungsten bar
烧结物 sinter
烧结物床层 sinter bed
烧结吸收剂 sintered absorbent
烧结纤维 sintered fiber[fibre]
烧结现象 sintering phenomenon
烧结限度 sintering limit
烧结相图 clinkerization diagram
烧结箱 sinter box
烧结性 agglutinating property
烧结性能 sintering character
烧结性试验 sintering test
烧结玄武岩 sintered basalt
烧结氧化铝 sintered alumina
烧结氧化铝车刀 <商品名> Sintex
烧结氧化铝刀具 <商品名> Sintex

S

烧结氧化铝坩埚 sintered alumina crucible
烧结氧化镁 sintered magnesia
烧结氧化铍坩埚 sintered beryllia crucible
烧结氧化物 sintered oxide
烧结氧化物磁铁 ceramics magnet
烧结因素 agglomerating agent; agglomerating factor
烧结阴极 sintered cathode
烧结硬质合金 cemented carbide; sintered hard alloy
烧结硬质合金材料 sintered hard metal material
烧结硬质合金钻头 sintered metal bit; sinter-set bit
烧结油页岩集料 sintered shale aggregate
烧结釉面砖 clinker tile
烧结元件 sintered component
烧结源 sintered source
烧结指数 agglomerating index
烧结制度 sintering schedule
烧结制品 sintered article; sintered product; sintered ware; sintering product
烧结质量 sintering quality
烧结周期 sintering cycle; sintering period
烧结轴承 sintered bearing
烧结轴承合金 sintered bearing metal
烧结砖 burned brick; burnt brick; Dutch brick; fired brick; sintered brick
烧结装置 agglomerating installation; agglomerating plant
烧结状态 burnt state
烧结钻头 sintered bit
烧结作用 agglomeration
烧尽 afterflaming; burn(ing)-out; burn-up; complete combustion; delayed burning
烧尽毒物 burn-out poison
烧净 burning-out
烧酒副产品 distiller's by-product
烧酒副产物 distiller's by-product
烧酒业下脚料 distillery refuse
烧酒糟 distiller's dried soluble; distiller's spent grains
烧酒糟残液 distiller's soluble
烧酒糟残液干燥物 dry distiller's soluble
烧酒糟饲料 disller's feed
烧开的 boiled
烧开水通知 boil notice
烧孔 hole burning
烧孔模型 hole-burning model
烧孔效应 hole-burning effect
烧矿法 calcination
烧垃圾的人 cremator
烧蓝 blu(e)ing
烧沥青工人 potman
烧炼 clinkering
烧炼清油 burnt oil
烧了起来 burn-up
烧料 frit
烧裂 decrepitate; fire cracking; firing crack
烧裂缝 firing crack
烧裂声 decrepitation
烧林垦地法 milpa system
烧硫炉气体 pot gas
烧硫灭菌 sulphuring
烧炉 firing drawing-chamber
烧铝土矿 burnt bauxite
烧绿石 chalcoamprite; niobpyrochlore; pyrochlorite; pyrrhite
烧绿石含量 pyrochlore content
烧绿石结构 pyrochlore structure
烧绿石碳酸岩矿床 pyroch ore-bearing carbonatite deposit
烧绿石型结构 pyrochlore type structure
烧绿岩 pyrrhite
烧落 <混凝土> thermal texturing
烧毛板 flamed slab
烧煤 coal burning; coal firing
烧煤仓 coal dust storage hopper
烧煤船 coal-burning vessel
烧煤的 coal-fired
烧煤的窑 coal fired kiln
烧煤粉的 dust fired; pulverised [pulverized] coal fired
烧煤工人 stoker
烧煤锅炉 coal-burning boiler; coal-fir boiler
烧煤集中采暖法 coal-fired central heating
烧煤炉 coal-fired furnace
烧明矾 dried alum
烧木锅炉 wood-fired boiler
烧耐火黏[粘]土 schamot(te)
烧黏[粘]土 baked clay; burned clay; burnt clay; burnt gault; fired clay
烧黏[粘]土道砟 burnt clay ballast
烧黏[粘]土路面 burnt clay pavement
烧黏[粘]土水泥 brick cement
烧盆 saggar[sagger]
烧坯 sintered compact; sintering briquet(te); sintering compact; sintering shape
烧坯孔隙度 sintered porosity
烧瓶 boiling flask; bulb; flask
烧瓶加热器 flask heater
烧瓶刷 flask brush
烧瓶(支)架 flask holder
烧气锅炉 gas-fired boiler
烧汽化液体燃料的窑 pregasified fuel oil fired kiln
烧砌沟管 masonry sewer
烧嵌 shrink(age) fit(ting); shrunk fit
烧嵌环 shrunk ring
烧青砖法 bluing
烧球 hot-bulb
烧球式柴油机 semi-diesel
烧球式点火 hot-bulb ignition
烧球式发动机 hot-bulb engine
烧去 burn-off
烧去旧漆 torching
烧却灰 ash from incineration
烧燃 combustion
烧热的 heated
烧熔 burn(a); burning-out; sweat(ing); scorification <试金法>
烧熔边缘 <焊接> pitting
烧熔结合剂 fused bond
烧入装饰 encaustic decoration
烧色 annealing colo(u)r
烧山 swale
烧伤 burn(a); burning; burnt; dieburn; fire injury
烧上 burn(ing)-in
烧勺伸缩柄 telescopic(al) dipper stick
烧生物蛋白石 burnt gaize
烧失量 burnable quantity; fire loss volume; ignition loss; loss on ignition <粉煤灰的一个指标>
烧失量百分率 percent ignition loss
烧失量曲线 ignition loss curve
烧石膏 baked plaster; bassanite; burnt gypsum; burnt lime; burnt plaster; calcined gypsum; calcium sulphate hemihydrate; dead-burnt gypsum; dierite; hard burnt; hard-burnt plaster; hard plaster; plaster of Paris
烧石膏灰浆 anhydrous calcium sulphate plaster
烧石灰 burned lime; calcined lime; plaster of Paris; burnt lime
烧石灰工人 limeburner
烧蚀 ablation; burn-through; erosion; loss on ignition
烧蚀保护层 ablative shielding
烧蚀表面 ablating surface
烧蚀材料 ablation material; ablative; ablative material
烧蚀层 ablative layer
烧蚀反应 ablative response
烧蚀防护 ablative protection
烧蚀防护法 ablative mode of protection
烧蚀防护罩 ablation shield
烧蚀复合材料 ablative composite material
烧蚀隔热材料 ablative insulating material
烧蚀机理 ablation mechanism
烧蚀剂 ablative agent; ablative material; ablator
烧蚀聚合物 ablative polymer
烧蚀绝热材料 ablation insulating material
烧蚀冷却 ablation cooling
烧蚀冷却材料 ablative cooling material
烧蚀率 ablating rate; ablativity
烧蚀热 heat of ablation
烧蚀热防护 ablative thermal protection
烧蚀闪光灯 ablative flashlamp
烧蚀试验 ablation test
烧蚀速率 ablation velocity
烧蚀特性 ablative characteristic
烧蚀体 ablator
烧蚀涂层 ablation coating
烧蚀涂料 ablation coating
烧蚀性材料 ablator
烧蚀性聚合物 ablation polymer
烧蚀性塑料 ablative plastics
烧蚀因数 ablation factor
烧蚀重量 ablated weight
烧死 death by burning
烧损 burning loss; fire waste; loss by burning; melting loss; overbake; overburning; overfiring; scaling loss; damaged by fire
烧损的 fire-damaged
烧损率 <建筑物烧损面积与总面积之比> ratio of burnt area
烧损木 burnt wood
烧缩 crawling; creeping; fire [firing] shrinkage
烧炭 carbon burning; chark(ing)
烧搪瓷 baked enamel
烧桶 ladle pot
烧头 burner
烧透 burn-through; burn through melt down; thorough burning
烧透的 double-burnt; hard-burned; hard-fired; well-fired
烧透的瓦 well-burnt tile
烧透的砖 hard-fired brick; well-fired brick
烧透石灰 burnt lime
烧透熟料 well-burned clinker
烧透砖 burned brick; burnt brick; well-burned brick
烧涂法 burning-on
烧土 soil burning
烧土制品 clay article
烧土砖墙 clay wall tile
烧脱 burning off
烧完 after-combustion; burn-out
烧纹电流 sintering current
烧析 ex(s)udation; sweating
烧箱 saggar[sagger]
烧箱土 saggar[sagger] clay
烧窑 kiln firing; kilning
烧窑辅助设备 kiln furniture
烧窑工人 kiln man
烧窑记录 kiln log
烧窑排出的余热 kiln waste heat
烧页岩 burned shale; burnt shale
烧页岩制品 burned shale product; burnt shale product
烧液体燃料的 liquid fired
烧硬 bake
烧硬的 hard-burned
烧油 fuel oil
烧油船 oil burner; oil fuel ship
烧油的 oil firing; oil-burning; oil-fired
烧油的炉火 oil-burning fire
烧油的燃烧器 oil-burning burner
烧油电厂 oil-fired station; oil-fired thermal plant
烧油锻炉 oil forge
烧油反射炉 oil-fired air-furnace
烧油干燥机 oil-burning drier[dryer]; oil-fired drier[dryer]
烧油供热装置 fuel-fired heating equipment
烧油锅炉 oil-fired boiler
烧油壶 oil-burning kettle
烧油回转窑 oil-fired rotary kiln
烧油炉子 oil-fired furnace
烧油烹饪器 oil-burning cooker
烧油系统 oil firing system
烧油窑炉 oil fired kiln; oil-fired furnace
烧釉炉 enamel stove
烧釉面砖的烘炉 biscuit oven
烧釉窑 saggar house
烧余残渣 ignition residue
烧渣 calcigenous; firing residue
烧胀板岩 expanded slate
烧胀板岩工厂 expanded slate factory
烧胀黏[粘]土骨料 expanded clay aggregate
烧胀页岩 expanded shale
烧赭石 burned ocher[ochre]; burnt ocher[ochre]
烧赭土 burnt sienna; burnt umber
烧针 head pin
烧震 combustion shock
烧制 burning; calcining; firing
烧制产品 burned product
烧制车间 furnace room
烧制成的灰 burnt dust
烧制黏[粘]土曲瓦 burned clay curved roofing tile
烧制品 burned ware; coctile <建筑用>
烧制前修饰加工砖坯 dressed brick
烧制熔块 fritting
烧制水泥用矿渣 cement slag
烧制陶器 clay body
烧制填料 burnt filler
烧蛭石 <可做建筑材料> zonolite
烧舟 boat; combustion boat
烧轴 overheating of axle-box
烧煮 cooking
烧煮温度 cooking temperature
烧砖 brick burning
烧砖台 brick cage
烧灼 burning; cauterize
烧灼残渣 ignition residue
烧灼剂 cauterantia; cautery
烧灼减量 weight loss by ignition
烧灼减重 loss on ignition
烧灼器 cauter
烧灼试验 ignition test
烧灼术 cauterization; cautery
烧灼样足 burning feet
烧棕土 burnt umber
烧钻 bit burnt out; burn a bit; burning of bit; burnt bit
烧嘴 burner nozzle; external burner lip
烧嘴标高 burner level
烧嘴堵塞 clogging of the burner
烧嘴风机 <热风炉> burner blower

烧嘴固定板 burner fixing plate
烧嘴管线 burner piping
烧嘴耐火砖 burner fire-brick
烧嘴逆火 flash back
烧嘴钳 burner pliers
烧嘴砖 burner block;nozzle brick

稍大的配件 loose fitting

稍带咸味的水 brackish water
稍黑的 blackish
稍加浓铀堆 near-natural uranium reactor
稍尖的 subacute
稍结工厂 sintering plant
稍慢渗透 moderately slow permeability
稍密 sparse
稍稍 in a way
稍湿 slightly humid
稍湿的 slightly damp;slightly wet
稍凸的檐壁 pulvinated frieze
稍微浑浊的 briefly cloudy
稍咸的 brackish
稍有溶蚀 slightly dissolution
稍有污损的纸 <英国在这种纸的包装上打 XX 的印记,美国打 R 印记> retree
稍有圆角的边棱 eased arris

艄 helm;rudder;stern

艄板 transom
艄型 buttock

勺柄 dipper handle;kettle holder

勺斗 bagger;bucket;digger;dipper
勺斗柄 dipper arm
勺斗刃口 dipper lip
勺斗式机铲 pan shovel
勺斗式装载机 scoop loader
勺轮 bucket wheel
勺轮式排出装置 scoop wheel distributor
勺皿 casserole
勺取 bail-out;dipper
勺取法 shovel method
勺取炉 bail-out furnace
勺扰土样 spoon sample
勺式进料器 scoop feeder
勺形叉 scoop fork
勺形铲 scoop shovel
勺形螺钻 posthole auger
勺形取土器 spoon soil sampler
勺形钻 bailer; gouge drill; posthole auger;bucket auger;spoon bit
勺形(钻孔)除渣器 auger cleaner
勺形钻头 gouge bit;pod bit
勺样检验 spoon test
勺状断层 spoon-like fault
勺子 scooper;shovel
勺钻 bucket auger; half-round drill; posthole auger; spoon bit; drilling bucket
勺钻取样 spoon sampling
勺钻土样 spoon sample
勺钻取样器 spoon sampler

芍药 peony

杓 dan;ladle

杓柄 dipper handle
杓铲 scoop shovel
杓斗柄 <挖掘机的> dipper stick; dipper arm
杓斗铲 dipper shovel
杓斗齿 dipper teeth
杓斗刃口 dipper lip
杓斗挖泥船 bagger
杓斗挖泥机 dipper dredge(r)
杓管 scoop tube
杓管控制的耦合器 scoop controlled coupling
杓管深度控制 scoop depth control
杓兰属【植】moccasin flower
杓轮 bucket wheel
杓轮回收设备 bucket wheel reclaimer
杓轮加载器 bucket wheel loader
杓式进料器 scoop feeder
杓式挖泥船 scoop dredge(r)
杓式挖泥机 scoop dredge(r)
杓形钻头 bailer bit;dowel bit
杓状聚光灯 scoop light
杓子 scoop

少保养蓄电池 low-maintenance battery

少报 underreport;understatement
少冰冻土 poor ice content frozen soil
少尘 <能见度不小于 2 千米> light dust
少沉淀的 mini-ash
少乘员车辆 single occupancy vehicle
少地震活动地区 quiet location
少废 waste minimization
少分配的成本及费用 underabsorbed burden
少分配的费用 under-allocated expenses;underapplied expenses
少分配制造费用 underabsorbed burden; underabsorbed overhead; underapplied manufacturing expenses; underapplied manufacturing overhead
少缝隙 little fissure
少付工资 underpay
少钢筋混凝土 under-reinforced concrete
少给 skimp
少耕法 minimum tillage
少耕制 minitillage system
少股合资 minority joint-venture
少灰拌和砂浆 lean mixed mortar
少灰的 low-ash;mini-ash
少灰混合料 lean mix(ture)
少灰混凝土 lean(mix) concrete;poor concrete
少灰碾压混凝土 poor mixing ratio rolled concrete
少灰砂浆 lean mortar
少灰水泥混凝土 lean cement concrete
少计总额 underfoot
少缴 shortage in delivery
少筋 less reinforcement
少筋的 under-reinforced
少筋混凝土 less reinforced concrete; plain concrete; rare reinforcement concrete
少筋混凝土结构 under-reinforced concrete structure;under-reinforcement concrete structure
少筋混凝土梁 under-reinforced concrete beam
少筋混凝土圬工 partially reinforced concrete masonry
少筋梁 low-reinforced beam; u-reinforced beam
少筋设计 under-reinforced design
少筋微弯板 under-reinforced slab with slightly curved bottom
少孔隙混凝土 concrete of low porosity
少矿渣水泥 low-slag cement
少量 drib(b)let; gleam; groat; handful; modicum; paucity; shred; small quantity
少量拆除 spot clearance
少量分散系统 patucidisperse system
少量购买 be purchased in small quantities
少量光吸收 low optic(al) absorption
少量沥青重铺面层 light asphalt resurfacing
少量配筋 light reinforcement
少量配筋(混凝土)路面 lightly reinforced pavement
少量气体逸出 little gas escape
少量剩余 shred
少量停车 light parking
少量污染源 small quantity pollution sources
少量细砂 trace fine sand
少量有机质 trace organic
少量元素 minor element
少量炸药 blown-out shot
少黏[粘]性土 less-cohesive soil
少碾压混凝土 lean rolled concrete
少排土桩 small displacement pile
少铅漆 <含氧化铅少于 5%> lead-restricted paint
少区近似 few-region approximation
少群分析 few-group analysis
少群扩散理论 few-group diffusion theory
少人值班的遥控电站 remotely controlled power station with skeleton attendance
少溶剂的 solventless
少熔渣水泥 low-slag cement
少色性 hypochromism
少砂拌合[和]物 undersanded mix(ture)
少砂大孔混凝土 hollow concrete with less sand
少砂混合料 <含砂过少的混合料> undersanded mix(ture)
少砂黏[粘]土 sand poor in clay;gumbo <密西西比河流域的一种细而少砂的黏[粘]土,湿则黏稠而极滑腻,干则裂成块者>
少收 shortage in collection
少收款项 amount under-collected
少熟料矿渣水泥 clinker-bearing slag cement
少数股控制 minority control
少数股权 minority holding; minority interest
少数股权股东 minority stockholders
少数控制网(络) oligarchic network
少数买主垄断 oligopoly;oligopsony
少数民族地区 territory of nationality
少数民族建筑 architecture of minority nationalities
少数民族居住地区 ghetto
少数派 minority
少数权益 minority interest
少数群 minority group
少数杂质 minority impurity
少数载流子 minority carrier
少数载流子发射极 minority-carrier emitter;minority emitter
少数载流子俘获 minority-carrier trapping
少数载流子注入 minority-carrier injection
少数种族 racial minority
少水泥混合料 lean mix(ture)
少水泥混凝土 lean cement concrete; lean mix concrete;poor concrete
少水年 dry year; low-flow year; year with low flow;year with low water
少算 under footing
少算费用 undercharge
少算款项 amount under-charged
少算运费 under-billing
少索体系 <斜拉桥> few stay system
少无氧化加热 scale-less of free heating
少先吊车 pioneer crane
少芯电缆 small capacity cable
少许空隙的 semi-porous
少循环湖 oligomictic lake
少盐的 oligohaline
少盐水 oligohaline
少音路面 <行车噪音较低> quiet(er) pavement
少英细晶岩 engadinite
少油断路器 oil minimum breaker; small oil volume circuit breaker
少油混合料 lean mix(ture)
少油警报 low oil alarm system
少油开关 low oil content circuit-breaker;oil minimum breaker
少油清漆 short-oil varnish
少油式测量用变压器 low oil content measuring transformer
少油式断路器 small oil volume circuit breaker
少云 partly clouded; partly cloudy; somewhat cloudy
少云的 briefly cloudy
少云天空 slightly clouded sky
少占空间的 space-saving

少年感化院 protectory

少年宫 children's hall;children's palace
少年教养院 detention home; juvey; juvie;protectory;reform school
少年之家 children's center[centre]; children's club;children's house
少年坐便器 juvenile water closet
少壮地形 topographic(al) adolescence;topographic(al) youth
少壮海岸 adolescent coast
少壮河 adolescent river
少壮河谷 adolescent valley
少壮期 <河流的> adolescence
少壮期的 <指河流> adolescent

邵德石 schoderite

绍雷配电制 Thury system

哨兵 sentinel;sentry

哨浮标 whistle buoy
哨声 squealing
哨声调制 whistle modulation
哨所 watch house
哨艇 vedette(boat)
哨音浮标 whistling buoy
哨音符号 whistle code
哨子 whistle

奢侈 extravagance;luxury

奢侈品 luxury

赊付 on-account payment

赊购 account purchase;bought for account;buy on credit;buy on tally; buy on the nod;buy on tick;buy on trust; credit buying; credit purchase; credit purchasing; hire purchase;purchase on credit
赊购发票 charge ticket
赊购费 account payable;on credit

S

赊购货物 buy goods on tick
赊购交易 credit transaction
赊购卡 access card
赊购账 charge account
赊购账户 charge account; credit account
赊购制度 credit system
赊货簿 pass book
赊货率 charge plate
赊买 account purchase
赊卖 tally system
赊卖价格 credit price
赊卖账户 account of credit sale
赊欠(交易) on credit; on tick
赊欠免言 no credit charge
赊欠清单 charge list
赊欠账项 mark-up
赊售 charge sales; credit business; sale on open book account; trust
赊售账户 charge account
赊销 account sale; charge sales; credit sale; deferred payment sale; payment on account; sale for account; sale on account; tally the trade
赊销发票 credit sale invoice
赊销法 tally system
赊销付款条件 credit terms
赊销合同 credit sale agreement
赊销价格 credit price
赊销金额 book credit
赊销贸易 tally trade
赊销商店 tally shop
赊账 account credit; account purchase; given credit; on account; on trust; open (book-) account; keep account
赊账付款 payable in account
赊账交易 credit transaction
赊账金额 amount of the credit; credit amount
赊账卡 charge card
赊账贸易 tally trade
赊账损失 credit loss
赊账条件 terms of credit
赊账支付 payment on account

舌板式铲运机 tongue scraper

舌板闸门 flap gate
舌瓣 flap(per); gate flap; hinged flash gate
舌瓣倾侧式闸门 flap gate
舌瓣闸板 oscillating flashboard
舌瓣闸门 bascule gate; oscillating flashboard
舌瓣状河漫滩 flood plain lobe
舌贝 tongue shell
舌槽 glossal canal; lingual groove
舌槽法兰 tongued and grooved flange
舌槽接缝 matched joint; offset; T and G joint; tongue and groove joint
舌槽接合 matched joint; match(ing) joint(ing); T and G connection; tongue and groove connection; tongue and groove joint
舌槽迷宫式 tongue and groove labyrinth
舌槽迷路 tongue and groove labyrinth
舌槽面 tongued and grooved surface
舌槽企口板 match boarding
舌槽式接合 tongued and grooved joint
舌槽榫 tongued and grooved timber
舌槽砖 notched brick
舌侧倾斜 linguoclination
舌颤动 tremulous tongue
舌点 linguale
舌阀 clack box; dart; dart valve; flapper; leaf valve
舌阀捞砂筒 auger with valve; dart-valve bailer
舌感试验 taste test
舌弧 lingual arch
舌簧 armature tensioning springs; reed; tongue
舌簧接点元件 reed switch
舌簧喇叭 armature loudspeaker; moving-armature loudspeaker
舌簧频率计 reed ga(u)ge
舌簧式电流表 tongue amperemeter
舌簧式继电器 reed relay
舌簧式受话器 reed-type receiver
舌簧式听筒 vibrating reed receiver
舌簧雾笛 reed fog horn
舌簧扬声器 armature loudspeaker
舌簧叶片 reed blade
舌尖 apex linguae; tongue tip
舌尖音 apical
舌接 tongue grafting; whip-and-tongue graft; whip grafting
舌门 flapper
舌片 foot lug; tongue(piece)
舌片垫圈 tab washer
舌片式槽楔 tongue wedge
舌钳 tongue forceps
舌饰【建】 tongue
舌榫 slip feather; tongue and groove
舌榫加工 matching and grooving
舌榫接合 cogged joint; tongue joint; tonguing
舌榫斜接 tongued miter[mitre]
舌榫斜拼合 tongue(d) miter[mitre]
舌头 back edge; backfin; full strip
舌形贝板层 Lingula flags
舌形部分 pecker
舌形扯破试验 tongue tear test
舌形阀 flap valve
舌形接口 lipped joint
舌形接头 lipped joint
舌形冷却管 serpentine cooler
舌形沙坝 <又称舌形砂坝> lingual bar
舌形沙坝坝沉积 lingual-bar deposit
舌形石 glossopetra
舌形试样法 tongue method
舌形试样撕破强力试验 tongue tear test
舌形撕破强力 tongue tear strength
舌形转辙器 tongue-type switch
舌焰 narrow flame
舌与槽的 tongued and grooved
舌状冰川 tongue-shaped glacier
舌状波痕 cuspate ripple mark
舌状大波痕 linguoid ripple mark
舌状的 lingulate; lobate; tongue-shaped; linguiform
舌状阀 cantilever valve
舌状分布 tongue-like distribution
舌状腹接 side tongue graft; side tongue-grafting
舌状海岸 lobate coast
舌状积沙 tongue-shaped sand flood
舌状流痕 lobate rill mark
舌状流域 tongue-like basin
舌状泥石流体 coulee
舌状盆地 tongue-like basin
舌状侵入体 ribbon injection
舌状丘 tongue-hill
舌状三角洲 lobate delta
舌状沙埋 sand cover of tongue shape
舌状沙洲 lingual bar; linguoid bar
舌状体 ligulate; liguliformcolulus
舌状物 tongue
舌状斜接 tongued miter[mitre]
舌状褶皱 tongue-like fold

蛇孢刺壳属 <拉> Ophiochaeta

蛇毒 snake poison; snake venom
蛇腹形 concertina
蛇腹形河曲 concertina meander; meander concertina
蛇腹形河弯 concertina meander; meander concertina
蛇腹形铁丝网 concertina wire
蛇腹状浮标 accordion buoy
蛇感受器 snake's sensor
蛇根木 snake wood
蛇管 coil(ed) pipe; coiler; flexible conduit; flexible pipe; helical coil; hose; hose pipe; pipe coil; spiral pipe; worm coil; worm pipe
蛇管过滤器 coil filter
蛇管换热器 coil heat exchanger
蛇管加热器 serpentine heater
蛇管夹套 coil jacket
蛇管接触室 serpentine contact chamber
蛇管冷凝器 coil condenser
蛇管式防毒面具 airline mask; airline respirator
蛇管式加热器 tubular heating coil
蛇管蒸发器 coil evapo(u)rator
蛇窖 snake pit
蛇节杆 articulated arm
蛇节连接板 articulated slab
蛇节(连接)方式 pinned system
蛇类展览馆 serpentarium
蛇链 pole chain
蛇笼堤 gabionade
蛇笼树脂 snake cage resin
蛇绿混杂体 ophiolitic melange
蛇绿榴辉岩 ophiolitic eclogite
蛇绿岩 ophiolite
蛇绿岩侵位 ophiolite emplacement
蛇绿岩套【地】 ophiolite suite; ophiolite complex
蛇麻草 hop clover
蛇麻草袋 hop pocket
蛇麻草田 hop-field
蛇麻草纤维织物 hop-fiber[fibre] cloth
蛇麻(草)园 hop-garden
蛇麻草枕头 hop pillow
蛇炮眼 snake hole
蛇皮 snake skin
蛇皮管 flexible conduit; flexible metal conduit; hosepipe; snake pipe
蛇皮绿 snakeskin green
蛇皮绿釉 green snakeskin glaze
蛇皮釉 snakeskin glaze
蛇丘 escar; sechar
蛇丘丛 esker knobs
蛇丘的 eskerine
蛇丘底碛 betalayers
蛇丘谷 esker trough
蛇丘脊 esker ridge
蛇丘三角洲 esker delta
蛇丘扇形地 esker fan
蛇丘状三角洲 osar delta
蛇曲 meander
蛇曲带 flood plain lobe
蛇曲河谷 meandering valley
蛇曲河(流) meandering river; meandering stream; snaking river; snaking stream; winding river winding stream
蛇伤 snake bite
蛇石 snakestone
蛇尾线【数】 ophiuride
蛇纹白云石 ophicalcite
蛇纹大理石 <暗绿色的> serpentinous marble; Maryland verde antique; serpentine marble
蛇纹大理岩 ophicalcite
蛇纹方解石 ophicalcite
蛇纹管 flexible conduit
蛇纹滑石 serpentine-talc
蛇纹粒玄斑岩 navite
蛇纹绿泥片岩 serpentine-chlorite schist
蛇纹木材 snake wood
蛇纹石 Scotch stone; snakestone; serpentine
蛇纹石板 serpentine plate; serpentine rock slab
蛇纹石大理岩 serpentine marble
蛇纹石化 serpentinization
蛇纹石棉 chrysotile asbestos
蛇纹石片岩 serpentine schist
蛇纹石墙 serpentine wall
蛇纹石石棉 serpentine asbestos
蛇纹石石棉矿床 chrysotile-asbestos deposit
蛇纹石砖 serpentine brick
蛇纹形石或砖墙 serpentine wall
蛇纹形砖石墙 serpentine wall
蛇纹岩 antigorite; ophiolite; serpentine rock; serpentinite
蛇纹岩化作用【地】 serpentinization
蛇纹玉 serpentine-jade
蛇纹状岩 ophite
蛇线轨迹 serpentine path
蛇行 serpentuate; snake motion; S-shaped motion
蛇行棒 snake
蛇行波 shake-motion wave
蛇行浮动 snake
蛇行纹 snake
蛇行运动 sinusoidal motion; snake motion; snake movement
蛇行振动特性 hunting behavio(u)r
蛇形摆动拉幅机 serpentine jigging tenter
蛇形斑点 snake mark
蛇形棒 snake
蛇形边 snaky edge
蛇形畴 snake domain
蛇形磁畴 serpentine domain; snake domain
蛇形导管 serpentine duct
蛇形导轨 snaky track
蛇形的 sinuous; snake-like; snaky
蛇形队搜索 snake formation search
蛇形飞行 snaking
蛇形飞行模式 snake flying mode
蛇形沟 spiral ditch
蛇形管 serpentine (coil); serpentine conduit; serpentine pipe; serpentine tube; hairpin coil; hose; hose pipe; loop; serpentuator; sinuous coil; worm(pipe) <蒸汽器的>
蛇形管过热器 multiloop superheater; zigzag superheater
蛇形管加热器 coil-tube heater; flexible heater
蛇形管冷却器 serpentine cooler
蛇形管省煤器 continuous loop (-type) economizer; loop economizer
蛇形管水封 loop seal
蛇形冷凝管 coiled condenser
蛇形冷却管 spiral coil cooling tube
蛇形冷却器 cooling coil(ed) pipe; serpentine cooler
蛇形泥浆泵 slurry snake pump
蛇形泥条 snakes
蛇形排管 serpentining
蛇形排列 serpentining
蛇形盘管 serpentine coil; serpentining; sinuous coil
蛇形盘管冷却器 serpentine cooler
蛇形墙 serpentine wall; crinkle-crankle <十八世纪>
蛇形丘 back furrow; serpentine kame; aesar; esker[eskar] <冰川沉积物堆积成的蛇曲形丘岗>
蛇形渠 spiral conduit
蛇形砂丘 seif(dune)
蛇形栅栏 snake fence
蛇形弹簧联轴器 Bibby coupling; Falk flexible coupling
蛇形弯管 snake bend

蛇形弯头 snake bend
蛇形纹 snake
蛇形线 serpentine
蛇形芯 snake core
蛇形移动 sinuous movement
蛇形引线 snake
蛇形运动 hunting movement; serpentine locomotion; snaking motion
蛇形蒸汽管 steam coil
蛇形钻 twist bit
蛇形钻杆 snake drill
蛇型捻股机 snake type strander
蛇穴(爆破)法 snakeholing method
蛇穴法二次爆破 <大块底部打眼爆破> snakeholing method
蛇穴式炮孔 snake hole
蛇穴式炮眼 snake hole
蛇穴式炮眼爆炸 snakehole shot
蛇穴形孔 snake hole
蛇窑 beehive kiln; rifle kiln
蛇状沙嘴 serpentine spit
蛇状压条 serpentine layer

舍

9 校验 cast-out-9-check

舍菲尔德平炉法 Sheffield process
舍菲尔环节函数 Shaffer stroke function
舍九法 casting-out nines
舍雷绿 mineral green
舍利夫拉杆 <一种开闭扇形窗或天窗的拉杆> sheriff
舍利格埃板桩 <由高密度聚乙烯材料制成> Schlegel sheeting
舍利(子)塔 <佛教的> Buddha shrine; domical mound; pagoda for Buddhist relics; Buddhist shrine; dagoba
舍零误差 rounding error
舍蒙阶 <晚泥盆世>【地】Chemungian stage
舍蒙群【地】Chemungian group
舍蒙统【地】Chemungian series
舍内越冬 in-wintering
舍弃 truncate
舍弃回收率 cut-off rate
舍弃率 rejection rate; rejection ratio
舍弃数 rejection number
舍弃说明 ignore specification
舍去【数】casting out; rounding; truncation
舍去零数 round off
舍去零数化成整数 radiusing
舍入 half adjust(ing); rounding-off
舍入常数 round-off constant
舍入成整数【数】round off
舍入的整数 rounded figure; rounding figure; rounding number
舍入法 rounding-off error; rounding-off method
舍入方式 rounding-off procedure
舍入符号 round-off symbol
舍入过程 rounding procedure
舍入进位 end-around carry
舍入精度 round-off accuracy
舍入开关【计】rounding switch
舍入区间运算 rounded interval arithmetic
舍入数 rounded number; rounding-off number; round-off figure
舍入位 rounding bit
舍入误差 end-around error; rounding error; round(ing)-off error
舍入误差积累 accumulation of rounding errors
舍入误差累加 round-off accumulating
舍入误差累加器 round-off accumulator
舍入移位 end-off shift
舍入指令 round-off order
舍入字符 separating character
舍饲场 yard system
舍特克利夫(机械)测深仪 Sutcliffe's sounding apparatus
舍尾移位 end-off shift
舍位 truncate; truncation
舍位法 truncation method
舍位误差 truncation error
舍温电磁振动器 Sherwin electromagnetic vibrator
舍项【计】truncation; truncate
舍项法 method of truncation; truncation method
舍项三角分解法 truncated triangular decomposition
舍项误差 truncation error
舍选技术 rejection technique

设

鞍褥于背 panel

设暗销的边缘 <水泥混凝土路面> dowel(l)ed edge
设堡 fortification
设备 equipment; apparatus; appliance; appointment; contrivance; device; equipage; facility; fitment; fixture; furnishing; furniture; gadget; installation; instrument; means; outfit; plant; upgrade; work horse; accommodation; aid
设备安全 safety equipment
设备安全要求 safety equipment requirement
设备安全证书 safety equipment certificate
设备安装 construction of fixtures; equipment arrangement; equipment assembly; equipment installation; erection; installation of equipment
设备安装地点 location of the machine to be installed
设备安装吊耳 lifting lug for equipment installation
设备安装费 cost of equipment installation; equipment installation cost
设备安装工 fitter; millwright
设备安装工程 fitter's work; plant engineering
设备安装工业 plant industry
设备安装工作 fitter's work; fitting work
设备安装公司 construction company for machinery and equipment installations
设备安装固定端 dead-end installation
设备安装和维护负责人 master mechanic
设备安装检查 facility installation review
设备安装检查员 installation inspector
设备安装使用工程 plant engineering
设备安装通知 equipment installation notice
设备安装与检修 equipment installation and checkout
设备安装重量 installation weight
设备保留 device reserve
设备保留字 device reserve word
设备保温 insulation of equipment
设备保险 equipment insurance
设备保养 corrective maintenance; equipment maintenance
设备报废 equipment scrap; retirement; written off equipment
设备备份 device backup
设备备用 equipment sparing
设备比较 facility comparison
设备变更明细表 equipment modification list
设备变卖损益 gain or loss on sale of equipment
设备标牌 name plate
设备标识 device identification
设备标识符 device identifier
设备标示器 device flag
设备标志 device flag; device identification
设备标准 equipment standard
设备表 equipment list
设备不能动作 device inoperable
设备不足 lack of equipment
设备布局 plant layout
设备布置 equipment arrangement; equipment layout; facility location; plant layout
设备布置图 layout of equipment; mechanical drawing
设备部分 environment division; equipment component
设备部件 equipment component; equipment unit
设备部件零件表 equipment component list
设备材料 equipment and materials; materials of equipment
设备材料供应 supply of equipment and materials
设备材料清单 list of equipment and materials; list of machinery and materials
设备参数 parameter of apparatus
设备参数(列)表 device parameter list
设备残值 remanent value of equipment
设备仓库 equipment depot
设备舱口 equipment hatch
设备操纵 device manipulation
设备操作规程 equipment regulation
设备操作顺序 equipment operational procedure
设备操作重量 operational weight of equipment
设备测试指令 device test instruction
设备层 apparatus floor; apparatus stor(e)y; equipment level; mechanical floor; mezzanine; technical floor
设备插口 appliance outlet
设备拆迁 equipment remove
设备产热 heat generated by equipment
设备超载 overflow
设备车间 equipment workshop
设备成本 equipment cost; facility cost
设备程序语言 device program(me) language
设备池管理 device pool management
设备尺寸 installation dimension
设备尺寸计算 sizing of equipment
设备充电 equipment charge
设备出厂证 name plate
设备出租 equipment renting; leasing of equipment
设备出租合同 equipment rental contract
设备储备 stock of equipment
设备处理程序 device handler
设备处理(单) disposition of equipment
设备处理控制 installation processing control
设备处理器 device handler
设备大修 equipment rebuilding
设备(代)码 device code
设备单 tables of equipment
设备单位重量 installation unit weight
设备单元 equipment unit
设备的安装修理工 millwright
设备的各部分构件及其相互联系图 graphical diagram
设备的工作容量 equipment volumetric capacity
设备的购置 equipment procurement
设备的技术陈旧率 obsolescence
设备的检查 check of equipment
设备的经济寿命 economic life
设备的壳体 body of equipment
设备的冷却 equipment cooling
设备的连接 attachment of devices
设备的生产效益 equipment productivity
设备的使用 employment of plant; plant operation
设备的卸下 detachment of devices
设备的重新安装 conversion of unit
设备的重新分配 deallocation of devices
设备的总体布置 general arrangement of the apparatus
设备登记 plant register
设备等待队列 device waiting queue
设备底座 device base; facility base
设备地脚螺栓一览表 list of anchor bolts for the equipments
设备地址 device address; unit address
设备电网 circuitry
设备电源 device power supply
设备电源板 facility power panel
设备吊顶 counter ceiling; false ceiling
设备调查 investigation of facilities
设备调换时间 equipment change-over time
设备定期维护 equipment periodic maintenance
设备定位 equipment siting
设备独立程序 device-independent program(me)
设备独立性 device independence
设备队列 device queue
设备对应表 device correspondence table
设备法兰 vessel flange
设备房 apparatus room; equipment room
设备废弃费用 obsolescence charges
设备费(用) facility cost; cost of equipment; cost of installation; equipment cost; equipment outlay; facility charges; installation cost; facility cost
设备费预算 plant cost estimating
设备分配 device allocation; facility assignment
设备分配程序 device allocation routine
设备分配信息 device assignment information
设备分摊费用 machine-rate charges
设备风扇 equipment blower
设备封存 equipment storing up
设备服务部 services & equipment division
设备服务程序 device service routine
设备服务任务 device service task
设备负荷测量 load meterage
设备负荷系数 apparatus load factor; plant load factor
设备负载因数 equipment capacity factor; plant load factor
设备附件 auxiliaries
设备赋值命令 device assignment command
设备改良 equipment modification
设备改造 equipment modification; equipment upgrading; revamping of equipment; scrap build
设备改装 conversion of unit; equipment modification; scrap build

设备隔间 plant compartment
设备隔振 equipment isolation
设备更改分析 design change analysis
设备更改申请 equipment change request
设备更换时间 equipment change-over time
设备更新 equipment renewal; equipment renovation; equipment replacement; re-equipment; renewal of equipment; renew and renovation of equipment; updating equipment
设备更新费用 cost of equipment replacement; replacement cost
设备更新改造自主权 decision right of equipment replacements and technical innovations
设备更新决策 facility replacement decision
设备更新模式 equipment replacement model
设备更新模型 equipment replacement model
设备更新问题 equipment replacement problem
设备更新政策 equipment replacement policy
设备工程 equipment engineering; equipment work; plant engineering
设备工程更改建议 facility engineering change proposal
设备工程师 equipment engineer; mechanical engineer; plant engineer
设备工程学 apparatus engineering; mechanical engineering
设备工作队列 device work queue
设备工作日志 equipment performance log
设备功率 equipment capacity; horsepower of equipment; installed power; plant capacity
设备供给仓库 equipment supply depot
设备供应 supply of equipment; technical supply
设备供应合同 supply of equipment contract
设备供应和安装合同 supply of equipment with erection contract
设备供应商 equipment supplier
设备供应源 equipment pool
设备购买订货单 facility purchase order
设备购置费 equipment cost; original equipment cost
设备故障 breakdown of equipment; equipment breakdown; equipment failure; equipment trouble
设备故障调查 equipment balk diagnosis
设备故障概率 probability of equipment failure
设备故障率 equipment failure rate
设备故障诊断 equipment fault diagnosis
设备管道 utility raceway
设备管理 equipment control; facility management; management of equipment
设备管理控制和时间调度程序 facility administration control and time schedule
设备管理区 equipment precinct
设备管理软件 facility management software
设备管理系统 facility management system
设备管路 utility run
设备管线 utility raceway; utility run
设备规范 equipment code
设备规格 equipment specification; specification of equipment
设备柜 equipment drawer; equipment locker
设备柜照明 equipment locker lighting; locker lighting
设备柜支架 equipment locker frame; locker frame
设备过剩 overcapacity; over-equipment
设备号 device number; equipment number
设备合格证 certificate of approval
设备合理使用年限 useful life of equipment
设备和工具 equipment and tool
设备和配件 equipment and supplies
设备荷载 equipment load
设备后援 device backup
设备互换(相容)性 equipment compatibility
设备换装 conversion of unit
设备机架报警 equipment rack alarm
设备机械故障失效 mechanical failure
设备机组 united equipment
设备基建成本 plant capital cost
设备基本控制块 device base control block
设备基础 apparatus foundation; device base; equipment foundation; facility base; machine foundation
设备基金 equipment funds
设备及物料的进出气闸 muck-lock
设备及材料明细表 list of equipment and materials
设备及附件 fittings and fitments
设备及工具选择专用卡片 check sheet
设备级 device level
设备计划 equipment planning; facility planning; plant planning
设备记录 plant register
设备记录簿 equipment recordkeeping
设备记录卡片 equipment record card
设备技术规定 equipment specification
设备技术规范 equipment specification
设备技术条件 equipment specification
设备寄存器 device register
设备加油 equipment charge
设备架 equipment frame
设备间 equipment room
设备兼容性 equipment compatibility; hardware compatibility
设备监督器 hardware monitor
设备检查 device check; equipment inspection; plant monitoring
设备检修 equipment overhaul
设备检修配电板 appliance panel
设备检验 unit check
设备检验费 equipment inspection expenses
设备检验装置 equipment test facility
设备鉴定 appraisal of equipment
设备交换 equipment interchange; exchange of equipment
设备交接单 <集装箱码头的> equipment interchange receipt; equipment receipt
设备校准程序 equipment calibration procedure
设备接地母线 equipment grounding
设备接地装置 equipment grounding conductor
设备接口 equipment interface; unit interface
设备接口模块 device interface module
设备接收的要求与检验 equipment acceptance requirements and inspections
设备接头 appliance outlet
设备接线盒 facility terminal cabinet
设备接线架 equipment patch bay
设备结构 structure of device; structure of equipment
设备结构变换 device reconfiguration
设备结构材料 plant constructional material
设备结果程序 device object program(me)
设备结束 device end
设备经济使用寿命 economic life of equipment; equipment economic life
设备净重 net weight of equipment
设备就绪 device ready
设备卡片 machine card
设备开关装置 device switching unit
设备抗震约束 seismic restraint of equipment
设备科 equipment division
设备壳体 body of equipment
设备可靠性 equipment dependability
设备可用率 device availability
设备可用性 device availability
设备空间 device space
设备空重 empty weight of equipment
设备控制 device control; equipment control; plant control
设备控制表 device control table
设备控制单元 device control cell
设备控制符号 device control character
设备控制寄存器 device control register
设备控制开关 equipment control switch
设备控制块 device control block
设备控制器 device controller; device control unit
设备控制台 facility control console
设备控制系统 appliance control system
设备控制中心 system control center [centre]
设备控制装置 device control unit
设备控制(字)符 device control character
设备库 equipment pool; equipment shed
设备库存量 equipment inventory
设备库管理 device pool management
设备跨越 device spanning
设备跨展 device spanning
设备扩充 equipment augmentation
设备老化更新 age replacement
设备类别 device class
设备类型 device type; type of equipment
设备类型逻辑部件 device type logical unit
设备冷却器 unit cooler
设备利用 disposition of equipment; equipment utilization; machine utilization; utilization of equipment
设备利用常数 capacity constant; equipment utility constant
设备利用定额 equipment utilization quota
设备利用率 capacity operating rate; device availability; equipment capacity factor; installation utilization rate; plant availability; plant factor; point availability; utility factor; utility factor of equipment; utilization rate of equipment; utilized efficiency of equipment
设备利用系数 plant factor; utility factor; utility factor of equipment
设备利用学 human engineering
设备连接 device attachment
设备连接前检查 precontact equipment check
设备联合经营的效果 effect of pooling facility
设备联结中继线 interfacility transfer trunk
设备流程图 equipment flowsheet
设备楼层 apparatus floor; apparatus stor(e)y
设备轮询 device polling
设备忙 device busy
设备忙碌控制 device busy control
设备描述块 device descriptor block
设备描述模块 device descriptor module
设备名(称) device name; equipment name; implementer [implementor] name
设备名字 device name
设备明细表 plant register
设备模拟程序 environment simulator
设备目标程序 device object program(me)
设备目录 list of equipment
设备内无样测量 sample-out count
设备能力 apparatus capacity; capacity of equipment; capacity of facility; equipment capacity; installation capacity; installed capacity; plant capacity
设备能力利用 plant capacity utilization
设备能力利用率 rate of utilization of equipment's capacity
设备能力利用系数 capacity factor
设备排队 device queue
设备排水能力测定单位 drainage fixture unit
设备配置 configuration; hardware configuration; equipment assortment
设备配置节 configuration section
设备配置平面图 disposition plan; disposition plan of equipment
设备配置图 layout of equipment
设备平均停运时间 average outage time of equipment; equipment average outage time
设备平面布置图 plan of equipment arrangement
设备破损 equipment breakdown
设备齐全车辆 self-contained vehicle
设备齐全的 self-contained; well-appointed; well-equipped
设备齐全的船坞 self-contained dock; well-appointed dock; well-equipped dock
设备齐全的活动房屋 self-contained mobile home
设备齐全的机组 self-contained unit
设备齐全的居住单元 self-contained flat
设备齐全的楼层 <旅馆或公寓的> flatels
设备齐全的泥浆(拌和)机 self-contained slurry machine
设备齐全的住房 self-contained dwelling
设备其他参数 other coefficient of equipment
设备起动 system start-up
设备清单 equipment list; list of equipment; schedule of accommodation; plant list; plant schedule
设备清洗 service cleaning
设备请求 facility request
设备区 central facility area
设备驱动程序 device driver
设备驱动机构 device driver
设备驱动器 device driver; device handler
设备缺乏 lack of equipment

设备确定 device resolution
设备容差 device tolerance
设备容积 volume of equipment
设备容量 apparatus capacity; equipment capacity; installation capacity; installed capacity; plant capacity
设备容量利用率 capacity factor
设备柔性接头 appliance flexible connection
设备软接头 appliance flexible connection
设备散热量 heat dissipating capacity of equipment; heat gain from appliance and equipment
设备散湿量 moisture gain from appliance and equipment
设备删除 unit deletion
设备商标 name plate
设备设计 design of plant; equipment design; facility design
设备设计标准 facility design criterion; installation design standard
设备设计标准文件 facility design criteria documents
设备设计坐标 facility design critical
设备生产调查 equipment production study
设备生产订货 installation production order
设备生产能力 operating rate
设备生产线 equipment product line
设备生产研究 equipment production study
设备失灵 equipment malfunction
设备失效 device failure
设备施用与装配技术 plant enrichment
设备时间利用率 rate of utilization of working time of equipment
设备识别符号 equipment identification code
设备史料 equipment-history record
设备使用 disposition of equipment
设备使用保养技术 terotechnology
设备使用费率 equipment rental rate
设备使用和维护工程师 facility engineer in use and maintenance
设备使用记录 equipment-history record
设备使用年限 asset life; duration of service; useful life of device
设备使用系数 plant factor
设备事故 equipment breakdown
设备试验装置 equipment test facility
设备试运转 equipment shakedown
设备室 equipment room
设备收据 equipment receipt
设备寿命 equipment life; life of equipment
设备输出格式 device output format
设备输出格式块 device output format block
设备输出口 equipment output port
设备输入队列 device input queue
设备输入格式 device input format
设备输入格式块 device input format block
设备输入噪声级 equipment input noise sound level
设备竖井 utility shaft
设备数据块 device data block
设备说明书 equipment specification
设备特性表格 device characteristics table
设备特性失真 equipment characteristic distortion
设备挑选 equipment selection
设备停用 equipment out-of-use
设备停用期的腐蚀 standstill corrosion
设备通风机 equipment blower
设备投资 equipment investment; installation cost
设备投资函数 equipment investment function
设备投资回收期 recoupment period of equipment investment
设备投资基金 fund for equipment investment
设备投资计划 equipment investment plan
设备图(纸) equipment drawing; mechanical drawing; service drawing
设备外壳 enclosure
设备外壳间距 shell-to-shell distance
设备(外壳)接地 equipment ground
设备外形尺寸 contour size of the unit(s)
设备外形检查 equipment configuration control
设备完好率 equipment availability; in good condition rate of equipment; proportion of equipment in good condition
设备完全的 well-found
设备完善的 self-contained; well-appointed; well-equipped
设备维护保养 maintenance of plant
设备维护不良 poor maintenance of equipment
设备维护管理系统 maintenance and management system of plant
设备维修 corrective maintenance; equipment maintenance; equipment service; maintenance of equipment; upkeep of equipment
设备维修场地 equipment repair yard
设备维修试验公差 maintenance test tolerance
设备维修小组 <土木工程承包的> black gang
设备温度 equipment temperature
设备窝工索赔 claim for idle plant
设备无故障时间 up-time
设备无依性 device independence
设备误差 equipment error
设备系列 equipment line; equipment train; range of equipment
设备系统 equipment system; furnishing system
设备细目 equipment item
设备线路 device line
设备限界 equipment ga(u)ge
设备限制 equipment constraint
设备相关程序 device-dependent program(me)
设备相关的 device-dependent
设备相容性 equipment compatibility
设备相位控制 set phase control
设备箱 equipment box
设备向量表 device vector table
设备项目 equipment item
设备消息处理程序 device message handler
设备小屋 equipment shanty
设备效率 device availability; device efficiency; plant efficiency
设备携带材料 on-equipment material
设备信托公司债券 equipment trust bond
设备信托债券 equipment trust bond
设备信息处理程序 device message handler
设备信息行 device line
设备型号 unit type
设备性能 equipment characteristic; equipment performance; unit performance
设备性能表 equipment performance log
设备性能试验报告 equipment performance report
设备修复 equipment restoration
设备修理场 equipment pool; equipment repair yard
设备修理方法 equipment repaired method; unit replacement system
设备修配工场 plant depot
设备虚址 virtual unit address
设备序列 device queue
设备选型 equipment selection; option of device
设备选择 equipment selection; option of device
设备选择逻辑 device selector logic
设备选择码 device select code
设备选择器 device selector
设备选择位 device selection bit
设备选择线路 device select line
设备选择校验 device selection check
设备询问 environment inquiry
设备延迟输出 equipment delay-output
设备要求 equipment requirement; facility request
设备要求规格 equipment requirement specification
设备页面 device page
设备一览表 equipment list; list of the equipment; plant register; equipment schedule
设备异常 unit exception
设备引线 fixture wire
设备应答 device acknowledge
设备营运维持费 operating maintenance of equipment
设备用房 <如通风房、电梯间、密闭水箱、水泵间等> mechanical room; building for equipment
设备用光缆 equipment optic(al) fibre cable
设备优先级 priority facility
设备与工艺研究 equipment and process research
设备与零备件 equipment and spare parts
设备与运输经理 plant and transport manager
设备与运输主任 plant and transport manager
设备元件 equipment component
设备元件明细表 equipment component list
设备源程序 device source program(me)
设备运输气闸 equipment air-lock
设备运行 equipment practice; operation activity
设备运行队列 device work queue
设备运行时间 unit run time
设备运营成本 plant running cost
设备运转技术 plant engineering
设备运转率 operation rate
设备运转状况记录簿 equipment performance log
设备在移动中进行的作业 on-the-fly
设备再组合 device reconfiguration
设备占用 hold facility
设备折旧 depreciation of equipment; equipment depreciation
设备折旧费 amortized installation cost; depreciation charges of equipment
设备折旧费摊销费及大修费 equipment depreciation charge apportion and over haul charges
设备折让 outfit allowance
设备诊断程序 device diagnostic program(me)
设备支承构架 equipment supporting deck
设备支持例行程序 device support routine
设备支架图 the equipment support drawing
设备制造 equipment manufacture
设备质量 equipment quality
设备质量成本 equipment quality cost
设备致动器 tooling actuator
设备中断 device interrupt
设备中断向量表 device interrupt vector table
设备中心 equipment center[centre]
设备种类 device category
设备重量 weight of equipment
设备专业 mechanical discipline; service discipline
设备转换装置 device switching unit
设备转移 transfer of equipment
设备装置目录 list of furniture and fixture
设备状况监视 equipment status monitoring
设备状况仪表板 equipment status panel
设备状态地址 device status address
设备状态寄存器 device status register
设备状态域 device status field
设备状态字 device status word
设备状态字段 device status field
设备状态字节 device status byte
设备准备好状态 device ready
设备资产估计 appraisal of plant assets
设备资金 equipment funds
设备资料 equipment information
设备字段 device field
设备字符控制 device character control
设备综合 equipment complex
设备综合工程学 terotechnology
设备综合利用率 rate of comprehensive utilization of equipment
设备总效率 overall plant efficiency
设备总重 gross weight of equipment
设备租费 equipment rental rate
设备租借费 equipment rental charges; equipment rental cost
设备租金 equipment rental
设备租赁 equipment lease[leasing]
设备租赁费 equipment rental charges; equipment rental cost
设备租用 equipment rental
设备租用率 plant-hire rate
设备组合程序 configurator
设备组装 equipment assembly
设备最大重量 maximum weight of equipment
设备最高生产能力 maximum plant capacity
设备最小工作参数 stalling work capacity
设备最小重量 minimum weight of equipment
设备最优化 equipment optimization
设备作业率 operation rate
设备坐标 device coordinate
设标 beaconing
设标船 dan layer
设标点 marked point; marked station; monumented[survey] point
设标里程 mileage of aids-to-navigation installed
设标密度 density of aids allocation; density of aids-to-navigation installed
设标水深 buoying depth; marking depth
设标者 buoy tender
设超高的线路 track with cant
设传力杆的板边 <水泥混凝土路面> dowel(l)ed edge
设船坞 dockize[dockise]
设船闸运河 canal with locks
设堤河段 embanked reach

S

设定变址指令 set index instruction
设定地震 scenario earthquake
设定点 set point
设定点控制 set point control
设定点调节器 set point adjuster
设定符 set symbol
设定股本 declared capital
设定航道系统 routing system
设定寄存器 set-up register
设定价值 declared value;stated value
设定目标 target setting
设定时间 set time
设定压力 setting pressure
设定运算数组 set operate stock
设定值 command;set point;setting value
设定值范围 range of set value
设定值跟踪 set point tracking
设定值控制 set point control
设定值调整 set point adjustment
设定值调整器 set value adjuster
设定资本 declared capital;stated capital
设法 contrive;excogitation;take measures
设法做到 contrive
设防 fortification;fortify
设防城市 fortified town;fortress-town
设防的 fortified
设防的井 fortress well
设防的居民区 fortified residence
设防地带 fortified zone
设防地区 bastion;fortified place
设防地震 precautionary earthquake
设防洞口 protected opening
设防房屋 < 苏格兰边境的 > bastle house
设防宫殿 fortified palace
设防教堂 fortified church;garnison church
设防烈度 design earthquake intensity;fortification intensity
设防农舍 bastel house;bastle
设防寺院 fortified monastery;fortress-monastery
设防要塞 fortified stronghold
设防住所 dun
设防走廊 protected corridor
设伏灯 ambush light
设浮标 buoying
设航标航道 demarcated channel
设横肋的 cross-ribbed
设计 chalk out;chalk to;contrive;cook;design;devise;engineering;excogitate;excogitation;layout;layout plan;machinery;outlay;planning;projecting;projection
设计安排 design programme
设计安全 design safety
设计安全系数 design factor of safety;design safety factor
设计安全限 design safety limit
设计岸边线 design waterline
设计坝顶标高 design crest level
设计坝顶高程 design crest level
设计板 layout board
设计保固期限 design life
设计保证值 design certified value
设计报告 design report
设计暴雪量 project storm
设计暴雨单位线法 unigraph design storm procedure
设计暴雨(径流量)design storm
设计暴雨量 project storm
设计比较 design comparison;design competition
设计比较方案 alternate design;design alternatives;alternative design
设计比较估算 comparative design estimate;contrivance
设计比例 design ratio
设计比赛 design contest
设计边坡 designed slope
设计编号 design number;drawing number
设计变动命令书 engineering change order
设计变更 design alteration;design change
设计变量 design variable;design change
设计标定功率 design power rating
设计标高 building grade;design(ed) elevation
设计标高标志点 designed elevation mark
设计标准 design criterion;design standard
设计标准船型 design vessel;significant shipform for dock and harbo(u)r
设计标准洪水 standard project flood
设计标准技术规范 design standard specification
设计标准手册 design standard manual
设计表 design table
设计表格 design schedule;design table
设计冰冻指数 < 10 年中最冷一个冬季的冰冻指数,或 30 年中三个最冷冬季的平均冰冻指数 > design freezing index
设计波参数 design wave parameter
设计波高 design wave height
设计波候 design wave climate
设计波况 design wave condition
设计波(浪)design wave
设计波浪标准 design wave criterion
设计波浪条件 design wave condition
设计波浪重现期 recurrence interval of design wave;return period of design wave
设计补充文件 design supplement
设计不符 different design
设计不良 poor design
设计不良建筑物综合征 sick-building syndrome
设计不全 missing design
设计布置 design layout;planning arrangement
设计步骤 design procedure;design step;design process
设计部门 design(ing) department;design section
设计采购建造 engineering-procurement-construction
设计采购与施工合同 engineering procurement and construction contract
设计参考 design reference
设计参考资料 design aids
设计参数 design condition;design (ed) parameter;design(ed) value;design variable
设计参数值 design value
设计草图 design brief;design draft;design outline;design plan;design sketch;draft design;rough plan;schematization design;scheme design;sketch plan
设计草图加工 design development
设计测量 planning survey
设计策略 design strategy
设计产品 design product
设计常数 design constant
设计潮位 design(ed) tide level;design(ed) sea level
设计车道 design lane
设计车辆 design vehicle
设计车辆尺寸 design vehicle dimension
设计车辆荷载 design vehicle loading
设计车辆荷载标准 loading standard for design vehicle
设计车辆或船舶速度 design speed for vehicles or vessels
设计车速 design speed;design speed of car;project speed
设计车速级差 design-speed interval
设计成本 design cost;projected cost
设计成分 designed composition
设计承受压力 design bearing pressure
设计程序 design approach;design procedure;design program(me);layout procedure
设计吃水 designed draft;design draught
设计吃水差 designed trim;initial trim
设计尺寸 design(ed) dimension;design size;projected dimension;nominal size
设计充满度 calculated filling;designed filling;designed full-load
设计出发点 design starting point
设计储量 design(able) reserves
设计储量失实 planning reserves existence
设计处 engineering and design department
设计船舶 reference ship;design ship
设计船时效率 design ship-hour efficiency
设计船型 design ship;design vessel type and size;dimensioning ship
设计船型尺寸 size of design vessel
设计错误 design fault
设计答案 design solution
设计代表 designated representative;design liaison
设计单位 design institute;design organization;design unit
设计挡水位 designed damming level
设计的 designed;engineered;rated
设计的安全系数不足的 undesigned
设计的波浪条件 design wave condition
设计的抽水量 designed pumping discharge
设计的分配 design distribution
设计的更改 change in design;design change
设计的购物中心 designed shopping centre[centre]
设计的合格性 design acceptability
设计的河槽 designed channel
设计的混凝土配合比 designed concrete mix
设计的基本原则 basis of design
设计的兼容性 design compatibility
设计的矿井 projected mine
设计的螺栓荷载 design bolt load
设计的内部条件 internal conditions for design
设计的人工成本 determining cost of labo(u)r
设计的水位降深值 value of drawdown designed
设计的统一性 unity of design
设计的外部条件 external conditions for design
设计的系统 designed system
设计的协调 < 建筑物各部分的 > unity of form
设计的真实性 frankness of design
设计的指导方针 guideline for design
设计等级 design category
设计低水位 design low water-level
设计底标高 intended bottom
设计地基反力模量 design subgrade modulus
设计地基反力系数 design subgrade factor
设计地面运动 design ground motion
设计地震 design earthquake
设计地震动参数 design parameters of ground motion
设计地震荷载 design seismic load
设计地震力 design seismic force
设计地震烈度 design earthquake intensity;design seismic degree;seismal degree for design;seismic degree for design
设计地震系数 design seismic coefficient
设计地震运动 design ground motion
设计点 design point
设计电路 design circuit
设计电压 design voltage;rated voltage
设计定案 design decision
设计定高程 design crest level
设计定型试验 design fixed type test
设计定义和保密 project definition and secrecy
设计定义和评述 project definition and survey
设计冬季温度 winter design temperature
设计动力荷载 design power loading
设计动态系统 dynamic(al) system of design
设计断面 design section;planned section
设计对工程区分的阶段 design phase
设计对基建、维护、营运总成本的估算 cost in use
设计对象 design object
设计吨位 builder's tonnage
设计额定功率 design power rating
设计额定量 design rating
设计二氧化碳度 designed carbon dioxide level
设计发展 design advance
设计反应谱 design response spectrum
设计范围 design latitude;scope of design
设计方案 design alternative;design approach;design concept;designing scheme;design option;design plan;design precept;design program(me);design proposal;design schematization;design scheme;design solution;layout of plan;project layout;contrivance;designing;contrivance
设计方案经济比较 economic comparison between design proposals;engineering economy
设计方案小组 project team
设计方法 design method;design procedure;method of design(ing)
设计方法学 design methodology
设计方面 design phase
设计方针 design guideline;design principle
设计费 design charges;design fee;expense on design works
设计分工 division of design
设计分析 design analysis
设计风暴 design storm
设计风力 design wind
设计风速 designed air velocity;design wind speed;design wind velocity
设计风压 design wind pressure
设计符号 design abbreviation
设计服务 design service
设计幅度 design latitude
设计辅助工具 design aids
设计负荷 designed load;design(ing) load;load rating
设计负荷点 designed load point
设计负荷系数 design load factor
设计负载 calculated load;design load;load rating
设计负责人 design head;package dealer;project manager

设计附件 design annex;design appendix;design supplement
设计复审 design review
设计复制投影 projection copying
设计改进程序 design improvement program(me)
设计概况 design profile
设计概念 design concept
设计概念的改变 design concept change
设计概念图 preliminary drawing;sketch plan
设计概算 budget estimate for construction design; designed estimation; design estimate; general estimate; preliminary computation; preliminary estimate
设计概算定额 design estimating norm
设计概算负荷 estimated design load
设计概算指标 design estimating index
设计概要 design brief
设计钢筋量 projecting reinforcement
设计高潮位 design high water-level
设计高程 design(ed) elevation
设计高度 design altitude
设计高水位 design high water-level
设计革新 design innovation
设计根据 design consideration
设计更改检验 design change verification
设计更改通知 design change notice; design revision notice
设计更改要求 design change request
设计更改一览 design change summary
设计更正证明文件 design change documentation
设计工程 design engineering
设计工程的检验 design engineering inspection
设计工程师 design(ing) engineer; project engineer
设计工程试验 design engineering test
设计工程一切险 engineering all risks
设计工程展览会(场) design engineering show
设计工具 design aids
设计工况 design condition;design point
设计工况效率 design power efficiency
设计工艺 design technology
设计工作 design effort;design operation;design work;design service
设计工作合同 contract for design service
设计工作计划 design programme
设计工作寿命 design working life
设计工作压力 design working pressure
设计工作站 engineering work station
设计公路桥的一种标准汽车荷载 <美> H-loading
设计公式 design formula
设计公司 design firm
设计功率 design capacity;design power;rated power
设计供暖负荷 design heating load
设计构思 design concept;design consideration
设计构想 design concept;design consideration
设计估算 design estimate;engineering estimate
设计估算的不可预见费 design contingency
设计骨架 design framework
设计故障 design error failure;design failure;design fault
设计顾问 design consultant
设计管径 design pipe diameter
设计管理 design administration; design control
设计管网分析 project network analysis
设计管线压力 internal design pressure
设计惯例 design practices
设计规程 design code;design specification
设计规定的参考尺寸 reference size
设计规范 design code;design criterion; design specification;design standard; project specification;specification for design
设计规格 project specification
设计规则 design code; design(ed) rule
设计规章 design code
设计过程 design procedure; design process
设计过程记录 design development record
设计过程线 design hydrograph
设计海面高程 design sea level
设计海啸 design tsunami
设计航道尺寸 designed dimension of channel
设计航道尺度 designed dimension of channel
设计航道高程 design channel elevation
设计航速 design(ed) speed
设计耗用量 design consumption
设计合同 contract of design; design contract
设计和建造 design-build
设计和建筑公司 design and construct firm
设计和施工合同 design and build contract
设计河底 design bottom; intended bottom
设计荷载 assumed load(ing); designed load; design(ing) load; load rating; specific rated load; specified (rated) load
设计荷载安全系数 design load safety factor;design load factor
设计荷载承受能力 design load capacity
设计荷载谱 design force spectrum
设计荷载水线 designer's load water line
设计荷载效应 design load effect
设计荷载因数 design load factor
设计荷载组合 design loading
设计荷载组合效应 design loading effect
设计洪峰 design flood peak
设计洪峰流量 design flood peak discharge
设计洪水 computed high water; design(ed) flood; project design flood;projected flood
设计洪水出现率 design flood occurrence
设计洪水过程线 design flood hydrograph; hypothetical flood hydrograph
设计洪水进流量 design flood inflow
设计洪水流量 design flood discharge;design flood flow
设计洪水频率 design(ed) flood frequency;design(ed) flow frequency
设计洪水入流量 design flood inflow
设计洪水位 design flood level;design flood stage
设计洪水重现期 design flood occurrence
设计洪水组合 design flood composition
设计后评价 post-design evaluation
设计厚度 design thickness
设计会议 design conference
设计混合料 design mix
设计混合料的成分平衡 balancing of proposed mix
设计混凝土保护层 nominal (concrete) cover
设计混凝土配合比 design concrete mix
设计活动 design activity
设计获准 planning consent
设计机构 project organization
设计基本地震 design basic [basis] earthquake
设计基本地震加速度 design basic acceleration of ground motion
设计基本事故 design basis accident
设计基本数据 basic design data
设计基本原理 design philosophy
设计基础 design basis;design foundation;design fundamental
设计基础数据 design basis data
设计基础资料 basic information for design
设计基准期 design reference period
设计基准强度 design basic strength
设计绩效标准 engineered performance standard
设计及建筑标准 standards of design and construction
设计极限 design limit
设计极限荷载 design limit load; design ultimate load;factored load
设计计划 design effort
设计计划表 design schedule
设计计划和预算体系 planning-programming-budgeting system
设计计划书 design program(me)
设计计算 design calculation; design computation
设计计算图 design diagram
设计记录 design memorandum
设计技术标准 design specification
设计纪要手册 project manual
设计剂量 calculated dose
设计加速度 design acceleration
设计加州承载比<又称设计的加利福尼亚承载比> design California bearing ratio[design CBR]
设计家 ornamentalist
设计假定 design assumption; design hypothesis
设计假设 design assumption; design hypothesis
设计兼施工 design build process;design construct
设计(兼)施工合同 design construct contract
设计检查 design check;design review
设计检查周期 design proof cycle
设计检验 design monitoring
设计检验规范 design control specification
设计检验图表 design control drawing
设计简图 sketch
设计-建设-运营-移交 design-build-operate-transfer
设计建议 design proposal
设计建造合同 design-build contract
设计建造师的现场代表 project representative
设计建筑师 design architect
设计鉴定试验 design evaluation test
设计降水量 design precipitation;project precipitation
设计降雨量 design precipitation;project precipitation
设计降雨强度 designing rainfall intensity
设计交通量 designed traffic volume; design flow;design traffic number; projected traffic volume; design volume <英>
设计交通流量<英> design flow
设计阶段 design(ing) stage;design period;design phase;drawing board;engineering grade;planning phase;project stage;stage of design
设计阶段的评价 assessment in time of design
设计结构 design framework
设计结果 design solution
设计截流量 design interception
设计截面 design section
设计进度表 design schedule; period of design;schedule of design
设计精度 design accuracy
设计井深 target depth
设计竞赛 design competition
设计竞争 design competition
设计举例 design example
设计矩阵 design matrix
设计距离 designed distance
设计决策 design decision
设计浚挖高程 design dredge level
设计卡片索引 planning card index
设计开采量 designed mining yield
设计开采深度 planning mining depth
设计开采时间 designed mining time
设计开挖线 B-line;pay line;tight spot <隧道的>
设计勘察合同 contract for carrying out design and survey works
设计抗剪强度 design shear strength
设计抗力 design resistance
设计考虑 design consideration
设计考虑事项 design consideration
设计科 drafting department
设计可靠度 designed reliability;reliability of design
设计可靠性 designed reliability;reliability of design
设计可能性 designability
设计客流量 designed passenger flow
设计空气动力学 design aerodynamics
设计空气冻结指数 design air freezing index
设计孔深 project depth
设计孔型 form a groove
设计控制 design control
设计控制表 planing control sheet
设计控制截面 principal design section
设计控制条件 design constraint
设计控制图表 design control drawing
设计枯水位 design low water-level
设计框架 design framework
设计理论 design theory;theory of design
设计力矩 design moment
设计立意竞赛 ideas competition
设计沥青用量 design asphalt content
设计连续梁的克劳斯法 Cross method of continuous beam
设计联络 design liaison
设计联络会议 design liaison meeting
设计联络组 design liaison group
设计联席会议 design liaison meeting
设计良好的 well-designed
设计烈度 design intensity;design seismicity
设计灵活性 design flexibility
设计流程 design cycle;design process
设计流量 computed discharge;design(ed) discharge; design(ed) flow; design rate;rated flow;design volume <美>
设计流量保证率 guarantee rate of designed discharge; guarantee rate of designed flow
设计流率 designed flow rate
设计流速 design current velocity;designed flow velocity; designed velocity;design velocity of flow
设计滤速 design filter rate

设计路段 design section
设计路拱高程 design crest level
设计路基强度值 design subgrade strength
设计轮廓 design profile
设计轮压 design wheel load
设计轮载 design wheel load
设计马力 designed horse-power
设计满载吃水 designed full load draft
设计满载吃水线 designed full load line
设计每小时耗热量 design heat consumption per hour
设计面积 planning area
设计秒流量 designed flow per second
设计模拟 design simulation
设计目标 design goal; design object; design objective
设计目的 design goal; design objective
设计内部压力 <管线的> internal design pressure
设计内压 internal design pressure
设计耐久性 designed durability
设计能力 design capability; design (ed) capacity; engineering capacity; projected capacity; rated capacity; contrivance
设计能量 design capacity
设计年度 design year
设计年限 design fixed number of year; period of design; design life
设计扭矩 design torque
设计浓度 design concentration
设计排放量 designed discharge
设计排水量 design displacement; designed discharge
设计配合比 proportioning; designed mix
设计配矿 design ingredient ore
设计偏差 design deviation
设计贫化 design dilution
设计贫化率 design dilution ratio
设计频率 design frequency
设计品质 quality of design
设计平面图 design plan
设计评价 design assessment
设计评审 design review
设计评审会议 design review conference
设计评选 selection of design
设计评议 design deliberation
设计坡度 designed grade
设计坡度线 designed gradient line
设计破碎波高 design breaker height
设计普查 design walk-through
设计谱 design spectrum
设计期 design period
设计期间 design period
设计期限 design (ed) period; period of design
设计起点 beginning of design
设计气候 design climatology
设计汽车 design vehicle
设计前的 predesign
设计前服务 predesign services
设计前期勘测 predesign survey
设计强度 design strength
设计清单 design checklist
设计曲面模型 designing curved surface model
设计曲线 design curve
设计趋向 engineering trend
设计权 design right
设计缺陷 design deficiency
设计热负荷 design heat load
设计热力负荷 design heating load
设计热效率 design thermal efficiency
设计人 designed by; designer; projector
设计人员 design personnel
设计任务 design problem; design task; task of design
设计任务书 design brief; design instruction; design order; design precept; design program (me); design prospectus; design specification; task letter of certificate
设计任务说明书 performance specification
设计日 design day
设计容量 design capacity
设计入库洪水流量 design flood inflow
设计商标 registered design
设计上的考虑 design consideration
设计深度 design (ed) depth; projected depth
设计审查 design examination; design review
设计审查意见 examine suggestions of design
设计审议 design deliberation
设计升力 design lift
设计生产规模 planning mine scale; planning production scale
设计生产率 designed output
设计生产能力 design duty; designed productive capacity
设计生产年限 production age of planning; mine age of planning【地】
设计师 architect; constructor; designer; engineer; fancier
设计师校核表 designer's check list
设计施工分离式承包 design, construction divided contracting
设计施工合同 design construction contract
设计施工合约 design construction contract
设计施工经验 design build experience
设计施工统包法 <美> design and build
设计施工小组 design construction team
设计施工最终阶段 final stage of design construction
设计湿度 design humidity; design moisture
设计湿球温度 design wet bulb temperature
设计(时采用的)负荷 designed loading
设计实践 design practice
设计使用年限 designed service life; designed working life
设计使用期(限) design life
设计使用寿命 design (service) life
设计使用周期 design life; design year
设计事务所 designing firm; design office
设计视距 design sight distance
设计试制车间 design development shop
设计室 design and drawing office; design (ing) department; designing room; design office; design section; design studio; drafting room; drawing office
设计室外气象条件 outdoor condition for designing
设计适应性 design adequacy
设计手册 design handbook; design manual; project manual
设计手段 technical engineering aid
设计寿命 designed (working) life; project life
设计疏浚标高 design dredge level
设计输出量 designed output
设计输入量 designed input
设计数据 project data; design data
设计数据单 design datasheet
设计数据校核表 information checklist for design
设计数值 design value
设计水流速度 design current speed
设计水面(高程) design water-level
设计水平年 design development year; design reference year
设计水深 design (ed) (water) depth; planned depth; project (ed) depth
设计水头 design (ed) (water) head; specified head
设计水位 design (ed) water-level; design (water) stage; projected water level; project water-level
设计水位保证率 guarantee rate of designed water-level
设计水位重现期 recurrence interval of designed water-level
设计水文过程线 design hydrograph; hypothetical hydrograph
设计水线(面) designed waterline
设计水质 designed water quality
设计说明 description of design; design specification
设计说明书 design book; design instruction; instruction of design; performance specification; design sheets
设计思路 design concept
设计思想 design concept; design philosophy; engineering philosophy
设计思想竞赛 ideas competition
设计速度 design (ed) speed; design (ed) velocity
设计损失 design losses
设计损失率 design losses ratio
设计所依据的自然条件 design basis natural event
设计特点 design feature
设计特征 design feature
设计特征周期 design characteristic period of ground motion
设计条件 design condition
设计停止指示 design stop order
设计通航洪水水位 navigation design flood stage
设计通航水位 designed navigable water-level
设计通行能力 design traffic capacity
设计图 constructional sketch; design plan; design scheme; plan sketch; project drawing; scaling drawing; scaling system; scheme
设计图案 layout (work)
设计图案处理 design processing
设计图表 design chart; design diagram; design table
设计图集 collected drawings; collective design drawings
设计图解法 designograph
设计图上的飞机 paper aircraft
设计图上切断桩头高程 cutoff elevation
设计图样 design draft; detail of design
设计图纸 design (ed) paper; design (ed) drawing
设计途径 design approach
设计推力 design thrust
设计吞吐量为 n 吨 designed for capacity of n tonnages
设计妥善的 well-designed
设计挖槽 design dredge-cut; projected dredge-cut
设计挖泥航速 designed dredging speed
设计挖深 designed cut depth; designed depth of dredging
设计挖深线 intended bottom
设计外力 design force
设计外载 design external load
设计弯沉值 design deflection value
设计弯矩 design bending moment; design moment
设计完备度 design adequacy
设计完美的 well-designed
设计网络 planning grid; planning net (work)
设计委托 design commission; design entrusting
设计位置 designed position
设计温差 design (ed) temperature difference
设计温度 design (ed) temperature
设计温度下的许用拉应力 allowable tensile stress at design temperature
设计文件 design document (ation); design file; design paper
设计文件复制 design paper's duplication
设计误差 design error; structural error
设计系列 design series
设计系数 design coefficient; design factor
设计系统 design system
设计系统程序包 design system package
设计系统化 systematization of design
设计细部 design details; details of design
设计细节 design details; details of design
设计细则 design details; details of design
设计现场 project site
设计线 design line
设计限值 limiting design value
设计相对湿度 design relative humidity
设计详图 design plan; detail drawing; detail of design
设计项目 design item; design project; project; projected undertaking; project item
设计项目倡议人 design promoter
设计项目等级 class of heaviness of planning mine
设计项目负责人 job captain
设计小时 <道路设计交通量的计时单位> design hour
设计小时交通量 design hour volume [DHV]
设计小时交通量的定向分布 directional distribution of DHV
设计小时交通量的方向分配 directional distribution of DHV
设计小时交通流量 design hourly volume
设计小时(控制)因素 design hour factor
设计小组 design team
设计效果 design effect
设计效率 design (ed) efficiency
设计协议 design agreement
设计心理学 psychology of design
设计新颖 ingenuity
设计(行车)速度 design speed
设计行动 design action
设计型号 designed type
设计性能 design performance; specified performance
设计修改 change in design; design alternation; design revision
设计许用应力 allowable design stress
设计蓄水位线 design storage level
设计旋转压实次数 design number of gyration
设计选择 design option
设计询问通知 engineering query note
设计压力 design (ed) pressure; engineered pressure
设计压实层厚 design lift
设计压缩比 design compression ratio
设计压缩变形 rated compression deflection
设计研制车间 design development shop
设计验证 design verification
设计验证试验 design verification test
设计扬程 design head

设计要点 design consideration; essentialities of design; main point of design
设计要求 design requirement; design specification
设计要求的图纸 design requirement drawing
设计要素 design element
设计业务 design operation; design profession
设计一览表 design summary
设计依据 basis of design; design basis; design consideration; design fundamental
设计依据地震 design basic [basis] earthquake
设计意图 design idea
设计因数 design(ing) factor
设计因素 design(ing) factor
设计因子 design(ing) factor
设计应力 design stress
设计应力法 assumed stress approach
设计盈满度 designed filling
设计用量 design quantity
设计用水量 designed duty of water; designed water consumption
设计优化 design optimization
设计有效荷载 design payload
设计余量过大的 over-designed
设计与估计 planning and estimating
设计与施工 design and construction; engineering and construction
设计与施工公司 design and construction firm
设计雨量 design rainfall; project rain
设计预算 design budget; detailed estimate
设计裕量过大的 over-designed
设计员 estimator
设计原理 design philosophy; design principle; principle of design
设计原型 prototype
设计原则 design philosophy; design precept; design principle; engineering philosophy; principle of design
设计院 designing institute; institute; institute of design
设计约束条件 design constraint
设计允许渗入量 infiltration design allowance
设计载重 design load
设计载重水线 designer's load water line; designed load water line
设计哲学 design philosophy
设计者 artificer; constructor; contriver; designer; deviser; fashioner; projector
设计者对承包商 proposal request
设计者发明人 artificer
设计震动力 design seismic force
设计震动系数 design seismic coefficient
设计政策 design policy
设计支承压力 design bearing pressure
设计值 designed value; specified value
设计职业 design profession
设计指标 design goal
设计指南 design guide(line)
设计指引 design guidance
设计制图 design drafting
设计制图学院 college of design
设计制图组长 job captain
设计制约 design constraint
设计制约条件 design constraint
设计制造一体化 integrated manufacturing system
设计制作万能钥匙型 master-keying
设计质量 design quality; quality of design
设计中的 on the drawing board
设计中考虑的问题 design consideration
设计中心 design center
设计终点 end of design
设计重量 design weight
设计重现期 design frequency; design recurrence interval; design return period
设计周到的 well-designed
设计周期 design cycle; design(ed) period; period of design
设计主任 design head
设计主题 design motif; motif
设计专家 design professional
设计专利权 design patent
设计专业人士 designated person
设计装配图 design assembly
设计状况 design situation
设计状态 design point
设计准确度 design accuracy
设计准则 design criterion; design guide(line); guidelines for design
设计咨询 design consultation
设计资料 design data; design document(ation); engineering data; planning information; project data
设计资料收集 design data gathering; engineering data gathering
设计资料图表 design datasheet
设计资用压力 <工作资用的压力值> design working pressure
设计自动化 design automation
设计自动化辅助工具 automation assistant
设计自动化工程 automated design engineering
设计自动化系统 design automation system
设计总方案 parti
设计总负责人 chief designer; chief engineer of the project; project director; project engineer; designer-in-charge
设计总工程师 chief design engineer; chief designer
设计总体组 overall design management
设计总造价 project cost
设计总重量 design gross weight
设计纵坡 designed longitudinal grade
设计纵倾 designed drag; designed trim
设计组 design group; design section; design team; project group; project team
设计组成 designed composition
设计组长 job captain
设计组装系统 design-it-yourself system
设计阻力 design resistance
设计钻孔直径 rated(bore) hole diameter
设计最大工作压力 maximum design working pressure
设计最大流量 design flood discharge
设计最大通航流量 designed maximum navigable discharge
设计最低水位保障率 guarantee rate of designed lowest stage
设计最低通航水位 designed lowest navigable stage; designed lowest navigable water-level
设计最高大流量重现期 recurrence interval of designed maximum discharge
设计最高水位重现期 recurrence interval of designed highest stage
设计最高通航水位 designed highest navigable stage; designed highest navigable water-level
设计最小接缝宽度 minimum design joint width
设计最小流量保障率 guarantee rate of designed minimum discharge
设计最小通航流量 designed minimum navigable discharge
设计坐标 design coordinate
设井栏 curb
设井圈 curb
设景 landscaping
设栏平交道 guard rail crossing
设老虎窗的屋顶 dormer(ed) roof
设立 establish; institution; set-up
设立边界 bordering
设立疆界 bordering
设立通信[讯]线路 setting-up the link
设立者 creator; institutor
设路缘设备 curbing machine
设路缘石 curb(ing)
设码头 dockize[dockise]
设施 device; facility; installation
设施部门 facility operation department
设施的补充 provision of facility
设施费 establishment charges
设施分散 facility dispersion
设施改进后按房地产临街部分多少加征税款 frontage assessment
设施更新 facility replacement
设施供应 installation supplying
设施和设备 installation and equipment
设施记录 facility record
设施利用率 facility usage ratio
设施迁改 utility diversion
设施审计 facility audit
设施疏开布置的机场 dispersed aerodrome
设竖井式进水口的涵洞 drop-inlet culvert
设水闸引水 sluicing
设算成本 imputed cost
设算利息 imputed interest
设围堤的 endyked
设围篱 pale
设圬工砌面的混凝土重力坝 concrete gravity dam with masonry facing
设屋顶窗的屋顶 dormer(ed) roof
设屋顶式闸门的坝 bear-trap dam
设陷阱 trap setting
设想 assume; assumption; conceive; contemplate; envisage; imagine; on the supposition that
设想的坝址 considered dam site
设想的方案 scenario
设想等高线 eoisohypse
设想方案 conceived proposal; hypothetical scenario
设想阶段 concept phase
设想全球气候 hypothetical global climate
设想误差 presumptive error
设想研究 opportunity study
设斜撑的墙 raker-braced wall
设熊阱式闸门的坝 bear-trap dam
设有便餐部的二等客车 buffet second
设有茶点部的公园 tea garden
设有单渡线的箭翎型编组线布置 <为编组多组和摘挂列车用> S-herringbone track layout; single type herringbone track layout
设有服务中心的建筑 building with service core
设有公用设施的竖井 shaft with installed services
设有海关的车站 customs station
设有基金之准备 funded reserve
设有驾驶台的动车 motorized A car
设有交叉渡线的箭翎型编组线 double type herringbone track
设有较多老虎窗的屋顶 dormer covering roof
设有空调房屋 air-conditioned building
设有两用沙发床的单人房间 single room with convertible couch-bed
设有墙帽的组合屋顶 built-up roofing with coping
设有市场的车站 market station
设有视听设备的教室 audio-visual classroom
设有调节水位装置的坝顶 controlled crest
设有下水道的地区 sewered area
设有邮件专间的守车 brake van with special compartment for mails
设有娱乐室的客车 club car; lounge car
设有重力传感器的计量料斗 weighing hopper with loadcell
设有住宅的建筑 building with dwellings
设于地板上的上射灯光 torchere
设于墙内侧的抗风化材料 weather back
设于舞台上的反射音响的大型壳体 orchestra shell
设月台 platform
设在岸上的 shore-based
设在柏林的新守卫室 new guard house at Berlin
设在地板上的燃烧装置 floor burning appliance
设在楼板上的燃烧装置 floor burning appliance
设在陆上的 land-based
设在墙上的吊钩 flashing hook
设在原料产地的 source-located
设闸的溢洪道顶 obstructed crest of spillway
设闸渠道 canal with locks; lock canal
设闸泄水堰 sluice weir
设闸溢洪道 gate spillway; obstructed spillway
设闸溢洪道顶 obstructed crest of spillway
设闸运河 canal with locks; lock(ed) canal
设栅 barricade
设栅栏 fencing
设站 refixation; stationing
设站点 occupied point; occupied station
设站(施测)河流 ga(u)ged stream
设站(施测)流域 ga(u)ged drainage; ga(u)ged drainage area; ga(u)ged drainage basin; ga(u)ged watershed
设站时间 station occupation time
设障 counterguard
设障沟 barrier ditch
设置 establish; installation; institution; locate; placing; plant
设置暗销的 dowel(l)ed
设置暗销的伸缩缝 dowelled expansion joint
设置标志牌 <客运站引导旅客用> sign posting
设置不全的 underplanted
设置测站 station siting
设置岔道 <道路> lay turn-out
设置成本 cost of set-up; set-up cost
设置传力杆 dowel(l)ing
设(置)传力杆的 dowel(l)ed
设置传力杆的横向伸缩缝 dowel(l)ed transverse expansion joints; dowel(l)ed transverse expansion or contraction joint
设置传力杆的伸(胀)缝 dowel(l)ed expansion joint

S

设置(道路)标志牌 signboarding
设置灯塔 beaconage
设置地下防污帷幕 setting-up underground protecting pollution curtain
设置断点 set breakpoint
设置方法 methods of installation
设置费(用) establishment charges; cost of installation
设置分岔 <道路> lay turn-out
设置浮标 buoy
设置浮标的航道 buoyed channel
设置浮筒 buoying; placing buoy
设置拱脚石 giving springer
设置购买费 equipment acquisition
设置轨底坡的钢轨 canted rail
设置航标 beaconage; setting beacons; signalization
设置航标的航道 buoyed channel
设置护道 berming
设置缓和曲线 rounding-off
设置基金期间 funded period
设置截污井排 setting-up well rows for protecting pollution
设置警戒线 cordon
设置路标 route marking
设置路缘石 curbing
设置模板 form placing
设置模壳 form placing
设置平面与坡度桩 <即平面与纵断面放样> set line-and-grade stakes
设置曲线 set-up curve
设置曲线超高 super-elevated a curve
设置伸缩缝 expansion jointing
设置伸胀缝 expansion jointing
设置时间 set-up time
设置挑出式脚手架用墙上预留孔穴 grappler
设置图 set-up diagram
设置信号 signalization
设置信号于 signalize
设置游隙 taking-up of play
设置在地面上的观测管道 survey pipes form the surface
设置在滑道前端底部的圆木 head log
设置噪声减低设施区 noise abatement zone
设置闸门的溢洪道 barrage type spillway
设置直达路线标志 through-route sign posting
设置直达路线号志 through-route sign posting
设雉堞 crenel(l)ate
设桩的 pegged
设总部 headquarters

社办企业 commune-run enterprise

社长 director; president
社队工业 commune/brigade run industry
社队企业 commune/brigade run enterprise
社会安全 public safety
社会边际成本 social marginal cost
社会边际费用 social marginal cost
社会福利 community welfare; public welfare; social welfare
社会福利部 <美> Community Welfare Department
社会工程 social engineering
社会基本设施 social infrastructure
社会基本设施成本 social overhead cost
社会基本设施费用 social overhead capital; social overhead cost
社会基础 substruction; substructure
社会基础设施 social infrastructure
社交场所 watering hole
社交关系图 sociogram
社交联谊楼 social hall
社交用户界面 social user interface
社交中心 social center[centre]
社区 community
社区参与 community participation
社区大学 community college
社区发展 community development
社区发展计划 community development plan
社区复兴 gentrification
社区改善计划 neighbo(u)rhood improvement scheme
社区共有的公寓项目 community apartment project
社区构成 community structure
社区购物中心 community shopping center[centre]
社区管理 community management
社区规划 community plan(ning)
社区规划师 community planner
社区及 EDI 系统 community and EDI system
社区间公路 intercommunity highway
社区建设 community development
社区建设样板 community development model
社区开发(工程) community development
社区开发公司 community development corporation
社区开发计划 community development program(me)
社区康乐设施 community amentities
社区设计 community design
社区设施 community service; community facility
社区设施规划 community facility plan
社区生活 community living
社区生活质量 quality of life
社区托儿所 community daycare housing
社区网络 portent
社区协会 community association
社区游憩设施 community recreational facility
社区增长管理 growth management
社区中心 community center[centre]
社区中心建筑 settlement house
社群建筑计划 social building program(me)
社群优势 social dominance
社所 bureau[复 bureau/bureaus]
社团 body of persons; college; commune; community; confraternity; league
社团财产 institutional property
社团成员 corporator
社团的 corporate; corporative
社团的成员公司 member company
社团的清真寺 collegiate mosque
社团地产 institutional property
社团法人 corporation aggregate; juridical person of an association
社团废物 institutional waste
社团公寓工程 community apartment project
社团会所 social club
社团建筑 institutional building
社团所有制 corporate ownership
社团土地利用 institutional land use
社团自动电话局 community dial office

射 throwing

射表 firing table
射表仰角 firing table elevation
射波测深法 acoustic(al) sounding
射波整流器 jet-wave rectifier
射程 actual range; amplitude; blow distance; flight; gunshot; jet distance; range of fire; reach; throw
射程表 range table
射程程控 preset range control
射程电离望远镜 range-ionization telescope
射程计算仪 rangekeeper
射程距离 range straggling
射程能量关系 energy-range relation
射程线 range line
射程遥测术 range-data telemetry
射出 effluence; ejection; jetting out; spray; stream
射出边 initial edge
射出的 ejected; emergent; radiate
射出点 discharge point
射出法 method of radiation
射出辐射 outgoing radiation
射出火焰 ejaculation of flame
射出角 angle of emergence
射出镜 outgoing mirror
射出水 injected water
射出速度 issuing velocity
射出物 effluence
射出者 ejaculator
射弹式取样器 projectile sampler
射氮 radioactive nitrogen
射的 projectile
射灯 reflector lamp
射碘 radioiodine
射电 radio
射电暴 radio storm
射电暴晕 halo of radio-burst region
射电场噪声比 radio field-to-noise ratio
射电窗口 radio window
射电刀 radio knife
射电等辐透 radio-isophote
射电等强线 radio-isophote
射电电子学 radionics
射电定位天文学 radio locational astrometry
射电洞 radio hole
射电多普勒效应 radio Doppler effect
射电辐射 radio emission; radio radiation
射电辐射流量 radio flux
射电复合线 radio recombination line
射电干扰 radio interference
射电干扰仪 radio interferometer
射电干涉测量法 radio interferometry; very long baseline interferometry
射电干涉仪 radio interferometer
射电光度 radio luminosity
射电光学 radio luminosity
射电光子 radio photon
射电轨 radio range
射电金相学 radiometallography
射电金属 radio metal
射电连续辐射 radio continuum
射电亮度 radio brightness
射电亮度分布 radio brightness distribution
射电亮温度 radio brightness temperature
射电六分仪 radiometric sextant
射电轮廓图 radio contour
射电脉冲 radio pulse
射电冕 radio corona
射电能量 radio energy
射电宁静太阳 quiet radio sun
射电抛物面天线 radio dish
射电喷流 radio jet
射电偏振测量 radio polarimetry
射电偏振计 radio polarimeter
射电频率 radio-frequency
射电频谱仪 radio spectrograph
射电谱 radio spectrum
射电谱斑 radio plage
射电谱线 radio line; radio spectral line
射电谱指数 radio spectral index
射电桥 radio bridge
射电全息照相术 radio holography
射电日象仪 radio heliograph
射电散射 radio scattering
射电闪烁 radio scintillation
射电食 radioeclipse
射电衰减 radio attenuation
射电双源 double radio source
射电太阳 radio sun
射电太阳单色图 radio spectroheliogram
射电太阳单色仪 radio spectroheliograph
射电天体测量学 radio astrometry
射电天文导航系统 radio celestial navigation system
射电天文观察 radio astronomy observation
射电天文台 radio astronomical observatory; radio astronomy station; radio observatory
射电天文卫星 radio astronomy satellite
射电天文学 radio astronomy
射电天文学测量法 radio astronomy measuring method
射电天文学家 radio astronomer
射电天文业务 radio astronomy service
射电天文站 radio astronomical station
射电透镜 radio lens
射电望远镜 radio telescope
射电荧光 radiofluorescence
射电源 radio source
射电源计数 radio source count
射电噪声 cosmic(al) noise; radio noise
射电噪声通量 radio noise flux
射电直径 radio diameter
射电指数 radio index
射钉 fired pin; stud shooting
射钉弹 bullet
射钉机 nailer; nail gun; stapler; stapling machine
射钉枪 brad setter; explosive-actuated gun; nail firing tool; nail gun; pistol-type stapler; powder-actuated fastening tool; power-actuated setting device; radio-frequency heating; radioheating; staple gun; stapler; stud gun
射钉丝规格 bullet ga(u)ge
射钉装置 nail shooting
射龟辐射 radio radiation
射击 gunning; shooting
射击安全区 no-fire area
射击安全线 no-fire line
射击场 gunnery range; proving ground; riffle range; shooting gallery; shooting range
射击挡板 firing hood
射击底座 firing base
射击电动协调装置 impulse generator; interrupter gear
射击方位角 firing azimuth
射击高度 operational height
射击基地 gunnery base
射击计划 scheme of fire
射击距离 firing range
射击孔 port
射击控制声呐 fire control sonar
射击控制象限仪 fire control quadrant
射击控制仪器 fire control instrument
射击区 field of fire
射击区域 sector of fire
射击散布图 shot group; shot pattern
射击式(侧壁)取样器 gun sampler

射击手 shooter
射击调平装置 firing jack
射击预习场 range preparation ground
射及区 swept area
射极电流 emitter current
射极电阻 emitter resistance
射极定时单稳电路 emitter timing monostable circuit
射极定时多谐振荡器 emitter timing multivibrator
射极负反馈 emitter degeneration
射极跟随放大器 emitter follower amplifier
射极跟随器的 emitter follower
射极跟随器逻辑 emitter follower logic
射极功率 emitter power
射极集极间距 emitter-collector separation
射极截止电流 emitter cut-off current
射极扩散层 emitter diffusion layer
射极扩散容量 emitter diffusion capacitance
射极脉冲 emitter pulse
射极耦合触发器 emitter-coupled trigger
射极耦合单元 emitter-coupled cell;emitter-coupled element
射极耦合的 emitter-coupled
射极区 emitter region
射极输出器 emitter amplifier
射极输出器的 emitter follower
射极特性 emitter characteristic
射极引线 emitter leg
射箭场 archery
射箭孔 <城堡上> balistraria
射角 quadrant angle;quadrant elevation
射角表 elevation table
射角差 elevation difference
射角的前置修正量 elevation prediction correction
射角分划 elevation scale
射角速率 elevation rate
射角指示器 elevation indicator
射解扩散模型 radical diffusion model
射解作用 radiolysis
射界 zone of fire
射界角 angle of traverse
射距散布 range spread
射孔 borehole springing;gun performation;perforation;meutriere <堡垒上的>
射孔冲头 perforative drift
射孔弹 explosive charge
射孔夹具 punch fixture
射孔器 well perforator;gun perforator <深钻钻工具>
射孔枪 firing gun;perforating gun
射孔枪装药器 gun-perforator loader
射孔完井 perforating completion
射孔装置 perforating equipment
射粒 radion
射量分析法 radiometric analysis
射量分析技术 radiometric technique
射流 efflux(ion);effusion;filament band;fluid;injector stream;jet current;jet efflux;jet flow;jet fluid;jet stream;shooting flow;stream flow;water efflux flow;water-jet
射流泵 eductor;efflux pump;ejector pump;injection pump;jet(flow) pump;jet injector;pump jet type;Shone ejector;water-jet pump
射流泵散弹钻进用钻头 jet-pump pellet impact drill bit
射流泵挖泥船 jet pump dredge(r)
射流比 jet ratio
射流边界 jet boundary;jet edge
射流波 jet wave
射流参数校正 jet parameter calibration
射流参数校准 jet parameter calibration
射流长度 jet length
射流冲击 impulse of jet;jet blow;jet impact
射流冲击范围 jet impact area;jet impact range
射流冲击力 jet force
射流冲击式滑车 jetting sled
射流冲击式钻机 jetting drill
射流冲刷 jet erosion
射流冲洗 jet wash
射流抽气泵 steam-jet ejector
射流抽吸式输送系统 suction-jet conveying system
射流出口 jet exit
射流穿孔 jet bit drilling;jet drilling;jet perforation
射流穿孔器 jet perforator
射流穿孔作业 jet perforating process
射流传感器 fluidic sensor
射流吹风冷却 jet cooling
射流吹洗 jet blasting
射流打桩法 water-jet method
射流打桩机 water-jet pile driver
射流导流槽 <发动机的> blast deflector
射流的 effusive;fluidic
射流电路 fluidic circuit
射流动力学 dynamics of jets;jet dynamics
射流发生器 fluidic generator
射流法钻进 drilling by jetting method
射流反馈 fluidic feedback
射流反循环钻进 jet reverse circulation drilling
射流反应 jet reaction
射流返回指令 break-point order
射流放大 fluid amplification
射流放大器 fluid(ic) amplifier
射流分离器 jet separator
射流分离现象 freeing of the nappe;jet separation;nappe separation
射流分散 jet diffusion;jet dispersion;jet spread
射流分散剂 jet disperser
射流分散器 jet disperser
射流风机 boaster fan;jet blower;jet fan;jet flow fan
射流风机阻挡 jet blocking
射流高度 height of jet;jet height
射流管 adjutage;jet(ting) pipe
射流管调节器 jet pipe regulator
射流管阀 jet action valve;jet pipe valve
射流管喷嘴 jet pipe tip
射流管式放大器 jet interaction amplifier
射流轨迹 jet path;path of jet;trajectory of jet
射流过渡 spray transfer
射流焊接 flow soldering
射流航迹 path of jet;trajectory of jet
射流互作用型元件 jet interaction element
射流几何 jet geometry
射流计算机 fluid computer
射流技术 fluerics;fluidics
射流继电器 fluidic relay;jet relay
射流加速度计 fluidic accelerometer
射流间歇计数器 fluidic batching counter
射流降水预压法 prepressing process of jet-drawdown
射流角 efflux angle
射流搅拌 jet mixing
射流截面 jet area
射流界限 edge of stream
射流紧缩 vena contracta[复 contractae]
射流进入点 jet entrance point
射流开关 fluidic switch;fluid valving
射流空化 jet cavitation
射流控制 efflux control;fluidic control
射流控制系统 jet control system
射流口 jet orifice
射流扩散角 jet divergence angle;spread
射流扩散器 jet diffuser[diffusor]
射流扩散装置 jet disperser
射流冷却 jet cooling
射流离散 separation of jet
射流力 jet force
射流流程 range of jet
射流流线 jet stack
射流路径 path of jet
射流脉冲整形器 fluidic pulse shaper
射流幕 jet curtain
射流黏[粘]度计 efflux visco(si)meter
射流喷口 efflux nozzle
射流喷射 jet injection
射流喷注 jet injection
射流喷嘴 ejector nozzle;jet nozzle
射流膨胀效应 jet swelling effect
射流偏导锥 efflux deflector cone
射流偏向器 thrust deflector
射流偏转 jet deflection
射流偏转舵 jetavator
射流曝气法 jet aeration
射流曝气气浮法 jet flo(a)tation
射流曝气器 efflux aerator;jet aerator
射流启动 jet priming
射流气浮法 jet flo(a)tation
射流器 ejector
射流器件 fluidic hardware
射流器式垂直起落飞机 ejector type vertical take-off and landing aircraft
射流器组 banks of ejectors
射流枪 bubble gun;range gun
射流切割 jet cutting
射流切割器 jetting cutter rod
射流区 forward flow zone
射流软管 jet hose
射流射程 range of jet
射流式步进电动机 fluidic stepping motor
射流式冲击器 fluidic type hydro-percussive tool
射流式轰炸机跑道 jet-bomber runway
射流式灰浆喷射机 jet flow mortar sprayer
射流式井点系统 ejector well point system
射流式喷气系统 jet aeration system
射流式喷射装置 hydraulic ejector
射流式喷水器 spray sprinkler
射流式喷头 ejector sprinkle head
射流式喷嘴 spray injector
射流式清理井眼螺钻 clean-out jet auger
射流式水轮机 free-jet-type turbine
射流式通风机 ejector type ventilator
射流式挖泥船 ejector type dredge(r)
射流式瓦斯喷燃器 jet gas burner
射流式闸门 jet flow gate
射流(束)收缩 contraction of jet
射流水功率 jet hydraulic horsepower
射流水马力 jet hydraulic horsepower
射流速度 efflux velocity;jet-stream velocity
射流体积流量计 fluidic volume flowmeter
射流调节器 jet regulator
射流挑出长度 jet trajectory length
射流通风 longitudinal ventilation with jet-blower
射流推进器 jet propeller;nozzle propeller
射流推力测定 jet thrust measurement
射流挖泥船 ejector dredge(r);jet dredge(r);jet suction dredge(r)
射流稳定器 fluidic stabilizer
射流问题 jet problem
射流涡轮机 radial-flow turbine
射流吸泥船 jet suction dredge(r)
射流稀释 jet dilution
射流系数 coefficient of efflux;efflux coefficient
射流系统 fluidic system
射流压力 jet pressure
射流相互作用型元件 stream-interacting element
射流相位检测回路 fluidic phase detection circuit
射流消能装置 jet disperser
射流效应 fluidic effect;jetting effect
射流学 fluerics;fluidics
射流有效高度 effective height of jet
射流元件 fluidic element
射流源 source of the jet
射流再压缩 jet recompression
射流胀大 jet exit
射流折流栅 blast fence
射流制动器 jet brake
射流中心 core of jet;jet-core region
射流轴(心) jet axis
射流轴心速度 jet axial velocity
射流转换 jet switching
射流转向器 jet deflector
射流装置 fluidic device;fluidizing system
射流状态 upper flow regime
射流阻尼 jet damping
射流组件 fluidic module
射流钻进 water-jet drilling;jet drilling
射流钻进作业 jet perforating process
射流钻井 jetting
射流最小断面点 <拉> vena-contracta
射流作用 jet(ting) action
射流作用挖沟机 jet action trencher
射脉菌属 <拉> Phlehia
射频 radio-frequency;radio wave frequency
射频保护率 radio-frequency protection ratio
射频边带均衡器 radio-frequency sideband equalizer
射频变换器 radio-frequency converter
射频变量器 radio-frequency transformer
射频变流器 radio-frequency current transformer
射频变频器 radio-frequency converter
射频变压器 radio-frequency transformer
射频标准信号发生器 radio-frequency standard signal generator
射频波 radio-frequency wave
射频波传播 radio wave propagation
射频波导(管) radio-frequency plumbing;radio-frequency waveguide;radio-frequency channel
射频波道 radio-frequency channel
射频波段 radio-frequency range
射频波反射 radio wave reflection
射频波接收 radio wave reception
射频波谱学 radio-frequency spectroscopy
射频部件 radio-frequency unit
射频差信号混频器 difference radio-frequency mixer
射频场 radio-frequency field
射频场强 radio field intensity;radio field strength
射频场强度 radio-frequency field intensity
射频场强噪声比 radio field-to-noise ratio

S

射频超导磁力仪 radio-frequency-squid magnetometer
射频成分 radio-frequency component
射频传感器 radio-frequency pick-up; radio-frequency sensor
射频传声器 radio-frequency microphone
射频传输线 radio-frequency line; radio-frequency transmission line
射频磁场 radio-frequency magnetic field
射频磁导计 radio-frequency permeameter
射频带宽度 radio-frequency bandwidth
射频灯 radio-frequency lamp
射频电磁场 radio-frequency electromagnetic field
射频电磁辐射 radio-frequency electromagnetic radiation
射频电极 radio-frequency electrode
射频电缆 radio-frequency cable; radio-frequency line
射频电流 radio-frequency current
射频电流传导面 radio-frequency carrying surface
射频电流计 radio-frequency ammeter
射频电路 radio-frequency circuit
射频电位 radio-frequency potential
射频电压 radio-frequency voltage
射频电阻 radio-frequency resistance
射频读出 radio-frequency reading
射频端 radio-frequency head
射频扼流圈 radio-frequency choke
射频发电机 radio-frequency alternator
射频发射 radio-frequency emission
射频发生器 radio-frequency generator; service oscillator
射频范围 radio-frequency region
射频放大 radio-frequency amplification
射频放大器 radio-frequency amplifier; radio-frequency gap; radio-frequency transmitter
射频放大器增益 radio-frequency amplifier gain
射频放大五极管 radio-frequency pentode
射频放电 radio-frequency discharge
射频放电检测器 radio-frequency discharge detector
射频放电器 radio-frequency gap
射频分光计 radio-frequency spectrometer
射频分离器 radio-frequency separator
射频分量 radio-frequency component
射频分配系统 radio-frequency distribution system
射频分析器 radio frequency analyser [analyzer]
射频封接 radio-frequency sealing
射频封接技术 radio-frequency sealing technique
射频伏特计 radio-frequency voltmeter
射频俘获 radio-frequency capture
射频辐射 radio-frequency radiation
射频(辐射)电源 radio-frequency power supply
射频辐射功率 radio-frequency radiation power
射频辐射记录器 radio-frequency radiation recorder set
射频干扰 radio-frequency interference
射频干扰测试 radio-frequency interference test
射频干扰场强 radio noise field intensity
射频干扰抑制装置 radio-frequency interference suppression equipment
射频干燥 radio-frequency drying; radio-frequency seasoning
射频感应 radio-frequency influence
射频感应加热 radio-frequency induction heating
射频感应加热器 radio-frequency induction heater
射频感应炉 radio-frequency induction furnace
射频高压电源 radio-frequency power supply
射频功率 radio-frequency power
射频功率放大器 radio-frequency power amplifier
射频功率耗散 radio-frequency power dissipation
射频功率损失 radio-frequency power loss
射频共振变压器 radio-frequency resonance transformer
射频共振法 radioresonance method
射频共振腔 radio-frequency cavity; radio-frequency liner; radio-frequency resonator
射频固化 radio-frequency cure; radio-frequency curing
射频故障探测 radio-frequency fault detection
射频管 radio-frequency tube
射频焊接 radio-frequency induction brazing
射频烘燥 radio-frequency drying
射频互调失真 radio-frequency intermodulation distortion
射频火花 radio-frequency spark
射频火花放电源 radio-frequency spark discharge source
射频火花离子源 radio-frequency spark ion source
射频(火)炬 radio-frequency torch
射频击出 radio-frequency knockout
射频击出共振 radio-frequency knockout resonance
射频击出技术 radio-frequency knockout technique
射频击出探针 radio-frequency knockout probe
射频击穿 radio-frequency arcing
射频激发 radio-frequency drive
射频激发器 radio-frequency driver
射频激射器 radio wave maser
射频级 radio-frequency stage
射频极谱法 radio-frequency polarography
射频技术 radio-frequency technique
射频加热 radio-frequency heating; radioheating
射频加热(技)术 radiothermics
射频加热器 radio heater
射频加速 radio-frequency acceleration
射频加速电压 radio-frequency accelerating potential
射频加速结构 radio-frequency accelerating structure
射频加速器 radio-frequency accelerator
射频加速隙 radio-frequency gap
射频监测接收机 radio-frequency check receiver
射频监控器 radio-frequency monitor
射频检测器 radio-frequency detector
射频溅射 radio-frequency sputtering
射频交流发电机 radio-frequency alternator
射频接收管 radio-frequency receiving tube
射频接收机 radio-frequency receiver
射频接收器 radio-frequency receiver
射频接转制 radio-frequency repeating system
射频结构 radio-frequency structure
射频介电加热 radio-frequency dielectric(al) heating
射频聚积 radio-frequency stacking
射频聚束 radio-frequency bunching
射频聚束器 radio-frequency buncher
射频开关 radio-frequency switch
射频开关脉冲倒置 radio-frequency switching pulse inversion
射频空腔 radio-frequency cavity
射频控制系统 radio-frequency control system
射频冷却试验 radio-frequency cold tests
射频灵敏度 radio-frequency sensitivity
射频漏泄 radio-frequency leakage
射频滤波器 radio-frequency filter
射频脉冲 radio-frequency impulse; radio-frequency pulse; radio pulse; wave packet
射频脉冲波形 radio-frequency pulse shape
射频脉冲发生机试验器 carpet tester
射频脉冲发生器 carpet tester; radio-frequency pulse generator
射频脉冲束 radio-frequency-pulsed beam
射频脉冲遥感 remote-sensing with radio-frequency pulse
射频脉塞 radio-frequency maser
射频密封 radio-frequency sealing
射频能量 radio-frequency energy
射频耦合环 radio-frequency coupling ring
射频耦合技术 radio-frequency coupling technique
射频耦合线圈 radio-frequency coupling loop
射频偏转器 radio-frequency deflector
射频偏转器系统 radio-frequency deflector system
射频漂移 radio-frequency shift
射频频谱 radio-frequency spectrum
射频频谱学 radio spectroscopy
射频屏蔽 radio shield(ing)
射频屏蔽装置 radio-frequency shielding fence
射频谱分析仪 radio-frequency spectrum analyser[analyzer]
射频全息摄影术 radio-frequency holography
射频全息(照相)术 radio-frequency holography
射频热核装置 radio-frequency thermonuclear
射频熔接 radio-frequency welding
射频设备 radio-frequency equipment
射频室 radio-frequency tank
射频适应器 radio-frequency adapter
射频输出探头 radio-frequency output probe
射频输入信号 radio-frequency input signal
射频衰减 radio-frequency attenuation
射频损耗角技术 radio-frequency loss-angle technique
射频损失 radio-frequency loss
射频探测器 radio-frequency probe
射频探雷器 radio-frequency mine detector
射频探针 radio-frequency probe
射频特性测试仪 radio-frequency characteristic measuring set
射频调谐 radio-frequency tuning
射频调谐变压器 tuned-radio-frequency transformer
射频调谐器 radio-frequency tuner
射频调制 radio-frequency modulation
射频调制器 radio-frequency modulator
射频通带 radio-frequency passband
射频通道 radio-frequency channel
射频头 radio-frequency head
射频图 radio-frequency chart
射频图像信号 radio picture signal
射频危害 radio-frequency hazard
射频微波激射器 radio-frequency maser
射频微观结构 radio-frequency microstructure
射频稳定度 radio-frequency stability
射频稳定区 radio-frequency stability region
射频稳态放电 radio-frequency steady state discharge
射频系统 radio-frequency system
射频线圈 radio-frequency coil
射频限制 radio-frequency confinement
射频相变 radio-frequency transition
射频谐波 radio-frequency harmonic
射频谐波序数 radio-frequency harmonic number
射频信标机 radio-frequency beacon
射频信道同步 radio-frequency channel synchronization
射频信号 radio-based signal; radio-frequency signal
射频信号发生器 radio-frequency signal generator
射频信号失落补偿器 radio-frequency dropout compensator
射频信号失落消除器 radio-frequency dropout killer
射频信号源 radio-frequency signal source
射频信扰比 radio-frequency wanted-to-interfering signal ratio
射频性能 radio-frequency performance
射频选择性 radio-frequency selectivity
射频抑止 radio-frequency confinement
射频抑制器 radio-frequency suppressor
射频预电离 radio-frequency preionization
射频预热 radio-frequency preheating
射频预选器 radio-frequency preselector
射频运用 radio-frequency operation
射频杂波消除 radio-frequency mute
射频载波 radio-frequency carrier
射频载波漂移 radio-frequency carrier shift
射频噪声 radio-frequency noise
射频噪声场强度 noise-field strength; radio noise field strength
射频增益控制 radio-frequency gain control
射频照相 radiograph
射频照相检查 radiographic(al) testing
射频振荡 radio-frequency oscillation
射频振荡发生器 radio-frequency alternator
射频振荡激发器 radio-frequency exciter
射频振荡器 radio-frequency generator; radio-frequency oscillator
射频振荡器型高压电源 radio-frequency oscillator high-voltage supply
射频直线加速器 radio-frequency linear accelerator
射频质谱计 radio-frequency mass spectrometer
射频质谱仪 radio-frequency mass spectrometer

射频周期 radio-frequency period
射频转换继电器 radio-frequency switching relay
射频转接 radio-frequency interconnection
射频转接器 radio-frequency adapter
射频自导引 radio-frequency homing guidance
射气 emanation;emanium
射气测量 emanation prospecting
射气测量法 emanometry
射气测量计 emanator
射气底数 base number of emanation
射气计 emanometer
射气技术 emanation technique
射气静电计 emanation electrometer
射气勘探 emanation prospecting
射气流密度 emanation flux density
射气率 emanating power;emanation rate
射气强度 emanation strength
射气热分析 emanation thermal analysis
射气试验氡 emanation test
射气系数 emanation coefficient
射气系数值 emanation coefficient value
射气箱 emanation chamber
射气验电器 emanation electroscope
射气仪 emanometer
射气仪格值 scale value of radiometer
射气仪类型 emanator type
射气源形态 form of emanation source
射气治疗院 emanatorium
射气总浓度 emanation total consistence
射汽抽气器 steam-jet ejector
射圈法 ring shooting
射入边 terminal edge
射入辐射 beam;incoming radiation
射入轨道 injection;orbital injection
射入口 entry portal
射入深度 penetration depth
射入斜角 angle of incidence
射杀比例 ratio kill
射杀量 hunter kill;sportman toll
射杀率 kill ratio
射沙紧实 shooting ramming
射声器 acoustic(al)horn
射声器测量 projector measurement
射声器功率响应 projector power response
射声头 projector head
射式 dual-scatter type
射手 shooter
射束 beam;pencil;streamer
射束摆动 beam swinging
射束摆动器 beam wobbler
射束包线 beam envelope
射束变向 beam switching
射束波导 beam waveguide
射束出射方向 beam-emergence direction
射束存储管 beam storage tube
射束存储器 beam storage
射束点 beam spot
射束电流 beam current
射束电流控制 beam current control
射束电流增益 beam current gain
射束对中 beam alignment
射束发散的 beam diverging
射束发生系统 beam generating system
射束反射器 beam reflector
射束方向 beam direction
射束分离层 beam splitting layer
射束分裂器 beam splitter
射束分配器 beam divider
射束干涉仪 beam interferometer
射束高度 beam height
射束光闸 beam shutter
射束横截面 beam area
射束会聚 beam convergence
射束记录 beam recording
射束加速电压 beam accelerating voltage
射束间距 beam spacing
射束监测器 beam monitor
射束角(度) angle of effluxion;angle of beam
射束角偏调 angular beam misalignment
射束校正 beam alignment;camera alignment
射束校正装置 beam alignment assembly
射束界限 beam boundary
射束阱 beam trap
射束聚焦 beam focusing
射束开关管 beam switching tube
射束开关频率 beam switching frequency
射束孔(径) aperture of beam;beam aperture
射束孔径角 beam angle
射束控制 beam control
射束宽度 beam width
射束扩展 beam spread
射束扩展函数 beam spread function
射束量子放大器 beam maser
射束灵敏度 beam sensitivity
射束零线 beam null line
射束脉塞 beam maser
射束密度 beam density
射束面 beam surface
射束面积 beam area
射束能 beam energy
射束能量 beam energy
射束黏[粘]附 < 重力仪上的 > beam sticking
射束排列 beam configuration
射束偏移 beam deflection
射束偏移因子 beam deviation factor
射束偏转 beam deflection
射束偏转管 beam deflection valve
射束偏转镜 beam deflecting mirror
射束剖析法 beam splitting
射束扫描法 beam-scanning method
射束扫描器 beam scanner
射束收集器 beam trap
射束衰减 beam attenuation
射束衰减器 beam attenuator
射束水平宽度 horizontal beam width
射束搜索 searchlighting
射束锁定 beam locking
射束探测器 beam finder
射束调制 beam modulation
射束调制存储器 beam-modulating memory
射束弯曲 beam bending
射束微摆 wobbulation
射束位移 ray displacement
射束位置调整 positioning of beams
射束熄灭 beam blanketing
射束形成光学 beam-forming optics
射束形成装置 beam-forming arrangement
射束形状 spray shape
射束型探测器 projected beam-type detector
射束旋转天线 rotary beam antenna
射束抑制 beam suppression
射束噪声 beam noise
射束遮拦 beam masking
射束直径 beam diameter
射束制导 beam rider guidance
射束中心 beam center[centre]
射束中心调整 beam centering[centring]
射束轴 beam axis
射束烛光 beam candlepower
射束柱 beam column
射束转换法 beam switching;lobe switching
射束转换管 beam switching tube
射束组态 beam configuration
射水 jet stream;jetting;water jet
射水泵 water-jet pump
射水沉没法 < 井点的 > jetting method
射水沉桩 jetted pile;jetting piling;pile water jet;sinking by jet piling;water-jet driving
射水沉桩法 jet method of pile-driving;jetting piling;pile jetting;sinking of pile by water jet;water-jet method;water-jet method of pile-driving;pile jetting(method);sinking by jetting;sinking pile by water jet
射水冲洗钻进法 jet ring drilling
射水抽气泵 water-jet air pump
射水抽气器 water-jet air ejector;water-operated ejector
射水除尘器 water-jet dust absorber
射水处理 water-blast;water blasting
射水打入的桩 jet(ted)pile
射水打桩 pile driving by water-jet
射水打桩法 jetting piling;sinking of pile by water jet;water-jet method of pile-driving;water-jet driving
射水打桩机 water-jet pile driver
射水发动机 hydromotor
射水法 jetting process;water jetting
射水杆(土壤)探测 jet probing
射水管 discharge jet;jetting lance < 水冲沉桩 >
射水进行表面清洁 abrasive jet cleaning
射水空气泵 hydraulic air pump
射水孔 injection orifice
射水口 < 井点的 > jetting orifice
射水螺旋桩 jetted screw pile
射水埋管装置 jet sled device for pipeline burial
射水器 water-jet
射水切割法 < 混凝土损坏部分的 > water-jet cutting
射水清除混凝土浮浆 sweep blasting
射水设备 jetting equipment;water jetting equipment
射水式挖泥船 water injection dredger
射水式钻机 wash drill
射水疏通【排】 jet clean
射水水管 jetting pipe
射水探测 jet probe
射水填土 jetting fill
射水推进 hydrojet propulsion
射水下沉法 < 井筒下沉的一种施工方法 > jetting process;sinking by jetting;sinking by water jet
射水钻进 wash drilling
射水钻井 jetting drill
射水钻孔法 wash boring method
射水钻探 jetting drilling
射塑 jet moulding
射塑喷射模塑法 jet moulding
射髓 < 木材的 > ray of wood
射投影纠正仪 orthophotoscope
射透比频谱 transmittance spectral
射微镜 microprojector
射水钻进 water jet drilling;water jetting
射吸式冲击器 jet vacuum type hydro-percussive tool
射线 beam;prong;ray
射线包迹 ray envelop
射线比 ray ratio
射线变压器 ray transformer
射线表示法 ray representation
射线病 radiation disease
射线薄壁组织 ray parenchyma
射线不透性 radiopacity
射线参数 ray parameter
射线参数方法 ray-parameter method
射线测厚 thickness measurement with ray
射线测井剖面 gamma-ray log
射线穿透管 penetron tube
射线穿透能力 radio transparency
射线传感器 radiation sensor;radiation transducer
射线导管间的 ray-vessel
射线导管间纹孔式 ray-vessel pitting
射线的数学研究 sciametry
射线底片 radiographic(al)film
射线底片对比度 radiographic(al)contrast
射线发光现象 radioluminescence
射线发散 ray divergence
射线法线 ray normal
射线反射 reflection of radiation
射线反行程 return trace
射线方程 ray equation
射线方法 method of radiation;ray method
射线防护塞 ray stopper
射线跟踪 gamma-ray tracking
射线固化涂料 radiation setting coating
射线故障检验法 radiographic(al)inspection
射线管 ray tube
射线光学 ray optics
射线轨迹 ray tracing;ray trajectory
射线轨迹图 ray tracing diagram
射线过滤板 filter
射线过滤器 ray filter
射线和加热混合杀菌法 radiopasteurization
射线化学 actinism;actinochemistry
射线活动摄影术 cineradiography
射线计 radiationmeter
射线记录 gamma-ray log
射线剂量传感器 radiation dose sensor;radiation dose transducer
射线校直 beam alignment
射线检测仪 radiation detector
射线检查法 radioscopy
射线检验 radial test;radiographic(al)inspection;radioexamination
射线检验法 radioscopy
射线交叉场 ray crossing
射线交点 ray intersection;ray intersection point
射线结构 ray structure
射线金相学 radiometallography
射线近似法 ray approximation(method)
射线晶体学 radiocrystallography
射线径迹 ray trace;ray tracing
射线径纹 ray fleck
射线聚焦 ray focusing
射线开关管 beam switching tube
射线开裂 ray check
射线可透过的 radiolucent
射线可透性 radioparency
射线孔 beam orifice
射线控制极 ray-control electrode
射线类别 ray type
射线理论 ray theory
射线理论地震图 ray theoretic(al)seismogram
射线理论方法 ray theoretic(al)method
射线立体测量学 radiographic(al)stereometry
射线路径 ray path
射线轮 ray wheel
射线脉冲 ray pulse
射线密度计 gamma radiation density meter

射线面 ray surface
射线偏移 ray deflection
射线偏转存储管 ray deflection type storage tube
射线谱 ray spectrum
射线枪 ray gun
射线强度的调制 modulation of beam intensity
射线曲率 ray curvature
射线曲面 ray surface
射线热发光 radio-thermoluminescence
射线杀伤 irradiation injury
射线筛管 ray sieve tube
射线伤害危险 radio hazard
射线摄影 radiography
射线深度偏移 ray depth migration
射线声学 geometric(al) acoustics;ray acoustics
射线示踪 ray tracing
射线收集器 beam trap
射线收注栅 beam catcher;beam trap
射线束 bundle of rays;ray bundle
射线束聚焦 beam focusing
射线速度 ray velocity
射线锁定装置 ray-locking device
射线探测仪 radiometer
射线探矿 radio prospecting
射线探伤 radiographic(al) inspection;ray detection
射线探伤率 rate of radiographic(al) examination
射线探伤试验 radiograph test
射线调制 ray modulation
射线体视术 radiographic(al) stereoscopy
射线透不过的 radiopaque
射线透度计 penetrameter[penetrometer]
射线透过性 radiolucency
射线透镜 radiographic(al) lens;radioscope
射线图 ray diagram
射线途径 ray path
射线危害 radio hazard
射线物理学 radiation physics
射线显迹法 autoradiography
射线显迹图 radioautograph
射线效应 ray effect
射线硬度测定计 radiochrometer
射线用量规定 rayage
射线源 radiographic(al) source
射线照片 radiograph;scotograph;skiagram;skiagraph
射线照射 radiation exposure
射线照相 radiogram;radiograph;radiographic(al) examination
射线照相部门 radiography department
射线照相的延米 linear metre of radiograph
射线照相等效因子 radiographic(al) equivalence factor
射线照相底版 radiographic(al) plate
射线照相法 radiography
射线照相检验 radiographic(al) examination
射线照相检验记录 record of radiographic(al) examination
射线照相胶片 radiographic(al) film
射线照相灵敏度 radiographic(al) sensitivity
射线照相乳胶 radiographic(al) emulsion
射线照相设备 radiographic(al) apparatus
射线照相试验 radiographic(al) testing
射线照相术 radiography
射线照相探伤法 radiographic(al) inspection
射线照相探伤检验 radiography examination
射线照相系数 radiograph factor
射线照相验收标准 acceptance standards for radiography
射线照相要求 radiographing requirements
射线照相纸 radiographic(al) paper
射线照相装置 radiographic(al) set-up
射线折射 refraction of ray
射线中心 ray center[centre]
射线转动 rotation of beam
射线追踪技术 ray trace technique
射线追踪偏移 ray tracing migration
射线灼伤 flash burn
射线自显迹 autoradiograph
射线纵散射效应 Plotnikow effect
射线作用 actinism
射线坐标 ray coordinates
射向与风后夹角 wind-fire angle
射芯 core shooting
射芯机 core shooter;core shooting machine;explosion type core blower
射型机 mo(u)ld shooter
射压造型机 shoot squeeze molding machine
射焰喷燃器 impact burner
射焰燃烧器 impact burner
射影 projection
射影A模 projective module A
射影保形几何 projective conformal geometry
射影变换 projective change;projective transformation
射影变换群 projective transformation group
射影变形 projective deformation
射影标架 projective frame;projective scheme
射影表示 projective representation
射影不变量 projective invariant
射影不变式 projective invariant
射影不变性 projective invariance
射影参数 projective parameter
射影测度 projective measurement
射影尺度 projective scale
射影次序公理 projective axioms of order
射影丛 projected bundle
射影簇 projective variety
射影代数簇 projective algebraic variety
射影代数的 projective algebraic
射影代数曲线 projective algebraic curve
射影的 projective
射影地恒等射影不变性原理 projectively identical
射影点列 projective ranges of points
射影度量 projective measurement
射影度量空间 projective metric space
射影对象 projective object
射影对应 projective correspondence
射影法 projective method
射影法线 projective normal
射影分解 projective resolution
射影覆盖 projective cover
射影干扰 projective rejection
射影关系 projective relation
射影函数 mapping function
射影画 projective drawing
射影极限 projective limit
射影极限空间 projective limit space
射影极限群 projective limit group
射影极小曲面 projective minimal surface
射影集 projective set
射影几何学 projection geometry;projective geometry
射影矩阵 projection matrix
射影空间 projective space
射影类群 projective class group
射影联络 projective connection
射影面积 plane of projection
射影面束 projective pencils of planes
射影纽结 projection knot
射影平面 projective plane
射影平坦的 projectively flat
射影平坦空间 projectively flat space
射影区间 projective interval
射影曲率张量 projective curvature tensor
射影群 projective group
射影双纽线 projective lemniscate
射影算子 projection operator
射影算子矩阵 projective operator matrix
射影同态 projection homomorphism
射影同调群 projective homology group
射影图形 projective figure
射影拓扑 projective topology
射影完全的 projectively complete
射影微分几何学 projective differential geometry
射影维数 projective dimension
射影系 projective system
射影线丛 projective line bundle
射影线束 projective pencils of lines
射影线元素 projective line element
射影辛群 projective symplectic group
射影性质 projective property
射影映射 projecting mapping
射影张量积 projective tensor product
射影直射变换 projective collineation
射影直线 projective line;projective straight line
射影值测度 projection-valued measure
射影中心 centre of projection
射影轴 axis of projection
射影柱 projecting cylinder
射影锥 projecting cone;projective cone
射影子空间 projective subspace
射影坐标 projective coordinates
射影坐标系 projective coordinate system
射油泵 oil injection pump;oil jet pump
射油器 oil ejector
射油系统 oil injection system
射油正时 injection timing
射针 dart
射阻长度 jet blocking length
射阻系数 jet blocking coefficient

涉渡口 ford

涉及的财务问题 financial implication
涉及的对象 referent
涉及的人或设施 <工程的> the parties
涉及地区 area affected
涉及多方面的 multilateral
涉及范围 area of coverage
涉及压力容器的法规 laws covering pressure vessels
涉禽 wading bird
涉水 wade;wading
涉水测流 wading measurement of discharge
涉水测流法 wading method
涉水测流杆 wading rod
涉水测深 wading measurement
涉水池 wading pool
涉水而行 squatter
涉水过河 ford a stream;wade across a river
涉水量测设备 wading equipment
涉水路 ford road
涉水深度 fording depth;wading depth
涉讼财产(或破产案产业)管理人的证书 receiver's certificate
涉外价格 price involving foreign countries
涉外经济合同法 foreign economic contract law
涉外民商事仲裁 arbitration of civil and commercial cases with foreign contact

摄测站 photogrammetric(al) station

摄动【天】 perturbation
摄动的 disturbed
摄动定理 perturbation theorem
摄动法 method of perturbation;perturbation method
摄动方程 perturbation equation
摄动方法 perturbation method
摄动分析 perturbation analysis
摄动符号 perturbation symbol
摄动轨道 perturbation orbit;track of perturbation
摄动函数 disturbing function;perturbative function
摄动阶 order of perturbation
摄动理论 perturbation theory;theory of perturbation
摄动力 disturbing force;perturbative force;perturbing force
摄动体 disturbing body;perturbing body
摄动问题 perturbed problem
摄动系数 perturbation coefficient
摄动项 perturbation term;perturbed term;perturbing term
摄动效应 disturbing effect
摄动因素 perturbation factor
摄动运动 disturbed motion
摄谱法 spectrography
摄谱分析 spectrography
摄谱分析法 spectrographic(al) analysis
摄谱轨道 spectrographic(al) orbit
摄谱鉴定 spectrographic(al) identification
摄谱控制分析 spectrographic(al) control analysis
摄谱术 spectrography
摄谱图 spectrogram
摄谱学 spectrography
摄谱仪 spectrograph
摄谱仪光栅 spectrograph grating
摄谱仪使用法 spectrography
摄谱照相机 spectrograph camera
摄取 intake;intussusception;uptake
摄取速率常数 uptake rate coefficient
摄全景 pan
摄全景动作 panoplay
摄入 ingestion
摄入率 uptake rate
摄食方法 food procuring contrivance
摄食洄游 feeding migration
摄食生物 consumer organism
摄食习性 feeding habit
摄氏 Celsius[C]
摄氏度(数) degree Celsius;degree Centigrade;degree of centigrade;centi-degree
摄氏热单位 centigrade heat unit
摄氏热量单位 centigrade thermal unit
摄氏温标 Celsius scale;centigrade (thermometric) scale;temperature degree of Centigrade
摄氏温标的 centigrade
摄氏温度 Celsius temperature;centigrade temperature;temperature degree of Centigrade;temperature in degree;degree of centigrade

S

摄氏温度标 Celsius thermometric scale; centigrade scale
摄氏温度表 Celsius thermometer; centigrade thermometer
摄氏温度度数 centigrade degree
摄氏温度计 Celsius scale; Celsius thermometer; centigrade temperature scale; centigrade thermometer
摄氏温度计的 centigrade
摄像 image pick-up; imagery; pick-up
摄像车操纵员 helmsman
摄像处理机 camera processor
摄像传感器 image sensor
摄像地质学 photogeology
摄像反差 image contrast
摄像管 camera tube; pick-up tube; television camera tube; pick-up
摄像管扫描图 camera-scan(ning) pattern
摄像管摄像机 vidicon camera
摄像管输出板 signal plate
摄像管余像测量仪 camera tube lag meter
摄像管预放器 pick-up tube preamp
摄像光谱特性 pick-up spectral characteristic
摄像机 camera; pick-up camera; telecamera
摄像机电缆 camera cable
摄像机调整表 camera sheet
摄像机放大器 head amplifier
摄像机监控器 camera monitor
摄像机控制台 camera control unit
摄像机物镜视角 camera angle
摄像机系统 camera chain
摄像机信道 camera channel
摄像机遥控装置 camera remote control unit
摄像机预放器 camera preamplifier
摄像角标度 angle scale
摄像镜头 pick-up lens
摄像雷达 imaging radar
摄像灵敏度 pick-up sensitivity
摄像排字机 photographic(al) typesetter
摄像屏 camera screen
摄像器 image pick-up device
摄像三原色 taking primaries
摄像调节器 frame hold
摄像头【计】 stylus
摄像系统 picture pick-up system
摄像执法【交】 camera enforcement
摄像装置 camera head; pick-up device
摄氧量 oxygen intake; oxygen uptake
摄氧速率 oxygen uptake rate
摄影 camera shot; filming; photo; photographing; picture-taking; shoot; shot; take a photograph
摄影暗盒 film pack
摄影凹版 photograve
摄影报道 photojournalism; photoreportage
摄影曝光 photoexposure
摄影比例尺 photographic(al) scale; scale of photography; taking scale
摄影波长 picture taking wavelength
摄影材料 photomaterial
摄影材料变形 photographic(al) material deformation
摄影参数 photographic(al) parameter
摄影舱口 camera port; window opening
摄影测绘 photomapping
摄影测绘设备 photomapping equipment
摄影测绘图 photogrammetric(al) map; photogrammetric(al) platting
摄影测绘学 photogrammetry
摄影测绘者 photogrammetrist
摄影测角仪 photogoniometer
摄影测量 photogrammetric(al) measurement; photogrammetric(al) measuration; photogrammetric(al) restitution; photogrammetric(al) survey; photogrammetry; photographic(al) survey
摄影测量草图 photogrammetric(al) sketch
摄影测量测图 photogrammetric(al) plotting; photogrammetric(al) restitution; photographic(al) plotting
摄影测量处理 photogrammetric(al) process
摄影测量的 photogrammetric(al)
摄影测量地图 photogrammetric(al) map
摄影测量定位 photogrammetric(al) fixing of position; photogrammetric(al) position finding
摄影测量法 photogrammetric(al) procedure; photogrammetry
摄影测量飞行 photogrammetric(al) flight
摄影测量分类 classification of photogrammetry
摄影测量工程师 photogrammetric(al) engineer
摄影测量过程 photogrammetric(al) process
摄影测量绘图器 photogrammetric(al) plotter
摄影测量绘图仪 aerocartograph; photogrammetric(al) plotter
摄影测量基线 photogrammetric(al) base
摄影测量畸变 photogrammetric(al) distortion
摄影测量计算机 photogrammetric(al) computer
摄影测量记录 photogrammetric(al) recording
摄影测量技术 photogrammetric(al) technology
摄影测量加密 photogrammetric(al) control
摄影测量加密点 photogrammetric(al) control point; photographic(al) point
摄影测量交会 photographic(al) resection
摄影测量精度 photogrammetric(al) accuracy
摄影测量镜头 photogrammetric(al) lens
摄影测量纠正 photogrammetric(al) rectification; photographic(al) rectification
摄影测量控制 photogrammetric(al) control
摄影测量控制点 photogrammetric(al) control point
摄影测量控制加密 photogrammetric(al) bridging
摄影测量立体摄影机 photogrammetric(al) stereocamera
摄影测量模型 photogrammetric(al) model
摄影测量内插 interpolation for photogrammetry; photogrammetric(al) interpolation
摄影测量设备 photogrammetric(al) surveying apparatus
摄影测量摄影机 photogrammetric(al) camera
摄影测量视差 photogrammetric(al) parallax
摄影测量术 photogrammetry; photographic(al) surveying; photometrology
摄影测量数字化系统 photographic(al) digitizing system
摄影测量数字获取系统 digital system of photographic(al) acquisition
摄影测量条件方程 photogrammetric(al) condition equation
摄影测量网 photogrammetric(al) network
摄影测量微分法 differential method of photogrammetry
摄影测量物镜 photogrammetric(al) lens; photogrammetric(al) objective
摄影测量系统 photogrammetric(al) system
摄影测量学 metrophotography; photographic(al) survey; photography; photogrammetry; photographic(al) surveying
摄影测量学的 photogrammetric(al)
摄影测量学(专)家 photogrammetrist
摄影测量仪(器) photogrammetric(al) apparatus; photogrammetric(al) instrument; photogrammetric(al) unit
摄影测量与遥感 photogrammetry and remote sensing
摄影测量员 air photographer; photogrammeter; photogrammetrist
摄影测量原图 map obtained by photogrammetric restitution; photographic(al) plot
摄影测量站 photogrammetric(al) station
摄影测量装置 photogrammetric(al) apparatus
摄影测量锥形法 <测定照片倾斜的一种解析方法> photogrammetric(al) pyramid
摄影测量自动化 photogrammetric(al) automatization
摄影测量作业员 plotting-machine operator
摄影测量坐标 machine coordinates; photogrammetric(al) coordinates
摄影测量坐标系(统) photogrammetric(al) coordinate system
摄影测量坐标仪 photogrammetric(al) coordinatograph
摄影测深法 photobathymetry
摄影测图 photogrammetric(al) mapping; photographic(al) mapping; photoplotting
摄影测图仪 photocartograph; photogrammetric(al) unit; photograph; photoplotting apparatus; photoplotting instrument; Photorestituteur <法国制造>
摄影测斜仪 photoclino-dipmeter; photoclinometer; photographic(al) inclinometer
摄影场 photostudio
摄影车 camera car
摄影车间 cinematographing department
摄影成果 photoproduct
摄影成图 photomapping
摄影成像 photographic(al) imagery
摄影处理 developing of photography; photographic(al) processing; photomechanical treatment; photoprocessing
摄影窗孔 camera window
摄影窗口 camera port; camera window
摄影存储器 photographic(al) storage
摄影打样图 photographic(al) layout drawing
摄影大地测量学 geodetic photogrammetry
摄影单位 photography organization
摄影的 photographic(al)
摄影的明暗部分 photo tonality
摄影底版 photographic(al) plate
摄影底片 photographic(al) plate; film of survey <测斜仪的>
摄影地层学 photostratigraphy
摄影地理学 photogeography
摄影地貌学 photogeomorphology
摄影地区代号 code of photography area
摄影地图 photographic(al) map
摄影地形测量 photographic(al) surveying
摄影地形测量学 phototopography
摄影地形图 photorelief map
摄影地质法 photogeology method
摄影地质工作者 photogeologist
摄影地质学 photogeology
摄影读数经纬仪 camera-read theodolite; photographic(al) theodolite; photographic(al) transit; photo-theodolite; phototransit
摄影反照率 photographic(al) albedo
摄影泛光 photoflood
摄影泛光灯 photoflood lamp
摄影方法 photographic(al) method; photographic(al) procedure
摄影方式 mode of photography
摄影方向 direction of optic(al) axis
摄影飞机 photographic(al) airplane; photoplane
摄影飞行 photographic(al) flight
摄影分辨率 photographic(al) resolving power; resolution of photography
摄影分区 flight block
摄影辐射点 photographic(al) radiant point
摄影负片 negative photograph
摄影复制 photographic(al) reproduction
摄影复制设备 photocopier
摄影覆盖地区 photographic(al) coverage
摄影改正器 photocorrector
摄影干版 photographic(al) plate
摄影干版的感光度 speed of photographic(al) plate
摄影高度 photo altitude
摄影跟踪 photographic(al) tracking
摄影跟踪经纬仪 kinetheodolite
摄影工程 photographic(al) engineering
摄影工艺 photographic(al) technology
摄影公式 photographic(al) formula
摄影观测 photographic(al) observation
摄影光度计 photographic(al) photometer
摄影光度学 photographic(al) photometry
摄影光圈 shooting aperture
摄影光束 taking bundle
摄影光速 photographic(al) bundle of rays
摄影光学 photographic(al) optics; photooptics
摄影光学系统 photographic(al) optic(al) system
摄影航带 photographic(al) strip
摄影航高 flying height of photography; photographic(al) flying height
摄影航迹 flight trace of photography
摄影航线 flight line of aerial photography; flying line of photography; photographic(al) strip; photo strip; pilot's trace <领航图上的>
摄影航线宽度与航高之比 width-height ratio
摄影航向 direction of flight

S

摄影红外 photographic(al) infrared
摄影化学 photographic(al) chemistry
摄影绘(轮廓)图 profiling
摄影绘图仪 photoplot
摄影绘制等高线法 photo-contour process
摄影火箭 photographic(al) rocket
摄影机 camera; photograph camera
摄影机暗箱 camera installation
摄影机曝光 camera recycle rate
摄影机操纵 handling of the camera
摄影机垂直摄全景 pandown
摄影机挡光板 dowser[douser]
摄影机的标志灯 camera marker
摄影机底座 camera base
摄影机吊舱 photographic(al) pod
摄影机分辨率 resolution of camera
摄影机俯摄 camera tilt down
摄影机俯仰运动 camera tilt
摄影机辅助装置 accessory camera attachment
摄影机几何参数 camera geometry
摄影机计时器 camera timer
摄影机记录 camera record
摄影机架 camera mounting
摄影机检定 camera calibration
摄影机检定场 camera calibration field
摄影机检定器 camera calibrator
摄影机检验 camera test
摄影机镜筒 camera cone
摄影机快门 camera shutter
摄影机框架 camera carrier
摄影机片门 picture gate
摄影机偏转 avertence of camera
摄影机倾斜 camera tilt
摄影机取景孔 camera eye
摄影机视角 camera angle
摄影机系统 camera array
摄影机悬挂装置 camera suspension; photographic(al) suspension
摄影机移动车 dolly car
摄影机轴 camera shaft
摄影机主光轴 photograph perpendicular; plate perpendicular
摄影机主距 principal distance of camera
摄影机主体 camera body
摄影机姿态 camera altitude
摄影机座架 camera mounting
摄影积分 photographic(al) integration
摄影基线 air base; photo base(line); photogrammetric (al) base line; photographic(al) base(line); photographing base(line)
摄影基线变形 photo-base distortion
摄影基线倾斜 base tilt
摄影基线水平长度 horizontal length of photo-baseline
摄影基准面 photographic(al) datum plane
摄影极谱仪 photographic(al) recording polarograph
摄影计时器 phototimer
摄影记录 histogram record; photographic(al) record; photographic(al) recording; photographic(al) registration; photolog
摄影记录读数 photographically recorded reading
摄影记录卡片 magnavue card
摄影记录气压计 photographic(al) barograph
摄影记录器 photographic(al) recorder; photorecorder
摄影记录系统 photographic(al) record system
摄影记时术 photochronography
摄影记时仪 photochronograph
摄影记者 pre-photographer
摄影纪录 photolog(ging)
摄影技巧 camera work
摄影技术 photographic(al) technique; shooting technique
摄影剂量仪 photographic(al) dosimeter
摄影剪辑 photograph montage
摄影检测 photodetection
摄影交会 photographic(al) intersection
摄影胶版印刷 photo-offset print
摄影胶带 photographic(al) tape
摄影胶卷 dry film; film band
摄影胶片 photographic(al) film
摄影胶片带 film band
摄影角度 camera angle; taking angle
摄影经纬仪 camera transit; field camera; photogrammeter; photo-theodolite; theodolite camera; transit camera
摄影经纬仪测量 photo-theodolite survey
摄影经纬仪导线 photo-polygonometric(al) traverse
摄影经纬仪跟踪站 photographic(al) tracking station
摄影经纬仪物镜 photo-theodolite objective
摄影镜头 camera lens; photographic(al) lens
摄影镜头延伸 camera extension
摄影距离 camera-to-subject distance; shooting distance
摄影聚光灯 photospot
摄影勘测 photographic(al) reconnaissance; photographic(al) surveying; photorecon; photoreconnaissance
摄影勘测员 photoreconnaissance pilot
摄影勘察 photographic(al) reconnaissance; photographic(al) surveying; photoreconnaissance
摄影拷贝 photocopy
摄影科学 photographic(al) science
摄影颗粒性 photographic(al) granularity
摄影刻图的 photoscribe
摄影刻印术 photolithographic(al) method
摄影宽度 latitude
摄影量角仪 photoangulator
摄影录音 photographic(al) recording; photographic(al) sound recording
摄影录音机 optic(al) sound recorder; photographic(al) sound recorder
摄影密度 photographic(al) density
摄影面积 coverage
摄影明胶 photogelatin
摄影模板 photographic(al) template
摄影排字机 photocomposer
摄影棚 film studio; photostudio; sound stage; studio
摄影偏转 avertence of camera
摄影拼接 photograph montage
摄影平面 photographic(al) plane; photoplane
摄影气压仪 photographic(al) barograph
摄影器材厂 photographic(al) equipment plant
摄影枪 camera gun
摄影倾斜角 photographic(al) tilt; photo tilt
摄影求积法 photo-planimetric(al) method
摄影取向角 photograph orientation angle
摄影日 photographic(al) day
摄影乳剂 photographic(al) emulsion
摄影乳剂层 photoemulsion layer; photographic(al) emulsion layer
摄影软片 film strip; photographic(al) film
摄影三角测量 photogrammetric(al) triangulation; phototriangulation
摄影三角测量仪 phototriangulator
摄影扫描器 photoscanner
摄影色调 photo tonality
摄影森林学 photo-forestry
摄影闪光弹 photoflash bomb
摄影闪光灯 photoflash lamp; photoflash light
摄影闪光混合物 photoflash composition
摄影师 cameraman; photographer
摄影时间 photography time
摄影实验室 photographic(al) laboratory; photo laboratory
摄影式日照计 photographic(al) sunshine recorder
摄影视场 picture angular field
摄影视场角 angle of photographic(al) coverage
摄影视距测量法 photo-tacheometry
摄影室 photographic(al) studio; studio
摄影室技术员 photographic(al) laboratory technician
摄影术 photography
摄影数据处理 photographic(al) data process
摄影速度 photographic(al) speed; photographic(al) velocity; picture rate; shooting rate; taking rate
摄影速率 photographic(al) speed
摄影缩小 photographic(al) reduction
摄影探测 photographic(al) detection
摄影天顶筒 photogrammetric(al) unit
摄影天顶望远镜 photographic(al) zenith telescope; photographic(al) zenith tube
摄影天体测量学 photographic(al) astrometry
摄影天体光度学 photographic(al) astrophotometry
摄影天体光谱学 photographic(al) astrospectroscopy
摄影天文学 photographic(al) astronomy
摄影透镜色差 photolens chromatism
摄影透视仪 photoperspectograph
摄影凸版 phototype
摄影网目版 photoscreen
摄影望远镜装备 phototelescopic technique
摄影卫星 camera satellite
摄影位置 air position; shooting position; taking position
摄影温度计 photographic(al) thermometer
摄影物镜 objective lens of photography; photographic(al) field lens
摄影物镜焦距 focal distance of photographic(al) lens
摄影系数 photographic(al) coefficient
摄影显微胶片 camera microfilm
摄影显微镜 photomicroscope
摄影箱 shot box
摄影新闻工作 photojournalism
摄影行迹 flying track of photography
摄影学 photography
摄影遥感系统 photographic(al) remote-sensing system; photographic(al) sensing system
摄影仪器 photographic(al) instrument
摄影影像 photographic(al) image
摄影员 cameraman; photographer
摄影原(像)片 photographic(al) original
摄影站 camera station; exposure station; photographic(al) station
摄影照准仪 photoalidade
摄影遮光 photographic(al) masking
摄影者 photog
摄影侦察 photorecon; photoreconnaissance
摄影侦察设备 photorecon equipment
摄影侦察卫星 photoreconnaissance satellite
摄影正片 photographic(al) transparency
摄影制版 heliotype; photochemigraphy; phototype
摄影制成等高线法 photo-contour process
摄影制图 photocharting; photogrammetry; photographic(al) map; photomap(ping)
摄影制图术 photocartography
摄影制图仪(器) photocartograph; photogrammetric(al) mapping instrument
摄影制锌版 heliozincograph
摄影制锌版术 heliozincography
摄影质量 photographic(al) quality
摄影轴 photograph axis
摄影轴方向 beam of exposing-axis; bearing of exposing axis
摄影主光距 principal axis of camera
摄影主光轴 photographic(al) perpendicular
摄影资料 photographic(al) data
摄影资料地区 photographic(al) coverage
摄影综合测图法 photo-planimetric(al) method
摄影纵距 longitudinal distance of photography; longitudinal photographic(al) distance
摄影坐标 <用坐标表示摄影实体位置> plate coordinates
摄远镜头 telephoto lens
摄远物镜 telephoto objective
摄云机 cloud camera
摄政时期风格 Regency style
摄政时期风格装饰 Regency ornament
摄政王式建筑 Regency style
摄制 mapping by photography
摄制机构 production agency
摄制图 photographic(al) map
摄制项 production area

麝

麝 musk deer

麝香 muskiness

申

申报 claim gamesmanship; declaration; declare; submit to the relevant authority

申报办公室 reporting room
申报表 declaration form; declaration list; return
申报承兑 acceptance declaration
申报单 declaration; declaration form; declaration list
申报股本 stated capital
申报关税 customs declaration
申报货载价值 declared value for carriage
申报价格 declared value
申报价值 declaration of value; declared value; value declared

申报金额 amount declared
申报进口 ship entry
申报进口日期 date of entry declaration
申报内容 declaration content
申报纳税身份 filing status
申报失实 misleading declaration
申报式保险契约 reporting contract
申报载重吨位 declare deadweight
申报重量 declared weight; said to weight
申报资本(额) declared capital
申克尔电压倍增器 Schenkel doubler
申克型过滤机 Schenk type filter
申明的转数/分钟 stated revolution per minute
申请 apply; make application; petition; proposal; propose; tender(out)
申请表 application
申请表格 application form; form of application
申请表项专利 application for a patent
申请步骤 procedure to apply
申请参数目录 request parameter list
申请参数目录串 request parameter list string
申请偿还 recourse back
申请撤回 application for withdrawal
申请撤销仲裁裁决 application to set aside an award
申请程序 application for credit; application procedure; requisition procedure
申请贷款 applying for loans
申请贷款人 loan applicant
申请单 request note; requisition
申请的工程 petitioned work project
申请登记 application for registration
申请抵押借款 apply for loan secured
申请地权 claim
申请调离 request for transfer
申请方法 method of application
申请费 anticipation fee; application fee
申请付款书 requisition for payment
申请复审 request reexamination
申请挂号人 applicant
申请(灌水)者 claimant
申请号 application number
申请计划 plan of application
申请检验 survey requested
申请借款 loan application
申请借款人 loan applicant
申请开采的国家 applicant country for exploitation
申请开采区域位置 applicant block position exploitation
申请领料单 requisition
申请绿灯信号 calling detector
申请赔偿的诉讼 action for reimbursement
申请赔偿付款 payment of claims
申请赔偿人 claimant; claimer
申请批准 application for approval; confirmation request
申请批准发展规划 planning application
申请批准放款 invite subscription for a loan
申请破产 bankruptcy petition; voluntary bankruptcy
申请人 applicant; claimant; declarant; claimer
申请人名单 waiting list
申请日期 applicant time
申请书 application; letter of application; requisition; written application
申请书格式 application form; form of application
申请特别承诺 application for special commitment
申请停办手续 enter a caveat
申请退还税金 application for drawback
申请退款 application for drawback
申请退税 claim for tax refund
申请外汇 application for foreign exchange
申请系统 requisition channel
申请显示 acquire display
申请(修建)工程 petitioned work
申请许可证的条件 licensing requirement
申请(宣告)合同无效诉讼 action for annulment of contract
申请要求 request
申请用地 acquisition of land
申请用图纸 application drawing
申请者 applicator; claimant
申请执照的条件 licensing requirement
申请中的专利 pending patent
申请仲裁 filling of the award
申请周期 requisition cycle
申请住房收入限额 income admission limits
申请注册 application for registration
申请专利 application for patent
申请装船单 application for shipment
申述意图函件 letter of intent(ion)
申诉 brief briefing; complain(t)
申诉程序 complaints procedure
申诉赔偿损失 action for damages
申诉人 complainant; declarant
申诉问题调查员 ombudsman
申诉与索赔 complaints and claims

伸

伸壁桥墩 jutting-off-pier
伸臂 cantilever; overhang
伸臂长度 boom-out; boom reach
伸臂大梁 cantilever girder
伸臂底座 cantilever footing
伸臂吊车 bracket crane
伸臂浮运架设安装 erection by protrusion and floating
伸臂极限长度 <起重机> boom-out
伸臂架设法 erection by protrusion
伸臂角度 boom angle
伸臂梁 overhanging beam
伸臂梁桥 cantilever bridge; cantilever timber beam bridge
伸臂末端 boom point
伸臂起重机 boom crane; jib crane
伸臂曲梁 overhanging curved beam
伸臂升降装置 boom lifting device
伸臂式变幅起重机 derricking jib crane
伸臂式打桩机 overhanging pile driver
伸臂式固定起重机 jib crane
伸臂式机具 boom rig
伸臂式链斗卸船机 dock leg elevator
伸臂式起重机 boom crane; boom hoist; gib crane; grab crane; jib crane
伸臂式塔吊 hammerhead crane
伸臂托架 arm extension bracket
伸臂系统 <起重机> boom reach
伸臂柱头 bracket capital
伸长 elongate; extension; outstretch; protend; protraction; stretch; stretch elongation
伸长百分率 percentage of elongation
伸长百分数 percentage of elongation
伸长比(值) elongation ratio; ratio of elongation
伸长臂 extension arm; extension boom
伸长变定 elongation set
伸长变形 extension strain
伸长标度 elongation scale
伸长表面 rectifying surface
伸长尺 extension rule
伸长的拉手 body grip
伸长电缆 extension cable
伸长度 elongation; extensibility; stretch elongation; tensile elongation
伸长端跨 spread span
伸长断裂 extension fracture
伸长反应 lengthening reaction
伸长方向 stretching direction
伸长杆 extension bar; extension rod
伸长管 extension pipe; extension tubing
伸长滑阀杆 extended slide valve rod
伸长机座 extension base
伸长计 elongation indicator; extens(i)ometer; extension meter; extensometer; strainometer; tautness meter; tensometer
伸长计读数 tensiometric measurement
伸长记录仪 elongation recorder
伸长接头 flanged expansion joint
伸长类变形 tensile type of deformation
伸长流动 elongational flow; extensional flow
伸长率 coefficient of elongation; coefficient of extension; elongation per unit length; extensibility; extension percent; percentage elongation; rate of elongation; rate of expansion; ratio of elongation; specific elongation
伸长率百分数 percentage elongation
伸长卵石 stretched pebbles
伸长模量 extension modulus; modulus of elongation
伸长内管 inner-tube extension
伸长黏[粘]度 elongational viscosity
伸长黏[粘]性 elongational viscosity
伸长片 extension piece
伸长破裂 tension fracture
伸长期 elongation stage; jointing stage; period of elongation
伸长器 lengthener; stretcher
伸长强度 strength of extension
伸长区 elongation areas; elongation zone
伸长曲面 rectifying surface
伸长生长 elongation growth
伸长式操雷头 extensible exercise head
伸长式测微计 extension micrometer
伸长式千分尺 extension micrometer
伸长式枢纽 prolonged junction terminal
伸长式支柱 expansion post
伸长试验 elongation test; extension test
伸长四点井网 elongated four-spot pattern
伸长速度 extension rate; extension speed
伸长速率 extension rate; extension speed
伸长弹性 elasticity of elongation
伸长网络 rectifier net
伸长系数 coefficient of elongation; coefficient of extension; elongation coefficient; elongation factor; extension coefficient; lengthening coefficient
伸长限度 allowable elongation
伸长斜顶式锅炉 extended wagon top boiler
伸长形变 elongation strain
伸长性 distensibility; extensibility; tensibility; tensility
伸长仪 extensometer
伸长仪装置 extensometer arrangement
伸长引线 extension lead
伸长应变 elongation strain; extensional strain
伸长应变速率 elongational rate
伸长运动 stretching motion
伸长张量 elongation tensor
伸长指示器 extension indicator
伸长指数 elongation index
伸长轴 axis of elongation; elongated axis
伸长作用 dilatation
伸出 beedle; booming(out); jut; outreach; overhang(ing); oversail; project(ing); protract; protrude; protrusion; put forth; stretch(-out); finger out <突堤等向水域>
伸出臂 cantilever arm; extension arm
伸出臂梁 overhanging beam
伸出边 outstanding leg
伸出部分 extension; outshot; shelf
伸出长度 extension elongation; outreach; overhanging length; projected length; reach; swing(ing) radius; stick-out <电极或焊丝的>
伸出窗外的工作平台 window jack scaffold
伸出的 projected; protrusive
伸出的钢挡 protruding bar
伸出的钢杆 protruding bar
伸出的钢筋 protruding bar
伸出的建筑 forebuilding
伸出的凉廊 outward opening; outwindow
伸出的竖直管道 rising duct
伸出的双头螺栓 pillar-bolt
伸出的危险物 outlying danger
伸出的支架 outstanding leg
伸出的支腿 extended outrigger
伸出底板的托盘 pallet with projecting floor
伸出地面 elevate above the soil
伸出阀箱 extension valve case
伸出杆 extension bar
伸出钢筋 projecting reinforcement; starter bar; stub rod
伸出键 extension key
伸出梁 outrigger; overhanging beam
伸出落下支腿 out-and-down outrigger
伸出施工缝的钢筋 tie bar
伸出式刮刀 expanding cutter; expansion cutter
伸出式桥墩 projecting abutment
伸出水面的 overhand
伸出腿 outstanding leg
伸出舷外的 outrigged
伸出舷外的栏杆 burton boom; outboard boom
伸出凿尖犁 barpoint plow
伸出凿尖犁铧 barpoint share
伸出支腿 outstanding leg
伸出支座轴承 free end bearing
伸出轴 projecting shaft
伸出锥度 extension taper
伸出座车床 extension lathe; extension machine
伸到框架或墙外的板 raised panel
伸得过长 overreach
伸顶通气管 roof extension; stack vent
伸放纸 overs
伸缝 stretching crack
伸幅机 tentering machine
伸角 angle of hade; hade
伸脚空间 <座位前> leg room
伸颈 crane
伸距 <起重机臂的> outreach
伸距长的 long-reach
伸锯 open
伸开 outstretch
伸开井架 extendible derrick
伸梁支架 boom brace
伸入岸中的土坝 ground sill rooted in the bank

伸入长度 built-in length
伸入管 dip nozzle
伸入海中的冰山舌 ice tongue
伸入基岩的截水墙 positive cut-off
伸入水域 finger out into water
伸入水中的圆形沙嘴 sand lobe
伸入向斜 reentrant syncline
伸手式控制 control within arm's length
伸缩 collapsing; come-and-go; dilation; expansion and contraction; stretching
伸缩案面台 extending table
伸缩案面桌 extension table
伸缩板 expansion plate
伸缩棒 extension stem
伸缩比 magnification ratio
伸缩臂 telescope arm; telescopic (al) arm; telescopic (al) boom; telescopic (al) jib; telescoping arm; pantograph
伸缩臂反铲挖掘机 extending dipper
伸缩臂杆 extension boom
伸缩臂架载人平台 extensible boom platform
伸缩臂起重机 telescopic (al) crane
伸缩臂汽车式起重机 telescoping truck-mounted crane
伸缩臂式起重机 telescopic (al) boom crane
伸缩臂式挖掘机 telescoping boom type excavation
伸缩臂式旋转起重机 retractable jib slewing crane
伸缩臂式装载机 telescopic (al) loader
伸缩臂液压反铲 Gradall
伸缩臂移动式高架工作平台 telescopic boom mobile elevating work platform
伸缩臂装置 < 起重机 > telescoping boom equipment
伸缩臂最大长度 maximum telescoping boom length
伸缩臂最小长度 minimum telescoping boom length
伸缩标尺【测】extension rod; sliding rod; sliding staff
伸缩柄 extension stem
伸缩波 expansion wave
伸缩薄膜 self-adhering film
伸缩补偿弯管 expansion bent; expansion loop
伸缩叉式起重机 reach forklift crane
伸缩叉式装卸车 reach forklift truck
伸缩插销 bullet bolt
伸缩铲挖机 telescopic (al) boom excavator
伸缩铲挖掘机 telescopic (al) boom excavator
伸缩长度 collapsing length
伸缩撑杆 telescopic (al) strut
伸缩撑螺 flexible stay bolt
伸缩撑螺栓 expansion staybolt
伸缩承口 telescopic (al) bellmouth
伸缩承座 expansion bearing
伸缩尺 extension rule; extension scale
伸缩传送带 telescopic (al) conveyer [conveyor]
伸缩大梁 extension girder
伸缩带 < 伸缩式输送机的 > telescopic (al) band
伸缩挡板 telescopic (al) baffle
伸缩导杆 telescopic (al) leader
伸缩导架 telescopic (al) guiding carriage
伸缩的 telescopic (al)
伸缩地板模板拱架 telescopic (al) floor shuttering center [centre]
伸缩地段 flexible segment
伸缩垫 expansion pad
伸缩垫盖 expansion pad cap
伸缩垫片 expansion washer
伸缩垫圈 expansion bead; expansion washer
伸缩吊 telescopic (al) hoist
伸缩吊臂 extensible boom
伸缩吊杆装置 < 灯光用 > lazyboy
伸缩度 dilatability
伸缩端 expansion end
伸缩(端)支座 expansion pedestal
伸缩对中座 telescopic (al) centering
伸缩法兰 creeping flange
伸缩缝 construction joint; control joint; dilatation joint; expanded joint; expansion; expansion gap; expansion opening; functional joint; gap allowed for expansion; movement joint; relief joint; shrinkage joint; sliding joint; slip joint; temperature joint; variator
伸缩缝材料 expansion joint material
伸缩缝盖板 dam; expansion joint cover
伸缩缝盖缝料 expansion joint sealing
伸缩缝盖缝条 expansion joint sealing
伸缩缝钢膨胀传力杆 < 水泥混凝土路面 > steel expansion dowel
伸缩缝接头钢筋 shrinkage bar
伸缩缝金属套筒 expansion joint cap strip
伸缩缝密封层 expansion joint sealant
伸缩缝密封体 expansion joint seal
伸缩缝嵌条 edge isolation; expansion strip
伸缩缝润滑销 < 无黏[粘]着钢筋的 > dowel lubricant
伸缩缝润滑油 dowel lubricant
伸缩缝填(充)料 expansion joint filler
伸缩缝狭带 expansion joint tape
伸缩缝止水层 expansion joint sealant
伸缩缝止水剂 expansion joint sealant
伸缩缝中的(止水)金属片 metal strip for expansion joint
伸缩缝中间的嵌缝板 filler board
伸缩缝装配 expansion joint assembly
伸缩附加力 additional temperature force; additional temperature load
伸缩杆 adjustable bar; extended stem; extension arm; extension rod; mast extension; retractable spear; telescoping arm
伸缩杆销 expansion link pin
伸缩缸筒 feed extension
伸缩缸总成 extension assembly
伸缩钢筋 expansion reinforcement
伸缩隔墙 telescope partition
伸缩关税 elastic tariff; flexible tariff
伸缩管 telescopic (al) duct; telescopic (al) pipe; telescoping tube; compensating pipe; draw tube; expansion pipe; expansion tube; extension tube; helically grilled tube; slip pipe
伸缩管端 gland joint
伸缩管接合 telescopic (al) joint
伸缩管连接 telescoped joint
伸缩管式连接 telescoping connection
伸缩管式联结 telescoping connection
伸缩管弯头 expansion pipe bend
伸缩管轴 slip-tube shaft
伸缩规 telescoping ga(u)ge
伸缩轨 telescoping rail
伸缩滚轴 expansion roller
伸缩滚轴支座 expansion roller
伸缩护罩 extensible cover
伸缩滑槽 extension chute
伸缩划料刀 knife extension blade
伸缩画图器 eidograph
伸缩环 expansion ring
伸缩回转梯 telescopic (al) revolving ladder
伸缩汇率 flexible rate
伸缩货叉 retractable shuttle
伸缩集电器 pantograph troll(e)y
伸缩集电器底座 pantograph base
伸缩计 ductilimeter; extensometer
伸缩继电器座 pantograph base
伸缩加成定价法 flexible markup practice
伸缩架 expansion bracket
伸缩间隙 expansion gap; gap allowed for expansion
伸缩剑杆 telescopic (al) rapier
伸缩键 slip key
伸缩交变应力 tension-compression cycling stress
伸缩脚手板 telescopic (al) scaffold board
伸缩铰链 expansion hinge
伸缩接缝 expansion joint; telescopic (al) joint; variator; expansion and contraction joint; running joint
伸缩接缝铁 expansion joint iron
伸缩接合 contraction joint; expansion and contraction joint; functional joint
伸缩接合套 slip-joint sleeve
伸缩接口 slip expansion joint
伸缩接头 expansion coupling; expansion joint; extension fitting; flanged expansion joint; flexible joint; gland joint; slide coupling; slip pipe; splice joint; telescopic (al) expansion joint; telescopic (al) line
伸缩接头单元 expansion joint unit
伸缩接头的修饰 expansion joint trim
伸缩接头横断面 expansion joint section
伸缩接头密封剂 expansion joint mastic (sealer)
伸缩接头黏[粘]胶 expansion joint mastic (sealer)
伸缩接头水密封 expansion joint waterstop
伸缩接头铁 expansion joint iron
伸缩接头外形 expansion joint profile
伸缩节 expanded joint; expansion bend; expansion joint; expansion knot; expansion loop; expansion piece; flexible connector
伸缩节安全阀 slip-joint safety valve
伸缩结 slip knot
伸缩结合 slip joint
伸缩进刀螺旋 telescopic (al) feed screw
伸缩井壁刮刀 expansion wall scraper
伸缩举重机构 telescopic (al) lifting mechanism
伸缩拉撑 expansion stay
伸缩拉杆 telescopic (al) drawbar
伸缩拉条 expansion brace
伸缩连杆 expansion link
伸缩连接 slip joint
伸缩连接件 extension link
伸缩连接器 extension link
伸缩联杆 telescopic (al) link
伸缩联轴器 extension coupling
伸缩梁 telescopic (al) girder
伸缩量规 telescoping ga(u)ge
伸缩裂缝 expansion crack(ing)
伸缩留量 expansion allowance
伸缩轮 expanding pulley; retractable wheel
伸缩螺钉 rigging screw
伸缩螺杆 turnbuckle
伸缩螺栓 retract bolt; slip bolt; expansion bolt; expanding bolt
伸缩螺丝 steel bottle screw; steel turnbuckle; stretching screw
伸缩螺丝旋凿 pump screwdriver
伸缩螺套 coupling screw; turnbuckle
伸缩螺纹接口 slip nipple
伸缩膜盒 flexible bellows
伸缩泡 contractile vacuole
伸缩皮腔 extension bellows
伸缩平衡锤 telescoping stabilizer
伸缩平衡重 telescoping outrigger; telescoping stabilizer
伸缩起落架 retractable landing gear; retractable under carriage
伸缩器 compensator; expansion bend; expansion loop; expansion piece; telescope piece
伸缩钳 extension tongs; slip-joint pliers
伸缩桥 traversing bridge
伸缩桥靴 expansion shoe
伸缩区 breathing zone
伸缩区长度 breathing length
伸缩曲线 expansion curve
伸缩驱动轴 telescopic (al) propeller shaft
伸缩圈 expanding ring; expansion loop; expansion ring
伸缩软管 bellows; expansion bellows
伸缩三脚架 sliding tripod; split leg tripod
伸缩升降车 reach truck
伸缩式臂杆 < 起重机的 > telescopic (al) boom; extensible boom
伸缩式臂架起重机 telescopic (al) jib crane
伸缩式标尺 adjusting latch; adjusting lath; telescopic (al) rod
伸缩式表耳簧 telescopic (al) style spring bar
伸缩式布料杆 telescopic (al) placing boom
伸缩式舱壁 telescoped bulkhead
伸缩式沉降测管 telescopic (al) tube settlement ga(u)ge
伸缩式沉降计 telescopic (al) settlement ga(u)ge
伸缩式吹灰器 retractable soot blower; telescopic (al) blower
伸缩式大梁 telescopic (al) girder
伸缩式带缆卷筒 extension warping barrel
伸缩式导向架 telescope lead
伸缩式的 extension-type
伸缩式吊臂 telescopic (al) hoist boom
伸缩式吊杆 telescopic (al) boom; telescoping jib
伸缩式吊杆起重机 telescopic (al) boom crane
伸缩式吊架 telescopic (al) spreader; telescoping spreader
伸缩式吊具 telescopic (al) spreader; telescoping spreader
伸缩式吊盘卡 retractable scaffold bracket
伸缩式动臂 telescopic (al) boom; telescoping boom
伸缩式斗柄 extendable dipperstick
伸缩式防撞装置 retractable fender system
伸缩式风钻 buzzy
伸缩式刚性梁 rigid extensible bar
伸缩式钢梁 retractable steel joist
伸缩式高架梯 extension trestle ladder
伸缩式隔水管 riser with slip joint
伸缩式拱架 telescopic (al) centering
伸缩式固定 telescopic (al) mount
伸缩式管线 telescopic (al) line
伸缩式锅炉 telescopic (al) boiler
伸缩式回转塔架 telescopic (al) rotary tower
伸缩式货叉 retractable fork
伸缩式机械手 telescopic (al) manipulator
伸缩式机罩 extension head
伸缩式集装箱起重机 reach stacker crane
伸缩式驾驶台 retractable pilot-house
伸缩式架座 telescopic (al) mount

伸缩式接管 telescopic(al) access tube
伸缩式接头 slip joint;telescopic(al) joint;telescoping joint
伸缩式进给 telescopic(al) feed
伸缩式进给装置 telescopic(al) feeder
伸缩式拉门 pantograph gate
伸缩式立柱 telescoping mast
伸缩式连接 slip joint;telescopic(al) joint
伸缩式联结装置 retractable drawbar
伸缩式溜槽 extensible pan; retractable chute;telescopic(al) chute
伸缩式履带挖掘机 telescoping crawler excavator
伸缩式落料管 telescopic spout
伸缩式埋头螺栓 extension flush bolt
伸缩式门 telescopic(al) door
伸缩式模板 telescopic (al) form (work);telescoping form(work)
伸缩式木三脚架 adjustable wooded legs;telescopic(al) wooden legs
伸缩式起吊卡车车身的吊车 telescopic hoist
伸缩式起重机 telehoist; telescopic (al) crane;telescopic(al) hoist
伸缩式起重机臂 telescopic jib
伸缩式气腿 retractable pusher-leg
伸缩式弃土区溢流口 <高低可调节的> telescoping weir
伸缩式潜望镜 retractable periscope
伸缩式切削臂 expanding cutting arm
伸缩式倾斜槽 retractable chute;telescopic(al) chute
伸缩式驱动轴 telescopic (al) drive shaft
伸缩式三脚架【测】 extension tripod; sliding tripod; adjustable legs; telescopic (al) tripod; telescoping tripod
伸缩式升降平台生 telescopic(al) lift truck
伸缩式输送管 extensible conveying pipe
伸缩式竖管 telescopic(al) riser
伸缩式双层装载槽 telescopic (al) loading trough
伸缩式塔身 telescopic(al) tower
伸缩式推进 telescopic(al) feed
伸缩式推进凿岩机 telescopic (al) feed hammer drill
伸缩式挖掘机 telescoping excavator
伸缩式挖掘机悬臂 telescoping excavator boom
伸缩式桅顶 telescopic(al) mast head
伸缩式桅杆 retractable mast; telescopic(al) mast;telescopic(al) pole
伸缩式桅杆起重机 telescoping derrick
伸缩式桅杆起重架 telescoping drilling derrick
伸缩式吸入管 sliding suction
伸缩式线盘卸料器 collapsible stripping spider
伸缩式卸槽 retractable chute; telescopic(al) chute
伸缩式卸料管 telescopic chute; telescoping chute
伸缩式卸料溜槽 extensible discharge trough
伸缩式心轴 collapsible mandrel
伸缩式袖珍钢卷尺 retractable steel pocket tape
伸缩式烟囱 telescopic(al) chimney
伸缩式液压顶杆 telescopic (al) hydraulic ram
伸缩式鹰架 telescopic(al) centering
伸缩式油缸 telescopic (al) cylinder; telescopic(al) ram
伸缩式运输机 extensible conveyer [conveyor]
伸缩式凿岩机 buzzy; hammer-drill stoper;stoper(drill);stoping drill
伸缩式造斜器 collapsible whipstock
伸缩式支臂 extensible arm
伸缩式支船架 telescopic(al) cradle
伸缩式支护 yieldable support
伸缩式支柱 expansion post
伸缩式中心钻头 expansive center bit
伸缩式助卷机辊 retractable wrapper rolls
伸缩式装货滑道 telescoping loading ramp
伸缩式装货斜道 telescoping loading ramp
伸缩式装料车 telescopic(al) bogie
伸缩式装载 telescopic (al) loading trough
伸缩式钻臂 telescoping boom
伸缩式钻机支柱 telescopic (al) drilling post
伸缩式钻架 telescopic(al) derrick
伸缩式钻井架 telescoping drilling derrick
伸缩式钻塔 telescopic(al) mast; telescopic(al) pole
伸缩式钻头 expansive bit
伸缩丝锥 collapsible tap
伸缩速度 telescopic(al) speed
伸缩塔架 telescopic(al) tower
伸缩塔式起重机 telescopic(al) tower crane
伸缩弹性 elasticity of elongation
伸缩弹指式捡拾滚筒 retracting spring-tooth pick-up cylinder
伸缩套杆 telescopic(al) bar
伸缩套管 telescopic (al) boom; telescopic (al) line; telescopic (al) shaft; telescoping casing; telescoping tube; telescopic (al) tube; expansion sleeve <传力杆的>
伸缩套管式天线 telescopic(al) mast
伸缩套管式天线杆 collapsible mast; dismountable mast
伸缩套筒 expansion sleeve;telescopic(al) shaft;telescopic(al) tube;telescoping tube
伸缩套筒叉 telescope fork;telescopic(al) fork
伸缩套筒铆锤 jam rivet(t)er
伸缩套筒式管接头 extension fitting
伸缩套筒式螺栓 expansion sleeve bolt
伸缩套筒液压缸 telescopic (al) hydraulic cylinder
伸缩梯 extending ladder; extension ladder;telescope ladder;telescopic(al) ladder
伸缩条 insulating strip
伸缩条款 adjustment clause; cost escalation clause; escalation clause; escalator clause
伸缩调节阀 expansion regulating valve
伸缩调节器 expanding device
伸缩调整轨 expansion adjusting rail; intermediate rail
伸缩调整件 expansion piece
伸缩调整器 compensator; expansion device
伸缩通风口 telescopic(al) vent
伸缩筒 retracting cylinder;telescopic(al) cylinder
伸缩筒式减震器 telescopic(al) damper;telescopic(al) shock absorber
伸缩筒式联轴节 telescopic (al) shaft coupling
伸缩筒型避震器 cylinder-type absorber;telescopic (al) type absorber;telescopic(al) type damper
伸缩腿架 telescope leg
伸缩弯管 expansion bend; expansion pipe bend
伸缩桅 housing mast;telescope mast; telescopic(al) top mast
伸缩箱式封盖 bellows seal gland
伸缩性 elasticity;retractility;slip joint
伸缩性发行法 elastic limit system
伸缩性供应 elastic supply
伸缩性汇率 flexible exchange rate
伸缩性价格 flexible price
伸缩性连接 concertina connection
伸缩性通货 elastic currency
伸缩性系数 elasticity coefficient
伸缩性张拉板 telescopic (al) tensile plate
伸缩旋风器 telescopic(al) cyclone
伸缩烟筒 telescopic(al) funnel
伸缩岩芯管 extension core barrel
伸缩液压缸 extension cylinder
伸缩仪 extens(i)ometer
伸缩应力 expansion or contraction stress
伸缩预算 flexible budgeting
伸缩钥匙 adjustable key;extension key
伸缩运动 stretching motion
伸缩闸瓦离合器 block clutch
伸缩债券 escalator bond
伸缩(胀)缝 expansion joint
伸缩罩 telescopic(al) cover
伸缩振动 shuttling; stretching vibration
伸缩支承 expansion bearing
伸缩支承杆 stretcher bar
伸缩支架 telescope support;telescopic(al) support
伸缩支架挖土机 telescopic(al) jibbed excavator
伸缩支柱 telescopic(al) leg;telescopic(al) prop;telescopic(al) support
伸缩支座 expansion and contraction bearing; expansion bearing; rocker bearing
伸缩指式捡拾器 retractable finger pick-up
伸缩轴 telescopic(al) shaft
伸缩轴传动 shaft drive
伸缩柱 telescopic (al) leg; telescopic(al) mast
伸缩柱塞 telescopic(al) plunger
伸缩爪 extendible and retractable dog
伸缩装卸机 telescopic(al) loader
伸缩装置 extension fitting; retractor device
伸缩桌 extension table; telescope table
伸缩自由的 retractile
伸缩钻臂 telescopic(al) boom
伸缩钻杆 bumper sub
伸缩钻头 expanding bit;expansion bit
伸头 peak with heading
伸腿的地方 <车、飞机等> leg room
伸线器 come-along
伸向【地】 hade;angle of hade
伸悬臂拱式大梁 cantilever arched bridge
伸展 elongation; extension; spread; stretch(ing);stretching run;stretch-out;unfolding
伸展的冰碛层 push moraine
伸展的根 spread root
伸展断层 extensional tectonics;extension fault
伸展多边形 open polygon
伸展范围 expanding reach
伸展钢筋 expansion reinforcement
伸展格栅 expanded joist
伸展构造 extensional fault
伸展构造型 extensional tectonic type
伸展机 flattener
伸展架设法 erection by launching; erection by protrusion
伸展结构 stretched out structure
伸展结晶 extension crystallization
伸展梁 spread beam
伸展率 percentage of elongation
伸展期 extension period
伸展绳 extension cord
伸展式收纸装置 extended delivery
伸展式锁块 expanding latch segment
伸展数 tensile figure
伸展速度 rate of stretch
伸展台 extending table
伸展台座 stretching bed
伸展梯 extending ladder
伸展性 extensibility
伸展支撑 expanding support
伸展组织 expansion tissue
伸展作用 extending
伸张 stretch(ing);tensile
伸张比 extension ratio
伸张次数 extension cycles
伸张阀 rebound valve
伸张杆 stretcher bar
伸张架式铰链 extension casement hinge
伸张力 expansion force
伸张木 tension wood
伸张器 stretcher;tensor
伸张器具 spreader
伸张强度 tensile strength
伸张索 <缆索铁路和架空索道> tension rope
伸张行程阻尼力 rebound resistance
伸张性 expansibility;expansibleness
伸张悬杆 expansion suspender
伸张仪 extens(i)ometer
伸胀 expansion
伸胀比 ratio of expansion
伸胀缝 variator
伸胀孔 expansion slot
伸胀力 expansive force
伸胀容许量 expansion allowance
伸胀套筒 expansion sleeve
伸胀悬杆 expansion suspender
伸胀支承 expansion bearing
伸直 unbend;unwind
伸直长度 stretched length
伸指现象 finger phenomenon;Souques' phenomenon

身边无子女的夫妇 empty nester

身份 identity;status
身份不明的 undisclosed
身份介绍信 letter of identification
身份说明 statement of identity
身份证 identification card[ID card];identity card; certificate of identification;certificate of identity
身份证件 identity documents
身份证明 identification paper; identity certificate
身份证明信 letter of identification; letter of identity
身份证书 letter of identification
身高 stature
身体健康检查 physical fitness test
身重量 tare

绅士锯 gent's saw

砷钯矿 arsenopallasinite

砷钡铝矾 weilerite
砷钡铀矿 heinrichite
砷铋矿 arsenobismite
砷铋镍钴矿 badenite
砷铋铅铀矿 asselbornite

砷铋石 rooseveltite
砷铋铜矿 mixite
砷铋铀矿 walpurgite
砷别洛夫石 arsenate-belovite
砷铂锇矿 osplatarsenite
砷铂矿 sperrylite
砷铂铱矿 iripatarsenite
砷车轮矿 seligmannite
砷碲锌铅石 dugganite
砷毒性麻痹 arsenical paralysis
砷矾铜矿 lindackerite
砷矾铅矿 endlichite
砷钙镁石 adelite
砷钙锰矿 rhodoarsenian
砷钙锰石 grischunit
砷钙钠铜矿 freirinite;lavendulan
砷钙石 haidingerite
砷钙铜矿 conichalcite;higginsite
砷钙铜石 conichalcite
砷钙锌矿 austenite
砷钙锌石 austinite
砷钙钇锰矿 retzianite
砷钙铀矿 arsenuranylite;uranospinite
砷铬铅矿 bellite
砷铬铜铅石 fornacite
砷汞钯矿 atheneite
砷钴钙石 roselite
砷钴矿 modderite;smaltine;tin-white cobalt
砷钴镍铁矿 westerveldite
砷硅铝锰石 ardennite
砷硅锰矿 schallerite
砷硅锌锰石 kraisslite
砷合金 arsenic alloy
砷华 arsenic bloom;arsenolite
砷化(合)物 arsenic compound
砷化镓 gallium arsenide;gallium arsenic
砷化镓半导体 gallium-arsenide semiconductor
砷化镓二极管 gallium-arsenide diode
砷化镓发光二极管 gallium-arsenide luminescence diode
砷化镓固态灯 gallium-arsenide solid-state lamp
砷化镓光电阴极 gallium-arsenide photocathode
砷化镓光源 gallium-arsenide light source
砷化镓红外发射器 gallium-arsenide infrared emitter
砷化镓激光器 gallium-arsenide laser
砷化镓结光源 gallium-arsenide-junction light source
砷化镓滤光片 gallium-arsenide optical filter
砷化镓面垒探测器 gallium-arsenide surface barrier detector
砷化镓探测器 gallium-arsenide detector
砷化镓自发红外光源 GaAs spontaneous infrared source
砷化铝 alumin(i)um arsenide
砷化钠 sodium arsenide
砷化镍 nickel arsenide
砷化镍结构 nickel arsenide structure
砷化氢 arsine
砷化氢中毒 arsine poisoning
砷化三氢 arsenic hydride
砷化锑 arsenical antimony
砷化铜 arsenical copper;copper arsenic
砷化物 arsenical;arsenide
砷化铟探测器 indium arsenide detector
砷黄 arsenic yellow
砷黄铁矿 arsenopyrite;mispickel
砷黄铁矿毒砂 arsenopyrite;mispickel
砷灰石 svabite
砷辉锑银矿 arsenomiargyrite
砷剂 arsenical
砷检出装置 arsenic apparatus
砷碱法脱硫 arsenic-caustic desulfuration
砷浸液 arsenic dip
砷离子 arsenic ion;arsenious ion
砷钌矿 ruthenarsenite
砷菱铅矾 beudantite
砷硫钒铜矿 arsenosulvanite
砷硫锑铅矿 geocronite
砷硫锑铜银矿 arsenpolybasite
砷硫铁铜矿 epigenite
砷硫铜铁矿 epigenite
砷硫银矿 pearceite
砷铝锰矿 synadelphite
砷铝石 mansfieldite
砷铝铜矿 ceruleite
砷铝铜石 luetheite
砷氯铅矿 finnemanite;georgiadesite
砷镁钙石 talmessite
砷镁锰石 manganeshoeresite
砷镁石 hoernesite
砷镁锌石 chudobaite
砷锰钙矿 caryinite
砷锰钙石 brandtite
砷锰矿 armangite
砷锰镁石 magnesium-chlorophoenicite
砷锰铅矿 caryinite;trigonite
砷钼铁钙矿 betpakdalite
砷钼铁钠石 sodium-betpakdalite
砷镍钯矿 majakite
砷镍钴矿 langisite;white nickel ore
砷镍矿 chloanthite;maucherite;nickel-skutterudite;temiskamite
砷镍铜矾 lindackerite
砷硼钙石 cabnite;cahnite
砷硼镁钙石 teruggite
砷铍钙石 asbecasite
砷铍硅钙石 asbecasite
砷铅合金 arsenical lead
砷铅矿 mimetite;petterdite;prixite
砷铅铝矾 hidalgoite
砷铅石 mimetite
砷铅铁矾 beudantite
砷铅铁矿 carminite
砷铅铁石 carminite
砷铅铀矿 huegelite
砷青铜 arsenical bronze
砷氢镁石 roesslerite
砷氢锰钙石 fluckite
砷热臭石 nelenite
砷受体 arsenoceptor
砷水锰矿 allactite
砷锶铝矾 kemmlitzite
砷锶铝石 arsenogoyazit
砷酸 acid;arsenic;ortho-arsenic acid
砷酸铋矿 atelestite;rhagite
砷酸滴定法 arsenometric titration
砷酸二硝基酚<木材防腐用> arsenate dinitrophenol
砷酸钙 calcium arsenate
砷酸根离子 arsenate ion
砷酸铬铜 copper-chrome arsenate
砷酸汞 mercuric arsenate;mercury arsenate
砷酸钴 cobalt arsenate
砷酸钾 Macquer's salt;potassium arsenate
砷酸镁 magnesium arsenate
砷酸钠 natrium arsenicum;sodium arsenate
砷酸镍 nickel(ous) arsenate
砷酸铅 lead arsenate
砷酸氢二钠 disodium hydrogen arsenate
砷酸铁 ferric arsenate
砷酸铜 copper arsenate
砷酸锌 zinc arsenate
砷酸亚铁 ferrous arsenate;iron arsenate
砷酸盐 arsenate
砷酸盐浓度 arsenate concentration
砷酸银 silver arsenate
砷酸酯 arsenate
砷钛钒石 tomichite
砷钛铁钙石 cafarsite
砷锑钯矿 mertieite
砷锑钙铜石 richelsdерfite
砷锑矿 allemontite;arsenical antimony;stibarsen
砷锑铁钙矿 stenhuggarite
砷铁矾 sarmientite
砷铁钙石 arseniosiderite
砷铁矿 symplesite
砷铁铝矿 liskeardite
砷铁铝石 liskeardite
砷铁镍矿 oregonite
砷铁铅矿 ludlockite
砷铁铅石 arsenbrackebuachite
砷铁铅锌石 jamesite
砷铁石 symplesite
砷铁铜石 chenevixite
砷铁锌铅石 tsumcorite
砷铜矾 parnauite
砷铜合金 arsenical copper
砷铜矿 domeykite
砷铜绿 parrot green
砷铜铅矿 duftite;plumbocuprite
砷铜铅石 duftite
砷铜钇矿 agardite
砷铜银矿 novakite
砷钍石 arsenthorite
砷污染 arsenic contamination;pollution by arsenic
砷污染物 arsenic contaminant;arsenic pollutant
砷硒银矿 rittingerite
砷硒黝铜矿 giraudite
砷锌钙矿 austinite
砷锌镉铜石 keyite
砷锌矿 reinerite
砷、锌、铜木材防腐剂 mineralized cell preservative
砷循环 arsenic cycle
砷氧化 arsenic oxidation
砷铱矿 iridarsenite
砷钇石 chernovite
砷钇铜石 agardite
砷铀铋矿 walpurgite;waltherite
砷铀矿 arsenuranylite;troegerite
砷黝铜矿 tennantite
砷真空镀层 arsenic vacuum coating
砷制剂 arsenical
砷中毒 arseniasis;arsenic(al) poisoning

深U形谷<冰川谷> yosemite
深暗的 hypermelanic
深暗颜色 sad colo(u)r
深凹槽 tectogene
深凹带 deep subsidence zone
深凹勾缝 deeply recessed joint pointing;hungry joint pointing
深凹灰缝 rustic joint
深凹接缝 racked joint
深凹式轮辋 drop center rim
深凹饰 bed mo(u)ld(ing)
深凹型方头钢锹 flat end deep rounded heel steel shovel
深凹圆线脚 quirk mo(u)lding
深奥的 abstruse
深拗槽 geotectogene;tectogene
深板大梁 deep plate girder
深泵井 deep pumped well
深变片麻岩 katagneiss
深变斜长片麻岩 kata plagioclase gneiss
深变正长片麻岩 kata-orthoclase gneiss
深变质带 catazone[katazone]
深变质带的 katazonal
深变质带岩石 kata-rock
深变质的 hypometamorphic
深变质作用 katametamorphism;metagenesis
深冰 deep ice
深播 deep seeding;under seeding
深播法 under seeded method
深播适宜 proper deep seeding
深不可测的 abysmal
深部 lower part
深部层位 deep level
深部产油层 deep pay
深部的 deep-seated
深部地壳变动 bathyderm
深部地热流体 deep-seated geothermal fluid
深部地热系统 deep-seated geothermal system
深部地下水 deep ground water;deep phreatic water
深部动力变质作用 hypokinematic metamorphism
深部断层 deep-seated fault
深部分散 deep-seated dispersion
深部分异作用 abyssal differentiation
深部缝隙<岩层中的> pit hole
深部干气带 deep dry gas zone
深部感受性 deep sensitivity
深部构造地球化学 deep tectono-geochemistry
深部构造线成矿说 metallogeny of deep lineaments
深部贯入 abysmal injection
深部含油砂层 deep sand
深部花岗岩 low-lever granite
深部化探 subsurface geochemical exploration
深部环境 deep-seated environment
深部缓流带 deep slow moving zone
深部级流岩溶 deep slow-flowage karst
深部开采 deep cut(ting)
深部勘探 deep prospecting
深部矿床 deep-seated deposit
深部类型 deep-seated type
深部砾石 deep gravel
深部煤层 underseam
深部黏[粘]土流 deep-seated clay flowage
深部穹隆 deep dome
深部区域变质作用 plutonic metamorphism
深部热储 deep reservoir
深部渗流 deep percolation;deep seepage
深部渗漏 deep percolation;deep seepage
深部渗水 deep percolation;deep seepage
深部渗透 deep percolation;deep seepage
深部水流系统 deep floe system
深部似境界剥采比 overburden ratio of deep approximate boundary limits
深部同化混染作用 abyssal assimilation and contamination
深部温度 deep temperature
深部循环热液 deep circulating hydrothermal solution
深部岩溶 deep karst
深部岩体温度 deep-rock temperature
深部异物定位器 profondometer
深部照明光纤 deep illumination optic(al) fiber
深部滞水层<湖的> monimolimnion
深采对比法 correlation between exploration and exploitation data
深舱壁 deep bulkhead

深舱堆装法 deep stowage
深舱货船 deep hold(cargo)ship
深舱容积 deep tank capacity
深舱水柜 deep tank
深舱油柜 deep tank
深藏的 deeplying
深藏矿床 deep-seated deposit
深槽 quirk;pool <航道>;deep channel;deep trench
深槽凹圆线脚 quirk bead
深槽冲刷 thalweg erosion
深槽的 deep-slot
深槽断面 deep channel section
深槽分级机 deep-pocket classifier
深槽感应电动机 current-displacement motor;deep-slot(induction)motor
深槽滚珠轴承 deep-groove(ball)bearing
深槽河 deep river channel;gut
深槽河道 incised river
深槽河段 deep pool section
深槽交错的浅滩 deeps-staggered shoal
深槽交错过渡浅滩 crossing-shoal with staggered pools; deeps-staggered crossing-shoal
深槽镘刀 quirk float
深槽木撑 timbering to deep trenches
深槽曝气 deep layer aeration
深槽曝气/气浮系统 deep tank aeration/ flo(a)tation system
深槽浅滩序列 pool-riffle sequence
深槽砂 channel sand
深槽式绕组 skin effect winding
深槽式转子 current throttling type rotor;skin effect rotor
深槽鼠笼式电动机 deep-bar motor; deep-slot squirrel-cage motor
深槽鼠笼式感应电动机 deep-slot squirrel-cage induction motor
深槽鼠笼式绕组 deep-bar cage winding
深槽鼠笼转子 extended bar rotor
深槽水沟 deep-furrow
深槽托辊 <皮带运输机用> deep-trough idler
深槽纹辊 crimping roll
深槽效应 deep-bar effect; deep-slot effect <电机的>
深槽轴承 deep-groove(ball)bearing
深侧向测井 deep lateral log
深测法的 bathymetric(al)
深测深度 depth of investigation
深测试验 deep test
深层 deep layer; deep lift; deep stratum
深层拌和法 deep admixture stabilization
深层拌和水泥处理地基 deep cement mixing
深层爆破 deep blasting
深层爆破压实 deep blasting compaction;explosive compaction
深层爆炸 deep blasting
深层采水 deep-casting
深层采样 in-deep sampling
深层沉降 deep settlement
深层沉降计 deep settlement ga(u)ge
深层沉降仪 deep settlement ga(u)ge
深层澄清池 depth clarifier
深层抽水 deep pumping
深层触探器 deep-sounding apparatus
深层触探设备 deep-sounding apparatus
深层触探试验 deep penetration test; deep-sounding test
深层次 deep level
深层粗粒岩 batholite
深层淬火面 deep hardened surface
深层带 bathypelagic zone
深层的 deep level; deeply buried; deep-seated
深层底土 deep subsoil
深层地铁车站 deep metro station
深层地下水 deep-bed ground water; profound groundwater
深层地下水位 deep-water table
深层地质填图 deep geologic(al) mapping
深层动物区系 bathypelagic fauna
深层冻结 deep freezing
深层发酵 submerged fermentation
深层反射 deep(er) reflex
深层分异运动 deep-seated different in movement
深层风化 deep-seated weathering
深层高强度沥青构造 deep-strength asphalt construction
深层高强度沥青路面 deep-strength asphalt pavement
深层耕作 layer mining ploughing
深层构造 deep lift construction;deep structure;infrastructure
深层构造图 infrastructure map
深层灌浆 deep-seated grouting
深层过滤 deep-bed filtration;in-depth filtration;depth filtration
深层过滤层 depth filter
深层过滤器 deep filter
深层滑动 deep(land)slide;deep slip
深层滑坡 deep slide;level landslide
深层化学拌和法 deep mixing method
深层环流 abysmal circulation; deep circulation
深层活动稳定性 slope stability
深层挤密法 deep extrusion method
深层加固 deep compaction;deep consolidation
深层加密法【岩】 deep compaction; deep densification
深层加实 deep densification
深层建筑 deep lift construction
深层搅拌 clay mixing consolidation
深层搅拌法 <软土地基加固的> deep mixing method;deep soil agitating process
深层搅拌法喷浆桩 deep bed stirring method gunite pile
深层搅拌桩 deep-mixed pile; deep mixing pile
深层开采量 deep production
深层开挖 deep(level)excavation
深层勘察 deep prospecting
深层勘探 deep prospecting
深层类型 deep-seated type
深层犁 deep-plough
深层沥青铺面 deep lift asphalt pavement
深层沥青施工 deep lift asphalt construction
深层流 deep(-water) current; deep-water flow
深层流动褶皱作用【地】 rheomorphic folding
深层煤矿 deep coal deposits
深层黏[粘]土 deep stratum of clay
深层排水井系统 deep well system
深层培养 deep culture; submerged culture
深层平板载荷试验 deep plate loading test
深层曝气 deep aeration tank;deep layer aeration;hypolimnetic aeration
深层曝气法 deep aeration system
深层气 deep gas
深层气富集带 deep gas abundance zone
深层气体测量 deep gas survey
深层取水 deep diversion; deep intaking;deep water intake
深层取样 deep sampling
深层取样器 deep sampler
深层燃烧 deep-seated burning
深层热水 deep thermal water
深层蠕动 deep creep;depth creep
深层深海的 abyssal benthic
深层渗流 deep percolation;deep seepage
深层渗漏 deep penetration;deep percolation;deep seepage
深层渗漏损失 deep percolation loss
深层渗滤 deep percolation
深层渗入量 quantity of percolation
深层渗水 deep percolation; deep seepage
深层渗透 deep penetration;deep percolation;deep seepage
深层生态学 deep ecology
深层石灰搅拌法 deep-lime mixing method
深层时窗长度 length of time window of deep layer
深层试验 deep test
深层水 deep phreatic water;deep water; water in deep layer; internal water
深层水井 deep-water well
深层水流 deep flow;deep-water current
深层水泥拌和 deep cement mixing
深层水团 deep-water mass
深层土(壤) deep soil;submerged soil
深层土壤水 deep subsoil water
深层稳定 deep stabilization
深层系数 deep coefficient; deep factor
深层咸水污染 deep salty water pollution
深层显影剂 depth development
深层压实 compaction of deep bed; deep compaction
深层压实法 deep compaction method
深层岩 deep-seated rock; plutonic rock
深层岩溶 deep karst
深层引水 deep diversion
深层原生水 internal primitive water
深层振动压实 vibrator-jetting deep compaction; deep vibratory compaction
深层振实 deep compaction
深层装舱系统 deep loading system of hopper
深层自记海流计 depth-current self-recording meter
深层自记海流器 depth-current self-recording recorder
深层钻探 deep boring; deep drilling; deep prospecting
深插 deep slotting
深长的 deep-drawn
深长的切痕 gash
深长海沟 foredeep
深长引道 <进入埃及古墓前的> dromos
深沉积盆地气 deep sedimentary basin gas
深成 subnate
深成变质作用【地】 katogenic metamorphism;plutonic metamorphism
深成的【地】 abysmal; abyssal; hypogenic; plutonic; hypogene; deep-seated
深成低温热液矿床 telethermal ore deposit
深成地震 plutonic earthquake
深成动力变质【地】 katogeno-dynamoetamorphism
深成动力花岗岩 hyperkinematic granite
深成高温热液的 hypothermal;katathermal
深成高温热液矿床 hypothermal deposit
深成贯入 abyssal injection
深成环境 abyssal
深成混合岩化方式 hypomigmatization way
深成活动性 hypogene mobility
深成火山 deep-seated volcano
深成角砾环状岩【地】 isothrausmatic rock
深成角砾石 plutonic breccia
深成角砾岩 plutonic breccia
深成矿床 hypogene deposit
深成矿物 plutonic mineral
深成理论 abyssal theory
深成脉状矿床 hypogene vein deposit
深成挠曲 plis du fond
深成侵入【地】 abyssal intrusion
深成侵入体 major intrusion; plutonic intrusion
深成热液 katathermal solution
深成热液矿床 katathermal ore deposit
深成渗透作用 hypofiltration
深成水【地】 plutonic water
深成说 abyssal theory
深成同化【地】 abyssal assimilation
深成现象 plutonism
深成相 abyssal facies;plutonic facies
深成岩 abysmal rock; abyssal rock; deep-seated rock; hypogene; hypogene rock;plutonic rock;plutonite
深成岩基带 hypobatholitic zone
深成岩浆 hypogene magma
深成岩脉 deep-seated dike[dyke]
深成岩侵入 abyssal intrusion
深成岩体【地】 pluton(e)
深成岩体含矿性评价 assessment of the productivity of plutons
深成油 deep oil
深成中温热液的 mesothermal
深成组构 mesogenetic fabric
深成作用 hypogene action
深成作用阶段 catagenetic stage
深橙色 copper
深吃水 deep-draft; heavy draft; deep-draught
深吃水船(舶) deep-draft vessel; deep-draught vessel; heavy draught vessel
深吃水船锚地 anchorage for deep draught vessel
深吃水船闸 deep-draft lock
深吃水航路 deep-draft route
深池 weel
深冲 deep-drawn;deep punching
深冲薄板 deep stamping sheet
深冲薄钢板 deep drawing sheet
深冲材料 deep-drawing material
深冲成型 <板材的> deep drawing
深冲钢 deep-drawing steel; deep punching steel
深冲沟 coulee; deep gulch; deep ravine;deep trench
深冲划伤 draw mark
深冲模 drawing die
深冲性 <板材的> deep-drawing quality;drawing quality
深冲压 deep drawing
深冲压成型 deep draw mo(u)ld(ing)
深冲压的薄板 deep-drawn sheet
深冲压模 deep draw mo(u)ld(ing)
深冲质量 drawing quality
深处 recess
深处水流 bathycurrent
深穿断裂 deep penetrating fault
深穿透 deep penetration
深床过滤法 deep-bed filtration
深床颗粒滤池 deep-bed granular filter

S

深床滤池 deep-bed filter
深床砂滤 deep-bed filtration
深脆沥青 kata-impsonite
深存水弯 <一种反虹吸存水弯> deep seal;deep-seal trap
深(大)断裂 deep fracture
深(大)梁 deep girder
深带变质作用 katazonal metamorphism
深到膝的 knee-deep
深的 bathymetric (al); profound; strong
深的格构大梁 deep lattice(d) girder
深低气压 deep depression
深低温 profound hypothermia
深低温研磨 freeze grinding
深低温治疗器 instrument for cryotherapy
深底带 profundal zone
深底的(湖、海) profundal
深地槽 tectogene
深地层处置 deep underground disposal
深地震测深 deep seismic sounding
深雕刻 deep engraving
深调 deep shadow
深碟式驾驶盘 deep dish steering wheel
深碟式路面铺设 deep dish paving
深碟式路面摊铺 <高速公路弯道路面超高布置形式,如深的碟子> deep dish paving
深冻 deep freezing
深冻处理 deep freezing
深冻货物 deep-frozen goods
深冻拖船 deep-freezing traverse
深洞 pot-hole
深斗板式输送机 deep pan apron conveyer[conveyor]
深斗裙式输送机 deep pan apron conveyer[conveyor]
深度 deepness; degree of depth; depth; fullness; hill-and-dale; vertical extent
深度X光治疗室 deep X-ray therapy room
深度报警装置 depth warning device
深度比 deep ratio;depth ratio
深度比例(尺) depth scale; scale of depth
深度编码器 depth encoder
深度变化指示器 depth deviation indicator
深度变换 deep conversion;depth conversion
深度标尺 depth scale
深度标引 depth indexing
深度标志 depth marker
深度表 deep meter; depth ga(u)ge; depth indicator;depth meter;depthometer
深度补偿换能器 depth compensated transducer
深度补偿湿式潜水服 depth compensated wet suit
深度参数 depth parameter
深度测定 deep test;depth determination
深度测量 depth measurement; measurement of depth
深度测量计 depth meter
深度测量误差 error in depth measurement
深度测微计 depth micrometer
深度尺 depth ga(u)ge;depth indicator;depthometer
深度抽提 drastic extraction
深度处理 advanced treatment; polishing process
深度处理法 advanced treatment process;deep treatment process
深度处理过程 advanced treatment process
深度锤 dip weight
深度淬火 deep-hardening
深度点 deep point; depth point; spot depth
深度点数 number of depth point
深度点位置 position of depth point
深度电导仪 bathyconductograph
深度电离 ionization in depth
深度冬眠 deep hibernation; hibernation
深度段 depth segment
深度队列 depth queue
深度法则 deep rule;depth rule
深度反向电流 deep return current
深度范围 deep range;depth range
深度放电 deep discharge
深度放松 deep disintegrating
深度废水处理 advanced waste (water) treatment
深度分辨率 depth resolution
深度分布 depth distribution
深度分带 deep zone;depth zone
深度分解 deep disintegrating
深度分类 depth classification
深度腐朽 advanced decay
深度改正 correction for depth
深度干涉 deep intervention
深度感 depth perception; perception of depth
深度贯穿 depth penetration
深度规 depth ga(u)ge;rule depth ga(u)ge
深度规测微器 micrometer depth ga(u)ge
深度规计 depth ga(u)ge
深度过滤 depth-type filtration
深度过滤器 deep filter
深度化整 depth round-off
深度回流 deep return current
深度基准面 depth datum;sea level datum;sounding datum;datum level
深度基准面保证率 assuring rate of depth datum; guaranteed rate of depth datum
深度棘轮调节装置 depth-ratchet setting
深度计 bathymeter; deep indicator; deep meter; depth ga(u)ge; depth indicator; depthometer; depth recorder;foot ga(u)ge
深度计仪器 depth ga(u)ge tool
深度记录 depth record
深度记录方式 fashion of depth record
深度记录器 deep recorder; depth recorder
深度记录线 <海洋> receiver line; receiving line
深度记录仪 deep recording device; depth recorder;depth recording device
深度剂量 depth dose
深度剂量分布 depth dose distribution
深度加固 deep consolidation
深度加氢裂化 overhydrocracking
深度剪取 depth clipping
深度浸蚀 deep etch;macroetching
深度净化 deep cleaning
深度卡规 depth calliper
深度可测的 fathomable
深度可调声呐 variable depth sonar
深度控制 depth control
深度控制器 deep controller; depth controller
深度控制装置 depth control device
深度宽度比 depth over width ratio
深度拉伸 deep drawing
深度累积式取样器 depth-integrating sampler
深度冷冻 cryogenic process; deep freeze;deep freezing; deep refrigeration
深度冷冻厂 intense cooling plant
深度冷冻器 deep freezer
深度冷却 copious cooling;deep cooling;deep refrigeration
深度-历时曲线 <降雨的> depth-duration curve
深度量规 depth ga(u)ge
深度裂化的 heavily cracked
深度裂化阶段 advanced stage
深度/密度剖面图 depth/density profile
深度密封存水弯 deep-seal trap
深度模式 <楼梯扶手弯曲部分> falling mo(u)ld
深度排序 depth sort
深度偏移 depth migration
深度偏移算子 depth migration operator
深度平衡树 height balanced tree
深度破坏 deep breaking up
深度剖面 deep profile
深度剖面图 depth section chart
深度千分表 dial depth ga(u)ge
深度千分尺 depth micrometer; micrometer depth ga(u)ge
深度千分卡规 micrometer depth ga(u)ge
深度清扫 deep cleaning
深度清洗 deep cleaning
深度曲线 deep curve;depth curve
深度-容积关系曲线 depth-volume curve
深度扫描声呐 depth scan(ning) sonar
深度渗入 frost penetration
深度渗碳 deep carburizing;jackmanizing
深度时间相关曲线 depth-time curve
深度数值 depth figure
深度数字转换器 depth digitizer;digital depth recorder
深度睡眠前时期 predormitium
深度-速度曲线 depth-velocity curve; vertical velocity curve
深度探头 deep probe;depth probe
深度梯度 deep gradient;depth gradient
深度调节螺杆 depth screw
深度调节器 depth adjuster;depth regulator
深度推断 depth interpretation
深度脱氮 advanced nitrogen removal
深度瓦解 deep disintegrating
深度温度计 bathythermograph;thermarine recorder; thermosounder recorder
深度温度联合记录仪 combined recording depth and temperature meter
深度温度仪 bathothermograph;bathythermograph
深度污水处理 advanced effluent treatment
深度污水处理后出水 polished effluent
深度污水处理塘 polishing pond
深度系数 deep coefficient;depth factor
深度细裂纹 crocodiling
深度线 depth line
深度限度 depth bound
深度效果 deep effect; depth effect; effect of depth
深度效应 deep effect; depth effect; effect of depth
深度信息脉冲 depth information pulse
深度压力计 depth manometer
深度-压力曲线 depth pressure plot
深度压实 deep compaction
深度氧化沥青 fully blown bitumen
深度遥测仪 depth telemeter
深度因数 deep factor;depth factor
深度因素 deep factor;depth factor
深度饮用水处理工艺 advanced drinking water treatment process
深度影响 effect of depth
深度优先 depth-first
深度优先过程 depth-first procedure
深度优先搜索 depth-first search
深度优先最小最大过程 depth-first minimax procedure
深度游标尺 depth vernier
深度域宽度 width of depth domain
深度遮檐 deep-throat coping
深度指标 depth index
深度指示器 deep indicator;depth indicator;depth recorder
深度指示仪 deep indicator;depth indicator;depth recorder
深度指数 depth exponent
深度轴比例尺 scale of depth axis
深度注记 depth figure; depth number;sounding mark
深度着色 deep tint
深度自记温度计 bathothermograph; bathythermograph
深度纵剖面 depth profile;longitudinal profile of depth
深断层 abyssal fault;profound fault
深断裂 profound fault
深断裂带 deep fracture zone
深断裂的标志 mark of deep fracture
深断裂类型 type of deep fracture
深断裂名称 name of deep fracture
深断裂系 deep fractures system
深断裂线性体 lineament related to deep faults
深断裂走向 trend of deep fracture
深翻 deep tillage
深翻耙齿 deep ripping shank
深分点荷载 <材料试验的> third point load(ing)
深粉红色 radiance
深风化 deep erosion
深扶手 deep handrail
深浮雕 high relief
深腐蚀 deep etch;etch back
深感应测井 deep investigation induction log
深感应测井曲线 deep investigation induction log curve
深橄榄绿 olive green deep
深镉橙颜料 cadmium deep orange
深镉红颜料 cadmium deep red
深铬黄 chrome yellow orange
深给进钻车 long-feed drill jumbo
深根吸水植物 mesophreatophyte
深根性作物 deep-rooted crop
深根植物 deep-rooted plant;phreatophyte
深根作物 deep-rooted plant
深耕 deep culture;deep till(age)
深耕法 deep digging
深耕犁 deep-plough;trench-plough
深耕土 deep-worked soil
深耕土壤 deep-worked soil
深耕细作 deep ploughing and intense cultivation;spade husbandry
深耕中耕机 grubber
深拱 rear vault
深沟 gulch;strong defence;sloot <南非大雨冲刷成的>;sluit <南非大雨冲刷成的>
深沟厕所 deep trench latrine
深谷 barranca; canyon; deep valley; glen; kloof; pronounced valley; ravine
深谷河流 canyoned river; canyoned stream

深管井 deep-tube well
深贯入 <用于沥青碎石路> complete penetration;deep penetration
深闺 innermost part
深硅铝层 bathyderm
深海 abysmal sea;abyss(al sea);blue sea;blue water;deep sea;deep water;exposed deep water
深海半深海序列 abyssal-semiabyssal succession
深海波 deep-sea wave;trochoidal wave
深海波浪计 deep-sea wave meter;deep-sea wave recorder
深海驳 deep-sea barge
深海簸脊圈闭 deep-winnowed-crestal trap
深海簸翼圈闭 deep-winnowed-flank trap
深海捕鱼 offshore fishing
深海采矿船 deep-sea mining ship
深海采矿环境研究 deep ocean mining environment study
深海槽 deep-sea canyon;deep-sea channel;deep-sea trough;oceanic trench;trough bend
深海测量 bathymetric(al) survey
深海测量法 bathymetry
深海测量学的 bathymetric(al)
深海测深 deep-sea sounding
深海测深锤 deep-sea lead;deep-sounding lead
深海测深机 deep-sounding machine
深海测深绳 deep-sea lead line;deep-sea line
深海测深仪 abysmal sea bathometer;bathometer;bathymeter;deep-sea bathometer
深海测深装置 deep-sea sounding apparatus
深海测温仪 bathythermograph
深海层浮游生物 bathy plankton
深海层生物 bathybic organism;bathypelagic organism
深海层鱼类 bathypelagic fishes
深海潮汐 deep-sea tide
深海沉积土 deep-sea sedimentary soil
深海沉积(物) abysmal deposit;abyssal deposit;abyssal sediment;bathyal sediment;deep-sea deposit;deep-sea sediment;pelagic(al) deposit;deposit(e) of the deep sea;thalassic deposit
深海沉积物层序 abyssal sediment sequence
深海沉积(物)取样器 benthos sediment sampler
深海沉积组合 abyssal deposits association
深海沉积作用 abyssal deposition
深海沉砂 deep-sea sand
深海传播 deep-sea propagation
深海带 bathypelagic zone;pelagic(al) zone
深海倒转温度表 deep-sea reversing thermometer
深海的 abyssal;archibenthic;bathybic;deep water;eunic;pelagic(al);thalassic;deep-sea
深海底 abysmal floor;abyssal depth;abyssal floor;benthos;deep-sea bottom;floor of deep ocean;profund(al) zone
深海底沉积环境 fondo
深海底沉积岩石 fondothem
深海底带 abyssal zone
深海底的 bathyal;bathybic
深海底地形 fondoform
深海底栖带 abyssal benthic zone
深海底栖群落 abyssal community
深海底栖生物 archibenthos;deep-sea benthos
深海地槽 thalassogeosyncline
深海地堑 deep-sea graben
深海地震 bathyseism;deep-sea earthquake;deep seism
深海地震测量 deep seismic survey
深海电缆 deep-sea cable
深海电视照相机 deep-eye
深海调查 deep-sea expedition
深海调查船 deep-sea research;deep submergence search vessel
深海调查潜水器 deep-sea research vehicle
深海定位 deep ocean location
深海动物区系 abyssal fauna;bathypelagic fauna;deep-sea fauna;pelagic(al) fauna
深海动物群 abyssal fauna;bathypelagic fauna;deep-sea fauna;pelagic(al) fauna
深海浮游生物 abyssopelagic plankton;bathypelagic plankton
深海沟 deep-sea furrow;deep-sea trench
深海构造 abyssal structure
深海谷 combe;coom;deep-sea channel
深海谷地沉积 abyssal valley deposit
深海海床 abyssal floor;deep-seabed
深海海底 deep-sea floor
深海海底的 abyssal benthic;abyssobenthic
深海海底电缆 deep-sea submarine cable
深海海底水听器装置 ocean bottom hydrophone
深海海面的 abyssal pelagic;abyssopelagic
深海海啸 deep-sea tsunami
深海航道 deep-sea channel
深海航路 deep-sea course;deep-sea lane
深海红土 abyssal red earth;red deep sea earth
深海环境 abyssal environment;deep ocean environment
深海环流 abyssal circulation;deep-sea circulation
深海火山 deep-seated volcano
深海机器人 deep-sea robot
深海技术 deep ocean technology;deep-sea technology
深海间隙水 deep ocean pore water
深海阶地 deep-sea terrace
深海开发技术 deep ocean exploitation technique
深海勘探器 bathyscaph(e)
深海考察(潜水)器 deep research vehicle
深海空间 deep space
深海矿藏 deep-sea deposit
深海矿物资源 abyssal mineral resources
深海矿物资源地质学研究 geologic(al) study of deep sea mineral resources
深海矿物资源勘查基础研究 basic study on exploration of deep sea mineral resources
深海矿物资源名称 name of abyssal mineral resources
深海扩散层 deep scattering layer
深海拉斑玄武岩 abyssal tholeiite
深海裂隙 abyssal gap
深海流 abyssal current;deep-sea current
深海流速仪 deep-sea velocimeter;deep-sea velocity meter
深海锚泊 deep-sea anchor dredge
深海锚泊地 deep-sea berth
深海锰结核 abyssal Mn nodule
深海模拟 deep-sea simulation
深海能源 deep ocean power source
深海泥 deep-sea mud
深海黏[粘]土 abysmal clay
深海抛锚绞车 deep-sea anchoring winch
深海盆地 abyssal basin;deep-sea basin
深海平原 abyssal plain;deep-sea plain
深海平原沉积 abyssal plain deposit
深海平原名称 name of abyssal plain
深海铅测锤 deep-sea lead
深海潜水 deep-sea diving
深海潜水船 deep submergence search vessel
深海潜水服 deep-sea diving dress
深海潜水工具 deep-sea diving outfit
深海潜水呼吸用的混合气体 <成分为氢、氦和氧> hydreliox
深海潜水帽 hard hat
深海潜水器 bathyscaph(e);bathyvessel
深海潜水球 bathysphere
深海潜水设备 deep-sea diving outfit
深海潜水员 deep-sea diver
深海浅层区 epipelagic region
深海丘陵 abyssal hill
深海球形潜水器 <能载人和仪器供研究深海动物的> bathysphere;benthoscope
深海球形摄影仪 benthograph;benthoscope
深海区 <大陆架以外,海面以下2000~6000米> abyssal zone;abysmal area;abyssal area;abyssal region;deep-sea zone;oceanic abyss;off sounding;pelagic(al) realm;profundal zone
深海取(海水)样 deep-casting
深海取芯钻进 deep-sea core drilling
深海群落 pontium
深海软泥 abyssal ooze;abyssoplite;deep dean ooze;deep ocean ooze;deep-sea ooze
深海散射层 deep scattering layer
深海沙丘 abyssal dune
深海砂矿 deep lead
深海山口 abyssal gap
深海扇 abyssal fan;deep-sea fan;submarine fan;turbidite fan
深海扇沉积 abyssal fan deposit
深海扇地貌 geomorphology of abyssal fan
深海扇名称 name of abyssal fan
深海上层动物区系 epipelagic fauna
深海摄影 deep-sea photography
深海摄影机 camera for deep-sea;deep-sea camera
深海深处 abyssal deeps
深海深度 abyssal depth
深海深水温度自记仪 deep-sea bathythermograph
深海生态学 abyssopelagic ecology;deep-sea ecology
深海生物 bathypelagic organism
深海生物散居层 deep scattering layer
深海生物相 bathymetric(al) biofacies
深海声波扩散层 deep sound scattering layer
深海声场 deep acoustic(al) field;deep sound field
深海声道 deep ocean channel;deep sound channel;deep sound duct
深海声道现象 deep soundpath phenomenon
深海声呐 deep depth sonar;deep-water sonar
深海疏浚 deep-sea dredging
深海水道 <海底峡谷> deep-sea channel
深海水力挖矿机 deep-sea hydraulic dredge(r)
深海水深温度自记仪 deep-sea bathythermograph
深海水域的 abyssal pelagic;abyssopelagic
深海碎屑沉积 deeper-marine clastic deposit
深海台地 deep-sea terrace
深海探测船 deep-sea detection ship;deep-sea surveying ship
深海探测器 bathyscaph(e);deep ocean survey vehicle
深海通道 abyssal gap
深海拖斗采矿船 deep-sea drag dredge(r)
深海拖缆系统 deep-towed system
深海拖网 deep-water trawl;pelagic(al) trawl
深海拖网绞车 deep-sea trawl winch
深海拖曳磁力仪 deep-towed magnetometer
深海温度 abyssal temperature
深海温度计 abyssal thermometer;deep-sea thermometer;sounding thermometer
深海温度记录图 bathythermogram
深海系泊 deep ocean mooring;deep-sea mooring
深海系留浮标 deep-sea moored buoy
深海系留仪器站 deep-moored instrument station
深海系统 deep-sea system
深海峡谷 deep-sea channel
深海狭缝 abyssal gap
深海下潜试验 deep-sea submergence test
深海相 abysmal facies;abyssal facies;abyssal phase;deep-sea facies
深海岩石 abysmal rock;abyssal rock;thalassic rock
深海岩芯 deep-sea core
深海盐度计 bathysalinometer
深海洋测深导航系统 bathymetric(al) navigation system
深海遥测设备 deep-sea robot
深海异重流盆地 alee basin
深海淤泥 abyssal ooze;radiolarian ooze
深海鱼类 abyssal fishes;abyssal pelagic fishes;deep-sea fishes
深海鱼族馆 oceanarium
深海渔业 deep-sea fishery
深海藻类 abyssal algae
深海照相机 camera of deep-sea;deep-sea camera
深海植物群丛 abyssal association
深海植物群落 bathyphytia
深海种群 abyssal population
深海锥 abyssal cone;deep-sea cone;submarine cone;submarine fan
深海浊积岩圈闭 deep-turbidite trap
深海资源 deep-sea resources
深海自记温度计 deep-sea bathythermometer
深海自记温度仪 deep-sea bathythermograph
深海自养浮游生物 autopelagic plankton
深海组合导航系统 integrated deep-water navigation system
深海钻机 deep-sea drill of deep-sea
深海钻探计划 deep-sea drilling project;deep sea drilling programme
深海作业 deep-sea operation;deep-sea working
深海作业船 deep ocean work boat;deep-sea work boat
深焊 deep welding
深河槽 deep channel;deep trench
深河谷 coulee

深河口 accul
深褐黑色 deep brown-black
深褐红色 deep brown-red
深褐黄色 deep brown-yellow
深褐灰色 deep brown-grey
深褐蓝色 deep brown-blue
深褐绿色 deep brown-green
深褐色 auburn; burnt sienna; chocolate; deep brown; mocha; sable; Vandyke brown
深褐色的 bistre[bister]
深褐色蓝图 sepia print
深褐色颜料 bistre[bister]
深褐紫色 deep brown-violet
深壑 clove
深黑的 jet black
深黑漆 black japan; Japanese lacquer
深黑色 aterrimus; blue black; jet black; pitch black
深黑色的 atrous; dark-colo(u)red; sable
深痕 gash
深衡薄板 deep sheet
深红 crimson; ponceau
深红的 cardinal; crimson; poppy
深红褐色 deep red-drown
深红黄色 deep red-yellow
深红灰色 deep red-grey
深红硫锑矿 dark red silver ore
深红色 cardinal red; cherry wine; crimson; garnet; poppy; royal pink; scarlet red; shrimp pink; wine
深红色布 cramoisy
深红色层 magenta layer
深红色的 cardinal(red); laky; scarlet; deep red
深红色的精制紫胶 garnet lac
深红色花岗岩<美国康涅狄格州产> Stony Greek granite
深红色滤光镜 deep red filter
深红色炻器 Bottger ware
深红色颜料 lake; red lake pigment
深红眼镜玻璃 dark red spectacle glass
深红银矿 dark red silver ore; dark ruby ore; pyrargyrite
深红赭色 almagra
深红紫色 deep red-violet
深泓抛泥 thalweg disposal
深泓线 axial channel; axis of channel; channel line; line of maximum depth; thalweg; valley line
深泓线冲刷 thalweg erosion
深厚反气旋 deep anticyclone
深厚(暖)高压 deep anticyclone
深厚水平沉积盆地 geobasin
深厚信风带 deep trades
深湖 deep lake
深湖沉积 deep lake deposit
深湖底 bathile
深湖底的<25米以下> bathile; euprofundal
深湖区 abime; abysm; abyss
深湖相 deep lake facies
深化 deepen
深环裂 deep shakes
深黄褐色 deep yellow-brown
深黄褐色硬木<产于中美洲> guambo
深黄红色 deep yellow-red
深黄灰色 deep yellow-grey
深黄绿色 deep yellow-green
深黄色 deep yellow
深黄色滤光片 dark yellow filter
深黄铀矿 becquerelite
深灰白色 deep grey-white
深灰褐色 deep grey-brown
深灰黑色 deep grey-black
深灰红色 deep grey-red
深灰黄色 deep grey-yellow
深灰蓝色 deep grey-blue
深灰绿色 deep grey-green
深灰色 dark-grey[gray]; dark-smoke; deep grey[gray]
深灰色泥灰 slaty marl
深灰紫色 deep grey-violet
深基 deep basement
深基础 deep foundation
深基础墩 deep pier
深基础码头 deep foundation pier
深基础桥墩 deep foundation pier
深基础施工法 deep foundation method
深基础柱 deep pier
深基础桩 deep pile
深基坑 deep building pit; deep excavation
深及踝部的泥泞 ankle-deep mud
深集水坑 deep sump
深挤压 cupping
深加工玻璃 deep-processed glass
深检修孔 deep manhole
深涧 clough
深胶 dark factice
深结扩散工艺 depth-diffusion process
深截式采煤机 deep web shearer
深金色 deep gold
深进水口 low-level intake
深浸式水翼装置 fully submerged hydrofoil system
深井 artesian well; borehole well; deep bore; deep shaft; draw well; deep well
深井泵 deep-well pumping unit; borehole pump; borehole turbine pump; deep-well working barrel; drowned pump; subsurface pump; sump pump; tube well pump; deep-well pump; Abyssinian pump; artesian well pump
深井泵房 deep-well pump house
深井泵光杆头平衡重量 polished rod head counterweight
深井泵光拉杆钩环 polish rod clamp
深井泵疏干法 deep-well pumping
深井泵吸水管 drop pipe
深井抽水 deep-well pumping
深井抽水机 deep-well pump
深井抽水机组 deep-well pumping unit
深井抽水影响面积 area of influence of well
深井处理 deep-well disposal; deep-well treatment
深井处曝气 deep aeration
深井处置 deep-well disposal
深井地震计 deep hole seismometer
深井地震仪 deep hole seismometer
深井点 deep-well point
深井点排水 deep well point drainage
深井吊梯 deep-well elevator
深井法 deep-well method
深井废水处理 deep-well wastewater treatment
深井废物处置 deep-well waste disposal
深井灌注 deep-well injection
深井回灌水 deep-well backwater
深井活塞泵 deep-well piston pump; deep-well plunger pump
深井基础的管片衬砌 segmental lining for deep shaft foundation
深井井管 drop pipe; drop pipe line
深井开采 deep mining
深井勘探 deep prospecting
深井离心泵 centrifugal deep-well pump
深井流量计测量 spinning survey
深井炉 soaking pit crane
深井滤水管 deep-well filter; screen of deep well
深井排水 deep-well dewatering; deep-well drainage; deep-well pumping
深井排水法 deep-well drainage method
深井排水系统 deep well drainage system
深井喷射泵 deep-well jet pump
深井喷水泵 deep-well jet pump
深井曝气 deep shaft aeration; deep-well aeration
深井曝气池 deep-well aeration tank
深井曝气法 deep-well aeration method
深井曝气活性污泥法 deep-well aeration activated sludge process
深井曝气生化法 deep-well aeration biochemical process
深井潜水泵 submersible deep well pump
深井取样器 deep-well sampling pump
深井升降梯 deep-well elevator
深井式泵站 well-sump type pumping plant
深井式抽水站 well-sump type pumping plant
深井式基础 deep shaft foundation
深井式进人孔 deep manhole
深井式人孔 deep manhole
深井式水泵 deep-well type pump
深井竖管 drop pipe line
深井双冲程水泵 double stroke deep well pump
深井水 deep phreatic water; deep-well water; phreatic water; water in deep layer
深井水泵 deep-well water pump
深井水泵排水 drainage by pumping from deep wells
深井水轮泵 deep-well turbine pump
深井水轮机 deep-well turbine
深井水位量测仪 deep-well water level measuring equipment
深井水污染 pollution of deep well water
深井台阵 deep-well array
深井提升 deep winding
深井提升机 deep-well elevator
深井透平泵 deep-well turbine pump
深井外施电流阴极保护系统 deep-well impressed-current cathodic protection system
深井涡轮泵 deep-well turbine pump
深井污水泵 deep-pit sewage pump
深井污水处理 deep shaft sewage treatment
深井系统 deep shaft system; deep-well system
深井油井水泥 deep-well oil well cement
深井照相机 deep-well camera
深井注入法 well injection
深井注水 deep-well injection
深井柱塞泵 deep-well plunger pump
深井钻进 deep-well drilling
深橘黄色的 luteous
深锯 deep cut(ting); deep-sawing
深掘松土器 chisel-tiller
深浚 deep dredging
深喀斯特 deep karst
深开挖 deep cut(ting); heavy cut; deep excavation; deep digging
深砍 slash; slash-cut
深壳的 bathydermal
深刻 deepening; profound
深刻槽 deep cut(ting)
深刻蚀 deep cut(ting)
深刻影响 profound effect
深坑 chasm; deep concavities; gulf; deep pit
深坑厕所 deep-pit latrine
深坑笼养鸡舍 deep-pit cage house
深坑曝气法 deep shaft aeration process
深空 deep space
深空跟踪网 deep space network
深空激光跟踪系统 deep space laser tracking system
深空激光通信[讯] deep space laser communication
深空探测设备 deep space instrumentation facility
深空探测系统 deep space instrumentation system
深孔 deep bore; long borehole
深孔爆破 deep hole blasting; long hole blasting; muffling; well-hole blasting
深孔爆破法 long hole method
深孔爆破技术 long hole blasting technique
深孔崩矿回采 blast-hole stoping
深孔崩落开采 blast-hole mining
深孔崩落开采法 blast-hole method; long blast hole work
深孔锤击式钻岩机 down-the-hole-hammer rig
深孔地震测探 deep hole seismic detection
深孔地震计 deep hole seismometer
深孔地震探测 deep hole seismic detection
深孔地震仪 deep hole seismometer
深孔伽马测量 deep hole gamma ray survey
深孔灌浆 deep hole grouting; long hole grouting
深孔开挖法 deep hole excavating method
深孔勘探 deep hole prospecting
深孔拉刀 broach for deep hole
深孔留矿法 long hole shrinkage method
深孔木钻 sash bit
深孔能谱测量 deep hole spectrum survey
深孔汽车式钻机 deep hole wagon drill
深孔扇形闸门 reservoir bottom sector gate
深孔石块 deep hole block
深孔镗床 deep hole borer
深孔梯段采矿法 long hole benching
深孔梯段回采法 long hole benching
深孔梯段掘进法 long hole benching
深孔凿 digging chisel
深孔凿岩 deep boring; deep drilling; deep hole(rock) drilling; long blast drilling; long hole blasting; long hole drilling
深孔凿岩工程 deep-drilling engineering
深孔凿岩机 blast hole drill
深孔凿岩设备 deep-drilling equipment
深孔炸弹 deep penetration bomb
深孔注水 deep hole infusion
深孔砖 vertical coring brick
深孔装药器 blast-hole charger
深孔钻 boring drill; gun drill; long-eye auger
深孔钻床 deep hole drilling machine; gun drilling machine
深孔钻刀头 solid boring head
深孔钻杆 solid boring bar
深孔钻机 deep hole drill; depth drill; deep-drilling rig; deep hole drilling machine; long hole drilling machine; long hole machine; longholing machine; downhole drill
深孔钻进 deep boring; long hole drilling; long holing; downhole drilling
深孔钻探 deep hole boring; deep hole drilling; deep hole prospecting
深孔钻探工程 deep-drilling engineering
深孔钻探设备 deep-drilling equipment
深孔钻头 deep hole drill; downhole bit; downhole drill

深孔钻凿 long hole drilling
深孔作业 deep hole work
深口<铸造表面缺陷> roke
深扣螺纹 recessed thread
深宽比 depth ratio;depth(-to)-width ratio;form ratio<河流的>
深拉 cupping;deep-drawn
深拉薄板 deep drawing sheet
深拉钢 deep-drawing steel
深拉模 press-through die
深拉伸 deep drawing
深拉试验 cup-drawing test
深拉试验机 cupping testing machine
深拉系数 cupping value
深拉压力机 cupping machine; cupping press
深拉延 deep drawing
深蓝 concentrated blue; ultramarine blue
深蓝褐色 deep blue-drown
深蓝黑色 blue black;deep blue-black
深蓝灰色 deep blue-grey
深蓝绿色 deep blue-green; green ultramarine
深蓝色 Berlin blue;blue black;bronze blue; dark blue; mazarine blue; navy(blue);Prussian blue;solid-blue
深蓝色料 mazarine blue
深蓝色石英 sapphire quartz
深蓝色颜料 bronze blue pigment; royal blue pigment
深蓝色鱼类 blue fish
深蓝釉 deep blue glaze
深蓝紫色 deep blue-violet
深肋T形梁板 deep-ribbed T-beam slab
深肋骨架 deep frame[framing]
深肋骨架可动颚板破碎机 deep frame swing jaw crusher
深冷 deep cooling
深冷泵 cryogenic pump
深冷超导体 cryogenic superconductor
深冷抽气面 cryoplate
深冷抽吸 cryopump
深冷的 cryogenic
深冷分离法 separation by deep refrigeration
深冷粉碎 cryogenic grinding; cryopulverization
深冷粉碎机 cryopulverizer
深冷干燥 cryodrying
深冷恒温器 cryostat
深冷化学 cryochemistry
深冷技术 cryogenics
深冷流体泵 cryogenic fluid pump
深冷面 cryogenic surface
深冷水等温层 hypolimnetic layer; hypolimnion
深冷水等温层曝气 hypolimnetic aeration;hypolimnion aeration
深冷水等温层区 hypolimnetic region;hypolimnion region
深冷水等温层缺氧 hypolimnetic anoxia
深冷水等温层缺氧量 hypolimnetic oxygen deficit;hypolimnetic oxygen depletion
深冷水等温层溶解氧消耗率 hypolimnetic dissolved oxygen consumption rate
深冷水等温层水流 hypolimnetic current
深冷水等温层水体 hypolimnetic waters
深冷水等温层需氧量 hypolimnetic oxygen demand
深冷温度 cryogenic temperature
深冷温度计 cryometer;frigorimeter
深冷吸附 cryosorption
深冷制冷器 cryogenerator
深冷致冷器 cryogenerator
深冷装置(系统)cryogenic system
深栗色 dark maroon
深梁 beam with high depth span ratio; deep bead;deep beam;wall beam
深梁矩柱系统 spandrel-wall girder-short column system
深梁弯曲 bending of deep beam
深裂 drastic crack
深裂的 cleft
深裂木材 cleft wood
深裂纹<深1/8英寸以下,1英寸=0.0254米> heavy torn grain
深流褶皱 rheomorphic fold
深流作用【地】rheomorphism
深六角螺母 deep hexagonal nut
深滤器 deep filter
深路堑 deep cut(ting);heavy cut
深绿褐色 deep green-brown
深绿黑色 deep green-black
深绿黄色 deep green-yellow
深绿灰色 deep green-grey
深绿辉石 fassaite
深绿磷灰石 carbonate-fluorapafite; staffelite
深绿色 bottle green;dark-green;deep green;invisible green;lime green
深绿色厚玻璃瓶 junk bottle
深绿颜料 deep green pigment
深绿玉髓 plasma
深轮辋 deep-well rim
深螺帽 deep nut
深螺纹 coarse thread
深螺纹喂料段 deep feed flight section
深埋 deep depth;deep embedment
深埋的 deeply buried; deeplying; deep-seated
深埋地下结构 deep underground structure
深埋钢管标 deep buried steel-pipe benchmark
深埋基础 deeply embedded foundation;deep(lying) foundation
深埋类型 deep-seated type
深埋裂纹 immerged crack
深埋双金属标【测】deep buried bi-metal benchmark
深埋隧道 deep-buried tunnel; deep-seated tunnel; deep tunnel; tunnel in deep length
深埋隧洞 deep-seated openings
深埋作用 deep burialism
深玫瑰红色 old rose
深密封 deep seal
深磨 deep cut(ting)
深墨色 raven
深谋远虑的筹划 design with proper forethought
深内陆海 deep inland sea
深奶色石<英国威尔特郡出产的> Hartham Park
深奶油色 deep cream
深能级 deep energy level;deep level
深能态 deep state
深碾裂缝 heavy roller check
深浓色的 lurid
深排水边沟 deep side drainage
深排水口 low-level outlet
深盘 deep plate
深刨法 deep digging
深炮眼 deep borehole;long hole
深炮眼崩矿法 deep hole method
深炮眼凿岩机 deep hole drill
深盆地热水垂直运移模式 migration model of thermal water upward from deep basin
深盆气圈闭 deep basin gas trap
深皮型重力构造 bathydermal type of gravitative tectonic
深撇泡沫 deep scraping
深破裂带 deep fracture zone
深曝气池 deep aeration tank
深漆木纹的工具 overgrainer
深漆木纹刷 overgrinding brush
深气升浮选机 deep air cell
深潜救生船 deep submergence rescue vessel
深潜救生艇 deep submergence rescue vehicle
深潜流洞 deep phreatic cave
深潜模拟器 deep-diving simulator
深潜器 barn yard;deep-diving submersible; deep-diving vehicle; hydrospace vehicle;hyperbaric vehicle
深潜潜水船 deep-diving vessel
深潜潜水器 deep-operating vehicle
深潜实验 deep-diving experiment
深潜试验船 deep-diving trial ship
深潜水器 bathyscaph(e)
深潜水位 deep-water table
深潜搜索器 deep submergence search vehicle
深潜系统 deep-diving system
深潜照相系统 deep submersible photographic(al) system
深潜装备 deep-diving device; deep-diving unit
深浅 dark-and-light
深浅计 densi(to)meter
深浅交映的油漆加工 ombré finish
深浅调节导管 depth control pipeline
深堑 deep cut(ting)
深堑侧壁<开挖的> high wall
深嵌的 deep-seated
深嵌类型 deep-seated type
深腔零件 deep-recessed part
深切 heavy cutting;incision;overdeepening
深切槽 deep slotting
深切的河流 entrenched stream
深切地区 enclosed country
深切沟 barranca
深切河(道)incised river;intrenched river
深切河谷 entrenched valley
深切河谷曲流 ingrown valley meander
深切河流 entrenched river;entrenched stream;incised river;incised stream; intrenched river;intrenched stream
深切河曲 entrenched meander;incised meander;inherited meander
深切河湾 entrenched meander;incised meander;intrenched meander
深切曲流 incised meander;entrenched meander;inclosed meander
深切融沟河(冻土区)beaded stream
深切弯曲河段 entrenched meander
深切削 deep cut(ting)
深切作用 incision
深侵蚀 deep-etching
深侵蚀河床 deeply eroded river bed
深青岩 greenstone
深倾 high oblique
深倾航测像片 high oblique
深倾航空摄影 high oblique aerial photography
深倾航空摄影测量 high oblique air survey
深倾交向摄影像片 high oblique convergent photograph
深倾摄影 high oblique photography
深倾斜航空摄影测量 high oblique air survey
深全息术 deep holography
深泉 deep(-seated) spring
深且狭窄的河谷 water gap
深染 engrain
深染的衬里纸 engrain lining paper
深染的墙纸 engrain wallpaper
深染的墙纸涂层 engrain wallpaper coat
深热源补给 abyssal heat recharge
深人孔 well hole
深熔暗包体 mianthite
深熔池 deep molten bath
深熔焊 deep penetration weld(ing)
深熔焊接 deep penetration welding; penetration fusion welding; penetration welding
深熔焊条 deep penetration electrode; deep penetration welding electrode
深熔花岗岩 anatectic granite
深熔混合岩化 anamigmatization
深熔角焊 deep fillet weld(ing)
深熔流动区 plutonic melting flow region
深熔岩浆 anatectic magma
深熔作用【地】anatexis
深熔作用方式 anatexis way
深入的 in depth;ingoing
深入调查 in-depth investigation; intensive investigation
深入发展 developing in depth
深入分析 in-depth analysis
深入干预 profound intervention
深入火场灭火战斗 entry fire fighting
深入检查 in-depth inspection
深入农村的小路 rural penetration
深入围岩的矿脉 extension of ore into wall
深入细微的 intensive
深入研究 examined in depth;in-depth study
深入振实法 deep compaction
深散射层 deep scattering layer;false bottom
深色 dark colo(u)r;deep colo(u)r; heavy-colo(u)r;high colo(u)r
深色安山岩 melanoandesite
深色斑 heavy stain
深色斑边材 heavy-stained sapwood
深色的 dark-colo(u)red;fuscous
深色钉砖 flare header
深色花岗岩<产于美国新罕布什尔州的> bear's den gray
深色火成岩 traprock
深色矿石 mafic
深色毛 black waxy soil
深色炻器 obsidian ware
深色土壤 dark soil
深色团 bathochrome
深色岩 melanocratic rock
深色油 dark oil
深山谷 barranca
深山糠草 bentgrass
深深感动的 piercing
深渗碳 deep cementing
深渗碳处理 jackmanizing
深渗(透)损失 deep percolation loss
深生的 hypogene;hypogenic
深时曲线 depth-duration curve
深蚀膏 deep-etching paste
深蚀糊 deep-etching paste
深蚀刻法 deep-etching
深蚀试验<用酸类深蚀试件,以便发现疵病> deep etch test
深式进水口 deep(water) intake;low(er) level intake
深式泄水孔 deep outlet; low-level outlet
深视力<目测前方对象之距离所需的视力> measuring vision
深试坑 deep trial pit
深室浮选机 deep-cell machine
深受气候影响的 much-weathered
深鼠笼 buried cage
深鼠笼条 deep bar
深鼠笼转子 deep-cage-bar rotor

S

深衰弱 deep fade
深水 deep water;bottom water
深水岸壁 deep-water quay
深水白头浪 comber
深水波 deep-sea wave; deep-water wave;deep wave
深水波长 wavelength in deep water
深水波高 deep-water wave height; wave height in deep water
深水波功率 deep-water power
深水波浪计 deep-sea wave meter; deep-sea wave recorder
深水波浪衰减 wave decay for deep-water area
深水波群速度 wave group velocity in deep water
深水波与浅水之间的波浪 intermediate wave
深水驳船 deep-draft barge
深水泊位 deep-sea berth;deep-water berth
深水采样器 deep-sea sampler;deep-water sampler
深水舱 deep tank
深水草本相 deep-water herbaceous facies
深水测量浮子 <一种浮子以电杆或竹竿等组成,用来测量水流的深水部分流动情况> logship
深水测深 deep-sounding
深水测深锤 deep-sea lead; deep-sounding lead;deep-water lead
深水测深器 deep-sounding apparatus
深水测深仪 deep-sounding apparatus
深水测听器 deep-water hydrophone
深水层 deep strata;deep-water layer; profund(al)zone
深水沉积物 pelagic(al)deposit; pelagic(al)sediment
深水沉积作用 pelagic(al)sedimentation
深水沉箱 deep-water caisson
深水承台 deep-water platform
深水池 deep pool
深水处 deep water place
深水船 deep-draft vessel
深水船舶 deep-draft vessel
深水船推荐航道 recommended track for deep draught vessels
深水带 aphytal; deep-water zone; profundal zone
深水的 abyssal;deep-draft;deep water
深水(底)冰 depth ice
深水底刺网 stab net
深水底栖生物 benthos
深水电缆 deep-sea cable
深水电视录像装置 deep telerecording unit
深水电子多普勒导航仪 NAVTRAK Doppler navigator
深水动物区系 deep-water fauna
深水墩 deep pier
深水阀压盖 tube gland
深水反转温度计 deep-sea reversing thermometer
深水防波堤 breakwater in deep water
深水放射测定器 deep underwater nuclear counter
深水封 deep seal
深水浮标 ball-and-line float; canister float;can(n)ister float;deep-water float;depth float;double float;loaded float
深水浮游生物 skotoplankton
深水浮子 depth float; loaded float; subsurface float
深水腐泥土 deep-water muck
深水复式采样器 deep-sea multiple sampler
深水港 deep-draft harbo(u)r;deep-draft port;deep-sea port;deep-water harbo(u)r;deep-water port; deep-sea harbo(u)r
深水港池 deep-water basin;deep-water slip
深水港坞 deep-water dock
深水光度计 bathyphotometer
深水柜 deep tank
深水过滤器 deep filter
深水海底 low bottom
深水海流计 deep-sea velocimeter;deep-water velocimeter
深水航槽 deep-water channel
深水航道 deep(-draft)channel;deep-draft navigation channel;deep-draft waterway;deep-water channel;deep-water fairway;deep-water slip;deep-water route
深水航线 deep-draft route;deep-water route
深水航运 deep-draft navigation
深水河槽 deep-draft channel
深水河道 ship canal; deep draught; deep draft waterway
深水湖泊 deep lake
深水湖盆 aphytal
深水湖区 bathyal region of lake
深水护面石 sub-surface stone
深水环流 deep-water circulation
深水换能器 deep-water transducer
深水混凝土结构 concrete deep-water structure;condeep
深水货轮 deep-draft vessel
深水基础 deep-water foundation
深水检波器 deep-water hydrophone
深水绞车 deep-sea winch
深水井 deep well
深水流速计 deep-sea velocimeter;deep-water velocimeter
深水码头 deep-water pier;deep-water quay;deep-water terminal;deep-water wharf;sea terminal
深水锚(泊)设备 deep-water moorings
深水锚地 deep-water berth;low bottom
深水盆地 deep basin
深水贫营养湖 deep oligotrophic lake
深水平台 deep-water platform
深水曝气 deep aeration tank; depth aeration
深水曝气活性污泥法 deep-water aeration activated sludge process
深水潜航器驾驶员 hydronaut
深水潜水 deep diving
深水潜水服 deep-diving suit
深水潜水衣 acqualung
深水潜水用呼吸剂 heliox
深水潜水员 deep diver
深水潜水钟 deep-water bell
深水桥墩 deep-water pier
深水情况 deep-water case
深水球形摄影仪 benthograph; benthoscope
深水区 abyssal region;deep-water area;deep-water zone;profundal zone
深水取样 deep-water sampling
深水取样器 deep-water sampler
深水泉 bahr[复 bahar]
深水入口 deep-water entrance
深水三角洲 deep-water delta
深水散射层 deep scattering layer
深水声呐 deep depth sonar
深水试验计划 deep-water test program(me)
深水水道 deep-draft channel; deep-draft water course;deep-draft waterway; deep-draught route; deep-draught waterway
深水水深 deep-water depth
深水水听器 deep-water hydrophone
深水探测 deep quest
深水探测器 deep sonde; deep-water hydrophone
深水探测仪 hydroscope
深水探视仪 hydroscope
深水探听器 deep-water hydrophone
深水通航工程 deep-draft project
深水砣 dipsea lead
深水湾 deep bay
深水望远镜的 hydroscopic
深水围堰 deep-water cofferdam
深水位条件 deep-water-table condition
深水温度计 bathythemometer; deep-water thermometer
深水无定形相 deep-water amorphous facies
深水习性 floating habit
深水系泊 deep-water mooring
深水系泊浮筒系统 deep-sea moor buoy array
深水系泊浮筒阵列 deep-sea moor buoy array
深水峡谷 deep-sea canyon
深水线 line of maximum depth
深水型冷却池 deep cooling basin; deep cooling pond; deep cooling pool
深水洋流 deep-sea current
深水油库 deep-water oil terminal
深水淤泥质波痕 abyssal mud ripple
深水域 deep-water area; deep-water zone
深水运河 ship canal
深水藻类 abyssal algae
深水炸弹 depth bomb; depth charge; diving torpedo
深水炸药包 depth charge
深水种类 deep-water species
深水驻波 deep water clapotis
深水桩支承围栏 <沉箱下沉时就位用> corral
深水浊流沉积 deep-water turbidity deposit
深水自记海流计 depth-current meter;depth-current recorder
深水自记流速仪 depth-current meter;depth-current recorder
深水钻井 deep-water drilling
深水钻井作业 deep-water drilling operation
深思熟虑的 well-advised
深松法 deep digging
深松土 deep digging
深松土器 deep digger
深松土中耕机 deep cultivator
深邃光泽 deep gloss
深潭 deep pool;weel
深潭泉水 pool spring
深探测电阻率测井图 deep investigation resistivity log plot
深探测感应测井图 deep investigation induction log plot
深掏槽 heavy cut
深填方 deep fill
深填土 deep fill
深同化作用【地】 abyssal assimilation
深头套筒 deep socket wrench
深挖 deep cut(ting); deepening; heavy cut
深挖方 heavy cut
深挖工作 heavy cutting
深挖沟 heavy cut
深挖掘 heavy cut
深挖坑 deep building pit
深挖取样 deep sampling
深挖式挖泥船 deep-digging dredge(r)
深挖土方 deep cut;high cut
深挖土机 deep digger
深挖削 deep breaking up
深挖型矿井 dig-down pit;sunken pit
深挖作业 heavy cutting
深外层空间 deep space
深弯 accul;deep bend
深弯度 deep camber
深弯月形 deep meniscus
深位的 deep-seated
深位火 deep-seated fire
深位汽泡 deep-seated blowhole
深温计 bathythermograph
深温图 bathythermogram
深温仪 bathythermograph
深纹饰混凝土 deep model(l)ed concrete
深吸水井 deep suction well
深吸水器 deep suction apparatus
深溪 dingle
深峡 barranca;clove
深峡谷 kloof
深峡海湾 geo;gio
深陷的 deep-set
深陷阱 deep trap
深相 depth facies
深向冲刷 deep erosion
深向侵蚀 deep erosion
深向渗流 deep percolation;deep seepage
深向渗滤 deep percolation;deep seepage
深向渗水 deep percolation;deep seepage
深向渗透 deep percolation;deep seepage
深型充气浮选槽 deep air cell
深休眠 deep dormancy
深压 deep drawing
深压模 deep-drawing die
深压载水舱 deep-water ballast tank
深延伸 deep drawing
深岩沟 bogaz
深岩浆 hypomagma
深眼爆破法 long hole method
深眼嘴子 long hole bit
深眼木钻 long-eye auger
深眼钻 long hole bit;long hole drill
深眼钻机 well borer
深腰船 deep waisted vessel
深夜不睡 stay up
深夜服务 late-night service
深夜噪声 midnight noise
深阴影 deep shadow
深印度红颜料 deep Indian red pigment
深油槽 deep sump
深油井水泥 deep-drilling cement
深渊 abime;abysm;abyss;chasm;deep pool;deep-sea basin
深渊带 abyssal pelagic zone
深渊的 abysmal;abyssopelagic
深渊底 abyssal depth;hadal
深渊底栖带 abyssal benthic zone
深渊底栖动物(区系) abyssal benthic fauna
深源 deep focus
深源的【地】 anatectic
深源地震 anatectic earthquake;bathyseism;deep-focus earthquake;palintectic earthquake;plutonic earthquake; deep-focus shake
深源地震面 deep-focus earthquake plane
深源喷气孔 deep-seated fumarole
深源气 deep source gas
深源岩 typhonic rock
深源震 plutonic earthquake
深源蒸汽 deep-seated steam
深远的 far-reaching; profound and lasting

深远的历史意义 far-reaching historical significance
深远的意义 profound significance
深远影响 far- profound impact; far-reaching impact
深沼草本群落 terrificientinerbosa
深沼泽 malezales
深照型漫射 narrow angle diffusion
深照型照明器 narrow angle lighting fittings
深罩型灯 deep shade type
深震 bathyseism; deep(-focus) earthquake; plutonic earthquake
深震带 deep seismic zone
深震源 deep-focus of earthquake
深中心 deep center[centre]
深朱红 scarlet vermilion
深注水井 deep injection well
深锥形浓缩池 deep-cone thickener
深紫褐色 deep violet-brown
深紫黑色 deep violet-black
深紫红色 deep violet-red; imperial purple; mulberry; petunia; prune; royal purple
深紫灰色 deep violet-grey
深紫蓝色 deep violet-blue
深紫色 amaranth; deep violet; grape; modena
深棕色 Cassel brown; chocolate-brown; nigger-brown
深棕色的 nut-brown
深棕色调色法 sepia toning
深棕色蓝图 sepia print
深棕色图纸 sepia
深棕色照片 sepia
深钻 deep boring; deep drilling
深钻孔 deep borehole; deep drill hole; gun drilling
深钻设备 deep-drilling equipment

神道神殿 <日本> Shinto shrine

神殿 shrine; prang <13 ~ 18 世纪泰国建筑中的>
神殿入口 monumental gateway
神羔像 <象征基督的一种标志> agnus dei
神户港【日本】Kobe Port
神经外科 neurosurgery
神经网络 <计算机中被视为类似神经细胞的网络，它增强了识别与吸收信息的能力> neural network; neuro
神经网络模拟 neuron network simulation
神经网络模型 neural network model
神经系统 nervous system
神经中枢 nerve center
神龛 aedicule; feretory; sacrarium; shrine
神龛顶部装饰 <礼拜堂> tabernacle work
神龛建筑 aedicular architecture
神龛式门窗 aedicule
神龛塔墩 <印度> stupa mound
神龛塔基 <印度> stupa-mound base
神龛塔神座 <印度> stupa shrine
神秘的事物 mystery
神秘性 mystique
神庙区 <古希腊> temenos
神社牌坊 <日本> torii
神圣的地区 sacred precinct
神圣的街道 sacred street
神圣的境界 sacred precinct
神圣的小树林 sacred grove of trees
神圣的岩石 Sacred Rock
神圣纪念碑 ecclesiastical monument
神圣祭坛 sacred alter
神圣十字架教堂 Holy Cross church
神使杖标志 caducei symbol
神坛 altar
神位圣舍 hieron
神仙葫芦 block and fall; block and tackle; block pulley; chain block; chain hoist; differential block; differential chain block; differential hoist; differential pulley purchase; differential tackle; duplex purchase; gun tackle; jenny; manual chain hoist; patent chain hoist; planetary hoist; pull lift; tackle
神仙鱼楔形器件 angelfish device
神像 cult image; cult statue; miraculous image
神学院 divinity school; seminary
神之羔羊饰 agnus dei

审查 auditing; examination; examine; investigation; review; scrutinization; scrutiny

审查报告 certificate of the auditor
审查程序 audit program(me)
审查程序设计 audit programming
审查的 examinatorial
审查范围 scope of examination
审查工程设计 engineering design review
审查进口项目 evaluating imported projects
审查经营方针 examining operational policy
审查评价文件 audit-review file
审查人 examinant; examiner; referee
审查提纲 check list
审查委员会 board of review; jury
审查文件 audit-review file
审查项目建议书 examine the project proposal
审查小组 review panel
审查账簿 book audit
审查账目 examination of account
审查者 investigator
审查追踪 audit trial
审订人 redactor
审定 approved; authorise[authorize]; check and decide; examine and approve
审定版本 authorized edition
审定的投标人名单 closed list of bidders
审定会计师 certified accountant
审定投标 examination of bids; screening of bids
审定投资项目 check investment item
审定图号 approved symbol
审定压力 authorized pressure
审稿 copy edit
审稿人 copy editor
审稿员 subeditor
审稿者 reader
审核 authorized by; examination and verification; examine and verify; examined and verified by
审核报告 audit report
审核跟踪 audit trial
审核通知书 advice of audit
审核投标 screening the bids
审核预算 examine and approve budget
审计 audit
审计报告 auditor's report; audit report; report on auditing and inspection
审计报告标准 reporting standard
审计部门 audit department
审计长 chief auditor; chief comptroller; comptroller
审计程序 audit procedure; audit program(me)
审计处 audit department
审计的统计估算法 statistic(al) estimation approach to audit
审计法 audit law; law of audit
审计范围 audit coverage
审计费 audit fee
审计跟踪 audit trial
审计官 commission of audit
审计合同 audit contract
审计机构 auditing body
审计机关 auditing offices
审计检查 examination of auditor
审计结果 findings of audit
审计结论 audit conclusion
审计解释部分 explanatory paragraph
审计局 Bureau of Audit
审计决算 audited accounts
审计客户表白书 letter of representation
审计客户律师表白书 legal representation letter
审计期 audit period
审计契约 audit engagement
审计权限 competence of auditor
审计人员 auditing officer
审计师 auditor; comptroller
审计手续 audit procedure
审计署审计长 auditor-general of auditing administration
审计委员会 auditor; board of audit-(ors); comptroller
审计小组 audit team
审计学 auditing
审计员 auditor; comptroller; auditing clerk
审计证明书 audit certificate; certificate of auditing
审计制度 audit system
审计主任 chief comptroller
审计准则声明 statement on audit standards
审校 proofing
审校参考图 correction copy; revised proof
审校符号 correction code; correction symbol
审校人 proofreader
审校样 examination copy
审校员 proofreader
审校纸 proof checking paper
审理 judge; trial
审理程序 inquisition procedure
审理货币条款 judgment currency clause
审理期限 duration of hearings
审理通知书 notice of hearing
审美的问题 aesthetic aspect
审美观 aesthetic standard
审美观念 aesthetic idea
审美距离 <用心理距离来解释审美现象的一种美学观点> aesthetic distance
审美考虑 aesthetic consideration
审美力 taste
审美能力 aesthetic judgment
审美意识 aesthetic consciousness; aesthetic sense
审判室 courtroom
审判员席 tribunal
审批 approval; examination and approval
审批单位 examination and approval authority
审批过程 approval process
审批机构 approval authority
审批阶段 validation phase
审批权 sign-off power
审批手续 approval procedure
审批意见 approving opinion
审片室 viewing room
审评小组 evaluation group
审慎的 deliberate
审慎的投资 prudent investment
审图 checking of drawings; check-up of drawings
审外局 controller's department
审讯室 judges room
审议 consideration; counsel; deliberation
审议次序 order of consideration
审议机构 deliberative organ
审音耦合矢量图 crosstalk coupling vector diagram
审阅中 under review
审制独立 judicial independence

肾硅锰矿 caryopilite

肾矿石 kidney(iron) ore
肾色油 kidney oil
肾铁矿 kidney ore
肾形石板 reniform slate
肾形线 nephroid
肾状构造 reniform structure
肾状矿石 reniform ore

甚长基线干涉仪 very long baseline interferometer

甚长线干涉测量 very long baseline interferometry
甚大规模集成电路 very large-scale integration
甚低密度采样 very low density sampling
甚低频 very low frequency
甚低频波 <频率 3 ~ 30 千赫兹，波长 100000 ~ 10000 米> very low frequency wave
甚低频电波传播 propagation of very low frequency
甚低频电磁仪 very low frequency electro-magnetic instrument
甚低频法测量剖面平面图 profile-plan figure of results very low frequency method
甚低频法测量剖面图 profile figure of results very low frequency method
甚低频高压试验 very low frequency high potential test
甚低频接收机 very low frequency receiver
甚低频人为干扰发射机 very low frequency jammer
甚低频视频信号 very low frequency video signal
甚低频通信[讯] very low frequency communication
甚短波 very short wave
甚高比特率数字用户线系统 very high bit rate digital subscriber line
甚高分辨率 very high resolution
甚高分辨率辐射计 very high resolution radiometer
甚高分辨率辐射仪 very high resolution radiometer
甚高分辨率扫描辐射计 very high resolution scan(ning) radiometer
甚高级语言【计】very-high-level language
甚高频 <30 ~ 300 兆赫兹> very high frequency
甚高频波 <频率 30 ~ 300 兆赫兹，波长 10 ~ 1 米> very high frequency wave
甚高频测向系统 very high frequency direction finder system
甚高频测向仪 very high frequency direction finder

甚高频传输 very high frequency transmission

甚高频传输线路 very high frequency link

甚高频带通滤波器 very high frequency bandpass filter

甚高频单波道接收机 very high frequency single channel receiver

甚高频单波道双向无线电设备 very high frequency single channel two-way radio equipment

甚高频导航系统 very high frequency navigation system

甚高频电波传播 propagation of very high frequency

甚高频电路 very high frequency circuit

甚高频电视接收机 very high frequency television receiver

甚高频电视接收设备 very high frequency television receiving equipment

甚高频电视频道选择器 very high frequency TV tuner

甚高频电视调谐机构 very high frequency tuning mechanism

甚高频电视中频部分 very high frequency television IF strip

甚高频电台 very high frequency(radio) station

甚高频定向 very high frequency direction finding

甚高频段 very high frequency band

甚高频发射机 very high frequency transmitter

甚高频固态电视 very high frequency solid-state TV

甚高频固态电视频道选择器 very high frequency solid-state TV tuner

甚高频率归航仪 very high frequency homing adapter

甚高频接收机 very high frequency receiver

甚高频脉冲控制的功率放大器 multitron

甚高频频道 very high frequency channel

甚高频频道选择器 very high frequency tuner

甚高频前置放大级 very high frequency preamplifier stage

甚高频全向无线电信标 very high frequency omnidirectional radio range

甚高频全向无线电指向标 very high frequency omnidirectional range

甚高频全向信标测距设备 very high frequency omnirange distance measuring equipment

甚高频全向指向标 very high frequency omnirange

甚高频人为干扰发射机 very high frequency jammer

甚高频数据线路干扰器 very high frequency data link jammer

甚高频天线阵 very high frequency antenna array

甚高频调幅收发两用机 very high frequency AM transceiver

甚高频调谐器 very high frequency tuner

甚高频通信[讯]站位置标记 very high frequency station location marker

甚高频无线电灯塔 very high frequency radio lighthouse

甚高频无线电电话设备 very high frequency radio telephone device; very high frequency radio telephone equipment

甚高频无线电话发射机 very high frequency radio telephony transmitter

甚高频无线电信标 very high frequency radio beacon

甚高频无线电指向标 very high frequency radio range

甚高频线性放大器 very high frequency linear amplifier

甚高频旋转式频道选择器 very high frequency turret tuner

甚高频与超高频测向器 very high frequency and ultra-high-frequency direction finder

甚高频振荡电路 oscillator circuit at very-high frequency

甚高频振荡器 very high frequency oscillator

甚高输出荧光灯 very-high-output fluorescent lamp

甚高速计算机 very high speed computer

甚高真空 very high vacuum

甚急闪 very quick flashing

甚疏浮冰群 very open pack ice

甚无差异类 thick indifference class

甚窄束接收器 very-narrow-beam receiver

胂凡纳明银盐 silver arsphenamine

胂基 arsino

渗斑 <混凝土表面的> efflorescence; white deposit

渗层 percolating bed

渗潮 transmission of humidity

渗潮性 moisture permeability

渗成脉 infiltration vein

渗出 bleeding; diffuse; distil(l); exfiltration; exude; ooze; oozing; outward seepage; perspiration; seep-off; spewing; sudation; sweat(ing); transude; weepage; seep

渗出的 bleeding; oozy; weeping

渗出沟处置法 seepage trench disposal

渗出沥青体 exsudatinite

渗出量 seepage discharge

渗出流 effluent seepage

渗出率 leaching rate

渗出面 seepage face; surface of seepage

渗出泉 filtration spring; weeping spring

渗出石油 seep oil

渗出试验 bleeding test

渗出水 rushing-out water; seep(age) water; seeping discharge; sweating; water seepage; outcrop water <地下洞井的>

渗出损失 seepage loss

渗出温度 exudating temperature

渗出物 diffusate; effusion; exudant; exudate; spew; ex(s)udation

渗出性指数 bleeding index

渗出液 diffusate; exudate; exudation; percolate; transudate

渗出液体 weepage

渗出晕 seepage halo

渗出作用 exudation; transudation

渗床 leaching bed; percolated bed; percolating bed; percolating filter; percolation bed

渗淡 qualify

渗氮 azotize; nitridation

渗氮表面 nitride(d) surface

渗氮层 nitration case; nitrided case; nitriding layer

渗氮的 nitrizing

渗氮钢 nitride steel; nitriding steel

渗氮工艺过程 nitriding process

渗氮合金 nitralloy; nitro-alloy

渗氮剂 nitriding medium

渗氮碳硬化处理 nitriding carbon case hardening

渗氮温度 nitriding temperature

渗氮硬化 nitride hardening; nitrogen hardening; nitriding

渗氮组织 nitriding structure

渗滴水 trickling water

渗豆 <铸造用语> sweat

渗毒率 leaching rate

渗钒 vanadinizing

渗铬 alphatizing; chromising[chromizing]; chromium impregnation; inchromizing

渗铬镀层 chromized coating

渗铬法 chromium implements

渗铬钢 chromized steel

渗沟 blind drain; leaching bed; leak ditch

渗管 leak pipe

渗灌 percolation irrigation; pitcher irrigation; subirrigation; subsurface irrigation

渗硅 ihrigizing; siliconising[siliconizing]

渗过 percolation

渗化 imbibition

渗化钢 calorized steel

渗浆纤维混凝土 slurry infiltrated fiber reinforced concrete

渗胶 bleed-through; strike-through

渗胶的 glue bleed through

渗截 seepage interception

渗金属 metallic cementation

渗金属法 metallic cementation process

渗金属塑料 metallized plastic

渗进作物内部 permeating the plant

渗井 cesspool; disposal well; dry well; negative well; percolation pit; percolation well

渗井排水 spray drain

渗径 flow path; leakage path; path of filtration; path of percolation; path of seepage; seepage path; seepage tracking

渗坑 leach pit; percolation pit; seepage pit; sink; sink hole; soak pit; swallow hole

渗坑排水 sinkhole drainage; sump drain

渗沥井 bleeder well

渗沥水处理工艺 treatment process of leachate

渗沥液特征 characteristics of leachate

渗磷 phosphatizing

渗流 seepage(flow); filtering flow; flow of seepage; infiltrated flow; infiltration flow; infiltration water; influent; influent seepage; interstitial flow; leak through liquid; percolation; porous flow; runoff in depth; underground flow; vadose

渗流槽 seepage tank

渗流测定 seepage measurement

渗流层 vadose zone

渗流场 infiltration field; seepage field

渗流场的类型 type of infiltration field

渗流场地 absorption field

渗流沉积 sieve deposit

渗流冲刷 seepage scour

渗流出口 seepage exit

渗流出逸点 seepage exit

渗流带 vadose zone

渗流的表示方法 expressive method of infiltration

渗流的实际流速 real velocity of seepage flow

渗流的主要特征 major characteristic of seepage

渗流地形 vadose feature

渗流电阻网模拟 resistance network analogue for seepage

渗流豆粒 vadose pisolite

渗流翻腾 seepage boil

渗流腐蚀 seepage corrosion

渗流沟 seepage trench

渗流固结 consolidation of permeability

渗流管 influent pipe; seepage pipe; seepage tube

渗流灌溉 irrigation by infiltration; ooze irrigation

渗流河 losing stream

渗流河床 seepage(river) bed

渗流湖 <与地下水有联系的> influent lake

渗流基本定律 basic law of seepage

渗流基本定律与公式 basic law and governing formula of seepage

渗流基床 seepage bed

渗流及水压力试验 seepage and water pressure test

渗流集水管道 <在地下埋设有孔管道> infiltration gallery

渗流集水廊道 infiltration gallery

渗流集水渠 infiltration channel

渗流计 seepage ga(u)ge; seepage meter[metre]

渗流计算模型 seepage calculation model

渗流截断 seepage interception

渗流井盖 cover for seepage pit

渗流坑 seepage pit

渗流控制 infiltration control; seepage control

渗流理论 seepage theory

渗流力 seepage force

渗流力学 permeation fluid mechanics

渗流连续性方程 continuity equation of seepage; seepage continuity equation

渗流量 osmotic flow; quantity of percolation; quantity of seepage; seepage discharge; seepage quantity

渗流量测设备 seepage measuring device

渗流流量 seepage flow; seepage discharge

渗流流速 seepage velocity

渗流路径 flow path; path of percolation; percolation path

渗流路线 flow path; path of percolation; percolation path

渗流面 seepage surface

渗流面积 seepage area

渗流模拟电阻网络 seepage analog(ue) resistance network

渗流爬径 line of credit

渗流爬径长度 creep path length

渗流抛物线 seepage parabola

渗流盆地 exudation basin

渗流破坏 seepage failure

渗流期 vadose epoch

渗流器 percolator

渗流侵蚀 seepage erosion

渗流区 seepage area; seepage zone; vadose area; vadose region; vadose zone

渗流泉 filtration spring; seepage spring; spring seepage; vadose spring; weeping spring

渗流实际流速 real velocity of seepage

渗流试验 seepage test

渗流水 buried stream; influent water; kremastic water; river bed water; seep(age) water; vadose water; wandering water

渗流水抽点检验 seepage water checking point

渗流水出流 vadose-water discharge

S

渗流水流量 vadose-water discharge
渗流水头 seepage head
渗流速度 seepage velocity; percolation velocity
渗流速率 seepage rate
渗流损失 infiltration loss; seepage loss water
渗流淘刷 seepage scour
渗流通道 seepage path
渗流途径 path of percolation
渗流网 flow net
渗流系数 coefficient of percolation; coefficient of seepage; seepage coefficient
渗流线 line of credit; line of creep; line of percolation; line of seepage; percolation line; phreatic line; seepage line
渗流线路 path of percolation; percoline
渗流形式 seepage pattern
渗流压力 percolation pressure; seepage pressure; exudation pressure; seepage force
渗流溢出点位置 position of vadose overflow
渗流溢出面 seepage face
渗流溢山点高程 height of vadose overflow
渗流应力 seepage stress
渗流运动形态分类 pattern classification of seepage movement
渗流运动要素 major factor of seepage movement
渗流运动主要公式 major formula of seepage flow
渗流折射定律 refraction law of seepage
渗流指数 infiltration index
渗硫 sulfurizing[sulphurizing]
渗硫处理 sulfinuz processing
渗硫混凝土 sulfur-infiltrated concrete
渗硫铁系含油轴承 ferro-porit bearing
渗漏 blow by; creep (ing); effluent seepage; exfiltration; influent seepage; leaching; leach out; leakage; percolation; seep; wastage; weeping; oozing
渗漏保险 insurance against leakage
渗漏报警器 leak alarm
渗漏材料 filtering material
渗漏草场 leaky lawn
渗漏测定 lysimetry
渗漏测定计 lysimeter
渗漏层 seepage course
渗漏出流量 seepage outflow
渗漏处 weep
渗漏带 percolation zone; water leakage zone; zone of percolation
渗漏带宽度 width of leakage zone
渗漏的 effluent; leaking; leaky
渗漏的船 leaky ship
渗漏滴水 trickling water
渗漏地基 leaky foundation
渗漏调查 seepage investigation
渗漏定位器 leakage locator
渗漏段位置 position of leakage sector
渗漏段岩性 rock type of leakage portion
渗漏范围 leakage area
渗漏管 leaky pipe
渗漏灌溉 replenishing irrigation
渗漏含水层 leaky aquifer
渗漏河 influent river; influent stream; seepage river; seepage stream
渗漏湖 seepage lake
渗漏回流 seepage reflection; seepage reflux(ion)
渗漏计 drainage ga(u)ge; percolation ga(u)ge
渗漏检测 leak detection
渗漏检测器 leak detector
渗漏检查 inspection for leakage; leak detection
渗漏接缝 leaking joint; leaky joint
渗漏井 leaking well; negative well; waste well
渗漏警告装置 seepage warning device
渗漏空气 leakage air
渗漏控制 seepage control
渗漏类型 type of leakage
渗漏量 amount of leakage; amount of seepage; filter loss; leakage water; percolating water; percolation water; quantity of leakage; quantity of percolation; quantity of seepage
渗漏量观测 leakage measurement
渗漏流量 leakage flow rate; seepage discharge
渗漏率 percolation rate; percolation ratio
渗漏面 leached surface; seepage face
渗漏能力 infiltration capacity; percolation capacity; transmission capacity
渗漏破坏 seepage failure
渗漏区 seepage area; seepage zone
渗漏试验 leak (age) test; lysimetric experiment; percolation test
渗漏水 fugitive water; influent water; leak (age) water; percolating water; percolation water; seepage water
渗漏水处 place of seepage and leakage
渗漏水地 seepy land
渗漏水排除 drainage of seepage
渗漏速度 <泥浆的> rate of penetration
渗漏速率 leakage rate
渗漏损失 infiltration loss; leakage loss; loss by percolation; percolation loss; seepage loss
渗漏探测器 leakage locator; leak detector
渗漏探测仪 leak detector
渗漏通道 channel of leakage
渗漏筒式榨油机 curb press
渗漏突变体 leaky mutant
渗漏途径 path of filtration
渗漏污水池 leaching cesspool
渗漏系数 coefficient of leaking; coefficient of percolation; coefficient of seepage; leakage coefficient; leakage factor
渗漏显示器 leak indicator
渗漏险 risk of leakage
渗漏线 filtration path; line of seepage; percolation path; seepage line
渗漏蓄水层 leaky aquifer
渗漏抑制 leakage reduction
渗漏因素 leakage factor
渗漏与渗透变形问题 problem of seepage and permeation deformation
渗漏晕 leakage halo
渗漏指示器 leakage indicator; leak detector
渗漏铸件 leaker
渗漏转换断层 leaky transform fault
渗滤 colation; diffusion; percolating; percolation; straining filtration
渗滤布 diffusing tissue
渗滤材料 filter material
渗滤槽 percolation tank
渗滤层 percolating bed; percolation bed; percolation layer
渗滤池 continuous filter; infiltration basin; percolating filter; percolation filter; percolator; straining filter
渗滤处理 percolation treatment
渗滤床 infiltration bed; percolation bed; seepage bed
渗滤带 zone of percolation
渗滤地下水 infiltration ground water
渗滤法 percolation process
渗滤法土地处理系统 infiltration land treatment system
渗滤沟 leaching trench
渗滤管 filter tube; infiltration pipe
渗滤基床 leach bed
渗滤计 diffusiometer
渗滤进水口 water intake with filter
渗滤井 infiltration well; percolation well
渗滤坑 adsorption pit; leaching pit; percolation pit; soakage pit; soakaway
渗滤廊道式进水口 water intake with filter gallery
渗滤量 amount of infiltration; amount of percolation; percolation ration; seepage
渗滤路径 path of percolation
渗滤路线 infiltration routing
渗滤率 percolation rate; percolation ratio
渗滤摩擦 percolation friction
渗滤摩阻力 percolation friction
渗滤片状径流溶蚀 percolate-flake runoff mixed corrosion
渗滤器 filter; percolating filter; percolation extractor; percolation filter; percolator; trickling filter
渗滤溶蚀 percolate corrosion
渗滤砂岩 filter sandstone
渗滤生物床 seepage biological bed
渗滤式进水口 water intake with filter gallery
渗滤式湿地生物生态系统 infiltrated wetland-biology ecology system
渗滤式污泥床 percolation-type sludge bed
渗滤试验 percolation test
渗滤水 infiltrate; infiltration water; leakage water; percolating water; percolation water
渗滤水水质 seepage water quality
渗滤速度 infiltration velocity; percolation velocity
渗滤速率 infiltration rate; rate of percolation
渗滤损失 seepage loss
渗滤脱沥青法 percolation deasphalting method
渗滤稀释 infiltration dilution
渗滤稀释活度系数 infiltration dilution activity coefficient
渗滤系数 coefficient of percolation; permeability coefficient
渗滤效应 infiltrating effect
渗滤压力 percolation pressure; seepage pressure
渗滤液 leaching solution; percolate
渗滤仪 percolation apparatus
渗滤异常 leakage anomaly
渗滤预防 percolation prevention
渗滤晕 infiltration halo; leakage halo
渗滤值 infiltration value
渗滤指数 infiltration index
渗滤作用 infiltration
渗铝 alumetizing; aluminising[aluminizing]; aluminium impregnation; alumin (i) um impregnation; calorizing; calorizing process
渗铝处理 calorize
渗铝法 <钢铁表面的> alitizing
渗铝钢 alumetized steel; aluminized steel; calorised[calorized] steel
渗率 percolation ratio
渗摩 osmol(e)
渗墨 ink bleed
渗泥压力 exudation pressure
渗排水防潮层 infiltrated dampproof course
渗硼 boriding; boronizing
渗硼硅 borosiliconizing
渗硼硅法 borosiliconizing
渗硼渗硅处理 borosiliconizing
渗气 permeation
渗气率 permeability to gas
渗汽空气 carburet(t)ed air
渗汽系数 coefficient of steam permeability
渗渠 infiltration canal
渗渠进水口 water intake with filter gallery
渗溶 vadose solution
渗入 imbibe; infiltrate; influent seepage; interfusion; leak-in; permeance; permeate; permeation; seep in; soaking in; transfusion
渗入补给 influent seepage
渗入池 infiltration pond
渗入处理 cementation
渗入的 inspersed
渗入的空气 infiltration air
渗入底物 influent substrate
渗入地下水 swallowed subsurface water
渗入剂 impregnating compound
渗入空气 entrainment of air
渗入空气量 quantity of air infiltration
渗入控制 infiltration
渗入廊道 infiltration gallery
渗入沥青 penetrating asphalt
渗入量 amount of percolation; infiltration capacity
渗入量曲线 infiltration capacity curve
渗入流量 infiltration flow; percolation flow; seepage flow
渗入面积 infiltration area
渗入能力 infiltration capacity
渗入盆地 infiltration basin
渗入强度 infiltration intensity; infiltration strength
渗入渠道 infiltration ditch
渗入热水的溪流 heated stream
渗入容量 infiltration volume
渗入时间 soaking time
渗入试坑深度 depth of pit of water infiltration test
渗入试验 infiltration test
渗入水 infiltrated water; infiltration water; influent water; water of infiltration; water of percolation; water seepage
渗入速度 infiltration rate; infiltration velocity
渗入速率 infiltration rate; infiltration velocity
渗入损失 infiltration loss
渗入通渠 infiltration gallery
渗入土壤 leaching into a soil
渗入土中 soak into a soil
渗入物 infiltrate; infiltration
渗入系数 infiltration coefficient
渗入型 infiltration type
渗入型水源地 infiltration water source
渗入性 accessibility
渗入性灌浆 permeable grouting; permeation grouting
渗入油灰 soaked to a putty
渗入雨水 infiltrating rainfall
渗入者 infiltrator
渗入指数 infiltration index
渗入重量 soaked weight
渗入注浆法 seep-in grouting method
渗入作用 seeping; infiltration
渗润线 seepage line
渗润作用 infiltration
渗色 bleeding

渗色试验 bleeding test
渗色污染的表面 bleeding surface
渗色颜料 bleeding pigment
渗失量 infiltration loss
渗湿性 humidity permeability; moisture permeability
渗湿异常 seepage anomaly
渗湿晕 seepage halo
渗水 creep(age); penetration of water; pervious to water; water; water creep; water ooze; water oozing; water penetration; water percolation; water seepage; weepage; bleed water
渗水比 specific permeability
渗水材料 permeable material; pervious material
渗水测定器 lysimeter
渗水层 permeable layer
渗水场地 leaching field
渗水池 infiltration basin; infiltration pond; percolation basin; percolation pond
渗水触气区 zone of aeration
渗水带 seepage belt; seepage zone; thief zone
渗水地层 water-bearing stratum; weeping formation
渗水点数 number of filter test
渗水丁坝 permeable dyke[dike]; permeable groin; permeable spur
渗水断层 dripping fault
渗水法 water permeability method
渗水防波堤 permeable breakwater
渗水缝 infiltration fissure; infiltration slit; permeable joint
渗水格笼堤 permeable crib dyke[dike]
渗水沟 infiltration ditch; seepage ditch; infiltration gutter; seepage gutter; weep drain
渗水沟管 weeper drain
渗水管 exfiltration pipe; infiltration pipe; porous pipe; weep drain; bleed pipe
渗水管沟 seepage trench; weeper drain
渗水管线 <地下排水系统> absorption line
渗水管引流 infiltration diversion
渗水管引水 <河底> infiltration diversion
渗水和土上涌 boil mud
渗水河 influent river; influent stream
渗水河床 filtration bed
渗水湖 seepage lake
渗水基岩 permeable bed
渗水计 lisimeter[lysimeter]
渗水计安装 lysimeter installation
渗水建筑物 permeable works; seepage structure
渗水接缝 permeable joint
渗水介质 permeable medium
渗水井 absorbing well; absorption well; blind catch basin; cesspit; dead well; drainage well; dung hole; filter well; leaching cesspool; leaching pit; negative well; percolation well; seepage pit; seepage well; waste well; weeping well; leaching well
渗水坑 leaching pit; seepage pit; soakaway; spreading pit; water sink; rummel <苏格兰语>
渗水坑井 soakaway
渗水孔 seepage hole; seep hole; weeper; weep hole; swallow-hole
渗水廊道 infiltration gallery
渗水量 leakage; seepage; water percolating capacity
渗水率 <混凝土> water rate
渗水面积 infiltration area; seepage area
渗水排除 drainage of leak
渗水曝气区 zone of aeration
渗水浅井 leaky shallow well
渗水强度 infiltration capacity
渗水渠 infiltration canal; infiltration ditch
渗水泉 seepage spring
渗水深度 water penetration
渗水试坑底面积 bottom area of pit of water infiltration test
渗水试验 infiltration test; in-situ permeability test; water seepage test
渗水试验方法 method of water infiltration test
渗水试验仪 infiltration test apparatus
渗水水头 infiltration head
渗水速度 seepage velocity
渗水速率 seepage rate
渗水隧道 infiltration tunnel
渗水通道 infiltration gallery; permeable channel
渗水土 permeable soil
渗水土层 permeable ground
渗水土路基 permeable soil subgrade; seeping soil roadbed
渗水洼地 leaky depression
渗水现象 <混凝土> water soaking
渗水性 phreatic permeability; water permeability; water seeping
渗水性注浆 permeability grouting
渗水压力试验 exudation pressure test
渗水压实作用 hydrocompaction
渗水岩层 permeable rock bed
渗水岩类 permeable rocks; weeping rocks
渗水岩石 permeable rock; weeping rock
渗水铸件 leaker
渗水作用 infiltration; influent action
渗速计 effusiometer; effusion meter
渗炭层 carburized layer
渗碳 acierage; acieration; brinelling; carbon pick-up; carbon pile-up; carburate; carburat(t)ing; carburetion; carburize; carburizing; case-harden(ing); cement carbon; cementing; face-harden
渗碳表面层 carburized case
渗碳层 carburized layer; cementation zone
渗碳层深度 carburized(case) depth; case depth
渗碳处理 carbonization
渗碳氮化 nicarbing; nitrocarburizing
渗碳的 cemented
渗碳法 acieration; cementation; cementation process
渗碳防护涂料 anti-carburizing paint
渗碳防止剂 anti-carburizer
渗碳钢 blister steel; carbonizing steel; carburizing steel; case-hardened steel; case-hardening steel; cementation steel; cemented steel; converted steel
渗碳钢软心 sap
渗碳过程 carburizing process; cementation process; cementing process
渗碳过量 over carburization
渗碳火焰 carburizing flame
渗碳剂 carburant; carburising agent; carburizer; cementation agent
渗碳加速剂 energizer
渗碳金属 carburized metal; cemented metal
渗碳炉 carburizing furnace; cementation furnace; cementing furnace
渗碳耐火砖 carbonized brick
渗碳黏[粘]土 carbonized clay
渗碳期 carburizing cycle
渗碳气氛 carburizing atmosphere
渗碳器 carburator
渗碳容器 carburizing container
渗碳烧结 carbusintering
渗碳深度 carbon penetration; carburizing depth
渗碳(碳素)钢 carbon carburizing steel
渗碳体 cementite; iron carbide
渗碳体的溶解度 cementite solubility
渗碳温度 carburizing temperature
渗碳细钢丝 <制特殊针用> pin bar
渗碳箱 cementing pot
渗碳样板钢 cemented templet steel
渗碳硬化层深度 carburated case depth
渗碳硬化法 cementing process
渗碳装置 cementing plant
渗碳组织 carburized structure
渗碳作用 carbur(iz)ation
渗铜钢 cupric cemented steel
渗铜铁 cemented iron
渗透 effusion; filtering; impenetrate; influent seepage; interfuse; leakage; leak through; penetrating; percolating; percolation; permeability; permeate; permeation; pervade; pierce; seepage; soakage
渗透保护层 penetrating sealer
渗透比降 percolation gradient
渗透变形 deformation due to seepage; seepage deformation
渗透变形防治措施 treatment measure of seepage deformation
渗透变形类型 type of seepage deformation
渗透变形试验 filtration erosion test; permeating deformation test
渗透薄膜 osmotic membrane
渗透补给 infiltration recharge
渗透参数 filtration parameter; infiltration parameter
渗透测定 infiltration measurement
渗透测定计 infiltrometer
渗透测定仪 permeameter(apparatus)
渗透测粒法 permeametry
渗透策略 penetrating strategy
渗透层 blotter coat; filter bed; filter layer; percolation bed; permeable stratum; pervious course
渗透长度 seepage length
渗透常数 permeability constant; permeation constant; transmission constant
渗透衬里炉 permeably-lined furnace
渗透池 osmotic cell; percolation pond; seepage basin
渗透冲击 osmotic shock
渗透出口 seepage exit
渗透处理 carburising[carburizing]; cementation; osmosis treatment; osmotic treatment
渗透床 seepage bed
渗透脆性 osmotic fragility
渗透带 percolation zone; permeable zone
渗透带直径 diameter of percolation zone
渗透当量 osmotic equivalent
渗透的 osmolar; osmotic; permeant
渗透的猝度 osmotic shock
渗透地板密封剂 <渗入木材的渍漆> penetrating floor(ing) sealer
渗透电场 infiltration electric(al) field
渗透电流 penetration current
渗透垫层 filter blanket
渗透定律 seepage law
渗透度 degree of penetration; osmolity; permeability
渗透法 osmose process; percolation method; percolation process
渗透方程 osmotic equation
渗透方程式 permeability equation
渗透方法 osmotic method
渗透分布 permeability distribution
渗透分离 osmotic separation; permeation separation
渗透分离设备 permeability separatory
渗透分析 dialysis
渗透缝 permeable joint
渗透浮力 seepage uplift
渗透干燥型油墨 penetration drying type ink
渗透隔膜 osmotic membrane
渗透公式 permeability formula
渗透功 osmotic work
渗透关系 osmotic relation
渗透管 osmosis tube; permeation tube
渗透灌浆 permeation grouting
渗透过滤床 percolating bed
渗透过滤器 percolating filter
渗透和渗透压 osmosis and osmotic pressure
渗透回流 seepage reflection; seepage reflux(ion)
渗透回水 return seepage
渗透回填 porous backfill(ing)
渗透基层 permeable base
渗透基底 permeable base
渗透极限 permeation limit
渗透计 osmometer; percolation ga(u)ge; permeability apparatus; permeability meter; permeameter(apparatus); permeator; seepage meter[metre]
渗透计算 seepage calculation
渗透剂 penetrant; penetrating agent; permeate agent
渗透剪力试验 drained shear test
渗透检查 penetrant inspection
渗透检验 penetrant inspection; penetrant test
渗透交代作用方式 infiltration metasomatism way
渗透交换 osmotic exchange
渗透接缝 permeable joint
渗透介质 permeating medium
渗透井 seepage well
渗透距离 penetration distance; seepage distance
渗透空气 infiltration air
渗透控制 infiltration control; percolation control
渗透控制率 controlled rate of penetration
渗透扩散法 <木材防腐处理> osmoses diffusion process
渗透冷空气量 permeating cold air
渗透理论 penetration theory
渗透力 osmotic force; penetrability; penetrating power; seepage force
渗透沥滤选矿法 percolation leaching
渗透量 amount of infiltration; infiltration capacity; permission through-out; quantity of seepage
渗透量曲线 infiltration capacity curve
渗透流 osmotic flow
渗透流量 discharge of permeability; seepage discharge
渗透滤层 percolating filter
渗透滤器 percolating filter
渗透路径 path of percolation; seepage path
渗透路线 path of seepage
渗透率 coefficient of percolation; coefficient of permeability; infiltration rate; intrinsic(al) permeability; penetration factor; percolation rate; permeability; permeability rate; permeation rate; rate of permeation; seepage rate; seepage rating; specific permeability

渗透率剖面 permeability profile
渗透率区域法 permeability-block method
渗透率曲线 infiltration rate curve
渗透率与饱和率 permeability-saturation
渗透弥散 pervasion
渗透密封 penetrating sealing
渗透面 pellicular front
渗透面积 infiltrating area;infiltration area
渗透膜 osmotic membrane;permeable membrane
渗透摩阻力 percolation friction
渗透能力 infiltration capacity;percolation capacity;transmission capacity
渗透浓度 osmotic concentration
渗透排斥力 osmotic repulsive pressure
渗透排水 filter drain;rubble drain
渗透排烟炉 permeably-lined furnace
渗透漂白 osmotic bleaching
渗透平衡 osmotic balance;osmotic equilibrium
渗透坡降 percolation gradient
渗透破坏 seepage failure
渗透剖面 seepage profile
渗透气化膜 devapo(u)ration membrane
渗透器 permeator
渗透潜能 osmosis potential
渗透强度 infiltration intensity
渗透清漆 penetrating varnish;penetration varnish
渗透清漆涂层 penetrating finish
渗透区 infiltration area;permeable zone;seepage zone;injection zone <污水注入井的>
渗透曲线 infiltration curve;penetration curve;permeability curve
渗透取水 infiltration diversion
渗透染色 penetration dy(e)ing
渗透染色剂 penetrating stain
渗透热损失 infiltration heat loss
渗透三轴试验 drained triaxial test
渗透色谱法 permeation chromatography
渗透上压力 seepage uplift
渗透深度 depth of penetration;length of penetration;penetration depth
渗透时间 penetration time;time of penetration
渗透时间滞后 permeation time lag
渗透式分离设备 permeability separatory apparatus
渗透式分离装置 permeability separatory apparatus
渗透式浸出器 percolation extractor
渗透式沥青路面 bituminous penetration road(pavement)
渗透势 osmotic potential;seepage potential
渗透势防渗 osmotic potential
渗透试验 infiltration experiment;percolation test;permeable test;seepage test
渗透试验器 osmoscope
渗透水 infiltration water;leakage water;osmotic water;percolating water;percolation water;permeate water;seemingly water;seep(age) water;water of infiltration;water of percolation
渗透水泵 seepage pump
渗透水带 percolating hose
渗透水量 infiltration volume;percolating volume;seepage volume
渗透水流 seepage flow
渗透水路 infiltration gallery
渗透水迁移 osmotic water transport
渗透水头 percolation head;seepage head
渗透水压 seepage pressure
渗透水运移 osmotic water transport
渗透速度 filtration velocity;percolation rate;seepage speed;seepage velocity;velocity of permeability
渗透速度-时间关系曲线 curve of percolating velocity-time
渗透速率 percolation rate;permeation rate;rate of percolation
渗透塑剂法 osmoplastic method
渗透塑剂防腐法 osmoplastic method
渗透损失 infiltration loss
渗透探伤 liquid penetrant test;oil whiting test;penetration inspection
渗透探伤法 permeation flaw detection
渗透探伤试验 chalk test
渗透梯度 percolation gradient
渗透天平 osmotic balance
渗透调节回游 osmoregulatory migration
渗透通径 path of percolation
渗透通量 permeate flux
渗透途径 filtration path;path of infiltration;path of percolation;path of seepage;percolation path;seepage path
渗透脱水 osmotic dehydration
渗透污垢 osmosis fouling
渗透物 permeate
渗透吸力 osmotic suction
渗透系数 coefficient of hydraulic conductivity;coefficient of infiltration;coefficient of penetration;coefficient of percolation;coefficient of permeability;filter factor;filtration coefficient;hydraulic conductivity;infiltration coefficient;osmotic coefficient;percolation coefficient;permeability coefficient;permeability factor;permeable coefficient;seepage coefficient;transmissibility;transmission coefficient
渗透系数测定 determination of permeable coefficient
渗透现象 osmotic phenomenon
渗透线 filter line;path of percolation;percolation line;percoline;seepage line
渗透限度 permeability limit
渗透效应 osmotic effect
渗透型水源地 penetration water source
渗透性 hydraulic conductivity;osmotaxis;osmotic permeability;penetrability;permeability;perviousness;osmosis
渗透性变化造成的圈闭 varying permeability trap
渗透性不良土壤 impervious soil
渗透性材料 permeable material
渗透性测定 permeability determination
渗透性测量 permeability survey
渗透性充填材料 pervious packfill material
渗透性冲击 osmotic shock
渗透性的 permeable
渗透性底漆 penetration priming
渗透性底油 penetration priming
渗透性地基土 pervious subsoil
渗透性方程 permeability equation
渗透性防腐 osmosis preservation
渗透性防腐剂 penetrability preservative
渗透性极低的面层 impervious coat
渗透性极低的岩床 impervious bed
渗透性漏失 seepage loss
渗透性密封材料 penetrating sealer
渗透性能 hydraulic permeability
渗透性清漆 penetrating finish
渗透性热储 permeable reservoir
渗透性试验 permeability test
渗透性试验杯 permeability testing cup
渗透性试验法 permeability method
渗透性试验机 permeability tester;permeability testing machine
渗透性试验仪 permeameter(apparatus)
渗透性土(壤) permeable soil
渗透性沿流向渐变含水层 aquifer of gradual change in hydraulic conductivity along flow direction
渗透性沿流向突变含水层 aquifer of sudden change in hydraulic conductivity along flow direction
渗透压 osmotic pressure;percolation pressure
渗透压测定法 osmometry
渗透压法 osmotic pressure
渗透压感受器 osmoreceptor
渗透压记录仪 osmograph
渗透压克分子 osmol(e)
渗透压力 osmotic pressure;percolation pressure;seepage force;seepage pressure
渗透压力测定法 osmometry
渗透压力法 osmotic method
渗透压力计 osmometer
渗透压力梯度 osmotic pressure gradient
渗透压摩尔 osmol(e)
渗透压强 osmotic pressure
渗透压梯度 osmotic pressure gradient
渗透压调节 osmoregulation;osmotic regulation
渗透压调节机制 osmoregulatory mechanism
渗透压头 infiltration head
渗透压效应 osmotic pressure effect
渗透压注射器 osmotic injector
渗透岩石 permeable rock;saturated rock
渗透盐剂 <一种木材防腐剂> osmosalt
渗透扬升力 seepage uplift
渗透扬压力 seepage uplift
渗透液 dialyzate;penetrating fluid
渗透液化 seepage liquefaction
渗透仪 infiltrometer;osmoscope;percolation apparatus;permeability testing machine;permeameter(apparatus);seepage meter[metre];drainage indicator <土的渗透系数测定仪>
渗透仪试样盒 permeability cell;permeameter cell
渗透因素 permeability factor
渗透应力 infiltration stress
渗透影响 osmotic effect
渗透涌气 seepage boil
渗透涌土 seepage boil
渗透用沥青 penetration asphalt
渗透油 penetrating oil
渗透有效面积 effective leakage area
渗透(雨)水 cut-off water
渗透运动 osmotic movement
渗透晕 leakage halo
渗透张量 permeability tensor
渗透蒸发 pervapo(u)rization
渗透值 osmotic value;penetration number
渗透指数 infiltration index;penetration index
渗透着色(剂) penetrating stain
渗透阻挡层 permeability barrier
渗透阻力 osmotic resistance;percolation resistance
渗透作用 infiltration;osmosis;osmotic effect;permeation;percolation
渗透作用和再分配 infiltration and redistribution
渗吸水 imbibition water;water of imbibition
渗吸速度 rate of percolation
渗析 dialyse;osmosis
渗析处理 dialysis treatment
渗析单元 dialysis unit
渗析分析 dialysis
渗析壳 dialyzing shell
渗析力 osmotic force
渗析膜 dialyser;dialysing membrane;dialysis membrane;dialytic membrane;dialyzator;dialyzer
渗析能 osmotic energy
渗析器 dialyser[dialyzer];dialyzator
渗析势 osmotic potential
渗析吸力 solute suction
渗析系数 osmotic coefficient
渗析现象 osmotic phenomenon
渗析压力 osmotic pressure
渗析压势 osmotic pressure potential
渗析压头 osmotic pressure head
渗析液 dialysate[dialyzate]
渗锡 stannize
渗锡处理 stannizing
渗压 consolidation;osmotic pressure
渗压比 consolidation ratio
渗压表 osmometer
渗压测定法 osmometry
渗压差 osmotic pressured difference;permeability pressure difference
渗压缓冲 osmotic buffering
渗压计 osmometer
渗压剂 osmoticum
渗压力 consolidation pressure
渗压曲线 consolidation curve;consolidation line
渗压势 potential of penetration pressure
渗压试验 oedometric test
渗压室 consolidation chamber
渗压梯度 permeability pressure gradient
渗压调节 osmoregulation
渗压系数 coefficient of consolidation
渗压仪 consolidation cell;consolidation(test) apparatus;consolidometer;oedometer
渗压状态 piezometric regime
渗液检验法 liquid penetrant inspection
渗液探伤 liquid penetrant inspection
渗液探伤试验 liquid penetrant test
渗油 greasing;oil seepage;spewing;suboiling
渗油性 greasiness
渗油性能 oil penetration
渗油性能试验 oil penetration test
渗渍盐剂 osmosalt

蜃

蜃景 looming;mirage

升

升 liter[litre]

升岸阶地 elevated shore face terrace
升板 jump form;lift slab
升板法 lift-slab method
升板法结构 lift-slab construction
升板法施工 lift-slab construction
升板滑模联合施工 lifting floor and sliding formwork combined construction
升板混凝土楼面板 lift-slab concrete floor
升板机 lifting slab machine
升板技术 lift-slab technique
升板建筑法 <屋顶或楼板在地坪上浇捣,然后用千斤顶顶升并搁在柱上> lift-slab construction

升板结构 lift-slab structure
升板卡圈 lift-slab collar
升板楼面 lift-plate floor
升板施工 lift construction
升板施工法 building construction with lifting method;lift-slab(method of) construction;lift-slab system
升板式装料机 lifting-gate feeder
升板体系 lift-slab system
升板支柱 lift-slab column
升标 raising of indices
升采样率 rising sample rate
升采样率处理 rising sample rate processing
升潮 flood tide;flood water
升车机 car lift
升沉【船】rising and sinking
升沉补偿器 < 海上钻探时,钻探船随波浪升沉而钻压不变 > heave compensator; wave motion compensator
升沉补偿设备 heave compensation equipment
升沉补偿设备规格 specification of heavy compensation equipment
升沉补偿设备类型 type of heavy compensation equipment
升沉补偿设备组成 component of heavy compensation equipment
升沉仪 heave meter
升程 < 悬挂装置的 > lift range
升程曲线 < 凸轮 > lifting curve
升程式止回阀 lift check valve
升程限制器 lift limiter;valve guard
升出幅角 rising amplitude
升船浮坞 lifting dock;offshore dock
升船滑道 patent slip; railway dry dock;shop railway; slip dock; marine(ship) railway < 缆车式的 >
升船机 barge lift; lift dock; ship elevator;shiplift;ship lifter;canal lift; elevating platform; marine lift; mechanical lift;;mechanical lift dock
升船机充水泄水装置 filling and emptying device of shiplift
升船机船箱 lock chamber
升船机的尺寸 dimensions of shiplift
升船机的能力 capacity of shiplift
升船机的水深 depth of shiplift
升船机的现场调查 site investigation for shiplift
升船机的选址 siting of shiplift
升船机额定提升能力 nominal lifting capacity of shiplift
升船机横向移船 side transfer
升船机密封装置 telescope sealing of shiplift
升船机平台 shiplift platform
升船机平台设计 shiplift platform design
升船机上船舶转移(上下船台) ship transfer
升船机转盘 turntable of shiplift
升船机纵向移船 longitudinal transfer
升船能力 lifting ship capacity; lifting ship power;lift capacity
升船起重机 ship jack
升船设备 barge lift
升船厢 lift chamber; ship chamber lock chamber;navigation chamber
升船斜面 inclined plane
升船斜坡 inclined plane
升船斜坡道 boat ramp
升船闸 lift lock
升达限幅 limit rise;limit up
升到顶 way aloft
升底窑 elevator kiln
升吊式闸门 lifting gate
升吊索 outhaul cable
升斗式排肥器 rising hopper feed
升端送纸系统 lifting head sheet-handling system
升/分钟 lit/min[liter per minute];lpm [liter per minute]
升符号 up symbol
升幅振荡 increasing oscillation
升负荷 load up
升负荷速度 loading rate
升杆阀 elevating valve
升杆棍 pike pole
升高 advance;ascend;heighten(ing); lifting; power-up; raising; rising; step up;upraise
升高比 ratio of rises
升高并移动零件的机构 lift-and-carry mechanism
升高的 elevated;rising;super-elevated
升高的压力 built-up pressure
升高的支座 analemma
升高电压 boosted voltage
升高度 ascent
升高二层底铺板 raised tank top
升高工作面 rise working face
升高海岸线 elevated shore line
升高荷载应力 advancing load stress
升高后甲板 raised quarter-deck
升高机 block-lifter
升高甲板 raised deck
升高检查设备 lifting and inspecting installation
升高肋板 rising floor
升高率 rate of rise
升高门铰 rising hinge
升高能力 climbing capacity
升高平台 raised platform
升高千斤顶 elevating jack
升高设备 elevating appliance
升高式高速公路 elevated freeway
升高试验 rise test
升高首甲板船 raised foredeck ship
升高首(楼)甲板 raised forecastle
升高速率 < 建筑混凝土的 > rate of placing
升高天线用摇柄 antenna elevation pawl
升高尾甲板船 raised quarter-deck ship
升高温度 elevated temperature; elevation of temperature
升高系数 step-up ratio
升高压的 step up
升高载荷应力 advancing load stress
升高转子装置 jack system
升高组 lift set
升格 upgrade;upgrading
升格销售 trading up
升弓弹簧 raising spring
升功率 output per litre; performance per litre[liter];power-to-volume ratio < 发动机的 >
升汞 corrosive sublimate; mercuric chloride;mercury bichloride; sublimate
升汞防腐 kyanize
升汞防腐法 Kyan's process; kyanising [kyanizing];kyanization
升汞防腐木材 kyanized wood
升汞浸渍处理 kyanising[kyanizing]
升谷机 grain elevator
升管 riser pipe
升轨 rail lift
升轨器 rail lifter;track lifter
升弧段 upward leg
升华 sublimation; sublime; subliming; vapo(u)rization without melting
升华白铅 sublimed white lead
升华泵 sublimation pump
升华冰 sublimated ice
升华沉淀作用 evaporation
升华的 sublimate;sublime
升华点 sublimation point
升华碘 sublimed iodine
升华法 sublimation method; sublimed method
升华干燥 freeze-dry(ing);lyophilization;sublimation drying
升华干燥机 freeze drier[dryer]
升华干燥器 freeze drier[dryer]
升华干燥添加剂 freeze drying additive
升华干燥装置 freeze drying plant
升华矿脉 sublimation vein
升华冷凝法 sublimation condensation
升华冷却 sublimation cooling
升华硫 sublimed sulfur
升华率 sublimation rate
升华皿 sublimation pot;subliming pot
升华能 sublimation energy
升华谱 sublimatogram
升华谱法 sublimatography
升华器 sublimator(y);sublimer
升华铅白 sublimed white lead
升华潜热 latent heat of sublimation
升华曲线 sublimation curve
升华热 heat of sublimation; latent heat of sublimation;sublimation heat
升华温度 sublimation temperature
升华物 sublimate
升华压力 sublimation pressure
升华仪 apophorometer
升华逸散作用 evaporation
升华印花 sublimation(transfer) printing
升华自铅 basic lead sulphate
升华作用 < 感情和态度转向其他对象的代替作用 > sublimation
升藿烷 homohopane
升级 preferment; promotion; stage; upgrade;upgrading
升级的 upgraded
升级适配器 staging adapter
升级数 ascending series
升级条款 escalation clause
升级训练 upgrading training
升计数器 up-counter
升降 fluctuation
升降百叶窗执手 blind lift
升降板 lift slab
升降臂 < 起重机 > lift(ing) arm
升降臂组件 lift arm assembly
升降变换齿轮 lifter change wheel
升降操纵杆托架 < 起重机 > hoist control lever bracket
升降叉车 forklift
升降车 lift truck;cage < 竖井内的 >
升降船闸 lift lock
升降窗玻璃的皮带 sash cord
升降磁铁 lifting magnet
升降挡板 adjustable end stop
升降刀 lifting knife
升降导杆 lifting guide pillar
升降导架 hoist mast
升降道 access trunk; downtake; fall way
升降底式炉 elevator furnace
升降电动机 elevating motor; elevator motor;hoisting motor;travel motor
升降电压调节器 buck-and-boost regulator
升降吊灯 rise-and-fall pendant
升降吊笼 elevator cage
升降调 rising-falling tone
升降顶 lifter roof
升降动作筒 elevator actuating cylinder
升降斗 cage lift(er);elevator scoop
升降段 dropping section
升降堆垛机 elevating piler
升降舵 elevating rudder; elevator; flipper;lifting plane
升降舵操纵杆 elevator stick
升降舵杆 elevator horn
升降舵控制 elevator control
升降舵控制杆 elevator control lever
升降舵控制索 elevator cable
升降舵偏转角 elevator angle
升降舵伺服系统 elevator servo
升降发射装置 elevator-launcher
升降法 rise and fall
升降风扇 < 气垫车的 > lift fan
升降副翼 elevon
升降盖(口) companion
升降杆 elevating lever; lifter(rod); lifting arm;lifting poker;lifting rod
升降杆送料炉 lift beam furnace
升降钢丝绳 hoist cable
升降钢索 wire halyard
升降高差 rise-and-fall
升降工作平台 aerial lift; aerial platform
升降工作台 rise-and-fall table
升降钩 drop hook;drop lifter
升降轨 lift rail
升降辊 depressing roll
升降滚轮 lift roller
升降合页 rising butt hinge; rising hinge;skew hinge
升降合作业 jacking operation
升降横杆 cross bar
升降滑板 lifter slide
升降滑架 lift bracket
升降回转台 coil lift-and-turn unit
升降机 cage lift(er);hoister;hoisting machine;lift(er);lifting tackle;mechanical lift; ram; retraction jack; elevator
升降机安全器 elevator car safety
升降机把手 hoisting lever
升降机保险装置 grip gear
升降机臂 elevator jib
升降机不停层开关 elevator nonstop switch
升降机操作器 door operator
升降机操作绳 elevator rope
升降机车库 lift-type car park
升降机车厢 elevator car
升降机衬垫 elevator liner
升降机大梁 hoist beam
升降机导轨 cage guide; elevator guide rail
升降机底滑车箱 boot of elevator
升降机(底)坑 elevator pit
升降机电缆 elevator cable; hoist cable
升降机电压开关 elevator potential switch
升降机动 up-and-down maneuver
升降机斗链 bucket paternoster
升降机额定荷载 lift rated load
升降机钢丝绳松弛开关 elevator slack-cable switch
升降机供电电缆 elevator cable
升降机构 elevating; hoist mechanism; lifting mechanism
升降机关节销 elevator wrist pin
升降机轨道 tilting track
升降机戽斗 elevator bucket
升降机滑轮固定梁 elevator sheave beam
升降机回动爪 reverse elevator pawl
升降机回合面 lift joint
升降机机房 elevator machine room
升降机机门接点 gate contact
升降机及回转器分动手把 hoist and drill head clutch lever
升降机棘轮 elevator ratchet wheel
升降机棘爪套 elevator pawl casing
升降机架 < 矿井 > shaft house
升降机间 hoistway
升降机脚手架 rigger's scaffold(ing)
升降机搅动器 elevator stirrer
升降机轿厢 elevator car

升降机轿厢底部超跑距离 bottom elevator car run-by

升降机轿厢动力关闭门 power-closed car door

升降机轿厢动力关闭闸门 power-closed car gate

升降机轿厢平层装置 elevator car leveling device

升降机井 hoistway

升降机井道 access well;lift trunk;well hole;elevator hoistway;elevator shaft;elevator well;lift shaft;lift well

升降机井道的门 hoistway door

升降机井顶楼 elevator tower

升降机井孔 lift well

升降机开口环 elevator split ring

升降机控制 elevator control;lift control

升降机控制阀接头 hoist control valve adapter[adaptor]

升降机控制器 elevator controller

升降机缆绳 elevator rope

升降机联锁装置 elevator interlock

升降机梁 elevator machine beam

升降机笼导轨 cage guide

升降机螺旋 elevator screw

升降机螺旋上端套 upper elevator screw bushing

升降机马达 elevator motor

升降机内控制 car-switch control

升降机平层区 elevator landing zone

升降机平层上限距离 elevator truck zone

升降机平衡 elevator balance

升降机平台 elevator landing;lift landing;lift platform

升降机平台空间 elevator platform space

升降机前厅 elevator hall

升降机曲柄 elevator crank

升降机扇形齿轮 elevator quadrant

升降机上墩下水设备 ship elevating plant

升降机设备(安装) elevator installation

升降机伸臂 elevator boom

升降机使用费 elevating charges

升降机竖井 lift pin;lift pit;lift shaft;lift well;elevator shaft

升降机速度 elevator speed

升降机塔(架) elevator tower

升降机塔柱 hoist mast

升降机停层装置 elevator landing stopping device

升降机停车场 lift-type car park

升降机停车装置 elevator parking device

升降机通道 lift trunk

升降机筒 cage hoist

升降机围壁 lift trunk

升降机衔接器 lift engager

升降机箱 elevator cage;elevator crib

升降机小齿轮 elevator pinion

升降机用绳 elevator rope

升降机罩衬板 elevator casing wearing plate

升降机支柱 cathead

升降机制造厂 elevator plant;lift maker

升降机轴衬 elevator shaft bush

升降机肘节销定位螺钉 elevator wrist pin keeper

升降机主动齿条 elevator driving rack

升降机爪簧 elevator pawl spring

升降机爪销 elevator pawl pin

升降机装置 elevator gear

升降机组成部分 elevator component

升降计数器 up-down counter

升降夹具 lift dog

升降架 crane;erector

升降交替的坡度 intermittent grading

升降绞车 hauling winch;lifter winch;lift(ing) winch

升降绞辘 <吊杆的> lift tackle

升降绞盘与钢丝绳结合 hoist winch and rope

升降绞索 holding line

升降铰链 lifting chain;skew butt

升降进给 vertical feed(ing)

升降井 access well;hauling shaft;manhole;shaft

升降距 pitch of rise and fall

升降锯 rise-and-fall-saw

升降锯机 rise-and-falling saw

升降开关 direction switch

升降孔 life span

升降孔盖 manway cover

升降控制 lift control

升降控制阀托架 <起重机> hoist control valve bracket

升降控制杆托架 <起重机> hoist control lever bracket

升降控制系统 lift control system

升降口 companion hatch(way);companion way;manhole

升降(口)扶梯 companion way

升降口盖 companion;manhole cover;manway cover;scuttle hatch(cover)

升降口花格 hatch grating

升降口门 hatchway door

升降口门连锁装置 hatchway door interlock

升降口梯 companion ladder

升降口围栏 hatchway enclosure

升降口围罩 companion

升降口闸门 hatchway gate

升降类型及其载重量 hoist types and capacities

升降链 hoisting chain

升降流 up-and-down welling

升降螺杆 lifting screw;lifting spindle

升降螺丝 up-and-down screw

升降螺旋 elevating screw

升降门 drop door;lifting gate;overhead door

升降门操作器 door operator

升降门附件 lifting door hardware

升降门铰链 cocking;rising butt hinge

升降门配件 lifting door fittings

升降门设备 lifting door furniture

升降平板车 elevating platform truck

升降平台 lifting platform

升降平台式移动钻井 elevating-deck-type mobile drilling rig

升降平台式钻孔台车 lift full-deck jumbo

升降坡度分级 classes of rise and fall;classes of grodient

升降起重臂 luffing boom

升降起重机 hoisting crane;lifting gear

升降起重联动装置 lifting gear

升降器 riser

升降桥 hoist bridge;lever draw bridge;lift(ing) bridge;vertical-lift bridge

升降趋势检验 test for upward or downward trend

升降伞形齿轮 <高低牙的> lifter bevel wheel

升降色谱法 ascending-descending chromatography

升降色谱分离 ascending-descending chromatograph

升降设备 crane;crane attachment;jacking equipment;lifting appliance;lift-on/lift-off equipment;vertical motion

升降绳 halliard;halyard

升降绳滑车 halyard block

升降绳索 hoist rope

升降时可透视景物的电梯 open-view lift

升降式 elevation type;over-and-under type

升降式搬运车 lift truck

升降式叉车搬运机 lift truck

升降式船坞 lift dock

升降式道口栏木 hanged barrier;lifting-gate barrier

升降式底阀 lifting bed valve

升降式吊灯 rise-and-fall luminaire

升降式吊货钩 Seattle pattern cargo hook

升降式浮顶罐 lifter roof tank

升降式感应圈 hoist-type induction coil

升降式工作平台 mast work platform

升降式刮土机 elevating scraper

升降式回转排气罩 lift-type rotary exhaust hood

升降式胶带运输机 liftable and lowerable belt conveyer[conveyor]

升降式脚手架 raise scaffold;raising scaffold

升降式接线器 panel switch

升降式开合桥 lift bridge

升降式栏木 lifting barrier;lifting boom;lifting gate

升降式排气罩 lift-type exhaust hood

升降式平路机 elevating grader

升降式平台 elevating platform;jack-up platform;raising platform

升降式平台车 platform lift truck

升降式潜望镜 elevator periscope

升降式桥孔 lift space;lift span

升降式停车场 lift park

升降式停车库 lift park

升降式挖掘铲 lifting share

升降式挖泥船 elevator dredge(r)

升降式挖泥机 elevator dredge(r)

升降式望远镜 elevator periscope

升降式悬臂栈桥 cantelcas

升降式闸门 drop gate

升降式止回阀 elevated type check valve;lift(type) check valve

升降式装卸车 elevator-type loading car;lift truck

升降式自动电话制 panel system

升降式钻机 jack-up rig

升降室 jack house

升降输送机 lift and conveyer[conveyor]

升降丝杆 elevating screw

升降速度 hoisting-lowering speed

升降索 hoist cable;halliard;haulyard <旗帆等的>

升降索卡环 haulyard shackle

升降索套 lifting sling

升降塔 elevation tower

升降塔架 elevator tower

升降塔架绞车 elevator tower hoist

升降台 cage assembly;elevating platform;elevating stage;elevating table;elevator platform;knee-and-column;lifting desk;lift table;raising platform;stage lift;knee <铣床的>;rise and fall table

升降台式炉 elevator furnace

升降台式铣床 column and knee type milling machine

升降台式小车 lift platform troll(e)y

升降台式载驳船 Seabee ship

升降台铣床 knee-and-column milling machine

升降台下水法 lifting platform launching

升降台型 knee type

升降弹簧 lifting spring

升降梯 companion ladder;escalator;lift ladder

升降梯口 port gangway

升降停车库 elevator garage

升降凸轮 lifter cam;lifting cam

升降凸轮轴 lifter motion cam shaft

升降托架 shears

升降托脚 lifter bracket

升降挖泥机 elevator dredge

升降尾门 elevating tail gate

升降蜗杆 elevating endless screw

升降屋顶 breather roof

升降舞台 drop stage;elevator stage;lift stage;stage lift

升降铣床 rise-and-fall milling machine

升降系统 hoisting system;jacking system

升降销 lifter pin

升降压变压器 step transformer

升降窑 elevator kiln

升降液压缸 lift cylinder;lift ram

升降液压缸管道 lift cylinder line

升降椅 chair lift

升降翼汽车 ram-wing vehicle

升降油泵托架 <起重机> hoist pump bracket

升降油缸 hoist cylinder

升降圆截锯 jump saw

升降圆锯机 jump saw

升降运动 lift movement;upland down movement

升降运动的限制开关 limit switch for hoist motion

升降运输车 coil buggy

升降闸门 lift(ing) gate

升降闸门堰 lift gate weir

升降支臂用滑轮组 <起重机的> pulley block luffing gear

升降支架 hoist frame

升降止动器 draft stop

升降指 lifting finger

升降制选择器 panel switch

升降制自动电话 panel telephone

升降轴 lifting shaft

升降轴臂 lifting shaft arm

升降轴锯 rising and falling saw

升降轴锁闭手柄 lift-shaft lock handle

升降装置 cage assembly;elevating gear;elevating installation;flipper;hoisting device;hoisting gear;jacking gear;lifting device;lifting gear;winding gear

升降钻具 hoisting lowering

升降作用 <地壳的> elevation and subsidence

升交点 ascending node;north bound node

升交点黄经 longitude of ascending

升交角距 argument of latitude

升角 <叉摆> lift angle

升井 riser shaft

升举 boost;elevate a turnable ladder

升(举)阀 lift valve

升举范围 lifting range

升举刮板 lifter flight

升举力 lifting force;raising force

升举能力 elevating capacity;lifting capacity;lifting power

升举试验 <装载机的> lifting test

升举速度 speed lifting

升举行程 lifting stroke

升举性能 lift efficiency

升举纸卷的滑车 reel lifting tackle

升空 levitation;lift-off

升空弹射器 launching catapult

升孔时间 uphole time

升孔速度 uphole velocity

升力 ascending power;ascensional force;elevating force;lift force

升力产生装置 lifting system

升力发动机吊舱 lift pod

升力发动机推力 lift thrust
升力分布 lift loading
升力分量 lift component
升力风扇 lift(ing) fan
升力换能器 lift transducer
升力减小 lift divergence
升力角 angle of lift
升力控制 lift control
升力拉条 lift bracing
升力理论 lift theory
升力力比 lift-drag ratio
升力力矩 lifting force moment
升力螺旋桨 lifting propeller
升力面 supporting plane
升力面面积 lifting surface area
升力扭转力矩 moment due to lift force
升力喷管 lift-device nozzle
升力曲线 lift curve
升力曲线斜率参数 lift-curve-slope parameter
升力特性 lift efficiency
升力体 lifting body
升力体再入 lifting-body reentry
升力尾翼 lifting tail
升力涡 lift(ing) vortex
升力系数 coefficient of lift; lift coefficient
升力系数指示器 velometer
升力线 lift(ing) line; lift wire
升力限度 margin of lift
升力卸减 lift dump
升力指示器 lift indicator
升力中心 center[centre] of lift
升力装置 lift(ing) unit
升力-阻力比 lift-drag ratio
升料器 filler hoist
升流 upflow; upwash
升流澄清池 upflow clarifier
升流床吸附法 expanded-bed adsorption
升流粗滤 upflow roughing filtration
升流粗滤池 upflow roughing filter
升流反应器 upflow reactor
升流复合生物滤池 upflow composite biofilter
升流固定床反应器 upflow fixed bed reactor
升流固着膜厌氧生物反应器 upflow fixed film anaerobic bioreactor
升流管 upspout
升流过滤工艺 upflow filtration process
升流流化床 upflow fluidized bed
升流膨胀中和滤池 upflow expansion neutralizing filter
升流曝气生物滤池 upflow biological aerated filter
升流曝气污泥床 aerated upflow sludge bed
升流生物过滤反应器 upflow biofiltration reactor
升流生物膜滤池 upflow biofilm filter
升流生物塔 upflow biotower
升流石灰床 upflow limestone bed
升流式池 upflow basin
升流式过滤 upflow filtration
升流式过滤法 upflow filtration process
升流式集中(采暖)炉 upflow-type central furnace
升流式接触澄清池 upflow contact clarifier
升流式连续蒸煮锅 upflow continuous digester
升流式卵石澄清池 upflow pebble bed clarifier
升流式凝聚作用 upflow coagulation
升流式砂滤池 upflow sand filter
升流式污泥床反应器 upflow sludge blanket reactor
升流式选粉机 upstream separator
升流式厌氧污泥床 upflow anaerobic sludge bed; upflow anaerobic sludge blanket
升流速度 upflow velocity
升流塔 upflow tower
升流填料床反应器 upflow pakced bed reactor
升流污泥床 upflow sludge bed; upflow sludge blanket
升流污泥床反应器 upflow sludge bed reactor
升流污泥床滤池 upflow sludge blanket filter
升流污水床滤池 upflow sludge bed-filter
升流斜板沉淀池 inclined plank settling tank of upflow
升流厌氧滤池 upflow anaerobic filter
升流厌氧生物滤池 upflow anaerobic biofilter
升流厌氧污泥床法 upflow anaerobic sludge bed process
升流厌氧污泥床反应器 upflow anaerobic blanket reactor; upflow anaerobic sludge bed reactor
升流厌氧污泥-固着膜生物反应器 upflow anaerobic sludge-fixed film bioreactor
升炉 kindling
升轮器 wheel lifting device
升每秒 liter per second
升幂 ascending power; increasing power; power to
升幂级数 ascending power series; series of increasing powers
升模 jump(ing) form; leaping formwork
升模施工 jumpforming; jumping shuttering
升膜浓缩 rising film condensation
升膜蒸发器 climbing-film evapo(u)rator; upward-flow evapo(u)rator
升频变换器 step-up frequency changer
升频器 up-converter
升频转换 up-conversion
升频转换激光器 up-converting laser
升坡 ascending grade; grade-up; plus grade; summit grading; track raising; upgrade; uphill gradient; uphill slope
升坡的 acclive; accliv(it)ous
升坡段 upgrade section
升坡皮带输送机 upgrade belt conveyer; uphill belt conveyer[conveyor]
升坡牵引 climbing power
升起 ascent; hoisting; lifting; uprise
升起部分 rising part
升起点 <天体> east point
升起吊杆 raise a boom; top the lift
升起时间 rise time
升起式平台 raising platform
升起速度 lifting velocity
升起特性 rising characteristic
升气管 air rising duct; riser; warm-air rising duct
升气管顶面水平偏差 difference in riser level
升气管中心距 center distance of riser
升热器 heat booster
升式 lift-type
升式阀 lift-type valve
升式阀旋塞 lift-type valve cock
升式盘 lift-type disk
升水 at premium; premium
升水泵 lift pump
升水沟筑物 water lifting structure
升水管 lift(ing) tube; water lifting conduit; water lifting pipe[piping]; water lifting tube[tubing]
升水管道阀 lift valve
升水机 water-raising engine
升水力 water furnishing ability
升水率 contango rate; rate of premium
升水头 elevation head
升水头渗透试验 rising-head permeability test
升送高度 delivery lift
升送链板护罩 <铲土机> elevator guard
升送器上口 top door
升送式铲运机 elevating scraper
升送式平地机 elevating grader
升送式挖掘机 elevator digger
升送式装载机 elevating loader; elevator-type loader
升送叶片 lifting blade
升速 raising speed; speed-up
升速器 speed increaser
升速曲线 climbing curve
升速特性 rising-speed characteristic
升速指示器 climb indicator
升酸器 acid elevator
升台 jacking up hull
升台高度 air gap
升台重量 jacking weight
升天线的摇柄 antenna elevation pawl
升艇机 boat lift
升托式浮船坞 depositing dock; depositing floating dock
升温 bringing-up; elevation of temperature; heating-up; rise in temperature; temperature rise; temperature up; warming; warming up
升温处理 hyperthermic treatment
升温等温淬火 step-up austempering
升温反应 temperature reaction
升温过程 temperature development
升温间 warming room
升温阶段 temperature rise period
升温结晶法 raising temperature crystallization
升温进样系统 temperature sample introduction system
升温控制 warm-up temperature control
升温裂纹 heating-up crack; up-quenching crack
升温期 heating-up period; temperature rise period; warming-up period
升温曲线 curve of temperature; heating-up curve
升温热损失 heating-up loss
升温时间 heating-up time; warming-up time
升温室 temperature rise room
升温速度 heating rate; heat up rate; heat-up speed
升温速率 heating rate; heat up rate
升温速率倒数曲线 inverse heating-rate curves
升温速率曲线 heating-rate curve
升温脱附 desorption by heating
升温效率 temperature rise efficiency
升温着色 heat tinting
升卧式平面闸门 lift flap plain gate; lift-lie plain gate
升线波脉 anacrotic pulse
升限 altitude capability; ceiling altitude; ceiling height; top of climb
升限指示器 ceiling height indicator
升/小时 liter per hour
升序 ascending order; ascending sequence
升序标号 ascending key
升序列 ascending chain
升序排列 ascending arrangement; ascending sort
升序序列 ascending sequence
升压 lifting pressure; pressure rise; boost(ing pressure); step-up of voltage【电】
升压泵 backing pump; booster pump
升压泵站 booster pumping station
升压比 step-up ratio
升压变电所 step-up substation
升压变电站 step-up substation
升压变流器 booster converter
升压变压器 booster; booster transformer; boosting transformer; step-up transformer; transformer booster
升压侧 <变压器> step-up side
升压充电 booster charge; boosting charge
升压抽水站 booster pumping station
升压的 air boosted
升压电 booster station
升压电池组 booster battery; boosting battery
升压电缆 booster cable
升压电路 booster circuit; step-up circuit
升压电容器 boost capacitor
升压电台 booster station
升压电阻 booster; boost resistor
升压二极管 booster diode; series-efficiency diode
升压阀 backup valve
升压放大器 booster amplifier
升压风机 booster(-type) fan
升压缸托架 booster cylinder bracket
升压鼓风机 booster blower
升压机 booster; positive booster
升压激励 step-up excitation
升压技术 pressure build-up technique
升压继电器 booster relay
升压加热器 booster heater
升压控制 boost control
升压脉冲变压器 step-up pulse transformer
升压器 booster; positive booster; step-up transformer
升压曲线 pressure rising curve
升压设备 booster device
升压时间 rise time
升压试验 pressure build-up test
升压水泵 booster pumping station
升压速度 rate of rise
升压误差 step-up error
升压线圈 step-up coil
升压压力计 boost ga(u)ge
升压压缩 compressor booster
升压压缩机 booster compressor
升压油泵 booster oil pump
升压增音电路 boosted boost circuit
升压站 booster installation; booster plant; booster station; step-up substation
升压直流变压器 step-up direct-current transformer
升压中心 anallobaric center[centre]; center[centre] of rise; pressure rise center[centre]
升压自耦变压器 step-up auto-transformer
升焰窑 up-draught kiln
升扬装药 lifting charge
升曳比 lift-drag ratio
升液泵 lift pump
升液斗 lift pot
升液量单位 liter
升液站 <污水> lift station
升鱼机 fish hoist(er); fish lift; fish elevator
升运带 lifting belt; lifting web
升运斗 elevator bucket
升运鼓轮 elevating drum; elevating wheel

升运和输送机械 elevating and conveying machinery
升运机 elevator;mechanical elevator
升运机构 elevating mechanism
升运机拉索 elevator cable
升运机皮带 elevator belt(ing)
升运机输送带 elevator conveyer belt
升运机用索 elevator cable
升运机罩 elevator casing
升运角 elevation angle
升运链 elevating chain;elevator chain
升运链板间距 elevator flight spacing
升运螺杆 elevating screw
升运螺旋 elevating screw
升运马达 elevator motor
升运能力 elevating capacity
升运皮带 lifting belt
升运器 lift conveyer[conveyor]
升运器槽 elevator channel
升运器底板 elevator boot
升运器底滑板 elevator bottom
升运器底滑脚 elevator boot
升运器斗 elevator cup
升运器帆布带 elevator apron;elevator canvas
升运器滑脚 elevator foot
升运器排出槽 elevator spout
升运器式捡拾装载机 elevator-type pick-up loader
升运器式装干草机 elevator hay loader
升运器喂料坑 elevator pit
升运器支架 elevator leg
升运式铲运机 elevated loading scraper;elevating scraper;elevator scraper
升运式平路机 elevating grader
升运式平土机 elevating grader
升运式挖掘机 elevator digger
升运式装载机 lifter-loader;elevator(-type)loader
升运塔 concreting tower
升运装载车 elevating wagon;lifter-loader
升运装置 lift and carry transfer
升值 appreciate;appreciation;revaluation;revalue;upvaluation;upward revaluation
升值机 upward revaluation
升职 promotion
升重 liter weight
升轴 lifting shaft
升轴托 lifting shaft bracket
升装铲运机 elevating scraper
升阻比 lift-drag ratio

生

生氨作用 ammonification
生柏油 crude tar
生板胶 sheet gum
生拌和料 raw mix
生表程序库 production library
生波板 wave paddle
生波机 wave generator;wave maker
生波历时 wave duration
生波器 <用于潮波模型试验> bore generator;wave generator
生波圆的半径 radius of generating circle
生材 green lumber;green material;green timber;green wood
生材比重 specific gravity in green
生材的 green
生材堆场 green lumber storage
生材链 green chain
生材(链条)分类台 green sorting deck
生材密度 green density
生材入口 green end
生材输送台 green chain
生财 stock-in-trade
生财资本 instrumental capital
生草 sward
生草丛 tussock;tussock grass
生草丛冻原 tussock tundra
生草丛植物 tussock plant
生草地 grass covered land
生草法 sod culture
生草禾本真草原 caespitoso-graminosa;steppa genuine
生草灰化 sward podzolic soil
生草灰化土 soddy podzolic soil;sodpodzol soil
生草灰化土区 sward illimerized soil area
生草灰化土试验 soddy podzolic soil test
生草山地草甸土 soddy-mountain meadow soil
生草土 soddy soil;turfy soil
生草休闲 late fallow
生产 childbirth;delivery;fabrication;labo(u)r;making;manufacture;pilot build;production
生产保全 productive maintenance
生产备件 running part;running spare
生产边界的平直性 flatness of production frontier
生产变量 manufacturing variable
生产标准 production standard
生产波动 fluctuation in production
生产剥采比 operational stripping ratio;production stripping ratio
生产薄弱环节 bottleneck operation;production bottle neck
生产不足 underproduction
生产布局 distribution of production;distribution of productive forces;location of production
生产部门 production sector
生产部门成本 producing department cost
生产部门费用 production department expenses
生产财富 productive wealth
生产财务计划 output and finance plan;plan of production and finance;production and finance plan;productive-financial plan
生产采场 active stope
生产参数 manufacturing parameter
生产残留物 production residues
生产残余物 production residues
生产操作 fabrication operation
生产操作能力 production operational capability
生产操作室 processing cell
生产测定 efficiency test;production test(ing)
生产层 pay sand;producing horizon;production formation
生产层底深 bottom depth of producing horizon
生产层顶深 top depth of producing horizon
生产层隔离法 zone of isolation
生产层深度 producing depth
生产层套管深 casing depth of producing zone
生产层有效因素 effective pay factor
生产产量 production output
生产常数 production constant
生产厂 manufacturing plant;producer;producer plant
生产厂的试验 manufacturer's test
生产厂家 manufacturer
生产厂家的授权信 letter of authority from manufacturer
生产厂家/供应商的保证书 manufacturer's/supplier's guarantee certificate
生产厂商保证书 manufacturer's bond
生产场地 production area
生产车间 fabricating bay;manufacturing shop;production works
生产车间费用 production department expenses
生产成本 conversion cost;cost of production;cost price;fabricating cost;fabrication cost;factory cost;first cost;initial cost;manufacturing cost;production cost;running cost;work cost
生产成本报告表 statement of production cost
生产成本比率 output cost ratio;production cost ratio
生产成本比重 specific production cost
生产成本表 production cost sheet;statement of cost of production;statement of production cost
生产成本差异 production cost variance
生产成本单 production order cost sheet
生产成本法则 law of cost
生产成本降低 cost decrease
生产成本降低一半 halve the production cost
生产成本提高 cost push
生产成本图 cost map
生产成本稳定 cost stability
生产成本预算 cost of production budget;production cost budget
生产成本增加 increasing cost
生产成本账户 production cost account
生产成本指数 index of production cost
生产成本中心 production cost center[centre]
生产成本最低的车型 base vehicle
生产成本最低化 cost minimization
生产城市 manufacturing city;producer-city
生产城市煤气的炼焦厂 coke-oven plant for town gas production
生产程序 construction procedure;production program(me);production routine;production sequence;productive program(me)
生产程序的程序 pure generator
生产程序图 flow sheet
生产抽水井 production well
生产储备 production reserves
生产储备资金 production stock capital
生产储层边界圈定 production reservoir delineation
生产大纲 production program(me)
生产单位 production department;production unit
生产单位成本 unit cost of production
生产单位法 unit-of-output method
生产单位折旧法 unit of production depreciation method
生产单元 manufactured unit
生产道路 production road
生产的 manufacturing;productive
生产的对偶性 duality in production
生产的矩形图 box diagram of production
生产的可能性 manufacturing feasibility
生产的溶解有机碳 produced dissolved organic carbon
生产的实物单位 unit of produced kinds
生产的因素 factor of production
生产地点 place of production
生产地理集中 geographic(al)concentration
生产地市场 local market
生产地质指导 control of production geology
生产调度 production despatching[dispatching];production management;production scheduling
生产定单 production order
生产定单号 number of production order
生产定额 job rate;job ration;output quota;performance standard;production quota;production standard;stream factor
生产定型检验 production approval inspection
生产定型试验 production-type test
生产动力的 power-producing
生产动态法 production performance method
生产队 production team
生产对人类有用的畜产品 produce animal products useful to man
生产多功能的(路面)面层材料工厂 multipurpose coating plant
生产额 capacity;delivery capacity
生产法 working system
生产方法 fabricating method;manufacturing method;methods of production;producing method;production method;production process
生产方式 mode of production
生产放款 productive loan
生产废料 processing waste;process scrap;production waste;scrap
生产废气 process gas;production waste
生产废水 factory effluent;factory sewage;manufacturing wastewater;nonpolluted industrial wastewater;production wastewater
生产废水处理场 treatment yard for productive wastewater
生产废液 manufacturing waste
生产废渣 production residues;production waste
生产费(用) cost of production;expenses of production;operating cost;production charges;production cost;production expenses;factory overhead <工厂的>;operation cost;running cost
生产费用表 statement of production expenses
生产费用和总运输费 production cost and freight
生产费用要素 element of production expenses
生产费用预算 budget of production expenses
生产分析管理技术 production analysis control technique
生产辅助建筑物 auxiliary buildings for production
生产概念 production concept of orientation
生产干线【道】 productive arterial road
生产各种食物 to produce various food
生产各种需要的木材 the production of wood for various uses
生产更多的粮食 to produce more food
生产工厂 manufacture plant;production manufacture plant;production plant
生产工程 production engineering
生产工程师 production engineer
生产工具 implements of production;instrument of production;means of production

S

生产工具和设备 production tool and equipment
生产工人 direct labo(u)r;productive worker
生产工人工资 wage of productive workers
生产工人劳动效率 labo(u)r efficiency of productive worker
生产工时法 productivity labo(u)r hours method
生产工艺 production technology;techniques of production
生产工艺过程 technical process of production;production process
生产工艺热负荷 process heating load
生产工艺学 process technology
生产工资 productive wage
生产工作 production work
生产工作面 active stope;forefield;working face
生产工作区 active workings
生产功能 working function
生产供不应求 underproduce
生产供水 industrial water supply;production water supply
生产供水系统 production water supply system
生产故障 fabricating defect;production breakdown;production difficulty;production holdup
生产关系 relationship of production
生产观点 production point of view;production view
生产管理 manufacturing management;plant supervision;production control;production management;productive management
生产管理费用 manufacturing overhead
生产管理系统 management operating system
生产管理制度 production management system
生产管制 output control;production control
生产规定 manufacturing specification
生产规范 manufacturing specification
生产规划 production plan(ning);production programming
生产规模 production capacity
生产规模厂 full-scale plant
生产规模的 production-scale
生产规模要求 manufacturing specification request
生产国标志 country of origin
生产过程 fabricating process;industrial process;manufacturing operation;procedure in production;process of production;production run;productive process;technologic(al)process
生产过程比例性 proportionality of production process
生产过程的检验 work-in-process inspection
生产过程的设计 process design
生产过程分析 process analysis
生产过程工程 process engineering
生产过程简图 outline of process
生产过程均衡性 production process equilibrium
生产过程空间布置组织 organization for productive process space
生产过程空间组织 organization for productive process space
生产过程控制 industrial process control;process control
生产过程控制计算机 process control computer
生产过程控制系统 industries process control system;process control system
生产过程时间接合组织 organization for productive process time
生产过程时间组织 organization for productive process time
生产过程图解 flow chart;flow sheet;flow process diagram
生产过程用检测仪 process instrumentation
生产过程用热 process heat
生产过程中产生的缺陷 in-process defect
生产过程中的加料 in-process addition
生产过程中的检查 in-process inspection
生产过程中的检验 in-process inspection
生产过程中的瓶颈现象 production bottle neck
生产过程自动化 process automation
生产过程组合 process mix
生产过剩 overproduce;overproduction
生产过剩理论 overproduction theory
生产含水层 production aquifer
生产函数 production function
生产合格证 product certification
生产合格证书 certification of construction
生产合同 production contract
生产合作 productive cooperation
生产合作社 producers cooperative
生产和技术指标 production and technical index;production and technical indication
生产和维护成本 operating and maintenance cost
生产和效率 production and efficiency
生产后的维修 post production service
生产后作业 post-production activity
生产环境 production environment
生产会议 conference on production
生产货物 producer goods
生产机械 manufacturing machinery
生产机制 production mechanism
生产积累 production accumulation;productive accumulation
生产基地 construction home base;construction support base;production base
生产基金 production fund
生产集的不可逆性 irreversibility for a production set
生产集的可加性 additivity for a production set
生产集的连续性 continuity for a production set
生产集的凸性 convexity for a production set
生产集约化 production intensification
生产集中 concentration of production
生产计划 fabricating program(me);operating plan;production plan(ning);production program(me);operating schedule
生产计划表 manufacturing plan sheet
生产计划的灵活性 flexibility of production planning
生产计划的评价 evaluation of product planning
生产计划和控制 production planning and control
生产计划科 design production department
生产计划控制系统 production planning control system
生产计划问题 production planning problem
生产计划与分析 planning and analysis for production
生产计划员 scheduler
生产记录 operating data;production record;record of production
生产技能 production skills
生产技术 fabricating technique;manufacturing technique;producing practice;production engineering;production technique;technology;technology of production
生产技术参数 technologic(al)production norm
生产技术构成 composition of production technologies
生产技术管理 productive technology management
生产技术规程 technical regulations for production
生产技术说明 operating technical statement
生产技术条件 production specification
生产技术系数 technical coefficient of production
生产技术要求 manufacturing requirement;
生产技术指标 production and technical indication
生产技术指导 production and technical guidance
生产技术准备 productive and technical preparation
生产加速度的力 force acceleration
生产价格 cost price;price of production;producer price;production price
生产价值 productive value
生产间接费用 manufacturing overhead
生产监督 manufacturing supervision;plant supervision
生产检验 production inspection
生产建设投资 productive construction fund
生产建设性支出 portion of state expenditures for production and construction
生产建设资金 production and construction funds
生产阶段 production phase;production stages
生产节拍 tact of production
生产节拍时间计算 tact timing
生产结构 pattern of production;productive structure
生产结构图 production structure diagram
生产进度 manufacturing schedule;production scheduling
生产进度表 production schedule
生产经济学 economics of production;production economics
生产经济原则 economic principles of production
生产经理 production manager
生产经营成本支出审计 productive and operating cost audit
生产经营性基建投资 investment in capital construction projects for commercial and productive purposes
生产井 field well;producing well;well in operation;well production
生产井测试 production well testing
生产井的钻探 production well drilling
生产井段 producing interval;production range
生产净额 net product
生产净能 net energy for production
生产竞赛 production drive;production emulation
生产勘探 production exploration
生产勘探工程系统 system of production engineering
生产考核 performance test
生产科 production department
生产可能性 production possibility
生产可能性曲线 production possibility curve
生产孔 production hole
生产控制 production control
生产控制程序 production control process
生产控制图表 production control graph
生产快报 express summary of operation
生产宽容度 tolerance level in production
生产矿井 production well
生产矿井测量 production shaft survey
生产矿井调查 investigation of productive mine
生产矿量 production reserves
生产矿量保有期限 period of retention ore reserves for production
生产矿量变动 variation of production reserves
生产矿区 being produced area
生产亏损 manufacturing loss
生产垃圾 production waste
生产劳动成本计算法 productive labo(u)r cost method
生产劳动小时计算法 productive labo(u)r hours method
生产类用具 commercial-type appliance
生产力 give out;producing power;productive capacity;productive force;productiveness;productive power;productivity;yield capacity;yield power
生产力比率 productivity ratio
生产力标准 productive forces criterion;standard of productive forces
生产力布局规划 planning of distribution of productive forces
生产力测定 productivity measurement
生产力的利用程度 utilization of capacity
生产力过剩 overcapacity
生产力计量 productivity measurement
生产力金字塔 pyramid of productivity
生产力控制指数 productivity rating index
生产力配置论 allocation theory
生产力趋势 productivity trend
生产力数量规模 productivity quantitative scale
生产力水平 productivity level
生产力弹性 elasticity of productivity;elasticity of the production function
生产力增长率 rate of growth of productivity
生产力指数 index of productivity
生产利润分配 production profit sharing
生产量 output;output capacity;production capacity;production quantity;productive output;productivity;quantity of production;throughput;volume of production
生产林 production forest
生产流 production current;withdrawal current

生产流程 manufacturing process; process flow; production flow(line); production process
生产流程分析 production flow(line) analysis; productive flow analysis
生产流程模型 production flow pattern
生产流程设计 process design
生产流程图 process flow sheet; production flowchart; production flow diagram
生产流水线 process line; production (flow) line
生产流水作业设备 production run equipment
生产率 capacity of production; capacity rating; manufacturing rate; output capacity; output rate; production coefficient; production rate; productiveness; productive rate; productivity; productivity rate; rate of production; rate of working; throughput; throughput rate
生产率变动范围 spread of productivity
生产率不足 undercapacity
生产率测定 efficiency testing; productivity measurement
生产率差异 productivity difference; productivity differential
生产率低的农业劳力 labo(u)r of low productivity in agriculture
生产率递减 diminishing productivity
生产率估算 production estimating
生产率会计 productivity accounting
生产率系数 productivity factor
生产绿地 productive plantation area
生产绿化地带 <城市> productive green
生产轮询 production poll
生产煤巷 active entry
生产煤样 coal sample for checking production
生产煤样编号 production coal sample number
生产每吨干物质 per ton of dry matter produced
生产密度 production density
生产面 production surface
生产面积 plant space; production area
生产命令 fabrication order; manufacturing order; production order
生产模型 production model
生产目标 production goal
生产耐用品的部门 durable goods sector
生产能 productive energy
生产能力 capacity; capacity production; delivery capacity; earning capacity; output capacity; production capability; production capacity; productive capability; productive capacity; productive power; productivity; throughput (capacity); work(ing) capacity
生产能力比率 capacity ratio; productivity ratio
生产能力薄弱环节 capacity bottleneck
生产能力差异 capacity variance
生产能力成本 capacity cost
生产能力费用 capacity expenses
生产能力估计 estimate of output capacity; estimate of production capability; estimate of productivity
生产能力过剩 excess production capacity; excess productive capacity; in excess of production capacity; overcapacity
生产能力建成率 ratio of completion under productive capacity
生产能力利用率 capacity utilization rate
生产能力试验 output test
生产能力水平 capacity level
生产能力投资 capacity investment
生产能力指数 index of production capacity
生产能量比率 capacity ratio
生产能量成本 productive capacity cost
生产年份 productive year
生产年限 productive life; productive life length
生产配方 factory formula; practical formulation; production formula
生产批 batch
生产批号 production run number
生产批量 production lot
生产品 product
生产平台 production platform
生产凭单 fabrication order; manufacturing order; production order
生产铺路小方石的机器 paving sett making machine
生产普查 census of production
生产期费用 production period cost
生产期限 period of production
生产企业 production enterprise
生产气候预测 ecoclimate forecasting
生产气体 process gas
生产气体炭黑过程 black process of production gas
生产前成本 preproduction cost
生产前基建费用 preproduction capital expenditures
生产前试验 preproduction-type test
生产潜力 latent productive capacity; potential of production; potential productive force; production potential; productive potentiality
生产强度 production intensity
生产情报 production information
生产情报和控制系统 production information and control system
生产区 production bay; production quarter
生产区域 production area
生产曲线 production curve
生产热负荷 production heat load
生产人工时数法 productive labo(u)r hours method
生产人员 operating personnel; production people
生产任务单 manufacture order
生产任务书 manufacture order
生产日记 production log
生产日期 date of manufacture
生产容量 <线路容量和运行速度的乘积> productive capacity
生产容量的分派 allocation of production capacities
生产溶液 plant solution
生产商品食糖 commercial production of sugar
生产上试用 pretest production
生产上限 upper limit of production
生产设备 income-producing equipment; machinery of production; manufacturing facility; manufacturing installation; production equipment; production facility; production unit
生产设备有效使用寿命 production life
生产设计 production design
生产设施 manufacturing facility
生产生态学 production ecology
生产声道 production track
生产剩余 surplus of production
生产石油 produce petroleum
生产时间 production time; productive time; time of production; working time
生产时间中的无效时间 non-productive working time
生产时间中的有效时间 productive working time
生产实践 producing practice; productive practice
生产食品和纤维的原料 raw materials to produce food and fiber
生产使用 plant use
生产事故 industrial accident
生产试验 factory test; pilot production; production test(ing)
生产试验船 production and testing ship
生产收益 proceedings from production
生产手段 means of production
生产术语 shop term
生产数据 operating data; operational data
生产数量折旧法 service output depreciation method
生产数字 production figure; production file
生产衰减率 ratio productive decreasing
生产衰减主要因素 principal factors of productive decreasing
生产衰退 decline in production
生产水果和蔬菜 produce fruits and vegetables
生产水平 production level; work level
生产说明书 production program(me)
生产损耗 production loss
生产损失 loss in production
生产所得支出 income-producing expenditures
生产台时 productive machine-hour
生产套管 production casing; production string
生产套管线 <石油> production casing string; production string of casing
生产条件 condition of production; plant condition; production condition; working condition
生产调节剂 growth regulation
生产调整 production adjustment
生产调整准备成本 setting-up cost
生产调整准备费用 setting-up expenses
生产停滞 damp production
生产通知单 factory order; production order
生产统计 manufacturing statistics
生产投资 productive investment
生产图 shop drawing
生产图解 production diagram
生产图纸 fabricating drawing; working drawing
生产维修 productive maintenance
生产温度 production temperature
生产污水 polluted industrial wastewater; production sewage; production waste
生产污水处理装置 process waste (water) treatment plant
生产物 product
生产物资 producer goods
生产误差故障 dependent failure; production error failure
生产系数 coefficient of production
生产系统 production system
生产下降曲线 production-decline curve
生产纤维 produced fiber[fibre]
生产现场 production field
生产现状 manufacturing status
生产线 assembly line; line of production; processing line; production chain; production line
生产线布置 layout of production line
生产线的评价 production line evaluation
生产线工艺 production line technique
生产线平衡 line balancing
生产线平衡分析 line balance analysis
生产线上设备 built-in unit
生产线损益计算书 production line profit and loss statement
生产线性能考核 production line performance test
生产线盈利能力 production line profitability
生产线职能 line function
生产线中热成型机 in-line type thermoforming machine
生产线装配法 assembly line principle
生产线装配作业 line-assembly work
生产限制因素 production constraint
生产巷道 active workings
生产消费 productive consumption
生产消费限制 limit for production and consumption
生产小时成本比较表 comparative statement of cost per productive hour
生产效率 job efficiency; production efficiency; production output
生产效能 productiveness
生产效用 utility of production
生产新树的方法 method of producing a new tree
生产信息处理系统 production information processing system
生产信息控制系统 production information and control system; production information control system
生产行为 act of production
生产形式 production model
生产型 production configuration
生产型铲斗 production version bucket
生产型号 production model; production version
生产型设备 plant-size equipment
生产型铣床 manufacturing miller
生产性 producibility
生产性查询 productive poll
生产性程序 production routine; productive routine
生产性毒物 productive poison
生产性工作时间 productive working time
生产性规范研究 full-scale plant study
生产性货物 capital goods
生产性积累 productive accumulation
生产性建设 construction for production purposes; construction for productive purposes
生产性建筑 productive building
生产性开发 commercial exploitation
生产性劳动 productive labo(u)r
生产性冷库 productive cold store
生产性逻辑测试机 production logic tester
生产性能 production performance
生产性能测定 performance test; record of performance
生产性能的性状 performance trait
生产性能登记 production registry
生产性能登记制 production registry system
生产性能随机测定 random sampling performance test
生产性能特征 performance trait
生产性能指数 performance index
生产性企业 production enterprise
生产性热储 productive reservoir
生产性任务 productive task
生产性森林 production forest
生产性设备 full-scale plant
生产性试验 factory testing; full-scale test(ing)
生产性停歇 productive downtime

S

生产性外伤 industrial injury
生产性研究 manufacturing research
生产性运输 productive transportation
生产性运行 production run
生产性支出 productive outlays
生产性助铲 < 铲土机 > production pushing
生产性装置 full-scale device
生产性状 production trait
生产性资产 productive assets
生产许可证 manufacturing certificate; manufacturing permit; production certificate; production license [licence]; production permit
生产循环 manufacturing cycle
生产压差 differential pressure under production; pressure difference of production; pressure difference under production
生产压力 production pressure
生产研究 production study
生产掩模 production mask
生产样品 production sample
生产要求 production requirement
生产要素 elements of production; factors of production; production factor
生产要素成本 production factor cost
生产要素得最低成本 least factor-cost for producing
生产要素组合 factor mix
生产冶金学 process metallurgy; production metallurgy
生产业务 business of producing; product service
生产一体化 productive unification
生产因素 factors of production
生产用地 productive land
生产用电解槽 production-scale cell
生产用房屋 production quarter
生产用固定资产 fixed assets used in production; productive (fixed) assets
生产用机床 production tool
生产用建筑 production building
生产用汽 process steam
生产用燃气轮机 processing gas turbine
生产用设备 plant-scale equipment
生产用水 process water; production water supply; water of productive us(ag)e
生产用水的供应 production water supply
生产用水井 production well
生产用水系统 production water supply system
生产用推土机 PAT dozer; production application tractor dozer
生产用压机 production press
生产用杂费 industrial sundry expenses
生产用蒸汽 process steam
生产(油)井 production well
生产油气比 gas-oil ratio; producing gas-oil ratio
生产与成本管理主任工程师 chief production control and costing engineer
生产预算 manufacturing budget; production budget
生产预制屋面料 ready roofing manufacture
生产原型 production prototype
生产运行 production run
生产凿岩 production drilling
生产责任制 production responsibility system; responsibility system for production; system of production responsibility
生产增长率 ratio of productive increasing
生产增长主要因素 principal factors of productive increasing
生产者 manufacturer; producer
生产者的风险 producer's risk
生产者价格 producer price
生产者剩余 producer's surplus
生产者协会 Producers Council
生产者需求 producer demand
生产者责任 producer responsibility
生产支出 productive expenditures
生产支线【道】 productive branch road
生产职能 production function
生产指标 output target; production index; production quota; production standard; production target
生产指导 operation instruction
生产指挥人员 line boss
生产指示图表 process chart
生产指数 production index; productive rating index
生产指数单位成本 production index unit cost
生产制造 production-manufacturing
生产制造工程 production-manufacturing engineering
生产质量检验 production quality test
生产中的废料 production waste
生产中断 production break
生产周期 fabricating cycle; period of production; producing period; production cycle; production period
生产专业化 productive specialization; specialization of production
生产装置 process unit
生产准备成本 set-up cost; starting load cost
生产准备费 production preparation fee
生产准备时间 lead time; make ready time; set-up time
生产准时制 just-in-time system
生产资本 capital producing; producer capital; production capital; productive capital; working capital
生产资本密集程度 capital intensity of production
生产资财 productive property
生产资金 producer capital; production capital; productive capital; working capital
生产资金定额 production capital quota
生产资料 capital equipment; capital goods; consumer goods; consumption goods; industrial goods; means of production; operating data; producer goods; production goods
生产资料补给 supply of means of production
生产资料转移价值 transferred value of the means of production
生产资源 resources for production
生产自动化 production automation
生产综合费用 synthetic(al) expenses
生产总成本 total production cost
生产总额 aggregate output
生产总值 total output value; total value of production
生产组织 factory management; organization of production
生产组织计划 organization and planning of production
生产组织技术 industrial engineering
生产钻进 production drilling
生产钻井 production drilling
生产作业能力 production operational capability
生产作业线 production line
生潮机 tide generator
生潮装置 < 水力模型的 > tide generator
生尘 dusting
生尘性 dustiness
生辰石 birthday stone
生衬 undercoating
生成 formation; generation
生成变换 generating transformation
生成程度 degree of maturity
生成程序 generating program(me); generating routine; generation routine; generator
生成程序段 generation phase
生成出现 generation occurrence
生成代码 generating code
生成的 generative
生成的单调类 generated monotone class
生成的可传类 generated hereditary class
生成的理想 generated ideal
生成的子群 generated subgroup
生成地址 calculated address; generated address; synthetic(al) address
生成点 generic point
生成点滴 formation of droplets
生成对 generating pair
生成多项式 generator polynomial
生成反应 reaction of formation
生成泛函 generating functional
生成非标准标号 creating nonstandard label
生成符号 generated symbol
生成概率 generating probability
生成规则 generating rule; generative rule; production rule
生成过程 generative process
生成函数 generating function
生成函数的处理 manipulation of generating function
生成焓 enthalpy of formation
生成环 generated ring
生成集 generated set
生成矩阵【数】 generated matrix
生成空间 generated space; span
生成力 generative power
生成例程 generating routine
生成脉冲 production burst
生成模式 generating scheme
生成母图 spanning supergraph
生成能力 generative capacity
生成气体 gassing
生成热 formation heat; generation of heat; heat of formation
生成日期 creation date
生成软件 generative software
生成森林 spanning forest
生成设备程序 creation facility program(me)
生成时间 rise time
生成式 production
生成式宏程序 generative type macro
生成式(计算机)制图法 generative graphics
生成寿命 generation lifetime
生成输入流 generation input stream
生成树【数】 spanning tree; generation tree
生成数 generation number
生成数据组 generation data group
生成水垢 scale formation; scale forming; scaling
生成算法 < 计算机的 > key generation algorithm; generating algorithm
生成算子 generating operator
生成态晶体 "as-grown" crystal
生成天体 generating celestial body
生成添印 generation printing
生成条件 formation condition
生成图式 generative scheme
生成土 genetic(al) soil
生成位错 product dislocation
生成文法 generative grammar
生成文件 spanned file
生成物 outgrowth; product; resultant
生成物理块记录 S-mod record
生成误差 < 使用不精确参数、公式引起的 > generated error
生成线 generating line
生成型式 generated form
生成型宏程序 generative type macro
生成序次 succession of generation
生成氧化膜 sull coat
生成语句 generated statement
生成元 generator
生成元的极小集 minimal set of generators
生成元素 generating element
生成运行 generating run
生成转换文法 generative transformational grammar
生成锥 generating cone
生成子集 generating subset
生成子图【数】 spanning subgraph
生成自同构 generating automorphism
生成自由能 free energy of formation
生成作用文法 generative action grammar
生虫的 wormy
生储盖组合 association of source reservoir and cap rock
生存 subsist(ence); survival
生存带 living zone
生存概率 probability of survival; survival probability
生存函数 survivorship function
生存环境 living environment
生存活动 survival movement
生存竞争 struggle for existence
生存空间 living room; living space; vivosphere
生存率 fraction; survival rate
生存能力 viability
生存期 duration; life cycle; life time
生存期间 life span
生存潜能 survival potential
生存曲线 survival curve; survivorship curve
生存时间 life span; survival time
生存条件 existing condition; life condition; survival condition
生存线对策 lifeline game
生存性 survivability; viability
生存性准则 survivability criterion[复 criteria]
生存战略 survival strategy
生的 raw; unslaked < 石灰 >
生底辟 salt diapir
生地 raw land
生地蜡 ader wax
生动建筑艺术 animation architectural art
生动实例 speaking example
生发丝裂缝 feather check(ing)
生废水 crude wastewater; fresh wastewater; raw wastewater
生粉筒仓 raw meal silo
生风区域 generation area
生根 rootage
生根带 rooting zone
生垢 scale formation
生垢率 rate of scale formation
生垢水 scale-production water
生垢因数 fouling factor
生骨料 raw aggregate
生硅藻土 raw diatomite
生褐斑的 < 书页等 > foxed
生弧电刷 arcing brush
生化变异 biochemical variation
生化参数 biochemical parameter

S

生化池 biochemical tank
生化池截留污水处理过程 lagooning
生化出水 biochemical effluent
生化处理 biochemical treatment
生化处理城市污水 biochemically treated municipal wastewater
生化处理装置 biochemical treatment facility
生化的 biochemical
生化法 biochemical process
生化反应 biochemical reaction
生化反应动力学 biochemical reaction kinetics
生化反应时间 biochemical reaction time
生化分类学 chemotaxonomy
生化过程 biochemical process
生化耗氧量 biochemical consumption of oxygen
生化环境 biochemical environment
生化混凝沉淀法 biochemistry-coagulation sedimentation method
生化机理 biochemical mechanism
生化甲烷势 biochemical methane potential
生化假设 biochemical hypothesis
生化降解 biochemical degradation
生化净化 biochemical purification
生活垃圾 domestic garbage;domestic waste;sanitary waste
生化滤层 bacteria bed;biologic(al) filter;contact bed;continuous filter
生化培养基 biochemical medium
生化燃料电池 biochemical fuel cell;bio-fuel cell
生化生态学 biochemical ecology
生化势能 biochemical energy potential
生化试验 biochemistry test
生化试验室 biochemistry laboratory
生化特征 biochemical characteristic
生化突变 biochemical mutation
生化突变体 biochemical mutant
生化污染标准 biochemical pollution criteria;biochemical pollution criterion
生化污水处理 biochemical sewage treatment
生化污水容量 biochemical capacity of wastewater
生化吸附法 biochemical absorption
生化相促作用 allelocatalysis
生化效应 biochemical effect
生化需氧量 biologic(al) oxygen demand
生化需氧量测定 biochemical oxygen demand determination;biochemical oxygen demand test
生化需氧量测定仪 biochemical oxygen demand analyser[analyzer]
生化需氧量沉浮 biochemical oxygen demand settling and rising
生化需氧量沉浮系数 biochemical oxygen demand settling and rising coefficient
生化需氧量反应动力学 biologic(al) oxygen demand reaction kinetics
生化需氧量负荷 biochemical oxygen demand load(ing)
生化需氧量缓和剂 biochemical oxygen demand moderator
生化需氧量浓度 biochemical oxygen demand concentration
生化需氧量曲线 biochemical oxygen demand curve
生化需氧量试验 biochemical oxygen demand test
生化需氧量污泥负荷 biochemical oxygen demand sludge loading
生化需氧量自记仪 biochemical oxygen demand automatic recorder;biochemical oxygen demand meter
生化需氧量总量负荷 biochemical oxygen demand mass loading
生化絮凝 biochemical flocculation
生化血指数 biochemical blood index
生化氧化作用 biochemical oxidation
生化抑制 biochemical inhibition
生化指示剂 biochemical indicator
生化指示物 biochemical indicator
生化指数 biochemical index
生化转化 biochemical conversion
生化自净化作用 biochemical self-purification
生化作用 biochemical action
生荒地 lay land;new soil;primitive soil;reclaimable virgin soil;sterile soil;virgin ground;virgin land;virgin soil
生荒地土壤 virgin soil
生黄麻 raw jute
生黄铜 cast brass
生灰斑 fish eye
生混合料 raw mix
生混凝土 green concrete
生活安宁 domestic tranquility
生活本能 life instinct
生活必需品 consumer goods;daily necessaries[necessities];essential goods;essentials of life;necessities of life
生活标准 living standard;standard of living
生活补贴 subsistence allowance
生活补助费 subsistence allowance
生活舱 living chamber
生活舱室 living accommodation
生活场所 life area
生活车 living car
生活出行 non-work trip
生活单元楼板面积 living unit floor space
生活的第一需要 life's prime want
生活抵抗力 vital hardiness
生活吊【机】accommodation crane
生活方式 life form;life style;style of living;way of life;commuterization <往返城市和郊区住所的>
生活房屋 accessory building
生活废水 domestic wastewater;house(hold) wastewater;sanitary wastewater;wastewater from living
生活废物 domestic waste;sanitary waste
生活费(用) cost of living;living cost;living expenses;subsistence cost;alimony;subsistence expenses
生活费用的提高 cost of living escalation
生活费用价格总指数 total price index of living cost
生活费用条款 cost of living clause
生活费用指数 cost of living indexes
生活费指数 cost of living indexes
生活粪便污水 domestic fecal sewage
生活服务设施 personal service establishment;service facility
生活福利建筑 welfare facility
生活福利设施 welfare facility
生活辅助建筑物 auxiliary buildings for living
生活给水关闭 domestic water shut-off
生活给水系统 domestic water system
生活工资 living wage
生活功能 vital function
生活供电线路 electric(al) power lifeline
生活供水 domestic water supply;service water
生活供水工程 domestic supply sewage
生活供水水质 domestic water supply quality
生活供水系统 domestic water supply system
生活固体废物 domestic solid wastes
生活固体垃圾 household waste
生活规律 law of life
生活过程 life process
生活耗水量 domestic water consumption
生活耗用量 domestic consumption
生活化粪池污水 domestic septic tank effluent
生活环境 bag;habitation;living environment;surrounding;surrounding environment;habitat
生活环境条件 living environmental condition
生活回用水系统 domestic water reuse system
生活活动力 life activity
生活机能 vital function
生活基础设施 <住宅、交通、上下水道、电气、煤气、道路、公园绿化等> basic life-related facility
生活基地 construction home base
生活间 changing room;locker room;rest quarter;welfare building;welfare quarter
生活津贴 living allowance;subsistence;subsistence allowance
生活居住区 living-dwelling zone
生活空间 life space;livehood space;living space
生活垃圾 consumer waste;domestic garbage;domestic refuse;home scrap;household garbage;house(hold) refuse
生活垃圾处理法 treatment of domestic refuse;house(hold) refuse disposal
生活垃圾处置系统 household disposal system
生活垃圾的处置 domestic refuse disposal
生活垃圾焚化炉 domestic incinerator
生活垃圾压实机 domestic refuse compactor
生活力 viability;vigour;vital force
生活炉灰 household cinder;household clinker
生活目标 life goal
生活能力 viability
生活排污 domestic discharge;sanitary drainage
生活期 life stage
生活气候学 domestic climatology
生活强度 life intensity
生活情况 existing condition
生活区 living area;living quarter;living zone;residential district;rest quarter;utility area
生活圈 life circle;living sphere;sphere of life;watersphere
生活设施 convenience[conveniency];domestic installation
生活时期 life period
生活史 life history
生活史表解 life cycle diagram
生活史的研究 studies of the life histories
生活室 sitting room
生活水平 cost of living;estate;level of living;living level;living standard;quality of life;standard of living;subsistence level
生活水平指标 index of living standard
生活水平指数 index of living standard
生活水源 domestic water supply
生活素质极高化 maximization of living standard
生活条件 living condition
生活维持费 subsistence
生活污泥 domestic sludge
生活污染 domestic pollution
生活污染源 domestic pollution sources;sanitary sources of pollution
生活污水 domestic sew(er)age;domestic wastewater;household sewage;household wastewater;house sewage;housing sewage;sanitary sewage;sanitary waste;sanitary wastewater;wastewater from living;wastewater of domestic use
生活污水泵 sanitary pump
生活污水产生指数 coefficient of domestic sewage production
生活污水处理 domestic sewage treatment;domestic wastewater treatment;residential sewage treatment
生活污水处理厂 residential sewage treatment plant
生活污水低温预处理 low-temperature pre-treatment of domestic sewage
生活污水管(道) house drain;sanitary sewer
生活污水(管道)系统 sanitary sewer system
生活污水管网 domestic sewer network;domestic sewer system
生活污水管系统 domestic sewer system
生活污水灌溉 household wastewater irrigation
生活污水后生物处理 domestic wastewater post-biotreatment
生活污水回收 domestic wastewater reclamation
生活污水窨井 sanitary sewer manhole
生活污水立管 soil stack
生活污水量 volume of domestic wastewater
生活污水流量 residential sewage flows
生活污水排放系统 sanitary drainage
生活污水排水管 house drain
生活污水排泄 building sanitary drain
生活污水排泄系统 sanitary drainage
生活污水深度处理 advanced treatment of domestic wastewater
生活污水窨井 sanitary sewer manhole
生活物质 living material;living substance
生活习惯 habits and customs
生活习性 life habit
生活现象 vital phenomenon
生活消费量 domestic consumption
生活消费品价格 consumer price
生活消耗 domestic demand
生活小区 biotope
生活型 living form
生活型级 life-form class
生活型谱 life-form spectrum
生活型式 life form
生活型系统 life-form system
生活型优势 life-form dominance
生活性乘车出行 non-work trip
生活需水量 domestic water demand
生活需要 domestic demand;living needs;living requirement
生活因素费用 cost of living factor
生活饮用水 domestic potable water
生活饮用水水质标准 potable water quality standards;quality standard of drinking water

S

生活饮用水卫生标准 sanitary standard for drinking water
生活用电电路 utility circuit
生活用电负荷 domestic load
生活用电量 domestic load
生活用电设施 domestic installation
生活用电需量 household demand
生活用锅炉 boiler for domestic use
生活用平台 living terrace
生活用气 domestic utilization of gas
生活用水 community consumption; domestic use of water; domestic water; domestic water use; living water use; service water; water for domestic usage; water for domestic use; water for human consumption; water for living
生活用水加热器 service water heater
生活用水量 domestic consumption; domestic water consumption; residential water consumption
生活用水水库 service water reservoir
生活用水水源 sources of domestic water
生活用水消毒 disinfection of domestic water
生活用烟囱 building chimney
生活有机体 living organism
生活再用 domestic reuse
生活在其他动物体内 living in the body of another animal
生活噪声 domestic noise
生活照明 living light
生活指数 cost of living indexes; index of living; living index
生活指数津贴 cost of living allowance
生活制度 daily regime; regime(n)
生活质量 quality of life
生活质量法 quality method of life
生活周期 growth cycle; life circle; life cycle
生活状况 living condition
生活资料 consumer goods; consumption goods; livelihood; living allowance; means of livelihood; means of living; means of subsistence
生活资料物价指数 consumer price index
生活组合 life assemblage
生活作用 vital action
生火 lighting-off
生火工具 fire irons
生货 raw goods
生机 vital force
生集料 raw aggregate
生计 livelihood; subsistence; sustenance
生姜 ginger
生浆制备 grouting
生胶混炼机 Banbury mixer
生胶块 raw rubber block
生胶料 green glue stock; green stock
生胶片 sheet rubber
生胶质 collagen; ossein(e)
生节 sound knot
生境 biotope
生境保护 habitat conservation; habitat preservation; habitat protection
生境变化 habitat deterioration
生境变化性 habitat variability
生境多样化 habitat diversification
生境多样性 beta-diversity; habitat diversity
生境恶化指数 habitat degradation index
生境范围 habitat range
生境分离 habitat segregation
生境分析 habitat analysis
生境改善 habitat modification
生境隔离 habitat isolation
生境管理 habitat management
生境混合作用 habitat complex
生境集合 habitat grouping
生境结果多样性 structural diversity of habitat
生境龛 ecologic(al) niche
生境(类)群 habitat group
生境评价法 habitat evaluation procedure
生境破坏 habitat destruction; habitat dislocation
生境区 biochore
生境容量 habitat volume
生境适当的 site suited
生境适宜指数 habitat suitability index
生境梯度 habitat gradient
生境条件 habitat condition; site condition
生境条件指数 habitat condition index
生境图 habitat mapping
生境稳定性 habitat stability
生境习性 habitability
生境形态 habitat form
生境型 habitat type
生境型式 habitat form
生境选择 habitat selection
生境因素 factor of the habitat; habitat factor; site factor
生境因子 factor of the habitat; habitat factor; site factor
生境质量 habitat quality
生境质量指数 habitat quality index
生境状况 habitat status
生境总体 habitat complex
生聚系数 source-accumulation coefficient
生来的 connatural
生理裂纹 <树木> quagginess
生理疲劳 physiologic(al) fatigue
生理气候学 physiologic(al) climatology
生理研究所 institute of physiology
生理盐水 normal saline
生力面 carrying plane
生沥青 crude asphalt
生辆位置的不断补充清单 perpetual inventory for car location
生料 crude material; slurry
生料拌和 raw mix
生料层 batch blanket
生料成分控制 raw meal composition control
生料成球 raw meal nodulizing
生料的 in the rough
生料粉 raw meal
生料(粉)均化 homogenization of raw meal
生料粉磨车间 raw meal grinding plant
生料粉磨设备 raw meal grinding plant
生料灰 kiln feed dust
生料混合库 blended meal slurry
生料浆 raw slurry; though slurry
生料浆磨 raw slurry mill
生料搅拌 raw meal blending
生料搅拌库 blended meal silo; raw blending silo; raw meal blending silo
生料接收站 pig receiver station
生料进给率比 ratio of raw feed rate
生料均匀性 raw meal homogeneity
生料库 raw meal silo; raw mix bin
生料块 briquet(te)
生料量 amount of raw mix
生料磨 raw mill; slurry grinding mill
生料配料 raw meal proportioning; raw mix proportioning
生料配料计算 calculation of raw mix proportions
生料球 granules; nodulized raw meal; raw meal granule; raw meal nodule; raw meal pellet
生料球的热稳定性 thermal stability of raw meal nodules
生料取样 raw meal sampling
生料圈 meal ring
生料熔成率 molten rate of batch
生料熔尽的 batch-free
生料/熟料比 raw material to clinker ratio
生料调配 raw meal blending
生料溢流 raw meal overflow
生料釉 raw glaze
生料预热器 raw meal preheater; raw mix preheater
生料云母 crude mica; raw mica
生料制备 raw mix preparation
生料质量控制系统 raw quality control system
生料中掺煤共同粉磨 coal intergrinding
生料助磨剂 raw meal grinding aid
生裂缝 seam
生硫化物 sulphidisation
生铝 cast alumin(i)um
生绿锈 patination
生麻 raw hemp
生满石笋的 stalagmitic(al)
生煤焦油 crude tar
生面团 dough
生灭过程【数】birth and death process
生命 biosis
生命层 biosphere
生命的 biotic
生命分子 biomolecule
生命工程 life project
生命规律学 bionomy
生命科学 <包括生物学、医学、心理学、人类学、社会学等> life science
生命力 life-force; nature; vital force; vitality
生命粒子 biomone
生命起源 arohebient
生命损失 life loss
生命体 biomass
生命危险 danger to life
生命维持系统 life support system
生命系统 life system
生命线 lifeblood; life line
生命线地震工程 lifeline earthquake engineering
生命线地震工程技术委员会 Technical Council on Lifeline Earthquake Engineering
生命线系统 lifeline system
生命盐类 <生物体所需的溶解盐类> biogenic salts
生命支持系统 life support system
生命周期 life cycle
生木 green wood
生奈尔 <重量单位> centner
生(牛)皮 raw hide
生脓 pyogenesis
生坯 green body; green compact; green pressing; green ware; pressed green compact
生坯存放室 greenhouse
生坯加工 green machining
生坯进窑机 kiln feeding unit
生坯开裂 green cracks
生坯孔隙度 green porosity
生坯库 green material store
生坯密度 green density; unfired density
生坯强度 green strength; unfired strength
生坯上釉 green glazing
生坯体 green ware
生皮 fell; green hide; hide; pelt
生皮仓库 hide house
生皮打号锤 hide-marking hammer
生皮打印机 hide stamping machine
生皮检验局 hide inspection bureau
生皮料 raw hide
生皮轮 rawhide wheel
生皮切屑 hide trimmings
生皮手锤 rawhide hammer
生皮制的 raw hide
生皮专用集装箱 hide container
生漆 Cheshu lacquer; Chinese lacquer; Japanese lacquer; raw lacquer
生漆油 raw linseed oil
生气 vitality
生气化 vitalization
生气剂 blowing agent; inflating medium
生气泡 blister
生气潜量 gas-source potential
生气岩层 source bed; source rock
生铅 bullion lead; pig lead; work lead
生铅釉 raw glaze
生球(团矿) green ball
生热 calorigenic; development of heat
生热的 calorifacient; calorific; calorigenic; thermogenic; thermogenous
生热反应 pyrogenic reaction
生热性 heat generation; heat production
生热学 thermogenics
生热中枢 thermogenic centre
生热作用 thermogenesis
生日判定 birthday paradox
生软毛的 woolly
生色团 chromogen; chromophore
生砂 <翻砂用新砂> greensand
生烧 false burning; under-burning; under-firing
生烧的 insufficiently burnt
生烧或过烧 faulty burning
生烧建筑砖 place brick
生烧无光的 frosted
生虱子 pediculosis
生石膏 calcined lime; plaster rock; plaster stone; potter stone; pure gypsum; raw gypsum; uncalcined gypsum
生石膏存放 gypsum stor(ag)e
生石灰 anhydrous lime; burned lime; burnt lime; calcined lime; calcining; calcium lime; calcium oxide; calcium quicklime; callow rock; ca(u)lk; caustic lime; dehydrated lime; hard-burned free lime; lump lime; pellouxite; quicklime; unslaked lime
生石灰仓库 quicklime bunker
生石灰厂 quicklime manufacturing plant
生石灰块 lump quicklime
生石灰磨 quicklime mill
生石灰熟化 slaking of quick lime
生石灰研磨机 quicklime mill
生石灰柱 quicklime pile; unslaked lime pile
生石灰桩 quicklime pile
生石灰桩施工法 chemio-pile method
生石棉 asbestos crude; raw asbestos
生柿汁 kaki-shibu
生手 apprentice; greener; green hand; inexperienced operator; inexpert; neophyte; newcomer; new hand; novice; poor hand; raw hand; tyro
生手工人 unskilled labo(u)r
生疏的 unfamiliar
生树脂 green resin
生水 crude water; natural water; raw water
生水储水池 raw water storage basin
生水垢 furred
生水水网 raw water network
生丝检验所 conditioning house
生死比 birth-death ratio

生酸的 acidific
生苔的 mossy
生态安全 ecologic(al) security
生态安全状况 ecologic(al) security status
生态保护 conservation of biologic(al) life; ecologic(al) protection
生态悲观主义 ecologic(al) pessimism
生态背景 ecologic(al) setting
生态变化 ecologic(al) change
生态变异 ecocline
生态变种 ecologic(al) variety
生态变种反应 ecophene
生态变种选择 ecotypical selection
生态标记 eco-mark
生态标签 eco-label
生态标志 ecolabelling
生态表型 ecologic(al) indicator; ecophenotype
生态表征 ecologic(al) indicator
生态博物馆 eco-museum
生态薄弱地带 ecologic(al) ly fragile terrain
生态补偿与赔偿 ecologic(al) compensation and amends
生态不平衡 ecologic(al) imbalance
生态参数 ecologic(al) parameter
生态操纵 ecologic(al) manipulation
生态草业 eco-herbary
生态层 ecosphere
生态差别 ecologic(al) diversity
生态差型 ecocline
生态差异 ecocline
生态承载力 ecologic(al) carrying capacity
生态城市 eco-city
生态城市规划 eco-city planning
生态池湿地系统 ecologic(al) pond-wetland system
生态池塘 eco-pond
生态持续发展 economic sustainable development
生态冲击 ecologic(al) backlashes; ecologic(al) impact
生态刺激 ecologic(al) stimulation
生态脆弱性 ecologic(al) vulnerability
生态大气 eco-atmosphere
生态大灾难 ecocatastrophe
生态道德标准 ecologic(al) moral standard
生态道德原则 ecologic(al) moral principle
生态的保持 preservation of ecology
生态等位种 ecologic(al) equivalent species
生态等值 ecologic(al) efficiency; ecologic(al) equivalence
生态地层单位 ecostratigraphic(al) unit
生态地层分类 ecostratigraphic(al) classification
生态地层学 ecostratigraphy
生态地理分布 ecologo-geographic(al) distribution
生态地理趋势 ecogeographical divergence
生态地理群落学 biogeocoenology
生态地理图 ecologic(al) map
生态地理学 ecologic(al) geography; ecologo-geography
生态地(球)植物学 ecologic(al) geobotany
生态地质学 geoecology
生态帝国主义 ecologic(al) imperialism
生态递阶模型 ecosystem hierarchical model
生态调查法 ecologic(al) survey method
生态动力学 ecologic(al) kinetics
生态动物地理学 ecologic(al) zoogeography
生态毒理性测试 ecotoxicological test
生态毒理性效应 ecotoxicological effect
生态毒理性影响 ecotoxicological impact
生态毒理学 ecotoxicology
生态毒性 ecologic(al) toxicity
生态对策 ecologic(al) strategy
生态多度计算 ecologic(al) bonitation
生态多态型现象 ecologic(al) polymorphism
生态多样性 ecologic(al) diversity
生态发生 ecogenesis
生态发展 ecodevelopment
生态发展论 theory of ecologic(al) development
生态防治 bionomic control; ecologic(al) prevention and treatment
生态分布 ecologic(al) distribution; economic distribution
生态分类 ecologic(al) classification
生态分歧 ecologic(al) divergence
生态分析 ecologic(al) analysis
生态风险表征 ecologic(al) risk characterization
生态风险分析 ecologic(al) risk analysis
生态风险级别 ecologic(al) risk grade
生态风险评价 ecologic(al) risk assessment
生态风险指数 ecologic(al) risk index
生态浮床 ecologic(al) floating-floor
生态幅度 ecologic(al) amplitude; ecologic(al) range
生态复合体 ecologic(al) complex
生态复原 ecologic(al) rehabilitation
生态概况 ecologic(al) aspects
生态干扰 ecologic(al) disturbance
生态感应 ecologic(al) response
生态隔离 ecologic(al) isolation
生态更替 ecologic(al) displacement
生态工程 economic project
生态工程系统 ecoengineering system; ecologic(al) engineering system
生态工程学 ecoengineering; ecologic(al) engineering
生态工业园 eco-industry park
生态工艺 ecologic(al) technology
生态公园 economic park
生态孤立 ecologic(al) isolation
生态关联 environmental association
生态关系 ecologic(al) relationship
生态观 ecologic(al) outlook
生态管理及审核计划 eco-management and audit scheme
生态规划 ecologic(al) planning
生态过程 ecologic(al) process
生态过渡带 ecotone
生态河流水质 ecologic(al) river quality
生态后果 ecologic(al) consequence
生态化学 ecologic(al) chemistry
生态环境 biotope; eco-environment; ecologic(al) environment; ecology environment; ecotone; ecotope
生态环境安全 ecologic(al) environmental safety; safety of ecologic(al) environment
生态环境保护 eco-environment(al) protection; protection of ecologic(al) environment
生态环境变化 eco-environment(al) change; ecologic(al) environmental change
生态环境处理 ecologic(al) approach
生态环境功能 eco-environment(al) function; ecologic(al) environmental function
生态环境观点 ecologic(al) approach
生态环境规划 eco-environment(al) planning; ecologic(al) environmental planning
生态环境建设 eco-environment(al) establishment; ecologic(al) environmental establishment
生态环境水文地质学 ecologic(al) environmental hydrogeology
生态环境特征 eco-environment(al) feature; ecologic(al) environmental feature
生态环境效应 eco-environment(al) effect; ecologic(al) environmental effect
生态环境指标 eco-environment(al) index; ecologic(al) environmental index
生态环境重建 ecologic(al) environmental rebuilding
生态环境资料 ecologic(al) environmental data; ecologic(al) environmental information
生态环境综合治理 all-round improvement in the eco-environment
生态恢复力 ecologic(al) resiliency
生态混凝土 <用废料制成的混凝土> ecologic(al) concrete
生态活动 eco-activity; ecologic(al) activity; ecology movement; economic activity
生态活动家 <致力环境保护免受污染> eco-activist
生态极限 ecologic(al) threshold
生态集合体 ecologic(al) assemblage
生态技术 ecologic(al) technique; ecotechnique; ecotechnology
生态技术处理系统 ecology technology treatment system
生态技术学 ecologic(al) technology
生态价值 ecologic(al) value; economic evaluation
生态监测 ecologic(al) monitoring
生态监测对策 ecologic(al) monitoring strategy
生态监测机构 ecologic(al) monitoring unit
生态建材 ecologic(al) building materials
生态建设 ecologic(al) construction
生态建筑 ecologic(al) architecture
生态建筑物 ecologic(al) building
生态建筑学 arcology
生态健康 ecologic(al) health
生态渐变群 ecocline
生态礁 ecologic(al) reef
生态教育 ecologic(al) education
生态接续 ecologic(al) succession
生态结构 ecologic(al) structure
生态界 ecosphere
生态金字塔 ecologic(al) pyramid
生态进化 ecologic(al) evolution
生态经济规划 ecologic(al) economic planning
生态经济价值观 value view of ecologic(al) economics
生态经济平衡 ecologic(al) economic equilibrium
生态经济生产观 production view of ecologic(al) economics
生态经济系统 ecologic(al) economic system
生态经济需求观 demand view of ecologic(al) economics
生态经济学 ecoeconomics; ecologic(al) economics
生态经济综合效益 economic comprehensive benefit
生态竞争 ecologic(al) competition
生态距离 ecologic(al) distance
生态绝对命令 ecologic(al) absolute order
生态均等 ecologic(al) equivalence
生态菌株 ecologic(al) strain
生态考虑 ecologic(al) consideration
生态科学 ecologic(al) science
生态可行性 eco-feasibility
生态控制 ecologic(al) control
生态会计学 ecologic(al) accounting
生态昆虫学 ecologic(al) entomology
生态类群 ecologic(al) group; mores
生态类型 ecotype
生态林业 eco-forestry
生态临界 ecologic(al) threshold
生态旅游 eco-tourism
生态伦理学 ecologic(al) ethics
生态密度 ecologic(al) density
生态灭绝 ecocide
生态敏感地区 ecologically sensitive area
生态敏感经济技术指标 ecology-sensitive economical technique index
生态模拟 ecologic(al) simulation; simulation in ecology
生态模型 ecologic(al) model
生态内稳定现象 ecologic(al) homeostasis
生态耐寒性 ecologic(al) winter hardiness
生态耐性 ecologic(al) tolerance
生态能动论 ecoactivism
生态年代学 ecochronology
生态年龄 ecologic(al) age
生态农场 ecologic(al) farm
生态农业 eco-agriculture; eco-farming; ecologic(al) agriculture; ecologic(al) farm; environmentally friendly agriculture
生态农业系统 eco-agricultural system
生态农业指标体系 target system of eco-agriculture
生态浓集 ecologic(al) concentration
生态平衡 balance of nature; ecologic(al) balance; ecologic(al) equilibrium; economic balance; economic equilibrium; nature balance
生态平衡不稳定 c(o)enosis
生态平衡论 theory of ecologic(al) equilibrium
生态平衡失调 disturbed ecologic(al) equilibrium
生态评价 ecologic(al) assessment; ecologic(al) evaluation; economic evaluation
生态评价标准 ecologic(al) evaluation criteria
生态破坏 ecologic(al) damage; ecologic(al) disruption; ecologic(al) failure
生态气候 ecoclimate
生态气候适应 ecoclimate adaptation
生态气候学 ecoclimatology; ecologic(al) climatology
生态气候预测 ecoclimate forecasting
生态迁移 ecologic(al) succession
生态倾差 ecocline
生态区 ecoregion; ecotope; ecozone
生态区划 ecologic(al) regionalization
生态区系 ecosystem
生态圈 ecosphere
生态群 cline; ecogroup; synusium
生态群落 ecologic(al) association; ecologic(al) community
生态群体 ecology group; ecologic(al) group
生态容量 ecologic(al) capacity
生态容限 ecologic(al) tolerance
生态设计 ecologic(al) design
生态社会学 ecologic(al) sociology
生态生长效率 ecologic(al) growth efficiency
生态失调 dysbiosis; ecologic(al) disturbance

S

生态食物链 ecologic(al) food chain
生态史 ecologic(al) ecogenesis
生态史观 ecologic(al) conception of history
生态示范区 eco-demonstration region;environment-friendly region
生态势 ecologic(al) potential
生态(适宜)环境 ecologic(al) niche
生态适应 ecologic(al) adaptation
生态适应性 ecologic(al) suitability
生态水动力学 ecohydrodynamics
生态水力学 eco-hydraulics
生态水文学 ecohydrology;ecologic(al) hydrology
生态水文学管理方法 ecohydrological water management approach
生态死亡率 ecologic(al) mortality
生态损害 ecologic(al) damage
生态塘系统 ecologic(al) pond system
生态特性 ecologic(al) character;ecosystem characterization
生态替换 ecologic(al) replacement
生态同类群 ecodeme
生态统计分析 ecologic(al) statistical analysis
生态投资 economic investment
生态图 ecologic(al) map
生态图解 ecograph
生态土壤学 edaphology
生态退化 ecologic(al) degeneration;ecologic(al) degradation
生态危害 ecologic(al) hazard
生态危机 ecocrisis;ecologic(al) crisis
生态危险 ecologic(al) risk
生态位 ecologic(al) niche;niche
生态位泛化 niche generalization
生态位分离 niche separation
生态位宽度 niche breadth
生态位能 bioecologic(al) potential
生态位特化 niche specialization
生态位重叠 niche overlap
生态文化 ecologic(al) culture
生态稳定性 ecologic(al) stability
生态稳态 ecologic(al) home-ostasis
生态污染极限 ecologic(al) pollution limit
生态无害产品 ecologically safe product;ecologically sound product
生态无害技术 ecologically sound technology;ecotechnology
生态无害特性 ecologically sound characteristic
生态物宽度 niche width
生态系的物质循环 material cycle in ecosystem
生态系动力学 ecosystem dynamics
生态系复原 ecosystem rehabilitation
生态系观点 ecosystem approach
生态系结构与功能 ecosystem structure and function
生态系列 ecologic(al) series
生态系列法 ecologic(al) serial method
生态系模型 ecosystem model
生态系内能量流动 energy flow in ecosystem
生态系迁移率 ecosystem transfer rate
生态系生态学 ecosystem ecology
生态系数 ecologic(al) factor
生态系统 bioecologic(al) system;biogeocenosis;economic system;ecologic(al) system;ecosystem
生态系统安全性 ecosystem security
生态系统保护对策 ecosystem protection strategy
生态系统不确定性分析 ecosystem uncertainty analysis
生态系统的动态功能 dynamic(al) function of ecosystem
生态系统的过程 ecosystem processes
生态系统动力学 dynamics of ecosystem;ecosystem dynamics
生态系统多样性 ecosystem diversity
生态系统发展的战略 strategy of ecosystem development
生态系统反应器 ecosystem response unit
生态系统服务 ecosystem service
生态系统工程 ecologic(al) system engineering;ecosystem engineering
生态系统管理 ecosystem management
生态系统管理理论 ecosystem management theory
生态系统监测 ecosystem monitoring
生态系统结构 ecosystem structure
生态系(统)界面 ecosystem interface
生态系统开发对策 ecosystem development strategy
生态系统开发战略 ecosystem development strategy
生态系统类型 ecosystem type
生态系统力能学 ecosystem energetics
生态系统模拟 ecosystem modeling
生态系统模拟模型 ecosystem simulating model
生态系统模型 ecologic(al) model
生态系统平衡 ecosystem balance
生态系统评价 ecosystem assessment
生态系统评价模型 ecosystem assessment model
生态系统评价指标体系 ecosystem assessment index system
生态系统生产力 ecosystem productivity
生态系统生态学 ecosystem ecology
生态系统水文模拟器 ecosystem hydrologic simulator
生态系统梯度 gradient of ecosystem
生态系统条件 ecosystem condition
生态系统退化 ecosystem degradation
生态系统稳定性 ecosystem stability
生态系统效应 ecosystem effect
生态系统型 ecosystem type
生态系统修复 ecosystem restoration
生态系统演化 ecosystem development
生态系统演替 ecosystem succession
生态系统养护组 ecosystem conservation group
生态系统要素 ecosystem component
生态系统预测 ecosystem prediction
生态系统制图 ecosystem mapping
生态系统中的物质和能量 matter and energy in ecosystem
生态系统中生物地球化学循环 biochemical cycle in ecosystem
生态系预报 ecosystem prediction
生态限度 ecologic(al) limit
生态相 ecologic(al) facies;ecologic(al) phase;environmental facies
生态相当 ecologic(al) equivalence
生态相互作用 ecologic(al) interaction
生态小环境 ecologic(al) niche
生态小生境 ecologic(al) niche
生态效率 ecologic(al) efficiency
生态效益 ecologic(al) benefit;ecologic(al) effect;economic benefit;economic effect
生态效应评价 ecologic(al) effect assessment
生态形态学 ecologic(al) morphology
生态型 ecologic(al) type
生态型林沼 swamp ecotype
生态性物种形式 ecologic(al) speciation
生态性状 ecologic(al) character
生态修复 ecologic(al) restoration
生态修复技术 ecologic(al) restoration technique
生态需求 ecologic(al) demand
生态序列 ecologic(al) succession
生态畜牧业 eco-animal husbandry
生态学 bioecology;bionomics;bionomy;ecology
生态学的和地理的测量 economic and geographical surveys
生态学的和地理的调查 ecologic(al) and geographical surveys
生态学的许可 ecologic(al) approval
生态学方法 ecologic(al) approach
生态学方面的利益 ecologic(al) benefit
生态学工作者 ecologist
生态学规律 ecologic(al) law
生态学家 ecologist
生态学谬论 ecologic(al) fallacy
生态学评价 ecologic(al) evaluation
生态学世界观 global outlook of ecology
生态学数学模式 mathematic(al) models of ecology
生态学思维 thoughts of ecology
生态学相关 ecologic(al) correlation
生态学效应 ecologic(al) effect
生态学哲学 philosophy of ecology
生态循环 ecocycle;ecologic(al) cycle
生态循环规律 ecocycle rule
生态演替 ecologic(al) succession
生态遥测术 ecotelemetry
生态遗传学 ecogenetics;ecologic(al) genetics
生态艺术 ecologic(al) art
生态意识 ecologic(al) consciousness
生态因素 ecofactor;ecologic(al) factor
生态因子 ecofactor;ecologic(al) factor
生态银行 ecology bank
生态影响 ecologic(al) consequence;ecologic(al) effect;ecologic(al) impact
生态影响评价 ecologic(al) impact assessment
生态影响研究 ecologic(al) effect study
生态用水 ecologic(al) water
生态优势 ecologic(al) dominance
生态与微量元素污染物分析 ecology and analysis of trace contaminant
生态预测 ecologic(al) forecasting
生态域 ecosphere
生态阈限 ecologic(al) threshold
生态原因 ecologic(al) consideration
生态运动 ecology movement
生态灾害 ecologic(al) disaster
生态灾难 eco-catastrophe;ecologic(al) disaster;environmental disaster
生态灾区 ecologic(al) disaster area
生态哲学 ecologic(al) philosophy
生态织物 ecologic(al) fabric
生态植物地理学 ecologic(al) plant geography
生态植物解剖学 ecologic(al) plant anatomy
生态植物学 ecologic(al) botany
生态指标植物 ecologic(al) indicator plant
生态指示物 ecologic(al) indicator
生态质量指数 biotic index
生态种 ecospecies
生态种发生 ecogenesis
生态种群 mos
生态种组 ecologic(al) species group
生态主导因素 ecologic(al) master factor
生态砖 <空心砖内置草种,成长后覆于砖面,绿化环境> ecologic(al) brick;economic brick
生态状况 ecologic(al) regime
生态宗(族) ecologic(al) race
生态综合 ecologic(al) synthesis
生态综合评价 ecologic(al) comprehensive assessment
生态综合体 ecologic(al) complex
生态综合指数 ecologic(al) comprehensive index
生态族 ecologic(al) race
生态组合 ecologic(al) assemblage;environmental association
生态最适度 ecologic(al) optimum
生态作物地理学 ecologic(al) crop geography
生碳化合物 carbonific
生碳酸钙的 calciferous
生陶瓷 green ware
生体机械学 bionics
生铁 cast-iron;foundry ingot;foundry iron;foundry pig;pig(iron)
生铁槽 pig bed
生铁打碎机 pig breaker
生铁锭 pig
生铁废钢平炉炼钢法 Martin process
生铁分支管 cast-iron lateral
生铁管 cast-iron pipe
生铁机 pig-casting machine
生铁集管 cast-iron header
生铁浇铸机 pig-casting machine
生铁搅拌器 pig-iron mixer
生铁块 cast-iron block;pig
生铁矿石法 pig and ore process
生铁连接管 cast-iron adapter[adaptor]
生铁喷雾粒化法 Roheisenzunder process
生铁破碎机 pig breaker
生铁套圈 cast-iron drive ferrule
生铁迂回管 cast-iron offset
生铁增碳 pigging(-up);pig up
生铁支柱座板 bloom base plate
生铁铸件 iron casting
生铁铸造 iron casting;iron founding
生桐油 raw tung oil
生铜 pig copper
生土 immature soil;mineral soil;new soil;raw soil;subsurface soil;virgin soil
生土建筑 earth building
生土结构房屋 raw soil structure
生土墙 cob walling
生土住宅 earth-sheltered home
生团块 green briquet(te)
生污泥 fresh sludge;green sludge;primary sludge;raw sludge;undigested sludge
生污水 crude wastewater;raw sewage
生物 creature;living beings;organism
生物安全 biosafety
生物安全用水 ecologically reliable water
生物胺 bio-amine
生物搬运 biologic(al) transport of organism
生物半衰期 biologic(al) half-life
生物包粒 biocoated grain
生物保护 biologic(al) protection;biologic(al) shielding
生物保护法 conservation law
生物保护区 biologic(al) reserve
生物暴露量 biologic(al) exposure
生物暴露指标 biologic(al) exposure indicator
生物被动吸收 passive adsorption of organisms
生物变化 biologic(al) transformation
生物变数 ecologic(al) variable
生物变异 biomutation
生物变质 biodeterioration
生物标本 biologic(al) sample
生物标本培育箱 cabinet
生物标定辐射当量 radiation-equivalent manikin calibration

生物标度 biologic(al) scale
生物标志物 biomarker
生物标准 biologic(al) standard
生物标准物质 biologic(al) standard substance
生物表面活性剂 biosurfactant
生物病虫害防治 biologic(al) pest control
生物病因 biologic(al) pathogen
生物玻璃 bioglass
生物剥蚀 biologic(al) denudation
生物薄膜 biologic(al) film
生物捕积岩 bafflestone
生物不能降解的物质 non-biodegradable substance
生物材料 biologic(al) material; biomaterial
生物材料监测 biologic(al) specimen monitoring
生物参数 biologic(al) parameter
生物参数不确定性 biologic(al) parameter uncertainty
生物残毒 biologic(al) residual toxin
生物残毒测定 biologic(al) residual toxin measurement
生物残骸分解的有机物 organic of biolysis
生物舱 biopak
生物操纵 biomanipulation
生物测定 bioassay; biologic(al) measure
生物测定参数 bioassay parameter
生物测定法 bioassary method; biologic(al) assay way
生物测定评价 bioassay evaluation
生物测定特性曲线 bioassay response
生物测定系统 bioassay system
生物测定学 biometrics; biometry
生物测量学 biometrics
生物测试 bioassay
生物层 biosphere; biostrata; biostrome
生物层闭 biostrome trap
生物层次 biotic level
生物层段 biomere
生物层灰岩 biostromal limestone; biostromic limestone
生物层面印痕 biostromic print
生物差距 biotal distance
生物差异 biologic(al) diversity
生物柴油 bio-diesel
生物产品 body products
生物产生的氯 biogenic chlorine; natural chlorine
生物沉淀 biogenous sediment; biologic(al) precipitation; bio-precipitation
生物沉淀法 biologic(al) purification
生物沉积 biogenetic deposit; biogenic sediment
生物沉积的 organosedimentary
生物沉积物 biogenic deposit; biogenic sediment; organic deposit
生物沉积作用 biologic(al) deposition
生物沉降 biologic(al) settlement
生物沉降法 bio-precipitation process
生物沉降过程 bio-precipitation process
生物陈列室 biology collection room
生物成分 biotic component; organic composition
生物成因标志 marker of biogenesis
生物成因的 biogenetic; biogenic
生物成因分化作用 biogenetic fractionation
生物成因分散 biogenic dispersion
生物成因构造 biogenic structure
生物成因泥沙 biogenic origin sand
生物成因气 biogenic gas
生物成因气带 biogenetic gas zone
生物成因气量 amount of biogenetic gas
生物成因燧石 biogenic chert
生物成因异 biogenic anomaly
生物池塘 biologic(al) pool
生物池氧化法 biologic(al) pond process
生物尺度 biologic(al) scale
生物冲淋 bioblooding
生物初始面 biont starting plane
生物初始面位置 position of biont starting plane
生物除草 biologic(al) weed control
生物除磷 biologic(al) phosphorus removal
生物除磷工艺 biologic(al) phosphorus removal process
生物除磷酸盐 biologic(al) phosphate removal
生物除磷脱氮系统 anaerobic-anoxic/oxic system
生物除磷系统 biologic(al) phosphorus removal system
生物除锰 biologic(al) manganese removal
生物除锰滤池 biologic(al) manganese removal filter
生物除铁除锰 biologic(al) removal of iron and manganese
生物除铁除锰池 biologic(al) iron and manganese removal filter
生物除污 biologic(al) depollution
生物除氧法 biologic(al) method for oxygen removal
生物处理 biologic(al) process; biologic(al) treatment; biotreatment
生物处理厕所 <其中之一种>"clivus multrum" toilet
生物处理池 biolytic tank
生物处理催化铁内循环工艺 biologic(al) treatment-catalytic iron inner cycle process
生物处理法 biologic(al) treatment method
生物处理工艺 biologic(al) treatment process
生物处理过程 biologic(al) treatment process
生物处理过的生活污水 biotreated domestic wastewater
生物处理过的污泥 biotreated sludge
生物处理过的污水出水 biologically treated sewage effluent
生物处理机理 biologic(al) treatment mechanism
生物处理系统 biologic(al) treatment system
生物处理站 biostation
生物传播 biotransmission
生物传感器 biosensor
生物创建 biopoiesis
生物磁效应 biomagnetic effect
生物磁性 biomagnetism
生物磁学 biomagnetism
生物刺激作用 biostimulation
生物促进剂 bio-energizer
生物催化剂 biocatalyst; biologic(al) catalyst
生物大分子 biomacromolecule
生物大气层 biosphere; ecosphere
生物大气单元 bio-atmospheric unit
生物代谢作用 biologic(al) metabolism; biotic metabolism
生物带 biozone; life belt; life zone; range zone of organism; biocycle <包括陆地、海洋和淡水域>
生物带生态学 life zone ecology
生物单位地层划分 biostratigraphic(al) classification
生物单元处理法 biologic(al) unit process
生物蛋白土 gaize
生物导航 bionavigation
生物的 biologic(al); biotic
生物的反硝化作用 biologic(al) denitrification
生物的机械搬运 physical transport of organism
生物的生存环境条件 living environmental conditions of organisms
生物的水质分析 biologic(al) water analysis
生物的物理搬运 physical transport of organism
生物堤 biologic(al) dike[dyke]
生物滴滤 biologic(al) trickling filtration
生物滴滤池 biologic(al) trickling filter
生物地层 biostratum[复 biostrata]
生物地层带 biostratigraphic(al) zone
生物地层单位 biostratigraphic(al) unit
生物地层对比 biostratigraphic(al) correlation
生物地层区 biostratigraphic(al) zone
生物地层学 biostratigraphy
生物地理分布 biogeographic(al) realm
生物地理气候区 bioeoclimate zone
生物地理区 biogeographic(al) province; biotic province
生物地理圈 biogeosphere
生物地理群落 biogeocoenosis; biogeocoenosium
生物地理群落学 biogeocoenology
生物地理图 ecologic(al) map
生物地理学 biogeography; chorology
生物地球化氮循环 biogeochemical nitrogen cycle
生物地球化学 biogeochemistry
生物地球化学参数 biogeochemical parameters
生物地球化学测量 biogeochemical survey
生物地球化学的 biogeochemical
生物地球化学富集 biogeochemical enrichment
生物地球化学过程 biogeochemical process
生物地球化学模拟 biogeochemical modeling
生物地球化学生态学 biogeochemical ecology
生物地球化学省 biogeochemical province
生物地球化学探测 biogeochemical prospecting
生物地球化学探矿 biogeochemical prospecting
生物地球化学性疾病 biogeochemical disease
生物地球化学循环 biogeochemical cycle
生物地球化学异常 biogeochemical anomaly
生物地球化学作用 biogeochemical processes
生物地球学障 biogeochemical barrier
生物地志 biogeography
生物地质化学 biogeochemistry
生物地质年代表 biologic(al) time scale
生物地质特征 biogeology
生物地质学 biogeology; biologic(al) geology
生物地质状况 biogeology
生物电 bioelectricity
生物电池 biobattery; biocell
生物电发生 bioelectrogenesis
生物电化学 bioelectrochemistry
生物电化转换 bioelectrochemical conversion
生物电极 bioelectrode
生物电控制 bio-control; bioelectric(al) control
生物电流 biocurrent; bioelectric(al) current
生物电流电池 biogalvanic battery
生物电模型 bioelectric(al) model
生物电势 bioelectric(al) potential; biopotential
生物电试验 electrobiologic(al) test
生物电位 bioelectric(al) potential
生物电学 electrobiology
生物电子学 bioelectronics
生物调查船 biologic(al) survey ship
生物动力学 biodynamics; biokinetics
生物动力学常数 biokinetics constant
生物动力学农业 biodynamic agriculture
生物豆状岩 biopisolite
生物毒理学 biologic(al) toxicity; biotoxicity
生物毒料 biocide
生物毒素 biotoxin
生物堆积灰岩 bio-accumulated limestone
生物堆积岩 hydrobiolite
生物堆积作用 bio-accumulation
生物堆置灰岩 bafflestone
生物对外来物质的保护 biologic(al) reactor shield
生物多类状态 biologic(al) diversity
生物多样性 biodiversity; biologic(al) diversity; biologic(al) heterogeneity
生物多样性的保护 conservation of biologic(al) diversity
生物多样性公约 Conversion on Biologic(al) Diversity
生物多样性和保护区 biologic(al) diversity and protected area
生物多样性基准 biodiversity criteria
生物多样性生态系统 biologically diverse ecosystem
生物惰性陶瓷 bioinert ceramics
生物发光 noctilucence
生物发光的 bioluminescent
生物发光菌 bioluminescent bacterium
生物发光现象 bioluminescence
生物发光指示菌 bioluminescent reporter bacterium
生物发霉 microorganism corrosion
生物发生 biogenesis
生物发生律 biogenetic law
生物法 biologic(al) process; bioprocess; biologic(al) method
生物法去污 biologic(al) depollution
生物反馈 biofeedback
生物反馈控制 biofeedback control
生物反硝化系统 biologic(al) denitrification system
生物反应 biologic(al) reaction
生物反应池 biologic(al) reactor
生物反应谱 biologic(al) response spectrum
生物反应器 biologic(al) reactor; bioreactor; microbial film reactor
生物方法 biologic(al) measure
生物防护 biologic(al) protection; biologic(al) shield; biologic(al) shielding
生物防护铅屏 lead biologic(al) shied
生物防治 biologic(al) protection; biologic(al) shielding
生物防治措施 biologic(al) control measure
生物防治法 biologic(al) control
生物防治污染 prevention and control of pollution with organism

生物放大因数 biologic(al) amplification factor; biologic(al) magnification
生物放大作用 bio-amplification; biologic(al) magnification
生物放射性废物 biologic(al) radioactive wastes
生物肥料 bio-fertilizer
生物废物处理 biologic(al) waste treatment
生物分布学 chorology
生物分带性 biozonation
生物分解池 biolytic tank
生物分解的 biodegradable
生物分解性的有机成分 biologically decomposable organic component
生物分解作用 biologic(al) decomposition; biologic(al) degradation; biolysis
生物分类上的等级 biologic(al) value
生物分类学 biologic(al) classification; biosystematics
生物分泌物 bioeffluent
生物分析法 bioanalysis; biologic(al) analysis
生物分子 biomolecule
生物分子反应 biomolecular reaction
生物风化 biologic(al) weathering; organic weathering
生物浮选 biologic(al) flo(a)tation
生物福利设施 public welfare facility
生物辐射防护混凝土 biologic(al) shielding concrete
生物辐射防护墙 biologic(al) shielding wall
生物辐射防护装置 biologic(al) shielding block
生物腐败 biologic(al) decay
生物腐蚀 biodeterioration; biologic(al) corrosion
生物负荷 biologic(al) load
生物负荷率 biologic(al) loading rate
生物附着 biofouling
生物复合体 bioplex
生物复制 clone
生物富集 bioenrichment; biologic(al) concentration; biologic(al) enrichment; biologic(al) magnification
生物富集系数 biologic(al) concentration factor
生物富集作用 biologic(al) enrichment function
生物感应 biologic(al) response
生物高分子 biopolymer
生物高分子水合作用 biopolymer hydration
生物高分子絮凝剂 biopolymer flocculant
生物隔离 biologic(al) isolation
生物个性 biont
生物耕作 biologic(al) husbandry; organic farming
生物工程 biologic(al) engineering; biotechnology
生物工程下游技术 downstream technique of biotechnology
生物工程学 bio-engineering; biologic(al) engineering; bionics
生物工艺学 biotechnology
生物公害 biohazard
生物功能性 biofunctionability
生物构架石灰岩 framestone
生物古地理区划 division of palaeobiogeographic(al) province
生物骨架结构 bioframework texture; organic framework texture
生物骨架孔隙 growth-framework pore
生物固氮 biologic(al) nitrogen fixation
生物固沙 sand fixation with biologic
生物固体 biologic(al) solid; biosolid
生物固体停留时间 biologic(al) solid retention time
生物挂膜 biofilm colonization
生物过程 biologic(al) process; bioprocess
生物过滤 <处理污水的方法> biofiltration; biologic(al) filtration
生物过滤层 biological filter
生物过滤池 biofilter; biologic(al) filter
生物过滤法 biofiltration method; biofiltration process; biologic(al) filtration process
生物过滤器 biofilter; biologic(al) filter
生物过滤设施 biologic(al) filter
生物海岸 biologic(al) coast
生物海洋学 biologic(al) oceanography
生物耗氧量 biotic oxygen demand
生物合成作用 biologic(al) synthesis; biosynthesis
生物痕迹 biogenic imprint
生物互扰 allelopathy
生物化石群 oryctocoenose
生物化学 biochemistry; biologic(al) chemistry
生物化学变化 biochemical change
生物化学沉积 biochemical deposit
生物化学沉积成矿作用 biochemical-sedimentary ore-forming process
生物化学沉积矿床 biochemical-sedimentary ore deposit
生物化学沉积作用 biochemical deposition
生物化学处理 biochemical treatment; biologic(al) chemical treatment
生物化学的 biochemical
生物化学的相互作用 biochemical interaction
生物化学法 biochemical process
生物化学反应 biochemical reaction
生物化学方法 biochemical method; chemical biologic(al) method
生物化学风化 biologic(al) chemical weathering
生物化学技术 biochemical technique
生物化学剂 biochemicals
生物化学降解作用 biochemical degradation
生物化学联用法 integrated biochemical process
生物化学煤化作用 biochemical coalification
生物化学凝胶化作用 biochemical gelefication
生物化学品 biochemical
生物化学生态学 biochemical ecology
生物化学损伤 biochemical lesion
生物化学污染标准 biochemical pollution standard
生物化学需氧量 biochemical oxygen demand
生物化学需氧量自动记录仪 biochemical oxygen demand automatic recorder
生物化学循环 biochemical cycle
生物化学岩 biochemical rock; biochemigenic rock
生物化学岩储集层 biochemical reservoir
生物化学战 biochemical warfare
生物化学作用 biochemical action
生物还原 biologic(al) reduction
生物环 biocycle
生物环境 biologic(al) environment; biotic environment
生物环境的 bioenvironmental
生物环境化学 bioenvironmental chemistry
生物环境调节实验装置 biotron
生物环境污染物 biologic(al) environment pollutant
生物环境指数 index of living environment
生物灰岩 biolimestone; biolithite
生物挥发 biovolatilization
生物混合深度 biologic(al) mixing depth
生物混合系统 biologic(al) hybrid system
生物混凝 bioflocculation; biologic(al) coagulation
生物活动 biologic(al) activity
生物活度 bioactivity
生物活化 bioactivitation
生物活性 bioactivity; biologic(al) activity
生物活性高分子 bioactive polymer
生物活性化合物 bioactive compound
生物活性剂 bioactivator; biologically active agent
生物活性滤池 biologically activated filter
生物活性炭 bioactivated carbon; biologically active carbon; biologic(al) activated carbon
生物活性炭法 biologic(al) activated carbon process
生物活性炭过滤 biologic(al) activated carbon filtration
生物活性炭滤池 biologic(al) activated carbon filter
生物活性炭污泥 biologic(al) activated carbon sludge
生物活性炭纤维 bioactivated carbon fiber[fibre]; biologic(al) activated carbon fiber[fibre]
生物活性陶瓷 bioactive ceramics
生物活性物质 biologically active substance
生物活性絮体 biologically active floc
生物活性指标 bioactivity index; index of bioactivity; index of biologic(al) activity
生物机理 biologic(al) mechanism
生物机体 living organism
生物机械学 bionics
生物积聚 bio-accumulation
生物积累 bio-accumulate; bio-accumulation
生物积累比 bio-accumulation ratio; biologic(al) accumulation ratio
生物积累系数 bio-accumulation factor; biologic(al) accumulation factor
生物积累效率 bio-accumulation efficiency; biologic(al) accumulation efficiency
生物积累元素 bio-accumulation element; biologically accumulated element
生物基准 biocriterion[复 biocriteria]; biologic(al) criterion[复 criteria]
生物基准点 biologic(al) bench mark
生物基准面 biodatum
生物激活剂 biostimulant
生物集群 biocoenose
生物记录中心 biologic(al) record center[centre]
生物技术 biologic(al) technique
生物技术安全 biotechnological safety
生物技术的社会经济影响 socioeconomic impact of biotechnologies
生物技术问题 biotechnological issues
生物技术学 biotechnology
生物剂量 biologic(al) dose
生物剂量测定法 biodosimetry
生物剂量学 biologic(al) dosimetry
生物剂量指示剂 biologic(al) dosimeter
生物剂浓度 biogenic substance concentration
生物甲基化作用 biologic(al) methylation
生物甲基烷势 biologic(al) methane potential
生物监测 <废液污染度的> biomonitoring; biologic(al) monitoring
生物监测工作组 biologic(al) monitoring working party
生物监测器 biomonitor
生物监控 biologic(al) monitoring
生物监视 biomonitoring
生物检测 biologic(al) analysis; biologic(al) detecting; biologic(al) detection; biologic(al) examination; biologic(al) test(ing)
生物检测法 biologic(al) detection method
生物检定 bioassay
生物检定参数 bioassay parameter
生物检定法 bioassay
生物检定系统 bioassay system
生物检验 bioassay; biologic(al) analysis; biologic(al) assay; biologic(al) examination
生物检验体 biologic(al) monitor
生物碱 alkaloid; natural base
生物碱毒素 alkaloid toxicant
生物建造海岸 biogenic coast
生物鉴别 biologic(al) discrimination
生物鉴定 bioassay; biologic(al) assay
生物降解 biodegrade
生物降解玻璃 biodegradable glass
生物降解材料 biodegradable material
生物降解的 biodegradable
生物降解动力学 biodegradation kinetics
生物降解法 biodegradation process
生物降解反应 biodegradation reaction
生物降解废水 biodegradation wastewater
生物降解化学物的能力 biodegradability of chemicals
生物降解技术 biodegradation technique; biologic(al) degradation technique
生物降解模量 biodegradation module
生物降解能力 biodegradability
生物降解势 biodegradation potential
生物降解速度 rate of bio-degradation
生物降解(速)率 rate of bio-degradation
生物降解途径 biodegradation pathway
生物降解性 biodegradability
生物降解有机物 biodegradable organic substance
生物降解作用 biodegradation; biologic(al) degradation
生物交互作用 biotic interaction
生物胶体 biocolloid
生物礁 bioherm; organic reef; klint
生物礁沉积 organic reef deposit
生物礁沉积模式 sedimentation model of organic reef
生物礁的岩屑 reef detritus
生物礁环境 organic reef environment
生物礁环形体 circular features of bioherms
生物礁灰岩 bioherm limestone
生物礁类型 type of reef
生物礁圈闭 bioherm trap
生物礁石 biogenic reef
生物礁相 organic reef facies
生物接触 biologic(al) contact oxidation process
生物接触法 bio-contact method; biologic(al) contact process

生物接触反应器 bio-contact reactor; biologic(al) contact oxidation reactor
生物接触器 bio-contactor; biologic-(al) contactor
生物接触氧化 bio-contact oxidation; biologic(al) contact oxidation
生物接触氧化法 bio-contact oxidation process; biologic (al) contact oxidation process
生物接触氧化滤池 bio-contact oxidation filter
生物接触氧化塘 bio-contact oxidation pond; biologic(al) contact oxidation pond
生物接触氧化-絮凝沉淀-过滤-消毒工艺 biologic (al) contact oxidation-flocculation sedimentation-filtration-disinfection process
生物接触氧化装置 biologic (al) contact oxidation unit
生物接触指数 biologic (al) exposure index
生物节律 biorhythm
生物洁净室 bio-cleaning room
生物结构 biogenetic texture
生物结晶学 biocrystallography
生物解毒 biodetoxification
生物界 biologic (al) universe; biosphere; biotic community; living nature; organic sphere; organic world; organ sphere
生物金字塔 biotic pyramid
生物进化 biologic (al) evolution; organic evolution
生物经济学 bioeconomics
生物景谱 biologic(al) spectrum
生物净化池 biologic (al) purification tank
生物净化(处理)厂 biologic(al) clarification plant
生物净化法 biologic(al) clarification; biologic (al) cleaning; biologic (al) purification
生物净化工艺 biologic (al) purification process
生物净化剂 bio-cleaning agent
生物净化作用 biologic (al) purification
生物静力学 biostatics
生物静力抑制 biostatic inhibitation
生物聚合物 biologic (al) polymer; biopolymer
生物聚合物水合作用 biopolymer hydration
生物聚集 bio-accumulation; biogenic accumulation
生物绝灭 extinction
生物砍屑结构 bioclastic texture
生物抗污染性 organism resistance to pollution
生物抗性 biologic(al) resistance
生物可处理性 biologic (al) treatability
生物可分解物 biodegradable
生物可降解废水 biodegradable waste; biodegradable wastewater
生物可降解性 <可以被微生物分解转化的性质> biodegradability
生物可利用磷 bioavailable phosphorus; biologically available phosphorus
生物可利用性 bioavailability
生物可利用指数 bioavailability index
生物控制 bio-control; biologic (al) control; biotic control
生物控制论 biocybemetics
生物矿化作用 biologic (al) mineralization
生物老化 biologic(al)ag(e)ing
生物类群 biologic(al) group
生物累积 bio-accumulation; biologic-(al) accumulation
生物离子交换法 biologic (al) ion exchange process
生物力学 biomechanics
生物立体量测学 biostereometrics
生物利用率 bioavailability
生物沥浸 bioleaching
生物沥滤 bioleaching
生物砾屑灰岩 biocalcirudite
生物链 biologic(al) chain
生物亮晶砾屑灰岩 biosparitic calcirudite; biosparrudite
生物亮晶石灰岩 biosparite
生物量 biologic(al) mass; biomass
生物量度 biologic(al) metric
生物量金字塔 organism quantity pyramid; pyramid of biomass
生物量指标变换 biomass index transformation
生物淋滤 bioleaching
生物淋滤情况 bioleaching condition
生物磷 biologic(al) phosphorus
生物流分类法 biologic (al) stream classification
生物流化床 biologic(al) flow bed; biologic(al) fluid(ized) bed
生物流化床法 biologic (al) fluidized bed process
生物流化床反应器 biologic (al) fluidized bed reactor
生物流化床技术 biologic (al) fluidized bed technique
生物流体力学 biofluid mechanics
生物硫酸浸蚀 biogenic sulfuric acid attack
生物滤池 bacteria bed; biofilter <特指回流式生物滤池>; biologic (al) filter; living filter; percolating filter; sprinkling filter; trickling filter; trickling filter bed
生物滤池表面水力负荷 hydraulic surface loading of biologic(al) filter
生物滤池表面有机负荷 organic surface loading of biologic(al) filter
生物滤池底板 biologic(al) filter floor
生物滤池堵塞 ponding in biologic-(al) filter
生物滤池法 biologic (al) filter process; trickling filter process
生物滤池反硝化系统 biofilter denitrification system
生物滤池辅助措施 biologic (al) filter appurtenances
生物滤池负荷 biofilter loading; biologic(al) filter load(ing)
生物滤池固定布水器 fixed distributor of biologic(al) filter
生物滤池固体负荷 solid loading of biologic(al) filter
生物滤池灰绳 psychada alternate
生物滤池灰蝇 sewage fly
生物滤池回流比 recirculation rate in biologic(al) filter
生物滤池回流量 recirculation flow in biologic(al) filter
生物滤池活动固定布水器 movable distributor of biologic(al) filter
生物滤池积水 biologic(al) filter ponding
生物滤池控制系统 biologic (al) filter control system
生物滤池馈水池 dosing tank in biologic(al) filter
生物滤池滤料 biologic (al) filter media
生物滤池滤料炉渣滤料 biologic (al) filter media-slag media
生物滤池滤料塑料滤料 biologic (al) filter media-plastic media
生物滤池滤料碎石滤料 biologic (al) filter media-rock media
生物滤池滤料阻塞 biologic (al) filter media plugging
生物滤池漫灌 biologic (al) filter flooding
生物滤池偶联工艺 biofilter coupled process
生物滤池填料 biologic(al) filter media
生物滤池填料阻塞 biologic (al) filter media plugging
生物滤池通风 ventilation in biologic-(al) filter
生物滤池脱氮系统 biofilter denitrification system
生物滤池脱膜 sloughing of biological filter
生物滤池仪表 biologic (al) filter instrument
生物滤池自然通风 nature ventilation in biologic(al) filter
生物滤床 bacteria bed; biologic (al) filtration bed; biology filter bed
生物滤器 bacteria filter; biofilter; biologic(al) filter; trickling filter
生物埋藏群 taphocoenose
生物密度扩大 biologic(al) magnification
生物敏感性指数 biologic(al) sensitivity index
生物模拟 biologic(al) simulation; biosimulation
生物膜 biofilm; biologic(al) film; biologic (al) membrane; biologic (al) slime; biomembrane
生物膜产率系数 biofilm yield coefficient
生物膜处理 biologic (al) film treatment
生物膜电极反应器 biofilm-electrode reactor
生物膜法 biologic (al) membrane method; biologic (al) membrane process; biologic(al) slime process; biomembrane process
生物膜反应器 biofilm reactor
生物膜厚度 biofilm thickness
生物膜技术 biofilm technique; biologic(al) membrane technique
生物膜监测器 biofilm monitor
生物膜密度 biofilm density
生物膜膜反应器 biofilm-membrane reactor
生物膜膜生物反应器 biofilm-membrane bioreactor; biologic (al) film-membrane bioreactor; biomembrane-membrane bioreactor
生物膜培养 biofilm culturing
生物膜膨胀床 attached-film expanded bed
生物膜生长 biofilm growth
生物膜生成 biofilm formation
生物膜生成机理 mechanism of biofilm formation
生物膜生物 biofilm organism
生物膜损耗率 biofilm loss rate
生物膜脱落 unloading; unloading of biotic film
生物膜脱落机理 biofilm detachment mechanism
生物膜稳定性 biofilm stability
生物膜厌氧流化床 biofilm anaerobic fluidized bed
生物膜载体 biofilm carrier
生物耐力 biotolerance
生物能 bioenergy; organic energy
生物能含量 bio-content
生物能系统 biologic (al) energy system
生物能源 biologic(al) energy source
生物泥屑灰岩 biocalcilutite
生物泥质结构 bio-argillaceous texture
生物年代地层单位 biochronostratigraphic(al) unit
生物年代学 biochronology
生物黏[粘]膜 biologic(al) slime
生物黏[粘]泥 biologic(al) fouling; biologic(al) slime; bioslime; slime
生物黏[粘]泥量 biologic (al) slime content
生物黏[粘]泥评估 bioslime evaluation
生物农耕 biologic(al) land-farming
生物农药 biologic(al) pesticide; biotic pesticide
生物浓集 biologic (al) concentration; biologic(al) magnification
生物浓缩 bio-concentrating
生物浓缩系数 bio-concentration factor
生物浓缩作用 bio-concentration; biologic(al) concentration
生物排斥说 exclusive relationship theory
生物排水 biologic(al) drainage
生物排泄物 biotic excreta
生物培养基 biologic(al) medium
生物喷洒 biosparging
生物品种 biologic(al) species
生物平衡 biobalance; biologic(al) balance; biologic(al) equilibrium; biotic balance
生物评价 bioassessment
生物屏蔽 biologic(al) shield
生物曝气作用 bio-aeration; biologic-(al) aeration
生物起源 biologic(al) origin
生物起源说 biogenesis
生物气 biologic(al) gas; fermentation gas
生物气候 bioclimate
生物气候变化 bioclimatic change
生物气候的 bioclimatic
生物气候界线 bioclimatic frontier
生物气候室 biotron
生物气候图 bioclimatic chart; bioclimatograph
生物气候图解 bioclimatograph
生物气候学 bioclimatics; bioclimatology; phenology
生物气体 biogas
生物气象学 biometeorology
生物迁移 biogenic migration; biologic-(al) transport
生物前处理 biologic(al) pretreatment
生物潜能 biopotential
生物潜穴孔隙 burrow porosity
生物强化 bioaugmentation; bioenhancement; biointensifying; biologic-(al) enhancement
生物强化工艺 bioaugmentation process
生物强化活性污泥反应器 bioaugmentation activated sludge reactor
生物强化技术 bioaugmentation technology; bioenhanced technology
生物强化颗粒活性炭 biologically enhanced granular activated carbon
生物强化序批间歇式膜生物反应器 bioaugmented sequencing batch membrane bioreactor
生物强化真菌 bioaugmentation fungus
生物侵蚀 biologic (al) attack; biologic-(al) erosion
生物侵蚀作用 bioerosion
生物丘 bioherm
生物球粒亮晶(石)灰岩 biopelsparite
生物球粒微晶(石)灰岩 biopelmicrite

生物球粒微亮晶灰岩 bispelmicrite
生物区 biologic(al) division
生物区系 biota
生物区系省 biotic province
生物区系影响 <对动植物生活> biota influence
生物区域界线 biochore
生物去除剩余磷 biologic(al) excess phosphorus removal
生物去除营养物 biologic(al) nutrient removal
生物去除营养物装置 biologic(al) nutrient removal plant
生物圈 biosphere; organic sphere; vivosphere; watersphere
生物圈保护区 biosphere protection area; biosphere reserve
生物圈保护行动计划 action plan for the biosphere reserves
生物圈地球化学 geochemistry of biosphere
生物圈平衡 biosphere balance
生物圈污染 biosphere contamination
生物圈演化 biosphere evolution
生物圈中心主义 biosphere centralism
生物群 biologic(al) group; biologic(al) population; biota
生物群丛 association of organisms
生物群落 bioc(o)enose; biocoen; bioco(e)nosis; biocommunity; biome; biotic community; cenosis; coen(osis)
生物群落地理学 geobiocenosis
生物群落多样性 diversity of biocommunities
生物群落法 biocommunity method; biologic(al) community method
生物群落关联 biocoenotic connection
生物群落区 biome
生物群落生境 biotope
生物群落效应 biocoenological effect
生物群落学 biocoenology; bioecology; bionomy
生物群迁移 biota transfer
生物群生态特征 biota ecologic(al) feature
生物群系 biome; biotic formation
生物群系型 biome-type; formation type
生物群转移 biota transfer
生物燃料 biologic(al) fuel
生物燃料电池 biologic(al) fuel cell
生物染色剂 biologic(al) stain
生物扰动构造 bioturbate structure
生物扰动结构 bioturbate texture
生物扰动岩 bioturbite
生物扰动作用 biologic(al) perturbation; bioturbation
生物乳化剂 biologic(al) emulsifier
生物软泥 bioslime
生物杀虫剂 bioactivator; biotic pesticide; living insecticide
生物杀伤剂 biologic(al) agent
生物砂滤池 <又称生物沙滤池> biologic(al) sand filter
生物砂滤系统 biologic(al) sand filtration system
生物砂屑灰岩 biocalcarenite
生物砂屑岩 bioarenite
生物筛选试验 biologic(al) screening test
生物上陆事件 organisms disembarkation
生物摄取 biologic(al) uptake
生物渗透 bio-osmosis
生物生产力 biologic(al) productivity; bioproductivity
生物生长 biologic(al) growth
生物生长阶段 biostage
生物生存环境 habitable environment
生物生境 biologic(al) habitat
生物生态地理学 ecologic(al) biogeography
生物生态型 biotic ecotype
生物生态学 bioecology; biologic(al) ecology
生物声学 bioacoustics
生物尸积群 thanalocoenosis
生物失调 biologic(al) disturbance
生物石灰岩 biogenetic limestone
生物时 biochron; biologic(al) time
生物试验 biologic(al) examination; biologic(al) test(ing)
生物试验法 method of bioassay
生物试验室 biology laboratory
生物试样 biologic(al) material; biologic(al) sample
生物适应性 biocompatibility
生物受害 hazard of organism; hazard to organism
生物数据 bio-data
生物数学 biologic(al) mathematics; biomathematics
生物衰变 biologic(al) decay
生物衰变常数 biologic(al) decay constant
生物水平指标 index of living standard
生物水文学 biohydrology
生物水污染 water pollution by organism
生物顺序 biotic succession
生物素 biotin
生物素砜 biotinsulfone
生物碎屑 bioclast; biodetritus
生物碎屑的 bioclastic
生物碎屑灰岩 biocalcarenite; bioclastic limestone
生物碎屑岩 bioclastic rock; bioclastics
生物损伤 biologic(al) damage
生物塔 biologic(al) tower
生物炭 biocarbon; biologic(al) carbon
生物炭池 biologic(al) carbon tank
生物碳酸 biogenic carbonic acid
生物塘 biologic(al) pond; biologic(al) pool; bio-oxidation pond; pond
生物塘法 organism pond process
生物陶瓷(学) bioceramics
生物陶粒滤池 biologic(al) ceram; site filter
生物特征 biologic(al) characteristic
生物梯度 biologic(al) gradient
生物体 living body; organism
生物体的意外释放 accidental release of organisms
生物体污染 organism pollution
生物体污染物负荷量 body burden of pollutant
生物体效应风险特征 risk characterization for organism level effect
生物体效应模型 organism level effect model
生物添加剂 bio-supplement
生物填料塔 biologic(al) packing tower
生物条件 biotic condition
生物条件指数 biotic condition index
生物铁工艺 bio-ferric process
生物铁技术 biology iron technology
生物铁强化活性污泥法 biologic(al) iron strengthening activated sludge process
生物通风 bioventing
生物通气 bio-aeration; bioventing
生物统计调查 biostatistical investigation
生物统计区 biostatistical area
生物统计学 biologic(al) statistics; biometrics; biometry; biostatistics
生物统计学方法 biometrical method
生物突变 sudden changing in biology
生物退化 biodeterioration; catagenesis
生物脱氮 bio-denitrification
生物脱氮系统 anoxic/oxic system; biologic(al) denitrification system
生物脱氮作用 biologic(al) denitrification
生物脱色 biodecolo(u)rization
生物外加剂 biologic(al) additive
生物完整性 biologic(al) integrity; biotic integrity
生物完整性指数 biologic(al) integrity index; biotic integrity index; index of biologic(al) integrity
生物危害 biohazard; biologic(al) damage; biologic(al) hazard
生物微电解法 biologic(al) microelectrolysis
生物微晶灰岩 biomicrite
生物微晶砾屑灰岩 biomicrudite
生物微晶石灰岩 biomicrite
生物微颗粒沉淀 bioparticle deposition
生物微亮晶灰岩 biomicrosparite; biosparite
生物微亮晶石灰岩 biomicrosparite
生物稳定化 biologic(al) stabilization; bio-stabilization
生物稳定剂 biostabilizer
生物稳定塘 biologic(al) stabilization pond; bio-stabilization pond
生物稳定土壤技术 biotechnical stabilization
生物稳定性 biologic(al) stability; biostability
生物稳定作用 biologic(al) stabilization
生物污泥 biologic(al) sludge; bioslime
生物污泥固体 biologic(al) sludge solid
生物污泥评价 bioslime evaluation
生物污染 bio-contamination; biofouling; biologic(al) contamination; biologic(al) pollution; biotic pollution
生物污染评价 biologic(al) pollution assessment
生物污染图 biotic pollution map
生物污染物 biologic(al) pollutant
生物污染源 biologic(al) sources of pollution
生物污染指数 biologic(al) index of pollution
生物污染综合防治 integrated control of biologic(al) pollution
生物污染综合控制 integrated control of biologic(al) pollution
生物污水 biologic(al) sewage
生物污水净化处理 biologic(al) sewage treatment
生物污着 biofouling
生物武器 biologic(al) weapons
生物物理风化 biologic(al) physical weathering
生物物理化学 biophysical chemistry
生物物理学 biophysics
生物雾 biofog
生物吸附 biologic(al) adsorption; biosorption
生物吸附法 biosorption process
生物吸附机理 biosorption mechanism
生物吸附剂 biosorbent
生物吸附曝气 biosorption aeration
生物吸附曝气氧化法 adsorption bio-oxidation process
生物吸附系数 bioadsorption of coefficient; biologic(al) absorption coefficient
生物吸附序列 sequence of biologic(al) absorption
生物吸着法 biosorption process
生物吸着作用 biosorption
生物系 biology department
生物系统 biologic(al) system; biosystem
生物系统学 biosystematics
生物细胞组织 biologic(al) cell tissue
生物细菌堵塞 biologic(al) bacteria blocking
生物显微镜 biomicroscope
生物相 biofacies
生物相带 biofacial zone
生物相古地理图 biofacies-paleotopographic(al) map
生物相容性 biocompatibility
生物相图 biofacies map
生物相组合带 biofacies-assemblage-zone
生物硝化脱氮作用 biologic(al) nitrification denitrification
生物小区 biotope
生物小群落 microcommunity
生物小循环 small biologic(al) cycle
生物小种 biologic(al) form
生物效应 bio-effect; biologic(al) effect
生物屑微晶灰岩 bioclastic-micritic limestone
生物形态 biologic(al) form
生物型 biotype
生物型谱 biologic(al) spectrum
生物性地方病 biologic(al) endemic disease
生物性暴露指示装置 biologic(al) exposure indicator
生物性气溶胶 biologic(al) aerosol
生物性损伤 biologic(al) injury
生物性污染 biologic(al) pollution
生物性药物 biochemical
生物性有毒物质 biologic(al) toxic material
生物修复 biologic(al) remediation; biologic(al) repair; bioremediation
生物修复技术 bioremediation technology
生物需氧量 biologic(al) oxygen demand
生物需氧量测定 biologic(al) oxygen demand test; BOD [biochemicaloxygen demand] determination
生物需氧量负荷 biologic(al) oxygen demand load
生物需氧量浓度 biologic(al) oxygen demand concentration
生物需氧量去除率 biologic(al) oxygen demand removal rate
生物絮凝沉淀 biologic(al) flocculation and precipitation
生物絮凝过程 bioflocculation process
生物絮凝和沉降作用 biologic(al) flocculation and precipitation
生物絮凝剂 bioflocculant
生物絮凝剂产生菌 bioflocculant producing bacteria
生物絮凝体 biologic(al) floc
生物絮凝物 biologic(al) floc
生物絮凝作用 biofloculation; biologic-(al) flocculation
生物蓄积 biologic(al) accumulation
生物悬浮物 bioseston
生物选择器 biologic(al) selector
生物学 biology; bionomy
生物学标度 biologic(al) scale
生物学标志 biologic(al) indication; marker of biology
生物学除污染 biologic(al) depollution
生物学的最小型 biologic(al) minimum size
生物学方法 biologic(al) method
生物学和医学的 biomedical
生物学化验 biologic(al) assay

S

生物学家 biologist
生物学检定 bioassay; biologic(al) assay
生物学鉴定 biologic(al) assessment; bioassay
生物学鉴定法 biologic(al) assay
生物学平衡 biologic(al) balance
生物学评价 biologic(al) assessment
生物学去污 biologic(al) depollution
生物学(上)的 biologic(al)
生物学试验 biologic(al) examination
生物学同位素效应 biologic(al) isotope effect
生物学污染指数 biotic index
生物学指标 biologic(al) indicator
生物学致死温度 biologic(al) zero point
生物学终点 biologic(al) end point
生物学最低温度 biologic(al) minimum temperature
生物循环 biocycle; biologic(al) cycle
生物循环流化床 biologic(al) circulating fluidized bed
生物岩 biogenetic rock; biogenic rock; biogenous rock; biohermite; biolite; biolith; organic rock; organolite
生物岩礁 bioherm
生物研究安全性 biosafety
生物衍生材料 biologic(al) derived material
生物演化 organic evolution
生物演替 biologic(al) succession; biotic succession
生物氧化 bacteria oxidation; biologic(al) oxidation process; bio-oxidation
生物氧化处理 biologic(al) oxidation treatment; bio-oxidation treatment
生物氧化法 biologic(al) oxidation process
生物氧化分量 biologic(al) oxidation component
生物氧化工艺 biologic(al) oxidation
生物氧化沟渠 bio-oxidation channel
生物氧化塘 bio-oxidation pond
生物氧化塘法 biologic(al) oxidation pond process
生物氧化塘工艺 biologic(al) oxidation pond process
生物氧化预处理 biologic(al) oxidation pre-treatment
生物氧化作用 biologic(al) oxidation
生物氧稳定化 bio-oxygen stabilization
生物样本 biologic(al) sample
生物样本分析 biologic(al) sample analysis
生物样品 biologic(al) sample
生物样品分析 biologic(al) sample analysis
生物样性状态 biologic(al) diversity
生物遥测(术) biotelemetry
生物医学工程 biomedical engineering
生物仪传感器 biometer sensor
生物遗迹 biogliph[bioglyph]
生物异常 biogenic anomaly
生物因素 <影响环境的> biotic factor; biologic(al) agent; biologic(al) factor
生物因子 biologic(al) factor; biotic factor
生物印痕 biogliph[bioglyph]
生物营养物 biotic nutrient
生物影响 biotic influence
生物用水 biologic(al) water
生物用水处理工艺 biologic(al) water treatment process
生物用水水质指标 biologic(al) water quality indices
生物油岩 oil source rock
生物有机化学 bioorganic chemistry
生物有机降解作用 biologic(al) organic degradation
生物有机体 biologic(al) organism
生物有效率 bioavailability; biologic(al) availability; biologic(al) effectiveness
生物有效性 biologic(al) effectiveness
生物有效有机碳 biologically available organic carbon
生物诱导除磷 biologic(al) induced phosphorus removal
生物淤泥 bioslime
生物与气候效应研究 biologic(al) and climate effects research
生物预处理 biologic(al) pretreatment; bio-pretreatment
生物域 ecosphere
生物阈限值 biologic(al) threshold limit value
生物元素 biologic(al) element
生物灾害 biohazard
生物载体 biomass support particle
生物再生 biologic(al) regeneration; bio-regeneration
生物再造 biogenetic reworking
生物早起警告 biologic(al) early warming
生物噪声 biologic(al) noise
生物渣滓 biologic(al) sludge
生物炸弹 bio-bomb
生物指标 biologic(al) indicator
生物指示法 bioindication method; biologic(al) indication method; biologic(al) indicator method
生物指示剂 bio-indicator; biologic(al) indicator
生物指示品种 bio-indicator; biologic(al) indicator species
生物指数 bioindex; biologic(al) index; biotic index
生物指数评价 biologic(al) index assessment
生物志 <包括动物志和植物志> biota
生物制剂 biologic(al) agent; biologic(al) preparation
生物制品 biologic(al)
生物制品研究所 institute of biologic(al) products
生物制氢 biohydrogen production
生物制药废水 biologic(al) pharmaceutical wastewater
生物治草 biologic(al) weed control
生物治虫 biologic(al) insect control
生物治理 <利用微生物对污染水进行处理> bioremediation
生物质 biomass
生物质废水 biomass wastewater
生物质焚烧发电 biomass incineration power
生物质过滤 biomass filtration
生物质累积系数 biomass accumulation efficient
生物质量标准 biologic(al) quality standard
生物质量指标 indication of bioquality
生物质密度 biomass concentration; biomass density
生物质能 biomass energy; biotic energy
生物质能发电 biomass power generation
生物质气化 biomass gasification
生物质群聚生物反应器 biomass aggregation bioreactor
生物质燃料 <如甲烷> biomass fuel
生物质生物降解 biomass biodegradation
生物质物理化学处理 biomass physicochemical treatment
生物质吸附剂 biomass adsorbent
生物质转化系数 conversion factor in biomass
生物致毒作用 biotoxication
生物致劣 biodeterioration
生物致死温度 biologic(al) zero point
生物智能 biologic(al) intelligence
生物滞积灰岩 bafflestone
生物中毒 biotoxication
生物终结面 biont terminal plane
生物终结面位置 position of biont terminal plane
生物钟 biochronometer; biologic(al) clock; living clock
生物钟氧化法 biologic(al) clock
生物种 biologic(al) species
生物种多样性指数 species diversity index
生物种群 biologic(al) population
生物周期 biologic(al) cycle
生物注气 biosparging
生物转化 bioconversion; biologic(al) conversion; biologic(al) disintegration; biotransformation
生物转化率 biotransformation rate
生物转化器 biotransformer
生物转化作用 biologic(al) transformation
生物转换面 biont transform plane
生物转换面位置 position of biont transform plane
生物转盘 biodisc; biologic(al) disc; biologic(al) rotating disc[disk]; rotating biologic(al) contactor; rotating biologic(al) disk; rotating contactor
生物转盘处理系统 biologic(al) disk treatment system
生物转盘法 rotating biologic(al) disc process
生物转盘废水处理法 fixed biological surfaces waste-water treatment method
生物转盘滤池 biodisk filter
生物转筒 biologic(al) drum
生物转运 biologic(al) transport; biotransport
生物状况 biologic(al) state
生物资源 biologic(al) resources; biotic resources; living resources
生物资源保护 protection of biologic(al) resources; protection of biotic resources
生物资源保护法 biologic(al) resource protection law
生物自净化作用 biologic(al) self-purification
生物组合 biologic(al) association
生物组合法 method of biologic(al) association
生物作用 biologic(al) action; biologic(al) agent
生物作用带 zone of biologic(al) effect
生物作用光谱 biologic(al) action spectrum
生雾温差 mist interval
生息贷款 interest-bearing load
生息资本 interest-bearing capital
生橡胶 caoutchouc; crude rubber; raw rubber
生橡胶粉 powdered raw rubber; raw rubber powder
生橡胶乳化涂料 raw rubber emulsion paint
生橡胶乳化液 raw rubber emulsion
生硝 caliche
生效 become effective; become operative; come into effect; come into force; come into operation; coming to force; enter into force; entry into force; go into force; take effect; validation
生效保险单 insurance policies in force
生效的专利 unexpired patent
生效日期 availability date; date of entry into force; effective date
生效温度 kick-off temperature
生锌釉 Bristol glaze
生新芽 tiller
生锈 formation of rust; oxidization [oxidisation]; oxidize; rust (formation); rust forming; rustiness; rusting; scaling; staining; tarnish
生锈部分 rusty part
生锈程度 degree of rusting
生锈的 iron stained; oxidized; rusty
生锈地区 <指美国东北部和中西部日益衰落的传统工业地区> rust bowl
生锈钢轨 rusty rail
生锈机理 rusting mechanism
生锈险 risk of rust
生压木 compression wood
生压坯 green pressing
生亚麻籽油 raw linseed oil
生烟强度单位 smoke unit
生盐的 saliferous
生阳极糊 green paste
生氧光合作用 oxygenic photosynthesis
生氧细菌 oxyphoto bacteria
生页岩 raw shale
生硬的 abrupt; harsh; raw; buckram
生油窗底界 base of source oil window
生油窗顶界 top of source oil window
生油岩层 source rock; source bed
生油岩分析仪热解分析 rock-eval pyrolysis analysis
生油岩潜量 source tock potential
生釉 raw glaze
生于河中的 fluvial; fluviatile
生育率 fertility rate
生育年龄组 child boarding group
生原体 biophore
生源说 biogenesis
生在岩隙中的 petrosus
生长 germination; grow(ing); growth; sprout; spurt; upspring; vegetate
生长凹坑 growth pit
生长背斜【地】 growth anticline
生长比率 growth ratio
生长比速 specific growth rate
生长变量 growth variable
生长波 growing wave
生长波管 growing wave tube
生长参数 growth parameter
生长层 growth lamellae; growth layer; growth striation; ring layer
生长常数 growth constant
生长迟缓 growth retardation
生长带 zone of growth
生长的 vegetal
生长点 growing point
生长动力学 growth dynamics
生长断层 growth fault; contemporaneous fault; flexure fault
生长断层发育特征 development feature of growth fault
生长断层-滚动背斜圈闭 growth fault-roll-over anticlinal trap
生长断层活动期 growing period of fault
生长伐 increment felling
生长反应 growth response
生长方程 growth equation
生长方法 growing directions
生长分界面 growth interface
生长峰 growth peak
生长高大的 tall growing
生长格子 growth lattice

S

生长构造 growth structure
生长管 growth tube
生长管盘 growth coil;growth lattice
生长规律 growth rhythm
生长过程 process of growth
生长过速 overgrowth
生长环带 growth zoning
生长环境 habitat
生长缓慢 poor growth
生长辉纹 growth striation
生长基 growth media
生长极限 limit to growth
生长计 auxanometer
生长记录器 auxograph
生长季节 growing season; season of growth
生长阶(梯)growth step
生长节律 growth rhythm
生长结 grown junction;growth semiconductor junction
生长结光电池 grown-junction photocell
生长结型晶体管 growth junction transistor
生长快的 large-growing
生长轮 growth layer; growth ring; ring layer
生长轮界 growth ring boundary
生长霉菌 fungi growth
生长面 growth surface
生长谱 auxanograph
生长期 duration of growing season; growing period; growing season; growth period
生长器 grower
生长丘 growth hillock
生长曲线 growth curve
生长缺陷 growth defect;growth fault
生长扇面边界 growth sector boundary
生长受阻 stunt
生长衰减期 period of declining growth
生长双晶 growth twin
生长速度 growth rate;rate of growth
生长(速)率 growth rate; rate of growth
生长速率起伏 growth rate fluctuating
生长苔藓的(海洋)浅滩 grassy ridge
生长条纹 growth striation
生长位置 growth site
生长习性 growth habit; habit of growth
生长系数 growth factor
生长狭温性 vegetative stenothermy
生长狭盐性 vegetative stenohaline
生长线 accretion line;growth line
生长限制因素 growth limiting factor
生长相 growth phase
生长箱 vegetation tank
生长向量 growth vector
生长小锥 growth hillock
生长效应 growth effect
生长学 auxology
生长仪 auxanograph
生长抑制剂 growth inhibitor
生长于牧场的 pastural
生长韵律 growth rhythm
生长在山区的 montane
生长在石灰岩上的 calcicolous
生长指数 growth index
生长周期 growth cycle
生长锥 cone of growth
生长组构 growth fabric
生长钻 increment borer
生赭石色 cyprus umber
生植物油 raw oil
生殖隔离 reproductive isolation
生殖过程【数】multiplicative process
生殖期 breeding period
生殖狭温性 reproductive stenothemy
生殖狭盐性 reproductive stenohaline
生纸浆 unbeaten pulp
生砖 raw brick

声

声霸卡 sound bias
声板 acoustic(al)board
声饱和 acoustic(al)saturation
声保真度 acoustic(al)fidelity
声爆 sonic boom
声爆影响区 <喷气式飞机的> bang-zone
声比抗 specific acoustic(al)reactance
声比阻抗 specific acoustic(al)impedance;unit-area acoustic impedance
声笔 sonic pen
声变化 sound variation
声变化器 acoustic(al)transfer
声变换器 acoustic(al)transformer
声表面波 surface acoustic(al)wave
声表面波编码器 surface acoustic(al) wave coder
声表面波波导 surface acoustic(al) waveguide
声表面波传感器 surface acoustic(al) wave sensor
声表面波带通滤波器 surface acoustic-(al)wave bandpass filter
声表面波放大器 surface acoustic(al) wave amplifier
声表面波换能器 acoustic(al)surface wave transducer; surface acoustic-(al)wave transducer
声表面波技术 surface acoustic(al) wave technique
声表面波卷积器 surface acoustic(al) wave convolver
声表面波滤波器 surface acoustic(al) wave filter
声表面波式压力传感器 surface acoustic(al)wave pressure sensor
声表面波相关器 surface acoustic(al) wave correlator
声表面波延迟线 acoustic(al)surface wave delay line
声表面波移相器 surface acoustic(al) wave phase shifter
声表面波振荡器 surface acoustic(al) wave oscillator
声表面脉冲压缩滤波器 surface acoustic(al) wave pulse compression filter
声波 acoustic(al)wave;sonic wave; sound wave
声波比 acoustic(al)ratio
声波比较仪 sonic comparator
声波标记图 acoustic(al)signature
声波波长 walk length of sound
声波测点数 number of sonic logging test
声波测定表 sonic wave gage
声波测定仪 sound locator[locater]
声波测高计 sonic altimeter;sound altimeter;sound ranging altimeter
声波测井 acoustic(al)log(ging); sonic log(ging);sound log(ging)
声波测井长度 total length of sound wave logging well
声波测井成果图 sonic log plot
声波测井法 sonic logging method
声波测井记录 sonic logging record
声波测井曲线 acoustic(al)log curve
声波测井仪 sonic tool
声波测距 sound location;sound measurement;sound ranging
声波测距定位 sound ranging location
声波测距法 ranging method of sound wave
声波测距基线 sound ranging base
声波测距控制 sound ranging control
声波测距器 sound radar
声波测距台 sound ranging station
声波测距仪 sound ranger
声波测量 acoustic(al)survey(ing); sonic survey
声波测漏 sonic leak detection
声波测漏仪 sonic leak tester
声波测深法 acoustic(al)depth sounding; acoustic(al) sounding; sonic sounding
声波测深器 sounder;sonic depth finder
声波测深仪 sonic depth-finding instrument;sounder
声波测位法 sonic location method
声波测位和测距 sound finding and ranging
声波测向器 phonozenograph
声波长度 wavelength of sound
声波除尘 dust collection by sound wave; sonic dedusting; sonic dust removal;sonic precipitation
声波除尘器 sonic precipitator
声波除灰 sonic dedusting;sonic dust removal
声波传播 sonic propagation; sound propagation
声波传播速度 acoustic(al)wave velocity
声波传输系数 acoustic(al)transmissivity
声波打桩机 sonic pendulum; sonic pile driver;sound pile driver
声波大气探测 acoustic(al)sounding of atmosphere
声波导 acoustic(al) duct; acoustic-(al)waveguide
声波导航 sonar navigation;sonic navigation;sound homing
声波导航和测距法 sonar method
声波的 acoustic(al); acoustomotive; sonic
声波的干涉 interference of sound wave
声波的减弱 decay of sound
声波的衰减 decay of sound
声波地层学 acoustic(al)stratigraphy
声波电视测井成果图 sonic borehole televiewer plot
声波电视测井(曲线)borehole acoustic(al)televiewing
声波电视测井仪 borehole acoustic-(al)televiewing logger
声波定位 sound location
声波定位和测距 sonar[sound navigation and ranging]
声波定位器 acoustic(al)locator; acoustic(al)detector;audio-locator; audiolocator; phonozenograph; sonic detector;sound locator[locater]; sound radar
声波定位仪 acoustic(al)locating device; sonar [sound navigation and ranging];sound locator[locater]
声波定位与测距(系统)sofar;sound fixing and ranging
声波定向器 direction-listening device
声波杜普勒流速剖面仪 acoustic Doppler current profiler
声波多普勒测流计 acoustic(al) Doppler current meter;acoustic(al) Doppler current recorder
声波多普勒水流断面测绘仪 acoustic (al)Doppler current profiler
声波发光 sonoluminescence
声波发光机 phonophote
声波发射定位 pinger location
声波发射监测法 acoustic(al)emission monitor(ing)
声波发射器 pinger;sonic transmitter; transmitter;acoustic(al)pinger
声波法 acoustic(al)wave method;sonic method;sonic technique;sound wave method
声波法测井 sonic well logging
声波法测探 acoustic(al)logging
声波反射 sonic reflection; sound reflection
声波反射云 acoustic(al)cloud
声波放大器 acoustic(al)wave amplifier
声波分析 acoustic(al)wave analysis
声波分析器 sonic analyser[analyzer]
声波分析数据 sound analysis data
声波分析仪 sonic analyser[analyzer]; sound-analyzing equipment
声波风速计 sonic anemometer
声波幅度测井 acoustic(al)amplitude log
声波幅度测井曲线 acoustic(al)amplitude log curve
声波幅度测井仪 acoustic(al)amplitude logger
声波辐射 sound emission;sound radiation;sound wave radiation
声波辐射照相术 sonoradiography
声波干扰 acoustic(al)interference
声波干涉仪 acoustic(al)interferometer
声波跟踪 sound tracking
声波海流计 acoustic(al)ocean-current meter
声波和密度测井曲线 curve of acousilog and density log
声波核晶过程 sonic nucleation
声波化学分析器 sonic chemical analyser[analyzer]
声波换能器 acoustic(al)transducer
声波换能器阵列 acoustic(al)transducer array
声波集尘 sonic precipitation
声波记录仪 phonautograph
声波记振仪 phonautograph
声波检波器 aural detector
声波检验 sonic testing
声波接合器 sonic bonder
声波井下电视 seisviewer
声波警报系统 sonic alarm system
声波勘探 acoustic(al)exploration
声波孔隙度 sonic porosity
声波雷达 acoustic(al)radar
声波理论 acoustic(al)wave theory
声波料位指示器 sonic level indicator
声波脉冲 sound(im)pulse
声波密度仪 acoustic(al) density probe
声波模数 sonic modulus
声波凝聚 acoustic(al) conglomeration
声波凝聚作用 acoustic(al) coagulation
声波喷嘴 acoustic(al)nozzle
声波疲劳测试 sonic fatigue measurement
声波频率分析器 sound analyser[analyzer]
声波剖面仪 acoustic(al)profiler
声波清洁器 sonic cleaner
声波清洗 sonic cleaning
声波曲射 flexure of sound wave
声波全波列测井成果图 full acoustic-(al)wavetrains recording
声波全波列测井仪 variable-density logger
声波全息术 acoustic(al)holography; sonoholography
声波全息学 acoustic(al)holography; sonoholography
声波全息照相术 acoustic(al)holography;sonoholography
声波散射 acoustic(al)scattering
声波筛分机 sonic screening machine

声波摄影学 sound wave photography
声波深度记录器 sonic depth recorder
声波时差 sonic wave interval transit time
声波示波器 acoustic(al)oscillograph
声波式细度仪 sonic fineness tester
声波视地层因数 sonic apparent formation factor
声波试验 sonic test
声波束 acoustic(al) beam; beam of sound; sound beam
声波衰减 attenuation of sound
声波速度 sonic speed; sonic velocity
声波速度测井 acoustic(al) velocity log
声波速度测井曲线 acoustic(al) velocity log curve
声波速度记录 acoustic(al)velocity log
声波速率 rate of sonic wave
声波锁眼机 sonic button hole machine
声波探测 acoustic(al) detection; acoustic(al) exploration; sonic prospecting; sonic sounding; sound spotting
声波探测器 acoustic(al) detector; sonic detector; sound probe
声波探测仪 sonic detector; soniscope; sound locator[locater]
声波探查 sonoprobe; sonic prospecting <探查海底地层>
声波探漏 sonic leak detection
声波探伤 sonic flaw detection
声波探伤器 acoustic(al) crack detector; acoustic(al) flaw detector; sonic detector
声波探伤仪 acoustic(al) detector; sonic analyser[analyzer]; sonic detector; soniscope
声波探索 sound spotting
声波透射法检测 crosshole sonic logging
声波透视 acoustic(al) method across-ing borehole
声波透视法 cross-hole sonic logging
声波透视曲线 acoustic(al) curve acrossing boreholes
声波图 acoustic(al) pattern; acoustic(al) picture; audiogram; audiograph; sonogram
声波温度计 sonic thermometer
声波纹 <仪器记录下来的> voice-print
声波显示器 phonodeik
声波显示仪 phonodeik
声波絮凝 sonic flocculation
声波寻的 sound homing
声波压强 pressure of sound wave
声波衍射 diffraction of sound wave
声波遥测法 acoustic(al) remote measurement
声波液位计 sonic liquid-level meter
声波仪 acoustic(al) wave hydrophone; sonic apparatus; soniscope; sonograph
声波应变计 acoustic(al) strain ga(u)ge; sonic strain ga(u)ge
声波应变仪 acoustic(al) resistance ga(u)ge; acoustic(al) strain ga(u)ge; sonic ga(u)ge
声波照射 acoustic(al) illumination
声波照相 sound wave photography
声波照相法 phonophotography
声波照相记录 phonophotogram
声波照相术 phonophotography
声波折射 sound deflection
声波振荡 acoustic(al) oscillation
声波振动 acoustic(al) vibration; sonic vibration
声波振动记录仪 phonautograph
声波振幅传输系数 sound amplitude transmission coefficient
声波中质点速度 sound particle velocity
声波装置 sonic device
声波自动记录仪 <记录振动波形> phonoscope; phonograph
声波阻抗 acoustic(al) impedance
声波钻头 <浅孔用的> sonic drill
声箔 sound foil
声薄膜存储器 sonic-film memory; sonic-film storage
声不连续性 acoustic(al) discontinuity
声布置 sound arrangement
声参量阵 parametric (al) acoustic(al) array
声参数 acoustic(al) parameter
声残余 acoustic(al) hangover
声测标图板 sound ranging plotting board
声测分排 sound ranging section
声测高度计 sonic altimeter
声测角计 phonozenograph
声测井 acoustic(al) well logging; sound log(ging); acoustic(al) emission well
声测距 sound ranging
声测距定位 sound ranging location
声测距基线 sound ranging base
声测距技术 sound ranging technique
声测距控制 sound ranging control
声测距(离)调整 sound ranging adjustment
声测距设备 sound ranging set
声测距仪 acoustic(al) ranger; phonotelemeter
声测距站 sound ranging station
声测连 sound ranging company
声测量 acoustic(al) measurement
声测量技术 sound maintenance
声测量系统 acoustic(al) measurement system
声测流量计 sonic flowmeter
声测排 sound ranging platoon
声测情报 sound information
声测设备 sound equipment
声测深度 sounding
声测深器 sonoprobe
声测术 acoumetry
声测所 sound ranging observation post
声测探器 sounding device
声测线 wave measure line
声测线完整性函数值 the function value of wave measure line's integrity
声测向 acoustic(al) bearing
声测仪 <测弹性模量用> soniscope
声测仪器 sound ranging instrument
声层析法 acoustic(al) tomography
声场 acoustic(al) field; sound field
声场标绘器 sound field plotter
声场分布 sound field distribution
声场校正 field calibration
声场校准 sound field calibration
声场起伏 acoustic(al) field fluctuation; sound field fluctuation
声场调整 field adjustment
声衬板 acoustic(al) form board
声称精度 claimed accuracy
声成像 acoustic(al) imaging; ultrasonic imaging
声程 sound path
声冲击吸收器 acoustic(al) shock absorber
声储存器 acoustic(al) memory; acoustic(al) storage
声处理 sound-proofing
声处理的马达 sound-proof motor
声处理结构 acoustically treated construction
声穿透 insonification
声穿透声透射 insonify
声传播 acoustic(al) propagation; sound propagation; sound transmission
声传播常数 acoustic(al) propagation constant; sonic propagation constant; sound propagation constant
声传播降低 sound transmission reduction
声传播系数 acoustic(al) propagation coefficient
声传导 sonic propagation; sound conduction; sound propagation
声传递 acoustic(al) transmission
声传感器 sonic transducer
声传输 acoustic(al) transmission; transmission of sound
声传输系统 acoustic(al) transmission system; sound transmission system
声传输线 acoustic(al) line
声传输性 acoustic(al) transmissivity
声传输质量 sound transmission qualities
声传输装置 acoustic(al) transmission system
声窗 acoustic(al) window
声磁电效应 acoustomgneto-electric(al) effect
声存储器 acoustic(al) memory; sonic storage
声达 <声定位和测距仪> sodar; sound radar
声达距离 audible range; range of audibility; range of hearing
声带 vocal cords
声导 acoustic(al) conductance; acoustic(al) conduction; conductance
声导航 sonic navigation
声导率 acoustic(al) conductivity
声导纳 acoustic(al) admittance
声导体 sound conductor
声导引系统 acoustic(al) homing system
声道 acoustic(al) channel; audio track; aural channel; sound channel; sound duct; sound track; speech channel <电视>
声道波 channel wave
声的侧传 flanking of sound
声的迟滞 acoustic(al) hangover
声的重发 reproduction of sound; sound reproduction
声的传播 sonic propagation; sound propagation
声的传递 transmission of sound
声的反射 sonic reflection
声的改正 acoustic(al) correction
声的干涉 interference of sound
声的过滤 acoustic(al) filtering
声的立体感 acoustic(al) perspective
声的能量密度 energy density of sound
声的衰减 acoustic(al) attenuation
声的痛阈 painfully loud sound
声的投射 sound projection
声的透射 transmission of sound
声的折射 refraction of sound
声的指向性 directivity of sound
声底数值 noise background
声电变换器 electroacoustic(al) transducer
声电变换效率 acoustoelectric(al) index
声电变换指数 acoustoelectric(al) index
声电波 acoustoelectric(al) wave
声电材料 acoustoelectric(al) material
声电的 acoustoelectric(al); electroacoustic(al)
声电功率比 acoustoelectric(al) power ratio
声电化学 sonoelectrochemistry
声电换能器 acoustic(al)-electrical transducer; electroacoustic (al) transducer
声电火花 sound spark
声电流计 sonic-electronic flow meter
声电平 vocal level
声电系数 acoustoelectric(al) coefficient
声电效应 acoustoelectric(al) effect
声电学的 acoustoelectric(al)
声电振荡器 acoustoelectric(al) oscillator
声电指数 acoustoelectric(al) index
声电子学 acoustoelectronics
声调放大器 note amplifier
声调升高的 upward
声调学 tonetics
声定位 acoustic(al) fix; acoustic(al) fix ranging; acoustic(al) positioning; sonic fix; sound location
声定位参考系统 acoustic(al) position reference system
声定位器 sonic locator; sounder
声定位仪 acoustic(al) locating device; sonic locator; sound locator [locater]
声定位装置 acoustic(al) locating device; acoustic(al) locating installation
声定向器 sonic locator
声动电效应 <声波与电荷的互作用> acoustodynamic(al) effect
声动力的 acoustodynamic(al)
声动力效应 acoustodynamic(al) effect
声断路开关 sound-off switch
声对抗 acoustic(al) countermeasure
声发 <水中测声器> sofar; sound fixing and ranging
声发波道 sofar channel
声发传播 sofar propagation
声发动机 phonomotor
声发射 acoustic(al) emission; sound emission
声发射波 acoustic(al) emission wave
声发射测井计 acoustic(al) piezometer
声发射测量系统 acoustic(al) emission system
声发射测压计 acoustic(al) piezometer
声发射法 acoustic(al) emissive method; sound emission method
声发射分布分析仪 distribution analyser of acoustic(al) emission
声发射分析技术 acoustic(al) emission analysis technology
声发射技术 acoustic(al) emission technique; acoustic(al) emission technology
声发射监测 acoustic(al) emission monitor(ing)
声发射监测系统 acoustic(al) emission monitoring system
声发射检测 acoustic(al) emission test
声发射接收器 sound emission microphone
声发射率 acoustic(al) emission rate
声发射频谱 acoustic(al) emission spectrum
声发射试验 acoustic(al) emission testing
声发射探测系统 acoustic(al) emission detection system
声发射无损探伤法 sound emission for non-destructive testing
声发射信号 acoustic(al) emission signal
声发射仪 sound emission indicator
声发射应变计 sonic strain ga(u)ge

S

声发射源 acoustic(al)emission source
声发生器 sound generator
声发系统 sofar system
声法测井 sonic log(ging)
声法拉第旋转 acoustic(al) Faraday rotation
声反馈 acoustic(al)feedback;acoustic(al) regeneration;throwback < 扩声系统的扬声器反施于传声器的声强 >
声反馈抑制 acoustic(al) feedback suppression
声反射 acoustic(al)reflex;sonic reflection;sound reflection
声反射板 sound mirror;sound reflection board
声反射比 acoustic(al)reflectivity
声反射镜 acoustic(al)mirror
声反射率 acoustic(al) reflectivity;sound reflection efficiency;sound reflectivity
声反射器 sound reflector
声反射系数 acoustic(al) reflection coefficient;acoustic(al) reflection factor;acoustic(al) reflectivity;pressure reflection coefficient;sound reflection coefficient;sound reflection factor
声反射性 acoustic(al)reflectivity
声反射因数 sound reflection factor
声放射 sound radiation
声分布 distribution of sound;sound distribution
声分布测量 sound distribution measurement
声分布记录 recording of sound distribution
声分路放大器 sound branch amplifier
声分配放大器 sound distribution amplifier
声分析 sound analysis
声分析仪 sound analyser[analyzer]
声缝隙 sound slit
声浮标 acoustic(al)buoy
声浮标指示设备 sonobuoy indicator equipment
声幅测井 amplitude log
声辐射 acoustic(al)radiation;emanation of sound;sound radiation
声辐射导航 sound homing
声辐射级 sound exposure level
声辐射计 acoustic(al) radiometer;sound radiometer
声辐射器 acoustic(al)radiator;radiator of sound;sound radiator
声辐射线 sound ray
声辐射压 acoustic(al)radiation pressure;sound radiation pressure
声辐射阻抗 sound radiation impedance
声负荷 acoustic(al)load
声负载 acoustic(al)load
声复版 sound-print
声干扰 acoustic(al)dazzle;acoustic(al)disturbance;acoustic(al)jamming;interference of sound;sonic disturbance;sound interference;sound reverberation
声干扰探测 acoustic(al)intrusion detector
声干扰消除器 acoustic(al)clarifier
声干扰效应 sound disturbance effect
声干涉 acoustic(al) interference;sound interference
声干涉探测器 acoustic(al)intrusion detector
声干涉仪 sonic interferometer
声感 phonoreception
声感单位 sensation unit
声感检车器 < 由经过车辆所产生的声波来引动 > sound-sensitive vehicle detector
声感觉量 sensation quantity of sound
声感抗 acoustic(al)inertance
声感图像输入板 acoustic(al)tablet
声隔离 sound isolation
声跟随器 sound follower
声跟踪 acoustic(al)tracking
声工程学 acoustic(al)engineering
声功率 acoustic(al) power;sound power
声功率传输系数 sound power transmission coefficient
声功率反射系数 sound power reflection coefficient
声功率级 acoustic(al) power level;sound power level
声功率计算 sound power calculation
声功率密度 acoustic(al)power density;sound power density
声功率输出 acoustic(al)power output
声功率吸收系数 sound power absorption coefficient
声功率值 sound power value
声共鸣 acoustic(al)resonance
声共振 acoustic(al)resonance
声共振器 acoustic(al)resonator
声管 sound pipe
声管共振 pipe resonance
声惯量 acoustic(al)inertance;acoustic(al)inertia
声光报警装置 sound and light alarm device
声光变换元件 acousto-optic(al)cell
声光玻璃 acousto-optic(al)glass
声光布喇格衍射 acousto-optic(al) Bragg diffraction
声光材料 acousto-optic(al)material
声光材料优值 figure of merit of acousto-optic(al)material
声光测距 sound and flash ranging
声光成像 acousto-optic(al)imaging
声光催化 sonophotocatalysis
声光催化反应器 sonophotocatalytic reactor
声光催化分解 sonophotocatalytic decomposition
声光催化技术 sonophotocatalytic technology
声光催化降解 sonophotocatalytic degradation
声光催化破坏 sonophotocatalytic destruction
声光的 acousto-optic(al)
声光浮标 acousto-optic(al) buoy;lighted sound buoy
声光光量开关 acousto-optic(al)quantity switch
声光盒 acousto-optic(al)cell
声光化学反应器 sonophotochemical reactor
声光化学降解 sonophotochemical degradation
声光化学破坏 sonophotochemical destruction
声光介质 acousto-optic(al)medium
声光晶体 acoustic(al)-optic(al) crystal;acousto-optic(al)crystal
声光可调滤波器 acousto-optic(al) tunable filter
声、光、空调等设施组合的吊顶 service integrated ceiling
声光滤波器 acousto-optic(al)filter
声光脉冲调制器 acousto-optic(al) pulse modulator
声光偏转 acousto-optic(al)deflection
声光偏转器 acousto-optic(al)deflection device;acousto-optic(al) deflector
声光品质因数 acousto-optic(al)quality factor
声光腔 acousto-optic(al)cavity
声光扫描 acousto-optic(al) scan(ning)
声光扫描器 acousto-optic(al) scanner
声光栅 acoustic(al)grating
声光摄影术 sonophotography
声光时差 flashbang
声光束控制偏转器 acousto-optic(al) beam steering deflector
声光水听器 acousto-optic(al)hydrophone
声光锁模 acousto-optic(al) modelocking
声光锁模倍频器 acousto-optic(al) modelocker frequency double
声光陶瓷 phonooptic(al)ceramics
声光调谐激光器 acousto-optically tuned laser
声光调制 acousto-optic(al) modulation
声光调制器 acousto-optic(al)cell;acousto-optic(al) modulation device;acousto-optic(al)modulator
声光(相互)作用 acousto-optic(al) interaction
声光像转换器 acousto-optic(al)image converter
声光效应 acoustic(al)-optic(al) effect;acousto-optic(al)effect
声光信号系统 acousto-optic(al)signal system;audio and visual signal system
声光性能指数 index of acousto-optic(al)property
声光学 acousto-optics
声光折射效应 acoustophotorefractive effect
声过调量 sound overshooting
声焊 sonic soldering
声航速仪 acoustic(al)marine speedometer
声号 sound signal
声号导航信标 talk and listen beacon
声耗能元件 acoustic(al) dissipative element
声合成 phonosynthesis;synthesis of sound
声互作用 acoustic(al)interaction
声化反应 phonochemical reaction
声化学 phonochemistry;sonochemistry
声化学发光 sonic chemiluminescence
声化学反应器 sonochemical reactor
声化学降解 sonochemical degradation
声化学破坏 sonochemical destruction
声化学氧化序批间歇式反应器活性污泥法 sonochemical oxidation-sequencing batch reactor activated sludge process
声画 sound picture
声画编辑机 sound moviola
声环境设计 acoustic(al)environment design;sound environment design
声环境试验室 acoustic(al) environment laboratory
声换能器 acoustic(al) transducer;sound transducer
声换能器阵列 acoustic(al) transducer array
声回授 acoustic(al)feedback
声昏迷 acoustic(al)dazzle
声混合 sound mixing
声火花定位器 sonic spark locator
声火花室 sonic spark chamber
声迹 audio track;sound path;sound track
声迹超前 sound track advance
声迹复版 sound transfer
声迹基准磁平 reference surface induction for the sound track
声迹解调器 sound track demodulator
声迹刻纹机 sound track engraving apparatus
声迹速率 sound track speed
声迹信号 track signal
声基宽 sound stage
声基阵 acoustic(al)array
声畸变 acoustic(al)distortion
声激波 sonic shock
声激励 acoustic(al)excitation;sound excitation
声激励器 sound driver
声激励声码器 voice-excited vocoder
声级 level of sound;sound level
声级标定仪 sound level calibrator
声级测定仪 sound level recorder
声级测量 sound level measurement
声级测量法 method of sound level measurement
声级测量设备 sound level measuring device
声级差 level difference;sound level difference
声级计 level meter;noise meter;psophometer;sound(level)meter
声级计数据 sound level meter data
声级计衰减器 sound level meter attenuator
声级校准器 sound level calibrator
声级控制 sound level control
声级评定 rating of sound level
声级信号 sound level signal
声级仪 acoustimeter
声级中值 medium of sound level
声级自动记录仪 sound level automatic recorder
声极限 sound limit
声加热 acoustic(al)heating
声监测系统 acoustic(al) monitoring system
声监控 acoustic(al)monitoring
声监控器 sound monitor
声监控系统 acoustic(al) monitoring system
声监听 acoustic(al)monitoring
声监听仪 acoubuoy
声减弱指数 sound-reduction index
声剪辑 sound cutting
声检波器 acoustic(al)detector
声检测 sound detection
声检测器 acoustic(al)detector;sound detector
声检索 sonic search
声渐显 fade in
声渐隐 fade out
声降系数 acoustic(al) reduction factor
声接收 sound reception
声接收机 acoustic(al)receiver;sound receiver
声接受器 acoustic(al)receiver;sound receiver
声接头 sound splice
声解作用 phonolysis
声介质 acoustic(al) medium;sound intermediate
声劲 acoustic(al)stiffness
声劲度抗 acoustic(al)stiffness reactance
声经纬仪 acoustic(al)theodolite
声阱 noise trap;sound trap
声景象 auditory perspective
声镜 acoustic(al) mirror;sound mirror
声聚 sonic agglomeration
声聚焦 acoustic(al) focus(ing);sound focus(ing)

声聚焦试验片 sound focus test film
声绝缘 acoustic(al) insulation;sound-insulating;sound insulation
声绝缘层 sound-insulating layer
声绝缘能力 sound insulation capability
声绝缘系数 acoustic(al) reduction coefficient;sound reduction coefficient
声觉 sound sensation
声均匀分布 uniform sound distribution
声卡【计】sound card
声开关 acoustic(al) switch
声勘探 sonic prospecting
声抗 acoustic(al) impedance;acoustic(al) reactance
声抗率 specific acoustic(al) reactance
声空化 acoustic(al) cavitation
声孔径 sound aperture
声控 acoustic(al) control;sound control
声控玻璃 acoustic(al) glass
声控材料 sound-control material
声控机器人 televox
声控技术 sound-control technique;voice control technique
声控靠泊码头 acoustically approach docking control
声控磨机喂料 audio mill feed control
声控室 sound-control room
声控释放式浮标系统 sono controlled type buoy system
声控遥测系统 acoustic(al) control and telemetry system
声控增益调节器 voice-operated gain adjusting device
声控增益调整器 vogad
声控制 voice control
声控蛭石粉刷 vermiculite sound-control plaster
声控蛭石灰泥 vermiculite sound-control plaster
声扩散 acoustic(al) diffusion;sound diffusion
声扩散率 sound diffusivity
声浪 sonic wave;sound wave
声雷达 acoustic(al) radar;sodar;sound radar
声垒 sonic barrier;sound barrier
声力 sound power
声力电话 sound-powered telephone
声力电话机 sound-powered telephone set
声力电话系统 sound-powered telephone system
声力级 intensity level of sound
声链轮 sound sprocket
声量 sound level;sound volume;volume(of sound)
声量单位 volume unit
声量范围 sound volume range;volume range;volume sound range
声量计 volume unit meter
声量检查 audition
声量控制 sound volume control;volume control
声量扩大 volume of expansion
声量扩展 volume expansion
声量扩展器 automatic volume expander;volume expander
声量能通 sound energy flux
声量失真 volume distortion
声量效率 volume efficiency
声量压缩器 automatic volume compressor;volume compressor
声量指示器 level indicator;volume indicator
声列 tonic train
声列信号 tonic train signal(l)ing
声流 acoustic(al) streaming;sound stream(ing)
声流量计 sonic flowmeter
声流吸声器 sound stream absorber
声漏 sound leakage
声滤 filtration of sound
声滤波器 acoustic(al) filter;acoustic(al) wave filter
声路 sound travel
声马赫计 acoustic(al) Mach meter
声码器 sounder;vocoder;voice coder
声脉冲 acoustic(al) impulse;acoustic(al) pulse;pulse of sound;sound(im) pulse
声脉冲传播 acoustic(al) pulse propagation
声脉冲定位器 acoustic(al) pinger locator
声脉冲发射器 pinger
声脉冲发射器测深系统 pinger system
声脉冲发送器 acoustic(al) pinger;pinger
声脉冲回声探测系统 pinger proof echo-sounding system
声脉冲衰变 sound-pulse decay
声脉冲速度 sonic pulse velocity
声脉冲信号 ping
声脉冲照相 sound-pulse photography
声脉塞 acoustic(al) maser
声盲区 zone of silence
声媒质 sound bearing medium;sound intermedium
声媒质体积速度 sound volume velocity
声迷宫 acoustic(al) labyrinth
声密度 sound density
声密封 acoustic(al) seal
声敏探测器 sound-sensitive detector
声敏元件 acoustic(al) sensor
声明 affidavit;announcement;assert;avow;declaration;dictum[复 dicta/dictums];manifest(o);proclaim;proclamation;pronouncement;protest;statement;testify
声明价格 declared value
声明书 declarative statement;statement
声明条约无效 denunciation
声模 acoustic(al) mode
声模散射 acoustic(al) mode scattering
声模式 sound pattern
声幕 sound screen
声呐 acoustic(al) susceptance;asdic gear;dipping asdic;sound radar;mountain goat;sonar[sound navigation and ranging];sound locator[locater]
声呐背景噪声 sonar background noise
声呐本底噪声 sonar background noise
声呐标图 sonar plot
声呐兵 ping jockey
声呐参数 sonar parameter
声呐操纵员 sonar operator
声呐操作员教练器 sonar operator trainer
声呐测距控制 sound ranging control
声呐测量 sonar survey
声呐测量成像的海底 ensonification field
声呐测深器 sonoprobe
声呐测深设备 sonar sounding set
声呐测试装置 sonar test set
声呐测速仪 sonar speed measuring set
声呐测位器 sonar position plotter
声呐传输 sonar transmission
声呐窗 sonar window
声呐单边带通信[讯]设备 sonar single sideband communication
声呐导航 sonar navigation
声呐导流罩 sonar dome;sonar housing;sonodome
声呐吊舱 <反潜飞机上> sonar nacelle
声呐定位 asdic spotting;radio acoustic(al) position finding
声呐定位器 mountain goat
声呐对抗与诱骗 sonar countermeasures and deception
声呐发射机 sonar transmitter
声呐发射器 sonar projector;sonar transmitter
声呐发送器 sonar projector
声呐法 asdic method;sonar method
声呐反潜艇探测 sonar anti-submarine detection
声呐反射物标 sonar target
声呐方程 sonar equation
声呐放大器 sonar amplifier
声呐分解器 sonar resolver
声呐浮标 radio sonobuoy;sonar buoy;sonobuoy
声呐浮标参考系统 sonobuoy reference system
声呐浮标舱 sonar-buoy compartment
声呐浮标测量 sonobuoys survey
声呐浮标降落伞 sonobuoy parachute
声呐浮标接收机 sonobuoy receiver set
声呐浮标屏障 sonobuoy barrier
声呐浮标群 sonobuoy pattern
声呐浮标设备 sonobuoy equipment
声呐浮标信号接收波道 sonobuoy channel
声呐浮标折射系统 sonobuoy refraction system
声呐浮标指示设备 sonobuoy indicator equipment
声呐复示器 asdic repeater
声呐干扰 sonar countermeasures
声呐干扰和假象 sonar countermeasures and deception
声呐攻击指挥中心 sonar attack center[centre]
声呐合格试验 sonar certification test
声呐盒 sonar capsule
声呐环境噪声 sonar background noise
声呐换能器 sonar projector;sonar transducer
声呐换能器阵 sonar transducer array
声呐回声探测装置 sonar echo sounder set
声呐记录器 sonar recorder
声呐技术 sonar technique;sonar technology
声呐技术师 sonar technician
声呐技术性能 sonar technical characteristic
声呐监视系统 sonar surveillance system
声呐鉴定站 sonar evaluation station
声呐校准站 sonar calibration station
声呐教练器 sonar trainer
声呐接收机 sonar receiver
声呐坑 sonar dome pit
声呐控制室 sonar control room
声呐控制系统 sonar control system
声呐脉冲 ping
声呐脉冲测距系统 sonar pinger system
声呐脉冲发射器 sonar pinger
声呐模拟器 sonar simulator
声呐目标 sonar target
声呐情报中心 sonar information center[centre]
声呐入水 sonar dunking
声呐扫床 sonar sweep(ning)
声呐扫海 sonar sweep(ning)
声呐扫描器 sonar scanner
声呐扫描设备 sonar scan(ning) set
声呐扫描仪 side scan sonar photo;sonar scanner
声呐设备 sonar[sound navigation and ranging];sonar equipment;sonar set
声呐射束 sonar beam
声呐声波发生器 sonar thumper unit
声呐声源级 sonar source level
声呐室 <附在军舰船底的> sonar dome
声呐试验驳 sonar test barge
声呐收听装置 sonar listening set
声呐数据 sonar data
声呐数据记录器 sonar data recorder
声呐搜索 asdic search
声呐搜索扇面 sonar search arc
声呐探测法 asdic method
声呐探测器 asdic gear
声呐探测仪 sonar sounding set
声呐探测与测距仪 sonar detecting-ranging set
声呐听音仪 sonar listening set
声呐通信[讯] sonar communication
声呐投吊设备 sonar hoist set
声呐投吊与换能器组合 sonar hoist and transducer group
声呐透声窗 sonar window
声呐图 sonar chart
声呐拖曳系统 sonar towing system
声呐位置传感器 sonar position sensor
声呐系统 sonar system
声呐显示器 sonar control indicator;sonar display
声呐信标 sonar beacon
声呐信号 sonar signal
声呐信号处理 sonar signal processing
声呐信号处理机 sonar signal processor
声呐信号处理计算机 sonar signal processing computer
声呐信号模拟器 sonar signal simulator
声呐性能计算机 sonar performance computer
声呐训练舰 sonar school ship
声呐仪探测 sonar instrumentation probe
声呐引伸仪 sonar extensometer
声呐应答系统 sonar transponder system
声呐预报 sonar prediction
声呐早期警戒 sonar early warning
声呐噪声记录器 sonar noise recorder
声呐站测距分机 ranging set
声呐站自记测波仪 sonar wave recorder
声呐罩 sonar dome
声呐阵测深系统 sonar array sounding system
声呐阵列 sonar array
声呐指示系统 sonar indicator system
声呐装置 sonar equipment;sonar gear
声呐自噪声 sonar self-noise
声呐作用距离 asdic range;sonar range
声能比 acoustic(al) energy ratio
声能比较仪 sonic comparator
声能表 acoustometer
声能反射系数 sound energy reflection coefficient
声能级 power level;sound power level
声能量 acoustic(al) energy;sonic energy;sound energy
声能流密度 acoustic(al) energy flow density;sound energy flow density
声能密度 <单位为焦耳/立方米> acoustic(al) energy density;sound energy density

S

声能疲劳 sonic fatigue
声能通量 sound energy flux
声能通量密度 sound energy flux density
声能通信[讯]系统 sound-powered communication system
声能学 sonics
声能应用设备 sonic applicator
声能锥 <回声测深仪的> cone of acoustic(al) energy
声欧(姆) acoustic(al) ohm
声偶极子 acoustic(al) dipole; acoustic(al) doublet
声耦合剂 <用于材料的超声无损检验> acoustic(al) coupling agent
声耦合器 acoustic(al) coupler; acoustic(al)(data) coupler
声盘 audio disc[disk]; disk record
声盘放送器 record player
声配置 acoustic(al) colo(u) ration
声疲劳试验 acoustic(al) fatigue test
声偏转 acoustic(al) deflector
声偏转电路 acoustic(al) deflection circuit
声频 audible frequency; audio frequency; note frequency; sonic frequency; sound frequency
声频保真度 audio fidelity
声频保真控制 audio-fidelity control
声频报警器 audible alarm
声频变换 audio-frequency change
声频变压器 audioformer; audio-frequency transformer
声频波动因数 audio-frequency variation factor
声频插入通道 audio cue channel
声频带 audio band; sonic frequency band
声频带宽 sound bandwidth
声频电流 audio-frequency current
声频电路 audio circuit
声频电路测试器 audio chanalyst
声频扼流圈 audio-frequency choke
声频发电机 audio generator
声频发生器 acoustic(al) frequency generator; audible frequency generator; audio-frequency generator
声频法 audio method
声频范围 range of audibility; sound frequency range
声频方式 audio system
声频放大器 acoustic(al) frequency amplifier; audible frequency amplifier; audifier; audio-frequency amplifier
声频分析器 sound frequency analyser [analyzer]
声频工程学 audio engineering
声频功率 audio(-frequency) power
声频管 audio tube
声频回音装置 audio response unit
声频混频器 audio mixer
声频基准磁平 reference audio level
声频级 audio-frequency stage
声频记录仪 sound range recorder
声频检波器 aural detector
声频鉴频器 sound discriminator
声频控制放大器 audio fader amplifier
声频滤波器 sceptron
声频率 acoustic(al) frequency
声频率计 audio-frequency meter
声频内插振荡器 audio interpolation oscillator
声频频段 audible region
声频频率 audio-frequency rate
声频频谱计 sound spectrometer
声频谱 noise spectrum
声频谱仪 acoustic(al) spectrometer
声频区 audible region
声频散 acoustic(al) dispersion; sound dispersion
声频散射 acoustic(al) dispersion
声频扫描器 audio scanner
声频筛 sonic sifter
声频上频率 super-audio frequency
声频设备 audio-frequency apparatus
声频声子 acoustic(al) phonon
声频视频程序转换器 audio routing switcher; video routing switcher
声频输出 audio-frequency output
声频衰减器 audio attenuator
声频探头 sound probe
声频特性图示仪 audio-frequency characteristic oscillograph
声频调制 voice modulation
声频调制-解调器 acoustic(al)(data) coupler
声频通道 audio channel
声频通信[讯]线中路 sound line
声频图 audiogram
声频图像 audio image
声频推挽放大器 push-pull audio amplifier
声频下频率 subaudio frequency
声频线 audio line
声频线路 audio line
声频陷波器 sound trap
声频响应信息 audio response message
声频响应装置 audio response unit
声频信号 acoustic(al)(frequency) signal; audible (frequency) signal; audio (-frequency) signal; sound signal
声频信号编码器 audio coder
声频信号接线台 audio patch bay
声频信号雷达站 Emma
声频信号陷波器 sound trap
声频信号中频放大器 sound IF amplifier tube
声频选择器 <滤去低频声波> phono-selectoscope
声频询问 audio inquiry
声频压缩扩展器 audio compressor expander
声频仪 audio equipment
声频远程通信[讯]线路 audio telecommunication line
声频载波 aural carrier
声频噪声 audible noise; audio-frequency noise
声频斩峰器 audio peak chopper
声频振荡器 audio-frequency oscillator; audio-oscillator
声频振动钻进 drilling with sound vibration
声频振铃 voice frequency ringing
声频支 acoustic(al) branch
声频支路 acoustic(al) branch
声频指示器 audio indicator
声频终端 audio terminal
声屏 acoustic(al) shed
声屏蔽 acoustic(al) screen; acoustic(al) shielding
声屏障 acoustic(al) barrier; sound barrier
声剖面图 acoustic(al) profile
声谱 acoustic(al) spectrum; audible spectrum; audio spectrum; sound spectrum
声谱测量 sound spectrum measurement
声谱测量学 sonometry
声谱分析 sound spectrum analysis
声谱分析器 sound analyser [analyzer]; sound spectrum analyser [analyzer]
声谱分析仪 sound spectrum analyser [analyzer]
声谱图 sonagraph[sonograph]
声谱显示仪 direct translator; sonolator
声谱仪 acoustic(al) spectrograph; acoustic(al) spectrometer; acoustic(al) spectroscope; audio spectrometer; sonograph; sound spectrograph
声气冷却系统 sonic gas cooling system
声腔 vocal cavity
声强 flux density; intensity of sound; speech volume; strength of sound
声强比 acoustic(al) ratio
声强测量 sound intensity measurement
声强测量法 phonometry
声强测量计 acoustimeter[acoustometer]
声强测量器 acoustimeter[acoustometer]
声强测量仪 sound intensity measuring device
声强度 acoustic(al) intensity; sound intensity; sound energy flux density
声强度级 sound intensity level
声强度计 phonmeter
声强反射系数 sound intensity reflection coefficient
声强级 acoustic(al) intensity level; intensity level; sound intensity level
声强计 acoustimeter; phonemeter; phon(o) meter; sonometer; sound intensity meter
声强起伏 sound intensity fluctuation
声强衰减 sound intensity decay
声强速度定律 noise-velocity law
声强透射系数 sound intensity transmission coefficient
声强仪 acoustimeter; sound intensity meter
声桥 acoustic(al) bridge; sound bridge
声区 sound area
声圈 voice coil
声全息法 acoustic(al) holography
声全息图 acoustic(al) hologram
声全息图的再现 reconstruction of acoustic(al) hologram
声全息系统 acoustic(al) holographic system
声全息照片 acoustic(al) hologram
声全息照相术 acoustic(al) holography
声权级 acoustic(al) weighted level
声染色 sound colo(u) ration
声扰 interference of sound
声扰动 acoustic(al) disturbance
声绕射 sound diffraction; sound round-about radiant
声容 acoustic(al) capacitance
声容单位 acoustic(al) capacitance unit
声容抗 acoustic(al) compliance
声入侵防护 acoustic(al) intrusion protection
声入侵探测器 acoustic(al) intrusion detector
声散射 scattering of sound
声散射层 sonic scattering layer
声散射体 sound scatterer
声散射物 sound-scattering object
声扫描滚筒 sound-scan(ning) drum
声扫描设备 sound scan(ning) device
声栅 acoustic(al) grating; sound grating
声射 sound reflection
声射线摄影系统 sonaradiography system
声射线照相术 sonoradiography
声失配 acoustic(al) mismatch
声失真 sound distortion
声湿氧化 sonication wet oxidation
声湿氧化技术 sonication wet oxidation technique
声识别输入 acoustic(al) recognition input
声室 sound chamber
声室式传声器 sound cell microphone
声舒适度指数 acoustic(al) comfort index
声输出 acoustic(al) output; sound output
声输出功率 acoustic(al) output power
声输入 vocal input
声束 sound beam
声束偏转器 acoustic(al) beam deflector
声束效应 beaming effect
声衰减 attenuation of sound; sound deadening
声衰减测量 sound attenuation measurement
声衰减常数 acoustic(al) attenuation constant; sound attenuation constant
声衰减器 sound attenuator
声衰减系数 acoustic(al) attenuation coefficient; sound attenuation coefficient; sound attenuation factor
声衰减指数 sound-reduction index
声衰减装置 sound fading device
声衰落 sound fading
声顺 acoustic(al) compliance
声搜 sonar surveillance system
声速 speed of sound
声速测井 acoustic(al)(velocity) log(ging); sounding log
声速测井仪 acoustic(al) velocity logger
声速测量仪 sound velocimeter
声速测深器 sonic sounder
声速垂直分布 sound speed profile
声速怠速化油器 sonic-idle carburetor
声速的 sonic
声速度 acoustic(al) speed; acoustic(al) velocity; sonic speed; sonic velocity; sound speed; sound velocity; velocity of sound
声速法 sound velocity method
声速范围内的 transonic
声速分布 sound velocity distribution
声速改正 correction for sound velocity
声速化油 sonic carburetion
声速化油器 sonic carburetor
声速计 velocimeter
声速校准 sound velocity calibration
声速泥面指示器 sonic sludge level indicator
声速喷射微粉机 micronizer[micronister]
声速喷嘴化油器 sonic-nozzle carburetor
声速射流化油器 sonic discharge carburetor
声速梯度 sound velocity gradient
声速误差 sound velocity error
声速线 sonic line
声速相关器 acoustic(al) velocity correlator
声速压缩 transonic compression
声速仪 sound velocimeter; sound velocity meter
声速应变仪 acoustic(al) strain ga(u)ge
声损伤 acoustic(al) trauma
声损失 acoustic(al) loss
声锁 sound lock; sound trap
声弹性 acoustoelasticity
声弹性法 acoustoelasticity
声探测 sonic survey; sonoprobe; sounding

声探测剖面 sonoprobe profile
声探测剖面线 sonoprobe line
声探测器 sonic locator; sound detector
声探测仪 sonograph
声探伤仪 reflectoscope
声探头 sonic probe; sound probe
声探针 acoustic(al) probe
声特性 acoustic(al) property
声特性阻抗 acoustic(al) characteristic impedance; acoustic(al) intrinsic impedance
声特征 acoustic(al) signature
声调制解调器 acoustic(al) modem
声通量 sound flux
声通路 sound travel
声通信[讯] acoustic(al) message
声投射 sound projection
声透镜 acoustic(al) lens; sound lens
声透镜定向 acoustic(al) lens guide
声透射 insonification; sound transmission
声透射比 acoustic(al) transmissivity
声透射损失 sound transmission loss; transmission loss
声透射损失因数 sound-reduction factor
声透射系数 acoustic(al) transmission factor; acoustic(al) transmissivity; sound transmission coefficient
声透射性 acoustic(al) transmissivity
声透系数 acoustic(al) transmission factor
声图 sonogram
声图鉴定 voice recognition
声图判读 interpretation of echogram
声图判译 interpretation of echogram
声图形 acoustic(al) figure
声望 prestige
声微弱 tenuity
声位移探测器 acoustic(al) displacement detector
声位置线 sonic line of position
声纹 sound-groove; vocal print
声稳定性 sound stability
声污染 sonic pollution; sound pollution
声雾 acoustic(al) fog
声吸收 acoustic(al) absorption; sound absorption
声吸收率 acoustic(al) absorptivity; sound absorptivity
声吸收损耗 acoustic(al) absorption loss
声吸收系数 acoustic(al) absorption coefficient; acoustic(al) absorptivity; sonic absorption coefficient; sound absorption coefficient
声系统 acoustic(al) system
声下的 infra-acoustic(al)
声显微镜 acoustic(al) microscope; sonomicroscope
声线 sound ray
声线轨迹 sound ray trace
声线轨迹仪 sound ray tracking plotter; sound ray tracking set
声线列阵 acoustic(al) array
声线图 ray picture; sound ray diagram
声相关计程仪 acoustic(al) correlation log
声相容性 acoustic(al) compatibility
声相(位) acoustic(al) phase
声相移网络 network of sound phase-shift
声响 noise background
声响标志 acoustic(al) signal
声响标准 acoustic(al) standard
声响测高计 acoustic(al) altimeter
声响测距 sonic ranging
声响测流仪 acoustic(al) flow meter
声响测探器 aural detector
声响车辆探测器 sound-sensitive vehicle detector
声响处理 acoustic(al) correction
声响处理合适的会堂 acoustically satisfactory auditorium
声响传导系数 acoustic(al) transmission factor
声响传感器 acoustic(al) strain ga(u)ge
声响船位 acoustic(al) fix; sonic fix
声响磁性水雷 acoustic(al) magnetic mine
声响导航法 acoustic(al) navigation; sonic navigation
声响的 audio
声响灯浮标 combination buoy
声响度计 phon(o) meter
声响反射 sound reflecting
声响反射板 abat-voix
声响方位 acoustic(al) bearing; sonic bearing
声响分析 sound analysis
声响浮标 bell whistle buoy; sound buoy
声响改善 acoustic(al) correction
声响跟踪系统 acoustic(al) tracking system
声响功率 acoustic(al) power
声响火(灾报)警系统 fire alarm sounding system
声响计 acoustometer
声响假目标 acoustic(al) decoy
声响降低因素 sound-reduction factor
声响焦点 sound foci
声响警报 aural alarm
声响警报器 acoustic(al) alarm (unit); audible alarm
声响警报装置 sound alarm unit
声响控制 sound control
声响控制式栏木 microphone-operated barrier
声响能 acoustic(al) power
声响频率 audio frequency; sound frequency
声响器 squealer
声响式流速计 acoustic(al) current meter
声响衰减管 sound attenuation duct
声响水雷 acoustic(al) mine; listening mine
声响探测仪 sound-detecting and ranging
声响探向器 acoustic(al) direction-finder
声响探向仪 acoustic(al) direction-finder
声响通信[讯] sound communication
声响同步器 acoustic(al) synchronizer
声响位置线 acoustic(al) line of position; sonic line of position
声响污染 acoustic(al) pollution; sound pollution
声响系数 acoustic(al) coefficient; coefficient of acoustics
声响消减系数 acoustic(al) reduction factor
声响消震器 acoustic(al) shock absorber
声响效果滤波器 sound effect filter
声响信号 acoustic(al) signal; audible signal
声响信号器材 sound signal(l)ing appliance
声响性能 acoustic(al) property
声响应变计 acoustic(al) strain ga(u)ge
声响折减系数 acoustic(al) reduction factor
声响整步器 acoustic(al) synchronizer
声响指示器 squealer
声响助航系统 audible aid to navigation
声像 acoustic(al) image; sound image
声像变换器 acoustic(al) image converter
声像的 audio-visual
声像法 <超声检测的> sound image method
声像间隔 sound-to-picture separate [separation]
声像情报 audio-visual information
声像同步录制系统 double system sound recording
声像图 ultra-sonogram
声像显示 acoustic(al) imaging
声像转换管 acoustic(al) image converter
声像转换器 acoustic(al) image converter
声消除 sound elimination
声消衰 sound decay
声效率 acoustic(al) efficiency; sound output
声信号 audible signal; aural signal
声信号处理 acoustic(al) signal processing
声信号设备 acoustic(al) signal(l)ing; acoustic(al) signal(l)ing equipment
声信号识别 acoustic(al) signal recognition
声信号数字 sound signal digital
声信号通道 acoustic(al) channel
声信号载波 voice carrier
声信息 acoustic(al) message
声性质 acoustic(al) property
声悬浮 acoustic(al) levitation
声学 acoustics; phonics
声学倍频带 octave-band
声学不稳定性 acoustic(al) instability
声学材料 acoustic(al) material; sound-absorbent material
声学测波仪 acoustic(al) wave ga(u)ge
声学测高计 acoustic(al) altimeter
声学测距 acoustic(al) range
声学测量 acoustic(al) measurement
声学测量单元 acoustic(al) measuring unit
声学测量的最佳频率 preferred frequencies for acoustic(al) measurement
声学测量系统 acoustic(al) measuring system
声学测量仪 acoustic(al) indicator
声学测量装置 acoustic(al) measuring unit
声学测流测深装置 sonic flow-measuring assembly
声学测流量系统 acoustic(al) flow measuring system
声学测深设备 depth-finder sonic equipment
声学测试仪器 sound testing instrument
声学测温法 acoustic(al) thermometry
声学层深度 sonic layer depth
声学处理 acoustic(al) treatment; sound treatment
声学处理建筑材料 acoustic(al) construction material
声学处理金属箔 acoustic(al) foil
声学处理面层 acoustic(al) facing
声学处理施工法 acoustic(al) construction method
声学处理纤维板 acoustic(al) fiber[fibre] board
声学传输系统 acoustic(al) transmission system
声学单位 acoustic(al) unit
声学导航 acoustic(al) navigation
声学的 acoustic(al); electroacoustic(al)
声学点阵振动 acoustic(al) lattice vibration
声学调查 acoustic(al) survey(ing)
声学定位 acoustic(al) positioning
声学定位和测距 acoustic(al) fixing and ranging; acoustic(al) positioning and ranging
声学定位系统 acoustic(al) positioning system
声学多普勒定位系统 acoustic(al) Doppler fixing system
声学多普勒技术 acoustic(al) Doppler technology
声学法测压 pressure determination by acoustic(al) means
声学方法 acoustic(al) method
声学分析 acoustic(al) analysis
声学风速表 acoustic(al) anemometer; sonic anemometer
声学风速计 acoustic(al) anemometer; sonic anemometer
声学风速仪 acoustic(al) anemometer; sonic anemometer
声学刚性边界 sound-hard boundary
声学高度计 acoustic(al) altimeter; sonic altimeter
声学工程 acoustic(al) engineering; sound engineering
声学工程师 acoustician; acoustics engineer
声学工作者 acoustician
声学构造 acoustic(al) construction
声学顾问 acoustics consultant
声学过滤 <基岩面上覆盖层对地震能量传递的作用> acoustic(al) blanking
声学海洋学 acoustic(al) oceanography
声学互易定理 acoustic(al) reciprocity theorem
声学环境 acoustic(al) environment
声学机理 acoustic(al) mechanism
声学基准系统 acoustic(al) reference
声学技术 acoustic(al) technique; acoustic(al) technology
声学家 acoustician
声学建筑模型 acoustic(al) architectural model
声学结构 acoustic(al) construction
声学警报系统 sound alarm system
声学粒度分析仪 acoustic(al) particle size analyser[analyzer]
声学量测量 measurement of acoustic(al) quantity
声学量度 acoustic(al) measurement
声学流量计 acoustic(al) flow meter
声学流速廓线仪 acoustic(al) theodolite
声学脉冲收发机导航系统 acoustic(al) transponder navigation system
声学模型试验 acoustic(al) model test
声学破裂速度 sonic rupture velocity
声学气体分析计 sonic gas analyser [analyzer]
声学气体分析器 acoustic(al) gas analyser[analyzer]
声学全息摄影术 acoustic(al) holography
声学全息术 acoustic(al) holography
声学全息照相术 acoustic(al) holography
声学软边界 sound-soft boundary
声学扫描器 acoustic(al) scanner
声学设备 acoustic(al) equipment
声学设计 acoustic(al) design
声学矢量平均海流计 acoustic(al) vector averaging current meter

S

声学示波器 acoustic(al) oscillograph
声学试验 acoustic(al) experiment; acoustic(al) investigation; acoustic(al) test
声学试验室 sound camera
声学数据 sound data
声学水雷 sonic mine; sound mine
声学探测 acoustic(al) investigation; acoustic(al) sounding
声学特性 acoustic(al) behavio(u)r; acoustic(al) characteristic
声学特征 acoustic(al) characteristic
声学条件因素 acoustic(al) condition factor
声学温度计 acoustic(al) thermometer; sonic thermometer
声学系统 acoustic(al) system; sound system
声学显示器 acoustic(al) indicator
声学效应 acoustic(al) effect
声学性能 acoustic(al) behavio(u)r; acoustic(al) property
声学寻的装置 acoustic(al) homing device
声学遥测 acoustic(al) telemetering
声学遥测(技)术 acoustic(al) telemetering; phonotelemetry
声学遥测系统 acoustic(al) telemetry system
声学遥感 sound remote sensing
声学遥控 acoustic(al) remote control
声学仪表 acoustic(al) instrument; sonic device; sonic instrument; sound device
声学仪器 acoustic(al) instrument; sonic device; sonic instrument; sound device
声学异常 acoustic(al) anomaly
声学因素 acoustic(al) factor
声学应变仪 acoustic(al)(type) strain ga(u)ge; sonic strain ga(u)ge
声学应答器导航系统 acoustic(al) transponder navigation system
声学应答器系统 acoustic(al) transponder system
声学与噪声控制 acoustics and noise control
声学元件 acoustic(al) element; sound cell
声学振动 acoustic(al) vibration
声学指示器 acoustic(al) indicator
声学制导 acoustic(al) guidance
声学驻波 standing acoustic(al) waves
声学专家 acoustics expert
声学自记推移质测定器 acoustic(al) bed-load recorder
声循环 sound looping
声压 acoustic(al) pressure; acoustic(al) radiation pressure; acoustomotive pressure; sound pressure
声压补偿法 sound pressure compensation method
声压测量传声器 pressure-measuring microphone
声压场 acoustic(al) pressure field; sound pressure field
声压传感器 sound pressure sensor
声压带式传声器 pressure ribbon microphone
声压电压比 acoustoelectric(al) pressure ratio
声压反射率 pressure reflection ratio
声压反射系数 sound pressure reflection coefficient
声压幅度 pressure amplitude
声压级 acoustic(al) pressure level; pressure level; sound pressure level
声压级测量 sound-pressure-level measurement
声压级分贝标度 decibel scale for sound pressure level
声压计 acoustic(al) pressure meter; sound pressure meter
声压校正 sound pressure correction
声压校准 pressure calibration; sound pressure calibration
声压均衡 sound pressure equalization
声压连续性 continuity of acoustic(al) pressure; continuity of sound pressure
声压灵敏度 pressure response; pressure sensitivity
声压式传声器 pressure microphone
声压式话筒 pressure microphone
声压式水听器 sound pressure hydrophone
声压水平 sound pressure level
声压水听器 pressure hydrophone
声压图像说 pressure pattern theory
声压响应 pressure response; sound pressure response
声压振荡 oscillation of sound pressure
声压值 acoustic(al) pressure value
声压转换 pressure transformation
声延迟 acoustic(al) delay; sound lag
声延迟器 sonic delay device
声延迟线 sonic delay-line
声延迟线存储器 acoustic(al) delay-line memory; acoustic(al) delay-line storage
声言模糊 inarticulation
声衍射 acoustic(al) diffraction; sound diffraction
声衍射栅 acoustic(al) diffraction grating
声掩蔽 sound masking
声掩蔽听力图 noise audiogram
声洋流计 acoustic(al) ocean-current meter
声遥测海水温度计 acoustic(al) telemetry bathythermometer
声遥测(技)术 acoustic(al) telemetry; phonotelemetry
声遥感 acoustic(al) remote sensing
声抑制 sound rejection; sound suppression
声抑制电容器 anti-hum condenser
声音保真度 acoustic(al) fidelity
声音报警 audible alarm
声音报警装置 audible alarm unit
声音逼真度 acoustic(al) fidelity
声音编码器 vocoder
声音变量 wobble
声音辨识器 speaker-dependent
声音拨号电话 sound dial telephone
声音操作开关 voice-operated switch
声音操作控制 voice-operated control
声音操作调节器 voice-operated regulator
声音侧面传播 flanking transmission of sound
声音测距法 acoustic(al) ranging
声音测量法 acoustic(al) survey(ing)
声音测量器 audiometer[audiometre]
声音传播 propagation of sound; sound transmission; transmission of sound
声音传播区 insonified zone
声音传播速度 velocity of propagation of sound
声音传播系数 sound transmission coefficient
声音传递 sound transmission; transmission sound
声音传递损失 sound transmission loss
声音传递因数 acoustic(al) transmission factor
声音传输线 sound line
声音磁迹 magnetic sound track
声音存储器 acoustic(al) memory; acoustic(al) storage
声音单元 voice unit
声音的 sonic; tonal
声音的传导体 conductor of sound
声音的混响 reverberation of sound
声音的识别 sound recognition
声音的数字编码 digital coding of voice
声音电平 sound level
声音叠复 overlapping of sounds
声音反射 reflection of sound
声音反射器 sound baffle
声音反射系数 sound reflection coefficient; sound reflection factor
声音方位 sound bearing
声音方向 direction of sound
声音放大器 audio amplifier; sound intensifier
声音辐射 sound radiation
声音复叠 overlapping of sounds
声音干扰 acoustic(al) noise
声音感觉 sound perception
声音工作器 voice-operated device
声音工作增益调整装置 voice-operated gain adjusting device
声音合成器 sound synthesizer; voice synthesizer
声音和数据同时传输协议 digital simultaneous voice and data
声音很轻的 noiseless
声音恢复系统 sound retrieval system
声音回答 audio inquiry
声音回答计算器 audio response calculator
声音绘制 sound rendering
声音机构 acoustic(al) mechanism
声音激活系统 voice-activation system
声音计时 acoustic(al) timing
声音计时机构 acoustic(al) timing machine
声音监视 acoustic(al) surveillance
声音建模 sound model(l)ing
声音降低指数 sound reduction index
声音焦点 focal area for sound
声音校正器 sound corrector
声音接收终端 voice acceptance terminal
声音警报 audible alarm
声音卡 sound bias
声音控制室 sound-control booth
声音控制装置 sonic device
声音扩散场 diffuse sound field
声音理解 speech understanding
声音耦合 acoustic(al) coupling
声音耦合器 acoustic(al)(data) coupler
声音频带 voiceband
声音强度 intensity of sound; sound intensity; sound pressure level
声音清晰度 distinctness of sound; sound articulation
声音全息记录器 holophone
声音扰频器 voice scrambler
声音绕射 diffraction of sound
声音散播 sound splitting
声音失真 acoustic(al) distortion; distortion of sound
声音识别 speech recognition
声音识别系统 sound recognition system
声音输出扫描器 voice output scanner
声音输出(装置) voice output
声音输入 voice input
声音输入装置 acoustic(al) input device; audio input device
声音数字自动交替 automatic alternate voice/data
声音衰变曲线 sound decay curve
声音衰减 sound attenuation; sound decay
声音衰减管道 attenuation duct
声音衰减率 rate of decay
声音衰减器 sound attenuator
声音随时间衰减 sound attenuation
声音通路 sound channel
声音通信[讯] audio communication
声音投映示波器 phonoprojectoscope
声音透射系数 sound transmission coefficient
声音吸收 absorption of sound; sound absorption
声音吸收系数 sound absorption coefficient
声音响度 speech volume; volume of sound
声音响应器 audio response unit
声音响应装置 audio response unit
声音消除 sound elimination
声音消衰 sound decay
声音信号 acoustic(al) signal; aural signal; sound signal
声音信号混合器 sound mixer
声音信息 sound message
声音行程 travel of sound
声音询问 audio inquiry
声音压缩 voice compression
声音延迟线 acoustic(al) delay line
声音掩蔽作用 masking effect of sound
声音抑制器 sourdine
声音音程 sound interval
声音应答 audio response
声音应答控制 audio response control
声音应答器 audio response unit
声音应答信息 audio response message
声音应答终端 audio response terminal
声音应答装置 audio response device; audio response unit; voice answer back
声音迂回传播 flanking transmission of sound
声音载频 sound carrier frequency
声音增强系统 sound-reinforcement system; sound-reinforcing system
声音折射 refraction of sound
声音振动 sound oscillation
声音震颤 wabble
声音直接传播 direct sound transmission
声音中断 sound breakdown
声音终端 audio terminal; voice terminal
声音自记器 volume recorder
声音最高水平 maximum sound level
声引导系统 acoustic(al) homing system
声引信 acoustic(al) fuse
声引信水雷 acoustic(al) mine
声印 voiceprint
声印象 sound impression
声应变仪 acoustic(al) strain ga(u)ge
声应答器 acoustic(al) transponder
声影 sound shadow
声影区 <声波不能到达的地区> acoustic(al) shadow; sound shadow; sound shadow region; sound shadow zone
声影显示的 visual aural
声影显示范围 visual-aural range
声邮系统 audio-mail system
声与光的 acousto-optic(al)
声誉 odo(u)r
声元件 sound component
声源 acoustic(al) driver; acoustic(al) source; radiator of sound; sonic source; sound event; sound genera-

tor;sound source;source of sound
声源定位 auditory localization;localization of sound source;location of sound sources
声源方位 sound bearing
声源辐射功率 radiation power of sound source
声源功率 sound power of a source
声源级 source level
声源校正 acoustic(al) calibrator
声源勘定器 sound locator[locater]
声源强度 acoustic(al) source strength; sound source intensity; strength of sound source
声源室 <用以判断声级 D 的> room containing the source;sound source room;source room
声源输出功率 sound power of a source
声乐 vocal music
声再生 acoustic(al) regeneration
声噪声 acoustic(al) noise
声噪声测量 acoustic(al) noise measurement
声闸 sound lock
声障 acoustic(al);sonic barrier;sonic threshold;sound barrier
声障板 acoustic(al) baffle; sound screen
声照射 insonification
声照射器 sound irradiator
声罩 acoustic(al) shell
声折射 acoustic(al) refraction; refraction of sound;sound refraction
声折射系数 sound deflection coefficient
声振 sound vibration
声振荡 sound oscillation
声振动 sound vibration
声振动频率法 audiosonometry
声振动特征长度 characteristic length of acoustic(al) vibration
声振器 boomer;sonic vibrator
声振试验 acoustic(al) noise test
声振音响效应 sonic boom effect
声震 acoustic(al) shock; booming; sonic bang;sonic boom
声震层 sonic boom carpet
声震仪 sonic boom carpet
声支撑 acoustic(al) suspension
声支路 acoustic(al) branch
声指令 vocal command
声制导 acoustic(al) guiding;acoustic(al) homing
声制导系统 acoustic(al) guiding system;acoustic(al) homing system
声质量 acoustic(al) inertance;acoustic(al) mass
声质量抗 acoustic(al) mass reactance;mass reactance
声质设计 acoustic(al) quality design
声致电离 acoustic(al) ionization
声致断裂 sound-induced breakdown
声致故障 sound failure
声致冷光 sonoluminescence
声致疲劳 sonic fatigue; acoustic(al) fatigue
声致眩晕 acoustic(al) dazzle
声致振动 sound-induced vibration
声中继器 acoustic(al) relay
声子 <晶体点阵振动能的量子> phonon
声子参量放大器 phonon parametric(al) amplifier
声子发射 phonon emission
声子反射光栅 phonon reflecting grating
声子激发 phonon excitation
声子密度 phonon density
声子模型 phonon model
声子耦合能级 phonon coupled level
声子谱 phonon spectrum
声子谱仪 phonon spectrometer
声子热导 phonon conductance
声子热导率 phonon thermal conductivity
声子物理学 phonoy physics
声自导系统 acoustic(al) homing system;sonic homing system
声自导引 sonic homing
声自导装置 acoustic(al) homing device
声自动引导头 acoustic(al) homing head
声阻 acoustic(al) resistance; phonon drag;sound resistance
声阻检验 acoustic(al) resistance inspection
声阻抗 acoustic(al) impedance
声阻抗比 acoustic(al) impedance ratio
声阻抗率 specific acoustic(al) impedance
声阻抗因子 acoustic(al) impedance factor
声阻率 specific acoustic(al) resistance
声阻尼 acoustic(al) damping
声阻尼器 sound damper
声阻应变仪 acoustic(al) resistance ga(u)ge

牲

牲口船 cattle carrier; horse boat; cattle ship
牲口地下(通)道 stock creep; stock pass;stock subway
牲口耳号 earmark
牲口栏 livestock enclosure; pinfold; pound
牲口排泄物 cattle waste
牲口棚 barn; livestock shed; stock barn
牲口圈 pen
牲口税 <在路上通行的> earmarked tax
牲口血块 gore
牲口运输装置 cattle fitting
牲畜 cattle;livestock
牲畜保险 livestock insurance
牲畜舱单 cattle manifest
牲畜场废料 stockyard waste
牲畜车 cattle carrier; cattle wagon; livestock car;livestock wagon
牲畜大农业 livestock ranching
牲畜单位 animal unit
牲畜单位月 animal-unit month
牲畜道交叉 cattle crossing
牲畜吊具 cattle sling
牲畜防护网栏 cattle grid
牲畜肥育业 feeding industry
牲畜肥育站 feeding center[centre]
牲畜分布图 stock map
牲畜粪 animal manure
牲畜粪便 excrement of animals; waste material dropped by farm animals
牲畜粪便利用 use of animal waste
牲畜副产品 animal by-products
牲畜给水栓 livestock hydrant
牲畜供水停车站 watering halt
牲畜管理设备 stock handling facility
牲畜灌药器 drencher
牲畜过路处 cattle crossing
牲畜护栏 cattle guard;stock guard
牲畜货柜 pen container
牲畜集装箱 livestock container; pen container
牲畜集装箱船 cattle-container ship
牲畜供水车站 watering station
牲畜交易税 domestic animal trade tax
牲畜结构 livestock production structure
牲畜栏 cattle fitting; cattle stall; corral;pen;stock corral
牲畜码头 cattle pier;cattle wharf
牲畜排出物污染 pollution by livestock effluent
牲畜棚 animal house
牲畜评价 livestock judging
牲畜破坏程度 disturbance by livestock
牲畜栖息场 bed ground
牲畜圈 cattle pen
牲畜市场 live market;stock market
牲畜市场税 toll turn
牲畜数量 livestock number
牲畜饲喂场废水 feedlot wastewater
牲畜饲喂场排水 feedlot runoff
牲畜饲喂场饮水池 feedlot lagoon
牲畜饲喂站 feeding station
牲畜饲养 animal husbandry; livestock breeding;stock raising
牲畜饲养场 dry lot;feedlot
牲畜塘 stock pond
牲畜跳板 stock chute
牲畜通道 stock pass
牲畜围栏 pen;penfold;stock pen
牲畜污染 stockbreeding pollution
牲畜污水污染 pollution by livestock effluent
牲畜小道 <铁道或公路下面的> cattle creep; cattle pass; cattle walkway
牲畜斜槽 inclined chute
牲畜需水量 livestock requirement
牲畜押运人 drover
牲畜饮用 drinking water for livestock
牲畜运输 livestock traffic
牲畜运输保险 livestock transit insurance
牲畜运输船 cattle-carrying ship; cattle carrier
牲畜运输路程单 livestock waybill
牲畜运载工具 <如运送活牲畜的特殊集装箱等> cattle carrier
牲畜折合系数 livestock conversion coefficient
牲畜装车斜台 cattle loading ramp; livestock loading ramp
牲畜装车站台 livestock loading platform
牲畜装卸斜坡台 cattle loading ramp
牲畜装卸站台 cattle dock
牲畜装运船 cattle carrier
牲畜总增率 gross rate of increase of domestic animals
牲畜(走的土)路 drove road

绳

绳 twist;rope <英制长度单位, =20英尺>

绳槽 rope groove
绳槽卷筒 grooved hoist drum
绳测 string measurement
绳测功器 rope brake
绳插座 rope socket
绳铲挖掘机 cable excavator
绳车 hawser reel;reel
绳尺 string tape
绳传动 rope drive
绳传动装置 rope gearing
绳锤 cord weight
绳带 rope belt
绳带式输送机 cable belt conveyer [conveyor];rope and belt conveyer [conveyor]
绳带式运输机 cable belt conveyer [conveyor];rope and belt conveyer [conveyor]
绳道 rope race
绳的韧性 stiffening of rope
绳的下端 bottom rope
绳吊索 rope sling
绳斗电铲 dragline
绳渡 rope ferry
绳端结 crown knot;wale knot
绳端套孔 becket
绳端心环 heart shape thimble; pear-shaped thimble;thimble
绳段 ran
绳断股 stranded
绳割 line cut
绳沟鼓筒 grooved drum
绳沟滚筒 grooved drum
绳钩 cord hook;rope lug;rurp <爬山攀冰用>
绳钩法 line fishing;lining
绳钩锚链钩 cable hook
绳股 strand
绳股捻成的钢丝绳 strand-laid rope
绳股捻成的缆索 cable-laid rope
绳股捻距 <钢丝绳的> strand pitch
绳股与股丝不同捻向的钢丝绳结构 regular lay
绳股与股丝同捻向的钢丝绳结构 lang lay
绳罐道 cable guide
绳盒 twine carrier
绳滑车 rope tackle block
绳环 becket;bight;parrel;rove
绳回环 loop;sling
绳夹 cable clamp; rope clip; twine holder mechanism; twine keeper; twine retainer
绳夹板 twine holder plate
绳夹的夹头 <打结器的> retainer clip
绳夹清洁器 twine holder cleaner
绳架 rope support
绳绞导线 rope-lay conductor
绳绞接 rope splice
绳绞盘 rope reel
绳接头 cable joint;rope coupling
绳结 bends and hitched;knot
绳卷的中空部位 tier
绳卡 Babcock socket; rope cappel; stringing grip
绳卡中穿绳的孔 wood pecker hole
绳孔 <滑车上> swallow
绳扣 bight;bale sling <装卸件货的>
绳扣成组货件 wire sling package
绳拉条 cord stay
绳拉卸车法 cable-pullout unloading method
绳缆缠包机 cable binding tool
绳缆搓绞机 strander
绳缆端头 rope socket
绳缆用油 batch oil
绳链制动器 snubber
绳路 cord circuit
绳路记发器交接电路 <入中继、出中继> register chooser for cord circuit
绳路架 link frame
绳路信号灯 cord lamp
绳轮 cable pulley; loading sheave; rope sheave
绳轮托臂 jack boom
绳毛垫 shag mat;thrum mat
绳毛绒 foxes;thrum
绳帽 roller-ratchet socket
绳扭转方向 rope lay
绳碰垫 mat fender;rope fender
绳圈 bail;deck quoit;fake;quoit
绳圈合扎 eye seizing;throat seizing
绳圈碰垫 grommet fender
绳绒地毯 chenille carpet
绳绒线 chenille

绳绒织物 chenille
绳式蛤壳形抓斗 cable clamshell
绳式股芯 independent wire rope core
绳式股芯钢丝绳 independent wire rope core wire rope
绳式过滤机 string filter
绳式喷头 rope-type head
绳式钻具 cable rig
绳束 strick
绳栓 lading band anchor;lading strap anchor
绳速 rope speed
绳索 cable line; rope; cord(ing); horse;marlin(e);patch cord;purchase; rigging; roping; stranding wire;string;winding rope;overwinding <为吊重物而在滚筒上的>; cordage <总称>
绳索包皮 snake wire
绳索编接接头 married joint
绳索操纵 cord control;rope control
绳索操纵的 rope-operated
绳索操作的 cord operated
绳索测力计 rope dynamometer
绳索插脚 rope spear
绳索抄平器 rope levelling unit
绳索冲击 surge
绳索传动 cord drive;rope drive[driving];rope transmission;rope working
绳索垂度 rope curve
绳索打滑 rope slipping
绳索单缀 single lacing
绳索导轮 wire-line truck
绳索导向装置 wire-line guide
绳索的捻距 lay of rope
绳索的捻向 lay
绳索的扭曲 squirm
绳索的纽结 hockle
绳索的散端 fag-end
绳索的下垂 sag of rope
绳索的最大偏角 <绞车的> fleet angle
绳索吊的屋顶 rope roof
绳索吊环 cordage sling
绳索定线 string lining
绳索飞轮 rope-pulley flywheel
绳索覆盖 cord covering
绳索和小盘输送机 rope-and-button conveyer[conveyor]
绳索滑车 fairlead;rope pulley;rope sheave
绳索滑轮 rope pulley
绳索滑轮式 rope and pulley type
绳索滑落 rope slip
绳索滑座 cable slide
绳索环圈 rope strap
绳索机 crosser
绳索急弯处 nip
绳索夹具 rope catching device
绳索检验 examination of ropes
绳索绞盘 rope winch
绳索绞纹 cantline
绳索接长 rope extension
绳索接头 wire-line sub
绳索卷筒 line drum
绳索卷扬机 tugger hoist
绳索卡尺 rope ga(u)ge
绳索控制的 rope-controlled
绳索控制的机械式起落机构 rope-controlled mechanical lift
绳索控制灌溉 cablegation irrigation
绳索捆绑端 dead-end hitch
绳索拉紧 rope stretch
绳索拉力 line pull;rope pull
绳索联结 rope coupling
绳索楼梯 string stair(case)
绳索轮 rope pulley;rope sheave
绳索密封 cord seal(ing)
绳索模板 lacing board
绳索磨损 rope wear
绳索挠度 cable sag
绳索捻距 rope turn
绳索扭结 kink;nip
绳索牵引机 cable-towed machine
绳索牵引犁 cable plough
绳索牵引能力 rope capacity
绳索牵引式铲运机 cable scraper
绳索牵引式刮泥刀 rope-hauled scraper blade
绳索强度 cord strength
绳索强度试验机 rope testing machine
绳索桥 rope suspension bridge
绳索倾斜计 rope angle ga(u)ge
绳索取芯 wire-line coring
绳索取芯工具 wire-line tool
绳索取芯管 wire-line core barrel
绳索取芯绞车 wire-line core reel; wire-line drum; wire-line hoist; wire-line reel;wire-line winch
绳索取芯金刚石钻进 wire-line diamond drilling
绳索取芯内管 wire-line cable
绳索取芯内管接头 inner-tube head for wire line
绳索取芯器 wire-line coring device
绳索取芯装置 cable core tube assembly;wire-line coring apparatus
绳索取芯钻杆 wire-line drill pipe; wire-line drill rod
绳索取芯钻杆接头 wire-line drill rod coupling
绳索取芯钻进 wire-line core drilling
绳索取芯钻具 wire-line coring system
绳索取芯钻头 wire-line core bit; wire-line coring bit
绳索取芯作业 wire-line work
绳索绕缠法 reeving
绳索绕缠系统 reeving system
绳索绕组节距 winding pitch
绳索润滑剂 rope lubricant
绳索润滑油 cordage oil
绳索散端 loosed end
绳索散开 unravel
绳索式冲击钻进 cable drilling
绳索式穿孔机 churn drill
绳索式抖动器 cable shaker
绳索式分级机 rope grader
绳索式起落机构 rope-operated lift
绳索式挖泥机 cable dredging machine
绳索式线脚 torsade
绳索式凿岩器 churn
绳索式真空过滤器 string vacuum filter
绳索式钻孔机 wire-line drill(ing machine)
绳索式钻头 wire-line bit
绳索收紧器 tension toggle operator
绳索输送机 cable conveyer[conveyor];rope conveyer[conveyor]
绳索双缀 double lacing
绳索死结 builder's knot
绳索松弛部分 slatch
绳索松弛停止器 slack rope stop
绳索松股 long jawed
绳索松脱 rope slip
绳索速度 speed of rope
绳索锁紧装置 cable locking system
绳索套环 rope thimble
绳索梯 ratlin(e)
绳索铁道 cog railway
绳索推进系统 <在编组线连挂车辆>【铁】rope propulsion system
绳索推进(装置) rope crowd(ing)
绳索拖动的 rope-powered
绳索纹饰线脚 rope mo(u)lding
绳索小车 rope trolley
绳索运输 rope haulage
绳索运输道 rope drift
绳索运输机 rope conveyer[conveyor];rope haulage machine
绳索扎结 <用绳把电缆固定到吊线上> marline tie
绳索支承鞍座 <悬索桥塔上的> cable saddle
绳索织物 cord woven fabric
绳索制动器 rope brake
绳索(终端)固定 rope dead end
绳索装罐机 rope cager
绳索装饰嵌条 rope mo(u)lding
绳索组节距 winding pitch
绳索钻具法【岩】cable tool method
绳索钻钻具 cable tool
绳套 bail;becket loop;lashing;rope sling;rope socket
绳套曲线 loop curve;loop rating
绳套形水位流量关系曲线 rating loop
绳梯 gangway; jack ladder; Jacob's ladder; jumping ladder; pair of steps;rope ladder;man rope <上船用的>
绳梯棍 ladder rod;ladder rung
绳梯横杆 rundle
绳梯横索 ratlin(e);ratlings
绳梯中央绳 concluding line
绳条 strand;yarn
绳条结 marline knot;rope yarn knot
绳条填料 rope yarn packing
绳筒 rope roll;twine can
绳头 pigtail
绳头编帽 <防碰撞用> thrum cap
绳头插接 dogging
绳头缠扎 common whipping; palm whipping;plain whipping
绳头承窝 new era socket
绳头冲击装置 rope attachment
绳头打捞筒 tongue socket
绳头合扎 palm and needle whipping
绳头夹 bulldog grip;wire clip
绳头结 lanyard knot;wale knot;wall knot
绳头扣紧座 rope socket
绳网吊索 western ocean plant
绳网区 net and rope making yard
绳尾 tail
绳尾编成尖形 cross pointing
绳纹 rope figure;rope pattern;ropiness
绳纹壁缘 twisted rope frieze
绳纹型取土器 <日本古土器> Jomon-type pottery
绳纹装饰 cord-mark decoration
绳屑 foxes;thrum;junk <填缝隙用的>
绳芯 center[centre] of a rope;core; core of a rope;rope heart
绳芯麻条 core yarns
绳型聚四氟乙烯石棉填料 cord type teflon asbestos packing
绳眼 cable eye
绳油 rope grease
绳针结 admiralty hitch;marlin spike hitch
绳整计算器 string-line calculator
绳正法整正曲线 string lining of curve
绳织轮胎 cord tire[tyre]
绳状花纹 rope figure
绳状嵌缝条 rope caulk
绳状熔岩【地】ropy lava;corded lava;paha lava;pahoehoe;pahoehoe lava
绳状熔岩构造 pahoehoe lava structure
绳状熔岩流 ropy lava flow
绳状体 corpus restiformis;restiform body
绳状物 ropy
绳状洗布机 winch machine
绳状洗涤机 rope scourer;rope-scouring machine;rope washer
绳状轧漂机 rope chemicking machine
绳状轧水机 rope squeezer
绳子盘 rim pulley
绳子式的 ficelle

省道 provincial road; provincial (trunk) highway

省道网 provincial highway system
省的规划 provincial planning
省地图 provincial map
省电灯丝 dim filament
省电管 dull emitter tube;low-filament drain tube
省费用 cost saving
省干线公路 <省道> provincial trunk highway
省工 labo(u)r saving;saving of labo(u)r;underwork
省工法 labo(u)r-saving device
省工设备 labo(u)r-saving device
省工装备 labo(u)r-saving device
省公路局 provincial highway commission
省公路系统 provincial highway system
省会 provincial capital
省级干线公路 provincial trunk highway
省计划 state planning
省际高速公路 interprovincial expressway;interprovincial freeway
省际公路 interprovincial highway
省界 provincial boundary
省力 labo(u)r-saving
省力操纵阀 low-effort control valve
省力的 effortless
省力的变速装置 effortless shifting
省力滑车(轮)组 pulley block for a gain in force
省力有效法则 <认知机能> stinginess rule
省力装置 work saving device
省略法 ellipsis
省略符号 ellipsis;ignore character
省略规则 default rule
省略时解释 default interpretation
省略时页面 default page
省略误差 error of omission;omitted error
省略因数 omission factor
省略值 default value
省煤器 coal economizer;economizer
省煤器管 economizer tube
省煤器护板 economizer casing
省煤器灰斗 economizer hopper
省煤器汽化 economizer steaming
省煤器受热面 economizer surface
省钱 save money
省缺 default
省热器 economizer;heat saving apparatus
省时(的) timesaving
省时分析 time-saving analysis
省时商品 time-saving commodities
省时装置 time-saving apparatus; time-saving device;time-saving unit
省水池 thrift basin;water saving basin
省水船闸 saving lock;storage thrift lock;thrift lock;water-saving lock
省水灌溉 water-saving irrigation
省水器 water economizer[economiser]
省水闸室 water-saving chamber
省水装置 water-saving installation
省位拨号盘 space-saver dial
省辖市 municipality of provincially administered;provincially administered municipality
省行政(管理)办公大楼 provincial administration building

省医院 provincial hospital
省银行 country bank
省油阀 economizer valve
省油汽化器 economy-type carburettor
省油器 fuel economizer;oil economizer
省油器本体 economizer body
省油针阀 <汽化器内> needle valve of economizer
省油轴承 oil saving bearing
省余压力 positive pressure
省展览会 provincial exhibition
省政府 provincial government
省政府大楼 provincial government building

圣·彼得拉博格标准 <木材标准尺度> St. Peteraburg standard

圣·罗马广场 <位于威尼斯> The Piazza of S. Mark
圣·乔治十字架 St. George's cross
圣·保罗大教堂 S. Paul's Cathedral
圣·彼得大教堂 <公元 1506 至 1626 年建于罗马> S. Peter Cathedral
圣·马克广场 St. Mark Square
圣·尼古拉教堂 S.Nichola's church
圣·索菲亚教堂 S. Sophia Cathedral
圣爱尔摩火 Saint Elmo's fire
圣安德列斯断层 <美> San Andreas fault
圣安德列斯入字型构造 San Andress-type structure
圣安德列斯转换断层 San Andress transform fault
圣安德鲁交叉砌合 <英国式交叉砌合> Saint Andrew's cross bond
圣安德鲁十字架 Saltire cross
圣安娜风 Santa Anna
圣杯 Holy Grail
圣彼得-圣保罗的荷兰巴罗克式大教堂 Dutch Baroque Cathedral of St Peter and St Paul
圣壁 iconostasis
圣饼烘房 host bakery
圣餐饼陈列器 monstrance
圣餐房 bethlehem
圣餐盒 ostensory
圣餐具橱 almary[almery];a(u)mbry
圣餐面包房 host bakery
圣餐室 prothesis
圣餐台 altar;communion table
圣城 holy city
圣诞节 Christmas
圣诞树 Christmas tree;English holly
圣地 bethel;hagiasterium;sacred place;shrine;ambitus <教堂周围的>
圣地围墙 peribolos
圣地亚哥 <智利首都> Santiago
圣地岩窟 chaitya cave
圣殿 sanctuary
圣殿或庙宇 <古罗马> delubrum
圣殿围墙 peribolos
圣多伦火山灰 Santorin ash
圣多伦水泥 Santorin cement
圣多伦土 Santorin earth
圣多美和普林西比 São Tome and Principe
圣多明戈红木 San Domingo mahogany
圣多明各 <多米尼加首都> Santo Domingo
圣非石 santafeite
圣佛利斯符号 Schoenflies symbol
圣佛利斯-国际符号 Schoenflies-international symbol
圣弗朗西斯科河 Sao Francisco River
圣胡安淡褐色硬木 <产于中美洲> San Juan
圣胡安港 <波多黎各> Port San Juan
圣卡杰坦教堂 St. Kajetan church
圣龛 mihrab;prayer niche
圣库 altar of repose
圣劳伦斯河 St Lawrence River
圣栎 holm(e)(oak)
圣灵教堂 church of the Holy Spirit
圣铃 sanctus bell
圣柳丛 salt ceders
圣楼 tribune
圣卢西亚 St Lucia
圣马力诺 <圣马力诺首都> San Marino
圣母玛利亚教堂 church of the Miraculous Virgin
圣母玛利亚礼拜堂 Lady Chapel
圣母堂 Lady Chapel
圣母院 Lady Chapel
圣牛角【给】sacred cows
圣盘 Holy Grail
圣器 <古代教堂中的> diaconicum
圣器室 sacristy;sceuophylacium;diaconicon <古代教堂中的>
圣人礼拜堂 saint's church
圣人墓 saint tomb
圣萨尔瓦多 <萨尔瓦多首都> San Salvador
圣三一教堂 church of the Holy Trinity
圣十字教堂 church of the Holy Cross
圣水 hyssop
圣水池 <意大利教堂的> pila
圣水喷的墩柱 <教堂中的> pillar piscina
圣水盆 benitier;holy-water basin
圣水器 aspersorium
圣水石盆 <教堂门口的> holy-water stone
圣水体 <教堂入口> stoup
圣所 abaton; innermost part; inner sanctum
圣台背面突出物 gradin(e)
圣台顶上的石板 altar slab
圣台隔扇 altar screen
圣泰非拖挂运输公司 <美> Santa Fe Trail Transportation Company
圣坛 bema;chancel
圣坛侧廊 chancel aisle
圣坛的台面 mensa
圣坛地毯 altar carpet
圣坛拱顶 chancel arch
圣坛后高架 refable
圣坛栏杆 altar rails
圣坛旁的空间 <教堂的> presbyterium;presbytery
圣坛屏风 altar screen
圣坛屏饰 chancel screen
圣坛前挂饰 altar front(al)
圣坛上面装饰性平顶 ceilure
圣坛台面 altar slab;altar stone
圣坛围栏 chancel rail
圣堂 abaton;great altar;sacrarium
圣体容器 pix(is)
圣徒遗物存放箱 chasse
圣维南(等效力)公式 Saint Venant's equation
圣维南方程 Saint Venant equation
圣维南模型 Saint Venant model
圣维南扭力 Saint Venant torsion
圣维南扭矩常数 Saint Venant's torsion constant
圣维南扭转 free torsion
圣维南塑性体 Saint Venant plastic body
圣维南弹性原理 Saint Venant principle of elasticity
圣维南物体 Saint Venant body
圣维南原理 <即等力载原理> principle of Saint Venant; Saint Venant principle
圣维特大教堂 <10 世纪初捷克布拉格> Chram St. Vita
圣物匣 reliquary
圣物箱 kist
圣像 cult statue;miraculous image
圣像屏 iconostasis
圣像周围的光轮 mandorla
圣像周围的光晕 mandorla
圣衣会教堂 Carmelite church
圣油瓶龛 chrismatory
圣约瑟 <哥斯达黎加首都> San Jose
圣障 iconostasis
圣者形像 image of a saint
圣钟 sanctus bell

胜地 famous scenic spot

胜负概率比 odds ratio
胜过 outbalance; out perform; outstrip;outweigh;prevail against;prevail over;surpass;transcend
胜利拱门 cancel arch
胜利纪念碑 victory memorial
胜利柱 <古罗马> column of triumph
胜任人员 competent person

盛冰期 pleniglacial period

盛产粮食地带 granary
盛潮河段 maritime section of stream
盛大的 pompous
盛废物容器 waste container
盛钢桶 tap ladle
盛钢桶衬砖 ladle brick
盛钢桶吊运车 ladle crane trolley
盛钢桶倾注装置 ladle tilter
盛开 bloom
盛盘 catch basin
盛盆期 widespread period of basin
盛器 dispenser
盛燃时间 burning period
盛饰 adornment
盛饰建筑 decorated architecture;ornamental architecture;ornamented architecture
盛饰建筑形式 <英国哥特式建筑之第二阶段> decorated style
盛饰罗马风 heavy Romanesque(style)
盛饰砌合 decorative bond
盛饰时代 decorative period
盛饰时期 decorative period
盛衰 ebb and flow;ups and downs;vicissitude
盛水杯 drainage cup
盛水试漏 full water test;water filling test
盛酸器 acid receiver
盛夏 high summer
盛行波(浪)dominant wave;predominant wave;prevailing wave
盛行潮流 predominant tidal flow
盛行的 prevailing
盛行的顺岸漂沙 dominant longshore drift
盛行方向 predominant direction
盛行风 dominating wind; predominant wind; prevailing wind; dominant wind
盛行风向 prevailing wind direction;reigning winds direction;prevailing direction of wind
盛行海流 predominant current
盛行浪 predominant wave
盛行流 predominant current;prevailing current
盛行能见度 prevailing visibility
盛行漂沙 predominant drift
盛行入射波 prevailing incident wave
盛行西风带 prevailing westerlies
盛行性污染物 prevalent pollutant
盛压缩气体的钢瓶 pressure bottle
盛液池 holding pond
盛液盘 drip pan
盛油盘 drip pan
盛油器 oil catcher
盛油容器 oil-holder
盛雨斗 rain ga(u)ge bucket
盛雨筒 rain ga(u)ge bucket
盛渣罐 slag receiver
盛渣桶 sludge ladle
盛装【建】emblazon;finery
盛壮年谷 full mature valley
盛壮年期 full mature stage;full maturity

剩差向量 vector of residuals

剩磁 remanence; remanent magnetism; residual magnetism; retentive magnetism; retentiveness; retentivity
剩磁测定 residual magnetism measurement
剩磁场中平面 remanent median surface
剩磁笛簧接线器 remreed switch
剩磁电感 residual induction
剩磁法 residual field method;residual method
剩磁感应 remanence
剩磁感应式继电器 remanence type relay
剩磁化 residual magnetization
剩磁激励 residual excitation
剩磁继电器 remanence relay
剩磁间隙调整 residual setting
剩磁励磁 residual magnetism excitation
剩磁平剖图 remanence field profile on plane
剩磁通 residual flux
剩磁效应 magnet after effect; remanence effect; residual magnetism effect
剩磁仪 remanent magnetometer
剩弹计数器 round counter
剩电 residual electricity
剩簧继电器 remreed relay
剩簧接线器 remreed matrix;remreed switch
剩货 leavings
剩留水分 residual moisture
剩下 remain
剩下未做完的工程 arrear of work
剩药 unfired explosive
剩油回输(管)线 excess-oil return line
剩油量 allowance;innage
剩油量测定 ga(u)ging by innage
剩余 leavings;overmeasure;redundant;remain; residual; residuary; surplus;wasting
剩余百分率 residual percentage
剩余饱和度 residual saturation
剩余保险费 unearned premium
剩余变动性 residual variability
剩余变化 residual variation
剩余变量 remaining variable;surplus variable
剩余变位 permanent deflection
剩余变形 residual deformation
剩余变异性 residual variability
剩余标准差 residual standard error
剩余波纹 residual ripple
剩余部分 remainder;rest
剩余材料 stub bar;surplus material
剩余财产 residual property
剩余财产分配 distribution of residual

property
剩余财产索偿权 residual claim
剩余财产账 equity account
剩余参数 rest parameter
剩余策略 remaining strategy
剩余产品 residual product; surplus product
剩余产权 remainder
剩余长度 over-length; residue length
剩余偿金 residual settlement
剩余场曲 residual field curvature
剩余沉降 residual settlement
剩余承载力 residual carrying capacity
剩余冲击谱 residual shock spectrum
剩余冲击响应谱 residual shock response spectrum
剩余臭氧 ozone residual
剩余纯收入 residual net income
剩余磁场 anomalous field; remanent field; residual field
剩余磁感应 remanence; residual induction; remanent induction
剩余磁感应强度 residual magnetic flux density
剩余磁化强度 remanence; remanence magnetization; remanent magnetization; residual magnetization
剩余磁化强度类型 type of remanent magnetization
剩余磁矩 remanent magnetic moment
剩余磁力仪 remanent magnetometer
剩余磁密 remanent flux density
剩余磁扰 residual disturbance
剩余磁通 remanent flux; residual magnetic flux
剩余磁通量密度 residual flux density; residual magnetic induction
剩余磁通密度 remanence; residual magnetic flux density
剩余磁性 remanent magnetism
剩余磁性的自反转 self-reversal of remanent magnetization
剩余磁异常 residual magnetic anomaly
剩余大气阻力 residual atmospheric drag
剩余代码 residue code
剩余氮时间 residual nitrogen time
剩余导磁率 residual permeability
剩余的 leftover; remnant; residual; residuary; surplus
剩余的零钱 odd money
剩余地区 residual sector
剩余点群 residual point-group
剩余电导 residual conduction
剩余电感 residual induction
剩余电荷 residual charge; residual electric(al) charge
剩余电荷放电装置 earth switch
剩余电离 residual ionization
剩余电力 dump power; surplus power
剩余电量 dump energy
剩余电流 after current; residual current
剩余电流继电器 residual current relay
剩余电路 residual circuit
剩余电容 residual capacitance
剩余电压 residual voltage
剩余电子区域 N-region
剩余电阻 residual resistance
剩余电阻比 residual resistance ratio
剩余定理 residue theorem
剩余动力 excess power; surplus power
剩余度 degree of redundancy; redundancy
剩余额 surplus(age)
剩余二次发射 residual secondary emission
剩余发电量 surplus generation
剩余发射 residual emission
剩余法 method of residuals; method of residues
剩余反射 residual reflection
剩余反射率 residual reflectance
剩余反应性 built-in reactivity; excess reactivity
剩余方差 residual variance
剩余方程 residual equation
剩余放电 residual discharge; soaking-out
剩余放射 residual emission
剩余放射尘 residual radioactive dust
剩余放射性 residual activity
剩余放射性强度 residual activity level
剩余放射性污染 residual radioactive contamination
剩余废气 remainder of exhaust gases; residual gas
剩余费用 remaining cost
剩余服务寿命 remaining service life
剩余浮力 residual buoyancy; surplus buoyancy
剩余辐射 residual radiation
剩余辐射滤光板 reststrahlen plate
剩余付款 residual payment
剩余干扰 residual disturbance
剩余感应 surplus induction
剩余高度 residual altitude
剩余耕地 residual earth
剩余工程 remaining works
剩余公因子方差 residual communality
剩余功率 dump power; surplus power
剩余构造应力 residual tectonic stress
剩余购买力 discretionary purchasing power
剩余固定资本 net stock of capital
剩余光线 residual ray
剩余含水量 residual moisture content
剩余函数 survival function
剩余航程 range-to-go
剩余核 residual nucleus
剩余核辐射 residual nuclear radiation
剩余荷载 surplus load
剩余厚度 residual thickness
剩余灰浆堆 cold pile
剩余恢复降深 residual recovered drawdown
剩余回潮率 residual regain
剩余回归法 residual regression method
剩余活度 residual activity
剩余活性 residual activity
剩余活性污泥 excess activated sludge; surplus activated sludge
剩余货物 residue cargo
剩余畸变 residual distortion
剩余激发 residual excitation
剩余极化 residual polarization
剩余极化强度 remanent polarization
剩余集 residual set
剩余计数 residual count
剩余价力 residual valence force
剩余价值 remaining value; surplus value; residual value
剩余价值学说 surplus value theory
剩余间隙 residual gap
剩余剪抗 residual shear resistance
剩余剪切强度 residual shear strength
剩余检查 residue check
剩余降深 residual drawdown
剩余降水 precipitation excess
剩余交通量 surplus traffic volume
剩余校验 redundancy check; residual check; residue check
剩余经济寿命 remaining economic life
剩余景观 residual landscape
剩余径向速度 residual radial velocity
剩余静水压力 excess hydrostatic(al) pressure
剩余静稳性面积 area under residual static stability curve
剩余静校正 residual static correction
剩余矩阵 residual matrix
剩余距离显示器 range-to-go display
剩余均方 residual mean square
剩余抗剪强度 residual shear strength
剩余可采储量 residual recoverable reserves
剩余可理解度 residual intelligibility
剩余空气 excess air; residual air
剩余空隙 residual void
剩余孔隙水压力 excess pore water pressure; residual pore water pressure
剩余孔隙压力 residual pore pressure
剩余拉力 residual tensioning force
剩余拉伸 residual stretch
剩余劳动力 spare labo(u)r force; surplus labo(u)r
剩余类 residue class
剩余类代数 residue class algebra
剩余类环【计】 residual class ring; residue class ring
剩余类空间 residue class space
剩余类群 residue class group
剩余类域 residue class field
剩余累积曲线 residual mass curve
剩余累积(曲线)图 residual mass diagram
剩余离子 residual ion
剩余理论 surplus theory
剩余力 excess force; remaining force
剩余力矩 excess torque
剩余力量 surplus energy; surplus power
剩余利润 surplus profit
剩余量 surplus
剩余流量 residual discharge; residual flow; surplus flow
剩余流速水头 residual velocity head
剩余硫代硫酸盐含量 residual thiosulphate content
剩余氯 residual chlorine; residue chlorine
剩余马力 excess horsepower
剩余码 residue code
剩余脉冲 after impulse; afterpulse; residual impulse
剩余煤气 off-gas
剩余煤气火力发电站 excessive gas recovery power plant
剩余面积 excess area
剩余模 residual mode
剩余内张力 residual stressing force
剩余能力【铁】 surplus seats and berthes on train; surplus capacity
剩余能量 dump energy; idle capacity; surplus energy; excess energy
剩余黏[粘]度 residual viscosity
剩余扭矩 surplus torque
剩余农产品 farm surplus
剩余耦合 residual coupling
剩余偏差 offset; residual deflection; residual deviation; residual discrepancy
剩余偏移 residual deviation; residual migration; residual offset
剩余票 stand-by ticket
剩余平方和 residual sum of squares
剩余谱 residual spectrum
剩余气分析器 residual gas analyser [analyzer]
剩余气泡 residual bubble
剩余气(体) residual gas
剩余气体成分 residual gas composition
剩余气体散射 residual gas scattering
剩余气体原子 residual gas atom
剩余气隙 armature residual gap; residual gap
剩余强度 residual strength
剩余倾角 remain inclination
剩余球差 residual spheric(al) aberration
剩余球面像差带 zone of residual spheric(al) aberration
剩余权限 residual competence
剩余燃料 residual fuel
剩余扰动 residual disturbance
剩余热 delayed heat; residual heat
剩余容积 residual volume
剩余容量 residual capacity; surplus capacity
剩余色差 residual achromatic aberration
剩余色散 residual dispersion
剩余闪烁 residual flicker
剩余设备 waste appliance
剩余射程 residual range
剩余射线 residual ray; reststrahlen; reststrahlen ray
剩余射线滤光板 reststrahlen plate
剩余射线滤光器 reststrahlen filter
剩余射线谱带 reststrahlen band
剩余射线区 reststrahlen region
剩余伸长 residual elongation; residual extension
剩余伸张力 residual stretching force
剩余失衡 residual unbalance
剩余失真 residual distortion
剩余湿度 residual humidity
剩余湿陷量 residual collapse settlement
剩余石灰 residual lime
剩余时间 excess time; remaining time
剩余使用年限 <建筑物的> remaining durable years; remaining service life
剩余使用寿命 remaining operational life
剩余视差 residual parallax
剩余视向速度 residual radial velocity
剩余收缩 residual shrinkage
剩余收益 surplus profit
剩余收益法 remaining-benefit method
剩余寿命 remaining life
剩余受弯能力 residual moment capacity
剩余输出 residue output
剩余束流 remainder of beam
剩余数 remainder
剩余数系 residue number system
剩余数系统 residue system
剩余衰减 residual attenuation
剩余衰减畸变 residual attenuation distortion
剩余水 excess of water; super-abundance water; surplus water
剩余水环境容量 water environment residual capacity
剩余水灰比 remained water-cement ratio
剩余水量 surplus capacity; surplus content; residual flow
剩余水量蓄存 storage of surplus water
剩余水体 residual water mass
剩余水头 excess head; residual head; residual water head
剩余水位 residual water height; residual water-level; residual water stage
剩余水压力 residual water pressure
剩余水蒸气 residual water vapo(u)r
剩余速度 remaining velocity; residual velocity
剩余速度损失 residual velocity loss
剩余酸度 residual acidity
剩余随机变量 residual random variable
剩余随机过程 stochastic residual

process
剩余损耗 residual loss
剩余缩率 residual shrinkage
剩余弹性 residual elasticity
剩余特征 residue character
剩余通量 residual flux
剩余图 residual map
剩余图像信息 remaining video information
剩余土地 residual land;remnant
剩余土方 surplus earth;surplus soil
剩余未平衡摩擦系数 remaining unbalanced friction factor
剩余位 residual potential
剩余位移 residual displacement
剩余污泥 excess(ive) sludge; sludge wasting; super-abundance sludge; surplus sludge; vestigial sludge; waste sludge
剩余污泥产量 excess sludge production
剩余污泥处理 excess sludge treatment
剩余污泥减量 excess sludge reduction
剩余污泥量 quantity of excess sludge; quantity of residual sludge
剩余污泥浓度 excess sludge concentration; residual sludge concentration
剩余无线电自差 remainder radio deviation
剩余物 fag-end; leftover; remainder; remains; residua; residual deposit; residual material;surplus;residue
剩余物品拍卖 rummager sale
剩余物资 surplus material
剩余物资处 surplus materials division
剩余误差 remainder error; remaining error;residual error
剩余误差比 residual error ratio
剩余吸收 residual absorption
剩余系 residue system
剩余现款 spare cash
剩余相关 residual correlation
剩余相互作用 residual interaction
剩余响应 residual response
剩余向量 residual vector
剩余项 remainder term
剩余相位误差 residual phase error
剩余像差 residual aberration
剩余像散 residual astigmatism
剩余消毒剂浓度指数值 excess disinfectant concentration index value
剩余效果 residual effect
剩余效益法 remaining-benefit method
剩余效应 residual effect
剩余效用 surplus utility
剩余楔 residual wedge
剩余信号 residual signal
剩余形变 residual deformation
剩余压差调节器 differential pressure controller
剩余压力 excessive pressure; overpressure; pressure excess; residual compression; residual pressure; super-pressure; surplus pressure
剩余压力降低 residual pressure drop
剩余压力调节器 differential pressure controller
剩余氧计算法 computing method of residual oxygen
剩余异常 residual anomaly
剩余异常法 residual anomaly method
剩余因子 residual factor; surplus factor
剩余引力场 residual gravitational field
剩余应变 residual strain
剩余应变释放 remnant strain release
剩余应力 internal stress; residual stress; steady stress
剩余硬度 residual hardness
剩余油 surplus oil
剩余雨量 rainfall excess
剩余域 residue field
剩余原料 surplus stock
剩余运输工具 surplus transportation
剩余杂散光 residual stray light
剩余杂质 residual impurity
剩余噪声谱 residual noise spectrum
剩余债款 debt remain
剩余折射改正 residual refraction correction
剩余振动 residual vibration
剩余振幅 residual amplitude
剩余值 surplus value
剩余种群 residual population
剩余重力异常 residual gravity anomaly
剩余重力异常图 residual gravity anomaly map
剩余重力值 residual gravity
剩余资本 surplus capital
剩余资本主权 residual equity
剩余资产 residual assets; residuary estate; surplus assets
剩余资金 surplus fund
剩余资金吸收 absorption of surplus funds
剩余子波确定 defining residual wavelet
剩余子波消除 residual wavelet cutback
剩余自差 remaining deviation; residual deviation
剩余自由气 residual free gas
剩余阻力 residual resistance; residuary resistance
剩余最大量 residual peak

尸

尸骨安置所 columbarium
尸骨仓 bone house
尸骨存放处 charnel house; charnel house
尸骨存放所 <公墓小教堂里的> carnary
尸骨停放房 <墓葬挖掘出的> shullhouse
尸体 cadaver;carcass;corpse
尸体板 death board
尸体陈列处 mortuary block
尸体床 deathbed
尸体检验 postmortem examination
尸体解剖 postmortem
尸体解剖室 autopsy room
尸体剖检 autopsy

失

失败百分率 percent failure
失败次数 mortality
失败的 unsuccessful; abortive; lost ground
失败的执行 unsuccessful execution
失败分析 failure analysis
失败概率 probability of failure; failure probability
失败呼叫 unsuccessful call
失败计划 unsuccessful plan
失败率 failure rate;mortality
失败模式的效应和鉴定分析 failure mode effect and criticality analysis
失败模式和效应分析 failure modes and effect analysis
失败模型 failure model
失败区域 failure go-to field
失败射孔 <管壁未穿透的> flush perforation
失败试验 water haul
失败效应 underdog effect
失败异常方式 failure exception mode
失败原因 cause of failure
失辨觉能 agnosia
失辨色能 colo(u)r agnosia
失步 de-synchronizing; locking out; losing step;losing synchronism;loss of synchronization; pulling out of synchronism;stepout【无】
失步保护 loss of synchronism protection
失步操作 out-of-step operation
失步功率 pull-out power
失步继电器 stepout relay
失步间隔 hold-off interval
失步开合 out-of-step switching
失步试验 pull-out test
失步运行 out-of-step operation
失步转矩 pull-out torque
失策的 ill-advised
失常 aberration
失常效应 wiggle effect
失潮 vanishing tide
失磁 loss of excitation;loss of field
失磁保护(装置) loss of excitation protection
失磁继电器 excitation-loss relay;field loss relay;deenergized relay
失磁接点 deenergized contact
失电量概率 probability of energy loss
失电子【物】 betatopic
失访偏性 withdrawal bias
失光 lost of gloss
失光材料 deadening
失衡 disbalance; out of trim; overbalance;unbalance
失衡的 unbalanced
失衡电压 out-of-balance potential
失衡载重 out-of-balance weight
失火警钟 fire bell
失火危险 fire hazard
失火自动报警器 automatic fire alarm
失聚 misconvergence
失均衡 overbalance
失控 incontrollable;out of control
失控车辆 errant vehicle
失控船 vessel not under command
失控导叶 free guide vane
失控的工资率 runaway rate
失控点 out-of-control point
失控堆 runaway reactor
失控反应 runaway reaction
失控洪水 wild flooding
失控井孔 wild bore
失控料流 feed on rush;feed surge
失控流动 uncontrolled flow
失控潜水 uncontrolled diving
失控上升 <反应堆功率> runaway
失控数据 off-control data
失控速度 runaway speed
失控调节器 runaway governor
失控温度 runaway temperature
失控效应 runaway effect
失控信号灯 breakdown light;non-under command light
失蜡造型 <铸造> investment casting
失蜡铸件 investment casting;lost wax casting
失蜡铸型 investment mo(u)ld
失蜡铸造 dewaxing casting; investment casting;lost wax casting
失蜡铸造法 cire-perdue method;cireperdue process; lose-wax process; lost wax process
失利的 lost ground
失灵 abort;breaking down;conk;dysfunction;failure;false;malfunction; Maloperation;not work
失灵概率 malfunction probability; probability of malfunction
失灵制动器 unserviceable brake
失绿病 chlorosis
失明 blindness
失模铸造 lost pattern casting
失能 deenergization
失能的 deenergized
失能继电器 deenergized relay
失能位置 deenergized position
失能周期 deenergized period
失配 maladjustment;misalignment;mismatch(ing);mistermination;unmatch
失配参数 mismatch parameter
失配重叠系数 mismatch contact ratio
失配的 maladjusted
失配电路 dismatched circuit
失配电压 misalignment voltage
失配发送器 error pick-off
失配角 displacement angle
失配开槽线 mismatch slotted line
失配树 none tree
失配衰减 mismatch attenuation
失配损耗 mismatch(ing) loss
失配损耗测量器 return loss measuring set
失配条件 mismatch condition
失配误差 mismatch error
失配信号传感器 error pick-up
失配信号发生器 mismatched generator
失配因数 mismatching factor
失配指示器 mismatch indicator
失强时间 zero strength time
失强温度 zero strength temperature
失窃货物 stolen goods
失去 atrophy
失去安定性 loss of stability
失去表皮 loss of epidermis
失去博览光泽 devitrification;devitrify
失去操纵 loss of control; run-out of controls
失去操纵的 out-of-operation
失去操纵的状态 uncontrollable condition
失去操纵性 loss of controllability
失去跟踪 mistrack
失去功能 dysfunction
失去功用的 disfunctional
失去光泽 devitrification;fogging;sleepiness;staining;frost <玻璃等>
失去光泽的渣 devitrified slag
失去航行能力的船 disabled ship
失去机动能力 immobilization
失去机动性 loss of mobility
失去交通作用的道路 obsolescent road;obsolete road
失去均衡 out-of-balance
失去控制 not under command
失去控制的 out of control
失去控制的船 vessel not under command
失去控制的动作 uncontrollable maneuver
失去控制的紧急情况 uncontrolled flight condition emergency
失去控制的速率 runaway speed
失去控制的(自流)井 wild well
失去控制市场能力 lose ability to control the market
失去联锁 loss of interlocking
失去列车分路 loss of shunt
失去买卖机会 lose one's market
失去能力 incapacitation
失去膨压 lose turgor
失去平衡 disequilibrium;outbalance;overbalance;unbalance;imbalance
失去平衡的 off-balance;out-of-balance
失去润滑 grease worms
失去生气 withering-away
失去时效的 stale
失去时效的债权 barred claim
失去时效的债务 barred obligation (debt)

S

失去时效权利 barred right
失去市场 loss of market
失去水分 drying out
失去速度 lose speed
失去弹性橡胶 impoverished rubber
失去调节 disaccommodation
失去通货资格 demonetization
失去同步 break step; dropout of step; fallout of step; loss of synchronization
失去无线电控制 loss of radio control
失去信用 loss credit standing
失去意义 dwindle
失去圆度 out-of-roundness
失去真空 loss of vacuum
失去正常形状的 out of shape
失去植被的地区 <因旱灾等造成的> dust bowl
失去作用 out of action; stall
失去座位 unseating
失却控制 get out of control
失热 loss of heat
失容 volume loss
失散 defocussing
失色 sleepiness
失时效 time barred
失时效的债务 outlawed debt; stale debt
失实记录 blind entry
失实数字 blind figure
失事 accident; crash; distress; fatal accident; shipwrecked; wreck(age) <船只、火车、飞机等的>
失事船(舶) shipwreck; wreck(ed ship); wrecked vessel
失事船船员 shipwrecked crew
失事船的漂浮物 floating wreckage; sea wreck
失事船浮出的货物 flotsam
失事船只 wrack
失事船只中的货物 wreck
失事飞机定位信标 crash-locator beacon
失事飞机中的货物 wreck
失事浮标 release buoy
失事火车中的货物 wreck
失事紧急救援处 emergency office
失事救援车 <飞机> crash ambulance; crash truck; crash wagon
失事救援方位信标 recovery locator beacon
失事率 wreck rate
失事潜舱 wrecked submarine
失事抢救车 wrecker
失事区 disaster area
失事跳伞 bail-out
失事险 accidental damage
失事信标 distress beacon
失事信号 emergency signal; wreck signal
失事应急发动机 abort engine
失事预测 failure prediction
失事折损 breakage
失事指示信号 failure indication
失水 dehydration; loss water; water loss
失水量 fluid loss volume; water loss; filter loss <泥浆的>
失水量降低剂 <泥浆> fluid loss reducing agent
失水量添加剂 <泥浆> fluid loss additive
失水量调节剂 <泥浆> water-loss control agent; filter loss agent
失水率 rate of water loss
失水事故 loss-of-coolant accident
失水试验 loss-of-coolant experiment
失水收缩 water-loss shrinkage
失水收缩极限 water-loss shrinkage limit
失水收缩裂缝 water-loss shrinkage crack
失水收缩裂纹 water-loss shrinkage crack
失水收缩情况 water-loss shrinkage behavio(u)r
失水收缩曲线 water-loss shrinkage curve
失水收缩性能 water-loss shrinkage behavio(u)r
失水收缩应力 water-loss shrinkage stress
失水收缩值 water-loss shrinkage value
失水速率 rate-of-loss of coolant
失水性 filtration quality
失速 loss of speed; speed loss; speed reduction; stallout; lug down; stall(ing)
失速边际 stall margin
失速变矩系数 stall torque ratio
失速颤振 <飞机飞行时的空气动力学现象> stall(ing) flutter
失速冲角 angle of stall(ing)
失速传播 stall propagation
失速的严重程度 <发动机> stall severity
失速点 stall(ing) point; stall spot
失速度 stall speed; stall velocity
失速范围 stalling range
失速改出操纵机构 stall recovery control unit
失速改出系统 stall recovery system
失速改出系统工作指示灯 stall recovery operating lamp
失速改出系统故障警告灯 stall recovery failure warning lamp
失速告警的 stall-warning
失速攻角 angle of stall(ing)
失速荷载 stalling load
失速后俯冲 stall-dive
失速继电器 stalling relay
失速角 angle of stall(ing); burble angle
失速界限 stalling limit
失速警告灯 stall warning light
失速警告器 stall warning device
失速警告系统 stall-warning system
失速警告指示器 stall warning indicator
失速警告装置 stall-warning unit
失速力矩 stalling moment
失速流量 <涡轮> stall flow
失速马赫数 stalling Mach number
失速模拟 stall simulation
失速扭矩 stall torque
失速判别 stall recognition
失速平坠着陆 pancking
失速起步 stall start
失速气流 stalled flow
失速区 stalled area; stalling region; stall zone
失速区内的流体 stalled fluid
失速试验 stalling trial
失速速度 <液力变矩器等的> stall(ing) speed
失速探测器 prestall detector
失速探测器阀 standstill detector valve
失速特性 stall performance
失速条件 stall condition
失速下滑 stalled glide
失速下降 stalled descent
失速线 <涡轮性能曲线> stall limit; stall line
失速信号器 stallometer
失速型螺母扳手 stall-type nut runner
失速延迟 stall delay
失速叶片 stalled blade
失速仪 stallometer
失速预测器 stall detector; standstill detector
失速预防系统 stall protection system
失速指示器 stall indicator
失速转矩 stall torque
失速转矩比 stall torque ratio
失速转弯 stalling turn
失速转速 stalled speed
失速状态 stall(ed) condition; stalling behavio(u)r; stall mode
失速状态试验 stalling run
失速着陆 stalled landing
失塑裂纹 ductility-dip crack
失算 miscalculate; miscalculation
失锁 losing lock
失锁指示器 unlock indicator
失弹性橡胶 umpoverished rubber
失调 disadjust; disaccommodation; disarrangement; disorder; disturbance; loss of adjustment; maladjustment; malfunction; misadjustment; misalignment; mismatch; mistermination; out of adjustment; stepout; detuning【电】
失调传感器 error signal transmitter
失调的 maladjusted; off(set)-tune
失调电流 offset current
失调电路 mistuned circuit
失调电容器 detuning capacitor
失调电压 misalignment voltage; offset voltage
失调负荷 regulation drop-out; regulation pullout
失调继电器 stepout relay
失调角 angular displacement; error angle
失调时差 out-of-step
失调现象 detuning phenomenon[复 phenomena]; imbalance
失调帧 out-of-frame
失调指数 index of detuning
失调准 misalignment
失听危险标准 damage risk criterion
失头 loss of head
失透 blinding; devitrification
失透玻璃 devitrified glass; dewdrop glass
失透石 devitrite
失透性焊剂 devitrifying solder
失透釉 devitrification glaze
失土 loss of soil
失望 disappointment
失位线圈 misplaced winding
失稳 destabilization; local buckling
失稳断裂 rupture in buckling
失稳荷载 collapsing load
失稳扩展 non-stable propagation
失稳裂纹扩展 unstable crack growth; unstable crack propagation
失稳破坏 failure due to instability
失稳情况 state of instability
失稳危险 risk of buckling
失稳载荷 collapsing load
失稳指数 buckling index
失稳状况 buckling behavio(u)r
失稳作用 destabilizing effect
失物清单 list of missing articles
失物招领处 lost and found department; lost and found office; lost property office
失物自理 not responsible for anything lost
失误 bust; fluff; lapse; pitfall
失误代价 miss cost
失误电平 missing level
失误概率 miss probability
失误脉冲 missed pulse
失误位 missing bit
失吸现象 loss of suction
失向 misorientation
失相 loss of phase; out-of-phase
失效 break down; cease to be available; cease to be in effect; desuetude; failure; inactivation; incapacitation; lapse; out of run; out-of-work; shrivel
失效保护 fail safe
失效保护安全联锁 fail-safe safety interlock
失效保险 fail safe
失效保险单 lapsed policy
失效保险控制 fail-safe control
失效保险系统 fail-safe system; fail system
失效标准 failure criterion[复 criteria]
失效材料 sterile material
失效存储器 dead file
失效单位 failure unit
失效的 defunct; non-serviceable; out-of-operation; out-of-service; out-of-work
失效的刮路机 fault drag
失效的海绵铁 spent iron sponge; spent oxide
失效的路刮 fault drag
失效的石灰 dead lime
失效的水权 forfeited water rights
失效地址 fail address
失效点 fail point
失效电子管 inactive valve
失效分布 failure distribution
失效分析 failure analysis
失效分析程序 failure analytical procedure
失效概率 failure probability; probability of failure; probability of structural failure
失效故障平均间隔期 mean time between failures
失效轨枕 defective sleeper
失效(后)复元 failure recovery
失效机构 failure mechanism
失效机理 failure mechanism
失效级 failure level
失效记录 failure logging; failure record
失效接缝 <黏[粘]结不良的> starved joint
失效节点 failure node
失效晶体 dead crystal
失效绝缘子 deteriorated insulator
失效客票 invalid ticket; non-valid ticket
失效控制 fail control; fall control
失效控制线 lose line
失效类别 fail category
失效率 failure rate; rate of failures
失效率的观测值 observed failure rate
失效率等级 failure rate level
失效率加速度 failure rate acceleration
失效率加速系数 failure rate acceleration factor
失效率平均函数 failure rate average function
失效率曲线 failure rate characteristic
失效密度 failure density
失效面 failure surface
失效模式 failure mode
失效模式分布 failure mode distribution
失效模式分析 failure mode analysis
失效模式与影响分析 failure model and effect analysis
失效模型 failure model
失效判据 failure criterion
失效炮孔组 lost round
失效炮眼 failed hole; missed hole
失效平均时间 mean time to fail
失效契约 void contract
失效日期 expiry date
失效容限 fault tolerant
失效弱化 fail soft

失效石灰 killed lime
失效时间 out-of-service time
失效寿命 burn-out life;failure life
失效树分析 failure tree analysis;fault tree analysis
失效数据 fail data
失效弹簧 dead spring
失效提单 stale bill of lading
失效条款 lapse provision
失效位置 dead position
失效温度 fail temperature
失效物理 failure physics
失效显影液 spent developer
失效预测 failure prediction
失效枕木 defective sleeper;defective tie
失效证券 nullified bond
失效支票 out-of-date check;stale check
失效终端 dead terminal
失效周期 cycle failure
失效状态 failure state
失效准则 failure criterion
失谐 detune;detuning;disarrangement;mismatch;mistermination;mistune;mistuning;tune out
失谐的 disadjust
失谐电路 detuned circuit
失谐电压 misalignment voltage
失谐短截线 detuning stub
失谐式鉴别器 off-tune type discriminator
失谐天线 dumb antenna
失信 breach of faith
失信用 break faith;loss credit
失信用的 lost ground
失形 disfigure(ment)
失形的 outmoded
失性石灰 <无消化能力的石灰> dead lime
失修 disrepair
失修道路 unmade road
失修的 in bad repair;out-of-repair
失修的因素 <车辆的> deadline factor
失压 decompression;loss of pressure;de-voltage【电】
失压保护 no-voltage protection
失压继电器 no-voltage relay;under-voltage relay
失压力矩 decompression moment
失压脱扣器 no-voltage release
失业 out of employ;out-of-work;unemployment
失业保险 insurance for unemployment;unemployment insurance
失业保险费 unemployment insurance expenses
失业保险救济 unemployment insurance benefits
失业保险税 unemployment insurance tax
失业保险统计 unemployment insurance statistics
失业补偿 unemployment compensation
失业补偿金 unemployment compensation funds
失业补贴 unemployment compensation;unemployment pay
失业补助 unemployment compensation
失业补助金 unemployment benefit
失业储备金 unemployment reserves
失业的 disemployed;jobless;unemployed;workless
失业范围 incidence of unemployment
失业工人 idle labo(u)r
失业工人救济金 labo(u)r relief
失业基金 unemployment fund
失业津贴 unemployment dole
失业救济 unemployment relief;unemployment compensation
失业救济工程 relief works
失业救济金 unemployment benefit;unemployment pay
失业救济委员会 unemployment compensation board
失业均衡 unemployment equilibrium
失业恐慌 fear of unemployment
失业劳动力 unemployed labor force
失业理论 theory of unemployment
失业率 jobless rate;rate of unemployment;unemployment rate
失业率的差别 dispersion of unemployment
失业期限 duration of unemployment
失业人数 unemployment
失业守恒规律 law of conservation of unemployed
失业严重地区 development district
失业者(总称)the workless
失约 delinquency
失泽 tarnishing
失泽作用 tarnishing action
失真 anamorphoser;anamorphosis;distortion;skewness
失真变压器 distortion transformer
失真拨号 <不正确拨号> distorted dial(l)ing
失真波 distorted wave;wave of distortion
失真波形 distorted waveform
失真补偿 distortion compensation
失真测试器 distortion set
失真的 fuzzy
失真的现金余额 distorted cash balance
失真的形象 anamorphoser
失真度 degree of distortion;distortion factor
失真度表 distortion factor meter
失真度测量仪 distortion factor meter
失真度测试仪 distortion meter
失真范围 distortion range
失真放大器 distorting amplifier
失真分析器 distortion analyser[analyzer]
失真光学仪 anamorphoser
失真和偏移 distortion and bias
失真极限 distortion limit
失真交变电流 distorted alternating current
失真胶片 jam
失真脉冲码 distortion pulse code
失真区 distorted region
失真声 distorted sound
失真衰减量 klirr-attenuation
失真衰减器 distortion pad
失真透镜 anamorphoser
失真图像 fault image;fuzzy image;scrambled image
失真网络 distorting network
失真误差 distortion error
失真系数 distortion factor
失真系数测试器 distortion factor meter
失真限制作用 distortion limited operation
失真消除 distortion elimination
失真效应 distortion effect;effect of distortion
失真信号 distorted signal
失真性质 nature of distortion
失真延迟 distortion delay
失真因数 distortion factor
失真影像 fault image;scrambled image
失真噪声电干扰 distortion noise
失真指示器 distortion indicator
失知觉的 unconscious
失职 breach of duty;delinquency;dereliction of duty;neglect of duty;negligence of duty;omission
失职者 delinquent;derelict <美>
失中 disalignment
失重 agravic;loss in weight;loss of weight;loss on weight;weightlessness;weight loss;zero gravity
失重补偿给料器 loss-in-weight feeder
失重的 weightless
失重环境 zero-gravity environment
失重阶段 zero-gravity period
失重率 rate of weight loss
失重模拟器 null-g simulator
失重黏[粘]性土(壤) weightless cohesive soil
失重情况 agravic;agravity
失重曲线 weightlessness curve
失重时间 zero-gravity time
失重试验 zero-gravity test
失重试验塔 zero-gravity tower
失重条件 zero-gravity condition
失重系数 coefficient of weight loss
失重效应 weightlessness effect;zero-gravity effect
失重装置 zero-gravity facility
失重状态 null-gravity state;weightlessness;weightless state;zero-g(ravity)behavio(u)r;zero-g(ravity) condition
失主 loser
失准 misaim
失准值 misalignment
失踪 disappearance
失踪被保险人 lost policy holder
失踪船(舶)missing ship;lost ship
失踪日期 date of loss

师范学校 normal school;teachers school;training school

师范学院 normal college;teachers college;teacher's training college;training college

虱 crawler;louse[复 lice]

虱病 pediculosis
虱子 <俚语> crumb

诗歌立架 <东方教堂中的> analogion

诗条石 poem-engraved stone slab

施彩 <青花瓷釉上加彩饰> bedye

施彩作用 decoration
施测(测)站 occupied station
施测单位 organization of surveying
施测断面 demarcated section
施测年代 age of surveying
施测站点 occupied station
施堆肥机 compost applier
施恩罗克棱镜 Schonrock prism
施恩罗克偏光半荫仪 Schonrock half shade
施恩罗克自准直目镜 Schonrock autocollimating eyepiece
施珐琅设备 enamel(l)ing equipment
施珐琅用金属 enamel(l)ing metals
施钒铅铁石 cechite
施防水剂 doping
施肥 apply fertilizer;fertilization;fertilizer application;manurial application;manuring;soil;spread manure
施肥附加装置 fertilizer attachment
施肥沟 dressing furrow
施肥灌溉 fertile irrigation;fertilizer irrigation;fertilizing irrigation;manuring irrigation
施肥过多 excessive use of fertilizer
施肥机 fertilizer applicator;fertilizer drill
施肥机具 fertilizing equipment
施肥机械 fertilizer machinery
施肥价值 fertilizing value
施肥开沟器 applicator boot
施肥量调节器 fertilizer quantity regulator
施肥器具 fertilizing equipment
施肥勺 manure ladle
施肥条播机 combined fertilizer-and-seed drill
施肥装置 fertilizer drill unit
施感磁化 inducing magnet
施感电荷 inducing charge
施感电流 inducing current
施感电路 inductance circuit
施工 construct(ion);execution(of work);operation
施工安全 construction safety;safety during construction
施工安全分析 job safety analysis
施工安全管理 construction safety management
施工安全规则 construction safety regulation
施工安全条例 constructional safety regulation
施工安装 construction and installation
施工安装图 construction and erection drawing
施工安装用的机械及工具 machines and tools for construction and erection
施工扳手 construction wrench
施工保证 construction warranty
施工保证书 completion bond
施工报表 construction forms
施工报告 construction report;job report
施工便道 builder's road;construction road;pioneer road;service gangway
施工便桥 construction bridge;pathway;service bridge;temporary bridge for construction;accommodation bridge
施工便线 construction detour;work track
施工标志 construction marker;construction sign
施工标志浮标 marker buoy
施工标桩 grade stacker;construction stake
施工标准 standard of construction
施工标准化 construction standardization
施工驳船 construction barge
施工不便 construction inconvenience
施工不良 poor construction
施工布置 construction layout
施工布置图 construction plan;draft of construction
施工步道 catwalk
施工步骤 construction procedure;construction sequence;construction step;site procedure;construction process
施工部门 construction department
施工部门经理 field engineering manager
施工部署计划 work organization plan
施工参考标点 reference object
施工仓 <水泥混凝土路面的> con-

S

struction bay
施工操作 construction operation
施工操作与安全 construction operations and safety
施工操作中的加热 heating for construction operations
施工测量 constructional measurement;construction survey(ing)
施工层面 working level
施工差错 construction fault
施工场地 builder's yard;construction area; construction plant; construction site; construction yard; site of construction;work-yard
施工场地布置 layout of construction work
施工场地额外费用 construction site premium;construction yard premium
施工场地清理 site cleaning
施工场地升降机 construction site hoist
施工场地图 construction map
施工场所 job location
施工超期 construction time overrun
施工车辆 construction vehicle
施工成本 cost of work
施工成本计算 cost for construction work
施工成本审计 construction cost audit
施工成型 job-shaped
施工承包 construction contracting
施工(承包)风险(费) construction risk
施工承包人 construction contractor
施工承包商 construction contractor
施工承包者 construction contractor
施工程序 construction operation; construction procedure; construction program(me); construction sequence; construction step; execution program(me) for works; working sequence
施工程序设计 programming of construction
施工程序图 process chart
施工程序网络图 project network
施工程序摘要说明书 memorandum of procedure
施工尺寸 working dimension
施工抽水 pumping during construction
施工筹资 construction financing
施工出图 drawing issued for construction
施工初期阶段 initial period of construction
施工处 construction department
施工船舶 construction vessel; workboat
施工船坞 <英> construction dock
施工措施计划 construction procedure
施工错误 construction fault
施工大桥轴线 proposed construction bridge axis
施工大样图 construction details
施工贷款 construction loan
施工单 construction sheet;job sheet
施工单位 builder;construction organization; construction unit; crew in charge of construction; unit in charge of construction
施工单位竣工决算 last account of contractor
施工挡水板 construction waterstop
施工导洞 construction heading; construction shaft
施工导流 construction diversion; river diversion during construction; river handling
施工导流布置 river diversion arrangement
施工导流隧洞 approach adit
施工导线 construction traverse
施工道路 access road; construction way
施工的建设项目 project under construction
施工地点 construction location; construction site;job location;working place
施工地平 construction horizon
施工吊车 construction crane; construction hoist;crane hoist
施工吊篮 work boat
施工吊椅 boatswain's chair
施工调度 construction dispatching
施工调度计划 construction dispatching plan; dispatching and scheduling of drilling job【岩】
施工订货(单) construction order
施工定额 building quota;construction norm;construction quota
施工定位销 construction dowel
施工段 construction section
施工队 constructional force; construction brigade; construction team; bull gang <普通的>
施工队伍 construction crew; work force
施工队伍调遣费 construction crew shifting costs
施工范围 scope of construction work
施工方案 arrangement and method for construction;arrangement and method statement;construction plan;construction program(me);construction scheme; executive project; statement of arrangement and method for construction
施工方法 construction method; execution method;job practice;method of application; method of construction; method of execution;ways to apply
施工方法说明 description of construction method;description of execution method; description of job practice
施工方格控制 square control network for construction
施工方式 application form; form of construction work;survey mode
施工防护平台 catch platform
施工妨碍 construction interference
施工放线槽板 batter board
施工放样 setting out(for construction survey);construction enlargement; construction layout
施工放样测量 layout survey; setting-out survey
施工放样认可 acknowledgement of construction enlargement
施工放样详图 layout plan
施工废弃物 construction waste
施工费(用) construction cost;cost of construction; engineering cost; executive cost; expenditure on construction;project cost
施工费用的固定限制 fixed limit of construction cost
施工费用估算报告书 <设计单位向业主提交的> statement of probable construction cost
施工费用指数 construction cost index; expenditure on construction; project cost
施工分期 construction phase; construction stage
施工分期付款 process payment; progress payment
施工粉笔画线 chalk line
施工封闭线路 line occupation for works
施工缝 cold joint; construction gap; construction joint; isolation joint; lift joint; movement joint; running joint; stoppage joint; work(ing) joint; stop-end joint <混凝土>; cold joint <新旧混凝土之间的>
施工缝处混凝土的堵头 kicker
施工缝防水构造图 structural drawing of construction joint waterproofing
施工缝小木条 chamfer strip
施工缝粘贴带 construction joint tape
施工服务工作 constructional service
施工辅助工厂 construction(al) plant
施工辅助设施 construction aid
施工附属企业 auxiliary factory for construction
施工概况 general conditions of construction
施工概率 probability of construction
施工概要 general features of construction
施工干扰 construction interference
施工港湾 construction harbo(u)r
施工高度 construction depth
施工工长 masterbuilder
施工工程 construction(al) works; construction project
施工工程安装 construction engineering installation
施工工程设计 construction engineering design
施工工程师 construction engineer; operating engineer
施工工程学 construction engineering
施工工地 construction site
施工工具 construction tool
施工工棚 construction camp
施工工期 construction period
施工工艺 construction technique; construction technology
施工工种 construction trade
施工工作 construction(al) work; construction operation
施工工作船 construction craft
施工公司 construction company
施工构件 working element
施工估价 construction estimate
施工骨料总量 total aggregate
施工固体分 application solid
施工挂篮 <预应力梁悬臂法> travel(l)ing carriage
施工管理 construction administration; construction control; construction management; construction supervision; execution control; works management
施工管理费 construction management cost; construction overhead; job overhead cost
施工管理公司 construction management firm
施工管理合同 construction management contract
施工管理合约 construction management contract
施工管理量测 construction control measure
施工管理研究 construction management research
施工规程 construction procedure; construction specification
施工规范 code of practice; construction specification; construction standard; job specification
施工规范协会 Construction Specification Institute
施工规划 construction(al) planning; construction(al) programming
施工规模 scope of construction item; size of construction
施工轨道 contractor track
施工过程 construction procedure; construction process; course of construction
施工行业 construction trade
施工合同 construction contract
施工合同的公证 notarization of construction contract
施工合同的签证 identification of construction contract
施工合同管理 administration of construction contract; administration of the construction contract
施工合作体 cooperation of contractors
施工和拆毁的废料 construction and demolition waste
施工核准程序 construction authorization procedure
施工荷载 construction(al) load(ing); erection load; operating toad;working load;site load
施工横坑道 construction adit
施工横巷 construction adit
施工洪水 construction flood
施工后(的) post-construction
施工后的痕迹 construction scar
施工后检查 inspection after construction
施工后评估 post-construction evaluation
施工后水深测量 post-bathymetric survey
施工滑道 building bent
施工环境影响 environmental impact of construction
施工机构迁移费 construction organization shifting costs
施工机具 builder's equipment; constructional appliance; constructional equipment; constructional installation; constructional plant; contractor's plant; plant for construction; working equipment
施工机桥 construction trestle
施工机械 construction(al) machinery; construction(al) mechanism; construction(al) plant; work(ing) machine; worksite equipment
施工机械表格式 equipment formats
施工机械场地 contractor's spread
施工机械的记录 equipment records
施工机械的租金 equipment rental
施工机械费 cost of construction equipment; cost of constructor's mechanical plant
施工机械费预算 equipment cost estimate
施工机械化 mechanization of building operation; mechanization of construction; mechanization of earth-moving; mechanized construction
施工机械化系数 coefficient of construction mechanization
施工机械排放物 construction machine emission
施工机械设备 constructor's mechanical plant; mechanical plant for construction
施工机械使用费 construction plant costs
施工机械使用核算流程图 equipment flow chart of information
施工机械使用效率 plant efficiency rate
施工机械台时费 equipment charge out rates

施工机械噪声 construction machinery noise
施工机械振动 construction machinery vibration
施工基坑 construction pit
施工基面 construction level
施工基面标高 construction level;formation level
施工基线 construction base line
施工基准面 construction datum;datum plane for construction;working datum
施工基准线<路基的> zero grade line
施工计划 building scheme;constructional program(me);construction plan(ning);construction plant;construction program(me);job plan(ning);schematization of execution;working plan
施工计划条例 construction regulation
施工计划图 execution scheme drawing
施工计划系统 construction planning system
施工记录 case history;construction journal;construction notes;construction record(ing)
施工记录纪念碑 monument construction and records
施工记录图 record drawings
施工技术 construction(al)technique;construction technology;engineering technique
施工技术标准 construction specification
施工技术财务计划 financial plan for construction technology
施工技术管理 construction technology management
施工技术规格书 specification for construction
施工技术要求 construction specification
施工季节 construction season;working season
施工加强计划 work enforcement plan
施工间接费 construction overhead
施工监测 construction monitoring
施工监督 construction supervision;monitoring;supervision;supervision of construction
施工监督官员 construction supervising authority
施工监控 construction supervision
施工监理 construction supervision;construction inspection;site inspection
施工监理单位 construction supervising authority
施工监理员 construction supervisor
施工检查 inspecting of construction;inspection of construction
施工检查员 construction inspector;inspector
施工简图 sketch of job
施工交通 construction transportation
施工交通路线 construction access road
施工绞车 builder's winch
施工阶段 construction period;construction phase;construction stage;stage of construction;phase of construction
施工阶段-施工合同执行的管理 construction phase-administration of the construction contract
施工截流 river closure
施工截水沟 French drain
施工截水墙 construction waterstop
施工进场道路 construction access road
施工进场坡道 construction access ramp
施工进程 progress of construction works
施工进度 construction progress;construction work progress;rate of progress
施工进度报告 construction progress report
施工进度表 construction schedule;erection schedule;Gantt chart;job plan;phased program(me) of works;schedule of construction;schedule of operations;time schedule(chart);work-schedule
施工进度量测 progress measurement
施工进度管理 construction project time management
施工进度横道图 bar chart
施工进度计划 construction scheduling;construction schedule;schedule of construction;detailed construction schedule;job schedule;program(me);progress plan;progress schedule
施工进度条线图 bar chart for progress
施工进度图 progress schedule
施工进度线图 progress chart
施工进展 construction progress
施工进展(定期)报告 construction progress report;construction status report;progress report;project status report
施工经济学 construction economics
施工经理 construction manager
施工经验 construction experience
施工井字架 construction shaft
施工卷尺 builder's tape
施工卷扬机 builder's hoist;builder's winch;contractor winch
施工决算 final account;final settlement
施工卡 operation job card
施工卡车 construction truck
施工开始 beginning of construction;commencement of construction work
施工勘察 exploration for construction;investigation during construction
施工科 construction department
施工可能性 constructability;construction possibility;probability of construction
施工坑 construction pit
施工控制 construction control;job control
施工控制网 construction control network
施工控制性进度 mandatory construction schedule
施工廊道 construction gallery
施工劳动力 construction labo(u)r
施工劳工 builder's labo(u)rer
施工类别 construction type
施工力矩 applied moment
施工力量 construction manpower;work force
施工联合体 consortium of contractors
施工量 quantity of work;volume of construction
施工量平衡 balancing of construction volume
施工列车 construction train
施工临时房屋 construction camp
施工临时荷载 temporary loading during construction
施工临时路线 contractor's track
施工临时螺栓 construction bolt
施工临时平台<英> staging
施工临时支撑 falsework
施工灵活性 construction flexibility
施工零杂工 builder's handyman
施工流程图 layout chart
施工流水作业法 construction streamline method;streamlined method construction
施工路段交通管理 construction section traffic management
施工路面 working surface
施工路线 construction access road
施工码头<工程施工临时用的> construction dock;construction wharf
施工慢行 running below normal velocity for construction
施工面积 floor space under construction
施工面宽度 working width
施工目标 construction goal
施工目的 construction goal
施工难度 construction difficulty
施工能力 capacity of execution;construction capacity;execution capacity
施工排水系统 construction drainage system
施工配合图 coordination drawings
施工棚 construction shed
施工皮尺 builder's tape
施工贫化 construct dilution
施工贫化率 construct dilution ratio
施工平接 construction butt joint
施工平面(布置)图 construction plan
施工平台 work deck;temporary staging
施工凭证 building certificate
施工剖面图 constructed profile
施工期 construction stage;course of construction
施工期测量 survey work during construction
施工期抽水 pumping during construction
施工期度汛 flood protection during construction
施工期高水位 construction flood
施工期荷载 temporary load during construction
施工期间 construction period
施工期交通 construction traffic
施工期孔隙压力 construction pore pressure
施工期利息 interest during construction
施工期绕行路计划 detour plan
施工期绕行平面图 detour plan
施工期通风 ventilation during construction
施工期通航 navigation during construction period
施工期限 construction term;period of construction
施工期涨价 escalation during construction
施工期中的建筑设备 building equipment in construction period
施工企业 construction enterprise;construction firm
施工企业工程项目审计 project audit of construction enterprise
施工起重机 construction crane;construction hoist
施工起重塔架 construction tower;erection tower
施工前 preconstruction
施工前测量 prior building survey
施工前估算 preconstruction estimate
施工前规划 pre-project planning
施工前解冻法 preconstruction thawing
施工前期(阶段) preconstruction stage
施工前期咨询服务 preconstruction advisory service
施工前试验 preconstruction trial
施工前水深测量 pre-bathymetric survey
施工前预算 preconstruction estimate
施工前(准备)阶段 preconstruction stage
施工桥 construction bridge
施工轻便铁路 field railway
施工区 built-up area;construction area;construction zone;work(ing) area;work(ing) zone
施工区段 building section;construction zone
施工区域 construction zone
施工缺陷 constructional defect;constructional deficiency
施工(绕行)便道 construction detour
施工人员 builder;constructer;construction crew;construction gang;construction party;construction personnel;house builder;work force;constructor
施工人员便道 builder's road
施工人员驻地 camp;camp ground
施工认可书 necessary preconstruction approvals
施工任务 construction job
施工任务单 construction job sheet
施工日 working day
施工日报表 daily construction report
施工日程表 construction calendar;construction schedule
施工日记 builder's diary;construction diary;construction journal;daily record of construction
施工日志 builder's diary;construction diary;construction journal;daily record of construction
施工日志进度表 stunt-head
施工容许误差 construction fit;construction tolerance
施工设备 builder's equipment;construction(al) equipment;construction(al) installation;construction(al) plant;contractor's plant;plant for construction;work(ing) equipment
施工设备布置图 layout of construction plant
施工设备厂 construction plant
施工设备场地 contractor's spread
施工设备费 cost of construction equipment;cost of construction plant
施工设备和技术的发展 construction equipment technical development;construction progress
施工设备及防护措施 construction equipment and protection
施工设备手册 construction equipment manual
施工设备项目 construction equipment item
施工设备性能 construction equipment performance
施工设计 detailed engineering;final design;construction design
施工设计人员 construction person;detailer
施工设计者 construction designer
施工设施 construction facility;preparatory facility;preparatory works
施工深度 construction depth
施工升降机 builder's lift;construction elevator;crane hoist
施工生产基地 construction support base

施工生活基地 camp; camp site; construction camp
施工时的临时线路 contractor's track
施工时含水量 placement water content
施工时间 construction time; execution time
施工时期 construction period
施工实践 construction practice
施工事故 construction accident
施工手册 construction manual
施工手段 construction means; construction way
施工术语 construction term
施工竖井 construction shaft; working shaft
施工水平 construction horizon
施工水位 construction water-level; construction water-stage; stage for construction; working water-level; working water-stage
施工水准仪 builder's level
施工顺序 construction sequence; sequence of construction; sequence of trades
施工说明卡片 card of work order
施工说明(书) construction specification; general description of construction; contract specification; operation sheet; closed specification <严格做法和材料的>
施工说明书阶段 construction documents phase
施工说明书摘要 streamlined specifications
施工死亡事故 fatal construction mishap
施工速度 construction velocity; rate of construction; rate of performance; rate of progress; stage of execution works
施工宿营车 camp car
施工损失 construct losses
施工损失率 construct losses ratio
施工索 erection cable
施工索赔 construction claims
施工台架 builder's staging
施工(台)跳板 runway board
施工套管 construction casing
施工天窗【铁】 construction "interval"; possessive interval for construction
施工条件 construction condition; condition for construction; working condition
施工铁路 constructor's railway; contractor rail
施工通道 construction access road; working gallery; access window
施工通风 construction ventilation
施工通路 construction access road; working gallery
施工通知 notice to proceed
施工图 construction chart; construction diagram; construction drawing; contract drawing; detail (ed) drawing; layout chart; shop drawing; work drawing; working graphics; working map; working plan
施工图估算 construction estimate
施工图绘图员 detailer
施工图阶段 stage of constructive map
施工图阶段预算 budget of construction drawing stage
施工图设计 construction(al) drawing design; construction documents design; detailed design; execution design; working design
施工图设计阶段 construction documents design phase; design stage of constructive map
施工图设计审查 evaluation of construction design; examination of construction design
施工图(图纸)目录 list of working drawings
施工图预算 budget according to working drawing; budget making for working design; budget of construction drawing project; construction drawing budget; construction estimate; detailed estimate (based on working drawing); working drawing estimate; working drawings based on estimate
施工图预算的单价法 unit price method in working drawing budget
施工图预算的实物法 material object method in working drawing budget
施工图预算审计 audit of working drawing budget
施工图纸及技术说明书 plan and specification
施工土坞 construction basin
施工、完工和修补缺陷 execute, complete and remedy defects
施工网络进度 construction network schedule
施工文件 <包括施工图及说明> construction documents
施工稳定范围 application temperature range
施工误差 construction error; inaccuracy of erection
施工细节 detailed engineering
施工现场 building field; building site; camp site; construction field; construction site; fabricating yard; job location; job site; placement area; work site
施工现场安全措施 construction site security
施工现场安装 job site installation
施工现场采暖器 construction heater
施工现场的物料搬运设备 material handling device for construction sites
施工现场活动厕所 chick sale
施工现场机械工 job's housekeeping' mechanic
施工现场机修工 job's housekeeping' mechanic
施工现场加热器 construction heater
施工现场交通 site traffic
施工现场经验 as-constructed experience
施工现场调查 site investigation
施工现场图 construction map
施工现场围栏 high rise
施工现场信号 working site signal
施工现场用的居住拖(挂)车 house trailer for construction sites
施工现场用的卧拖车 sleeping trailer for construction sites
施工现场运输 transportation on the site
施工现场准备 site preparation
施工现场总建筑师 architect in charge
施工现场总平面布置图 overall site layout
施工现场总体布置图 overall layout of construction
施工线 building line; construction line; formation line
施工限制线 neat line; net line
施工详图 construction(al) details; detail of construction; production drawing; working details; constructional drawing; working diagram; working drawing; shop drawing
施工详图尺寸 figuring
施工项目 construction item; constructive project; item under construction; project under construction
施工项目编号 construction item reference number
施工项目管理 construction project management
施工效率 efficiency of construction
施工斜坡道 access ramp
施工信道 point of access
施工性能 application property; workability
施工需电量 construction demand of electricity
施工许可证 builder's license[licence]; construction license; construction permit; permissible certificate of construction
施工序列 construction sequence
施工压力 construction forcing
施工研究 construction research
施工演示 construction show
施工验收技术规范 technical code for work and acceptance
施工样板金属构件 templet hardware
施工一切险 construction all risks
施工依据 manufacture bases
施工遗迹 construction scar
施工营地 camp; camp building; camp site; construction camp; housing of staff
施工营业执照 builder's license [licence]
施工应力 construction stress; erection stress; temporary stress
施工用道路 construction road
施工用电 construction demand of electricity
施工用扶梯 builder's ladder
施工用钢栈桥 construction steel trestle
施工用航道 channel for working; working channel
施工用架空索道 construction ropeway
施工用角尺 builder's square
施工用脚手架 builder's scaffold; builder's staging
施工用卷扬机 builder's hoist
施工用螺栓 construction bolt
施工用千斤顶 builder's jack
施工用强力胶 construction adhesive
施工用手推车 builder hand cart
施工用水准仪 builder's level; builder's tape
施工用塑料 construction plastics
施工用踏步梯 builder's ladder
施工用挑架 builder's jack
施工用油毛毡 construction felt
施工用栈桥 construction trestle
施工预报 forecast(ing) for construction
施工预付款 advance for construction
施工预算 construction budget; construction estimate; detailed estimate (of construction cost); estimates of construction
施工预算书 written estimate
施工员日志 builder's diary; constructor's diary
施工月 construction month
施工运输 construction traffic
施工运输道路 haul road
施工运输方式 transporting means for building construction
施工运输工具 transporting means for building construction
施工杂费 construction overhead
施工杂志 construction journal
施工造价 construction cost
施工噪声 construction noise
施工炸药 explosives for construction
施工窄轨铁路 constructor's narrow railroad
施工栈架 viaduct
施工帐篷 construction camp
施工照明 construction lighting
施工整地 pioneering
施工支洞 access adit; access tunnel; service tunnel
施工支架 falsework
施工织物 construction fabric
施工执照 builder's license; building permit; consent to build; construction license [licence]; construction permit; necessary preconstruction approvals; necessary preconstruction permits
施工止水条 construction waterbar
施工指标 construction index
施工指令 construction order; directives of construction
施工质量 construction quality; working quality; workmanship
施工质量低劣 poor workmanship
施工质量控制 construction quality control
施工质量认可 acknowledgement of construction quality
施工中安全和健康调节 safety and health regulation for construction
施工中重复操作 repetitive operations in building
施工中的 under construction
施工中的工程 construction in progress; work-in-progress; work underconstruction
施工中的建筑物 structural under construction; structure under construction
施工中间阶段 intermediate constructional stage
施工中楼板临时开的运输材料的孔洞 chase hole
施工中有节拍的操作 repetitive operations in building
施工周期 construction cycle; construction period
施工主管人员 superintendent of construction
施工注意事项 construction caution
施工准备 site preparation; construction preparation; preliminary work for construction; preparations for construction; preparatory plan of construction
施工准备工作 preparatory works
施工准备工作计划 program(me) of preparatory works
施工准备计划 preparatory plan; preparatory plan of construction
施工准备阶段 construction preparation stage; preconstruction stage
施工准备期 construction preparation period
施工准则 construction guide
施工准则线 <路基的> zero grade line
施工总布置图 construction site general layout; overall construction site plan
施工总分包 general, sub-contracting of construction
施工总工期 construction period; construction duration
施工总进度 general construction schedule; general progress of construction; general schedule of construction; construction master schedule; construction overall schedule

施工总进度计划 master construction schedule plan; overall construction plan
施工总平面布置 layout of construction work; general layout of construction
施工总平面图 construction general layout; construction general plan; general arrangement plan of construction; overall construction site plan
施工总说明 general description of construction
施工总体布置 layout of construction work; construction general layout
施工总体积 total volume of construction
施工总则 general conditions of construction; general description of construction
施工组织 construction organization; organization of construction
施工组织计划 work organization plan
施工组织设计 constructional organization design; construction management plan(ning); construction organization; design of organizing construction; design of working organization; organizational arrangement for construction; preparation of construction plan; construction planning
施工组织设计平面图 construction plan
施工组织设计优选法 critical path diagram; critical path method; critical path technique
施工钻探 construction drilling
施工作业 construction operation; construction works
施工作业参考面 surface of operation
施工作业计划 work element construction program(me)
施工作业模拟 simulation of construction operation
施工坐标 construction coordinates
施工坐标系 construction coordinate system
施骨肥 boning
施灌人员 irrigator
施焊部分 welding portion
施惠人 obliger[obligor]
施混合肥料 composting
施加 application; applied to; apply; imposition
施加薄雾涂层 apply a mist coat
施加的负荷 applied loading
施加的功 applied work
施加的扭力矩 applied torque
施加的推力 applied thrust
施加的外力 externally applied force
施加的转矩 applied torque
施加点 point of application
施加负荷 application of load; applied load
施加荷载 application of load; applied load; load application; imposed load <区别于自重>
施加力 apply force
施加力矩 application moment; applied moment
施加黏[粘]合剂 application of binder
施加黏[粘]合料 binder application
施加黏[粘]结剂 application of binder
施加速率 application rate
施加外力速度 rate of head movement
施加压力 exert(ing) pressure; forcing; twist a tail
施加压重 ballasting up
施加应力 application of stress; applied stress; stressing
施加应力的速度 rate of stress application
施加应力的预应力钢索 stressed tendon
施加影响 impose
施加预加应力设备 stressing equipment
施加预应力 prestress
施加预应力端 stressing end
施加预应力构件的弹性缩短 elastic shortening of prestressed members
施加制动 application of the brake; brake application
施加重力 weight application
施胶辊 furnishing roll
施胶装置 size applicator
施救泵 salvage pump
施救费用 sue and labo(u)r charges
施救费用理算 adjustment of salvage loss
施救(整理)条款 sue and labo(u)r clause
施控磁场 controlling magnetic field
施控系统 controlling equipment; controlling system
施控装置 controlling equipment
施拉车间 tensioning plant
施拉盖电动机 Schrage motor
施拉力次序 tensioning order
施拉力方法 tensioning method
施拉力工场 tensioning yard
施拉力楔形物 tensioning wedge
施乐克 <一种连接金属管的粉末> Selek
施力 application of force; force application
施力点 application point; application point of force; force point; point of application; point of application of force; point of force application
施力角 angle of application
施力线 line of action
施利仑光阑 schlieren stop
施利仑投影透镜 Schlieren projection lens
施利仑投影系统 Schlieren projection system
施利伦照相法 Schlieren-method of photography
施铃格拉姆电磁勘探法 Slingram method
施卢姆贝格尔倾角测量仪 Schlumberger dipmeter
施卢姆贝格尔热长度试验仪 Schlumberger apparatus
施卢姆贝格尔热梳片式长度试验仪 Schlumberger comb sorter
施卢姆贝格尔台阵 Schlumberger array
施锚机械 bolting machine
施锚设备 anchorage device
施镁量 amount of magnesium applied
施密特触发电路 Schmidt toggle circuit
施密特锤 Schmidt hammer
施密特定律 Schmidt law
施密特反射镜系统 Schmidt mirror system
施密特反射望远镜系统 Schmidt-type optical system
施密特反应 Schmidt reaction
施密特非球面校正板 Schmidt aspheric(al) corrector plate
施密特光学投影系统 Schmidt projection optics
施密特光学系统 Schmidt (optical) system; Schmidt system; Schmidt optics
施密特回弹锤 Schmidt rebound hammer
施密特活塞液压电动机 Schmidt motor
施密特极限 <核磁矩的上下限> Schmidt limit
施密特校正版 Schmidt correction plate; Schmidt corrector
施密特校正镜 plano-aspheric corrector
施密特经纬仪 Schmidt theodolite
施密特棱镜 Schmidt prism
施密特摄影机 Schmidt camera
施密特式热流计 Schmidt heat flow meter
施密特试锤 Schmidt hammer
施密特投影 Schmidt projection
施密特投影电视机 Schmidt television projector
施密特投影机 Schmitt projector
施密特投影箱 Schmidt camera
施密特透镜 Schmidt lens
施密特透镜系统 Schmidt lens system
施密特图 Schmidt plot
施密特网 equal-area net; Schmidt net
施密特望远镜 Schmidt telescope
施密特线 Schmidt line
施密特照相机 Schmidt camera
施密特甄别器 Schmidt discriminator
施密特正交化法 Schmidt orthogonalization
施密特转矩表 Schmidt torque meter
施奈德型立窑 Scheider kiln
施喷 <混凝土> placement of shotcrete
施铅封 place under seal
施羟镍矿 theophrastite
施舍物 alms
施石灰机 lime spreader
施氏矿物 Schwertmannite
施氏网络 Schmidt net
施式网 Schmidt net
施水 watering
施水率 water application rate
施水效率 water application efficiency
施塔赫分类 Stach's microlithotypes of coal
施特伦茨石 strunzite
施体【化】donor
施调河流 donor stream
施调流域 donor basin
施调区 donor region
施挖前后水深 water depth before and after dredging
施瓦茨希尔德坐标 Schwarzschild coordinates
施瓦尔兹不等式 Schwarz's inequality
施魏德勒大梁 Schwedler's girder
施魏德勒穹隆 Schwedler's dome
施温福特绿颜料 Schweinfurt(h) green
施行 administer; apply; dispensation; execute; go into effect; implement; perform; put in force; rendition; take effect
施行的 operative
施行减速制动 drag brake application
施行细则 enforcement rules; rules for implementation
施行制动 application of the brake; brake application
施行制动机的阶段制动 make a graduated application of the brake
施行中的项目 operational project
施压 pressing; pressurization
施压铲土的浅铲斗 positive-action low-bowl
施压阀 pressure valve
施压活塞 pressure piston
施压集团 pressure group
施药方法 application method
施业案编制员 working plans officers
施业案初步报告 preliminary working plan report
施业案管理区 working plans circle
施业案面积 working plan area
施业案总监 working plans conservator
施业规划 working scheme
施业计划 plan of management
施业控制 control
施业控制表 control form
施业期 working plan period
施业区 circle; working circle
施业区划 working division
施以不透明的釉 opaque glazing
施用除虫剂 application of insecticide
施用的最适宜时间 the optimum of application
施用方法 application process
施用方式 method of application
施用规范印章 stamp for application of code
施用量 application rate
施用量必须加大 a height dosage must be used
施用日期 date of application
施用石灰 liming
施用时间 time of application
施用同样剂量 apply at the same rate
施釉 glazing
施釉不足 short glaze
施釉法 thrown glaze
施釉坩埚 glazed pot
施釉工 dipper
施釉柜 glazing booth
施釉机 glazing machine; machine for glazing
施釉色料 glazing colo(u)r
施釉陶瓷 glazed ceramics; glazed porcelain
施予者 dispenser
施闸牌 apply brake board
施张车间 tensioning plant
施张力次序 tensioning order
施张力方法 tensioning method
施张力工场 tensioning yard
施张力楔形物 tensioning wedge
施照度 illuminance
施照器 illuminant; illuminator
施照体 illuminant; illuminator
施赈所 almonry; almshouse
施主 donator; donor
施主掺杂剂 donor dopant
施主离子 donor ion
施主密度 density of donor; donor density
施主密度减小 donor depletion
施主能级 donor level; donor site
施主浓度 donor concentration
施主物质 <半导体> donor material
施主徙动 donor migration
施主杂质 donor impurity
施主杂质含量 donor impurity level

狮

狮大门 lion gate
狮面具 lion-mask
狮面用花岗岩 granite for ornament
狮面装饰 lion-mask
狮身人面像 androsphinx
狮身羊头像 criosphinx
狮头羊身蛇尾饰 chimera
狮子山阅江楼 <中国> Yangtze Yue Jiang pavilion on lion hills
狮子院南厅 <西班牙阿尔罕伯拉宫> Hall of Abencerrages

湿

湿 hygro; soaking
湿拔 wet drawing
湿斑 water stain
湿板 wet-plate

S

湿板胶接 wet-cemented;wet glued
湿板效率 wet plate efficiency
湿版法 wet-plate process
湿版摄影术 wet-plate photography
湿版洗印 wet process
湿版照相 wet-plate photography
湿拌 wet-mixing
湿拌步骤 wet-mix process
湿拌法 wet-mix process
湿拌和 moist mixing;wet mix(ture)
湿拌和机 wet mixer
湿拌和料 wet mix(ture)
湿拌混合料 wet mix(ture)
湿拌混凝土 wet mix(ture);wet-mix concrete
湿拌混凝土骨料 wet-mix concrete aggregate
湿拌混凝土集料 wet-mix concrete aggregate
湿拌机 moistening and mixing machine
湿拌级配碎石 wet mix(ture)
湿拌级配碎石基层 wet-mix base; wetted graded stone base
湿拌喷射混凝土 wet-mix shotcrete
湿拌碎石 premixed water-bound macadam;wet-mix macadam <有级配的>
湿拌碎石混合料 wet-mix macadam mix(ture)
湿拌碎石基层 wet mix macadam
湿拌碎石路 wet-mix macadam
湿拌重量 wet-mix weight
湿拌作业 wet-mix process
湿饱和蒸汽 moist saturated steam; wet saturated steam
湿背式锅炉 wet-back boiler
湿比热 humid heat
湿比热容 moisture specific heat
湿比容 humid volume
湿比体积 moisture specific volume
湿壁画 buon fresco
湿壁降膜吸收塔 falling film type absorber
湿壁塔 wetted wall column; wetted wall tower
湿边 live edge;wet edge
湿边保持性 wet-edge retention
湿边剂 wet-edge agent
湿边时间 wet-edge time
湿边增充剂 wet edge extender
湿变电阻器 hygristor
湿变形 green deformation
湿变性土 udent;udert
湿表面浸涂漆 water dip lacquer
湿表面施工涂料 underwater coating compound
湿补 wet patch
湿布卷条 rag doll
湿部 wet end
湿部化学 wet end chemistry
湿部配比 wet end furnish
湿部牵引力 wet draw
湿部添加剂 wet end additive
湿部效率 wet end efficiency
湿部装饰 wet end finish;wet finish
湿擦 wet rubbing
湿擦拭 wet smear
湿材 green lumber; green timber; green wood; unseasoned lumber [timber/wood]; water-core; wet-wood;water soak <含水率高的木材>
湿材料 wet stock
湿材(栖)白蚁 damp-wood termite
湿材重 green weight
湿舱型载驳货船 wet-cell type barge carrier
湿藏 store moist
湿槽 wet sump
湿槽润滑 wet sump lubrication
湿槽润滑系统 wet sump lubricating system
湿草地群落 telmathium
湿草甸 aquiprata;marsh;moist meadow;wet meadow
湿草甸黑土 moist meadow black soil
湿草甸土 wet meadow soil
湿草原群落 hygrophorbium
湿草原土 brunizem; planosol; prairie (-forest) soil
湿层合法 wet laminate process
湿差电感光计 panzer actimometer
湿差应力 moisture difference stress
湿掺和 wet blending
湿抄机 presser-pate; wet board former;wet machine
湿抄机操作工 wet machine tender
湿潮土 wet meadow soil
湿沉积(物) wet deposition
湿沉箱 wet caisson
湿衬垫蒸发冷却器 wetted pad evaporative cooler
湿成岩 humid rock
湿成岩石成因论 humid lithogenesis
湿澄清法 wet liming
湿冲击 hygral shock
湿冲积土 sansouire
湿稠灰浆 green plaster
湿出铁口 green hole
湿(储)藏法 wet storage
湿储法 wet stowage
湿处理 wet process;moist after-treatment
湿触点 wet contact
湿船坞 closed dock; enclosed dock; wet basin;wet dock
湿催化 moist catalysis
湿存水 hygroscopic moisture; hygroscopic water
湿搓纹的 wet-grained
湿打浆 wet beating
湿打磨 wet sanding
湿大气 damp atmosphere
湿代谢作用 hygrometabolism
湿袋法等静压成型 wet bag isostatic pressing
湿单位重 wet unit weight
湿挡板室 wet baffle chamber
湿捣法 moist tamping; moist rodding <试件制备的>
湿捣碎 wet stamp
湿捣碎法 wet stamping
湿的 humid;moist;wet
湿的层压法 wet-laminating process
湿的跑道 juicy runway;wet runway
湿滴损失 wet loss
湿底电集尘器 wet bottom precipitator
湿底发生炉 wet bottom producer
湿底炉渣 wet bottom boiler slag
湿底燃烧炉 wet bottom furnace
湿地 causeway; dismal; everglade; glade;marsh(al land);marshy area;marshy ground;marshy land;morass; nunja; slash; slough; swamp(ed land); swampy land; waterlogged land; wet ground; wetland; wet soil;wetted land
湿地白杨 swamp cottonwood
湿地草本群落 humidiherbosa
湿地草原 water grass
湿地冲积扇 humid fan
湿地处理系统 wetland treatment system
湿地腐殖土 marsh muck;marsh podzol
湿地灌丛 marsh scrub
湿地灰壤 marsh podzol
湿地基地 wet foundation
湿地履带板 crawler swamp shoe; swamp shoe
湿地排水 bog drainage
湿地槭 red maple; scarlet maple; swamp maple
湿地群落 mesophytia
湿地热田 wet geothermal field
湿地生境 wetland habitat
湿地生态系统 wetland ecosystem
湿地坍方 wet landslide
湿地坍坡 wet landslide
湿地土 marsh soil
湿地土崩 wet landslide
湿地土壤 marshy ground;wetland soil
湿地推土机 swamp bulldozer
湿地系统 wetland system
湿地修复 wetland restoration
湿地用推土机 swamp bulldozer
湿地沼泽 marsh
湿点 wet point
湿电池 wet cell
湿电池组 wet battery
湿电解电容器 wet-electrolytic capacitor
湿淀粉 wet starch
湿定形 wet setting
湿定形效应 wet-memory effect
湿动态 hygrokinesis
湿冻原土 wet tundra soil
湿度 degree of humidity; degree of moisture; degree of wetness; humidity; humidness; hydrometer; moistness;moisture capacity;moisture content; moisture degree; moisture-holding capacity;moisture level; dampness; water content; wetness(fraction)
湿度百分率 percentage(of) humidity;percentage of moisture
湿度百分数 percentage(of) humidity;percentage of moisture
湿度比 humidity ratio;moisture content ratio;psychrometric ratio
湿度比降 humidity gradient
湿度比学 hydrostatics
湿度变动带深度 depth of moisture variation zone
湿度变化曲线 moisture variation curve
湿度变化引起的翘曲 moisture warping
湿度变形 moisture movement
湿度表 hygrometric table;hygrometer <指仪表>
湿度表格 glaisher's table
湿度测定 humidity determination;humidity test;measurement of humidity;moisture determination
湿度测定法 hygrometry;hygroscopy; psychrometric method;psychrometry
湿度测量 moisture measurement
湿度测量法 psychrometric method
湿度测量计 humidity meter;moisture meter
湿度测量控制仪 hygrostat
湿度测量探头编号 number the probe of moisture measurement
湿度测量探头埋设深度 buried depth of moisture measuring probe
湿度测量探状数目 number of moisture measuring probes
湿度测试器 moisture detector
湿度测头 humidity sensor
湿度查算表 psychrometric table
湿度差 moisture deficiency;psychrometric difference
湿度场 moisture field
湿度陈化 moisture ag(e)ing
湿度传感器 humidity sensor;humidity transducer
湿度大 high humidity
湿度当量 centrifuge moisture equivalent;moisture equivalent
湿度动态 moisture regime(n)
湿度对跳火电压特性曲线 moisture and flash over voltage curve
湿度范围 moisture regime(n)
湿度防护 humidity protection;moisture protection
湿度分析 psychrometric analysis
湿度分析器 humidity analyser[analyzer]; moisture analyser[analyzer]; water extractor
湿度腐蚀试验 humidity test;moisture corrosion test
湿度公式 psychrometric formula
湿度关系 humidity relation
湿度盒 humidity box
湿度恒定器 hygrostat
湿度混合比 humidity mixing ratio
湿度混合率 humidity mixing rate
湿度积分器 moisture integrator
湿度计 drimeter; humidiometer; humidity ga(u)ge; humidity meter; hygrograph; hygrometer; hygroscope;moisture apparatus;moisture content meter[metre]; moisture ga(u)ge;moisture indicator;moisture instrument; moisture meter; moisture tester; psychrometer; succulometer
湿度计常数 psychrometer constant
湿度计的 hygroscopic
湿度计的干球 dry bulb
湿度计算尺 humidity slide-rule
湿度计算器 psychrometric calculator
湿度计算式 psychrometric formula
湿度计算图 psychrometric chart
湿度记录器 humidity recorder; hygrograph;recording hygrometer
湿度记录仪 hygrograph; moisture recorder
湿度检测 humidity measurement
湿度检测器 humidity detector
湿度检定箱 hygrostat
湿度降低 dehumidification
湿度校正系数 humidity correction factor
湿度结合比 moisture combined ratio
湿度界限 humidity limit
湿度控制 humidity control(ling); moisture(content) control
湿度控制器 humidity controller; humidity regulator
湿度廓线 moisture profile
湿度量 humidity quantity
湿度密度 humidity density
湿度密度关系 moisture-density relationship
湿度密度计 humidity densimeter;humidity density meter
湿度密度计法 density-hygrometer method
湿度密度曲线 moisture-density curve
湿度密度试验 moisture-density test
湿度敏感元件 humidity sensor
湿度摩阻应力 <指混凝土路面板因湿度变化而发生的> moisture-friction stress
湿度逆增 moisture inversion
湿度平衡 humidity equilibrium; hygrometric equilibrium; moisture balance;moisture equilibrium
湿度器 hygroscope
湿度翘曲应力 <混凝土路面板与底层因湿度差而引起的> moisture-warping stress
湿度区 humidity province
湿度势 humidity potential; moisture potential

湿度试验 humid test; moisture test(ing); wet test
湿度试验柜 humidity-testing cabinet
湿度试验器 moisture tester
湿度试验容器 humidity chamber
湿度试验室 humidity cell; humidity chamber
湿度室 humidity cabinet
湿度受感元件 humidity sensitive element
湿度水液控制 controlling humidity level
湿度损失 wetness loss
湿度梯度 gradient of moisture; humidity gradient; moist(ure) gradient
湿度调节 humidity control(ling); humidity regulation; moisture control
湿度调节板 humidostat
湿度调节器 humidistat; humidity controller; humidity regulator; humidostat; hygrostat; psychrometric regulator
湿度调节仪 humidistat[humidostat]
湿度调整 moisture adjustment
湿度图 humidity chart; humidity diagram; hygrogram; psychrometric chart
湿度温度计 hygrothermograph
湿度稳定土法 moisture stabilization
湿度吸收 moisture absorption
湿度吸收试验 moisture absorption test
湿度系数 humidity coefficient; moisture factor
湿度箱 humidity cabinet; humidity chamber; sweat box
湿度效应 humidity factor
湿度学 hygrology
湿度养护 damp storage
湿度养护密封剂 moisture curing sealant
湿度养护黏[粘]结剂 moisture curing adhesive
湿度仪 humidity detector; hygronom; moistograph; moisture meter
湿度移动 moisture movement
湿度因素 humidity factor
湿度因子 humidity factor
湿度造成的胀缩 moisture movement
湿度增加 humidity increase
湿度胀缩 moisture movement
湿度指示器 humidity indicator; moisture indicator
湿度指数 humidity index; moisture index; wetness index
湿度中值 medial humidity
湿度转移 <混凝土> moisture migration
湿度自动控制器 automatic humidity controller
湿度自动调节 automatic moisture control
湿度自记曲线 hygrogram
湿端 green end
湿段旋风分离器 wet-stage cyclone
湿断面 wetted cross-section
湿吨 latent heat load; moisture ton; wet ton
湿发动机 wet engine
湿发射场 wet pad
湿法 wet method; wet way; wet process
湿法安装玻璃 wet glazing
湿法包膜法 wet coating procedure
湿法闭路系统细粉磨 wet closed-circuit fine grinding
湿法薄毡 wet-laid mat; wet-process mat
湿法缠绕 wet winding
湿法长回转窑 long wet-process rotary kiln
湿法长窑 long wet process kiln
湿法成网 wet-laying
湿法成网非织造织物 wet-laid nonwoven fabric
湿法成网工艺 wet laying process
湿法成网机 wet-web former
湿法成型 wet mo(u)lding
湿法冲洗 wet flushing
湿法除尘 wet dedusting; wet dust collection; wet dust extraction; wet dust removal
湿法除尘器 hydrofilter; scrubber
湿法除灰 wet ash removal
湿法除泥 desliming
湿法除氧化皮 wet-scale disposal
湿法处理 wet handling; wet-processing; wet treatment
湿法磁选 wet cobbing; wet magnetic dressing; wet magnetic separation
湿法磁选机 wet magnetic separator; wet separator
湿法打眼 hydraulic drilling; wet drilling
湿法镀锌 wet galvanizing
湿法纺丝 wet spinning
湿法纺丝机 wet-spinning frame; wet-spinning machine
湿法分级 wet classification
湿法分级器 wet classifier
湿法分离 wet separation
湿法分析 analysis by wet way; humid analysis; wet analysis; wet assay(ing)
湿法分选 wet split
湿法分选机 wet separator
湿法粉磨 wet grinding; wet-milling
湿法粉碎 water ground; waterproof pulverization
湿法粉碎磨 wet crushing mill
湿法敷(涂)层 wet layup
湿法复制 liquid duplicating
湿法改干法 wet to dry process conversion; wet to modification
湿法干法混合筛分技术 wet-dry sieving technique
湿法工艺 wet process
湿法工艺调制泥浆过程 slip process
湿法工艺窑 wet-process kiln
湿法鼓式磁选机 wet drum cobber
湿法管磨 wet grinding tube mill
湿法过滤 wet filtration
湿法过滤中的工业纤维 industrial fabrics in wet filtration
湿法过筛的 wet-screened
湿法化学 wet chemistry
湿法化学分析 wet chemical analysis
湿法还原 wet reduction
湿法灰化 wet ashing
湿法回转钻进 wet rotary drilling
湿法混合 <带有液态燃油的混合气> wet-mixing (method); wet-mixing process
湿法混料器 wet mixer
湿法集尘器 wet dust arrester
湿法集尘器系统 wet-type system of dust control
湿法加工 wet process
湿法加工装置 wet-process installation
湿法胶接强度 wet bonding strength
湿法搅拌批量 wet mixing batch rating
湿法精选 wet cleaning; wet concentration; wet-milling
湿法净化 wet cleaning; wet purification; wet purifying
湿法开采 wet mining
湿法颗粒分析 wet mechanical analysis
湿法拉拔 wet drawing
湿法拉丝 wet wire drawing
湿法拉制 liquor finish draw
湿法粒径分析 wet mechanical analysis
湿法炼油 wet rendering
湿法磷酸 phosphoric acid by wet process
湿法煤气净化器 wet gas purifier
湿法磨 wet grinding mill; wet mill
湿法磨盘 wet pan grinder
湿法磨碎垃圾 wet milling
湿法黏[粘]合 wet bonding
湿法黏[粘]土砖 soft mud brick
湿法碾磨 wet grinding
湿法盘磨 wet pan grinding
湿法抛光 wet polishing; wet tumbling
湿法喷补 wet spraying
湿法喷砂 wet blast
湿法喷丸清理 wet blast cleaning
湿法破碎 wet crushing
湿法破碎的 wet crushed
湿法破碎机 wet crushing mill
湿法铺放 wet layup
湿法铺毡 wet-felting
湿法铺装 wet-felting
湿法起皱 wet crepe
湿法气体净化器 wet gas purifier
湿法气体洗涤器 wet gas scrubber
湿法清洁 <石棉污染物的> wet cleaning
湿法清理 wet blasting
湿法清刷机 wet cleaner
湿法清洗器 wet cleaner
湿法球窑 wet nodule kiln
湿法取样 wet sampling
湿法燃气净化 wet purification of combusting gas
湿法燃烧 wet combustion
湿法熔剂镀锌 wet galvanizing
湿法筛分 wet screening; wet sieving
湿法筛分分析 wet screen analysis; wet sieve analysis
湿法烧制转窑 wet-process rotary kiln
湿法生产 <水泥的> wet process
湿法生产的硬质纤维板 wet-process hardboard
湿法生产装置 wet-process installation
湿法生料磨 raw slurry mill
湿法施工 wet construction
湿法收集尘埃器 wet collector
湿法水泥回转窑 wet-process rotary cement kiln
湿法水泥生产 wet-process cement production
湿法水泥生产工艺 wet-process of cement manufacture
湿法搪瓷 wet-process enameling
湿法提纯 wet purification
湿法提浓 wet concentration
湿法填装 wet-filling; wet packing
湿法涂搪 wet-process enameling
湿法脱除细粒 de-slurrying
湿法脱氮 wet denitrification
湿法脱硫 liquid purification; wet desulphurization
湿法喂料窑 wet feed kiln
湿法无纺布 wet nonwoven fabrics
湿法无介质磨 hydrofall mill
湿法物理处理 wet physical processing
湿法吸尘器 wet cap collector
湿法熄焦 wet quenching
湿法洗涤 wet scrubbing
湿法洗涤器 wet cleaner
湿法洗气 wet gas cleaning
湿法洗选机 wet washer
湿法细碎 wet grinding
湿法消尘 wet method of dust-suppression
湿法选矿 wet dressing; wet-mill concentration
湿法选矿厂 wet-milling plant
湿法选矿后的水分容差 wet wash allowance
湿法压热处理的 wet-autoclaved
湿法压制砖 wet mud brick
湿法烟气脱硫 wet-process of fume gas desulfurization
湿法研磨 wet grinding
湿法养生 moist curing
湿法氧化脱臭 wet oxidation deodorizing
湿法窑 wet-process kiln
湿法冶金 hydrometallurgy; wet metallurgy; wet-process metallurgy
湿法预裂 wet preliminary splitting
湿法原料闭路系统粉磨 wet-saw closed-circuit grinding
湿法再生 wet reclamation
湿法再生砂 washed sand
湿法凿岩 hydraulic drilling; wet drill cutting
湿法制备的磷酸 wet phosphoric acid
湿法制瓷 wet-process porcelain
湿法制矿棉板 wet-process for making mineral wool slab
湿法制造 wet lay system; wet manufacture
湿法制造混凝土构件 wet cast process concrete member
湿法制毡机 wet-laid tissue machine
湿法贮存污染 wet storage stain
湿法转化工艺 wet conversion process
湿法转筒抛光 wet barrel tumbling
湿法转窑 wet-process-type rotary kiln
湿法装填 wet filling filter; wet packing
湿法自磨 wet autogenous mill
湿法钻进 wet boring; wet drilling
湿法钻井 wet well cuttings
湿法钻探 wet drilling
湿法作业 wet method operation; wet operation; wet work
湿法(作业)回转窑 wet-process rotary kiln
湿反应 wet reaction
湿方强度 wet cube strength
湿房间 wet room
湿纺 wet spinning
湿飞弧电压 wet flashover voltage
湿飞弧试验 rain arc-over test
湿废鞣料 wet tan
湿分含量 moisture content
湿分绞棒 wet dividing rod
湿分离 wet split
湿分析 wet(mechanical) analysis
湿分析法 wet analysis method
湿粉 wet-milling
湿粉料 wet mash
湿粉碎机 wet crushing mill
湿粉研磨机 wet powder grinder
湿封皮 moisture cover
湿锋 water front
湿缝 wet joint
湿敷裹 wet dressing
湿浮筒 wet float
湿腐 damp rot; moist rot; water rot
湿腐朽 <木材的> wet rot
湿负荷计算 moisture load calculation
湿附着试验 wet adhesion test
湿副热带 subhumid tropics
湿干联合筛分法 wet-and-dry screening
湿干两用磨削机 wet-and-dry combination grinding machine
湿干试验 wetting-and-drying test
湿干循环 humid-dry cycling
湿格筛 wet sieve
湿铬鞣革 wet blue
湿工况 wet cooling condition
湿工序 wet trade
湿供暖系统 wet heating system
湿构造分隔墙 wet construction partition(wall)

S

湿毂式转子 wet hub rotor
湿骨料搅拌 wet-aggregate process
湿鼓风 wet blasting
湿固化 curing in moisture; moisture curing; steam set
湿固化氨基甲酸酯 moisture-cured urethane
湿固化氨基甲酸酯涂料 moisture-cured polyurethane coating
湿固化聚氨酯涂料 wet cured polyurethane paint
湿固化漆 moisture-hardening varnish
湿固着油墨 moisture-set ink
湿管喷水灭火系统 wet-pipe sprinkler system
湿管喷水器系统 wet-pipe sprinkler system
湿管式冷凝 wet condense pipe
湿管系统 wet-pipe system
湿辊磨机 wet pan
湿辊磨机研磨 wet pan milling
湿过程显影 wet-process development
湿害 wet injury
湿含量 humidity ratio
湿含量梯度 moisture gradient
湿寒 raw
湿旱生的 tropophilous
湿旱生植物 trop(op)hyte
湿旱生植物群落 tropophytia
湿黑干球温度指标 wet-bulb globe temperature
湿后松弛回缩 relaxation shrinkage on wetting
湿弧距离 wet arcing length
湿滑警告标志 slippery when wet sign
湿滑抗力 wet-skid resistance
湿滑区 humid zone
湿滑试验 wet-skidding test
湿滑阻抗 wet-skid resistance
湿化 air slacked
湿化崩解 slaking
湿化含水量 slaking water content
湿化含水率 slaking water content
湿化粒径分析 wet mechanical analysis
湿化热 heat of wetting
湿化时间 time of slaking
湿化试验 slaking test
湿化学法 wet chemical process
湿化学腐蚀 wet chemical etching
湿化学技术 wet chemical technique
湿簧继电器 mercury-wetted relay
湿灰浆 wet mortar
湿回水管 wet return line
湿回水系统 wet return system
湿混 wet blending
湿混合 blunging; moist mixing
湿混合时间 wet-mixing time
湿混凝土分批搅拌站 wet-concrete batching plant
湿混凝土混合料 wet concrete mix
湿混凝土(混合料的)塑性稠度 wet consistency of concrete
湿货 wet cargo
湿击法 <试件制备的> moist tamping
湿机阶段 soak period
湿积冰 slush icing
湿基 wet base
湿基成分 wet basis component
湿基分析 wet basis analysis
湿基准 wet basis
湿集料法施工 wet sand-binder construction
湿集料搅拌 wet-aggregate process
湿季 raining season; wet season
湿季风 <夏季西南季风> wet monsoon
湿加工 wet-processing
湿加工尺寸 green sized
湿架式凿岩机 wet drifter

湿剪强度 green shear strength
湿浆离解机 wet disintegrator
湿交联法 wet cross-linking process
湿浇制法 wet cast process
湿浇注管 wet cast pipe
湿胶合的 wet-cemented
湿搅拌 moist mixing; wet blending; wet mix(ture)
湿接成型 wet sticking process
湿接缝 wet connection
湿接合 wet hookup
湿接头 wet joint
湿进料 wet feeding
湿进料混合器 wet feed mixer
湿精矿 wet concentrate
湿精矿仓 wet concentrate bin
湿精炼 wet puddling
湿井 wet well
湿境土 humid soil
湿锯木屑 damp sawdust
湿绝热 condensation adiabat; saturation adiabat
湿绝热变化 moist adiabatic change; wet adiabatic change
湿绝热不稳定性 wet adiabatic instability
湿绝热的 wet adiabatic
湿绝热过程 moist adiabatic process; saturation-adiabatic process
湿绝热温度差 moist adiabatic temperature difference; wet adiabatic temperature difference
湿绝热线 moist adiabat; wet adiabat
湿绝热直减率 moist adiabatic lapse rate; saturation-adiabatic lapse rate; wet adiabatic lapse rate
湿绝缘 moist adiabat
湿开挖 wet cut(ting); wet excavation
湿龛 wet niche
湿抗张强度 wet tensile(strength)
湿抗张强度试验 wet tensile test
湿抗张强度仪 wet tensile tester
湿空气 air vapo(u)r mixture; humid air; moist air; soft air; wet air
湿空气泵 moist air pump
湿空气比容 humid volume
湿空气比体积 humid volume
湿空气的湿含量 humidity ratio of moist air
湿空气电化学氧化 wet electrochemical oxidation
湿空气过氧化物氧化 wet peroxide oxidation
湿空气焓湿图 enthalpy-humidity diagram
湿空气回流 recirculation of wet air
湿空气冷却器 wet air cooler
湿空气量 enthalpy of moist air
湿空气流 moist air stream
湿空气密度 density of moist air
湿空气热容 humid heat capacity of air
湿空气压缩机 wet air compressor
湿空气氧化法 pressurized aqueous combustion; wet air oxidation
湿空气氧化技术 wet air oxidation technology
湿空气氧化塔 wet cooling tower
湿空气氧化装置 wet air oxidation unit
湿孔 wet hole
湿块 wet lump
湿块云母 hygrophilite
湿拉力 wet tensile(strength)
湿拉强度 green tensile strength
湿牢度性能 wet fastness property
湿老成土 udult
湿冷 wet cold
湿冷拉钢丝 wet-cold-drawn wire
湿冷却 wet cooling

湿冷却条件 wet cooling condition
湿篱笆 wet fence
湿砾石 wet gravel
湿连接 wet connection
湿帘 <用以降低室内气温,保持潮湿的门帘或窗帘> tatty
湿量 moisture content; moisture quantity
湿量密度 moisture density
湿料 wet material; wet mix(ture); wet piece
湿料储箱 wet hopper
湿料分批法 <混凝土> wet batch method
湿料计盘 <按混凝土湿料盘数计算拌和机生产率> wet batch rating
湿料计盘法 wet batch method
湿料加料器 wet feeder
湿料搅碎机 wet masher with tap
湿料炉 wet-process kiln
湿料碾碎机 wet masher
湿料喷浆 wet-mix gunite
湿料喷枪 wet-mix gun
湿料喷射法 wet spraying
湿料输送 wet feed
湿料运送法 wet batch method
湿裂 hygrochase
湿临界 wet criticality
湿临界的 wet critical
湿淋的 dripping wet
湿流取样 wet stream sampling
湿硫熏漂白 wet stoving
湿路附着力 wet grip
湿路堑 wet cutting
湿轮碾机 wet wheel mill
湿罗盘 fluid compass
湿麻袋养护 wet sacks curing
湿麻袋养生法 wet burlap method
湿霾 damp haze; misty
湿煤气净化法 wet gas purification
湿煤器 sprinkler
湿密度 moisture density; wet compacted weight; wet density
湿密度控制 moisture-density control
湿密封 moisture seal
湿密控制 moisture-density control
湿面 wetted surface
湿面分馏塔 wetted surface column
湿面筋 wet gluten
湿面轮廓测定仪 wet surface profilometer
湿面系数 wetted surface coefficient; wetted surface factor
湿敏电容 humicap
湿敏电容器 humidity sensitive capacitor
湿敏电阻 humistor; hygristor
湿敏电阻材料 moisture sensitive resistance material
湿敏电阻器 humidity-dependent resistor
湿敏陶瓷 humiceram; humidity sensitive ceramics; hygrometric sensitive ceramics
湿敏效应 humidity sensitive effect
湿敏元件 dew cell; humidity sensitive element; moisture sensor
湿模制砖 slop mo(u)lded brick; slop mo(u)lding; slop mo(u)lding brick
湿模砖 water-struck brick
湿膜 moisture film; wet film
湿膜测厚仪 wet film ga(u)ge; wet film thickness ga(u)ge
湿膜刮涂器 drawdown bar; drawdown blade
湿膜厚度 wet film thickness
湿膜厚度计 film thickness ga(u)ge
湿膜黏[粘]聚力 moisture-film cohesion

湿膜养护剂 <混凝土的> wet membrane curing agent
湿磨 rasp mill; wet comminution; wet-milling
湿磨白茬 sponge sanding
湿磨白坯 sponge sanding
湿磨白云母 wet ground muscovite mica
湿磨操作 wet grinding operation
湿磨车间 wet grinding plant
湿磨处理 <废物的> wet-milling
湿磨法 wet grinding; wet method
湿磨干烧 wet grinding and dry process burning
湿磨机 wet comminuter; wet grinder; wet grinding machine; wet mill; wet pulverizer
湿磨精加工 wet finishing
湿磨矿渣硅酸盐水泥 Trier cement
湿磨木浆 wet mat; wet-milling timber pulp
湿磨设备 wet grinding plant
湿磨碎机 wet pan
湿磨削装置 wet grinding attachment
湿木白蚁 damp-wood termite
湿木板 green lumber
湿木材 moist wood
湿耐磨强度 wet rub strength
湿耐磨性能 wet rub quality
湿泥饼 wet cake
湿泥芯 greensand core
湿泥砖 wet mud brick
湿年 wet year
湿黏[粘]法 damp and stick method
湿黏[粘]合 wet adhesive bonding
湿黏[粘]合强度 wet bond strength
湿黏[粘]胶合板 wet-glued plywood
湿黏[粘]结强度 green bond
湿黏[粘]土的密度 bulk density of wet clay
湿黏[粘]土的密度孔隙度 density porosity of wet clay
湿黏[粘]土的视声波孔隙度 apparent sonic porosity of wet clay
湿黏[粘]土的中子孔隙度 neutron log porosity of wet clay
湿黏[粘]性 wet adhesion
湿捻 wet doubling; wet twisting
湿碾混合机 edge runner mixer
湿碾混凝土 wet-rolled concrete
湿碾机 edge runner-wet mill
湿碾矿渣 wet grinding slay(concrete); wet ground slag
湿碾矿渣混凝土 slag slurry concrete; slurry slag concrete; wet-rolled granulated slag concrete
湿碾磨细 wet reduction
湿碾碎 wet reduction
湿暖气候 humid mesothermal climate; humid temperature climate
湿排气总管 wet exhaust manifolds
湿抛光 wet polishing
湿刨 hit-and-miss
湿跑道 wet runway
湿跑道刹车距离 wet braking distance
湿泡温度计 wet-bulb thermometer
湿泡砖 wet brick
湿配料 moistened batch; wet batch
湿喷 wet-mix shotcreting <混凝土>; wet blast(ing) <含有磨蚀剂的水>; wet shot <喷火器不经点燃而喷射胶状燃料>
湿喷清理 wet blast cleaning
湿喷砂 wet abrasive blasting; wet blast
湿喷砂法 wet blasting process
湿喷砂机 wet abrasive blasting machine
湿膨胀 hygral expansion; moisture expansion; wet-expand; wet expansion
湿膨胀率 hygroexpansivity

S

湿膨胀指数 hygral expansion index
湿碰湿(工艺) wet-on-wet
湿碰湿涂装 wet-on-wet coating
湿坯 green body
湿坯表面光洁层 self-slip of green body
湿皮 wet pelt
湿铺工艺 wet layup technique
湿期 wet spell
湿起动 wet start
湿起飞力 <以水喷射作为助力的起飞力> wet takeoff power
湿起毛 wet raising
湿起绒 wet raising
湿气 combination gas;humid gas;humidness; moist air; moisture;sweat;wet;wet gas;humidity
湿气泵 wet air pump;wet vacuum pump
湿气传导 moisture conductance
湿气带 wet gas zone
湿气分离器 mist extractor;mist separator
湿气腐蚀 humidity corrosion
湿气干燥法 moisture air method
湿气固化聚氨酯 moisture-cured urethane
湿气固化密封膏 humidly cured sealant
湿气管 wet air duct
湿气盒 moisture cell
湿气阶段产烃率 hydrocarbon productivity of wet-gas stage
湿气浸透 dampness penetration
湿气模糊 humidity blushing
湿气凝珠 <在表面凝结水加工> sweat
湿气排除效率 moisture-excluding efficiency
湿气迁移 humidity migration
湿气渗透 moisture vapo(u)r transmission;pervious to moisture
湿气生成底界 base of wet gas generation
湿气生成峰 peak of wet gas generation
湿气室 moist(ure) closet;moist(ure) chamber;moist(ure) room
湿气室的防潮 moist room damp-proofing
湿气梯度 gradient of moisture;moisture gradient
湿气透射率 moisture vapo(u)r transmission rate
湿气团 moist air mass
湿气吸收 moisture absorption
湿气箱 wet box
湿气消失带 wet gas phase-out zone
湿气移动 humidity migration;humidity movement
湿气移动途径 humidity passage
湿气致白(漆病) humidity blushing
湿气重的 humid
湿气转移 migration of humidity
湿汽 moist steam
湿汽缸套结构 wet-sleeve construction
湿汽膜 moisture film
湿汽蒸 wet steaming
湿砌 wet installation
湿强度 green strength; wet state strength;wet strength
湿强度保留率 wet strength retention
湿强度剂 wet strength agent
湿强纸袋 wet strength paper sack
湿墙加色漆 fresco
湿墙上的白斑 saltpeter rot
湿墙上画壁画 fresco
湿墙涂漆 painting damp wall
湿氢 wet hydrogen
湿氢技术 wet hydrogen technique
湿球 wet bulb
湿球冷却温度表 wet cooling thermometer;wet katathermometer
湿球磨耗试验 wet shot rattle test
湿球磨 wet ball mill
湿球气流计 wet gas flow meter
湿球气体流量计 wet gas meter
湿球位温 wet-bulb potential temperature
湿球温差 wet-bulb depression
湿球温度 wet-bulb globe temperature;wet-bulb temperature
湿球温度差 wet-bulb depression
湿球温度法 wet-bulb temperature measuring method; wet-bulb thermometry
湿球温度计 wet-bulb thermometer
湿球温降 wet-bulb depression
湿球系数 wet-bulb coefficient
湿球研磨机 wet ball grinding mill
湿球装置 wet sphere device
湿曲率 wet buckling
湿燃料 wet fuel
湿燃烧法 wet combustion process
湿燃室 wet combustion chamber
湿燃室锅炉 wet combustion chamber boiler
湿绕 <涂有湿树脂的线圈> wet-wound
湿热 humid heat;moist heat;sensible heat
湿热比 sensible heat factor
湿热处理 hydrothermal treatment
湿热带 humid tropical belt; humid tropical region;humid tropics
湿热带性 humid tropicality
湿热的 muggy
湿热的天气 sticky weather
湿热地区 humid tropical region
湿热定形 wet heat setting
湿热多雨气候 tropic(al) rain climate
湿热负荷 sensible heat load;wet ton
湿热环境 hot-moist environment
湿热季候风 muggy monsoon
湿热交变试验 alternate humidity and temperature test
湿热锯 warm saw
湿热灭菌法 moist heat sterilization;moist heat sterilizer
湿热气候 hot humid climate; humid tropics climate
湿热气候带 humid tropics climate zone
湿热容 humid heat capacity
湿热杀菌 wet heat sterilization
湿热试验 damp heat test
湿热效应 hygrothermal effect
湿热养护 hydrothermal curing
湿热养护法 <混凝土> hydrothermal method of curing
湿热因子 sensible heat factor
湿热值 wet calorific power;wet calorific value
湿热转换器 moist heat transfer
湿热状态 humid tropical condition
湿容积 humid(ifying) volume; wet volume
湿容量 moisture capacity; wet compacted weight;wet volume capacity
湿容重 wet(bulk) density; wet compacted weight;wet unit weight
湿乳胶技术 wet-emulsion technique
湿入口 moisture entry
湿软土 udoll
湿软土上的底脚 mud slab
湿润 humectation;humidification;moisten;wetness;wetting
湿润表面 wetted surface
湿润薄膜 wetting film
湿润不稳定 wet instability;wet unstable
湿润程度 sogginess
湿润处理 moist curing;water finish
湿润处理的 moist-cured
湿润大陆性气候 moist continental climate
湿润带 humid zone;wetting zone
湿润的 soggy
湿润地区 humid area;moist region
湿润度 wettability
湿润断面 wetted cross-section
湿润范围 wetted perimeter
湿润分散剂 wetting and dispersing agent
湿润锋 wetting front
湿润灌溉 moistening irrigation;wetting irrigation
湿润滚筒 dampener
湿润过渡生物带 humid transition life zone
湿润剂 humectant; wet-out agent; wetting additive; wetting agent; surface active agent <一种酸洗添加剂>
湿润剂摆动传递辊 damper vibrator roller
湿润剂施加辊 damper applicator roller
湿润加工 damping
湿润角 <毛细管的> wetting angle
湿润界面 wet(ting) front
湿润控制 damping control
湿润宽度选定器 damping width selector
湿润率 wetting rate
湿润面 wet surface
湿润面积 wetted area;wetted cross-section
湿润摩擦阻抗 wet rub
湿润能力 wettability;wetting ability;wetting capacity;wetting power
湿润黏[粘]合强度 wet strength of adhesion
湿润期 humid period
湿润气候 humid(mesothermal) climate;moist climate
湿润器 dampener; damper; humidifier;moistener
湿润区 humid region
湿润区沿海沙丘 fossil dune
湿润热 heat of wetting;wetting heat
湿润热带地区 region of humid tropics
湿润热带地区的平整土地方法 tillage methods in humid tropics
湿润容器 wet pan
湿润色 wet colo(u)r
湿润时间 wet time
湿润时期 wetting period
湿润室 dampening chamber
湿润树脂的视密度 apparent wet density
湿润土地 springy land
湿润土(壤) humid soil;soggy soil
湿润温带森林地区 the forest in the humid temperate zones
湿润温和气候 humid temperate climate;humid temperature climate
湿润温室 humidified greenhouse
湿润系数 humidity coefficient; humidity factor;moisture coefficient
湿润线 line of saturation; wetting front
湿润箱试验 humidity cabinet test
湿润效应 humidifying effect
湿润性 humidification; wet ability; wettability;wet(ting) property
湿润压辊 damping pressure roller
湿润养护 moist curing;wet curing
湿润养生 moist curing;wet curing
湿润液浓度 fountain-solution density
湿润因素 moisture factor
湿润指数 humidex; humidity index; index of moisture; index of wetness;moisture index
湿润周边 wetted perimeter
湿润装置 damping device;damping unit;humidifying installation;wetting apparatus;wetting mechanism;wetting unit
湿润状态 wet condition;wet state
湿润阻尼 damping
湿润作用 humidification
湿色料 wet colo(u)r
湿砂 damp sand;green sand <铸造用>
湿砂拌和 wet sand mix
湿砂处理 wet sand treatment
湿砂打磨 wet sanding
湿砂浆 wet mortar
湿砂黏[粘]结施工 wet sand-binder construction
湿砂喷射 wet sand blasting
湿砂喷射器 wet sand blaster
湿砂胎模 greensand match
湿砂芯 green core
湿砂型 greensand mo(u)lding
湿砂型芯 cod;greensand core;green core
湿砂养护 <混凝土的> wet sand cure;wet sand curing
湿砂养护法 <混凝土> wet sand process
湿砂养生法 wet sand process
湿砂(造)型 greensand mo(u)lding
湿砂铸法 greensand casting
湿砂铸造 greensand casting
湿筛法 <筛分的> wet screening (method);wet sieve method
湿筛分 moist screening; wet screening;wet sieve;wet sieving
湿筛(分)的 wet-screened
湿筛砂浆 wet-screened mortar
湿筛析 wet screening;wet sieve;wet sieving
湿筛系数 <筛分的> wet screening factor
湿闪弧距离 wet flashover distance
湿闪络电压 wet flashover potential
湿闪络试验 wet flashover test
湿闪试验 wet flash test
湿烧法 wet combustion method
湿舌 moist tongue
湿伸长率 wet expansion
湿生的 ombrogenous
湿生动物 hygrocole;mesocole
湿生植物 hydrophyte;hygrophyte
湿绳改正 wet-line correction
湿绳校正 <水文测验的> wet-line correction
湿绳梯 wet net
湿绳梯训练 wet-net training
湿施工 in the wet;wet
湿石粉抛光 ashing
湿石灰 wet lime
湿石灰涤汽法 wet limestone scrubbing
湿石灰涤汽器 wet limestone scrubber
湿蚀 wet corrosion
湿式拌和器 slurry-type seed mixer
湿式报警 wet-type alarm system
湿式并流低位冷凝器 wet parallel flow low lying condenser
湿式操作破碎机 wet-worked crusher
湿式车闸 oil-immersed brake
湿式尘埃分离器 wet dust separator
湿式冲击(取样)器 wet impinger
湿式除尘 wet dust collection; wet dust extraction;wet dust removal
湿式除尘器 gas scrubber; scrubber; wet cap collector; wet cleaner; wet collector; wet collector of particulate matter;wet dust collector;wet dust separator; wet scrubber; wet separator;wet-type dust collector

S

湿式除尘装置 wet dust collection device
湿式储气罐 water-sealed(gas) holder
湿式储气柜 fluid seal gas holder
湿式触尘装置 wet dust collection equipment
湿式磁粉 wet magnetic particle(powder)
湿式磁选 wet magnetic separation
湿式磁选机 wet magnetic separator
湿式粗选磁选机 wet magnetic cobber
湿式带型磁选机 roche separator
湿式导轨式凿岩机 wet drifter
湿式捣矿机 wet stamp mill
湿式等离子体 humid air plasma
湿式涤气器 wet scrubber
湿式笛簧继电器 wet-reed relay
湿式电集尘器 wet-type electric precipitator
湿式电气除尘器 wet electric(al) dust precipitator
湿式防护 wet preservation
湿式分级 wet classification
湿式分离机 wet separator
湿式焚化法 Zimmerman process
湿式焚烧 wet combustion
湿式粉磨机 wet powder grinder
湿式风镐 wet pick
湿式腐蚀 wet corrosion
湿式干燥器 humidity drier
湿式缸套 wet cylinder lining; wet sleeve
湿式缸套气缸 wet-sleeve cylinder
湿式钢盘离合器 oil-disc clutch
湿式给料水泥回转窑 wet-fed cement kiln
湿式给药机 wet reagent feeder
湿式给药器 wet reagent feeder
湿式光电池 wet-type photocell
湿式光泽机 wet calender stack
湿式过滤器 irrigated filter; wet filter
湿式过氧化物法 wet peroxide oxidation
湿式海底采油树 wet subsea X-mass tree
湿式焊接技术 wet welding technique
湿式和干式混合破碎法 mulling
湿式化学分离法 wet chemical separation method
湿式回火 wet tempering
湿式回水 wet return
湿式回水系统 wet return system
湿式回水蒸汽供暖系统 steam-heating system with wet return
湿式混合 wet-mixing
湿式混合酪朊胶 wet-mix casein glue
湿式集尘器 wet collector; wet dust collector
湿式集尘器设备 wet washing plant
湿式集尘装置 wet dust collection device
湿式挤压成型 wet extrusion molding
湿式拣选 wet sorting
湿式胶合板 wet-glued plywood
湿式搅拌机 wet mixer
湿式净气塔 tower washer
湿式静电除尘器 wet cottrell
湿式静电集尘器 irrigated electrostatic precipitator
湿式开松机 wet opener
湿式空气电池 wet-type air cell
湿式空气过滤器 wet-type air filter
湿式空气净化设备 wet-type air cleaner
湿式空气净化装置 wet-type air cleaner
湿式空气冷却器 wet-type air cooler
湿式空气滤清器 wet-type cleaner
湿式空气氧化法 wet air oxidation process
湿式空气氧化系统 wet air oxidation system
湿式矿灯 wet-cell caplight
湿式垃圾管道 wet-type refuse channel
湿式拉丝机 wet-type wire drawing machine
湿式冷风机 wet-type air cooler
湿式冷凝管 wet condense pipe
湿式冷凝器 wet condenser
湿式冷凝水回水管 wet condensate return
湿式冷却器 wet-type cooler
湿式冷却塔 wet cooling tower
湿式冷却系统 wet-type cooling system
湿式离合器 wet-type clutch
湿式离心除尘器 wet centrifugal dust scrubber
湿式离心分级器 centriclone
湿式力矩电动机 immersed torque motor
湿式力矩马达 immersed torque motor; oil-immersed torque motor
湿式立管系统 wet standpipe system
湿式砾磨机 wet grinding pebble mill
湿式量气表 wet gas meter
湿式量水计 wet-type water meter
湿式滤清器 scrubber; wet cleaner
湿式轮碾(粉碎)机 wet-edge runner mill; wet pan
湿式罗盘 liquid compass
湿式锚杆安装机 wet-type bolting machine
湿式煤气表 wet gas meter; wet meter (of gas)
湿式煤气洗涤器 gas washer
湿式灭火机 water type extinguisher
湿式磨矿机 wet grinding machine
湿式磨削 wet grinding
湿式凝结水管 wet return pipe
湿式排渣炉 wet bottom furnace
湿式盘式制动器 wet disc brake
湿式抛光法 wet lapping
湿式喷粉 wet dusting
湿式喷粉机 dew duster
湿式喷砂 wet blast; wet sand blast(ing)
湿式喷射方法 wet-mix method
湿式喷水系统 wet-pipe sprinkler system
湿式膨胀法 wet inflation
湿式碰撞器 <测定空气中灰尘含量的> water impinger; wet impinger
湿式气表 wet test meter
湿式气缸套 wet liner; wet-type cylinder liner
湿式气镐 wet pick
湿式气柜 dish gas holder; wet storage holder(s); wet-type gasholder
湿式气体流量计 wet gas(flow) meter
湿式气体洗涤器 gas washer
湿式汽缸套 wet cylinder liner
湿式潜水 wet diving
湿式潜水服 wet(diving) suit
湿式潜水服压缩性 compressibility of wet suit
湿式潜水器 wet submersible
湿式潜水器系统 wet vehicle system
湿式潜水钟 wet diving bell
湿式清理滚筒 wet tumbler; wet tumbling barrel
湿式球磨机 wet ball grinding mill; wet ball mill
湿式燃烧法 wet combustion
湿式燃烧过程 wet combustion process
湿式润滑剂 wet lubricant
湿式升船机 wet ship lift
湿式石膏法 wet gypsum process
湿式试验气表 wet test meter
湿式收尘器 wet dust collector; wet precipitator
湿式收集器 wet collector
湿式手持式凿岩机 wet sinker drill
湿式双曲线面冷却塔 wet-type hyperbolic(al) cooling tower
湿式水表 wet-type water ga(u)ge; wet-type(water) meter
湿式水加热潜水服 hot-water wet suit
湿式水洗涤器 wet cleaner
湿式伺服阀 wet-type servo-valve
湿式塔式磨粉机 wet-type tower abrasion mill
湿式填充物过滤器 wet filling filter
湿式通风器 wet ventilator
湿式温度计 wet thermometer
湿式无介质磨 wet autogenous mill
湿式雾滴收集器 wet mist collector
湿式吸风机 wet suction fan
湿式洗涤器 wet scrubber
湿式洗气 wet washing
湿式洗气塔 scrubber wash tower
湿式下向凿岩机 wet sinker drill
湿式纤维过滤器 wet fibre filter
湿式消防喷水系统 sprinkler system-wet pipe
湿式蓄电池 wet storage battery
湿式悬轮混料机 Muller mixer
湿式旋风除尘器 wet cyclone; wet rotoclone collector
湿式旋风分离器 wet cyclone
湿式旋风集尘器 wet rotoclone collector
湿式旋风收尘器 wet rotoclone collector
湿式旋转凿岩 wet rotary drilling
湿式旋转钻眼 wet rotary drilling
湿式压缩制冷循环 wet compression refrigeration cycle
湿式烟道气脱硫系统 wet flue gas desulfurization system
湿式烟气处理 wet flue gas treatment
湿式研磨 wet lapping
湿式氧化 wet oxidation
湿式氧化法 wet oxidation method; wet oxidation process
湿式摇纱 wet reeling
湿式油底壳 wet sump
湿式圆筒磁选机 wet drum magnetic separator
湿式圆筒选矿机 wet drum cobber
湿式凿岩 water-fed rock drilling; wet boring; wet drilling
湿式凿岩机 wash boring drill; water-feed rock drill; water hammer drill; water injection air hammer; water injection drill; wet drill; wet-type machine
湿式凿岩钻 wet-type rock drill
湿式凿岩钻头 injection drill-bit
湿式真空泵 wet vacuum pump
湿式纸板机 wet board machine
湿式纸型用纸 wet flong
湿式重选方法 wet-gravity concentration method
湿式转鼓抛光 wet tumbling
湿式转运救生 wet transfer rescue
湿式撞击式滤尘器 wet impinger
湿式自动喷水灭火系统 wet automatic sprinkler system
湿式自动喷水系统 automatic sprinkler system; automatic wet-pipe sprinkler system
湿式钻眼 wet drilling
湿式作业 wet method operation; wet-processing
湿视 wet look
湿视有光膜 wet look gloss film
湿试样 wet sample
湿试仪表 wet test meter
湿室 humid room; wet room
湿室泵房 wet well pumping house
湿室电气设备 wet room fixture
湿室防潮 wet room dampproofing
湿室隔墙 wet room partition(wall)
湿室(内)条件 <温度25℃,相对湿度75%> moist room condition
湿室事务 wet room service
湿室照明装置 wet room light fitting; wet room light(ing) fixture; wet room luminaire(fixture)
湿熟石灰 wet slaked lime
湿水法张紧 <绘图纸用> stretch
湿水硅钙石 nekoite
湿水货 damp goods
湿水沼土 wet tundra soil
湿松弛复原 wet relaxation
湿送料法 wet feed method
湿塑缝 <水泥混凝土的> wet formed joint
湿碎法 wet crushing
湿碎磨 wet crushing mill; wet stamping mill
湿损 <货被水损> wet damage
湿损坏 moisture damage
湿损货 sea damaged goods
湿损事故 wet damage accident
湿损纸 wet broke
湿塔 wet tank; wet tower
湿态 hygrometric state
湿态层贴材料 wet layup material
湿态层贴胶合法 wet layup method of bonding
湿态刮平 wet screed
湿态加油 wet fat-liquoring
湿态喷漆法 wet-on-wet painting
湿态强度 <型砂的> green strength; wet(state) strength
湿态容积 wet bulk
湿态性能 green property
湿碳化 wet carbonization
湿套 moisture cover
湿套印性 trapping
湿体积 humid volume
湿体系抽空 <用喷水空气泵抽空的装置> wet system of evacuating
湿体重 wet volume weight
湿天然气 wet gas; wet natural gas
湿贴 wet combining
湿通风管道 wet vent
湿通路 moisture entry
湿投设备 wet feed facility
湿投装置 wet feed device; wet feed equipment; wet feed unit
湿透 drench(ing); drip; drop; sopping wet <衣物等>
湿透的 logged; sloppy; soggy
湿透气性 green permeability
湿涂层量 wet pick-up
湿涂喷膜厚度 wet film thickness
湿土 moist earth; moor; wet soil
湿(土)崩坍 <雪、土等的> wet avalanche
湿土堆雪崩 wet avalanche
湿土块 bat
湿土囊 <填方中的> pockets of swamp material
湿土器 soil soaker
湿土坍坡 wet landslide
湿褪色 wet fading
湿挖 <基坑的> in the wet
湿挖方 <基坑的> wet cut; wet excavation
湿弯曲试验 wet flex test
湿喂 slop feeding
湿喂法 wet feeding
湿温计 hygrothermograph
湿温气候 humid mesothermal climate
湿稳性 hygro-stability
湿污泥 liquid sludge; wet sludge
湿污泥池 wet sludge tank

湿污染 wet soiling
湿坞 non-tidal basin;wet dock
湿坞闸门 guard lock;lock of dock;entrance lock
湿物料 wet stock
湿物料干燥机 wet stock dryer
湿雾 wet fog
湿吸量 hygroscopic moisture content
湿洗 laundering;wet cleaning
湿洗器 wet scrubber;wet washer
湿显色 wet development
湿线校正 wet-line correction
湿陷 collapse;hydrocompaction;hydroconsolidation;settlement by soaking;slump
湿陷沉降 slumping type settlement
湿陷量 collapse settlement;slumping type settlement
湿陷量分级 collapsibility grading index
湿陷起始压力 initial collapse pressure
湿陷起始压力试验 initial collapsing pressure test
湿陷系数 coefficient of collapsibility;wet subsidence coefficient
湿陷系数试验 coefficient test of collapsibility
湿陷性 collapsibility;subsidability
湿陷性沉降 slumping type settlement
湿陷性分级指标 collapsibility grading index
湿陷性黄土 collapsible loess;loess collapsible slumping;slumping loess;wet-settling losses
湿陷性黄土的地基承载力 bearing capacity of loess with subsidability
湿陷性土 collapsible soil;collapsing soil;slumping soil;water sensitive soil
湿陷压力 collapse pressure;collapsing pressure
湿相 wetting phase
湿箱 < 用以贮存液体 > wet tank
湿消化法 < 固体废物厌氧分解处理 > wet digestion method
湿芯 greensand core
湿芯材 wetwood
湿芯静电耗散器 wet-wick static dissipator
湿新成土 udent
湿型离心铸造 greensand spinning
湿型砂 greensand
湿型涂料 greensand facing
湿性胶合法 wet gluing
湿修坯体 water finishing
湿朽 cellar rot;wet rot
湿选 water concentration;wet concentration;wet treatment < 矿石 >
湿选法 wet cleaning;wet cleaning method;wet cleaning process;wet preparation;wet screening;wet washing
湿雪 cooking snow;moist snow;snezhura;snow slush;water snow;wet snow
湿雪崩 damp-snow avalanche;wet avalanche
湿雪崩坍 damp-snow avalanche
湿压成型 plastic pressing;wet pressing
湿压法 ram pressing
湿压光 moist calendering
湿压机 wet press
湿压力滚子 moistening pressure roller
湿压强度 green compression strength;wet-compressive strength
湿压湿喷漆法 < 前一道漆未干时就喷后一道漆 > wet-on-wet
湿压湿涂漆法 wet-on-wet painting
湿压湿印刷 wet-on-wet
湿压实重量 wet compacted weight
湿压损失 wet press(ing) loss;wet pressure loss
湿压缩 wet compression
湿压仪 < 测试受湿与未受湿试件抗压强度的仪器 > pressure-wetting equipment
湿压硬质纤维板 wet pressed hardboard
湿压榨 wet pressing
湿压纸浆 wet pressed pulp
湿亚热带 subhumid tropics
湿岩生动物区系 hygropetrical fauna
湿岩生物 hygropetrobios
湿研 wet grinding
湿研矿渣 wet ground slag
湿研磨 wet grind
湿研磨分离法 discrete process of wet
湿研盘 wet pan mill
湿盐 < 含镁钙等杂质的 > wet salt
湿盐土 wet solonchak
湿验定 wet assay(ing)
湿养护 < 混凝土 > moist after-treatment;moist cure;moist curing;water curing;wet curing
湿养护的 moist-cured
湿养护混凝土 moist-cured concrete;water cured concrete;wet-cured concrete
湿养护室 damp-storage closet;water curing room;fog room
湿养黏[粘]合剂 moisture curing adhesive
湿养膨胀 expansion wetting
湿养生 moist cure;moist curing;water curing;wet curing
湿养室 moist room;moist cabinet
湿氧化法 wet oxidation method;wet oxidation process
湿氧化土 udox
湿样品 wet sample
湿引入 moisture entry
湿应力 < 木材的 > green stress
湿用胶 wet-use adhesive
湿用胶粘剂 wet-use adhesive
湿游移法 wet migration process
湿原 moor
湿原料 wet feed material
湿源 moisture source;moisture supply
湿运 lifting with water;lifting with wetted chamber;wet-lifting
湿运升船机 wet ship lift
湿凿(碎)石 wet rock cuttings
湿渣 wet slag
湿渣粉 wet ground slag
湿涨率 < 砂的 > bulking factor
湿胀 bulking;moisture expansion;water swelling
湿胀量 swelling capacity
湿胀率 bulking factor;percentage bulking < 砂的 >
湿胀曲线 < 砂的体积 > bulking curve
湿胀容许量 allowance for bulking;bulking allowance
湿胀砂 bulking sand
湿胀系数 bulking factor;coefficient of bulking;coefficient of swelling
湿胀现象 bulking phenomenon [复 phenomena]
湿胀效应 bulking effect
湿胀性 bulking;bulking effect;swelling property
湿胀性材料 swelling material
湿胀因素 bulking factor
湿胀应力 swelling stress
湿胀指数 swelling index
湿沼泽 damp marsh
湿沼泽地 carr;quag;quagmire
湿沼泽地带 carr-lands
湿罩光 fogged coat
湿折皱恢复 wet without-dry recovery
湿蒸硫化 wet steam cure
湿蒸馏 boiling down
湿蒸呢 wet decating;wet decatizing
湿蒸呢机 wet decatizer
湿蒸气 moisture vapo(u)r;wet vapo(u)r
湿蒸气参数 quality of wet vapo(u)r
湿蒸汽 moist steam;prime steam;wet steam < 含水分的蒸汽 >
湿蒸汽干燥度 dryness fraction of wet steam
湿蒸汽冷却堆 wet steam cooled reactor
湿蒸汽汽轮机 wet steam turbine
湿蒸汽田 wet steam field
湿蒸汽养护 < 混凝土 > wet steam curing
湿整理 dressing
湿枝 wet shoots
湿纸边 wet trim
湿纸幅 wet web
湿纸幅张力测定仪 pressductor
湿纸浆 wet pulp
湿纸耐破度 wet mullen
湿纸耐破强度 wet bursting strength
湿纸蒸发表 Piche atmometer;wet paper atmometer
湿制剂 wet preparation
湿制陶瓷 wet-process porcelain
湿治 moist curing;water after-treatment;water curing;wet curing
湿治薄膜 curing membrane
湿治混凝土 moist-cured concrete;water cured concrete
湿治室 < 养护混凝土用 > damp-storage closet;moist cabinet;moist(ure) chamber;moist(ure) closet;moist(ure) room
湿治水 curing water
湿治水养护 water curing
湿致胀缩 moisture movement
湿重 humid weight;natural weight;wet weight < 加满油、水后空车重 >
湿周(边) wetted perimeter
湿皱纹 wet wrinkle
湿装配 wet installation
湿渍 stained by moisture;wet-stained
湿阻 moisture resistance
湿组分 wet basis
湿钻法 wet drill method
湿钻孔 wet drill hole;wet drilling
湿钻探 wet drilling
湿钻屑 wet well cuttings
湿作业 wet trade
湿作业隔断 wet construction partition(wall)
湿作业施工 wet construction
湿作用温度 humid operative temperature

十

十-二进制转换 decimal-to-binary conversion

十八点统计 eighteen-o'clock statistics
十八斗 connection block
十八甲基八硅氧烷 catadecamethyloctosiloxane
十八进制的 octodenary
十八开本 eighteenmo;octodecimo
十八酸 octadecanoic acid;stearic acid
十八碳三烯酸 calendic acid;el(a)eostearic acid
十八碳酸 n-octadecanaic acid
十八烷 octadecane
十八烷酸 n-octadecanaic acid
十八烯醇 oleic alcohol
十八信息移频电源 eighteen-code frequency-shift power supply
十磅表压力 ten-pounds ga(u)ge pressure
十磅汽油 ten-pounds gasoline
十磅重的东西 tenpounder
十倍 decade;tenfold
十倍的 decuple;tenfold
十边形 decagon
十仓混和机 ten-fold mixer
十打 < 计数单位 > great hundred;long hundred
十大功劳树 Chinese manhonia
十单元平板车 < 美国圣泰菲铁路首创,以关节接合十个骨架结构而成,用于背驮式运输 > Ten-Pack
十导线 ten-wire
十的 decadal;denary
十的幂 power of ten
十冬腊月 the cold months of the year
十度级二甲苯 ten degree grade xylene
十二胺 lauryl amine
十二边 twelve edge
十二边形 d(u)odecagon;twenty-sided polygon
十二边形螺丝 dodecagon screw
十二边形螺丝头 dodecagon-head-(ed) screw
十二齿小蠹 ips sexdeutatus
十二穿孔 twelve punch
十二醇 lauryl alcohol;l-dodecanol
十二的 duodecimal;duodenary
十二点球 twelve point sphere
十二点扫描器 twelve point scanner
十二方套筒 twelve point socket
十二方套筒扳手 twelve point socket wrench
十二分配器 divide-by-two circuit
十二分频电路 divide-by-two circuit
十二分算 duodecimal
十二分音符法 twelve note music
十二分之几的 duodenary
十二分之一的 duodecimal
十二缸 twelve-cylinder
十二根钢丝的 twelve-wire
十二股(金属丝)缆绳 twelve-wire cable
十二海里界限 twelve-mile limit
十二回线示波器 Hathaway oscillograph
十二基 lauryl
十二级风 force twelve wind;hurricane;wind of Beaufort force twelve
十二角形 dodecagon
十二脚管座 duodecimal base
十二节环 twelve-membered ring
十二进位算法 duodecimal system
十二进(位)制 duodecimal system
十二进位制的 decimal binary;duodenary
十二进制的 duodecimal
十二进制记数法 duodecimal notation
十二进制数字 duodecimal number;duodecimal numeral
十二开本 duodecimo;twelvemo
十二开间 twelve-bay
十二开纸 duodecimo
十二菱面体 rhombododecahedron
十二路混合设备 twelve-direction mixing unit
十二路载波设备 twelve-circuit carrier system
十二路载波制 twelve-circuit carrier system
十二罗 < 计数单位,等于 1728 个 > great gross
十二螺栓通风潜水装置 twelve bolt ventilation type diving equipment
十二脉冲整流 twelve pulse rectifying

十二面体 dodecahedra;dodecahedron
十二面体的 dodecahedral
十二面体基苯 dodecahedron
十二面体金刚石 dodecahedral stone
十二面体形状的 dodecahedron-shaped
十二氢戊搭烯并戊搭烯 staurane
十二醛 lauraldehyde;lauric aldehyde
十二双酸 dodecanedioic acid
十二(碳)烯丁二酸酐 dodecenyl succinic anhydride
十二碳烯基琥珀酸 dodecenyl succinic anhydrite
十二碳烯基琥珀酸酐 dodecenyl succinic anhydride
十二碳烯双酸 dodecenedioic acid
十二碳脂肪酸 lauric acid
十二烷 dihexyl;dodecane;dodecyl
十二烷胺盐酸盐 laurylamine hydrochloride
十二烷醇 lauryl alcohol
十二烷基 dodecyl
十二烷基苯 detergent alkylate;dodecylbenzene
十二烷基苯磺酸钠 neopelex;sodium dodecyl benzene sulfonate
十二烷基苯磺酸盐 dodecyl benzene sulphonate
十二烷基磺化乙酸酯 laurylsulfoacetate
十二烷基硫酸钠 lauryl sodium sulfate;sodium lauryl sulfate
十二烷基硫酸盐 lauryl sulfate
十二(烷)酸 dodecanoic acid;lauric acid
十二烷酸丙二醇酯 propylene glycol laurate
十二烷酸二羟丙基酯 lauricidin
十二烷酸戊酯 amyl laurate
十二烯 laurylene
十二烯酸 lauroleic acid
十二旬风 bad-i-sad-o-bistroz;wind of 120 days;Seistan < 一种强劲的北风 >
十二指肠溃疡 duodenal ulcer
十二柱式 duodecastyle
十二柱式门廊 dodecastyle portico
十二柱式神庙 dodecastyle temple
十二纵标格式 twelve-ordinate scheme
十翻二 decimal-to-binary
十分安全 foolproof
十分安全的方法 foolproof way
十分宝贵的 of great price
十分标 scale-of-ten
十分标电路 scale-of-ten circuit
十分度盘 ten's dial
十分度取样 ten sampling intervals per decade
十分肥沃的土地 rank soil
十分黑的物质 atrament
十分清洁的 bioclean
十分位距 decile interval
十分位数 decile
十分之九 nine-tenths
十分之六的因数 six-tenths factor
十分之六准则 six-tenths rule
十分之七定则 seven-tenths rule
十分之一倍频程 one-tenth octave
十分之一大波 one-tenth maximum wave
十分之一当量浓度 tenth-normal solution
十分之一当量浓度的 tenth-normal
十分之一功率点宽度 tenth-power width
十分之一毫克 decimilligram
十分之一厚度值 tenth-thickness value;tenth-value thickness
十分之一秒经纬仪 one-tenth second theodolite
十分之一值层 tenth-value layer
十分之一最大波高 one-tenth maximum wave
十分之一最大波高平均值 mean of the highest one-tenth of wave heights; average highest one-tenth wave height
十分钟最大功率额定值 ten-minute rating
十分注意 all the attention
十个单元关节式薄片轻质平板车所组成的节油背驮式运输列车 < 美国圣泰非铁路首创每单元平车装运十台,整列车共计装运 100 台重载挂车,来回一次可节省燃料 6000 加仑,约合 22712.47 升 > ten-pack fuel foiler train
十个桥跨的 ten-bay
十个一组 decade
十公斤 myriagram(me)
十公里 myriameter[myriametre]
十环 tenth ring
十级风 storm; whole gale; wind of Beaufort force ten
十角形 decagon
十角形的 decagonal
十节环 ten-membered ring
十进安培秤 deka-ampere balance
十进制图像数据 decimal picture data
十进制系统 decimal index system; decimal system
十进制小数 decimal fraction
十进制小数部分 decimal part
十进制转换 decimal conversion
十九进制的 novendenary
十九碳烯酸 jecoleic acid
十九烷 nonadecane
十卷 decade
十开间的 ten-bay
十克 decagram(me)[dekagram(me)]
十框蜂箱 ten-frame-hive
十立方公尺 dekastere
十立方米 decastere[dekastere]
十联固结仪 ten-in-one odeometer
十六胺盐酸盐 hexadecyl amine-hydrochloride
十六醇 hexadecanol;hexadecyl alcohol
十六醇胺 hexadecylol amine
十六分标 scale of sixteen
十六分之一日分潮 sixteenth-diurnal constituent
十六分之一弯头 one-sixteenths bend
十六氟氧基环辛烷 hexadecafluorooxy-cyclooctane
十六基 hexadecyl
十六基磷酸盐 hexadecyl phosphate
十六基硫醇 hexadecyl mercaptan
十六基亚磺酰胺 hexadecyl sulfenamide
十六极矩 hexadecapole moment
十六进位法 hexadecimal notation
十六进位计数器 hexadecimal counter
十六进制【计】 hexadecimal system; sexadecimal(number) system
十六进制表示法 hexadecimal notation
十六开本 dicimo-sixto;sextodecimo
十六开(的)纸 decimosexto; sextodecimo
十六内酯 hexadecanolide
十六酸 hexadecanoic acid;hexadecoic acid;hexadecylic acid
十六酸十六酯 hexadecyl palmitate
十六碳二烯二酸 hexadecadienedioic acid
十六碳二烯酸 hexadecadienoic acid
十六碳炔 hexadecine[hexadecyne]
十六碳炔二酸 hexadecylene diacid
十六碳炔二羧酸 hexadecylene dicarboxylic acid
十六碳炔酸 hexadecynoic acid
十六碳三烯酸 hexadecatrienoic acid
十六碳烯 hexadeeene
十六碳烯二甲酸 hexadecylene dicarboxylic acid
十六碳烯二酸 hexadecendioic acid; hexadeeene diacid
十六碳烯二羧酸 hexadeeene dicarboxylic acid
十六碳烯酸 hexadecylenic acid
十六烷 cetane;hexadecane
十六烷二甲酸 hexadecane dicarboxylic acid
十六烷二酸 hexadecandioic acid (hexadecane diacid)
十六烷基胺 hexadecylamine
十六烷基硫酸 hexadecyl hydrosulfate
十六烷酸 palmic acid; palmitinic acid;palmitate
十六(烷)酸铝 alumin(i)um palmitate
十六烷值 cetane number;cetane value
十六烷值试验机试验 CFR[cooperative fuel research committee] engine test
十六烯值 cetene number
十六元数 sedenion
十轮卡车 ten wheeler
十米 decameter[decametre]; dekameter[dekametre]
十米波 decametre wave
十面体 decagon;decahedron[复 decahedrons/decahedra]
十面体的 decahedral
十年 decade;decennium;ten years
十年的 decennary;decennial
十年间 decade
十年间的 decadal
十年期 decade
十年一次的 decennial
十年一遇洪水 once-in-ten-year flood; ten-year(recurrence interval) flood; unusual flood
十年一遇之洪水 decimal flood
十年责任期 decennial liability
十牛顿 decanewton
十七醇 heptadecanol
十七基 heptadecyl
十七进制的 septendecimal
十七酸 heptadecanoic acid
十七碳烯 heptadecene
十七烷 heptadecane
十七烷胺 heptadecylamine
十七烷基磷酸 heptadecyl phosphoric acid
十七酰 heptadecanoyl
十氢化萘 decahydronaphthalene; decaline
十氢化四硅 tetrasilicane;tetrasilicon decahydride
十氢化四硼 tetraboron decahydride
十取一抽样 one-in-ten sampling
十三胺 tridecyl amine
十三个 long dozen
十三进制的 terdenary
十三块 tridecyne
十三陵 < 中国明朝帝王墓 > Ming Tombs of Emperor
十三硫酸盐 tridecyl sulfate
十三醛 tridecylic aldehyde
十三酸 tridecanoic acid
十三烷 tridecane
十三烷二酸 tridecandioic acid;tridecane diacid
十三烷二羧酸 tridecane dicarboxylic acid
十三烷基苯 tridane;tridecyl benzene
十三烷基膦酸 tridecane phosphonic acid
十三烯 tridecylene
十三烯二酸 tridecylendioic acid; tridecylenediacid
十三烯二羧酸 tridecene dicarboxylic acid
十三烯酸 tridecylenic acid
十三酰 tridecanoyl
十升 decaliter[decalitre];dekaliter[decalitre]
十水硫酸钠 Glauber salt
十水碳酸钠 sal soda
十四脚管底 diheptal base
十四进制的 quaterdenary
十四腈 tridecyl cyanide
十四日 < 两星期 > fortnight
十四烃基硫酸钠 natrium tetradecylsulfuricum
十四烷 tetradecane
十四烷酸 myristic acid
十通道多谱扫描仪 ten-channel multispectral scanner
十通阀 ten-way valve
十万万 billion;milliard < 英 >
十位 decade;ten's place
十位分量 ten-bit component
十位分指示盘 tens of minutes indicator
十位累加器 decimal accumulator
十位数测微器 ten-digit micrometer
十位跳杆 tens jumper
十位跳杆隔片 tens jumper distance piece
十五边形 pentadecagon
十五点移动平均 fifteen points moving average
十五个大气压下的湿度百分数 < 相当枯萎百分数 > fifteen atmosphere percentage
十五进制的 quindenary
十五迷宫 fifteen puzzle
十五酸 pentadecanoic acid; pentadecylic acid
十五(碳)烷 pentadecane
十五烷醇 pentadecanol
十五烷酮 pentadecanone
十五烯二酸 pentadecendioic acid; pentadecene diacid
十五烯酸 pentadecylenic acid
十五酰 pentadecanoyl
十弦螺线 ten-chord spiral
十弦螺旋曲线 ten-chord spiral
十项道路试验 ten point road test
十星螺钉 tens star screw
十溴二苯醚 decabromodiphenyl ether
十溴联苯醚 decabromodiphenyl ether
十一胺 undecylamine
十一倍性 hendecaploidy
十一边形 hendecagon;undecagon
十一乘八厘米像片 quarter plate
十一醇 undecylic alcohol
十一点移动平均 eleven points moving average
十一行穿孔 eleven punch
十一基 hendecyl;undecyl
十一基苯酚 undecyl phenol
十一基聚氧乙烯醚醇 undecyl-polyoxyethylene-ether-alcohol
十一基硫代硫酸盐 undecyl thiosulfate
十一基噻吩烷 undecyl thiophane
十一级风 storm gale;wind of Beaufort force eleven; violent storm < 风速 28.5 ~ 32.6 米/秒 >
十一角形 hendecagon
十一脚管底 magnal base
十一脚管座 magnal socket
十一进制的 undecimal
十一腈 undecanonitrile
十一硫醇 undecyl mercaptan
十一面体 hendecahedron
十一醛 hendecanal; undecanal; undecylic aldehyde

十一炔 undecyne
十一酸 undecanoic acid; undecylic acid
十一碳二烯酸 undecandienoic acid
十一碳炔 hendecyne
十一碳炔二酸 undecyndioic acid;undecyne diacid
十一碳炔二羧酸 undecyne dicarboxylic acid
十一碳炔酸 undecynic acid
十一碳烯 undecylene
十一碳烯醇 undecylenic alcohol
十一碳烯二酸 undecendioic acid;undecene diacid
十一碳烯二羧酸 undecene dicarboxylic acid
十一碳烯基 undecenyl
十一碳烯酸钠 sodium undecylenate
十一碳烯酸盐 undecylenate
十一烷 hendecane;undecane
十一烷的 undecanoic
十一烷酸 hendecanoic acid
十一烯 hendecene
十一烯酸 undecap;undecylenic acid
十一酰 undecanoyl
十一酰胺 undecanoic amide
十亿 kilomega;billion <美>;milliard <英>
十亿电子伏特 <美> billion electron-volts;giga-electron-volts
十亿吨 gigaton
十亿分率 parts per billion
十亿分之几 parts per billion
十亿分之一 <10^{-9}> billionth; part per billion[ppb]
十亿加仑/天 billion gallons per day
十亿年【天】billion years;aeon
十亿位 billibit;kilomegabit
十英尺见方的小屋侧房 ten-footer
十有八九 ten to one
十元环形脉冲计数器 ring-of-ten circuit
十趾吊 hang ten
十中取码 one out often code
十中取一码 one-out-of-ten code
十重 tenfold
十重的 tenfold
十周年 <美> decennial
十柱式房屋 decastyle
十柱式建筑 decastyle building;decastylos
十柱式门廊 decastyle;decastyle portico
十柱式庙宇 decastyle temple
十柱式柱廊 decastyle
十柱式走廊 decastyle
十柱柱列式 decastyle
十字 cross-shaped budding
十字凹口螺丝 cross recess(ed) screw
十字把手 crossbar handle
十字板 cross-shaped plate
十字板拌和机 criss-cross mixer
十字板常数 vane borer constant
十字板贯入仪 vane penetrometer
十字板剪力强度 vane shear strength
十字板剪力试验 vane shear test; crosshead shear test
十字板剪力试验点数 number of vane shear test
十字板剪力试验机 vane shear apparatus
十字板剪力试验仪 vane shear test apparatus
十字板剪力仪 vane apparatus;vane shear borer;vane shear tester;vane shear apparatus
十字板剪切强度 vane shear strength
十字板剪切试验 <土壤的> crosshead shear test; vane shear test; vane test
十字板剪切试验点数 number of vane shear test
十字板剪切试验机 vane shear apparatus
十字板剪切仪 vane apparatus;vane shear apparatus;vane shear borer;vane shear tester
十字板抗剪强度 vane shear strength;vane strength
十字板抗剪强度值 vane strength value
十字板试验 <土壤的> auger vane test
十字板头 crosshead
十字板头长度 length of vane
十字板头直径 diameter of vane;vane diameter
十字板仪 four-bladed vane;vane borer
十字半搭接 cross-halved joint
十字绷带 crucial bandage
十字臂 cross arm
十字臂沉降管 cross arm settlement ga(u)ge
十字臂沉降计 cross arm settlement ga(u)ge
十字臂沉降仪 cross arm settlement ga(u)ge
十字臂分层沉降管 cross arm settlement ga(u)ge
十字臂固结管 cross arm settlement ga(u)ge
十字扁铲 cross bit;four-point(ed) bit
十字标 retic(u)le
十字标记 cross mark
十字柄 feed handle
十字部 hip cross
十字槽 cross-notch
十字槽螺钉 cruciform slot screw; Phillips screw
十字槽螺钉头 Phillips head screw
十字槽螺母 Phillips head
十字槽螺丝帽 cruciform slot head
十字槽埋头螺钉 Phillips recessed-head screw
十字槽头螺钉 cross recess(ed) screw; cross recess head screw; Phillips head; Phillips screw; plus screw
十字槽头螺钉用的螺丝起子 screw driver for cross slot head screw
十字测天仪 cross staff;fore staff
十字叉 four-arm spider; spider; joint spider <万向接头的>
十字叉衬套 spider bushing
十字叉传动 spider drive
十字叉连接 circular spider type joint
十字叉线 hairs
十字叉线校正环 reticle adjusting ring
十字岔管接头 double junction
十字铲 bull point
十字撑 X-brace;cross bracing
十字撑架 X-bracing
十字撑条 cross brace
十字冲(击)钻 cross-mouthed chisel
十字锤 pick hammer
十字锤式破碎机 cross type hammer crusher
十字锤头 cross-peen; cross-peen hammer
十字搭接 crosslap;crosslap joint
十字刀撑 cross brace
十字的 geneva
十字阀 cross valve
十字沸石 garronite
十字分划对准目标 reticle-on-target
十字分划线 cross division line
十字分配料道 cross-cap
十字缝 cross seam;cross stitch;herringbone
十字缝法 herringboning
十字符号 tick mark
十字斧 pick ax(e)
十字斧柄 pickaxe handle
十字杆 <直角器的> cross staff
十字杆件 crossbar
十字镐 pick ax(e);pick(er);drag bit; hack iron; kirk; mandrel [mandril]; mattock; moil; moyle; navvy pick;pecker
十字格 crossed grid
十字格版框 crossbar
十字格栅 cross grid
十字跟踪 cross-track(ing)
十字拱 cross vault
十字沟槽联轴节 cross-slot coupling; Oldham's coupling
十字骨针 stauractine
十字管 cross branch; cross tee; pipe cross
十字管道 crossduct
十字管接头 cross joint(ing); pipe cross;cross pipe
十字管气泡水准仪 cross-bubble
十字管头 cross piece
十字光 cross light
十字光标 crossing of light; tracking cross
十字规矩线 corner mark; corner sticks; corner tick; register mark; register sticks; register tick; registration sticks;tick;tick mark
十字焊 cross-wire weld
十字焊缝试样 cruciform test specimen
十字号 criss-cross
十字合扎 <绳结的> cross seizing; right-angle seizing
十字痕迹 cross mark
十字横臂 cross arm
十字横梁 cruciform girder
十字花饰 dogtooth
十字花纹的斜边 cross bevel
十字滑轨 cross slide
十字滑块联轴器 cross slide coupling; double slider coupling;Oldham coupling;slider coupling
十字划痕 cross scribe
十字环 cross ring
十字回转门 turnstile
十字混合 criss-cross mixing
十字脊屋顶 cross-ridge roof
十字记号 crux[复 cruxes/cruces]
十字架 cross piece;cross staff head; crucifix; dog back; manhole dog; rood;spider;strong back
十字架底的窗 pede window
十字架隔屏中央的拱 rood arch
十字架警告标志 <平交道口> warning crossing
十字架上面的花格挑顶 ceilure;celure
十字架式万向节 cardan joint;Hooke's joint;universal joint
十字架饰柱 cruciferous column
十字架坛 presbyterium
十字架形曲线 cruciform curve
十字架支承梁 rood beam
十字架柱 cruciferous column; cruciform column
十字架状的教堂 cruciform church
十字尖锤 cross-peen;cross-peen sledge hammer
十字尖头锤 cross-peen hammer
十字交叉 cross-nailed material; decussation;right-angled intersection; interstitium <教堂平面中的>
十字交叉道 square crossing
十字交叉点 Latin cross
十字交叉焊缝 cross weld
十字交叉刷涂法 crossing brush coating
十字交叉线 cross wires
十字交汇 cross type intersection
十字铰链 cross-garnet butt;cross tail butt
十字接管 double tee
十字接头 backflow connection; cross connection;cross joint(ing);cruciform joint; four-way box; pipe cross; X-connection; four-way union <连接管子的>
十字接头体 cross body
十字节点板 cruciform gusset plate
十字结 stauros
十字结构 criss-cross structure
十字结构星形轮 geneva cam
十字结联轴节 crosshead
十字镜 <测定光在晶体中偏振平面方向的仪器> stauroscope
十字军城堡 crusader castle
十字军式建筑 crusader architecture
十字军式教堂 crusader type church
十字开关 cross-switch
十字空心钳 double hollow bit tongs
十字孔板【船】cloverleaf plate
十字孔板用锥头 cone for clover leaf plate
十字框架 cross frame
十字拉杆 saltier cross bars
十字拉线 crossing guy
十字蓝晶 staurolite cyanite
十字立体交叉 cross interchange
十字链 cross chain
十字梁 cross beam; cross member; rood beam;rood screen
十字梁基础 grillage
十字路 cross nailed material;cross-over road; cross road; four cross-way;four way;four wont way
十字路口 carrefour; crossing; cross road;four corners;crossway
十字路口的交通指挥装置 silent cop
十字路口画有黄线的路段 <车辆应在其一侧停下,除非在显示红灯以前能行驶到另一侧> box junction
十字路口交通拥堵 grid lock
十字轮机构 geneva gear; geneva mechanism
十字轮胎 cross-ply tire[tyre]
十字螺钉头的螺丝刀 screwdriver for cross-head screw
十字螺帽的螺旋插座 screwdriver socket for cross-head screw
十字螺帽的螺旋套筒 screwdriver socket for cross-head screw
十字螺丝刀 cross driver;Philips driver;Phillips screwdriver
十字螺丝起子 cross driver;plus driver
十字瞄准线 cross hairs
十字排列 cross spread
十字配筋 cross-section iron; cruciform iron
十字平接头 crosslap joint
十字砌合层 broken course
十字锹 pike
十字桥台 cross-shaped abutment
十字穹顶 cross vault
十字圈 cross in circle
十字山石 pleysteinite
十字石 granitite;staurolite
十字石白云母片岩 staurolite-muscovite-schist
十字石二云母片岩 staurotite dimicaceous schist
十字石云母片岩 staurotile
十字式砌叠 <砖的> cross bond
十字式钻头 silot
十字手柄 capstan handle;cross handle
十字束节 cross socket
十字双晶 crossed twinning

十字丝 cross-hair retic(u)le; cross lines; cross web; cross wire; filar cross(ing); graticule; hair cross(ing); retic(u)le crossing; spider lines; stadia hairs; retic(u)le <光学仪器的>
十字丝板 cross-hair retic(u)le
十字丝测微计 crossbar micrometer
十字丝测微器 cross-wire micrometer
十字丝横丝 horizontal crosshair
十字丝环【测】 cross-hair ring; retic(u)le ring
十字丝间距 thread interval
十字丝校正环 retic(u)le adjusting ring
十字丝校正螺旋 retic(u)le adjustment screw
十字丝接头抗裂试验 cruciform cracking test
十字丝距 interval between graticule wires
十字丝目镜 cross-filar eyepiece; cross-hair eyepiece; filar eyepiece
十字丝片 cross-hair retic(u)le; diaphragm
十字丝平面 cross-hair plane
十字丝视差 parallax of cross-hairs; parallax of reticule; parallax of wires
十字丝竖丝 vertical cross-hair
十字丝网目板 cross-ruling screen; retic(u)le screen
十字丝细线条 hairline
十字丝线 vertical hair
十字丝照明 reticle illumination
十字丝照明装置【测】 cross-hair illuminating attachment
十字丝中心 center of reticule
十字铁 cross-section bar; cross-section iron; cruciform iron
十字通风 cross ventilation
十字头 crosshead; crossbar; cross fitting; cross head(ing); cross-over tee; cross point; four-way tee; joint center [centre]; joint cross pinblock; pinblock; rinse cross head; joint cross <万向接头的>
十字头臂 crosshead arm
十字头边瓦 crosshead side shoe
十字头扁栓 crosshead gib
十字头扁销 crosshead key
十字头衬垫 crosshead lining
十字头导板 crosshead guide plate
十字头导承 connecting rod guidance
十字头导杆 crosshead guide bar
十字头导框 crosshead guide
十字头端 crosshead end
十字头发动机 engine with outside guide
十字头关节销 crosshead wrist pin
十字头观察口盖 crosshead inspection head
十字头烘箱 crosshead oven
十字头滑板 crosshead shoe; slide bar
十字头滑槽 crosshead slipper
十字头滑槽衬垫 crosshead gib
十字头滑块 crosshead(slipper) shoe; crosshead slipper
十字头加热器 crosshead heater
十字头夹板 crosshead plate
十字头连杆 crosshead link
十字头联杆 crosshead beam
十字头螺钉 Allen screw; recessed head screw
十字头螺丝刀 Phillips screwdriver
十字头铆钉 crosshead rivet
十字头上瓦 crosshead top shoe
十字头伸长杆 crosshead extension rod
十字头式发动机 crosshead engine
十字头式活塞 crosshead piston
十字头式行动机构 crosshead type walking mechanism
十字头尾杆 crosshead arm
十字头系船柱 cruciform bollard
十字头下瓦 crosshead bottom shoe
十字头销 crosshead cotter; cross (head) pin; gudgeon pin
十字头销衬套 wrist-pin bearing
十字头销套 crosshead pin bushing; wrist-pin collar
十字头销轴承 bearing of the cross head pin
十字头压板 crosshead wrist pin washer
十字头圆销 crosshead wrist pin
十字头圆销螺母 crosshead wrist pin nut
十字头凿 cross-mouthed chisel
十字头凿子 diamond jumper
十字头轴承 crosshead bearing
十字头钻 cross-mouthed drill
十字凸花纹 raised crossover rib
十字涂覆法 cross coating
十字围栏 cross rail
十字艉 cruciform stern
十字纹 cross pattern
十字纹基床 cross-bedding
十字纹孔对 crossed pits
十字五通接头 side outlet crossing
十字系缆桩 cross bitt; crosshead bollard; cross-shape bollard
十字系索 bunt gasket; cross gasket
十字纤维 cross fiber[fibre]
十字线 cross curve; cross lines; hairline; retic(u)le; hair cross <测量仪器目镜中的>
十字线对准 reticle alignment
十字线分划板 cross-line graticule
十字线夹 crossing clamp
十字线目镜 criss-cross eyepiece; cross-hair eyepiece
十字线片 diaphragm
十字线平面 cross-line plane
十字线视差 parallax of reticule
十字线竖丝【测】 vertical hair
十字线中心 center of reticule
十字销头 cross pin head
十字形 cross nailed material; cross shape; cross type
十字形暗销 cross type dowel
十字形凹槽 cross groove
十字形播种 cross-sow
十字形布置圆顶教堂 ambulatory church
十字形槽 cross bath; cross slot
十字形齿轮钻头 cross-roller bit
十字形大厦 cruciform block
十字形的 criss-cross; crossed; cross-like; cross-shaped; cruciform; crucishaped; decussate
十字形底层平面 cross-shape ground plan
十字形底层平面图 cruciform ground plan
十字形底架 cruciform ground frame
十字形地窖 cruciform crypt
十字形丁坝 cruciform groin; cruciform groyne
十字形定向耦合器 cross-guide coupler
十字形对称 cruciform symmetry
十字形对出叶 decussate leaf
十字形墩 cross pier
十字形舵 cruciform of rudders
十字形耳堂【建】 transept
十字形二歧分枝式【植】 cruciate dichotomy
十字形防波堤 cruciform breakwater
十字形房屋 cross block
十字形分色镜 dichroic cross
十字形钢钻头 four-winged steel bit
十字形格子砖 cruciform checker
十字形工具 spider kit
十字形工字钢组合柱 larimer column
十字形拱顶教堂 cross-domed church; cruciform-domed church
十字形构件 dagger
十字形构造 decussate structure
十字形管 cross pipe; four-way pipe
十字形管子接头配件 pipe cross
十字形横挡 crossbar; X-stretcher <家具的>
十字形横梁 cruciform member
十字形横梁框架 cruciform frame
十字形横移挖泥法 cross-shaped transverse dredging method
十字形花纹 <花格窗中> cross quarters
十字形花样 criss-cross pattern
十字形滑道 spider slip
十字形换乘 cross-shaped transfer
十字形机构轮 geneva wheel
十字形机架 X-frame
十字形记号 Christ-cross
十字形建筑 cruciform building
十字形建筑平面交叉的方形部分 transept square
十字形交叉 cross intersection; four-leg intersection; square crossing; cross road【道】
十字形交叉口 cross intersection
十字形交叉中心线 square crossing center[centre] line
十字形教堂 cross church; cruciform centrally planned church; cruciform church; cruciform centralized church <主室在中央>
十字形教堂的双耳堂 double transept
十字形教堂的袖廊 double transept
十字形教堂翼部 transept
十字形接合件 joint cross
十字形接头 geneva cross
十字形结构 cross texture; cruciform construction
十字形截面 cruciform cross-section; star section
十字形卷线轴 spider reel
十字形控制杆 cross control rod
十字形框架 X-frame
十字形拉力试验 cross tension test
十字形拉条 diamond brace
十字形缆桩 cross bitt; cruciform bollard
十字形连接件 cross connector
十字形联轴器 cross-linked coupling
十字形螺丝刀 plus driver
十字形螺丝起子 cruciform screw driver
十字形铆接 cross-riveting
十字形摩天楼 cruciform skyscraper
十字形泥芯 cruciform core
十字形黏[粘]土冲切器 cruciform clay cutter
十字形平交路 cross roads
十字形平面房屋 cross building
十字形平面图 cross-shaped plan; cruciferous plan; cruciform plan
十字形钎头 cross type bit; cruciform bit; four-pattern bit; four-wing bit
十字形桥梁 cruciform bridge
十字形切割器 cross-section cutter
十字形热交换器 cross system of heat exchanger
十字形伞 cross parachute
十字形式样 cruciform pattern
十字形枢纽 cross type junction terminal
十字形水准器 cross level
十字形榫槽 cross slot
十字形弹簧 spider spring
十字形体 cruciform mass
十字形体的四肢 cross limb
十字形铁芯 cruciform core
十字形铁圆管 tubular cross bar
十字形图案 criss-cross; criss-cross pattern; cruciform pattern
十字形弯 cross brake
十字形屋体 cruciform mass of a building
十字形无岩芯钻头 cross plug bit
十字形系船柱 cruciform bollard
十字形效果 cruciform effect
十字形形心 cross-shape core; cruciform core
十字形牙轮钻头 cross-roller bit; cross-section cone bit; cross-section cutter
十字形圆顶教堂 ambulatory church
十字形凿刀 cross-cut chisel
十字形支墩 cross-shape pier
十字形支柱 cross-shape pier; cross-shape support; cruciform support
十字形制模板 four-entry pallet; four-way pallet
十字形轴头 pinblock
十字形柱 cross pier; cross(-shape) column; cross support
十字形柱墩 cruciform pier
十字形组合横臂 cross-sectional arm
十字形四辊万能轧机 sack mill
十字形钻 cross-mouthed drill; star drill
十字形钻具 cross-edged drill
十字形钻头 cross bit; cross chisel; square bit; cross-bladed chisel bit; cross-cut bit; cross drilling bit; cross-edged bit; cross matched drill; cruciform bit; star-type bit; X-bit; X-chisel; cross-chopping bit <破碎孔底岩柱用>
十字碹 crossed arch; crossed crown
十字靴 cross-over shoe
十字影线 cross hatching
十字晕 sun cross
十字晕渲 cross hatching
十字凿 cross chisel
十字栅门 crossbar slings
十字照准丝 cross hairs; cross wires
十字支撑 cross bracing
十字支座 cross beam
十字轴 center[centre] crossing; center [centre] piece; cross(axle); cross pin; spider
十字轴传动 quarter-turn drive
十字轴式起落架 cross-axle undercarriage
十字轴式套管扳手 cross limb wrench
十字轴形节头 cross pin type joint
十字轴中心 spider center[centre]
十字肘材 dagger knee
十字柱 cruciferous column
十字转盘 two-way turntable
十字状 crosswise
十字状地 criss-cross
十字准线 cross hairs; cross wire
十字准线光标 cross-hair cursor
十字准线光标数字转化器 digitizer
十字纵丝 vertical hair
十字钻 cross bit
十字钻探钻头 four-point(ed) bit
十字钻头 cross bit; cross drill bit; cross-mounted bit; cutaway mine bit; four-bladed bit; four-edged bit; four-wing rotary bit; square bit; star-pattern drilling bit <冲击钻用>
十字钻头刃口 cross bit cutting edge
十足工作天数 clear working days
十足满载 <船舶载量> full and down
十足密封 positive seal
十足重量 lumping weight
十足准备计划 one-hundred percent

什锦锉 key file

什锦角尺 combination square
什锦香精花园挂毯 millfleurs garden tapestry
什锦组锉 Swiss pattern file
什切青港 <波兰> Port Szczecin
什卫道夫模型 Schwedoff model

石鞍 saddle reef

石岸 rock bank
石暗沟 stone subdrain
石凹面 sunk face
石坝 boulder dam; masonry dam; stone dam
石坝堆脚 rock toe
石斑【地】 lithosporic
石斑木 Raphiolepis indica
石板 ashlar; ashlar piece; broad stone; flag(stone); greystone slate; plate stone; slabstone; slat(e); slate board(ing); stone flag; tabula rasa; tile shoe; tilestone; flagging stone <铺路用>; stone slate
石板扒钉 slate cramp
石板板壁 slate siding
石板边缘磨光机 edge polisher
石板衬砌 stone slab revetment
石板垂直面 slate hung
石场粗加工的方形石料 block stone
石板道 flagstone
石板的匠称 <石工按尺寸命名的石板> wippet; wivet
石板的匠名 <石工按尺寸命名石板> tant
石板的棱锥顶 pyramidal crown of stone slabs
石板底层 bed of a slate
石板地板 slab floor
石板地面 slab floor; slate floor cover(ing); stone slab floor(ing); flagstone flooring <室内>
石板地面终饰 slate floor(ing) finish
石板斧 slate axe; sax
石板覆层 stone slab cladding
石板覆盖的 slate-covered
石板橄榄绿色 slate olive
石板工 rockman
石板工锤 slater's hammer
石板工人 slater
石板工用锤 slater's hammer
石板工用脚手架 slater's scaffold
石板固定钉 slate fixing nail
石板固定器 slate fixer
石板褐色 slate tan
石板黑(色) slate black
石板护岸 stone slab lining; stone slab revetment
石板灰 slate gray[grey]
石板剪断机 slate shears
石板蓝(色) slate blue
石板楼板 slab floor
石板楼地面 slate flooring; slate slab flooring
石板楼面 stone slab floor(ing)
石板路 flagstone walk; sett-paved road; stone road
石板路面 flagstone pavement; flagstone paving; stone slab floor(ing); flagging
石板绿(色) slate green
石板锚碇法 veneer stone anchoring
石板面层 slate clad
石板面配电盘 slate switch panel
石板劈裂机 slate cleaving machine
石板铺的地板 slatted floor
石板铺的石面 slatted floor
石板铺面 pavement of flagstones; paving with flags; rag paving
石板铺砌 flagstone pavement; flagstone paving; flagstone pitching; paving with flags; slab pavement; slab paving
石板砌合 ragwork
石板砌面 stone slab
石板砌筑 ragwork
石板嵌板 slate panel
石板墙板 slate wall panel
石板桥 cladded bridge; stone slab bridge
石板切割具 slate cutter
石板色 slate
石板贴面 slate clad; stone slab facing; stone veneer(ing)
石板贴面做法 veneer stone facing system
石板瓦 backer; rag(stone); roofing slate; slate; stone shingle; stone slate; healing stone; peggies < 10 英寸 × 10 英寸 >; duchess < 12 英寸 × 24 英寸 >; marchioness < 22 英寸 × 12 英寸 >
石板瓦钉 slate nail; slating nail
石板瓦盖 greystone slate; stone slate
石板瓦工 slater
石板瓦工程 slate work; slating
石板瓦工锤 slate hammer
石板瓦工的脚手架 clippie
石板瓦挂钩 slate hook
石板瓦挂瓦条 slate batten
石板瓦两边搭接 Dutch lap
石板瓦前缘 front edge
石板瓦墙面 slate hanging; weather slating
石板瓦双燕尾榫 slate cramp
石板瓦贴面 hung slating
石板瓦屋脊 slate ridge; slate roll
石板瓦屋面 slate roofing; stone slab roofing
石板瓦修换 tab
石板瓦准尺 slating ga(u)ge
石板瓦钻孔 holing of slab tile; holing of slates; thirling
石板屋顶 slate(d) roof
石板屋顶排水沟 slate valley
石板屋顶望板 slate roof sheathing
石板屋脊 slate roll
石板屋面 stone slab roofing
石板屋面瓦用钉 slater's nails
石板镶面 stone slab facing
石板压顶 slate coping
石板栅栏 slat fence
石板整边工具 whittle
石板质的 slaty
石板装饰 stonework decoration
石板状的 slaty
石板桌面 slate table
石板紫色 slate violet
石版 lithograph; lithostone
石版复制术 autography
石版印刷 lithographic(al) printing
石版印刷术 lithography
石版印刷纸 litho printing
石版油墨 lithographic(al) ink
石棒 stone bar
石宝塔 stone pagoda
石宝座 stone throne
石碑 stela [复 stelal/steles]; stele; stone tablet
石碑圈 peristalith
石崩 avalanche of rock
石笔杆 styiolith
石笔片麻石 pencil gneiss
石笔石 pencil stone
石壁 stone wall(ing)
石壁倒塌 rock wall failure
石壁破坏 rock wall failure
石边饰 <建筑物上的> stone curb; stone kerb
石标 brancher; stoneman <用作界标的石堆>
石冰川 rock glacier
石冰块 block glacier
石驳 stone barge
石布道坛 <教堂中> stone pulpit
石材 stone material
石材表面装饰加工 stonework decorative finish; stonework enrichment; stonework ornamental finish
石材的风化 weathering of stone
石材地板 stone floor(ing)
石材地面覆盖板 stone floor cover(ing)
石材垫板 stone filler
石材垫片 stone filler
石材雕带 stone frieze
石材二次破碎机 fine stone crusher
石材分离机 stone splitter
石材骨料混凝土 stone-aggregate concrete
石材加工 stone-working
石材加工机械 stone-working machine
石材建筑材料 stone construction(al) material
石材建筑方法 stone construction(al) method
石材(建筑)立面 stone facade
石材(建筑)正面 stone facade
石材楼梯间 stone stair(case)
石材磨床 rock mill
石材磨光机 stone polishing machine
石材抛光 stone polishing
石材劈裂机 stone splitter
石材破碎装置 stone crushing installation
石材强度 stone strength
石材强度等级 strength grading of stone
石材切割法 stone cutting method
石材切割机 stone cutter; stone cutting machine
石材商品分类 trade grouping of rocks
石材饰面 stone facing; stone finish
石材填料 stone filler
石材贴面板 stone facing slab
石材贴面的混凝土板 stone-faced concrete panel
石材弯曲件 stone curving
石材铣磨成型工作 stone shaping work
石材铣磨工作 stone hewing work; stone milling work
石材细缝锯 jointer saw
石材(修筑的)喷水池 stone font
石材修琢法 stone dressing
石材研磨机 stone grinding machine
石材硬度 hardness of stone
石材斩劈工作 stone hewing work
石材整形工作 stone shaping work
石材支架 stone support
石材制品 stone; stone products
石材装修 stone finish
石菜花矿物 agaric mineral
石蚕属 germander
石槽 rock-cut ditch
石侧面 flank
石层 <天然的> stone layer
石层风化带 ruptured zone
石层缝 commissure
石层裂缝方位 seam orientation
石层之上的土层 <采石场> top cap
石长凳 stone bench
石肠坝 sausage dam
石厂 crushing plant
石场板岩 stonesfield slate
石场废块 quarry waste
石场废料 quarry refuse; refuse of pile
石场弃石 quarry refuse; quarry waste; refuse of pile
石场石屑 quarry fines
石场碎石 quarry refuse; refuse of pile
石场未分选料 pit-run; quarry run
石尘 crushed dust
石衬(砌) stone lining; steening
石撑壁 stone counterfort
石川 stone stream
石川岛播磨装卸方式 IHI system
石川石 ishikawaite
石窗盘 stone window sill
石窗台 stone sill
石窗台板 stone sill of window; stone window sill
石锤 crandall; stone hammer
石锤修饰 bush-hammered dressing
石锤修琢 bush-hammered dressing
石锤琢面 bush-hammered face
石莼(属) green laver
石带 stone stripe
石丹哈德 <一个厂商牌号,出售防水材料、填充料、硬化剂、防腐剂等建筑产品> Stonhard
石挡墙 mortar rubble retaining wall
石刀推挤机 slate scudding machine
石导流堤 stone training wall
石的 calculary
石的粗琢 scapple
石的形成 lithogenesy
石灯笼 <常见于日本式庭园中> stone lantern
石堤 rock bank; rock embankment; stone dike [dyke]; stone embankment; stone levee
石滴水檐 dripstone
石底 stony bottom
石底河川 hard-bed stream
石底河流 hard-bed river; hard-bed stream
石底基础 stone footing
石底脚 stone footing
石底座 stone footing
石地 rocky ground
石地群落 phellium
石垫 stone template; stone templet
石垫层 bedding stone; stone bedding; stone matting; stone packing
石雕 cameo; carved stone; sculpture in stone; stone carving
石雕花格窗 stone tracery window; tracery window
石雕刻 stone carving
石雕刻术 stereotomy; stone sculpture
石雕刻物 stone sculpture
石雕男像 stone male figure
石雕女像 stone female figure
石雕塔 stone-carved pagoda
石雕像 rock-cut statue
石雕琢机 stone-dressing machine
石雕坐像 rock-hewn seated statue
石吊兰 Lysionotus pauciflorus
石丁坝 stone built groyne
石丁头 stone header
石顶盖板 stone capping slab
石洞 stone cavern; pit hole <储存气体用>
石洞门 stone portal
石堆 bourock; cairn; heap of rubble; moraine; piled rock; rock windrow; rubble pile
石堆标记 carn
石堆堤 riprap
石碓 stamp mill
石墩 stone counterfort; stone pillar
石盾 cave shield; cave palette
石盾厚度 thickness of stone shield
石盾直径 diameter of rock shield
石垛 puck
石垛带 rock wall
石垛工 backman; waller
石垛节间 <未填充部分> open set

S

石垛平巷 dummy road
石蛾 caddis fly
石法券 stone arch
石帆 sea fern
石翻犁 right-hand plough
石方 cubic(al) meter of stone; loose yards
石方爆破 stonework explosion
石方爆炸 stonework explosion
石方车厢 quarry body
石方除尘机 rockdusting machine
石方工程 rock excavation; rock works; stoneworks
石方机械 rock machine
石方开挖 rock cut
石方量 compacted yard
石方路基 rock subgrade
石方用格栅铲斗 skeleton rock bucket
石方用加深花纹 rock extra deep tread
石方用轮胎 rock tire
石方用轮胎花纹 rock tread
石方用特种花纹 rock-extra tread
石方用重型铲斗 heavy duty rock bucket
石方凿工 squarer
石方装载机 rock loading shovel
石方作业 rock excavation
石防波堤 stone breakwater; stone built groyne; stone jetty; stone pier
石舫 marble boat; stone boat
石分界(围)墙 stone boundary wall
石坟墓 stone tomb
石粉 crusher (rock) dust; crushed sand; cut stone; filler; fine dust; gritty dust; ground rock powder; meal; mineral filler; mountain flour; mountain meal; pulverised [pulverized] stone; quarry dust; rock dust; rock flour; rock meal; rock powder; slate powder; stone dust; stone powder; stone sand
石粉厂 stone mill
石粉称量斗 filler weigh hopper
石粉储藏装置 filler storage unit
石粉废水 stone wastewater
石粉工厂 stone mill
石粉工场 stone mill
石粉计量器 filler scale
石粉沥青 asphalt with rock flour
石粉提升机 filler elevator
石粉筒仓 filler silo
石缝握裹木条 ranging bond
石扶壁 stone counterfort
石扶垛 stone counterfort
石扶手 stone balustrade; stone railing
石斧 chopper; slate knife; slater's axe; tomahawk; whittle; zax
石盖顶 stone coping
石膏 calcium sulfate [sulphate]; hydrated sulphate of lime; mineral white; parge(t); plaster stone; salt lime; salt lime calcium sulphate; sulphate of lime
石膏白云岩建造 gypsum-dolomite formation
石膏白云岩亚建造 gypsum-dolomite sub-formation
石膏斑 gypsum stains
石膏板 gypsum building material; gypsum panel; gypsum plate; gypsum plank; gypsum tile; plaster panel; plaster slab; plaster tablet; sheetrock; Thistle board < 排挡间的 >
石膏板成型站 plasterboard forming station
石膏板钉 plasterboard nail
石膏板顶棚 dry ceiling; gypsum board; plate ceiling
石膏板干燥机 plasterboard drier
石膏板隔断 gypsum lath partition; gypsum slab partition
石膏板隔墙系统 gypsum board enclosure system
石膏板护面纸 boardcarton
石膏板木龙骨隔断 plasterboard stud partition
石膏板墙 gypsum board wall; gypsum drywall
石膏板天棚 dry ceiling
石膏板条 board lath; gypsum board lath; long lath; plaster's lath(ing)
石膏板条的起始固定件 starter clip
石膏板(条)钉 gypsum lath nail
石膏板围护系统 gypsum board enclosure system
石膏拌料 gypsum stuff
石膏保温 gypsum insulation
石膏绷带 plaster bandage; plaster of Paris bandage
石膏壁板 gypsum plasterboard; gypsum wallboard
石膏边缘 casing bead
石膏变形 gypsum transformation
石膏标准稠度 standard consistency of gypsum
石膏冰垢 gypseous scale
石膏薄板 gypsum sheet
石膏薄板顶棚 gypsum sheet ceiling
石膏薄板隔墙 gypsum sheet partition
石膏材料 gypsum block
石膏仓 gypsum bin; gypsum hopper
石膏仓库 gypsum stor(ag)e
石膏层 gypsum horizon; gypsum layer; plastering
石膏掺和浆 gypsum ga(u)ging plaster
石膏炒锅 gypsum calcinating kettle
石膏沉积物 playa deposit
石膏衬板 backing gypsum; gypsum backer board; gypsum backing board; gypsum board sheathing; gypsum sheathing
石膏稠浆 gypsum slurry
石膏床 plaster bed; plaster of Paris bed
石膏打底的 gypsum based
石膏打底灰 gypsum bond(ing) plaster
石膏打底抹灰 gypsum backing(mixed) plaster; gypsum basecoat (mixed) plaster; hard wall
石膏大板 plasterboard
石膏大理石 gypsum marble
石膏单元 gypsum unit
石膏底板 backing board; gypsum base board
石膏底抹灰 gypsum undercoat(mixed) plaster
石膏底子 gesso
石膏地形模型 plaster relief model
石膏地形模型塑造装置 relief-milling device
石膏垫板 plaster bat
石膏雕刻 gypsograph; plaster carving
石膏雕刻术 gypsography
石膏雕模 plaster carving
石膏吊顶 gypsum ceiling
石膏吊顶板 gypsum ceiling board
石膏煅烧 gypsum calcination
石膏堆栈 gypsum stor(ag)e
石膏矾土膨胀水泥 gypsum aluminate expansive cement
石膏防火盖板 gypsum fireproofing
石膏防火盖面 gypsum fireproofing
石膏房 plaster room
石膏肥料 gypsum
石膏粉 cement plaster; gess(o) < 雕塑用 >; ground gypsum; gypsum meal; gypsum powder; land plaster; powdered gypsum; terra alba
石膏粉面材料 plastic-coated material
石膏粉面层 plaster of Paris
石膏粉饰 gypsum plaster(ing); Murite
石膏粉饰的光板 plain sheet of gypsum plaster board
石膏粉刷 gypsum plaster(ing)
石膏粉刷饰面 gypsum trowel finish
石膏粉碎器 gypsum breaker; gypsum crusher
石膏复合板 gypsum composite panel
石膏盖板 gypsum board sheathing; gypsum sheathing
石膏隔墙板 gypsum panel
石膏隔墙砌块 gypsum partition block
石膏隔墙砖 gypsum partition tile
石膏工厂 gypsum plant
石膏构件 gypsum unit
石膏含量 gypsum content
石膏和硬石膏矿床 gypsum and hydrite deposit
石膏糊团 bads
石膏花 anthodite
石膏花饰 perget
石膏化 gypsification
石膏化铁炉渣水泥 cupola slag sulphated cement; gypsum iron slag cement
石膏缓冷水泥 gypsum-retarded cement
石膏缓凝剂 calcium sulphate retarder; gypsum retarder; gypsum retarding agent; plaster retarder
石膏缓凝水泥 gypsum-retarded cement
石膏灰 tiling plaster
石膏灰缓凝剂 keratin
石膏灰浆 calcium sulfate plaster; ga(u)ge stuff; gypsum mortar
石膏灰胶纸柏板 gypsum plasterboard
石膏灰泥 gypsum cement; gypsum marl; gypsum plaster
石膏灰泥板 gypsum plasterboard
石膏灰泥板片 plasterboard sheet
石膏灰泥(薄)板隔墙 plasterboard sheet partition wall
石膏灰泥纤维板 gypsum base board
石膏灰泥组合板 gypsum plasterboard composite
石膏灰岩角砾岩 brockram
石膏混凝土 gypsum concrete; plaster concrete
石膏混凝土拌和料 gypsum concrete mill mixture
石膏基快速硬化灰 granite plaster
石膏基料砂浆 gypsum based mortar
石膏激发 gypsum activation
石膏夹心板 gypsum cassette slab
石膏夹心纸板 sheetrock
石膏剪 plaster shears
石膏建材 gypsum building material
石膏浆 calcium plaster; gypsum paste; gypsum plaster; gypsum putty; plaster slip; sulfate plaster
石膏浆混合器 plaster mixer
石膏浆抹灰基层 hard wall
石膏浆凝结 plaster set
石膏浇注 plaster casting
石膏胶结料 selenite cement
石膏胶结物 gypsum cement
石膏胶凝材料 calcium sulfate cement; gypsum plaster
石膏胶凝灰 gypsum cement plaster
石膏胶粘剂 gypsum adhesive
石膏搅拌机 plaster mixer
石膏搅拌器 plaster mixer
石膏浸渗墙布 gypsum-coated wall fabric
石膏晶石 gypseous spar; gypsum spar
石膏晶体 gypsum crystal
石膏净粉饰 gypsum neat plaster
石膏镜盘 plaster block
石膏聚积层 gypseous horizon
石膏绝热材料 < 隔墙等用的 > insulex
石膏开采 gypsum quarry
石膏开裂监视标志 plaster pad
石膏壳模 case mo(u)ld
石膏空心墙板 gypsum hollow wall panel
石膏块 lump gypsum; plaster block; rock gypsum
石膏块材 gypsum block; gypsum block cement; gypsum slag cement; supersulphated slag cement
石膏块培养基 gypsum block medium
石膏块土壤湿度计 gypsum block soil moisturemeter
石膏矿床 gypsum deposit
石膏矿石 gypsum ore
石膏矿渣板 gypsum slag board
石膏矿渣水泥 gypsum slag cement; Kuhl cement; slag-gypsum cement; slag-sulphate cement; supersulfated (slag) cement; supersulphated slag cement
石膏粒 gypsum sand
石膏连续炒锅 gypsum continuous kettle
石膏料浆站 plaster slurry mixing station
石膏裂缝 plaster crack
石膏龙骨 gypsum stud
石膏墁灰制品 stick and rag work
石膏面砖 gypsum tile
石膏模 gypsum pattern
石膏模联合成型机 plaster mixing and pouring machine
石膏模造法 plaster mo(u)ld casting
石膏模真空注浆法 plaster mo(u)ld vacuum casting
石膏模注型法 plaster mo(u)lding
石膏抹底墙 hard wall
石膏抹灰底板 gypsum lath(ing); gypsum plank; rock lath(ing)
石膏抹灰顶棚 gypsum plaster ceiling
石膏抹灰混合料 gypsum plaster mixture
石膏抹灰面 calcium sulphate plaster
石膏抹灰墙板 gypsum plasterboard panel
石膏抹灰饰面 gypsum finish
石膏抹料 gypsum plaster
石膏抹面 hard finish
石膏抹面灰浆 calcium sulphate plaster
石膏抹面装饰 gypsum trowel finish
石膏模板 gypsum form board
石膏模型 gypsum model; gypsum mo(u)ld; plaster cast; plaster model; plaster mo(u)ld; plaster pattern; print < 扁平装饰品的 >
石膏模型法 plaster mo(u)ld method
石膏模型真空注浆生产线 plaster mo(u)ld vacuum casting line
石膏木丝灰浆 gypsum wood-fibered [fibred] plaster
石膏木丝灰泥 gypsum wood-fibered [fibred] plaster
石膏泥 cement plaster
石膏泥板 gypsum plaster board
石膏泥浆 gypsum mud; gypsum slurry
石膏腻子 finish plaster
石膏黏[粘]结剂 gypsum cement
石膏刨花板 gypsum wood board; gypsum wood slab
石膏膨胀 expansion due to gypsum
石膏片 gypsum plate
石膏破碎车间 gypsum crushing plant

石膏破碎机 gypsum crusher;gypsum crushing plant
石膏砌块 gypsum block;gypsum building material;gypsum unit
石膏砌块隔断 gypsum block partition
石膏砌体 gypsum masonry
石膏嵌缝膏 gypsum joint filler
石膏嵌缝料 gypsum joint filler
石膏墙板 gypsum block;gypsum unit;gypsum wall baseboard; gypsum wallboard;gypsum plasterboard
石膏墙粉 calcium sulphate plaster
石膏墙粉整平板 calcium sulphate plaster screen
石膏墙筋 gypsum stud
石膏墙面板 gypsum sheathing board; gypsum sheathing plasterboard
石膏墙面抹灰 gypsum wall plaster
石膏侵 gypsum cutting
石膏球 gypsum sphere
石膏人造石 protean stone
石膏(三合土)路 gypsum road
石膏纱布 gypsum gauze
石膏砂浆 gypsum-sand mortar
石膏砂抹灰 gypsum-sand plaster
石膏烧锅 gypsum kettle
石膏(烧)盘 gypsum pan
石膏石灰拌料 gypsum-lime stuff
石膏石灰灰浆 gypsum-lime plaster
石膏石灰浆 gypsum-lime mortar
石膏式矿物 gypsoide
石膏试板 selenite plate
石膏试块 gypsum pat
石膏饰板 veneer plaster
石膏饰面基层 veneer base
石膏双向密肋板 gypsum waffle slab
石膏水泥 gypsum cement;selenite cement;selenitic lime
石膏水泥灰泥 cement plaster
石膏胎模 plaster match
石膏天花板 gypsum ceiling board; gypsum plaster ceiling; plastic ceiling panel
石膏天花镶板 plaster ceiling(panel); plaster ceiling(slab)
石膏填孔 gypsum filling
石膏填孔料 gypsum filler(block)
石膏填块 gypsum filler(block)
石膏条板 board lath; plaster lath(ing);gypsum board;gypsum lath(ing);gypsum plank
石膏条板钉 gypsum lath nail
石膏图案装饰 graffito
石膏土 gypsum earth
石膏瓦 gypsum roof(ing) tile;gypsum tile
石膏外模 gypsum mo(u)ld
石膏望板 gypsum sheathing; gypsum sheathing board
石膏微粉 mineral white
石膏为主要成分的岩石 gyprock
石膏圬工 gypsum masonry
石膏屋面板 gypsum roof(ing) board; gypsum roof(ing) plank; gypsum roof(ing) slab
石膏吸声板 absorptive backing; acoustic(al) perforated gypsum board;gypsum acoustic(al) board
石膏吸收法 absorption of gypsum method;gypsum absorption method
石膏纤维板 sheetrock
石膏纤维板墙 plasterboard wall
石膏纤维灰浆 fibered[fibred] plaster
石膏纤维混凝土 gypsum fiber[fibre] concrete
石膏线脚 plaster fillet
石膏线脚灰 gypsum mo(u)lding plaster
石膏线条粉刷 gypsum mo(u)lding plaster
石膏镶板 backing board
石膏镶饰灰 gypsum veneer plaster
石膏像 plaster figure;plaster statue
石膏屑 gypseous sand
石膏芯 gypsum core
石膏芯板 gypsum core board
石膏芯墙板 gypsum panel
石膏型 gypsum mo(u)ld;plaster cast
石膏型加压整铸铝模板法 pressure match plate process
石膏型铸造 plaster casting
石膏需用量 gypsum requirement
石膏压条 plaster stop
石膏岩 gyprock;gypsum rock;selenolite
石膏岩溶 gypseous karst
石膏窑 plaster kiln
石膏医疗室 plaster sand
石膏印模 plaster impression
石膏硬块 batting block
石膏硬石膏矿石 gypsum-anhydrite ore
石膏原矿 run-of-bank gravel;run-of-mine gypsum
石膏罩面灰泥 gypsum ga(u)ging plaster
石膏珍珠岩灰浆 gypsum perlite plaster
石膏珍珠岩灰泥 gypsum perlite plaster
石膏珍珠岩墙板 gypsum perlite wall board
石膏真空处理机 vacuum gypsum treatment machine
石膏植物 gypsum plant
石膏纸板抹灰底 rocking lath(ing)
石膏指示植物 indicator plant of gypsum
石膏制品 gypsum article; gypsum product;plastering
石膏制品的模具 section mo(u)ld
石膏制酸法 anhydrite process
石膏制造硫酸(和水泥)的方法 gypsum-sulfuric acid process
石膏制造石 protean stone
石膏(质人造)石 protean stone
石膏蛭石灰浆 gypsum-vermiculite plaster;vermiculite gypsum plaster
石膏注件 plaster cast
石膏砖 gypsum brick
石膏装防火隔声板 calcium sulphate incrustation
石膏装饰板 calcium sulphate incrustant; fresco; gypsum plasterboard
石膏状的 gypseous
石膏状灰泥 gypseous marl
石膏状坚石 gypseous solid rock
石镐 stone pick
石戈壁植被 rock pavement vegetation
石格栅 stone grille
石格子窗 stone grille
石隔板 stone lining
石工 block chopper; brancher; lapicide; mason; rock header; rock works;scabbler;square man;squarer; stone cutter; stone dresser; stoneman;stone mason
石工保存覆盖层 preservative coating for stone work
石工标记 banker mark;mason's mark; stonemason's mark
石工标志 mason mark
石工厂 stonework
石工衬砌 stone masonry lining
石工程 masonry
石工齿凿 claw tool
石工初刻作业 broached work
石工锤 axhammer; cavel; pane hammer;stone hammer; stonemason's hammer;stone sledge
石工粗刻 broach
石工的 masonic
石工短锤 mash hammer
石工肺 chalicosis
石工斧 ax(e)[复 axes];axhammer; jedding axe;kevil
石工镐 quarryman's pick;quarry pick
石工工长 master mason
石工工具 stone cutter's tool
石工尖斧锤 kevel
石工锯 helicoidal saw; stonemason's saw
石工刻线槽 broached work
石工宽凿 broach
石工连接件 masonry fixing
石工腻子 mason's putty
石工平凿 drove chisel
石工手锤 stone cutter's hand hammer
石工小锤 acisculis;scutch
石工小屋 mason lodge; stonemason's lodge
石工斜角缝 mason's stop
石工修饰工作 dragged work
石工修整 rock work dressing
石工用斧子 zax
石工用锯 masonry saw
石工用宽凿 tooler
石工用凿刀 stonecutter's chisel
石工凿(刀) stonecutter's chisel; drove;waster
石工作台 stone bench
石拱 rock arch;stone arch
石拱的锁石 sagitta
石拱顶 stone vault
石拱顶棚 stone-vaulted ceiling
石拱廊 stone arcade
石拱门 rock arch door
石拱桥 masonry arch(ed) bridge; rock arch bridge;stone arch bridge
石拱桥拱顶石 keystone at the crown of stone bridge
石拱桥木拱架 timber centering for stone arch bridge
石拱圈石块间的铁榫 iron key to connect the stone blocks and arch ring
石拱圬工 rock arching;stone arching
石拱座 stone shoulder
石沟 bergfall furrow
石构交叉拱(肋) stone built groyne
石骨架结构 stone skeleton construction
石鼓丘 rock drumlin
石棺 sarco(phagus);stone coffin
石棺上的神坛 altar over sarcophagus
石棺上的祭坛 pulpit over sarcophagus
石辊 cylindric(al) stone;stone roll(er)
石辊式磨粉机 drag-stone mill
石滚 rock burst
石滚磨 drag-stone mill
石滚筒 stone roll(er)
石椁坟墓 mastabah tomb
石过梁 lintel stone;stone lintel
石海 block field; rock block field; stone field
石海绵目 lithistida
石涵(洞) masonry-stone culvert; stone culvert
石河 rock river; rock stream; stone-river
石荷叶 cave lotus leaf
石横档 <门窗> stone transom
石弧线 stone curving
石护墙 <用来保护海塘免遭海水冲刷> stone apron
石花 cave flower
石花菜 agar
石花台 stone flower bed
石华 onyx
石华大理石 onyx marble
石化 petrify
石化产品 petrochemicals
石化的 petrochemical
石化的树林 fossil forest
石化废水 petrochemical wastewater; wastewater from petrochemical
石化废水处理 petrochemical wastewater treatment
石化废水回用 petrochemical wastewater reuse
石化工业 petrochemical industry
石化工业废水 petrochemical industry waste(water); wastewater from petrochemical industry
石化公司 petrochemical corporation
石化木 lithoxyl(e);lithoxylite;petrified wood;woodstone
石化松香 fossil resin
石化土 petrified soil
石化液 petrifying liquid
石化有机废水 petrochemical industry organic wastewater
石化作用 lithification;petrification
石划 scutch
石环 stone circle;stone ring
石灰 calcium oxide; calx[复 calces/calxes]; caustic lime; kalk; lime; stone lime
石灰白 lime white
石灰斑 lime blast; lime stain; liming stain
石灰板天花板 dry ceiling
石灰拌和机 lime mixer
石灰包 <炸药> lime cartridge
石灰包裹体 concretions of limestone
石灰饱和比 lime saturation ratio
石灰饱和的 lime-saturated
石灰饱和度 degree of lime saturation;lime saturation
石灰饱和器 lime saturator
石灰饱和系数 lime saturation coefficient;lime saturation factor
石灰饱和值 lime saturation value
石灰爆 lime pops
石灰焙烧 lime roasting
石灰崩落开采法 lime caving
石灰比(例) lime ratio
石灰标线 chalk line marking
石灰标准值 <控制水泥适当石灰含量的系数之一> lime standard value
石灰饼 lime cake
石灰玻璃 lime(-silicate) glass
石灰采石场 lime quarry
石灰仓 lime bin
石灰槽 lime trough;liming tank
石灰测定器 calcimeter
石灰产浆量 yield of lime
石灰厂 lime plant
石灰场 lime yard
石灰场废水 lime yard waste(water)
石灰沉淀法 lime precipitation
石灰沉积 adarce
石灰沉积物 <在盆地边缘> rimstone
石灰池 banker;lime banker;lime pit
石灰虫 Spirorbis
石灰储存槽 lime silo
石灰处理 lime treatment;liming;lime coating <钢丝的>
石灰处理的 limed
石灰处理的骨料 lime-treated aggregate
石灰处理的集料 lime-treated aggregate
石灰处理的泥浆 lime-treated mud
石灰处理的松香 calcium resinate
石灰处理下水道污泥 lime-treated sewage sludge
石灰纯碱软化法 lime-sodium carbonate softening method

石灰粗骨料混凝土 lime coarse aggregate concrete
石灰粗集料混凝土 lime coarse aggregate concrete
石灰打底的 lime based
石灰袋 lime bag
石灰氮 lime nitrogen
石灰的 calcic
石灰的残效 residual value of lime
石灰的烧制 burning of lime
石灰的熟化 slake of lime
石灰的水化物 hydrate of lime
石灰的未消化残渣含量试验 unslaked residue test of lime
石灰的相容性 lime compatibility
石灰的消化作用 lime hydration
石灰的亚硫酸氢盐 bisulfite of lime
石灰的氧化镁含量试验 magnesium oxide test of lime
石灰的有效氧化钙含量试验 effective calcium oxide content test of lime
石灰的有效氧化钙和氧化镁含量试验 effective CaO and MgO content test of lime
石灰的种类和等级 type and grade of lime
石灰灯 lime light
石灰煅烧 lime burning
石灰煅烧过度的 hard-burned
石灰(煅烧)窑 lime kiln
石灰堆 lime pile
石灰(堆)场废物 lime yard waste
石灰二次煅烧 lime recalcining
石灰法 lime base process; lime process
石灰矾土水泥 lime aluminous cement
石灰防锈涂层 lime rust-coating
石灰分解 lime disintegration
石灰粉 calcareous slack; flour lime (stone); lime ash; limestone powder; powder (ed) lime; pulverised [pulverized] lime; selected lump lime; lime powder
石灰粉尘 lime dust
石灰粉炼加固法 lime-flyash stabilization
石灰粉煤灰 lime and fly-ash
石灰粉煤灰混合料 <俗称二灰> lime-flyash mixture
石灰粉煤灰混凝土 <含粗集料> lime-flyash concrete
石灰粉煤灰矿渣 lime and fly-ash slag
石灰粉煤灰沙砾基层 lime-flyash-crushed sand gravel base(course)
石灰粉煤灰水泥 lime-flyash cement
石灰粉煤灰碎石 lime-flyash treated broken stone
石灰粉煤灰碎石基层 lime-flayash-crushed stone base(course)
石灰粉煤灰土 lime and fly-ash soil; soil-lime flyash
石灰粉煤灰土基层 lime-flyash soil base
石灰粉煤灰土结碎石路 lime-flyash-soil bound macadam
石灰粉煤灰稳定法 lime-flyash stabilization
石灰粉末 lime powder
石灰粉喷射 lime injection
石灰粉砂岩 calcisiltite
石灰粉饰 lime plaster(ing)
石灰粉刷 lime paste; lime plastering; lime wash; lime whiting; plaster lime
石灰粉刷爆裂 pitting of plaster
石灰粉刷砂浆 lime plaster
石灰风化 lime disintegration; lime efflorescence
石灰改良土壤 lime-improved soil
石灰改善(土)层 lime-modified layer
石灰改善土壤 lime-modified soil
石灰坩埚 lime crucible
石灰膏 cream of lime; lime paste; lime plaster; lime putty
石灰膏抹灰 hard finish
石灰膏乳化沥青 emulsified asphalt with lime paste; lime plaster emulsified asphalt
石灰膏涂层 setting coat (plaster); skimming coat
石灰膏罩面层 putty coat
石灰隔热抹灰层 lime insulating plaster
石灰垢 lime scale
石灰固化 lime solidification
石灰灌浆 lime grout(ing)
石灰光 lime light
石灰硅石 lime dinas
石灰硅酸盐石头 calcareous petrosilex
石灰硅酸盐水泥 Portland limestone cement
石灰过多的水泥 over-limed cement
石灰过剩 excess of lime
石灰含量 lime status
石灰含量测定 lime content determination
石灰(含量)测定器 calcimeter
石灰含量过多的水泥 over-limed cement
石灰含砂量 sand-carrying capacity of lime
石灰和水泥撒料器 lime and cement spreader
石灰核 <老冲积层下> kalar
石灰红 lime red
石灰花 lime bloom
石灰华 adarce; calcareous sinter; calcareous tufa; calcarious sinter; calcareous tufa; calc-sinter; calc-tufa; sinter; tiburitine; travertine; tufa; tuff
石灰华帷幕 draperies; drip curtain; drip drapery
石灰华柱 travertine column
石灰化 calcify(ing)
石灰化合能力 lime combing capacity
石灰化作用 calcification
石灰还原浴 lime vat
石灰黄 lime yellow
石灰灰掺和料 pozzolan(a) admixture
石灰灰泥 lime plaster
石灰回收 lime recovery
石灰回转炉 rotary lime kiln
石灰混合器 prelimer
石灰混凝法 lime coagulation
石灰混凝土 <石灰、砂、砾石混合物> lime concrete
石灰活化性 lime reactivity
石灰活性氧化钙 active calcium oxide of lime
石灰火山(灰)混合料 lime pozzolan
石灰火山灰浆 lime trass mortar
石灰火山灰稳定 lime-pozzolan stabilization
石灰火山灰稳定土法 lime-pozzolan stabilization
石灰火山灰质混合物 lime-pozzolan mix
石灰火山灰质混凝土 lime-pozzolan concrete
石灰火山灰质水泥 lime pozzolanic cement
石灰或水泥稳定灰渣 lime or cement stabilized ash
石灰基底 lime base
石灰基底层 lime basecoat
石灰基底油脂 lime base grease
石灰基恩水泥 <高温焙烧白石膏粉与石灰膏拌成的面层粉刷材料> lime-Keene's cement
石灰及苏打粉软化水方法 lime and soda-ash process
石灰激发 lime stimulation
石灰极限含量 maximum lime content
石灰剂量 lime dosage
石灰加固 lime treatment
石灰加固法 lime stabilization
石灰加固土 lime-stabilized soil
石灰加固效果 effectiveness of lime treatment
石灰加料器 lime feeder
石灰间 lime yard
石灰碱石头 calc-alkali rock
石灰碱釉 lime alkali glaze
石灰浆 calcimine; caustic mud; caustic sludge; cream of lime; ga(u)ged stuff; lime cream; lime milk; lime slurry; lime white; milk of lime
石灰浆底 kalsomine
石灰浆粉刷层 milk of lime coat
石灰浆缓凝剂 keratin
石灰浆面层 milk of lime coat
石灰浆黏[粘]结能力 lime binding capacity
石灰浆墙面变色块 butterfly
石灰浆刷 lime brush
石灰浆投加量 milk of lime feeder
石灰浆涂层 milk of lime coat
石灰浆涂刷 lime wash
石灰胶 lime paste
石灰胶结多孔混凝土 lime-bound aerated concrete
石灰胶结料 lime cement
石灰角页岩 calciferous petrosilex
石灰脚病 scaly leg
石灰搅拌机 lime agitator; lime mixer
石灰搅拌器 lime agitator; lime mixer; lime raker; lime stirrer
石灰结合的 lime-bound
石灰结合的硅质耐火材料 lime bonded silica refractory
石灰结合的硅砖 lime bonded silica brick
石灰结合的建筑构件 lime-bound building component
石灰结核 lime nodule
石灰结节 lime knot
石灰结碎石 lime bound macadam
石灰结碎石路 lime bound macadam
石灰浸物质 liming material
石灰精陶 lime earthenware
石灰净化 lime purification; lime purifying
石灰拘酸蒸馏器 lime lee still
石灰聚合松香 limed poly-pale
石灰开采漏斗 lime caving bin
石灰苛性沉淀 lime caustic precipitation
石灰坑 cloup; dolina[doline]; lime pit; pit lime; swallow hole
石灰库 lime bin; lime storage
石灰块 lump lime
石灰快凝的 fast-setting to lime
石灰矿 lime quarry
石灰矿渣水泥 lightweight slag cement; lime-slag cement
石灰蓝 lime blue
石灰类别 sort of lime
石灰类黏[粘]结料 calcareous cement
石灰累积作用 lime accumulation
石灰立窑 vertical lime kiln
石灰砾石 lime gravel
石灰砾石粉煤灰混合料 gravel-flyash-lime mix
石灰砾石凝灰岩混合料 gravel-tuff-lime mix
石灰硫黄合剂 lime sulphur(solution)
石灰硫酸铜液 Bordeaux mixture
石灰(滤)饼废物 lime cake waste
石灰滤渣 lime cake waste
石灰滤渣废水 lime cake wastewater
石灰绿 lime green
石灰落水洞 lime sink
石灰煤矸石水泥 lime-spoil cement
石灰煤渣 lime and breeze; lime cinder
石灰煤渣灰浆 black ash mortar; black mortar
石灰煤渣混合料 lime-cinder mixture
石灰煤渣矿渣 lime and breeze slag
石灰镁氧比例 lime-magnesia ratio
石灰密封油背 lime putty
石灰面 powder lime
石灰模数 lime modulus
石灰抹面层 lime finish coat
石灰耐火材料 lime refractory
石灰泥 lime slurry
石灰泥处理 lime mud disposal
石灰泥灰岩 lime marl(ite)
石灰泥浆 caustic mud; lime mud; milk of lime
石灰泥岩 calcilutite
石灰腻子 lime putty
石灰黏[粘]合 binding of lime
石灰黏[粘]合料 calcareous cement
石灰黏[粘]结的砂岩 lime cemented sandstone
石灰黏[粘]结料 binding of lime
石灰黏[粘]结碎石(路) lime-bound macadam
石灰黏[粘]泥 calcareous mud
石灰黏[粘]土 lime clay
石灰碾碎机 lime pulverizer
石灰凝结沉淀 lime coagulation sedimentation
石灰凝结法 lime coagulation
石灰耙 lime raker
石灰盘 calcareous pan
石灰磐 lime pan
石灰泡砂岩 lime sand rock
石灰配比 cement-aggregate compatibility
石灰配料 raw meal prepared from lime
石灰片岩 limestone schist
石灰漂液 lime bleach liquor
石灰品种 sort of lime
石灰气纯化 lime gas purification
石灰气化 lime boil
石灰砌合的 limed
石灰侵蚀的 lime aggressive
石灰侵蚀性 lime aggressivity
石灰溶液喷嘴 lime wash nozzle
石灰融合分析法 lime-fusion method; lime ignition method
石灰鞣革 limed hide
石灰乳 lime milk; lime weeping <混凝土裂缝处分泌出的>; milk of lime
石灰乳槽 liming still
石灰乳室 liming chamber
石灰乳液 cream of lime; lime milk; milk of lime
石灰乳浊液 cream of lime; lime cream; lime emulsion
石灰软化法 lime softening; softening by lime
石灰软化设备 lime softening plant; lime water softening plant
石灰软水厂 lime water softening plant
石灰软水设备 lime water softening plant
石灰软水试验 lime water softening test
石灰软水装置 lime softening plant; lime water softening plant
石灰撒布机 bulk lime spreader; lime distributor; lime sower; lime spreader
石灰撒施机 lime sower; lime spreader

S

石灰三合土 lime concrete;lime sand blocked brick concrete; lime sand-broken brick concrete
石灰砂浆 calcium plaster; lime mortar;lime plaster;statumen <古罗马用于铺路的>
石灰砂浆地板 lime mortar flooring
石灰砂浆粉刷 lime mortar plastering
石灰砂浆抹面二遍作法 two-coat lime plaster
石灰砂砌面砖 calcium silicate facing brick
石灰砂砂浆 lime sand mortar
石灰砂石 lime sandstone
石灰砂岩 lime sandstone
石灰砂岩碎片 arenaceous limestone chip(ping)s
石灰砂岩屑 arenaceous limestone chip(ping)s
石灰砂砖 lime sand brick
石灰砂桩 lime pile with sand; lime sand pile
石灰筛 lime screen
石灰烧黏[粘]土水泥 lime burnt clay cement
石灰深层搅拌法 deep lime mixing method
石灰渗坑 lime sink
石灰生物处理 lime biologic(al) treatment
石灰生物污泥 lime biologic(al) sludge
石灰施用 lime application
石灰石 chalkstone; dolomite; flux lime; fossil rock; lime rock; limestone;raw limestone
石灰石柏油碎石路 limestone tar macadam
石灰石板 balatte
石灰石采石场 limestone quarry
石灰石采石工 limestone quarry operator
石灰石仓库 limestone storage; limestone store
石灰石掺和料 limestone addition
石灰石厂 limestone plant
石灰石沉积 limestone deposit
石灰石粗骨料混凝土 limestone coarse aggregate concrete
石灰石粗集料混凝土 limestone coarse aggregate concrete
石灰石大理石 limestone marble
石灰石道砟 limestone ballast
石灰石的处理 milling of limestone
石灰石的粉磨 milling of limestone
石灰石的需要 limestone needs
石灰石地层 limestone stratum [复 strata]
石灰石法 limestone-based process
石灰石反应器 limestone reactor
石灰石粉 agstone; calcite flour; ground limestone; limestone flour; limestone powder
石灰石粉末 limestone dust;powdered limestone
石灰石粉碎机 lime mill
石灰石膏工艺 lime gypsum process
石灰石膏灰浆 lime gypsum mortar
石灰石骨料 limestone aggregate
石灰石褐土 limestone brown loam
石灰石红土 limestone red earth
石灰/石灰石法烟气脱硫 flue gas desulfurization with lime and/or limestone
石灰石混合料 limestone addition
石灰石集料 limestone aggregate
石灰石计量喂料机 limestone weighing feeder
石灰石纪念碑 monument of limestone
石灰石块 limestone block
石灰石矿 limestone quarry
石灰石矿粉 limestone filler
石灰石玛琋脂 limestone mastic
石灰石凝块 limestone concretion
石灰石喷入法脱硫 desulfurization by limestone injection process
石灰石破碎机 limestone breaker; limestone crusher
石灰石破碎设备 limestone crushing plant
石灰石千枚岩 limestone phyllite
石灰石清洗法 limestone scrubbing (process)
石灰石清洗机 limestone washer
石灰石清洗污泥 limestone scrubber sludge
石灰石湿式洗涤器 limestone wet scrubber
石灰石塔 lime rock tower
石灰石土松土机 limestone ripper
石灰石外形 calcareous facies
石灰石圬工 limestone masonry(work)
石灰石吸收塔 limestone tower
石灰石相 limestone facies
石灰石屑 chicken grit;limestone chip(ping)s;race
石灰石屑填料 limestone filler
石灰石岩 limestone rock
石灰石岩屑 limestone debris
石灰石页岩 limestone slate
石灰石油灰 limestone putty
石灰石中和 limestone neutralization
石灰石中和处理 limestone neutralization treatment
石灰石助熔剂 limestone flux
石灰熟化 lime conditioning; lime slaking
石灰熟化机 lime hydrating machine
石灰熟化器 lime slaker; liming apparatus
石灰熟化时间 time of slaking
石灰竖窑 lime shaft kiln
石灰刷 lime plaster
石灰刷白 lime (white) wash; lime whiting
石灰刷浆 lime wash
石灰水 aqua calcis;calcareous water; lime wash;lime water;whitewash
石灰水淬渣(混合料) lime granulated-slag(mixture)
石灰水粉刷 lime whitewash
石灰水化 liming
石灰水化法 wet slaking process
石灰水化机 lime hydrating machine
石灰水化热 hydration heat of lime
石灰水浆涂料 lime washing
石灰水泥 grappler cement; lime cement
石灰水泥粉煤灰混合料 <用作路面的基层> lime cement flyash mix
石灰水泥粉煤灰混合物 lime cement flyash mix
石灰水泥粉煤灰(三合)混凝土 lime cement flyash concrete
石灰水泥粉刷 lime cement mixed plaster
石灰水泥灰浆抹面 lime cement finish
石灰水泥灰泥 cement temper
石灰水泥混凝土 cement lime concrete
石灰水泥拉毛粉饰 lime-cement stucco
石灰水泥面层 lime cement finish
石灰水泥砂浆 compo mortar; lime and cement mortar
石灰水泥柱 lime-cement column
石灰水泥桩 lime-cement pile
石灰水煮练 lime boil
石灰松香 hardened rosin;limed rosin
石灰松香清漆 gloss oil; limed rosin varnish
石灰苏打处理 lime soda treatment
石灰苏打法 lime and soda process;lime soda method;lime soda process
石灰苏打灰软化法 lime soda ash softening
石灰苏打软化 lime soda softening
石灰苏打软化法 lime soda softening method;lime soda softening process
石灰苏打软水法 lime soda softening method
石灰碎砾岩 lime rubble rock
石灰碎裂 lime disintegration
石灰碎砖三合土 lime sand-broken brick concrete
石灰碳酸钠软化法 lime-sodium carbonate softening method
石灰陶渣灰泥 lime chamotte mortar
石灰陶渣砂浆 lime chamotte mortar
石灰添加量 lime addition
石灰添加料 lime addition
石灰筒仓 lime silo
石灰投量 dose of lime;lime dose
石灰涂层 lime coat(ing);lime plaster
石灰涂料 calcicoater;lime paint
石灰土 calcareous soil;lime clay;lime (stone) soil;soil-lime
石灰土(或水泥)搅拌机 temperer
石灰土基层 lime-soil base(course)
石灰土路面 lime-soil pavement
石灰土稳定法 lime-soil stabilization
石灰土植物 limestone plant
石灰脱碳酸化作用 lime decarbonization
石灰洼地 calc-pan
石灰外表面抹灰泥 lime external plaster
石灰外粉刷 lime external rendering
石灰位测定降水酸碱度的标准 lime potential
石灰位减少表示酸性增加 lime potential
石灰稳定的 lime-stabilized
石灰稳定法 lime stabilization
石灰稳定骨料 lime-treated aggregate
石灰稳定灰渣 lime stabilized ash
石灰稳定集料 lime-treated aggregate
石灰稳定砂土路面 lime-stabilized sand pavement
石灰稳定土 lime stabilization; lime-stabilized soil; soil stabilization with lime
石灰稳定(土壤)法 stabilization with lime
石灰污泥 lime sludge
石灰污泥处理 lime sludge treatment
石灰污泥处置 lime sludge disposal
石灰污泥中的镁 magnesium in lime sludge
石灰无光釉 lime matt
石灰误差法 lime deviation method
石灰系数 lime factor
石灰细度 fineness of lime
石灰相 calcareous facies
石灰消和池 pudding bin
石灰消化 liming;lime slaking
石灰消化槽 lime slaking tank; lime slaking trough
石灰消化池 banker
石灰消化分离器 lime classifier
石灰消化鼓 drum slaker
石灰消化机 lime hydrating machine; lime slaking machine
石灰消化坑 lime slaking pit
石灰消化器 lime slaker
石灰消化速度 slaking rate of quick lime
石灰消化筒 lime slaking drum
石灰消化作用 lime slaking
石灰消解器 lime slacker
石灰性 calcareous
石灰性冲积土 calcareous alluvial soil
石灰性钙 calcareous earth
石灰性红土 red calcareous soil
石灰性黏[粘]土 calcareous clay
石灰性土(壤) calcareous soil; limy soil;terra calcis
石灰性土壤表层施用 surface applicants to calcareous soil
石灰需要量 lime requirement
石灰悬浮体 suspension of lime
石灰悬浮液 lime suspension
石灰悬浊液 lime suspension
石灰循环系统 lime recycling system
石灰压碎机 lime crusher
石灰亚硫酸氢盐 bisulfite of lime
石灰亚硫酸盐 bistagite of lime
石灰岩 chalkstone; lime rock; Portland stone
石灰岩参差面 clint
石灰岩测井 limestone log
石灰岩大理石 limestone marble
石灰岩地沥青 <一种天然岩沥青> limestone rock asphalt
石灰岩地形 limestone terrain
石灰岩电极系 limestone sonde
石灰岩洞 limestone cavern
石灰岩发育的土壤 limestone soil
石灰岩沟 lapie
石灰岩和铁矿石等矿床的上部地层 weald-clay
石灰岩夹渣 concretions of limestone
石灰岩建造 limestone formation
石灰岩结核 limestone concretion
石灰岩矿床 limestone deposit
石灰岩类型 limestone type
石灰岩砾岩 limestone conglomerate
石灰岩裂隙地 fissure lime stone land
石灰岩落水洞 limestone sink
石灰岩密度 density of limestone
石灰岩盆地 cockpit
石灰岩溶洞 carst; carst river; limestone cave
石灰岩砂 limestone sand
石灰岩渗坑 limestone sink
石灰岩生产者 lime producer
石灰岩石 Bath stone
石灰岩水泥 dolomite cement
石灰岩松石机 limestone ripper
石灰岩松土机 limestone ripper
石灰岩碎石 limestone rubble
石灰岩岩脊 clint
石灰岩岩溶 calcareous karst
石灰岩质 limestone
石灰岩砖 limestone brick
石灰盐 lime salt
石灰氧化硅水系统 lime silica-water system
石灰氧化期 boiling period
石灰窑 limeburner;lime burning kiln; limestone kiln
石灰窑厂 lime burning plant
石灰窑气 lime kiln gas
石灰液 lime liquor
石灰因素 lime factor
石灰引起的起霜现象 lime efflorescence
石灰硬磐 lime(hard)pan
石灰油漆 lime paint
石灰釉 calcareous glaze;lime glaze
石灰浴 lime bath
石灰运输设备 lime-handling equipment
石灰载体 <如水泥熟料等> lime carrier
石灰再(煅)烧 lime recalcining
石灰在空气中熟化的 air slaked
石灰藻 calcareous algae
石灰皂 lime soap
石灰造成的 calcigenous
石灰渣 grappier;lime cinder;lime mud
石灰渣处理 lime mud disposal

石灰渣水泥 grappier cement
石灰胀裂 < 建筑砖缺陷 > lime blowing;popping
石灰脂 lime base grease;lime grease
石灰制品 lime product
石灰质 calc
石灰质白坯陶瓷 calcareous whiteware
石灰质板岩 lime slate
石灰质材料 calcareous material
石灰质沉积物 lime deposit
石灰质沉凝灰岩 calcareous tuff
石灰质粗砂岩 calcareous grit
石灰质大理石 calcitic marble
石灰质的 calcareous;calcitic;limy
石灰质骨料 calcareous aggregate
石灰质含量 lime content
石灰质集料 calcareous aggregate
石灰质结核 lime concretion
石灰质结块 limy concretion
石灰质结硬 concretion of lime
石灰质精陶 calcareous whiteware
石灰质精陶器 lime earthenware
石灰质矿渣 calcareous slag
石灰质砾石 calcareous gravel; limestone gravel
石灰质卵石砾岩 limestone pebble conglomerate
石灰质泥灰岩 calcareous marl(ite)
石灰质黏[粘]板岩 calcareous clay slate
石灰质黏[粘]土 calcareous clay;limy clay
石灰质黏[粘]土板岩 calcareous clay slate
石灰质凝灰岩碎屑 clastic lime tuff
石灰质凝结物 limy concretion
石灰质泉 calcareous spring
石灰质砂 calcareous sand;malm;lime sand
石灰质砂岩 calcareous sandstone; limy sandstone;lime sandstone
石灰质石膏 calcareous alabaster
石灰质石英 calcareous silex
石灰质松石 travertine
石灰质燧石 calcareous silex
石灰质陶器 calcareous earthenware
石灰质陶器材料 calcareous earthenware type material
石灰质铁石 calcareous iron-stone
石灰质铁质结核 lime iron-concretion
石灰质土 calcareous earth;calcareous soil;puttsand;putty sand
石灰质土壤 limy soil
石灰质岩 calcareous rock
石灰质岩石 calcareous stone
石灰质页岩 calcareous shale
石灰质淤泥 calcareous ooze
石灰质藻 calcareous alga
石灰中和处理 neutralization with lime
石灰柱 lime column;lime pile;quicklime pile
石灰砖 lime brick
石灰桩 lime column; lime pile; lime stake;quicklime pile
石灰桩法 lime column method
石混凝土 ballast concrete
石基础 stone foundation
石级 stone steps
石级风化 kilt
石级间搭接扣 pien(d) check
石脊 ledge
石祭坛 stone altar
石祭坛块件 stone altar piece
石祭坛屏障 stone altar reredos;stone altar retable;stone altar screen
石架间 aft-bay
石尖顶 stone spire
石尖塔 stone spire
石建筑 stone construction
石建筑凸出部分 stone shoulder
石建筑突额 stone shoulder
石讲堂 < 教堂中 > stone pulpit
石匠 brancher; rock header; stone dresser;stoneman;stone mason
石匠锤 club hammer
石匠的平凿 drove;drove chisel
石匠工场 stonework
石匠型板 mason's mo(u)ld
石焦油 rock tar
石铰接 stone hinge
石阶 perron;sarn;stone steps
石阶码头 gha(u)t
石阶踏步背榫 back joint
石阶梯 stone stair(case)
石街 stone row
石节点 stone hinge
石结构 stone construction; stone structure;stonework
石结构房屋 stone structure house
石结砾石路 stone-bound gravel road
石界标 stone bound
石精整机 stone-dressing machine
石井 stone shaft
石井栏 stone curb
石矩形涵洞 stone box culvert
石锯 grub saw;rock saw;stone saw
石坎 cross wall
石刻 lithoglyph; stone carving; stone inscription
石坑 stone pit
石坑水 quarry water
石坑岩层含水 quarry sap
石孔隙 pores of stone
石控制台 < 教堂中 > stone pulpit
石扣 stone cramp
石窟 cave(rn);cave temple;dolmen; grotto; rock cave; rock-cut building;rock temple
石窟建筑 rock architecture
石窟陵墓 rock-cut tomb
石窟庙宇 rock-cut temple
石窟墓 rock sepulchre
石窟墓室 < 古埃及 > speos
石窟神庙 speos
石窟圣堂 rock sanctuary
石窟寺 rock-cut temple; rock-hewn temple
石窟寺院 chaitya cave; rock-hewn monastery
石库门 stone portal
石块 block of rock; block of stone; chump;gobbet;quarry stone;quarry waste;rock chunk;rock lump; stone; stone block; stone mass; stone cut;ground apples < 俚语 >
石块爆破 blistering on the stone
石块崩落 displacement of rock
石块表面的凹陷 lewis hole
石块衬砌 stone masonry lining
石块齿凿面 tooth chisel finish of stone
石块粗加工 knobbling
石块打毛机 sett roughening machine
石块大小 stone size
石块的搬运 rock handling
石块的处理 rock handling
石块吊升夹具 stone lifting tongs
石块防波堤 block breakwater;quarrying rock breakwater
石块分离机 stone cleaner;stone eliminator;stone extractor;stone separator
石块分离器 stone retarder
石块分选机 stone-separating mill
石块基层 stone bedding
石块间接缝 abre(a)uvior
石块间细缝 abre(a)uvior
石块间隙缝 < 即拱石或石砌体缝 > abre(a)uvior
石块接缝填料 sett joint sealing
石块流动 rock flowage
石块路 sett-paved road
石块路基边缘 stone shoulder
石块路面 pavement of stone blocks; sett paving;stone block pavement; stone pavement
石块路面用玛琋脂填缝料 sett paving mastic joint sealer
石块面层的岸壁 quay wall with stone facing
石块面层的岸墙 quay wall with stone facing
石块排除器 stone releaser
石块铺底 bottoming; pitching; stone bottoming
石块铺路面 stone block pavement; stone block paving
石块铺面 pavement of stone blocks; paving in setts; paving in stone blocks;stone block pavement;stone pavement
石块铺面工程 sett paving works
石块铺面工作 sett paving work
石块铺面夯实机 sett paving rammer; sett paving tamper
石块铺面振动器 sett paving vibrator
石块铺砌 block pavement; bottoming; paving in setts; penning; set block paving;sett paving
石块铺砌路面 stone block pavement
石块铺砌路面的砾石 gravel for sett pavings
石块铺砌面连接 jointing of sett paving
石块切割机 block chopper
石块清除器 rock ejector
石块试验机 sett tester
石块受压强度 block compression strength
石块输送吊斗 sett-handling bucket
石块填充料 pierrotage
石块铁夹 < 砌筑用的 > metal cramp
石块围堰 rock dike[dyke] cofferdam
石块消除器 rock ejector
石块牙砌 tusses
石块缘琢 margin draft
石块制作 stone blockwork
石块坠落 falls of stone
石笸 gabion;rock crib
石笸垫层 < 保护桥梁免遭水流冲刷用 > gabion mat
石矿 quarry
石矿层 quarry bed
石矿体 stone deposit(e)
石腊油 petrolatum
石蜡 ceresin(e) wax;mineral wax;ozocerite[ozokerite]; paraffin(e); paraffin(e) wax; petroleum wax; wax;earth wax
石蜡沉积 deposition of paraffin
石蜡打底 paraffin(e) embedding
石蜡的 paraffinaceous
石蜡的熔点 melting point of paraffin(e) wax
石蜡底子 paraffin(e) base
石蜡电容器 paraffin(e) condenser
石蜡二甲苯溶液 paraffin(e) xylol
石蜡发动机 paraffin(e) motor
石蜡发汗 paraffin(e) sweating
石蜡防射线屏蔽墙 paraffin(e) radiation shielding wall
石蜡防水法 paraffin(e) waterproofing
石蜡防水帆布 paraffin(e) duck
石蜡分馏 wax fractionation
石蜡封盖 paraffin(e) coating
石蜡敷糊 pliable paraffin
石蜡垢 paraffin(e) dirt
石蜡刮削器 paraffin(e) scraper
石蜡含量 paraffin(e) content;paraffinicity; paraffinicity curing compound
石蜡和油乳液 paraffin(e) wax and oil emulsion
石蜡环烷型石油 paraffin(e) naphthenic oil
石蜡基 paraffin(e) base
石蜡基混凝土养护合成物 paraffin(e)-base concrete curing compound
石蜡基混凝土养护剂 paraffin(e)-base concrete curing compound
石蜡基沥青 paraffin(e)-base asphalt
石蜡基沥青混合基原油 paraffin(e)-asphalt petroleum
石蜡基润滑油 paraffin(e)-base lubricating oil
石蜡基石油 paraffin(e)-base oil;paraffin(e)-base petroleum; paraffinic-base(crude) petroleum
石蜡基石油精 < 一种溶剂 > paraffin (e) naphtha
石蜡基原油 paraffin(e)-base crude oil;paraffin(e) crude oil;paraffinic-base crude(oil);paraffinic crude
石蜡浸透探伤 paraffin(e) test
石蜡浸纸 wax-impregnated paper
石蜡浸注材 paraffin(e)-impregnated wood
石蜡浸渍法 paraffin(e) immersion method
石蜡浸渍混凝土 wax impregnated concrete
石蜡绝缘线 paraffin(e) wire
石蜡块 paraffin(e) block;paraffin(e) mass
石蜡矿 paraffin(e) deposit
石蜡沥青基石油 paraffin(e)-asphalt petroleum
石蜡馏分 paraffin(e) distillate;paraffin(e) pressed distillate; pressed distillate
石蜡瘤 oleogranuloma;oleoma;paraffinoma
石蜡滤波器 paraffin(e) filter
石蜡氯化装置 paraffin(e) chlorination unit
石蜡煤 paraffin(e) coal
石蜡密封度试验 sealing strength test
石蜡切片 paraffin(e)(-cut) section
石蜡切片法 paraffin(e) method
石蜡醛 paraffin(e) aldehyde
石蜡燃料 parol
石蜡染色剂 paraffin(e) stain
石蜡溶液 paraffin(e) solution
石蜡熔点试验 paraffin(e) wax melting point test
石蜡乳剂 paraffin(e) emulsion
石蜡乳液 paraffin(e) solution
石蜡乳液浸润剂 paraffin(e) emulsion sizer
石蜡乳状液 wax emulsion
石蜡软布 nujol mull
石蜡润滑 paraffin(e) lubrication
石蜡闪点测定仪 Abel tester
石蜡酸 paraffin(e) acid
石蜡烃的射解作用 radiolysis of alkanes
石蜡烃煤油 paraffin(e)
石蜡桶 slack barrel
石蜡涂层 paraffin(e) coating
石蜡脱脂 paraffin(e) degreasing
石蜡系 paraffin(e) series
石蜡系碳氢化合物 paraffin(e) hydrocarbon
石蜡系烃 paraffinic hydrocarbon
石蜡型石油 paraffinic oil
石蜡压滤机 paraffin(e) press
石蜡氧化 paraffin(e) wax oxidation
石蜡页岩 paraffin(e) shale

石蜡油 liquid paraffin;liquid petrolatum;oil of paraffin;paraffin(e) oil;paraffinic oil;paraffinum liquidus;white mineral oil
石蜡油膏 paraffin(e) jelly
石蜡浴 paraffin(e) bath;wax bath
石蜡皂 paraffin(e) soap
石蜡渣油 paraffin(e) flux
石蜡蒸馏液 wax oil
石蜡纸 paraffin(e) paper
石蜡质的 paraffinaceous
石蜡质量试验 paraffin(e) wax quality test
石蜡中间基石油 paraffin(e) intermediate crude
石蜡铸造 investment cast
石蜡族的 paraffinaceous;paraffinic
石蜡族酸 paraffinic acid
石蜡族烃 paraffin(e) hydrocarbon
石拦河堰 stone weir
石栏杆 stone balustrade;stone railing
石栏杆小柱 stone baluster;stone banister
石篮坝 basket dam
石勒板 <古建筑的> orthstat
石肋 stone rib
石肋拱顶 stone-ribbed vault
石肋拱桥 stone-ribbed arch bridge
石肋脚 <古典神庙下部的护墙石板或古建筑的> orthostat
石肋穹隆 stone-ribbed dome
石肋圆屋顶 stone-ribbed cupola
石类爆裂 lime popping
石篱 stone hedge
石理 vein
石沥青 asphaltite
石栗 belgaum walnut; candle berry tree;candlenut;Chinese varnish tree
石栗果油 anda-assu oil;nogueria oil
石栗子油 candlenut oil;lumbang oil
石砾 chad
石砾道砟 shingle ballast
石砾质河床 dry wash
石砾状杂层 <松软岩石中的> boulder-like inclusions
石粒喷涂板 aggregate-coated panel
石莲 <三叠纪的海百合>【地】encrinite
石莲灰岩 encrinal limestone
石梁 ledge(of rocks);rock bar;rock beam;stone beam
石梁桥 stone beam bridge
石梁托 stone corbel
石料 building stone;rag;rock material;stone;stone material;stony material
石料凹框 sunk draft
石料保持性 stone retention
石料表面加工 scutching
石料表面加工的 faced
石料表面修整工程 boasted work
石料采掘工作 rock excavation work
石料仓 stone bin
石料场地质平面图 geological plan of rock quarry
石料场地质剖面图 geological section of rock quarry
石料称量器 stone batcher
石料尺寸 stone size
石料冲击值 lashed stone value
石料次品 wasting
石料粗加工 angle dunting; drove; stugged;scapplingling
石料粗面的 quarry-faced
石料粗琢 boasting
石料打磨机 dresser
石料大件 stone bull
石料的初琢 scrappling
石料的短槽纹修琢 drove finish of stone
石料的磨光性 polishing characteristics
石料的天然含水量 quarry sap
石料等级 <路用> gradation of stone
石料(翻斗)车 stone skip
石料防腐剂 stone preservative;stone preserving agent
石料分级 stone classification; stone gradation
石料分级粉碎 stage crushing
石料封顶 stone capping
石料浮沉冲洗法 sink-float process
石料覆盖 stone cladding
石料供料定量器 stone batcher
石料供应列车 stone supply train
石料骨架 stone skeleton
石料裹覆试验 stone coating test
石料护岸 rock revetment
石料护面 stone surfacing; stone veneer(ing)
石料护坡的岛 pitched island
石料基层 stone packing
石料计量箱 stone batcher
石料加工锤 knapping hammer
石料加工砂 stone sand
石料加速磨光机 accelerated stone polishing tester
石料加速磨光仪 accelerated stone polishing tester
石料坚固性的试验 soundness test
石料结构 stone structure
石料锯 masonry crosscut saw; rock saw
石料开采 cutting of stones; quarrying;winning of natural stone
石料料堆 rock windrow
石料路面 stone pavement
石料磨光机 stone rubbing machine
石料磨光系数 polished stone coefficient
石料磨光值 <一种表示防滑性能的指标> polished stone value
石料磨耗试验 rattle test
石料磨损值 polished stone value
石料磨砖试验 rattle test
石料耐久性 rock durability
石料排水沟 stone drain
石料破碎 rock breaking;stone crushing
石料破碎厂 rock breaking plant
石料破碎船 rock breaker vessel
石料破碎机 reduction stone crusher
石料破碎筛分厂 combined crushing and screening plant
石料破碎设备 rock breaking plant
石料铺底 bottoming
石料铺砌路面 stone paving
石料砌成的房屋 stone built building
石料砌面 stone dressing;stone(sur) facing;stone veneer(ing)
石料砌面的 stone-faced
石料砌体上下层嵌接 tabled joint
石料褥层 stone mattress
石料撒布箱 spreader box
石料筛分厂 miner(al) separation plant;stone sizing plant
石料筛分装置 mineral separation plant
石料饰面 stone dressing;stone(sur) facing;stone veneer(ing)
石料顺层理面垂直放置并垂直于墙面的砌石法 edge bedding
石料顺层理面垂直放置并平行于墙面的砌石法 face bedding
石料顺层理面放置并平行于接缝的砌石法 joint bedding
石料摊铺机 aggregate paver; stone spreader
石料天然面 natural face of stone;quarry-faced stone;quarry face of stone
石料填筑的长堤 rock causeway
石料填筑的人行道 rock causeway
石料贴面 stone cladding
石料统一法 method of unifying the rock materials
石料挖斗 rock dipper
石料镶面 stone-pitched facing; stone (sur)facing;stone veneer(ing)
石料镶面工程 stone cladding work
石料修饰 tooled finish;tooling
石料修琢 dressing
石料压盖 stone capping
石料压碎强度 stone crushing strength
石料研磨机 stone grinder;stone rubbing machine
石料凿毛 stugging
石料凿磨量规 honing ga(u)ge
石料轧碎 stone granulating
石料逐级破碎 stage breaking (of stone);stage reduction(of stone)
石料抓斗 rock grapple
石料琢边富余量 drafted margin
石料琢面 stone dressing
石料自动磨光机 automatic stone polisher
石林 <一种岩溶景观> stone forest; rock forest;hoodoos
石流 rock flow; rock glacier; rock storm;rock stream;stone stream
石硫合剂 calcium polysulfide; lime sulfur
石榴【植】garnet;pomegranate
石榴变粒岩 garnet granulite
石榴长英麻粒岩 garnet felsic granulite
石榴虫胶 garnet lac
石榴二辉麻粒岩 pirigarnite
石榴橄榄岩 garnet peridotite
石榴硅卡岩 garnetiferous skarn
石榴红色 garnet
石榴铰链 <T形门铰> garnet hinge
石榴皮 granatum
石榴色紫胶 garnet lac
石榴闪辉麻粒岩 garnet amphibole pyroxene granulite
石榴石 carchedonius;carchedony
石榴石白云母石英片岩 garnet muscovite quartz schist
石榴石虫胶片 garnet(shel)lac
石榴石的 garnetiferous
石榴石二辉橄榄岩 garnet lherzolite
石榴石二云母片岩 garnet dimicaceous schist
石榴石二云母石英片岩 garnet two mica quartz schist
石榴石粉砂纸 garnet paper
石榴石橄榄岩 garnet peridotite
石榴石黑云母石英片岩 garnet biotite quartz schist
石榴石化合物 garnet compound
石榴石辉石岩 garnet pyroxenite
石榴石激光器 garnet laser
石榴石晶体 garnet crystal
石榴石绢云母千枚岩 garnet-sericite phyllite
石榴石绿帘石矽卡岩 garnet epidote sharn
石榴石磨料 garnet abrasive
石榴石片麻岩 garnet gneiss
石榴石千枚岩 garnet-phyllite
石榴石砂 garnet sand
石榴石砂纸 garnet paper
石榴石石英片岩 garnet-quartz schist
石榴石透辉石矽卡岩 garnet diopside sharn
石榴石矽卡岩 garnet sharn
石榴石型铁氧体 garnet type ferrite
石榴石型颜料 garnet type pigment
石榴石岩 garnet rock; garnetyte; granitite
石榴石云母片岩 garnet-mica schist
石榴石紫胶 garnet(shel)lac
石榴石棕 <红棕色> garnet brown
石榴树 pomegranate
石榴酸 punicic acid
石榴透辉钙长片麻岩 garnet diopside anorthite gneiss
石榴透辉角闪斜长麻粒岩 garnet diopside amphibole plagioclase granulite
石榴斜长片麻岩 garnet plagioclase-gneiss
石榴硬柱蓝闪片岩 garnet lawsonite glaucophane schist
石榴子石 garnet
石榴子石含量 garnet content
石榴子石化 garnetization
石榴子石矿床 garnet deposit
石榴子石岩 garnetite
石榴子岩 garnet rock
石榴紫苏透辉斜长麻粒岩 garnet hypersthene diopside plagioclase granulite
石榴紫苏斜长麻粒岩 garnet hypersthene plagioclase granulite
石龙 stone roll(er)
石笼 crib;cribwork filled with stone; gabion; pannier; rock crib; rock-filled gabion;rock-filled wire gabion;stone basket;stone-filled crib; stone-mesh
石笼坝 crib dam;gabionade;gabion dam; stone-case dam; stone-filled crib dam;wire dam
石笼沉排 reno mattress
石笼挡栅 gabion boom
石笼堤 gabion wall
石笼护岸 gabion revetment
石笼结构 stone-mesh construction; gabion structure
石笼筐 gabion basket
石笼筐子 pannier
石笼框垛 cribwork
石笼拦河埂 gabion boom
石笼排 gabion mattress
石笼墙 gabion wall
石笼围堰 stone cage cofferdam
石笼组 gabionade
石楼厅 stone balcony
石路面 sarn
石路牙 stone curb
石绿 green earth; malachite; mineral green;mountain green
石绿颜料 green basic copper carbonate
石麻 amianthus;mountain flax
石马赛克 stone mosaic
石幔 curtain
石锚 killagh;kill(i)ck
石楣窗 stone transom
石煤 bone coal;stone(-like) coal
石门 cross adit; cross-entry; jack hole; rock crosscut; cross-cut <采矿>
石门窗侧壁 jamb post;jamb stone
石门架 stone portal
石门槛 stone door threshold; stone sill of door; stone threshold; stone sill
石门掘进 cross-measure drifting
石门框槽口 giblet check
石门枢 stone hinge
石幂 stone mulch
石棉 asbestos;amiant(h);amianthin(it)e;amianthoide;amianthus;amiantos; asbestus; carystine; earth flax; fossil flax; mountain cork; mountain flax; mountain leather; rock leather;rock silk;rock wool; salamander's wool;salamander wool; rock cork

S

石棉板 asbestos board; asbestos plate; asbestos sheet(ing); asbest sheet; sheet asbestos
石棉板瓦 asbestos tile
石棉包覆金属 asbestos-covered metal
石棉包线 asbestos-covered wire
石棉保护 asbestos protection
石棉保护板 asbestos board shield
石棉保护金属屋顶 asbestos-protected metal roofing
石棉保温板 asbestos insulating board; asbestos insulating sheet
石棉保温层 asbestos blanket; asbestos clothing; asbestos insulation
石棉壁板 asbestos siding
石棉编包管 asbestos braided tubing
石棉标准检验筛 asbestos standard testing machine
石棉波纹板 asbestos-corrugated board; asbestos-corrugated panel
石棉波形板板桩 corrugated asbestos sheet pile
石棉薄板 asbestos sheet(ing); atlasite <表面呈大理石花纹或木纹的>
石棉薄片 flake asbestos
石棉布 asbestos canvas; asbestos cloth; asbestos fabric; asbestos filter cloth; asbestos woven fabric
石棉擦拭夹 asbestos press
石棉采矿场 asbestos quarry
石棉层 asbestos layer
石棉层压板 asbestos laminate; asbestos plywood
石棉插件 asbestos insert(ion)
石棉尘 asbestos dust
石棉沉着病 amianthosis; asbestosis
石棉衬垫 asbestos lining
石棉衬垫板 asbestos liner sheet
石棉衬里黄铜垫圈 brass asbestos-lined gasket
石棉衬里铜垫圈 copper asbestos-lined gasket
石棉衬圈 asbestos card liner
石棉衬网 asbestos wire gauze
石棉粗纱 asbestos roving
石棉带 asbestos ribbon; asbestos tape
石棉挡板 asbestos apron
石棉的 asbestic; asbestous
石棉底板 asbestos apron
石棉地沥青 asbestophalt
石棉垫 asbestos cushion; asbestos insert(ion); asbestos mat; asbestos pad
石棉垫层支座 asbestos pad bearing
石棉垫料 asbestos packing
石棉垫片 asbestos filler; asbestos gasket
石棉垫圈 asbestos gasket; asbestos ring; asbestos washer
石棉堵漏丝 control wool
石棉短料 asbestos shorts
石棉短衣 asbestos jacket
石棉对角瓦 asbestos diagonal tile
石棉防护层 asbestos protection
石棉防火幕 asbestos fire curtain
石棉防火毡 asbestos fire blanket
石棉防水 asbestos waterproofing
石棉防水板 asbestos waterproofing slab
石棉防水布 asbestos waterproofing cloth
石棉纺丝堆放试验 array test for asbestos spinning fiber[fibre]
石棉纺丝阵列试验 array test for asbestos spinning fiber[fibre]
石棉纺织品 asbestos textile fabric
石棉肺 <肺的石棉吸入病> asbestosis; pulmonary asbestosis
石棉分段汽管套 asbestos sectional steam pipe covering
石棉粉 asbestine; flake asbestos; pulverised[pulverized] asbestos
石棉粉尘 asbestos dust
石棉粉末 asbestos flour; ground asbestos; asbestos powder
石棉粉饰 asbestos plaster
石棉风选 pneumatic concentration of asbestos
石棉服 asbestos clothing; asbestos suit
石棉覆盖板 asbestos millboard
石棉覆盖层 asbestos covering
石棉盖 asbestos cover
石棉盖板 asbestos apron
石棉钢片 asbestos steel slice
石棉隔膜电解槽 asbestos diaphragm electrolytic cell
石棉隔热层 asbestos insulation
石棉构造的 asbestiform
石棉箍 asbestos ring
石棉管 asbestos conduit; asbestos pipe; asbestos tube
石棉管道 asbestos duct
石棉硅酸钙薄板 asbestos calcium silicate board; asbestos calcium silicate sheet
石棉硅酸盐水泥 Portland cement asbestos
石棉硅藻土 asbestos-diatomaceous earth; asbestos-diatomite
石棉辊 asbestos roll
石棉锅炉罩 asbestos boiler clothing
石棉裹腿 asbestos legging
石棉过滤器 asbestos filter
石棉红纸板 red compressed asbestos sheet
石棉护层 asbestos protection
石棉护层金属屋顶 asbestos-protected metal roofing
石棉护面具 asbestos mask
石棉缓冲垫 asbestos washer
石棉灰 asbestos ash; asbestos plaster
石棉灰浆 asbestos mortar; asbestos plaster
石棉灰泥 asbestos plaster
石棉混凝土 asbestos concrete
石棉混凝土板 asbestos concrete slab
石棉混凝土管 asbestos concrete pipe
石棉基的 asbestos-based
石棉基地沥青油毛毡 asbestos-based asphalt felted fabric
石棉基地沥青油毛纸 asbestos-based asphalt paper
石棉基沥青油毛纸 asbestos-based asphaltic-bitumen paper; asbestos-based bitumen paper
石棉基沥青油毡 asbestos-based asphalt felt; asbestos-based bitumen felt
石棉(基)油(毛)毡 asbestos-based asphalt felt; asbestos-based felt
石棉加工 asbestos dressing; asbestos manufacturing
石棉加强板 asbestos-reinforced panel
石棉夹心胶合板 asbestos-veneer plywood
石棉检验筛 asbestos testing sieve
石棉建筑平板 asbestos flat building board; asbestos flat building sheet
石棉胶板 asbestos rubber sheet
石棉胶布板 asbestos textolite
石棉胶合板 asbestos-cement board; asbestos plywood
石棉角形屋脊板 asbestos angular ridging
石棉接口浇道 asbestos joint runner
石棉接头浇道 asbestos joint runner
石棉结构屋面 asbestos structural roofing
石棉聚氯乙烯 asbestos PVC
石棉聚氯乙烯板 vinyl-asbestos tile
石棉聚氯乙烯塑料地板 asbestos PVC floor(ing)
石棉聚氯乙烯塑料地面砖 asbestos PVC floor(ing) tile
石棉聚氯乙烯塑料楼面砖 asbestos PVC floor(ing) tile
石棉卷筒 asbestos reel
石棉绝热板 asbestos insulating board
石棉绝热薄板 asbestos insulating sheet
石棉绝热材料 asbestonite; asbestos heat-insulation material
石棉绝热体 asbestos insulation
石棉绝缘 asbestos insulation
石棉绝缘板 asbestos insulating slab
石棉绝缘材料 asbestonite; asbestos insulating material
石棉绝缘电缆 asbestos insulated cable
石棉绝缘纸 insulating asbestos paper
石棉开采 asbestos mining; asbestos quarry
石棉颗粒 asbestification particle
石棉裤 asbestos trousers
石棉块 asbestos block
石棉矿床 asbestos deposit
石棉矿(井) asbestos mine
石棉矿开采 asbestos mining
石棉矿石 asbestos mineral
石棉矿石的地质品位 geologic(al) tenor of asbestos ore
石棉矿石的石棉地质含量 geologic(al) content of asbestos in the ore
石棉矿石矿山品位 mine tenor of asbestos ore
石棉矿物 asbestos mineral
石棉矿渣棉 asbestos slag wool
石棉沥青 asbestos-bitumen; asbestumen
石棉沥青板 asbestos-bitumen sheet; semi-rigid asbestos-bitumen sheet
石棉沥青层 asbestos asphalt layer
石棉沥青浆 asbestos asphalt paste
石棉沥青热塑板 asbestos-bitumen thermoplastic sheet
石棉沥青填塞 asbestos-bitumen fill
石棉沥青屋面材料 astos
石棉沥青油毛毡 asbestos-based asphalt bitumen felted fabric
石棉沥青油毡 asbestos-based bitumen felt
石棉沥青制品 asbestos asphalt products
石棉两指宽口大手套 asbestos thumb gloves with gauntlets
石棉两指手套 asbestos thumb gloves
石棉滤布 asbestos filter cloth
石棉滤池 asbestos filter
石棉滤(油)器 asbestos filter
石棉滤纸 asbestos filter
石棉麻丝板 asbestos millboard
石棉毛毡管套 asbestos and hair felt pipe covering
石棉帽 asbestos cap
石棉密封 asbestos packing
石棉密封垫片 asbestos sheet packing
石棉密封罩 asbestos encapsulation
石棉密封制品 asbestos sealing product
石棉模板 asbestos formboard
石棉摩擦材料 asbestos friction material
石棉摩擦片 asbestos friction sheet
石棉幕帘 asbestos curtain
石棉耐火屋顶板 asbestos roof shingle
石棉耐火砖 asbestos firebrick
石棉耐油橡胶板 resistant asbestos packing sheet
石棉泥 asbestos clay; asbestos composition
石棉碾 chaser
石棉碾磨 asbestos milling
石棉盘根 asbestos packing
石棉泡沫混凝土 asbestos foam(ed) concrete
石棉劈分性 cleavability of asbestos
石棉片 asbestos sheet(ing); sheet asbestos
石棉平板 asbestos flat board; flat asbestos sheet
石棉嵌料 asbestos insert(ion)
石棉墙板 asbestos siding; asbestos wallboard; asbestos wall sheet; asbestos wood
石棉清除 asbestos removal
石棉圈 asbestos ring
石棉绒 amiant(h); amianthin(it)e; amiant(h)us; asbestos fiber[fibre]; asbestos float; asbestos wool; asbestos yarn
石棉绒的 amiathine
石棉乳胶板 emulsion asbestos sheet
石棉褥 asbestos mattress
石棉软管 asbestos hose
石棉纱 asbestos yarn
石棉刹车带 asbestos brake lining
石棉砂浆 asbestos mortar
石棉绳 asbestos cord; asbestos rope; rat joint runner <用于嵌缝>
石棉绳编织的覆盖层 braided asbestos cord covering
石棉绳缠层 asbestos cord covering
石棉绳防潮层 asbestos cord covering
石棉绳绝缘层 asbestos cord covering
石棉绳填密法 asbestos rope packing
石棉石毛 <含石棉纤维> rock wool
石棉试验机 asbestos tester
石棉饰面胶合板 asbestos-veneer plywood
石棉手套 asbestos gloves; asbestos mitten
石棉书皮纸板 asbestos millboard
石棉水泥 asbestos cement; cement asbestos; fiber[fibre] cement
石棉水泥板 asbestos-cement board; asbestos-cement panel; asbestos-cement sheet; board of asbestos cement; cement asbestos board; cementitious sheet; Eternit slab; transite
石棉水泥板顶棚 asbestos-cement board ceiling; ceiling of asbestos cement sheets
石棉水泥板覆面 asbestos-cement cladding
石棉水泥板结构屋面 asbestos-cement board structural roofing; asbestos-cement structural roofing
石棉水泥板平顶 asbestos-cement board ceiling; ceiling of asbestos cement sheets
石棉水泥板条 asbestos-cement slate
石棉水泥板瓦 asbestos-cement slate
石棉水泥板岩 Eternit slate
石棉水泥半波瓦 asbestos-cement semi-corrugated shingle
石棉水泥保温板 asbestos-cement insulation board
石棉水泥波瓦标准张 nominal asbestos cement corrugated shingle(sheet)
石棉水泥波纹板 asbestos-cement corrugated board; asbestos-cement corrugated panel; asbestos-cement corrugated sheet(ing); asbestos-cement fluted board; eternity corrugated sheet
石棉水泥波纹屋顶 asbestos-cement corrugated roof cladding
石棉水泥波纹屋面板 asbestos-cement corrugated roof(ing) board

石棉水泥波形瓦 asbestos-cement corrugated sheet(ing); corrugated-asbestos cement sheet;fibrotile
石棉水泥波形瓦屋面 asbestos-cement corrugated sheet roofing;corrugated-asbestos cement sheet roofing
石棉水泥材料 asbestone
石棉水泥材料通风装置 asbestos-cement ventilator
石棉水泥槽纹板 asbestos-cement fluted board;asbestos-cement fluted sheet
石棉水泥厂 asbestos-cement factory
石棉水泥衬板 asbestos-cement lining sheet;asbestos-cement roof sheathing
石棉水泥衬垫 asbestos-cement lining
石棉水泥衬里 asbestos-cement lining
石棉水泥抽风器 asbestos-cement extractor
石棉水泥抽风装置 asbestos-cement extract(ion) ventilation unit
石棉水泥抽气机 asbestos-cement extractor
石棉水泥抽气装置 asbestos-cement extract(ion) ventilation unit
石棉水泥窗台板 asbestos-cement window sill
石棉水泥导管 asbestos-cement duct
石棉水泥地板 asbestos-cement floor(ing)
石棉水泥地板材料 asbestos-cement flooring material
石棉水泥电气绝缘板 asbestos-cement electric(al) insulating board
石棉水泥垫块 asbestos-cement spacer
石棉水泥定型板 asbestos-cement profile(d) board
石棉水泥方形屋顶天沟 cement asbestos box roof gutter
石棉水泥防火板 transite
石棉水泥防雨板 asbestos-cement siding
石棉水泥防雨壁板 asbestos-cement siding
石棉水泥防雨墙板 asbestos-cement siding
石棉水泥房屋 <一种预制好的> cemestos
石棉水泥废水管 asbestos-cement refuse water pipe
石棉水泥粉面 asbestos-cement facing;asbestos-cement front
石棉水泥封闭 asbestos-cement closure
石棉水泥封口 asbestos-cement closure
石棉水泥封檐板 asbestos-cement lining sheet
石棉水泥盖屋板 asbestos-cement shake
石棉水泥隔板 asbestos-cement distance piece; asbestos-cement separator
石棉水泥隔墙 asbestos-cement partition
石棉水泥构件 asbestos-cement member
石棉水泥挂墙板 asbestos-cement siding
石棉水泥管 asbestos-cement duct;asbestos-cement pipe; cement asbestos pipe; cement asbestos tube; eternit pipe;transite
石棉水泥管标准米 nominal meter of asbestos-cement pipe
石棉水泥管腐蚀 asbestos-cement pipe corrosion
石棉水泥管接头 asbestos-cement pipe joint
石棉水泥管配件 asbestos-cement pipe fitting
石棉水泥管瓦 cranked sheet
石棉水泥管(铸铁)配件 asbestos cement fitting
石棉水泥花盆 asbestos-cement flower container
石棉水泥化粪池 asbestos-cement septic tank
石棉水泥集管 asbestos-cement header
石棉水泥脊瓦 asbestos-cement ridge capping; asbestos-cement ridging tile
石棉水泥建筑构件 asbestos-cement building component; asbestos-cement building member;asbestos-cement building unit
石棉水泥接头 asbestos-cement joint
石棉水泥绝缘板 asbestos-cement insulation board
石棉水泥绝缘薄板 asbestos-cement insulating sheet
石棉水泥落水管 asbestos-cement conductor; asbestos-cement downcomer; asbestos-cement downpipe; asbestos-cement downspout; asbestos-cement fall pipe
石棉水泥门面 asbestos-cement facade
石棉水泥面 asbestos-cement face
石棉水泥面层 asbestos-cement facing;asbestos-cement floor(ing)
石棉水泥模板 asbestos-cement form board;asbestos-cement formwork
石棉水泥模型 asbestos-cement mo(u)ld
石棉水泥暖气管 asbestos-cement flue
石棉水泥排水管 asbestos-cement discharge pipe; asbestos-cement drain(age) pipe;asbestos-cement leader; asbestos-cement refuse water pipe
石棉水泥喷泉池 asbestos-cement fountain basin
石棉水泥喷水池 asbestos-cement fountain basin
石棉水泥平板 asbestos-cement fat siding;asbestos-cement flat board; asbestos-cement flat (run) sheet; plane asbestos cement slate
石棉水泥平瓦 asbestos-cement roofing shingle; asbestos-cement shingle;asbestos-cement slate roofing
石棉水泥铺面 asbestos-cement facade
石棉水泥器皿 asbestos-cement ware
石棉水泥墙板 asbestos-cement sheet; asbestos-cement shingle; asbestos-cement siding shake; asbestos-cement wallboard; asbestos-cement wall panel; asbestos-cement wall shingle;poilite
石棉水泥墙面板 asbestos-cement siding
石棉水泥柔性板 asbestos-cement flexible board
石棉水泥砂浆 asbestos-cement mortar;cement asbestos mortar
石棉水泥实板 asbestos-cement solid board
石棉水泥实心板 asbestos-cement solid board
石棉水泥饰面 asbestos-cement facade; asbestos-cement face; asbestos-cement(sur) facing
石棉水泥饰面板 asbestos-cement facade slab; asbestos-cement surfacing sheet
石棉水泥饰面层 asbestos-cement facing
石棉水泥水槽 asbestos-cement cistern
石棉水泥天沟 asbestos-cement gutter
石棉水泥填充片 asbestos-cement filler strip
石棉水泥填充墙 asbestos-cement cladding
石棉水泥挑口板 <用于平屋面的> asbestos-cement fascia board (for flat roof)
石棉水泥通风管 asbestos-cement vent(ilating) pipe
石棉水泥通风机 asbestos-cement extractor
石棉水泥通风器 asbestos-cement extractor
石棉水泥通风装置 asbestos-cement extract(ion) ventilation unit
石棉水泥瓦 asbestos-cement shingle; asbestos-cement tile;cement asbestos tile;poilite
石棉水泥瓦标准张 nominal asbestos cement sheet
石棉水泥瓦楞板 asbestos-cement corrugated panel; asbestos-cement fluted board
石棉水泥瓦楞薄板 asbestos-cement corrugated sheet(ing)
石棉水泥瓦楞屋顶 asbestos-cement corrugated roof cladding
石棉水泥瓦楞屋面板 asbestos-cement corrugated roof(ing) board
石棉水泥瓦屋面 asbestos-cement tile roofing
石棉水泥外挂板 asbestos-cement cladding
石棉水泥外墙护板 asbestos-cement cladding
石棉水泥望板 asbestos-cement roof sheathing
石棉水泥围栏 asbestos-cement fence
石棉水泥污水管 asbestos-cement foul water pipe;asbestos-cement refuse water pipe; asbestos-cement sewage pipe; asbestos-cement sewer pipe
石棉水泥屋顶 asbestos-cement roofing
石棉水泥屋顶板 asbestos roofing
石棉水泥屋顶防水层 eternit roof sheathing
石棉水泥屋顶箱形天沟 asbestos-cement box roof gutter
石棉水泥屋面 asbestos-cement roofing;eternit roofing
石棉水泥屋面板 asbestos-cement roofing board; asbestos-cement roofing panel; asbestos-cement roofing sheet;asbestos-cement roof sheathing; asbestos-cement roofing shingle
石棉水泥屋面覆盖层 asbestos-cement roof cladding; asbestos-cement roof cover(ing)
石棉水泥屋面平瓦 asbestos-cement roofing slate
石棉水泥屋面石板瓦 asbestos-cement slate
石棉水泥屋面瓦 asbestos-cement roofing shingle; asbestos-cement roofing tile
石棉水泥屋面檐沟 asbestos-cement roof gutter
石棉水泥纤维素 asbestos-cement cellulose
石棉水泥镶面板 asbestos-cement flat (run) panel; asbestos-cement flat (run) sheet;asbestos-cement sheeting
石棉水泥橡胶瓦 asbestos-cement rubber tile
石棉水泥斜沟槽 asbestos-cement valley gutter
石棉水泥斜沟底板 asbestos-cement valley board
石棉水泥泄水管 asbestos-cement downcomer
石棉水泥型板 asbestos-cement form board
石棉水泥蓄水池 asbestos-cement cistern
石棉水泥压脊瓦 asbestos-cement ridge capping tile; asbestos-cement ridge covering tile
石棉水泥压力管 asbestos-cement pressure pipe
石棉水泥烟道 asbestos-cement flue
石棉水泥烟道衬管 asbestos-cement flue liner
石棉水泥檐沟 asbestos-cement eaves gutter;asbestos-cement eaves trough
石棉水泥檐口板 <用于平屋面的> asbestos-cement fascia board (for flat roof)
石棉水泥异型板 asbestos-cement profile(d) board
石棉水泥雨水管 asbestos-cement rainwater conductor
石棉水泥雨水落水管 asbestos-cement conductor; asbestos-cement leader
石棉水泥雨水檐沟 asbestos-cement rainwater gutter
石棉水泥栅栏 asbestos-cement fence
石棉水泥毡 cementitious sheet
石棉水泥支管 asbestos-cement lateral
石棉水泥制品 asbestos-cement article; asbestos-cement composition; asbestos-cement product; asbestos-cement ware;handcraft;Durobestos <一种专利产品>
石棉水泥砖 asbestos-cement tile
石棉丝 asbestos silk;mineral flux
石棉丝沥青胶 bitumastic
石棉松解度 asbestos fibrillation
石棉塑料地板 asbestos plastic floor(ing)
石棉塑料楼面板 asbestos plastic floor(ing)
石棉塑料镶块 retinacs insert
石棉毯 asbestos mat;asbestos blanket
石棉套袖 asbestos sleeve
石棉填充 asbestos packing
石棉填充料 asbestos filler
石棉填充片 asbestos-cement filler strip
石棉填缝浇口 asbestos joint runner; pouring runner
石棉填缝绳 <灌铅接头的> pouring rope
石棉填料 asbestos filler;asbestos gasket; asbestos jointing; asbestos packing;asbestos wadding
石棉条 asbestos sliver;asbestos strip
石棉涂料 asbestos paint
石棉瓦 asbestic tile;asbestiform tile; asbestos sheet(ing);asbestos shingle;asbestos tile;mineral fiber tile
石棉瓦厂 asbestos tile works
石棉瓦尾部的铆钉 <防止风掀起> crampo(o)n
石棉瓦屋顶 asbestos roofing;asbestos tile roofing
石棉外套 asbestos lagging
石棉外罩 asbestos encasement
石棉围裙 asbestos apron
石棉尾矿 asbestos tailing
石棉污染 asbestos pollution;pollution by asbestos

石棉屋顶瓦 asbestos roof shingle
石棉屋面板 asbestos roofing sheet; asbestos(roof)shingle
石棉屋面材料 asbestos roofing; asbestos roofing material
石棉屋面瓦 asbestos roof shingle
石棉屋面檐沟 asbestos roof gutter
石棉屋面雨水天沟 asbestos rainwater gutter; asbestos roof gutter
石棉吸选法 aspiration method of asbestos processing
石棉膝垫 asbestos knee pad
石棉纤维 asbestos fiber[fibre]; rock wool; salamander wool
石棉纤维板 asbestos fiber board
石棉纤维包管绳 asbestos fibrous pipe lagging
石棉纤维保温 asbestos fiber insulation
石棉纤维长度 length of asbestos fiber
石棉纤维分级 classification of asbestos fiber
石棉纤维粉 asbestos mortite
石棉纤维灰浆 asbestos fiber plaster
石棉纤维加筋的 asbestos fiber-reinforced
石棉纤维加筋墙板 asbestos fiber-reinforced sheet(ing)
石棉纤维加劲聚酯 asbestos fiber-reinforced polyester
石棉纤维绝热 asbestos fiber insulation
石棉纤维绝热材料 heat insulation material of asbestos fibre
石棉纤维素 cellulose-asbestos
石棉纤维填缝 asbestos fiber filling
石棉纤维砖 asbestos fiber tile
石棉线 asbestos string; asbestos thread; asbestos wire; asbestos yarn
石棉镶板 asbestos sheet(ing)
石棉橡胶板 asbestos rubber sheet; asbestos sheet with insertion rubber; it-plate; paronite
石棉橡胶垫片 asbestos-packing gasket
石棉橡胶密封垫片 asbestos rubber gasket
石棉橡胶瓦 asbestos rubber tile
石棉小体 asbestos body
石棉鞋 asbestos shoes
石棉鞋盖 asbestos gaiter
石棉屑 ground asbestos
石棉心铁丝网 asbestos center gauze
石棉性的 asbestine
石棉絮 asbestos wadding
石棉选矿 asbestos dressing
石棉压盖填料 asbestos gland packing
石棉岩 asbestos rock
石棉岩板 asbestos slate
石棉衣 asbestos suit
石棉乙烯地板面层 asbestos-vinyl floor cover(ing)
石棉乙烯楼板面层 asbestos-vinyl floor cover(ing)
石棉乙烯饰面板 asbestos-vinyl tile
石棉乙烯饰面砖 asbestos-vinyl tile
石棉乙烯制品 asbestos-vinyl compound; asbestos-vinyl composition
石棉乙烯(制品)材料 asbestos-vinyl mass; asbestos-vinyl material
石棉油灰 asbestos putty
石棉油毡 asbestos felt
石棉疣 asbestos wart
石棉有机物的 asbestos-organic
石棉雨水沟 asbestos rainwater gutter
石棉雨水天沟 asbestos gutter
石棉云母 asbestos mica; asbestos mire
石棉增强板 asbestos-reinforced panel
石棉增强塑料 asbestos-reinforced plastics
石棉增强效率 reinforcing efficiency of asbestos
石棉渣道砟 asbester ballast
石棉闸衬【机】asbestos brake lining
石棉毡 asbestos blanket; asbestos felt; mineral flax
石棉罩 asbestos clothing
石棉罩衫 asbestos overall
石棉织构 texture of asbestos
石棉织构指数 texture index of asbestos
石棉织品 asbestos cloth; asbestos fabric
石棉织物 asbestos cloth; asbestos fabric; asbestos woven fabric
石棉纸 asbestos paper; asbestos sheet-(ing)
石棉纸板 asbestos fiber[fibre]; compressed asbestos sheet; asbestos cardboard
石棉纸带 asbestos paper tape
石棉纸垫片 asbestos paper gasket
石棉纸油毡 asphalt-saturated asbestos felt; saturated bitumen asbestos felt
石棉制动衬片 asbestos brake lining; woven asbestos brake lining
石棉制动带 asbestos brake belt
石棉制动摩擦片 asbestos brake facing
石棉制动制品 asbestos friction product
石棉制绝热材料 asbestonite
石棉制离合器摩擦片 asbestos friction product
石棉制品 asbestos article; asbestos product
石棉制品厂 asbestos product factory
石棉质的 asbestine
石棉致癌性 asbestos carcinogenesis; carcinogenicity of asbestos
石棉主体纤维 constituent fiber[fibre] of asbestos
石棉砖 asbestos brick
石棉状 asbestiform
石棉状的 asbestine; asbestoid
石棉状矿物 asbestiform mineral
石棉阻燃金属屋顶 asbestos-protected metal roofing
石面 plain work; stone surfacing; stone veneer(ing)
石面凹缝 rustic joint
石面凹坑 pitchhole
石面板 stone slab cladding
石面半砍平工作 half-plain work
石面半砍平细工 half-plain work
石面锤琢 hammered finish of stone
石面打光 polished finish of stone
石面的 stone-faced
石面拱 stone-faced arch
石面截槽 jad
石面刻槽工作 broached work
石面刻线槽 punched work
石面磨光 honing; polished finish; polished finish of stone; polished work
石面磨光机 dunter machine; surfacer
石面磨平工作 plain work
石面抛光 polished finish of stone
石面抛光机 dunter
石面砌体 stone-faced masonry
石面曲线槽 broached work
石面圬工 stone-faced masonry
石面修琢 picking
石面修琢锤 patent(bush)hammer
石面修琢凿 stone-dressing stem
石面凿 daubing
石面凿毛 da(u)bbing
石面斩凿 stone granulating
石面植物 exochomophyte
石面砖 ashlar brick; rock-faced brick
石面斫平(工作)plain work
石面琢边 margin draft
石面琢新 regrate; regrating skin
石明矾 rock alum
石磨 buhr(stone)mill; burr mill; stone grinder
石磨成精光表面 stone finish
石磨机 burrstone mill
石末沉着病 chalicosis; silicosis
石末肺 chalicosis
石漠 hammada; rock desert; stone desert
石墨 black lead(ore); carbon; graphitic carbon; meta-anthracite; mineral black; mineral carbon; plot lead; plumbagine; plumbago; pot lead; wad
石墨板 graphite bearing; graphite cake; graphite plate
石墨板材 graphite sheet
石墨板岩 graphite slate
石墨棒 graphite rod; graphite solid cylinder
石墨棒电炉 graphite bar electric(al) furnace
石墨棒炉 graphite-rod furnace
石墨棒阳极 graphite-rod anode
石墨棒阴极 graphite-rod cathode
石墨包衬 graphite ladle liner
石墨杯雾化器 graphite cup atomizer
石墨杯脂 graphite cup grease
石墨捕捉器 kish collector
石墨槽 graphite cuvette
石墨槽原子化器 graphite cuvette atomizer
石墨层 graphite layer
石墨层间化合物 lamellar compound of graphite
石墨巢孔 <铸件缺陷> graphite nest
石墨尘肺 graphite pneumoconiosis; graphosis
石墨坩锅 graphite crucible
石墨沉淀 aquadag
石墨衬垫 graphite backing; graphite packing
石墨衬里 graphite lining
石墨衬套 graphite bushing; graphite insert; graphite liner
石墨尺寸 graphite size
石墨处理的编织填料 graphited braided asbestos packing
石墨粗大疏松组织 graphite nest
石墨大理岩 graphite marble
石墨单晶 graphite single crystal
石墨的 graphitic; plumbaginous
石墨灯丝原子储存器 carbon filament atom reservoir
石墨电极 graphite electrode; graphite resistor rod
石墨电极块 graphite electrode slab
石墨电极炉 carbon bar furnace
石墨电解槽 graphitic cell
石墨电刷 electrographite brush; graphite brush
石墨电阻 graphite resistance
石墨电阻棒 graphite resistor rod
石墨电阻加热器 graphite resistance heater
石墨电阻炉 carbon resistor furnace; graphite heater furnace
石墨电阻器 graphite resistor
石墨垫料 graphitized packing
石墨端包 graphite ladle
石墨堆 graphite stack
石墨反射层 graphite reflector
石墨反应堆 graphite reactor
石墨防腐涂料 graphite paint
石墨分离 kish
石墨粉 dag; flake graphite; graphite powder; ground graphite; plumbagine
石墨粉尘 graphite dust
石墨腐蚀 graphite corrosion
石墨钙基润滑脂 graphite lime base grease
石墨干润滑剂 graphite dry lubricant
石墨坩埚 black-lead crucible; carbon crucible; graphite crucible; graphite pot; plumbago crucible; plumbago pot; graphite cup
石墨坩埚原子化法 atomization in graphite crucible
石墨感应环 graphite susceptor ring
石墨高温计 graphite pyrometer
石墨膏 graphite grease; oildag
石墨隔板 graphite septum
石墨管 graphite pipe; graphite tube
石墨管炉 graphite-tube furnace
石墨管炉原子化法 atomization in graphite tube furnace
石墨含量 graphite content
石墨合成炉 graphitic synthetic(al) furnace
石墨合金 graphite alloy
石墨(和)碳化硅热电偶 graphite to silicon-carbide couple
石墨黑云碳酸岩 graphite sovite
石墨黑云斜长片麻岩 graphite biotite plagioclase gneiss
石墨滑环 graphite slip ring
石墨滑水 aquadag
石墨滑水涂料 aquadag
石墨化程度 degree of graphitization
石墨化处理 graphitizing
石墨化电极 graphitized electrode
石墨化分子筛 graphitization molecular sieve
石墨化腐蚀 graphitic corrosion
石墨化钢 graphitizable steel; hybrid metal
石墨化工设备 graphitic chemical equipment
石墨化剂 graphitizer
石墨化炉 graphitizating furnace
石墨化炭黑 graphitized black; graphitized carbon black; graphon
石墨化碳 graphitizating carbon; graphitized carbon; graphitizing carbon
石墨化碳丝 graphitized carbon filament
石墨化退火 graphitizing annealing
石墨化阳极块 graphitised anode block
石墨化元素 graphite element; graphitizing element
石墨化作用 graphitic action; graphitization; graphitizing
石墨环面 graphite annulus
石墨换热器 graphite heat exchanger
石墨黄铜混合料 graphite-brass composition
石墨灰 dag
石墨活塞 graphite piston
石墨活塞环 graphite piston ring
石墨活性区 graphite core
石墨火花技术 graphite spark technique
石墨基复合材料 graphite-base composite material
石墨基合成材料 graphite-base composite material
石墨集结 kish lock
石墨计量泵 graphite metering pump
石墨剂 graphite preparation
石墨减速剂 graphite moderator
石墨浆涂料 darmold
石墨胶层 aquadag coating
石墨胶合剂 carbon cement
石墨胶态溶液 cohydrol
石墨搅拌器 graphite agitator
石墨接合剂 graphite jointing compound
石墨接受器 graphite susceptor

石墨结构 graphite structure
石墨结合耐火材料 graphite bonded refractory
石墨金刚石相平衡线 graphite-diamond-equilibrium line
石墨金刚石液相三相点 graphite-diamond-liquid triple point
石墨金属层间化合物 graphite-metal lamellar compound
石墨径向密封片 graphite apex seal
石墨块 graphite block;graphite cake
石墨块挡风圈 graphite block air seal
石墨块密封圈 graphite block air seal
石墨块体 graphite block body
石墨矿床 graphite deposit
石墨矿石 graphite ore
石墨类型 graphite type
石墨流槽 graphite spout
石墨炉 graphite furnace;graphite oven
石墨炉雾化器 graphite furnace atomizer
石墨炉原子吸收光谱法 graphite furnace atomic absorption spectrometry
石墨慢化棒 graphite moderator stringer
石墨慢化堆 graphite-moderated reactor
石墨慢化热中子反应堆 thermal graphite reactor
石墨慢化栅格 graphite-moderated lattice
石墨密封垫 graphite packing
石墨密封环 carbon seal;graphite seal
石墨密封圈 carbon seal ring
石墨密封条 carbon shoe
石墨密封制品 graphite sealing products
石墨模具 graphite jig; graphite mo(u)ld
石墨模具车间 graphite mo(u)ld department
石墨耐火材料制品 graphite refractory product
石墨(耐火)砖 graphite brick
石墨黏[粘]土 graphite clay
石墨黏[粘]土砖 graphite-clay brick
石墨盘根 graphite packing
石墨喷管衬套 carbon nozzle insert
石墨喷射管 graphite sparge tube
石墨喷雾器 graphite atomizer
石墨片 graphite cake;graphite flake; graphite sheet
石墨片麻岩 graphite gneiss
石墨片岩 graphite schist;grapholite; plumbaginous schist
石墨片阻加热器 graphite resistance heater
石墨偏转板 graphite deflector
石墨漂浮 graphite flo(a)tation
石墨屏蔽 graphite shielding
石墨气冷反应堆 gas-graphite reactor
石墨铅基合金 graphite metal
石墨铅油 graphite lead paint
石墨青铜 graphite bronze;graphitized bronze
石墨青铜轴承 graphite bronze bearing
石墨青铜轴承合金 oilite
石墨球 graphite nodule;graphite pebble
石墨球化 spheroidization of graphite
石墨球状组织 spheroidal structure
石墨热柱 graphite thermal column
石墨容器 crystal vessel;graphite container;graphite vessel
石墨润滑棒 graphite lubricating rod
石墨润滑的无油轴承 graphited oilless bearing
石墨润滑法 graphite lubrication
石墨润滑剂 black lead lubrication; graphite lubricant;oildag
石墨润滑器 graphite lubricator
石墨润滑油 graphite lubricating oil; oildag;graphite oil
石墨润滑脂 graphited oil; graphite grease
石墨润滑轴承 oilless bearing
石墨润滑轴瓦 oilless bushing
石墨烧舟 graphite boat
石墨熟料坩埚 graphite grog crucible
石墨水泥 graphite cement
石墨塑料 graphite plastic
石墨酸 graphite acid
石墨碳 graphite carbon
石墨陶瓷纤维 graphite-ceramic fibre reinforced phenolic resin
石墨淘洗水 graphite water
石墨套管 graphite bush; graphite sheath;graphite sleeve
石墨体 graphite body
石墨体氮化钢 graphitic nitralloy
石墨(体)钢 graphitic steel
石墨填充酰胺纤维 <商品名> Nylatron
石墨填料 graphite packing
石墨填料密封环 graphite sealing pad
石墨填密 graphite packing
石墨通气器 graphite breather
石墨涂层 graphite coating
石墨涂料 black-lead paint; foundry facing; graphite paint; plumbago dressing
石墨推承环 carbon thrust ring
石墨洼勺 graphite ladle
石墨微粉 graphite micropowder
石墨窝 <铸件缺陷> blacking hole
石墨吸收器 graphitic absorber
石墨析出 graphite precipitation
石墨细棒 graphite spine
石墨纤维材料 graphitized filamentary material
石墨纤维加强塑料 graphite fiber[fibre] reinforced plastic
石墨纤维(增强剂) graphite fiber[fibre]
石墨纤维增强水泥 carbon fiber[fibre] reinforced cement
石墨纤维织物 graphite fabric
石墨楔 graphite wedge
石墨形态 graphite form
石墨形状 graphite shape
石墨型 graphite mo(u)ld
石墨型结构纤维 graphite type structure fibre
石墨须晶 graphite whisker
石墨悬浮液 aquadag
石墨悬浮油剂 oildag
石墨压型管 graphitic pressed pipe
石墨岩 graphitic rock
石墨阳极 graphite anode
石墨阳极篮 graphite anode basket
石墨氧化膜 graphite oxide membrane
石墨氧化物 graphite oxide
石墨引燃阳极 graphite ignition anode
石墨油膏 gredag
石墨(油)漆 graphite paint
石墨(油)脂 graphite grease
石墨铀堆 graphite-uranium pile
石墨(原子)减速反应堆 graphite-moderated reactor
石墨圆筒 graphite cylinder
石墨云母 graphitic mica
石墨杂质 <金刚石晶体中的> carbon spot
石墨增强预应力桩 graphite prestressed pile
石墨渣 kish slag
石墨罩 graphite wool
石墨制垫圈 graphite washer
石墨制品 graphite product
石墨质耐火材料 plumbago refractory
石墨质耐火黏[粘]土 plumbago clay
石墨质黏[粘]土 graphitic clay
石墨质岩 graphocite
石墨舟 graphite boat
石墨轴承 graphite bearing
石墨砖结构 graphite brick work
石墨状材料 graphite-like material
石墨状的 graphitoid
石墨状结构 graphite-like structure
石墨状软锰矿 plumbagolike pyrolusite
石木雕像 acrolith
石墓 sepulcher
石墓室 dolmen;mastaba <古埃及>
石幕墙 stone curtain wall
石南 heath;rosebay
石南花 heath bell
石南灰壤 heath podzol
石南林 heath forest
石南泥炭 heath peat
石南属的植物 heather
石楠灌木群落 Ericifruticeta
石楠荒地 moss heath
石楠荒漠沙土 belisand
石楠荒原 breckland;heath land
石楠木本群落 Ericiliosa
石楠属 <拉> Photinia
石脑溶剂 naphtha solvent
石脑油 benzin(e); naftha; naphtha; paraffin(e) naphtha; petroleum naphtha
石脑油残渣 naphtha residue
石脑油重整的汽提塔 naphtha reformer stripper
石脑油二甲苯当量 <一种指示沥青材料不等质程度的指标> naphtha-xylene equivalent
石脑油脚 naphtha residue
石脑油聚合重整 naphtha polyforming
石脑油馏分 naphtha cut
石脑油炉 naphtha furnace
石脑油气 naphtha gas
石脑油气回馏过程 naphtha gas reversion
石脑油溶剂 diluent naphtha
石脑油塔 naphtha column
石脑油英 naphthein
石脑油蒸馏 naphtha distillation
石脑油重整器 naphtha reformer
石黏[粘]土芯 stone-clay core
石碾 stone roll(er)
石牛腿 <从墙上伸出的> stone bracket;stone corbel
石耙 rock rake;stone rake
石排水槽 stone discharge gutter; stone drain(age) gutter
石牌坊 dolmen
石泡 lithophysa
石泡构造 lithophysa structure
石泡熔结凝灰岩 lithophysa welded tuff
石棚屋 stone hut
石片 ashlar piece; chippings; chip stone;flake;rock chip(ping)s;rock slice;scabblings;slat(e);stone chip(ping)s;stone fragment
石片滤床 slate bed
石片面墙圬工 bastard masonry
石片压碎机 chip-breaker
石拼花地砖 stone mosaix
石平巷 stone drift
石平整机 stone-dressing machine
石坪 rock plateau
石屏障 stone curtain wall;stone screenings
石铺道路 stone course
石铺渠道 stone-lined canal
石瀑 stone fall
石瀑高度 height of stone fall
石瀑宽度 width of stone fall
石栖动物 petrocole
石漆 mineral varnish
石旗 cave flag
石砌 <井壁的> steining
石砌暗沟 stone drain
石砌八字墙 wing masonry
石砌层 stone course
石砌道路 sett-paved road; stone-pitched road
石砌的 stone-pitched
石砌的小屋与圆塔围成建筑体系 <史前撒丁岛独有的> nurag(h)e
石砌堤 stone dike[dyke]
石砌滴水 drip mo(u)ld(ing)
石砌防波堤 quarrying rock breakwater
石砌房屋 stone building
石砌扶壁 stone counterfort
石砌拱顶状屋顶 stone-vaulted roof
石砌沟渠 rubble drain
石砌涵洞 stone culvert
石砌合 stone bond
石砌护岸 stone revetment
石砌护坡 stone pitching
石砌护坡的底层 subbase of stone pitching
石砌护坡堤 stone levee
石砌建筑物 stone building;stonework
石砌教堂 stone church;stone temple
石砌窖 kist
石砌矩形涵洞 stonework box culvert
石砌矩形水道 stonework box drain
石砌块 stone block
石砌块层 assize
石砌块工程 stone blockwork
石砌块工作 stone blockwork
石砌楼梯 stone stair(case)
石砌陆标 cairn
石砌路肩 stone shoulder
石砌路面 stone-surfaced road
石砌盲沟 stone drain
石砌面 stone veneer(ing)
石砌明水沟 Irish bridge
石砌墓 kist
石砌排水沟 blind drain;French drain; Irish bridge;rubble drain;stone discharge gutter; stone drain; stone drain(age) gutter
石砌桥墩 stone masonry pier
石砌人行道 stone sidewalk
石砌人行道石板 stone sidewalk paving flag
石砌水槽 stone gutter
石砌寺庙 stone temple
石砌隧道拱顶 stone tunnel vault
石砌台阶 stone steps
石砌体 stone(-faced) masonry;stonework
石砌体齿形接口 toothing of stone
石砌体定位销 bed dowel
石砌体灰缝 abre(a)uvior
石砌体或金属制品制出浮雕 abate
石砌体交叉缝 interlocking joint
石砌体结合 binding of stones
石砌体留齿插口 toothing of stone
石砌体锚固件 stone anchor
石砌体预留齿形接口 toothing of stone
石砌体(中的)暗销 bed dowel
石砌桶形拱顶 stone barrel vault
石砌筒形拱顶 stone wagon vault
石砌外墙面线 ashlar line
石砌围墙 stone fence;stone hedge
石砌圬工 stone(block) masonry
石砌箱形涵洞 stonework box culvert
石砌箱形水道 stonework box drain
石砌翼墙 wing masonry
石砌圆形拱顶 stone tunnel vault
石砌圆柱体拱顶 stone barrel vault

石器 stone article;stone vessel;stoneware
石器时代 anthropolithic age;Lithic;Stone Age
石钳工台 stone bench
石墙 boulder wall;stone wall(ing)
石墙墩 stone pier
石墙粉刷 stone wall plaster
石墙缝隙填石灰 garret
石墙面砖 stone wall tile
石桥 clapper bridge;stone bridge
石桥墩 rock abutment;stone pier
石桥门 stone portal
石桥施工 stone bridge construction
石桥台 rock abutment
石青 azure copper ore;azure stone;azurite;azurite malachite;blue malachite;chessy copper;chessylite;mountain blue
石青蓝 <灰绿蓝色> azure blue;chessylite blue;blue carbonate of copper
石穹 stone vault
石穹屋顶 stone-vaulted ceiling
石球 flint pebble
石球式热风炉 pebble stove
石球藻属 Lithococcus
石渠 rock-cut ditch;stone canal
石圈 stone circle
石券 stone arch
石阙 stone watchtower
石燃性的 asbestine
石绒 amiant(h);amianthin(it)e;amianthus;asbestos[asbestus]
石软木 agglomerated cork
石蕊 lacca coerula;lacmus
石蕊蓝 lichen blue
石蕊色素 litmus
石蕊(试)纸 litmus paper
石蕊试纸试验 litmus test
石蕊素 azolitmin
石砂 crushed dust;crusher screenings;stone sand
石砂比 coarse-to-fine-aggregate ratio
石山表示法 rock drawing
石珊瑚 cave coral;madrepore;stony-coral;hexacoral;scleractinian
石扇 rock fan
石生演替系列 lithosere
石生植物 <如苔藓等> chomophyte;lithophytes;petrophytes;saxicolous
石生植物群落 cremnion
石十字架 <坟墓用> stone cross
石实验台 stone bench
石饰面 stone finish
石饰面板 stone facing slab
石饰面拱 stone-faced arch
石饰面圬工 stone-faced masonry
石室 rock chamber
石首石肢木身雕像 acrolith
石首鱼 croaker
石束带层 stone string course
石竖框窗 stone mullion window
石栓 stone bolt
石水槽 raggle
石顺坝 stone training wall
石松 club moss;stone pine;wolf's claw
石松孢子 lycopodium spore
石松纲 Lycopodiatae
石髓 chalcedony
石笋 dripstone;stalagmite
石笋高度 height of stalagmite
石笋直径 diameter of stalagmite
石笋状的 stalagmitic(al)
石榫 stone bolt
石塔 pinnacle;stone pagoda
石台阶 stone altar;stone perron
石台式祭坛 stone table altar
石台座 perron
石滩 groundsel;ground sill;rock patch
石炭 stone coal
石炭二叠过渡期【地】Permo-Carboniferous
石炭二叠纪 Anthracolithic period;Permo-Carboniferous period
石炭二叠纪冰期 carboniferous-Permian glacial stage
石炭二叠系 Anthraeolithic system;Perinea-carboniferous system
石炭化泥炭 charred peat
石炭极 baked carbon
石炭纪【地】Carbonic period;Carboniferous period;carbon period
石炭纪的 Carboniferous
石炭纪灰岩 Carbonic limestone;Carboniferous limestone;mountain limestone
石炭纪石灰岩 Carboniferous limestone
石炭石 carbonate of lime
石碳 lithocarbon
石碳酸 benzophenol acid;carbolic;carbolic acid;phenol;phenol acid;phenylic acid;acidum carbolicum <拉>
石碳酸灌木 creosote bush;Larrea divaricata
石碳酸酒精 carbol-alcohol
石碳酸品红染剂 carbolfuchsin stain
石碳酸品红溶液 carbolfuchsin solution
石碳酸系数 phenol coefficient
石碳酸盐 carbolate;phenate
石碳酸液 cresylic acid
石碳酸中毒 carbolism(us);phenol poisoning
石碳系【地】Carboniferous system
石碳与二氧化碳的平衡 carbonate balance
石填料 hard core;stone filling
石条 slate
石铁陨石 siderolite;stony-iron meteorite
石亭 stone pavilion
石头 rock;stone
石头布道坛 <教堂中的> natural stone pulpit
石头的 stony
石头点缀 stone ornament
石头防腐处理 <一种处理石头方法> church's baryta process
石头接合 stone hinge
石头铺路工人 base pav(i)er
石头相互咬合 stone-stone interlock
石头之间嵌锁 stone-stone interlock
石头支承面 bearing stone
石头钻机 rock drill
石突堤式码头 stone jetty
石托臂 <从墙上伸出的> stone bracket
石托架 <从墙上伸出的> stone bracket
石托肩 stone corbel
石瓦 masonry tile;stone tile;tile shoe;tilestone
石弯道 stone curving;stone sweep(ning)
石王位 stone throne
石围栏 stone fence;stone railing;stone wall(ing)
石围裙 stone apron
石围堰 rock cofferdam
石纹涂装法 marble figure coating
石窝 tafoni
石窝混凝土 stone pockets of concrete
石窝水 quarry water
石圬工 rubble;stone masonry;stonework
石圬工挡土墙 stone masonry retaining wall
石圬工涵(洞) masonry-stone culvert
石屋 stone building;stone house
石屋顶 stone roof
石屋顶覆盖层 stone roof cover(ing)
石屋顶尖塔 stone roof spire
石屋尖顶 stone roof spire
石屋面 stone roof cover(ing)
石屋面瓦 stone roof(ing) tile
石细胞 sclereid
石细厂 stonework
石细工 stonework
石隙层 zone of rock fracture
石隙流层 zone of rock flowage
石隙滤层 zone of rock flowage
石隙植物 chasmophytes
石纤维 mineral wool
石线 stone line
石线脚 stone string course
石香肠 boudin(age)
石香肠长度 boudin length
石香肠断面形态 section shape of boudin
石香肠厚度 boudin thickness
石香肠宽度 boudin width
石箱形涵洞 stone box culvert
石镶面圬工 stone-faced masonry
石巷 hard heading
石像 rock statue
石像动物 stone animal
石小径 stone footpath
石小屋 stone hut
石楔 stone cutting wedge
石楔及垫片 feathers and plug
石屑 aggregate chip(ping)s;attle;chip(ping)s;gal(l)et;ground rock;knocking;mineral filler;quarry dust;quarry refuse;quarry rubbish;quarry waste;riddlings;rock chip(ping)s;rock dust;rock screenings;scabblings;scalping;stone chip(ping)s;stone dust;stone fragment;stone screenings;tailings;blinding <填缝用>
石屑层 layer of chip(ping)s
石屑垫层 chip(ping)s carpet
石屑堆 scree;talus material
石屑堆放棚 grits storage hangar;gritting material hangar
石屑堆栈 gritting material store
石屑肺 lithosis
石屑粉刷拌合[和]物 stuc mixture
石屑封层 screening seal coat
石屑敷面胶合板 aggregate plywood
石屑敷面转移法 aggregate transfer
石屑骨料 chip(ping)s aggregate
石屑混合物 chip(ping)s compound
石屑混凝土 chip(ping)s concrete
石屑集料 chip(ping)s aggregate
石屑冷拌沥青 Damman cold asphaltic concrete
石屑沥青混合料 chipping compound
石屑料仓 grit bin
石屑路面 gritty finish
石屑碾辊附件 gritting roll attachment
石屑凝灰岩 liter tuff
石屑铺面 chipped surface;chip(ping)s surfacing;gritting surface dressing;gritty finish;gritty surface dressing;gritty surface finish
石屑铺面琢石面 chipped stone surface
石屑铺撒车 grit spreading vehicle;gritting lorry
石屑铺撒工 hand chip(ping)s spreader
石屑铺撒机 hand chip(ping)s spreader
石屑撒布机 chip distributor;chip(ping) spreader;road gritting machine;stone spreader;gritter
石屑、石渣等制造的石料 reconstructed stone
石屑摊铺机 chip distributor
石屑填充料 stone chip filler
石屑选分器 chip rejector
石屑压碎机 crushing rolls for chip(ping)s production
石屑毡层 chipping carpet
石屑罩面 chip seal;covering of screenings;cover of stone chip(ping)s
石屑贮仓 gritting material bin
石修整机 stone-dressing machine
石修琢机 stone-dressing machine
石压顶 cut-stone coping;stone capping;stone cope[coping]
石芽 clint;stony sprout
石研磨机 stone miller
石盐 halite;rock salt
石盐地质学 salt geology
石盐镁矾岩 kieseritite
石盐岩 halite rock
石堰 stone weir
石阳台 stone balcony
石液酸废液 phenol waste(liquor)
石印 lithograph(y)
石印板(岩) lithographic(al) slate
石印的 lithographic(al)
石印灰岩 lithographic(al) limestone
石印胶滚 lithography roll
石印结构 lithographic(al) texture
石印蜡毛 lithographic(al) crayon
石印蓝 lithosized blue
石印墨 lithographic(al) ink
石印盘 litho master
石印清漆 lithographic(al) varnish
石印石 lithographic(al) chalk;lithographic(al) slate;lithographic(al) stone;lithograph stone
石印石结构 lithogration texture
石印石灰岩 lithographic limestone
石印术 lithography;polyautography
石印油 lithographic(al) oil;litho oil
石印油画 oleograph
石印原版 litho master
石印毡 litho felt;printing blanket
石印照相制版 lithophotogravure
石印纸 litho-paper
石印转印墨辊 nap roller
石英 rock crystal
石英安粗岩 quartz latite
石英安山岩 dacite;quartz andesite
石英摆 quartz pendulum
石英摆倾斜仪 quartz pendulum tiltmeter
石英斑岩 granite porphyry;quartz porphyry
石英板 quartz plate
石英板岩 quartz slate
石英棒 quartz bar;quartz(push) rod
石英棒检术 quartz rod technique
石英棒伸缩仪 quartz bar extensometer
石英棒照明系统 quartz-rod illumination system
石英包层光纤 silica cladded fiber[fibre]
石英保护管 quartz protecting tube
石英倍长岩 quartz-bytownite
石英壁钟 wall quartz clock
石英变换控制 quartz control
石英表 quartz watch
石英表机芯 quartz crystal movement
石英表校表仪 quartz-timer;quartz watch tester
石英波长计 crystal wavemeter;quartz wavemeter
石英玻璃 quartz glass;silex(glass);silica glass;vycor glass
石英玻璃布 quartz cloth
石英玻璃灯 quartz glass lamp
石英玻璃坩埚 quartz glass pot
石英玻璃观察孔 silica window
石英玻璃结构单元 vitron
石英玻璃滤光片 fused silica filter

石英玻璃瓶 quartz bottle
石英玻璃器皿 quartz glass ware
石英玻璃弹簧 quartz glass spring
石英玻璃温度计 quartz glass thermometer
石英玻璃纤维 quartz glass fibre
石英玻璃纤维材料 Refrasil
石英玻璃延迟线存储器 quartz delay-line memory
石英玻璃砖 quartz glass brick
石英材料 silica material
石英测力感传器 quartz force transducer
石英长石斑岩 quartz feldspar porphyry
石英长石云母斑岩 quartz feldspar mica porphyry
石英衬底 quartz substrate
石英储存管 quartz container tube
石英磁力仪 quartz magnetometer
石英次杂砂岩 quartzose subgraywacke
石英粗面斑岩 quartz trachyte porphyry
石英粗面粒玄岩 quartz-banakite
石英粗面岩 liparite;quartz trachyte
石英大理岩 quartz marble
石英的生长 growth of quartz
石英灯 quartz lamp
石英灯泡 quartz bulb
石英碘灯 quartz iodine lamp
石英碘钨灯 quartz iodine tungsten lamp
石英电气石岩 roche moutonnee rock
石英电钟 crystal electric(al) clock; quartz crystal electric(al) clock
石英电子表 quartz electronic watch
石英电子手表 crystal-oscillator watch; quartz crystal watch
石英电子组件 quartz electronic modules
石英二长斑岩 quartz-monzonite porphyry
石英二长石 adamellite
石英二长细晶岩 quartz-monzonite-aplite
石英二长岩 adamellite; normal granite; quartz monzonite
石英发送机 quartz transmitter
石英反应室 quartz reaction chamber
石英方解石脉型金矿石 gold ore of quartz-calcite type
石英霏细岩 quartz felsite
石英分光测光仪 quartz spectrophotometer
石英分光光度计 quartz spectrophotometer
石英分光镜 quartz spectroscope
石英玢岩 quartz porphyrite
石英粉 ground quartz; powdered quartz; quartz flour; quartz powder; silex; silica dust; silica flour
石英粉尘 silica dust
石英粉尘吞噬作用 quartz-dust phagocytosis
石英粉绝缘 quartz powder insulation
石英粉涂料 silica flour wash
石英浮渣 silica scum
石英千枚岩 quartz-phyllite
石英杆 quartz rod
石英杆应变地震仪 quarts rod strain seismometer
石英坩埚 quartz crucible; silica crucible
石英感传器 quartz transducer
石英橄榄粒玄岩 quartz olivine dolerite
石英钢片谐振器 quartz-steel resonator
石英汞灯 quartz mercury lamp
石英钩 quartz hook
石英骨料 quartz aggregate
石英管 quartz ampoule; quartz capsule; quartz tube; silica tube
石英管定硫试验 quartz-tube test for sulfur
石英管反应器 quartz-tube reactor
石英管辐射器 quartz-tube radiator
石英管红外线加热器 quartz-tube infrared heater
石英管温度计 quartz-tube thermometer
石英管芯 <熔模中> quartz core
石英光纤 silica fibre
石英光楔 quartz wedge
石英光楔补偿器 quartz wedge compensator
石英光学平面 quartz optic(al) flat
石英光学声子激射器 quartz optic(al) phonon maser
石英过滤砂 quartz filter sand
石英黑电气岩 schorl rock
石英恒温器 crystal thermostat
石英化 quartzification
石英化合漆 quartz compound
石英换能器 quartz transducer
石英辉长岩 quartz gabbro
石英辉绿岩 quartz diabase
石英基片 quartz substrate
石英加热器 quartz heater
石英加速度计 quartz accelerometer
石英夹持簧 quartz retaining spring
石英架 quartz frame
石英碱流岩 quartz pantellerite
石英渐变型光纤 silica graded fibre
石英胶接剂 quartz cement
石英角斑玢岩 quartz keratophyre
石英角斑岩 baschtauite; quartz keratophyre
石英角斑岩质玻璃 quartz keratophyre glass
石英角闪片岩 quartz hornblende schist
石英校准的 crystal checked
石英结晶 growth of quartz
石英结晶作用 quartz crystallization
石英进样管 quartz sample
石英晶体 piezoid; quartz crystal
石英晶体薄膜监控器 quartz crystal thin film monitor
石英晶体单色仪 quartz crystal monochromator
石英晶体单元 quartz crystal unit
石英晶体的参考轴 X-axis
石英晶体电子表 quartz crystal electronic watch
石英晶体监视器 quartz crystal monitor
石英晶体控制 quartz crystal control
石英晶体控制间歇振荡器 quartz crystal controlled blocking oscillator
石英晶体控制接收机 quartz crystal controlled receiver
石英晶体控制调频接收机 quartz crystal controlled FM receiver
石英晶体滤波器 quartz crystal filter; quartz crystal wave filter
石英晶体频率振荡器 quartz crystal frequency oscillator
石英晶体切割机 quartz crystal cutter
石英晶体时钟 quartz crystal clock
石英晶体水下传声器 quartz crystal subaqueous microphone
石英晶体稳频器 quartz crystal stabilizer
石英晶体谐振子 quartz crystal resonator
石英晶体压力变换器 quartz crystal pressure transducer
石英晶体元件 crystal element
石英晶体振荡器 quartz crystal oscillator; quartz oscillator
石英晶体振荡器装配面 quartz crystal oscillator side
石英晶体振子 quartz crystal unit
石英晶质玻璃 rock crystal
石英巨砾 <岩石中的> bastard quartz
石英聚光镜 quartz condensing-lens
石英绢云母千枚岩 quartz-sericite phyllite
石英绝缘子 quartz insulator
石英颗粒 quartz grain
石英控制 quartz pilot
石英控制发射机 crystal-controlled transmitter; quartz-controlled transmitter
石英控制振荡器 quartz controlled oscillator
石英矿脉 quartz reef
石英拉斑玄武岩 quartz tholeiite
石英类岩 quartziferous rocks
石英棱镜 quartz prism
石英棱镜滤波器 quartz prism filter
石英棱镜仪 quartz prism instrument
石英砾石 quartz conglomerate; quartz gravel
石英粒玄岩 quartz dolerite
石英卤灯 quartz halogen lamp
石英卤气灯 quartz halogen lamp
石英卤素灯 tungsten quartz halogen lamp
石英卤素钨灯 quartz-halogen-tungsten lamp
石英滤波器 quartz filter; quartz wave filter
石英绿泥片岩 quartz chlorite schist
石英绿泥千枚岩 quartz-chlorite phyllite
石英螺旋线秤 quartz spiral balance
石英螺旋线天平 quartz spiral balance
石英脉 quartz reef; quartz vein
石英脉金 reef gold
石英脉矿 quartz mine
石英脉型金矿石 gold ore of quartz vein type
石英猫眼石 quartz cat's-eye
石英毛 quartz wool
石英毛细管 quartz capillary
石英蒙瓦克岩 quartz mengwacke
石英锰榴岩 gondite
石英棉 quartz wool; silica wool
石英面砖 quartz(ite) tile
石英皿 silica dish
石英膜 quartz film
石英膜真空计 quartz membrane ga(u)ge
石英磨机 quartz mill
石英钠长千枚岩 quartz-albite phyllite
石英钠长岩 quartz-albitite
石英闹钟 quartz alarm clock
石英耦合器 quartz coupler
石英盘 quartz disk
石英泡沫玻璃 quartz foam glass
石英喷灯 quartz burner
石英劈 quartz wedge
石英片 piezoid; quartz plate
石英片盒 quartz plate holder
石英片麻岩 quartzite gneiss
石英片岩 quartz-schist
石英偏振单色仪 quartz-polaroid monochromator
石英频率 quartz frequency
石英频率检查仪 quartz-printer
石英汽车钟 quartz digital car clock
石英器皿 silica ware
石英千枚岩 quartzose phyllite
石英切割 quartz cutting
石英侵入体 quartzitic intrusion
石英青盘岩 quartz-prophyry
石英球悬吊型喷淋头 quartzoid bulb pendant type sprinkler
石英燃烧管 quartz burner
石英刃口 quartz knife edge
石英容器 quartz container
石英溶剂 quartz flux
石英纱 quartz yarn
石英砂 arenaceous quartz; crystal silica sand; gan(n)ister sand; quartz grain; quartzite; quartz sand; silica sand
石英砂矿床 quartz sand deposit
石英砂水分 quartz moisture
石英砂屑岩 quartz arenite
石英砂岩 quartzose sandstone; quartzy sandstone; silicarenite; silica silicarenite
石英砂岩建造 quartz sandstone formation
石英砂岩矿床 quartz sandstone deposit
石英闪长玢岩 quartz-diorite porphyrite
石英闪长岩 quartz-bearing diorite; quartz diorite; tonalite
石英闪长岩侵入 quartz-diorite intrusion
石英闪光管 quartz flash tube
石英烧杯 quartz beaker
石英摄谱仪 quartz spectrograph
石英石灰水泥 silica-lime cement
石英石金矿脉 quartz mining
石英石晶体 quartz rock crystal
石英拾音器 piezoelectric(al) pick-up
石英视场致平器 quartz field flattener
石英试验仪表 quartz testing instrument
石英手表测试仪 quartz tester
石英双单色仪 quartz double monochromator
石英双历表 quartz day-date
石英水晶 Brazilian pebble
石英水泥 quartz cement
石英水平磁强计 quartz horizontal magnetometer
石英水银灯 quartz mercury lamp; silica lamp
石英水银电弧灯 quartz mercury arc lamp
石英水银弧光灯 quartz mercury arc lamp
石英水银气灯 quartz mercury vapo(u)r lamp
石英丝 quartz fiber[fibre]
石英丝摆 quartz-fibre pendulums
石英丝剂量计 quartz-fibre dose meter; quartz-fibre dosimeter
石英丝静电计 quartz-fibre electrometer
石英丝黏[粘]滞规 quartz fibre ga(u)ge
石英丝弹簧秤 quartz-fibre balance
石英丝微量天平 quartz-fibre microbalance
石英丝压力计 quartz-fibre manometer
石英丝验电器 quartz-fibre electroscope
石英碎屑 quartz chip(ping)s
石英弹簧 quartz spring
石英弹簧秤 quartz spring balance
石英弹簧式重力仪 quartz spring gravimeter
石英套筒 quartz sleeve
石英梯 quartz ladder
石英天文钟 crystal chronometer
石英条 quartz bar
石英条带矿 quartz-banded ore
石英透镜 quartz lens
石英透镜方法 quartz-lens method
石英透镜摄影 photography with quartz lens
石英瓦克岩 quartzwacke
石英外壳 quartz container

石英微晶闪长岩 quartz microbiorite
石英微晶正长岩 quartz microsyenite
石英温度计 quartz crystal thermometer;quartz thermometer
石英纹理漆 quartz paint
石英稳定器 crystal unit;quartz stabilizer
石英稳定性 quartz stabilization
石英稳频 crystal control
石英稳频的 crystal-controlled
石英稳频法 frequency stabilization by quartz resonator
石英稳频器 quartz frequency stabilizer
石英细脉 quartz stringer
石英细砂 silica fine sand
石英细屑 silica sand fines
石英纤维 quartz fiber[fibre]
石英纤维板 quartz wool sheet
石英纤维秤 quartz-fibre balance
石英纤维悬浮液 quartz-fibre suspension
石英纤维增强酚醛塑料 quartz-fibre reinforced phenolics
石英纤维阵列 quartz-fibre array
石英小卵石 quartz pebbles
石英楔片检偏器 quartz wedge analyser
石英楔(子) quartz wedge
石英斜长岩 quartz-anorthosite
石英谐振器 quartz resonator
石英玄武岩 quartz basalt
石英压电变送器 quartz-piezoelectric-(al) transducer
石英压力传感器 quartz pressure sensor;quartz pressure transducer
石英延迟线 quartz delay line
石英延迟线分析器 quartz delay-line analyser[analyzer]
石英岩 aposandstone;quartzite;quartz rock;rock quartzite;silica rock
石英岩薄板 quartzite slab
石英岩材料 quartzite material
石英岩层 quartzite bed;billy
石英岩道砟 quartzite ballast
石英岩花砖 quartzite tile
石英岩矿床 quartzite deposit
石英岩砾岩 quartzite conglomerate
石英岩玉石 quartzite jade
石英岩砖 quartzite brick
石英岩状板岩 quartzite slate
石英岩状砂岩 quartzite sandstone; quartzitic sandstone
石英页岩 quartzose shale
石英仪器 quartz apparatus
石英应变仪 quartz strain ga(u)ge
石英萤石矿石 quartz-fluorite ore
石英萤石透镜 quartz-fluorite achromat;quartz-fluorite lens
石英元件 quartz element
石英云母片岩 quartz mica schist
石英杂砂岩 quartz graywacke
石英再生生长胶结物 quartz regrowth cement
石英真空管 silica valve
石英真空微量天平 quartz vacuum microbalance
石英振荡器 quartzite oscillator
石英振动片 piezoelectric(al) vibrator
石英振子 quartz crystal vibrator; quartz oscillator;quartz resonator
石英蒸馏器 quartz still
石英正长岩 birkremite;quartz syenite
石英支架 quartz holder
石英指示器 quartz indicator
石英指数 quartz index
石英制品 quartz ware
石英质 quartz
石英质沉积物 quartz sediment
石英质的 quartziferous;quartzose;quartzy
石英质灰岩 quartzose limestone
石英质砾岩 quartzose conglomerate
石英质泥沙 quartz sediment
石英质砂坑 quartz pit sand
石英质砂石 quartzy sandstone
石英质砂岩 quartzite sandstone; quartzitic sandstone; quartz (ose) sandstone
石英质土 cherty soil;quartzitic soil
石英质岩 quartziferous rock
石英质岩石 caple
石英质中长石 dacite
石英钟 crystal clock;crystal electric(al) clock; quartz chronometer; quartz clock
石英钟检验仪 quartz clock tester
石英舟 quartz boat; railboat; silica boat
石英舟架 quartz holder for boat
石英主控振荡器 quartz master oscillator
石英砖 quartz block; quartz (fire) brick;quartzite brick
石英转变 quartz inversion
石英紫外光 quartz ultraviolet light
石英紫外光人工老化试验机 quartz ultraviolet light weatherometer
石英紫外激光器 quartz ultraviolet laser
石拥壁 stone counterfort
石油 fossiloil; mineral oil; bergol; black gold; earth oil; petrol; petroleum oil;rock oil;stone oil
石油柏油 petroleum tar
石油伴生气 gas associated with crude oil
石油饱和的砂岩 oil-saturated sandstone
石油苯 petrobenzene;petroleum benzene
石油泵 petroleum pump
石油标价 oil posted price;petroleum posted price
石油丙烷 petrogas
石油驳船 oil hulk;petrol barge
石油捕集器 <炼油厂下水道的> salvage sump
石油参考价格 oil reference price
石油残油 petroleum residue
石油残渣 black oil;petroleum residue
石油残渣油 petroleum tar
石油测定器 petroleum tester
石油测井结果解释(方法) interpretation of petroleum log
石油产地使用费 oil royalty; royalty petroleum
石油产品 oil products;petroleum liquids;petroleum products
石油产品包装容器 oil package
石油产品的精制 refining of petroleum product
石油产品计量记录仪 pneumeractor
石油产品质量管理 quality control of oil products
石油产区 oildom;petroleum province
石油车用润滑油 petroleum motor oil
石油成因 genesis of petroleum; oil genesis; origin of oil; origin of petroleum
石油成因说 petroleum origin theory
石油赤字 oil(-induced) deficit
石油重整 petroleum reforming
石油除草剂 oil herbicide
石油储备 petroleum accumulation
石油储藏量 oil reserves; petroleum reserves
石油储藏量估计 estimates of petroleum reserves
石油储存 oil storage
石油储存罐 field tank
石油储罐阻火器 flame arrester for petroleum tank
石油储量 oil deposit;prospective oil
石油处理 petrolization
石油处理法 petrolage
石油醇酸 oil alkyd
石油醇酸树脂 oil alkyd resin
石油醇酸油漆 oil alkyd paint
石油代用品 petroleum substitute;petroleum supplement
石油贷款 oil facility
石油的饱和烃 saturated hydrocarbon of petroleum
石油的比重 specific gravity of petroleum
石油的不饱和烃 unsaturated hydrocarbon of petroleum
石油的不足 oil shortage
石油的初沸点 initial boiling point of petroleum
石油的初疑点 initial set of petroleum
石油的电阻率 resistivity of petroleum
石油的沸点 boiling point of petroleum
石油的痕量元素 trace element of oil
石油的留度 density of petroleum
石油的黏[粘]度 viscosity of petroleum
石油的凝固点 temperature of freezing of petroleum
石油的膨胀系数 coefficient of expansion of petroleum
石油的皮囊装运 oil skinning
石油的气味 odo(u)r of oil
石油的氢稳定同位素组成 stable hydrogen isotope composition of petroleum
石油的热演化 thermal alteration of petroleum
石油的收缩率 shrinkage of petroleum
石油的收缩因子 oil shrinkage factor
石油的溯色 colo(u)r of petroleum
石油的碳稳定同位素组成 stable carbon isotope composition of petroleum
石油的体积系数 volume coefficient of petroleum
石油的消耗 consumption of petroleum
石油的旋光性 optic(al) activity of oil
石油的萤光色 fluorescence colo(u)r of petroleum
石油的元素比 element ratio of oil
石油的运移 migration of oil
石油的折光指数 refractive index for petroleum
石油的主要元素 principal element of oil
石油的属性 nature of petroleum
石油地沥青 oil asphalt;petrol(eum) asphalt
石油地球化学 petroleum geochemistry
石油地球化学分析及同位素地球化学 petroleum geochemical analysis and isotopic geochemistry
石油地震勘探 petroleum seismic prospecting;shoot for oil
石油地质 oil geology
石油地质学 petroleum geology
石油地质钻孔 petroleum geologic(al) hole
石油动力 oil power
石油动态 petroleum situation
石油冻 petroleum jelly;vaseline
石油对水生生物的影响 oil effect on aquatic life
石油发动机 petroleum engine
石油发酵 petroleum fermentation
石油法 petroleum law;kerosene method <测土密度用>
石油法案 petroleum legislation
石油芳烃 aromatic oil;petroleum aromatics
石油防喷器 blow(out) preventer
石油放泄弯管 petrol trap
石油非烃化合物 petroleum non-hydrocarbon compound
石油废气 petroleum exhaust
石油废水 petroleum wastewater
石油废物 petroleum waste
石油废液 petroleum waste
石油分解微生物 petroleum degrading microorganism
石油分类 classification of petroleum
石油分离器 petroleum separator
石油分馏物 petroleum fraction
石油分析 petroleum analysis
石油富集带 oil abundance zone
石油干性油 petropon
石油港(口) petroleum port; petroleum harbo(u)r
石油膏 petroleum butter; petroleum jelly
石油工程 petroleum engineering
石油工程师 petroleum engineer
石油工业 oil industry; petroleum industry
石油工业发展趋势 oil industry trends
石油工业废水 oil industry wastewater; petroleum industry wastewater
石油工业用管材 oil country tubular goods
石油公司国际航运论坛 Oil Companies International Marine Forum
石油公司海上油污赔偿协会 Oil Companies Institute for Marine Pollution Compensation Ltd.
石油供应过剩 oil glut
石油构造井 oil structure well
石油固体 petroleum solid
石油管道 oil pipeline;petroleum pipeline
石油管路系统 petroleum pipeline system oil pipeline system
石油管线 oil pipeline;petroleum pipeline
石油管线工地 firing-line
石油灌装栈桥 oil-loading rack
石油锅炉燃料 boiler oil
石油含硫量分析器 petroleum sulfur analyser[analyzer]
石油和天然气勘探开采租约 oil and gas lease
石油和天然气遥测 oil and gas telemetry
石油湖 <海湾战争时科威特油田受破坏形成的> oil lake
石油化工 petrochemical engineering; petrochemical industry
石油化工产品 petrochemical derivatives; petrochemical products; petrochemicals
石油化工厂 petrochemical plant; petrochemical works; petroleum chemical plant
石油化工废水 petrochemical wastewater; wastewater from petrochemical industry
石油化工废物 petrochemical waste
石油化工工业 petrochemical industry
石油化工工艺装置 petrochemical process unit
石油化工公司 petrochemical complex; petrochemical corporation
石油化工联合工厂 petrochemical complex
石油化工设计院 petrochemical engineering(designing) institute
石油化工原料 petrochemical materials
石油化工制品 petrochemicals; petroleum chemicals

石油化工装置 petrochemical plant
石油化工总厂 general petrochemical works;petrochemical complex
石油化工总公司 petrochemical complex;petrochemical corporation
石油化学 petrochemistry;petroleum chemistry
石油化学产品 petrochemicals;petroleum chemicals
石油化学产品的储藏 petrochemical storage
石油化学产品的储藏器 petrochemical storage vessel
石油化学产品的气味 petrochemical odo(u)r
石油化学产品的生产 petrochemical manufacture
石油化学的 petrochemical
石油化学反应 petrochemical reaction
石油化学废料处置 petrochemical waste disposal
石油化学废弃物 petrochemical waste
石油化学废物处置 petrochemical waste disposal
石油化学工厂 petrochemical plant
石油化学工业 petrochemical industry
石油化学工业废水 wastewater from petrochemical industry
石油化学工业局 petrochemical industrial bureau
石油化学工业总公司 petrochemical industry corporation
石油化学化合物 petrochemical compound
石油化学加工 petrochemical processing
石油化学加工装置 petrochemical processing plant
石油化学品的储存容器 petrochemical storage vessel
石油化学制品 petrochemicals;petroleum chemicals
石油化学中间产品 petrochemical intermediate
石油化学总公司 petrochemical complex;petrochemical corporation
石油环烷 naphthene
石油磺酸 mahogany acid
石油磺酸酸铝泡沫剂 alumin(i)um sulfonate foamer;petroleum
石油磺酸盐 mahogany petroleum sulfonate;mahogany sulfonate;petroleum sulfonate;petronate
石油磺酸油 petroleum sulfonate
石油磺酸皂 mahogany soap
石油挥发气 petroleum vapo(u)r
石油挥发气体 petroleum gas
石油挥发油 benzin(e);petroleum benzene
石油回收 petroleum recovery
石油火焰 oil-fired flame;oil fires
石油机车 petroleum locomotive
石油机械厂 petrolic machinery plant
石油积聚 oil accumulation
石油基地 oil base
石油基防锈剂 petroleum base rust preventive
石油基能源产品 petroleum-based energy product
石油基泥浆 oil-based mud
石油基添加剂 petroleum base additive
石油基钻泥 oil-based mud
石油及天然气成因 petroleum and natural gas origin
石油及天然气资源 oil and gas resources
石油计量器 oil meter
石油计量仪器 petroleum metering instrument
石油计税标价 petroleum posted price
石油加工 oil processing;petroleum processing
石油加工产品 refinery petroleum products;refinery products
石油加工厂 oil refinery(plant);processing plant
石油加工厂废水 petroleum refinery wastewater
石油加工厂废物 petroleum refinery waste
石油加工废料 petroleum refinery waste
石油加工废(弃)物 petroleum processing waste
石油加工过程流程 petroleum processing flow
石油加工化学 chemofining
石油加工装置 oil processing unit
石油加热膨胀 petroleum expansion
石油加压处理法 Marietta process
石油价格模拟模型 oil price simulation model
石油简易分析 simple analysis of petroleum
石油碱 petroleum base
石油焦 petrol coke;petroleum oil coke;refinery coke
石油焦沥青 asphaltic pyrobitumen
石油焦炭 oil coke;petroleum coke
石油焦油 <石油裂化后蒸馏残渣> petroleum tar
石油焦油沥青 petroleum tar
石油脚 petroleum residue
石油接收码头 crude receiving terminal
石油截留设施 oil interceptor
石油进口税 oil duties
石油进口限额 oil imports quota
石油浸染岩石 oil-impregnated rock
石油禁运 oil embargo
石油精 naphtha;petroleum spirit
石油精炼废水 waste from petroleum refinery
石油精炼工厂废水 wastewater from petroleum refinery
石油精炼和石油化工厂 refinery and petrochemical plant
石油精馏器 oil rectifier
石油井 oil-spring;oil well
石油井用旋转钻头 oil field rotary bit
石油聚集 oil accumulation
石油聚酯树脂 oil alkyd resin
石油开采 oil exploitation;oil production;petroleum extraction
石油开采权 oil concession
石油开发 oil development
石油开发装置 oil development equipment
石油勘探 oil exploration;oil prospecting;petroleum prospecting
石油勘探船 petroleum surveying ship
石油勘探地质学家 rock hound
石油勘探数据库 oil exploration data base
石油勘探者 oil prospector
石油科学 naphthology;petroleum science
石油可燃性组分 inflammable constituent of petroleum
石油可溶性组分 soluble constituent of petroleum
石油矿藏耗减 oil depletion
石油矿藏耗减优惠 oil depletion allowance
石油矿床 oil deposit;petroleum deposit
石油矿浆矿石三用船 oil-slurry-ore carrier
石油矿石运输船 oil ore carrier
石油扩散 petroleum migration
石油蜡 petroleum wax
石油类柏油脂 petroleum pitch
石油类产品 petroleum product
石油类焦油脂 <如裂化石油沥青硬渣,天然硬沥青> petroleum pitch
石油类农药 oil pesticide
石油沥 petrobitumen
石油沥青 asphalt(um);petroleum (asphaltic) bitumen;petroleum asphalt;petroleum pitch;petroleum tar;semi-asphaltic flux
石油沥青泵 asphalt pump
石油沥青材料 asphalt material
石油沥青掺和剂 asphalt addition
石油沥青衬里 asphalt lining
石油沥青底层 asphalt-base course
石油沥青底子 asphaltic base
石油沥青地面 asphalt floor(ing)
石油沥青垫层 asphalt mattress
石油沥青防潮层 asphalt damp-proof course
石油沥青粉末 asphalt powder
石油沥青骨料混合料 asphalt-aggregate mixture
石油沥青灌浆 asphalt grouting
石油沥青灌注料 asphalt grout
石油沥青混合料 asphalt mixture
石油沥青(混合)摊铺机 asphalt paver
石油沥青混凝土 asphalt concrete
石油沥青基料 asphalt binder
石油沥青加热锅 asphalt heater
石油沥青浇面 asphalt topping
石油沥青胶熔制锅 mastic asphalt cooker
石油沥青焦炭 petroleum pitch coke
石油沥青焦油沥青化合物 bitumen-tar blend;bitumen-tar mixture
石油沥青焦油沥青混合料 <一般指含煤沥青较多的> bitumen-tar blend
石油沥青接缝 bleeding joint
石油沥青冷底子油 asphalt-base oil
石油沥青路 asphalt road
石油沥青路面 pavement of asphalt
石油沥青玛琋脂 asphalt(ic) mastic
石油沥青煤 asphalt coal
石油沥青面层切割机 asphalt cutter
石油沥青膜 asphalt membrane
石油沥青黏[粘]层 asphalt tack coat
石油沥青喷洒机 asphalt distributor
石油沥青片瓦 asphalt shingle
石油沥青溶解于四氯化碳中的比例 <测定石油沥青所含杂质> proportion of bitumen soluble in carbon tetrachloride
石油沥青熔渣 asphalt(ic) clinker
石油沥青乳液 asphaltic emulsion
石油沥青砂浆 asphalt mortar
石油沥青砂胶 asphalt mastic;mastic asphalt
石油沥青填料 asphalt filler
石油沥青透层 asphalt prime coat;asphalt primer
石油沥青涂层 coat of asphalt
石油沥青涂料 asphalt paint
石油沥青稳定 asphalt stabilization
石油沥青烯 asphaltene
石油沥青岩 asphaltite
石油沥青油毡 asphaltic felt;asphalt saturated organic felt
石油沥青毡 asphalt felt;bituminous felt
石油沥青质 asphaltene
石油沥青砖 asphaltic acid
石油联合加工 combined oil-processing
石油炼厂 petroleum refinery
石油炼油厂 oil refining factory
石油炼制 petroleum refining
石油炼制厂 oil refinery(plant)
石油炼制废水 petroleum refining wastewater
石油炼制工业 petroleum refining industry
石油炼制工业污水 petroleum refining industry sewage
石油炼制炉 petroleum refining furnace
石油炼制能力 petroleum refining capacity
石油炼制燃料 oil fuel
石油炼制蒸馏釜 oil refining still
石油裂化 cracking of petroleum;petroleum cracking
石油裂化过程 petroleum cracking process
石油裂解 cracking of petroleum;petroleum cracking
石油裂解气 cracking gas of petroleum
石油淋 petrolin(e)
石油流动问题 oil flow problem
石油馏出物 petroleum distillate
石油馏分 fraction of petroleum;oil distillate;petroleum cut;petroleum fraction
石油码头 oil handling wharf;oil jetty;oil pier;oil terminal;oil wharf;petroleum jetty;petroleum pier;petroleum terminal;petroleum wharf;tanker terminal
石油埋藏量 oil deposit
石油媒染剂 oil mordant
石油美元 oil dollar;petro-dollar
石油美元再循环 recycling of petrodollars
石油醚 benzinam;benzin(e);ligarine;light petrolene;ligroin(e);petroleum benzin;petroleum ether;petrolic ether
石油醚不溶物 petroleum ether insoluble
石油醚提出物 petroleum ether soluble matter
石油密度计 hydrometer
石油灭火系统 petrochemical extinguishing system
石油耐久性 kerosene resistance
石油黏[粘]度 oil viscosity
石油农业 petro-agriculture
石油浓缩物 mineral oil concentrate
石油牌价 oil post price
石油配给 oil rationing
石油喷射器 petroleum feeder
石油品种 grade of oil
石油破坏蒸馏得到的烃气 oil gas;oil-gas from petroleum
石油破坏蒸馏制造烃气的过程 oil-gas process
石油普查 oil finding;oil search
石油企业家 petroleum producer
石油起源 oil genesis;oil origin;petroleum origin
石油气 commercial rock gas;natural gas;oil gas;petrogas;petroleum gas;petrol vapo(u)r
石油气分离 petroleum gas separation
石油气体 petroleum gas
石油气体油 petroleum gas oil
石油气味 petroleum odour
石油气压缩机 petroleum gas compressor
石油气族 gas family
石油迁移 petroleum migration
石油青灌注碎石路面 asphalt-grouted surfacing
石油青灌注碎石面层 asphalt-grouted surfacing
石油轻溶剂油 ligroin(e)
石油情报 petroleum intelligence

石油区 petroleum region
石油区间勤务处 petroleum intersectional service
石油圈闭 petroleum trap
石油泉 oil-spring
石油燃料 oil fuel;petroleum fuel
石油燃料代用品 alternate fuel
石油燃料油 petroleum fuel oil
石油燃料造成的火灾 oil fires
石油燃烧 oil combustion
石油热裂解 thermal cracking of petroleum
石油溶剂 mineral turpentine;petroleum solvent
石油溶剂油 mineral spirit; mineral turps;petroleum spirit;white spirit
石油溶剂油容忍度 mineral spirit tolerance
石油溶于天然气 solubility of oil in gas
石油乳剂 kerosene emulsion;petroleum emulsion
石油乳液 kerosene emulsion;paraffin(e) emulsion;petroleum emulsion
石油软膏 petrol ointment;petrosapol
石油软化剂 oil-extender; petroleum flux
石油软制剂 petroleum flux
石油润滑剂 petroleum grease;petroleum lubricant
石油润滑脂 petroleum grease;petroleum lubricating grease
石油洒播法 petrolization
石油散货矿石三用船 oil-bulk-ore carrier
石油散货两用船 combination tanker
石油色谱 petroleum chromatography
石油设备 oil equipment
石油设施 petroleum facility;petroleum installations
石油生产 oil production
石油生产国 petroleum production country
石油生产及输出国家 Oil Producing and Exporting Countries
石油生成 oil genesis;petroleum formation
石油生成峰 peak of oil generation
石油省 petroleum province
石油石蜡 petroleum paraffin
石油时代 petroleum times
石油使用 oil appliance
石油试验法 petroleum testing method
石油试钻井 wildcatting
石油收集 oil gathering
石油收入 oil money
石油收入税 petroleum revenue tax
石油输出国组织 <欧佩克> Organization of Petroleum Exporting Countries
石油输送 oil delivery;oil-transferring
石油输送管线 crude line
石油束 filament of oil
石油树脂 hydrocarbon resin;petroleum resin;petropol;petroresin
石油水煤气 oil-water gas
石油水煤气焦油 oil-water gas tar
石油/水泥浆/矿砂三用船 oil/slurry/ore ship
石油水溶液 oil in water solution
石油税 petrol duties
石油酸 petroleum acids
石油损耗 oil loss
石油损失控制 oil losses control
石油炭黑 oil black;petroleum black
石油炭黑厂 oil-black plant
石油探测器 oil finder
石油碳 petrocarbon
石油提炼 petroleum extraction;petroleum refining
石油体积收缩率 volumetric(al) shrinkage of oil
石油天然气 natural petroleum gas
石油天然气勘探开采租约 oil and gas lease
石油天然气钻探 oil and gas drilling
石油添加剂 petroleum additive
石油烃 petroleum hydrocarbon
石油烃类化合物 petroleum hydrocarbon compound
石油烃树脂 petroleum hydrocarbon resin
石油烃油 petroleum hydrocarbon oil
石油通货 petrocurrency
石油统计 oil statistics
石油桶 petrol barrel
石油突堤码头 oil jetty;oil pier
石油脱硫 petroleum sweetening
石油脱轻 <蒸去轻质油> topping of petroleum
石油危机 oil crisis;oil shock
石油微生物 petroleum microorganism
石油微生物学 petroleum microbiology
石油尾气 paraffinic gas
石油稳定性 kerosene resistance
石油污染 oil pollution; petroleum contamination;petroleum pollution
石油污染残留(物) oil pollution residue
石油污染的海岸 oil polluted seashore
石油污染的水域 oil polluted waters
石油污染的遥感分析 remote-sensing analysis of oil pollution
石油污染法令 Oil Pollution Act
石油污染防治 oil pollution control
石油污染监视系统 oil pollution surveillance system
石油污染检测 oil pollution detection
石油污染紧急情况 oil pollution emergency
石油污染控制 oil pollution control
石油污染事故 oil pollution emergency
石油污染水域 oil-polluted waters
石油污染损害 oil pollution damage
石油污染物 petroleum contaminant;petroleum pollutant
石油污染遥感系统 oil pollution remote-sensing system
石油物理性质 physical property of petroleum
石油烯 petrolene
石油稀释剂 petroleum flux;petroleum thinner;white spirit <油漆的>
石油系合成干性油 hydrocarbon drying oil;petroleum drying oil
石油系油类 petroleum oil
石油消费国 oil consumption country
石油消耗量 petrol consumption
石油卸货码头 crude unloading terminal
石油信贷 <为石油购买者提供的信贷> oil facility
石油形成 evolution of petroleum;petroleum genesis
石油性质 nature of oil
石油需要量 oil demand
石油絮凝脱水法 flocculation oil removing method
石油学会试验法 <英> induced polarization test methods
石油学院 petroleum institute
石油岩资源 oil-shale resources
石油衍生的燃料 petroleum-derived fuel
石油衍生烃 petroleum-derived hydrocarbon
石油衍生物 oil derivative;petroleum derivative;petroleum-derived product
石油野外钻探车车身 oil field body
石油业 oil interest
石油业者 oilman
石油仪器分析 apparatus analysis of petroleum
石油移动 oil migration
石油乙炔 petroacetylene
石油溢出 oil spillage
石油溢出的预防控制及清洁措施 oil spill prevention control and clean-up
石油盈余 oil surplus
石油荧光 petroleum bloom
石油硬沥青 petroleum pitch;thermal asphalt;thermal pitch
石油用品 oil appliance
石油油料 petroleum oil
石油与水蒸气的混合物 oil-steam mixture
石油元素组成 element of petroleum
石油原油 petroleum crude oil
石油运输 oil transport; petroleum transportation
石油运输船 oil tanker;petroleum ship;tanker
石油运移 oil migration;petroleum migration
石油蕴藏量 oil reserves
石油杂酚油配合液 <木材防腐> creosote-petroleum solution
石油再生 petroleum recovery
石油再生厂 oil regeneration plant
石油皂 petroleum soap
石油渣油 dead oil;petroleum residual oil
石油炸药 axite
石油窄馏分 narrow cut petroleum fractions
石油蒸发损失 oil shrinkage loss
石油蒸馏 petroleum distillation
石油蒸馏残余物 petroleum distillation residue;petroleum tailings
石油蒸馏釜 petroleum still
石油蒸馏锅 crude still
石油蒸馏炉的炉管支架 oil-still tube supports
石油蒸馏器 oil rectifier
石油蒸馏塔 petroleum oil column
石油蒸馏液 petroleum distillate
石油蒸馏在煤油后的馏分 solar oil
石油蒸馏泵 oil refinery pump
石油蒸气 petroleum vapo(u)r
石油直馏 reducing of crude oil
石油制品 oil product; petroleum product
石油质 melthene
石油致冷剂 petroleum refrigerant
石油中间产品 petroleum intermediate product
石油重油 black petroleum product;black product
石油贮存 oil storage
石油专用列车 oil unit train
石油砖 petroleum briquet
石油转移 travel of oil
石油转运港 oil transshipment port
石油装船码头 oil-loading terminal
石油装卸码头 petroleum jetty
石油装载 oil shipment
石油装载许可证 oil permit
石油资源 oil resources;petroleum resources
石油资源保护 oil conservation;petroleum resources conservation
石油资源调查 petroleum resources survey
石油资源节约 petroleum conservation
石油综合利用 overall utilization of petroleum;utilization of petroleum
石油组成 component of petroleum;petroleum composition
石油钻采工程 petroleum engineering
石油钻机 oil rig
石油钻井 petroleum drilling
石油钻井废液 oil drilling waste
石油钻井平台 oil drilling platform;oil platform
石油钻井用钢丝绳 petroleum well drilling wire rope
石油钻井装置投资 oil drilling equipment
石油钻塔 oil rig
石油钻探 boring for oil;oil drilling;petroleum drilling
石油钻探平台 drill ship;oil drilling island;oil drilling platform
石油钻探人工岛 oil drilling island
石油作业 petroleum operation
石榆木 rock elm
石缘琢边 margin draft;margin draught
石陨石 aerolite;aerolith;stony meteorite
石錾 chipping chisel;stone chisel
石凿 chipping chisel;puncheon;rock chisel; stone chisel; stone pick;comb chisel;plain chisel
石凿平 plain work
石造挑水坝 stone dike[dyke]
石造物 stonework
石渣 ballast(aggregate);break stone;tailing; quarry muck; scappling;chip;crushed rock ballast;crushed-stone ballast; debris; quarry dust;rock ballast;rock chunk;rock spoil;stone ballast; tailings; quarry rubbish <采石场的弃料>
石渣坝 rock debris dam
石渣仓 stone bin
石渣场 quarry
石渣车 ballast wagon
石渣传送带 muck conveyer[conveyor]
石渣传送机 muck conveyer[conveyor]
石渣道床 ballast bed
石渣底料 crushed rock base material
石渣垫层 scappling
石渣斗车 bunker car
石渣混凝土 ballast concrete
石渣加工厂 crushing plant
石渣犁 spreader
石渣列车 ballast train
石渣溜槽 muck chute
石渣漏底车 ballast car;ballast truck
石渣漏斗车 ballast hopper wagon
石渣路 ballast road;ballast road bed
石渣路基 ballast bed
石渣路床 ballast road bed
石渣路肩 ballast shoulder
石渣滤水层 ballast filter
石渣滤水池 ballast filter
石渣滤水床 ballast filter
石渣煤 ballast coal
石渣面层 surface of ballast
石渣耙 stone fork
石渣铺面的 ballast - surfaced
石渣撒布车 ballast spreader
石渣筛 ballast screen
石渣升降机 hoisting muck car
石渣线 quarry siding
石渣运输系统 muck handling system
石渣抓斗 muck fork
石渣装载机 ballast loader
石轧砂 crushed-stone sand
石辗 stone roll(er)
石遮帽 stone coping
石针 lithostyle;tentaculocyst
石枕 stone block tie
石支托 stone corbel
石支座拱顶 stone cradle vault
石枝【地】 helictite
石直棂窗 stone mullion window

石制代型 stone die
石制面板 stone hedge
石制品 rock product;stoneware
石制品工业 rock industry
石制品下水管 stoneware drain pipe
石制镶板 stone hedge
石质岸 petrous coast;rocky shore
石质层 <土壤岩石接触层> lithic contact
石质的 lithical;petrean;petrous
石质底土 rocky subsoil
石质地 rock land
石质多边形土 stone polygon
石质粉尘 stone dust
石质构造 rock fabric
石质构造土 stone structure soil
石质海岸 rocky coast;rocky shore
石质海绵属 Petrostroma
石质华 lithoid tufa
石质荒漠 hammada;rock desert;stone desert;stony desert
石质荒漠群落 rupideserta
石质基础 soling
石质介质 rock medium
石质矿石 lithic ore
石质拉沟 rock cutting
石质路 stone road
石质路基 rock subgrade
石质路堑 rock cutting
石质路缘石 stone curb
石质腻子黏[粘]结 stone putty cementation
石质凝灰岩 lithic tuff
石质浅滩 rock(y)shoal
石质沙漠 rock desert; stone desert; hamada
石质砂岩 lithic sandstone
石质扇形地 rock fan
石质石膏 rock gypsum
石质隧洞 rock bore
石质土 chisley soil; lithosoil; rocky soil; skeletal soil; skeleton soil; stony soil
石质土壤 rocky soil
石质纤维 <即石棉> earth flax;rock wool
石质硬底子 flinty ground
石质油灰 stone putty
石质油灰黏[粘]结 stone putty cementation
石质杂砂岩 lithic graywacke
石质中梃 mullion
石质中梃窗 stone mullion window
石竹 China pink; dianthus chinensin; pink
石竹淡红色 carnation rose
石竹红色 carnation red
石竹花 pink
石竹花园 dianthus garden
石竹科 Caryophyllaceae
石竹色的 pink
石竹萜烯 caryophyllene
石竹园 pink garden
石竹属 carnation
石柱 hoodoos; pedestal stone; pedestal rock; peristele; pillar; stack; stalacto-statagmite; stela[复 stelal/steles]; stone column; stone pillar; stalactite column
石柱长度 length of karstic stony column
石柱上架石梁的纪念门 trilith(on)
石柱碎石桩 stone column
石柱型 stone order
石柱直径 diameter of karstic stony column
石柱中楣 stone frieze
石筑结构 stone structure
石筑楼梯 stone stair(case)
石桩 bootleg; monument; rock pile; stone mark;stone pillar
石状黏[粘]土 leck;Clay spar <商品名>
石锥 stone prick;coup-de-poing
石子 gravel;rounded pebble
石子混凝土 stone concrete
石子露出 <水磨石饰面的> reveal lining
石子路 cobbled road; cobblestone street; stone road; stoneway; dirt road
石子热涂(沥青)厂 hot-coated-stone plant
石子热涂(沥青)车间 hot-coated-stone plant
石子韧性试验 toughness test of gravel
石子筛 stone screen
石子受料箱 stone receiving box
石子滩 shingle beach
石梓属 bushbeech
石钻 churn drill;stone drill
石钻头 aiguille
石嘴 rocky point;spur
石作安装 stone fixing
石作安装工程 stone fixing work
石作(工程) stonework
石作锚碇 anchoring of stonework
石作锚碇法 anchor method of stonework
石坐凳 bench table

时【地】chron

时变 time-variation;time-varying
时变变阻抗 time-variable impedance
时变参数模型 time-varying parameter model
时变场 time-varying field
时变的 time-dependent;time-variant
时变的时间比例尺 time-varying time scale
时变动态(分析) time to day dynamic
时变功率谱 time-varying power spectrum
时变化 hourly fluctuation; hourly vibration
时变化系数 hourly variation coefficient;hourly variation factor
时变化因数 hourly variation factor
时变角 angle of distortion
时变控制 time-dependent control
时变控制系统 time-variant control system
时变力 time-varying force
时变滤波 time-varying filtering
时变模型 time-varying model
时变时延散布信道 time-varying channel with delay spread
时变梯度 time-varying gradient
时变通量 time-changing flux;time-varying flux
时变图 time-varying figure
时变网络 time-varying network
时变系数 time-varying coefficient
时变系统 time-dependent system;time variant system;time-varying system
时变线性过滤器 time-varying linear filter
时变线性滤波器 time-varying linear filter
时变性 time dependence
时变性能 time behavio(u)r
时变正弦曲线梯度 time-varying sinusoidal gradient
时变坐标 time-varying coordinate
时标 completion cycle; hour-index; time mark(er); time marking; time reference;time scale
时标触发器 clocked flip-flop
时标道 clock track
时标的相对论性扩展 relativistic dilation of time-scale
时标的准确度 accuracy of time scale
时标电路 timing circuit
时标电容器 timing capacitor; timing condenser
时标电阻 timing resistor
时标定时 timing signal
时标抖动 time jitter
时标读出元件 timing mark sensor element
时标发生器 time marker; time mark generator
时标换算系数 time scale factor
时标寄存器 marker register
时标校验 time scale check; timing mark check
时标脉冲 timing pulse
时标脉冲发生器 time pulse generator;timing pulse generator
时标盘 hour circle
时标输入 clock input
时标速率 clock rate
时标同步 timing signal
时标图 clock plot
时标系统 time system;timing system
时标信号 timing signal
时标序列 timing sequence
时标因子 time scale factor
时标振荡器 signal generator; time mark generator; timing generator; wave generator
时不变 time invariance;time-invariant
时不变滤波 time-invariant filtering
时不变模型 time-invariant model
时不变系统 time-invariant system
时差 hour rate;moveout <地震波的>
时差定位法 space-time processing; time-of-arrival location
时差反应 jetlag
时差方程式 equation of time
时差角 parallactic angle
时差校准带 skew master tape
时差率 time-preference rate;equation of time
时差曲线 curve of time equation; moveout curve
时差调整 rate adjustment
时差转盘 <设定绿灯信号开始时间的> offset dial
时常发生的 frequent
时常汇报 keep one advised of
时程法 time-history method
时程反应 time-history response
时程分配 temporal distribution
时程分配型式 temporal pattern
时程分析 time-history analysis
时程分析法 time-history analysis method
时程记录 time record(ing)
时程记录仪 time-history recorder
时程曲线 time-history curve; time-travel curve;travel-time curve
时迟 time lag
时迟系统 time-lag system
时齿条 hour rack
时畴 time domain
时畴显示 time domain display
时出力 hourly output
时窗 time window
时窗长度 time window length
时窗起始时间 window start time
时窗移动间隔 shift interval of time window
时窗终了时间 window end time
时锤 hour hammer
时锤操纵杆簧 hour hammer operating lever spring
时锤操纵杆偏心销 hour hammer operating lever eccentric
时锤杆簧 hour hammer spring
时锤杆压片 hour hammer cover
时锤杆桩 hour hammer stud
时锤中心杆 hour hammer intermediate lever
时代 age;era;times
时代地层单位【地】time-stratigraphic(al) unit
时代地层相 time-stratigraphic(al) facies
时代对比 time correlation
时代分类 time classification
时代风格 period style
时代间断【地】time break
时代式样 period style
时代性 temporal spirit
时带 chronozone
时点数列 specified time series
时段 time interval
时段钟 time-of-day clock
时断时续 intermittent disconnection; on again off again;on-and-off
时断时续的 on-off
时断时续地 off-and-on
时断信号 time break
时法 hour system
时分 hour minute; minute of time; time division
时分保密机 time division scrambler
时分操作 one-at-a-time operation
时分的 time-shared
时分电路 time division circuit
时分电子电话交换机 time division electronic telephone switching system
时分定序 base-timing sequencing
时分多工技术 time multiplex technique
时分多路传输 time division multiplex
时分多路传输系统 time division multiplex transmission system
时分多路存取 time division multiple access
时分多路访问 time division multiple access
时分多路复用 time division multiplex
时分多路复用技术 time division multiplexing
时分多路复用系统 time multiplexed system
时分多路调制器 time division multiplexer
时分多路通信[讯] channel(l)ing time dividing;channel(l)ized time dividing;time dividing channel(l)ing
时分多路转换 time division multiplexing
时分多路转换器 time division multiplexer
时分多路转接 time division multiplexing
时分多路总线 time division multiplex bus
时分多址存取 time division multiple access
时分多址联结方式 time division multiple access
时分多址通信[讯]系统 time division multiplex access message system
时分复用 time division multiplexing; time multiplexing
时分复用技术 time division multiplexing technique
时分激光器 time-sharing laser
时分交换 time division switching
时分交换网 time division switching network
时分开关 time division switching
时分控制通路 time division control access

S

时分量 real component
时分数据链路 time division data link
时分双工 time division duplex
时分通道 time-derived channel
时分系统 time division system
时分信道 time-derived channel
时分指令同步方式 time division command synchronization
时分制 time division system
时分制电报 time division telegraph
时分制多路传输 time division multiplex transmission; time multiplex; time multiplex transmission
时分制通信[讯] time division system
时分制遥测系统 time division telemetry system
时分总控制 time division common control
时负荷 hourly load
时高剖面 time-height section
时号 hour mark; time signal
时号归算 reduction of time signals
时号校正 correction to time signal
时号接收机 time receiver
时号自记仪 ondulateur
时机 conjuncture; opportunity
时机成熟 high time
时机代价 opportunity cost
时机心托限制环 hour mechanism support banking ring
时基 time base
时基波形 time base waveform
时基颤动 time flutter
时基充电电容器 time base charging capacitor
时基重复频率 time base repetition
时基电路 time base circuit
时基电容器 timing capacitor; timing condenser
时基电压 time base voltage
时基抖动 time base flutter
时基发生器 time base generator
时基分解 time base resolving
时基功率放大器 time base power amplifier
时基光学测距仪 time based optic(al) range finder
时基基准信息 time base reference information
时基计数电路 time base counter chain
时基校正 time base correction
时基控制 time base control
时基扩展 time base expansion
时基滤波 time base filtering
时基频率 time base frequency
时基曲线图 time base diagram
时基扫描 time base sweep; time sweep
时基扫描波束 time reference scan(ning) beam
时基扫描多谐振荡器 time base sweep multivibrator
时基调整器 time base corrector
时基图 time base diagram
时基稳定 time base stabilizing
时基误差 time base error
时基误差校正器 time base error correction
时基协调器 time based coordinator
时基信号 time code
时基信号发生器 time base generator
时基压缩 time base compression
时基压缩式频谱分析仪 time base pressing type spectrum analyser [analyzer]
时基振荡器 time base generator
时基装置 time base unit
时基准确度 time base accuracy
时计 chronometer; horologe; hour meter; time keeper; time piece; timer; timekeeper < 钟表等 >
时计差 chronometer correction
时计齿轮 horological gear
时计电路 chronometer circuit
时计改正 chronometer correction
时计校正 chronometer correction
时计数附加中间轮 hour counter additional intermediate wheel
时计数轮 hour-counting wheel
时计数轮夹板 hour-counting wheel bridge
时计数轮摩擦簧 hour-counting wheel friction spring
时计数轮套 hour-counting wheel tube
时计数轮桩 hour-counting wheel stud
时计数器传动齿轴 hour counter driving pinion
时计数器传动轮 hour counter driving wheel
时计数器过轮 hour counter setting wheel
时计数器活动杆 hour counter operating lever
时计数器活动杆构压片 hour counter operating lever hook
时计数器活动杆簧 hour counter operating lever spring
时计数器连接桩 hour counter coupling stud
时计数器锁杆 hour counter lock
时计数器锁杆簧 hour counter lock spring
时计数器跳杆 hour counter jumper
时计数器跳杆垫 hour counter jumper seating
时计数器跳杆偏心轴 hour counter jumper eccentric
时计数器中间轮 hour counter intermediate wheel
时计数器轴 hour counter spindle
时计数凸轮 hour counter cam
时计数指示器 hour counter indicator
时计数柱轮 hour counter column wheel
时计数柱轮跳杆 hour counter column wheel jumper
时计误差 chronometer error
时记 time mark
时记录 hour recording
时际变率 interhourly variability
时价 current cost; current price; market value; present price; prevail(ing) price; quotation; ruling price
时价发行 tap issue
时价会计 current cost accounting; current-value accounting
时价率 average rate
时价行情 tap rate
时价支出法 market price at time of issue method
时价总额 aggregate value of listed stock
时间安排 scheduling
时间安排表 schedule; timing
时间安排不当 mis-timing
时间摆 time balance
时间比 time ratio
时间比尺 < 模型 > time scale ratio
时间比较仪 chronocomparator; time comparator; time interval comparator
时间比可控的斩波器 time-ratio-controlled chopper
时间比控制 time rate control
时间比例 time scale
时间比例标度变化 memomotion
时间比例控制器 time proportioning controller
时间边缘效应 time edge effect
时间编码 temporal coding; time code; time(en) coding
时间编码读出器 time code reader
时间编码发生器 time code generator
时间编码器 clock coder; time coder
时间编码转换器 time code translator
时间变更 time change
时间变化 temporal change
时间变化边界条件 time-varying boundary condition
时间变化标程 time-varying level
时间变化控制 time-variation control
时间变化数据 time-variable data
时间变化增益 time-varied gain
时间变换参数 time-varying parameter
时间变换器 time converter; time transformer
时间变量 time-variable; time-variant
时间变量因素 time-dependent factor
时间变率 time rate; time rate of change
时间变慢效应 time dilation
时间变数 time-variable
时间变异 time variance
时间标度 airsecond; time scale
时间标度的放大 time magnifying
时间标法 time scale calibration method
时间标记 time mark; timer
时间标记装置 time-marking device
时间标志 timing mark
时间标准 time standard
时间标准发生器 time standard generator
时间表 program(me); time bill; time chart; time profile; time schedule; timetable; timing schedule
时间表编辑 timetable editor
时间表格文件 timetable file
时间表管理 timetable handling
时间表和违约赔偿 time schedule and liquidated damages
时间表控制 time schedule control; time schedule variable control
时间表控制器 time schedule controller
时间表问题 scheduling problem
时间波形 time waveform
时间补偿 time bias
时间不变量 time-invariant
时间不变系统 time-invariant system
时间不变性 time invariance
时间不变性运算 time-invariant operation
时间不变原理 principle of time invariance
时间不同 asynchronism; asynchronization
时间不同的 asynchronous
时间步长 time step
时间步长比 time-step ratio
时间步进的 time-stepping
时间采样 time sampling
时间采样定理 temporal sampling theorem
时间参考尺度 reference time scale
时间参考信号 timing reference signal
时间参考坐标 time reference coordinate
时间参数 time reference; time parameter
时间参数空间 time parameter space
时间槽模式 time slot pattern
时间测定法 chronometry
时间测量 time measurement
时间测录器 counter timer
时间层序 temporal sequence
时间差 time difference
时间差别 < 运输高峰时间与非高峰时间客运票价的差别 > time differential
时间差分法 time differencing method
时间差价买卖 time spread
时间差距 lag time
时间常量 characteristic time; lag coefficient; time constant
时间常数 characteristic time; time constant
时间常数法 time constant method
时间常数项 time constant term
时间场校准 time field calibration
时间超前 time lead
时间车间距 time spacing of vehicles
时间车速分布 time-speed distribution
时间沉降曲线 time-settlement curve; time-settlement graph; time-subsidence curve
时间成本 < 按时间计算的费用成本 > time cost
时间成本估计 time cost estimates
时间成本权衡法 time cost trade-off
时间程序 time-program(me)
时间程序控制 time pattern control; time program(me) control
时间程序指令 time program(me) command
时间迟滞 time lag
时间持续 time remaining
时间尺度 time scale
时间重叠 time-interleaving
时间抽样 time sample
时间抽样管 time-sampling tube
时间传感器 control timer; timer; time sensor
时间传感元件 time-sensing element
时间窗 time window
时间窗口法 time window method
时间错误 timing error
时间代码 time code
时间带 chronozone
时间带宽积 time bandwidth product
时间单位 time unit; unit of time; time cell; time quantum
时间当量分钟 equivalent minute
时间导数 time derivation
时间的 horary; temporal
时间的风险 time risk
时间的宽限 time allowance
时间的损失 leeway
时间的推移 lapse of time
时间的消逝 lapse
时间的延长 extension of time
时间等温线 chr(on)oisotherm
时间地层单位 time-rock unit
时间地层相 time-stratigraphic(al) facies
时间地点集聚性 time-place clustering
时间地址 time address
时间电动机 clockwork motor
时间电流临界值 time-current threshold
时间电流试验 time-current test
时间电流特性 time-current characteristic
时间电路 time circuit
时间电码 time code; timing code
时间电码控制系统 time code control system
时间电码控制制 time code control system
时间定常系统 time non-variant system
时间定额 time norm; time value; standard piece time < 工件 >
时间定额标准 time quota standard
时间定购货单 time compulsory purchase order
时间动作分析 time and motion analysis
时间动作研究 time and motion study; time and movement study
时间毒性曲线 time-toxicity curve
时间读数 time read; time taking
时间读数精度 time accuracy

时间度数 time degree
时间段 period; slot; time quantum; time slice
时间段号 slot number
时间段排序 slot sorting
时间段组 slot group
时间断面 time section
时间队列 time queue
时间对比 time comparison
时间对数法 logarithm-of-time method
时间对数拟合法 logarithm-of-time fitting method
时间对数配算法 logarithm-of-time fitting method
时间反差指数曲线 time-contrast-index curve
时间反射 time reflex
时间反响 time response
时间反演 time reflection; time reversal
时间反应 time response
时间反应谱 time response spectrum
时间反转不变性 time reversal invariance
时间范围 time frame; time limit(ation); time range
时间方程式 equation of time
时间方位 time azimuth
时间方位表 time azimuth table
时间方位显示 time bearing display
时间方位显示器 time bearing display
时间-方向-温度图 chronoanemoisothermal diagram
时间防护 protection of time
时间放大倍数 time magnification
时间费用 time cost
时间费用权衡 time cost trade-off
时间分辨 time resolution
时间分辨 X 射线衍射 time-resolved X-ray diffraction
时间分辨变像管 time-resolving image converter
时间分辨的 time-resolved
时间分辨发射光谱 time-resolved emission spectrum
时间分辨分光光度计 time resolving spectrometer
时间分辨干涉光谱学 time-resolved interference spectroscopy
时间分辨光谱学 time-resolved spectroscopy
时间分辨力 temporal resolution; time sense
时间分辨率 temporal resolution; time resolution
时间分辨谱 time-resolved spectrum
时间分辨摄谱仪 time-resolved spectrograph
时间分辨显微照相术 time-resolved microphotography
时间分辨荧光 time-resolved fluorescence
时间分布 temporal distribution; time distribution
时间分步法 time-step method
时间分段 time slice; time slicing
时间分段信号 time tick
时间分割 time division; time sharing
时间分割乘法器 time division multiplier
时间分割电路 time discriminating circuit; time division circuit; time-sharing circuit
时间分割多路传输 time division multiplex transmission
时间分割多路传输系统 time-shared multiplexing system
时间分割多路调制 multiplier time division modulation
时间分割多路通信[讯] time dividing channeling; time division multiplex communication
时间分割(多路通信[讯])检波器 time-shared detector
时间分割多路通信[讯]制 multiplex time division system
时间分割分品复多路传输 time multiplex
时间分割管 time share tube
时间分割激光器 time-sharing laser
时间分割开关 time division switching
时间分割式分压 voltage division by time division
时间分割式无线电中继系统 time division radio relay system
时间分割系统 time division system
时间分割制 time division system
时间分割制交换系统 time division switching system
时间分割综合控制 time division common control
时间分隔 time division; time separation; timing separation
时间分隔乘法器 time division multiplier
时间分隔多路电报制 time division multiplex telegraph system
时间分隔法 time slot
时间分隔控制存取 time division control access
时间分隔脉冲(多路)通信[讯]制 impulse time division system; pulse time division system
时间分隔制 time split system
时间分集 time diversity
时间分解变像管 time-resolving image converter
时间分离法 time discretization method
时间分路器 time de-multiplexer
时间分配 temporal distribution; time distribution; time sharing
时间分配的 time-shared
时间分配法 time-sharing method
时间分配放大器 time-shared amplifier
时间分配分析器 time-distribution analyser
时间分配系统 time-sharing system
时间分配信号方式 time-assigned signal(l)ing
时间分配执行系统 time-shared executive system; time-sharing executive system; time-sharing operating system
时间分片 time slicing
时间分析器 time analyser[analyzer]
时间分析图 time analysis chart; time analysis diagram
时间分享 time sharing
时间分序 time sequencing
时间分选 time sorting
时间分支信道 time-derived channel
时间风险 time risk
时间服务 time service
时间复杂性 time complexity
时间改变的 time-stepping
时间改正 time correction
时间改正因子 time-correction factor
时间干涉 temporal interference
时间高度部分 time-height section
时间隔离 temporal isolation
时间工作 time service
时间共享 time share
时间固结(关系)曲线 time-consolidation curve
时间关联 association in time; time correlation
时间关系 time relationship
时间关系分析 time-history analysis
时间关系曲线 history; time curve
时间关系曲线图 time history
时间关系式 time-history form
时间观象台 time observatory
时间管理 time management
时间光通量曲线 time-light curve
时间归一化法 time normalization
时间过程 time course; time process; duration <统筹方法中完成一个活动所需时间>
时间海侵单位 time-transgressive unit
时间函数 time function
时间函数发生器 time function generator
时间函数系数 <随时间变化的系数> time-dependent coefficient; time-varying coefficient
时间合成法 time-synthesis schema
时间和频率标准 time and frequency standard
时间和事件记录 time and events record
时间和温度效应 effect of time and temperature
时间核算统计 time accounts statistics
时间荷载曲线 time-load curve
时间恒定的 time-invariant
时间恒定调整器 time-invariant regulator
时间恒定系统 time-invariant system
时间衡量单位 time measurement unit
时间互换 time tie
时间互换测验 time reversal test
时间互换性 time reversibility
时间划分 time division; time sharing
时间划分倍增装置 time division multiplier unit
时间划分乘法器 time division multiplier
时间划分乘法装置 time division multiplier unit
时间划分多路传输 time division multiplexing
时间划分法 time-sharing method
时间划分方案 time-sharing scheme
时间划分计算机 time-share computer
时间划分制 time-sharing scheme
时间换算 time conversion
时间恢复误差 time recovery error
时间积分 time integral
时间积分比尺 integral time scale; time integral scale
时间积分光谱 time-integrated spectrum[复 spectra]
时间积分级 time-integration stage
时间积累采样器 time-integrating sampler
时间积累法 time-integration method
时间基点 time origin
时间基线 time baseline; time reference line <用于信号系统>
时间基线视差法 time baseline parallax method
时间基准 time base; time reference
时间基准标记 timing reference
时间基准脉冲 time reference pulse
时间基准系统 time reference system
时间及日期 time and date
时间极限 time limit; time period
时间极性 time polarity
时间极性控制 time polarity control
时间极性控制码 time polarity control code
时间计量方法 method of time measurement
时间计数分布 time counting distribution
时间计算 time calculation
时间记号 time mark
时间记录 time keeping; time record(ing)
时间记录法 chronography
时间记录分类 timekeeping classification
时间记录器 chronograph; time recorder; time regulator
时间记录系统 timekeeping system
时间记录员 time checker; timekeeper
时间记忆 time memory
时间纪录员 time keeper
时间继电器 time delay device; timed-relay; time element relay; timer; time relay; timing ranger; timing relay; timing unit
时间寄存器 time register
时间加权暴露量 time-weighted exposure level
时间加权的 time-weighted
时间加权绝对误差 time-weighted absolute error
时间加权平方误差 time-weighted squared error
时间加权平均浓度 time-weighted average concentration
时间加权平均值 time-weighted average
时间价格差 time-price differential
时间价值 temporal value; time value; trend of the times; value of time
时间价值因素 time value factor
时间间隔 floor-to-floor time; interval; time break; time cell; time interval; time piece; time range; time separation; time spacing; time span; timing separation
时间间隔闭塞系统 time interval block system
时间间隔闭塞制 time interval block system
时间间隔测量 time interval measurement
时间间隔测量器 intervalometer; time interval indicator
时间间隔测量仪 time interval measuring instrument
时间间隔错误 time interval error
时间间隔地图 time interval map
时间间隔发送器 clock multivibrator
时间间隔法 interval method; time interval method
时间间隔分布 interval distribution
时间间隔计 intervalometer
时间间隔计(数)测量器 counter timer
时间间隔计数器 time interval counter
时间间隔记录器 time interval recorder
时间间隔铃 time interval bell
时间间隔平均法 time interval method
时间间隔调节器 interval timer
时间间隔系统 time interval system
时间间隔行车 time interval running; train interval running【铁】
时间间隔行车制 time interval system; time separation system
时间间隔原理 time interval principle
时间间隔运行 time interval running
时间间隔指示器 time interval indicator
时间间隔制 time interval system
时间间距 time interval
时间间隙 time interval; time slot
时间检验 time check
时间鉴别 time discrimination
时间鉴别电路 time discriminating circuit
时间鉴频器 time discriminator
时间降深曲线 time drawdown curve
时间角 time-angle
时间校验 time check
时间校正 time correction
时间校正电路 time correction circuit
时间校准 time calibration

S

时间阶梯 time step
时间结构 time structure
时间解调 time demodulation
时间界限 time line
时间界限图灵机 time-bounded Turing machine
时间紧急 time constraint
时间进度表 progress chart;time schedule;schedule
时间经历 time history
时间经历反应 time-history response
时间经历分析 time-history analysis
时间精度因子 time dilution of precision
时间警告信号 time alarm
时间距 time interval
时间距离 time distance
时间距离(关系)曲线 curve time distance; time-distance curve; time-path curve;time-travel curve
时间距离关系图 time-distance graph
时间距离迹线 time-distance trajectory;time-space trajectory
时间距离继电器 time-distance relay
时间距离图 time-distance diagram; time-space diagram
时间距离选择器开关 time or distance selector switch
时间距离原则 time-distance basis
时间角色 time case
时间均衡器 time equalizer
时间均匀过程 time-homogeneous process
时间均值 time average
时间开关 clock switch
时间可逆性 time reversibility
时间可用度 time availability
时间刻度 time record(ing); time scale
时间、空间和速度相互关系的研究 time-space-speed study
时间空间问题 time-space problem
时间空间运行图 time-space diagram
时间控制 duration control;time control;timing-control
时间控制标记 timing control reference
时间控制继电器 control timer
时间控制码 time and control code
时间控制器 time controller
时间控制装置 timing device
时间跨度 time space;time span
时间宽度 time width
时间宽度调制 duration time modulation
时间宽限 <如对行车时分> time margin
时间框架 time frame
时间扩展 time dilation; time expansion
时间累积采样器 time-integrating sampler
时间累积法 time-integration method
时间累积悬移质采样器 time-integrating suspended sampler
时间历程 time history
时间历程文件 time-history file
时间利用 time utility
时间利用率 availability factor; time availability
时间利用系数 availability factor;time factor
时间利用效率 time efficiency
时间连接 time connection
时间量程 time scale
时间量化 time quantization; time sampling;timing sampling
时间量化连续信息 sampled analog(ue)data;sampled data
时间量级 time frame
时间量准 time scale
时间量子 time quantum
时间量子法 time quantum method
时间量子化 time quantization
时间临界进程 time critical process
时间零点 time zero;zero time
时间流变换 time stream transformation
时间流结构 time flow mechanism
时间流量曲线 time-discharge curve
时间流逝 time lapse
时间流逝变慢效应 time dilatation
时间滤波 time filtering
时间滤波器 temporal filter
时间路径 time path
时间率 time rate
时间落后 time lag
时间码 time code
时间码重放放大器 time code playback amplifier
时间码重放增益 time code playback gain
时间码磁迹检测电平 time code track detecting level
时间码磁头 time code head
时间码定时 time code timing
时间码发生器 time code generator
时间码检测器 time code detector
时间码静噪 time code muting
时间码控制 time code control
时间码录制电平 time code record level
时间码录制放大器 time code record amplifier
时间码信号 time code signal
时间码选通脉冲 time code gate pulse
时间码指示灯驱动器 time code lamp driver
时间码字 time code word
时间码阻塞 time code disable
时间脉冲 time(im)pulse
时间脉冲倍增器 time division multiplier
时间脉冲分配器 time pulse distributor;timing pulse distributor
时间门 time gate
时间密码 time code
时间-面积法 time-area method
时间面积降水深度曲线 <暴雨的> time-area-depth curve
时间-面积曲线 time area curve
时间-面积图 time-area diagram
时间模拟 time simulation
时间模式 time pattern
时间目标 time target
时间目标合同 target time contract
时间浓度曲线 time-concentration curve
时间配合法 time fitting
时间配置 time setting
时间配准 temporal registration
时间膨胀曲线 time swelling curve
时间匹配 temporal match
时间片 quantum[复 quanta]; sliced time;time slice
时间片端 time slice end
时间片分时 time slice time-sharing
时间片间隔 time slice interval
时间片轮转 round-robin
时间偏差 time deviation
时间偏好 time preference
时间偏好的评估 assessment of time preference
时间偏好率 rate of time preference
时间偏向参数 time bias parameter
时间频率 time frequency
时间频率避碰系统 time-frequency collision avoidance system
时间频率滤波 time-frequency filtering
时间频率谱 temporal frequency spectrum
时间频率域 temporal frequency domain;time-frequency domain
时间平方根定律 root-time law
时间平方根拟合法 <一种土固结试验结果的整理法、压法> square root of time fitting method
时间平均 time average
时间平均产额 time-averaged yield
时间平均车速 time mean speed
时间平均乘积阵 time-average product array
时间平均电流 time-averaged current
时间平均法全息干涉测量术 time-average holographic interferometry
时间平均干涉量度学 time-average interferometry
时间平均干涉仪 time-average interferometer
时间平均强度 time-average intensity
时间平均全息术 time-average holography
时间平均全息图 time-average hologram
时间平均全息照片 time-averaged hologram
时间平均输出 time-average output
时间平均束流 time-average beam
时间平均速度 time-average velocity; time mean speed;time-mean velocity
时间平均速率 time mean speed
时间平均条纹图样 time-average pattern
时间平均样 time-average sample
时间平均阴影条纹法 time-average shadow-Moire method
时间平均值 time average
时间平稳过程 time stationary process
时间破坏预测技术 time-deterioration prediction technique
时间剖面 time profile;time section
时间剖面对比 time sections correlation
时间剖面图 time cross-section
时间谱系 time hierarchy
时间期限 time limit
时间齐次的【数】 temporal homogeneous
时间起伏 time jitter
时间前进间隔 time interval of advancement
时间-强度(关系)曲线 time-intensity curve
时间-强度关系曲线图 <混凝土> time-intensity graph
时间求经度 longitude by chronometer
时间区带法 time band method
时间区分 time division; time share [sharing]
时间区分多路传输 time division multiplier
时间曲线 hour-out line; time curve; time front;timetable
时间曲线图 time graph;time plot
时间趋势 temporal trend;time trend
时间趋势分析 time-trend analysis
时间燃烧温度曲线 time-fire temperature curve
时间冗余 time redundancy
时间冗余度 temporal redundancy
时间蠕动曲线 time-creep curve
时间扫掠器 time sweep unit
时间扫描 time scan(ning)
时间扫描双稳态多谐振荡器 time sweep flip-flop
时间上的叠加 temporal summation
时间上的协同 time coordination
时间深度曲线 time-depth curve
时间深度图 time-depth chart; time-depth graph
时间深度图分析 time-depth plot analysis
时间深度转换 time-depth conversion
时间矢量 time vector
时间矢量分析法 time vector method
时间矢量图 phasor diagram;time diagram
时间是本合同的基本要素 time is of the essence of the contract
时间事件分析 time-event analysis
时间视差 temporal parallax
时间适应 temporal adaptation
时间收缩曲线 time-shrinkage curve
时间受控装置 time unit
时间输出循环 time-output cycle
时间输出指令 time-output command
时间数列 time series
时间数列模型 time series model
时间数列趋势 time series trend
时间数列相关 correlation of time series
时间数列预测法 time series forecasting
时间数字变换器 time digital converter
时间数字转换 time-to-digit(al) conversion
时间水平 time horizon
时间水位曲线 time stage curve
时间顺序 time sequence;time sequencing
时间瞬态 time transient
时间速度插值间隔 time-velocity interpolation interval
时间速度对数 numbers of time-velocity couple
时间速度图 velocity-time diagram
时间损耗 time loss
时间损失 leeway; loss of time; lost time;time loss;time penalty
时间损失津贴 allowance for lost time
时间损失系数 coefficient of time lost
时间缩短法 time condensation method
时间锁闭 time locking
时间锁定 time lock
时间套利 time arbitrage
时间特性 time behavio(u)r; time character;time response
时间梯度 time gradient
时间梯度系数 time gradient coefficient
时间提前 time advance
时间调节 time adjustment
时间调节范围 timing range
时间调节器 time regulator; timing-control
时间调整 time adjustment;time phasing
时间调整技术 time-align technique
时间调整收益率 time-adjusted rate of return
时间调整装置 time adjusting device
时间调制 time modulation
时间调制技术 time modulation technique
时间调制束 time-modulated beam
时间跳动 time jitter
时间贴现 time discounting
时间通道 timing channel
时间同步 clock synchronization;time lock;time synchronization
时间同步器 synchrotimer
时间图 diagram on the plane of the celestial equator;time diagram
时间图表 time chart
时间推移 lapse;passage of time;time lapse;time phase
时间推移干涉量度术 time-lapse interferometry
时间推移指示器 time-lapse indicator
时间拖延过长 drawn-out
时间微比尺 microtime scale

时间为零的活动 <统筹方法> zero time activity
时间维度 time dimension
时间位移【数】time shifting
时间位移曲线 time-displacement curve
时间位置码 time and location code
时间温度控制 time-temperature control
时间温度曲线 time-temperature curve
时间温度图 time-temperature diagram
时间温度显示器 time-temperature indicator
时间温度形态转变曲线 time-temperature-transformation curve
时间温度指数 time-temperature index
时间温度转变曲线 time-temperature-transformation diagram
时间稳定性 time stability
时间误差 time error
时间误差校正 time error correction
时间系列 chronological series; time series
时间系数 time factor
时间下沉曲线 time-subsidence curve
时间显示单元 time display unit
时间显示控制开关 time display control switch
时间显示控制仪 time display controller
时间显微镜 time microscope
时间线 time vector
时间线性控制 time proportional control
时间限度 time dimension;time limitation
时间限制 time bar; time limit; time slicing
时间限制电路 gate
时间相干光 temporally coherent light; time-coherent light
时间相干光束 temporal coherent beam
时间相干性 temporal coherence;time coherence
时间相关 association in time; temporal correlation; time correlation; time dependence
时间相关变形 time-dependent deformation
时间相关传播 time-dependent propagation
时间相关法 computation of wagon turnround time on the basis of correlating time spent;time correlation method
时间相关观测 time-correlated observation
时间相关回跳 time-dependent rebound
时间相关机制 time-dependent mechanism
时间相关挠度 time-dependent deflection
时间相关扰动 time-dependent perturbation
时间相关透镜效应 time-dependent lensing effect
时间相关系统 time correlation system
时间相角 time phase angle
时间相量 time phasor
时间相邻场 time-adjacent fields
时间相位 time phase
时间相位不均衡 phase inequality of time
时间相位角 time phase angle
时间相位散布 time phase dispersion
时间响应 temporal response;time response
时间向量 time arrow
时间消耗 expenditure of time
时间消逝 efflux;lapse of time
时间消隐 time blanking
时间效应 time(-dependent)effect
时间效应关系 time-effect relationship
时间效应曲线 time-effect curve
时间效用 time utility
时间协同 time coordination
时间谐波 <电压和电流波形的> time harmonic
时间信号 time code;time signal;timing signal
时间信号发生器 time signal generator
时间信号塞孔 time jack
时间行程曲线 time-travel curve
时间行程图 time-travel diagram
时间形变曲线 time deformation curve
时间修正值 time corrected value
时间序列 temporal series; time sequence;time series
时间序列的季节效应 seasonal effect in time series
时间序列法 time series method
时间序列分布 time series distribution
时间序列分析 analysis of time series; time series analysis
时间序列数据 time series data
时间序列向量 time series vector
时间序列预测 prediction of time series;time series prediction
时间选择 time selection
时间选择电路 time selection circuit
时间选择器 time gate;time selector; time sorter
时间选择衰落 time selective fading
时间选择性干扰 time division jamming
时间选择性衰落 time selective fading
时间选择性作用 time selective action
时间寻址信号方式 time-address signal(l)ing
时间压力曲线 time-pressure curve
时间压缩 time-compression;time lapse
时间压缩编码 time-compression coding
时间压缩(多路)复用 time-compression multiplex(ing)
时间压缩关系曲线 time-compression curve
时间压缩式单边带系统 time-compressed single sideband system; time-compression single sideband system
时间压缩相关器 compressed-time correlator
时间延长 time expand
时间延迟 time delay; time lag; time postpone
时间延迟鉴频雷达应答器 interrogated time offset frequency agile racon
时间延迟校正方法 time delay correction means
时间延迟失真 time delay distortion
时间延迟调节器 timer
时间延误 time delay
时间研究 time study
时间演变 time domain
时间要素 element of time
时间一致 time correlation
时间依从 temporal dependence
时间依赖 time-dependent
时间依赖性流体 time-dependent fluid
时间移动 time moving
时间已过 time-out
时间因次 temporal dimension; time dimension
时间因数 time factor
时间因素 time element;time factor
时间因素变形 time-dependent deformation
时间因子 time factor
时间因子曲线 time factor curve
时间引信 time fuse
时间应变曲线 time deformation curve
时间应变实验 time-strain experiment
时间应变图 time-strain diagram
时间映射 time mapping
时间映射异步模拟 time-mapping asynchronous simulation
时间硬度 time hardness
时间硬化理论 time hardening law
时间优惠 time preference
时间优惠率 rate of time preference
时间游标尺 time vernier
时间与地点 time and venue
时间与高度方位 time-and-altitude azimuth
时间与固结关系曲线 time-consolidation curve
时间与空间 time and space
时间与事件记录仪 time and events recorder
时间预分配 time prearranged assignment;time preassign
时间预分配多址 time-preassigned multiple address
时间预算 time budget
时间域 time domain;time field
时间域处理 time domain processing
时间域的直接解 direct solution in time domain
时间域电磁法 time domain electromagnetic method
时间域分析法 time domain analysis
时间域激电仪 induced polarization instrument in time domain
时间域激发极化法 time domain induced polarization method
时间域宽度 width of time domain
时间域滤波法 time domain filtering
时间增量 time increment
时间增量指数 time increment index
时间增益 temporal gain
时间增益控制 temporal gain control; time gain control
时间占有率 time occupancy
时间障碍 time penalty
时间整定 time setting
时间帧 time frame
时间知觉 time perception
时间指标 time target
时间指标合同 target time contract
时间指示 time indication
时间指示器 time-indicating device; time marker
时间滞差 lag of time;time lag;time of delay
时间滞后 lag of time;time lag;time of delay
时间滞后式 time-lagged type
时间置位 time setting
时间置位电路 time-setting circuit
时间中心 time center[centre]
时间中心差法 leapfrog method
时间周期 period of time;time-cycle; time period
时间轴 <潮汐曲线> axis of time;time axis;time base
时间轴线 axis of time
时间注记 time figure
时间转换 time changeover; time reversal
时间装定 time set(ting)
时间准确度 time accuracy
时间组合 time combination
时间最佳行驶方向 time optimal driving direction
时间最小化 time-minimization
时间最优的 time optimal
时间最优控制 time optimal control; time optimum control
时间最优控制系统 time optimum control system
时间最优问题 time optimal problem
时间作用 time action;time effect
时间坐标 time base;time coordinates
时件工资 time and piece rate
时降 time-fall
时交通量 hourly capacity
时交通量变化图 diagram of hourly traffic variation
时角 meridian angle;hour angle
时角差 hour angle difference;meridian angle difference
时角赤道坐标系统 dependent equatorial coordinate system
时角赤纬轴座架 hour angle-declination axis mount
时角方位图 hour angle hyperbola
时角基本量 basic increment of hour angle
时角求方位法 time azimuth
时角系统 hour angle system
时角增量 increment of hour angle
时角(坐标)系 hour angle system of coordinates
时距 chronometric distance;time distance;time interval
时距工资 travel time
时距关系曲线 hodograph; time-distance curve
时距扩大法 time-range extending method
时距曲线 hodograph;time curve;time-distance graph;time-path curve;time-travel curve;travel-time curve
时距曲线的平均斜率 average slope of the travel time curve
时距曲线向量图 time plot
时距探空仪 time interval radiosonde
时距图 time-distance graph; time-space diagram;time-space graph
时距原理 <按时距图组织道路施工的> time-distance basis
时距照相机 time-lapse camera
时均流速 time-mean velocity
时均全息照像 time-average holography
时均速度 time averaged velocity
时均温度 time averaged temperature
时均值 hourly mean value
时刻表 schedule;timetable
时刻表比较程序 timetable comparing program(me)
时刻表编辑工作站 workshop for timetable editor
时刻表编辑器 timetable editor
时刻表规定的定期列车 regular train; regulation train;scheduled train
时刻表规定的会让站点 schedule meeting or passing point
时刻表规定的货物列车 scheduled freight train
时刻表规定的列车会让站 scheduled passing station
时刻表规定的列车交会站 scheduled meeting station
时刻表规定的列车运行 scheduled service
时刻表规定的时间 booked time
时刻表规定的时刻 scheduled time
时刻表规定的速度 scheduled speed
时刻表规定的停车 booked stop;regular stop;scheduled stop
时刻表规定的行车优先权 timetable superiority
时刻表规定的运转时分 scheduled running time
时刻表框 timetable holder

时刻表协调 timetable coordination
时刻表座架 timetable holder
时刻度环 hour circle
时刻准备好的 ready to operate
时空 space time
时空变化 temporal and spatial variation
时空变换 space-time transformation
时空标架 frame of space and time
时空不变式 invariant in space-time
时空布朗运动 space-time Brownian motion
时空产率 space-time yield
时空处理 space-time processing
时空代数 space-time algebra
时空单位 space-time unit
时空的 spatiotemporal
时空点 space-time point
时空度量 space-time metric
时空多变性 spatiotemporal variability
时空方程式 space-time equation
时空分布 spacio-temporal distribution
时空概念 space-time concept(ion)
时空关系 time-space relationship
时空函数 space-time function
时空间隔 time-space interval
时空结构 space-time structure
时空扩展 spatiotemporal spread
时空类型 space-time pattern
时空连续体 space time; space-time continuum
时空连续域 space-time continuum
时空流形 space-time manifold
时空滤波器 space-time filter
时空律 space-time law
时空模型 temporally spatial model
时空配合 space-time processing
时空迁移 spacio-time migration
时空曲率 space-time curvature
时空曲线 space-time curve
时空随机过程 space-time random process; space-time stochastic process
时空特征 spacio-temporal characteristic
时空图 space-time diagram; time-space diagram
时空网络 time-space network
时空相关 temporal and spatial correlations
时空相关函数 space-time correlation function
时空相关图 space-time correlogram
时空相关性 spatial and temporal dependence
时空因数 temporal and spatial factors
时空预测 spacio-temporal prediction
时空元 space-time element
时空折衷 time-space made-off; time-space trade-off
时空坐标 space-time coordinates
时空坐标系 space-time coordinate system
时控的 timed; time-dependent
时控恒温计 chronotherm
时控开关 automatic clock switch
时控应力累积作用 time-dependent stress accumulation
时控应力松驰作用 time-dependent stress relaxation
时控预测 spacio-temporal prediction
时控装置 time device; timed unit
时拉紧的 tense
时令的 in season
时令河 intermittent river; intermittent stream; seasonal river; seasonal stream
时令湖 seasonal lake
时流量 hourly rate of discharge
时路 time channel
时率法 rate law method
时轮传动轮 hour wheel driving wheel
时轮刀具 hour wheel cutter
时轮垫 hour wheel seating
时轮垫圈 hour wheel washer
时轮簧夹 hour wheel spring-clip
时轮摩擦簧 hour wheel friction spring
时轮配合记号 timing mark
时轮桩 hour wheel stud
时轮组件 hour wheel module
时码读出器 time code reader
时码信号 time code signal
时髦的 go-go
时髦建筑师 fashionable architect
时面深曲线 time-area-depth curve; depth-area-duration curve <降雨的>
时盘 hour indicator
时偏 time bias
时偏积分器 time bias integrator
时偏整定 time bias setting
时频地址矩阵 time-frequency address matrix
时频相调制 time-frequency-phase modulation
时平均速度 temporal mean velocity
时期 length of time; period; time period
时期数列 periodic(al) series
时齐的【数】 temporal homogeneous; time homogeneous
时齐过程 temporally homogeneous process
时区 hour zone; time belt; time zone; zone of time
时区(标)号 zone description
时区病 time-zone disease
时区范围 zone boundary
时区时间 zone time
时区图 map of time zones; time-zone chart
时区针 time-zone hand
时区子午线 time meridian
时圈 circle of declination; circle of right ascension; hour circle
时日 time zone
时深面 <降雨的> duration-depth-area
时深面值 date-depth-area value
时深转换 time-depth conversion
时深转换空间归位值 space restoration value of time depth transform
时时 ever and again
时式 fashion
时式序列 sequence of tense
时事 current event
时事评论员 publicist
时事通讯 newsletter; newssheet
时速 speed per hour
时速曲线 speed-time curve
时速图 velocity-time diagram
时随变形 time-dependent deformation
时随行为 time-dependent behavio(u)r
时损率 hourly loss rate
时锁杆 hour locking lever
时态 tense
时桃轮 hour heart
时桃轮簧 hour heart spring
时跳 time-hopping
时统中心 timing center[centre]
时为 beam date
时温迭加 time-temperature superposition
时温移动因子 time-temperature shift factor
时间特性 time characteristic
时蜗形凸轮 hour snail
时系列 time series
时系图 diagram on the plane of the celestial equator; diagram on the plane of the equinoctial; time diagram
时隙 time slot
时隙号码 time slot number
时隙互换 time slot interchange
时隙内信号方式 in-slot signal(l)ing
时现时隐效应 hit-and-miss effect
时线 time line
时限 time bar; time limit(ation); time period
时限低电压保护装置 time-undervoltage protection
时限电磁继电器缓放线圈 slow releasing coil of time delayed magnetic relay
时限电动机 timing motor
时限电路 time circuit
时限调度 deadline scheduling
时限法 time limit method
时限方式 time limit system
时限服务 time service
时限附件 time limit attachment
时限过电流 time overcurrent
时限过电流保护 time-lag over-current protection; time-overcurrent protection
时限过电流继电器 time-overcurrent relay
时限过载继电器 time-lag overload relay
时限计 quantum clock
时限继电器 time-lag relay; time limit relay; timing relay
时限减速 chronotropic(al) deceleration
时限警报 time alarm
时限特性 selectivity characteristic
时限调正装置 time adjusting device
时限元件 timing element
时限元件继电器 time element relay
时限装置 time adjusting device; timing device
时限自动开关 time cut-out
时相 time phase
时相阶段 temporal stage
时项法 time-term method
时像 time image
时效 ag(e)ing effect; seasoning; statute of limitations; time effect
时效变化 secular change; secular distortion change; secular variation
时效变形 secular distortion
时效不足 under-ag(e)ing
时效成型 age forming
时效抽样检验法 age sampling
时效处理 ag(e)ing treatment; seasoning; time-effect treatment
时效脆性 embrittlement by aging
时效电阻 ag(e)ing resistance
时效钢 aged steel; ag(e)ing steel
时效规范 ag(e)ing condition
时效过程 ag(e)ing process
时效过的 aged
时效合金粉末 aged alloy powder
时效化 ag(e)ing
时效劣化 aged deterioration
时效裂纹 ag(e)ing crack
时效龄期硬化 age hardening
时效破裂 season cracking
时效期 length of the limitation period; prescriptive period
时效期开始(日期) commencement of the limitation period
时效期限 limitation period
时效强化 ag(e)ing strengthening
时效情况 time-dependent behavio(u)r
时效柔韧性 aged flexibility
时效软化 age softening
时效试验 ag(e)ing test; age test
时效衰变 time-aged deterioration
时效特性 time-dependent behavio(u)r
时效稳定性 ag(e)ing stabile
时效效应 ag(e)ing behavio(u)r; time behavio(u)r
时效行为 ag(e)ing behavio(u)r
时效因素 aged factor
时效应力分析 time-dependent stress analysis
时效硬化 age hardening; ag(e)ing hardening
时效硬化不锈钢 age hardening stainless steel
时效用水权 water rights acquired by prescription
时效终止 cease to run(of limitation period)
时效阻力 ag(e)ing resistance
时新的 modern
时薪制 stab
时星 clock star; ten-day star; time star
时星轮衬圈 hour star ring
时星轮座 hour star support
时星形轮 hour star
时星形轮跳杆 hour star jumper
时星形轮跳杆偏心销 hour star jumper eccentric
时序 sequence[in time]; time course; time sequence[sequencing]; timing
时序编码 sequential coding
时序变换器 sequential transducer
时序标记 gomma
时序标记图 sequential marked graph
时序表 time scale
时序波瓣法 sequential lobing
时序操作 sequential operation
时序层次 sequential hierarchy
时序层次片段 sequential hierarchy segment
时序乘积 time-ordering product
时序处理 sequential processing
时序存取 sequential access
时序的 sequential; time sequential
时序电路 sequence circuit; sequential circuit
时序调度 sequential scheduling
时序调度系统 sequential scheduling system
时序堆栈作业控制 sequential stacked job control
时序放大 sequential maximization
时序分析 sequential analysis; time sequential analysis; time series analysis
时序分组 date time group
时序工作 sequential working
时序关系 sequential relationship
时序关系矩阵 sequential relationship matrix
时序函数 sequential function
时序机 sequential machine
时序机理论 sequential machine theory
时序极大化 sequential maximization
时序计数器 sequential counter
时序计算 sequential computation
时序计算机 consecutive-sequence computer; sequential computer
时序检验 logrank test; sequence check
时序校正元件 sequential correcting element
时序结构 sequential organization
时序截断脉冲发射 stutter shooting
时序开关联锁 sequence switch interlocking; sequential switch interlocking
时序开关网络 sequential switching network
时序控制 sequence[sequencing] control; sequential control; time-oriented sequential control
时序控制发送机自动起动 sequentially controlled automatic transmitter start

时序控制机构 sequential control mechanism
时序控制计数器 sequence control counter
时序控制寄存器 sequence control register
时序控制器 sequential controller; time schedule controller
时序控制算法 sequential control algorithm
时序链环 sequential link
时序列 time series
时序逻辑 sequential logic
时序逻辑方程 sequential logic equation
时序逻辑网络 sequential logic network
时序逻辑系统 sequential logic system
时序逻辑元件 sequential logic element
时序码 sequence code
时序脉冲 time sequential pulse
时序脉冲发生器 sequence timer
时序平均数 chronological average
时序时钟 sequence timer
时序式发电机 chronometric tach(e)ometer
时序收缩 time-ordered contraction
时序图 sequence chart
时序网络 sequential network
时序系统 sequential system
时序险态 sequential hazard
时序线路 sequential circuit
时序相关 correlation of time series
时序相关段 sequential dependent segment
时序相似性检测算法 sequential similarity detection algorithm
时序信号 sequence signal
时序元件 sequential element
时序运算 sequential operation
时序质隐含式 sequential prime implicant form
时序组织 sequential organization
时延 delay time;time delay;time lag; time lapse
时延差 delay inequality
时延常数 delay constant
时延电路 time delay circuit
时延分析器 time delay analyser[analyzer]
时延畸变 delay distortion
时延计数器 delay counter
时延均衡 delay equalization
时延均衡器 delay equalizer
时延容许量 delay allowance
时延调整范围 range of time-lag settings
时延网络 time delay network
时延因子 time delay factor
时延作用 time-lag action
时岩单位 time-rock unit
时岩对比【地】 time rock correlation
时样 fashion style
时移键控 time-shift keying
时移算符 time displacement operator
时移信号 shifted signal
时疫 epidemic;plague
时用比 hourly utilization ratio
时用率 hourly utilization ratio
时用水量 hourly capacity;hourly water consumption
时域 time domain
时域编码 time domain coding
时域波形 time domain waveform
时域多重访问系统 time domain multiple access system
时域法 time domain method
时域反射计 time domain reflectometer
时域反射仪 time domain reflectometer
时域分析 analysis in time domain; time domain analysis
时域分析法 time domain analysis method
时域解 time solution
时域局部构造图 local structural map in time domain
时域均衡器 time domain equalizer
时域区域构造图 regional structural map in time domain
时域失真 time domain distortion
时域疏样 decimation in time
时域响应 time domain response
时域研究 time domain study
时域与频域 time domain and frequency-domain
时域振幅剩余曲线 amplitude residual curve in time domain
时域转换 time domain conversion
时增殖率 hourly growth rate
时闸 time gating
时针 hour counter hand;hour hand
时(针)轮 hour wheel
时针式喷灌 center pivot sprinkler irrigation
时针式喷灌机 center pivoted sprinkler
时值 chronaxie【生】; chronaxy; time value
时值测量 chronaximetry
时值的 time characteristic
时值会计法 current-value accounting
时值计 chronaxie meter
时指示器传动簧 hour indicator spring driver
时指示器传动轮 hour indicator driving wheel
时指示器传动轮盖 hour indicator driving wheel cover
时指示器锁杆 hour indicator lock
时指示器托板 hour indicator maintaining plate
时指示器座 hour indicator support
时指示器座垫 hour indicator support seating
时指示器座压片 hour indicator support cover
时制 time system
时致误差 time error
时滞 time lag
时滞电路 time-lag network
时滞电容器 time delay condenser; time-lag condenser
时滞环节 time-lag element
时滞继电器 time-lag relay
时滞器件 time-lag device
时滞熔断丝 time-lag fuse
时滞系统 time-lag system
时滞需量计 lagged-demand meter
时滞因数 time-lag factor
时滞引起的扰动 time-lag disturbance
时滞域 lag domain
时滞元件 time-lag element
时滞装置 time limit attachment
时滞作用 time-lag action
时钟 clock;time clock;timing clock
时钟比较器 clock comparator
时钟笔头 clock pen
时钟单位 clock-unit
时钟调度 clock time scheduling
时钟读数 clock indication
时钟对比 clock comparison
时钟发生器 clock generator
时钟发生驱动器 clock generator and driver
时钟改正 clock correction
时钟更新 clock transaction
时钟功能 time clock feature
时钟过程 clock procedure
时钟和信号模块 clock and tones module
时钟花坛 flower clock
时钟恢复 clock recovery
时钟机构 clock mechanism; clockwork
时钟计数器 clock counter
时钟计数时间 counting clocktime
时钟校准 clock error;clock set
时钟节拍频率 clock frequency
时钟进程 clock process
时钟控制 clock control;timer control
时钟控制模块 timer control module
时钟控制系统 clock control system; time-controlled system
时钟链 clock chain
时钟逻辑 clocked logic
时钟脉冲 clock pulse; master clock; sprocket pulse;time clock
时钟脉冲边沿 clock edge
时钟脉冲产生器 gate generator;time pulse generator
时钟脉冲重复频率 clock repetition rate
时钟脉冲道 clock track
时钟脉冲电路 clock pulse circuit
时钟脉冲发生器 clock pulse generator; restore-pulse generator; sprocket pulse generator
时钟脉冲放大器 clock amplifier
时钟脉冲分配 time-impulse distribution
时钟脉冲分配器 time pulse distributor
时钟脉冲进入顺序 clock run-in sequence
时钟脉冲宽度 clock pulse width
时钟脉冲门 clock gate
时钟脉冲频率 clock pulse frequency; clock rate
时钟脉冲驱动器 clock driver
时钟脉冲输入 clock in;clock input
时钟脉冲系统 clock system
时钟脉冲限制器 clock qualifier
时钟脉冲相位差 clock skew
时钟脉冲源 clock source
时钟脉冲终端 clock terminal
时钟脉冲周期 clock cycle;clock period
时钟脉动 clock pulse
时钟频率 clock frequency
时钟起停功能 timer start/stop function
时钟驱动(信号)控制机 time clock actuated controller
时钟日差 clock rate
时钟生成器 clock generator
时钟时间 clock time
时钟时序电路 clocked sequential circuit
时钟式自记潮位计 tidal clock
时钟收音机 clock radio
时钟速率 clock rate
时钟同步器 clock synchronizer
时钟图 clock diagram
时钟歪斜 clock skew
时钟位 clock bit
时钟相位差 clock skew
时钟信号 clock signal
时钟信号测试 clock test
时钟信号发生器 clock signal generator
时钟信号频率 clock signal frequency
时钟佯谬 clock paradox
时钟移位寄存器 clock shift register
时钟振荡器 clock oscillator
时钟值 timer value
时钟中断 clock interrupt;timer interrupt
时钟周期 clock period
时钟装置 clock installation
时钟字码板式行车时刻表 <如市郊列车每一刻钟开行一次的时刻表> clock-face timetable
时钟字盘 hour plate
时钟组件 clock module
时轴 time axis;time base
时轴偏移 time-axis shift
时装店 dressmaker
时装用品小店 boutique

识别 characteristic mark; discriminate; identification; identify(ing)

识别板 identification plate
识别被呼叫线路设备 called line identification facility
识别被呼叫线路信号 called line identification signal
识别标记 identifying mark
识别标签 identification tag
识别标志 distinguishing mark; dog tag; identification beacon; identification division; identification index; identification mark; identifying feature; recognition mark
识别部分 identification division
识别部位 identification site; recognition site; recognization part; recognization site
识别差 recognition differential
识别产品 identification product
识别成本法 identified cost method
识别程序 recognition procedure; recognition program(me); recognizer
识别代号 identification number
识别带 identification tape
识别带头 identification leader
识别带尾 identification trailer
识别单 identification form
识别导航 navigation by recognition
识别灯 identification lamp; identification light
识别灯标 recognition light
识别地址 identification address
识别点 identification point
识别电路 identify circuit; identifier circuit
识别调用 identifying call
识别段 identification burst
识别方法 identification method; recognition method
识别符 identifier
识别符表 identifier list
识别符长度 identifier length
识别符号 character recognition; distinguishing mark; distinguishing symbol; identification code; identification mark; identification sign; identification symbol; identify code
识别符计数 identifier count
识别符说明 identifier declaration
识别符指示字 identifier pointer
识别符属性 identifier attribute
识别概率 identification probability
识别功能 recognition function
识别关键字 identification key
识别规则 recognition rule
识别过程 identify procedure; recognition process
识别函数 recognition function
识别和控制处理机 recognition and control processor
识别呼号 identification call letter
识别呼叫 identification call letter
识别呼叫线路设备 calling line identification facility
识别呼叫线路信号 calling line identification signal
识别机 recognizing machine

S

识别记号 identification sign
识别记录 identification record
识别技术 identification technique
识别尖旗 distinguishing pennant
识别接收机 identification receiver
识别矩阵 recognition matrix
识别卡(片) identification card
识别卡阅读器 identification card reader
识别空段 identify dummy section
识别控制段 identify control section
识别率 discrimination
识别码 authentication code; heading code; ID code; identification code; identify code
识别脉冲 identification(im) pulse
识别门 recognition gate
识别密码 recognition code
识别名 identification name
识别目标 recognized target
识别能力 discernibility; recognition capability
识别牌 identifying plate
识别票号 ticket identification
识别期 identification phase; recognization phase
识别器 detector; discriminator; identifier; recognizer
识别染料 identifying dye
识别闪光 identification blink
识别声呐 identification sonar
识别绳条 identification thread; Rogue's yarn
识别数据 identification data
识别数(码) identifying number; identification number
识别数字 character recognition
识别顺序 recognition sequence
识别速率 recognition rate
识别算法 identification algorithm; recognition algorithm; recognizer
识别条件 identification condition
识别图像 recognition image
识别网络 recognition network
识别危险 recognition hazard
识别危险菱形图 recognition hazardous diamond
识别卫星 identification satellite
识别问题 identification problem
识别无线电发射机 identification set
识别误差 identification error
识别系统 identification system; recognition system; recognization system
识别项 identification item
识别信标 identification beacon
识别信号 call(ing) signal; distinguishing signal; identifiable signal; identification signal; identifying signal; recognition signal
识别信号灯 marker lamp; recognition light
识别信号发生器 identification generator
识别信号源 identification source
识别性 identity
识别性截击 recognition intercept
识别序列号 identification sequence number
识别仪器 identification instrument
识别音 discriminating tone
识别用照明信标 landmark beacon
识别阈限(极限) identification threshold
识别载波 discriminatory carrier
识别制 recognition system
识别装置 identification device; identification equipment; judging device
识别字母 identification letter
识别综合系统 integrated identification system
识别总数 recognized sum
识模认 pattern recognition
识阈 threshold of consciousness

实

实棒岸墩 gravity ability

实保金额 amount of insurance carried
实报实销 actual cost payment; get full payment for all spending
实报实销合同 cost-plus contract
实报实销加管理费合同 cost-plus-fee agreement; fee plus expense agreement
实边三角刮刀 triangular solid scraper
实变函数论 theory of functions of real variable
实变量 physical argument; real variable
实变量函数 function of real variables; real variable function
实变量数据 real variable data
实变数 actual argument; real variable
实变(数)函数 real variable function
实变元 actual argument; real argument
实补码 true complement
实部【数】 active component; real part
实部符号 real part of symbol
实部算子 real part operator
实部条件 real-part condition
实参【计】 actual parameter
实参数 real parameter
实测 actual measurement; field measurement; practical survey
实测暴雨类型 observed storm pattern
实测比重 measured specific gravity
实测标 half-mark
实测波高 observed wave height
实测潮位 observed tidal height; observed tide height
实测的 measured
实测等高线 instrument(al) contour
实测地图 field map; survey map
实测地质剖面图 field-acquired geologic(al) profile; surveyed geologic(al) profile
实测点 eyeball
实测点据 measured points
实测读数 observation reading
实测反馈 measured feedback
实测方向 observed direction
实测风 measured wind; observed wind
实测干扰流量 measuring interference flow
实测工时 observed time
实测功率 measured power; observed power
实测过程线 recorded hydrograph
实测过度调量 measured overshoot
实测含沙量 measured load
实测河流 ga(u)ged river; ga(u)ged stream
实测荷载 measured load
实测洪水 measured flood; observed flood; recorded flood
实测洪水流量 measured flood discharge; observed flood discharge
实测记录 actual observation record; observation report
实测精度 accuracy of observation
实测径流 measured runoff; measurement runoff
实测距离 ga(u)ged distance
实测历年水位 actual surveyed water-levels over the years
实测流量 measured discharge; observed current; observed discharge; observed flow
实测流速 measured flow velocity; observed flow velocity
实测流域 ga(u)ged drainage area; ga(u)ged watershed
实测漏泄量 actual leakage
实测落差 measured fall; observed fall
实测密度 measured density
实测面 effective surface
实测浓度 actual concentration; measured concentration; practical concentration of measurement
实测排放污染物负荷 observed discharge pollutant loading
实测频率分布 observed frequency distribution
实测平均值 mean of observation
实测剖面图 practical measured profile
实测强度 observed strength; situ strength
实测曲线 measured curve
实测深度 observed depth
实测时间 elapsed time
实测输沙率 measured sediment discharge
实测数据 actually measured data; measured data; observation(al) data; observed data
实测数量 measured amount
实测数值 measured value
实测水力梯度 actual hydraulic gradient
实测水深 measured water depth; observed water depth; sounding depth
实测水位 measured stage; observed stage; observed water-level
实测水文过程线 recorded hydrograph
实测同位素成分 determining isotope composition
实测图 measured drawing; surveyed drawing
实测位置 observed position
实测温度 observed temperature
实测系列 observed series
实测巷道 gallery measured
实测项目 observed items
实测效率 efficiency by input/output test
实测辛烷值 rated octane number
实测雪水当量 measured water equivalent of snow
实测应力 measured stress
实测英里 measured mile
实测油水过渡带厚度 measuring thickness of oil-water transition zone
实测原图 field map; field sheet; field survey sheet; original plot; plane-table map; survey sheet
实测震中 instrumental epicenter[epicentre]
实测值 actual measured value; measured value; observed value; recorded value
实测资料 actually measured data; field(-surveyed) data; measured data; observed data; recorded value
实测自重湿陷量 actually tested self-weight
实测纵断面 measured longitudinal profile; observed profile; measured profile
实测纵剖面 measured longitudinal profile; measured profile; observed profile
实测最大的 maximum-recorded
实测最大洪峰 maximum-recorded flood peak
实测最大洪峰流量 maximum-recorded flood peak discharge
实测最大洪水 maximum experienced flood
实测最大降水量 maximum observed precipitation
实测最大流量 maximum-recorded flow
实测最低的 minimum recorded
实测最低水位 lowest recorded level; lowest recorded stage; measured lowest stage; minimum recorded stage
实测最高水位 highest recorded stage; highest recorded(water) level; maximum-recorded stage; measured highest stage
实测最小流量 lowest recorded discharge; measured lowest discharge; minimum recorded discharge
实长 true length
实常数 real constant
实场 real field
实车试验场试验 proving ground test
实尺放样 full-scale lofting
实尺锯切的 country-cut
实尺模型 full-scale model; full-size model
实尺设计图 full-size design
实尺图 full-size drawing
实齿 pleodont; solid tooth
实齿圆锯 solid-plate circular saw; solid-tooth circular saw
实出价 firm bid
实出勤工日数 actual man-days in attendance
实船试验 full-scale ship test; prototype test(ing)
实船适航性试验 full-scale seakeeping trials
实船丈量 taking-off
实磁芯 solid pole
实磁性 real magnetism
实存 actual balance; real storage
实存储器 real storage
实存储页表 real storage page table
实代码 true code
实单模群 real unimodular group
实得 net proceeds
实得工资 real wage; take-home pay <扣除指税等以后的>
实得还款数目 liquidated sum
实得率法 effective yield method
实得平均数 obtained mean
实得平均值 obtained mean
实得摊销 effective-yield amortization
实得增值 realized appreciation
实底面积 solid surface
实底燃烧室 solid-bottom combustion chamber
实地 real-world
实地板 solids
实地辨认 identify on the ground
实地测量 in-situ measurement
实地查勘 field reconnaissance
实地尺寸 ground dimension; natural size; terrestrial dimension
实地的 on-scene
实地地区 real address area
实地调查 field inquiry; field investigation; field work; on-the-spot inquiry; on-the-spot investigation
实地调查的 fact finding
实地调查法 field method
实地调查研究 field research
实地调查员 field worker
实地调查者 fact finder
实地定界 demarcate on the ground
实地定位工作 field location work
实地定线 field location work
实地方案 field program(me)
实地放样 field layout

实地观察 autopsy;field observation

实地计数 physical count

实地检测 revise in field;revise in the field; revise on ground; revise on the ground; walk-through inspection

实地检查 spot-check

实地勘察 autopsy

实地考察 first hand; first investigation;on-the-spot inspection;on-the-spot investigation

实地考察报告 field report

实地快速定向能力 eye for ground

实地流速 effective velocity;field velocity

实地描绘 field sketching

实地目测 eye for ground

实地盘存 actual inventory; physical inventory

实地盘存簿 physical inventory book

实地盘存法 physical inventory method

实地盘存控制法 physical inventory control

实地盘存制 physical inventory system

实地球 real earth

实地试验 field test(ing); in-place test;site test;site trial;spot testing

实地收集 field collection

实地图 actual map

实地训练 practical training

实地研究 field study

实地预测雾情方法 rule-of-thumb in forecasting of fog

实地运行实验 field running test

实地资源 in-situ resource

实点 hard dot;real point

实顶枝【植】fertile telome

实动工时 actual working hours

实锻 solid forging

实堆积 close piling;solid piling

实垛法 bulk stacking

实二次型 real quadratic form

实发工资 take-home pay

实发功率 power developed

实发数量 actual quantity issued

实发性故障频率 instantaneous failure rate

实方 bank measure;solid yardage

实方黏[粘]土 bank clay

实方体积 compacted volume

实方土的计算方法 bank measure

实方装载 loading from bank

实放热量临界温度 recalescent point

实废活性污泥 real waste activated sludge

实费承包工程 cost-plus-fee contact

实分量 real component

实分量剖面平面图 profiling-plan figure of real component

实分量相对异常曲线 relative anomaly curve of real component

实分区 real partition

实分析 real analysis

实符号 real symbol

实付费用 out-of-pocket cost

实付工资 net pay

实付工资率 wage rate paid

实付税款 duty actually paid

实腹板桁架 solid-web truss

实腹板梁 solid-web plate beam

实腹板桥 solid-web bridge

实腹板式桁架梁 solid-wed girder

实腹大梁 steel solid web girder

实腹大梁结构 structure of solid web girders

实腹复合结构的屈服点安全 yield safety of solid-web composite structures

实腹刚架 solid-web rigid frame

实腹钢格栅 solid-web steel joist;solid-web joist

实腹钢梁 solid-web steel joist; steel plain web beam;solid-web joist

实腹钢梁桥 steel plain web beam bridge

实腹工字梁 solid-web beam

实腹拱 solid-spandrel arch;solid-web arch;spandrel-filled arch

实腹拱桥 arch bridge with filled spandrel; filled spandrel arch bridge; solid arch bridge; solid-spandrel arch bridge;spandrel-filled arch bridge

实腹横梁框架 girder bent

实腹结构 solid web construction

实腹肋拱 solid-rib(bed)arch

实腹梁 plain girder;solid-web girder

实腹木大梁 wooden plain webbed girder

实腹木梁 wooden plain webbed beam

实腹石拱桥 filled spandrel stone arch bridge

实腹式桥 solid-web bridge

实腹式石拱桥 filled stone arch bridge

实腹式柱 solid-web(bed)column

实腹组合钢梁桥 steel plain web composite girder bridge

实覆梁 plain beam

实覆梁桥 plain beam bridge

实干 <涂层> actual drying; hard (through-)dry;through-drying

实干时间 hard drying time

实感 materialism

实感温度 sensory temperature

实格式项 real format item

实根【数】real root

实拱肩拱 solid-spandrel arch

实股 real stock

实过度调量 <用于自动化> real overshoot

实函数 real function

实函数扩充 extension of real function

实滑距 actual slip;real slip;true slip

实货柜 loaded container

实机 real machine

实积 bulk piling;solid cubic(al)content

实积比 solidity ratio

实积形数 reducing factor

实绩 actual performance;performance

实绩差异 performance variance

实绩列车运行时刻表 <列车调度员记录的> train sheet

实绩统一换算 equivalent performance

实集的 closely spaced

实计利息 exact interest

实际安全荷载 actual safe load

实际安全剂量 virtually safe dose

实际搬迁 physical relocation

实际半径 real radius

实际半毛重 real demi-gross weight

实际包络线解调器 practical envelope demodulator

实际保护率 effective rate of protection

实际报酬率法 effective yield method

实际报价 effective offer

实际比例尺 real scale

实际比率 effective rate

实际比推力 actual specific impulse

实际比重 actual specific gravity

实际边际税率 effective marginal tax rate

实际边界剪切应力 actual boundary shear stress

实际边坡 actual slope

实际编码 actual coding

实际贬值 effective depreciation

实际变动 actual change

实际变量 actual argument;actual variable

实际变量参数 actual variable parameter

实际变现值 realization value

实际变异性 true variability

实际变元 actual argument

实际变元阵列 actual argument array

实际变元值 actual argument value

实际标高 actual elevation;actual level

实际标准 actual standard

实际标准成本 practical standard cost

实际标准成本法 actual normal cost system

实际并行 actual pairing

实际波浪 actual wave

实际玻璃成分 actual glass composition

实际玻璃组成 actual glass composition

实际部件 physical unit

实际材积 <不扣除缺陷部分> full-scale

实际材积量度 true volume measure

实际材料图 actual material map;map of primitive data;practical material figure; primitive data map; primitive data plan

实际财产 realized property

实际财产的转移 actual delivery

实际财产主体 actual corpus

实际参数 actual parameter

实际残留极限 practical residue limit

实际操作 actual operation

实际测量 actual measurement

实际测试数据 actual test data

实际差异 physical variance

实际产出 actual output

实际产量 active output;actual output;actual production;actual throughput;effective output;net output;real output

实际产率 true yield

实际产权 material equity

实际产业 real property

实际产值 real product

实际长 actual(tube)length

实际长度 actual length;effective length;physical length;virtual length

实际偿还 actual discharge

实际偿还能力 actual capacity to repay

实际车道车流 substantial lane flow

实际车流曲线 actual flow curve

实际车速 prevailing speed

实际沉降-时间曲线 actual settlement-time curve

实际陈废率 actual rate of obsolescence

实际成本 actual cost;bona fide cost;effective cost; out-of-pocket cost; real cost;true cost

实际成本法 actual cost method

实际成本惯例 historic(al)cost rules (for revenue recognition)

实际成本会计制度 actual cost accounting system

实际成本计算 actual cost calculation; actual costing

实际成本贸易条件 real cost terms of trade

实际成本曲线 outlay curve

实际成本账 cost incurred account

实际成本制 actual cost system

实际成荒率 practical quarry-stone yield

实际成绩 actual achievement actual performance

实际承包人 actual tender

实际承运人 actual carrier

实际尺寸 actual dimension; actual measurement; actual size; dressed size;full size;natural size;physical dimension; real dimension; real size;trim size;whole size;manufactured size

实际尺度 actual dimension; actual size;physical size

实际冲角 actual angle of attack

实际冲刷量 true erosion

实际出发时间 actual time of departure

实际出力 actual output;true output

实际出勤人日数 actual man-days in attendance

实际出生率 realized natality

实际出水量 actual water output

实际储藏量 actual reserve

实际储存量 actual stock

实际储蓄 real saving

实际储蓄总额 aggregate realized savings

实际船型 actual ship form

实际存储单元 physical memory location

实际存储空间 physical memory space

实际存在的十进制小数点 actual decimal point

实际大小 actual size

实际代价 actual cost

实际代码 actual code

实际单耗 actual volume of consumption per unit

实际单位 effective unit;equivalent unit

实际导程 <打桩时的> actual lead; real lead

实际到达能力 actually realized productive capacity

实际到达时间 actual time of arrival

实际的 concrete; effective; practicable; practical; pragmatic; realistic; substantial

实际的标准成本 practical standard cost

实际的光学系统 practical optic(al) system

实际的系统 actual system

实际登记项 actual entry

实际地下水流速 field groundwater velocity

实际地形 actual landform

实际地址 actual address;physical address;real address

实际地址区 physical address block; real address area

实际电路 actual circuit

实际电容量 actual capacitance

实际电压 virtual voltage

实际顶点 physical vertex

实际读数 actual reading

实际断层移距 real fault displacement

实际断线 actual cord break

实际煅烧率 actual degree of calcinations

实际对地航程 distance over the ground

实际发生额会计 actual basis accounting

实际发生时间 actual occurrence time

实际范围 practical framework

实际方案 action plan

实际放牧期 actual use range

实际放牧月 actual use range

实际飞越时间 actual time over

实际废料 actual garbage

实际费用 actual charges;actual cost; actual expenses;out-of-pocket cost

实际费用账 cost incurred account

实际费用制度 actual cost system

实际分辨率 true resolution

实际分解率 actual degree of calcinations; actual degree of decarbonation

实际分选比重 Tromp cut-point

S

实际峰荷 actual peak load
实际缝隙 actual gap
实际服务 physical service
实际付款条款 effective payment clause
实际付税人 actual taxpayer
实际负担率 actual burden rate
实际负荷 actual load(ing);practical duty
实际负荷试验 actual loading test
实际负荷因素 actual load factor
实际负载 actual load(ing);net load
实际负载试验 actual loading test
实际负载因素 actual load factor
实际负债 actual debt;actual liability;effective liabilities;real liability
实际附加水头 actual supplementary head
实际干燥过程 actual drying process;practical drying process
实际感觉噪声级 effective perceived noise level
实际高程 actual elevation;actual height;true height
实际高度 actual height;true height;virtual height
实际给料 actual feed
实际工况 actual condition
实际工况测定 actual performance measurement
实际工期 as-built schedule
实际工时 actual man-hour;actual working hours
实际工资 actual wage;real wage;take-home pay
实际工资率 actual labo(u)r rate;effective pay rate;real wage rate
实际工资收入 real wage income
实际工资指数 index number of real wages;real wage index
实际工作 real work
实际工作吨公里 <包括车辆自重> virtual ton-kilometers [kilometres] worked
实际工作负荷 <起重机的> practical working load
实际工作活荷载 service live load
实际工作经验 real experience
实际工作能力 hands-on background
实际工作人日数 actual man-days
实际工作人员 actual personnel;actual staff;real staff
实际工作时差值 total float
实际工作时间 actual working time;real work time;running time
实际工作台日数 actual working machine-days
实际工作天数 actual working days;effective days;effective working days
实际工作压力 actual working pressure
实际公差 actual allowance
实际功率 actual(horse) power;real (horse) power;true(horse) power;true watt;useful power
实际功率图 actual power chart
实际功能 practical duty
实际攻角 actual angle of attack
实际供给 actual supply
实际供给价格 true supply price
实际供水量 real quantity of water supply
实际购买 actual purchase
实际购买力 real purchasing power
实际购入价格 actual purchase price
实际购置成本法 actual acquisition cost basis
实际故障 physical fault
实际故障容限 physical fault tolerance
实际关键码 actual key
实际关键字 actual key
实际关税税率 effective tariff;effective tax rate
实际观测时间 actual time of observation;real observation time
实际观测条件 real observation condition
实际观测值 actual observed value
实际管理机构 effective management organization
实际管理机构所在地 place of effective management
实际光路追迹 actual ray trace
实际规范 actual specification
实际规格尺寸 actual size
实际规划 physical planning
实际规模 physical size;real size
实际国民生产总值 actual gross national products
实际过程 real process
实际过失和知情 actual fault and privity
实际海底 actual bottom
实际海平面 actual sea level
实际含灰量 actual lime content
实际焊喉深度 actual throat depth
实际行 actual line
实际航程 actual distance;distance made good
实际航速 speed made good
实际航向 actual travel direction
实际耗汽率 actual steam rate
实际耗蚀率 actual rate of wastage
实际耗损与陈旧率 actual rate of wastage and obsolescence
实际耗氧量 actual oxygen requirement
实际和预计收入对照表 statement of actual and estimated revenue
实际和预计支出对照表 statement of actual and estimated expenditures
实际荷载 actual load(ing);real load;working load
实际荷载试验 actual loading test
实际荷重法 real triaxial process
实际横断面 actual section
实际横断面面积 actual cross-section area
实际横截面 actual section
实际横剖面面积 actual cross-section area
实际洪峰记录 records of actual floods
实际洪水记录 records of actual floods
实际厚度 actual thickness;real thickness
实际花费 out-of-pocket
实际换出【计】physical swap-out
实际回流 actual reflux;finite reflux
实际回收率 practice recovery
实际汇价 effective rate;effective rate of exchange
实际汇率 actual exchange rate;actual rate of exchange;real rate of exchange
实际汇率指数 effective exchange rate index
实际会聚 practical convergence
实际混合气标准循环 real mixture standard cycle
实际货币 real money
实际货币持有额 actual money holding
实际货币余额 actual balance of money;real money balance
实际货物 actuals;physicals
实际获利 yield
实际获利率 yield rate
实际基地址 physical base address
实际级联 practical cascade
实际极限 actual limit;practical limit
实际计算面积 actual calculating area
实际计算数据 real arithmetic(al) data
实际记录 physical record;record block
实际记录长度 physical record length
实际家庭规模 actual family size
实际价格 actual cost;actual price;effective price;net cost;realized price;real price;true price
实际价值 actual value;practical value;real value;tangible value;true value;virtual value;realistic value <与名义价值相对>
实际价值法 actual value method
实际价值计算基础 actual value basis
实际驾驶行为 actual driving behavio(u)r
实际间接制造费用 actual manufacturing overhead
实际间隙 actual allowance;actual gap;actual play
实际检查 actual inspection
实际建议 factual proposal
实际建造时间 real construction time
实际建筑工期 actual construction time
实际建筑期限 actual construction time
实际建筑施工的准备工作 preliminary work for construction
实际降雨量 actual amount of rainfall
实际交付 actual delivery;actual handing over
实际交货 actual delivery;physical delivery
实际交通(容)量 practical (traffic) capacity;practice(traffic) capacity
实际交易 real transaction
实际交易值 actual quotation
实际胶质 existent gum
实际教训 object lesson
实际接点温度 virtual junction temperature
实际接合 actual interface
实际结构 actual structure;real structure
实际结果 actual result;real result
实际结束 physical end
实际截面 real cross-section
实际金额 actual amount of money
实际金融资产量 actual banking assets
实际进动 induced precession
实际进度 actual advance
实际进度与计划进度对比 progress actual vs scheduled
实际进路 route actually followed
实际经验 physical experience;practical experience
实际经由进路 route actually followed
实际晶粒度 actual grain size
实际晶体 actual crystal;real crystal
实际净空 actual clearance
实际净利润 real net profit
实际净载重量 actual payload
实际净重 actual net weight
实际径流量 actual run-off
实际静水压力值 real value of safe water pressure
实际拒付 actual dishono(u)r
实际距离 actual distance;true distance
实际浚挖高程 actual dredging level
实际竣工 substantial completion
实际开采量 real mining yield
实际开工日期 actual start date
实际开航时间 actual time of departure
实际开价 actual quotation
实际开始施工 being actual construction
实际开支 actual expenses
实际开支成本 out-of-pocket cost
实际靠岸时间 actual time of berthing
实际可靠性 achieved reliability
实际空速 true empty-running speed
实际空速计 true airspeed meter
实际空隙率 effective porosity;true porosity
实际空运飞机数 prime airlift
实际孔径 effective aperture;real aperture;true aperture
实际孔隙率 actual porosity;practical porosity
实际控制 working control
实际控制的范围 the extent of actual control
实际库存 actual stock;physical inventory
实际矿物成分 <岩石的> mode
实际亏损 actual loss
实际拦截时间 actual time of interception
实际劳动成本 real labo(u)r cost
实际离岸时间 actual time of unberthing
实际离差 actual deviation
实际里程计算法 method of computing wagon kilometers actually run
实际力矩 actual moment
实际力学强度 actual technical strength
实际利率 actual rate of interest;effective availability;effective interest rate;effective rate of interest;real interest rate;true rate of interest
实际利润 actual profit;realized profit;true profit
实际利息 effective interest
实际利息法 effective interest method
实际利息收入 actual interest income
实际利息收益 actual interest income
实际利益 actual benefit
实际例行程序的数据区 actual routine data area
实际粒径 actual grain size
实际亮度 intrinsic(al) brilliance
实际列车运行图 actual graph of train running;actual train running graph
实际领导 <非正式领导,但因其个人条件而有实际影响力> actual leader
实际流动 actual flow
实际流量 actual discharge;actual flow
实际流量系数 actual discharge coefficient
实际流速 actual velocity;real velocity;true velocity
实际流速测定法 determination of real flow velocity
实际流体 actual fluid;real fluid
实际流通额 effective circulation
实际履行 actual performance
实际绿灯时间【交】controller green time;displayed green time
实际螺距 actual pitch
实际马力 actual horsepower;real horsepower;real output;true horsepower
实际媒体 physical medium
实际密度 actual density;true density
实际面积 actual area;real area
实际模拟 realistic simulation
实际磨损 physical wear and tear
实际内摩擦角 angle of true internal friction
实际能力 activity capability;real capacity;virtual rating
实际能力因素 practical factors
实际能量 actual energy
实际能率 practical duty
实际年代学 parachronology
实际年利率 actual annual interest rate;effective annual interest rate

实际年龄 actual age;physical age
实际年息 true annual interest
实际年限 physical life
实际黏[粘]度 practical viscosity
实际黏[粘]度测定 practical viscosity measurement
实际欧姆 true ohm
实际排量 actual displacement;practical flow rate
实际排气速度 actual exhaust velocity
实际排水量【船】actual displacement
实际盘存 actual inventory;physical inventory
实际炮数 actual shot number
实际皮重 actual gross weight;actual tare;particular tare;real tare
实际偏差 actual allowance;actual deviation
实际偏心 actual eccentricity
实际票面价值 effective par
实际频度 actual frequency
实际平衡 real balance
实际平均螺距 experimental mean pitch
实际平均压力 actual mean pressure
实际平均值 true mean value
实际平均质量 average outgoing quality
实际平面 physical plane
实际坡度 actual grade;effective grade;real gradient
实际破断荷载 actual breaking load
实际破断力 actual breaking force
实际破坏荷载 actual breaking load
实际破坏应力 actual stress at fracture;actual stress by fracture
实际起飞飞行航道 net take-off flight path
实际起飞时间 actual time of departure
实际气体 actual gas;imperfect gas;real gas
实际气体状态方程 equation of state of real gas
实际气温 real air temperature
实际气隙 physical air gap
实际气压 actual pressure
实际汽耗 actual steam consumption
实际牵伸 resultant draft
实际前角 true rake
实际潜水深度 actual diving depth
实际强度 actual strength;technical strength
实际切力 actual shear force
实际切削 actual cut
实际侵蚀量 true erosion
实际情况 physical circumstance;practical consideration;practical situation
实际曲线 real curves
实际屈服点 actual yield point
实际屈服应力 actual yield stress
实际取得成本 actual acquisition cost
实际全损 absolute total loss;actual total loss
实际全损理算 adjustment of actual total loss
实际全损调整 adjustment of actual total loss
实际权数 actual weight
实际热效率 actual thermal efficiency
实际人口 actual population;de-facto population
实际人员(情况)actual personnel;actual staff
实际人员数量 real personnel
实际任意数据项 actual derived data item
实际容差 actual allowance
实际容积 actual volume
实际容量 actual capacity;practical capacity;real capacity
实际容许量 real allowance
实际溶液 real solution
实际蠕变 true creep
实际入口 actual entry
实际筛分 <粒状材料的> actual grading curve
实际筛分粒度 effective screen cutpoint
实际筛下回收率 true undersize recovery
实际山体压力 genuine mountain-pressure
实际商品 actuals
实际上 actually;in effect;materially
实际上升限度 operating ceiling;operational ceiling
实际上限 actual upper bound
实际设备 actual device;physical device
实际设备表 physical device table
实际设备的分配 allocation of real devices
实际设备坐标 actual device coordinates
实际设置 actual setting
实际射速 usable rate of fire
实际深度 actual depth
实际渗流速度 actual seepage velocity
实际渗透率 effective permeability
实际渗透系数 field permeability coefficient
实际升力 actual lift
实际升限 operating ceiling;operational ceiling
实际生产成本 real cost of production
实际生产量 effective output
实际生产率 practical productivity
实际生产能力 effective capacity;practical production capability;actual capacity;practical capacity
实际生产时间 actual productive time
实际生活标准 physical standard of living
实际生活费指数 actual cost of living index
实际生活水平 real standard of living
实际生态位 realized niche
实际生息率 actual yield
实际声源位置 actual sound position
实际剩余 real surplus
实际失效率 acceptable failure rate
实际施工程序 actual construction procedure;actual construction sequence
实际施工期 actual construction time
实际(十进制)小数点 actual decimal point
实际时间 actual time;material time;real-time
实际实现 actual implementation
实际使用 in-service use
实际使用带宽 utilized bandwidth
实际使用年限 actual(service)life;physical life;practical life
实际使用期限 actual life
实际使用权 beneficial occupancy
实际使用日期 date of actual use
实际使用时间 time used
实际使用试验 actual service test
实际使用寿命 actual service life;physical life;practical life
实际使用条件 actual service condition
实际驶离车流图式 out flow pattern
实际市场汇率 market rates of exchange in effect
实际示功图 actual indicator card
实际示意图 actual indicator;actual indicator card;real indicator
实际试验 field test(ing)
实际试验数据 actual test data
实际试验压力 actual testing pressure
实际收费载重 actual payload
实际收回 actual eviction
实际收入 real earning;real income
实际收入指数 real income index
实际收益 actual income;real income;returns in real term;true earning
实际收益率 actual yield;effective yield
实际寿命 activity life;device lifetime;effective life(time);physical life;practical life;actual life
实际输出 actual output
实际输出功率 real output power
实际输量 actual throughput
实际输入输出 physical input-output
实际输水量 actual output
实际输送吨公里 <不包括车辆自重> virtual ton-kilometers[kilometres] hauled
实际数据传输率 actual data transfer rate
实际数据块处理程序 actual block processor
实际(数)量 actual quantity
实际数值 actual numeric(al)value
实际数字 actual figure;real figure
实际数字地址 physical numeric(al) address
实际水管理条件下的作物产量 shadow yield
实际水平 practical level;true level
实际水深 actual depth;actual water depth
实际水头 actual head;real head
实际水头试验装置 real-head test rig
实际水位高度 actual height of water-level
实际水位降深值 real value of drawdown
实际水位下降值 real water-level decrease
实际水温 actual water temperature
实际税率 effective rate of tax;effective tax rate
实际顺序存取 physical sequential access
实际死亡率 actual death rate;actual mortality;effective death rate;effective mortality rate
实际送达 actual service
实际速度 actual speed;actual velocity;opening speed;physical speed;speed made good;virtual velocity
实际速率 actual speed
实际酸度 actual acidity
实际损害赔偿金 actual damages
实际损耗率及废弃率 actual rate of wastage and obsolescence
实际损失 actual loss;real loss
实际损失率 actual loss ratio
实际损益 realized gains and losses
实际所得 real income
实际所有权 beneficial ownership
实际坍落度 true slump
实际体积 true volume
实际天顶距 actual zenith distance;true zenith distance
实际填充深度 actual filling depth
实际填土 actual earth fill
实际条件 field condition
实际贴现率 true discount
实际停放量 actual parking volume
实际通道 physical channel
实际通过量 actual throughput
实际通过能力 practical throughput capacity;practical tonnage capacity
实际通行能力 real traffic capacity
实际投资 actual investment;real investment
实际透过的 actually transmitted
实际涂布率 practical spreading rate
实际土壤密度 actual density of soil
实际土压 true earth pressure
实际推力 actual thrust
实际弯矩 actual moment
实际完成的工作 actual performance
实际完成工作量 actual cost of work performed
实际完成情况 actual performance
实际完成日期 physical completion date
实际完成时间 actual finish time;actual time of completion
实际完工日期 date of substantial completion;actual finish date
实际完工证书 certificate of practical completion
实际网络配置 physical network configuration
实际违约 actual breach
实际维数 actual dimension
实际维修时间 active repair time
实际位移 actual displacement;real displacement
实际位置 actual position;physical location;real position
实际温度 actual temperature;true temperature
实际温度等值线 thermoisohyp
实际文件名称 physical file name
实际问题 physical problem;practical consideration;practical problem;practical question
实际无用数据 actual garbage
实际误差 actual error;practical error
实际吸附 actual absorption;true adsorption
实际吸附密度 actual absorption density;true adsorption density
实际吸收 actual absorption;true adsorption
实际系统时间 physical system time
实际细度 actual size
实际下界 actual lower bound
实际现场情况 actual field condition
实际现金价值 actual cash value
实际现金余额 actual cash balance;real cash balance;real money balance
实际现金支出 out-of-pocket cost;out-of-pocket costs or expenses
实际线路通过能力 actual track capacity
实际限度 actual limit
实际像差 actual aberration
实际像点 actual image point
实际消费 actual consumption
实际消耗功率 actual power consumption
实际销售 effective sale
实际销售量 actual sale
实际效果 actual result;real effect
实际效率 actual efficiency;practical efficiency
实际效益 tangible advance
实际效益估算 ex post measurement of benefit
实际效用 actual utility
实际斜距离 operational slant range
实际薪资率 effective pay rate
实际信用 real credit
实际行动 actual performance
实际行驶时间 time of traveling;travel(l)ing time
实际形状 true form
实际性能 actual performance
实际性状 actual behavio(u)r
实际修理时间 active repair time
实际需求 actual demand;effective demand;physical demand
实际需氧量 real oxygen requirement

S

实际需要 actual demand;actual need
实际需要量 actual requirement
实际雪线 actual snowline
实际循环 actual cycle
实际训练 hands-on
实际压力 active pressure;actual pressure;operating pressure;real pressure
实际压力角 operating pressure angle
实际压力限度 practical pressure limitations
实际压缩 <在绝热和等温压缩之间> actual compression
实际烟囱高度 real stack height
实际延迟 actual relay;true relay
实际验证 physical verification
实际验证过的结构 field-tested design
实际扬程 actual head
实际扬水量 actual delivery
实际业绩 actual performance
实际业绩与规定完成业绩的偏差 performance deviation
实际液体 real fluid;real liquor
实际应变 actual strain;true strain
实际应力 actual stress;field stress;true stress
实际应用 practical application
实际应用课程 implementation classes;implementation courses
实际盈余 realized surplus
实际用户 actual user
实际用水量 real quantity of water consumption
实际油温 actual oil temperature
实际有效汇率 real effective exchange rate
实际有效坡度 virtual slope
实际余额效应论 real-balance-effect theory
实际预期标准成本 actual expected standard cost
实际预算 physical budget
实际预算标准成本 actual expected standard cost
实际预算赤字 physical budget deficit
实际阈 practical threshold
实际约束 physical constraint
实际运动 actual motion
实际运费率 actual freight rate
实际运输能力 physical transport capacity
实际运输时间 actual transportation time
实际运行 actual motion;proper motion
实际运行程序数据区 actual runtime data area
实际运行时间 actual run time
实际运行时数据区 actual runtime data area
实际运载能力 actual carrying capacity
实际运转 actual operation;actual play;real-world operation
实际载荷 actual load(ing);practical load
实际载货量 actual cargo capacity;actual payload
实际载重吨位 actual deadweight
实际载重量 real loading capacity
实际暂驻存储区 physical transient area
实际造价 actual cost
实际增长率 actual rate of growth;real growth rate
实际增长水平下降 decline of real growth rate
实际增益 actual gain
实际增值 realized appreciation
实际债款 actual debt
实际债务 actual debt;effective debt;real debt
实际占有的动产 choses in possession
实际占有(权)actual possession
实际占有人 actual holder;occupant
实际长势 actual growing situation
实际折旧 actual depreciation;physical depreciation;realized depreciation
实际折旧的原因 effective depreciation cause
实际折扣 true discount
实际辙叉心 actual nose of crossing
实际蒸发量 actual evapo(u)ration
实际蒸发蒸腾量 actual evapotranspiration
实际蒸发作用 real evapo(u)ration
实际蒸腾 real evapotranspiration
实际整合 paraconformity
实际正常成本 actual normal cost
实际支出 actual expenditures;actual outlay;out-of-pocket expenses
实际支出成本 outlay cost
实际支出总额 total actual spending
实际支付率 rate actually paid
实际支付数额 net amount of payment;net amount paid
实际知识 know-how;practical knowledge
实际执行 actual execution
实际(执行)订货单 actual order
实际执行合同 actual order
实际执行预算 effective working budget
实际值 actual value;real value
实际职能 physical function
实际职务 active duty
实际植被 real vegetation
实际指令 actual instruction
实际制动距离 active braking distance
实际制冷效果 net refrigerating effect
实际制造成本 actual manufacturing cost
实际制造费用分配率 actual burden rate
实际置信度 actual degree of belief
实际中心距(离)operating center distance
实际重力 actual weight
实际重量 actual weight
实际贮量 actual reserves
实际转角视角 actual corner sight angle
实际转头 effective head
实际转氧效率 field oxygen transfer efficiency
实际装斗物料 bucket payload
实际装货量 actual cargo capacity
实际装卸工作日数 man days actual used in loading/discharging
实际装载量 actual weight of load
实际状况 actual state;existing circumstance;real state
实际状态 actual state;existing circumstance;real state;virtual condition
实际准备装船数量 actual quantity to be shipped
实际准确度 available accuracy
实际资本 actual capital;real capital
实际资本流动 real capital flow
实际资本收益率 yield of real capital
实际资本系数 actual capital coefficient
实际资产 actual assets;net assets;physical assets
实际资金利润率 actual profile rate of funds;effective profit rate on funds
实际资源 real resources
实际子女数 actual family size
实际综合税率 effective aggregate tax rate
实际总成本 total cost used
实际总坡度 total effective grade
实际总收入 effective gross income
实际总重量 actual gross weight
实际总阻力 total effective grade
实际租金 actual rent
实际最大供水管路 actual full supply line;actual maximum flow line;actual maximum supply line
实际最大损失 absolute maximum loss
实际作业半径 actual working radius
实际作业工作日 actual man-days at work
实际作业天数 effective days;effective working days
实际作用部位 the actual site of action
实价 firm offer;intrinsic(al) value;net cost;net price;real price
实价保险 value policy
实价率 value number
实价评估 good faith estimate
实肩拱 filled spandrel arch;spandrel-filled arch
实肩式拱桥 spandrel-filled arch bridge
实践 experience
实践的系统工程师 practicing system engineer
实践结果 practical result
实践经验 hands-on;practical experience
实践科学 work science
实践性 practicality
实践知识 know-how
实践智力 practical intelligence
实践智能 concrete intelligence;practical intelligence
实箭头 solid arrow
实焦点 real focus;true focus
实焦面 real focal plane
实铰 real hinge
实缴 paid in
实缴部分 paid-in portion
实缴股本 capital paid-in;capital stock paid in;contributed capital;paid-in capital;share premium account
实缴股本总额 total paid-in capital
实缴金额 amount paid in
实缴税收会计核算 flow-through accounting
实缴资本 contributed capital
实井 real well
实景显示 realistic display
实矩阵 real matrix
实距 actual distance
实空间 real space
实孔径雷达 real aperture radar
实库 thesaurus
实况 truth
实况报告 factual record
实况电视广播 live [TV] television broadcast(ing)
实况调查 fact finding
实况广播 live program(me);running commentary
实况广播技术 live studio technique
实况广播间 commentator's booth
实况广播节目 live program(me)
实况广播节目电路 outside broadcast programme circuit
实况广播室 live studio
实况节目插播 live insert
实况录像 outside broadcast recording
实况录音 outside broadcast recording
实况转播 field pick-up;outside broadcast(ing)
实况转播车 outside broadcast vehicle
实况转播解说员 commentator
实矿石 positive ore
实扩散函数 real spread function
实肋 solid rib
实肋板 plate floor;solid floor
实肋拱 solid-rib(bed) arch;solid rid arch
实肋拱桥 solid-ribbed arch bridge
实力薄弱的开发商被淘汰掉<经济危机中> shakeout
实力地位 position of strength
实利函数 utility function
实利计息 exact interest
实利人 rational-economic man
实利主义 materialism
实例 case;example;illustration;instance;living example;sample
实例查询 actual query
实例程序图 example program(me) graph
实例分析 case analysis;case study
实例记载 case history
实例历史包 case history package
实例研究 case study
实梁 carrier beam;real beam;solid beam
实量 real quantity
实量需求 manifest demand
实路对幻路的串话 side-to-phantom crosstalk
实路对实路的串音 side-to-side crosstalk
实码头接岸端 root
实面积 solid area
实模式 real mode;real pattern
实模型 real mould
实模铸造法 cavityless casting
实模撞击成型 solid bossed
实能 actual energy
实盘 firm offer;firm quotation;offer with engagement
实盘交易 firm bargain
实膨胀 real expansion
实片英尺 super-foot true
实拼门 barred door;batten door;ledged door
实频率轴 real frequency axis
实平面 real plane
实奇点 real singularity
实砌(楼梯)旋栓 solid newel
实砌墙 solid wall
实墙 blank wall;blind wall;dead wall
实桥面 solid bridge floor
实情调查 facts survey
实球体 solid sphere
实区间 real interval
实曲线 full curve;solid curve line;solid line curve
实驱动 real drive
实驱动器 real driver
实容重 solid unit weight
实容重法 real unit weight process
实生矮林 seedling coppice
实生林 seeding crop
实生苗 seedling
实生苗种子园 seedling seed orchard
实生树 seedling-plant;seedling tree
实生树林带 seedling strip
实生树植区 seedling clearing
实生树桩 seedling stump
实声源 real source
实施 accomplishment;administer;bring into effect;carry into execution;carry into practice;come into operation;enforcement;execution;go into operation;implement-(ation);put into effect;put into execution;putting into operation;staging;take effect
实施标准 standard of performance
实施步骤 implementation steps
实施程序 implementation procedure;operational procedure
实施的核查 check practice

实施法规 code of practice
实施方法 meaner of execution
实施费用 executive cost
实施管理 execution control
实施规程 code of practice
实施规则 enforce a rule
实施计划 action plan;execution plan;implementation plan;starting plan
实施阶段 feasibility stage;implementation phase
实施可能性 operational feasibility
实施期 effective date
实施日期 date of enforcement
实施设计 execution design
实施手册 enforcement manual
实施细则 implementation regulation;rule for implementation
实施线网 execution road network
实施项目保函 performance bond
实施与监督 implementation and supervision
实时 actual time;current time;intrinsic(al)time;true time
实时安培容量 real-time ampacity
实时保护 real-time guard
实时保护方式 real-time guard mode
实时编码 real-time coding
实时辨识 real-time identification
实时并行操作 real-time concurrency operation
实时并行性 real-time concurrence
实时波形 real-time waveform
实时采集处理系统 real-time acquisition and processing system
实时操作 real-time operation; real-time working;true time operation
实时操作方式 real-time operational mode
实时操作联机 real-time operational on-line
实时操作系统 real-time operating system;real-time operational system
实时测量 real-time measurement;real-time survey
实时查询 real-time remote inquiry
实时差分接收机 real-time differential receiver
实时成批处理 real-time batch processing
实时程序 real-time program(me)
实时程序编制 real-time programming
实时程序编制系统 real-time production monitoring system; real-time program(me)development system
实时程序仿真 real-time program(me)simulation
实时程序开发系统 real-time program(me)development system
实时程序设计 real-time programming
实时处理 real-time processing
实时处理机 real-time processor
实时处理机设计 real-time processor design
实时处理控制系统 real-time processing control system
实时处理通信 real-time processing communication
实时处理系统 real-time processing system
实时传递系统 real-time transmission system
实时传输 live transmission;real-time transmission
实时传输和非实时传输 real-time transmission and non-real-time transmission
实时传输系统 real-time transmission system
实时传输线 real-time link
实时传送系统 real-time transmission system
实时磁盘操作系统【计】real-time disk operating system
实时的 real-time;time-bound
实时地址 real-time address
实时电视 real-time television
实时电子监视 real-time electronic surveillance
实时定址 real-time addressing
实时动态测绘 real-time kinematic surveying and mapping
实时动态放映显示器 real-time dynamic(al)projection display
实时动态分析 real-time dynamic
实时读出 real-time readout
实时度标 real-time scale
实时多带图灵机 real-time multitape Turing machine
实时多光谱观察 real-time multispectral viewing
实时多普勒成像系统 real-time Doppler imaging system
实时多重计算 real-time multicomputing
实时范围 real-time range
实时方式 real-time mode
实时仿真 real-time simulation
实时仿真器 real-time simulator
实时分时 real-time time-sharing
实时分析 real-time analysis
实时分析器 real-time analyser[analyzer]
实时分析系统 real-time analysis system
实时分析仪 real-time analyser[analyzer]
实时服务程序 real-time and service software routine
实时复合计算机 real-time computer complex
实时干涉测量法 real-time interferometry
实时干涉量度学 real-time interferometry
实时干涉仪 real-time interferometer
实时更新系统 real-time revision and correction system
实时工作 real-time operation; real-time working
实时工作比 real-time working ratio
实时故障 real-time fail
实时观点 real-time point of view
实时管理 real-time executive
实时光学补偿滤光器 real-time optic(al)matched filter
实时光学相关 real-time optic(al) correlation
实时光学信号处理器 real-time optic(al)signal processor
实时过程控制系统 real-time process control system
实时过程语言 real-time process language
实时缓冲器 real-time buffer
实时灰阶显示 real-time grey level display
实时激光线路 real-time laser link
实时计时器 real-time clock
实时计算 real-time computation
实时计算方法 real-time computing technique
实时计算复合体 real-time computing complex
实时计算机 real-time computer;real time machine
实时计算机程序 real-time computer program(me)
实时计算机系统 real-time computer system
实时计算机中心 real-time computer center[centre]
实时计算控制 real-time computing control
实时计算中心 real-time computing center[centre]
实时计算综合装置 real-time computer complex
实时记录 real-time recording
实时监测仪 real-time monitor
实时监督程序 real-time monitor
实时监督器 real-time monitor
实时监控程序 real-time monitor
实时监控器 real-time monitor
实时减压法 real-time decompression
实时交通感应控制系统 real-time traffic responsive control system
实时校正 real-time correction
实时接口 real-time interface
实时进程 real-time process
实时精确声级测量 real-time precision level measurement
实时距离处理 real-time range processing
实时控制 real-time control
实时控制板 real-time option board
实时控制(例行)程序 real-time control routine
实时控制器 real-time controller
实时控制输入输出 real-time control input/output
实时控制系统 real-time control system
实时控制应用 real-time control application
实时联机操作 real-time on-line operation
实时联机系统 real-time on-line system
实时列车编组程序 real-time train assembler
实时流量演算 real-time flow routing
实时录音 real-time recording
实时滤波器 real-time filter
实时模拟 real-time simulation
实时模拟计算机 real-time analogue computer
实时模拟器 real-time simulator
实时模拟数字计算 real-time-analog-digital computation
实时模拟装置 real-time simulator
实时模式识别 real-time pattern recognition
实时模型 real-time model
实时排队 real-time queue
实时配位 real-time coordination
实时频率分析 real-time frequency analysis
实时频率分析仪 real-time frequency analyser[analyzer]
实时频谱分析器 real-time spectrum analyser[analyzer]
实时谱分析 real-time spectral analysis
实时谱分析仪 real-time spectrum analyser
实时谱分析员 real-time spectrum analyser
实时前台任务 real-time foreground task
实时情报系统 real-time information system
实时全数字频谱分析仪 real-time all-digital spectrum analyser[analyzer]
实时全息法 real-time holography
实时全息干涉测量(术)real-time holographic interferometry
实时全息术 real-time holography
实时全息图 real-time hologram
实时全息照相再现 real-time holographic reconstruction
实时任务 real-time task
实时软件 real-time software
实时三维显示 real-time three-dimensional display
实时摄影测量 real-time photogrammetry
实时摄影术 real-time photography
实时生产监视系统 real-time production monitoring system
实时时钟 time-of-day clock;real-time clock
实时时钟分时 real-time clock time-sharing
实时时钟计数字 real-time clock count word
实时时钟记录 real-time clock log
实时时钟模件 real-time clock module
实时时钟诊断程序 real-time clock diagnostics
实时时钟中断 real-time clock interrupt
实时示波器 real-time oscilloscope
实时输出(信号)real-time output
实时输入 real-time input
实时输入输出 real-time input-output
实时输入输出控制 real-time input-output control
实时输入输出控制器 real-time input-output controller
实时输入输出译码器 real-time input-output translator
实时数据 real-time data
实时数据处理 actual time data processing; real-time data processing;real-time data reduction
实时数据传输 real-time data transmission
实时数据控制系统 real-time data control system
实时数据整理 real-time data reduction
实时数字控制器 real-time digital governor
实时数字频谱分析器 real-time digital spectrum analyser[analyzer]
实时算法 real-time algorithm
实时算法试验器 real-time algorithmic tester
实时探测 real-time detection
实时条纹 real-time fringe
实时调试程序 real-time debug program(me)
实时调试例行程序 real-time debug routine
实时通道 real-time channel
实时通信[讯] real-time communication;real-time traffic
实时通信[讯]处理 real-time communication processing
实时通信[讯]系统 real-time communication system
实时图像处理 image processing in real-time
实时外推法 real-time extrapolation
实时微计算机 real-time microcomputer
实时微计算机系统 real-time microcomputer system
实时卫星计算机 real-time satellite computer
实时位置 real-time position
实时污染综合监测仪 integrated real-time contamination monitor
实时系统 event driven system; real-time system
实时系统软件 real-time system software
实时系统特性 real-time system characteristic
实时系统微处理机 real-time system microprocessor

S

实时显示 real-time display
实时显示程序 real-time display program(me)
实时显示器 real-time display device
实时线性探测系统 real-time linear detection system
实时相对寻址 real-time relative addressing
实时相关 real-time correlation
实时相互作用情报检索系统 real-time interactive reference retrieval system
实时响应 real-time response
实时协议 real-time protocol
实时信号处理 real-time signal processing
实时信号分析 real-time signal analysis
实时信息 real-time information
实时信息理论 real-time information theory
实时信息系统 real-time information system
实时选择板 real-time option board
实时样机分析器 real-time prototype analyser[analyzer]
实时遥测 real-time telemetry
实时遥测处理系统 real-time telemetry processing system
实时遥测发射机 real-time telemeter transmitter
实时遥测线 real-time telemetry link
实时译码器 real-time decoder
实时应用 real-time application
实时用户 active user
实时语言 real-time language
实时预报 real-time forecast(ing);real-time prediction
实时预测 real-time forecast(ing);real-time prediction
实时预定座席 real-time seat reservation
实时远程询问(站) real-time remote inquiry
实时运算 real-time operation; real-time working;true time operation
实时运行(操作)模式 real-time operation mode
实时在线操作 real-time on-line operation
实时执行操作系统 real-time executive operating system
实时执行程序 real-time executive routine
实时执行程序系统 real-time executive routine system
实时执行系统 real-time executive system
实时指定 real-time assignment
实时指令 real-time command
实时质量控制 real-time quality control
实时中央处理 real-time batch processing; real-time central processing
实时中央处理机 real-time central processor
实时装置 real-time clock
实时资料传输 real-time data transmission
实时资料收集 real-time data collection
实时自动数字光学跟踪器 real-time automatic digital optic(al)tracker
实时最优化程序 real-time optimization program(me)
实时作业 real-time job
实示需氧器 anoxyscope
实事求是的 down-to-earth;pragmatic
实事求是 come down to earth
实视场 real field of view
实收 net receipt
实收保险费 earned premium
实收股本 paid-up capital;paid-up stock
实收基础 receipts basis
实收款项 proceeds of sale
实收率 extraction yield;recovery efficiency
实收数量 actual quantity received
实收资本 capital paid-in;paid-in capital;paid-up capital
实收资本利润率 profit ratio of paid-up capital
实收资本增加 paid-in capital increase
实售价格 realized price
实数【数】 actual number; real; real number;real quantity
实数标志 real denotation
实数部分 active component; energy component;power component;real component;real part
实数的绝对值 absolute value of a real number;numeric(al)value of a real number
实数分析法 actual number analysis
实数付现 net cash
实数加 real add
实数减 real substract
实数据项 real data item
实数类型 type real
实数累加器 real accumulator
实数连续统 arithmetic(al)continuum;continuum of real numbers
实数平面 number plane
实数商 quotient of real numbers
实数系【数】 system of real numbers
实数域 real number field
实(数)运算 real arithmetic
实数直线【数】 number line
实数值场强度 real valued intensity
实数制 numeric(al)rating system
实数轴 axis of reals; real axis; real number axis
实素几何学 real geometry
实算术型常数的精度 precision of real arithmetic(al)constants
实胎轮辋 rim for solid type
实体 embody;entity;materiality;noumenon[复 noumena];quintessence;solid body;solid mass;stereo;subsistence;stereotomy <尤指石头>
实体岸壁 solid wall
实体岸壁建筑界线 bulkhead line
实体岸墩 gravity abutment
实体坝 gravity dam;solid dam;solid gravity dam
实体板 solid panel
实体辨别 stereognosis
实体标识符 entity identifier
实体表面位 potential of materiel surface
实体材积 solid volume
实体测定法 body measurement method
实体测量图 stereometric(al)map
实体存储器 physical memory
实体错觉 haptic illusion
实体挡土墙 solid retaining wall
实体的 entitative;physical
实体地面 solid floor
实体地形 stereotopography
实体电路 physical circuit
实体丁坝 solid groin;solid groyne
实体短木段 solid billet
实体断面 solid section
实体堆积的 close-piled
实体法 material law;substantive law
实体法性质的规定 substantive rules
实体方位码 bank cubic(al)yard
实体仿真 partial simulation;physical simulation
实体分隔设施 physical divider
实体分配 physical distribution
实体分配管理 physical distribution management
实体粉刷工程 solid plasterwork
实体扶壁 solid buttress
实体腹板 solid web;solid-web plate
实体腹板梁 solid-web girder
实体腹板桥 solid-web bridge;solid-web plate bridge
实体干船坞 solid dry dock
实体杆件 solid member
实体感 perception of solidity
实体隔墙 solid partition
实体拱 solid arch
实体鼓壁式离心机 solid wall bowl centrifuge
实体关系 entity relationship
实体关系分析 entity relationship analysis
实体关系模型 entity relationship model
实体关系图 entity relationship diagram
实体观测测波仪 stereo wave meter
实体规划 physical planning
实体化 objectivization
实体幻灯机 stereopticon
实体绘制【计】 entity draw
实体混凝土 realcrete
实体混凝土地面 solid floor
实体混凝土块 solid block
实体积 solid volume
实体积比率 solid volume percentage
实体积的 bulked down
实体集 entity set
实体集合模型 entity set model
实体记录 entity record
实体记录集 entity record set
实体剪力墙 solid shear wall
实体浇铸 solid casting
实体浇铸法 solid casting method
实体浇铸门 solid cast door
实体胶轮 solid rubber wheel
实体铰刀 solid reamer
实体结构 massive construction;massive structure;solid structure
实体结构体系 solid structural system
实体镜画 stereograph
实体矩形梁 solid rectangular beam
实体块 solid block
实体框架 solid frame work
实体类型 entity type
实体立方码 <爆破岩石> bank cubic(al)yard
实体联系法 entity relationship approach
实体梁 solid beam
实体楼板 solid floor
实体轮胎 solid tire[tyre]
实体码头 solid pier
实体(码头)岸壁 solid quay wall
实体码头前沿线 bulkhead line
实体门 solid door
实体密度 in-situ density
实体名字 physical name
实体模拟 partial simulation;physical simulation
实体模型 dummy; full-scale mock-up;material model;mock-up;physical model
实体抹灰 solid plasterwork
实体偏心轮 solid eccentric sheave
实体平面 physical plane
实体砌筑单位 solid masonry unit
实体墙 solid masonry wall;solid wall
实体墙结构 structure with solid face
实体桥墩 solid pier
实体桥面 solid bridge floor
实体桥面板 solid bridge deck; solid deck
实体桥台 solid abutment
实体权利 substantive right
实体商品 physical commodity
实体摄影 stereophotograph
实体摄影法 solidography
实体世界 physical world
实体式 solid type
实体式船坞构造 solid type of dock construction
实体式吊机臂 solid jib
实体式结构 closed construction;solid construction
实体式码头 dock of closed construction;dock of solid construction;solid pier; solid (type) quay; solid (type)wharf;bulkhead quay wall
实体式码头结构 solid type of dock construction
实体式起重机臂 plate jib
实体式顺岸码头 solid marginal wharf;solid wharf
实体式转子 solid type rotor
实体视觉 stereovision
实体数量加权法 weighting method of physical quantity
实体填筑码头 solid-fill quay
实体填筑式突堤码头 solid-fill type pier
实体突堤 fill(ed)-in pier;solid jetty;solid pier
实体图 stereogram
实体外转鼓 outer solid bowl
实体围篱 solid fence
实体圬工 solid masonry(work);solid unit masonry
实体圬工单元 solid masonry unit
实体圬工挡土墙 solid masonry retaining wall
实体圬工块 solid block
实体物切割术 stereotomy
实体系统 physical system
实体系统时间 problem time
实体线 physical line
实体斜坡道 paving sloping way;solid sloping way
实体型混凝土块体 solid concrete block
实体性 materiality
实体栅(栏)solid fence
实体账户 real account
实体支墩 solid buttress
实体支墩基础 solid pier foundation
实体重力坝 solid gravity dam
实体重力式岸壁 solid gravity wall
实体轴承 solid bearing
实体柱 solid column;substantial pillar
实体砖石单元 solid masonry unit
实体锥柄绞刀夹头 solid taper shank reamer chuck
实体资产 real assets
实填土防止基础或土方滑动 buttress fill
实铁芯 solid iron core
实投资本 paid-in investment
实土方量 net volume of earthwork
实伪变量 real pseudo-variable
实位置 real location
实文件 real file
实务 actual practice
实物 entity;matter;real(istic)object;substance;true object
实物标准 material standard
实物补偿 physical make-up;recovery in kind
实物测量 actual measurement
实物尺寸 full-scale;full size;full-size measurement;natural-size drawing
实物大模型 mock-up
实物大小 natural scale;natural size
实物大小的 full-sized;lifesize(d)
实物单位 physical unit;unit in kind
实物担保 real guarantee;real security
实物地租 rent in kind

实物电路图 pictorial diagram
实物调查 inventory survey
实物定额 norm in kind
实物分隔 physical separated
实物分配 physical distribution
实物付税 tax in kind
实物工程量 project volume in physical units
实物工资 natural wage; truck; wage and salary in kind; wages in kind
实物工资制 tommy; truck system
实物股息 dividend in kind
实物幻灯机 epidiascope
实物货币 commodity money
实物基准 material standard
实物计划 physical plan
实物计量 physical measure
实物价格 physical price
实物交割 delivery settlement
实物交换经济 barter economy
实物交易 barter; barter business; barter trade
实物缴纳 pay-in kind
实物教学 object lesson; object teaching
实物津贴 allowance in kind
实物净重 net weight
实物捐献 contribution in kind
实物捐助 contribution in kind
实物空间 real space
实物库存 physical holding of stock
实物栏杆 physical barrier
实物量 physical quantity; quantity of goods produced
实物量度 measure in kind; quantity measure
实物量具 material measure
实物量指标 indicator of physical output
实物流动 real flow
实物模 natural pattern
实物模拟 physical analog(ue)
实物模型 physical model
实物模铸型 reproduced model mo(u)ld(ing)
实物纳税 tax payment in kind
实物赔偿 reparations in kind
实物品质 actual quality
实物商品 physical commodity
实物市场 actuals market; physical market
实物试验 actual loading test; actual test; full-scale experiment; full-scale test; full-sized test
实物试验风洞 full-scale tunnel
实物试验台 full-scale test bench
实物收入 income in kind
实物数量 physical volume
实物水价 <以小麦棉花若干斤计价> water price in kind
实物税 real tax
实物体积 physical volume
实物投入 input in kind
实物投入产出 physical input-output
实物投资 investment in kind
实物投资估价过高 overestimation of investment in kind
实物土地税 land tax collect in kind
实物先张的 full-scale pretensioned
实物消耗定额 norm of material consumption
实物信用 real credit
实物形态 physical form
实物样机 hardware prototype
实物样品 hardware prototype
实物一半大小的 half life-size
实物预拉的 full-scale pretensioned
实物预算 physical budget
实物造型 reproduced model mo(u)ld-(ing)
实物账户 real account
实物折旧 physical depreciation
实物征用 requisition in kind
实物支持 material supply
实物支付 in-kind payment; payment in kind
实物指标 index in kind; index in terms of material products; physical indicator; target in kind
实物资本 physical capital; real capital
实物资本财物 physical capital goods
实物资产 assets in kind; physical assets; physical capital goods; real assets; tangible assets
实物自然大小 natural scale
实物走私 physical smuggling
实物坐标 real-world coordinate
实习 exercise; exercitation; practice; perform on-the-job training
实习班 seminar; workshop
实习车间 practice workshop
实习船 school ship; training ship
实习工厂 factory attached to school
实习工程师 engineer-in-training; intern engineer; student engineer
实习绘图员 drawing office apprentice
实习驾驶室 apprentice officer; deck cadet
实习建筑师 architect-in-training; intern architect
实习教室 practical room
实习课 exercise class; training course
实习楼 practical room block; practical room building
实习轮机员 apprentice engineer; cadet engineer; engineer cadet
实习生 apprentice; cadet; improver; probationer
实习水手 apprentice sailor; apprentice seaman; seaman apprentice
实习台 practice position
实习型钢 solid bar
实习医生 intern(e)
实习引航员 apprentice pilot
实显交通需求 manifest demand
实现 accomplish; effectuate; implementation; make real; materialize
实现保护贸易政策 putting into effect the policy of trade protection
实现标准成本 attainable standard cost
实现的非唯一性 nonuniqueness of realization
实现的利润 realized profit
实现反应 realization response
实现分部制 divisionalization
实现过程 implementation procedure
实现机械化 mechanize
实现计划 execute a plan
实现技术 implementation technique
实现价格 realized price
实现价值 realized value
实现可持续发展 maintain sustainable development
实现利润 bring about profit
实现利益 profits realized
实现目标的方法 aim-oriented approach
实现收入 revenue realization
实现算法 realization algorithm
实现损益准备 reserve for realized profits and losses
实现系统 implement system
实现性模块 implementation module
实现需求 implementation requirement
实现盈余 realized surplus
实现语言 implementation language
实现者 implementor
实现者名 implementor name
实现周期 performance period
实线 actual line; continuous line; drawn line; full line; hard-wire; physical wire; real line; solid line; unbroken line
实线电路 physical circuit; side circuit
实线箭头 solid arrow
实线路 audio line
实线曲线 block curve
实线线路加感线圈 side circuit loading coil
实箱 <集装箱> full unit
实向量空间 real vector space
实巷 dead-end street
实像 real image
实像模型 iconic model
实像平面 real image plane
实像全息立体模型 real image holographic(al) stereomodel
实像全息图的扩散函数 real hologram spread function
实效 actual effect; effectiveness; practical result; substantial results; useful effect
实效尺寸 virtual size
实效出行 <为达到指定目的的出行> utilitarian trip
实效工资 efficiency wage
实效状态 virtual condition
实心 solid core
实心坝 solid dam
实心板 solid flat plate; solid panel; solid plate; solid sheet; solid slab
实心板拱 solid-barrel arch
实心板拱桥 solid-barrel arch bridge
实心棒 solid rod
实心棒材 solid bar
实心玻璃微球体 solid glass microsphere
实心玻璃砖 solid glass block
实心部分 solid section
实心舱内支柱 solid hold pillar
实心层积隔墙 solid laminated partition
实心长方形踏步 solid rectangular step
实心长方形台阶 solid rectangular step
实心长方形梯级 solid rectangular flyer
实心长方形悬臂台阶 solid rectangular cantilevered step
实心车轮 solid wheel
实心车轴 solid axle
实心车轴轴承 solid axle bearing
实心车轴轴箱 solid axle box
实心冲头 solid punch
实心船模 solid-block model
实心窗格 blank tracery
实心窗花格 bland tracery; blind tracery
实心窗架 solid frame; solid window arcade; solid window frame
实心窗框 solid frame; solid window arcade
实心磁轭 solid yoke
实心磁极 solid magnetic pole
实心导辊 solid supporting roller
实心导线 solid conductor
实心的 stuffed
实心的窗框 solid window frame
实心地板 solid floor
实心电缆 solid(type) cable
实心端 solid end
实心断面 solid(cross-) section
实心法 solid-core method
实心方钢筋 solid square bar
实心方石阶步 square step
实心腹板 solid web
实心腹板大梁 steel solid web girder
实心腹板钢梁 steel solid web beam
实心杆 pole spar; solid spar
实心钢筋混凝土楼板 solid reinforced concrete floor(slab)
实心钢门 solid steel door
实心钢钎头 steel solid bit
实心钢钻头 steel solid bit
实心高炉泡沫矿渣混凝土砌块 solid foamed slag concrete block
实心隔墙 solid partition
实心拱 blank arch; blind arch; massive arch
实心拱廊 blank arcade; blind arcade; blind gallery; blind triforium
实心构件 solid element
实心管柱 <填以混凝土> filled pipe column
实心硅酸钙砖 solid calcium silicate brick
实心辊轧 solid roll
实心滚子 solid roller
实心滚子轴承 solid roller bearing
实心焊丝 solid welding wire
实心横隔墙 solid cross wall
实心护圈 solid case
实心环 solid torus
实心簧眼 spring solid eye
实心灰泥 solid plasterwork
实心灰砂墙 solid lime sand wall
实心灰砂砖 solid sand-lime brick
实心混凝土 solid concrete
实心混凝土板 solid concrete slab
实心混凝土方块 solid block; solid concrete block
实心混凝土梁 non-voided concrete beam; solid concrete beam
实心混凝土砌块 concrete brick; solid concrete block
实心活塞 solid piston
实心集 solid set
实心挤压型材 solid extruded shape
实心件 solid piece
实心胶轮 solid rubber wheel
实心胶胎 solid rubber tire[tyre]
实心接线柱 solid post
实心截面 solid section
实心截面柱 column of solid section
实心金刚石钻头 solid diamond bit
实心金属垫片 solid flat metal gasket
实心金属反射器 solid metal reflector
实心卷轴 solid roll
实心绝缘子 solid insulator
实心块 solid slug
实心块材 pier block
实心块材模子 solid-block mo(u)ld
实心块材砌筑墙 solid-block masonry (work)
实心块材试验机 solid-block tester
实心块材制机 solid-block machine
实心栏杆 blind balustrade; solid balustrade
实心连拱 arcature
实心连拱廊 blank arcade; blind arcade; wall arcade
实心连环拱廊 wall arcade
实心连接 solid connection
实心梁 solid beam; solid(-web) girder
实心亮点 solid center[centre] spot
实心流束喷嘴 solid-stream nozzle
实心楼板 solid floor
实心楼板电气供热 solid-floor heating by electricity
实心楼板供热 solid-floor heating
实心楼板供热电线 solid-floor heating cable
实心轮胎 band tire; block tire; rubber band tire; solid tire[tyre]
实心铆钉 solid rivet
实心煤渣砖 solid cinder tile
实心门 solid(-core) door
实心门框 solid door frame; solid frame
实心模锻机 solid die forging machinery
实心抹灰工程 solid plasterwork
实心木船模 block model
实心木地板 plank-on-edge floor; solid

wood floor
实心木垛 solid chock;solid cribbing;solid wooden chock
实心木垛支护 solid crib timbering
实心木料 solid wood
实心木楼板 solid wood floor
实心黏[粘]土砌块 solid clay unit
实心黏[粘]土砖 solid clay brick
实心女儿墙 bahut
实心刨花板木门 solid wood flake door
实心膨胀性炉渣混凝土砌块 solid expanded cinder concrete block
实心坯 solid billet
实心平板 solid slab
实心平板楼板 solid flat plate floor
实心平面门 solid-core flush door
实心砌块 solid masonry unit; solid block
实心砌体 solid masonry(work)
实心砌体墙 solid masonry wall
实心砌体圬工 solid unit masonry
实心钎杆 solid drill steel
实心钎钢 solid steel
实心墙 solid wall
实心墙板隔墙 solid wallboard partition
实心桥墩 solid abutment;solid pier
实心球 solid sphere
实心曲轴 solid crankshaft
实心圈 solid rim
实心全板门 solid door
实心熔渣砌块 solid clinker block
实心升降臂 solid lift arm
实心石膏板 solid gypsum board
实心石膏板隔墙 solid gypsum partition
实心石膏建筑板 solid gypsum building board
实心石膏屋顶砌块 solid gypsum roof(ing) block
实心石灰砂砖 solid lime sand brick
实心石泡构造 sincere lithophysa structure
实心竖框窗 solid mullion window
实心塔 solid pagoda
实心踏步 solid step
实心炭棒 solid carbon
实心炭刷 solid carbon brush
实心(体)护栏 high containment parapet
实心填充物 solid-core packing
实心铁芯 solid iron core
实心砼梁 blind concrete beam; non-voided concrete beam
实心铜线 solid copper wire
实心图形 solid object
实心圬工 solid masonry(work)
实心屋顶 solid roof
实心现浇板 solid in-situ cast slab
实心橡胶轮胎 solid rubber tire[tyre]
实心芯板 <胶合板的> solid core
实心型材 solid shape
实心旋梯中柱 solid newel post
实心堰 solid weir
实心阳极 heavy anode; massive anode;solid anode
实心翼梁 solid spar
实心圆花窗 blank rosette; blind rosette
实心圆柱形转子 solid cylindrical rotor
实心载体 solid-core support
实心轧辊 solid roll
实心支承物 solid-core support
实心支柱 solid pillar;solid strut
实心中框窗 solid mullion window
实心中柱旋梯 solid newel stair(case)
实心轴 solid axis[复 axes];solid axle;solid shaft(ing)
实心注浆 solid casting
实心柱 solid column
实心铸件 solid casting
实心砖 solid brick
实心砖砌拱顶 solid brick vault
实心砖砌体 solid brickwork
实心砖砌圬工 solid brick masonry
实心砖砌烟囱 solid brick chimney
实心砖墙 solid brick wall
实心砖墙填充墙 solid masonry infill wall
实心砖塔 solid brick pagoda
实心转子 solid rotor
实心桩 solid pile
实心锥喷嘴 solid-cone(type) nozzle
实心钻法 solid drilling
实心钻孔 solid-core boring
实心钻深 solid-core boring
实心钻头 boring cutter;solid bit;solid crown bit
实心钻头凿岩 solid drilling
实心座 solid seat
实信号 real signal
实行 carry into effect;carry into execution; bring into effect; carryout; execute; execution; perform; practice;prosecute;prosecution;put into practice;putting into operation
实行的规则 <建筑部门、权威机构> approved rule
实行封锁 enforce a block
实行节约 practise economy
实行紧缩 belt tightening
实行可能性 practicability
实行冷冻法 put into practice freezing method
实行者 executant
实行指令 imperative instruction
实行制裁 apply sanction
实型 real type;solid pattern
实型表达式 real expression
实型数据 real data
实型说明 real type specification
实型铸造 cavityless casting; full-mo(u)ld casting process
实验 experiment;test;trial
实验靶场 proving range
实验班 workshop
实验包交换业务 experimental packet switching service
实验报告 laboratory report
实验爆破 practical shot
实验备忘录 experimental memo
实验变差函数 experimental variogram
实验变差函数值 experimental variogram value
实验变数 experimental variation
实验变数效果 effect of experimental variable
实验布置 experimental arrangement; experimental set-up
实验步骤 experimental procedure;laboratory procedure
实验材料 guinea-pig
实验测定 experimental determination
实验策略 experimental strategy
实验常数 empiric(al) constant;experimental constant
实验厂 trial plant
实验场 experimental station
实验车 experimental vehicle
实验车间 experimental shop
实验城市 experimental city
实验程序 experimental procedure;experimental sequence
实验处 experimental station
实验传质系数 experimental mass-transfer coefficient
实验大纲 test program(me)
实验单位 experimental unit
实验的 developmental; empiric(al); experimental
实验的不确定性 experimental uncertainty
实验地貌学 experimental geomorphology
实验地球化学 experimental geochemistry
实验地球化学图 experimental geochemistry diagram
实验地震学 experimental seismology
实验地质学 experimental geology
实验点 experimental point
实验电路 experimental circuit
实验电台 experimental station
实验电信卫星 experimental telecommunications satellite
实验电压 experimental voltage
实验调查法 <市场调查的方法之一> experimental survey
实验定价法 experimental pricing
实验定律 experimental law
实验定则 experimental law
实验动力反应堆 experimental power reactor
实验动物 experimental animal
实验段长度 length of test sector
实验法 experimentation
实验反应堆 reactor experiment
实验范围 scope of experiment(ation)
实验方案 experimental program(me)
实验方法 experimental method; experimental technique
实验费用 experimental expenses
实验分布 experimental distribution
实验分析 experimental analysis
实验服 lab-gown
实验浮选槽 experimental test cell
实验工厂 pilot plant;semi-work;trial plant
实验工作 cut-and-try work; experimental;experimental work
实验工作服 lab coat
实验工作面 experimental face
实验工作台 laboratory bench
实验公式 empiric(al) equation; experimental formula
实验构造地球化学 experimental tectono-geochemistry
实验构造地质学 experimental structural geology
实验观察 experimental observation
实验管道 experimental channel
实验规划 design of experiment
实验(规)律 experimental law
实验规则 experimental law
实验过程 experimentation
实验函数发生器 empiric(al) function generator
实验函数生成程序 empiric(al) function generator
实验合金 technic metal
实验河槽 laboratory channel
实验核物理学 experimental nuclear physics
实验化学 experimental chemistry
实验回路 experimental loop
实验回路栅元 experimental cell
实验混凝土 labcrete
实验机车 experimental locomotive
实验计划 experimental arrangement; experimental set-up
实验技术 experimental technique
实验检查 experimental check
实验校验 experimental check
实验阶段 experimental phase;experimental stage
实验结果 experimental result;laboratory result;result of experiment
实验结果处理 observation reduction
实验结果分析 interpretation
实验结果归纳 experimental results reduction
实验结果整理 interpretation
实验介子谱学 experimental meson spectroscopy
实验(经验)成果 empiric(al) result
实验剧场 experimental theater[theatre]
实验聚变动力反应堆 experimental fusion power reactor
实验坑 experimental pit
实验孔 experimental port
实验孔道 experimental hole
实验矿井 experimental mine
实验矿物学 experimental mineralogy
实验力学 experimental mechanics
实验流体力学 experimental fluid mechanics
实验流域 experimental basin; experimental watershed
实验楼 laboratory block; laboratory building
实验路 experimental road
实验论文 experimental papers
实验螺距 experimental pitch
实验模拟 test simulation
实验模拟法 experimental analogic method
实验模拟理论 experimental analogy theory
实验模型 experimental prototype; pilot model; predictable pattern; research model
实验模型分析 experimental model analysis;laboratory model analysis
实验模型结构 laboratory model construction
实验农场 experimental farm
实验平均螺距 experimental mean pitch
实验气象学 experimental meteorology
实验潜水 experimental diving
实验强度 laboratory strength
实验区 experimental area
实验曲线 experimental curve
实验任务 experimental duty
实验筛 testing sieve
实验设备 experimental equipment;experimental facility;experimental installation;experimental plant
实验设备用计算机 laboratory instrument computer
实验设计 design of experiment; experimental design;experimental design work
实验生态系统 microcosm(os)
实验生物学 experimental biology
实验生物研究所 Institute of Experimental Biology
实验师 experimentalist
实验时间 experimental period;experimental time
实验式 empiric(al) formula
实验试验 full-scale test
实验室 experimental depot; experimental laboratory;lab(oratory)
实验室拌和器 laboratory mixer
实验室标准 laboratory standard
实验室标准传声器 laboratory standard microphone
实验室标准筛网目 laboratory sieve-mesh
实验室表现 laboratory performance
实验室波谱测量仪器 laboratory spectral measurement instrument
实验室玻璃器皿 laboratory glassware
实验室参考标准器 laboratory reference standard
实验室测量 laboratory measurement
实验室测量法 laboratory surveying
实验室测试 laboratory test

实验室程序 laboratory procedure
实验室瓷砖 laboratory tile
实验室打浆机 experimental beater
实验室的辅助管线 laboratory service
实验室的频率标准 laboratory standard of frequency
实验室发动机试验法 laboratory test engine method
实验室方法 laboratory method
实验室分样机 laboratory sample divider
实验室风化法 laboratory exposure
实验室干集料 laboratory-dry aggregate
实验室工作服 laboratory clothing
实验室固结试验 laboratory consolidation test
实验室挂车 laboratory trailer
实验室规模 bench scale; laboratory scale; lab-scale
实验室规模的 bench scale
实验室规模分析 bench-scale analysis
实验室过滤机 laboratory filter
实验室和温室 laboratory and greenhouse
实验室环境 laboratory environment
实验室级器械 laboratory-scale apparatus
实验室级用水 laboratory-grade water
实验室计算机 laboratory computer
实验室计算机体系 laboratory computer hierarchy
实验室技术 laboratory technique
实验室技术员 laboratory technician
实验室加速试验 accelerated lab(oratory)
实验室间 interlaboratory
实验室间的试验比较 interlaboratory test comparisons
实验室间偏倚 between laboratory bias
实验室间试验 interlaboratory trial
实验室检验 laboratory inspection
实验室鉴定 laboratory qualification
实验室搅拌器 laboratory stirring device
实验室搅拌装置 laboratory stirring device
实验室可靠性试验 laboratory reliability test
实验室模拟 laboratory simulation
实验室模拟降质像 laboratory-simulated degraded imagery
实验室模拟试验 laboratory simulation test
实验室模型 laboratory model
实验室内的 intralaboratory
实验室内快速试验 accelerated laboratory test
实验室内燃机试验 laboratory engine test
实验室评定 laboratory assessment; laboratory evaluation
实验室评定者 laboratory assessor
实验室铺面砖 laboratory tile
实验室气蚀法 laboratory exposure
实验室器皿 laboratory ware; labware
实验室潜水舱 laboratory diving tank
实验室染色机 laboratory dyeing machine
实验室认证 laboratory certification
实验室熔炼产品 laboratory melts
实验室筛选设备 laboratory sieving equipment
实验室设备 laboratory apparatus; laboratory equipment; laboratory furniture
实验室渗透系数 laboratory coefficient of permeability; standard coefficient of permeability
实验室实习 laboratory practice
实验室实验 laboratory experiment
实验室使用 laboratory use
实验室试剂 laboratory reagent
实验室试验 bench-scale experiment; in-house test; laboratory experiment; laboratory test; small test
实验室试验的发动机 laboratory test engine
实验室试验规模 laboratory test scale
实验室试验寿命 laboratory life
实验室试验土样 laboratory-tested soil specimen
实验室试验研究用发动机 laboratory engine
实验室试验装置 laboratory testing rig
实验室试样 laboratory sample
实验室数据库 laboratory database
实验室台面砖 laboratory bench tile
实验室条件 laboratory condition
实验室(土的)十字板试验 laboratory vane test
实验室土壤木块培养基 laboratory soil block culture
实验室土样拌和器 laboratory soil mixer
实验室拖车 laboratory trailer
实验室万能挤出机 universal laboratory extruder
实验室吸滤机 laboratory suction filter
实验室系 laboratory frame; laboratory system
实验室系速度 laboratory velocity
实验室小型计算机 laboratory minicomputer
实验室辛烷值 laboratory octane number
实验室辛烷值测定法 laboratory knock-testing method
实验室信号处理机 laboratory signal processor
实验室信号处理设备 laboratory signal processing instrument
实验室型反应器 laboratory size reactor
实验室型浊度计 lab-type turbidimeter
实验室研钵研磨机 laboratory mortar grinder
实验室研究 laboratory investigation; lab study
实验室研究方法 laboratory procedure
实验室验证 laboratory proofing
实验室样号 laboratory sample number
实验室窑炉 laboratory furnace
实验室业务 laboratory service
实验室仪表 laboratory instrument
实验室仪器 laboratory appliance; laboratory-type instrument; laboratory apparatus
实验室用车 laboratory vehicle
实验室用电炉 laboratory electric furnace
实验室用堆 laboratory reactor
实验室用混砂机 laboratory muller
实验室用火箭发动机 laboratory rocket
实验室用具 laboratory paraphernalia
实验室用炉 laboratory furnace
实验室用磨 laboratory grinder; laboratory mill
实验室用磨浆机 laboratory refiner
实验室用砂浴 laboratory sandbath
实验室用水 laboratory water
实验室用显微镜 laboratories microscope
实验室用窑 laboratory furnace
实验室振荡器 laboratory oscillator
实验室蒸煮锅 laboratory digester
实验室纸机 laboratory paper machine
实验室质量控制 laboratory quality control
实验室质谱仪 mass-spectrometer in laboratory
实验室中间生产研究 laboratory pilot plant study
实验室转子磨 laboratory rotor mill
实验室装置 laboratory installation
实验室锥形磨浆机 laboratory Jordan
实验室桌面砖 laboratory bench tile
实验室资格证明 laboratory accreditation
实验室自动化系统 laboratory automation system
实验室钻探试验 laboratory drilling
实验手册 laboratory exercise; laboratory manual
实验术 experimentation
实验束流 experimental beam
实验数据 experimental data; experimental findings; laboratory data; laboratory findings; observed data
实验数据处理系统 experimental data processing system
实验数据记录 laboratory data record
实验数据库 experimental data base
实验数值 empiric(al) value
实验数字电视 experimental digital television
实验水槽 experimental flume; experimental tank
实验水池 experimental basin
实验思路 experimental considerations
实验所 experimental depot
实验塔 laboratory tower
实验台 experimental bench; experimental table; experiment rig
实验台上试验法 method of static testing
实验弹性力学 experimental elasticity
实验套管 experiment thimble
实验特性 experimental feature
实验条件 experimental condition; test condition
实验铁路线 experimentable line; line used for experiments
实验厅 experimental room
实验通信[讯]卫星 experimental communications satellite
实验同步卫星 experimental synchronous satellite
实验卫星通信[讯]地面站 experimental satellite communication earth station
实验温度 experimental temperature
实验物理学 experimental physics
实验误差 experimental error
实验吸回效应 experimental resorption effect
实验细节 experimental detail
实验线路 experimental line
实验相关图 experimental correlogram
实验小区 experimental plot; field plot
实验形态学 experimental morphology
实验型磨矿机 laboratory mill
实验型无线电接收机 experimental radio receiver
实验性的 brassboard; pilot; prototype
实验性的调查结果 experimental findings
实验性的房地产 experimental estate
实验性多谱段扫描仪 experimental multispectral scanner
实验性反应堆 experimental reactor; preliminary pile assembly
实验性沸水反应堆 experimental boiling water reactor
实验性(工程)项目 experimental project
实验性过程模拟 experimental process simulation
实验性过热反应堆 experimental superheat reactor
实验性集水区 experimental watershed
实验性检验 experimental test
实验性建筑 experimental building
实验性鉴定 experimental verification
实验性雷达设备 experimental radar equipment
实验性模型 experimental model
实验性能 experimental performance
实验性气冷反应堆 experimental gas cooled reactor
实验性消退 experimental extinction
实验性研究 experimental study
实验性有机冷却反应堆 experimental organic cooled reactor
实验性增殖反应堆 experimental breeder reactor
实验性住房抵押贷款保险 experimental housing mortgage insurance
实验性住宅 case study house
实验学科 laboratory course
实验岩石力学 experimental rock mechanics
实验岩石学 experimental petrology
实验研究 experiment investigation
实验研究法 experimental investigation; experimental methodology
实验研究计划 experimental research program(me)
实验研究卫星 experimental research satellite
实验验证 experimental verification
实验样品 laboratory sample
实验仪器 experimental apparatus; laboratory apparatus; laboratory instrument
实验仪器配置 experimental layout
实验仪器组 experiment packet
实验应力分布法 empiric(al) stress distribution method
实验应力分析 experimental stress analysis
实验用打浆机 experimental beater
实验用地 practice ground
实验用模型 empiric(al) model; sample mo(u)ld
实验员 assayer; experimentator; laboratory technician
实验站 experimental establishment; experiment(al) station
实验者 experimenter
实验证据 experimental evidence; experimental proof
实验证明 experimental confirmation; experimental verification; result verification
实验值 experimental data; experiment value; result value
实验制图设备 experimental cartographic facility
实验中学 experimental middle school
实验周期 experimental period
实验转炉 experimental converter
实验装置 experimental apparatus; experimental device; experimental facility; experimental installation; experimental mounting; experimental provision
实验准备 experimental preparation; experimental set-up
实验资料 experimental data; laboratory findings; observed data
实验组件 experiment package
实验最大干密度 laboratory maximum dry density
实样模 natural pattern
实业管理 industrial administration

实业家 businessman; entrepreneur; industrist
实业界 business career; business community
实业界巨头 tycoon
实业银行 industrial bank
实业制度 industrial system
实异常 practical anomaly
实硬化曲线 real hardening curve
实用 utility
实用饱和度 practical degree of saturation
实用比 utilization ratio
实用编目 practical cataloguing
实用标准 practical standard
实用操作小品 practical tips
实用超高度 practical superelevation
实用车辆 serviceable vehicle
实用程序 service program(me); utility program(me)
实用程序包 utility package
实用程序控制设施 utility control facility
实用程序设计员 utility program(me)
实用冲压式空气喷气发动机 practical athodyd; practical ramjet
实用单位 practical unit
实用单位制 practical system; practical system of units
实用的 practical
实用等效电路 practical equivalent circuit
实用地层学 practical stratigraphy
实用地震学 practical seismology
实用电单位 practical electric(al) unit
实用电容率 practical specific capacity
实用毒剂 economic poison
实用度盘表 practical circle-setting table
实用对话 utility session
实用法律 applicable law
实用房屋 functional building
实用分类 practical classification
实用分析 practical analysis; proximate analysis; rough analysis
实用服务设施程序 utility facility program(me)
实用负荷 useful load
实用负载 live load; useful load
实用公式 practical formula
实用功能 utility function
实用故障诊断电路 practical fault diagnostic circuit
实用光学 practical optics
实用规范 code of practice
实用荷载 practical load; service load
实用化试验台 development test station
实用化学 practical chemistry
实用级配 practical grading
实用计算机 practical computer
实用计算性 practical computability
实用记录系统 utility logger system
实用技术 operative technology
实用价格 applied cost
实用价值 practical value; use value
实用建筑 functional architecture; utility architecture
实用建筑主义 functionalism
实用建筑主义者 functionalist
实用解法 pragmatic solution
实用经济计量模型 practical econometric models
实用经济学 applied economics
实用经济准则 pragmatic economic criterion
实用静噪器 actual noise silencer
实用控制设备 utility control facility
实用例行程序 utility routine
实用列类法 practical collocation
实用临界温度 working critical temperature
实用率 utility ratio
实用美术 applied fine arts
实用美术学校 school of applied art
实用模型 utility model; working model
实用模型专利 utility model patent
实用目的区划 applied divisional plan
实用耐磨指数 service wear index
实用年限 physical life
实用农业化学 < 化学工业和农业一体化的近代技术 > chemurgy
实用配方 practical formulation
实用容许应力 proof stress
实用软件 utility software
实用软件包 utility software package
实用上的耐冲击性 service-shock resistance
实用上升限度 practical ceiling height
实用设备表 utility device list
实用设备程序 utility facility program(me)
实用设计 utility design
实用升限 practical ceiling; service ceiling
实用升限高度 service ceiling height
实用时间 utility time
实用试验 in-service test(ing); in-use testing; operational test
实用手册 application manual
实用寿命 physical life; practical life
实用水力学 practical hydraulics
实用水平 realistic scale
实用水质指标 water quality economic index
实用弹限应力 proof stress
实用弹性极限 proof limit
实用弹性极限应力 proof stress
实用特伦特水质模型 Trente economic water quality model
实用天文学 practical astronomy
实用调试 utility debug
实用调试程序 utility debugger
实用烃产率曲线 practical hydrocarbon productivity curve
实用通信[讯]卫星 telecommunications satellite
实用通行能力 < 根据现有条件,在不致造成不合理的延滞或使驾驶人感到操纵困难的情况下,一条道路或指定车道通过的最大车辆数 > practical traffic capacity
实用土力学 practical soil mechanics
实用无线电发射机 utility radio transmitter
实用无线通信[讯] utility radio communication
实用稀释试验 use-dilution test
实用系统 utility system
实用项目 off-the-shelf item
实用卸货时间 loading time used; time of discharge used
实用新型 utility model
实用型 utility type
实用型变压器 < 从公用事业供电给企业的 > utility transformer; utility transformer
实用型中小马力牵引车 utility tractor
实用性 applicability; practicability; practicality; usefulness; utility
实用性结构 functional structure
实用研究 applied research
实用盐标 practical salinity scale
实用盐度 practical salinity
实用盐度计 practical salinity unit
实用验收试验 operational acceptance test
实用艺术 applied art
实用艺术工艺品 artware
实用溢洪道 ogee spillway
实用溢洪堰 ogee weir
实用原则 functional principle
实用载荷 practical load
实用噪声抑制器 actual noise silencer
实用证书 utility certificate
实用支援装备 operational support equipment
实用知识 practical knowledge
实用织物 utility cloths
实用值 practical value
实用主义 pragmaticism
实用主义的 pragmatic
实用柱温 practical column temperature
实有功率 actual horsepower
实有荷载 service load
实有马力 actual horsepower
实有人数 actual number of persons
实有数量 outturn
实有元件 actual element
实域 real domain; real field
实域局限 real restriction
实元 actual argument
实圆点 darkened circle
实跃迁 real transition
实运算常数 real arithmetic(al) constant
实载货吨位 cargo deadweight tonnage
实载率 actual loading factor; actual load rate; carrying ratio; rate of real loading capacity
实载试验 actual loading test
实在 actuality; earnest; noumenon[复 noumena]
实在边界条件 essential boundary condition
实在参数 actual parameter
实在参数表 actual parameter list
实在参数部分 actual parameter part
实在参数结合 actual parameter association
实在参数显示 actual parameter display
实在存储器 real storage; real memory
实在的 actual; intrinsic(al); substantive
实在地形 actual landform
实在电流 actual current
实在法 positive law
实在工作 net work
实在行标行 actual row of rower
实在论 realism
实在气体 real gas
实在权衡曲线 actual trade-off curve
实在上界 actual upper bound
实在识别算法 actual recognizer
实在数组 actual array
实在数组变元 actual array argument
实在数组说明符 actual array declarator
实在说明词 actual declarer
实在说明符 actual declarator
实在贴现 true discount
实在稳定性 actual stability
实在下界 actual lower bound
实在序列 actual sequence
实在压力 real pressure
实在增量 actual delta
实在折扣 true discount
实在字数 number of actual word
实账 real account
实账户 < 即资产账户负债账户及资本账户 > permanent account; real account
实正交群 real orthogonal group
实证法学派 positivist
实证概率 positive probability
实证古生物学 actuopaleontology
实证经济学 positive economics
实证论 positivism
实证研究 empiric(al) research; empiric(al) study
实证哲学 positivistic philosophy
实证主义 positivism
实证主义法学 positivist jurisprudence
实证主义者 positivist
实支成本 outlay cost
实支成本账户 cost applied account
实支的 out-of-pocket
实值 actual worth; intrinsic(al) value; real value
实值的【数】 real valued
实值低于账面的资产 watered asset
实值函数 real-valued function
实值函数序列 sequence of real-valued functions
实值随机过程 real-valued stochastic process
实值项目 real-value item
实值资产 real-value assets
实(址)方式 real mode
实址区 real address area
实指数 real exponent
实质 parenchyma
实质本原蕴涵 essential prime implicant
实质成本 < 即以实物数量计算成本 > real cost
实质错误 material fallacy
实质的 essential; substantial; tangible
实质等价 essentially equivalent
实质多重输出本原 essential multiple output prime implicant
实质工资 real wage
实质含意 material implication
实质互惠 material reciprocity
实质化 substantiation
实质记录 physical record
实质利率 real interest rate
实质票据 real bill
实质上 essentially; in substance; materially
实质上的损害 essential quality of damages
实质审查 examine as to substance
实质事项 matter of substance
实质收益 economic income; real earning
实质违约 material breach
实质险态 essential hazard
实质协议 substantive agreement
实质性 essentiality; materiality; tangibility
实质性的结论 substantive finding
实质性公约 substantive convention
实质性损害 physical damage
实质性损失 material damage
实质性条款 material term; substantive provision
实质性问题 question of substance; substantive issue
实质性协议 substantive agreement
实质因素 physical cause
实质有效汇率指数 real effective exchange rate index
实质折旧 economic depreciation
实中心柱式螺旋楼梯 solid newel stair(case)
实重 absolute weight; actual weight; real weight; true weight
实轴 material axis; real axis; solid axis [复 axes]
实装机容量 real installation capacity
实自变量 actual argument
实足年龄 age at last birthday
实作台日数 actual working machine-days

拾

拾波 pick-up
拾波电路 pick-up circuit

拾波环 pick-up loop
拾波器 adapter; pick-up device; pick-up unit
拾波线圈 pick-up coil; pick-up device; pick-up loop
拾波因数 pick-up factor
拾草捆机 bale collector
拾荒人 bone-grubber
拾件钳 pick-up tongs
拾垃圾的人 scavenger
拾墨度 degree of pick
拾起 pick-up
拾取 pick(-off)
拾取标识符 pick identifier
拾取窗口 pick-up window
拾取刀 picker knives
拾取电压 pick-up voltage
拾取耦合 pick-off coupling
拾取器 pick-up arm
拾取绕组 pick-up winding
拾取设备的操作方法 pick mode
拾取设备状态 pick device state
拾取时间 pick-up time
拾取速度 pick-up velocity
拾取线圈 pick-up winding
拾取信号天线 pick-up antenna
拾取装置 pick-up device
拾声 adapterization
拾声器 acoustic(al) pick-up
拾声器头 pick-up cartridge
拾声器心座 pick-up cartridge
拾声头 soundhead
拾声线圈 pick-up coil
拾脱机构 uncoupling gear
拾纬 filling pick-up
拾音 adapterization
拾音插头 phonoplug
拾音器 acoustic(al) pick-up; adapter [adaptor]; mechanical reproducer; pick-up; pick-up device; pick-up head
拾音器电路 pick-up circuit
拾音器盒 pick-up cartridge
拾音器声道分隔 pick-up channel separation
拾音器心座 cartridge
拾音设备 sound pick-up equipment
拾音时间 pick-up time
拾音头 pick-up head
拾音头去磁器 audio head demagnetizer
拾音系数 pick-up factor
拾音线路 pick-up line
拾振器 oscillation pickup; vibration detector; vibration pickup
拾振仪 vibration pickup
拾震器 geophone; shock-adapter
拾震器臂 pick-up arm
拾震器的输出 output of pick-up
拾震器墩子 seismometer pier
拾震器线圈 pick-up coil

炻

炻器 semi-porcelain; stoneware; toki

炻器墙地砖 stoneware tile
炻器釉 stoneware glaze

蚀

蚀斑 plaque

蚀斑检定法 plaque assay method
蚀斑形成单位 plaque forming unit
蚀本 loss capital
蚀本出售 sell at a loss; selling at less than cost
蚀本生意 back bargain; lose money in a business; loss maker
蚀边结构 margination texture
蚀变【地】 alter
蚀变白榴石 metaleucite
蚀变产物 alteration product
蚀变带 alteration envelope; alteration halo; alteration zone; altered aureole; alternation zone
蚀变带控制点 control point of alteration zone
蚀变点 alteration point
蚀变辉长石 allalinite
蚀变辉长岩 allalinite
蚀变火山岩 altered volcanic rock
蚀变火山岩类 altered volcanic rocks
蚀变假像 alternation-pseudomorphism
蚀变交代实验 rock alteration simulating
蚀变矿物 altered mineral
蚀变矿物学 alteration mineralogy
蚀变砂矿物 altered sand mineral; alterite
蚀变钛铁矿 weathered ilmenite
蚀变系数 index of alteration
蚀变玄武岩 melaphyre
蚀变岩石 alteration rock; altered rock
蚀变样品 altered sample
蚀变作用【地】 alteration
蚀残掩冲体 klippe[复 klippen]
蚀穿 eating-through; pit-through < 汽体的 >
蚀船虫 ship-worm
蚀掉 eating away
蚀顶 deroofing
蚀顶喷发 deroofing eruption
蚀洞 scoring
蚀防护罩 ablation shield
蚀割 etch cut
蚀沟 etched groove
蚀痕 etch(ing) pit
蚀痕角 etch angle
蚀坏的 eating
蚀积滨线 graded shoreline
蚀积台地 cut-and-built platform
蚀刻 acid embossing; etch; incise
蚀刻病毒 etch virus
蚀刻玻璃 etching glass
蚀刻材料 etching material
蚀刻槽 etched groove; etching bath
蚀刻槽纹 etched line
蚀刻测斜法 etch method
蚀刻成凹凸 etch into relief
蚀刻处理 etching treatment
蚀刻刀 etching knife
蚀刻的 acid etching
蚀刻的树脂铸模 etched resin cast
蚀刻底漆 etch primer
蚀刻电路 etched circuit
蚀刻法 etching method; etching pit
蚀刻反射镜 etched mirror; etched reflector
蚀刻废液 etching waste liquor
蚀刻粉 etching powder
蚀刻管 etch tube
蚀刻光栅 etched grating
蚀刻辊 etcher roll
蚀刻痕迹 etching mark
蚀刻环线 etch ring
蚀刻黄铜 dipped brass
蚀刻机 etching machine
蚀刻技术 etching technique
蚀刻剂 etchant; etching(re) agent
蚀刻坑 etch pit
蚀刻坑计数法 etch-pit counting method
蚀刻孔 etch hole
蚀刻抛光 etch-polish(ing)
蚀刻片 etch slide
蚀刻器 etcher
蚀刻腔 etched cavity
蚀刻清漆 etching varnish
蚀刻染色 etching dye
蚀刻溶液 etching solution
蚀刻溶液浓度 etching solution concentration
蚀刻溶液温度 etching solution temperature
蚀刻师 etcher
蚀刻时间 etching period; etching time
蚀刻试验 etching test
蚀刻术 etching
蚀刻图 etch figure
蚀刻图像 etched image; etching pattern
蚀刻图章 stamp etching
蚀刻线 etch line
蚀刻线路板 etched circuit board
蚀刻型 etch pattern
蚀刻掩模 etching mask
蚀刻印刷 etch printing
蚀刻用石蜡 etcher's wax
蚀刻用针 etcher's needle
蚀刻装饰 etched decoration
蚀刻装置 etching device
蚀坑 corrosion pit; pit
蚀孔 etching pit
蚀面 abrasion
蚀木虫 woodworm; xylophagan
蚀丘 etched hill; etch hillock
蚀图 corrosion figure
蚀退滨线 shoreline of retrogradation
蚀洗(用)涂料 wash primer
蚀像 etch(ed) figure; etching figure
蚀穴 blowhole
蚀余硅质岩 rotten stone; terra cariosa
蚀余红土 terra rossa
蚀余山 old mountain; relic(t) mountain
蚀余柱 outlier
蚀原作用【地】 plain denudation

食

食 <日月的> eclipse

食槽 manger
食草动物 grazing animal; grazing herbivore; herbivorous animal
食虫植物 carnivore; insectivorous plant
食橱 kas
食道 swallow
食底泥动物 bottom feeder
食底栖生物者 benthos eater; benthos feeder
食浮游生物的 piscivorous; plankton-eating
食具柜 dresser
食木的 xylophagous
食年 eclipse year
食品 food(stuff); food substance
食品包冰衣 glazing of food
食品包装纸 food wrapper
食品包装纸用涂料 grease resistant paper coating
食品保藏 food preservation
食品保温器 hot food server
食品保温箱 food warmer
食品备用室 service pantry
食品补助费 foodstuff subsidization
食品厂 bakery and confectionery; food products factory
食品厂废水 food-mill wastewater
食品储藏 food storage
食品储藏室 buttery; food room; pantry; provision room
食品储藏性能 food stability
食品处理者 food-handler
食品店 bakery; canteen; food stores
食品冻结 freezing of food
食品法典委员会 Codex Alimentarius
食品防腐剂 food preservative
食品工程 food engineering
食品工业 food industry; foodstuff industry
食品工业废水 food industry wastewater; food processing wastewater
食品工业废水处理 wastewater treatment of food processing
食品工业用橡胶 foodstuff rubber
食品工艺学 food technology
食品工艺杂志 food technology
食品公司 food company
食品供应 food supply
食品罐头厂 packing house; packing plant
食品罐头清漆 food can varnish
食品柜 ambry; food locker
食品化学 food chemistry
食品加工 food handling; food processing
食品加工厂 food processing plant; packing house; packing plant
食品加工厂废水 packing house waste (water)
食品加工场 food processing factory
食品加工副产物 packing house by-product
食品加工工业 food processing industry
食品加工和制造 food processing and manufacturing
食品监测室 food monitoring house
食品检测 food detection
食品检查 food inspection
食品检验证明书 food inspection certificate
食品结构原则 principles of food composition
食品津贴 foodstuff subsidization
食品科学 food science; sitiology
食品科学学报 Journal of Food Science
食品库 provision store
食品冷冻间 chill room; food refrigerating room
食品铺 bake house
食品商 grocer
食品商店 commissary; provisions shop
食品生产指数 index number of food production
食品室 ambry; covey; larder; pantry; provision(s) room
食品输送车 dinner wagon
食品输送架 < 有脚轮的 > dinner wagon
食品添加剂 food additive
食品箱 food box
食品小卖部 buttery; canteen
食品行业 food service industry
食品与药物管理局 Food and Drug Administration
食品原料 raw-food material
食品运输车 food-handling truck
食品杂货店 grocery; grocery store; groceteria < 美 >
食品增补剂 food supplement
食品质量管理 quality control of food
食品中毒 alimentary toxicosis; sitotoxism(us)
食品组成 food composition
食器柜 < 文艺复兴时期 > credence
食肉动物 carnivore; carnivorous animal
食肉性生物 carnivore
食宿费用收据 accommodation bill
食堂 canteen; dining room; dining saloon; eating room; mess(canteen); mess hall; mess house; mess room; cenaculum < 古罗马住宅 >; refectorium < 教堂、神学院的 >; souper-y < 美 >; refectory < 又称职工饭堂 >

S

食堂发售食物窗口 buttery hatch
食堂礼堂 cafetorium
食梯 dinner lift; dumb waiter; food conveying elevator; food lift
食梯井 dumbwaiter shaft
食物 nourishment; nurture; sustenance; susten(ta)tion; victuals
食物保温箱 hot cupboard; warming oven
食物不足 alimentary deficiency
食物废物 food waste
食物废渣粉碎机 food waste grinder
食物垃圾 food waste
食物类污垢 food soil
食物链 food chain; food cycle
食物链的降解 degradation of food chain
食物网 food web
食物-微生物量比 food-microorganism ratio
食物污染分析 food pollution analysis
食物污染监测器 food contamination monitor
食物循环 food cycle
食物质量比 food/mass ratio
食物中毒 alimentary toxicosis; sitotoxism
食性 feeding habit; food habit
食盐 common salt; sodium chloride
食盐出口 salt export
食盐架 <家畜的> salt rack
食盐水防冻剂 cryohydrate
食盐稳定道路 salt-stabilized road
食盐稳定法 salt stabilisation[stabilization]
食用成熟度 edible ripeness
食用粮 food grain
食用伞菌 champignon
食用水舱 culinary water tank
食用油 edible oil; sweet oil
食用油污染 edible oil contamination
食用园林布置 edible landscaping
食用植物 food plant
食用作物 alimentary crop
食油罐车 vegetable oil tank car
食余残渣 food waste
食鱼的 piscivorous
食源体系 food cycle
食指 index finger

史册 annals

史达塔克斯 <一种专利纤维板> Startex
史丹尼弗莱斯 <一种专利屋面防水材料> Stoniflex
史丹尼特 <一种专利墙面材料> Stonite
史蒂芬尼头饰 Stephane
史迹岩层的类型 pattern of historic formations
史赖伯全组合测角法 Schreiber's method in all combination
史赖伯虚拟观测方程 Schreiber's fictitious observation equation
史料 evidence; historic(al) materials; historic(al) source
史密森标准地球 Smithsonian standard earth
史密森协议 Smithsonian Agreement
史密森星表 Smithsonian astrophysical observatory star catalogue
史密斯导纳圆图 Smith admittance chart
史密斯(电磁)离合器 Smith's coupling
史密斯防火地板 Smith fireproof floor
史密斯高温电热线合金 Smith alloy
史密斯工作法 Smith's work
史密斯关联 Smith correlation
史密斯滚筒式染色机 Smith drum type machine
史密斯河柳 Smith river willow
史密斯近似式 Smith's approximate formula
史密斯柳 Smith willow
史密斯磨矿机 Smith kominuter
史密斯强力试验机 Smith's strength tester
史密斯曲线 Smith curve
史密斯三轴法 <试验沥青混凝土用> Smith triaxial method
史密斯式(沥青)混合料配合三轴试验设计法 <美> Smith triaxial method of mix-design
史密斯铁基合金 Smith's alloy
史密斯型无杆锚 Smith's stockless anchor; Smith anchor
史密斯圆图 Smith chart
史密斯折射仪 Smith's refractometer
史密斯枕座轴承 Smith saddle bearing
史密斯织物耐磨试验仪 Smith's cloth abrasion tester
史密斯阻抗圆图 Smith impedance chart
史普劳尔 <一种脚手架牌号> Sprowl
史前的 prehistoric
史前地震 prehistoric earthquake
史前归化植物 prehistoric naturalized plant
史前纪念石碑 mensao
史前巨石墓 dolmen; table stone
史前期土器 prehistoric pottery
史前人类学 protohistory
史前时代 prehistoric age
史前时期 protohistory
史前时期用石块架成的纪念物 dolmen
史前史 prehistory
史前世界 <石器时代> prehistoric world
史前学 prehistory
史前有机物遗存时期的放射性碳 carbon dating
史前之石棺或石墓 cistvaen
史坦塞牌专利风门 <用于机械通风> Stayset damper
史特劳格尔式 Strowger type
史特劳格尔式出中继第二级寻线机 Strowger system outgoing secondary line switch
史特劳格尔式继电器 Strowger type relay
史特劳格尔式自动电话制 Strowger automatic system
史特劳格尔选择器 Strowger selector
史特劳格尔制 Strowger system
史特劳格尔自动电话机键 Strowger mechanism

矢【数】 sagitta

矢板 <隧道中的> forepole; forepoling(board)
矢板梁 <隧道中的> forepoling girder
矢车菊 corn-flower
矢点图 vector point diagram
矢端曲线 hodograph
矢耳石 sagitta
矢幅角 argument of vector
矢高 arrow height; bilge; bowed height; height of arc; rise; rise of arch; vector height
矢高小的石拱 trimmer arch
矢函(数) vector function
矢积 cross product; outer product; vector product
矢径 radii vectors; radius vector; vector
矢径法 radius vector method
矢径线 line of radius vector
矢距尺 ordinate staff
矢跨比 ratio of rise to space; ratio of rise to span; rise-span ratio; rise to span ratio
矢量 complexor; phasor; vector(quantity)
矢量比 vector ratio
矢量表 vectorial representation
矢量表示法 vector representation
矢量波 vector wave
矢量波动方程 vector wave equation
矢量波动函数 vector wave function
矢量玻色子 vector boson
矢量差 phasor difference; vector difference
矢量产生器 vector generator
矢量长度 vector length
矢量场 vector(ial) field
矢量乘法 vector multiplication
矢量乘积 vector product
矢量处理 vector processing
矢量代数 vector algebra
矢量导纳 vector admittance
矢量到光栅 vector to raster
矢量的 vectorial
矢量的环流量 circulation of a vector
矢量的散度 divergence of a vector
矢量的梯度 gradient of a vector
矢量电流 vector current
矢量电压表 vector voltmeter
矢量定理 vector theorem
矢量对消器 vector canceller
矢量多边形 vector polygon
矢量发生器 vector generator
矢量法 vector method
矢量方程 vector equation
矢量方向 direction of a vector
矢量分隔 vector separation
矢量分解 vector separation
矢量分量 component of a vector
矢量分析 resolution of vectors; vector analysis
矢量分析法 time vector method
矢量分析计算器 complex plane analyser[analyzer]
矢量分析器 vectorlyser[vectorlyzer]
矢量伏特计 vector voltmeter
矢量幅角 argument of vector
矢量跟踪系统 vector servomechanism
矢量公式 vector equation
矢量功率 vector power
矢量功率图 geometric(al) power diagram
矢量关系 vectorial relation
矢量管 vector tube
矢量光电效应 vectorial photoelectric(al) effect
矢量轨道常数 vectorial orbital constant
矢量轨迹 vector locus
矢量轨迹法 vectorial track method
矢量过程 vector process
矢量函数 phasor function; vector function
矢量函数发生器 vector function generator
矢量合成 addition of vectors; composition of vectors; vectorial resultant
矢量和 vector addition; vector sum
矢量和激励线性预测 vector sum excited linear prediction
矢量化 vectorize
矢量环 vector loop
矢量绘图 vector(point) plotting
矢量绘图法 vector drawing method
矢量机构 vector mechanism
矢量积 cross product; gross product
矢量集 set of vectors; vector set
矢量计 vector meter
矢量计算 vector calculus
矢量计算机 vector computer
矢量记号 vector notation
矢量加法 addition of vectors; vector addition; composition of vectors
矢量加速度 vector acceleration
矢量角 polar angle; vector angle; vectorial angle
矢量解法 vector solution
矢量解释法 vectorial interpretation method
矢量解算 resolution of a vector
矢量解析 vector analysis
矢量矩阵 vector matrix
矢量空间 vector space
矢量控制 vector control
矢量控制器 vector controller
矢量力 vector force
矢量粒子 vecton
矢量瞄准器 vector sight
矢量模的基底 basis of a vector module
矢量模型 vector model
矢量内积 inner product of a vector
矢量逆变分量 contravariant component of a vector
矢量耦合系数 Clebsch-Gordan coefficient; vector coupling coefficient
矢量平衡 vector balancing
矢量平均海流计 vector-averaged current meter
矢量平均值 vectorial mean
矢量强度法 vectorial intensity method
矢量强度解释法 vector intensity method
矢量倾角法 vector inclination method
矢量三角形 vectorial triangle
矢量扫描 vector scan
矢量示波器 vector oscilloscope
矢量式 vector expression
矢量势 vector potential
矢量算子 vector operator
矢量特性 vectorial property
矢量天体测量 vectorial astrometry
矢量通量 vector flux
矢量投影图 vector projection
矢量图 arrow diagram; phasor diagram; space diagram; vectogram; vector(ial) diagram; vectorial representation
矢量位 vector potential
矢量位移 vector displacement
矢量问题 vector problem
矢量显示 vector display
矢量显示器 vectorscope
矢量相关 vector correlation
矢量相互作用 vector interaction
矢量形式 vector form
矢量性质 vector property
矢量旋度 curl of a vector
矢量旋转 vector rotation
矢量应力 vector stress
矢量应力图 vector stress diagram
矢量有效结构体系 vector-active structure system
矢量原点 origin of vector
矢量圆图 circle of vector diagram; clock diagram
矢量运算 vector action; vector operation
矢量运算子 vector operator
矢量指令 vector instruction
矢量制图数据 vector plotted cartographic(al) data
矢量阻抗 vector impedance
矢量阻尼 vector damping
矢量组 vector group
矢量组的数标 numeric(al) index of

the vector group
矢轮 girth gear
矢面积 vector area
矢圈 girth gear
矢算 vector calculus
矢算子 vector operator
矢通量 flux of vector
矢线 edge
矢线图 arrow diagram
矢形感器 sensillum sagittiforme
矢形尖卷窗 lancet window
矢形天线 arrow antenna
矢形图像 sagittal image
矢性介子 vector meson
矢性空间 vector space
矢性目标函数 vector-valued objective function
矢性判据 vector-valued criterion
矢元 vector element
矢状的 sagittal
矢状缝 sagittal suture
矢状沟 sagittal groove
矢状径 sagittal diameter
矢状平面 sagittal plane;sagittal section
矢状饰【建】dart;lancet
矢锥【岩】tap
矢锥吃口 gripping action
矢锥扭紧 gripping action

使安全 secure

使安稳 soften
使暗 blacken
使凹 cavern;dent
使凹处隆起 <筑路> raising of sags
使凹进 indent
使白 whiten
使摆动 vibrate
使摆脱 extricate;rid
使败废 blight
使饱和 saturate
使保持铅直 fetch up plumb
使抱偏见 prejudice
使爆发 set-off
使爆炸 fulminate
使崩溃 crumple
使绷紧 tightening
使泵起动 priming the pump
使闭气 choke
使变暗 darken;tarnish
使变白 blanch
使变成粉 pulverize
使变得陡峭 steepen
使变得清楚 <电视、电码等> unscramble
使变得世故 sophisticate
使变短 short out
使变黑 blacken;nigrify
使变化 variate
使变坏 penalizic
使变色 discolo(u)r
使变位 dislocate
使变稀薄的 attenuant
使变形 distort;strain;transfigure;transform
使变形的 deformative
使变性 denaturalize;denature;modify
使变硬 consolidate
使标准化 calibrate;normalize
使表面变硬 face-hardened
使表面粗糙 toothing
使表土层肥沃 enrich the top-soil layer
使冰川化 glaciate
使波动 undulate
使玻璃化 vitrify
使玻璃失去光泽和透明 devitrify
使剥蚀 denude
使不安 disconcert
使不安定 unsettle
使不纯的 adulterant
使不方便 discommode
使不规则化 randomization;randomizing
使不合格 dishabilitate;disqualify
使不活动 deactivated
使不利 penalize;penalizic
使不连接 disconnect
使不平衡 disproportion
使不生效的 infirmatory
使不适 discomfort
使不适当 unfit
使不受风化 weather away
使不同 variate
使不透水 seal;waterproof
使不稳定 destabilize
使不显著 overshadow
使不相称 disproportion
使不相宜 unfit
使不朽 perpetuate
使布满麻点 pock mark
使布满麻面 pock mark
使参数量化 parametrized
使残废 crippling
使层化 stratify
使产权文书完备 perfecting title
使颤动 vibrate
使超荷 hypercharge
使车子颠簸的道路凹凸不平处 <俚语> thank-you-madam
使沉淀 depositing;precipitate
使沉淀成胶态 peptisation [peptization]
使成奥氏体 austentize
使成八字形 splay
使成比例 proportionate;proportioning
使成冰状 glaciate
使成波纹 engrail
使成波状 corrugate
使成层 stratify
使成虫蚀状的装饰 vermiculate
使成大块 bulking
使成洞 cavern
使成非彩色 achromatize
使成粉状 powder
使成蜂窝状 cellulate;honeycomb
使成格言 proverb
使成格子式 checker
使成拱状 roach
使成沟渠 channelize
使成惯例 routinization
使成棍棒形 club
使成糊状 levigate
使成话柄 proverb
使成急斜面 escarp
使成浆 pulp
使成胶状 agglutinate;gelatinize
使成胶状物集合 agglutination
使成蓝色 blu(e)ing
使成立 call into being
使成粒状 granulate;granulation
使成流线型 streamline
使成螺旋形 spiral
使成螺旋状 corkscrew
使成平板状 tabulate
使成平面 tabulate
使成棋盘格状 checker
使成穹顶 domed
使成球形 conglobate
使成乳剂 emulsify
使成乳色 opalize
使成三倍 treble
使成三角形 triangulate
使成扇形 scallop
使成事实 materialize
使成熟 refining
使成水平 level(l)ing
使成四倍 quadruplicate
使成四重 quadruplicate
使成似联式 duplicate
使成套 unitization
使成同时 synchronization
使成网状 ramify
使成为可塑 plasticize
使成为四倍 quadrupling
使成为微小粒子 micronize
使成为有理数 rationalize
使成五倍 quintuple
使成细胞状 cellulate
使成星状 stellify
使成一个单位 unitization
使成一体 unify
使成原子 atomize
使成圆角 radiusing
使成圆形 radiusing;rounding;round(ing)-off
使成杂色 counterchange;mott(e)
使成针状结晶 needle
使成直角相交 square up
使成指数 exponentiate
使成锥形 cone
使成紫色 purple
使成组 grouping
使呈气态 aerify
使呈杂色 mottle
使承担 entail
使城(都)市化 urbanize
使吃饱 satiate
使充满 impregnate;inundate
使稠 stiffen
使稠化 inspissate
使臭氧化【化】ozonize
使出汗 sweat
使出入相抵 make both ends meet
使穿 clothe
使传动 drive round
使船搁浅 beach;grounding;run ashore
使船横风 get beam to wind
使船后退 aback;back ship;lay aback
使船急转前进 heave a ship about ahead
使船倾侧 careen
使船倾斜 careening;heave down
使船头更近风向 <帆船> haul up
使船头离开风 box off
使船向一侧倾斜 <小船修理时> hove down
使船只倾侧以便修理 heave down
使垂直 plumb
使醇化 alcoholize
使从属 subordination
使粗 thicken
使粗糙 fret;roughen
使脆化 embrittle
使存货过剩 glut the market
使错开 cross-set
使达顶点 climax
使大气变性的产物 atmospheric transformation product
使带放射性 radioactivate
使带酸性 acidulate
使淡化 freshen
使得瘟疫 plague
使等零 nullification
使等同 level(l)ing
使等于零 nullify
使滴 dribble
使滴下 dribble;trickle
使抵销 counterweigh
使地区化 regionalization
使颠倒 invert
使电缆特性曲线均衡 cable equalization
使电脑化 cybernate
使电子计算机化 computerise;cybernate
使调换 transpose
使跌价 depress
使定位 bind in position
使定形 concretize;formalize
使动 actuate
使动摇 destabilize;unhinge
使动作 bring into action;call into play
使动作的 actuating
使冻结 glaciate
使短路 short out
使断裂 decouple
使对称 symmetrize
使对立 counterpose;contrapose
使钝 dull
使钝化 immunizing;passivate【冶】
使多产 fecundate
使恶化 deteriorate
使发动机空转 <确定发动机的工作规范> race an engine
使发光 brighten
使发黑光 japanned
使发黑亮 japanned
使发酵 leaven
使发亮 brighten
使发生 breed
使发香 perfume
使发皱 crimp
使反弯 recurve
使反向 invert
使访问 make to access
使肥 tallow
使肥沃 fecundate
使分化 differentiate
使分开 unhinge
使分离 disengage;rend
使分裂 disrupt;rend
使分裂开 break apart
使分馏 dephlegmate
使分路 shunt
使分凝 dephlegmate
使分支 ramify
使丰饶 fecundate
使服从 submit
使浮起 buoy up;refloat(ing)
使腐败 taint
使腐烂 decompose;putrefy
使负担过重 over-tax
使负担债务 encumber
使负重担 burden;tax
使复活 reactivate;resurrect
使复苏 <如用表面处治法复苏已老化的沥青路> rejuvenate;rejuvenation
使复新 refresh
使复原 reinstate;unscramble
使复杂 tangle
使改道 rechannel;redirect
使改方向 redirect
使干涸 dry-up
使干乳 dry off
使干燥 dehumidify;dry(-up);exsiccate;through-dry
使干燥的 exsiccative;siccative
使革命化 revolutionize
使隔开 space
使隔离 sequester;sequestrate
使更完善 make something of
使更新 rejuvenate
使更有用 make something of
使工程师满意 to the satisfaction of the engineer
使工地清洁 site cleanup
使工作 bring into service
使公式化 formularize
使供应市场 commercialize
使共振 syntonizer
使固定 consolidate;fixate;root(ing)
使固定不变 stereotype
使固定的 stuck
使固结 consolidating
使固体变成颗粒状 prill
使馆 diplomatic mission;embassy
使馆区 diplomatic row;legation quar-

S

ter
使馆全体人员 legation
使光滑 burnish; sleek; slick; smoothing adjustment
使硅化 silicify
使轨道电路继电器落下的分路电阻 drop shunt of a track circuit
使过饱和 super-saturated
使过度冷(却) subcool
使过热 overheat
使过湿的 overwet
使含糊 obscure
使含沥青 bituminize
使好 making good
使合标准 calibrate;standardize
使合并 consolidate
使合法 legalize;regularize
使合法化 decriminalize
使合格 qualify
使合理 rationalize
使合理化 rationalize
使合同无效 vitiate contract
使合于某种风格 stylized
使和解 reconcile
使荷载进入塑性阶段的荷载 plastic load
使黑 denigrate
使红 redden
使厚 thicken
使互相关系 correlate
使滑 slip
使滑动 sliding
使还原 decompose;hydrogenate
使环行 circulate
使缓和 defuse;lull;mitigate
使缓解 respite
使换气 ventilate
使恢复 reinstate
使恢复翻新 rejuvenation
使恢复水分 moisturize
使恢复元气 revitalize
使恢复原状 revest;undo
使恢复正常灵敏度 decohere
使挥发 volatilize
使回火 temper draw
使混合 mingle
使混乱 confuse;perplex
使(混)乱随机化 randomize
使混同 confuse
使混杂 interlard
使活动 call into play
使活跃 rev
使(火车)转轨 shunt
使火成岩内分为多少有规则岩层的节理 <因冷缩等原因而形成> absondering
使机器坚固 ruggedize
使(机械等)人性化 hominize
使机械化 mechanize
使激冷 chill
使极化 polarize
使挤入 bulging in
使记起 remind
使寂静 still
使夹杂 interlard
使尖 point
使坚固 fortify
使减到最小 minimize
使减弱 dampen
使减小 minify
使减压 depressurize
使碱化 alkalize;basify
使渐细 taper
使浆与水面平行 feather
使僵持 deadlock
使降低保密等级 declassify
使降级 down grade;demote <美>
使交错 counterchange
使交错货物核对员 checker
使焦 parcel

使绞结 kink
使矫直 straighten
使节厅 <西班牙阿尔罕伯拉宫> Hall of Ambassadors
使结合成一体 tie-in
使结晶 crystallizer
使解除 relieve from
使解调 demodulate
使解冻 unfreeze
使金属化 metalate
使(金属)回火 attemper
使紧急制动无效 overriding of an emergency brake application
使紧密 tighten
使紧密接合 knit
使紧张 strain;tension
使劲干 blaze away
使劲关门 slam
使劲关上 slam
使劲拉 yank
使近代化 updating
使精洁 purify(ing)
使精通 familiarise[familiarize]
使井喷出泥砂等物质 blowing a well
使痉挛 convulse
使静 silence
使静止 still
使纠结 kink
使具体化 concretize;crystallizer;embody;materialize;substantiate
使具有隔音性 deaden
使具有货币性质 monetization
使具有力量 muscle
使具有形式 formalize
使具有一定形状 configurate
使具有正位移 guide positively
使具有资格 qualify
使剧烈震动 convulse
使聚焦 bring in focus
使卷曲 crimp;crinkle;curl
使卷曲口 crimp
使绝热 athermalize
使绝缘 isolation
使均等 equalize
使均衡 equalize;proportioning
使均匀 homogenize
使均值 equibalance
使均重 equiponderate
使卡嗒响 click
使开动 put into operation
使抗锈 rust-proof(ing)
使空出来 empty
使空气变香 scent
使空气以气泡状存在于混凝土中 entrain
使空闲 vacate
使枯萎 blight
使快干 flash dry
使扩张 augmenting;distend
使拉紧 tightening
使劳动过度 overlabo(u)r
使冷却 chill;refrigerate
使离合器分离的安全装置 declutching safety device
使离开 beam off;detach
使离开原定进程 derailer
使离原位 delocalization
使离中心 decentre[decenter]
使粒状材料变成流体 prill
使连接 <用键> feather
使联机 vary on-line
使两端相接 make both ends meet
使两极分化 polarize
使两履带相对反转 counterrotate track
使另一方不受损害 hold harmless
使流出 debouch;let out
使流态化 fluidize
使流通 circulate
使硫化 sulphurize;vulcanize
使隆起 haunch up

使露出 denude
使路面不利 weakening
使轮廓模糊 dislimn
使轮流 rotate
使满足 satiate;suffice
使冒泡 to bubble up
使门关上 slam
使蒙受 entail
使弥漫 interfuse
使迷失方向 disorient
使密 thicken
使密合不漏 tighten
使面对 confront
使苗条 slenderize
使明了 clear up
使明晰 clarify
使命 mission;vocation
使模糊 blur
使模糊不清 bedim
使模块化【计】modularize
使木质化 lignify
使内接【数】inscribe
使内向弯曲 incurvate;incurve
使耐火 fire-proofing
使耐用 ruggedize
使挠曲 deflect
使能够 enable
使能缓冲器 enable buffer
使能信号 enable signal;enabling signal
使泥沙充塞了河流 filling rivers with sand and mud
使泥土水分蒸发 evapotranspire
使黏[粘]稠 toughen
使黏[粘]结 agglutinate
使凝固 concretize
使凝集 agglutinate
使凝结 coagulate
使凝聚成团 conglomerate
使浓 thicken
使浓厚 inspissate
使浓缩 inspissate
使排出 sweat
使排列成列 bring into line
使喷出 spurt
使蓬松 bush
使膨胀 distend
使偏差 deflect
使偏斜 deflect
使偏心 decentre
使偏振 polarize
使飘荡 waft
使贫化 leaning
使平 adequation;flatten
使平衡 counter-balance; counterpoise; counterweigh;equibalance;equipoise; equiponderate;poise
使平滑 plane;smoothing
使平缓 flattening
使平静 mollifying
使平均 counterpoise;equate
使平齐 flush
使平息 becalm
使平整 level off flush
使平直 line up
使破产 bankrupt
使破坏 carry to failure
使破裂 flaw
使普及 popularize
使起波纹 ripple
使起涟漪 riffle
使起微波 ripple
使起重机吊杆起落 luff
使起皱纹 corrugate;crimp
使气泡居中 set bubble
使气体化 aerify
使气体与碳氢化合物混合 carburat(t)ing;carburet(ion)
使浅 shoal
使强韧 toughen
使(墙等)不漏声 deafen

使翘曲 curl;torture
使氢化 hydrogenate
使倾斜 raking
使清洁 detersive;sweeten
使清洁爆炸声 detersive
使清洁的 smectic
使清新 refresh
使曲线光滑 adjust the angles by curves
使确定不变 bind
使燃烧 inflame
使人极度紧张的 high-pressure
使人口减少 unpeople(d)
使人一时失明的眩光 disability glare
使容易 facilitate;make easy
使容易做 smooth the way
使溶液的盐分增大 salt slug
使柔和 soften;tone-down
使柔顺 supple
使乳化 emulsify
使软化 mollifying;soften
使润滑 lubricate
使弱 dampen;soften;weaken
使散屑 decohere
使色变深 sadden
使色彩柔和的技术 scumbling technique
使伤残 disable
使商品化 commercialization; commercialize
使商业化 commercialize
使上升 upheave
使摄动【天】perturb
使深刻 deepen
使渗出 sweat
使渗入 interfuse
使渗透 osmose
使升华 sublimate;sublime
使升值 revaluation
使生根 root
使生皮软化 bate
使生效 validate
使生锈 tarnish
使生振动的 vibratory
使声音向下传播的装置 abatsons
使失步 pull out
使失去爆炸性 defuse
使失去玻璃光泽 devitrify
使失去法律效力 outlaw
使失去光泽 deaden;flatten
使失去平衡 overbalance
使失去时效 outlaw
使失去自然性能 denature
使失效 deactivated;deactivation;neutralization;neutralize
使失效平衡 neutralize
使失谐 disarrange
使湿 dampen
使湿润 lubricate
使湿透 drown;shower
使十字丝对焦清晰 diopter adjustment
使石料表面光滑的 faced
使石墨化 graphitize
使时间一致 synchronization
使实验条件完全相同 duplicating experimental condition
使市场不稳定的投机 destabilization speculation
使适合 quadrate
使适应 acclimate; adapt; condition; proportionate
使适应气候条件 weatherizing
使收缩 astringe
使收支相抵 make both ends meet; make buckle and tongue meet; meet one's expenses
使艏斜桅倾斜 steeve
使受冰川影响 glaciate
使受超高频声波的作用 insonate
使受催化作用 catalyze
使受电子计算机控制 cybernate
使受法律约束 bind

使受力过大 over-stress
使受审查 subject to examination
使疏松 fluff
使熟悉 familiarise[familiarize]
使数字化 digitalize
使衰弱 emaciate;wither
使松动 loosen;unset
使松碎 loosen
使松脱 unthread
使苏醒 resurrect
使酸 sour
使酸的 acidific
使酸化 acidize
使随机化 randomization
使私营化 denationalise[denationalize]
使损坏 endamage
使缩到最小 minimize
使缩小 minify
使索赔无效 extinguish a claim
使索赔有效 validate a claim
使太拥挤 overcrowd
使套入 nest
使体系化 systematize
使甜 sweeten
使听不见 deafen
使停止不动 becalm
使停止流溢 stanch
使通导 turn-on
使通电 energization
使通风 ventilate
使通过 navigate
使通货膨胀 inflate
使通脉冲 enable[enabling] pulse
使通俗 popularize
使通俗化 familiarise[familiarize]
使通晓 familiarise[familiarize]
使同步 bring into step;timing
使同等的 coordinative
使同高 level(l)ing
使同相 bring in phase
使同相位 bring in phase
使同样 unify
使痛苦 discomfort
使凸起 relieving
使突出 underline
使图像轮廓鲜明 crispen
使(土壤)消瘦 emaciate
使团 mission
使褪色 destain;discolo(u)r
使脱机 vary off-line
使脱开 unlink
使脱离 disjoint
使脱色 bleach
使脱水 desiccate;evaporate
使瓦解 disorganize
使外接 circumscribe
使外切 circumscribe
使弯 bend;ramp
使弯道外侧超高 bank up
使弯曲 curl;incurvate;incurve;inflect;quirk;scrunch
使完全变形 transmogrify
使完全硬化 harden right out
使微型化 miniaturize
使为零 zero fill;zeroize
使为难 bewilder;discommode;perplex;stumble
使萎缩 rivel
使位移 detrude
使文明 civilize
使吻合 dovetail
使紊乱 disorganize
使污损 blur
使无光泽 matt finish
使无害 defuse
使无能 disable
使无色 achromatize
使无效 blank out;invalidation;nullification;nullify;override;quash
使无效的 diriment
使无用 cripple
使无资格 disable
使物料成型的铲斗 profiled bucket
使雾化 atomize
使吸湿 imbue
使稀薄 rarefy
使稀疏 rarefy
使熄灭 blank out;quench
使习惯于 wont
使系数相等 equating coefficient
使系统化 systematize
使系统化者 systematizer
使细长 slenderize
使下垂 droop;weigh
使下水【船】launch
使显著 heighten(ing)
使现代化 bring up to date;update;updating
使现裂纹 craze
使线划光滑 line cleaning
使线条光滑 line cleaning
使陷入困难 swamp
使相称 equipoise;proportionate
使相当 proportionate
使相等 equate
使相等命令 equate directive
使相互参照 cross reference
使想起 remind
使向东 orientate
使向后靠 recline
使向上 uplift
使向外张开 flare
使向下倾斜 tilt down
使消根 rationalize
使消色 achromatize
使消失 evaporate
使硝化 nitrify
使小型化 miniaturize
使协调 attune
使协调的 coordinative
使斜削 splay
使斜倚 recline
使谐振 syntonizer
使新机器开得顺利 run-in
使新生 regenerate;revitalize
使新鲜 freshen
使休息 repose
使蓄电池完全放电 running the battery down
使旋转 rotate;slew
使循环 circulate;cycle
使压入 bulging in
使延长 lengthen
使延迟的 retardant
使岩石裸露 denude
使颜料增量的能力 pigment extending capability
使氧化 oxidize
使液化 fluidize
使一同沉淀 coprecipitate
使一致 quadrate;reconcile;uniform;unify
使依附于 attach to
使仪器对零 adjust to zero
使移动 unset
使乙酰化 acetylated
使阴暗 dull;overcast
使硬 stiffen(ing)
使硬化 hardened;stiffen;vulcanize
使拥塞 back-up
使永存 perpetuate
使用安全性 safety of use
使用半径 blow distance
使用保养手册 service and maintenance manual
使用保养说明标签 hangtag
使用保养说明书 service and maintenance manual
使用比汽油重的燃料的发动机 oil-burning engine
使用标准 technical standard
使用不当的土地 misuse land
使用不当造成的故障 abuse failure
使用不可靠的 unserviceable
使用不可靠性 unserviceability
使用部门 customer
使用材料 materials used;used material
使用财产 goods in use
使用参数 operation parameter
使用操作系统的程序 librarian
使用测试 performance test;test in service
使用长度 serviceable length;working length
使用长期车票 commute
使用车【铁】used train
使用车计划 plan of wagons utilized for loading;transshipment and recording
使用车数 number of cars used;rolling stock amount of using
使用成本 cost of use;use-cost
使用成本差 use-cost difference
使用程序 application program(me);service program(me);service routine
使用存储器方法 paging system
使用大量劳动力的 labo(u)r-intensive
使用带宽 utilized bandwidth
使用带宽比 utilized bandwidth ratio
使用贷款 utilization of a loan
使用单位 building user;building using party
使用单位主管 user management
使用单元面积 area of unit occupancy
使用导则 application guide
使用的委任 authorization use
使用等级 service rating
使用点 point of application
使用电力 electrification
使用电力的设备 electric(al) appliance
使用电视接收机的人 televisor
使用电压 operating voltage;service voltage
使用电子技术的 technetronic
使用定额 service rating
使用动力 working power
使用多种语言的 multilingual
使用法规 appropriation law
使用泛黄 use-yellowing
使用范围 blow distance;field of use;range of use;scope of use;serviceable range
使用方案 operational version
使用方法 application method;method of application;methods employed
使用方式 usage mode;use-pattern
使用费汇款的最高限额 ceiling on remittance of royalties
使用费率 rate of royalty
使用费(用)charges for use;working cost;cost of operation;operating cost;operation cost;running cost;tariff;use fee;user charges
使用分级法令 use classes order
使用分区 use zoning
使用分区图 use zone plan
使用分区制 use zoning
使用负荷 applied load
使用负荷频谱 service load spectrum
使用负载 live load;occupancy load;service load;traffic load;working load
使用公差 operational limit
使用功率 service power
使用规程 operating specification;service instruction
使用规格 technical standard
使用规划 service regulation
使用规则 service regulation
使用国 user state
使用过程中的调整 service adjustment
使用过程中损坏 service failure
使用过度 excessive manipulation;overuse;overwork
使用过度的 overriding
使用焊剂自动焊接 automatic welding with flux
使用焊接性 overall weldability;service weldability
使用合格的制品 performance-proved product;use-proved product
使用和维修手册 operation and maintenance manuals
使用荷载 live load;occupancy load;service load;traffic load;working load
使用荷载设计法 working load design
使用荷载应力 service load stress
使用后调查 post-occupancy survey
使用后评估 post-occupation evaluation
使用厚度 application thickness;used thickness
使用货盘的货物 palletized cargo
使用机器 put a machine into service;set-up a machine
使用机械的时间分配 machine time
使用及试验的限制 service and test restrictions
使用计数 usage count;use count
使用计算机的地面模型 digital ground model
使用计算机规划 using computers to plan
使用记录 service history
使用技术 use technology
使用价值 exchange value;service value;usable value;use(ful) value;value in use;value of exchange
使用价值量 magnitude of use-value
使用价值总和 aggregate use-values
使用检验 in-service inspection
使用建筑面积 <办公楼等> useable floor area
使用阶段 phase of use
使用介质 working medium
使用借贷合同 contract of loan for use
使用借款 loan for use
使用可焊性 overall weldability;service weldability
使用可靠 operational safety
使用可靠时间 serviceable time;uptime
使用可靠性 serviceability;use reliability
使用可靠性系数 serviceability ratio
使用可能性 use capability
使用控制机构 access control mechanism
使用控制台 utility control console
使用劳动力 use of labor
使用链的溢出 overflow with chaining
使用两种燃料的发动机 dual-fuel engine
使用率 activity ratio;application rate;occupating coefficient;occupation coefficient;rate of utilization;use percentage;use rate;utilization ratio
使用麻醉区(域)anesthetizing area
使用码头税 <英> lastage
使用码头延期罚金 penalty dockage
使用面 face
使用面积 area of use;floorage;usable area;usable floor area;useful area;useful floor area;utilization area
使用命令 utility command

使用磨损 service wear
使用目的 purpose of use
使用内链的溢出 overflow with internal chaining
使用耐久性试验 life test(ing)
使用能力 serviceability;usability;use capacity
使用能力极限状态 serviceability limit state
使用年限 age limit;durable years;duration of service;length of life;life length;life space;life span;operational life;operational time;serviceability life;service(able) life;service age;service period after completion;tenure of use;term of expectancy;useful time;working life;life time <设备或建筑物>
使用年限比较 life-comparison
使用年限表 <固定资产> life table
使用年限法 service-life method
使用年限积点折旧法 years-digit depreciation method
使用年限因素 life factor
使用农业劳动力 use of agricultural labor
使用频带 us(e)able frequency band
使用频繁的楼板 heavy-duty floor(ing)
使用频率 usable frequency
使用品质 in-serve behavio(u)r
使用品质记录 service-behavio(u)r record
使用评价 operational evaluation
使用期 length of life;life time;period of use;usage time;working life;tenure <不动产的>;life cycle <软件等的>
使用期成本 life-cycle cost
使用期费用 cost in use
使用期满时残值 end-of-life salvage value
使用期弃水 operational waste
使用期试验 end-use test(ing);life tester
使用期特性 life performance
使用期维护 physical maintenance
使用期限 durable life;duration of service;length of life;length of vitality;life durability;life span;life time;live time;operating life;operation life;period of service;period of usage;service life;term of life;time limit of service;working life;serviceable life;service age;term of service;useful life
使用期限保险 life insurance
使用期限测定值 lifetime measurement
使用期限测试器 life detector
使用期限长的 long life
使用期限的特性曲线 life curve
使用期限概率 lifetime probability
使用期(限)试验 life test(ing)
使用期限质量控制 quality control of life
使用期性能 end-use performance
使用期修理费用 life repair cost
使用契约 guaranty bond
使用器具 instrumentation
使用前的调试工作试车 shakedown operation
使用前的调整 debugging
使用前对系统的整体性试验 integrity test before use;operational test before use
使用前调试 shakedown
使用强度 characterisation strength;characteristic strength
使用清洁能源 burn clean fuel
使用情况 condition of service;in-serve behavio(u)r;in-service behavio(u)r;service condition;service performance
使用权 appropriative rights;easement;right of usage;use rights;usership;tenure <土地的>
使用权出租 operating lease
使用权证明书 deed of appropriation
使用人员 user of service
使用日数 days in use
使用杀虫剂 use of insecticide
使用上的困难 operational difficulty
使用上的限制 limitation on usage
使用设备表 utility device list
使用申请 request of utilization
使用升限 service ceiling
使用时间 durable hours;hours of use;life length;occupation period;service time;time of commissioning;time of usage;usage time
使用时间累加器 elapsed time totalizer
使用时期 length of life;service period;use age
使用试验 performance test;service test;service trial
使用试验机 service test machine
使用收益权 usufruct
使用手册 instruction manual;service manual
使用寿命 age limit;applicational life(span);depreciable life;in-service life;length of life;life space;life span;life time;longevity;longevity of service;long life;operating life;operating time;operational life span;operation life;period of service;serviceable life;service life(time);useful life;useful physical life;useful time;working life
使用寿命保证周期 operational life proof cycle
使用寿命降低 loss of life
使用寿命期 useful life period
使用寿命期内维修成本 lifetime maintenance cost
使用寿命期限 service life period
使用寿命试验 life test(ing)
使用寿命周期成本 life cycle cost(ing)
使用数据 service data
使用水平 service level
使用税 use tax;royalty <版权专利权等的使用税>
使用说明(书) instruction manual;operating instruction(manual);operating specification;operator's handbook;direction for use;shop instruction;instruction book;operating instruction;operation book;operation instruction;operation sheet;operation specification;run book;handling instruction
使用速度范围 operating speed range
使用损耗 service deterioration;service wear
使用缩图器的 pantographic(al)
使用缩图器绘成的缩图 pantographic(al) reduction
使用特性 character of service;operating performance;operational performance
使用特性曲线 operating characteristic curve
使用特征 operational characteristic
使用条件 condition of use;operating condition;service condition;working condition
使用调和 proportioning
使用图表 service chart
使用外资收益率 rate of return on the use of foreign capital
使用完好率 operational availability
使用完好周期 operational availability
使用维护工程师 utilization engineer
使用维护规程 service instruction
使用位 usage bit;use bit
使用温度 application temperature;serviceability temperature;serviceable temperature;temperature of use;usage temperature;working temperature
使用温度范围 service temperature range
使用稳定性 service durability;stability in use
使用物料 spent material
使用物品的商品 commodity considered as a use-value
使用系数 availability factor;coefficient of efficiency;coefficient of performance;coefficient of utility;coefficient of utilization;maintenance factor;occupation coefficient;utilization coefficient
使用细菌的过程 bacteriological process
使用细菌进行净化 bacteriological purification
使用细则 service instruction;service manual
使用先进技术 using advance technology
使用限度 service limits
使用限制 service restriction
使用限制权 deed restriction
使用效率 coefficient of efficiency;service efficiency
使用信息论、计算机科学和多种事实 systems engineering
使用性 <焊条的> operating characteristic
使用性磨耗 service wear
使用性磨损 service wear
使用性能 behavio(u)r in service;character of service;functional performance;operating performance;performance in service;serviceability;service performance;service property;use characteristic;use property;handling quality
使用性能范围 serviceability limitation
使用性能级别 <路面> serviceability rating
使用性能极限状态 serviceability limit state
使用性能试验 usability test
使用性试验 fleet testing
使用性质会改变的土地价值 speculative value
使用修磨过的钻头 rerun a bit
使用须知 working direction
使用须知标签 care label
使用许可税 occupancy permit
使用许可证 certificate of occupancy;permit occupancy
使用压力 service pressure;working pressure
使用要求 operating requirement;operation(al) requirement;serviceability requirement;use requirement;on the road requirements 求 <对于汽车、拖车的>
使用液体燃料的燃气轮机 oil-fired gas turbine
使用一般设备 using ordinary equipment
使用一次的小气瓶 disposable cartridge
使用仪器 instrumentation;instrument of use
使用因素 service factor;usage factor
使用因子 usage factor
使用应力 applied stress;operating stress;service stress
使用有准备 ready for use
使用鱼叉 gig
使用与维护 operation and maintenance
使用与占据保险 use and occupancy insurance
使用预制构件建造的 preengineered
使用月票经常来往 commutation
使用再生纸 use recycle paper
使用者 applicator;consumer;employer;user
使用者得益经济分析 user-benefit economic analysis
使用者亮度旋钮 operator's brilliance control
使用者平均等待时间 <一种衡量道路服务水平的指标> mean user-waiting time
使用者最优分配原则 user-optimized assignment principle
使用者坐标 user coordinate
使用真空度 useful vacuum degree
使用证 letter of credit[L/C]
使用支腿时的定额"on outrigger" load rating
使用执照 occupancy permit
使用指南 operating guide;operating manual
使用指示书 operating instruction manual;operating manual
使用质量 functional quality(of pavement);service property;service quality;use quality
使用中 in service
使用中的固定资产 fixed assets in use
使用中的故障 failure in service
使用中的机车 locomotive in service
使用中的跑道 active runway
使用中的油罐 tank in use
使用中磨损 wear and tear in use
使用中试验 in-service test(ing)
使用中损坏 service failure
使用重量 operating weight;service weight;working weight
使用属性 use attribute
使用专款授权 obligation authority
使用装置 operative installations
使用状况 behavio(u)r in service;service behavio(u)r;service condition;utilization condition;working condition
使用状态 service condition;service state
使用状态阶段 serviceability stage
使用状态挠度限值 deflection limit for serviceability
使用状态试验 service-type test
使用状态说明 status note
使用资本 use capital
使用资本额 capital employed
使用资金 disposable fund
使用资源收益率 ratio of return on resources employed
使用总年限折旧法 sum of expected life method of depreciation
使用总资本收益率 rate of earnings on total capital employed
使用租赁 operating lease
使用租约 operating lease
使有差别 differentiate
使有差异 differentiate
使有导电性 metallization
使有规则 regularize
使有机化 organize
使有机物变腐烂 causing organic material to rot

S

使有棱角 corner
使有深槽 quirk
使有文化 civilize
使有系统 systematize
使有效 validate
使有新的活力 revitalize
使有用 utilization;utilize
使有褶痕 crease
使有秩序 regularize;systematize
使有组织 regularize;systematize
使淤塞 silt up
使与沥青混合 bituminizing
使与水银混合 amalgamate
使与碳化合 carbonize;carburate;carburet
使与碳酸化合 carbonate
使与溴化合 bromate
使与原来熟悉的环境隔离 deracinate
使与轴线平行 collimate
使匀称 symmetrize
使运行 bring into service;put into operation;put into service
使再浮起 refloat(ing)
使再符合标准 restandardize
使再活化 reactivate
使再结合 reintegrate
使再生 regenerate;revivify
使再现 resurrect
使暂停 respite
使增加 multiply
使增加密度 densify
使张紧 tightening
使者 emissary;envoy
使真实 substantiate
使振动 vibrate
使镇静 mitigate
使蒸发 evaporate
使蒸气化 vapo(u)rize
使正常化 normalize
使正规化【数】normalize
使正态化 normalize
使直动【机】translate
使植物冻死 winter kill
使制度化 institutionalize
使窒息 stifle;suffocate
使中间透光 <剪树> keeping the centre of tree open
使中立化 neutralize
使重新完整 reintegrate
使逐步减缩 de-escalate
使逐步降级 de-escalate
使转弯 swerve
使转向 divert;put over
使桩顶开花 broom
使桩顶篷裂 broom
使浊 thicken
使资源利用达到最大限度 maximize the use of resources
使自动化 automatize
使自己适应于 adjust oneself to
使最佳化 optimalize
使作废 invalidation
使作用的 actuating
使坐 seat

始

始白垩层【地】Eocretaceous

始爆剂 primer detonator
始爆器 primer
始变质作用 eometamorphism
始成土 cambisol;inceptisol
始成岩阶段 eogenetic stage
始地台 eoplatform
始点 beginning point;departure;initial point;initial vertex;origin;point of origin
始点荷载 initial load
始点时间 time of origin
始点事项 <统筹方法中标志网络中一个或多个活动的开始的一个事项> beginning event
始锭 starting ingot
始动点 incipient point
始动电流 pick-up current
始动电压 pick-up voltage
始动继电器 initiating relay
始动调节作用 antecedent regulation
始动值 pick-up;pick-up value
始端按钮 entrance button;entry button
始端继电器 entrance relay
始端系统 origin system
始对象 source object
始发编组列车技术作业过程 operating procedure of outbound train
始发端 sending end
始发方式 originating mode
始发港 original port;port of departure;port of origin;terminal port
始发交通 originating traffic
始发井 launching shaft;starting shaft
始发量 initiator
始发列车 originating train
始发流水号 original sequence number
始发旅客列车 originating passenger train
始发旅客人数 passengers originated
始发枢纽 initial terminal
始发站 initial station;originating station
始发直达货物列车 through freight train made up at one loading station;through goods train made up at one loading station
始发直达列车 through train originated from one loading train
始发作业费 operation charges of start off
始发作业支出 starting operation expenditures
始法列车 outgoing train
始泛大陆【地】proto-Pangaea
始沸点 bubble point;initial boiling point
始寒武纪【地】Eocambrian period
始寒武纪冰川作用 Eocambrian glaciation
始寒武统【地】Eocambrian;Infracambrian
始航 departure
始航船位 departure position
始航点 departure point
始航向 departure course;initial course
始极槽 starting sheet cell
始极片 starting sheet
始极片母板 starting sheet blank
始极片种板 starting sheet blank
始加速度 starting acceleration
始铰纲【地】Eoarticulate
始结点 beginning node
始裂荷载 load at first crack
始凝点 clouding point
始盘类 eodiscid
始膨点 original dilation point
始倾角 primary dip
始曲点 beginning of curve
始生代【地】Archean;Eozoic(era)
始生界 Eozoic group
始生物【地】eozoon
始石器 eolith
始石器时代 Eolithic Age
始石器时代的 Eolithic
始石器时期 Eolithic Period
始态 primary state
始拖 commencement of towing
始线 initial line
始效值 threshold value
始新纪石油 Eocene era oil
始新世【地】Eocene;Eocene epoch
始新世海浸 Eocene transgression
始新世末期事件 terminal Eocene event
始新世黏[粘]土 Eocene clay
始新世气候 Eocene climate
始新统【地】Eocene series
始新系【地】Eocene system
始因 incipient cause
始站 from-depot;from-tank farm
始值 initial value
始值保持量 threshold retention
始终 all along;alpha and omega;from beginning to end;throughout;top and tail
始终保持安全浮泊 always safely afloat
始终保持浮泊 always afloat
始终标记 sentinel
始终如一的置换 consistent substitute
始终赢利的企业 trend bucker
始重 starting weight
始锥 cone of origin

驶

驶出 drive out;put out;ride-out;steam off;take departure

驶出车道 exit lane
驶出车流 pulling-out of vehicle
驶出车行道 exit roadway
驶出船闸 leaving the dock
驶出道路 exit road
驶出的 off bound
驶出路外事故 runoff-road accident
驶出率 rate of discharge
驶出坡道 exit ramp;off bound ramp;off ramp
驶出时间 exit time
驶出速度 exit speed;exit velocity
驶出匝道 exit ramp;off ramp
驶出状态 discharging state
驶出左转弯 left exit turn
驶(船) conn
驶帆 under canvas;under sail
驶帆架 horse rail
驶帆术语 sailing term
驶帆训练船 sail training vessel
驶回 pullback;resail
驶回本国 homeward bound
驶近 approach;beam in with
驶近(交叉口)速率 approach speed
驶近陆地 make for land;make the land;making the land
驶近陆地初见灯光 landfall light;making light
驶近速度 approaching speed;approaching velocity;velocity of approach
驶进 drive-in
驶进港湾 embay
驶进速度 approach speed
驶离 beam off;get off;sheer off;stand-off
驶离泊位时间 deberthing lime
驶离车流图式 go flow pattern
驶离港 leave port
驶离路外车辆 off-road vehicle
驶离码头 undock
驶离时间 departure time
驶离图式(形态) departure pattern
驶骂曲线 exit curve
驶入 ride-in
驶入车流 pulling-in of vehicle
驶入车行道 entrance roadway
驶入的 on bound
驶入和驶出信号 entering and leaving signals
驶入坡道 entrance ramp;entry ramp;on ramp
驶入曲线 entrance curve
驶入时间 entry time
驶入驶出(交通)调查 input-output study
驶入、驶出坡(匝)道 ro/ro ramp[roll-on/roll-off ramp]
驶入匝道 entrance ramp;entry ramp;on bound ramp;on ramp
驶往……的车道 bound lane
驶往国外的 outward bound
驶向 beam in with
驶向港 port of destination
驶向海岸 stand in to land
驶向海面 take an offing
驶向下风 bear up to

氏

氏方法(地下水力学) Theis method

世

世 <=10^9年> eon

世代 generation;life cycle
世代材料的成熟期 generation material for maturity
世代长度 generation length
世代号 generation number
世代间隔 generation interval;generation length
世代交替 alternation of generation;digenesis;metagenesis
世代年龄曲线 generation age curve
世代时间 generation time
世代数 generation number
世代数据集 generation data set
世代数据组 generation data group
世代文件组 generation file group
世代演替 succession of generation
世代指数 generation index
世代(周)期 generation period;generative period
世纪 century
世纪变化 secular variation
世纪的 secular
世纪年 centurial year
世间月 karma month
世界霸权 world domination
世界包装组织 World Packaging Organization
世界保护监测中心 World Conservation Monitoring Center[Centre]
世界标准地震网 world-wide standardized seismic network
世界标准地震仪台网 world-wide standard seismograph network
世界标准日 World Standard Day
世界表层环流图 Chart of the World General Surface Current Circulation;General Surface Current Circulation of the World
世界博览会 World's Fair
世界测地系统 world geodetic system
世界产量 world-wide production
世界潮汐假说 global tide hypothesis
世界车速记录 world speed record
世界臭氧层行动计划 world plan of action on the ozone layer
世界臭氧监测网 world ozone network
世界臭氧资料中心 world ozone data center[centre]
世界出口价格指数 world export prices indices
世界出口贸易公平比率 equitable share of world export trade
世界出口贸易总额 world's total trade volume
世界储量 world resources
世界大地测量坐标系(统) world geodetic coordinate system

世界大地网 world geodetic network
世界大气 world's atmosphere
世界大气基金 World Atmosphere Fund
世界大洋 world ocean
世界大洋航路 ocean passages for the world
世界大洋水深图 bathymetric(al) chart; general bathymetric(al) chart of the oceans; ocean bathymetric(al) chart
世界大自然基金 World Wide Fund for Nature
世界岛 < 即欧亚非大陆 > world island
世界岛中心 < 指亚非欧中心 > heartland
世界道路会议 World Road Conference
世界的 international
世界地磁图 the world geomagnetic chart
世界地理 world geography
世界地理基准系 world geographic(al) reference system
世界地图 international map of the World; map of the world; spheric(al) map; world map
世界地图集 atlas of the world; world atlas
世界地震分布简图 sketch map of earthquake of the world
世界地震图 seismic world map
世界地震资料 world-wide earthquake data
世界地质图委员会 Commission for the Geological Map of the World
世界第一流水平的 world-class
世界点 world point
世界定点海洋观测站 fixed oceanographic(al) stations of the world
世界动力会议 World Power Conference
世界动力学 world dynamics
世界动物保护联合会 World Federation for the Protection of Animals
世界多圆锥投影坐标方格 world polyconic(al) grid
世界范围 worldwise
世界范围的 world-wide
世界范围的耕地损失 the world loss of cropland
世界方格坐标 world grid
世界肥料经济 world fertilizer economy
世界肥料使用量 world use of fertilizer
世界分布 world distribution
世界辐射地图 world radiation map
世界辐射监测网 world radiation network
世界港口指南 World Port Index
世界工业和技术研究组织协会 World Association of Industrial and Technological Research Organization
世界公民 cosmopolite
世界关税同盟 international custom union
世界观 world outlook; world view
世界海事大学 World Maritime University
世界海洋 world ocean
世界海洋环流实验 world ocean circulation experiment
世界海洋日 world oceans day
世界海洋组织 World Ocean Organization
世界航空图 world aeronautical chart
世界航空线 world air route
世界环境 world environment
世界环境和资源委员会 World Environment and Resources Council
世界环境日 world environment day
世界环境学会 World Environment Institute
世界环境研究所 World Environment Institute; Worldwatch Institute < 美 >
世界环境与发展委员会 World Commission on Environment and Development
世界环境状况 state of the world environment
世界货币 universal money; world currency; world money
世界纪录 world's record
世界降雨纪录 world rainfall record
世界交通图 world-wide communication network
世界金融市场 world financial market
世界经济 world economy
世界经济概览 world economic survey
世界巨型油气田 world giant oil and gas field
世界开发预算 world development budget
世界科学工作者协会 World Federation of Scientific Workers
世界科学数据中心 World Data Centre
世界劳工联合会 World Confederation of Labo(u)r
世界历 world calendar
世界粮库 world food bank
世界粮食计划署 < 联合国机构，设在美国纽约 > World Food Programme
世界粮食日 World Food Day
世界旅游与汽车组织 World Touring and Automobile Organization
世界贸易 world commerce; world embracing commerce; world trade
世界贸易矩阵 world trade matrix
世界贸易量 volume of world trade
世界贸易模式 pattern of world trade
世界贸易中心 World Trade Center
世界贸易中心联合会 World Trade Centres Association
世界贸易中心塔楼 World Trade Centre Towers
世界贸易总量 quantum of world trade
世界贸易组织 World Trade Organization
世界模型 world model
世界能源会议 World Energy Conference
世界农业发展指示计划 Indicative World Plan for Agricultural Development
世界配额 world quota
世界气候 world climate
世界气候极端值 world weather extremes
世界气候研究计划 world climate research programme
世界气象日 world meteorological day
世界气象中心 World Meteorological Center
世界气象资料中心 World Data Centres for Meteorology
世界气象组织 < 联合国机构，设在瑞士日内瓦 > World Meteorological Organization
世界气象组织天气码 World Meteorological Organization Code
世界情报系统交流中心 World Information Systems Exchange
世界区域预报系统 world area forecast system
世界人口年 world population year
世界人类遗产公约 World Heritage Convention
世界日 world day
世界三大谷类作物 world's three-major cereals
世界森林目录 world forest inventory
世界森林资源 world forest resources
世界商船队 world merchant fleet
世界商船总吨位 world merchant tonnage
世界商业 world commerce
世界生产模式 world production pattern
世界生产总值 gross world product
世界石油 world oil
世界石油产量 global oil production
世界石油市场 world oil market
世界时 Greenwich civil time; Greenwich mean time; universal time; world(-wide) time
世界时区 world time zone
世界时区图 Chart of Time Zones of the World
世界市场 international market; world(-wide) market
世界市场参考价格 reference world market price
世界市场价格 international market price; world price; world market price
世界市场税率 world market rate
世界疏浚会议 World Dredging Conference
世界疏浚协会 World Dredging Association; World Organization of Dredging Association
世界疏浚组织 World Dredging Organization
世界输出总量 volume of world exports
世界数据查询系统 world data referral system
世界数据传输业务 world data transmission service
世界数据中心 world data center[centre]
世界水 < 月刊 > World Water
世界水平 international level; world level; world standard
世界水日 world water day
世界水资源 world's water resources
世界天气 world weather
世界天气监测网 world weather watch
世界天气监视网 world weather watch
世界通货 world currency
世界通用时间 universal time
世界统一信号系统 world-wide uniform signal(l)ing system
世界图组 world map series
世界土壤地图 world soil map
世界土壤退化和危害图 world map of soil degradation and hazards
世界微生物资料中心 world data centre on microorganism
世界卫生组织 < 联合国机构，设在瑞士日内瓦 > World Health Organization
世界卫生组织水安全计划 Water Safety Plans of World Health Organization
世界系统模型 world system model
世界先进水平 advanced world level; highest world standard; world advanced level
世界线 world-line
世界销售量 world-wide sales
世界协调时 universal time coordinated
世界协调时间 world concordant time
世界性大港 world port
世界性的 cosmopolitan
世界性的传染病 pandemic
世界性的电信网络 world-wide telecommunication network
世界性都市 < 假设整个世界为一个城市 > ecumenopolis
世界性通货膨胀 world inflation
世界养护监视中心 World Conservation Monitoring Center
世界野生动物基金会 World Wildlife Funds
世界遗产 world heritage
世界遗产保护地 world heritage site
世界银行 International Bank for Reconstruction and Development; World Bank < 即国际复兴开发银行 >
世界银行采购指南 Guideline for Procurement of I-BRD
世界银行环境部 World Bank Environmental Ministry
世界银行集团 World Bank Group
世界银行经济发展研究所 Economic Development Institute
世界邮政联盟 < 联合国 > Universal Postal Union
世界运费率 world scale
世界运价表 world scale
世界战略 world strategy
世界知识 world knowledge
世界知识产权组织 < 联合国机构，设在瑞士日内瓦 > World Intellectual Property Organization; World Organization of Intellectual Property
世界钟 world clock; world time clock
世界种 cosmopolitan species
世界重力测量系统 world gravimetric system
世界重力基点 world-wide gravimetric basic point
世界主要盆地名称 world's major basin name
世界专利索引 world patents index
世界资源储备状况 situation of world mineral reserves
世界资源研究机构 < 华盛顿 > World Resources Institute
世界资源研究所 world resources institute
世界自然保护基金会 World Wild Fund for Nature
世界自然保护中心 World Conservation Monitoring Center
世界自然资源保护大纲 world conservation strategy
世界自由贸易 free world trade
世界坐标 world coordinates
世界坐标系 world coordinate system
世俗巴利卡 civic basilica
世俗的长方形会堂 < 非宗教性的 > secular basilica
世俗哥特式风格 profane Gothic style
世俗哥特式建筑 civic Gothic(style)
世俗建筑学 profane architecture
世袭不动产 freehold
世袭财产 estate of inheritance; hereditary estate; hereditary property; heritage; patrimony
世袭地的保有权 freeholder
世系任务 ancestral task
世系图 lineal chart

市

市办公共设施 city-owned utilities
市办公共事业 city-owned undertakings
市边缘区 urban fringe
市长 city manager
市场 market(hall); market place; market stead; mart; baza(a)r; Rialto; tekram; trading floor; tryst
市场办公室 market cross
市场饱和 market saturation
市场饱和度 percent saturation of mar-

ket
市场饱和价格 satiety price
市场报告 market report
市场本能 market instinct
市场比率 market ratio
市场变化 turn of the market
市场波动 market fluctuation
市场博弈 market game
市场不完全性 imperfections in markets
市场不稳定 market unhealthy; market unsettled
市场不正常 market unhealthy
市场采购 market purchasing; purchase from the market
市场操纵 market dominance; market manipulation
市场测验 market test(ing)
市场策略 market(ing) strategy
市场层次 market level; tiers in market
市场尺寸 <商品> market size
市场充分供应 satiety
市场存货过多 glut the market; market overstocked
市场呆滞 heavy market
市场导向工业 market-oriented industry
市场导向机制 market-oriented mechanism
市场导向经济 market-directed economics
市场道路 market road
市场的景况 tone(of the market)
市场的类型 category of market
市场的敏感性 market sensitiveness
市场的潜在倾向 undertone
市场的透明度 transparency of market
市场地位 market position
市场调查 market investigation; market research; market survey
市场调查报告名称 name of marketing surveying report
市场调查程序 market survey procedure
市场调查公司 market research company
市场调查资料 market feedback
市场调研 market study
市场跌落 falling market
市场定单 market order
市场定购 market order
市场定价 market-set prices
市场动力 market forces
市场动态 market trend; movement in the market
市场对策 market game
市场对称 market symmetry
市场对象 market target
市场多样化 market diversification
市场发育 growth of the market; market growth
市场发展 market development
市场发展与培育 promotion of market development
市场法规 market regulation
市场繁荣 brisk market
市场反应能力 ability to react to market conditions
市场范围 market area
市场房屋 market building
市场分割 market segmentation
市场分类 classification of market; market classification; market grouping
市场分配 market share; market sharing; share of market
市场分配协定 market-sharing arrangement
市场分析 market analysis
市场份额 market share; portion of marketing
市场份额矩阵 market share matrix
市场风险 market risk
市场封闭 market closure
市场概况 market profile
市场干预 market intervention
市场割据 market separation
市场工资 market wage
市场工作 marketing
市场公开价格 open market value
市场功能 marketing function
市场供给表 market supply schedule
市场供求 market supply and demand
市场供求情况 market supply-and-demand situation
市场供需平衡 market equilibrium
市场供应 supply of the market
市场供应量 apparent availability
市场购买可能性 commercial availability
市场估价 market assessment; market valuation
市场观念 marketing concept
市场管理 market administration; market control; market management
市场管理所 market house
市场广场 market square
市场规划 city planning
市场规模 market scale; market size
市场国库券 market bill
市场过剩 glut the market
市场和计划经济组合 combined market and planning economy
市场和销售系统 market and marketing system
市场花园 market garden
市场化 general adoption of the market principle
市场化水土保持 marketing conservation
市场划分 market segmentation
市场环节分析 analysis of market links
市场回稳 rally
市场汇率 market exchange rate; market rate of exchange
市场活动 marketing activity
市场活动调查 marketing research
市场活跃 active market; brisk market; buoyant market
市场机构 market mechanism
市场机制 market mechanism
市场机制调节作用 regulation by the market mechanism
市场计划 marketing plan
市场价差保证金 market difference margin
市场价(格) market price
市场价格变动准备金 reserve for market fluctuation
市场价格的下跌 market depreciation
市场价格趋势 tendency of market
市场价格政策 marketing price policies
市场价格总额 aggregate market value
市场价值 commercial value; par value; realized value; trading value; market value
市场价值法 market value method
市场价值评估法 market value approach
市场坚挺 strong market
市场坚稳 steady market
市场交易 market deal; market transaction
市场交易费用 market transaction cost
市场交易税 hallage
市场交易折扣 discount allowance; market discount allowance
市场结构 market mechanism; market structure
市场结算 market clearing
市场借债 market borrowing
市场金额 market amount
市场金价的不稳定 instability of gold market prices
市场金融 market finance
市场紧张 market shortages
市场经济 exchange economy; market(-based) economy; market-directed economy; market economics; market(-oriented) economy
市场经济结构 market economy structure
市场经营战略 market management strategy
市场经营组合 marketing mix
市场景况 business cycle marketing
市场竞争 market competition
市场竞争对策 game of market competition
市场竞争力 marketability
市场竞争前景 anticipating of marketing competition
市场开发 market development
市场开放政策 open market policy
市场考察 market exploration
市场可供商品量 availability in market supplies
市场控制商 control commerce
市场狂热 superheat(ing)
市场扩张 market expansion
市场劳动力价格 market price of labo(u)r
市场力量 market forces
市场力量控制 control through market forces
市场利率 market interest rates; market rate
市场利率贷款 market rate loan
市场利率低 easiness of money market
市场联合 market combination
市场流动性 market liquidity
市场垄断 corner on the market; monopolized market
市场买卖 marketing
市场买卖人 marketer
市场卖主 marketeer
市场名称 <木材> market term
市场明朗度 market transparency
市场模型 market model
市场目标 market objective
市场疲跌 bearish market
市场疲软 bearish market; corner on the market; sluggish market; soft market; weak market
市场平衡 market equilibrium
市场气氛 market sentiment
市场前景 market prospect
市场潜力 market potential
市场潜在需求量 market potential
市场清算价格 market cleaning price
市场情况 market condition
市场区分 market segmentation
市场区(域) market area
市场趋势 market trend
市场渠道 market channel
市场缺口 market has openings
市场确定 market identification
市场扰乱 market disruption
市场容量 capacity of the market; market absorption capacity
市场容纳量 absorption of market
市场扫描 market profile
市场商品 marketing
市场商业中心区 market place
市场上的钢材 black bar
市场上的买空卖空 speculation on the rise and fall of the market
市场上各种买卖的混合体 marketing mix
市场上可买到的 commercially available
市场上可以买到的代用品 commercially available substitute
市场深度 depth of the market
市场审核 market audit
市场渗入 market penetration
市场渗透 market penetration
市场渗透定价 market-penetration pricing
市场生产 market production
市场实际容量 actual market volume
市场实现 market realization
市场试销 test marketing
市场收入额 geologic(al) market receipt charges
市场收入增长率 increasing rate of geologic(al) market receipt
市场收益率 market rate of return
市场寿命期 marketing life
市场衰退 market failure
市场税 lastage[lestage]
市场损失 loss of market
市场所在地 market area; market place
市场索赔 market claim
市场探测法 market survey method
市场体系 marketing system
市场调节 market regulation; regulation by market; regulation through the market
市场调节器 regulator of the market
市场调节职能 market mechanism
市场贴现利率 market rate of discount; market discount rate
市场位置 location of markets
市场污染权建立 market-creation of pollution rights
市场污染许可证 marketed pollution permit
市场吸收能力 market absorption
市场限值 market threshold
市场销路 access
市场销售 market sale; outlet sale
市场销售量 annual marketing
市场效率 market efficiency
市场信息 market(ing) information
市场行情 market quotation
市场行为 market behavio(u)r
市场性 marketability
市场休业 market off
市场需求 market demand; market requirement
市场需求价格 market demand price
市场需求曲线 market demand curve
市场需要 market demand
市场需要量 market requirement
市场需要预测 market demand forecasting
市场研究 market research
市场研究与分析销售潜力 market research and analyze sales potential
市场要求的质量 market quality
市场业务 marketing function
市场议价标准 market appraisal standard
市场营销学 market marketing
市场营业费 marketing cost
市场影响 market influence
市场用语 <木材> market term
市场优势 market preference
市场预测 market forecast(ing); market prediction
市场预测年限 limited in years of market forecasting
市场原则 marketing principle
市场再调节 market readjustment
市场展望 marketing anticipating; market outlook; market prospect
市场占有率 market share
市场战略模拟 marketing strategy simu-

S

lation
市场涨价 appreciation
市场政策 marketing policy
市场支配力 market power
市场指数 market index
市场质量情报 market quality information
市场中心 market center[centre]
市场主体 marketing main body
市场专家 marketing expert
市场走势 market behaviour
市场租价 market rent
市场足量供应 saturation
市场族群 market segment
市场阻塞 < 指市场挤迫现象 > market congestion
市尺 < 中国长度单位 > chi
市寸 < 1 市寸 = 1/30 米 > cun
市电 commercial power
市电电源 mains supply
市电频率 commercial electricity frequency; commercial frequency; power frequency
市电收音机 electric(al) set; socket-powered set
市电停电 interruption of mains supply
市房管局 municipal housing management bureau
市废物 town waste
市肺 < 人口稠密,交通拥挤市区的小花园、绿地 > lung
市管县 municipally affiliated county
市花 city flower
市话长话混合接续 combined local and toll operation
市话电缆 city cable; local cable; telephone exchange area cable
市话分局 < 多局制情况下 > local telephone office; urban telephone office
市话分局交换 centerx
市话汇接局 switching office; tandem office
市话交换机 local switch
市话交换网 local exchange network
市话局 local central office; local exchange center[centre]; local telephone office; urban telephone office
市话网 local telephone network; public telephone network; urban telephone network
市话业务 local service
市话占线 local busy condition
市话支局 local telephone branch office
市级道路系统 city system
市集 baza(a)r; expo; market town
市际 intercity
市际乘客 intercity passengers
市际出行 interurban trip
市际道路 intercity road; interurban road
市际的 intercity; intertown; interurban
市际电气化铁路线 intercity electric railway line
市际高速铁路 high-speed intercity rail
市际公共汽车 < 长途汽车 > intercity bus
市际公共汽车交通 interurban bus service
市际公共汽车运输 intercity bus transportation
市际活动网络优化模型 model for optimizing the network of intercity activities
市际交通 intercity communication; intercity traffic; interurban traffic
市际街道 interurban street
市际客运业务 city-to-city passenger service; inter-city passenger business
市际快速交通服务 express intercity service
市际旅客 intercity passengers
市际旅客列车 intercity train
市际汽车货运 intercity trucking
市际汽车交通 interurban service
市际汽车运输吨数 intercity truck tonnage
市际铁道 interurban railroad[railway]
市际铁路 intercity rail; interurban railroad[railway]
市际通勤交通 intercity commuting
市际通信[讯] intercity communication
市际(运输)计划 intercity program(me)
市价 current cost; current price; market; market price; prevailing rate; rule price
市价比较法 relative market values method
市价变动 fluctuation
市价表 price current
市价单 market order
市价跌落 loss of market
市价订购 market order
市价法 < 指联合成本的分配 > market price method
市价上涨 appreciation of market price
市价突然下跌 break-in the market
市价下跌 market write-down
市价正上涨 market advancing
市价正下降 market declining
市间长途运输 interurban long haul service
市间公共汽车 interurban bus
市间公共汽车交通 interurban bus service
市间交通 interurban service; interurban traffic
市间街道 interurban street
市间铁路 interurban railroad[railway]
市郊 faubourg; outer city; outskirt; subtopia; suburb; suburban area; urban fringe
市郊穿梭旅客列车 commuter push-pull(train)
市郊道路 suburban road
市郊的 suburban
市郊的大商业中心 supercenter[centre]
市郊的扩大 suburb sprawl
市郊电话线 suburban telephone line
市郊电气铁路 suburban electrified railway
市郊定期车票票价 commutation fare
市郊定期客票 periodic(al) suburban ticket
市郊定期客票发售机 commuter ticket issuing machine
市郊定期票 commutation ticket
市郊(定期票)客运 commuter passenger traffic; commuting passenger traffic
市郊房地产 suburban estate
市郊富人住宅区 excurb
市郊高级住宅区 cocktail belt
市郊高速公路 urban clearway
市郊工业区 industrial suburb
市郊公路 suburban highway
市郊关节列车组 articulated suburban train set
市郊化 suburbanization; suburbanize
市郊火车站 suburban train station
市郊交通 suburban traffic
市郊界线 suburban line
市郊经常客流 commuter movement
市郊景观 suburbscape
市郊居民 exurbanite
市郊居住区 suburban populated area; suburban residential area
市郊客车 commuter car; suburban coach; suburban passenger car
市郊客流 suburban passenger flow
市郊客流图 suburban passenger flow diagram
市郊扩大 suburb dispersal
市郊列车 suburban train
市郊列车运行 suburban service
市郊旅客 suburban passenger
市郊旅客列车 commuter train; suburban passenger train
市郊绿化区 suburban green area
市郊蔓延 suburban dispersal
市郊内运输业务 inner suburban service
市郊农业 suburban agriculture
市郊贫民窟 < 北非 > bidonville
市郊贫民区 < 俚语 > slurb(ia)
市郊区 suburb; suburban district
市郊区道路 suburban road
市郊区服务业 suburban service
市郊区列车 commuter railroad
市郊商业区 outlying business district; shopping center[centre] < 围绕大型停车场的 >
市郊上下班往返旅客运输 suburban commuters travel
市郊铁路 commuter railroad; suburban line; suburban railroad; suburban railway
市郊铁路系统 suburban railway system
市郊铁路线 commuting line; suburban line
市郊停车场 fringe parking
市郊线路 suburban line; suburban track
市郊选择通信[讯]业务 optional extended area service
市郊延伸 suburb dispersal
市郊园林化 garden suburb
市郊月季票 commutation ticket
市郊月季票票价 commutation fare
市郊运输 commuting traffic; suburban traffic
市郊站 suburban station
市郊至市郊线路 suburb-to-suburb route
市郊住宅区 dormitory suburb
市郊转车点 suburban transfer point
市街 city street; town road; town street
市街电车 street tramway
市街进站口 street approach
市街铁道 street railroad track
市街有轨电车道 electric(al) street railroad
市街运量观测 street survey; street traffic survey
市界 city line
市斤 < 1 市斤 = 0.5 千克 > jin; catty
市况 market condition
市况萧条 big bad
市面 conjuncture; market condition
市面冷淡 dull market
市面灵通的人 Hipster
市面萧条 dull market
市面兴旺 swimming market
市面租金 open market rent
市民 townsman
市民参与 citizen participation
市民广场 citizen's square
市民会堂 citizen hall
市民集会所 < 古罗马 > comitium
市民团体 citizen organization
市亩 < = 1/15 公顷 > mu
市内 city proper; downtown(area); urban district
市内草地广场 mall
"市内长途"转换开关 local-distance switch
市内车站 in-town station; town station
市内乘车出行 internal trip
市内的 intra-city; intraurban
市内地下铁道车站 metropolitan railroad station
市内地下铁道钢轨 metropolitan railroad track
市内电报 local telegram
市内电车 street car; tramcar
市内电话 city telephone; local telephone; urban telephone; local call
市内电话分局 local office of city telephone; local telephone office; urban telephone office
市内电话服务区 local service area
市内电话交换机 local telephone exchange
市内电话交换系统 city telephone switching system
市内电话局 local exchange; local station; local telephone exchange; local telephone office; local telephone switching system; urban telephone office
市内电话区 telephone exchange area
市内电话网 district telephone network; local exchange area; local plant; local telephone network; public telephone network; urban telephone network
市内电话系统 local telephone system
市内(电话)线路图 town distribution scheme
市内电话支局 local telephone branch office
市内电话中心局 local central office
市内电缆 city cable; exchange area cable
市内分局 local station
市内分区道路 distributor road
市内公共汽车 local service; local service urban bus; town car; urban bus
市内和长途电话合用选择器 combined local and trunk selector
市内呼叫 exchange call; local call
市内呼叫按扣 local call push-button
市内呼叫按钮 local call push-button
市内话务员 local operator
市内价郊区通信[讯]业务 extended area service
市内交换 local exchange
市内交换机 local switch board
市内交通 intra-city traffic; local traffic; street traffic; urban transport
市内街道 city street; municipal block; town street
市内就业条例 local employment act
市内流动 intra-urban mobility
市内绿地广场 mall
市内汽车 town car
市内人口迁移 intra-urban migration
市内商业区车站 downtown station
市内疏散通路 inner relief road
市内铁道 metropolitan railroad; metropolitan railway
市内铁道车站 metropolitan railroad station
市内铁道钢轨 metropolitan railroad track; metropolitan railway track
市内铁路 intramural railway; metropolitan railway
市内停车场 city parking lot

市内通话 local connection
市内通勤交通 intra-city commuting
市内通信[讯] local service
市内线路 local line
市内线路制 local line system
市内相互作用 intra-urban interaction
市内小车灯照明 dimmed illumination
市内行车束光 town driving beam
市内行车小光灯 town driving beam
市内行驶里程 urban-mileage
市内行政区 <美> borough
市内选择机 local selector
市内业务 local service
市内音频电缆 local audio cable
市内运输 city service; intra-urban transportation; urban transport
市内运送 city terminal service
市内运行 intra-city travel
市内直穿铁路线 diametric(al) in-town line
市内中继线 local junction; local trunk
市内中继线路 local central office; local junction circuit
市内中心局 local central office
市内终接器 local connector
市内住宅区 inner residential area
市内转换中心 local exchange center [centre]
市区 built-up area; city area; city district; commune; municipal area; municipality; town (ship); urban district; urban section
市区搬运· town cartage
市区办公室 town office
市区边缘地带车辆停放处 fringe parking
市区超速道路 urban clearway
市区道路 city street; municipal road; road; town road; urban road; urban area
市区道路铺设 municipal surface
市区的 urban
市区电车道 street tramway
市区电话 urban telephone
市区定期票旅客 urban commuter
市区发展规划 planned urban development
市区范围 city limits
市区房地产 urban property
市区改造 urban renewal
市区干路 urban arterial highway
市区高速铁路系统 urban rail transit
市区公共交通 urban mass transportation
市区公共交通管理局 <美> Urban Mass Transportation Administration
市区公共汽车 municipal bus
市区公路 urban highway
市区广场 downtown plaza
市区规划 urban area planning; urban district planning
市区化 <主要指农用地逐渐变成住宅用地> urbanization
市区化控制区 urbanization control area
市区化区域 <规划成为市区的地区> urbanization promotion area
市区化调整区域 <规划中对建设加以抑制的地区，即非市区化地区> urbanization control area
市区环境 urban area district environment
市区火车站 municipal station
市区货运站 off-track freight station
市区间的有轨电车 interurban tramway
市区建设 urban sprawl
市区建筑 street building
市区交通 urban traffic
市区交通通达度 traffic accessibility in city
市区交通运输系统 urban mobile system; urbmobile system
市区街道 urban street
市区街道交通流量控制 urban street traffic flow control
市区街面道口 street crossing
市区界 city boundary; city limits; municipal boundary
市区开拓 municipal development
市区扩展 municipal extension; urban area extension; urban district sprawl; urban extension; urban sprawl
市区旅馆 municipal hotel
市区面积比 urban area to city area
市区浓度 urban area concentration
市区膨胀 urban expansion; urban explosion
市区人口 urban population
市区散漫扩展地区 sprawled area
市区桑拿浴室 municipal sauna bath
市区伸延 urban sprawl
市区树林 municipal forest
市区隧道 urban tunnel
市区铁道 street railroad
市区铁路 city railway; street railroad; street railway; urban railway
市区通话(业务) local traffic
市区(土)地 urban land
市区土壤流失率 urban erosion rate
市区外的 outbound city; out-city
市区污染 urban area pollution
市区污染源 urban area pollution sources
市区无规划扩展区 sprawled area
市区无规划漫伸地区 sprawled area
市区线网 urban road network
市区拥挤 urban area congestion; urban congestion
市区用地 urban area land
市区游泳池 municipal swimming pool
市区有轨电车道 street railroad track
市区园林 urban garden
市区运输 urban traffic; urban transportation
市区中心 municipal center[centre]
市区周围地带 peri-urban area
市区住宅 urban dwelling
市容 city aesthetics; city appearance; townscape; urban aesthetics; urban appearance
市容规划 townscape plan(ning)
市售包装 consumer package
市售商品 articles on free market
市售水泥 marketed cement
市售涂料 consumer paint; trade coating; trade sales paint
市外的 extramural
市外商业机构办公区 <远离市中心的> executive park
市辖县 municipally administered county
市行政官 city manager
市议会 city council
市银行 city bank
市营发电厂 municipal power plant
市用煤气 city gas
市有地 <美> city estate
市缘存车场 perimeter car park; peripheral car park
市缘停车场 perimeter car park; peripheral car park
市远郊的 exurban
市债 city bond
市镇 bourg; town
市镇白天亮度 town sunniness
市镇道路 township road
市镇道路系统 township system
市镇范围 township corner
市镇范围界 <美国土地测量> range line
市镇公路 township highway
市镇间的 interborough
市镇建筑规范 township building code
市镇街道 town road
市镇界 civil township line
市镇界线 township line
市镇轮廓 township corner
市镇排水 town drainage
市镇铁路 town railway
市镇中心 town center[centre]
市政 civil administration; municipal administration
市政(城市)财政及预算 municipal finance and budgeting
市政大楼 municipal building
市政大厅 urban hall
市政大厦 municipal building; urban hall
市政当局 municipality
市政当局控制的港口 municipal-controlled harbo(u)r
市政的 municipal
市政法规 municipal code
市政服务(工作) municipal service
市政府 municipal government; municipality
市政府(管理的)码头 municipal dock
市政府拥有的港口 municipally owned port
市政工程 civic work; municipal (civil) engineering; municipal works; public works; utility
市政工程部 Ministry of Public Works
市政工程测量 municipal engineering survey; public engineering survey; public works survey
市政工程承包人 public works contractor
市政工程承包商 public works contractor
市政工程承包者 public works contractor
市政工程处 department of public works; department public works
市政工程局 Bureau of Public Works; Department of Public Works; Municipal Works Bureau; Public Works Department
市政工程设计 public works project
市政工程设施 municipal engineering facility
市政工程师 municipal engineer
市政工程学 municipal engineering; urban engineering
市政工程研究所 <日本> Public Works Research Institute
市政公务地图 official map
市政公用设施 public utility
市政公债 municipal bond; municipal loan
市政固体废料 municipal solid waste
市政管理 municipal administration; municipal management
市政管理图 administrative map
市政规程 municipal code
市政规划 urban planning
市政合质监视 municipal compliance monitoring
市政会计 municipal accounting
市政基础设施 municipal infrastructure; public infrastructure
市政基金 municipal fund
市政监督 municipal supervision
市政建筑(物) civic building; municipal building
市政剧院 municipal theatre
市政垃圾 municipal refuse; municipal rubbish
市政排水工程 municipal sew(er)age
市政排水管道 public sewer
市政区 ward
市政设施 public facility; urban facility
市政收集垃圾系统 municipal refuge collection system
市政税 municipal tax
市政隧道 municipal service tunnel
市政隧洞 municipal service tunnel
市政厅 city council; city hall; municipal building; municipal hall; municipal office; town hall; prytaneion <古希腊的>
市政厅建筑群 city hall complex; town hall complex
市政图书馆 municipal library
市政图形信息系统 municipal graphic information system
市政委员会 Board of Aldermen
市政卫生 municipal sanitation
市政卫生掩埋场 municipal sanitary landfill
市政污水系统 public off-site sanitary sewage disposal
市政现状图 official map
市政项目 municipal project
市政学 civics
市政业务 municipal service
市政预算 municipal budget
市政再用 municipal reuse
市政债券 municipal bond
市政中心 municipal center[centre]; civic center[centre]
市值 current cost
市制 Chinese system of weights and measures
市中较高处 uptown
市中较高处的 uptown
市中心 central city; city center[centre]; town center[centre]
市中心存(停)车场 central area park
市中心功能 function of civic center [centre]
市中心规划 planning of city center [centre]
市中心交通区 central traffic district
市中心较高处 uptown
市中心区 corduroy city; core city; metropolitan area; urban core; downtown (area); urban center [centre]
市中心区规划 civic centre planning
市中心区重新规划 remodel(l)ing of city center[centre]
市中心商业街 commercial center[centre]
市中心商业区 downtown(area)
市中心停车区 central parking district
市中心外围地区 fringe area
市自治机构 municipal cooperation

示波 oscillography

示波比较法 oscilloscopic comparison
示波测量术 oscillometry
示波的 oscillometric
示波滴定 oscillometric titration
示波法 oscillographic method; oscillography
示波分析器 scope analyser[analyzer]
示波管 oscillogram tube; oscillographic tube; oscilloscope tube; oscillotron
示波积分器 oscillographic integrator
示波极谱法 oscillographic polarography; oscillopolarography; oscilloscopic polarography
示波计 oscillometer
示波记录 electrographic recording
示波记录读码器 oscillograph record reader device
示波记录器 visigraph recorder
示波器 electrograph; hymograph; ondoscope; oscillograph; oscillometer;

S

oscilloscope
示波器测试头 oscilloprobe
示波器打印机 oscillograph printer
示波器带上的记录曲线 oscillograph tracing
示波器迹线 oscillograph trace; oscilloscope trace
示波器记录 oscillograph record(ing); oscilloscope record
示波器记录装置 oscillograph recording system
示波器校正 scope calibration
示波器亮度控制 beam control
示波器描迹 oscilloscope trace
示波器偏转极性 deflection polarity of an oscilloscope
示波器扫描 scope sweep
示波器试验 oscilloscope test
示波器试验主 oscillograph test wagon
示波器水平扫描 scope horizontal sweep
示波器探测器 oscilloscope-detector
示波器探头 oscilloprobe
示波器探头电容 oscilloscope probe capacitance
示波器探针 oscilloscope coupling
示波器显示检测 oscilloscope display test
示波器响应 scope response
示波器响应记录法 response recording
示波器荧光屏 oscilloscope screen
示波器用摄影机 oscillorecord camera; oscilloscope camera
示波器照相 oscilloscope photograph
示波器照相机 oscillograph camera; oscilloscope camera
示波器照像(术) oscillograph photography
示波器振子 oscillograph vibrator
示波色谱法 oscilloscopic chromatography
示波筲 oscillotron
示波术 oscillography; oscilloscopy
示波图 oscillogram; oscillograph trace; oscilloscope pattern; oscilloscope trace
示波图读出器 oscillogram trace reader
示波图记录照相机 oscillorecord camera
示波图摄影机 oscillorecord camera
示波显示 oscillographic display
示波线 depression line; slope line
示波仪 oscillograph; oscillometer; oscilloscope
示差测光 differential photometry
示差测压术 differential manometry
示差的 differential
示差发光二极管指示器 differential light emitting diode indicator
示差分隔间 differential compartment
示差分析 differential analysis
示差光谱 difference spectrum
示差流速计 differential ga(u)ge
示差膨胀仪 differential dilatometer
示差热分析 differential thermal analysis
示差热膨胀计 differential dilatometer
示差扫描热量计 differential scanning calorimeter
示差水头 differential head
示差温度计 differential thermometer
示差压力计 differential manometer
示差液体压力计 differential manometer
示差折光计 differential refractometer
示差折光检测器 differential refraction detector
示差折射计 differential refractometer
示潮器 tidal indicator; tide indicator
示滴仪 disdropmeter
示点器 station pointer
示顶底构造 geopetal structure
示顶底组构 geopetal fabric
示动器滚筒 indicator drum
示读装置 reading device
示度 reading; register
示范 demonstration; show-how <技术、工艺等>
示范表演 demo
示范表演空中喷药 put on a demonstration of aero-spraying
示范布置 typical layout
示范车 demonstration plant
示范城市建设计划 demonstration cities program(me)
示范厨房 demonstration kitchen
示范典型 demonstration model
示范电站 demonstration plant
示范堆 demonstration reactor
示范法规 model statute
示范工厂 demonstration plant
示范工程计划 demonstration project
示范工程(项目) demonstration project
示范工程项目的拨款 demonstration project grant
示范工艺线 modular technologic(al) line
示范馆 demonstration building
示范合同 model contract
示范计划 demonstrative project
示范建设项目的拨款 demonstration project grant
示范建筑 demonstration building
示范建筑规范 model building code
示范(建筑)项目 demonstration project
示范决策分析 parablgm decision analysis
示范林 demonstration forest
示范农场 demonstrate farm
示范区 demonstration area; demonstration plot; pilot area; representative area
示范试件 representative sample
示范试验 demonstration test
示范试验计划 demonstration-cum-trial scheme
示范试验区 pilot-testing project
示范试验项目 pilot-testing project
示范试样 representative sample
示范塘 demonstration pond
示范图片 model photo
示范效果 demonstration effect
示范效益 demonstration effect
示范性试验 demonstration test
示范性住房 model home
示范照片 model photo
示范者 demo; demonstrator
示范装置 demonstration plant; demonstration unit
示范自行车路线 demonstration bicycle route
示范作品 demo
示功卡 indicator card
示功阀 indicator valve; indicator cock
示功器 engine indicator; ergograph; indicator; monograph; power indicator
示功器笔 indicator pencil
示功器传动机构 indicator drive
示功器管件 indicator fitting
示功器驱动装置 indicator driver
示功器塞 indicator plug
示功器弹簧 indicator spring
示功器通道 indicator-channel
示功器旋塞 indicator cock
示功器用纸 indicator paper
示功图 diagram of work; ergogram; indicator card; indicator chart; indicator diagram; indicator diagram of work
示功图因数 diagram factor
示功仪缩动器 indicator reducing motion
示构分析 rational analysis
示构合成 rational synthesis
示构式 rational formula
示号器 annunciator; numerator
示号台 call indicator board
示号信号系统 call indicator system
示滑油流量计 visible oil flow ga(u)ge
示教板 demonstrator
示教方式 teach mode
示教器 demonstrator
示教图 working diagram
示教显示器 instruction display
示教再现式机器人 teaching and playback robot
示教者 demonstrator
示教状态 teach mode
示警灯 warning light
示警浮标 danger buoy
示警气压计 alarm manometer
示警温度汁 alarm thermometer
示警柱 warning post
示警桩 warning post
示警装置 telltale; warning device
示距天体 distance indicator
示廓灯 flank indicator; side indicator
示力涂料 stress coat
示例 instantiation; note example; paradigm
示量等高线 index contour
示零器 null indicator
示流灯 current indicator lamp
示流器 current indicator
示漏气体 probe gas
示漏器 leak indicator
示忙电路 busy indicating circuit
示忙喀音 click
示忙器 busy indicator
示忙闪烁 busy flash
示扭器 torsion indicator
示频率 frequency indicator
示坡线 depression line; fall line in terrain; slope hachure
示热油漆 thermoindicator paint
示色温度 colo(u)r temperature
示深管 <测探仪的> chemical pipe
示声波器 acoustic(al) oscillogram
示湿涂料 humidity-indicating paint
示数管 inditron
示数盘 dial
示数器 numerator; numeroscope
示数仪表 cyclometer
示速计 speed indicator
示速器 speed indicator; stroboscope
示速悬球 speed ball
示酸色 acid colo(u)r
示图 pictorial view
示威游行 demo
示位标 position indicating mark
示位标识别数据 beacon identification data
示位灯 clearance light
示位器 position(al) indicator
示位信标 <机场> location beacon; locator
示温棒 thermoscopic bar
示温法 tempil pellets
示温片 chameleon thermometer
示温漆 heat indicating paint; heat-sensitive paint; temperature indicating paint; temperature sensitive paint; thermal paint; thermindex; thermoindicator paint
示温器 temperature indicator
示温熔锥 fusible cone; pyrometric cone
示温熔锥比值 pyrometric cone equivalent
示温熔锥当量 pyrometric cone equivalent
示温熔锥估价 pyrometric cone evaluation
示温色 thermocolo(u)r
示温涂层 temperature indicating coating
示温涂料 chameleon paint; colo(u)r-changing temperature indicating paint; heat indicating paint; temperature indicating paint; temperature sensitive paint; thermindex; thermocolo(u)r; thermoindicator paint; thermopaint
示温物质 heat-sensitive material
示温颜料 heat indicating pigment; temperature indicating paint; temperature indicating pigment; thermocolo(u)r
示误三角形 triangle of doubt; triangle of error
示误三角形定点法 triangle-of-error method
示误三角形后方交会法 resection by inverted triangle of error
示闲灯 idle-indicating lamp
示闲灯电路 free lamp circuit
示闲信号 idle indicating signal
示向模型 pilot model
示向器 direction indicating device
示向信号 directional signal
示像改变 aspect change; aspect modification
示像回声测深仪 videograph echo sounder
示像名称 aspect name
示性分析 rational analysis
示性函数 indicative function
示性曲面 characteristic surface
示性式【化】 rational formula
示性运动 indicial motion
示性指标 characterisation index
示性指示 characteristic index
示序标志 geopetal criterion
示序组构 geopetal fabric
示压螺栓 load-indicating bolt
示压温度计 pressure ga(u)ge thermometer
示意布局 diagrammatic layout
示意布置图 diagrammatic arrangement
示意草图 sketch map; sketch plan
示意的 schematic
示意断面图 constructed profile; schematic section
示意符号 coding
示意概略 schematic summary
示意平面图 diagrammatic plan
示意剖面图 constructed profile; schematic cross-section
示意图 schematic drawing [illustration/map/representation/diagram]; abridged general view; conventional diagram; delineation; diagrammatic chart [drawing/illustration/sketch/view]; freehand drawing; illustrative diagram; outline; pictorial diagram; presentation drawing; rough draft [drawing/sketch]; schemation diagram; scheme; simplified version; skeleton drawing (map); sketch drawing(map)
示意图粗线 heavy line
示意图实线 heavy line
示意性布置 diagrammatic arrangement
示意摘要 schematic summary
示油规 sight oil indicator

示油器 oil indicator
示园 forest garden
示振计 vibrating meter; vibrograph; vibroscope
示振器 kaleidophone; vibrating meter; vibrograph; vibroscope
示振仪 vibrating meter; vibrograph; vibroscope
示值 indicating value
示值读数 readout
示值读数装置 reading off device
示值范围 indication range
示值分布 spread of indications
示值精度 indicating accuracy
示值偏差 deviation of reading
示值误差 indicating error; indication error
示职叙词 docuterm
示质路基 sandy subgrade
示重计 weight indicator
示重图表 weight indicator chart
示踪 label(l)ing; probe; tagging; tracing
示踪标志 tracer label
示踪玻璃 tracer glass
示踪材料 tracer material
示踪测定 tracer determination
示踪测流法 tracer flow method; water tagging; water tracing
示踪程序 trace program(me); tracer
示踪钉 spike
示踪法 tagging method; tracer method; tracer technique
示踪法试验 tracer test(ing)
示踪分析 tracer analysis
示踪分子 tagged molecule; tracer molecule
示踪化合物 tracer compound
示踪化学 tracer chemistry
示踪畸变 tracing distortion
示踪技术 tracer technique; tracing technique
示踪剂 tracer(material)
示踪剂本底值 background value of tracer
示踪剂测流法 tracer flow method
示踪剂到达时间 time of tracer arrival
示踪剂的弥散角 dispersive angle of tracer
示踪剂毒性 poison of tracer
示踪剂法 method of tracer
示踪剂放射性强度 radioactive intensity of tracer
示踪剂峰值浓度 peak concentration of tracer
示踪剂价格 cost of tracer
示踪剂检出灵敏度 sensitivity of tracer detection
示踪剂接收探头 tracer receive probe
示踪剂量 tracer dose
示踪剂脉动法 tracer pulse technique
示踪剂浓度 concentration of tracer
示踪剂浓度变化过程曲线 graph of tracer concentration change
示踪剂浓度稀释速度 velocity of concentration dilution of tracer
示踪剂曲线 tracer curve
示踪剂试验 tracer experiment
示踪剂随水运移性能 function of tracer migration with water
示踪剂投放方法 method of tracer putting in
示踪剂投放含水层岩性 lithologic(al) characters of aquifer where tracer is put into
示踪剂投放器 tracer releaser
示踪剂投放时间 time of tracer drop
示踪剂稳定性 stability of tracer
示踪剂稀释技术 tracer dilution technique
示踪剂用量 quantity of tracer used
示踪剂种类 type of tracer
示踪剂注入量 quantity of tracer injection
示踪剂注入深度 depth of tracer injection
示踪剂注入速度 velocity of tracer injection
示踪颗粒 label(l)ed particle; tracer grain
示踪扩散实验 tracer atmospheric diffusion experiment; tracer experiment
示踪扩散系数 tracer diffusion coefficient
示踪量 trace level; trace quantity
示踪量萃取 tracer-scale extraction
示踪量级 tracer level
示踪量级放射性 tracer-level activity
示踪能力 traceability
示踪气体 search gas; tracer gas
示踪气体测定仪 tracer gas instrument
示踪气体法 tracer gas technique
示踪器 tracer
示踪染料 tracer dye
示踪溶液 spiked solution; tracer solution
示踪沙 sediment tracer
示踪实验 tracer experiment
示踪实验室 tracer laboratory
示踪试验 tracer experiment; tracing experiment
示踪试验法 method of tracing test
示踪探测器 tracer detector
示踪碳 carbon tracer
示踪同位素 label(l)ed isotope; spiked isotope; tracer isotope
示踪物标 tracer-label(l)ing
示踪物的放射性 tracer activity
示踪物量 tracer scale
示踪物实验室 tracerlab
示踪物试验 tracer experiment
示踪物质 probe material; tracer material; tracer
示踪物置换法 tracer-displacement technique
示踪物稀释测流量法 dilution method
示踪细菌 bacteria(l) tracer
示踪元素 indicator element; tracer; tracer element
示踪元素法 tracer method
示踪元素试验 tracer test(ing)
示踪元素研究 tracer study
示踪原子 chemical tracer; isotopic tracer; label(l)ed atom; radioactive tracer; tagged atom; tracer atom
示踪原子法 label(l)ed atom method
示踪原子分析 tracer atom analysis
示踪原子扩散 tracer diffusion
示踪装置 tracer
示踪装置加热 tracer heating

式符 formal

式样 format; hue; mode; model(l)ing; style
式样的 modal
式样的纯净 purity of style
式样法 model method
式样翻新成本 retrofit cost
式样过时的 outmoded
式样开发 stylistic development
式样特征 stylistic feature

事故 accident; breakdown; catastrophe; distress; emergence; emergency; incident; injury cases; malfunction; mishap; out-of-order

事故安全阀 guard valve
事故按钮 emergency button
事故棒<原子堆> emergency rod; safety element; safety member
事故保护 safe control
事故保护定值器 trip setting
事故保护停堆 reactor trip
事故保险 accident insurance; insurance against accident
事故保险阀 guard valve
事故报废 retirement through accident
事故报告 accident report; reporting of accidents; reporting of an accident
事故报警信号 accident warning signal; emergency alarm signal
事故备用 emergency duty; incidental reserve
事故备用泵 emergency pump
事故备用灯 emergency lamp
事故备用电源 emergency auxiliary power
事故备用电站 emergency power plant
事故备用发电机 emergency generator
事故备用过滤器 emergency filter
事故备用容量 emergency capacity
事故备用设备 casualty spare
事故备用线路 emergency line
事故备用蓄电池室 emergency battery room
事故表示灯 accident indication lamp; accident indication light; emergency light
事故补偿金 hazard bonus
事故操纵 emergency control
事故插座 emergency consent
事故厂用变压器 emergency station service transformer
事故出口 emergency door; emergency exit
事故储罐 emergency tank
事故处理 accident disposal; accident treatment; settlement of accident; trouble removal
事故处理工具【岩】 fishing tool
事故处理一般工具 general fishing tools
事故处理专用工具 special finishing tools
事故的危险 accident hazard
事故的应变计划 planning for incidents
事故灯 obstruction lamp
事故登记簿<美> blotter
事故地点 accident location; accident spot
事故地点发射机 emergency locator transmitter
事故地点图 accident spot map
事故电流 fault current
事故电梯 emergency lift
事故电源 emergency power; emergency power source; emergency power supply
事故电源线 emergency feed line
事故电源装置 emergency power supply unit
事故调查 accident investigation; accident-prone investigation; accident survey(ing); damage survey
事故叠梁闸门 emergency stoplog
事故断电时间 break time on fault
事故多的 accident-prone
事故多的地点 black spot
事故多段 accident-prone section
事故多发的 accident-prone
事故多发地段【交】 high-accident location; trouble spot
事故多发点 black spot
事故多发性 accident proneness
事故发生 accidental occurrence
事故发生地点 accident site
事故发生地点平面图 plan of the accident site
事故发生率 accident toll; accident rate
事故发生时间 troubles happening time
事故发生原因及规程 accident-cause code
事故阀 emergency valve; guard valve
事故防护 accident protection; safety control
事故防护安全限度 safety trip level
事故防护程度 safety trip level
事故防护制动 safety control brake application
事故防止 accident prevention
事故防止条款<海上保险> sue and labo(u)r clause
事故放水口 emergency relief
事故放油阀 dump valve
事故费用 accident cost
事故分级<坝的> hazard rating
事故分析 accident analysis; fault analysis
事故分析报告 failure analysis report
事故分析图 accident analysis diagram; collision diagram; collision strut; analysis diagram
事故风扇 emergency fan
事故风险 accident risk
事故负伤 accidental injury
事故感受器 emergency cut-off receiver
事故给水 emergency water supply
事故工况 emergency condition
事故供电网 emergency network
事故关闭 emergency shut-off; emergency closure
事故关闭设备 emergency closure device
事故关闭装置 emergency closing device; emergency closure device
事故管理 accident management
事故管理系统 accident management system; risk management system
事故过电压 abnormal overvoltage
事故荷载 contingency loading; accidental load
事故黑点 accident black spots
事故后 post-accident
事故后的紧急服务 post-accident emergency service
事故后果 fallout
事故后火灾 post-accident fire
事故滑动门 emergency slide gate
事故及健康保障计划 accident and health plan
事故及停待时间 trouble and waiting time
事故及医疗保险 trouble and health insurance
事故记分 accident points
事故记录 accident record
事故记录卡片 trouble chart
事故继电器 fault relay
事故监督 fault control
事故减压法<出高压沉箱的方法> decanting
事故检测 accident detection; incident detection
事故检查 damage survey; incident detection
事故检修 accident maintenance; emergency repair
事故鉴定 accident survey(ing); damage survey
事故降低率 accident reductive rate
事故接线 emergency connection

S

事故解案时间 troubles free time
事故紧急关闭 emergency shut-off
事故警报 accidental alarm
事故警告标志 incident warning sign
事故警告灯 emergency warning lamp
事故救护车 crash truck
事故救护医院 accident hospital
事故救护站 accident ambulance station
事故救援 accident rescue
事故开关 emergency switch
事故可能性 accident probability
事故控制 accident control;emergency control;fault control
事故控制对比 accident control ratio
事故控制率 accident control ratio
事故类型 accident pattern
事故冷却器 emergency cooler
事故历史 case history
事故例数 case-load
事故率 failure rate;fatality rate;outage;possibility of trouble
事故门槽 outer stop
事故密度 density of accident
事故苗子 accident exposure;dangerous accident
事故苗子指数法 accident exposure index method
事故敏感 accident-prone
事故磨铣 emergency milling-out;emergency milling-up
事故凝汽器 dump condenser
事故(排)出口 emergency outlet
事故排放 accidental discharge
事故排放阀 dump valve
事故排放口 emergency outlet
事故排风 emergency air exhaust
事故排气 emergency exhaust
事故排汽 emergency dump steam;emergency exhaust
事故判定【交】 accident conviction
事故旁路 emergency by-pass
事故旁通 emergency by-pass
事故赔偿 accident indemnity;indemnity of accident
事故配电盘 emergency panel
事故喷淋器 emergency shower
事故频率 accident frequency
事故起重车 breakdown crane wagon
事故起重机 <铁路的> breakdown crane
事故前预案 preemergency planning
事故抢救 first aid
事故切断 emergency switching-off
事故倾向性 accident proneness
事故情况 accident condition
事故区 emergency area
事故日期 casualty date
事故伤亡图 casualty map
事故失效 accident failure
事故时高照射 emergency high exposure
事故时间 accident time
事故手动装置 emergency hand-drive
事故树 event tree;fault tree
事故树分析 fault tree analysis
事故甩负荷 emergency load dump
事故水 water used for emergency case
事故死亡 accidental death
事故死亡保险 accident death insurance
事故死亡保险赔偿费 accidental death benefit
事故死亡率 accident(al) death rate;accident involvement rate;death rate of accident
事故死亡人数 accident fatalities
事故速报 accident circular telegram
事故损坏 accident damage;accident defect;accident fault;damage caused by accident
事故损失 accident toll
事故损失费 accident cost
事故损失时间 troubles waste time
事故索赔表 occurrence form
事故探测 incident detection
事故条件分析图 condition diagram
事故跳闸电气系统 emergency trip wire system
事故停产时间 break time on fault
事故停车 disastrous shutdown;emergency shutdown
事故停车道 emergency stopping lane
事故停电 forced outage;unscheduled interruption
事故停堆 breakdown
事故停堆棒 emergency shutdown member
事故停工 accidental shutdown
事故停工时间 mechanical downtime
事故停机 emergency outage;emergency shutdown;forced outage
事故停机监察器 shut-down monitor
事故停机率 forced outage rate
事故停运 forced shutdown
事故通风 emergency ventilation
事故通风井 emergency ventilation shaft
事故通风系统 emergency ventilation system
事故通信[讯] emergency communication
事故统计 accident record
事故统计学 accident statistics
事故统计资料 accident statistics
事故图 accident map
事故图表 trouble chart
事故危险(性) accident risk;risk of failure
事故维修 accidental maintenance;breakdown maintenance;emergency maintenance
事故现场 emergency scene;scene of accident
事故现场情况图 accident map
事故消息 exception message
事故效应 damaging effect
事故泄放 accidental discharge
事故心理【交】 accident psychology
事故信号 accident signal;emergency signal;fault signal(l)ing;obstruction signal;trouble back signal
事故信号灯 emergency light
事故信号发送器 emergency signal transmitter
事故信号系统 alarm signal system
事故信号装置 emergency signal(l)ing device
事故性辐照 accidental irradiation
事故性接触 accidental contact
事故性排放 accidental release
事故性神经病 accident neurosis
事故性水污染风险 accident water environment risk
事故性停工时间 outage time
事故性污染 accident(al) contamination
事故性溢漏 accidental spill
事故性质 character of accident;failure property
事故性中毒 unintentional poisoning
事故性转储 disaster dump
事故性转贮 disaster dump
事故修理 damage repair;emergency repair
事故讯号装置 emergency signal(l)ing device
事故严重程度 severity of injuries
事故严重度 <死伤与非死伤事故之比> accident involvement severity
事故严重度指数 severity index
事故严重率 <人身伤亡的> accident severity rate
事故严重性 accident involvement severity
事故易发生 liability to accidents
事故溢洪道 emergency spillway
事故因素 contingency factor
事故引导按钮 emergency call-on button
事故隐患 accident threat
事故应急泵 emergency pump
事故用水池 emergency pit
事故有形损伤 accidental bodily injury
事故与工具 troubles and tool
事故预测 accident forecast
事故预防 accident averting;accident prevention;accident protection;prevention of accidents
事故预防规程 accident prevention regulation
事故预防计划 accident prevention program(me)
事故预防员 accident prevention officer
事故原因 cause of accident;offender;source of damage
事故运行 accidental operation
事故运行方式 accidental operation mode
事故责任 accident liability;accident risk;liability to accidents
事故责任局 administration at fault
事故闸 emergency closure;emergency lock;escape lock
事故闸门 emergency(closure)gate;stop gate;guard gate
事故闸门槽 emergency gate slot
事故闸门关闭 emergency gate cut-off
事故摘要 incident summary
事故照明 emergency lighting
事故照明灯 safety lamp
事故照明配电设备 emergency lighting distribution equipment
事故照明配电箱 emergency lighting distribution box
事故照明器 emergency light
事故照明切换盘 accident lighting transfer board
事故照明线路 emergency lighting wiring
事故照射 accident exposure
事故征候 accident-cause factor;accident proneness;potential accident cause
事故支出 accident cost
事故指挥中心 emergency command center[centre]
事故指示信号 trouble blinking
事故指示装置 alarm annunciator
事故制动器 emergency brake
事故中子剂量测定法 neutron accident dosimetry
事故种类 type of trouble
事故贮油设施 emergency oil storage installation
事故转换设备 emergency throw-over equipment
事故状态 accidental state;emergency condition
事故状态稳定性 stability under accident conditions
事故资料 accident data
事故资料记录器 accident data recorder
事故自动检测系统 automatic incident detecting system;automatic incident detection system
事故自动切断阀 self-closing valve
事故自动刹车装置 deadman device
事故阻塞 accident congestion;accident jam
事后报关 post entry
事后编辑 post edit(ing)
事后标定模式 post-calibration mode
事后补正 nunc pro tunc
事后测定 after-only measurement
事后偿付 ex post payoff
事后成效 post result
事后处理程序 post-processor
事后的 postmortem
事后的认识 hindsight
事后的思考 after-thought
事后对策的 remedial
事后对策管理 remedial management
事后分析 after-the-fact analysis;back analysis;ex post analysis;postmortem
事后付款系统 post billing system
事后概率 a posteriori probability
事后概率法则 posterior probability law
事后估计 a posteriori estimate
事后估值 a posteriori estimate
事后计算 ex post calculation
事后监督 subsequent supervision;supervision afterwards
事后检查 postmortem
事后校正模式 post-calibration mode
事后模拟 ex post simulation
事后剖析 postmortem
事后剖析程序 postmortem routine
事后设计 <企业管理> after-only design
事后审计 post audit
事后试验 post-testing
事后适应 postadaptation
事后投资 ex post investment
事后维修 correction maintenance
事后预测 ex post forecast
事后预测误差 ex post forecast error
事绩 deed
事件 case;circumstance;episode;event;incident;occurrence;transaction
事件报告 event report;incident report
事件报告系统 event reporting system
事件编写 event numbering
事件变量 event variable
事件标记 event flag
事件标记笔 event marker
事件标记组 event flag cluster
事件标志 event flag
事件标志束 event flag cluster
事件表 event list
事件表达 event representation
事件不出现 <概率论> event's failing
事件不可简化 case irreducibility
事件层理 event stratification
事件产生过程 event-generating process
事件程序 event routine
事件出现 event's happening
事件处理 event processing
事件存储与分配部件 event storage and distribution unit
事件大小 event size
事件代数 event algebra
事件的必然 logic(al) of events
事件的高潮 climax
事件的交换性 event commutativity
事件的最早发生时间 earliest event occurrence time
事件登记 event posting
事件地层学 event stratigraphy
事件调度程序 event schedule
事件队列 event queue
事件分时控制 event-spaced time control
事件分析 accident analysis

事件概率 probability of happening; probability of occurrence
事件概率回归 event probability regression
事件概率回归估计 regression estimation of event probability
事件感知卡 event-sensing card
事件跟踪程序 event tracer
事件规模 event size
事件号 incident number
事件号码 event number
事件计数器 event counter
事件记录 event record; incident record; logging; logout
事件记录器 event recorder
事件记录系统 event recording system
事件记入 event posting
事件监督 event monitoring
事件检索 fact retrieval
事件建立 event establishment
事件节点网 event node network
事件控制块 event control block
事件控制器 event controller
事件控制用信息组 block event control
事件类型 incident type
事件链 chain of events; event chain
事件流 flow of event
事件率 incident rate
事件描述 incident description
事件名 event name
事件内函数 event built-in function
事件配置 event posting
事件频度 event frequency
事件频率 event frequency
事件平均浓度 event mean concentration
事件驱动 event-driven
事件驱动系统 event driven system
事件驱动执行程序 event driven executive
事件任选 event option
事件任选项 event option
事件扫描 event scan(ning)
事件扫描机构 event-scan(ning) mechanism
事件识别时间 event identification time
事件树 event tree
事件数据 event data
事件水平线 event horizon
事件松弛时间 event slack time
事件探测器 event detector
事件同步 event synchronization
事件同时性 event synchronization
事件伪变量 event pseudo-variable
事件位置 event location
事件文件 incident file
事件序列 sequence of events
事件循环 cycle of event
事件延迟 event delay
事件属性 event attribute
事件自报式系统 event reporting system
事件最早时间 earliest event time
事界 event horizon
事例 case history; instance
事例调查 case survey
事例研究 case study
事前 beforehand
事前分布 a priori distribution
事前分析 ex ante analysis; preanalysis
事前概率 a priori probability; theoretic(al) probability
事前估计(值) advance estimate; a prior estimate
事前计划 pre-incident plan
事前技术调研 prior-art search
事前监督 supervision in advance
事前警告 advance warning
事前审计 preaudit
事前误差界 a priori error bound
事前预测 ex ante forecast(ing)
事前预防管理 preventive management
事前折旧利润 predepreciation profit
事前准备 advanced preparation
事前资料 preliminary information
事实 deed
事实报告 factual report
事实陈述书 factum[复 facta/factums]
事实的证据 factual evidence
事实记录 statement of facts
事实检索 fact retrieval
事实检索系统 fact retrieval system
事实前提 factual premise
事实人口 de-facto population
事实上 actually; as a matter of fact; in actual fact
事实上的承兑 virtual acceptance
事实上的承认 de facto recognition
事实上接受 virtual acceptance
事实相关 tact correlation
事实信息 factual information
事实真相的系统记录 reporting
事态 condition of affairs; conjuncture; course of things; state of affairs; status of affairs
事务 affair; business; office work; pursuit
事务标识器 transaction identifier
事务部 purser department
事务部门 administration section; back office
事务策略 business game
事务长【船】 purser
事务长室【船】 purser room
事务处 business office
事务处理 transaction
事务处理程序 transaction processing program(me); transaction program(me)
事务处理机 business machine
事务处理机定时 business machine clocking
事务处理记录 transaction logging
事务处理记录标题 transaction record header
事务处理命令保密性 transaction command security
事务处理数据组记录块 transaction data set record block
事务处理系统 transacter; transaction processing system
事务处理业务 transaction service
事务代码 transaction code
事务带 transaction tape
事务对策 business game
事务费(用) clerical cost; office expenses
事务工作自动化 office automa(tiza)tion
事务官 commissioner
事务管理 office management
事务管理的数据处理 administrative data processing
事务管理系统 transaction management system
事务荷载平衡 transaction load balancing
事务计算机 office machine
事务记录 transaction journal; transaction log; transaction record
事务记录号 transaction record number
事务卡 transaction card
事务控制程序 transaction control program(me)
事务控制系统 transaction control system
事务清单 transaction listing
事务确认 transaction validation
事务日志 transaction journal
事务室 clerk's office
事务数据 transaction data
事务数据处理 business data processing
事务数据集 transaction data set
事务所 business premises; commercial premises; office
事务网络业务 transaction network service
事务文件 detail file; transaction file
事务系列 flow
事务显示 transaction display
事务性的 clerical
事务性工作 clerical operation; desk-side exercise
事务用计算机 business machine; office machine
事务用计算机时钟 business machine clocking
事务员 clerical staff; office clerk
事务终端系统 administrative terminal system
事务主义者 routineer
事务自动化 business automation
事物的内部规律 inherent laws of things
事物的主体 corpus[复 corpora]
事物管理 regulation of affairs
事物管理机 business machine
事物两面的较显著一面 obverse
事先安排的预算 planned budget
事先编辑 preediting
事先测试 before test
事先偿付 ex ante payoff
事先承兑 preacceptance
事先磋商 preliminary consultation; prior consultation
事先防止损失 take measures to prevent losses in advance
事先估定 prior estimation
事先估计 prior estimation
事先估计值 advanced estimate
事先计划好的机动性 plan-ahead flexibility
事先监督 supervision in advance
事先批准 prior approval
事先批准的分包人 pre-approved subcontractor
事先平衡 planned balance
事先审查 precensor
事先试验 pretest(ing)
事先通知制度 prior informed consent procedure
事先宣布 preannounce
事先选定的噪声标准曲线 preferred noise criterion curve
事先知情同意 prior informed consent
事先准备 advance preparation
事先作出(想法) preconceive
事项 particulars; proceedings; transaction
事项标准差离 <统筹方法中关于事项预计日期散布度的一个度量> standard deviation of an event
事项处理 transaction
事项处理程序 transaction program(me)
事项处理服务 transaction service
事项处理服务程序 transaction service
事项(处理)记录 transaction record
事项(处理)记录头标 transaction record header
事项处理均衡装入 transaction load balancing
事项处理码 transaction code
事项(处理)命令安全性 transaction command security
事项处理日志 transaction journal
事项处理显示 transaction display
事项处理选择表 transaction selection menu
事项处理重新启动 transaction restart
事项登记 transaction log
事项复原 transaction backout
事项检索 fact retrieval
事项进度表 event schedule
事项码 transaction code
事项情报检索系统 factorial information retrieval system
事项数据集 transaction data set
事项数据集记录块 transaction data set record block
事业 cause; enterprise; task; undertaking
事业单位 institution
事业费 funds for public undertaking; operation expenses; undertaking expenditures
事业机构 institutional organization
事业机构会计 institutional accounting
事业经费预算 budget of undertaking expenditures
事业收入 income from undertakings; undertaking revenue
事业支出 enterprise expenses
事主 client; clientele

势伴流 potential wake; streamline wake

势差 difference of potential; potential difference
势差现象 potentiation
势场 potential field
势的叠加 superposition of potential
势的叠加规则 superposition rule of potential
势电极 potential electrode
势电解质 potential electrolyte
势陡坡 potential gradient
势分布 potential distribution
势峰 potential hump
势谷 potential trough
势函数 potential function
势降 potential drop
势阱 potential barrier; potential hole; potential well
势垒 barrier potential; potential barrier; potential hill
势垒层 barrier region
势垒场 barrier field
势垒穿透 barrier penetration
势垒电容 barrier capacitance
势垒高度 barrier height
势垒宽度 barrier width; potential barrier width
势垒能 barrier energy
势垒区 barrier region
势垒栅 barrier gate
势垒效应 barrier effect
势力 force; influence; power
势力范围 perisphere; sphere of influence; zone of influence
势流 potential flow; potential motion of a fluid
势论 potential theory
势密度 potential density
势能 energy of position; latent energy; potential energy; static energy
势能槽 potential trough
势能分布 potential energy contribution
势能理论 potential theory
势能曲线 curve of potential energy
势能梯度 gradient of potential energy
势坪 potential plateau

势曲线 power curve
势散射 potential scattering
势水头 potential head
势态分析器 situation analyser[analyzer]
势态控制 situation control
势态识别 situation recognition
势弹性散射 elastic potential scattering
势梯度 potential gradient
势头 position head; potential head; potential water head; momentum [复 momenta]
势透过率 potential transmittance
势位线 potential line
势温度 potential temperature
势吸收率 potential absorptance
势限 potential boundary
势像 potential image
势穴 potential hole
势移 potential shift
势源 potential source
势越二极管 barrier injection and transit time diode

视

视半径 apparent radius; apparent semidiameter
视杯 optic(al)cup
视背景光 bias light compensation
视奔赴点 apparent vertex
视比重 apparent gravity;apparent specific gravity;bulk specific gravity
视比重煤样 coal sample for determination of apparent specific gravity
视比阻 apparent resistivity
视扁率 apparent flattening
视变量 apparent variable
视变数 apparent variable
视标 sighting post;sighting rod;sighting target
视表面 apparent surface
视表速 apparent surface velocity
视柄 optic(al)stalk
视波高 apparent wave height
视不整合 apparent unconformity
视测光度法 sight photometer method
视测积存管颗粒分析器 visual accumulation tube size analyser[analyzer]
视测计数技术 visual counting technique
视测误差【测】observation error
视测仪 scope
视测浊度(测定)法 scopometry
视测浊度计 scopometer;scotometer
视层状 bedways
视察 general view;overlook
视察法 method of inspection
视察飞行 view flying
视察工程 observation of works
视察工作 observation of the work
视察孔 sight
视察人员 inspectorate
视察团 inspectorate
视察现场 inspection of site
视察员 overlooker
视察者 examining officer;inspector;visitor
视差 error in viewing;optic(al)parallax;parallax(effect);parallax error;reading error
视差变换器 parallax converter
视差表 parallactic table;parallax table
视差补偿 parallax compensation
视差补偿机构 parallax offset mechanism
视差补偿器 parallax compensator
视差不等 parallactic inequality;parallax inequality
视差不等潮龄 age of parallax inequality
视差测定 parallax determination
视差测高程 parallax heighting
视差测量 parallactic measurement;parallax measurement
视差测图镜 stereometer
视差差异 parallax difference
视差潮龄 age of parallax tide;parallax age of tide
视差尺 parallactic bar;parallax arm;parallax bar
视差导线 parallactic polygon;subtense traverse
视差的 parallactic
视差等线测量 parallactic polygonometry
视差动 parallactic motion
视差法测距 subtense method distance measurement;taping by parallactic method
视差方程 parallax equation
视差改正 adjustment for parallax;parallactic correction;parallax correction
视差改正量 parallax correction;parallax in altitude
视差杆 parallactic bar;parallax bar;parallax measurer;parallax rod
视差格网 parallactic grid
视差光楔 parallax wedge
视差轨道 parallactic orbit
视差滑尺 parallactic slide;parallax slide
视差环节 scheme of parallactic distance measurement
视差基线 parallactic base;subtense base
视差计 parallaxometer
视差计算 parallax computation
视差角 angle of convergence[convergency];angle of parallax;angular parallax;convergence angle;parallactic angle;subtense angle
视差角(测距)法 subtense method
视差矫正 optic(al)refinement
视差校正 correction for parallax;parallax correction
视差校正发射机 parallax range transmitter
视差校正计算机 parallax computer
视差校正器 parallax offset mechanism
视差校正取景器 parallax correcting finder
视差较【测】parallax difference;differential parallax
视差镜术 parallactoscopy
视差量测杆 parallax bar
视差量测仪 parallactic instrument;parallax instrument;parallax measurer
视差螺旋 parallax screw
视差偏移 parallactic shift
视差偏转 apparent parallactic shift
视差三角形 parallactic triangle
视差式彩色荫罩管 parallax colo(u)r tube
视差数据 parallactic data
视差天平动 parallactic libration
视差调节 parallax adjustment
视差椭圆 parallactic ellipse
视差网 parallactic net
视差网格 parallactic grid
视差位移 parallactic displacement;parallactic shift;parallax displacement
视差误差 parallactic error;parallax error
视差显微镜 parallactic microscope;parallax microscope
视差现象 parallax effect
视差像片 parallactic photograph
视差消除 elimination of parallax
视差效应 parallax effect
视差楔 parallactic wedge;parallax wedge
视差星 parallax star
视差修正 parallactic correction
视差修正瞄准器 no-parallax viewfinder
视差移动 parallactic motion;parallactic movement;parallactic shift
视差因子 parallax factor
视差与蒙气差 parallax and refraction
视差与折光差 parallax and refraction
视差运动 parallactic motion
视差照片 parallactic photograph
视差照相测量术 parallax photogrammetry
视差折光差 parallactic refraction
视差折射计 parallax refractometer
视差周视体视图片 parallax panoramagram
视差自动检测 automatic parallax detection
视场 angular coverage;field coverage;field of view;field of vision;range of response;view(ing)field;visual field;field of view
视场背景 surround of a comparison field
视场表面 apparent surface
视场的明净 clearness of field
视场范围 field range
视场光阑 field diaphragm;field stop
视场角 angle of coverage;angle of view;angle of visual field;angular coverage;field angle;viewing angle;vision angle;visual angle;angular field;covering power <摄影机的>
视场亮度 field luminance
视场深度 depth of field
视场数 field of view number
视场外监听 blind monitoring
视场像 field image
视场仪 field of view meter
视场遮光 field stop;stop field of view
视场直径 field number
视场致平器 field flattener
视场中心的渐晕 vignetting at the centre of the field
视超光速 apparent superluminal velocity
视超光速运动 apparent superluminal motion
视车镜 enoscope
视程 range of visibility;range of vision;visual range
视程圈 circle of visibility
视程障碍 obstruction to vision
视尺寸 apparent size
视赤道坐标 apparent equatorial coordinates
视赤经 apparent right ascension
视赤纬 apparent declination
视充电率 apparent chargeability
视储量 visible reserves
视窗 viewport;window
视窗操作系统【计】Windows
视窗环境 windows environment
视窗随机存取存储器【计】Windows RAM
视垂线 apparent vertical;dynamic(al)vertical
视磁化率 apparent susceptibility
视磁化率计算 computing of apparent susceptibility
视磁化率图 apparent susceptibility chart
视磁化率值 apparent susceptibility value
视错觉 optic(al)illusion;visual illusion
视带 optic(al)zone
视等时年龄 apparent isochron age
视地层常数 apparent formation constant
视地层厚度 apparent bed-thickness
视地层间隔 apparent stratigraphic(al)interval
视地层离距 apparent stratigraphic(al)separation
视地层水电阻率 apparent resistivity of formation water
视地层位移幅度 apparent heave slip
视地层因数 apparent formation factor
视地面速度 apparent surface velocity
视地平 eye-level;natural horizon;sensible horizon;topocentric horizon
视地平俯角 apparent depression of the horizon;dip of the horizon
视地平距离 distance of visibility;distance of visible horizon;radius of visibility
视地平面 local horizon
视地平圈 sensible horizon
视地平纬度 apparent altitude
视地平线 apparent horizon;visible horizon;visual horizon
视地下水流速 apparent groundwater velocity
视地下水面 apparent groundwater table
视点 eye point;perspective center[centre];point of sight;point of vision;reference point;sight point;viewport
视点移动 viewpoint movement
视电荷 apparent charge
视电阻 apparent resistance
视电阻率 apparent resistivity;apparent specific resistance;apparent specific resistivity
视电阻率测井曲线 apparent resistivity log curve
视电阻率测量精度 survey precision of apparent resistivity
视电阻率拟合 comparison in the apparent resistivity
视电阻率曲线 apparent resistivity curve
视电阻率曲线的横向比例尺 horizontal scale of apparent resistivity
视电阻率图 figure of apparent resistivity
视迭覆 apparent superposition
视动 apparent motion;apparent movement
视动反应 optokinetic response;optomotor response
视读数 visual reading
视度 degree of visual;diopter[dioptre];visual degree
视度差 diopter difference
视度分划圈 diopter dividing ring
视度计 dioptrometer
视度圈 diopter ring
视度调节 diopter regulation;visual accommodation
视度筒 diopter cylinder
视度透镜 diopter lens
视度系数 visibility factor
视度仪 visibility meter
视度值 diopter value

视度转螺 diopter rotating screw
视断层活动 apparent movement of faults
视断距 apparent fault displacement; apparent slip
视发水雷 observation mine
视翻正反射 optic(al) righting reflex; visual-righting reflex
视方位 aspect
视方位角 apparent azimuth(angle)
视方向 apparent direction
视仿红星等 apparent photo-red magnitude
视仿视星等 apparent photovisual magnitude
视放大率 visual amplification
视丰度 apparent abundance
视风 apparent wind; relative wind
视风向 apparent direction of wind
视辐射点 apparent radiant
视辐射星等 apparent rediometric magnitude
视辐线 optic(al) radiation
视干线 <透视图上的> eye line
视杆 retinal rod
视感测色 visual colo(u)rimetry
视感控器 perception
视感控制器 perceptron
视高程 apparent altitude; apparent height; eye-level; rectified altitude
视高度 apparent altitude; apparent height; eye-level; rectified altitude
视高改正 height of eye correction
视高校正 height of eye correction
视公转 apparent revolution
视功率 apparent power
视光标志 optic(al) signal
视光点直径 apparent spot diameter
视光诱导 optic(al) guidance
视光轴角 apparent optic(al) axial angle
视规 visual ga(u)ge
视轨道 apparent orbit; apparent path
视海岸线 apparent coastline; apparent shoreline
视海底 apparent bottom
视航向 apparent course
视荷载 apparent load
视恒星时 apparent sidereal-time
视横断距 apparent heave
视红星等 apparent red magnitude
视厚度 apparent thickness; apparent width <地层的>
视厚聚焦 visual focusing
视滑距 apparent slip
视滑脱 apparent slip
视环形火山 apparent crater
视幻觉 visual hallucination
视黄基 retinyl
视黄经 apparent longitude
视黄纬 apparent latitude
视活化能 pseudo-activation energy
视基板 optic(al) placode; optic(al) plate
视激电率 apparent polarization
视级 scale of visibility; visibility scale
视极化率 apparent chargeability
视极性 apparent polarity
视极移 apparent polar-wandering
视极移轨迹 apparent polar-wander-(ing) path
视极移曲线 apparent polar-wander-(ing) path
视极移速率 rate of apparent polar-wandering
视见变换 viewing transformation
视见参考点 view reference point
视见度 visibility
视见函数 visibility function
视见函数修正玻璃 visibility function modification glass
视见平截头体 viewing frustum
视见区 <图形的> viewport
视见曲线 visibility curve
视见限度 limit of velocity
视见因数 visibility factor
视见约束体 view volume
视交叉 optic(al) chiasma
视交叉的 opticochiasmatic
视焦(点) apparent focus
视角 angle of sight; angle of view; angle of vision; apparent angle; look-(out) angle; sight angle; view angle; vision angle; visual angle
视角板 sight angles board
视角场 angular field
视角法 visual angle method
视角范围 angular field; angular field of view
视角面 ocular surface
视角因数 view-factor
视角阈值 angular threshold of eye
视界 angular coverage; angular field; coverage; field coverage; field of view; field of vision; purview; range of sight; range of visibility; range of vision; scope; sight; sight distance; viewing field; visibility; visual field
视界半径 radius of visibility
视界大的 pancreatic; panoramic
视界大的透镜 pantoscope
视界对准 visual alignment
视界范围内的远处海面 offing
视界弧 arc of visibility; sector
视界角度 angle of visibility; aspect angle; visual angle
视界校准误差 field alignment error
视界宽度 width of coverage
视界内的远处海面 offing
视界平淡 flatness of field
视界缩小 contract horizon
视界条件 visibility condition
视界图 visibility chart
视界效应 horizon effect
视界(以)外 out of sight
视界障碍 obstacle to visibility
视金属因素 apparent metal factor
视近点角 apparent anomaly
视进动 apparent precession; apparent wander
视井径 apparent wellbore radius
视井径比 apparent wellbore ratio
视景线 vista
视径 visual path
视镜 sight glass; sight level glass; viewing mirror
视镜护罩 sight glass shield
视距 line-of-sight coverage; range of visibility; seeing distance; sight distance; sight length; stadia distance; view(ing) distance; vision clearance; visual distance; visual range; apparent distance
视距比 viewing ratio
视距标尺 distance ga(u)ge; sight rod; stadia scale; tachymeter staff
视距标尺截距 stadia intercept
视距标杆 stadia rod
视距表 stadia table
视距不良的弯道 concealed bend
视距测高 stadia level(l)ing
视距测距术 tach(e)ometric(al) surveying; tach(e)ometry
视距测距仪 stadia (metric) range finder
视距测距指示器 stadiametric range indicator
视距测量 stadia measurement; stadia shot; tach(e)ometer measurement; tach(e)ometer shot; tach(e)ometric(al); survey(ing)
视距测量法 stadia method; tach(e)ometric(al) surveying; stadia; stadia survey(ing); tach(e)ometry
视距测量工作 stadia work
视距测量光楔 distance measuring wedge
视距测量手簿 tach(e)ometric(al) book; tach(e)ometric(al) record book
视距测量术 tach(e)ometric(al) surveying
视距测量学 tach(e)ometry; tachymetry
视距测量照相机 stadiametric camera
视距测平 stadia level(l)ing
视距测站 stadia station
视距常数 stadia constant; stadia factor
视距乘常数 stadia multiplication constant; stadia ratio
视距尺 flag pole; range pole; range rod; ranging pole; ranging rod; stadia(rod)
视距尺截距 stadia intercept
视距传播 line of light propagating
视距导线 stadia traverse; tacheometer traverse; tach(e)ometric(al) polygon; tach(e)ometric(al) traverse; tachometric(al) polygon
视距点 stadia point; stadia station; tach(e)ometric(al) station
视距读数 stadia reading
视距多边形 tach(e)ometric(al) polygon
视距法 stadia method; subtense method; subtense technique; tach(e)ometric(al) method; tachymetry
视距法测距 taping by tach(e)ometric(al) method
视距改正 stadia reduction
视距改正表 table of stadia reduction
视距改正图 stadia reduction diagram
视距公式 stadia formula
视距光楔 stadia wedge; tach(e)ometric(al) prism attachment
视距弧 stadia arc; stadia circle
视距换算表 table of stadia reduction
视距计 stadimeter
视距计算 stadia computation
视距计算表 stadia table; tach(e)ometer table; tach(e)ometric(al) table; tachymetric(al) table
视距计算尺 stadia slide ruler; tach-(e)ometric(al) ruler
视距计算机 stadia computer
视距计算盘 stadia computing disk [disc]
视距计算器 stadia computer
视距记录簿 tachymetric(al) (record) book
视距加常数 additive constant; stadia additive constant
视距间隔 stadia interval
视距间距 stadia interval
视距校正图表 stadia correction diagram
视距经纬仪 distance measuring theodolite; stadia theodolite; stadia transit; tach(e)ometer theodolite; tachymeter-transit
视距经纬仪测量 stadia transit survey; transit-and-stadia survey
视距经纬仪导线测量手薄 transit stadia traverse field book
视距镜 anallatic lens
视距快速测定术 tach(e)ometry
视距离 anallatic distance; apparent distance
视距离模数 apparent modulus
视距量角器 tach(e)ometric(al) protractor
视距列线图 tach(e)ometric(al) nomogram
视距瞄准 tach(e)ometer shot
视距诺谟图 tach(e)ometric(al) nomograph
视距平板仪测量 stadia plane-table survey
视距三角 <交叉口> sight triangle
视距三角线 clear sight triangle
视距手提水准仪 stadia hand level
视距水平 stadia level(l)ing
视距水准测量 stadia level(l)ing
视距水准仪 stadia level; tach(e)ometric(al) level
视距丝 stadia hairs; stadia lines; stadia reticule; stadia wire
视距丝乘常数 stadia lines multiplication factor
视距丝环 stadia wire ring
视距丝式经纬仪 transit with stadia wires; theodolite with stadia wires
视距通信[讯] horizon communication
视距通信[讯]系统 line of sight
视距图表 stadia diagram
视距外通信[讯]方式 over-the-horizon system
视距外无线电通信[讯] radio communication beyond the horizon
视距望远镜 anallatic telescope; stadia telescope; stadimetric telescope
视距微波无线中继站(系统) line-of-sight microwave radio relay station
视距误差 defect in vision
视距线 stadia hairs; stadia lines; stadia point
视距信号 visual signal
视距信号接收 visual distance reception
视距仪 apomecometer; cross-wire meter; stadia; stadia range finder; stadi(o)meter; tach(e)ometer; tachymeter
视距仪标尺 tach(e)ometer staff
视距仪测量 tachymeter surveying
视距仪分度器 tach(e)ometric(al) protractor
视距仪制动装置 tach(e)ometer stop gear
视距因数 stadia factor
视距照准仪 tach(e)ometric(al) alidade; tachometric(al) alidad(e); tachymetric(al) alidade
视距准尺 stadia
视觉 sight-shot; visible sensation; visual perception
视觉报警 visual alarm
视觉比较 visual comparison
视觉边限 peripheral vision
视觉变量 visual variable
视觉辨别 visual discrimination
视觉标示 visual indication
视觉表示装置 visual indication device
视觉不稳定 visual instability
视觉测光 visual photometry
视觉持续时间 duration of vision
视觉尺寸 visual size
视觉刺激物 visual stimuli
视觉的 ocular
视觉读数 visible reading; visual reading
视觉对比 visual contrast
视觉对象设计 <指图表、阵列、标志、包装等的设计> visual design
视觉二重性理论 duplicity theory of vision
视觉法 visual method

S

视觉反馈 visual feedback
视觉反射率 visual reflection factor
视觉反射系数 visual reflection factor
视觉反应 visual impact; visual response
视觉分辨理 visual resolving power
视觉分辨率 visual resolution
视觉分析 <街景设计的> visual analysis
视觉改正 optic(al)correction
视觉感 visual sensation
视觉功能 visual performance
视觉过程 visual process
视觉航标 visual aids; visual beacon; visual navigation mark
视觉呼号 visual call sign
视觉环境 visual environment
视觉汇合 visual fusion
视觉机器人 vision robot
视觉极限 visual threshold
视觉检查 sight control
视觉警报 visual alarm
视觉警报信号 visual alarm signal
视觉警告 visual warning
视觉警示灯 visual warning lamp
视觉宽度 width of coverage
视觉亮度 visual brightness
视觉灵敏度 eye response; foveal sensitivity; visual sensitivity
视觉灵敏度特性 visual sensitivity characteristic
视觉密度 optic(al)density
视觉敏度 visual acuity
视觉敏锐度 acuteness of vision; visual acuity
视觉敏锐角 visual acuity
视觉敏锐性 visual acuity
视觉疲劳 eye strain; visual fatigue
视觉平衡 visual balance
视觉平面 visual plane
视觉器官 ocular system
视觉清晰度 sharpness of sight; sharpness of vision; visual sharpness
视觉情报 visual information
视觉区 visual perception area; visual-sensory area
视觉曲线 eye sensitivity curve
视觉权利 visual rights
视觉缺失 ablepsia; blindness
视觉缺陷 defect of vision; visual defect
视觉锐度 sharpness of vision
视觉色度计 visual colo(u)rimeter
视觉上的不稳定性 visual instability
视觉上的稳定性 visual stability
视觉式指示器 visual detector
视觉势态 visual sense modality
视觉适应 visual adaptation
视觉舒适 visual adaptation
视觉舒适概率 visual comfort probability
视觉舒适感 visual comfort
视觉双性理论 duplicity theory of vision
视觉速度 speed of vision
视觉特性 visual characteristic
视觉特征 visual characteristic
视觉通信[讯] visual communication
视觉通信[讯]系统 visual communication system
视觉通知 visual annunciation
视觉外形 visual form
视觉稳定 visual stability
视觉问题 visual problem
视觉污染 sight pollution; visual pollution
视觉误差 collimation error; defect in vision; visual error
视觉细胞 visual cell
视觉显示 visible display; visual display
视觉显示装置 visual display unit
视觉像 visible image; visual image
视觉效果 visual effect
视觉效应 visual stimuli
视觉信号 optic(al)marking; optic(al)signal; visible signal; visual signal
视觉信号机【铁】 visible signal
视觉信息 visual information
视觉掩蔽 visual masking
视觉艺术 visual arts
视觉因素 visual factor
视觉印象 visual impression
视觉影响 visual impact
视觉阈 threshold visibility
视觉运动伸缩性 visuo-motor flexibility
视觉暂留 perpetual of vision; persistence of vision; retentivity of vision; visual perpetual; visual persistence
视觉暂留时间 retentivity time of eye
视觉质量 visual quality
视觉中枢 optic(al)center[centre]; visual center[centre]
视觉中心 fovea
视觉助航设施 visible aids to navigation
视觉驻留 persistence of vision
视觉追踪 visual tracking
视觉走廊 visual corridor
视觉阻尼 optic(al)damping
视抗剪强度 apparent shear-strength
视空间 visual space
视孔 eye hole; eyelet; eye pit; oillet(te)
视孔盖 wicket
视孔口 sight size
视孔隙率 apparent porosity
视口 viewport
视拉长 apparent elongation
视累积 apparent accumulation
视力 eyesight; seeing; sight; vision; visual sense modality
视力表 test chart
视力测定法 optometry
视力差 weak-eyed
视力的阻碍 obstruction of vision
视力范围 range of vision
视力分辨率 vision resolution; visual resolution
视力计 optometer
视力减退 hypopsia
视力检查 check by sight
视力检定表 test types of sight
视力检验表 eye chart
视力检验星 eyesight tester
视力近点 near point of vision
视力控制 visual control
视力敏锐度 eyesight acuity; visual acuity
视力模糊 blurred vision
视力疲劳 asthenopia
视力锐敏 oxyopia
视力适应性 adaptation of vision
视力衰退 failing vision
视力损害 impairment of vision
视力训练 visual training
视力训练辅助器材 visual training aids
视力增强 vision enhancement
视力障碍 blurred vision
视力作用范围 covering power
视立体图 visual relief map
视亮度 apparent brightness; apparent luminance; visual acuity
视流速 apparent velocity
视流质 apparent fluidity
视螺距 apparent pitch
视落差 apparent throw
视煤气发生率 pseudo-productivity of gas
视密度 apparent density
视密度计 volumenometer
视面积 apparent area
视敏读 visual acuity
视敏度 vision
视模数 apparent modulus; secant modulus
视模型 perceived model
视摩擦角 apparent angle of friction
视目视星等 apparent visual magnitude
视内聚力 apparent cohesion
视内摩擦角 angle of apparent internal friction
视内切 apparent interior contact
视泥浆滤液电阻率 apparent resistivity of mud filtrate
视年龄 apparent age; apparent life
视黏[粘]度 apparent viscosity
视黏[粘]滞系数 apparent coefficient of viscosity
视黏[粘]滞性 apparent viscosity
视宁度扰动 seeing disturbance
视宁像 seeing image
视宁圆面 seeing disc[disk]
视凝聚力 apparent cohesion
视盘 optic(al)disc[disk]
视膨胀 apparent expansion
视膨胀系数 coefficient of apparent expansion
视偏差分力 apparent deviation component
视漂移 apparent precession; apparent wander
视频 image frequency; video frequency exchanger; vision frequency; vision mixer; visual frequency
视频包络 video envelope
视频保护带 video guard band
视频壁 video wall
视频编码技术 video coding technique
视频变换器 video converter
视频变频器 video converter
视频变压器 video transformer
视频表 videometer
视频表校准 videometer calibration
视频表驱动 videometer drive
视频波道 video link
视频波段 video band; video frequency band
视频波形 video waveform
视频彩条选择器 video colo(u)r bar selector
视频参数 video parameter
视频测试 video measurement
视频测试信号发生器 video test signal generator
视频插孔板 video jackfield
视频插入 video insertion
视频插座 video socket
视频成像仪 video mapper
视频处理 video processing
视频处理器 video processor
视频传递特征 video transfer characteristic
视频磁头方位角 video head azimuth angle
视频磁像仪 video magnetograph
视频存储器【计】 video memory
视频带宽 video bandwidth
视频单片坐标量测仪 video/mono comparator
视频的 video
视频地图 video map
视频电报 videotext; videtex
视频电话 visual telephone
视频电流 picture current; video current
视频(电视)图像处理【计】 video image processing
视频电文 videotext
视频电压放大器 video voltage amplifier
视频定位器系统 video point locator system
视频定位设施 video position device
视频发射机 video transmitter
视频发送 video transmission
视频返回(地球)数据 video-returned data
视频放大 video amplification
视频放大器 video amplifier
视频分配放大器 video frequency distribution amplifier
视频分配开关单元 video distribution switching unit
视频符号 video sign
视频光检波器 video optical detector
视频光源 video optical source
视频获取系统 video acquisition system
视频间隙 video gap
视频解调器 picture demodulator
视频矩阵 video crowbar
视频开关 video switching
视频控制信号 video control signal
视频录像机 video recording
视频滤波器 video filter
视频率 video frequency
视频片 video plate
视频频谱 visual spectrum
视频频域 video domain
视频切换 video switching
视频切换器 video switch(er)
视频切换台 cutbank
视频散率 percent frequency effect
视频时逝录像机 video character generator
视频示波器 envelope widescope
视频输入 video input
视频数据显示系统 video data display system
视频数字传输 digital video transmission
视频随机存取存储器【计】 video RAM
视频调制 video modulation
视频通道 video channel
视频图像 video image; video picture
视频图形适配器 video graphics array
视频显示 video presentation; video display
视频显示器 video display; video display unit
视频显示终端 video display terminal
视频线路放大器 videoline
视频相关器 video correlator
视频信道 video channel; vision channel
视频信道中的干扰限制器 video interference limiter
视频信号 video signal
视频信号包线 outline of video signal
视频信号编码器 video coder
视频信号波形 video waveform
视频信号传输线对 video pair
视频信号磁带记录器 video tape recorder
视频信号电路 video circuit
视频信号混合 video mixing
视频信号畸变 video distortion
视频信号极性 polarity of picture signal
视频信号记录 video recorder
视频信号接收 video reception
视频信号频率响应 video response
视频信号调制深度百分率 picture modulation percentage
视频信号载波 video vision carrier
视频信号增益 video gain

视频信息 video information
视频信息处理机 video processor
视频杂波测试仪 video noise meter
视频终端 video terminal
视频装置 video unit
视平超距 apparent horizontal overlap
视平错 apparent heave
视平面 plane of vision;view plane
视平面距离 view plane distance
视平线 horizon
视坡度 apparent slope
视前区 preoptic(al) region
视强度 apparent strength
视倾伏 apparent plunge
视倾角 apparent angle of dip;apparent dip
视倾向 apparent dip
视倾斜 apparent dip;apparent tilt
视倾斜投影法 false dip projection method
视情况维护 condition-based maintenance
视区 viewport;vision area;zone of vision
视热星等 apparent bolometric magnitude
视日出 apparent sunrise;visible sunrise
视日没 apparent sunset;visible sunset
视容积 apparent volume
视容积效率 apparent volumetric(al) efficiency
视容量 apparent capacity
视熔 visual fusion
视熔频率 visual-fusion frequency
视如圆木屋的外墙披叠板 log cabin siding
视色亮度 brightness
视上超 apparent onlap
视设置 apparent setting
视渗透速度 apparent seepage velocity
视时 apparent solar time;true solar time
视束损害 optic(al) tract lesion
视衰减度 apparent attenuation degree
视衰减时 apparent decay time
视双星 optic(al) double;optic(al) pair
视水平 apparent level
视水平超距 apparent horizontal overlap
视水平离距 apparent horizontal separation
视水平线 apparent horizon
视水位 apparent water-level;apparent water stage;apparent water table
视速度 apparent velocity
视速度对比 apparent velocity correlation
视速率 apparent rate;apparent speed
视速仪 tachistoscope
视岁差 apparent precession
视损失 apparent loss
视太阳 apparent sun;true sun
视太阳年 apparent solar year
视太阳日 apparent solar day;true solar day
视太阳时 apparent solar time;true solar time
视太阳时角 hour angle of apparent sun
视弹性极限 apparent elastic limit
视弹性限度 apparent elastic limit
视体 view volume
视体积 apparent volume
视天顶距 apparent zenith distance
视天空亮度 apparent sky brightness
视天平动 apparent libration
视听材料 audio-visual material
视听超声波检验设备 audio-visual ultrasonic testing equipment
视听的 audio-visual
视听教材 audiovisual
视听教具 audio-visual aids;audiovisual facility
视听教育中心 audio-visual center[centre]
视听警告系统 audio-visual warning system
视听空间 auditory-visual space
视听器材 audiovisual
视听情报 audio-visual information
视听设备 audiovisual
视听图书 audio-visual book
视听文献 audio-visual documents
视听无线电指向标 visual-aural radio range
视听系统 audio-visual system;video system
视听协调 auditory-visual integration
视听整合 auditory-visual integration
视听资料 audiovisual
视同现金支票 check as cash
视突起 apparent relief
视图 view
视图菜单 view menu
视图重构 view restructuring
视图点 viewpoint
视图分类 view classification
视图建模 view modeling
视图面 view surface
视图模型化 view modeling
视图数据 viewing data
视图文件绘图 view file plotting
视图正视方向 view-up
视外切 apparent exterior contact
视外区域 parafoveal region
视外线 ultraphotic ray
视网膜 retina[复 retinas/retinae]
视网膜的 retinal
视为重要 count for much
视位置 apparent place;apparent position
视温度 apparent temperature
视午 apparent noon
视物显多症 polyopia
视误差 apparent error;visible error
视吸收 apparent absorption
视晰度 visual acuity;visual sharpness
视下落断距 apparent downthrow
视线 eye axis;line of sight;sight(ing) line(of vision);transit line;viewing line;vision line;visual line;visual line of vision;visual ray
视线被挡的座椅 blind seat
视线标高 elevation of line of sight
视线不良的交叉口 blind crossing
视线传播 line-of-sight propagation
视线法 line-of-sight system
视线范围 horizon range from an object;range of sight;sight-line coverage
视线干扰 visual intrusion
视线高 height of vision
视线高程 <即仪器高程>【测】elevation of sight line;elevation of sight
视线高度 eye height;height of sight-(ing) line
视线角 angle of sight;aspect angle
视线距离 horizon range;line-of-sight distance;line-sight distance;optic-(al) range;seeing distance;sighting distance
视线距离测量 measurement of line of sight distance
视线棱锥 visual ray pyramid
视线理 apparent lineation
视线链路 line-of-sight link
视线内的传播 line-of-sight propagation
视线锐度试验 acuity test
视线升高 eye-level rise
视线速度【天】radial velocity
视线条件 sighting condition
视线引导 optic(al) guidance;visual guidance
视线引导标志 sight-line induction sign
视线诱导栽 sight-line indication planting
视线张角 viewing angle
视线障碍物 sight obstruction;visual obstruction
视线遮蔽 vision screen(ing)
视线走廊 visual corridor
视相对运动 apparent relative movement
视相速度 apparent phase velocity
视向 line of vision
视向分量 line-of-sight component
视向速度 line-of-sight velocity;radial velocity;sight-line velocity
视向速度光谱仪 radial velocity spectrometer
视向速度曲线 radial velocity curve
视向速度扫描仪 radial velocity scanner
视向速度仪 radial velocity meter
视像 visible image
视像曝光计 visual exposure meter
视像管 staticon;vidicon(tube)
视像管望远镜 vidicon telescope
视像检测系统 video detector system
视像文本系统 videotext
视消融 apparent ablation
视消色差 visual achromatism
视效率 apparent efficiency
视星等 apparent magnitude
视形 view
视压缩波 apparent compressional wave
视压缩性 apparent compressibility
视眼 oillet(te)
视摇摆 apparent rolling
视野 angular field;distance vision;eye reach;eyeshot;eyesight;field coverage;field of view;field of vision;free view;ken;outlook;range of visibility;range of vision;scope;sight;viewing field;visual field
视野测量法 perimetry
视野范围 field range
视野归算 sight reduction
视野归算表 sight reduction table
视野计 perimeter
视野检查 perimetry
视野角 angle of visual field;viewing angle
视野镜 cycloscope
视野可达 180°的 fish eye
视野轮廓测定器 schematograph
视野深度 depth of field;depth of scene
视野缩小 contraction of visual field
视野图 visual field diagram
视野狭小 narrowed visual field
视野仪表 perimeter
视野以外的 unsighted
视野张角 visual angle
视应变 apparent strain
视应力 apparent stress
视迎角 apparent angle of attack
视影 seeing image
视影圆面 seeing disc[disk]
视硬度 apparent hardness
视油板 sight glass
视油规 sight oil indicator
视有闪色效应 goniochromatic effect
视有效场 apparent effective field
视预固结压力 virtual preconsolidation pressure
视域 eye;sighting line;field of view
视域大小 field of view
视域光阑 field diaphragm
视域宽度 <窗的> sight size
视阈 visual threshold
视原基 optic(al) rudiment
视运动 apparent motion;apparent movement;relative motion
视在半径 apparent radius
视在比电阻 apparent resistivity;apparent specific resistance
视在齿密度 apparent tooth density
视在磁阻 apparent resistance
视在的 apparent
视在电动势 apparent electromotive force
视在电感 apparent impedance
视在电容 apparent capacity
视在电阻 apparent resistance
视在电阻率 apparent resistivity;apparent specific resistance
视在方位 apparent bearing
视在方位角 apparent azimuth(angle)
视在分辨率 apparent resolution
视在负载 apparent load
视在高度 apparent height;virtual height
视在功率 apparent output;apparent power;total volt-ampere
视在黑色 apparent black
视在进动 apparent precession
视在距离 apparent range
视在距离变化率 apparent range rate
视在抗剪强度 apparent shear(ing) strength
视在力 apparent force
视在亮度 apparent brightness;apparent intensity
视在螺距 apparent pitch
视在落差 apparent head
视在马力 indicated horsepower
视在摩擦角 apparent angle of friction;virtual friction-angle
视在目标 virtual target
视在内摩擦角 angle of apparent internal friction;apparent angle of internal friction
视在强度 apparent intensity
视在清晰度 apparent resolution
视在容量 apparent capacity
视在色 apparent colo(u)r
视在渗透系数 apparent permeability
视在输出 apparent output
视在瓦特 apparent watt
视在位置 apparent place;apparent position
视在误差 apparent error
视在效率 apparent efficiency
视在旋进 apparent precession
视在颜色 apparent colo(u)r
视在仰角 apparent elevation angle
视在预固结压力 apparent preconsolidation pressure
视在值 apparent value
视在质量 apparent mass
视在阻抗 apparent impedance
视正断层 apparent normal fault
视正午 apparent noon
视知觉 visual perception
视直径 apparent diameter;visual diameter
视直线 line of collimation
视质量 virtual mass
视中心 visual center[centre]
视重量 apparent weight
视周日路径 apparent diurnal path
视轴 axis of sight;axis of visual cone;boresight;eye axis;optic(al) axis;sight axis
视轴测定立体镜 haploscope
视轴线 visual axis
视烛光 apparent candle power
视柱式高温计 high-temperature visi-

ble column thermometer
视锥 collimation cone;retinal cone
视锥轴 axis of visual cone
视准 collimate;collimation
视准标 sight vane
视准标杆 collimating rod;collimating staff
视准标志 collimation mark
视准差 error of collimation
视准差改正 collimation correction error
视准差校正 collimation error correction
视准常数 collimation constant
视准尺架 sight bracket
视准点 observed point;sighting point
视准法 collimation adjustment;collimation method
视准改正 collimation correction
视准高度 height of collimation
视准管 collimator;tubular collimator
视准管棱镜 collimator prism
视准轨 <用于校核路沟深度的> sight rail
视准检景器 collimator-finder
视准校正 collimation adjustment;collimation correction
视准精度 collimation accuracy
视准面 collimation plane;plane of collimation;plane of sight;sighting plane
视准器 vane
视准设备 collimation system
视准误差【测】error of collimation;collimation error
视准系统 <大地测量> collimation system
视准线 collimating line;collimating ray;line of collimation;line of sight;observation line;observing line;pointing line;collimation line
视准线法 <变形观测法之一> collimation line method;method of alignment
视准线观测 collimation line observation
视准仪 alidade;alignment collimator;collimator;sight alidade;sighting rule
视准正午 apparent noon
视准轴 aiming axis;axis of collimation;collimation axis;line of collimation;sight axis;visual axis
视准轴偏心 eccentricity of collimation axis
视准轴线 line of collimation
视紫 visual violet
视紫红 visual purple
视自转 apparent rotation
视总体压缩系数 pseudo-bulk compressibility
视纵断面 apparent profile;apparent throw
视阻抗 apparent impedance
视阻力 apparent drag;apparent resistance
视阻力系数 apparent drag coefficient
视最大高度 apparent maximum altitude
视坐标 apparent coordinates

试

试安装 trial erection
试板 accessory plate;test panel;test plate
试板材料 material of test piece
试拌 trial mix;trial batch
试拌法 trial-batch method
试拌法设计 trial mix design
试拌和 trial mix
试拌和组次 trial mix series
试拌混合料 trial mixture
试拌混合料设计 trial mix design
试拌校正法 trial-and-error method
试拌配合比 trial mix
试拌配料(比) trial-batch mixing proportion
试拌配料法 trial-batch method
试棒 coupon;test rod
试爆 trial blasting
试泵间 drafting pit
试泵箱 pumper pit
试笔 test probe
试编年度报表 tentative annual report
试编预算 preliminary budget;tentative budget
试标 test-object
试饼 <水泥安定性试验的> pat
试饼染迹试验 <鉴定沥青混凝土含油量> pay stain test
试饼试验 pat test
试饼试验法 pat test method
试播 trial seeding
试采 preproduction
试采船 experimental mining ship
试采地点 experimental mining site
试采方法和设备 experimental mining method and device
试采国 experimental mining country
试采阶段 preproduction stage
试采井 development test
试采区 preproduction area
试采区水深 depth of experimental mining area
试采时间 experimental mining time
试操试验 trial test
试操纵 trial maneuver
试操纵航向 trial course
试操纵速度 trial speed
试操作 dry run;dummy run;manual simulation
试操作时间 time to maneuver
试槽 test trench;trial trench
试槽法 trial trench method
试测点 trial point
试测和误差修正定线法 range-in
试测线 random line
试查 audit trial
试产品检验 preproduction inspection or test
试车 trial drive;trial run;breaking-in;commissioning;commissioning test;green test;initial start-up;pre-operation(al) test;road test;run-in;run-up;shakedown run;start-up test;test;test drive;test run(ning);test working
试车Ⅵ板路 corrugated road;washboard road
试车场 proving ground;skid pad;test ground
试车船坞 wet slip
试车道 proving road
试车功率 trial output
试车轨道 run-up track
试车驾驶员 test driver
试车架 running support
试车间 test cell
试车间检测仪器设备 test cell instrumentation
试车马力 trial horse-power
试车码头 quay for propeller thrust trial
试车跑道 test course;test road
试车期 breaking-in period;start-up period
试车浅水池 shallow water splash
试车圈 pace lap
试车时期 breaking-in period;trial period
试车事故 <发动机> running-up accident
试车试验 <道路> driving test
试车数据 <发动机> firing test data
试车水潭 water splash
试车速度 trial speed
试车速跑道 speed-testing runway
试车台 test bay;test bed
试车台点火 test bed firing
试车台结构 stand structure
试车台试车 test cell run
试车台运转 test stand operation
试车线 train test line
试车行程 trial race;trial trip
试车用粉 bonamite powder
试车执照 trial license
试车周期 start-up period
试车状态 trial condition
试充电 trial charging
试抽 bailing test
试除 trial division
试除数 trial divisor
试凑步骤 trial-and-error procedure
试凑 cut-and-trial;cut-and-try
试凑法 cut-and-try method;method of trials and errors;trial-and-error method;try-and-error method
试凑工作 cut-and-try work
试凑解法 trial-and-error solution
试错法 trial-and-error method;trial-and-error procedure
试错模型 trial-and-error model
试错曲线 trial curve
试错设计法 trial-and-error design method
试打桩 indicator pile
试点 experimental work in selected point
试点测量 orientation survey
试点单位 pilot agency
试点调查 pilot investigation;pilot survey
试点工程 pilot project
试点区 pilot area
试点试验 experimental test
试点田 conduct tests at selected plot
试点项目 pilot project
试点研究 pilot study
试电笔 electroprobe;electroscope;screwdriver with voltage tester;test pencil
试电器 live line tester
试定舱位 tentative booking
试抖(配料)法 trial-batch method
试舵 testing of steering gear
试飞 flight testing;test flight;test-fly(ing)
试飞发动机 flight test engine
试飞工作单 test flight worksheet
试飞驾驶员 test pilot
试飞跑道 test runway
试飞任务 test mission
试飞用跑道 test runway
试风 air test;compressed-air test
试杆 test rod
试割 test tapping
试给架 test-tube rack
试工 labo(u)r test
试沟 discovering trench
试购 sample order;trial order
试管 testing tube
试管机 pipe-testing machine
试管夹 test-tube clamp;test-tube holder
试管架 test-tube rack;test-tube stand;test-tube support
试管离心机 test-tube centrifuge
试管凝集试验 tube agglutination test
试管区 in vitro
试管试验 tube test
试管刷 test-tube brush
试管斜面 test-tube slant
试焊 test weld
试焊接 test welding
试航 sea trial;shakedown cruise;ship trial;test sailing;trial run;trial voyage;trial strip【航空】
试航船长 trial captain
试航航程 trial voyage
试航排水量 trial displacement
试航速度 trial speed
试航委员会 trial board
试航预测图 trial predication diagram
试航状态 trial condition
试合法 method by trial
试荷载 trial load
试荷载分析 trial load analysis
试换法 trial substitution
试机器 try engine
试极纸 pole paper
试记录 trial record
试剂 agent;reactant;reagent;reductant
试剂纯 reagent's purity
试剂等级 reagent grade
试剂分配器 reagent distributor
试剂规格 regent specification
试剂级别 reagent grade
试剂加入器 reagent feeder
试剂鉴定 reagent identification
试剂接收孔 tracer receiving hole
试剂空白 reagent blank
试剂类化学药品 reagent chemicals
试剂配制 dosing(of) reagent
试剂瓶 reagent bottle
试剂溶液 reagent solution
试剂特性 specificity of reagent
试剂投放孔 tracer injecting hole
试剂投配 dosing(of) reagent
试剂选择性 selectivity of reagent
试剂种类 reagent type
试驾驶 trial drive
试件 sample piece;specimen;test piece;test specimen
试件包封 specimen encasement
试件材料 material for test
试件采样 sampling
试件的表面效应 surface effect in test specimens
试件放射性 sample activity
试件封顶 capping of specimen
试件封套 specimen encasement
试件高度 height of sample
试件个数 number of test-pieces
试件几何学 <高与直径之比> specimen geometry
试件检验 sample inspection
试件面积 area of sample
试件模 specimen mold
试件(模)盒 specimen carrier
试件模具 test piece mould
试件栅 specimen grating
试件养护 curing of test specimen
试件制备 specimen preparation;test material preparation
试件准备 specimen preparation
试搅拌 test mixing
试搅拌法设计 trial mix design
试接 wiped joint
试金 assay
试金吨砝码 assay ton weight
试金砝码 button weight
试金分析法 fire assay
试金坩埚 scorifier
试金石 basanite;lydianite;touchstone;test stone
试金实验室 assay laboratory
试金室 assay laboratory
试金天平 assay balance

试金物 assay
试金者 assayer
试金值 assay value
试井 testing of well;well test
试井车 truck-mounted slick line reel unit
试井记录 test boring record
试镜架 trial frame
试镜头 screen test(ing)
试卷 test paper
试掘 sinking;test drill;test pitting
试开车 test working
试坑 bore pit; pilot hole; prospect hole;prospect pit;test pit;trial pit
试坑尺寸 dimension of test pit
试坑抽水渗透试验 pumping-out test from trial pit
试坑调查 adit test
试坑荷载试验 pit loading test
试坑记录 log of test pit
试坑浸水试验 pit immersion test
试坑勘探 prospecting by trial pits
试坑深度 depth of test pit
试坑渗水法 pit permeability method
试坑数 number of trial pits
试孔 prospect hole; test boring; test hole;test well;trial hole
试孔钻探 test hole drilling
试孔钻探工作 test-hole work
试块 briquet(te); sample piece; test cube;testing block;test piece
试块拉力试验 briquet(te) tension test
试块联合模型 briquet(te)gang mould
试块试验 pat test
试块张拉试验 briquet(te) tension test
试量 weighted portion
试料 assay;test portion;trier
试料制备 test material preparation
试模 testing mo(u)ld;specimen mold <作试件用的>
试模环口 collar extension
试镍剂 reagent for nickel
试拍 trial exposure
试配逼近法 spline fit approximation
试配法 spline fit method;trial-and-error method;trial method
试配法配合混凝土成分 proportioning by trial method
试配法配料 proportioning by trial method
试配合 trial batch;trial mix
试配沥青含量 trial asphalt content
试配料法 trial-batch method
试配强度 <混凝土等的> target strength
试配曲线 spline fit curve
试配体积的最大密度配料法 box method
试匹配 trial match
试片 coupon;test block;test piece
试片电影院 show case
试拼装 tentative assembly; trial assembly
试瓶 trial jar
试起动 experimental starting
试切 trial cut
试驱动 trial drive
试染 trial dyeing
试烧 burning test
试设计 trial design
试生产 pilot production; preproduction;preproduction trial;producing test;trial produce;trial production;trial run
试生产材料 pilot material
试生产测量 pilot survey
试生产费用 pilot production expenditures;start-up cost
试生产机械 preproduction machine
试生产检验 initial production test
试生产线 pilot production line
试生产铸件 prototype casting
试生产装置 pilot device;pilot plant;pilot unit
试收 acceptance
试收量规 acceptance ga(u)ge
试输电 trial transmission
试水阀 pet valve
试水器 water diviner;water finder
试水位旋塞 ga(u)ge cock
试送电 trial line charger
试速航行 speed trial trip
试算 dibble-dabble;pilot calculation
试算逼近法 trial-and-error method
试算表 daily trial balance;trial balance;trial table
试算成本 pro forma cost
试算尺寸 trial dimension
试算法 cut-and-trial method;cut-and-try method; cut-and-try process; method by trial; method of trial; method of trials and errors; trial-and-error method; trial-and-error procedure; trial-and-error process; trial method
试算费用 pro forma cost
试算分析法 budget analysis
试算荷载法 trial-load method
试算解 trial solution
试算解法程序【计】heuristic routine
试探 feel;heuristic;heuristic approach;pricking;probe
试探靶 probe target
试探标号 tentative label
试探步骤 heuristic approach
试探查找技术 heuristic search technique
试探程序 heuristic program(me); heuristic routine
试探穿刺 exploratory puncture
试探电极 electric(al) probe; probe; probe electrode
试探法 cut-and-try method; cut-and-try process;error method;heuristic method;heuristics;method of trial;method of trials and errors; seamount method
试探规划法 heuristic programming
试探规则 heuristic rule
试探函数 tentative function
试探和误差法 method of trials and errors
试探解 trial solution
试探井 trial-test well
试探孔 probe inlet
试探控制方式 tentative control regime
试探粒子 probe particle
试探器 experimental probe
试探矢量 trial vector
试探试验 trial-and-error test;trial test
试探搜索 heuristic search
试探线圈法 search coil method
试探向量 trial vector
试探信号 probe signal
试探性策略 tentative strategy
试探性的控制策略 tentative control strategy
试探性观测 trial-and-error observation
试探性计算 scouting-type calculation
试探性试验 running test;trial test
试探性桩 preliminary pile
试探钻头 test bit
试填 test filling
试填充 trial fill(ing)
试填土 trial fill(ing)
试听播音室 audition studio
试听室 audiometric rooms; sound-shield enclosure
试图 endeavour
试图成为 attempt to be
试涂层 trial coat
试推销上市 test-market
试挖 test-pit digging;trial excavation;trial dredging【疏】
试挖泥 trial dredging
试挖横坑道 test adit
试位迭代法 regular false interaction
试位法 false position;method of false position;regular false method
试误法 trial-and-error method
试线杆 test pole
试线夹 test clip
试线器 line tester
试线区段 test section
试线用触排 private bank
试线用线弧 private bank
试销 approval sales;on approval;sales on approval;test marketing;trial sale
试销的 semi-commercial
试销书 approval copy
试行 probation;trial implementation
试行本 advance(d)copy
试行标准 tentative specification;tentative standard
试行的 tentative
试行规定 trial regulation
试行(技术)规范 tentative specification
试行条例 proposed regulation
试行推销 on approval
试行运转 preliminary operation
试行章程 experimental regulation
试选样品 pilot model;pilot sample
试压 pressure test(ing)
试压板 test plate
试压保压时间 holding time at test pressure
试压泵 hydraulic test pump
试压部位 position to be tested
试演 tryout
试验 assay; experiment; paternity test; proofing; proofness; prove; proving; put on trial; put to test; tentative; test; test by trial; trial; trial run;tryout
试验T形管 test tee pipe;test tee tube
试验安排 test arrangement
试验按钮 testing button
试验靶场跟踪 test-range tracking
试验坝 trial embankment
试验板 breadboard;test board;testing panel;test slab;test plate
试验板模型 breadboard model
试验板调制盘叶片 checkerboard reticle blade
试验板座 breadboard socket
试验拌合[和]物 test mix(ture)
试验棒 proof stick;test bar
试验报告 report of test;test information;test report
试验报告编号 test report No.
试验报告表 test report table
试验报告单 testing report sheet
试验曝光 test exposure
试验爆破 trial blast;trial shot
试验杯 test glass
试验泵 test pump
试验笔 test pencil
试验变差 experimental variation
试验变压器 testing transformer
试验标志 proof mark
试验标准 test criterion;test norm
试验表 testing table;trial table
试验表格 test card;test table
试验表面 testing surface
试验波浪 test wave
试验不合格 below proof
试验不确定性 experiment uncertainity
试验布置 test(ing) arrangement;testing configuration
试验步骤 test(ing) procedure; test operating procedure
试验部分 test portion;test section
试验部门 testing department
试验材料 test material
试验参数 experimental parameter; test parameter
试验操作程序 test operating procedure
试验槽 experimental tank; test cell; test tank;test trough
试验测定 test determination
试验测定曲线 data curve
试验测量 test measurement
试验层数 number of test strata
试验插塞 test plug
试验插座 testing socket
试验长度 test length
试验常规 testing routine
试验常数 test constant
试验厂 demonstration plant; demo plant;trial plant
试验厂规模 pilot plant scale
试验场 experimental plat; proof ground;proving ground;test ground;testing field; testing ground; trial ground
试验场的基底噪声 field baseline noise
试验场地 experiment field;testing site
试验场观测项目 observation items on testing field
试验场规格 test site specification
试验场建立日期 date of testing ground established
试验场设施 proving ground facility
试验场试验 proving ground test
试验场所 test site
试验场位置 location of testing site
试验车 instruction carriage; lab-on-wheel; laboratory vehicle; testing car;test station truck;test wagon;trial vehicle
试验车测试间 measurement compartment
试验车车身 testing body
试验车间 development shop; testing department; testing plant; testing shop
试验车速跑道 speed-testing runway
试验尘 test dust
试验成本 experimental cost; experimentation cost
试验成果 experimental result;test result
试验成果推广 spread of results
试验成果总表 soil test data sheet
试验程序 check program(me); experimental sequence; procedure of test;testing program(me); testing sequence;test procedure;test routine
试验池 test basin;test(ing)tank
试验持续期 test duration
试验持续时间 duration of test runs
试验抽水 trial pumping
试验出水率 tested capacity
试验触点 test contact
试验船 experimental ship;test vessel
试验床 test bed;test board
试验次数 order number of tests;test frequency;test number
试验错误 experimental mistake
试验打桩 test driving; trial driving; trial piling
试验大纲 test(ing) program(me); test(ing)schedule

S

试验单 testing sheet; test sheet
试验单位 experimental unit; testing agency
试验单元 test unit
试验导洞 test adit
试验得出的 provable
试验的 developmental; testing
试验的技术要求 test requirement
试验的限制 test limitation
试验的一批 test batch
试验的应用 test application
试验灯 test burner; test lamp
试验等级 test class
试验堤 test embankment; trial embankment
试验地 experimental field; sample plot(ting); study plot; test plot
试验地点 test(ing) site
试验地锚 test anchor
试验点 test(ing) point
试验点分布区 test point scattering
试验点火 test firing
试验点密度 density of test points
试验点散布图 scatter diagram
试验点数 number of test points
试验点阵 trial array
试验电池 test cell
试验电荷 test charge
试验电解槽 experimental cell
试验电缆 test cable
试验电流 test current
试验电路 breadboard circuit; hookup
试验电路板 breadboard
试验电气仪表 electric(a) instrument testing device
试验电压 testing tension; testing voltage
试验电站 experimental power station
试验调查 pilot survey
试验丁坝 test groin; test groyne
试验订单 trial order
试验洞 test heading
试验读数 test reading
试验段 experimental section; measuring section; trial lot; trial section
试验段层位 position of experimental part
试验段横截面面积 working section area
试验段深度 depth of experimental part
试验段数 number of test segments
试验段岩性 lithologic(al) characters of experimental part
试验断面 test profile; test section
试验堆 test(ing) reactor
试验队 testing crew
试验发动机 test engine
试验发动机用的样机 motor test vehicle
试验发射井 test silo
试验发射装置 test(-type) launcher
试验阀 pet valve; test(ing) valve
试验法 cut-and-trial method; cut-and-try method; cut-and-try process; method of testing; set and try method; testing method; test procedure
试验法规 test code
试验法配料 proportioning by trial method
试验范围 scope of experimentation; scope of testing; scope of tests; testing range; trial stretch
试验方案 test scheme
试验方法 experimental method; method of testing; test(ing) method; testing procedure; experimentation
试验方式 test mode
试验房屋 experiment building
试验放大器 try amplifier
试验飞机 testing aircraft
试验飞行 trial flight
试验费 cost of testing; experimental expenses; testing expenses
试验分析报告 test analysis report
试验风洞 test air tunnel
试验辅助设备 test accessories; test accessory; test support equipment
试验负荷 proof load; testing load
试验负载 proof load; testing load
试验负责人 test supervisor
试验附加装置 test adapter
试验复现性 reproducibility of tests
试验干线 test trunk
试验杆 check bar; test bar
试验缸 test cylinder jar; test jar
试验杠杆 testing lever
试验根据 experimental evidence
试验工厂 semi-plant; semi-work; test-tube plant
试验工厂研究 pilot plant work
试验工程 demonstrative project
试验工程布置数据 data of test engineering arrangement
试验工程师 testing engineer
试验工地 testing field
试验工具 test apparatus; test instrument
试验工况的稳定性 constancy of test condition
试验工人 experimental labour
试验工作 test work; trial work
试验工作面 test face; trial face
试验公差 test tolerance
试验沟 pilot ditch
试验观测点号 number of observation point for test
试验观测点数 observation point number for test
试验观测记录 observation record for test
试验观测孔数 observation well number for test
试验观测台 model test platform
试验观测线编号 number of the observation line for test
试验观测线长度 length of observation line for test
试验观测线方位 direction of observation line for test
试验观测线条数 observation line number for test
试验观察 test observation
试验观察孔号 number of observation well for test
试验管 test glass; test tube
试验灌浆 test grouting
试验罐 test tank
试验规程 testing regulation; testing specification; test instructions; test procedure; test protocol
试验规范 test code; test specification
试验规格 test specification
试验规划 test project
试验规则 test code; test regulation; test rule
试验轨道 test track
试验锅炉 pilot boiler
试验过程 test procedure
试验过的 exd
试验焊缝 control seam; pilot seam; practice weld
试验焊接 test welding; trial weld
试验号(码) test number
试验合格 bear the test; pass test; stand test
试验合格的分数 upcheck
试验合格证书 test certificate
试验河段 test reach
试验荷载 finder charge; proof load; roof load; test(ing) load; trial load
试验盒 test box
试验滑轨 railroad test track
试验环 proving ring
试验环道 test loop road; test track
试验环境 test environment
试验环线 test circuit; test loop; test ring
试验环形天线 test loop antenna
试验换热器 test heat exchanger
试验灰浆 testing mortar
试验回路 test loop
试验混合器 test mixer
试验混合物 test mix(ture)
试验混凝土 test concrete
试验火 test fire
试验火锥 test cone
试验机 tester; test(ing) machine; trier; probe aircraft <有实验设备的飞机>
试验机测程 capacity range
试验机的准确性 accuracy of testing machine
试验机底板 bed plate
试验机夹具 grip of testing machine
试验机架 test rack
试验机器 testing machine
试验机数据 test engine data
试验机组 test crew
试验基地 proving ground; test base; test site; experimental plot
试验及维修车 test and maintenance truck
试验计 tester
试验计划 pilot project; pilot scheme; test(ing) program(me); test plan; test plot
试验计量 test measurement
试验记录 record of test; test record; trial sheet
试验记录表 testing record sheet
试验记录单 test record sheet
试验记录卡 testing record sheet
试验技术 experimental technique; testing skill; testing technique
试验技术规范 test requirement specification
试验继电器 test relay
试验加荷 test loading
试验夹具 test fixture
试验驾驶 <测汽车性能> test drive
试验架 proving frame; proving stand; rig for test; test console; test frame; testing jig; testing rig; testing site; test stand
试验间 test bay; test building; test chamber; test cubicle
试验间断时间 interrupted period of test
试验监测器 test monitor
试验检查 test check
试验检查人 test examiner
试验检定(证)书 testing certificate
试验鉴定 test evaluation
试验阶段 experimental stage; test period; test phase; test session; trial-and-error phase
试验接缝 experimental joint
试验结构 test structure
试验结果 findings of test; outcome of test; results of tests; testing findings; testing result
试验结果的复验性 reproducibility of tests
试验结果的调整 adjuster of results
试验结果绘制线图 plotting of the results
试验结果曲线的示意图 test profile
试验结果再现性 reproducibility of tests
试验结论 conclusions of testing
试验结束 off-test
试验结束时间 finishing time of test
试验截面 test section
试验介质 testing medium
试验进度表 test schedule
试验井 experimental well; pilot well; test shaft; test well
试验井出水率 tested capacity of well
试验矩阵 test matrix
试验距离 test distance
试验卡片 test card
试验开关 test switch
试验开始 on-test
试验开始时间 initial time of test
试验开挖 test cut; test excavation
试验可靠性 testing reliability
试验坑 test trench
试验孔 test holing; testing drill hole; trial bore; trial hole
试验孔布置型式 test bore pattern
试验孔号 number of test wells
试验孔深度 depth of test well
试验孔数 test well number
试验孔组号 group number of test well
试验控制人员 test control officer
试验控制台 test console
试验块 test block
试验廊道 testing gallery
试验类别 kind of test
试验里程 test mile
试验例程 test routine
试验粒度 testing size
试验粒子 test particle
试验连接 test splice
试验梁 test beam
试验梁模 test beam mo(u)ld
试验量测装置 experiment measuring device
试验料制备 test material preparation
试验列车 test train
试验龄期 testing age
试验领导人 test director
试验流化速度 test fluidized velocity
试验流域 experimental catchment; experimental watershed
试验炉 trial furnace
试验路 experimental road; test road
试验路堤 test embankment
试验路段 trial lot; trial pavement section
试验路段法 trial-section method
试验路断面 experimental section
试验路面 experimental surface; trial pavement
试验轮 test wheel
试验码 test code
试验脉冲 test pulse
试验模拟器 test simulator
试验模式 test mode
试验模型 experimental model; mock-up; prototype; test dummy; test model; test pattern
试验木炭 <用于测定天然气中的含油量> charcoal test
试验目的 test objective
试验能力 test(ed) capacity
试验农场 agronomy farm; experimental farm; testing farm
试验排列 experimental arrangement
试验盘 <用于试验材料抗压力> test plate; test board
试验盘试板 test panel
试验跑道 test(ing) track; test lane
试验配方 test formulation; test prescription; test recipe
试验配合 laboratory proportioning; trial mix
试验配合比 trial proportioning
试验配合法 proportioning by trial method; trial mixing method

试验配料法 proportioning by trial method;trial mixing method
试验批料 test batch
试验片 test film;test piece
试验偏差 experimental variation
试验偏倚 test bias
试验品系 testing line
试验平均值 test average
试验平台 testing platform
试验平巷 testing gallery
试验评价 test evaluation
试验期 test period
试验期中 in-test
试验气体 test gas
试验汽车 test car
试验汽车耐久性的砾石环道 gravel durability loop
试验汽车耐久性的砾石路 gravel durability road
试验器 effector;exerciser;testing set
试验前处理 pretest treatment
试验前的 pretest
试验前的调整 pretest conditioning
试验前检查 pretest inspection
试验前准备 pretest treatment
试验腔 test cavity
试验强度 strength of testing; test strength
试验墙 experimental wall;test wall
试验桥 experimental bridge; prototype bridge;prototype bridge
试验区 demonstration area; experiment block; experiment plot; pilt area;pilt zone;sample plot(ting); study plot;test area;test zone
试验区段 test section
试验区用脱粒机 nursery separator
试验曲线 assay curve;test diagram; testing curve;trial curve
试验取样 test sampling
试验取值点 data point
试验燃料 test fuel
试验(燃)气 test gas
试验人员 crew member
试验日期 date of test(ing);date tested;test date
试验容量 test capacity
试验容器 test vessel
试验溶液 test solution
试验熔断器灯 fuse test light
试验筛 test sieve
试验筛板 testing sieve plate
试验筛布 testing sieve cloth
试验筛布系列 testing sieve cloth series
试验筛的标准化 testing sieve standardization
试验筛机械振动器 mechanical sieve shaker
试验筛网 testing sieve cloth
试验筛网标准 testing sieve cloth standard
试验筛网系列 testing sieve cloth series
试验筛析 test sieve
试验筛振动器 testing sieve shaker
试验筛组的单位筛 nester
试验设备 experimental equipment; experimental rig;experiment facility; pilot installation; research instrument;test facility;testing appliance; testing equipment; testing plant;testing rig;test outfit;test unit
试验设备的检查 check of test equipment
试验设备工程学 test equipment engineering
试验设备误差分析报告 test equipment error analysis report
试验设计 design of experiment; test project
试验设计分析【数】 design of experiment analysis
试验生产设备 pilot production unit
试验声盘 test record
试验时程 <汽车> test mile
试验时的损坏 test failure
试验时间 test duration;testing time; time duration of test;time of test(ing)
试验时龄期 age at test
试验时数 time of test(ing)
试验矢量 trial vector
试验室 lab;laboratory;test bay;test cell;test chamber;test house;testing laboratory;testing room
试验室规模 laboratory scale
试验室排烟橱 laboratory flume hood
试验室配合(比) laboratory proportioning
试验室配合比设计 laboratory mix design
试验室染色机 lab dyer
试验室人员 laboratory staff
试验室设备 laboratory equipment
试验室湿筛分法 laboratory decantation method
试验室试剂 laboratory reagent
试验室试验 laboratory test
试验室试验述评 review of laboratory tests
试验室试样 laboratory sample
试验室研究 laboratory research; laboratory study
试验室用混合器 lab mixer
试验室用小型混合机 blender
试验室用仪器 laboratory instrument
试验室资料 lab data
试验输入端每分钟转数 test input rpm[rotations per minute]
试验术 experimentation
试验竖井 test(ing)pit
试验数据 experimental data;supporting laboratory data;tentation data; test data;test figure
试验数据报告 test data report
试验数据单 test data memorandum
试验数据发生 test data generator
试验数据换算 test data reduction
试验数据汇总表 test data summa sheet
试验数据记录 test data record;test data sheet
试验数据记录器 test data recorder
试验数据统计表 test data summa sheet
试验数据系统 test data system
试验数字 test figure;trial number
试验栓 test cock
试验水表 test water meter
试验水槽 experimental tank; test flume;laboratory flume
试验水池 experimental tank
试验水头 test(ing)head
试验水准 test level
试验顺序 series of test;test sequence
试验速度 test speed;trial speed
试验速度的记录 test speed record
试验速率 test rate
试验隧道 test tunnel
试验台 bedstand; bedstead; experimental bench; laboratory bench; observation desk; stand; test bed; test bench; test block; test board; testing rig; testing site; test panel; test platform;test rack;test stage
试验台测试仪器 test bed instrument; test bed instrumentation
试验台车 test ring
试验台的平台 test rig platform
试验台滚轮 test bench roller
试验台架 test bed;testing bench;testing block
试验台校验 bench calibration
试验台控制板 test rig panel board
试验台面 test bed;test floor
试验台上试验 rig test(ing)
试验台设备 test bench installation
试验台试验 bench run; bench test; shore trial <船用机械的>
试验台试验的发动机 laboratory test engine
试验台仪表 test rig instrument
试验台座 test(ing)floor;test(ing) bed;test(ing)stand
试验弹簧 test spring
试验天线 test antenna
试验田 study plot;testing field;test plot
试验填方 test embankment;test(ing) fill
试验条 test bar
试验条件 condition of experiment;experimental condition;test condition
试验调制 test modulation
试验通知 test information
试验统计资料 test statistics
试验图 test chart
试验图表 test figure
试验图像 test pattern
试验图像发生器 test pattern generator
试验涂层 trial coat
试验土工学 experimental soil engineering
试验土力学 experimental soil mechanics
试验土样名称 name of testing soil sample
试验土样组数 group number of testing soil sample
试验网 experimental network;try-net
试验位置 testing position;try state
试验温度 service temperature
试验文件 test documentation
试验物 trier
试验误差 experimental error;test error
试验系列 test series
试验系统 pilot system; proven system;testing system;trial system
试验系统的柔度 testing system flexibility
试验细节 test details
试验细则 testing specification
试验现场 experimental field; experimental site;testing site
试验线 test wire
试验线路 hookup;test track
试验线路板 breach board
试验限度 test limit
试验箱 testing box
试验项目 experimental project; pilot project;test item;trial heading
试验小车 test truck
试验小区 experimental plot;test plot
试验效率 test efficiency
试验心轴 test mandrel
试验芯样 test core
试验信号振荡器 test signal oscillator
试验性 testability
试验性拌和 trial batch
试验性布置 tentative layout
试验性采样 survey sampling
试验性测量 trial measurement
试验性充水 dummy fill
试验性导弹 test missile
试验性的 cut-and-trial; cut-and-try; developmental;tentative;way out
试验性的规格 tentative specification
试验性的气体 probe gas
试验性调查 experimental investigation
试验性反应堆 pilot reactor
试验性方案 tentative schematization
试验性方案(技术)规范 tentative scheme
试验性俯冲 test dive
试验性工厂 pilot plant
试验性工程 experimental engineering
试验性公路 experimental highway
试验性观测台 test observatory
试验性规模 pilot scale
试验性过程 pilot process
试验性海洋资料系统 pilot ocean data system
试验性混凝土 experimental concrete
试验性混凝土路面 experimental concrete surface
试验性结构 testing structure
试验性开发 experimental exploitation
试验性开挖 trial excavation
试验性孔 test hole
试验性模型 pilot model
试验性模型系统 pilot model system
试验性模型研究 pilot model study
试验性能 experimental performance; test performance
试验性配料 trial concrete mix
试验性破坏荷载 test failure load
试验性上升 test ascent
试验性设备 pilot plant;pilot unit
试验性设计 experimental design
试验性设计规范 tentative design criterion
试验性设计要求 tentative design criterion
试验性生产 pilot plant scale production; pilot production; preproduction;test production
试验性生产流水线 pilot production line
试验性时间表 tentative schedule
试验性试验 proving ground test
试验性疏浚 trial dredge; test dredging;trial dredging
试验性数据 tentative data
试验性提议 tentative
试验性填土 experimental banking
试验性挖泥 trial dredge;test dredging
试验性挖泥船 experimental dredge(r)
试验性微量筛分设备 pilot microscreening unit
试验性系统 pilot system
试验性现代化 tentative modernistic
试验性压水 trial injecting water
试验性研究 pilot study
试验性预应力混凝土路 experimental prestressed concrete road
试验性运转 experimental run
试验性整治工程 trial regulation works
试验性筑堤 trail embankment
试验性着陆 test landing
试验性钻孔 trial hole;trial holing
试验性钻探 test boring
试验需用时间 required test time
试验许可证 sanction for test
试验旋塞 test cock;try cock
试验学会 testing institute
试验循环 test cycle
试验压力 test(ing) pressure; test pressing
试验压力计 test pressure ga(u)ge
试验压力计的校核 calibration of test ga(u)ge
试验压头 test head
试验研究 advanced development;experimental investigation; experimental study
试验研究费 experiment and research expenses
试验研究工程 testing and research engineering

S

试验研究及开发费用 experiment research and development expenses
试验研究院 testing institute
试验研磨机 assay mill
试验样板 plaques;test panel;test piece
试验样机 development type;experimental prototype;test vehicle;trial-and-error model
试验样机车轮荷重 test weight
试验样机的重量 test weight
试验样品 development type;test piece model;test sample;test specimen
试验样品规格 specification of experimental sample
试验样品模具 test sample mo(u)ld
试验样张 test chart
试验窑 experimental kiln
试验窑炉 pilot furnace;pilot kiln
试验要求 test requirement
试验要求手册 test requirement manual
试验液体 testing liquid
试验仪表 test-meter;trier
试验仪表板 test instrument board
试验仪表灵敏度 sensitivity of tester
试验仪表要求 test ga(u)ge requirement
试验仪器 test(ing) apparatus;test(ing) instrument;tester
试验仪器的校准 calibration of testing instrument
试验仪器的准确性 accuracy of testing apparatus
试验仪器清单 test equipment list
试验仪器支架 test bed;test board
试验引线 test lead
试验应力 proof stress
试验用 on trial
试验用按钮 test push button
试验用变压器 testing transformer
试验用测力计 test dynamometer
试验用测温锥 test cone
试验用车辆样车 prototype vehicle for test purposes
试验用地锚 test anchor
试验用地面 test surface
试验用反应器 test reactor
试验用飞机 testing aeroplane
试验用粉尘 test dust
试验用附属装置 test paraphernalia
试验用功率吸收风扇 test fan
试验用环行天线 test loop
试验用夹子 test clamp
试验用架子 test stand
试验用降落伞 test parachute
试验用接线 test harness
试验用开关 test switch
试验用流体 test fluid
试验用路面 test surface
试验用闷头 <钢管的> test head
试验用配电板 testing switchboard
试验用染料 test dye
试验用筛 testing screen;test(ing) sieve
试验用筛穿孔板 perforated plate for test sieves
试验用设备 experimental unit;lab-size equipment
试验用水 test water
试验用水泥 testing cement
试验用铜版 brassboard
试验用线圈 test coil
试验用旋塞 ga(u)ge tap
试验用压出机 laboratory size extruder
试验用液体 test fluid
试验用重量单位 test weight
试验优先次序 testing priority
试验有效载荷 experimental payload
试验与操作 test and handling
试验与操作群 test and operation group
试验员 experimenter;investigator;test clerk;tester;trier
试验原理 testing principle
试验允差 test tolerance
试验运转 test run(ning);test working
试验载荷 trial load
试验凿岩 test drilling
试验责任表 test duty table
试验站 experimental station;field station;pilot depot;test house;testing station;trial plant;trial station
试验站网 experimental station system
试验者 tester;trier
试验证据 experimental evidence
试验证(明)书 certificate of proof;testing certificate;certificate of test
试验值 experimental value;tested value;test figure;trial value
试验纸条 testing strip;trial slip
试验指标 test index[复 indices]
试验指导人 test conductor
试验指导书 test code
试验指导小组 test direction team
试验指示器 test(ing) indicator
试验制备 test specimen preparation
试验质料检存 information retrieval and documentation of experiments
试验中 on trial;under test
试验中心 test center[centre]
试验周期 period of test(ing);test cycle;test(ing) period;trial period
试验注入 implant test
试验柱 testing column
试验转炉 experimental converter
试验桩 test(ing) pile[piling];trial pile
试验装备 rig for testing;testing outfit;test rig;test set-up
试验装药 trial charge
试验装载 test loading
试验装置 analyser[analyzer];bedstead;experimental rig;experiment(al) set-up;pilot installation;semiplant;test assembly;test facility;testing apparatus;testing unit;test set;trial installation
试验准备 test arrangement;test setup
试验桌 experimental bench;experimental desk;test desk
试验资料 test information;testing data;model data
试验字表 logatom
试验总表 test summary sheet
试验总延续时间 total continuous period of test
试验组 test team
试验组次 test
试验组件 test assembly;test module
试验组数 group number of test
试验钻车 test ring
试验钻进 test drilling
试验钻孔 test bore(hole);test boring;test hole;trial bore-hole
试验钻孔平面布置图 planar distribution map of testing borehole
试验钻孔数 number of test wells
试验钻头 test bit
试验作业 experimentation
试样 exemplar;handsel;proof sample;representative sample;sample(piece);sampling material;specimen;test coupon;test sample;test specimen;batch sample <道路材料>;advance print <指印刷>
试样斑 sample spot
试样板 sample board
试样棒 coupon
试样保存 preservation of samples
试样壁排水条件 side drain of sample
试样编号 sample indexing
试样编录 sample record
试样标记 sample mark(ing)
试样冰冻(试验) sample freezing
试样侧壁排水条 <三轴试验的> side drain
试样陈列室 sampler room
试样尺寸 sample size;size of specimen;specimen size
试样尺寸码 sample size letter
试样尺寸效应 effect of specimen
试样处理量 sample throughput
试样的定量组成 quantitative composition of samples
试样的定性组成 qualitative composition of samples
试样的制备 test run(ning)
试样断面 sample section
试样对准 specimen alignment
试样翻新 retrofit
试样分流器 sample splitter
试样分取器 sample splitter;sample splitting device
试样分散性 sample dispersion
试样分析 specimen analysis;specimen assay
试样分析器 sample analyser[analyzer]
试样粉碎 sample reduction
试样罐 sample carrier
试样光束 sample light beam
试样规格 dimension of sample
试样号 test piece number[test pc No.]
试样化学成分 chemical composition of samples
试样划痕器 sample streaker
试样回收 sample recovery
试样混合 sample loss
试样夹具 specimen holder
试样夹具套 stressing head
试样架 specimen mounting
试样检验 test check
试样件 coupon
试样件试验 coupon testing
试样降解 sample degradation
试样块 coupon
试样冷凝 sample condensation
试样量 sample size
试样量损耗 sample size loss
试样流出阀 effluent sample
试样炉 assay furnace
试样帽 top cap
试样面积比 <取土器的> sample area ratio
试样描述 description of sample
试样名称 sample name
试样模(子) test specimen mo(u)ld;sample mo(u)ld
试样浓缩 sample concentration
试样浓缩器 sample concentrator
试样盘 sampling tray
试样抛光机 sample polishing machine
试样批成粒机 sample-batch granulator
试样劈裂器 sample splitter;sample splitting device
试样品位 sampled grade
试样平均值 sample mean
试样破碎机 sample crusher
试样汽化 sample evapo(u)ration
试样前处理 pretreatment of samples
试样染色 test dyeing
试样容器 sample carrier;sample container
试样溶液 sample solution
试样试验 specimen test
试样收集 sample collection
试样收集器 sample collector
试样输送 sample transport
试样数量 sample quantity
试样说明 test description of sample
试样损耗 sample loss
试样缩分 sample divider pass
试样缩分器 riffler
试样探棒 sample probe
试样探针 sample probe;sample throughput
试样提出的顺序效果 order effect
试样调理时间 sample conditioning time
试样筒 test specimen tube
试样涂污 sample smear
试样温度 sample temperature
试样污染 sample contamination
试样相关系数 sample correlation coefficient
试样协变性 sample covariance
试样压制机 briquetter
试样岩心 test drill core
试样研磨 buck
试样预处理 sample pretreatment
试样支持器 sample holder
试样织工 pattern weaver
试样制备 preparation of specimen;sample preparation
试样制备方法 sample preparation method
试样制备时的缩分 division during sample preparation
试样种类 sample type
试样注射 sample injection
试样注射口 sample injection port
试样抓取器 sample grabber
试样转移操作 sample transfer operation
试样准备 sample preparation
试样组(件) specimen assembly
试药 reagent
试液 test solution
试液径旋塞 ga(u)ge cock;ga(u)ging cock
试印样 first print;preliminary sheet
试用 on probation;probationary appointment;shakedown;trial-on;tryout
试用版 proof edition
试用标准 tentative;tentative standard
试用材料 trial material
试用程序 hello program(me)
试用除数 trial divisor
试用道路 trail road
试用的 tentative
试用灯标 experimental light
试用地点 trial site
试用地沥青 trial asphalt
试用函数 trial function
试用混合料 trial mix
试用交换局 trial exchange
试用阶段 trial period
试用路线研究 trial road study
试用农药杀虫剂 probation agricultural insecticide
试用配电盘 test board
试用坡度线 tentative grade line;trial grade line
试用期 driving cycle;during probation;probation(ary) period;trial period
试用人员 person on probation;probationer
试用试验 proof test
试用数据 tentation data;tentative data
试用值 tentative value;test value;trial value
试用中 on trial
试用资料 tentation data
试油 test oil
试油器 oil tester
试油水泥塞 testing plug
试运销 sample shipment
试运行 commissioning;commissioning test;pilot run;preliminary operation;preoperation;preoperation(al) test;running-in;test operation;test run(ning);trial operation;trial run(ning)

试运行时间 commissioning time
试运行试验 running-in test; commissioning test
试运转 breaking-in; commissioning; green run; green test; preoperation; run(ning)-in; service trial; shake-down run; start-up test; test; test run(ning); trial operation; trial run(ning) test; trial trip
试运转费用 start run cost; trial run cost; trial running expense
试运转功率 trial trip rating
试运转过程 breaking-in process
试运转列车 test run train; trial trip train
试运转期 break-in period
试运转时间 time of commissioning
试运转时期 running-in period
试运转顺序 start-up trial run procedure
试运转样机 running-in machine
试运转装置 running-in machine
试载 test load; trial load
试载法 <拱坝应力分析的> trial-load method
试造车间 laboratory shop
试闸 brake test; brake trial
试针 test point
试值法 false position
试纸 indicating paper; indicator paper; reagent paper; test paper
试纸(检测)法 test paper method
试制 advanced development; pilot build; pilot production; preproduction; trial-manufacture
试制车间 development shop; experiment(al) department; laboratory shop
试制发动机 preproduction engine
试制费 experiment manufacturing cost
试制工作 development work
试制基金 <新产品> trial production fund
试制间 experimental shop
试制阶段 development stage
试制模型 preproduction model
试制品 preindustrial prototype
试制品试验 development test; prototype test(ing)
试制前试验 preproduction test
试制涂层 trial coat
试制样品 pilot sample; preproduction model; prototype unit
试制铸件 pilot casting
试置荷重 trial load
试注水 pilot flood
试转 preliminary operation; run(ning)-in
试转速度 running-in speed
试桩 preliminary pile; pre-piling; test pile; test piling; trial driving; trial pile
试桩荷载 pile test load
试桩强度 ultimate resistance
试桩台架 cribwork
试装配 trial assembly; trial erection; trial making-up
试钻 exploratory boring; pilot boring; probe boring; prospecting drill(ing); scout boring; test boring; test drill
试钻洞 trial bore hole; trial borehole
试钻工具 trial boring tool
试钻井 prospecting shaft
试钻孔 test bore(hole); trial(bore) hole; trial boring
试钻探 trial boring; test boring
试钻钻孔 exploratory(bore) hole

饰

饰板 ornamental plate; plaque(tte)
饰边 cording; selvage; selvedge; trim-edge
饰边门洞 cased opening; trimmed opening
饰边盆景 pattern edged plate
饰边植物 edging plant; skirting plant
饰编织物状的线脚 platted mo(u)lding
饰变 modification
饰窗花格 window trim
饰带 braid; frieze; ribband; ribbon; ribbon band
饰带板 ribbon board
饰带层 ribbon course
饰带压条 plate facing
饰钉 ornamental nail; stud
饰顶 surmount
饰革 dressing leather
饰花钢条 <围墙用> ornamental steel bar
饰花字头 initial ornamental
饰件 <脊瓦上的> fleur
饰金 gilding; gold plating
饰锯齿形的线脚 saw-tooth mo(u)lding
饰孔 pink
饰框 escutcheon; escutcheon plate
饰面 decorative lamination; decorative overlay; facing; facing surface; fair-faced finish; finish(ing); ground finish; ornamental surface; overcoating; plastering; surface finish; veneer
饰面板 decorative laminate; face slab; face veneer; facing board; facing panel; facing slab; veneer; wood veneer
饰面板等级 veneer grade
饰面板胶粘剂 veneer bonding adhesive
饰面板黏[粘]结剂 veneer cementing agent
饰面拌和料 face mix; facing mix
饰面玻璃 decoration glass
饰面箔 facing foil
饰面薄板 decorative(wood) veneer
饰面材料 coating material; decorative material; facework material; facing mass; finishing material; finish mass; surfacing material; facing material
饰面材料成分 facing composition
饰面层 facing layer; finishing coat
饰面层厚度的标定粉刷 <俗称出拓饼> ga(u)ging plaster for finish(ing)
饰面瓷砖 clay facing tile; facing tile
饰面的钢筋混凝土 fair-faced reinforced concrete
饰面钉 casing nail; finishing nail
饰面粉刷 finish plaster
饰面复合材料 facing compound
饰面工(程) face work
饰面(工程)类别 type of finishing
饰面工程用钉 finishing nail
饰面工程用油灰 filling putty
饰面工具 facing kit
饰面工作 air-faced work; facing work; veneering work
饰面构件 face component; face member; face unit; facing member
饰面构造 veneered construction
饰面骨料 facing aggregate
饰面烘干 drying of screeds
饰面环 finishing ring
饰面混合材料 face mix; facing mix(ture)
饰面混凝土 face concrete; fair-faced concrete; finished concrete
饰面混凝土板 faced concrete panel; fair-faced concrete panel; fair-faced concrete slab
饰面混凝土房屋正面部件 fair-faced concrete facade building unit
饰面混凝土骨料 fair-faced concrete aggregate
饰面混凝土集料 fair-faced aggregate; fair-faced concrete aggregate
饰面混凝土浇注 fair-faced concrete cast(ing)
饰面混凝土结构 fair-faced concrete texture
饰面混凝土梁 fair-faced concrete beam
饰面混凝土楼梯 fair-faced concrete stair(case)
饰面混凝土模板 fair-faced concrete forms; fair-faced concrete shuttering
饰面混凝土墙 fair-faced concrete wall
饰面混凝土柱 fair-faced concrete column
饰面基层 veneer
饰面集料 facing aggregate
饰面加强效果 stiffening effect of cladding
饰面建筑构件 face building component; face building member; face building unit
饰面胶带布 <装修技术中的> adhesive masking tape
饰面胶合板 decorative glued plywood; facing plywood; fancy plywood; veneered plywood
饰面金属箔 facing foil
饰面门框 cabinet jamb
饰面模板 face formwork; shell formwork
饰面抹灰条板 veneer plaster lath
饰面黏[粘]土砖 face clay brick
饰面黏[粘]土砖砌体 fair-faced clay brickwork
饰面喷浆 facing grouting
饰面漆 finishing paint
饰面砌块 decorative block; decorative faced block; face(d) block; facing block; fair-faced block
饰面砌体 masonry veneer
饰面砌筑砖 facing clay brick
饰面墙 faced wall; veneered wall
饰面轻质混凝土 fair-faced lightweight concrete
饰面轻质混凝土砌块 fair-faced lightweight concrete block
饰面轻质混凝土砖 fair-faced lightweight concrete tile
饰面清漆 finishing varnish
饰面石 face stone; facing stone
饰面石板 facing stone slab; plain ashlar
饰面熟石灰 finishing hydrated lime
饰面水泥 finish cement
饰面说明 description of finishes
饰面碎料板 dressed particle board
饰面涂层 finishing coat
饰面涂漆 finishing system
饰面圬工 veneering masonry(work)
饰面系统 system of facing
饰面细工 elaborate decorative coating; encrustation; incrustation
饰面型防火涂料 finishing fire retardant paint
饰面用大理岩 marble for decoration
饰面用的混凝土组合件 fair-faced cast compound unit
饰面用的混预制混凝土构件 fair-faced cast-concrete component
饰面用的预制混凝土组合件 fair-faced cast-concrete compound unit
饰面用石制件 facing stone ware
饰面用油灰 face putty; lacquer putty
饰面找平 screed finish
饰面纸浆水泥板 decorated pulp cement brick
饰面砖 air-faced tile; face brick; face tile; facing brick; facing tile; rustic brick; tapestry brick; veneer brick; wall tie; wall tiling
饰面砖隔墙 facing tile partition(wall)
饰面砖工 facing brickwork
饰面砖砌筑 facing brick bond
饰面砖砌筑的隔墙 facing block partition
饰面砖石砌筑 fair-faced masonry(work)
饰品 adornment; decoration; frilled organ
饰墙布用滚筒 roll of wall covering
饰墙后底层图迹在面层上透视 photographing
饰墙花毡 arras
饰球 passementerie
饰山墙 attached gable
饰双锥体线脚 double cone mo(u)lding
饰条制造工 trim fabricator; trim maker
饰物 adornment
饰线 mo(u)ld line
饰线单元 mo(u)lding unit
饰以花彩的圆窗 garland circular window
饰以锯齿形的 embattled
饰以锯齿形的女儿墙 embattled parapet wall
饰以锯齿形的桥上护墙 embattled bridge parapet
饰以小枝 sprig
饰用黄铜 nu-gild; nu-gold
饰有布褶的雕刻板 napkin pattern
饰有纹章的盾 scutcheon
饰缘植物 edging plant
饰针 fibula
饰砖 Dutch tile; ornamental brick

S

室

室对 cell pair
室对电阻 cell pair resistance
室法铅白 chamber white lead
室干 <木材> room-dry
室干集料 laboratory-dry aggregate
室光装入 room-light loading
室炉 room furnace
室内 interior
室内阿尔发卡法 indoor Alpha card survey
室内矮平台 estrade
室内安全电梯 interior escape stair(case)
室内安全梯 interior escape stair(case)
室内安装 indoor location; interior location
室内本底 room background
室内壁板装饰 cabinet finish
室内壁骨 interior stud
室内变电所 indoor substation
室内变电站 indoor substation
室内变水头渗透性试验 falling-head permeability test in laboratory
室内变压器 indoor substation
室内表面装饰 interior surfacing
室内玻璃安装 interior glazing
室内玻璃窗 interior glazing
室内玻璃门 indoor glass door; inside glass door; interior glass door
室内不准吃喝 eating drinking prohibited within
室内布线 house wiring; indoor wiring; interior wiring; plumbing work

室内布线用导线 interior wire
室内布置 inner layout;inside layout;interior decoration;interior layout
室内财产火灾保险 contents rate
室内采光 inner lighting; inside lighting;internal lighting
室内采暖 space heating
室内采暖炉 room air heating furnace
室内采暖炉火 room air heating
室内采暖器 room heater;space heater;space heating appliance
室内采暖设备 space heating plant
室内参数 room parameter
室内侧玻璃 indoor pane;inside pane
室内测量法 indoor measuring technique
室内常数 room constant
室内潮气 interior moisture
室内尘埃 house moss
室内陈设玻璃 interior furnishing glass
室内尺寸 interior dimension
室内初学者泳池 indoor learner's pool
室内厨房 kitchenet(te)
室内船台 covered in berth; covered slip
室内窗 borrowed light; interior window
室内窗槛 inner window cill[sill]
室内窗台 indoor window cill[sill]
室内吹干 dark air-curing
室内垂直偶极天线 lounge antenna
室内存车场 indoor parking space
室内大便器 close-stool
室内大理石 indoor marble
室内道路模拟试验装置 indoor road tester
室内的 indoor
室内的闷浊空气 fug
室内灯 indoor lamp; room lamp; room light
室内灯光喷泉 illuminated indoor fountain
室内地板 flooring
室内地面 floor finish
室内地面高程 building ground elevation;elevation of floor level
室内地面材料 flooring;interior flooring
室内地面混凝土层 surface concrete
室内地面门后夹 flooring door catch
室内地面平面 floor level
室内地坪高程 building ground elevation;elevation of floor level
室内(电话)交换系统 house exchange system
室内电话系统 house telephone system
室内电话线 house telephone circuit
室内电缆 house cable;inside(plant) cable
室内电路 interior circuit;room circuit
室内电气装置 electric(al) fixture of a room
室内电信线路分布系统 internal cable distribution system
室内电影院 indoor theatre[theater]
室内吊车 cabin heater
室内吊顶 suspended ceiling
室内顶面 interior face
室内斗技场 indoor arena
室内分析方法 laboratory analytical method
室内风干的 room-dry
室内风俗画 conversation piece
室内封闭式燃具 room-sealed appliance;type C appliance
室内敷管 indoor piping
室内敷线工 indoor wire layer; narrow back <俚语>
室内干球 room dry bulb

室内感触 feeling of space
室内高尔夫球场棚 indoor golf range shelter
室内隔热隔声 interior insulation; internal insulation
室内工程 interior works
室内工程行业 trade for interior work
室内工程用的型材 unit for interior work
室内工作 indoor work;inside work;interior works;office work
室内供暖 space heating
室内供热 interior heat
室内供水管系 interior wet standpipes
室内供水系统 indoor water supply system
室内管道 interior conduit; interior piping
室内管道工程 plumbing;internal pipework
室内管道工程固定装置 plumbing fixture
室内管道工程系统 plumbing system
室内管道架设 roughing-in
室内管道系统 indoor pipe system
室内管工 indoor plumbing worker
室内管工系统 plumbing system
室内管沟 indoor pipe trench;indoor trench
室内管沟系统 plumbing system
室内管网 house pipe system; plumbing network
室内管系 house connection; indoor pipe system
室内管线布置 service layout
室内广播 broadcast studio; studio broadcast(ing)
室内广播节目 live program(me)
室内规划 inner layout;inside layout;interior layout
室内滚筒式底盘测功器 indoor chassis rolls dynamometer
室内锅炉 indoor boiler
室内恒水头渗透性试验 constant head permeability test in laboratory
室内恒温计 room thermostat
室内恒温器 indoor thermostat; room thermostat
室内恒温调节器 room thermostat
室内护墙板 interior panel
室内护墙栏杆 wall protector
室内花格饰 interior tracery
室内花园 indoor garden;interior garden
室内画 interior picture
室内环境 indoor environment; internal ambience
室内环境工程 building services
室内混凝土柱 interior concrete column
室内混响 room reverberation
室内火电厂 indoor thermal power plant
室内火力发电站 indoor thermal power plant
室内给排水工程 plumbing
室内给排水平面图 indoor layout of plumbing system
室内给排水系统图 indoor plumbing system drawing
室内计算 calculation in office;office calculation;office computation
室内加工 office operation
室内加工车间 indoor processing shop
室内嫁接 grafting
室内间隔 interior partitioning
室内检核 office control
室内建筑板 indoor building board;interior building board;interior building panel
室内建筑钢板 indoor building sheet

室内建筑学 interior architecture
室内建筑用板 internal(building) panel
室内交通线 circulation route; route of circulation
室内胶合板用黏[粘]结剂 intermediate adhesive
室内胶乳半光泽涂料 interior latex semigloss paint
室内胶乳光泽涂料 interior latex gloss paint
室内角落尘埃 house moss
室内角隅 inside corner; interior corner
室内脚手架 inner scaffold; inside scaffold;internal scaffold(ing)
室内教练池 indoor teaching pool
室内街道 <罗马建筑艺术中的> colonnaded street
室内结霜 frost in houses
室内景观设计 interior landscape design
室内景物 indoor scene
室内净高 ceiling height;floor-to-ceiling height
室内净面积 room enclosing area
室内开关 cubicle switch;snap switch
室内开关所 indoor switching station
室内开关站 switchgear room
室内康乐中心 indoor recreation center[centre]
室内靠墙的一圈长椅 podium[复 podiums/podia]
室内空间 indoor space;interior space
室内空气 interior air
室内空气计算参数 indoor air design condition
室内空气加热燃烧器 room air heating burning appliance
室内空气冷却器 room air cooler
室内空气流速 indoor air velocity
室内空气调节器 room air conditioner
室内空气污染 indoor air pollution
室内空气污染物 indoor air pollutant
室内空气质量 indoor air quality
室内空气质量标准 indoor air quality standard
室内空气状况 state of indoor air; state of room air
室内空调器 room air conditioner
室内空调系统 room air conditioning system
室内快速试验 accelerated laboratory test
室内冷气机 indoor air cooler; inside air cooler;interior air cooler
室内冷却器 room cooler
室内立管 house riser;riser pipe;service riser
室内立面 internal elevation
室内立面图 interior elevation
室内亮度 room brightness;room brilliancy
室内列柱 columnar interior
室内流通的空气 entrained air
室内溜冰场 ice palace; indoor ice rink; indoor skating rink; roller-skating rink
室内楼梯 access stair(case);indoor stair(case);internal stair(case)
室内楼梯间 interior stair(case)
室内楼梯栏杆 interior stair-rail
室内露点 room dew point
室内录音 indoor recording
室内路线 interior wiring
室内绿化 indoor planting
室内氯化橡胶涂料 indoor chlorinated rubber paint
室内煤场 indoor coal yard
室内门窗侧壁砖 interior door jamb tile

室内门执手 inside handle
室内闷热 frowst
室内弥散试验 method of dispersion experiment in laboratory
室内面积 room area
室内模型 room model
室内末端装置 room terminal unit
室内木材涂色 interior wood stain
室内木器 interior woodwork
室内木条百叶窗 indoor slatted blind
室内黏[粘]结剂 interior cement(ing agent)
室内暖气 room heater
室内拍摄 indoor shot
室内排水 house drainage
室内排水管 building drain; interior drain pipe;protected waste pipe
室内排水管存水弯 house trap
室内排水系统 building drainage system;house sewage system;interior plumbing system
室内抛光油漆 indoor clear varnish; indoor gloss varnish
室内跑道 indoor track
室内配电 interior distribution; internal distribution
室内配电亭 kiosk
室内配电线 house wiring
室内配管 interior piping
室内配线 house service
室内配线用瓷绝缘子 porcelain insulator for interior wiring
室内喷泉 indoor fountain
室内喷湿式加湿器 room spray-type humidifier
室内喷湿式增湿器 room spray-type humidifier
室内频率响应 room response
室内平面布置 room-layout
室内铺地砖 indoor tile
室内漆器 indoor lacquer
室内气氛 internal ambience; internal atmosphere
室内气候 indoor climate;indoor weather;interior climate;room climate
室内气流 indoor air flow; room air draft;room air motion
室内汽轮机 indoor turbine
室内砌筑分隔墙 interior masonry dividing wall
室内墙 room wall;interior wall
室内墙角保护层 corner guard
室内墙顶涂层 inside coat
室内墙面 interior face; interior wall surface
室内墙面处理 inner surfacing; internal surfacing
室内清漆 indoor varnish
室内球场 inner court; sphaeristerium <古罗马>
室内取暖器 cabin heater
室内全密闭牵引变电所 indoor totally-enclosed substation
室内全密闭牵引变电站 indoor totally-enclosed substation
室内燃气管道 building gas pipe; indoor pipe;service pipe
室内热环境 indoor thermal environment
室内热量 indoor heat; interior heat; room heat
室内热源 indoor heat source
室内热增量 indoor heat gain;interior heat gain
室内人工照明 artificial indoor illumination;artificial indoor lighting
室内人造气候学 conditional climatology
室内乳化油漆 indoor emulsion paint
室内乳剂油漆 indoor emulsion paint

室内散射 room scattering
室内散射中子 room-scattered neutron
室内扇风机 room blower
室内墙上安装的室温调节器 room thermostat
室内取暖炉 room heater
室内上光清漆 interior gloss varnish
室内设备 indoor equipment; indoor facility; indoor installation
室内设计 indoor design; interior design
室内设计师 interior decorator; interior designer
室内设计温度 indoor design temperature
室内设施 inside plant
室内射击练习枪弹 gallery practice ammunition
室内摄影 indoor photography; indoor shot; live pick-up
室内升高的地面 halfpace
室内声场 room sound field
室内声响绝缘 room sound insulation
室内声响衰减 room sound attenuation
室内声学 room acoustics
室内声学条件 acoustic(al) colo(u)ration
室内施工 interior works
室内湿度 indoor humidity; indoor moisture; inside humidity; interior humidity; interior moisture; room humidity; room moisture
室内十字板试验 laboratory vane test
室内市场 indoor market; market hall
室内式发电站 indoor-type power station
室内式水电站 indoor hydro electric(al) station
室内试验 indoor test; laboratory experiment; laboratory test
室内试验场 indoor proving ground
室内试验分析项目 terms of experiment analysis at lab(oratory)
室内试验装置 laboratory testing rig
室内饰面 inside facing
室内饰面油漆 indoor finish(ing) paint
室内收集器 chamber collector
室内舒适水平 interior comfort level
室内水质自净试验 self-cleaning test of water quality in laboratory
室内送话器 booth microphone
室内太平梯 interior escape stair(case)
室内踢脚板供暖 domestic baseboard heating; domestic base plate heating
室内体育馆 indoor stadium
室内(天然)干燥 indoor seasoning
室内天线 indoor aerial; indoor antenna; inside antenna; internal antenna; room antenna
室内条件 indoor condition
室内庭园 indoor garden
室内庭院 indoor court; interior court; internal court
室内停车场 indoor parking area; parking garage
室内通风 indoor air ventilation; room air draft; room ventilation
室内通风系统 indoor ventilation system
室内透明清漆 indoor clear varnish
室内透视 interior perspective
室内凸出烟囱 chimney chest
室内涂层 indoor coating; internal coating
室内涂料 indoor coating; internal coating
室内涂色 interior stain
室内土工试验 laboratory soil test
室内外抱 indoor reveal; interior reveal
室内外标高差 difference of elevation between indoor and outdoor; difference of elevation between inside and outside
室内外计算参数 indoor and outdoor design condition
室内外两用地毯 indoor-outdoor carpet(ing)
室内外设计参数 indoor and outdoor design condition
室内外下水道连接管 slant
室内网球场 indoor tennis court
室内网球设施 indoor tennis facility
室内卫生设备安装 roughing-in
室内温度 indoor temperature; inside room temperature; interior temperature; room temperature
室内温度及湿度舒适标准 comfort standard
室内温度控制 indoor temperature control; interior temperature control
室内温度探测器 indoor temperature sensor; room temperature sensor
室内温度调节 indoor temperature control; interior temperature control
室内温湿度基数 indoor reference for air temperature and relative humidity
室内温湿度允许波动范围 allowed indoor fluctuation of temperature and relative humidity
室内圬工墙 indoor masonry wall
室内污染级 indoor pollution level
室内污水管 building sewer
室内戏院 indoor theatre[theater]
室内系统 in-house system; in-plant system
室内系统布置 inside system layout
室内细木工 inner joinery; inside joinery
室内线 house line; indoor line; indoor wire; interior wire
室内线缆 inside plant
室内线路 inside wiring; interior wiring
室内相对湿度 indoor relative humidity; interior relative humidity
室内消防 interior fire protection
室内消防栓 indoor fire hydrant
室内消防系统 indoor fire extinguishing system
室内消火栓 indoor fire hydrant; wall hydrant
室内小摆设 accessories
室内小棚 <教士听忏悔的> confessional
室内小气候 cryptoclimate[kryptoclimate]; house microclimate
室内小气候学 cryptoclimatology [kryptoclimatology]
室内小眺台 minstrel gallery
室内型变压器 indoor transformer
室内修整装饰工作 profile for interior work
室内悬吊脚手架 interior hung scaffold
室内悬挂脚手架 interior hung scaffold
室内烟道 internal flue
室内样品 indoor sample
室内仪表用 dry instrument
室内音质评价 room acoustic(al) assessment
室内用的扬声器 cabinet speaker
室内用建筑塑料 building plastic for internal use
室内用胶 interior glue
室内用胶合板 interior(-type) plywood
室内用胶粘剂 interior adhesive
室内用胶着剂 interior glue
室内用煤气炉 gas-fired floor furnace
室内用黏[粘]合剂 interior adhesive; interior bonding agent
室内用配电箱 cubicle switch
室内用乳化涂料 interior emulsion paint
室内用涂料 interior paint
室内用油漆 interior paint
室内用油性着色剂 interior oil stain
室内用釉面砖 glazed interior tile
室内用装修材料 interior finish board
室内油漆 indoor paint
室内游泳池 bath accessory; indoor pool; indoor swimming pool; natatorium[复 natatoriums/natatoria]; swimming bath
室内游泳训练池 training pool hall
室内有色清漆 indoor pigmented varnish
室内隅角处金属条 <防止抹灰粉裂纹> cornerite
室内雨水管 building storm sewer
室内浴池 alveus
室内越冬 indoor wintering
室内运输系统 indoor transportation system
室内栽培 indoor growing
室内噪声 indoor noise; inside noise; interior noise; room noise
室内噪声标准曲线 indoor noise criterion curve; inside noise criterion curve; interior noise criterion curve; room noise criterion curve
室内噪声级 room noise level
室内噪声谱 room noise spectrum
室内噪音 room noise
室内增热 interior heat gain
室内照明 cabin lighting; indoor illumination; indoor lighting; inside illumination; interior illumination; interior lighting; internal illumination; room illumination
室内照明安装高度 interior mounting height
室内照明器附件 indoor fittings
室内照明设备 indoor light fitting
室内照明设备分级 luminaire classification
室内照明装置 indoor luminaire
室内照相 indoor photography
室内遮篷 interior awning
室内蒸汽养护 room steam-curing
室内支柱 interior stanchion; interior support
室内植物 house plant; indoor plant
室内植物景观 interior plantscape
室内纸张工具箱 house journal; house magazine
室内柱 internal column
室内柱子 interior stanchion
室内砖 interior tile
室内转鼓试验台 indoor road tester
室内装管工人 internal plumber
室内装潢 upholster(ing); upholstery
室内装潢设计者 house decorator; house designer
室内装潢型材 shape for interior work
室内装潢杂志 house journal; house magazine
室内装饰 indoor decorating; indoor decoration; interior decorating; interior decoration; interior finish
室内装饰布置 decor
室内装饰材料 furnishing material; upholstery
室内装饰件 interior trim
室内装饰抹灰 cement stucco
室内装饰品 <台布、窗帘、地毯等> upholstery
室内装饰师 interior decorator
室内装饰业 upholstery
室内装饰植物 house plant; indoor decorative plant
室内装饰专业 upholstery
室内装修 interior finish; interior trim(ming); internal finish; internal trimming
室内装修墙纸 companion paper
室内装修商 upholsterer
室内装修设计 interior design
室内装修设计者 interior designer
室内装修与设施 inner work
室内装置 indoor fixture; interior fixture <尤指灯具>; indoor installation; indoor location; indoor set; inner fixtures; inside fixtures
室内浊气 fug
室内总管 rising main
室内走廊 interior corridor
室内作业 office operation; office work
室内作业法 paper method
室内做法表 room finish schedule
室式干燥机 cell drier[dryer]; chamber drier[dryer]; room dryer[drier]
室式干燥器 cabinet dryer[drier]; cell drier[dryer]; chamber drier[dryer]; room dryer[drier]
室式干燥室 cabinet dryer[drier]
室式炉 batch-type furnace
室式炉退火 batch annealing
室式收集器 chamber collector
室式温度检测器 space temperature-pulse transmission
室式窑 chamber kiln
室数 number of chamber
室外 outside
室外安装 exterior installation; outdoor location
室外安装地点 outdoor location
室外暴露 outdoor exposure
室外暴露试验 atmospheric exposure test ourdoor exposure test
室外避雷器 outdoor arrester
室外变电所 open air substation; outdoor substation
室外变电站 outdoor substation
室外变压器 outdoor transformer
室外标志 outdoor sign
室外布线 outdoor wiring; outside run; outside wiring
室外餐台 dining terrace
室外操作 field operation
室外厕所 jake; outdoor closet; outdoor toilet; privy
室外场地完成面标高 finished ground level
室外冲洗水 outside flush(ing) water
室外抽风机 exterior fan
室外出线端子 outdoor termination module
室外厨房 outside kitchen
室外储存 outdoor storage
室外打靶场 outdoor shooting area
室外大气压力 outside atmospheric pressure
室外的 outdoor
室外灯插座 outdoor lamp-socket
室外灯柱 exterior lamp post; outdoor lampholder
室外灯座 outdoor lampholder
室外底漆 outside primer
室外地板 external floor slab
室外地面 grade
室外地面标高 elevation of levelled ground
室外地坪标高 ground elevation
室外地下室墙 external cellar wall

室外电缆 external cable; outside cable
室外电路 exterior circuit
室外电影院 open air cinema
室外吊斗提升机 exterior ship-hoist
室外堆垛储存 stored in a stack out-of-door
室外堆积 outdoor pile
室外发电机 outdoor type generator
室外发射机 outside transmitter
室外防火梯 exterior escape stairway
室外防御工事 outer defense
室外分配装置 outdoor switchgear
室外风机 outdoor air fan
室外风压 outside wind pressure
室外风栅 outdoor air grille
室外扶手 external handrail; outer handrail; outside(hand) rail
室外干燥 open air drying; open air seasoning; outdoor seasoning
室外隔声 exterior noise insulation
室外工程 external works; exterior component
室外工作 field work; outdoor work(ing); outside work
室外供暖 outdoor heating
室外供热 outdoor heating
室外沟槽 outer ditch
室外管道 external piping; outdoor piping
室外管道预留接头 outdoor prepiping
室外管道综合图 general layout of outdoor pipelines
室外管沟 outdoor pipe trench; outdoor trench
室外管系 outdoor piping
室外管线 outdoor pipeline; outdoor piping
室外管线综合图 combination of outdoor pipelines
室外管线总平面图 general plan of outdoor pipelines
室外广播 nemo; outside broadcast(ing)
室外广播节目 outdoor program(me)
室外广告标志 outdoor advertising sign
室外广告建筑 outdoor advertising structure
室外锅炉 open air boiler; outdoor boiler
室外弧光灯 open arc lamp
室外户外用途 outdoor use
室外环境 outdoor environment
室外灰膏集料 exterior plaster aggregate
室外灰泥集料 exterior plaster aggregate
室外集合的风雨棚 outdoor assembly shelter
室外集会场所 place of outdoor assembly
室外挤奶台 outdoor milking bail
室外给排水总平面图 general layout of outdoor water supply and drainage system
室外(给水)立管 outside standpipe
室外给水栓 outside hydrant
室外计算日平均温度 outdoor design mean daily temperature
室外家具 outdoor furniture
室外架线 outside wiring
室外建筑 outdoor architecture
室外讲台 external pulpit
室外交换装置 outdoor switchgear
室外教学游泳池 outdoor leaner's pool
室外绝缘 outdoor insulation
室外开关装置 outdoor type switch gear
室外空地 open air space; outdoor space
室外空间 exterior space; outdoor space
室外空气 exterior air; fresh air; outdoor air; outside air
室外空气补偿控制系统 control system with outside temperature compensation
室外空气处理机组 outside air unit
室外空气计算参数 outdoor air design condition
室外空气进风道 outside air intake duct
室外空气进口 fresh-air intake; outside air intake; outside air opening
室外空气进入(口) outside air intake
室外空气入口 outside air intake
室外空气温度 outdoor air temperature; outside air temperature
室外空气温度传感装置 outdoor air temperature sensing device
室外空气吸入风道 outside air intake duct
室外空气吸入口 outdoor air intake
室外控制台 outside pulpit
室外栏杆 outside rail
室外临界照度 outdoor critical illuminance
室外龙头 outside tap
室外楼板 outside floor slab
室外楼梯 exterior stair(case); outdoor stair(case); outside stair(case); outside stairway; perron
室外楼梯扶手 outside stairrail
室外氯化橡胶漆 outer chlorinated rubber paint
室外马赛克面层 outdoor mosaic finish
室外(门)把手 outside handrail
室外面板 external panel
室外明楼梯 unenclosed exterior stair(case)
室外耐久性 exterior durability; outdoor durability
室外牛栏 cow pen
室外跑道 open air track; outdoor track
室外配电 outside distribution
室外(配电)变电所 outdoor substation
室外配电盒 outdoor distributing box
室外配电箱 outdoor distributing box
室外配电装置 switch yard
室外平台 outdoor platform; terrace
室外曝晒试验 outdoor exposure test
室外起居空间 outdoor sitting space
室外起居平台 outdoor living area; outdoor living patio
室外气干法 open air seasoning; outdoor seasoning
室外气候 external climate
室外气象参数 outdoor meteorological parameter
室外清漆 outer varnish(ing)
室外穹隆 outer dome
室外商业公告装置 outdoor commercial advertising device
室外设备 field equipment; open air installation; outdoor equipment; outdoor facility; outdoor installation; outdoor location; outdoor plant
室外设计温度 outdoor design temperature
室外设计温湿度 outdoor design temperature and humidity
室外设施 open air facility
室外摄像 outdoor pick-up
室外摄影 field pick-up; outdoor photography; outdoor shot
室外声学 free acoustics
室外湿度 outdoor humidity; outside humidity
室外湿度探测器 outdoor humidity sensor
室外市话电缆 outdoor public telephone cable
室外室内传声损失 outdoor-indoor transmission loss
室外疏散路线 external escape route; external evacuation route
室外刷漆 outdoor painting
室外台阶 exterior steps; outdoor steps; perron
室外台阶踏板 perron step
室外太平梯 exterior escape stairway
室外梯级 perron
室外天线 exterior aerial; exterior antenna; external antenna; open aerial; open antenna; outdoor aerial; outdoor antenna; outside antenna
室外条件 outdoor condition; outside condition
室外调相机 outdoor phase modifier
室外调谐线圈 outdoor tuning coil
室外庭院 exterior yard
室外通道 outside walk
室外通信[讯]设备 outside plant
室外通用地毯 anywhere carpet
室外涂料 outdoor coating
室外网球场 outdoor tennis court
室外温度 exterior temperature; external temperature; field temperature; open air temperature; opening temperature; outdoor temperature; outside temperature
室外温度传感器 exterior temperature sensing device
室外温度传感设备 outdoor temperature sensing device
室外温度探测器 field temperature sensor; outdoor temperature sensor; outside temperature sensor
室外稳定水压塔 outside standpipe
室外下水三通接头 house slant
室外线 outside wire
室外线用套管 outdoor type bushing
室外相对湿度 outside relative humidity
室外镶板 external panel
室外消防龙头 open air hydrant
室外消防洒水系统 drencher system
室外消防栓 open air hydrant; outdoor hydrant; outside hydrant
室外消火栓 outdoor fire hydrant
室外新风 outside supply air
室外新鲜空气 outside fresh air
室外信号 outdoor signal
室外信号机【铁】 outdoor signal
室外型电动机 outdoor motor
室外阳台 outdoor corridor
室外养护 field curing
室外应力龟裂寿命 outdoor stress crack life
室外用防水黏[粘]结剂 exterior adhesive
室外用建筑漆 exterior house paint
室外用建筑塑料 building plastic for external use
室外用胶合板 exterior(-grade) plywood; exterior type plywood; weatherproof plywood
室外用胶粘剂标准板 standard with exterior glue
室外用清漆 outdoor varnish
室外用乳胶涂料 exterior latex paint
室外用涂料 exterior coating
室外用油漆 exterior paint; outdoor paint; outer paint
室外油画 outdoor painting
室外油漆 external oil paint
室外油漆工程 external painting (work)
室外油质涂料 exterior oil paint
室外油质油漆 exterior oil paint
室外游泳池 open air swimming pool; outdoor swimming pool
室外有光泽漆 outer gloss paint
室外余地 open air space
室外浴场 open air bath
室外浴池 outdoor bath
室外院子 outer court
室外运动设施 outdoor sports facility
室外栽培 outdoor planting
室外噪声 outdoor noise
室外照明 exterior illumination; exterior lighting; external lighting; outdoor illumination; outdoor lighting; outside illumination; outside lighting
室外照明装置 exterior lighting unit
室外遮阳篷 outside awning blind
室外住宅装饰 outdoor home decoration
室外装饰 outside trim; upholstery
室外装修 exterior finish; exterior trim
室外装置 open air installation; outdoor location; outdoor plant
室外自然老化试验 natural outdoor weathering test
室外综合温度 sol-air temperature
室外总平面布置图 plot plan
室外走道 outdoor corridor
室外最后一层粉刷 external final rendering
室温 ambient temperature; room temperature
室温操作 ambient operation
室温超导体 room temperature superconductor
室温储藏 room temperature storage
室温传感器 room temperature sensor
室温电导率 room temperature conductivity
室温辐射 room temperature radiation
室温固化 ambient cure; room curing; room temperature cure
室温固化胶粘剂 room temperature setting adhesive
室温固化黏[粘]合剂 room temperature setting adhesive
室温光照条件下 at room temperature in daylight
室温寄存 storage in ambient temperature
室温控制 room temperature control
室温控制系统 room temperature control system
室温硫化 room temperature vulcanising; room temperature vulcanization
室温硫化硅橡胶 room temperature setting adhesive
室温敏感元件 room temperature sensor
室温黏[粘]结 room temperature gluing
室温凝结黏[粘]结剂 room-temperature setting adhesive
室温强度 room temperature strength
室温蠕变 cold flow
室温乳化清洗 cold emulsifiable cleaning
室温施工煤焦沥青涂料 cold-applied coal-tar coating
室温施工涂料 cold-applied coating
室温时效 room temperature ag(e)ing
室温舒适曲线 room temperature comfort curve
室温稳定性 room temperature stability
室温性能 room temperature property
室温性质 room temperature property
室温硬化 cold-setting; room temperature setting
室温硬化环氧树脂 epoxy resin hardened at room temperature

室温噪声 room noise;room temperature noise
室温指示灯 thermal light
室温自动调节器 room type thermostat
室吸水剂 water absorption
室形指数 room index
室压 < 三轴试验 > cell pressure
室用恒温器 room thermostat
室噪声 room noise

拭wipe

拭布 wiping cloth;wiping rag
拭擦效应 wiping effect
拭接 wiped joint
拭镜纸 lens paper
拭子 swab
拭子拉杆 < 管壁清扫 > swabbing line

是非法 yes-no-method

是否法试验 go-no-go test

柿红 persimmon juice; persimmon red

柿浆 kaki-shibu
柿木 kaki;calamander < 产于东印度 >
柿漆 persimmon juice
柿漆饰面 persimmon juice work
柿色炻器釉 kaki
柿涩酚 shibuol
柿液 persimmon juice
柿油 kaki-shibu;persimmon juice
柿汁丹宁 shibuol; tannin of persimmon
柿属 ebony; persimmon; Diospyros < 拉 >
柿子树 persimmon

适暗曲线 scotopic curve

适暗眼 scotopic eye
适被淹没的 awash
适草地的 poic
适草性 phytophily
适氮植物 nitrophile
适当曝光 correct exposure; right exposure
适当补偿(费) just compensation
适当场所 niche;suitable place
适当尺寸 reasonable size
适当存货 optimum inventory
适当措施 adequate measure
适当担保 justifying bail
适当的安排 proper arrangement
适当的产量 optimum yield
适当的场合 suitable occasion
适当的工具 proper implement
适当的公共机关 appropriate public authority
适当的供给 moderate supply
适当的机动灵活 proper maneuvering
适当的机会 convenience [conveniency]
适当的技术 appropriate technology
适当的例子 a case in point
适当的票据 eligible paper
适当的屏蔽 adequate shielding
适当的强度 adequate strength
适当的润滑作用 adequate lubrication
适当的设备 adequate provision
适当的时机 opportune moment
适当的条件 moderate condition
适当的委付通知 reasonable notice of abandonment
适当的位置 niche
适当的细化程度 appropriate level of detail
适当的要求 moderate demand
适当的租金 adequate rent
适当放养度 proper stocking
适当管理 adequate management
适当规范 aptitudal station
适当规模集成电路 right scale integration
适当计划的 well-planned
适当价格 reasonable price;right price
适当降价 modest price reduction
适当块状 sizable lump
适当利率 appropriate interest rate
适当利润 reasonable profit
适当平均(数) fair average
适当权限 due authority
适当润滑 proper lubrication
适当设计的垃圾收集车辆 suitably designed collection vehicle
适当时候 due course;due time
适当使用免耕法 using free-cultivation properly
适当数量 right quantity
适当调和 just compromise
适当位置 suitable position
适当温度 moderate temperature
适当稳定的 moderately stable
适当稀疏 < 树木的 > selective thinning
适当线向 proper alignment
适当相位接入 suitable phase switching
适当性 appropriation;eligibility;fitness
适当性能 proper property
适当涨价 modest price increase
适当制裁 appropriate sanction
适当(中)规模集成电路 right scale integrated circuit
适点法 method of curve fitting
适度保护关税 moderate protective duty
适度从紧原则 appropriately stringent financial policy
适度的报酬 measured consideration
适度的经济增长 appropriate economic growth
适度的人口 contributory population
适度的雾 moderate fog
适度地 moderately; within limit; within measure
适度放牧 conservative grazing
适度放牧量 proper stocking
适度管理 moderate management
适度规模经营 moderate scale management
适度烘成砖 medium-baked brick
适度积累率 appropriate accumulation rate
适度剂量 moderate dose
适度价格 moderate price
适度冷却 comfort cooling
适度利用 proper use
适度内压 proper inflation
适度烧成的瓦 medium-baked tile
适度烧成的砖 medium-baked brick
适度疏伐 moderate thinning
适度退化 graceful degradation
适度消费 moderate consumption
适度修剪 moderate pruning
适度需求 moderate demand
适度循环优化 moderate loop optimization
适度植物 moderate plant
适逢(日期) fall on
适腐的 saprophilous
适腐动物 saprophile
适腐殖质的 oxygeophilus
适钙植物 calcicole; calcipete; calciphile
适干旱的 xerophytic
适耕地 arable land; cultivable land; tillable land;tillage land
适耕地生物 agrophilous
适耕土壤 workable soil
适光的 photophilic;photophilous
适光范围 photopic range
适光曲线 photopic curve
适寒植物 hekistotherm
适旱变态 xeromorphosis
适旱变态的 xeroplastic
适旱的 xerophilous
适旱性植物 exrophreatophyte
适旱植物 xerophile;xerophilous plant
适航包装 seaworthy packing
适航的 airworthy;seaworthy
适航等级 grade of fit
适航(能)力 seaworthiness; seakeeping
适航区测量 navigation area survey
适航水深 nautical depth
适航水域 navigable waters
适航性 seakeeping; seaworthiness; navigability; sea-going ability; seagoing capacity【船】; seakeeping ability;airworthiness < 指飞机 >
适航性良好的船 heavy weather vessel
适航性能【船】 seakeeping characteristics;seaworthiness
适航性认可条款 seaworthiness admitted clause
适航性试验 seakeeping trial
适航证书【船】 seaworthiness certificate
适航状态 navigable condition; seaworthy
适航纵倾 sailing trim
适合 congruity;match up
适合标度 proper scale
适合捕捞 eumetric fishing
适合产量 eumetric yield
适合程度 appropriateness
适合当地的气温条件 be suited to local weather
适合当地行车与气候条件的公路 "tailor-made" highway
适合的 adaptive;congruous
适合的备择方案 adaptive alternatives
适合的速度信号 suitable speed signal
适合度 fitness; grade of fit; goodness of fit
适合度测定 goodness of fit test(ing)
适合度检测 goodness of fit test(ing)
适合度系数 coefficient of conformity
适合度因数 fitness figure
适合短日照的地区 adaptation to short-day region
适合反应堆用的石墨 reactor-grade graphite
适合海上运输包装 seaworthy packing
适合环境 keep with surroundings
适合人类的环境 for human environment
适合商销 in merchantable condition
适合湿度 workable moisture
适合时机 seasonable
适合食用的 edible
适合条件 condition of compatibility
适合条件带 zone of optimum condition
适合卫生的 wholesome
适合性 compatibility; suitability; versatility
适合性测定 test of goodness of fit
适合性检验 compatibility test
适合于循环操作的 suitability for cycle operation
适合运行的 fit for running
适合值 fit value
适合种植油橄榄 be suited to grow olive
适合装箱货 suitable containerable cargo
适货 cargo-worthy
适碱的 basophilous
适碱植物 alkaline plant
适筋梁 under-reinforced beam
适筋破坏 balanced failure
适浪性 weatherly quality
适量 proper quantity;right amount
适量沉淀 acclimatement; acclimation;acclimatization
适量收缩 approximate shrinkage
适林的 nemoral
适锚性 fitness for anchorage
适内微小气候 indoor microclimate
适配差分脉码调制 adaptive differential pulse code modulation
适配单元 adaptation unit
适配度 degree of adaptability
适配功能 adaptation function
适配环 adapter ring
适配控制 adaptive control
适配器 adapter;receptor
适配器检查 adapter check
适配器控制块 adapter control block
适配裙部 adapter skirt
适配箱 adapter junction box
适配性 suitability
适配性试验设备 compatibility test unit
适潜潜水员 fit diver
适热生物 thermophilous organism
适热微生物 thermophilic microorganism
适热细菌 thermotropic(al) bacteria
适热型 thermophilic form
适日植物 heliophilous plant
适沙丘的 thinophilus
适砂植物 psammophile
适湿植物 mesophyte
适石地的 petrodophilus
适时的 timely;tim(e)ous;updated
适时校正法 updating technique
适时进口 optimum import
适时修正 update;updating
适时修正法 updating technique
适树的 dendrophilous
适酸的 acidophil(e);acidophilic;acidophilous;oxyphilous
适酸植物 oxylophyte
适贴配合 snug fit
适土性 adaptability of soil
适温 thermotactic optimum
适温带 mesothermophilous
适温的 thermophilic;thermophilous
适温活性污泥法 thermophilic activated sludge process
适温细菌 thermophilic bacteria
适温性 thermophily
适温植物 thermophyte
适稀树干草原群落的 psilic
适线法 curve fit(ting); curve fitting method;method of curve fitting
适箱货 containerizable cargo; containerizable commodity; container load freight
适箱货物 goods fitting for container transport
适销产品 marketable goods; marketable products
适销对路的产品 products that have a ready market
适销价格 marketable price
适销性 marketability;merchantability
适销质量 good merchantable quality
适形线栅 conformal wire grating
适雪植物 chionophilous
适压细菌 barophilic bacteria

适淹 awash
适淹礁 rock awash
适岩地的 phellophilus
适盐的 drimophilous
适盐生物 halobiont;halophile
适盐微生物 halophile
适盐植物 halophile
适阳的 heliophilous
适阳植物 heliophile;heliophilous plant
适宜包装 proper packing
适宜采用桩基的 pile fit
适宜产量 optimum yield
适宜的环境 adapted circumstance
适宜的气候 good growing weather
适宜的硬件设计 proper hardware design
适宜的运输途径 due course of transit
适宜的噪声标准曲线 preferred noise criterion curve
适宜范围 comfort zone
适宜各种气候 all weather
适宜各种用途 all-purpose
适宜含水量 suitable moisture content
适宜含水率 optimum water content
适宜交收情形 deliverable state
适宜居住的场地 habitable space
适宜居住的房间 room-habitable
适宜面积的确定 estimates of optimum plot size
适宜抛锚的底质 holding ground for anchors
适宜品种 adapted varieties
适宜生境 adequate habitat; suitable habitat
适宜生长条件 optimal growth condition
适宜室温 comfort zone
适宜速度 convenient speed
适宜条件 suitable condition
适宜停放重型车辆或设备的地面 hardstanding
适宜温度 preference temperature; preferential temperature
适宜相位 proper phase
适宜性分析 suitability analysis
适宜于多种用途的 all-purpose
适宜于压力输送系统的液体 pneumatic fluid
适宜运行通量 suitable running flux
适宜种 preferential species
适阴的 heliophobic;heliophobous
适阴(的植物) umbrosus
适阴植物 ombrophyte;sciophile
适应 acclimation; acclimatization; acclimatize; accordance; adapt; adjustment; aptitude; conformability; conformation;match up
适应变化的能力 adaptability to change
适应变异性 adequate variability
适应不良 <个体与环境之间> maladjustment
适应测试 adaptive testing
适应的 adap(ta)tive;commensurate
适应的技术 appropriate technology
适应的线性动态及二次判据 adaptive linear dynamics and quadratic criterion
适应点 adaptation point
适应电平 adaptation level
适应电阻 adaptation resistance; adapter resistance
适应度 sufficiency
适应反馈控制 adaptive feedback control
适应范围 accommodation;rangeability
适应峰 adaptive peak
适应赋色 adaptive colo(u)ration
适应干扰控制器 disturbance-accommodating controller
适应干热环境的 xerothermic
适应各种气候的 all weather
适应各种气候的可靠性 all-weather reliability
适应关系 conformity relation
适应规范的要求 meet the requirement of specifications
适应过程 adaptation process; adaptive process
适应河 adjusted river;adjusted stream
适应环境 acclimatization; environmental adaptation
适应活动 adaptative activity
适应机制 adaptive mechanism
适应计 adaptometer
适应技术 adaptive technique
适应阶段 adaptation period; adaptive phase
适应类型 adapted types;adaptive form
适应理论 adaptation theory; theory of adaptation
适应力 adaptability
适应亮度 adaptation brightness; adaptation level; adaptation luminance; field luminance
适应流量 competent discharge
适应滤波器 adaptive filter
适应耐受性 adaptation tolerance
适应能力 acclimatization; adaptability; adaptation level; adaptive capacity; adaptive faculty
适应品种 adapted breed
适应期 <木材在建筑物内短期储存,以获得理想的内部温度> acclimatization
适应气候 accustom to climate; climatize
适应迁移 adaptive migratory
适应区段 <车辆在隧道口前使驾驶员适应光度变化的过渡区段> adaptation zone
适应趋同 epharmonic convergence
适应趋异 adaptive divergence
适应生存 survival of the fittest
适应时间 adaptation period
适应市场经济的计划 plan-oriented market economy
适应市场经济计划体系 plan-oriented market economy system
适应崎岖地形使用的施工设备 rough terrain equipment
适应式控制系统 adaptive control system
适应式路径选择 adaptive routing
适应式通信[讯] adaptive communication
适应水平 adaptation level
适应水系 adjusted drainage
适应算法 adaptive algorithm
适应特殊生态的 ecotopic
适应条件 condition of compatibility
适应退化 adaptative regression; adaptive regression
适应系数 accommodation coefficient; optitude factor <材料的>
适应现象 adaptation
适应线路条件的制动 braking to suit line conditions
适应消退 adaptive regression
适应行车 accommodate the traffic
适应行为 adaptive behavio(u)r
适应形状 adapted form
适应型【生】 ecad
适应性 adaptability; applicability; compatibility; flexibility; plasticity; versatility
适应性处理 compatible treatment
适应性的变化过程 process of adaptation
适应性放射 adaptive radiation
适应性辐射 adaptive radiation
适应性机能 homeostatic mechanism
适应性技术 appropriate technology
适应性检定 <包括交通安全及结构强度等> sufficiency rating
适应性建模 adaptive modelling
适应性建模方法 adaptive modelling approach
适应性进化 adaptive evolution
适应性决策 adaptive decision
适应性控制 adaptive control
适应性扩散 adaptive dispersion
适应性强的物种 adaptable species
适应性试验 aptitude test; compatibility test
适应性通道倍增器 flexible channel multiplier
适应性系统 adaptive system
适应性研究 adaptation study
适应性再使用 adaptive re-use
适应循环 adaptive-coping cycle
适应要求 fit the bill
适应于要求 adequate to the demand
适应预报 adaptive prediction
适应元件 adaptive element
适应照明 adaptation lighting; adapting lighting
适应值 adaptive value
适应自然的设计 design with nature
适应最优化 adaptive optimization
适用标准 applied code; code requirement
适用船用技术 marinization
适用的 rated; viable
适用的法律 applicable law; governing law
适用的水 suitable water
适用的水质标准 applicable water quality standards
适用地下水水质 groundwater quality suitable
适用地下水水质带 zones of groundwater quality suitable
适用度 relevance grade; relevance weights
适用范围 applicability; extent to which applied; rangeability; range of applicability; scope of application; sphere of application
适用范围与专业 scope and field of application
适用轨型 suitable rail type
适用技术 appropriate technology; suitable technology
适用价格 blanket price
适用量 dosage
适用率 relevance factor
适用期 available period; pot life; shelf life; storage life; usable life; usual life
适用区 zone of application
适用时间 <漆料> spreadable life
适用寿命 shelf life
适用条件 applicable condition
适用条款 applicable notes; applicable provision
适用铁矿石 usable iron ore
适用系数 proper-use factor
适用形式 service form
适用性 adaptability; applicability; fitness for use; performance; serviceability; suitability; usability
适用性检查 suitability inspection
适用性试验 employment and suitability test; operational suitability test; operational test
适用性试验的鉴定相配的 suitability test evaluation
适用性系数 applicability parameter; suitability number
适用性准则 serviceability criterion
适用烟道 appliance flue
适用样品 adequate sample
适用于定区放牧 applicable for set stock grazing
适用于各种情况的规则 blanket rule
适用于各种情况的焊条 all-position electrode
适用于机械挖掘的土壤 ground suitable for mechanical excavation
适用于空间 space-rated
适用于某一具体地点的模型 site-specific adaptive model
适用于热带地区种植 be suitable for grow in tropical area
适用于人的 man-rated
适用于石灰性土壤 suitable for limed
适用于水下的 subaquatic; subaqueous
适用于住宅炉灶的烟囱 residential appliance type chimney
适用于最小的农村远端集成器和最大的国际关口局 full range
适用于做地板的木料 floor board
适用资料 relevant information
适于步行的 walkable
适于承载 loadable
适于城市的运输系统 city oriented transport system
适于放牧 fit for depasturing
适于耕种的 arable
适于耕种的地 arable land
适于航海的 snug
适于航行 navigability
适于居住 habitability
适于居住的 fit for human habitation; inhabitable
适于居住的房间 livable room
适于居住的楼面积 livable floor area
适于居住环境 habitable environment
适于利用计算机的操作体制 computer-oriented operational system
适于拍摄电视的 telegenic
适于气候的汽油 climatic gasoline
适于食用的 suitable for eating
适于试验的 fit for testing
适于水下的 subaquatic; subaqueous
适于水下使用 submersed
适于夏季的 aestival
适于以物易物(或小额贸易)的商品 truck
适于用灰刀填塞的填缝物 <有别于用刷的填缝物> knifing filler
适于远洋航行的 sea going
适于赠送的 presentable
适于装入程序的形式 loader-compatible form
适雨天植物 hydrochimous
适雨植物 ombrophile; ombrophyte
适者生存 survival of the fittest
适中 moderation
适中的 temperate
适中的格式 medium format
适中的数量 moderate quantity
适中的温度 moderate moisture
适中的雨量 moderate rainfall
适中的植距 intermediate spacing
适中改变试验计划 moderate change test project
适中密度 optimum density
适中束 reasonable beam
适装证书 certificate of fitness

铈

铈玻璃 cerium glass
铈的 ceric
铈钙钛矿 knopite
铈硅钙球墨铸铁 oz cast iron
铈硅磷灰石 britholite
铈硅石 cerite
铈合金 mischmetal

铈褐帘石 cerorthite
铈矿 cerium ore
铈镧稀土合金 mischmetal
铈磷硅钍石 cerphosphorhuttonite
铈磷灰石 britholite
铈钠闪石 chiklite
铈铌钙钛矿 loparite
铈热还原剂 cerium-thermic reducer
铈烧绿石 cerian pyrochlore;ceriopyrochlore;ceruranopyrochlore;marignacite
铈石 ceria;cerianite
铈酸盐 cerate
铈铁 ferrocerium
铈铁白云石 cerium-ankerite;codazzite
铈钍吸气剂 cerium-thorium getter
铈钨华 cerotungstite
铈烯土 cerium mischmetall
铈钇矿 yttrocerite
铈钇钛铁矿 kalkowskite
铈铀钛铁矿 davidite;ferutite
铈铀铁钛矿 ufertite

释出 liberation

释出的矿物 released mineral
释出能量 released energy
释出中子 released neutron
释放 deliverance; delivery; disconnect;drop-away;given off;release;releasing;relieve;set free;tripping;unbind;uncage;uncaging;unhitch;unlatch;unloose(n);cast off
释放安匝 release ampere turn
释放按钮 release button; release-push; releasing button; trip push button
释放保护信号 release guard signal
释放报警 release alarm
释放比 release ratio;release to birth ratio
释放柄 release handle;releasing handle
释放参数 dropout value
释放操作手柄 relief lever
释放叉 discharging pallet
释放叉瓦 unlocking pallet
释放持续时间 duration of release
释放触点 releasing contact
释放的继电器 relay deenergized; relay released
释放点 release point
释放电磁铁 release magnet; releasing magnet
释放电流 drop-away current; dropout current; release current; releasing current
释放电流值 dropping current strength
释放电路 release circuit
释放电压 drop-away voltage; release voltage
释放电压的正常值 normal dropout voltage
释放定时器 release timer
释放轭 release yoke
释放阀 relief valve
释放风压 release air pressure
释放杆 discharge lever;release lever; release link
释放杆簧 unlocking lever spring
释放杆簧夹 unlocking lever spring clip
释放杆螺母 unlocking lever nut
释放杆套 unlocking lever tube
释放钩 release hook
释放钩架 release hook bracket
释放挂钩 release shackle
释放规律 release pattern
释放滚轴 discharging roller
释放簧 slack spring;trip spring
释放机构 release gear;release mechanism; releasing gear; releasing mechanism; tripping gear; tripping mechanism
释放机械 release machinery
释放基本数据 release of basic data
释放棘爪杆 unlocking click yoke
释放棘爪杆簧 unlocking click yoke spring
释放棘爪簧 unlocking click spring
释放剂 releaser;releasing liquid
释放继电器 release relay
释放键 release key
释放角 separation angle
释放接点 deenergized contact
释放进路 releasing of a route
释放禁止操作指示器 release-quiesce indicator
释放开关 release-push
释放矿物 released mineral
释放力 release [releasing] force; release power
释放力矩 release moment
释放连接杆 release link
释放联轴节 releasing coupling
释放量 burst size
释放疗法 release therapy
释放灵敏度 release sensitivity
释放率 discharge rate
释放轮 unlocking wheel
释放轮操作杆 unlocking wheel operating lever
释放轮操作杆簧 unlocking wheel operating lever spring
释放轮棘爪 unlocking wheel click
释放轮夹持螺钉 unlocking wheel holder screw
释放轮摩擦簧 unlocking wheel friction spring
释放轮片 unlocking wheel plate
释放轮压板隔片 unlocking wheel plate distance piece
释放脉冲 releasing impulse;releasing pulse
释放面积 disengagement area
释放命令 release command
释放能 let-loose energy
释放扭力 torque release
释放期间 < 继电器的 > deenergized period
释放气回收 purge gas recovery
释放汽 disengaged vapo(u)r
释放器 releaser
释放球阀 ball relief valve
释放热量 heat release;release of heat
释放热能 release thermal energy
释放容许量 detachment allowance
释放时间 down-away time;drop-away time;release time;take-down time
释放闩 release bar
释放速度 liberation velocity
释放速率 rate of release;release rate
释放踏板 release the pedal
释放弹簧 release spring; retracting spring
释放条件 release requisition
释放停止指示符 release-quiesce indicator
释放凸轮 release[releasing] cam;trip cam
释放凸轮簧 unlocking cam spring
释放位置 off position;position 0
释放温度 releasing temperature
释放线 release wire;releasing wire
释放线圈 release[releasing] coil
释放信号 release signal;unlock signal
释放信号寄存器 disconnection register
释放信息素 release pheromone
释放压力 release pressure
释放压力控制阀 relief pressure control valve
释放延迟 hangover;releasing lag
释放延迟时间 hangover delay
释放因子 release[releasing] factor;releasing hormone
释放语句【计】free statement
释放占用 release busy
释放值 drop-away value;dropout value;release[releasing] value
释放指令 release command; release order
释放中继线 release trunk
释放周期 deenergized period;release period
释放爪 release piece;unlocking click;unlocking finger
释放爪簧 unlocking finger spring
释放装置 releaser;release unit;releasing arrangement;releasing device;releasing gear; releasing mechanism;tripping device
释放状态 open state; release condition;released state
释放阻力 unlocking resistance
释放作用 release action
释负 uncharge
释荷阀 unloading valve
释荷面积 unloading area
释荷阳极 relieving anode
释荷装置 relief mechanism
释磷 phosphorus release
释磷速率 rate of phosphorus release
释能过程 exoergic process
释能密度 power density
释气 outgas;outgassing
释气剂 bubble-release agent
释气聚合物 gas-yielding polymer
释气率 outgassing rate
释去(负担)relieve of
释热 heat generation
释热率 rate of heat release
释热率量热计试验 heat release rate calorimeter test
释热元件 fuel element
释热元件损伤探测器 burst slug detector
释热元件外壳 fuel element jacket;slug can
释热元件组件 stack of fuel elements
释热组件 cluster of fuel elements
释水度 specific storage
释水系数 storativity
释压 relief of pressure; pressure release
释压带 relieved zone
释压阀 pressure-relief valve
释压接口 pressure-release coupling
释压节理作用【地】pressure-release jointing
释压裂隙 release fracture
释压面 pressure-release surface
释压试验 pressure-relief test
释压装置 pressure-release assembly
释抑 hold off
释抑电路 hold-off circuit
释重节理 lift joint;release joint

嗜粪生物 coprophyte

嗜铬染色 chromaffin
嗜铬体 chromaffin body
嗜光的 photophilic
嗜寒的 cryophilic
嗜碱微生物 alkalophilic microorganism
嗜碱性 basophilla
嗜碱植物 alkaline plant
嗜菌体学 protobiology
嗜冷微生物 psychrophilic microorganism
嗜冷细菌 psychrophilic bacteria
嗜冷厌氧废水处理 psychrophilic anaerobic wastewater treatment
嗜硫细菌 thiophilic bacteria
嗜眠病 sleeping sickness
嗜热的 thermophilic
嗜热范围 thermophilic range
嗜热好氧生物废水处理 thermophilic aerobic biological wastewater treatment
嗜热菌 thermophile bacteria; thermophilic bacteria
嗜热生物 thermophile
嗜热微生物 thermophilic microorganism
嗜热消化 thermophilic digestion
嗜酸的 acidophilic;acidophilous
嗜酸染色的 oxychromatic
嗜酸染色质 oxychromatin
嗜酸乳杆菌 Boas-Oppler bacillus; Lactobacillus acidophilus
嗜酸生物 acidophole
嗜酸微生物 acidophilic microorganism
嗜酸细菌 acidophilic bacteria
嗜酸植物 acidophilic plant
嗜铁细菌 iron bacteria
嗜温的 mesophilic
嗜温生物 mesophile
嗜温消化 mesophilic digestion
嗜温消化范围 mesophilic range
嗜温性细菌 mesophilic bacteria
嗜污水真菌 lymaphile
嗜盐球菌 halococci
嗜盐微生物 halophile; halophilic microorganism
嗜盐细菌 halophilous bacteria; salt loving bacteria
嗜氧细菌 halophilous bacteria
嗜银性 argyrophilia
嗜有机质的 metatrophic
嗜中性 neutrophilia
嗜中性的 neutrophil(e); neutrophilic;neutrophilous
嗜中性颗粒 neutrophilic granule
嗜中性粒 neutrophil(e)granule

噬菌体 bacteriophage;phage

噬硫杆菌 thiobacillus
噬铁细菌氧化剂 ferrobacillus ferrooxidant

螫sting

收板装置 catcher

收报 code reception;telegraph reception
收报处 receiving office
收报穿孔机 reperforator
收报船 ship of destination
收报地址 destination address
收报分配器 reception distributor
收报复凿机 receiving reperforating apparatus
收报机 radio telegraph receiver; receiver; recorder; telegraphic register; telegraph receiver; telephone register;ticker
收报人 addressee
收报凿孔机 reperforator
收臂后长度 < 起重机 > retracted length
收编程序 librarian
收标期限 bid time
收舱阀 stripping valve

收藏 collect;house;store up;stow
收藏起来 lock in
收藏室 <皮箱、行旅箱等的,英国> boxroom
收车时间 off-running time
收尘 dust collection;dust recovery
收尘袋 dust bag;dust-collecting sleeve;soot pocket
收尘器 dedusting filter;dust allayer;dust collector;dust extractor;dust precipitator;grit arrestor;precipitator
收尘设备 dust-collecting installation
收尘系统 dust-collecting system
收尘效率 dust collection efficiency
收尘效率参数 precipitation rate parameter
收尘旋风筒 dust-collecting cyclone
收尘装置 dust-arrester installation;dust-collecting installation
收尘总面积 gross filter area
收成 crop;harvest
收成保险 crop insurance
收成保险计划 crop insurance scheme
收成比例地租 crop share rent
收成不佳 poor crop
收成估计 crop prospect
收成预测 crop forecast
收尺日 measuring day
收锤标准 condition for stopping hammering
收存人 consignee
收大于支 revenues are over expenditures
收带盘 machine reel;take-up reel;take-up spool
收到 come to hand;receipt;received
收到材料 purchasing delivery receipt
收到材料报告单 return of purchases
收到出口信用证 export letter of credit received
收到的贷放款项 funds received for participation in loans
收到的脉冲 received impulse
收到付款 charge collect
收到和发出 received and forwarded
收到货后付款 pay(able) on receipt
收到基 as received basis
收到马力 delivered horse power
收到日期 date received
收到时间 time of receipt
收到数量 quantity received
收到通知 acknowledge
收到效果 payoff
收到许可证条款 receipt of license clause
收得率损失 yield loss
收点 shipping destination
收电人 addressee
收电子注册 collector
收赌注处 <赌场的> pool room
收兑金银外币 buy and change gold;silver and foreign currencies
收兑债券 call a bond
收发 transceiving
收发报机 receiver-transmitter;send-receiver
收发报汽车 telecar
收发电文 code-practice material
收发端机 transceiver
收发断路接点 send-receiving-break contact
收发分置雷达 bistatic radar
收发分置声呐 bistatic sonar
收发管 anti-transmit-receive tube;transmit-receive tube
收发合置声呐 monostatic sonar
收发话机 voice transmitter-receiver
收发话器 headphone;monophone
收发混合运行 composite working
收发机保护放电管 transmit-receive tube
收发键盘装置 receive, send keyboard set
收发接地转换开关 send-receive-ground-switch
收发距 distance between transmitter and receiver
收发开关 transmit-receive switch
收发两用机 composite set;receiver-transmitter;receiver-transmitter unit;transmitter-receiver
收发两用无线电设备 two-way radio
收发器 transceiver;transmitter-receiver
收发器数据传送装置 transceiver data link
收发设备 transceiver
收发室 office for incoming and outgoing mail;receiving and dispatching room;receiving room;dispatcher's office
收发天线 dual mode antenna
收发信防卫度 near end crosstalk ratio between two directions of transmission of a circuit;receiving-transmitting level difference
收发信机 transceiver
收帆绞辘 garnet purchase
收帆索 downhauler;tripping line
收方 debit(or);debit side;receipt side
收放机构 retraction jack
收放式电缆卷筒 retractable cable reel
收放式发射器 retractable launching box
收放式发射装置 retractable launcher
收放式扶梯 retractable stairway
收放式减速板 retractable air brake
收放式起落架 extendible landing gear
收放式散热器 retractable radiator
收放式跳板 retractable ramp
收放式尾轮 retractable tail wheel
收放式着陆灯 retractable landing lamp;retractable landing light
收放支腿时间 retractable time
收费 charge;fee;toll
收费比例 scale of charges
收费便宜的公共汽车 jitney bus
收费标准 fee scale
收费表 rate sheet;tariff <旅馆或公用事业的>
收费财政 toll financing
收费厕所 pay toilet
收费长途电话 toll call
收费车道 toll lane
收费车道计算机控制机 toll lane computer controller
收费车道检测器 toll lane detector
收费车道栏杆 toll lane rail
收费处 toll house
收费处路障 toll barrier
收费错误 error in charging
收费岛 toll island
收费道路 <收通行费的道路> toll road;tollway
收费的 charged;pay
收费的高速公路 turnpike
收费的运输 charged traffic
收费低廉 moderate cost
收费电话 charged call
收费电缆网络 pay cable network
收费电视 fee television;pay-as-you-see;pay-as-you-see television;pay-as-you-view;pay television;subscriber television;subscription television
收费电视广播节目 subscription television broadcast program(me)
收费额 toll revenues
收费额的制定 <公用事业的> rate-making
收费法 fee method
收费方式 toll model
收费高速公路 toll motorway;turnpike;turnpike road
收费公路 toll highway;tollway
收费公路公债 toll road bonds
收费估计 fee estimate
收费估算法 <按人工、开支加权的> multiple of direct personnel expense
收费管理 toll administration
收费广场 toll plaza
收费过多 overcharge
收费过高的 exorbitant
收费计时器 chargeable time counter
收费计算机 charge computer
收费记录处理器 toll transaction processors
收费监督控制台 toll supervisors console
收费鉴定人 fee appraiser
收费交通 toll traffic volume
收费距离 charged distance
收费卡口 toll gate
收费拦门 toll bar
收费里程 charged distance
收费立体交叉 toll-type interchange
收费路 toll road
收费率表 schedule of terms and conditions
收费率区分器 tariff zoner
收费门 pike;toll bar;toll gate
收费浓缩 toll enrichment
收费棚 toll canopy
收费评估人 fee appraiser
收费器 rater
收费桥 toll bridge
收费人 toll collector
收费少的小公共汽车 jitney
收费设备 toll collection
收费设施 toll facility
收费收入 toll revenues
收费水平 tariff level
收费送货 charge send
收费隧道 toll tunnel
收费所 toll booth;toll house
收费弹性 toll elasticity
收费亭【道】 toll booth;toll house
收费亭设施 toll-settings
收费停车 fee-parking
收费停车场 commercial parking area;commercial parking facility;commercial parking lot;parking area with charges
收费停车处 commercial parking space
收费通话 chargeable call
收费通行券 toll pass ticket
收费图书馆 lending library;pay library
收费系统软件 toll system software
收费显示器 toll display
收费项目 charge collectable
收费用电磁铁 <投币电话机> collecting magnet
收费用水量 billing demand
收费运输 revenue-earning traffic
收费载重 payload
收费载重率 payload ratio
收费栅 pike
收费站 fare collecting station;toll (collection) station
收费者 toll collector;toller
收费支票户 fee checking account
收费指数 charge index
收费制(度) rate system;rating system;tariff system
收费制式 toll model
收费中心 toll center[centre]
收费重量 chargeable weight
收费自动调整条款 escalation clause
收分 batter(ing) <指墙>;contracture;entasis[复 entases] <指柱>
收分法 method of entasis
收分墙 baterred wall;talus wall
收分曲线微凸线 <西方古典式柱身的> entasis[复 entases]
收付平衡 balance
收付实现基础 cash basis
收付实现会计制 account(ing) on cash basis
收付实现制 accounting on the cash basis
收付实现制会计 cash basis accounting
收复 recapture;reclaim;reclamation;regain
收干草的耙 hayrack
收割 cropping;harvesting;reap
收割机 cropper;harvester;harvesting machine;reaper
收割者 reaper
收工 cease work;knock off;pack up;stop work for the day
收工集合 packing-up and assembly;gathering after completion of dredging work【疏】
收购 procurement;purchase
收购、仓储及运输 purchase, storage and transportation
收购额 value of purchase
收购股权 tender offer
收购国 country of purchase
收购和拆卸废船的承包人 ship-breaker
收购机构 procurement agency
收购基数 purchase cardinals
收购价(格) procurement price;purchase price;purchasing price
收购价格指数 procurement price index
收购借贷储存方案 purchase-loan storage program(me)
收购量 volume of purchase
收购农副产品 purchase farm produce and sideline products
收购手续费 acquisition commission
收函通知书 form of acknowledgment
收话器 telephone receiver
收回 call in;drawback;drawing(-in);payback;recede;recover;regain;repossession;resume;resumption;retract;retraction;retrieve;revoke;revulsion;withdraw;retirement <成本、通货等的>
收回报价 revoke an offer;withdraw an offer
收回财产 eviction;evict property
收回产权诉讼 ejectment
收回产业 repossessed property
收回成本 cost recovering;retire
收回出租的房地产的诉讼 writ of attachment
收回待修产品 call-back
收回贷款 call in a loan;recall loan;recover loan
收回的票据 retired bill
收回的请求 reclamation
收回抵押权 take back a mortgage
收回非法扣留动产 replevin
收回费用 recoup oneself
收回分期付款货物 repossession
收回股份基金 stock redemption fund
收回坏账 bad debt recovery
收回坏账收益 income from recoveries of bad debts
收回或取消建筑许可证 cancellation of building licence
收回基建费 capital recovery cost
收回(价)值 recover value
收回旧欠 collect outstanding account
收回票据 retire
收回期 payback period;retirement period

收回权 re-entry
收回使用 reclamation service
收回提单 surrender bill of lading; withdrawal of bill of lading
收回投资 recouping the capital outlay
收回投资费 capital recovery cost
收回已售产品 product withdraw
收回优先股票准备 reserve for redemption of preferred stock
收回摘要 resumption
收回债券 retire bonds
收回证件 decertify
收回注销 surrender for cancellation
收回资本 capital recovery
收回资金的时间标准 recovery guideline
收回租地 eviction
收回租借 re-entry
收回租屋 eviction
收货报告单 received report
收货备运 received for shipment
收货处 receiving office
收货单 cargo receipt; cargo sheet; receiving note
收货单位 consignee; receiving unit
收货地点 place of receipt
收货费用 receiving expenses
收货付款 collection on delivery; collect on delivery
收货估价单 receiving quotation
收货候装 received for shipment
收货回单 acknowledgement
收货检验 inspection test
收货码头 receiving terminal; reception terminal
收货棚 receiving shed
收货票据 goods received note
收货凭单 consignment sheet
收货凭证 receipt voucher
收货人 cargo receiver; consignee; recipient
收货人保证书 consignee's letter of guarantee
收货人仓库 consignee's warehouse
收货人地址 consignee's address
收货人拒收货物 refused freight
收货人要求变更货车到达站 redirection of wagon by consignee
收货人指示 order of consignee
收货时付款 cash on receipt of merchandise
收货室 receiving office
收货收据 <领货收据> consignee's receipt
收货通知 receiving note
收货物税 excise
收货与交货 receiving and delivery
收货员 goods clerk; receiving clerk
收货站 receiving station
收获表 yield table
收获表标准地 yield sample plot
收获残余物 crop residue
收获(测定)法 harvest method
收获的农作物 cropper
收获递减律 law of diminishing returns
收获堆垛机 harvester stacker
收获谷物记量器 grain register
收获机 cropper; harvester
收获机的输送器 harvester conveyer [conveyor]
收获机具 harvesting equipment
收获机用输送器帆布带 harvest duck
收获季节 harvest season
收获量 crop; harvest yield; take; yield
收获量调节 yield regulation
收获率 recovery; yield; yielding capacity
收获前干燥剂 pre-harvest desiccant
收获用附加栏板 harvest ladder
收获用冷藏装置 harvest cooler
收获原理 cropping principle
收获运装机 loader-harvester
收获者 harvester
收获装置 harvesting apparatus; harvesting mechanism
收集 acquisition; collect; balling up; call in; catch; entrap; gathering; trap
收集板 catch tray; collecting board
收集槽 collecting gutter; collecting vat; holding tank; receiving tank
收集场 collecting field
收集车 collecting vehicle
收集沉沙设施 grit catcher
收集程序 collection procedure
收集单元 collector unit
收集到的许多品系 many of the strains in the collection
收集的 collecting
收集地点 collecting site
收集点 bleeding point
收集电极 collector electrode; passive electrode
收集电流 collection of current
收集电流比 collected-current ratio
收集电流系统 system of current collection
收集电势 collecting potential
收集电位 collecting potential
收集堆运螺旋 gathering auger
收集方法 collection method
收集方向 collecting direction
收集工商业废料的箱子 lugger body
收集管【给】collecting sewer
收集管线 gathering line
收集罐 drip tank
收集辊 collecting drum
收集过滤器 entrainment filter
收集或验收方法 <轧材的> take-up method
收集极 collector
收集极开路的反相器 inverter open collector
收集极圆筒 collector cylinder
收集技术资料 engineering data search
收集角 collection angle
收集井 incoming well
收集阱 collection trap; incoming barrel; incoming compartment trap
收集坑 collecting sump; collection sump
收集空气中污染颗粒系统 air-polluted particles capture system
收集垃圾专营权 franchise collection
收集漏料的漏斗 spillage collection hopper
收集螺栓 tie bolt
收集螺旋 gathering screw
收集牛奶用奶罐车 bulk milk collection lorry
收集盘 catch basin; catch tray; drip pan
收集频率 collection frequency
收集瓶 receiving flask
收集器 accumulator; catcher; collector; drip cup; gatherer; interceptor; receiver; scoop; trap
收集器效能 collector performance
收集栅网 collector mesh
收集设备 collecting device
收集设计资料 engineering data search
收集时间 acquisition time; collection time
收集桶 gathering barrel
收集土壤溶液 collecting soil solutions
收集网横向钢筋 longitudinal reinforcement of the collecting net
收集物 collection
收集系统 collecting system; collection system; gathering system
收集箱 collecting box
收集效率 collection efficiency
收集信息 acquisition of information
收集性能试验 trap test
收集液 drips
收集溢油 containing overflows
收集有用废物 scavenge
收集站 collection center [centre]; gathering station; collection stop
收集者 collector
收集中心 collection center[centre]
收集装置 gathering attachment; gathering unit
收集资料 collect data; collection of data; data gathering; data search; glean; swap data
收件局 receiving office
收件人 addressee; consignee; receiver
收浆 set of mortar
收截面门梃 <通常在上部装玻璃处收小> diminishing stile
收紧 hauling; tighten; backsetting <砌墙等工作>
收紧背隙 take-up the backlash
收紧间隙 take-up the slack
收紧辘绳 round up
收紧锚(索) bring home; fetch home
收紧器 tension toggle operator; tightener
收紧绳索 tail up on a rope; take-up slack
收紧索 laniard; lanyard
收紧橡胶带装置 <输送机> take-up device
收紧游隙 take-up the lost motion
收进 run-in; take in; backsetting; setback <砌墙等工作>
收进舰首水平舵 rig in bow planes
收进式扶垛 setback buttress
收进线 setback line
收颈 collaring; necking down
收据 acknowledgement; acknowledgement of receipt; quittance; receipt; receipt note; voucher
收据簿 receipt book
收据存根 stub of receipt
收据票据 collecting note
收卷 stow
收卷辊 wind-up roll
收卷松脱 doffing
收卷装置 wrap-up
收孔 <拉丝模模孔的> batter(ing)
收口 <空心铸件或管件的> closing in
收口部分 mouth
收口匣形天沟 <女儿墙后的箱形水槽> tapered parapet gutter
收口用去角七分头 three-quarter closer
收款 collection; collection of payment; make collections; receipt of payment
收款本票 collection note
收款比率 collection ratio
收款便条 receipt slip
收款处 cash department; receiver's office
收款传票 receipt voucher
收款代理商 collecting agent
收款汇票 collecting bill
收款率 <指账款回收率> collection rate
收款凭单 receiving voucher; warrant
收款清单 collection schedule; schedule of collection
收款人 beneficiary; payee; remittee
收款人现金账 receiver's cash account
收款入账 account payee
收款书 covering warrant
收款银行 due bank
收款员 collector; deposit(e) teller; receiving cashier
收敛 astringe; constriction; contraction; waist; convergence
收敛凹凸透镜 converging meniscus
收敛半径 convergence radius; converging radius; radius of convergence
收敛比 convergence ratio
收敛不一致的 non-uniformly convergent
收敛不足 misconvergence
收敛部分 contraction section; contractor
收敛槽 converging channel
收敛测量 convergence [convergency] measure(ment)
收敛差 convergence[convergency] error
收敛常数 convergence constant; convergency constant
收敛程度 degree of convergence[convergency]
收敛磁场 convergent magnetic field
收敛的 astringent; convergent
收敛的阶 order of convergence
收敛点 convergence point; converging point
收敛迭代程序 convergent iterative procedure
收敛度 degree of convergence [convergency]
收敛罚函数 convergence penalty function
收敛法 convergence method
收敛反应 convergent response
收敛反应堆 convergent reactor
收敛分支系统 convergent branching system
收敛管道内的流动 contracting duct flow
收敛横坐标 abscissa of convergence
收敛级数 convergence series; convergent series
收敛极限 limit of convergence
收敛计 convergence ga(u)ge; convergence indicator; convergent ga(u)ge; convergometer
收敛计算 convergence calculation; convergency calculation
收敛剂 astringent; astringent
收敛检验 test for convergence
收敛角 angle of convergence[convergency]; convergence angle; convergent angle; theta angle; mapping angle <子午线的>
收敛近似 converging approximation
收敛矩阵 convergent matrix
收敛控制 convergence control; convergent control
收敛扩散 convergent-divergence
收敛扩散的 convergent-divergent
收敛扩散形喷管 con-dinozzle; convergent-divergent nozzle
收敛理论 convergence theory
收敛幂级数环 ring of convergent power series
收敛年龄 convergent ages
收敛判据 convergence criterion [复 criteria]; criterion [复 criteria] of convergence
收敛喷嘴 convergent nozzle
收敛频率 convergence frequency
收敛区间 interval of convergence
收敛区(域) convergence region; region of convergence
收敛射线 converging ray
收敛失效 misconvergence
收敛时间曲线 convergence time curve
收敛式燃烧室 constrictor
收敛试验 convergence test
收敛室 necked-down chamber

收敛束 convergent beam
收敛水流 convergent flow; converging flow; reduced flow
收敛速度 contractive velocity; rapidity of convergence; rate of convergence; speed of convergence
收敛速率 convergence rate; rate of convergence; speed of convergence
收敛酸 styphnic acid
收敛算法 convergence algorithm
收敛条件 condition of convergence
收敛图 circle of convergence; convergence map
收敛尾部 tapered tail
收敛系统 collective system; convergence system
收敛响应 convergent response
收敛项 convergent
收敛形 convergent contour
收敛形管道 contracted channel
收敛形进气道 contracted air duct
收敛形喷管 constrictor nozzle
收敛形喷嘴 convergence nozzle
收敛形状 contraction shape
收敛型进气管 effuser
收敛性 astringency; convergence property; convergence【数】
收敛性的 astringent
收敛性加速 convergence acceleration
收敛性检验 test of convergence
收敛性判别准则 convergence criterion; converging criterion
收敛性判定 convergence test
收敛性判定准则 convergence criterion
收敛性判据 convergence criterion; converging criterion
收敛性物质 astringent substance
收敛序列 convergence sequence; convergent sequence
收敛仪 <地下工程观测用的> convergence indicator
收敛因素 funnel(l)ing factor
收敛因子 contractive factor; convergence constant; convergence factor
收敛域 contractive domain; convergence domain; domain of convergence
收敛原理 contractive principle
收敛原则 convergence principle
收敛圆 circle of convergence
收敛正向修剪 convergence forward pruning
收敛值 convergency value
收敛指数 contractive index; convergence exponent
收敛轴 axis of convergence
收敛属性 convergence attribute
收敛准则 contractive criterion; convergence criterion
收敛子 convergent
收敛作用 astringency; converging action
收链使船前进 heave a ship ahead
收料报告 material receipt; material report
收料差异 materials received variance
收料单 material receipt sheet; material received sheet; materials received note; materials received sheet; receipt of material
收料单位 material receiving unit
收料斗 receiving hopper; receive hopper
收料汇总表 summary of materials received
收料试验 acceptance test
收料员 receiving clerk
收领货物通知书 advice of receipt
收拢 retraction
收拢位置 retracted position
收拢状态 retracted position
收录 embody; enrol(l)
收录两用机 radio-recorder
收录时间 time of receipt; time of recording
收率 yield
收码器 code receiver; receiver
收买 bribe; buy off; buy over; purchase; suborn; tampering
收买权 preemption
收锚复绞辘 fish tackle
收锚杆 fish boom; fish davit; half davit
收锚滑车索 fish fall
收锚进孔 house the anchor
收锚时链与睡眠成锐角的状态 astay
收锚索 ring rope
收纳 pick-up
收能的 endoergic
收能反应 endergonic reaction; endoergic reaction
收泥器 dirt excluder; dirt trap
收盘 closing; closing quotation
收盘出价和要求 closing bid and asked price
收盘大减价 liquidating sale
收盘汇率 closing rate
收盘价(格) closing price; closing rate
收盘降价 closing low
收盘叫价 closing call
收盘上涨 closing high
收盘行市 closing quotation
收片盒 take-up magazine; take-up spool
收票员 gateman; taker; ticket-collector; lobby man <戏院、剧场的>
收票员报告 ticket collectors report
收起 bowse away; lay in; retracting
收起起落架 cleaning up undercarriage
收气 gettering
收气剂 getter
收气剂溅散 getter flash
收气器 getter
收讫 paid; payment received; received; received in full
收讫戳记 receipted stamp
收讫待运提单 received for shipment bill of lading
收讫发票 receipt invoice
收讫通知书 acknowledgement
收讫章 receipt stamp; received stamp
收汽箱 receiver tank
收清 received in full
收取的利率 interest charges
收取点 bleeding point
收(取)费用 collection of charges
收取附加费 collection of additional charges; imposition of surcharge
收容 house
收容机场 recovery airfield
收容所 asylum; collecting post; home for the homeless; sheltering post; xenodochium
收入 earnings; income; means; proceeds; revenue
收入表 account of receipts
收入不平衡 income inequality
收入材料撮总表 abstract of materials received
收入财政 revenue financing
收入仓库 warehouse
收入查核 income audit
收入差距 income differential
收入传票 collection voucher; receipt voucher; receiving slip
收入的边际效用 marginal utility of income
收入动态 income behavio(u)r
收入额 revenue position
收入额高限 income ceiling
收入费用表 statement of income and expenses
收入分布 income distribution
收入分类 income bracket; income group
收入分类账 receipt ledger
收入分配 distribution of income; income distribution
收入分配账户 distribution of income account
收入概算 estimated receipts; estimate of incomes; estimates of income
收入估计减少数 estimated decrease in income
收入管理信息系统 income management information system
收入和成本间的比较 trade-offs of the benefit
收入和盈余账户 income and surplus accounts
收入和支出 revenue and expenditures
收入核算 revenue accounting; revenue calculation
收入汇总 income summary
收入获得能力 revenue-yielding capacity
收入集中存储 revenue pool
收入记录 record of earnings
收入仅敷支出的 marginal
收入来源 source of revenue
收入利润率 the income and profit ratio
收入利息 interest received
收入量 volume of receipts
收入流程 revenue stream
收入流量 flow of income; income stream
收入能力 earning capability
收入平均数 income averaging
收入凭单 receipt documents; voucher for receipts
收入凭证簿 book of original document for receipts
收入实现 realization of revenue
收入受益人 income beneficiary
收入水平 income level
收入速度 income velocity
收入弹性 income elasticity
收入梯度 income gradient
收入调节税 regulatory income tax
收入调整 adjustment to income
收入现金 cash received
收入线 income line
收入限值 present worth of income; time-adjusted revenue
收入向量 revenue vector
收入项目 item of receipt
收入消费曲线 income consumption curve
收入效应 income effect
收入需要 revenue requirement
收入因子 income multiplier
收入预测 income projection
收入预算 revenue budget
收入再分配 income redistribution
收入增殖作用 multiplier effect
收入债券 income bond
收入债务比率 debt coverage ratio
收入账户 account of receipts; receiving account
收入账目 account of receipts; account to receive; revenue accounting
收入折旧法 depreciation revenue method
收入政策 income policy
收入周期 income cycle
收入总额 gross income; gross receipt; gross revenue; total receipt
收入最高轮伐期 rotation of the highest income
收湿的 hydroscopic
收湿器 desiccator
收湿物 hydroscopic substance
收湿效应 hydroscopic effect
收湿性 hygroscopicity; moisture-absorptivity
收湿性材料 hygroscopic material
收湿性降低 hygroscopic depression
收湿作用 hydroscopic effect
收时 time receiving
收拾 pack up
收拾干净 cleaning up
收市报价 closing quotation
收市订单 market on close
收市汇率 closing rate
收市价 closing price
收市行情 closing quotation
收市执行 execution at the close; on close
收受存款银行 deposited bank
收受故障通知的座席 position for reception of fault notices
收受器 receptacle
收受人 receiver; recipient
收受时间 time of acceptance
收束导堤 converging jetty
收束谷 hour-glass valley
收束管嘴 converging nozzle
收束渠道 converging channel
收束水流 converging flow
收束形管嘴 converging nozzle
收束形双导流堤 convergent jetties
收水器 drift eliminator
收税 levy duties on; receive tax; tax collection; toll
收税单 duty-paid certificate
收税道路 pike; tollway; turnpike road
收税的实施 enforcement of tax collection
收税高速公路 turnpike
收税过重 over-tax
收税卡 toll gate
收税路 turnpike; turnpike road
收税马力 taxable horsepower
收税人 collector; tax collector; tax receiver; toll man
收税亭 toll booth
收税员 collector; tax collector; tax receiver
收税栅 pike; turnpike
收缩 astriction; contract; curtailment; deflation; narrowing; retract(ion); shortening; shrink(ing); take-up; drawdown【化】; cissing <油漆由于黏[粘]结不足的收缩>
收缩百分比 shrinking percentage
收缩百分率 contraction percentage
收缩板分级机 constriction-plate classifier
收缩棒 pining rod
收缩包装 shrink wrapping
收缩比 contraction ratio; shrinkage ratio
收缩比例 contraction proportion
收缩比例尺 shrinkage rule
收缩边界 convergent boundary
收缩变换 retracting transformation; shrinking transformation
收缩变形 contraction distortion; shrinkage deformation; shrinkage distortion; shrinkage strain
收缩标志 settle mark; shrinkage mark
收缩波 shrinkage wave; shrinking wave
收缩薄膜 shrinkable film
收缩补偿 shrinkage-compensating; shrinkage compensation
收缩补偿混凝土 shrinkage-compensating concrete
收缩补偿水泥 shrinkage compensates cement; shrinkage-compensating cement

收缩部分 constriction;tapered section

收缩槽缝 <缝两边的> shrinkage groove

收缩测定器 mirror-apparatus

收缩测定仪 shrinkage meter

收缩差(别) shrinkage difference; difference in shrinkage

收缩沉降量 shrinkage settlement

收缩承口 convergent mouthpiece

收缩程度 shrinkage degree;shrinkage level

收缩尺 contraction ga(u)ge

收缩齿 tapered tooth

收缩处理 contract disposal

收缩的 astringent; contractive; convergent

收缩的工作面 shrinkage stope

收缩电阻器 pinch resistor

收缩定律 shrink rule

收缩度 degree of shrinkage;shrinkage degree;shrinking measure

收缩度控制 shrinkage control

收缩端<管的> serrated end

收缩段 constricted section; contracted section;contraction

收缩段距离 contraction distance

收缩段水头损失 loss of head due to contraction

收缩段压头损失 loss of head due to contraction

收缩断层 contraction fault

收缩断裂破损 shrinkage fracture distress

收缩断面 constricted section; contracted cross-section; contracted section; necked-down section; section of contraction; shrinkage section;vena contract;vena contracta [复 contractae <拉>]

收缩断面水深 depth of constricted section;depth of vena contraction

收缩发纹 shrinkage crack(ing)

收缩法兰 shrink flange

收缩翻边 shrink flanging

收缩翻边模 shrink-flanging die

收缩分度 shrink graduations

收缩缝 coating space; contractive joint; cooling joint; cooling space; shrinkage joint;contraction joint

收缩缝灌浆 control joint grouting

收缩缝压力灌浆 contraction joint grouting under pressure

收缩钢筋 shrinkage bar;shrinkage reinforcement;shrinkage rod

收缩根 contractile root

收缩公差 shrinking tolerance

收缩管 collapsible tube; contraction pipe

收缩管道 constriction; convergent conduit;converging duct

收缩管压头损失 loss of head due to contraction

收缩管嘴 convergent mouthpiece

收缩光栅 shrinking raster

收缩和温度钢筋 shrinkage and temperature steel

收缩和细化 shrinking and thinning

收缩河段 contracting reach

收缩核 retract

收缩痕迹 shrink mark

收缩环 retraction ring

收缩回采工作 shrinkage stoping

收缩机 shrinker

收缩级联 reduction cascading

收缩极限 contraction limit; contractive limit;shrinkage limit

收缩夹紧头 shrink grip tool joint

收缩假说 contraction hypothesis

收缩间期 intersystole

收缩渐变段 converging transition; converging transition

收缩校正作用 shrinkage correcting action

收缩阶段 contraction phase

收缩接缝 shrinkage joint;contraction joint

收缩接口 convergent mouthpiece

收缩节理【地】 joint of retreat;contraction joint;shrinkage joint

收缩截面 contracted section

收缩筋 shrinkage rib

收缩卷曲 shrinkage crimping

收缩开裂 cracking due to shrinkage; green cracks; shrinkage crack(ing);shrinkage-induced cracking

收缩开裂钢环试验法 ring test for shrinkage cracking

收缩孔 shrinkage hole; shrink hole; sink hole;tapered hole

收缩孔洞 contraction cavity

收缩孔口 contracted orifice

收缩孔隙 shrinkage porosity

收缩宽度 contracted width

收缩扩大 converging-diverging

收缩扩张管 convergent-divergent channel

收缩拉力 shrinkage tension

收缩拉应力 shrinkage tensile stress

收缩拉应力裂缝 tensile shrinkage stress crack

收缩冷铁模 contracting chill

收缩力 retractile force; retraction; shrinkage force

收缩量 amount of contraction; amount of shrinkage; amount of shrinking; contraction; shrinkage; shrinkage mass;wring

收缩裂缝 check crack; contracting crack;contraction crack(ing);contraction fissure;contractive crack; plastic crack(ing);plastic cracking of concrete;shrinkage crack(ing); shrinkage fracture

收缩裂纹 check crack;contraction crack; contraction fissure; shrinkage crack(ing)

收缩裂隙 shrinkage crack(ing);shrinkage gap

收缩流 contracted flow

收缩留量 allowance for contraction; allowance for shrinkage; shrinkage allowance

收缩率 contractibility rate; contraction ratio;degree of shrinkage;rate of contraction; rate of shrinkage; shrinkage factor; shrinkage rate; shrinkage ratio

收缩率总校正 process shrinkage

收缩码 punctured code

收缩脉冲电路 narrowing circuit

收缩脉动 contraction pulsation

收缩模量 shrinkage modulus

收缩模数 shrinkage modulus

收缩黏[粘]土 contractive clay;shrinkage clay

收缩泡 contractile vacuole

收缩配合 shrinkage fit(ting)

收缩喷注<水或气的> contracted jet

收缩喷嘴 contracting nozzle;convergent nozzle;converging nozzle

收缩膨胀 come-and-go; convergent-dilatation

收缩膨胀的 contractive-dilative;convergent-divergent

收缩偏析 shrinkage segregation

收缩破裂 contraction crack

收缩企口 contraction groove

收缩期 contraction phase

收缩期回缩 systolic retraction

收缩器 compress;constrictor;retracter;retractor

收缩前膨胀 preshrinkage expansion

收缩区 shrinking zone

收缩曲线 shrinkage curve

收缩趋势 shrinkage tendency

收缩渠道 convergent channel

收缩圈 shrink ring

收缩热 contraction heat

收缩容差 shrinkage allowance

收缩容许量 shrinkage allowance

收缩容许值 shrinkage allowance

收缩蠕变关系 shrinkage-creep relationship

收缩射流 contracted jet

收缩式测向器环形天线 retractable direction finder loop

收缩式发射器 retractor launcher

收缩式发射装置 retractable launching device

收缩式进气道 effuser

收缩式天线 retractable antenna

收缩式系缆柱 retractable bollard

收缩式闸 contracting brake

收缩式栅栏门 folding lattice gate

收缩试验 contractive test; shrinkage test

收缩试验仪(器) shrinkage test apparatus

收缩疏松 shrinkage porosity

收缩水 shrinkage water

收缩水道 contracted waterway

收缩水流 contracting current;convergent flow;converging flow

收缩说 contraction theory

收缩速度 rate of shrinkage

收缩速率 shrinkage rate

收缩损失 contraction loss; shrinkage loss;loss of shrinkage <预应力的>

收缩损失应力 shrinkage loss stress

收缩缩短 shrinkage shortening

收缩特性 shrinkage character;shrinkage property

收缩梯度 shrinkage gradient

收缩体积 shrinkage volume

收缩通道 contract channel; converging passage

收缩头 feeder head;piped end;shrinkage head

收缩土 contractive soil

收缩微 differential shrinkage

收缩微差 shrinkage difference

收缩为主要趋势的脉动 pulsation but in which the essential trend is contracting

收缩温度 shrinkage temperature

收缩稳定区 contracted stable region

收缩物 constrictor

收缩系数 coefficient of contraction; coefficient of shrinkage;contraction coefficient; contraction ratio; contractive coefficient;shrinkage coefficient;shrinkage factor

收缩线 shrinkage front;shrinkage line

收缩限度 shrinkage limit

收缩相 contraction phase

收缩效应 blockage effect; pinch; pinch effect;shrinkage effect

收缩效应电流 pinch current

收缩形变 shrinkage strain

收缩型短管 converging short tube

收缩型管道 convergent pipe

收缩型喷嘴 converging nozzle

收缩型卸料通道 converging discharge channel

收缩性 constringence[constringency]; contracti(bi)lity;contractibleness; shrinkage

收缩性大易裂开和扭曲的木材 tension wood

收缩性骨料 shrinkage aggregate

收缩性集料 shrinkage aggregate

收缩性锚固 shrinkage anchoring

收缩性能 shrinkage behavio(u)r

收缩性试验器 shrinkage apparatus

收缩性土 contractive soil

收缩性纤维 retractable fiber[fibre]

收缩修正 constriction correction

收缩修正系数 shrinkage correction factor

收缩穴 shrinkage vug

收缩压力 contract pressure

收缩堰 contracted weir; weir with contraction

收缩仪 retractometer; shrinkage apparatus

收缩因数 contraction factor; shrinkage factor

收缩因数增减率 shrinkage gradient

收缩引起的拉应力 tensile shrinkage stress

收缩引起的预应力损失 prestressing loss due to shrinkage

收缩应变 contraction strain;contractive strain;shrinkage strain

收缩应力 contraction stress;contractive stress; retraction stress; shrink(age) stress

收缩影响 effect of contraction

收缩映射 shrinking mapping

收缩映象 contraction mapping

收缩油罐 collapsed storage tank

收缩釉 shrinkage glaze

收缩余量 allowance for shrinkage; shrinkage allowance

收缩与温度钢筋 shrinkage and temperature reinforcement

收缩宇宙模型 contracting model

收缩预热的 pinch-preheated

收缩允许量 allowance for shrinkage; shrinkage allowance

收缩障碍物 constriction obstacle

收缩褶缘 shrink flanging

收缩值 shrinkage value

收缩指数 contractive index;shrinkage index <即塑限和缩限的差值>

收缩皱纹 shrinkage mark

收缩柱 pinch column

收缩装配 shrink fit

收缩装置 retractor

收缩状态的卷筒 <卷取机的> collapsed mandrel

收缩嘴 constricting nozzle

收缩作用 contraction; shrinkage action

收条 acknowledgement; receipt; voucher

收听 sound reading;tune into;tuning

收听方式 listening mode

收听干扰 impedance of listening

收听广播 listening-in

收听器 listener; listening appliance; listening device

收听时间 listening period

收听装置 listening appliance;listening device

收听阻抗 impedance of listening

收通行费的(道路)设施【道】 toll facility

收通行费地点【道】 toll-booth area

收通行费公路 toll highway

收通行费立体交叉 toll-type interchange

收通行税地区【道】 toll area

收通行税卡【道】 toll bar

收头半砖 half header

收头棒 termination bar

收头线脚 stop mo(u)lding

收图室 incoming divisor

收妥通知 acknowledgement of receipt

收尾 alpha and omega

收尾凹槽 stop-chamfer
收尾程序 epilog(ue)
收尾的半砖 half header
收尾工场 terminal yard
收尾工程 arrears;tailing-in works
收尾工作 arrears;tailing-in work
收尾过程 epilog
收尾时期 tail-out period
收尾速度 terminal velocity
收下的粉尘 collected dust
收线架 <钢丝绳> take-up stand
收线卷筒 take-up block
收线轮 take-up pulley
收线装置 take-up
收像 reproduced image
收像对比度 reproduced image contrast
收像分辨力 reproduced image resolution
收像清晰度 reproduced image resolution
收像细节 reproduced image fineness
收屑器 cuttings chute;cuttings pit
收信地址 address
收信放大管 receiving amplifier
收信放大器 reception amplifier
收信管 receiver tube;receiving tube
收信机 receiver machine
收信人信箱 mailbox
收信台 receiving station
收信效率 receiving efficiency
收信站 receiving station
收益 earnings;gains;income;profit;revenue
收益报表 earnings report
收益报告 earnings report
收益比率 earnings ratio;income ratio
收益变动性 earnings variability
收益标准 earnings standard
收益表 earnings statement;income sheet;income statement;statement of earnings
收益表比率 income sheet ratio
收益表账户 income statement account
收益偿债能力比率 earning coverage ratio
收益成本比 benefit-cost ratio
收益成本分析 benefit-cost analysis;benefit-cost study
收益重分配 income redistribution
收益的支出 expenditure on revenue
收益递减 decreasing profit;diminishing return
收益递减律 law of diminishing returns
收益递增 increasing return
收益掉期法 yield pick-up swap
收益对风险比率 reward-to-variability ratio
收益额 earning capacity
收益法 capitalization approach
收益分成抵押 equity stake mortgage
收益分配 division of earnings;division of income;income apportionment;income distribution
收益分配计划 gain sharing plan
收益分享 revenue sharing
收益股利比率 earnings-dividend ratio
收益观察 earning observation
收益和亏损 profit and loss
收益汇总表 summary of earnings
收益及支出明细表 statement of income and expenditures
收益减除数 income deduction
收益井 earning well
收益净值 net earning
收益决定 income determination
收益流通速度 income velocity
收益留存 earning retained
收益率 earning rate;earning ratio;earning yield;potential return;rate of return;revenue position;yield rate
收益率差异 yield variance
收益率计算 computation of rate of return
收益能力 earning capacity;earning power
收益期 payback period
收益曲线 yield curve
收益审计 income audit
收益受益人 income beneficiary
收益损失 loss in revenue
收益投资比 benefit-cost ratio
收益营运日 <桥梁或结构开始有收益营运的日期> revenue operation date
收益营运日后 post-revenue operation date
收益与资产总价之比 income-to-total assets
收益预测 earnings forecast(ing)
收益预算 income audit
收益再投资 reinvestment of earnings
收益债券 income bone
收益账(户) income account
收益折现法 time-adjusted-return method
收益之实现 revenue realization
收益支出(费用) income expenditures;charge against revenue;income charges;revenue expenditures
收益质量 earning quality
收益状况 earnings position
收益资本化方法 earnings capitalization method
收益资本化价值 earnings capitalized value
收益资产 earning assets;revenue assets
收音电唱机 radio phonograph
收音电唱两用机 autoradiogram;radiogram;radiogramophone;radiophonograph
收音机 radio;radio receiver;radio set;receiver;receiving set;wireless set
收音机的报时声 squeak
收用 expropriation
收油车间 receiving house
收窄河段 contracted river reach
收债人 debt collector
收账 collect account
收账报告书 collection report
收账部门 collection department
收账程序 account-receivable program(me)
收账费用 collection expenses
收账流动值 collection float
收账期 collection period
收账人 bill collector;collection clerk;collector
收账员 bill collector;collection clerk
收账政策 collection policy
收针板 narrowing combing
收针杆 narrowing rod
收针装置 narrowing attachment
收支 expense and receipts;income and expenses;incomings and outgoings;revenue and expenditures
收支比 income-expense ratio
收支比值 default ratio
收支表 statement of income and expenditures
收支不平衡 payment in balance;payment imbalance
收支差额 balance of payment;gap between revenues and expenditures
收支差额补助 balance-of-payment assistance
收支赤字 deficit balance
收支范围 division of the revenue and expenditures boundaries
收支费用 marginal cost
收支概算 estimated expenditures and revenue
收支关系 expenditure-income relation
收支汇总表 summary of receipts and expenditures;summary statement of receipts and expenditures
收支计算书 account of business
收支决算 final accounting of revenue and expenditures
收支款项期报 cash flow statement
收支困难 payments problem
收支两平的收入水准 break-even level of income
收支两讫 account balanced
收支明细账 receipt and disbursement statement
收支逆差 balance of payments deficit
收支平衡 balance between income and expenditures;balanced budget;balanced receipts and payment;balance expenditures with income;balance of income and outlay;balance of payment;break-even
收支平衡表 balance sheet
收支平衡并有盈余的预算 balance budget with surplus
收支平衡点 break-even point;default point;receipts and disbursements breakeven point
收支平衡定价法 break-even pricing
收支平衡分析 break-even analysis
收支平衡图 break-even chart
收支平衡营运率 break-even rate of operation
收支平衡政策 balance-of-payments policy
收支凭证 receipts and payment documents
收支清单 income and expenditure statement
收支顺差 active balance of payments;balance of payment surplus;payments balance with surplus
收支相等 pay-as-you-earn
收支相抵 break-even;clear expenses;expense and receipts in balance;expense balance receipts;revenues and expenditures are in balance
收支一览表 statement of the income and expenditures
收支预算 budget for revenues and expenditures
收支摘要本 abstract book of receipts and payments
收支摘要簿 abstract book of receipts and payments
收支账户 account of receipts and payments
收支账目 revenue and expenditure account
收支状况 payment position
收脂 dipping
收脂工 dipper
收纸 collection
收纸叼纸牙 fly hand
收纸滚筒 collecting cylinder
收纸架 form receiving tray
收纸系统 delivery system
收皱 gathering
收注栅 catcher
收注栅空间 catcher space

手

手按 hand push
手按弹簧定位装置 spring ga(u)ge
手拔 hand-lifting
手把 grip handle;hand hold;holder;lug
手把操作 lever-operated
手把吊链 handle hanging chain
手把杆 handle stem;push rod
手把杆弹簧 push rod spring
手把杆套 handle stem holder
手把给进岩芯钻机 hand feed core drill;lever feed core drill
手把取出位 handle-off position
手把闩 handle latch
手把闩弹簧 handle latch spring
手把套口 handle socket
手把托架底板 handle bracket patch plate
手把型千斤顶 lever-type jack
手把支点 hand fulcrum
手把支柱 hand fulcrum
手摆式拉力滑车 lever block
手扳冲床 arbor press
手扳冲刀 hand punch
手扳道岔 hand-operated switch;manually operated points;manually operation switch;manual switch
手扳葫芦 lever block
手扳螺纹攻 hand tap
手扳丝锥 hand tap
手扳压床 hand-arbo(u)r press;arbor press
手扳压机 arbor press
手扳铡断机 hand shear cutting machine
手扳钻 ratchet drill
手搬大钳回转 rotate with a wrench
手搬钻机 clack mill
手拌法 hand mix procedure;hand mixing
手拌和料 hand mixture
手拌混凝土 hand-mixed concrete;mix batch concrete
手拌料 hand mix
手边 at hand;handy;on hand
手编编译程序 hand-coded compiler;hand-written compiler
手编程序 machine language program(me);manual code programming
手编分析程序 hand-coded analyser
手编绒线 hand-knitting yarn
手编数字化器 manual editing table
手编无错程序 star program(me)
手标本 hand specimen
手表 watch;wrist watch
手表带 watchband
手表调整台 watch holder
手表误差 watch error
手表油 oil for watch
手柄 control lever;grip;hand control lever;hand hold;hand knob;hand(le)grip;hand lever;handset;hand shank;holder;knob;lever;lock-on;rein;shank;joystick lever <一杆多用的>
手柄把 lever handle
手柄保险 hand control lock
手柄表示灯 lever lamp;lever light
手柄操舵 on-off steering
手柄操纵 handle operation;stick control
手柄操纵员 leverman
手柄操纵转向 hand lever steer
手柄槽 cavity at handle
手柄带指环的剪刀 bulldog snip
手柄导板 handle guide plate
手柄端头 tiller grip
手柄防松螺母 handle lock nut
手柄杆 handle bar;handle lever;hand-operating lever
手柄杆导管 handle bar guide tube
手柄杆螺栓 handle lever fastening bolt
手柄固定板 handle setting plate

手柄固定弹簧 handle setting spring
手柄固定轴 handle setting shaft
手柄剪机 hand lever shearing machine
手柄接点 lever contact
手柄开关 bat-handle switch; lever switch
手柄空位 lever space
手柄控制 hand grip control; lever control; stick control
手柄末端 bottom level
手柄木 handle blank
手柄设定旋钮 handle setting knob
手柄伸长部 handle extension
手柄式 lever type
手柄式方形钢制蝶阀 square steel butterfly valve with handle
手柄锁定器 hand control lock
手柄锁紧 handle locking
手柄锁紧螺母 handle lock nut
手柄弹键盖螺钉 lever latch cap screw
手柄套管 lever collar
手柄调整 hand lever adjustment
手柄停机 manual stop
手柄停止 manual stop
手柄位置 handle position
手柄窝 handle socket
手柄销 handle pin
手柄型保险丝 handle-type fuse
手柄支点 arm pivot
手柄组件 handle sub-assembly
手拨号码机 numberer
手拨轮 hand gear
手薄编号 numbering of field book
手薄类别 classification of field book
手部污染监测器 hand contamination monitor
手簿格式 form of note; note form
手采岩样 hand specimen
手操测距仪 stadimeter
手操舵柄 hand tiller
手操器 manual actuator
手操推进器 hand-operated propeller
手操纵 hand control
手操纵带式打磨器 hand belt sander
手操纵的 hand-operated; manual-operated
手操纵的吊车 hand crane
手操纵渡线 hand throw crossover
手操纵阀 hand control valve
手操纵进刀 hand feed
手操纵栏木 hand-operated barrier; hand-operated gate; manly operated gate
手操纵力 manual steering force
手操纵喷灌器 hand-controlled sprinkler
手操纵砂带磨机 hand belt sanding machine
手操纵式危急信号 hand distress signal
手操纵脱轨器 hand throw derail
手操纵弯折机 hand bender
手操纵制动阀 hand-operated brake valve
手操纵转向 hand steer(ing); manual steering
手操作 manual drive
手操作的 manually operated
手操作杆 hand-operating lever
手操作混凝土拌和器 sweat board
手操作空间范围 hand reach
手操作千斤顶 hand-operated jack
手操作位置 manual position
手操作仪表板 manual panel
手册 companion; directory; ench(e) iridion[复 ench(e) iridia/ ench(e) iridions]; handbook; manual book; reference book
手册分编 manual breakdown
手册和专著 manuals and monographs
手册资料 manual data
手测量 hand measurement
手测深锤 hand lead
手叉 hand fork
手铲 hand shovel
手铲刀 putty knife
手长 hand length
手抄本 hand-written copy; manuscript
手抄电报 manual telegraphy
手抄纸 hand-made paper
手车 barrow; cart; lorry; trundle
手车采矿 barrow excavation
手车触头 truck contact
手车道板 barrow runner
手车工 wheeler
手车工人 barrow man
手车轨道 lorry rail; lorry track
手车式撒布机 barrow-type spreader
手车式撒布碎屑机 barrow-type chip(ping)s spreader
手车跳板 barrow runner
手扯长度法 hand-drawing method
手扯强力 handle strength
手扯纤维长度测定法 hand-measured staple length; hand stapling
手沉锤 hand sinker
手持 hand hold
手持步谈机 handcart set
手持测角仪 hand goniometer
手持测速器 hand speedometer
手持测向器 hand goniometer
手持打夯机 hand tamper
手持导轨式两用凿岩机 converted drifter
手持捣实棒 hand-operated compacting beam; hand-operated tamping beam
手持的小型装置 handset
手持电动打磨机 orbital sander
手持电焊机 hand welder
手持电焊面罩 hand shield
手持电焊器 hand welder
手持电话机 hand telephone set
手持电掘凿器 electric(al) digger
手持动力锯 power hand saw
手持对讲机 hand held radio system
手持盾牌 manual shield
手持泛光灯 hand-held flood light
手持放大镜 hand lens; hand magnifier; pocket lens
手持风锤 jack
手持风镐 <凿岩用> air hand hammer rock drill
手持风速表 hand anemometer
手持风速计 hand anemometer
手持风钻 hand-held drill
手持高频对讲机 hand held high frequency
手持高频式凿岩机 hand-held high-frequency drill
手持工件磨光 off-hand grinding
手持工具 hand tool
手持刮刀 pull scraper
手持焊接护目罩 hand shield
手持焊枪 hand torch
手持夯实器 jumping jack
手持红外报警器 handhold infrared alarm
手持护目罩 <焊工的> hand shield
手持灰泥板 mud pan
手持混凝土振捣器 handheld concrete vibrator
手持机动铲 clay digger
手持机动工具 power-driven hand-tool; powered hand tool
手持机动锯 power hand saw
手持激光夜视仪 hand laser night vision device
手持计程仪 hand log
手持计算器 hand-held calculator
手持架式凿岩机 hand-held drifter
手持件 handpiece
手持卷扬机 hand hoist
手持控制器 hand-held control unit
手持量角计 hand goniometer
手持铆钉锤 hand rivet(ing) hammer
手持铆钉枪 hand rivet(ing) machine
手持煤油炉 primus stove
手持磨削 freehand grinding
手持黏[粘]土铲 clay digger
手持刨子 planing machine
手持喷枪 hand gun
手持喷洒管 manual spray hose
手持气动锤 jack hammer
手持热电偶探测器 hand-held thermocouple probe
手持热像仪 hand-held thermal imager
手持润滑油枪 hand grease gun
手持摄影 hand-held exposure; hand-held photography
手持摄影机 hand-held camera
手持声呐 handheld sonar
手持湿度测量仪 hand-held moisture meter
手持式 hand-held
手持式曝光表 hand-held meter
手持式步话机 handie-talkie; handy-talkie
手持式步谈机 handie-talkie
手持式彩色电视摄像系统 hand-held colo(u)r TV camera system
手持式冲击凿岩机 hand-held hammer drill
手持式传声器 hand-held microphone
手持式打结器 hand-held knotting device
手持式打孔器 hand punch
手持式的 hand-held
手持式电动工具 hand electric(al) tool
手持式电脑 hand-held PC
手持式电钻 hand-held electric(al) drill
手持式风速仪 hand-held anemometer
手持式风钻 hammer hand-held drill; hand drill(ing machine); hand hammer drill; jack hammer; unmounted drill
手持式风钻钎子 jack bit
手持式钢钎 hand drill steel
手持式个人计算机 hand-held personnel computer
手持式焊工面罩 hand shield
手持式话筒 hand microphone
手持式回转风动凿岩机 hand-held self-rotating air-hammer drill
手持式计算机 hand-held computer
手持式接受机 hand receiver
手持式(淋浴)莲蓬头 hand shower
手持式挠度仪 portable deflectometer
手持式喷淋器 hand shower
手持式喷枪 hand spray
手持式气动锤 jack hammer
手持式气动工具 portable air tool
手持式气动钻机 jack hammer
手持式轻便凿孔机支架 old man
手持式砂轮机 hand grinder
手持式受话器 hand receiver
手持式双目镜 hand-held binocular
手持式水平仪 hand level
手持式送受话器 hand microtelephone
手持式形变仪 portable deflectometer
手持式岩粉撒布器 hand gritter
手持式凿岩机 breast drill; hammer; hand drill(ing machine); hand hammer; hand-held drilling machine; hand-held(rock) drill; hand-held sinking drill; jack hammer drill; sinker; sinker drill; unmounted drill; jack hammer; pusher
手持式凿岩机工 jack hammer man
手持式凿岩头 jack bit
手持式凿岩钻 hand drill(ing machine); hand-held drill
手持式自动电话机 telephone of hand-set-type
手持式钻机 breast drill; unmounted drill
手持收发话筒 handset
手持受话器 hand telephone
手持数字化器 manual digitizer
手持数字热电偶高温计 hand-held thermocouple probe
手持水平仪 Abney level
手持水准 monocular hand level
手持水准仪 Abney(hand) level; hand level
手持送受话器 hand combination set; handset; hand set(ting)
手持弹簧传动旋转底质取样器 hand-line spring driven rotary-bucket bed material sampler
手持探照灯 hand-held flood light
手持提升器 hand hoist
手持听筒 hand receiver
手持头戴送受话器 hand-headset assembly
手持弯管器 conductor bender
手持望远镜式激光测距机 telescopic(al) hand-held laser rangefinder
手持信号圆盘 hand signal disc
手持液体罗盘 liquid hand compass
手持凿岩机 nager; rock drill hammer
手持振动器 hand vibrator
手持终端 handle held terminal
手持转数计 hand revolution counter
手持转速表 hand tach(e)ometer
手冲床 hand punch
手冲击钻 drill hammer; hand hammer drill
手冲钻 hand(-held) sinker
手冲钻机 drill hammer; hand hammer drill
手抽泵 hand-operated pump
手锄 hand hoe
手穿孔卡片 manual card
手传动 hand drive
手传动装置 hand-operated gear
手吹风机 hand drier[dryer]
手槌 <非金属的> lump hammer
手锤 engineer's hammer; hand hammer; bishop; club hammer; common ram; hand(le) hammer; hand tamp; lump hammer; mall hammer; wooden hammer; flat lump hammer <非金属大锤>
手锤布氏硬度计 hand Brinell's hardness tester
手锤测深 handlead sounding; handlead survey
手锤敲击硬化 hand peening
手锤球 hand lead
手锤选矿 cob walling
手锤凿岩 hand hammer drilling
手锤组件 hammer sub-assembly
手锤钻眼 hand hammer drilling
手磁铁 hand magnet
手锉 arm file; hand file
手打眼钢钎 jupper
手打油杯 hand primer
手打褶裥 hand-run tucks
手大小的矿石标本 hand specimen
手带动 hand motion
手挡 forehand
手导镜 cheiroscope
手捣 hand-tamped
手捣管 hand-tamped pipe
手灯 lantern; portable lamp
手灯电池 lantern battery

手点焊 poke weld(ing)
手电动机 hand motor
手电容 hand capacity
手电筒 electric(al)torch;flash;flash(ing)lamp;flash(ing)light;pocket flashlight;pocket lamp;torch
手电筒灯泡 pocket lamp bulb
手电筒电池 flashlight battery
手电钻 electric(al)portable drill;motor drill
手雕 handcarving
手钓 hand lining
手钓船 hand-liner
手钓母船 dory-hand-liner
手钓丝 hand line
手钓艇 handline boat
手定则 <右手或左手定则> hand rule
手动 hand drive; hand movement; hand power;manual drive;manually operation
手动按钮 manual button
手动摆动泵 semi-rotary hand pump
手动摆移柄 jitterbug
手动搬道机 hand switch box
手动板材剪切机 hand plate shears
手动板材切割机 hand plate shears
手动拌和机 manual blending mixer
手动棒 manual rod
手动饱和度调整 manual saturation (colo(u)r) control
手动保安装置 manual tripping device
手动报警按钮 manual alarm button
手动报警系统 manual alarm system
手动备用控制系统 manual backup control system
手动备用调节装置 manual backup
手动背负式喷雾器 hand-powered knapsack sprayer
手动泵 hand-operated pump; hand pump;manual pump
手动闭环过程控制 manual closed-loop process control
手动闭环控制系统 manual closed-loop control system
手动闭路控制 manual closed control
手动闭锁 manual block
手动编辑 manual editing
手动变速 manual shift
手动变速器 <安装在汽车底板上的> stick shift
手动变速箱 manual transmission
手动变速型 manual shift transmission
手动变速装置 manual gear shifting
手动标志板 sign paddle
手动并联【电】manual paralleling
手动补偿 manual compensation
手动操舵室 hand steering room
手动操舵装置 hand power steering gear; hand steering gear; manual steering equipment
手动操纵 manual control;manual manipulation;manual operation
手动操纵阀 manually operated valve
手动操纵轮 hand wheel
手动操纵系统 hand operated system; manual control system
手动操纵振动精整机 manually operated vibrating finisher
手动操作 hand operation;manual operation;manual manipulation
手动操作的 manually controlled
手动操作杆 manual lever
手动操作搅拌装置 manual operating batch plant
手动操作配电盘 hand-operated distribution panel
手动操作印刷机 hand press
手动操作站 manual station
手动操作装置 hand-operating device
手动测读仪 manual readout
手动测角计 hand goniometer
手动测向仪 manual direction finder
手动叉式装卸车 hand-fork truck
手动车 hand car
手动车床 turns
手动车装置 hand winding device
手动程序 manual program(me)
手动秤 manual weight batcher
手动齿轮(装置) hand travel(l)ing gear
手动充气救生衣 manually operated gas inflatable lifejacket
手动冲床 toggle puncher
手动冲击钻杆 spring pole
手动冲击钻架 spring pole rig
手动冲击钻进 spring-pole drilling
手动冲孔机 hand punch(ing machine)
手动冲压机 hand punch
手动冲钻 hand churn drill
手动重调 manual-reset adjustment
手动重调保险控制 manual-reset safety control
手动抽水机 hand pump;manual pumping unit
手动抽油机 manual pumping unit
手动除灰 manual removing of ashes
手动穿孔 <在穿孔卡上用手揿穿孔机校正个别卡的错误>【计】spot punch
手动穿孔机 hand punch
手动传动箱 manual(shift) transmission
手动传动装置 manual take-over drive
手动窗开关器 hand window (control) gearing
手动窗孔卡片安装机 hand aperture card mounter
手动吹风器 bellows
手动粗切锥铰刀 roughing hand taper reamer
手动打气泵 hand-operated air pump
手动打闸 hand off
手动打桩机 hand-operated driver
手动单梁桥吊 hand-operated overhang crane
手动单向离合器 manual-control freewheeling
手动挡水板 handstop
手动刀架 hand slide rest
手动道岔 hand-operated points;hand throw switch; manually operated points
手动的 hand-actuated; hand-driven; hand-operated;manual;manual acting; manually operated; manumotive;unpowered
手动低速挡 manual low gear
手动地 manually
手动地址开关 manual address switch
手动点火锅炉 manually lighted boiler
手动电磁阀 hand control solenoid
手动电路控制器 electric(al)plunger; hand circuit controller
手动电压控制 manual voltage control
手动电钻 hand-operated electric(al) drill
手动吊车 hand-operated crane
手动吊杆 hand-driven batten
手动吊绳冲击式钻眼 hand churn drilling
手动蝶阀 butterfly valve with lockable actuator;manual butterfly valve
手动定量给进阀 measuring valve
手动堵塞计 manual plugging meter
手动断路 hands-off;manual off
手动断路流阀 manual shutoff valve
手动断续器 hand interruptor
手动多叶调节阀 hand-operated multivane regulating valve
手动舵柄 hand tiller
手动颚式破碎机 hand-operated jaw crusher;manual jaw crusher
手动阀(门) hand control valve;hand (-operated) valve; manually operated gate;manually operated valve; manual-operated valve;manual valve
手动阀门定位器 manual valve positioner
手动阀位控制器 manual valve positioner
手动翻斗车 band tip-cart
手动翻转装置 hand tilting device
手动反向机构 hand reversing gear
手动方式 manual mode
手动仿形铣床 routing machine
手动飞溅润滑系统 manually operated spray system
手动分类卡片 hand-sort card
手动风阀 damper with lockable actuator
手动风门 manual damper
手动风门杆 hand throttle lever
手动风箱式喷粉器 midget duster; puff duster
手动辅助定位 manual assisted positioning
手动复归 hand resetting
手动复归信号器 hand-restoring indicator
手动复位 hand reset;hand-restoring; manual reset
手动复位继电器 hand reset relay
手动复位式 manual-reset type
手动复印机 manual driven duplicator
手动复原 hand reset;hand-restoring
手动复制机 manual duplicator
手动干油站 manual grease station
手动杆式夹桩器 manual lever chuck
手动钢筋剪断机 hand steel shears
手动钢筋切断器 hand bar cutter
手动钢筋弯曲机 manual steel bar bender
手动钢绳冲击钻机 hand churn drill
手动钢绳冲击钻进 hand churn drilling
手动钢丝绳冲击式钻机 hand churn drill
手动钢丝索绞车 hand wire cable winch
手动钢索绞车 hand steel cable winch;hand steel rope winch
手动钢索卷扬机 hand steel cable winch;hand steel rope winch
手动杠杆扩胎器 hand lever tyre expander
手动杠杆式上油泵 hand plunger grease pump
手动杠杆制动器 hand lever brake
手动割草机 hand turf cutter
手动给进 hand-operated feed
手动给进钻机 hand feed
手动跟踪 hand tracking; manual tracking
手动跟踪转速计 manual-tracking tach(e)ometer
手动工具 hand tool
手动功能 manual function
手动攻丝 hand tapping
手动刮板 hand scraper
手动刮水器 hand wiper
手动刮土机 drag loader shovel
手动关闭阀 manual shutoff valve
手动关断阀 hand stop valve
手动贯入仪 hand penetrometer
手动光圈光阑 manual iris
手动焊割炬 hand blowpipe; hand torch
手动焊接机 hand-held welder
手动焊钳点焊 poke weld(ing)
手动夯 manually operated rammer
手动和自动(转换)开关 manual-automatic switch
手动横向进给机构 hand traverse gear
手动葫芦 chain block; chain hoist; handy lift hoist block;manual chain hoist;pulling jack
手动葫芦门式起重机 gantry crane with chain hoist
手动戽斗 hand bucket
手动滑板送料 hand-operated slide feed
手动滑车 hand pulley
手动滑车组 hand-operated pulley block
手动滑轮吊车 hand block(and tackle)
手动滑轮组 hand-operated block(and tackle);hand-operated pulley block
手动滑枕反向 manual ram reverse
手动还原继电器 hand reset relay
手动环程序控制 manual closed-loop process control
手动换挡杆 hand gear shift lever
手动换向把 manual reversing handle
手动换向阀 hand-operated direction valve
手动灰浆喷枪 hand-operated mortar gun
手动回路管制器 electric(al)plunger; hand circuit controller
手动回转泵 manual rotary (drum) pump
手动混合阀 hand mixing valve
手动混凝土搅拌机 handy concrete mixer
手动活塞泵 hand-operated piston pump;syringe
手动火警报警器 hand fire alarm(device);manual fire alarm;manually operated fire alarm
手动火警系统 manual fire alarm system
手动火焰清理 hand scarfing
手动火灾报警按钮 manual fire alarm system
手动货油阀 manual control of cargo valve
手动机 hand machine
手动机构 hand-operating mechanism
手动机械 hand machine
手动棘轮 hand ratchet
手动挤焊 poke weld(ing);push weld(ing)
手动挤泥条机 wad box
手动计算器 manual calculating unit
手动计重器 manual weight batcher
手动加荷 manual loading
手动加油站 manual grease station
手动加注润滑脂 manual greasing
手动夹盘 hand-operating chuck
手动兼自动的 manual-automatic
手动监控系统 manual monitored control system
手动剪草机 hand-controlled grass cutter
手动剪切机 hand shears
手动检查 manual check; manual examination
手动键 manipulated key;manipulating key
手动交换 manual exchange
手动交通信号 manual traffic signal
手动焦油喷洒机 hand tar spraying machine
手动绞车 hand cable winch;hand-operated winch
手动绞刀 hand reamer
手动搅拌齿轮(装置) hand-operated stirring gear
手动搅动器 hand agitator
手动节风板 manual control damper

手动节气阀 hand throttle
手动节气门 manual mixture control
手动节气门按钮 hand throttle button
手动截止阀 hand stop valve; manual globe valve
手动解脱 master trip
手动金刚石钻机 hand diamond drill
手动紧急停堆 <反应堆> manual scram
手动进给 follow-up hand feed; manual feed
手动进给机构 hand feeding mechanism
手动进给钻床 sensitive drilling machine
手动卷百叶 hand-operated rolling shutter
手动卷缠弯曲机 manually powered draw bending machine
手动卷绕机 manual hoist; manual winder
手动卷扬机 hand winch
手动均衡器 manual equalizer
手动卡盘 hand chuck
手动开关 hand-operated switch; manually operated switch; manual switch
手动开关器 hand operator
手动开环控制 manual open-loop control
手动靠模铣床 routing machine
手动控制 finger control; hand control; handling operation; manual (-operated) control
手动控制板 manual control panel
手动控制的 hand-controlled; manually controlled; hand-operated
手动控制的调车场 hand-controlled yard
手动控制阀 hand control valve; hand valve
手动控制器 hand controller; manual controller
手动控制台 hand-held console
手动控制系统 hand control system; hand operated system; manual control mode
手动控制站 manual control station
手动扩孔钻 hand reamer
手动拉杆 hand lever
手动栏木 hand-operated gate; manly operated gate; manually operated gate
手动冷镦机 hand cold header
手动离合器杆 hand clutch lever
手动立式液力压榨机 hand-operated vertical hydraulic press
手动链滑车 hand chain block
手动链滑车卷扬机 hand chain block hoist
手动链轮起重机 manual chain hoist
手动链式触探机 hand chain sounding machine
手动列车控制装置 manual train control device
手动螺杆 hand screw
手动螺纹梳刀 hand chaser
手动螺旋挤泥条机 dod box
手动螺旋压机 hand power screw press; swing press
手动螺旋压孔机 hand power screw punching press
手动螺旋压力机 hand-operated screw press; hand screw press
手动螺旋钻 hand-operated auger
手动铆(钉)机 hand riveter
手动煤气关闭阀 manual gas shut-off valve
手动门 manually operated door
手动模刀 former
手动能量 manual capacity regulator
手动泥炭采掘机 hand turf cutter
手动拈线机 hand twiner
手动碾磨机 quern
手动扭锁 hand-engaged twist-lock
手动排气 manual air vent
手动盘车装置 hand barring; hand turning gear; manually operated turning gear; manual shaft turning device
手动旁通阀 hand by-pass valve
手动喷布机 manual spray hose
手动喷粉机 hand gun duster
手动喷粉器 hand-operated duster
手动喷枪 manual spray gun
手动喷洒器 hand sprayer; hand water sprayer
手动喷射泵 hand-operated spray(ing) pump
手动喷射器 hand-operated spreader; hand syringe
手动喷水器 hand boom; hand sprayer; hand water sprayer
手动喷雾泵 hand-operated spray(ing) pump
手动喷雾机 manual spray hose
手动喷雾器 hand sprayer; manual sprayer; manual spray hose
手动膨胀阀 hand expansion valve; manual expansion valve
手动平板机 arbor press; hand-driven screw press
手动平衡重升降机 manually operated counterweight lift
手动破碎机 hand crusher
手动启闭机 hand lift
手动启闭门 hand-operated door; manually operated door
手动启闭闸门 manually operated gate
手动启动器 hand-operated starter; manual starter
手动启动注水 manual priming
手动启门机 hand lift
手动起动 manual starting
手动起动泵 hand primer
手动起动开关 manual starting switch
手动起动器 hand-operated starter
手动起锚机 hand-powered capstan
手动起锚绞盘 hand-powered capstan
手动起重机 hand crane; hand lift; hand-operated crane; manual crane
手动起重机滑轮组 hand block (and tackle)
手动起重器 hand jack (screw); hand screw
手动气泵 hand air pump
手动气力喷雾机 manual pneumatic sprayer
手动气门研磨器 hand valve grinder
手动气体进样阀 manual gas sampling valve
手动千斤顶 hand jack (screw); hand screw
手动嵌缝枪 hand-operated ca(u)lking gun
手动桥式吊车 hand-operated travel(l)ing bridge crane
手动桥式起重机 manual overhead crane
手动桥式行车 hand-operated travel(l)ing bridge crane
手动切板机 hand lever shears
手动切刀 hand-powered cutter
手动切断 hand off
手动切断机 hand-powered cutter
手动切割机 hand-operated cutting machine
手动-切换-自动选择开关 manual-off-automatic selector switch
手动轻型岩石钻 lightweight hand rock drill
手动倾倒装置 manual tilting device
手动倾翻传动装置 hand-operated tipping drive
手动倾卸设备 hand tilting device
手动曲柄 hand crank
手动驱动 hand-operated gear
手动驱动器 hand-operated driver; manual driver
手动取样器 manual sampler
手动乳液喷洒器 emulsion hand sprayer
手动润滑器 hand lubricator; manual lubricator
手动润滑油站 manual pumping unit
手动撒布器 hand-operated spreader
手动洒水装置 drencher system; drenching installation
手动扫掠 manual search
手动色调控制 manual toning control
手动砂轮整形工具 handwheel-dressing tool
手动筛 hand screen; riddle
手动筛分振动器 manual sieve shaker
手动筛网 manually operated screen
手动上条表 hand winding watch
手动设备备份 manual device backup
手动神仙葫芦 hand-operated block (and tackle); hand chain block
手动升降机 hand (power) elevator; handworked lift; manlift elevator
手动升压机 manual booster
手动石屑铺撒器 hand-operated gritter
手动石屑铺砂机 hand-operated gritter
手动示向器 hand direction indicator
手动式 manual type
手动式拌和机 handy mixer
手动式电位计 manual potentiometer
手动式管井提水工具 mosti
手动式静电喷漆机 hand-operating electrostatic sprayer
手动式灭火器 hand fire extinguisher
手动式涂料无气喷射机 manual non-pneumatic paint sprayer
手动式旋压 manual spinning
手动式装修吊篮 manual basket
手动式装修平台 manual lifting platform
手动释放 manual release
手动释放钮 manual discharge button
手动释放装置 hand release; manual tripping device
手动输入 manual input
手动输入装置 manual input device
手动数据输入 manual data input
手动双雾角 hand-operated dual horn
手动水泵 hand water pump
手动水表 manual-controlled water meter
手动水电站 manual hydroelectric(al) station
手动水泥喷枪 hand-operated gun
手动送料刨板机 hand planing machine
手动搜索 manual search
手动台式磨床 hand bench grinder
手动弹涂机 hand paint catapult
手动陶车 hand-driven pottery's wheel
手动淘汰盘 hand buddle
手动套管挂 manual casing hanger
手动提捞 hand bailing
手动提前点火 hand advance
手动提升 hand hoisting
手动提升机 hand-operated lifting machine; manual hoist
手动填装 hand charged
手动调节 finger control; hand adjustment; manual adjustment; manual regulation
手动调节器 hand (-operated) regulator; manual regulator
手动调速 manual speed adjustment
手动调压 manual voltage regulation
手动调整 hand regulating; hand regulation; hand reset; manual regulation; manual setter
手动调整器 hand-controlled regulator
手动跳汰机 hand(-operated) jig
手动跳闸 hand trip (ping); manual trip (ping)
手动跳闸把手 hand trip control
手动跳闸断路器 fixed trip circuit breaker
手动跳闸开关 fixed trip switch
手动铁鞋安置机 manual skate machine
手动停机(装置) manual shut-down
手动停油杆 manual cut-out lever
手动同步变速箱 synchro manual transmission
手动土壤螺旋(麻花)钻 hand soil auger
手动土壤消毒器 hand-operated soil disinfector
手动推断送进 hand-operated push feed
手动脱扣 hand trip(ping)
手动脱扣器 hand trip gear
手动脱扣装置 manual tripping device
手动挖泥机 hand dredge(r)
手动弯曲辊 hand-operated bending roll
手动弯曲机 hand bender
手动位置 manual position
手动无线电测向仪 manual direction finder; manual radio direction finder
手动铣床 hand miller; hand milling machine
手动系统 manual system
手动限位器 finger stop
手动限制器 manual limiter
手动消防泵 manual fire pump; manually controlled fire pump
手动小车升降机 hand-operated troll(e)y hoist
手动小打样机 proving press
手动信号 hand signal
手动信号机【铁】 manual-controlled signal
手动信号设备 hand signal(l)ing device
手动信号设施 hand signal(l)ing device
手动修整 hand dressed
手动续纸器 hand feeder
手动悬挂式单轨系统 hand-operated suspended monorail system
手动旋臂起重机 hand-slewing crane
手动旋钮 manual knob
手动旋转开关 manual rotary switch
手动旋转式切削工具 hand-operated rotary cutting tools
手动选择性调整 manual selectivity control
手动询问 manual interrogation
手动压床 hand press
手动压尖机 hand pointer
手动压接钳 mechanical hand pressing pliers
手动压力机 mandrel press; manually operated press
手动压力螺钉 manually operated screw
手动压下装置 hand screwdown gear
手动压油泵 manual oil pressure pump
手动摇臂 hand-operated rocker arm; manual rocker
手动遥控 remote manual control
手动遥控按钮 manual remote control
手动遥控操纵杆 manual remote control

S

手动遥控键 manual remote control
手动遥控设备 manual remote control
手动叶轮泵 manual vane pump
手动液面计 hand-operated (level) ga(u)ge
手动液压泵 manual hydraulic pump
手动液压操舵装置 hand-hydraulic steering gear
手动液压机 hand-operated hydraulic press
手动液压式钢筋剪切机 hand-hydraulic bar shears
手动液压式钢筋切断器 hand jack type bar shears
手动液压托盘搬运车 manual hydraulic tray pushcart
手动液压弯管机 hand-hydraulic pipe bender
手动易拧紧接头 handy connection
手动音量控制 hand volume control
手动应急控制 emergency hand control
手动应急运转装置 manual emergency running device
手动油泵 hand oil pump
手动油门 hand throttle
手动油门按钮 hand throttle button
手动油门杆 hand throttle lever
手动油枪 hand gun
手动有轨巷道堆垛起重机 manual S/R[storage/retrieval] machine
手动预调控制 manual preset control
手动运算 hand operate
手动运行 hand operation
手动杂务梯 hand-operated service lift; hand power service lift
手动增压泵 < 内燃机的 > hand bypass pump
手动增益控制 manual gain control
手动闸 hand brake
手动闸门 hand door; manual-operated gate
手动振捣器 hand-manipulated vibrator
手动振动磨光机 hand-operated vibrating finisher
手动振动抹面机 hand-operated vibrating finisher
手动振动抛光机 hand vibrating finisher
手动振动器 hand-manipulated vibrator
手动整平板 hand-operated screed
手动正时 manual timing
手动止动阀 manual stop valve
手动制 manual system
手动制动器 hand brake; manual brake; service brake
手动制动释放装置 manual brake release device
手动制轮杆 hand ratchet
手动制轮机 hand ratchet
手动重量称量器 manual weight batcher
手动主截流阀 manual main shutoff valve
手动主轴制动器 hand spindle brake
手动助力装置 manual assist
手动注油装置 hand primer
手动柱塞式液压泵 hand plunger pump
手动柱塞涂料泵 hand plunger paint pump
手动转换 manual shift
手动转换开关 manual changeover switch
手动转轮 < 启门的 > hand wheel
手动转向 manual steering
手动转向机构 handy steering mechanism
手动转向力 hand force; manual steering effort
手动转向装置 hand steering gear
手动转辙器 hand-operated points; hand-operated switch; hand throw switch; manually operated points
手动装卸车 hand-lift truck
手动装置 hand appliance; hand gear; hand priming device; manual unit
手动追踪瞄准 manual tracking
手动自动 manauto
手动自动变换 manual-automatic change-over
手动自动的 manual-automatic
手动自动开关 manual-automatic switch
手动自动起动选择器 manual auto run-up selector
手动自动转换继电器 manual-automatic relay
手动阻风阀 hand choke
手动组合开关 manual switchgroup
手动钻机 hand borer
手动钻进 hand drill(ing machine)
手动钻头 hand borer
手动作业 manual work(ing)
手洞门 handhole door
手段 claim gamesmanship; craft; gateway; instrumentality; means
手段与目的分析 means-ends analysis
手锻 hand forging
手锻炉 smith forging furnace
手锻模 hand die
手堆的 hand-placed
手法 maneuver; manuduction; sleight; tactics; technicist
手法主义 <16 世纪末，意大利古典建筑风格 > mannerism
手放的 hand-placed
手风门 hand throttle
手风琴 < 一种六角形的 > concertina; accordion
手风琴式触点簧片 accordion contact
手风琴式隔墙 concertina partition
手风琴式连接 concertina connection
手风琴式折叠门 accordion folding door; concertina folding door
手风箱 hand bellows; hand blower; handle air blower
手风钻 jack hammer
手缝接缝 hand-sewn seam
手缝线迹 hand stitch
手扶 hand-held
手扶采煤机 hand coal cutting machine
手扶叉车 hand truck lift truck; pedestrian-controlled fork(lift) truck
手扶-乘坐两用叉车 motorized hand/rider truck
手扶打夯机 hand tamper
手扶单轮压路机 single-drum roller
手扶单轮振动压路机 single-drum vibratory roller
手扶电动工具 electric(al) hand-held tools
手扶动力工具 hand-held power tool
手扶感应笔 hand-held pointer
手扶割草机 walk type mower
手扶跟踪 manual tracing
手扶跟踪数字化法 manual followed digitizing
手扶跟踪数字化器 manual followed digitizer; manual tracing digitizer
手扶跟踪数字化仪 manual followed digitizer; manual tracing digitizer
手扶刮土小车 manually guided drag skip
手扶夯实机 hand-held tamper
手扶机动叉式搬运车 motorized hand truck
手扶机械铲 hand scraper; manual scraper
手扶平路机 hand grader
手扶式除雪机 hand-guided snow remover
手扶式单轮压路机 hand-guided single drum(roller); walk behind single drum
手扶式混凝土振捣修整机 manually guided vibrating concrete finisher
手扶式沥青洒布机 walking asphalt sprayer
手扶式马铃薯挖掘机 walking potato digger
手扶式农具 home-garden type
手扶式升降搬运车 walkie fork lift truck
手扶式弹齿中耕机 walking spring-tooth cultivator
手扶式推土机 push bulldozer
手扶式拖拉机 walking tractor
手扶式压路机 pedestrian roller; walk-behind
手扶式振动压路机 walk behind vibration roller
手扶数字化器 manually assisted digitizer; manually operated digitizer
手扶拖拉机 garden tractor; hand tractor; push bulldozer
手扶拖拉机配装式喷雾机 mounted sprayer for walking tractor
手扶拖拉机用旋转式灌木切除机 rotary brush cutter for walking tractor
手扶旋耕机 walking rotary cultivator
手扶旋转电刨 hand rotary electric(al) planer
手扶压路机 walk-behind roller
手扶园艺拖拉机 autogardener; walk-behind garden tractor
手扶振动辊 vibrating hand-roller
手扶震动碾 walk-behind roller
手扶中耕机 walking cultivator
手辐轮 hand spoke wheel
手斧 hand axe; hatchet; small axe
手斧石 hatchet stone
手杆 hand spike; holder
手杆柄 handle bar grip
手杆衬套 handle bar bushing
手杆螺栓 handle bar bolt
手杆式开关 hand lever shifter
手杆枢轴 hand lever pivot shaft
手杆闩 hand lever latch
手感 hand feeling; hand handle; handle; hand property; hand touch
手感测试仪 hand tester
手感柔软 soft hand
手感柔软涂层 coating with soft hand
手感柔软涂料 soft-feel coating
手感试验 hand test
手感舒适 hand feel and drape
手感土壤 feeling the soil
手感重量 heft
手高 < 堆货高度 > hand high
手稿 autograph; handwriting; manuscript
手稿复制品 autograph
手镐 pick ax(e)
手工 handling work; manual labo(u)r
手工安装 hand fitting
手工凹印 hand gravure
手工搬运 manual transportation
手工板金加工 hand plate working
手工半自动焊接 manual semi-automatic welding
手工拌法 hand mixing
手工包装的 hand-packed
手工包装作业线 packing shift
手工编程序 manual programming
手工编图 manual compilation
手工玻璃成型 chair work
手工裁切 manual cut(ting)
手工采集试样 hand specimen
手工采掘 hand-got
手工采掘工 hand(pick) miner
手工采掘泥炭 hand-cut peat
手工采矿面 hand-mined face
手工采样 hand sampling; manual sampling
手工采运机 hand logger
手工彩绘 hand painting
手工彩饰 hand decoration
手工操纵 hand control; lever control; manual control; manual manipulation
手工操纵的 hand steered
手工操纵的货运电梯 hand-operated elevator
手工操纵的货运升降机 hand-operated elevator
手工操纵转车台 hand-operated traverser
手工操作 hand operation; hand-handling; hand labo(u)r; handworked hand labo(u)r; manual handling; manual manipulation; manual operation
手工操作程序 manual procedure
手工操作穿孔卡片 hand-operated punched card; manually operated punched card
手工操作传动链 hand-operated chain drive
手工操作刀具 manual-operated cutter
手工操作的 hand-operated; manually operated
手工操作的动力 manual power
手工操作的粉刷 manual rendering
手工操作的复位时间 manually reset time switch
手工操作的混合阀门 manual mixing valve
手工操作的铆接 manual riveting
手工操作的配电间 manual switchroom
手工操作的水拌和机 manual water blending mixer
手工操作的水混合阀门 manual water mixing valve
手工操作的水喷灌混合机 manual water shower mixer
手工操作的提升机滑轮 hand-operated hoist block
手工操作的弯钢筋器 hand-operated bar bender
手工操作工具 hand tool
手工操作铆接机 manual riveting machine
手工操作设备 hand operation equipment; manually operated equipment
手工操作通风柜 handworked hood
手工操作装置 manipulative device
手工测微器 hand micrometer
手工产品 hand product
手工铲掘 hand spading
手工铲料 hand shovel(l)ing; hand spading
手工铲土 hand shovel(l)ing; hand spading
手工铲削清理 hand chipping
手工铲装 hand shovel(l)ing
手工车刀 hand turning tool
手工称量 manual proportioning
手工成型 artificial forming; hand finish(ing); freehand shaping; hand forming; hand mo(u)lding; manual assembly
手工成型的 hand mo(u)lded
手工成型法 hand-mo(u)lding press
手工程序 manual program(me)
手工充填 hand stowing
手工充填焊丝 manually feeding fillered

手工冲击钻 kirner
手工冲印 <坯面上的> hand cutter
手工重调 hand reset
手工出灰 hand poking
手工出灰气体发生器 hand-poked producer
手工出渣 hand lashing
手工除尘 hand ashing
手工除尘气体发生炉 hand-ashed producer
手工除锈 hand cleaning;hand rust removing
手工处理法 manipulative device
手工穿孔机 hand feed punch
手工吹风器 hand bellows
手工吹筒法 hand cylinder method
手工锤击 hand bumping
手工粗削的 hand-hewn
手工存放的 hand-placed
手工搓纹 hand graining
手工打结 hand tying
手工打麻 hand scutching
手工打入工具 hand driver
手工打眼 manual drilling
手工打样样张 hand proof
手工打桩机 hand driver
手工单工系统 manual simplex system
手工刀具 hand cutting tool
手工捣棒 hand compacting beam
手工捣锤 hand rammer
手工捣固法 hand rodding
手工捣实 compacting by hand;hand ramming;manual consolidation
手工的 manipulative;manual
手工电弧堆焊 manual arc welding;manual electric (al) arc pile up welding
手工电弧焊 manual electric(al) arc welding;covered arc welding;hand arc welding;hand electric (al) arc welding
手工电路 manual circuit
手工电渣焊 manual electro-slag welding
手工垫版 hand-cut overlay
手工雕刻 hand engraving
手工短刨 bench plane
手工锻炉 hand calciner
手工锻造 blacksmithing;hand forging
手工锻制(的) hand(icraft) forging
手工锻制铁工 hand-forged ironwork
手工堆垛工 hand stacker
手工方法 hand (labo(u)r) method;manual method
手工纺织呢 homespun
手工放样 manual lofting
手工分类 hand-sort
手工分类卡 hand sorted card
手工分离牙边法 hand scalloping
手工分色 hand-colo(u)r separation
手工分选的 hand sorting
手工粉刷 hand rendering
手工粉刷石膏 gypsum hand plaster
手工缝边 hand felling;hand hemming
手工工具 hand tool;manipulative device
手工工具架 hand rest
手工工人 manual labo(u)r;trade(s) man
手工工作 manual work
手工刮除 manual scraper
手工刮具 pull scraper
手工刮涂 knife applied
手工管接头 hand tight
手工管子切割器 hand pipe tool;hand tube cutter
手工过筛 hand sieving
手工焊机 manual welding machine
手工焊(接) hand welding;manual welding
手工焊接设备 manual welding installation
手工焊条 stick electrode
手工夯 hand tamp
手工夯击 hand tamping
手工夯具 bishop
手工夯实 hand tamping
手工夯实的 hand-packed
手工呼叫 manual call(ing)
手工画边 hand lining
手工换位 hand transposition
手工绘草图 freehand sketch
手工绘原图 hand-drawn original
手工绘制瓷砖 hand-painted picture tile
手工积层法 hand lay-up method
手工基本动作 manual element
手工极谱仪 manual polarograph
手工计数 manual counting
手工记录 hand-kept
手工技能 handicraft
手工加工 hand finishing;handwork
手工加工的 handworked
手工加工的矩形石 hand-cut random rectangular ashlar
手工加料气体发生炉 hand-fed producer
手工加煤 hand stoked
手工加煤机 hand stoker
手工加油 hand oiling
手工加载 hand-loaded
手工加脂法 hand stuffing
手工剪 hammer-shears;hand shears
手工剪板机 hand shearer
手工剪修 hand clipping
手工检查 desk checking
手工检索 manual searching
手工浇铸 hand teem(ing)
手工搅拌 hand mixing
手工搅拌器 hand stirrer
手工接合器 manual bonder
手工金属电弧焊 manual metal-arc welding
手工进料 hand feed
手工进料搅拌机 hand loaded mixer
手工精削 hand finishing
手工精修 hand finish
手工精整加工 hand finish
手工精装本 extra bound
手工具 small tool
手工具盒 glove compartment
手工具箱盖 glove compartment cover
手工锯 hacksaw frame
手工开采工作面 hand won face
手工开挖 hand excavation
手工刻图 hand engraving;hand scribing;manual engraving;manual scribing
手工控快门 hand shutter
手工控制 hand control;manual control
手工拉坯 hand throwing
手工劳动 hand labo(u)r;manual labo(u)r
手工镘刀 hand trowel;hand float
手工镘平 manual floating
手工铆钉 hand rivet(ing)
手工铆接 hand rivet(ing);riveting by hand
手工密实填充 solid hand packing
手工描花地毯 hand-painted carpet
手工描绘 hand drawing
手工灭火队领班 hand crew boss
手工模板印花织物 block print
手工模版印花 hand block printing
手工模版印花织物 hand-blocked fabric
手工模版印制 hand blocking
手工模型 hand form block
手工模压机 hand-mo(u)lding press
手工模制 hand mo(u)ld
手工磨床 hand grinder
手工磨光 hand polishing;off-hand grinding
手工磨光器 hand grinder;hand jigger
手工磨刻 freehand grinding
手工磨平地板 flogging
手工磨碎 hand grinding
手工磨削 hand grinding
手工抹灰 hand applied(mixed) plaster;hand plastering;hand rendering
手工抹平混凝土工具 hand finisher
手工抹子 hand float
手工木制瓦 hand-made shingle
手工拈线机 hand twiner
手工捏练 hand kneading
手工捏练成型 <耐火砖的> pinching
手工捏塑 hand forming
手工排版 handset;manual composition
手工排字 handset
手工抛光的 hand-polished
手工刨皮 hand shaving
手工配合 manual fitting
手工喷淋 hand spraying
手工喷雾 hand spraying
手工劈开 hand split
手工劈裂 hand split
手工劈制的木片瓦 hand split wood shingle
手工劈制的木质板壁 hand split wood siding shingle
手工劈制的木质屋顶板 hand cleft wood shingle
手工劈制的屋顶板 hand split wood shingle
手工品 hand-built product;hand-made article
手工平板切割机 hand plate shears
手工剖幅 hand slitting
手工铺路队 hand laying gang
手工铺路组 hand laying gang
手工铺撒 hand spreading
手工铺撒工 hand spreader
手工铺撒机 hand spreader
手工铺撒器 hand spreader
手工铺撒石屑 hand gritting
手工铺砂 hand spreading
手工铺设 hand placement;hand placing
手工启动 hand starting;manual starting
手工启动器 hand starter;manual starter
手工起动曲柄 manual starting crank
手工起动摇把 manual starting crank
手工气割 manual gas cutting
手工气焊 manual gas welding
手工砌石 hand-packed rockfill
手工砌筑块石 hand rubble
手工砌筑毛石 hand rubble
手工器具 manipulative device
手工钎子 jumper
手工嵌线 hand insertion
手工敲铲除锈 hand chipping and scraping
手工切割 hand cutting;manual cut(ting)
手工切片机 hand microtome
手工切削刀具 hand cutter
手工清除筛 hand-cleaned screen
手工清底找平 bottoming
手工清理 manual finishing
手工取样 grab sample;hand sample;hand sampling;manual sampling
手工取样器 hand sample cutter;manual sampler
手工燃烧 hand-firing
手工揉泥楔块 hand wedging
手工润滑 manual application of lubricant
手工撒布修补法 manual spray patching
手工筛 hand sieve
手工筛分 hand sieving
手工筛网印花织物 hand screen-printed fabric
手工上料的 handling loaded;hand-loaded
手工烧火 hand-firing;hand stoking
手工烧火的 hand-fired
手工生产 hand-handling;manual operation
手工石膏 manual applied plaster
手工石屑铺撒器 hand-operated chip(ping)s spreader
手工梳理 hand carding;hand combing
手工梳麻 hand dressing;hand hackle
手工梳麻台 hackle
手工输入 manual input
手工输入穿孔机 hand feed punch
手工输入装置 manual input unit;manual word generator
手工输送穿孔机 hand feed punch
手工数值化 hand digitized
手工刷浆 hand coat;hand swabbing
手工刷漆 hand brushing
手工刷色 hand swabbing
手工刷涂 hand brushing
手工丝网印刷法 hand screening
手工塑性制砖法 slop mo(u)lding
手工缩样 sample hand reducing
手工台 hand bench
手工摊铺 hand placement;hand spreading
手工掏槽 handhole;hand holing
手工套扣 hand tapping
手工挑选 hand pick
手工调整 hand adjustment;manual fitting;manual setting
手工筒式绞车 hand drum winch
手工筒式卷扬机 hand drum winch
手工涂布机 hand applicator
手工涂刷 hand rendering
手工涂刷和缠绕 manual coating and wrapping;manual laying up
手工涂搪 manual enamelling
手工推光 hand glassing
手工推挤 hand scudding
手工推平 hand set(ting);hand slicking
手工挖掘 hand digging
手工挖掘井 hand-dug well
手工弯筋机 hand bending machine
手工弯曲机 hand bending machine
手工弯折机 hand bending machine
手工喂料轧制的方孔型 hand square pass
手工喂料轧制的圆孔型 hand round pass
手工钨极惰性气体保护焊 manual tungsten inert gas welding
手工钨极交直流氩弧焊机 manual AC and DC tungsten-pole argon arc welding machine
手工洗污渍 swealing
手工下线 hand insertion
手工线测仪 handline ga(u)ge
手工线脚 hand mo(u)lding
手工线条 hand mo(u)lding
手工镶嵌 hand set(ting)
手工镶嵌金刚石钻头 hand set bit
手工小地毯 <中东和远东的> oriental rug
手工修改 hand completion;manual correction;manual update
手工修整 hand chipping;hand finish(ing)
手工修正 manual amendment
手工选择穿孔卡片 hand-operated punched card;manually operated punched card

手工压机 hand press
手工压浆棒子 jitterbug
手工压接 manipulative compression joint;manipulative joint
手工压模 hand mo(u)ld
手工压模的 hand mo(u)lded;hand blocked
手工压实的 hand-packed
手工研磨 freehand grinding;hand lapping;off-hand grinding;hand grinding
手工研磨机 hand grinder
手工样品 hand sample;hand specimen
手工业 handicraft(industry)
手工业方式 rule-of-thumb
手工业区 domestic industrial district
手工艺 handicraft;handicraft art;manual craft
手工艺工人 handicraftsman
手工艺工作者 artist-craftsman
手工艺品 articles of handicraft art;handicraft;handiwork
手工艺室 craft room
手工艺型花格窗 handicraft-type metal grille
手工艺型金属栅 handicraft-type metal grille
手工印花木模 printing block
手工印花墙纸 hand-blocked(wall) paper;hand-printed(wall) paper
手工印坯 hand mo(u)lding
手工印刷 hand printing
手工印刷的(糊)墙纸 hand-printed (wall) paper
手工印制 hand printing
手工印制的 hand blocked
手工用棒捣实 hand rodding
手工晕渲法 manual hill shading;manual hill shading technique
手工錾面 hand tooled finish
手工凿掘 hand picking
手工凿孔 hand boring
手工凿孔机 hand perforator;punch perforator
手工凿密 hand ca(u)lking
手工凿面 hand tooled finish
手工凿平 hand chipping
手工凿岩 hand hammer drilling;manual rock drilling
手工凿子 cold chisel
手工造型法 hand mo(u)lding
手工造型灰铁铸件 hand-mo(u)lded grey iron casting
手工造纸作坊 hand-made paper mill
手工针织 hand knitting
手工针织品 hand knitted hosiery
手工振动式整平板 <混凝土路面用> hand-operated vibrating [vibration] screed
手工整平板 hand-operated screed
手工整修 hand finishing;hand tooled finish
手工整修的 hand-finished
手工整修机 <平整混凝土的> hand finisher
手工整修器 hand finisher
手工纸 hand-made paper
手工制(成)的 hand-wrought;handmade;home-made
手工制管鞋 turn-back
手工制品 handiwork;product of handiwork
手工制榫眼机 hand mortising machine
手工制型瓦 hand-mo(u)ld tile
手工制造的 hand-built
手工制造增强塑料膜 hand lay-up
手工制砖 hand-mo(u)lded brick
手工制作 hand labo(u)r;shanty method
手工置放 hand placement;hand placing
手工铸模 hand mo(u)ld
手工砖 hand(-made) brick
手工转动的 hand-rotated
手工转绘 hand transferring
手工转写 hand transferring
手工装订 hand bookbinding
手工装配 hand assembly;hand fitting;manual fitting;manual setting
手工装饰工作 hand coating job
手工装岩 hand mucking
手工装载工作 hand loading job
手工装载工作面 hand-filled face
手工着色 hand colo(u)ring;hand-paint
手工着色的 hand-colo(u)red
手工着色地图 hand-colo(u)red map
手工琢石 dressed;hand dressed
手工组装 hand fitting;manual fitting
手工钻孔机 hand drill(ing machine)
手工钻探 hand boring
手工最后加工 hand finish
手工作图 manual plotting
手工作业 handwork
手工作业噪声 hand finishing noise
手弓 handbow
手弓锯 coping saw
手弓锯框 hand hack saw frame
手弓锯片 hand hack saw blade
手勾草图 freehand sketch
手勾等高线 sketched contour
手勾剖面 freehand profile
手钩 hand hook;stevedore hook
手钩扯破包皮 cover torn by hand hooks
手钩破洞 hook holes
手估 gross estimate
手刮刀 hand scraper
手焊 hand welding;manual weld
手焊烙铁 hand soldering bit
手焊设备 hand welding installation
手撼试验 dilatancy test;shaking test
手夯 bishop;commander;common ram;hand ram(mer);pommel;set ram
手夯槌 hand ram
手夯锤 hand ram;hand beetle
手夯实 hand ram
手夯硪 punner
手夯斜切面 Bishop's miter[mitre]
手弧焊 manual(metal-) arc welding;shielded metal manual electric arc welding
手糊成型 hand lay-up
手虎钳 hand vice [vise];vice [vise]-grip
手护 hand saver
手换挡 hand shifting
手绘草图 freehand sketch
手绘图 freehand drawing
手绘原图 hand-drawn original
手击发 manual firing
手机 cellular telephone;hand microtelephone;hand set(ting);mobile phone
手机送话器 hand set transmitter
手迹 handwriting
手迹图形输入板 script graphics tablet
手加速杆 hand accelerator
手加油器 hand oiler
手加油(器)盖 hand oiler cap
手夹 hand file
手监测器 hand monitor
手拣 hand cleaning;hand-sort;hand sorting
手拣富矿 rich hand-picked ore
手拣矸石 hand-picked reject
手拣金砾 reef picking
手捡 hand picking
手剪 snippers
手剪机 hand shears
手检 hand-sort
手检查 hand inspection
手键 telephone key
手浇包 hand ladle;shank ladle
手绞车 hand winch;pull lift
手绞盘 jack roll
手铰孔 hand reaming
手节流杆 hand throttle lever
手结地毯 hand-knotted rug
手结栽绒地毯 knotted pile carpet;knotted rug
手巾 hand towel
手巾架 horse
手紧螺母 hand nut
手进刀 hand feed
手进刀平面刨床 hand feed surfacer
手进给穿孔机 hand feed punch
手进给轮 hand feed wheel
手精选的生石灰 best hand-picked quicklime
手举过肩的 overhand
手锯 arm saw;back saw;hand back saw;hand(-operated) saw;miter [mitre] saw;pad saw;plunging saw;quarter rip saw;single-handled saw;tenon saw
手锯柄 grip
手锯锉 hack file
手锯胶合板 hand sawing plywood
手锯条 arm saw blade;hand saw blade
手掘井筒 hand-dug pit
手掘(竖)井 hand-dug shaft
手(开)动 hand motion
手开闸 hands-off
手孔 <钢结构焊接拼装或养护时便于伸手入内操作的孔洞> handhole
手孔板 handhole plate
手孔板垫密片 handhole plate gasket
手孔板螺栓 handhole plate bolt
手孔轭 handhole yoke
手孔盖 handhole cover;handhole plate
手孔和人孔开口 handhole and manhole openings
手孔接头盖 handhole nipple cap
手孔凸缘螺纹接套 handhole flange nipple
手控安全开关 manual safety switch
手控按比例操纵系统 manual proportional control system
手控靶 hand-controlled target
手控曝光 manual exposure
手控闭环控制系统 monitored control system
手控便利 ease of handling
手控拨叉 manual selector fork
手控操纵 hand-controlled manipulation
手控测程仪 hand log
手控铲运车的绞车 hand scraper winch
手控超越控制 manual override control
手控齿轮换挡变速器 manual gearshift transmission
手控穿孔机 manual punch
手控传输 manual transmission
手控存储开关 manual storage switch
手控的 manipulative;manual;manually controlled
手控点火提前装置 manual spark advance
手控电动振动压路机 hand-guided power-propelled vibrating roller
手控断油开关 manual shutoff valve
手控发报机 manual telegraph transmitter
手控阀 hand-operated valve;manual (-operated) valve;selector valve
手控阀盖 hand control valve cover
手控方式 manual mode
手控防火门 manually operated fire resisting door
手控复位 manual reset
手控杆 hand(control) lever
手控跟踪 manual following
手控刮铲车的绞车 hand scraper winch
手控刮铲小车 hand scraper
手控刮铲卸货车 hand scraper unloader
手控航空摄影机 hand-photogrammetric camera
手控黄油枪 hand grease gun
手控火警报警器 manual fire station
手控机 manual controller
手控计数器 hand counter
手控计算器 hand calculator
手控监控系统 manual closed-loop control system;monitored control system
手控检查 hand-on inspection
手控交通信号 manually controlled traffic signal
手控开关 manual switch
手控开关存储器 manual-switch storage
手控开环 manual control open loop
手控可变焦距透镜 manual zoom lens
手控离合器 manually operated clutch
手控目标选择器 manual target selector
手控配平轮 hand trim wheel
手控起动开关 hand starting button;manual starter switch
手控起重机 hand control hoist
手控气阀 hand air valve
手控桥式吊车 hand-propelled travel(l)ing crane
手控入口 manual entry
手控撒播机 <播农作物种子用> hand sprayer
手控色调 manual hue
手控摄影机 hand-operated camera
手控升降器 hand-controlled lifting device
手控式 manual mode
手控式水枪 hand-controlled branch;hand-controlled nozzle
手控输入装置 manual input device
手控数码发生器 manual number generator
手控数字指令 manual digital command
手控数字转换器 manual digitizer
手控双轮推土机 hand-guided two-wheel dozer
手控随动操舵 hand follow-up steering
手控调节器 manual governor
手控调谐 manual tuning
手控同步 manual synchronization
手控透镜 manual lens
手控凸轮 manual selector cam
手控微调 manual fine tuning
手控稳住的 manually held
手控消防泵 manual fire pump
手控信号 hand control signal
手控信号台 hand control signal box
手控选择开关 manual selector switch
手控音量控制 manual volume control
手控油门 hand throttle control
手控增量键 manual-increment key
手控增益 manual gain
手控闸杆 brake hand control lever
手控振动刮板 hand vibrating screed
手控振动混凝土抹面机 hand-guided vibrating concrete finisher
手控帧 manual frame
手控制 hand control;manual control
手控制递开弹簧 control graduating spring

S

手控制动发动机点火 manual retrofire
手控制动杆 brake hand control lever
手控制阀活塞 hand control valve piston
手控制阀体 hand control valve body
手控周期 effort-controlled cycle; manually controlled work
手控转速计 manual tachoscope
手控自控 manauto
手控自控联合控制 man-machine control
手拉车 hand car(t);Jak-tung truck; man drawn truck;hand buggy
手拉粗丝 hand-drawn coarse fiber; starting waste fiber[fibre]
手拉的 man-drawn
手拉吊挂 pulley block
手拉滚筒 hand-drawn roller; hand-guided roller
手拉葫芦 chain block
手拉滑车组 hand pulley block
手拉加速杆 hand throttle
手拉紧 finger-tight;hand tight
手拉缆 dead rope
手拉链 hand chain
手拉链驱动 hand chain drive
手拉平口刨 planishing knife
手拉撒砂器传动杆 hand sander reach rod
手拉示功图 hand-pushed indicator diagram
手拉式点火提前杆 manual-advance lever
手拉式手制动器 pull-out hand brake
手拉锁 drawback lock
手拉弹簧秤 hand spring scale
手拉小车 hand cart
手拉行车 hand power truck crane
手拉压路机 hand-drawn roller
手拉样板 hand pulled template;hand template
手拉闸 hands-off
手拉镇压器 hand roller
手拉制动器 hand brake
手拉钻 fly drill
手喇叭 hand-actuated horn
手栏杆托架 handrail bracket
手缆 hand rope
手缆绞车 hand rope winch
手冷錾 hand cold chisel
手力 hand power
手力传动装置 hand(-operated) gear
手力复原 hand-restoring
手力杠杆闸 hand lever brake
手力割草机 hand mower
手力夯 hand tamper
手力机 hand machine
手力唧筒 hand pump
手力剪机 hand snips
手力接合 hand engagement
手力进料 hand feed
手力起道器 hand jack(screw)
手力起重器 hand jack(screw)
手力千斤顶 hand jack(screw);hand screw
手力轻便铆钉器 hand portable riveter
手力驱动 hand gear
手力压剪机 hand punching and shearing machine
手力移动式起重机 hand power track crane
手力制动鼓 hand brake drum
手力钻机 hand driller
手链小车 manual chain-driven carrier
手捩加紧 hand tight
手榴弹 bomb
手炉的 open-hearth
手录计数板 manual counting board

手轮 flier[flyer]; hand wheel; pilot wheel;derrick wheel < 在司钻台操纵动力机加速器的 >
手轮柄 handwheel handle
手轮操纵阀 handwheel valve
手轮操作阀 handwheel valve
手轮垫圈 handwheel washer
手轮开关 handwheel switch
手轮螺母 handwheel nut
手轮式水枪 pistol-type branch
手轮锁紧螺钉 wheel screw
手轮弹簧圈 handwheel coil
手轮推进的振动抹面机 handwheel propelled vibrating finisher
手轮推进的振动抹面器 handwheel propelled vibrating finisher
手轮托 < 水温调整阀 > handwheel seat
手轮压下装置 handwheel screw down
手轮轴 handwheel shaft
手轮轴盖 handwheel shaft cap;handwheel shaft cover
手螺母 hand nut
手螺旋夹 hand screw clamp; hand screw holdfast;screw clamp
手镘 hand floating
手镘板 hand float
手镘刀 hand float trowel
手镘干硬性混凝土 hand floating dry concrete
手镘抹面 hand float finish
手铆 hand rivet(ing)
手铆(钉)锤 hand riveted hammer; hand rivet(ing) hammer
手铆(接)机 hand rivet(ing) machine
手铆铆钉 hand-driven rivet
手磨 hand polishing
手磨机 hand mill
手磨气门摇臂 hand crank grinding tool
手木槌 hand-mallet
手捏 hand pinching
手捏变速操纵装置 hand gear control
手捏吸气式 hand-aspirated type
手拧螺丝 thumb screw
手耙 < 清理拦污栅用 > hand rake
手耙的 hand raked
手帕 napkin;pocket-handkerchief
手刨 hand plane
手刨床 hand planer;hand planing machine
手喷器 hand distributor;hand sprayer
手喷枪 hand distributor; hand lance; hand sprayer
手喷雾器 hand sprinkler
手劈木墙板片 riven wood siding shingle
手劈木瓦片 hand cleft wood shingle; hand split wood shingle
手劈外墙面板 hand cleft wood siding shingle; hand split wood siding shingle
手劈再锯木瓦板 hand split and resawn
手铺的 hand laying;hand-placed
手铺法 hand lay-up process
手铺片石 hand-placed riprap
手铺砌石(块) pitching stone
手铺石 hand-placed rock
手铺碎石 hand-pitched broken stone
手铺小块石底层 hand-pitched base
手铺小块石基层 hand-pitched base
手旗 hand flag;semaphore flag
手旗或手臂通信[讯] signal(l)ing by hand flags or arms
手旗通信[讯] semaphore signal
手旗信号 flag wagging; hand signal; semaphore signal
手起动 hand start;hand cranking
手起动泵 hand priming of pump

手起动装置 hand starting arrangement
手起落杆 hand lifting lever
手起落式犁 hand-lift plow
手砌的 hand-laid
手砌基础 hand-laid foundation
手砌块石 hand-placed rock
手钎子 hand bit
手钳 combination pliers; hand tongs; hand vice[vise]; nippler pliers; pincers; pliers[plyers]
手枪 pistol;shooter < 美 >
手枪射击场 pistol range
手枪式把手 pistol grip
手枪式钉器 pistol-type stapler
手枪式多穿孔 revolver type multiple punch
手枪式喷枪 pistol lance
手枪式喷雾器 spray pistol
手枪式握把 pistol grip
手枪形砖 pistol brick
手锹式碰锁 thumb latch
手橇 toboggan
手揿泵 hand pump
手揿插销 lift latch
手揿点货机 hand tally
手倾车 hand tip-cart; hand-tipping-barrow
手清洁度计数器 hand counter
手球场 handball court; handball playground
手球法 ball in hand
手驱动的 hand-driven
手绕 winding by hand
手绕的 hand-wound
手绕法 hand winding
手撒的 hand laying
手洒水器 hand sprinkler
手塞规 hand plug ga(u)ge
手刹车 grip(per) brake; hand brake; hand lever brake
手刹车操纵杆 hand brake lever
手刹车杆 lever brake
手筛 hand sieving
手上试验 hand test
手烧锅炉 hand-fired boiler
手烧炉 manually operated furnace
手烧炉膛 hand-fired furnace
手勺 hand ladle;hand scoop
手勺取样 spoon test
手绳操作 hand rope operation
手示信号 hand signal
手势信号 flag signal;hand signal
手饰铜 gilding metal
手书 holograph
手书件 handwriting
手术房屋 treatment building
手术观摩室 amphitheater[amphithcatre]
手术间 surgical suite
手术示教室 operating theatre[theater]
手术室 operating chamber; operating room; operation room; operative room
手术台 operating table
手术治疗 operative treatment
手刷 hand brush;hand scrubber
手水准(仪) Abney level; hand level; monocular hand level
手顺的 flowing
手丝锥 hand plug tap
手撕试验 hand tear test
手送碎木机 hand feed grinder
手算 hand calculation; hand computation; manual computation
手算法 hand calculation method
手抬泵拖车 portable pump trailer
手抬砂箱 one-man flask
手抬式钢轨钳 hand carrying type rail tongs
手抬消防泵 portable fire pump

手套 gloves;hand sleeves
手套衬里 glove liner
手套革 glove leather
手套机 glove knitting machine; glove machine;glove port
手套棉法兰绒 glove flannel
手套纱线 glove yarn
手套式操作箱 < 真空设备用 > glove box
手套式感觉缺失 glove anesthesia
手套箱 glove box;glove compartment
手套箱串列 glove-box train
手套箱盖锁扣 glove-box door latch
手套箱屏 glove box shield
手套箱线路 glove-box train
手套箱照明灯 glove-box lamp
手套压烫机 glove presser
手套制作 gloving
手提 tote
手提包 handbag; hand baggage; hand baggage grip;holdall;gripsack < 美 >
手提泵 portable pump
手提布氏硬度计 hand Brinell's hardness tester
手提步话机 handset
手提操作台 hand-held console
手提测读装置 portable readout unit
手提测试仪 portable testing set
手提测图摄影机 hand mapping camera
手提测斜仪 portable inclinometer
手提秤 portable scale
手提的 manpack;portable
手提灯 hand lamp; hand lantern; hand light; lantern; portable lamp; portable light;service lamp
手提灯变压器 hand lamp transformer
手提灯玻璃灯罩 hand lantern glass globe
手提地震计 portable seismometer
手提地震仪 portable seismograph
手提电锤和电钻 electric (al) hand hammer and(rock) drill
手提电动工具 electric(al) hand tool; hand electric(al) tool
手提电话 portable telephone
手提电剪刀 unishear
手提电锯 electric(al) hand saw
手提电磨机 electric(al) grinding machine
手提电刨 electric(al) hand shaper
手提电视摄像机 hand-held TV camera
手提电钻 electric (al) hand drill; hand-held electric(al) drill;handset
手提动力工具 hand power tool
手提泛光照明灯 hand float light
手提放大镜 hand magnifier
手提风速表 < 福斯风速表 >【气】 hand anemometer
手提风钻 hammer hand drill
手提封包机 portable bag closer
手提工具箱 portable toolbox
手提工作灯 portable lamp
手提鼓风机 portable blower
手提灌油壶 hand pouring pot
手提灌注器 hand injector
手提航拍摄影机 hand-held aerial camera
手提航摄影机 hand mapping camera
手提航行灯号 portable lamp;portable light
手提厚度计 portable thickness ga(u)ge
手提活塞式底质取样器 hand-held piston-type bed material sampler
手提火号 hand flare
手提火焰切割器 hand flame cutter
手提击实仪 portable compacter
手提机枪 portable machine gun

S

手提计算器 hand calculator
手提加油器 hand oiler
手提胶片显影剂 portable film developer
手提净水器 portable water purifier
手提刨孔机 portable router
手提扩音器 hailer
手提离心式水泵 hand-carry centrifugal pump
手提裂缝计 portable crack ga(u)ge
手提罗经 box and needle;hand bearing compass; hand compass; prismatic(al)compass
手提螺钻 hand auger
手提面罩 <焊工用> hand face shield
手提模 portable mo(u)ld
手提泥沙取样器 hand sediment sampler
手提喷洒器 hand sprayer
手提喷油壶 hand pot
手提皮箱 keister;suitcase
手提气压喷雾器 pneumatic hand-sprayer
手提润滑器 hand lubricator
手提闪频转速计 tach(e)oscope
手提摄影机 hand-held camera;hand-operated camera
手提升式刮土铲运机 hand-lift scraper
手提式 hand-held;man-portable
手提式摆式仪 portable pendulum tester
手提式报话机 hand-held
手提式步话机 handie-talkie
手提式测绘摄影机 hand mapping camera
手提式测角器 hand goniometer
手提式测试器 portable testing set
手提式测向器 hand goniometer
手提式超高速喷水系统 portable ultra-high-speed water spray system
手提式超声波焊机 cavitron
手提式吹风装置 hand blower
手提式吹灰器 hand lance
手提式磁带录音机 portable tape recorder
手提式磁力探伤仪 portable magnetic flaw detector
手提式的 hand-held;portable
手提式灯具 portable luminaire
手提式点焊机 gun welding machine
手提式电动泵 portable electric(al)pump
手提式电动工具 portable electric(al)tool
手提式电烘箱 portable electric(al)oven
手提式电锯 portable electric(al)saw
手提式电喇叭 hailer
手提式电视摄影机 walkie-lookie
手提式电子实验线路板设计 breadboard design
手提式电钻 electric(al)portable drill;hand electric(al)drill; portable electric(al)drill
手提式放射能指示器 portable radiation instrument
手提式分光光度计 hand spectrophotometer
手提式分光镜 hand spectroscope
手提式分析仪器 portable analytic(al)instrument
手提式风钻 hand-held drill
手提式辐照器 portable irradiator
手提式高温计 hand pyrometer
手提式高压蒸汽消毒器 high-pressure steam sterilizer of portable type
手提式工具 portable tool
手提式工具箱 portable toolbox
手提式光谱计 portable spectrometer
手提式轨枕捣固机 portable tie tamper
手提式夯击压实机 portable impact compactor
手提式夯实仪 portable compacter
手提式火焰喷射器 portable flame thrower
手提式火灾报警装置 hand-operated fire alarm installation
手提式计数器 portable counter
手提式计算机 hand-held computer; portable computer
手提式静电喷枪 electrostatic hand gun;electrostatic hand spray gun
手提式锯 portable saw
手提式卡规 microsnap ga(u)ge
手提式抗滑试验仪 portable skid-resistance test device
手提式刨孔机 portable router
手提式扩音器 bull-horn;loud-hailer
手提式灭火机 hand(fire)extinguisher; portable (fire) extinguisher; manual extinguisher
手提式灭火器 hand(fire)extinguisher; portable (fire) extinguisher; manual extinguisher
手提式灭火设备 hand-held extinguishing appliance
手提式灭火装置 hand extinguishing appliance
手提式磨光机 portable grinder
手提式泥沙采样器 hand sediment sampler
手提式泡沫发生器 portable foam (generating)device
手提式泡沫喷口 portable foam applicator
手提式喷壶 pouring can
手提式喷枪 hand spray gun
手提式喷燃器 portable burner
手提式喷洒设备 aerostyle
手提式器械 portable apparatus
手提式轻便电力工具 hand-held portable electric(al)tool
手提式倾斜仪 portable tiltmeter
手提式曲颈甑 portable retort
手提式砂轮机 portable grinder
手提式摄影机 gun camera;hand camera
手提式声级计 portable hand-held sound level meter
手提式收发台 hand transmitter receiver
手提式水磨石机 hand terrazzo grinder
手提式水准仪 flying level;hand level
手提式缩微胶卷阅读器 portable microfilm reader;hand viewer
手提式弹簧秤 hand spring scale
手提式弹簧吊秤 hand spring balance
手提式往复锯 portable saber saw
手提式微型电话机 hand microtelephone
手提式污染物采集器 portable contaminant collector
手提式无线电电话机 handie-talkie
手提式无线电设备 portable radio apparatus;portable radio equipment
手提式吸尘器 portable dust cleaner
手提式系统 portable system
手提式压力试验机 portable compression testing machine
手提式仪表 portable appliance
手提式仪器 portable appliance;portable instrument
手提式圆锯 portable circular saw
手提式阅读器 portable reader
手提式凿岩机 jack hammer
手提式振动抹平刀 vibratory hand float;vibratory hand trowel
手提式锥形贯入试验 portable cone penetration test
手提式资料记录器 portable data recorder
手提试验设备 portable testing set
手提数字记录器 portable digital recorder
手提桶 hand bucket
手提无线电步话机 handle-talker
手提无线电话设备 portable radio-telephone equipment
手提析光计 hand refractometer
手提箱 holdall
手提箱式高压乙炔发生器 hand case high pressure acetylene generator
手提信号灯 portable signal light
手提行李 personal effects;hand baggage;hand luggage <旅客随身携带>
手提油灯玻璃罩 hand oil lamp glasses
手提凿岩机 handhold drill; jack hammer
手提照相机 handie-lookie
手提终端机 hand-held terminal
手提转速表 tach(e)oscope
手提转速计 hand tach(e)ometer; portable tachoscope;tachoscope
手提转速器 hand tach(e)ometer
手提自给矿工灯 miner's hand lamp
手调弧光灯 hand-regulated arc lamp
手调铰刀 hand adjustable reamer
手调(节)hand adjustment;hand regulation; finger control; hand setting); manual adjustment; manual set
手调节膨胀阀 hand-regulating expansion valve
手调螺钉 manual control screw
手调螺距螺旋桨 manual control pitch propeller
手调式送经装置 hand-regulated warp let-off device
手调斜度切割器 manually angled cutter
手调谐 hand tuning
手调旋钮 manual adjusting knob
手调整 manual regulation; manual setting
手调装置 hand adjustment device
手头备用现金 cash on hand
手头装饰条 casing bead
手涂焊条 manually coating welding electrode
手推车 barrow truck;block truck;bogie[bogey/ boggy]; buggying; cart; dilly; go-cart; go-devil; hand barrow; hand car; hand(push)cart; hand wagon; larry; manual (handling)truck; manumotor; perambulator;pram;push car(t);push section car;ranking bar;sack barrow; troll(e)y; wheel barrow; wheeled pusher;hand buggy;hopper <有倾斜斗的>;carbunk <运木材往炉内烘干用的>
手推车磅秤 wheel-barrow scale
手推车道 barrow run
手推车道板 barrow runner
手推车队 barrow gang
手推车工人 barrow men
手推车轨道 lorry rail;lorry track
手推车过磅秤 wheel-barrow scale
手推车马道 barrow run
手推车式混凝土搅拌机 wheel-barrow concrete mixer
手推车式机动喷雾机 power wheel-barrow sprayer
手推车式喷雾机 wheel-barrow sprayer
手推车式喷雾器 push-type sprayer
手推车式撒播机 wheel-barrow broadcast seeder
手推车跳板 barrow runner
手推车狭桥 wheel-barrow run
手推车小路 wheel-barrow run
手推车小铁轨 troll(e)y rail
手推车运料 barrowing
手推车运输 barrowing;carting
手推车装载量 cartload
手推床 wheeled stretcher
手推带锯 hand feed band-saw
手推单辊压路机 manually propelled single-drum roller
手推的 hand-propelled;manumotive
手推点焊 poke weld(ing)
手推堆垛机 hand stacker
手推翻斗车 hand tipple;tippling-barrow
手推杠杆式放水龙头 push tap
手推割草机 hand mower
手推割捆机 push binder
手推刮铲 mixer-drive shovel
手推管道运输车 pipe hand truck
手推夯击机 hand-propelled compacting machine
手推夯实机 hand-propelled compacting machine
手推货车 hand truck
手推剪草机 hand mower
手推绞盘 hand capstan
手推垃圾车 refuse cart
手推冷乳剂喷洒器 pedestrian-controlled cold emulsion sprayer
手推犁 hand plow
手推两轮货车 rear car
手推磨 quern
手推墨辊 brayer;hand roller
手推车木板路 <工地临时铺设的> barrow run
手推平车 push section car
手推起重机 hand travel(l)ing crane
手推倾卸车 hand tip-cart
手推润滑油枪 hand push grease gun
手推洒水车 sprinkling cart
手推式钢轨超声波探伤机 hand ultrasonic detecting cart for rails
手推式构造深度仪 minitexture meter
手推式(火焰)发生器 <路面加热用> hand-propelled generator
手推式剪草机 hand lawnmover
手推式沥青 hand sprayer
手推式牛油枪 push grease gun
手推式喷雾车 spray barrow
手推式振动碾(压机)hand-propelled vibrating roller
手推双轮混凝土车 rickshaw
手推土车 navvy bar
手推托盘车 hand pallet truck
手推小车 decauville tub; decauville wagon;hand cart;travel(l)ing troll(e)y; troll(e)y; wheel barrow; hand-propelled troll(e)y
手推旋转起重机 hand-slewing crane
手推压实机 hand-propelled compacting machine
手推移动的振动压路机 pedestrian-controlled vibrating roller
手推移动式脚手架 manually propelled mobile scaffold
手推油墨辊 brayer roll
手推运货车 hand trolley; wagon; hand truck
手推运料车 barrow truck
手推运料车轨道 lorry rail
手推运输 barrowing
手推凿岩机车 drill carriage
手推振动辊压机 hand-propelled vibrating roller
手推振动磨光机 hand-propelled vibro-finisher
手推振动抹面机 hand-propelled vibro-finisher

手推振动压路机 hand-propelled vibrating roller
手推中耕器 push hoe
手推钻 push drill;well auger
手砣站台 horse block;sounding platform
手挖沉箱 hand-dug caisson
手弯 hand bending
手腕 wrist
手纹 hand print
手握定点器 mouse
手污防护板 <门上的> finger(-nail) plate
手铣 hand milling
手掀点货机 hand tally
手掀式喷雾器 manual sprayer
手掀锁的插销 jumbo bolt
手箱 casket
手镶(金刚石)钻头 hand set bit
手效应 hand effect
手挟式单辊压路机 walk behind single drum roller
手携灭火器 hand fire extinguisher
手携式转数计 hand tach(e)ometer
手写注记 hand lettering
手卸车 hand tip-cart; hand-tipping-barrow
手信号灯 glim;hand lantern
手信号旗 hand flag
手性的 chiral
手修 retouching
手修法 hand-trimming method
手续 course; formality; procedure; process
手续费 commission(charges);brokerage;charges for trouble;commitment fee;factorage;front-end fee;handing charge;handling cost;handling fee;percentage;rack off;service charges;service cost;service fee;servicing cost;handling charges;handling expenses
手续费及其他费用 commission fees and other charges
手续费率一览表 schedule of commission charges
手续费账目 commission account
手续纸印刷机 sheet-fed press
手旋紧面平均直径 average diameter of handtight face
手旋硫化机 hand-screw vulcanizing press
手旋螺钉 hand screw
手旋螺丝夹 <木工夹具> hand screw
手旋起重机 hand-slewing crane
手旋钻 rotary hand drill
手选 hand cleaning; hand picking; hand preparation; hand selection; hand sort(ing)
手选的 hand-picked
手选法 hand sorting method
手选分离 hand sorting separation
手选矸石 handpicking stone
手选工(人) conveyer picker; hand picker;ore picker
手选胶带输送机 picking conveyer [conveyor]
手选块铜 barrel copper
手选矿石块 cobbing
手选皮带 conveyer picker
手选取样 grab sampling
手选取样机 grab sampling machine
手选台 hand sorting table;picking table;sorting table;strake
手选优质石灰 best hand-picked lime
手选运输机 conveyer picker
手压泵 hand(force) pump; hand lance;hand-operated pump;manual pump;spraying lance
手压床 arbor press
手压点焊 push weld(ing)
手压罐式润滑油枪 hand-tank grease gun
手压磙 hand roller
手压机 hand-operated press; hand power press;manual press
手压井 man-pumped well
手压开关 petcock
手压沥青撒布器 hand distributor; hand sprayer
手压刨床 surface planer
手压喷洒器 hand-operated sprinkler
手压千斤顶 hand jack(screw)
手压潜水供气泵 diver's pump
手压撒布机 hand distributor
手压式牛油枪 push grease gun
手压式喷雾器 manual sprayer
手压油泵 oil hand pump
手压油枪注油器 hand gun loader
手压注油枪 hand gun
手压钻 press drill
手压钻床 sensitive drilling machine
手眼机器 eye-hand machine; hand-eye machine
手摇把 crank handle;hand crank
手摇把式转辙机 point machine with hand crane
手摇半转泵 hand semi-rotatory pump
手摇报警器 hand emergency signal-(l)ing apparatus
手摇泵 hand force pump; handling operated pump; handling pump; hand (-operated) pump; rower pump; stirrup hand pump; wing pump;wobble pump
手摇泵接头 manual pump disconnect
手摇舭水泵 hand bilge pump
手摇扁钻 nose bit
手摇玻璃窗 crank-operated window
手摇播种机 hand seeder
手摇舱底水泵 hand bilge pump
手摇舱面泵 hand deck-pump; handy billy
手摇测深仪 patent sounding machine
手摇叉动车 pallet truck
手摇车 handcar; hand(power) trolley;hand winch;troll(e)y;block truck;go-devil <铁路用>
手摇车床 hand lathe
手摇车底座【铁】troll(e)y base
手摇冲击钻 percussion hand drill
手摇冲击钻孔 percussion hand boring
手摇传送泵 manual transfer pump
手摇磁电机 handle magneto
手摇磁石发电机 magneto
手摇磁石发电机呼叫 magnetocall
手摇带阀活塞泵 bucket pump
手摇刀架 hand rest
手摇的 hand-operated;hand-powered
手摇吊包 geared crane ladle
手摇顶车器 jack with handle
手摇抖动器 hand shaker
手摇发电机 hand(-driven) generator;hand-dynamo; inductor; magneto-generator
手摇发电机系统 hand generator system
手摇发电机制 electric(al) hand generator system;hand generator system
手摇方向机 hand traversing mechanism
手摇方向机小齿轮 hand traverse pinion
手摇方向机制动联动装置 hand traversing brake linkage
手摇方向机轴 hand traverse shaft
手摇纺车 hand reeling machine
手摇风箱式喷粉器 hand bellowed duster
手摇封罐机 hand seamer
手摇干湿度表 sling psychrometer
手摇干湿度计 sling psychrometer
手摇干湿球湿度计 sling psychrometer
手摇杆弹簧 hand priming lever spring
手摇工具车床 hand tool lathe
手摇供电式紧急发信机 hand-powered emergency transmitter
手摇(鼓)风机 hand blower
手摇观察泵 sight pump
手摇惯性起动机 hand inertia starter
手摇轨道车 hand car
手摇呼叫 manual call(ing); manual ringing
手摇滑脂增压器 hand lever type grease pump
手摇回转泵 rotary hand pump
手摇火警器 manual fire alarm box
手摇机器 hand machine
手摇计算 hand computation
手摇计算机 calculator;hand computer;manual calculating machine
手摇计算器 hand(-held) calculator;hand tally
手摇计算装置 hand calculator
手摇绞车 hand cable winch; hand hoist;hand(-operated) winch;hand putter;monkey winch
手摇绞车拉紧装置 hand-winch take-up
手摇绞车提升 hoisting by hand-operated gear
手摇绞盘 hand winch
手摇搅拌机 manual blending mixer
手摇警报器 manual alarm
手摇警报装置 manual alarm system
手摇卷线车 hand-operated reel
手摇卷扬机 hand hoist; hand lifting winch;hand windlass;manual hoist
手摇快开人孔 manual quick opening manhole
手摇缆车 pain trolley;plain trolley
手摇缆索 <绞盘上的> hand cable winch
手摇缆索绞车 hand cable winch
手摇离心机 hand centrifuge
手摇立式螺旋压榨机 hand-operated vertical screw press
手摇链式回柱机 <采矿用> sylvester
手摇链条炉箅 hand chain grate
手摇铃 handbell
手摇炉箅 hand grate
手摇轮钻 wheel brace drill
手摇螺丝扳手接长节 ratchet wrench extension
手摇螺旋桨舢板 hand propelling boat
手摇螺旋钻 augering by hand; hand-operated auger
手摇螺钻机 hand-operated auger; ratchet bit brace
手摇马铃薯种切块机 hand potato cutter
手摇码头起重机 hand wharf crane
手摇灭火泵 stirrup(hand) pump
手摇喷粉机 crank duster;hand duster;manual duster
手摇喷粉器 hand(-operated) duster
手摇喷雾器 hand sprayer
手摇平板车 hand pallet truck
手摇破碎机 hand crusher
手摇起动 hand cranking; hand starting
手摇起动柄 hand-operating crank
手摇起动磁电机 hand starting magneto
手摇起动机 hand-driven starter;hand (-operated) starter
手摇起动手柄 manual cranking handle
手摇起锚机 pump brake windlass; pumping windlass
手摇起重 hand cranking
手摇起重机 hand crane;hand lift
手摇起重机车 hand-lift truck
手摇起重绞车 hand-driven crab;hand lifting winch;reel purchase
手摇起重装置 hand-lift device;hand-lift unit
手摇千斤顶 hand jack(screw);jack with handle
手摇切片机 rotary microtome
手摇切碎机 hand-operated chopper
手摇切纸机 hand paper cutter
手摇曲柄 crank handle;cranker;hand crank
手摇曲柄绣花机 crank handle embroidery
手摇曲柄钻 brace; brace and bit; brace drill; hand brace drill; brace bit;hand brace;crank brace
手摇燃料泵 hand fuel pump
手摇撒播机 hand sower
手摇砂轮 hand grinder;hand grinding machine
手摇砂轮机 hand-driven grinder;hand grinding wheel
手摇筛 hand screen
手摇上向凿岩机 hand rotation stoper
手摇升压器 hand booster
手摇湿度计 sling psychrometer
手摇式 hand-rotated;shaking
手摇式长钻 churner
手摇式火警装置 manual fire alarm system
手摇式计数器 hand tally
手摇式绞盘 hand capstan;windlass
手摇式捻度计 manual twist tester
手摇式起重机 hand lifting winch
手摇式轻型静力触探 hand light static cone penetration test
手摇式种子拌药机 hand seed dresser
手摇水果榨汁机 hand-operated fruit press
手摇碎矿机 hand crusher
手摇跳汰机 hand jig
手摇土钻 manual soil auger
手摇推进 hand feed
手摇推进导轨式凿岩机 hand feed drifter
手摇推进式倾料器 hand-propelled tripper
手摇腿 hand cranking leg
手摇脱粒机 hand threshing machine
手摇蜗轮传动计数器 worm-geared hand counter
手摇污水泵 hand bilge pump
手摇消防泵 hand fire pump
手摇小绞车 dolly winch; monkey winch
手摇旋转 manual traverse
手摇旋转机 ratchet brace; ratchet drill
手摇旋转手柄 manual traversing handle
手摇旋钻机 ratchet brace; ratchet drill
手摇研磨机 hand-operated grinder
手摇研磨器 hand-driven gear transmission bit;hand-driven gear transmission drill
手摇印刷机 hand press
手摇油泵 hand oil pump
手摇轧碎机 hand crusher
手摇振铃器 manual ringer
手摇转 hand traverse
手摇转速计 tach(e)oscope
手摇转向机构 manual changeover
手摇装置 manual setter
手摇走刀 follow-up hand feed

S

手摇钻 bit brace; bit stock; crank auger; crank brace; crank hand brace; drilling jig; hand bore; hand drill(ing machine); hand-operated drill; road auger; rotary hand drill; twist gimlet; wimble; straight hand drill
手摇钻的钻头承窝 pod
手摇钻弓柄 drill bow
手摇钻机 hand drill machine; manual auger; hand boring machine
手摇钻夹具 steel puller
手摇钻孔机 hand-operated boring machine
手摇钻台架 drill stand
手摇钻探 hand auger boring; hand boring
手摇钻头 pin bit
手摇钻土器 hand earth auger
手移起重机 hand power travelling crane; hand travel(l)ing crane
手艺 craft; craftsmanship; hand(i) craft; handicraftsman; mystery; useful arts; workmanship
手艺低劣 poor workmanship
手艺(工)人 artisan[artizan]; craftsman; handicraftsman; tradesman
手艺师 <欧洲中世纪的> magister
手艺知识 artisan knowledge
手印 finger mark; finger print; hand print
手用槽刨和舌刨 hand grooving and tonguing plane
手用测程器 log-ship
手用吹风器 bellows; hand bellows; handle air blower; pair of bellows
手用大槌 forehammer; mall hammer
手用大锤 forehammer; mall; sledge; sledgehammer; uphand sledge (hammer)
手用电钻 electric(al) hand drill
手用钢铆钉模 hand steel snap
手用管牙丝锥 hand tap for pipe thread
手用环柄弓锯 circular handle hack saw
手用螺母丝锥 hand nut tap
手用螺丝攻 hand tap
手用螺旋尖钻 gimlet bit
手用螺旋钻 hand auger
手用喷枪 <喷雾时用> hand gun
手用平底丝锥 bottoming hand tap
手用塞型丝锥 plug hand tap
手用水平仪 hand level
手用水准仪 hand level
手用丝锥 hand screw tap; hand tap; set tap
手用修理工具袋 hand refinishing kit
手用油石 handstone
手用圆刃楔形锤 hand fuller
手用锥形丝锥 hand taper tap
手用锥形销孔铰刀 hand taper pin reamer
手油壶 hand lubricator; hand oiler
手油门 hand acceleration
手油门拉杆盖支架 hand throttle link cover bracket
手油门拉线 hand throttle wire
手圆凿 hand gouge
手栽幼苗 hand-plant
手凿 hand-held sinker
手凿子 hand bit; hand chisel
手轧棉 hand-ginned cotton
手轧圆钢 hand rounds
手闸 hand brake
手闸操作杆 hand brake operating lever
手闸操作杆联杆 hand brake operating lever link
手闸操作杆联节销 hand brake operating lever link pin
手闸操作杆瓦 hand brake operating lever shoe
手闸衬片铆钉 hand brake lining rivet
手闸传力杆 hand brake transfer lever
手闸带 hand brake band
手闸带碇夹弹簧 hand brake band anchor clip spring
手闸带锚定杆调整螺钉 hand brake band anchor bar adjusting screw
手闸定位托架 hand brake locating bracket
手闸放松弹簧 hand brake release spring
手闸杆 hand brake lever; hand brake lever arm
手闸杆操作弹簧 hand brake lever operating spring
手闸杆操作止点 hand brake lever operating stop
手闸杆轭 hand brake lever yoke
手闸杆接头 hand brake lever adapter
手闸杆拉杆轭 hand brake lever pull rod yoke
手闸杆扇形齿轮 hand brake lever sector
手闸杆闩 hand brake lever latch
手闸杆弹键杆钮 hand brake lever latch rod knob
手闸杆托架 hand brake lever bracket
手闸杆握柄弹簧 hand brake lever grip spring
手闸杆爪 hand brake lever pawl
手闸杆爪弹簧盖 hand brake lever pawl spring cup
手闸杆爪销 hand brake lever pawl pin
手闸箍端托架 hand brake band anchor bracket; hand brake band bracket
手闸横轴 hand brake cross-shaft
手闸棘爪杆 hand brake lever pawl rod; hand brake support tie rod
手闸控制 hand brake control
手闸调整螺栓托架 hand brake adjusting bolt bracket
手闸调准螺栓 hand brake adjusting bolt
手闸调准螺栓簧 hand brake adjusting bolt spring
手闸凸轮 hand brake cam
手闸凸轮杆 hand brake cam lever
手闸压缩弹簧 hand brake compression spring
手闸支架 hand brake support
手掌 palm
手砧 dressing stake; hand anvil
手征 chirality
手征的 chiral
手征对称性 chirality
手征性 chirality
手织簇绒地毯 hand-knotted pile carpet
手织地毯 hand-knotted rug
手执的钩丝 hand line
手执计数器 tally
手纸 bumf
手纸盒 paper-holder
手纸架 paper-holder; toilet roll holder
手纸匣 paper-holder
手指操纵 finger-tip control
手指刮花 finger sgraffito
手指规 hand director
手指画 finger painting
手指拧紧 <使螺丝达到所能拧紧的程度> finger-tight
手指梳花 finger combing
手指输入装置 finger-input system
手指套 cot
手指印 finger mark
手指状喷涂图案 tails
手制 hand-made
手制草图 cartographic(al) sketching
手制的 hand-made
手制钉 hand-made nail
手制动 hand braking
手制动导架 hand brake guide
手制动阀 hand brake valve
手制动杠杆链 hand brake lever chain
手制动杠杆支点托 hand brake lever fulcrum bracket
手制动盒 hand brake housing
手制动机 hand brake
手制动棘轮 brake ratchet; hand brake ratchet wheel
手制动棘轮掣子 brake dog; hand brake ratchet pawl
手制动棘轮掣子锤 brake pawl weight; hand brake ratchet pawl weight
手制动棘轮掣子托 brake pawl carrier
手制动棘爪拉杆 hand brake pawl rod
手制动接合指示灯 hand brake telltale
手制动拉杆 brake chain connecting rod; hand brake pull rod
手制动拉杆导架 hand brake pull rod guide; hand brake rod guide
手制动拉杆吊 hand brake pull rod guide anchor
手制动拉杆及拉链 hand brake connection
手制动拉杆链 hand brake pull rod chain
手制动链 hand brake chain
手制动链导板 hand brake chain guide
手制动链导架 hand brake chain carrier
手制动链滑轮 hand brake chain roller
手制动链轮 hand brake chain sheave; hand brake chain wheel
手制动链轮托 hand brake chain roller bracket
手制动链筒 brake chain drum
手制动轮 hand brake wheel
手制动轮掣棘子 brake finger
手制动盘 hand brake disc[disk]
手制动器 hand(operated) brake; lever brake
手制动器驱动杆调整叉 hand brake drive rod adjusting fork
手制动器已接合 hand brake on
手制动器支架 hand brake bracket
手制动手把盖 hand brake handle cap
手制动踏板 brake foot board; brake step; end platform; hand brake step
手制动踏板托 brake step bracket; hand brake step bracket
手制动台 hand brake platform
手制动瓦销 hand brake shoe pin
手制动蜗杆 brake chain worm
手制动轴 brake mast; brake shaft; brake staff; hand brake mast; hand brake shaft
手制动轴臂 brake shaft arm
手制动轴承拉条 brake shaft step brace
手制动轴导架 brake shaft bearing; brake shaft guide; hand brake mast support; hand brake shaft guide
手制动轴棘轮掣子 brake shaft pawl
手制动轴卡板 hand brake guide plate
手制动轴拉杆 brake shaft connecting rod
手制动轴链 brake shaft chain
手制动轴链轮 brake shaft chain sheave; hand brake sprocket
手制动轴上导架 upper brake shaft bearing
手制动轴套 brake shaft sleeve; brake spool; hand brake shaft sleeve
手制动轴套管或导架 brake shaft thimble
手制动轴托 brake shaft bracket; brake shaft step; hand brake master step
手制动轴罩 brake shaft casing
手制动轴止挡 hand brake shaft stopper
手制动轴装置 hand brake shaft rigging
手制动爪杆 hand brake latch rod
手制动装置 hand brake
手制锻件 hand forging
手制节点 hand-formed joint
手制陶器 black shape
手制砖 hand-formed brick; hand-made brick
手掷靶装置 handtrap
手置模片辐射三角测量 hand-templet radiation plot; hand-templet radiation triangulation
手置式移栽机 hand-fed transplanter
手注油器 hand oiler
手注油器阀 hand oiler valve
手注油器簧 hand oiler spring
手注油器体 hand oiler body
手筑 hand-tamped
手爪模件 grip module
手转伸缩式凿岩机 hand rotation stoper
手转铁水包 hand shank; shank ladle
手转铁水桶 shank ladle
手转陀螺 teetotum
手转压力机 hand screw press
手转钻井法 hand rotary drilling
手转钻井机 hand rotary drilling machine
手装开关 flush switch
手锥 <测定磨具硬度工具> grading tool; gimlet
手钻 auger screw; gimlet; hand auger; hand boring; hand drill(ing machine); jumper(drill); jumping drill; push drill
手钻机 hand drill(ing machine)
手钻尖头 screwdriver bit
手钻尖头握柄 screwdriver bit holder
手钻炮眼 churn
手钻子 hand bit
手钻钻头 gimlet bit; gimlet point; hand boring bit
手座钳 vise grip pliers

守车 brakeman's caboose; brake van; cabin car; caboose(car); guard's van; hack; monkey house; van(car); crummy <货客运列车的>; dog house; zoo <俚语>; way car <列车尾部的>; shanty; conductor's car <美>

守车边灯 <向司机显示白灯,向后方列车显示红灯> side lamp
守车发电机 caboose generator
守车阀 caboose valve
守车瞭望顶棚 cupola
守车瞭望顶棚标志灯 cupola marker lamp
守车瞭望顶棚扶手 cupola hand rail
守车瞭望顶棚脚蹬 cupola inside step
守车瞭望顶棚信号灯 cupola signal lamp
守车停留线 brake van track; caboose track
守车线 caboose parking track; caboose track
守车员 car rider
守车员车厢 car rider's box
守车员室 car rider's box
守车制动阀 guard's van valve
守合同 hono(u)r contract
守恒 conservation

守恒定理 conservation theorem
守恒定律 conservation law
守恒方程 equation of conservation
守恒关系(式)conservation relation
守恒过程 conservation process
守恒力 conservation force
守恒凝固 conservative freezing
守恒派建筑 Mannerist architecture
守恒散射 conservative scattering
守恒矢量流 conserved vector current
守恒系(统)conservation system;conservative system
守恒性 conservation property
守恒性质 conservative property
守恒原理 principle of conservation
守恒轴矢流 conserved axial current
守护 guardianship
守护程序 demons
守护电话 guarder telephone
守护进程 guardianship process
守护神 daemon
守旧派建筑 <十七世纪意大利的> Mannerist architecture
守门工人【港】porter
守门人 door keeper
守桥人 bridge ward
守时 time keeping
守时刻的 punctual
守时性能 timekeeping performance
守时仪器 timekeeper
守听时间 hour of watch
守望 lookout
守望处 <西班牙建筑中的> mirador
守望所 watch house
守望台 crow's nest;watch tower
守望亭 watch box
守望员室 watchman's room
守望着 lookout man
守卫 caretaker;guard
守卫室 clock-house;warder house
守卫一个目标 guarding a target
守卫者 guarder
守位浮标 watch buoy
守信用 maintain commercial integrity
守夜 vigilance
守夜人的职位 <古罗马> excubitorium;watchman shed
守夜人员 dark horse
守夜者 night watchman
守约 abide by;follow treaty
守约的一方 observant party
守约方 observant party
守约重义 hono(u)ring contract and acting in good faith
守则 regulations;rule
守住某物标方位 keep one's bearings

首

首岸推 bow cushion
首笔现付金额 down payment
首标 header(label);heading
首标卡 header card
首标开始 start of header
首波 head wave
首波到达 first wave arrival
首部标签 header label
首部缓冲器 header buffer
首部记录 leader
首部结构 header
首部链轮 head sprocket
首部式布置 <指水电站> upstream station arrangement
首部双层底 double bottom forward
首部水槽 head flume
首部调压井 headrace surge tank
首部舷墙 bow bulwark
首部信息 header message
首部闸门 crown gate;head gate
首部闸门灌水率 head-gate duty of water
首部装卸设备 bow loading and discharge arrangement
首采区 initial mining district;primary mining area
首侧推半速向右 bow thrust half to starboard
首侧推半速向左 bow thrust half to port
首侧推器 bow thruster
首侧推全速向右 bow thrust full to starboard
首侧推全速向左 bow thrust full to port
首侧推停车 bow thrust stop
首层 first floor <美>;ground floor <英>
首层楼层 above-grade subfloor
首层楼面 above-grade subfloor;above-ground subfloor
首层平面 first floor plan <美>;ground-floor plan <英>
首层涂色 primary colo(u)r;primitive colo(u)r
首吃水大于尾吃水 trim by bow;trim by head;trim by stern
首创 initiate;originate;pioneer;take the initiative
首创建造师 pioneering architect
首创精神 creative initiative;initiative;pioneering spirit
首垂线 fore perpendicular;forward perpendicular
首次报价 first bid
首次逼近 first approximation
首次拨款 initial appropriation
首次衬砌 primary lining
首次淬火 early hardening
首次分簇 primary clustering
首次分析价值 head value
首次付款 down payment
首次更新 initial update
首次故障前平均时间 mean time to first failure
首次故障周期 mean time to first failure
首次灌满渠道 priming
首次航行 virgin voyage;maiden voyage
首次火警 first alarm
首次剂量 initial dose
首次检验 first inspection
首次浇筑 first pour
首次近似 first approximation
首次经过时间 first passage time
首次拷贝时间 first copy-out time
首次可采销量 primary recovery
首次来压 first coming pressure
首次满足 first fit
首次满足法 first fit method
首次模型 proplasm
首次碾压 breakdown rolling
首次碰撞 initial collision
首次偏移概率 first excursion probability
首次屈服 first yield
首次熔炼 pill heat
首次适合 first fit
首次送电 initial power receiving
首次损坏 primary failure
首次显影 first developing
首次要求 first demand
首次要求即付保函 guarantee on the first demand
首次有界区间度量 first bounded interval measure
首次阵风 first gust
首次置业房子 starter home
首次注册税 first registration tax
首倒缆 fore spring line;forward spring
首灯 bow light
首跌落【船】dropping
首顶肘板 breast hook
首都 capital;metropolis
首都的 metropolitan
首都地区 capital territory;metropolitan area
首都行政区 metropolitan district
首端绞车 head winch
首端轴 head shaft
首发台 station of origin
首犯 prime culprit
首符 first symbol
首府 capital;metropolis
首行进页 first line form advance
首行满排余均缩排 hanging indention
首航 first open water;virgin voyage
首(号)卡 header card
首横缆 forward breast line
首环功能部件 first loop feature
首货门 bow port
首尖舱 forepeak
首检波点线起始桩号 start stake number of the first receiver line
首检波点线终了桩号 end stake number of the first receiver line
首件 initial workpiece
首角投影法 first-angle projection
首阶 starting step
首阶梯楼梯踏步 first step
首盔 bow visor
首缆 bow fast;headline
首力矩 primary moment
首列 begin column
首领 captain;chief
首楼【船】forecastle
首楼后端 break of forecastle
首轮 head pulley;head roller;head sheave
首买权 right of first refusal
首锚 anchor bower;bow anchor;bower
首锚链 bower chain
首门【船】bow door;bow shell door
首脑会议 summit conference
首年追加折旧费 additional first year depreciation
首炮号 head shot number
首喷水转向舵 bow jet rudder
首批 first batch
首批订货 initial order
首偏荡幅度 yawing amplitude
首期付款 down payment;initial payment
首期工程(建筑)费 initial construction cost
首期建设 priority construction
首期建筑 priority construction
首期现款 <购置房地产的> original equity
首倾【船】by the head;down by the head;trim by the bow;trim by the head;trim by the stern
首曲线 index contour;intermediate contour;intermediate curve;mediate contour;mediate curve;principal contour;standard contour;standard curve
首取路由 first-choice route
首绳 head rope
首饰店 jeweler's shop
首数【数】characteristics of logarithm
首水尺修正 stem correction
首塔 <缆式起重用> head tower
首涂材料 prime material;priming material
首涂层 primary coat(ing);prim(r) coat(ing);prime membrane;priming coat
首涂色料 priming colo(u)r
首涂油 primer
首涂油漆 primer coat(ing)
首尾 alpha and omega
首尾不稳定性 head-tail instability
首尾不一致 change about
首尾吃水不等 uneven keel
首尾吃水差 trim;vessel trim
首尾吃水相等 even keel
首尾导标 double crossing leading mark;fore-and-aft range mark;head and stern leading marks
首尾机构 terminal organs
首尾两天均包在内 both days inclusive
首尾同型船 double-ended ship;double ender
首尾系泊 fore-and-aft mooring;moor fore and aft;moor head and stern;mooring by the head and stern;mooring fore and aft
首尾衔接地 end-to-end
首尾线 fore-and-aft line;keel line
首尾相接 nose-to-tail
首尾相接的钢索 endless rope
首尾相接的轨道组合 <输送机> endless track assembly
首尾相连的 end-to-end
首尾向的 fore and aft
首尾循环的 end-around
首尾重线 full ended
首位城市 primate city
首位度 primacy
首席代表 chief delegate;chief representative
首席法官 chief justice
首席工程师 lead engineer
首席顾问 adviser head;chief adviser
首席经理 top manager
首席科学家 chief scientist
首席领航员 chief pilot
首席销货员 star salesman
首席执行官 chief executive officer
首席仲裁员 umpire
首席专业顾问 principal professional adviser
首先进入市场 first-to-market
首先考虑 first consideration
首先留置权 first lien
首先起火的建筑物 fire building
首先是 above all
首先显影 first developing
首先选择时延最小的输出话路 first free channel select
首先要考虑的事 first consideration
首舷方位 bow bearing
首舷角 bow angle
首向标线 heading index
首向陀螺仪 heading gyroscope
首向指示 heading indication;heading line
首项 leader;first term【数】
首项编目 first item list
首项表 first item list
首项附注 preliminary note
首项系数 leading coefficient
首项系数为一的多项式 monic polynomial
首斜缆 forward bow spring
首选 first choice
首选电路 first-choice circuit
首选方案 preferred option
首摇角 angle of yaw
首要保护标准 primary protection standard
首要化学品 priority chemical
首要生态系统 priority ecosystem
首要损害事件 first harmful event
首要条款 paramount clause
首要污染物 priority pollutant
首要物质清单 priority substances list

首要因素 overriding factor; paramount factor
首要原则 first principle
首要准备金 primary reserves
首页标题 headline
首页指示器 first page indicator
首一 monic
首一多项式 monic polynomial
首因 primacy
首站 head bulk depot;head bulk plant; head depot;head tank farm
首站设施 head facility;head installation
首长 header;paramount
首长的职位 masterhood
首长府 < 意大利威尼斯 > Doge's Palace
首长身份 mastership
首震 preliminary shock
首支架【船】fore poppet
首支架压力【船】fore poppet pressure
首枝渐近线 asymptote of the first part of the curve
首柱板 stem plate
首柱脚 < 龙骨前端 > stem foot
首柱木 head spar(tree)
首柱破浪材肘板 cheek knee
首柱倾角 angle of bow rake
首柱曲率 stem curvature
首柱上端 stem head
首柱头 stem head
首子午线 basis meridian;first meridian; initial meridian; prime meridian;principal meridian;zero meridian
首字母缩略词 acronym;initialism
首纵剖面 bow and buttock plane

艏 bow

艏波 bow wave
艏部桥架 bow ladder
艏部系泊 bow-mooring
艏舱 bow compartment
艏舱壁 collision bulkhead; forepeak bulkhead
艏侧推器机械舱 bow thrust machinery compartment
艏沉 dipping
艏沉没 founder by the head
艏沉深度 dipping height
艏吃水 fore draft;forward draft
艏垂线 forward perpendicular
艏导缆孔 bow-chock mooring pipe; stem chock
艏导缆器 bow fairleader
艏导缆钳 bull-nose
艏倒缆 fore spring;forward spring
艏舵 bow rudder
艏机型船 bow-engined ship
艏基阵 bow array
艏尖舱 collision compartment;forepeak
艏尖舱壁 forepeak bulkhead
艏缆 bow fast;bowline;headline
艏缆方驳 bowline scow;head line scow
艏肋骨 bow frame
艏楼 forecastle
艏楼安装 forecastle erection
艏螺旋桨 bow propeller
艏落 dropping
艏锚 bower
艏锚绞车 bow winch; head anchor winch
艏门 bow shell door
艏喷水转向舵 bow jet rudder
艏气封 bow seal
艏桥楼 forward bridge
艏倾 head trim;trim by bow;trim by head;trim by stern
艏声呐导流罩 bow(asdic)dome
艏推进器 bow propeller
艏弯曲部 luff
艏尾吃水相等 on an even keel
艏尾系泊 moor head and stern;mooring fore and aft
艏艉线 keel line
艏系缆 bow fast;head fast
艏向 heading
艏向监测器 heading monitor
艏向误差 error of heading
艏向线 heading line
艏向向上 < 雷达显示 > head up
艏斜桁仰角 steeve
艏斜缆 forward bow spring
艏摇 yawing
艏踵 forefoot
艏柱 stem
艏转向推进器 bow steering propeller

寿

寿命 durability; length of life; life length;life period;life span;life time;livability;longevity;span of life;viability

寿命保险 life insurance
寿命比……长 outlast
寿命表 life table
寿命测定 biometrics
寿命测定计 life meter
寿命长的 long-lasting
寿命短建筑物 limited life structure
寿命分布 age distribution
寿命分散系数 factor of life scatter
寿命估计 life expectance[expectancy]
寿命极限 lifetime limitation
寿命计算公式 life formula
寿命末期 end of life
寿命年限 length of life
寿命曲线 life curve
寿命三角形 life-span triangle
寿命时期 life time
寿命试验 length-of-life test;life test-(ing);longevity test
寿命试验台 life-test rack
寿命水平 life level
寿命缩短 loss of life
寿命特性 life performance
寿命系数 life factor
寿命性能 life performance
寿命性能曲线 life performance curve
寿命延长 lifetime dilation
寿命要求 life requirement
寿命已到的 condemned
寿命预测 lifetime prediction
寿命预期 life expectancy
寿命终止 < 可靠性术语 > end of life
寿命周期 life cycle
寿命周期成本 life cycle cost(ing)
寿命周期费用 < 初期费用,经营费用及维修费用之和 > life cycle cost-(ing)
寿命周期分析法 life cycle analysis method
寿命周期管理 life-cycle management
寿命周期价格 life cycle cost(ing)
寿命周期经济分析 life cycle economic analysis
寿命周期评估 life-cycle assessment
寿命锥体 life-span taper
寿山石 agalmatolite;pagodite
寿限保险 insurance till death
寿限折旧法 depreciation-age-life method

受

受碍转动 hindered rotation

受保护存储 protected storage
受保护单元 protected location
受保护的电力线路 protected power circuit
受保护的混凝土板角 protected corner
受保护的计算机系统 protected computer system
受保护的物种 protected species
受保护的野生动物 protected wild animal
受保护的自然景观 protected landscape
受保护国 protectorate
受保护机器 protected machine
受保护空间 protected space
受保护区域 protected field
受保护植物 attacked plant
受保护资源 locked resource
受保人 insurant
受暴风颠簸的 storm tossed
受暴风停阻的 storm stayed
受暴风雨打击的 storm beat(en)
受变质矿床 metamorphosed(ore) deposit
受变质片岩 metamorphosed schist
受冰川作用的 glaciated;glacierized
受冰河作用的 glaciated;glaciered
受冰影响水域 ice affected waters
受波浪袭击的 wave-swept
受补偿支付房价的账户 earned home payments account
受补井 recharge well
受补贴的出口 subsidized export
受补贴的进口 subsidized import
受补贴的运价率 subsidized rate
受补贴住房 subsidized housing
受补助的 grant-aided
受补助者 grantee
受侧限试样 laterally confined specimen
受侧限压缩试验 laterally confined compression test
受侧向荷载 laterally loaded
受潮 be affected with damp; damping;exposed to water;wetting
受潮变质 deterioration through moisture
受潮地区 tide land
受潮腐烂 ret
受潮面积 wetted area
受潮受热险 risk of sweating and-or heating
受潮水冲刷的 tide washed
受潮水影响的海岸码头(或指形码头)tidewater pier
受潮水影响的码头铁路 tidewater railroad
受潮水涨落而被交替淹没或露出的泥沼 tidal flat
受潮水阻碍 tide-bound
受潮汐影响的部分 < 河流 > tideway
受潮汐影响的海岸 tidal-water coast
受潮汐影响的河道 tidal river
受潮线间漏电 weather cross
受潮性 wettability
受潮岩 tide rock
受尘埃污染 contamination from dust
受冲岸 caving bank
受处理机限制 processor-limited
受锤顶 drive cap
受锤桩帽 driving cap;driving helmet
受存储器限制的 memory-bound
受挫的 crossed;frustration
受带限制 tape limited
受单剪螺栓 bolting single shear
受到季风影响的国家 monsoon affected country
受到控制的易变性 controlled variability
受到破坏的水系 deranged drainage
受到人类冲击的资源 human-impacted resources
受到严格限制 subject to severe limitations
受到严格限制的化学品 severely restricted chemical
受到中等限制的土壤 soils subject to moderate lamination
受抵押人 mortgagee;pledge;wortgagee
受点 current collection
受电变压器 relay transformer
受电电平 incoming level
受电端 incoming end; power user end;receiving end;relay end
受电端电压 relay end voltage
受电端阻抗 receiving-end impedance
受电杆 boom;troll(e)y pole
受电杆轴销 troll(e)y-pivot
受电杆座 troll(e)y base
受电弓 bow current collector; current-collector bow;pantograph
受电弓安装 installation to pantograph
受电弓车 pantograph coach
受电弓承槽 pantograph well
受电弓动态接触力 dynamic(al) contact force
受电弓弓头 pantograph head
受电弓滑板 collecting pan; pantograph pan
受电弓检修间 pantograph workshop
受电弓降低后长度 length of lowered pantograph
受电弓降弓装置 pantograph dropping device
受电弓接触压力 pantograph adherence pressure; pantograph contact pressure
受电弓控制 pantograph control
受电弓控制机构 pantograph control mechanism;pantograph cylinder
受电弓框架 bow frame
受电弓框架自然振荡 natural oscillation of the pantograph frame
受电弓离线 pantograph bounce
受电弓配置 pantograph arrangement
受电弓偏摆量 pantograph sway
受电弓气动控制装置 pneumatic pantograph control
受电弓气路塞门 air pipe cock of pantograph
受电弓汽缸 pantograph cylinder
受电弓试验台 testing device for pantograph
受电弓提升过多 excessive pantograph lift
受电弓悬挂装置 bow suspension
受电弓阻尼器 pantograph damper
受电器 current collector
受电器侧的桥架 collector side bridge
受电器杆支枢 troll(e)y-pivot
受电设备 power receiving equipment
受电头 current collector; troll(e)y head
受电靴 third rail collector shoe
受电装置 current-collecting gear
受钉板 nailable plate
受钉槽钢 nailing channel
受钉的 nailing
受钉的墙角小木条 wood ground
受钉混凝土 < 可打钉的混凝土 > nail-(ing) concrete; nailable concrete; nailcrete
受钉基板 nailing ground
受钉金属块 metal nailing plug
受钉块 fixing block;fixing brick;nailable block; nailing block; nailing plug
受钉块钉 nailer
受钉木块 nailing block;nog

受钉木塞 nailing plug;wooden nailing plug
受钉木丝板 woodwool nailing slab
受钉木条 nailer;wood nailing strip
受钉木砖 anchor brick
受钉刨花板 woodwool nailing slab
受钉砌块 fixing block
受钉嵌条 fixing fillet;fixing pad;nailing strip
受钉性 nailability;nail holding;nailing property
受钉压缝条 fixing pad
受钉砖 dowel brick;fixing brick;nailable brick;nailing brick
受钉砖块 fixing brick
受动负载的梁 dynamically loaded beam
受冻掉角 frost-bitten corner
受恶劣天气所迫 stress of weather
受反冲力而动作的 recoil-operated
受范体力学 plasticity
受范性 plasticity
受方 licensee
受风的一面 windward side
受风距离 distance of suffering wind
受风浪冲击而转向的 weather cocking
受风流漂动的船 roader;roadster
受风率 wind resistance
受风面积 wind(age) area;windward area
受风区(域) wind-swept area;fetch region
受风影响 direction of exposure
受辐射的金刚石 irradiated diamond
受辐射作用的水 activated water
受辐照表面 irradiated surface
受辐照气体 irradiated gas
受抚养人 dependent
受腐蚀混凝土 attacked concrete
受缚电子 fixed electron(ic)
受干扰样品 disturbed sample
受感反应时间 reaction time
受感器 pick-up;sensor
受高压的 highly pressed
受工业污染而废弃的土地 derelict land
受股人 transferee of stock
受雇(就业)能力 employability
受雇人员酬金 compensation of employees
受雇者 employee
受雇做(家庭)杂事的人 home handyman
受管制的车流 restricted traffic
受光不褪色的漆 fast-to-light paint
受光的 photic
受光伐 removal cutting;secondary felling
受光镜 illuminated mirror
受光墙 illuminated wall
受光区 photic zone
受规章限制 regulation limit
受规章限制的 regulatory
受过照射的 irradiated
受海水冲刷的 washed by the sea
受海水作用的混凝土 concrete exposed to sea water
受害 detriment
受害初期最明显的迹象 the most conspicuous early sign of attack
受害方 aggrieved party;damaged party;injured party
受害方本身过失 act of negligence
受害方的过失 contributory negligence
受害方损坏 damaged party
受害国 injured state
受害苗 affected seedling
受害人 victim
受害于 labo(u)r under
受害者 victim
受含硫水作用而蚀变的 alunitized
受寒指数 chill factor
受旱后的复原 recovery from drought stress
受焊金属试件 base metal test specimen
受荷 load bearing
受荷龄期 age of loading
受荷面积 area of loading
受荷期龄 age of hardening
受荷载的骨架 loaded skeleton
受荷载的机理 loaded mechanism
受荷载的空间结构 loaded space structure
受荷载的框架 loaded frame
受荷载的三维结构 loaded three-dimensional structure
受横向荷载的 transversely loaded
受横向荷载桩 M 值法 subsoil reaction modulus M method for laterally loaded pile
受洪水淹没的 flooded
受护陆架 protected shelf
受话号码 called number
受话器 receiver;telephone(receiver)
受话装置 receiver unit
受欢迎的 well-accepted
受环境所迫 stress of circumstances
受贿 accept(ance) bribe;bribery
受惠国 benefit country
受惠人 obligee
受惠者 beneficiary
受惠者偿还原则 beneficiary repayment
受惠者负担原则 beneficiary pay principle
受火严重性试验 fire exposure severity test
受货人 consignee;recipient of goods
受货员 receiving clerk
受击面 batted surface
受激波 excited wave
受激布里渊漫射 stimulated Brillouin diffusion
受激布里渊散射 stimulated Brillouin scattering
受激布里渊效应 stimulated Brillouin effect
受激场 excited-field
受激磁铁 excited magnet
受激单能态 excited singlet state
受激电平 excited level
受激电子 excited electron
受激发射 stimulated emission
受激发射机 driven transmitter
受激发射探测 stimulated emission detection
受激发射探测器 stimulated emission detector
受激发射系数 stimulated emission coefficient;B-coefficient
受激发射跃迁几率 stimulated emission transition probability
受激分子【化】 excited molecule
受激辐射 stimulated radiation
受激辐射可调电子放大器 Teaser
受激辐射式光频放大器 light amplification by stimulated emission of radiation
受激复合辐射 stimulated recombination radiation
受激共振 excited resonance
受激共振腔 excited cavity
受激光场 stimulating optical field
受激光发射 stimulated light emission
受激光照射 lase
受激光子 stimulated photon
受激光子释放 stimulated photon liberation
受激康普顿散射 stimulated Compton scattering
受激拉曼发射 stimulated Raman mission
受激拉曼散射 stimulated Raman scattering
受激拉曼效应 stimulated Raman effect
受激离子 excited ion
受激磷光粉 excited phosphor
受激络合物 excited complex
受激能带 excited band
受激能级 excited energy level
受激瑞利光散射 stimulated Rayleigh scattering
受激瑞利效应 stimulated Rayleigh effect
受激瑞利翼散射 stimulated Rayleigh wing scattering
受激散射 stimulated scattering
受激态簇 manifold of excited state
受激吸收 stimulated absorption
受激吸收跃迁几率 stimulated absorption transition probability
受激乙烯 excited ethylene
受激有效质量 equivalent excited effective mass
受激阈 stimulated threshold
受激原子 excited atom
受激跃迁 excited transition;induced transition;stimulated transition
受激振荡 stimulated oscillation
受激振荡频率 forced frequency
受激振动 excited vibration
受激中心 excited center[centre]
受激状态 excited state
受激作用 stimulated action;stimulating action
受计算机限制 computer-limited
受计算量限制的 compute bound;compute-limited
受计算量限制的计算 compute-bound computation
受岬角包围的海区 closed sea
受剪 in shear
受剪承载能力 shear capacity
受剪对角斜杆 diagonal shear member
受剪杆件 membering shear;member in shear
受剪构件 member in shear;shear member
受剪加固筋比 ratio of shear reinforcing bar
受剪节点 shear joint
受剪螺栓 shear bolt
受剪铆钉 rivet in shear
受剪面 shearing face
受剪面积 shearing area
受剪敏感的 shear-susceptible
受剪破坏 fail in shear
受剪墙 shear wall
受剪区 shear zone
受剪弹性模量 modulus of elasticity due to shear
受剪贴角焊缝 shear fillet
受剪销钉 shear(ing) pin
受剪斜杆 diagonal shear member
受剪状态 shear behavio(u)r
受检查人 examinee
受检音源 tone to detect
受溅地带 splash zone
受浆器 pulp catcher
受奖者 awardee
受叫方 called party
受津贴者 <尤指政府雇员> payroller
受浸后的复原 recovery from submergence
受浸蚀性 erodibility
受矩节点 node holding moment
受卡盒 <穿孔卡计算机的> stacker
受空气污染的地热流体 aerated geothermal fluid
受控爆破技术 controlled blasting technique
受控闭锁 controlled block
受控变量 controlled variable;regulated variable
受控变量给定值 set value of controlled variable
受控参数 controlled parameter
受控掺杂 controlled doping
受控重写 controlled rewriting
受控词汇表 controlled vocabulary
受控淬火 controlled quenching
受控存储器 controlled storage
受控存储区分配 controlled storage allocation
受控存取通路 controlled access path
受控存取性 controlled accessibility
受控的耗用量 controlled consumption
受控的计算试验 controlled computational experiment
受控的系统过程入口 controlled entry to system procedure
受控对象 controlled plant
受控发送器 controlled sender
受控观念 ideas of influence
受控光束扫描 directed beam scan
受控焓流 enthalpy-controlled flow
受控航行器 maneuvering craft
受控核聚变 controlled fusion
受控混凝土 controlled concrete
受控价 controlled valence
受控检错程序 controlled postmortem program(me)
受控件 controlled member
受控接口模块 supervised interface module
受控结构聚合物 controlled texture polymer
受控紧急制动 controlled emergency (brake) application
受控聚变堆 controlled fusion reactor;controlled thermonuclear reactor
受控聚合速度 controlled polymerization rate
受控可逆式电码轨道电路 controlled reversible coded track circuit
受控扩散反应 diffusion-controlled reaction
受控扩散反应速率 diffusion-controlled reaction rate
受控扩散释放消毒剂 diffusion-controlled release disinfect
受控冷却 controlled cooling
受控量 controlled quantity
受控料流均化库 controlled flow silo
受控流率【交】 controlled rate
受控漏泄 controlled leak
受控炉气 controlled atmosphere
受控落下 controlled drop-away;controlled drop-out
受控气氛 controlled atmosphere
受控倾卸法 controlled tipping
受控情报 controlled information
受控取消 controlled cancel
受控燃烧 controlled burning
受控热分解 controlled thermal decomposition
受控溶耗性聚合物 controlled depletion polymer
受控扫描 directed scan
受控生态系统污染试验 pollution experiment in model ecosystem
受控生物膜硝化滤池 biofilm-controlled nitrifying trickling filter
受控释放 controlled release
受控水体 controlled waters
受控速度台从锁相 controlled rate genlock

受控条件 controlled condition
受控网络 controlled network
受控涡流 controlled vortex flow
受控系统 controlled system
受控信号 controlled signal
受控蓄水 controlled storage
受控雪崩整流器 controlled avalanche rectifier
受控氧化作用 controlled oxidation
受控载波调制 controlled carrier modulation
受控载波系统 controlled carrier system
受控制的 controlled;slave
受控制的产品 controllable products
受控制的空气 controlled atmosphere
受控制的施工 controlled construction
受控制公司 controlled corporation
受控制人行道横道 controlled pedestrian crosswalk
受控装置 controlled device
受控状态 controlled state
受控籽晶法 controlled seeding
受款人 accepter;beneficiary;payee
受款人账户 payee's account
受矿(煤)仓 receiving bunker
受拉边 tension flange;tension side
受拉变形 tension deformation
受拉部分 advancing side;part in tension
受拉侧(一边)tension side
受拉承载能力 tensile capacity
受拉大梁 stretch girder
受拉带件 tension strap
受拉的 tensile
受拉段 tight side
受拉断面 tensile section
受拉多边线 funicular tension line
受拉多边形 tension polygon
受拉法兰 tension flange
受拉分裂的力 tensile splitting force
受拉杆(件)tension member;bar in tension;rod in tension;membering tension;member in tension;tensile member;tension element;tensioning rod
受拉钢筋 positive reinforcement;tensile bar;tensile reinforcement;tensile steel;tension bar;tension(ing) reinforcement;tension steel
受拉钢筋断面面积 cross-sectional area of tensile reinforcement
受拉钢筋束 tensioning tendon
受拉钢丝 tensioned wire
受拉构件 component under tension;member in flexure;member in tension;tension element;tension member
受拉过程 tensioning process
受拉焊缝 tension fillet;tension weld
受拉荷载 tensile load
受拉滑轮 tension pulley
受拉剪力裂缝 tensile shear crack
受拉绞合线 tensioning strand
受拉接点 tensile joint;tension joint
受拉接合 tensile joint;tension joint
受拉接头 tensile joint;tension joint
受拉劲化 <混凝土> tension stiffening
受拉开裂 crack in tension;tensile capability;tensile cracking;tension cracking
受拉缆索 tension cable
受拉肋 tension rib
受拉力钢 tensioning steel
受拉力钢缆 tensioning cable
受拉力块 tensioning block
受拉力梁 tension beam
受拉连接 tension connection
受拉链 tension chain
受拉梁 stretched beam
受拉梁翼 tension flange
受拉裂缝 tensile crack;tension crack
受拉裂缝深度 depth of tension crack
受拉铆钉 rivet in tension;tensioned rivet
受拉面 tension(ed) face;tension(ed) side
受拉面积 tensile area;tension(ing) area
受拉膜破损 tensile membrane failure
受拉能力 tensile power;tensile quality
受拉皮带的铰支座 belt tensioning hinged support
受拉疲劳强度 limit of fatigue in tension
受拉拼接 tension splice
受拉破坏 fail in tension;tensile failure;tension failure
受拉牵条 tension member
受拉区(域)tensile region;tensile area;tensile zone;tension area;tension region;tension(ing) zone
受拉屈服点 tensile yield point;tension yield point
受拉屈服强度 tensile yield strength
受拉伸长 elongation due to tension;elongation in tension;extension elongation
受拉伸裂 tension crack
受拉试件 tensile test piece;tension specimen
受拉试样 tensile coupon
受拉套管 tension sleeve
受拉贴角条 tension fillet
受拉系杆 tension tie
受拉系统 tensile system;tensioning system
受拉纤维 stretched fiber[fibre];tension fiber[fibre]
受拉弦杆 tension chord
受拉斜杆 diagonal in tension;tension diagonal
受拉斜桩 tension raker
受拉行态 tensile behavio(u)r
受拉翼缘 tension flange
受拉应变 tensile strain;tension strain;tensive strain
受拉应力 tension stress
受拉预应力 tensile prestress
受拉张裂 tensile capability
受拉振动 forced vibration
受拉主应力轨迹线 tensile trajectory
受拉柱 joggle post
受拉桩 tension pile
受拉桩的下拉力 dragdown force
受拉状态 state of tension
受涝的 watersick
受镭作用 radiumize
受冷面积 area of cooling
受理 acceptance of carriage
受理上诉程序 appellate procedure
受理上诉的 appellate
受力 stress(ing)
受力部件 supporting parts
受力齿面 approach side
受力的 weighted
受力点 working point
受力范围 field of load
受力杆件 stress(ed) member
受力钢筋 active reinforcement;principal bar;main reinforcement
受力构件 bearing carrier;stress(ed) member
受力滑车 fall block
受力环 compression ring
受力夹层板 stressed sandwich panel
受力结构 framing;supporting structure
受力筋 bearing bar;carrying steel;main bar;main reinforcement
受力缆索 bearer cable
受力梁 bearing beam
受力零件 supporting parts
受力螺栓 tension bolt
受力锚 riding anchor;working anchor
受力蒙皮 stressed covering
受力面 carrying surface;stress surface;thrust(sur) face
受力面层结构 stressed-skin construction
受力面积 force area
受力绳(索)bearer cable;carrying rope
受力双折射 mechanical birefringence;stress birefringence
受力索 carrying rope
受力图 force diagram;reciprocal diagram
受力外包层 stressed skin
受力弦杆 stress chord
受力限度 limit of strength
受力状况 stress state
受力状态 state of stress;strain condition;stress state
受力状态分析 stress state analysis
受料仓 receiver bin;receiver hopper;receiving bin;receiving bunker;receiving pocket
受料槽 charge chute;receptacle trough
受料场所 receiving point
受料地点 receiving point
受料斗 receiving hopper;pump hopper <泵送混凝土的>
受料格栅 receiving skid;stationary grate
受料罐 receiving bucket
受料胶带输送机 receiving belt conveyer[conveyor]
受料孔 receiving opening
受料口 receiving mouth;receiving opening
受料漏斗 receiving cone;receiving hopper
受料浅井 receiving pit
受料区 material receiving region
受料容器 receiving container
受料输送机 receiving conveyer[conveyor]
受料台架 receiving skid
受料站 receiving station
受领 take delivery
受拢林 disturbed forest
受氯化物损害的混凝土 chloride contaminated concrete
受磨部件 wearing parts
受磨零件 wearing parts
受纳水道 receiving watercourse
受纳水体 receiving(body of) waters
受纳水体标准 receiving water standard
受纳体【化】receptor
受纳污水 receive effluent
受难者 victim
受挠杆件 member in flexure
受挠构件 member in flexure
受挠弯曲 flexural bending
受泥船 mud barge
受黏[粘]着限制的牵引力 tractive effort at adhesion limit
受扭承载能力 torsional capacity
受扭杆 torsion rod
受扭杆簧 torsion rod spring
受扭杆件 member in torsion;torque member;torsion member
受扭构件 member in torsion;torque member;torsion member
受扭晶体 twister
受浓雾所困的 fogbound
受盘公司 acquiring company
受盘人 offeree
受劈性 cleavability
受偏心倾斜荷载的承载力 bearing capacity under eccentric and inclined load
受票人 drawee
受迫波 forced wave
受迫采暖 forced heating
受迫采暖系统 forced heating system
受迫采暖装置 forced heating installation
受迫对流 forced convection
受迫发射 induced mission
受迫复合 induced recombination
受迫供热 forced heating
受迫供热系统 forced heating system
受迫供热装置 forced heating installation
受迫鼓热风炉 forced warm air furnace
受迫康普顿散射 induced Compton scattering
受迫偶极辐射 induced dipole radiation
受迫偶极子跃迁 forced-dipole transition
受迫热风采暖 forced warm air heating
受迫弯曲波 forced flexural wave
受迫吸收 forced absorption
受迫谐振 forced harmonic vibration
受迫循环集中供热 forced circulation central heating
受迫跃迁 forced transition
受迫章动 forced nutation
受迫振荡 forced oscillation
受迫振荡电流 forced alternating current
受迫振动 forced vibration
受迫振动互谱分析 forced vibration cross-spectrum analysis
受迫振动频率 forcing frequency
受迫转动 forced rotation
受迫组合散射 induced combination scattering
受启发的 enlightened
受气候影响的延误 weather delay
受汽管 <复涨式机车> receiver pipe
受契约束缚的人 bondsman
受器 cup
受器压力 receiver pressure
受浅滩或暗礁包围的港 locked harbo(u)r
受切接合部件 shear connector
受侵蚀地面 erosion surface;planation surface
受侵蚀植物 attacked plant
受屈曲构件 member subject to buckling
受权 be authorized
受权调查范围 term of reference
受权签字人 authorized signator
受权人 attorney
受权一方 entitled party
受让方 transferee
受让人 alienee;assignee;endorsee;grantee;indorsee;surrenderee;transferee
受让者 alienee;assignee;endorsee;grantee;indorsee;surrenderee;transferee
受扰 0 输出信号 disturbed zero output signal
受扰 1 输出信号 disturbed one output signal
受扰单元 disturbed cell
受扰的零电压 disturbed voltage zero
受扰回(声)波 disturbing echo
受扰面 perturbed surface
受扰破裂区 disturbed fractured zone
受扰区【物】disturbed area
受扰日 disturbed day
受扰射流 disturbed jet
受扰体 disturbed body
受扰响应电压 disturbed response voltage
受扰响应信号 disturbed response signal

受扰元件 disturbed cell
受扰运动 disturbed motion;perturbed motion
受扰周期势场 perturbed periodic(al) potential field
受热 be affected by the heat;be heated;heating
受热变脆 heat embrittlement
受热变色 heat discolo(u)red
受热变形 temperature distortion
受热变质 heat damage
受热表面 heat accepting surface
受热剥落 thermal spalling
受热分解的 thermolabile
受热环流器 thermic syphon;thermic syphon tube
受热介质 heat accepting medium
受热开裂 pinching
受热面 heat-absorbing surface;heating surface;heat receiving surface;hot area;hot face
受热面的积灰 fouling of heating
受热面负荷 heat absorption rate of heating surface
受热面积 heat absorption area;heated area;heating(surface) area;hot area
受热面积单位蒸气负荷 rate of evapo(u)ration per unit heating surface
受热面积灰 fouling of heating surface
受热面积碳 fouling of heating surface
受热面积蒸发率 rate of evapo(u)ration per unit heating surface
受热面面积 heating supply area;heating surface area
受热面蒸发率 evaporation rate of heating surface
受热破裂 <玻璃的局部> thermal breakage
受热器 heat receiver;heat sink
受热区 heat-affected zone
受热色变 temperature colo(u)ration
受热时间 heating time
受热时状况 behavio(u)r under heat
受热体 heating body;heat receiver
受热系数 coefficient of heat perception
受热延长 thermal elongation
受热作用 heat perception
受入塞孔 "receiving in" jack
受伤薄壁组织 traumatic parenchyma;wound parenchyma
受伤的 traumatic
受伤年轮 traumatic ring
受伤射线管胞 traumatic ray tracheid
受伤树脂道 traumatic resin canal
受伤心材 traumatic heartwood
受摄根数 perturbed elements
受摄轨道 disturbed orbit
受摄开普勒运动 perturbed Keplerian notion
受摄体 disturbed body
受摄运动 perturbed motion
受摄运动方程 perturbed equation of motion
受摄坐标 disturbed coordinates
受审核方 auditee
受声室 <隔声测量时受声的房间> sound receiving room
受湿性 wettability;wetting property
受湿由货主负责 owner's risk of wetting
受时效限制的债务 barred debt;barred obligation(debt)
受时效限制丧失索赔权 claim barred by reason of limitation
受试人 examinee
受输出限制的 output limited
受输入输出限制 input-output limited
受输入输出限制的计算 input-output-bound computation
受输入输出限制的作业 input-output-bound job
受双剪螺栓 bolt in double shear
受霜冻的局部地区 <被围地区、阻塞区、小海湾等> frost pocket
受水 exposed to water
受水池 receiving basin;reception basin
受水河 recipient stream
受水涝的 watersick
受水面积 catchment area;intake area
受水盆地 reception basin
受水区 intake area;reception area;reception basin;water catchment area
受水箱 reception tank
受丝卷绕装置 pick-up device
受速度限制的设备 speed limiting device
受酸影响的 acid-affected
受损 damaged;under average
受损财产 damaged property
受损船 damaged ship
受损反射 frustrated reflection
受损后价值 damaged value
受损后市价 damaged market value
受损货物 damaged cargo;damaged goods
受损货物报告书 damaged cargo report
受损货物的市价 damaged market value
受损情况 damaged condition
受损伤 corrode
受损(失)程度 extent of damage
受损时间 debatable time
受损事故 broken down
受损状况 condition in damaged
受锁电路 lock-on circuit
受锁副载波 locked subcarrier
受弹簧力作用 spring-loaded
受体 acceptor;receptor
受体部位 acceptor site
受体面积 receptor area
受体模型 receptor model
受调放大器 modulated amplifier
受调节水流 regulated stream
受调振荡器 modulated oscillator
受投资需求拉动 driven by investment demand
受托保管的财产 fiduciary property
受托程序 trusted program(me)
受托代购人 indentee
受托的 fiduciary
受托管理人 <财产、业务等的> trustee
受托管之房地产 trust estate
受托监护人 fiduciary guardian
受托期限 mandatory period
受托人 attorney;bailee;consignee;depositary;executioner;executor;fiduciary;referee;trustee
受托人代客保险 bailee's customers' insurance
受托人条款 bailee clause
受托人投资 trustee investment
受托人资格 fiduciary capacity
受托者 assignee;executor
受弯部分 part in bending
受弯承载能力 flexural capacity
受弯的剪切裂缝 flexural shear crack
受弯的梁 girder subjected to bending
受弯的黏[粘]附力试验 flexural bond test
受弯杆件 member in bending;member in flexure
受弯钢筋 compression bar;flexure reinforcement;flexural reinforcement
受弯钢筋截面积 area of flexural steel
受弯构件 bending member;flexural member;member in bending;member in flexure;membering bending
受弯环接头 flexural loop joint
受弯建筑物 flexural building
受弯结构体系 structure system in bending
受弯力臂 bending arm
受弯力筋 deflected tendon
受弯耐力 bending endurance;repeated flexural strength
受弯疲劳 bending fatigue
受弯破坏 bending failure;fail in bending
受弯曲构件 member subject to buckling
受弯曲荷载的 transversely loaded
受弯试件 bend test piece
受弯试验 bending test
受弯引起的黏[粘]附力破坏 flexural bond
受弯引起的屈曲 flexural buckling
受弯引起的失稳 flexural buckling
受危害 jeopardise[jeopardize]
受威胁群体和资源 stock-at-risk
受威胁物种 threatened species
受委托 be authorized
受委托人 assignee;bailee;entrusted person;mandatory person
受委托者 assignee;bailee;entrusted person;authorized person
受温度变化的影响 be acted upon by temperature changes
受污染的 contaminated
受污染的浚挖物质 contaminated dredged soil
受污染河流 polluted river
受污染人口金字塔 pyramid of polluted population
受污染水 polluted water
受污染土地 contaminated land
受污染土(壤) contaminated soil
受污染物质的覆盖 capping of contaminated material
受污水污染 polluted by sewage
受下水道污染的 contaminated by sewage
受限变量 bound variable
受限分机线 restricted extension line
受限可加 summable bounded
受限类型 restricted type
受限名 qualified name
受限模糊自动机 restricted fuzzy automaton
受限盆地 barred basin;restricted basin;silled basin
受限射流 jet in a confined space;restricted jet
受限图灵机 restricted Turing machine
受限项 limited entry
受限于 subject to
受限运动 restrained motion
受限制 bound
受限制的 limited;obligatory
受限制的股票 restricted stock
受限制的留存收益 restricted retained earnings(or surplus)
受限制的视野 restricted view
受限制的投标 restricted tender
受限制的现金 restricted cash
受限制的资产 restricted assets
受限制区域 restricted area
受限制水域 restricted waters
受限制映射 restricted mapping
受限制姿态 trapped state
受薪者 payroller
受信机 receiving instrument
受信托人销售 trustee's sale
受信用 be credited
受训练者 exerciser
受训人 trainee
受训者 trainee
受压 compress;incompression;under compression
受压板 compressed plate
受压边 compression side;compressive side
受压边缘 compression edge;pressed edge
受压变色 piezochrom(at)ism
受压变形 compression deformation;compressive deformation
受压不漏气的 pressure-tight
受压部件 pressure-containing parts
受压残留变形 compression residue set
受压侧 compression face;compression side
受压侧面 compression side
受压层 compressional layer;compressive layer
受压层深度 <地基计算的> compression zone depth;depth of compression zone
受压承载能力 compressive capacity
受压磁层 compressed magnetosphere
受压带 compression zone;compressive zone
受压的 compressive
受压的初始破坏 <钢筋混凝土的> primary compression failure
受压的加筋材料 compressed reinforcing
受压的加劲材料 compressed reinforcing
受压的梁 compressed boom
受压的弦杆 <桁架中的> compressed chord
受压的翼缘 <指梁中的> compressed flange
受压的柱 compressed column
受压地下水 artesian groundwater;confined groundwater
受压点 compression point
受压断面 compression cross-section
受压范围 compression region;compressive range;compressive region
受压腹板开裂 compressive web breaking
受压腹板破坏 compressive web failure;compressive web rupture
受压腹板破裂 compression web breaking
受压腹杆 compression web member
受压杆 bar in compression;compressed bar;compressed rod;compressive boom;rod in compression
受压杆件 compression(al) member;compression bar;member in compression
受压钢筋 compressed bar;compressed reinforcement;compression(al) bar;compression reinforcement;compression steel;compressive reinforcement;rod in compression
受压钢筋对接套筒 end-bearing sleeve
受压钢筋截面积 area of compression steel
受压钢筋面积 compression steel area
受压拱 compression arch
受压构件 component under compression;compressed element;compressed member;compressional member;member in compression;compression member;membering compression
受压关闭弯嘴水龙头 compression bib cock faucet
受压横断面 cross-section under compression
受压护舷 compressed fender
受压环 compression ring
受压环理论 ring compression theory
受压混凝土 compression concrete
受压接头 compression joint

S

受压结构接缝 compressed bearing joint
受压截面 compressive cross-section
受压壳层 pressure shell
受压快干油墨 pressure-set ink
受压宽度 compressed width
受压肋 compression rib
受压立方体(试块) compression cube
受压连杆 compression link
受压连接杆 pressure pitman
受压联系件 strut bracing
受压梁 strut beam
受压流体 pressure fluid
受压面 compression(al) face; pressure side; pressure surface
受压面积 area of pressure; compression(al) area; load area
受压木 compression wood; overtopped
受压黏[粘]度 pressure viscosity
受压拼接 compression splice
受压破坏 compression failure; compression rupture; compressive failure; fail in compression
受压期间 pressure period
受压起重臂 compression boom
受压气密性 compression
受压区 compressed zone; compression(al) zone; compression region; compressive region
受压区高度 < 地基计算的 > compression zone depth; depth of compressive zone
受压屈服点 compressive yield point
受压人孔 pressure manhole
受压容积 compressive volume
受压容器 pressure vessel
受压容器 X 光探伤 X-ray flaw detection for pressure vessel
受压容器应力测试 stress testing for pressure vessel
受压伸臂 strut boom
受压试件 compression specimen; compression test piece; compressive specimen
受压试件开裂 cracking in compression specimen
受压试验 compression test; compressive test
受压试验与气密性试验 hydrostatic and pneumatic pressure test
受压室压力 < 三轴试验的 > cell pressure
受压室压力控制器 cell pressure control
受压手孔 pressure handhole
受压水 compressed water; confined water
受压损坏 compression damage; fail in compression
受压特性 compression property
受压通风 blowing ventilation
受压突出 pressure bump
受压桅杆 < 起重机的 > compression boom
受压稳流 confined steady flow
受压物体 pressurant
受压下 under-pressure
受压先破坏 primary compression failure
受压纤维 compression fiber[fibre]
受压弦杆 compression(al) chord; compression boom; compression(al) check; compressive chord
受压斜杆 compressed diagonal; compression(al) diagonal; diagonal in compression; diagonal strut
受压斜支柱 compressing diagonal; compression diagonal
受压斜桩 compression raker; compressive raker
受压液体 pressure fluid
受压翼板 compressed plate
受压翼缘 compression(al) flange; compressive flange
受压翼缘压屈 compression flange buckling
受压应变 compressive strain
受压应力 compression stress; compressive force; compressive stress
受压油液 oil under pressure
受压元件 pressure element
受压圆柱体(试块) compression cylinder
受压支撑 compression strut
受压支杆 compression strut
受压支柱 compression strut
受压桩 compression pile; non-uplift pile
受压自紧密封件 pressure-energized seal
受压最强面 extreme compression fiber[fibre]
受押人 mortgagee
受淹地区 flooded land; inundated land
受淹耕地 inundated cultivated land
受淹面积 flood(ed) area; inundated area
受淹人口 inundated population
受盐损害的 salt damaged
受衍射限制镜头 diffraction-limited lens
受验者 subject
受洋流影响 feeling the current
受氧体 oxygen acceptor
受邀投标者 invited bidder
受遥控的信号楼 remotely controlled tower
受要约人 offeree
受液器 accumulator
受遗赠人 devisee
受遗赠者 legatee
受抑反射 frustrated reflection
受抑全反射 frustrated total reflection
受抑全反射滤光器 frustrated total reflection filter
受抑水舌 depressed nappe
受抑制出行 suppressed trip
受抑制的 suppressed
受抑制反应 inhibited reaction
受抑制氧化 inhibited oxidation
受益 fertilization
受益比 benefit ratio
受益成本比 benefit-cost ratio
受益成本比法 benefit-cost ratio method
受益方 beneficiary party; benefited interest(party); benefited party
受益费 benefited-user charge
受益分析 benefit analysis
受益理论 < 由于道路的修建或改善而使邻接和附近财产受益的理论 > theory of benefits
受益利息 beneficial interest
受益率 benefit factor; benefit ratio
受益权 beneficial right
受益权益 beneficial interest
受益群体分析 benefited group analysis
受益人 beneficial owner; beneficiary; recipient
受益人的声明 beneficiary's statement
受益人的要求 beneficiary's demand
受益使用权 beneficial occupancy
受益所有人 beneficial owner
受益所有者 beneficial owner
受益信托 benefit trust
受益原则 benefit principle
受益者 beneficiary; obligee; stakeholder
受益者负担(费) beneficiary charge
受益者角色 beneficiary case
受益者权利 rights of beneficiaries
受益征税 inclusion fee
受益值 benefit value
受应力的薄膜 stressed membrane
受应力的接头 stressed connection
受应力的黏[粘]土 stressed clay
受应力钢筋腱 stressing tendon
受应力面积 stressing area
受应力影响的模量 stress-dependent modulus[复 moduli]; stress modulus
受影响的区域 affected area; involved area
受影响面 area of infection
受影响区 area of infection
受油腐蚀 corroded by oil
受油机 refueled aircraft
受油机接近加油机 pushing of the receiver
受油接头 nozzle-sinker device
受油器 lubricating head; oil catcher; oil distributor; oil receiver; oil supply head; sinker
受有开孔墙约束的大梁 girder restraint provided by walls with openings
受有特权的 concessionary
受雨器 rain ga(u)ge receiver
受雨区 catchment area
受原子爆炸污染的 atom-stricken
受援国 aid-receiving nation; benefit country; recipient country
受约人 promisee
受约束 bound; entering
受约束参数 constrained parameter
受约束的地下水 confined groundwater
受约束的防潮层 constrained damping layer
受约束回归模型 constrained regression model
受约束混凝土 confined concrete
受约束机构 constrained mechanism
受约束流 confined flow
受约束税率 bound rates of duty
受约束系统 constrained system
受约束振动 constrained vibration
受约束最优化 constrained optimization
受灾地区 disaster area; distress area; stricken area
受灾害地区 calamity area
受灾率 damage ratio
受灾土地 devastated land
受灾住房的抵押保险 disaster housing mortgage insurance
受载 on-load(ing); stand under load
受载变形特性 load deformation characteristic
受载变形图 load-deformation diagram
受载轨道 loaded track
受载焊缝 strength weld; strong seam
受载机理 loaded mechanism
受载历时 duration of load
受载期间 lay days
受载时间挠度曲线 load-time deflection curve
受载弹簧 spring-loaded
受载弦杆 loaded chord
受载状态 loaded state
受赠人 donee
受张地带 tension zone
受张分裂的力 tensile splitting force
受张杆件 tensioning rod
受张钢筋 tensioning reinforcement
受张钢筋束 tensioning tendon
受张过程 tensioning process
受张绞合线 tensioning strand
受张肋 tension rib
受张力的面积 tensioning area
受张力钢 tensioning steel
受张力钢缆 tensioning cable
受张力块 tensioning block
受张连接 tension connection
受张能力 tensile power; tensile quality
受张区(域) tensioning zone; tensile zone
受张系杆 tension tie
受张系统 tensioning system
受张性能 tensile property
受张应变 tensile strain
受障碍的错误呼叫 mutilated call
受照面 plane of illumination
受照栅极 illuminated grid
受照体 illuminated body
受振点 receiving point
受震程度 seismicity
受震区 disturbed area
受政府补贴的住房 subsidized apartment; subsidized dwelling; subsidized house-building
受支配 subject
受脂法 cupping setting up
受指导于 under the auspices of
受制车流 forced flow
受制渗气式汽化器 restricted air bled carburetor
受制约的价格 forced price
受制障碍 controlling obstacle
受制转弯 restricted turns
受质子溶剂 proton acceptor solvent
受中子辐照的 neutron-irradiated
受重负的 heavy-duty
受重负的楼板 heavy-duty floor(ing)
受重力作用 gravitate; gravitating
受重力作用下降 gravity drop
受主 acceptor; acceptor material; donee
受主掺杂 acceptor doping
受主掺质 acceptor dopant
受主电流 < 半导体中 > hole current
受主反应 acceptor reaction
受主光谱 acceptor spectrum
受主结合能 acceptor binding energy
受主能带 acceptor band
受主能级 acceptor level
受主浓度 acceptor concentration; acceptor density
受主缺陷 accepter defect; acceptor imperfection
受主态 acceptor state
受主调整晶体 acceptor adjusted crystal
受主物质 acceptor material
受主杂质 acceptor impurity
受主杂质能级 acceptor impurity level
受主中心 acceptor center[centre]
受装索 pass line
受资助城市 entitlement city
受租让者 concessionaire
受阻胺 hindered amine
受阻波浪 damped wave
受阻沉降 hindered settling
受阻沉降比例 hindered-settling ratio
受阻沉降速率 rate of hindered settling
受阻船 hampered vessel
受阻的交通 stalled traffic
受阻酚 hindered phenol
受阻酚抗氧剂 hindered phenol antioxidant
受阻码 barred code
受阻收缩 hindered contraction
受阻旋转 hindered rotation
受阻延误 delay through obstructions
受尊敬的 prestigious
受作用的 exposed

狩

狩猎保留地 game preserve
狩猎场 game land

狩猎地 chase
狩猎区 game area; hunting field; hunting reserve
狩猎小屋 shooting box

兽的长牙 tusks

兽功率 animal power
兽骨炭 animal black
兽角形杆 horn-shaped pole
兽栏 box stall
兽力车 animal drawn traffic
兽力车过道路标 cattle pass marker
兽力运输 teaming
兽毛毡 animal hair felt
兽皮 animal skin; fell
兽皮集装箱 hide container
兽皮纸 animal parchment
兽炭 animal charcoal
兽炭黑 animal black; bone black
兽头形装饰 catshead
兽图 animal
兽瘟 murrain
兽纹 animal motif; animal pattern
兽形的 zoomorphic
兽形柱 beast column; zoomorphic column; zoophoric column
兽形装饰 animal-shaped ornament; zoomorphic ornament; zoomorphism in ornament
兽穴 burrow; den
兽医站 veterinary station
兽医诊所 veterinary clinic
兽医总医院 veterinary general hospital
兽用 for livestock

售出发票 invoice outward

售出过多 over-sell
售出租回合约 sell-and-leaseback agreement
售方赊贷 seller financing
售后服务 after-sale service; service after sale
售货场 sale(s) room
售货处 sale(s) room
售货单 sales note
售货点终端 point-of-sale terminal
售货柜台 sales counter
售货合同 sales contract
售货合约 bill of sale
售货回扣 sales rebate and allowance
售货机 vender[vendor]; vending machine
售货竞争 sales contest
售货面积 <商店> sale area
售货清单 bill of sales
售货区 sales area
售货确认书 sale confirmation
售货条件 sales term
售货员 salesman; shopman; sales clerk
售货员室 salesman room
售货正式文件 bill of sales
售价 sales price; selling price; trading value; selling value <美>
售价高 come high
售价公道 reasonable price
售价公道的 reasonableness
售价总数 gross proceeds
售料 sale of material
售料单 sales order
售票 booking; sale of tickets; ticketing; ticket selling
售票处 booking hall; booking office; box office; issuing office; ticket counter; ticket office
售票窗(口) ticket window; booking window; guichet; booking office window
售票大厅 ticket hall
售票代理处 ticket agency
售票当日有效 valid on date of issue
售票房 ticket booth
售票柜台 ticket counter
售票机 booking office machine
售票局 issuing administration
售票口 wicket
售票日期 date of issue
售票日期盖戳机 ticket dating machine
售票室 box office; ticket room
售票台 ticket counter
售票台窗口 opening at ticket counter
售票铁路 issuing carrier
售票厅 booking hall; ticket lobby
售票问讯处 ticketing information area
售票系统 ticket-selling system
售票员 booking clerk; ticket issuing clerk; troll(e)y man
售票员显示 operator display
售票站 issuing station
售票自动化 automatization of ticket sales
售前服务 pre-sales service
售缺 sell out
售完 close out
售主备件分类 vendor provisioning parts breakdown
售主部件更改 vendor parts modification
售主部件号 vendor parts number
售主部件索引 vendor parts index
售主部件一览表 vendor parts list
售主装货单据 vendor's shipping documents
售主装运说明书 vendor shipping instruction

授标 award of contract

授标函 letter of award
授标决定 award decision
授标名单 list of award
授标前会议 pre-award meeting
授标通知 award notification; bill notification; notification of award
授标通知书 letter of acceptance
授标准则 award criterion
授粉 pollination
授给物 grant
授奖典礼 commencement
授课 oral communication
授权 authorization; authorize; delegation of authority; entitle; investiture; mandate; warranty
授权程序分析报告 authorized program(me) analysis report
授权出口例行程序 authorization exit-routine
授权代表 authorized representative
授权代理人 authorized agent
授权代码 authorization code
授权的 facultative
授权的检验人 authorized surveyor
授权的权力 authority to delegate; authorized signature
授权的限制 restrictions on authority
授权的验船师 surveyor authorized
授权地方维护的道路 deligated road
授权法案 enabling act; enabling legislation
授权范围 scope of authority
授权付款 authority to pay
授权给 empower
授权和撤回 delegation and revocation
授权机构 authorized agency
授权检查员 authorized inspector
授权借记 authorization to debit
授权开立汇票 authority to draw
授权控制 authorization control
授权批准书 note of authorisation[authorization]
授权签署 authorized signature
授权签字 authority to sign
授权人 authorized person
授权人签字 authorized signature
授权人员 authorized officer
授权申明 enabling declaration
授权声明 note of authorisation[authorization]
授权书 power of attorney; certificate of authority; commission of authority; letter of authorization
授权通知书 note of authorisation[authorization]
授权投资 authorized investment
授权文件 authorization file
授权文书 letter of attorney; power of attorney
授权信 letter of authority
授权信号 enabling signal
授权信息 authorization message
授权行为 act of authorization
授权议付 authority to negotiate
授权银行 authorized bank; instruct a bank
授权原则 principle of delegation
授权者将财产契据或保证书交给第三方保存 escrow
授权证 warrant
授权证书 certificate of authorization; certificate of entitlement; letter of authority
授权支出 authorizing of expenditure
授时 issue the official calendar; time service; time transmission
授受机柱 <行车凭证> exchanger post
授信额度 facility amount
授信人 credit giver
授以勋章 decorate
授予 vesting
授予合同 awarding contract; award of contract
授予口粮 ration
授予全权 invested with full authority
授予权利 vesting of rights
授予权责 delegation of authority and responsibility
授予委托书 invest…with power of attorney
授予者 granter[grantor]
授与 confer; grant
授与者 conferrer

瘦长型船 slender ship

瘦地槽【地】 leptogeosyncline
瘦风帆 asleep
瘦尖形抓 sand fluke
瘦焦煤 lean coking coal
瘦煤 black jack; blind coal; dry burning coal; lean coal; meagre coal
瘦黏[粘]砂 antiquated sand
瘦黏[粘]土 adobe; lean clay; meagre clay; mild clay; sandy clay
瘦弱的 slight
瘦砂 lean mo(u)lding sand
瘦体船 thin ship
瘦土 infertile soil
瘦削 thin out
瘦削型船 fine cut stern
瘦型船 fine shaped ship; fine ship
瘦型砂 lean mo(u)lding sand; mild sand; weak sand
瘦渣 lean slag

书报亭 kiosk; news stall; newsstand

书背 backbone; back freight
书橱 book closet
书店 bookseller's; book shop; bookstore
书堆 bookstall
书堆组构 bookhouse fabric
书法 calligraphy; chirography
书法家 penman
书法文体 pencraft
书房 den; sanctum; study room; scriptorium <修道院的>
书柜 bookcase
书柜式干燥器 bookcase drier
书柜 bookstand
书画电传机 autotelegraph
书籍的末页 colophon
书籍封面烫压装饰 tooling
书籍卷头插图 frontispiece
书籍压平机 book smashing machine
书籍正文后面附录的资料 backmatter
书脊 backbone; back freight
书脊扒圆机 book back rounding machine
书脊背衬料 backlining
书脊护舌 <精装书的> head cap
书脊棱带 raised bands
书脊黏[粘]衬 backlining
书脊黏[粘]纱布机 book backbone lining up machine
书脊清漆 book binder's varnish
书脊上胶机 back gluer
书记官长 <国际法院> registrar
书架 bookrack; book shelf; bookstack
书架导标 range guide
书架号码 call number
书架右侧板 range end
书架桌 book table
书架左侧板 range front
书卡袋 pockets on the cover of the work
书刊地图 map in books and periodicals
书库 bibliotheca; book repository; bookstack room; book vault; stack room; stacks
书库便梯 library steps
书库叠层方式 <书库与阅览室分开设置的> stack system
书库楼 stack block
书库内小阅览凹室 carrel(l)
书库内小阅览角 carrel(l)
书面保证 written undertaking
书面报告 reading report; written report
书面材料 written material
书面陈述 written statement
书面答复 answer in writing; rewrite; written reply
书面单据 written documents
书面的 writing
书面订货 written order
书面沟通 written communication
书面合同 contract in writing; letter contract; literal contract; written contract
书面记录 written statement
书面教材 written material
书面结清单 written discharge
书面联络法 written liaison method
书面批复 written approval
书面批准 approval in writing; written approval
书面凭据 written confirmation
书面凭证 written confirmation
书面契约 letter agreement; written contract
书面确认 confirmation in writing; written confirmation

书面申请 written application
书面声明 poop sheet;written declaration;written statement
书面烫印标记 titling
书面提问 written inquiry
书面通知 written notice
书面同意 written approval; written consent;written permission
书面同意在发生工程变更 consent of surety
书面文件 escript;written documents
书面协议 written agreement;letter agreement
书面信托 express trust
书面许可 written permission
书面羊皮 basan
书面要求 written request
书面要式合同 specialty contract
书面证据 documentary evidence;evidence documentary
书面证明 certification;written confirmation;written documents;written evidence
书面证明的合同 contract evidenced by a writing
书面纸 cover paper
书面质询 written inquiry
书面(资料)工作 paper work
书名标题 caption title
书名全称 full title
书末的版权页 colophon
书目 bibliography
书目编纂者 bibliographer
书目耦合 bibliographic(al) coupling
书目提要 bibliography
书皮纸 paper for covering books
书皮纸板 mill board
书室-卧室 study-bedroom
书摊 bookstand
书套 forel;slip case;slip cover
书帖编号 numbering of sections
书帖标记 designation marks
书亭 book-kiosk;bookstall
书写品 black and white
书写纸 writing paper
书信电报 letter telegram
书信复写器 letter press
书信合约 letter agreement
书形空心砖 <带凹凸边的> book tile
书型排架法 physical arrangement
书页构造 book structure
书页黏[粘]土 book-leaf clay
书页式电容器 book capacitor
书页式构造 <岩石> book structure
书页式拼板 book matching
书页式拼装法 book matching
书页岩【地】 bookstone;page stone
书页云母 book mica;mica book
书页折角 dog ear
书有阴影线的面积 hatched area
书院 lyceum
书斋 inner sanctum
书证 documentary evidence
书志目录 biblio
书中所附供比较的材料 apparatus criticus
书状电容器 book capacitor
书桌 analogion;davenport;desk;writing desk;writing table
书桌锁 desk lock

叔胺【化】 tertiary amine

叔胺值 tertiary amine value
叔醇 tertiary alcohol
叔丁醇 tert-butyl alcohol;tertiary butanol
叔丁基次氯酸盐法 tertiary butylhypochlorite process
叔丁基过氧化氢 tertiary hydroperoxide
叔丁基过氧化物 tert-butyl peroxide
叔丁基化过氧氢 tert-butylhydroperoxide
叔丁基邻苯二酚 tert-butyl catechol;tert-butylpyrocatechol
叔碳酸乙烯酯 vinyl ester of versatic acid;vinyl versatate
叔碳原子 tertiary carbon atom
叔硝基化合物 tertiary nitro compounds

枢 center[centre];hub;pin;pivot

枢板 pin plate
枢承 pin bearing;pivot socket
枢承推土机 tilting bulldozer
枢承支座 pin support
枢齿 armature tooth
枢杆 hinged arm;hinged bar;pivoted arm
枢机 helm
枢铰 pivot hinge
枢接 pin-connected; pin-keyed; pinned joint
枢接构架 pin truss
枢接合 pin(-connected) joint
枢接合的 pin-jointed
枢接桁架 pin-connected truss
枢接桁架桥 pin-connected truss bridge
枢接框架 pin-connected frame
枢接柱 pin-joint column
枢臼轴承 angular bearing
枢孔垫片 pin packing
枢孔填板 pin packing
枢孔填料 pin packing
枢木 Aralia
枢纽 hinge;junction;pivot;terminal;terminus[复 termini];hub
枢纽变电所 pivotal substation
枢纽变电站 load-center substation
枢纽布置图 layout of hydrojunction;project layout;terminal layout;layout of hydroproject
枢纽侧伏角 pitch of hinge
枢纽电气化 electrification of terminal
枢纽调车场 terminal yard
枢纽调车场平面图 plan of terminal yard
枢纽断层 hinge fault;pivotal fault
枢纽港 main port;pivotal harbo(u)r;pivotal port;terminal port;center[entre] port; direct-call port; hub port;hub and load center[centre] port
枢纽工程 key project
枢纽环道 junction roundabouts
枢纽环线 terminal loop
枢纽机场 key-airport
枢纽交叉 <联动信号系统的> key junction
枢纽进站线路疏解 approach junction terminal tracks de-crossing
枢纽联轨站 junction
枢纽联络线 terminal connecting line
枢纽内部调车 intra-terminal switching
枢纽内调车 terminal shunting;terminal switching
枢纽内线路 intra-terminal track
枢纽内行程 terminal run
枢纽内运行 terminal movement
枢纽前方线路所 block post before hub
枢纽前方站【铁】 station in advance of terminal
枢纽倾伏角 plunge angle of hinge
枢纽倾伏向 plunging of hinge
枢纽区 terminal district;terminal zone
枢纽区段站 junction terminal district station
枢纽区域环行线 belt line
枢纽铁路公司 terminal company
枢纽通过能力 terminal capacity
枢纽位置 key position
枢纽线 hinge line
枢纽限界容量 terminal clearance capacity
枢纽小运转列车 junction terminal transfer train
枢纽信号 junction signal
枢纽遥控 centralized traffic control at terminal
枢纽迂回线 terminal roundabout line
枢纽闸站 lock station
枢纽站 hub station; interchange station; junction centre station; junction station;railroad terminal
枢纽直径线 diametric(al) line of junction terminal
枢纽周围各区段 surrounding districts of terminal
枢纽走向 strike of hinge
枢栓 pintle
枢销 pivot pin;trunnion
枢销盖 pivot pin cap
枢销锁销 pivot pin lock pin
枢销外壳 vertical pivot pin housing
枢销悬挂式反铲挖掘机 fixed pivot back-hoe
枢心 pivoting point
枢芯清漆 core varnish
枢形夹座 pivot holder
枢支承 pivot bearing
枢支座 tip bearing
枢制动器 pivoted detent
枢轴 gudgeon; king journal; knuckle pivot; pintel; pintle; pivot (pin); pivot shaft; trunnion (axis); weigh bar shaft
枢轴凹座 <船闸闸门> hollow quoin
枢轴掣子 pivoted detent
枢轴承 gantry post; pivot(ing) bearing;tip bearing
枢轴承平旋桥 center bearing swing bridge;pivot-bearing swing bridge
枢轴的 pivoted
枢轴点 pivotal point
枢轴垫 pivot pad
枢轴吊斗提升输送机 pivot-bucket conveyor-elevator
枢轴斗式输送器 pivoted bucket conveyer[conveyor]
枢轴杠杆锁 pin tumbler lock
枢轴高压 pivoting high
枢轴关节 trochoid
枢轴合页 pin hinge
枢轴环 pivot cup
枢轴继电器 pivoted relay
枢轴架 trunnion bracket
枢轴间隔 pivotal interval
枢轴绞链接合 pivot knuckle joint
枢轴铰链 loose pin hinge
枢轴接合 skewback hinge
枢轴颈 pivot journal;vertical journal
枢轴梁 pivoting beam
枢轴螺钉 pivot screw
枢轴螺栓 pivot bolt
枢轴螺旋 pivot-point screw
枢轴抛光机 pivot polisher
枢轴三角架 pivot tripod
枢轴式发射机 pivot transmitter
枢轴式犁辕锁定器 pivot beam lock
枢轴式皮带运输机 pivoted conveyer[conveyor]
枢轴式擒纵机构 pivoted detent
枢轴式天文钟擒纵机构 pivoted detent escapement
枢轴位置 pivot location
枢轴舷窗 opening side light;pivoting side light
枢轴衔铁 clapper; hinge(d) armature;pivoted armature
枢轴线 pivotal line;pivot axis
枢轴销 pivot pin
枢轴旋动舷窗 pivoted scuttle
枢轴研磨工具 jacot tool
枢轴研磨机 pivot burnisher
枢轴仰开桥 trunnion bascule bridge
枢轴因素 pivoted factor
枢轴元素 pivot(al) element
枢轴元素方程 pivotal element equation
枢轴支承 pivot(al) bearing;pivot suspension
枢轴支承发动机 pivoted engine
枢轴支点 pivot point
枢轴支座 trunnion bearing
枢轴值 pivotal value
枢轴轴承 trunnion bearing
枢轴柱 quoin post
枢轴状 trochoid
枢轴座 <旋开窗的> socket
枢转断层【地】 pivotal fault
枢转起重机 kingpin slewing crane;pivot slewing crane
枢转式通风口 pivoted ventilation window
枢转性 pivotability
枢转轴 pivotal axis
枢桩 key pile
枢椎 axis[复 axes]

梳辫子 queue-up

梳尺 comb scale
梳齿 comb dent
梳齿板 comb plate; coxcomb; fish back
梳齿刀 rack-type cutter
梳齿导流 comb diversion
梳齿滚筒 hackle drum
梳齿函数 comb function
梳齿机 rack shaper;rack shaping machine
梳齿接 corner locked joint
梳齿结合 cornerlock joint
梳齿滤波器 comb filter
梳齿转筒 comb dresser
梳齿状滤波器 <多通带滤波器> comb filter
梳出的麻屑 flax tow
梳刀 carding tool;chasing tool
梳刀盘 chaser
梳粉机 carding machine
梳辊式并条机 rotary drawing frame
梳机 comb
梳集板 collecting comb
梳荚凹板 carding bottom
梳解机 breaker beater
梳理 carding;pectination;tease
梳理的粗杂材 combed fascine raft
梳理滚筒 comber cylinder
梳理机 carding machine;hackling machine
梳理机废料清除工 fettler
梳理机针布 card clothing
梳理下脚 scutcher waste
梳轮 comb
梳麻导板 hackle guide
梳麻台 hackle bench
梳麻装置 rippling device
梳毛厂 wool-carding mill
梳棉机 hackle
梳式 herringbone pattern
梳式布置 <港池> comb system

梳式导丝器 comb guide
梳式滑道 comb type slipway
梳式加料 strip filling
梳式接缝 steel comb joint
梳式结合 combed joint
梳式码头布置 herringbone wharf layout
梳式排水系统 herringbone drain(age) system
梳刷 card;combing;strip
梳刷滚筒 picking drum
梳刷轮 combing beater
梳刷面饰 combed finish;dragged finish
梳刷转筒 comb dresser
梳刷装置 rippling device
梳松机 teaser
梳条 sliver
梳形板 comb joint;comb plate;spacer
梳形避雷器 comb arrester; comb lightning arrester; lightning comb protector
梳形电极 comb poles
梳形电路 comb type circuit
梳形集电极 comb collector
梳形间隔 side spacer
梳形接缝 comb joint
梳形接合 finger joint
梳形聚合物 comb polymer
梳形卡子 side spacer
梳形滤波器 comb filter
梳形耙斗 rake scraper
梳形刨刀 tooth plane iron
梳形膨胀缝 finger-type expansion joint
梳形双光束光谱仪 comb type double beam spectrometer
梳形天线 comb antenna
梳形温莎椅 comb-back chair
梳形纹木材 comb-grained wood
梳形物 coxcomb
梳形系索耳 comb cleat
梳形线圈 lattice coil
梳形张力器 gate tensioner
梳形支架 <窑具> comb rack
梳凿 comb
梳摘滚筒 carding cylinder; picking drum
梳摘装置 stripper unit
梳针定位 set of wires
梳制花纹 combing
梳妆架 <盥洗室> comb rack;toilet rack
梳妆室 dressing cubicle
梳妆台 dresser; toilet table; dressing table
梳妆台式洗面器 vanity basin
梳妆用具 toilet articles;toilet set
梳状 comb-like;honeycomb
梳状变换 combescure transformation
梳状波发生器 comb generator
梳状剥落 comby
梳状测云器 comb nephoscope
梳状测针 rake probe
梳状打底 comb rendering
梳状刀片 comb
梳状的 comby
梳状电极 comb-shaped electrode
梳状粉刷 comb plaster
梳状蜂窝 comb
梳状构造 comb-like structure
梳状刮刀 cockscomb
梳状管 Pitot traverse
梳状函数 comb function
梳状横向扩展全加器 comb-shaped transverse spreading adder
梳状结构 Comanchic structure;comb structure;comb-like texture【地】
梳状静压管 Pitot-static rake
梳状裂纹 reeds
梳状滤波器 comb filter
梳状抹灰 comb rendering
梳状黏[粘]合胶粘剂 close-contact adhesive
梳状皮托管 comb pitot
梳状山脊 comb-shaped ridge
梳状探头 rake probe
梳状探针 rake probe
梳状纹 coxcomb
梳状物 comb;pectination
梳状线 pectinate line
梳状褶皱 comb-shaped fold
梳状轴承 corrugated bearing

疏

疏冰 open ice
疏冰群 open pack
疏波 wave of rarefaction
疏部 part of rarefaction
疏成帧 loose framing
疏承地板梁 <保温车> floor rack stringer
疏出 drain away;drain out
疏丛草 bunchgrass
疏导 divert;drainage;training for discharge
疏导工程 discharge engineering
疏导坍塌工程 work for deflecting avalanche
疏电(子)的 electrophobic
疏堆积 loosely packed;loose packing
疏伐 improvement cutting
疏伐周期 thinning cycle
疏干 dewatering;drainage;drying out;sewer;unwater(ing)
疏干半径 dewatering radius
疏干表面 drainage face
疏干工程排水量 discharge of dewatering engineering
疏干井水 drainage by desiccation
疏干开采量 depletion yield;unwatering yield
疏干孔 dewatering borehole
疏干面积 drainage area
疏干排水 drainage by desiccation
疏干排水系统 dewatering system
疏干区 dewatering area
疏干时间 dewatering time
疏干水量体积 dewatering water volume
疏干水平 dewatering level
疏干体积 dewatering volume
疏干系数 depletion coefficient;dewatering coefficient;unwatering coefficient
疏干系统 unwatering system
疏干巷道 dewatering lane; drainage gallery
疏干类型 dewatering type
疏干因素 dewatering factor
疏港 evacuation of cargo from port; intensive dispatch of port cargo
疏管距布置 open tube spacing
疏灌丛 shrub land
疏忽错误 missing error
疏忽条款 inchmaree clause; negligence clause
疏解 untwine;untwining
疏解车站咽喉 untwine a station bottleneck
疏解机 fluffer
疏解进站线路 untwine the approach lines
疏浚 drag; dredging; hydraulic excavation; scour; training; training for depth
疏浚标高 dredge level
疏浚标志 dredging mark
疏浚材料 dredged material; dredging material
疏浚操纵轮 dredging wheel
疏浚操作 dredging operation
疏浚槽 dredge cut
疏浚测量 dredging survey
疏浚测图 dredging survey sheet
疏浚铲斗 dredging shovel
疏浚车轮 scoop wheel
疏浚承包人 dredging contractor
疏浚船(舶) dredging vessel;dispersal vessel;dredge
疏浚的泊位 dredged berth
疏浚的航槽 dredged channel
疏浚的航道 dredged channel
疏浚的环境方面 environmental aspects of dredging
疏浚地点 dredging site
疏浚地段 dredging site
疏浚方法 dredging method
疏浚废料 dredging spoil
疏浚改善航道 channel improved by dredging
疏浚高程 dredge elevation;dredge level
疏浚工 dredge
疏浚工程 conservancy engineering; deepening project; dredging engineering; dredging project; dredging works
疏浚工程测量 dredge engineering survey
疏浚工程对环境与海上安全的影响 effect of the dredging works on the environment and on maritime safety
疏浚工程辅助船舶 ancillary craft for dredging works; auxiliary craft for dredging works
疏浚工程竣工验收 completion acceptance of dredging works
疏浚工程排泥点或吹填点 discharge point
疏浚工程配套船舶 associated craft for dredging works
疏浚工程图 dredging project chart
疏浚工地 dredging site
疏浚工况 dredging site condition
疏浚工人 dredger
疏浚工艺 dredging process; dredging technique;dredging technology
疏浚工作 dredge work; dredging operation
疏浚公司 dredging firm;dredging company
疏浚管道 dredging pipe(line)
疏浚轨迹 dredging track
疏浚过程自动化 automation of dredging process
疏浚航槽 improved channel
疏浚航道 improved channel
疏浚河道 train river channel
疏浚河底 deepening of the stream floor;dredging bottom
疏浚戽斗 dewatering bucket
疏浚机 dredge(r);hydraulic dredge(r); hydraulic excavator; mud drag;mud dredge(r);mud drum
疏浚机链斗架 dredging ladder
疏浚机械 dredging machinery
疏浚技术 dredging technique
疏浚界 dredging community
疏浚经济 dredging economics
疏浚开挖 dredge excavation
疏浚勘测 dredging exploration survey
疏浚量 dredging quantity;volume of dredging
疏浚能力 dredging capacity
疏浚泥沙倾倒区 disposal area
疏浚排泥填岸 artificial replenishment
疏浚频率 dredging frequency
疏浚评估 dredging assessment
疏浚弃泥区 dumping ground
疏浚弃土 dredged material; dredged spoil matter; dumped material; dumping dredged silt
疏浚强度图 dredging intensity chart
疏浚区 area to be dredged; dredged area
疏浚区浮标 dredging buoy
疏浚任务 dredging task
疏浚容许误差 dredging tolerance
疏浚设备 dredging equipment; dredging plant
疏浚设备类型 type of dredging plant
疏浚深度 dredging depth
疏浚施工 dredging construction; dredging execution;dredging operation
疏浚施工标志 dredging symbol
疏浚施工负责人 dredgemaster
疏浚时间利用率 time utilization rate in a project
疏浚水道 channel dredging
疏浚水道蓄水池 scouring basin;sluicing pond
疏浚填筑 dredging and reclamation
疏浚土 dredging material; dredging matter; dredging muck; dredging soil;spoil
疏浚土处理 disposal of dredged material;dredged material disposal;spoil disposal
疏浚土处置 dredged material disposal;spoil disposal
疏浚土的有效利用 beneficial use of dredged material
疏浚土分级 gradation of soils for dredging purposes
疏浚土分类 classification of dredging soil;classification of soils and rocks to be dredged
疏浚土管理 management of dredged materials
疏浚土评价框架 dredged material assessment framework
疏浚拖刮取土器 dredge sampler
疏浚维护工作 <河道,港口的> conservancy matters
疏浚污泥 dredge muck
疏浚污染 pollution by dredging
疏浚系统 dredging system
疏浚项目 dredging project
疏浚允许超深 allowance for overdepth dredging
疏浚作业 dredging; dredging operation
疏浚作业中的测量 during dredging survey
疏开 deploy
疏开布置的跑道 dispersed runway
疏开队形 loose formations
疏拦污栅 coarse rack
疏篱 hurdle
疏林 light forest
疏林草地 park land
疏林草原 park land;veld(t) <南非>
疏林地 open wood land
疏密波 compressionally dilatational wave; compression-dila(ta)tion wave; dilatational wave; push-pull wave; wave of condensation and rarefaction
疏密(抽头)变换器 rough and fine tap changer
疏密度 degree of closeness;degree of density;laxity
疏密构造 dense-sparse structure
疏密制 variable-density system
疏木纹 coarse grain;open grain
疏木纹的 open-grained
疏年轮木材 wide-ringed timber
疏耦合 weak coupling
疏排 white-out

疏铺柴排 loose brush mattress
疏铺瓷砖 open slating
疏铺梢料 loose brush
疏铺石板（瓦）open slating; spaced slating; half slating
疏铺望板 spaced boarding
疏渠管用的清除轮 clearing wheel
疏绕线圈 spaced winding
疏溶剂色谱 solvophobic chromatography
疏散 decentralization; decentralize; dispersal; diversity; evacuate; evacuation
疏散背板 open sheathing
疏散车辆道路 relief road
疏散程序 evacuation procedure
疏散出口 emergency exit
疏散道路 diversion road
疏散方向 direction of evacuation
疏散滑槽 evacuation chute
疏散滑门 evacuation slide
疏散机场 aircraft dispersal area
疏散计划 evacuation plan
疏散交通 direction of traffic; diversion of traffic; relief of traffic
疏散警报 evacuation alarm
疏散口 exit; escape
疏散流量 discharge value
疏散楼梯 emergency stair (case); escape stair (case); fire escape; fire escape stair (case)
疏散路线 escape route; evacuated route; evacuation route; route of escape
疏散路线标志 escape route sign
疏散路线照明 escape route lighting
疏散路线指示标志 evacuation route marker
疏散率 discharge rate
疏散能力 evacuation capability
疏散爬梯 escape ladder; safety ladder
疏散坡道 escape ramp
疏散器材 evacuation equipment kit
疏散区 <机场的> dispersal parking area; evacuated area
疏散人口 depopulation
疏散设施 means of egress; means of escape
疏散绳 escape rope
疏散时间 egress time; evacuation time
疏散通道 evacuation passageway; exit passageway
疏散通道照明装置 escape path lighting system
疏散网络计算机模型 evacuation network computer model
疏散信号 evacuation signal
疏散性 diversity
疏散演习 evacuation drill
疏散用地 dispersal area
疏散照明 escape lighting
疏散照明系统 escape lighting system
疏散照明装置 escape lighting unit
疏散支柱 open sheathing
疏散指示灯 evacuation strobe light
疏散走廊 escape corridor; exit corridor
疏扫描 coarse scan (ning)
疏上胶层 open coat
疏梢褥护岸 loose brush revetment
疏树草地 park land
疏树草原 boschveld; park land
疏水 dewatering; drain (age); drain off; easying; lyophobic
疏水泵 drainage pump; flood pump; return-drain pump
疏水表面 hydrophobic surface
疏水材料 hydrophobic material
疏水槽 drain (age) tank; drain trough; drip trough
疏水的 hydrophobic; water-hating; water-rejecting
疏水地板 <保温车> floor rack
疏水地板条 floor rack slat
疏水斗 effluent hopper
疏水阀 bleeder; draining valve; steam trap
疏水分数 hydrophobic fraction
疏水浮石 hydrophobic pumice
疏水管 drainage connection; drain (age) pipe
疏水管路水封 condensate drain loop
疏水罐 drain trap
疏水化合物 hydrophobic compound
疏水化学污染物 hydrophobic chemical pollutant
疏水活化 hydrophobic activation
疏水基 hydrophobic group
疏水基相互作用 hydrophobic interaction
疏水剂 anti-hydro; hydrophobic admixture; hydrophobing agent; water-repeller
疏水键 hydrophobic bond
疏水键合 hydrophobic bonding
疏水胶体 hydrophobe; hydrophobic colloid; hydrophobic gel
疏水冷却器 drain (age) cooler
疏水冷却区 drainage cooling zone
疏水链 hydrophobic chain
疏水膜 hydrophobic membrane
疏水内芯 hydrophobic core
疏水膨润土 organophilic bentonite
疏水膨胀器 flash tank
疏水气体电极 hydrophobic gas electrode
疏水器 condensate trap; condensate water discharge; condensation trap; drip trap; moisture trap; standpipe; steam trap; trap
疏水器臂 trap arm
疏水器出口 trap outlet
疏水器的最低部位 trap dip
疏水器的最高部位 trap weir
疏水器浮子室 drainer float chamber
疏水溶胶 hydrophobic sol
疏水溶质 hydrophobic solute
疏水塞子 hydrophobic plug
疏水隧洞 drainage culvert
疏水(调节)器 drainage controller
疏水微孔膜 hydrophobic microporous membrane
疏水物 hydrophobe; hydrophobic material
疏水物质 hydrophobic substance
疏水物种 hydrophobic species
疏水吸附剂 hydrophobic adsorbent
疏水吸附质 hydrophobic adsorbate
疏水系统 <船坞> drainage channel and pump for the maintenance of working condition
疏水箱 blow tank; drain trap
疏水橡胶 hydrophobic rubber
疏水型过热器 drainable superheater
疏水性 hydrophobicity; hydrophobic nature
疏水性缔合 lyophobic association
疏水性粉尘 hydrophobic dust; lyophobic dust
疏水性胶体 lyophobic colloid
疏水性颗粒 hydrophobic particle
疏水性网格膜滤池 hydrophobic grid membrane filter
疏水性相互作用 hydrophobic interaction
疏水性指数 hydrophobicity index
疏水旋塞 clean-out plug; clearing plug
疏水油 hydrophobic oil
疏水有机化合物 hydrophobic organic compound
疏水有机化学物 hydrophobic organic chemicals
疏水有机污染物 hydrophobic organic contaminant; hydrophobic organic pollutant
疏水中空纤维膜 hydrophobic hollow fiber[fibre] membrane
疏水装置 drain system
疏松 rarefaction; shatter; porous spot <铸件缺陷>
疏松拌和机 pulvimixer
疏松表面 the loose surface
疏松冰 lolly ice
疏松材料 bulk material
疏松层 loosened layer; tectorium; unconsolidated layer
疏松沉淀物 loose deposit; unconsolidated material
疏松充填 loose fill
疏松的 chessom; incoherent; loose (ned); porous
疏松的母质层 regolith
疏松的素烧瓷体 porous
疏松的氧化皮 loose scale
疏松的植物纤维 uncompressed vegetable fiber[fibre]
疏松底板 loose bottom
疏松底部的平巷拱 <采矿> loose-footed roadway arch
疏松地面 loose ground
疏松点 <测定沥青土中土的最佳含水量用> fluffy point
疏松度 fraction void; gas porosity; porosity
疏松堆块 loose blocks
疏松粉末 loose powder
疏松稿秆 loose straw
疏松骨料 loose aggregate
疏松焊缝 porous weld
疏松回填 unconsolidated backfill
疏松混合料 open-textured mix
疏松级配 open-textured grading
疏松集料 loose aggregate
疏松节疤 <木材> loose knot
疏松结构 open structure; porous structure
疏松结合 loose combination
疏松界面模型 loose-boundary model
疏松块堆 loose blocks
疏松鳞锈 loose scale
疏松煤 loose coal
疏松膜 loose membrane
疏松石蜡 slack wax
疏松水垢 loose scale
疏松体积 loose volume
疏松填充 loose packing
疏松填料 aerated filler; bulking filler; loose filler
疏松填石 loose rock-fill
疏松土粒 hold the particles apart
疏松土(壤) free soil; light earth; loose soil; mellow soil; porous soil; spongy soil; chessom
疏松纹理 loosened grain
疏松物质 unconsolidated material
疏松性 friability; sponginess
疏松压制 loose compaction
疏松岩石 loose rock
疏松盐土 loose saline soil
疏松氧化皮 loose scale
疏松质地 loose texture
疏松重量 loose weight
疏松状态 loose state
疏通 opening up; unplug; unstop
疏通沟渠 opening-up notches and gutters
疏通积水 away the collected water
疏涂(饰)层 open coat
疏纹木材 open-grained timber; wide-ringed timber
疏相 lean phase
疏橡胶性 rubber-phobe
疏液的【化】lyophobic
疏液胶体 lyophobe colloid; lyophobic colloid
疏液胶体溶液 lyophobic colloidal solution
疏液器 liquid trap
疏液溶胶 lyophobic sol
疏液体 lyophobe
疏液物 lyophobe
疏液物料 lyophobic material
疏液性 lyophobicity
疏液性介晶现象 lyotropic mesomorphism
疏油的 oleophobic
疏油胶体 oleophobic colloid
疏油整理剂 oleophobic finisher
疏油作用 oleophobic effect
疏有机性的 organophobic
疏于保养 failure to maintain
疏于检查 failure to inspect
疏于维护 failure to maintain
疏运 dispatch
疏运工具 dispatching tools; dispatch means; means of dispatch
疏运能力 <港口的> evacuation capacity of harbo (u) r; evacuation capacity of port
疏运设备 facility for dispatch
疏枝法 thinning
疏植 wide planting
疏质子的 protophobic
疏质子介质 aprotic media
疏质子溶剂 aprotic solvent; phobic solvent
疏柱式的 areostyle; wide-spaced
疏柱式建筑寺庙 araeostyle temple
疏柱式建筑物 <柱距等于柱径的四倍或四倍以上> ar (a) eostyle; wide-spaced building
疏柱式庙 wide-spaced temple
疏柱式寺庙 araeostyle temple

舒

舒尔茨公式 Schulz formula
舒尔茨锌基轴承合金 Schulz alloy
舒缓波状的 gently undulated
舒缓波状断层 gently wavy fault
舒缓波状断裂 sine-curve fault
舒缓波状构造带 sine-curve tectonic belt
舒勒摆 Schuler pendulum
舒利莱恩光学系统 Schlieren optic (al) system
舒曼顶棚 Schumann's ceiling
舒曼金属骨料砂浆覆盖层 Schumann metal (lic) aggregate mortar cover (ing)
舒适采暖系统 comfort heating system
舒适带 comfort line; comfort zone
舒适的地方 cubbyhole
舒适的减速率视距 comfortable deceleration
舒适的停车视距 comfortable stopping sight distance
舒适度 comfortable degree
舒适度计 eupatheoscope
舒适度曲线 comfort curve
舒适度图 comfort chart
舒适范围 comfort standard; comfort zone
舒适感 comfortable feeling; feeling of comfort
舒适感线图 comfort chart
舒适焓湿图 comfort chart; comfort psychrometric chart
舒适减速 comfort deceleration
舒适降温 comfort cooling

舒适空(气)调(节)comfort air conditioning;air-conditioning comfort
舒适控制 comfort control
舒适冷气系统 comfort cooling system
舒适冷气装置 comfort cooling unit
舒适冷却 comfort cooling
舒适冷却系统 comfort cooling system
舒适气流 comfort current;comfort stream
舒适气流速度 comfort stream velocity
舒适区(域)comfort zone;zone of comfort
舒适权 amenity right
舒适设备 <客车的> amenity facility
舒适设施 amenities;amenity facility
舒适湿度 comfort humidity
舒适速度 comfort speed
舒适条件 comfort condition
舒适通风 comfort ventilation
舒适图 comfort chart
舒适温度 comfort temperature
舒适温度区 comfort temperature line
舒适温湿图 comfort psychrometric chart
舒适纤维 <一种平织用的高张力聚酯纤维> comfort fiber[fibre]
舒适线 comfort line
舒适小房间 cubby
舒适型结构 relaxed and loose structure
舒适性 amenity;comfortability
舒适性标准 amenity standard
舒适性空(气)调(节)comfort air conditioning
舒适性频率 ride-frequency
舒适性设计 amenity design
舒适性指数 comfort index
舒适压力 comfort pressure
舒适要求 comfort requirement
舒适指标 comfort index
舒斯特机制 Schuster mechanism
舒斯特问题 Schuster problem
舒威赫港 <科威特> Port Shuwaikh
舒张期 diastole

输

输冰桥 overline bridge for ice transportation

输出 fan out;lead-out;output
输出版 trade edition
输出保险 export insurance
输出报告 output report
输出泵 rear pump
输出比例 export ratio
输出比例尺 output scale
输出编辑 output edit
输出变化 output variation
输出变换器 output translator
输出变量 output variable
输出变量器 output transformer
输出变数 output variable
输出变送器 output transducer
输出变压器 output transformer
输出标记 output identification;output token
输出标题 output header
输出表 output list;output meter
输出表接合器 output-meter adapter
输出波 outgoing wave;output wave
输出波长 output wavelength
输出波导 output waveguide
输出波动 output pulsation
输出补偿 output back-off
输出不变区 dead band;dead zone
输出部分 output part
输出部件 output block;output unit
输出参数 output parameter
输出参数地址 output parameter address
输出操作 output function
输出操作杆 output control lever
输出测度 output measure
输出测量表 output meter
输出插孔 receptacle outlet
输出插座 output socket
输出常数 output constant
输出成果 outputting result
输出程序 output program(me);output routine;output writer
输出齿轮 output gear
输出处理机 output handler;output processor
输出穿孔机 output punch(er)
输出传递函数 output transfer function
输出传动轴 take-off propeller shaft
输出传送 output transfer
输出窗 output window
输出存储区 output block
输出错误 output error
输出打印机 output printer
输出带 throw-out spiral
输出贷款 export loan
输出单元 output element;output unit
输出导纳 output admittance
输出的 deferent
输出的差错率 output error rate
输出的搜索损失 output hunting loss
输出地点 export point
输出电导 output conductance
输出电极 output electrode
输出电抗 output reactance
输出电缆 output cable
输出电流 current output;emergent current;output current
输出电流额定值 output current rating
输出电路 outgoing circuit;output channel;output circuit;output end
输出电路阻抗 output circuit impedance
输出电平 output level
输出电平不稳定性 output level instability
输出电平自动调节放大器 fading amplifier
输出电容 output capacitance
输出电刷 output brush
输出电信号 electric(al)output signal
输出电压 output voltage
输出电阻 output reactance;output resistance
输出动量 leaving momentum
输出动脉 efferent artery
输出端 delivery end;mouth;outfan;outlet(end);output end;output lead;output port;coil-out
输出端导纳 output admittance
输出端扼流圈 output choke;output choke coil
输出端反射镜 output end mirror
输出端复接电路 multiple output circuit
输出端干扰 output disturbance
输出端回程 output flyback
输出端极间电容 output interelectrode capacitance
输出端极性 output polarity
输出端接触器 output contactor
输出端接口器 output interface adapter
输出端脉冲 reproduced pulse
输出端面镜 output end mirror
输出端数 fan out
输出端衰减器 output pad
输出端损耗 exit loss
输出端损失 exit loss
输出端信号极性 output polarity
输出端指示电压 output indicating voltage
输出端子 carry-out terminal;lead-out terminal;outlet terminal;output terminal
输出队列 output queue
输出对输出的串音 output-to-output crosstalk
输出多路调制器 output multiplexer
输出额定值 output rating
输出扼流圈 output choke
输出阀 delivery valve;outlet valve
输出法兰盘 output flange
输出反馈 output feedback
输出反射系数 output reflection coefficient
输出范围 output area;output range
输出方程 output equation
输出放大级 output amplifier stage
输出放大器 output amplifier
输出放大器阻抗 output amplifier impedance
输出分布处理 output distributed processing
输出分程序 output block
输出分拣 output sort
输出分解 output resolution
输出分类 output category
输出分路总管 delivery manifold
输出幅度电平 output amplitude level
输出辐射 output radiation
输出负荷 output load(ing)
输出负载 output load(ing)
输出负载率 output loading factor
输出负载因数 output loading factor
输出干扰 output disturbance
输出干扰声 output ripple
输出港 delivery port;export harbo(u)r;outport;port of origin
输出格式 output form(at)
输出格式程序 output formatter
输出格式说明 output format specification
输出功 delivery work;output work
输出功率 capacity power;delivered horse power;delivered power;developed power;kilowatt output;output;output(horse)power;power output;wattage output
输出功率的确定 determination of output
输出功率计 output power meter
输出功率密度 output power density
输出功率谱 output power spectrum
输出功率试验 output test
输出功率调节器 output governor
输出功率透平 output turbine
输出功率稳定度 output power stability
输出功率与输入功率比 output-input ratio
输出功率增量 incremental delivered power
输出功率值 output rating
输出共振器 output resonator
输出管 delivery pipe;efferent;efferent duct;output tube
输出管接头处流量 discharge connection delivery
输出光阑 output diaphragm
输出光束 output beam
输出光束功率 output beam power
输出光束能量 output beam energy
输出光束耦合 output beam coupling
输出光通量 output light flux
输出光子 output photon
输出辊道 run-out table
输出国 country of exportation;exporter
输出过程 output procedure;output process
输出函数表 plotting board;plotting table
输出合格证 certificate for export
输出环节 output element
输出缓冲存储器 output buffer storage
输出缓冲器 output buffer
输出缓冲区 output block
输出灰岩 donor limestone
输出回路 output circuit
输出回描 output flyback
输出回输 output feedback
输出机 output unit
输出机构 output element;output gear;output mechanism
输出机能 efferentation
输出基准 output reference
输出激励器 output driver
输出级 output level;output stage
输出极限 output limit
输出集 output ensemble
输出计 output meter
输出计附加器 output-meter adapter
输出计算机变量 output machine variable
输出记录 output record
输出记录程序 output writing program(me)
输出记录机 outscriber
输出记录器 output printer;output recorder;output writer
输出技术 export technique
输出继电器控制板 output relay panel
输出寄存器 output register
输出尖峰 output spike
输出监督程序中断 output monitor interrupt
输出监视器 output monitor
输出监听器 actual monitor
输出键 run-out key
输出奖励金 export bounty
输出交叉控制 output interleaving control
输出交流声电平 output hum level
输出胶片 output film
输出胶片幅面 output film format
输出角 output angle
输出角分布 angular distribution of output
输出接卡箱 output stacker
输出接口 output interface
输出接口器 output interface adapter
输出接线 output connection
输出接线端 outlet terminal
输出接线盒【电】extension terminal box
输出节点 output node
输出结构 export structure
输出解耦零点 output decoupling zero
输出解调器 output demodulator
输出借位 output borrow
输出进程 output process
输出进位 output carry
输出禁止器 output inhibitor
输出井 output well
输出镜 outgoing mirror
输出镜透射率 output mirror transmission
输出矩阵 output matrix
输出开关 output switch
输出可控性 output controllability
输出客流 output passenger flow
输出空腔谐振器 output cavity
输出空运港机上交货价 ex plane
输出孔口 delivery orifice
输出控制 outgoing control;output control
输出控制台 output panel
输出口 delivery outlet;outcome
输出馈路 outgoing feeder

输出馈线 outgoing feeder
输出扩展电路 out-expander
输出劳动力 export labo(u)r power
输出力 output force
输出力矩 output moment;output torque
输出立管 export riser
输出例行程序 output routine
输出连接 outgoing junction; output connection
输出量 aggregate output; output; output capacity; output quantity; output variable; pump capacity; send-out
输出量的希望值 desired output
输出列表格式 output listing format
输出流道 delivery channel
输出流量 delivery flushing flow; output flow
输出流量范围 delivery range; output flow range
输出流量控制 output traffic control
输出流量试验 output test
输出漏斗 output hopper
输出滤波器 output filter
输出路径 outgoing route
输出率 output factor; output rating; output ratio; rate of discharge; specific output
输出螺旋线 throw-out spiral
输出马力 delivered horse power; output horsepower
输出脉冲 output(im)pulse
输出脉冲幅度 output pulse amplitude
输出脉冲记录器 outpulsing register
输出脉冲宽度 output pulse width
输出脉冲频率 output pulse frequency
输出脉动 output ripple; ripple in output
输出媒体 output medium
输出门 <电路的> outgate
输出面 output face
输出模块 output module
输出模块阀 output module valve
输出目标制 export target
输出内阻抗 internal output impedance
输出能力 fan-out capability; output capability; output capacity
输出能量 output energy
输出能量滤波 output energy filtering
输出能量稳定度 output energy stability
输出泥沙总量 total sediment outflow
输出扭矩 output torque; torque output
输出扭矩分配器 output torque divider
输出耦合 output coupling
输出耦合系数 output coupling factor
输出耦合装置 output coupling device
输出排队 output queue; output work queue; work output queue
输出喷管 delivery jet pipe
输出喷嘴 delivery cone; delivery nozzle; discharge nozzle
输出片 output chip
输出偏振 output polarization
输出频率 load frequency; rated frequency
输出频率调节 load frequency control
输出品 export products
输出屏 output screen
输出器 follower; output unit
输出千瓦 kilowatt output
输出腔 outlet chamber
输出强度 output intensity
输出强制信号 output control signal
输出倾销 export dumping
输出区 output area; output block; output section
输出曲线 curve of output
输出曲线描绘台 output table
输出绕组 output winding
输出热量 heat output; quantity of heat given up
输出任选 output option
输出入连锁制 import-export link system
输出入相抵协定 import-export offset agreement
输出软管 delivery hose; outlet hose
输出塞孔 outgoing jack; output jack
输出三极管 output triode
输出设备 out-device; output device; output equipment; output unit
输出设备结构 output architecture
输出生产 export production
输出声功率 output acoustic(al)power
输出失业 exporting unemployment
输出时间系列 output time series
输出时序 output timing
输出矢量 output vector
输出收卡机 output stacker
输出输入控制程序 input output control system
输出束 output beam; output bundle
输出数据 outgoing data; output data; dataout
输出数据结构 output data structure
输出数据选通 output data strobe
输出数据组 output block
输出数位 output digit
输出数字 output digit
输出数字数据 output digital data
输出衰减 output attenuation
输出衰减器 output pad
输出水头 delivery head
输出税率 export tariff
输出顺序号 output sequence number
输出伺服机构 output servo-mechanism
输出速度 exit velocity; output speed
输出速率 output speed
输出台 output table
输出弹簧 delivery spring
输出探测器 output detector
输出套管 downstream end
输出特性 output characteristic
输出条 output bars
输出调节 output regulation
输出调节器 output controller; output governor; output regulator
输出调节器问题 output regulator problem
输出调整 regulations of output
输出通道 output channel
输出通量 efflux
输出推力 thrust output
输出瓦数 wattage output
输出外围控制 output peripheral control
输出网路 output network
输出网络 output network
输出网络节 output link
输出微分反馈 output derivative feedback
输出为先的对开信用证 back-to-back credit export first
输出位 carry-out bit
输出文件 out(put) file
输出稳定度 output stability
输出误差 output error
输出系数 discharge coefficient; output coefficient; output factor
输出系统 delivery system; output system
输出隙 output gap
输出狭缝 output slit
输出下限 bottoming
输出显示器 output display unit
输出显示区 output display area
输出显示设备 output display unit
输出线(路) outlet line; output line
输出线路滤波器 outlet line filter
输出线圈 output winding
输出线性部分 output linear group
输出限额 export quota
输出限制 export restriction
输出限制器 output limiter
输出效率 delivery efficiency; output efficiency
输出谐振电路 output resonant circuit
输出谐振腔隙 catcher gap
输出信道 delivery channel; output channel
输出信号 outcoming signal; outgoing signal; output signal; outlet signal
输出信号变化 output drift
输出信号波形 signal output waveform; signal waveform
输出信号的峰间电压 peak-to-peak output signal voltage
输出信号位数 bit of output signal
输出信号振幅 output amplitude
输出信息 output information
输出信息块 output block
输出信息量控制 output traffic control
输出信息组 output block
输出信用保险 export credit insurance
输出行程 delivery stroke
输出形式 output form
输出许可证 export license[licence]; export permit
输出选择器 outlet selector
输出压力 delivery pressure; discharge pressure; outlet pressure; output pressure
输出压力表 delivery ga(u)ge
输出压力转速传感器 speed-to-pressure transducer
输出延迟 output delay
输出扬程 outlet watershed
输出因数 output factor
输出因素 output factor
输出引线 output lead
输出印刷装置 output printer
输出语句【计】 output statement
输出元件 output element
输出运行时间 outgoing run time
输出载波 outgoing carrier
输出凿孔机 summary puncher
输出噪声 output noise
输出增量 output increment
输出增益级 output gain stage
输出者 exporter
输出阵列 output array
输出振幅 output amplitude
输出执行 output executive
输出值 output; readout【计】
输出值不稳定 output disturbance
输出纸带 output tape
输出纸带穿孔机 output tape punch
输出指令 output command; output instruction; output order
输出指示器 output indicator
输出制导信号 guidance output
输出中断 output break
输出中断寄存器 output interrupt register
输出中继线 output trunk
输出中止 output disable
输出种类 output class
输出轴 output axis; output shaft; take-off
输出轴防泥护罩 mud shield of output shaft
输出轴调速器 output shaft governor
输出轴凸缘 output flange
输出属性 output attribute
输出转矩 output torque; pull-out torque
输出转数 output revolutions
输出转速 output speed
输出装置 output device; output equipment; output unit
输出装置终装配 final assembly
输出子程序 output subroutine
输出字符串 output string
输出总线 output bus; storage-out bus
输出总线驱动器 output bus driver
输出总线校验 output bus check
输出阻抗 output impedance
输出组 output block
输出作业流 output job stream
输带辊 capstan roller
输带机构 paper-feeding mechanism
输带机组 conveyer train
输带轴 feed spool
输导 transfusion
输导薄壁组织 conducting parenchyma
输导砂设施 sand guiding and regulating device
输导系统 carrier system
输导组织 transfusion tissue
输导组织束 transfusion strand
输导作用 translocation
输电 electric(al) distribution; electric(al)(power) transmission; power transmission; transmission of electricity
输电成本 transmission cost
输电带 charge-carrying belt; charging belt
输电电杆 transmission mast for power
输电电缆 feeder cable; live wire
输电电压 transmission pressure
输电端 sending end
输电费用 transmission cost
输电干线 electric(al) main; power main
输电杆 transmission mast; transmission pole
输电(杆)塔 transmission tower
输电环 contact ring
输电及通信[讯]线路图 plan of power transmission and telecommunication
输电技术 electric(al) power transmission technique
输电链 charging chain
输电能力 transmission capacity; transmitting capacity; carrying capacity
输电频率 transmission frequency
输电曲线 transmission curve
输电容量 transmission capacity; transmitting capacity
输电设备 transmission equipment; transmission facility; transmission plant
输电设施 transmission facility
输电丝 transmission line
输电损耗 power transmission loss; transmission loss
输电损失 power transmission loss; transmission loss
输电损失增量 incremental transmission loss
输电塔 electric(al) transmission pole tower; power transmission tower
输电塔架 power transmission tower; transmission line pylon; transmission line tower
输电塔桅 power transmission tower; transmission line pylon; transmission line tower
输电铁塔 electric(al) power pylon
输电网 grid system; power transmission network; transmission grid; transmission network

输电系统 power transmission system; transmission system
输电线 conveyor cable; hot wire; live wire; living wire; power cable; power current transmission line; power line; power main; power transmission line; supply cable; electric (al) conductor; electricity transmission line
输电线的短分支 spur line
输电线电压 power-line voltage
输电线杆 transmission mast; transmission pole
输电线路 electric (al) transmission line; power line; power transmission line; transmission line
输电线路保护 line protection
输电线路保护装置 line protection
输电线路并行布置 parallel arrangement of transmission lines
输电线路测量 route survey for power transmission line
输电线路常数 line constant
输电线路干扰 power-line interference
输电线路塔 transmission line tower; transmission tower
输电线牌号 transmission line trademark
输电线塔架 power-line tower; transmission mast
输电线网 gridiron
输电线载波 power-line carrier
输电效率 efficiency of transmission; transmission efficiency; transmitting efficiency
输电仪表板 transmission panel
输电总开关 service entrance switch
输格 <电传机纸页向前> space feed
输给系统 logistics
输灰管真空破坏阀 ash vacuum breaker
输浆管路 grout pipe line
输浆器 slurry feeder
输金点 gold point
输砾河流 shingle carrying river
输料槽 trough conveyer[conveyor]
输料斗 charging magazine
输料管内径 internal diameter of transporting material tube
输料软管 conveying hose; material hose <受压混凝土在软管中输送>
输煤管 coal chute
输煤机 coal conveyer[conveyor]; coal transporter
输煤溜槽 coal chute
输煤设备 coal handling installation; coal handling plant
输煤设备耗电量 power to prepare coal
输煤装置 coal handling plant
输墨系统 inking system
输奶泵 milk lifter
输奶软管管路 flexible milk line
输奶系统 milk-transfer system
输泥管(道) slurry pipe; hydraulic fill pipeline; shore pipe; split line; split pipe; spoil line; mud pipe-line
输泥管浮体 mud pipe floater
输泥管接头 floating mud-pipe joint
输配电 transmission and distribution
输配电线 transmission and distribution line
输配水 water distribution and transmission
输片 transport film
输片齿轮 sprocket; sprocket gear
输片惰性轮 sprocket wheel
输片杆 film advance lever
输片鼓轮 film drum
输片机构 film transport
输片轮 film advancing wheel
输片速度 film velocity
输片系统 film transporting system
输片爪 pulldown claw
输漆软管 paint hose
输气管 air conduit; air manifold; appendix[复 appendices/appendixes]; gas pipeline
输气管道润滑器 boll-weevil lubricator
输气管规格 gas pipeline standards
输气管进入孔 appendix man hole
输气管丝锥 gas pipe tap
输气管调压阀 gas pressure regulator valve
输气管线 gas main
输气管线系统 transmission system
输气机 aerophore
输气孔 transfer port
输气量 carrying capacity
输气喷管 feed air nozzle
输气软管 air hose
输气损耗 line loss; transmission loss
输气系统 air delivery system; gas transmission system
输气压力 distribution pressure
输气英马力 air horsepower
输气支管 delivery pipe branch
输热 admission of heat
输热系数 heat-transfer coefficient
输入 carry-in; fan-in; impact; importation; infan; lead-in; loading; transfuse; import <指进口物品>
输入按钮 input button; load button
输入白噪声 input white noise
输入包 input packet
输入笔 stylus
输入边 input side
输入编辑 input editing
输入变换器 input translator
输入变量 input variable
输入变量名 input variable name
输入变数 input variable
输入变压器 input transformer
输入变压器式 input transformer type
输入变元 input argument
输入标题 input header
输入表 input list
输入表示码 incoming indication code
输入波 input wave
输入波动 input disturbance
输入补偿 input back-off
输入补偿电流 input offset current
输入补偿电流漂移 input offset current drift
输入补偿电压 input offset voltage
输入补偿电压漂移 input offset voltage drift
输入不足 under inlet
输入部件 input block
输入参考值 input reference
输入参数 input parameter
输入操作 input operation
输入侧 input side
输入测度 input measure
输入差错 loading error
输入产出模型 input-output model
输入程序 input program(me); input routine; loader routine; loading routine
输入程序片 load module
输入冲击脉冲 input stroke
输入储存 input store
输入储卡箱 input magazine
输入处理 input process(ing)
输入处理机 input processor
输入穿孔机 receiving punch
输入传感器 <发动机示波器等的> input pick-up
输入传输 incoming transmission
输入传送 input transfer
输入串 input string
输入窗口 input window
输入存储器 input storage
输入存储区 <存储器的> input block
输入错误 input error
输入代码 input code
输入带 input tape
输入带密度 input tape density
输入带排序 input tape sorting
输入带型号 input tape format
输入单元 input element
输入导纳 input admittance
输入的 inward
输入第一选择器 incoming first selector
输入点 enter point; entry point; input point; load point
输入电波 incoming wave
输入电导 input conductance
输入电极 input electrode; receiving electrode
输入电抗 input reactance
输入电流 entering flux; incoming current; input current; received current
输入电流曲线 arrival current curve
输入电路 incoming circuit; input channel; input circuit
输入电路噪声 input circuit noise
输入电平 incoming level; input level
输入电容 grid capacitance; input capacitance
输入电位计 input potentiometer
输入电压 input voltage
输入电压范围 input voltage range
输入电压滤波 input voltage filtration
输入电源绕组 input power winding
输入电阻 input resistance; input resistor
输入动量 entering momentum
输入端 input end; carry-in terminal; incoming terminal; input channel; input port; lead-in; receiving end; terminal; coil in; infan
输入端的等效噪声 equivalent input noise
输入端等效噪声电平 input absolute noise level
输入端电路 inlet circuit
输入端放大器 input amplifier
输入端封接 lead seal
输入端接口器 input interface adapter
输入端接线图 input wiring diagram
输入端绝对噪声电平 input absolute noise level
输入端数 fan-in
输入端阻抗 sending-end impedance
输入断言 input assertion
输入队列 input queue; input rank
输入队列溢出 input queue overflow
输入多路调制器 input multiplexer
输入额 input
输入扼流圈 input choke
输入阀 transfer valve
输入法规 law of import
输入翻译程序 input translator
输入反射系数 input reflection coefficient
输入反应性谱 input reactivity spectrum
输入方程 input equation
输入方式 load mode
输入放大器 input amplifier
输入分辨能力 input resolution
输入分布处理 input distributed processing
输入分程序 input block
输入分类 input category
输入分析 input analysis
输入符号 incoming symbol; input symbol
输入负荷 accepted load
输入负载率 input loading factor
输入感测 input sensing
输入钢筋直径 feeding diameter
输入港 port of entry
输入港船边交货价格 free overboard; free overside
输入港水上船边交货 overside delivery
输入格式 input format
输入格式控制 input format control
输入工作 input service; input work
输入工作队列 input work queue
输入工作排队 input work queue
输入功 input work
输入功率 input power; power input; receiving power
输入功率曲线 input power curve
输入功率绕组 input power winding
输入功率调节装置 fading unit
输入共存程序 input symbiont
输入共模电压摆幅 input common-mode voltage swing
输入共模范围 input common-mode range
输入共模抑制比 input common-mode rejection ratio
输入共振器 input resonator
输入管 admitting pipe; inlet pipe; inlet tube; input tube
输入管道 access duct
输入管道系统 input duct system
输入管线 inlet line
输入光束 input beam; input light beam
输入光学系统 fore optics system
输入规格 input specification
输入辊道 approach table; run-in table
输入过程 input procedure; input process(ing)
输入荷载 accepted load
输入呼叫识别 incoming call identification
输入环 input loop
输入环节 input element
输入缓冲程序 inbuffer
输入缓冲寄存器 input buffer register
输入缓冲器 input buffer
输入缓冲区 input block
输入灰岩 receptor limestone
输入回描 input flyback
输入机 input unit
输入机构 feed mechanism
输入机制 feed mechanism
输入基元 input primitive
输入基准值 input reference
输入基准轴 input reference axis
输入畸变 input skew
输入级 input stage
输入极限 input limit
输入集 input ensemble
输入计算机 input computer
输入计算机变量 input machine variable
输入继电器 input relay
输入寄存器 incoming register; input register
输入加工 input process(ing)
输入加载 input loading
输入加载式前置放大器 input loading type preamplifier
输入加载因数 input loading factor
输入价值 import value
输入键 enter key; entry key; load key
输入键盘 input keyboard
输入键请求 enter key request
输入奖励金 import bounty
输入交叉控制 input interleaving control
输入交流声电平 input hum level
输入角 input angle

输入角分布 angular distribution of input
输入校验设备 input checking equipment
输入阶跃 input-step
输入接卡箱 input stacker
输入接口 input interface
输入接线 input connection
输入接线端 outlet terminal
输入接线柱 lead-in clamp
输入节点 input node
输入结构 import structure
输入结束 end of input
输入解耦零点 input decoupling zero
输入解调器 input demodulator
输入借位 input borrow
输入进程 input process(ing)
输入进位 input carry
输入禁止器 input inhibitor
输入晶体 input crystal
输入井 input well
输入镜 input mirror
输入镜透射率 input mirror transmission
输入就业 import employment
输入矩阵 input matrix
输入卷数 number of input traces
输入卡片 input card
输入卡片箱 card hopper
输入开关 in switch
输入可控性 input controllability
输入客流 input passenger flow
输入空间 input space
输入空气 air-in;entrained air
输入空腔 input cavity
输入空腔谐振器 input cavity
输入控制 input control
输入控制器 input controller; input control unit
输入控制移位寄存器 input control shift register
输入口 input port
输入口岸 port of entry
输入库 input magazine
输入框 input mask
输入馈电线 incoming feeder
输入馈线 incoming feeder
输入扩展电路 input expander
输入扩展轨道 input expansion path
输入扩展路径 input expansion path
输入扩展器 input expander
输入劳动力 imported labo(u)r; import labour power
输入力矩 input moment;input torque
输入例程 input routine
输入例行测试 incoming routine test
输入例行程序 input routine
输入连接 inlet connection;input connection
输入连接点 incoming junction
输入连线 input link
输入量 input quantity;intake
输入量的希望值 desired input
输入量曲线 input curve
输入量与输出量的关系特性 input-output characteristic
输入列表格式 input listing format
输入流 inlet flow
输入流道 input channel;input duct
输入流类程 input stream class
输入流量 inlet flow rate; input flow rate
输入流量范围 input flow range
输入流量控制 input traffic control
输入漏斗 input hopper
输入漏泄 input leakage
输入滤波器 input filter
输入旅客 inward passenger
输入率 input rating;input ratio
输入逻辑 input logic
输入逻辑变量 input logic variable
输入马力 input horsepower
输入脉冲 incoming pulse; input-impulse;input pulse
输入脉冲幅度 input pulse amplitude
输入脉冲宽度 input pulse width
输入脉冲频率 input pulse frequency
输入脉动 input ripple;ripple in input
输入媒体 input medium
输入门 ingate;input gate
输入面 input face
输入模块 input module
输入模块阀 input module valve
输入模块库 load module library
输入目标制 import target
输入能 energy input
输入能力 input capability; input capacity
输入能量 input energy;intake
输入能量稳定度 input energy stability
输入能源效率比值 energy efficiency ratio
输入泥沙总量 input total sediment inflow;total sediment inflow
输入扭矩 input torque
输入浓度 input concentration
输入耦合器 input coupler
输入耦合因数 input coupling factor
输入排队 input(work)queue
输入喷嘴 input nozzle
输入片 input chip
输入偏振 input polarization
输入品 import
输入品需求函数 input demand function
输入屏 input screen
输入器 input unit;loader
输入千瓦 kilowatt input
输入前编辑 preinput editing
输入腔 inlet chamber
输入强度 input intensity
输入倾销 import dumping
输入区 input area
输入曲线 arrival curve; curve of input
输入绕组 input winding
输入热量 admission of heat; heat input; ingress of heat; quantity of heat supplied
输入任务 incoming task
输入任选 input option
输入入连锁制 import import link system
输入入相抵协定 import import offset agreement
输入软管 inlet hose
输入塞孔 input jack
输入扫描 input scan
输入设备 input device; input equipment;input unit
输入设备结构 input architecture
输入声功率 input acoustic(al)power
输入失配损耗 power mismatch loss
输入失调电流 input offset current
输入失业 importing unemployment
输入时间 input time
输入时间常数 input time constant
输入时间系列 input time series
输入时钟 input clock
输入识别 input identification
输入事务处理 input transaction processing
输入手续 process of import
输入受限系统 input-limited system
输入输出 input-output
输入输出标准接口 input-output standard interface
输入输出部件 input-output unit
输入输出操作 input-output operation
输入输出程序库 input-output library
输入输出程序系统 input-output programming system
输入输出处理机 input-output processor
输入输出处理器 input-output processor
输入输出存储区 input-output storage
输入输出单元 input-output unit
输入输出电传机 input-output teletype unit
输入输出电路板 input-output circuit board
输入输出端口 input-output port
输入输出对 input-output pair
输入输出方式 input-output mode
输入输出分析 input-output analysis
输入输出共同文件 combined file
输入输出关系 input-output relation
输入输出过程 input-output process
输入输出盒 in-out box
输入输出缓冲区 input-output buffer
输入输出缓冲通道 buffered input-output channel
输入输出记录介质 input-output recording medium
输入输出技术 input-output technique
输入输出寄存器 input-output register
输入输出接口 input-output interface
输入输出接口模块 input-output interface module
输入输出接口系统 input-output interface system
输入输出介质 input-output medium
输入输出控制 input-output control
输入输出控制程序 input-output control program(me)
输入输出控制器 input-output controller;input-output unit
输入输出控制系统 input output control system
输入输出控制装置 input-output control unit
输入输出块 in-out box
输入输出命令的链接 chaining of input/output commands
输入输出模型 input-output model
输入输出器 input-output unit
输入输出请求字 input-output request word
输入输出区 input-output area
输入输出设备 input-output device; input-output equipment
输入输出设备控制器 input-output device controller
输入输出适配器 input-output adapter
输入输出数据通道 input-output data channel
输入输出速度 input-output speed
输入输出特征 input-output characteristic
输入输出通道 input-output channel
输入输出相互关系定理 input-output cross correlation theorem
输入输出箱 in-out box
输入输出效率之比 input-output efficiency
输入输出语句 input-output statement
输入输出执行程序 input-output executive
输入输出指令 input-output command; input-output instruction
输入输出中断 input-output interrupt
输入输出终端 input-output termination
输入输出重叠 input-output overlap
输入输出装置 input-output device; input-output equipment; input-output medium;input-out unit
输入输出总线 input-output bus
输入束 input beam;input bundle
输入数 input number
输入数据 incoming data; input data; input information;data-in【计】
输入数据处理机 input data processor
输入数据的检验 input data proof
输入数据合法性确认 input data validation
输入数据集 input data set
输入数据结构 input data structure
输入数据库 load-data base
输入数据块 input block
输入数据误差 error of input data
输入数据系统 input data system; input system
输入数据选通(脉冲)input data strobe
输入数据指令 input data instruction
输入数据转换器 input data translator
输入数据总线 input data bus
输入数据组 input block
输入数字 input digit
输入数字数据 input digital data
输入衰减 input attenuation
输入衰减器 input attenuator;input pad
输入水 imported water
输入顺序号 input sequence number
输入速度 input speed;input velocity
输入速率 input rate
输入台 input table
输入探测器 input detector
输入特性 input characteristic
输入条 input bars
输入调节阀 input control valve
输入调制 input modulation
输入通带 input pass-band
输入通道 input channel
输入通量 influx;inward flux
输入通信[讯]量 incoming traffic
输入同步脉冲 incoming sync(hronization)pulse
输入头 input head
输入图像 input image;input picture
输入图形 tablet pattern
输入图元 input primitive
输入瓦数 wattage input
输入外围控制 input peripheral control
输入外资 incoming of foreign capital; introduce foreign capital
输入网络 input network
输入网络节 input link
输入微程序 load micro-program(me)
输入微分反馈 input derivative feedback
输入为先的对开信用证 back-to-back credit import first
输入位【计】carry in bit
输入位流 incoming bit stream
输入文件 input document;input file
输入文件标号处理 input file label handling
输入文件代码 input file code
输入稳定度 input stability
输入物镜 input objective
输入物种 input species
输入误差 input error;loading error
输入系数 input factor
输入系统 input system
输入隙 input gap
输入隙电压 input gap voltage
输入狭缝 input slit
输入下限 bottoming
输入衔接器 input adapter
输入显示设备 input display unit
输入线(路)incoming line;inlet line; input line;lead in wire
输入线路滤波器 inlet line filter
输入线圈 input winding
输入线性部分 input linear group
输入限额 import quota
输入限制 import restriction

输入限制电路 input-restricted channel
输入限制器 input limiter
输入相关 input correlation
输入向量 input vector
输入项目 entry item
输入小齿轮 input pinion
输入效率 input efficiency
输入谐振电路 input resonant circuit
输入谐振腔隙 buncher gap
输入信道 input channel
输入信号 incoming signal;input signal
输入信号变化 input drift
输入信号波形 signal input waveform
输入信号电平 input signal level
输入信号电压 input signal voltage
输入信号发生器 input signal generator
输入信号反相 input inversion
输入信号振幅 input amplitude
输入信号转换器 monitoring element
输入信息 incoming information;input information
输入信息块 input block
输入信息量控制 input traffic control
输入信息组 input block
输入信用保险 import credit insurance
输入压力 inlet pressure;input pressure
输入压气 incoming air stream
输入引导 bootstrap
输入引导子程序 bootstrap loader
输入语句【计】 input statement
输入元件 input element;receiver
输入源<信息> input well
输入运动 input motion
输入载波 incoming carrier
输入站【计】 reading station;transactor
输入值 initial value;input value
输入指令 entry instruction
输入指令码 input instruction code
输入中继电路 incoming trunk circuit
输入中继线 incoming trunk
输入终端 entry terminal
输入终端交换局 incoming terminal exchange
输入重新排列 input rearrangement
输入轴 input axis;input shaft;intake shaft
输入轴护罩 guard of input shaft
输入诸元 incoming data;input data
输入属性 input attribute
输入转换器 input converter
输入转速 input speed
输入装置 headend system;input device;input equipment;input unit
输入状态 input state
输入资料 input file
输入子模块 input submodule
输入子系统 input subsystem
输入字符 input character
输入总线 input bus;storage in bus
输入阻抗 feed impedance;input impedance
输入组 input block
输入组件 input module
输入作业队列 input job queue;input work queue
输入作业流 input job flow;input job stream
输入作用 input action
输沙<又称输砂> sanding;sediment transmission; sediment transport; silt transport; sand transport; silt-carrying
输沙比 sediment delivery ratio
输沙比率 specific rate of sediment transport
输沙措施 measure of translating sand
输沙措施装置 sand guiding and regulating device
输沙公式 sediment transport formula
输沙管 sand line;sand transport pipe
输沙过程 sediment delivery process; sediment transport process; transportation process
输沙河道 shingle carrying river
输沙机理 mechanism of sediment transport
输沙计 sand transport meter
输沙记录 sediment record
输沙廊道 gallery for sediment transport;gallery of sediment transport
输沙粒级 grain-size of sediment discharge
输沙量 amount of sediment transport; discharge of solids; flow of solid matter; nature load; sediment discharge (quantity); sediment flux; sediment load; sediment outflow; sediment runoff; silt discharge;solid discharge;solid flow; silt load;fine sediment load
输沙量测定 sediment discharge measurement
输沙量过程线 sediment hydrograph
输沙量连续相似 continual similarity of sediment discharge
输沙量流量关系 sediment-discharge rating;silt-discharging rating
输沙量流量关系曲线 sediment-discharge rating curve
输沙量率定 silt discharge rating
输沙量模数 modulus of silt discharge
输沙量曲线 sediment discharge curve; sediment-rating curve
输沙量相似 similarity of sediment discharge
输沙量沿程变化相似 similarity of sediment discharge
输沙量与流量的关系 sediment discharging rating
输沙-流量关系曲线 sediment-rating curve
输沙率 rate of sand motion; rate of sand movement; rate of sediment discharge; nature load; sand discharge;sediment delivery rate;sediment discharge rate; sediment flux; sediment flux rate; sediment rate; sediment transport concentration; sediment transport rate; silt discharge;silt load;solid discharge; solid flow;solid load-discharge ratio; specific rate of sediment transport
输沙率比例 scale of sediment transport
输沙率过程线 sediment hydrograph
输沙率流量比 load-discharge ratio
输沙率流量关系曲线 sediment-discharge rating curve
输沙模量 sediment runoff modulus
输沙能力 sediment transport capacity;sediment transport competence; sediment transport competency;silt carrying capacity;transport capacity; transport competence; transport competent
输沙浓度 sediment transport concentration;transport concentration
输沙平衡 balancing of sediment transportation; equilibrium of sediment transportation
输沙破坏研究 sediment transport balance study
输沙强度 sediment discharge intensity
输沙设备<绕道的> by passing plant;sand by-passing
输沙设施 sand discharge facility;sediment discharge facility
输水 conveyance;transmission of water;transport of water;water delivery; water-distribution; water transfer
输水比 conveyance ratio
输水成本 delivered cost
输水道 aqueduct; conduit; conveyer way; driving channel; head race; raceway;water conduit;water conveyance
输水道断面 conduit section
输水道拱 aqueduct arch
输水道孔 aqueduct arch
输水的 hydrophobic
输水洞工程地质勘查 engineering geologic (al) investigation of water tunnel
输水渡槽 canal aqueduct
输水阀(门) filling and emptying valve; water delivery valve;delivery valve
输水方法 water-carrier method
输水费用 delivered cost; handling charges
输水干管 delivery main;transmission main;water main;main water line
输水干管水表 mainline meter
输水高度 delivery lift;water delivery lift
输水沟 field ridge
输水管 water line;water pipe
输水管道 aqueduct; conduit; conduit pipe; delivery conduit; delivery pipe;raceway
输水管道末端压力水池 terminal reservoir
输水管道系统 conduit system
输水管阀 conduit valve
输水管接头 spud
输水管路 hydraulic(pipe)line
输水管配件 conduit fittings
输水管系统 water pipe line system
输水管线 conveyance system; delivery pipe line;transmission line
输水管之间的相互接通 cross connection
输水涵洞 conveyance culvert; filling culvert
输水建筑物 conveyance structure; water conveyance structure
输水胶管 water hose
输水结构物 conveyance structure
输水廊道 culvert;culvert gallery;filling and emptying culvert; water gallery
输水量曲线 demand curve
输水流道 flow passage
输水率 conveyance;conveyance factor
输水能力 carrying capacity; conveyance capacity; conveyance power of water;conveying capacity;transmissivity
输水桥 canal aqueduct;conduit bridge
输水渠(道) canal aqueduct; carrier drain; conveyance canal; conveying canal; conveying channel; delivery canal; driving channel; head race; transfer canal;conveyance channel
输水软管 delivery hose;water hose
输水设备 water delivery facility
输水设施 handling installation; water conveyance; water conveyance facility;water-conveying facility
输水时间曲线图 histograph
输水水头 delivery head;delivery lift; water delivery head;water delivery lift
输水隧道 conveyance tunnel; conveying tunnel; delivery tunnel; water-carriage tunnel;water carrying tunnel; water delivery tunnel; water (way) tunnel
输水隧道入口 water tunnel portal
输水隧洞 conveyance tunnel;convey(ing) tunnel; delivery tunnel; water carrying tunnel;water delivery tunnel;water(way) tunnel
输水隧洞入口 water tunnel portal
输水损失 conveyance loss; delivery loss;water delivery loss
输水损失率 rate of conveyance loss
输水系数 conveyance factor;conveyance ratio
输水系统 conveyance system; filling and emptying system; water-carriage system
输水线路长度 length of transporting water line
输水因素 conveyance factor
输水支管 delivery pipe branch
输水中断 interruption of throughput
输水主管 delivery main;water main
输水装置 humidifier
输水总管 delivery main
输送 conduction; convey (ing); deliver; entrain; feed (ing); movement; teaming; transfer; transport (tation)
输送板 delivery board
输送包 bull ladle
输送泵 delivery pump; discharge pump;transfer pump
输送泵安全阀 transmission pump relief valve
输送泵车间 transfer pumping unit
输送泵减压阀 transmission pump relief valve
输送标准 entrainment criterion
输送部件 transport unit
输送槽 conveying chute; conveying trough;conveyor trough
输送车 delivery wagon;transfer car
输送冲程 delivery stroke
输送船 transport ship
输送带 conveyer apron; conveyer [conveyor] band; conveyer[conveyor] belt (ing); conveyer draper; conveying belt; feed apron; feed belt;load transfer device;travel(l) ing apron
输送带秤 conveyer-type scale
输送带挡板 belt stop plate
输送带的保护器 conveyor belt protector
输送带地道 conveyor tunnel
输送带分批称量 proportioning by conveyor belt
输送带架 apron carrier
输送带接头 belt lacing
输送带控制器 conveyor belt control device
输送带宽度 belt width
输送带配料称量 proportioning by conveyor belt
输送带启动器 conveyor belt actuator
输送带倾角 delivery angle
输送带清扫器 belt cleaner
输送带式布面洗选机 cleaning apron
输送带式淬火槽 conveyer-type quench tank
输送带式干燥机 conveyer drier[dryer]
输送带式加热炉 conveyor type furnace
输送带式连续炉 conveyer-type continuous furnace
输送带式炉 conveyer furnace;travel(l)ing furnace
输送带式卸载机 canvas-apron unloader
输送带式逐稿器 raddle raker;raddle straw rack

输送带式装干草机 endless apron hay loader
输送带式装载机 conveyer loader
输送带速度 belt speed
输送带调位器 belt aligner
输送带停转机构 draper stop
输送带托辊 belt idler; carrier ilder; conveyer idler
输送带系统 belt system
输送带张紧装置 conveyer stretcher
输送带支承滚筒 belt supporting roller
输送带支持滚轮 conveyer idler roller
输送带支架 conveyer bridge
输送带装载机 conveyor loader
输送带阻塞地点 conveyor jam location
输送的 deferent
输送斗 conveyor bucket; transfer hopper
输送端螺栓和螺母 delivery bolt and nut
输送端阻抗 sending-end impedance
输送段 barrow-way; conveying track
输送阀 delivery valve; discharge valve; transfer valve
输送范围 range of throughput
输送范围换位杆 transmission range selector lever
输送范围换向杆 transmission range selector lever
输送方法 means of delivery
输送方式 model of transportation
输送方向 direction of throughput
输送费 pipage; transport(ation) expenses
输送风机 conveying fan
输送干管 delivery main
输送高度 discharge head; height of delivery
输送格框 crate
输送工具 means of delivery; transfer means
输送功率 transporting power
输送沟 duct
输送固体管线 pipeline to move solids
输送刮泥器 conveyor type scraper
输送管(道) conveying conduit; convey(ing) tube[tubing]; delivery conduit; delivery pipe; delivery tube; feeding conduit; running piping; transfer piping; conveying duct; conveying pipe(line); line; transmission pipeline; transport conduit; conveyer pipe; deferent; deferent duct; delivery fitting; duct; feed pipe; flow pipe; gathering line; supply pipe; tube conveyer[conveyor]
输送管的外涂层 exterior coating of pipe line
输送管路 delivery line; transfer line
输送管路图 piping layout
输送管线 conveying pipe line; delivery pipe line; line; line of pipes; pipeworks; transportation pipeline
输送管线断裂 line break; line check
输送管压力 line pressure
输送轨道 delivery track; feed track
输送辊道 <横行的> rollgang; travel(l)ing roller table
输送滚筒 feed roll
输送滑槽 delivery chute
输送机 conveyer[conveyor]; reclaimer
输送机变向装置 conveyer reverse unit
输送机承料槽 conveyer trough; conveying trough
输送机传动小齿轮 conveyer drive pinion
输送机传动轴衬套 conveyer drive shaft bush(ing)
输送机带支架 conveyor belt frame
输送机导向装置 conveyor guide means
输送机道 loft
输送机的搬运机构 transfer mechanism of conveyor system
输送机的头轮 head pulley
输送机端承 conveyer end bearing
输送机端承盖 conveyer end bearing cover
输送机端承盖衬套 conveyer end bearing cover bushing
输送机房 conveyor house; junction house
输送机构 conveying mechanism
输送机滚筒 conveyor roller
输送机滚柱 conveyer roller
输送机滑行托 conveyer slide support
输送机化 conveyerisation[conveyerization]
输送机回程皮带惰轮 return idler
输送机回动盖 conveyer reverse cover
输送机回送系统的自动起动装置 self-actuating return conveyor system
输送机回行扇形齿轮 conveyer reverse quadrant
输送机机架 belt transom
输送机棘轮 conveyer ratchet wheel
输送机棘爪盖 conveyer pawl casing cover
输送机棘爪回动法 conveyer pawl reverse
输送机棘爪回动销 conveyer pawl reverse pin
输送机计时装置 conveyor timing mechanism
输送机架腹杆 body rod for conveyer mounting
输送机胶带 travel(l)ing apron; travel(l)ing belt
输送机胶带托辊 idler
输送机进料板 feed apron
输送机廊道 conveyer gallery; transfer gallery of conveyer[conveyor]
输送机连接调整装置 conveyor junction regulating mechanism
输送机链 conveyer[conveyor] chain
输送机螺旋轴 conveyer screw shaft
输送机螺旋轴联轴节 conveyer screw shaft coupling
输送机螺旋主动轴 conveyer screw drive shaft
输送机煤斗支座 conveyer hopper support
输送机门 conveyor door
输送机挠性主动轴 conveyer flexible drive shaft
输送机式喷砂机 conveyer-type sand blast machine
输送机式提升机 conveyor elevator
输送机塔架 loading tower
输送机停止按钮 conveyor stop
输送机托辊 conveyor idler
输送机托辊组合 conveyor idler assembly
输送机窝球节 conveyer ball joint
输送机系统 conveyer system
输送机线路 conveyer line; conveyor route
输送机卸料器 belt tripper; conveyer belt tripper
输送机械 conveying machinery
输送机用减速机 reducer for conveyer[conveyor]
输送机张紧滚筒 tension drum of conveyer[conveyor]
输送机支架 conveyer bridge
输送机支架结构 conveyor frame structure
输送机主动棘轮 conveyer drive ratchet wheel
输送机主动轴 conveyer drive shaft
输送机主动轴端盖 conveyer drive shaft end cover
输送机装料漏斗 conveyer loading hopper
输送机装卸 cargo-handling by conveyer[conveyor]
输送机自动步道 travelator
输送机组 conveyer train
输送计划 displacement plan
输送检验 feed check
输送浇包 transfer ladle
输送校正器调节机 doctor
输送介质 conveying medium; pumped medium
输送井 delivery shaft
输送距离 conveying distance; transporting distance
输送孔 sprocket hole; transfer port; feed hole【计】
输送孔距 feed pitch
输送口 delivery port
输送冷石油 running of cold oil
输送链 carrier chain; conveyer chain; elevating chain
输送链保持装置 conveyor chain hold-down means
输送链和轨道的组合体 conveyor chain and track-way assembly
输送链式饲料分送器 bunk feeder with chain conveyer[conveyor]
输送链式挖掘机 web-conveyor digger
输送链装料站 chain loading station
输送量 conveyed quantity; conveying capacity; pumpability; transport capacity
输送量调节装置 proportioner
输送料斗 conveyer[conveyor] bucket; conveying bucket
输送(料斗)索道 conveying line
输送溜槽 conveying chute
输送流速 transportation velocity; transporting velocity
输送流体 conveyance fluid
输送率 rate of delivery; rate of transport; transfer rate
输送螺旋 conveying screw; delivery auger; endless screw
输送螺旋止推板 conveyer screw thrust plate
输送煤斗承板 conveyer hopper bearing plate
输送模型 transportation model
输送能力 carrying capacity; conveying capacity; deliverability; delivery capacity; delivery value; displacement capacity; displacement transmissivity; traffic (-carrying) capacity; transmissivity; transport competence[competency]; transporting capacity; transporting power
输送能量 convey(er) energy; conveying energy
输送黏[粘]油的内螺纹管 <约有三米长的螺纹> rifled pipe
输送皮带架 conveyer boom
输送起重机 transshipment crane
输送器 conveyer[conveyor]; feeder; forwarder; towveyor; transporter; transport unit; transveyer
输送器传动轴 conveyer drive shaft
输送器传动装置 conveyer drive
输送器回动爪 reverse conveyer pawl
输送器架 conveyer boom
输送器漏斗 conveyer hopper
输送器皮带轮 conveyer pulley
输送器前进爪 forward conveyer pawl
输送器式干燥炉 apron drier[dryer]; conveyer drier[dryer]
输送器型刮泥机 conveyor type scraper
输送器轴 conveyer axle
输送器爪式链节 lug link
输送情况 state of delivery
输送渠 duct
输送渠道 delivery conduit
输送热风 heat air delivery
输送容积 delivery volume
输送容器 transport box
输送软管 conveying hose; delivery hose; material hose <输送灌浆、砂浆等材料的>; delivery hose <喷射混凝土、泵送混凝土或砂浆的>
输送设备 conveyance device; conveyancer; conveyer[conveyor]; conveyer equipment; conveying equipment; conveying facility; conveying plant
输送式刮泥机 conveyor type scraper
输送水量 delivery discharge
输送水头 delivery head; delivery lift
输送速度 conveying speed; delivery rate; transporting velocity
输送隧道 convey tunnel; delivery tunnel
输送隧洞 convey tunnel; delivery tunnel
输送损失 conveyance loss
输送提升联合机 convelater[convelator]
输送提升系统 transportation hoist system
输送托辊 conveyor idler
输送蜗杆 conveying worm
输送物 deferent
输送系统 conveyer system; conveying system; delivery system
输送线 conveying line; delivery line
输送线路图 flow plan
输送相 delivery phase
输送斜槽 conveying chute
输送卸槽 feed chute
输送压力 delivery pressure; discharge pressure; feed pressure; head pressure
输送压头 delivery head lift; discharge head
输送延迟部件 transport delay unit
输送延迟装置 transport delay unit
输送液体性质 pumped fluid characteristics
输送引进水体 delivering imported water
输送油 transferring oil
输送轧碎机 conveyer crusher
输送站 dispatch station; flowing plant
输送制动 travel brake
输送中的产品 product in transit
输送主管 flow main
输送装置 conveying appliance; delivery mechanism; feedway; handler; transport unit
输送状态 feed status
输送总管 delivery main
输索绞盘 messenger winch
输氧呼吸器 oxygen-breathing apparatus
输液管 perfusion tube
输移比 delivery ratio
输移沉积物 transported deposit
输移冲刷流速 transporting erosive velocity
输移率 transfer rate; transport rate
输移模型 <废水、污染质等的> transport model; transshipment model

输移滞后距离 transport relaxation distance
输油 oil transportation
输油泵 cargo oil pump; oil-line pump; oil transfer pump; transfer pump
输油泵房 oil transfer pump house
输油臂 <油码头> oil-loading arm
输油管 oil conveying pipe; oil delivery pipe; oil duct; oil pipe(line); petroleum pipeline; pipe line
输油管泵 oil-line pump
输油管道 oil conveying pipeline; oil pipeline; pipe line
输油管道测量 petroleum pipeline survey
输油管伸缩接头 expansion pipe joint
输油管水面浮运法 <铺设海底管线的> surface floatation method
输油管网 flow circuit
输油管系(统) cargo pipe line; oil conveying piping system; oil transport system
输油管线 oil-feed line; oil(pipe) line; transmission line
输油管信道 <码头上的> pipeway
输油管压力 flowline pressure
输油加热器 suction heater
输油胶管 cargo oil hose
输油轮 shuttle tanker
输油潜水艇 oil-carrying submarine; refueling submarine
输油软管 flexible cargo hose; oil (conveying) hose; shore installation hose; oil loading hose
输油设备 oil transfer equipment
输油输气管道 oil and gas pipeline
输油塔 distributing column
输油系统 oil transfer system
输油线 carrier line
输油循环总管 main circulating loop
输油证 certificate of clearance; clearance of certificate
输油终点站 terminal station
输运 haulage
输运方式 mode of transport
输运过程 transport process
输运平均自由程 transport mean free path
输运气体 carrier gas; transport gas
输运因数 transport factor
输渣水 <运送矿渣用的水> transportation water
输纸 paper feed
输纸机 paper guide
输纸机构 paper advance mechanism; paper transport mechanism
输纸孔 feed hole
输纸轮 tape feed spindle
输纸器 paper transport
输纸系统 paper transport system
输纸装置 extension delivery; paper moving device
输种管 seed spout

蔬

蔬菜 garden shed; garden stuff; green stuff; truck crop; vegetable wax; vegetal

蔬菜地窖 vegetable cellar
蔬菜碟 plant bowl
蔬菜废料 vegetable waste
蔬菜废物 vegetable waste
蔬菜分级机 vegetable sorter
蔬菜耕作 <美> truck farming
蔬菜加工 vegetable processing
蔬菜加工间 vegetable preparation room
蔬菜加工业 vegetable industry
蔬菜蜡 vegetable wax
蔬菜农场 truck farm; truck garden
蔬菜配制室 vegetable preparation room
蔬菜水果店 green grocery
蔬菜水果加工废物 vegetable and fruit processing wastes
蔬菜挖掘器 vegetable lifter
蔬菜洗涤池 vegetable preparation tank
蔬菜园艺 <美> market gardening
蔬菜栽培研究所 Vegetable Farming Institute
蔬菜作物 vegetable crop

赎

赎单 retire a bill; retire shipping documents

赎换代理 replacement agent
赎回 call; redeem; redemption; retirement
赎回抵押品 redeem mortgage
赎回抵押权 equity of redemption
赎回典契 redemption of mortgage
赎回合同 contract of ransom
赎回价格 call price
赎回价值 redemption value
赎回扣押权的费用 release price
赎回票据 bills retired
赎回期 redemption period
赎回溢价 call premium; redemption premium
赎回债券溢价 call premium on bonds
赎买权 right of redemption
赎票 retire a bill; retiring a bill; take-up a bill
赎票率 retirement rate
赎票申请书 application of retiring bill
赎债基金 redemption fund
赎债价格 call price

熟

熟白云石 magnefer

熟材 imperfect heart wood; imperfect ripe wood; ripe wood
熟材树种 tree with imperfect heartwood
熟成法 curing process
熟成鼓 ripened drum
熟成过程 <沥青乳胶> curing process
熟成能力 <沥青乳液> curing power
熟成期 <沥青乳胶> curing time; curing period
熟成速率 <沥青乳胶> curing rate
熟床【给】 ripe bed
熟大漆 boiled urushi
熟的 boiled
熟地 cultivated soil; long-cultivated field; long cultivated land; old arable soil; old land
熟地翻耕 ploughing of cultivated land
熟地犁铧 stubble share
熟读的 deep read
熟粉 prepared powder
熟腐殖质 mull
熟钢 wrought steel
熟钢管 wrought-steel pipe
熟化 age; ag(e)ing; conditioning; cure; curing; fattening up; maturing; ripening; slaking <石灰的>
熟化槽 digester; digestion tank
熟化层 anthropic epipedon
熟化程度 amount of cure; curing degree
熟化程序 degree of cure
熟化池 maturation pond; maturing bin; slaking pan; boiling tub <石灰>
熟化的 slack
熟化法 maturation process; maturing process
熟化方法 cure system
熟化规程 cure schedule
熟化过程 ag(e)ing process
熟化过滤器 ripened filter
熟化后 <指石灰> after slake
熟化混凝土 matured concrete
熟化机理 curing mechanism
熟化剂 curing agent
熟化减慢 cure retardation
熟化黏[粘]度 aged viscosity
熟化期 ag(e)ing period; conditioning period; curing time; maturation period
熟化器 ager; digester[digestor]; rapid steam ager
熟化区 curing area
熟化石灰 slake
熟化石灰螺旋输送器 slaking screw
熟化时间 ag(e)ing time; cure time; curing time; time of slaking
熟化室 curing chamber; curing room
熟化树脂 cured resin
熟化速率 <沥青乳胶> curing rate; cure rate
熟化塘 maturation pond
熟化条件 ag(e)ing condition
熟化土层 agric
熟化土壤 tilth top soil; vegetable soil
熟化温度 curing temperature; maturing temperature
熟化污泥 ripe sludge
熟化箱 maturing bin
熟化性能 slaking behavio(u)r
熟化周期 cure cycle
熟化装置 cure system
熟化阻滞剂 cure retarder
熟荒地 idle land
熟黄铜 wrought brass
熟焦油 boiled tar
熟练 dexterity; proficiency; skill
熟练程度 degree of skillfulness; qualification; skill level
熟练的 boiled; experienced; proficient; ripe; skilled; skil(l)ful; workmanlike
熟练的程序设计员 skilled programmer
熟练的管理人员 expert manager
熟练的混凝土工人 skilled concrete worker
熟练的手艺 expert worksmanship
熟练电工 journeyman electrician
熟练工(人) experienced worker; full-fledged worker; hot short; old hand; skilled labo(u)r(er); skilled worker; veteran worker; journeyman; skilled man
熟练工作 skilled work
熟练和半熟练劳动力 skilled and semi-skilled labo(u)r
熟练技工 journey-man; skilled manpower; skilled technical personnel; skilled worker
熟练技能 skilled craft
熟练技师 master mechanic
熟练技艺 skilled craft
熟练劳动(工作) skilled occupation
熟练劳动力 skilled labo(u)r(er)
熟练沥青工 potman
熟练泥瓦工 master mason
熟练炮工 shotfirer
熟练人员 skilled man; skilled staff
熟练石工 stone mason
熟练水平测试 proficiency testing
熟练水手 able(-bodied) seaman
熟练圬工 master mason
熟练行业 expertise[expertize]
熟练油漆工 journeyman painter
熟练者 practitioner
熟炼油 boiled oil
熟炼油漆 boiled oil paint
熟料 burned filler; clinker; clinker aggregate; grog
熟料标号 strength grading of clinker
熟料层 clinker bed
熟料产量 clinker output; yield of clinker
熟料成球 balling of clinker
熟料秤 clinker scale
熟料出口 clinker outlet
熟料储库 clinker storage(building)
熟料大块 caked clinker
熟料带走热 clinker
熟料的快速冷却 rapid cooling of clinker
熟料的升重测定 liter weight test of clinker
熟料的易磨性 grindability of clinker
熟料斗 clinker hopper
熟料煅烧 clinker burning
熟料堆场 clinker pile; clinker stockpiling plant
熟料堆存设施 clinker stockpiling facility
熟料堆棚 clinker storage shed
熟料对煤灰的吸收 absorption of ash
熟料粉尘 clinker dust
熟料粉磨 clinker grinding
熟料粉磨机 clinker grinding mill
熟料粉磨试验 clinker grinding test
熟料过筛 grading of grog
熟料含量 <水泥中的> clinker content
熟料含率 cullet adding
熟料活性 reactivity of clinker
熟料活性曲线 clinker activity curve
熟料基体 clinker matrix
熟料急冷 quenching of clinker
熟料结大块 clinker sausaging
熟料结块 caked clinker
熟料结粒 clinker grain
熟料结圈 clinker ring
熟料坑 clinker pit
熟料孔隙度 clinker porosity
熟料库 clinker storage hall
熟料快速冷却 rapid clinker cooling
熟料矿物 clinker mineral
熟料拉链机 clinker drag chain conveyer[conveyor]
熟料冷却机 clinker cooler
熟料冷却机效率 clinker cooler efficiency
熟料冷却器 clinker cooler
熟料料斗 clinker hopper
熟料溜槽 clinker chute
熟料螺旋输送机 clinker screw
熟料磨 clinker mill
熟料磨料机 grog mill
熟料磨碎机 grog mill
熟料耐火材料 grog refractory
熟料耐火黏[粘]土 grog fireclay mortar
熟料排出温度 exit temperature of clinker
熟料棚 clinker storage hall
熟料漂白 clinker bleaching
熟料破碎 clinker reduction
熟料破碎机 clinker breaker; clinker crusher
熟料破碎机坑 clinker crusher pit
熟料球 clinker nodule; nodule of clinker
熟料圈 hot zone ring
熟料圈的圈料 materials of clinker ring
熟料热耗 heat consumption of clinker
熟料散布器 clinker spreader
熟料砂 chamot(te) sand
熟料烧成 clinker burning; clinker sintering

熟料升重 liter weight of clinker
熟料升重自动测量仪 automatic measuring instrument for liter-weight of clinker
熟料生成热 heat of clinker formation
熟料输送机 clinker conveyer[conveyor];clinker handling conveyer[conveyor]
熟料水化形成的氢氧化钙 portlandite
熟料水泥 clinker cement
熟料水泥砖 clinker cement brick
熟料松密度 clinker bulk density
熟料显微结构 microstructure of clinker
熟料相 clinker phase
熟料卸出带走的热损失 clinker discharge loss
熟料卸料闸门 clinker discharge gate
熟料形成 clinker formation
熟料形成带 clinker-forming zone
熟料形成反应 reaction of clinker formation
熟料循环过程 clinker recirculating process
熟料颜色 colo(u)r of clinker
熟料液相 clinker liquid
熟料余热 clinker waste heat
熟料中熔融物 liquid in clinker
熟料骤冷 quenching of clinker
熟料骤冷冷却机 quenching clinker cooler
熟料贮库 clinker pit;clinker storage (building)
熟料转窑 chamot(te) rotary kiln
熟料组分 clinker constitute
熟铝 wrought alumin(i)um
熟铝合金 alumin(i)um wrought alloy;wrought alumin(i)um alloy
熟耐火黏[粘]土 chamot(te)
熟耐火土 chamot(te)
熟黏[粘]土 burned clay
熟膨胀率 thermal expansibility
熟石膏 calcined gypsum; calcined plaster; calcium sulphate hemihydrate;ga(u)ging plaster;hemi-hydrate;Paris plaster;plaster(of Paris)
熟石膏拌合[和]物 plaster stuff
熟石膏仓库 plaster store
熟石膏粉 dry hydrate
熟石膏灰泥 hemi-hydrate plaster; plaster of Paris mortar
熟石膏浆 hemi-hydrate plaster
熟石膏浆找平层 screed of plaster of Paris
熟石膏平缘 fillet of plaster of Paris
熟石膏饰面 plaster of Paris finish
熟石灰 calcium hydrate;calcium hydroxide; caustic lime; drowned lime;ga(u)ging plaster;hydralime; hydrated lime;hydrate of lime;hydrous lime; lime white; limoid; slack lime;slaked lime;water lime; white chalk lime;white lime
熟石灰粉 dry hydrate
熟石灰空气消化石灰 air slaked lime
熟食店 delicatessen;delicatessen store
熟手 old hand
熟树脂 resinoid
熟思 deliberate
熟苏籽油 boiled perilla oil
熟塘 maturate pond
熟铁 bushelled iron; dug-iron; forge(d) iron;knobbled iron;puddle(d) iron; puddle steel; soft iron; wrought iron
熟铁板 boster;wrought iron plate
熟铁棒材 < 搅炼炉产品 > puddled bar
熟铁扁条 muck;muck bar
熟铁扁条粗轧机 muck mill
熟铁扁条束 muck pile
熟铁扁条轧辊 muck roll
熟铁扁条轧机 muck-rolling mill
熟铁插销 wrought iron bolt
熟铁成球 balling
熟铁初轧扁条 muck bar
熟铁初轧条 millbar
熟铁窗栅 wrought iron window grille
熟铁吹炼法 bloomery process
熟铁吹炼炉 bloomery
熟铁锤炼 nobbing
熟铁粗轧坯 < 挤制熟铁轧成品 > muck bar
熟铁吊钩 wrought iron hook
熟铁废料 wrought iron scrap
熟铁管 forged pipe;wrought iron conduit;wrought iron pipe;wrought iron piping;wrought iron tube;wrought iron tubing;wrought pipe
熟铁光车刀 wrought iron finishing turning tool
熟铁块 bloomery iron;loup
熟铁块吹炼法 bloomary process
熟铁链条 wrought iron chain
熟铁坯挤压 spindling;spingling
熟铁坯小头 ancony
熟铁片式锅炉 wrought iron sectional boiler
熟铁条 plated bar;muck bar < 熟铁搅炼炉的最初产物 >
熟铁弯头 wrought iron bend
熟铁斜三通 wrought iron wye
熟铁渣 hearth cinder
熟铁轧机 puddle rolling mill
熟桐油 boiled oil; boiled tung oil; boiled wood oil; heat-bodied tung oil;tung stand oil; wood oil stand oil
熟桐油打底漆 primer oil
熟铜 wrought brass;wrought copper; Muntz(metal)
熟土 anthropic soil; cultivated soil; fermented soil;mature soil;mellow earth;mellow soil;ripe soil
熟污泥 ripe sludge
熟污渣 ripe sludge
熟悉 acquaintanceship; acquaintance with; acquaint oneself with; conversance;up in(on)
熟悉船上帆缆业务 know the rope
熟悉汇兑的人 cambist
熟亚麻仁油 boiled linseed oil;boiled oil
熟亚麻子油 bodied linseed oil
熟油 boiled oil;kettle-boiled oil;polymerized oil;stand oil
熟油光泽涂料 stand oil gloss paint
熟油含量 stand oil content
熟油壶 stand oil kettle;stand oil pot
熟油漆 stand oil paint
熟釉 fritted glaze
熟釉瓷砖 frit-glaze tile
熟渣 matured slag
熟枝插条 mature cutting
熟制耐火材料 grog refractory

暑

暑季混凝土施工 hot-weather concreting

暑天混凝土施工 hot weathering concreting
暑天浇灌混凝土 hot application of concrete
暑湾 sinus aestuum

黍

黍 panic grass

署

署名 affix one's signature; autograph;signature

署名人 the undersigned;undersigned

鼠

鼠 rat

鼠标(器) mouse
鼠巢式测辐射热计 rat's nest type bolometer
鼠道 mole channel
鼠道排水管 mole channel
鼠道式地下排水道 mole drain
鼠道式排水(沟) mole drainage
鼠洞 mouse hole
鼠洞管道排水 mole pipe drainage
鼠洞机 moling machine
鼠洞犁 mole plough
鼠洞排水 mole drainage
鼠洞式孔 coyote hole;gopher hole
鼠洞式排水道 mole channel
鼠洞式排水管 mole drain
鼠灰色 slate gray[grey]
鼠忌避剂 rat repellent;rodent repellent
鼠李属 buckthorn
鼠笼 rotor cage
鼠笼电枢 short-circuited armature
鼠笼端环 cage ring
鼠笼绕组 squirrel-cage winding
鼠笼式 squirrel-cage
鼠笼式打捞器 mouse-trap
鼠笼式灯丝 squirrel-cage type filament
鼠笼式电动机 cage motor; mouse-mill motor; squirrel-cage motor; squirrel motor
鼠笼式电枢 squirrel-cage armature
鼠笼式粉碎机 squirrel-cage disintegrator
鼠笼式风机 squirrel-cage fan
鼠笼式风扇 sirocco fan;squirrel-cage type fan
鼠笼式感应电动机 squirrel-cage induction motor
鼠笼式感应型 squirrel-cage induction type
鼠笼式鼓风机 squirrel-cage blower
鼠笼式磨碎机 squirrel-cage mill
鼠笼式绕组 cage winding; squirrel-cage winding
鼠笼式三相异步牵引电动机 three-phase squirrel-cage asynchronous traction motor
鼠笼式碎解机 squirrel-cage disintegrator
鼠笼式推斥电动机 squirrel-cage repulsion motor
鼠笼式转子 cage rotor;squirrel-cage rotor
鼠笼天线 cage antenna
鼠笼条 squirrel-cage bar
鼠笼形平衡机 squirrel-cage balancing machine
鼠笼形栅极 squirrel-cage grid
鼠色 mouse;mouse-colo(u)r
鼠尾 rat tail
鼠尾锉 rat tail(ed) file
鼠尾喷灯 rat-tail burner
鼠尾弹簧 door closer-spring
鼠尾形 squirrel tail
鼠尾形管子连接器 squirrel-tall pipe jointer
鼠穴 mouse hole
鼠咬 damaged by rats
鼠疫 black death;pestilence;plague

蜀

蜀黍 sorghum

蜀柱 kingpost

曙

曙粉红色 dawn pink

曙光 morning twilight
曙光的 auroral
曙光宫 Palace of the Dawn
曙光红 < 一种淡红色染料 > eosine; tetrabromofluore scein
曙光石 eosphorite
曙光效应 dawn effect
曙红 bromeosin
曙红色淀 eosine lake
曙暮光 twilight
曙暮光发射 twilight emission
曙暮光改正 twilight correction
曙暮光光谱 twilight spectrum
曙暮光弧 arch twilight
曙暮光亮度 twilight brightness
曙暮光区 twilight zone
曙暮辉 crepuscular ray
曙暮辉光 crepuscular ray
曙暮辉弧 crepuscular arch
曙暮色 dusk

属

属 genus

属的【生】 generic
属地 dependency
属地管辖 territorial jurisdiction
属具 accessory; appendage; appurtenance;attachment
属模标本 genotype;lygotype
属人职权 personal competence
属性 attribute
属性抽样 attribute sampling;sampling of attributes
属性单元 template;templet
属性的 attributive
属性方案模型【交】 attribute-alternative model
属性关系 relation on attributes
属性检测器 property detector
属性检验法 method of attributes
属性取样 cluster sampling
属性缺省(规划) default for attribute
属性设计模型 attribute-based model
属性特征 attribute property
属性统计 statistics of attributes
属性文法 attribute grammar
属性因子分解 attribute factoring
属性值 property value
属于水底的 demersal
属于银合成的 argental

术

术语 scientific terminology; buzz word;jargon;onomasticon;technicality; technical language; technical term; technical word; technics;technology

术语标准 terminology standard
术语标准化 standardization of terminology
术语表 nomenclature
术语汇集 technology
术语学 terminology

束

束比 beam ratio

束箔光谱学 beam-foil spectroscopy
束捕获栅 beam catcher
束捕集器 beam catcher
束不稳定性 beam instability
束槽 shackle
束柴 fascine bundle
束柴作业 fascine work
束成型 beam shaping
束存储器 < 光束或电子束 > beam stor-

age
束带 bridle;lace;spanner band
束带层 oversailing course; string course;tabling;belt course <墙上凸出的装饰层>
束带的 belted
束带滑车 rigger
束带式惯性导航设备 gimballess inertial navigation equipment;strapped-down inertial navigation equipment
束带式铰链 band and gudgeon hinge;band and hook hinge; hook-and-band hinge
束带饰 taenia
束带柱 laced column
束道 beam trace
束的角宽 angular beam width
束电流 beam current
束电流比 ratio of gun currents
束电流杂波 beam noise
束电位 fasciculation potential
束发划小蠹 <拉> Scolytoplatypus supersiliosus
束发散 misconvergence of beams
束发散角 beam divergence angle
束反向 beam reversal
束沸石 desmine;radiated zeolite;stilbite
束分离器 beam separator
束缚 bondage;captivity;colligate;colligation; crimp; girt(h); restrain;stricture;tether;yoke;tie-down
束缚波 forced wave
束缚电荷 blowout electricity;bound charge;bound electricity
束缚电子 bound electron
束缚辐射 imprisoned radiation
束缚钢筋 bound reinforcement
束缚孔隙度 bound porosity
束缚模 bound mode
束缚能 binding energy;binding power;bound energy
束缚散射 bound scattering
束缚生长激素 bound auxin
束缚试验 lock test
束缚水 adsorption water; adsorptive water; attached water; bound water;combined water;fixed groundwater;held water;hygroscopic water
束缚水饱和度 irreducible bound water saturation
束缚水分 fixed moisture
束缚态 bound state
束缚涡流 bound vortex
束缚线 lashing wire
束缚向量 bound vector; localized vector
束钢 bridle iron;fag(g)ot steel
束箍 tie hoop
束谷机 gleaner
束管 beam tube
束光阑 beam diaphragm
束光缆 bundle cable
束会聚 beam convergence
束集 beam convergence;constriction
束几何 beam geometry
束校正 beam alignment
束节式取土器 ringed-line barrel sampler; thin-walled shoe and barrel sampler
束结装订 kettlestitch
束截面 area of beam
束筋 bundle reinforcement
束紧 astringe;lacing
束紧的金属线 lacing wire
束紧皮绳 lacing leather
束紧如瓶 bottle tight
束紧用具 lacer
束径迹 beam trace
束聚焦 beam focusing
束聚焦电压 beam-focus voltage
束开关 beam switch
束孔 beam hole
束孔径 beam orifice
束控 beam control;shading
束控换能器 shaded transducer
束口防波堤 converging breakwater;converging mole;converging pier
束口模 reducing die
束口式调压塔 restricted entry type of surge tank; restricted orifice surge tank
束宽 beam width
束扩展望远镜 expanding telescope
束梁 gird
束亮度 beam brightness
束磷钙铀矿 phurcalite
束流 beam current
束流坝 training wall
束流残像 beam current lag
束流成分 beam component
束流放大器 beam current amplifier
束流分离器 beam separator
束流分配装置 beam switch yard
束流负载周期 beam duty cycle
束流刚度 beam rigidity
束流光学 beam optics
束流光学装置 beam-optics arrangement
束流计量学 beam dosimetry
束流监测变量器 beam-measuring transformer
束流交叉角 beam crossing angle
束流截面的长半轴 semi-major axis of beam cross-section
束流径向控制 radial control of beam
束流径向位置 radial beam position
束流控制 beam handling;beam steering
束流控制系统 beam-control(led) system
束流脉冲 beam burst
束流脉冲重复率 beam pulse rate
束流模型 beam model
束流喷射 beam spraying
束流偏转管 beam deflection valve
束流品质 quality of beam
束流破坏 beam destructure
束流扫描器 beam scanner
束流试验 beam practice
束流收集器 beam dump
束流输送系统组件 beam component
束流输运系统 beam guide;beam-optics arrangement
束流输运线 beam corridor
束流调定 setting-up of beam
束流调制 modulation of beam
束流显示装置 beam sensing system
束流限制器 beam clipper
束流消失 beam breakup
束流性质 nature of beam
束流中心 beam center[centre]
束流重心 beam centroid
束流坐标 beam coordinate
束木 bond timber
束木基础 fascine foundation
束木支架 fascine cradle
束内观察 intrabeam viewing
束偏转 beam deflection;beam deflexion;beam steering
束平行光管 beam collimator
束剖面图 beam profile
束强度 beam intensity
束强监测器 beam monitor
束倾角 beam tilt angle
束热速度 beam thermal velocity
束锐化 beam sharpening
束射功率管 beam power tube;novar
束射五极管 beam pentode
束绳钩 rope binding hook
束石 bonder;bond stone
束石层 course of bondstones
束式光缆组件 bundle cable assembly
束式斜拉桥 bundle(type) cable-stayed bridge
束手无策 shrivel
束熟铁扁条 muck bar piling
束树 fascine wood
束水坝 contracting dam
束水堤 side pier
束水工程 constriction project; contraction works
束水攻沙 clearing sands with converged[converging] flow; scouring by contraction works; sediment flushing by contraction works
束水归槽 converging flow into main channel
束水孔 contracted opening
束缩堰 weir with contraction
束套式钢丝绳吊具 bridle wire rope sling
束调制百分数 beam modulation percentage
束调制度 beam modulation percentage;percentage beam modulation
束铁 fag(g)ot;fag(g)oted iron
束通量 beam flux
束筒 bundled tube;modular tube
束筒体系 bundled tube system
束位移 beam displacement
束稳定化 beam stabilization
束狭冲刷 constriction scour
束狭段 constricted section
束狭工程 contraction works
束狭河道 contracted channel; restricted channel
束狭急流 neck current
束狭渠道 contracting duct;restricted channel
束狭渠道水流 contracting duct flow
束狭水道 restricted waterway
束线 wire harness
束线机 buncher;bunching machine
束限制孔径 beam-limiting aperture
束消隐 beam blanking;beam suppression
束效应 <声、光等的> beaming effect
束形放电 bunch discharge
束腰 girdling
束腰板 frieze panel
束腰螺栓 waisted bolt
束腰竖沟 glyph
束扎 bundle
束窄的航道 constricted channel;contracted channel;contracted fairway
束窄段 constricted reach;constricted section
束窄工程 constricting works;contraction works
束窄河床 contracting riverbed
束窄设施 constricting works
束窄水道 constricted channel; contracted channel; contracted waterway
束窄水流 constricted flow;contracted flow
束支 bundle branch
束直径 beam diameter
束住 engird(le)
束柱 clustered columns
束柱带 cimbia
束砖 bonding brick;tie brick
束装运输 containerized transport
束状 fascicular;fasciculate
束状变晶结构 bunchy blastic texture
束状波 beam wave
束状电缆 bunched cable
束状接头 cluster joint
束状结构 bunchy texture; sheaflike structure
束状马氏体 packet martensite
束状纹 bundle finishing
束准直仪 beam collimator
束着屏误差 beam-landing screen error

述

述录音机 dictaphone[dictophone]

树

树疤 burl
树白蚁 <拉> Neotermes sinensis
树槽 tree wall
树草漂浮物 sudd
树层 tree layer
树叉木 crotchwood
树杈 crotch
树查廊道 inspection gallery
树池保护格栅 tree grate
树丛 boscage; bosk(et); bosquet; clump; coppice (wood); copse; grove;lucus;spinn(e)y
树丛繁茂的 scrubby
树丛剪修 selective thinning
树丛群落 alsium
树丛植物 alsad
树丛滞流 <河道> tree retards
树带 tree belt
树的遍历 traverse of tree;tree traversal
树顶 tree top
树墩 stump;tree stump
树多项式 tree polynomial
树分类【计】tree classification
树蜂 giant wood wasp; horntails; wood wasp
树复制 tree copy
树干 bole; branch of tree; raddle; stock;tree trunk;trunk
树干保护套栏 tree guard
树干材 stem timber
树干的 truncal
树干的表皮 outer covering of tree trunk
树干地面直径 diameter at ground level
树干风裂 wind shake
树干腐朽病 trunk tree
树干火 stand fire
树干级木 stem class
树干绿化 trunk greening
树干梢径 top diameter
树干形数 stem form factor
树干枝条 trunk shoot
树杆制矿用梯子 monkey ladder
树高 height of tree;tree height
树高测定 hypsometric(al) measurement
树高测定器 hypsometer
树高测量 tree height measurement
树高度 height of tree
树高公式 tree height formula
树高级木 height class
树根 stub
树根表 table of values
树根薄板 butt veneer
树根和块石抓机 root and rock rake
树根和石块清除器 root and rock rake
树根掘除 stumping off
树根孔洞 root hole
树根耙 root rake
树根清理工 grubber
树根挖掘器 root topper
树根桩 root pile;stump
树挂 air hoar;rime
树冠 coma;crown;tree crown;tree top
树冠矮化层 dwarfed crown
树冠层 crown canopy;crown layer
树冠长度 crown length

树冠长势过重 top heaviness
树冠滴水 crown drip
树冠覆盖 crown cover
树冠覆盖度 crown cover degree
树冠覆盖密度 density of canopy
树冠覆盖面 tree canopy
树冠厚(度) crown depth;length
树冠火 crowning fire
树冠级 crown class
树冠级干 crown class
树冠截留 canopy interception;crown interception
树冠宽度 width of tree
树冠率 crown percent(ratio)
树冠茂密的树木 canopy tree
树冠密度 crown density
树冠面 crown cross-section
树冠面积指数 crown-area index
树冠喷灌机 overtree sprinkler
树冠平均直径 crown mean diameter
树冠蔷薇 tree rose
树冠疏枝 setting free the crown
树冠线 skyline
树冠效应 foliage effect
树冠郁闭 canopy;crown closure
树冠郁闭度 crown density
树冠遮蔽度 crown closure
树冠直径量尺 tree crown scale
树号 blaze
树基材 butt log
树胶 gum;gum resin;liquid resin;vegetable glue;wood gum;xylan(e)
树胶桉 kino eucalyptus
树胶斑纹 gum vein
树胶板 resin plate
树胶产品 gum naval stores
树胶带 gum belt
树胶道 gum passage
树胶的 resinous
树胶的采集 gumming
树胶分泌 gumming
树胶管 gum canal;gum duct
树胶含量 gum content
树胶结合剂 gum cement; resinoid bond
树胶精油 gum spirits of turpentine; gum turpentine
树胶脉纹 gum vein
树胶密封层 gummed sealing tape
树胶囊 gum pocket
树胶黏[粘]合剂 resinoid bond
树胶黏[粘]条 gum tape
树胶圈 gutta-percha ring
树胶熔炼 gum running;gum-solution
树胶软管 gum hose
树胶塞 gum plug
树胶树脂 gum resin;resinoid
树胶树脂产品 gum naval stores
树胶水彩画法 gouache
树胶水彩画颜料 gouache
树胶水彩颜料 gouache paint
树胶松香 resin
树胶条纹 gum streak
树胶印画法 gum process
树胶汁 gutta percha
树胶制的 gummy
树胶质 gumminess
树胶状 gumminess
树胶状残渣 gummy residue
树胶状沉淀 gummy residue
树胶状的 gummy;gummy appearance
树胶状结构 gummy formation
树节 <木材缺陷> branched knot; burl;knop
树节点 tree node
树节洞 hollow knot
树景 treescape
树景大理岩地 forest marble
树景整治 vista clearing
树径 tree-walk
树径尺 tree caliper;tree cal(l)pers
树径记录仪 dendrograph
树矩阵 tree matrix
树锯 tree saw
树坑 tree well
树蜡 vegetable wax
树篱(笆) hedgerow;brush hurdle;espalier;green fence;hedge;live enclosure; live fence; quick fence; quick hedge;quickset;raddle;tree wall;green fence
树篱带 screen planting strip
树篱修剪器 hedge trimmer
树立 uprear
树立标志 sign posting
树立觇标 erect beacon;signal erection
树立者 erector
树列深景 vista
树林 coed;forest;grove;hurst;woodland;woods
树林的 woodland
树林培植 arboriculture
树林石 polymnite
树林植被 forest cover
树龄 tree age
树瘤 burl;tree wart;wart
树瘤材 burr wood;knaggy wood
树瘤纹 brittle heart;burl;knot
树轮测年法 tree ring dating method
树码 tree code
树码译码 tree code decoding
树码字 tree code word
树苗 plant; sapling; seedling; tree seedlings;young plant
树苗床 tree nursery
树苗和小树 seedlings and small trees
树苗圃 tree nursery
树苗挖掘机 tree balling machine
树苗育种 tree breeding
树模 tree mo(u)ld
树模式 tree schema
树木 crop of formed trees; trees; woody plant
树木矮化法 nanigation
树木保护 greenery conservation;protection of tree
树木保护装置 tree guard
树木被车辆撞断阻力 tree resistance
树木被砍光的 cut-over
树木被砍光的空地 cut-over
树木病毒 arbovirus
树木病害 tree disease
树木剥皮 debark
树木不整形期 formless stage
树木材积表 tree volume table
树木打枝剪 dresser
树木档案库 tree archive
树木的 arboreal;arboreous
树木的髓心 pith
树木地面直径 ground level diameter
树木调查 tree census
树木顶梢枯死 dieback
树木毒杀剂 arboricides;silvicide
树木多的 silvan
树木繁茂的 wooded
树木分类 tree classification
树木根部膨大 root swelling
树木根部特大的 swell-butted
树木根围 rhizosphere
树木刮痕器 scratcher
树木挂淤法 <控制河槽时用> tree retard method
树木很密的地 heavily wooded area
树木护栏 tree guard
树木护理 tree maintenance
树木化石 dendrolite
树木环形带 girdling
树木基端防腐处理 butt-end treatment
树木夹皮 ingrown bark
树木间隔 spacing of trees
树木剪切 tree shears
树木剪枝 lop
树木节瘤 plethora
树木界限 tree limit
树木景观 treescape
树木老化 ag(e)ing of trees
树木类型辨认 tree type identification
树木茂盛的 arboreous;woody
树木苗圃 nursery for trees
树木耐阴性 tolerance
树木年代学 dendrochronology
树木年轮 annual ring(of timber);annular ring;growth ring;tree ring
树木黏[粘]料 tree-glue
树木配植 arrangement of trees and shrubs
树木群落 xylium
树木生长 arboreal growth
树木生长限界 <指海拔高限界> forest line
树木生态观察 tree survey
树木石南 tree heath
树木实积 solid content
树木示范小区 demonstration plot
树木碎裂 tree burst
树木挖根清出场地 clearing and grubbing
树木围栅 tree grate
树木线 timber line;tree line
树木线以上地区 alpestrine
树木限界 <纬度限界> timber limit; tree limit
树木心材腐烂 heart rot
树木形数 tree form-factor
树木修剪术 topiary art
树木修整 reshaping of trees
树木学 dendrology;dendrometry
树木学家 arborist
树木硬节 knurl;knur(r)
树木有用材价值部分 merchantable
树木园 arboretum
树木栽培学 arboriculture
树木在一个生长季节长成的圈(层) growth ring
树木轧碎机 tree crusher
树木支柱 tree prop
树木枝条 branches of trees
树木注射 tree injection
树木注射器 tree injector
树木阻塞 tree entanglement
树内皮 bass
树年轮 tree ring
树配电方式 tree system
树棚 living-tree pergola
树皮 bark;cortex[复 cortices/cortexes];rind
树皮剥刀 barking iron
树皮剥脱 bark slip
树皮布 barkcloth;tapa cloth
树皮残植煤 bark liptobiolith
树皮堵漏处理 tree bark
树皮粉尘 bark dust
树皮割痕 hack
树皮光滑的 smooth-barked
树皮规 bark ga(u)ge
树皮画 bark picture
树皮焦油 wood bark tar
树皮扣除率 bark allowance
树皮拉毛饰面 flattened stipple finish
树皮煤 bark coal
树皮内的直径 diameter inside bark
树皮鞣革 bark
树皮鞣革厂 bark tannery
树皮鞣料 bark tan(ner);tanbark
树皮上粗糙带鳞状的表面 ross
树皮碎裂机 bark breaker
树皮碎屑堵漏剂 <商品名> Control Fiber
树皮脱离 bark slipping
树皮外的直径 diameter outside bark
树皮纤维堵漏剂 cedar seal
树栖的 arboreous
树墙 espalier
树群 woodlot
树删除程序 tree deletion
树上的苔藓 tree moss
树上附生植物 epiphta arboricosa
树上生活的 arboreal
树上雾滴 fog drip
树梢 forest fascine;tree top
树梢坝 brush dam
树舌 shelf fungus
树舌属 <拉> Ganoderma
树身 bole
树石松 tree clubmoss
树石竹 tree pink
树搜索 tree search
树算法 tree algorithm
树髓 pitch
树塔形的 excurrent
树图 tree diagraph;tree graph
树文法 tree grammar
树下矮灌丛群落 sleganochamephytium
树下矮灌木 underplant
树下草地 under growth
树下喷水 under-tree sprinkling
树下植被 under growth
树线 timber line;tree line
树心 pith
树心板 center[centre] plank
树心节 pith knot
树形 tree form
树形笔石目 Dendroidea
树形布局 tree topology
树形层次 tree hierarchy
树形重叠 overlay tree
树形登记项 tree entry
树形电路 tree circuit
树形发生电压 tree initiation voltage
树形分支结构 tree and branch architecture
树形结构 tree-like structure
树形结构计算 tree computation of height
树形结构简化算法 tree height reduction algorithm
树形矩阵 tree matrix
树形决策图 decision tree
树形毛巾架 towel tree
树形排序 tree ordering;tree sorting
树形配线 wiring tree
树形数据库管理系统 tree database management system
树形图 arborescence; dendrogram; tree derivation;tree diagram;trellis diagram
树形网络 tree network
树形网络分派信息处理机 tree allocated processor
树形系统 tree system
树形相位差型式 tree-offset pattern
树形寻优法 tree search method
树穴 bark pocket; inbark; ingrown bark;plant pit
树叶 leaf(age)
树叶覆盖 foliage cover
树叶枯死 leaf necrosis
树叶茂盛的 leafy
树叶状装饰 branched work
树液 sap
树液置代法 <木材防腐> boucherizing
树液置换处理 <木材防腐> sap displacement treatment
树艺学 arboriculture
树荫 umbrage
树荫处 bower
树罂粟 tree poppy
树元组团 collection of tuples of trees

S

树汁色 sap green
树枝 branch(wood);twig
树枝编织的 pleached
树枝虫胶 stick lac
树枝防护网 brush guard
树枝分枝 arborization
树枝覆膜砂 resin coated sand
树枝挂油法 tree retard method
树枝挂淤 tree retards
树枝建筑<一种以树为屋脊,枝为屋面的原始建筑> cruck construction
树枝晶 fir-tree crystal
树枝配电方式 tree system
树枝梢捆 brush roll
树枝石 dendrite
树枝石的 dendritic
树枝式(水管)网 ramified system
树枝式系统 tree system
树枝束 kid
树枝条立式沙障 twing mattress sand-break
树枝席排 mat of bush
树枝形 dendritic pattern
树枝形的 dendriform;dendritic
树枝形晶体 arborescent crystal;dendrite
树枝形排水系统 dendritic drainage;dendritic drainage system
树枝形配电方式 tree system
树枝扎捆装置 plant binder
树枝状 arborization;dendroid
树枝状冰川 dendritic glacier
树枝状冰晶 dendritic crystal
树枝状的 arborescent;dendritic;dendroidal;tree-like
树枝状地下水系 branch work
树枝状洞穴 branch work
树枝状断层 diverging fault
树枝状粉末 dendritic powder
树枝状高分子 dendrimer
树枝状构形 dendritic topology
树枝状构造 dendritic structure
树枝状谷 dendritic valleys
树枝状管网 branch-off pipeline network;ramified pipe system;tree-branch pipeline;tree pipe system
树枝状管网系统 tree system
树枝状河谷 dendritic(al)valley
树枝状河网 arborescent river network;tree-like drainage
树枝状河网系统 dendritic(al)drainage(system)
树枝状混合岩 ramification migmatite
树枝状击穿 treeing breakdown
树枝状结构 dendritic structure
树枝状结晶 dendritic crystallization
树枝状晶体 arborescent crystal;dendrite;dendrite crystal;dendritic crystal;fernleaf crystal;fir-tree crystal;pine-tree crystal;treeing;tree-like crystal
树枝状晶体簇 cluster of dendrites
树枝状巨晶<钢锭结构缺陷> ingotism
树枝状模式 dendritic drainage pattern
树枝状排水系(统) dendritic drainage system;arborescent drainage pattern
树枝状配水系统 tree-type distribution system
树枝状偏析 dendritic segregation
树枝状石灰华 dendritic tufa;dendroid tufa
树枝状水系 arborescent drainage;dendritic drainage;dendritic drainage system;dendritic river system
树枝状图 arborescence
树枝状装饰 dendritic decoration;dendritic dressing;rinceau
树枝状组织 dendritic structure;treeing
树脂 colophonium;colophony;jaffaite;mastic;peucine;pitch;rosin;wood gum
树脂柏油(沥青)resinous tar
树脂斑 pitch streak(ing)
树脂板 resin panel;resin plate
树脂包封 resin-encapsulate
树脂包裹颜料颗粒 resin encased pigment particle
树脂饱和的 resin saturated
树脂变质 resin depletion
树脂表面处理 resinous surface treatment
树脂表面处治 resinous surface treatment
树脂表面处治法 resinous surface treating method
树脂表面木材 resin-treated wood
树脂表面墙纸 resin-faced building paper
树脂表面形成的乙醇 resin-forming alcohol
树脂玻璃 Plexiglass;resin glass
树脂玻璃界面 resin-glass interface
树脂玻璃救生船 resin-glass lifeboat
树脂薄膜 resin film;resin sheet
树脂薄膜罩面层 resin sheet overlay
树脂薄片 resin flake
树脂捕捉器 resin(ous)trapper
树脂采集器 resin tapper
树脂采收 resin tapping
树脂残渣 gum residue
树脂残植煤 resin liptobiolith
树脂层 resin bed;resin layer
树脂产品 naval stores
树脂衬里 resin-lining
树脂成分 resin constituent;resinous composition
树脂处理胶合板 resin treated plywood
树脂处理木材 resin-treated wood
树脂传递模塑成型 resin transfer mo(u)lding
树脂醇 resin alcohol;resinol
树脂醇类 resin alcohols
树脂丹宁醇 resinotannol
树脂道 resin canal;resin duct;resin passage
树脂的 gummy;resinaceous
树脂滴点 resin dropping point
树脂点滴试验 resin spot test
树脂点滴试验法 resin spot test method
树脂电 resinous electricity
树脂度 resinousness
树脂凡立水 resin clear varnish
树脂反应釜 resin kettle
树脂反应锅 reaction vessel
树脂反应器 resin reaction vessel
树脂防水胶粘剂 resinous waterproof adhesive
树脂分批人浆法 resin-in-pulp process
树脂分散体 resin dispersion
树脂酚 resinol
树脂粉末 resin powder
树脂缝 pitch seam;resin seam
树脂覆膜 resin coating;resin precoating
树脂改良水泥 resin-modified cement
树脂改性的水泥混凝土 resin-modified cement concrete
树脂改性混凝土 resin concrete
树脂改性木材 resin modified wood
树脂改性水泥 resin-modified cement
树脂膏 resin plaster
树脂工艺学 resin technology
树脂沟道 resin channel
树脂固定销 resin-anchored bolt
树脂固化玻璃棉 uncured glass wool
树脂固化剂 resin type curing agent
树脂管 pitch tube;resin tube
树脂灌浆 resin grout(ing)
树脂锅 resin kettle
树脂含量 proportion of resin present;resin content
树脂合金 resin alloy
树脂褐煤 resinous lignite
树脂痕 pitch seam
树脂烘焙机 resin curing machine
树脂糊 resin paste
树脂化合物 resin compound
树脂化混凝土 resinification concrete
树脂化剂 resinifying agent
树脂化作用 resinification
树脂灰泥 resin putty
树脂混合料 resin compound
树脂混凝土 resin concrete
树脂混配 resin compounding
树脂机械损耗 physical resin loss
树脂基 resin base
树脂基薄膜养护 resin-based membrane curing
树脂基薄涂层 thin resin-base coating
树脂基的 resin-based
树脂基灰浆 resin-based mortar
树脂基灰泥 resin-based putty
树脂基混凝土 resin-based concrete
树脂基混凝土养护剂 resin-based concrete curing agent;resin-based concrete curing compound
树脂基胶结料 artificial bonding adhesive;artificial bonding agent
树脂基胶粘剂 resin-based adhesive;resin-based cement
树脂基腻子 resin-based putty
树脂基黏[粘]合剂 resin-based adhesive;resin-based binder
树脂基黏[粘]结料 artificial bonding adhesive;artificial bonding agent
树脂基嵌缝料 resin-base ca(u)lking compound
树脂基涂膜 resin-based liquid membrane
树脂基养护剂 resin-base curing compound
树脂集聚 resin-rich spot
树脂加工 resin treatment
树脂浆 paste resin
树脂浆液 resin grout
树脂降解 resin degradation
树脂交换剂 resinous exchanger
树脂交换容量 resin exchange capacity
树脂交换柱 resin-column
树脂交联度 cross-linking of resin
树脂浇注型变压器 resin molded transformer
树脂胶合 resin bonding;resinoid bond;resinous bond
树脂胶合板 campoboard;compo board;resin-bonded plywood;resin-bonded slab
树脂胶合的 resin-bonded
树脂胶合剂 resin cement
树脂胶合塞 resin plug
树脂胶化批量 monkey
树脂胶夹衬板 rosin sized sheathing
树脂胶结层 resin bonding layer
树脂胶结木屑板 resin-bonded chipboard;resin chip board
树脂胶结软木嵌缝料 resin-bound cork filler
树脂胶结软木填缝材料 resin-bound cork filler
树脂胶结砂轮 resinoid wheel
树脂胶结碎木板 resin-bonded chipboard;resin chip board
树脂胶泥芯 resin mortar core
树脂胶凝材料 bakelite adhesive;bakelite cement
树脂胶粘剂 resin adhesive
树脂搅动罐 resin surge pot
树脂节 pitch knot
树脂结合的 resin bonding
树脂结合剂 resinous bond;resinous cement
树脂结合剂砂带 resin belt
树脂结合剂砂轮 resin bond wheel
树脂结合耐火材料 resin-bonded refractory
树脂结合砂轮 resin-bonded wheel;resinoid wheel
树脂浸透纸面胶合板 resin-impregnated paper-faced plywood
树脂浸析 resin leading
树脂浸制多孔陶瓷 resin-impregnated porous ceramics
树脂浸制陶瓷 resin-impregnated ceramics
树脂浸注木材 impreg;resin-impregnated wood
树脂浸渍的 resin-impregnated
树脂浸渍木 resin-treated wood
树脂浸渍木材 impreg;resin-impregnated wood
树脂浸渍湿增强材料 wet layup
树脂浸渍洗槽 resin impregnation bath
树脂浸渍线圈 resin-impregnated coil
树脂浸渍增强材料 resin-impregnated reinforced material;layup
树脂精 resin spirit
树脂阱 resinous trap
树脂镜质体 resocollinite
树脂聚合物防潮层 pitch-polymer damp course
树脂绝缘绕组 resin-insulated winding
树脂菌类体 resino-sclerotinite
树脂科 terebinthaceae
树脂颗粒 resin particle
树脂壳模铸造 resin shell mold casting
树脂孔脂孔 pitch pocket
树脂矿浆吸附法 resin-in-pulp process
树脂蜡 resin wax
树脂类 resinoid
树脂类材料 resinous material
树脂类稳定剂 resinous stabilizer
树脂沥青 antracen
树脂沥青黄麻皮电缆 pitch bitumen jute cable[PBJ cable]
树脂裂 pitch shake
树脂硫 resin sulfur
树脂硫化 resin cure
树脂瘤 resin gall;resinous cancer
树脂锚杆 resin bolt
树脂锚固杆 resin anchor
树脂锚固锚杆法 resin-anchored bolting method
树脂锚固锚杆法 resin-anchored bolt
树脂锚固栓 resin anchor
树脂煤 resinous coal
树脂煤素质 resinite
树脂密封元件 resin cast component
树脂棉板 resin-bonded board
树脂棉毡 resin-bonded felt
树脂膜电极 resin-membrane electrode
树脂磨损 resin wearing
树脂母料 resin concentrate
树脂木 candle wood
树脂木屑板 resin chip board
树脂囊 black check;black streak;pitch pocket;resin duct;resinocyst;resin pocket;resin cyst<木材的>
树脂腻子 resin cement;resinous putty;resin putty
树脂黏[粘]度 resin viscosity
树脂黏[粘]合 resin bonding
树脂黏[粘]合的 resin-bound

树脂黏[粘]合剂 resin binder
树脂黏[粘]合螺栓 resin bonding bolt
树脂黏[粘]合线圈 resin-bonded coil
树脂黏[粘]胶 resin glue
树脂黏[粘]接剂 resin glue
树脂黏[粘]结 resin bond
树脂黏[粘]结的层压板 Resweld
树脂黏[粘]结的磁铁 resin-bonded magnet
树脂黏[粘]结剂 resin glue; resinoid bond; resin adhesive
树脂黏[粘]结砂轮 resinoid wheel
树脂配合石膏灰浆 resin-gypsum plaster
树脂配合体 resin-ligand
树脂喷枪 resin gun
树脂片 chips
树脂坡莫合金 resin permalloy
树脂漆 lacquer type organic coating; resin (-emulsion) paint; resin lacquer
树脂腔 resin cavity
树脂清漆 resin clear varnish; resin varnish
树脂容量 resin capacity
树脂溶剂 resin solvent
树脂溶液 resin solution
树脂溶胀 resin swelling
树脂乳胶漆 resin-emulsion paint
树脂乳漆 resin emulsion paint
树脂乳液 resin latex; resinous emulsion
树脂乳液涂料 resin-emulsion paint
树脂乳状液 resin emulsion
树脂润滑脂 rosin grease
树脂砂 resin-bonded sand
树脂砂衬离心铸管法 Monocast process
树脂砂覆砂造型 permanent-backed resin shell process
树脂砂浆 resin mortar
树脂砂浆胶合剂 binder for resin mortar
树脂砂芯黏[粘]结剂 plastic core binder
树脂渗出 bleed-out
树脂渗出条痕 resin streak
树脂渗溢 resinosis
树脂生产废水 resin manufacturing wastewater
树脂石 retinite
树脂石膏胶结材料 resin-gypsum cement
树脂式锚杆 bolt of resinification
树脂适应期 resin pot life; resin working life
树脂收集器 resin trap
树脂寿命 resin life
树脂水泥 resin cement
树脂素 resene
树脂酸 resin(ous) acid
树脂酸苄酯 benzyl resinate
树脂酸钙 calcium resinate
树脂酸铝 alumin(i)um resinate
树脂酸锰 manganese resinate
树脂酸钠 sodium resinate
树脂酸铅 lead resinate
树脂酸色淀 resin acid lake
树脂酸树脂 resinolic acid resin
树脂酸铁 ferric resinate; iron resinate
树脂酸铜 copper resinate
树脂酸锌 zinc resinate
树脂酸盐 resinate
树脂酸盐光泽彩 resinate lustre
树脂酸盐类 resinates
树脂酸钴 cobalt resinate
树脂髓线 pitch ray
树脂碎木板 resin chip board
树脂羧基铁粉 poly-iron
树脂陶瓷结合剂 resinoid and vitrified bond
树脂提浓物 resin concentrate
树脂体 resinite
树脂添加剂掺和机 resin additive blender
树脂填充 resin filling; resin impregnation
树脂填充料 resin matrix
树脂条痕 pitch streak(ing)
树脂条纹 pitch streak(ing); resin streak
树脂涂层 resin coat; resin coating
树脂涂敷 resin coating
树脂涂覆辊 resin applicator roller
树脂涂料 cold coating; resin-emulsion paint
树脂味 resinous odo(u)r
树脂纹 pitch streak(ing)
树脂污染 resin fouling
树脂吸附工艺 resin adsorption process
树脂吸附剂 resin adsorbent
树脂吸附剂芬顿试剂氧化法 resin adsorbent-Fenton reagent oxidation process
树脂系腻子 plastic putty
树脂系清漆 resinous varnish
树脂系统 resin system
树脂系涂料 resinous varnish
树脂纤维齿轮 fiber gear
树脂纤维护面层胶合板 overlaid plywood
树脂显微照相 resinography
树脂屑 resin-lint
树脂心节 pitch knot
树脂形成 gum formation
树脂型胶粘剂 resin adhesive
树脂型锚杆 resin rock bolt
树脂型黏[粘]合剂 resin adhesive
树脂型酸 resinoid acid
树脂型物 resinoid
树脂型芯黏[粘]结剂 plastic core binder; resin core binder
树脂型增塑剂 resinous plasticizer
树脂性 resinousness
树脂性瓷漆 varnish paint
树脂性的 peucinous; resinaceous
树脂悬浮液 resin suspension
树脂悬融 resin suspension
树脂压力注射成型 resin injection process
树脂压力注射成型机 resin injection machine
树脂颜料比 resin-pigment ratio
树脂油 resineon; resin oil
树脂油灰 resinous putty
树脂油墨 resin ink
树脂油桐 gumlactree
树脂淤积 resin pocket
树脂原 resinogen
树脂再生 resin regeneration
树脂皂 resin soap
树脂增量剂 resin extender
树脂增塑剂 resin plasticizer
树脂整理 resin finish(ing)
树脂支承 resin support
树脂酯 resin ester
树脂质的 resinous
树脂质(煤岩) resinite
树脂质油 resinous oil
树脂致白 gum blushing; resin blushing
树脂致雾 gum bloom
树脂中毒 resin poison
树脂注射模塑 resin injection mo(u)lding
树脂贮存期 resin shelf life; resin storage life
树脂柱式吸附法 resin-in-column
树脂转模拼装工艺 resin transfer molding fabrication process
树脂状沉淀物 resin
树脂状断口 resinous fracture
树脂状多元醇 resinous polyol
树脂状光泽 resinous luster[lustre]
树脂状聚合物 resinous polymer
树脂状物质 resinous material; resinous matter; resinous substance
树脂状杂质 resinous impurity
树脂状增塑剂 resinous plasticizer
树脂组 resinoid group
树指示字 tree pointer
树质 arborescence
树种 tree species
树种规划 planning of trees and shrubs
树种试种小区 species trial plot
树种选择 tree species selection
树桩 stub; stump; tree stump
树桩拔除机 stumper; stump puller; stump remover; tree stumper
树桩基础 stump foundation
树桩脉序 dendroid venation
树桩木 buttwood
树桩饰面板 stump veneer
树桩撕碎机 stump splitter
树桩挖除机 stump-clearing digger
树桩挖掘机 stub puller
树桩围培 stump fence
树状 arborescence
树状表达式 tree representation
树状的 arboreal; arboreous; arboroid
树状二元型 tree binary form
树状分级 dendrachy
树状分析图 relevance tree
树状隔丝 dendrophysis
树状构形 tree topology
树状结构 tree structure
树状结构的 dendriform; dendritic
树状晶体 tree crystalization
树状框视图 treeframed view
树状水系 tree-like drainage
树状搜索法 star method search method; star-search method
树状图(表) dendrogram
树状图解 tree diagram
树状转移表 tree-like transition table
树状自动机 tree automaton

竖

竖板 mullion; riser(board); vertical plate

竖板高度 riser height
竖板型造波机 vertical board type wave generator
竖板栅栏 pale fence; paling fence
竖板桩 vertical sheeting
竖背式送话器 solid back transmitter
竖比例 scale of height
竖臂式挖沟机 vertical boom ditcher; vertical boom trenching machine
竖标距 ordinate
竖标示牌 sign post
竖波 vertical wave
竖槽隔墙 chase partition
竖槽间 interglyphe
竖槽接槎 chase bonding
竖槽式跌水井 stand chute type drop-manhole; stand chute type drop-well
竖槽凿 scribing gouge
竖插销 vertical bolt; vertical cat bar
竖撑 soldier; sway brace
竖撑条 vertical batter stud
竖尺 vertical bar; vertical staff
竖尺员 rodman
竖冲舂粉机 vertical impact pulverizer
竖窗 vertical window
竖窗框 stanchion
竖挡板 day joint; stunt end
竖的 endway; upright; vertical
竖地旗杆 ground set flagpole
竖吊杆 vertical hanger; hip vertical < 指承受局部横梁荷载者 >
竖吊门 vertically suspension door
竖钉对接护墙板 upright panel
竖钉墙板 vertical siding
竖锭式卷绕机 spindle winder
竖斗式拌和机 vertical-bowl mixer
竖斗式搅拌机 vertical bucket mixer
竖堆法 end stacking
竖耳 prick ear
竖枋木 soldier; soldier pile
竖缝 perpendicular joint; perpends; perps; side joint; vertical seam; vertical joint
竖缝板 plate for standing seams
竖复板 vertical sheathing
竖杆 kingpost truss; upright post; vertical; vertical bar; vertical member; vertical post; montant < 嵌板的 >
竖杆机 post driver
竖杆件 vertical strut member
竖杆坑钻掘机 posthole drill
竖杆孔挖坑机 posthole borer
竖杆联结系 bracing with verticals
竖杆式挖沟机 vertical boom type trenching machine
竖高改正 orthometric correction
竖隔流板式反应池 vertical flow baffled reaction basin
竖隔流板式混合池 vertical flow baffled mixing basin
竖隔膜电池 vertical diaphragm cell
竖沟 chase; flute; vertical drain(age)
竖沟环割 frill girdling
竖沟切口 frill cuts
竖固掩盖 < 防空洞上 > bursting layer
竖挂木板瓦 vertical shingling
竖观覆盖图 elevation coverage diagram
竖管 standpipe; riser(pipe); stack pipe; standing gutter; standing pipe; standing tube; upright conduit; upright pipe[piping]; upright tube[tubing]; vertical pipe[piping]; vertical shaft; vertical tube[tubing]; driving tube < 自流井的 >
竖管阀门 riser valve
竖管裹接 upright wiped joint
竖管过水能力 stack capacity; stack content; stack power
竖管连接件 rising connector
竖管连接器 rising connector
竖管流水能力 stack(ed) capacity
竖管排烟能力 stack capacity; stack content; stack power
竖管配件 stack fittings
竖管人孔 standpipe manhole
竖管式测压计 open pipe piezometer; standpipe piezometer
竖管式跌水井 standpipe type drop-manhole; standpipe type drop-well
竖管式冷水箱 cold water tank with vertical pipes
竖管式水箱 standpipe tank
竖管式悬移质采样器 standpipe suspension load sampler; vertical pipe suspension load sampler
竖管式蒸发器 vertical tube evapo(u)rator
竖管水位计 standpipe piezometer
竖管通风 stack venting
竖管通风口 stack vent
竖管通气孔 stack vent
竖管透气 stack vent
竖管下部分排放垃圾 continuous waste and vent
竖管栅栏 pale fence; paling fence
竖管罩 shaft-cover
竖管支架 riser support
竖罐炼锌 zinc vertical retorting
竖罐炼锌厂 vertical retort zinc smelter

竖罐熔炼 vertical retort smelting
竖罐提炼法 New Jersey retort process;vertical retort process
竖罐蒸馏法 vertical retort method
竖光柱 sun pillar
竖滚柱 capstan roller
竖焊 vertical weld(ing)
竖焊缝 cross joint(ing)
竖荷载空间框架 vertical-load-carrying space frame
竖活塞 stand cock;vertical piston
竖活栓 <厕所的> stand cock
竖或斜过道 shaft way
竖基尺视差法 subtense method with vertical staff
竖基尺视距仪 depression position finder;depression range finder
竖架挖沟机 vertical beam ditcher
竖角 elevation angle
竖铰链窗 casement window
竖铰链窗框架 casement frame
竖铰链窗扇 casement;casement sash
竖铰链气窗 casement light
竖接缝 finger joint
竖截槽 vertical kerf
竖截面 vertical section;comb cut <椽子顶端与脊檩的接合面>
竖筋砖 vertical-fiber[fibre] lug brick
竖劲杆 vertical stiffener
竖井 shaft(way);shaft well;aven;cenote;coal pit;drilled shaft;fall way;perpendicular well;pot-hole;rising shaft;silo;tube well drainage;vertical bar;vertical shaft;vertical well;well shaft;drive pipe <自流井的>
竖井采矿 shaft mining
竖井测量 shaft survey
竖井测量仪(器) shaft survey instrument
竖井沉基 sunk shaft
竖井沉陷 shaft sinking
竖井衬壁 steining
竖井衬层 shaft lining
竖井衬砌 shaft lining;shaft support
竖井抽水泵 shaft-sinking pump
竖井垂锤 pilot bob
竖井垂球静止位置 zero position of shaft plummet
竖井垂线定向法 shaft plumbing
竖井垂线钢丝 shaft plumbing wire
竖井垂线井下定向工具 shaft plumbing tool
竖井垂直轴线 silo centreline
竖井导管 shaft column
竖井导轨 guide rail
竖井的建设 construction of shaft
竖井的首部建筑 shaft headworks
竖井的仪器垂测 instrumental shaft plumbing
竖井底板 shaft bottom;shaft floor
竖井底部 bottom of shaft
竖井吊锤线 shaft plumbing wire
竖井吊桶 kibble
竖井跌水式溢洪道 drop-inlet spillway
竖井顶机房 shaft house
竖井顶架 headframe of shaft
竖井定向测量 shaft orientation survey
竖井定向法 three point problem
竖井反掘 shaft raising
竖井反掘法 raising method
竖井反掘工作 raise work
竖井封闭 shaft enclosure
竖井盖 silo door
竖井钢索 <输送工人及物品的> shaft cable
竖井高程传递 elevation transmission for shaft
竖井隔板 shaft partition wall
竖井构造 shaft construction
竖井罐笼 shaft cage
竖井光学垂准法 method of optic(al) shaft plumbing
竖井荷载 load on shaft
竖井横截面 cross-section through shaft
竖井环板壁 cribbing
竖井基础 shaft base
竖井基座 shaft seat
竖井箕斗 shaft skip
竖井激光指向 laser guide of vertical shaft
竖井加高 shaft raising
竖井间壁 hoistway enclosure
竖井阶梯 shaft stair(case)
竖井结构 shaft construction
竖井截面 shaft cross-section
竖井进水口 shaft intake
竖井井管 well casing
竖井井架 headframe of shaft;lifting tower;shaft jumbo
竖井掘进 shaft excavation
竖井掘进盾构 shaft jumbo
竖井掘进机 shaft jumbo;shaft well digger
竖井开采 shaft mining
竖井开拓 vertical development
竖井开挖 shaft excavation
竖井开挖爆破 shaft-sinking blasting
竖井口 fore shaft
竖井联系测量 shaft connection survey
竖井炉 vertical kiln
竖井盲井联合开拓 shaft-blind slope development
竖井密封 shaft sealing
竖井模板 shaft formwork;shaft shuttering
竖井磨煤机 pi mill
竖井木框板壁 cribbing
竖井内大型吊斗 sinking bucket
竖井内径 silo inside diameter
竖井排水 drainage by vertical well;shaft drainage;vertical drainage;well drainage
竖井平台 shaft platform;sollar
竖井铅垂定位 shaft plumbing
竖井铅垂定位金属丝 shaft plumbing wire
竖井铅锤 shaft plummet
竖井铅直线 shaft plumbing wire
竖井墙 hoistway enclosure;well wall
竖井墙壁 shaft wall
竖井墙系统 shaft wall system
竖井清洁工 pimp
竖井深度 depth of a shaft
竖井深度测量 shaft depth measurement
竖井深度计 depth indicator of shaft
竖井施工 shaft construction
竖井施工测量 shaft construction survey
竖井式 vertical shaft type
竖井式泵站 shaft pumping station
竖井式抽水站 shaft pumping station
竖井式船闸 high lift lock;lock with high lift;shaft(navigation) lock
竖井式电炉 electric(al) shaft furnace
竖井式电站 shaft power station
竖井式进水口 shaft intake
竖井式磨煤机 impact mill;mine mill
竖井式生物反应器 vertical well biologic(al) reactor
竖井式水轮机 pit-type turbine
竖井式溢洪道 glory-hole spillway;shaft spillway
竖井式鱼闸 shaft fish lock
竖井数 number of vertical shaft
竖井双垂线定向 double plumbing of shaft
竖井提升 shaft winding
竖井提升机 shaft hoist;vertical shaft winder
竖井调压水槽 shaft surge tank
竖井调压水箱 shaft surge tank
竖井通道 shaft chimney
竖井通风 shaft ventilation
竖井通风管 silo duct
竖井通风机 shaft fan
竖井挖掘机 shaft excavation;shaft jumbo;shaft well digger
竖井圬工 shaft masonry(work)
竖井下沉 shaft sinking;sunk shaft
竖井信号 shaft signal(l)ing
竖井信号系统 shaft signal(l)ing system
竖井延深 shaft deepening
竖井溢洪道 overflow standpipe
竖井运输机 shaft carrier
竖井闸 shaft inland lock
竖井闸门 chamber interceptor
竖井支撑 shaft support;shaft timbering
竖井支护 lining of shaft
竖井中心线 silo centreline
竖井重锤 pilot bob
竖井抓岩机 shaft grab
竖井装载 shaft loading
竖井自卸式吊箱 battle ship
竖井总深度 total depth of vertical shaft
竖井钻车 shaft jumbo;sinking jumbo
竖井钻机 shaft borer
竖井钻架 shaft jumbo
竖井钻进 shaft boring;shaft drilling
竖井钻进法 shaft boring method
竖井钻进机 shaft boring machine
竖距 vertical distance
竖锯 jig saw;scroll saw;vertical frame saw
竖坑 vertical shaft
竖坑底 shaft bottom
竖坑口 brace
竖坑缆 shaft cable
竖孔 vertical hole;vertical perforation
竖孔空心砖 V-brick;vertically perforated brick
竖孔砖 end construction tile
竖框 mullion[munnion];mull on;parting rail;stile【建】
竖框窗 munnion window
竖框梃 stile
竖拉百叶门 vertical sliding shutter door
竖拉窗扇 vertical sash;vertical sliding sash;vertical sliding sash-window;vertical sliding window sash
竖拉杆 suspension member;suspension rod
竖拉卷帘式铁门 vertical sliding steel shutter door
竖拉门 vertical slide door
竖立 apeak;erection;horning;rear up;stand;up-end
竖立安放 to be kept upright
竖立板 stand sheet
竖立标石 monumentation
竖立冰 ropac[ropak];turret ice
竖立船 floating instrument platform
竖立的 upright
竖立的东西 upright
竖立的石块 orth(o)stat
竖立电缆 vertical rise cable
竖立杆 vertical rod
竖立拱模 erect centering
竖立构件 upstand
竖立拐肘 pedestal crank;vertical crank
竖立拐肘座 vertical crank stand
竖立货舱护板 vertical sparring
竖立集水布置 vertical water-collecting layout;vertical water-collection layout
竖立脚手架 erect scaffold
竖立砌层 upright course
竖立式板 upright board
竖立式容器 vertical vessel
竖立影像 upright image
竖立在屋顶上的旗杆 roof set flagpole
竖立柱子 erection column;plant column
竖立着 upright
竖立钻塔 run derrick
竖链 riser chain
竖链浮筒 riser-type buoy
竖梁 vertical beam
竖菱形系数 vertical prismatic coefficient
竖流 upflow
竖流表面曝气器 vertical surface aerator
竖流沉淀池 ventilating flow sedimentation tank;vertical flow sedimentation tank;vertical sedimentation basin;vertical sedimentation tank;vertical settling basin
竖流池 vertical flow basin
竖流滤池 vertical filter
竖流滤法 vertical filtration
竖流凝聚作用 upflow coagulation
竖流式沉淀池 upward-flow sedimentation tank;vertical flow settling tank;vertical flow tank;vertical settling tank
竖流式澄清池 upward flow clarifier
竖流式接触澄清池 upflow contact clarifier
竖流涡轮曝气器 vertical turbine aerator
竖流折流式絮凝池 vertical tableflap flocculation tank
竖龙骨 vertical keel;centre keel <单底的>
竖龙头 stand cock
竖炉 well furnace
竖炉熔化 vertical fusion
竖炉熔炼 vertical fusion
竖码 <盘类制品竖着装窑> rearing
竖码入窑 rearing
竖脉 rake vein
竖锚 vertical anchor
竖面 vertical plane
竖面车钩 vertical plane coupler
竖面浅槽饰 glyph
竖模 vertical mo(u)ld
竖木 quebracho;shore
竖幕 vertical curtain
竖排水管 vertical chain
竖盘水准器 altitude level
竖盘指示差 index error of vertical circle;vertical index error
竖盘制动 <经纬仪> vertical clamp
竖片 riser
竖瓶积深式采样器 vertical bottle depth-integrating sampler
竖坡面 talus
竖剖面 vertical section
竖铺木瓦 weather shingling
竖起 bristle;stack
竖起的 erective;vertical
竖起力矩电动机 erection torque motor
竖起钮 cocking button
竖砌 on end laying
竖砌池壁结构 full-depth construction
竖砌池壁砖 full-depth block;palisade block
竖砌大砖 soldier block
竖砌空斗墙 rowlock wall
竖砌砖 brick laid on edge;brick-on-end;rolock;rolok;rowlock

竖砌砖层 rowlock course; soldier course
竖砌(砖)拱 rowlock arch
竖砌砖拱砌合 rowlock bond of arch
竖砌(砖)空心墙 rowlock cavity wall
竖墙 straight wall
竖切 <尖轨头部垂直切去轨头及底缘> straight cut
竖切式尖轨的啮合 joggle for straight-cut switches
竖琴 harp
竖琴式(布置) harp arrangement
竖琴式钢丝带 steel wire harp-type screen
竖琴式管子结构加热炉 harp
竖琴式筛 harp type screen
竖琴式斜拉桥 harp cable-stayed bridge; harp-shaped cable-stayed bridge; harp-shaped stay cables bridge; harp type cable-stayed bridge
竖琴式斜缆桥 harp type cable-stayed bridge
竖琴形 harp configuration
竖琴形拉索 hard-type stay cable; harp type stay cable
竖琴形索 harp cable
竖琴形斜拉桥 harp-shaped stay cables bridge; harp type cable-stayed bridge
竖琴形卸扣 harp-shaped shackle
竖琴样结构 lyra
竖琴座 <星座名> harp
竖曲线 round(ing)-off curve; vertical curvature; vertical curve
竖曲线半径 radius of the vertical curve; rounding-off radius; vertical curve radius
竖曲线测设 vertical curve location
竖曲线顶点 hump; summit of vertical curve; top point of vertical curve
竖曲线起点 point of vertical curvature; point of vertical curve
竖曲线切线 profile tangent
竖曲线位置 seat of vertical curve
竖曲线最低点 low point of vertical curve
竖曲线最小长度 minimum length of vertical curve
竖曲线最优化设计 <用电子计算机设计> optimization of vertical curves
竖绒 <使绒毛竖起> pile setting
竖升开启桥 bascule bridge
竖式百叶(窗) vertical louver[vouvre]
竖式百叶门 upright ventilating slit door
竖式窗框 vertical sash
竖式吹炉 vertical converter
竖式挡板混合池 vertical baffled mixing basin
竖式导井 perpendicular shaft
竖式二进制代码 column binary code
竖式二进制方式 column-binary mode
竖式二进制卡片 column binary card
竖式二进制(数) column binary; Chinese binary
竖式干燥器 vertical drier[dryer]
竖式工业泵 vertical industrial pump
竖式固定百叶窗 vertical jalousie
竖式滑窗 vertical sash
竖式混合器 shaft mixer
竖式拉窗 vertical sliding window
竖式缆索铁路 ascensor
竖式离心脱水器 carpenter centrifuge
竖式连续泵 vertical in-line pump
竖式联锁箱 vertical locking box
竖式料斗提升机 vertical bucket elevator
竖式炉 shaft furnace; tower furnace
竖式嵌接 edge scarf; vertical scarf
竖式燃烧炉 vertical shaft furnace
竖式石灰窑 shaft lime kiln; vertical lime burner
竖式锁簧床 vertical locking box
竖式梯形泄水槽 vertical trapezoidal sluice
竖式小便器 stall-type urinal
竖式絮凝池 vertical flocculator
竖式旋转冲击钻机 vole drill
竖式窑 shaft kiln
竖式叶轮泵 vertical turbine pump
竖式移动泵 vertical moving pump
竖式移动床 vertical moving bed
竖式移动甑 vertical transportable retort
竖式造纸机 verti-former
竖式甑 vertical retort
竖式蒸罐【化】 vertical retort
竖式蒸馏釜 vertical still
竖式蒸馏器 vertical retort
竖式蒸瓶焦油沥青 vertical retort tar
竖式蒸瓶煤沥青 vertical retort tar
竖式中悬窗 vertical centre hung pivot window
竖式转炉 Great Falls converter
竖式转锥磨机 colloid mill
竖式装配 vertical assembly
竖式钻孔机 vertical boring machine
竖视线 vertical height
竖栓 vertical bolt
竖水管 riser water pipe
竖丝【测】 vertical(cross-) hair; vertical thread; vertical wire
竖锁 upright lock
竖锁闭 longitudinal locking
竖锁条 longitudinal locking bar
竖梯 vertical ladder
竖梯井孔 ladder well
竖天线 vertical aerial; vertical antenna
竖天效应 vertical aerial effect
竖条 column bar
竖条测试 column bar test
竖条宽度 width of vertical strip
竖条纹饰两槽间的平面 <最古希腊建筑中的> meros
竖条栅栏 picket fence
竖通风道 air chimney
竖筒 column
竖筒坩埚炉 shaft crucible furnace
竖筒式干燥 column dried
竖筒式干燥机 column drier
竖筒式自动分批干燥机 automatic batch column dryer
竖筒形海洋研究船 spar ship
竖凸沿缸砖 vertical-fibre lug brick
竖凸沿砖 vertical-fibre lug brick
竖瓦 tile hanging; vertical tiling
竖线 vertical line
竖线锯 jib saw; jig saw
竖线条 vertical mo(u)lding
竖线条建筑 vertical accent
竖向 V 形开挖 vertical wedge cut
竖向百叶(窗) vertical louver[vouvre]; vertical blind
竖向摆 vertical pendulum
竖向板 <墙面边沿上的> fascia[复 fa(s)ciae/fa(s)cias]
竖向板条百叶窗 vertical slatted blind
竖向比尺 vertical scale
竖向比尺增大的模型 vertically exaggerated model
竖向比例 vertical scale
竖向变态模型 vertically exaggerated model
竖向变形 vertical deformation
竖向布置 perpendicular layout
竖向布置图 vertical planning
竖向槽式运输机 vertical tray conveyors
竖向测量 vertical survey
竖向尺寸放大比率 vertical exaggeration ratio
竖向冲刷 vertical scour
竖向出口 vertical exit
竖向磁力运输机 vertical magnetic transporter
竖向错列的墙或剪力墙建筑 staggered-wall-beam building
竖向错位 vertical dislocation
竖向打桩反力 vertical piling reaction
竖向单管系统 vertical one-pipe system; vertical one-tube system
竖向的 vertical
竖向地面运动 vertical ground motion
竖向地球速率 vertical earth rate
竖向地震系数 vertical seismic coefficient
竖向地震仪 vertical component seismograph
竖向定位 vertical layout
竖向定线 vertical alignment
竖向定线角【数】 vertical alignment of road
竖向动态位移量 vertical dynamic(al) load displacement
竖向断层 vertical fault
竖向断层崖 vertical fault scarp
竖向断距 vertical slip
竖向断面图 vertical section
竖向堆积 vertical accretion
竖向防护层 vertical mulching
竖向放大比 vertical amplification ratio; vertical exaggeration ratio
竖向分布 vertical distribution
竖向分割 vertical division
竖向分量 vertical component
竖向分区 vertical division block
竖向风(荷)载 vertical wind load
竖向风速计 vertical anemometer
竖向缝 vertical joint
竖向缝焊 vertical seam welding
竖向腹板加劲肋 vertical web stiffener
竖向覆盖层 vertical mulching
竖向刚劲式桥 vertically stiff bridge
竖向钢板衬砌 vertical steel lining
竖向钢板护面 vertical steel facing
竖向钢板镶面 vertical steel surfacing
竖向钢筋 vertical bar; vertical reinforcement; vertical reinforcing steel
竖向钢壳装置 vertical metal spiral setting
竖向构件 vertical member
竖向箍筋 vertical stirrup
竖向固结系数 coefficient of vertical consolidation
竖向挂瓦 vertical tiling
竖向管道 duct riser
竖向规划 site engineering; vertical planning
竖向轨道鼓曲 vertical track buckling
竖向航行净空 vertical navigational clearance
竖向荷载 vertical load(ing)
竖向荷载系统 vertical load system
竖向桁架 vertical truss
竖向护墙板 vertical shingling; vertical wainscot
竖向滑动窗 vertical sliding window
竖向滑动模板 vertical slip form
竖向滑距 vertical dip slip
竖向滑模 vertical slip form
竖向缓和曲线 transition curve between gradients; vertical easement curve
竖向缓和曲线半径 transition radii on gradients
竖向混合式防波堤 vertical type composite breakwater
竖向活动框格窗 balanced sash window
竖向激励 longitudinal excitation; vertical excitation
竖向集水布置 vertical water-collecting layout; vertical water-collection layout
竖向加劲 vertical stiffening
竖向加劲材 vertical stiffener
竖向加劲杆 vertical stiffener
竖向加劲肋 vertical stiffener
竖向加速度 vertical acceleration
竖向加速度计 vertical accelerometer
竖向加速度均方根 root mean square vertical acceleration
竖向剪(力) normal shear; vertical shear
竖向剪切 vertical shear
竖向剪切法 vertical shear method
竖向剪切梁模型 vertical shear beam model
竖向剪切模型 vertical shear model
竖向铰接 vertical articulation
竖向搅拌机 vertical mixer
竖向接触应力 vertical contact stress
竖向接缝 vertical abutment joint
竖向接缝面 vertical joint face
竖向截水体 <坝工> chimney drain
竖向介曲线 vertical easement curve
竖向净空 overhead clearance; vertical clearance
竖向静态位移量 vertical static load displacement
竖向控制 vertical control
竖向扩散系数 vertical diffusion coefficient
竖向联斗挖沟机 vertical ladder ditcher
竖向流 vertical flow
竖向流速曲线 vertical velocity curve
竖向流速梯度 vertical velocity gradient
竖向盲沟 vertical blind ditch
竖向挠度 vertical deflection
竖向泥沙浓度分布 vertical sediment concentration distribution
竖向排气管 upcast shaft
竖向排水 vertical drain(age)
竖向排水管 drop connection <连接检查井的>; vertical bleeder drain <减压排水坞底板下的>
竖向排水砂井 vertical sand drain
竖向排水系统 chimney drain system; drainage curtain <混凝土坝体>
竖向(配)筋 vertical reinforcement
竖向膨胀 vertical expansion
竖向偏斜 vertical skew
竖向平面 vertical plane
竖向起动器 vertical starter
竖向气闸管 vertical air-lock tube
竖向墙板 vertical siding
竖向曲度 vertical curvature
竖向缺口 vertical cut
竖向上浮 vertical ascent
竖向设计 economic schematization; elevation planning; vertical design; vertical planning; design of elevation; elevation scheme; elevation design <城市道路的>
竖向摄影 vertical photograph
竖向伸缩缝 vertical expansion joint
竖向升降桥 vertical-lift bridge
竖向施工缝 vertical construction joint
竖向石砌体 face-bedded stone
竖向石砌筑 face-bedded
竖向视距 vertical sight distance
竖向收缩率 vertical shrinkage
竖向枢轴式百叶窗 vertically pivoted shutter
竖向疏散口 vertical escape exit; vertical exit
竖向双管系统 vertical two-pipe system; vertical two-piping system

S

竖向水流速度 vertical water velocity
竖向速度 vertical velocity
竖向塑料纸板排水 vertical wick drain
竖向弹簧支枢铰链 vertical spring-pivot hinge
竖向弹力常数 vertical spring constant
竖向调高器 extension device
竖向调高装置 extension device
竖向贴砖 vertical tiling
竖向通道 vertical exit;vertical opening;vertical passageway
竖向土压力 vertical earth pressure
竖向推拉门 vertical sliding door
竖向挖方 vertical cut
竖向挖心黏[粘]土砖的砖石工程 masonry work of vertical coring clay bricks
竖向弯沉 vertical deflection
竖向弯挠运动 vertical deflection movement
竖向弯曲运动 vertical bending motion
竖向位错 vertical dislocation
竖向位移 perpendicular displacement;vertical displacement;virtual displacement
竖向纹理 vertical grain
竖向纹理砌筑 <沉积岩类等层状石材的> face-bedded
竖向稳定度 vertical stability
竖向稳定性 vertical stability
竖向涡流 vertical eddy
竖向线位 vertical alignment
竖向悬臂单元 vertical cantilever element
竖向压力 vertical pressure
竖向烟囱 se-duct
竖向延伸 vertical extension;vertical extent
竖向延伸量 vertical extent
竖向移动模板 vertical travel(l)ing formwork
竖向移动式转臂吊机 vertically travel(l)ing derrick crane
竖向淤积 vertical accretion
竖向预(加)应力 vertical prestress(ing)
竖向约束 vertical restraint
竖向运动 vertical motion;vertical movement
竖向振动 vertical vibration
竖向震动 vertical shock
竖向震度 vertical seismic coefficient
竖向支撑(杆) sway bracing;sway strut;sway brace;vertical batter stud;vertical bracing;vertical strut
竖向支距 vertical offset
竖向轴 vertical axis
竖向驻波 standing gravity wave
竖向柱 plumb post
竖向桩 plumb pile
竖斜杆桁架 N-truss
竖斜杆型桁架(梁) Pratt truss
竖斜面【建】 talus
竖斜面(挡土)墙 talus wall
竖芯 vertical grille
竖形式 vertical format
竖形圆筒式加热炉 updraft type heater
竖旋翅 bascule leaf
竖旋活动桥 bascule bridge
竖旋孔 bascule space;bascule span
竖旋桥 balance bridge;bascule bridge;leaf bridge;weigh-bridge
竖旋桥的双翼 bascule
竖旋桥的翼 leaf of bascule bridge
竖旋桥孔 bascule spalling
竖旋桥桥墩 bascule pier
竖旋桥桥跨 bascule spalling
竖旋桥桥翼 leaf of bascule bridge
竖旋桥支柱 anchor column
竖旋式路栏 bascule barrier
竖旋翼 bascule leaf
竖旋闸门 bascule gate
竖旋闸门堰 shutter weir
竖穴坟墓 shaft grave;shaft mastaba <古埃及一种长方形平顶斜坡的>
竖窑 shaft furnace;updraft kiln
竖窑石灰 shaft kiln lime
竖窑水泥 shaft kiln cement
竖窑原料 shaft kiln stone
竖于基地上的标志 on-premises sign
竖圆柱坐标系 right circular cylinder coordinate
竖胀潜量 potential vertical rise
竖支撑 vertical bracing
竖直 erect;plumb
竖直摆 vertical pendulum
竖直比长器 vertical comparator
竖直比例尺 vertical scale
竖直补偿信号 vertical shading signal
竖直场强磁力仪 vertical field balance;vertical field magnetometer;vertical force magnetometer
竖直大炮眼的钻孔 toe-to-toe drilling
竖直带隙 direct band gap
竖直的 edgewise
竖直的碰冒头 vertical meeting rail
竖直地 vertically
竖直度 verticality
竖直度盘 altitude circle
竖直度盘精度【测】 vertical accuracy;accuracy of vertical circle
竖直断面 vertical section
竖直对缝砌 vertical bond
竖直反力 vertical reaction
竖直方向 vertical direction
竖直分辨力检验楔 vertical resolution wedge
竖直分布 vertical distribution
竖直分力 vertical component
竖直分量 vertical component
竖直风道 vertical air duct
竖直缝 vertical joint
竖直箍筋 vertical stirrup
竖直管道 straight duct
竖直航空摄影 vertical aerophotography
竖直航摄像片 vertical aerial photograph;vertical air photograph
竖直航摄照片 vertical photograph
竖直荷载 vertical load(ing)
竖直滑距 vertical dip slip;vertical slip
竖直回描 vertical retrace
竖直回水管 vertical return
竖直加劲杆 vertical stiffener
竖直加强杆 vertical stiffener
竖直剪切 vertical shear
竖直浇筑砖镶板 vertical cast brick panel
竖直角 upright angle;vertical angle
竖直角测量 measurement of vertical angle
竖直铰链 drag hinge
竖直结构 vertical structure
竖直筋 vertical bar
竖直距离 vertical distance
竖直锯框 vertical saw-frame
竖直刻度盘 vertical circle
竖直力 vertical force
竖直溜管 vertical chute
竖直螺栓 foot bolt
竖直面 vertical(sur)face
竖直挠曲振动 vertical bending vibration
竖直排水 vertical drain
竖直偏转电流 vertical yoke current
竖直剖面 sectional elevation
竖直铺墙面板 vertical wall slab
竖直气闸 vertical lock
竖直强度磁力仪 vertical intensity variometer;Z variometer
竖直切削 vertical cut
竖直圈 vertical circle
竖直三角测量 vertical triangulation
竖直扫描 vertical sweep
竖直扫描发生器 vertical scanning generator
竖直摄谱仪 vertical spectrograph
竖直摄影相片 vertical print
竖直升降机部件 vertical cantilever element
竖直式连续摄影机 vertical strip camera
竖直式圆锯 vertical saw
竖直视差螺旋 vertical parallax screw
竖直衰落 vertical fading
竖直水尺 vertical ga(u)ge
竖直丝 vertical wire
竖直速度曲线法 vertical velocity curve method
竖直速度梯度 vertical velocity gradient
竖直挑出的招牌 vertical projecting sign
竖直同步 vertical synchronization
竖直同步脉冲 vertical synchronizing pulse
竖直同步脉冲分离器 vertical separator
竖直同步调整 vertical-hold control
竖直投影 vertical projection
竖直拖车 trailer-erector
竖直位移限位器 elevating stops
竖直位置 edgewise placing;upright position;vertical position
竖直稳定性 vertical stability
竖直析像力 vertical definition;vertical resolution
竖直狭通气孔 loop window
竖直线偏差 deviation of the vertical
竖直线性调节 vertical linearity control
竖直镶板 vertical panel
竖直性 verticality
竖直旋涡 vertical eddy
竖直叶片 vertical blade
竖直运动 vertical motion
竖直闸门堰 draw door weir
竖直障碍声呐 vertical obstacle sonar
竖直褶皱 vertical fold
竖直支撑 vertical bracing;vertical shoring
竖直制动螺旋 vertical clamp screw
竖直中心调节 vertical centering[centring] control
竖直中悬窗 vertical center[centre] hung window casement
竖直轴倾斜误差 error due to inclination of vertical axis
竖直砖铺路面 edgewise clay brick paving
竖直转弯 vertical turn
竖直转轴 vertical axis of revolution
竖直坐标 vertical coordinate
竖轴 mill spindle;standing axis;vertical shaft
竖轴窗扇 vertical pivoted sash;vertical pivoted window;vertical pivot hung window
竖轴风机 vertical axis wind turbine
竖轴复式汽轮机 vertical compound steam turbine
竖轴干船坞排水泵 vertical spindle dry dock drainage pump
竖轴弧形闸门 radial gate with vertical axes
竖轴环流 sagittal-axis circular current
竖轴混凝土蜗壳 vertical concrete spiral
竖轴混流泄水泵 vertical spindle mixed-flow dewatering pump
竖轴铰接弧形闸门 vertically-hinged sector gate
竖轴离心泵 vertical spin-die centrifugal pump
竖轴流速仪 vertical axial current meter;vertical shaft current meter
竖轴磨床 vertical shaft grinder
竖轴扇形闸门 vertical axis sector gate
竖轴式布置 vertical shaft arrangement
竖轴式沉淀池 vertical flow sedimentation tank
竖轴式沉砂池 vertical flow grit chamber
竖轴式金属蜗壳装置 vertical metal spiral setting
竖轴式汽轮机 vertical shaft water turbine;vertical turbine
竖轴式水轮机 vertical shaft water turbine;vertical turbine
竖轴式透平 vertical shaft water turbine;vertical turbine
竖轴式涡轮机 vertical shaft water turbine;vertical turbine
竖轴式旋窗 vertically pivoted window
竖轴弹簧铰链 vertical spring-pivot hinge
竖轴误差 standing axis error
竖轴旋涡 eddy with vertical axis;standing eddy;vertical whirl
竖轴旋转 <门轴的> vertical hinge revolving
竖轴叶片搅拌机 vertical spindle paddle mixer
竖轴支承 crapaudine
竖轴轴流整周进水式水轮机 vertical shaft axial flow full-admission turbine
竖轴转轮型风机 vertical axis rotor-type wind turbines
竖柱 upstand;vertical post
竖砖 brick-on-end;row-block
竖砖层 brick soldier course
竖砖缝 collar joint;head joint
竖砖拱 brick-end arch;brick-on-end soldier arch;soldier arch
竖砖空斗墙 rat-trap bond
竖砖平拱 soldier arch
竖砖铺砌 rolock paving
竖砖砌层 soldier course
竖砖券 rowlock arch
竖桩 soldier beam;soldier pile;vertical pile
竖桩板 vertical sheathing
竖锥形弹簧悬置 vertical volute spring suspension
竖准器 vertical collimator
竖着 on end
竖着放 stand on end
竖坐标 ordinate

数

数百千米 several hundred kilometers
数百万 multimillion
数变词 numeric(al)variable
数变项 data item;numeric(al)variable
数标 number scale;numeric(al)index
数表 numeric(al)table
数表示法 number representation
数不清的 incomputable;innumerable;innumerous;unnumbered
数测定计 phakometer;phakoscope
数乘 scalar multiplication
数传机 data set
数传机同步 data set clocking
数串 string

数错 miscount
数得清 numerable
数的 numeric(al)
数的标准分布 standard layout
数的表示 numeric(al) representation
数的串行表示 serial representation
数的阶部分 exponential part of number
数的逆 inverse of a number
数的配置 digit layout
数的上整数 ceiling of number
数的舍入法 rounding-off method
数的显示 presentation of number
数的指数部分 exponential part of number
数地址码 number address code
数额未定贷款 open-end loan
数发生器 number generator
数符号 numeric(al) symbol
数符交通控制 digital traffic control; traffic digital control
数格式 number format
数股钢丝绳合成的左旋粗钢丝绳 cable-laid wire rope
数贯 sequence; suite
数函项 numeric(al) function
数和 scalar sum
数环 ring of numbers
数积 scalar product
数基 number base; number system base
数记 number scale
数记号 number token
数阶法 step spectrum method
数解法 algebraic(al) method; analytic(al) method
数据 data; datum; digital data; digital information; evidence; finding; numeric(al) data; quantitative data
数据安排形式控制 format control
数据安全 data safety
数据安全性 data security
数据包 data package; data packet
数据包标题 data packet header
数据包容量直方图 data packet size histogram
数据包应答 data packet acknowledg(e)ment
数据保持【计】data hold
数据保持装置 data hold
数据保存方式 data save mode
数据保护 data protection; data safety guard
数据保密 data privacy; data security; security of data
数据保密通信[讯] data security communication
数据报 datagram
数据报告 data report
数据报文 data message
数据报文交换机 data message switching system
数据报文交换系统 data message switching system
数据报业务 datagram service
数据比较仪 data comparator
数据编号 data number
数据编辑 data editing; data edition
数据编辑系统 data editing system
数据编码器 data encoder
数据编码系统 data code; data encoding system
数据编译 data origination
数据编址存储器 data-addressed memory
数据变化表 data change table
数据变化系统 data translating system
数据变换 converted data; data conversion; data reduction; data transformation; transformation of data
数据变换接收机 data conversion receiver
数据变换器 data converter; data reducer; data transducer
数据变换系统 data conversion system; data translating system
数据标绘器 data plotter
数据标绘仪 data plotter
数据标记 data token
数据标记包 data token packet
数据标记机制 data token mechanism
数据标题 data header
数据表 data book; data table
数据表达装置 data presentation device
数据表结构 data list structure
数据表块 data table block; table block
数据表示法 data(re) presentation
数据并合 merging
数据补偿程序 data compensation routine
数据捕捉 data capture
数据不收敛不能定位 no fix data
数据不足 inadequate of data
数据布局 data layout
数据部分 data division
数据采集 data acquisition; data collection
数据采集和监测设备 data acquisition and monitoring equipment
数据采集和控制系统 data acquisition and control system
数据采集记录器 data acquisition instrument; data-logging instrument
数据采集控制系统 data acquisition control system
数据采集平台 data collection platform
数据采集设备 data acquisition equipment; data acquisition unit
数据采集手段 data collection mean
数据采集探测 data acquisition probe
数据采集系统 data collecting system; data gathering system; data acquisition system
数据采集与处理 data acquisition and processing
数据采集与控制 data acquisition and control
数据采集照相机 data acquisition camera
数据采样开关 data sampling switch
数据采样控制系统 sampled-data control system
数据彩色表示方法 colo(u)red display of data
数据彩色系统 data colo(u)r system
数据仓库【计】data warehousing
数据操纵 data manipulation
数据操作 data manipulation
数据操作寄存器 data manipulation register
数据操作语言 data manipulation language
数据测量系统 data measuring system
数据层次 data hierarchy
数据插入 data inserter
数据插入变换器 data insertion converter
数据插入程序 data inserter
数据插入系统 data insertion system
数据查询 data challenge
数据差错率 data error rate
数据常数 data constant
数据撤出 data withdrawal
数据成分 data component; data element
数据程序振荡器 digitally programmable oscillator
数据重定位 data realignment
数据重发器 data repeater
数据重返主存储器 roll-in
数据重排列 data rearrangement
数据重现 data reproduction
数据重新定位 alignment of data
数据抽出 data withdrawal
数据抽象 data abstraction
数据抽象语言 data abstraction language
数据抽样 data sampling
数据初(始)加工 data origination
数据储存 data banking
数据储存和检索 data storage and retrieval
数据储存器 data storage
数据处理 handling data; data conversion; data handling; data management; data manipulation; data processing; data reduction; data treatment; digital data handling; informatics; process data; processing; processing of data; reduction of data; treatment of data
数据处理部件 data processing unit; data processor
数据处理程序 data-handling program(me); data-handling routine; data processor
数据处理单元 data processing element
数据处理方法 data processing method
数据处理和解释推断 data processing and interpretation
数据处理活动 data processing activity
数据处理机 data processor; data processing machine; datatron <十进制计算机中的设备>
数据处理计算机 data processing computer
数据处理记录器 data processing recorder
数据处理技术 data processing technique; data technology
数据处理阶段 data processing stage
数据处理开始 beginning of data handling
数据处理能力 data-handling capacity
数据处理盘 data panel
数据处理器 data processor
数据处理清单 data processing inventory
数据处理容量 data-handling capacity
数据处理设备 data-handling equipment; data-handling system; data processing equipment; data processing installation; data processing unit
数据处理设备公司 datamation
数据处理时间 data processing time
数据处理算法 data processing algorithm
数据处理系统 data processing system; digital data handling system; digital data processing system; data-handling system
数据处理序列 data processing sequence
数据处理业 datamation
数据处理与分布 data reduction and distribution
数据处理元件 data-handling component
数据处理中心 data-handling center [centre]; data processing center [centre]; data reduction center[centre]
数据处理终端设备 data processing terminal equipment
数据处理装置 data-handling equipment; data processing complex; data processing device; data processing equipment; data processing unit
数据处理自动化 datamation
数据传递 data transfer
数据传递率 data transfer rate
数据传递速度 data transfer rate
数据传递速率 message data rate
数据传递系统 data transmission system
数据传感器 data pick-up; data sensor; data transducer
数据传接口 data transmission interface
数据传输 data communication; data transfer; data transmission
数据传输电脑化 computerization of data transmission
数据传输端站 data link terminal; data transmission terminal
数据传输发信机 data link communicator
数据传输反回装置 data transmission echoing unit
数据传输分配板 data panel
数据传输服务 data-transmission service
数据传输公路 data-transmission highway
数据传输机 data set
数据传输机适配器 data set adapter
数据传输阶段 data phase
数据传输接口 date transmission interface
数据传输接收设备 data link receiving set
数据传输控制 data link control
数据传输控制器 data traffic director
数据传输块 data transmission block
数据传输设备 data communication equipment; data set
数据传输视频显示器 data transmission video display unit
数据传输速度 data transmission speed
数据传输速率 message transmission rate
数据传输网 data communication network; data transmission network
数据传输系统 data link system; data transmission system
数据传输线 data line; data link; data transmission line
数据传输线路 data transmission link
数据传输线路终端 data link terminal
数据传输效率 data transmission efficiency; data transmission ratio; data transmission utilization measure
数据传输协议 data transmission protocol
数据传输信道 data transmission channel
数据传输信号平滑 data smoothing
数据传输与转换 data transmission and switching
数据传输终端 data link terminal; data transmission terminal
数据传输终端设备 data transmission terminal equipment
数据传输终端装置 terminal installation for data transmission
数据传输转换器 data set
数据传输装置 data transmission device; data transmission package; data transmission setup; data transmission unit
数据传送 data transfer; data transmission
数据传送的自动同步系统 selsyn-data system

数据传送方式 data mode
数据传送放大器 data repeater
数据传送分配板 data panel
数据传送换码 data link escape
数据传送率 data transfer rate;rate of data signal(l)ing
数据传送速度 data transfer rate
数据传送速率 data transfer rate
数据传送系统 data transmission system
数据传送系统测试器 data transmission testing set
数据传信率 data signal(l)ing rate
数据窗口 data window
数据(磁)道 data track
数据簇 aggregate of data
数据存储 information storage;record storage
数据存储检索系统 data storage and retrieval system
数据存储块 data store block
数据存储器【计】 data storage;data accumulator;data-setting box;data storage device
数据存储区 data storage area
数据存储设备 data storage equipment
数据存储桶 data bucket
数据存储系统 data storage system;record storage system
数据存储与检索系统 data storage and retrieval system
数据存储制 data storage system
数据存储装置 data storage device;data storage unit
数据存储自动记录 data storage register
数据存取 data access
数据存取程序 data access program(me)
数据存取法 data access method
数据存取方式【计】 data access method
数据存取寄存器 data access register
数据存取时间 data time
数据存取速度【计】 access speed
数据存取速率 data access rate
数据存取通路 data access path
数据存取系统 data access system
数据存取语言【计】 data access language
数据存取装置 data access arrangement
数据存入和记录设备 data-logging-recording equipment
数据打印格式 data layout
数据打印器 data printer
数据代理终端 data agent
数据代码 data code
数据代码翻译 data code translation
数据代码转换 data code conversion;data code translation
数据单 data sheet
数据单段 data sheet field
数据单元 data element;database;data cell;data location;data unit
数据单元存储器 data cell storage
数据单组 data sheet field
数据的隔离 data privacy
数据的计算机化评价 computerized evaluation of data
数据的逻辑视图 logic(al) view of data
数据的修匀法 graduation of data
数据的修整法 graduation of data
数据的预先加工 predigestion of data
数据登记项 data entry
数据登录系统 data-logging system;data recording system
数据地址 data address
数据地址存储 data-addressed memory
数据点 data point
数据点标绘器 digital point plotter
数据电报服务 datagram service
数据电传 data telex
数据电传打字机 data teletypewriter set
数据电话电路 data telephone circuit
数据电话机 dataphone
数据电话拾音器 dataphone adapter[adaptor]
数据电话适配器 dataphone adapter[adaptor]
数据电话数字系统 dataphone digital system
数据电话业务 dataphone service
数据电话转换器<一种将数据用电话线路从一个远存取点达到中央计算机运算的转换器>【计】 dataphone adapter[adaptor]
数据电路 data circuit
数据电路端接设备 data circuit terminating equipment
数据电路交换机 circuit switching system for data
数据电路交换系统 data circuit switching system
数据电路设备 data circuit equipment
数据电路透明性 data circuit transparency
数据电路终接设备 data circuit terminating equipment
数据电视检阅 viewdata
数据电信 data communication
数据电源系统 data power
数据定界符 data delimiter
数据定位 data positioning
数据定向传输 data directed transmission
数据定向输出 data-directed output
数据定义 data definition
数据定义方式 data definition mode
数据定义符 data delimiter
数据定义名字 data definition name
数据定义语句 data definition statement
数据定义语言 data definition language
数据动向系统 data trend
数据读出器 data reader
数据读出设备 data-taking equipment
数据读出装置 data readout setup
数据读数系统 data reading system
数据独立存取模型 data independence access model
数据独立性 data independence
数据段 data segment
数据段缓冲器 segment data buffer
数据对齐 alignment of data;data alignment
数据对数目 number of data-pairs
数据多次重新定义 multiple redefinition of data
数据多路复用器 data multiplexer
数据多路复用系统 data multiplexing system
数据多路系统 data multiplexing system
数据发射 data transmission
数据发生器 number generator
数据发送部件 data unit
数据发送机 data transmitter;data unit
数据发送器 data source
数据发送设备 data transmitting equipment
数据发送装置 data source
数据翻译系统 data translating system
数据反馈 data feedback
数据反演 data inversion
数据范围 data area
数据访问法 data access method
数据访问寄存器 data access register
数据访问路径 data access path
数据访问系统 data access system
数据分隔符 data delimiter
数据分级 data staging
数据分级结构 data hierarchy
数据分拣设备 data sorting system
数据分块 de-blocking
数据分类 data qualification
数据分离器 data extractor
数据分配表 data distribution list
数据分配计划 data distribution plan
数据分配系统 data distribution system
数据分配中心 data distribution center[centre]
数据分区界 compartment boundary
数据分析 analyze the data;data analysis
数据分析技术 data analysis technique
数据分析控制台 data analysis console
数据分析器 digital analyser[analyzer]
数据分析设备 data analysis facility
数据分析系统 data analysis system
数据分析中心 data analysis center[centre]
数据分支 data path
数据分组 data clustering;data package;data packet
数据服务单元 data service units
数据浮标 data buoy
数据浮标系统 data buoy system
数据辅助环 data-aided loop
数据复示 data recall
数据复杂性 data complexity
数据复制 data reproduction
数据复制系统 digital data reproduction system
数据副本 data copy;data transcription
数据改变环 modification loop
数据改正 adjustment of data
数据高速通路 data highway
数据格式 data format;data layout
数据格式编排 data formatting
数据格式的错误 data formatting error
数据更改建议 data change proposal
数据共享 data sharing
数据构模 data modeling
数据构形 data configuration
数据估计 data evaluation
数据管理 data management
数据管理程序 data management program(me)
数据管理公共代码 data management common code
数据管理功能 data management function
数据管理系统 data management system
数据管理与分析系统 data management and analysis system
数据管理员 data administrator
数据广播网络 data broadcasting network
数据归纳 data conversion;data reduction;reduction of data
数据规则 data rule
数据过量运行 data overrun
数据过滤 data filtering
数据航行 data sailing
数据合并 data packing
数据核实 data validation
数据互换【计】 data interchange
数据化 datamation
数据化教学设备 digitizing trainer
数据化字符 digital character
数据化自动编图 digital automatic map compilation
数据还原 reduction of data
数据缓冲器区 data buffer
数据换算 data conversion;data reduction;reduction of data;revaluation of data
数据换算程序 data exchange program(me)
数据换算硬件 data conversion hardware
数据恢复 data recovery
数据回放系统 data playback system
数据汇编 data compilation
数据汇集 data collection(mean);data gathering
数据汇集板 data gathering panel
数据汇集器 data concentrator
数据汇接器 data sink
数据绘图仪 data plotter
数据获得系统 data acquisition system;data collecting system;data gathering system
数据获取 data acquisition;data capture
数据获取电路 data acquisition circuit
数据获取级 data acquisition stage
数据获取平台 data acquisition platform
数据获取系统 data acquisition system
数据机 data unit
数据机制 data mechanism
数据积分法 value integration
数据积累 data accumulation
数据基 database
数据基准面 datum plane;datum surface
数据级 data level
数据级存取 data level access
数据集 data set
数据集标号 data set label
数据集存取 data set access
数据集地址分配 allocation of data set
数据集分配 data set allocation
数据集合 data collection
数据集合控制器 data acquisition controller
数据集合设备 data acquisition equipment
数据集结构 data set organization
数据集结束 end of data set
数据集结尾出口程序 end-of-data-set exit routine
数据集就绪 data set ready
数据集控制 data set control
数据集连接器 data set coupler
数据集名 data set name
数据集迁移 data set migration
数据集入口 data set entry
数据集识别 data set identification
数据集实用程序 data set utility
数据集输出入 data set in-out
数据集顺序号 data set sequence number
数据集信息 data set information
数据集序列号 data set serial number
数据集引用号 data set reference number
数据集指令 data set instruction
数据集中 data concentration
数据集中处理 centralized data processing
数据集中(分配)器 data concentrator
数据集中格式 data concentration formatting
数据集组 data set group
数据计数 data counts
数据计算机中心 data computer center[centre]
数据记录 data logging
数据记录表 data logger;data sheet
数据记录长度 data record size

数据记录程序 data logger
数据记录处理 data recording processing
数据记录带 data tape
数据记录方式 fashion of data record
数据记录放大器 data recording amplifier
数据记录格式 data record format
数据记录工作台 data logger operation(al) desk
数据记录机 data-logging machine
数据记录技术 data-logging technique
数据记录介质 data carrier
数据记录媒体 data carrier; data medium
数据记录能力 data recording capability
数据记录器 data inscriber; data logger; data printer; data recorder
数据记录器和核对器 data logger and verifier
数据记录头 data recording head
数据记录载体 data carrier; data medium
数据记录者 recorder
数据记录装置 data-logging device; data-logging equipment; data recording device; data chamber <航摄仪镜箱中记录辅助数据的部分>
数据寄存器 data register
数据加工 data smoothing; treatment of data; data processing
数据加密 data encryption
数据加密标准 data encryption standard
数据驾驶台 data bridge
数据驾驶仪 data pilot
数据间距误差 data spacing error
数据间隙 data slit
数据兼容性 data compatibility
数据监控系统 data monitoring system
数据检查表 data check list
数据检查台 data inspection station
数据检索 data retrieval
数据检索系统 data retrieval system
数据检验 data verification
数据简化 data conversion; data reduction; reduction of data
数据简化程序 data reduction program(me)
数据简化和处理装置 data reduction and processing
数据简化器 data reducer
数据简化输入程序 data reduction input program(me)
数据简化系统 data reduction system
数据简缩 data reduction
数据鉴别处理系统 discrimination data processing system
数据鉴定 data identification
数据键 data key
数据键盘 number generator
数据交换 data exchange; data interchange; data-switch(ing)
数据交换机 data exchanger; data-switching exchange
数据交换接口【计】data exchange interface
数据交换器 data exchanger
数据交换设备 data exchange unit
数据交换通道 data exchange channel
数据交换系统 data exchange system; data interchange system
数据交换业务 data exchange service
数据交换用标准码 standard code for interchange
数据交换中心 data-switching center [centre]
数据交换装置 data exchange unit
数据胶片 data film
数据校验 data check; data validation
数据校验程序 data verifying program(me)
数据校正 adjustment of data; data correction
数据阶段 data phase
数据接收标志 data accepted flag
数据接收机 data receiver
数据接收器 data sink
数据接收速度 information rate
数据接收站 data-acquisition statement
数据接收终端机 data sink
数据结构 data hierarchy; data structure
数据结构变换 data structure mapping
数据结构存储器 data structure memory
数据结构化系统开发方法 data structured system development method
数据结构树 data structure tree
数据结构图 data structure diagram
数据结构选择 data structure choice
数据结束标志 end of data mark(er)
数据结束指令 data end command
数据结尾 end of data
数据介质 data medium
数据界面 data interface
数据紧缩 data compaction
数据紧缩技术 data compaction technique
数据进入项 data entry
数据进位 data migration
数据精化 data purification
数据精简 data compaction; data reduction
数据精练 data purification
数据就绪标记 data ready flag
数据矩阵 data matrix
数据卷宗 data volume
数据卡(片) data card
数据可靠程度 degree of accuracy of datum
数据可靠性 data reliability
数据控制 data management
数据控制程序的检验 format check
数据控制间隔 data control interval
数据控制块 data control block
数据控制器 recording controller
数据控制系统 data control system
数据控制员 control clerk
数据控制装置 data control unit
数据库【计】database; data library; date pool; data bank
数据库安全系统 database safety system
数据库安全性 database security
数据库保护 database protection
数据库保密 database privacy
数据库标识符 database identifier
数据库参数 data bank parameter; database parameter; data library parameter
数据库操作 database manipulation
数据库程序 data bank handler; database handler; database procedure; data library handler
数据库抽象化 database abstraction
数据库初始化 database initialization
数据库创建 database initialization
数据库存储器 archival memory; database storage
数据库存取语言 database access language
数据库单元 database location
数据库定义语言 database definition language
数据库翻译 database interpretation
数据库服务程序 database server
数据库服务器 database server
数据库更新 database updating
数据库关键码 database key
数据库管理 database management
数据库管理程序 database administrator
数据库管理系统 data bank management system; database management system
数据库管理系统计算机 database management system computer
数据库管理员 database administrator; database manager
数据库过程 database procedure
数据库后备 database back-up
数据库恢复 database recovery
数据库绘图软件 database graph software
数据库机 database machine
数据库集 data bank
数据库计算机 database computer
数据库结果向导 database result wizard
数据库可靠性 database reliability
数据库可移植性 database portability
数据库控制程序 database control
数据库逻辑组织 database logical organization
数据库码项 database key item
数据库描述 database description
数据库描述项 database description entry
数据库模型 database model
数据库目录 database directory
数据库设计 database design
数据库设计员 database practitioner
数据库生成程序 database generator
数据库实用程序 database utility
数据库数据存储器 data bank
数据库数据名 database data name
数据库数据模型 database data model
数据库体系结构 database architecture
数据库统计 statistics of database
数据库物理组织 database physical organization
数据库系统 database system
数据库性能 database performance
数据库修饰符 database modifier
数据库一致性 database consistence [consistency]
数据库异常条件 database exception condition
数据库引擎 database engine
数据库应用系统 database application system
数据库有效率 database availability
数据库预置 database initialization
数据库源 data pool
数据库制图 database mapping
数据库子系统 database subsystem
数据块 data block; block data
数据块传送结束符 end-of-transmission block character
数据块语句 block data subprogram(me)
数据块子程序 block data subprogram(me)
数据宽度 data window
数据扩充块 data extend block; data extent block
数据来源 data origination; data source
数据雷达 data radar
数据类 data class
数据类目 data element
数据类型 data type; number type
数据类型缺席规则 default for data type
数据类型说明(书) data type specification
数据类型转换 data type conversion
数据累积 data accumulation
数据累积网络 data accumulation network
数据累加器 data accumulator
数据连接 data connection
数据联锁 data locking
数据联网 data networking
数据联系 data link
数据链 chain data; data chain
数据链标记 chain data flag
数据链接 data chain(ing)
数据链路 data link
数据链路层 data link layer
数据链路级协议 data link level protocol
数据链路控制 data link control
数据链路控制程序 data link control program(me)
数据链路连接标识 data link connection identifier
数据链路连接器 data link connector
数据链路配制 data link configuration
数据链取 data chaining
数据量 data size; data volume; quantity of data
数据列表 data list
数据流【计】data stream; data flow
数据流操作符 data flow operator
数据流程图 data flowchart; data flow diagram; data flow graph
数据流程序 data flow program(me)
数据流传输 stream transmission
数据流过程 data flow procedure
数据流计算机 data flow computer
数据流巨型计算机 data flow supercomputer
数据流模型 data flow model
数据流说明 stream data specification
数据流通量 data traffic
数据流向寄存器 data direction register
数据流语言 data flow language
数据流指令 data flow instruction
数据滤波 data filtering
数据滤波网络 data filtering network; data smoothing network
数据路径选择 data routing
数据率 data link
数据码 numeric(al) data code
数据码检索 data code indexing
数据脉冲 data pulse
数据媒体 carrier
数据描述 data description
数据描述程序 data description program(me)
数据描述符 data descriptor
数据描述(记入)项 data description entry
数据描述语言 data description language
数据敏感性故障 data sensitive fault
数据名(称) data name
数据模块 data module
数据模拟 data modeling; data simulation
数据模式 data pattern
数据模型 data model
数据拟合 data fitting
数据耦合器 data coupler
数据排序 data ordering
数据配置 data layout
数据批处理 batched data processing
数据平滑 data smoothing
数据平滑网络 data smoothing network
数据平均挡 data average
数据平移 translation of data
数据评估 data evaluation

数据评价法 data evaluation technique
数据剖面 data profile
数据谱系关系 data ownership
数据启动控制 data-initiated control
数据起点 data origination
数据器 data set
数据迁移 data migration
数据前送 data forwarding
数据清除 data dump
数据请求 data demand
数据区 data bit;data field
数据区编号 data area number
数据区分 data separation
数据区分符号 data separator
数据区域 data area;data realm
数据驱动的 data driven
数据驱动计算机 data-driven machine
数据取出 data fetch
数据取样系统 sampled-data system
数据取样仪 data sampling unit
数据缺口 data gap
数据确定 data determination
数据群集 packing of data
数据容量 data capacity
数据容器 data capsule
数据融合(技术) data fusion
数据冗余性 data redundance
数据入口 data entry
数据扫描装置. data scanner
数据筛选 data filtering
数据设计 data design
数据设计方案 data design layout
数据摄影记录 photographic(al) data recording
数据摄影记录装置 photographic(al) storage
数据时间片 data slot
数据时间序列法 sequential method of data
数据识别 data identification
数据实际结构独立性 physical data independence
数据实体 data entity
数据实体化 data materialization
数据使用标识符 data use identifier
数据适配器 data adapter unit
数据收发两用机 data transceiver
数据收发装置 data source and sink
数据收发准备状态 data ready
数据收集 data acquisition;data aggregation; data capture; data collection;data gathering
数据收集方法 data collection method;data gathering method
数据收集浮标站 data buoy
数据收集功能 data collection function
数据收集和分析 data collection and analysis
数据收集和整理系统 data acquisition and interpretation system
数据收集计算机 data acquisition computer
数据收集记录系统 data acquisition logging system
数据收集控制器 data acquisition controller
数据收集盘 data gathering panel
数据收集平台 data collecting platform;data collection platform
数据收集器 data collecting device; data collector
数据收集日期 data of datum logging
数据收集设备 data collecting equipment;data gathering set
数据收集卫星 data collection satellite
数据收集系统 data capture system; data collecting system;data gathering system
数据收集形式 data acquisition form; data collection form
数据收集与处理系统 data acquisition and processing system
数据收集与分配 data acquisition and distribution
数据收集与记录系统 data acquisition and recording system
数据收集与简化系统 data acquisition and reduction system
数据收集与控制系统 data acquisition and control system
数据收集站 data acquisition platform;data collection platform
数据收集中心 data acquisition center [centre]
数据收集终端 data collector terminal
数据收集装置 data acquisition unit
数据手册 data(hand)book
数据输出 data output;presentation of information
数据输出电平 data output level
数据输出分压器 data potentiometer
数据输出校验器 data logger checker
数据输出门 data output gate
数据输出器 data logger
数据输出装置 output medium
数据输入 data entry;data-in;data input
数据输入端 data input pin
数据输入法 data input procedure
数据输入分类法 data inputting methodology
数据输入监控器 data input supervisor
数据输入键盘 data entry keyboard
数据输入器 data-in;data input unit; data inserter;datin
数据输入条件 entry condition
数据输入系统 data entry system
数据输入显示 data input display
数据输入语句 data input statement
数据输入装置 data entry unit;datasetting box;track ball
数据输入总线 data input bus
数据输送介质存储器 data carrier storage
数据输送模件 data transmission module
数据数量 data bulk;data quantity;data volume
数据数率 data rate
数据数字处理 data digital handling
数据说明错误【计】 data-declaration error
数据说明卡边印刷 end printing
数据说明语句 data declaration statement
数据送话器 dataphone
数据搜集 data compilation
数据搜索系统 data acquisition system
数据速度 data speed
数据宿 data sink
数据索引 data index
数据探测法 data snooping
数据淘汰 data-rejection
数据套 nest
数据剔除 data-rejection
数据提纯 data purification
数据条件 data qualification
数据条件码 data condition code
数据条目 data entry
数据调整 data alignment
数据调整网络 data alignment network
数据调整属性 data alignment attribute
数据调制调解器 data modem
数据通道 data channel
数据通道带宽 data channel bandwidth
数据通道多路复用器 data channel multiplexer
数据通道复用器 data channel multiplexer
数据通道行式打印机 data channel line printer
数据通道中断 data channel interrupt
数据通道周期 data channel cycle
数据通道周期挪用 data channel cycle stealing
数据通路【计】 data highway; data path
数据通路交换系统 data message switching system
数据通路模型 data way model
数据通信[讯] data communication
数据通信[讯]处理机 data communication processor
数据通信[讯]多路复用器 data communication multiplexer
数据通信[讯]换码 data link escape
数据通信[讯]换码字符 data link escape character
数据通信[讯]监督程序 data communication monitor
数据通信[讯]监督端 data communication monitor
数据通信[讯]监控程序 data communication monitor
数据通信[讯]监视器 data communication monitor
数据通信[讯]交换机 data communication exchange;data exchange
数据通信[讯]设备 data communication equipment
数据通信[讯]输入缓冲器 data communication input buffer
数据通信[讯]速度 data signal(l)ing rate
数据通信[讯]速率透明性 data signal(l)ing rate transparency
数据通信[讯]网(络) data communication network
数据通信[讯]系统 data communication system
数据通信[讯]线路 data communication link;data link;tie line
数据通信[讯]协议 data communication protocol
数据通信[讯]信道 data communication channel
数据通信[讯]业务 data communication service
数据通信[讯]站 data communication station
数据通信[讯]终端 data communication terminal
数据通信[讯]转接器 data communication adapter unit
数据通信[讯]自由化 data communication liberalization
数据同步 data synchronization
数据同步传输电缆 data cable
数据同步器 data synchro; data synchronization unit
数据同步装置 data synchronization unit
数据头 data head
数据图 datagram;data map
数据图表 tabulated data
数据图形 datagraphics
数据图形输入板 data tablet
数据外存储器 data file
数据完整性 data integrity
数据网(络) data network
数据网络码 data network identification code
数据卫星转播 satellite relay of data
数据位【计】 data bit
数据位置 data location
数据文件 data file
数据文件结束 end of data file
数据文件类程 data file class
数据文件生成 data file generation
数据文件转换 data file converter
数据稳定 data stabilization
数据误差 data error;error in data
数据系统规范 data system specification
数据系统集合 data system integration
数据系统接口 data system interface
数据细目 data item
数据显示 data display;data presentation
数据显示技术 data presentation technique
数据显示控制器 data display controller
数据显示器 data display equipment; data display unit
数据显示区 display field
数据显示系统 data display system
数据显示中心 data display center [centre]
数据显示终端 video data terminal
数据线 data wire
数据线路 data circuit
数据线路接口 data line interface
数据线路透明性 data circuit transparency
数据线路终端 data line terminal
数据线路终端设备 data line terminating equipment
数据限定规程 data qualification procedure
数据限制(条件) data qualification
数据相关 data correlation
数据相关信息 data association message
数据相关性 data dependency
数据响应 data response
数据向量 data vector
数据向上延拓法 upward continuation method of data
数据向下延拓法 downward continuation method of data
数据项 data element;data item
数据项表 data item table
数据项级 data item level
数据项列表 data dictionary
数据项验证 data item validation
数据项再分组 regrouping data item
数据项组 data item group
数据消除 data take-off
数据消减 data degradation
数据小区组 blockette
数据信道 data channel
数据信道转换器 data channel converter
数据信号 data signal
数据信号传输 data signal(l)ing
数据信号传输率 data signal(l)ing rate
数据信号传输速率 rate of data signal(l)ing
数据信号传送率 data signal(l)ing rate
数据信号发生器 data information signal generator
数据信号放大器 data repeater
数据信号速率 data signal(l)ing rate
数据信息处理器 data handler
数据信息交换机 message switching system for data
数据信息交换系统 data message switching system
数据信息通路 data highway
数据信息线路 data information line
数据信息信号发生器 data information signal generator
数据修匀 data smoothing;graduation

S

of data
数据修正 data revision
数据需求 data requirements
数据序列 data sequence
数据旋转 rotation of data
数据选取速度 access speed
数据选通脉冲 data strobe pulse
数据选择时间 access time
数据巡回检测 data-logging
数据巡回检测器 data logger
数据巡回检测装置 data logger;data-logging equipment
数据压缩 data compaction;data compression;data condensation
数据压缩程序 data reduction program(me);data reduction subsystem
数据压缩技术 digital data compaction technique
数据压缩算法 data condensation algorithm
数据压缩通信[讯] data compression communication
数据压缩子程序 data reduction subroutine
数据压缩子系统 data reduction subsystem;reduction subsystem
数据验证 data verification
数据遥测(术)data telemetry;remote data telemetry
数据遥测系统 data telemetry system
数据遥示器 remote data indicator
数据要求 data demand
数据要求协议 data demand protocol
数据一览表 data sheet
数据一致性 data consistency[consistence]
数据异常 data exception
数据译码系统 data encoding system
数据易变性 data volatility
数据因子分析法 factor analysis method of data
数据引用错误【计】data-reference error
数据引用方法 method of data reference
数据映象 data mapping
数据优化计算机 data optimizing computer
数据有效性 data validation;data validity
数据右移 data shift right
数据预处理 data preprocessing
数据预(处理)加工【计】predigestion of data
数据预先加工 data origination;data preparation;predigestion of data
数据预置语句 data initialization statement
数据元件 data cell;data element
数据元素 data element
数据元字典 data element dictionary
数据源 data pool;data source
数据约束 data constraint
数据阅读器 data reader
数据载波 data carrier
数据载波故障检测器 data carrier failure detector
数据载波检波器 data carrier detector
数据载体存储器 data carrier storage
数据载子 data carrier
数据再构成 data restructuring
数据再生 data reproduction
数据暂时存储区 temporary area
数据栈 data stack
数据站 data station
数据真实性 data validity
数据阵列 array of data;data array
数据整理 data ordering;data reduction
数据整理系统 data reduction system
数据整删 data reduction
数据帧 data frame
数据值范围检查 range check
数据指定 data specification
数据指针 data pointer
数据质量监控系统 data quality monitoring system
数据中继 data relay
数据中继器 data link
数据中继卫星 data relay satellite
数据中继卫星系统 data relay satellite system
数据中心 data center[centre]
数据中转 data relay
数据终端 data terminal
数据终端就绪 data terminal ready
数据终端设备 data terminal equipment
数据终端设备标准接口 EIA-RS 232 interface
数据终端信息自动编码和传输系统 data terminal message compiler and transmission system
数据终端装置 data terminal unit
数据属性 data attribute
数据属性表 data attribute list
数据转储 data dump
数据转发卫星 data relay satellite
数据转换 data conversion;data dump;data-switch(ing);data transfer;data transition;data translation
数据转换变形 transformation
数据转换程序 data converter
数据转换接口控制器 data transfer interface controller
数据转换接收器 data conversion receiver
数据转换器 data converter
数据转换实用程序 data conversion utility
数据转换特性 data conversion feature
数据转换系统 data conversion system;data transfer system
数据转换线 data conversion line
数据转换元件 data-handling component
数据转换中心 data-switching center[centre]
数据转换装置 data transfer unit
数据转接 data-switch(ing)
数据转接器 data adapter;data adapter unit
数据转接装置 data adapter unit
数据转录 data synchronization;data transcription
数据转录设备 data transcription equipment
数据转送中心 data relay center[centre]
数据转移 data transfer
数据装置 data device;data set;data unit
数据装置标记 data set label
数据装置就绪 data set ready
数据装置同步 data set clocking
数据状态 data mode
数据状态字 data status word
数据准备 data preparation
数据资料 data file
数据资料处理负责人 data processing manager
数据资料记录 data-logging
数据资料库 data bank
数据资料仪器操作 data instrumentation
数据子集 data subset;subset data
数据子句 data clause
数据子模型 data submodel
数据子语言 data sublanguage
数据字 data word
数据字典 data dictionary
数据字节 data byte
数据自动采集与控制系统 automated data acquisition and control system
数据自动抽出 automated data extraction
数据自动处理 datamation
数据自动处理机 autodata processor
数据自动处理器 data handler
数据自动处理系统 data automation system
数据自动传输器 data link
数据自动传输装置 communication link;information link;tie line;tie-link;data link
数据自动传送介质 automated data medium
数据自动化 datamation
数据自动记录 data record(ing);data-logging
数据自动记录系统 data-logging system;data recording system
数据自动记录仪 data logger
数据自动记录装置 automatic data recording device
数据自动检查 automatic inspection of data
数据自动交换系统 automated data interchange system
数据自相关 autocorrelation of data
数据自整角机 data synchro
数据总库 data bank
数据总线 data bus
数据总线接通 data bus enable
数据总线耦合器 data bus coupler
数据组 case;data array;data field;data packet;data set
数据组传送终了符 end-of-transmission block
数据组传送终了符号 end-of-transmission block character
数据组定时 data set clocking
数据组混合 amalgamation of the data sets
数据组结束 end of data block
数据组末信号 end of block signal
数据组世代 generation data group
数据组项目 group item
数据组织 data organization
数据最佳计算机 data optimizing computer
数据作业控制综合设施 data operation control complex
数均 number average
数均分子量 number average molecular weight
数控 digital control;numeral system;numeric(al)control
数控车床 numeric(al) controlled lathe
数控冲床 numeric(al)control press
数控带 numeric(al)control tape
数控带的信道 column of tape
数控电视 digital television
数控电子进给磨床 numeric(al)controlled and electronic-infeed grinder
数控绘图 automatic data plotting;computer support mapping;digital controlled drawing;digital drawing;digital plot;plot data;plot the data
数控绘图机 automatic data plotter;automatic numeric controlled plotter;cartographic(al)digitizing plotter system;digital plotter;digital tracing machine;digital tracing table;numeric(al) control drawing machine;numeric(al) plotter;Calcomp <商品名,美国>
数控绘图系统 cartographic(al)digitizing plotter system
数控绘图桌 digital control table;digital tracing table
数控机床 numeric(al) control machine;numerically controlled machine;numerically controlled machine tool
数控机器人 numeric(al) control robot
数控机械手 numeric(al) control robot
数控计算机 digital control computer;numeric(al)control computer
数控切割 numeric(al)control cutting
数控切割机 numeric(al)control cutting machine
数控切削 numeric(al)control cutting
数控绕线机 numeric(al)control filament winder;numerically controlled filament-winding machine
数控设备 numeric(al)control device
数控深孔钻床 numeric(al) control deep hole drilling machine
数控调直钢筋切断机 Nc steel straight-cutting machine
数控铣床 numeric(al)control milling machine
数控系统 numeric(al) control;numeric(al)control system
数控显示 digital display
数控氧气切割 numeric(al)controlled oxy-cutting
数控振荡器 number control oscillator
数控正射投影纠正仪 numeric(al) control orthophotoscope
数控中心 numeric(al)control center[centre]
数控装配机 numeric(al) assembling machine
数控钻机 control figures drill
数控坐标量测仪 numerically controlled comparator
数控坐标磨削 numeric(al)controlled jig grinding
数理测量学 mathematical geodesy
数理差异 quantity variance
数理地理学 mathematic(al) geography
数理发生法 mathematic(al)genetics
数理分析 mathematic(al)analysis
数理符号逻辑 logistic
数理概率 mathematic(al)probability
数理规划法 mathematic(al)programming
数理经济学 mathematic(al) economics
数理逻辑 logistic;mathematic(al) logic;symbolic logic;logic(al) for mathematicians
数理名学 mathematic(al) logic;symbolic logic
数理气候 mathematic(al)climate;solar climate
数理气象学 mathematic(al) meteorology
数理人口学 mathematic(al)demography
数理适宜 mathematic(al) optimization
数理天文 mathematic(al)astronomy
数理统计分析 mathematic(al) statistic(al)analysis
数理统计分析法 mathematic(al) statistic(al)method
数理统计相关 statistic(al)correlation
数理统计学 mathematic(al)statistics

S

数理统计学理论 mathematic(al) theory of statistics
数理统计预报方法 forecasting by mathematic(al) statistics
数理学派 mathematic(al) school
数理语言学 mathematic(al) linguistics
数理哲学 mathematic(al) philosophy
数例 numeric(al) example
数粒法 number grain method
数量 amount; deal; number; numeric(al) measure; quantity
数量变化 change in value; quantitative change; quantitative variation; quantity change
数量变量 quantitative variable
数量标志 quantity mark
数量标准 quantitative criterion; quantitative standard
数量表 bill of quantities; quantity sheet; table of quantities
数量表达式 quantitative expression
数量表示 quantitate
数量表示法 quantitative representation
数量不确定因素 nonquantifiable factor
数量不足 quantity not sufficient
数量查定 quantitative assessment
数量成本 volume cost
数量成熟期 quantitative maturity
数量大 high number
数量待定的合同 open-end contract
数量待定的契约 open-end contract
数量单位 quantity unit; unit of quantity
数量的 quantitative; scalar
数量等级 quantitative rating
数量调查 quantity survey(ing)
数量多 high number
数量方法 quantitative approach; quantitative method
数量分布 quantity distribution
数量分配 quantity allocation
数量分滩 quantity allocation
数量分析 quantitative analysis
数量符号 quantitive attribute
数量复验 check weight
数量概算 quantity survey(ing)
数量概算法 < 用于房地产评估 > quantity survey method
数量估计 quantity survey(ing)
数量估算 quantity survey(ing)
数量关系 quantitative relationship
数量管理 quantitative management
数量管制 quantitative restriction
数量合同 quantity contract; quantum contract
数量和百分比 number and percentage
数量化理论 theory of quantification
数量化理论程序 theory of quantification program(me)
数量化理论模型 theory model of quantification
数量积 dot product; scalar product
数量级 order(of magnitudes)
数量级相关 order-dependence
数量计算 computation of quantities
数量记录 quantity recording
数量减少 depletion of magnitudes
数量检验 inspection of quantity
数量检验书 inspection certificate of quantity
数量界限 quantitative limit
数量金字塔 pyramid of numbers; quantitative pyramid
数量控制 quantity control
数量控制器 quantity controller
数量密度 number density
数量明细表 bill of quantities; schedule of quantities
数量平均分子量 number average molecular weight
数量平均聚合度 number average degree of polymerization
数量清单 bill of quantities; schedule of quantities
数量试验模型 scale experimental model
数量索赔 quantity claim
数量特征 quantitative character; quantitive attribute; scalar property
数量条件 term of quantity
数量调节 quantity control
数量限制器 quantity limiter
数量性状 quantitative character
数量有某人决定 quantity at one's option
数量预报 quantitative forecast(ing)
数量增加 quantities uplifted
数量增减 quantity change
数量增减范围条款 more or less terms
数量折扣 quantity discount; volume discount
数量证明书 certificate of quantity
数量证书 certificate of quantity
数量指标 quantitative index (number); quantity index(number)
数量指数 quantitative index (number); quantity index(number)
数量质量流程图 quantitative-qualitative flow graph
数量属性 quantitative attribute
数量锥体 pyramid of numbers
数量资料 quantitative data
数列 sequence; sequence of numbers
数路供电用电池组 banked battery
数轮 numeral wheel
数论【数】 number theory; theory of numbers
数论法 number-theoretic method
数论泛函 number-theoretic function
数论函数 number-theoretic function
数码 code number; digit; digital code; figure; numeral
数码表示的图幅 figure shown by number
数码代号 identification code number
数码带射流阀 coded tape valve
数码发生器 number generator
数码管 nixie display indicator; nixie light; nixie tube
数码管读出装置 nixie readout
数码管枪 charactron gun
数码管译码器 nixie decoder
数码键 figure key
数码键位 figure shift
数码孔 numeric(al) aperture
数码控制的 numerically controlled
数码控制式机床 numeric(al) control machine
数码面积 digit area
数码模型 digital model
数码目标特性 digital target characteristic
数码钮 digital knob
数码示像 numeric(al) aspect
数码式图像 picture in digital form
数码显示器 nixie display
数码信号示像 numeral signal aspect
数码压模 number stamp set
数码掩模 cipher mask
数码影像光盘 < 数码多功能光盘 > digital versatile disc
数码影音光碟 digital video disc
数码影音光盘 digital video disc
数码遮掩装置 figure mask
数码总线 number bus
数码组合 grouping of figure
数密度 number density
数模 digifax
数模电路 digilogue circuit; digital-to-analog(ue) circuit
数模换能器 digital-to-analog (ue) converter
数模信道 digilogue channel; digital-to-analog(ue) channel
数模译码器 digital-to-analog(ue) decoder
数模转换 digital-to-analog(ue) conversion
数模转换阶梯 digital-to-analog(ue) ladder
数模转换器 digital-to-analog (ue) converter
数模转模拟器 digital-to-analog(ue) simulation
数目 amount; number; population
数目表 numeric(al) statement
数目词 entry word
数目极小的钱 dime
数目计算 numbering
数目顺序 numeric(al) order
数目字 numeric(al) word
数筛 number sieve
数式 numeric(al) expression
数算符 number operator
数突变 numeric(al) mutation
数位 digit order number; numeric(al) digit
数位波形 digit wave form
数位电极 digit electrode
数位叠加和 sideways sum
数位分配常数 layout constant
数位计数器 digit counter
数位减缩 digit compression
数位列 digit column
数位流传输 stream bit transmission
数位脉冲 digit impulse
数位面 digit plane
数位面驱动器 digit plane driver
数位配置 digit layout
数位配置参数 digit layout parameter
数位容量 digit capacity; digit spacing
数位时隙 digit time slot
数位位置 bit location; digit position
数位线 digit line
数位压缩 digit compression
数位延迟 digits delay
数位总和 hash total
数系 number system; numeral system; numeric(al) system
数显千分尺 digimatic micrometer
数行列式【数】 Jacobian
数形算术项 numeric-pictured arithmetic item
数序(列) number sequence; sequence of numbers
数学 mathematics
数学摆 mathematic (al) pendulum; perfect pendulum; simple pendulum
数学编译程序 mathematic(al) compiler
数学变换 mathematic(al) manipulation
数学表示法 mathematic(al) representation
数学表征 mathematic(al) characterization
数学参数 mathematic(al) parameter
数学残数 mathematic(al) residues
数学常数 mathematic(al) constant
数学程序 mathematic(al) routine
数学程序编制 mathematic(al) programming
数学程序设计 mathematic(al) programming
数学程序设计系统 mathematic(al) programming system
数学处理 mathematic(al) treatment
数学船模 mathematic(al) ship model
数学大地测量学 mathematic(al) geodesy
数学的反演 mathematic(al) inversion
数学的或然律 mathematic(al) law of probability
数学的期望(值) mathematic(al) expectation
数学地理学 mathematic(al) geography
数学地形模拟算法 mathematic(al) terrain analog(ue) computation
数学地形模型 mathematic(al) terrain model
数学地质(学) geomathematics; mathematic(al) geology
数学定律 mathematic(al) law
数学定义 mathematic(al) definition
数学对象 mathematic(al) object
数学法 mathematic(al) method
数学方程 mathematics equation; numeric(al) equation
数学方法预报天气 numerical forecasting
数学方法预测气候 numerical weather prediction
数学放样 mathematic(al) lofting
数学分解 mathematic(al) decomposition
数学分析 mathematic(al) analysis
数学符号 mathematic(al) symbol
数学公理 mathematic(al) axiom
数学固有函数 mathematic(al) built-in function
数学关系 mathematic(al) relation
数学归纳法 complete induction; mathematic(al) induction
数学规划 mathematic(al) programming
数学规划模型 mathematic(al) planning model
数学函数 mathematic(al) function
数学函数程序 mathematic(al) function program(me)
数学函数库 mathematic(al) function library
数学和统计学 mathematics and statistics
数学化 mathemati(ci)zation
数学换算 mathematic(al) conversion
数学机器理论 mathematic(al) machine theory
数学基础 basic mathematics; mathematic(al) basis; mathematic(al) foundation
数学计算 mathematic(al) computation
数学计算过程 mathematic(al) proceeding
数学计算装置 mathematic(al) equipment
数学家 mathematician
数学家海山 mathematicians seamount
数学检验 mathematic(al) check
数学简化 mathematic(al) reduction; mathematic(al) simplification
数学交通模型 traffic model
数学校核 mathematic(al) check
数学校验 mathematic(al) check
数学结构 mathematic(al) structure
数学解(答) mathematic(al) solution
数学解释 mathematic(al) treatment
数学解析 math analysis; mathematic(al) analysis
数学近似法 mathematic (al) approach; mathematic(al) approximation
数学经验 mathematic(al) experience
数学矩阵 mathematic(al) matrix
数学控制 numerical control

S

数学控制方式 mathematic(al) control mode
数学控制理论 mathematic(al) control theory
数学控制模型 mathematic(al) control model
数学理论 mathematic(al) theory
数学理想化 mathematic(al) idealization
数学例行子程序 mathematic(al) subroutine
数学论述 mathematic(al) treatment
数学论证 mathematic(al) argument
数学逻辑 mathematic(al) logic
数学面 mathematic(al) surface
数学模拟 mathematic(al) model(l)ing; mathematic(al) shaping; mathematic(al) simulation
数学模拟的方法 method of mathematic(al) simulation
数学模拟的类型 type of mathematic(al) simulation
数学模拟实验 mathematic(al) model experiment
数学模式<解决科学技术问题的数学形式> mathematic(al) mode
数学模数 mathematic(al) module
数学模型 mathematic(al) model
数学模型法 mathematic(al) model method
数学模型试验 verification of mathematic(al) model
数学平均 mathematic(al) mean
数学期望公式 mathematic(al) expectation formula
数学期望值 mathematic(al) expectation value
数学曲线拟合 mathematic(al) curve fitting
数学上的危险 mathematic(al) danger
数学设备 mathematic(al) instrument
数学生态学 mathematic(al) ecology
数学式 mathematic(al) expression
数学式计算机 digital computer
数学适合 mathematic(al) optimization
数学收敛系统 mathematically convergent system
数学术 mathematicasis
数学算符 mathematic(al) operator
数学体系 mathematic(al) system
数学通则 mathematic(al) generalization
数学统计或说明 numeric(al) statement
数学投影 mathematic(al) projection
数学推理 mathematic(al) induction
数学推论 mathematic(al) inference
数学推算 mathematic(al) derivation
数学物理方程 equation of mathematic(al) physics
数学物理模型 mathematic(al) physics model
数学物理学 mathematic(al) physics
数学系统理论 mathematic(al) system theory
数学形式 mathematic(al) form
数学形式体系 mathematic(al) formalism
数学样条函数 mathematic(al) splines
数学仪器 mathematic(al) instrument
数学应用 digital operation
数学用表 mathematic(al) table
数学优化 mathematic(al) optimization
数学优选法 mathematic(al) optimization method
数学语义学 mathematic(al) semantics
数学预测 mathematic(al) prediction
数学运算 mathematic(al) manipulation
数学造型 mathematic(al) model(l)ing
数学阵列 mathematic(al) arrays
数学制图学 mathematic(al) cartography
数学转换 mathematic(al) transformation
数学作图 mathematic(al) construction
数样本 numeric(al) example
数以千计 count by thousands
数域 number field
数域描述符 numeric(al) field descriptor
数域数据 numeric(al) field data
数域说明符 numeric(al) field descriptor
数域形式 numeric(al) field form
数值 magnitude; noise background; number; numeric(al) number; numeric(al) value; value
数值逼近 numeric(al) approximation; digital approximation
数值比较 numeric(al) comparison
数值比较装置 number comparing device
数值比例 numeric(al) proportion
数值编辑字符 numeric-edited character
数值编码 numeric(al) coding
数值变化 change in value; numeric(al) change
数值变位码 numeric(al) conversion code
数值表 numeric(al) table; numeric(al) tabular
数值表达式 numeric(al) expression
数值表示 numeric(al) representation
数值表示法 numeric(al) expression
数值不稳定性 numeric(al) instability
数值部分 mantissa; numeric(al) part
数值差量 dispersion of numbers
数值常数 numeric(al) constant
数值处理 numeric(al) processing; numeric(al) treatment
数值处理方法 numeric(al) procedure
数值传输线 number transfer bus
数值错误 numeric(al) fault
数值打字杆 numeric(al) type-bar
数值代数 numeric(al) algebra
数值单位 numeric(al) unit; value of unit
数值当量 numeric(al) equivalent
数值的 numeric(al)
数值的离散 dispersion of numbers
数值的稳定性 numeric(al) stability
数值的逐步逼近法 numeric(al) step by step method
数值等级 numeric(al) class; value class
数值等价的 numerically equivalent
数值点 numeric(al) point
数值读出 numeric(al) read-out
数值法 numeric(al) representation; numeric(al) method
数值法分析 numeric(al) method analysis
数值法说明 explanation of numeric(al) method
数值法演题步骤 operation of numeric(al) method
数值范围 number range; numeric(al) range; range of values
数值方法 numeric(al) approach; numeric(al) method
数值分布曲线中频率最高的数值 mode
数值分段法 numeric(al) paneling
数值分类学 numeric(al) taxonomy
数值分析 number analysis; numeric(al) analysis; numeric(al) calculus
数值估计 number estimate; numeric(al) estimate
数值关系 numeric(al) relation; numeric(al) relationship
数值函数 numeric(al) function
数值积分 numeric(al) integration; number integration; numeric(al) quadrature
数值积分法 numeric(al) integrating; numeric(al) methods of integration
数值积分器 digital differential analyser[analyzer]
数值积分误差 numeric(al) integration error
数值级数 numeric(al) series
数值计算 number calculation; number computation; numeric(al) calculation; numeric(al) calculus; numeric(al) computation; numeric(al) evaluation
数值计算法 method of numeric(al) calculation; number method; numeric(al) method
数值计算过程 number calculation process
数值计算误差 error in numeric(al) calculation
数值计算研究机 number cruncher
数值计算用软件 mathematic(al) software
数值计算有关参数 parameter related to digital calculation
数值记录表 data logger
数值技术 numeric(al) technique
数值校验 numeric(al) check
数值解 arithmetic(al) solution; numeric(al) calculation
数值解法 numeric(al) solution
数值解水质模型 digital solution of water quality model
数值近似法 digital approximation; number approximation; numeric(al) approximation
数值精度 numeric(al) accuracy; numeric(al) precision
数值距离 numeric(al) distance
数值科学 numeric(al) science
数值刻度 numeric(al) scale
数值孔径 numeric(al) aperture
数值孔径计 apertometer
数值控制 numeric(al) control
数值控制机 numeric(al) control machine
数值控制设备 numeric(al) control device
数值控制语言处理机 numeric(al) control language processor
数值控制装置 numeric(al) control device
数值口径 numeric(al) aperture
数值口径计 apertometer
数值离差 dispersion of numbers
数值离散性 scatter of value
数值例 numeric(al) example
数值流观测 numeric(al) flow visualization
数值流型方法 numeric(al) manifold method
数值码 numeric(al) code
数值模拟 numeric(al) model(l)ing; numeric(al) simulation
数值模拟方法 numeric(al) analog(ue) method
数值模拟计算 mathematic(al) model calculation
数值模拟流 numerically stimulated flow
数值模拟验证 verification of mathematic(al) model
数值模拟中心 numeric(al) modeling center[centre]
数值模式 numeric(al) mode
数值模型 numeric(al) model
数值内插法 numeric(al) interpolation
数值偏心(率) numeric(al) eccentricity
数值评价 numeric(al) evaluation
数值求面积 numeric(al) quadrature
数值求面积法 numeric(al) quadrature
数值三角学 numeric(al) trigonometry
数值色差 numeric(al) colo(u)r difference
数值实验 numeric(al) experiment
数值式 digital
数值试验 numeric(al) experimentation
数值数据代码 numeric(al) data code
数值数学 numeric(al) mathematics
数值衰减因子 numeric(al) damping factor
数值搜索 numeric(al) search
数值特性 numeric(al) characteristic
数值特征 numeric(al) characteristic
数值特征曲线 numeric(al) characteristic curve
数值天气预报 numeric(al) weather forecasting; numeric(al) weather prediction
数值调整 figure adjustment
数值图解 numeric(al) graph
数值图解法 numeric(al)-graphic(al) method
数值图像变量 numeric(al) pictured variable
数值图像格式项 numeric(al) picture format item
数值图像说明 numeric(al) picture specification
数值微分法 number differentiation; numeric(al) differentiation
数值文字 numeric(al) literal
数值稳定性 numeric(al) stability
数值问题 numeric(al) problem
数值误差 numberical fault; numeric(al) error; numeric(al) fault
数值向量 numeric(al) vector
数值项 numeric(al) item; numeric(al) term
数值信号 numeric(al) signal
数值岩土力学 numeric(al) geomechanics
数值研究 numeric(al) study
数值因子 numeric(al) factor
数值预报 mathematic(al) forecast(ing); numeric(al) prediction; quantitative forecast(ing)
数值预报法 numeric(al) forecast(ing)
数值张量 numeric(al) tensor
数值转换代码 numeric(al) conversion code
数值字 numeric(al) word
数值字符数据 numeric(al) character data
数值自动规格化 automatic number normalization
数制 base notation; number representation system; number system; numeric(al) system
数周期 one number time
数轴 number axis
数轴及记录设备 axle counting and recording device
数轴器 axle counter
数轴装置 axle-counting apparatus
数属性 number attribute
数转换 number conversion

数转换器 number converter
数字 digital number; figure; cipher; number; numeric(al); numeric(al) character; numeric(al) digit
数字安置盘 dial for number setting
数字白板 digital white board
数字包裹 digital packet
数字倍减器 digital demultiplier
数字比较 numeric(al) comparison
数字比较器 digital comparator
数字比例尺 digital scale; natural scale; numeric(al) scale; representation fraction
数字比例量 proportional
数字比率表 digital ratiometer
数字比时同步指示器 digital synchronometer
数字编程资料自动记录仪 programmable digital data logger
数字编号 <车站、调车场、货物品名> numeric(al) code
数字编辑 digital editing
数字编辑项 numeric(al) edited item
数字编辑字符 numeric(al) edited character
数字编码 digital code; digital coding; figure code; numeric(al) coding
数字编码的字符集 numeric(al) coded character set
数字编码器 digital(en) coder
数字编码器手册 digital encoder handbook
数字编码声音 digit-coded voice
数字编码系统 digital encoding system
数字编码指令 numerically coded instruction
数字变换 digital conversion; mathematic(al) manipulation
数字变换接收机 digital conversion receiver
数字变换码 numeric(al) conversion code
数字变换器 digital converter; digital translator; digit(al) izer
数字变量 numeric(al) variable
数字变元指示符 numeric(al) argument indicator
数字标度盘 digital dial
数字标号 number designation
数字标记 figure notation; number symbol; numeric(al) symbol
数字表 numeration table
数字表达 numeric(al) expression
数字表达式 mathematic(al) expression
数字表面模型 digital surface model
数字表示灯 numeral light; numeral light indicator
数字表示法 digital representation; discrete representation; number representation; number representation system; numeric(al) representation
数字表示器 numeral indicator
数字表示制 number representation system
数字拨号 digit dial(l) ing
数字拨号锁 numeral type dial lock
数字拨号音频 data dial tone
数字波面测量干涉仪 digital wavefront measuring interferometer
数字不稳定性 numeric(al) instability
数字步进式记录器 digital stepping recorder
数字部分 numeric(al) part; numeric(al) portion
数字采集系统 digital acquisition system
数字操作数 numeric(al) operand
数字操作系统 digital operation system
数字测井仪 digital logger
数字测距装置 digital range unit
数字测量仪 digital measuring apparatus
数字测图 digital restitution; numeric(al) restitution
数字测图系统 digital mapping system
数字测图仪 digital restitution instrument
数字测微计 digital micrometer
数字插入装置 digital insertion unit
数字常数 numeric(al) constant
数字乘法器 digital multiplier; digital multiplier unit
数字乘法装置 digital multiplier
数字乘积 product of numbers
数字程控交换机 digital program(me) controlled switch
数字程序化的伺服机构 digitally programmable servo
数字程序控制 digital process control
数字程序自动控制 prodac control; programmed digital automatic control
数字迟延信号发生器 digital delay generator
数字冲模 numbering die
数字冲压机 figure punch
数字除法器 digital divider
数字储存器 digital memory
数字处理 digital handling; digital processing; numeric(al) process
数字处理器 digital processing unit
数字处理系统 digital processing system
数字处理装置 digital processing unit
数字穿孔(机) digit(al) punch; numeric(al) punch
数字传感器 digital sensor; digital transducer
数字传声器 digital microphone
数字传输 digital transmission
数字传输系统 digital communication system; digital transmission system
数字传送干线 number transfer bus
数字传送总线 number transfer bus
数字传信 digital signal(l) ing
数字垂线偏转 digital vertical deflection
数字磁带格式 digital-tape format
数字磁鼓 digital drum
数字次序 numeric(al) order
数字存储器 digital memory; digital storage; number storage
数字存储系统【计】 digital storage system
数字错误 numeric(al) error
数字打印机 digital printer; numeric(al) printer
数字大气模型 digital atmospheric model
数字代号 digital name; numeric(al) name; digital code
数字代码 numeric(al) code
数字代码盘 digital code wheel
数字代码输入 numeric(al) code input
数字带 digit(al) tape; number tape; numeric(al) tape
数字倒置 transposition; transposition of figures
数字捣弄 number crunching
数字道 digit path; digit track
数字道路地图 digital road(street) map
数字道路拓朴图 digital roadway topological map
数字的 digital; numeric(al); numeral
数字的尺度 numberical scale
数字的二进制 binary system of figures
数字的密度 numberical density
数字的最左位 high order position
数字低通滤波 digital low-pass filtering
数字低通滤波器 digital low-pass filter
数字地面模型 digital terrain model; digital ground model
数字地图 digital map; numeric(al) map
数字地图信息处理 dealing with digitized cartographic(al) data
数字地图修测 digital map revision
数字地图制图 digital cartography
数字地形模数 digital terrain module
数字地形模数程序 digital terrain module program
数字地形模型 digital surface model; digital terrain model
数字地震计 digital seismometer
数字地震(记录)系统 digital seismic system
数字地震检波器 digital geophone
数字地震图 digital seismogram
数字地址 numeric(al) address
数字地址码 number address code
数字电法仪 digital instrument for electric(al) method
数字电话网 digital telephone network
数字电路 digital circuit
数字电路系统 digital circuitry
数字电路族 digital circuit family
数字电码 figure code
数字电码变换器 digital code converter
数字电码组合 numeric(al) character
数字电脑 digital computer
数字电视 digital television
数字电视监视器 digital television monitor
数字电视检测器 digital television monitor
数字电视系统 digital television system
数字电视转换器 digital television converter
数字电位计制式 digital potentiometer system
数字电压(编)码 digital voltage code
数字电压译码器 digital voltage encoder
数字电子地图 digital electronic map
数字电子读数机 electron digital counter
数字电子计算机 digital computer; digital electronic brain; electronic digital computer
数字电子计算机工作程序 digital computer routine
数字电子学 digital electronics
数字电阻应变仪 digital resistance strain indicator
数字调度电话车站分系统 station sub-system of the digital dispatching phone
数字调度电话主系统 main system of the digital dispatching phone
数字定向 digital orientation; numeric(al) orientation
数字读出 numeric(al) read-out
数字读出光度计 digital readout photometer
数字读出辉光管 digitron
数字读出机 digitiser
数字读出计时器 digital readout timer
数字读出示波器 digital readout oscilloscope
数字读出系统 digital readout system
数字读出仪 digital readout device
数字读出装置 digital readout
数字段 digital section
数字断面数据 digital profile data
数字多波束基阵 digital multibeam steering array
数字多波束扫描声呐 digital multibeam scan(ning) sonar
数字多路同步器 digital multiplexing synchronizer
数字多用表 digital multimeter
数字耳机 digital earphone
数字二进位脉冲 digit bit pulse
数字发射器 digit emitter
数字发生器 number generator
数字发送器 digital transmitter; digit emitter
数字法畸变校正 numeric(al) distortion correction
数字番号 numeric(al) designation
数字反馈 digital feedback
数字反射全息图 digital reflection hologram
数字方程 digital equation
数字方程式 numeric(al) equation
数字方法 digital method
数字仿真 digital simulation
数字仿真模拟计算机 digital simulated analog(ue) computer
数字分辨率 digital resolution
数字分接 digital demultiplexing
数字分接器 digital demultiplexer
数字分类 digital sorting
数字分类学 numeric(al) taxonomy
数字分配架 digital distribution frame
数字分频器 digital frequency divider
数字分析 numeric(al) analysis
数字分析仪 digital analyser[analyzer]
数字分压器制式 digital potentiometer system
数字粉尘测量仪 digital dust measuring apparatus
数字符产生器 digit-symbol generator
数字符号 digital character; numeric(al) character; numeric(al) notation; numeric(al) symbol; symbol of numeral; digit sign
数字复接 digital multiplexing
数字复接设备 digital multiplex equipment
数字复用设备 digital multiplex equipment
数字复用体系 digital multiplex hierarchy
数字复用系列 digital multiplex hierarchy
数字伽马测量 digital gamma ray survey
数字改正 numeric(al) correction
数字高程模型 digital elevation model
数字高斯计 digital Gaussmeter
数字格式 number format
数字格形式 digital format
数字跟踪系统 digital servosystem
数字公路网 computerized highway network
数字功能 digital function
数字固态液晶显示仪表 solid-state liquid crystal digital
数字管 nixie tube
数字管驱动器 nixie driver
数字管线地图 digital cable map
数字光滑函数 digital smoothing function
数字光偏转器 digital indexed light deflector
数字光学跟踪装置 digital optical tracking set
数字滚筒 number roll
数字过程控制 digital process control
数字过程控制系统 digital process control system; numeric(al) process control system

数字海岸发生器 digital coast line generator
数字海底电缆 digital submarine cable
数字海流计 digital current meter;digital current recorder
数字函数发生器 digital function generator
数字盒式磁带 digital cartridge
数字盒式磁带机 digital cassette
数字盒式磁盘 digital cartridge
数字后备电路 digit backup
数字后援 digital back-up
数字化 digit(al)ization;digitize;numeralization;quantization
数字化版图 digitized layout
数字化比例尺 digitizing scale
数字化彩色信号 digitized colour signal
数字化测图 digital mapping;digitized mapping;measurements in digital;numeric(al)mapping
数字化的设计图 digitized layout
数字化地形数据 digitized terrain data
数字化点数据 digitized point data;quantized point data
数字化多媒体广播 digital multimedia broadcasting
数字化方法 digitization method
数字化分类程序 digital classification program(me)
数字化跟踪头 digitized
数字化过程 digitized processing;digitizing process
数字化盒式磁带加速度仪 digital cassette accelerograph
数字化回放系统 digitized playback system
数字化技术 digital technique
数字化加速度 digitized acceleration
数字化加速度图 digitized accelerogram
数字化阶段 digitizer stage
数字化精度 digitizing accuracy
数字化控制器 digital controller
数字化面 digitized surface
数字化模型 digitized model
数字化平台 digitized platform;digitized table
数字化器 digigrid;digitising[digitizing] board;digitising table;digitizing equipment;digitizing machine;quantizer
数字化器模拟设备 simdig;simulator of the digitizer
数字化器型号 digitizer format
数字化强震仪 digitized strong seismometer
数字化设备 digitizer
数字化摄影 digitized photography
数字化摄影测图 digital photogrammetric plotting
数字化视频信号 digitized video
数字化数据 digitized data
数字化台 digital ramp;digitized ramp;digitizing station;solid-state table
数字化图像 digitized image
数字化图像处理 digitized image processing
数字化图形输入板 digital picture input tablet
数字化文件 digitized file
数字化误差 digitalization error;digitized error
数字化系统 digital system
数字化小组 digital team
数字化信息 digitized information;digitized signal
数字化信息线路 digital message link
数字化仪 digitizer
数字化仪板 digitizing tablet
数字化仪点方式 point mode of digitizer
数字化仪器 digitized instrumentation
数字化仪原图定向 map orientation of digitizer
数字化影像 digitized image
数字化语言技术 digitized speech technology
数字化制图 digitized cartography;digitized mapping
数字化制图技术 digital mapping technique
数字化制图系统 system for digital mapping
数字化转换器 digitizer
数字化装置 digit(al)izer
数字化字符表示 digital character representation
数字化自动驾驶仪 digital autopilot
数字话路扩容系统 digital circuit multiplication system
数字话音插入 digital speech interpolation
数字话音内插信道 digital speech interpolation channel
数字环路载波 digital loop carrier
数字换挡 figure shift;numeric(al) shift
数字回声测深仪 digital echo sounder
数字回声抵消器 digital echo canceller
数字回声抑制器 digital echo suppressor
数字回溯法 digital back-up
数字回转打印机 number revolving stamp
数字绘图 digital plot
数字绘图机 digital plotter
数字绘图系统 digital plotting system
数字绘图仪 digital plotter
数字货币 digital money
数字机 digital exchanger
数字积分 digital integration;numeric(al)integration
数字积分电路 digital integrating circuit
数字积分机 digital integrator
数字积分计算机 digital integration computer
数字积分器 digital integrator
数字基带 digital baseband
数字基群 digital group;digroup;primary digital group
数字基准 numeric(al)reference
数字激光视盘 digital video disc
数字激光束偏转器 digital laser beam deflector
数字激励器 digit driver
数字集成电路 complex digital circuit;digital integrated circuit
数字集成电路元件 digital integrated circuit element
数字计数 digit count
数字计数器 digital counter
数字计算 digital computation;numeric(al)computation
数字计算法 numeric(al)procedure
数字计算机 digital computer;digital machine
数字计算机程序 digital computer program(me)
数字计算机程序编制 digital computer programming
数字计算机控制 numeric(al)control
数字计算机全息术 digital computer holography
数字计算机系统 digital computer system
数字计算机中心 digital computer center[centre]
数字计算机自动化 digital automatization
数字计算交通控制 digital traffic control
数字计算器 digital calculator
数字计算系统 digital computing system
数字记录 digital recording
数字记录打印机 number record printer
数字记录地震仪 digital seismograph
数字记录过程 digital recording process
数字记录盒式磁带机 data cartridge
数字记录回声测深 digitized echo-sounding
数字记录器 digital recorder;numeroscope
数字记录器件 numeric(al)recording device
数字记录设备 digital recording equipment
数字记录式岸用验潮仪 digitally recording offshore tide ga(u)ge
数字记录系统 digital recording system
数字记录仪 digital recorder
数字继电器 digital relay
数字寄存斑记 digit register spot
数字寄存点 digit register spot
数字加法器 digital adder
数字加工 digital processing
数字家用网 digital home network
数字间间隔 blank
数字间无间隔记录 non-return recording
数字间有间隔的记录 return recording
数字监控 digital supervision
数字减法器 digital subtracter
数字检波 digital detection
数字检测 digital detection
数字检定 digital detection
数字检验 character check
数字键 ten key
数字键(控)穿孔 numeric(al)keypunch
数字键控孔机 digital keypunch;numeric(al)keypunch
数字键盘 digital keyboard;numeric(al)keyboard;ten-key board
数字键盘装置 numeric(al)keyboard device
数字交叉连接设备 digital cross connect equipment
数字交换 digital switching
数字交换程序 digital exchange program(me)
数字交换机 digital switching system
数字交换模块 digital switching module
数字交换网络 digital switching network
数字交通控制 <用数字计算机的> digital traffic control
数字角位置 digital angular position
数字校验 digit check
数字校正 numeric(al)correction
数字阶部分 exponent part of number
数字接地模型 digital ground(terrain)model
数字接口 digital interface
数字接收站 digital receiving station
数字结果 digital result
数字结果记录 recording of digital result
数字解码器 digital decoder
数字解算器 digital resolver
数字解析测图仪 numeric(al)analytic compiler
数字近似值 digital anemometer
数字经济 digital economy
数字经纬仪 digital theodolite;digital transit
数字警报传送器 digital alarm transmitter
数字纠正 digital rectification
数字纠正放大机 numeric(al)rectifying enlarger
数字卷筒 digital roll;numeric(al)roll
数字卡尺(规) digital caliper
数字开关 digital switch;number switch;numeric(al)switch
数字刻度 digital scale;numeric(al)scale
数字孔 <卡片上的> digit punch
数字控制 digital control;numeric(al)control
数字控制测量机 numerically controlled measuring machine
数字控制冲床 numerically controlled punching machine
数字控制的 numerically controlled
数字控制工作台 numerically controlled table
数字控制划线机 numerically controlled layout machine
数字控制回转头压力机 numeric(al)controlled turret punch press
数字控制机 digitally controlled machine;numerically controlled machine
数字控制机床 numeric(al)control machine;numerically controlled machine tool
数字控制机床群 line-up of numeric(al)control machine tools
数字控制机器人 digital control robot
数字控制计算机 numeric(al)control computer;digital control computer
数字控制剪床 numerically controlled shears
数字控制磨床 numerically controlled grinding machine
数字控制普通车床 numerically controlled engine lathe
数字控制绕线机 numeric(al)control filament winder
数字控制设备 numeric(al)control device
数字控制镗床 numerically controlled boring machine
数字控制凸轮磨床 grinding machines cam numeric(al)control
数字控制卫星 digital controlled satellite
数字控制铣床 numerically controlled milling machine
数字控制系统 numeric(al)control system
数字控制折弯机 numerically controlled benders
数字控制转塔车床 numerically controlled turret lathe
数字控制装置 numeric(al)control device
数字控制钻床 numerically controlled drilling machine
数字扩展系统 digital expansion system
数字栏 numeric(al)field
数字雷达测高计 digiralt[digital radar altimeter]
数字雷达地面模拟系统 digital radar landmass simulation
数字类 numeric(al)class
数字类测试 numeric(al)class test
数字立体测图 digital stereocompilation;numeric(al)stereocompilation
数字例题 numeric(al)example
数字链接 digital connection
数字链路 digital link
数字量测仪 digital ga(u)ge

数字量读入 read digital input
数字量规 digital ga(u)ge
数字量化器 digital quantizer
数字流 digital stream
数字流量计 digital flowmeter
数字滤波 digital filtering
数字滤波法 digital filter method
数字滤波器 digital filter; numeric(al) filter
数字率 digit rate
数字轮 digital drum; digit wheel
数字逻辑 symbolic logic; digital logic
数字逻辑线路 digital logic routine
数字逻辑演算装置 digital logic trainer
数字码 cipher code; numeric(al) code
数字脉冲 digit(al)(im)pulse
数字脉冲持续时间 digital pulse duration
数字脉冲宽度 digit duration
数字脉冲序列 digital pulse sequence
数字面 digit plane
数字面积光度计 digital area photometer
数字描绘器 digital plotter
数字模拟 digital analog(ue); digital simulation; numeric(al) analogy; numeric(al) model(l)ing; numeric-(al) simulation
数字模拟变换器 digital-analog(ue) converter
数字模拟乘法器 digital-analog(ue) multiplication
数字模拟计算机系统 digital simulator computer system
数字模拟技术 digital simulation technique
数字模拟信息转换装置 digiverter
数字模拟型 digital-analog(ue) type
数字模拟译码器 digital-analog(ue) decoder
数字模拟语言 digital simulation language
数字模拟转变 digital-to-analog(ue) conversion
数字模拟转换 digital(-to)-analog-(ue) conversion
数字模拟转换电路 digital-analog-(ue) conversion circuit
数字模拟转换阶梯信号发生器 digital-to-analog(ue) ladder
数字模拟转换器 digital-analog(ue) converter
数字模拟装置 digital simulator
数字模片 digital template; number template
数字模式 digital model; figure pattern; numeric(al) mode
数字模型 digital model; numeric(al) model
数字模型试验 digital model test
数字内插 digital interpolation; numeric(al) interpolation
数字盘 dial; figure disk[disc]; number disc; number plate
数字配位 digital coordination
数字偏差 digital deflection
数字拼读法 figure of mark pronunciation
数字频率 numeric(al) frequency
数字频率测深仪 digital frequency sounding instrument
数字频率计 counter-type frequency meter
数字坡度模型 digital slope model
数字谱分析 digital spectral analysis
数字期望值 mathematic(al) expective value
数字旗 numeral pennant
数字启动 digital enable
数字气象站 numeric(al) weather facility
数字器 digitiser[digitizer]
数字千分表 digital dial ga(u)ge
数字签名 digital signature
数字区段 digital block
数字区分符 digit specifier
数字区域 numeric(al) area
数字驱动脉冲 digit drive pulse
数字驱动器 digit driver
数字全天候相机 digital all-sky camera
数字全息图 digital hologram
数字热电偶装置 digital thermocouple unit
数字热敏电阻装置 digital thermister unit
数字容量 numeric(al) capacity
数字扫描 digital scan
数字扫描器 digital scanner
数字筛选作用 digital filtering
数字上 numerically
数字舍入误差 error in rounding number
数字设备公司 digital equipment corporation
数字摄影 digital photography
数字摄影测量学 digital photogrammetry; numeric(al) photogrammetry
数字声音文件 digital audio
数字十进制计数器 digital decade counter
数字时间 digit time
数字时间发播 digital time dissemination
数字时钟 digital clock
数字识别系统 numeric(al) recognition system
数字示度 numeric(al) indication
数字示像 numeral aspect
数字式保护装置 digital protection device
数字式变阻器 digital rheostat
数字式测厚仪 digital thickness ga(u)ge
数字式测量仪 digital measuring instrument
数字式测斜计 digital inclinometer
数字式磁性强度仪 digital magnetic intensity instrument
数字式导航设备 digital navigation set
数字式地震记录系统 digital seismic system
数字式电路 digital circuitry
数字式电位计 <分压器> digital potentiometer
数字式电压表 digital voltmeter; digivolt(meter)
数字式电液控制系统 digital electric-hydraulic control system
数字式电子通用计算机 digital electronic universal computing engine
数字式仿拟 digital simulation
数字式分类 digital sort
数字式伏特计 digital voltmeter
数字式功能部件 digital function
数字式惯性导航系统 digital inertial navigation system
数字式海底地震仪 digital ocean-bottom seismograph
数字式计数器 digital counter
数字式记录 digital record(ing)
数字式记录分析器 digital recording analyser[analyzer]
数字式记录器 digital data recorder
数字式架空牵引网保护装置 numeric-(al) protection for overhead contact lines
数字式校表仪 visotest
数字式控制设计语言 digital control design language
数字式控制台 digital console
数字式控制系统 digital control system
数字式力传感器 digital force transducer
数字式模拟器 digital simulator
数字式欧姆表 digital ohmmeter
数字式频率分析器 digital frequency analyser[analyzer]
数字式频率机 digital frequency counter
数字式频率计 digital frequency meter
数字式频率监视器 digital frequency monitor
数字式频率显示 digital frequency display
数字式设备 digital equipment
数字式摄像机 digital camera
数字式深层地震仪 digital deep seismograph
数字式声呐 digital sonar
数字式石英校表仪 quartz-timer
数字式调速(器) digital transmission shift
数字式通知装置 digital announcement device
数字式同步语音数据 digital simultaneous voice data
数字式透射密度计 digital transmission densitometer
数字式万用表 digimer; digital multimeter
数字式微波接力通信[讯]系统 digital microwave relay system
数字式无线电仪表 digital radio meter
数字式误差监控分系统 digital error monitoring subsystem
数字式系数单元 digital coefficient unit
数字式显示 digital display
数字式延迟发生器 digital delay generator
数字式遥控装置 digital remote unit
数字式液晶显示管 signatron
数字式仪表 digital instrument
数字式应变仪 digital strain indicator
数字式远距离测量 digital telemeter; digital telemetering
数字式转矩计 digital type torquemeter
数字式转速计 digital tachometer
数字式自动化 digital autom(atiz)ation
数字式自动频率控制 digital automatic frequency control
数字式自动装置 digital automation device
数字式阻塞倒相放大器 digital block inverting amplifier
数字式阻塞双稳态触发器 digital block flip-flop
数字事件记录器 digital event recorder
数字视频 digital video
数字视频带宽 digital video bandwidth
数字视频网 digital visual frequency network
数字视频显示(器) digital-to-video display
数字视听委员会 Digital Audio Visual Council
数字输出 digital output
数字输出计时器 digital output timer
数字输入 digital input
数字输入数据 digital input data
数字数据 digital data; numeric(al) data
数字数据包 digital data packet
数字数据变换器 digital data converter
数字数据处理 digital data process
数字数据处理机 digital data processor
数字数据处理器 digital data processor
数字数据处理设备 digital data processing equipment
数字数据处理系统 digital data processing system
数字数据传输 digital data transmission
数字数据传输器 digital data transmitter
数字数据存储装置 digital data storage unit
数字数据带 digital data tape
数字数据发播系统 digital data broadcast system
数字数据发送器 digital data transmitter
数字数据分配器 digital data distributor
数字数据计算机 digital data computer
数字数据记录器 digital data recorder
数字数据记录系统 digital data recording system
数字数据接收机 digital data receiver
数字数据库 digital input base; numeric(al) data base
数字数据群 digital data group
数字数据收发机 digital data transceiver
数字数据收集系统 digital data acquisition system
数字数据输出变换单元 digital data output conversion element
数字数据通信[讯] digital data communication
数字数据网 digital data network
数字数据系统 digital data service; digital data system
数字数据显示系统 digital data display system
数字数据项 numeric(al) data item
数字数据消除 digital data wash
数字数据信道 digital data channel
数字数据终端 digital data terminal
数字数据自动传输线 digital data link
数字-数字数据变换器 digital-to-digital data converter
数字衰减系统 digital attenuator system
数字水位计 digital water-stage recorder
数字水准测量 digital leveling
数字水准仪 digital level
数字顺序 numeric(al) order
数字顺序控制 digital process control
数字伺服机构 digital servomechanism
数字伺服系统 digital servo
数字速度表示器 <司机室> digital speed indicator
数字速率 digital rate
数字算符 mathematic(al) operator
数字算子 numeric(al) operator
数字锁相环 digital phase-locked loop
数字台秤 digital bench scale
数字天气预报 numeric(al) weather prediction
数字填充 digital filling
数字调频 frequency shift keying
数字调谐 digital tuning
数字调制 digital modulation
数字调制解调器 digital modem
数字通带滤波器 digital bandpass filter
数字通道 digital path
数字通信[讯] digital communication
数字通信[讯]设备 digital communication set
数字通信[讯]系统 digital communication system
数字通用光盘 digital versatile disc

S

数字同步 digilock; digit synchronization
数字图解形式 graphic(al) digital form
数字图像 digital image; digital picture; numeric(al) picture
数字图像编码装置 digital image coding device
数字图像变换器 digital-to-image converter
数字图像处理 digital image processing
数字图像处理产品 digital image processing products
数字图像处理方法 digital image processing technique
数字图像处理技术 digital image processing technique
数字图像处理系统 digital image processing system
数字图像的重现 restoration of digital imagery
数字图像的恢复 restoration of digital imagery
数字图像广播 digital video broadcast
数字图像记录仪 digital image recorder
数字图像加工 digital image processing
数字图像扫描记录系统 digital image scanning and plotting system
数字图像相关 digital image correlation
数字图像帧结构 digital framing structure
数字图形处理 digital graphic(al) processing
数字网络 digital network
数字网络分析器 digital network analyser
数字网络体系 digital network architecture
数字微波电路 digital microcircuit
数字微波通信[讯] digital microwave communication
数字微波无线电 digital microwave radio
数字微波中继系统 digital microwave relay system
数字微电路 digital microcircuit
数字微分 numeric(al) differential; numeric(al) differentiation
数字微分分析机【计】digital differential analyser[analyzer]
数字微型组件 digital module
数字未记【计】dropout
数字位 digit(al) bit; numeric(al) bit; figure shift <电传打字机的>
数字位计算 layout count
数字位数据 numeric(al) bit data
数字位置 digit place; digit position
数字位置控制 numeric(al) positional control
数字位置模型 digital situation model
数字位置信息 digital position information
数字温度指示器 <其中的一种> pyrodigit
数字温盐深测量仪 digital salinity temperature and depth instrument
数字稳定性 numeric(al) stability
数字无绳电话 digital cordless telephone
数字无线电进局线路 digital radio entrance link
数字无线段 digital radio section
数字无线通道 digital ratio path
数字无线系统 digital ratio system
数字吸收 digit absorption
数字吸收器 digit absorber
数字吸收选择器 digit-absorbing selector
数字系数 digital coefficient; numeric(al) coefficient
数字系统 digital logic system
数字系统量化误差 digital system quantizing error
数字系统设计自动化 design automation of digital system
数字显示 digital present(ation); display in digital form; numeric(al) display
数字显示秤 circular dial-type scale
数字显示传感器 digital display sensor
数字显示单道地震仪 single trace seismograph with digital display
数字显示电路 digital display circuit
数字显示法 number telling method
数字显示管 numeric(al) indicator tube
数字显示计数器 digital-presentation counter
数字显示计算器 digital-presentation counter
数字显示器 digital display; digital indicating system; digital indicator; nixie display; numeric(al) indicator; numeroscope
数字显示器件 numeric(al) display device
数字显示示波器 digital display scope
数字显示式电子表 digital watch
数字显示式计量仪器 cyclometer counter
数字显示天平 circular dial-type scale
数字显示系统 digital indicating system
数字显示信号发生器 digital display generator
数字显示压力指示器 digital readout pressure indicator
数字显示仪 digital indicator
数字显示指示器 nixie display indicator
数字显示终端【计】alphanumeric video terminal
数字显示装置 digital readout; digital display device; digital display unit; numeric(al) display device; numeric(al) display unit; readout
数字显字法 number telling method
数字显字器 digital indicating system
数字线化图 digital line graph
数字线路 digital circuit
数字线路微型组件 digital circuit module
数字相关 digital correlation
数字相关器 digital correlation kit; digital correlator
数字相机 digital camera
数字相位计 digital phasometer
数字相位检测器 digital phase detector
数字镶嵌 digital mosaic
数字项 numeric(al) item
数字像管 digicon; digital image tube
数字像片镶图 digital photomosaic
数字协处理器 numeric(al) coprocessor
数字信道连接 digital channel link
数字信号 digital signal; numeric(al) signal
数字信号处理器 digital signal processor
数字信号传送总线 digit transfer bus; digit transfer trunk
数字信号发生器 digital signal generator
数字信号复接器 digital multiplexer
数字信号系统 digital signal(l)ing system
数字信号显示 numeral signal aspect
数字信号音 digital signal(l)ing tone
数字信号周期 digit signal(l)ing period; digit signal(l)ing time
数字信号转换器 digital signal converter
数字信令 digital signal(l)ing
数字信息 digital data; digital information; numeric(al) information
数字信息按专题分类 digital thematic classification
数字信息处理 digital information processing; numeric(al) information processing
数字信息处理系统 digital information processing system; numeric(al) information processing system
数字信息检查 digital information detection
数字信息输入系统 digital message entry system
数字信息输入装置 digital message entry device
数字信息调制系统 digital data modulation system
数字信息显示 digital information display
数字信息显示器 digital information display(unit)
数字信息显示系统 digital information display system
数字信息业务 digital data service
数字信息终端 digital message terminal
数字行列式 digital determinant; numeric(al) determinant
数字形式 digital form(at); figuration
数字形式的 figurative
数字序列 number sequence
数字序列完整性 digital sequence integrity
数字选择器 digit selector; numeric(al) selector
数字选择通信[讯] digital selective communication
数字选择性呼叫 digital selective calling
数字选择性呼叫设备 digital selective call equipment
数字选组单元 digital switching element
数字学 numerology
数字雪盖地图 digital snow map
数字寻呼系统 digital paging system
数字循环移动 digital loop movie
数字压缩 digit compression
数字延迟元件 digit delay element
数字遥测 digital telemeter(ing)
数字遥测寄存器 digital telemetry register
数字遥测术 digital telemetry
数字遥测系统 digital telemetry system
数字遥测装置 digital telemetry unit
数字移动 numeric(al) move
数字移相器 digital phase shifter
数字译码器 digital decoder
数字溢出计数器 digital down counter; digital up counter
数字因数 numeric(al) factor
数字印模 number stamp
数字印字机 digital printer
数字影像 digital image
数字用户环 digital subscriber loop
数字用户模块 digital subscriber module
数字用户终端 digital subscriber terminal
数字有线通信[讯]系统 digicom
数字与控制装置 digital processing and control unit
数字语言插入 digital speech interpolation
数字语言系统 digital voice system
数字语音编码 digital speech coding
数字域 numeric(al) field
数字域数据 numeric(al) field data
数字元地面模型 digital ground model
数字元模型 digital model
数字云图 digital cloud map
数字运算 digital operation
数字运算电路 digital operation(al) circuit
数字运算技术 operational digital technique
数字运算中心 digital arithmetic(al) center[centre]
数字再生器 digital regenerator
数字凿孔机 numeric(al) perforator
数字噪声 digital noise
数字增量绘图仪 digital incremental plotter
数字增强 digital enhancement
数字展示 digital readout
数字照相机 digital camera
数字值 digital quantity; numeric(al) bit; numeric(al) value
数字指令 digital command
数字指令系统 digital command system
数字指示 numeric(al) display; numeric(al) indication
数字指示灯 digital indicating lamp
数字指示管 digitron
数字指示器 digital indicator
数字指示系统 digital indication system
数字制 numeration system
数字制表机 numeric(al) tabulator
数字制图软件 digital mapping software
数字制图学 digital cartography
数字中继模块 digital trunk module
数字中继线 digital trunk
数字终端 digital terminal
数字终端局 digital end office
数字周期 one-digit time
数字属性 number attribute
数字注记线 digital line; numeric(al) line
数字转换 digital conversion; digitization; numeric(al) inversion; quantization
数字转换板 digitizing tablet
数字转换代码 numeric(al) conversion code
数字转换器 box car; digital converter; digital quantizer; digital transducer; digiverter; quantizer; quantizing encoder
数字转换型计算机 quantized computer
数字转接干线 digit transfer bus
数字转锁 trick lock
数字装置 digital device
数字资料 digital data; numeric(al) data
数字资料交换网 digital data exchange
数字资料收集和处理系统 digital data acquisition and processing system
数字资料收集和整理系统 digital data acquisition and processing system
数字子集 digital subset
数字字符集 numeric(al) character set
数字字符数据 decimal picture data; numeric(al) character data
数字字符子集 numeric(al) character subset
数字字符组 numeric(al) character set
数字字面文字 numeric(al) literal
数字字母发生器 character generator
数字字母符 numeric(al) alphabetic

数字字母显示器 character display
数字自动跟踪与测距 digital automatic tracking and ranging
数字自动化装置 digital automatization
数字自动机 digital automat
数字自动聚焦 digital automatic focus
数字自记器 digital recorder
数字自记水位仪 digital water-stage recorder
数字自适应技术 digital adaptive technique
数字综合电报 numeric(al) summary message
数字总和法<美国租赁业一种折旧法> sum-of-the-year digits
数字总线 number bus
数字组 blockette;digital group;subblock
数字组合器 digital combiner
数字作图 digiplot
数字坐标 digitized coordinates;numeric(al) coordinates
数字坐标输出 digital coordinate output
数组 array of data;digit group;group of numbers
数组变量 array variable
数组标识符 array identifier
数组标志 block mark
数组表 array list
数组表达式 array expression
数组操作 array manipulation
数组长度 block length
数组成分 array component
数组重新定义 array redimension
数组处理 array manipulation;array processing
数组处理机 array processor;arrester processor
数组处理计算 array processing computation
数组处理内部函数 array manipulation built-in function
数组处理器 array processor
数组代数 array algebra
数组的表示 representation of arrays
数组地址 block address
数组定义 array defining
数组段 array segment
数组多路信道 block multiplexer channel
数组分量 array component
数组分配 array allocation
数组符号表 array symbol table
数组复写子程序 array copy subroutine
数组赋值 array assignment
数组划分 array partitioning
数组计算机 array computer
数组间隙 block gap
数组间运算 array-array operation
数组校验 array verify
数组结构 array architecture
数组界 array bound
数组距 array pitch
数组块 array box
数组宽度 array extent
数组连接字段 array linkage field
数组流 array stream
数组逻辑 array logic
数组名字表 array name table
数组模块 array module
数组取消符号 block cancel character;block ignore character
数组说明 array declaration
数组说明符 array declarator
数组说明语句 array declarator statement
数组算法 array algorithm
数组维数 array dimension
数组文件 array file
数组下标 array index;array subscript
数组向量 array vector
数组项 array item
数组信息 dope vector
数组形式参数 array formal parameter
数组元素 array element;element in array
数组元素后继函数 array element successor function
数组元素名 array element name
数组元素下标 array element subscript
数组元素引用 array element reference
数组运算 array operation
数组转置 array transpose
数罪并罚 concurrence of offences
数罪合计 several count

刷 badger;brushing

刷白 blanch;lime wash;whiten(ing);whitewash(ing);whiting
刷白表面 whitened surface
刷白层 white coat(ing)
刷白面 whitened face
刷帮 scallop;wall breaking
刷帮爆破 brushing shot
刷帮工序 slabbing operation
刷帮加宽的煤巷 slab entry
刷帮截煤机 slabbing machine
刷臂 brush arm
刷边 edge polishing
刷柄销 brush holder pin
刷薄<油漆过厚> pick-up sage
刷布 brushing;napping
刷彩 washbanding
刷层 brush coat
刷尘器 dust scrubber
刷大白浆 whitewashing
刷大角 reaming angle
刷大巷道 brush
刷底漆 priming
刷底子油(沥青) priming
刷垫板机 brushing machine for caul plate
刷掉 brush off;brushout
刷斗 brush hopper
刷镀 brush plating
刷方减载 cutting for balance of ground
刷仿木纹 brush graining
刷粉 brush dust
刷敷填配料 spackling compound for brush application
刷敷涂层 brush-applied coat(ing);brush coating
刷敷涂料 brush coat
刷杆 brush holder stud
刷钢丝刷 wire brushing
刷管器 tube scaler
刷光 brushing;brush up;satin finish;scratch brushing
刷光机 brush(ing) machine
刷光剂 brush polish
刷光胶合板 brushed plywood
刷光饰面 scratch-brushed finish
刷光涂料 brushing compound
刷光性 brushing property
刷过白漆的 white-painted
刷盒 brush box
刷痕 brush mark;brush trace;ropiness
刷弧 brush arc
刷花 brushing decoration
刷花纹的油漆刷子 mottler
刷环 brush ring
刷灰 brush dust;brushing
刷混凝土墙涂料 concrete paint
刷混凝土墙用浆 concrete paint
刷迹圈 brush track
刷夹 brush rigging
刷架 brush bracket;brush yoke
刷尖放电 brushing
刷浆 distempering;mopping;plastering;swabbing
刷浆并铺贴在墙上 pasting and applying to wall
刷浆法 mopping method
刷降河川 degrading stream
刷金属板机 plate brushing machine
刷金属底漆 metal priming
刷净 outwash
刷卡 wipe through
刷亮 brushing
刷了黏[粘]结剂等候黏[粘]合的时间 open time
刷了石灰的 limed
刷料 brush coat
刷鳞机 brusher
刷拢到一起的东西 brushing
刷拢来的东西 brushing
刷路搭接处 lap
刷轮 brush(ing) wheel
刷毛 brush of kernel;comb;brush finish<混凝土的>
刷毛混凝土 brushed concrete;scrubbed concrete
刷毛混凝土板 scrubbed concrete slab
刷毛混凝土花槽 scrubbed concrete flower trough
刷毛混凝土面层 scrubbed concrete facing
刷毛机 carder;carding engine
刷毛面<混凝土路面的> brushed finish;brushed surface
刷毛饰面 brushed surface finish
刷毛纹理<一种防滑措施> brush texture
刷毛行程<混凝土路面刷毛机的一次> brush stroke
刷面 broom finish
刷面层 brushed surface
刷面处理 brush(ed) finish
刷面麻布(带)<水泥混凝土路面的> burlap drag
刷模 brush cast
刷磨光 brush polishing
刷末层粉漆 skim
刷木纹状漆面 brush graining
刷平流淌(油漆) pick-up sage
刷坡坡面 scouring slope
刷漆 brushing lacquer;lacquering;painting
刷漆样板 brushout
刷墙粉 kalsomine;powdered distemper;wall plaster;wall stuff;whitewash
刷墙粉料 kalsomine
刷墙粉渣 kalsomine
刷墙石灰浆 calcimine
刷墙石灰水 limewash
刷墙水粉 calcimine;kalsomine[calsomine]
刷墙水浆涂料 distemperature;distemper paint
刷墙水浆涂料用色料 distemper colo(u)r
刷墙鬃刷 kalsomine
刷清 brush off
刷清路面 broom finish
刷清漆 varnishing
刷去 brush down;brush(ing) off
刷染法 staining
刷色 colo(u)r wash;swabbing
刷色辊 brayer roll
刷深 degradation;degrading
刷深程度<河床的> upper of degradation;magnitude of degradation
刷深河槽 degrading channel
刷深河床 degrading stream
刷深河流 degrading river;degrading stream;downcutting river;downcutting stream
刷深河湾 incised meander;entrenched meander;inherited meander
刷石灰 lime wash
刷石灰的 limed
刷石灰浆 lime whiting
刷石灰浆排刷 limewash brush
刷石灰浆用排刷 limewash brush
刷石灰水 lime wash;whitewash
刷式电极 brush electrode
刷式刮路机 brush drag
刷式滚筒 brush cylinder
刷式机构 brush gear
刷式捡拾器 brush pick-up
刷式锯齿轧花机 brush saw gin
刷式排种器 brush feed
刷式排种装置 brush feed mechanism
刷式曝气器<污水处理> brush-type aerator
刷式清理机 brusher
刷式清石机 brush stone separator
刷式清选机 brush cleaner
刷式石块分离机 brush stone separator
刷式石块清石机 brush stone separator
刷式涂布机 brush spreader
刷式轧花机 brush cotton gin
刷式摘棉铃机 brush-type stripper
刷式阻种器 brush cutoff
刷饰面 brushed finish
刷饰木纹状饰面【建】 brush graining
刷水 swabbing
刷水笔 swab
刷水泥浆 cement wash
刷水色<上清漆前的工序> overgraining
刷损量<稳定土磨耗试验用> brushing loss
刷损试验<试验稳定土的耐久性> brushing test
刷贴商标标签 affixing trademark tags
刷涂 brush coat(ing);brushing;brush-on;brooming<沥青>
刷涂层 brush coat
刷涂底漆 brushing primer
刷涂堆漆 piling
刷涂法 brush coating method;brush painting;spread coating
刷涂挥发性漆 brushing lacquer
刷涂机 brush coater
刷涂料 painting;searing
刷涂时拉力 drag during brushing
刷涂梯(形条)纹 ladders in painting
刷涂涂层 brush-applied coat(ing)
刷涂涂装 brush-applied coat(ing)
刷涂性 brushability;brushing property
刷涂用腻子 brush filler
刷涂用漆 brushing-on paint
刷涂阻力 brush drag
刷位移 brush displacement
刷纹面层 brush finish
刷纹饰面 brush finish
刷握 brush carrier;brush holder;brush rigging
刷握臂 brush holder rod
刷握柄 rock(er) arm;rocking arm
刷握架 brush holder yoke
刷握弹簧 brush spring
刷握弹簧调整器 brush spring adjuster
刷握支柱 brush holder stud
刷洗车 brush truck
刷洗干燥机 scrubbing-and-drying unit
刷洗机 scrubbing unit
刷洗性 scrubbability

刷下 brushing down;clean down
刷新 brushing;brush up;face-lift;furbish;refreshen
刷新存储器 refresh memory
刷新的 regenerate
刷新控制器 refresh controller
刷新驱动 refresher driving
刷新式显示器 refresh tube
刷新速率 refresh rate
刷新周期 refresh cycle
刷形充气器 brush aerator
刷形触点 brush contact
刷形挡风雨条 brush weather strip
刷形电弧 brush arc
刷形电晕 brush corona
刷形放电【电】brush discharge
刷形开关 laminated-brush switch
刷形联轴节 brush coupling
刷形密封条 brush weather strip
刷形曝气器 brush aerator
刷型构造 brush-type structure
刷移角 brush displacement
刷印 brush mark
刷印的装载限制 stenciled load limit
刷用钢丝 brush handle wire
刷油脱模 drip strip
刷油作业 brush application
刷釉 brushing glazing
刷蒸机 brushing and steaming
刷状痕 brush mark;rub;scrub mark;scuff mark
刷子 badger;brush;scrubber;swab
刷子处理的混凝土表面 brushed surface
刷子工 brushman
刷子拉毛饰面 brush finish
刷字模板 stencil

衰变 decaying;degradation

衰变波 decaying wave
衰变不稳定性 decay instability
衰变参数 decay parameter;degradation parameter
衰变产物 daughter product;decay daughter;decay product;descendant;disintegration product
衰变常数 decay constant;destruction constant;disintegration constant
衰变迟滞 decay lag
衰变电子 decay electron
衰变定律 decay law
衰变方程式 decay equation
衰变方式 decay mode;disintegration mode;mode of decay
衰变分支比 decay branching ratio
衰变辐射剂量 decay radiation dose
衰变光子 decay photon
衰变过程 decay process
衰变毫居 millicurie-destroyed
衰变计 decremeter
衰变减量 attenuation
衰变角 decay angle
衰变类型 decay mode;decay type
衰变冷却器 decay heat cooler
衰变链 decay chain;disintegration chain
衰变率 decay ratio
衰变脉冲 decaying pulse
衰变模量 modulus of decay
衰变能 disintegration energy
衰变能量 decay energy
衰变期 decay period;decay time
衰变曲线 decay curve
衰变曲线分析 decay curve analysis
衰变热 decay heat
衰变时间 decay time
衰变速率 decay rate;rate of decay
衰变特性 decay characteristic
衰变图 decay mode;decay scheme;disintegration mode
衰变系 decay series
衰变系列 decay sequence
衰变系数 attenuation coefficient;coefficient of attenuation;coefficient of decay;decay coefficient;decay module;disintegration coefficient
衰变相 decay phase
衰变相互作用 decay interaction
衰变型 disintegration mode
衰变性能 decay property
衰变修正值 attenuation corrected value
衰变序列 decay sequence
衰变因素 decay factor
衰变振荡 decay oscillation
衰变指数 degradation index
衰变子体 decay daughter
衰变族 decay series
衰耗 pad control
衰耗表 attenuation meter
衰耗常数 attenuation constant;decay constant
衰耗电缆 attenuating cable
衰耗均衡器 attenuation equalizer
衰耗控制 pad control
衰耗器 attenuation network;attenuator;pad
衰耗器盒 attenuation box
衰耗器继电器 pad relay
衰耗调整 damping control
衰耗系数 decay coefficient
衰化 degradation
衰化参数 degradation parameter
衰化误差 ag(e)ing error
衰化指数 degradation index
衰坏 failure
衰坏百分率 percent failure
衰坏面 failed surface
衰坏伸长 elongation at failure
衰减 attenuate;attenuation;build-down;damp(en)ing;deadening;decay(ing);degeneration;fade down;fall(ing) off;weakening;deamplification
衰减安全系数 fading safety factor
衰减板 attenuating plate
衰减半导体 degenerate semiconductor
衰减倍数 damping factor
衰减比 attenuation ratio;ratio of attenuation;rejection ratio;specific damping
衰减比测仪 attenuation comparator
衰减标准 attenuation criterion[复 criteria]
衰减波 damped wave;decadent wave;decaying wave
衰减波变压器 jigger
衰减波群 jig
衰减补偿 attenuation compensation
衰减补偿器 attenuation equalizer
衰减材料 attenuating material
衰减参数 damping parameter
衰减测定装置 attenuation measuring device
衰减测量 attenuation measurement
衰减测量计 attenuation ga(u)ge
衰减测量器 decremeter
衰减测试 attenuation test
衰减层 damping course
衰减长度 relaxation length
衰减常量 attenuation constant;damping constant;decay constant
衰减常数 attenuation constant;damping constant;decay constant
衰减程度 attenuation range
衰减传导电流 decaying conduction current
衰减带 attenuation band;attenuation region;stop band
衰减的 damped;decadent;degenerative
衰减的振动 damped vibration
衰减电流 damped alternating current
衰减电路 attenuator circuit;damping circuit
衰减电压 evanescent voltage
衰减电子振荡器 damped electron oscillator
衰减电阻 damping resistance
衰减定律 fall-off law;decay law
衰减度 extent of damping
衰减段距离 attenuation distance
衰减多次倾斜叠加处理 attenuation multiple slant stacking processing
衰减法 decay method
衰减范围 attenuation range;range of attenuation
衰减方程 attenuation equation
衰减方程式 decay equation
衰减分级指示灯 attenuation step pilot lamp
衰减关系式 attenuation relation
衰减管 attenuator tube
衰减过程 attenuation process
衰减横摇运动 roll subsidence mode
衰减畸变 attenuation distortion
衰减激波 decaying shock wave
衰减极限 fading margin
衰减计 decremeter
衰减交流电 damped alternating current
衰减角 loss angle
衰减角频率 damped angular frequency
衰减阶段 declining phase;waning stage
衰减截面 attenuation cross section
衰减矩 damping moment
衰减距离 attenuation distance;decay distance <波浪的>;distance of degeneration
衰减均衡 attenuation equalization
衰减均衡器 attenuation equalizer
衰减库 decay reservoir
衰减力 damping force
衰减量 decrement;decrement of damping
衰减量测量 decrement measurement
衰减率 attenuation rate;decay rate;numeric(al) decrement;rate of attenuation;rate of decay attenuation;rate of fall;ratio of attenuation
衰减媒质 attenuating medium
衰减模量 modulus of decay
衰减模数 modulus of attenuation
衰减能量 damping capacity
衰减频带 attenuation band
衰减频率失真 attenuation-frequency distortion
衰减频率特性 attenuation-frequency characteristic
衰减期间 period of damping out
衰减器 attenuating pad;attenuation pad;attenuator;pad;resistance network;resistive network
衰减器控制 pad control
衰减器谐振 rejector resonance
衰减前向波 decreasing forward wave
衰减区 attenuation band;attenuation region;decay area;fading region
衰减曲线 attenuation curve;decay curve;die away curve;reduction curve
衰减曲线法 decay curve method
衰减全发射谱 attenuated total reflection spectroscopy
衰减全反射 attenuated total reflectance;attenuated total reflection
衰减全反射比 attenuated total reflectance;frustrated internal reflectance;internal reflectance spectroscopy
衰减全反射光谱 attenuated total reflectance spectroscopy
衰减剩余时差 attenuate residual moveout
衰减失真 attenuation distortion
衰减时间 attenuation time;decay time;die-away time;relaxation time;storage time
衰减试验 die-away test
衰减试验器 fading machine
衰减瞬变过程 damping transient
衰减瞬变量 damping transient
衰减瞬态 decaying transient
衰减速率 decay rate;rate of decay
衰减损失 attenuation loss
衰减特性 attenuation characteristic;decay characteristic;fading characteristic
衰减特性曲线 attenuation characteristic curve
衰减特征 attenuation characteristic
衰减条件 damp condition
衰减网路 attenuation network
衰减网络 damping network
衰减尾部 tailing
衰减无线电信号的低大气层 substandard surface layer
衰减系数 attenuation coefficient;attenuation factor;coefficient of attenuation;coefficient of decay;coefficient of extinction;damper coefficient;damping coefficient;damping ratio;decay coefficient;decay factor;extinguishing coefficient;modulus decay;reduction factor;subsidence ratio
衰减系数值 attenuation coefficient value
衰减限制孔 fading choke
衰减限制作用 attenuation limited operation
衰减相 declining phase
衰减相位 decay phase
衰减箱 attenuator box
衰减响应 convergent response
衰减项 attenuation term
衰减消失 dying out
衰减信号 deamplification signal;fading signal
衰减性质 attenuation property
衰减因数 attenuation factor;damping factor;decay factor;quadripole attenuation factor
衰减因子 attenuation factor;damping factor;decay factor
衰减云母陶瓷 attenuating mica ceramic
衰减运动 attenuation motion;convergent mode of motion;damp motion
衰减增殖期 period of declining growth
衰减振荡 convergent oscillation;damped oscillation;dying oscillation
衰减振荡成分 damped oscillatory component
衰减振动 attenuation vibration
衰减正弦量 damped sinusoidal quantity
衰减值 pad value
衰减指示器 fade indicator
衰减指数 attenuation index;damped exponential;decaying exponential
衰减指数定律 exponential law of attenuation
衰减周期 attenuation period;decay period

衰减装置 damping device
衰减阻抗 damped impedance
衰减作用 attenuation; attenuative effect; damping effect; fading
衰竭 exhaustion; failure; prostration
衰竭井 stripper well
衰竭土 senile soil
衰竭性 exhaustibility
衰老 dote; senescence; senility
衰老比 senescent ratio
衰老变化 senile change
衰老的 decrepit; senile
衰老河 senile river; senile stream
衰老湖 senescent lake
衰老期 death phase
衰老失效 wear-out failure
衰老试验 long-term ag(e)ing test
衰老现象 senilism
衰落 declining; fading; falling off; freak; labefac(ta)tion; turn down
衰落边际 fade margin; fading margin
衰落补偿装置 anti-fading device
衰落储备 fading margin
衰落带宽 fading bandwidth
衰落的 decadent; down grade
衰落分布 fading distribution
衰落峰值 fading peak
衰落幅度 amplitude of fading
衰落控制 fading control
衰落频谱 fading spectrum
衰落期 winter
衰落区 fade area; fade zone; fading area
衰落区图 fade chart
衰落群体 depauperate colony
衰落(深)度 fading depth
衰落时间 decay time; die-away time
衰落试验器 fading machine
衰落速度 rapidity of fading
衰落损失 fading loss
衰落现象 fade out
衰落信道 fading channel
衰落信号 fading signal
衰落迁腐 < 城市 > blight and flight
衰落余量 fading margin
衰落增长 fading rise
衰落指示器 fade indicator
衰落中的生物分类群 decreasing taxa
衰落周期 fading period
衰落装置 fading unit
衰灭振荡 dying oscillation
衰弱 asthenia; weakness; withering-away
衰弱的 faint; worn
衰弱期 adynamic(al) stage
衰弱效应 devitalizing effect
衰退 atrophy; declining; degeneracy; degeneration; diminution; downbeat; drop-off; ebb(tide); fail; fall(ing) off; recession; regression; slow up; slump; fading【电】
衰退倒闭 degenerated bankruptcy
衰退的 downcast; retrogressive; downhill
衰退范围 fading range
衰退过程 degenerative process
衰退和恢复试验 fade-and-recovery test
衰退记忆 fading memory
衰退阶段 downhill
衰退控制 fader control
衰退邻里 declining neighbo(u)rhood
衰退率 fading rate
衰退期 decline phase; declining stage; waning stage
衰退曲线 decline curve
衰退速度 decline rate
衰退特性 fade characteristics
衰退效应 fading effect
衰退性损坏 degradation failure
衰退再结晶作用 degenerative recrystallization; degradation recrystallization
衰退支 failed arm
衰亡裂谷 failed rift
衰亡期 decline phase
衰萎 deadness
衰朽阶段 age of decline

摔倒 tumble down

摔毁后起火 postcrash fire
摔跤场 wrestling ring
摔土率 scuffler
摔土器 scuffler
摔下 spill

甩摆动作 wobbling action

甩板 flail knife; knockout plate
甩板式刀片 flail-type blade
甩布架 plaiting apparatus
甩出机 kick-off
甩出预探井 extension test
甩刀 flail knife; free swinging knife
甩刀固定螺栓 flail hanger retainer bolt
甩刀式割草机 flail mower
甩刀式滚筒 flail rotor
甩刀式茎秆切碎机 knife type shredder
甩刀式切碎机 flail cutter
甩刀式切碎装置 flail
甩刀式清垄器 flail row cleaner
甩刀式旋转切碎机 flail rotor
甩刀装置 flail mechanism
甩掉 throw off
甩动 whip(ping)
甩负荷 load rejection; load-shedding; load thrown off; rejection of load; relief load; relieved load
甩负荷试验 governor test; load dump test; load rejection test; load shutdown test; load throw-off test
甩负荷装置 load-shedding equipment
甩疙瘩面 spatter dash
甩灰打底 spatter dash
甩击 whipping
甩溅 throwing
甩胶 whirl coating
甩开 ditching
甩开钻井 outstep drilling
甩客 denial of passenger
甩力 knockout press
甩链式撒肥机 rotor spreader
甩链子 throwing the chain
甩卖 on-sale
甩满负荷 rejection of full load; total load rejection
甩蜜机 honey centrifuge; honey extractor
甩炮 squib
甩漆布 paint harling
甩漆饰 paint harling
甩石涂装法 paint harling
甩水环 throw ring; water deflector
甩水机 hydroextracting cage
甩水墙 watershot walling
甩水圈 thrower
甩丝法 spinning
甩套器 kicker
甩涂 centrifugal finishing
甩土转速 rotation speed for cast
甩尾 whipping
甩油 splashing of oil
甩油齿轮 oil gear
甩油环 lubricating disc; oil slinger; oil splash ring; oil thrower ring; thrower; throw ring
甩油盘 disc; oil thrower wheel
甩油圈 oil slinger; slinger
甩油装置 oil thrower
甩子 hollow swage; swage; swaging hammer

闩柄 bolt handle; bolt lever

闩承座 latch retainer
闩的定位器 latch retainer
闩钩 latch dog
闩卡体 < 绳索取芯工具 > latch body
闩块 lock block
闩门杠 shutting post
闩上突出的键 latch lug jaw
闩上突出的锁 latch lug jaw
闩式挡料装置 latch stop
闩锁 breech lock; latch
闩锁电路 latch circuit
闩锁继电器 latched relay; latching relay; latch-in relay; lock(ing) relay; lock-up relay; relay with latching
闩锁寄存器 latch register
闩锁器 latch unit
闩锁全加器 latching full adder
闩锁式前机头 < 钻机的 > latch-type front head
闩锁位测试 latch bit test
闩锁型极化继电器 polarized-latching type relay
闩锁译码器 latch decoder
闩锁状态 latch mode
闩体 breechblock
闩头导槽 bolt guide
闩托 latch bracket
闩销 bolt; latch bolt; rod bolt
闩销插 latch holder
闩爪 bar claw
闩座 latch holder

拴 hitch

拴绳锁环 pull-iron
拴住脚手架和脚手板的钩子或捆索 scaffold hitch

栓 cotter; cross lock; forelock; pin; plug; stopper; stopple; tether

栓板【机】 key plate
栓槽 croze
栓承 pin bearing
栓船桩 make fast
栓道 keyway
栓钉 bolt; male pin; peg; stop pin; stud
栓钉垫板 pin filler
栓钉端 pin end
栓钉开关 peg switch
栓钉孔 pin-hole
栓钉螺帽 pin nut
栓钉錾 cant chisel
栓钉支承 pin bearing
栓定浮筏 captive float; reefing float
栓端支杆 pin-ended strut
栓端柱 pin-ended column
栓固能力 restraint capability
栓固强度 stacking strength
栓固试验 restraint test
栓固元件 securing attachments
栓焊 high-strength bolt-welded
栓焊钢梁 bolted and welded steel girder or truss
栓焊钢桥 bolted and welded steel bridge; welded and(high strength) bolt connected steel bridge; welded and high strength bolted steep bridge
栓化的 suberized
栓化纤维素 adipo-cellulose
栓化作用 suberization
栓簧 bolt spring
栓剂 suppository
栓礁 plug reef
栓接 bolting
栓接板 pin plate
栓接的 pin-connected
栓接端 pinned end
栓接法兰 bolt flange; screw flange
栓接分部式浮(船)坞 bolted sectional dock
栓接腹板 bolted web
栓接钢桥 bolted steel bridge
栓接构架 pin truss
栓接桁架 pin-connected truss
栓接节点 pin-connected joint
栓接结构 articulated structure
栓接框架 pin-connected frame
栓接上桅 housing topmast
栓接榫凿(块石) bolting iron
栓接压力 bolted compression coupling
栓接翼缘接头 bolt-on-flange joint
栓紧 chocking-up
栓紧装置 toggler
栓菌属 < 拉 > Trametes
栓考克 stop cock
栓孔 key hole; key holing
栓孔锯 keyhole saw
栓牢 make fast
栓连接 fix with plugs
栓连接分段拼装式船坞 bolted sectional dock
栓流式气力输送装置 plug-flow type pneumatic conveyer[conveyor]
栓流输送 plug-flow conveying
栓螺栓隔片 stud spacer
栓马环 horse tying ring
栓木 kalopanax
栓内层 phelloderm
栓皮 cork; phellem
栓皮加工板 < 一般为隔热、隔音用 > corkwood plank
栓皮栎 cork bark; corkoak; corktree; oriental oak; phellem
栓塞 embolism; ramming piston
栓塞阀 ground key valve; plug valve
栓塞(封闭)灌浆法 packer(method of) grouting
栓塞锯 dowel saw; plug saw
栓塞流动 plug flow
栓塞式保险丝 plug fuse
栓塞式排水阀 plug drain valve
栓塞型阀 corporation stop
栓锁 bolt-lock
栓锁带 latchstring
栓锁法 bolted method
栓体 key
栓体长度 length of bolt shank
栓位 bolting
栓窝式 peg and socket
栓系 tie-down
栓系点 tie-down point
栓系设备 < 栓系装在货车上的木料、汽车等 > tie-down equipment
栓系图 tie-down diagram
栓销 coak; peg dowel; pin; plug
栓与栓的间隔 peg
栓轴 stud shaft
栓轴承 spigot bearing
栓住 fastening
栓住螺母 captive nut
栓柱焊接 peg welding
栓转电闸 peg switch
栓桩 deadman
栓锥感器 sensillum styloconicum
栓子 embolus

S

栓座 key seat

涮洗池 rinse tank

涮洗水 rinse water

双D 盒回旋加速器 two-dee cyclotron

双 D 盒系统 two-dee system
双 L 形楼梯 double L stair(case)
双 T 板 pi slab
双 T 地板 double T floor slab
双 T 接头 double T junction
双 T 框架 double tee frame
双 T 滤波器 twin-T filter
双 T 形板 double tee[T] plate; double tee[T] slab
双 T 形电路 twin-T network
双 T 形构件 double T unit
双 T 形接头 double T fitting
双 T 形结构 double T frame
双 T 形框架 double T frame
双 T 形梁 double tee[T] beam
双 T 形梁板 double tee [T] floor (slab)
双 T 形模板 double tee form(work)
双 T 形模壳 double T formwork
双 T 形铁 double tee[T] iron
双 T 形网络 twin-T network
双 T 形屋面板 double tee roof slab
双 T 形型滤波器 twin-tee filter
双 T 形预制大梁 double tee[T] prefabricated girder
双 T 支座 double T bed
双 T 字形屋面板 double T roof(ing) slab
双 U 形膨胀接头 double offset U bend
双 V 粒子 two-vee
双 V 形槽 double vee[V] gutter
双 V 形的 double vee[V]
双 V 形接头 double V-butt joint
双 V 形天线 X-antenna
双安培滴定法 biamperometry
双安全阀 double safety valve
双岸式布置 twin layout
双岸式水电站 twin power plant
双胺坚牢紫 diamine fast violet
双胺精染料 diaminogen dye
双胺染料 diamine
双凹槽砖 double-frogged brick
双凹的 biconcave; concave-concave; double concave
双凹镜 biconcave lens
双凹面的 biconcave
双凹面透镜 biconcave lens
双凹透镜 bicone lens; concavo-concave lens; double concave glass; double concave lens
双凹凸式密封面 double male and female
双凹形 biconcave; concavo-concave; double concave
双凹形的 concavo-concave
双凹圆 double intended circle
双八面体 dioctahedron
双八面体的 dioctahedral
双八字结 stevedore knot
双百叶片快门 two-blade shutter
双摆 double pendulum
双摆动支座 double tumbler bearing
双摆法 two-pendulum method
双摆门 double-swing(ing) door
双摆门的门框 double swing frame
双摆式地震仪 duplex pendulum seismograph
双摆式支座 double pendulum bearing
双摆四连杆机构 double swing lever mechanism
双摆重力仪 two-pendulum gravimeter
双班乘务组 double crew
双班运行 two shifts run
双班制 double shift(ing); watch and watch
双板 biplate
双板舱壁 double plate bulkhead
双板层 double course; eaves course
双板分粒机 two-deck classifier
双板犁 butting plow
双板梁 double plate girder
双板列预热器 twin line preheater
双板面式房屋 double slab-type building
双板面式建筑 double straight-line block; double straight-line building
双板式建筑 double slab-type block
双板条 double batten
双板瓦层 doubling tile course
双板压榨机 two-daylight press; two-plate press
双板音响器 double plate sound
双板桩 double sheet pile; twin sheet piles
双半波滤光片 double halfwave filter
双半径半圆头铆钉窝模 double radius button head snap
双半面晶形 disphenoid
双半内圈轴承 split inner race bearing
双半日潮 double half-day tide
双半式连接卸扣 Kenter shackle
双半筒型海底取样抓斗 Peterson grab
双半外圈轴承 split outer race bearing
双半轴承 axle bearing in two parts
双拌式混凝土拌和摊铺机 twin-batch pav(i)er
双拌式混凝土摊铺机 twin-batch pav(i)er
双拌式搅拌筒 twin-batch mixing drum
双拌载重车 <可装两拌混凝土> two-batch truck
双瓣 bivalve
双瓣阀 double flap valve
双瓣料斗 clamshell scoop
双瓣扇形斗门 clamshell gate bucket
双瓣式抓斗 clamshell grab; two-jaw grab; two-piece clamshell bucket
双瓣卧式闸门 bear-trap gate
双瓣阳极 split anode
双棒浇注装置 double stopper arrangement
双包层光学纤维 doubly clad optical fiber[fibre]
双包层条形介质波导 double bedded slab dielectric(al) waveguide; doubly cladded slab dielectric(al) waveguide
双包装 twin pack; two-pack system
双包装铝粉漆 ready-to-mix alumin(i)um paint
双包装涂料 two-pack coating
双孢霉属 <拉> Didymosporium
双保护垫的滚珠轴承 two-shields ball bearing
双保价 dual evaluation
双保价条款 dual evaluation clause
双保险锁 double-throw lock
双报警 double knock
双抱 <用钢带止动鼓的> double embrace
双抱钩 clamp hook; clipper hooks; clove hook; match hook; sister hooks; double hook
双曝光干涉度量学 double exposure interferometry
双曝光全息法 double exposure holography
双曝光全息干涉法 double exposure holographic interferometry
双曝光全息摄影术 double exposure holography
双曝光全息术 double exposure holography
双曝光全息图 double-exposed hologram; double exposure hologram
双北向采光框架 twin northlight frame
双贝克线 biback line
双背齿轮装置 <车床的> double back gear
双倍 diplex; double
双倍长常数 double length constant
双倍长乘法 double length multiplication
双倍长度的 double length
双倍长度数 double length number
双倍长工作 double length working
双倍长寄存器 double length register
双倍长累加 double length accumulation
双倍长累加器 double length accumulator
双倍长数 double length number; double precision number
双倍长运算 double length operation
双倍长字 double length word
双倍尺寸 double size
双倍充电 double charges
双倍大的砖 double-sized brick
双倍带电离子 doubly charged ion
双倍的 duplex
双倍递减余额折旧 double-declining balance depreciation
双倍递减余额折旧法 double-declining balance depreciation method
双倍读出 double read out
双倍负载 double charges
双倍复滑车 two-fold purchase
双倍工资 double time
双倍工作单元 double length working
双倍滑轮组 two-fold purchase; two-fold tackle
双倍计时表 doubling time meter
双倍接收 double reception
双倍精度常数 double precision constant
双倍精度复数型 double precision complex type
双倍精度工作 double length working
双倍精度量 double precision number; double precision quantity
双倍精度数 double precision number
双倍精度硬件 double precision hardware
双倍精度运算 double precision arithmetic; double precision operation
双倍精密度 double precision
双倍宽度标准的 <耐火砖> double standard
双倍率仪器 dual-power instrument
双倍密度 double density
双倍赔偿 double indemnity
双倍赔偿条款 double indemnity clause
双倍器 duplex
双倍收费 double charges
双倍四滑轮滑车 two-fold purchase
双倍体 diploid
双倍同步速度电动机 bisynchronous motor
双倍危险角 double danger angle
双倍位运算 double length arithmetic
双倍用料的构架 double framing
双倍余额递减法 double-declining balance method
双倍直线法折旧率 twice the straight-line depreciation rate
双倍字长 double length
双苯胺 dianiline
双苯胺青染料 dianilazurin(e)
双苯胺枣红染料 dianil bordeaux
双苯绕蒽酮绿 dibenzanthron green
双泵合流 duplex pump feeding
双泵轮变压器 double pump roll torque converter
双泵轮液力变矩器 double impeller torque converter
双比例量测法 double proportionate measurement
双比特转移 dibit transition
双比重计法 <土的分散度或团粒度试验的> double hydrometer method
双笔电位计式连续记录器 two-pen potentiometer continuous recorder
双笔记录器 two-pen recorder
双笔记录装置 twin pen recorder
双笔图表记录器 two-pen chart recorder
双闭磁路继电器 double shunt field relay
双闭箍 double closing hoop
双闭管 double closed tube; doubly-closed tube
双闭合接点 double make contact
双闭合褶皱 double zigzag fold
双闭塞机 double block instrument
双壁板桩围堰 double wall sheet pile cofferdam
双壁冲击管 <锤击钻进用的> double wall driving pipe
双壁储罐 double integrity tank
双壁分隔 double wall separation
双壁钢围堰钻孔桩基础 double wall steel cofferdam and bored pile foundation
双壁管 double-skin duct; tube within a tube
双壁罐 double-walled tank
双壁航天器 twin-walled spacecraft
双壁换热器 double wall heat exchanger
双壁井 double-walled well
双壁开沟铲 lister shovel
双壁开沟铲式犁体 middle breaker sweep bottom
双壁开沟犁 buster plow; ditcher; double breasted plough; lister <干旱地区用>
双壁开沟型 buster
双壁开沟作垄犁 lister plough
双壁壳 double-walled shell
双壁冷却器 double-walled heat exchanger
双壁犁 lister; ribbing plow
双壁楼梯 dog-legged stair(case)
双壁起垄犁 buster plough; double mould-broad ridging plough
双壁取土器 soil sampler of double wall
双壁式掺和机 twin-shell blender
双壁式楼梯 staircase of dog-legged type
双壁围堰 double wall cofferdam
双壁支撑 double wall buttress
双壁柱 accouplement of pilasters; coupled pilasters; paired pilasters
双壁钻杆 dual concentric drill pipe; dual-wall drill pipe
双臂板下象限臂板信号机【铁】 double arm lower quadrant semaphore signal
双臂板信号机【铁】 double-arm semaphore; double blade semaphore
双臂采矿台车 two-boom stope-jumbo
双臂操作旋压 level-tool spinning
双臂存取 dual access
双臂电桥 double bridge
双臂轭 two-arm yoke

双臂杠杆 double arm lever
双臂搅拌机 two-arm kneader
双臂捏合机 double arm kneader
双臂捏练机 double arm kneader
双臂铺轨机 twin boom tracklayer; twin jib tracklayer
双臂曲柄 bell crank
双臂曲柄杆 bell crank lever
双臂曲轴 double-throw crankshaft
双臂式开合桥 double-arm draw bridge
双臂式旋转喷射冲洗装置 dual-arm rotating jet washer
双臂式旋转喷射装置 dual-arm rotating jet device
双臂式仰开桥 double draw bridge
双臂受电弓 crossed-arm pantograph; double arm pantograph
双臂隧道掘进机 two-arm tunnel(l)ing machine
双臂旋涡图样 two-armed spiral pattern
双臂凿岩台车 twin-boom carriage
双臂支架 two-armed spider
双臂钻车 twin-boom drill rig
双边 doubling piece
双边 K 形坡口 double K groove
双边 Z 变换 two-sided Z-transform
双边齿轮驱动压力机 twin-geared press
双边传动 double-sided gear drive
双边带 both sideband; twin sideband
双边带传输 double-sideband transmission
双边带传输制 double-sideband transmission system
双边带的 double-sideband
双边带电话 double-sideband telephone
双边带发射机 double-sideband transmitter
双边带发射载波 double-sideband emitted carrier
双边带减幅载波 double-sideband reduce carrier
双边带接收机 double-sideband receiver
双边带调幅电话 amplitude modulation telegraphy double sideband
双边带调制 double-sideband modulation
双边带调制法 double-sideband modulation system
双边带无线电话 double-sideband radiophone
双边带信号 double-sideband signal
双边带信号发生器 double-sideband signal generator
双边带抑制载波 double-sideband suppressed carrier
双边带制 double-sideband system
双边贷款 bilateral loans
双边单位元 two-sided identity
双边的 bilateral
双边电路 bilateral circuit
双边对流 bilateral flow
双边发展贷款 bilateral development loan
双边法律关系 bilateral legal relations
双边钢领 doubled flanged ring
双边公约 bilateral convention
双边供电 two-way power feeding
双边关系 dyad; two way relationship
双边关系原则 <国际贸易> bilateralism
双边函数 double-sided function
双边合伙 bilateral partners
双边合同 bilateral contract
双边和多边经济合作 bilateral and multilateral economic cooperation
双边会谈 dyad
双边会议 bilateral
双边货物 two-sided goods
双边接收线圈 two-side receiving coil
双边进口配额 bilateral import quotas
双边经济援助 bilateral economic aid
双边靠船码头 twin jetties
双边拉普拉斯变换 bilateral Laplace transformation; two-sided Laplace transform
双边理想 two-sided ideal
双边零 two-sided zero
双边流形 two-sided manifold
双边垄断 bilateral monopoly
双边贸易 bilateral trade; bilateral trade exchange; two-way trade
双边贸易(及)支付协定 bilateral trade and payment agreement
双边面积 bilateral area
双边模 two-sided module
双边磨边机 double edge grinding machine
双边逆元 two-sided inverse
双边配额 bilateral quota
双边破图廓 bi-margin format
双边谱 two-sided spectra
双边契约 bilateral contract
双边器件 two-sided unit
双边倾卸车 double-side tipping wagon
双边清算 bilateral clearing
双边饰门 double-margined door
双边税收抵免协定 bilateral agreement on tax credit
双边谈判和协定 bilateral negotiations agreements
双边套利 bilateral arbitrage
双边替换 two-sided alternative
双边条约 bilateral treaty
双边调制声道 bilateral-area track
双边铁鞋 double edged skate
双边突堤 twin jetties
双边限额 bilateral quota
双边箱形截面 two-bay box section
双边协定 bilateral agreement; bilateral convention
双边协定配额 bilateral quota
双边协调标准 bilaterally harmonized standard
双边协议 bilateral agreement
双边信贷互惠协定 bilateral swap agreement
双边性投资 bilateral commitments of capital
双边研磨机 double edge grinder; twin edge grinder
双边有价证券投资 bilateral portfolio investment
双边援助 bilateral aid; bilateral assistance
双边缘 double selvedge
双边约束 bilateral constraint
双边枕底清筛机 double line sleeper bed sieve machine
双边支撑 double triangulated system
双边支付协定 bilateral payments agreement
双边支付协议 bilateral payments agreement
双边支重轮 double flange track roller
双边中期贷款 bilateral midterms loan
双边专约 bilateral convention
双编号图幅 binominal sheet
双编结 double becket bend; double bend; double sheet bend
双扁杆 double flat rod
双便池 twin bowls
双变 bivariate
双变法 double variation method
双变换制 double conversion system
双变量 bivariate
双变量的 bivariant; bivariate
双变量多项式 bivariate polynomial
双变量分布 bivariate distribution
双变量分析 bivariate analysis
双变量函数 bivariate generating function
双变量函数发生器 function generator of two variable; two-variable function generator
双变量回归 bivariate regression
双变量计算机 two-variable computer
双变量记录器 X-Y recorder
双变量矩阵 two-variable matrix
双变量母函数 bivariate generating function
双变量随机过程 bivariate stochastic(al) process
双变量系统 bivariant system
双变量正态分布 bivariate normal distribution
双变量正态曲面 bivariate normal surface
双变频超外差 double superhet(erodyne)
双变频超外差式接收机 double superheterodyne receiver; dual-conversion superheterodyne receiver
双变式 bivariant
双变数计算机 two-variable computer
双变体系 bivariant system
双变性转变 enantiotropic inversion
双变质带【地】 paired metamorphic belt
双变状态 bivariant state
双标尺法 <检验水准仪> two-staff method
双标度 two-scale
双标度法 two-scale method
双标度仪表 double scale instrument
双标准接收机 dual standard receiver
双标准经线等角圆柱投影 transverse cylindric(al) conformal projection with two standard meridians
双标准内插法 double standard interpolative method
双标准纬线等积圆锥海图 conic(al) projection with two standard parallels; secant conic(al) chart
双标准纬线等积圆锥投影 secant conic(al) projection
双标准纬线圆锥投影 conic(al) projection with two standard parallels; secant conic(al) projection
双表轨计时器 double dial timer
双表面处理(路面) dual surface treatment
双表盘 double dial
双丙酮 diacetone
双丙酮丙烯酰胺 diacetone acrylamide
双丙酮醇 diacetone alcohol
双柄 twingrip
双柄剥皮刀 timber shave
双柄侧向气钻 two-handed offset drill
双柄高脚薄釉酒杯 kantharos[kantharus]
双柄刮皮刀 two-handled knife
双柄锯 double handed saw
双柄拉刨 draw knife
双柄铁水包 double shank ladle
双柄细颈瓶 amphora
双饼铁滑车 iron double blocks
双并方位标 two-bearings and run between
双并双串压气机 duplex tandem compressor
双波瓣天线系统 double lobe system
双波长薄层色谱扫描器 dual-wavelength thin layer chromatography scanner
双波长分光光度法 dual-wavelength spectrophotometry
双波长分光光度计 dual-wavelength spectrophotometer
双波长激光器 dual laser
双波长染料激光器 two-wavelength dye laser
双波长双光束紫外可见分光光度计 dual-wavelength-double beam ultraviolet-visible spectrophotometer
双波长显微镜 two-wavelength microscope
双波长显微术 two-wavelength microscopy
双波道 duplex channel
双波道的 twin-channel
双波段 dual range; two-waveband
双波段红外传感器 dual waveband infrared sensor
双波段接收机 two-band receiver; two-waveband radio set
双波发送 double transmission
双波束技术 dual-beam technique
双波纹板回转式预热器 double undulated rotational preheater
双波纹管差压计 double waved tube differential pressure ga(u)ge
双波纹铁皮 double-corrugated sheet
双波形测距系统 dual mode range acquisition system
双波性 double ripple
双玻色子 diboson; two-boson
双不变平均 two-sided invariant mean
双不变伪度量 two-sided invariant pseudometric
双不变线性泛函 two-sided invariant linear functionals
双不动中心问题 problem of two fixed centers[centres]
双不胶透镜 air-spaced doublet
双步梁 two-step cross beam
双裁边锯 double edger; twin edger
双彩颜料 Duocreme pigment
双参数 double parameter; two-parameter
双参数控制 two-parameter control
双参数流动 two-parameter flow
双参数自适应控制系统 two-parameter adaptive control system
双仓泵 twin pressure vessel conveyer[conveyor]; twin vessel conveyer[conveyor]
双仓斗 double hopper
双仓复式磨机 two-compartment compound mill
双仓磨 two-chamber mill; two-compartment mill
双仓球磨机 two-compartment ball mill
双仓输送泵 double pressure-vessel conveyer[conveyor]
双仓输送机 twin pressure vessel conveyer[conveyor]
双舱储槽 two-compartment tank
双舱化粪池 two-compartment septic tank
双舱耙吸式挖泥船 trailing suction double hopper dredge(r)
双舱式潜水器 two-compartment submersible
双操作数指令 double operand instruction
双槽 twin tanks
双槽板 <突出的装饰面,带双槽> diglyph
双槽齿 double steps
双槽串珠线脚 double-quirk(ed) bead
双槽定影 two-bath fixation; two-solution fixation
双槽断面 double channel section

双槽阀 double sheave pulley
双槽路 two-circuit
双槽面 diglyph
双槽排挡 diglyph
双槽式拌和机 double trough type mixer
双槽凸轮 double track cam
双槽凸圆线脚 double-quirk(ed) bead
双槽系统 twin pot system
双槽钻头 two-flute bit
双侧 bilateral
双侧板式除雪机 double cheek snow plough
双侧壁导洞法 twin-side heading method
双侧臂 double-sided arm
双侧城镇通道 two-sided town gateway
双侧齿板 double-sided toothed plate
双侧电抗联结器 double reactance bonds
双侧检验 two-sided test; two-tailed test
双侧浇口 double branch gate
双侧进气压缩机 double-sided compressor
双侧锚式挖泥机 double-sided anchor dredge(r)
双侧门 double wing door
双侧平衡铊式吊窗 double hung window
双侧谱密度 two-sided spectral density
双侧倾卸车 double-side tipping wagon
双侧曲面 two-sided surface
双侧人行道 sidewalks provided on both sides
双侧水柜机车 side tank locomotive
双侧托架柱 double-bracket post
双侧系统 bilateral system
双侧向测井 dual laterolog
双侧向测井曲线 dual laterolog curve
双侧向量空间 two-sided vector space
双侧向破裂 bilateral rupture
双侧堰溢流 double-side weir overflow
双侧正则表示 two-sided regular representation
双侧指数 two-sided exponential
双侧指数分布 two-sided exponential distribution
双侧制动 clasp braking
双侧阻抗联结器 double impedance bond
双侧作用继动器 double-acting relay
双测距交会法 range-range intersection
双测量六分仪 double sextant
双测压管测流(速)计 sympiezometer
双层 bilayer; double ply; double skin; two-tier; double coat
双层安全玻璃 double-pane safety glass
双层暗盒 double plate holder
双层百叶风口 double deflection register
双层百叶风门 double deflection grille
双层板 double course; doubling plate; two-piece panel
双层板梁 double plate girder
双层板桩 double sheet piling
双层板桩围堰 double wall cofferdam; double-walled sheet-pile cofferdam
双层保护 double shielding
双层保护层 double layer coat
双层比色计 bicolo(u)rimeter
双层壁 cavity wall
双层壁断面 double-walled profile
双层壁锅炉 double-walled boiler
双层壁挤压件 double-walled extrusion
双层壁柱正面 two-tiered pilastered facade
双层编线包皮 double braid covering
双层编织物 double layer of braid
双层扁股钢丝绳 double flattened strand
双层表面处理 double surface treatment; two-pass surface treatment
双层表面处治 <路面> double surface treatment
双层表面位 surface potential of double layers
双层丙伦尼龙缆 double braided polyester nylon rope
双层波阵面推进 wavefront advance in two layers
双层玻璃 double glass; pair glass; pair of glasses
双层玻璃安全视镜 safety sight glass with double layered glasses
双层玻璃板 double glazing glass; double glazing unit
双层玻璃窗 dehydrated window; double glass window; double glazed [glazing] window; double glaze unit; double glazed[glazing] unit; dual glazing
双层玻璃窗的室内侧玻璃 inside pane
双层玻璃窗框 double glazed window frame
双层玻璃单元 double glazing unit
双层玻璃空隙 interspace of double glazing
双层玻璃气密封接 double glass seal
双层薄壳 double shell
双层采油井 dual-pumping well
双层舱壁 double bulkhead
双层草垫 double reed mat
双层测标 double target
双层层压制品 two-layer laminates
双层超速干道 double deck freeway
双层车 double deck vehicle; two-decker
双层车库 two-level garage; two-stor(e)y garage
双层车辆 double decker
双层车箱 double deck coach
双层车站 double deck station
双层车振动筛 double deck vibrating screen
双层沉淀池 double layer settling tank; Emscher filter; Imhoff tank; two-stor(e)y sedimentation tank; two-stor(e)y settling tank
双层沉降槽 two-stor(e)y sedimentation tank
双层衬里 dual layer lining
双层衬砌 double layer lining
双层衬砌法 double-lining method
双层承面 double deck
双层冲沙闸(门) double sluice gate
双层重合环状接合 gill
双层出油井 dual zone well
双层储箱 two-stor(e)y tank
双层穿廊客车 gallery double-deck car
双层船 double decker
双层船底 double bottom
双层船底框架 double bottom cellular
双层船壳 double hull
双层船壳的 double-hulled
双层船壳结构 double hull structure
双层船体 double hull
双层窗 combination window; double-framed window; storm window; winter window
双层窗的外窗 storm window
双层床 bunk bed; double deck bed; double decker
双层次 double layer
双层次操作 two-level operation
双层次施工 two-level operation
双层次问题 two-layered problem
双层大桥 <铁路公路两用的> double decked bridge
双层单动取土器 swivel-type double tube core barrel
双层道路 double deck road
双层道砟 two-layer ballast
双层的 double decked; double-walled; two-course; two-layered; two-stor(e)y; two-storied
双层的度体系 two-decked
双层灯 jacketed lamp
双层底 water bottom【船】; double bottom
双层底边板 boundary plank; margin plate
双层底舱 double bottom compartment; double bottom tank
双层底舱顶 crown of a double bottom
双层底船 double bottom ship
双层底顶部 inner bottom; tank top
双层底炉 double hearth furnace
双层底内底板 double bottom tank top plating
双层底清洁组 double bottom party
双层底水密试验 double bottom test
双层底水压载 bottom water ballast
双层底外侧肘板 margin bracket; margin plate bracket
双层底翼板 bilge bracket; tankside bracket; tankside knee; wing bracket
双层底之间的间隔 double bottom cells
双层底组合肋板 double bottom open floor
双层地板 access floor(ing); double floor(ing); raised floor(ing)
双层地板式 access flooring system
双层地板饰面 double floor finish
双层地壳模型 double layer crust model
双层地下室 double vault
双层电车 double decker
双层电光机 duplex Shreiner calender
双层电极 two-layer electrode
双层电离室 twin ionization chamber
双层电梯 double deck elevator; double deck lift; tandem elevator; tandem lift
双层叠合绝缘玻璃板 two-piece laminated insulating glass
双层叠绕组 double layer lap winding
双层叠瓦 double deck bus; double decker; double lap tile
双层堵塞器 twin packer
双层镀膜 two-layer coating
双层多机钻车 two-deck jumbo
双层多晶硅 double level polysilicon
双层多丝丙伦缆 double braided polypropylene filament hawser
双层筏式基础 double raft foundation
双层飞机 double deck airplane
双层分级机 double deck classifier
双层分离式隔墙 staggered-stud partition
双层粉饰 double coating
双层风门 air lock door
双层风闸门 air lock
双层浮顶 double deck floating roof
双层幅板车轮 double plate wheel
双层干燥床 two-floor kiln
双层干燥机 two-deck drier[dryer]
双层坩埚 double crucible
双层刚性单剑杆 twinned single rigid rapiers
双层钢 clad steel; composite steel
双层钢筋 <钢筋混凝土中的> double layer of reinforcement
双层钢围囹 double-walled steel waling
双层高速公路 double deck freeway
双层高踢脚板 double skirting
双层割面 two-stor(e)y faces
双层隔仓板 double layer diaphragm plate; double wall partition
双层隔断 double-leaf party wall; double partition
双层隔墙 double partition
双层隔音地板 double sound-boarded floor
双层公共汽车 double deck(ed) bus; double decker
双层拱 double(deck) arch
双层共用隔墙 double-leaf party wall
双层沟瓦 double gutter tile
双层构架的 double-framed
双层构造 double layer construction; two-layer construction
双层关节货车 articulated double deck car
双层馆构筑法 two-course method
双层管 bimetallic tube
双层管板 double tube sheet
双层管道 double-skin duct
双层罐笼 two-decker cage
双层轨道表示盘 double deck track diagram
双层滚轴输送机 overlapping roller conveyer[conveyor]
双层过滤层 dual layer filter; drum filter
双层过滤池 double layer filter
双层过滤器 double layer filter
双层海崖 two-stor(e)y cliff; two-storied cliff
双层含水层 double layer aquifer; two-stor(e)yed aquifer
双层河 <喀斯特> both cauce
双层桁架桥 double decked bridge; two-storied truss bridge
双层烘缸 two-deck drier[dryer]
双层化粪池 two-stor(e)y septic tank
双层灰条 brandering; counter lathing; cross furring; double lathing
双层混凝土地面 two-concrete floor; two-course concrete floor
双层混凝土路面 two-course concrete pavement
双层混凝土铺面 two-course concrete pavement
双层活性滤池 double active filter
双层火车 double decker; double decker train
双层货垫 cross dunnage; double dunnage
双层机壳式 double casing
双层机壳式电机 double casing motor; sing machine
双层集装箱车 double stack(ed) container car
双层集装箱货车 double-stack container wagon
双层集装箱列车 double-stack container train
双层夹沥青包装纸 union kraft
双层甲板船 double decker; tweendeck ship; two-decked ship; two-decker
双层假平顶 double false ceiling
双层假天棚 double false ceiling
双层减反射膜 double layer antireflection coating
双层建筑 double layer construction
双层胶合板 two-ply
双层胶粘 double size
双层搅拌储存库 double deck blending and storage silo

双层教堂 double church;two-stor(e)yed church
双层接触滤池 double contact filter
双层街道 double deck street
双层结构 double deck(er);two-course construction;two-layered texture
双层结构的 double deck
双层结构矿物 two-sheet mineral
双层介质滤池 double media filter;dual-media filler
双层界面晶体生长理论 double layer interface theory of crystal growth
双层金属板 bimetal plate
双层金属带 bimetallic strip
双层金属闪光漆 two-layer metallic effect paint
双层金属铸造 composite casting
双层晶格 two-layer lattice
双层晶片单元 bimorph cell
双层晶体 bimorph crystal
双层晶体元件 bimorph cell
双层卷边机 double seamer
双层绝缘材料 double insulation
双层绝缘芦苇垫 double insulating reed mat
双层绝缘线 double insulated conductor
双层开沟 double trenching
双层开关窗 double casement window
双层铠装电缆 double armo(u)red cable
双层壳体 <采用立体构架的> double-layered space frame shell
双层客车 bi-level(rack)car;double deck car;double deck coach;double decker(carriage);double decker coach;double decker passenger coach;two tier coach
双层空间壳架 two-layered space frame shell
双层控制盘 double deck control panel
双层库 double stor(e)y silo;two-stor(e)y silo
双层跨运车 two-high straddle carrier
双层矿物 double level mineral;two-layer mineral
双层框架 double frame
双层框架隔墙 double tier partition
双层框架间壁 double tier partition
双层框架楼板 double-framed floor
双层拉床 by-level broaching machine
双层拉杆 double deck pull rod
双层拉丝作业 double level fibre forming;double level operation
双层蜡焊管法 bundyweld
双层廊 two-stor(e)yed corridor
双层离子交换 stratified bed ion exchange
双层犁 double cut plough;double deck plough;double depth plough
双层立交 two-level junction
双层沥青表面处治 double bituminous surface treatment
双层沥青混凝土路面 Warrenite-bitulithic pavement
双层连续抹灰(作业)double up
双层磷光体 double layer phosphor
双层菱形天线 two-tier rhombic
双层流 two-layer flow
双层龙骨地板 double-joisted floor;framed floor
双层楼板 access floor(ing);double(-skin)floor
双层楼板饰面 double floor finish
双层楼的 two-floored
双层楼面 double floor
双层漏斗 double funnel
双层芦苇垫 double reed mat
双层炉 double decked oven;double deck furnace
双层炉壁 double furnace wall
双层滤池 double bed filter;double filter
双层滤池滤料 double bed filter material
双层滤料 two-layer filter medium
双层滤料滤池 dual-media filler
双层滤膜工艺 dual membrane process
双层滤器 duplex strainer
双层路面 two-course pavement
双层旅客车队 double deck fleet
双层铝窗 alumin(i)um double window
双层轮胎 two-ply tyre[tire]
双层螺旋式线圈 two-layer spiral coil
双层铆接 two-ply riveting
双层煤气发生器 double zone gas producer
双层门 combination door;two-storied gate
双层面层做法 double back
双层膜 double layer coating;duplex film
双层抹灰 two-coat plaster;two-coat work
双层木百叶窗 two layers of timber louvres
双层木板墙 double plank wall
双层木板桥面 two-ply bridge floors
双层木格栅 two layers of wood(en) joists
双层木铺板 double deal
双层木条 doubling piece
双层幕墙结构 double-skin curtain wall construction
双层内罩炉 double muffle furnace
双层尼龙缆 double braided nylon hawser
双层黏[粘]合安全玻璃 two-piece laminated safety sheet glass
双层浓密机 two-tray type thickener
双层排架 two-stor(e)y bent
双层盘管 duplex coil
双层皮带 double belt;two-ply belt
双层平开窗 double window
双层平台 double decked platform
双层平纹织物 double plains
双层屏蔽罩 double shield enclosure
双层铺 bunk bed;double berth;double bunk
双层铺板法 double coursing
双层铺面 two-course pavement
双层铺砌法 <路面> two-coat work;two-course method
双层铺筑法 two-coat work;two-course method
双层汽车道路 double deck motorway
双层汽车库 two-floor garage
双层汽车运输车 double deck car-carrying wagon
双层汽缸 double shell casing
双层砌体 diamicton
双层铅包皮 double-lead covering
双层墙 cavity wall;double wall;jacketed wall
双层墙板 double layer wallboard
双层墙结构 double-wall construction
双层桥(梁) double decked bridge;double(-leaf) level bridge;two-layer level bridge
双层桥(梁)的下层桥面 low deck of bridge
双层桥楼【船】two-tier bridge
双层桥面 double deck(er)
双层穹顶 bi-vault;double vault
双层取芯管 double tube core barrel
双层取芯器 double core barrel;double tube core barrel
双层取样器 double tube sampler
双层绕组 double layer winding;two layer winding;two-position winding
双层绒头地毯 face-to-face pile carpet
双层溶液 dual layer solution
双层三层架货车 <铁路用> rack car
双层三角针道 twin cam tracks
双层三角座 two-tier cambox
双层纱架 two-height creel
双层砂集水池 dual sand catching basin
双层筛 double classifier;double deck screen
双层烧结钟罩 double-walled bell jar
双层伸缩卸料嘴 double bellow type spout
双层升降窗 double-sashed window
双层牲畜车 stock car double deck
双层施工 two-course construction;two-layer construction
双层施工体系 two-level building operational system
双层实心叠层的 double solid laminated
双层市场 two-tier market
双层式【数】two-layer equation;two-layer formula
双层式Y形立体交叉 two-level fork junction
双层式表面处理 double course surface treatment
双层式表面处治 double course surface treatment;dual course surface treatment;double coating;double-layered treatment;double seal coat;double surface dressing
双层式道路 double level road
双层式高速干道 double deck freeway
双层式环形立体交叉 two-level roundabout;two-level roundabout junction
双层式环形立体枢纽 two-level roundabout junction
双层式立体交叉 two-level junction
双层式滤尘器 two-stage type filter
双层式(路面设计)公式 layer equation;layer formula;two-layer equation[formula];two-layer formula
双层式(污水处理)池 two-stor(e)y tank
双层数 double layer
双层双动取土器 rigid-type double tube core barrel
双层双动岩芯管 core barrel of double tube-rigid
双层双面波纹纸板 double wall corrugated board
双层丝包铜线 double silk covered copper wire
双层丝包线 double silk-covered wire
双层丝绝缘的 double silk covered
双层丝绒织机 double velvet loom
双层塑模 two-level mold
双层弹簧垫圈 double spring washer
双层提升水闸 double sluice gate
双层提升(闸)门 double lifting gate
双层体理论 <用于柔性路面设计的理论> double layer theory;two-layer theory
双层体系 double-layered system
双层体系表面垂直位移 vertical displacement of double layered system surface
双层屉烤炉 two-deck drawplate oven
双层天车 double deck type of crown block
双层天花 double ceiling
双层天幕 double awning
双层天线阵 two-stacked array
双层铁线铠装电缆 double wire-armo(u)red cable
双层同心钻杆柱 dual concentric drilling string
双层同心钻管 dual concentric drill pipe
双层筒 double cylinder;two-layered cylinder
双层透镜 double layer lens
双层涂布 double coat
双层涂层 double coating
双层涂料纸 double coated paper
双层涂漆 two-coat work
双层外底 two-sloe
双层外壳 double casing
双层完成 dual completion
双层完成井 dually completed well
双层碗橱 press cupboard
双层网架 double layer grid
双层网络 double layer grid
双层位 double layer potential
双层屋顶 <上层排水,下层吊顶> double roof(ing);double-skin roof;weathering
双层舞台 two-floor stage;two-level stage;two-stor(e)y stage
双层系统 two-layer system
双层纤维板 double wall fireboard
双层箱形托盘 box pallet with a double floor
双层橡皮垫圈接头 double rubber-gasket joint
双层消音地板 double sound-boarded floor
双层小公寓 double decker
双层效应 double layer effect
双层泄水闸(门)double sluice gate
双层型 two-layer type
双层畜舍 bank-type barn;loft barn
双层絮凝沉淀池 two-level flocculation and sedimentation tank
双层岩芯管 double core barrel;double tube core barrel;double tube corer
双层岩芯管拆卸专用钳 Parmalee wrench
双层掩盖型 double covered type
双层摇床 double deck table
双层摇动筛 double shaking screen
双层药皮焊条 double coated electrode
双层荧光屏 double layer phosphor
双层油罐 two-compartment oil tank
双层油漆 doubling back;two-coat paint
双层运输机 two-deck transport
双层增透膜 double layer antireflection coating
双层轧辊 sleeved roll
双层轧制 pairing
双层闸门 double-leaf gate;double-leave gate
双层摺门 double folding door
双层支管砖 dual lateral tile
双层织物 double cloth;two-layer fabric
双层织造 two-layer weave
双层织造的织物 double-woven
双层织造网筛布 double-woven-wire cloth
双层纸 duplex paper
双层纸面石膏板 double layer wallboard
双层滞水池 two-stor(e)y aquifer
双层中空玻璃 double glazing unit;pair glass
双层柱头 double capital
双层柱状节理 two-layer columnar joints
双层筑路法 tow-course construction
双层铸石 two-layer cast stone
双层转塔 dual level turret
双层转筒燃烧器 double level rotary burner;Teclu burner

双层转子 two-level rotor
双层装置 double decker
双层纵向木壳板 double fore-and-aft planking
双层钻车 two-deck jumbo
双层作业线 double level geometry
双叉触点簧片 bifurcated contact
双叉的 biramous
双叉分接盒 bifurcating box
双叉挂车夹 fork
双叉管 bifurcation;double branch
双叉轨 double switch;double turnout
双叉河口 double distributary estuary
双叉接插件 bifurcated contact
双叉口 bifurcation
双叉口扳手 double open end spanner;double open end wrench
双叉溜槽 twin chute;two-way chute
双叉式钢绳捞矛 two-prong rope grab
双叉式支撑 double forked strut;K-strut
双叉爪钩 two-prong claw hook
双插大小头 spigot and spigot reducer
双插法 double interpolation
双插渐缩管 two-plug-in reducing pipe[piping];two-plug-in reducing tube[tubing]
双插口箱形托盘 box pallet with two apertures
双插入式箱形托盘 two-entry box pallet
双插塞塞绳 double-ended cord
双插销座 two-pin socket
双插芯式调谐器 double slug tuner
双插座 double socket;duplex outlet;duplex receptacle;twin socket
双插座接线盒 duplex outlet
双差动 double differential
双差固定解 double differential fix resolution
双差频效应 double super effect
双差速器 double differential
双掺杂 codope
双缠绕头 twin winding head
双产品试验 two-product test
双铲斗刮土机 two-bowl scraper
双铲口门框 double rebated frame
双铲口门膛板 double rebated lining
双超外差接收 double superheterodyne reception;triple detection
双超外差接收机 double superheterodyne receiver
双超重火石玻璃 double extra dense flint
双潮 agger;double tide;gulder
双车车库 two-car garage
双车串联铲运机组 two-bowl tandem scraper unit
双车串联翻车机 two-car tandem dumper
双车道 double lane;double stripe;double way;dual carriageway (road);dual drive way
双车道单幅车行道 two-lane single carriage
双车道道路 double lane road;double track;two-lane road(way)
双车道的 dual lane;two-lane
双车道浮桥 treadway bridge
双车道公路 double-land highway;double lane highway;dual highway;two-lane highway;two-lane road
双车道公路隧道 dual carriageway road tunnel
双车道交通 double lane traffic;two-lane traffic
双车道汽车路 double carriageway motorway
双车道桥(梁) double lane bridge;two-lane bridge
双车道隧道 two-lane tunnel
双车道隧管 two-lane tube
双车舵船 twin-screw and single-rudder ship
双车交通 two-lane traffic
双车三舵船 twin-screw and triple-rudder ship
双车试验 two-car test
双车双舵船 twin-screw and twin-rudder ship
双车头牵引式列车 double header
双车行道汽车路 dual carriageway motorway
双掣索结 double stopper hitch
双衬砌 double lining
双称端的 double-bellied
双称多谐振荡器 symmetric(al) multivibrator
双撑杆 anchor tower;pole with strut;pull-off pole;stayed pole;strutted pole
双成分粒子 two-component particle
双成分熔断器 dual-element fuse
双成分纤维 biconstituent fiber[fibre]
双成型头型热成型机 twin former type thermoforming machine
双承大小头 bell and bell reducer;transition of double-socket
双承丁字管 sanitary tee
双承渐缩管 two-socket reducing pipe
双承口管 double socket;double socket pipe
双承力索吊弦线夹 dropper clip for twin catenary wire;twin catenary wire dropper clip
双承力索滑动吊弦线夹 sliding dropper clamp for twin catenary wire
双承力索锚固线夹 twin catenary wire anchor clamp
双承力索双耳线夹 put-off clamp for twin catenary wire
双承曲管接头 both ends bell bend
双承式屋盖 double-framed roof
双承弯管 bend with double hub
双承弯头 double bell bend
双承一插丁字(三通)管接 both ends bell and spigot T
双承异径管 transition of double socket;two-socket reducing pipe
双乘法器 paired multiplier
双程 double pass(age);double passing
双程测量法 double run method;double run procedure
双程插销 double-throw bolt
双程出入境签证 two-way exit-entry visa
双程单色仪 double pass monochromator
双程分光计 double pass spectrometer
双程航路 two-way route
双程冷凝器 two-pass condenser
双程时间对齐 two-way time aligned
双程时间对齐叠加 stack presentation with two-way times aligned
双程式 two-pass type
双程水准测量 double run level(l)ing;double run line;duplicate level line;reciprocal level(l)ing
双程水准(测量路)线 double run line;duplicate level line
双程台卡导航系统 two-range Decca
双程往返观测 two-way observation
双程运输 two-way loading transport
双池(式)布置 <潮汐发电> two-pool plan
双池窑炉 twin tank furnace
双齿轨 double rack
双齿辊破碎机 double-geared roller crusher;double toothed crusher
双齿履带板 double grouser shoe
双齿轮轴承壳 compound gear bearing shell
双齿铣刀 straddle gear cutter
双翅管 finned duplex tube
双充气轮胎的车轮 pneumatic dual tired wheel
双充气轮胎的轮子 pneumatic dual tired wheel
双冲程的 double stroke;two-stroke
双冲程发动机 two-stroke engine
双冲程深井泵 double stroke deep well pump
双冲程式 double stroke
双冲构造 duplex structure
双冲轨道 two-impulse trajectory
双冲燃烧 opposed firing
双冲压喷气发动机靶机 twin-ram-jet target
双重 doubling;doubly;duplexing;duplication
双重板条 double lath
双重板桩式码头 double sheet pile type wharf
双重包装 double pack
双重保险 double insurance
双重保险单 double insurance policy
双重曝光 double exposure
双重贝塔衰变 double beta decay
双重背书 double endorsement
双重比例尺作图 drawing with two scales
双重比例控制 dual-ratio control
双重闭锁塞门 double locking cock
双重壁阶 dual offset
双重标志 double marking;duplication of marking
双重标准 double standard
双重表 dual meter
双重操纵 dual operation
双重操纵道岔 dual-operated switch
双重操纵系统 twin controls
双重差拍 double beat
双重车道 dual carriageway(road)
双重沉积 duplex deposit
双重承管 double pipe
双重抽样 double sampling
双重处理 double handling
双重传动 dual drive
双重刺网 trammel net
双重淬火 double quenching
双重错误 double error
双重大门 double gateway
双重代表权 dual representation
双重代理人 dual agency
双重氮化合物 bis-diazo compound
双重的 diploid;duplex;duplicate;twifold;double;dual;two-fold;two-ply
双重的Y形管 double Y branch
双重低潮 double low water
双重底 double bottom
双重点的 bifocal
双重点火 dual ignition
双重电路校验 dual circuitry check
双重电信 duplex transmission
双重淀积 duplex deposit
双重叠(代) double overlap
双重读数放大器 dual sense amplifier
双重堵缝 double ca(u)lking
双重对称 two-fold symmetry
双重对冲 double hedging
双重阀 double valve
双重法 double method
双重反馈 double reaction
双重反馈放大器 duplex feedback amplifier
双重反应 double reaction
双重反应调节器 double response controller
双重方法 dual mode
双重方位投影 double azimuthal projection
双重防洪堤系统 double levee system
双重防水节点 two-stage joint
双重防锈 dual weather protection
双重放大 dual amplification
双重放大镜 double magnifier
双重放大器 dual amplifier
双重费率制度 dual-rate system
双重分布【数】 double distribution
双重分次群 bigraded group
双重分级 double grading
双重分集接收机 dual diversity receiver
双重分类 two-fold-classification
双重分配通路 dual system
双重否定 double denial
双重否决 double veto
双重浮标 <一浮于水上,一浮于水中,测流速用> twin float
双重符合 two-fold coincidence
双重负债 double liability
双重复合带头 binary combination head
双重腹杆系桁架 double-webbed truss
双重钢筋 <受拉和受压> double (layer of) reinforcement
双重钢筋梁 double reinforced beam;doubly reinforced beam
双重高潮 double height water
双重隔电子 double petticoat insulator;double shed insulator;petticoat insulator
双重隔墙 double partition;two-withe [wythe]
双重拱门 double archway
双重构架墙 double-framed wall
双重估价 dual estimate
双重故障 twin failure
双重关节的 double-jointed
双重关税 double tariff;dual tariff
双重管式换热器 dual tubes heat exchanger
双重光谱投射器 double spectroprojector
双重广播 dual broadcast(ing)
双重国籍 double nationality;dual nationality
双重过滤 double filtration
双重过滤层 double filtration layer
双重过滤介质 dual filtration media
双重号码 double number
双重合开关 double coincidence switch
双重核共振磁力仪 double nuclear resonance magnetometer
双重喉管 double Venturi
双重后座装置 double recoil system
双重厚壁管 double extra heavy pipe;double extra strong pipe
双重护面 double armo(u)ring
双重滑架 dual carriage
双重滑移窗 twin sliding window
双重回代 double back-substitution
双重回转运动 double rotation motion
双重汇价制 two-tier exchange system
双重汇率 dual exchange rate;two-tier exchange rate
双重汇率制度 tow-tier exchange system
双重混凝 double coagulation
双重活动性 double activity
双重活度 double activity
双重活塞型 ram with ram type
双重货位检测装置 bin detection device
双重机构 dual mechanism
双重机壳 double hulled
双重机械联锁 duplex mechanical interlocking

双重机制 dual mechanism
双重积分陀螺 double integrating gyro
双重基点制 dual basing-point system
双重激发态 doubly excited state
双重极化天线 dual-pole antenna
双重计划 dual plan
双重记录密度 dual density
双重加固的 doubly reinforced
双重价格 double price;two-tier price
双重价格基础 double price basis
双重价格条款 dual valuation clause
双重监视系统 double check system
双重减速 double reduction
双重检波 double detection
双重检查 duplication check
双重检验量 duplication check
双重交叉 double crossing
双重交叉桁架 double intersecting truss
双重交换 double exchange
双重交换作用 double exchange interaction
双重交替 double alternate
双重胶片 bipack
双重校对 double check;twin check
双重校验 double check; duplication check;twin check
双重接缝 double seam
双重接头 double connector
双重接头连接 cross connection
双重结构 double texture; doublet structure
双重结构系统 dual structural system; dual structure system
双重介质渗流模型 binary medium flow model
双重金属 duplex metal
双重进位 double carry
双重经济结构 dual structure of economy
双重晶粒结构 duplex grain size
双重精度法 double precision
双重卷边 double seam
双重卷边接缝 double seaming
双重绝缘 double insulation
双重绝缘的 double insulated
双重绝缘的设备 double insulated appliance
双重绝缘电动机 double insulated motor
双重绝缘子 double insulator
双重均匀信道 doubly uniform channel
双重铠装 double armo(u)ring
双重(课)税协定 double tax treaty
双重空腔 duplex cavity
双重控制 <自动和人工> double(-model)control;dual mode control; duplicate control; dual operation; dual control
双重控制道岔 dual-control(led) switch;dual-operated switch
双重控制式动力转辙机 dual-control power switch machine
双重控制系统 dual control system
双重控制制 dual control system
双重控制转辙机构 dual switch mechanism
双重馈电 duplex feeding
双重扩散 double diffusion
双重扩散处理 double diffusion treatment
双重拉伸试验 double tension test
双重冷凝式离心热泵冷冻机 double condenser type centrifugal heat pump chiller
双重冷却的 dual-cooled
双重利润 double profit
双重连接列表 doubly inked list
双重连续扩散处理 <木材防腐> double diffusion treatment
双重联结环状列表 doubly linked circular list
双重联结线性列表 doubly linked linear list
双重梁 brace summer;double beam
双重螺旋起重器 telescoping jack
双重螺旋千斤顶 telescoping jack
双重螺旋弹簧 nest spring
双重慢沙滤 double slow sand filtration
双重铆钉接合 double riveted joint
双重门 double door
双重门船闸 double gate lock
双重门道 double gateway
双重门拱廊 double archway
双重门过道 double archway
双重密封 double(lip)seal;dual-seal
双重密封人井盖 double seal manhole cover
双重母线 double bus
双重目的的 dual-purpose;double edged
双重内外密封 dual inside and outside seal
双重纳税年度 dual-status tax year
双重凝聚 double coagulation
双重配筋 double reinforcement
双重配筋混凝土 double reinforced concrete
双重配筋混凝土梁 double reinforced concrete beam
双重配筋梁 double reinforced beam
双重喷嘴式喷射头 duplex nozzle
双重偏移 dual offset
双重频率变换 double frequency conversion
双重频率谱密度函数 double frequency spectral density function
双重平焊接 double butt welded
双重破碎机 duplex breaker; duplex crusher
双重气体保护焊 dual gas-shield welding
双重牵引的 double heading
双重嵌封 two-stage seal
双重墙式防波堤 double wall breakwater
双重擒纵机构 duplex escapement
双重穹隆 double vault
双重曲度的凹陷部分 circular-circular sunk
双重绕组 duplex winding
双重热机械处理 double thermomechanical treatment
双重人格 double personality; dual personality
双重熔点 double melting point
双重三星轮重力式擒纵机 double three-legged gravity escarpment
双重散热片二极管 double heat sink diode
双重散射激光测速计 dual scatter laser velocimeter
双重设计准则 dual design criterion
双重十字轴式等速万向节 double jaw joint
双重时间测量计 doubling time meter
双重收缩 double contraction;double shrinkage
双重收缩齿 duplex tooth taper
双重收缩模 double contraction pattern
双重税率制 double tariff system
双重税(收)double taxation
双重榫 two-stepped tenon
双重锁 duplex lock;two-bolt lock
双重锁闭器 duplex lock
双重锁定 double-locked
双重态 doublet
双重探询 double-polling
双重特性 double grading
双重特性天线 dual antenna
双重体系 dual system
双重调幅乘法器 double amplitude-modulation multiplier
双重调节 double regulation;dual regulation
双重调节阀 double regulation valve
双重调节接力器 double regulation servomotor
双重调节器 double governor
双重调速器 dual regulation governor;double governor
双重调制 double modulation
双重通风罩 double hood
双重通信[讯]方式 duplex communication system
双重投票 plural vote
双重投影法 double projection
双重投影系统 double projection system
双重透镜 doublet lens
双重透镜系统 two-lens system
双重透镜制 two-lens system
双重透视圆柱投影 double perspective cylindric(al)projection
双重凸凹榫接 double rabbet
双重图像 ghost
双重图像效应 ghost effect
双重涂层 double coat
双重退火 double annealing
双重托架 dual carriage
双重椭圆系 double elliptical system
双重外汇市场 dual exchange market; two-tier foreign exchange market
双重外壳 pair case
双重外门 double outdoor
双重位置式信号 two-position signal
双重温度控制 dual temperature control
双重稳定性能点 bistable performance point
双重屋顶 double roof(ing)
双重稀释法 double dilution(method)
双重系统 dual system;duplex system
双重夏令时 double summer time
双重显示 dual display
双重线 doublet
双重线法 <矿山测量传递方向> double plumbing method
双重线精细结构 doublet-fine structure
双重线圈灯丝 double coil filament
双重线项 double state
双重限止器 <上、下都限制> double limiter
双重像纸 double weight paper
双重性 dual nature
双重悬链线 compound catenary
双重选择权 double option
双重循环 double circulation
双重压力 dual-pressure
双重压制 double compression
双重延迟扫描 dual delayed sweep
双重研磨的平板玻璃 twin ground plate glass
双重衍射 two-fold diffraction
双重咬口 double seam
双重音汽笛 double blast
双重用途 double duty
双重用途包装 dual-use packaging
双重用途的 dual-purpose
双重余额递减折旧法 depreciation-double declining balance method
双重圆顶 double vault
双重圆筒沉箱式防波堤 dual cylindrical caisson breakwater
双重钥匙锁 duplex lock
双重运费制 <运费同盟的两价运费制> dual-rate system
双重责任 double liability
双重真空处理 double vacuum treatment
双重振荡器 two-mass oscillator
双重征税 double taxation
双重支撑 dual bracing
双重支撑系统 dual bracing system
双重支承楼板 double-framed floor
双重执照 dual chartering
双重纸 two-ply stock
双重转运 double handling
双重组合透镜 doublet combination lens
双重罪行 double criminality
双重作业 double freight operations; dual operations of wagon
双重作用 dual function
双重作用控制器 double response controller
双重作用说 duplicity theory
双重作用调节器 double response controller
双抽气式汽轮机 double pass-out turbine
双抽运 double pumping action
双畴模型 two-domain model
双出口厕所 two-holer
双出口齿轮泵 double discharge gear pump
双出口阀 twin valve
双出口挤泥机 twin die
双出口料斗 double-outlet cone
双出口煤斗 double-outlet cone
双出口消火栓 double hydrant
双处理机主机 dual processor main frame
双触点 collateral contact;double contact;twin contacts
双触点电键 double contact key
双触点断电器 two-point breaker
双触点滑动开关 double contact sliding switch
双触点配电器 alternate distributor
双触发 doublet trigger
双触键 double contact key; double touch
双触排开关 double bank switch
双穿孔 double punch(ing)
双穿孔机 double punch
双传动皮带 twin-drive belt
双传动式 double drive
双传动输送机 dual drive conveyer [conveyor]
双传动选粉机 dual drive separator
双传动(装置)double transmission; twin drive
双船身式飞机 twin-boat-type airplane
双船身式水上飞机 twin-hull flying boat
双船体 twin hulls
双船坞 twin docks
双船闸 twin locks
双船作业 dual-ship operating
双串式 dual tandem
双串式(飞机)着陆装置 dual tandem landing gears
双串式轮 twin-tandem wheel
双串式着陆轮架装置 dual-in-tandem landing gear assembly
双窗 two-light window
双窗办公室 two-window office
双窗铰接预制墙板 two-window glue-jointed two-piece panel
双窗帘 double shutter curtain
双窗墙板 two-window panel
双床【给】double bed
双床加氢脱硫 two-bed hydrodesulfurization
双床间客房 twin beds guest room; two-bed guest room
双床水软化器 two-bed demineralizer

S

双床脱矿质器 two-bed demineralizer
双床吸附减湿器 dual-bed dehumidifier
双床装置 two-bed system
双垂尾布局 twin-finned layout
双垂尾飞机 twin-finned airplane
双垂线 double plumbing
双垂线法 double plumbing method
双垂直平衡滑动窗 vertically sliding balanced sash
双垂直尾翼的 twin-finned
双锤破碎机 double hammer breaker
双锥式拌筒 double cone drum
双唇扣环 double lip retaining ring
双唇油封 two lip seal
双磁场异步电机 two-field induction machine
双磁道纠错法 double error correction scheme
双磁电机点火 dual ignition
双磁迹录音机 half-track recorder
双磁极电机 bipolar machine
双磁极受话器 bipolar receiver
双磁路 double magnetic circuit
双磁铁引出系统 two-magnet extraction system
双磁头卷带方式 two-headed wrap
双磁头通电磁化法 <磁粉探伤> prod magnetizing method
双磁芯 bimag
双磁芯存储方式 two-core per bit
双磁芯开关 two-core switch
双磁滞回线 double hysteresis loop
双雌接口 double female
双次验收抽样法 double sampling plan
双存储位 double bucket
双搭板对接 butt joint with double straps
双搭接焊缝 double strap seam
双搭瓦 double lap tile
双打包机 dual packer
双打拿管 duodynatron
双打蒸汽机桩锤 double-acting steam pile hammer
双大裁 double cap;double large
双大梁 double girders; dual girders; twin girders
双大邮裁 double large post
双代号网络 activity on branch;arrow network
双代号网络计划 network plan of arrow
双代号网络图 arrow diagram
双代森锰 manoc
双代森锌 zinoc
双带 duo-tape
双带锯机(制材厂)twin band mill
双带锯制材厂 double bandmill;double band sawmill
双带眶棘鲈 two-striped scolopsis
双带滤波器 two-band filter
双带模型 two-band model
双带式升降机 double-belted elevator
双带式输送机 double belt conveyer [conveyor]; twin-belt conveyer [conveyor]
双带式淘洗机 twin-belt vanner
双带式压机 double belt press
双带图像 two-band picture
双带耀斑 two-ribbon flare
双带运输装置 twin tape transporter
双袋库式磨木机 two-pocket magazine grinder
双单色仪 double monochromator; double monochrometer
双单色仪装置 double monochromator arrangement
双单射 bijection
双单元 twin pack
双单元爆裂防护器 double unit type blow-out preventer
双单元组水力泵 double unit hydraulic pump
双弹子锁 double cylinder lock
双挡冲洗系统 dual flushing system
双挡大便器 dual-cycle toilet
双挡导轨 double guiding rail
双挡链环 double bar link
双刀开关 double pole on-off switch
双刀单掷 double pole single-throw
双刀单掷开关 double pole single throw switch
双刀的 double pole
双刀割草机 double knife mower
双刀辊式切纸机 synchro-fly cutter
双刀开关 double point switch;double pole switch
双刀口密封 seal by double knife-edges double knife-edge seal
双刀口狭缝 bilaterally variable slit; bilateral slit
双刀盘切齿法 tangear(method)
双刀刨 double plane iron
双刀刨齿机 two-tool machine
双刀双掷 dpdt [double pole double throw]
双刀双掷开关 double pole double throw switch
双刀头刀夹 twin-head holder
双刀旋坯机 double template jigger
双刀转换开关 double pole change-over switch; two pole change-over switch
双导板泵阀 dual guided slush service valve
双导洞掘进 double heading
双导管 dual duct
双导架式升降机 twin-mast hoist
双导流堤 twin jetties
双导流尾鳍 twin-skeg stern
双导轮转向架 pony truck
双导频 double pilot
双导体 two-conductor
双导线 double conductor
双导线臂板信号机【铁】double wire semaphore signal
双导线传输线 twin line
双导线的 twin-lead
双导线调整器 double wire compensator
双导线系统 double wire system
双导线线路 double wire line; two-conductor line
双导线信号设备 double wire signal-(l)ing device
双导线信号握柄 double wire signal lever
双导线制 double wire system
双导线转辙机构 double wire point mechanism
双导向式立柱 double-side leader
双岛 twin islet
双倒锤定线法 working in on line
双倒镜法 <延长直线的> double line method;double traverse method
双倒镜直线延长法 <在固定点上延伸直线的方法> wiggling-in on line
双道传送 dual-path transmission
双道分析 dual channel analysis
双道夫搓条机 double doffer condenser
双道管道 bipass
双道火焰光度控制器 dual channel flame photometric detector
双道记录器 two-channel recorder
双道记录仪 two-canal recorder
双道矿块分选机 two-channel lump sorting machine
双道冷凝器 two-pass condenser
双道门 double(up)door;fancy door
双道旁侧声呐 dual channel side scan sonar
双道天线收发转换开关 two-channel duplexer
双道涂料 double coating
双道威扫描棱镜 double Dove scan prism
双道装置 dual channel device
双的 diploid;dual;duo;duple(x)
双灯丝的 bifilar
双灯丝灯泡 bilux bulb; double filament bulb
双灯座 twin-lamp socket
双登插管 double offset
双等离子体 ambiplasma
双等离子体不稳定性 ambiplasma instability
双等离子体发射器 duoplasmatron
双等离子(体)管 duoplasmatron
双低潮 double low water;tidal double ebb
双堤式防波堤 double breakwater
双滴电极 double dropping electrode
双底水准仪 double bottom level
双地址 double address;double bucket;two-address
双地址栈 dual address stack
双点 two point
双点插入法 double point interpolation
双点点焊接头 duplex spot weld
双点吊悬挂式脚手架 two-point suspended scaffold
双点法 <测定控制点> reference pair method
双点焊 duplex spot weld
双点焊机 duplex spot welder
双点焊接 duplex spot welding
双点划线 double dot dash line;two-dot chain line
双点画线 two-dot chain line
双点火 double ignition
双点记录器 two-point recorder
双点内插法 double point interpolation
双点啮合 double point toothing
双点调焦 two-point focus setting
双点调谐 double spot tuning; repeat point
双点显示 double dot display
双点显字器 double dot display
双点压力机 two-point press
双电层 double electric (al) layer; double electrode layer;double layer;electric (al) double layer; electrostatic double layer
双电动机 double motor
双电动机车 two-motor car
双电动机机组 double unit motor
双电动机驱动 twin drive
双电杆 twin pole
双电感式电压调整器 double induction regulator
双电荷层 dipole layer
双电荷离子 double charge ion
双电弧隔板 double arc chute
双电弧技术 double arc technique
双电弧熄弧沟 double arc chute
双电机组换速 speed-change with multimotors
双电极火花塞 two-point spark plug
双电极室 two-electrode chamber
双电极系统 two-electrode system
双电缆系统 twin-cable system
双电缆制 twin-cable system
双电离室 double ionization chamber
双电离压力计 double ionization ga-(u)ge
双电流 double current
双电流电炉 double current furnace
双电流发电机 double current generator
双电流制 double current system
双电流制式电力机车 double current locomotive
双电路 dual circuit
双电路的 double circuit
双电路法 two-circuit method
双电路微处理机 two-circuit microprocessor
双电平读出 dual level sensing
双电平工作 bi-level operation
双电平逻辑 two-level logic
双电容器式电动机 dual-capacitor motor
双电势静电透镜 bipotential electrostatic lens
双电枢 double armature; two-armature
双电枢电动机 double armature motor
双电枢发电机 two-armature generator
双电位 bipotential
双电位法测井曲线 dual normal log curve
双电位透镜 bipotential lens
双电压 dual voltage
双电压电力机车 dual-voltage electric locomotive
双电源 duplicate supply
双电源插座 duplex outlet
双电源电路 two-supply circuit
双电源供给 dual-power supply
双电致伸缩继电器 capaswitch
双电子复合 dielectronic recombination
双电子供体 two-electron donor
双电子谱线 two-electron spectrum
双电子枪存储管 metrechon
双电子枪示波器 two-gun oscillograph
双电子束 dual beam; twin-cathode ray beam
双电子束放大器 double stream amplifier
双电子束射线管 double stream tube
双电子束示波器 double beam oscilloscope; double oscillograph; double oscilloscope; dual-trace oscilloscope;dual channel oscilloscope
双电子束示波术 double beam oscillograph
双电子束同步示波器 dual-beam synchroscope
双电子系 two-electron system
双电子转移 two-electron shift
双凋 dual regulation
双吊架 twin-lift spreader
双吊杆联动绞辘 schooner stay
双吊杆联合起吊系统 married gear;union purchase
双吊杆联合作业法 burtoning system; married fall system
双吊杆装卸 two-derrick cargo handling
双吊钩 double hung
双吊货套链 double chain sling
双吊笼吊车 twin cage hoist
双吊式吊具 twin-lift spreader
双吊式集装箱装卸桥 twin-lift transporter container crane
双吊索 double sling;two-lay sling
双叠板簧 double spring
双叠波 double wave
双碟轮平面磨床 double disk grinder
双蝶滚刀 twin-disk cutters
双丁梁 double tee beam
双丁支座 double T bed
双丁字梁 double T-beam
双钉合 double nailing
双顶梁 dual head girder
双顶圆锥轧碎机 double crusher

双定位器准直线 locator dual collimating line
双定心法<用于直线延长> double centering[centring]
双定转子电动机 two-stator-rotor motor
双定子感应电机 two-stator induction machine
双动冲床 double-action press
双动冲压水压机 double-action hydraulic press
双动锤 double-acting hammer
双动打桩锤 double-acting pile-driver;double-acting(pile) hammer
双动道岔 double-acting points;double-working switches
双动道岔乙端【铁】 other end of double working turnout
双动的 double-acting
双动电开关 double-acting switch
双动发动机 double-acting engine
双动风箱 double-action bellows
双动夯 double acting ram
双动混合器 double motion mixer
双动活塞 double-acting piston;opposed piston
双动活塞柴油机 opposed-piston diesel engine
双动机械泵 duplex pump
双动机械压力机 double-action mechanical press
双动(棘)爪 double-acting pawl
双动减震器 double-acting damper
双动桨式混合器 double motion paddle mixer
双动桨叶拌和机<即向相反方向> double motion paddle mixer
双动铰链 double-acting hinge
双动搅拌机 double motion mixer;gyromixer
双动搅拌器 double motion agitator
双动进料器 double-acting feeder
双动卷扬机 double-acting hoist
双动力的自载铲运机 twin-powered self-loading scraper
双动力机车 dual-powered locomotive
双动力曝气塘 dual-power aerated lagoon
双动力驱动 twin-powered drive;two-motor drive
双动力式多格塘 dual-power multicellular lagoon
双动力式多格塘系统 dual-power multicellular lagoon system
双动力式机械 dual horse power
双动力水平塘系统 dual-power level lagoon system
双动力挖土机 twin-powered excavator
双动力型汽车 hybrid vehicle
双动门 double-acting door
双动模 double-acting die;double-action die
双动片电容器 twin rotor capacitor;twin rotor condenser
双动式泵 double-acting pump
双动式测力计 double-acting forcemeter
双动式打桩机 double-acting pile hammer
双动式的 double-acting;double action
双动式精梳机 double-acting comber
双动式空气压气机 double-acting air compressor
双动式空气压缩机 double-acting air compressor
双动式控制阀 double-acting control van
双动式弥雾机 double-acting atomizer
双动式喷雾机 double-acting sprayer
双动式气动打桩锤 double-acting air hammer
双动式汽锤 double-acting steam hammer
双动式手摇泵 double-acting hand pump
双动式水泵 double-acting pump;double-action pump;return-acting pump
双动式压缩机 double-acting compressor
双动式轧花机 double-acting gin
双动式蒸汽锤 double-acting steam hammer
双动式蒸汽机 double-acting steam engine
双动双层取芯器 core head of double tube-rigid type;double tube rigid barrel;double tube rigid type core barrel;rigid double tube;rigid type core barrel;rigid type double tube core barrel;stationary inner-tube core barrel
双动双层岩芯管 core head of double tube-rigid type;double tube rigid barrel;double tube rigid type core barrel;rigid double tube;rigid type core barrel;rigid type double tube core barrel;stationary inner-tube core barrel
双动双管取芯钻具 rigid type double tube core barrel
双动弹簧铰链 double-acting spring butt;double-action spring butt
双动选择器 bank-and-wiper switch
双动压床 double-acting press;double-action press
双动压力机 double-action press
双动压气机 double-acting compressor
双动压缩机 double-acting compressor
双动压制 double-acting compression
双动液压千斤顶 double-acting hydraulic jack
双动与弹簧回位型 double-acting and spring return type
双动蒸汽锤 double-action steam hammer;holding-up hammer
双动柱模 toggle draw die
双动转辙器 double slip switch
双动桩锤 double-acting ram
双动自止门 swing(ing) door
双动作继电器 double-acting relay
双动作选择器 two motion switch
双斗铲运机 two-bowl scraper
双斗的 twin-bucket
双斗给料机 double scoop feeder
双斗刮土机 two-bowl scraper
双斗式挖土机 double scoop type excavator
双斗提升机 double bucket elevator
双读表 dual meter
双读游标<向两方向读数的> folded vernier
双独立地图编码 dual independent map encoding
双堵塞器 double packer
双渡线 double crossover;intersecting crossover;scissors crossing;scissors crossover
双渡越速调管 double transit oscillator
双端 double pointed
双端的 double-end(ed)
双端法兰管 spool
双端干船坞 canal dock
双端功率发射管 double-ended power transmitting tube
双端管 bitermitron
双端规 go-and-no-go ga(u)ge
双端机车 double-ended locomotive
双端尖的 fusiform
双端开关 two-terminal switch
双端控制 double-ended control
双端口存储器 dual-ported memory
双端馈电 duplex feeding
双端馈电的 dual feed
双端螺栓 double-ended bolt;stud (bolt)
双端螺柱 stud bolt
双端面机械密封 double mechanical seal
双端面密封 double mechanical end face seal
双端面磨床 parallel surface grinding machine;twin surface grinder
双端磨床 double-ended grinder
双端汽缸 double-ended cylinder
双端塞绳电路 double ended cord circuit
双端式镀膜线 double-ended coating line
双端头车 double ender
双端凸缘管 spool;spool piece
双端铣刀 double end mill
双端系统 double-ended system
双端卸载 end-to-end discharge
双端型热成型机 twin former type thermoforming machine
双端修饰的 double end trimmed
双端引线 double end
双端引线型 double-ended type
双端制榫机 double end tenoner
双端装药爆破筒 double-ended charge
双短截线调谐器 two-stub tuner
双短线调谐器 double stub tuner
双段法 two-section method
双段接头 double steps
双段破碎机 two-stage crusher
双段衰减器 double attenuator
双断 double break
双断断路器 double break circuit breaker
双断接点 double break contact;twin-break contact
双断接头 double break contact
双断开关 double break jack;double break switch
双断口灭磁开关 two-pole field circuit breaker
双断路点开关 double break switch
双断面 dual cross-section;twin cross-section
双断塞孔 double break jack
双断双闭触点 double break double-make contact
双对 biconjugate
双对称 disymmetry;double symmetry
双对称大梁 bisymmetric(al) girder
双对称的 bisymmetric(al);double symmetrical
双对称断面 double symmetrical cross-section
双对称平衡给料器 duplex balanced feeder
双对称式修饰 bisymmetric(al) mo(u)lding
双对数 double logarithm
双对数标绘 log-log plot
双对数频谱比 log-log spectral ratio
双对数斜率 log-log slope
双对数增加 double logarithm increase
双对数诊断图版法 log-log diagnostic plot
双对数坐标 double logarithm coordinate;log-log coordinates;log-log scale
双对数坐标图 log-log plot
双对数坐标系统 log-log grid
双对数坐标纸 log-log paper
双墩 twin-pier
双墩肘形尾水管 two-pier elbow draft tube
双多孔密封膏背衬材料 bicellular sealant backing
双多面型 double brilliant
双舵 double vane
双舵船 twin-rudder ship
双额电表 two-rate meter
双额定电压 double voltage rating
双轭通气管 dual yoke vent
双颚式破碎机 twin-jaw crusher
双颚式轧碎机 twin-jaw crusher
双颚式抓斗 clamshell bucket;clamshell grab
双耳鞍子 catenary wire suspension clamp with hook
双耳长颈瓶 amphora
双耳的 binaural
双耳定位 binaural localization
双耳机受话器 bi-telephone receiver
双耳结 double ear knot
双耳连接器 cross link(age)
双耳连接头 guy clevis eye-bolt
双耳联板 dual-lug yoke plate
双耳受话器 biphone;double headphone
双耳双管线夹 twin-tube clamp with double eye
双耳线夹 clevis end clamp;pull-off clamp
双耳效应 biaural effect
双耳楔型线夹 clevis end wedge-type clamp
双耳轴磨机 two-trunnion mill
双二次 biquadratic;quartic
双二次的【数】 biquadratic
双二次绕组变压器 double secondary transformer
双二分法 double dichotomy
双二辊式平整机 uni-temper mill
双二辊轧机 double duo-mill
双二极管 duodiode;twin diode
双二极管单极性箝位电路 one-way double-diode clamp
双二极管限幅器 double-diode limiter
双二极三极管 double diode-triode;duodiode-triode
双二极五极管 double diode pentode;duodiode-pentode;duplex diode pentode
双二进制 duobinary system
双二进制的 duobinary
双二进制码 duobinary code
双二进制数 duobinary number
双发动机 bi-motor;duplex engine;twin engine
双发动机舱 two-engine nacelle
双发动机车辆 two-engined vehicle
双发动机的 twin-engined
双发动机飞机 twin-engined plane;two-motored airplane
双发动机公共汽车 twin-engined bus
双发动机火磁电机 twin spark magneto
双发动机货机 twin-engined freighter
双发动机减振回转起重机 two-engine rubber-mounted slewing crane
双发动机减振伸缩臂起重机 two-engine rubber-mounted telescopic(al) crane
双发动机减振悬臂起重机 two-engine rubber-mounted slewing crane
双发动机进场着陆 two-engine approach
双发动机汽车 twin engine vehicle
双发动机汽车吊 two-engine rubber-mounted crane
双发动机汽车挖土机 two-engine rubber-mounted excavator
双发动机驱动 double-engine drive;twin engine;twin-engined drive
双发动机驱动式 double engine

S

双发动机式铲运机 dual engine scraper;twin-engine scraper
双发动机式平地机 two-engine grader
双发动机系统 two-engine system
双发动机型 tandem powered
双发动机着陆 two-engine landing
双发射极晶体管 two-emitter transistor
双发生器 dual producer
双阀混凝土泵 two-valve concrete pump
双阀开关 open and shut valve
双法兰 Z 形管件 double-flanged zed
双法兰短管 flanged spool piece
双法兰渐缩管 double-flanged reducer
双法兰克反射镜 double Frank's mirror
双法兰空心轴 two-flanged hollow shaft
双法兰外移管件 double-flanged zed
双法兰轴 two-flanged shaft
双翻铲机 twin rocker shovel
双翻斗 double skip
双翻式翻车机 two-car tandem dumper
双反冲洗法 double recoil method
双反光立体镜 double reflecting stereoscope
双反面编织 links-links knitting
双反面横机 links-links flat bar knitting machine
双反面提花滚筒 links-links pattern drum
双反面提花三角 links-links jacquard cam
双反面提花装置 links-links patterning mechanism
双反面图案 links-links design
双反面线圈 links-links stitch
双反面针织物 pearl fabrics;purl fabrics
双反面织物 links-links fabric
双反射 bireflection;double reflection
双反射分布函数 bidirectional reflectance distribution function
双反射镜 double mirror
双反射镜法 double mirror technique
双反射镜系统 two-mirror system
双反射率 bireflectance;bireflectivity
双反射器天线 Cassegrain(ian) antenna
双反像平面镜系统 double reversing mirror system
双反应电机理论 two-reaction machine theory
双反应法 double reaction;two-reaction method
双反应论 two-reaction theory
双方(当事人) both;both parties;by and between
双方的利益 common interest;mutual interest
双方过失碰撞条款 both to blame collision clauses
双方话终拆线 last-party release;last-subscriber release
双方解石片 double calcite plate
双方解约 mutual recession
双方面体 dihexahedron
双方同意 by mutual consent;mutual consent
双方头传动 double end square drive
双方位计算机 two-bearing computer
双方向的 two-directional
双方向平衡 duplex balance
双方向运行 double directional working(on single track line)
双方协定 mutual agreement
双方议定 come to an understanding
双方有利 win-win
双方有责任碰撞条款 both to blame collision clauses
双方之间的责任 liability between the parties
双防爆膜 dual rupture disks
双房间客房 double room
双放大率测距仪 two-magnification rangefinder
双放电稳定 double discharge stabilization
双放音带 double play tape
双扉门 bi-part door
双扉(竖铰链)窗 double casement window
双扉闸门 double-leaf gate;double-leave gate
双沸系统设备 two-boiling system
双费率 double fee rate
双分叉式楼梯 staircase of bifurcated type
双分道采样器 dual channel sampler
双分道岔 double turnout
双分集偶极天线阵 space-dipole array
双分雷达 bistatic radar
双分离对称消像散镜头 air-spaced double anastigmat
双分量电磁计程仪 two-component electromagnetic log
双分裂导线线路 twin bundled lines
双分流 double split flow
双分楼梯 bifurcated stair(case);bifurcated type stair(case);staircase of bifurcated type
双分平行楼梯 double-return stair(case)
双分式对折楼梯 double-return stair(case)
双分透镜 air-spaced doublet
双分子 dimolecular
双分子层 bilayer;bimolecular layer
双分子反应 reaction of the second order
双分子膜 bimolecular film
双分子碰撞 bimolecular collision
双分子终止(反应) bimolecular termination
双酚 A bisphenol-A;dian;diphenylol propane
双酚 A 环氧化物 bisphenol epoxide
双酚 A 环氧树脂 bisphenol A epoxy resin;dian epoxy resin
双酚 A 型环氧树脂 diphenylol propane epoxy resin
双酚 A 型聚酯树脂 bisphenol A polyester resin
双酚 F 环氧树脂 bisphenol F epoxy resin
双酚酸 diphenolic acid
双份的 duplicate
双份估价 dual estimate
双风泵 duplex air compressor
双风泵调节器 duplex compressor governor
双风道 dual duct
双风道方式 dual duct system
双风道混合箱 double duct mixing box;mixing unit
双风道空调系统 dual duct air conditioning system
双风道喷嘴 dual channel burner
双风道燃烧器 dual channel burner
双风道系统 double duct system;dual duct system
双风管 dual duct
双风管混合装置 dual duct blender unit
双风管空气调节系统 dual duct air conditioning system;dual duct system
双风化层 double layer weathering
双风门化油器 twin-choke carburetor
双风扇 double fan
双风扇清选 two-fan cleaning
双风向标 bivane
双风巷通风系统 leakage intake system
双封(闭)层 double seal(coat)
双封隔器 twin packer
双封型油管接头 dual-seal tubing joint
双封制 <向业主提交报价的一种方式> two envelops
双封轴承 double seal(ed) bearing
双峰 double peaks;split-blip
双峰暴雨 twin storms
双峰波 double humped wave
双峰潮汛 double tidal flood
双峰的 bimodal;double hump
双峰(地形)的纵断面 longitudinal profile of double incline
双峰对数分布图 double peak logistic distribution map
双峰反应谱 response spectrum with double peaks
双峰分布【数】 bimodal distribution
双峰共振 crevasse;double humped resonance
双峰洪水 twin floods
双峰火山作用 bimodal volcanism
双峰粒度分布 bimodal size distribution
双峰裂距 doublet peak split distance
双峰脉冲 doublet impulse
双峰曲线 bimodal curve
双峰热 double quotidian fever
双峰式(泥沙级配)分布 bimodal distribution
双峰索道 bimodal distribution
双峰态的 bimodal
双峰态性 bimodality
双峰驼 bactrian camel;two-humped camel
双峰效应 double hump effect
双峰谐振 crevasse;double humped resonance
双峰型粒径曲线 bimodal curve
双峰性 bimodality
双峰异常 double peak anomaly
双峰云母组构 bimodal mica fabric
双锋 two-edged
双缝磁头 dual-gap head
双缝干涉 two-slit interference
双缝干涉仪 double slit interferometer;two-slit interferometer
双缝(隙) double slit
双缝衍射 two-slit diffraction
双伏特计法 two voltmeter method
双浮标 <测流用的> double float
双浮标系泊 double buoys mooring;securing to buoys fore and aft;two-point mooring
双浮标系统 two-part buoy system
双浮离泊 slipping from head-and-astern buoys
双浮筒 twin float
双浮筒式水上飞机 double buoys seaplane;double float seaplane;pontoon seaplane;twin-float seaplane
双浮筒系泊 double buoys mooring;securing to buoys fore and aft;two-point mooring
双符合谱仪 double coincidence spectrometer
双幅 double amplitude;total amplitude
双幅度峰值 double amplitude peak
双幅二车道公路 dual two-lane highway
双幅联画雕刻 diptych
双幅路 dual carriageway(road)
双幅门 double-margined door
双幅三车道公路 dual three-lane highway
双幅式车行道 dual carriageway(road)
双幅式车行道高速公路 dual carriageway motorway
双幅式(道)路 <相对行车方向之间有中央分隔带,俗称两块板道路> dual carriageway(road)
双幅式公路 dual highway
双幅式平行滑行道 dual parallel taxiways
双幅式双车道高速公路 dual two-lane motorway
双幅式行车道公路 divided highway;dual carriageway road
双幅折叠画板 diptych
双幅值 peak-to-peak value
双幅字盘架 double frame
双辐射 biradial
双负载电路 twin load circuit
双负重轮 dual load wheel
双负阻管 duodynatron
双复磷 toxogonin
双复面透镜 bitoric lens
双复式(四车道)公路 <用中央分隔带与边缘分隔带分成四个专用车道的道路,道路中部为过境车道,两侧为地方车道> dual-dual highway
双复式着陆架 twin-twin landing gear
双副 two-pack
双副边变压器 double secondary transformer
双副载波系统 two-subcarrier system
双副载波制 twin subcarrier;two-subcarrier system
双腹板 T 形梁 double-webbed T-beam;twin-webbed T-beam;two-webbed T beam
双腹板大梁 two-webbed plate girder
双腹板的 twin-webbed;two-webbed
双腹板工字梁 double-webbed I-beam
双腹板拱 double-webbed plate arch
双腹板拱梁 double-webbed plate arch(ed) beam
双腹板(桁)梁 through-shaped girder;twin-webbed(plate) girder
双腹板门 double-panelled door;two-panel(led) door
双腹梁 double-webbed beam;two-webbed beam
双腹式断面 double-webbed section
双盖板搭接接头 double lap joint;double strap lap joint
双盖板对接 double cover butt joint
双盖板接头 double strap joint
双盖表 hunter
双盖缝 double-lock seam;double weld(ed seam)
双盖覆瓦状 quincuncial aestivation
双盖立缝 double standing welt
双盖立缝式镀锌板屋面 double standing welt-type zinc roof
双盖条立缝 double standing seam
双盖条立缝式镀锌板屋面 double standing seam-type zinc roof
双盖瓦 double lap tile
双干板离合器 double dry plate clutch
双干管 double main
双干管式 double main system;dual main system;two-main system
双干管系统 double main system;dual main system;two-main system
双干管制 double main system;dual main system;two-main system
双干涉仪 dual interferometer
双干树 twin geminate;twin planting
双干线编索 railway sennit
双干线系统 double main system
双干燥鼓 double drying drum
双甘油 dialycerin;diglycerol

双杆 double pole
双杆保险杠 twin-bar bumper
双杆变压器 two-stub transformer
双杆划线规 double bar ga(u)ge
双杆机构 double lever mechanism
双杆件的 two-pack;two-part
双杆控制 dual control
双杆联吊装卸方式 union purchase system
双杆气压计 dual traverse barograph
双杆式弹性道钉 double shafted spring spike
双杆弹性道钉 twin-shafted resilient spike
双杆提引器 dual rod puller
双杆线路 H-pole line
双杆应变计(仪) sister bar strain ga(u)ge
双杆支脚手架 double pole scaffold
双杆装卸法 cargo-handling with double boom; cargo-handling with married fall
双坩埚法 double crucible method; double crucible process
双感应八侧向测井 dual induction-laterolog 8
双感应八侧向测井曲线 dual induction-laterolog 8 curve
双感应测井 dual induction log
双感应测井曲线 dual induction curve
双感应测井图 dual induction log plot
双感应聚焦测井 dual induction-focused log
双感应聚焦测井曲线 dual induction-focused log curve
双感应聚焦测井图 dual induction-focused log plot
双感应球形聚焦测井 dual induction-spheric(al) focused log
双感应球形聚焦测井曲线 dual induction-spheric(al) focused log curve
双缸泵 double pump;dual pump;twin-(-cylinder) pump; two-cylinder pump
双缸泵调节器 duplex pump governor
双缸柴油机 twin diesel
双缸的 twin-cylinder
双缸端面进气转子发动机 twin rotor side-intake-port engine
双缸发动机 twin-cylinder engine
双缸风冷式发动机 twin-cylinder air-cooled engine
双缸复动泵 duplex double-acting pump
双缸复合机车 cross compound locomotive;two-cylinder compound locomotive
双缸复涨式机车 cross compound locomotive
双缸活塞泵 duplex piston pump; twin-cylinder piston pump
双缸机车 two-cylinder locomotive
双缸机试验 two-cylinder test
双缸气泵 duplex air pump
双缸气锤 compound steam hammer
双缸曲柄泵 duplex power pump
双缸式发动机 double cylinder engine
双缸式涡轮机 two-cylinder turbine
双缸式液压操舵装置 two-ram hydraulic steering gear
双缸双作用泵 double-acting duplex pump;duplex double-acting pump
双缸体 twin rotor housing
双缸往复式泵 duplex reciprocating pump
双缸压缩机 duplex cylinder compressor
双缸液压触探机 double cylinder hydraulic sounding machine
双缸液压混凝土泵 Putzmeister pump
双缸液压千斤顶 jack pair
双缸蒸汽机 double cylinder steam engine;twin steam engine
双缸蒸汽往复泵 duplex steam piston pump
双缸柱塞泵 duplex plunger pump
双缸转子发动机 twin rotor engine
双缸转子内燃机 rotary combustion engine;tworotor
双钢板桩套入锁口并焊接定位后的打桩法 double sheets
双钢带铠装 double tape armo(u)r
双钢筋 twinned bars
双钢梳式接缝 double steel comb joint
双杠 parallel bars
双杠杆 double lever;dual lever
双高潮 double high tide;double high water;tidal double flood
双高斯镜头 double Gauss lens
双高斯式物镜 double Gauss type lens
双高斯透镜 Biotar lens;double Gauss lenses
双高斯型镜头 Biotar lens
双高斯型物镜 double Gauss objective;Gauss double type object-lens
双高斯型物镜变形 double Gauss derivatives
双割锯 double cut saw
双格储槽 two-compartment tank
双格窗 double-pane sash
双格窗框 two-pane sash
双格化粪池 two-compartment septic tank
双格栅楼板 double-joisted floor
双格斜井 two-compartment slope
双隔 double partition
双隔间化粪池 dual compartment septic tank
双隔膜泵 dual diaphragm pump
双隔膜跳汰机 twin-diaphragm jig
双隔膜圆锥分级机 double diaphragm cone
双膈膜泵 twin-diaphragm pump
双膈膜夹具跳汰机 twin-diaphragm jig
双根<由两根钻杆组成的> couplet
双根平行变形 parallel texturing
双根钻杆 double length of drill pipe
双工帮电机 duplex repeater
双工闭塞机 sosin block instrument
双工拨号 duplex dialling
双工拨号盘 duplex dial
双工博多(电报)机 baudot duplex apparatus
双工操作 duplex operation
双工传输 duplex transmission;double transmission
双工传真机 duplex facsimile equipment
双工的 duplex
双工电报 duplex operation
双工电报电路 duplex connection
双工电话 duplex telephone
双工电路 duplex circuit;duplex line
双工发送 double transmission;duplex transmission
双工仿真电路 duplex artificial circuit
双工仿真线路 duplex artificial line
双工放大器 bilateral amplifier
双工高速数据 duplex high speed data
双工工作 diplex operation; duplex working
双工机 duplexer
双工计算机 duplex computer
双工接收 double reception
双工控制 duplex control
双工况运行 dual operation
双工滤波器 combining filter
双工平衡 duplex balance
双工平衡线 duplex artificial line
双工器 diplexer[duplexer]
双工桥接电报 duplex bridge telegraph
双工设备 diplex system;duplex apparatus
双工通报 duplex working
双工通信[讯] duplex communication; duplex operation;duplex working
双工无线电传输 diplex radio transmission
双工系统 duplex system
双工信道 duplex channel
双工信令 duplex signal(l)ing
双工运行 duplexing; duplex operation;duplex transmission
双工振荡器 diplex generator
双工制 diplex system; duplex operation
双工质锅炉 binary cycle boiler
双工质燃烧器 dual burner
双工质系统 two-fluid system
双工质循环 binary cycle
双工转发器 duplex repeater
双工装置 duplex apparatus
双工资 double time
双工字钢柱 double H steel column
双工作触点 double make contact
双工作方式干扰机 noise and deception jammer
双工作轮水轮机 twin-runner(type) turbine;two-runner turbine
双工作轮涡轮机 twin-runner(type) turbine;two-runner turbine
双弓板弹簧 elliptic(al) spring
双弓弹簧 double laminated spring
双弓弦桁架 double bowstring truss
双弓形折流板 double segmental baffle
双公扣大小头 double pin sub
双功能催化剂 dual-catalyst
双功能键 double function key
双功圆盘耙 double-action disc[disk] harrow
双供电系统 dual supply system
双供水系统 dual supply system
双拱顶池窑 double crown tank furnace
双拱炉膛 double arch furnace
双拱隧道 double vault tunnel
双拱挑檐 double arched corbel table
双拱屋顶 bi-vault
双拱窑炉 double crown furnace
双拱圆顶 double vault
双拱支承的平板<支承平板的牛腿呈拱形> double arched corbel table
双共轭 biconjugate
双共振 double resonance;doubly resonance
双共振光参量振荡 double resonance optic(al) parameter oscillation
双共振光参量振荡器 double resonant optic(al) parameter oscillator
双共振摄谱仪 double resonance spectrograph
双共振微波激射 two-resonance maser
双共振振荡器 doubly resonant oscillator
双沟道 double channel
双沟道型 dual channel type
双沟型 sycon
双钩 sister hooks
双钩吊车 twin hock crane
双钩吊钩 double hook
双钩法兰螺扣 hook and hook turnbuckle
双钩环 double shackle
双钩起重机 double crane
双钩伸缩螺扣 hook and hook turnbuckle
双构架 sandwich frame
双构件的 two-component
双构件系统 two-component system
双箍闭合 double-hoop closing
双谷模型 two-valley model
双股 bifilar;double ply;two-ply
双股冰川 transection glacier;twinned glacier
双股并线络筒 two-end cheese winding
双股导线 twin conductor
双股的 bifilar; double wound; double strand
双股电缆 duplex cable; pair(ed) cable;twinax cable
双股吊索 brothers
双股钢丝绳 twin cable
双股滚柱链 two-strand roller chain
双股滚子链 double strand roller chain
双股绞车 winch with double barrel
双股绞钢丝束 twin-twisted wire strand
双股绞合 twinning
双股金属丝网布 double gauze wire cloth
双股聚合物 double stranded polymer
双股绝缘电线 ripcord
双股溜槽 bifurcated chute
双股流型预热器 twin stream preheater
双股螺纹钢筋 twin-twisted bar reinforcement;twin-twisted bars
双股螺旋 bifilar helix
双股螺旋线 double helix structure
双股扭合钢筋 twin-twisted bar reinforcement;twin-twisted bars
双股扭绞的 twin-twisted
双股扭绞钢丝束 twin-twisted wire strand
双股扭绞四芯电缆 multiple twin quad
双股扭转钢丝束 twin twisted wire strand
双股软线 twin flexible cord
双股纱线捻度试验仪 two-thread yarn twist tester
双股绳 twine
双股丝 doubled folded yarn; folded filament yarn
双股四芯电缆 multiple twin quad
双股线 twine; twin wire; two-fold yarn;two-ply yarn
双股油麻绳 marlin(e)
双鼓的 double-drum;dual-drum
双鼓分类法 twin drum sorting
双鼓卷纸机 two-drum reel;two-drum winder
双鼓式混凝土拌和摊铺机 twin pav(i)er
双鼓式混凝土搅拌摊铺机 twin pav(i)er
双鼓式(混凝土)铺路机 dual-drum pav(i)er;twin pav(i)er
双鼓式转鼓 dual-drum tumbler
双鼓筒锅炉 bi-drum boiler
双鼓筒混凝土搅拌机 double-drum (concrete) mixer
双鼓筒搅拌机 double-drum mixer
双鼓筒卷扬机装置 double-drum hoisting gear
双鼓筒铺路机 double-drum pav(i)er
双鼓真空离心过滤机 double bowl vacuum centrifuge
双固化机理 dual cure mechanism
双故障 double failure
双刮刀钻头 two-way bit
双胍 biguanide
双挂车列车 twin trailer train
双挂车汽车列车 twin-trailer(road) train
双挂钩砖 double tuckstone

S

双挂汽车列车 double trailer train
双拐结点 biflecnode
双拐曲柄 two-throw crank
双拐曲轴 double-throw crankshaft
双拐弯拉张器 double pull-off
双关测量计 on-off ga(u)ge
双关对称振幅 dual symmetric amplitude
双观测 double measurement; pair observation; pairs of observation
双观测组 group of two observations
双官能团的 bifunctional
双冠形结节 double crown knot
双管 double pipe; duplex tube
双管伴随加热法 double pipe heat tracing
双管采暖系统 double pipe heating system; two-pipe heating system; two-pipe system
双管簇冷凝器 double bundle condenser
双管单封隔器 double tube single-packer
双管道排水系统 two-pipe system
双管道向上送热供暖系统 < 下部分配集中供暖 > two-pipe rising system
双管道向下供暖系统 < 上部分配集中供暖 > two-piece drop system
双管的 double barrel(l)ed
双管法 double tube method
双管废气锅炉 double tube waste heat boiler
双管封闭式孔隙水压力仪 double seal pore water piezometer
双管供暖系统 double pipe heating system; two-pipe heating system; two-pipe system
双管供热系统 two-pipe heat supply system
双管供水系统 two-pipe water supply system
双管沟 dual duct
双管灌浆法 double tube grouting method
双管涵洞 double pipe culvert; two-pipe culvert
双管焊接 double jointing; double welding
双管焊接厂 double jointing plant
双管焊接场地 double jointing site; double jointing yard
双管焊接机 double jointing machine
双管化学灌浆法 two-shore method
双管火箭发射架 twin missile carrier
双管加热器 twin-tube heater
双管绞刀 double-lead screw; double screw conveyer[conveyor]
双管绞刀输送机 twin-screw conveyer[conveyor]
双管绞刀喂料机 twin-screw feeder
双管卡 double pipe clamp; double pipe clip
双管冷却管(道) coiled cooling pipe
双管立式喷雾器 duplex vertical sprayer
双管路液压制动系统 dual hydraulic circuit brake system
双管螺旋输送机 double screw conveyer[conveyor]; twin-screw conveyer[conveyor]
双管枪 double barrel
双管取芯钻具 double tube core barrel
双管取样器 double tube sampler
双管燃烧管 double tube burner
双管热交换器 double pipe heat exchanger
双管热水供暖 two-pipe hot water heating
双管上给式热水供暖系统 two-pipe hot water upfeed system
双管伸缩式钻塔 duplex design tubular derrick
双管渗透计 double tube permeameter
双管式测压计 hydraulic twin tube piezometer
双管式孔隙压力计 hydraulic twin tube piezometer
双管式(岩)芯管 double tube core barrel
双管输送机 twin-tube conveyer[conveyor]
双管双封隔器 double tube double packer
双管隧道 two-tube tunnel
双管围油栏 double tubed containment boom
双管卫生工程管道 two-pipe plumbing
双管卫生设备 two-stack plumbing
双管卧式喷雾器 duplex horizontal sprayer
双管系统 double pipe system; dual system; twin-pipe system
双管下给式热水供暖系统 two-pipe hot water downfeed system
双管线 double(pipe) line; twin pipeline
双管线夹 twin-tube clamp
双管型隧道 twin-tube tunnel
双管旋转型表面喷射岩芯钻筒 double tube swivel-type face ejection core barrel
双管仪 two-pipe meter
双管直接回水(供暖)系统 twin-pipe direct return system; two-pipe direct return system
双管制动系统 double pipe brake system
双管制蒸汽热网 two-pipe steam heat-supply network
双罐厚浆混合机 twin heavy-duty paste mixer
双罐笼提升 two-cage hoisting
双罐式气动灰浆泵 double pot mortar pump
双光灯泡 two-light lamp
双光点技术 double spot technique
双光度计 dual photometer
双光缆 dual fiber[fibre] cable
双光谱复印法 dual spectrum
双光气 diphosgene
双光前大灯 two-light headlamp
双光栅 double grating
双光栅单色仪 double grating monochromator
双光栅电视系统 dual raster television system
双光栅摄谱仪 double grating spectrograph; twin grating spectrograph
双光束 double beam; dual beam
双光束车头灯 dual-beam headlight
双光束电零点红外分光计 double beam electric-null infrared spectrometer
双光束斐佐干涉条纹 double beam Fizeau fringes
双光束分光光度计 double beam spectrophotometer
双光束干涉 two-beam interference; two-ray interference
双光束干涉测量学 two-beam interferometry
双光束干涉分光计 two-beam interference spectrometer
双光束干涉量度学 two-beam interferometry
双光束干涉图 two-beam interference pattern
双光束干涉显微镜 two-beam interference microscope
双光束干涉显微术 two-beam interference microscopy
双光束干涉仪 double beam interferometer; dual-beam interferometer
双光束光度计 double beam photometer
双光束光栅分光光度计 double beam grating spectrophotometer
双光束记录分光光度计 double beam recording spectrophotometer
双光束技术 double beam technique
双光束密度计 double beam densitometer
双光束偏振器 double beam polarizer
双光束散斑干涉度量法 dual-beam speckle interferometry
双光束摄谱仪 two-beam spectrograph
双光束系统 double beam system
双光束显微分光光度计 double beam microspectrophotometer
双光束原理 double beam principle
双光束远红外傅立叶光谱仪 double beam far-infrared Fourier spectrometer
双光眼镜 bifocal eye glass; bifocals; bifocal spectacle
双光子全息术 two-photon holography
双光子吸收 two-photon absorption
双光子荧光法 two-photon fluorescence method
双光子荧光监测器 two-photon fluorescence monitor
双轨 double line; two-track; pair of tracks
双轨导向器 flat double guides
双轨道 double track
双轨道桥 double track bridge
双轨的 double track; dual track
双轨地面铁路 two-rail surface track
双轨电动起重机 double rail motor crab; double rail motor hoist
双轨翻斗绞车 double track skip hoist
双轨刮料机 two-rail scraper
双轨滑道 twin rail runway
双轨记录 twin-track recording
双轨距车辆 vehicle with dual-ga(u)ge wheelsets
双轨距铁路 dual ga(u)ge railway; three-rail railway
双轨距线路 double ga(u)ge line; dual ga(u)ge line; three-rail track
双轨录音机 dual-track recorder
双轨贸易 two-way trade
双轨输入 double rail input
双轨用车调车场 double track swing parting
双轨隧道 double tracking tunnel
双轨条 double-rail
双轨条单车缓行器 two-rail single car retarder
双轨条轨道电路 double rail circuit; double track circuit
双轨条式轨道电路 double rail track circuit; two-rail track circuit
双轨条式缓行器 double rail retarder
双轨条式减速器 double rail retarder
双轨条双车减速器 two-rail double-car retarder
双轨铁道 double track railroad; double track railway; three-rail track; two-rail railway
双轨铁道桥 double track bridge
双轨铁路 double track railroad; double track railway; duorail; three-rail track; two-rail railway
双轨铁路桥 double track bridge
双轨推拉门 double track sliding door
双轨系统 two-rail system
双轨线路 double track line; double track route
双轨销售 dual distribution
双轨枕 twin sleepers; twin ties
双轨枕夹接 double sleeper joint
双轨枕接头 double sleeper joint
双轨制 double track system
双辊变距压滤机 twin-roll vari-nip press
双辊磁选机 two-roll machine
双辊干燥机 twin-drum drier[dryer]; twin-roll drum drier[dryer]
双辊辊压机 double roller press
双辊滚动烘干机 twin-roll drum drier [dryer]
双辊环 cross collar
双辊混合机 double shaft mixer
双辊混合加温磨机 two-roll mixing warm-up mill
双辊混合器 double roller mixer
双辊挤压机 two-roll press
双辊加压 twin-roll press
双辊搅拌机 double shaft mixer
双辊拉伸机 twin roller stretching machine
双辊连续塑化混合机 twin rotor continuous fluxing mixer
双辊链斗提升机 bucket elevator with double roller type chains
双辊路面破坏机 double drum breaker
双辊磨 two-roller
双辊碾磨机 mill of edge runner type
双辊破碎机 double roll breaker; double roll crusher; twin-roll breaker; twin-roll crusher
双辊驱动 dual drive
双辊筛 two-roll screen
双辊式浸润器 doubled-roller applicator
双辊式碾碎机 edge runner mill
双辊式轧机 double high rolling mill
双辊涂漆机 roll kiss coater; two-roll coater
双辊涂油 double roller size applicator
双辊喂料机 twin-roll feeder
双辊压光机 single nip calender
双辊压花机 two-roll embossing machine
双辊压力机 double roll press
双辊压滤机 twin-roll press
双辊压延玻璃 double-rolled glass
双辊压延机 two-bowl calender; two-roll calender
双辊研磨机 double roller mill; edge runner mill; end-runner mill
双辊轴 double roller
双辊轴筛 two-roll grizzly
双滚动面轮胎 duo-tread tyre
双滚轮导缆钳 double roller chock
双滚轮式 twin roller type
双滚轮振动压路机 twin-vibratory roller
双滚擒纵机构 double roller(lever) escapement
双滚筒 double roll; dual roller; twin-roll
双滚筒干燥机 double-drum drier[dryer]; twin-drum drier[dryer]
双滚筒绞车 double-drum winch
双滚筒搅拌机 dual-drum mixer
双滚筒卷扬机 two-drum hoist
双滚筒联合采煤机 double-ended machine
双滚筒磨耗试验 Deval abrasion test
双滚筒耙矿机 two-drum slusher
双滚筒破碎机 double roll breaker; double roll crusher; dual roll breaker; dual roll crusher; two-roll crusher
双滚筒铺路机 dual-drum pav(i)er

双滚筒式轧花机 double battery gin
双滚筒碎石机 double roll breaker; double roll crusher; dual roll breaker; dual roll crusher
双滚筒缩绒机 compound milling machine
双滚筒提升机 double-drum hoist
双滚筒压路机 tandem roller
双滚筒运输机 double-drum haulage engine
双滚筒轧碎机 double crusher roll; double roll crusher
双滚折线屋面瓦 double roll mansard tile
双滚轴破碎机 two-roll crusher
双滚柱 twin roller
双滚柱链 duplex chain
双滚柱轴承 two-roller bearing
双滚子链 duplex roller chain
双磙 double roller
双锅锅炉 double cylindrical boiler
双锅筒式锅炉 double-drum boiler
双锅系统 two-pot system
双过道客舱 two-aisle cabin
双过滤器 duplex filter
双过轮 double back gear
双涵(洞)double culvert
双焊 butt weld
双焊道搭接接头 double bead lap joint
双夯实梁修整机 double tamping beam finisher
双夯实梁整面机 double tamping beam finisher
双行 duplicate rows
双行的 twin-row
双行定植法 plant two-rows method
双行翻拍法 duo method
双行滚珠轴承圈 twin-row ball-bearing slewing ring
双行铆钉 double rivet; double row rivet
双行铆钉搭接缝 double riveted lap joint
双行铆钉对接缝 double riveted butt joint
双行铆接 double riveting; double row riveting
双行铆接的 double riveted
双行铆接缝 double row riveted joint
双行铆接头 double (row) riveted joint
双行停车 double parking
双行挖掘升运器 two-row elevator-digger
双行星齿轮组 dual planetary gears
双行星式混合机 double planetary-mixer
双行影像系统 duo
双行圆盘耙 tandem disc[disk] harrow
双行作业机具 two-row machine
双行座 double row; double row seating
双航道无线电指向标 two-course radio range
双航道运河 two-lane canal
双航路 dual airway; twin-track airway
双航路航线飞行规则 twin-track procedure
双航向信标 two-course range
双号 double number
双号筒辐射器 double horn radiator
双号筒扬声器 double horn loudspeaker
双合 pairing
双合唱诗班 double choir; double quire
双合唱诗班教堂 double-quire church
双合放大镜 doublet magnifier
双合分式楼梯 double-return stair(case)
双合结 cow hitch; Lark's head
双合套 cow hitch; Lark's head
双合套结 dead eye hitch
双合透镜 double lens; doublet; two-component lens
双合物镜 doublet objective
双合页门 double-hinged gate
双核络合物 binuclear complex
双荷重竖井投影 two-weight shaft plumbing
双荷子【物】dyon
双桁架梁龙门起重机 double girder truss gantry crane
双横承力索 twin headspan wire
双横承力索线夹 twin headspan wire clamp
双横挡十字架 Lorraine cross
双横杆十字架 double cross
双横滚路缘石 double roll curb tile
双烘筒上浆机 two-cylinder dresser
双红灯显示系统 double-red indication system
双红灯显示制 double-red indication system
双喉管燃烧器 double burner; double throat burner; duplex burner
双后轮胎 dual rear tire[tyre]
双后桥半挂车 tandem semitrailer
双后轴(轮架)bogie[bog(e)y]
双厚 double thickness
双弧长短幅外旋轮线气缸 two-lobed epitrochoidal bore
双弧度 bending in two direction
双弧工作室 two-lobe chamber
双弧焊 dual arc weld
双弧焊接 dual arc welding; twin arc welding
双弧外旋轮线缸体 two-lobed epitrochoidal housing
双弧外旋轮线工作室 two-lobed epitrochoidal working chamber
双弧外旋轮线旋转机械 two-lobed epitrochoidal rotary mechanism
双弧现象 double arcing
双弧线 double camber
双弧自动焊机 twin arc automatic welding machine
双弧钻头 double arc bit
双壶系统 two-pot system
双互换制度 double reversal system
双户房屋 duplex house
双户连体住宅 semi-detached dwelling
双户住宅 duplex apartment; duplex dwelling; duplex-type house; dwelling duplex; two-family house
双护盖 two-piece calotte
双戽斗式提水工具 two-bucket lift
双花大绳接结 double carrick bend
双花钩结 double blackwall hitch; stunner knot
双花菱形结节 double diamond knot
双花青 dicyanine
双花绳端结 double wall knot
双花筒复合提花机 two-cylinder compound jacquard
双花筒提花机 double cylinder jacquard
双华伦式桁架 double Warren truss
双铧犁 double bladed plough; double bladed plow; double furrow plough; double shared plough
双滑车 sister block
双滑车之一 sister block
双滑程 double slippage
双滑触变阻器 double slide rheostat
双滑道下水 double way launching; two-way launching
双滑动臂调谐线圈 two-side receiving coil
双滑阀 double slide valve
双滑阀调节 with two sliding valve governing
双滑块曲柄链系 double slider crank chain
双滑轮 sister block; sister hooks
双滑轮导缆器 double roller chock
双滑轮吊架 <起重机> sling block
双滑轮组 twin pulley block
双滑台式压机 double sliding table press
双滑线电桥 double slide wire bridge
双滑线电势计 double slide wire potentiometer
双滑移 duplex slip
双化合价 bivalence
双化油器系统 twin carburetor system
双画面同轴电缆 double screen coaxial cable
双环 forked loop; spectacle
双环板 twin bracket
双环存储装置 two-ring storage system
双环道三次线 two-circuited cubic
双环阀 double ring valve
双环杆 double eye end rod
双环管进样阀 two-loop sampling valve
双环海百合 dicyclic(al) crinoids
双环己基甲烷 bicyclohexyll-methane
双环己基乙烷 bicyclohexyll-ethane
双环库 double ring silo
双环流 bicirculation; double gyre
双环壬烷 bicyclononane
双环伸缩螺丝 eye and eye turnbuckle
双环渗水法 method of water infiltration with dual ring
双环渗透仪 double ring infiltrometer
双环式船尾框架 spectacle stern frame
双环式船尾轴架 spectacle type shaft bracket
双环式活塞 two-ring piston
双环适配器 dual ring adaptor
双环天线 double loop antenna
双环亭 linked circular kiosk
双环烷烃 bicycloalkane
双环温度控制器 two-loop temperature controller
双环戊二烯 dicyclopentadiene
双环戊基甲烷 dicyclopentylmethane
双环衔铁 four ring snaffle bit
双环辛烷 bicyclooctane
双环信号控制机 dual ring signal controller
双环形偏转线圈 dual toroidal yoke
双环氧乙烷 bioxirane; butadiene dioxide
双缓冲 double buffer
双缓冲数据传送 double buffered data transfer
双换流器 double converter
双换位线棒 double Roebel bar
双换向器电动机 double commutator motor
双黄线 double amber lines
双簧车钩 with two spring draw gear
双簧管 oboe
双簧类 double mechanical reed
双簧门 double wing door
双簧牵引装置 with two spring draw gear
双灰条铺砌 shell bedding
双回波 double echo; paired echo
双回波校准法 double bounce calibration
双回程锅炉 two-pass boiler
双回火 double tempering
双回廊 double ambulatory
双回流阀 double-return valve
双回路 two-circuit; double circuit
双回路对消 echo cancellation
双回路供电电源 double-return circuit power supply
双回路静液压传动 dual-path hydrostatic drive
双回路气动系统 dual circuit air system
双回路伺服机构 two-loop servomechanism
双回路调谐器 two-circuit tuner
双回路线 double circuit line
双回路线架 double circuit tower
双回路线塔 double circuit tower
双回路液力耦合器 twin circuit coupling
双回收回路 double recovery circuit
双回行楼梯 double-return stair(case)
双回旋曲线 biclothoid
双回转反击式汽轮机 Stahl turbine
双汇流排 double bus
双汇流条调整 two-busbar regulation
双活门可调喷管 clamshell nozzle
双活塞 double piston
双活塞泵 double piston pump
双活塞对动发动机 double oppose engine
双活塞发动机 duplex engine
双活塞杆的活塞 two-rod piston
双活塞升杆 two ram-lifting rod
双活塞制动缸 duplex brake cylinder
双活性区 two-core
双火管锅炉 double fired boiler; double flue boiler
双火花间隙 double gap
双火花塞点火 dual ignition; twin ignition
双火花塞分电器 two-spark distributor
双火星塞点火 twin spark ignition
双火焰检测器 dual flame detector
双货棚屋顶 butterfly roof
双货厢卡车 two-batch truck
双击镦锻 double blow heading
双击冷镦机 double header
双击面锤 double-faced hammer
双击式水轮机 Banki turbine; cross-flow turbine; Michell-Banki turbine
双击式蒸汽锤 double-acting steam hammer
双机 duplex machine; two-shipper
双机并联推土机 side-by-side (bull) dozer
双机并行起重机 twin travel(l)ing crane
双机操作 two-machine operation
双机车 double locomotive
双机单轴式 twin-engined single-shaft system
双机粉磨流程 two-mill type of grinding; two-unit grinding
双机跟进滑行 two-by-two taxiing
双机壳电机 double casing machine
双机目标 twin target
双机坡度【铁】double heading grade; double heading gradient; helper grade
双机牵引 double heading
双机牵引的列车 double-headed train
双机驱动 dual-motor drive
双机热备 hot standby
双机身布局 twin-fuselage configuration
双机身飞机 double fuselage plane
双机台车 two-machine jumbo
双机头自动发报机 double head automatic transmitter
双机系统 dual system
双机械接头连接件 double mechanical joint connecting piece
双机械抓手 double hand grip

S

双机运行 duplex running; assisted running <列车头部有两台机车>
双机制 double unit system
双机着陆 pair landing
双机组 pair assembly; two-aircraft element; two-ship element; two unit <起动电动机和充电发电机组成的机组>
双机钻车 dual drill rig; twin-drill jumbo; two-drill jumbo; double drill jumbo
双机座型发电机 double-framed generator
双机座轧机 double stand rolling mill; twin-stand mill
双迹法 two-trace method
双迹放大器 dual-trace amplifier
双迹示波器 dual-trace oscilloscope
双积 dyadic product
双积的 dyadic
双积等式 dyadic equality
双积分法 <梁的挠度静定解法> double integration(method)
双积分器系统 two-integrator system
双积分陀螺仪 double integrating gyroscope
双积和 dyadic sum
双基测距计 two-base range finder
双基的 biradical
双基地雷达 bistatic radar
双基点法(汉森法) reference pair method
双基火箭推进剂 double base rocket propellant
双基火药 bipropellant; double base powder
双基极二极管 double base diode
双基极光电晶体管 double base phototransistor
双基极晶体管 double base transistor
双基坑吊车 double foundation pit hoist
双基推进剂 double base propellant
双基线(气压测高)法 two-base method
双基炸药 <硝化甘油与硝化纤维> double base powder
双箕斗 double loop pattern
双箕斗提升系统 twin-skip system
双激波进口 double shock intake
双激发复合 recombination with double excitation
双激光器 duolaser; twin laser
双激元激光器 two-exciter laser
双级 twin-stage
双级泵 close-coupled pump; double-stage pump; dual-section pump; two-stage pump
双级操纵 dual-ratio steering
双级齿轮传动 double geared
双级除盐装置 two-demineralization system
双级船闸 double lift dock
双级的 two-stage
双级电吸尘器 two-stage precipitator
双级方案 two-level scheme
双级方程曲线 oval of Cassini
双级管 two-stage tube
双级归零记录系统 two-level return system
双级活塞 two-diameter piston
双级级联起电机 two-stage cascade generator
双级减速 double reduction
双级减速齿轮传动装置 double reduction gear device
双级节流制冷器 compound throttling refrigerator
双级空气喷淋室 two-stage air washer
双级空气压缩机 two-stage air compressor
双级控制 two-stage control
双级冷却恒温器 double-stage cooling thermostat
双级粒状活性炭处理 dual-stage granular activated carbon treatment
双级喷管 two-step nozzle
双级喷射系统 two-step injection system
双级膨胀 double-stage expansion
双级破碎机 two-stage crusher
双级燃料喷嘴 duplex fuel nozzle
双级式 double-stage
双级式码头 double decked wharf
双级式燃气轮机 two-stage gas turbine
双级式透平 double-stage turbine
双级塑炼机 two-stage plasticator
双级调节器 two-stage regulator
双级凸轮 double lift cam
双级涡流旋风除尘器 two-stage vortex cyclone
双级涡流旋风分离器 two-stage vortex cyclone
双级涡轮增压 two-stage turbocharging
双级涡轮增压器 twin turbo-charger
双级系统 two-bed system
双级显微镜 two-stage microscope
双级显微术 two-step microscopy
双级消化池 double-stage digester
双级旋风除尘器 double-stage cyclone
双级旋风分离器 double-stage cyclone
双级循环锅炉 dual-circulation boiler
双级压缩 compound compression; two-stage compression
双级压缩机 compound compressor; double-stage compressor
双级压缩机涡轮喷气发动机 compound turbo jet
双级压缩式制冷循环 refrigeration cycle of two stage compression
双级液力变距器 two-stage converter
双级音频放大器 double-note amplifier
双级增压器 double-stage super-charger
双级真空挤泥机 two-stage de-airing extruder
双级蒸发装置 two-effect evapo(u)rator system
双级质谱计 two-stage mass spectrometer
双级注水泥 two-stage cementing
双级最终传动【机】 double reduction final drive
双级作用 two-step action
双极 dipole
双极 n 投开关 double n-way throw switch
双极变换器 bipolar converter
双极层模 dipole-layer mode
双极场 ambipolar field; bipolar field
双极场效应晶体管 bipolar transistor; nesistor
双极存储器 bipolar memory; bipolar storage
双极单点焊 gap weld
双极单投开关 double pole single throw switch
双极的 ambipolar; bipolar; double pole; two-polar
双极电磁驱动单元 bipolar magnetic driving unit
双极电动机 bipolar electric(al) motor
双极电机 bipolar machine
双极电极 bipolar electrode
双极电解 bieletrolysis
双极电解槽 bipolar cell
双极电路 bipolar circuit
双极电门 two-pole switch
双极电枢绕组 bipolar armature winding
双极电源 bipolar power supply
双极断流器 double pole cut-out
双极发电机 bipolar dynamo; bipolar generator
双极放大器 bipolar amplifier
双极高压直流输电系统 bipolar high voltage direct current system
双极黑子 bipolar sunspot
双极化 dual polarization
双极化波共用天线 duplex polarization antenna
双极化处理器 bipolar processor
双极集成电路 bipolar integration circuit
双极继电器 double pole relay
双极晶体管存储器 bipolar
双极晶体管开关时间 switching time of bipolar transistor
双极静电计 binary electrometer
双极开关 double pole on-off switch; double pole switch; two-pole switch
双极扩散 ambipolar diffusion
双极拉线开关 double pole pull cord type switch
双极屏蔽总线 bipolar mask bus
双极器件 bipolar device
双极器件制造 bipolar fabrication
双极枪 two-electrode gun
双极群 bipolar group
双极绕组 bipolar winding
双极熔断器 double pole fuse
双极施压法 double pressure system
双极式发电机 bipolar dynamo
双极双断法 double-pole double-break method
双极双断防护法 double-pole double-break protection
双极双投开关 double pole double throw switch
双极水流玫瑰图 bipolar current rose
双极碳弧焊 twin carbon arc welding
双极微控制器 bipolar microcontroller
双极微型控制器 bipolar microcontroller
双极位片微机 bipolar bit-slice microcomputer
双极型集成电路 bipolar integration circuit
双极性 bipolarity
双极性编码 bipolar coding
双极性层 double sheath
双极性的 bipolar
双极性等离子体 ambiplasma
双极性电极 bipolar electrode
双极性电流法 double current method
双极性分布 bipolar distribution
双极性扩散数 bipolar diffusivity
双极性脉冲 bipolar pulse
双极性脉冲调制 double-polarity pulse-amplitude modulation
双极性破坏点 bipolar violation
双极性剖面 dipolar section
双极性器件 bipolar device
双极性调制 bipolar modulation
双极性信号 bipolar signal
双极性有效迁移率 ambipolar effective mobility
双极性栅 bipolar gate
双极掩膜总线 bipolar mask bus
双极元件 bipolar cell
双极闸刀开关 two-pole knife switch
双极振荡器 dipole oscillator
双极正面连接 double pole front connection
双极坐标 bipolar[dipolar] coordinates
双极坐标系 bipolar coordinate system
双集水沟 double gulley
双给水系统 dual water supply system
双脊屋顶 double gable roof
双计数法 double counting method
双计数器计算机 two-counter machine
双计算机背对背系统 two-computer back-to-back system
双计算机系统 duplex computer system
双记录校验 double record check
双记名票据 double name paper
双剂联用灭火系统 dual-agent system
双季稻 double crop(ping) rice; double harvest rice; two-crop-paddy
双季冻土 pereletok
双季对流混合湖 dimictic lake
双季栽培制 double cropping system
双加厚管 double extra strong pipe
双季回水湖 dimictic lake
双加料夹丝玻璃成型机 double glass wired glass machine; double pass machine
双加热器体系 twin-heater system
双夹板对接 butt joint with double traps
双夹瓣三角 two-point cam
双夹钳 twingrip
双颊板式雪犁 double cheek snow plough
双甲板船 double decked ship
双甲亚胺 bisazomethine
双价电度表 double tariff(system) meter
双价电度累计装置 two-rate register
双价离子 divalent ion
双价率两部电度预付计 two-rate two-part prepayment meter
双介质孔隙率过滤器 dual-media porosity filter
双价制 double tariff system
双架钻机 twin-boom drilling rig
双尖窗 <中间有竖框的> double lancet window
双尖钉 double-pointed tack
双尖顶窗 double lancet window
双尖翻换松土铲 reversible shovel
双尖镐 mandrel[mandril]; tubber
双尖拱 two-cusped arch
双尖料堆 double cone pile
双尖松土锄铲 reversible steel
双尖头 split-blip
双尖头镐 quarry pick
双尖圆形 <古建筑哥特式的> vesica-piscis
双尖嘴无杆锚 Dunn anchor
双间隔 binary stage
双间隔化粪池 dual compartment septic tank
双间回水盒 alternating return trap
双间回水箱 alternating return trap
双间圆木屋 double cabin
双肩垫板 double shoulder(ed) tie plate
双减速比机构 dual-ratio reduction
双减速变速箱 double reduction
双减速齿轮【机】 double reduction gear
双减速终传动 double reduction drive
双减振阀 double damper valve
双剪力 double shear
双剪力撑 double bridging
双剪力铆钉 rivet in double shear
双剪螺栓 bolt in double shear; double shear bolt
双剪铆(钉)接头 double shear riveted joint
双剪铆接 double shear riveted joint
双剪(切)铆钉 rivet in double shear; riveting double shear; double shear rivet

双检波接收机 double detection receiver; dual-detector receiver
双检测器 double detector; dual detector
双检测器单向阀 double detector check
双筒卷扬机 double-drum winch
双碱光电阴极 bialkali photocathode
双碱流程 dual alkali process of flue gas desulfurization
双碱洗涤法 double alkali scrubber process
双碱效应 double alkali effect
双件模 two-piece mo(u)ld
双键 double bond
双键结合 doubly linked bond
双键离解 paired-bond dissociation
双键位变异构体 metaisomer
双键位移 double bond shift
双键悬挂法 double catenary suspension
双键异构 double bond isomerism
双桨捏合机 double blade kneader
双桨轻划艇 sculler
双桨艇 pair oar boat
双桨舟 pair oar boat
双交 double cross; two-way cross
双交叉 double crossover; double junction
双交叉撑 double bridging
双交代作用 dimetasomatism
双交代作用方式 dimetasomation way
双交替 double crossing over
双交替配时方案 double alternate
双交替式 bialternant
双交圆 crossed circles
双浇 double teem(ing)
双胶合轮廓投影物镜 doublet profile lens
双胶轮手推车 buggy
双胶轮运输车 wheel trailer
双胶水流模式 double water flow model
双胶透镜 cemented doublet
双胶物镜 doublet objective
双焦点 bifocus
双焦点的 bifocal
双焦点干涉显微镜 double focus interference microscope
双焦点干涉仪 double focus interferometer
双焦点管 bifocal tube
双焦点片 bifocal segment
双焦点透镜 bifocal lens; bifocals
双焦法 double focus method
双焦镜 bifocal lens
双焦镜片 bifocals
双焦距测定表 hyperfocal chart
双焦面 bifocal segment; bifocal surface
双焦取景器 double finder telescope; two-position viewfinder
双焦透镜 bifocal lens
双焦眼镜 bifocal eye glass; bifocal spectacle
双焦装置 double focus arrangement
双角T字钢 double angle tee
双角测定仪 station pointer
双角测向法 two-bearings and run between
双角的 bicorn
双角度针 angle pin
双角缝舟藻 Raphoneis amphiceros
双角钢 double angle bar
双角钢腹杆 double angle web
双角钢杆件 double angle
双角钢构成的系杆 double angle tie
双角钢构件 double angle
双角剪力 double angle shear
双角器 biangler
双角探头 twin angle probe
双角铁 double angle; double angle iron
双角铣刀 double angle (milling) cutter
双角形鱼尾板 double angle fishplate
双角型钢 double angle section
双角锥(体) bipyramid
双绞 multiple twin
双绞电缆 pair(ed) cable; twin transposition cable
双绞钢筋 twin-twisted reinforcement
双绞股线 twin-twisted strand
双绞屏蔽电缆 twisted shielded pair cable
双绞曲柄卡盘 double toggle chuck
双绞软线 twisted cord
双绞式馈线 twisted-pair feeder
双绞线 twisted-pair cable; twisted-pair line; twisted-pair wire
双绞圆钢筋 twin-twisted round bars
双脚扳手 spanner wrench
双脚规 cal(l)ipers; divider
双脚规形夹 cal(l)iper splint
双脚架 bipod
双脚式起重机 shear-leg derrick crane
双脚桅 bipod mast
双脚支柱 flat bearing
双铰承窝 two-pin socket
双铰单跨框架 two-hinged frame of one bay; two-hinged frame of one span
双铰的 two-hinged
双铰方框架 double-hinged rectangular frame; double-pinned rectangular frame
双铰刚架 two-hinged rigid frame; two-hinge frame; two-pin rigid frame
双铰钢筋混凝土拱 two-hinged reinforced concrete arch
双铰弓弦拱 two-hinged braced arch
双铰弓形大梁 double-hinged segmental arch(ed) girder; double-pinned segmental arch(ed) girder
双铰拱 arch hinged at ends; double-hinged arch; double hinged body; two-hinged arch; two-hinged body; two-pinned arch
双铰拱大梁 double-hinged arch(ed) girder
双铰拱桁架 two-hinged arch truss
双铰拱架 double-hinged arch(ed) frame; double-linked arch(ed) frame; double-pinned arch(ed) frame
双铰拱桥 double-hinged arch bridge; two-hinged arch bridge
双铰拱桥梁 two-pinned arch bridge
双铰拱形框架 two-articulated arched frame
双铰构架 two-hinge frame
双铰接的 two-linked; two-pinned
双铰接方框架 double-articulated rectangular frame
双铰接弓形拱大梁 double-articulated segmental arched girder
双铰接拱架 double-articulated arch(ed) frame
双铰接平拱大梁 double-articulated flat arch(ed) girder
双铰接平抛物线拱大梁 double-articulated flat parabolic arch(ed) girder
双铰接人字架 double-articulated gable(d) frame
双铰接转向机构 double-articulated steer
双铰接转向式平地机 double-articulated steer grader
双铰结构 two-hinged structure
双铰矩形框架 double-hinged rectangular frame
双铰框架 double-hinged frame; double-linked frame; double-pinned frame; two-hinge frame; two-pin frame
双铰链 double-strand chain
双铰链滑车 double purchase pulley
双铰链提引器 double gate elevator
双铰门式刚构桥 two-hinged portal frame bridge
双铰门式刚架 two-hinged portal frame
双铰抛物线拱大梁 double-hinged parabolic arch(ed) girder; double-pinned parabolic arch(ed) girder; two-articulated parabolic arched girder
双铰抛物线拱形梁 double-hinged parabolic arch(ed) girder
双铰平拱大梁 double-hinged flat arch(ed) girder; double-linked flat arch(ed) girder; double-pinned flat arch(ed) girder; two-articulated flat arched girder
双铰平拱形梁 double-hinged flat arch(ed) girder
双铰平抛物线拱大梁 double-hinged flat parabolic arch(ed) girder; double-linked flat parabolic arch(ed) girder; double-pinned flat parabolic arch(ed) girder; two-articulated flat parabolic arched girder
双铰人字架 double-hinged gable(d) frame; double-pinned gable(d) frame
双铰人字形框架 two-articulated gable(d) frame
双铰软线 twisted cord
双铰山墙式框架 two-articulated gable(d) frame
双铰扇形拱大梁 two-articulated segmental arched girder
双铰式 two-hinged type
双搅拌器 double agitator
双阶齿榫 double steps
双阶燃烧发动机 two-stage combustion engine
双阶式混凝土拌和站 double-stage concrete mixing plant
双阶式搅拌机 double-stage mixing plant
双阶撕碎/成粒机 two-stage shredder/granulator
双阶凸轮 double lift cam
双阶作用 two-step action
双接触的简单悬挂 twin simple catenary
双接触线 twin contact wire
双接触线的简单链形悬挂 simple catenary equipment with twin contact wire
双接触线吊弦线夹 dropper clip for twin contact wire
双接触线定位线夹 twin contact wire clip
双接触线接头线夹 twin contact wire clamp
双接触线链形悬挂 compound catenary equipment; double link suspension
双接点 double contact; twin contacts; two point
双接点聚光灯 bipost lamp
双接缝 double joint(ing)
双接缝罐 double seam can
双接键 double contact key
双接近系统 two-approach system
双接近制 two-approach system
双接(口) double nip
双接口罐 double seam can
双接力器 twin servomotors
双接门 bride-door
双接收器 dual channel collector; dual channel receiver
双接收制 two-bin system
双接贴板 double butt strap
双接头圆规 double-jointed compasses
双节 binodal; two-section
双节臂 twin lever
双节波动 binodal seiche
双节柴油机车 double-unit diesel locomotive
双节底盘 two-piece chassis
双节点垂直振动 two-noded vertical vibration
双节点振动 two-noded vibration
双节扼流圈 two-section choke coil
双节公共汽车 trailer bus
双节固溶曲线 binodal curve
双节横波 binodal lateral wave
双节假潮 binobal seiche; dicrotic seiche
双节铰接车 twin articulated vehicles
双节距 bipitch
双节卡车 trailer truck
双节抗流圈 two-section choke
双节流阀 twin throttle
双节滤波器 two-mesh filter; two-section filter
双节螺纹 double thread
双节水面波动 binodal seiche
双结 double knot
双结合 double jointing
双截顶圆锥密封 seal by double conical frustum
双截门 stable door
双截墙裙 double skirting
双截头圆锥密封 double truncated cone seal
双截头指数分布 double truncated exponential distribution
双介质过滤器 dual-media filler
双介质滤层 dual-media filler
双介质摄影测量学 two-media photogrammetry
双界面 biface
双界张力 bifacial tension
双金属 composite metal; duplex metal; laminated metal; thermometal
双金属摆轮 bimetallic balance
双金属板 <单面或双面覆层的> inlay clad plate; bimetallic plate; bimetal sheet
双金属棒材 composite metal rod
双金属表盘温度计 bimetallic dial thermometer
双金属补偿片 bimetallic compensation strip
双金属补偿器 bimetal compensator
双金属材料 bimetallic material
双金属层板 inlay clad plate
双金属衬筒 bimetallic lined barrel
双金属尺 bimetallic ruler
双金属尺法 bimetal ruler method
双金属带 bimetallic strap; bimetallic strip
双金属(刀形)开关 bimetal blade
双金属刀形闸 bimetal blade
双金属导体 bimetallic conductor
双金属导线 bimetallic conductor
双金属的 bimetallic
双金属灯 bimetal lamp
双金属断路器 bimetal release
双金属腐蚀 bimetallic corrosion; couple corrosion
双金属杆件 bimetallic bar
双金属感光计 bimetallic actinometer
双金属钢 clad steel
双金属管 bimetal tube
双金属恒温控制器 bimetallic thermostat

S

双金属恒温器 bimetal thermostat
双金属基线测量器械 bimetal base (line) (measuring) apparatus; duplex base-line apparatus
双金属及多层板材轧制 sandwich rolling
双金属挤出原料筒 bimetallic barrel
双金属挤压 coextrusion
双金属挤压的 co-extruded
双金属继电器 bimetal relay
双金属浇注置换 double pouring displacement technique
双金属(接)触点 bimetal contact
双金属开关 bimetal release
双金属控制器 bimetal pilot
双金属离心铸管法 dual metal casting
双金属密封 bimetallic (spring cone) seal
双金属坯块 bimetal compact
双金属片 bimetallic blade; bimetallic element; bimetallic strip
双金属片补偿 bimetallic strip compensation
双金属片触点 bimetal blade
双金属片传感器 bimetallic strip ga(u)ge
双金属片电容器 bimetallic capacitor; bimetallic condenser
双金属片断路器 bimetal release
双金属片盖 bimetal cover
双金属片过热断路装置 bimetallic strip thermal device
双金属片恒温控制器 pilotherm
双金属片继电器 bimetallic strip relay
双金属片开关 bimetal release
双金属片温度计 bimetallic thermometer
双金属侵蚀 bimetallic corrosion; couple corrosion
双金属圈温度计 bimetal coil thermometer
双金属热电偶 bimetallic thermocouple
双金属热膨胀开关 thermoswitch
双金属日射表 bimetallic actinometer
双金属熔丝 bimetal fuse
双金属式度盘温度计 bimetal type dial thermometer
双金属式仪表 bimetallic instrument
双金属式指针温度计 bimetal type dial thermometer
双金属水准标 bimetal benchmark
双金属丝法 bimetallic fiber method
双金属弹簧 bimetallic spring
双金属套片式翅片管 duplex continuous integral fin tube
双金属条 bimetal strip
双金属调节器 bimetallic regulator
双金属温度计 bimetallic thermometer; differential thermometer
双金属温度记录器 bimetallic thermograph
双金属温度调节器 bimetallic temperature regulator; bimetallic thermostat
双金属线 bimetallic wire; copper-clad steel wire; composite wire <铜包钢线>
双金属延时继电器 bimetal time delay relay
双金属仪表 bimetallic instrument
双金属元件 bimetallic element
双金属真空计 bimetallic vacuum ga(u)ge
双金属制品 bimetallic article
双金属轴承 bimetallic bearing
双金属铸件 composite casting; compound casting
双金属铸造 bimetal casting
双金属转子 bimetallic rotor
双金属组合条 bimetal strip
双筋 double reinforcement
双筋混凝土 <具有受压钢筋和受拉钢筋的> doubly reinforced concrete
双筋截面 double reinforced section; doubly reinforced section
双筋梁 beam with double reinforcement; double reinforced beam; double reinforcement beam; beam with compression steel <有受压钢筋>
双筋梁配筋梁 double-reinforced concrete beam
双筋履带板 double grouser
双筋履带板的履带 double grouser track
双进风离心通风机 double inlet centrifugal fan
双进口风机 double inlet fan
双进路 either directional route; either route; up-and-down train routes
双进气阀 double intake valve
双进水口离心泵 double centrifugal pump
双进线【铁】 either directional route
双进样系统 double inlet system
双经 double ends; double warp
双经双纬格子花纹 two-and-two check
双经纬仪定位法 two-transit fix method
双经纬仪观测 double-theodolite observation; two-theodolite observation; two-transit observation
双茎式装饰 twining stem mo(u)lding
双晶 bicrystal; bimorph; compound crystal; twin(ned) crystal; twinning
双晶带 <金属中的> twin band
双晶的 dimorphic
双晶法 twin method
双晶分光计 double crystal spectrometer
双晶缝合线 partition line
双晶固溶体 solid solution with twinning
双晶滑棱面 twin gliding plane
双晶滑移 twin gliding
双晶间界 twin boundary
双晶界面 twin boundary
双晶类型 type of twinning
双晶律 twinning law
双晶面 twin(ning) plane
双晶面反射 twin plane reflection
双晶片 twin lamella
双晶摄谱仪 double crystal spectrograph
双晶石 eudidymite
双晶体 bicrystal; twin crystal
双晶体分光计 two-crystal spectrometer
双晶体管 bipolar transistor; pair transistor; twin transistor
双晶体检波器 combination-crystal detector; perikon detector
双晶体三极管 tandem transistor
双晶条纹 twin striation
双晶位错 twining dislocation
双晶纹 twin striation
双晶锡石 visor tin
双晶现象 dimorphism; twinning
双晶消除 detwinning
双晶形成 twin formation
双晶衍射计 double crystal diffractometer
双晶中心 twin center[centre]
双晶轴 twin(ning) axis
双精度操作 double precision operation
双精度存储 store double precision
双精度的 double precision
双精度浮点宏指令 double precision floating macro-order
双精度加法 special add
双精度阶部分 double precision exponent part
双精度实数 double precision real data
双精度实数型 double precision type
双精度实型 double precision
双精度数 double precision numeral
双精度值 double precision value
双精馏器 birectifier
双精确度运算 double precision arithmetic
双井 twin well
双井定向法 two-shaft orientation method
双井开采 dual well completion
双井联动抽油设备 back crank pumping units
双井式调压井 double shaft surge tank
双井筒钻进 double barrel(l)ed drilling; simultaneous drilling
双阱式活动堰 automatic double trap weir; double trap movable weir
双颈瓶 two-neck bottle
双警报器 twin electric(al) horn
双径传输 dual-path transmission
双径滚筒绞盘 Chinese capstan; double-drum windlass
双径向 biradial
双镜 bimirror
双镜测量摄影机 twin photogrammetric camera
双镜反像镜 double reversing mirror
双镜立体摄影机 binocular stereocamera
双镜连续摄影机 twin serial camera
双镜摄影机 twin camera
双镜式电流计 double mirror galvanometer
双镜头 twin lens; two-shot
双镜头测量摄影机 twin photogrammetric camera
双镜头连续摄影机 twin serial camera
双镜头拍摄 two-shot
双镜头倾斜航摄照片 twin oblique (air) photograph; two-oblique photograph
双镜头倾斜摄影术 two-oblique photography
双镜头摄影机 bicamera; twin camera
双镜头摄影照片 twin photograph
双镜头照相机 bicamera; twin-lens camera
双镜物镜 two-mirror objective
双镜消像散镜 two-mirror anastigmat
双镜直角器 double optic(al) square
双镜装置 double mirror device
双鸠尾榫键 double dovetail key; dovetail feather joint
双桔酸 digallic acid
双距离 dual range; two-range
双距离式台卡导航系统 two-range Decca
双距螺旋桨 controlled pitch airscrew; two-pitch airscrew
双锯齿声迹 duplex sound track
双锯齿形布置 double saw-tooth system
双锯齿形框架 twin saw-tooth frame
双锯面对开材 crown-tree
双聚光器 double condenser
双聚焦 double focusing
双聚焦系统 double focus system
双聚焦质谱计 double focusing mass spectrograph
双聚焦质谱仪 double focusing mass spectrograph; double focusing mass spectrometer
双卷暗盒 double magazine
双卷边山墙突瓦 double roll verge tile
双卷边瓦 double roll under-ridge tile
双卷筒柴油绞车 double-drum diesel winch
双卷筒铲运提升机 two-drum drag scraper hoist
双卷筒打桩绞车 double-drum pile driving winch
双卷筒的 dual-drum
双卷筒绞车 double-drum winch; double purchase winch; dual-drum winch
双卷筒卷扬机 double-drum winch; twin drum winch
双卷筒缆索控制装置 double-drum cable control unit
双卷筒起网机 double-drum winch
双卷筒提升机 double-drum hoist
双卷筒蒸汽绞车 double-drum steam winch
双卷筒蒸汽提升机 two-drum steam hoist
双卷筒抓斗起重机 dual-drum grabbing crane
双卷卫生纸配出器 twin-roll toilet tissue dispenser
双卷位开卷机 double coil holder arrangement
双卷圆簧组 double coil nest spring
双均压环 double strap
双卡 dual card
双卡模块 two-card module
双卡瓦打捞器 bulldog double slip spear
双卡瓦打捞筒 <可循环洗井的> double slip casing bowl
双卡瓦卡盘 two-jaw(ed) chuck
双开玻璃摆动门 glass swing door
双开窗 double sash casement window
双开阀 double beat valve
双开扶手宽梯 double-cleat ladder
双开活动扳手 central-adjustable handle-adjustable wrench
双开活络扳手 coach wrench
双开口扳手 double open end wrench
双开拉门 biparting door
双开路接点 double break contact
双开门 double-acting engine; double door; double leaves door; double up door; fancy door
双开门吊卡 double gate elevator
双开门门挡 double gate stop
双开门门梃 meeting stiles
双开门止门器 double gate stop
双开幕 travel(l)er curtain
双开平旋桥 double swing-span bridge
双开铺面砖 double format pavior
双开式料筒 split barrel
双开式弹簧铝门 alumin(i)um swing door
双开式弹簧门 double-acting door; swing(ing) door
双开式旋转铰链 helical hinge
双开弹簧门附件 swing door fittings
双开弹簧门关闭器 swing door closer
双开弹簧门管式手柄 swing door tubular grip handle
双开弹簧门管式握手 swing door tubular grip handle
双开弹簧门设备 swing door furniture
双开弹簧门手柄 swing door grip handle
双开弹簧门推板 swing door push handle
双开弹簧门推柄 swing door push handle
双开弹簧门推梗 swing door push handle
双开弹簧门握手 swing door grip handle
双开弹簧门小五金 swing door hardware

双开弹簧门硬件 swing door hardware
双开弹簧门装配 swing door furniture
双开弹性门 flexible door
双开挖面掘进 double heading
双开摇摆弹簧门 swing door
双铠装 double armature
双壳拌和器 twin-shell blender
双壳层地板 double-skin floor
双壳层结构 double-skin construction
双壳的 double shell
双壳动物地理区 bivalve faunal province
双壳结构 double casing construction
双壳类 bivalve
双壳式吸收制冷机 double shell type absorption refrigerating machine
双壳艇 double-skin boat
双壳外墙 external masonry wall of double-leaf cavity construction
双壳型 double crust type
双刻度 double scale;duplicate scale;two-scale
双刻度法 two-scale method
双刻度经纬仪 double circle theodolite; double circle transit; double reading theodolite
双刻度(盘)读数 double circle reading
双刻度周波计 two-scale frequency meter
双空格点 divacancy
双空腔缝 double cavity joint; joint with double cavities
双空腔谐振器 duplex cavity
双空位 divacancy;double vacancy
双孔 binary cell
双孔板 spectacle plate
双孔板式涵洞 twin slab culvert
双孔插座出线口 duplex receptacle outlet
双孔的 binocular;double opening
双孔电磁波法 two-boreholes electromagnet wave method
双孔电子透镜 two-aperture electronic lens
双孔定向耦合器 two-hole directional coupler
双孔构架 spectacle frame
双孔管道 two-way duct;dual duct
双孔涵洞 double conduit; twin-cell culverts;twin culvert
双孔涵管 twin-cell culverts
双孔和无孔检测 double punch and blank-column detection
双孔交通隧道 two-barrel transit tube
双孔空心黏[粘]土砖 double hollow clay(building) block
双孔空心砌块 two-core block
双拉杆钢板桩岸壁 steel sheet pile wall with anchors at two levels
双孔连接线夹 cable lug with two connecting holes
双孔联板 twin pin clevis link; two-pin strap
双孔耦合器 two-hole coupler
双孔砌块 two-cavity block;two-core block
双孔三维显示 binocular three dimensional display
双孔式喷嘴 double hole type nozzle
双孔水道 double conduit
双孔隧道 eye tunnels; twin(-bore) tunnel
双孔透镜 double aperture lens; two-aperture lens
双孔线圈 binocular coil
双孔楔子 two-hole wedge
双孔旋流片 two-hole whirl plate
双孔重载 two-span heavy load
双控开关 double control switch; on-off switch;two-way switch
双控制极 double gate
双口 double port;two-port
双口锻工钳 double pick-up tongs
双口阀 double port valve
双口给料器 two-gate feeder
双口供料器 two-gate feeder
双口管瓦管 double vent pipe tile
双口接头 duplex adapter
双口喷枪 double-head(ed) spray gun;double nozzle spray gun;two-component(spray)gun
双口喷枪喷涂 catalyst spraying
双口式绕组 doubly re-entrant winding
双口油桶 two-spout oil can
双口装载溜槽 two-gate loading chute
双扣吊链 chain sling with two ears
双跨 double space;two-span
双跨大梁 two-span girder
双跨底链的系船浮筒 double span moorings
双跨度 double spans;twin spans;two-bay
双跨间 double bay
双跨缆索吊桥 two-span cable supported bridge
双跨连续梁 beam with central prop
双跨连续桥 two-span continuous bridge
双跨梁 two-span beam
双跨轮 double back gear
双跨排架 double bent
双跨桥 two-span bridge
双跨桥门架 two-bay portal frame
双跨人字形框架 two-bay gable(d) frame
双跨山墙式框架 two-bay gable(d) frame
双跨双坡屋面 saddle-back roof;saddle roof
双跨屋架结构 double span trussed roof structure
双块混凝土轨枕 dual block tie;twin-block concrete sleeper
双块式 twin block type
双块式混凝土轨枕 concrete sleeper of twin-block design
双块制动器 double block brake
双快水泥 quick-setting quick-hardening cement
双矿物法 dual mineral method
双框架陀螺仪 two-frame gyroscope
双框锯厂 twin-frame(saw)mill
双框锯机制材厂 twin-frame(saw) mill
双馈 double feed
双馈串联电动机 doubly fed series(or repulsion)motor
双馈电式异步电机 double fed asynchronous motor
双馈路 duplicate feeder
双馈天线 double fed antenna
双馈同步电动机 double-fed synchronous motor
双馈推斥电动机 double fed repulsion motor
双馈线 twin feeder
双扩散法 <木材防腐法> double-diffusion method
双扩散晶体管 double-diffused transistor
双扩散面 double diffusive interface
双括板平整器 twin-screed finisher
双括板整平机 two-screed finisher
双括板整平器 two-beam finisher
双括号 double brackets
双括棱镜 biprism;duplicating prism
双拉力 double tension
双拉螺栓 stud
双拉门 biparting door
双拉盘式轨道衡 <车辆在运行中不论连挂与否,两端分别测定重量而后加总的方法> two draft scale
双喇叭扬声器 double horn loudspeaker
双蜡烛烛台 two-light candle stick
双栏式分类账账户 double column ledger account
双栏式排版 half measure
双栏税则 double column tariff
双缆架空索道 double cable ropeway
双缆坡道 double-cleat ladder
双缆式架空索道 double rope aerial cableway
双缆索道 bi-cable ropeway; double cable ropeway; double cable suspension line;double ropeway
双缆索起重机 double cable crane
双缆柱 double bitt
双捞牙 <捞钢丝绳用> double corkscrew
双乐音门铃 double chime unit
双雷管起爆 double priming
双肋 double rib
双肋的 double-webbed
双肋式 double web
双肋式桥梁 double-rib bridge;double-webbed bridge
双累积曲线 double mass curve
双棱 double rib
双棱(角)混凝土砌块 double corner block
双棱镜 biprism;double prism
双棱镜垂准 double prism plummet
双棱镜单色仪 double prism monochromator;duo-prism monochromator
双棱镜分离器 twin-prism separator
双棱镜干涉 biprism interference
双棱镜光学直角镜 double prism optical square
双棱镜射谱仪 double prism spectrograph
双棱镜摄谱法 double prism spectrography
双棱镜望远镜 double prism field glass
双棱镜仪 two-prism device
双棱镜直角器 double prism(atic) square; twin prism square; two-prism square
双棱柱(体)biprism
双棱锥(体)bipyramid[dipyramid]
双离合器 double clutch
双离子束的同时接收 simultaneous collection of two-ion beams
双离子束探测器 dual ion beam detector
双力 double force
双力绞车 double purchase winch
双力矩 bimoment
双力偶 double couple;two-couple
双力偶法 double couple method;two-couple method
双力偶模型 double couple model
双力作用的 duo-servo
双力作用制动器 duo-servobrake
双历 day-date
双历手表 calendar day and date watch
双历瞬跳杆 calendar unlocking yoke
双历压片 combined maintaining plate
双历装置 day-and-date device
双立方体 double cube
双立方体室 double-cube room
双立管 double riser pipe
双立管系统 double-stack system;dual riser system
双立体镜 dual stereoscope
双立筒预热窑 double shaft preheater kiln;twin-shaft preheater kiln
双立柱 double column vertical boring machine
双励磁线性电机 dual exciting linear motor
双沥青纯碱纸 double pitch sodium paper
双粒级煤 doubles
双粒序构造 double graded structure
双连 double linking
双连壁橱 close-coupled closet
双连厕所 close-coupled closet
双连画 diptych
双连接表 doubly linked list
双连绝缘子串 double strings;twin insulator strings
双连开关 tumbler switch; two-way tumbler switch
双连开关接线法 strapping wire
双连式房屋 semi-detached building
双连式喷油嘴 duplex fuel nozzle
双连式住宅 semi-detached house
双连锁西班牙瓦 double-interlocking Spanish tile
双连通图 biconnected graph
双连销瓦 tile with double interlock
双连续 bicontinuous
双连续拌和叶片式搅拌机 twin-shaft continuous-mix pugmill
双连续映象 bicontinuous mapping
双联 two-fold-gang
双联安全阀 dual relief valve
双联板排水涵洞 twin slab drainage culvert
双联本印刷 fore and aft
双联泵 duplex pump; double(-section) pump; dual pump; split-flow pump;tandem pumps;twin pumps;two-way flow pumps
双联操纵机构 coupled control
双联铲运法 duplex earth shoveling process
双联乘法器 paired multiplication
双联齿轮 compound gear; duplicate gear
双联齿轮泵 tandem gear pump
双联传动装置 dual drive
双联船闸 twin ship lock
双联单 double draft
双联单缸泵 twin simplex pump
双联单作用泵 twin-single pump
双联的 dual; duplicate; siamesing; two-gang;duplex
双联电容器 double capacitor;double condenser;dual capacitor;two-gang condenser
双联吊车 duplex crane
双联订单 indent
双联发动机 dual engine;twin engine
双联发射装置 twin launcher; twin launching device
双联阀 double valve;dual valve
双联法 duplexing;duplex practice;duplex process
双联风扇 twin fan
双联浮标 twin float
双联杆 double bar link
双联固结仪 two-in-one odeometer
双联管 twin tubes
双联管式涵洞 twin-pipe culvert
双联光电管 twin photocell
双联合同 indenture
双联滑轮组 twin pulley block
双联环规 twin ring ga(u)ge
双联幻灯片析像器 dual slide scanner
双联汇票 draft in duplicate
双联会合 double junction
双联混凝土泵 double concrete pump
双联火道 twin flue
双联机枪装置 twin-gun attachment

双联计算机 duplex computer; twin calculating machine
双联夹 double couplers
双联角形止动垫圈 twin horn lock washer
双联结点 double junction
双联晶 bicrystal
双联绝缘子串 twin insulator strings
双联开关 coupled twin switch; ganged switch; tumbler switch; two-way switch; two-wire switch
双联可变电容器 two-gang variable capacitor
双联空气泵 dual air pump
双联孔 twin opening
双联孔人行隧道 twin opening subway
双联控制器 twin controller
双联炼钢法 duplex process for steel making
双联炉 duplex furnace
双联滤器 duplex filter; twin strainer
双联模 double cavity mo(u)ld
双联起重机 twin crane
双联汽化器 duplex carburetor
双联桥式起重机 twin travel(l)ing crane
双联擒纵机构 duplex escapement
双联球轴承 duplex ball bearing
双联曲柄轴 double-throw crankshaft
双联燃料泵 twin fuel engine; twin fuel pump
双联燃气发生器 twin type gasifier
双联熔炼 duplex melting
双联式泵 duplex pump
双联式分级机 duplex(rake) classifier
双(联式)计算机 duplex computer
双联式夹紧装置 duplicate clamping arrangement
双联式交通管理 coupled control
双联式链 duplex chain
双联式气泵 duplex pump
双联式旋转压气机 rotating duplex compressor
双联式住宅 duplex house; two-family house
双联梳棉机 duocards
双联锁 duplex lock
双联锁瓦 double-interlocking tile
双联调节器 duplex regulator
双联通风管 dual vent
双联同轴操作 gang-operated
双联推土机 double dozer; yo-yo <俚语>
双联拖车 double bottom trailer combination; doubler; doubles
双联拖车方式 double bottom trailer system; doublers
双联拖拉机 twin tractor
双联箱(横水管)锅炉 double header boiler
双联箱形刚架 twin-box-shaped rigid frame
双联羊脚碾 double sheepfoot roller
双联闸室 twin chamber
双联直角止动垫圈 twin vertical horn lock washer
双联住宅 two-family duplex; two-family dwelling(duplex)
双联助推器 paired booster
双联铸铁 duplex iron
双联装置 twin installation
双炼 duplex
双炼法 duplex process
双炼钢 duplex steel
双炼铁 double refined iron
双链 double chain
双链拔丝机 dual chain bench
双链斗式升运器 twin-chain elevator
双链钩 chain legs
双链刮板制动输送机 double chain retarder
双链接技术 double chained technique
双链接树 doubly chained tree
双链式吊桥 double chain suspension bridge
双链式拉拔机 double chain draw bench
双链式线性表 double chained linear list
双链式悬索桥 double chain suspension bridge; suspension bridge with double cable system
双链式运输带 endless chain conveyer [conveyor]
双链输送机 double chain conveyer [conveyor]
双链形悬挂【铁】 compound catenary equipment
双链转环 mooring shackle; mooring swivel
双链装载抓斗 double chain rehandling grab
双梁 twin beams
双梁吊车 double beam crane
双梁机翼 double longeron wing
双梁结构 twin-spar construction
双梁门架 double beam mast
双梁门式起重机 double girder gantry crane
双梁桥 double girder bridge
双梁桥式起重机 double girder overhead crane; twin beam bridge crane
双梁桥式抓斗起重机 twin beam bridge type grabbing crane
双梁上楼板挑出 double T-beam
双梁通用机架 open frame tool carrier
双梁系楼盖 double-framed floor
双亮度信号灯 two-level signal
双量程 double range; dual range
双量程的 double dial
双量程伏特计 double range voltmeter
双料 double strength
双料玻璃 double-strength glass
双料窗玻璃 double-strength window glass; double thickness sheet glass; double thickness window glass
双料轮胎 dual tire[tyre]; twin tire [tyre]
双料门 deadlocking lath
双料配制的灰泥 two-part putty
双料水玻璃 double water glass
双料运载车 <可载两种混凝土> two-batch truck
双料振荡器 two-mass oscillator
双料振动器 two-mass vibrator
双料钟 double charging bell
双列 biserial; double row
双列包装机 double track packing machine
双列比利时式轧机 double Belgian mill
双列布置 <机电> double row layout
双列缠绕机 double torsion machine
双列车流 double file stream
双列触发器 dual rank flip-flop
双列的 twin-row; two-bank
双列分行机 dual liner
双列辐射型发动机 double row radial engine
双列(滚珠)轴承 double row bearing
双列合并旅客列车 double length jointed passenger train
双列寄存器 dual rank register
双列径向滚珠轴承 double row ball journal bearing
双列刻度水准(标)尺 double scale rod
双列螺旋弹簧缠绕机 double torsion machine
双列内向牛舍 faced-in; facing cows in
双列(汽缸)发动机 two-bank engine
双列球轴承 two-row ball bearing
双列蛇形管圈 double loop
双列式牛舍 double row cowshed
双列式悬浮预热器 double string suspension preheater
双列式猪舍 double row hog house
双列套筒滚子链 twin-roller chain
双列停车 double parking
双列网眼 two-course net
双列系数 biserial coefficient
双列相关比率 biserial ratio of correlation
双列相关系数 bi-serial coefficient of correlation
双列向心短滚柱轴承 double row radial short cylindrical roller bearing
双列向心鼓面滚柱轴承 double row radial spherical roller bearing
双列向心滚珠轴承 double row ball bearing
双列向心球轴承 double row radial ball bearing
双列预分解炉窑 separate line calciner kiln
双列预热器 twin preheater; twin stream preheater
双列预热器塔 twin towers
双列圆盘耙 double-action disc[disk] harrow; tandem disc[disk] harrow
双列圆盘式路耙 double-action disc [disk] harrow
双列支柱 double pillar
双列直插 dual-in-line
双列直插式 dual inline type
双列直插式封装 dual inline package
双列直插式外壳 dual inline package
双列直插式组件 dual inline package
双列柱廊的 dipteral
双列柱廊式建筑 dipteral building
双列柱廊式神庙 dipteral temple
双列柱子 double colonnade
双列自动调心球面球轴承 self-aligning double row ball bearing
双列自动调心轴承 self-aligning double row bearing
双列组合列车 double train
双菱形(互通式)立体交叉 double diamond interchange
双菱形天线 double rhombic antenna
双零值静切力胶体 <起始和十分钟均为零的> zero-zero gel
双溜放【铁】 double humping(of two trains); dual humping (of two trains); simultaneous humping of two trains
双溜放设备 double humping facility; dual humping facility
双溜放驼峰 double rolling hump
双流 dual flow
双流不稳性 two-stream instability
双流程堆芯 two-flow core
双流程冷凝器 two-flow condenser
双流程选矿厂 two-stream preparation plant
双流程蒸汽过热器 two-pass superheater
双流传号 double current mark
双流单工 double current simplex
双流道叶轮 double channel impeller; two channel impeller
双流的 double flow
双流电键 double current key; double tapper key
双流电缆 two-condition cable
双流电缆码 double current cable code
双流电力牵引 dual mode electric(al) traction
双流发电机 double current generator
双流分配 double distribution
双流归零制 double current return to zero system
双流浇铸 twin casting
双流滤池 reverse-current filter
双流路 dual flow path
双流排汽 double flow exhaust
双流桥接式双工 double current bridge duplex
双流式 double current method
双流式继电器 double current relay
双流式冷却器 double stream cooler
双流式凝汽器 two-pass condenser
双流式汽轮机 double flow turbine; twin water turbine; two-runner turbine
双流式球磨机 double-ended ball mill
双流式水轮机 double flow turbine; twin turbine
双流式透平 double flow turbine; twin water turbine; two-runner turbine
双流式涡轮机 double flow turbine; twin water turbine; two-runner turbine
双流式油喷燃器 dual-flow oil burner
双流水线码 double current cable code
双流塔板 double pass(flow) tray
双流体 two-fluid
双流体反应堆 two-fluid reactor
双流体模式 two-fluid mode
双流体模型 two-fluid model
双流体喷雾器 dual-fluid atomizer
双流体喷嘴 two-fluid spray nozzle
双流体雾化器 dual-fluid atomizer
双流体系统 binary fluid system; two-fluid system
双流体循环 binary fluid cycle
双流体压力计 two-fluid manometer
双流通报 double current signal(l)ing; double current working
双流线形 <钻杆接头> double streamline
双流向地段 two-way flow zone
双流形汽轮机 twin turbine
双流形涡轮机 twin turbine
双流诱导式通风 double flow induced-draft
双流制 double current system
双流制单工电路 double current simplex circuit
双流作用计 electrodynamometer
双硫磷 abate; biothion
双硫腙分光光度法 spectrophotometric(al) method with dithizone
双硫腙提取法 dithizone extraction method
双六(单元)天线阵 double six array
双六缸发动机 twin-six engine
双六角的 dihexagonal
双六角棱镜 dihexagonal lens
双六面体 dihexahedron
双六汽缸 twin six cylinders
双笼钢筋 double-cage reinforcing
双笼升降机 twin cage hoist
双笼式钢筋架 double-cage reinforcement
双陇槽 diglyph
双炉胆锅炉 double flue boiler; Lancashire boiler
双炉门式锅炉 double-ended boiler
双炉膛 double bed burner; double furnace; double hearth burner; twin furnace
双炉膛锅炉 double furnace boiler; twin-furnace boiler
双炉选择性裂化 two-coil selective cracking
双滤料 dual-medium
双滤网 duplex strainer

双滤油器 twin filter
双路 twin-channel; two-circuit; two-way
双路操纵的 two-control
双路测距装置 two-path distance measuring system
双路传动 dual-path drive
双路的 twin-channel
双路电路 two-path circuit
双路叠绕法 two-in-hand winding
双路动圈传声器 two-way dynamic microphone
双路反馈 two-path feedback
双路分配器 two-way distributor
双路干燥器 double pass drier[dryer]
双路供气 parallel gas feed
双路共振吸声器 two-circuit resonant absorber
双路管道 two-way duct
双路混合器 fader potentiometer
双路基道路 separated formation road
双路进线针织机 two-feed knitter
双路开关 two-circuit switch; two-way switch; two-way tap
双路控制器 dual channel controller
双路馈电 dual feed; duplex feeding; duplicate supply
双路冷凝器 two-pass condenser
双路冷却 two-pass cooling
双路喷燃器 duplex burner
双路燃油喷嘴 duplex burner
双路绕组 two-circuit winding
双路式发动机 two-spool engine
双路式燃气轮机 two-spool gas turbine
双路式涡轮喷气发动机 by-pass engine
双路调节器 dual channel regulator
双路通路 duplex channel
双路通信[讯] duplex communication
双路拖运机 double strand drag conveyer[conveyor]
双路系统 double path system
双路移频制 four frequency diplex; frequency shift twinplex; twinplex
双路预付电度计 two-circuit prepayment meter
双路运输带 double strand conveyer [conveyor]
双路制 twinplex
双路转发器 forked repeater
双路转接器 duplex adapter
双率计 double rate
双率税则 bilinear tariff; double tariff
双氯甲醚 dichloromethyl ether
双氯烷基醚 bischloromethyl ether
双氯乙基硫 yperite
双轮 coupled wheel; double round; dual wheel; twin wheel
双轮铲运机 two-wheel scraper
双轮串联式压路 two-wheel tandem roller
双轮单铧犁 gallows plow
双轮的 dual-drum
双轮底盘 bicycle undercarriage; two-wheeled chassis
双轮反循环钻机 hydrofraise; hydromill
双轮反循环钻机造孔法 hydrofraise method
双轮负荷 twin-wheel loading
双轮复滑车 two-part line tackle
双轮钢滑车 double sheave steel block
双轮沟槽压路机 dual-compression trench roller
双轮刮路机 two-wheel scraper
双轮荷载 dual - tire [tyre] loading; dual wheel loading
双轮后推助力器 two-wheel trailer booster
双轮滑车 double block; sister(-sheave) block; two(-sheave) block
双轮滑轮 double block; sister(-sheave) block; two(-sheave) block
双轮机动车 autoped
双轮架式犁 two-wheel frame plow
双轮犁 double wheel plow; two-wheel plough
双轮路碾 dual-drum roller
双轮轮胎货车 dual tire[tyre] truck
双轮内燃机式压路机 tandem road roller
双轮盘涡轮转子 double disk turbine rotor
双轮炮车 artillery cart
双轮起落架 <飞机> dual wheel under carriage; two-wheel landing gear; twin landing gear
双轮气胎电车 two-wheel rubber-tired[tyred] troll(e)y
双轮前起落架 twin-wheel nose gear
双轮驱动式播种机 two-wheel drive planter
双轮式(飞机)着陆架 bicycle landing gear
双轮式碾压机 tandem roller
双轮手推车 barrow; hand barrow; hurl barrow; two-wheel handcar
双轮手推翻斗车 rocker-dump handcart
双轮双铧犁 double shared double wheeled plow; double-share wheeled plough
双轮双喷嘴式水轮机 twin-wheel twin nozzle type turbine
双轮胎 double tires[tyres]; twin tires [tyres]
双轮胎可拆式轮辋间隔圈 rim spacer
双轮胎碾 dual-tyre roller
双轮胎胎压平衡器 pressure equalizer for twin tyres
双轮同时移动起重机 twin travel(l)ing crane
双轮拖车 semi-trailer; semi-trailer van; semi-van
双轮拖车式反应类道路平整度仪 two-track RTRRMS [reaction-type road regularity machining system]
双轮拖拉机 two-wheeled tractor
双轮拖拉机牵引式铲运机 two-wheel tractor-pulled scraper
双轮小车 dandy
双轮斜击式水轮机 double Turgo turbine
双轮压路机 dual-drum roller; tandem roller; two-axle tandem roller
双轮移动起重机 twin travel(l)ing crane
双轮运圆木车 sulky
双轮振荡压路机 double vibratory roller
双轮振动碾 tandem vibrating roller
双轮振动压路机 double drum vibrating roller; double vibrating roller; double vibratory roller; tandem vibrating roller; twin-vibratory roller
双轮支架 two wheel cradle
双轮制动器 two-wheel brake
双轮转环滑车 double pivoting block
双轮转向装置 dual wheel steering gear
双轮装配 dual wheel assembly
双轮钻探车 two-wheel wagon drill
双罗贝尔线棒 double Roebel bar
双罗盘系统 twin compass system
双螺带式搅拌器 double ribbon agitator
双螺杆 queen bolt
双螺杆泵 double screw pump; Quimby pump; two-screw pump
双螺杆顶车机 double worm pusher
双螺杆混合机 twin-screw mixer
双螺杆挤出复合 twin-screw extrusion compounding
双螺杆挤出机 cotruder; double screw extruder; dual worm extruder; twin-screw extruder
双螺杆聚合器 twin-screw polymerization reactor
双螺杆叶片式拌和机 twin-screw pugmill
双螺杆与螺帽 twin-screw and nut
双螺管阀门 two solenoid valve
双螺帽 double nuts
双螺母 double nuts
双螺栓 double bolt
双螺栓杆 queen bolt
双螺丝直尾车床鸡心压头 double screw straight tail lathe dog
双螺纹的 twin-twisted
双螺纹接套 double nipple
双螺纹螺杆 double thread screw; two-flight screw
双螺纹面钢筋 twin-twisted round bars
双螺线 bifilar helix; bipitch; compound clothoid
双螺线灯丝 coiled coil
双螺线加热器 coiled-coil heater
双螺线慢波电路 bifilar helix slow-wave circuit
双螺形水轮机 twin spiral water turbine
双螺旋 double helix
双螺旋泵 twin-screw pump
双螺旋齿轮 double helical (spur) gear; herringbone gear
双螺旋打捞器 double corkscrew
双螺旋的 bipitch
双螺旋分级机 duplex-spiral classifier
双螺旋分砂机 twin-screw sand classifier
双螺旋混合机 double worm mixer
双螺旋挤泥机 double wing pusher
双螺旋桨船 twin-screw ship; twin-screw vessel
双螺旋桨的 twin propeller; twin screw
双螺旋桨混合器 double helical mixer
双螺旋桨轮船 twin-screw steamer
双螺旋搅拌机 double worm mixer
双螺旋犁 twin-screw propeller
双螺旋楼梯 double spiral stair(case)
双螺旋坡道 double helical ramp
双螺旋式输送机 double-lead screw; double spiral screw conveyer
双螺旋送料散装水泥拖车 twin-screw bulk cement trailer
双螺旋喂料器 two-screw chip feeder
双螺旋洗矿机 double screw washer
双螺旋线 bifilar helix
双螺旋线灯丝 coiled-coil filament; double helical heater
双螺旋线千分尺 bifilar micrometer
双螺旋斜坡道 double spiral ramp
双螺旋形的双螺旋 double helical double helix
双螺旋坐标量测仪 two-screw comparator
双螺柱式钻车 double screw column drill mounting
双落潮流 double ebb; double tidal ebb
双落料机 double drop machine
双麻袋 double jute bag
双马达 bi-motor
双马达传动 duplex set
双马来亚胺树脂复合材料 bimaleimide resin composite
双马牵引的四轮车 two-horse chariot
双马蹄焰池窑 double horse-shoe flame tank furnace
双码 dicode
双码信号 dicode signal
双买主垄断 duopsony
双卖主垄断 duopoly
双脉冲 dipulse; two-pulse
双脉冲编码 twin pulse code
双脉冲触发器 double trigger
双脉冲触发信号 double trigger
双脉冲电台 double pulsing station
双脉冲定时器 two-pulse timer
双脉冲对消器 two-pulse canceler
双脉冲发射台 double pulse station
双脉冲方式 dipulse system
双脉冲激励 double pulse excitation
双脉冲计数电路 two-pulse counting circuit
双脉冲记录法 double pulse recording
双脉冲记录方式 double pulse recording mode
双脉冲馈给控制 two-element feed control
双脉冲列发射 double pulsing
双脉冲码 dicode; dipulse code
双脉冲调制 double pulse modulation
双脉冲选择 double pulse selection; binary pulse-code modulation
双脉码调制 binary pulse-code modulation
双漫射密度 doubly diffuse density
双镘整平机 two-screed finishing machine
双盲法 double-blind method
双锚泊 moor; riding at two anchors; riding to two anchors
双锚泊改单锚泊 unmoor
双锚单交叉 cross chain; half elbow
双锚定桩靴 double grouser track shoe
双锚碇板墙 double anchored sheet wall
双锚碇墙 double anchored wall
双锚碇墙的设计 design of double anchored wall
双锚杆锚碇设施 double tie-rod anchorage
双锚杆墙 double-anchored wall
双锚锚泊 mooring with two anchors; two-anchor mooring
双锚锁环 mooring swivel
双锚停泊 mooring with two anchors
双锚系泊 mooring with two anchors; two-anchor mooring
双帽钉 double-headed nail; form nail; scaffold nail
双帽螺栓 standing bolt
双没食子酸 digallic acid
双玫瑰花型 double rose
双镁合金板 Magclad
双门 double gate; two-door
双门插销 double door bolt
双门敞车 sport-tourer
双门敞篷轿车 roadster
双门车身 two-door(type) body
双门道 pair of gateways
双门的大门 dipylon
双门家用冰箱 double door household refrigerator
双门廊 double portio
双门篷车 roadster
双门轻便小汽车 pony car
双门扇 double door leaves
双门式船闸 double gate lock
双门小客车 two-door car
双(门)叶闸门 double-leaf gate
双米点法 double meter point
双密度 dual density
双密度编码 double-density encoding
双密度格式 double-density format
双密度装置 dual density device
双密封 twin seal

双密封穿孔套管 twin-seal tapping sleeve
双密封垫 double seal
双密封器 twin packer
双面 diprosopy
双面 U 形坡口 double U groove
双面 V 形对焊 double V-butt welding
双面 V 形对焊接 double V-butt weld joint
双面 V 形对接 double V-butt joint
双面 V 形对接焊 double V-butt joint weld
双面 V 形平焊接 double V-butt weld
双面安装 back-to-back arrangement
双面凹的 biconcave
双面拔气罩 double-sided hood
双面靶 two-sided target
双面板 dual platen
双面板阀门 double-skin plate valve
双面包夹板门 single-measure door
双面包铁皮门 door covered with sheet iron on both sides
双面包锡双金属轧制耐蚀铝板 Zinnal
双面标尺 reversible rod; reversible staff
双面标志 double face sign
双面波 double wave
双面槽 double groove
双面层隔墙 double-skin partition (wall)
双面层(空心)瓦 double shell tile
双面差厚镀锡 differential plating
双面差厚涂镀 differential coating
双面差厚涂法 differential coating
双面铲斗 double-acting bucket
双面撑杆 anchoring pole
双面齿 double cog
双面齿条 double-sided rack
双面齿钥匙 double-bitted key
双面处理 double measure; double surface treatment
双面锤 <砌砖石用的> club hammer; lump hammer; mash hammer
双面锉刀 double ender
双面搭焊接 double fillet lap joint
双面单密度软磁盘 two-sided single-density diskette
双面刀 two-edged knife
双面刀盘 alternate blade cutter; spread-blade cutter
双面的 bidirectional; bifacial; double-faced
双面的硬木板 duo-faced hardboard
双面等厚镀层镀锡薄钢板 straight electrolytic(tin) plate
双面地槽 bilateral geosyncline
双面地弹簧 double-acting floor spring; double-action floor spring
双面点焊 direct spot welding; direct welding
双面调车信号机【铁】 double-sided shunting signal; front-and-back shunting signal; signal for shunting front and back
双面丁字形焊接 double fillet welded T-joint
双面镀锡薄板 dual-coat plate
双面缎 double-faced satin
双面缎子织物 double-faced satin
双面对焊接缝 double-welded butt joint
双面颚式破碎机 double face jaw crusher
双面阀 double face valve
双面浮雕饰 bossed(on) both sides
双面辐射式加热炉 equiflux heater
双面复制页 duopage
双面盖板 doubling plate
双面盖板缝 double-lock seam
双面盖板接头 double strap web joint; double strapped joint
双面盖缝 double welted seam
双面盖条立缝 double welted standing seam
双面钢筋 double reinforcement
双面隔墙 double-sided partition(wall)
双面隔墙砖 double-faced partition wall tile
双面工具 biface tool
双面光的门 single-measure door
双面光整处理 <板的> standard bright finish
双面规尺 <整修混凝土板边缘的> double edging tool
双面辊涂机 double coater
双面焊搭接接头 double-welded lap joint
双面焊对接接头 double-welded butt joint
双面焊缝 double fillet welded joint; double-side welding joint; double-welded joint
双面焊(接) both sides welding; two-side welding; welding by both sides; double-side welding
双面焊接接头 double fillet welded joint; double-welded joint
双面荷兰式砌合 double Flemish bond
双面横棱缎 peau de sole
双面灰板条 double lath
双面活动钳 always-ready wrench
双面机动镘刀 double mechanical trowel
双面机架 double-sided rack
双面积调制声迹 dulateral track
双面畸形 double distortion; diprosopia
双面加工 surfaced on both sides
双面加热 sandwich heating
双面加压 compaction by double-action; double-action pressing
双面剪床 double shears
双面剪力铆钉 rivet in double shear
双面剪切 double shear
双面剪切法 double shear method
双面剪切试验 double shear test
双面建筑 double-leaf construction
双面交叉说 two-plane theory of chiasma
双面胶带 double-faced adhesion band
双面角焊缝 twin fillet welt
双面角焊缝搭接接头 double fillet lap joint
双面接触式冻结装置 double contact freezer
双面结构 bilateral structure
双面金属型 reversing gravity die; reversing permanent mould
双面进料 twin feed; dual feed
双面进路表示器 double-sided route indicator
双面进气离心式压气机 double entry centrifugal compressor
双面进气离心式压缩机 centrifugal double sided compressor
双面进气压气机 double entry compressor
双面进气叶轮 double entry impeller
双面经二重织物 reversible warp backed weave
双面井底车场 two-sided bottom switchyard
双面可折铰链 reversible hinge
双面冷床 double cooling bank; double cooling bed
双面离心式压缩机 bilateral centrifugal compressor
双面联胎 janiceps
双面链钳 reversible chain tong; reversing chain tong
双面列车流 double file stream
双面临街院落 double frontage lot yard
双面模板 built-up plate; double-sided pattern plate; double-side shuttering; match plate; split-plate pattern
双面模(板模) match plate pattern
双面磨粉机 duplex mill
双面抹灰泥 plastered both sides
双面啮合齿轮检查仪 dual flank gear rolling tester
双面啮合检查仪 two-flank gear rolling tester
双面抛光 twin polishing; two-sided finish
双面抛光设备 twin polisher
双面刨光机 double surfacer; double thicknesser
双面平衡 two-plane balancing
双面坡 double pitch
双面坡口 double groove
双面坡口焊缝 double groove weld joint
双面坡口接头 double groove joint
双面嵌镶幕 double-sided mosaic; two-side mosaic
双面墙 double face wall
双面墙结构 <木框架结构夹衬板,两边覆以胶合板的墙体结构> conventional wall construction; double wall construction
双面撬锤 double-faced sledge hammer
双面燃烧式锅炉 double-ended boiler
双面刃端铣刀 two-lip end milling cutter
双面绒布 two-faced plush cloth
双面绒头地毯 union carpet
双面三角皮带 double angle V-belt
双面上光 double glazing
双面式 two-plane type
双面受剪 double shear
双面书架 double-faced book rack; double-faced stack; double-sided book shelves
双面甩车调车场 two-sided swing parting
双面双密度软磁盘 two-sided double-density diskette
双面水冷壁 dividing waterwall; division wall
双面水准(标)尺 double-sided staff; double target level(l)ing rod
双面水准管 reversible level tube
双面说 two-plane theory
双面索 cables in double plane
双面弹簧插销 double-action spring bolt
双面弹簧铰链 double-acting spring butt hinge; double-action spring butt hinge
双面弹簧门 double-acting spring door; double swing door
双面搪瓷器 double face ware
双面掏槽 two-sided cutting
双面提花地毯 ingrain carpet
双面提花黄麻地毯 ingrain jute carpeting
双面提花墙纸 ingrain wallpaper
双面提花墙纸覆盖层 ingrain wallpaper cover
双面凸的 biconvex
双面凸轮 double track cam
双面凸起镶板 double raised panel
双面凸缘轮 double-flanged wheel
双面涂布 coated both sides; double face coating; double-sided coating; double spread; two-coat; two-sided coating
双面涂布带 double coated tape
双面涂层 surfaced on both sides
双面涂胶 double spread; double spread gluing
双面涂蜡 two-sided waxing
双面托盘 double deck pallet
双面温床 double-sloping bed
双面吸气罩 double-side-draft hood
双面铣刀 straddle cutter; straddle mill
双面线脚门头板 double-faced architrave
双面楔 double wedge
双面斜边 miter bevel both sides
双面斜棱边 bevel on both surfaces
双面信号(机)【铁】 back-to-back signal; two-way signal
双面信号机构 back-to-back signal heads
双面型板 match plate; reversible pattern plate
双面型板分型面 matched parting
双面型叠箱造型法 stack mo(u)lding
双面型箱造型 stack mo(u)lding
双面型芯撑 double head chaplet
双面修饰 double measure
双面修整 dressed two sides; two-side dressing
双面压刨 double thicknesser
双面研磨 twin grinding
双面研磨玻璃板 twin ground plate
双面研磨的镜玻璃 twin ground plate
双面研磨法 Pilkington twin process; twin-plate process
双面研磨机 twin grinder
双面摇摆筛 swinging double sieve
双面摇门 double-acting door; double swing door
双面摇纱机 double reeling frame
双面要涂层的 coated both sides
双面一顺一丁砌合 double Flemish bond
双面印刷电路 eircuitron
双面印刷机 perfect(ing) press; perfector
双面有线脚的木工活 double measure
双面釉墙面砖 wall tile glazed on both sides
双面闸瓦 clasp brake shoe
双面粘贴黏[粘]合剂 two-way stick adhesive
双面针织物 double knit; double knit fabric; two-sided knit goods
双面整理 full-finish
双面织物 double-faced; reversible cloth
双面注浆 double casting
双面砖 double-faced tile
双面装修 double measure
双面子钟 double-faced secondary clock
双面自行车架 double-sided bicycle stand
双秒机构 split-second mechanism
双秒轮 split-second wheel
双秒推扭 split-second pusher
双灭火剂灭火 dual agent attack
双名 binomen
双名法 binomial nomenclature
双名汇票 two-party draft
双名命名法 binomial nomenclature
双模 double mode; double module; dual mode; twin mo(u)ld
双模拔丝机 double-dies wire drawing machine
双模理论 two-film theory
双模量 reduced modulus
双模壳 dual mo(u)ld
双模式编码 dual mode coding
双模式火成活动 bimodal igneous activity

双模式火山杂岩 bimodal volcanic complex
双模式裂谷火山活动 bimodal rift volcanism
双模数 double modulus
双模态的 bimodal
双模行波管 dual mode travelling wave tube
双模型 double model
双模型立体模片 double model stereotemplate
双模制砖机 two-mold machine for forming brick
双膜电极 double membrane electrode
双膜法 double membrane process
双膜厚涂布器 duplex film applicator
双膜理论 two-film theory
双膜气压系统 double membrane pneumatic system
双膜式泵 double diaphragm pump
双膜式土压力盒 double diaphragm pressure cell; double diaphragm pressure ga(u)ge
双末级传动 double final drive
双母接头钻铤 double box collar
双母扣管子 box to box pipe
双母扣(接头)box to box
双母扣钻杆 box to box pipe
双母线 double busbar;duplicate-busbar
双母线结线 double bus connection
双母线曲面 doubly ruled surface
双母线调整 two-busbar regulation
双母线系统 double-busbar system
双母线系统联络开关 coupler bay of a duplicate-busbar system
双木边梁 double header
双木浮式消波堤 twin-log floating breakwater
双木过梁 double timber lintel
双木塞注水泥法 two-plug cementation
双木桅杆 twin timber mast; twin wood(en) mast
双木柱 twin timber mast;twin wood(en) mast
双目衬比 binocular contrast
双目垂直角差 dipvergence
双目的 binocular
双目观测 binocular observation
双目观察 binocular viewing; binocular vision
双目观察头 binocular viewing head
双目观察系统 binocular viewing system
双目观察装置 binocular viewing device
双目光度学 binocular photometry
双目混色 binocular colo(u)r mixing
双目检眼镜 binophthalmoscope
双目镜 binoculars;paired eyepiece
双目镜观察 binocular view
双目镜管 binocular tube
双目镜头 binocular head
双目立体显微镜 stereoscopic(al) microscope
双目配色 binocular colo(u)r matching
双目实体显微镜 binocular microscope
双目视差 binocular parallax
双目视场 binocular visual field
双目视觉 binocular vision
双目视系统 binocular visual system
双目视野 binocular field; binocular field of view
双目手持式水准仪 binocular hand level
双目双像 binocular diplopia
双目调整 binocular adjustment
双目望远镜 binocular
双目望远镜棱镜 binocular prism
双目显微镜 binocular microscope
双内插 double interpolation
双能谷模型 two-valley model
双能量过渡过程 double energy transient
双能源机车 battery trolley locomotive;combination locomotive
双泥条绞合手柄 double interlaced handle
双逆流预热器 Dopol;Polysius double stream counter flow preheater
双逆止阀装置 double check-valve assembly
双年轮 double annual ring
双碾盘混砂机 multimull(er)
双碾盘连续混砂机 multimill
双捏拌和机 twin pug mill
双啮误差 radial composite error
双扭钢筋 twin-twisted bar reinforcement
双扭线电缆 paired cable
双扭线流量计 double-torsion-line flowmeter
双扭转的 twin-twisted
双纽曲线 lemniscate curve
双纽线【数】lemniscate
双纽线电缆 twisted-pair cable
双纽线方向性图接收 lemniscate reception
双纽线函数 lemniscate function
双纽线坐标 lemniscate coordinates
双纽型缓和曲线 lemniscate transition curve
双钮话筒 double-button transmitter
双偶氮 bisazo
双偶氮化合物 bisazo compound;tetrazo compound
双偶氮黄颜料 bisazo yellow
双偶氮甲苯 bisazo benzil
双偶氮颜料 bisazo pigment
双偶极 quadripole
双偶极天线 double doublet; double doublet antenna
双偶极子 double dipole
双偶极子 H 天线 lazy H antenna
双偶然性体系 two-contingency system
双偶效应 double double effect
双偶自动机 dual-automation
双爬犁 double sledge
双耙侧向搂草机 two-raker
双耙湿式粗选磁选机 rake-type duplex wet magnetic cobber
双拍 double beat
双拍加法器 two-bit-time-adder
双排 double row;two-bank;two-tier
双排板桩堤 rock-filled double row sheet-pile breakwater
双排板桩墙结构 two-wall sheet-pile structure
双排板桩填充突堤 rock-filled jetty
双排板桩围堰 double row pile cofferdam;double wall cofferdam;double wall sheet pile cofferdam;two-wall sheet-piling cofferdam
双排板桩箱形围堰 two-walled sheet-pile cofferdam
双排标准组合插件 dual inline package
双排布置炮眼 double row spacing
双排车流 double file stream
双排的 two-bank
双排吊车 double hoist
双排钉 double fastening
双排多斗挖泥船 double ladder dredge(r)
双排钢板桩 double-wall sheet pile
双排钢板桩结构 double-wall sheet piled structure
双排钢板桩墙 double steel sheet pile wall
双排钢板桩突堤码头 sheet pile tied wall pier
双排钢板桩围堰 double wall steel sheet pile cofferdam
双排股线 two-wire-wide strands
双排滚珠止推轴承 double row ball thrust bearing
双排滚子链 double row roller chain
双排话传电报交换台 <16 座席以上> double tier equipment
双排环 double-style hanger
双排混凝土桩防渗墙 two-ranks of concrete pile curtains
双排建筑用吊车 double builders hoist
双排脚手架 independent-pole scaffold-(ing);double pole scaffold
双排阶段泵 two-bank stage pump
双排锯 two-gang saw
双排链 duplex chain
双排链斗卸车机 dual buckct unloader
双排量涡壳式水轮机 double discharge spiral water turbine; double spiral turbine
双排铆接头 double riveted joint
双排排样 double row layout
双排排样冲模 double punch layout
双排炮孔爆破法 two-row method
双排气管 twin exhaust pipes
双排汽 double flow
双排汽缸 two-row cylinder
双排汽口汽轮机 double-ended turbine
双排墙 double wall
双排绕组 two-range winding
双排式柴油机 double bank engine; double row engine
双排式内燃机 double bank engine; double row engine
双排式圆盘耙 double-action disc [disk] harrow
双排水 double drain
双排丝散光灯 double broad
双排四轮压路机 twin double drum roller
双排梯架 double bank
双排外夹环滚动轴承 double cup bearing
双排星形发动机 double row radial engine
双排星型发动机 double bank radial engine; twin-row radial engine; two-bank radial engine
双排圆盘耙路机 double-action disc [disk] harrow; double gang disk harrow
双排轴承 double row
双排柱 paired columns
双排柱脚手架 double pole scaffold
双排桩 double piles
双排桩横编柴梢式拦沙坝 double post row crosswise brush type check dam
双排桩梢料坝 double post row brush dam
双排座四人小客车 surrey
双盘传动 twin-disk transmission
双盘打浆机 duplex beater
双盘电泳法 double disc electrophoresis
双盘丁字管 two-collar tee fitting
双盘防逆阀 dual disc check valve
双盘弯头 both end flanged bend
双盘离合器 twin-plate clutch
双盘磨机 burr mill
双盘耙路机 double disc harrow
双盘强烈混合机 twin-pan intensive mixer
双盘石磨 buhr(stone) mill;burr mill
双盘式流变仪 disk-disk rheometer
双盘式磨碎机 attrition mill
双盘式抹光机 double plate trowelling machine
双盘式砂纸打磨机 double disc type sand papering machine
双盘弹性传动器 twin-disc resilient drive
双盘型线圈 double disk winding
双盘研磨机 double plate grinding machine;two-lap lapping machine
双盘液力变扭器 twin-disk hydraulic torque converter
双盘液力耦合器 twin-disk hydraulic coupling
双盘一插丁字管接 both ends flange and spigot T
双盘张力器 two-disc tensioner
双判决问题 two-decision problem
双抛物线大梁 double parabolic girder
双刨刀 double plane iron
双跑道机场 two-runway field
双跑楼梯 double flight stair(case); dual-flight stair(case);half turn
双跑平行楼梯 <指无梯井的双跑平行楼梯> dog-legged stair(case); narrow U stair(case);U-stair
双跑直楼梯 straight stair(case) with landing
双配筋的 double reinforced
双配筋梁 double reinforced beam
双配水(配电)系统 dual distribution system
双配位吸附质 bidentate adsorbate
双喷管化油器 two-jet carburetor
双喷口喷枪 two-nozzle gun
双喷气发动机 twinjet
双喷油(柴油机)dual fuel(l)ing
双喷嘴 double jet;twin-jet nozzle
双喷嘴吹灰器 dual nozzle blower
双喷嘴挡板阀 twin flapper-and-nozzle valve;two-jet flapper valve
双喷嘴水轮机 double shot turbine; two-jet-type turbine
双盆洗涤池 double bowl sink
双皮带层压机 double belt laminator
双皮带引出机 twin-belt take-off machine
双片 biplate;unmarried print
双片齿轮 split gear
双片胶粘板 two-piece glue-jointed panel
双片距冷却器 double spacing finned cooler
双片摄影机 double camera
双片式径向密封片 two-piece apex seal
双片式离合器 twin-plate clutch;two-plate clutch
双片式滤波器 bilithic filter
双片式轮辋 two-piece rim
双片式摩擦离合器 twin dry plate clutch
双片式热变电阻桥 two-disk thermistor bridge
双片衰减器 double vane attenuater [attenuator]
双片照相机 process camera
双偏流测风法 double drift
双偏心 double eccentricity
双偏心连接器 double eccentric connector
双偏心塔式篦子 dual eccentric tower type grate
双偏振 dual polarization
双偏振荡 dual-polarization oscillation
双偏振环形激光器 dual-polarized ring laser
双偏振激光器 dual-polarization laser

S

双偏振接收 reception of double polarization
双偏置 double offset
双偏置管 <卫生管道的> jump over
双偏置极化继电器 polarized double-biased relay
双偏置螺丝起子 double offset screwdriver
双偏转指示管 double deflection indicator tube
双漂移雪崩二极管 double drift region avalanche diode
双拼板 <木结构> double butt strap
双频 double frequency
双频拨号 two-frequency current selection; two-frequency dialling; two-frequency selection
双频拨号设备 two-frequency dialling equipment
双频测深仪 dual-frequency sounder; two-frequency sounder
双频长途拨号制 two-frequency long-distance selection system
双频带 double band
双频带扬声器 twin loudspeaker
双频道 two-way
双频道发声 dual channel sound
双频道激电仪 double frequency induced polarization instrument
双频道接收机 two-way receiver
双频道数字激电仪 digital induced polarization instrument with dual frequency
双频道扬声器 two-way speaker
双频电流选择 two-frequency current selection
双频电码 double frequency code; dual frequency code
双频段 two-band
双频多普勒系统 Doppler system with double frequency
双频发电机 double frequency generator
双频发生器 diplex generator
双频感应器 double frequency inductor
双频回声测深仪 dual-frequency echo-sounder; two-frequency echo-sounder
双频间隔 separation of two frequencies
双频率 bifrequency
双频率编码 double frequency coding
双频率电动机 dual-frequency motor
双频率电力机车 dual-frequency locomotive
双频率法 <用于超声测试> two frequency method
双频率双边带传输系统 double frequency both sideband transmission system
双频率坦声波测试法 two frequency method
双频率通路 two-frequency channel
双频谱 bispectrum
双频气体激光器 two-frequency gas laser
双频双工制 two-frequency duplex
双频顺序信号制 double frequency sequence signal(l)ing system
双频信标 double frequency beacon; dual-frequency beacon; two-frequency beacon
双频信号 two-frequency signal
双频信号发生器 <同时产生两个不同频率信号的电子管振荡器> diplex generator; two-frequency signal generator
双频信号制 diplex
双频鱼探仪 two-frequency fish finder
双频振荡 double frequency oscillation
双频振荡器 double frequency oscillator
双频直流变换器 dual-frequency converter; two-frequency converter
双频制 bifrequency system; two-fold frequency system
双频制信号 two-frequency signal(l)ing
双频制信号接收机 two-frequency signal receiver
双平顶大石锤 double flat face stone sledge
双平干矮树 double horizontal cordon
双平衡混频器 double balanced mixer
双平衡调制器 double balanced modulator
双平接板 double butt strap
双平开窗插销 double casement fastener
双平面 biplanar; biplane; diplane
双平面波全息图 two-plane-wave hologram
双平面回转 biplanar revolution
双平面建筑 double slab-type block
双平面镜 bimirror
双平面耦合器 biplanar coupler
双平面显像管 biplanar image tube
双平巷 twin drift; two-heading entry
双平巷采煤法 double entry method
双平巷掘进系统 two-entry system of mining
双平行煤巷 double headings
双平行偶极子天线 lazy H antenna
双平行平巷布置 twin entry
双平行四边形 double parallelogram
双平行四边形连接 double parallelogram linkage
双平行圆锯制材厂 scrag mill; skrag mill; twin-circular mill
双平型 amphiplatyan
双屏极三极管 twin-plate triode
双屏伸缩隔墙 two-screen expanding wall
双坡 double way gradient
双坡层屋顶 double pitch roof; double roll roof
双坡的 double pitched; gavel
双坡顶山墙 hip gable
双坡盖顶 saddle-back capping; saddle-back coping
双坡梁 double pitched beam
双坡面屋顶 double pitched roof
双坡内倾屋顶 double pent
双坡式屋顶 double-framed roof
双坡天窗 double pitched skylight
双坡屋顶 comb roof; couple(d) roof; duo-pitched roof; gable roof; saddle-back roof; double lean-to roof <中间有天沟的>
双坡屋面 double pitch roof; gable roof; span roof
双坡压顶 double splayed coping
双铺舱 two-berth room
双谱分光双星 double line spectroscopic binary; two-spectra binary
双谱线结构 doublet structure
双齐边机 double sizer
双齐伯天线 double Zepp antenna
双企口门框 double-rabbeted frame; double rebated; double rebated frame
双企口瓦 double gutter tile
双起重柱 double king post
双气阀喷嘴 dual-valve jetting tip
双气管 twin tubing
双气化器 duplex carburetor
双气轮胎 dual pneumatic tyre
双气循环 binary vapo(u)r cycle
双汽包锅炉 bi-drum boiler; double-drum boiler
双汽缸 twin cylinders
双汽缸泵 duplex cylinder pump
双汽缸蒸汽机 two-cylinder steam engine
双汽泡锅炉 double cylindrical boiler
双汽循环 binary cycle
双汽源汽轮机 combined main and exhaust steam turbine
双千斤顶 jack pair
双千斤顶法 double jack process
双钳导缆钳 double tongued chock
双浅倾转换 twin low-oblique transformation
双嵌线平顶 double-headed ceiling
双戗堤截流 double dike closure
双戗堤立堵截流法 end-tipped closure with two banks
双腔泵 dual chamber pump; dual element pump
双腔的 pair
双腔放大管 two cavity amplifier tube
双腔管 twin barrel tube
双腔化油器 twin barrel carburetor
双腔滤波器 two-chamber filter
双腔式中棂 double boxed mullion
双腔速调管 double transit oscillator; two-cavity klystron; two-resonator klystron
双腔速调管放大器 two-resonator klystron amplifier
双腔谐振电路 duplex cavity
双腔谐振器 double resonator
双腔型板 duplicate cavity plate
双腔液力耦合器 two-space fluid coupling
双腔振荡器 two-cavity oscillator
双强钢管 double extra strong steel pipe
双强坠陀杆 balance weight bar
双墙板层 double wall board layer
双墙的 double-walled; two-walled
双墙木板桩泥土填心围堰 double wall timber piling cofferdam filled with puddle
双墙式结构 double wall structure
双墙式围堰 double wall cofferdam
双敲铆法 pin riveting
双敲铆接法 pin riveting
双橇 two-sled
双桥墩 twin-pier
双桥梁 double plate girder
双桥梁摇座 double rocker bearing
双桥驱动 dual drive
双桥驱动式铲运机 two axle drive scraper
双桥驱动式后卸卡车 two-axle drive rear-dump
双桥全驱动的底卸卡车 two axles and drive bottom-dump
双桥式铲运机 two axles scraper
双桥式触探仪 double bridge-type penetrometer
双桥探头 double bridge probe
双桥轴驱动 two-axle drive
双桥轴式 twin axle reaction
双桥组 tandem axles
双桥组横梁 bogie beam
双桥组驱动 tandem axle drive
双桥组驱动式底卸卡车 tandem axle drive bottom-dump
双桥组中心距 tandem spacing
双桥组中心支点 tandem center point
双切齿锯 peg-raker saw
双切断塞门 double cut-out cock
双切线 bitangent
双倾的 bilateral
双倾伏褶皱 doubly plunging fold
双倾后代 amphiclinous hybrids; amphiclinous progeny
双氰胺 dicyandiamide
双穹隆 bi-vault
双球程磨机 two-race mill
双球浮子 double float; twin(-ball) float
双球脚阀 double ball foot valve
双球扣栓 double ball catch
双球连接杆臂 double ball pitman arm
双球面摆 double spheric(al) pendulum
双球面透镜 bispheric lens
双球式电阻温度计 two-bulb type resistance thermometer
双球式浮标 twin-ball float
双区 two-region
双区测光 dual-pattern light measurement
双区大牵伸 two-zone high drafting
双区电除尘器 two-stage precipitator
双区堆 two-region reactor
双区堆芯 two-zoned-core
双区反应堆 core-blanket reactor
双区管式炉 two-zone tubular furnace
双区空调系统 two-zone air-conditioning system
双区拉伸体系 two-zone drawing system
双区逆搅拌系统 dual zone counterflow mixing system
双曲板 double parabolic slab
双曲蚌线 hydrostatic conchoid
双曲臂和面机 double Z-arm mixer
双曲扁壳 double curvature shallow shell; double curved shell; doubly curved shell
双曲变换 hyperbolic(al) transformation
双曲柄 coupling crank; double crank
双曲柄闭塞器 double crank lock
双曲柄机构 double crank mechanism
双曲柄连杆机构 drag link
双曲薄壳 double curvature shell; doubly curved shell; shell of double curvature
双曲薄壳穹顶 doubly curved shell dome
双曲薄壳屋顶 double bent shell roof; hyperbolic(al) shell roof
双曲测度 hyperbolic(al) measure
双曲船位线 hyperbolic(al) position line
双曲磁镜透 hyperbolic(al) magnetic lens
双曲丛 hyperbolic(al) bundle
双曲代换 hyperbolic(al) substitution
双曲的钉格板 double curved spike grid
双曲点 hyperbolic(al) point; neutral point
双曲度量 hyperbolic(al) metric
双曲对合 hyperbolic(al) involution
双曲非球面 hyperboloid aspheric(al) surface
双曲工 <立面和平面都是曲线> circle-on-circle
双曲拱 double curvature arch; two-way curved arch
双曲拱坝 cupola dam; dome(d)(arched) dam; double curvature arch dam; double curved arch dam; hyperbolic(al) arch dam; hyperboloid arched dam; orange-peel shape dam; vaulted dam
双曲拱顶 double curved vault roof
双曲拱渡槽 double curvature arch aqueduct; double curvature arch flume
双曲拱桥 arch bridge with two-way curvature; double arch bridge; double convex arch bridge; double curved arch bridge; hyperbolic(al) arch; two way curved arch bridge

双曲拱形坝 orange-peel shape dam
双曲拐叶片搅拌机 double sigmoid-blade mixer
双曲轨组合式钝角辙叉 built-up obtuse crossing with both rails curved
双曲轨组合式普通辙叉 built-up common crossing with both rails curved
双曲奇点 hyperbolic(al) singular point
双曲几何 hyperbolic(al) geometry
双曲接合杆 double offset joint; double offset rod
双曲肋 double curved rib; sweeping rib
双曲连拱坝 multiple dome dam
双曲流形 hyperbolic(al) manifold
双曲流(运动)方程 hyperbolic(al) flow equation
双曲率 bicurvature; double curvature; hyperbolicity
双曲率平移薄壳 double curvature translation(al) shell
双曲率弯曲的 doubly curved
双曲率叶片 double curvature vane
双曲螺线 hyperbolic(al) conchoid; hyperbolic(al) spiral; reciprocal spiral
双曲面 double camber; hyperboloid【数】
双曲面玻璃 double curved glass
双曲面薄壳 hyperboloidal shell
双曲面齿轮 hypoid gear
双曲面齿轮传动轴 hypoid axle
双曲面齿轮系 skew axis gear [复 skew axes gears]
双曲面次镜 hyperboloidal secondary mirror
双曲面的 hyperboloidal
双曲面的叶 sheet of hyperboloid
双曲面反射镜 hyperboloidal mirror
双曲面镜 hyperbolic(al) mirror
双曲面冷却塔 hyperbolic(al) cooling tower
双曲面壳体 double curvature shell; doubly curved shell; hyperboloidal shell
双曲面透镜 bitoric lens
双曲面图 hyperbolic(al) chart
双曲面位置 hyperboloidic position
双曲抛物面【数】 hyperbolic(al) paraboloid
双曲抛物面(薄)壳 hypar shell; hyperbolic(al) paraboloidal shell; shell of hyperbolic(al) paraboloid
双曲抛物面壳顶 hyperbolic(al) paraboloidal roof
双曲抛物面壳体 hyperbolic(al) paraboloidal shell; parabolic(al) hyperboloid shell; shell of hyperbolic(al) paraboloid
双曲抛物面伞形薄壳屋顶 hyperbolic(al) paraboloid umbrella shell
双曲抛物面屋顶 hyperbolic(al) paraboloid roof
双曲抛物挠线 cubic(al) hyperbolic(al) parabola
双曲抛物线 hyperbolic(al) parabola
双曲偏转板 hyperbolic(al) deflector
双曲平面 hyperbolic(al) plane
双曲平面度量几何 hyperbolic(al) metric geometry in a plane
双曲平凸透镜 hyperbolic(al) plano-convex lenses
双曲壳体 vaulted shell
双曲区域 hyperbolic(al) region
双曲三角学 hyperbolic(al) trigonometry
双曲扇形 hyperbolic(al) sector
双曲透镜 hyperbolic(al) lens
双曲透射 hyperbolic(al) homology
双曲位置面 hyperbolic(al) surface of position
双曲线 hyperbola; hyperbolic(al) curve
双曲线贝塞耳函数 hyperbolic(al) Bessel functions
双曲线笔 double ruling pen; hyperbolic(al) pen; railway pen; railroad pen; railroad; road-pen <绘铁路道路等图的>
双曲线变移器 hyperbolic(al) inversor
双曲线波 hyperbolic(al) wave
双曲线薄壳 hyperbolic(al) shell
双曲线测位制 distance-difference measurement
双曲线齿轮 hyperbolic(al) gear; hyperboloidal gear
双曲线齿轮传动 hyperbolic(al) wheel drive
双曲线齿轮油 hypoid gear oil
双曲线导航 hyperbolic(al) navigation
双曲线导航定位系统 hyperbolic(al) navigation positioning system
双曲线导航图 hyperbolic(al) navigation chart
双曲线导航制 hyperbolic(al) navigation system
双曲线倒数 hyperbolic(al) inverse
双曲线的 hyperbolic(al)
双曲线电子导航系统 hyperbolic(al) electronic navigation system
双曲线定律 hyperbolic(al) law
双曲线定位 hyperbolic(al) fix; hyperbolic(al) position; hyperbolic(al) position finding
双曲线定位系统 differential distance system; hyperbolic(al) position fixing system; hyperbolic(al) positioning system
双曲线定位线 hyperbolic(al) position line
双曲线定位仪 hyperbolic(al) position finder
双曲线定位仪用显示管 Loran indicator
双曲线定位制 differential distance system
双曲线对数 hyperbolic(al) logarithm
双曲线多瓦浦测位测速器 hyperdop
双曲线法 hyperbolic(al) method
双曲线反光镜 elliptic(al) glass reflector
双曲线方程 hyperbolic(al) equation
双曲线幅度 hyperbolic(al) amplitude
双曲线覆盖范围 hyperbolic(al) area coverage
双曲线拱 hyperbolic(al) arch
双曲线拱桥 curved cross-section arch bridge
双曲线规 hyperbolograph
双曲线轨道 hyperbolic(al) orbit; hyperbolic(al) trajectory
双曲线轨迹 hyperbolic(al) locus; hyperbolic(al) trace
双曲线函数【数】 hyperbolic(al) function
双曲线函数波 hyperbolic(al) wave
双曲线号筒 hyperbolic(al) horn
双曲线级数 hyperbolic(al) series
双曲线(家具)腿 ogee bracket foot
双曲线矫直辊 hyperboloid straightening roll
双曲线距离 hyperbolic(al) distance
双曲线空间 hyperbolic(al) space
双曲线拉平 hyperbolic(al) flare out
双曲线喇叭 hyperbolic(al) horn
双曲线喇叭天线 hyperbolic(al) horn; hyperbolic(al) horn antenna
双曲线雷达系统 hyperbolic(al) radar system
双曲线冷却塔 hyperbolic(al) cooling tower
双曲线流动方程 hyperbolic(al) flow equation
双曲线脉冲雷达系统 hyperbolic(al) pulsed radar system
双曲线模型 hyperbolic(al) model
双曲线抛物面 four gable form; hyperbolic(al) paraboloid
双曲线抛物面圆锥体 hyperbolic(al) paraboloid conoid
双曲线偏光图 bicurve polarization figure
双曲线切线 hyperbolic(al) tangent
双曲线区 hyperbolic(al) zone
双曲线三角学 hyperbolic(al) trigonometry
双曲线扫描发生器 hyperbolic(al) sweep generator
双曲线栅格 hyperbolic(al) grill
双曲线式喇叭 hyperbolic(al) horn
双曲线式螺杆泵 hyperbolic(al) screw pump
双曲线衰减定律 hyperbolic(al) decay law
双曲线速度 hyperbolic(al) velocity
双曲线塔 hyperbolic(al) tower
双曲线弹性插孔 hyperbolic(al) contact
双曲线特性 hyperbolic(al) characteristic
双曲线体 hyperboloid
双曲线天线 hyperbolic(al) antenna
双曲线铁炮 hyperbolic(al) cone
双曲线图 hyperbolic(al) chart
双曲线网格 hyperbolic(al) pattern lattice; hyperbolic(al) system
双曲线网格海图 hyperbolic(al) lattice chart
双曲线微分方程 hyperbolic(al) differential equation
双曲线位置线 hyperbolic(al) line of position
双曲线无线电船位 hyperbolic(al) fix
双曲线无线电导航 hyperbolic(al) radio navigation
双曲线无线电导航设备 hyperbolic(al) radio navigation aids
双曲线无线电导航系统 hyperbolic(al) radio navigation system
双曲线系统 hyperbolic(al) system
双曲线卸料口 hyperbolic(al) outlet
双曲线形磁轴 hyperbolic(al) magnetic axis
双曲线形喇叭筒 hyperbolic(al) horn
双曲线形室 hyperbolic(al) chamber
双曲线形梯度 hyperbolic(al) grading
双曲线型 hyperbolic-type
双曲线型偏微分方程式 hyperbolic(al) partial differential equation
双曲线应力分布 hyperbolic(al) stress distribution
双曲线优 hyperboloid
双曲线运动 hyperbolic(al) motion
双曲线助航 hyperbolic(al) navigational aids
双曲线助航法 hyperbolic(al) navigation aids
双曲线柱面 hyperbolic(al) cylinder
双曲线族 hyperbolic(al) pattern
双曲线坐标 hydrostatic coordinates
双曲形线脚 talon mo(u)lding
双曲型二次曲面 hyperbolic(al) quadratic surface
双曲型方程 hyperbolic(al) equation
双曲型非欧几里得几何学 hyperbolic(al) non-Euclidean geometry
双曲型黎曼曲面 hyperbolic(al) Riemann surface
双曲型衰减 hyperbolic(al) decline
双曲型算子 hyperbolic(al) operator
双曲型微分方程 hyperbolic(al) differential equation
双曲型域 hyperbolic(al) domain
双曲性的圆束 hyperbolic(al) pencil of circles
双曲性线汇 hyperbolic(al) congruence
双曲性直射 hyperbolic(al) collineation; hyperbolic(al) projectivity
双曲性直线 hyperbolic(al) lines
双曲悬链线 hydrostatic catenary; hyperbolic(al) catenary
双曲叶片 double bent blade
双曲余割 hyperbolic(al) cosecant
双曲余切 hyperbolic(al) cotangent
双曲余切函数 hyperbolic(al) cotangent function
双曲余弦 hyperbolic(al) cosine
双曲余弦函数 hyperbolic(al) cosine function
双曲元 hyperbolic(al) element
双曲圆顶薄壳 double curved cupola shell
双曲正割 hyperbolic(al) secant
双曲正切 hyperbolic(al) tangent
双曲正弦 hyperbolic(al) sine
双曲正弦函数 hyperbolic(al) sine function
双曲轴发动机 twin crankshaft engine
双曲轴驱动筛 screen with twin crank-shaft drive
双曲轴压力机 double crank press; duplex crank press
双曲柱面 hyperbolic(al) cylindrical surface; hyperbolic(al) cylinder
双曲柱面坐标 hyperbolic(al) cylindrical coordinates
双曲坐标 hyperbolic(al) coordinates
双曲坐标系 hyperbolic(al) coordinate system
双驱动(器) dual drive; twin drive
双驱轮 dual drive wheel
双渠 dual channel; twin-channel
双渠道调节器 dual channel regulator
双渠道推销 dual channels of distribution
双取效应 binaural effect
双圈变阻器 dual rheostat
双圈测角器 two-circle goniometer
双圈反射测角仪 two-circle reflecting goniometer
双圈环管 double loop
双圈式晶体界面角测量计 two-cycle goniometer
双圈锁 double turn lock
双圈转台 revolving stage with disc and outer ring
双全息图 dual hologram
双全息图法 two-hologram method
双缺口破断试验 nick-break test
双缺圆折流板 center-to-side baffles
双裙瓷绝缘子 double petticoat porcelain insulator
双裙隔电子 double petticoat insulator; double shed insulator
双裙绝缘子 double petticoat insulator
双裙式瓷绝缘子 double shed porcelain insulator
双群常数 two-group constant
双群处理 two-group treatment
双群扩散理论 two-group diffusion theory
双群理论 two-group theory
双群临界质量 two-group critical mass
双群模型 two-group model
双群箱 twin hive
双燃弧 double arcing
双燃机 dual-fuel engine
双燃料柴油机 dual-fuel diesel

双燃料发动机 double fuel engine
双燃料系统 bifuel system; dual-fuel system
双燃料窑 dual-fuel fired furnace
双燃烧管锅炉 double combustion chamber boiler; double fired boiler; double flue boiler; double furnace boiler
双燃烧和除氮炉法 dual combustion and denitration furnace process
双燃烧器 <可用两种燃料的> dual burner
双燃烧室 double combustion chamber
双燃式发动机 dual combustion engine
双燃循环 dual combustion cycle
双燃用户 dual-fuel customer
双绕变压器 bifilar transformer
双绕的 double wound
双绕地段 detouring section of both line
双绕扼流圈 bifilar choke
双绕无感线圈 bifilar coil
双绕线圈 double wound coil
双绕组变压器 two-circuit transformer; two-winding transformer
双绕组串励伺服电动机 split series servomotor
双绕组电枢 double winding armature
双绕组继电器 double winding relay
双绕组交流发电机 double winding alternator
双绕组伺服电动机 split series servo-mechanism
双绕组同步发电机 double winding synchronous generator
双人舱 two-berth room
双人操纵搅拌站 two man plant
双人操作锯 two-handed saw
双人乘员组 two-man crew
双人乘员组位置 two-man crew station
双人床 double bed
双人床房 double room
双人床客房 double bed(ded) room; double bed guest room
双人大锯 pit saw
双人贷款 piggyback tandem mortgage
双人房间 <有两张单人床的> twin(-bedded) room
双人钢架掩蔽部 two-man steel shelter
双人工作 two-handed work
双人公寓 two-person flat
双人光学瞄准系统 two-men optic(al) system
双人横切锯 whip saw
双人教练飞行 dual instruction flight
双人锯 double handed saw; two-man cross-cut saw; two-man saw
双人锯木 pit sawing
双人(客)房 double guest room; two-bed guest room
双人散兵坑 two-man foxhole
双人沙发 love seat
双人手工打孔 double hand drilling
双人手工打眼 double hand drilling
双人双桨赛艇 racing skiff
双人双桨小艇 funny
双人特写镜头 two-shot
双人卧室 two-bed room; two-bed sleeping room
双人椅子 tete-a-tete
双人用成套咖啡具 cabaret
双人用竖拉大锯 pit saw
双人制 two-man rule
双人住舱 twin berth
双人字纹 double herringbone
双人字屋顶 double gable roof
双人自行车 tandem bicycle
双人座车 two-seater
双人座高速度汽车 speedster
双人座四轮敞篷马车 victoria
双人座椅 double chair; dual seat; love seat; twin car seat; two-bench-type seat
双刃带锯 double cutting bands; double cutting bandsaw
双刃刀具 double point tool
双刃刀锯 double edge knife saw
双刃的 double edged
双刃端面铣刀 two-lip end mill
双刃斧 brick ax(e); double(-bitted) axe
双刃刮路机 two-blade drag
双刃铰刀 spade reamer
双刃锯 reversible saw
双刃开槽锯 double edge grooving saw
双刃刨 double iron plane
双刃凿 double chisel
双刃钻头 double chisel; two-bladed bit; two-point bit; double cutting drill
双日 even-numbered days
双日潮 double day tide
双容罐 double containment tank
双溶剂过程 two-solvent process
双溶剂精炼 double solvent refining
双溶剂润滑油精制过程 duo-sol process
双溶剂梯度 two-solvent gradient
双溶剂提取 duo-sol extraction
双溶质体系 bisolute system
双熔锭 double melted ingot
双熔块配釉 double fritting
双熔块釉 double frit glaze
双肉豆蔻醚 dimyristyl ether
双(入口大)门 dual portal
双入口卡片 double entry card
双塞棒钢包 twin stopped ladle
双塞点火 dual ignition
双塞灌浆 double packer grouting
双塞孔 double jack; twin jack
双塞绳式中继台 double cord system trunk board
双塞绳制 dicord system
双三层截流井 double sand catching basin
双三叉体 amphitriaene
双三醇 farnesol
双三次 bicubic
双三极管 double triode; double-triode tube; double-triode valve; dual triode; duotriode; duplex triode; pentatron; twin triode
双三角结构 double triangulated system
双三角孔桁架 double triangular truss
双三角形 ditrigon
双三角形桁架 double Warren truss
双三角形接线法 delta-delta connection
双三进 biternary
双三进制码 biternary code
双散射 double scattering
双散射密度 doubly diffuse density
双散射实验 double scattering experiment
双扫描过程 double sweep procedure
双扫气泵 twin scavenge pump
双色 two-tone colo(u)r
双色版 bichromatic plate; duotone
双色版印刷 double tone printing
双色版油墨 bitone ink; double tone(ink); duotone ink; duplex colo(u)r ink
双色比色的 bicolo(u)rimetric
双色比色计 bicolo(u)rimeter
双色玻璃 dichroic glass
双色彩汽车 two-tone vehicle
双色重现 two-colo(u)r reproduction
双色的 bichrome; duotone; duple
双色调的 duotone; two-tone
双色调革 two-tone leather
双色调染色 two-tone dyeing
双色调涂饰 two-toning
双色调涂饰革 two-tone finished leather
双色调涂饰剂 two-tone finish
双色调网点印版 duotype
双色调油墨 double tone(ink)
双色调制盘 two-colo(u)r reticule
双色分束镜 beam splitting dichroic mirror
双色分析仪 bichromatic analyser[analyzer]
双色复制法 two-colo(u)r process
双色高温计 two-colo(u)r pyrometer
双色跟踪器 two-colo(u)r tracker
双色跟踪装置 two-colo(u)r tracker
双色管 two-colo(u)r tube
双色光波长 dual optic(al) wavelength
双色过程 two-colo(u)r process
双色航摄胶片 bicolo(u)r aerial film
双色激光测距仪 two-colo(u)r EDM [electronic distance measurement] instrument; two-colo(u)r electronic distance measuring instrument; two-colo(u)r laser distance finder; two-colo(u)r laser distancer; two-colo(u)r laser range finder
双色激光测距仪 <英国制造> Georan
双色胶片 dichromated gelatin film
双色结子花线 two-colo(u)r knop yarn
双色(喷)漆 two-tone finish
双色染料 dichromatic dye
双色软线 dichromatic cable
双色散 double dispersion
双色散光谱 doubly dispersed spectrum
双色摄影术 two-colo(u)r photography
双色水下光度计 dual-filter hydrophotometer
双色探测器阵列 two-colo(u)r detector array
双色套印法 duotone
双色条纹 two-tone stripe
双色图 bicolo(u)rable graph
双色涂层瓷砖 graffito tile
双色网线板 duograph
双色温度计 two-colo(u)r pyrometer
双色无线电定位 two-colo(u)r radiolocation
双色系统 bicolo(u)r system
双色显像管 two-colo(u)r tube
双色像 two-colo(u)r image
双色油漆 two-tone paint
双色再现 two-colo(u)r reproduction
双色照片 duotone
双色指示剂 two-colo(u)r indicator
双色制版法 two-colo(u)r process
双色注塑 double shot mo(u)lding
双色着色器 dual-purpose tinter
双沙嘴 double spits
双纱包线 double cotton-covered wire
双刹车控制 dual brake-control
双山墙屋顶 double gable roof; M-roof; trough roof
双山头屋顶 double gable roof; M-roof; trough roof
双山形梁 trough beam
双栅 double grid
双栅场效应晶体管 dual-gate field-effect transistor
双栅的 two-grid
双栅管 double-grid tube; dual-grid tube; space charge tetrode
双栅极 bigrid
双栅整流器 twin-grid rectifier
双闪黄光灯 <上下二黄光灯交替闪光,用以保护学童安全横穿街道> twin flashing yellow lights
双扇窗 casement window; combination window; coupled window; folding casement; two-casement window; two-sashed window
双扇窗户 double sash window
双扇对开门 double-hinged door
双扇隔板 two-piece bulkheads
双扇弧形门 double clam gate
双扇滑门 biparting door
双扇滑门安装 double sliding door installation
双扇货物隔板 two piece load divider
双扇拉门 double plug door; double sliding door
双扇门 biparting door; combination door; double door; double-leaf door; double wing door; folding door; leaf door; twin door; twin gates
双扇门零件 double door hardware
双扇门零配件 double door fittings
双扇门设备 double door furniture
双扇门中先开启的门扇 opening door; active leaf
双扇双向门 double-action double door
双扇弹簧门 double-swinging door
双扇弹簧门缝嵌条 split astragal
双扇推拉门 biparting sliding door
双扇小圆窗 winged bull
双扇形掏槽 double fan cut
双扇形天线 di-fan antenna
双扇形闸门 double-leaf hook-shaped gate; double-leaf hook-type gate
双扇摇门五金件 swinging-pair hardware
双扇移门 double sliding door
双扇折门 double jackknife door
双上下圆锯制材厂 double circular mill; double circular sawmill
双勺喂料机 double scoop feeder
双舌结合 double tongued joint
双舌锁 two-bolt lock
双蛇杖标志 caducei symbol
双设计点飞机 two-design point aircraft
双射 bijection
双射电源 double radio source
双射极开关 dual-emitter switch
双射流 double jet
双射流干粉炮 dual-stream dry chemical nozzle
双射频溅射 double radio frequency sputtering
双射束模型 twin beam model
双射线 dual beam
双射线示波器 double beam oscilloscope; double oscilloscope; double trace scope; dual-beam oscilloscope; dual channel oscilloscope; dual-trace oscilloscope
双射映射 bijective mapping
双身半潜式起重船 twin-hull semi-submersible derrick barge
双身半潜式钻井船 twin-hull semi-submersible derrick barge
双砷硫铅矿 rathite
双升程凸轮 double lift cam
双升船机 twin lift
双升萜烷 bishomohopane
双升降船机 twin lifts
双生斑 twin spot
双生暴雨 twin storms
双生冰川 transection glacier

双生波 twin wave
双生侧面河 twin-lateral river; twin laterals; twin-lateral stream
双生产线 dual-product line
双生的 geminate; twined
双生河流 twin river
双生洪水 twin floods
双生叶 binate leaves; twin leaves
双生支流 twin laterals
双声道 double track; dual track
双声道/4 声道立体声接收机 stereo/4-channel receiver
双声道磁带录音机 double track tape recorder
双声道立体声 stereophony
双声低音雾号 two-tone diaphone
双声轨 <录音磁带> double track
双声迹的 dual track
双声迹录音 sound-on-sound recording
双声迹录音机 dual-track recorder
双声源 double sound source
双声子激发 two-phonon excitation
双绳吊桶 two-line grab (bucket)
双绳多瓣式抓岩机 two-line clamshell
双绳卷筒 twin rope drum
双绳提升 two-line string up
双绳抓斗 double rope grab; twin rope grab; two-line grab (bucket)
双绳抓斗绞车 two-rope grab hoist
双绳抓岩机 two-line clamshell
双十字符 diesis
双十字头系缆柱 double cross head bollard
双十字线 paired-line cross
双十字形系缆柱 double ridding bollard
双石英棱镜光谱辐射计 double quartz prism spectroradiometer
双石英片 biquartz
双时间仪 doubling time meter
双时效 double aging
双时元 chronon
双实线 solid double line
双实心轮胎 twin solid tyres
双实型【计】double precision
双矢的 bivectorial
双矢旋量 motor
双矢坐标 bivectorial coordinates
双示踪技术 double tracer technique
双式混合机 duplex mixer
双式混料机 duplex mixer
双式控制 dual mode control
双式跳汰机 duplex jig
双式振荡器 double oscillator
双视 double vision
双视像管 bivicon
双试验段风洞 duplex wind tunnel
双室拌和筒 two-compartment mixing drum
双室差动式调压室 gallery simple differential surge chamber
双室池窑 two-compartment tank
双室船闸 twin lock; twin navigation lock; two-chamber lock; double lock
双室床【给】double chamber bed
双室电解槽 two-compartment cell
双室焚化炉 dual chamber incinerator
双室风缸 double chamber reservoir
双室化粪池 double compartment septic tank
双室加热炉 double deck furnace
双室静电除尘器 twin chamber electrostatic precipitator
双室立舱 two-compartment silo
双室料箱 two-compartment bin
双室炉(膛) double chamber furnace
双室平衡罐 compound surge tank
双室气化器 two-barrel carburetor
双室气门 biforous spiracle
双室式捡拾压捆机 dual chamber baler
双室式磨机 two-compartment mill
双室式隧道窑 double chamber tunnel kiln
双室式调压井 double chamber surge tank
双室式调压室 double chamber surge shaft
双室式制动器 two-chamber brake
双室调压池 compound surge tank
双室跳汰机 duplex jig
双室箱形截面 double box section
双室闸门 tow-navigation lock; twin lock; two-chamber lock
双室真空炉 two-chamber vacuum furnace
双室贮气筒 two-chamber air reservoir
双手把浴盆 twingrip bath (tub)
双手操纵 two-hand control
双手操作法 conjoined manipulation
双手锤 cross-peen sledge
双手大锤 double jack
双手检查 bimanual examination
双手控制器 two-handed control
双手轮打的大锤 sledgehammer
双手使用的大锤 sledgehammer
双手作业钻进 double hand drilling
双枢杆 double-hinged arm; double-jointed arm
双枢框架 two-pin frame
双枢直流电动机 two-armature direct current motor
双枢轴拱 two-pinned arch
双枢轴闸门 double pivoted gate
双梳式接缝 double steel comb joint
双输出五极管 double output pentode
双输泥箱 duplicate sluice box
双输入混频器 double input mixer
双输入加法器 two-input adder
双输入减法器 two-input subtracter
双输入开关 two-input switch
双输入控制器 dual input controller
双输入门 two-input gate
双输入伺服系统 two-input servo
双鼠笼感应电动机 double squirrel-cage inductor motor
双鼠笼绕组 double squirrel-cage winding
双鼠笼式电动机 Boucherot (squirrel-cage) motor; double squirrel-cage motor
双鼠笼转子 double squirrel-cage rotor; two-cage rotor
双束 two-beam
双束存储环 two-beam storage ring
双束电子感应加速器 dual-beam betatron
双束分光光度计 double beam spectrophotometer
双束分光计 double beam spectrometer
双束干涉显微镜 two-beam interference microscope
双束观测 dual-beam observation
双束管 two-beam tube
双束光学系统 double beam optic (al) system
双束加速器 two-beam accelerator
双束脉冲加速器 two-beam pulsed accelerator
双束全息术 two-beam holography
双束射线管 double beam tube
双束射线示波器 double beam oscillograph
双束示波器 double beam oscilloscope; dual channel oscilloscope; twin beam oscilloscope
双束双聚焦质谱计 double beam double focusing mass spectrometer
双束双聚焦质谱仪 double beam double focusing mass spectrometer
双束衍射 two-beam diffraction
双束阴极射线管 double beam cathode-ray tube
双束制 dual-beam system
双竖杆桁架 queen post; queen-post truss
双竖沟环割 double frill girdle
双竖索面 two-vertical cable planes
双竖索面斜拉桥 double-vertical plane cable-stayed bridge
双数 dual; even
双数的 even number
双漱洗座 double wash (ing) stand
双刷刮水器 tandem wiper
双栓塞(封闭)灌浆装置 tube-a-manchette
双栓塞灌浆法 double packer grouting
双栓锁 two-bolt lock
双双射电源 double double radio source
双水泵 double pump; duplex pump; twin pumps
双水分子 dihydrol
双水化(镁)石灰 <石灰中氧化镁及氧化钙全部水化> dihydrate dolomitic lime
双水级船门 double lift
双水级(船闸) double lift
双水内冷 double water internal cooling
双水内冷发电机 double water internal cooling generator
双水内冷汽轮发电机 turbogenerator with inner water-cooled stator and rotor
双水位泵 duplex-headed pump
双水相萃取 two-aqueous phase extraction
双水源给水方案 dual water scheme
双水源给水计划 <地面水地下水联合使用的> dual water scheme
双水源供水 two-source supply
双水源系统 two-source system
双水准仪法 double level method
双顺序作用阀 dual sequence valve
双顺砖 double stretcher
双丝 double thread; double wire
双丝包的 double silk covered
双丝并列埋弧焊 twin-wire parallel power submerged arc welding
双丝测微计 bifilar micrometer
双丝串联埋弧焊 twin-wire series power submerged arc welding
双丝灯 twin-filament lamp
双丝灯泡 double filament lamp; two-filament bulb
双丝法 double thread method
双丝弓 twin-wire arch
双丝照准 pointing by sandwiching
双丝重力仪 bifilar gravimeter
双四边形 ditetragon
双四极管 duo-tetrode
双四面体 ditetrahedron
双送风管 double blast pipe
双速齿轮变速机 two-speed gearbox
双速齿轮变速装置 two-speed gear
双速齿轮逆转装置 two-speed reversing gear
双速传动 two-speed drive
双速的 two-speed
双速电动机 two-speed motor
双速功率输出轴 twin-speed power take-off
双速供料 two-speed feed
双速鼓风机 two-speed blower
双速后桥 dual performance rear axle
双速滑环式电动机 two-speed slipring motor
双速减速 dual-ratio reduction
双速减速机 two-speed gearbox
双速接头杆 two-speed adapter lever
双速控制器 two-speed controller
双速离合器 two-speed clutch
双速水泵水轮机 two-speed pump-turbine
双速天轴 two-speed overhead countershaft
双速主减速齿轮 two-speed final gear
双随机矩阵 doubly stochastic matrix
双隧道 twin tunnels
双榫 divided tenon; twin tenons
双榫接 bridle joint
双榫接合 double tenon joint
双榫舌 double tenon; twin tenons
双榫舌检修孔盖板 double-seal manhole cover
双榫头 double tenon
双梭口提花机 double shed jacquard
双缩脲反应 biuret reaction
双索单钩吊货系统 burtoning system; union purchase system
双索吊钩吊索系统 married couple; married fall
双索吊梁 connecting traverse
双索架空索道 bi-cable ropeway
双索卷扬机 double cable winch; double rope winch
双索空中缆车 double-rope tramway
双索链钩 two-leg sling
双索面 double cable plane; double-plane cable
双索面竖琴形斜拉桥 double plane harp cable stayed bridge
双索面斜拉桥 double plane cable-stayed bridge
双索索道 <一来一往> jig back; bi-cable ropeway
双索悬索桥 double cable suspension bridge
双索抓斗 double chain grab; double rope grab
双索抓斗起重机 double chain grab crane; double rope grab crane
双索转载抓斗 double rope rehandling grab
双锁边 double seaming
双锁盖条 double-lock welt
双锁卷边接合 double-lock seam
双锁口 H 桩 double-interlocking H-pile
双锁装置 double-locking device
双塔 twin towers
双塔板铆钉对接 joint with double strap
双塔的 double towered; twin-towered
双塔的正面 dual tower facade
双塔建筑物正面 two-tower building facade
双塔空气分离系统 double column air separation system
双塔立面 dual tower facade
双塔连续接触床 dual-bed continuous contactor
双塔轮式前端 double nose
双塔门楼 twin-towered gatehouse
双塔式城楼 double-towered gatehouse
双塔式大门 double-towered gatehouse
双塔式建筑 paired towers; twin towers
双塔式建筑物正面 twin-towered facade
双塔式立面(建筑) double-towered facade
双塔式悬浮预热器 dual tower suspension preheater
双塔式闸室 double-towered gatehouse
双塔寺 <中国泉州开元寺> Two-Pagoda

双塔斜拉桥 two pylon cable stayed bridge
双塔正面 double-towered facade
双踏板操纵 two-pedal operation
双踏板控制 two-pedal control
双踏步木接点 double steps
双胎车轮 dual tire[tyre] wheel
双胎后轮 dual rear wheel
双台成型机 two-table machine
双台电梯 dual elevator
双台分粒器 two-deck classifier
双态 bifurcation;two-state
双态变量 binary-state variable
双态变数 binary-state variable
双态操作 dual operation
双态磁芯 bimag
双态过程 two-state process
双态控制 two-state control
双态逻辑元件 binary logic element
双态码 two-condition code
双态通信[讯] binary signal(ling)
双态陀螺罗经 double state compass
双态系统 two-state system
双态元件 binary element;toggle
双态装置 two-state device
双探头扫描机 twin-detector scanner
双探头探伤仪 two-transducer flaw detector
双探针系统 double probe system
双探针雪密度计 twin-probe snow-density ga(u)ge
双探针装置 two-probe arrangement
双碳极电弧 twin carbon arc
双碳极弧光灯 double carbon arc lamp;twin carbon arc lamp
双碳精盒传声器 double-button carbon microphone
双碳窑炉 double crown furnace
双搪(瓷)double cover-coat enamel
双糖 disaccharide
双掏槽 double cut
双套阀 double sleeve valve
双套管接头 double cap casing head
双套管式换热器 dual tubes heat exchanger
双套管温度计 enclosed scale thermometer
双套管转轴 double tubular shaft
双套减速齿轮(装置)double reduction gear
双套结 bowline on a bight; double bowline; Portuguese bowline; two-bowline
双套系统 duplexed system
双套线 double detour
双套信号钟装置【铁】twin chime unit
双套针管 double sheathed needle
双套装置 duplexed system; duplicating device
双套钻 two-pocket bore
双特征射线 bicharacteristic ray
双特征性 bicharacteristics
双梯段剥离 two-bench stripping
双梯段楼梯 double flight stair(case)
双梯厢组 two-car bank
双提升吊钩 double lifting hooks
双蹄片式 double shoes
双蹄式制动 two-shoe brake
双蹄式制动器 double shoe brake
双蹄外缩式制动器 two-shoe external contracting brake
双体 binary
双体半潜式起重船 twin-hull semi-submersible derrick barge
双体半潜式钻井船 twin-hull semi-submersible derrick barge
双体半潜式钻探船 twin-hull semi-submersible derrick barge
双体驳船 catamaran barge
双体船 catamaran;double canoe;twin hull boat;twin hull craft;twin hull ship;twin-hull vessel
双体趸船 twin pontoon
双体筏 catamaran
双体浮座护舷 catamaran fender
双体海船 sea-going catamaran
双体混凝土输送泵 double concrete pump
双体平台 catamaran platform; twin-hull unit
双体起吊重件驳船 catamaran heavy-lift barge
双体气垫船 captured air bubble craft
双体推轮 catamaran towboat; catamaran tug
双体拖网(渔)船 catamaran trawler; twin-hull trawler
双体挖泥船 twin-hulled dredge(r); catamaran dredge(r)
双体卫星 two-body satellite
双体载驳船 straddler
双体载驳货船 barge aboard catamaran;barge-carrying catamaran
双体载驳货船子驳 Bacat barge;barge aboard catamaran barge
双体钻探船 catamaran driller; catamaran drilling vessel
双天线同步卫星 two-aerial synchronized satellite
双天线消噪声接收机 barrage receiver
双填充子 di-interstitials
双填隙(子)di-interstitials
双条单向道 one-way pair
双条件门 biconditional gate
双条圈条器 twin coiler
双条杉天牛 <拉> Semanotus bifasciatus
双条双轭法 double bar and yoke method
双条纹 double fringe
双调幅 double-amplitude modulation
双调归并 bitonic merge
双调归并网络 bitonic merging network
双调和的 biharmonic
双调和方程 biharmonic equation
双调和函数 biharmonic function
双调和算子 biharmonic operator
双调节 double regulation
双调节阀 double damper valve;double regulating valve
双调节角阀 double regulating angle valve
双调节片可调喷管 clamshell nozzle
双调节器 double governor
双调节悬挂系 double adjustment suspension
双调谐 double response;double tuning
双调谐的 double-tuned
双调谐电路 double-tuned circuit
双调谐电路接收 two-circuit reception
双调谐电路接收机 double circuit receiver;two-circuit receiver
双调谐电路耦合 double-tuned coupling
双调谐放大器 double-tuned amplifier
双调谐检波器 double-tuned detector
双调谐天线 multiple tuned antenna
双调序列 bitonic sequence
双调整卡普兰式水轮机 double regulated Kaplan turbine
双调制 dual modulation
双跳板 twin ramp
双跳电路 double-hoop circuit
双贴层压机 two-ply laminating machine
双贴角搭焊 two-fillet lap weld
双贴面绝热制品 double-faced insulation
双萜 diterpene
双萜类 diterpenoids
双铁线式铠装电缆 double wire-armo(u)red cable
双铁芯调压变压器 two-core voltage regulating transformer
双停歇期运动 two-dwell motion
双通 bipass
双通传感器 bilateral transducer
双通道 twin-channel; two-channel; two-path
双通道伴音电路 dual channel sound
双通道伴音方式 dual channel sound
双通道存取【计】dual port access
双通道单工 double channel simplex
双通道单脉冲雷达 two-channel monopulse radar
双通道的 twin-channel;two-pass;two-channel
双通道放大器 dual channel amplifier
双通道辐射计 two-channel radiometer
双通道接收机 two-channel receiver
双通道控制器 dual channel controller;dual port controller
双通道热红外扫描仪 dual channel thermal infrared scanner
双通道双工 double channel duplex
双通道系统 dual channel system
双通道旋转变压器 two-channel resolver;two-speed resolver
双通道噪声系统 two-channel noise figure
双通道自动驾驶仪 two-axis autopilot
双通道自整角机 two-channel selsyn; two-speed synchro
双通的 bilateral
双通电路元件 bilateral circuit element
双通阀 two-way valve
双通风管系统 double-stack system
双通供料溜槽 two-way feeding chute
双通活门 two-way valve
双通接头 duplex fitting
双通开关 bilateral switching; two-throw switch;two-way cock
双通路 binary channels;diplex
双通路活性区 two-flow core; two-path core
双通路开关 dual duct switch
双通路卧式蓄热室 two-pass horizontal regenerator
双通面 bilateral surface
双同步通信[讯] binary synchronous communication
双同步协议 bisync protocol
双同轴电缆 twin axial cable
双桶存储器 double bucket
双筒 double cylinder
双筒拌和机 twin drum mixer
双筒泵 duplex pump
双筒比色法 bicolo(u)rimetric method;bicolo(u)rimetry
双筒比色计 bicolo(u)rimeter
双筒步行式压路机 double roll pedestrian-controlled roller
双筒赤道仪 double equatorial
双筒的 binocular; double barrel(l)ed;double-drum;dual-drum
双筒电缆控制单元 double-drum cable control unit
双筒电子透镜 two-cylinder electron lens
双筒放大镜 binocular loupe;binocular magnifier
双筒分离式卷扬机 capstan winch
双筒观测镜 binocular viewer
双筒管 twin barrel tube
双筒锅炉 double cylindrical boiler
双筒回转烘干机 double shell rotary drier
双筒回转窑 double cylinder type rotary kiln
双筒混合机 twin-cylinder mixer; twin-shell blender;twin-shell mixer
双筒混凝土摊铺机 twin-drum type concrete paver
双筒绞车 double-drum hoist;double-drum rope winch; twin drum winch;two-drum winch
双筒井 dual well
双筒镜 binocular body; glasses; prismatic(al)binoculars
双筒卷扬机 double-drum hoist;double-drum rope winch; twin drum winch; two-drum crab; two-drum winch
双筒棱镜望远镜 binocular prism telescope
双筒立体摄影机 binocular stereocamera
双筒立体显微镜 orthostereoscope
双筒联生 twin cylinders
双筒联体 twin cylinders
双筒密集钻井 dual bore cluster drilling
双筒瞄准具 binocular sight
双筒磨耗试验机 Deval(abrasion testing)machine;Dever abrasion tester
双筒泥炮 double barrel-mud gun
双筒起重机 double-drum hoist
双筒潜望镜 binocular-type periscope
双筒枪 double barrel
双筒驱动 twin drum drive
双筒式柴油卷扬机 double-drum diesel hoist;double-drum diesel winch
双筒式锅炉管 field tube
双筒式灰浆搅拌机 double-drum type mortar mixer;twin mortar mixer
双筒式磨耗试验 Deval abrasion test
双筒手持水准仪 binocular hand level
双筒手动泵 double barrel hand pump
双筒手提水准仪 binocular hand level
双筒太阳单色光照相仪 double spectroheliograph
双筒提升机构 two-drum crab
双筒天体摄影仪 double astrograph
双筒天体照相仪 double astrograph
双筒透镜 two-cylindrical lens
双筒望远镜 binocular prism telescope; binoculars; binocular telescope; field glass(es); prismatic(al)binoculars
双筒望远镜玻璃 bifocals
双筒望远镜观察 binocular vision
双筒望远镜遮片 binocular mat
双筒显微镜 binocular microscope; dual microscope; double microscope
双筒线圈 binocular coil
双筒羊角碾 sheep-horn double barrel roller
双筒咬口瓦 double roll interlocking tile;double Roman tile
双筒仪器 binocular instrument
双筒振动(路)碾 double-drum roller; dual drum vibratory roller;twin-vibratory roller
双筒振动式打桩锤 twin vibratory pile hammer
双筒振动压路机 double-drum roller; dual drum vibratory roller;twin-vibratory roller
双头 bipitch
双头扳手 double ender;double end(ed) spanner; double end wrench; double head wrench;spanner double ends

双头泵 bull pump
双头插销 double male; stacking adapter; stacking fitting
双头串联电弧自焊机 twin-head tandem arc welding machine
双头串列弧焊机 twin-head tandem arc welding machine
双头锤 club hammer
双头导轨 double-headed rail
双头捣锤 duplex tamper
双头的 bifurcate; double-end(ed)
双头的套筒扳手 double-ended box spanner; double-ended box wrench
双头丁坝 double head spur dike [dyke]
双头钉 double-headed nail; dual head nail; duplex-headed nail; duplex nails; scaffold nail
双头对击锤 automatic hammer with two opposite working rams
双头鳄口扳手 double end alligator type wrench
双头斧 double-headed axe
双头腹壁牵开器 double-ended abdominal retractors
双头杆螺纹 double stem threads
双头钢轨 double head rail; I-rail
双头公扣接头 pin-to-pin coupling
双头钩针 twin needle
双头管 double heading pipe
双头管扳手 double end alligator type wrench
双头管形灯(泡) double-ended tubular lamp
双头轨 bull-head(ed) rail; double head rail
双头轨座 bull-head(ed) rail chair
双头滚刀 double thread hob
双头锅炉 double-ended boiler
双头环节链 stud link chain
双头环形扳钳 double-ended ring spanner
双头环形扳手 double-ended ring wrench
双头活塞杆 through rod
双头机车 double-ended locomotive
双头加压法 <用两个活动压头从相对方向同时压缩试件成型> double-plunger compaction procedure
双头尖钉 double-pointed nail
双头绞盘 double capstan
双头接点 bifurcated contact; split contact; twin contacts
双头进料螺杆 double flight feed crew
双头浸入式电弧焊机 twin-head submerged arc welding machine
双头卡规 double end snap ga(u)ge
双头开关 two-way switch
双头开关器 two-knob lock
双头开口扳手 double open end wrench
双头垄断 <即市场某一商品由两家卖主垄断> duopoly
双头垄断市场 duopsony
双头鹿角形 twin-head type sloping lobe
双头螺杆 double thread screw
双头螺栓 double-end(ed) bolt; double-pointed nail; double screwed bolt; double-start screw; handrail bolt; pin screw; stud; stud bolt; stud pin; stud screw; threaded stud; through bolt
双头螺栓连接 studded connection
双头螺栓螺纹 stud bolt thread
双头螺栓拧出器 stud extractor; stud remover
双头螺栓拧入器 stud setter
双头螺栓柱螺栓 bolt stud; screw stud; stud bolt
双头螺丝 double threaded screw
双头螺丝钉 dowel screw
双头螺纹 double-start thread; double thread; double thread screw
双头螺纹螺杆 two-thread worm
双头螺旋 twin-feed spiral
双头螺旋加料器 double flight feed crew
双头螺旋螺旋泵 double helix screw pump
双头螺旋钻机 twin-head auger
双头螺钻机 twin-head auger
双头幕式涂布 double header curtain coating
双头内螺纹接合器 double female adaptor
双头配件 double male
双头喷灯 twin torch
双头喷枪 double head spray gun
双头喷嘴 double swivel nozzle
双头平刮刀 double end flat scraper; double head flat scraper
双头牵开器 double-ended retractors
双头切口扳手 double key
双头球磨机 double-ended ball mill
双头燃烧器 duplex burner
双头摄影机 double-headed camera
双头式(大圆头) bull head
双头式堆装件 double male stacking fitting
双头式钢轨 bull-head(ed) rail
双头式夹板 tapered step joint bar
双头式列车 double header
双头水平螺旋钻机 twin-head horizontal rotary drill(ing machine)
双头水枪 double revolving branch
双头死扳手 double-ended wrench
双头缩小三通 tee reducing on both runs
双头探针 double-ended probes
双头镗床 double-ended boring machine
双头调压器 double top air compressor governor
双头通电磁化法 prod magnetizing method
双头同性接口 double adapter
双头弯脖梅花扳手 double hexagon opening double offset box socket wrench
双头弯管机 double-headed tube bender
双头蜗杆 double thread worm
双头铣床 duplex head milling machine
双头线夹 double-ended clamp
双头小艇 peapod dinghy
双头斜榫 double bevel scarve
双头鸭嘴笔 railroad; road-pen
双头研磨机 double head grinder; dual head grinder; twin-head grinder; two-head grinder
双头圆钉 double-headed nail
双头圆锯 twin-circular saw
双头胀大钢管 both ends swelled steel pipe
双头转体扳手 double-ended swivel wrench
双头自动弧焊机 two-head automatic arc welding machine
双投 double throw
双投触点 double-throw contact
双投断路器 double-throw circuit breaker
双投隔离开关 double-throw disconnecting switch
双投接点 two-way contact
双投开关 double-throw switch; two-way switch
双投药池 twin dosing tanks
双投影绘图仪 double projection plotter; twin projection plotter
双投影立体绘图仪 double projection stereo-plotter
双投影器 dual projector; twinplex projector
双投影器测图仪 double projection plotter; twinplex plotter
双投影器绘图仪 twinplex plotter
双投影仪 twinplex projector
双投影直接观测立体测图仪 double projection direct-viewing stereo-plotter
双透镜 two-lens
双透镜放大镜 double lens magnifier
双透镜目镜 two-lens ocular
双透镜物镜 double objective; lens doublet; two-lens objective; two-lens ocular
双透镜组件 doublet component
双透射法 double transmission method
双透射效应 double transmission
双凸板摄影机 lenticular plate camera
双凸瓣式水泵 two-lobed pump
双凸窗 double projected window
双凸的 biconvex; convexo-convex; double convex
双凸角凸轮 twin-lobe cam
双凸块凸轮 two-lobe cam
双凸轮旋转泵 two-lobed rotary pump
双凸轮旋转活塞 twin-lobe rotary piston
双凸轮轴的链传动装置 twin camshaft chain drive
双凸轮轴式 twin cam shaft type
双凸面的 biconvex; convexo-convex
双凸面透镜 biconvex lens
双凸面圆锯 double conical saw; taper-ground saw
双凸式线脚 double bolection mo(u)lding
双凸榫 double tenons; twin tenons
双凸透镜 biconvex lens; convexo-convex lens; double convex glass; double convex lens
双凸透镜状 lenticular
双凸透镜状胶片 lenticular film
双凸形 convexo-convex
双凸形配件 double male fitting
双凸型 double convex type
双凸缘 double flange
双凸缘排气管 double flange
双凸缘压路机 double flange roller
双突瓣式泵 two-lobed pump
双突堤式码头 double finger pier
双图 digraph
双图像 double image
双图像回波 double image echo
双涂层 double layer of coating
双涂层 X 射线照相片 double coated X-ray film
双涂层屏 double coated screen
双涂层荧光屏 double screen
双推单溜 single rolling on double pushing track
双推电路 push-push circuit
双推放大器 push-push amplifier
双推进器 twin propeller; twin screw
双推进器冲击破碎机 double impeller impact breaker
双推进器船舶 twin-screw steamer
双推进器船尾肋骨 spectacle stern frame
双推进器架 spectacle type shaft bracket
双推进器轴毂 spectacle bossing
双推力 dual thrust
双推力发动机 dual-thrust motor
双推双溜【铁】 double humping and double rolling; double rolling on double pushing track
双推挽 double push-pull
双腿的 bicrural
双腿管制钻塔 two-leg pipe derrick
双腿井架 double pole mast
双腿木钻塔 double rig
双腿起重爪 two-legged lewis
双腿式起重机 bipod; bipod crane; shear-leg crane
双托梁 double trimmer
双拖网渔船 hull trawler; pair trawler; two-boat trawler
双拖网渔轮 hull trawler; pair trawler; two-boat trawler
双脱氧法 dideoxy
双驼峰 double hump; double incline; dual hump; dual humping facility
双驼峰砂 one-screen
双椭圆 bielliptic(al)
双椭圆率 biellipticity
双椭圆腔 double elliptical cavity
双椭圆弹簧 double elliptical spring
双椭圆形水箱 double ellipsoidal tank
双瓦特计法 two-wattmeter method
双瓦闸 double block brake
双瓦制动器 double shoe brake; two-shoe brake
双瓦自动闸 duo-servobrake
双外层水带 double jacket hose
双外差法 double heterodyne system
双外壳斗式提升机 double casing bucket elevator
双外廊布局 double exterior-corridor layout
双外螺丝 barrel nipple; long nipple
双外延法 double epitaxial method
双弯凹凸弧形线脚 ressaut
双弯边对接焊缝 double-flanged butt weld
双弯边对接接头 double-flanged butt joint
双弯薄壳(屋)顶 double bent shell roof
双弯存水弯 running trap
双弯钩栓锁 double hook bolt lock
双弯管 double bend; twin elbow
双弯虹吸管 double-return siphon
双弯接头 double-bend(ing) fitting; twin bend; twin elbow
双弯流型放大器 elbow amplifier
双弯曲 double curved; tangent bend
双弯曲玻璃 double bent glass
双弯曲的 doubly curved
双弯曲的桥梁 double curved bridge
双弯曲嚼子 ogee bit
双弯曲刨刀 ogee plane(iron)
双弯曲刨铁 ogee plane(iron)
双弯曲线 ogee curve
双弯曲线脚 lesbian cyma(tium)
双弯曲线形铲斗 bucket of ogee
双弯曲形的 ogee
双弯曲轴 double-throw crankshaft; two-throw crankshaft
双弯刃剪 crossing curved blades snips
双弯三通接头 twin elbow; twin ell
双弯头 double bend; twin elbow; twin ell
双弯头 T 形接头 double sweep tee
双弯头半椭圆形钢板弹簧 semi-elliptic(al) scroll spring
双弯头管 double elbow pipe
双弯头三通 double sweep tee
双弯头水封 S-trap
双弯头水落管 swan-neck downpipe
双弯线条 ogee mo(u)lding
双弯形屋顶 ogee arch; ogee roof
双弯形溢洪道 ogee spillway
双弯形溢流堰 ogee spillway
双碗式绝缘子 double cup insulator

S

双碗形 double bell
双万向节 double Hooke's joint;double universal joint
双万向节传动 two-joint drive
双万向节传动轴 double-jointed propeller shaft
双万向联轴节 double universal coupler
双腕臂底座 joint clevis for twin tube
双腕臂底座连接架 connector for twin cantilever bracket
双腕臂上底座 upper bracket for double track cantilever
双腕臂下底座 lower bracket for double track cantilever
双网板装备 dual otter board system
双网成型装置 twin former;two-wire former
双网虹吸管 double-return siphon
双网型浸渍机 double screen saturator
双网造纸机 two-wire paper machine
双桅船 two-master
双桅杆起重设备 twin-derrick system
双桅横帆船 brig
双桅平底船 piragua
双桅纵帆船 <加拿大的> jake
双尾撑飞机 twin-boom aircraft
双尾船体 catamaran-stern hull
双尾检验 two-tailed test
双尾气排放管 double exhaust pipe
双尾引送 double pick insertion
双位 dibit
双位变距螺旋桨 two-pitch propeller;two-position propeller
双位不连续温度控制 two-position discontinuous temperature control
双位测定 two-site assay
双位差隙 two-position differential gap
双位差隙控制 two-position differential gap control
双位差隙作用 two-position differential gap action
双位错误 double bit error
双位的 on-off
双位电平控制 high low level control
双位电平调节器 high low level control
双位动作 bang-bang action
双位二进制 diad
双位阀 on-off valve; two-position valve
双位继电器 two-position relay
双位夹具 twin-station fixture
双位进位 double carry
双位卷取机 duplicate recoiler
双位开关 on-off (control) switch; open and shut valve
双位开关恒温器 on-off thermostat
双位开关控制器 on-and-off controller
双位控制 off-on control; on-off control;stop-go control
双位控制动作 two-step action
双位控制器 on-off controller; two-position controller
双位控制伺服机构 on-off control servomechanism
双位控制作用 two-position action; two-step action
双位式控制器 on-and-off controller
双位式信号机【铁】two-position signal
双位调节 high low level control; on-off control; snap action; two-position control
双位调节器 two-position controller
双位外底压合机 two-station sole press
双位温度控制器 two-position temperature controller
双位无回程系统 two-level nonreturn system
双位线 double digit wire
双位信号 two-position signal
双位液面控制 high low level control
双位液面调节器 high low level control
双位有回程系统 two-level return system
双位制 two-position mode; two-position system
双位置 two-position
双位置控制 on-off control; two-position control
双位置控制动作 on-off control action
双位置原理 double site principle
双位置作用 two-position action
双温度恒温器 dual setting thermostat
双温分离 dual temperature separation
双温冷柜 dual temperature refrigerator
双温区炉 two-zone furnace
双温水系统 dual temperature water system
双(温)盐水系统 dual temperature brine system
双纹 double cut;twinned grooves
双纹锉 cross-cut file;double cut file
双纹锉刀 double cut file
双纹螺栓 double threaded screw
双纹螺丝 double threaded screw
双稳触发电路 bistable trigger circuit; stable trigger circuit;toggle
双稳触发器 bistable trigger;flip-flop
双稳存储器 bistable memory
双稳(定)的 bistable
双稳定性 bistability
双稳定状态 two-stable position
双稳多谐振荡器 bistable multivibrator
双稳态 bistable state
双稳态部件 bistable unit
双稳态触发电路 flip-flop circuit
双稳态触发器 flip-and-flop generator;flip-flop multivibrator
双稳态触发(器)翻转时间 flip-flop transition time
双稳态触发器线路 bistable trigger circuit
双稳态存贮显示器 distable storage tube
双稳态电路 lockover circuit; two-state circuit;bistable circuit
双稳态多频振荡器 Eccles-Jordan multivibrator
双稳态多谐振荡电路 flip-flop circuit;flopover circuit
双稳态多谐振荡管 flip-flop tube
双稳态多谐振荡器 bistable multioscillator; flip-and-flop; flip-flop; flip-flop generator; flip-flop multivibrator
双稳态多谐振荡器电路 Eccles-Jordan circuit
双稳态多谐振荡器翻转时间 flip-flop transition time
双稳态多谐振动器 bistable multivibrator
双稳态反射率 bistatic reflectivity
双稳态工作 bistable operation
双稳态光电子元件 bistable optoelectronic element
双稳态光发射体 bistable light emitter
双稳态光学元件 bistable optical element
双稳态光学装置 bistable optical device
双稳态激光装置 laser bistable device
双稳态继电器 bistable relay
双稳态控制 flip-flop control
双稳态逻辑元件 binary logic element
双稳态器件 bistable device
双稳态状 bistable
双稳态性能 bistable behaviour
双稳位置 two-stable position
双稳线路 bistable circuit
双稳型 double stable type
双稳性 bistability
双稳元件 bistable component; bistable element;bistable unit
双涡轮 dual turbine
双涡轮变矩器 double turbine torque converter
双涡轮式增压器 dual turbocharger
双涡轮液力变矩器 double turbine torque; twin-turbine torque converter
双蜗杆 twin worm
双蜗壳 double volute
双蜗壳式泵 twin volute pump
双蜗壳(式)水轮机 twin spiral(water)turbine;double casing turbine
双蜗壳水泵 twin-volute water pump
双握索结 manrope knot
双污泥序批间歇式反应器工艺 two-sludge sequencing batch reactor process
双屋面 double roof(ing)
双五点网 double five-spot
双五角棱镜 double pentagonal prism
双五进制码 biquinary code
双坞室式船坞 dock with intermediate gate
双戊 dicyclopentadiene
双戊烯 bipentene
双物镜摄影机 double camera
双物镜显微镜 double objective microscope
双吸管架挖泥船 double ladder dredge(r)
双吸口 double suction
双吸(口)泵 double suction pump; close-coupled pump; two-throw pump
双吸口叶轮 double suction impeller
双吸离心泵 double suction centrifugal pump;double suction pump
双吸入式叶轮 double suction impeller
双吸式 double suction
双吸式泵 double suction pump
双吸式单级螺旋泵 double suction single stage volute type pump
双吸式单级蜗壳式泵 double suction single stage volute type pump
双吸式离心泵 double centrifugal pump
双吸式立管 double suction riser
双吸式叶轮 double-sided impeller
双吸收 biabsorption
双吸压缩机 double suction compressor
双吸叶轮 double entry impeller
双烯酮 diketen(e)
双稀释剂 double spike
双洗碟盆水盘 twin dishwashing sink bowls
双系列 twin stream
双系列悬浮预热器 suspension preheater with two strings
双系统 dual system
双系统供电 two-system power supply
双系斜杆 double cancellation
双隙缝谐振腔 double gap cavity
双隙速调管 two-gap klystron
双隙头 two-split head
双匣窗梃 double boxed mullion
双狭缝 double aperture slit
双下标变量【数】double subscripted variable
双酰胺蜡 bisamide wax
双弦法 <校正铁道> two-chord method
双弦式方格桁架 Town lattice truss
双衔铁 double armature
双衔铁继电器 double armature relay; two-armature relay
双显示器 dual display
双线 bifilar;double stripe;two-wire
双线T形陷波电路 bifilar T-trap circuit
双线摆 bifilar pendulum
双线半自动闭塞 double line semi-automatic block
双线包缝机 two-thread overlock machine
双线保险丝插塞 two-way fuse plug
双线保险丝插座 two-way fuse socket
双线笔 double line pen
双线臂板信号机【铁】double wire operated semaphore
双线变压器 bifilar transformer
双线标板 paired-line target
双线布线 two-conductor wiring
双线操纵 twin-wire control
双线插入段 double track insert; double track interpolation; short lengths of double track
双线插头 two-pin plug
双线车道 dual carriageway(road)
双线出入线 access line for double track
双线传送装置 double strand conveyer[conveyor]
双线船闸 double lines of locks;double(way) lock;dual lines of lock;duplicate lock; twin lock; two-way lock
双线导向线 double track guideway
双线道道路 double track line
双线道路 double track line; double track;two-track road
双线道路系统 two-way road system
双线的 bifilar;double tracking;double wound;dual line;two-wire
双线等距线夹 feeder clamp; twin feeder wire
双线地物 double feature; double line feature
双线电磁示波器 bifilar electromagnetic oscillograph
双线电缆 paired cable
双线电路 duplicate line;two-wire circuit;two-wire system
双线电路放大器 two-wire amplifier
双线电路制 two-wire circuit system
双线电梯 twin lifts
双线电位计 bifilar potentiometer;dual linear potentiometer
双线吊弦 double line hanger
双线段 double line;dual line
双线多级船闸 twin flight locks;twin multi-stage locks
双线扼流圈 bifilar inductor
双线法 double wire method; duplex-wire method
双线帆布 twisted thread canvas
双线反向行车 train movement against the current of traffic
双线反向运行 running against current of traffic;running-in reverse order
双线方式 two-wire system
双线(飞机)跑道 dual runway
双线分离度 doublet separation
双线符号 double line symbolization
双线改为单线行车 <一条线路进行大修或发生故障时> conversion to single track working; single-line working of double track line
双线改为单线行车的命令 single lining order
双线干道 arterial pair
双线干管 duplicate main

双线干线 double track main line
双线割胶制度 two-cut tapping system
双线公里里程 double track kilometrage
双线公里数 double-tracked kilometrage
双线公路 double driveway
双线光谱 doublet spectrum
双线轨道 double track(guideway)
双线航道 double lane channel;double way channel; dual lane; two-lane channel;two-way channel
双线航道的 double lane
双线化 double tracking; doubling of the track
双线回路 two-wire circuit
双线回路端子 two-wire loop terminal
双线加工作业线 two-strand line
双线架空吊车索道 bi-cable aerial tramway
双线架空索道 bi-cable aerial; twin-cable aerial tramway; twin-cable ropeway
双线间隔 doublet interval
双线检流计 bifilar galvanometer
双线交通 double line traffic;two-way traffic
双线接地 two-line ground
双线接地故障 double ground fault
双线静电计 bifilar electrometer
双线刻刀 double line cutter;double line scriber
双线刻度 <一种度盘刻度方法,为了减少刻度误差,刻成双线> double line graduation
双线刻针 double line point
双线控制 twin-wire control
双线快门开关 double cable release
双线馈线线夹 twin feeder clamp
双线缆 twin wire
双线缆道 bicable tramway;jig back
双线联络线 double line connection
双线临时改为单线行车 temporary single-line working
双线流向指示器 bifilar current indicator
双线路 duplex line;twin line;double line road【道】
双线路液力制动系 two-circuit hydraulic braking system
双线逻辑 double rail logic
双线螺丝 double-start thread screw
双线螺纹 double thread
双线螺旋线 bifilar helix
双线码头 twin-pier
双线耦合空腔谐振器 two-line cavity
双线屏蔽电缆 two-conductor shielded cable
双线铺设 double lay;dual lay
双线期 diplotene
双线起动 two-wire start
双线桥 double line bridge; double-track bridge
双线区段 double line section;double track section
双线区间 double track section
双线区间分段反向行车 reversible road working
双线圈 double coil;twin coil
双线圈变压器 two-winding transformer
双线圈磁感应检测器 two-coil magnetic detector
双线圈灯丝 double coil filament
双线圈电流继电器 two coil current relay
双线圈法 two-coil method
双线圈方法 double coil method
双线圈继电器 double-coil(ed) relay; double wound relay; two-element relay
双线圈仪表 two-coil instrument
双线绕法 bifilar winding
双线绕式电灯 double-coiled lamp
双线绕式转子 double wound rotor
双线绕组 bifilar winding; two-wire winding
双线塞绳 double flex
双线三级船闸 two-way three-step ship lock
双线扫描法 two-trace method
双线栅变频器 grating converter
双线上单向行车 unidirectional working
双线上二闭塞区间三示像的显示 two-block three-aspect indication
双线上三闭塞区段四显示的表示法 <前进、注意、限速、停车> three-block four aspect indication
双线升船机 double lift;double ship lifts;twin lifts
双线示波法 double beam oscillograph
双线示波器 bifilar oscilloscope;dual-beam system; two-beam oscillograph
双线式馈电线 twin feeder
双线式缆道 bi-cable system
双线式轧机 twin-strand mill
双线输送机 double strand drag conveyer[conveyor]
双线水准测量 double line level(l)ing
双线隧道 double line tunnel;double track tunnel;twin-track tunnel
双线隧洞 double line tunnel;double track tunnel
双线索道 bi-cable tramway;jig back
双线锁缝机 union special double lockstitch machine
双线套口机 two-thread linking machine
双线梯级船闸 twinned flight locks
双线天线 two-wire antenna
双线铁道 double-tracked railroad; double-tracked railway; double track line
双线铁道隧道 double-tracked railway tunnel
双线铁路 double-track(ed) railroad; double-track(ed) railway; double track line
双线铁路隧道 double-tracked railway tunnel
双线铁路隧洞 double line railway tunnel
双线同步示波器 dual-beam synchronoscope
双线(同心)圆 double circle
双线图 double line plan
双线图案 paired-line patterns
双线退火和酸洗作业线 two-strand anneal and pickle line
双线无感绕法 bifilar winding
双线无感线卷 bifilar winding;double winding
双线无感线圈 bifilar choke
双线系统 double wire system
双线线路 double(wire) line; two-wire route;double track
双线形式 bilinear form
双线型滞回曲线 bilinear hysteretic curve
双线性 bilinearity
双线性变换 bilinear transformation
双线性变换分析 bilinear transformation analysis
双线性材料 bilinear material
双线性插值 bilinear interpolation
双线性程序设计 bilinear programming
双线性的 bilinear
双线性函数 bilinear function
双线性矩阵式 bilinear form;bilinear matrix form
双线性抗力函数 <动力作用下延性材料> bilinear resistance function
双线性理想滞变回线 bilinear idealized hysteresis loop
双线性理想滞变曲线 bilinear idealized hysteresis loop
双线性律 bilinear law
双线性模型 bilinear model
双线性式 bilinear expression
双线性试验函数 bilinear trial function
双线性算符 bilinear operator
双线性算子 bilinear operator
双线性弯矩挠度关系 bilinear moment deflection relationship
双线性系数 bilinear coefficient;bilinear factor
双线性系统 bilinear system
双线性形式 bilinear form
双线性形式的秩 rank of a bilinear form
双线性映射 bilinear mapping
双线性硬化模型 bilinear hardening model
双线性运算符 bilinear operator
双线性运算号 bilinear operator
双线性运算子 bilinear operator
双线性展开式 bilinear expansion
双线性滞变弹簧 bilinear hysteretic spring
双线性滞变型弹簧 bilinear hysteresis-type spring
双线悬吊 bifilar suspension
双线悬吊滑轮 two-track suspension pulley
双线悬挂 bifilar suspension
双线悬置 bifilar suspension
双线掩模 double wire mask
双线圆锯制材厂 double circular sawmill
双线运河 two-lane canal; two-way canal
双线运行 double line working
双线载波制 two-wire carrier system
双线增音器 two-wire repeater
双线轧机 double strand mill
双线闸室 twin chamber
双线占线路总里程的百分率 percentage of double tracks to total route kilometers[kilometres]
双线制 double trolley system;double wire system
双线制帮电机 two-wire translator
双线制线路【铁】 dual system track
双线中继线 two-wire junction
双线自动闭塞系统 double track automatic block system
双线自动闭锁 double line automatic block
双限抽样检验方案 two-sided sampling plans
双相 double phase
双相变压器 two-phase transformer
双相的 biphase;diphase;two-phase;quarter phase
双相电流 two-phase current
双相电位 diphasic potential
双相分界线 coexistence border
双相共格衍射图 two-phase coherent diffractogram
双相合金 two-phase alloy
双相继电器 two-phase relay
双相接地故障 double earth fault
双相冷却剂致冷机 polyphase freezer
双相脉冲 diphasic pulse
双相喷嘴 two-fluid nozzle
双相平衡 biphase equilibrium; two-phase equilibrium
双相清洗 diphase cleaning
双相区 coexistence region;two-phase region
双相输出 complementary output
双相水流 two-phase flow
双相死亡现象 diphase morality;two-phase morality
双相位 quarter phase
双相性的 diphasic
双相整流 biphase rectification
双相整流器 biphase rectifier
双厢船坞 double dock
双厢船闸 double lock
双箱吊具 twin-lifting spreader
双箱角锥分级机 two-box spitzkasten
双箱式 double box
双箱式大梁 double box girder
双箱式集装箱起重机 twin-lift container crane
双箱式梁桥 double bowstring truss; double box girder bridge
双箱形截面大梁 twin-box girder
双箱型泵 dual-tank type pump
双箱制 two-bin system
双箱制动器 Carpenter brake
双箱桩 double box pile
双箱锥形选粒器 two-box spitzkasten
双镶板式门 two-panelled door
双镶板线脚 double bolection mo(u)lding
双向 bidirection; both-way; two-direction;two-way
双向凹形交叉 double-sag crossing
双向把手 reversible handle
双向百叶送风口 double deflection supply grille
双向摆动门 double swing door
双向板 <具有直角双向钢筋> flat slab;two way slab
双向板楼板 two-way flat slab floor
双向板体系 two-way slab system
双向半导体开关 ovonic switch
双向半双工电路 two-way half-duplex circuit
双向拌泥机 twin pug
双向帮电机 two-action translator
双向报价 two-way price quotation
双向报键 bug key
双向爆破 two-way shot
双向泵 two-way pump
双向逼近法 undirectional approach
双向闭塞 both-way block
双向编组场 <上下行各设一个编组场>【铁】 two-way classification yard; two-way marshalling yard; double type marshalling yard
双向编组站【铁】 double type marshalling station
双向变换器 bidirectional transducer; bilateral transducer; reversible transducer
双向变量泵 two-way variable displacement pump
双向变速箱 reverse gearbox
双向并行 two-way simultaneous
双向波谱 two-dimensional wave spectrum
双向波形 bidirectional waveform
双向操作 bidirectional operation; two-way operation
双向操作机构 double-acting power unit
双向测角仪 double Gauss goniometer
双向测试口 bidirectional test port
双向测线 reciprocal sight line
双向测振器 two-component pallograph
双向插座 double-sided socket

S

双向差额套头 spreads
双向缠绕 wound in both directions
双向场致发光 ovonic electroluminescence
双向车道 reversible flue
双向车辆 dual-way vehicle
双向车行道 reversible traffic lane
双向乘式犁 two-way sulky plow
双向齿锯 double cut saw
双向齿轮泵 double direction gear pump
双向出入口 two-way inlet
双向触点 two-way contact
双向传动 two-way drive
双向传动轮系 reverted train
双向传感器 bidirectional transducer; bilateral transducer
双向传声器 bidirectional microphone
双向传输 bidirectional transmission; bipolar transmission; two-way transmission
双向传输图像的电视信道 reversible television channel
双向传输线(路) reversible link; reversible line; two-action line
双向船舶行驶航道中间的安全富余航道 ship clearance lane
双向船闸 two-way lock
双向垂直压缩波 two-way vertical compressional wave
双向单工操作 two-way simplex operation
双向单工电路 two-way duplex circuit; two-way simplex circuit
双向单工连接 two-way simplex connection
双向单射 bijection; bijective mapping
双向单稳电路 bidirectional one-shot
双向道路 two-way road
双向的 bidirectional; bipartite; both-way; duo; duplex; either direction; either-way; ovonic; reversible; two-dimensional; two-way; up-down
双向等距平行肋型楼板 waffle slab
双向地面运动 bi-dimensional ground motion
双向电机 bipolar machine; either-rotation motor
双向电缆 bidirectional cable
双向电缆道 two-way cable conduit; two-way conduit tile
双向电流 bidirectional current; duo-directional current
双向电流传送 double current transmission
双向电流脉冲 double current impulsing
双向电路 bilateral circuit; circuit worked alternately; two-way cable conduit; two-way circuit
双向电码化信号线路电路 either-direction coded signal line circuit
双向电视 two-way television
双向电视通道 reversible TV channel
双向电视中继系统 two-way television relay
双向调度集中 either-direction centralized traffic control
双向定位 bilateral bearing
双向定位系统 two-dimensional positioning system
双向动作的桩锤 double-action pile hammer
双向读角系统 double-sided angle-reading system
双向渡线 facing point crossover
双向短接管 double nipple
双向对称加速器 two-way symmetric accelerator
双向对话通信[讯]系统 communication system to provide two-way conversation
双向对接铰 double-acting butt
双向多载客车车道 bidirectional HOV [heavy operated vehicle] facility
双向二级管开关 bilateral diode switch
双向二极管 bilateral diode
双向发射电视信道 reversible television channel
双向发送 two-way transmission
双向阀 double beat valve; two-way valve
双向法 two-dimensional method
双向翻斗车 two-way dumper; two-way tipper
双向翻倾 two-way dumping
双向反射分布函数 bidirectional reflectance distribution function
双向泛光照明 twin-flood lighting
双向放大器 bilateral amplifier; two-way amplifier
双向分布 two-way distribution
双向分岔 bilateral splitting
双向分类 two-way sort
双向风向标 bivane
双向风钻 <狭小地方用的> reversible close-quarter pneumatic drill
双向封闭灌浆 <钻孔内> double packer grouting
双向复制 bidirectional replication
双向干线 both-way trunk line
双向刚性增音机 rigid bidirectional repeater; rigid two-way repeater
双向钢格构(楼)面板系统 battle deck(bridge) floor
双向钢结构(楼)面板 <房屋和桥梁用> battle deck(bridge) floor
双向钢筋 bilateral bar; two-way reinforcement
双向钢筋地基(础) two-way reinforcement footing; two-way reinforced footing
双向钢筋混凝土板 two-way concrete slab; two-way reinforced concrete slab
双向钢筋网 bidirectional wire mesh
双向格构网架 two-way lattice(d) grid
双向格栅 two-way joist
双向格栅构造 two-way joist construction
双向格栅建筑 two-way joist construction
双向格栅矩形板建筑 waffle-plate construction
双向格式网架楼板 diagrid floor
双向工作的天线转换装置 duplex system
双向公差 bilateral tolerance
双向公差制 bilateral system of tolerance
双向公式 dyadic formula
双向供水 multiway service pipe system
双向沟通渠道 two-way communication channel
双向谷 double valley
双向固结 two-dimensional consolidation
双向固结试验 two-dimensional consolidation test
双向刮路机 two-way drag
双向管理图 two-way control chart
双向光缆 bidirectional cable
双向硅对称开关 biswitch
双向过滤 biflow filter
双向过闸 double way lockage; two-way transit
双向航道 double way channel; two-way channel; two-way route
双向航道交通 two-way traffic
双向航行 two way ship traffic
双向毫安表 double reading milliammeter
双向合页 double-acting hinge; double-action hinge
双向荷载 biaxial load; two-directional load(ing)
双向桁架系统 two-way truss system
双向横列式编组站 bidirectional transversal type marshalling station
双向滑动杆 thrust arm
双向滑动门 biparting door
双向滑门 double sliding gate
双向环流 double circular current
双向换挡装置 shift shuttle
双向换能器 bidirectional transducer; bilateral transducer
双向回弹阀门 double check valve
双向混合式编组站 bidirectional combined type marshalling station
双向活动杆 double-throw lever
双向击穿二极管 bidirectional breakdown diode
双向基础 two-way footing
双向激振 two-directional excitation
双向棘轮扳手 reversible ratchet handle
双向计数器 backward forward counter; bidirectional counter; forward backward counter; reversible counter; up-down counter
双向记忆开关 ovonic memory switch
双向继电器 bidirectional relay; duo-directional relay
双向加筋 two-way reinforcement
双向加筋板 slab with two-way reinforcement
双向加速度表 dual accelerometer; two-component accelerometer
双向加速度计 dual accelerometer; two-component accelerometer
双向加速度仪 dual accelerometer; two-component accelerometer
双向加速器 two-way accelerator
双向夹紧装置 counter clamp; counter cramp
双向减震器 double-acting shock absorber
双向剪切变形 double shear deformation
双向检波器 dual detector
双向浆叶拌和机 double motion paddle mixer; double shaft paddle mixer
双向浆叶搅拌机 double motion paddle mixer; double shaft paddle mixer
双向交叉桁架 double intersecting truss
双向交错层理 bimodal cross-bedding
双向交换 two-way alternate
双向交流信息 two-way flow of message
双向交替通信[讯] two-way alternate communication
双向交通道路 two-way road
双向交通公路 two-way road
双向交通量 bidirectional traffic; two-way traffic
双向角柱 two-way column
双向铰链 bilateral hinge; double-acting hinge; two-acting hinge
双向搅拌机 double agitator
双向接触器 two-way contactor
双向进给阀 cross feed valve
双向晶体管 bidirectional transistor
双向径向扇加速器 two-way radial sector accelerator
双向镜 two-way mirror
双向距离显示器 back-by-back display
双向聚焦 double directional focusing; two-directional focusing
双向开关 change-over switch; double-throw switch; throw-over switch; two-way switch
双向开关半导体器件 ovonics
双向开关元件 sidac
双向开门框 double egress frame
双向可逆侧犁 reversible side plough
双向可调操纵装置 two-way control
双向可调的 bilaterally adjustable
双向可调司机座椅 two-way seat
双向可调谐振器 two-way stretch resonator
双向刻度 center[centre] zero scale
双向空间钢网架 two-way space-framed steel structure
双向空间网架 two-way space grid
双向控制 bilateral control
双向控制阀 <凿岩机> double kicker port valve
双向控制继电器 non-recycling control relay
双向控制器 reversible controller
双向跨越 spanning in two direction
双向块形基脚 two-way block footing
双向快速转换开关 fast bi-directional switch
双向馈电 two-way feed
双向扩散 bilateral diffusion; double diffusion
双向扩散沉淀试验 double diffusion precipitation test
双向拉伸 biaxial tension; two-way stretch
双向拉伸应变 biaxial tensile strain
双向缆道 double cableway
双向肋平板 two-way ribbed slab
双向离合器 bidirectional clutch
双向离子变频器 <交流电源用> cycloinverter; cycloconverter
双向犁 alternate plow; hillside plough; spinner-type plow; swivel plough
双向犁翻转机构 plow reverser
双向连接 <线路的> both-way junction; double way connection
双向链接列表 bidirectional chained list
双向链路 two-way link
双向梁 two-way beam; two-way joist
双向梁格结构 beam and girder construction
双向列表 bidirectional list
双向掠挡装置 quick shift shuttle
双向淋洗技术 two-way elution technique
双向流 two-dimensional flow; two-way flow; bidirectional flow
双向流程线 bidirectional flow
双向流动 bidirectional flow
双向流滤池 biflow filter; by flow filter
双向龙骨 two-way joist
双向龙骨构造 two-way joist construction
双向龙头 two-way cock
双向楼板 two-way floor
双向滤池 bidirectional filter; biflow filter; two-way filter
双向螺杆 reverse-flighted screw
双向螺纹套管 double screw connector
双向螺旋桩 screw pile with two turns
双向脉冲 bidirectional pulse
双向脉冲传输 two-way pulse transmission
双向脉冲序列 bidirectional pulse train
双向镘灰板 two-way pallet
双向门 double-acting door; double-action door
双向门挡 emergency door stop
双向门框 double-acting frame

双向密肋 two-way ribs
双向密肋板 waffle slab
双向密肋楼板 cassette floor; waffle floor
双向密肋模板 waffle slab form
双向膜片活塞式空气弹簧 reversible diaphragm piston type pneumatic spring
双向木楔法 counter wedging
双向内部通话装置 two-way intercommunication system
双向内部通信[讯]系统 two-way intercom system
双向挠曲格栅 double deflection grille
双向能力 bi-directional capability
双向扭矩 double direction twist moment
双向耦合器 bidirectional counter
双向排列 two-way spread
双向排水 double(-slope) drainage; two-way drainage
双向刨 double end block plane
双向配合模 counter locked die
双向配件基础 two-way reinforced foundation
双向配筋 cross reinforcement; two-way reinforcement
双向配筋板 two-way flat slab; two-way reinforced slab
双向配筋的 two-way reinforced
双向配筋的钢筋混凝土 two-way reinforced concrete
双向配筋法 two-way system of reinforcement
双向配筋混凝土 two-way reinforced concrete
双向配筋混凝土板 two way slab
双向配筋基础 two-way reinforced footing
双向配筋基脚 two-way footing; two-way reinforced footing
双向配筋体系 two-way system of reinforcement
双向配筋系统 two-way system of reinforcement
双向匹配 bilateral matching
双向匹配衰减器 bilaterally matched attenuator
双向偏心距 biaxial eccentricity
双向偏转 bilateral deflection
双向平板 slab spanning in two directions; two-way flat slab
双向平板楼面 two-way flat slab floor
双向平衡运行 <船闸或航道> two-way balanced traffic
双向平面调车场 double flat yard
双向坡 double pitch
双向坡道 two-way ramp
双向坡度 two-way slope
双向破裂 bilateral rupture
双向启动可逆电动机 externally reversible motor
双向起动器 reversing starter
双向气腿 double-acting air leg
双向迁移 diorientation migration
双向钳位 two-way clamp
双向箝位电路 bidirectional clamping circuit
双向桥 two-way bridge
双向切割锯 double cut saw
双向切削龙门刨 double cut planer
双向倾卸车 two-way dumper; two-way tipper
双向倾卸车身 two-way dump body
双向倾卸挂车 two-way dump trailer
双向倾卸货车 two-way dump truck
双向倾卸拖车 two-way dump trailer
双向取向 dual orientation
双向人字闸门 bidirectional retaining mitre gate
双向三车道道路 three-lane dual carriageway road
双向三次曲线 bipartite cubic
双向三级六场纵列式编组站 three-stage-six-yard bidirectional longitudinal marshalling station
双向三极管开关 bilateral triode switch
双向扫描 bilateral scan(ning)
双向扫描术 two-dimensional scan(ning) technique
双向筛选筛 screen for two screening directions
双向渗透固结 two-dimensional consolidation
双向升降机 dual action lift
双向实时通信[讯] two-way real-time communication
双向式空气压气机 double-acting(air) compressor
双向式液压操纵 two-way hydraulic control
双向视频 two-way video
双向视线 reciprocal sight line
双向试探 bidirectional heuristic
双向试验 biaxial test
双向收缩 two-dimensional constriction
双向受控式传送 two-way controlled transmission
双向受力基脚 two-way footing
双向(受力)强度 biaxial strength
双向输出 duo-directional output
双向输电高压直流系统 reversible high-voltage direct current system
双向树 undirected tree
双向双工电路 two-way duplex circuit
双向双工制 double current system
双向水流玫瑰图 bimodal current rose
双向水流洗涤方式 two-water action
双向水听器 bidirectional hydrophone
双向水头船闸 double turn lock
双向水准测定 bilateral level(l)ing
双向水准测量 bilateral level(l)ing
双向伺服机构 bilateral servo-mechanism
双向隧道 dual tunnel
双向索道 double cableway; double ropeway
双向锁 reversible lock
双向锁定 boxed anchor; two-way anchor
双向弹簧铰(链) double-acting spring butt; double-acting spring hinge; helical hinge; two-way spring butt
双向弹簧门 swing door
双向套筒 bipolar socket
双向体系 two-way system
双向天顶距 reciprocal zenith distance
双向天线 bidirectional antenna; bilateral antenna; Janus
双向调风器 double deflection register
双向调制器 bidirectional modulator
双向停车标志 <次要道路与高级道路交叉时,在次要道路的双向上同时树立停车标志,要求停车等候高级干道车流空档> two-way stop sign
双向挺杆 positive tappet
双向通道 dual channel; duplex channel; two-way channel; two-wire channel
双向通话电路 mutual circuit
双向通信[讯] both-way communication; two-way communication; bilateral communication
双向通信[讯]业务 two-way traffic
双向同步加速器 two-way synchrotron
双向同时传输 simultaneous two-way transmission
双向同时交互作用 two-way simultaneous interaction
双向同时通信 two-way simultaneous communication
双向透镜 double image lens
双向图 two-dimensional plot
双向推拉门 double sliding gate
双向推力球轴承 double direction thrust ball bearing
双向推力轴承 two-directional thrust bearing
双向推土机 two-way dozer
双向拖拉机 reversible tractor
双向驼峰编组场【铁】 double hump yard; up-and-down hump yard
双向外伸轴 double extended shaft
双向弯曲 biaxial bending; compound bending
双向弯曲薄壳 shell curved in two directions
双向网格 two-way grid
双向网络 bilateral network
双向往复泵 double displacement pump
双向无线电话 duplex radiophone
双向无线电设备 two-way radio
双向无线电通信[讯] transmitting and receiving service; two-way radio communication
双向无线电通信[讯]系统 two-way radio system
双向系数 bi-dimensional coefficient
双向系统 bilateral system; two-way system
双向下倾褶曲 doubly plunging fold
双向先断后合触点 two-way break-before-make contact
双向线 bidirectional lines
双向线脚 double-faced
双向线路 bidirectional lines; two-way cable conduit; two-way circuit
双向限幅 double limiting
双向限幅器 clipper-limiter; double limiter
双向限位开关 two-way limit switch
双向削波电路 bidirectional clipping circuit
双向楔牢 counter wedging
双向斜坡 double slope
双向(泄)水闸 double beat sluice
双向卸料 two-way chute
双向卸料槽 two-way spout; two-way chute
双向信标 two-course beacon
双向信道 both-way channel; duplex channel; two-way channel
双向信号 both-way signal(l)ing; either-direction signal(l)ing
双向信号显示 either-direction signal indication
双向信号显示法 both-way signal indication
双向信号装置 both-way signal(l)ing
双向行车 either direction running; either direction working; two-way traffic
双向行车线路 track for either direction working
双向行车装置 bilateral gear
双向型 double deflection model
双向性 bilateral
双向性硅钢片 cubex
双向(絮凝)沉淀池 cyclofloc settling tank
双向旋塞 two-way cock
双向旋转泵 reversible pump
双向(旋转)电动机 reversing motor
双向旋转马达 reversing motor
双向旋转门 swinging door
双向选组级示闲器 idle two-way selector stage indicator
双向雪犁 two-way snow plow
双向循环阀 reversible circulation valve
双向压力压机 dual-pressure press
双向压模 double-acting die
双向压缩 biaxial compression; dual compression
双向压缩法 two-dimensional compression method
双向压制 compaction by double-action; double-action pressing; two-directional pressing
双向烟道 revertible flue
双向摇摆大门 swing gate
双向摇摆门 swing door
双向业务线 service both-way circuit
双向液压减振器 two-way hydraulic shock absorber
双向移门 double sliding door
双向移圈花纹 two-way stitch transfer
双向移相器 two-way phase switcher
双向因子 biaxiality factor
双向应力 biaxial stress; two dimensional stress
双向应力部位 biaxial stress area
双向应力状态 biaxial state of stress; two-dimensional state of stress
双向影线 cross hatch(ing)
双向油缸 two-way cylinder
双向游振 two-directional excitation
双向预应力 double prestressing; two-dimension prestress(ing)
双向预应力板 two-way prestressed slab
双向预应力混凝土 doubly prestressed concrete
双向阈值开关 ovonic threshold switch
双向运动 bidirectional movement
双向运河 two-way canal
双向运输机 two-way conveyer[conveyor]
双向运行 bidirectional traffic; double direction running; either-direction operation; either direction traffic; either-way operation; either-way traffic; two-way operation
双向运行动车列车 reversible motor-coach train
双向运行自动闭塞【铁】 automatic block with double direction running; double direction automatic block
双向运转的曲轴 cranks working in opposite directions
双向匝道【道】 two-way ramp
双向择一通信[讯] either-way communication
双向增强夹层板 bidirectional laminate
双向增音机 two-way repeater
双向增音器 two-way repeater
双向闸刀开关 double-throw knife switch
双向闸机 reversible gate
双向展开法 two-dimensional development method
双向张力 biaxial tension
双向折弯 dogleg
双向阵 binomial array
双向振动 bi-dimensional shaking motion
双向振动磨 vibroenergy mill
双向振动台 bi-axial shaking table
双向振幅 <垂直的以厘米为单位,旋转的以度为单位> double amplitude
双向振铃中继线电路 two-way ring down trunk circuit
双向支座 bilateral support
双向直线步进电动机 two-directional linear stepping motor

S

双向止流阀 double branch stop valve
双向止推轴承 double-thrust bearing
双向纸层析 two-dimensional paper chromatography
双向纸色谱 two-dimensional paper chromatography
双向纸上色层分析法 two-way paper chromatography
双向制 bilateral system
双向中继器 two-way repeater
双向中继线 both-way trunk(line); two-way trunk(line)
双向中继线继电器组 bothway trunk relay set
双向柱 two-way column
双向转变 enantiotropic transformation; reversible transformation
双向转发器 two-action translator
双向转换 bilateral switching
双向转换开关 two-way reversing switch; bidirectional switch
双向转换器 bidirectional transducer
双向装车机 dual loader
双向装置 two-way device
双向自动闭塞【铁】both-direction automatic signal(l)ing; double direction automatic block; either-direction signal(l)ing; either-direction traffic automatic block
双向自动闭塞号志 automatic block signal in both directions
双向自动闭塞系统 double direction automatic block system; either-direction traffic automatic block system
双向自动闭塞信号机【铁】automatic block signal in both directions
双向自动闭塞制 double direction automatic block system; either-direction traffic automatic block system
双向自动调整止推滚珠轴承 two-directional self-aligning ball thrust bearing
双向自动专用线路 two-way automatic private line
双向自卸车 shuttle dump truck
双向总线 bidirectional bus
双向总线驱动器 bidirectional bus driver
双向纵列编组站【铁】bidirectional longitudinal type marshalling station; three-stage/six yard marshalling station
双向阻抗 bilateral impedance
双向左转车道 two-way left-turn lane
双向作用 bidirectional; bilateral
双向作用的 double-acting
双向作用发动机 double-acting engine
双向作用水力千斤顶 double-acting hydraulic jack; double-acting hydraulic ram
双向作用液压马达 bidirectional hydraulic(al) motor
双向作用元件 bilateral element
双像 bifurcate image; diplopia; diplopy; echo image; split image
双像测距机 double image range finder
双像(测距)棱镜 double image prism
双像测距仪 coincidence range finder; concrete-image telemeter; double image tach(e)ometer; double image tachymeter; double image telemeter; double image range finder
双像测量【测】double image measurement
双像测图 stereorestitution
双像测微器 double image micrometer
双像重合测距仪 double image overlap tacheometer
双像对应 matching
双像法 double image method
双像法光学垂准 double image method optic(al) plumbing
双像分离 twin image separation
双像幅照相机 double frame camera
双像棱镜 doubling prism
双像量测理论 theory of double image measurement
双像量角仪 photogoniometer for pairs of photographs
双像目镜 double image eyepiece; double image ocular
双像全息照片 dual hologram
双像摄影测量学 double image photogrammetry; two-image photogrammetry
双像视距仪 Bosshardt-Zeiss reducing tacheometer; double image tach(e)ometer
双像视仪 double image tachymeter
双像天顶仪 double image zenith telescope
双像投影 double projection
双像投影测图仪 double projector
双像投影立体测图仪 double projection stereoscopic plotting instrument
双像投影器 double projector; binocular projector
双像投影仪 double projection instrument; double projection stereoscopic plotting instrument
双像效应图像 double image effect picture
双橡胶纤维片 twin fabric disk
双削齐边锯 chipping edger
双消毒 dual disinfection
双消像散透镜 double anastigmat
双硝炸药 dualin
双销拨盘 double driver plate
双销钉 double pin
双销擒纵机构 two-pin escapement
双销锁 two-bolt lock
双小裁 double small
双小车 double trolley
双小齿轮 double pinion
双效补偿 double offset
双效成型 double-action forming
双效冲模 double-action die
双效的 double effect【化】; double service; dual-purpose
双效抵消 double offset
双效开关 open and shut valve
双效控制设备 dual effect control device; dual effect controlling equipment
双效溶剂萃取(油脂)法 dual solvent process
双效调节器 duplex regulator
双效吸收式制冷剂 double effect absorption refrigerating machine
双效溴化锂吸收式制冷机 double effect lithium-bromide absorption-type refrigerating machine
双效压塑 compaction by double-action
双效压缩 dual compression
双效压缩机 dual effect compressor
双效压制 double(-acting) compression
双效蒸发器 double effect evapo(u)rator
双楔 folding wedges
双楔垫老虎钳 double-wedged clamp
双楔固定平行垫板 taper parallel
双楔式锚杆 sliding wedge bolt
双楔式切挖 double-wedge cut
双楔榫镫接<屋架中央主木与横木间的> cottered joint
双楔掏槽 double-wedge cut; W-cut
双楔形 double taper; double wedge
双楔形榫键 dovetail feather joint
双斜 T 形接头 double bevel T joint
双斜边 double bevel; double-beveled edge
双斜边肋骨 double cant frame
双斜边坡口焊缝 double bevel groove weld; double V groove weld
双斜槽口 double skew notch
双斜的 diclinic
双斜丁字接头 double bevel T joint
双斜杆系统<桁架梁的> arrow point bracing
双斜杆支撑系统<桁架梁的> arrow point bracing
双斜接合 double oblique junction
双斜量角器 double protractor
双斜率模/数转换器 dual slope analog/digital converter
双斜面 double bevel(ling); double cant; double incline; double inclined plane
双斜面的 bibevel(l)ed
双斜面钻 double chambered drill
双斜劈面木瓦板 tapersawn; tapersplit
双斜坡口焊 double bevel groove weld; double V groove weld
双斜(坡)屋顶面(顶棚) duo-pitch roof
双斜刃钎头 double taper bit
双斜刃钻头 double taper bit
双斜式 double bevel
双斜凸角系船柱<鹿角形的> twin-head sloping lobe bollard
双斜线 double bevel; double slashes
双斜楔 double taper wedge
双斜楔块 double taper wedge
双谐(波)的 biharmonic
双谐振 double resonance
双谐振动器 double harmonic oscillator
双泄水箱 duplicate sluice box
双卸料板 double stripper
双榍 bisphenoid
双心体 diplosome
双芯 diplocardia; twin-core
双芯扁电缆 flat twin cable
双芯扁软线 flat-braided cord
双芯变压器 two-core transformer
双芯导线 duplex wire; twin-core conductor
双芯的 twin-cored
双芯灯 duplex lamp
双芯电缆 duplex cable; paired cable; twin-core cable; two-core cable; two-conductor cable
双芯电流互感器 twin-core current transformer
双芯电器引线 two-core fixture wire
双芯电线<非同轴的> duplex cable; paired cable; twin conductor; twin-core cable
双芯拱 two-centered[centred] arch; two-centered body
双芯光缆 twin fiber cable
双芯焊条 twin electrode
双芯绞合电缆 twisted pair; twisted-pair cable
双芯铰合线 twisted pair wiring
双芯扭在一起的绝缘电缆 twisted pair
双芯起爆电缆 twin-core shot firing cable
双芯软绳 twin flexible cord
双芯塞绳 two-way cord
双芯塑料电线 double core plastic wire
双芯同轴电缆 twin-concentric cable
双芯头 double core point
双芯线 two-core cable
双芯线终端套管 bifurcating box
双芯引线 twin-lead
双信伴传机 diplexer
双信道 binary channels
双信道的 twin-channel
双信道电路 two-channel circuit
双信道跟踪接收机 two-channel tracking receiver
双信道开关 two-channel switch
双信道制 twin channel system
双信道转换开关 two-channel duplexer
双信号闭塞制 composite block system
双信号测试法 two-signal test method
双信号电报键控 double signal telegraph keying
双信号法 two-signal method
双信号码的自动识别装置 automatic number identification
双信号区 bi-signal zone
双信号同时同向传送 diplex
双信号选择性 two-signal selectivity
双信号振铃 composite ringer; composite ringing
双信回路 set composite
双信接收 double reception
双信息处理机 dual processor
双信息处理机系统 dual processor system
双信息处理系统 dual processing system
双信息系统 double message system
双星 binary star; dicoaster; double star
双星导航 two stars navigation
双星等高测时法 method of time determination by two stars in equal altitude; time by equal altitude of two stars
双星点 double star point
双星点衍射像 diffraction image of double star point
双星绞电缆 double star guad cable
双星绞结构 double star-quad construction
双星绕转 rotation of binaries
双星天文惯性制导系统 two-star stellar inertial guidance system
双星团 double cluster
双星形接法 star-star connection; Y-Y connection
双星形连接整流器 double Y connected rectifier
双星型接法 Y-Y connection
双星座天文惯性导航 two-star inertial guidance
双形晶 double formed crystals
双形现象 dimorphism
双型成像系统 biform imaging system
双型分布 bimodal distribution
双型劈理 bimodal cleavage
双型腔压型 two-cavity die
双型腔铸型 two-cavity mould
双性反应 amphoteric reaction
双雄接口 double male
双雄榫 double tenons; twin tenons
双虚线 double line of pecks
双悬臂 double cantilevers; twin cantilevers
双悬臂薄壳 double cantilever shell
双悬臂堆料机 twin-boom stacker
双悬臂夹式引伸计 double cantilever clip-in displacement meter
双悬臂梁 beam with overhanging ends; double cantilever beams
双悬臂梁桥 double cantilever beam bridge; double cantilever girder bridge

双悬臂梁试样 double cantilever beam specimen
双悬臂龙门吊车 double cantilever gantry crane
双悬臂锚孔 double cantilever anchor space; double cantilever anchor span
双悬臂门式起重机 double cantilever gantry crane
双悬臂起重机 double cantilever crane
双悬臂壳体 double cantilever shell
双悬臂式旋桥 double cantilever swing bridge
双悬臂塔柱 double cantilever pylon
双悬臂屋架 double cantilever roof truss
双悬臂卸船机 double link unloader
双悬臂旋开桥 double cantilever swing bridge
双悬臂支撑塔杆 mast with double cantilevers; mast with twin cantilevers
双悬臂装卸桥 double cantilever gantry crane
双悬窗 double hung(sash) window
双悬窗的锤箱 weight pocket
双悬窗的弹簧配重 spiral balance
双悬窗分隔木条 parting stop
双悬窗框底边线沟槽 check stop
双悬窗平衡锤滑框 sash run
双悬窗扇中梃 parting bead; parting slip; parting strip
双悬窗系吊带的铁件 sash-cord iron
双悬链线结构 double catenary construction
双旋窗的顶椗子 window yoke
双旋大门 swing gate
双旋光 bi-rotation
双旋回海岸 two-cycle coast
双旋回海崖 two-stor(e)y cliff
双旋流器 duplex cyclone
双旋网捕集器 double cyclone
双旋涡 vortex pair
双旋涡饰 double roll
双旋翼式布局 twin rotor layout
双旋翼系统 twin rotor system
双旋转磁场法 two-revolving field method
双旋转法 swing-swing method
双旋转钩 double swivel hook
双旋子阀门 dual rotor valve
双选通 double gate
双碹 double arches; recessed arch
双碹顶池窑 Amco type tank furnace
双穴模 double cavity mo(u)ld
双循环 bicirculating; dual circulation
双循环地热电站 binary cycle geothermal power generator
双循环电路 two-cycle scheme
双循环反应堆发电厂 dual-cycle reactor plant
双循环方案 two-cycle scheme
双循环沸水堆 dual-cycle boiling-water reactor
双循环谷 two-cycle valley; two-stor(e)y valley; valley-in-valley
双循环混合制冷剂天然气液化工艺流程 dual-mixed refrigerant cycle liquefaction process
双循环解法 dicyclic(al) solution
双循环两相生物处理 binary cycle two-phase biologic(al) process
双循环系统 binary cycle system
双压氨制冷 double compression ammonia refrigeration
双压泵 dual-pressure pump
双压电动机 dual-voltage motor
双压电晶片 bimorph
双压电晶片元件 bimorph cell
双压盖磨边机 two-cup edging machine
双压杆桁架 double strutted frame
双压杆桁架梁 double strut trussed beam
双压杆框架 double strutted frame
双压挤熟铁 double fag(g)ot iron
双压控制 double pressure control
双压力计 double pressure ga(u)ge
双压力容器输送机 double pressure-vessel conveyer[conveyor]
双压料锯 double swage saw
双压路机 double roller
双压区复式压榨 twin press
双压塑模 floating chase mould
双压缩式接头 double compression fitting
双压透平 mixed pressure turbine
双压选择器 duplex pressure selector
双压油路 dual-pressure circuit
双压制动 double pressure control
双牙轮钻头 double cone bit; twin-cone bit; two-cone bit; two-roller bit
双烟道 twin flue
双烟道锅炉 double flue boiler
双岩心管 double tube barrel
双沿口 twin bead
双盐(混合物) double salt
双檐层木板瓦 double eaves course
双檐瓦层 double eaves course; doubling course
双檐钻头 double-fluted bit
双衍射 double diffraction
双衍射介质 double diffracting medium
双眼 binoculus
双眼单筒镜 binocular
双眼观察头 binocular head
双眼集合 binocular integration
双眼检眼镜 binophthalmoscope
双眼镜照相 stereograph
双眼螺钉 eye bolt
双眼视差 binocular parallax
双眼视轴交会 binocular fixation
双眼望远镜 binocle; field glass(es); telestereoscope
双眼仪器 binocular instrument
双眼照相镜 stereoscope
双演出场的剧院 twin theatre[theater]
双堰孔溢洪道 two-bay spillway
双燕尾键 double dovetail key; dovetail feather
双燕尾榫口 dovetail feather
双燕尾榫扣 double dovetail key; dovetail cramp
双羊角 twin horn cleat
双阳极 binode; double anode
双阳极的 bianode
双阳极负阻管 negat(r)on
双阳极管 binode
双氧化物铁电体 two-oxide ferroelectrics
双氧气 oxozone; oxydol
双氧水 hydrogen peroxide; oxyful
双氧稳定剂 stabilizer of hydrogen peroxide
双氧铀(根) uranyl
双样板 double screed
双样板横向修整机 double screed transverse finisher
双样板横向整面机 double screed transverse finisher
双样板修机 two-screed finishing machine
双样板修整机 double screed finisher
双样板整面机 double screed finisher
双样本检验 two-sample test
双样固结试验 double specimen oedometer test
双样试验机 <同时能进行两个试样的试验> two-unit tester
双摇摆门 double swing gate
双摇摆支座 double pendulum bearing
双摇动支座 double rocker bearing
双摇杆机构 double connecting rod mechanism
双摇杆碎石机 compound toggle lever stone crusher
双摇枕转向架 double transom truck
双摇座 double rocker bearing; double tumbler bearing
双咬口 double-locked seam; double-locked welt
双咬口立缝 double-locked standing seam
双咬口平接缝 double-locked flat seam
双咬口十字缝 double-locked cross welt
双叶瓣饰 double feathering
双叶长短轴圆外旋轮线形状 two-lobed epitrochoidal shape
双叶垂直提升闸门 double-leaf vertical lift gate
双叶的 two-bladed
双叶阀门 butterfly valve
双叶仿星器 two-turn stellarator
双叶风扇 two-bladed fan
双叶覆盖 two-sheeted covering
双叶拱 lenticular arch
双叶拱桥 lenticular arch bridge
双叶戽斗 double lobe bucket
双叶轮 bilobed wheel
双叶轮式冲击粉碎机 double impeller impact mill
双叶轮式冲击破碎机 double impeller impact breaker
双叶轮式浮选机 double impeller flo(a)tation cell
双叶螺母 butterfly nut
双叶螺栓 butterfly bolt; wing-headed bolt
双叶螺旋加料器 double flight feed crew
双叶螺旋桨 two-blade propeller
双叶农用风车 Fales-Stuart windmill
双叶刨槽铣刀 two-wing slotting cutter
双叶片 double vane; twin blades
双叶片分批分散混合机 double blade batch dispersion mixer
双叶片混料机 double zone(pugmill) mixer
双叶片水泵 two-lobed pump
双叶片真空泵 double vane exhauster
双叶片转轮 twin-blade runner
双叶饰 two-leaf
双叶竖旋桥 twin-leaf bascule bridge
双叶双曲面 biparted hyperboloid; hyperboloid of two sheets
双叶锁风翻板阀 air-tight double tipping valve
双叶推进器 two-bladed propeller
双叶线【数】 double folium
双叶形线 double folium
双叶旋转双曲面 hyperboloid of two sheets of revolution
双叶转子鼓风机 two-lobe blower
双页 pair of pages
双页地图 two-sheet map
双页地图集 double atlas
双页复印法 duplex method
双页黏[粘]土砖饰面 double-leaf clay brick veneer
双页拍摄法 duo method
双页缩拍法 duo method
双液冲洗 rapidoprint
双液淬火 double quenching; interrupted hardening
双液单系统注浆 twin liquid single system injection
双液电池 double fluid cell
双液法 two-fluid process; two-shot method <化学灌浆的>
双液灌浆 double shot grouting; two-shot grout(ing); two-shot solution grout(ing)
双液灌浆法 <化学灌浆的> two-fluid process
双液硅化法 Joosten process
双液硅化加固法 double shot silification
双液化学灌注 <多指用水玻璃和氯化钙灌入土中,亦称硅化法> two shot chemical grouting
双液化学加固土壤法 two-fluid chemical consolidation process
双液离心机 two-liquid centrifuge
双液面流量计 two-fluid manometer
双液双系统注浆 twin liquid twin system injection
双液显影 divided developer; split developer; two-bath developer; two-solution developer
双液型环氧底漆 double fluid epoxy primer
双液压的 double hydraulic
双液压千斤顶 twin hydraulic jack
双一次性 bilinear
双一烷基马来酸二丁锡 dibutyltin bis monoalkylmaleate
双乙酸联苯酯 diphenol diacetate
双乙酸盐 diacetate
双乙酰 dimethyl diketone
双乙酰芳胺黄 diarylide yellow
双乙字管 double offset
双异丁烯酸化合物 twin-part methacrylate compound
双异旋光 bi-rotation
双异质结构 double heterostructure
双异质结激光器 double heterojunction laser
双翼 double vane; twin blades
双翼波形瓦 double flap pantile
双翼采油树 dual wing Christmas tree
双翼船台 two-wing building berth
双翼的 double-leaf
双翼对角式通风 ventilation of two-way and opposite angles
双翼飞机 biplane
双翼衡重式仰开桥 twin-leaf bascule bridge
双翼衡重仰开桥 twin-leaf bascule bridge
双翼回采工作面 two-winged stope
双翼机 biplane
双翼岬 winged headland
双翼开合桥 double draw bridge
双翼螺母 wing nut
双翼门 double-leaf door
双翼黏[粘]土波形瓦 double flap clay pantile
双翼平旋桥 double-leaf swing bridge
双翼式蝴蝶阀门 biplane butterfly valve
双翼竖旋桥 double-leaf bascule bridge; twin-leaf bascule bridge
双翼水上飞机 hydrobiplane
双翼仰开桥 double draw bridge
双翼铸钢钻头 cast-steel 2-wing bit
双翼钻头 two-way bit
双因贸易条件 double factorial terms of trade
双因素方差分析 two-factor variance analysis
双因素理论 two-factor theory
双阴极 twin cathode
双阴极充气三极管 pulsatron
双阴极管 bicathode tube
双音 diphonia
双音多频 dual-tone multi-frequency; dual-tone multiple frequency
双音多频键控脉冲 dual tone multi-frequency key pulsing

双音发生器 two-tone diaphone
双音工作 two-tone working
双音检波器 two-tone detector
双音喇叭 two-tone horn
双音频率 dual-tone frequency
双音圈扬声器 double coil loudspeaker
双音系统 two-tone system
双音信号发生器 double-tone signal oscillator
双音信号法 two-tone method
双引出线波导管连接 two-port waveguide-junction
双引擎 twin engine
双引擎单翼机 twin-engined monoplane
双引擎拖拉机 twin-engine tractor
双引线 double lead
双印 mackle
双荧光层的 double-fluorescence-layered
双赢 win-win
双影 double image
双影像处理系统 dual image processing system
双硬度橡胶圈 dual-hardness rubber ring
双用 dual mode
双用测定器体系 dual ga(u)ge system
双用的 double service
双用封隔器 straddle packer
双用角尺 double L-square
双用刨 compass plane
双用燃料发动机 dual-fuel engine
双用熔炉 dipout furnace
双用通气管 dual vent
双用途的 double duty
双用途机器 dual-purpose machine
双用线 dual-use line
双用钻头 combined bit; combined drill
双优先 double priority
双优先式 double priority mode
双油封 double oil seal
双油管完井 dual completion well; dual string completion
双油路布置 dual path arrangement
双油路喷嘴 duplex burner
双油路设计 dual path arrangement
双油楔轴承 two-wedge bearing
双游标 double vernier
双游离基 biradical
双有机阳离子膨润土 dual organic cation bentonite
双釉面瓷器 double-faced ware
双语的 bilingual
双浴法 two-bath process
双浴显影 two-bath development
双浴显影剂 two-bath developer
双阈检测 double threshold detection
双阈检定 double threshold detection
双阈(值) dual threshold
双元 binary cell; double base; unit doublet
双元传声器 two-element microphone
双元发动机 binary engine
双元干涉仪 double interferometer; two-element interferometer
双元件 double element
双元件电子系统 two-element electronic system
双元件馈给控制 two-element feed control
双元件调节器 two-element regulator
双元脉冲 unit double(t) impulse
双元燃料发动机 binary engine
双元推进剂 bireactant
双元系统 tow-part system
双元循环 binary cycle
双元蒸气发动机 binary vapo(u)r engine
双原子的 diatomic
双原子分子的形成 dimerization
双原子价 bivalence
双原子气体单质 diatomic gas
双原子气体化合物 diatomic gas
双圆板耙路机 double disk harrow
双圆测角仪 two-circle goniometer
双圆垂直荷载 dual circle vertical load
双圆光轮 vesica-piscis
双圆弧形 Gothic
双圆角 double nose
双圆角式柱 double bull type column
双圆角铣刀 double circular cornering cutter
双圆角砖 double(bull-) nose brick
双圆锯裁边机 double edger; twin edger
双圆锯机 double circular mill
双圆盘保险装置 double roller safety action
双圆盘滚轮 double disk roller
双圆盘划行器 double disc [disk] marker
双圆盘机构 roller mechanism
双圆盘开沟器 double disc[disk] opener; twin-disc[disk] colter
双圆盘耙 double disc[disk] harrow
双圆盘擒纵机构 double roller(lever) escapement
双圆盘轴承 twin-disc[disk]-bearing
双圆盘钻头 two-disc[disk] bit
双圆片离合器 double plate clutch
双圆投影 double circular projection
双圆凸瓦 double roll tile
双圆线脚 double bead
双圆柱 bicylindrical
双圆柱共振器 bicylindrical resonator
双圆柱形楼梯 double cylinder stair(case)
双圆转动单晶相机 double circle rotating crystal camera
双圆锥的 biconic(al)
双圆锥活动连接器 biconic(al) connector
双圆锥破碎机 twin cone crusher
双圆锥式空气分级机 double cone air classifier
双圆锥形扬声器 duo-cone
双缘钢轨轮 double-flanged rail wheel
双缘轮 channel wheel; double flange wheel
双缘轮胎 double head tyre
双源 double source; duplicate source
双源独立点火 two-independent ignition
双源供电 dual service
双源静寂时间测定法 two-source dead time determination
双源距密度测井 dual densilog
双源距密度测井曲线 dual densilog curve
双源频率键控 two-source frequency keying
双月桂基硫醇二丁锡 dibutyltin bis
双月刊 <期刊名> Alternate Months; Bimonthly
双运算器 twin arithmetic(al) units
双运算数操作 dyadic operation
双运行图制度 <一个适用于列车对数多的期间,一个适用于列车对数少的期间> dual diagram system
双匝 double turns
双匝线组 two-turn winding
双载波损耗 two-carrier loss
双载流子晶体管 bipolar transistor
双再用系统 dual reuse system
双錾头钻头 double chisel bit; double chisel drill bit
双择检测 binary detection
双择检验 binary detection
双择判定 binary decision
双择判决 binary decision
双渣法 double slag process
双渣熔炼 two-slag practice
双扎线圈 bincoculars coil
双轧机 twin roller
双闸 double gate
双闸板式防喷器 double ram type preventer; preventer of double ram type
双闸板型防喷器 double ram type preventer; preventer of double ram type
双闸刀开关 double bladed knife switch
双闸孔溢洪道 two-bay spillway
双闸门给料机 two-gate feeder
双闸门减压舱 two-lock recompression chamber
双闸室船闸 dual chamber lock; double lock
双闸瓦 double block; double brake shoe
双闸瓦制动 double block brake; double shoe brake
双占用 glare
双站雷达 bistatic radar
双张丝悬吊 bifilar suspension
双涨潮流 double flood; double tidal flood
双胀式蒸汽机 double expansion engine
双账式资产负债表 double account form of balance sheet
双账制 double account system
双找平修整器 double screed finisher
双照射 double irradiation
双遮光器快门 opposed shutter
双折板 steps
双折边对接 double-flanged butt joint
双折边缝 double-lock welt
双折布抛光轮 double buff
双折窗(扇) two-fold window; two-leafed window
双折的 two-leaf
双折返线 double turn-back track
双折铰接百叶门 two-leaf hinged shutter door
双折楼梯 dog-leg stair(case)
双折门 bifolding door; double folding door; two-fold door; two-leafed door
双折上翻门 two-part up and over door
双折射 birefraction; birefringence; double refraction
双折射板 birefringent plate
双折射板的慢轴 slow axis of a birefringent plate
双折射波导 birefringent waveguide
双折射补偿板 birefringence compensating plate
双折射补偿器 birefringent compensator
双折射单色计 birefringent monochrometer
双折射的 birefringent
双折射法 birefringence method <量测岩石残余应力的>; method of double refraction <研究边界层流态用>
双折射反光器 double refraction reflector unit
双折射方解石 double refraction calcspar
双折射干涉显微镜 birefringent interference microscope
双折射干涉仪 birefringent interferometer
双折射各向异性 birefringence anisotropy
双折射光楔 birefringent wedge
双折射检查仪 birefringencemeter
双折射角 double refraction angle
双折射解调器 birefringent demodulator
双折射介质 birefringent medium
双折射晶体 birefringence crystal; birefringent crystal; doubly refracting crystal
双折射棱镜 birefringent prism
双折射连锁滤光器 birefringent chain filter
双折射滤光器 birefringence filter; birefringent filter; monochromatic filter
双折射率 birefringence
双折射媒质 birefringent medium; doubly refracting medium
双折射片 birefringent plate
双折射偏振器 birefringent polarizer
双折射器 doublet refractor
双折射热光系数 birefringence thermooptic(al) coefficient
双折射双片 doubly refracting biplate
双折射贴片 birefringent coating
双折射贴片的加强效应 reinforced effect of birefringent coatings
双折射透明方解石 calc-spar
双折射图案 birefringence pattern
双折射望远镜 Cassegrain(ian) telescope
双折射效应 birefringent effect
双折射性能 birefringence property
双折射延迟器 birefringent retarder
双折弹簧门 two-leaf flexible swing door
双折梯 pair of steps; steps
双折推拉百叶门 two-leaf sliding shutter door
双折屋顶 gambrel roof; mansard roof
双折线屋顶 double pitched roof
双折隐梯 double fold disappearing stair(case)
双摺 duplication
双摺窗(扇) two-fold window; two-leafed window
双摺门 bi-fold door; bifolding door; two-door; two-fold door; two-leafed door
双辙叉 double frog
双针 crosspointer
双针床 two-needle bar
双针风表 duplex air ga(u)ge
双针计 duplex ga(u)ge
双针记录式温度计 two-pen recording thermometer
双针秒表 split seconds chronograph
双针气压计 duplex air ga(u)ge
双针式测量仪表 crosspointer instrument
双针式指示器 crosspointer indicator; dual indicator
双针停表 double stop watch
双针压力计 duplex pressure ga(u)ge
双真空排气台 two-ply vacuum exhaust station
双枕梁 double body bolster
双振荡模式 double oscillation mode
双振荡型 double moding
双振捣器 double vibrator
双振动 double vibration
双振动器 double vibrator; twin vibrator
双振幅 peak-to-peak amplitude
双震带 double seismic zone
双整流子电动机 double commutator motor
双正交的 biorthogonal
双正交关系 biorthogonality relation
双正交函数集 biorthogonal sets of functions

双正交天线阵 binomial antenna array
双正则 biregular
双支撑 double bracing
双支撑的 double beat
双支撑桁架梁 double strut trussed beam
双支撑架 double strutted frame
双支撑体系 dual bracing system
双支撑系统 double bracing system
双支承 double base; twingrip
双支承的悬挂式脚手架 float scaffold
双支承舵 double bearing rudder; double supported rudder
双支承砂轮无心磨床 twin grip type centerless grinder
双支的 biramous
双支点 colon
双支点椽条 double jack rafter
双支墩 double buttress
双支墩坝 double buttress dam
双支管(件) double Y; double branch pipe
双支河流 twin river
双支架 double cradle; double jack bar
双支脚尾轴架 two-leg strut
双支距(测量) double offset
双支渠分水闸 bifurcation headgates of laterals
双支枢式 double pivoted pattern
双支枢式闸门 double pivoted gate
双支腿龙门吊 full arch gantry crane
双支腿门座 full arch gantry
双支弯头 double branch bend; double branch elbow
双支下锚横梁 double branch anchor beam
双支柱 double jack
双枝三次曲线 bipartite cubic
双枝形天线 twin branch type antenna
双肢吊杆 double member
双肢钢柱 battened steel column
双肢剪力墙 coupled shear wall
双肢起重杆 shear-leg derrick crane
双肢柱 battened columns; coupled columns
双直轨组合式钝角辙叉 built crossing up obtuse crossing with both rails straight
双直轨组合式普通辙叉 built-up common crossing with both rails straight
双直角(定线)棱镜 double right-angle prism
双直角四边形 birectangular quadrilateral
双直角形 birectangular
双直纹曲面 doubly ruled surface
双直线的 bilinear
双直线式房屋 double straight-line building
双直线式建筑 double straight-line block; double straight-line building
双值 diadic[dyadic]; two-value
双值变量 two-valued variable
双值电容式电动机 two-value capacitor motor
双值函数 two-valued function
双值监控程序 di-value monitor
双值逻辑元件 binary logic element
双值群 double group
双值输出信号 two-valued output signal
双值算子 diadic operator
双值条件 binary condition
双值限幅 slicing
双值性 ambiguity; double valuedness
双值性传感器 ambiguity sensor
双值性函数 ambiguity function
双值性判定 sense research
双值运算 dyadic operation
双值状态 binary condition; two-valued condition
双止回阀 double-return valve; dual check valve; duplex check valve
双纸盒扬声器 duo-cone
双纸盆扬声器 double cone loudspeaker
双指标法 double index method
双指数分布 double exponential distribution
双指向标志 double arrow sign
双指压板 double finger clamp
双指针自动测向仪 dual automatic direction-finder
双趾弹簧扣件 double-toe rail fastener
双制动爪 double dog
双制计算器 duplex calculator
双制冷系统 double refrigerant system
双质命名法 <土壤颗粒的> contrasting texture
双质子 diproton
双质子放射性 two-proton radioactivity
双质子衰变 two-proton decay
双掷 double(t) throw
双掷开关 double-throw switch
双掷旋转开关 double-throw rotary switch
双中殿教堂 double naved church
双中间轴式结构 twin-layshaft configuration
双中频超外差接收法 double superheterodyne reception
双中心规整性 ditactic
双中心经纬仪 double-center[centre] theodolite; double-center [centre] transit
双中心壳模型 two-centre shell model
双中柱 distyele
双中子 dineutron
双钟罩压差计 double bell differential manometer
双众数粒度分布 bimodal size distribution
双周 fortnight
双周波发电机 two-cycle generator
双周抵押 biweekly mortgage
双周对数纸 two-cycle log paper
双周分潮 fortnightly component; fortnightly tide
双周检车库 two-week inspecting shed
双周刊 biweekly; fortnightly
双周期信号 double cycled signal
双周期(信号交叉口) double cycling
双周期性状态 biperiodical regime
双周-三月检车库 two-week and three months inspection shed
双轴 double shaft; tandem; tandem axles; twin axles
双轴半挂车 two-axle semitrailer
双轴拌和机 twin pug
双轴比 dual axle ratios
双轴布置 cross compound arrangement
双轴缠绕法 biaxial winding
双轴铲运车 two-axle scraper
双轴车 bogie truck
双轴车架 two-axle carrier
双轴车轮 tandem wheels
双轴承 duplex bearing
双轴承筛分机 twin-bearing screen
双轴承型电机 two-bearing machine
双轴传动机构 two-axle gear
双轴串联式压路机 two-axle tandem roller
双轴锤式粉碎机 two-shaft hammer mill
双轴锤式破碎机 double shaft hammer crusher; double spindle hammer breaker; double spindle hammer crusher
双轴锤式碎石机 double spindle hammer crusher
双轴锤式轧碎机 double spindle hammer crusher
双轴磁芯存储器 biased memory; biaxial memory
双轴磁芯元件 biax magnetic element
双轴的 biaxial; twin-shaft
双轴等效电路 two-axis equivalent circuit
双轴定向 biaxial orientation
双轴对称 double-axis symmetry
双轴对称截面 double symmetric cross-section
双轴对合 skew involution(in space)
双轴发动机 dual-shaft engine; two-shaft engine
双轴干涉图 biaxial interference figure
双轴各向异性 biaxial anisotropy
双轴挂车 two-axle trailer
双轴光弹(性)(应变)计 biaxial photoelastic(strain)ga(u)ge
双轴光弹仪 biaxial photoelastic(strain) ga(u)ge
双轴光学对称 biaxial optical symmetry
双轴辊轴筛 Ross roll grizzl(e)y
双轴滚齿机 duplex gear hobber
双轴荷载 biaxial load; tandem axle load
双轴混合机 double shaft mixer
双轴混合/挤出机 double arm mixer/extruder
双轴混凝土捏拌机 two-shaft pug-mill concrete mixer
双轴货车 two-axle truck
双轴货挂车 two-axle truck trailer
双轴机组 cross compound unit
双轴激光陀螺仪 two-axis laser gyroscope
双轴尖式 double pivoted pattern; double pivoted type
双轴桨板搅拌机 twin-shaft paddle mixer
双轴胶片显影槽 rewind tank
双轴搅拌混合器 double axial mixer
双轴搅拌机 biaxial-shaft mixer; double shaft pug mill; twin-shaft (paddle) mixer
双轴结构 cross compound arrangement
双轴晶光率体 biaxial indicatrix
双轴晶体 biaxial crystal
双轴抗拉试验 biaxial tensile test
双轴抗压 biaxial compression
双轴抗压强度 biaxial compression strength
双轴练泥机 double arm kneader
双轴两轮驱动车辆 four-by-two vehicle
双轴流式水轮机 double axial flow turbine
双轴挠曲 biaxial bending
双轴耙运机 two-axle scraper
双轴破碎机 double shaft crusher
双轴牵引车 two-axle motor unit
双轴强度 biaxial strength
双轴强制拌和器 twin-shaft compulsory pugmill mixer
双轴桥 bogie beam
双轴切丝机 double bolt cutter
双轴倾斜仪 biaxial tiltmeter
双轴驱动 tandem drive
双轴驱动刮刀 tandem bowl scraper
双轴驱动链 tandem drive chain
双轴驱动式铲运机 two axle drive scraper
双轴驱动式后卸卡车 two-axle drive rear-dump
双轴驱动输送机 tandem drive conveyer[conveyor]
双轴驱动轴套 tandem drive housing
双轴取向 biaxial orientation
双轴全挂车 two-axle full trailer
双轴燃气轮机 dual-shaft gas turbine (engine); two-shaft gas turbine
双轴燃涡轮机 dual-shaft gas turbine engine
双轴筛 twin-shaft screen
双轴筛式捏合机 screen kneader with two shafts
双轴伸长型应变椭圆 biaxial extension type of strain ellipse
双轴伸缩应变 biaxial longitudinal strain
双轴式铲运机 two axles scraper
双轴式的 twin-shaft
双轴式发动机 two-axle engine
双轴式桨叶拌和机 double motion paddle mixer; double shaft paddle mixer; twin-shaft paddle mixer; twin-shaft pug mill
双轴式桨叶搅拌机 double motion paddle mixer; double shaft paddle mixer
双轴式搅拌机 double shaft mixer
双轴式搅拌转子 double shaft mix rotor
双轴式汽轮机 cross compound turbine
双轴式强制搅拌器 double shaft agitator
双轴式拖拉机 four-wheel tractor
双轴式涡桨搅拌机 twin-shaft paddle mixer
双轴式压路机 double axle roller
双轴式装载机 four wheel loader
双轴手纸架 double roll toilet tissue fixture
双轴受拉 biaxial tension
双轴水轮机 two shaft turbine
双轴四轮驱动车辆 four-by-four vehicle
双轴缩短型应变椭圆 biaxial compression type of strain ellipse
双轴探头 biaxial probe
双轴镗床 two-spindle borer
双轴厅 <在克里特的克诺索斯宫中的> Hall of the Double Axes
双轴同步电机 two-axis synchronous machine
双轴拖挂货车 jeep trailer
双轴陀螺稳定平台 biaxial gyrostabilized platform
双轴弯曲 biaxial bending
双轴弯曲应力 biaxial bending stress
双轴涡轮机 cross compound turbine
双轴涡轮喷气式发动机 double compound turbojet
双轴铣床 double milling machine; duplex milling machine
双轴系统 biaxial system
双轴向变形 biaxial deformation
双轴向侧力 biaxial lateral load
双轴向负荷 biaxial loading
双轴向荷载 biaxial loading
双轴向拉伸 biaxial stretch
双轴向拉伸成型机 biaxial stretch-forming machine
双轴向黏[粘]性 biaxial viscosity
双轴向倾斜仪 biaxial inclinometer
双轴向应力 biaxial stress
双轴向应力系统 biaxial stress system
双轴向影响 <泊松比的> biaxial effect(of Poisson's ratio)
双轴向张力 biaxial tension
双轴斜床身卡盘机床 chucker two-axis slant bed machine
双轴斜床身万能机床 universal two-axis slant bed machine

双轴性 biaxiality
双轴悬挂装置 two-axis hitch;two-axis linkage
双轴压力 biaxial compression
双轴压气机 two-spool compressor
双轴压缩 biaxial compression
双轴叶式拌和机 double shaft paddle mixer
双轴仪 biaxial apparatus
双轴应力场 biaxial stress fields
双轴应力应变关系 biaxial stress-strain relationship
双轴应力状态 biaxial state of stress;biaxial stress state
双轴元件 biax element
双轴圆盘锯床 double cut-off saw
双轴载荷 tandem axle load
双轴振动台 biaxial shaking table
双轴主驱动 dual final drives
双轴转向架 two-axle bogie
双轴自由燃气轮机 two-shaft free power turbine engine
双轴坐标 bi-axial coordinates
双肘板 double toggle
双肘板颚式破碎机 brake crusher;double toggle crusher;double toggle jaw crusher
双肘杆颚式破碎机 blade type jaw crusher;Blake-type jaw crusher
双肘管 twin elbow
双肘节 double toggle
双肘破碎机 double toggle crusher
双肘形弯管 double elbow pipe
双绉 crepe de chine
双皱纹农用滚筒 double-corrugated farm roller
双珠饰的 double bead
双主动齿轮单出轴齿轮 twin-input single-output gear
双主动齿轮单出轴减速齿轮 twin-pinion single-output reduction gear
双主动齿轮减速齿轮 twin-input reduction gear
双主动脉球 double aortic knob
双主应力 <三个主应力中的一个等于零> biaxial stress
双主轴箱机床 dual center[centre]
双主柱桩 twin king pile
双助势蹄制动器 two-leading shoe brake
双注电子波管 permactron
双注管 double beam tube;double stream tube
双注射模塑 double shot mo(u)lding
双柱 column pair;double strut;geminated columns;paired columns;pair of columns;twin columns;delphinorum columnae <古罗马竞技场内有雕像的>
双柱插头 two-pin plug
双柱导架 one leader with two stay
双柱底脚 combined(column)footing
双柱吊架 gallows bit(ts)
双柱墩 twin-pier;twin-shaft pier
双柱法注水泥 cementing between two moving plugs
双柱高速压力机 double-sided high speed press
双柱桁构加强梁 queen-trussed beam
双柱桁架 double hanging(roof) truss;queen truss
双柱桁架撑梁 queen-trussed beam
双柱桁架撑木 straining sill
双柱桁架的分柱木 straining sill
双柱桁架的立柱 queen post
双柱桁架加强梁 queen-trussed beam
双柱桁架立杆 prick post;queen post
双柱桁架梁 double strut trussed beam
双柱桁架螺栓 queen bolt
双柱桁架竖杆 queen bolt;queen rod
双柱桁架屋顶 queen-post roof
双柱桁梁 queen-post girder
双柱桁条 queen-post purlin(e)
双柱架 queen post
双柱架与填充墙 queen-post and wind filling
双柱架中的一个柱 queen post
双柱径间距柱廊 systyle
双柱缆桩 double post bollard
双柱廊建筑 dipteral building;dipteros
双柱立式镗床 double column vertical boring machine
双柱联体 twin columns
双柱梁 queen girder
双柱檩屋架 purlin(e) roof with queen post;purlin(e) with queen post
双柱龙门刨床 double column planing machine;two-column planer
双柱龙门铣床 double column planer milling machine
双柱(冒)头 double capital
双柱门廊 in antis
双柱面 bicylindrical
双柱面透镜 bicylinder;bicylinder lens
双柱木桁架 queen-post wooden
双柱木桁架桥 queen-post timber truss bridge
双柱暖气片 two-column radiator
双柱刨床 double column planer
双柱汽车举升器 two-post car lift
双柱人字形起重机 breast derrick
双柱塞摆动缸 rotary actuator with two pistons
双柱塞泵 duplex plunger pump
双柱塞式防喷器 preventer of double ram type
双柱塞压机 double ram press
双柱散热器 two-column radiator
双柱色谱法 dual column chromatography
双柱上撑式大梁 queen-post girder
双柱上撑式桁架 double hanging (roof) truss;queen-post truss
双柱上撑式桁架拉杆 straining beam
双柱上撑式桁架屋顶 queen-post truss roof
双柱上撑式桁架下弦拉杆 straining sill
双柱上撑式桁架柱间弦杆 strutting piece
双柱上撑式梁 queen girder;queen-post beam;queen-trussed beam
双柱上撑式梁桥 queen-post trussed beam bridge
双柱上撑式屋架 queen-post truss roof
双柱升降机 twin-post lift
双柱式 double column;double column type;dyostyle;queen post
双柱式变压器 two-legged transformer
双柱式标志牌 double post sign
双柱式车身底架支撑型举升器 twin-frame lift
双柱式锤 double-framed hammer
双柱式单相变压器 two-legged single-phase transformer
双柱式的 distyle
双柱式墩 twin-shaft pier
双柱式桁架 queen truss
双柱式机床 double column machine
双柱式剪切机 gate shears
双柱式檩条 queen-post purlin(e)
双柱式檩支屋顶 purlin(e) roof with queen post
双柱式龙门刨床 double housing planer
双柱式门廊 distyle in antis;portico with two columns
双柱式汽车举升器 twin-post auto lift
双柱式桥墩 coupled column pier;distyle pier;double column pier;twin-shaft pier;two-columned pier;two-shaft pier
双柱式桥塔 distyle pylon
双柱式升车机 twin-post lift
双柱式塔架 double jack bar
双柱式屋顶 queen-post roof
双柱式屋顶桁架 queen-post roof truss
双柱式压床 straight-side press
双柱式钻架 double jack bar
双柱双锥式卷筒 bi-cylindro-conic-(al) drum
双柱头螺栓 stud
双柱头系缆柱 double head bitt
双柱托架 queen-post truss
双柱托梁 queen-post girder
双柱桅 goal post mast;pair masts;queen mast;queen post
双柱窝式机械手 castle manipulator
双柱屋顶桁架 double post roof truss
双柱屋架 double post roof;queen post
双柱洗脸盆 two-leg lavatory
双柱系船柱 double bitt;double bollard
双柱系船桩 double bitt;double bollard;double piles;mooring bollard;timberheads
双柱系缆柱 double bitt;double bollard
双柱系缆桩 double bitt;double bollard;double piles;mooring bollard;timberheads
双柱下撑式桁架 hog chain truss
双柱型灯头 bipost base
双柱型铁芯 two-column core
双柱型系船柱 double-head bollard;double bitt
双柱型有轨巷道堆垛起重机 double mast S/R [storage/retrieval] machine
双柱压床 straight-side press
双柱压力机 double arm press
双柱液压刨床 double column hydraulic planer
双柱支架 double jack;twin columns support
双柱钻架 double drill column;double jack
双爪夹头 union chuck
双爪卡盘 box chuck;union chuck
双爪锚 double-fluked anchor
双爪钳 vulsellum forceps
双爪式挠性联轴节 double claw flexible coupler
双砖 double brick;twin brick
双转点 <抄平中> double turning point
双转点法【测】double rodded method;double-turning point method;duplex turning point method
双转点水准测量 bilateral level(l)ing;double(line)level(l)ing
双转点水准测量路线 simultaneous double(level)line
双转点(水准)路线 double rodded line
双转鼓干燥器 double-drum drier [dryer]
双转鼓式干燥机 twin-roll drum drier [dryer]
双转鼓锁 duplex lock
双转环套筒 double swivel barrel
双转换呼叫 double switch call
双转轮泵 bi-rotor pump
双转轮式水轮机 twin(-runner type) turbine;two-runner turbine
双转轮式涡轮机 twin(-runner type) turbine;two-runner turbine
双转轮水轮机 twin-runner turbine;twin water turbine
双转盘台 double deck rotary
双转式双滚筒 contra-rotating drum
双转筒研磨机 two-compartment tube mill
双转向 double steering
双转向装置 twin-steer
双转轴细矿机 two-log washer
双转子 dual rotors
双转子摆式罗经 two-gyro pendulous gyrocompass
双转子泵 bi-rotor pump
双转子冲击式破碎机 double rotor impactor
双转子锤式磨 twin rotor hammer mill
双转子锤式破碎机 double impeller breaker;double rotor hammer crusher;impact breaker;twin rotor hammer crusher
双转子的 birotor;two-spool
双转子电动机 spinner motor
双转子发动机 birotary engine;two-spool engine
双转子反击式磨机 double impeller impact mill
双转子反击式破碎机 double impeller impact crusher;double rotor impact pulverizer
双转子鼓风机 Root's blower
双转子辊式破碎机 twin rotor roll crusher;double rotor crusher
双转子连续塑化混合机 twin rotor continuous fluxing mixer
双转子破碎机 twin rotor crusher
双转子燃气发生机 two-spool gas generator
双转子燃气轮机 birotary turbine
双转子涡轮螺桨发动机 compound turboprop
双转子压气机 dual compressor;two-spool compressor
双转子压缩机 two-spool compressor
双桩 double piles
双桩联体 twin columns
双桩帽 double pile helmet
双装药 double load
双锥 bipyramid;duo-cone
双锥标 double tower
双锥垫密封 seal by duplex conical gasket
双锥反转出料式搅拌机 double cone invert discharging type mixer
双锥鼓 dual-cone drum
双锥管接合 cone joint
双锥辊挤浆机 fibercone press;Messing cone press
双锥滚筒式球磨机 double-ended ball mill
双锥环的最大半径间隙 radial gap of double cone ring
双锥混合机 double cone mixer
双锥离合器 double cone clutch;double vee clutch
双锥密封 seal with double cone
双锥密封的螺栓力计算 calculation for bolt load of double cone seal
双锥密封环 double cone seal ring
双锥面垫圈 double cone gasket
双锥面旋轮 <旋压用> double-tapered roller
双锥区 double cone
双锥式混合制粒机 twin-cone mixer-granulator
双锥式混凝土搅拌机 double cone concrete mixer
双锥式卷筒 twin-cone drum
双锥套式加油 two-drogue refueling
双锥体 bipyramidal
双锥体料堆 double cone pile
双锥体倾斜混凝土混合器 duo-cone

S

tilting type mixer
双锥天线 biconic(al)antenna
双锥形 double cone shape
双锥形掺和机 double cone blender
双锥形反射镜 biconic(al)reflector
双锥形混合机 double cone blender
双锥形桨式混合机 double cone impeller mixer
双锥形可倾斜拌和机 duo-cone tilting type mixer
双锥形喇叭 biconic(al)horn
双锥形粒子分级 double cone classifier
双锥形密封 duo-cone seal
双锥形破碎机 double cone crusher
双锥形天线 biconic(al)antenna
双锥形自落式混凝土搅拌机 biconic-(al)gravity concrete mixer
双锥型滚柱轴承 double-tapered roller bearing
双锥型密封牛 du-cone seal
双锥型旋风筒 double cone type cyclone
双锥选粉机 double cone classifier
双锥牙轮钻头 two-cone bit
双锥转鼓 double conical rotary vessel
双锥钻头 double taper bit
双缀 latticing
双缀条 double lacing;double lathing;double latticing
双子浮标 loaded float;twin float
双子叶树材 dicotyledonous wood
双子叶植物 dicotyledon
双字 double word
双字长运算 double length arithmetic;double length operation
双字典 double-dictionary
双字母组 eigram
双自由度陀螺仪 two degrees of freedom gyroscope
双踪扫描 two-trace sweep
双踪示波器 double beam oscilloscope; double-trace oscillograph; double traces oscilloscope
双总线 dual bus
双纵向舱壁油轮 twin-bulkhead tanker
双走道布局 double corridor layout
双走廊房屋 two-aisle building
双阻力理论 two-resistance theory
双阻塞部件 double block
双阻触点式继电器 double pole relay
双组定片电容器 double-stator capacitor;tandem capacitor
双组分 dual pack; two-component; two-package
双组分玻璃纤维 bicomponent glass fiber
双组分催化聚氨酯涂料 two-package catalyst urethane coating
双组分多元醇聚氨酯涂料 two-package polyol urethane coating
双组分复合纤维 bicomponent composite fiber[fibre]
双组分混凝土 two-component concrete
双组分火箭燃料 diergol
双组分挤压纺丝头 bicomponent extruded spinning head
双组分甲基丙烯酸酯化合物 twin-component methacrylate compound
双组分胶粘剂 two-component glue;two-part adhesive
双组分聚氨酯涂料 bicomponent polyurethane coating
双组分流 two-component flow
双组分密封膏 two-component sealant;two-part sealant
双组分密封胶 two-part sealant
双组分密封料 two-part sealant
双组分黏[粘]合剂 two-component adhesive
双组分黏[粘]接剂 two-part adhesive
双组分黏[粘]结剂 two-component glue
双组分配方 two-package formulation
双组分喷枪 two-component spray gun
双组分喷涂 two-component spraying
双组分喷涂机 two-component spray apparatus
双组分燃料 bireactant
双组分式 two-pack system
双组分衰变 two-component decay
双组分体系 two-component system
双组分涂料 two-can coating; two-component coating;two-part paint
双组分系 two(-pack)component system
双组分纤维纺丝装置 bicomponent spinning device
双组分显微煤岩类型 bimaceral microlithotype
双组分预均化堆场 double component blending bed
双组件 twin pack
双组空气抽逐器 two-element air ejector
双组犁 two-gang plough; two-gang plow
双组系统 two-unit system
双组线 twisted-pair line
双组信号 two-way signal
双组液力制动器 dual hydraulic(al) brake
双组元推进剂 bipropellant
双钻 twin-drill
双钻机 dual drill rig;twin-drill
双嘴包装机 double bag filling machine;two-spout unit packing machine
双嘴锄 double pick
双嘴镐 double pick
双嘴割炬 two-tip torch
双嘴喷枪 dual nozzle spray gun
双嘴汽化器 auxiliary jet carburetor;double jet carburetor
双嘴勺 two-lip ladle
双嘴手杓 two-lip hand ladle
双嘴砧 bickern;bick-iron
双作动式制动器 twin-action brake
双作用铲斗 double-acting bucket
双作用锤 double-action hammer
双作用打桩锤 double-acting piling hammer
双作用单水池 double-acting single basin
双作用的 double action;double-acting
双作用的端铣刀 double-action end cutter
双作用阀 double-acting valve
双作用管 double purpose valve
双作用回转振击器 double-acting rotary jar
双作用活塞 double-acting piston
双作用继电器 double-acting relay
双作用铰链 double-acting butt hinge
双作用接力器 double-acting servomotor
双作用进料器 double-acting feeder
双作用离合器 double-action clutch
双作用离心泵 double-acting centrifugal pump
双作用链条 twin duty chain
双作用气腿 double-acting air leg
双作用(汽)缸 double-acting cylinder
双作用汽(桩)锤 <蒸汽和空气> double-acting hammer
双作用千斤顶 double-acting jack;double-action jack;double-function jack
双作用式 difunctional;double-acting
双作用式泵 double-acting pump
双作用式测力计 double-acting forcemeter
双作用式发动机 double-acting engine
双作用式煤气机 double-acting gas engine
双作用水池 double-acting basin
双作用压路机 dual roller
双作用压缩机 double-acting compressor
双作用压缩汽缸 double-acting compressed air cylinder
双作用液压成型法 hydromatic process
双作用液压缸 double-acting hydraulic(al)cylinder;double-acting ram;double-action hydraulic cylinder;two way ram
双作用液压件 double-acting hydraulics
双作用油缸 two-way cylinder
双作用蒸汽锤 double-acting steam hammer
双作用蒸汽机桩锤 double-acting steam pile hammer
双作用止回阀 double check valve
双作用轴箱 dual-system axle box
双坐标外形加工 two-axis contouring
双座敞篷轿车 roadster
双座舱 two-seat cabin
双座单翼机 two-seater monoplane
双座的 two-seat(ed)
双座阀 double beat drop valve;double seat valve;dual-seat valve
双座飞机 two-seater airplane
双座高速敞篷汽车 speedster
双座激光陀螺仪 two-frame laser gyroscope
双座轿车 grand touring car
双座截击机 two-place interceptor
双座跑车 sport coupe
双座平衡阀 double seat balanced valve
双座汽车 two-person car;two-seater
双座轻型汽车 sport car
双座球阀 double seated ball valve
双座式【机】double seater
双座式主汽阀 double seated main stop valve
双座四轮轿式马车 coupe
双座隧道 twin tunnels
双座小(型)汽车 bug;voiturette
双座型 two-seat(ed)version
双座自行车 tandem
双座座舱 two-crew cockpit

霜

霜拔 frost heave[heaving]
霜白表面 frosted finish
霜白花测试 efflorescence test
霜白花盐 efflorescent salt
霜斑 <混凝土表面的> white deposit
霜带 frosty-zone
霜袋地 frost pocket
霜点 frost point
霜点方法 eight dimensional technique;frost-point technique
霜点湿度表 frost-point hygrometer
霜点湿度计 frost-point hygrometer
霜冻 frostbite;frost(ing)
霜冻标准的评定 frost-criteria evaluation
霜冻剥落 frost scaling
霜冻地带 <美国北部的> frost belt
霜冻点 frost point
霜冻防护 frost protection
霜冻感受性 susceptibility to frost
霜冻害 frost damage
霜冻毁坏 damage by frost
霜冻开裂 frost cracking
霜冻孔 frost hole
霜冻裂缝 frost fracture
霜冻轮 frost ring
霜冻敏感的 frost susceptible
霜冻敏感性 frost susceptibility;non-frost susceptibility
霜冻耐受性 susceptibility to frost
霜冻破坏 damage by frost
霜冻区 frost zone
霜冻融化时间 frost-melt period
霜冻砂子摊铺机 frost gritter
霜冻伤疤 frost rib
霜冻渗入地带 frost penetration zone
霜冻渗入深度 frost penetration depth
霜冻损害 damage by frost;frost damage
霜冻天气 frosty weather
霜冻土丘 agollissartog
霜冻危险 frost hazard
霜冻穴 <树木的> frost hollow
霜冻指数 <冷季连续日平均负气温的最大总合值> frost index
霜冻准则 frost criterion
霜冻作用 action of frost;frost action
霜冻作用土壤层 frost-active soil
霜度 degrees of frost
霜沸土 frost boil soil
霜高 frost level
霜害 freeze injury;frostbite;frost injury;injury by frost
霜花 frost flower; ice flower; frost work <水蒸气凝结成的>
霜花玻璃 frosted glass
霜花粗面装饰 frosted rustic work
霜花面饰 frosted finish
霜花纱 frosted yarn
霜花饰 frost work
霜花图案 frost patterns
霜花(纹)装饰 <银器等的> frost work
霜华 salt efflorescence
霜化玻璃 frosted glass
霜化槽 frosting bath
霜化面 frosted face
霜灰色 chateau grey
霜晶 hoar crystal
霜晶石 pachnolite
霜棱或霜脊 frost rib or ridge
霜裂 frost cleft;frost crack
霜瘤 frost canker
霜面化 frosting
霜棚 frost-shed
霜期 frost period; frost season; frost stage
霜日 frost day
霜蚀作用 nivation
霜凇 hard rime;hoar-frost;air hoar;white frost
霜凸地 frost boil(ing)
霜洼 frost hollow;frost pocket
霜纹 frosting
霜雾 frost mist;frost smoke
霜线 frost line
霜形冰 rime ice
霜烟 frost smoke
霜盐 frosting salt
霜灾 frost hazard
霜肿 frost rib
霜柱 ice pillar

爽

爽身粉 talc powder

谁

谁投资谁受益论 theory that those who has invested should be benefited
谁污染谁治理 one who pollutes should be responsible for treating
谁污染谁治理原则 principle of keeping pollution controlled by whomever making

水埃洛石 endellite; hydrated halloysite; hydrohalloysite; hydrokaolin

水安装 water fitting
水铵长石 buddingtonite
水岸 water's edge
水胺硫磷 isocarbophos
水坝 dam; hydraulic dam; hydraulic fill dam; hydro dam; water dam
水坝边缘 dam shoulder
水坝顶 hydro dam crest
水坝观测 dam measurement
水坝和船闸 lock and dam
水坝建筑 construction of dams
水坝溃决 dam break(ing); dam-bursting; dam failure
水坝面板斜坡度 taper of face slab
水坝破坏 dam break(ing); dam failure
水坝失事 failure of dam
水坝式发电站 dam type power plant
水坝梯级 barrage steps; dam steps
水坝体 hydro dam body
水坝重量 weight of dam
水白铅矿 hydrocerussite; plumbonacrite
水白色的 water white
水白色高级燃料油 water-white high-grade burning oil
水白色矿物油 water-white mineral oil
水白色煤油 water-white kerosene
水白色石蜡 water-white paraffin(e) wax
水白色油 water-white oil
水白酸 water-white acid
水白云母 damourite; hydromuscovite
水白云石 hydrodolomite
水百分含量 content of water
水柏油 bitumastic solution
水斑 water mark; water spotting; water stain; water spots
水斑铀矿 ianthinite
水板铅铀矿 richetite
水半管裹接 underhand wiped joint
水半球 water hemisphere; oceanic hemisphere
水包 water drum
水包油 oil in water
水包油乳化泥浆 oil-in-water emulsion mud
水包油乳化液 emulsion oil in water; oil-in-water emulsion
水包油型 oil-in-water type
水包油型乳化剂 oil-in-water type emulsifier
水包油型乳化液 oil-in-water emulsion
水包油型乳剂 oil-in-water emulsion
水包油型乳液 water-phased emulsion
水饱和 saturate with water
水饱和的 water-impregnated; water-saturated
水饱和地层 water-saturated bed; water-saturated formation
水饱和曲线 water-saturated curve
水饱和试验 water saturated test; water saturation test
水饱和土壤 water-saturated soil
水饱和系数 water saturation coefficient
水饱和(岩)层 water-saturated bed; water-saturated formation
水饱和样品 water-saturated sample
水饱和值 water saturation value
水保持 water conservation
水保持论 hydro-conservatism
水保护 water conservation
水保险阀 hydraulic back-pressure valve
水保险器 <焊接> hydraulic back-pressure valve
水爆法 hydroblast(ing)
水钡锶烧绿石 pandaite
水钡铀云母 bergenite
水泵 compressed water pump; flush pump; pump; raw water pump; water(proof) pump; laigh lift <泵组中最下面的>
水泵安置 pump arrangement
水泵安装 installations of pump; pump setting
水泵比速 pump specific speed; specific speed of pump
水泵表 flush pump ga(u)ge
水泵不灌水起动的 self-priming
水泵厂 water pump works
水泵衬毡 water pump felt
水泵充水 priming of pump; pump priming
水泵抽水量 pumpage; pump delivery; pump output
水泵抽压洗井 well cleaning trough the agency of pumping and compressing of pump
水泵出力能力 water pump capacity
水泵出水管 water pump outlet pipe
水泵出水管道 pumping line
水泵出水管阀 discharge valve of pump
水泵出水量 discharge of pump
水泵出水率 capacity of pump; pump capacity
水泵传动挡油圈 water pump drive oil thrower
水泵传动链 water pump driving chain
水泵传动链轮 water pump drive sprocket
水泵传动轴 water pump drive spindle
水泵的活塞余隙 pumping clearance
水泵的水封 pump seal
水泵底座 pump column
水泵垫密片 water pump gasket
水泵动力端 power end
水泵额定流量 capacity of pump
水泵阀 pump valve
水泵房 pump building; pump chamber; pump house; pump plant; pump room; water pump house
水泵房出水池 header tank of pumping house
水泵封垫 water pump gland
水泵封圈推力弹簧 water pump seal thrust spring
水泵负荷 pump load
水泵盖板 water pump cover plate
水泵杆 pump rod
水泵缸套 hoghead
水泵缸(体) pump bowl
水泵工程 pump works
水泵工况运行 pumping operation
水泵固定螺钉 water pump fixing screw
水泵关死扬程 nodischarge pump head
水泵管垫圈 pump ring
水泵规格 ordination of pump
水泵荷载 pump load
水泵护圈 water pump seal
水泵护圈弹簧 water pump seal spring
水泵护圈锁环 water pump lock ring
水泵滑车及风扇 water pump pulley and fan
水泵活塞 water piston
水泵活塞杆的密封 pump packing
水泵机械效率 pump mechanical efficiency; wire-to-water efficiency
水泵机组 pumping unit
水泵基础 pumping unit
水泵级 pump stage
水泵集水井 pump suction well; pump sump
水泵间 pump box; pump house; pump room
水泵间格 pumpway
水泵接合器 fire department pumper connection; pump adapter; pumper connection
水泵结合器 water pump combiner
水泵进水管口滤网 pump strainer
水泵进水花管底阀 suction basket; suction rose
水泵进水软管夹 water pump inlet hose clamp
水泵净吸入高度 net pump suction head
水泵净扬程 net lifting head of pumps
水泵绝缘装置 pump insulation
水泵壳(体) water pump housing; pump housing; water pump cover
水泵坑 pump pit
水泵空气室 pump chamber
水泵空吸状态 <不进水> on air
水泵口径 pump caliber
水泵拉杆拧紧工具 fence jack
水泵廊道 pump gallery
水泵理论排水量 theoretic(al) pump displacement
水泵连杆 string rod
水泵连接器 water pump coupling
水泵莲蓬头 pump basket
水泵联结器 water pump coupling
水泵链动齿轮 water pump chain driving gear
水泵零流量输入功率 nodischarge pump input
水泵零流量扬程 nodischarge pump head
水泵流量 pump capacity; pump discharge; pump flow; pump output
水泵流量扬程曲线 pump capacity head curve; pump discharge head curve
水泵轮毂拆卸器 water pump hub puller
水泵密封 water pump seal
水泵密封弹簧导套 water pump seal spring guide
水泵密封圈螺帽 water pump gland nut
水泵密封填料 water pump packing
水泵名称 name of pump
水泵排出管 pump water conductor
水泵排(出)量 delivery of pump; pump delivery; pump output; pump volume
水泵排水 pumping drainage
水泵排水沟 pump drain
水泵排水连接管 water pump outlet fitting
水泵排水量 delivery of pump; discharge of pump; pump delivery; pump drainage
水泵盘根 water pump packing
水泵旁通弯头 water pump by-pass elbow
水泵配电箱 subdistribution box for pump
水泵皮带轮 water pump pulley
水泵皮碗 pump bucket; pump cup (leather); pump leather
水泵歧管 pump manifold
水泵启动水 automation priming
水泵气蚀 cavitation in pump; cavitation of pump
水泵前垫密片 water pump front gasket
水泵容积效率 pump volumetric(al) efficiency
水泵润滑剂 water pump lubricant
水泵润滑脂 water pump grease
水泵设备 pumping equipment; pumping outfit
水泵深度 pump setting
水泵生产率 delivery of pump
水泵试验 pump test(ing)
水泵室 pump room
水泵输出端 delivery side of pump
水泵输入功率 pump input
水泵输水管 delivery pipe of pump
水泵输送管 pump piping
水泵水封垫圈 water pump seal retainer washer
水泵水封环 water pump gland ring
水泵水封皮碗 water pump seal cup
水泵水封压盖凸缘 water pump gland flange
水泵水封座环 water pump seal retainer
水泵水轮机 pump-turbine; reversible turbine
水泵水头 pump(ing) head
水泵水头损失 water head loss of pumps
水泵弹簧 water pump spring
水泵套垫 water pump bushing packing
水泵特性 characteristics of pump; pump characteristic
水泵特性曲线 pump characteristic curve
水泵体 pump body; water pump body; water pump casing
水泵调节器 pump governor
水泵拖车 trailer pump
水泵外壳 pump case
水泵网 pumping system
水泵吸入管头 snore piece
水泵吸入扬程 suction head
水泵吸收管头 snore piece
水泵吸水高度 pump suction head
水泵吸水管 water pump suction pipe
水泵吸水管底止回阀 foot valve
水泵吸水管阀 suction valve of pump
水泵吸水管滤网 pump basket
水泵吸水孔 pump intake
水泵吸水龙头 pump snoring box
水泵系列 row of pump
水泵系统 pumping system
水泵下置深度 depth of pump installation
水泵效率 efficiency of pump; pump efficiency
水泵型号 model of pump
水泵性能 pump performance
水泵压头 head-on pump
水泵淹没深度 pump submergence
水泵扬程 height of delivery; lifting capacity; lifting head of pumps; lift of pump; pump delivery head; pump head; pump lift
水泵叶轮 impellers of pumps; pump impeller; water pump impeller; water pump vane
水泵叶轮片 pump impeller blade
水泵叶轮气蚀 cavitation of pump impellers
水泵叶轮弹簧 water pump impeller spring
水泵溢流 discharge of pump
水泵用手钳 water pump pliers
水泵油杯 water pump oiler
水泵油膏 water pump grease
水泵有效吸入扬程 net pump suction head
水泵站 lift station; pumping plant; pump(ing) station; pump works; water lodge
水泵站集水井 wet well

S

水泵止推环 water pump thrust washer
水泵轴 pump axis; pump rod; water pump shaft; water pump spindle
水泵轴承 bearings of pumps; water pump bearing
水泵轴齿轮 water pump shaft gear
水泵轴隔套 water pump spindle distance piece
水泵轴抛油环 water pump shaft slinger
水泵轴轴承 water pump shaft bearing
水泵主动轴 water pump driving shaft
水泵主动装置 water pump driving gear
水泵铸铁外壳 water pump cast housing; water pump casting
水泵转子 impeller
水泵装置 pumping arrangement
水泵状态触点 pump status contact
水泵总水头 total pump head
水泵总效率 wire-to-water efficiency
水泵总扬程 gross pump head; total pump(ing) head
水比理论 water ratio theory
水碧 amethyst
水壁炉膛 water-wall furnace
水壁炉子 water-wall furnace
水壁式垃圾焚化炉 water-wall refuse incinerator
水壁式炉 water-wall furnace
水壁式气垫艇 water-wall air-cushion vehicle; water-wall craft
水篦 shutter grate
水边 brim; water front; water edge; waterside
水边车站 waterside station
水边的 aquatic; riparian
水边的陆地、建筑物或市区 water front
水边低沙丘 fore dune
水边地 waterside land
水边地区 riparian zone; waterfrontage
水边公园 waterfront park
水边构件 waterfront component
水边建筑(物) waterfront construction; waterfront structure; waterside construction; waterside structure
水边开发 waterside development
水边砾石 waterside gravel
水边绿地 waterfront green
水边绿化 waterfront green
水边设施 waterfront facility
水边填筑 riparian fill
水边土地业主 riparian owner
水边土地业主的权利 riparian rights
水边围护工程 riparian enclosure
水边线 encroachment line; water-edge line; water front; water edge
水变质作用 aqueous metamorphism; hydrometamorphism
水变阻器 hydrorheostat; water rheostat
水变阻器起动 liquid-rheostat starting
水标尺 water ga(u)ge; draft meter; immersion scale; tide ga(u)ge; tide staff; river ga(u)ge; stream ga(u)ge < 河、溪流的 >
水标高轴线 axis of heights
水标准物质 water standard substance
水表 cock meter; flow ga(u)ge; water flow ga(u)ge; water ga(u)ge; water meter
水表玻璃 ga(u)ge glass; water soda glass
水表玻璃管 water ga(u)ge glass tube
水表玻璃砖 water ga(u)ge glass plate
水表层漂浮生物 neuston
水表齿轮系 meter gear train
水表的流通能力 flow capacity of water meter
水表灯 water ga(u)ge lamp
水表阀 meter stop
水表盖 water meter lid
水表盒圈和盖 meter box ring and cover
水表护套 water ga(u)ge shield
水表计量 water(meter) metering
水表记录负载系数 water meter load factor
水表井 water meter chamber
水表坑 meter pit
水表孔 ga(u)ge cock hole
水表控制的流量 metered flow
水表控制的配水系统 metered system
水表口径 bore of water meter
水表面 water surface
水表面负荷系数 water meter load factor
水表塞门 ga(u)ge cock
水表收费率 meter rate
水表水滴 ga(u)ge cock dripper
水表系统 water meter system
水表箱 delivery box; meter box; water meter box
水表修理工场 meter repair shop
水表旋塞 water ga(u)ge cock
水滨 water edge
水滨的 waterside
水滨地 water front
水滨公园 waterfront park; waterside park
水滨建筑物 waterfront structure; waterside structure
水滨生活的 riparian
水冰 water ice
水兵草 water soldier
水病 hydrophobia
水波 water wave
水波到达时间 water-wave arrivals
水波能 water-wave energy
水波箱法 ripple tank method
水波折射 wave refraction
水波中月影 moonpath
水玻璃 liquid glass; potassium silicate; silicate of soda; sodium silicate; soluble glass; water glass; water soda ash glass; water soda glass
水玻璃二氧化碳硬化砂法 silicate process
水玻璃混合料 soluble glass mix(ture)
水玻璃混合物 water-glass mix(ture)
水玻璃混凝土 sodium silicate concrete; sodium silicon concrete
水玻璃浆 soluble glass paste
水玻璃胶 water-glass coat; water-glass mastic
水玻璃胶结料 sodium silicate cement
水玻璃胶泥 sodium silicate mastic
水玻璃胶凝材料 sodium silicate cement
水玻璃结合耐火浇注料 water-glass-bonded refractory castable
水玻璃矿渣砂浆 water-glass slag mortar
水玻璃玛琋脂 soluble glass mastic; water-glass mastic
水玻璃模数 modulus of water glass
水玻璃耐火混凝土 water-glass refractory concrete
水玻璃耐酸混凝土 water-glass acid-proof concrete
水玻璃腻子 soluble glass putty
水玻璃黏[粘]结的砂轮 vitrified-bonded grinding wheel
水玻璃黏[粘]结剂 sodium silicate adhesive; sodium silicate binder; water-glass mastic
水玻璃黏[粘]结砂轮 silicate-bond wheel
水玻璃嵌料 water-glass putty
水玻璃溶液 soluble glass solution; water-glass solution
水玻璃砂 sodium silicate bonded sand; water-glass bonded sand
水玻璃搪瓷 water-glass enamel
水玻璃涂层 soluble glass coat; water-glass coat
水玻璃涂料 silicate paint; water-glass paint
水玻璃颜料 water-glass colo(u)r; water-glass paint
水玻璃油灰 soluble glass putty; water-glass putty
水玻璃(油)漆 water-glass paint
水玻璃油漆涂层 water-glass paint coat
水玻璃注浆材料 water-glass grout
水玻璃自硬砂 self-setting silicate process
水播植物 hydrochore
水薄荷油 poco oil
水薄膜表面张力 surface tension of moisture films
水簸分析 elutriation analysis
水簸机 hydraulic jig
水簸箕 splash block; splash pan
水不可溶聚电解质 water-insoluble polyeletrolyte
水不溶物 insoluble residue in water
水不溶物含量 insoluble residue content
水不溶物质 water-insoluble matter
水不溶性防胶剂 water-insoluble gum inhibitor
水不溶性切削油 water-insoluble cutting oil
水布植物 hydrochore; hydrosporae
水采法 hydraulicking
水采工人 hydraulic miner
水采样 water sampling
水采样器 water sampler
水采样系统 hydrophone system
水采样装置 water sampling equipment
水采作业 hydraulic mining operation
水彩 tempera; water colo(u)r
水彩壁画 < 抹灰面上的 > secco
水彩画 painting in watercolo(u)r; tempera painting; wash drawing; water colo(u)r painting
水彩画法 aquarelle
水彩画色 water colo(u)r
水彩画颜料 moist colo(u)r; tempera colo(u)r; water colo(u)r
水彩画纸 water colo(u)r paper
水彩溶液 aqueous vehicle
水彩颜料涂白法 gouache
水彩油漆 cold water paint
水参考点 water reference point
水仓 drain pit; sump
水舱 water chamber; water compartment; water lodge; water tank
水舱泵 sump pump
水槽 water channel; basin; fluid passage; flume; gullet; gutter; killesse; launder; pentrough; raggle < 尤指屋檐下的 >; sink; water-bath; water cistern; water groove; water trough; waterway; porthole < 钻头的 >
水槽车 bowser; cistern car; tank car; water truck
水槽地 tidal flat
水槽挂车 tank trailer
水槽壶 kettle
水槽孔道 specus
水槽流线型化 streamlining of water way
水槽模拟 model(l)ing with water tank; water tank model(l)ing
水槽模型实验 water tank model(l)ing test
水槽汽车 tanker; tank truck
水槽去除 water weed removal
水槽生长 water weed growth
水槽式洒水机 flume distributor
水槽试验 flume experiment; flume test; tank test
水槽数目 < 钻头 > number of water way
水槽艇 bowser boat
水槽檐槽 gutter
水槽运输 hydraulic flume transport
水槽驻波 standing wave
水草 aquatic; aquatic growth; aquatic plant; aquatic weed; marshy weeds; water grass; water plant; water weed
水草丛生的河流 weedy river
水草丛生的水体 weediness of waters
水草丛生的水域 weedy waters
水草地 flow meadow; tidal flat; water meadow
水草控制 aquatic weed control
水草切除机 water weed cutter
水草切除器 aquatic weed cutter
水草生长 aquatic growth
水草酸钙石 whewellite
水草滩 water meadow
水草田 pioases; oasis[复 oases] < 沙漠中的 >
水草在水体中大量繁殖 vegetal invasion of water
水册 cadaster[cadastre]
水侧 water side
水侧腐蚀 waterside corrosion
水侧污垢热阻 waterside fouling resistance
水侧污垢系数 waterside fouling factor
水厕 reverse trap; water closet; water lock; reverse trap water closet < 带存水弯的 >
水厕污水 sewage from water closet
水测定评估 water test evaluation
水测高温计 hydropyrometer
水层 aqueous layer; hydrosphere; sheet of waters; water-bearing bed
水层带 pelagic(al) zone
水层单位 hydrostratigraphic(al) unit
水层动物区系 pelagic(al) fauna
水层混合 destratification
水层深度 depth of water layer
水层生活 pelagic(al) mode of life
水层生物 pelagic(al) organism; pelagos
水层食物链 pelagic(al) food chain
水层温度差异区 mesolimnion zone
水层下沉去油法 sinking method
水插 water cutting
水插接 water cutting grafting
水差计 differential manometer
水差压变送器 water different pressure transducer
水掺和黏[粘]土 blunge
水产 marine products
水产动物 aquatic livestock
水产工业 marine products industry
水产海洋学 fisheries oceanography
水产加工车间 fish handling and processing shed
水产加工船 factory ship; factory vessel
水产加工业区 processing industry area of aquatic products

水产冷藏制罐船 refrigerating and cannery ship
水产品(货) marine products; aquatic products
水产品加工厂 aquatic product processing factory
水产学院 fisheries institute; marine products institute
水产养殖 aquaculture; aquatic farm; breeding of aquatic products; fish farming
水产养殖学 aquiculture
水产业 aquatic products industry; fishery industry; marine industry
水产用水 fisheries water; marine products water
水产用水标准 standard of fisheries water
水产植物 aquatic plant
水产专家 fisheries expert
水产资源 aquatic resources; fishery resources
水产资源保护 aquatic resource conservation; conservation of aquatic resources; fishery conservation; fishery protection
水产资源保护法 law of conservation of aquatic resources
水产资源繁殖保护条例 regulations of breeding and conservation of aquatic resources
水产资源开发 development of aquatic living resources
水厂 clarification plant; water plant; water supply works; water works
水厂的运转 waterworks operation
水厂工程 water project; water works
水厂工程师 waterworks engineer
水厂规划 waterworks planning
水厂进水口 intake of water works
水厂铝 waterworks alumin(i)um
水厂取水口 intake of water works
水厂运作 water operation
水厂自用水 water consumed in water works; water consumption in water-works
水场 flow field
水车 aqueous vehicle; Chinese wheel; current wheel; mill wheel; water mill; water-wheel; water cart; water truck; water wagon <运水用的>
水车出水槽 mill tail
水车的水道 mill race
水车的尾水渠 mill tail
水车工 millwright
水车进水槽 mill race
水车轮 water-wheel
水车螺旋叶轮 helical runner
水车尾水渠 mill tail
水车引水道 leat; mill stream <磨坊矿或农场动力的>
水车用的水流 mill stream
水车用水流 mill run
水沉材料 water-laid material
水沉积 water sedimentation
水沉积物 water-deposited material
水成变质岩 para-nocks; pararock
水成变质岩类 pararocks
水成层 aqueous horizon; hydromorphic horizon
水成沉淀岩 hydrolith
水成沉积层 water-borne deposit; water-formed deposit
水成沉积(物) aqueous deposit; hydrogenetic sediment; water-borne deposit; water-borne sediment; water-deposited material; water-deposited matter; water-formed deposit; water-formed sediment; hydatogen sediment
水成成因的 aquiferous
水成纯砂岩 hydrosilicarenyte
水成的【地】 hydatogenous; aqueous; hydatogenic; hydatomorphic; hydrogenetic; hydrogenic
水成电极 water-borne electrode
水成分散 hydromorphic dispersion
水成过程 hydromorphous process
水成交错层理构造 aqueous cross-bedding structure
水成角砾岩 subaqueous breccia
水成矿物 hydatogenous mineral
水成论 neptunianism; neptunian theory; neptunism
水成论的 neptunian
水成论者 neptunian; neptunist
水成锰结核 hydrogenous Mn nodule
水成磨损 water worn
水成片麻岩 paragneiss
水成片岩 paraschist
水成说 neptunianism
水成塑性 hydroplasticity
水成碎屑 hydroclast
水成碎屑的 hydroclastic
水成碎屑岩 hydroclastic rock; hydroclastics
水成碳酸盐碎屑岩 hydrolith
水成土 aquatic soil; aqueous soil; aquic taxa; hydrogenetic soil; hydrogenic soil; hydromorphic soil
水成物质 hydrogenous material; water-borne material
水成型态 hydromorphy
水成岩 aqueous rock; hydatogenous rock; hydrogenic rock; hydrogenous rock; hydrolith; katogene rock; neptunic rock; sedimentary rock
水成岩薄层剥离 peel thrust
水成岩层理面 bedding plane
水成岩的岩化过程 lithification
水成岩盖 hydrolaccolith
水成岩构造 dish structure
水成岩内小褶皱 intraformational fold
水成岩墙 neptunic dike[dyke]; sedimentary dike[dyke]
水成盐 water-borne salt
水成异常 hydromorphic anomaly
水成有机污染物 aqueous organic pollutant
水成者 neptunist
水成组分 hydrogenous constituent
水成作用【地】 hydatogenesis; hydrogenesis
水程 water lead; water path
水程航行 voyage
水程里格 marine league
水澄明度仪 water clarity meter
水澄清剂 water clarifier
水澄清器 water clarifier
水澄清作用 water classification
水池 basin; cistern; collecting basin; holding pond; pond; pool; water basin; water chamber; water cistern; water pit; water pool; water pot; water tank
水池成粒法 pit process
水池处理 tank treatment
水池的日调节容量 daily pondage
水池防渗衬砌 reservoir seepage-proof lining
水池灌溉 tank irrigation
水池龙头 <厨房> sink bib
水池排水管 sink drain; tank sewer
水池墙 wall of a basin
水池容量 tank content
水池渗漏 seepage of tank
水池式反应堆 aquarium reactor
水池式鱼道 pool(-type) fish pass
水池试验 tank test
水池水淬法 pit process
水池水位 pool elevation
水池位置 location of tank
水池型堆 open-pool reactor
水池养护 curing by ponding
水池溢流量 tank overflow
水池指示器 water tank indicator
水尺 depth ga(u)ge; draft mark; ga(u)ging rod; hydrometric(al) ga(u)ge; staff ga(u)ge; tide pole; tide staff; water ga(u)ge; water stake
水尺读数 ga(u)ge reading; staff reading
水尺高程 water ga(u)ge level
水尺高度 ga(u)ge height
水尺计重 checking weight by draft
水尺检量 draught survey
水尺检验 draught survey
水尺井 ga(u)ging well; stilling well
水尺零点 ga(u)ge datum; datum plane of ga(u)ge; ga(u)ge zero; zero of ga(u)ge; zero point of rod
水尺零点标高 elevation of ga(u)ge datum; elevation of ga(u)ge zero
水尺零点高程 elevation of ga(u)ge zero; level of ga(u)ge zero; level of zero of ga(u)ge
水尺零点以上高程 height above datum
水尺零点以上高度 height above datum
水尺水位 water ga(u)ge level
水尺特定读数 <对应某一给定流量> specific ga(u)ge reading
水尺误差 staff error
水尺钟 ga(u)ge clock
水尺组 multiple tide staff
水赤铁矿 hydrohaematite; turgite[turjite]
水赤铜矿 hydrocuprite
水充满的 water-impregnated
水冲 flashing; sluicing; water flush; water jet(ting)
水冲采掘 jetting process
水冲厕座 water closet
水冲沉桩 hydraulic jet piling; water-jet piling; water jetting of pile; pile jetting
水冲沉桩法 jetting piling; pile jetting (method); sinking by jet piling; sinking by jetting; sinking (of) pile by water jet
水冲成粒法 jet process
水冲锤击(混合打桩)法 combined flushing and ramming
水冲打桩法 water-jet driving; water-jet method of pile-driving
水冲打桩机 water-jet pile driver
水冲法 jetting process; water-jetting(method)
水冲法沉排抛石护岸 falling apron
水冲法沉桩 pile jetting(method)
水冲法打桩 flushing and ramming
水冲法取样 jet probe
水冲法填土 jetting fill
水冲击 rush of water; water slug
水冲击力 hydraulic impact
水冲击探测器 water impact detector
水冲螺旋桩 jetted screw pile
水冲马桶 water closet
水冲耙头 water jet draghead
水冲筛分析 washed sieve analysis
水冲蚀作用 water erosion
水冲式厕所 water closet
水冲式分选机 water-impulse separator
水冲式沟渠 water-borne sewerage
水冲式贯入仪 wash-point penetrometer
水冲式钻进 wash drilling
水冲式钻探 wash boring; wash drilling
水冲刷 water erosion; water scouring
水冲刷排驱机理 water flushing expulsion mechanism
水冲碎石机 water-cannon rock-breaker
水冲探测 wash probing
水冲填土法 jetting fill
水冲污水 water-borne sewerage
水冲洗 water flushing; water wash(ing)
水冲系统 water-carriage system
水冲穴 <岩溶地区的> light hole
水冲压力 hydraulic impact
水冲岩屑钻进 self-cleaning drilling
水冲圆的 water-rolled
水冲渣 flush slag
水冲桩 jet(ted) pile
水冲状脉 water hammer pulse
水冲钻 water hammer drill
水冲钻进 wash drilling; wash boring
水冲钻进法 wash boring method
水冲钻孔 hydroauger hole; self-cleaning drilling; wash boring
水冲钻孔法 hydrauger hole method; wash boring method; water-boring method
水冲钻孔机 water hammer drill
水冲钻探 self-cleaning drilling; water flush boring; water flush drilling; washboring
水冲钻探法 hydrauger method
水冲钻探孔 hydrauger hole
水冲作用 water hammer
水抽出物 water extract
水抽子 ejector pump; hydroejector; water-operated ejector
水臭氧化作用 aqueous ozonation
水除臭 water deodo(u)ring
水除氟 water defluorination
水除气 water degassing
水除气设备 water deaerating equipment
水除色 water decolo(u)ring
水除霜 water defrosting
水除味 water taste control
水储存 water storage
水储器 drumwall
水处理 aqueous treatment; treatment of water; water conditioning; water processing; water treating
水处理厂 water conditioning plant; water purification plant; water treatment plant
水处理厂废水 water treating plant waste (water); water treatment plant wastewater
水处理厂污泥 water treatment plant sludge
水处理单元 water processing element; water treatment entity; water treatment unit
水处理单元过程费用函数 cost function of unit process of water treatment
水处理单元过程(最)优化设计 optimization design of unit process of water treatment
水处理单元过程作业优化设计 optimization design of unit operation in water treatment
水处理工程 water treatment works
水处理工艺 water treatment process; water treatment technology
水处理工艺流程 water treatment technological process
水处理缓蚀剂 water treatment corrosion inhibitor
水处理技术 water treatment technology

水处理剂 water treatment agent; water treatment chemicals
水处理料剂 water treatment medium and reagent
水处理器 water conditioner
水处理清洗剂 water treatment cleaning agent
水处理杀菌剂 water treatment biocide
水处理设备 water conditioning plant; water purification structure; water treating equipment; water treatment equipment; water treatment facility
水处理设施 water treatment facility
水处理室 water treatment room
水处理系统 water disposal system; water handling system; water treatment system
水处理系统优化设计 optimization design of water treatment system
水处理项目 item of water treatment
水处理絮凝剂 water treatment flocculant
水处理药品 water treating chemicals
水处理运作 water treatment operation
水处理质量 quality of treated water
水处理终端过滤器 polisher
水处理装置 water treating equipment; water treatment equipment; water treatment facility; water treatment plant; water treatment unit
水处治 water cure
水传病原体 water-borne pathogen
水传播病 water-borne disease
水传播病原生物 water-borne pathogenic organism
水传播的污染物 water-borne contaminant
水传播放射性 water-borne radioactivity
水传动 water transmission
水传染 water-borne infection
水传(染)疾病 water-borne disease
水传染细菌性疾病 water-borne bacterial disease
水传声音 water-borne sound
水传时疫 water-borne epidemic
水传送 water transmission
水传物 water-borne body
水船 water barge
水串时清扫效率 breakthrough sweep efficiency
水锤 hammer blow; water hammer; impingement; knocking; water ram
水锤泵 hydraulic ram pump; ram pump; water hammer pump
水锤泵站 ram station
水锤波 water hammer wave
水锤测流法 pressure-time method
水锤冲击 hammer shock
水锤的再现期 resurge phase of water hammer
水锤法 pressure-time method
水锤防护装置 water hammer arrester [arrestor]
水锤腐蚀 impinging corrosion
水锤击 hammering
水锤击吸收器 water hammer shock absorber
水锤间隔时间 interval of water hammer
水锤试验机 water hammer tester
水锤现象 water hammer
水锤消除器 water hammer arrester [arrestor]
水锤压力 water hammer press(ing); water hammer pressure
水锤扬水机 hydraulic ram; water ram
水锤噪声 water hammer noise
水锤制止器 water hammer arrester [arrestor]
水锤作用 water hammer action; water hammering
水锤作用消除器 water hammer eliminator
水磁 hydromagnetic
水磁化处理 water magnetization treatment
水磁铁矿 hydromagnetite
水磁学 hydromagnetism
水从表面向下冻结 water freezes from the surface downwards
水葱 great bulrush
水淬 drag ladle; quench in water
水淬槽 watch-quench tank
水淬成粒法 hydraulic granulation; water granulation
水淬的 water-quenched
水淬多孔矿渣 expanded slag
水淬法 jet process; water quench(ing)
水淬法(制)碎玻璃 drag-ladled cullet; quenched cullet; shredded cullet; draghead cullet
水淬方法 slaking process
水淬钢 water-hardened steel
水淬高炉(矿)渣 granulated blast furnace slag; quench blast furnace slag; water quench blast furnace slag
水淬火 water hardening; water quench(ing)
水淬火槽 water quench tank
水淬矿渣 slaking slag
水淬粒状高炉矿渣 water-quenched granulated blast-furnace slag
水淬粒状矿渣 water-granulated slag
水淬粒状熔渣 water-granulated slag
水淬裂纹 water crack
水淬炉渣 foamed blast-furnace slag; quenched blast furnace slag
水淬煤渣烧结制品 water-quenched cinder sintered product
水淬泡沫法 <熔渣的> water foaming method
水淬碎玻璃 dragaded cullet
水淬硬化 water hardening; water quench(ing)
水淬硬化法 water hanger
水淬硬盘条 water-hardened wire rod
水淬渣 <又称粒状渣> water-quenched slag; granulated slag; granulating slag
水萃取 water extraction
水萃取法 water extraction method
水存法 water storage
水存木材制材厂 wet mill
水打磨 wet sanding
水带 hardwall hose; water band
水带包布 hose bandage; hose gaiter
水带泵车 hose laying lorry; hose tender
水带比阻 water hose resistance
水带编织层 hose carcass
水带操作员 hose operator
水带储藏架 hose storage rack
水带储藏室 hose storage room
水带的带束叠装 flake
水带的马蹄形叠装 hoseshoe hose load
水带的铺设 hose laying
水带的弹性 elasticity of hose
水带吊带 hose becket
水带吊钩 hose hoist
水带吊索 hose sling
水带堵头 hose cap
水带断流器 hose strangler
水带房 hose house
水带分水器 hose manifold
水带负载 hose load
水带附件 hose accessory; hose tools
水带覆盖层 hose casing
水带挂钩 hose strap
水带滚筒 hose roller
水带烘房 hose drier[dryer]
水带护桥 hose bridge; hose jumper; hose ramp
水带接口扳手 hose coupling spanner; hose coupling wrench
水带接口螺纹 hose thread
水带卷盘 hose roll
水带螺塞 hose plug
水带三脚架 hose control
水带塔 hose tower
水带提带 hose belt
水带跳动 hose pulsation
水带外层 hose jacket
水带线泵送 pumping in-line
水带箱 hose bed; hose compartment; hose tank
水带异径接口 hose adapter
水袋 water bag
水单硫铁矿 hydrotroilite
水胆矾 brochantite; warringtonite
水蛋白石 hydrophane
水当量 <雪溶后的水深> water equivalent
水道 navigation water course; water course; conduit; flow channel; pass way of water; racecourse; race track; rhine; stream course; water channel; water lane; water passage; water passway; water race; water road; water tunnel; waterway; raceway <美>; klong <泰国>; leat; gullet
水道保护 channel protection
水道边浅滩 middle ground
水道变迁 shifting of waterway
水道标 sign for water pipe【给】; mark for waterway【航海】
水道标示器 channel marker
水道测量 hydrographic(al) measurement; hydrographic (al) survey(ing); sound ranging
水道测量部 hydrographic(al) office
水道测量船 hydrographic(al) ship; hydrographic(al) survey ship; hydrographic(al) survey vessel; hydrographic(al) vessel
水道测量的 hydrographic(al)
水道测量公报 hydrographic(al) bulletin
水道测量精密扫描回声测深仪 hydrographic (al) precision scan(ning) echo sounder
水道测量局 hydrographic(al) office; hydrographic(al) service
水道测量六分仪 hydrographic(al) surveying sextant
水道测量人员 hydrographer
水道测量术 hydrography
水道测量数据处理系统 hydrographic(al) data processing system
水道测量数字定位与深度自记系统 hydrographic(al) digital positioning and depth recording system
水道测量艇 surveying boat
水道测量图 hydrograph
水道测量学 hydrography; hydrographic(al) surveying
水道测量学家 hydrographer
水道测量与成图系统 hydrographic(al) survey and charting system
水道测量与制图系统 hydrographic(al) survey and charting system
水道测量员 hydrographer; nautical surveyor
水道测量者 hydrographer; nautical surveyor
水道测量资料 hydrographic(al) data
水道测量资料收集系统 hydrographic(al) data acquisition system
水道测深 hydrographic(al) sounding
水道测深(定)线 alignment of sounding
水道充填交错层理【地】 channel fill cross-bedding
水道出口损失 outfall loss
水道床 water course bed; waterway bed
水道导流工程 channel training works
水道定线 waterway alignment
水道堵塞 channel block
水道断面面积 cross-sectional area of waterway
水道发展管理局 <美> Waterway Development Authority
水道阀门 conduit valve
水道分汊(处) bifurcation; channel bifurcation
水道分汊点 <对进口船而言> bifurcation point
水道浮标 channel buoy
水道改变 channel change
水道改道 channel change
水道干线 trunk line waterway
水道钢丝拖运器 wire drag
水道港口海岸与海洋工程学报 <美国土木工程学会季刊> Journal of Waterway, Port, Coastal and Ocean Engineering
水道工程 waterway engineering; waterway works
水道横截面 cross-section of waterway
水道横向变化 lateral change of channel
水道汇合点 confluence of channels; junction of channels
水道基床 water course bed
水道及港口分会会刊 Journal of the Waterway and Harbo(u)r Division
水道加固建筑物 waterway stabilization structure
水道建设工程 waterway project
水道降水 channel precipitation
水道交叉 stream crossing
水道交叉地点 stream crossing site
水道交通 waterway traffic
水道截面积 area of waterway
水道进口 conduit entrance
水道开发 waterway development
水道开挖 channel excavation
水道勘测 water course survey; waterway survey
水道宽度 channel width
水道扩大段 expansion in channel
水道类型 channel pattern
水道流向 flowing alignment
水道轮廓 channel configuration
水道面积 waterway area
水道模 channel cast
水道摩阻力 conduit resistance
水道、汽车联运运价率 joint water-motor rate
水道区段 level reach
水道容量 water-carrying capacity
水道入口 channel entrance; channel intake; entrance of a channel
水道设立航标 marking of a channel
水道渗透能力 water percolating capacity
水道实验站 waterway experiment station
水道试验站 <美国陆军工程兵团> Waterways Experiment Station

水道输水 channel conveyance
水道刷深 channel degradation
水道隧道 waterway tunnel
水道隧洞 waterway tunnel
水道损失 channel loss
水道、铁路联运价率 joint water-rail rate
水道、铁路、汽车联运价率 joint water-rail-motor rate
水道通过能力 waterway capacity
水道通行权 right to watercourses
水道图 hydrographic(al) chart; hydrographic(al) map
水道弯曲段 channel bend; channel with a bend
水道网 hydrographic(al) net(work); waterway net(work)
水道稳定措施结构 waterway stabilization structure
水道污染 water courses pollution
水道系统 channel system
水道斜坡进口 mitered inlet
水道学 fluviology
水道移动 channel shift
水道淤积 channel aggradation
水道运输驳船协会 <美> Water Transport Barge Association
水道运输业者 water carrier
水道整治工程 channel regulation works; channel training works
水道指示灯 channel light
水道中的湍急水流 quick-water in water channel
水道中沙洲 middle-ground shoal
水道中心线 channel center [centre] line
水道铸型 channel cast
水道状况 channel regime(n)
水道自动测量系统 hydrographic(al) automated system
水道阻力 resistance of water course
水稻 aquatic rice; lowland rice; paddy; paddy rice; water seeded rice; wet rice
水稻地灌溉系统 rice-based irrigation system
水稻联合收割机 rice combine
水稻田 paddy; paddy field; rice paddy
水稻田用农药 paddy field pesticide
水稻土 paddy soil; rice paddy soil
水稻秧田 rice nursery; rice seed bed
水的 aquatic; water-borne; watery
水的宝贵特征之一 one of valuable property of water
水的饱和 water saturation
水的保持 water conservation
水的保护与采集 conservation and collection of water
水的比重 density of water
水的表面膜 surface film of water
水的表面张力 surface tension of water
水的冰点 water freezing point
水的补充 provision of water
水的补给 recharging
水的不合理使用 unreasonable use of water
水的不渗透性 water imperviousness
水的侧向冲刷作用 lateral erosion by water action
水的成本 water cost
水的成浆作用 puddling action of water
水的澄清 water clarification
水的澄清度 water clarity
水的冲击 wave shock
水的冲刷作用 erosion by water action
水的重复利用 reuse of water; water recycling; water reuse
水的重复利用率 rate of water reuse
水的臭味 water odo(u)r
水的臭氧处理 ozone treatment of water
水的臭氧化作用 water ozonization
水的除臭 deodo(u)rization of water
水的除盐 de-ionization; desalting of water
水的除氧 deactivation of water
水的处理 disposal of water; water treatment
水的传输 water transmission
水的磁化 water magnetization
水的磁性处理法 magnetic water treatment
水的大肠杆菌群值 coli-group titre of water
水的大肠杆菌指数 coli-group index of water
水的淡性 freshness of water
水的地下储存 ground storage of water
水的电导度 electric(al) conductivity of water
水的电加热装置 electric(al) water heating appliance
水的电解 electrolysis of water
水的电离 ionization of water
水的动黏[粘]滞度 kinematic(al) viscosity of water
水的动黏[粘]滞性 kinematic(al) viscosity of water
水的毒物指数 toxicological index of water
水的钝化 deactivation of water
水的法规 water code
水的翻转 turnover of water
水的反复利用 repeating use of water
水的放射性 radioactivity of water
水的非碱性硬度 non-alkaline hardness
水的分布 water distance
水的分层 stratification of water
水的分解 decomposition of water
水的分类 water classification
水的分配 allocation of water
水的分析 water analysis
水的丰度 abundance of water
水的浮力 buoyancy of water
水的辐射消毒 radiation sterilization of water
水的腐蚀性 corrosiveness of water; corrosivity of water
水的腐蚀作用 corrosive action of water
水的复原 restoration of water
水的富养分化 water enrichment
水的感官性状 aesthetic properties of water
水的感官质量 water aesthetic quality
水的高潮 high tide of water
水的供给 provision of water
水的固定 fixation of water
水的固定残渣 fixed residue of water
水的管理 water management
水的规划 water plan
水的过程线 hydrograph
水的过滤 filtration of water; water filtration
水的耗量 water consumption
水的后处理 water after-treatment
水的化学成分 chemical composition of water
水的化学成分含量 chemical content of water
水的化学分析 chemical water analysis
水的化学结构 water structure
水的化学结构理论 water structure theory
水的化学净化 chemical water purification
水的化学特征 chemical characteristics of water
水的化学性质 chemical properties of water
水的化学性状 chemical properties of water
水的缓冲强度 buffer intensity of water
水的缓泻效应 laxative effect of water
水的回收 recovery of water
水的回用 reuse of water; water reuse
水的混床除盐 mixed bed deionization
水的混凝 coagulation of water
水的混浊度 turbidity of water; water turbidity
水的集中处理 central water treatment
水的加氯消毒 water chlorination
水的加氯消毒法 chlorination of water
水的加热软化法 water softening by heat(ing)
水的间接再用 indirect-use of water
水的减活 deactivation of water
水的减活化作用 deactivation of water
水的碱性硬度 alkaline hardness
水的检定 water detection
水的简易消毒处理 simplified water disinfection
水的碱度 alkalinity of water
水的截流 interception of water
水的进入 ingress of water
水的净化 decontamination of water; depollution of water; water decontamination; water purification; water treatment
水的净化处理 water treatment
水的净化与再利用 water purification and reuse
水的净化作用 water clarification
水的开发 water development
水的可口性 palatability of water
水的可压缩性 water compressibility
水的可饮用性 potability of water
水的控制运用 water control
水的矿化 mineralization of water
水的矿化度 mineral content of water
水的离解 dissociation of water
水的离子积常数 ionization product constant of water
水的利用 utilization of water; water use; water utilization
水的利用含量 utilizing water content
水的利用量 utilizing water capacity
水的利用效率 water use efficiency
水的两相聚合分离 aqueous two-phase polymer separation
水的临界深度 critical depth of water
水的流入 water inrush
水的硫化物 sulfide of water
水的硫酸盐 sulfate of water
水的氯化处理 chlorination of water; water chlorination
水的氯化物 chloride of water
水的氯化消毒法 chlorination disinfection of water
水的毛细上升 capillary ascension of water
水的密度 density of water; water mass density; moisture density <土中的>
水的面积 water area
水的黏[粘]度 viscosity of water; water viscosity
水的凝聚 coagulation of water
水的农药污染 aquatic pesticide contamination
水的排泄 water excretion
水的漂失 drift loss
水的坡度 hydraulic grade
水的侵蚀性 aggressiveness of water
水的侵蚀作用 erosion by water action
水的氢稳定同位素值 value of stale hydrogen isotope in water
水的氢稳定同位素组成 stable hydrogen isotopic composition of water
水的清洁标准 clean water standard
水的去污染 depollution of water
水的全分析 complete water analysis
水的扰动 water disturbance
水的热处理 treatment of hot water; treatment of thermal water
水的热容量 heat capacity of water
水的热性质 thermal property of water
水的容重 unit weight of water
水的溶解度 aqueous solubility
水的溶解性固体 soluble solids of water
水的蠕升 water creep
水的入侵 encroachment of water
水的软化 demineralization of water; water softening; hardness removal; mitigation; softening; softening of water; water correction; water demineralizing
水的软化剂 deincrustant
水的软化装置 apparatus for water softening
水的三相点温度 triple point of water
水的色度 colo(u)r of water
水的色度和浊度 colo(u)r and turbidity in water
水的上升运动 upward movement of water
水的射解 radiolysis of water
水的审计 water audit
水的渗漏 water seepage
水的渗入 seepage of water
水的渗透 water filtration; water intrusion; water permeation
水的生化污染指标 biologic(al) index of water-pollution
水的生物污染指数 biologic(al) index of water-pollution
水的试验 water test
水的输送 water conveyance; water transport(ation)
水的输送设备 water conveyance facility
水的酸度 acidity of water; water acidity
水的酸化 aquatic acidification
水的酸碱值 pH value of water
水的损失 water loss
水的碳酸盐硬度 carbonate(d) hardness of water
水的特殊成分分析 special composition quality of water
水的特性 characteristics of water
水的提取 abstraction of water
水的提取应用 abstractive use of water
水的提升 lifting of water; raising of water
水的提升高度 static-elevation difference
水的体积热膨胀系数 thermal expansivity of water volume
水的调质 water conditioning
水的通道 passage of water
水的通过量 throughput of water
水的同位素 water isotope
水的透明度 transparency of water; water clarity; water transparency
水的突变性 mutagenicity of water

S

水的推移力 traction force; tractive force
水的脱矿化 demineralization of water
水的脱盐 desalting of water; water demineralizing; water desalting
水的脱盐作用 water desalination
水的微生物污染 microbial contamination of water
水的温度波动 temperature fluctuation in water
水的温度分级 grade of water temperature
水的稳定处理 vaccination of water
水的污染 water contamination; water pollution
水的污染含量 water contaminant level
水的物理检测 physical examination of water
水的物理检验 physical examination of water; physical quality of water
水的物理特性 physical property of water
水的物理性质 physical property of water
水的吸收 water uptake
水的吸收能力 water sorptivity
水的细菌分析 water analysis of bacteria; bacterial composition analysis of water
水的细菌检验 bacterial examination of water; bacteriological examination of water
水的细菌净化 bacterial purification of water
水的显微镜检验 microscopic examination of water
水的相对渗透率 relative permeability to water
水的消毒 disinfection of water; water disinfection; water sterilization
水的消耗 consumption of water; expenditure of water
水的性质 water property
水的需求 water demand
水的需求量 water requirement
水的悬浮固体 suspensible solid of water
水的循环 circulation of water; water circulation; recycling of water
水的循环利用率 rate of water circulation and water utilization
水的循环使用 <已处理的> water recycling
水的压力 water pressure
水的亚型 water sub-type
水的颜色 water colo(u)r
水的养分富化措施 water enrichment
水的氧稳定同位素值 value of stale oxygen isotope in water
水的氧稳定同位素组成 stable oxygen isotopic composition of water
水的液相离子平衡 aqueous phase ionic equilibrium of water
水的硬度 alkaline hardness; water hardness; hardness of water
水的硬化 hardening in running water
水的永久硬度 permanent hardness of water
水的有效电阻率 effective water resistivity
水的有效利用 beneficial use of water; water benefit use
水的有效渗透率 effective permeability to water
水的有效性 availability of water
水的有益利用 beneficial use of water
水的运动 water movement
水的运动黏[粘]度 kinematic(al) viscosity of water
水的再利用 water reuse
水的再曝气 water re-aeration
水的再生和再用 water renovation and reuse
水的再用 reuse of water
水的蒸馏 water distillation
水的蒸气压 vapo(u)r pressure of water
水的质量密度 mass density of water
水的重金属离子污染源 heavy metal ion source of water pollution
水的重力循环系统 water circulation gravity system
水的专用 private use of water
水的专用权 water appropriation
水的浊度 turbidity of water
水的自由流动层 free sheet of water
水的综合管理 integrated water management
水的综合利用 multipurpose use of water
水的总固体 total solids of water
水的总硬度 total hardness of water
水的阻力 water resistance
水的组分 water composition
水的最终用途 ultimate use of water
水滴 dripper; drip(ping); drop; tear drop; water drip; water drop(let)
水滴冲刷 water drop scavenging
水滴大小 drop size
水滴大小分布 drop size distribution
水滴挡板 drift eliminator
水滴定法 water titration(method)
水滴花纹玻璃 water drop glass
水滴花样玻璃 dewdrop glass
水滴集电器 water dropper
水滴冷却 cooling through water droplets
水滴破碎理论 breaking-drop theory
水滴清除 <云中尘埃被形成的> rain-out
水滴渗透试验 drop penetration test(ing)
水滴声 dripping
水滴式冷却塔 drop cooling tower
水滴试验 water drop test
水滴形油罐 Horton spheroid; spheroid
水滴直径 drop diameter; drop size
水滴状气球 teardrop balloon
水底 submarine; sunken; water bottom
水底爆炸 underwater blasting
水底波 bottom wave
水底采泥机 bag dredge
水底采样袋 bag dredge
水底采样器 bottom sampler; dredge(r)
水底参考电缆 ocean bottom reference cable
水底沉积(物) benthal deposit; benthal sediment; subaqueous deposit; bottom deposit
水底粗粒有机物 coarse benthic organic matter
水底大型植物 submerged macrophyte
水底的 benthal; benthic; subaquatic; subaqueous; submarine; sunk
水底等高线 bottom contour; submarine contour
水底地形 bottom configuration; bottom contour; bottom relief; bottom topography
水底地形图 bathyorographic(al) map; bottom relief drawing; hypsobathymetric(al) map
水底电报 submarine telegraphy
水底电缆 stream cable; subaquatic cable; subaqueous cable; submarine cable; underwater cable
水底电缆电报系统 submarine cable telegraph system
水底电缆浮标 cable buoy; telegraph buoy
水底电缆故障信号铃 submarine call bell
水底工程 submarine works
水底公路 underwater highway
水底管 submerged pipe
水底管形隧道 immersed tube tunnel
水底滑动 subaqueous glide
水底环境传感系统 bottom-environmental sensing system
水底混凝土 subaqueous concrete
水底拉曳法 bottom pull method
水底流 bed current
水底锚链浮标 ground cable buoy; ground chain buoy
水底锚链浮筒 ground cable buoy; ground chain buoy
水底煤气发生炉 water-bottom gas producer
水-底泥界面 water-sediment interface
水底爬行(观测)器 creeper
水底坡度 underwater slope; underwater gradient
水底浅层剖面 subbottom profile
水底浅层剖面测量 subbottom profiling
水底浅层剖面仪 subbottom profiler
水底区域 bottom region
水底取土器 bottom sampler
水底取土钻 bottom sampler
水底取样器 bottom probe; bottom sampling device; bottom sampler
水底取样设备 bottom sampling device
水底群落 bottom community
水底人行地道 underwater pedestrian tunnel
水底设施 submarine installation
水底摄影 bottom photography
水底深处的 benth(on)ic
水底生物 benthon; benthos; bottom-dwelling organism; bottom organism
水底生物带 benthos belt; bottom belt
水底实验室 habitat
水底水雷 submarine mine
水底隧道 immersed tunnel; subaqueous tunnel; submarine tunnel; submerged tunnel; underwater tunnel
水底探测器 bottom probe
水底特性 bottom characteristic; nature of the bottom
水底铁路隧道 chunnel
水底停留时间 bottom time; duration of staying on bottom
水底通道 underway
水底土壤 underwater soil
水底拖网 bottom trawl
水底拖曳计程仪 <浅水航行测定航向与速度的仪器> ground log
水底挖泥抓斗 bottom grab
水底微生物 benthic microorganism
水底物质密度 bed density
水底细粒有机物 fine benthic organic matter
水底线路 submarine line
水底研究室 habitat
水底样品 benthic sample
水底游动植物 nectonic benthon
水底有机物 benthic organic matter
水底有机物沉积 benthal deposit
水底整平船 knife holder ship
水底植物 phytobenthon; plant benthon
水地球化学测量 geochemical water survey
水碲氢铅石 oboyerite
水碲铁矿 mackayite
水碲铜石 graemite
水碲锌矿 zemannite
水点 water droplet
水点侵蚀 erosion
水碘钙石 brueggenite
水碘铜矿 bellingerite
水电 hydroelectric(al) power; hydroelectricity; hydropower; water power
水电坝 hydroelectric(al) dam
水电比拟(法) electrohydrodynamic(al) analog(ue); conductive liquid analog(ue)
水电部门支出 water and electricity sector expenses
水电厂 hydraulic power plant; hydroelectric(al) plant; hydroelectric(al) power plant; hydroplant
水电传动装置 electrohydraulic actuator
水电的 hydroelectric(al)
水电洞室 power-house cavern
水电段 water and power supply section
水电工程 engineering scheme hydroelectric(al) engineering; hydroelectric(al) power project; hydroelectric(al) project; hydroelectric(al) scheme; hydroengineering; hydropower project; hydro scheme; water power engineering; hydraulic and hydroelectric engineering
水电工程方案 hydroelectric scheme
水电工程师 hydroelectric(al) engineer
水电工程师协会 Institution of Water and Power Engineers
水电工程项目 hydraulic power project
水电管线 utility line
水电管线廊道 tunnel for utility mains
水电规划方案 hydroelectric(al) scheme
水电规划枢纽 hydroelectric(al) scheme
水电计划 hydroelectric(al) project
水电计划的基底负荷 base-load hydroelectric(al) project
水电计划的基底负载 base-load hydroelectric(al) project
水电建设 hydroelectric(al) development
水电建筑 hydroelectric(al) construction
水电解质 aqueous electrolyte
水电解质代谢紊乱 water-electrolyte imbalance
水电局 hydroelectric(al) board
水电开发 hydraulic power development; hydroelectric(al) development; hydropower development; water power development
水电开发工程 hydraulic development project; hydroelectric(al) development project; hydropower development project
水电开发计划 hydroelectric(al) scheme
水电开发利用 hydroelectric(al) exploitation
水电开发许可证 hydroelectric(al) license[licence]
水电来源 hydroelectric(al) resources
水电模拟法 electrohydrodynamic(al) analog(ue)
水电能 hydroelectric(al) energy
水电潜力 hydroelectric(al) potentiali-

ty
水电施工 hydroelectric(al) construction
水电系统 hydrosystem
水电项目 hydroelectric(al) project
水电项目议定书 <条约草案,会谈记录> hydroelectric(al) protocol
水电效应破岩 rock fragmentation by electrohydraulic effect
水电要素 element of water power
水电引水渠 power canal
水电蕴藏量 hydroelectric(al) potentiality; hydroelectric(al) power; hydroelectric(al) power potential; hydropower potential; potential output; power potential
水电站 hydraulic power plant; hydroelectric(al) generating station; hydroelectric(al) generator; hydroelectric(al) plant; hydroelectric(al) power plant; hydroelectric(al) power station; hydroelectric(al) station; hydroplant; hydropower house; hydropower plant with reservoir; hydropower station; water generating station; water power plant; water power station
水电站厂房 hydroelectric(al) power house; powerhouse of hydropower station
水电站地址 water power site
水电站分期开发 multistage water power development
水电站和有关工程 hydroelectric(al) power facilities and related works
水电站建设 construction of hydroelectric(al) station
水电站进水口 intake of hydropower station
水电站满负荷泄流量 full-plant discharge
水电站满荷宣泄流量 full-plant discharge
水电站双岸式布置 twin layout
水电站隧洞 hydropower tunnel
水电站梯级开发 multistage water power development
水电站调压塔 surge tank
水电站尾水池 afterbay
水电站尾水渠 tailrace
水电站系统 hydroelectric(al) scheme
水电站引水工程 diversion system of hydropower station
水电站引水设备 diversion system of hydropower station
水电站引水隧道 power tunnel
水电站引水系统 diversion system of hydropower station; headrace system of hydropower station
水电站运行 hydroelectric(al) power station operation
水电站站址 water power site
水电资源 hydroelectric(al) resources; hydropower resources; power potential; water electric resources
水电阻 water resistance
水电阻器 water resistor; water rheostat
水垫 water bottom; water cushion; water-pillow
水垫层 water-bearing
水垫床 water-bed
水垫式活动发射台 water cushion movable launch platform
水垫塘 plunge pool
水垫消力池 cushion pool
水垫压力 water cushion pressure
水垫原理 water cushion principle
水顶压逆流再生 water blanket conversion current regeneration; water hold down of conversion current regeneration
水定时取样器 water-timed sampler
水定时式取样机 water-timed sampler
水冬瓜油 Morinda citrifolia oleum
水动 hydrodynamic(al)
水动比拟 hydrodynamic(al) analogy
水动滑阀 water slide valve
水动机 water motor
水动力 hydrodynamic(al) force
水动力参数 hydrodynamic(al) parameter
水动力冲击 hydrodynamic(al) impact; hydrodynamic(al) shock
水动力传导性 hydrodynamic(al) conductivity
水动力传动 hydrodynamic(al) fluid drive; hydrokinetic fluid drive
水动力粗糙面 hydrodynamically rough surface
水动力的 hydrodynamic(al)
水动力地 hydrodynamically
水动力调查 hydrodynamic(al) investigation
水动力分离器 hydrodynamic separator
水动力光滑面 hydrodynamically smooth surface
水动力混合特性 hydrodynamic(al) mixing characteristic
水动力计 hydrodynamometer
水动力计算 hydrodynamic(al) computation
水动力扩散 hydrodynamic(al) diffusion; hydrodynamic(al) dispersion
水动力扩散系数 hydrodynamic(al) dispersion coefficient
水动力扩散张量 hydrodynamic(al) dispersion tensor
水动力理论 hydrodynamic(al) theory
水动力力矩 hydrodynamic(al) moment
水动力流场 hydrodynamic(al) flow field
水动力弥散 hydrodynamic(al) dispersion
水动力弥散机理 mechanism of hydrodynamic(al) dispersion
水动力弥散系数 dispersion coefficient of hydrodynamics; dynamic(al) dispersivity
水动力模拟 fluid flow analogy; hydrodynamic(al) analogy
水动力平滑面 hydrodynamic(al) smooth surface
水动力圈闭 hydrodynamic(al) trap
水动力升力 hydrodynamic(al) lift
水动力条件 hydrodynamic(al) condition
水动力稳定性 hydrodynamic(al) stability
水动力系数 hydrodynamic(al) coefficient
水动力现象 hydrodynamic(al) phenomenon
水动力相似 hydrodynamic(al) similitude
水动力效应 hydrodynamic effect
水动力性能 hydrodynamic(al) performance
水动力学 hydrodynamics; hydrokinetics
水动力学的 hydrodynamic(al)
水动力学方程 hydrodynamic(al) equation
水动力学基本方程 basic equation of hydrodynamics
水动力学模型 hydrodynamic(al) model
水动力学微分方程 hydrodynamic(al) differential equation
水动力学研究 hydrodynamic(al) research
水动力压力 hydrodynamic(al) pressure; moisture hydrodynamic(al) pressure
水动力压头 hydrodynamic(al) head
水动力延滞 hydrodynamic(al) lag
水动力影响 hydrodynamic(al) effect
水动力油藏 hydrodynamic(al) pool
水动力噪声 hydrodynamic(al) noise
水动力遮挡 hydrodynamic(al) barrier
水动力质量 hydrodynamic(al) mass
水动力质量系数 hydrodynamic(al) mass coefficient
水动力滞后 hydrodynamic(al) lag
水动力阻力 hydrodynamic(al) drag; hydrodynamic(al) resistance
水动力阻滞 hydrodynamic(al) lag
水动型海面升降 eustatism; hydrocratic eustacy
水动闸门 hydraulic operate gate
水冻土 frozen soil
水冻岩溶 cryokarst
水洞 dolina [doline]; pot-hole; water hole; water sink; water tunnel <空化试验用>
水洞试验 water tunnel test
水都 water city; watery city
水斗 bailer [bailor]; bucket; fountain; impeller vane; sink(grating); water skip; runner bucket <冲击式的>
水斗辊 ductor roller
水斗联结(装置) <水轮机上的> bucket attachment
水斗式水轮机 jet type impulse turbine; nozzle type turbine; Pelton's turbine; scoop-type turbine; tangential turbine
水斗式转轮 tangential wheel
水斗转轮 bucket wheel
水度环境质量评价 reservoir environmental quality assessment
水端 water end
水短柱石 penkvilksite
水碓 water-driven stamp mill
水多硅鳞云母 hydropolylithionite
水舵 <水上飞机的> water rudder
水二次冷却系统 water recooling system
水阀 faucet; hydrovalve; water injection valve; water valve
水阀壁龛 faucet alcove
水阀杆 water valve stem
水阀帽 water stuffing box; water valve bonnet
水阀帽螺母 water valve bonnet nut
水阀体水门 water valve
水阀填密螺母 water valve packing nut
水阀压盖 water valve gland
水阀座 water valve seat
水法 water act; water decree; water law
水法规 water by-laws; water code
水法规咨询 water by-laws advisory
水法后处理 <核燃料等的> aqueous reprocessing
水矾石 aloite; diaspore
水矾土 diaspore clay; diasporite
水矾性砖红壤 bauxitic laterite
水钒钡石 gamagarite
水钒钙石 rossite
水钒铝矿 steigerite
水钒铝石 steigerite
水钒镁矿 hummerite
水钒镁钠石 huemulite
水钒锰铅矿 brackebuschite
水钒钠钙石 crantsite
水钒钠石 barnesite
水钒镍锌矿 kolovratite
水钒锶钙矿 delrioite
水钒酸铝石 alvanite
水钒铁矿 fervanite
水钒铜矿 usbekite [uzbekite]; volborthite
水钒铜铀矿 sengierite
水钒铀矿 ferghanite
水繁缕 water chickweed
水反射率 water albedo
水反射系统 water-reflected system
水反洗 water backwash
水反应堆 water reactor
水反应器 water reactor
水泛地 bottom land
水泛滥 flush of water
水方解石 hydrocalcite; hydroconite
水方钠石 hydrosodalite
水方硼石 hydroboracite
水防护屏 water shield
水房 water house
水放射性测量计 water radioactivity meter
水放射性计 water activity meter
水放射性记录器 water monitor
水放射性监测器 water monitor
水放射性污染源 radioactive sources of water pollution
水费 charges for water; water charges; water cost; water fee; water price; water rate; water tax
水费补贴 water fee subsidy
水费率 water rate
水费一览表 water service schedule
水费征收 cost recovery
水分 degree of moisture; moistness; moisture; moisture capacity; wet; wetness
水分百分率 percentage of moisture
水分饱和 water saturation
水分饱和亏欠度 water saturation deficit
水分保持 moisture conservation; water conservation; water-retention
水分保持的 water-retentive
水分保持量 water retaining capacity; water retentivity
水分保持能力 water retaining capacity
水分保持值 water-retention value
水分比 moisture ratio
水分变化 moisture movement; moisture migration
水分补偿 moisture compensation
水分捕集器 moisture trap
水分不定 moisture deficiency; water deficiency
水分不稳定性 hydrolability
水分不足 hydrologic(al) deficit; water deficiency; water deficient; water deficit
水分不足以生长作物 barely enough moisture to produce crops
水分测定计 drimeter
水分测定箱 moisture-box
水分测定仪 drimeter; moisture meter; moisture teller; water determination apparatus
水分常数 moisture constant; water constant
水分储存能力 moisture storage capacity
水分储存器 moisture storage
水分传递阻止膜 non-moisture transferring membrane
水分大的料浆 fluid slurry
水分代谢 water metabolism
水分带出量 primage
水分当量 moisture equivalent

水分的变化 moisture variation
水分的传导度 water conductivity
水分的存在状态 existent state of moisture
水分的分泌 weepage;weeping
水分的内扩散 interior diffusion of water
水分的渗流 weepage
水分动态 moisture regime(n);water regime(n)
水分动移 water movement
水分多 washiness
水分分布 moisture-distribution
水分分离器 moisture-catcher;moisture separator;water catcher;water separator
水分分散 aqueous dispersion
水分分析 moisture analysis;water analysis
水分供给 moisture supply
水分固体关系 moisture-solid relationship
水分关系 water relation
水分管理 water management
水分管理和保持 water management and conservation
水分过多 excess moisture;excess water
水分过剩 excess moisture
水分含量 amount of water;humidity;moisture charge;moisture content
水分含量测定 determination of moisture content
水分耗损 water consumption
水分活性 water activity
水分积累 moisture accumulation
水分计 moisture meter
水分检查机 conditioning machine
水分结合比 moisture combined ratio
水分结合力 water combining power
水分控制 humidity control;moisture content control;moisture control;water control
水分快速测定仪 moisture teller
水分亏缺 moisture deficit
水分亏损 water deficit
水分来源 moisture source
水分离器 water trap
水分离指数 water separation index
水分利用率 efficiency of water application
水分临界点 critical moisture point;critical water point
水分流失 water loss
水分内循环 internal water circulation
水分凝结作用 hydrogenesis
水分排除效率 moisture-excluding efficiency
水分配 water-distribution
水分配站 water dispense
水分配政策 water distribution policy
水分平衡 hydrologic(al) balance;moisture balance;moisture equilibrium;water equilibrium
水分平衡循环 hydrologic(al) balance
水分迁移 moisture migration;moisture transfer;water translocation
水分欠缺 water deficit
水分侵入 moisture ingress
水分取样 moisture sampling
水分散法 water dispersion
水分散剂 water dispersible liquid
水分散区 region of dispersed water
水分散特种环氧树脂 special water dispersion epoxy resin
水分散体 aqueous dispersion;water dispersion
水分散系 aqueous dispersion
水分散性催干剂 water dispersible drier
水分散性烘漆 water dispersion baking paint
水分散性环氧 water dispersible epoxide
水分散性涂料 water dispersible paint
水分散性硬脂酸锌 water dispersible zinc stearate
水分散性油 water dispersible oil
水分散液 water dispersible liquid
水分散晕类型 water dispersion halo types
水分上升运动 upward movement of water
水分渗入 ingress of water
水分渗入地面深度 moisture penetration depth of ground
水分渗入率 moisture penetration
水分渗入深度 moisture penetration depth
水分渗透 moisture penetration;penetration of dampness
水分渗透率 rate of water infiltration
水分渗透能力 water percolating capacity
水分渗透指数 moisture permeability index
水分渗透作用 penetration of dampness
水分渗透作用和再分配 water infilitration and redistribution
水分渗移 moisture migration
水分试验 moisture test(ing)
水分试样 moisture sample;water sample
水分收集器 moisture trap
水分收支 water household
水分输入 moisture charge
水分输送 moisture transfer;moisture transport
水分数值 moisture figure
水分水平渗漏 water contour percolation
水分损耗 loss of water
水分损失 moisture loss
水分梯度 gradation of moisture;moisture gradient
水分调节 humidity control;water adjustment
水分调节作用 moisture regulating function;water regulating function
水分调整 moisture adjustment
水分通量 moisture flux
水分吸力 moisture suction
水分吸收 moisture absorption;water absorption
水分吸收很慢 slow to absorb water
水分吸收率 rate of water absorption
水分吸收能力 water-absorbing capacity
水分吸胀后 after imbibition of water
水分析 water analysis
水分析成果 results of water analysis
水分析法 water analysis method
水分析样品 sample for water analysis
水分析资料整理方法 processing methods of water analytical data
水分现状 moisture status
水分效应 hydrologic(al) effect;moisture loss
水分修正 moisture adjustment
水分修正系数 moisture adjustment factor
水分循环 hydrologic(al) cycle;water circulation;water cycle
水分压榨机 moisture expeller
水分一览表 water service schedule
水分移动 moisture movement;moisture transfer
水分引力 moisture attraction;water attraction
水分应力 <土内> moisture stress
水分诱出量 primage
水分运动 moisture movement
水分运动到土壤内 movement of water into soil
水分张力 moisture tension
水分张力入渗计 tension infiltrometer
水分蒸发 evaporation of moisture;evaporation of water;water evapo(u)ration
水分蒸发期 <烧窑初期> water smoking period
水分蒸发强度 evaporating intensity of water
水分蒸腾计 evapotranspirometer
水分直减率 hydrolapse
水分指示计 moisture indicator
水分指数 moisture index
水分贮藏及利用 moisture storage and use
水分转移 moisture migration
水分状况 moisture regime(n);water regime(n)
水分资料 hydrologic(al) data
水分子 water molecule
水分子的黏[粘]吸 adhesive attraction of water
水分子偶 pairs of water molecule
水分子在一次涨落潮循环中运动的水平距离 tidal excursion
水分子族 water cluster
水粉比 water-powder ratio
水粉画 opaque water-colo(u)r painting;size distemper;tempera;water colo(u)r painting
水粉画颜料 distemper
水粉涂料 water paint
水粉颜色 size colo(u)r
水风筒【冶】trompe
水风信子 water hyacinth
水封 aquaseal;hydraulic seal;hydroseal;intercepting trap;liquid pack;seal;seal water;water-base;water block;water bosh;water closing;water lock;water lute;water packing;water seal;water shutter;water siphon;water-tight seal;water trap;interceptor <烟气进口与联箱间的>
水封标准 trap standard
水封槽 water seal tank
水封池 water-sealed basin;water-sealed tank
水封储气罐 water-sealed gas holder;wet gas holder
水封袋 <代替炮泥用> water-filled stemming bag
水封的 water-sealed
水封堵塞袋 water-stemming bag
水封发生器 water luted producer
水封阀 water shutter
水封防爆阀 water-sealed gas valve
水封防爆门 water-sealed explosion door
水封防尘连接 liquid-tight dustguard joint
水封高度 depth of water seal
水封隔间 flashback chamber
水封隔器 water packer
水封管 liquid trap;sealing water pipe;stink trap;syphon trap;water sealing pipe
水封管接头 sealing water pipe connection
水封罐 water-sealed tank
水封护罩 shock absorber expander
水封环 seal cage;seal loop;water sealing ring
水封接缝 water seal joint
水封接头 water joint
水封井 water-sealed well
水封煤气发生炉 water-bottom gas producer
水封煤气发生器 water bosh producer
水封气表 water-sealed gas meter
水封气体发生器 water bosh producer
水封器 liquid packing;liquid seal
水封圈 water sealing ring
水封入口 sealing-liquid duct
水封设备 water-sealed equipment
水封深度 depth of trap seal;depth of water;trap seal
水封式阀 water-sealed valve
水封式煤气发生炉 wet bottom gas producer
水封式气柜 water sealing gas holder
水封式杀菌机 hydrolock(retort)
水封式砂泵 hydroseal sand pump
水封试验 leak test by filling water
水封水管 seal water pipe
水封水箱 seal water tank
水封套 hydraulic gland;water-sealed gland
水封通气管 water trap vent
水封退火炉 water-sealed annealing furnace
水封弯管中的静水面 crown weir
水封弯头 peat trap
水封屋面 water-filled roof
水封箱 flashback tank
水封压盖 water gland
水封阴井 trap pit
水封用水箱 seal water tank
水封油库 water-sealed oil storage
水封运输机 water-seal conveyer[conveyor]
水封轴承 water-sealed bearing
水封装置 water seal arrangement;water-sealed packing
水氟钙铈矾 chukhrovite Ce
水氟钙钇矾 chukhrovite
水氟硅钙石 bultfonteinite
水氟化 water fluorination
水氟磷铝钙石 morinite
水氟铝钙矿 yaroslavite
水氟铝钙石 carlhintzeite
水氟铝镁矾 wilcoxite
水氟铝锶石 tikhonenkovite
水氟硼石 johachidolite
水氟碳钙钍矿 thorbastnaesite
水浮浪幼体 hydroplanula
水浮力 buoyancy of water;water buoyancy
水浮莲 water lettuce
水腐蚀 aqueous corrosion;water corrosion
水负峰 water dip
水负荷 water load(ing)
水负载 water load(ing)
水负载功率计 water load power meter
水附着的 water-clogged
水复钒矿 corvusite
水覆盖因子 water-covering factor
水改良剂 water conditioner
水钙钒铀矿 rauvite
水钙沸石 abrazite;gismondine;gismondite
水钙磷镁石 hautefeuillite
水钙榴石 hydrougrandite
水钙铝榴石 hydrogrossular
水钙芒硝 hydroglauberite
水钙镁铀矿 swartzite
水钙锰榴石 henryptite
水钙霞石 hydrocancrinite
水钙硝石 nitrocalcite
水钙铀矿 calxiouranoite
水干 water-drying
水高岭土 hydrokaolinite
水膏比 water gypsum ratio

S

水锆石 hydrozircon;malacon(e);orvillite
水隔器 water shutter
水铬镁矾 redingtonite
水铬铅矿 iranite
水耕法 water culture;water tillage
水工程 < 广义的水利工程 > water engineering
水工程师协会 Institute of Water Engineers
水工程师学会会刊 < 双月刊 > Journal of the Institution of Water Engineers
水工程序控制 hydraulic sequence control
水工大模型 hydraulic mockup
水工的 hydraulic;hydrotechnic(al)
水工堤坝 hydraulic embankment
水工地面试验机 hydraulic ground-testing machine
水工钢结构 hydraulic steel construction;hydraulic steel structure;steel engineering hydraulics
水工工程 hydraulic engineering works
水工工程试验室 hydraulic engineering laboratory
水工构造物 hydraulic structure;water constructional works
水工构筑物 hydraulic structure;water constructional works
水工管道部件 hydraulic conduit sections
水工混凝土 hydraulic(engineering) concrete
水工混凝土泵 hydraulic concrete pump
水工机械 civil engineering hydraulics equipment
水工基础工程 foundation engineering of hydraulic structure
水工计算 hydraulic calculation
水工技术 hydrotechnics
水工建筑 hydraulic structure
水工建筑技术 hydraulic architecture
水工建筑物 hydraulic engineering structure; hydraulic engineering works;hydraulic structure;hydrostructure;marine construction;marine structure; maritime construction; maritime structure; maritime works;water works
水工建筑物地点 marine site
水工建筑物荷载 hydraulic structure load
水工建筑物级别 grades of water conservancy engineering
水工建筑物建筑界限 bulkhead line
水工建筑物模型 hydraulic structure model; hydrostructure model; marine construction model; marine structure model
水工建筑学 hydraulic architecture
水工结构 hydraulic engineering structure; hydrostatic structure; hydrostructure
水工结构模型 hydraulic structure model;hydrostructure model
水工结构模型试验 hydraulic structure model test; hydrostructure model test
水工结构物 hydraulic structure;water constructional works
水工模型 hydraulic model
水工模型几何变态 geometric(al)distortion in hydraulic model
水工模型试验 hydraulic model testing)
水工模型研究 hydraulic model research;hydraulic model study
水工漆 underwater paint
水工上的抛石 clean-dumped rockfill
水工实验室 hydraulic laboratory
水工试验 hydraulic test
水工试验模型 hydraulic model test(ing)
水工试验室 hydraulic laboratory
水工试验堰 notch plate
水工试验用放水室 outlet chamber
水工隧道 aqueduct tunnel;hydraulic tunnel
水工隧洞 hydraulic tunnel
水工隧洞构造 components of hydraulic tunnel
水工隧洞类型 classification of hydraulic tunnel
水工缩尺模型 hydraulic scale model
水工缩尺模型试验 hydraulic scale model research
水工缩尺模型研究 hydraulic scale model research
水工学 hydraulic engineering;hydrotechnics
水工学的 hydrotechnological
水工研究 hydraulic engineering research;hydraulic engineering study; hydrotechnic(al) research; hydrotechnic(al)study
水工业 water and wastewater industry;water industry
水工业污染源 industrial sources of water pollution
水工艺学 water technology
水公用事业 water utility
水公用事业管理 water utility management
水公用事业协会 Water Utility Council
水功率 water horsepower
水功率转换系数 hydraulic conversion coefficient
水攻法 < 采油 > water-flooding
水供给 water supply
水供应来源 source of water supply
水供应质量 quality of water supply
水拱 water arch
水共振 water resonance
水沟 canal;ditch;drain;gole;gutter;killesse; leat; lode; mill stream; moat;runnel;scupper;stank;water ditch;water furrow;land waste < 沿岸冰面上的 >
水沟挡板 sluice dam
水沟的阶梯形断面 stepped section of ditch
水沟防臭水封 gull(e)y trap
水沟盖板 bridge stone;drain cover slab
水沟涵洞 brook culvert
水沟交叉 trench crossing
水沟井 gull(e)y trap
水沟排水 water channel drainage
水沟排污阱 gull(e)y trap
水沟清理机 ditch cleaner
水沟堰 sluice weir
水垢 cloud;deposit(e) of scale;dirt and mud;fur;incrustant;incrustation;incrustation scale;lime deposit;scale;scale deposit;scale incrustation;scum;sediment
水垢槽 scale pit
水垢层 scale crust
水垢沉积(物) deposit(e) of scale;scale deposit
水垢防止剂 scale preventive
水垢覆盖面 scale coated surface
水垢及沉积 scale and sediment deposit
水垢净化器 scaler
水垢清除工具 boiler scaling appliance
水垢清除器 scale breaker
水垢热阻 scale heat-resistance
水垢热阻系数 scale heat-resistance coefficient
水垢溶化 disincrustation
水垢溶剂 scale solvent
水垢生成 formation of scale
水垢形成 scale formation
水垢影响 scale effect
水垢增率 scaling rate
水谷线 talweg
水股射程 range of jet
水钴矿 heterogenite;stainierite
水钴铀矾 cobalt zippeite
水固比 water-solid ratio;water/solids ratio
水(固)结的 water-consolidated
水固结作用 hydroconsolidation
水固体堆 solid water reactor
水管 conduit(pipe);raceway;water conduit; water hose; water line; water pipe; water piping; water supply pipe;water tube;water tubing
水管板 water-tube plate
水管壁 water wall
水管柄 water hose stem
水管插口 spigot joint;spigot of conduit;spigot of pipe;spigot of tube
水管车 < 消防用 > hose car
水管冲洗 washing of pipe;washing of water pipe
水管出口 water outlet
水管串联 series of conduits;series of pipes;series of tubes
水管道 water conduit;water line
水管吊托 water pipe hanger
水管定线 flowing alignment
水管阀门 water valve
水管防冻塑性材料 kyljack
水管敷设权 water piping right
水管腐蚀 waterline attack
水管干线 hydraulic main
水管管件 water pipe fittings
水管锅炉 water-tube boiler
水管环路 water-tube circuit
水管火管合并式锅炉 fire-tube-water tube boiler
水管加热温室 pipe heated hotbed
水管检查井 water pipe manhole;water piping manhole
水管交接 cross connection
水管接件 water supply fittings
水管接口套筒 ferrule
水管接头 water connection;water pipe head
水管接头的封铅接合 blown joint
水管结合配件 water connection;water connection fitting
水管进孔 water pipe manhole
水管绝缘 pipe insulation
水管口盖 water connection cap
水管冷凝器 water-tube condenser
水管冷却器 water-tube type cooler
水管理措施 water management measure
水管理单位 water management unit
水管理对策 water management strategy
水管理法规 water management regulation
水管理法令 Water Control Act;Water Management Act
水管理方案 water management option
水管理方法 water management method
水管理规划 water management outline planning
水管理活动 water management activity
水管理机构 water management agency
水管理计划 water management plan;water management program(me)
水管理局 water authority
水管理决策 water management decision
水管理决策法 water management decision-making process
水管理模型 water management model
水管理目标 water management objective
水管理情况 water management scenario
水管理区 water management area;water management district
水管理区界 water management arena
水管理人员 water manager
水管理实践 water management practice
水管理系统 water management system
水管理协议 water management agreement
水管理政策 water management policy
水管理指标 water management index
水管理准则 water management criterion
水管理综合体 water management complex
水管连通 < 饮用水与工业水串通 > cross connection
水管路 water piping
水管内径 water hose inner diameter
水管配件 pipe fittings
水管喷口 epistomium
水管歧管 water lines manifold
水管气体冷凝器 water-tube gas condenser
水管桥 aqueduct;water-conduit bridge;water pipeline bridge
水管清垢器 water-tube cleaner
水管软管夹 water pipe hose clip
水管软接管 water pipe hose
水管塞 plug for water pipe
水管式比较仪 fluid displacement comparator
水管式沉降计 hydrostatic settlement cell
水管式沉降仪 water level settlement ga(u)ge
水管式锅炉 centre boiler
水管式火箱 water-tube firebox
水管式水准仪 hydrostatic level(l)ing apparatus
水管栓 tap cock
水管(水流)比降 pipe flow gradient
水管套座 pipe-insert
水管突扩大处流速水头损失 Borda loss
水管外罩 waterline bonnet
水管网 waterline network
水管温度探测器 pipe temperature sensor
水管系 hydraulic pipe line;hydraulic piping
水管系统 water vascular system
水管下置深度 setting depth of water pipe
水管线(路) water(pipe) line;water piping
水管线路施工设备 waterline construction equipment
水管小龙头 nose cock
水管削角处接头 angle collar
水管旋塞 nose cock
水管旋转接头 water swivel
水管窨井 water pipe manhole
水管因素 pipe factor
水管支墩 anchorage; buttress; thrust

block of pipe
水管直径 diameter of water pipe
水管制局 Water Control Board
水管转角处铸铁削角接头 angle collar
水管嘴 nozzle
水灌溉 irrigation by channelling water
水罐 ewer;hydria;jug;pitcher
水罐车 cistern car;ester tank wagon;water tank;water tank car;water tank lorry;water truck
水罐区面积 water tank area
水罐拖车 water trailer
水罐消防车 fire-extinguishing tanker;fire-tank wagon
水罐指示器 water tank indicator
水光 water glaze
水规划区 water planning region
水硅钡锰石 verplanckite
水硅钡石 krauskopfite
水硅钒钙石 cavansite
水硅钒锌镍矿 kurumsakite
水硅钙锆石 armstrongite
水硅钙钾石 reyerite
水硅钙锰石 kittatinnyite
水硅钙石 hillebrandite;okenite
水硅钙铜石 stringhamite
水硅钙铀矿 haiweeite
水硅锆钾石 kostylcvite
水硅锆钠钙石 loudounite
水硅锆钠石 terskite
水硅铬石 rilabdite
水硅灰石 foshallasite;radiophyllite
水硅钾铀矿 weeksite
水硅磷钙石 viseite
水硅铝钙石 roggianite
水硅铝钾石 lithosite;shilkinite
水硅铝石 hydralsite
水硅铝钛镧矿 karnasurtite
水硅锰钙铍石 chiavennite
水硅锰钙石 ruizite
水硅锰镁锌矿 ga(u)geite
水硅锰石 dosulite
水硅钠锰石 raite
水硅钠铌石 epistolite
水硅钠石 kanemite
水硅铌钛矿 labuntsovite
水硅镍矿 connarite;refdanskite
水硅硼钠石 searlesite
水硅铍石 beryllite
水硅铅铀矿 orlite
水硅石 silhydrite
水硅钛锰钠石 tisinaite
水硅钛钠石 murmanite
水硅钛锶石 ohmilite
水硅铁锰矿 sturtite
水硅铁石 hisingerite;sjogrenite
水硅铜钙石 kinoite
水硅铜石 gilalite
水硅铜铀矿 cuprosklodowskite
水硅钍铀矿 enalite
水硅锌钙钾石 minehillite
水硅锌钙石 junitfoite
水硅铀矿 coffinite
水柜 storage tank;tank;water cistern;water tank;cistern;deposit(e)
水柜车 bowser;water tank car
水柜底 tank water bottom
水柜顶 crown of a tank
水柜定位铁 tank lug
水柜阀滤清器 tank valve strainer
水柜防漏检查 tank test
水柜盖 tank lid
水柜火钩架 tank fire tools bracket
水柜角钢 tank angle
水柜均衡管 tank equalizing pipe
水柜(开关)阀 tank valve
水柜拉条 tank brace
水柜滤水器 tank hose strainer
水柜旁腿 tank water legs
水柜容量 tankage
水柜式机车 tank locomotive
水柜水尺 tank ga(u)ge;tank scale
水柜水垢 tank scale
水柜梯 tank ladder
水柜艇 bowser boat
水柜拖车 tank trailer
水柜污水水深登记黑板 sounding board
水柜注水孔 tank filling hole
水辊 dampener;dampening roller;damper;water roll
水滚 hydraulic roller
水锅炉 homogeneous solution-type reactor;water boiler
水锅炉反应堆 water boiler;water-boiler reactor
水果 garden shed;garden stuff
水果表面涂层 coating on fruit fare
水果仓库 fruit shed;fruit stock
水果船 fruit carrier
水果店兼冷饮店 fruit parlour
水果和蔬菜干燥机 fruit and vegetable drier[dryer]
水果和蔬菜加工厂 fruit and vegetable processing plant
水果和蔬菜加工废物 fruit and vegetable processing waste
水果摊 fruit stand
水果运输船 fruit carrier;fruit ship;maritime fruit carrier
水果专用船 fruit carrier
水过滤 water filtration
水过滤厂 water filtration plant
水过滤网 water sieve
水过滤装置 water filtration plant
水害 flood damage;water trouble
水害冲毁 washout
水含盐的 brackish
水旱地两用拖拉机 amphibious machine
水撼法 method of water rammer
水航学<有关水动力学及水面,水下推进航行的学问> hydronautics
水耗率 unit water quantity
水合白石灰 hydrated white lime
水合层 hydration sphere
水合成 hydrosynthesis
水合催化剂 catalyst;hydration catalyst
水合的 aquo;hydrated;hydrating;hydrous
水合电子 aqueous electron;hydrated electron
水合电子剂量测定法 hydrated electron dosimetry
水合电子剂量计 hydrated electron dosimeter
水合多水高岭石 ablykite
水合多水高岭土 hydrohalloysite
水合二氧化钛 hydrated titanium dioxide
水合发光 lyo-luminescence
水合反应器 hydration reactor
水合封闭处理 sealing by hydration
水合高岭土 hydrous kaolin
水合铬绿 hydrated chromium green
水合光泽处理 hydration polishing
水合硅酸钙 afwillite;hydrous calcium silicate
水合硅酸铝 hydrated alumin(i)um silicate
水合硅酸铝镁 hydrated magnesium alumin(i)um silicate;trisimint
水合硅酸铝土 allophane soil
水合硅酸镁 hydrated magnesium silicate;hydrous magnesium silicate
水合硅酸镁石棉 chrysotile
水合硅酸盐 hydrosilicate;hydrous silicate
水合硅酸盐石棉 chrysotilite
水合过程 hydrate process;process of hydration
水合过度 overhydration
水合过氧化物 hydrated peroxide
水合碱 hydrated basic
水合金属离子 hydrated metallic ion
水合精磨法 hydration polishing
水合肼 hydrated hydrazine;hydrazine hydrate
水合肼类废水 hydrazine hydrate wastewater
水合离子 aquated ion;aquo ion;hydrated ion;solvated H-ion
水合离子半径 hydrated ion radius
水合磷酸铀铣钙 hydrated calcium uranyl phosphate
水合硫酸钙 hydrated calcium sulphate;hydrous calcium sulfate[sulphate]
水合硫酸钙硬壳 gypcrete
水合硫酸铝 hydrated alumin(i)um sulfate
水合率 hydration rate
水合氯 chlorine hydrate
水合氯化铝 alumin(i)um chlorohydrate
水合氯醛 chloral hydrate;crystalline chloral hydrated chloral;hydral;noctec;novochlorhydrate
水合氯氧化铜 oxivor
水合能力 hydratability
水合黏[粘]土 hydrated clay
水合抛光 hydration polishing
水合器 hydrator
水合氢 hydronium ion;hydroxonium
水合氢离子 hydrogen ion;hydronium ion;oxoniu mion
水合氢离子铁矾 hydronium jarosite
水合氢氧化铝 algeldrate
水合热 heat of hydration;hydrate heat;hydration heat
水合乳化树脂 aqueous resin emulsion
水合润滑脂 hydrated grease
水合三氯乙醛 chloral hydrate
水合三溴乙醛 bromal hydrate
水合式 hydrated form
水合树脂乳化液 aqueous resin emulsion
水合数 hydration number
水合水 hydrate water;hydration water;water of hydration
水合碳酸钾 hydrated potash
水合特性吸附物种 hydrated specifically adsorbed species
水合同分异构 hydration isomerism;hydratisomery
水合五氧化二锑 hydrous antimony pentoxide
水合物 hydrate;hydrated matter
水合物碱度 hydrate alkalinity
水合物碱量 hydrate alkalinity
水合物生成 cimolite formation
水合物抑制剂 hydrate inhibitor
水合纤维素 hydrated cellulose
水合纤维纸板 fiber[fibre] paper
水合硝基酸 nitroic acid
水合硝酸 nitric hydrate
水合硝酸汞 hydrated mercurous nitrate
水合性 hydrability;hydratability
水合盐 hydrated salt;hydrous salt;salt hydrate
水合颜料 water colo(u)r;water colo(u)r pigment
水合阳离子 hydrated cation
水合氧化沉淀物 hydrous oxide precipitate
水合氧化铬 hydrated chromium oxide;emerald oxide of chromium;hydrated chromium oxide green;smeraldino
水合氧化铝 alumina hydrate;hydrated alumina
水合氧化锰 hydrous manganese oxide
水合氧化膜 hydrous oxide membrane
水合氧化钛 hydrous titanium oxide
水合氧化铁 ferric oxide hydroxide;hydrous iron oxide
水合氧化物 hydrous oxide
水合衣 hydration mantle
水合乙醛酸 glyoxylic acid
水合抑制剂 hydrate inhibitor
水合茚三酮 ninhydrin;triketohydrindene hydrate
水合皂 hydrated soap
水合值 hydration value
水合质子 proton hydrate
水合装置 hydration plant
水合作用 aquation;aquotization;hydrate;hydration
水和不溶物含量的测定 sediment and water test
水和沉渣含量 water and sediment content
水和沉渣含量测定 water and sediment test
水和非水相液体混合流 simultaneous water and non-aqueous phase liquid flow
水和胶结料之比 water-binder ratio
水和空气可以通过的空隙 small spaces through which water and air may pass
水和稀浆传运法 water slurry transportation
水荷载 water ballasting;water load(ing)
水鹤 water crane;standpipe【铁】
水鹤臂 water crane arm;water crane jib
水鹤标志 water crane indicator;water crane sign
水鹤表示器 water crane indicator
水鹤操纵台 operating bench of water crane
水鹤室 water crane well
水鹤值班房 on-duty room for water crane operator
水鹤柱 water crane column;water crane stand
水黑氯铜矿 hydromelanothallite
水黑锰矿 hydrohausmannite
水黑铜矿 hydrotenorite
水黑稀金矿 hydroeuxenite
水黑云母 hydrobiotite
水黑蛭石 eukamptite
水痕 wash marking;water mark;water stain<陶瓷缺陷>
水恒工况 constant duty
水恒温器 water thermostat
水衡重 water ballast
水红砷锌石 koettigite
水候学<研究影响水中生物的化学物理环境的科学> hydroclimatology
水呼吸动物 water-breathing animal
水壶柄 kettle holder
水湖流域 water catchment area
水葫芦 water hyacinth
水糊剂 water suspension paste
水戽(斗叶)bucket-type wheel
水戽式水轮机 impulse water turbine;Pelton's turbine;Pelton(water)wheel
水花 blowing spray;lipper;water bloom;water flowers;water spray
水花飞溅 splash
水花区 bloom
水花四溅地划桨 row wet
水华 algal bloom;bloom;flowering;

water bloom;algae bloom <水体富氧化引起的藻类繁茂生长>
水华病 bloom disease
水华控制 water bloom control
水华现象 water bloom
水滑 hydroplaning
水滑大理石 pencatite
水滑道 <运木材用> flume;wet slide
水滑结晶灰岩 brucite marble;pencatite;predazzite
水滑石 brucite; gavite; hydrotalcite [hydrotalkite]
水滑速度 <降雨时车轮在路面上产生漂浮滑移时的速度> hydroplaning speed
水滑现象 <在高速道路上降雨时,车轮产生严重漂浮滑移的现象> hydroplaning phenomenon
水滑作用 solifluxion
水化白云石灰 hydrated dolomitic lime;hydrated magnesium lime
水化半径 hydrated radius
水化变色 hydration discolo(u)ration
水化层 hydrated layer; hydrated sheath;hydrated shell
水化层法年代测定 hydration dating
水化产物 hydration product
水化程度 degree of hydration;extent of hydration;hydration degree
水化的 hydrated;hydrating;hydrous
水化的天然骨料 hydrated natural aggregate
水化的天然集料 hydrated natural aggregate
水化低热 low heat of hydration
水化动力学 kinetics of hydration
水化度 slaking value
水化法储存 hydrate storage
水化反应 hydration reaction
水化方法 slaking process
水化粉石灰 hydrated lime powder
水化构造 hydration structure
水化硅铝酸钙 calcium aluminate silicate hydrate; calcium silicoaluminate hydrate
水化硅酸二钙 dicalcium silicate hydrate
水化硅酸钙 calcium hydrosilicate; calcium silicate hydrate; hydrated calcium silicate;tobermorite
水化硅酸三钙 tricalcium silicate hydrate
水化硅酸盐 hydrosilicate;silicate hydrate
水化硅氧 hydrated silica
水化过程 hydration process;process of hydration
水化含镁石灰 dolomitic hydrate
水化合器 hydrator
水化(合)作用 aquation
水化化学 hydration chemistry
水化黄长石 gehlenite hydrate
水化混凝土 hydrated concrete
水化活性 hydration activity
水化碱性铜碳酸盐 hydrated basic carbonate of copper
水化浆料 hydrated stock
水化胶束 hydrated micelle
水化精磨机 hydrafiner
水化矿物 hydrated mineral
水化硫铝酸钙 calcium aluminate sulphate hydrate;calcium ferrite sulphate hydrate; calcium sulfoaluminate hydrate
水化铝酸二钙 dicalcium aluminate hydrate
水化铝酸钙 calcium aluminate hydrate
水化铝酸镁 riopan
水化铝酸三钙 tricalcium aluminate hydrate
水化铝酸四钙 tetracalcium aluminate hydrate
水化铝酸一钙 monocalcium aluminate hydrate
水化率 hydration rate
水化氯铝酸钙 calcium aluminate chloride hydrate
水化氯铁酸钙 calcium chloroferrite hydrate
水化膜 hydration sheath; sheath of hydration
水化能 hydration energy
水化黏[粘]土 slaking clay
水化期 hydration period
水化器 hydrator
水化潜能 latent hydraulic power
水化热 heat of hydration; hydration heat;setting heat;heat of hardening <水泥的>
水化热的消散 dissipation of heat of hydration
水化热散逸 dissipation of heat of hydration
水化热造成的开裂 thermal cracking
水化热造成的裂缝 thermal cracking
水化三铝酸四钙 tetracalcium trialuminate hydrate
水化三氧二铬绿颜料 Veronese green
水化熵 entropy of hydration
水化深度 hydration depth
水化生石灰 hydraulic quicklime
水化石灰 autoclaved lime; callow rock;hydrated lime;slaked lime
水化石灰法 hydrated lime process; lime process
水化石灰膏 hydrate putty
水化石灰膏泥 hydrated lime putty
水化石灰硅酸盐 hydrated lime silicate
水化式 hydrated form
水化试验 slaking test
水化水 hydration water;water of hydration
水化水泥 hydrated cement
水化水泥和石灰成分 hydraulic cement and lime composition
水化水硬石灰 hydrated hydraulic lime
水化碳铝酸钙 calcium aluminate carbonate hydrate; calcium carboaluminate hydrate
水化碳酸钙镁 hydrated dolomitic lime;hydrated magnesium lime
水化铁铝酸钙 calcium aluminate ferrite hydrate; calcium alumo-ferrite hydrate
水化铁酸钙 calcium ferrite hydrate
水化温度 hydration temperature;temperature(of)set(ting) <水泥的>
水化物 hydrate;hydration product
水化物转化 hydrate conversion
水化系数 coefficient of hydration
水化纤维素 cellulose hydrates
水化硝酸纤维素 hydrated nitrocellulose
水化性 hydrability
水化性能 hydrating capacity
水化学 aquatic chemistry; aqueous chemistry
水化学边界条件 hydrochemical boundary condition
水化学变化 hydrochemical variation
水化学采样 hydrochemical sampling
水化学测井 hydrochemical logging
水化学测量 hydrochemical survey
水化学成分 water chemical component
水化学成分测定 measurement of chemical composition of water
水化学成分分类法 method of hydrochemistry composition classifications
水化学成果图 hydrochemical resultant plot
水化学处理 chemical water treatment
水化学垂直分带 hydrochemical vertical zoning
水化学的 hydrochemical
水化学调查 hydrochemical investigation;hydrochemical survey
水化学动力学 hydrochemical dynamics
水化学分层 hydrochemical stratification
水化学分层采样 hydrochemical stratification sampling
水化学分带 hydrochemical zoning
水化学分带性 hydrochemical zonality
水化学分区图 map of hydrochemical division
水化学分析 chemical analysis of water;hydrochemical analysis
水化学复杂程度 complicate degree of water chemistry
水化学环境 hydrochemical environment
水化学基 hydrochemical basis
水化学监测网 hydrochemical network
水化学检测 chemical examination of water
水化学检验 chemical examination of water
水化学径流模数法 hydrochemical runoff modulus method
水化学勘探 hydrochemical prospecting
水化学类型 hydrochemical type;water chemical type
水化学类型图 map of hydrochemical type
水化学模型 hydrochemical model
水化学剖面图 hydrochemical profile
水化学水平分带 hydrochemical horizontal zoning
水化学特征 hydrochemical characteristic; hydrochemical property; water chemical property
水化学图 hydrochemical chart;hydrochemical map
水化学系统 water chemistry system
水化学相 hydrochemical facies
水化学形成的影响因素 effective factors of hydrochemical formation
水化学形成作用 water chemical formation
水化学研究 hydrochemical research
水化学演化 hydrochemical evolution
水化学异常 hydrochemical anomaly
水化学找矿标志 hydrochemical mark of mineral deposits prospecting
水化学指标 hydrochemical regime
水化学柱状图 hydrochemical columnar section
水化学状况 hydrochemical regime
水化学资料 hydrochemical data
水化学组分 hydrochemical component; hydrochemical constituent; water chemical component
水化学组分测定 measurement of chemical composition of water
水化阳离子 hydrated cation
水化阳离子含量 content of hydrobase-exchange
水化阳离子种类 kinds of hydrobase-exchange
水化氧化铝 alumina hydrate
水化氧化铁 hydrated ferric oxide;hydrated iron oxide
水化云母 hydrous mica
水化值 slaking value
水化质子 hydrated proton
水化阻滞剂 hydration retarder
水化作用 hydrate;hydration
水化作用减速剂 hydration retarder
水化作用水 water of hydration
水化作用速率 rate of hydration
水化作用温差 hydration temperature range
水化作用温度 temperature of hydration
水桦 water birch
水还原剂 water-reducing agent
水环 <混凝土喷枪中的> water ring
水环泵 water ring pump
水环境 hydroenvironment;water environment
水环境安全 water environment security
水环境安全评价 water environment security evaluation
水环境保护 water environmental protection
水环境保护功能区 functional district of water environment protection
水环境本底值 background of water environment
水环境参数 water environmental parameter
水环境承载能力 water environmental carrying capacity
水环境承载容量 water environmental carrying capacity
水环境毒物指示评价 toxic index evaluation of water environment
水环境恶化 water environment worsening
水环境风险 water environment risk
水环境功能区 functional district of water environment
水环境管理 management of water environment; water environment management
水环境管理工作者 water environmental manager
水环境管理模型 water environmental management model
水环境规划 water environmental planning
水环境化学 aquatic environmental chemistry; water environmental chemistry
水环境极限 water environment limit
水环境决策支持系统 decision support for water and environmental system
水环境模拟 hydroenvironmental simulation;water environmental simulation
水环境评价 water environmental assessment; water environmental evaluation
水环境容量 carrying capacity of water environment;environmental capacity of water;water environment capacity
水环境特征 water environmental characteristic
水环境污染 water environmental pollution
水环境污染预测 water environmental pollution forecasting
水环境系统 water environmental system
水环境系统工程 water environmental system engineering
水环境信息 water environmental information
水环境信息系统 water environmental information system

水环境修复 aquatic environment restoration;rehabilitation of water environment; water environmental restoration
水环境压力 water environmental pressure
水环境压力指数 water environmental pressure index
水环境样本 water environmental sample
水环境治理 water environmental treatment
水环境质量 water environmental quality
水环境质量标准 water environmental quality standard
水环境质量模型 model of water environmental quality
水环境质量评价 water environmental quality assessment
水环境质量指数 water environmental quality index
水环空气泵 water ring air pump
水环流器 water circulator
水环式压缩机 liquid piston compressor;liquid ring compressor
水环式真空泵 water ring vacuum pump
水缓冲器 weed buffer
水换热器 water-to-water heat exchanger
水荒 water famine; water scarcity; water shortage
水黄长石 juanite
水灰比 water-cement factor; water-powder ratio
水灰比定律 Abram's law; water-cement ratio law
水灰比法配合混凝土 proportioning by water-cement ratio
水灰比法配料 proportioning by water-cement ratio
水灰比高的混凝土 wet concrete
水灰比理论 water-cement ratio theory
水灰比例 cement-water ratio; water(-to)-cement ratio
水灰比配合<混凝土> proportioning by water-cement ratio
水灰比配料法 proportioning by water-cement ratio
水灰比恰好的混凝土 earth-damp concrete;earth moist concrete
水灰比强度定律 water-cement ratio strength law
水灰比值 water cement value
水辉石 hectorite
水回火 water tempering
水回路 water loop
水回收 recycled water; water reclamation;water recovery
水回收厂 water reclamation plant
水回收设备 water recovery apparatus
水回收系统 water reclamation system;water recovery system
水回收装置 water reclamation plant; water recovery apparatus
水毁 flood damage; washout (by flood);washout of flood
水混合的 water-mixing
水混合阀 water mixing valve
水混合喷嘴 water-mixing nozzle
水混连接 water-mixing direct connection
水混凝 water coagulation
水混凝土搅拌厂 floating concrete mixing plant
水混溶性 water miscibility
水混性剂 water miscible formation
水混油 water-in-oil
水活性炭处理法 activated carbon process in water treatment
水火成的 aqueo-igneous;hydatopyrogenic;hydroplutonic
水火成岩 aqueo-igneous rock;hydroplutonic rock
水火电混合替代方案 mixed hydrothermal alternative
水火电力系统 hydrosteam system
水击 hammer blow;hydraulic shock; line shock; pipe hammer; water hammer
水击波 water hammer wave
水击吸收器 shock pressure absorber; water hammer absorber; water hammer detector
水击压力 water hammer pressure
水击作用 water hammering
水机浮筒 aerofloat
水迹印 water line;water mark
水积岸砾石 water-deposited bank gravel
水积(泥)土 water-deposited soil
水基冲洗液 water-base fluid
水基的 water-base(d)
水基封口胶 water-base compound
水基工业用涂料 water-based industrial finish
水基混合物 water-based vehicle
水基胶粘剂 water-based adhesive
水基胶乳分散 water-base latex dispersion
水基胶乳漆 water-base latex paint
水基泥浆<钻泥> water-base mud
水基黏[粘]结剂 water-base adhesive
水基泡沫 water-base foam
水基清洁剂 water-base cleaner
水基润滑剂 water-base lubricant
水基石墨润滑剂 colloidal graphite mixed with water
水基铁矾 hydroglockerite
水基涂料 water-base(d) paint;waterborne coating;water-carried paint
水基涂饰剂 water-base dressing;water-base finish
水基洗井液 water-base bore fluids
水基性矾 hydrobasaluminite
水基液体 water fluid;water liquid
水基油洗井乳化泥浆 water-base oil emulsion mud
水基载色剂 water-based vehicle
水基质系统 water-base system
水基着色剂 water stain
水基钻泥 water-base mud
水激波 water shock
水激活电池组 water-activated battery
水及煤气接头 water and gas connection
水级铁率<即铝率=%三氧化二铝/%三氧化二铁> iron modulus
水极 water electrode
水急冷器 water-quencher
水计量罐 water measuring tank
水剂 aqua; aqueous liquid; water aqua;water solution
水加到电石中的型式 water to carbide type
水加氯(处理) water chlorination; aqueous chlorination
水加热 water heat(ing)
水加热盘管 water heating coil
水加热器 water heater
水加热系统 water heating system
水加注器 water dispenser
水夹带式垃圾焚化炉 water-wall refuse incinerator
水夹套 jacket of water;water jacket; water leg<锅炉下部的>
水夹套式垃圾焚化炉 water-wall refuse incinerator
水钾玻璃 water-pearl-ash glass; water-potash glass
水钾钙霞石 sacrofanite
水钾铊矾 monsmedite
水价 cost of water;water price;water rate;water tariff
水监测网 water surveillance network
水监测系统 water monitoring system
水监视网 water surveillance network
水减速反应堆 water-moderated reactor
水检测 water detection
水检眼镜 orthoscope
水检验 water testing
水碱 furred;incrustation;scale;thermonatrite
水碱黄铜矿 orickite
水溅飞雾 bonfire
水鉴定法 water detection
水浆 cream
水浆管理 water-mud management
水浆涂料 calcimine;distemper;slurry coating
水浆研磨机 magazine grinder
水浆颜料 pulp colo(u)r
水浆状颜料 water slurry of pigment
水降 fall(ing)(head) of water
水交换 water exchange
水交替 water exchange
水浇地 irrigated fields;irrigated land
水浇地面积 irrigable area
水胶 hydrogel; joiner's glue; splicing cement;water gel
水胶比 water-binder ratio
水/胶结质之比 water/cementitious ratio
水胶涂料 glue water paint
水胶炸药 water gel explosive
水窖 water cellar
水接触角 water contact angle
水接头 water joint(ing);water swivel
水节霉 Leptomitus lacteus
水结的 water-bound
水结底基层 hydraulically bound subbase
水结合剂 water-binding agent
水结合力 water bind(ing)
水结合料之比 water-binder ratio
水结路基 water-bound base
水结碎石<无结合料的碎石路> plain macadam; water-bound broken stone;water-bound hoggin;water-bound macadam
水结碎石路 macadam road compacted with water;water bond macadam; water-bound broken stone road;water-bound macadam road
水结碎石路面 water-bound broken stone road surface; water-bound macadam; water-bound macadam pavement; water-bound macadam surface
水结碎石面层 water-bound surfacing
水解【化】 hydrolytic decomposition; hydrolization [hydrolyzation]; hydrolyze
水解变质 hydrolytic spoilage
水解槽 hydrolytic tank
水解产物 hydrolysate [hydrolyzate]; hydrolysis product
水解产物污水 hydrolyzate effluent
水解常数 hydrolytic constant
水解沉积物 hydrolyte;hydrolyzate
水解池 hydrolytic tank;Travis tank
水解池出水 hydrolysis basin effluent
水解处理 hydrolytic treatment
水解催化剂 hydrolyst
水解的 hydrolytic
水解电离 hydrolytic dissociation
水解电量计 water coulometer
水解电流 hydrolysis current; hydrolytic current
水解电压计 water voltmeter
水解淀粉 hydrolyzed starch
水解丁基浮选剂 hydrolysis butyl flo(a)tation aids
水解度 degree of hydrolysis
水解法 water disintegrating
水解反应 hydrolytic reaction
水解反应器 hydrolysis reactor; hydrolytic reactor
水解分裂 hydrolytic scission
水解罐 hydrolytic tank
水解过程 hydrolytic process
水解好氧工艺 hydrolysis-aerobic process
水解好氧循环序批间歇式反应器工艺 hydrolysis-aerobic loop sequencing batch reactor process
水解剂 hydrolytic reagent
水解降解 hydrolytic degradation
水解胶体 hydrocolloid
水解接触氧化气浮混凝生物炭工艺 hydrolysis-contact oxidation-air flo(a)tation-biological carbon process
水解秸秆 hydrolyzed straw
水解晶种 hydrolysis nuclei
水解聚丙烯腈 hydrolysis polyacrylonitrile
水解聚丙烯酰胺 hydrolyzed polyacrylamide
水解聚合反应 hydrolysis polymerization;hydrolytic-polymeric reaction
水解类鞣质 hydrolyzate
水解离解 hydrolytic dissociation
水解率 degree of hydrolysis;hydrolysis yield
水解马来酸酐 hydrolyzed polymaleic anhydride
水解酶 hydrolytic enzyme
水解耐久性试验 slake-durability test
水解黏[粘]土 slaking clay
水解平衡 hydrolysis equilibrium
水解曝气生物滤池硝化工艺 hydrolysis-biological aerated filter nitrification process
水解器 hydrolyzer
水解生物接触氧化工艺 hydrolyzation and biological contact oxidation process
水解石灰 hydrolytic slaked lime; water slaked lime
水解试验 slaking test
水解水 water of hydrolysis
水解速率 hydrolysis rate
水解酸化 hydrolysis acidification; hydrolytic acidification; hydrolyze acidification
水解酸化池 hydrolysis acidification pool;hydrolytic acidification basin
水解酸化好氧工艺 hydrolytic acidification-oxic process
水解酸化好氧氧化工艺 hydrolytic acidification-oxic oxidation process
水解酸化接触氧化混凝沉淀工艺 hydrolytic acidification-contact oxidation-coagulation sedimentation process
水解酸化缺氧生物工艺 hydrolytic acidification-anoxic biological process
水解酸化升流式曝气生物滤池工艺 hydrolytic acidification-upflow biological aerated filter process
水解酸化生物接触氧化工艺 hydrolytic acidification-biological contact oxidation process
水解酸化序批式活性污泥法 hydrolytic acidification-sequencing activated sludge process
水解酸化厌氧好氧生化工艺 hydro-

lytic acidification-anaerobic-aerobic biochemical process
水解稳定性 hydrolytic stability; stability to hydrolysis
水解吸附 hydrolytic adsorption
水解细菌 hydrolytic bacteria
水解纤维素 hydrated cellulose; hydrocellulose
水解纤维素硝酸酯 hydrocellulose nitrate
水解效应 hydrolytic effect
水解性氮 hydrolyzable nitrogen
水解性酸度 hydrolytic acidity
水解氧化 hydrolytic oxidation
水解液 digest; hydrolysate [hydrolyzate]
水解脂肪 hydrolyzed fat
水解值 slaking value
水解质 hydrolyte
水解紫胶 hydrolyzed shellac
水解作用 aquolysis; hydrolytic action; hydrolysis; hydrolytic dissociation
水介质 aqueous medium
水介质旋流器 water-only cyclone
水介质液压给进机构 water hydraulic-feed mechanism
水界 hydrosphere; water boundary
水界读数 water reading
水界面 water termination
水金云母 hydrophlogopite
水襟翼 hydroflap
水进式层序 transgressive sequence
水进韵律 transgressive rhythm
水浸 water infusion; water-logging; water mark; water encroachment <开采石油时>
水浸安定度试验 <评价沥青混合物耐水性的试验方法> immersion stability test
水浸岸 backshore
水浸冰厚 depth of immersed ice
水浸冰碛 water-soaked till
水浸出 aqueous leaching; water-leach
水浸出物 water extraction
水浸出液 water leach liquor
水浸地 water-logged farmland
水浸法 soaking process; water-seasoning; water soaking process; immersed method <一种探伤方法>
水浸牢度 fastness to water
水浸泡过的砂 inundated sand
水浸区间 bilged compartment
水浸试验 water soaking test
水浸提法 water extract method
水浸提液 water extract
水浸透 water-logging
水浸透测定 water penetration test
水浸透的 water-soaked
水浸物镜 water-immersion objective
水浸系数 water encroachment coefficient
水浸系统 water-immersion system
水浸岩芯 water wet core
水浸养护混凝土法 pounding method of curing concrete
水浸装置 water immersion
水浸渍 soaking in water; water retting; water immersion; water-logging
水经济学 water economics
水晶 crystal; Irish diamond; lake-george diamond; mountain crystal; pebble; quartz (crystal); rock crystal
水晶斑岩 quartz porphyry
水晶玻璃 crystal glass; cut glass
水晶的 crystalline
水晶灯 crystal lamp; quartz lamp
水晶吊灯 crystal chandelier
水晶宫 crystal palace; Glass Palace
水晶棺 crystal sacrophagus
水晶胶 clean plastic; clean polyester
水晶矿床 mountain deposit
水晶劈 quartz wedge
水晶片 quartz plate; quartz slate
水晶切型 quartz crystal cutting
水晶石 kryolith
水晶体 crystalline; crystalline lens
水晶天 crystalline heaven
水晶楔 quartz wedge
水晶自动楼梯 crystalator
水精处理装置 water polishing unit
水精灵 <一种自航浅水微型挖泥船> water witch
水井 water bore; water well; well
水井半径 well radius
水井泵 well-point pump
水井布局的工程参数 engineering parameter of well arrangement
水井抽排 pumped well drain
水井出水量 capacity of well; water-well yield; water yield of wells
水井出水量测定设备 water-well yield test equipment
水井出水率 capacity of well
水井的引用半径 substitute radius of well
水井的增量 development of a well
水井底部与地下蓄水层连接的暗渠 kanat
水井电视检查 TV well inspection
水井吊泵 pump suspended in well
水井调查 investigation of well
水井调查点 observation point of well
水井分布范围 range of well distribution
水井复原 well rehabilitation
水井更新 well rehabilitation
水井工程病害 hazard of well engineering
水井工程病害类型 type of well engineering hazard
水井工艺 water-well technology
水井过滤层 well screen
水井合理深度 rational depth of well
水井积垢 well incrustation
水井结构 well construction
水井结构数据 data of well structure
水井框支架 curbing
水井力学 water-well mechanics
水井排数 number of well rows
水井使用年限 service life of well
水井竖管 drop pipe; drop pipe line
水井数目 well number
水井损失 well resistance
水井维修 water-well maintenance
水井卫生保护 sanitary well protection
水井污染 well atmosphere pollution
水井性能 well performance
水井影响半径 radius of influence of well
水井影响面积 area of influence of well
水井影响区 area of influence of well
水井注入法 water injection of well
水井钻机 hydrogeologic (al) and water well drilling rig; water-well driller; water-well drill (ing machine); water-well drilling rig; water-well rig; well drill
水井钻架 water-well drilling rig
水井钻进 water-well drilling; well drilling
水井钻探 water-well drilling
水井最大出水量 maximum water yield of well
水井最大水位降深 maximum water-level drawdown in well
水景观 aqua landscape
水景(庭)园 water garden; aqua garden
水警艇 police boat; police launch
水净化 aquatic purification; water softening
水净化厂 water conditioning plant; water purification plant
水净化池 water purifying tank
水净化工程 water purification project
水净化剂 water purification agent; water scavenging agent
水净化器 water clarifier; water purifier
水净化设备 water correction plant; water purification plant
水净化生物预处理 water purification bio-pretreatment
水净化系统 water purification system
水净化效率 water purification efficiency
水净化站 water purification station
水净化装置 water purification plant; water purifier
水净化作用 water purification; water treatment
水静电计 hydroelectrometer
水静力学 hydrostatics
水静力学基本方程 basic equation of hydrostatics
水静力运动器 hydrostatic vehicle
水静压测深棒 hydrostatic depth ga(u)ge
水静压式杀菌 hydrostatic sterilization
水静压式杀菌机 hydromatic retort; hydrostatic cooker; hydrostatic sterilizer
水纠纷 water disputes
水纠纷法庭 water disputes tribunal
水炬 water torch
水掘法 hydraulicking
水均衡 water balance
水均衡方程 equation of hydrologic equilibrium
水均衡预测 hydrologic(al) budget
水均匀 hydrologic(al) budget
水均匀反应堆 aqueous homogeneous reactor
水菌 water bacteria
水卡计 water calorimeter
水开发政策 water-development policy
水开关 cock tap
水科学 hydroscience
水壳 water hull
水可进入的孔隙 water permeable voids
水可流动性系数 aquatic mobility coefficient
水可湿性 water-wettable
水坑 pond (ing); pool; puddle; slop; spoil pool; water hole; water pit; water trap
水坑格栅 sink grating
水坑沟槽 puddle trench
水空两用飞机 amphibian
水空气换热器 water-air heat exchanger
水空气热泵 water to air heat pump
水空气系统 water to air system
水孔隙 water pore; water void
水控阀 water control valve
水控装置 water control device
水口 water gap; water slot; water spout; waterway; weir
水口钮 sprue button
水口深度 <钻头> depth of kerf
水口闸板 nozzle gate
水口砖 nozzle brick
水口座砖 pocket block
水库 reservoir; storage reservoir; water (storage) reservoir; artificial lake; barrage lake; conservation pool; hydraulic reservoir; impounded basin; impounded water; impounding pond; impounding reservoir; impound water; ponded lake; water supply; weirage
水库安全规定法 reservoir safety provision act
水库岸防护林 reservoir bank protection forest
水库岸坡稳定性 stability of reservoir slope
水库岸线 reservoir shoreline
水库坝系统 reservoir dam system
水库边坡稳定性 stability of reservoir slope
水库边缘 reservoir rim; rim
水库表土剥离 reservoir stripping; stripping of reservoir
水库表土清除 reservoir stripping; stripping of reservoir
水库不稳定三角洲 unstable reservoir delta
水库操作 reservoir operation; storage operation
水库测量 reservoir survey
水库沉积(物) reservoir deposit; reservoir sediment; sediment distribution in reservoir
水库沉积作用 reservoir sedimentation; sedimentation of reservoirs
水库沉泥 reservoir silting
水库衬里 reservoir lining
水库衬砌 reservoir lining
水库充水 reservoir filling
水库冲沙 reservoir sediment flushing; reservoir sediment washout
水库出流量 outflow from reservoir
水库出水量 outflow from reservoir
水库储存 reservoir storage
水库储存容积 storage volume
水库储沙能力 sediment storage capacity of reservoir
水库储水量 reservoir storage
水库触发地震 reservoir triggered earthquake
水库堤 reservoir embankment
水库底沉陷 subsidence of reservoir bottom
水库地震 reservoir (-induced) earthquake
水库地震频度 frequency of reservoir earthquake
水库地震效应 reservoir seismic effect
水库地震与蓄水关系 relation between storage water and reservoir earthquake
水库地质图 geological map of reservoir
水库调查 reservoir investigation; storage investigation
水库调度 reservoir operation; reservoir regulation
水库调度表 reservoir operation chart
水库调度曲线 operating rule curve of reservoir; rule curve
水库调度图 reservoir operation chart
水库调度指导曲线 reservoir operation guide curve
水库断面 reservoir range; reservoir section
水库堆沙容积 sediment storage capacity of reservoir
水库防洪库容 flood retention capacity
水库防护林 reservoir protection forest
水库防渗铺盖 reservoir seepage-proof lining
水库放空 emptying of reservoir; reservoir emptying

水库放水 draft from reservoir; draft from storage
水库放水口 reservoir outlet
水库放水量 outflow from reservoir
水库放水流量 reservoir release rate
水库放水曲线 reservoir depleting curve
水库分类 classification of reservoir
水库负载 reservoir loading
水库港 reservoir port
水库搁沙率 reservoir trap efficiency
水库工程 reservoir engineering; storage works
水库工程地质勘察 engineering geologic(al) investigation of reservoir
水库工作库容 working storage in reservoir
水库工作深度 reservoir drawdown
水库供水量 reservoir draft; reservoir yield
水库观测 reservoir observation
水库观测记录 reservoir-inspection record
水库管理 reservoir operation; storage operation; reservoir management
水库荷载 reservoir load
水库洪水演算 reservoir routing; storage routing
水库湖面波动 seiche oscillation
水库环境 pond environment
水库回水 reservoir backwater
水库回水变动区 fluctuating backwater area of reservoir
水库回水计算 backwater computation of reservoir
水库基础 reservoir foundation
水库减淤积渠 reservoir by-wash channel
水库截留(泥沙)效率 reservoir trap efficiency; trap efficiency of reservoir
水库截留食料 <鱼类的> entrapment of nutrients by reservoir
水库截沙效率 reservoir trap efficiency; trap efficiency of reservoir
水库紧急放空 emergency emptying of reservoir
水库进水过程线 reservoir inflow hydrograph
水库经营政策 reservoir operating policy
水库净蒸发量 net reservoir evapo(u)ration
水库径流 flow pass reservoir
水库静水头 static head of reservoir
水库可用年限 reservoir life
水库空库时合力线 line of resultants of reservoir empty
水库控制蓄水 controlled filling of a reservoir
水库库岸 reservoir shore
水库库区航道 channels within reservoir region
水库库容 reservoir capacity; storage capacity of reservoir
水库库容曲线 reservoir capacity curve; storage capacity curve
水库库址 reservoir site
水库来水量 reservoir inflow
水库来水预报 forecasting of reservoir inflow
水库拦沙效率 reservoir trap efficiency
水库拦蓄 reservoir detention
水库拦蓄量 reservoir holdout
水库拦鱼器 jigger
水库累积水质效果评价 evaluation of cumulative reservoir water quality impact
水库冷却 reservoir-cooling
水库利用 reservoir utilization
水库流出水 reservoir outflow
水库旅游业 reservoir recreation
水库满库时合力线 line of resultants of reservoir full
水库面积 reservoir area; water reservoir area
水库面积-库容(关系)曲线 reservoir area-capacity curve
水库末端 reservoir head
水库末端淤积 uppermost deposition
水库泥沙 reservoir deposit; reservoir sediment; reservoir silt
水库泥沙淤积速率 reservoir sediment accumulation rate
水库排出口工程 reservoir outlet works
水库排泄口 outlet from reservoir
水库前淤积 aggradation above reservoir
水库墙 reservoir wall
水库清基 reservoir clearing; reservoir stripping; stripping of reservoir
水库清理 cleaning of reservoir; reservoir cleaning; reservoir stripping
水库清淤 evacuation of sediment; reservoir desilting
水库区 reservoir basin
水库区测绘面积 geologic(al) mapping area of reservoir region
水库区地质测绘 geology mapping of reservoir area
水库全消落时期 period of full drawdown
水库群 multistorage system
水库容积 capacity of reservoir; volume of reservoir
水库容积曲线 capacity curve
水库容量 capacity; reservoir capacity; reservoir volume; volume of reservoir
水库容量计算 calculation of reservoir capacity
水库容量曲线 reservoir capacity curve
水库容量纵剖面 reservoir capacity profile
水库入流 reservoir inflow
水库入流过程线 reservoir inflow hydrograph
水库三角洲 reservoir delta
水库上游低堰 low weir upstream of reservoir
水库设计洪水 reservoir design flood
水库设计洪水位 design flood level
水库设计容量 reservoir design capacity
水库设计水位 reservoir design level
水库渗漏的水 fugitive water
水库渗漏量 amount of reservoir leakage; amount of reservoir seepage; leakage from reservoir; reservoir leakage; reservoir seepage; seepage from reservoir
水库渗涌 reservoir leakage
水库生境 reservoir habitat
水库生态学 reservoir ecology
水库实际供水量 reservoir draft
水库实验水文站 reservoir experiment hydrological station
水库使用年限 life of reservoir; reservoir life
水库式发电厂 reservoir type power plant
水库寿命 life of reservoir; reservoir life
水库水井 reservoir water shaft
水库水量平衡 reservoir water balance
水库水面波动 seiche
水库水塔 reservoir water tower
水库水位 pool level; reservoir level; reservoir stage; storage level; water-level of reservoir
水库水位变化幅度 amplitude of water-level of reservoir
水库水位降落 drawdown
水库水位泄降 drawdown
水库水位指示器 telltale
水库水文测验 hydrometry for reservoir
水库水文站 reservoir hydrological station
水库水源 dam spring
水库水质 reservoir water quality
水库水质控制 reservoir pollution control
水库水质模型 reservoir water quality model; water quality model of reservoir
水库水质条件 reservoir water quality condition
水库死库容 dead storage of reservoir
水库死水位 top of inactive storage; dead water level of reservoir
水库坍岸 bank ruin of reservoir; reservoir bank caving; reservoir bank failure; shoreline erosion
水库坍岸阶段 stage of bank ruin of reservoir
水库特性 reservoir characteristics
水库提取水量 reservoir draft
水库调洪演算 flood routing through reservoir; reservoir flood routing; reservoir routing; storage routing
水库调节 reservoir regulation
水库调容演算 storage routing
水库调蓄 reservoir regulation
水库统计学 reservoir statistics
水库位置 reservoir site
水库污染 pollution of reservoir; reservoir pollution
水库系统 multistorage system; water reservoir system
水库下游沿程冲刷 degradation along downstream reach of reservoir
水库消落 drawdown of reservoir; reservoir drawdown
水库消落范围 available depth of reservoir
水库消落期 drawdown period of reservoir
水库泄放 draft from storage
水库泄放速度 reservoir release rate
水库泄放速率 reservoir release rate
水库泄降 drawdown of reservoir
水库泄降深度 drawdown
水库泄水 reservoir sluicing; flash
水库泄水道 afterbay
水库蓄水 filling of reservoir; reservoir filling; reservoir raise; water conservation
水库蓄水变化 change of reservoir storage; variation of reservoir storage
水库蓄水量 pondage; reservoir storage; reservoir storage capacity; storage capacity of reservoir; storage in reservoir
水库蓄水水质 reservoir storage quality
水库蓄水位 reservoir pool level
水库蓄水位线及高程 storage leave and elevation of reservoir
水库压力 reservoir pressure
水库压力梯度 reservoir pressure gradient
水库淹没 reservoir flowage; reservoir inundation
水库淹没界线测量 survey of reservoir inundation line; reservoir inundation line survey
水库淹没区 inundated area of reservoir; reservoir basin; reservoir flowage area; reservoir inundation area
水库淹没损失 reservoir inundation damage; reservoir inundation loss
水库淹没线测量 reservoir flooded line survey
水库演算 reservoir routing
水库养鱼 fish culture in reservoir
水库异重流 density current in reservoir
水库溢洪道 reservoir spillway; spillway for reservoir
水库用途 purposes of reservoir
水库有效(工作)深度 available depth of reservoir
水库有效库容 effective capacity of reservoir; reservoir live storage
水库有效容量 effective capacity of reservoir; reservoir live storage
水库有效蓄水量 reservoir live storage
水库诱发地震 reservoir filling-triggering earthquake; reservoir impounding induced seismicity; reservoir-induced earthquake; reservoir triggered earthquake
水库诱发地震监测 monitoring of induced earthquake of reservoir
水库诱发地震特点 character of reservoir induced earthquake
水库淤积 filling of reservoir; reservoir deposition; reservoir sedimentation; reservoir silting; sedimentation of reservoirs; silting of reservoir
水库淤积测量 reservoir accretion survey; reservoir deposition survey; reservoir sedimentation survey
水库淤积观测 reservoir deposition observation; reservoir sedimentation observation
水库淤积速度 reservoir sediment accumulation rate
水库淤积物 reservoir deposit; reservoir sediment
水库淤沙库容 storage capacity for deposition; storage capacity of sedimentation
水库淤填 filling of reservoir; reservoir deposition; reservoir sedimentation
水库渔业 reservoir fishery
水库源流 dam spring
水库运行 storage operation
水库运行表 reservoir operation procedure
水库运用 reservoir operation; storage operation
水库运用表 reservoir operation procedure
水库运用曲线 operating rule curve of reservoir
水库运用图表 reservoir operation chart
水库运用指导曲线 reservoir operation guide curve
水库运作 reservoir operation; storage operation
水库运作表 reservoir operation chart; reservoir operation procedure
水库运作曲线 operating rule curve of reservoir
水库运作指导曲线 reservoir operation guide curve
水库征地 land confiscation on reservoir area
水库蒸发 reservoir evapo(u)ration
水库正常(蓄)水位 normal pond level; normal pool level; normal reservoir elevation
水库中的回水 backwater in reservoir
水库中蓄积的水量 accumulated water
水库中蓄水量 accumulated water

S

水库周边 reservoir perimeter
水库轴线 axis of reservoir
水库(贮)水 reservoir water
水库综合调度 reservoir integrated regulation
水库综合利用 reservoir comprehensive utilization
水库总库容 reservoir volume; total reservoir storage
水库总容积 reservoir volume
水库总容量 total reservoir storage; total storage(capacity)
水库最低水位 minimum pool level
水库最低运行水位 top of inactive storage
水库最高蓄水位 top water level
水扩张树脂 water-extended polyester;water-extended resin
水蜡树 privet
水拦污 water screening
水拦油栅 floating boom
水蓝 saxe blue
水蓝宝石 aquamarine
水蓝方石 hydrohauyne
水蓝晶石 hydrocyanite
水蓝色 aqua blue
水镧铈石 calkinsite
水廊 corridor on water;water gallery
水浪 water spray
水浪翻花 boil
水老鼠 underwater pump;water-rat
水涝 water-log(ging)
水涝地 chaor; saturated ground; submerged land; water-lodging field; water-lodging land; water-logged (farm) land;water-logged ground
水涝害 flooding damage; flooding injury
水涝灾 water-logging disaster
水勒斯 <一种防水液> Waterex
水雷 mine;mo(u)ldy;torpedo
水雷区 mine area;mine field
水雷区浮标 submarine mining buoy
水雷区危险 mine risk
水雷扫清 mine removal
水雷危险区 areas dangerous due to mines
水雷障碍 mine barrage
水雷阵 mine barrage
水冷壁 water(-cooled) wall; water screen; boiler wall; furnace wall; spaced tube wall
水冷壁吹灰器 wall blower
水冷壁管 water screen tube; water-wall tube
水冷壁管屏 panelized-tube-wall section
水冷壁回路 wall circuit; water-wall circuit
水冷壁联箱 water screen header;water-wall header
水冷壁炉膛 water-evapo(u)rating furnace
水冷壁面 wall-cooling surface;water-wall surface
水冷壁屏 water-wall panel
水冷壁受热面 water-wall surface
水冷壁悬挂装置 wall hook
水冷壁中的水循环 wall circulation
水冷变压器 water-cooled transformer
水冷槽 quenching bath
水冷柴油机 water-cooled diesel engine
水冷冲天炉 water jacket cupola
水冷(处理) hydrocooling
水冷淬火 cold quench
水冷的 water cooling
水冷底板 water-cooled base; water-cooled bottom plate
水冷电磁铁 water-cooled electro-magnet
水冷电动机 water-cooled motor
水冷电极 water-cooled electrode;water-cooled electrode bar
水冷电极夹 water-cooled electrode clamp
水冷电接头 water-cooled contact
水冷电子管 water cooled vacuum tube
水冷电阻 water-cooled resistance
水冷电阻器 water-cooled resistor
水冷顶板 water-cooled head plate
水冷定型模 water-cooled former
水冷堆 water-cooled reactor
水冷发电机 water-cooled generator
水冷发动机 water-cooling engine
水冷发火管 water-cooled ignitron
水冷发射台 wet emplacement
水冷阀 water-cooled valve
水冷法 water cooling
水冷反应堆 water-cooled reactor
水冷分隔器 water-cooled spacer
水冷分凝管 knockout coil
水冷风口 water-cooled tuyere
水冷服装 water-cooled suit
水冷辐射源 water-cooled source
水冷坩埚 cold crucible;water-cooled crucible
水冷坩埚电弧熔炼 cold-mo(u)ld arc melting
水冷高频引线 water-cooled high frequency lead
水冷沟 water-cooling groove
水冷管 water-cooled tube; water-cooling tube
水冷管式铜坩埚 water-cooled copper tube mo(u)ld
水冷辊 water-cooled rolls
水冷焊炬 water-cooled welding torch
水冷黄铜接头 water-cooled brass connection
水冷活塞 water-cooled piston
水冷加热器 water-cooled heater
水冷夹头 water-cooled clamp
水冷空气冷却器 water to air cooler
水冷拉丝卷筒 water-cooled capstan
水冷流槽 water-cooled launder
水冷炉篦焚化器 water-cooled grate incinerator
水冷炉壁 water-cooled furnace wall
水冷炉箅 grate with water circulation;water grate
水冷炉底 water-cooled base
水冷炉盖 water-cooled cover
水冷炉壳 water-cooled furnace body; water-cooled shell
水冷炉排片 water bar
水冷炉墙 water-cooled furnace wall
水冷炉膛 water-cooled furnace
水冷炉条 water bar
水冷炉栅 water-cooled grating
水冷门拱 chill arch
水冷密封 water-cooled seal
水冷模 water-cooled mo(u)ld
水冷磨 water-cooled mill
水冷耐火壁 bailey wall
水冷凝器 water condenser
水冷盘管 water-cooled coil
水冷汽缸 water-cooled cylinder
水冷汽轮发电机 water-cooled turbo-generator
水冷腔 waterway
水冷强制油循环式变压器 water-cooled forced-oil transformer
水冷却槽 water-cooling tank
水冷却成套设备 water chiller package
水冷却的 water-cooled
水冷却法 water cooling
水冷却剂 water coolant
水冷却炉箅 water-cooled grate
水冷却慢化堆 water-cooled and water-moderated reactor
水冷却内壁 water-cooled wall
水冷却盘管 water-cooling coil
水冷却器 hydrocooler; water chiller; water-cooler
水冷却器放水阀 water cooler valve
水冷却器后火桥 water back
水冷却栅格 water-cooled lattice
水冷却设备 water chiller equipment
水冷却塔 water-cooling tower
水冷却套 water-cooled jacket;water-cooling jacket
水冷却系统 water-cooling system
水冷却压缩机 water-cooled compressor
水冷却诱导通风塔 water-cooling induced draft tower
水冷却原子反应堆 swimming pool reactor
水冷却轴承 water-cooled footstep
水冷却装置 water chiller;water chilling unit
水冷热交换器 oil-to-water heat exchanger
水冷韧化处理 water toughening
水冷闪烁计数器 water-cooled scintillation counter
水冷式 water-cooled type
水冷式测针 water-cooled probe
水冷式磁控管 water-cooled magnetron
水冷式的 water-cooled
水冷式电动机 motor with water cooling
水冷式电机 water-cooled machine
水冷式发动机 liquid cooled engine; water-cooled engine
水冷式阀门 water-cooled valve
水冷式负载电阻 water-cooled load resistance
水冷式绞车 water-cooled hoist
水冷式开炼机 water-cooled rolls
水冷式空调器 water-cooled air conditioner
水冷式空气压缩机 water-cooled air compressor
水冷式冷凝器 water-cooled condenser
水冷式排气歧管 water-cooled exhaust manifold
水冷式汽缸 water-cooled cylinder
水冷式汽轮机 water-cooled turbine
水冷式壳体 water-cooled housing
水冷式刹车卷筒 water-cooled brake drum
水冷式设备 water-cooled equipment
水冷式套筒 cooling water jacket;water-cooled jacket
水冷式涡轮机 water-cooled turbine
水冷式消音器 water-cooled silencer
水冷式叶片 water-cooled blade
水冷式引燃管 water-cooled ignition fuse
水冷式余热锅炉 water-tube waste-heated boiler
水冷式再冷凝器 water-cooled after-condenser
水冷式照明装置 water-cooled luminaire
水冷式整体空调机 water-cooled packaged air conditioner
水冷式制动器 water-cooled brake
水冷式制冷剂冷凝器 water-cooled refrigerant condenser
水冷式转鼓 water-cooled drum
水冷式自动焊枪 water-cooled automatic electrode holder
水冷室 water-cooling chamber
水冷受热面 wall-cooling surface
水冷水缸套 cold water cylinder liner
水冷塔 water-cooling tower
水冷套 water collar; water-cooled housing;water jacket
水冷铁铸件 tymp
水冷铜电极顶杆 water-cooled copper electrode ram
水冷铜坩埚熔化台 water-cooled copper melting platform
水冷铜极靴 water-cooled copper shoe
水冷退火 water annealing
水冷托梁 water back
水冷外壁 bailey wall
水冷往复式冷水机组 water-cooled reciprocating water chiller
水冷稳定板 water-cooling stabilizer panel
水冷钨尖电极 water-cooled tungsten tipped rod
水冷旋管 water-cooled coil
水冷阴极 water-cooled cathode
水冷引燃管 water-cooled ignitron
水冷用空腔 water cavity
水冷油浸变压器 water-cooled oil-immersed transformer
水冷油绝缘变压器 water-cooled oil-insulated transformer
水冷油冷却器 water-to-oil cooler
水冷渣挡 water back
水冷针形阀 water-cooled needle valve
水冷真空管 water cooled vacuum tube
水冷轴承 water-cooled bearing
水冷注塞 water-cooled stopper
水冷转子 water-cooled rotor
水冷装置 water chiller;water-cooling plant
水离解反应 water dissociation reaction
水离子化 water ionization
水理机 water dresser
水理图 hydrologic(al) map
水理学 hydrology
水理学家 hydraulician
水力 hydropower; water power; hydraulic power
水力半径 hydraulic mean depth; hydraulic mean path;hydraulic radius
水力半径比例 hydraulic radius scale
水力伴流 hydraulic wake
水力爆破 hydraulic blasting
水力爆破工作面 water blasting face
水力爆破筒 hydraulic cartridge
水力崩落 hydraulic loosening
水力泵 hydraulic pump unit
水力泵喷雾器 hydraulic pump sprayer
水力比降 hydraulic gradient; hydraulic slope; specific hydraulic slope
水力比率 hydraulic ratio
水力比拟 hydraulic analog(ue)
水力边界条件 hydraulic boundary condition
水力变幅 hydraulic luffing
水力变力矩器 hydrodynamic(al) moment transformer
水力表 hydraulic ga(u)ge; hydraulic scale
水力波 bore;hydrodynamic(al) wave
水力剥离 hydraulic stripping; strip with water
水力剥离法 water stripping
水力剥皮 hydraulic barking; jet barking
水力剥皮机 hydrabarker; hydraulic barker; hydraulic slab debarker; hydrobarker;jet barker
水力播种 hydroseeding
水力播种机 hydroseeder
水力捕金器 hydraulic trap
水力采矿 hydraulic mine; hydraulic

mining;hydroextraction
水力采煤 hydraulic coal mining; hydrocoal mining
水力采砂船 hydrojet dredge
水力参数 hydraulic parameter
水力操纵 hydraulic control;hydraulic steering
水力操纵的 hydraulic operated
水力操纵器 hydraulic manipulator
水力操作 hydraulic operation
水力操作的 hydraulic operated
水力操作条件 hydraulic operating condition
水力糙度 hydraulic roughness
水力糙率 hydraulic roughness
水力漕运 hydraulic flume transport
水力测功机 hydraulic brake;hydraulic dynamometer;hydrodynamometer
水力测功计 hydraulic dynamometer; hydrodynamometer
水力测功率 hydraulic dynamometer; hydrodynamometer
水力测功器 hydraulic dynamometer; hydrodynamometer; hydromonitor; water brake
水力测力计 hydraulic dynamometer; hydrodynamometer
水力测量截面 hydrometric(al)(cross-)section
水力测量学 hydraulic measurement; hydrometry
水力测试器 hydroscope
水力测压计 hydrodynamic(al) piezometer
水力超载 hydraulic overload(ing)
水力潮汐河 hydraulic tidal stream
水力车轮压装机 hydraulic wheel press
水力沉积的岩石碎屑 hydroclastics
水力沉降计 hydraulic settlement ga(u)ge
水力沉降值 hydraulic subsiding value
水力沉桩法 jetting piling; sinking of pile by water jet;jetting
水力迟缓器 hydraulic retarder
水力充填 controlled-gravity stowing; flow stowing; hydraulic gobbing; hydraulic mine filling; hydraulic stowing;jetting fill
水力充填坝 hydraulic fill dam
水力充填法 hydraulic fill method; water-jetting process
水力充填管 flushing pipe
水力充填土 hydraulic fill soil
水力冲锤 hydraulic ram
水力冲灰沟 ash sluice way
水力冲灰沟器 hydraulic ash ejector
水力冲灰器 ash ejector; clinker ejector
水力冲毁堤岸 hydraulic spoil bank
水力冲击 hydraulic blow; hydraulic hammer; hydraulic impact; hydraulic shock; hydrostatic lock; water hammer
水力冲击剥皮机 hydraulic jet barker
水力冲击机工人 nozzleman
水力冲击钻进法 hydraulic percussion drilling method
水力冲击钻探法 hydraulic percussion drilling method
水力冲积堤 hydraulic dam; hydraulic fill dam; hydraulic levee
水力冲孔机 hydraulic punching machine
水力冲矿机 hydraulic giant
水力冲裂法 hydraulic fracturing
水力冲泥 hydraulic dredging; hydraulic flushing of sediment
水力冲泥管 beach pipe
水力冲泥机 hydraulic dredge(r); hydraulic excavator
水力冲沙法 hydraulic sluicing
水力冲砂 hydraulic flushing
水力冲射沉桩 pile sinking with the water jet
水力冲射泥石 hydraulic mining
水力冲射器 hydraulic giant; hydraulic gun
水力冲蚀 water abrasion
水力冲刷 ablation; hydraulic scour; water scour
水力冲刷浮土法 slickens
水力冲刷勘探 costean
水力冲刷砾石 hydraulic gravel
水力冲刷挖泥船 hydraulic erosion dredge(r)
水力冲刷钻进 drilling by jetting method
水力冲填 float fill; hydraulic embankment; hydraulic fill(soil); hydraulic slushing; sluiced fill; slushing; water-laid deposit; wet laid deposit
水力冲填坝 hydraulic fill dam; pond filling dam
水力冲填堤 hydraulic levee
水力冲填法 hydraulic fill method; hydraulic fill process; hydraulic percussion(drilling) method; water-jetting process
水力冲填管道 hydraulic fill pipeline; shore pipe
水力冲填土 hydraulic fill(soil)
水力冲挖 hydraulic excavation; hydraulic extraction; hydroextraction
水力冲挖机 hydraulic excavator
水力冲挖土管 pipe for hydraulicking
水力冲洗 elutriation by water; hydraulic flushing
水力冲洗泵 <钻探用> hydroboring pump
水力冲洗机 hydraulic giant
水力冲洗装置 jetting device
水力冲泄 sluice
水力冲压机 hydraulic punching machine; hydraulic ram
水力冲淤法 hydraulic sluicing
水力冲运 sluicing
水力冲渣法 water conservancy residue sluicing method
水力抽气泵 water suction pump
水力抽气器 water-jet exhaust
水力出碴 hydraulic mucking
水力出坯机构 hydraulic knock-out mechanism
水力除尘 hydraulic dust removal; hydraulic type dust removal
水力除尘器 wet precipitator; wet-type dust collector
水力除灰 hydraulic ash removal; hydraulic ash sluicing; water soaking
水力除灰法 ash pumping
水力除灰沟道系统 ash trench system
水力除灰灰浆 ash slurry
水力除灰系统 ash removal system by slurry pump
水力除鳞喷嘴 water descaling sprayer
水力除鳞设备 hydraulic jet descaler
水力除鳞系统 hydraulic descaling system
水力除鳞装置 water descaling unit
水力除土 sluicing
水力除渣 water sluice
水力除渣系统 water sluicing system
水力储蓄器 hydraulic accumulator
水力处理器 hydrotreater
水力传导度 hydraulic conductivity
水力传导度系数 coefficient of hydraulic conductivity
水力传导率 hydraulic conductivity
水力传导系数 coefficient of hydraulic conductivity; fluid conductivity; hydraulic conductivity
水力传导性 hydraulic conductivity
水力传动 hydraulic drive; hydraulic transmission; hydrodynamic(al) drive; hydromatic drive
水力传动的平旋桥 hydrostatic swing bridge
水力传动控制 hydraulic control; hydraulic steering
水力传动图 hydraulic chart
水力传动装置 hydraulic transmission gear
水力传输系数 hydraulic conductivity
水力传送 hydraulic feed
水力传送带 hydraulic belt
水力吹风管 exhauster[exhaustor]
水力吹填 hydraulic fill(soil)
水力吹填法 hydraulic fill process
水力锤 hydraulic hammer
水力粗糙的 hydraulically rough
水力粗糙区 hydraulically rough zone
水力打大包机 hydraulic baling press
水力打浆 hydrabeating
水力打桩 hydraulic pile driving
水力打桩机 hydraulic pile driver
水力当量 hydraulic equivalent
水力捣矿机 water-driven stamp mill
水力的 liquid operated
水力的动力供应 hydraulic power supply
水力等价粒径 hydraulically equivalent size
水力等效 hydraulic equivalence
水力等效颗粒 hydraulically equivalent particles
水力等效粒径 hydraulically equivalent size
水力地面试验机 hydraulic ground-testing machine
水力定型机 hydraulic calibrating press
水力动力波 H-wave; hydrodynamic(al) wave
水力动力法 hydrodynamic(al) means
水力断裂 hydraulic fracture; hydraulic fracturing; hydrofracture
水力断裂试验 hydraulic fracture test
水力断裂作用 hydraulic fracturing
水力断流阀 hydraulic shutoff valve
水力断面 hydraulic profile; hydraulic section
水力锻压机 hydraulic forging press
水力锻造 hydraulic forging
水力发电 hydraulic power supply; hydroelectric(al) generation; hydroelectric(al) power; hydroelectric(al) power generation; hydroelectricity; hydroelectricity generation; hydrogeneration; hydropower; water power
水力发电坝 hydroelectric(al) dam; water power dam
水力发电策略 hydro policy
水力发电厂 hydraulic power plant; hydroelectric(al) generating station; hydroelectric(al) generator; hydroelectric(al) generator plant; hydroelectric(al) power plant; hydroelectric(al) power station; hydroplant; hydropower plant with reservoir; water generating plant; water generating station; water power plant
水力发电厂容量 <正常水头满流情况下,电厂最大发电量> capacity of a hydroelectric(al) plant
水力发电厂址 hydraulic power site; hydropower site; water power site
水力发电的 hydroelectric(al)
水力发电方针 hydro policy
水力发电工程 hydraulic power project; hydroelectric(al) engineering; hydroelectric(al) power development; hydroelectric(al) project; water power development; water power engineering; water power project; water power scheme; water power works
水力发电工程方案 hydraulic power project; hydroelectric(al) power project
水力发电工程师 hydroelectric(al) engineer
水力发电规划 hydroelectric(al) power planning
水力发电河流 power river
水力发电机 hydroalternator; hydroelectric(al) generator; hydrogenerator
水力发电机组 hydroelectric(al) generating set; hydroelectric(al) generator set
水力发电计划 hydroelectric(al) project; water power project; water power scheme
水力发电建设 hydroelectric(al) development
水力发电建设项目 hydroelectric(al) project
水力发电建筑 hydroelectric(al) construction
水力发电开发 hydroelectric(al)(power) development
水力发电开发方案 hydraulic power scheme
水力发电开发利用 hydroelectric(al) exploitation and utilization
水力发电控制 hydraulic power control
水力发电联合企业 hydroelectric(al) complex
水力发电能 hydroelectric(al) energy
水力发电设备 hydroelectric(al) installation
水力发电施工 hydroelectric(al) construction
水力发电实践 hydroelectric(al) practice
水力发电枢纽 hydraulic power scheme; water power development
水力发电隧洞 hydropower tunnel
水力发电系统 hydraulic power system
水力发电项目 hydroelectric(al) project
水力发电用水 hydroelectric(al) use of water
水力发电与坝工建设 <英国月刊> Water Power & Dam Construction
水力发电与坝工建设手册 <英国年刊> Handbook-Water Power and Dam Construction
水力发电站 hydroelectric(al) plant; hydroelectric(al) power station; hydropower station; water power station
水力发电贮水池 hydraulic power pool
水力发电装置 hydroelectric(al) installation; hydro installation
水力发电资源 hydropower resources
水力发动机 hydraulic engine; hydraulic motor; hydromotor; water engine; water motor; water power engine; water pressure engine
水力发射机 hydraulic giant
水力发送器 hydraulic power transmitter
水力阀 fluid valve; hydraulic valve
水力法掘进 hydraulic sluicing

S

水力法排土 hydraulic sluicing
水力法钻探 jetting
水力反馈 hydraulic force feedback
水力防波堤 hydraulic breakwater
水力防松螺帽 hydraulic check nut
水力纺纱机 water frame
水力分级 hydraulic classification; hydraulic separation; hydroclassification; hydroseparation
水力分级机 hydration classifier; hydraulic classifier; hydrosizer
水力分级器 hydroseparator
水力分类法 hydroclassifying
水力分离(法) hydroseparation; hydraulic separation
水力分离机 hydroseparator
水力分离器 hydraulic separator; hydroseparator
水力分离作用 hydroseparation
水力分粒法 hydroseparation
水力分粒机 hydraulic classifier; hydraulic sizer
水力分粒作用 hydroseparation
水力分砂器 hydraulic sand sizer
水力分析 hydraulic analysis
水力分选 hydraulic classification; hydraulic sorting
水力分选槽 water scalping tank
水力分选机 hydroseparator; desilter
水力分选器 hydraulic classifier
水力封隔器定位器 hydraulic packer holddown
水力浮槽 hydrobowl
水力浮槽分级机 hydrobowl classifier
水力浮选 hydraulic flo(a)tation
水力辅助设备 hydraulic power assistance
水力辅助装置 hydraulic assist(ance) device
水力负荷 hydraulic load(ing)
水力负荷率 hydraulic loading rate
水力工程 hydraulic works; hydraulic engineering
水力工程计划 hydraulics engineering project
水力工程师 hydraulic engineer
水力工程学 hydraulic engineering; hydroengineering; hydrotechnics
水力工程学报 <美国土木工程学会月刊> Journal of Hydraulic Engineering
水力工程学的 hydrotechnological
水力工况 hydraulic condition; hydraulic regime(n)
水力工艺学的 hydrotechnological
水力功率 hydraulic power
水力功率计 hydraulic dynamometer; hydrodynamometer
水力固结作用 hydroconsolidation
水力管 hydraulic tube; hydraulic pipe
水力管道 hydraulic pipe line
水力管道冲泥 hydraulic pipe line dredge(r)
水力管路 hydraulic pipe line
水力管路冲泥 hydraulic pipe line dredge(r)
水力管线 hydraulic pipe line
水力管线冲泥 hydraulic pipe line dredge(r)
水力管线开挖机 hydraulic pipe line dredge(r)
水力灌浆 hydraulic slushing
水力灌砂 hydraulic placement of sand
水力光滑的 hydraulically smooth
水力光滑管 hydraulically smooth pipe
水力光滑流态 hydraulically smooth regime
水力光滑区 hydraulically smooth zone
水力过渡过程 hydraulic transient
水力过负(荷) hydraulic overloading
水力夯锤 hydraulic ram; water ram
水力荷载 hydraulic load(ing)
水力荷载系统 hydraulic loading system
水力缓冲器 hydraulic buffer
水力回填 hydraulic backfilling
水力回填材料 hydraulic fill material
水力回旋钻探 hydraulic rotary drilling
水力回转方法 hydraulic rotary method
水力回转钻探 hydraulic rotary drilling
水力混汞捕汞器 hydraulic trap
水力混合料 hydraulic mix(ture)
水力混合器 hydraulic mixer
水力活塞 hydraulic piston
水力机 hydraulic engine; hydraulic machine; water engine
水力机操作者 hydraulicker
水力机械 hydraulic machinery; hydromachine; water power machinery
水力机械采矿法 hydromechanical mining method
水力机械的 hydromechanical
水力机械的水道 race
水力机械化 hydromechanisation[hydromechanization]
水力机械化矿井 hydromine
水力机械间 hydromechanical cell
水力机组 water power set
水力积分仪 hydraulic integraph
水力激波 <开闸门时的冲击波> hydraulic bore
水力集中联锁 hydraulic interlocking
水力几何形态 hydraulic geometry
水力几何形状变化 hydraulic-geometry change
水力几何形状关系 hydraulic geometry relationship
水力几何学 hydraulic geometry
水力挤压机 hydroextractor
水力计 aquameter; potamometer
水力计算 hydraulic calculating; hydraulic computation
水力计算值 computing figure of hydraulic power
水力继动阀膜 hydraulic relay valve diaphragm
水力继动阀体 hydraulic relay valve body
水力夹头 hydraulic chuck
水力监测 hydraulic monitor
水力减震器 hydraulic shock absorber
水力剪切扩散设备 hydraulic shear diffuser
水力剪切力 hydraulic shear
水力检测 hydraulic measurement
水力检测仪 hydromonitor
水力检查计 hydromonitor
水力建筑物 waterfront structure
水力交换 hydraulic exchange
水力绞车 hydraulic jigger; hydraulic power winch
水力绞刀 hydrofraise
水力绞盘 hydraulic capstan
水力绞滩 hydraulic rapids-heaving; hydraulic rapids-warping; rapids warping by water power
水力搅拌设备 hydraulic stirring device
水力结合器 hydraulic binder
水力解棉机 hydraulic defibering machine
水力进刀 hydraulic feed
水力进给 hydraulic feed
水力经济断面 hydraulically profitable section
水力井下泵 hydraulic bottom hole pump
水力警铃 hydraulic alarm
水力静压头 elevation head
水力卷轴机 hydraulic beamer
水力掘进 hydraulicking
水力掘进冲挖 hydraulic mining; hydraulicking
水力掘进开采 hydraulicking; hydraulic mining
水力掘进挖土 hydraulicking; hydraulic mining
水力掘土 outlet sluice
水力开采 hydraulic excavation; hydraulic extraction; hydraulicking; hydraulic mining; hydraulic stripping; hydroextraction
水力开采法 hydraulic mining system; hydraulic sluicing; hydromining
水力开采矿井 hydromine
水力开采水枪 hydraulic mining giant
水力开动 hydraulic actuation
水力开发 hydraulic development; hydrodevelopment; water power development
水力开发地点 water power site
水力开发调查 water power survey
水力开发区划 waterworks planning
水力开沟铲 hydraulic trench-forming shovel
水力开沟锄 hydraulic trenching
水力开沟机 hydraulic trencher
水力开沟耙 hydraulic trenching
水力开土机 hydraulic bulldozer
水力开挖 hydraulic excavation
水力开凿 jet cutting
水力勘察 water power survey
水力勘探 hydraulic prospecting
水力空化 hydrodynamic(al) cavitation
水力控制器 hydraulic controller; hydroman
水力离析 hydraulic separation
水力离析器 hydraulic separator
水力离心除渣器 hydraulic centrifugal cleaner
水力离心分级器 hydrocyclone
水力离心分离器 hydrocyclone
水力离心分粒器 hydrocyclone
水力离心机 hydroextractor
水力利用 utilization of water power
水力连系 hydraulic connection
水力联合系统 hydraulic power pool
水力联锁 hydraulic interlocking
水力联系 hydraulic connection; hydraulic interrelation
水力联轴器 hydraulic clutch; hydraulic coupling
水力流动 hydraulic flow
水力流网 hydraulic flow net
水力龙门起重机 hydraulic gantry crane
水力滤器 hydraulic filter
水力螺旋钻 hydrauger
水力落差 hydraulic drop
水力马力 water horsepower
水力铆钉机 hydraulic riveting machine
水力铆接机 hydraulic riveter; hydraulic riveting machine
水力模量 hydraulic modulus
水力模拟 hydraulic analogy; hydraulic similitude
水力模拟盘 hydraulic analog(ue) table
水力模拟水槽 hydraulic analog(ue) water channel; hydraulic analogy water channel
水力摩擦 hydraulic friction
水力摩阻 flow resistance; hydraulic friction; hydraulic resistance
水力摩阻系数 hydraulic friction coefficient
水力磨粉机 water mill
水力磨蚀 hydraulic erosion
水力磨损 hydraulic wear
水力囊 hydraulic capsule
水力能源 hydraulic energy; source of hydropower
水力泥砂充填 hydraulic silting
水力黏[粘]合剂 hydraulic binding medium
水力黏[粘]合料 hydraulic cementing agent; hydraulic cementing material; hydraulic matrix
水力浓缩槽 hydrotator-thickener
水力耙式分级机 hydraulic rake classifier
水力排管式挖掘船 hydraulic pipe line dredge(r)
水力排料 hydraulic discharge
水力排泥泵 hydraulic ejector
水力排土场 hydraulic dump
水力跑道装置 hydraulic race track device
水力泡沫塑料射送器【救】 ejector pump for injecting plastic foam
水力泡沫塔 hydraulic foam tower
水力喷砂法 sand jet method
水力喷射 hydrojet; water-jet
水力喷射搅拌机 hydraulic jet mixer
水力喷射开采 hydraulic jet mining
水力喷射器 hydraulic ejector; hydroejector; water(-operated) ejector
水力喷射式挖掘机 hydrojet dredge(r)
水力喷射式挖泥船 hydrojet dredge(r)
水力喷射装置 jetting device
水力喷射钻进 jet drilling
水力喷射钻井 jetting
水力喷射钻孔 jet hole
水力喷雾器 hydraulic sprayer
水力喷洗 hydraulic jetting
水力喷嘴 hydrojet nozzle
水力劈裂 hydraulic fracture; hydraulic fracturing; hydraulic splitting; hydrofracture; hydrofracturing
水力劈木机 hydraulic splitter
水力劈石机 hydraulic splitter
水力平衡 hydraulic balance; hydraulic equilibrium
水力平均半径 hydraulic mean radius
水力平均比例 hydraulic mean ratio
水力平均深度 hydraulic mean depth
水力平路机 hydraulic bullgrader
水力屏障 hydraulic barrier
水力坡度 hydraulic grade; hydraulic gradient; hydraulic inclination; hydraulic slope; water grade; water gradient; water inclination; water slope
水力坡度线 hydraulic grade line
水力坡降 hydraulic drop; hydraulic gradient; hydraulic slope; hydraulic grade
水力坡降率 specific hydraulic slope
水力坡降线 hydraulic grade line; hydraulic gradient; hydraulic gradient line; piezometric line; slope of hydraulic grade line
水力破坏 hydraulic damage; hydraulic fracture; hydraulic collapse
水力破裂 hydraulic fracturing
水力破裂法 hydraulic shattering; hydrofracturing method
水力破裂试验 hydraulic fracture test
水力破碎 hydraulic breaking; hydrofracture
水力破碎法 means of hydraulic fracturing; hydrofracturing method

S

水力启动系统 hydraulic starting system
水力启门机 hydraulic-operated gate lifting device
水力起锚机 hydraulic capstan
水力起重船坞 hydraulic lift dock
水力起重机 hydraulic crane;hydraulic hoist;hydraulic lift;hydrocrane;water crane
水力起重器 hydraulic jack
水力千斤顶 hydraulic jack
水力牵引 hydraulic drag; hydraulic traction
水力强度 hydraulic strength
水力桥式起重机 hydraulic gantry crane
水力切变充气机 hydraulic shear aerator
水力切割器 hydraulic cutter
水力侵蚀 water erosion
水力清除 hydraulic stripping
水力清除法 hydraulic cleaning
水力清基 hydraulic stripping
水力清焦 hydraulic decoking
水力清焦法 hydraulic method of coke removal
水力清理 hydroblast(ing)
水力清扫 water soaking
水力清砂 hydraulic blast;hydroblast(ing)
水力清洗 hydraulic cleaning; hydrocleaning
水力情况 hydraulic regime
水力求积仪 hydrointegrator
水力区域化 hydraulic regionalization
水力驱动 liquamatic
水力驱动的平旋桥 hydrostatic swing bridge
水力驱动旋转钻进 hydraulic rotary drilling
水力软管 hydraulic hose
水力筛分设备 hydraulic grading outfit
水力上经济的断面 hydraulically profitable section
水力上行压机 hydraulic up-stroke press
水力设备 hydraulic equipment; hydraulic set(ting)
水力设计 hydraulic design
水力设计准则 hydraulic design criteria
水力射流 hydraulic jet
水力射流搅拌机 hydraulic jet mixer
水力深度 <等于过水面积除以水面宽度,通常用 D 表示> hydraulic depth
水力渗透率 hydraulic permeation rate
水力渗透性 hydraulic permeability
水力升船机 hydraulic shiplift
水力升降车 hydraulic lifting truck
水力升降船坞 hydraulic lift dock
水力升降机 hydraulic lift
水力生物膜负荷 hydraulic biofilm load
水力失调 hydraulic disorder;hydraulic misadjustment
水力失调度 hydraulic imbalance
水力实验室 hydraulic laboratory
水力式测压计 hydraulic piezometer
水力式防水试验 hydrodynamic(al) water-resistance test
水力式分级机 hydroscillator classifier
水力式封隔器 hydraulic packer
水力式(土壤)水分蒸腾计 hydraulic evapotranspirometer
水力式土壤蒸腾计 hydraulic evapotranspirometer
水力势能 hydraulic potential
水力试验 hydraulic test(ing);hydrostatic test(ing)
水力试验槽 hydraulic flume
水力试验机 hydraulic tester
水力试验模型 hydraulic model
水力试验室 hydraulic laboratory
水力试验箱 hydraulic test box
水力枢纽 hydraulic development
水力枢纽布置 hydraulic project planning
水力疏浚管道 hydraulic pipe line dredge(r)
水力疏浚机 hydraulic dredge(r)
水力疏通 hydraulic clean
水力输灰装置 hydraulic ash conveyer [conveyor]
水力输泥管 <水力冲填坝用的> beach pipe
水力输送 hydraulic conveying;hydrotransport;water transport(ation)
水力输送的 water-borne
水力输送法 hydraulic transport(ation)
水力输送管系 water-carriage pipe system
水力输送机 hydraulic conveyer[conveyor]
水力输送设备 hydraulic-transporting device
水力输送装置 hydraulic conveying device
水力输渣体系 hydraulic transport system
水力双管压力计 hydraulic twin tube piezometer
水力水电工程规模 engineering scale of water conservancy facility and hydroelectric(al) station
水力水头 hydraulic head
水力(水下喷水)防波堤 hydraulic breakwater
水力瞬变过程 hydraulic transient
水力松土机 hydraulic ripper
水力碎浆机 aquapulper;hydrabrusher;pulper
水力碎石机 hydraulic rock breaker
水力损失 hydraulic loss;hydraulic resistance
水力缩尺模型试验 hydraulic scale model research
水力缩尺模型研究 hydraulic scale model research
水力弹性模型 hydroelastic model
水力弹性稳定性 hydroelastic stability
水力镗床 hydraulic boring machine
水力淘析器 lavoflux
水力淘选 hydraulic jigging
水力特性河段 hydraulic reach
水力特性曲线 hydraulic characteristic curve
水力特征 hydraulic characteristic;hydraulic property
水力特征曲线 hydraulic characteristic curve
水力梯度 hydraulic grade; hydraulic gradient
水力梯度分析 hydraulic gradient analysis
水力梯度线 hydraulic gradient line
水力提取 hydraulic extraction;hydroextraction
水力提升 hydraulic hoisting; water winding
水力提升船坞 hydraulic dock; hydraulic lift dock
水力提升工作台 hydraulic lifting work platform
水力提升机 hydraulic elevator; hydraulic hoist;hydraulic lift
水力提升能力 hydraulic lifting capacity
水力提升平台 hydraulic lifting platform
水力提升器 hydraulic elevator;jet elevator
水力提升器排砂 grit discharge with hydraulic elevator
水力提升系统 hydrolift system
水力填充 open hydraulic fill
水力填方 hydraulic fill(soil);sluiced fill
水力填砂 hydraulic placement of sand;hydraulic sand filling;sand fill
水力填土坝坝心 core pool
水力填土法 hydraulic filling;slushing
水力填土管道 hydraulic fill pipeline
水力填土围堰 hydraulic earth-fill cofferdam
水力填筑 hydraulic fill(soil)
水力填筑坝 hydraulic fill dam
水力填筑的 hydraulically filled
水力填筑法 hydraulic fill process
水力填筑路堤 hydraulic fill embankment
水力填筑曲线 hydraulic fill curve
水力条件 hydraulic condition
水力调节 hydraulic governing
水力调节器 hydraulic governor; hydraulic pressure regulator; hydraulic regulator
水力跳汰机 hydraulic jig
水力停留时间 hydraulic detention time;hydraulic residence time;hydraulic retention time
水力通沟 hydraulic flushing of sewers
水力桶状升降机 hydraulic barrel lifter
水力透平 hydraulic turbine
水力透水性 hydraulic permeability
水力推进机 hydraulic propeller
水力推进器 hydraulic propeller; hydromotor propeller
水力推移 hydraulic traction
水力拖耙 hydraulic rake
水力拖曳 hydraulic traction
水力脱模 hydraulic sloughing
水力挖掘 hydraulic excavation
水力挖掘机 hydraulic excavator;jetting gear
水力挖泥船 hydraulic dredge(r);hydraulic suction dredge(r)
水力挖泥机 hydraulic drag;hydraulic dredge(r);hydraulic navvy
水力挖土 hydraulic excavation; hydraulicking
水力挖土机 hydraulic excavator;hydraulic navvy
水力挖凿机 hydraulic excavator;hydraulic navvy
水力弯管机 hydraulic bender
水力弯轨机 hydraulic rail bender
水力网 hydraulic grid;hydraulic net
水力尾砂充填 hydraulic tailing fill
水力位能 hydraulic potential; potential hydroenergy
水力位势 hydraulic potential; potential hydroenergy
水力稳定器 hydraulic stabilizer
水力稳定性 hydraulic stability
水力涡轮发电机 hydraulic turbine generator
水力涡轮机 hydraulic turbine;water turbine
水力涡轮机发电机 hydraulic turbine generator
水力污泥排放管 hydraulic sludge withdrawal pipe
水力污泥提升管 hydraulic sludge withdrawal pipe
水力吸泵挖矿机 hydraulic suction dredge(r)
水力吸泥机 hydraulic dredge(r)
水力吸气器 hydroaspirator
水力洗管器 hydraulic pipe cleaner
水力系数 hydraulic coefficient
水力系统 hydraulic system;hydrosystem
水力下沉值 hydraulic subsiding value
水力下行压机 hydraulic down-stroke press
水力纤维离解机 hydroflaker
水力相似 hydraulic analog(ue);hydraulic similitude
水力相似律 hydraulic similarity
水力相似性 hydraulic similarity
水力相似性定律 law of hydraulic similitude
水力相似原理 principle of hydraulic similarity
水力消能器 hydraulic buffer
水力效率 hydraulic efficiency
水力形状 hydraulic form; hydraulic shape
水力型试验 hydraulic model test(ing);model test
水力型芯打出机 hydrocore-knock-out machine
水力性能 hydraulic behavio(u)r;hydraulic performance
水力性质 hydraulic property
水力絮凝 hydraulic flocculation
水力悬臂起重机 hydraulic slewing crane
水力悬浮 hydraulic suspension
水力旋臂起重机 hydraulic slewing crane
水力旋风器 hydracyclone
水力旋流分离过程 Hydrocyclone process
水力旋流分离器 cyclone classifier; hydroclone separator; hydrocyclone;hydraulic cyclone;liquid cyclone
水力旋流器 centriclone; hydrocyclone; liquid solid cyclone; water cyclone;wet cyclone
水力旋流淘析器 hydraulic cyclone elutriator
水力旋式澄清池 hydraulic and circular clarifier
水力旋转方法 hydraulic rotary method
水力旋转分粒机 hydraulic cyclone
水力旋转凿井机组 hydraulic rotary drilling rig
水力旋转钻进机组 hydraulic rotary drilling rig
水力旋转钻井机 hydraulic rotary drilling machine
水力旋转钻探 hydraulic rotary drilling
水力旋转钻探机组 hydraulic rotary drilling rig
水力学 hydraulics;hydromechanics
水力学测量 hydraulic measurement
水力学的 hydraulic
水力学法流量演算 hydraulic routing
水力学(方法)洪水演算 hydraulic routing
水力学公式 hydraulic formula
水力学计算 hydraulic calculation;hydraulic computation
水力学计算尺 hydraulic slide rule
水力学家 hydraulician
水力学理论 hydraulics theory
水力学模拟 hydraulic analog(ue)
水力学模拟模型 hydraulic analog(ue)model
水力学模型 hydraulic model
水力学模型试验 hydraulic model test-

S

(ing)
水力学模型研究 hydraulic model study
水力学设计准则 hydraulic design criterion
水力学试验 hydraulic(s) test
水力学试验模型 hydraulic testing model
水力学数学模型 mathematic(al) models of hydraulic
水力学特性 hydraulic characteristic; hydraulic performance; hydraulic property
水力学相似原理 principle of hydraulic similitude
水力学研究 hydraulics research
水力学研究站 hydraulics research station
水力学与气体力学 <美国月刊> Hydraulics and Pneumatics
水力循环澄清池 circulator clarifier; circulator pressure type clarifier; hydraulically circulated clarifier; hydraulic circulating clarifier; hydraulic recirculation clarifier
水力循环给水器 hydraulic circulating head
水力循环加速澄清池 circulating acceleration clarifier
水力压接 rubber gasket connection
水力压紧 liquid packing
水力压具 hydraulic holddown
水力压力盒 hydraulic load cell
水力压裂 hydraulic fracturing; hydrofracture
水力压裂处理 waterfrac treatment
水力压裂法 method of hydraulic pressure fracture
水力压裂工艺 fracturing technology
水力压裂试验 hydraulic fracture test
水力压轮机 hydraulic wheel press
水力压气机 hydraulic air compressor
水力压缩机 hydraulic compressor
水力压头 hydraulic head
水力压弯机 hydraulic press brake
水力压型机 hydraulic mo(u)lding press
水力延缓 hydrodynamic(al) lag
水力岩石回填 hydraulic rock fill
水力岩石压力测量传感器 hydraulic load cell
水力岩提断器 hydraulic core extractor
水力岩芯提取器 hydraulic core extractor
水力岩压测量传感器 hydraulic load cell
水力研究 hydraulic study
水力研究所 hydraulic research station
水力研磨机 hydrofraise
水力扬射泵 water-jet-lift pump
水力摇床洗选 hydraulic rocking
水力遥测术 hydraulic telemetry
水力要素 element of water power; hydraulic constituent; hydraulic element
水力冶金 hydrometallurgy
水力叶轮泵 hydraulic vane pump
水力曳引 hydraulic traction
水力液 hydraulic liquid
水力液体罐 hydraulic liquid tank
水力液体箱 hydraulic liquid tank
水力液压夯锤 hydraulic ram
水力翼型 hydrodynamic(al) profile
水力翼栅 hydrofoil cascade
水力因数 hydraulic factor
水力因素 hydraulic element; hydraulic factor
水力应变法 hydroforging
水力硬化作用 hydraulic hardening
水力淤填堤 hydraulic levee
水力淤填法 hydraulic fill process
水力元素 hydraulic element
水力原动机 hydraulic prime motor
水力原动力 hydraulic prime mover
水力圆网成型器 hydraulic former
水力圆锥体 hydraucone
水力圆锥网成型器 hydraulic former
水力运输 hydraulic haulage; hydrotransport; hydraulic transport
水力运输设备 hydraulic-transporting device; hydrotransporting device
水力运输设施 hydraulic transport system
水力运送 hydraulic haulage; hydrotransport
水力运土 hydraulic soil transportation
水力运土方 ground sluicing
水力蕴藏量 potential power; potential water-power resources; water power potential
水力蕴藏率 specific water power potential; specific water power resources
水力凿岩 hydraulic rock cutting
水力凿岩机 hydraulic rock drill
水力增压器 hydraulic intensifier
水力闸 hydraulic brake; water brake
水力闸阀 hydraulic gate valve
水力闸门 hydraulic shutter
水力闸式测功器 water brake
水力张拉 hydraulic tensioning
水力折弯机 hydraulic bender; hydraulic bending machine
水力真空泵 water-operated vacuum pump
水力振荡器 hydraulic oscillator; hydroscillator
水力振动 hydraulic vibration
水力直径 hydraulic diameter
水力值 hydraulic value
水力指数 hydraulic exponent
水力指数法 hydraulic index method
水力制 hydraulic system
水力制动机 cataract
水力制动器 cataract; hydraulic brake; water brake
水力致裂 hydraulic fracturing
水力致裂法 method of hydraulic pressure fracture; hydrofracturing method
水力致裂法原位应力测试 in-situ stress test by hydraulic fracturing method; in-situ stress test by hydrofracturing method
水力滞后器 hydraulic retarder
水力重力梯度 hydraulic gravity gradient
水力轴线 hydraulic axis
水力助推 hydraulic assist(ance)
水力柱 hydraulic post
水力转矩 hydraulic torque
水力装料 hydraulic load(ing)
水力装岩机 hydromucker
水力装载 hydraulic load(ing)
水力装置 hydraulic set(ting)
水力状况 hydraulic regime(n)
水力状态 hydraulic regime(n)
水力锥式尾水管 hydraucone tube; hydraucone type draft tube
水力资源 hydroelectric(al) resources; water power reserves; water power resources
水力资源分布图 water power map
水力资源开发 development of hydroelectric(al) resources
水力资源蕴藏量 potential hydroelectric(al) resources; potential water-power resources
水力资源综合利用 multipurpose water utilization
水力纵剖面 hydraulic profile
水力阻抗性 hydraulic resistivity
水力阻力 hydraulic drag; hydraulic resistance
水力阻力系数 hydraulic resistance coefficient
水力钻机 hydraulic drill; jetting drill; wash boring drill
水力钻进 jet drilling
水力钻进法 jet method of drilling
水力钻进设备 wash boring rig
水力钻井 jetted well; jetting
水力钻具 hydrauger; hydraulic thrust boring machine; hydro-drill
水力钻具钻进的钻孔 hydrauger hole
水力钻探 hydraulic drilling; jetting drilling
水力钻探法 hydrauger hole method
水力最佳断面 best hydraulic cross-section
水力最优断面 optimum hydraulic cross-section
水力作用 hydraulic action; water action
水力作用产生的碎屑 hydroclastics
水利 water conservation; water conservancy
水利部 Ministry of Water Resources
水利部门 water sector
水利产业 water industry
水利措施 water conservancy measures; water control
水利电力部 Ministry of Water Conservancy and Electric(al) Power; Ministry of Water Resources and Electric(al) Power
水利发电站 natural flow station
水利法规 water code; water conservancy law and regulation
水利工程 conservancy works; conservation project; hydraulic project; hydraulic works; hydroengineering; hydrologic(al) project; hydroproject; riparian works; water conservancy; water conservancy project; water conservancy works; water project; water resource project; water utilization project
水利工程测量 hydrographic(al) engineering survey
水利工程等级 grade of hydraulic projects
水利工程管理 hydraulic project management
水利工程建设项目 hydraulic engineering project
水利工程设计 hydraulic design; hydroengineering design
水利工程设施 hydraulic works
水利工程师 hydraulician; hydraulic(s) engineer; water engineer
水利工程师协会 Institution of Water Engineers
水利工程实践 hydraulics engineering practice
水利工程系统 water resources system
水利工程学 hydraulic engineering; water engineering; hydrotechnics
水利工程运行 hydraulic project operation
水利工程闸门 hydraulic engineering gate
水利官员 water commissioner
水利管理 water management
水利管理机构 water board; water conservancy
水利管理局 water authority; water conservancy
水利规范 water code
水利规划 water plan
水利化 adequate irrigation
水利会议 water conservancy conference
水利机械化 hydromechanisation[hydromechanization]
水利及森林资源保护 viver and forestry conservation
水利计划 water project
水利技术 hydrotechnics
水利技术研究 hydrotechnic(al) research
水利建设 water conservancy construction
水利经济 water economy
水利经济分析 hydroeconomic analysis; water economic analysis
水利经济学 economics of water conservancy; water economics; hydroeconomics
水利局 catchment board; hydrotechnic(al) bureau; water board
水利开发 hydraulic development; water development
水利勘测 water survey
水利科学 water science
水利科学研究 hydrotechnic(al) research
水利领域 water sector
水利农业发展 hydroagriculture development
水利权 rights of water
水利设施 water conservancy facility; water resource facility; hydrotechnical facilities
水利设施综合体 hydroengineering complex
水利实验室 hydraulics laboratory
水利史 water conservancy history
水利枢纽 hydraulic complex; hydraulic development; hydraulic engineering complex; hydrocomplex; hydroengineering complex; hydrojunction; hydroproject
水利枢纽布置 hydrojunction layout
水利枢纽工程 key water-control project
水利枢纽工程测量 engineering survey of hydraulic hinge
水利水电发展管理局 water power development authority
水利水电工程地质勘察 engineering geologic(al) investigation of water conservancy facility and hydroelectric(al) station
水利水电技术 water resources and hydropower technique
水利水力开发机构 water and power development authority
水利隧道 irrigation tunnel
水利土壤改良 hydromelioration
水利委员 conservator
水利委员会 water conservancy commission
水利卫生 hydraulic sanitation
水利系统 hydraulic engineering system; hydrocomplex; hydrosystem
水利项目 hydraulic project
水利协会 water users association
水利用 <英国月刊> Water Services
水利债券 irrigation bonds
水利政策 water policy
水利志 water conservancy annals
水利专家 irrigationist
水利专员 water commissioner
水利资源 hydropower resources; water power resources; water re-

sources
水利资源工程 water resources engineering;water resources project
水利资源管理 water resources management
水利资源规划 water resources planning
水利资源开发 water resources development
水利资源利用 water conservation
水利资源清账 inventory of water resources
水利资源系统 water resources system
水利资源研究 water resources research
水利资源蕴藏量 potential waterpower resources
水利资源综合开发 comprehensive development of water resources
水利资源综合利用 multipurpose water utilization
水利资源综合利用规划 integrated water resource planning
水沥青 bituminous grout
水沥青铀矿 hydronasturan;hydronitchblende
水栎 barren oak;black jack;water oak
水粒纹 water grain
水帘 water screen;water sheet
水帘洞 water curtain cave
水帘管 granulating screen
水梁衡器 water weighing device
水亮漆的 japanned
水量 quantity of water; volume of water
水量安全 water quantity security
水量表 water(flow)meter
水量表旋塞 water ga(u)ge cock
水量不足 deficiency in water;water shortage
水量不足的河流 misfit river;misfit stream
水量不足地区 water deficient area; water deficit region; water-short area
水量的测量 measurement of water quantities
水量动态 regime(n)of water quantity
水量分布 water-distribution
水量分流 diverting water
水量分配 water allocation
水量丰富 abundant in water; abundant of water quantity
水量供需平衡 budget of water demand and water supply;water supply and demand balance
水量管理模型 model of water quantity management
水量过剩河段 surplus section
水量极丰富 great abundant of water quantity
水量计 moisture meter;water(flow) meter
水量计算 water balance;water budget;water quantity calculation
水量记录计 <涨潮时的> floodometer
水量减小系数 coefficient of water decrease
水量交换 water exchange
水量均衡法 water resource balance method
水量亏耗过程线 depletion hydrograph
水量模拟 water quantity analogy
水量模型 model of water quantity
水量配比 water proportioning
水量配置 water allotment
水量贫乏 poor of water quantity
水量平衡 hydrologic(al)accounting; hydrologic(al) balance; hydrologic(al) equilibrium; water household; water balance
水量平衡表 hydrologic(al)budget
水量平衡法 water balance method; water budget method
水量平衡方程 hydrologic(al) equation
水量平衡核算 water balance accounting
水量平衡预算 water balance budget
水量热计 water calorimeter
水量热器 water calorimeter
水量收支 hydrologic(al)budget
水量收支平衡 water budget
水量收支预算 water budget
水量水质管理 water management
水量水质模拟模型 water quantity-quality simulation model
水量水质综合规划 integrated water quantity and quality planning
水量水质综合模拟 integrated water quantity-quality modeling
水量损失量 water loss
水量调节 water regulation
水量调节阀 water regulating valve
水量调节器 water regulator
水量系统 water quantity system
水量盈余区 water surplus region
水量增加 flow augmentation
水量中等 middle of water quantity
水疗 water cure
水疗场 watering-place
水料比 water-solid ratio
水裂 water crack
水裂解 water-splitting
水淋冷凝器 drip condenser; drip cooler;trickling cooler
水淋冷却器 surface spray cooler; trickling cooler;water drip cooler
水淋式空调设备 humidifying and air conditioning equipment
水淋式冷却器 drip cooler;trickle cooler
水磷铵镁石 hannayite
水磷钡铁矿 ruaskovite
水磷钒铝矿 gutsevichite
水磷钒铝石 schoderite
水磷钒铁矿 rusacovite[rusakovite]
水磷钙钾石 englishite
水磷钙铝钙石 lehiite
水磷钙锰矿 robertsite
水磷钙铍石 uralolite
水磷钙石 isoclasite
水磷钙铁石 calcioferrite
水磷钙钍矿 brockite
水磷钙铀矿 tristramite
水磷灰石 hydroapatite
水磷钾铝石 minyulite
水磷钪石 kolbeckite;sterrettite
水磷铝矾 sanjuanite
水磷铝钙钾石 englishite
水磷铝钙镁石 ovencerite
水磷铝钙石 overite
水磷铝钾石 minyulite
水磷铝碱石 millisite
水磷铝镁锰石 lungokite
水磷铝镁石 suozalite
水磷铝锰石 sinkankarite
水磷铝钠石 wardite
水磷铝铅矿 plumbogummite;plumboresinite;schadeite
水磷铝石 senegalite
水磷铝铜石 zapatalite
水磷铝铀云母 moreauite
水磷镁石 newberyite
水磷镁铁石 ushkovite
水磷镁铜石 nissonite
水磷锰石 reddingite
水磷钠石 natrophosphate
水磷钠锶石 nastrophite
水磷铍钙石 hydroherderite
水磷铍锰石 roscherite
水磷铍隅石 fransoletite
水磷铅钍石 grayite
水磷氢钠石 dorfmanite
水磷铈矿 rhabdophane
水磷铁钙镁石 segelerite
水磷铁矿 kertschenite
水磷铁镁石 garyansellite
水磷铁锰石 switzerite
水磷铁钠石 cyrilovite
水磷铁铅石 drugmanite
水磷铁石 delvauxene;delvauxite;giniite;phosphoferrite
水磷铁锶矿 lusungite
水磷钍铀矿 kivuite
水磷锌铝矿 kehoeite
水磷钇矿 churchite
水磷铀矿 ningyoite
水磷铀铅矿 dumontite;przhevalskite
水菱钙石 rabbittite
水菱镁矿 hydromagnesite
水菱钇矿 tengerite
水菱铀矿 sharpite
水流 convection current; current (flow);current of water;effluent; efflux; flow stream; fluent; stream current; water course; water current;flow of current
水流摆动 current shifting
水流比能 specific energy of current; specific energy of flow
水流变形 flow distortion
水流标桩 current stake
水流表 water current meter
水流表面坡度 water surface gradient
水流波痕 aqueous current ripple mark; current ripple mark
水流不规则性 flow irregularity
水流不均匀性 flow irregularity;flow non-uniformity
水流不连续面 surface of discontinuity current
水流不连续性 discontinuity of current
水流不稳定性 flow instability
水流参数 flow parameter
水流测定 current measurement;flow measurement;hydrometry
水流测定设备 current measuring device;current measuring plant
水流测定装置 current measuring equipment
水流测量 hydrographic(al) survey(ing)
水流测量图 hydrograph
水流层理 current bedding
水流产生的钻孔 borehole producing by flow
水流场 water flow net
水流成分 flow component
水流尺度 stream dimension
水流冲涤 water elutriation
水流冲击 rush of current; water shock
水流冲击力 water flow impact pressure
水流冲击试验 hose-stream test
水流冲积堤 hydraulic levee
水流冲刷 current scour; erosion by current;fluvial abrasion;water erosion
水流抽气管 aspirator
水流出 throughput of water
水流出露口 outcrop of water
水流出逸口 outcrop of water
水流传播时间曲线 water travel (time) curve
水流传感器 water flow sensing unit
水流导向 current deflecting
水流导向设备 current deflector
水流道 flow passage
水流的共轭水位 alternate stages of flow
水流的连续条件 continuity condition for flow
水流的连续性 continuity of flow
水流的水力学特性 hydraulic property of current
水流的影响 effect of current
水流底部收缩 bottom contraction
水流动 water flow
水流动力学 flow dynamics
水流动力轴线 axis of channel; dynamic (al) axis of flow; dynamic (al) axis of water flow; flow dynamic(al) axis; hydrodynamic (al) axis;line of maximum depth
水流动量 flow momentum; momentum of flow
水流动量方程 flow momentum equation
水流动能 flow kinetic energy
水流动曲线 flow-through curve
水流动态 flow regime
水流断面 active cross-section; flow area
水流堆积物 water flow deposit
水流反应器 flow reactor
水流方向 direction of stream; set of stream; water flow direction; current direction
水流方向角 angle of current
水流分布 flow distribution; water spread
水流分类 flow classification
水流分离 flow separation; streamflow separation
水流分离区 separation area
水流分离现象 flow separation phenomenon
水流分配均衡的输水系统 balanced flow system
水流分散 water spreading
水流分析 flow analysis
水流分选 separation in streaming current
水流复形 flow distortion
水流改道 flow diversion
水流改向 diversion of flow
水流干扰 interference of current
水流干扰覆盖层 flow obstruction cover
水流刚离开量水堰下游的平均流速 velocity of retreat
水流功率 stream power
水流沟渠 water-borne sewerage
水流鼓风器 water-blast
水流观测 current observation; current survey;flow observation
水流观测站 current observation station
水流轨迹 water trajectory;streak line
水流过程 water flow process
水流过滤装置 flow filter
水流荷载 current load(ing)
水流痕(迹) current mark
水流横截面 area of flow;sectional area of flow
水流环量 number of swirls of water flow
水流回路 flow circuit
水流畸变 flow distortion
水流集中 flow convergence
水流几何学 flow geometry
水流计 current meter
水流计及变送器 flow meter and transmitter
水流继电器 water flow relay
水流间断面 surface of discontinuity current

S

水流监测器 water flow monitor
水流剪力比例 scale of flow shear force
水流交叉 stream crossing
水流交错波痕 current cross ripple mark
水流角度 angle of current
水流接点 water contact
水流结构 flow structure
水流截断 interception of water
水流截面积 area of waterway
水流进出口 < 闸板控制的 > paddle hole
水流静止区 keld
水流均衡 flow equalization
水流均一性 flow uniqueness
水流均匀性 flow uniformity
水流开关 flow switch
水流开展区 flow expansion area
水流空蚀 water cavity
水流控制 flow control
水流控制措施 water flow control measures
水流控制设施 hydraulic monitor; water control facility
水流宽度 flow width; spread of flow
水流扩散 flow diffusion; flow expansion
水流扩展 spreading of stream; water spreading
水流扩张 flow expansion
水流雷诺数 Reynold's number of flow
水流力 < 作用于水工建筑物的 > drag force; current force; flow force; hydraulic drag
水流力系数 coefficient of drag; current force coefficient; drag coefficient
水流力学 flow mechanics; river mechanics < 河流的 >
水流连续方程 continuity equation of flow
水流量 discharge of water; throughput of water; flow of water
水流量测量 water flow measurement
水流量计 water flowmeter; water ga(u)ge
水流量警报系统 water-flow-alarm system
水流量热器 water flow calorimeter
水流量调节器 water flow regulator
水流流程 course of currents; water circuit
水流流动时间 flow-through period
水流流量 volume of flow; water discharge
水流流速 drift
水流流线 flow stream-line
水流流向 course of currents
水流流向角 angle of current; angle of flow
水流落差 fall of stream; fall of water
水流脉动 flow fluctuation; flow pulsation
水流玫瑰图 current rose
水流密度频率 channel frequency; stream frequency
水流模拟 flow analogy; flow modeling; flow simulation; water flow simulation
水流模式 current pattern; flow pattern
水流模型 flow model
水流模型试验 flow model test
水流摩擦 flow friction
水流摩擦坡 friction(al) slope
水流摩阻(力) flow friction
水流能量 flow strength; stream energy; water energy
水流能头 energy head
水流逆向 flow reversal
水流偏转 flow deflection
水流漂移 current shifting
水流频率(玫瑰)图 current rose
水流平面 flow plane
水流平面图 plan of flow filaments
水流坡度 flow gradient; slope of water
水流坡降 fall of stream; flow inclination
水流破坏 flow failure
水流起伏 flow fluctuation
水流起源 flow initiation
水流气味控制 < 给排水工程的 > odo(u)r control of water flow
水流强度 competence of stream; flow intensity; flow strength; intensity of current
水流强度指数 flow intensity index
水流清除固体垃圾系统 water borne solid waste removal system
水流情况 regime(n) of flow; stream condition; flow regime
水流情况改变 flow modification
水流情势 flow regime; regime(n) of flow
水流区 zone of flow
水流曲线 current curve
水流驱动泵 stream driven pump
水流权 right to watercourses
水流扰动 disturbance of flow
水流扰动传播 passage flow disturbance
水流绕过结构物引起的严重冲刷 outflanking
水流日射强度表 water flow pyrheliometer
水流入 water influx
水流入速度 water influx rate
水流深度 depth of flow; flow depth
水流深宽比 form ratio
水流失的河流 losing stream
水流时间 time of flow
水流式计程仪 submerged screw log
水流式冷却器 open surface cooler
水流式煤气热水器 instantaneous gas water heater
水流势能 flow potential; streaming potential
水流试验接头 flow test connection
水流事故 flow failure
水流收缩 jet contraction; flow contraction
水流输灰道 ash sluice
水流输渣道 ash sluice
水流数据 flow data
水流水力参数 hydraulic parameters of flow
水流水力特性 hydraulic behavio(u)r of flow
水流水力阻抗 hydraulic resistance of flow
水流水头损失 resistance loss
水流速度 current velocity; flow velocity; stream velocity; velocity of flow; water flow velocity; water velocity
水流速度单位 < 经过 1 平方英寸洞口的流水速度,用于美国西部 > miner's inch
水流速率 rate of flow
水流损失高度 < 指降雨与流量的高差 > height of loss of water
水流特性 flow characteristic; flow property; flow behaviour
水流特性系数 flow characteristic coefficient
水流体积含沙量浓度 spatial concentration of sediment
水流体系 flow regime
水流条件 current condition; flow condition
水流条件的稳定状态 steady-state of flow condition
水流条件改变 flow modification
水流调节 flow regulation
水流调节阀 water flow regulating valve; water regulating valve
水流调节器 flow regulator
水流通道 flow channel; hydraulic duct
水流图 current chart
水流图表 water flow chart
水流图解 current diagram
水流湍急处 quick water
水流湍急的 fast-flowing
水流推移力 current tractive force; flow tractive force
水流拖曳 current drag
水流拖曳痕迹 drag mark
水流纹理 current lamination
水流纹线 thread of stream
水流紊动 flow turbulence; stream turbulence
水流紊动激励器 tripping device
水流紊动条件 flow turbulence condition
水流紊动现象 flow turbulence phenomenon[复 phenomena]; flow turbulent phenomenon[复 phenomena]
水流稳定性 flow stability
水流问路 flow by-pass
水流吸收系统 disposal field
水流系数 stream factor
水流系统 flow system
水流下沉 downwelling
水流舷角 relative bearing of the current
水流线 elementary stream; filament of water; flow(stream) line
水流线理 current lineation; parting lineation
水流相似准则 flow similarity criterion
水流挟带泥沙 water-borne sediment
水流挟带作用 stream traction
水流挟沙能力 sediment-carrying capacity of flow; sediment transport capacity of flow; silt-carrying capacity of flow
水流形式 flow pattern
水流型式 current type
水流循环回路 water flow circuit
水流循环系统 water circuit system; water flow circulation system; water circulation system
水流压力 current pressure; stream-flow pressure
水流压力传感器 stream pressure probe
水流压力系数 coefficient of current pressure
水流溢出口 outcrop of water
水流溢顶 topping
水流引起的振荡 flow-induced oscillation
水流迎面阻力 current drag force
水流影响 current effect
水流涌入 inrush of water
水流诱导的振动 flow-induced vibration
水流诱发振动 flow-induced vibration
水流遇到障碍向后喷射 displacement current
水流约束 confinement of flow
水流运动 flow motion; flow movement; water motion; water movement
水流运动示踪剂 water movement tracer
水流运动相似 similarity of flow motion; similarity of flow movement
水流运动指示剂 water movement tracer
水流运土 hydraulicking
水流噪声 flow noise
水流展宽 spread of flow
水流障碍 discharge obstacle
水流整治 regulation of flow
水流指示灯 water flow indicator light
水流指示器 current indicator
水流中心线 midstream
水流轴线 axis of flow; flow axis; stream centerline; stream center [centre] line
水流转角 angle of current; angle of turning flow; turning angle of flow
水流转移 diversion of flow
水流状况 current regime; flow regime; regime(n) of flow; stream condition
水流状态 current regime; flow regime; regime(n) of flow; stream condition
水流状态图 current state diagram
水流资料 flow data
水流自动掺气试验 experiment on self aerated flow
水流自由表面 free surface of flow
水流总量 volume of flow
水流纵剖面曲线 flow profile
水流纵剖面图 flow profile; water surface profile
水流阻力 flow resistance; resistance to flow; resistance to water-flow; drag force < 作用在水工结构上的 >
水流阻力系数 coefficient of flow resistance; flow resistance factor; drag force coefficient
水流阻力系数比例 scale of flow resistance factor
水流作用 effect of current; action of current
水硫碲铅石 schieffelinite
水硫砷铁石 zykaite
水硫酸铝石 alunogen
水硫酸铁铬矿 redingtonite
水硫碳钙镁石 tatarskite
水硫锑矿 klebelsbergite
水硫铁钠石 erdite
水硫硝镍铝石 hydrombobomkite
水硫铀矿 uranopilite
水馏分 aqueous distillate
水榴石 hibschite; hydrogarnet
水柳 water willow; weeping willow
水龙凹龛 water alcove
水龙带 fire hose; hose; hose line; hose pipe; rotary hose; rubber hose; spray hose; water hose; fireman hose < 消防员用的 >
水龙带编结机 hose braider
水龙带不稳定性 fire hose instability
水龙带阀门 hose valve
水龙带管接头 hose union
水龙带滚筒 hose reel
水龙带架 hose handling frame
水龙带接头 hose union
水龙带接头螺纹 hose thread
水龙带接嘴 outlet hose nozzle
水龙带卷筒 hose drum; hose reel
水龙带龙头 hose bib; hose cock; sill cock
水龙带喷嘴 hose nozzle
水龙带射口 hose nozzle
水龙带箱 hose box; hose cabinet
水龙带旋塞 hose bib; hose cock; sill cock
水龙带转动头 Butterworth head
水龙管 hose
水龙接嘴 outlet hose nozzle

水龙卷 twist(er);water spout
水龙螺纹 hose thread
水龙软管 hose pipe
水龙射水试验 <对热的隔墙或门作耐力试验> hose-stream test
水龙头 bib(b)(cock);bib valve; cock tap; faucet; hydrant; hydrovalve; plug water valve; stop cock; water cock; water faucet; water tap; rotary swivel; water swivel【岩】
水龙头把手 cock handle
水龙头侧轴颈 swivel eye
水龙头承插口 faucet
水龙头大钩销钉 swivel hook pin
水龙头阀(门) bib valve;faucet valve
水龙头管茎 hydrant stem
水龙头接管 faucet pipe
水龙头接管夹具 service clamp
水龙头接头 faucet joint
水龙头坑 hydrant pit
水龙头软管 swivel arm;swivel hose
水龙头提环 swivel bail
水龙头外壳 swivel body;swivel shell
水龙头异径接头 swivel sub
水龙头中心管 swivel stem
水龙头最大负荷 maximum load of swivel
水龙头最大转速 maximum rotational speed
水漏 <计时用> hour-glass
水漏斗 hyponome;water funnel
水陆的 terraqueous
水陆航空港 combined water and land air base
水陆机场 combined water and land air base
水陆建筑场地 amphibious site
水陆交接点车站 terminal
水陆交接点码头 terminal
水陆联合保险 mixed sea and land risk
水陆联用列车 boat train
水陆联运 combined land/sea; combined through transport; combined water-and-land transport; connexion; land-and-water coordinated transport; portage; through transport by land and water; water-land transshipment
水陆联运车 flexi-van
水陆联运的 dual mode
水陆联运集装箱 amphibious container
水陆联运集装箱系统 <英> freightliner
水陆联运计划 rail-water through goods transport plan
水陆联运列车 boat train
水陆联运设备 harbo(u)r terminal facility;port terminal facility
水陆联运提单 overland bill of lading
水陆联运站 ferry terminal; rail and water terminal
水陆联运转换点 overland common point
水陆两栖机械 amphibious plant
水陆两栖消防车 fire-fighting amphibian
水陆两用车(辆) amphibian vehicle; amtrack; duck; alligator; amphibian truck; amphibious vehicle; amphicar; canoe automobile; combined sea/land; duck truck; terrapin; transport
水陆两用车辆江河模拟试验装置 river simulator for amphibian test
水陆两用单斗挖掘机 amphibious shovel
水陆两用的 amphibian; amphibious; land-and-water
水陆两用飞机 amphibian;amphibious aircraft;amphibious plane;duck
水陆两用工程机械 amphibian construction equipment
水陆两用机械铲 amphibious mechanical shovel
水陆两用吉普车 amphibian jeep;sea jeep;seep
水陆两用摩托车 duck
水陆两用起落架 amphibian(landing) gear;amphibious landing gear
水陆两用起重机 amphibious crane
水陆两用气垫船 amphibious hovercraft
水陆两用汽车 amphibian automobile; amphibious car; amphibious vehicle;canoe automobile
水陆两用牵引车 amphibian tractor; amphibious tractor;swamp buggy
水陆两用轻型汽车 jeep
水陆两用坦克 alligator; amphibian; amphibian tank; amphibious tank; buffalo
水陆两用推土机 amphibious bulldozer
水陆两用拖拉机 buffalo
水陆两用挖掘机 land-and-water excavator
水陆两用挖泥船 amphibian dredge(r);amphibious dredge(r)
水陆两用载重(汽)车 amphibian truck;amphibious truck;amph-trk
水陆两用装甲车 sea jeep
水陆两用自动车 weasel
水陆履带牵引车 amphtrac
水陆码头 ferry terminal
水陆平底军用车 alligator
水陆牵引车 amphibious tractor; swamp buggy
水陆生态系统 aquatic and terrestrial ecosystem
水陆枢纽 rail-water terminal
水陆坦克 amphibious tank
水陆形成的 terraqueous
水陆运费 freight and cartage
水陆运输 shipping and traffic;surface movement; water and land transportation
水陆直达联运 through transport by land and water
水陆中转货棚 transit shed;transship shed
水陆转运点 navigation head
水陆转运站 navigation head
水滤器 hydrofilter;water filter
水滤网 water screen
水路 water course;waterway;aquage; nullah;passage;passageway;gut <河、海、湖沼中适于航运的>;lode <英国方言>
水路测量术 hydrography
水路尺寸 waterway size
水路出版物 hydrographic(al) publications
水路的 hydrographic(al)
水路的中段 mid-channel
水路改道 avulsion of water course; diversion of water course; diversion of waterway
水路交通图 shipping-line map
水路竞争 water competition
水路设施 aqueduct facilit
水路铁路联运 joint water-and-rail transportation
水路图【测】 hydrographic(al) chart
水路运输 water(-borne) transport(ation);waterway transport
水路运输的 water-borne
水路转铁路运输 water-to-rail
水铝氟石 prosopite
水铝钙石 hydrocalumite
水铝高岭石 dillinite
水铝铬石 knipovichite
水铝黄长石 straetlingite
水铝矿 gibbsite; hydrargillite; hydroscarbroite
水铝矿耐火材料制品 gibbsite refractory product
水铝镍石 takovite
水铝石 boehmite;diaspore;gibbsite
水铝石耐火材料制品 diaspore refractory product
水铝氧 gibbsite;hydrargillite
水铝英石 allophane;allophanite
水铝铀云母 uranospathite
水绿矾 melanterite
水绿榴石 hydrogrossular;hydrogrossularite
水绿皂石 griffithite
水氯碲铜石 tlalocite
水氯钙石 sinjarite
水氯化动力学 aqueous chlorination kinetics
水氯化铝 water alumin(i)um chloride
水氯铝铜矾 aubertite
水氯镁石 bischofite
水氯镍石 nickelbischofite
水氯硼钙镁石 chelkarite
水氯硼钙石 hilgardite
水氯硼碱铝石 satimolite
水氯硼镁石 shabynite
水氯铅矿 fiedlerite;laurionite
水氯铅石 fiedlerite
水氯羟锌石 simonkolleite
水氯铁镁石 iowaite
水氯铜矾 antofagastite
水氯铜矿 anthonyite;eriochalcite
水氯铜铅矿 pseudoboleite
水氯铜石 eriochalcite
水轮 hydraulic wheel;water-wheel
水轮泵 isogyre pump turbine;turbine pump;water turbine pump
水轮泵站 turbine-pump station
水轮槽 water-wheel pit
水轮发电机 hydraulic generator;hydrogenerator; turbine generator; waterwheel generator
水轮发电机轴 turbo-generator shaft
水轮发电机组 hydroelectric(al) generating set; hydrogenerating unit; hydroturbine generator unit;turbogenerator set;water turbine generator set
水轮刮削机 hydraulic wheel scraper
水轮戽斗 bucket of a water wheel
水轮机 hydraulic prime mover; hydraulic turbine; hydroturbine; water turbine
水轮机安装 turbine setting
水轮机安装高程 turbine setting
水轮机保护阀 turbine guard valve
水轮机保证效率 guaranteed turbine efficiency
水轮机本体 turbine proper
水轮机补气 turbine venting
水轮机补气装置 turbine aerator
水轮机操作盘 turbine operating board
水轮机层 <水电站的> turbine floor
水轮机常数 turbine constant
水轮机超开度容量 over-gate capacity
水轮机出力 turbine output
水轮机大轴 turbine main shaft
水轮机导叶 guide vane of turbine; turbine wicket gate
水轮机导轴承 main guide bearing
水轮机的埋深降低 deep setting
水轮机顶盖或底环 turbine lid
水轮机阀 turbine valve
水轮机阀门 turbine gate
水轮机仿真器 turbine simulator
水轮机盖 turbine cover; turbine lid; turbine lip
水轮机给定功率 turbine demand
水轮机工况 turbine operation
水轮机工况运行 turbining
水轮机固定导叶环 stay vane ring
水轮机管路系统 turbine piping; turbine tubing
水轮机过渡现象 transient turbine phenomenon
水轮机过流量 turbine discharge
水轮机机壳 turbine case
水轮机机坑 turbine chamber;turbine pit
水轮机机坑衬砌 turbine pit liner
水轮机检查平台 turbine inspection platform
水轮机进口阀 valve at turbine inlet
水轮机进气管 turbine air vent pipe
水轮机进水阀 turbine inlet valve
水轮机进水弯管 turbine inlet bend
水轮机净出力 net turbine power
水轮机空转 turbine idling
水轮机控制盘 turbine control panel
水轮机廊道 turbine gallery
水轮机旁通管 turbine bypass
水轮机润滑 hydroturbine lubrication
水轮机式离心泵 centrifugal pump of turbine type
水轮机试验规范 turbine test code
水轮机室 turbine chamber; turbine room
水轮机输出功率 turbine output
水轮机输入功率 turbine input
水轮机水槽 tailrace
水轮机水斗 bucket of a water wheel
水轮机送水弯管 turbine inlet bend
水轮机特性 turbine characteristic
水轮机调节 turbine regulation
水轮机调速器 hydrogovernor
水轮机调速系统 water turbine regulation system
水轮机通气 turbine venting
水轮机通气管 turbine air vent pipe
水轮机尾水道 wheel race; turbine draft tube
水轮机蜗壳 spiral case
水轮机吸收能力 absorption capacity of turbine
水轮机系列 series of turbine
水轮机效率 efficiency of turbine;turbine efficiency
水轮机泄流量 wheel discharge capacity
水轮机型 type of turbine
水轮机性能曲线 performance diagram
水轮机叶片 blade of water turbine; blade of water wheel;turbine blade
水轮机油 water turbine oil
水轮机闸门 turbine gate
水轮机罩 turbine case
水轮机轴 turbine shaft
水轮机轴密封套 turbine shaft gland
水轮机轴向推力 turbine axial thrust
水轮机主轴 turbine main shaft
水轮机转轮 turbine runner
水轮机自调因数 turbine self-regulation factor
水轮机最大过流量 maximum turbine discharge capacity
水轮机最大过水能力 maximum turbine discharge capacity
水轮机座环 turbine stay ring
水轮计程仪 submerged screw log
水轮交流发电机 water-wheel alternator
水轮坑 water-wheel pit;wheel pit
水轮驱动布水器 water-wheel driven

distributor
水轮驱动发电机 water turbine driven generator
水轮式流量计 turbine(flow)meter
水轮式水表 water-wheel meter
水轮伺服电动机 turbine servomotor
水轮同步发电机 hydroelectric(al) synchronous machine
水轮叶板 paddle board of water wheel
水轮(制动)机 Froude brake
水轮制动器 hydraulic wheel brake
水络合离子 aquo-complexed ion
水络物 water complex
水落斗 cistern head; cistern hopper; conductor head; conductor hopper; gutter spout funnel; hopper head; leader head; leader hopper; rainwater head; rainwater hopper; spitter
水落管 downfall pipe; downflow pipe; down pipe; downspout; fall pipe; gullet; leader(pipe); rain conductor; rain(fall) leader; rain pipe; rain spout; rainwater leader; rainwater pipe; spouting; stair spout; water spout
水落管槽 boot
水落管出口的散水 kickout
水落管箍 leader hook
水落(管)固定带 leader strap
水落管接长部分 leader extension
水落管接头 drop-pipe connection
水落管卡 U-iron
水落管口 downpipe shoe
水落管滤池 downpipe filter; fall-pipe filter; gutte pipe filter
水落管弯头 elbowed leader
水落管下导水砌块 splash block
水落管鞋 leader shoe
水落管圆筒 leader drum
水落环管 drip loop
水落饰件 gutter member
水落管水平出水口 rainwater shoe
水落铁 <俗称> U-iron
水落铁卡 fastener
水落头 rainfall water head; rainwater head
水马力 water horsepower
水码头 water terminal
水脉 water vein
水螨 water mite
水慢化反应堆 water-moderated reactor
水慢化剂 water moderator
水漫顶 water overtopping
水漫土 water-logged soil
水媒传布 hydrochory
水媒传染 water-borne infection
水媒传染病 water-borne disease; water-borne epidemic
水媒疾病 water-borne disease
水媒介物 aqueous vehicle
水媒流行病 water-borne epidemic
水媒污染物 water-borne contaminant
水媒细菌性疾病 water-borne bacterial disease
水媒植物 hydrophilae
水煤气 blue gas; generator; water gas
水煤气柏油防腐油 water-gas-tar creosote
水煤气成套设备 water-gas set
水煤气催化剂 water-gas catalyst
水煤气搭焊钢管法 water-gas lapweld process
水煤气电池 water-gas cell
水煤气发生炉 blue gas generator; water-gas generator
水煤气发生器 steam-gas generator; water-gas apparatus; water-gas generator; water-gas machine
水煤气发生装置 carburet(t)ed water gas installation
水煤气反应 water-gas reaction
水煤气管 blue gas pipe; water-gas pipe
水煤气管线 water-gas pipeline
水煤气过程 water-gas process
水煤气焊 water-gas weld
水煤气焊接 water-gas welding
水煤气合成油 cogasin
水煤气焦炭 water-gas coke
水煤气焦油 water-gas tar
水煤气焦油沥青 water-gas tar pitch
水煤气焦油乳状液 water-gas tar emulsion
水煤气焦油脂 water-gas tar pitch
水煤气冷凝器 water-gas condenser
水煤气沥青 water-gas pitch
水煤气沥青乳胶 emulsion of water-gas tar
水煤气煤焦油 blue gas tar
水煤气硬沥青 water-gas tar pitch
水煤气转换 water-gas shift
水煤气装置 blue gas set; water-gas set
水霉菌 water mould
水镁大理石 brucite marble
水镁矾 kieserite
水镁方解石 hydromagnesite
水镁橄榄石 hydrofosterite
水镁铬矿 barbertonite
水镁铝矾 seelandite
水镁铝石 manasseite
水镁石 brucite; texalite
水镁石大理岩 pencatite; brucite marble
水镁石石棉 brucite asbestos
水镁石岩 brucitite
水镁铁石 sjogrenite
水镁硝石 nitromagnesite
水镁铀矾 magnesium zippeite
水门 clough; gole; penstock; sluice; water gap; water gate; water valve
水门城 <美国华盛顿的俚称> Watergate City
水锰矾 mallardite
水锰橄榄石 hydrotephroite
水锰辉石 neotocite
水锰矿 gray manganese ore; manganite; newkirkite
水锰磷铁矿 kryzhanovskite
水弥散 aqueous dispersion
水密舱 compartment
水密舱壁 water-tight bulkhead
水密舱口 water-tight hatch
水密舱口盖 water-tight hatch cover
水密插头 weatherproof receptacle
水密的 liquid-tight; waterproof; water-tight
水密灯 water-tight lamp
水密灯罩 water-tight lobe
水密电缆 water-tight cable
水密垫料 watertight gasket
水密垫圈 watertight gasket
水密度增加速度 densimetric velocity
水密分舱 water-tight compartment; water-tight subdivision
水密分舱区划 compartmentation; subdivision
水密分段 water-tight subdivision
水密分线盒 water-tight branch box
水密封 clearance leakage; water stop; water sealing
水密封材料 water-bar material; water-stop material
水密封垫 water sealing gland
水密封口 packing water seal; water-tight seal
水密封屋面 water-filled roof
水密盖 water-tight cover
水密隔舱 water-tight bulkhead
水密隔舱骨架 water-tight frame
水密隔堵 water-tight bulkhead
水密工程 water-tight work
水密工作 water-tight work
水密关闭 water-tight closure
水密关闭记录 water-tight closure log
水密关闭物 watertight closure
水密柜 water-tight tank
水密海图筒 waterproof chart container
水密(环保型)两瓣抓斗 watertight clamshell bucket
水密甲板 water-tight deck
水密接缝 water-tight joint
水密接合 water-tight joint
水密接头 water-tight connector; water-tight joint
水密结构 water-tight structure
水密结合 water-tight joint
水密开关 water-tight switch
水密空气箱 water-tight air case
水密肋板 water-tight floor
水密肋骨 water-tight frame
水密铆合 water-tight riveting
水密铆接 water-tight riveting
水密铆距 water-tight pitch
水密门 water-tight door
水密门试验 flooding test; water-tight door test
水密配件 water-tight fitting
水密区 bilged compartment
水密容器 breaker; water-tight vessel
水密施工缝 water-tight construction joint
水密试验 waterproof test; water-tight test
水密室 water-tight compartment
水密双层底救生艇 decked lifeboat; pontoon lifeboat
水密水箱 water-tight box
水密填料 stopwater
水密外壳 water-tight case
水密完整性 water-tight integrity
水密性 water impermeability; waterproofing quality; waterproofness; water-tightness
水密性表壳 water-tight watch case
水密性材料 water-tight material
水密性的混凝土平屋顶 water-tight concrete flat roof
水密性地下室 water-tight basement
水密性混凝土 water-tight concrete
水密性试验 waterproofness test; water test; water-tightness test
水密性试验装置 water test unit
水密袖口 waterproof cuff
水密闸门 water-tight sluice door
水密纸张包装 water-tight paper packing
水密钻孔 tight
水面 sheet of waters; surface of water; water-level; water plane; water spread
水面安全航行区 area safe for surface navigation
水面凹陷 depression of level
水面般的光泽 water glaze
水面爆炸 water surface burst
水面比降 slope of water surface; surface gradient; surface slope; surface slope of water; water surface gradient; water surface slope
水面比率 water surface slope
水面表层冰 surface ice
水面冰层 surface ice
水面波 surface wave
水面波动 surface beat(ing); water-level fluctuation; water surface fluctuation
水面波动幅度 <水库、湖泊或内海由地震或大风等引起的> amplitude of seiche
水面波浪起伏的 lumpy
水面薄冰层 sheet ice
水面测流浮标 surface buoy
水面测流浮子 surface float
水面层流 laminar surface of flow
水面充氧潜水装置 surface demand diving apparatus
水面船舶 above-water craft; surface craft
水面船舶活动锚泊地 moving surface ship haven
水面船迹 wake
水面船只 surface craft; surface ship
水面垂直速度 vertical velocity of water surface
水面大浪 water surface roller
水面的油污 slick
水面等高线 contour of water table; contours of water table
水面法 <测流的> surface method
水面反射 water-reflected
水面反应 water surface response
水面(飞)机场 seadrome
水面飞行器 surface craft
水面风 surface wind
水面风速 overwater wind speed
水面浮标 surface float; water surface float
水面浮标(测流)法 surface float method
水面浮标瓶 surface drift bottle
水面浮标系数 surface float coefficient
水面浮油 oil slick; oil spill; slick
水面浮油封栏法 surface containment method
水面浮油刮集装置 sea-skimming equipment
水面浮油回收船 oil recovery vessel
水面浮游生物 pleyston; surface plankton
水面浮子 surface float; water surface float
水面俯仰起重机 level-luffing crane
水面附近的岸边活性沉积物 active stream sediment near water surface
水面覆盖力 water-covering capacity
水面高程 surface elevation; water-level elevation; water surface elevation
水面(高程)纵剖面 surface profile
水面供气式潜水装具 Hookah type diving apparatus; Hookah type diving device; Hookah type diving equipment; Hookah type diving unit; surface-supplied diving apparatus; surface-supplied diving device; surface-supplied diving equipment; surface-supplied diving unit
水面管线 floating(pipe)line
水面光泽 water glaze
水面滚浪 surface roller
水面航行 surface navigation
水面航行船 surface boat
水面耗层 water surface
水面核爆炸 water surface nuclear burst
水面核舰艇 nuclear surface ship
水面横比降 transverse slope of water surface
水面滑行艇 hydroskimmer; hydroski vehicle
水面滑走快艇 hydroplane
水面积负荷 <沉淀池面积与流入量之比> surface load(ing)
水面急剧涨落 rapid fluctuation of water-level

水面计 water column
水面间隔时间 surface interval
水面减压 surface decompression
水面舰艇 surface craft;surface ship
水面降落 drawdown
水面降落范围 drawdown range
水面降水 channel precipitation
水面静止的渠道 sleeping canal
水面静止的运河 sleeping canal
水面控制元件 level control element
水面宽度 level breath; spread of flow;water surface width;width of water-level
水面涟漪 current rips
水面流速 surface velocity
水面掠行艇 <包括滑行艇、水翼艇、气垫船等> hydroskimmer
水面落差 fall of water surface; stream fall
水面面积 water surface area
水面目标标图板 surface plot
水面跑道 <水上飞机的> fairway
水面漂(浮)物刮集船 surface dredge(r)
水面平衡涵洞 level(l)ing culvert
水面平衡涵洞出水口 level(l)ing culvert outlet
水面平衡涵洞进水口 level(l)ing culvert intake
水面坡度 surface slope;water surface gradient; water surface slope; water-table slope
水面坡降 hydraulic inclination; slope of water surface;surface slope;water surface gradient; water surface slope
水面破碎 water-break
水面剖面 hydraulic surface profile
水面剖面图 water surface profile
水面起伏 surface undulation
水面气候 hydroclimate
水面气候因素 hydroclimatic factor
水面清污器 scummer
水面曲线 surface curve;surface profile;water surface profile
水面曲线法 surface curve method; water surface profile method
水面散热系数 coefficient of water surface heat exchange
水面上的 above water;over water
水面上的浮油 slick
水面上升曲线 rising surface curve
水面上升速度 <船闸灌水时> level rise velocity
水面上形状 shape above water
水面设备 floating equipment
水面升高 <螺旋桨等引起的> mound
水面升降(波动) water surface fluctuation
水面双峰波 binodal seiche
水面双节驻波振荡 dicrotic seiche
水面水平 surface level;water surface level
水面随波浮标 surface following buoy
水面提斗式温度计 surface bucket thermometer
水面提高 raising of water
水面微波 riffle
水面伪装 water surface disguise
水面未封冻季节 open-water season
水面温度 water surface temperature
水面温度测定系统 water surface temperature measuring system
水面温度分布 water surface temperature distribution
水面稳定设备 hydrostabilizer
水面吸氧减压 surface oxygen decompression
水面下浮桥 underwater buoyant bridge
水面下浮子 sub-surface float
水面下降 down surge; drawdown; lowering of water level; negative surge; negative wave; top contraction; water-table decline
水面下降试验 drawdown test
水面下漂浮生物 hyponeuston
水面下水层 subsurface
水面线 flowage line; hydraulic grade line; hydraulic line; water surface curve; water surface profile
水面泄降 negative wave
水面行驶速度 water speed
水面以上坡度 slope above the water
水面油膜 oil slick
水面油污传感器 surface oil pickup
水面油污自动平衡刮集装置 self-level(l)ing unit for removing pollution
水面油栅 oil slick boom
水面遮盖力试验 water coverage test
水面折射 overwater refraction
水面蒸发 surface evapo(u)ration
水面蒸发量 amount of surface evapo(u)ration; evaporation from water surface
水面植物堆积 sudd
水面植物汇集 sudd
水面植物漂浮物 <阻碍航行的> sudd
水面至堤顶的高度 crest freeboard
水面中心流速 central surface velocity
水面自由降落 free fall
水面综合散热系数 heat-transfer coefficient
水面纵剖面 stage profile
水面最大高程 maximum water surface elevation
水灭火系统 water fire-extinguishing system
水灭藻 algae removal
水敏 water-sensitive
水敏(感)性 water sensitivity
水敏性地层 water sensitive formation
水敏性砂岩 water sensitive sandstone
水敏性土 water sensitive soil
水敏性页岩 water sensitive shale
水明晰度计 water clarity meter
水模型 water model
水膜 aqueous film; moisture film; water film; water membrane
水膜不破表面 water break-free surface
水膜残迹 water-break
水膜除尘器 water dust scrubber; water-film cyclone; water-film deduster; water-film separator; water scrubber
水膜垫 water-film bearing
水膜分离器 water-film separator
水膜集尘器 water-film dust collector
水膜静电吸尘器 water-film electrostatic precipitator
水膜冷却 cooling through water film
水膜理论 water-film theory
水膜力 film force
水膜流 film flow
水膜内聚力 moisture-film cohesion
水膜黏[粘]聚力 moisture-film cohesion
水膜破裂 water-break
水膜上漂浮生物 supraneuston
水膜深度表 water-film depth ga(u)ge
水膜式冷却塔 film type cooling tower
水膜吸尘器 water-membrane scrubber
水膜吸收器 wetted wall absorber
水膜系数 water-film coefficient
水膜形成泡沫 aqueous film forming foam
水膜压力 film pressure
水磨 levigation; liquid honing; water milling
水磨工序 grinding process
水磨机械 millwork
水磨匠 millwright
水磨沥青地面 terrazzo asphalt tile
水磨砾石 <被水磨光的砾石> water-worn gravel
水磨料喷射切割 water-abrasive jet cutting
水磨砂布 waterproof abrasive cloth
水磨砂纸 waterproof abrasive paper
水磨砂纸打光 sandpapering with water
水磨石 terrazzo; granolith; granolithic concrete; pelikanite; rubbed concrete; waterstone
水磨石凹圆挑檐 terrazzo cove
水磨石板 terrazzo block; terrazzo tile; Venetian mosaic
水磨石车间 terrazzo plant
水磨石冲洗台 terrazzo sink drop
水磨石窗台 terrazzo cill[sill]
水磨石打磨机 terrazzo grinding machine
水磨石挡板 terrazzo coping
水磨石的底层 setting bed
水磨石的罩面处理 terrazzo finish
水磨石地板面饰面 terrazzo floor(ing) finish
水磨石地板坪饰面 terrazzo floor(ing) finish
水磨石地面 terrazzo finish; terrazzo flow; terrazzo surface; terrazzo floor(ing)
水磨石(地面)分格条 divider strip
水磨石地面覆盖层 terrazzo floor cover(ing)
水磨石地坪 terrazzo finish; terrazzo floor(ing)
水磨石顶层 terrazzo cope
水磨石分格条 dividing strip for terrazzo
水磨石分块 terrazzo panelling
水磨石覆盖 terrazzo capping
水磨石盖顶 terrazzo coping
水磨石隔板 terrazzo dividing panel
水磨石工 terrazzo layer
水磨石工厂 terrazzo plant
水磨石工场 terrazzo trade
水磨石工程 terrazzo works
水磨石工作 terrazzo work
水磨石拱 terrazzo cove
水磨石骨料 terrazzo aggregate
水磨石护壁板 terrazzo dado
水磨石混合料 terrazzo mix(ture)
水磨石混凝土 terrazzo concrete
水磨石机 terrazzo grinder; terrazzo machine; wet surface grinder
水磨石基础 terrazzo base
水磨石集料 terrazzo aggregate
水磨石颗粒 terrazzo grain
水磨石块 terrazzo panel
水磨石楼梯踏板 terrazzo stair tread
水磨石路面 traffic deck surfacing
水磨石马赛克 mosaic terrazzo
水磨石面板 terrazzo slab
水磨石面层 palladiana
水磨石面混凝土 concrete terrazzo
水磨石磨光 grinding of terrazzo
水磨石平板 terrazzo slab; terrazzo tile
水磨石平面 terrazzo tile
水磨石铺地砖 terrazzo flooring tile
水磨石铺面 terrazzo pavement
水磨石铺嵌工 terrazzo layer
水磨石器皿 terrazzo ware
水磨石嵌条 dividing strip for terrazzo
水磨石墙顶盖 terrazzo wall capping; terrazzo wall cope
水磨石墙帽 terrazzo(wall) cope
水磨石墙面板 terrazzo wall tile
水磨石墙面砖 terrazzo wall tile
水磨石墙裙 terrazzo dado
水磨石砂浆垫层 setting bed
水磨石上光 terrazzo polishing
水磨石石屑 terrazzo chip(ping)s
水磨石饰面 finish with terrazzo; terazzo-finish; terrazzo dressing; terrazzo finish; terrazzo finished surface
水磨石饰面砖 terrazzo tile
水磨石手工业 terrazzo trade
水磨石水槽落水 terrazzo sink drop
水磨石踏板 terrazzo tread
水磨石踢脚线 terrazzo skirting
水磨石罩面 terrazzo topping
水磨石整体罩面 monolithic terrazzo
水磨石制品 terrazzo ware
水磨石制造机 terrazzo tile machine
水磨石砖 terrazzo tile
水磨石(砖)厂 terrazzo tile press
水磨石作业 terrazzo work
水磨蚀 water abrasion
水磨土 cimolite; lemnian earth
水磨云母 wet ground mica
水磨砖压机 terrazzo tile press
水沫 blowing spray
水沫线 waterfront line
水墨画 ink and wash; wash painting
水墨渲染 rendering with water and ink; water-and-ink rendering
水母 jellyfish; medusa
水母毒素 physaliatoxin
水母类 Tubularia crocea
水母期 jellyfish stage
水钼铁化矿 molybdic ocher
水钼铁矿 ferrimolybdite
水幕 water curtain; water screen; water ring <防尘用>; water wall <即水冷壁>【港】
水幕保护系统 water sprinkling protection system
水幕除尘器 screen collector; water curtain collector
水幕喷漆橱 water curtain spray booth
水幕喷头 drencher head
水幕喷嘴 water spray nozzle
水幕式灭火器 drencher fire extinguisher
水幕系统 drencher system
水幕装置 drencher installation; drenching installation
水内冰 frazil ice; needle ice; submerged ice; underwater ice
水内冷 internal water cooling
水内冷水轮发电机 water internal cooling hydraulic generator
水内声速计 velocimeter
水内小冰块 frazil ice; submerged ice
水钠矾石 alumian
水钠钙锆石 lemoynite
水钠铝矾 mendozite
水钠镁矾 uklonskovite
水钠锰矿 birnessite
水钠铀矾 sodium zippeite
水钠铀矿 clarkeite
水囊 pockets of water; water pocket
水能 hydraulic energy; hydraulic power; hydroenergy; water power
水能动力学 hydrodynamics
水能分布图 water power map
水能力 carrying capacity
水能利用 hydroenergy utilization; hydropower development; water power utilization
水能权 water power rights

S

水能溶解的 water miscible
水能梯度 energy gradient; energy slope
水能线 energy gradient;energy line
水能运作阀 hydraulic valve
水能资源 hydroelectric (al) resources; potential water-power resources; water energy resource; water power resources
水泥 cement;hydraulicity
水泥 B 盐 belite
水泥安定性 cement soundness; sounding of cement; soundness of cement
水泥安定性试验 cement soundness test;test for soundness
水泥八字角 < 又称水泥填角 > cement fillet;weather fillet
水泥板 cement board; cement flag; cementitious sheet; cement plate; cement tile;xylolith
水泥板盖屋顶 cement board roof covering
水泥板护岸 cement slab revetment; concrete slab revetment
水泥板铺面 cement flag paving
水泥板(铺砌)路面 cement flag pavement;cement tile pavement
水泥拌和机 cement mixer
水泥拌和土 soil-cement
水泥拌和土护岸 soil cement revetment
水泥拌和土砖块护岸 soil-cement block revetment
水泥包转向器 sack deflector
水泥包装 encasure
水泥包装车间 cement packing plant
水泥包装机 cement packer; cement packing machine
水泥包装设备 cement packing plant
水泥贝壳基层 cement-shell base course
水泥泵 cement pump
水泥扁饼 cement pat
水泥标号(强度等级) brand of cement;cement brand;cement grade; cement mark;cement strength number;grade of cement; quality mark of cement; strength grading of cement;strength of cement
水泥标准规范 cement standard specification
水泥标准筛 cement standard screen
水泥标准试件 briquet(te)
水泥表层 crust of cement
水泥表皮层 cement skin
水泥饼 pat;pat of cement
水泥玻璃纤维板 cement board
水泥薄板 cement sheet
水泥薄板屋面 cement sheet roofing
水泥薄浆 cement-water grout; weak cement grout;wet cement grout
水泥薄浆拌和机 grout mixer
水泥薄浆罐 ga(u)ge pot
水泥薄浆壶 ga(u)ge pot
水泥薄浆搅拌机 grout mixer
水泥薄膜 cement film;cement skin
水泥薄屋面板 cement sheet roof cladding
水泥不能从储料仓或铁路车辆中流出 pack set cement
水泥采样器 cement sampler
水泥仓库 cement shed; cement storage;cement store;shed for cement; cement hangar
水泥仓筒鼓风设备 cement silo aeration
水泥测井 cement log
水泥(测)针 < 测定水泥凝结时间用的 > cement needle
水泥层 cement bed;layer of cement
水泥掺和剂 cement admix(ture)
水泥掺和料 filler for cement
水泥掺土混合料 soil-cement mixture
水泥铲 cement shovel
水泥厂 cement factory;cement mill; cement plant;cement works
水泥厂废水 wastewater from cement mill; wastewater from cement plant; wastewater from cement product
水泥厂自动控制系统 automatic controlling system of cement plant
水泥超塑化剂 cement superplasticizer
水泥车 cement wagon
水泥车泵压 pump pressure of cementing unit
水泥尘肺 cement pneumoconiosis
水泥尘土 cement dust
水泥衬层 cement lining
水泥衬里 cement lining
水泥衬里管(道) cement-lined pipe
水泥衬里套管 cement-lined casing
水泥衬砌 cement lining
水泥衬砌的 cement-lined
水泥衬砌管(道) cement-lined pipe
水泥称量 cement weighing
水泥称量机 cement weighing machine
水泥称量器 cement weigher
水泥称量系统 cement dosing system
水泥称量装置 cement batching plant; cement weigh-batching device; cement weigh-batching unit
水泥称料斗 cement bin
水泥称重给料器 cement weighing hopper
水泥成分 cement constituent;cement composition;cement compound
水泥成品 finished cement
水泥成球 balling-up of cement
水泥成团 balling-up of cement
水泥承包商 cement contractor
水泥承托器 cement retainer
水泥秤 cement meter;cement weigher
水泥冲筋 cement screed
水泥抽筒 cement dump
水泥稠度 consistency of cement
水泥初凝时间 thickening time of cement
水泥初凝时间测定 thickening time test of cement
水泥储仓 cement bin
水泥储藏室 cement bin; cement bunker;cement silo;cement storage silo
水泥储槽 cement storage tank
水泥储存站 cement shed
水泥储库 cement storage silo
水泥储棚 cement shed
水泥处理的 cement treated
水泥处理的承重层 cement treated supporting layer
水泥处理基层 cement treated base
水泥处理基底 cement treated base
水泥处理(土)路 cement treated road
水泥处理土壤 cement modified soil
水泥处治的 cement treated
水泥处治的承重层 cement treated supporting layer
水泥处治骨料 cement modified aggregate
水泥处治基层 cement treated base
水泥处治集料 cement modified aggregate
水泥处治粒 cement treated granular base
水泥处治铝红土 cement treated laterite
水泥船 concrete boat; concrete vessel;reinforce concrete vessel
水泥纯浆 cement paste
水泥次品 substandard cement
水泥粗粒 over-size grains in cement
水泥促凝剂 cement accelerator; set accelerator(for cement)
水泥窜槽 cement channeling
水泥打底涂层 cement-based coating
水泥大理石 cement marble
水泥代替料 cement replacement
水泥代用材料 cement replacement material
水泥代用品 cement substitute; cement replacement
水泥袋 cement bag;cement sack
水泥袋提升机 sack elevator
水泥袋装发运 dispatch of bagged cement
水泥袋装载机 bag loading machine
水泥当量系数 < 混凝土混合材料的活性相当于水泥活性的比数 > cement equivalent factor
水泥的 cementatory
水泥的比表面积 specific surface of cement
水泥的变质 deterioration of cement
水泥的成熟度 maturity of cement
水泥的非正常凝结 abnormal setting cement
水泥的横向张骨 hoop
水泥的假凝 false set of cement paste
水泥的铝率 alumina ratio of cement
水泥的气动输送 pneumatic cement handling
水泥的气动运输 pneumatic cement handling
水泥的石灰饱和系数 lime saturation degree;lime saturation factor of cement
水泥的水凝性 hydraulicity
水泥的细度 fineness of cement
水泥的延迟膨胀 delayed expansion of cement
水泥等级 cement grade
水泥底卸料专用设备 special equipment for silo discharge
水泥地板 cement floor(ing);cement mortar flooring
水泥地沥青砂浆 cement asphalt mortar
水泥地面 cement floor(ing);cement mortar flooring
水泥地面肥育场 paved feedlot
水泥地面肥育圈 paved lot
水泥地面坚硬材料 floor hardener
水泥地面砖 cement floor tile
水泥地坪 cement floor(ing)
水泥电杆 concrete pole
水泥垫块 cement cushion block
水泥钉 cement nail; concrete nail; masonry nail
水泥堵漏法 cement box method
水泥堵漏箱 cement box
水泥煅烧 cement burning
水泥煅烧方法 cement burning process
水泥煅烧工艺 cement burning process
水泥煅烧过程 cement burning process
水泥堆 cement dump
水泥堆料车 bulker
水泥趸(船) concrete barge
水泥多孔板 cement hollow slab
水泥垛 cement dump
水泥惰性掺和料 filler for cement
水泥发运中转站 cement distribution terminal
水泥翻沫 laitance
水泥翻沫层 laitance layer
水泥返高 return top of slurry
水泥泛浆 < 混凝土表面 > bleeding cement
水泥泛水 cement flashing
水泥防潮存放处 weather cement shack
水泥防水材料 cement waterproofer
水泥防水粉 cement waterproofing powder
水泥防水涂层 cement waterproofing coating
水泥仿大理石 cement imitation marble
水泥仿石粉刷 stuke
水泥费 cement cost
水泥分布机 cement distributor
水泥分配机 cement distributor
水泥分配器 cement spreader
水泥分批进料秤 cement batch weigher
水泥分批进料计量器 cement batcher;cement weigh batcher
水泥分散剂 cement dispersing agent
水泥粉 cement flour
水泥粉尘 cement dust
水泥粉光 smooth cement finish
水泥粉光面 cement-based glazed finish
水泥粉煤灰碎石桩 cement flyash gravel pile
水泥粉磨工艺 cement finishing process
水泥粉磨控制 cement grinding control
水泥粉磨细度 fineness of grinding cement
水泥粉磨站 cement grinding station
水泥粉饰 cement plaster; hardwall plaster
水泥粉饰面 cement plaster finish
水泥粉饰墙 hardwall plaster wall
水泥粉刷 cement dressing; cement plaster(ing);cement rendering
水泥粉刷料 cement slurry
水泥(粉刷)硬化剂 cement hardener
水泥风动输送机 cement pneumatic conveyer[conveyor]
水泥风化 aeration of cement
水泥封闭 cementation;cement-bond; cement seal;grout sealing
水泥封孔 cement sealing hole;injected hole
水泥封涂 cement lute;lute;putty
水泥敷面 cement dressing
水泥浮船坞 concrete floating dock
水泥浮浆 bleeding cement;laitance
水泥浮浆(表)层 laitance coating; laitance layer;surface laitance
水泥改良路基 cement modified subgrade
水泥改良土(壤) cement modified soil
水泥改善路基 cement modified subgrade
水泥改善土(壤) cement modified soil
水泥改性剂 < 一种改善水泥制品性质的液体 > cemprover
水泥盖层 coat(ing) of cement
水泥盖面 cement covering; cement facing
水泥杆菌 < 硫酸盐对水泥侵蚀的 > cement bacillus(e);Candlot's salt; ettringite
水泥高压釜试验 < 测定水泥安定性 > autoclave test for cement
水泥高压蒸养膨胀(率) autoclave expansion
水泥工厂粉尘 dust from cement factory
水泥工尘肺 cement worker's pneumoconiosis
水泥工业 cement industry

水泥工艺 cement technology
水泥骨料比(率) cement-aggregate ratio; ratio of cement of aggregate
水泥骨料间的相互适应性 cement-aggregate compatibility
水泥骨料间反应 cement-aggregate reaction
水泥骨料间粉结 cement-aggregate bond
水泥固定处理法 < 软地基 > cement stabilization
水泥固定法 tile cement fixing method
水泥固定性 soundness of cement
水泥固化 cementation; cement solidification
水泥固化的 cement-solidified
水泥固结砾石 cement stabilized gravel
水泥固结卵石 cement stabilized gravel
水泥固结土 soil cement
水泥固井 cementing off
水泥馆 < 1939 年苏黎世博览会马耶设计的薄壳建筑 > Cement Pavilion
水泥管 cement pipe; cement tube
水泥管接头 cemented socket joint
水泥管椭圆率 ellipticity of cement pipe
水泥管异型件 cement pipe fittings
水泥贯透法 cement penetration method
水泥灌浆 additive soil stabilization; cementation; cement grout; cement injection
水泥灌浆泵 cementing pump; grout pump
水泥灌浆材料 cement grouting
水泥灌浆车 cementing truck
水泥灌浆的 cement-grouted
水泥灌浆法 cementation process; cement grouting
水泥灌浆工具 cementing tool
水泥灌浆工艺 cementation process
水泥灌浆固结法 < 软地基 > cementation process
水泥灌浆机 cement grouter; cement injector
水泥灌浆加固 cement stabilization
水泥灌浆孔 cementing hole
水泥灌浆料 cement slurry
水泥灌浆区 cementation zone
水泥灌浆设备 cementing outfit
水泥灌浆填缝 cement grout filler
水泥灌浆填料 cement grout filler
水泥灌浆压力头 cementing head
水泥灌浆硬化法 cementation process
水泥灌注 cement injection
水泥罐 cement dump; cement silo; cement tank
水泥罐车 cement tanker
水泥光泽涂层 cement glazed coat(ing)
水泥裹砂(法) < 一种喷混凝土的施工工艺 > cement enveloped sand; sand enveloped with cement
水泥含碱量 alkali content of cement
水泥含量 cement content; cement factor
水泥含量滴定试验 cement content titration test
水泥含量高的混凝土 rich concrete
水泥含量试验 cement content test
水泥和混凝土协会 < 英 > Cement and Concrete Association
水泥和水搅拌 mixing of cement
水泥和添加剂混合时无痕迹 absence of streaks
水泥和添加剂混合时无痕线 absence of streaks
水泥核 cement core
水泥候凝时间 waiting on cement; waiting on cement time
水泥护面层 cement coat(ing); cement covering; cement facing; cement finish
水泥花砖 cement decorated floor tile; cement tile
水泥滑道 cement slide
水泥化合物的形成 formation of cement compounds
水泥化学 cement chemistry; chemistry of cement
水泥环顶深度 top depth of cemented annulus
水泥缓凝剂 cement retarder; retarder for cement
水泥灰 cement dust
水泥灰饼 cement screed
水泥灰浆 cement mortar
水泥灰浆灌注 cement mortar grouting
水泥灰浆体积收缩 volume contraction
水泥灰浆注射 cement mortar injection
水泥灰泥 cement plaster
水泥灰色 cement gray
水泥灰岩 cement rock
水泥回填 back-stuffing with cement
水泥回转窑 cement rotary kiln; rotary cement kiln
水泥回转窑计算机控制 computer control of rotary kiln
水泥回转窑窑衬 cement rotary kiln lining
水泥混合材料 addition of cement
水泥混合料 cement admix(ture); cement mix(ture)
水泥混合物 cement compound
水泥混凝土 cement concrete
水泥混凝土板相对转动刚度 relative stiffness of slab
水泥混凝土板相对转动劲度 relative stiffness of slab
水泥混凝土泵 cement concrete pump; concrete pump
水泥混凝土标号 cement concrete mark
水泥混凝土补强层 concrete overlay
水泥混凝土衬砌 cement concrete lining
水泥混凝土(道)路 cement concrete road
水泥混凝土骨料 cement concrete aggregate
水泥混凝土裹砂 sand enveloped with cement concrete
水泥混凝土混合料 cement concrete mixture
水泥混凝土(混合料)拌和设备 cement concrete mixing plant
水泥混凝土混合料泵 cement concrete mixture pump
水泥混凝土混合料摊铺机 cement concrete mixture pav(i)er
水泥混凝土浇注机 concrete placer
水泥(混凝土)接榫 cement joggle
水泥混凝土锯缝机 concrete saw
水泥混凝土路面 cement concrete pavement
水泥混凝土路面加厚层 concrete overlay
水泥混凝土路面切割机 cement concrete pavement cutter
水泥混凝土路面修复 concrete pavement restoration
水泥混凝土路面修整机 concrete finisher
水泥混凝土跑道 cement concrete runway
水泥混凝土配合比 proportioning of cement concrete
水泥混凝土喷枪 cement concrete gunite machine
水泥混凝土铺面 cement concrete pavement
水泥混凝土嵌缝板 deformed plate
水泥混凝土强度等级 cement concrete mark
水泥混凝土切缝机 concrete cutter; concrete joint cutter
水泥混凝土切割机 concrete cutter
水泥混凝土清缝机 concrete joint cleaner
水泥混凝土摊铺机 cement concrete spreading machine
水泥混凝土填缝机 concrete joint sealer
水泥混凝土镶边 < 道路上的 > haunching
水泥混凝土养护用棉花毡 cotton mat for curing
水泥混凝土预制块路面 concrete block pavement
水泥混凝土振捣器 concrete vibrator
水泥混凝土整面机 cement concrete finisher
水泥活性 activity of cement
水泥积灰 concrete spillage
水泥基 cementing matrix
水泥基础 cement base
水泥基防水面层 cement-base waterproof coating
水泥基防水涂层 cement-based waterproof coating
水泥基防水涂料 cement-base waterproof coating
水泥基料 cement matrix
水泥基黏[粘]合剂 cement-based adhesive
水泥基黏[粘]结剂 cement-based adhesive
水泥基涂料 cement-based paint
水泥基制品 cement-based product
水泥及土混合料 soil cement
水泥级配磅秤 cement proportioning scale
水泥级配螺旋输送机 cement proportioning screw
水泥级配装置 cement proportioning plant
水泥集料比(率) cement-aggregate ratio
水泥集料的实体积 < 即绝对体积 = 材料重比重 × 水单位重 > solid volume of cement and aggregate
水泥集料间的相互适应性 cement-aggregate compatibility
水泥集料间反应 cement-aggregate reaction
水泥集料间黏[粘]结 cement-aggregate bond
水泥挤入管柱作业 bradenhead squeeze job
水泥挤压 cement squeeze
水泥计量 cement batching
水泥计量机 cement weighing machine
水泥计量螺旋输送机 cement measuring screw
水泥计量设备 cement measuring plant
水泥加固材料 cement stabilized material
水泥加固的 cement stabilized
水泥加固的承载层 cement treated base
水泥加固的基础 cement treated subgrade
水泥加固的路面 cement treated base
水泥加固法 cementation process
水泥加固土 cement soil stabilization; cement stabilized soil; soil cement; cement treated soil
水泥加固作用 cement stabilization
水泥检验 cement test(ing)
水泥浆 bonding paste; cement(ing) paste; fluid cement grout; grout; liquid cement; mastic cement; slurry cement; water cement slurry; wet paste
水泥浆拌和机 cement slurry mixer; grout mixer
水泥浆泵 cement slurry pump
水泥浆比重 gravity of slurry
水泥浆表面露纹处理 graffito
水泥浆操平 cement screeding
水泥浆层 < 新旧混凝土之间的 > knitting layer
水泥浆稠度 cement consistency; grout consistency[consistence]
水泥浆袋 grout baffling
水泥浆的含量 paste content
水泥浆封固套管 grout casing
水泥浆附加量 excess slurry volume
水泥浆骨料结合力 paste aggregate bond
水泥浆固着式岩石锚杆 Perfo-type rock bolt
水泥浆含量 < 混凝土 > paste volume
水泥浆集料结合力 paste aggregate bond
水泥浆计算诺模图 grout mix computation chart
水泥浆剂 cement paste
水泥浆加固 consolidation grouting; consolidation with cement grouting
水泥浆搅拌机 cement mixer; grout mixer
水泥浆金属界面 paste metal interface
水泥浆扩散 travel of grout
水泥浆料 cement stuff
水泥浆流动性能 cement slurry fluid property
水泥浆流度锥 flow cone
水泥浆流化剂 grout fluidifier
水泥浆滤出 leaching of cement paste
水泥浆锚固 cement grout anchors
水泥浆面找平 cement screeding
水泥浆沫 < 混凝土表面 > bleeding cement
水泥浆沫层 laitance layer
水泥浆沫上浮 laitance
水泥浆喷枪 cement blower; cement gun
水泥浆喷射法 cement gun shooting
水泥浆砌毛石圬工 cement rubble masonry
水泥浆嵌缝 mortar fillet
水泥浆渗漏 < 由酸溶液等使水化物分解 > leaching of cement paste
水泥浆渗散试验锥 grout flow cone
水泥浆渗透法生产的钢纤维混凝土 slurry-infiltrated fibre reinforced concrete
水泥浆收缩 paste shrinkage
水泥浆输送管 cement grout discharge hose
水泥浆刷面 cement wash
水泥浆体 cement paste
水泥浆体积收缩 < 用长颈瓶测定的 > volume flask contraction; volume flask contraction of shotcrete
水泥浆体膨胀 paste expansion
水泥浆体试饼 pat of cement-water paste
水泥浆涂层 dope coat
水泥浆涂料 cement(-water) paint
水泥浆涂刷 cement wash
水泥浆吸入速度 grout-acceptance rate
水泥浆液 cement grout; cement slurry

水泥浆用量 amount of slurry; consumption of cement slurry
水泥浆注入器 cementing injector; grout injector
水泥胶 cement gel; cement glue
水泥胶合料 cement mastic
水泥胶结 cementation; cement-bond; cementing
水泥胶结材料 cement-bound material
水泥胶结测井 cement-bond log
水泥胶结测井曲线 cement-bond log curve
水泥胶结测井图 bond log
水泥胶结的破碎岩石 cement treated crushed rock
水泥胶结料 cement-bound material; cement matrix
水泥胶结木屑板 wood cement particleboard
水泥胶结碎石 cement penetration
水泥胶结碎石路 cement-bound macadam
水泥胶乳 cement latex
水泥胶体 cement gel
水泥焦渣 cement cinder
水泥焦渣预制块 cement cinder block
水泥绞刀 screw conveyer for bulk cement
水泥搅拌机 cement batcher; grouting machine
水泥搅拌器 cement mixer
水泥搅拌柱 cement column
水泥窖灰 cement flue dust
水泥接缝 cement for joints; cement(ed) joint
水泥接合 cement(ed) joint
水泥接口 cement for joints; cement joint
水泥接榫 cement joggle
水泥接头 cement joint
水泥节约剂 cement economiser[economizer]
水泥结的 concrete bound
水泥结地面 cement paving floor
水泥结构船 concrete ship; plastered ship
水泥结合的 cement-bound
水泥结合料 cement-bound material; cement matrix
水泥结合碎石路面 concrete-bound macadam; concrete-bound pavement
水泥结颗粒材料 cement-bound granular material
水泥结壳 coating of cement
水泥(结)砾石 gravel cement
水泥结砾石路面 cement-bound road
水泥结(粒料)基层 cement-bound base
水泥结粒料路面 cement-bound road
水泥结粒料(石料) cement-bound granular material
水泥结(粒料、碎石或砾石)路面 cement-bound road
水泥结石 harden grout film; set cement
水泥结碎石 cement-bound macadam; concrete-bound macadam; macadam-cement(with Vibro-cem process) <用振动灌浆法>
水泥结碎石基层 cement macadamix base
水泥结碎石路 concrete-bound macadam
水泥结碎石路面 cement-bound macadam pavement; concrete-bound pavement; cement-bound road; cement-bound surfacing
水泥结碎石路面施工法 cement macadamix method
水泥结碎石面层 cement-bound surface; cement-bound surfacing
水泥结团 balling up of cement
水泥结硬块 air-set lamps of cement
水泥晶粒 cement grain
水泥井 cement
水泥井管 cement well pipe
水泥净浆 cement paste; cement-water mixture; fresh paste; neat cement mortar; neat cement paste; neat paste; slurry of neat cement
水泥净浆标准稠度 normal consistency for cement paste
水泥净浆标准稠度需水量 water requirement for normal consistency for cement paste
水泥净浆体透明磨片 thin-section of cement paste
水泥聚乙烯酸酯乳液混凝土 cement-polyvinyl acetate emulsion concrete
水泥聚乙烯酸酯乳液浆料 cement-polyvinyl acetate emulsion
水泥绝对体积 solid volume of cement
水泥抗压强度 cement compression strength
水泥抗折机 concrete anti-breaking machine
水泥颗粒 cement grain; cement particle
水泥颗粒分布 network of cement particles
水泥空隙比 void-cement ratio
水泥空隙比理论 cement void ratio theory
水泥空心砌块 cement hollow block
水泥孔隙比 void-cement ratio; cement-space ratio
水泥库 cement bin; cement bunker; cement silo; finished storage silo
水泥库侧卸料器 lateral unloading valve of cement silo
水泥库底卸料阀 bottom unloading valve of cement silo
水泥块 clinker cement
水泥快凝的 fast-setting to cement
水泥矿料 cement mineral
水泥矿物的生成 formation of cement compounds
水泥矿渣 cement slag
水泥拉力标准试块 briquet(te)
水泥拉力试块成型机 briquet(te) press
水泥拉毛粉刷 cement stucco
水泥拉毛抹灰 cement stucco
水泥篮 cement basket
水泥类型 cement type; type of cement
水泥冷却剂 cement cooler
水泥冷却器 cement cooler
水泥冷却设备 cement cooling plant
水泥立窑 cement shaft kiln; cement vertical kiln; vertical cement kiln; vertical kiln
水泥沥青 cement asphalt
水泥沥青浆 cement-bitumen grout
水泥砾石混合料 gravel cement mixture
水泥量与骨料孔隙比(率) cement-space ratio
水泥量与集料孔隙比(率) cement-space ratio
水泥料筒 cement basket
水泥炉衬 cement lining
水泥(路)面 cement surface
水泥路面切缝机 concrete joint cutter
水泥路面清缝机 concrete joint cleaner
水泥路铺筑机械 concrete paving machine
水泥路铺筑设备 concrete paving equipment
水泥螺旋的陡坡输送机 cement screw conveyer for steep conveying
水泥螺旋输送器 cement screw conveyer[conveyor]
水泥螺旋喂料机 cement screw feeder
水泥玛琋脂 mastic cement
水泥码头 cement quay; cement terminal; cement wharf
水泥镘刀 cement trowel
水泥毛石挡土墙 cement rubble retaining wall
水泥毛石的 cement rubble
水泥毛石砌体 cement rubble masonry
水泥毛石圬工 cement rubble masonry
水泥弥散剂 cement dispersing agent
水泥密度 density of cement
水泥密封 cement seal
水泥面层 cement finish; cement topping
水泥面层的镘平 cement floating
水泥面层镘平 cement floating
水泥面层抹面 cement floating
水泥面道路 cement-bound road
水泥磨 clinker grinding mill; clinker mill
水泥磨光板 finish blade
水泥磨机 cement mill; finish grinding mill
水泥磨碎机 cement grinding mill
水泥磨细(分散)附加剂 cement dispersion admixture
水泥抹光 float with cement
水泥抹灰 cement plaster finish; cement plaster(ing)
水泥抹面 cement coat(ing); cement covering; cement facing; cement finish; cement flo(a)tation; cement floating; cement plaster finish; cement plastering; cement rendering
水泥抹面防水 cement plaster-coat waterproof(ing)
水泥抹面灰泥 cement plaster
水泥抹面砂浆 cement stuff
水泥抹平 float coat
水泥木椽板 wood cement board
水泥木料地面处理 cement-wood flooring
水泥木丝板 cemented excelsior board; wood cement board; wood-wool cement plate
水泥木丝建筑板 cement-bound excelsior building slab
水泥木屑地面 cement-wood floor
水泥木屑建筑板 cement-bound excelsior building slab
水泥木屑楼板 cement-wood floor
水泥木屑砌块 wood cement block
水泥钠矿 franconite
水泥(泥)浆 cement grout; cement slurry
水泥黏[粘]固性 cement fastness
水泥黏[粘]合砖 cementitious brick
水泥黏[粘]浆 cement-based adhesive
水泥黏[粘]结层 bonding layer
水泥黏[粘]结混凝土罩面层 cement bonded concrete overlay
水泥黏[粘]结剂 cement gel; cement glue
水泥黏[粘]结力测井记录 cement-bonded log
水泥黏[粘]结碎石路面 concrete-bound pavement
水泥黏[粘]结岩芯 grout core
水泥黏[粘]土分段回填 sectional back-stuffing with clay and cement
水泥黏[粘]土灌浆 cement clay grouting
水泥黏[粘]土矿石 cement clay ore
水泥黏[粘]土砂浆 cement clay mortar
水泥啮合 cement joggle
水泥凝固 cement setting; set of cement
水泥凝固测针 cement setting needle
水泥凝固时间测定针 cement needle
水泥凝固速度 rate of setting
水泥凝胶 cement gel; cement glue
水泥凝结 cement setting; set of cement
水泥凝结试验 cement setting test
水泥凝结硬化理论 theory of cement setting and hardening
水泥凝结指数 hydraulicity Vicat index
水泥凝硬作用 pozzolanic reaction
水泥排出管 grout outlet hose
水泥排水管 cement sewer pipe
水泥牌号 brand of cement; cement brand; cement mark
水泥刨花板 cement-bonded particleboard; cement chipwood; cemented chip board; cemented excelsior board; cement particleboard; wood cement particle board; wood-shaving-cement plate
水泥泡沫混凝土 foamed cement concrete
水泥配料仓 cement batching bin
水泥配料秤 cement batching scale
水泥配料计量器 cement batcher
水泥配料螺旋输送器 cement batching screw
水泥配料器 cement batcher
水泥配料试验 cement mixing test(for emulsified asphalt)
水泥配料页岩 shale for cement burden
水泥配料用红土 laterite for cement burden
水泥配料用黄土 loess for cement burden
水泥配料用泥岩 mudstone for cement burden
水泥配料用黏[粘]土 clay for cement burden
水泥配料用砂岩 sandstone for cement burden
水泥配料有黏[粘]土矿床 clay deposit for cement
水泥配料装置 cement batching plant
水泥喷补枪 grouter
水泥喷浆 cementation; cement grouting; cement injection; cement throwing jet; jet grouting with cement
水泥喷浆衬砌 gunite lining
水泥喷浆灌浆 gunite lining
水泥喷浇工作 cement gun work
水泥喷枪 air cement gun; cement injector; cement jet; cement throwing jet; concrete cement gun; concrete gun; spray gun; throwing jet; cement gun; gunite
水泥喷射灌浆 gunite lining
水泥喷射灌浆衬砌 gunite lining
水泥喷射枪 wonder gun
水泥喷射涂层 spray cement coating
水泥膨润土乳浆 cement bentonite milk
水泥膨润土团矿 cement bentonite pellets
水泥膨胀珍珠岩制品 expanded pe(a)rlite cement product
水泥平衡涵洞 cement level(l)ing-culvert
水泥平衡涵洞出水口 cement level(l)ing-culvert outlet
水泥平衡涵洞进水口 cement level-

(l)ing-culvert intake
水泥平瓦 cement plain(roofing) tile; flat cement(roofing) tile
水泥平整器 cement finisher
水泥屏障 <在一排孔中灌水泥形成的> grout curtain
水泥砌合的 limed
水泥砌块 cement block
水泥砌体 cement mason
水泥嵌缝 cement ca(u)lked joint
水泥枪 air cement gun; grouting apparatus
水泥枪喷涂混凝土 cement gun concrete
水泥强度 set strength of cement; strength of cement; cement strength
水泥强度预测 prediction of cement strength
水泥侵蚀 cement cut(ting); cement scouring
水泥取样器 cement sampler
水泥染料 cement colo(u)rs
水泥人造大理石 cement artificial marble
水泥容器 cement vessel
水泥容重计 cement density ga(u)ge
水泥熔碴 cement clinker
水泥熔渣 cement clinker
水泥柔性地面 cement rubber latex flooring
水泥乳浆 cement milk
水泥乳液混凝土 cement latex
水泥乳浊液混合物 cement emulsion mixes
水泥软练法 mushy consistence[consistency]
水泥软练试验 test by wet mortar
水泥撒布机 cement injector; cement spreader
水泥塞长度 cement plug length
水泥塞挡圈 cement baffle collar
水泥塞底深度 bottom depth of producing horizon
水泥塞顶深度 top depth of cementing plug
水泥散货散装 bulk loading of cement
水泥散水板 cement flashing
水泥散装设备 bulk loader
水-泥沙-流量关系曲线 water-sediment-rating curve
水泥砂 cement-bonded sand
水泥砂垫层 cement and sand cushion
水泥砂灰浆 cement-sand mortar
水泥砂混合料 cement-sand mix(ture)
水泥砂浆 cement mortar; cement grout; cement slurry; compo; grouting mortar; sand-cement grout; sand-cement mortar; sanded cement grout; sand grout; slush; colgrout <用于预填骨料混凝土的>
水泥砂浆板 reinforced cement mortar board
水泥砂浆拌和机 cement mortar mixer; cement-sand grout mixer
水泥砂浆保护层 cement mortar protective course; protective course of cement mortar
水泥砂浆标号 cement mortar mark; grade of cement mortar
水泥砂浆层 cement-sand bed; cement-sand grout
水泥砂浆掺和料 mortar admixture
水泥砂浆衬里 cement mortar lining
水泥砂浆衬砌 cement mortar lining
水泥砂浆衬砌沟渠 cement-lined ditch
水泥砂浆打底 dash-bond coat
水泥砂浆底层 base course of cement mortar; cement mortar base course
水泥砂浆底涂层 scratch coat
水泥砂浆地面 cement mortar surface; cement screed
水泥砂浆垫层 cement mortar bedding cushion; cement mortar pad; cement-sand cushion
水泥砂浆垫块 cement spacer; cement washer
水泥砂浆防(泛)水条 mortar fillet; weather fillet
水泥砂浆防水做法 cement mortar waterproofing system
水泥砂浆粉面 cement mortar rendering
水泥砂浆粉面层 dash-bond coat
水泥砂浆粉刷 cement mortar plaster(ing)
水泥砂浆粉刷分格 cement mortar plaster sectioned
水泥砂浆勾缝 cement mortar pointing; pointing with cement mortar
水泥砂浆贯入仪 <测定水泥砂浆硬化速率的仪具> mortar penetrometer
水泥砂浆灌浆 cement mortar grouting; mortar grouting
水泥砂浆护面 cement covering; cement facing
水泥砂浆混合料 cement mortar mixture
水泥砂浆加防水剂 cement mortar with waterproof additive; cement mortar with waterproof compound
水泥砂浆搅拌器 sand-cement grout mixer
水泥砂浆接缝 cement mortar joint
水泥砂浆接口 cement mortar joint
水泥砂浆接头 cement mortar joint
水泥砂浆拉伸试块 briquet(te)
水泥砂浆锚栓 cement grouted bolt
水泥砂浆面层 cement screed
水泥砂浆抹面 cement mortar coating; cement mortar plaster(ing); cement plaster
水泥砂浆内衬管 cement lining pipe
水泥砂浆黏[粘]结层 <混凝土施工缝> bonding layer
水泥砂浆喷浆 cement mortar whitewashing
水泥砂浆喷枪 shotcrete gun; shotcreting gun; cement gun
水泥砂浆喷射机 shotcrete machine
水泥砂浆喷涂 slush grouting
水泥砂浆砌片石 cement-water glass grout
水泥砂浆嵌(齿合)缝 cement joggle
水泥砂浆强度 strength of cement mortar
水泥砂浆强度等级 cement mortar mark; grade of cement mortar
水泥砂浆填缝碎石路 cement-bound macadam
水泥砂浆填料 cement grout filler
水泥砂浆贴角条 cement fillet
水泥砂浆涂层 cement mortar coating; dash-bond coat; spatterdash
水泥砂浆外加剂 additive to cement grout
水泥砂浆需水比 water requirement ratio of cement mortar
水泥砂浆需水量 water requirement of cement mortar
水泥砂浆找平 cement mortar screeding
水泥砂浆找平层 cement mortar screed; cement screed
水泥砂浆找坡 cement mortar screeding to falls
水泥砂浆罩面 cement mortar covering; cement mortar finish
水泥砂浆中夹入气泡 capillary space
水泥砂胶 cement mastic; mastic cement
水泥砂砾 cement-sand-gravel
水泥砂预拌料 premixed plaster
水泥砂造型 cement-bonded mo(u)lding; cement-sand molding
水泥砂整平板 cement-sand screed
水泥烧结块 cement clinker
水泥烧粒 cement grit
水泥少的混合料 poor mix(ture)
水泥射浆法 <压力射入水喷浆的黏[粘]结方法> cementation process
水泥深层搅拌法 cement deep mixing method
水泥渗水仪 <一种测定水泥浆或水泥砂浆渗水速率的仪器> cement bleeding apparatus
水泥升降机 cement bucket elevator; cement elevator
水泥生产设备 cement making plant
水泥生产用石灰岩 cement rock
水泥生磨 cement raw meal
水泥石 cement stone; hardened cement; hardened cement paste; paste matrix
水泥石粉浆 bumicky
水泥石膏灰泥 cement plaster
水泥石灰灰泥 cement temper
水泥石灰混凝土 cement lime concrete
水泥、石灰、砂混合料 cement lime sand mix(ture)
水泥石灰砂浆 cement lime (sand) mortar; ga(u)ged mortar; lime and cement mortar; compo mortar
水泥石灰砂浆粉刷 cement lime sand mortar plaster; lime and cement mortar plaster
水泥石孔隙 pores in set cement paste
水泥石棉板 cement asbestos board; cement asbestos slate; uralite
水泥石棉边沟 cement asbestos gutter
水泥石棉薄板 cement asbestos sheet
水泥石棉地板 cement asbestos floor(ing)
水泥石棉定距块 cement asbestos distance piece; cement asbestos separator; cement asbestos spacer
水泥石棉废水管 cement asbestos refuse water pipe
水泥石棉管 cement asbestos pipe; cement asbestos tube; cement block
水泥石棉建筑构件 cement asbestos building member
水泥石棉面 cement asbestos face
水泥石棉排水管 cement asbestos discharge pipe
水泥石棉配件 cement asbestos fittings
水泥石棉器皿 cement asbestos ware
水泥石棉墙板 cement asbestos wall panel
水泥石棉墙面板 cement asbestos siding
水泥石棉实心板 cement asbestos solid board
水泥石棉水落管 cement asbestos conductor
水泥石棉天沟 cement asbestos(roof) gutter
水泥石棉涂层 cement asbestos coating
水泥石棉瓦 cement asbestos roof covering; cement asbestos roofing shingle
水泥石棉瓦楞板 cement asbestos corrugated board
水泥石棉污水管 cement asbestos sewage pipe
水泥石棉(屋)脊瓦 cement asbestos ridge capping tile
水泥石棉屋面板 cement asbestos roofing board; cement asbestos shingle
水泥石棉屋面瓦 cement asbestos roofing slate
水泥石棉烟道 cement asbestos flue
水泥石棉雨水排水制品 cement asbestos rainwater article
水泥石棉栅栏 cement asbestos fence
水泥石棉制品 cement asbestos article; cement asbestos ware
水泥石屑拌和料 cement stone-dust mixture; face mix
水泥试饼 cement cake; pat of cement; cement pat
水泥试件养护室 moist room
水泥试块 briquet(te); cement briquet(te)
水泥试块成型机 briquet(te) press
水泥试块拉力试验机 briquet-testing machine
水泥试块压制机 briquet(te) press
水泥试条 bar of cement
水泥试验 cement test(ing)
水泥试验机 cement test(ing) machine
水泥试验器 cement tester
水泥试验室 cement lab(oratory)
水泥试验台 test bed of cement
水泥试验用标准砂 cement-testing sand; standard cement testing sand
水泥试样 sample of cement
水泥饰粉 hardwall plaster
水泥饰面 cement dressing; cement finish; cement veneer
水泥饰面涂料 cement paint
水泥收缩仪 cement contraction ga(u)ge
水泥受料斗 cement receiving hopper
水泥输送管 cement conveying pipe
水泥输送机 cement conveyer[conveyor]
水泥输送设备 cement handling installation
水泥输送索道 cement cableway; cement supply ropeway
水泥熟料 cement clinker; clinker
水泥熟料成分 cement clinker composition; clinker composition; clinker constituent
水泥熟料的转桶研磨 vortex-chamber grinding of cement clinker
水泥熟料化学 cement clinker chemistry; clinker chemistry
水泥熟料结圈 clinker ring
水泥熟料颗粒 cement clinker grain; clinker grain
水泥熟料库 clinker store
水泥孰料快速冷却器 rapid clinker cooler
水泥熟料矿物 clinker mineral of cement
水泥熟料矿物质 clinker material
水泥熟料磨 clinker grinding mill
水泥熟料相 cement clinker phase
水泥熟料形成 clinker formation
水泥熟料研磨 cement clinker grinding; clinker grinding
水泥熟料中的矿物成分 felite
水泥熟料状态 clinker phase
水泥熟料组成 clinker composition
水泥刷 cement paint brush
水泥刷面 cement rendering
水泥水玻璃双浆 cement-water glass grout
水泥水合产品 cement hydration product
水泥水合作用 hydration of cement
水泥水化试验 cement hydration test
水泥水化温度 temperature of cement hydration
水泥水化物 cement-hydrate
水泥水化作用 cement hydration; hy-

S

dration of cement; cement-hydrate
水泥水凝性 hydraulic activity
水泥送料计量器 cement batcher
水泥送料器 cement feeder
水泥速凝剂 cement hardener; rapid setting agent for cement
水泥塑料冷涂釉料 cement-plastic cold glaze
水泥塑料冷涂釉质墙面 cement-plastic cold-glazed wall coat(ing)
水泥塑料透明涂层 cement-plastic vitreous surfacing
水泥碎木板 cement chipwood
水泥碎石混合料基层 cement macadamix base
水泥碎石混合料路面施工法 cement macadamix method
水泥碎石路面 cement macadam
水泥榫接合 cement joggle joint
水泥摊铺机 cement distributor
水泥碳化铁体 cement cementite
水泥糖浆砂造型法 cement-sand molasses process
水泥套管头 cement casing head
水泥套管靴 cement casing shoe
水泥提升机 cement bucket elevator
水泥体积安定性 cement volume soundness
水泥体系 cement system
水泥添加剂 cement additive; cement agent
水泥填缝料 cement filler
水泥填缝料浆 cement filler grout
水泥填缝稀浆 cement filler slurry
水泥填角 cement fillet; weather fillet
水泥填料 filler for cement
水泥调和剂 cement temper
水泥调浆 cement treated grout
水泥调质处理 functional addition
水泥调质添加剂 functional addition
水泥铁 cement iron
水泥筒仓 cement silo; cement storage silo
水泥筒仓风动设备 cement silo aeration; cement silo aerator
水泥头 cement head
水泥涂层 cement coat(ing); cement paint; coating of cement
水泥涂料 cement wash; cement paint
水泥涂面钉 cement-coated nail
水泥土 cement stabilized soil; soil-cement(mixture); soilcrete
水泥土衬砌 soil-cement lining
水泥土底基层的路面 soil-cement pavement
水泥土和石灰土中的掺和 additive in soil-cement and soil-lime
水泥土混合料 soil cement
水泥土基层 cement-soil base; soil-cement base
水泥土加固 cement soil stabilization
水泥土加固法 soil-cement processing
水泥土搅拌法 cement deep mixing method
水泥土浆 cement treated soil grout(ing); soil-cement slurry
水泥土泥浆 cement treated soil slurry
水泥土壤 soil cement
水泥土稀砂浆 cement treated soil slurry
水泥团 balling up of cement
水泥团块 cement balls
水泥拖车 cement trailer unit
水泥瓦(管) cement tile; concrete tile
水泥外粉剂 cement external plaster
水泥外加剂 addition of cement; cement additive; cement admix(ture)
水泥完善性 soundness of cement
水泥帷幕 cement curtain
水泥稳定 cement stabilization
水泥稳定大路 cement stabilized road
水泥稳定的 cement stabilized
水泥稳定的路面 cement stabilized pavement
水泥稳定法 cement stabilization; stabilization with cement
水泥稳定骨料 cement modified aggregate
水泥稳定积砂路面 cement stabilized sand pavement
水泥稳定集料 cement modified aggregate
水泥稳定砾石底基层 cement stabilized gravel subbase
水泥稳定粒料 cement-bound granular material
水泥稳定砂砾底层 cement treated sand-gravel base
水泥稳定砂土 cement stabilized soil; sand cement
水泥稳定土 soil cement
水泥稳定土处理 soil-cement treatment
水泥稳定土处治 soil-cement treatment
水泥稳定土法 soil-cement processing
水泥稳定土基层 soil-cement base
水泥稳定土路 cement stabilized road; soil-cement road
水泥稳定土(壤) cement stabilization soil; cement stabilized soil; soil cement; cement soil stabilization
水泥稳定土壤法 <路面基层土壤加 4% ~15% 水泥与适量水分压实使土壤稳定> stabilization with cement
水泥稳定液施工法 <防护钻孔壁面用的> stabilized liquid method
水泥稳固白垩 cement stabilized chalk
水泥污水管 cement sewer pipe
水泥圬工 cement mason
水泥屋顶瓦 cement roof(ing) tile
水泥屋顶瓦制造机 cement roofing tile machine
水泥屋面瓦 cement roof(ing) tile
水泥(屋)瓦 cement roof(ing) tile
水泥物理 cement physics
水泥物理学 physics of cement
水泥吸收速度 rate of grout-acceptance
水泥稀浆 cement filler grout; cement slurry; cement suspension
水泥系数 cement coefficient; cement factor
水泥系数法 fixed-cement factor method
水泥系涂料 cement family coating material
水泥细度 cement fineness
水泥细度试验 cement fineness test; Blaine test
水泥细砂浆 fine sand cement mortar
水泥隙灰比 void-cement ratio
水泥下水管 cement sewer pipe
水泥纤维板 cement fibrolite plate; cement fibrous plate; fiber cement board
水泥相 cement phase
水泥橡胶乳胶 cement rubber latex
水泥橡胶乳胶混合料 fleximer
水泥消耗量 cement consumption
水泥泻水 cement flashing
水泥卸料设备 cement unloading equipment
水泥卸载设备 cement unloading equipment
水泥型 cement mo(u)ld
水泥型块 cement block
水泥性能的 cementitious
水泥修面 cement finish
水泥悬浮液 cement suspension
水泥靴 cement shoe
水泥压缝条 cement fillet
水泥压力灌浆 cement grout
水泥压力输送泵 Fuller Kinyon pump
水泥压气输送机 pneumatic cement conveyer[conveyor]
水泥压线条 cement fillet
水泥研磨 cement grinding
水泥研磨机 cement grinding mill; cement mill
水泥研磨细度 grinding fineness of cement
水泥颜料 cement pigment
水泥颜色 cement colo(u)r
水泥掩体 block house
水泥样品 cement sample
水泥窑 cement kiln
水泥窑的链条串联系统设施 chain system installation in cement kiln
水泥窑灰尘 cement kiln dust
水泥窑喂料 cement kiln feed
水泥要求 cement requirement
水泥硬固检验针 cement needle
水泥硬化 hardening of cement
水泥硬化动力学 hardening kinetics of cement
水泥硬化率 cementation index
水泥硬化指数 cementation index
水泥硬结块 air-set lump
水泥硬练试验 test by dry mortar
水泥用粗面岩 trachyte for cement
水泥用大理岩 marble for cement
水泥用户 cement user
水泥用灰岩 limestone for cement; cement rock
水泥用辉绿岩 diabase for cement
水泥用量 amount of cement; cement content; cement factor
水泥用量多的混凝土 fat concrete
水泥用凝灰岩 tuff for cement
水泥用漆 cement paint
水泥用石灰岩 cement rock; cement stone
水泥由于水化作用的体积变化 autogenous volume change
水泥与骨料的相容性 compatibility of cement aggregate
水泥与集料的相容性 compatibility of cement aggregate
水泥与石灰制造 <双月刊> Cement and Lime Manufacture
水泥与细骨料比值 cement fine ratio
水泥预制块 cement block
水泥原料 cement raw material
水泥原料矿产 cement raw material commodities
水泥圆角线 cement fillet
水泥圆线脚 cement fillet
水泥运输 cement handling
水泥运输车 cement delivery truck
水泥运输船 cement carrier
水泥运输工具 cement haulage unit
水泥运输滑道 cement slide
水泥运输机 cement conveyer[conveyor]
水泥在包装中结块 pack set
水泥早强促进剂 promotor of initial resistance of cement
水泥增强剂 cement temper
水泥增硬剂粉刷 cement hardener rendering
水泥胀圈 cement basket
水泥找平材料 cement screed material
水泥找平层 cement screed-coat
水泥罩面 cement covering; cement dressing; cement facing; cement finish; cement overlay
水泥真空输送 vacuum handling of cement
水泥蒸压器 cement autoclave
水泥蒸养膨胀(率) autoclave expansion
水泥整平抹光机 cement finisher
水泥止回阀 cement cutoff valve
水泥止水 water sealing with cement
水泥止水材料 cement waterproofer
水泥制品 cement article; cement manufacture; cement product
水泥制造的大理石 cement manufactured marble
水泥制造工厂 cement manufacturer
水泥制造者 cement manufacturer
水泥质的 cementitious
水泥质含量 cementitious content
水泥质量 cement quality
水泥质量等级 cement quality class; grade of cement; grade of cement quality
水泥蛭石 cement vermiculite
水泥中转站 cement terminal
水泥终凝 final set of cement
水泥种类 type of cement
水泥重量秤 cement weighing scale
水泥重量投配料斗 cement weighing batcher
水泥重量投配器 cement weighing hopper
水泥注浆 cement grout; cement injection
水泥注浆管头 cement casing shoe
水泥贮仓 cement bunker; cement silo; cement storage silo
水泥贮存仓 cement bin
水泥柱用棘轮 tension wheel assembly for concrete pole
水泥专用船 cement barge
水泥砖 cement brick
水泥砖铺地 cement-bound road; cement tile pavement
水泥砖铺(路)面 cement-bound road; cement tile pavement
水泥桩 cement pile
水泥装袋 sacking cement
水泥装袋机 cement bagging machine
水泥装饰品 <在石膏模中模制的> cement pressing
水泥装卸 cement handling
水泥装卸设备 cement handling installation
水泥状 cement-like
水泥状化合物 cement-like mixture
水泥着色 pigmentation of cement
水泥着色剂 cement colo(u)rant
水泥着色耐碱颜料 limefast pigment for colo(u)ring cement
水泥着色颜料 pigment for colo(u)ring cement
水泥自动称量器 automatic cement batcher
水泥自动配料器 automatic cement batcher
水泥综合生产 combined production process of cement
水泥组成物 cement constituent
水泥最小用量 minimum cement content
水泥座垫 concrete base pad
水铌钽石 niohydroxite
水铌钇矿 hydrosamarskite
水年 water year
水黏[粘]的 water-clogged
水鸟 aquatic bird; water bird; waterfowl
水镍铀矾 nickel zippeite
水镍钴矾 moorhouseite
水柠檬钙石 earlandite
水凝(固) hydraulic set(ting)
水凝灰浆 hydraulic mortar
水凝混凝土 hydraulic concrete
水凝活性 hydraulic activity

水凝胶 aquagel[aquogel];hydrogel
水凝黏[粘]合剂 hydraulic cementing agent
水凝砂浆 hydraulic mortar
水凝石灰砂浆 water lime mortar
水凝水泥 hydraulic cement;water cement
水凝水泥涂料 water-cement paint
水凝性 hydraulic activity;hydraulicity
水凝性指数 hydraulicity Vicat index
水凝硬性 hydraulicity
水牛 buffalo
水扭 water twist
水暖 hot-water heating; water heat(ing)
水暖电设备的装配部件 mechanical unit
水暖工 fitter for heating installations;pipe fitter;plumber
水暖工助手 plumber's mate
水暖管配件 plumbing fixture
水暖气片 water-filled radiator
水暖器 hot-water stove;water radiator
水暖散热器 hot-water radiation
水暖系统 hydronic heating; water heating system
水排泄 water sluice
水盘 water pond
水蟠管 water-coil
水畔车站 waterside station
水畔栽植 waterside planting
水旁带 waterfront zone
水泡 bleb; blister; bubble cell; bulb; pocking;water-bubble
水泡铋矿 hydrobismutite
水泡式气压机 bubble type pneumatic ga(u)ge
水泡液 vesicle fluid[liquid/liquor]
水泡音 bubble
水疱疹 vesicle
水培(养)water culture
水配重 water ballast
水喷淋系统 system of water sprinkler;water-drench system
水喷洒 atomized water spray
水喷洒管 water spray pipe
水喷射 injection of water;water-jet blast
水喷射泵 water ejector;water stream injection pump
水喷射器 water ejector; water-jet blower;water-jet injector
水喷射(真空)泵 water-jet pump
水喷雾 atomized water spray
水喷雾灭火系统 water spray extinguishing system
水喷雾器 water sprayer
水喷真空<用喷射泵能达到的真空> water-jet vacuum
水喷嘴 operating water nozzle;water injection nozzle
水盆 water basin;water tub
水盆地渗入补给 infiltration from water basin
水硼铵石 ammonioborite
水硼钙石 froloyite;ginorite;hayesite;hayyesenite; hydroboracite; hydroborocalcite
水硼钙锶石 kurgantaite
水硼钾石 santite
水硼铝钙矾 charlesite
水硼氯钙钾石 ivanovite
水硼镁石 admontite;kaliborite;paternoite
水硼钠镁石 rivadavite
水硼钠石 sborgite
水硼铍石 berboritae
水硼锶石 veatchite
水片硅碱钙石 hydroodelhayelite
水漂生物 pleuston
水漂植物 pleuston
水瓢 baler;scoop;skeet
水瓢罐 baillierod
水瓢戽斗 bailer;baler
水平 diapason;horizon
水平 V 形天线 horizontal vee antenna
水平安定面 horizontal fin;horizontal stabilizer
水平安全出口 horizontal exit
水平安装 horizontal application;horizontal setting
水平鞍座 horizontal saddle
水平暗渠 kanat
水平百叶窗 horizontal louvers
水平摆 horizontal pendulum
水平摆动 horizontal hunting
水平摆动角 angle of horizontal swing
水平摆动式悬架【机】horizontal suspension
水平摆动悬架 horizontal suspension
水平摆锻锤头 horizontal forming shoe
水平摆角 angle of horizontal swing
水平摆式起动器 horizontal pendulum starter
水平搬运机械 horizontal handing machinery
水平板 horizontal plane; horizontal plate;level(l)ing plate
水平板防波堤 horizontal plate breakwater
水平板条 binding piece
水平板桩 interpit sheeting (sheathing)
水平棒 horizon bar
水平爆破 horizontal blasting
水平被动土压力 passive lateral pressure
水平比测计 horizontal comparator
水平比测器 horizontal comparator
水平比长仪 horizontal comparator
水平比尺 horizontal scale
水平比较仪 horizontal comparator
水平比例 horizontal scale
水平闭合差 horizon closure
水平篦式冷却器 horizontal grate cooler
水平臂 horizontal jib
水平变幅 level luffing
水平变幅臂架式起重机 level-luffing jib crane
水平变幅门座式起重机 level-luffing gantry crane
水平变幅伸臂式起重机 level-luffing jib crane
水平变化 horizontal variation
水平变化值间的最大差 maximum difference between level change values
水平变位 horizontal displacement
水平变形 horizontal deformation
水平标尺 horizontal scale; level(ling) rod;level(ing) staff;level(l)ing pole
水平标度盘 horizon dial
水平标杆 level(l)ing rod;level(l)ing staff;level(l)ing pole
水平标志 level(l)er;level mark
水平标注 horizontal dimensioning
水平标桩 level peg
水平表 water glass
水平表面 horizontal surface
水平表面荷载 horizontal surface load
水平柄轴 horizontal arbor
水平并联板 horizontal parallel strap
水平波 horizontal wave;horizon wave
水平波束宽度 horizontal beam width
水平玻璃(安装)horizontal glazing
水平玻璃格条 lay bar
水平玻璃观察孔 horizon glass
水平玻璃管 level vial
水平玻璃压条 lay bar
水平不平顺 irregularity of cross level
水平不透水层 horizontal impervious layer
水平布里季曼法 horizontal Bridgman method
水平布置燃烧器 horizontal burner
水平部分 horizontal component
水平参考误差 level(l)ed reference error
水平参数 horizontal parameter
水平侧倾 horizontal sway
水平侧枝 horizontal lateral
水平测定 level determination
水平测动仪 horizontal movement ga(u)ge
水平测杆 boning rod
水平测量 differential level(l)ing; horizontal level(l)ing; horizontal survey; level; level survey; measurement of level;rigging
水平测量人员 level(l)er
水平测量图 rigging diagram
水平测量误差 error in level(l)ing; horizontal error
水平测量仪 leveling instrument
水平测试 horizontal checkout
水平层 flat bed;horizontal layer;level(l)ing course;level(l)ing layer
水平层堆料 horizontal layer stacking
水平层堆料机 horizontal layer stacker
水平层缝 coursing joint
水平层积木 horizontally laminated wood
水平层理 horizontal bedding; horizontal stratification; level(l)ing bedding
水平层理构造 horizontal bedding structure
水平层流 horizontal laminar flow
水平层流式洁净室 cross-flow clean room
水平层流通风橱 laminar flow cupboard
水平层面 horizontal bedding
水平层砌方块工程 coursed blockwork
水平层砌方石 coursed ashlar
水平层砌体 coursed masonry
水平层岩溶 horizontal karst
水平层状介质 horizontally layered medium
水平差 difference in level;difference of level;inclination;level error
水平参考误差 leveled reference error
水平场强磁力仪 horizontal field magnetometer;horizontal force magnetometer
水平场强图 horizontal field strength diagrams
水平沉淀 horizontal precipitation; horizontal sedimentation
水平沉淀池 horizontal flow tank; horizontal sedimentation basin;horizontal sedimentation tank;horizontal settling tank
水平撑 horizontal shore; horizontal strut;strut;horizontal bracing
水平撑接头 strut joint
水平撑木 straining piece
水平撑柱 dog shore
水平成层砂岩 horizontally stratified sandstone
水平池槽 horizontal tank
水平尺 air level; boning rod; grade rod;level bar; level(l)ing board; level ruler;surveyor's rod;surveyor's staff;track level bar
水平尺寸 horizontal size; horizontal dimension
水平尺垫 staff plate
水平尺调整 lining for water-level ga(u)ge
水平齿轮 horizontal gear
水平冲断层 horizontal thrust fault
水平冲击破碎机 horizontal impact crusher
水平冲击钎子 horizontal thrust borer
水平冲击式凿眼机 horizontal thrust borer
水平重合控制 horizontal registration control
水平抽出的铠装开关装置 horizontal draw-out metal-clad switchgear
水平出口 horizontal exit
水平出挑【建】horizontal overhung
水平储罐 horizontal vessel
水平触点 water-level contact
水平触发器 horizontal trigger
水平传播 horizontal transmission
水平传动 horizontal fine motion drive
水平传感器 horizontal sensor; level sensor
水平传送 horizontal transfer; horizontal transmission
水平传送机 horizontal conveyer[conveyor]
水平船台 horizontal building berth
水平床身式铣床 horizontal bed type milling machine
水平垂直尺 plumb and level
水平垂直偏转线圈 horizontal-vertical deflection coil
水平磁秤 horizontal variometer
水平磁感应 horizontal induction
水平磁化 horizontal magnetization
水平磁力 horizontal magnetic force
水平磁力强度 horizontal magnetic intensity
水平磁力梯度仪 magnetic horizontal gradiometer
水平磁力仪 horizontal field balance; horizontal magnetometer; horizontal variometer
水平磁强变感器 horizontal intensity variometer;horizontal variometer
水平磁强计 horizontal intensity variometer
水平磁强仪 horizontal force instrument; horizontal magnetometer; horizontal vibrating needle
水平磁针 horizontal magnetic needle
水平错动 horizontal dislocation
水平错距 horizontal separation;offset
水平错开 horizontal offset;horizontal separation
水平错位断层 horizontal separation fault
水平搭接<钢筋> horizontal lap
水平打桩反力 horizontal piling reaction
水平带锯 horizontal band saw
水平带宽 horizontal bandwidth
水平带式滤机 horizontal belt filter
水平单干形 horizontal cordon
水平单管采暖系统 one-pipe loop circuit heating system; single-pipe loop circuit heating system
水平单管供暖系统 one-pipe loop circuit heating system
水平单管系统 horizontal one-pipe system
水平单火管锅炉 kornish boiler
水平挡土板<开挖基槽的> horizontal sheeting
水平挡土木板<开挖基槽的> horizontal timber sheeting
水平刀架 horizontal tool head

水平导管 horizontal duct
水平导向磁铁 horizontal steering magnet
水平导向力 level guidance force
水平倒飞 level inverted
水平道路行驶阻力功率 level road horsepower
水平的 aclinal; aclinic; aflat; azimuthal; horizontal
水平的汽水界面 horizontal gas-water interface
水平的油水界面 horizontal oil-water interface
水平的中线 horizontal center[centre] line
水平等高线 contours of water table
水平等温线 horizontal isotherm
水平等效(值) horizontal equivalent
水平低通滤波作用 horizontal low-pass filtering action
水平底盘 horizontal chassis
水平地层 <未经褶皱的> acline; aclinic structure
水平地层离距 horizontal stratigraphic(al) separation
水平地带 horizontal zone
水平地面 horizontal ground plane; level ground
水平地球速率 horizontal earth rate
水平地温梯度 horizontal geothermal gradient
水平地震系数 horizontal seismic coefficient; seismal coefficient; seismic coefficient
水平地震仪 horizontal component seismograph; horizontal seismograph
水平地质剖面 geologic(al) plan
水平递止阀 horizontal check valve
水平电场 horizontal component of electric(al) field
水平电极 horizontal electrode
水平电流密度分量 horizontal current density component
水平电子束熄灭 horizontal blanking
水平垫杆 bearer bar
水平吊盖人孔 manhole with horizontal hanging cover
水平吊杆 backbar; backing bar; horizontal jib; horizontal boom <美>
水平(吊杆)起落 level luffing
水平吊杆塔吊 horizontal jib tower crane
水平吊杆塔式吊车 horizontal boom tower crane
水平调车【铁】 flat shunting
水平调车场 flat shunting yard; horizontal shunting yard
水平叠加 horizontal stacking
水平叠加剖面 horizontal stack section
水平顶撑 horizontal shore[shoring]
水平顶管 horizontal pipe jacking
水平定距块 horizontal spacer
水平定距器 horizontal spacer
水平定起角 lateral jump
水平定位片 horizontal spacer
水平定线 horizontal alignment; horizontal alinement; horizontal layout
水平定心电位计 horizontal centering potentiometer
水平定心放大器 horizontal centering amplifier
水平定中心放大器 horizontal centering amplifier
水平锭料 horizontal ingot
水平动态幅度 horizontal dynamic amplitude
水平动态会聚 horizontal dynamic convergence
水平动态会聚校正 horizontal dynamic convergence
水平动态聚焦 horizontal dynamic focusing
水平动态振幅控制 horizontal dynamic amplitude control
水平动物区系范围 horizontal faunal area
水平冻胀力 horizontal heave force
水平洞穴 cueva
水平度 degree of level; horizontality; levelness
水平度盘 azimuth plateau; azimuth scale; graduated horizontal circle; horizontal circle; horizontal limb; lower plate; limbus <包括与其相连的部分>
水平度盘安置螺旋 circle setting screw
水平度盘比(例)尺 horizontal scale
水平度盘定位螺钉 horizontal circle setting screw
水平度盘水准器 horizontal circle level; surface level
水平度盘水准仪 horizontal circle level
水平度盘直径 diameter of horizontal circle
水平段 horizontal section; horizontal segment; level section
水平断层 horizontal dislocation; horizontal fault
水平断错 horizontal dislocation
水平断错距离 offset
水平断距 heave
水平断路开关 horizontal break switch
水平断面 horizontal profile; horizontal section; level cross-section; plane section; profile in elevation
水平断面法 horizontal section method
水平断面曲线 horizontal section curve
水平对称校正 horizontal symmetry correction
水平对称天线 horizontal double antenna
水平对流 advection
水平对置式发动机 pancake engine
水平对置式十二缸 twin six
水平对置式十二缸发动机 twin-six engine
水平对置式四缸发动机 horizontally opposed flat-four engine
水平对中控制 horizontal centering control
水平对准 horizontal alignment
水平多干枝 horizontal palmette
水平多级蒸发器 horizontal tube multistage evapo(u)ration
水平多式压条 horizontal multiple layering
水平舵 diving rudder; fin; horizontal rudder; hydrovane; planes at zero
水平舵上浮转角 rise angle
水平舵转舵机械 fin-tilting machinery
水平耳轴安装架 horizontal trunnion mount
水平二阶导数 second horizontal derivative
水平发动机额定推力 thrust level
水平发散度 horizontal divergence
水平发散喷嘴扩散泵 horizontal divergent-nozzle pump
水平发射 horizontal launching
水平发射度 horizontal emittance
水平阀 horizontal valve
水平法兰面 horizontal flange
水平翻门 trap door
水平反向器 horizontal inverter
水平反作用分力 horizontal component of reaction
水平反作用力 horizontal reaction
水平范围 horizontal extent
水平方格 horizontal grid
水平方框支架采矿法 horizontal square-set method
水平方向 horizontal direction
水平方向成型法 horizontal direction formation
水平方向观测手簿 horizontal angle of book
水平方向或坐标 horizontal direction or coordinates
水平方向气化 lateral gasification
水平方向清晰度 resolution in line direction
水平方向危险角 danger angle; horizontal danger angle
水平方向性 horizontal directive tendency
水平(方向)旋窗 horizontally pivoted window
水平方向最大拉伸应力 maximum horizontal tensile stress
水平防潮层 horizontal damp-proof course
水平防火壁 horizontal firewall
水平防挠材 horizontal stiffener
水平防渗 horizontal prevention leakage
水平放大 horizontal exaggeration
水平放大率 horizontal magnification factor
水平放大器 horizontal amplifier
水平放电管 horizontal discharge tube
水平飞行 horizontal flight; level flight; level off <飞机降落前的>
水平飞行指示器 horizontal flight indicator; level flight indicator
水平分辨力 horizontal definition; horizontal resolution
水平分辨率 horizontal resolution
水平分辨率极限 horizontal resolution limit
水平分辨能力 horizontal resolution
水平分辨力检验楔 horizontal resolution wedges
水平分布 horizontal distribution
水平分层 horizontal slice
水平分层充填采矿法 horizontal cut and fill stoping
水平分层方框支架采矿法 horizontal square-set system
水平分层开采法 horizontal slicing
水平分层现象 <岩石等的> horizontal stratification
水平分带性 horizontal zoning
水平分动量 horizontal momentum
水平分度头 index table; plain dividing head
水平分解力 horizontal resolution
水平分解力测试条 horizontal resolution bars
水平分界线 <标准穿孔卡上下面积的> curtate
水平分离 horizontal separation
水平分离器 horizontal separator
水平分力 horizontal component
水平分量 horizontal component
水平分量地震仪 horizontal component seismograph
水平分量电磁响应值 electromagnetic response of horizontal component
水平分量记录 horizontal component seismogram
水平分裂 horizontal split
水平分析 horizontal analysis
水平分应力 horizontal component of stress
水平风系 horizontal wind system
水平风向 horizontal wind direction
水平缝 horizontal joint
水平敷设 horizontal laying
水平浮筒式升船机 shiplift with horizontal floats
水平幅度 horizontal amplitude
水平幅度控制 horizontal amplitude control
水平幅度调整 horizontal amplitude adjustment; horizontal size control; line amplitude adjustment
水平辐射状炮眼组圈 horizontal ring
水平辐射状钻孔 horadiam drilling; horizontal ring drilling
水平辐射状钻孔金刚石钻进 horizontal radial diamond drilling
水平俯仰起重机 horizontal level luffing crane; level-luffing crane
水平复式压条 horizontal multiple layering
水平复位 horizontal reset
水平复位时钟 horizontal reset clock
水平干扰加速度 horizontal disturbing acceleration
水平干扰系数 coefficient of horizontal interference
水平杆 horizontal bar; horizontal shaft; level bar
水平杆件 horizontal member
水平刚度 horizontal stiffness
水平钢化 horizontal tempering
水平钢筋 horizontal bar; horizontal reinforcement
水平杠杆 horizontal lever
水平高度 level spacing
水平格筛 horizontal grizzly
水平格栅撑 horizontal bridging
水平格田灌溉 level basin irrigation
水平隔 dissepiment
水平隔板 horizontal baffle; horizontal diaphragm
水平隔墙 horizontal bulkhead
水平隔水底板 horizontal impervious bottom bed
水平隔行扫描 horizontal interlace
水平隔行扫描图形 horizontally interlaced pattern
水平根 horizontal branch; horizontal root
水平埂 terrace
水平供水支管 horizontal supplying branch; horizontal supply pipe
水平拱单元 horizontal arch element
水平拱元件 <拱坝的> horizontal arch element
水平共面装置 horizontal common-plane array
水平沟 contour ditch; contour furrow; contour trench; horizontal ditch; horizontal trench
水平沟槽的挡板和拉条 horizontal trench sheeting and bracing
水平沟槽的挡板和支撑 horizontal trench sheeting and bracing
水平构件 horizontal member; top rail <框架顶上的>
水平箍带 string course
水平固结系数 coefficient of horizontal consolidation
水平固结压力 horizontal consolidation pressure
水平挂衣杆 hangrod
水平关联树 horizontal relevance tree
水平管(道) horizontal branch; horizontal pipe[piping]; horizontal tube [tubing]
水平管段 horizontal run
水平管多效 horizontal tube multiple-effect
水平管井 horizontal pipe well
水平管式振动器 horizontal tube type

vibrator
水平管式蒸发器 horizontal tube evapo(u)rator
水平管束单元 horizontal tube bundle units
水平管吸收器 horizontal tubular absorber
水平罐 horizontal retort
水平光 horizon light
水平光程透射率 horizontal path transmittance
水平光分量 horizontal light component
水平光弧 horizontal sector
水平光调节螺钉 horizontal light beam adjusting screw
水平规 level ga(u)ge;level indicator
水平轨道 level track
水平轨距尺 combination level-board;combination level-ga(u)ge
水平轨枕垫板 horizontal tie plate
水平辊道 level roller runway
水平辊锻机械手 horizontal roll forging mechanical hand
水平辊机座 horizontal roll stand
水平辊磨 horizontal roller mill
水平滚动条 horizontal scroll bar
水平滚轮导缆器 fairleader with horizontal roller
水平滚轴碾磨机 horizontal roller grinding mill
水平滚轴碎石机 horizontal roller grinding mill
水平过扫描 horizontal overscan
水平海曼 horizontal apron
水平焊 horizontal weld;level weld
水平焊缝 horizontal seam;side seam
水平焊接 downhand welding;horizontal welding
水平航空像片 vertical aerial photograph
水平合力 horizontal resultant
水平合轴 horizontal centering[centring]
水平和垂直奇偶检验码 horizontal and vertical parity check code
水平和垂直坑道系统 horizontal-vertical system of exploring opening
水平和垂直坑钻结合系统 horizontal-vertical system exploring opening-drilling
水平和垂直钻探系统 horizontal-vertical drilling system
水平荷载 horizontal load(ing)
水平荷载反力 horizontal load reaction
水平荷载反应 horizontal load reaction
水平荷载反作用 horizontal load reaction
水平哼声干扰条纹 horizontal hum bars
水平桁 horizontal girder
水平桁架 horizontal truss
水平横波 SH-wave
水平横波反射 reflection of SH-wave
水平横撑 horizontal brace
水平横洞 horizontal adit
水平横断面 level section
水平横杆 horizontal crossbar
水平横截面 level cross-section
水平横梁 horizontal crossbar
水平横楣 lintel-tol
水平横向支撑(系) horizontal lateral bracing
水平衡 water budget;water balance
水平衡场 runoff plat;water balance plat
水平衡干管 water balancing main
水平后方交会法 horizontal resection
水平厚度 horizontal breadth;horizontal thickness
水平护板 horizontal sheeting
水平护垣 horizontal apron
水平滑车 fleeting tackle
水平滑动 horizontal slip;horizontal slide[sliding]
水平滑动窗 horizontal sliding window
水平滑动式入口 horizontal slide type entrance
水平滑杆 horizontal sliding bar;traverse rod <挂窗帘等的>
水平滑距 horizontal slip
水平滑移 horizontal slide[sliding]
水平滑移模板 horizontal travel(l)ing formwork
水平滑座式热锯 horizontal sliding-frame hot saw
水平划分 horizontal division
水平环辊式磨机 horizontal roller mill
水平环形钻进 horizontal ring drilling
水平簧 horizontal spring
水平晃动 horizontal jitter;sloshing
水平灰缝 bed joint
水平回归周期 horizontal retrace period
水平回描 horizontal retrace
水平回描消隐 horizontal blanketing;horizontal blanking
水平回描消隐时期 horizontal blanking period
水平回扫 horizontal flyback;horizontal flyback sweep
水平回扫变压器 horizontal flyback transformer
水平回扫时间 horizontal flyback time
水平回扫时间比 horizontal retrace ratio
水平回扫消隐时间 horizontal blanking period
水平回扫周期 horizontal flyback period
水平回声探测仪 horizontal projection echo sounder
水平回填 horizontal backfill
水平回转圆盘式真空过滤机 horizontal rotary disk type vacuum filter
水平会聚 horizontal convergence
水平会聚电流 horizontal convergence current
水平会聚电路 horizontal convergence circuit
水平会聚线圈 horizontal convergence coil
水平会聚形状控制 horizontal convergence shape control
水平混合式防波堤 horizontally composite breakwater
水平混合式直立堤 horizontal composite breakwater
水平活动遮阳板 movable horizontal louver shading device
水平活荷载 horizontal live load
水平火焰燃烧器 inshot burner
水平或倾斜的滚辊叠片 horizontal or inclined rollstack
水平奇偶校验 horizontal parity check
水平基础梁 grade beam
水平基床 horizontal bed
水平基线 horizontal datum
水平基线法 horizontal base-line method
水平基桩 level(l)ing peg;level peg
水平基准面 horizontal reference;level reference
水平畸变 horizontal skew
水平激励 horizontal fine motion drive
水平极化 horizontal polarization
水平极化接收天线 horizontally polarized receiving antenna
水平极化天线 ground plane antenna;horizontally polarized antenna
水平极化天线阵 horizontally polarized array
水平极化无方向性天线 horizontally polarized non-directional antenna
水平集合 level set
水平集水布置 horizontal collecting water plant;horizontal water-collecting layout
水平集水廊道 horizontal gallery
水平挤出机 horizontal extruder
水平计 chinometer;level ga(u)ge
水平记录仪 level recorder
水平迹线 horizontal trace
水平加荷 horizontal load(ing)
水平加筋 horizontal reinforcing
水平加劲 horizontal stiffening
水平加劲杆 horizontal stiffener
水平加劲肋 horizontal stiffener
水平加劲支撑 sash brace
水平加强 horizontal reinforcing
水平加强筋 horizontal stiffener
水平加速度 horizontal acceleration
水平加速度峰值 horizontal peak acceleration
水平加压叶滤机 horizontal pressure leaf filter
水平夹角定位 fixing by horizontal angles
水平夹角法 horizontal sextant angles method
水平间垂高 lift between two levels
水平间隔 horizontal interval;horizontal spacing
水平间隔指示器 horizontal separation indicator
水平间距 horizontal distance;horizontal separation;horizontal spacing;level interval;level spacing
水平间隙 horizontal clearance
水平剪刀撑 spanner;span piece
水平剪力 horizontal shear;in-plane shear
水平剪力撑 cross brace;horizontal cross brace;spanner;span piece
水平剪力法 horizontal shear method
水平剪力机 horizontal shear machine
水平剪力仪 horizontal shear machine
水平剪切 horizontal shear;in-plane shear
水平剪切法 horizontal shear method
水平剪切机 horizontal shear machine
水平剪切速率 rate of horizontal shear
水平剪切仪 horizontal shear machine
水平剪切应力 horizontal shear(ing) stress
水平剪切作用 horizontal shearing
水平检波器 horizontal pick-up
水平检查孔指示器 water sight level indicator
水平检查指示器 water sight level indicator
水平检验 horizontal check;horizontal parity check
水平建造 horizontal erection
水平渐近线 horizontal asymptote
水平键控脉冲 horizontal keying pulse
水平浆缝 horizontal grout joint
水平交叉 level crossing
水平交叉口 horizontal crossing;level-(l)ing crossing
水平交错点 horizontally interlaced dot
水平交会 horizontal intersection
水平浇注 pouring on flat
水平浇注系统 <铸造> parting-line gating system
水平胶带取料机 flat-running reclaiming belt
水平胶带输送机 flat-running belt conveyer[conveyor];horizontal belt conveyer[conveyor]
水平胶合木梁 horizontal glued-laminated timber beam
水平胶结叠层木梁 horizontal glued-laminated timber beam
水平礁石 horizontal ledge
水平角 horizontal angle
水平角闭合差【测】 error of closure of horizon;closure error of horizon
水平角测量 measurement of horizontal angle
水平角方向观测法 direction method of measuring horizontal angles;direction method of observation;method of measuring horizontal angles
水平角观测 horizontal angle observation
水平角焊 horizontal fillet weld(ing)
水平角焊缝 fillet weld in the flat position
水平角全测 close of horizon;closure of horizontal angle
水平绞线器 level-wind
水平铰接窗 horizontally pivoted window
水平铰链 falling hinge
水平校验 horizontal check
水平校验装置 flush mounting
水平校整线圈 horizontal alignment coil
水平校正 horizontal adjustment;level correction
水平校正块 level(l)ing block
水平校准 horizontal alignment
水平校准层 level regulating course
水平阶地 contour terrace
水平接缝 horizontal joint;horizontal seam;curb joint;curb roll <复折形屋顶变坡处的>;collar joint <圬工墙的>
水平接管 horizontal nozzle
水平接合 horizontal joint
水平接头 horizontal joint;horizontal seam
水平节理【地】 bathroclase;horizontal cleavage;horizontal joint
水平节理花岗岩采石场 sheet quarry
水平截面 horizontal(cross-) section;horizontal profile;level cross-section;plane cross-section
水平筋 horizontal
水平进给 horizontal feed
水平进料削片机 horizontal feed chipper;horizontal spout chipper
水平精度几何因子 horizontal dilution of precision
水平井 horizontal well
水平净空 horizontal clearance
水平径流 horizontal flow
水平径向的散发 radius diffusion
水平静会聚磁铁 horizontal static convergence magnet
水平静态会聚 horizontal static convergence
水平镜 horizon glass;horizon mirror
水平镜地平 artificial horizon
水平居中调整 horizontal centering control
水平距离 distance reduced to the horizontal;ground range;horizontal distance;horizon(tal) range;horizontal equivalent <等高线间的>
水平距离比例尺 scale of horizontal distance
水平距离测量 measurement of horizontal distance
水平距离或坐标 horizontal distance

S

or coordinates
水平距离图 ground plane plot
水平距离显示 ground plane plot
水平锯齿波 horizontal sawtooth
水平锯齿波电流 horizontal sawtooth current
水平锯齿波发生器 horizontal sawtooth former
水平锯齿波缓冲 horizontal sawtooth buff
水平锯机 splitter;splitting machine
水平聚焦 horizontal focus(ing)
水平聚焦扇块 horizontally focusing sector
水平聚焦四极透镜 horizontally focusing quadrupole
水平决策矩阵 horizontal decision matrix
水平掘孔机 horizontal thrust borer
水平均衡调查 metastasy
水平开采 horizon mining
水平开关 level switch; transversal switch
水平开挖 horizontal cut; horizontal excavation;level cutting
水平勘探 horizontal exploration
水平勘探法 horizontal exploration method
水平勘探巷道支护 horizontal exploratory tunnel
水平抗风撑架 horizontal wind bracing
水平抗剪强度 horizontal shear strength
水平刻度盘 horizontal circle; horizontal plate < 经纬仪的 >
水平坑道 adit; foot rill; horizontal tunnel;level adit;level drift
水平坑道口 adit opening
水平坑道系统 horizontal system of exploring opening
水平坑钻结合系统 horizontal system of combined exploratory opening drilling
水平空腔 horizontal cavity; horizontal cell
水平孔 horizontal hole;horizontal orifice
水平孔测斜仪 clinoscope
水平孔径校正 horizontal aperture correction
水平孔隙度 horizontal porosity
水平孔砖 horizontal coring brick
水平孔钻机 horizontal drill(er); sidewall drill(er)
水平孔钻进 horizontal drilling
水平控制 horizontal control; level control
水平控制网 horizontal control network
水平口 horizontal orifice
水平矿层 horizontal seam
水平矿浆流分级机 horizontal pulp-current classifier
水平矿脉 horizontal lode;level seam
水平矿柱 level pillar
水平框架 horizontal frame
水平馈送 horizontal feed
水平扩散 horizontal diffusion
水平扩散段 horizontal diffuser
水平扩散器 horizontal diffuser
水平扩散系数 horizontal diffusion coefficient
水平拉杆 bawk; girt(h); horizontal tie(back);horizontal tie beam; lateral tie beam; lunding beam; tie beam; horizontal brace [bracing]; collar beam < 屋顶框架的 >
水平拉管法 Danner process
水平拉晶技术 horizontal pulling technique
水平拉力 horizontal pull
水平拉梁 horizontal brace
水平拉索 horizontal tieback
水平拉条 horizontal bracing
水平拉线 horizontal guy; horizontal stay
水平拉制机 horizontal drawing machine
水平喇叭口扩角 angle of horizontal flare
水平垒结 horizontal constitution
水平肋板 horizontal rib
水平棱镜 horizon(tal) prism
水平棱柱 horizontal prism
水平犁沟 contour furrow
水平犁头混合机 horizontal plough mixer
水平力 horizontal force
水平力分布系数 horizontal force distribution coefficient
水平力矩 horizontal moment
水平连杆 waling stripe
水平连杆型颚式破碎机 horizontal pitman jaw crusher
水平连杆型破碎机 horizontal pitman crusher
水平连接 horizontal joint
水平连接处 < 烟道、烟囱等的 > breeching
水平连接器 < 集装箱 > flush deck socket
水平帘子 horizontal lattice
水平联结 horizontal connection
水平联系 horizontal bracing
水平联系杆 collar beam
水平联系木条 < 间墙柱间的 > nogging
水平联轴节 horizontal coupling;horizontal joint
水平梁 horizontal beam
水平料封 horizontal material seal
水平裂缝 horizontal fracture
水平临空面 horizontal free face
水平临时支撑 horizontal temporary prop
水平菱形天线 horizontal rhombic aerials; horizontal rhombic antennas
水平零位 horizontal zero
水平流 horizontal current
水平流层 horizontal fluid layer
水平流沉淀池 horizontal flow settling basin
水平流程图 horizontal flow chart
水平流动 horizontal delivery
水平流动型喷管 horizontal flow nozzle
水平流速 horizontal velocity
水平流速梯度 horizontal velocity gradient
水平流液洞 level throat; straight throat
水平楼板 horizontal floor
水平炉箅 horizontal fire grate
水平滤层 horizontal filter
水平滤水井 horizontal filter-well
水平滤纸色谱 horizontal filter paper chromatography
水平路堑 level cutting
水平轮廓校正器 horizontal contour corrector
水平轮廓信号 horizontal contour signal
水平轮组【机】 horizontal guiding wheels
水平螺条掺和机 horizontal ribbon blender
水平螺旋钻 horizontal auger
水平螺翼式干式水表 horizontal spiral type dry water meter
水平脉冲 horizontal pulse
水平脉冲定时 horizontal pulse timing
水平脉冲宽度调整 horizontal pulse width set
水平脉冲箝位 horizontal clamping
水平脉冲同步 horizontal pulse timing
水平锚碇 horizontal anchorage
水平锚碇板 horizontal plate anchorage
水平锚碇墙 horizontal wall anchorage
水平锚固 horizontal anchorage
水平锚具 horizontal anchorage
水平贸易 < 发展水平相近国家间的贸易 > horizontal trade
水平煤层 horizontal coal seam; horizontal seam
水平面 horizon(tal) plane; level (face);level of water;level plane; level surface;water-level
水平面标高 height of level
水平面布置图 horizontal planning
水平面测量 level measurement
水平面的实线 bold horizontal line
水平面定线 horizontal alignment
水平面浮子 water-level float
水平面高程 elevation of water surface
水平面高度 elevation of water surface
水平面航行 horizontal navigation
水平面回扫脉冲 horizontal flyback pulse
水平面铰接 horizontal articulation
水平面控制标石 horizontal control monument
水平面控制(测量)网 horizontal control network
水平面控制界牌 horizontal control monument
水平面控制面 horizontal control section
水平面面积力矩 moment of area of water plane
水平面内的辐射图 horizontal radiation pattern
水平面调节 level control
水平面调节器 water-level regulator
水平面投影 horizontal projection
水平瞄准 horizontal sight;pointing in direction
水平瞄准稳定 horizon-seeking stabilization
水平瞄准修正 lateral pointing correction
水平模板支撑 horizontal form(work) support
水平模型 horizontal model;level model
水平母线 horizontal bus
水平木 byatt;waling
水平木板 horizontal timber plank
水平木板壁 horizontal boarding wall
水平木板挡板 horizontal timber sheeting
水平木板栅 horizontal timber sheeting
水平木支撑 bar timbering
水平内芯 horizontal core; horizontal coring
水平挠度 horizontal deflection
水平挠曲振动 horizontal bending vibration
水平能见度 horizontal visibility
水平逆止阀 horizontal check valve
水平扭曲 horizontal deformation
水平偶极测深 horizontal dipole sounding
水平偶极天线阵 horizontal dipole curtain antenna
水平偶极子 horizontal dipole
水平偶极子天线 horizontal doublet antenna
水平排水暗管 blind level; horizontal blind
水平排水垫层 horizontal drainage gallery
水平排水沟 horizontal drain; level gutter
水平排水管 horizontal drainage pipe
水平排水井 horizontal filter-well
水平排水井系统 horizontal well system
水平排水铺盖 horizontal drainage blanket
水平排水支管 horizontal(drainage) branch
水平盘 horizontal plate
水平盘管 horizontal tube coil
水平盘式破碎机 horizontal shaft disc crusher
水平抛物波 horizontal parabola
水平抛物波直流控制电压 horizontal parabola DC control voltage
水平抛物控制 horizontal parabola control
水平刨削面 planning horizontal surface
水平炮孔 flat hole
水平炮眼 flat hole;horizontal hole
水平炮眼掏壶 snake hole springing
水平泡 spirit bubble
水平配管 horizontal distribution pipe;horizontal distributor
水平配光曲线 horizontal distribution
水平喷吹法 horizontal blowing process
水平披叠板 horizontal siding
水平偏差 horizontal deflection; horizontal departure; horizontal deviation
水平偏角 angle of horizontal swing
水平偏向角 horizontal angle of deviation
水平偏斜 horizontal skew
水平偏移 lateral drift; lateral deviation < 钻孔的 >
水平偏移校正 horizontal shift control
水平偏振 horizontal polarization
水平偏振波 horizontally polarized wave
水平偏振电磁波 horizontally polarized electromagnetic wave
水平偏振横波 horizontal polarized shear wave
水平偏振天线 horizontally polarized antenna
水平偏转 horizontal deflection
水平偏转板 horizontal deflection plate
水平偏转波型 horizontal deflection waveform
水平偏转电极 horizontal-deflecting electrode;horizontal deflection electrode
水平偏转电流 horizontal yoke current
水平偏转电路 horizontal deflecting circuit
水平偏转放大管 horizontal deflection amplifier(tube)
水平偏转换向器 horizontal inverter
水平偏转激励器 horizontal driver
水平偏转级 horizontal deflection stage
水平偏转输出电路 horizontal deflection output circuit
水平偏转输出管 horizontal deflection tube
水平偏转调整 horizontal deflection control
水平偏转系统 horizontal deflection

system
水平偏转线圈 horizontal deflection coil;horizontal deflection yoke
水平偏转振荡器 horizontal deflection oscillator
水平偏转准确度 horizontal deviation accuracy
水平漂移 horizontal drift
水平平衡 horizontal equilibrium
水平平面图 level plan
水平平台 level(l)ing bench
水平平行机构 horizontal parallel holing
水平坡 near-level grade
水平坡度 horizontal gradient
水平剖分 horizontally split
水平剖分式多级泵 horizontal-split multi-stage pump
水平剖分式横向(分流)廊道 <船闸> horizontally split cross culvert
水平剖分式机壳 horizontally split casing
水平剖面测量 horizontal profiling
水平剖面图 horizontal profile;horizontal section;profile in elevation;sectional plan(e)
水平畦灌 level border irrigation
水平起动器 horizontal starter
水平起飞 horizontal take-off
水平起落吊杆齿轮 level-luffing gear
水平起落吊杆顶杆 level-luffing ram
水平起落吊杆滑车 level-luffing gear
水平起落吊杆索 level-luffing line
水平起落吊杆塔式起重机 level-luffing tower crane
水平起落吊杆支架 level-luffing jib
水平起重杆 horizontal jib
水平起重机 luffing crane
水平气流 advection
水平气流分级机 horizontal current air classifier
水平砌缝 bed joint
水平砌合 horizontal bonding
水平砌筑法 horizontal setting
水平器 water level
水平钳位 horizontal clamping
水平潜流人工湿地 horizontal subsurface flow constructed wetland
水平强度 horizontal intensity
水平强度磁变仪 horizontal intensity variometer
水平强度等磁变线图 horizontal intensity isoporic charts
水平强度等磁力线图 horizontal isomagnetic chart
水平强度分布图 horizontal intensity chart
水平桥接 horizontal bridging
水平桥接操作 horizontal bridging operation
水平桥接作业 horizontal bridging operation
水平切片 dropping cut slice
水平切线 horizontal tangent
水平倾滑 horizontal dip slip
水平倾斜 horizontal tilt
水平清晰度 horizontal definition;horizontal detail;horizontal resolution
水平球轴承 horizontal ball bearing
水平区分 <电视雷达导航系统> horizontal separation
水平区格板 horizontal panel
水平区域熔炼 horizontal zone melting
水平曲度 horizontal curvature
水平曲率 horizontal curvature
水平曲桥 horizontally curved bridge
水平曲线 horizontal curve
水平曲线法 horizontal curve method
水平燃烧 horizontal combustion;horizontal firing
水平人孔 horizontal manhole
水平韧性剪切带【地】horizontal ductile shear zone
水平容器 horizontal vessel
水平溶蚀带 horizontal corrosion zone
水平熔烛 shelving
水平入射 glancing incidence
水平软百叶 horizontal blind
水平软铁磁感应 horizontal bar induction
水平散度 horizontal divergence
水平散热筋 horizontal fin
水平扫描 horizontal scan(ning);horizontal sweep;line spectrum[复spectra];searching lighting
水平扫描不够 horizontal under-scan
水平扫描点的水平速度 horizontal velocity of the scan(ning) spot
水平扫描电路 horizontal-scan(ning) circuit
水平扫描电平箝位 horizontal clamping
水平扫描电压 horizontal sweep voltage
水平扫描多谐振荡器 horizontal multivibrator
水平扫描发生器 horizontal-scan(ning) generator
水平扫描放大器 horizontal sweep amplifier
水平扫描非线性 horizontal scan nonlinearity
水平扫描幅度调整 horizontal size control
水平扫描隔行扫描技术 horizontal-interlace technique
水平扫描激励器 horizontal sweep driver
水平扫描间隔 horizontal-scan(ning) interval
水平扫描频率 horizontal frequency;horizontal line frequency;horizontal repetition rate;horizontal-scan(ning) frequency
水平扫描器 horizontal scanner
水平扫描时间 horizontal-scan(ning) interval
水平扫描输出电路 horizontal-scan(ning) output circuit
水平扫描同步制导 synchroguide
水平扫描系统 horizontal sweep system
水平扫描线 horizontal-scan(ning) line
水平扫描消隐时间 horizontal blanking interval
水平扫描效率 horizontal efficiency
水平扫描信号 horizontal timebase
水平扫描信号发生器 horizontal timebase generator
水平扫描信号放大器 horizontal amplifier
水平扫描选择 horizontal sweep selection
水平扫描振荡 oscillations in the horizontal scan
水平扫描振荡电路 horizontal oscillator circuit
水平扫描振荡器 horizontal oscillator;horizontal sweep generator
水平扫描振荡器用变压器 horizontal oscillating transformer
水平扫描装置 horizontal-scan(ning) device
水平色谱法 horizontal chromatography
水平色散 horizontal dispersion
水平筛 horizontal screen;level screen
水平筛架 horizontal screen chassis
水平筛井 horizontal screen well
水平射程 horizontal range
水平射击区 horizontal field of fire
水平射角 quadrant elevation
水平射流 horizontal jet;horizontally projected jet
水平射束宽度 horizontal beam width
水平摄像片 horizontal photograph
水平摄影【航测】horizontal photograph
水平伸缩平台 telescopic(al) level carriage
水平渗(流)井 horizontal filter-well
水平渗透率 horizontal permeability
水平渗透试验法 horizontal seepage test method
水平渗透系数 horizontal permeability coefficient
水平升降吊杆 level-luffing boom
水平升降吊杆缆索 level-luffing cable
水平失真 horizontal skew
水平施工缝 horizontal construction lift;horizontal construction joint
水平石坑道 rock drift
水平时基 horizontal timebase
水平式 horizontal type
水平式臂架 horizontal jib
水平式采样器 horizontal sampler
水平式差分表 horizontal difference table
水平式城市 horizontal city
水平式打光机 horizontal-bed glazing jack
水平式继电器 horizontal type relay
水平式联系 horizontal linkage;lateral linkage
水平式流液洞 level throat
水平式滤池 horizontal filter
水平式平板筛 horizontal flat screen
水平式摄影机 <航摄> horizon camera
水平式施胶压榨 horizontal size press
水平式天窗 flat skylight
水平式楔形陶槽 horizontal wedge-cut
水平式压缩机 horizontal type compressor
水平式闸槛 level-sill
水平式装载机 level loader
水平式组织图表 horizontal organization chart
水平视差 horizontal parallax
水平视差滑尺 horizontal parallax slide
水平视差校正 horizontal parallax correction
水平视差螺旋 horizontal parallax screw
水平视角 horizontal view(ing) angle
水平视距 horizontal sight distance
水平视距标尺 subtense bar
水平视线 horizontal line of sight
水平视野 horizontal field of view
水平视准差 horizontal collimation error
水平试验器 level trier
水平试样 horizon sample
水平收敛调整 horizontal convergence control
水平收敛位移量测 horizontal convergence measurement
水平收缩 horizontal contraction
水平受力构件 horizontal load carrying member
水平枢轴 trunnion axle
水平输出 horizontal output
水平输出变压器 horizontal output transformer
水平输出管 horizontal output tube
水平输出级 horizontal output stage
水平输出偏转电路 horizontal output deflection circuit
水平输送 horizontal feed
水平输送距离 horizontal delivery distance
水平束 horizontal beam
水平束的截面大小 horizontal beam size
水平束孔 horizontal beam hole
水平竖直偏转线圈 horizontal-vertical deflection coil
水平甩链式除茎叶器 chain-type horizontal stripper
水平甩链式茎杆切碎机 horizontal flail slasher
水平双滑系统 plane compound-slide system
水平双频振荡器 twice-horizontal frequency oscillator
水平水流 horizontal flow
水平丝 horizontal hair
水平四周 all-round the horizon
水平搜索 horizon scan
水平搜索方式 horizontal searching mode
水平速度 horizontal velocity
水平速度分布曲线 horizontal velocity distribution curve
水平速度梯度 horizontal velocity gradient
水平塑料带排水 strip drain
水平损失 loss in level
水平锁闭 horizontal locking
水平锁床 horizontal locking bed
水平锁相 horizontal lock
水平台 horizontal stand
水平太阳望远镜 horizontal solar telescope
水平弹簧铰链 horizontal spring hinge
水平炭化炉 horizontal retort
水平探测器 level detector
水平探坑 horizontal drilling pit
水平探鱼仪 horizontal fish finder
水平掏槽 horizontal cut;level cutting
水平梯 horizontal ladder
水平梯度 horizontal gradient
水平梯田 bench terrace;contour terrace;level terrace
水平梯形波发生器 horizontal trapezoid generator
水平梯形畸变校正 horizontal keystone correction
水平梯形密封垫 horizontal ladder gasket
水平提升杆 horizontal lifting beam
水平体长 level body length
水平天顶仪 horizontal zenith telescope
水平天线 horizontal antenna
水平挑檐 horizontal cornice
水平条 horizontal bar
水平条控制 H-bar control
水平条信号发生器 horizontal bar generator
水平条信号控制 horizontal bar control
水平调节 level(l)ing
水平调节扳手 face spanner
水平调节杆 level(l)ing lever
水平调节机构 level(l)ing mechanism
水平调节起落杆 level(l)ing lift rod
水平调节器 level governor;regulator of level
水平调节设备 level(l)ing equipment
水平调节手柄 level(l)ing control lever
水平调节装置 level(l)ing device
水平调整 horizontal adjustment;horizontal trim(ming)

S

水平调正 horizontal adjustment
水平调准 level adjustment
水平停准度【机】 precision of horizontal positioning
水平通道 horizontal adit; horizontal channel; horizontal gallery; horizontal passage; horizontal exit
水平通过式气流 horizontal through flow
水平通路 horizontal opening
水平同步 horizontal hold; horizontal synchronization
水平同步比特 horizontal sync(hronizing) bit
水平同步标志 horizontal sync(hronizing) marker
水平同步的 horizontal simultaneous
水平同步电路 horizontal sync(hronizing) circuit
水平同步鉴别器 horizontal sync(hronizing) discriminator
水平同步控制 horizontal sync(hronizing) control
水平同步脉冲 horizontal sync(hronizing)(im) pulse
水平同步脉冲分离器 horizontal separator; horizontal synchronization pulse separator
水平同步脉冲基底电平 horizontal sync(hronizing) pedestal
水平同步时间 horizontal sync(hronizing) time
水平同步调节电位器 horizontal sync(hronizing) potentiometer
水平同步调节器 horizontal sync(hronizing) regulator
水平同步信号 horizontal drive signal; horizontal synchronizing signal
水平同步信号发生器 horizontal sync(hronizing) generator
水平同步信号后延时间 back-back porch
水平同步装置 horizontal synchronizing device
水平桶形失真 horizontal barrel distortion
水平投影 horizontal plan; horizontal projection
水平投影面 horizontal plane of projection; horizontal projection plane
水平投影面积 horizontal projected area; horizontal projection area; plan area
水平投影图 ground plan
水平透水性 transverse permeability
水平图像 horizontal image; horizontal picture
水平土摩擦力 horizontal earth friction force; horizontal soil friction force
水平土压力 horizontal earth pressure
水平湍流扩散 horizontal turbulent diffusion
水平推动 horizontal fine motion drive
水平推动篦式冷却器 horizontally driven grate cooler
水平推动级 horizontal drive stage
水平推动控制 horizontal drive control
水平推动脉冲 horizontal driving impulse
水平推动信号 horizontal driving signal
水平推剪试验 horizontal-push shear test
水平推拉窗 horizontal sliding window; sliding sash; sliding window
水平推拉门 horizontal sliding door; slide door
水平推拉门滑槽 slide door groove
水平推拉栓锁 cabinet latch
水平推力 horizontal pushing force; horizontal thrust
水平推力装置 center shift
水平托辊 flat idler; flat roller
水平挖方支撑 horizontal sheeting; horizontal sheeting for excavation
水平挖掘 <挖掘机> level cut
水平外挂墙板 horizontal cladding
水平外墙面板 horizontal siding
水平外延反应管 horizontal epitaxial reactor
水平弯曲构件 horizontally curved member
水平危险角 horizontal danger angle
水平微程序 horizontal microprogram(me)
水平微程序设计 horizontal microprogramming
水平微动螺旋 horizontal drive screw; horizontal fine motion drive; horizontal fine motion screw; horizontal tangent screw
水平微动钮 horizontal slow motion knob
水平微动装置 horizontal slow motion device
水平围堰 horizontal cofferdam
水平位错 horizontal dislocation
水平位牙槽纤维 horizontal dental slot fiber[fibre]
水平位移 horizontal displacement; horizontal movement; lateral shift <岩石断层的>
水平位移变幅机构 level-luffing gear
水平位移变幅起重机 level-luffing crane
水平位移变幅起重装置 level-luffing gear
水平位移测量 horizontal displacement measurement
水平位移观测 horizontal displacement observation
水平位移计 horizontal movement ga(u)ge
水平位移阶段 horizontal displacement stage
水平位置 horizontality; horizontal location; horizontal position; level attitude; level position
水平位置焊接 horizontal position welding
水平位置基准点 horizontal datum
水平位置面上 horizontal position dial up
水平位置面下 horizontal position dial down
水平位置调整 horizontal centering control; horizontal positioning
水平位置指示器 horizontal situation indicator
水平喂料机 level feeder
水平温差法 horizontal gradient technique
水平纹层 horizontal lamination
水平紊动扩散 horizontal turbulent diffusion
水平稳定器 horizontal stabilizer; level(l)ing shoe
水平稳定性 horizontal stability; level stability
水平涡流交换 horizontal eddy exchange
水平屋面天沟 level roof gutter
水平物 horizontal objective
水平误差放大因子 <GPS 定位> horizontal dilution of precision
水平析出 horizontal precipitation
水平析像能力 horizontal resolution
水平熄灭 horizontal blanking
水平铣 horizontal milling
水平系杆 horizontal rigid tie bar; intertie
水平系统 horizontal system
水平纤维 horizontal fiber[fibre]
水平弦杆桁架 flat-chord truss
水平线 contour line; horizon(tal); horizontal line; level line; sea line; water line
水平线比例尺 horizontal scale; X-scale <在倾斜照片上的>
水平线冰光 glare
水平线传导 horizontal line transfer
水平线定向 horizontal alignment
水平线附近云隙中射出 underbright
水平线跟踪系统 horizon scanner; horizon tracker
水平线及水平面 level line and surface
水平线降低 lowering of horizon
水平线框法相对磁异常剖面平面图 magnetic anomaly of horizontal loop method
水平线框法相对磁异常剖面图 profile figure of relative magnetic anomaly of horizontal loop method
水平线路 horizontal alignment
水平线圈 horizontal coil
水平线圈法 horizontal loop method
水平线位 horizontal alignment
水平线向 horizontal alignment
水平线形 horizontal alignment
水平线性调整 horizontal linearity control
水平线性调整线圈 horizontal linearity coil
水平线状元素 horizontal linear element
水平相干加强 horizontal coherent enhancement
水平相位 horizontal phase
水平相位调整 horizontal phasing control
水平相位控制 horizontal phasing control
水平箱柜 horizontal tank
水平镶板 horizontal panel; lay panel
水平响应计数器 flat-response counter
水平向摆式基础 horizontal pendulum footing
水平向地面运动 horizontal ground motion
水平向地震仪 horizontal component seismogram; horizontal component seismograph; horizontal motion seismograph
水平向滑模 horizontal slipforming
水平向基床反力系数 coefficient of horizontal subgrade reaction
水平向挤出 horizontal extrusion
水平向静态位移量 horizontal static load displacement
水平向静载位移 horizontal static load displacement
水平向徐变 horizontal creep
水平巷道 adit; level course; drift
水平巷道掘进 drifting
水平像片 horizontal picture
水平像散补偿器 horizontal astigmatism compensator
水平像散校正器 horizontal astigmatism corrector
水平消除 horizontal wipe
水平消隐 horizontal blanking
水平消隐电平 horizontal blanking level
水平消隐多谐振荡器 horizontal blanking multivibrator
水平消隐间隔 horizontal blanking interval
水平消隐脉冲 horizontal blanking impulse; horizontal blanking pulse
水平消隐脉冲周期 horizontal blackout period
水平消隐时间 horizontal blanking time
水平消隐信号 horizontal blanking signal
水平楔式开挖 horizontal wedge-cut
水平楔形消除 horizontal wedge wipe
水平斜撑 angle brace; angle tie; brace; horizontal skew; horizontal diagonal bracing; sway brace
水平斜拉杆 horizontal diagonal bracing
水平卸料车 horizontal discharge vehicle
水平卸载垃圾车 horizontal discharge vehicle
水平芯头 parting-line print
水平信号 horizontal signal
水平行程 horizontal throw; travel
水平形变 horizontal deformation
水平型洒水喷头 horizontal sprinkler
水平性质 horizontality
水平修边 horizontal trim(ming)
水平序列晶格空间 horizontal-series lattice spacing
水平悬臂半径 horizontal reach
水平悬臂梁 cantilever boom
水平悬臂伸距 horizontal reach
水平悬挂 horizontal suspension
水平旋台 horizontal face plate
水平旋转起重机起落吊杆 level-luffing slewing crane
水平压板 horizontal pressure foot
水平压块 level(l)ing block
水平压力 horizontal pressure
水平压力变化 variation of horizontal pressure
水平压力(分布)图 diagram of horizontal water pressures
水平压力机 level press
水平压强梯度 horizontal pressure gradient
水平压缩效应 horizontal compressional effect
水平压缩应力 horizontal compressional stress
水平烟道 breech(ing); horizontal flue
水平烟道炼焦炉 horizontal-flue coke oven
水平烟道炉 horizontal-flued oven
水平岩层【地】 horizontal bed; horizontal seam; horizontal stratum[复 strata]
水平岩层构造地貌 landform of horizontal stratigraphic(al) structure
水平岩块 block field
水平掩覆 horizontal overlap
水平眼钻车 strip borer
水平窑 horizontal kiln
水平摇杆破碎机 horizontal pitman crusher
水平一阶导数 first horizontal derivative
水平仪 level(l)er; level instrument; alignment ga(u)ge; gradienter; level(ling) ga(u)ge; level(l)ing instrument; level measuring set; level meter; level sensor; niveau; surveyor's level; water-level indicator; plumb level <有铅垂线的>
水平仪玻璃管 bubbler tube
水平仪高 height of instrument
水平仪气泡 bubble cell; bubbler of level; bulb of level; level vial
水平仪式比测仪 level comparator
水平仪指示器 level meter

S

水平移动 horizontal movement；horizontal scroll；horizontal shift(ing)；parallel motion
水平移位 horizontal displacement
水平移相器 horizontal phase shifter
水平以上 above-ground level
水平以下 below ground level
水平翼缘角钢 horizontal flange angle iron
水平阴影 horizontal shading
水平阴影信号 horizontal shading signal
水平引伸 horizontal extent
水平应变 horizontal strain
水平应变模量 horizontal strain modulus
水平应力 horizontal stress
水平应力线 line of horizontal stress
水平油槽 horizontal oil groove
水平余隙 horizontal clearance
水平鱼探仪 horizontal fish finder
水平隅撑 horizontal angle brace
水平雨水沟 level rainwater gutter
水平预裂爆破 horizontal presplitting blast
水平圆环色谱 horizontal circular chromatography
水平圆柱 horizontal cylinder
水平允许偏差 horizontal tolerance
水平运动 horizontal motion；horizontal movement；tangential movement
水平运动地震仪 horizontal motion seismograph
水平运动粒子 horizontal particle
水平运动主导性 predominance of horizontal movement
水平运距 1 英里的 1 立方码材料 mile yard
水平运输 horizontal transportation
水平运输机 level conveyer[conveyor]
水平运输平巷掘进 horizontal crosscut；horizontal fine motion drive
水平运行高低速度【机】 high low horizontal travel(l)ing speed
水平运行高中速度【机】 high medium horizontal travel(l)ing speed
水平运行机构 horizontal travel(l)ing mechanism
水平运行加速度 horizontal travel(l)ing acceleration
水平运行时间 horizontal travel(l)ing time
水平运行速度【机】 horizontal travel(l)ing speed
水平轧辊 horizontal roll
水平轧辊机座 horizontal roll stand
水平轧线 horizontal path
水平展开 horizontal development
水平张节理 lift joint
水平丈量 drop chaining；horizontal chaining
水平障碍声呐 horizontal obstacle sonar
水平找平设备 level(l)ing equipment
水平照度 horizontal illumination
水平照明 horizontal illumination；horizontal lighting
水平照相机 horizontal camera
水平照准线 horizontal collimation；horizontal wire
水平遮阳 horizontal sunshade
水平遮阳板 horizontal overhung；horizontal sunshading board；shading by horizontal baffles
水平遮阳装置 horizontal type shading device
水平折光 horizontal refraction
水平折光差 horizontal refraction error
水平折射 horizontal refraction
水平褶皱 horizontal fold
水平枕形畸变校正 horizontal pincushion correction
水平阵 horizontal array
水平振摆因子 horizontal pendulum factor
水平振荡器 horizontal oscillator tube
水平振动 horizontal vibration
水平振动筛 horizontal vibrating screen；level vibrating screen
水平振实 horizontal compaction
水平震度 horizontal seismic coefficient
水平整流设备 horizontal flow compensation
水平整修机 level(l)ing finisher
水平整枝 horizontal training
水平正方波 horizontal square wave
水平之下 below normal
水平支 horizontal branch
水平支撑 dog shore；horizontal shoring；horizontal shuttering support；horizontal strut(ting)；lateral bracing；horizontal brace[bracing]；lacing <支撑构件的>
水平支撑板沟槽 trench with lateral lagging
水平支撑法 box sheeting；horizontal timbering
水平支撑架 horizontal shore
水平支承桁架 horizontal shore
水平支承架 horizontal shore
水平支承梁 horizontal shore
水平支杆 horizontal bridging；horizontal strut
水平支管 horizontal branch pipe
水平支恒星 horizontal branch star
水平支肋 horizontal stiffener
水平支索 triatic stay
水平支销 horizontal pivot pin
水平支座 horizontal seat(ing)
水平枝 horizontal branch
水平枝撑 bar timbering
水平直规 horizontal straight edge
水平直线 horizontal straight line
水平(直)线性 horizontal linearity
水平止动装置 lateral stops；rubber
水平指示器 level indicator；level meter
水平指引 horizontal indexing
水平制表 horizontal tabulation
水平制表符号 horizontal tabulation character
水平制动螺钉 horizontal clamping screw
水平制动螺旋 horizontal clamp
水平致偏放大器 horizontal deflection amplifier
水平中分面 horizontal flange；horizontal joint；horizontal split
水平中分面法兰 horizontal joint flange
水平中心控制 horizontal centering control
水平中心调整 horizontal centering [centring]；horizontal centering alignment
水平中心调准 horizontal centering alignment
水平中心位置电位器 horizontal centering potentiometer
水平中心线 horizontal center[centre] line
水平中轴旋窗 horizontal center-pivot-hung sash window
水平种植 contour cultivation
水平重锤拉紧装置 horizontal gravity take up
水平重力梯度 horizontal gradient of gravity
水平舟 horizontal boat
水平舟区熔提纯 horizontal boat zone refining
水平舟形坩埚 horizontal boat-shaped crucible
水平轴 cross axis；horizontal shaft；transit axis；transverse axis；traverse axis；trunnion axis
水平轴泵 horizontal spindle pump
水平轴换置装置 reversing apparatus
水平轴混凝土拌和机 <即非倾侧式或筒式的> horizontal axis concrete mixer
水平轴离心泵 horizontal shaft centrifugal pump
水平轴离心消防泵 horizontal shaft centrifugal fire pump
水平轴流速仪 horizontal axis current meter；horizontal shaft current meter[metre]
水平轴平面磨床 horizontal spindle surface grinding machine
水平轴倾斜误差 error due to inclination of horizontal axis
水平轴式拌和机 horizontal axis mixer
水平轴式搅拌机 horizontal axis mixer
水平轴式水轮机 horizontal shaft water turbine
水平轴式压榨机 horizontal spindle press
水平轴误差 error of horizontal axis
水平轴线 horizontal axis
水平轴型动力厂 horizontal shaft type power plant
水平轴型动力装置 horizontal shaft type power plant
水平轴型发电厂 horizontal shaft type power plant
水平轴转换装置 horizontal reversing apparatus
水平主动压力 active-lateral pressure
水平主平面 horizontal principal plane
水平主轴 horizontal spindle；level(l)ing spindle
水平柱 horizontal strutting
水平砖窑 horizontal brick kiln
水平转刀式割草切碎机 horizontal rotary-knife shredder-mower
水平转动焊接 horizontal-rolled position welding
水平转盘 <起重机的> bull wheel
水平转盘送料 horizontal dial feed
水平转弯 horizontal turn
水平转移 horizontal transfer
水平转轴搅拌机 horizontal shaft mixer
水平桩 grade stake；level(l)ing peg；level stake
水平桩反力系数 coefficient of horizontal pile reaction
水平装配 horizontal assembly；horizontal erection
水平装饰带 <阳台栏杆下的，中世纪建筑> king-table
水平装卸法 ferry traffic；horizontal traffic；roll on/roll off system；roll on/roll off traffic；ro/ro system
水平装卸方式 horizontal handling type
水平装卸设施 roll on/roll off facility
水平装置 horizontal setting
水平状态 horizontality；level condition；level position
水平撞击记录 seismogram of horizontal weight drop
水平着陆 horizontal landing
水平子午环 horizontal transit circle
水平自动导航仪 horizon autonavigator
水平自动调节吊具 self-level(l)ing spreader
水平自行车架 horizontal bicycle stand
水平总应力 horizontal total stress
水平纵向永久磁性 permanent longitudinal magnetism
水平阻力 horizontal resistance
水平阻尼 horizontal damping
水平组装 horizontal assembly；horizontal erection
水平钻洞 horizontal boring
水平钻洞机 horizontal boring machine
水平钻机 horizontal boring rig；horizontal boring unit；horizontal drilling rig；sidewall drill(er)
水平钻孔 horizontal borehole；horizontal boring；horizontal (drill) hole；horizontal drilling
水平钻孔穿孔机 horizontal strip borer
水平钻孔断面 horizontal borehole profile
水平钻孔机 horizontal boring unit；horizontal drilling machine；strip borer
水平钻孔排水 horizontal borehole drain
水平钻孔压力记录器 horizontal borehole pressure recorder
水平钻孔钻机 horizontal borer；horizontal drill(er)；horizontal strip borer <露天采矿用的>
水平钻探 horizontal drilling
水平钻探机 horizontal drilling machine
水平钻探坑 horizontal drilling pit
水平钻探系统 horizontal drilling system
水平坐标 horizontal coordinate
水平坐标系 horizontal system of coordinates
水平坐标系统 horizontal coordinate system
水屏(蔽) water shield(ing)
水屏式加热器 water-wall panel heater
水瓶 pitcher；water bottle
水瓶容量 water capacity of bottle
水坡升船机 water slope
水坡式升船机 water slope；water slope shiplift；water wedge shiplift
水坡式升船设施 water slope shiplift；water wedge shiplift
水破碎 hydrofracturing
水曝气动力学 water aeration kinetics
水曝清砂 hydraulic blast
水栖哺乳动物 aquatic mammal
水栖地 aquatic habitat
水栖寡毛类 aquatic oligochaeta
水期变化 water-stage fluctuation
水气 hydrosphere；sweat(ing)
水气比 air-water ratio；water-air ratio
水气并动的 hydropneumatic
水气并用的 hydropneumatic
水气不透性 weather tightness
水气采样 vapo(u)r sampling
水气储压器 hydropneumatic accumulator
水气分界面 liquid vapo(u)r surface
水气腹 hydraeroperitoneum
水气含量 humidity
水气灰比 water-air-cement ratio
水气混合比 vapo(u)r mixing ratio
水气混合物 water-air mix(ture)
水气计 hydroscope
水气扩散 damp diffusion
水气来量 moisture inflow
水气率 water rate
水气密度 vapo(u)r density
水气凝结体 hydrometeor

水气凝结物 hydrometeor
水气浓度 vapo(u)r concentration
水气式沉降盒 hydropneumatic settlement cell
水气输送 vapo(u)r transport
水气提取 whiz
水气体系 water-air regime
水气通量 vapo(u)r flux
水气透过(速)率 water vapo(u)r transmission rate
水气微波激射 water vapo(u)r maser
水气尾迹 vapo(u)r trail
水气温室效应反馈 water-greenhouse feedback
水气现象 hydrometeor
水气心包 hydropneumopericardium
水气压法 vapo(u)r pressure method
水气压力 vapo(u)r pressure
水气压力差 vapo(u)r pressure deficit
水气压梯度 vapo(u)r pressure gradient
水汽 aqueous vapo(u)r; moisture; smoke; steam; water vapo(u)r; green water <植物蒸腾的>
水汽饱和度 water vapo(u)r saturation
水汽饱和(率) water vapo(u)r saturation
水汽比 water ratio
水汽波导 hydroduct
水汽不敷 vapo(u)r deficit
水汽层 water vapo(u)r layer
水汽垂直梯度 vertical water vapo(u)r gradient
水汽的紊流交换 turbulent interchange of moisture
水汽放大 moisture maximization
水汽分光镜 water vapo(u)r spectroscope
水汽改正因子 moisture adjustment factor
水汽供应 moisture supply
水汽含量 moisture charge; moisture content; water vapo(u)r content
水汽化点 steam point
水汽化学分析 chemical analysis of vapo(u)r
水汽极大化 moisture maximization
水汽亏缺 vapo(u)r deficit
水汽(来)源【地】moisture source
水汽冷凝 condensation of moisture
水汽连续吸收 water vapo(u)r continuum
水汽流 moisture flux; vapo(u)r flow
水汽密度 water vapo(u)r density
水汽密封 water vapo(u)r seal
水汽凝结体 hydrometeor
水汽凝结物 hydrometeor
水汽浓度 vapo(u)r concentration
水汽泡 water vapo(u)r bubble
水汽迁移 migration of vapo(u)r
水汽圈 water atmosphere
水汽缺乏 moisture deficit
水汽容量 water vapo(u)r capacity
水汽入流量 moisture inflow
水汽输入 moisture charge
水汽输送 water vapo(u)r transfer; water vapo(u)r transport
水汽输移 transport of moisture
水汽调整 moisture maximization
水汽通量 moisture flux
水汽尾迹 contrail
水汽吸收 water vapo(u)r absorption
水汽学 atmology
水汽循环 circulation of water vapo(u)r
水汽压力 aqueous vapo(u)r pressure; water vapo(u)r pressure
水汽张力 water vapo(u)r tension

水汽状况 water-air regime
水迁移标志化合物 water-borne typochemical compound
水迁移标志元素 water-borne typomorphic element
水迁移能力 aquatic migration capacity
水迁移系数 aquatic migration coefficient; coefficient of aqueous migration
水铅铀矿 sayrite
水铅中毒 hydrosaturnism
水枪 giant; hydraulic giant; hydraulic gun; hydraulic monitor; jetting gear; water gun; water syringe; water crane <救火用>
水枪工 nozzleman
水枪喷嘴 atomizing jet; hose nozzle; hydraulic giant nozzle; monitor nozzle; monitor spout; water giant nozzle
水枪清洗 water lancing cleaning
水枪射流 giant jet
水枪压头 monitor head
水枪转向装置 monitor deflector
水墙 water wall
水羟碲铁石 sonoraite
水羟硅铝钙石 vertumnite
水羟硅钠石 kenyaite
水羟磷铝矾 hotsonite
水羟磷铝钙石 gatumbaite
水羟磷铝石 vashegyite
水羟磷铝锌石 kleemanite
水羟铝矾 zaherite
水羟铝矾石 hydrobasaluminite
水羟氯铜矿 claringbullite
水羟络离子 aquo-hydroxo complex ion
水羟锰矿 vernadite
水羟镍石 jamborite
水羟砷铝石 bulachite
水羟砷锰石 hemafibrite
水羟砷锌石 legrandite
水羟碳铜石 georgeite
水橇 hydroski
水亲合(能)力 water affinity
水侵 water invasion
水侵量 water influx
水侵前缘推进 encroaching water advance
水侵入 intrusion of water
水侵蚀 erosion by water; water erosion
水侵蚀曲线 curve of water erosion
水侵事故 water troubles
水芹绿色 cress green
水禽 aquatic bird; waterfowl
水禽类 natatores
水青槲 aquatic blue oak
水清洁法修正案 Clean Water Act Amendments
水情 flood regime (n); hydrologic(al) information; hydrologic(al) regime; hydroregime; regime(n); water regime(n)
水情传递 transmission of water regime information; transmitting of water regime information
水情符号 ice code
水情记录 river record
水情预报 river forecast(ing)
水情站 flood-reporting station; reporting station
水情站网 reporting network
水情状态系数 <洪峰时最大年径流与最小年径流的比率> coefficient of regime(n)
水情自动测报 hydrologic(al) regime automatic monitoring and forecasting
水蚯蚓 aquatic earthworms

水球 water polo
水球计时器 water polo timer
水区 pool; water area
水区扩张 hydrocratic
水曲柳 ashtree; Manchurian ash; tamo
水曲柳黏[粘]土 ashtree clay
水曲油田 water-controlled field
水驱 water drive
水驱剂 water-flooding agent
水驱面积 flood coverage
水驱特征曲线法 characteristic curve method of water drive
水驱添加剂 water-flooding additives
水驱油 water-oil displacement
水驱油藏 water-driven pool; water-driven reservoir
水驱油层 water-controlled reservoir; water-driven reservoir
水驱油曲线 brine-into-oil curve
水驱油实验 flood pot test
水渠 canal; ditch; water channel; water conduit; penstock <美>; raceline <水车的>; specus
水渠衬砌 canal liner
水渠底 canal bottom
水渠关闭 canal closure
水渠汇合点 channel junction
水渠进水口 channel intake
水渠桥 water-conduit bridge
水渠受益税 canal advantage rate
水渠隧道坡度 grades for penstock tunnel
水渠隧洞 canal tunnel
水渠网 network of canals
水渠中通道 colluviarium
水取样器 water sampler
水圈 hydrosphere; water ring; watersphere
水圈地球化学 geochemistry of hydrosphere
水圈地球化学异常 hydrogeochemical anomaly
水圈环境 water environment
水权 water rights
水权的公款条例 common-money-rule for water rights
水权的管理 administration of water rights
水权的价值 value of water rights
水权的民法条例 civil-law-rule for water rights
水权的民用条例 civil-law-rule for water rights
水权的专有 appropriation of water rights
水权法规 water code for water rights
水权翻案禁条 estoppel of water rights
水权价值 water rights value
水权判决 adjudication of water rights
水权体制 water rights system
水权享用者 appropriator of water rights
水权拥有者 appropriator of water rights
水权专有 appropriation of water rights
水权专有者 appropriative water rights; appropriator of water rights
水权专有准则 doctrine-of-appropriation of water rights
水泉 head spring; spring
水泉消退 backwashing
水缺乏 oligohydria
水染剂 water stain
水热伴生气体 hydrothermal gas
水热爆炸 hydrothermal explosion
水热爆炸口 hydrothermal explosion crater
水热爆炸口湖 hydrothermal crater lake

水热爆炸口系统 hydrothermal crater system
水热变质 hydrometamorphism; hydrothermal metamorphism
水热变质作用 hydrometamorphism
水热处理 hydrothermal treatment
水热法 hydrothermal method
水热反应 hydrothermal coefficient
水热工艺 hydrothermal process
水热合成 hydrothermal synthesis
水热合成法 dynamic(al) hydrothermal process
水热活动喷气 hydrothermal emissions
水热活化 hydrothermal activation
水热交换器 water-heat exchanger
水热硫化物系统 hydrothermal sulfide system
水热喷发的产状分类 occurrence classify of hydrothermal eruption
水热喷发形式 type of hydrothermal eruption
水热喷汽孔 hydrothermal fumarole
水热平衡 hydrothermal balance
水热区 fluid area; fluid site; fluid zone; hydrothermal area
水热溶液 hydrothermal solution
水热生长 hydrothermal growth
水热蚀变【地】hydrothermal alteration
水热蚀变带 hydrothermal alteration zone
水热蚀变地面 hydrothermal altered ground
水热蚀变矿物 hydrothermal alteration minerals
水热蚀变区 region of hydrothermal alteration
水热蚀变热储 hydrothermal altered reservoir
水热蚀变图 map of hydrothermal alteration
水热蚀变岩石 hydrothermal rock
水热蚀变晕 hydrothermal alteration halos
水热条件 hydrothermal condition
水热污染源 thermal source of water pollution
水热系数 hydrothermal coefficient
水热系统 hydrothermal system
水热显示点 hydrothermal features
水热型地热田 hydrothermal geothermal field
水热养护 hydrothermal curing
水热仪 hydrothermograph
水热因素 hydrothermal factor
水热增压 aquathermal pressure
水热蒸汽 hydrothermal steam
水热状况 hydrothermal condition
水热作用的 hydrothermal
水人工污染源 man-made source of water pollution
水容积 water capacity; water space
水容量 moisture-holding capacity; water capacity; water volume
水容器 water receptacle
水溶抽提 hydrotropic extraction
水溶度 aqueous solubility; water capacity; water solubility; water-soluble fraction
水溶剂 hydrosolvent; water solvent
水溶剂碱 aquo-base
水溶剂酸 aquo-acid
水溶剂增感剂 water-base photosensitizer
水溶胶 hydrogel; hydrosol; lyosol; aqua <拉>
水溶蓝 water blue
水溶黏[粘]合剂 water-borne adhesive
水溶农药 water-borne pesticide

水溶漆 water paint
水溶气 dissolved gas in water
水溶溶液 hydrotropic solution
水溶色料 water stain
水溶态空气 water-soluble air
水溶态污染物 water-soluble state pollutant
水溶体 hydrosol
水溶物 hydrotrope;water-soluble matter
水溶型天然气矿床 natural gas deposit of dissolved-in-water type
水溶性 miscibility with water;water solubility
水溶性氨基树脂 water-soluble amino resin
水溶性丙烯酸树脂 water-soluble acrylic resin
水溶性材料 water-base paint
水溶性产物 water-soluble product
水溶性醇酸树脂 water-soluble alkyd resin
水溶性催化剂 water-soluble catalyst
水溶性的 water miscible;water-soluble
水溶性防腐剂 water-borne preservative; water-borne-type preservation;water-soluble preservative
水溶性酚醛树脂 water-soluble phenol resin
水溶性粉(剂) water-soluble powder
水溶性腐殖质 water-soluble humus
水溶性高分子聚合物 water-soluble polymer
水溶性烘漆 water-soluble baking paint
水溶性活性染料 water-soluble active dye
水溶性基团 water-soluble group
水溶性胶 water-soluble glue
水溶性金属离子 water-soluble metal ion
水溶性聚合物 water-soluble polymer
水溶性聚乙烯醇纤维 water-soluble polyvinyl alcohol fiber[fibre]
水溶性颗粒 water-soluble granule
水溶性磷 water-soluble phosphate
水溶性磷肥 water-soluble phosphate fertilizer
水溶性磷酸 water-soluble phosphoric acid
水溶性木材防腐剂 water-soluble wood preservative;water-borne wood preservative
水溶性木素 hydrotropic lignin
水溶黏[粘]结剂 water borne adhesive
水溶性脲醛树脂涂料 water-soluble urea resin coating
水溶性农药 water-soluble pesticide
水溶性硼 water-soluble boron
水溶性漆 aqueous soluble paint;water-soluble paint
水溶性气干型漆 water-soluble drying paint
水溶性汽油分数 water-soluble gasoline fraction
水溶性切削液 water-soluble metalworking liquid
水溶性染料 water-soluble dye(stuff); water stain
水溶性乳化油 water emulsifiable oils
水溶性乳化脂膏 water emulsifiable paste
水溶性乳剂 water-soluble emulsion
水溶性润滑剂 soluble oil;water-soluble lubricant
水溶性色素 water colo(u)r
水溶性石油磺酸 water-soluble petroleum sulfonic acid
水溶性试验 miscibility with water test
水溶性树脂 water-soluble resin
水溶性树脂涂料 water-soluble resin coating
水溶性酸和碱 water-soluble acid and alkali
水溶性碳 water-soluble carbon
水溶性碳试验 water-soluble carbon test
水溶性碳水化合物 water-soluble carbon hydrate
水溶性填料 water filler
水溶性涂料 distemper; water-based paint; water soluble paint; water based coating
水溶性涂饰着色剂 water coating colo(u)r
水溶性维生素 water-soluble vitamin
水溶性无机物 water-soluble inorganic substance
水溶性物质 water-soluble material
水溶性盐 water-soluble salt
水溶性盐类 water-soluble salts
水溶性颜料 washable water paint; water-soluble dye (stuff); water-soluble paint
水溶性颜料涂饰剂 water pigment finishes
水溶性颜料着色剂 water pigment colo(u)r
水溶性阳离子型聚合物 water-soluble cationic polymer
水溶性氧化淀粉 water-soluble oxidized starch
水溶性药物 water-soluble pharmaceutics
水溶性油 water-soluble oil
水溶性油灰 <用来填塞裂缝或小孔> water putty
水溶性油墨 water-base ink
水溶性(油)脂 water-soluble grease
水溶性有机润滑剂 carbowax
水溶性有机物 water-soluble organic substance
水溶性制剂 water miscible formulation
水溶性重金属离子 water-soluble heavy metal ion
水溶性自干涂料 water-soluble air drying paint
水溶盐含量 content of water-soluble salts
水溶液 aqueous solution;aqueous solution of ammonia; liquor; water-(y) solution
水溶液的含氢指数 hydrogen index of water solution
水溶液电解槽 aqueous solution electrolytic cell
水溶液法 aqua-solution method
水溶液共聚合作用 aqueous solution co-polymerization
水溶液降温法 aqueous solution cooling method
水溶液聚合作用 aqueous solution polymerization
水溶液生长法 aqueous solution growth method
水溶液温差法 aqueous solution temperature differential method
水溶液蒸发法 aqueous solution evapo(u)ration method
水溶油 <冷却切削刀具用> aqueous soluble oil
水溶增溶剂 hydrotropic solubilizer
水溶助长性 hydrotropy
水溶助剂 hydrotrope; hydrotropic agent
水融 aqueous fusion
水融霜 water defrosting
水乳化漆 aqueous emulsion paint; water emulsion paint
水乳化抑制剂 water emulsion inhibitor
水乳化淤渣 water emulsion sludge
水乳剂 aqueous emulsion
水乳胶 water emulsion
水乳胶蜡 water wax
水乳胶漆 aqueous emulsion paint
水乳消毒剂 water emulsifiable disinfectant
水乳型丙烯酸涂料 latex water-thinned acrylic
水乳型防水涂料 emulsive waterproof paint
水乳型聚醋酸乙烯涂料 latex water-thinned polyvinyl acetate
水乳型聚氯乙烯涂料 latex water-thinned polyvinyl chloride
水乳液 aqueous emulsion; emulsion water;water miscible liquid
水乳状液 aqueous emulsion
水乳(浊状)胶体 aqueous emulsion
水褥 Arnott's bed; hydrostatic bed; water-bed
水软管 water-tube hose
水软化 softening of water;water softening
水软化工厂 softening plant; water softening plant
水软化剂 water softening agent
水软化器 demineralizer; water demineralizer
水软化设备 water softening plant
水软化装置 water softening apparatus
水润滑的 water-lubricated
水润滑给水泵 water-lubricated feed pump
水润滑膜 lubricating film of water
水润滑液膜轴承 water-lubricated fluid-film bearing
水润滑轴承 water-lubricated bearing
水润滑轴瓦 water-lubricated bushing
水润湿的 water-wet(ted)
水润湿的磨石 water-wetted sharpening stone
水润湿砂 water wet sand
水弱化作用 hydration weakening
水塞 water plug
水塞及链 water plug and chain
水塞体 water cock body
水散布的 water dispersal
水散热的 water-cooled
水散热器 water radiator
水散热系统 water-cooling system
水色 water colo(u)r;water stain
水色分级 water colo(u)r scale
水色计 water colo(u)r meter
水沙充填 flow stowing;hydraulic silting
水沙冲填料 hydraulic fill(soil)
水沙关系曲线 sediment-discharge rating curve
水沙混合体 water-sediment mixture
水沙混合物 water-sediment mixture
水沙空间变化 spatial variation in water and sediment
水沙条件 water-sediment condition
水沙条件的改变 change of water and sediment condition
水沙沿程变化 spatial variation in water and sediment
水沙运动 water-sediment motion; water-sediment movement
水沙状况 water-sediment regime
水沙资料 flow and sediment data
水刹车 hydromatic brake
水砂充填 controlled-gravity stowing; hydraulic (back) filling; hydraulic flushing; hydraulic gravel fill; hydraulic stowing;sand fill
水砂充填法 sand filling method
水砂充填管道 hydraulic stowing pipe
水砂粉饰 sand plaster
水砂混合体 water-sediment complex; water-sediment mixture
水砂抛光(处理) wet blasting; liquid honing
水砂藓 Rhacomitrium aquaticum
水砂纸 waterproof abrasive paper; wet abrasive paper
水砂资源 water-sand resources
水筛 grating;water sieve
水筛机 hydraulic classifier
水筛炉底 bottom water screen;floor screen
水山核桃 water hickory
水杉 dawn redwood; metasequoia; redwood
水杉属 <拉> metasequoia
水闪石 hydroamphibole
水上安定器 hydrospring;hydrostabilizer
水上靶场 water range
水上拌和厂 batcher plant barge
水上保持实验站 experiment station of water and soil conservation
水上边坡 slope on water
水上标志 floating mark
水上部分破损 above-water damage
水上仓库 floating shed
水上沉桩 floating piling;pile driving over water
水上城市 aquapolis;floating city;hydropolis;watery city
水上出租船 water taxi
水上储量 overwater reserves
水上储木场 lumber storage basin; timber storage basin;wood storage basin
水上船舶 floating craft
水上打桩 marine piling;pile driven over water;pile driving over water
水上打桩机 floating driver; floating pile driver; pontoon pile driver; pontoon pile driving plant
水上打桩架 floating pile-driving rig
水上打桩设备 driving plant;floating pile driving plant; floating piling plant;pontoon pile driving plant
水上单翼机 hydromonoplane
水上的 above water;aquatic
水上的机械发烟器 water-base mechanical generator
水上吊机 floating derrick
水上定位 location above water;position above water
水上动力站 floating power barge
水上发电站 floating power barge
水上房屋 pile building
水上飞机 aeroboat; aquaplane; float plane;hydro(aero)plane;hydroairplane;seaplane;water plane
水上飞机场 seadrome
水上飞机船身 seaplane hull
水上飞机的水上性能 rough water behavio(u)r
水上飞机登陆处 seaplane landing area
水上飞机地面拖车 seaplane dolly
水上飞机浮筒 seaplane float
水上飞机供应舰 seaplane depot ship
水上飞机基地 seaplane base
水上飞机降落场 seaplane alighting area
水上飞机降落在水面 seaplane alights lands on the water
水上飞机离水面起飞 seaplane takes off from the water
水上飞机锚地 anchorage for seaplanes

S

水上飞机锚地浮标 aeronautical anchorage buoy
水上飞机母舰 seaplane mother ship
水上飞机起落区域 seaplane basin
水上飞机牵引艇 seaplane boat
水上飞机前机身 forebody
水上飞机勤务支援船 seaplane tender
水上飞机试验水槽 seaplane tank
水上飞机停泊场 seaplane anchorage
水上飞机停泊处 seaplane berth
水上飞机停泊区 anchorage for seaplanes
水上飞机系留区 anchorage for seaplanes
水上飞机专用水域 reserved seaplane area
水上飞艇 marine aircraft
水上飞行 overwater flight
水上飞行快艇 hydroplane speedboat
水上浮动花园 floating island
水上浮动软管 floating hose
水上浮木 drift wood
水上浮式起重机 floating crane
水上浮油吸聚剂 oil herder
水上高度 air draught
水上工厂 workshop barge
水上工程 maritime engineering
水上公安机关 harbo(u)r police
水上(观测)仪器平台 floating instrument platform
水上管道 floating pipeline
水上管线 floating pipeline
水上过驳 lighterage on water; midstream transfer
水上航标制 buoyage
水上航道浮标系统 <设在航道的左侧或者右侧的浮标> lateral system
水上花园 water garden
水上滑翔机 hydroglider
水上滑行 hydroplane
水上滑行艇 hydro(air)plane
水上混凝土拌和厂 floating concrete factory; pontoon-mounted concreting plant
水上货运 water transport(ation)
水上机场 marine airfield; marine airport
水上机场降落灯 seadrome contact lights
水上机场锚地浮标 aeronautical anchorage buoy
水上机具 floating rig
水上机械设备 floating plant
水上机械施工法 floating plant system
水上稽查员 jerquer
水上集鱼灯 above-water fish lamp
水上加油驳 marine bunkering service barge
水上加油装置 marine bunkering unit
水上减载 lightening on water
水上检修 afloat repair
水上交通 marine traffic; water-borne traffic
水上交通管理 marine traffic control
水上交通管理系统 marine traffic control system
水上交通系统 water-transit system
水上接力泵站 floating booster station
水上节日赛艇运动 regatta
水上景观 supra aqua landscape
水上警察 water police; customs guard; customs officer; marine police; water guard
水上救生的狭条浮板 paddle board
水上救生设备 water survival gear
水上救生训练 water survival training
水上开采 working above the water
水上勘察 offshore exploration; overwater exploration; overwater investigation
水上勘探 offshore exploration; overwater exploration; overwater investigation
水上康乐活动 water recreation area
水上快艇 glider
水上旅店 aquatel
水上旅馆 bo(a)tel
水上贸易 water-borne commerce
水上木材分类场 timber sorting pond
水上排泥管 floating discharge pipe
水上排泥管线 floating discharge pipeline; floating pipeline
水上排水量 water-borne displacement
水上抛填 dumping and filling on water
水上跑道 <水上飞机起落的跑道> water lane; sea runway
水上棚屋 shanty boat; shanty town
水上拼接 mating afloat
水上坡角 natural slop angle above water surface
水上起吊 lift-on the water
水上起落架 water landing gear
水上起落跑道 water lane
水上起重机 boat derrick; crane ship; floater crane; floating steel crane; pontoon crane; pontoon derrick
水上砌石护面 pitching above water
水上人口 floating population
水上三角洲 subaerial delta
水上设备 marine equipment
水上升降机 floating elevator
水上生态毒理学 aquatic ecotoxicology
水上施工 overwater construction; construction afloat
水上施工机械 floating plant; overwater construction machinery
水上施工时间 working time over water
水上施工围堰 water cofferdam
水上试验站 test bed boat
水上适航性 seaworthiness
水上输油管线 floating oil(pipe)line
水上搜索 overwater search
水上探测 overwater search
水上探验 offshore exploration
水上提升机 floating elevator
水上庭园 water garden
水上挖泥船 marine dredge(r)
水上稳定性 water stability
水上屋 lake dwelling; pile dwelling
水上舾装 outfit afloat
水上小修【船】 light repairs afloat
水上卸货 discharge afloat
水上行动无线电话设备 maritime mobile radio-telephone equipment
水上行动业务 maritime mobile service
水上修理车间 floating workshop
水上悬臂起重机 floating jib crane
水上巡逻工作 water-borne patrol service
水上溢油刮集臂架 spill boom
水上营房 barracks ship
水上鱼雷发射管 above-water torpedo; deck torpedo tube
水上娱乐 aquatic recreation
水上运动 aquatics; aquatic sports; water sports
水上运动场 aquatic sport area; aquatic sport waters; water sport area
水上运动建筑物 aquatic building(s)
水上运输 marine shipping; marine traffic; water-carriage; water transport(ation); water borne transport
水上运输工具 marine equipment; marine media; water(-borne) carrier
水上运输机 water-base transport
水上运输设备 hydromotive equipment
水上运输线 water route; water transport line
水上战斗机 seaplane fighter
水上障碍物 water obstacle
水上蒸馏 water and steam distillation
水上职业 watermanship
水上指航标 boom
水上制砂工厂 floating sand plant
水上住房 house boat
水上住户 lake-village
水上住家 house boat
水上住宅 floating house; houseboat
水上贮木 water storage
水上贮木场 holding ground; river depot; timber basin; timber pond
水上抓斗起重机 grabbing floating crane
水上桩屋 pile building
水上装卸 cargo-handling on water
水上装卸锚地 cargo-handling anchorage
水上状态 surface condition
水上钻机 rock drill barge; rock drill vessel
水上钻井 pier drilling
水上钻探 drilling on water region; drilling on waterways; overwater boring[drilling]; boring over water
水上作业 overwater work
水烧绿石 hydropyrochlore
水勺 bailer
水舌 nappe; water tongue; free sheet of water <堰口的>
水舌掺和齿坎 <溢流堰顶的> nappe interrupter
水舌分离 nappe separation
水舌轮廓线 nappe profile
水舌面 nappe face
水舌上缘线 upper nappe profile
水舌收缩 jet contraction
水舌贴附 adherence of nappe
水舌脱离 freeing of the nappe; nappe separation; separation of jet
水舌下的真空 vacuum under nappe
水舌下通气的溢流堰 aerated weir
水舌下缘线 lower nappe profile
水舌形成 coning
水舌型溢洪道顶 nappe-shaped crest of spillway
水舌折流器 nape deflector
水蛇 water snake
水蛇纹石 deweylite; gymnite; hydroserpentine
水射沉桩 pile sinking with the water jet
水射沉桩法 jetting method
水射抽气泵 water-jet aspirator
水射打桩机 water-jet driver
水射机 hydraulic giant
水射空气泵 ejector water air pump
水射流 water-jet
水射流喷射器 water-jet injector
水射流破碎岩石 rock fracture by water-jet impact
水射流破岩 rock breaking by water jet
水射流切割 water-jet cutting
水射喷射泵 water-jet pump
水射曝气器 water-jet aerator
水射器 eductor; ejector
水射式沉桩机 water-jet pile driver
水射通风器 water-jet ventilator
水射吸尘器 water-jet dust absorber
水砷钙锰石 wallkilldellite
水砷钙石 sainfeldite
水砷钙铁矿 yukonite
水砷钙铁石 yukonite
水砷钙铜石 shubnikovite
水砷钾铀矿 abernathyite
水砷铝铜矿 liroconite
水砷镁石 brassite
水砷锰矿 arsenoclasite[arsenoklasite]
水砷铍石 bearsite
水砷氢铁石 kaatialaite
水砷氢铜石 lindackerite
水砷铁石 kankite
水砷铁铜石 arthurite
水砷铜矿 freirinite; lavendulan
水砷铜石 strashimirite
水砷锌矿 adamite
水砷锌铅石 helmutwinklerite
水砷锌石 koettigite
水砷钴石 cobaltkoritnigite
水砷钴铁石 smolianinovite
水深 depth of water; flow depth; water depth
水深比例尺 <水工模型的> depth scale ratio
水深变化 depth variation
水深变浅的影响 shoal water influence
水深标志 water depth marker
水深表面 hydrographic(al) datum
水深波长比 relative water depth
水深不确实 sounding doubtful
水深不足 <航运> deficient draft; depth deficiency
水深测量 bathymetric(al) measure; bathymetric(al) survey; sounding; sounding survey; water depth measurement
水深测量器 bathometer; bathymeter
水深测量详细程度 bathymetric(al) detail
水深测深器 bathometer; bathymeter
水深测试 sounding test
水深吃水比 water depth/ship draft ratio; water depth-to-ship draft ratio
水深尺 hydrobarometer
水深的 bathymetric(al)
水深地形图 bathyorographic(al) map; hypsobathymetric(al) map
水深范围 depth range
水深改正 correction of depth; correction of soundings
水深归算 reduction of soundings
水深基准 sounding datum
水深基准点 chart datum
水深基(准)面 hydrographic(al) datum; sea level datum; sounding datum
水深及扫海勘测 sonar and sweep survey
水深及扫海勘查 sonar and sweep survey
水深计 altitude ga(u)ge; fathometer
水深记录器 depth recorder
水深加密测量 additional sounding
水深流量关系 depth-to-discharge relation
水深-流速关系曲线 depth-velocity curve
水深剖面仪 bathymetric(al) profiler
水深摄影测量 photobathymetry
水深手表 diver's wrist depth ga(u)ge
水深水温自记仪 bathothermograph
水深探测 diving
水深探测锤 sounding bob
水深突变 mutation of water depth
水深图 bathygram; bathymetric(al) chart; depth chart; fathogram; water depth diagram; water depth figure; water depth map
水深维护 maintenance of depth
水深温度自记仪 bathythermograph
水深线 water depth line
水深信号 water depth signal; water signal
水深信号标 depth signal mark
水深信号杆 depth signal pole; depth

signal rod;depth signal staff
水深仪 fathometer
水深仪测绳 string of bathometer;string of fathometer
水深与流量关系曲线 rating curve
水深值的改正 reduction of soundings
水深指示器 depth indicator
水渗 water-logging
水渗导性 hydraulic conductivity
水渗漏 water percolation
水渗透 pervious to water;water filtration;water penetration
水渗透率 specific permeability of water;water permeability
水渗透能力 water percolating capacity
水渗透性试验 water permeability test
水升高 <未凝固混凝土的> water gain
水升水流分级机 hydrotator
水生贝壳类动物 shellfish
水生薄荷 water mint
水生哺乳动物 aquatic mammal
水生不定根 aquatic adventitious roots
水生草本群落 aquiherbosa;aquiprata
水生沉积层 water-borne deposit;water-borne sediment;water-formed deposit;water-formed sediment
水生沉积物 water-borne deposit;water-borne sediment;water-formed deposit;water-formed sediment;water-home sediment
水生成 hydrogenesis
水生大型植被(群落) aquatic macrophyte vegetation
水生大型植被修复 aquatic macrophyte vegetation restoration
水生大型植物(群落) aquatic macrophyte
水生的 aquatic;hydric;hydrogenous;hydrophytic;water-borne;water grown
水生丁香蓼 water seedbox
水生动物 aquatic animal;hydrocole
水生动物区系 aquatic fauna
水生动物生理生态学 physiologic(al) ecology of aquatic animal
水生动植物 aquatic
水生动植物栖息地 aquatic habitat
水生毒理学 aquatic toxicology
水生毒素 aquatic toxicant
水生毒性 aquatic toxicity
水生毒性试验 aquatic toxicity test
水生浮游生物 hydroplankton
水生腐殖质 aquatic humic substance
水生附着生物 peryphyton
水生富里酸 aquatic fulvic acid
水生根 water root
水生害虫 aquatic pest
水生害虫防治 aquatic pest control
水生害虫控制 aquatic pest control
水生花卉 aquatic flower
水生化学 aquatic chemistry
水生环境 aquatic environment;aquatic habitat
水生活污染源 domestic source of water pollution
水生甲壳类动物养殖场 shellfish bed
水生菌 aquatic bacteria
水生科学和渔业资料系统 aquatic science and fisheries information system
水生昆虫 aquatic insect;water inhabitating insect
水生绿肥 aquatic green manure
水生爬行类 aquatic reptiles
水生栖息地 aquatic habitat
水生气候 hydroclimate
水生群落 aquatic community
水生生长物 aquatic growth
水生生境 aquatic habitat
水生生境度 amount of aquatic habitat
水生生境适宜性 aquatic habitat suitability
水生生境资源 aquatic habitat resources
水生生态毒理学 aquatic ecotoxicology
水生生态环境 aquatic eco-environment
水生生态环境风险评价 aquatic eco-environment environmental risk assessment
水生生态环境环境质量 aquatic eco-environment quality;aquatic ecologic environmental quality
水生生态系(统) aquatic ecosystem
水生生态系统污染 aquatic ecosystem pollution
水生生态系统要素 aquatic ecosystem element
水生生态修复 remediation of aquatic ecology;restoration of aquatic ecology
水生生态学 aquatic ecology;hydroecology
水生生态因素 hydroecologic(al) factor
水生生态综合指数 aquatic ecological comprehensive index
水生生物 aquatic;aquatic biota;aquatic creature;aquatic life;aquatic organism;hydrobiont;hydrobios
水生生物边界条件 hydrobiologic(al) boundary condition
水生生物测定 aquatic organism determination
水生生物测试 aquatic biologic(al) test(ing)
水生生物毒性 aquatic life toxicity
水生生物反应 hydrobiologic(al) response
水生生物机体的有机物 organic of aquatic biotic body
水生生物急性毒性试验 acute toxicity test of aquatic organism
水生生物监测 hydrobiologic(al) monitoring
水生生物检定 aquatic organism determination
水生生物鉴定 assay for aquatic organism
水生生物慢性毒性试验 chronic toxicity test of aquatic organism
水生生物模拟 hydrobiologic(al) simulation
水生生物模拟模型 hydrobiologic(al) simulation model
水生生物农药污染 aquatic pesticide contamination
水生生物栖息地 water habitat
水生生物群 aquatic biota
水生生物声 aquatic bioacoustic(al)
水生生物试验 aquatic biologic(al) test(ing)
水生生物受害 hazard of aquatic organism;hazard to aquatic organism
水生生物死亡率 rate of mortality of aquatic life
水生生物特性 hydrobiologic(al) characteristic
水生生物相 aquatic bioda
水生生物学 aquatic biology;hydrobiology
水生生物亚急性毒性试验 submature toxicity test of aquatic organism;sublethnal effect test of aquatic organism
水生生物研究 hydrobiologic(al) research
水生生物样本 hydrobiologic(al) sample
水生生物状况 hydrobiologic(al) regime
水生生物资源 living aquatic resource
水生生物组分 hydrobiotic constituent
水生生息环境 aquatic habitat
水生食虫植物 water insectivorous plant
水生食肉类 hydradephaga
水生食物链 aquatic food chain
水生食物网 aquatic feed web
水生态模型 water ecologic model
水生态系(统) hydro ecosystem
水生态现象 hydrobiologic(al) phenomenon[复 phenomena]
水生态学 water ecology
水生土 aquatic soil
水生微生物 aquatic microorganism
水生微生物生态系统 aquatic microbial ecosystem
水生微生物学 aquatic microbiology
水生维管束植物 aquatic vascular plant
水生污染 aquatic pollution
水生无脊椎动物 aquatic invertebrate;stream invertebrate
水生物 aquatic life
水生物体 water-borne body
水生物学 water biology
水生物质 water-borne material
水生系统 aquatic system
水生细菌 aquatic bacteria;water(-borne) bacteria
水生形态 hydromorphism
水生序列 hydrosere
水生岩 hydrogenic rock;hydrolith
水生岩类 hydroliths
水生演变 hydrarch succession
水生演替 hydrarch succession
水生演替的 hydrarch
水生演替系列 hydrarch sere;hydrosere
水生一年生植物 hydrotherophyte
水生有机物 aquatic organic matter
水生有机物分析仪 organics in water analyser[analyzer]
水生有壳类动物 shellfish
水生园艺 water gardening
水生杂草 aquatic weed;water weed
水生杂草防除 water weed control
水生杂草防治 aquatic weed control
水生杂草控制 aquatic weed control
水生栽培植物 aquaculture plant
水生藻类 hydrobiontic algae
水生真菌 aquatic fungus[复 fungi]
水生植被 aquatic macrophyte;aquatic vegetation;hydrophytic vegetation;underwater vegetation
水生植物 aquaplant;aquatic growth;aquatic life;aquatic plant;aquatic vegetation;hydrophyte[hygrophyte];underwater vegetation;water plant;weed <尤指海藻>
水生植物丛 under growth
水生植物好氧塘 aerobic pond with water plant
水生植物区系 aquatic flora
水生植物群落 aquatic plant community
水生植物生物学 aquatic plant biology
水生植物塘 hydrophyte pond
水生植物叶(形饰) water leaves
水生植物园 aquatic plant garden;water garden
水生资源 aquatic resources
水生资源保护 aquatic resource conservation
水生紫罗兰 water gillyflower;water violet
水生组分 hydrogenous constituent
水生作物 aquatic crops;aquatic plant
水声 hydroacoustics;underwater sound;water-borne sound
水声测距 subaqueous sound ranging
水声测距仪 echometer;hydroacoustic(al) range finder
水声测距指示器 acoustic(al) ranging indicator
水声测量 underwater sound measurement
水声测试设施 underwater acoustic(al) test facility
水声测位仪 pinger;sonar[sound navigation and ranging]
水声测向仪 hydroacoustic(al) bearing indicator
水声定位 hydrolocation;underwater positioning
水声定位器 phonozenograph
水声定位仪 hydrolocator
水声定向器 sonic locator
水声法 asdic method
水声方位 asdic bearing;hydrophone bearing
水声浮标 sonobuoy
水声换能器 underwater acoustic(al) transducer;underwater sound transducer
水声换能器基阵 underwater transducer array
水声计程仪 acoustic(al) log
水声接触点 hydroacoustic(al) contact
水声接收基阵 hydroacoustic(al) receiving base
水声接收器 nautical receiving set
水声率 hydroacoustic(al) rate
水声脉冲 ping
水声全息系统 acoustic(al) holography system
水声设备 underwater listening device;underwater sound equipment;underwater sound gear
水声设备控制室 asdic control room
水声搜索 acoustic(al) sweep
水声速度 hydroacoustic(al) speed;hydroacoustic(al) velocity
水声探测法 underwater acoustic(al) exploration
水声探测器 hydrophonic detector;sonar[sound navigation and ranging]
水声通信[讯]机 underwater acoustic(al) communication apparatus;underwater acoustic(al) communication set
水声通信[讯]设备 underwater acoustic(al) communication set
水声通信[讯]系统 underwater sound communication system
水声吸收器 water sound absorber
水声系统 underwater acoustic(al) system;underwater sound system
水声信号 submarine sound signal
水声学 hydroacoustics;marine acoustics;underwater acoustic
水声学测量 underwater acoustic(al) measurement
水声研究船 acoustic(al) research vessel
水声研究室 underwater sound laboratory
水声遥测器 underwater acoustic(al) telemeter
水声仪器 sound gear
水声仪器监听海区 ensonified area
水声应答器 acoustic(al) responder
水声悦耳的流水 musical waters
水声悦耳的泉水 musical waters
水声站 hydroacoustic(al) station
水声振动器 hydroacoustic(al) vibrator
水声振幅分析器 hydroacoustic(al)

amplitude analyser
水声综合测量仪 underwater sound integrated measuring set
水剩余物 water surplus
水湿 wet with by water
水湿地 meadow bog;meadow moor
水湿模砖 water-struck brick
水湿性 hydroscopicity
水石层 waterstone
水石流 waterstone flow
水石榴石 hydrogarnet
水石墨 aquadag
水石盆景 rock and water miniature garden
水蚀 abrasion;erosion by water;water abrasion;water erosion
水蚀残沙 aqueo-residual sand
水蚀的 water worn
水蚀度 turbidity of water
水蚀沟 rain channel
水蚀曲线 gradient curve
水蚀石灰洞 karst
水市场 water marking
水事工程 water works
水事纠纷 dispute over water affairs;water affairs dispute
水事宪章 water charter
水势 water potential
水势能的估计 estimating water potential
水势梯度 hydraulic potential gradient;water potential gradient
水试验压力 water test pressure
水铈铀磷钙石 lermontovite
水释涂料 water thinned paint
水收集器 water arrester;water collector
水收支 water economy
水手 blue jacket;boatman;crew;deck hand;hands;jack tar;mariner;sailor(man);sand scratcher;seafarer;sea faring man;seafolk;seaman;shipman;swabber;waterman
水手餐袋 garland
水手餐盘 kid
水手舱 crew's accommodation;crew's quarter;crew's space;seamen's quarter
水手厕所 seamen's head
水手长 boatswain
水手长仓库 boatswain's locker;boatswain's store
水手长口笛 boatswain's pipe;boatswain's whistle
水手袋 duffle bag;housewife;sea bag
水手袋系绳 bag lanyard
水手刀 jackknife;sailor pocket knife;sea knife
水手刀系带 knife lanyard
水手的 sea going;seaman's
水手短外套 pea coat;pea jacket
水手工作服 sailor-working suit
水手柜 sailor-chest;sea chest;seaman's chest
水手计划 mariner program(me)
水手技术 watermanship
水手结 double bowline;double caulker
水手用箱 chest;kit locker
水手长 bosun
水输管道 hydraulic pipe line
水输污泥道 water carrying sewerage
水输污泥系统 water-carried sewerage system
水输系统 water-carriage system
水刷 bosh;spreader brush
水刷表面处理 hydraulic finish
水刷石 exposed aggregate;granitic plaster;rustic finish;rustic terrazzo;scrubbed granolithic finish;washed granolithic plaster
水刷石面层 washed finish
水刷石墙面 granitic stucco coating
水刷石饰面 exposed aggregate finish;fair-faced granite aggregate finish;rustic(or) washed finish;scrubbed finish
水刷石饰面法 green cut clean-up
水刷洗机 hydrobrusher
水衰减 water fade
水水动力堆 water-water energetic reactor
水水堆 water-water reactor
水水换热器 water(-to-) water heat exchanger
水水热交换器 water-water heat exchanger
水水式 water-water type
水水式换热器 water-water type heat exchanger
水水式加热器 water-water type heater
水水系统 water-to-water system
水水预热器 water-water preheater
水丝菌 water mould
水丝绿铁石 souzalite
水丝铀矿 studtite
水松 china-cypress;water pine;yew
水松木材 yew
水松属 china-cypress
水松紫杉 yew
水送系统 water-carriage system
水送制 <粪便等> water-carriage system
水苏玻璃 water soda glass
水速 water rate;water speed;water velocity
水速计 hydrodynamometer;stream measurer
水塑比 water-plastic(ity) ratio
水塑的 hydroplastic
水塑性 hydroplasticity
水碎槽 water granulator
水碎流槽 granulation launder
水碎炉渣 disintegrating slag
水碎渣 granulated slag
水隧道下 immersed tunnel
水损 <货物受水渍损坏> damage by water;seawater damage;water damage(d)
水损害 <沥青路面的> moisture induced damage
水损耗 lost water
水损条款 sea damaged terms
水损调整横倾装置 damage control locker
水塔 elevated reservoir;elevated(gravity) tank;elevated water tank;head tank;overhead tank;tank on tower;tank tower;tower;water surge tower;water tank;water tower
水塔地基勘察 exploration of water tower foundation
水獭 beaver
水獭堤 beaver dam
水苔 sphagnum
水苔泥炭 sphagnum peat
水态雾 water fog
水态云 water cloud
水钛锆石 oliveiraite
水钛铌钇铈矿 scheteligite
水钛铁矿 kleberite
水潭 swag;water hole;water sink
水碳钙镁铜矿 callaghanite
水碳钙镁铀矿 rabbittite
水碳钙铀矿 urancalcarite;wyartite
水碳锆锶石 weloganite
水碳铬镁石 barbertonite
水碳硅钙石 scawtite
水碳钾钙石 buetschliite
水碳镧铈石 calkinsite
水碳铝钡石 dresserite
水碳铝钙石 alumohydrocalcite
水碳铝镁石 manasseite
水碳铝铅矿 dundasite
水碳铝铅石 dundasite
水碳铝锶石 strontiodresserite
水碳镁钙石 sergeevite
水碳镁钾石 baylissite
水碳镁矿 barringtonite
水碳镁铝石 indigirite
水碳镁石 hydromagnesite;nesquehonite
水碳镍矿 hellyerite
水碳硼(钙镁)石 carboborite
水碳氢钙石【地】 earlandite
水碳铁镁石 sjogrenite
水碳铁镍矿 reevesite
水碳铜矾 nakauriite
水碳铜镁石 callaghanite
水碳钇石 lokkaite
水碳铀矿 sharpite
水塘 aquatic pond;aquatic pool;holding pond;pond;pool;puddle;water cistern
水塘入口处 entrance of the pond
水淘洗 water elutriation
水淘选 water elutriation
水套 jacket(of) water;jacket space;splash jacket;water chamber;water-cooled jacket
水套安全阀 water jacket safety valve
水套存水加温器 jacket water heater
水套的 water-jacketed
水套顶盖板 water jacket top cover
水套放水阀 jacket drain cock;jacket water cock
水套放水管 water jacket drain pipe
水套放水旋塞 water jacket drain cock
水套盖板 water jacket cover plate
水套供暖 jacket heating
水套管 jacket pipe
水套加热 jacket heating
水套加热烘箱 water-jacketed oven
水套加热炉 indirect fire heater
水套金属发火管 water jacketed metal ignitron
水套金属引燃管 water jacketed metal ignitron
水套进口 jacket inlet
水套空间 jacket space
水套冷凝器 water-jacketed condenser
水套冷却 jacket type cooling
水套冷却的螺旋输送机 water-jacketed screw conveyer[conveyor]
水套冷却金属点燃管 water-jacked metal ignition
水套冷却炉 water-jacketed furnace
水套冷却器 water-jacketed cooler
水套冷却式发动机 jacket-cooled engine
水套冷却室 water-jacketed chamber
水套冷却水 jacket cooling water
水套煤气发生炉 water-jacketed producer
水套内冷却水温度 <发动机汽缸> jacket temperature
水套气体发生器 water-jacketed producer
水套腔 water jacket space
水套墙 jacketed wall
水套塞 water jacket plug
水套式化油器 water-jacketed carburettor
水套水泵 jacket water pump
水套水冷器 jacket water cooler
水套损失 jacket loss
水套通道 jacket passage
水套外壁 water jacket outer wall
水套温度 jacket water temperature
水套型芯 jacket core
水套装置 water-jacketing
水藤 water vine
水梯级 water cascade
水锑铝铜石 cualstibite
水锑铅矿 bindheimite;moffrasite
水锑铜矿 partzite
水锑银矿 stetefeldtite
水提取法 water extraction method
水体 body of water;mass of water;water body;water mass;waters;water substance
水体保护 conservation of water;water protection
水体边界 boundary of water body
水体边缘 coastal waters
水体病原体污染 pathogen(e) contamination of water body;pathogen pollution of water body
水体穿透能力 water penetration
水体的富营养化 eutrophication
水体的化学净化过程 chemical purity process of water body
水体的生物净化过程 biologic(al) purity process of water body
水体的物理净化过程 physical purity process of water body
水体的植物营养物质污染 floristic nutrient pollution of water body
水体的自然污染物 corollary pollutant
水体底泥污染源 sources of sediment contamination of waters
水体毒性生物状况评价 assessment of toxicological state of water body
水体二次污染 secondary pollution of water body
水体翻转 overturn;overturning of water body
水体放射性污染 radioactive pollution of water body
水体非点源污染模型 non-point source pollution model of water body
水体粪(便)污染指示菌 indicator bacteria of water fecal pollution
水体复原 restoration of waters
水体富营养化 eutrification of water body;eutrophication of water body;eutrophication of waters
水体富营养化相关因素 eutrification-related factor of water body
水体覆盖区 water-covered area
水体工业污染 industrial pollution of water body
水体荷载 loading of water
水体化学污染 chemical contamination of water body
水体环境容量 water body environmental capacity
水体环境条件 water body environmental state
水体环境条件评价 water body environmental state assessment
水体环境值 environmental value of water
水体积压缩系数 elastic coefficient of water
水体净化 self-purification of waters
水体静止部分 keld
水体类别 water body classification
水体类型 type of water body
水体历史描述 hydrognosy
水体连续性 water mass continuity
水体流动 body current
水体名词学 hydronymy
水体农药污染 water body pesticide pollution
水体农业 hydroagriculture
水体农业污染 agricultural pollution of water body
水体迁移 mass transport
水体清洁法 Clean Water Act

水体扰动 water disturbance
水体热污染 thermal pollution of water body; water body heat pollution; water body thermal pollution
水体热污染源 water body thermal pollution source
水体人为非点污染源 anthropogenic non-point pollution source of water body
水体生活污染 domestic pollution of water body
水体生物净化 biologic(al) purification of water body
水体生物污染 biologic(al) pollution of water body
水体生物污染控制 water body biological pollution control
水体生物修复 biologic(al) remediation of water body
水体史 hydrognosy
水体水质标准 water body water quality standard
水体酸化 acidification of water body
水体特性 characterization of water body
水体同化能力 assimilative capacity of water body; water body assimilative capacity
水体同化容量 assimilative capacity of water body; water body assimilative capacity
水体透光层 euphotic zone
水体透光带 euphotic zone
水体微生物污水自然净化 natural purification of wastewater by water microorganism
水体污染 water body pollution; water pollution
水体污染调查 water pollution survey
水体污染控制 water body pollution control
水体污染控制标准 water body pollution control standard
水体污染物 pollutant in water body
水体污染物混合 pollutants mixing in water body
水体污染吸附 adsorption of pollutant in water body
水体污染源 pollution source of water body; source of pollutant in water body; water body pollution source
水体污染源管理 management of water body pollution
水体污染源管理措施 management measures of water body pollution sources
水体污染源控制和管理 control and management of pollution of water body; control and management of water body pollution
水体污染源评价模型 assessment model of water body pollution sources
水体无机物污染 inorganic pollution of water body
水体物理化学条件 physico-chemical condition of water body
水体物理净化 physical purification of water body
水体物理污染 physical pollution of water body
水体稀释 dilution of water bodies
水体细菌标准 bacteriological standards for water body
水体修复 remediation of water body
水体有机污染 organic pollution of water body
水体杂锦 breaking of the meres
水体重金属离子污染 heavy metal ion pollution of water body
水体重金属污染 heavy metal pollution of water body
水体贮水池 body of water
水体转移 mass transport
水体状况 state of water body
水体浊度 water body turbidity
水体自净(化作用) self-purification of water body; self-purification of waters; water body self-purification
水体自净能力 capacity of water body self-purification
水体自净生物工艺 biologic(al) process of water body self-purification
水体自净指数 indices of self-purification of water bodies
水体自然老化过程 eutrophication
水天然气系统接触角 contact angle of water-gas system
水天然气系统界面张力 interfacial tension of water-gas system
水天然污染源 natural sources of water pollution
水天线 sea horizon; sea line; skyline
水田 irrigated field; irrigated land; marshy field; paddy field
水田(机耕)船 paddy field boat
水田犁铧 rice share
水田耙 paddy field harrow
水田梯田 level terrace
水田土 paddy soil
水田土壤化学 the chemistry of submerged soils
水田作物 wet crop
水填坝水池 hog box
水填心墙 hydraulic core
水条线 filament line
水调节阀 water regulating valve; water valve
水调节剂 water modifier
水调节器 water regulator
水(调)腻子 water putty
水调色浆 pulp colo(u)r
水铁矾 szomolnokite
水铁矿 ferrihydrite
水铁联运码头 rail-water terminal; water-rail terminal
水铁镁石 muskoxite
水铁镍矾 hydrohonessite
水铁蛇纹石 hydrophite
水铁盐 hydromolysite
水听器 hydrophone; hydrophonic detector; pressure detector
水听器基阵 hydrophone array
水听器列 hydrophone array
水通量 water flux
水通深度 depth in channel
水铜铅钒 woodwardite
水铜氯铅矿 pseudoboleite
水桶 bucket; pail; water barrel; water bosh; water butt; watering bucket; water pot
水桶温度计 bucket thermometer
水筒 ajutage
水头 delivery head; flow head; head of pressure; head of water; hydraulic pressure head; pressure head; water head
水头摆动曲线 head oscillation curve
水头泵 head pump
水头边界条件 head boundary condition
水头变化范围 range of head
水头波动曲线 head oscillation curve
水头测量总单位压力 total water ga(u)ge
水头差 difference of water head; differential head; head difference; head differential
水头的单位损失 specific loss of head
水头范围 head range
水头方程式 head equation [H-equation]
水头分布 head distribution
水头高程 headwater elevation
水头恢复 gain of head; head recovery
水头计 head meter
水头降落 head fall
水头降速场 velocity field of water head drawdown
水头静压 static head
水头控制 headwater control
水头控制闸门 head control(ling) gate
水头历时曲线 head duration curve
水头流量关系 head-discharge relation
水头流量关系曲线 head-discharge relation curve
水头落差 drop in head; falling head; head fall
水头摩擦 head friction
水头摩擦损失 friction(al) head; friction(al) loss of (water) head
水头摩阻 hydraulic friction
水头排出(输送)压力 head pressure
水头坡降 water head gradient
水头-容量曲线 <水泵的> head-capacity curve
水头损耗 fall loss
水头损失 head loss; hydraulic friction; loss(in) head; loss in level; loss of head; lost head; pressure gradient
水头损失记录仪 head loss recorder
水头损失(量测)计 loss of head ga(u)ge
水头损失率 <指单位长度的水头损失> specific head loss; rate of head loss; rate of loss
水头损失指示器 head loss indicator; loss of head indicator
水头梯度 gradient of head; gradient of water head; pressure gradient; water head gradient; water-table gradient
水头梯度场 gradient field of water head
水头调节 headwater control
水头下降 <水库或水槽的> dropping head
水头压力 head pressure
水头压位差 head pressure
水头要求 head requirements
水头英寸数 inches of head
水头再分配 redistribution of (water) head
水头增长 gain of head
水头增流器 head increaser
水头阻力 resistance to head
水透不过去的 impervious
水透明度表 water transparency recorder
水透明度测量 water transparency measurement
水透明度测量装置 water transparency measurement device
水透式洗毛法 water-through wool scouring
水土百分比 soil water percentage
水土保持 conservancy; conservation of soil and water; soil and water conservancy; soil and water conservation; soil conservation; water-and-soil conservation
水土保持坝 gull(e)y-control dam; soil and water conservation dam
水土保持措施 erosion control; water-and-soil-conservation measure
水土保持的进展 conservation progress
水土保持地区 conservancy area
水土保持法 law of soil and water conservation; law of water and soil conservation; soil conservation law
水土保持放牧 conservative grazing
水土保持耕作 conservation farming; conservation till(age)
水土保持工程 conservancy engineering; soil and water conservation engineering; water-and-soil-conservation engineering
水土保持工程师 stream conservancy engineer
水土保持工作条例 regulation of water and soil conservation work
水土保持规划 planning of soil and water conservation; soil and water conservation planning; soil and water conservation program(me)
水土保持技术 conservation technology
水土保持科技市场营销 marketing conservation
水土保持林 forest for soil and water conservation; forest for water and soil conservation
水土保持区 conservation area; conservation district; land consolidation area; soil and water conservation district; soil and water conservation region
水土保持区划 soil and water conservation zoning
水土保持实践 soil and water conservation practice
水土保持实施计划 soil and water conservation program(me)
水土保持实验站 experiment station of soil and water conservation
水土保持塑料网 erosion control plastic net
水土保持学报 <美国双月刊> Journal of Soil and Water Conservation
水土保持学家 conservationist
水土保持研究部 <美> Soil and Water Conservation Research Division
水土保持杂志 <美> Journal of Soil, Water Conservation
水土保护地区 conservation area; soil conservation area
水土比例 ratio between water and soil
水土病 acclimation fever; water-borne and soil-borne disease
水土病因 disease caused by soil and water; soil and water pathogenic factor
水土病因水文地质调查 hydrogeologic(al) survey of disease caused by soil and water
水土病因源 soil and water source of disease
水土不服 not acclimatized; not accustomed to the climate
水土防护 earth water proofing
水土交界线 mud line
水土开发 land-and-water development
水土利用 soil and water utilization
水土流失 erosion loss; land erosion; loss by run-off of soil; loss of soil; soil and water loss; soil erosion; water-and-soil erosion; water loss and soil erosion
水土流失控制 soil erosion control
水土评估方法 soil and water assessment tool
水土适应 acclimation; acclimatization
水土植物关系 water-soil-plant relations
水土资源平衡 balance of water and land resources
水钍石 hydrothorite

水团 mass of water;water mass
水团尺寸 patch size
水团大小 patch size
水团分析 water mass analysis
水团类型 water type
水团移动 mass movement
水推进器 water propeller
水推力 hydraulic thrust
水推力损耗 hydraulic thrust loss
水腿 water leg
水退 falling of water;subside
水退韵律 regressive rhythm
水豚虱 Asellus aquatics
水脱矿物质 water demineralization
水脱坯砖 water-struck brick
水脱盐 water desalination
水脱氧 deactivation of water
水砣 head lead;lead;sounding lead
水砣测深 handlead sounding; handlead survey
水砣测深手 leadsman
水砣测深台 sounding platform
水砣长度记号 lead link marks;marks and deeps
水砣长度记录 lead link marks;marks and deeps
水砣滑车 lead block
水砣黏[粘]样剂 arming
水砣绳 handlead line;lead line
水砣绳标记 leadline mark
水砣手 leadsman
水砣手安全带 leadsman gripe; leadsman's gripe
水砣手防湿围裙 leadsman's apron
水砣台 leadsman's platform
水洼 charco;puddles of water;water hole <在干河床上的>
水洼地 moss-land
水洼地双壁开沟耕作法 basin listing
水外冷空气循环式电机 water-air-cooled machine
水完全回用 complete water reuse
水网 water net; drainage scheme; flow net;net of water courses;river system;stream system;water reticulation system;water system;anastomosis[复 anastomoses]
水网稻田地 paddy fields with dense hydrographic(al)system
水网类型 stream channel pattern; stream drainage pattern
水网系统 water reticulation system
水危机 water crisis
水微波激射 water maser
水微生物 water microorganism
水位 elevation of water;height of water;level of water;level stage;river level;water-level;water plane;water stage;water-table
水位保持时间 stage duration
水位报警(器)high and low water alarm;water-level alarm
水位比 water-table ratio
水位比降曲线 stage slope curve
水位比率 stage ratio
水位比率法 stage-ratio method
水位变程 stage fluctuation range
水位变动 fluctuation in stage;fluctuation of water table
水位变动带 belt of fluctuation of water;fluctuation belt of water table
水位变幅 amplitude of stage; amplitude of water-level; difference of level;difference of water-level variation;range of stage; stage amplitude of variation; stage fluctuation range;water-level amplitude
水位变幅降雨量比值法 method of water-level variation-precipitation ratio
水位变化 denivellation;fluctuation of water-level;stage fluctuation;variation in water level; variation in water table; water-level fluctuation; water-level variation; water-table fluctuation
水位变化范围 range of stage
水位变化曲线 stage-duration curve
水位变化趋势 tendency of stage
水位标 river ga(u)ge;water-level ga(u)ge
水位标玻璃 water-level ga(u)ge glass
水位标尺 ga(u)ge staff;stage ga(u)ge; stream ga(u)ge;water ga(u)ge;water-level scale;water-level staff;water post;water-level ga(u)ge
水位标记 water(-level)mark
水位标记仪 index limnograph
水位标示仪 water-level measuring post;water-level indicator
水位标线间面积 flat
水位标志 water indicator; waterline target;water mark
水位表 ga(u)ge glass;nilometer;water ga(u)ge;water-level ga(u)ge; water-level indicator; water-stage register
水位表保护罩 water ga(u)ge protector
水位表玻璃(管)water ga(u)ge glass
水位表汽管 water ga(u)ge steam pipe
水位表塞门 water ga(u)ge cock
水位波动 stage fluctuation;water-level fluctuation; water-table fluctuation;surging <调压塔内的>
水位波动带 belt of fluctuation of water table
水位波动范围 stage fluctuation range
水位波动缓冲器 water-level depressor
水位波动曲线 variation curve of water table
水位玻璃管 glass ga(u)ge;water column with ga(u)ge glass;water ga(u)ge glass;water glass
水位玻璃管保护装置 protector ga(u)ge glass
水位玻璃托【机】 ga(u)ge glass bracket
水位测定 stage measurement; water-level measurement;water-level measuring
水位测量 stage measurement; water-level measurement; water-level measuring
水位测量表 position water meter
水位测量曲线图 stage-hydrograph
水位测流量法 stage discharge method
水位测站 hydrometric(al)ga(u)ge
水位测针 needle ga(u)ge;point ga(u)ge
水位差 difference in water levels; difference of level; difference of water level;head of water;hydraulic gradient; potential pressure head; pressure head; range of stage;water head;pressure head
水位持续曲线 <显示一年间的水位变化> water-level duration curve
水位尺 depth ga(u)ge; staff ga(u)ge;stage rod;tide staff;water ga(u)ge; water-level measure post; water-level measuring post;water-level scale; water-level staff; foot ga(u)ge;river ga(u)ge
水位尺读数 staff reading
水位传导系数 coefficient of water-level; coefficient of water-level conductivity
水位传递器 water stage transmitter; water teleindicator
水位传递试验法 method of water-level transfer
水位传感器 water-level receiver;water stage transmitter
水位传感装置 level-sensing device
水位传送器 level transducer; stage transmitter; water-level transmitter;water stage transmitter
水位达到最高点的 high water
水位的多年变幅 perennial fluctuation of water-level
水位的年变幅 annual fluctuation of water-level
水位的日变幅 daily fluctuation of water-level
水位等高线 contours of water table; water-level contour
水位地质观测点密度 density of hydrogeologic(al) observation points
水位动态 regime(n)of groundwater-level
水位读数 ga(u)ge reading
水位读数装置 ga(u)ge readout
水位发送器 level transducer;water-level transmitter
水位反应性系数 water-level reactivity coefficient
水位分带改正 correction of water zoning
水位浮标 telltale float
水位浮球 water-level float
水位浮子 water-level float
水位改正 correction of water-level
水位高程 height of water-level;water elevation;water-level
水位高度 level of water; ga(u)ge height
水位关系曲线 stage relation curve; stage-relationship curve
水位观测 ga(u)ge observation;observation of stage; stage observation;stream-ga(u)ging;water-level observation
水位观测断面 stage measurement cross-section
水位观测精度 accuracy of water-level observation
水位观测频率 frequency of water-level observation
水位观测时间 time of water-level observation
水位观测室 ga(u)ge chamber
水位观测所 water ga(u)ge station
水位观测仪器 instrument of ga(u)ge observation
水位管 sighting conduit; sighting pipe;sighting tube
水位过程线 hydrograph of water (tide) level; level hydrograph; stage-hydrograph;stage-time curve
水位过程线零点 hydrographic(al) datum
水位过程线上升段 rising limb
水位含砂量关系过程线 sediment hydrograph
水位痕迹 water-level mark
水位后退 retrogression of water stage
水位恢复法 water-level recovery method
水位恢复曲线 recovery curve of water table
水位恢复试验的水井公式 well formula of water-level recovery test
水位恢复值 recovery value of water-level
水位回跌 lowering of the water level
水位回升曲线 regeneration curve
水位回升值 water-table in picking up
水位基点 ga(u)ge mark
水位基面 hydrographic(al)datum
水位基准面 ga(u)ge datum
水位急降 sudden drawdown
水位计 fluviograph; glass water ga(u)ge; hydrologic(al) ga(u)ge; level ga(u)ge; nilometer; niloscope; position water meter; sight ga(u)ge; tide ga(u)ge; water column;water(-level) ga(u)ge; water-level recorder; water-stage ga(u)ge
水位计玻璃 water ga(u)ge glass
水位计井 water-stage recorder well
水位计平衡器 subtank
水位计旋塞 water ga(u)ge cock
水位记录 water-level record; water-stage record
水位记录表 water-stage register
水位记录传递器 water stage transmitter
水位记录器 water-level recorder;water-stage recorder;water-stage register
水位记录仪 mareograph; marigraph; water-level recorder
水位记录站 water stage record station
水位记录装置 level-sensing device
水位监控器 water monitor
水位降低 decline of water table; drawdown;lowering of water level;stage reduction;subsidence;water lowering
水位降落 falling of water table;fall in water-level;fall of stage;fall of water-level;fall of water table
水位降落漏斗 depression cone
水位降落区 area of influence of well
水位降落曲线 curve of depression; drawdown curve
水位降落时间 time of fall
水位降落线 hydraulic grade line
水位降深 drawdown
水位降深次数 frequency of drawdown
水位降深历时曲线 drawdown-time curve
水位降深值 value of drawdown
水位降速场 field of water-level falling velocity
水位揭示牌 watermark board
水位截面图 profile of water table
水位井 ga(u)ging well
水位警报器 water alarm
水位警报装置 water-level alarm
水位径流数曲线 stage versus Q/F curve
水位绝对拟合误差 absolute fitting error of water table
水位控制 control of water-level;level control;water-level control
水位控制阀 altitude control valve;altitude valve
水位控制和止回阀 altitude and check valve
水位控制器 level controller; level sensor; water-level controller; water monitor
水位控制网 level control network; water-level control network
水位控制系统 water-level control system
水位-库容曲线 water-level capacity curve; elevation-capacity curve; stage capacity curve; stage storage curve
水位宽度 water-level width

水位累积下降值 accumulated decrease of water table
水位历时(关系)曲线 water-level duration curve; duration curve of stage; duration curve of water-level; stage-duration curve
水位量测 water-level measurement
水位零点 datum water level; ga(u)ge datum; ga(u)ge zero; hydrographic-(al) datum; zero water level
水位流量关系 discharge-stage relation; flow stage relation; ga(u)ge correlation; ga(u)ge-discharge relation(ship); stage-discharge relation(ship); ga(u)ge relation <河流测站的>
水位流量关系表 discharge table
水位流量关系公式 stage-discharge formula
水位流量关系环线 loop rating curve; loop stage-discharge relation; rating loop
水位流量关系曲线 discharge rating curve; discharge-stage curve; discharging curve; ga(u)ge-discharge curve; raring curve; stage-discharge curve; stage-discharge relation curve; water-level discharge relation curve; flow rating curve; rating curve
水位流量关系曲线的延长 extending [extension] of rating curve
水位流量关系曲线的外延 extrapolation of rating curve
水位流量关系绳套曲线 loop rating curve
水位流量关系稳定性 stability of rating curve; stability of stage-discharge relation
水位流量环形关系曲线 loop rating curve
水位流量记录 stage-discharge record
水位流量模数曲线 stage versus Q/F curve
水位流量年表 annual record of water-stage and discharge
水位流量曲线 discharge rating curve; stage-discharge curve; water-level discharge curve
水位流量图 stage-discharge diagram
水位流量相关 ga(u)ge correlation
水位-流速曲线 depth-velocity curve
水位落差 height of water
水位落程的顺序号 number of drawdown sequence
水位面积法 stage-area method
水位面积曲线 elevation-area curve
水位面积曲线法 stage-area curve method
水位频率 frequency of stage
水位频率曲线 stage frequency curve; water-level frequency curve
水位坡降 downdraw; drawdown
水位器开关 ga(u)ge cock
水位曲线 stage-hydrograph; water-level curve
水位曲线图 stage-hydrograph
水位日际变化 interdiurnal water-level change; interdiurnal water-level variation
水位容积曲线 capacity-elevation curve; storage-elevation curve; water-level capacity curve; elevation-capacity curve
水位上升平均高度 mean rise of water; rise of mean water-level
水位上升期 period of water-level raising
水位上升速率 raising rate of water-level
水位上涌 positive surge
水位上涨 rise of level; rise of river; rise of stage; rise of water-level
水位上涨高度 height of rise of water-level; rise of the river
水位上涨率 rate of level rise
水位上涨速度 level rise velocity; velocity of rise of water
水位上涨速率 rate of rise of water
水位升高 rise of water
水位升高速度 level rise velocity
水位时间曲线 stage-duration curve
水位输沙量关系曲线 sediment discharge curve; sediment runoff curve
水位水量曲线 water-level capacity curve
水位水深关系曲线 level-depth relation curve; stage-depth relation curve
水位探测头 water-level probe
水位梯度场 water-level gradient field
水位提高 raising of water
水位调节 level control; regulations of level; water-level control
水位调节阀 level control valve
水位调节器 water-level regulator
水位调整器 water-level governor
水位统测井点数目 number of simultaneous observation
水位统测时间 time of simultaneous observation
水位突变 mutation of water-level
水位突(然泄)降 sudden drawdown
水位图 hydrograph; stage diagram; stage-hydrograph; water-level diagram; water-table map
水位图示仪 graphic(al) water-stage register
水位退落 recession of water-level; recession of water stage
水位位置 deep-water table; location of water-level
水位下降 decline of water-level; decline of water table; degradation of level; depression of level; downdraw; drawdown; falling of water table; fall of stage; fall of water-level; lowering of water level; recession of level; water-level depression; water-level lowering
水位下降比 drawdown ratio
水位下降漏斗 cone of water table depression
水位下降期 decline period of water table; falling stage; period of water-level decline
水位下降曲线 drawdown curve of water-level; drawdown curve of water-stage; water-level descent curve; water-level lowering curve
水位下降时间 duration of fall
水位下降试验 <地下水> test lowering
水位下降速率 dropping rate of water-level
水位下降线 line of water-level decline
水位下落 decline of water table
水位线 ga(u)ging line; water(-level) line; water mark
水位线法 ga(u)ging line method; waterline method
水位相对拟合误差 relative fitting error of water table
水位相关法 stage relation method
水位相关曲线 correlation curve of water-level; line of corresponding stage; relation curve of water-level; stage correlation curve; stage-relationship curve
水位削减值 reducing value of water-level
水位消落 drawdown of level; drawdown of water level; draw-in; retrogression of level
水位消退 drawdown of level; drawdown of water level; draw-in; retrogression of level
水位泄降 <水库的> drawdown
水位泄降范围 drawdown range
水位泄降流量曲线 drawdown-discharge curve
水位泄降区 drawdown zone
水位信号 water-level signal
水位旋塞阀 ga(u)ge cock valve
水位延时曲线 water-level duration curve
水位演算 level routing; stage routing
水位遥测器 remote water-level controller
水位遥测仪 telemetering device of water-level; telemetering device of water stage
水位遥控器 remote water-level controller
水位仪 fluviograph; level ga(u)ge; water level ga(u)ge
水位以上毛细水边缘 capillary fringe
水位异常 water-level anomaly
水位影响 influence of water-level; influence of water table
水位壅高 raising of water level
水位预报 stage forecast(ing)
水位站 ga(u)ging station; staff ga(u)ge; stage-ga(u)ging station; water-level measuring post; water-level station
水位站基准线 ga(u)ge line
水位站网 ga(u)ge station network
水位涨落 fluctuation in stage; stage fluctuation; water-level fluctuation; water-table fluctuation
水位涨落规律 regularity of fluctuation in stage
水位针尺 water-level needle ga(u)ge
水位指示尺 dip rod
水位指示浮标 ga(u)ge float; water-level float
水位指示浮子 ga(u)ge float; water-level float
水位指示器 dipper stick; dip rod; dipstick; ga(u)ge glass column; level indicator; level meter; water-level ga(u)ge; water-level indicator; water-level tell-tale
水位指示系统 water level indicating system
水位指示装置 water locating device
水位轴 <潮汐曲线的> axis of heights
水位骤落 sudden drawdown
水位骤(然泄)降 rapid drawdown
水位状况 water regime(n)
水位状况曲线 water-level duration curve
水位资料 water-stage data
水位自动记录仪 water-stage recorder; water stage transmitter
水位自动调节器 water-level automatic regulator
水位自动调节装置 water-level automatic regulating device
水位自记仪 graphic(al) water-stage register
水味 taste of water
水温 water temperature
水温表 water temperature ga(u)ge; water temperature indicator; water thermometer
水温表指示器 water-thermometer dash unit
水温测量 water temperature measurement
水温测量仪器 instrument for measuring water temperature
水温超限开关 excessive water temperature switch
水温成层 water temperature stratification
水温递减速度 rate of water temperature progressively decreasing
水温递增速度 rate of water temperature progressively increasing
水温度调节 water conditioning
水温感传器 water temperature sensor
水温计 thermometer for water temperature; water temperature ga(u)ge; water temperature indicator; water thermometer
水温继电器 water temperature relay
水温开关 water temperature switch
水温控制 water temperature control
水温控制装置 water temperature control unit
水温模型 water temperature model
水温水深曲线 bathythermogram
水温调节器 water temperature regulator
水温调整阀 hot water regulating valve
水温异常 water temperature anomaly
水温自动控制器 water temperature shut-off
水温自动调节器 aquastat
水文暴雨 hydrologic(al) storm
水文比拟 hydrologic(al) analogy
水文变化过程 hydrographic(al) process; hydrologic(al) process
水文标志 hydrographic(al) markers
水文表 hydrographic(al) table
水文参数 hydrographic(al) parameter
水文参证 hydrologic(al) benchmark
水文参证点 hydrologic(al) benchmark
水文测船 hydrometric(al) boat; hydrometric(al) dingey
水文测井 hydrologic(al) well logging
水文测井结果解释 interpretation of hydrogeologic(al) log
水文测量 ga(u)ging; hydrographic(al) measurement; hydrographic(al) survey(ing); hydrologic(al) survey; hydrometric(al) measurement; hydrometric(al) surveying; water survey
水文测量船 hydrographic(al) vessel; ga(u)ging vessel
水文测量浮标 hydrographic(al) buoy; survey buoy
水文测量回声测深仪 hydrographic echo sounder
水文测量基准面 <深度为零> hydrographic(al) datum
水文测量记录 hydrographic(al) record; stream-flow record
水文测量局 hydrographic(al) office
水文测量六分仪 hydrographic(al) sextant
水文测量声呐 hydrographic(al) sonar
水文测量师 hydrographic(al) surveyor
水文测量艇 hydrographic(al) launch; sounding vessel; survey vessel
水文测量网络 hydrometric(al) network
水文测量学 hydrography; hydrometry; hydrographic(al) surveying
水文测量员 hydrographer
水文测量站 ga(u)ge[ga(u)ging] station
水文测量者 hydrographer
水文测流缆道 measuring cableway

水文测桥 bridge for streamflow measurement; bridge for stream-ga(u)ging; hydrometric(al) bridge
水文测绳倾斜角和方位角指示器 hydrographic(al) wire slope and azimuth indicator
水文测速仪 hydrometric(al) flowmeter; hydrometric(al) propeller
水文测艇 hydrometric(al) dingey
水文测验 ga(u)ging; hydrographic(al) measurement; hydrologic(al) observation; hydrologic(al) survey; hydrometric(al) measurement; hydrometry
水文测验便桥 ga(u)ging foot bridge
水文测验处 hydrographic(al) office
水文测验船 hydrographic(al) ship; hydrographic(al) vessel; hydrometric(al) boat
水文测验断面 demarcated section; hydrometric(al) profile; hydrometric(al) section
水文测验断面基柱 ga(u)ge line pillar
水文测验断面线 ga(u)ge line
水文测验浮标 hydrometric(al) float
水文测验浮子 hydrometric(al) float
水文测验工作者 hydrometrist
水文测验记录 hydrometric(al) records
水文测验架空缆道 hydrometric(al) aerial ferry
水文测验缆索 hydrographic(al) wire
水文测验实测资料 hydrometric(al) records
水文测验图 hydrographic(al) map
水文测验旋桨流速仪 hydrometric(al) propeller
水文测验学 hydrography; hydrometry
水文测验站网 hydrometric(al) network; stream-ga(u)ging(station) network
水文测站 ga(u)ging station; hydrometric(al) station
水文测站常规调查 normal investigation of hydrometric(al) station
水文测站控制 hydrometric(al) station control
水文查勘 hydrologic(al) exploration; hydrologic(al) survey; water survey
水文传感器 hydrologic(al) sensor
水文档案 hydrologic(al) documents
水文等值线图 hydroisopleth chart; hydroisopleth map; hydrologic(al) map
水文地层单元 hydrostratigraphic(al) unit
水文地理测量六分仪 hydrographic(al) sextant; sounding sextant; surveying sextant
水文地理的 hydrogeographic(al); hydrographic(al)
水文地理(分区)图 hydrographic(al) map
水文地理工作者 hydrographer
水文地理声呐 hydrographic(al) sonar
水文地理特征 hydrographic(al) feature
水文地理图 hydrographic(al) chart
水文地理学 hydro(geo)graphy
水文地理学家 hydrographer
水文地理资料 hydrographic(al) data
水文地球化学 hydrogeochemistry
水文地球化学带 hydrogeochemical zone
水文地球化学的 hydrogeochemical
水文地球化学分带 hydrogeochemical zoning
水文地球化学分带和分区 hydrogeochemical zoning and division
水文地球化学分区 hydrogeochemical division
水文地球化学分区依据 basis of hydro-geochemical division
水文地球化学环境 hydrogeochemical environment
水文地球化学环境分类 classification of hydrogeochemical environment
水文地球化学环境指标 hydrogeochemical index
水文地球化学勘探 hydrogeochemical prospecting
水文地球化学区 hydrogeochemical district
水文地球化学省 hydrogeochemical province
水文地球化学图 hydrogeochemical map
水文地球化学循环 hydrogeochemical cycle
水文地球化学亚区 hydrogeochemical sub-district
水文地球化学异常种类 hydrogeochemical abnormality types
水文地球化学找矿 hydrogeochemical prospecting
水文地区 hydrologic(al) region
水文地质比拟法 analogy method of hydrogeology
水文地质标志 geohydrologic(al) marker
水文地质参数 hydrogeologic(al) parameter
水文地质参数分区图 map of hydrogeologic(al) parameter division
水文地质草图 primary hydrogeologic(al) map
水文地质测绘 hydrogeologic(al) mapping; hydrogeologic(al) survey (ing and mapping)
水文地质测绘方法 method of hydrogeologic(al) mapping
水文地质初步勘察 preliminary hydrogeologic(al) investigation
水文地质单元 geohydrographic(al) unit; hydrogeologic(al) unit
水文地质单元编号 number of hydrogeologic(al) unit
水文地质地球物理勘探 hydrogeologic(al) prospecting
水文地质点数 number of hydrogeologic(al) point
水文地质调查 hydrogeologic(al) survey; hydrogeology survey
水文地质调查成果 result of hydrogeologic(al) investigation
水文地质调查种类 type of hydrogeologic(al) survey
水文地质方法 hydrologic(al) method
水文地质分区 groundwater province
水文地质分区图 map of hydrogeologic(al) division
水文地质工作者 hydrogeologist
水文地质观测点 observation point of hydrogeology
水文地质观测孔 hydrogeologic(al) observation hole
水文地质化学 geohydrologic(al) geochemistry; hydrogeochemistry
水文地质环境 geohydrologic(al) environment
水文地质基础性图件 basic maps of hydrogeology
水文地质计算模型 model of hydrogeologic(al) calculation
水文地质勘察 hydrogeologic(al) investigation
水文地质勘察孔 hydrogeologic(al) prospect hole
水文地质勘探 hydrogeologic(al) exploration
水文地质勘探孔 exploration drill hole for hydrogeology
水文地质勘探试验设备 equipment of hydrogeologic(al) exploration test
水文地质控制 hydrogeologic(al) control
水文地质评议 hydrogeologic(al) appraisal
水文地质剖面图 hydrogeologic(al) profile; hydrologic(al) profile
水文地质普查孔 general survey drill hole for hydrogeology
水文地质师 hydrogeologist
水文地质试验 hydrogeologic(al) test; hydrologic(al) test
水文地质试验孔 hydrogeologic(al) test hole
水文地质数据库 hydrogeology data base
水文地质水井钻机 hydrogeologic(al) well drilling machine
水文地质特性 hydrogeologic(al) characteristic; hydrogeologic(al) nature
水文地质特征 hydrogeologic(al) characteristic
水文地质条件 hydrogeologic(al) condition; hydrologic(al) and geologic(al) condition
水文地质条件程度等级 degree grade of hydrogeologic(al) condition
水文地质条件复杂 mixture conditions of hydrogeology
水文地质条件复杂程度 complicate degree of hydrogeologic(al) condition
水文地质条件概化图 generalized map of hydrogeologic(al) condition
水文地质通用名词术语 universal terms on hydrogeology
水文地质图 hydrogeologic(al) chart; hydrogeologic(al) map
水文地质详细勘察 detailed hydrogeologic(al) investigation
水文地质学 geohydrology; groundwater geology; water geology; hydrogeology; hydrologic(al) geology
水文地质学基础 principles of hydrogeology
水文地质学家 hydrogeologist; water geologist
水文地质资料 hydrogeologic(al) information; hydrogeologic(al) materials
水文地质钻孔 hydrogeologic(al) (bore) hole; hydrogeologic(al) drill-hole
水文地质钻孔类别 type of hydrogeologic(al) borehole
水文地质钻探 hydrogeologic(al) drilling
水文调查 hydrographic(al) survey(ing); hydrologic(al) inquiry; hydrologic(al) investigation; hydrologic(al) survey; water investigation; water survey
水文调绘 hydrographic(al) annotation; hydrologic(al) annotation
水文断面(图) hydrometric(al) section
水文反应动力学 kinetics of hydrologic(al) reactions
水文方程 hydrologic(al) equation
水文分割法 method of hydrograph separation
水文分界线 hydrologic(al) divide
水文分类 sea condition type
水文分区 hydrologic(al) regionalization; hydrologic(al) zonality
水文分水界 hydrologic(al) divide
水文分水岭 hydrologic(al) divide
水文分析 hydrologic(al) analysis
水文风暴 hydrologic(al) storm
水文锋面 hydrologic(al) front
水文服务机构 hydrologic(al) service
水文服务事业 hydrologic(al) service
水文工程地震勘探 hydrogeology and engineering seismic survey
水文工程地质学 hydrologic(al) and engineering geology
水文工程方法 hydrologic(al) engineering method
水文工作点 point of hydrology
水文工作者 hydrologist
水文关系 water relation
水文观测 hydrographic(al) observation; hydrologic(al) observation; stream-flow ga(u)ging; stream(-flow) measurement; stream-ga(u)ging
水文观测点 hydrologic(al) post
水文观测点记录表 record table of hydrogeologic(al) observation point
水文观测断面 hydrometric(al) section
水文观测井 ga(u)ge well; ga(u)ging well
水文观测站 anchor station; stream-flow measurement station; ga(u)ge station; ga(u)ging station; hydrographic(al) station; hydrologic(al) station; hydrometric(al) station
水文过程量平衡 hydrologic(al) balance; water balance
水文过程模拟 hydrologic(al) process modelling
水文过程模型试验 hydrologic(al) process modelling test
水文过程线 <特指流量过程线> hydrograph
水文过程线的退水段 falling limb of hydrograph
水文过程线的涨水段 rising limb of hydrograph
水文过程线分割 hydrograph separation; separation of hydrograph
水文过程线分割法 hydrograph separation method
水文过程线分析 hydrograph analysis; separation of hydrograph; hydrograph separation
水文过程线划分 hydrologic(al) separation
水文过程线形式 shape of hydrograph
水文过程线形状 hydrograph shape
水文和气象固定电台 hydrologic(al) and meteorologic(al) fixed station
水文和气象移动电台 hydrologic(al) and meteorologic(al) mobile station
水文化学 chemical hydrology; hydrochemistry
水文化学调查 hydrochemical survey
水文化学研究 hydrochemical research
水文基本测站 hydrometric(al) basic station; base hydrometric(al) station
水文基准面 hydrographic(al) datum
水文极端事件 <洪涝旱碱等> hydrologic(al) extreme events
水文极值 hydrologic(al) extreme
水文计算 ga(u)ging computation; hydrologic(al) accounting; hydrologic(al) calculation; hydrologic(al) computation; hydrologic(al) design; water balance; water budg-

et;water calculation
水文计算模拟程序 hydrocomputation simulation program(me)
水文记录 hydrologic(al) record
水文检测 hydrologic(al) measurement
水文绞车 hydrologic(al) winch
水文精密扫描回声测深仪 hydrographic(al) precision scan(ning) echo sounder
水文警报 hydrologic(al) warning
水文局 hydrographic(al) department;hydrographic(al) service
水文勘测 hydrographic(al) reconnaissance; hydrologic(al) survey; water survey
水文勘查 hydrologic(al) exploration
水文科学 hydroscience;scientific hydrology
水文科学学报 <英国季刊> Hydrologic(al) Sciences Journal
水文亏缺 hydrologic(al) deficit
水文缆车 hydrometric(al) cable car
水文缆道 hydrometric(al) cableway
水文历时线 duration hydrograph
水文流速仪 hydrometric(al) current meter;hydrometric(al) flowmeter;hydrometric(al) propeller
水文流域 hydrologic(al) basin
水文模拟 hydrologic(al) analog(y); hydrologic(al) simulation
水文模型 hydrologic(al) analogue; hydrologic(al) model
水文内循环 internal water circulation
水文年(度) climatic year; hydrographic(al) year; hydrologic(al) year; water year
水文年鉴 hydrologic(al) almanac; hydrologic(al) year; hydrologic(al) yearbook; water year; water year book
水文盆地 hydrologic(al) basin
水文频率分析 hydrologic(al) frequency analysis
水文频率计算 hydrologic(al) frequency calculation; hydrologic(al) frequency computation
水文频率曲线 hydrologic(al) frequency curve
水文平衡 <指降雨量、蒸发量、土壤渗透量和地面径流量的平衡> hydrologic(al) balance; hydrologic(al) budget
水文剖面 hydrologic(al) section
水文气候 hydroclimate
水文气候变异性 hydroclimatic variability
水文气候学 hydroclimatology
水文气候因素 hydroclimatic factor
水文气象 hydrometeor
水文气象的 hydrometeorologic(al)
水文气象调查 hydrometeorologic(al) survey
水文气象调查船 hydrometeorologic(al) ship
水文气象工作者 hydrometeorologist
水文气象关系 hydrometeorologic(al) relation; hydrometeorologic(al) relationship
水文气象(观测)站网 hydrometeorologic(al) network
水文气象条件 hydrologic(al) and meteorologic(al) condition
水文气象委员会 <属世界气象组织> Commission for Hydrometeorology
水文气象学 hydrometeorology
水文气象学家 hydrometeorologist
水文气象仪 hydrometeorologic(al) instrument
水文气象预报 hydrometeorologic(al) forecasting
水文气象预报所 hydrometeorologic(al) forecasting center[centre]
水文气象预报系统 hydrometer system
水文气象预报中心 hydrometeorologic(al) forecasting center[centre]
水文气象站 hydrometeorologic(al) service; hydrometeorologic(al) station
水文清点 hydrologic(al) inventory
水文情报 hydrologic(al) information
水文情况 hydrographic(al) regime; hydrologic(al) regime; hydroregime; water regime(n)
水文情势 hydrographic(al) regime; hydrologic(al) regime; hydroregime; water regime(n)
水文情态变化 hydrologic(al) modification
水文情态模拟 hydrologic(al) analogy
水文区划 hydrologic(al) regionalization; hydrologic(al) zonality
水文区域 hydrologic(al) region
水文曲线 hydrograph; hydrographic(al) curve
水文趋势预报 basic hydrologic(al) forecast(ing)
水文设计 hydrologic(al) design
水文施测 hydrographic(al) cast
水文施测器 trawl
水文施测站址 demarcated site
水文十年 hydrologic(al) decade
水文实测资料 hydrologic(al) record
水文实验 hydrologic(al) experiment
水文实验室 hydrologic(al) laboratory
水文实验站 hydrologic(al) experiment(al) station
水文事件 hydrologic(al) event
水文试验 hydraulics model test
水文手册 hydrologic(al) manual
水文数据 hydrologic(al) data
水文数据库 hydrologic(al) data bank; hydrologic(al) database
水文数据遥测 telemetering of hydrologic(al) data
水文数理统计 mathematic(al) statistics of hydrology
水文水资源计算 hydrology and water resources calculation
水文损失量 hydrologic(al) abstraction
水文所 hydrologic(al) service
水文探测设备 hydrologic(al) sensor
水文特点 hydrographic(al) feature
水文特性 hydrographic(al) feature; hydrologic(al) characteristics; hydrologic(al) property; water regime(n); hydrologic(al) response
水文特征 hydrography feature
水文特征调查 hydrologic(al) characteristic survey
水文特征值 hydrologic(al) characteristic
水文条件 hydrologic(al) condition
水文统计参数 hydrologic(al) statistical parameter
水文统计特征(值) hydrologic(al) statistical characteristic
水文统计学 hydrologic(al) statistics
水文图 hydrograph; hydrologic(al) map
水文图表 hydrochart; hydrometric(al) scheme
水文图集 hydrologic(al) atlas
水文土壤调查 hydropedological survey
水文拖测器板 trawl board
水文拖测网 trawl head
水文网疏密度 texture of drainage network
水文网(站) hydrographic(al) net(work)
水文卫星跟踪与记录 hydrographic(al) satellite tracking and recording
水文文明 hydrologic(al) civilization
水文物理的 hydrophysical
水文物理定律 hydrophysical law
水文物理过程 hydrophysical process
水文物理现象 hydrology
水文物理学 hydrophysics
水文物探 geophysical prospecting of hydrology
水文物探测井曲线 curve of hydrogeophysical logging
水文系列 hydrosequence
水文系统 hydrologic(al) system; hydrosystem
水文现象分带性 zonality of hydrologic(al) phenomena
水文现象分区性 zonality of hydrologic(al) phenomena
水文相变图 hydrophase diagram
水文相似 hydrologic(al) analogy
水文响应单元 hydrologic(al) response unit
水文响应(曲线) hydrologic(al) response
水文形态学的 hydromorphological
水文型 hydrologic(al) type
水文性质 hydrographic(al) feature; hydrologic(al) property
水文序列 hydrosequence
水文学 hydrology
水文学报 <荷兰不定期刊> Journal of Hydrology
水文学的 hydrologic(al)
水文学方程 hydrologic(al) equation
水文学方法 hydrologic(al) method
水文学家 hydrologist
水文学准备工作 preliminary hydrologic work
水文循环 circulation of water; hydrologic(al) cycle; water circulation; water cycle
水文研究 hydrologic(al) investigation; hydrologic(al) research
水文演算 hydrologic(al) routing; routing; stream-flow routing
水文遥测技术 hydrologic(al) remote sensing technique
水文遥感 remote-sensing in hydrology
水文要素 drainage; hydrographic(al) detail; hydrographic(al) feature; hydrography feature; hydrologic(al) element; hydrologic(al) factor; water feature
水文要素变化 hydrologic(al) modification
水文要素名 hydrographic(al) name
水文要素综合表 synthetic(al) table of hydrologic(al) elements
水文一致性 hydrologic(al) homogeneity
水文仪器 hydrologic(al) apparatus
水文异常 hydrologic(al) anomaly
水文因素 hydrographic(al) factor; hydrologic(al) factor
水文因素分离 separation of hydrographic(al) components
水文用水准点 hydrologic(al) benchmark
水文与气象测验 hydrologic(al) and meteorologic(al) survey
水文与气象调查 hydrologic(al) and meteorologic(al) survey
水文预报 hydrograph forecast; hydrographic(al) forecast(ing); hydrologic(al) forecast(ing); hydrologic(al) prognosis
水文预报所 hydrologic(al) forecasting centre
水文预报中心 hydrologic(al) forecasting centre
水文预测 hydrographic(al) prognosis; hydrologic(al) prognosis
水文预算 hydrologic(al) budget
水文站 stream-ga(u)ging station; stream measurement station; ga(u)ging station; measuring station
水文站基面 hydrometric(al) station datum(plane)
水文站设备 ga(u)ging station equipment
水文站网 hydrologic(al) network
水文站网规划设计 hydrologic(al) network planning and design
水文站网密度 density of ga(u)ging station; density of hydrologic(al) network; ga(u)ging station density
水文站资料 ga(u)ge station record
水文正常值 normal hydrologic(al) value
水文制图 hydrographic(al) cartography
水文制图学 hydrographic(al) cartography
水文周期 hydrologic(al) cycle; hydrologic(al) period; hydroperiod
水文属性 hydrologic(al) property
水文专用测站 hydrometric(al) station for special purposes
水文状况 hydrologic(al) condition; hydrologic(al) regime; water regime(n)
水文资料 hydrologic(al) data; hydrologic(al) documents
水文资料表 hydrographic(al) table
水文资料采集 hydrographic(al) data acquisition; hydrographic(al) data collection; hydrographic(al) data gathering; hydrologic(al) data acquisition; hydrologic(al) data collection; hydrologic(al) data gathering
水文资料插补 hydrographic(al) data interpolation; hydrologic(al) data interpolation
水文资料抽样 hydrologic(al) data sampling
水文资料处理 hydrographic(al) data processing; hydrologic(al) data processing
水文资料分析 analysis of hydrologic data; hydrographic(al) data analysis; hydrologic(al) data analysis
水文资料复原 restoration of hydrologic(al) data
水文资料还原 hydrologic(al) data restoration; restoration of hydrologic(al) data
水文资料库 hydrographic(al) database; hydrologic(al) data bank
水文资料系列代表性 sequence representation of hydrographic(al) data; sequence representation of hydrologic(al) data
水文资料修补 restoration of hydrologic(al) data
水文资料修复 hydrologic(al) data restoration; restoration of hydrologic(al) data
水文资料序列 hydrosequence
水文资料选样 hydrologic(al) data sampling
水文资料站 hydrographic(al) data station; hydrologic(al) data station
水文资料整编 compilation of hydrologic(al) data; hydrographic(al) data compilation; hydrologic(al) data compilation; processing of hydrologic(al) data

S

水文资料整理 compilation of hydrologic(al) data; hydrographic(al) data compilation; hydrologic(al) data compilation; processing of hydrologic(al) data
水文资料中心 hydrologic(al) data station
水文自动测报系统 automatic system of hydrologic(al) data collection and transmission
水文自动记录仪 hydrological recording ga(u)ge
水文综合过程线 complex hydrograph
水文钻探 hydrogeologic(al) drilling
水纹 drag mark; water crack; water thread; water wave
水纹病 watermark disease
水纹路 crazy paving path
水纹面饰 water finish
水纹图样 Moiré pattern
水稳定场 water-stable area
水稳定场上限 high limit of water stable area
水稳定场下限 low limit of water stable area
水稳定的 moisture-proof; moisture-resistant
水稳定等离子弹射器 water stabilized plasma thrower
水稳定剂 water stabiliser[stabilizer]
水稳定性 water stability
水稳定性系数 coefficient of water stability
水稳定指数 stability index of water
水稳性 hydrologic(al) stability
水稳性骨料 water-stable aggregate
水稳性团聚(体) water-stable aggregate
水问题 water problem
水(问题)研究科学委员会 Scientific Committee on Water Research
水涡流测功器 water vortex brake
水涡轮 hydroturbine
水涡轮泵 water turbo-pump
水涡喷射发动机 hydroturbojet
水窝 drain pit; water pocket
水窝泵 <排除油池积水用> sump pump
水污斑 water stain
水污泥干化 sewage sludge drying
水污染 aquatic pollution; vitiate of water; water contamination; water pollution
水污染标定 standardization of water pollution
水污染测定 measurement of water pollution; measuring of water pollution
水污染常规分析指标 index of routine analysis for water pollution
水污染成因 origin of water pollution
水污染处理 water pollution treatment
水污染的生物测试 bioassay of water pollution
水污染点源 point source of water pollution
水污染调查 sanitary survey; water pollution survey
水污染对策 water pollution control strategy
水污染对健康的影响 hazardous health effects of water pollution
水污染防治 control and prevention of water pollution; water pollution control(ing); water pollution prevention and control
水污染防治法 law of water pollution control; law of water pollution prevention
水污染防治工程 control engineering of water pollution; water pollution control engineering
水污染防治监督管理机构 organs of supervision and management of water pollution prevention and control
水污染防治监督管理机关 organs of supervision and management of water pollution prevention and control
水污染防治控制法 water pollution prevention and control law
水污染防治站 water pollution control station
水污染防治政策 water pollution control policy
水污染放射性源 radioactive source of water pollution
水污染费 charge on water pollution
水污染风险 water pollution risk
水污染负荷 water pollution load
水污染负荷指数 water pollution load index
水污染工业源 industrial sources of water pollution
水污染固定监测站 stationary monitoring station for water pollution
水污染固定源 stationary source of water pollution
水污染化学 water pollution chemistry
水污染监测 water pollution monitoring
水污染监测系统 water pollution detection system; water pollution monitoring system
水污染监测仪 water pollution monitor
水污染监控器 water pollution monitor
水污染交通源 transportation source of water pollution
水污染纠纷 water pollution dispute
水污染控制 water pollution control
水污染控制厂 water pollution control plant
水污染控制对策树 game tree of water pollution control
水污染控制法规 water pollution control law; water pollution control regulations
水污染控制法令 Water Pollution Control Act
水污染控制工程 water pollution control engineering
水污染控制管理局 Water Pollution Control Administration
水污染控制规划 water pollution control planning
水污染控制局 Division of Water Pollution Control; Water Pollution Control Board
水污染控制立法 water pollution control legislation
水污染控制联合会 Water Pollution Control Federation
水污染控制联合会会刊 Journal of the Water Pollution Control Federation
水污染控制联合会学报<月刊> Journal of the Water Pollution Control Federation
水污染控制目标 water pollution control goal
水污染控制目标树 target tree of water pollution control
水污染控制条例 Water Pollution Control Act
水污染控制系统 water pollution control system
水污染控制系统规划 water pollution control system planning
水污染控制系统设计 water pollution control system planning
水污染控制协定 water pollution control compact
水污染控制协会 Institute of Water Pollution Control
水污染控制学会 Institute of Water Pollution Control
水污染控制杂志 Journal of Water Pollution Control
水污染控制站 water pollution control station
水污染控制政策 water pollution control policy
水污染控制中心 water pollution control center[centre]
水污染控制装置 water pollution control plant
水污染连续自动监测系统 continuous and automatic monitoring system for water pollution
水污染流动监测站 mobile monitoring station for water pollution
水污染面源 area source of water pollution
水污染模式 water pollution mode
水污染模型 water pollution model
水污染农业源 agricultural source of water pollution
水污染排放许可证 water pollution discharge permit
水污染排放许可证制度 water pollution discharge permitting system
水污染潜在能力 water pollution potential
水污染潜在源 potential source of water pollution
水污染区 region of water pollution
水污染生态效应 ecologic(al) effect for water pollution
水污染生物测定 bioassay for water pollution; water pollution bioassay
水污染生物监测 biological monitoring of water pollution
水污染生物学评价 bioassessment of water pollution
水污染生物指示法 biologic(al) indication of water pollution; biologic(al) indicator of water pollution
水污染生物指数 biologic(al) index of water-pollution; biotic index of water pollution
水污染事故 water pollution accident
水污染水平 water contamination level
水污染图 water pollution map
水污染危害 water pollution hazard
水污染问题 water pollution problem
水污染物 aquatic pollutant; water contaminant; water pollutant
水污染物毒性生物评价 biologic(al) assessment of water pollutant toxicity
水污染系统 water pollution system
水污染线源 lineal source of water pollution; linear source of water pollution
水污染消除 abatement of water pollution
水污染效应 effect of water pollution
水污染型农药 water-pollution-prone pesticide
水污染修复 remediation of water pollution
水污染研究 water pollution research
水污染研究实验室 water pollution research laboratory
水污染研究厅 Water Pollution Research Board
水污染遥感 remote-sensing for water pollution
水污染预测 prediction of water pollution
水污染预防 water pollution prevention
水污染预防控制法 water pollution prevention and control law
水污染源 sources of water pollution; water pollution sources
水污染指标 index of water pollution; water pollution index
水污染指示生物 indicating organism for water pollution; indicator organisms for water pollution; water pollution indicating organism
水污染指数 index of water pollution; water pollution index
水污染重金属离子源 heavy metal ion source of water pollution
水污染追踪 water pollution tracing
水污染自然源 natural sources of water pollution
水污染综合防治 comprehension prevention and control of water pollution; comprehensive prevention and control of water pollution; comprehensive water pollution control; integrated control of water pollution
水污染综合防治规划 planning of comprehensive water pollution control
水污染综合控制 comprehensive water pollution control; integrated control of water pollution; integrated water pollution control
水污染综合控制规划 integrated planning of water pollution control; planning of comprehensive water pollution control
水污染综合治理 integrated treatment of water pollution
水钨华 hydrotungstite; meymacite
水钨铝矿 anthoinite
水钨铁铝矿 mpororoite
水坞 basin
水物质运动学 hygrokinematics
水雾 spray; water smoke; water smoking
水雾化 atomized water; water atomization
水雾化器 water atomiser[atomizer]
水雾灭火系统 water fog sprinkler
水雾喷洒 atomized water spray
水雾喷洒法 water fog spray method
水雾喷射 atomized water jet
水雾喷射系统 water spray injector system
水雾喷头 spray nozzle
水雾喷头有效射程 effective range of spray nozzle
水雾喷嘴 atomized water jet; hydraulic spray nozzle; spray nozzle
水雾系统 water spray system
水雾凿岩 water mist drilling
水雾锥 water spray cone
水吸附曲线 water sorption curve
水吸管 suction pipe
水吸式通风罩 updraft hood
水吸收法控制含氟废气 control of fluorine gases by absorption with water
水吸收法脱氮 control of NO_x by adsorption with water
水析 elutriation
水析分级法 elutriation method
水析器 wet elutriator
水硒镍石 ahlfeldite
水硒铁石 mandarinoite
水硒钴石 cobaltomenite
水晰 newt

S

水稀释比 dilution ratio of water
水稀释的 water thinned
水稀释漆 water paint
水稀释性 water dilutable
水稀释性面漆 water-reducible finish
水稀释性漆 water thinnable paint; water thinned paint
水稀释性树脂 water-reducible water-resin
水稀释性涂料 water-reducible coating; water thinnable paint; water thinned paint
水锡矿 hydroromarchite
水锡石 hydrocassiterite; varlamoffite
水螅 tubularia; hydra
水螅变形虫 hydramoeba
水螅(虫)纲 hydrozoa
水螅虫类 hydroidea
水螅珊瑚目 hydrocorallina
水螅水母 hydromedusa
水螅体 polypus
水洗 washing; water scrubbing
水洗槽 flushing tank; washing tank; wash-water tank
水洗除尘器 water scrubber
水洗除浆处理 aqueous desizing
水洗(涤)器 water scrubber
水洗法空气除尘装置 air-borne-dust elutriator
水洗法脱臭 washing method deodorizing
水洗废气净化装置 water exhaust conditioner
水洗浮雕 wash of relief
水洗浮石骨料喷射混凝土 concrete-spraying with washed pumice gravel
水洗机 scouring machine
水洗砾石 wash gravel
水洗砂 washed-out sand; masonry aggregate <用于砂浆拌和料的>
水洗筛 wet washing sieve
水洗湿式脱硫器 scrubber type wet desulfurizing equipment
水洗石饰面 rustic finish; washed finish
水洗时间 washing period; washing time
水洗式废气净化器 water exhaust conditioner
水洗式喷漆室 water wash spray booth
水洗塔 water scrubber; water scrubbing tower
水洗型尘埃收集系统 wet-scrubbing type collection system
水洗装置 water washing device
水系 aquo-system; channel net; channel system; drainage (channel); drainage system; hydrographic(al) net(work); hydrography feature; net of water courses; river system; stream net(work); stream system; water course; water reticulation system
水系版 blueprinting plate; cyan printing plate; drainage board; drainage plate; drainage separation; water plate
水系表示法 hydrography representation; water representation
水系布局 drainage pattern
水系测量 drainage survey
水系沉积物 stream sediment
水系沉积物采样 sampling of stream sediment
水系沉积物采样环镜 stream sediment sampling environment
水系沉积物采样位置 stream sediment sampling locality
水系沉积物测量 stream sediment survey
水系沉积物地球化学测量 geochemical stream sediment survey
水系沉积物样品 stream sediment sample
水系沉积物样品类型 type of stream sediment samples
水系沉积物异常 drainage sediment anomaly(stream sediment anomaly)
水系地物 drainage feature
水系调查 drainage survey; water system survey
水系对选线的关系 relation of location to drainage
水系恶化 water system deterioration
水系发育的流域 well-drained stream basin
水系法 drainage method
水系分布型式 drainage distribution pattern
水系格式 drainage pattern
水系规划 water system planning
水系级别 stream order
水系碱 aquo-base
水系类型 channel pattern; stream channel pattern; stream drainage pattern; drainage pattern
水系密度 drainage density
水系名称 hydronymy
水系模式 drainage pattern
水系排水网 drainage net(work)
水系腔 hydrocoel
水系生态学 ecology of river system
水系水异常 stream water anomaly
水系酸 aquo-acid
水系特征 characteristics of drainage; drainage characteristic
水系统 distributed system; water system
水系统竖向分区 vertical zoning of water system
水系图 drainage map; hydrographic(al) chart; hydrographic(al) map; water system map
水系退化 water system deterioration
水系网 drainage net(work)
水系网分析 drainage-network analysis
水系线性体 lineament along drainage system
水系型式 drainage pattern
水系形状 drainage pattern
水系异常 drainage anomaly
水系原图 drainage drawing
水细菌 water bacteria
水细菌成分分析 bacteria composition quality of water
水细菌学 water bacteriology
水细鳞白云母 damourite
水隙比 water void ratio
水隙舵 hydrogap rudder
水隙中的峰值 water-hole peaking
水峡 water gap
水霞石 hydronephelite; ranite
水下 submarine
水下安全预警 underwater security advance warning
水下安装 underwater installation
水下岸坡 inshore; shoreface
水下坝砂体圈闭 subaqueous bar trap
水下保养 underwater maintenance
水下爆扩 underwater blasting
水下爆破 submarine blasting; underwater blasting; underwater explosion; water shooting
水下爆破弹 limpet; limpet mine
水下爆破队 underwater demolition team
水下爆破胶质炸药 underwater blasting gelatin(e)
水下爆破孔 underwater blasting hole; underwater explosion-making hole
水下爆破切割 underwater blasting cutting; underwater explosion cutting
水下爆破形成的大浪 base surge
水下爆炸 subsurface burst; underwater blasting; underwater burst; underwater explosion; underwater shooting
水下爆炸伤 underwater blast injury; underwater explosion injury
水下爆炸研究 underwater blasting research; underwater explosion research
水下爆炸中心 underwater zero
水下贝塔伽马测量 beta-gamma survey under water
水下泵 submersible pump
水下比重 submerged specific weight
水下边界 submerged margin
水下边坡 slope in water; underwater slope
水下边滩 shoreface
水下冰 submerged ice; underwater ice
水下冰山 ram
水下兵器定位器 underwater ordnance locator
水下兵器研究试验艇 underwater ordnance research boat
水下波 subsurface wave; underwater wave
水下波导 underwater wave guide
水下剥蚀作用 dereption
水下部分 submerged body; submerged portion
水下部分船体 submerged hull; underwater hull
水下部分破损 submerged damage; underwater damage
水下采掘岩石 subaqueous rock excavation
水下采矿船 submerged mining ship; underwater mining ship
水下草本群落 submersiprata
水下草测 subaqueous running survey
水下侧面面积 lateral plane underwater
水下测距 undersea ranging; underwater range finding
水下测量 subaqueous survey; underwater acoustic(al) measurement; underwater survey
水下测量作业 underwater survey work
水下测声设备 acoustic(al) underwater survey equipment
水下测听器 hydrophone
水下测音器 hydrophone
水下拆除 underwater demolition; underwater dismantlement
水下柴埽垫 subaqueous mat(tress)
水下超声波接受器 submarine ultrasonic receiving set
水下超声波探测系统 sonar[sound navigation and ranging]
水下超压焊接 hyperbaric welding
水下车库 underwater garage
水下沉船 submerged wreck; sunken wreck
水下沉管 underwater tube
水下沉积倾斜岩层 clinothem
水下沉积(物) submerged deposit; submerged sediment; underwater deposit; underwater sediment
水下沉排 water fascine
水下冲刷 subsurface erosion
水下冲压式喷射发动机 hydroduct
水下出口 underwater outfall
水下出水口冲刷 submerged jet scour
水下除污 underwater cleaning
水下储罐 underwater storage tank
水下储量 underwater reserves
水下储能器 subsea accumulator
水下触探 subsurface sounding
水下穿道 underway
水下穿越 marine crossing
水下传感器 underwater sensors
水下传声器 underwater microphone
水下传音的 hydroacoustic(al)
水下船身电视摄像机 underwater hull TV camera
水下船体 underbody; underwater body
水下窗 <游泳训练或电视装置用的> underwater window
水下窗间墙 apron wall
水下磁测 magnetic survey under water
水下打捞 underwater salvage
水下打桩 underwater pile driving
水下打桩锤 underwater pile hammer
水下袋内浇进混凝土 depositing underwater concrete in bags
水下袋装混凝土工程 underwater bagwork
水下单位重量 submerged unit weight
水下挡墙 underwater revetment
水下导弹发射舰 underwater missile firing ship
水下导弹试验与鉴定中心 underwater missile test evaluation center [centre]
水下导洞法 top and bottom pilot tunneling method
水下导管浇注混凝土 tremie tube concrete
水下导航 underwater navigation
水下导缆器 underwater fairleader
水下导引头 underwater self-homing device
水下倒泥 underwater dump
水下的 subaquatic; subaqueous; subfluvial; submarine; submerged; under water
水下灯 underwater lamp
水下等高线 submarine contour; subsurface contour
水下等离子体焊接法 underwater plasma welding process
水下堤 subaqueous levee; submerged wall
水下堤坝 submerged groin
水下底槛 submerged sill
水下地层剖面勘探 subbottom profiling exploration
水下地基 wet foundation
水下地貌 subaqueous geomorphy
水下地形 bottom profile; bottom relief; submarine relief; submarine topography; underwater topography
水下地形测量 bathymetric(al) survey; bathyorographic(al) surveying; underwater topographic(al) survey
水下地形图 subaqueous topographic map; underwater topographic(al) map
水下地震 underwater earthquake
水下地震检波器 pressure detector
水下地震检波器漏水检查器 hydrophone
水下地震勘探 underwater seismics
水下地震勘探法 hydrosonde
水下地震映象探测法 underwater seismic image detecting
水下电割 underwater arc cutting
水下电焊 under water welding
水下电弧焊机 submerged arc welding machine
水下电弧焊(接) submerged arc weld-

S

ing; underwater arc welding
水下电话 underwater telephone
水下电话机 underwater sonic communication gear; underwater sound communication set
水下电缆 estuary cable; subaqueous cable; submarine cable; underwater cable
水下电缆浮标 telegraph buoy
水下电视 undersea television[TV]; underwater television[TV]
水下电视观测 underwater television [TV] observation
水下电视摄像机 underwater television[TV] camera
水下电信电缆浮标 submarine telegraph buoy
水下电氧切割 underwater position fixing
水下吊机 underwater crane
水下定位 underwater positioning
水下定位装置 underwater positioning apparatus; underwater positioning device; underwater positioning tool; underwater positioning unit
水下定置网 midwater trap net
水下独立开合双扇门 heck
水下段 subaqueous section
水下断流阀 underwater stop valve
水下断面 immersed section; wet(ted)(cross-)section
水下断面面积 area of wetter cross-section; wet cross-sectional area
水下堆石坝 underwater rockfill dam
水下堆石护坡 falling apron; launching apron
水下多孔扩散器排放 submerged multiport diffuser discharge
水下多普勒导航系统 underwater Doppler navigation system
水下舵 hydroflap
水下发电厂 submerged power plant
水下发射 underwater emission; underwater firing
水下方块安装 underwater block erection; underwater block setting
水下防波堤 submerged breakwater; underwater breakwater
水下防腐涂层 subaqueous antifouling coating
水下防护 underwater protection
水下防扩散覆盖层 underwater capping); subaqueous cap(ping)
水下防污漆 underwater antifouling coat
水下防御 underwater defence
水下放射性检测仪 underwater radiac set
水下废水场 submerged wastewater field
水下废物场 submerged waste field
水下分段油漆 underwater individual coat
水下分流河道沉积 submarine distributary channel deposit
水下风化作用 metharmosis
水下封层 tremie seal; underwater seal
水下封底 tremie seal; underwater seal
水下伏护 underwater revetment
水下浮标 ball-and-line float; depth float; loaded float; submerged buoy; subsurface float
水下浮动滑道【船】underwater floating launching way
水下浮容重 submerged unit weight
水下浮筒 <测定流速用> subsurface float; submerged float
水下浮子 ball-and-line float; subsurface float
水下辅助装置 underwater servicing device; underwater servicing unit
水下腐蚀 subaqueous corrosion; submerged corrosion
水下附体 underwater appendage
水下钢管 submarine steel pipeline
水下钢管隧道 immersed tube tunnel
水下钢丝防喷阀 subsea lubricator valve
水下钢丝防喷阀管缆绞车 hose reel unit for subsea lubrication valve
水下高压箱里干施工电焊 hyperbaric welding
水下割草机 underwater cutter
水下割条 electrode for underwater cutting
水下跟踪系统 underwater tracking system
水下工程 submarine works; underwater construction; underwater engineering; underwater works
水下工程地质取样管 hydroengineering geologic(al) sampler
水下工程机械 underwater construction equipment; underwater construction machinery
水下工程基础的修整 submerged grading of foundation
水下构件尺寸 dimensions of submerged member
水下构筑物 submerged structure; submerged work; underwater structure; underwater works
水下古三角洲 submerged paleo delta
水下固化涂料 underwater setting coating
水下刮土机 underwater scraper
水下观测 underwater observation
水下观测板 trawl board
水下观察 underwater observation
水下管 subaquatic pipe; subaqueous pipe; subaqueous tube; submaritime conduit; submaritime pipe; submaritime tube; submerged pipe; underwater pipe
水下管道 subaqueous pipeline; submarine pipeline; submerged pipeline; undersea pipeline; underwater pipeline
水下管道的加重包裹层 <防止管道上浮> weight coating
水下管道抵消浮力的保护层 negative buoyancy coating
水下管沟 underwater pipe ditch; underwater trench
水下管沟开挖 excavation of underwater pipe ditch; underwater pipe ditch excavation; underwater trench excavation
水下管浇混凝土 tremie tube concrete
水下管路 marine pipeline; offshore pipeline; submarine pipeline; subwater pipeline; undersea pipeline; underwater pipeline
水下管式隧道 submerged tube tunnel
水下管线 marine pipeline; offshore pipeline; submarine pipeline; subwater pipeline; underwater pipeline
水下管柱基础 submarine pipe column foundation
水下灌注混凝土 concreting in water; subaqueous concreting; submerged concreting; underwater concreting
水下灌注混凝土导管 tremie pipe
水下灌注混凝土顶面 level of tremied concrete
水下灌注混凝土基础 tremied concrete base
水下灌注混凝土榫槽 tremie concrete dowel
水下光度计 bathyphotometer; hydrophotometer; submarine photometer
水下光合作用带 euphoric zone
水下光缆 underwater optic(al) fiber [fibre] cable
水下焊割两用焊枪 underwater arc cutting welding torch
水下焊接 submerged welding; underwater welding
水下焊炬 underwater torch
水下焊条 underwater electrode
水下航行 submarine navigation; underwater navigation
水下河槽 submerged channel
水下核爆炸 underwater nuclear burst
水下横断面测量 underwater cross-section survey
水下呼吸器 rebreather; scuba; underwater breathing apparatus
水下呼吸生物 underwater breathing organism
水下呼吸装置 underwater breathing apparatus
水下弧焊 submerged arc welding
水下弧形闸门 submersible tainter gate
水下护脚 lower apron; underwater apron
水下护坡 underwater revetment
水下护坦 lower apron; underwater apron
水下滑道 underwater slipways-end
水下滑动 subaqueous slumping
水下滑坡 subaqueous landslide; submerged slide; underwater landslide
水下滑塌 subaqueous slump
水下滑塌沉积 subaqueous slump deposit
水下环礁 drowned atoll
水下环境 underwater environment
水下环境噪声 underwater ambient noise
水下换能器 underwater transducer
水下回声测深仪 underwater sound gear
水下回声地震剖面仪 hydrosound underwater seismic profiler
水下回声探测法 subsurface echo sounding
水下回填的压实 consolidation of underwater fill
水下混凝土 subaqueous concrete; submerged concrete; tremie concrete; underwater concrete
水下混凝土导管 tremie; tremie pipe; tremie tube
水下混凝土法 tremie method
水下混凝土封底 tremie concrete seal
水下混凝土工程 underwater concrete works
水下混凝土浇筑 underwater concreting
水下混凝土浇筑箱 underwater concreting box
水下混凝土连接 <沉埋隧管的> tremie concrete connection
水下混凝土配合比 underwater concrete mix
水下混凝土(配合)混合料 underwater concrete mix
水下混凝土平台 condeep platform
水下混凝土施工 subaqueous concreting; underwater construction
水下混凝土止水层 tremie concrete seal
水下混凝土桩灌注 cast-in-situ of underwater concrete pile
水下机床夯实 underwater bed tamping
水下机床整平 underwater bed level(l)ing
水下机器人 underwater robot
水下机械军士 underwater mechanic
水下机械手 underwater manipulator
水下基槽开挖 excavation of underwater foundation trench; subaqueous foundation trench cutting; underwater foundation trench excavation
水下基础 subaqueous foundation; underwater foundation; foundation under water
水下基础修整 submerged grading of foundation
水下基床夯实 compaction of underwater bedding; underwater foundation-bed tamping
水下基床整平 underwater foundation-bed level(l)ing
水下激光测量系统 underwater laser surveying system
水下激光电视 underwater laser television
水下激光雷达 undersea laser radar; underwater laser radar
水下激光器 underwater laser
水下激光通信[讯] laser underwater communication
水下激光装置 underwater laser device
水下极限坡角 limit angle of slope under water
水下集成声呐系统 submarine integrated sonar system
水下挤淤爆破 underwater squeezing explosion
水下计程仪 submerged log
水下尖礁 underwater rock pinnacle
水下监视 undersea surveillance; underwater surveillance
水下监视系统 underwater-surveillance system
水下监听站 underwater listening post
水下剪切强度测试器 submersible shear strength tester
水下检查 underwater inspection
水下检验 in-water survey
水下建筑(物) submarine structure; submarine works; submerged structure; submerged work
水下舰艇 undersea ship; underwater craft
水下交通隧道 subaqueous vehicular tunnel
水下浇灌 placing under water; tremie
水下浇灌混凝土 concreting in water; placing under water; subaqueous concreting; submerged concreting; underwater concreting
水下浇注混凝土 concreting in water; subaqueous concreting; submerged concreting; underwater concreting
水下浇筑混凝土 concreting in water; subaqueous concreting; submerged concreting; underwater concreting
水下礁(丘) nab
水下阶地 shoreface terrace
水下阶地减压法 underwater stage decompression
水下结构 underwater structure
水下结构或构件的宽度或直径 width or diameter of submerged structure or member
水下截割 under water cutting
水下井口装置 subsea equipment; wellhead cellar
水下景观 subaqual landscape
水下静物摄像机 underwater still camera
水下居住舱 underwater habitat
水下锯割 underwater sawing

水下锯机 submarine saw
水下开采 offshore mining; subaqueous mining; undersea mining; working under the water
水下开采岩石 subaquatic rock excavation
水下开挖 cutting under water; excavate underwater; wet cut (ting); underwater excavation
水下开挖法 wet excavation
水下勘测 subaqueous reconnaissance survey; subaqueous survey; underwater survey
水下控制盒 subsea control pod
水下控制系统 subsea control system
水下孔口 submerged orifice
水下快干混凝土 underwater fast-hardening concrete
水下矿石运输船 submarine ore carrier
水下窥视窗 underwater viewer
水下扩散器 submerged diffuser; underwater diffuser
水下扩音器 hydrophone
水下雷管 subaqueous detonator; submarine detonator; underwater detector
水下雷区浮标 submarine mine area buoy
水下立模 underwater formwork
水下立体电视 underwater stereoscopic television
水下砾石浅滩 gravel bank
水下连接 underwater connection
水下链传动涵洞 < 船闸开闭闸门的 > chain culvert
水下流速 subsurface velocity
水下隆起 submarine uplift
水下露头 submerged outcrop
水下录像 underwater video-recording
水下裸爆装药 surface charge
水下裸露爆破 underwater dobie blasting
水下埋管隧道 submerged tube tunnel
水下脉动式喷射发动机 hydropulse
水下模拟系统 undersea simulation system
水下目标模拟器 underwater target simulator
水下目标识别与延时系统 underwater target identification and delay system
水下目标探测器 underwater object detecting set
水下挠性管 underwater flexible pipe
水下能见度 underwater visibility
水下泥泵 submerged dredge pump
水下泥流 underwater solifluction
水下泥石流 subaqueous debris flow
水下泥石流沉积 subaqueous debris deposit
水下黏[粘]结剂 underwater adhesive
水下排泥管 submarine discharge pipe; underwater discharge pipe
水下排泥管线 submarine discharge pipeline; submerged discharge pipeline; underwater discharge pipeline
水下排气 submerged exhaust; underwater exhaust
水下排水沟 submerged valley
水下排污口 submerged outfall; submerged outlet
水下排淤爆破 underwater desilting explosion
水下抛泥区 submerged disposal area
水下抛石护坦 stone apron; launching apron
水下抛石基础 enrockment
水下抛填 underwater rockfill
水下培养 submerged culture
水下平板振动器 underwater vibratory plate-compactor
水下平台 underwater platform
水下坡度 underwater gradient
水下坡角 nature slope angle under water
水下破冰机 submerged ice cracking engine
水下铺管驳船 lay-barge
水下铺管船 submarine pipe-laying vessel
水下曝气器 underwater aerator
水下栖所 underwater habitat
水下栖息地 underwater habitat
水下起爆器 subaqueous detonator
水下气电割炬 gas electric(al) submarine torch
水下气管线 gas submarine pipeline
水下气体保护电焊 underwater air shield arc welding
水下器具 subsea equipment
水下潜槛 subaqueous sill
水下浅层剖面勘探 subbottom profile exploration
水下浅层剖面探测 subbottom profile exploration
水下浅滩 underwater bank
水下墙 underwater wall
水下切割 < 氢氧焰 > cutting under water; underwater cut(ting)
水下切割割炬 underwater cutting blowpipe
水下切割设备 underwater cutting equipment
水下侵蚀 subaqueous corrosion; submerged corrosion; subsurface erosion
水下清查【疏】 underwater removal of pretreated rock
水下清除岩石 subaqueous rock excavation
水下清洗和维护平台 submerged cleaning and maintenance platform
水下清淤 underwater desilting
水下清渣 underwater debris clearing
水下丘 subaqueous dune
水下全息摄影 underwater holography
水下泉 subaquatic spring; submarine spring
水下区 < 海工建筑物在低潮及设计波谷以下地带 > underwater zone; submerged zone
水下扰动 submerged disturbance
水下容重 submerged unit weight
水下熔化极水喷射切割 underwater consumable-electrode water-jet cutting
水下软管束 subsea hose bundle
水下弱透光带 disphotic zone
水下三角洲 subaquatic delta; submerged delta; undersea delta
水下三角洲平原 subaqueous delta plain; submerged delta plain
水下散射计 underwater scattering meter
水下沙坝 underwater bar
水下沙脊 subaqueous sand ridge; submarine bar crest; underwater sand ridge
水下沙槛 submerged sill
水下沙丘 underwater dune
水下沙滩 sandbank
水下沙洲 submerged bar
水下筛 underwater screen
水下扇层序 submarine fan sequence
水下设备 undersea device
水下射流 subaqueous jet; submerged jet
水下摄食者 midwater feeder
水下摄像机 underwater video camera
水下摄影 submarine photography; underwater photography
水下摄影测量术 underwater photogrammetry
水下摄影测量学 underwater photogrammetry
水下摄影机 fish-eye camera; submarine camera; underwater camera
水下摄影检查 underwater photographic(al) inspection
水下摄影术 submarine photography; underwater photography
水下升降机 underwater crane
水下生境 underwater habitat
水下生理学 underwater physiology
水下生物 submerged aquatics
水下声 underwater acoustic; underwater sound
水下声波 underwater sound wave
水下声波定位器 sonar[sound navigation and ranging]
水下声道 underwater acoustic(al) waveguide; underwater sound channel
水下声发射器 underwater sound projector
水下声脉冲转发器 subsea transponder
水下声信号 underwater sound signal(l)ing
水下声压式传声器 underwater sound-pressure microphone
水下声音传播区 ensonified zone; insonified zone
水下施工 subaquatic work; subaqueous construction; subaqueous execution; subaqueous work; underwater execution; underwater construction
水下施工涂料 underwater coating compound; underwater paint
水下湿电焊 wet welding
水下石埂 rock bar; ledge of rock
水下实验室 submarine laboratory
水下拾音器 hydrophone
水下视觉 underwater vision
水下试验 underwater test
水下试验靶场 underwater test range
水下试验室 sea-lab(oratory)
水下试样 subaqueous sample; underwater sample
水下收报机 submarine receiving set
水下树墩 planter; submerged stump
水下竖井 underwater shaft
水下双浮标 canister float; can(n)ister float
水下水雷 submarine mine
水下水生植被 submerged aquatic vegetation
水下水生植物 submerged aquatic plant
水下水声换能器 underwater sound transducer
水下水声装置 underwater sonic gear
水下水质测定仪 submersible water quality monitor
水下搜索设备 underwater search equipment
水下速凝砂浆 hydraulic mortar
水下碎石机 subaquatic rock breaker
水下隧道 subaquatic tunnel; subaqueous tunnel; submarine tunnel; submerged tunnel; underwater tunnel
水下隧道工人 sand hog
水下隧洞 subaquatic tunnel; subaqueous tunnel; submerged tunnel; underwater tunnel
水下索具 underwater rigging
水下塌方 submarine slide
水下探测 underwater detection
水下探测船 undersea exploration ship
水下探测器 asdic gear; detectoscope; hydrophonic detector
水下探测设备 underwater detection equipment
水下探测仪 asdic; sonar[sound navigation and ranging]
水下探测与分析系统 underwater detection and classification system
水下探伤 underwater defect detecting; underwater fault detection
水下探油船 underwater oil search ship
水下躺焊 underwater fire-cracker welding
水下梯道 companion way
水下体积 immersed volume
水下天然堤沉积 submerged natural levee deposit
水下天线 submerged antenna; underwater antenna
水下填方 underwater fill
水下填石坝 underwater rockfill dam
水下填土 underwater fill
水下填筑 underwater fill
水下铁丝网 boom net
水下听声器校准器 hydrophone calibrator
水下听声器响应 hydrophone response
水下听声器噪声 hydrophone noise
水下听音器 hydrophone
水下停车库 underwater parking building
水下通信[讯] underwater communication
水下通信[讯]和探测系统 undersea communication and detection system
水下通信[讯]系统 underwater communications system
水下投射体 underwater projectile
水下透明度计 underwater transparency meter
水下突堤 submerged jetty
水下图像 underwater picture
水下土溜作用 subsoilfluction
水下土样 subaqueous sample; underwater sample
水下推进器噪声 underwater propulsion noise
水下推土机 submarine(bull)dozer; underwater(bull)dozer; submerged bulldozer
水下拖测器 submarine sentry
水下拖曳器 underwater towed craft; underwater tow vehicle
水下挖方 underwater excavation; wet excavation
水下挖掘 underwater digging; underwater excavation
水下挖掘机(械) marine excavation; underwater excavator
水下挖泥机 submarine(bull)dozer; underwater(bull)dozer
水下挖石 rock dredging
水下挖土 cutting under water; underwater excavation; wet excavation
水下完成 submerged completion
水下完井 subsea completion
水下完井系统 subsea completion system
水下望远镜 water telescope
水下位置 position underwater; submarine site
水下温度计 submarine thermometer
水下温度突变层 thermocline layer
水下温度与氧记录仪 submersible temperature and oxygen recording equipment
水下纹影 underwater schlieren
水下涡轮喷射发动机 hydroturbojet
水下武器站 underwater weapons station
水下武器指挥室 underwater battery

plotting room
水下物体 immersed body
水下物体寻觅及打捞工具 solaris
水下雾钟 submarine fog bell
水下吸泥泵 submerged pump; ladder pump
水下吸氧减压法 underwater oxygen decompression
水下显示 underwater display
水下线 underwater line
水下线船宽 beam at water line
水下线面饱满系数 waterline coefficient
水下卸泥 dumping in water
水下信号 subaquatic signal; subaqueous signal; submarine signal
水下信号接收装置 submarine signal receiving apparatus
水下信号器 submarine signal apparatus
水下信号装置 subaqueous signal(l)ing device
水下形成的 subaquatic; subaqueous
水下性能 underwater performance
水下休止角 <土的> underwater angle of repose
水下修补 underwater patching
水下修理 underwater repair
水下选法 wet analysis
水下压力表 underwater barometer
水下淹没区 submerged zone
水下岩颈 underwater plug
水下岩石 hidden rock
水下岩石采掘 subaqueous rock excavation; submerged rock excavation
水下岩石开挖 subaqueous rock excavation; submerged rock excavation
水下岩芯钻探 submarine core drill
水下扬声器 pallesthesiometer; subaqueous loudspeaker; underwater loudspeaker
水下遥测(术) underwater telemetry
水下遥测装置 remote underwater detection device
水下遥控 remote underwater manipulation
水下遥控继续装置 underwater robot
水下液压铰刀驱动装置 underwater hydraulic-operated cutter drive gear
水下液压挖掘机 underwater hydraulic excavator
水下仪器 bottom instrument; submarine instrument; underwater instrument
水下异重流 underflow-density current
水下翼 hydroflap
水下音测 underwater ranging
水下音测队 underwater ranging battery
水下音响 underwater sound
水下音响测距发音站 submarine sound ranging station
水下音响电话 underwater sound telephone
水下音响通信[讯]系统 underwater acoustic(al) communication system
水下音响信号 submarine sound signal
水下音响装置 underwater sound device
水下引爆器 underwater detector
水下油管 submarine pipeline
水下油罐 submerged storage tank
水下(油)井 submerged well; underwater well
水下油漆 underwater coat(ing)
水下油渗推测器 underwater petroleum seep sensor
水下元件切断机 underwater cartridge chopper
水下凿岩机 submarine rock-cutter; submarine rock drill
水下凿石钻机 underwater rock drill
水下造粒 underwater pelletizing
水下噪声 underwater noise
水下闸门 submerged gate; submersible gate
水下栅栏 boomer
水下炸胶 subaqueous blasting gelatin [gelatine]
水下炸礁 reef blasting underwater; underwater reef blasting
水下障碍物 submerged bearing; submerged obstacle; sunken danger; underwater danger; underwater obstacle
水下照明 subaquatic illumination; submarine light(ing); underwater illumination; underwater lighting
水下照明灯 underwater light
水下照(明)度 underwater illumination
水下照相机 fish-eye camera; underwater camera
水下运载工具 underwater vehicle
水下障碍物测量 underwater obstruction survey
水下振荡器 submarine oscillator
水下振动 subaqueous vibration
水下振动器 subaqueous vibrator
水下振砂器 terra probe
水下震动 subaqueous shock
水下蒸发器 submerged evaporator
水下整平定高度板 strike-off template
水下支承 submerged bearing
水下植被 submarine vegetation
水下植物 subaquatic plant; submarine plant
水下重力流沉积 subaqueous gravity flow deposit
水下重量 submerged weight
水下洲滩 underwater bank
水下轴承 submerged bearing
水下住所 underwater habitat
水下贮存 underwater storage
水下贮罐 underwater tank
水下贮油罐 underwater oil tank
水下贮油库 underwater oil tank
水下抓捞斗 underwater bucket
水下桩墩 submerged stump
水下装备 underwater kit
水下装油臂 submarine loading boom
水下装油管 submarine loading line
水下装油塔系统 submerged turret loading system
水下状态 submerged condition
水下浊度计 underwater nephelometer
水下自持呼吸器的潜水员 scuba diver
水下自动装置 underwater robot
水下自然地理学 submarine physiography
水下纵断面测量 underwater longitudinal-section survey
水下钻车 underwater jumbo
水下钻机 underwater drill(ing machine)
水下钻进 underwater drilling; wet drilling
水下钻井 underwater drilling; wet drilling
水下钻井设备 subsea drilling equipment
水下钻具控制系统 control system of subsea drilling equipment
水下钻孔爆破 underwater bore-hole blasting
水下钻孔船 blue water vessel
水下钻孔炸礁船 underwater drilling and blasting ship
水下钻探 underwater drilling; wet drilling
水下钻探船 underwater boring vessel; underwater drilling vessel
水下钻探设备水下钻具 subsea drilling equipment
水下作业 subsea operation; underwater operation; underwater works
水下作业大队 underwater operation group
水下作业灯 underwater task light
水下作用的黏[粘]合值 binding value under the action of water
水仙菖蒲 corn-flag
水仙花(色的) daffodil(e)
水仙黄色 daffodil(e) yellow
水仙属 narcissus
水纤菱镁矿 artinite
水纤蛇纹石 pelhamine
水纤铁矿 hydrogoethite
水纤铁矿质学 hydrogoethite
水显影 water development
水险 marine insurance; marine risks
水险保单 marine insurance policy; marine policy
水险保险单 marine insurance policy
水险法 marine insurance act
水险公司 marine underwriter
水险掮客 ship broker
水险投保单 marine insurance application
水藓 sphagnum; sphagnum moss; water moss
水藓泥炭 sphagnum peat
水藓沼泽 muskeg; sphagnum bog; sphagnous swamp
水线 water line
水线标志 water mark
水线标志牌 subaqueous cable marker
水线长 length on water line; waterline length; length on the load water line
水线带 paint line; waterline topping
水线带漆 boot topping paint; water-line paint
水线登陆电缆 land cable
水线电报 cablegram
水线电话 submarine telephone
水线定界线的平面 plane surface circumscribed by the water line
水线斗容 bucket capacity at waterline; waterline bucket capacity
水线端点凹陷 hollow ended
水线房 subaqueous cable house
水线附近 water line
水线角 angle of entrance
水线码 cable code
水线面 plane of flo(a)tation; water plane
水线面积 waterplane area; area of water plane
水线面积系数 coefficient of water plane; waterplane area coefficient
水线面浸水部分中心 centre of water plane
水线面宽 beam on waterline
水线面面积 area of water plane; waterplane area
水线面面积曲线 curve of water plane area
水线面系数 load water-line coefficient; water plane area coefficient
水线面中心 centre of flo(a)tation
水线模型 waterline model
水线平面 <水工建筑物等> water plane
水线平面面积 <水工建筑物等> waterplane area
水线漆 boot topping paint; topping paint; water flo(a)tation paint; waterline paint
水线上部建筑 dead works; upper works
水线上部舷侧用黑漆 topside black
水线上侧面积 lateral area above water
水线涂料 boot topping paint; boot water-line paint; waterline paint
水线涂漆 boot topping paint; boot water-line paint; waterline paint
水线下船体纵剖面面积 lateral underwater area
水线以上部分 topside
水线以上船体 upper works
水线以下的 under water
水线终端房 subaqueous cable house
水线周长 waterline girth
水线最大宽度 maximum waterline beam
水乡景色 waterscape
水乡景象 waterscape
水相对渗透率 water relative permeability
水相溶性 water compatibility
水箱 header; radiator; storage cistern; water block; water box; water chamber; water drum; water tank
水箱残渣 after-flush
水箱盖 radiator cap
水箱管线 subaqueous pipeline
水箱加热器 tank heater
水箱降低 tank lowering
水箱龙头 tank cock
水箱滤器 tank strainer
水箱模型 <水文预报的> tank model
水箱塔架 tank tower
水箱台高 tank raising
水箱旋塞 tank cock
水箱溢流管 <抽水马桶> warning pipe
水箱溢水示警装置 telltale
水箱罩 radiator hood
水箱中扩散管 diffuser tube in tank
水相 aqueous phase; hydrofacies; water phase
水相摩尔分数 aqueous phase mole fraction
水相日光催化消毒工艺 aqueous phase solar photocatalytic detoxification process; aqueous phase solar photocatalytic disinfection process
水相调节剂 water-phase modifier
水象学 hydrography
水消毒法 water disinfection method
水消毒膜技术 membrane technology of water disinfection
水消费率 water rate
水消耗量 water consumption
水硝碱镁矾 humberstonite
水斜硅镁石 hydroclinohumite
水携带的 water-borne
水泄漏量 tank leakage
水榭 riverside pavilion; waterside pavilion
水锌矾 gunningite
水锌矿 calamine; hemimorphite; hydrozincite[hydrozinkite]; marionite; zinc bloom
水锌锰矿 hydrohetaerolite
水锌铀矾 zinc zippeite
水信息学 hydroinformatics
水星地质学 geology of Mercury
水星计划 Mercury program(me); project Mercury
水行政法规 administrative regulations of water
水行政复议 administration reconsideration of water
水行政管理 administration of water

水行政规章 administrative rules of water
水行政立法 administrative legislation of water
水行政司法 administrative of justice in water
水行政行为 administrative action of water
水行政执法 administration of law in water
水行政主管部门 department of water administration
水型 water type
水型集尘器 water type dust collector
水型灭火系统 water type fire extinguishing system
水型图 water type diagram
水型油泥 water type sludge
水性 aquosity
水性地面涂料 water-borne floor paint
水性多尿 hydrodiuresis
水性分散液 aqueous dispersion
水性建筑涂料 water-borne architectural coating
水性绝缘漆 insulative water paint
水性(可塑)拉毛漆 water tex
水性(可塑)拉毛涂装法 water tex finishing
水性凝胶 hydrogel
水性排出物 watery discharge
水性漆 water(-base) paint
水性漆料 aqueous vehicle
水性清漆 water-varnish
水性溶胶 aquosol
水性水泥涂料 cement-water paint
水性涂料 aqueous coating; water (-base) coating; water (-base) paint; water-borne coating; decatone
水性稀释剂 aqueous flux
水性颜料 water dispersible pigment
水性油墨 water colo(u)r ink
水性着色剂 water stain
水锈 fur(ring); incrustation; scale incrustation
水锈层 scale crust
水需要量 water requirement
水絮凝 water flocculation
水蓄冷 chilled water storage
水悬浮 aqueous suspension
水悬浮体 water suspension
水悬浮物 water slurry
水悬浮液 water slurry; water suspension
水悬剂 aqueous suspension
水旋板 hydrospire
水旋滚 hydraulic roller
水旋器 hydrocyclone
水旋塞 water cock
水旋转澄清池 screw motion clarifier
水旋转接头 gooseneck; swivel neck; water swivel
水漩涡力 water eddy force
水漩穴 water sink
水选 elutriate; elutriation; selection with water; water concentration; water gravity selection; wet cleaning
水选材料 water-sorted material
水选法 wet cleaning process
水选箱 water scalping tank
水学 hydrotechnics
水雪 water snow
水熏作用 water smoking
水循环 moisture cycle; water recycle
水循环泵 circulating water pump; water circulating pump; water circulation pump
水循环供暖系统 drop system
水循环管 water circulating pipe
水循环管线 water circulation pipeline
水循环检测器 water circulation detector
水循环冷却处理系统 cooling treatment system for water circulation
水循环器 water circulator
水循环区 water cycle area
水循环速度 water circulation velocity
水循环系数 water circulation coefficient
水循环系统 water circulating system; water-cycling system
水压 fall of water; hydropressure; hydrostatic pressure; water pressing
水压爆煤筒 coalburster
水压爆破试验 hydraulic bursting test; hydrostatic burst testing
水压泵 hydraulic pump unit
水压比降 gradient of piezometric head
水压变动率 water pressure regulation
水压表 hydraulic scale; water pressure ga(u)ge
水压剥皮器 hydraulic peeler
水压舱 water ballast
水压测深 water pressure type sounding
水压差 differential water pressure; potential pressure head
水压沉井凿井 forced drop shaft
水压成型 hydraulic forming
水压承插接头 hydrostatic joint
水压承接头 hydrostatic joint
水压冲除 <用高压水拆除损坏的混凝土> hydrodemolition
水压传动 hydraulic transmission
水压锤 hydraulic hammer
水压的 hydraulic; hydrodynamic(al)
水压电阻型测波仪 pressure type variable resistance wave meter
水压陡度 hydraulic gradient
水压锻机 hydraulic forging press
水压锻造 hydraulic forging
水压发动机 hydromotor
水压阀 hydraulic valve; piezograph valve; water valve
水压法 hydrostatic process; water pressure test
水压法洞室试验 pressure chamber test
水压法试验 <测定岩石变形性能的> water loading test
水压分布 water pressure distribution
水压封闭泵 Vacseal pump
水压缸 hydraulic cylinder
水压管道 pressure conduit
水压柜 hydraulic tank
水压荷载 water pressure load
水压缓冲器 hydraulic pressure snubber
水压混凝土泵 water hydraulic concrete pump
水压活塞取土器 hydraulic piston sampler; pressure tide gauge
水压机 hydraulic compressor; hydraulic engine; hydraulic forging press; hydraulic machine; hydraulic press; hydraulic ram; hydraustatic press; hydropress; hydrostatic press; liquid press; piezograph engine; water engine; water press; water pressure engine
水压机打包 hydraulic press packing
水压机过滤帆布 hydraulic press duck
水压机积贮器 hydraulic press accumulator
水压机模座 die carrier
水压机汽缸 ram cylinder
水压机下横梁 press bed
水压机械 hydraulic machinery
水压机压板 ram platen
水压挤出机 hydraulic extruder
水压计 hydraulic ga(u)ge; hydraulic indicator; hydraulic piezometer; hydraulic pressure ga(u)ge; hydrostatic ga(u)ge; piezometer; water ga(u)ge; water piezometer; water pressure ga(u)ge
水压计程仪 dynamic (al) pressure log; Pitot log; pressure log
水压计程仪的压力 Pitot pressure
水压技术 water hydraulics
水压继电器 water pressure relay
水压加载试验 water loading test
水压假说 hydraulic theory
水压降 water pressure drop
水压降低值 value of water pressure decrease
水压接力器 water pressure servomotor
水压接头 hydrostatic joint
水压紧密性试验 water pressure test for tightness
水压进给式钻机 auger with hydraulic feed
水压开关 hydraulic pressure switch; water switch
水压空化 hydraulic cavitation
水压块 hydraulic block
水压扩张术 hydrostatic dilatation
水压力 hydraulic pressure; water loading; water pressure(force)
水压力缓冲器 hydraulic pressure snubber
水压力计 piezometer
水压力液压 hydraulic pressure
水压联轴器 hydraulic coupling; hydrocoupling
水压裂缝 hydrofracture
水压流 hydraulic flow
水压硫化机 hydraulic vulcanizing press
水压铆机 hydraulic riveter; riveting ram
水压铆接 hydraulic riveting
水压面 piezometric level; piezometric surface
水压面下降 decline of piezometric surface
水压膜片 water diaphragm
水压逆止阀 hydraulic back-pressure valve
水压喷射 jet of water
水压平衡(管网) hydraulic balance
水压平面 piezometric level
水压坡度 hydraulic gradient; water banking
水压坡降线 hydraulic gradient; piezometric line; piezometric map
水压破碎机 hydraulic breaker (attachment)
水压起重机 hydraulic crane
水压强度和密封性试验 water pressure test for strength and tightness
水压强度试验 water pressure test for strength
水压驱动 water pressure drive
水压驱动的井 water dependent well
水压渗透性 hydraulic permeability
水压升降船坞 hydraulic dock
水压升降机 hydraulic elevator; hydraulic lift
水压式波高计 pressure meter type of wave recorder
水压式波高仪 pressure meter type of wave recorder
水压式成型器 hydroformer
水压式垂直升船机 hydraulic vertical shiplift
水压式的 hydrostatic
水压式漆膜抗张强度计 rumpometer
水压式起重机 water crane
水压式渗透仪 water pressure type leak detector
水压式引信 hydrostatic fuse
水压式自记验潮仪 hydraulic tide ga(u)ge
水压试验 hydraulic pressure test(ing); hydraulic test; hydrostatic test(ing); hydrotest; water pressure test; water test
水压试验泵 hydraulic test pump
水压试验合格证 hydrostatic inspection certificate
水压试验机 hydraulic testing machine; hydrostatic test(ing) machine
水压试验结果 result of hydraulic test
水压试验压力 hydraulic test pressure; non-shock fluid pressure
水压试验装置 water test unit
水压室 hydraulic chamber; hydraulic pressure cell
水压水试验 hydrostatic pressure test(ing)
水压水位 level of the hydraulic pressure
水压松散冲采法 infusion jet method
水压缩性 water compressibility
水压台 hydraulic block
水压探测器 pressure sounder
水压提升机 hydraulic lift; piezograph lift
水压调节器 hydrostat; water pressure regulator
水压调整器 water pressure regulator
水压调整装置 water pressure regulating device
水压头 hydraulic head; hydraulic ram; liquid pressure; pressure head
水压图 pressure diagram
水压图形 water pressure pattern
水压推进式钻机 auger with hydraulic feed
水压网格 water pressure grid
水压温度计 combination altitude ga(u)ge and thermometer
水压系统 hydraulic system
水压线 hydraulic grade; hydraulic grade line; hydraulic gradient line
水压线坡度 slope of hydraulic grade line
水压线圈 piezograph
水压线图 piezograph
水压蓄力器 hydraulic oil accumulator
水压循环疲劳试验 hydrostatic cyclic fatigue test
水压压头 hydraulic pressure head
水压验潮仪 pressure ga(u)ge
水压验证试验 hydrostatic proof test
水压扬汲机 hydraulic ram
水压引信 pressure mechanism
水压应力 hydrodynamic(al) stress
水压用水 hydraulic water
水压载 water ballast
水压载水衡重 <镇船> water ballast
水压载系统 water ballast system
水压轧光机 hydraulic calender
水压轧机 hydraulic mangle
水压轧水机 hydraulic mangle
水压闸 hydraulic brake
水压胀裂法 hydraulic fracturing; hydrofracturing
水压制动器 water brake
水压致裂法 hydraulic fracturing; hydrofracturing(method)
水压致裂试验 hydraulic fracture test
水压钟 hydrostatic bell
水压主管 hydraulic main
水压贮存系统 aqua system
水压钻进式钻机 auger with hydraulic feed

水压钻探机 hydraulic thrust boring machine
水压作用 hydraulic action
水涯线 hydrologic(al) line; line of water area
水淹地区 flood land; submerged land
水延迟线 water delay line
水岩 water-bearing rock
水/岩比值 water/rock ratio
水岩盖 cryolaccolith; hydrolaccolith
水研究 water research
水研究所 institute of water research
水研究委员会 Committee on Water Research
水研究文摘 water research abstract
水研究协会 water research association
水研究中心 water research centre
水盐度 salinity of water
水盐碱化 water salination; water salinization
水盐平衡 water and salt balance
水眼 <钻头的> nozzle outlet; water eye; water passage; water hole
水眼压力 <钻头的> nozzle pressure
水堰 mill-weir
水秧田 irrigated nursery
水扬程 water-raising capacity
水扬压力 hydraulic uplift pressure
水杨 bigcatkin willow
水杨醇 salicyl alcohol; saligenol
水杨酐 salicylic anhydride
水杨硫磷 salithion
水杨醛 salicyl aldehyde
水杨醛分光光度法 salicyl aldehyde spectrophotometry
水杨醛生产废水 salicyl aldehyde production wastewater
水杨酸 salicylic acid
水杨酸胺 salicylate amide
水杨酸苯汞 phenylmercuric salicylate
水杨酸苯基酯 salicylic acid phenyl ester
水杨酸苯酯 phenyl salicylate
水杨酸对辛苯酯 p-octylphenyl salicylate
水杨酸分工光度法 spectrophotometer method with salicylic acid
水杨酸内酯 salicylide
水杨酸钠 sodium salicylate
水杨酸十二烷酯 dodecyl salicylate
水杨酸戊酯 amylsalicylate
水杨酸盐 salicylate
水杨酸乙酯 ethyl salicylate; salethyl
水杨酰胺 salicylamide; salicylic amide
水杨酰替苯胺 salicylanilide
水养法 <保护层> water curing
水养护 water cure; water curing
水养护混凝土 water cured concrete
水氧化反应 water oxidation reaction
水氧钨矿 meymacite
水样保存 preservation of water samples
水样(本) water sample; aqueous humor; aqueous sample; sample of water
水样本采集 collection of water sample
水样采集 water sampling
水样采集装置 water sample sampling device
水样采取层位 sampling layer of water
水样采取的时间 sampling date of water
水样采取方法 sampling method of water
水样采取深度 sampling depth of water
水样采取时间 time of water sampling
水样采样点 water sampling point
水样储存 water sample storage
水样处理 sample processing; water sample processing
水样的保存 preserving of water sample
水样颠倒采样瓶 reversing water bottle
水样分析 analysis of water sample
水样化验单位 analysis unit of water sample
水样化验时间 analysis date of water sample
水样来源 source of water sample
水样类别 water sample kind
水样滤器 water sample filter
水样品类型 type of water sample
水样瓶 water sample bottle
水样取样瓶 water sampling bottle
水样取样器 hydrophore
水样全(指标)分析 complete water analysis
水样容器 sample container
水样数 number of water sample
水样数量 sampling quantity of water
水样稳定化 water sample stabilization
水样液 aqueous humor
水窑 water cellar
水冶 hydrometallurgy; wet metallurgy
水冶处理 wet metallurgical processing
水冶和火冶联合法 column of hydrometallurgy and pyrometallurgy
水冶设备 hydrometallurgical plant
水冶提取 hydrometallurgical extraction
水叶 hydrofoil; water foliage
水叶装饰 <柱头的树叶状装饰> water-leaf; lesbian leaf
水液除尘器 hydrofilter
水液深成的 hydroplutonic
水液体 <拉> aqua
水液相 fluid water phase
水液注入的 aqueous-injection
水一般的 watery
水衣 water jacket
水翼 hydrofoil(wing); hydroplane; hydrovane; water wing
水翼操纵系统 hydrofoil control system
水翼衬翼 annex(e) foil
水翼船 hydrofoil boat; hydrofoil craft; hydrofoil ship; hydroplane; wing boat
水翼船稳定装置 hydrofoil stabilization device
水翼渡船 hydrofoil ferry
水翼面积 area of hydrofoil
水翼式水陆两用车 hydrofoil amphibian
水翼式堰 hydrofoil weir
水翼艇 foil craft; hydrofoil boat; hydrofoil craft; hydrofoil vessel
水翼艇的双翼 hydrofoil
水翼消波器 hydrofoil subduer
水翼型 airfoil profile
水翼叶栅 hydrofoil cascade
水翼装置 hydrofoil unit
水因数 water factor
水音测漏装置 water phone
水银 mercury; quick silver
水银摆 mercurial pendulum
水银半波整流器 ignitron
水银保险装置 mercury cut-off
水银杯连接 mercury cup connection
水银泵 mercury pump
水银避电器 mercury discharge lamp
水银避雷器 mercury arrester
水银变频器 mercury frequency changer
水银标准 mercury standard
水银薄膜继电器 mercury film relay
水银槽 mercury bath; mercury cell; mercury pond; mercury pool; mercury tank; mercury trough
水银槽气压计 cistern barometer
水银槽阴极 mercury pool cathode
水银测孔仪 mercury porosimeter; mercury pressure porosimeter
水银插棒式继电器 mercury plunger relay
水银超增感 mercury hypersensitizing
水银沉降计 mercury settlement ga(u)ge
水银池 mercury pool
水银充满式 mercury-filled
水银充气整流管 mercury vapo(u)r tube
水银触点 mercury-wetted contact
水银触点继电器 mercury-contact relay
水银垂球 <测垂直度用> mercury bob
水银存储器 mercury memory; mercury memory storage; mercury storage
水银存储装置 mercury storage
水银的 mercurial
水银灯 finsen light; mercury light; mercury vapo(u)r lamp; mercury vapo(u)r light; quick silver lamp
水银灯泡 mercury vapo(u)r lantern
水银灯照明 mercury vapor illumination
水银地平 mercury horizon
水银地震计 cacciatore
水银电池 mercury battery
水银电动机式 mercury motor type
水银电动式安培小时表 mercury motor meter
水银电动式仪表 mercury motor meter
水银电度表 mercury meter
水银电弧整流器 mercury-arc rectifier
水银电极 mercury electrode
水银电解槽 mercury electrolytic cell
水银电解电度计 mercury electrolytic meter
水银电解质换能器 mercury-electrolyte transducer
水银电阻器 mercury resistor
水银断路器 mercury break(er)
水银断续器 mercury interrupter
水银阀 mercury valve
水银法 mercury method
水银法转换 Hg conversion
水银反射器 mercury reflector
水银反向换流器 mercury inverter
水银防腐法 kyanising[kyanizing]; kyanization
水银放电灯 mercurous discharge lamp; mercury discharge lamp
水银铬化 merchromize
水银共振辐射 mercury resonance radiation
水银光谱 mercury spectrum
水银光谱灯 mercury spectral lamp
水银锅炉 mercury boiler
水银恒温器 mercury thermostat
水银弧光灯 Finsen lamp; mercury-arc lamp
水银缓冲器 mercury dashpot
水银换流器 mercury converter; mercury inverter
水银激光器 mercury laser
水银集流器 <电测扭矩用> mercury bath collector
水银记录式温度计 mercury recording thermometer
水银继电器 mercury relay
水银继电器脉冲发生器 mercury-relay pulse generator
水银继电器水银接点开关 mercury switch
水银加厚法 mercurial intensification; mercury intensification
水银检波器 mercury detector
水银接触器 mercury treadle
水银接触温度计 mercury-contact thermometer
水银接触装置 mercury-bath contact apparatus
水银接点 mercurial contact; mercury contact
水银接点继电器 mercury-wetted contact relay; mercury-wetted relay
水银接点开关 mercury-contact switch
水银开关 mercoid switch; mercury contact; mercury cut-off; mercury switch
水银开关联锁装置 mercury switch interlock
水银孔率法 <用水银测孔率的方法> mercury porosimetry
水银孔隙仪 mercury porosimeter
水银矿石 mercury ore
水银扩散泵 mercury air pump; mercury diffusion pump; mercury vapo(u)r air pump; mercury vapo(u)r diffusion pump
水银冷冻试验 mercury freezing test
水银冷凝器 mercury condenser
水银冷却阀 mercury cooled valve
水银冷却气门 mercury cooled valve
水银连接 <环形水银开关> mercury connection
水银脉冲发生器 mercury pulser
水银密封 mercury seal
水银密封釜 mercury seal pot
水银模 mercast pattern
水银模铸造 mercast
水银凝固点 mercury freezing point
水银盘地平 artificial horizon; mercury horizon
水银频率变换器 mercury frequency changer
水银气泵 mercury air pump
水银气体放电管 mercury gas discharge lamp
水银气压表 mercurial barometer; mercury barometer
水银气压计 mercurial barometer; mercury barometer; Torricellian barometer
水银汽轮机 mercury turbine
水银器连杆 ballistic arm
水银器连杆轴承 ballistic link bearing
水银器纬度修正盘 latitude level corrector; latitude level(l)ing dial
水银器轴承 ballistic pivot bearing
水银切断器 mercury cut-off
水银清洗设备 mercury cleaner
水银容器 mercury reservoir
水银渗透法 <测气孔率> mercury penetration method
水银湿簧片开关 mercury-wetted reed switch
水银示差温度计 mercury differential thermometer
水银式压差记录器 mercurial type differential pressure recorder
水银试验 <测应力腐蚀> mercury test
水银水准仪 mercurial level
水银踏板 mercury treadle
水银调节装置 mercury regulating unit
水银同位素灯 mercury isotope lamp

S

水银温度表 mercurial thermometer; mercury thermometer
水银温度计 mercurial thermometer; mercury (-filled) thermometer; mercury-in-glass thermometer
水银温度继电器 mercury type temperature relay
水银温度调节器 mercury thermoregulator
水银涡轮机 mercury turbine
水银钨丝灯 mercury-tungsten lamp
水银限时解锁器 mercurial time release
水银旋转计数器 mercurial rotational counter
水银压孔法 <测孔隙率的> mercury injection method
水银压力表 mercuric pressure ga(u)ge; mercury ga(u)ge; mercury manometer; mercury pressure ga(u)ge
水银压力计 mercurial ga(u)ge; mercuric pressure ga(u)ge; mercury ga(u)ge; mercury manometer
水银压入法 mercury penetration method
水银延迟线 mercury delay line
水银延时装置 mercury timing valve
水银遥控开关 mercury teleswitch
水银液 quick water
水银阴极 mercury pool
水银阴极辉点 mercury spot
水银荧光灯 <一种光度强用电省的现代道路照明灯> fluorescent mercury discharge lamp; fluorescent mercury lamp; mercury vapo(u)r lamp
水银油压力盒 mercury-oil pressure cell
水银真空泵 mercury (vacuum) pump
水银真空计 mercury vacuum ga(u)ge
水银蒸气泵 mercury vapo(u)r pump
水银蒸气灯 mercury vapo(u)r lamp
水银蒸气放电灯 mercury vapo(u)r discharge lamp
水银蒸气阻隔筒 arc guide
水银蒸煮器 mercury boiler; mercury controller; mercury converter
水银整流器 mercury-arc rectifier; mercury rectifier; vapo(u)r rectifier
水银直流变换器 mercury direct current transformer
水银指示器 <测量油舱油位用> mercury indicator
水银中毒 hydrargyria; hydrargyrism; mercurialism; mercurial poisoning
水银柱 column of mercury; measuring column; mercury column
水银柱高度 mercury pressure
水银柱气压表 mercury barometer
水银柱压力 mercury absolute pressure
水银柱英寸数 inch of mercury
水银转换开关 mercoid
水印 <纸张上的> water mark
水印横格线 water line
水应力 water stress
水英寸 <0.0254 米孔径管口 24 小时的放水量> water-inch
水映光 water sky
水硬 hydraulic set(ting)
水硬度 water hardness
水硬度滴定管 hydrotimetric burette
水硬度计 hydrotimeter; water hardness indicator
水硬度记录仪器 water hardness recording instrument
水硬度量瓶 hydrotimetric flask
水硬硅钙石 hydroxonotlite
水硬剂 hydraulic agent
水硬结合力 hydraulic binding quality
水硬率 hydraulic index; hydraulic rate; hydraulic modulus <水泥等>
水硬模量 hydraulic modulus
水硬黏[粘]结剂 hydraulic glue
水硬强度 hydraulic strength
水硬石膏 hydraulic gypsum
水硬石灰 calcareous cement; hydraulic lime; pozzolanic lime; blue lias lime <侏罗纪岩层石灰石中的>; water lime
水硬石灰砂浆 hydraulic lime mortar
水硬石灰石 hydraulic limestone
水硬水泥 calcareous cement
水硬水泥混凝土 hydraulic cement concrete
水硬水泥砂浆 hydraulic cement mortar
水硬系数 hydraulic module; Vicat index
水硬性 hydraulic binding quality; hydraulicity; hydraulic property; hydraulic activity <水泥>
水硬性材料 hydraulic material
水硬性掺和料 hydraulic admixture
水硬性大块 hydration induced lump
水硬性的 hydraulic hardening
水硬性硅酸盐 hydraulic calcium silicate
水硬性化石灰 hydraulic hydrated lime
水硬性灰浆 hydraulic mortar
水硬性混合料 hydraulic admixture
水硬性加气混凝土 air-entraining hydraulic cement
水硬性建筑材料 hydraulic material
水硬性胶 hydraulic glue
水硬性胶合剂 hydraulic binder; hydraulic binding agent
水硬性胶结剂 hydraulic binder; hydraulic binding agent
水硬性胶结料 hydraulic binding agent; hydraulic bonding agent
水硬性胶凝材料 hydraulic binder; hydraulic binding agent; hydraulic binding material; hydraulic cement; hydraulic cementing material; hydraulic cementitious material; water cement
水硬性结合的底基层 hydraulically bound subbase
水硬性结合剂 hydraulic bond
水硬性结合料 hydraulic binder
水硬性结合料罩面 hydraulically bond overlay; hydraulically bound overlay
水硬性模数 hydraulic modulus
水硬性抹灰 hydraulic plaster
水硬性耐火材料 hydraulic refractory
水硬性耐火泥浆 hydraulic setting mortar
水硬性耐火水泥 hydraulic refractory cement
水硬性黏[粘]合剂 hydraulic binder
水硬性黏[粘]合料 hydraulic binder; latent hydraulic binder
水硬性黏[粘]结料 hydraulic binder; hydraulic binding agent; hydraulic cementing material
水硬性砂浆 hydraulic (setting) mortar
水硬性石灰 hydraulic lime; water lime
水硬性石灰水泥 hydraulic plaster
水硬性石灰岩 hydraulic limestone
水硬性熟石灰 hydraulic hydrated lime
水硬性水化石灰 hydraulic hydrated lime
水硬性水泥 <耐高温达 1 300°C> Kestner cement; hydraulic cement; water cement
水硬性添加剂 hydraulic admixture
水硬性外加剂 hydraulic additive
水硬性物质 hydraulics
水硬性指数 hydraulic index
水硬性作用 pozzolanic reaction
水硬作用 <水泥的> hydraulic hardening
水涌 swelling
水涌入 water inrush
水油比 water-oil factor; water-oil ratio
水油比曲线法 water-oil ratio curve method
水油分离器 water and oil separator; water-oil separator
水油柜检查 tank survey
水油过渡带 water-to-oil area
水油乳化液 chocolate mousse
水油乳剂 water-oil emulsion
水油乳浊液 water-oil emulsion
水油污渍 water-oil stain
水铀矾 zippeite
水铀铅矿 <铅和铀的氧化物> masuyite
水淤泥 water sludge; water slurry
水与沉积物的分界面 water/sediment interface
水与废水处理 <英国月刊> Water & Waste Treatment
水与废水的生物学 biology of water and wastewater
水与结合料之比 water-binder ratio
水与裸皮重量比 water-to-pelt ratio
水与泥沙混合物 water-sediment mixture
水与泥沙状况 water-sediment regime
水与黏[粘]结料之比 water-binder ratio
水与污水处理厂 water and effluent treatment plant
水俣病 <1956 年发现在日本北九州熊本县的水俣市，系由 CHISSO 水俣工厂排水中的甲基水银所致> Minamata disease
水俣病防治 Minamata disease control
水俣病事件 Minamata disease event; Minamata disease incident
水俣湾 Minamata bay
水玉髓 lutecite
水芋绿色 calla green
水浴 bain-marie
水浴锅 water-bath
水浴冷却 water-bath cooling
水浴瓶 bubbler
水浴器 bain-marie; water-bath
水浴蒸发器 water-bath evapo(u)rator
水预处理 water pretreatment
水预冷池 water for cooling tank
水预冷器 water forecooler; water precooler
水预热器 water preheater
水域 marine area; marine sea; sheet of waters; water body; water region; waters
水域保护 water body protection; watershed protection; waters protection
水域除藻 weed removal from surface waters
水域地震映象探测法 water seismic image detecting
水域法 water area method
水域浮雕地球仪 hydrographic(al) relief globe
水域光学 hydrooptics
水域环境变化试验 aqua-environment variation test
水域环境特征 waters environment characteristic
水域环境卫生 water area environmental sanitation
水域环境卫生管理 water area environmental sanitary administration; water area environmental sanitary management
水域环境卫生水平 water area environmental sanitary level
水域环境污染 waters environment pollution
水域面积 water area; water space
水域前沿 water frontage
水域权 right-of-way
水域设施 water facility
水域生态系(统) ecosystem of water body; hydro ecosystem; watershed ecosystem
水域生态学 hydroecology
水域微生物学 water microbiology
水域污染 water pollution
水域污染生态学 pollution ecology of waters; waters pollution ecology
水域污水循环 recycling of water or sewage
水域油污染 oil pollution of waters
水缘 water front
水缘榴石 hibschite
水源 alimentation of river; fountain; fountain head; head of a river; head spring; headwater; head waters of river; source; source of water; spring head; water head; water resource; water source; water supply
水源保护 protection of water resources; source protection; water conservation; water resource conservation; water resource protection; water source conservation; water source protection
水源保护措施 water conservation measures; water protective measure
水源保护措施实力 efficiency of water protection measures
水源保护法 water conservation law
水源保护工程 water conservation engineering
水源保护管理 water protection management
水源保护活动 water protection activity
水源保护林 forest for conservation of water supply; protection forest for water conservation; water conservation forest
水源保护区 security zone of water source; water protection area
水源保护体系 water protective complex
水源泵站 source pump station
水源病 water-borne disease
水源补给 water recharging
水源测量 water resource survey
水源层 source bed
水源产量 hydraulic yield
水源持续产量 sustained yield of water source
水源传染 water-borne infection
水源地 water head site
水源地建设总投资 total investment of water source construction
水源地区 stream source area
水源地投产年限 period of water source operation
水源地投产时间 time of water source operation
水源地总出水量 total yield of water sources

S

水源调查 headwater survey; water source survey
水源防污 pollution prevention at the source
水源高程 headwater elevation
水源工程 water source project
水源工程设备 equipment for water supply works
水源管理 water management
水源涵养林 forest for conservation of water supply; protection forest for water conservation; water conservation forest
水源互相袭夺 interpiracy of water sources
水源环境卫生防护 sanitary protection of water source
水源环境卫生区 sanitary zone for water source
水源勘察图 headwater prospect map
水源勘探 water prospecting; water resources exploration
水源勘探图 water prospect map
水源控制 control of headwater; control of source; headwater control; water source control
水源利用 water source utilization
水源联合运用 <地表水和地下水的> conjunctive water use
水源盆地 headwater basin
水源区 catchment area; contributing region; donor region; watershed
水源区保护 water basin protection
水源区防污控制规划 water basin pollution prevention and control planning
水源热泵 water source heat pump
水源热泵系统 water source heat pump system
水源热泵装置 water source heat pump system
水源树林 headwater forest
水源水 source water
水源水质保护区 water source quality protection zone
水源图 water prospect map
水源卫生 sanitary of water source
水源污染 water source pollution
水源系统 water resources system
水源箱 suction pit
水源性疾病 water-borne disease
水源选择 selection of water source; water source selection
水跃 backwater jump; hydraulic jump; jump; pressure jump; water jump
水跃波 hydraulic jump wave
水跃波高 hydraulic jump wave height
水跃长度 distance of hydraulic jump; length of jump
水跃的能量损失 energy loss in hydraulic jump
水跃的相对高度 relative height of hydraulic jump
水跃高度 height of hydraulic jump; height of jump; jump height
水跃混合设备 hydraulic jump mixing device
水跃距离 distance of hydraulic jump; jump distance
水跃能量消耗 energy loss in hydraulic jump
水跃能头损失 energy loss in hydraulic jump
水跃特性 hydraulic jump property
水跃挖流槽 standing-wave flume
水跃尾部 water spring tail
水跃位置 location of jump
水跃消能 energy dissipation by hydraulic jump
水跃值 value of hydraulic jump
水云 water cloud
水云母【地】 hydromica
水云母埃洛石型高岭土 kaolin ore of hydromica-halloysite type
水云母高岭石型高岭土矿石 kaolin ore of hydromica-kaolinite type
水云母化作用 damouritization
水云母泥岩 hydromica mudstone
水云母黏[粘]土 hydromica clay
水云母片岩 damourite-schist
水陨硫铁 hydrotroilite
水运 water-borne traffic; waterway traffic; freight; marine conveyance; marine navigation; transport by water; water-borne transport; water transport(ation); waterway transportation
水运保险单 marine insurance policy
水运比重 rate of freight(passenger) traffic by water
水运船队 river fleet
水运代理人 shipping agent
水运的 water-borne
水运动学 hydrokinematics
水运废物 water-carried wastes
水运费 freight of shipping; waterage
水运腹地 the tributary area of water transport
水运工程定额 maritime engineering rating
水运工程概算 maritime engineering estimate
水运工程规划 maritime engineering programming
水运工程决算 final accounts of maritime engineering
水运工程设计 maritime engineering design
水运工程网络 network of maritime engineering
水运工程网络规划 network programming of maritime engineering
水运工程协会 Association of Maritime Engineering
水运工程学 marine traffic engineering; maritime engineering; water transport engineering
水运工程学会 Society of Maritime Engineering
水运工程岩土勘察 geotechnic(al) investigation for port and waterway engineering
水运工程预算 maritime engineering budget
水运工程质量验收标准 quality acceptance standard of maritime engineering
水运工具 watercraft; water transportation carrier
水运货流选择 selection of cargo flow
水运货物 waterborne freight
水运货运密度 density of freight traffic by water
水运经济调查 economic investigation for water transport; economic survey of maritime navigation
水运经济规划 economic planning for water transport
水运垃圾 water-carriage of garbage
水运量 water-borne commerce
水运量预测 forecasting of water-borne traffic
水运密度 density of water transport
水运起讫点 water lead
水运枢纽 marine terminal; waterfront terminal
水运输 water-carriage
水运条款 <指对陆上危险不保> water-borne clause
水运通过能力 water-borne throughput
水运通信[讯]网 communication network of water transportation
水运系统 water-carriage system; water(-borne) transportation system
水运系统管道 water-borne sewerage
水运业 water-borne commerce
水运业者 shipping agent
水运运费 freight charges for water transport; waterage
水运运价 freight rate of water transport
水运运量 freight volume of water transport; water-borne traffic
水运运输组织 organism of water transport
水运政策 water-borne policy
水运中的货物 cargo afloat
水运终点站 marine terminal
水杂淀积 basic sediment and water; bottom settlings; sediment and water
水灾 flood catastrophe; flood(damage); flood disaster; flood havoc; inundation
水灾保险 flood insurance; inundation insurance
水灾贷款 flood relief loan
水灾风险 flood hazard
水栽法 hydroponics
水栽培 hydrocultivation; hydroponics; water culture
水栽植物 hydroponic plant
水栽植物培养 hydroponic plant growth
水载 water load(ing)
水载废物 water-borne waste
水载泥土的填土法 hydraulic fill (soil)
水载污染物 water-borne contaminant
水载物 water-borne body
水再生 water refreshing; water renovation; water reuse; water treatment
水再循环 water recycle
水再循环处理系统 water treatment system by recirculation
水再用 water reuse
水再用系统 water reuse system
水蚤 daphnia; water flea
水藻 alga[复 algae]
水藻混浊度 algae turbidity
水渣 grain slag
水渣车 slag buggy
水闸 aboideau; dike drainage lock; flood gate; gole; lock; mill dam; penstock; sasse; shuttle; sluice; sluice dam; sluice gate; tide lock; water lock; water shutter
水闸坝 penstock dam
水闸坝坑道 penstock dam gallery
水闸坝瞭望塔 penstock dam gallery
水闸坝歧管 penstock manifold
水闸坝支管 penstock manifold
水闸板 water board
水闸操纵机构 gate operating gear; sluice control mechanism
水闸操纵机械 gate operating gear; sluice control mechanism
水闸操作机构 gate operating gear; sluice control mechanism
水闸操作机械 gate operating gear; sluice control mechanism
水闸阀 sluice valve
水闸高差 lockage
水闸高低度 lockage
水闸管道 penstock pipe
水闸管理人 lock-keeper
水闸管理员 lock keeper; locksman
水闸和发电站 station and lock unit
水闸滑动泄水门 sliding sluice gate
水闸滑动闸门 sliding sluice gate
水闸看守人 lock tender
水闸孔 sluice opening
水闸门 clough; gate closure; sluice door; stop gate; water shutter
水闸门槛 lock sill
水闸前后高低度 lockage
水闸上部门扉 upper leaf
水闸上游门扉 upper leaf
水闸室泄水 unwatering of lock chamber
水闸水 water seal
水闸通行税 lockage
水闸箱 chamber box
水闸泄水道 lock sluiceway
水闸泄水涵洞 lock culvert
水闸蓄水池 scouring basin; sluicing pond
水闸用材料 lockage
水闸闸槽 gate recess
水闸闸门 sluice gate; water gate; sluice paddle
水闸闸头 sluice head
水栅 boom set
水沾污白斑 water spotting
水站 water station; water supply station
水张力 water tension
水张力膜 skin of water
水涨落速度装置 water leg
水胀 water swelling
水胀氟云母 water-swelling fluoromica
水账 water budget
水沼地 aqua marsh
水沼土 aquamarsh soil
水照云光 blink; water sky
水锗钙矾 schaurteite
水锗铅矾 fleischerite
水针 <凿岩机的> water needle; flushing tube; water tube
水针硅钙石 mountainite
水针铁矿岩 hydrogoethite rock
水侦查 water detection
水镇重路碾 water ballast roller
水镇重压路机 <加水调节重量的压路机> water ballast roller
水震 <水下爆破引起的> water shock
水蒸发器 water vapo(u)rizer
水蒸馏驳 water distilling barge
水蒸馏器 water distilling apparatus
水蒸气 aqueous vapo(u)r; reek; steam(vapo(u)r); vapo(u)r; water vapo(u)r; smother
水蒸气保护电弧焊 water vapo(u)r arc welding; steam shielded arc welding
水蒸气保护焊(接) water vapo(u)r welding
水蒸气爆破筒 hydrox; hydrox blaster; hydrox cartridge; hydrox cylinder
水蒸气不渗透性 water vapo(u)r impermeability; water vapo(u)r imperviousness
水蒸气不渗透性能 water vapo(u)r resistance
水蒸气测定法 atmometry
水蒸气抽出泵 vapo(u)r extractor pump
水蒸气处理沥青 steam asphalt
水蒸气传递 water vapo(u)r transmission
水蒸气传递速率 water vapo(u)r transmission rate
水蒸气传递系数 water vapo(u)r transfer coefficient
水蒸气窗口 window of water vapo(u)r

S

水蒸气的入流 inflow of water vapo(u)r
水蒸气等离子体 water vapo(u)r plasma
水蒸气分压力 partial pressure of water vapo(u)r; partial vapor pressure
水蒸气隔离层 water vapo(u)r barrier
水蒸气管 water vapo(u)r pipe
水蒸气含量 water vapo(u)r content
水蒸气激光器 water vapo(u)r laser
水蒸气扩散 water vapo(u)r diffusance; water vapo(u)r diffusion; diffusion of water vapo(u)r
水蒸气量 water vapo(u)r quantity
水蒸气密度 vapo(u)r concentration; water vapo(u)r density
水蒸气凝结 water vapo(u)r condensation; water recovery
水蒸气浓度 water vapo(u)r concentration
水蒸气喷射泵 hydrosteam ejector pump
水蒸气迁移 water vapo(u)r migration
水蒸气清除装置 steam cleaning unit
水蒸气渗透 vapo(u)r penetration; water vapo(u)r transmission; water vapo(u)r permeance
水蒸气渗透率 water vapo(u)r permeability; water vapo(u)r transmission rate
水蒸气渗透系数 water vapo(u)r penetration coefficient
水蒸气渗透性 water vapo(u)r permeability
水蒸气试验 water vapo(u)r test
水蒸气释放 water vapo(u)r release
水蒸气提浓法 water vapo(u)r concentration
水蒸气透过性 water vapo(u)r permeability; water vapo(u)r transmission
水蒸气透湿性 water vapo(u)r transmission
水蒸气图表 steam chart
水蒸气雾 water vapo(u)r mist
水蒸气吸收 water vapo(u)r absorption
水蒸气循环 vapo(u)r cycle
水蒸气压 water vapo(u)r pressure
水蒸气压力降落 water vapo(u)r pressure drop
水蒸气压力图 water vapo(u)r pressure graph
水蒸气蒸馏 wet distillation
水蒸气蒸馏过程 wet distillation process
水蒸气阻塞 water vapo(u)r barring
水蒸气阻滞层 water vapo(u)r retarder
水蒸气学 atmology
水政 water administration
水政策 water policy
水政策评价 assessment of water policy
水政策审查 water policy review
水政监察 water administration supervision
水值 water number; water value
水指示现象 hydroindication
水质 quality of water; water quality
水质安全 water quality security
水质保持 water quality conservation
水质保护 water quality conservation; water quality protection
水质保险 water quality insurance
水质保险企业联合组织 Water Quality Insurance Syndicate
水质本底 water quality background
水质变动 fluctuation of water quality; water quality change; water quality fluctuation; water quality variation
水质变化 fluctuation of water quality; variation of water quality; water change; water quality change; water quality fluctuation; water quality variation
水质变混 water to be turbid
水质变量 water quality variable
水质辨识 water mass identification
水质标识指数 water quality identification
水质标识指数法 water quality identification index
水质标准 quality standard for water; water(quality) criterion[复 criteria]; water(quality) standard
水质标准的类型 type of water quality standard
水质标准框架 water quality standards framework
水质波动 fluctuation of water quality
水质采样 water quality sampling
水质采样网 water quality sampling network
水质参数 water quality parameter
水质参数区别 regionalization of water quality parameters
水质参数组 water quality parameter group
水质测点 water quality measuring point
水质测定 water quality determination
水质测定试剂序列 time series of water quality determination
水质测定系统 water quality measuring system
水质测定仪 water quality ga(u)ge; water quality instrument
水质测量 water quality measurement
水质测试试剂 water quality testing agent
水质测站 water quality measurement station; water quality measuring point; water quality measuring station
水质场 water quality field
水质超标区面积 area of water quality over standard
水质成分 water quality constituent
水质除臭 removal of odo(u)r from water
水质除铁 removal of iron from water
水质处理 water conditioning; water quality processing
水质处理区 water quality treatment district
水质传感器 water quality sensor
水质脆弱带 water quality vulnerability zone
水质脆弱带图 water quality vulnerability zone map
水质脆弱带总图 general water quality vulnerability zone map
水质的理化检验 physical and chemical examination of water quality
水质等级 water quality grade
水质点 water particle
水质点轨迹 particle path; path line
水质点速度 particle velocity
水质点运动 mass transport; water particle motion; water particle movement; motive of water
水质点运动速度 mass transport velocity
水质调查 investigation of water quality; water quality investigation; water quality research
水质调查计划 water quality research program(me)
水质动态 dynamics of water quality; regime(n) of water quality
水质恶化 degradation of water quality; deterioration of water quality; water quality decline; water quality degradation; water quality deterioration
水质法案 water quality act
水质法令 water quality act
水质分层 water quality stratification
水质分级标准 water grading standard
水质分类法 system of water quality
水质分析 analysis of water; examination of water quality; water analysis; water quality analysis
水质分析法 method for analysis of water quality; water quality analysis method
水质分析模拟程序 water quality analysis simulation program(me)
水质分析系统 water quality analysis system
水质分析项目 analysis item of water quality
水质分析仪(器) instrument for quality analysis of water; water quality instrument
水质分析指标 water quality analysis index
水质风险评价 water quality risk judgment
水质改进 water renovation
水质改善 improvement of water quality; water quality correction; water quality improvement; water renovation
水质改善法案 Water Quality Improvement Act
水质改善法令 Water Quality Improvement Act
水质改善条例 Water Quality Improvement Act
水质改正 water correction
水质钙硬度 calcium hardness of water quality
水质更新 water renovation
水质工程 water quality engineering
水质工程局 Water Quality Engineering Division
水质观测 water quality observation
水质管理 water quality management
水质管理规划 water quality management planning
水质管理及保护 water quality management and protection
水质管理局 Department of Water Quality; Water Quality Administration
水质管理模型 model of water quality management
水质管理目标 water quality management goal
水质管理设施 water quality management facility
水质管理中心 water quality management center[centre]
水质规划 water quality planning; water quality program(me); water quality programming
水质规划模型 water quality planning model
水质过程线 water quality hydrograph
水质化学 hydrochemistry
水质化学参数 chemical water quality parameter
水质化学成分 water quality chemicals
水质化学模型 water quality chemical model
水质化学需氧量 water quality chemical oxygen demand
水质化学指标 chemical water quality index
水质化验检出值 pick-up
水质环境变化 change of water quality environment
水质基准 water quality criterion
水质计 water quality analyser[analyzer]
水质监测 water(quality) monitoring; water quality surveillance
水质监测车 water quality monitoring van
水质监测船 water quality monitoring ship; water quality ship
水质监测计划 water quality monitoring planning; water quality monitoring program(me)
水质监测器 water monitor; water quality monitor
水质监测网 water quality monitoring network
水质监测系统 water measuring and monitoring system; water quality monitoring system
水质监测系统设计 design of water quality monitoring system
水质监测仪 water monitor; water monitoring equipment; water quality monitor
水质监测站 water quality monitor; water quality monitoring station
水质监视 water quality monitoring; water quality surveillance
水质监视网 water quality surveillance network
水质检测 examination of water quality
水质检查 examination of water; water examination; water quality examination
水质检验 examination of water; examination of water quality; water analysis; water examination; water quality examination; water test
水质简分析 simple water quality analysis
水质健康风险评价 water quality health risk assessment
水质鉴定 water quality appraisal
水质净化设备 water conditioning device
水质可修复性 water quality recoverability
水质控制 water quality control
水质控制断面 water quality control cross-section
水质控制河段 water quality control (river) reach; water quality control stream reach
水质控制区 water quality control area
水质快速检测仪 rapid water quality monitor
水质矿化 water quality mineralization
水质类别 water quality class
水质类型 type of water quality
水质理化参数 water quality physicochemical parameter
水质立法 water quality legislation
水质量 water quality
水质量标准 water quality standard
水质劣化 water quality deterioration
水质弥散试验 dispersion test of water quality
水质模拟 model of water quality; water quality analogy
水质模拟模型 water quality simulation model
水质模拟预测 water quality simulation and prediction

S

水质模式 water quality mode
水质模数 water quality index
水质模型 water quality model
水质模型标定和验证 calibration and verification of water quality model
水质模型参数 parameter of water quality model
水质模型参数估值 parameter estimation of water quality model
水质模型单参数估值 single parameter estimation of water quality model
水质模型多参数估值 multiparameter estimation of water quality model
水质模型多参数梯度搜索估值 multiparameter estimation of water quality model by gradient searching
水质模型多参数网络搜索法估值 multiparameter estimation of water quality model by network searching
水质模型基本方程 basic equation of water quality model
水质模型检证 verification of water quality model
水质模型校正 water quality model calibration
水质模型灵敏度分析 sensitivity analysis of water quality model
水质模型评价 water quality model evaluation
水质目标 goal of water quality; water quality goal; water quality objective; water quality target
水质目标管理 water quality target management
水质浓度 water quality concentration
水质判据 water quality criterion[复 criteria]
水质平衡 water quality balance
水质评定 water assessment; water quality assessment
水质评定标准 quality criterion for water
水质评价 assessment of water quality; water evaluation; water quality assessment; water quality evaluation
水质评价标准 water quality assessment standard
水质评价参数 parameters of water quality assessment
水质评价模型 water quality assessment model
水质评价指标 water quality assessment index
水质区划 water quality regionalization
水质全分析 complete examination of water quality; complete water quality analysis
水质容限 water quality limit; water quality tolerance
水质软化 water demineralization; water softening
水质软化剂 water softener
水质软化器 water softener
水质软化设备 water softening equipment
水质色度 water quality colo(u)rity
水质生物参数 biologic(al) parameters of water quality
水质生物测定 bioassay of water quality
水质生物检验 biologic(al) analysis of water quality
水质事件 water quality event
水质试验 quality test of water; water examination; water test
水质试验装置 water test unit
水质试样 water sample; water specimen
水质输送模数 water quality transport module
水质数据 water quality data
水质数学模型 mathematic(al) water quality model; water quality mathematic(al) model
水质水量综合管理 integrated management of water quality and quantity
水质水文学 water quality hydrology
水质水源 sources of water quality
水质酸化 water quality acidization
水质特征分析 water quality characteristic analysis
水质条例 water quality act
水质调节 water quality regulation
水质调理 water conditioning; water quality conditioning
水质调整 water conditioning
水质图 water quality chart; water quality map
水质委员会 Water Quality Commission
水质卫生 water hygiene
水质卫生标准 water quality sanitary standard
水质卫生控制 water hygiene control
水质卫生指标 water quality sanitary index
水质稳定化 stabilization of water; water stabilization
水质稳定技术 stabilization technology of water quality; technology of water quality stabilization; water quality stabilization technology
水质稳定剂 water quality stabilizer
水质污染 aquatic pollution; water pollution; water quality pollution
水质污染保护管理 water pollution control; water quality pollution control
水质污染标志 water quality pollution index
水质污染处理厂 water pollution control plant
水质污染防止 water pollution prevention
水质污染防治 water pollution control; water quality pollution control
水质污染管理法(规) water quality pollution control law
水质污染管理立法 water quality pollution control legislation
水质污染管制法(规) water quality pollution control law
水质污染管制立法 water quality pollution control legislation
水质污染监测 water quality pollution monitoring
水质污染监测仪 water pollution monitor; water quality pollution monitor
水质污染控制 water pollution control; water quality pollution control
水质污染物 water quality pollutant
水质污染研究实验室 water quality pollution research laboratory
水质污染源 source of water pollution; water pollution sources; water quality pollution source
水质污染征兆 pollution sign of water quality
水质污染指数 water quality pollution index
水质无污染 zero water pollution
水质物质 water quality substance
水质细菌学检验 bacteriological examination of water quality
水质限制 water quality limit
水质响应曲线 water quality response curve
水质消毒 water disinfection
水质协会 water quality association
水质修复 restoration of water quality; water quality restoration
水质修复工艺 water renovation process
水质样本 water quality sample
水质要求 water quality requirement
水质要素 element of water quality; water quality element
水质因子 water quality factor
水质应力 water quality stress
水质硬度 water hardness
水质硬化 water quality hardening
水质优度 water quality level; water quality volume
水质油彩或水粉画颜料 autocrat
水质预报 water quality forecast; water quality prediction
水质预测 water quality prediction
水质站网 water quality network
水质诊断 water quality diagnosis
水质值 values of water quality; water quality value
水质指标 water quality guideline; water quality index
水质指示剂 water quality indicator
水质指示器 constituents quality index; water quality indicator
水质指数 index of water quality; water quality index
水质专项分析 special chemical analysis of water quality
水质状况 water quality condition; water quality situation
水质状态 water quality state
水质准则 water quality criterion
水质资料 water quality data
水质自动监测器 automatic water-quality monitor
水质自动监测系统 automatic water-quality monitoring system
水质自净试验 test of water self-purification
水质综合管理 comprehensive water quality management
水质综合管理规划 comprehensive water quality management planning
水质综合评价 comprehensive water quality assessment; water quality synthetic evaluation
水质综合污染指数 comprehensive pollution index of water quality
水质组成 water quality constituent
水质组成物质 water quality constituent mass
水质最低标准 minimum water quality criterion[复 criteria]
水质最优化 water quality optimization
水致发白 <漆病> water blush(ing)
水致流行病 water-borne epidemic
水致侵蚀作用 delignification
水蛭 hirudo; leech
水蛭病 hirudiniasis
水蛭皮炎 leech dermatitis
水蛭石 jefferisite
水蛭素 hirudin
水置换法 hydrosubstitution
水置换法取样 sampling by water displacement
水置换剂 <用于防锈剂中> water displacement agent
水置换试验 water displacement test
水中暗流 underwater current
水中爆炸 water shooting
水中哺乳动物 aquatic mammal
水中草原 aquiprata
水中沉桩 pile sinking in water
水中称重 immersion weighing
水中称重法 <测土密度> weight in water method; weigh process in water
水中冲击波 underwater blast wave
水中氚含量 tritium[H_3] content in water
水中淬火 quenching-in water
水中淬火的 quenched in water
水中带汽 carry-under
水中导架 <打桩的> underwater guide
水中倒土坝 water dumping dam
水中的 aquatic; submerged; under water
水中的浮碎屑 tripton
水中的溶解氧和氯离子 dissolved oxygen and chlorine ion
水中地震检波器 hydrophone
水中定位 hydrolocation
水中定位测时计 hydrolocation chronoscope
水中定位仪 hydrolocator
水中毒 water intoxication; water poisoning
水中舵 float rudder
水中方块试验法 <一种沥青材料软化点试验法> cube-in-water method
水中放电成型法 electrohydraulic forming; hydrospark forming process
水中放电法 electrohydraulics
水中放射性 water-borne radioactivity
水中服 wet suit
水中感声器 hydrophone
水中工程基础的整修 submerged grading of foundation
水中固定标志 fixed mark on water
水中固化 curing in water
水中含土砂率 sediment concentration
水中焊接 underwater welding; welding in water
水中呼吸 aquatic respiration
水中呼吸器 <潜水用> aqualung
水中混凝土 subaqueous concrete
水中积砂器 sand catcher
水中基础工程 submerged foundation
水中基础工程维修 submerged grading of foundation
水中急冷 water quench(ing)
水中集砂器 sand catcher; sand grain meter
水中胶 water-borne glue
水中结构 submerged construction
水中径迹测量 track survey in water
水中开关 water circuit breaker
水中扩散 hydrodiffusion
水中离解 dissociation in water
水中篱笆 water fence
水中滤网 underwater screen
水中密度 immersed density
水中泥沙单位 spatial concentration of sediment
水中泥沙单位水流体积含沙量 spatial concentration of sediment
水中泥沙集中区 spatial concentration of sediment
水中农药污染 aquatic pesticide contamination
水中抛石 pierre perdue
水中平衡器 hydrostatic organ
水中平均波速 average water velocity
水中切割 underwater cut(ting)
水中溶解的氧量 dissolved oxygen in water
水中溶解的氧气 oxygenation in water
水中溶解度 solubility in water
水中溶解性 solubility in water
水中扫描法 immersion scanning; immersion scan(ning method)

水中扫描术 immersion scan(ning) technique;immersion technique
水中射气浓度值 emanation concentration value in water
水中砷的极限 limit for arsenic in water
水中生境 undersea habitat
水中生物的物理及化学环境 hydroclimate
水中生物噪声 submarine biologic(al) noise
水中声 underwater sound
水中声波 underwater acoustic(al) wave
水中手电筒 marine torch
水中熟化 curing in water
水中水生植物 submerged water plant
水中苔藓 aquatic moss
水中探测 hydrospace detection
水中探礁器 submarine sentry
水中探音器 detectoscope
水中特性 <水陆两用车的> water performance
水中填土坝 dam by damping soils into ponded water
水中调整 aquatic adjustment
水中听音器 hydrophone
水中推移质 hydraulic traction
水中望远镜 hydroscope
水中微生物的浓度 microbiologic(al) concentration in water
水中污染物 water contaminant
水中雾号 submarine fog bell;submarine fog signal
水中雾钟 submarine bell;underwater bell
水中卸泥 dumping in water
水中信号 submarine signal
水中悬垂生物 periphyton
水中悬浮冰针 frazil ice
水中悬移质 hydraulic suspension
水中养护 storage under water;water storage
水中养护的 immersion-cured
水中养护试件 water storage of specimen
水中氧溶解度 solubility of oxygen in water
水中氧气消耗 oxygen depletion
水中幺重 submerged unit weight
水中隐树 snag
水中硬化 hydraulic set(ting)
水中用电机 submersible machine
水中油 oil in water
水中油扩散 oil-in-water dispersion
水中油型 oil-in-water type
水中油型乳液 oil-in-water emulsion
水中有机碳 dissolved organic carbon
水中浴解氧 dissolved oxygen
水中元素存在形式 existence mode of dissolved element
水中元素的同位素 dissolved element isotopic
水中运动场 hydrogymnasium
水中杂质 impurities in water
水中栽培 aquaculture
水中栽植 water planting
水中噪声测向仪 hydrophone
水中照明 underwater illumination
水中照明度 underwater illumination intensity
水中照相机 fish-eye camera
水中振荡器 submarine oscillator
水中最大容许浓度 maximum permissible concentration in water
水中最大容许未鉴别出的放射性核素浓度 maximum permissible concentration of unidentified radionuclides in water
水钟 water clock
水肿 edema <美>;oedema
水踵泵 hydraulic ram pump
水重力孤波 solitary water gravity wave
水重磷镁石 phosphorroessierite
水重砷镁石 roesslerite
水珠 bead;globule
水珠灰玻璃 water-pearl-ash glass
水煮回收 water cooked reclaim;water digestion reclaim
水注 jet of water;water-jet
水注入 water influx
水注入速度 rate of water injection;water influx rate
水注射 water-jet
水柱 water column;column of liquid;column of water;gargoyle;Shaff;spout;water crane;water spout
水柱避雷器 water column arrester;water-jet arrester
水柱参考点 water-post reference point
水柱高 height of water
水柱高度 head of liquid;head of water;height of water;water head
水柱式恒温器 hydrothermostat
水柱式水泵 water-jet pump
水柱水位 static level
水柱需氧量 water column oxygen demand
水柱压测量仪 head meter
水柱压力 water column pressure
水柱压力计 water draught ga(u)ge;water manometer;water pressure ga(u)ge
水转分选机 hydrotator
水转式分级机 hydrotator classifier
水转式分选机 hydrotator classifier
水转向 recurve
水状的 hydrous
水锥 <流向指示器> fair water cone;vertical coning;water cone
水锥浸入油井 coning the well
水坠坝 sluicing-siltation dam;slurry-fall fill dam
水坠法 sluicing-siltation method
水准闭合差 error of closure in level(l)ing
水准闭合环 circuit;closed level circle
水准标尺 grade rod;level(ling) pole;level(ling) rod;level(ling) staff;staff ga(u)ge;surveying rod;staff
水准标尺常数 constant of level(l)ing staff
水准标尺垫 staff plate
水准标尺读数 rod note
水准标尺配对 pairing the rods
水准标尺前后视读数和 red sum
水准标尺误差 level(l)ing rod error
水准标点 bench mark
水准标点图说(侧) description of benchmark
水准标杆 stadia rod
水准标石 level(l)ing pillar
水准标石类型 type of benchmark
水准标志 level(ling) marker
水准标桩 foot pin for level(l)ing rod;level(l)ing peg;survey plug
水准补标点 supplementary bench mark
水准补点 supplementary bench mark
水准测段 segment of level(l)ing
水准测杆 level rod
水准测管 level tube
水准测量 differential level(l)ing;field level(l)ing;grading;level(l)ing;level(l)ing process;level(ling) survey(ing);run a level
水准测量闭合差 error of closure level circuit
水准测量标尺 graduated measuring rod
水准测量标尺垫板 leveling plate
水准测量不符值 level(l)ing discrepancy
水准测量尺垫 benching iron
水准测量等级 elevation order
水准测量队 level(ling) party
水准测量附加装置 level(l)ing attachment
水准测量附件 grading attachment
水准测量复测 rerunning o f level(l)ing
水准测量高差 level(l)ing increment;spirit-level(l)ed elevation difference
水准测量高差增量 level(l)ing increment
水准测量工作 level(l)ing operation;level(l)ing work
水准测量基准面 leveling datum
水准测量记录簿 level book
水准测量控制 level control;spirit level control
水准测量平差 level(l)ing adjustment
水准测量区段 link of levels
水准测量人员 level-man
水准测量手簿 level book;level(ling) note
水准测量误差 level(l)ing error
水准测量仪器 level(l)ing instrument
水准测量员 level(l)er
水准测量折光差 refraction in level(l)ing
水准测量整平 level out
水准测量中与主线垂直方向的高程 cross-level
水准测量桩 level(l)ing peg
水准测量作业 level(l)ing practice
水准差 denivellation
水准常数 level constant
水准程序 benchmark routine
水准尺 air-bubble level;bubble level;level bar;level(ing) rod;level(ing) staff;level(l)ing rule;sight(ing) rod;surveyor's rod;pogo stick <带弹簧的>
水准尺尺垫 rod support;staff plate
水准尺垫 turning plate
水准尺读数 rod note;staff reading
水准尺端 level shoe;pointed shoe
水准尺端(金属)底板 level shoe
水准尺上的觇板 rod target
水准尺上的水准器 rod level
水准尺桩 spike
水准尺座 foot plate
水准带尺 <水准标尺上的因瓦带> level(l)ing tape
水准地形测量 topographic(al) level(l)ing
水准点 benchmark;datum mark(of levelling);level(l)ing point;point of reference;reference point;spirit-level(l)ed benchmark;spot level
水准点标高 benchmark elevation;benchmark value
水准点成果表 benchmark list
水准点等级 class of benchmark
水准点地面 benchmark soil
水准点高差及高程表 height difference of benchmarks and height list
水准点高程 bench level
水准点高程测量 <基平> benchmark level(l)ing
水准点高程成果表 benchmark list
水准点水平面 bench level
水准点说明 benchmark description
水准点一览表 benchmark list
水准点之记 benchmark description
水准复照仪 horizontal camera
水准改正 bubble correction
水准干线 primary level line;principal level line
水准杆 level(l)ing rod;level pole
水准管 bubble tube;tubular level;level tube
水准管反射镜 bubble reflector
水准管检定器 level tester;level testing instrument;level trier
水准管框 level tube housing
水准管目镜 eyepiece of level tube
水准管气泡 level tube bubble
水准管轴 axis of level tube;bubble (tube) axis;level tube axis
水准环 level(ling) loop
水准环闭合差 closure error of level circuit
水准回线 level(l)ing loop
水准基点 basic benchmark;control point;datum benchmark;level(l)ing base;original bench mark;primary benchmark;bench mark
水准基面 level(l)ing base
水准基线 level(l)ing base
水准基座 level(l)ing pad
水准记录簿 level book;level notes
水准架 level(l)ing support
水准监测点 monitoring point for level
水准检验仪 level testing instrument
水准结点 junction point of level(l)ing line
水准结点图 junction detail of level(l)ing
水准经纬仪 level theodolite;level transit;transit level
水准卷尺 level(l)ing tape
水准控制 level control
水准联络点 connecting point for level(l)ing
水准灵敏度 level sensitivity
水准零点 datum water level;height datum
水准路线 level(ling) line;line of level(l)ing
水准路线的权 weight of leveling line
水准路线名称 name of leveling line
水准路线数 number of leveling course
水准路线图 level(l)ing route map;sketch of level(l)ing lines
水准螺丝 level(l)ing screw
水准螺旋 level(l)ing screw
水准面 datum water level;level face;level(l)ing base;level plane;level surface;water mark;zero surface;water-level
水准面标记 level mark
水准面差 difference of level surface
水准面点 level mark
水准面法 level-surface method
水准面平均曲率 mean curvature of level surface
水准面起伏 undulation of the geoid
水准泡 <测量仪器的> vial
水准瓶 level(l)ing bottle
水准气泡 bubble of level;level(ling) bubble;level vial
水准气泡对中 centering[centring] of level(ing) bubble
水准气泡灵敏度 level sensitivity;sensitivity of levelling bubble
水准气泡偏差 deviation of level bubble
水准气泡室 level chamber
水准气泡轴线 axis of level(l)ing bubble
水准器 balance level;bubble;bubble cell;bubble level;level bubble;level device;level ga(u)ge;level(l)ing instrument;level tube;level vial;main level;spirit level;water-level
水准器玻璃管 level vial;vial
水准器东 level east
水准器分划值 scale value of level
水准器划分 level graduation
水准器检定器 level trier

S

水准器检验器 level trier
水准器检验仪 bubble tester; level tester
水准器角值 scale value of level
水准器校正 bubble correction
水准器灵敏度 level sensibility; level sensitivity; sensibility of level
水准器偏差 spirit level wind
水准器气泡偏差 deviation of level bubble
水准器西 level west
水准器轴线 axis of level(l)ing bubble; axis of level; level tube axis
水准球管 level(l)ing bulb
水准区段 section of level(l)ing
水准圈 <道路排水口的> level(l)ing ring
水准容器 level(l)ing vessel
水准塞 level plug
水准式测角仪 level type angle ga(u)ge
水准手簿 level book; level notes
水准手册 level book
水准调节器 level regulator; water-level regulator
水准椭球 level ellipsoid
水准网 level(ling) net(work)
水准网平差 adjustment of level(l)ing network
水准稳定 level equalization
水准误差 level error; disorder of cross-level <轨道的>
水准线 datum level; datum line; level-(ling) line; line of levels
水准线路平差 adjustment of level(l)ing circuits
水准旋转椭球 level ellipsoid of revolution
水准仪 level(bubble); air level; balance level; engineer's level; gradienter; level ga(u)ge; level indicator; level instrument; level(l)ing bubble; sight level; spirit level; spirit-level (l) ing instrument; survey-(ing) level; surveyor's level; telescope level; plumb level <有铅垂线的>
水准仪常数 level constant
水准仪高 height of level
水准仪格值 bubble scale value
水准仪管 level tube
水准仪检验校正 test and adjustment of level
水准仪校正 level correction
水准仪灵敏度测量仪 level trier
水准仪气泡 bubble of level; level vial
水准仪气泡偏差 deviation of level bubble
水准仪器 level(l)ing instrument
水准仪器零点误差 zero error of level
水准仪三脚架 level(l)ing tripod
水准仪水泡管 bubbler tube
水准仪调节器 level compensator
水准仪调整 regulation of level
水准仪微倾螺丝 tilting level screw
水准原点 basic level(l)ing origin; datum mark (of levelling); level (l) ing origin; original bench mark; standard datum of leveling
水准照尺 level(l)ing rod
水准折算 reduction of levels
水准支线 spur; spur line of level(l)ing; spur line of levels
水准指示器 stage indicator; water-level indicator
水准指针 level ga(u)ge; level indicator
水准轴 bubble axis; level axis; axis of level tube
水准桩 level(l)ing peg
水准资料 level(l)ing data; level(l)ing information
水准组 level party; level team
水资料单位 water data unit
水资料机构 water data unit
水资源 water mines; water resources
水资源保护 conservation of water resources; conservation of waters; protection of water resources; protection of waters; water resources conservation; water resources protection
水资源保护的法律措施 lawful measurement of protecting water resources
水资源保护区 water resources conservation zone
水资源不足 deficiency of water
水资源承载力 water resource carrying capacity
水资源大会 <美> Water Resources Congress
水资源的展望 perspectives on water
水资源调查报告 water resource investigation report
水资源发展中心 water resources development center[centre]
水资源法 water resources act
水资源法的名称 name of water resources law
水资源法令 Water Resource Act
水资源费 water resources charges
水资源分类 classification of water resources
水资源分配 allocation of water resources
水资源供需平衡 water supply and demand balance
水资源管理 management of water resources; water conservancy; water management; water resources stewardship
水资源管理部门 water management institution
水资源管理国际培训中心 International Training Centre for Water Resource Control
水资源管理机构 water management institution
水资源管理体系 management system of water resources
水资源管理体制 management system of water resources
水资源管理者 water resources guardian
水资源规划 water resources planning; watershed planning
水资源规划法令 Water Resources Planning Act
水资源规划数据 data of water resources planning
水资源规划与管理 water resources planning and management
水资源规划与管理学报 <美国土木工程学会季刊> Journal of Water Resources Planning and Management
水资源和水土保持 water resources and water and soil conservation
水资源计算方法 method of water resources calculation
水资源计算模型 model of water resources calculation
水资源进展 <英国季刊> Advances in Water Resources
水资源开发 development of water resources; water development
水资源开发方案 program(me) of water resources development
水资源开发工程投资 investment of water resources developing engineering
水资源开发培训中心 <印度> Water Resources Development and Training Center
水资源开发区 area of water development
水资源科学情报中心 <美> Water Resources Scientific Information Center(USA)
水资源利用 utilization of water resources
水资源利用程度图 map of utilization degree of water resources
水资源利用方针 utilization policy of water resources; water resources policy
水资源配置 allocation of water resources
水资源评估 water resources assessment; water resources evaluation
水资源评价 assessment of water resources; water resources assessment; water resources evaluation
水资源评价法 water resources assessment methodology
水资源所有权 water resources ownership
水资源通报 <美国双月刊> Water Resources Bulletin
水资源投入产出表 water resources input-output table
水资源完全消耗系数 complete consumption coefficient of water resources
水资源委员会 Committee on Water Resources; Water Resources Council
水资源文献选摘 <美国月刊> Selected Water Resources Abstracts
水资源问题 water resources problem
水资源污染 water resource pollution
水资源系统 water resources system
水资源系统分析 analysis of water resources system; system analysis of water resources; systematical analysis of water resources
水资源协调委员会 Water Resources Coordinating Committee
水资源学报 <联合国 ESCAP 季刊> Water Resources Journal
水资源研究 water resources study; Water Resources Research <美国双月刊>
水资源遥感 remote-sensing of water resources
水资源一体化管理 integrated water resource management
水资源政策 water resources policy
水资源直接消耗系数 direct consumption coefficient of water resources
水资源质量 water resource quality
水资源综合管理 integrated water resource management
水资源综合利用 comprehensive utilization of water resource; multipurpose utilization of water resources
水紫树 cotton gum; tupelo gum; water tupelo; water tupelo gum
水字交汇 sheaf type intersection
水自动比例采样器 automatic proportional water sampler
水渍 water damage(d); water mark; water-stained
水渍保险单 water damage insurance policy
水渍地 water-logged area
水渍过程 hydromorphous process
水渍货 water damaged cargo
水渍器皿 morted ware
水渍试验 water spotting
水渍险 with particular average
水渍现象 <混凝土的> water soaking
水总管 water header
水租权 emphytheutic lease
水族馆 aguaria; aquarium; aquatorium
水族馆消毒 aquarium disinfection
水阻力 hydraulic drag; water resistance
水阻力平衡 hydraulic resistance balance
水阻力中心 center[centre] of water drag; center[centre] of water resistance
水阻器 water resistor
水阻塞的 water-clogged
水阻试验装置 <牵引动力装置的> liquid resistor rating plant; liquid resistor testing plant
水嘴 water faucet
水作燃料 aqueous fuel; water as fuel

税单 duty memo; grand list; tax roll

税道 <美> turnpike road
税道关卡看守人 pike man
税额 amount of tax to be paid; assessment
税法 tariff law; tax law; taxation regulation
税费 expense of taxation; taxes and duties
税费不保 free from duty
税负 burden of taxation; tax bearing; tax burden
税负率 tax bearing ratio; tax burden ratio
税负能力 tax bearing capacity
税后 net of tax
税后工资 take-home pay
税后净利(润) net profit after tax
税后净收入 net of tax income
税后可支配收入 disposable income
税后利润 after-tax profit
税后收入 after-tax income; disposable income
税后收益 after-tax yield
税后收益额 earnings after tax
税后所得 income after taxes
税后现金流量 after-tax cash flow
税级 tax bracket
税结构 tax structure
税金 duties; tax
税金在内的价格 price including tax
税捐 duties; tax
税捐减免 tax relief
税款 imposition; taxation; tax money; tax payment
税款的征收 collection of tax
税款付讫 assessment paid; duty paid
税款计算 tax computation
税款减免 tax relief
税款清算 settlement of tax
税款调节 tax timing
税款延期缴纳 postponement of tax
税款预算 taxes estimating
税款滞纳罚金 penalty tax
税款专用 earmarking of taxes
税率 assessment percentage; duty rate; rate of duty; rate of taxation; rate of taxes; rating; tariff; tariff rate; tax rate
税率表 scale of tax; table of rates; tariff schedule; tax rate schedule; tax schedule; tax table
税率的余数 complement of tax rate
税率等级 grades of tax rates; tax bracket
税率递减 degression
税率高的 high duty

税率结构 tax rate structure
税率限度 tax rate limit
税率优待反污染投资 tax allowances for anti-pollution investments
税目 category of taxes; heading; item of tax; taxable item; taxation categories; tax item; tax roll
税契 tax title
税前的 pretax
税前价格 price before tax
税前利润 <未经扣除所得税> before-tax profit; pretax profit
税前收入 before-tax income; net before taxes; pretax income
税前收益 before-tax income
税前收益额 earnings before tax
税前所得 income before tax
税前现金除以投资股本 cash-on-cash
税前现金流动 cash throw-off
税前现金流量 cash flow before tax
税前现金流量财产价值 before-tax cash flow to equity
税前账面收益 pretax accounting income
税收 dues; revenue; taxation; tax levy; tax revenue; duty
税收裁决 adjudication of tax
税收参考价格 tax reference price
税收差别 tax differentiation
税收挡避 tax shield
税收的可接受性 acceptability
税收等级 tax bracket
税收抵免法 tax credit method
税收分配 tax apportionment
税收负担 tax burden
税收共付票据 tax anticipation bills
税收管理 tax administration
税收合同 revenue bond
税收机构 revenue office
税收间接抵免 indirect tax credit
税收减免 reduction and exemption of tax
税收减免的纳税人 persons eligible for tax exemption
税收减让 <指对投资的鼓励> tax concession; tax incentive
税收金额 tax take
税收漏洞 tax loophole
税收率 tax burden ratio
税收审计 tax audit
税收收入 tax receipts; tax revenue
税收特惠 tax privilege
税收条例 tax regulation
税收调整 tax adjustment
税收限度 tax limitation
税收削减 tax cutback
税收印花 revenue stamp
税收优待 tax allowance
税收优惠 tax concession
税收优惠政策 preferential tax policy
税收站 choky
税收政策 tax policy
税收直接抵免 direct tax credit
税收滞纳 tax delinquency
税所 toll house
税务登记 taxation registration
税务地图 revenue map; tax map
税务功能 tax function
税务官员 revenue officer
税务管理 tax administration
税务会计 tax accounting
税务机构 tax agency
税务机关 tax authorities
税务缉私人员 customs guard; customs officer; water guard
税务计划 tax planning
税务监查 tax audit
税务检查 tax search
税务局 commissioner; customs bureau; taxation bureau
税务扣押权 tax lien
税务署 excise house; revenue
税务所 revenue office; tax office
税务员 landwaiter; revenuer; tax collector
税源 source of revenue; tax fund
税则 book of rate; customs tariff; tariff; tariff schedule; tax regulation
税则的精细分类 refinement of tariff classifications
税则分类 classification of tariff
税则类别 tariff nomenclature
税则目录 tariff nomenclature
税征 tax certificate
税制 taxation; tax system
税制改革 tax reform(ation)
税制修改 tax revision
税种 category of taxes; item of taxation; tax categories

睡

睡袋 flea bag; fleaking
睡莲池 water-lily pool
睡铺 sleeping bunk
睡椅 chesterfield; couch

顺

顺岸变化 alongshore variation
顺岸波 alongshore wave
顺岸泊位 marginal berth
顺岸导流堤 parallel training wall
顺岸导墙 parallel training wall
顺岸海底浅槽 longshore trough
顺岸海流 longshore current
顺岸流 alongshore current; littoral current
顺岸流能通量 longshore energy flux
顺岸码头 <与岸平行> marginal quay; marginal wharf; marginal berth; marginal utility; mooring wharf; parallel berth; parallel quay; parallel wharf; quay; wharf; marginal pier <突堤前端的>
顺岸码头泊位 quay berth
顺岸码头不紧靠水池 quay breastwork
顺岸码头宽度 wharf width
顺岸码头面高程 quay level
顺岸漂沙 longshore drift
顺岸沙埂 longshore bar
顺岸沙洲 longshore bar
顺岸式码头 marginal type wharf
顺岸式码头港池 extended quay dock
顺岸式栈桥码头 shore staging
顺岸输沙 longshore (sediment) transport
顺岸输沙改向区 nodal zone
顺岸输沙净值 net longshore transport
顺岸输沙率 longshore sand transport rate; transport rate longshore sand
顺岸输沙室 longshore transport rate
顺岸透空式码头 marginal false quay
顺岸栈桥 shore bridge
顺岸栈桥式码头 open-type wharf
顺坝 guide wall; longitudinal dam; longitudinal dike [dyke]; parallel dam; parallel dike [dyke]; parallel training wall; training dam; training dike [dyke]; training embankment; training levee; training wall
顺苯三酚 pyrogallol
顺边砖 stretcher
顺便 in passing
顺便伽马检查 following gamma examination
顺变 transient variation
顺变示踪浓度 transient tracer concentration
顺变作用 transient effect
顺槽 sublevel
顺层 bedding
顺层边坡 slope of strata dip toward excavation
顺层剥离 bedding fissility
顺层冲断 veneering thrust
顺层冲断层 bedding glide; bedding thrust
顺层磁化 bedding magnetization
顺层断层【地】 bedding (plane) fault; plane fault
顺层滑动【地】 bedding slip; bedding glide
顺层滑坡 bedding plane landslide; consequent landslide
顺层混合岩 epibolite
顺层节理 bedding joint
顺层裂开性 bedding fissility
顺层流 accordant junction
顺层面 bedding plane
顺层劈理 bedding cleavage
顺层片理【地】 bedding schistosity
顺层韧性剪切带 bedding ductile shear zone
顺层溶蚀带 bedding corrosion zone
顺层掩卧褶皱 bedding recumbent folds
顺差 active balance; excess of exports; export surplus; favo(u)rable balance; overbalance; overbalance of exports; positive balance; surplus; trade surplus
顺差触排 slipped bank; slipped multiple
顺差复接 slipped multiple
顺差复接分品法 multiple slip grading
顺差国家 surplus country
顺铲开挖法 sequential excavating process
顺潮 fair tide; following tide; with the current; with tide
顺潮而行 tide sailing
顺车 going
顺车流车道 concurrent flow lane
顺车流方向 with flow
顺车偏心轮 free eccentric wheel
顺车汽轮机 ahead turbine
顺串号字段 run number field
顺串记录 run record
顺串生成 run generation
顺磁标记 spin labeling[labelling]
顺磁弛豫 paramagnetic relaxation
顺磁的 sideromagnetic
顺磁电流 paramagnetic current
顺磁法掺杂的 paramagnetically doped
顺磁法拉第效应 paramagnetic Faraday's effect
顺磁法拉第旋转 paramagnetic Faraday rotation
顺磁分析法 paramagnetic analytical method
顺磁各向异性 paramagnetic anisotropy
顺磁共振 paramagnetic resonance
顺磁共振波谱 paramagnetic resonance spectrum
顺磁共振波谱法 paramagnetic resonance spectroscopy
顺磁共振法 paramagnetic resonance method
顺磁共振光谱学 electron spin resonance spectroscopy
顺磁共振谱参数 paramagnetic resonance spectrum parameter
顺磁共振谱仪 paramagnetic resonance spectrometer
顺磁贡献 paramagnetic contribution
顺磁光谱 paramagnetic spectrum
顺磁合金 paramagnetic alloy
顺磁化合物 paramagnetic compound
顺磁晶体 paramagnetic crystal
顺磁居里点 paramagnetic Curie point
顺磁居里温度 paramagnetic Curie temperature
顺磁矩 paramagnetic moment
顺磁矿石 paramagnetic ore
顺磁粒料 paramagnetic particle
顺磁屏蔽 paramagnetic shielding
顺磁塞曼效应 paramagnetic Zeeman effect
顺磁散射 paramagnetic scattering
顺磁松弛 paramagnetic relaxation
顺磁体 paramagnet
顺磁铁 paramagnetic iron
顺磁位移 paramagnetic shift
顺磁物质 paramagnetic substance
顺磁吸收 paramagnetic absorption
顺磁系统 paramagnetic system
顺磁线宽 paramagnetic linewidth
顺磁效应 paramagnetic effect
顺磁谐振 paramagnetic resonance
顺磁性 paramagnetism
顺磁性材料 paramagnetic material
顺磁性磁化率 paramagnetic susceptibility
顺磁性的 paramagnetic
顺磁性固体微波激射器 paramagnetic solid state maser
顺磁性矿物 paramagnetism mineral
顺磁性配合物 paramagnetic complex
顺磁性体 paramagnetic body; paramagnetic substance
顺磁性位移试剂 paramagnetic shift reagent
顺磁性物质 paramagnet(ic material)
顺磁性谐振吸收 paramagnetic resonance absorption
顺磁性氧分析仪 paramagnetic oxygen analyser[analyzer]
顺磁盐 paramagnetic salt
顺磁氧流计 paramagnetic oxygen analyser[analyzer]
顺磁影响 paramagnetic contribution
顺磁杂质 paramagnetic impurity
顺磁展宽 paramagnetic broadening
顺磁中心 paramagnetic center[centre]
顺磁中心的电子顺磁共振谱 electron paramagnetic resonance spectrum of paramagnetic center[centre]
顺次 due course; due time; following; in due course; in succession
顺次的 in order; orderly; sequential; serial
顺次调度系统 sequential scheduling system
顺次过程 sequential process
顺次呼叫 serial call
顺次加倍 successive doubling
顺次排列 seriate; seriation
顺次配列 gradate
顺次扫描 progressive scan(ning); sequential scan(ning)
顺次序寄存器 sequence register
顺次溢出法 progressive overflow method
顺搓钢丝绳 lang-lay wire
顺道的 pathwise
顺堤 longitudinal dike[dyke]; longitudinal embankment
顺地貌 consequent landform
顺电导性 paraconductivity
顺丁烯二酸 maleic acid
顺丁烯二酸改性乙烯(基)共聚树脂 maleic acid-modified vinyl copolymer resin
顺丁烯二酸改性油 maleic treated oil
顺丁烯二(酸)酐 maleic anhydride
顺丁烯二酸酐改性松香 maleic modified rosin
顺丁烯二酸酐化油 maleic treated oil

顺丁烯二酸酐松香加成物 rosin-maleic adduct
顺丁烯二酸化油 maleic oil
顺丁烯二酸树脂 maleic acid resin; maleic resin
顺丁烯二酸树脂清漆 maleic acid resin varnish
顺丁烯二酸型聚合物树脂 maleic polymer resin
顺丁烯二酸酯树脂 maleate resin
顺丁烯二酰胺 maleamide
顺丁橡胶 cis-polybutadiene rubber
顺读游标尺 direct vernier
顺二甲基四氢化萘 cis-dimethyleterahydronaphthalene
顺发雷管 instantaneous cap
顺发起爆管 instantaneous detonator
顺纺锯木 bastard sawing
顺风 by the wind; fair wind; favo(u)rable wind; free wind; helping wind; tail wind; toward breeze; with the wind
顺风边 down-wind leg
顺风波浪 lee wave
顺风潮 lee(ward) tide
顺风潮流 lee(ward tide) current; leeward tidal current
顺风的 down the wind; down-wind
顺风掉抢 wear
顺风航向 down-wind course
顺风航行 run before the wind; run free; with wind
顺风火 head fire
顺风裥 knife pleat
顺风进场着陆 tailwind approach
顺风距离 down-wind distance
顺风浪旋转的固定式外海建筑物 compliant offshore structure
顺风落地 down-wind landing
顺风驶帆 going free; running free; running large; sail broad; sailing free; sailing large
顺风污染源 down-wind source of pollution
顺风向 down-wind; down-wind direction
顺风行驶 scud
顺风转舵 veer
顺光线木条 fair batten
顺光照明 front lighting
顺轨道或旧辙行驶 tracking
顺轨误差 along-track error
顺航潮流 fair tide
顺河坝 longitudinal dam
顺河堤 parallel dyke[dike]
顺河围堰 < 围堰平行于河道部分 > lateral cofferdam
顺滑曲线 smooth curve
顺化 naturalization
顺汇 direct remittance; favo(u)rable exchange
顺挤法 direct extrusion
顺挤压模 straight extrusion die head
顺加作用 cis-addition
顺价 cantango; normal market
顺桨 feathering; full feathering
顺桨泵 feathering pump
顺桨桨距 feathering pitch
顺桨铰链 feathering hinge
顺桨螺距 feather pitch
顺桨螺线管 propeller feathering solenoid
顺桨螺旋桨 feathering airscrew; feathering propeller(runner)
顺桨螺旋桨透平 feathering propeller turbine
顺桨螺旋桨涡轮机 feathering propeller turbine
顺桨叶轮 feathering blade runner
顺绞缆 lang-lay rope
顺截河缏储木场 parallel holding ground
顺锯材 flat-sawn timber
顺锯法 flat sawing
顺锯木材 cleft timber; flat-sawed lumber
顺浪 following sea; following swell; stern sea
顺浪航行 running before the seas
顺利进行 succeed planning
顺利收报 armchair copy
顺利运行 trouble-free operation
顺列布置 in-line position; in-line arrangement
顺列布置泵组 in-line pumps
顺列布置管束 in-line bank
顺列堆料 in-line stockpile
顺列论 syntax
顺列式枢纽 longitudinal arrangement type junction terminal
顺列式停车 longitudinal parking; parallel parking
顺列延伸式枢纽 sequentially prolonged junction terminal
顺裂碎面【地】 synclastic
顺流 afloat; cocurrent; direct flow; fair current; favo(u)rable current; following current; following sea; forward flow; parallel flow; uniflow; downstream current
顺流持续坡度 sustaining slope
顺流船 downstream boat; downstream ship; downstream vessel
顺流导流叶片 uniflow guide vane
顺流的 concurrent; downstream
顺流堤 longitudinal embankment
顺流墩尖 downstream nose
顺流而下 downstream
顺流阀 straight-flow valve
顺流锅炉 concurrent boiler
顺流过热器 parallel flow superheater
顺流烘干 drying in parallel current; parallel flow drying
顺流烘干机 parallel flow drier[dryer]
顺流后退 drop back downstream
顺流洄游 downstream migration
顺流洄游路程 downstream course
顺流洄游鱼 downstream migrant
顺流混合器 concurrent flow mixer
顺流交通 flow-through traffic
顺流接触 cocurrent contact
顺流浸出 concurrent leaching
顺流靠泊 going alongside with the current; going alongside with the tide
顺流冷凝器 parallel flow condenser
顺流冷却 cocurrent cooling
顺流喷射系统 downstream injection system
顺流盆地 upgrade basin
顺流坡度 < 水力学上用 > sustaining slope
顺流清管 on-flow pigging; on-flow pigging operation; on-stream operation; on-stream pigging
顺流热交换器 downstream heat exchanger
顺流热交换悬浮预热器窑 parallel flow suspension preheater kiln
顺流施工 downstream dredging
顺流式阀 straight-flow valve
顺流式空心阀 direct flow hollow spool valve
顺流式离心机 concurrent centrifuge
顺流式连续加热炉 parallel flow heating furnace
顺流式热交换器 parallel flow heat exchanger
顺流式压缩机 uniflow type compressor
顺流式烟道采暖炉 upflow furnace
顺流式再生 downflow regeneration
顺流拖锚 dredging with the current
顺流挖泥 downstream dredging
顺流挖泥法 dredging with current
顺流向 downstream direction
顺流向的 currentwise
顺流向流动 downstream flow
顺流再生 cocurrent regeneration; straight regeneration
顺流振荡 in-line oscillation
顺流转筒式烘干机 uniflow rotary dryer
顺留 direct stationary
顺路出行 pass-by trips
顺路的 pathwise
顺路(非转向)链状出行 undiverted linked trips
顺螺旋方向 pro-spin direction
顺码头 marginal wharf
顺面 stretcher(face)
顺面流 plane flow
顺面渗透性 in-plate permeability
顺面水流试验 planar water flow test
顺木纹 along the grain
顺木纹受压 compression parallel to grain
顺木纹压力 compression parallel to grain
顺木纹压缩 compression parallel to grain
顺逆换向开关 tumbler switch
顺逆转换向控制箱 < 平地机刮刀的 > circle reverse drive shaft
顺逆转角 < 平地机刮刀的 > circle reverse
顺逆转驱动轴 < 平地机刮刀的 > circle reverse control housing
顺年轮干裂 ring shake; wind shake
顺捻 Albert lay; lang lay; parallel lay
顺捻钢丝绳 lang-lay wire
顺扭转 positive torsion
顺排 in-line arrangement
顺排存储器 sequential storage
顺排挡 normal file
顺排管组 in-line bank of tubes
顺盘绳法 coil all rope with the sun
顺喷 downstream spray pattern
顺坡 along-grade
顺坡长度 < 超高度的 > gradual-decrease distance
顺坡滑动 translational slide
顺坡浇筑混凝土方法 advance slope method
顺坡耕 straight plowing
顺坡输水道 grade aqueduct
顺坡下溜 coaster effect
顺气流 favo(u)rable current
顺气流系统 parallel current system
顺砌 facing bond; stretcher
顺砌层 stretcher bond
顺砌的砖地坪 rowlock paving
顺砌混凝土砌块 stretcher block
顺砌砌块 < 混凝土 > stretcher block
顺砌砖 stretcher; stretcher brick
顺砌砖层 stretching course
顺倾向的下降盘 downdip block
顺射天线阵 end-fire antenna; end-fire array
顺绳扭劲的眼环结 German eye splice
顺时人口净增长率 intrinsic(al) rate of increase
顺时斜角 clockwise angle
顺时针方向 clockwise direction; clockwise sense; right-hand-wise
顺时针方向编号 numbering in clockwise direction
顺时针方向的 clockwise; right-hand(ed)
顺时针方向地 deasil
顺时针方向多边形 clockwise polygon
顺时针方向回转 clockwise rotation
顺时针方向力矩 right-hand moment
顺时针方向偏移 clockwise drift
顺时针方向旋转 clockwise rotation; dextrorotation; right-handed rotation
顺时针方向旋转的 dextrorotary
顺时针方向运动 clockwise motion
顺时针方向转动 clockwise motion; right-hand rotation; clockwise rotation
顺时针角 clockwise angle
顺时针螺旋偏振 clockwise helical polarization
顺时针满旋 fully clockwise
顺时针偏振电磁波 clockwise polarized electromagnetic wave
顺时针翘倾摆动 dextrally tilting swing
顺时针旋涡 clockwise vortex
顺时针旋转的 dextrorotary[dextrorotatory]
顺时针循环 clockwise cycling
顺时针转动 with the sun
顺时针转螺旋桨 right-handed propeller
顺式 cis; maleinoid form; syn-form
顺式聚丁二烯 cis-polybutadiene
顺式聚丁二烯橡胶 cis-polybutadiene rubber
顺式异构化合物 maleinoid
顺式异构现象 cis-isomerism
顺式有规聚合物 cistactic polymer
顺手的 well-placed
顺水风【气】 following wind
顺水杆 scaffold ledger
顺水条 < 钉挂瓦条用 > counter batten; batten; rainwater lath
顺弯 longitudinal distortion < 木材 >; easement; easing < 使转折处平缓 >
顺位 cis-position; order; syn-position
顺位效应 cis-effect
顺纹 along the fiber[fibre]; along the grain; parallel to grain; rift grain; straight grain; with the grain; parallel to the grain < 木材的 >
顺纹层压板 parallel laminate
顺纹插接 round splice; sailmaker's splice
顺纹剪切 < 木材的 > detrusion
顺纹剪切强度 shearing strength parallel to the grain
顺纹接缝 edge joint
顺纹锯 resaw
顺纹锯材 resawed lumber; resawn lumber
顺纹锯的木板 resawn board
顺纹锯法 flat sawing
顺纹锯解 rip sawing
顺纹锯开 bastard sawing; flat cutting; ripping
顺纹锯开木材 deepening
顺纹锯木 bastard sawing; flat cutting; ripping
顺纹抗剪强度 shearing strength of wood along the grain; shearing strength parallel to grain
顺纹抗拉强度 tensile strength parallel to grain
顺纹抗碎强度 crushing strength parallel to grain
顺纹抗压强度 crushing strength parallel to grain; compression strength parallel to grain; compressive strength parallel to grain; parallel-to-grain compressive strength
顺纹拉力 tension parallel to grain
顺纹木抹子 straight float
顺纹劈开的木材 cleft timber

顺纹劈制栗木篱笆 cleft-chestnut fencing
顺纹劈制栗木栅栏 cleft-chestnut paling
顺纹压材 parallel laminate
顺纹压缩 compression parallel to the grain
顺纹压缩法 longitudinal compression process
顺纹眼环插接 sailmaker's eye splice
顺纹应力 parallel-to-grain stress
顺纹纸 grain-long paper
顺纹作用 cis-orientation
顺铣 climb cut(ting);climb milling;downcut(ting);down milling
顺纤维方向应力 fiber[fibre] stress
顺向 consequent;with flow
顺向滨线 concordant shoreline
顺向车道 concurrent flow lane
顺向冲采 longitudinal efflux
顺向重叠进路 converging route;routes with overlapped section in the same direction
顺向传导 orthodromic conduction
顺向道岔【铁】trailing points;turnout back-passing the point;turnout passing-passing the point
顺向的 catacline;pathwise;straight-forward
顺向定位 cis-orientation
顺向断层 synthetic(al) fault
顺向断层崖 consequent fault scarp
顺向发车 dispatch trains in straight direction
顺向分带 normal zoning
顺向分水岭 consequent divide
顺向公共汽车车道 with-flow bus lane
顺向谷 consequent valley
顺向辊涂 forward roller-coating
顺向滚铣 climb hobbing
顺向海岸 concordant coast
顺向海岸线 concordant coastline
顺向河谷 consequent valley
顺向河(流)<流向与岩层倾向一致的> accordant river;accordant stream;cataclinal river;cataclinal stream;consequent drainage;consequent river;original river;consequent stream
顺向湖泊 consequent lake
顺向挤压 forward extrusion
顺向计数 counting forward;forward counting
顺向计算 counting forward;forward counting
顺向交叉【铁】same direction intersecting
顺向接车 receive trains in humping direction
顺向排水(系)consequent drainage
顺向坡<岩层倾向与边坡坡向相同的> cosequent slope
顺向水系 consequent drainage;consequent drainage system
顺向位移 forward lead;forward shift
顺向先成河 antecedent consequent river;anteconsequent stream
顺向行驶 trailing
顺向性 orthodromic
顺向张力 forward play
顺斜道逆斜河 diaclinal river;diaclinal stream
顺斜谷 acclinal valley;acclivous valley
顺斜河 cataclinal river
顺斜轴 synclinal axis
顺行 direct motion;prograde motion;progressive motion
顺行的 posigrade
顺行轨道 direct orbit;prograde orbit
顺行联想 forward association
顺行推进的 posigrade
顺行推进火箭的 posigrade
顺序 order;positive sequence;subsequence;succession
顺序安装法 progressive erection method
顺序报警器 sequential alarm
顺序曝光 sequential exposure
顺序爆破 consecutive firing;sequence blasting
顺序逼近法 sequential approach method
顺序比较指数 sequential comparison index
顺序编号 ordinal numeration;serial numbering
顺序编码 sequential encoding
顺序编排 sequential organization
顺序编排文件 sequential organization file
顺序编制程序员 sequence programmer
顺序变化 gradualness
顺序变速装置 progressive gear
顺序表 sequence chart;sequence list;sequence table
顺序表示法 sequential notation
顺序波束定向 sequential lobing
顺序采样法 sequential sampling
顺序操作 consecutive operation;sequence control;sequential operation
顺序操作符 sequential operator
顺序操作计算机 consecutive computer;sequential computer
顺序测定 sequential determination
顺序测量 proceeding measurement
顺序测试 sequential testing
顺序成批处理 sequential batch processing
顺序程序 sequential program(me)
顺序程序控制方法 sequence program(me) control method
顺序程序设计 sequential programming
顺序冲模 progressive die
顺序抽样 sequential sampling
顺序处理 sequential processing
顺序传输 sequential transmission
顺序淬火 progressive hardening
顺序存储存贮器 sequential access storage
顺序存储单元 sequential memory location
顺序存储器 sequential memory
顺序存储装置 sequential storage device
顺序存取 sequential access;serial access
顺序存取方法 sequential access method
顺序存取文件 file of sequential access;sequential access file
顺序存取显示 sequential access display
顺序错误 sequence error
顺序代码 sequence code
顺序的 ordinal;progressive;sequential
顺序电路 sequential circuit
顺序调度 sequential scheduling
顺序调度程序 sequential scheduler
顺序调度系统 sequential scheduling system
顺序调用 sequence call(ing)
顺序定时(起爆)器 sequential timer;sequence timer
顺序动作阀 sequence valve
顺序动作继电器 sequence relay;sequential relay
顺序动作模 gang die
顺序队列 sequential queue
顺序阀 sequence valve;sequential valve
顺序法则 priority rule;sequential rule
顺序反馈人机对话检索系统 sequential feedback interactive retrieval system
顺序仿形车削 progressive copy turning
顺序分段焊接法 selective block sequence
顺序分隔符 sequence separator
顺序分级 ordinal scale
顺序分级粉磨 progressive classified grinding
顺序分解 sequential decomposition
顺序分类 series classification
顺序分类算法 sequential classification algorithm
顺序分配 order distribution;sequential allocation
顺序分析 sequential analysis
顺序分析仪 sequenator;sequence analyser[analyzer]
顺序符号 sequence symbol
顺序共聚物 sequenced copolymer
顺序故障诊断 sequential fault diagnosis
顺序关系 order relation;ordinal relation
顺序规划 sequential program(me)
顺序号 current number;ordinal number;sequence number;sequential number;serial number;series number
顺序号数票 serial ticket
顺序呼叫 sequence call(ing)
顺序换景器 sequential switcher
顺序回路 sequencing circuit
顺序机 sequence machine
顺序集 ordered set
顺序计数器 sequence counter
顺序计算 sequencing computation;sequential calculation;sequential computation
顺序记号 sequence token
顺序记录 journal;sequential record
顺序记忆制彩色电视 colo(u)r television
顺序继电器 sequence relay;sequential relay
顺序寄存器 sequence register
顺序间歇式反应器 sequencing batch reactor
顺序监测器 sequence monitor
顺序监督程序 sequence monitor
顺序监控 sequential monitoring
顺序监控器 sequence controller;sequence monitor
顺序监视 sequential monitoring
顺序检测 sequential measurement
顺序检查 ordinal inspection
顺序检索 sequence search;sequential retrieval;sequential search
顺序检索文件 indexed sequential file
顺序检验 sequence check(ing)
顺序检验程序 sequence checking routine
顺序检验系统 sequence control system
顺序检验字 sequence check word
顺序渐近法 step-by-step method
顺序渐进 progressive procession
顺序渐进施工法 progressive construction method
顺序渐进现绕法 progressive cast-in-place method
顺序阶梯 sequential steps
顺序接班 heel-and-toe
顺序接触 progressive contact
顺序结构 sequential organization;sequential structure
顺序结构键记录 sequently structured keyed record
顺序进程 sequential process
顺序进位 sequential carry;successive carry
顺序浸蚀 sequence etching
顺序决策过程 sequential decision process
顺序决定方法 sequential decision procedure
顺序开关系统 sequential switching system
顺序空间测角交会 sequential three-dimensional angle intersection
顺序控制 sequence[sequencing/ sequential] control
顺序控制计数器 sequence control counter
顺序控制计算机 program(me)-controlled sequential computer
顺序控制寄存器 sequence control register
顺序控制器 sequence controller
顺序控制图 ladder diagram
顺序控制系统 sequence[sequential] control system
顺序控制移动 sequence-controlling advance
顺序控制元件 sequence control element
顺序拉开检验 ordering bias
顺序拉土法 sequential earth pulling process
顺序理论 sequencing theory
顺序联结 sequence interlock
顺序连续记录 consecutive entry
顺序鳞剥 progressive scaling
顺序逻辑 sequential logic
顺序脉冲 sequential pulse
顺序模 follow die
顺序凝固 directional solidification;progressive solidification
顺序排列 series arrangement;systematic arrangement
顺序判定方式 sequential decision procedure
顺序判定问题 sequential decision problem
顺序配板 slip matching
顺序喷油系统 sequential fuel injection system
顺序匹配 sequence matching
顺序拼木法 slide matching;slip matching
顺序拼装法 progressive erection;progressive placing
顺序剖面 serial section
顺序剖析方案 sequential parsing scheme
顺序启闭断路器组 cascade breakers
顺序启动泵 succeeding pump
顺序起动 sequential firing
顺序起动和控制设备 sequence starting and control device
顺序取样 sequential sampling
顺序任务 serial task
顺序扫描 progressive scan(ning);sequential scan(ning)
顺序扫描光栅 sequential raster
顺序扫描系统 sequential scan system
顺序扫描显示 sequential access display
顺序色差信号 sequential colo(u)r difference signal
顺序闪光灯 sequence flash lights
顺序闪光回路 sequential flasher circuit

S

顺序闪光进近灯 sequenced flashing approach light
顺序伸缩 telescope in sequence
顺序生产 production in series
顺序施工 work in succession
顺序时间调节器 sequence timer
顺序时延 sequential time delay
顺序识别 sequence recognition
顺序输入数据集 entry sequenced data set
顺序数 serial number
顺序数据变换 sequential data change
顺序数据存储 sequential data storage
顺序数据集 sequential data set
顺序数据结构 sequential data structure
顺序数据位 sequential data bit
顺序数字伺服机构 sequential digital servomechanism
顺序算子 sequential operator
顺序随机规划 sequential stochastic programming
顺序锁闭 sequential locking
顺序弹射系统 sequenced ejection system
顺序特征判别机制 sequential feature decision mechanism
顺序同步传输 sequential synchronous transmission
顺序统计量 order statistic
顺序投标 sequential tendering
顺序图 precedence diagram
顺序图像编码 sequential picture coding
顺序脱扣 sequential tripping
顺序外的 out of turn
顺序微程序设计 sequential microprogramming
顺序文件 sequential file
顺序文件存取 sequential file access
顺序文件结构 sequential file structure
顺序误差 sequence error; sequential error; systematic error
顺序项目文件 entry sequenced file
顺序信号 sequential signal
顺序性 succession
顺序选路算法 sequential routing algorithm
顺序选择 sequential selection
顺序选择法 tandem method
顺序选择器 sequence selector
顺序寻道【计】track to track seek
顺序寻优 sequential estimation
顺序寻优法 sequential search
顺序寻址 sequential addressing
顺序演变 consequent succession
顺序溢出 progressive overflow
顺序优化 sequential optimization
顺序由小而大 ascending order
顺序运算 sequence operation; sequential operation
顺序运算符 sequential operator
顺序运行 consecutive operation
顺序占用解放系统 the sequential occupancy release system
顺序照相 sequence photography
顺序直线扫探雷达 sequentially lobed radar
顺序制 sequential system
顺序制彩色电视 sequential colo(u)r television
顺序属性 sequential attribute
顺序转换器 sequential transducer
顺序装配 mount in series
顺序准确性检验 ordering bias
顺序自动控制 sequential automatic control
顺序自动装置 sequence automatics
顺序组装 in-line assembly
顺序作业 sequential access
顺序作用 sequencing
顺岩石天然节理开采的石料 quarry-faced
顺延 postpone
顺应 adjustment
顺应不良 maladjustment
顺应不良的 maladjusted
顺应地形布置的道路网 topographic(al) street system
顺应反应 adjustment reaction
顺应式平台 compliant platform
顺应行为 adaptive behavio(u)r
顺游标【测】direct vernier
顺直岸线 regular coastline
顺直冲积性河槽 straight alluvial channel
顺直段 straight reach
顺直海岸 regular coast; regular shore; smooth coast; smooth shore
顺直海岸线 regular coastline; regular shoreline; smooth coastline; smooth shoreline
顺直航道 straight channel
顺直航向 follow course
顺直河槽 straight channel
顺直河道 straight river; straight stream
顺直河段 straight reach; straight stretch; straight waterway section
顺直河流 straight stream
顺直砌合 running bond
顺直水道 straight stream; straight channel
顺直型河段 straight river reach
顺直正常河流 straight regular stream
顺芷酸 tiglic acid
顺轴的 direct axis
顺轴电抗 direct axis reactance
顺轴电路 direct axis circuit
顺筑法 bottom-up construction method
顺砖 stretcher; stretcher brick
顺砖层 course of stretchers; coursing; stretching course
顺砖错砌 isodomum
顺砖面 long face
顺砖内角斜切砌合体 clip bond
顺砖皮 flat course
顺砖铺地 stretcher paving
顺砖铺砌式竖砖路面 stretcher-bond type rowlock paving
顺砖砌层 stretcher course
顺砖砌合 chimney bond; stretcher bond; stretching bond; running bond
顺砖圬工砌合 stretcher masonry bond
顺砖压缝砌合 running bond
顺砖压缝圬工砌合 running masonry bond
顺转 corotation; forward rotation; veering
顺转风 veering wind
顺桩号而下 downstation
顺着地形的河流 consequent stream
顺着风向 down the wind
顺自转发射 corotational departure
顺作法 bottom-up method

瞬变 transitional; transitory

瞬变棒 transient rod
瞬变比较镜 blink comparator
瞬变波 transient wave
瞬变部分 transient part
瞬变测量 instantaneous measurement; transient measurement
瞬变测深法 transient sounding method
瞬变产生 transient generation
瞬变场法 transient field method; transient method; transit field method
瞬变场法感应电势剖面图 profile figure of potential induced in transient field method
瞬变场法感应电势曲线 curve of potential induced in transient field method
瞬变成像 transit imaging
瞬变传输率 instantaneous transmission rate
瞬变单日历 instantaneously changing simple calender
瞬变地图 instant map
瞬变电磁法 transient electromagnetic method
瞬变电动势 transient electromotive force; transient internal voltage
瞬变电抗 transient reactance
瞬变电流 transient current
瞬变电压 transient voltage
瞬变反应 transient reaction
瞬变分析 transient analysis
瞬变峰调节 peak transient regulation
瞬变峰调整 peak transient regulation
瞬变负荷效率 transient response efficiency
瞬变工况 transient condition
瞬变轨道 transient trajectory
瞬变过程 system transient output response; transient process
瞬变过程解扣 transient trip
瞬变过程特性 frequency response; transient response characteristic
瞬变过程中的转速特性 transient-speed characteristic
瞬变过电压指示器 transient indicator
瞬变航迹 transient flight path
瞬变荷载 instantaneous load(ing); transient load
瞬变恢复电压 transient recovery voltage
瞬变激波 transient shock wave
瞬变激光响应 transient laser response
瞬变计算 transient calculation
瞬变技术 transient technique
瞬变阶段 transient phase
瞬变空气动力学 transient aerodynamics
瞬变空穴 transient cavity
瞬变流动 transient flow; transition flow
瞬变流透气率 transient flow permeability
瞬变脉冲 ringing pulse
瞬变扰动 transient disturbance
瞬变热点 transient hot spots
瞬变热聚焦 transient thermal self focusing
瞬变热散焦 transient thermal defocusing
瞬变热晕 transient thermal blooming
瞬变色 flash colo(u)r
瞬变升力 transitional lift
瞬变时间 transient period; transient time
瞬变时间常数 transient time-constant
瞬变示波器 memnescope
瞬变输入信号 ramp input signal
瞬变数据 transient data
瞬变特性 transient property
瞬变特性试验器 transient analyser [analyzer]
瞬变条件 transient condition
瞬变统计学 transient statistics
瞬变温度 transient temperature
瞬变显微比较镜 blink microscope
瞬变现象 transient phenomenon
瞬变响应 transient response
瞬变消除 transient elimination
瞬变胁强 transient stress
瞬变行为 transient behavio(u)r
瞬变压力 transient pressure
瞬变压力脉冲 transient pressure pulse
瞬变抑制电容器 surge suppressing capacitor; transient suppressing capacitor
瞬变抑制器 transient suppressor
瞬变运动 transient motion
瞬变振动 transient vibration
瞬变值 transient value
瞬变周期 transient period
瞬变状态 transient condition; transient regime
瞬变阻抗 transient impedance
瞬变作用 transient effect
瞬磁变 transient magnetic variation
瞬磁变化 transient magnetic variation
瞬地磁变 transient geomagnetic variation
瞬动触点 snap action contact
瞬动过电流继电器 instantaneous overcurrent relay
瞬动继电器 instantaneous acting relay; instantaneous relay
瞬动接点 wiper-type contact
瞬动开关 instant-on switch; quick-made-and-break switch; snap switch
瞬动时限复归制 instantly acting time-limit resetting system
瞬动吸铁继电器 instantaneous attracted armature relay
瞬动咬合器 snapper
瞬读距离 glance-legibility distance
瞬断 hit; intermittent disconnection
瞬断率 short interruption rate
瞬发 graze
瞬发爆发 instantaneous firing
瞬发爆破 instantaneous blast
瞬发贝塔 prompt beta
瞬发倍增 prompt neutron multiplication
瞬发氮氧化物 prompt NO_x
瞬发的 non-delay
瞬发点火 instantaneous ignition
瞬发电雷管 instantaneous electric(al) detonator
瞬发反应性 prompt reactivity
瞬发辐射 prompt radiation
瞬发负温度系数 prompt negative temperature coefficient
瞬发伽马 prompt fission gammas
瞬发伽马辐射 prompt gamma radiation
瞬发伽马能量 prompt gamma energy
瞬发伽马射线 prompt gamma; prompt gamma ray
瞬发伽马射线分析 prompt gamma-ray analysis
瞬发洪河流 flashy stream
瞬发径流 prompt runoff; quick flow
瞬发雷管 instantaneous cap; instantaneous detonator; short relay detonator; undelayed detonator
瞬发裂变产物 prompt fission product
瞬发临界 prompt critical; prompt criticality
瞬发临界的 prompt super-critical
瞬发临界堆 prompt critical reactor
瞬发临界瞬变过程 prompt critical transient
瞬发脉冲 prompt pulse
瞬发每代时间 prompt generation time
瞬发起爆导火索 fuse of instantaneous detonating
瞬发衰变 prompt decay

瞬发衰变率 prompt decay rate
瞬发态 prompt-mode
瞬发停堆系数 prompt shut-down coefficient
瞬发通量斜变 prompt flux tilt
瞬发温度系数 prompt temperature coefficient
瞬发谐波 prompt harmonic
瞬发引线 instantaneous fusing
瞬发引信 instantaneous fuse; non-delay fuse; quick fuze; sensitive percussion fuze
瞬发中毒 prompt poisoning
瞬发中子 instantaneous neutron; prompt neutron
瞬发中子短时间闪光辐照 prompt neutron burst irradiation
瞬发中子堆 prompt reactor
瞬发中子份额 prompt neutron fraction
瞬发中子活化分析 prompt neutron activation analysis
瞬发中子面积 prompt neutron area
瞬发中子能谱 prompt neutron spectrum
瞬发中子寿期 prompt neutron lifetime
瞬发中子衰变常数 prompt neutron decay constant
瞬发周期 prompt period
瞬发周期事故 prompt period accident
瞬发撞击引信 instantaneous impact fuze
瞬即效应 immediate effect
瞬间 breath; instantaneousness; wink
瞬间比热容 instantaneous specific heat
瞬间变定 immediate set
瞬间变化 short-term fluctuation
瞬间冰冻 instantaneous freezing
瞬间操作 transient operation
瞬间测量 instantaneous measurement
瞬间超电压 transitory over voltage
瞬间超载 momentary overload(ing)
瞬间磁链 transient flux-linkage
瞬间的 momentary; temporal
瞬间的核辐射 prompt nuclear radiation
瞬间等离子体 transient plasma
瞬间点焊 shot weld(ing)
瞬间动作 snap action
瞬间读数 immediate reading
瞬间断开时间 transient off time
瞬间断路 instantaneous trip
瞬间分布 temporal distribution
瞬间峰值 transient peak
瞬间符合曲线 prompt coincidence curve
瞬间干燥 flash drying; wink-dry
瞬间干燥系统 flash drying system
瞬间高压试验 flash test
瞬间荷载 flashing load; flashy load; instantaneous load(ing)
瞬间畸变 transient distortion
瞬间极限电流 carrying current
瞬间接触式按钮 momentary-type contact push button
瞬间接收图像 instant vision
瞬间接收制 instant on system
瞬间距离 present range
瞬间可视图像 instant vision
瞬间控制 instantaneous control
瞬间快门 instantaneous shutter
瞬间临界性 prompt criticality
瞬间流变 transient flow
瞬间目标 fleeting target; transitory target
瞬间黏[粘]结剂 instantaneous adhesive
瞬间偏移控制 instantaneous deviation control
瞬间平衡 transient equilibrium
瞬间气化 instant vapo(u)rization
瞬间切断 instantaneous trip
瞬间燃烧压力 blast pressure
瞬间杀菌机 flash pasteurizer
瞬间摄影照片 instantaneous photograph
瞬间识别距离 <交通标志> glance-legibility distance
瞬间水样 snap sample
瞬间弹性变形 immediate elastic deformation
瞬间物质 intermediate species
瞬间谐波 transient harmonics
瞬间形变 immediate set
瞬间游离基 intermediate radical
瞬间自动跳闸 instantaneous trip
瞬间最大转速 maximum momentary speed
瞬间作用 momentary action
瞬凝 <水泥浆的> flash set of cement paste
瞬凝矾土高铝水泥 flash-setting alumina cement
瞬凝石膏 flash-setting gypsum
瞬凝水泥 flash-setting cement
瞬熔 instantaneous fusing
瞬时安装法 instantaneous erection
瞬时按钮开关 momentary-contact push button
瞬时半径 instantaneous radius
瞬时曝光 instantaneous exposure
瞬时爆发 instantaneous firing; instantaneous outburst
瞬时爆破 instant(aneous) blast(ing)
瞬时备用 instantaneous stand-by; momentary reserve
瞬时比热 instantaneous specific heat
瞬时闭合 rapid closing
瞬时边界影响 temporary boundary effect
瞬时变定 instantaneous set
瞬时变化 temporal variation; transient variation
瞬时变换日历 instant-change calendar
瞬时变位 instantaneous deflection
瞬时变形 immediate deformation; instantaneous deformation; temporary deformation; temporary set; transient deformation; transient deflection
瞬时变形阶段 instantaneous strain stage
瞬时表面变形 transient surface deflection
瞬时表面张力 instantaneous surface tension
瞬时波 transient wave
瞬时波动 momentary fluctuation; temporal fluctuation
瞬时波面 instantaneous wave elevation
瞬时波速 instantaneous celerity
瞬时薄层色谱法 instant thin-layer chromatography
瞬时采样 instantaneous sampling
瞬时采样器 instantaneous sampler
瞬时测量 instantaneous measurement
瞬时测时器 chronoscope
瞬时产生的不稳定电流 electric(al) transient
瞬时颤动 transient vibration
瞬时场 instantaneous field
瞬时超载 momentary overload(ing)
瞬时车速 instantaneous speed; spot speed
瞬时沉降 instantaneous settlement; distortion settlement; immediate settlement; initial settlement; undrained settlement; undrain settlement
瞬时程序存取 instant program(me) access
瞬时迟发爆破 instantaneous relay blast; short relay blasting
瞬时迟发雷管 short delay detonation; short delay detonator; short delay fuse
瞬时迟发起爆法 short delay blasting method
瞬时迟发引信 short delay fuse
瞬时赤道 equator of date
瞬时冲头能量 instantaneous punch energy
瞬时重合闸 instantaneous reclosing
瞬时抽水试验 transient pumping test
瞬时出量 instantaneous output
瞬时传冲 instantaneous(im) pulse
瞬时传动比. instantaneous transmission ratio
瞬时传输率 instantaneous transmission rate
瞬时磁化 flash magnetization
瞬时存储器 immediate-access storage; instantaneous storage; volatile store
瞬时带宽 instantaneous band; instant bandwidth
瞬时单位水过程线 instantaneous unit hydrograph
瞬时得热量 instantaneous heat gain
瞬时的 instantaneous; transitory
瞬时地面分辨率 instantaneous ground resolution
瞬时点燃热阴极灯 instant-start hot cathode lamp
瞬时点速 instantaneous particle velocity
瞬时点源 instantaneous point-source
瞬时电点火管 instantaneous squib
瞬时电动势 instantaneous electromotive force; instantaneous emf
瞬时电抗 instantaneous reactance
瞬时电流 instantaneous current; momentary current; transient current
瞬时电流效率 instantaneous current efficiency; transient current efficiency
瞬时电起爆器 instantaneous electric(al) detonator
瞬时电热水器 instantaneous electric(al) water heater
瞬时电压 instantaneous pressure; instantaneous voltage; transient voltage
瞬时电压跟随器 current-voltage follower
瞬时电子扫描 instantaneous electronic scan(ning)
瞬时电子引爆剂 instantaneous electric(al) detonator
瞬时电子引信 instantaneous electric(al) detonator
瞬时定额 short-time rating
瞬时动作 instantaneous action; instantaneous operation; split-second response
瞬时动作的 instant acting
瞬时冻结 instantaneous freezing
瞬时读出 instantaneous readout
瞬时读数 instantaneous reading; instantaneous value
瞬时短接 instantaneous short circuit
瞬时短路 instantaneous short circuit
瞬时短路试验 instantaneous short-circuit test
瞬时断流 instantaneous break; minute break
瞬时断路 momentary interruption
瞬时断续接电 inching
瞬时钝化 immediate passivation
瞬时发电雷管 instantaneous electric(al) detonator
瞬时反馈 temporary feedback
瞬时反应 instantaneous response; transient response
瞬时反应收率 instantaneous reaction yield
瞬时反应速度 instantaneous velocity of reaction
瞬时反应性 instantaneous reactivity
瞬时放电 instantaneous discharge
瞬时飞行数据 transient flight data
瞬时分布 instantaneous distribution; temporal distribution
瞬时分段复位 transient-divided reset
瞬时分离机构 quick-release gear
瞬时分路 instantaneous shunt
瞬时分路不良 instantaneous loss of shunting
瞬时风速 instantaneous wind speed
瞬时峰荷 transient peak loading
瞬时峰值 instantaneous peak value; transient peak value
瞬时峰值电流 instantaneous peak current
瞬时俘获 immediate capture
瞬时俘获效率 immediate-capture efficiency
瞬时浮点增益控制 transient floating point gain control
瞬时幅度 instantaneous amplitude
瞬时辐射 instantaneous radiation
瞬时辐射计 temporal radiometer
瞬时辐照 transient irradiation
瞬时辐照损伤 transient radiation damage
瞬时辐照效应 transient radiation effects
瞬时腐蚀(速)率 instantaneous corrosion rate
瞬时负荷 flash(y) load; instantaneous load(ing); momentary load; transient load
瞬时负载 momentary load; transient load
瞬时复位 instant reset
瞬时复原时间 instantaneous return time
瞬时傅立叶谱 instantaneous Fourier spectrum
瞬时干扰 instantaneous disturbance; hit
瞬时高峰负荷 instantaneous peak load
瞬时高峰需要量 instantaneous peak demand
瞬时隔板 flash barrier
瞬时隔离罩 flash barrier
瞬时根数 instantaneous elements
瞬时工作源 instantaneous source
瞬时功率 instantaneous capacity; instantaneous output; instantaneous power; momentary duty; momentary output; momentary power; swing capacity
瞬时功率谱 instantaneous power spectrum
瞬时功率输出 instantaneous power output
瞬时功率增加 momentary high power effort
瞬时故障 transient fault
瞬时故障率 instantaneous failure rate
瞬时光谱 instantaneous spectrum
瞬时轨道 instantaneous orbit
瞬时轨道圆 instantaneous circle
瞬时过冲 transient overshoot(ing)

S

瞬时过电流 momentary excess current
瞬时过电压 instantaneous overvoltage
瞬时过电压计数器 transient overvoltage counter
瞬时过调量 transient overshoot
瞬时过压容量 instantaneous overvoltage capacity
瞬时过载 instantaneous overload; momentary overload(ing)
瞬时过载电流 transient-overload current
瞬时函数 transient function
瞬时荷载 flash(y)load; instantaneous load(ing); momentary load; passing load; short duration load; transient load; transitory loading
瞬时荷重冲击仪 short duration load striker
瞬时洪峰 instantaneous peak
瞬时洪峰流量 instantaneous peak discharge; momentary peak discharge
瞬时环境 <汽车交通控制系统中的信息> immediate environment
瞬时换日 instant data setting
瞬时换向 instant(aneous) reverse
瞬时黄道 ecliptic of date
瞬时恢复 instantaneous recovery
瞬时恢复电压 transient recovery voltage
瞬时恢复力 instantaneous restoring force
瞬时恢复时间 instantaneous recovery time
瞬时回变 instantaneous recovery
瞬时回镜机构 instant return mirror mechanism
瞬时回转方式 mo-slewing system
瞬时或同步起爆雷管 instantaneous cap
瞬时畸变 transient distortion
瞬时激励器 instantaneous exciter
瞬时即变的 tempolabile
瞬时极 instantaneous pole
瞬时极限电流 instantaneous carrying current
瞬时计 chronoscope; microchronometer
瞬时记录 instantaneous record
瞬时记忆 immediate memory
瞬时剂量 prompt dose
瞬时继电器 instantaneous acting relay
瞬时加热 instantaneous heating
瞬时加热电子管 instant-heating tube
瞬时加热器 instantaneous geyser; instantaneous heater
瞬时加速度 brief acceleration; instantaneous acceleration; transient acceleration
瞬时加速度中心 instantaneous center [centre] of acceleration
瞬时加载 immediate loading
瞬时夹紧卡盘 quick-acting chuck
瞬时间应变 instantaneous strain
瞬时检测概率 instantaneous probability of detection
瞬时建成桥 <一种快速建成的新型预应力装配式混凝土桥> instant bridge
瞬时交通荷载 <快速移动的车辆荷载> fast-moving traffic load; instantaneous traffic load
瞬时角频率 instantaneous angular frequency
瞬时角速度 instantaneous angular speed; instantaneous angular velocity
瞬时校准 transient calibration
瞬时校准脉冲 transient calibration pulse
瞬时接触 instantaneous contact; momentary contact
瞬时接触按钮 momentary contact push button
瞬时接触开关 momentary contact switch
瞬时接点 momentary contact
瞬时接入 instant-start
瞬时接通 momentary connection
瞬时接通制 instant on system
瞬时结构 temporal structure
瞬时解 transient solution
瞬时劲度 instantaneous stiffness
瞬时经度 instantaneous longitude
瞬时距离 instantaneous distance
瞬时聚合 instantaneous polymerization
瞬时开关 instantaneous switch; quick make
瞬时抗挠曲强度 wink bending flexure strength
瞬时抗弯强力 instantaneous bending strength
瞬时可用性 instantaneous availability
瞬时空气速度 instantaneous air speed
瞬时空穴 transient cavity
瞬时孔隙压力 instantaneous pore pressure
瞬时快门 instantaneous shutter
瞬时扩展器 instantaneous expander
瞬时离合器 split-second clutch(coupling)
瞬时理论产量 instantaneous theoretic(al) output
瞬时力 transient force
瞬时力矩 transient moment
瞬时力矩最大值 maximum torque phase
瞬时力矩最小值 minimum torque phase
瞬时力偶 instantaneous couple
瞬时临界的 prompt super-critical
瞬时临界系统 instantaneous assembly
瞬时临界性 prompt criticality
瞬时流动透气性 transient flow permeability
瞬时流量 instantaneous delivery; instantaneous discharge; instantaneous rate of flow; momentary discharge
瞬时流密度 instantaneous stream density
瞬时流速 instantaneous flow rate; instantaneous(flow) velocity; transient current velocity
瞬时流线 instantaneous streamline
瞬时隆起 immediate heave
瞬时路面挠度 transient surface deflection
瞬时路面弯沉 transient surface deflection
瞬时脉冲 instantaneous(im) pulse
瞬时脉冲群 burst
瞬时脉冲群发生器 burst generator
瞬时脉动 instantaneous pulsation
瞬时毛坯能量 instantaneous billet energy
瞬时冒险 transient hazard
瞬时煤气热水器 instantaneous gas water heater
瞬时描述 instantaneous description
瞬时模量 instantaneous modulus
瞬时模型 instantaneous model
瞬时摩尔浓度 instantaneous molal concentration
瞬时耐压容量 instantaneous overvoltage capacity
瞬时能量 prompt energy
瞬时凝固 <混凝土等的> flash setting
瞬时凝结 flash set(ting); grab set
瞬时浓度 instantaneous concentration
瞬时偶极矩 instantaneous dipole moment; transient dipole moment
瞬时耦合 instant coupling
瞬时耦合模量 instantaneous coupling modulus
瞬时排放 instantaneous discharge; transient discharge
瞬时排放污染源 pollution source with instantaneous discharge
瞬时排油率 rate of instant oil discharge
瞬时喷出 instantaneous outburst
瞬时偏移控制 instantaneous deviation control
瞬时偏移控制电路 instantaneous deviation control circuit
瞬时偏移指示器 instantaneous deviation indicator
瞬时偏应力 instantaneous deviator stress
瞬时频带 instantaneous band
瞬时频率 instantaneous frequency
瞬时频率度 instantaneous frequency stability
瞬时频率剖面 instantaneous frequency section
瞬时频率稳定度 instantaneous frequency stability
瞬时频率指示接收机 instantaneous frequency-indicating receiver
瞬时频偏指示器 instantaneous deviation indicator
瞬时平均流速 temporal average velocity; temporal mean velocity
瞬时平均速度 temporal average velocity; temporal mean velocity
瞬时屏蔽 flash barrier
瞬时破坏 instantaneous damage; instantaneous failure; instantaneous rupture
瞬时剖面法 method of instantaneous profile
瞬时曝气 instantaneous aeration; temporal aeration
瞬时曝气器 instantaneous aerator
瞬时起爆 instantaneous firing; instantaneous ignition
瞬时起爆雷管 instantaneous cap
瞬时起动灯 quick start lamp
瞬时起动荧光灯 instant-start fluorescent lamp
瞬时起燃 instant-start
瞬时气穴 transient cavity
瞬时牵引力 transient pull
瞬时强度 instantaneous intensity; instantaneous strength; ultimate resistance
瞬时切断 instantaneous break
瞬时切入深度 instantaneous depth of cut
瞬时取样 instantaneous sample; instantaneous sampling
瞬时取样系统 prompt sample system
瞬时全部弃荷 instantaneous complete rejection; instantaneous complete rejection of load
瞬时全关 <阀的> instantaneous total closure
瞬时燃烧 instantaneous combustion
瞬时扰动 ringing
瞬时扰动时间 ringing time
瞬时热传导方程 transient heat conduction equation
瞬时热传递 transient heat flow
瞬时热水 instantaneous warm water
瞬时热水器 instantaneous geyser; instantaneous(sink) water heater; single-point heater
瞬时热效应 transient thermal response
瞬时热学性能 transient thermal behavio(u)r
瞬时人口净长率 instantaneous rate of increase
瞬时日差 instantaneous daily rate; instantaneous rate
瞬时容量 instantaneous capacity; momentary output
瞬时熔化 transient melting
瞬时柔量 instantaneous compliance
瞬时蠕变 transient creep
瞬时杀伤因素 instantaneous killing factor
瞬时闪火 instantaneous flash
瞬时上升速度 momentary rise of speed
瞬时摄影 instantaneous shot; instant photography; one-step photography
瞬时摄影片 instantaneous oxygen demand
瞬时渗流形式 transient seepage form
瞬时生长 transient growth
瞬时生化需氧量 immediate biochemical oxygen demand
瞬时声能密度 instantaneous sound energy density
瞬时声压 instantaneous acoustic(al) pressure; instantaneous sound pressure
瞬时失效 transient failure
瞬时失效率 instantaneous failure rate
瞬时时间 instantaneous time
瞬时式安全钳 instantaneous safety gear
瞬时视场 instantaneous field of view
瞬时视频检验 instantaneous video check
瞬时试验机 flash testing equipment
瞬时试样 snap sample
瞬时释放 instantaneous release; transient release
瞬时释放时间 instantaneous reoperate time
瞬时输出 instantaneous output; momentary output
瞬时熟炼点 instantaneous annealing point
瞬时束流 prompt beam current
瞬时数据传输率 instantaneous data-transfer rate
瞬时数均聚合度 instantaneous number average degree of polymerization
瞬时衰变率 transient-decay rate
瞬时衰落 instantaneous fading
瞬时水面 wave level
瞬时水面线观测 instantaneous surface profile observation
瞬时水头 instantaneous head
瞬时水位 instantaneous water-level; instantaneous water stage; momentary water-level; momentary water stage
瞬时水位降低 instantaneous drawdown
瞬时水位下降 instantaneous drawdown
瞬时速度 instantaneous speed; instantaneous velocity; momentary speed variation
瞬时速度变动率 instantaneous speed change
瞬时速度调节 instantaneous speed regulation
瞬时速度向量 instantaneous velocity vector
瞬时速度斜率 slope of instantaneous velocity
瞬时速度中心 instantaneous center [centre] of velocity

瞬时速率 instantaneous rate;momentary rate
瞬时速率量测设备 spot speed survey device
瞬时锁宅 momentary caging
瞬时太阳热传递 instantaneous solar heat transmission;instantaneous solar transmission
瞬时弹性 instantaneous elasticity
瞬时弹性变形 immediate elastic deformation
瞬时弹性模量 instantaneous modulus of elasticity;modulus of instantaneous elasticity
瞬时弹性挠度 instantaneous elastic deflection
瞬时弹性应变 instantaneous elastic strain
瞬时探测器 instantaneous detector
瞬时掏槽 instantaneous cut
瞬时特性 frequency response;instantaneous behavio(u)r;transient behavio(u)r;transient performance
瞬时天文北极 instantaneous astronomical north pole
瞬时条件 instantaneous condition
瞬时跳闸断路器 instantaneous trip circuit breaker
瞬时停顿 instantaneous stopping
瞬时通信[讯] instantaneous communication
瞬时突变裂变反应 spontaneous avalanche-like fission reaction
瞬时突水 transient bursting water
瞬时退火点 instantaneous annealing point
瞬时弯沉 <路面在行车通过时的> transient deflection
瞬时完全相位测量 complete instantaneous phase measurement
瞬时纬度 instantaneous latitude
瞬时位移 immediate movement;transient displacement
瞬时位置 instantaneous position
瞬时温度 transient temperature
瞬时温度变化梯度 transient temperature gradient
瞬时稳定储备系数 transient stability margin
瞬时稳定区 transient stability region
瞬时稳定性 instantaneous stability
瞬时污染 momentary pollution
瞬时污染源 momentary pollution source
瞬时误差 instantaneous error;transient error
瞬时误差测定 instantaneous error measurement
瞬时吸收 endomomental
瞬时系统 instantaneous system
瞬时线速度 instantaneous linear velocity
瞬时相干性 transient coherence
瞬时相关矩阵 instantaneous correlation matrix
瞬时相位 instantaneous phase
瞬时相位电流 instantaneous phase current
瞬时相位剖面 instantaneous phase section
瞬时相位数据体切片 slice of transient phase data block
瞬时相位指示器 instantaneous phase indicator
瞬时响应 instantaneous response
瞬时响应函数 instantaneous response function
瞬时像 instantaneous image
瞬时效率 instantaneous efficiency;momentary efficiency
瞬时效应 instantaneous effect;short-term effect;transient effect
瞬时斜率 instantaneous slope
瞬时谐振 transient resonance
瞬时形变 instantaneous deformation;transient deformation
瞬时形变速率 instantaneous rate of deformation
瞬时形象记忆 iconic memory
瞬时性 instantaneity
瞬时性能试验 instantaneous performance test
瞬时性态 transient behavio(u)r
瞬时性质 transient nature
瞬时需氧量 instantaneous photograph
瞬时徐变 transient creep
瞬时悬移质取样器 instantaneous suspension sampler
瞬时旋转 instantaneous rotation
瞬时旋转中心 instantaneous center [centre] of rotation
瞬时压扩 instantaneous companding
瞬时压扩器 instantaneous compandor
瞬时压力 instantaneous pressure;transient pressure
瞬时压力降(落) momentary fall of pressure
瞬时压力升高 momentary rise of pressure
瞬时压力下降 momentary fall of pressure
瞬时压实质量控制 instant compaction control
瞬时压缩 immediate compression
瞬时压缩机 instantaneous compressor
瞬时压缩扩展 instantaneous companding
瞬时压缩扩展器 instantaneous compandor
瞬时延迟 transient delay
瞬时延迟起爆筒 split-second delay cap
瞬时延发雷管 instant delay cap
瞬时延伸 instantaneous extension
瞬时摇摆 transient swing
瞬时引爆 instantaneous firing
瞬时应变 immediate strain;instantaneous strain;transient strain
瞬时应力 instantaneous stress;transient stress
瞬时应力波 transient stress wave
瞬时应力应变曲线 instantaneous stress-strain curve
瞬时涌浪 instantaneous surge
瞬时油气比 instantaneous gas-oil ratio
瞬时有效加速度 brief acceleration
瞬时预应力损失 immediate losses of prestress
瞬时运动 transient motion
瞬时运行电器 instantaneously operating apparatus
瞬时运行试验 instantaneous performance test
瞬时载荷 passing load
瞬时载流 instantaneous carrying current
瞬时载重 instantaneous load(ing)
瞬时再动作时间 instantaneous reoperate time
瞬时噪声 instantaneous noise
瞬时增压 momentary rise of pressure
瞬时遮光板 flash barrier
瞬时振荡 transient oscillation
瞬时振动 instantaneous vibration;transient vibration
瞬时振幅 instantaneous amplitude
瞬时振幅剖面 instantaneous amplitude section
瞬时蒸发加热器 flash heater
瞬时正切模量 instantaneous tangent modulus
瞬时支点 instantaneous fulcrum
瞬时值 actual value;instantaneous magnitude;instantaneous value;momentary value;spurt value
瞬时止动-开动控制 instant stop-start control
瞬时制动力 instantaneous braking power
瞬时制动率 instantaneous braking ratio
瞬时质量 instantaneous mass
瞬时中间产物 transient intermediate
瞬时中心 instant(aneous) center[centre]
瞬时重量 instantaneous weight
瞬时轴 instantaneous axis;temporary axis
瞬时转动轴 instantaneous axis of rotation
瞬时转化 instantaneous conversion
瞬时转换 instantaneous switching
瞬时转换开关 low-duty-cycle switch
瞬时转矩 instantaneous torque;transient torque
瞬时装载能力 instantaneous loading capacity
瞬时状态 instantaneous state;transient behavio(u)r;transient state;transition condition
瞬时自动频率控制 instantaneous automatic frequency control
瞬时自动增益控制 instantaneous automatic gain control
瞬时自信息 instantaneous self-information
瞬时总死亡系数 instantaneous total mortality coefficient
瞬时纵断面 <潮波引起的> instantaneous profile
瞬时阻力 transient drag
瞬时阻尼电流 instantaneous damper current
瞬时阻遇 transient repression
瞬时钻速 instantaneous penetration rate
瞬时最大风(力) instantaneous maximum wind
瞬时最大风速 instantaneous maximum wind velocity
瞬时最大风压 maximum instantaneous wind pressure
瞬时最大速度变化率 maximum momentary speed variation
瞬时最大速度变量 maximum momentary speed variation
瞬时最大转速 maximum instantaneous speed
瞬时最高速度 transient maximum speed
瞬时最小流量 momentary minimum discharge
瞬时作用 snap;transient action
瞬时作用的电磁铁 momentary duty electromagnet
瞬时作用力 transient action force
瞬蚀 flash rusting
瞬逝波 evanescent wave
瞬逝场 evanescent field
瞬逝光波 evanescent light wave
瞬态 momentary state;transient condition;transient state;transition condition
瞬态饱和 transient saturation
瞬态变量感传器 rate of change sensor
瞬态波形 transient waveform
瞬态测试仪器 instantaneous measuring apparatus
瞬态超速 transient overspeed
瞬态超压 transient overvoltage
瞬态超越度 transient overshoot
瞬态传播延迟时间 transient propagation delay
瞬态地球资源现象 transient earth resources phenomenon
瞬态电抗 transient reactance
瞬态电流 transient current
瞬态电流补偿 transient current offset
瞬态电路 transient circuit
瞬态电压 transient voltage
瞬态电压调整 transient voltage regulation
瞬态电阻 transient resistance
瞬态短路 transient short-circuit
瞬态法 transient method
瞬态反相峰值电压 transient peak-inverse voltage
瞬态反应 transient response
瞬态反应堆试验装置 transient reactor test facility
瞬态放大器 instantaneous amplifier
瞬态放电 spark
瞬态沸腾 transient boiling
瞬态分布 transient distribution
瞬态分析 instantaneous analysis;momentary state analysis;transient analysis;transient state analysis
瞬态分析器 instantaneous analyser [analyzer];transient analyser[analyzer]
瞬态高斯过程 transient Gaussian process
瞬态工作特性 instant operating characteristic
瞬态功率 transient power
瞬态共振 instantaneous resonance;transient resonance
瞬态过程 transient process
瞬态过调量 transient overshoot
瞬态过载反向电流限制 transient-overload forward-current limit
瞬态过载反向电压限制 transient overload reverse voltage limit
瞬态激光特性 transient laser behavio(u)r
瞬态激振 transient excitation
瞬态级 transient stage
瞬态记录 transient recording
瞬态记录器 instantaneous recorder;transient recorder
瞬态结构 transient buildup
瞬态空泡 transient cavity
瞬态孔隙压力 transient pore pressure
瞬态力 transient force
瞬态励磁电流 transient field current
瞬态脉冲 transient pulse
瞬态脉冲电压 transient surge voltage
瞬态偏差 transient deviation
瞬态曲线 transient curve
瞬态热传导 transient heat conduction
瞬态热管因子 transient hot channel factor
瞬态热阻抗 transient thermal impedance
瞬态蠕动 transient creep
瞬态声 transient sound
瞬态失真 over-strain;transient distortion
瞬态时间 transient time
瞬态实验 transient experiment
瞬态寿命 instantaneous lifetime
瞬态数据 transient data
瞬态数字转换器 transient digitizer
瞬态衰减电流 transient-decay current
瞬态特性 instantaneous characteristic;step response;surge characteristic;transient characteristic;transient response;unit function response
瞬态特征 unit impulse response

瞬态条件 transient condition
瞬态凸极性 transient saliency
瞬态推力 transient driving force
瞬态温度分布 transient temperature distribution
瞬态温升 transient temperature rise
瞬态稳定 transient stability
瞬态稳定度 transient state stability
瞬态稳定功率极限 transient stability power limit
瞬态稳定极限 transient stability limit
瞬态问题 transient problem
瞬态误差 transient error
瞬态系统偏差 transient system deviation
瞬态相位失真 transient phase distortion
瞬态响应 transient response
瞬态响应分析 instantaneous response analysis;transient response analysis
瞬态响应模型 transient response model
瞬态响应数据 transient response data
瞬态谐振 dynamic(al)resonance
瞬态性能 transient performance
瞬态压力 transient pressure
瞬态研究 transient study
瞬态应变 transient strain
瞬态应力 transient stress
瞬态应力能模量 transient stress-energy modulus
瞬态运动 transient motion
瞬态噪声 transient noise
瞬态针入法 < 快速测定土的热传导性方法 > transient-needle method
瞬态振荡 transient oscillation
瞬态振动 transient state vibration; transient vibration
瞬态振幅 instantaneous amplitude
瞬态正交模式法 normalized mode process in transient
瞬跳 instantaneously changing
瞬跳变 prompt jump
瞬跳变近似 prompt jump approximation
瞬跳杆衬圈 unlocking yoke ring
瞬跳杆传动轮桩 unlocking yoke driving wheel stud
瞬跳杆簧 unlocking yoke spring
瞬跳杆簧压片 unlocking yoke spring maintaining plate
瞬跳杆棘爪簧 unlocking yoke spring click
瞬跳杆驱动轮 unlocking yoke driver
瞬跳杆凸轮 unlocking yoke cam
瞬跳杆桩 unlocking yoke stud
瞬跳换日 instant data setting calendar
瞬跳换日装置及调对和保险机构 correctors and safety systems
瞬跳机构 instantaneous movement; instantaneous unlocking mechanism
瞬跳日历 instant-jump calendar
瞬跳日历机构 instantaneous calendar;instantaneous date mechanism
瞬跳双历机构 instantaneous date and day mechanism
瞬微脉动信号 transient micropulsation signal
瞬息 ephemera; transitory; trice; twinkling;wink
瞬息故障 transient fault
瞬息色谱法 instant chromatography
瞬息数据 ephemeral data
瞬息图像 transient image
瞬息作用 non-time delay;snappy
瞬现区 ephemeral region
瞬现温度 flash temperature
瞬心 instant(aneous) center[centre]
瞬心轨迹 centrode
瞬锈 flash rusting
瞬压曲线 isochron(e)
瞬压指示器 transient indicator
瞬载 transient load
瞬轴面 axode
瞬子 instanton

说

说白区 < 古希腊剧场舞台上 > logeion

说梗概 outline
说话干扰级 speech interference level
说还原反应 water-reduction reaction
说明比例尺 statement scale
说明变量 explanation variable
说明部分 declarative
说明操作 declarative name
说明操作码 declarative operation code
说明词 declarer
说明单元 declaration cell
说明符 declarator
说明符定义 declarator definition; declarator name
说明符扩展部分 specifier extension
说明符名称 declarator name
说明符注脚 declarator subscript
说明附注 explanatory note
说明宏指令 declarative macro instruction
说明卡(片)instruction card; identification card;instruction sheet
说明理由的仲裁裁决 award containing reasons
说明理由令 order to show cause
说明牌 indexing plate
说明片 instruction sheet
说明手册 instruction manual
说明书 guidebook;handbook;description[descriptive] manual; explanatory memorandum;instruction book; instruction manual;instruction sheet; manual; operation sheet; prospectus;spec;specification(sheet); synopsis;technical regulation
说明书编号 specification serial number
说明书规范 specification limit
说明书号 specification number
说明书小册子 instruction pamphlet
说明数组 declarative array
说明提要 descriptive abstract
说明图 clarification drawing;illustration;key diagram
说明图表 instruction sheet
说明图例 explanatory legend
说明图纸 explanatory drawing; clarification drawing < 工程修改补充的 >
说明位 detail bit
说明文件 supporting paper
说明项 descriptive item; descriptive term
说明性操作 declarative operation
说明性的 declarative;illustrative
说明性文摘 intermative abstract
说明性资料 accountability information
说明异常 specification exception
说明与描述语言 specification and description language
说明语句 declarative statement;specification statement
说明摘要 descriptive abstract
说明者 demonstrator; exponent; illustrator
说明注记 description note; descriptive data; descriptive name; descriptive report; descriptive statement; descriptive text; explanatory text;type matter
说明注解 explanatory comment
说明资料 detail file
说明子程序 specification subprogram(me)
说明字幕 side title;telltale title
说明作用域 scope of declaration
说听合用键 combined listening and speaking key

朔

朔潮 inferior tide

朔料填料生物滤池 plastic-medium trickling filter
朔日潮 change tide
朔望 the first and the fifteenth day of the lunar month
朔望潮 meridional syzygy tide;spring tide;syzygial tide;syzygy tide
朔望大潮 spring tide
朔望大潮最低水位 lowest low water spring tide
朔望低潮间隙 low-water full and change of moon
朔望低潮平均间隙 low-water full and change
朔望低水位高潮 low-water spring tide
朔望高潮 high water full and change; high water spring
朔望高潮间隙 common establishment;high water full and change; vulgar establishment
朔望间隙 syzygial interval;age
朔望平均低潮位 mean low water springs; mean monthly lowest water level
朔望平均高潮位 mean monthly highest water level
朔望平均满潮位 mean high water springs
朔望日期变化 Metonic cycle
朔望生潮力 long period force
朔望位置 syzygy position
朔望月 lunar month; synodic(al) month
朔望月潮间隙时平均低潮位 average low water lunitidal interval at syzygy
朔望月潮间隙时平均高潮位 average high water lunitidal interval at syzygy
朔望周期 synodic period

硕

硕士学位 mastership

蒴

蒴果 capsule

丝

丝氨酸 serine

丝板 filament plate;thread cutter
丝包电线 silk-covered cord
丝包漆包线 enamel silk-covered wire
丝包线 silk-covered wire;silk-insulated wire;silk yarn covered wire
丝饼头 cheese head
丝缠管 filament-wound pipe
丝厂 silk mill
丝厂废水 silk-mill wastewater
丝虫病 filariasis
丝绸 silk(cloth)
丝绸薄片 silk shavings
丝绸地图 rayon map;silken map
丝绸工艺 silking
丝绸墙壁衬垫 silk wall lining
丝绸墙壁覆面 silk wall(sur)facing
丝绸墙(壁饰面)布 silk wall covering
丝绸之路 < 中国 > Silk Road
丝传送带 wire belt
丝丛 line complex;tuft
丝丛法观察 tuft observation
丝带 filament band; ribband; ribbon; silk ribbon
丝堵 pipe plug;plug;screw(ed)plug
丝堵三通 tee with plug
丝对 screw nipple;stud < 气钻 >
丝缝线 silk suture
丝杆 guide screw; lead screw; screw mandrel;screw shaft
丝杆吊钩 cup hook
丝杆检查仪 lead screw tester
丝杆螺母 feed screw nut
丝杆驱动的 screw-driven
丝杆调节楔 screw wedge
丝杆销子 < 暖汽调整阀 > link pin
丝杠 lead screw;screw rod
丝杠车床 leading screw lathe
丝杠机构中的绳卡 temper-screw clamps
丝杠交换齿轮 lead screw gear
丝杠螺母副 screw pair
丝杠式插床 screw-driven slotter
丝杠式牛头刨床 screw-driven shaper
丝杠式刨床 screw-driven planer
丝根冷却器 nozzle cooler; nozzle shield
丝攻 thread tap
丝攻夹头盘 tapping chuck
丝管 fiber[fibre] tube
丝管筛 reel
丝光 silking;streakiness
丝光白 silk white
丝光白云母 gilbertite
丝光玻璃 satin finish glass;velvet-finish glass
丝光处理 < 织物的 > mercerize;mercerizing; mercerisation [mercerization]
丝光处理机 mercerizing machine
丝光断口 silky fracture
丝光沸石 flokite;mordenite
丝光缝纫线 mercerized sewing thread
丝光光泽 silk-like sheen
丝光化 mercerisation[mercerization]
丝光棉布 mercerized cotton
丝光面酸蚀 satin etch
丝光木棉 silk cotton
丝光漆 silky
丝光纱线 mercerized yarn
丝光酸蚀面 satin finish
丝光涂层 silk finish
丝光纤维素 mercerized cellulose
丝光性 silkiness
丝光研光机 silk finishing calender
丝光云母 silk mica
丝光整理 mercerized finish
丝光助剂 mercerizing assistant
丝光作用 mercerisation [mercerization]
丝硅镁石 loughlinite
丝焊器 wire bonder
丝毫 vestige
丝痕 sleek
丝极 filament
丝极变阻器 filament rheostat
丝极长度 filament length
丝极电积物 filament deposit
丝极电渣焊 electroslag welding with wire electrode
丝极电阻器 filament resistor
丝极放射性 filament activity
丝极交流声 heater hum
丝架 guide frame
丝晶 filament crystal
丝距 flight lead;wire interval
丝绢 silky

丝绢橡胶树 Kickxia rubber
丝绝缘 silk insulation
丝绝缘带 silk insulation tape
丝孔 threaded hole
丝扣 screw thread
丝扣保护件 thread protector
丝扣车床 threading machine
丝扣齿合长度 thread tooth length
丝扣法兰 thread flange
丝扣高度 thread height
丝扣管堵 screw plug
丝扣管接头 handling tight
丝扣接头 threaded joint(ing)
丝扣连接 screw(ed) connection; threaded connection
丝扣连接套管 screw joint casing; threaded joint casing
丝扣密封油 stabbing salve
丝扣拧得过紧 overtonging
丝扣漆 thread dope
丝扣欠紧 undertonging
丝扣润滑油 antigalling compound; joint grease
丝扣套筒 threaded sleeve
丝扣套筒接头 threaded sleeve joint
丝扣旋出 swivel off
丝扣余扣 residual thread
丝扣折断 thread failure
丝扣锥度 thread taper
丝扣最小抗拉强度 thread minimum tensile strength
丝框 reel
丝兰花 yucca
丝栎 silky oak
丝栎属 <拉> Grevillea
丝漏清漆 screening varnish
丝漏(印刷)油墨 silk screen ink
丝毛交织物 boratto
丝毛釉 glaze of furry appearance
丝煤 fusain;mother of coal
丝米 <1 丝米 = 0.0001 米> decimillimeter[decimillimetre]
丝绵 floss-silk
丝膜菌属 <拉> Cortinarius
丝钠铝石 dawsonite
丝屏罩 silk screen
丝切砖 wire-cut brick
丝绒 plush;velour;velvet
丝绒地毯 splush carpet
丝绒制的 velvet
丝筛布 silk bolting cloth
丝砷铜矿 trichalcite;tyrolite
丝束 tow
丝束成条机 tow transformer
丝束切断机 roving cutter
丝束直接成条 tow-to-top
丝锁气压计 Naudet's barometer
丝炭 dant;fossil charcoal;fusain;fusite;motherham;mother of coal
丝炭层 fusain layer
丝炭化 fusinization
丝炭化浑圆体 fusinized circleinite
丝炭化基质体 fusinized groudmassinite
丝炭化树脂 fusinized resin
丝炭化树脂体 fusi-resinite
丝炭化真菌物质 fusinized fungal
丝炭化作用 inertinitization
丝炭亮煤质煤 fusion bright coal
丝炭煤素质 fusinite
丝炭体 fusinite
丝炭组 fusain group;fusite group
丝锑铅矿 monimolite
丝条 silking
丝条集束机 tow collecting machine
丝头 nipple;threaded end
丝网 reticle; screen; silk net; silk screen;web;wire-mesh screen
丝网法 silk screening
丝网辊 roller screen
丝网剪断机 wire-mesh shearing device
丝网屏蔽法 silk screening process
丝网漆 silk screen paint
丝网涂漆 screen painting
丝网印花法 silk screen method
丝网印刷 screen print(ing); screen process printing;silk-screen printing
丝网印刷的墙纸 screen-printed wallpaper
丝网印刷法 stencil process
丝网印刷墙面板 silk-screened wall panel
丝网印刷清油 silk screen lacquer
丝网印刷涂料 silk screen paint
丝网印刷油墨 silk screen ink
丝网遮蔽法 hand screening
丝纹 rippling;silking;streakiness
丝锌铝石 fraipontite
丝絮 <观察气流、水流流线用的> tuft
丝牙口管子 grooved pipe
丝印 screen
丝云母 sericite mica
丝藻属 ulothrix
丝织厂 silk weaving mill
丝织地毯 polonaise rug
丝织锦 silk tapestry
丝织品 silk; silk fabrics goods; silk knit goods;silk stuff;silk textile
丝织业 silk textile
丝质次结构体 fusinite-posttelinite
丝质结构体 fusinite-telinite
丝质类 fusinite
丝质体 fusinite
丝质组 fusinoid group
丝状的 filamentary;filar;filiform;nematic;threadlike;filamentous
丝状电极 wire electrode
丝状断口 silky fracture
丝状断裂面 silky fracture
丝状反应 faden reaction;Mandelbaum's reaction
丝状分裂 mitoschisis
丝状腐蚀 filiform corrosion
丝状腐朽 stringy rot
丝状光泽 silk-like sheen
丝状活性污泥 filamental activated sludge;filamentous activated sludge
丝状火花 filamentary spark
丝状激光发射 filament lasing
丝状结构 filament
丝状晶体 trichite
丝状菌污泥 filamentous bacterial sludge
丝状菌污泥膨胀 filamentous bacterial sludge bulking
丝状脉 filiform pulse;thready pulse
丝状耦合 filamentary coupling
丝状器 filiform apparatus
丝状钎料 brazing wire
丝状区 filamentary region
丝状炭 peaty fibrous coal
丝状体 filament; filamentous form; mitoplast
丝状条纹 silking
丝状微生物 filamentous microorganism
丝状物 filament;silk
丝状形成 filamentation
丝状液晶 nematic liquid crystal
丝状衣壳 filamentous capsid
丝状阴极 filament(ary) cathode
丝状原纤维 filamentary fibril
丝状藻 filamentous algae
丝状藻泥炭 conferva peat
丝状真菌 filamentous fungi
丝状植物 nematophyte
丝状植物纤维 vegetable silk
丝锥 outside tap;screw tap;tap;thread tap
丝锥扳手 screw stock;tap handle;tap wrench
丝锥扳牙两用夹头 tap die holder
丝锥柄 tap holder
丝锥刀具 tap cutter
丝锥和铰刀铣刀 tap and reamer cutter
丝锥及扳牙 tap and die
丝锥夹头 clutch tap holder;tap chuck; tapping clamp
丝锥铰杠 clutch tap holder
丝锥接套 ball;bell screw;bell socket;bell tap;die nipple
丝锥磨床 tap grinder; tap-grinding machine
丝锥前端刃磨机 tap-nose grinder
丝锥刃瓣 cutter blade
丝锥手柄 tap handle
丝嘴 fibre feed arm

司

司磅员 weighman;weigher
司泵工 pump assistant; pumper; pumpman
司泵员 pumper
司尺员【测】chainman;rodman;sight carrier;staffman;tapeman
司筹员 tally clerk
司丹康钢锚板 Stelcon steel anchor plate
司丹康工业地面覆盖物 Stelcon industrial floor cover(ing)
司丹康混凝土筏式基础 Stelcon concrete raft
司蒂吉斯变换 Stieltjes transform
司蒂吉斯积分 Stieltjes integral
司东尼式闸门 <其辊轮是独立的,可沿门槽上下移动,不和闸门或闸墩相连> Stoney gate
司法部 <美> Department of Justice
司法部门 judicial department
司法裁决 judicial award;judicial decision
司法程序 judicial process; judicial program(me)
司法对合同义务的延期执行权 moratorium
司法、公证、公安 judiciary, notary and public security
司法机关 judicial authority; judicial organ
司法鉴定 judicial expertise
司法解决 judicial settlement
司法仲裁法院 judicial arbitration court
司各契上浆工序 Scotch dressing
司罐工 lander
司光员 light keeper
司号员 signalman
司机 cab operator; car driver; carman; driver; engine attendant; engine driver; engine man; engine runner; equipment operator; motor man;motor mechanician;operator; runner
司机报警系统【交】driver-alert system
司机舱 cockpit;driver's cabin
司机操纵室 control cab
司机操纵台 controlling desk;driver's control desk
司机长 head driver;head engineer
司机乘务报单 driver's report
司机乘务登记簿 driver's roster
司机单人操纵的 driver-only-operated
司机导致的设备故障 driver-induced equipment failure
司机阀 driver's valve; engineer's valve;motorman's valve
司机反应距离【交】driver reaction distance
司机防护网 operator protection screen
司机后方视野 rearward perception
司机脚踏板 driver's footplate
司机靠臂垫 arm cushion
司机控制台 driver's control desk
司机鸣笛标 whistle post sign
司机鸣笛预告标 driver's whistle warning board
司机培训辅助设备 driver training aid
司机棚凸头 nose end in front of cab
司机棚质量 canopy mass
司机前挡板 dash
司机前仪表盘 dash
司机全面负责运转制 driver-only operation
司机十分舒适 total operator comfort
司机室 cab(in);cabinet;cab mass; cockpit;driver's cab;driver's cage; driver's compartment;driving cab; engineer's cab; engineman's cab; motorman's cab(in); operating cab;operator's cab;operators cabin
司机室安装带 cab strap
司机室布置 driving cab layout
司机室操纵装置 cab fitting
司机室窗 cab window
司机室带电线束 cab harness
司机室挡烟板 cab smoke deflector
司机室灯 cab lamp
司机室地板 foot plate
司机室地板梁 cab floor beam
司机室顶部 cab roof
司机室渡板 cab apron
司机室反光镜 cab mirror
司机室扶手 cab handhold
司机室扶梯 cab step
司机室过道 operator's walkway
司机室后置式拖拉机 operator rear tractor
司机室后置式装载机 operator rear loader
司机室护板 cab guard
司机室或司机棚质量 cab or canopy mass
司机室加压器 cab pressurizer
司机室结构 cab structure
司机室门 cab door
司机室内加压器电线束 cab pressurizer harness
司机室内信号设备 in-cab signal(l)ing
司机室配置 driving cab layout
司机室前窗 cab front window
司机室前伸的载货车 lorry with forward cab
司机室前置式拖拉机 operator front tractor
司机室前置式装载机 operator front loader
司机室强光灯 cab flood
司机室取暖电动机 cab heater motor
司机室取暖器 cab heater
司机室取暖器变阻器 cab heater rheostat
司机室通道 operator's walkway
司机室通风器 cab ventilator
司机室托架 cab brace;cab bracket
司机室托座 cab support
司机室系统 operator arrangement
司机室信号 cab signal
司机室檐沟 cab gutter
司机室仪表装置 cab instrumentation
司机室右后地板 right rear floor
司机室与发动机并排布置 cab alongside-engine
司机室在发动机上方的 cab over engine
司机室遮阳板 cab visor

司机室置于发动机之后的机器 cab-behind-engine
司机室自动控制按钮 automatic monitor button
司机室组件 cab module
司机室左后地板 left rear floor
司机室座舱 cab module
司机手册 operator manual
司机受噪声影响 operator noise exposure
司机台控制器 platform container
司机通道 walkway
司机位置的噪声级 operator noise level;operator sound level
司机位置的噪声水平 operator noise level;operator sound level
司机位置靠后的机械 operator rear
司机位置靠前的车辆 operator front
司机习惯性 driver habit
司机信息系统【交】driver information system
司机行车日志 driver's log
司机行车注意小贴条 driver's cautious-running slip
司机真空制动阀 driver's vacuum brake valve
司机支持系统【交】driver support system
司机执照 operator's license[licence]
司机制动阀杆 engineer's brake valve spindle
司机制动阀手柄 engineer's brake valve handle
司机制动阀托架 engineer's brake valve bracket
司机助手 helper
司机座 cab seat;driver seat;operator seat
司机座处高地高度 height of seat
司机座处离地高度 seat height
司机座靠后的单辊压路机 ride behind single-drum roller
司机座椅 seat
司祭席 <教堂的> presbytery
司可巴比妥 quinalbarbitone;secobarbital
司克莱龙(铝基)合金 Scleron
司库 treasurer
司链员 chainman
司令部 command;headquarters
司令舰 staff ship
司令桥楼 flag bridge
司令台 admiral's bridge;flag bridge
司令员 commander
司炉 chief stoker; fire man; firing; furnace attendant;rabbler
司炉的呼唤电门 <机车乘务员通知信号员的呼唤电门> fireman's call plunger
司炉工 fire man
司炉工具 firing tools
司炉工平台 stoker's platform
司炉工人 stoker
司炉工作台 fireman's platform
司炉工作站位 fireman's platform
司炉用具 fire tool
司门 door operator
司品德 <商品名> Spyndle
司旗员 flagman
司签人 tallyman
司膳人 butler
司水 watermaster
司太克分析 Stekelian analysis
司太立合金 Stellite
司太立合金表面硬化 Stellite hard facing
司太立合金阀 Stellited valve
司太立合金覆面阀 Stellite faced valve
司太立合金气门 Stellite faced valve
司汀姆森奈塑料 Stimsonite
司托顿侧吹转炉 Stoughton converter
司行轮 escape(ment)
司仪 officiate
司闸车 brake van
司闸员 <俚语> brake(s)man
司闸员守车 brakeman's caboose
司长 director general
司钻 clutcher;rig runner
司钻班报表 driller's tour report
司钻电控盘 electric(al) driller's panel
司钻记录 driller's log
司钻位置 driller's position
司钻助手 rotary helper
司钻酌定 driller's discretion

私

私产 private property

私产房 private housing
私道 by-lane;by-walk
私的 by(e)
私法 private law
私房 private house;private housing
私房自住者 owner-occupier
私货 contraband;run goods;smuggled goods
私家花园 private garden
私家园林 private garden
私立病院 nursing home
私立疗养所 nursing home
私立学校 pay school;private school
私掠船 privateer
私密性 privacy
私人财产税 private property tax
私人餐室 private dining room
私人餐厅 private dining room
私人厂牌 private brands
私人场地 private area
私人车库 private parking garage
私人车主 private car ownership
私人成本 private cost
私人出租住宅 private rented housing
私人储备的水 private provision of water
私人储藏室 private storage garage
私人存储器 private memory
私人贷款 private loan
私人担保 sponsion
私人的 private
私人电话系统 private telephone system
私人短期资本流动 private short-term capital movement
私人发展 private development
私人房产业 private housing estate
私人房屋 lodging house; private building
私人飞机库 private airplane hangar
私人费用 private cost
私人公司 private company
私人股份 private shares
私人顾问 private councillor
私人会议 conclave
私人货物 private cargo
私人机构 private entity
私人间接投资 indirect private investment
私人建房屋 private housing
私人建筑 private building
私人交通 private transit
私人俱乐部 private club
私人开发 private development
私人开业的建筑师 private architect
私人空地 private open space
私人旷地 private open area
私人矿井 privy pit
私人楼座 <剧院的> private balcony
私人码头 private dock
私人密码 private code
私人灭火力量 private fire-fighting force
私人民航机场 privately owned civil aerodrome
私人企业 individual enterprise; private enterprise
私人汽车 private car; jitney (bus) <可以电话叫乘的行驶在半固定线路上的>
私人汽车库 private garage
私人汽车停车场 private car park
私人汽车停车处 private car park
私人汽车停车地 private car park
私人契约 private contract
私人签署 private seal
私人入口 private access
私人数据 private data
私人所有物 personal effects
私人田地 individual field
私人条款 private terms
私人停车场 private parking area;self parking
私人停车地(点) private parking place
私人停车地段 private parking lot
私人停车地区 private parking lot
私人停车区 private parking area
私人停机场 air park
私人通道 private access;private corridor
私人通信[讯]处 private address
私人通信[讯]系统 personal signal(l)ing system
私人投资 private finance;private investment
私人物品 personal effects
私人下水道 private sewer
私人下水管道 <公路下的> sewer connection
私人乡村屋宇 private village housing
私人小教堂 private chapel
私人协议 private treaty
私人信托款 private trust-fund
私人阳台 private balcony
私人医院 nursing home;private hospital;proprietary hospital
私人拥有的港口 privately owned harbo(u)r
私人用飞机场 personal-type airport
私人用灰坑 privy pit
私人用水 private use of water
私人用小路 private path
私人游憩用地 private open space
私人游泳池 private swimming pool
私人浴室 private bathroom;balnearium <古罗马时>
私人运输业 private carrier
私人债务 private debt
私人占用 private occupancy
私人占有汽车的平均数 car ownership rate
私人账户 private account
私人账目 private account
私人支票 individual check; personal check
私人直接投资 private direct investment
私人住房 private dwelling house
私人住所 private residence
私人住宅 private dwelling; family mansion <罗马人的>
私人住宅电梯 private residence elevator
私人住宅斜坡升降机 private residence inclined lift
私人专利 private monopoly
私人资本 private capital
私人资金 private fund
私室 garderobe;growlery;inner sanctum;private room
私售 illicit sale
私吞 feather one's nest;misappropriation
私下出售的 under-the-counter
私下协商 private consultation
私营 private operation
私营报酬(利润)率 private rate of return
私营部门 private sector
私营部门业主 private sector client
私营船舶 private vessel
私营导航设备 <美> private aid-to-navigation
私营道路 private road
私营的 private
私营电力生产商 independent power producer
私营工程合同 contract of privately performed work
私营工商业 privately owned industrial and commercial enterprises
私营公司 private company; private corporation
私营公用事业 private utility
私营公用事业的 quasi-public
私营化 privatization
私营建筑师 architect in private practice
私营经济 self-employed and private business
私营林场 tree farm
私营码头 private dock;private quay;private wharf
私营贸易 private trading
私营企业 independent power producer;private enterprise;private interest;private sector
私营企业家 private entrepreneur
私营企业界 private sector
私营商船 private boat;sailing on her own bottom
私营商港 private port
私营收集垃圾公司 private collection firm
私营铁路 private(-owned) railway
私营污水处理及排放系统 private sewage disposal system
私营线共用 community of private siding
私营银行 private bank
私营助航设施 private aid-to-navigation
私用 appropriation;personal use
私用道路 private road
私用井 private well
私用空间 private space
私用楼梯 private stairway
私用面积 <室内外为每户保留的面积> private area
私用室 private room
私用水 private use of water
私用水龙头 private tap
私用园地 allotment garden
私用运输 private transportation
私有保护信号系统 proprietary protective signal(l)ing system
私有部分 private sector
私有财产 personal effects; personal estate; private domain; private property
私有岔线 private siding
私有产业周 private property week
私有车 private car
私有车辆 privately owned vehicle
私有船舶 privately owned ship
私有道路 accommodation road
私有的成套公寓 condominium
私有房产标志牌 private identification sign
私有房屋 private house
私有港(口) private port; privately owned port

私有给水工程 private water supply
私有公用事业 private utility
私有供水 private water supply
私有共管 <一种公寓产权的形式> condominium
私有管道式下水道 private conduit-type sewer
私有化 privatization
私有货车 private-owned wagon
私有货车公司 private car line
私有货车主 owner of private wagon
私有机车 private locomotive
私有集装箱 privately owned container
私有集装箱结构标准 structure standard of privately-owned container
私有监狱 <古罗马囚禁奴隶的> ergastulum
私有林控制法规 forest regulation
私有密钥【计】private key
私有配水系统 private water distribution system
私有水 private water
私有铁路 privately owned railway
私有通信[讯]技术协议 private communication technology protocol
私有土地 private ground
私有污水处理厂 private sewage disposal
私有系统 private system
私有信号系统 proprietary signal(l)ing system
私有银行 private bank
私有制 private ownership; system of private ownership
私有专用报警系统 proprietary alarm system
私运进口 smuggle goods into a port
私宅游泳池 private residential swimming pool
私章 signet
私自拨款 background paper
私自离船船员 deserter
私自起货 land goods without permit
私自下货 ship goods without permit

咝

咝声 singing

思

思想体系 ideology

斯

斯宾塞氏边坡稳定分析法 Spencer method of slope stability analysis
斯蔡斯玻璃 <一种有光彩的铅质玻璃> Strass
斯乘 <力的单位,1 斯乘 = 1000 牛顿> sthene
斯粗司钢 <一种高强度钢> Strux
斯德哥尔摩 <瑞典首都> Stockholm
斯德哥尔摩国际水研究所 Stockholm International Water Institute
斯德哥尔摩黑色原料 Stockholm black
斯德哥尔摩焦油 Stockholm pine tar; Stockholm tar
斯德哥尔摩水论坛 Stockholm Water Front
斯德兰斯基石 stranskiite
斯蒂尔-埃斯林根摆动吊运车 Still-Esslingen swing lift
斯蒂尔森扳手 Stillson wrench
斯蒂尔森管子钳 Stillson wrench
斯蒂尔斯法 Stiles method
斯蒂尔亚德含水量测定仪 Steelyard moisture meter
斯蒂范阶 <晚石炭世>【地】Stephanian
斯蒂费尔式轧机 Stiefel mill
斯蒂费尔自动轧管法 Stiefel process
斯蒂芬定律 Steffen's law
斯蒂芬法 Steffen's process
斯蒂芬集装箱起重装置 Stephen conjack
斯蒂芬逊减摩轴承合金 Stephenson's alloy
斯蒂芬逊连杆 Stephenson link
斯蒂格马 <长度单位,1 斯蒂格马 = 10^{-12}米> stigma
斯蒂克轧机 all-pull mill
斯蒂普利特玻璃 <一种轧有纹饰的不透明玻璃> Stippolyte
斯蒂文森百叶箱 Stevenson screen
斯蒂文思真空圆网纸机 Stevens former
斯蒂晓夫石 stishovite
斯丁洛特代数 Steenrod algebra
斯丁洛特平方 Steenrod square
斯笃尔特运动【地】Sturt orogeny
斯笃尔特万特型鼓风机 Sturtevent blower
斯笃尔特万特型离心分级机 Sturtevant whirlwind classifier
斯笃尔特万特型离心式选粉机 Sturtevant type air separator
斯菲尔德-斯科特模型 Sehofield-Scat model
斯芬克斯林荫大道 avenue of Sphinxes
斯芬克斯林荫路 avenue of Sphinxes
斯芬克斯门 sphinx gate
斯芬克斯(狮身人面像)sphinx
斯夫特克里特 <一种快凝水泥> Swiftcrete
斯硅钾钍钙石 screavyite
斯基道阶 <早奥陶世>【地】Skiddavian(stage)
斯金纳箱 Skinner box
斯卡尔恩体系建筑 Skarne
斯卡格雷红色石灰石 Scaglia
斯卡摩齐柱式 Scamozzi
斯开普顿-贝仑法 Skempton (and) Bjerrum's method
斯开普顿法 Skempton method
斯堪的纳维亚灰浆 Scandinavian plaster
斯堪的纳维亚建筑 Scandinavian architecture
斯堪的纳维亚石膏粉饰 Scandinavian plaster
斯堪的纳维亚油漆工艺师联合会 Skandinaviska Lackteknikers Forbund
斯堪的亚屋面砖 Scandia roofing tile
斯康诺 <一种预制木屋> Scano
斯柯帕龙镜头 Skoparon lens
斯科林克尔砂岩 Skrinkle sandstone
斯科舍板块 Scotia plate
斯科特-达西法 Scott-Darcy process
斯科特接线法 <把三相电变为二相电的> Scott's connection
斯科特拉伸强力试验仪 Scott tensile tester
斯科特黏[粘]度计 Scott's viscosimeter
斯科特容量计 Scott volumeter
斯科特式纱线强力仪 Scott system serigraph
斯科特试验机 Scott's tester
斯科特水泥 <石灰加 5% 石膏> Scott's cement
斯科特斯康法 <一种不锈钢表面硬化法> Scottsconizing
斯科特效应 Scott's effect
斯科特型杆式联结装置 Scott's jockey; Scott jockey
斯科特预制住房 Scottwood
斯科特预制住宅 Scottwood
斯科特真空圆网成型装置 Scott's former
斯科特蒸发器 Scott's evapo(u)rator
斯克莱布诺材积表 Scribner log rule
斯克列隆铝基合金 Skleron
斯寇夫窑 Scove kiln
斯库巴潜水 Scuba diving
斯库尔携带式硬度计 Sekur machine
斯库普法喷镀 Schoop process
斯库普喷镀法 Schoop process
斯拉格合金 <一种锌基轴承合金> Slage metal
斯拉兹曼试剂法 Slatzmen reagent method
斯莱夫陆核 Slave nucleus
斯莱特定则 Slater's rule
斯莱特斯 <一种沥青防水剂> Slatex
斯莱特行列式 Slater determinant
斯莱特行列式波函数 Slater-determinant wave function
斯勒格 Geepound; Slug <质量单位>【物】
斯里兰卡 <亚洲> Sri Lanka
斯里兰卡锆石 Ceylon zircon
斯里兰卡吉纳树胶 Ceylon kino
斯里兰卡楝木 Ceylon cedar; Ceylon mahogany
斯里兰卡石 srilankite
斯里兰卡天然石墨 Ceylon graphite
斯里兰卡铁力木 Ceylon ironwood
斯里兰卡乌木 Ceylon ebony
斯力克锻轧机 slick mill
斯利希特法 Slichter method
斯硫锑铅矿 sterryite
斯硫铜矿 spiomkopite
斯隆模式 <一种管理体制模式> Sloan's model
斯伦贝谢公司数控测井仪 Cyber service unit
斯马雷钢铁脱硫法 Smalley process
斯马利安法 Smalian's method
斯马利安公式 Smalian's formula
斯孟布法 SMB method; Sverdrup-Munk-Bretchneider method
斯米电池 Smee cell
斯密斯加热法 Smitherm process
斯木兹硅锰锆铁合金 Smz alloy
斯内克河 Snake River
斯奈德电感炉 Snyder induction furnace
斯奈德破碎法 Snyder crushing method
斯奈德取样器 Snyder sampler
斯奈德寿命试验 <一种绝缘油老化试验法> Snyder life test
斯涅耳定律 Snell's law
斯涅耳折射定律 Snell laws of refraction
斯涅伦表示法 Snellen notation
斯涅伦小数 Snellen fraction
斯诺克里特 <一种白色或彩色混凝土> Snowcrete
斯诺克效应 Snoek effect
斯诺森 <一种防水的彩色水泥涂料> Snowcem
斯帕克曼分类 Spackman's classification
斯潘德克斯弹性纤维 Spandex
斯潘司金属 Spence metal
斯佩尔空心板 Spair hollow panel
斯佩克吸收测定仪 Spekker absorptiometer
斯佩里电解法 Sperry process
斯佩里铅基轴承合金 Sperry's metal
斯佩里型圆形淘汰盘 Sperry buddle
斯皮兹柏金西斯山脉 Montes spitzbergensis
斯颇林运行(平稳性)指标 Sperlin ride index
斯普雷帕克填料 Spraypak packing
斯普利特的第渥克利欣宫 <南斯拉夫> Palace of Diocletian at Split
斯普隆公式 Sprounq's formula
斯普伦格尔泵 Sprengel pump
斯普伦格尔炸药 Sprengel explosive
斯羟铜矿 spertiniite
斯砷锰石 sterlinghillite
斯石英 stishovite
斯氏阀动机构 Stephenson valve motion
斯氏阀动装置 link motion
斯氏体 steadite
斯式加热器 <一种安装在墙上的预制暖气道> Seduct heater
斯式预制暖气道 se-duct
斯水氧钒矾 stanleyite
斯塔布斯-佩里系统 Stubbs-Perry system
斯塔德尔夹 Stader's splint
斯塔锻造铝基合金 Studal
斯塔尔-爱森矿渣 <炼钢生铁生成的矿渣> Stahl-Eisen slag
斯塔尔顿 Stahlton
斯塔尔顿式厂 Stahlton plant
斯塔尔顿式窗过梁 Stahlton prestressed window lintel
斯塔尔顿式地板 Stahlton prestressed floor
斯塔尔顿式工厂 Stahlton factory
斯塔尔顿式过梁 Stahlton prestressed lintel
斯塔尔顿式(空心楼)板 Stahlton plank
斯塔尔顿式梁 Stahlton prestressed beam
斯塔尔顿式门过梁 Stahlton prestressed door lintel
斯塔尔顿式墙板 Stahlton prestressed wall panel; Stahlton prestressed wall slab
斯塔尔顿式墙体单元 Stahlton wall unit
斯塔尔顿式屋顶 Stahlton prestressed roof
斯塔尔顿式屋面板 Stahlton prestressed roof slab
斯塔尔顿式砖板 Stahlton prestressed clay plank
斯塔尔顿式砖窗过梁 Stahlton plank window lintel
斯塔尔顿式砖楼板 Stahlton floor
斯塔尔顿式砖屋顶 Stahlton roof
斯塔尔顿式砖屋面板 Stahlton roof slab
斯塔尔汽轮机 Stahl turbine
斯塔弗郡蓝砖 Staffordshire blues
斯塔福阶 <晚石炭世>【地】Staffordian
斯塔克-爱因斯坦定律 Stark-Einstein law
斯塔克藩数 Stark number
斯塔克感生频移 Stark induced frequency shift
斯塔克光谱学 Stark spectroscopy
斯塔克劈裂 Stark splitting
斯塔克位移 Stark shift
斯塔克效应 Stark effect
斯塔克增宽 Stark broadening
斯塔林定律 Starling's law of the heart
斯塔梅试验 Stamey test
斯塔密卡邦法 Stamicarbon process
斯塔密卡邦公司 Stamicarbon
斯塔佩利火山岩统 Stapeley volcanic series
斯塔山诺电炉 Stassano furnace
斯塔斯弗特沉积物 Stassfurt deposits
斯塔斯弗特钾盐 Stassfurt potash salt
斯塔斯弗特盐 Stassfurt salt
斯塔特朗喷枪 Statron gun
斯太尔轴承合金 Star alloy
斯泰诺林压力水泥稠化试验仪 Stanolind pressure thickening time tester
斯坦弗水准仪 Stampfer level
斯坦福接头 <一种承插管接头,承端设斜口沥青圈使管子在一直线上> Stanford joint

斯坦海尔透镜 Steinheil lens
斯坦利紧松指示器 Stanley up and down work
斯坦利-肯特纤维 Stanley Kent's fibres
斯坦曼钉 Steinmann pin
斯坦纳林德压力试验 Stanolind pressure test
斯坦纳气泡黏[粘]度汁 Steiner bubble viscometer
斯坦山地体 Stanley mountain terrane
斯坦脱<放射性单位,1 斯坦脱 = 3.63 × 10^{-27}Ci > stat
斯忒藩-玻耳兹曼常数 Stefan-Boltzmann constant
斯忒藩-玻耳兹曼定律 Stefan-Boltzmann law
斯忒藩-玻耳兹曼辐射定律 Stefan-Boltzmann law of radiation
斯忒藩公式 Stefan's formula
斯忒藩数 Stefan number
斯特德菲尔特炉 Stetefeldt furnace
斯特德曼方式 Steadman system
斯特德曼填料 Steadman packing
斯特格撕裂试验 Stegers crazing test
斯特克尔式轧机 Steckel mill
斯特克尔轧制法 Steckel rolling
斯特拉基石 Stracekite
斯特拉尼姆镁铝合金 Stalanium
斯特拉斯堡松节油 Strasbourg turpentine
斯特兰米脱楼板 Stramit slab
斯特兰米脱木构架隔墙 Stramit partition(wall) in timber framing
斯特兰特方法<轻骨料烧结法> Strand process
斯特劳伯曲线 Straub curve
斯特劳哈数<流体力学中一个无量纲数> Strouhal number
斯特劳斯螺旋桩 Strauss bore(d) pile
斯特劳斯试验 Strauss test
斯特雷顿岩群 Stretton group
斯特里普特<一种专利除漆剂> Stript
斯特里特-费尔普斯方程 Streeter-Phelps equation
斯特里特锌白铜 Sterlite
斯特林发动机 Stirling's engine
斯特林公式 Stirling's formula
斯特林锅炉 Stirling's boiler
斯特林合金 Stirling's metal
斯特林黄铜 sterling metal
斯特林近似法 Stirling's approximation
斯特林精炼锌法 Stirling's process
斯特林炼锌电弧炉 Stirling's furnace
斯特林内插公式 Stirling's interpolation formula
斯特林数 Stirling's number
斯特林铜镍锌合金 Sterlin
斯特林外燃(发动)机 Stifling external-combustion engine
斯特林锡锌铝合金焊料 Sterling aluminium solder
斯特林循环 Stirling's cycle
斯特龙伯格不对称漂移 Stromberg asymmetrical drift
斯特龙伯格图 Stromberg diagram
斯特龙博利式 Strombolian
斯特龙根半径 Stromgren radius
斯特龙根测光 Stromgren photometry
斯特龙根球 Stromgren sphere
斯特龙根四色指数 Stromgren four colo(u)r index
斯特鲁斯高强度钢 Strux
斯特罗含铁锰黄铜 Sterro metal
斯特罗黄铜合金 Sterro alloy
斯特曼填料 Stedman packing
斯特梅尔型黏[粘]度计 Sturmer viscometer
斯特耶铝铜合金 Stay alloy
斯梯巴天线 Sterba antenna
斯梯巴帷形天线 Sterba curtain
斯梯尔弗雷德间冰阶 Stillfried b interstade
斯提里运动 Styrian orogeny
斯提奈房屋<一种商品装配式房屋> Steane house
斯通表示定理 Stone's representation theorem
斯通电路 Stone circuit
斯通定理 Stone's theorem
斯通-杰尔里效用函数 Stone-Geary utility function
斯通尼式提升平板闸门 Stoney vertical lift gate
斯图迪特焊条合金 Stoodite
斯图尔特风车 Stuart windmill
斯图尔特人口分布定律 Stewart's law
斯图尔特石 stewartite
斯图尔特铸造铝合金 Stewart alloy
斯图尔特桩<一种现场浇制混凝土桩> Stewart's pile
斯图加特粗糙度测定仪 Stuttgart roughometer
斯图姆定理 Sturm's theorem
斯图姆-刘维尔微分方程 Sturm-Liouville differential equation
斯图姆-刘维尔问题 eigenvalue problem;Sturm-Liouville problem
斯图姆-刘维尔系 Sturm-Liouville system
斯图姆式叶片液压电动机 Sturm motor
斯图姆序列 Sturm's sequence
斯托比电炉 Stobie furnace
斯托达特浅盘布水器 Stoddard tray distributor
斯托达特溶剂 Stoddard solvent
斯托多拉法 Stodola method
斯托耳兹光栏 Stolz stop
斯托克纳波 Stohesina wave
斯(托克斯)<旧运动黏[粘]度单位,现改用米^2/秒,即二次方米每秒> stoke
斯托克斯波 Stokes' wave
斯托克斯参数 Stokes' parameter
斯托克斯参数测量仪 Stokes' meter
斯托克斯定理 Stokes' theorem
斯托克斯定律 Stokes' law
斯托克斯定律诺谟图 Stokes law homograph
斯托克斯方程 Stokes' equation
斯托克斯辐射 Stokes' radiation
斯托克斯公式 Stokes' formula
斯托克斯函数 Stokes' function
斯托克斯积分 Stokes' integral
斯托克斯积分定理 Stokes' integral theorem
斯托克斯颗粒沉淀定律 Stokes law of particle sedimentation
斯托克斯-拉曼散射 Stokes-Raman scattering
斯托克斯粒径 Stokes' diameter
斯托克斯流动 Stokes' flow
斯托克斯流函数 Stokes' stream function
斯托克斯流体 Stokesian fluid
斯托克斯黏[粘]滞度 Stokes' viscosity
斯托克斯漂移 Stokes' drift
斯托克斯频移 Stokes' shift
斯托克斯谱线 Stokes' line
斯托克斯日照仪 Stokes' heliometer
斯托克斯数 Stokes' number
斯托克斯凸轮式压机 Stokes' press
斯托克斯椭球常数 Stokes' constant for ellipsoid
斯托克斯问题 Stokes' problem
斯托克斯跃迁 Stokes' transition
斯托克斯阻力定律 Stokes' law of resistance
斯托麦旋转黏[粘]度计 Stomer visco(si)meter
斯托默黏[粘]度计 Stormer visco(si)meter
斯托默黏[粘]性 Stormer viscosity
斯托普丝分类 Stopes' classification
斯托普丝-赫尔冷分茨 Stopes-Heerlen system
斯托沃克闸门 Stauwerke gate
斯托希-莫拉夫斯基试验 Storch-Morawsk test
斯脱劳贝尔滚轴竖旋桥 Stroke roller bascule bridge
斯脱劳贝尔滚轴仰开桥 Stroke roller bascule bridge
斯瓦德摇杆硬度计 Sward rocker
斯瓦德硬度 Sward hardness
斯瓦德-扎德莱尔摇杆硬度计 Sward-Zeidler rocker
斯威科振动筛 Sweco vibrating screen
斯威士兰<非洲> Swaziland
斯维可芬造山运动【地】Svecofennian orogeny
斯维拉正向极性亚带 Thvera normal polarity subzone
斯维拉正向极性亚时 Thvera normal polarity subchron
斯维拉正向极性亚时间带 Thvera normal polarity subchronzone
斯维特兰型叶片过滤器 Sweetland filter
斯铀硅矿 swamboite
斯皂石 stevensite

锶白 strontium white

锶测定法 strontium method
锶长石 slawsonite
锶单位 strontium unit;sunshine unit
锶沸石 brewsterite
锶硅钛铈铁矿 strontiochevkinite
锶黄 strontium chromate; strontium yellow
锶钾铀矿 agrinierite
锶矿石 strontium ore
锶矿异常 anomaly of strontium ore
锶帘石 hancockite
锶磷灰石 fermorite;strontium-apatite
锶龄<地质年代测定的年龄> strontium age
锶-铅同位素相关性 strontium-lead isotope correlation
锶砷磷灰石 fermorite
锶铈磷灰石 belovite
锶水泥 strontium cement; strontium silicate cement
锶水硼钙石 strontioginorite
锶铁钛矿 crichtonite
锶同位素增长线 strontium development line
锶氧同位素相关性 strontium-oxygen isotope correlation

嘶嘶声 hiss;fizz<漏气>

撕成长带 ribbon cut

撕成两半 tear in two
撕断 tear fracture
撕断力 tearout
撕断纸带接转方式 torn tape relay system
撕分 bursting
撕毁合同 scrap a contract
撕开 rive;tear
撕开盖 tear-off lid
撕开型裂纹 tearing mode of crack
撕力 tearing force
撕裂 avulsion; lacerate; laceration; peel;rack;rend;rip;tear rupture
撕裂(III 型)变形<用于断裂力学> mode(III) of anti-plane slide
撕裂斑 tear stains
撕裂不稳定性 tearing instability
撕裂的 tearing
撕裂度测定仪 tear tester
撕裂检验器 tearing tester
撕裂角 angle of tear
撕裂力 tearout
撕裂模 tearing mode
撕裂模不稳定性 tearing mode instability
撕裂能 tearing energy
撕裂破坏 tearing failure
撕裂强度 peel(ing) strength; resistance of tearing;tearing strength
撕裂强度试验 tearing-strength test
撕裂式扩展 mode of anti-plane slide
撕裂试验 tear test;peel test<测定橡胶与金属板黏[粘]附力>
撕裂试验机 tear tester
撕裂试验用夹头 tear jaw
撕裂速率 tear speed
撕裂纹<木材加工表面的> torn grain
撕裂线<包装袋> tear line
撕裂型裂缝 tearing crack
撕裂应变 tearing strain
撕裂应力 tear stress
撕裂状的 lacerated
撕膜 dye stripping; out and removal of coating; removal of coating; remove coating
撕膜材料 peel coat material;strip material
撕膜法 open-window process; strip mashing process
撕膜聚酯片基 strip-coated polyester sheet
撕膜蒙片 open-window mask
撕膜片 peel coat film; strip coating film;stripping mask
撕捏 masticate
撕捏机 masticator
撕破 tear(ing)
撕破强度 tearing resistance; tearing strength
撕破强度试验 tear test
撕破试验 tearing test
撕碎 lacerate;torn to pieces
撕碎辊 shredding roll
撕碎机 shredder; shredder machine; shredmaster
撕碎垃圾 refuse shredding
撕条 tear tape
撕贴两用墙纸 strippable paper
撕脱 avulsion;Evulsion
撕下 tearout
撕下纸带 torn tape
撕下纸带转发中心 torn-tape switching center[centre]

死坝 dead dam

死扳手 solid wrench
死板的 off the peg;off-the-rack<美>
死焙烧 dead burn(ing)
死冰川 dead glacier
死层 dead layer
死岔道 dead arm
死岔线 dead-end siding
死潮 dead tide
死城 necropolis
死冲程 dead stroke
死冲锤 dead-stroke hammer
死船 dead ship

死窗 dead window
死带 <航摄> dead area;dead zone
死导条 idle bar
死倒木 dead and down
死道数 number of death traces
死地被物 dead soil covering;litter
死点 dead center[centre];deadcenter[deadcentre];dead point;dead spot <接收机的>
死点发火 dead center ignition
死点记号 dead center mark
死点位置 middle gear;mid-gear
死点压力 dead-end pressure
死电路 dead circuit
死顶点 <车床的> dead center[centre];dead point
死顶尖 back dead center[centre]; dead center[centre]; deadcentre[deadcentre];fixed center[centre]
死顶尖车床 dead center lathe
死动物处理场 dead animal processing plant
死洞 dead hole
死洞穴 dead cave
死端 non-jacking end
死端焊 dead-end weld
死端头 dead end
死断层 dead fault;passive fault
死堆 deadpile
死堆场 dead stackyard;dead storage
死法兰 blank flange
死港 ghost harbo(u)r
死谷 dead valley
死拐角 dead corner
死海 Dead Sea
死海沥青 Dead Sea asphalt; Jews pitch
死河汊 dead branch of river; dead stream branch;dead water
死荷载 permanent load(ing); dead load
死荷载力矩 fixed load moment;fixed weight moment
死荷载应力 fixed weight stress
死胡同 blind alley;blind gallery;blind pass; closes; cul-de-sac(street); dead end;dead-end path;dead-end road; dead-end street; impasse; pocket;spur track
死湖 extinct lake;blind lake <无补给的湖>;lacus mortis <月球>
死滑坡 dead landslide
死火山 extinct volcano;inactive volcano;quiet volcano
死火山堆【地】 puy
死火山口 extinct volcano crater
死机预防 hang-up prevention
死间歇泉 dead geyser
死角 blind side; blind spot; corner pocket; dead angle; dead ground; dead space;dead volume
死角区 dead area;dead ground
死角燃烧 dead burn(ing)
死接合 closed coupling
死节 black knot; dead knot; encased knot
死结 builder's knot;dead knot;encase knot;hard knot;snarl knot
死晶体 dead crystal
死井 dead well
死静 death calm
死可燃物 dead fuel
死可燃物含水量 dead fuel moisture
死空间 dead space
死空气 dead air;stale air
死空隙 dead-air void
死孔 dead hole
死口焊口 flying-in weld;tie-in weld
死扣 <绳索使用不当造成的> kinky;coil buckling
死库容 dead storage; inactive storage;non-effective storage
死连拱廊道 dead arcade
死连接 dead joint
死联轴节 fast coupling
死裂谷 extinct rift valley
死炉 dead furnace
死路 impasse
死螺母 nut cap
死螺栓 dead bolt
死锚 dead-end termination
死面 dead front
死木寄生菌 saprogen
死木生物 saproxylobios
死排水沟 blind drain
死谱带 dead band
死起角 angle of deadrise
死钱 dead money
死墙 blank wall;blink wall
死区 blind spot; dead area; dead band;dead belt;dead ground;dead region; dead space; dead volume; gap band; inert zone; silent zone; stagnant wake <气流中的>
死区段 dead section;dead track section
死区非线性调节器 dead-zone nonlinear regulator
死区宽度 skip distance
死区时间【计】 dead time
死区调节器 dead-zone regulator
死区域 dead zone
死渠段 dead part of canal;dead reach (of canal)
死去 dying out
死泉 inactive spring
死容积 dead volume
死梢 dead-top
死烧 dead burn;dead roasting;sweet roasting
死烧白云石 magnefer
死烧白云石水泥 cement of deadburned dolomite
死烧的 dead burn(ing); dead-burned;double-burnt;hard-burned
死烧矾土 dead-burned bauxite
死烧镁砂 dead-burned magnesite
死烧耐火白云石 dead-burned refractory dolomite
死烧石膏 dead-burned gypsum
死烧氧化镁 dead-burned magnesia; hard-burned magnesia
死绳固定轮 deadline sheave
死绳拉力 deadline stresses
死绳锚定 deadline anchor
死石灰 killed lime
死时间 dead time
死时间改正 coincidence correction; dead time correction
死时间校正 corrected of dead time value
死书 <指放错位置而无法找到的书> dead book
死水 adherent water;backwater;dead water; fixed water; lentic water; non-moving water; slack water; stagnant water;standing water;unfree water
死水池 dub
死水的 len(i)tic
死水沟 dead furrow
死水河 billabong
死水湖 dead lake; drainless lake; mortlake
死水环境 lentic environment
死水末端 dead end
死水区 dead area;dead(water) zone; slough;stagnant area
死水潭 billabong
死水塘 stagnant basin;stagnant pond; stagnant pool
死水头 dead end
死水洼地 billabong
死水位 dead storage level;dead water-level; dead water stage; minimum operating water level【港】
死水域 dead space
死水支槽 dead channel
死水支管 dead leg
死锁 deadlight lock;deadlock
死锁避免 dead avoidance
死锁舌 dead bolt
死锁特性 deadlock property
死态 dead state
死膛发生炉 dead hearth generator
死体积 dead volume
死天窗 dead light
死条纹 kill strip
死通道 blind gallery
死头 dead end;deadhead
死头(干)管 dead-end main
死头街道 <即此路不通> dead-end street
死头路 dead-end road
死头式布管 dean-end layout
死头烟道 dead-end flue
死土 dead soil
死土层 dead horizon
死弯 dogleg
死亡 death; demise; ending; killing back
死亡百分率 per cent death loss
死亡保险 assurance on death; death insurance;life insurance
死亡比(值) death ratio
死亡财产自动转让 gift cause mortis
死亡登记 death registration
死亡点 death point
死亡抚恤金 death benefit
死亡控制 death control
死亡量 mortality
死亡裂谷 extinct block
死亡率 death rate;fatal rate;mortality(rate)
死亡率比 mortality ratio
死亡率的自然因素 natural factors of mortality
死亡率多元回归分析 mortality multiple regression analysis
死亡率曲线 mortality curve
死亡率系数 mortality coefficient
死亡频率 fatality frequency
死亡期 death phase
死亡人数 fatalities;fire deaths
死亡事故 accidental death;fatal accident;fatal crash;fatality
死亡事故报告系统制 fatal accident reporting system
死亡事故率 fatal accident rate
死亡数 death count;mortality
死亡速率 rate of death
死亡统计 necrology
死亡系数 coefficient of mortality
死亡宣告 declaration of death
死亡学 thanatology
死亡原因 cause of death
死亡诊断书 death certificate
死亡证 certificate of death
死亡专率 specific death rate;specific mortality
死亡组合 death assemblage
死位 dead position
死物寄生菌 necroparasite
死隙 dead space
死线 deadline
死巷 blind alley; blind pass; cul-de-sac(street);dead end
死巷道 dead tunnel
死巷式出入口 L shaped entrance and exit
死循环 endless loop;infinite loop
死因 cause of death
死因构成比 proportion of death cause
死因死亡率 specific mortality of cause
死因学 thanatology
死鱼 kill fish
死载荷应力条件 dead load stress condition
死债 dead loan
死者雕像 statue of a deceased
死止块 definite stop
死滞区 dead area
死资本 dead capital
死资产 dead assets

四 阿【建】 hip roof

四氨络物 tetrammine
四胺 fouramine
四百度制度盘分划 centesimal(circle)graduation
四百年的 quadricentennial
四百天钟 four-hundred-day clock
四百周年 quadricentennial
四摆仪 four-pendulum apparatus
四班轮值 quarter watch
四板门 four-panel door
四板式铲斗式支承架 four-plate loader tower
四瓣 pin tongs
四瓣戽斗挖土机 orange-peel excavator
四瓣花饰 quarterfoil
四瓣加重抓斗 reinforced four-piece clamshell bucket
四瓣玫瑰线 four-leaved rose curve
四瓣式戽斗 orange-peel bucket
四瓣式抓斗 four-piece clamshell bucket
四瓣弹簧夹 springer
四瓣眼形窗 quatrefoil oculus window
四瓣轴瓦的滑动轴承 box quarter
四瓣抓土机 four-bladed circular grab
四倍 fourfold
四倍长寄存器 quadruple length register
四倍长字 quad word
四倍乘数 quadrupler
四倍的 quadruple(x); quad(ruplicate);tetraploid
四倍地 quadruple;quadruply
四倍脉冲 pulse quadrupling
四倍频 quadruple
四倍(频)器 quadrupler
四倍数 quadruple
四倍体 tetraploid
四倍压器 quadrupler
四倍于 quadrupling
四倍重的 quad
四倍柱底径间距的柱廊 araeostyle
四苯并四氮杂卟吩 tetrabenzo-porphyrazine
四苯并紫菜嗪 tetrabenzo-porphyrazine
四苯基甲烷 tetraphernl methane
四苯硼酸钠 sodium tetraphenylborate;tetraphenylboron sodium
四苯乙烯 tetraphenyl ethylene
四边单元体 quadrilateral element
四边的 quadrangular;quadrilateral
四边夹紧的嵌板 closed sandwich type panel
四边简支 simply supported on four sides
四边结构密封镶嵌玻璃 four-sided structural sealant glazing
四边刨方板 square-edged board
四边切方 square cut
四边切方平瓦 square cut plain tile

四边体 quad
四边图 four-sided figure
四边线饰栏杆柱 square-turned baluster
四边线饰螺旋梯中柱 square-turned newel
四边形 four-sided figure; quadrangle; tetragon; quadrilateral
四边形层顶 quadrangle roof; quadrangular roof
四边形单元 quadrilateral element
四边形单元体特性 quadrilateral element characteristic
四边形的 dimetric; quadrangular; tetragonal
四边形的支承 quadrilateral support
四边形断面 quadrangular section
四边形机座 quadrangle frame
四边形建筑构件 quarrel
四边形缆 quad
四边形棱柱 quadrilateral prism
四边形联杆转向机构 parallelogram linkage steering
四边形列柱(廊)式 quadrilateral peristyle
四边形裂隙桩【测】 quadrilateral tension crack stake
四边形扭转悬挂 quadrifilar torsion suspension; quadrilateral torsion suspension
四边形四分体 tetragonal tetrad
四边形特性 quadrilateral characteristics
四边形网 network of quadrilaterals
四边形屋顶 quadrangle roof
四边形物 quadrilateral
四边有条板的棚车 <装运家禽家畜> skeleton wagon
四边支撑楼板 two-way flat slab
四边支承板 two way slab; slab supported on four sides
四边支承的 supported along four sides; supported on four sides
四边琢方石铺地面 square-edged flooring
四变数 quaternary
四变数系统 quaternary system
四标两角定位 split fix
四标三角法 four-point method
四标准英尺的单位 tetrapody
四丙烯苯磺酸盐 tetrapropylene-benzene sulfonate
四丙烯苯磺酸酯 tetrapropylene-benzene sulfonate
四饼滑车 fourfold block; quardruple block
四波段光谱数据 four-band multispectral data
四波混合 four-wave mixing
四波混频 four-wave mixing
四波节振动 four-node oscillation
四波求和混频 four-wave sum-mixing
四步法 four-step rule
四步热处理 four-step heat treatment
四部分 tetramerous
四部分组成的 quadripartite
四部分组合体 quadripartite
四部组成 quaternary
四参量模型 four parameter model
四槽钢柱 four-groove steel column; Gray column
四槽扩孔钻 four-groove drill
四槽锥柄心孔钻床 four-flute taper-shank core drill
四槽钻头 four-groove drill
四侧有窗车厢 four-light body
四层半导体开关管 binistor
四层半导体开关器件 binistor
四层带 quadriply belt
四层的 fourply
四层盾构 <隧道> four-deck jumbo
四层罐笼 four-decker cage
四层甲板船 four-decker
四层交叉 four-level crossing
四层接合缝 four-ply seam
四层开关 four-layer switch
四层立交构筑物 <道路交叉> four-level interchange
四层立体交叉口 four-level super cross-road
四层立体交叉路 four-level super cross-road
四层帘布层轮胎 four-ply tyre
四层器件 four-layer device
四层绕组 four-layer winding
四层筛 four-deck screen
四层筛分器 four-deck grader
四层筛网 four-deck screen
四层筛子 four-deck screen
四层式表面处治 quadruple(course) surface treatment
四层涂刷法 four-coat system
四层织物 four ply cloth; quadruple cloth
四层织物的第二层组织 upper inside weave
四层组合覆盖屋面 four-ply built-up roof cover(ing)
四叉抓斗 four-tine grapple
四插角的 four-limbed
四插座转接器 four socket adapter
四岔交叉 four-leg intersection
四岔路 four way; four wont way
四岔路交叉(口) four-leg intersection
四岔路口 four corners; four ways
四岔路立体交叉 four-leg interchange
四车道 four-lane
四车道道路 four-lane road
四车道分隔道路 four-lane divided roadway
四车道公路 dual-dual highway; four-lane highway; four-lane road
四车道桥梁 four-lane bridge
四车轴构造 four-axle configuration
四乘幂 biquadratic
四程 quadruple pass
四程机 four-stroke engine
四齿轮钻头 four cutter bit
四翅槐树 kowhai
四冲 <内燃机的> Otto cycle
四冲程柴油发动机 four-stroke diesel engine
四冲程柴油机 four-stroke-cycle diesel; four-stroke oil engine
四冲程船用柴油机 four-stroke marine diesel
四冲程的 four-stroke
四冲程发动机 four-cycle engine; Otto (cycle) engine; four-stroke engine
四冲程循环 four-stroke cycle; Otto cycle; four-cycle
四冲程循环柴油机 four-cycle diesel
四冲程循环的 four-cycle
四冲程循环内燃机 Otto(cycle) engine
四冲程引擎 four-stroke engine
四冲程原理 four-stroke-cycle principle
四重 fourfold; quad; quadruple; tetramerous
四重的 quadruple(x); quadruplicate
四重点 quadruple point
四重对称 fourfold symmetry
四重对称轴 fourfold axis of symmetry
四重分集运用 quadruple diversity operation
四重峰 quadruple peak
四重符合记录设备 quadrupole coincidence set
四重格 quadruple lattice
四重根 quadruple root
四重寄生物 quaternary parasite
四重简并 quadruple degeneracy
四重节 fourfold node
四重离子 quadruple ion
四重联拍测绘照相机 quadruple serial photogrammetric camera
四重联轴节 fourfold coupling
四重矢量积 quadruple vector product
四重式多辊矫直机 four-high roller level-(l)er
四重误差检测 quadruple error detection
四重线 quartet(te)
四重线号 quadruplex
四重向量积 quadruple vector product
四重性 quadrupleness
四重正交 quadruple orthogonal
四重轴 fourfold axis; tetragon
四重组 quartet(te)
四触点塞孔 four-point jack
四川红杉 masters larch
四川梓 ducloux catalpa
四窗 four-light
四床的 four-bed
四次不尽根 quartic surd
四次代数曲线 quartic
四次的【数】 quartic; biquadratic
四次对称晶 tetrad
四次对称轴 axis of tetragonal symmetry; fourfold axis of symmetry
四次方 biquadratic; quadruplicate
四次方程式【数】 biquadratic; quartic equation
四次方的 quadruplicate
四次方定律 <轴载换算的> fourth power law
四次方概念 <轴载换算的> fourth power concept
四次方公式 fourth power law
四次进尺深度 pull length
四次幂 biquadratic
四次配位 fourfold coordination
四次曲面 quartic surface
四次曲线 biquadratic curve; quartic curve
四次式 quarternary quantic
四次提升岩芯长度 pull length
四次圆纹曲面 bicircular surface; cyclide
四大地壳波系 four essential crustal-wave systems of the earth
四大稳 four large stabilizing
四带分类 four-tape sort
四单轮滑车组 bell purchase
四单位制 four figure system
四单元码 four-unit code
四单元制 four digit system
四氮杂萘并苯 tetraazaphenalene
四挡变速 four speed gear shift
四挡变速器 four speed transmission
四挡变速箱 four speed gear box
四挡速度 fourth speed
四刀边秤量机构 <混凝土搅拌厂设备> four-knife edge weigh(ing) gear
四的 quaternary
四等 fourth order
四等舱旅客 steerage passenger
四等分 quadrate; quarter(ing)
四等分的 quartered
四等分螺钉 quarter screw
四等分取样法 quartering sampling
四等三角点 fourth-order triangulation point
四等水准测量 fourth-order level(l)-ing
四等水准点 fourth-order benchmark
四等水准观测手薄 fourth-order leveling field book
四等体 quartet(te)
四地址 four-address; quadruple address
四地址计算机 four-address computer
四地址码 four-address code
四地址指令 four-address instruction
四地址指令格式 four-address instruction format
四地址指令码 four-address instruction code
四点安装法 four-point mounting
四点测角头 four-point nose
四点吊吊具 four-point suspension spreader
四点定位法 bow and beam bearings; four-point bearing
四点法 four-electrode system; four-point method
四点方位 four-point bearing
四点荷载 four-point load(ing)
四点接触球轴承 four-point contact ball
四点控制法 four-point control method
四点面坐标 quadriplanar plane coordinates
四点木垛 four-point pigsty(e)
四点起动箱 four-point starting box
四点曲轴压力机 four-point press
四点(三分点)梁式弯曲疲劳试验 four-point bending beam fatigue test
四点探针法 four-point probe method
四点小角衍射图 four-point small angle diagram
四点型六点 quadrangular set of six points
四点悬吊装置 four-point suspension mounting
四点悬挂法 four-point mounting
四点悬挂装置 four-point linkage
四点悬式冲床 four-point suspension
四点一列探针 in-line four-point probe
四电极系统 four-electrode system; four-point method
四叠板 lamina quadrigemina; quadrigeminal plate
四叠体 corpora quadrigemina; quadrigemina; quadrigeminal body
四叠体的 quadrigeminus
四叠体上臂 prebrachium
四叠体下丘 testis cerebri
四顶点定理 four-vertex theorem
四动道岔 quadruple-working switches
四度空间【数】 four-dimensional space
四端 tetrapolar; two-port
四端参量 two-port parameter
四端的 quadripolar
四端电路 four-pole circuit; quadripole
四端级际网络 four-terminal interstage network
四端器件 four-terminal device
四端网格 four-pole network; four-terminal network; quadripole
四端网络 quadripole network; two-terminal pair network
四端网络参量 two-port parameter
四端网络常数 four-terminal constant
四端网络传感器 two-port transducer
四端网络的 tetrapolar
四端网络的导纳 four-pole admittance
四端网络方程 four-terminal equation
四端网络节 section of a recurrent structure
四端网络累接传播常数 quadripole propagation constant
四端网络衰耗 four-terminal attenuation
四端网络衰减 four attenuation termi-

nal attenuation
四端网络衰减量 four-terminal attenuation
四端网络衰减因数 quadripole attenuation factor
四端网络振荡器 four-terminal oscillator
四端网络综合 two-port network synthesis
四端系统 transducer
四端循环器 four-terminal circulator
四段横肋拱顶 four-part cross-rib(bed) vault
四对称圆锥体钢筋混凝土管<防波堤用> vierbein
四对数增加 four-log increase
四对中之一 quadruplicate
四阀式液压系统 four valve hydraulics
四方 all quarters; square; cardinal point<即东南西北>
四方板 quarter panel
四方单锥 tetragonal pyramid
四方刀架 squaring head
四方的 tetragonal
四方复铁天蓝石 lipscombite
四方格 square lattice
四方铬铁矿 donathite
四方鼓轮 quadrangular prism; quadrangular tumbler
四方光面面砖 square cut plain tile
四方剪机 squaring shears
四方进刀架 square groove
四方晶格 tetragonal lattice
四方晶体 tetragonal crystal
四方晶(体)结构 tetragonal structure
四方晶系 pyramidal system; tetragonal system
四方硫砷铜矿 luzonite
四方硫铁矿 mackinawite
四方铆钉窝子夹头 square snap chuck
四方锰铁矿 iwakiite
四方面 quadrilateral
四方钠沸石 tetranatrolite
四方镍纹石 tetrataenite
四方偏方(三八)面体 tetragonal trapezohedron
四方嵌板 square panel
四方羟锡锰石 tetrawickmanite
四方羟锌石 sweetite
四方切削 square turning
四方双锥 tetragonal dipyramid
四方丝锥 core tap
四方四面体 sphenoid; tetragonal disphenoid
四方钽锡矿 staringite
四方锑铅矿 genkinite
四方体堆置 tetrahedral packing
四方头 square head
四方位可调座椅 four-way adjustable
四方纤铁矿 akaganeite
四方镶板 square panel
四方向左转信号相<即四引道都有左转信号相> quad left turn phasing
四方协定 quadripartite agreement
四方形的 square-edged
四方形地形图纸<美> topographic(al) quadrangle sheet
四方形排列 square array
四方形隧道 square-section tunnel
四方形瓦 quadrel
四方性 tetragonality
四方氧化锆多晶陶瓷 tetragonal zirconia polycrystal ceramics
四方圆穹亭<莎桑尼亚建筑的> Tchahar taq
四方柱 tetragonal prism
四方采样法 quaternary sampling
四分的 quadripartite
四分点<统计学中频率分布距一端为3/4,另一端为1/4的点> quartile; quarter point
四分点的 quartile
四分点荷载 quarter-point loading
四分点横(隔)梁 quarter-point diaphragm
四分点挠度 quarter-point deflection
四分度取样 four sampling intervals per decade
四分对称 tetartohedry; tetartosymmetry; tetrahedry
四分对称性 tetartohedrism
四分法 method of quartering; quartering; quartering method
四分法工艺过程 quartering process
四分法取样 quartering way
四分花窗格 quadripartite tracery
四分搅拌法 quadrant blending method
四分搅拌系统 quadrant blending system
四分开板材 vertical grain
四分量石英管应变计 four-component quartz tube strainmeter
四分面的 tetartohedral
四分面晶体 tetratohedral crystal
四分面体 tetartohedron
四分面像 tetartohedry
四分面形 tetartohedral form
四分切 quadrisection
四分穹顶 quadripartite vault
四分穹隆 four-part vault
四分取样法 quartation; quarter dividing method; quartering; method of quartering<选取代表性试验材料的方法>
四分日潮 quarter-diurnal tide
四分缩样法 sample quartering
四分体 quadrant; tetrad
四分体的 quadrantal
四分位差 interquartile range; quartile deviation; semi-interquartile range
四分位分法 quartile division
四分位量度 quartile measurement
四分位偏差 quartile deviation
四分位散度 quartile dispersion
四分位数 quartile
四分位数分隔 interquartile separation
四分位数间隔<统计的> interquartile range
四分位数间距 interquartile range
四分位系数 quartile coefficient
四分型 tetrad segregation type
四分仪 quad(rant); sector
四分音符 crotchet
四分圆 quarter circle
四分圆线脚 quadrant; quarter round; quarter-round mo(u)lding
四分枕木 quarter tie
四分之三 three quarter
四分之三闭砖法 three-quarter closer
四分之三壁龛 three-quarter niche
四分之三侧视 three-quarter view
四分之三长丁砖 three-quarter header
四分之三浮动轴 three-quarter floating axle
四分之三浮式车轴 three-quarter-floating axle shaft
四分之三浮式轴 three-quarters float(ing) axle
四分之三夹板表 three-quarter plate watch
四分之三浇筑 three-quarter mo(u)lding
四分之三块体 three-quarter block
四分之三露径壁柱<露出墙面> three-quarter column
四分之三履带车 three-quarter track
四分之三履带式的 three-quarter tracked
四分之三幂定律 four-thirds law
四分之三碰撞(责任)条款 three-fourth running down clause
四分之三设计车速法<一种设计弯道超高的方法> three-quarter design speed method
四分之三凸圆饰 bowtel(l)
四分之三瓦 three-quarter tile
四分之三小柱 three-quarter small column
四分之三倚柱 bowtel(l)
四分之三硬度 three-quarter hard
四分之三柱 three-quarter(attached) column
四分之三砖 king closer; three-quarter bat; three-quarter brick; three-quarter header; three-quarter tile
四分之一 one quarter; quarter(ly); quartern
四分之一凹圆线脚 cavetto mo(u)lding; quarter hollow
四分之一波长 quarter wavelength
四分之一波长变换器 quarter-wave transformer
四分之一波长变量器 one-quarter wave skirt
四分之一波长补偿器 quarter-wave plate compensator
四分之一波长层 quarter-wave layer
四分之一波长的接地偶极子 quarter-wave ground(ed) dipole
四分之一波长短线 quarter-wave stub
四分之一波长辐射器 quarter-wave radiator
四分之一波长跨接线 one-quarter wave skirt
四分之一波长滤波器 quarter-wave filter
四分之一波长膜 quarter-wave film
四分之一波长膜系 quarter-wave stack
四分之一波长判断标准 quarter-wave criterion
四分之一波长匹配 quarter-matching
四分之一波长衰减器 quarter-wave attenuater[attenuator]
四分之一波长天线 quarter-wave antenna
四分之一波长线 quarter-wavelength line
四分之一波长线段 quarter section
四分之一波长谐振 quarter-wave resonance
四分之一波长折射片 quarter-wave plate
四分之一波电压 quarter-wave voltage
四分之一波晶片 quarter-wave plate
四分之一波片 quarter-wave(length) plate
四分之一波线 quarter-wave line
四分之一波限 quarter-wave limit
四分之一车模拟 quarter-car simulation
四分之一车模型<描述路面平整度的> quarter car model
四分之一车指数 quarter-car index
四分之一寸厚木板 quarter stuff
四分之一大小的石块 stone of one-quarter brick size
四分之一的 demisemi; quarterly; subquadruple
四分之一段 quarter section
四分之一高跨比 quarter height-to-span ratio[1/4 height-to-span ratio]
四分之一高跨比斜坡<高1与跨4之比> one-fourth pitch; quarter pitch
四分之一管子弯头 one-quarter pipe bend
四分之一加法器 quarter adder
四分之一间距点 quarter point
四分之一交错锯木 alternate quarter sawn
四分之一接头 quarter joint
四分之一节 quarter section
四分之一截头锥体 quadrant of truncated cone
四分之一径向锯 quarter-sawed
四分之一径向锯木 quarter-sawn
四分之一锯开的木材 quarter-sawed lumber
四分之一开口铝管<用于路面排水> quarter split aluminium pipe
四分之一跨度点处弯矩 quarter-point moment
四分之一跨度点荷载 quarter(ed)-point loading
四分之一块砖 quarter bat
四分之一流量 quarter flow
四分之一码宽 quarter width
四分之一品脱 quartern
四分之一平方乘法器 quarter-square multiplier
四分之一平方英里 quarter section
四分之一日潮 quarter-diurnal tide
四分之一日分潮 quarter-diurnal constituent
四分之一矢高法<测设平缓圆弧的近似法> quadrant method
四分之一缩尺 quarter size
四分之一图幅 quarter section
四分之一弯头 one-quarter bend; quarter bend
四分之一小柱 quarter small column
四分之一行频间置 quarter-line offset
四分之一翼弦点 quarter-chord point
四分之一英寸的长度单位 digit
四分之一英寸地图 quarter-inch map
四分之一英寸厚板条 lath and a half
四分之一英寸厚木板 quarter stuff
四分之一圆 conge; quadrant; quarter circle
四分之一圆木 quartered log
四分之一圆铣刀 rounding over cutter; round-over bit
四分之一圆线脚 quarter-round mo(u)lding
四分之一圆形反射器 quadrant reflector
四分之一圆缘饰 quarter round
四分之一圆周 quadrant
四分之一周缘饰 ovolo mo(u)lding
四分之一昼间潮 quarter-diurnal tide
四分之一砖 one-quarter bat; one-quarter clay brick; quarter brick; quarter closer; quarter closure
四分之一砖砌合 one-quarter brick bond
四分之一桩<放在道路横断面的四分之一处> quarter peg; quarter-stake
四分之一字长 quarter word
四分直径 quartile diameter
四分值 quartile
四分中二<指四等分的中部二等分> middle half
四分周干围 quarter girth
四分周卷尺 quarter-girth tape
四分砖 quarter bat; quarter closer
四分锥 tetartopyramid
四氟化硅 silicon tetrafluoride
四氟化硅处理混凝土表面 ocrating
四氟化硅混凝土 ocrate concrete
四氟化硅气体 silicon tetrafluoride gas
四氟化甲烷 tetrafluoromethane
四氟化硫 sulfur tetrafluoride
四氟化铀 green salt; uranium tetrafluoride
四氟乙烯 eteline; perfuoroethylene; tetrafluoethylene
四氟乙烯板橡胶支座 Teflon-neoprene bearing
四氟乙烯板支座 Teflon plate bearing

S

四氟乙烯树脂 tetrafluoroethylene resin
四幅 fourth officer
四幅式公路 dual-dual highway
四幅式路 quadri-carriageway road
四副 fourth officer
四钙高铝水泥 tetracalcium alumina cement
四甘醇 tetraethylene glycol
四杆节点 four-member nodal point
四杆联动机构 four-bar linkage
四杆链 four-bar chain
四杆运动 four-bar motion
四杆转向联动装置 four-bar steering linkage
四缸操舵装置 four-cylinder steering gear
四缸复式涡轮机 quadruple compound turbine
四缸机车 four-cylinder locomotive
四缸汽轮机 four-cylinder turbine
四缸三胀式蒸汽机 four-cylinder triple-expansion engine
四缸直列 four-in-line
四缸直列发动机 four-in-line engine
四缸转子发动机 four-rotor engine
四格表 fourfold table
四个 tetrad
四个基本学科 four basic fields
四个来回路 four-turn road
四个一套 quartet(te);tetrad
四个一组 quartet(te);quaternion;tetrad
四个一组的 quaternary
四个一组系统 quaternary system
四个月无霜期 four-months frost-free period
四个主要垂直装置 four principal vertical position
四根交叉肋拱顶 quadripartite cross-rib(bed)vault
四工(的)quadruple(x)
四工电路 quadruplex circuit
S
四工电路系统 quadruplex system
四工铅 tetraethyl-lead
四工通报用电键 reverting key
四工系统 quadruplex system
四功能紧急闪光报警灯 four-way emergency flasher
四钩吊货索 bunch of fours;cargo bridle
四构件轮辋式车轮 four-piece wheel
四构件棚子 four-piece set
四构件式轮辋 four-piece rim
四构件支架 <包括帽梁、支柱及槛梁,用于软弱地层中井巷的超前开挖> four-piece set;four-section support
四股 four-cant
四股编绳法 four-strand round sennit
四股的 fourply
四股吊索 brothers
四股钢丝 four-strand rope
四股绳 four-cant;four-strand rope
四股油绳 lanyard stuff
四股正搓绳 shroud laid rope
四刮刀钻头 four-blade-drag bit;four-drag-blade bit
四管包装机 four-tube packing machine
四管扩散炉 four stack diffusion furnace
四管式空调系统 four-pipe air conditioning system
四管通风系统 four-pipe air system
四管制水系统 four-pipe water system
四光榴石 tetrahedral garnet
四轨 fourth engineer
四轨道录音机 four-track recorder
四轨录音放音 quadraphonics;quadraphony;quadrasonics
四辊粗轧机 four-high rougher
四辊可逆式带材冷轧机 four-high reversing cold strip mill
四辊磨 four-roll mill
四辊碾碎机 four-roll crusher
四辊式破碎机 four-roll crusher
四辊式轧机 four-high rolling mill;four-roller mill
四辊式装置 four-high setup
四辊压延机 four-bowl calender;four-roll calender
四辊轧机 four-high mill
四辊轧碎机 twin-dual roll crusher
四辊轴承 four-roller bearing
四辊钻头 four-roller bit
四滚筒球磨机 Hyswing ball mill
四滚筒蒸汽提升机 four-drum steam hoist
四国海盆 Shikoku basin
四行的 quadruple
四行键盘 four-row keyboard
四行交叉桁架 quadruple intersection truss
四行铆钉接合 quadruple riveted joint
四行铆接法 quadruple riveting
四行排齐封装 quad-in-line
四航道无线电导航台 four-course radio range station
四航向无线电信标 four-course radio beacon;four-course radio range
四航向中频无线电信标 four-course MF radio range
四毫米锁闭 check 4 mm opening
四合庭院 hypaethral
四合透镜物镜 four-lens objective
四合物镜 four-lens objective
四合星 quadruple star
四合(一)铲斗 four-in-one bucket
四合(一)抓斗 four-in-one bucket
四合院 courtyard with building on the four sides;four family dwelling;quad(rangle);quadrangle houses
四合院中敞开的庭院 open court of a quadrangle
四合院住宅 courtyard dwelling house;courtyard house;residential quadrangle
四合组 quarternary system
四喉管式 <化油器> quadrajet
四户自有公寓 quadrominium
四滑板式十字头 four guide crosshead
四滑道下水 four-way launching
四滑轮系统 four-part line
四滑轮游动滑车组 four sheave travel(l)ing trevelling block
四滑轮组 quadruple block
四环链 four-link chain
四环式立体交叉 cloverleaf crossing;cloverleaf interchange;cloverleaf junction
四簧片塞孔 four-point jack
四汇流条调整 four busbar regulation
四机单轴式 four-engine single-shaft system
四机台车 quadruple rig
四机凿岩台车 four-drill rig
四机钻车的轮臂 quadruple boom
四机座连续式带材轧机 four-stand tandem cold strip mill
四机座连续式轧机 four-stand tandem mill
四级泵 four-stage pump
四级侧生的 quadrilateral
四级成矿远景区 the fourth grade of minerogenetic prospect
四级成煤远景区 the fourth grade of coal-forming prospect
四级处理 quaternary treatment
四级串列加速器 four-stage tandem accelerator
四级的 four-stage;quarternary
四级电离 quaternary ionisation
四级反应 fourth-order reaction
四级风 moderate breeze;moderate wind;wind of Beaufort force four
四级公路 fourth class highway
四级管【电】quadrode
四级结构 quarternary structure
四级结构面 grade four discontinuity
四级结构体 grade four texture body
四级精度 rough grade
四级浪 force-four wave;moderate sea
四级逻辑 four level logic
四级锚链 grade 4 chain
四级能见度 mist;thin fog
四级渠道 feeder canal;quarternary canal
四级旋风分离器 tetra-cyclone
四级压缩 four-stage compression
四级压缩机 four-stage compressor
四级涌 moderate swell
四级油气远景区 the fourth grade of oil-gas prospect
四级预热器 four-stage preheater
四级增压器 four-stage blower;four-stage supercharger
四级振荡器 four-level generator
四级质谱法 four-stage mass spectrometry
四级质谱计 four-stage mass spectrometer
四极磁铁 quadrupole electromagnet;quadrupole magnet
四极磁透镜对 quadrupole doublet
四极的 four-pole;quadripolar;tetrapolar
四极电动机 four-pole motor
四极电矩 electric(al)quadrupole moment
四极电透镜 electric(al)quadrupole lens
四极多重线 quadrupole multiplet
四极法 four-electrode method
四极分量 quadrupole component
四极分裂 quadrupole split
四极功率放大管 screen-grid power tube
四极管 dynatron;four-electrode tube;screened plate tube;screen-grid tube;shielded-grid tube;shielded plate tube;tetrode
四极管放大器 tetrode amplifier
四极继电器 four-pole relay
四极晶体管 tetrode transistor
四极矩质谱法 quadrupole mass spectrometry
四极力 quadrupole force
四极滤质器 quadrupole mass filter
四极能谱 quadrupole spectrum
四极偶合常数 quadrupole coupling constant
四极耦合 quadrupole coupling
四极匹配 quadrupole matching
四极谱仪 quadrupole spectrometer
四极汽轮发电机 four-pole turbo-generator
四极三合透镜 quadrupole-lens triplet
四极三透镜组 quadrupole-triplet lens
四极射频场 quadrupole radio frequency field
四极剩余气体分析器 quadrupole residual gas analyser[analyzer]
四极双峰 quadrupole-double peak
四极双合透镜 quadrupole-doublet lens
四极四元数 quadruquaternion
四极调焦透镜 quadrupole focusing lens
四极透镜 quadrupole lens
四极透镜对 quadrupole lenses pair
四极透镜聚焦系统 quadrupole-lens focusing system
四极透镜孔径 quadrupole aperture
四极网络 four-pole network
四极吸收谱线 quadrupole adsorption line
四极系统 quadrupole system
四极效应 quadrupole effect
四极性 quadrupolarity;tetrapolarity
四极源 quadrupole source
四极跃迁 quadrupole transition
四极展宽 quadrupole broadening
四极真空管 tetrode
四极振荡 quadrupole vibration
四极质量分析器 quadrupole mass analyser[analyzer]
四极质谱计 quadrupole mass spectrometer
四极质谱仪 quadrupole mass spectrometer
四极子 quadrupole
四极子强度 quadrupole strength
四季不断的 perennial
四季可用的计量器 all-weather ga(u)ging device
四季厅 atrium[复 atria/triums]
四加四的双向管道连接器 double manifold with four plus four connectors
四甲铵 tetramethyl ammonium
四甲醇甲烷 pentaerythritol
四甲基乙二醇 pinacol
四甲铅 tetraethyl-lead
四甲氧(基)铬酸盐 tetroxychromate
四甲氧甲替苯代三聚氰二胺 tetrakis methoxymethyl benzoguanamine
四价 quadrivalency;tetravalence
四价铂的 platinic
四价的化 quaternary
四价钒的 vanadic
四价碱 tetra-atomic base
四价铅的 plumbic
四价铈的 ceric
四价酸 atomic acid;quadribasic acid;tetrabasic acid
四价钛的 titanic
四价硒的 selenic
四价锡的【化】stannic
四价铀的 uranous
四价元素 tetrad
四尖镀锌钢刺线 four-point galvanized steel barbed wire
四间的 four-bay
四件套浴室 four-fixture bathroom
四件一套 quartet(te)
四件一套的(东西)quadruplet
四件运动链 quaternary link
四角 four corners
四角板 quadrilateral plate
四角包 tetra-pack
四角包底水泥袋 block-bottom bag
四角的 quadrangular
四角方钢 quarter octagon steel
四角拱顶 crossing vault
四角拱墩(柱)cross impost
四角加柱的货板 post pallet
四角加柱的托盘 post pallet
四角架 quadrupod
四角晶系 tetragonal crystal system
四角空心方块 hollow square block
四角拉杆 four-cornered bar
四角棱镜 tetragonal prism
四角棱柱体 square prism
四角螺丝套 boss
四角木材 cant
四角嵌条 four-sided bar
四角清晰度高的图像 picture with high definition in the corners
四角燃烧 corner firing

四角燃烧锅炉 corner-fired boiler
四角人兽像柱头 protomai capital
四角三八面体 tetragonal trisoctahedron;trapezohedron
四角三四面体 deltohedron
四角数 quadrangular number
四角塔 four-cornered tower;four-sided tower
四角塔楼 crossing tower
四角稳定体 <防波堤用> stabilopode
四角斜桁帆 donkey topsail
四角斜纵帆 lug sail
四角形 quadrangle;tetragon
四角形窗花格 four-lobe tracery
四角形的 dimetric; quadrangular; quadrilateral
四角形公开法庭 open court of a quadrangle
四角形桁架 quadrangle truss; quadrangular truss
四角形凸轮 four lobe cam
四角型材 four-sided bar
四角型钢 four-cornered bar
四角圆顶小阁 cross-cupola; cross-dome
四角攒尖顶 pyramid roof
四角支柱 four-cornered column
四角柱 four-cornered column; four-sided column;quadrangular prism
四角转子 four-sided rotor
四角锥 quadrangular pyramid
四角锥空间构架 pyramidic space frame
四脚锥块体 tetrapod block
四角锥体 <防波堤用> quadripod
四角锥网架 square pyramid space grid
四脚插头 four-pin plug
四脚吊链 four-leg sling
四脚吊绳 four-leg sling
四脚防波石 tetrapod
四脚管底 four-pin base
四脚管座 four-pin base
四脚护堤块 tetrapod
四脚混凝土块 <用于护堤> tetrapod
四脚架 quadripod
四脚空心方块 hollow square block; tetrapod hollow block
四脚块 tetrapod
四脚块体 quadripod
四脚体 tetrapod
四脚铁插头 four-way iron plug
四脚桅 quadruped mast
四脚支座 tetrapod
四脚重力擒纵机构 four legged gravity escapement
四脚锥 quadripod
四脚锥体 tetrahedron; tetrapod <防波堤用>
四脚锥体堤 tetrapod wall
四脚锥体堆墙 tetrapod wall
四脚锥体浇铸模 tetrapod mo(u)ld
四脚钻塔 tetragonal derrick
四铰拱 four-hinged arch
四铰管涵 quadric-hinged pipe culvert;quadri-hinge-pipe culvert
四阶 quadravalence
四阶段法 four-stage method
四阶张量 tetradic
四接点开关 four-point switch
四接合 four-arm junction
四节点平矩形剪切面 four-node flat rectangular shear panel
四节环 four ring;tetra-atomic ring
四节云梯车 quadruple ladder truck
四进位制 four digit system; four figure system
四进制 quaternary system
四进制乘法 quaternary multiplication
四进制的 quaternary
四进制浮点增益控制 biquadratic floating point gain control
四进制逻辑 quaternary logic
四进制码 quaternary code
四进制数 quaternary number
四进制数系 quaternary number system
四晶 fourling
四晶的 tetramorphous
四晶现象 tetramorphism
四晶形 tetramorphous
四井采掘 quadruple well completion
四井开采 quadruple well completion
四镜头 four barrel
四镜头连续航测摄影机 quadruple serial air survey camera
四镜头摄影机 quadruple camera
四九墙 double brick wall
四聚体 tetramer
四聚物 tetramer
四聚乙醛 metaldehyde
四开 quartering; quarto; quarter saw <指开圆木>
四开本 quarto
四开薄木板 quartered veneer
四开薄木片 quartered veneer
四开材梁 quarter beam
四开裁 demy
四开单板 quartered veneer
四开道岔 half-open points; half-open switch
四开的 edge-grained; quartered; quarter-sawed; quarter-sawn; rift cut; rift-sawed; rift sawn
四开断面 quarter section
四开环 pendant collet
四开截面 quarter section
四开精选橡木材 quarter-sawed clear oak; quarter-sawed select oak
四开径切木料 quarter-sawed grain
四开锯木 quartered timber; quartering; quarter-sawed; quarter sawing; rift saw(ing)
四开锯木的 rift sawn
四开锯木法 vertical grained
四开木材 comb grain timber; edge grain; quarter-cut; quartered lumber; quarter sawed timber; quarter timber; rift-sawn timber; quarter-sawed; edge grain(sawed) timber
四开木材纹 quarter grain
四开木梁 quarter beam
四开木料 edge-grained lumber
四开木压条 quadrant mo(u)ld; quarter round
四开日期 fourth spudding date
四开填塞砖 quarter closer
四开图纸 demy; demy check folio; foolscap-third; sheet-and-half foolscap; bastard <16 英寸×20 英寸>; crown cap <英国 15 英寸×20 英寸, 美国 15 英寸×19 英寸>
四开位置 half-open position
四开无节疤橡木材 quarter-sawed clear oak
四开原木 quarter-sawed
四开圆材 quarter timber
四开圆木 quartered log; quarter sawed timber
四开纸 quarto paper
四孔插座引出线 quadruplex receptacle outlet
四孔插座出线口 quadruplex receptacle outlet
四孔的 quadripuntal
四孔卡片 quadripuntal
四孔联板 four-pin clevis link; four-pin strap
四孔鱼尾板 four-bolt splice bar
四跨的 four-bay;four-span
四跨度的 four-bay;four-span
四框架的 four gimbaled
四缆索起重机 four cable crane
四肋单箱式 single quadruple box
四肋拱顶 quadripartite rib vault
四肋式单箱截面 single quadruple box section
四类土壤 four types of soil
四棱刀 four-square tool
四棱角的 tetraquetrous
四棱平尺 tetrahedral toolmaker's straight edge
四棱体 frustum of rectangular pyramid
四棱支柱 four-sided column
四棱柱 quadrangular
四棱锥 quadrilateral pyramid; rectangular pyramid
四楞凸轮 four-boss breaker cam
四力矩定理 theorem of four moments
四立根高处的钻塔工作台 eighty-board
四连杆 parallelogram
四连杆机构 four bar linkage mechanism;quadric crank mechanism
四连杆结构 parallelogram type linkage
四连杆结构的集材机 parallelogram configuration skidder
四连杆式吊臂 <起重机> double lever jib
四连杆式裂土器 parallelogram ripper
四连杆式伸臂 double lever jib
四连杆式伸臂起重机 crane with double lever jib
四连杆式松土器 parallelogram ripper
四连杆系统 parallelogram system
四连闪光 quadruple burner; quadruple flashing light
四连轴式 four-wheel coupled
四联苯 quaterphenyl
四联等应变直剪仪 double link equivalent-strain direct shear system
四联杆机构 four-bar linkage; four-link mechanism
四联杆式起重机 double link type luffing crane
四联杆式卸船机 double link unloader
四联公式 four parts formula
四联式住宅 quadruplex house
四联体 quadruplet
四联压气机 quadruplex compressor
四联钻管 fourble
四列扁平封装 quad flat package
四列圆锥滚柱轴承 four-row tapered roller journal bearing
四列直插封装 quad-in-line package
四裂花瓣 quadrifid petal
四磷酸钠 sodium tetraphosphate
四六黄铜 Muntz metal
四六面体 tetrahexahedron; tetrakis hexahedron
四笼间 <女厕所的> powder room
四楼 <英> third floor;third stor(e)y
四楼房间 three-pair
四漏斗车 quadruple hopper
四卤化物 tetrahalide
四路 quadruple
四路传输系统 quadruple system
四路传输制 quadruple system
四路的 quadruple
四路多工电报 quadruplex telegraph
四路多工系统 quadruplex
四路多工制 quadruplex system
四路分集传输系统 quadruple-diversity system
四路分配 quadruple distribution
四路分配器 quadruple distributor
四路给油润滑器 four feed lubricator
四路交叉 four-leg intersection; four way
四路交叉的 quadrivial
四路接线箱 four-way junction box
四路开关 four-way switch
四路联箱 four-way joint box
四路时分多路复用 four-channel time division multiplex
四路调制器 quad multiplexer
四路通信[讯]制 multiplex tetrode system
四路无线电通信[讯]终端设备 four-channel radio terminal set
四路相交的交叉 four-leg intersection
四路(肢)交叉口 four-arms intersection
四路钟驱动器 quad clock driver
四路转换阀 four-way change-over valve
四路转换开关 four-channel switch
四氯 tetrachloro
四氯苯 tetrachloro benzene
四氯苯胺 tetrachloro aniline
四氯苯酚 tetrachlorophenol
四氯苯酞 rabicide
四氯(代)甲烷 tetrachloromethane
四氯(代)邻苯二甲酸酐 tetrachlorophthalic anhydride
四氯代萘 naphthalene tetrachloride
四氯酚 tetrachlorophenol
四氯庚烷 tetrachloroheptane
四氯硅烷 tetrachloro silicane
四氯化铂 platinum tetrachloride
四氯化碲 telluric tetrachloride; tellurium tetrachloride
四氯化二锑 antimony tetroxide
四氯化钒 vanadium tetrachloride
四氯化锆 zirconium tetrachloride
四氯化硅 silicon chloride; silicon tetrachloride
四氯化(合)物 tetrachloride; tetrachloro compound
四氯化铼 rhenium tetrachloride
四氯化钼 molybdenum tetrachloride
四氯化镎 neptunium tetrachloride
四氯化铅 lead tetrachloride; plumbic chloride
四氯化钛 titanic chloride; titanium tetrachloride
四氯化碳 carbon tester; carbon tetrachloride; phenixin; tetrachloromethane
四氯化碳活度 carbon tetrachloride activity
四氯化碳灭火机 carbon tetrachloride fire extinguisher
四氯化碳熔断器 carbon tetrachloride fuse
四氯化碳熔丝 carbon tetrachloride fuse
四氯化碳中毒 carbon tetrachloride poisoning
四氯化钨 tungsten tetrachloride
四氯化锡 butter of tin; tin tetrachloride
四氯化铱 iridic chloride; iridium tetrachloride
四氯化乙烯 ethylene tetrachloride
四氯化铀 uranic chloride; uranium tetrachloride
四氯化锗 germanic chloride
四氯联苯 tetrachlorobiphenyl
四氯酞酸 tetrachlorophthalic acid
四氯酞酸酐 tetrachlorophthalic anhydride
四氯硝基苯 tecnazene
四氯乙烷 cellon; ethylene tetrachloride; tetrachloroethane
四氯乙烯 perchloro-ethylene; perclene; tetrachloroethylene; perchlorethylene

四氯异吲哚啉酮 tetra chloroisoindolinone
四轮 four-wheel
四轮操纵的 four-wheeled steer(ing)
四轮叉车起吊卡车 four-wheeled fork lift truck
四轮铲运机 four-wheeled scraper
四轮车 brougham; four-wheeler; quadricycle
四轮车辆 four-axle vehicle; four-wheel car
四轮传动 four-wheel drive
四轮传动车辆 four-wheel drive vehicle
四轮大车 farm wagon
四轮导向的 four-wheeled steer(ing)
四轮的底盘 four-wheeled chassis
四轮的机架 four-wheeled chassis
四轮的蟹爪式起重机 four-wheeled crab
四轮防滑装置 four-wheel anti-skid system
四轮钢滑车 four-sheave steel block
四轮挂车 four-wheel trailer
四轮滑车 fourfold block; fourfold purchase
四轮机车转向架 four-wheel engine truck
四轮机动车辆 four-wheel motor vehicle
四轮卡车 <英> rulley
四轮括土机 four-wheeled scraper
四轮马车 buckboard; phaeton <一种轻快的>
四轮马车雕饰 four-horse(d) chariot
四轮平板车(搬运) platform carriage
四轮起落架 four-wheel landing gear
四轮汽车 four-wheel automobile; four-wheel(ed) vehicle
四轮牵引机 four-wheel tractor
四轮驱动 four-wheel drive
四轮驱动的拖拉机 four-wheeled drive tractor
四轮驱动的装载机 four-wheeled drive loader
四轮驱动货车 four-wheel drive lorry
四轮驱动牵引车 four-wheel drive tractor
四轮驱动式车辆 four-wheel drive vehicle
四轮驱动式反铲挖掘装载机 four-wheel drive backhoe-loader
四轮驱动式平地机 four-wheel drive grader
四轮驱动式拖拉机 four-wheel drive tractor
四轮驱动式装载机 four-wheel drive loader
四轮驱动台车 four-wheel drive bogie
四轮驱动拖拉铲运机 four-wheel-drive tractor shovel
四轮驱动载货汽车 four-wheel drive truck
四轮驱动自行车辆 four-wheel drive self-propelled vehicle
四轮刹车 four-wheel brake
四轮式平地机 four-wheel grader
四轮式拖拉机 four-wheel tractor
四轮式转向 four wheel steer
四轮式装载机 four wheel loader
四轮手推车 four-wheeled hand truck
四轮四座马车 barouche
四轮胎小客车 four-tire[tyre] car
四轮拖车 four-wheel trailer; four-wheel wagon
四轮拖拉机 four-wheeled tractor
四轮橡胶拖拉机 four-wheeled rubber-tyred roller
四轮小车式 <双轮双轴> dual in tandem
四轮修车起重器 four-wheel garage jack; four-wheel screw jack
四轮原地转向 four-wheel crab steering
四轮原动机 four-wheeled prime mover
四轮(运)货车 wagon
四轮运木挂车 four-wheel timber trailer
四轮运坯车 carousel
四轮闸 four-wheel brake
四轮振动压路机 four vibrating drum roller
四轮制动 four-wheel braking
四轮制动器 four-wheel brake
四轮制动设备 four-wheel braking
四轮转向 four-wheel steering
四轮转向架 bogie[bog(e)y]; four-wheel bogie
四轮转向式拖拉机 four-wheel steer tractor
四轮转向拖拉机 four-wheel steering tractor
四罗拉式大牵伸装置 four line high draft system
四螺旋桨船 quadruple screw motor ship; quadruple screw steamer
四马达移动起重机 four-motor travel-(l)ing crane
四马马车 four in hand
四马拖车雕饰 quadriga[复 quadrigae]
四马一套 four-abreast hitch
四马战车雕饰 quadriga[复 quadrigae]
四门工作台 four way entry
四门轿车 landaulet; sedan
四门篷车 phaeton
四面顶锤 tetrahedral anvil
四面陡坡屋顶 helm roof
四面对称曲线 tetrahedral symmetric curve
四面斧砍的 square hewn
四面函数 tetrahedral function
四面环海的 seabound; sea-girt
四面加压式砧模 tetrahedral anvil
四面剪力 quadruple shear
四面锯切 square cut
四面锯切的 square-sawn
四面开槽方块 Antifier block
四面块体 <空心的> tetrahedron block
四面块(体)护岸 tetrahedron block revetment
四面雷达监视 four-way radar surveillance
四面落水的天窗 hipped dormer(window)
四面落水屋顶 hipped roof
四面刨 matcher; sizer
四面刨床 four-sided planing; quadrilateral planning machine
四面刨削机 matcher
四面体 tetrahedroid; tetrahedron
四面体层 tetrahedral layer
四面体场 octahedral field
四面体丛 tetrahedral cluster
四面体簇 tetrahedral cluster
四面体单元 tetrahedral element
四面体单元体的特征 tetrahedral element characteristic
四面体的 tetrahedral
四面体的顶点 tessarace
四面体对称性 tetrahedral symmetry
四面体构型 tetrahedral configuration
四面体混合轨函数 tetrahedral hybrid orbital function
四面体假说 tetrahedral theory
四面体间隙 tetrahedral interstice
四面体键结构 tetrahedral bonding structure
四面体角 tetrahedral angle
四面体(晶)片 tetrahedral sheet
四面体孔 tetrahedral pore
四面体块 tetrahedral block
四面体块体 tetrahedron block
四面体棱镜 kaleidoscope prism
四面体排列 tetrahedral arrangement
四面体配位 tetrahedral coordination
四面体缺陷 tetrahedral defect
四面体群 tetrahedral group; tetrahedron group
四面体绕组 tetrahedral winding
四面体式 tetrahedral formula
四面体数组 tetrahedral array
四面体研究卫星 tetrahedral research satellite
四面体坐标系 tetrahedral coordinate system
四面通道 four-sided gateway
四面线汇 tetrahedral congruence
四面有门廊或柱廊的宫院 tetrastoon [复 totrastoa]
四面装(配)玻璃的瞭望台 glazed lookout deck
四面装(配)玻璃的平台 glazed observation platform
四面装(配)玻璃的月台【铁】 glazed viewing platform
四面装饰 therming
四面坐标 tetrahedral coordinates
四能级 four-level
四能级材料 four-level material
四能级发射体 four-level emitter
四能级系统 four-level system
四能级荧光固体 four-level fluorescent solid
四能级荧光晶体 four-level fluorescent crystal
四能级振荡器 four-level generator
四年时间 quadrennium
四扭编组 squaring
四耙分级机 quadruplex rake classifier
四耙式分级机 four-rake classifier
四排滚棒式轴承 four-race roller bearing
四排链条传动 quadruple chain drive
四排汽口汽轮机 quadruple-flow turbine
四排汽口涡轮机 quadruple-flow turbine
四配位数 tetrahedral coordination
四配位体 quadridentate
四配位体螯合物 quadridentate chelate
四喷嘴对喷式气流粉碎机 Blaw Knox jet mill
四硼酸 tetraboric acid
四硼酸锂 lithium tetraborate
四硼酸钠 borax; sodium tetraborate
四硼酸盐 tetraborate
四硼烷 tetraborane
四片层矿物 four-sheet mineral
四片对接镶嵌 four-piece butt matching
四片对拼法 four-piece butt matching
四频器 quadrupler
四频双工制 four frequency diplex
四频制 quadruple-frequency system
四坡顶【建】 hip roof
四坡顶垂脊脊瓦 arris hip tile
四坡顶山墙端 hipped end
四坡顶(直)脊瓦 arris hip tile
四坡孟莎屋顶 hip mansard roof
四坡面板屋顶 hipped slab roof
四坡木板屋顶 timber hipped-plate roof; wooden hipped-plate roof
四坡排水管屋顶 hip-and-valley roof
四坡平屋顶 deck-on-hip; hip and flat roof
四坡屋顶 clipped gable; hip and hip roof; hipped roof; hip roof; Italian roof; pavilion roof; quadrangle roof; whole hip
四坡屋顶面坡椽 hip rafter
四坡屋顶上的平屋顶 deck-on-hip
四坡屋顶中脊上的瓦或盖条 hip roll
四坡阴戗屋顶 hip-and-valley roof
四坡攒角屋顶 helm roof
四破木料 edge-grained lumber
四汽缸发动机 four-banger; four-cylinder engine
四汽缸机车 four-cylinder locomotive
四腔化油器 four-choke carburettor
四腔式 four barrel
四腔速调管 four cavity klystron
四羟甲基环己醇 tetramethylol cyclohexanol
四羟甲基氯化锑 tetrakis hydroxymethyl phosphonium chloride
四羟甲基替苯代三聚氰二胺 tetramethylol benzoguanamine
四桥式底卸载货车 four axles bottom-dump
四氢呋喃 tetrahydrofuran
四氢化吡咯 pyrrolidine; tetrahydropyrrole
四氢化呋喃甲醇 tetrahydrofurfuryl alcohol
四氢化邻苯二甲酸 tetrahydrophthalic acid
四氢化邻苯二甲酸酐 tetrahydrophthalic anhydride
四氢化萘 tetrahydronaphthalene
四氢化三甲基匹 tetrahydrotrimethypicene
四氢糠醇 tetrahydrofurfuryl alcohol
四氢糠醇丙烯酸酯 tetrafurfuryl acrylate
四氢萘 tetrahydronaphthalene
四氢噻唑 thiazolidine
四球机 four-ball apparatus
四球机负荷能力试验 four-ball load carrying capacity test
四球极压试验机 four-ball extreme pressure tester
四球角锥 four-ball top
四球摩擦机 four-ball machine
四球润滑剂性能测定仪 four-ball tester
四球式极压润滑剂试验机 four-ball extreme pressure lubricant tester
四球试验机 four-ball tester; four-ball test rig
四区穹隆 quadripartite vault
四曲柄 four-throw
四曲柄压力机 four-crank press
四曲拐曲轴压力机 four-point press
四取一 select 1 in 4
四人一组 foursome; quartet(te)
四刃钎头 four-point bit
四日市哮喘 yokkaichi asthma
四色测光 four-colo(u)r photometry
四色测光系统 four-colo(u)r photometry system
四色定理 four colo(u)r theorem
四色分色制版法 four-colo(u)r separation process
四色复制 four-colo(u)r reproduction
四色胶片机 four-colo(u)r offset press
四色金 four-colo(u)r gold
四色问题 four-colo(u)r problem
四色印刷 four-colo(u)r printing
四筛分级器 four-deck grader
四筛分选 four-deck screen
四山墙形式 four gable form
四扇式折叠推拉(遮蔽)门 four-leaf sliding folding shutter door
四扇式转角推拉(遮蔽)门 four-leaf around corner sliding shutter door
四扇式转门 four-wing revolving door
四扇一组的拉门 four-unit sliding door

四扇整体门 four-unit sliding door

四上五下牵伸 four over five draft

四舍五入 half adjust; round down; rounding; round off; round up

四舍五入法【数】rounding-off error

四射珊瑚绝灭 rugose coral extinction

四深裂的 quadripartite

四声道立体声 quadraphonics; quadraphony; quadrasonics

四声迹磁带 four-track tape

四绳索摩擦提升机 four-rope friction winder

四绳提升 four-rope winding

四十八根七毫米直径的钢丝绳 cable of 48 seven mm wires

四十八根七毫米直径的钢丝索 cable of 48 seven mm wires

四十九点移动平均 forty-nine points moving average

四十烷 tetracontane

四十五转唱片 forty-five record

四十五度 Y 形管 forty-four degree lateral

四十五度大半径弯头 long radius 45° bend

四十五度角尺 <木工用> miter[mitre] box

四十五度角扶手弯头 mitred knee

四十五度弯头 eighth bend; forty-five degree swan-neck; forty-four degree bend

四十五度屋面坡度 square pitch

四十五度削接头 forty-four degree miter[mitre] joint

四十五度斜角铁板 <抹灰工用> miter[mitre] rod

四十五度斜接缝 miter[mitre] joint

四十五度斜屋面 square pitch; square roof

四十五度斜削接头 miter[mitre] joint

四十五度支管 forty-four degree angle branch

四十五度纵横交叉 transversal crossing

四十五角斜接 mitre[miter]

四十五角斜锯小梁 mitered[mitred]-and-cut string

四十英尺提升管子钻塔 forty feet pull tubular steel derrick

四氏岩石分类法【地】C.I.P.W.[Cross, Iddings, Pirsson, Washington] system of rock classification

四示像三闭塞区段系统 four-aspect three-block system

四示像三闭塞区段制 four-aspect three-block system

四示像色灯信号 four-aspect colo(u)r light signal

四示像色灯信号机【铁】four-aspect colo(u)r light signal

四示像信号法 four-aspect signal(l)ing

四示像自动闭塞 four-aspect automatic block

四示像自动闭塞系统 four-aspect automatic block system

四示像自动闭塞制 four-aspect automatic block system

四室分级机 four-spigot classifier

四室公寓 four-room(ed) flat

四室骨料喂送器 four-compartment aggregate feeder

四室料仓 four-compartment bin

四室炉 four hearth furnace

四室式磨碎机 four-compartment mill

四室套房 four-room(ed) flat

四栓鱼尾板 four-bolt splice bar; 4-bolt splice bar

四水白铁矾 rozenite

四水合酒石酸钾钠 Rochelle salt

四水合物 tetrahydrate

四水镁矾 leonhardtite; starkeyite

四水锰矾 ilesite

四水钠硼石 biringuccite

四水硼钙石 nobleite

四水硼砂 kernite; rasorite

四水硼锶石 tunnellite

四水铜铁矾 guildite

四水泻盐 starkeyite

四水锌矾 boyleite

四水钴矾 aplowite

四素组分组 tetrad grouping

四素组效应 tetrad effect

四速变速箱 four speed gear box

四速齿轮 four speed gear

四速挡的 four-speed

四速动力换挡 four-speed power shift transmission

四速快门 four speed shutter

四速无声变速箱 four silent-speed gearbox

四酸 tetracid

四酸价碱 tetracid base

四酸式盐 tetrahydric salt

四羧酸 tetrabasic carboxylic acid

四羧酸酯 tetrabasic ester

四索杆式抓斗 four-rope rod type grab

四索木材抓斗 four-rope log grapple

四索悬挂抓斗 four-rope suspension grab(bing)

四索抓斗 four-rope grab

四锁存器 quad latch

四台凿岩机的伸梁 <钻车上> quadruple boom

四台凿岩机的钻车 four-drill rig

四台凿岩机钻车 quadruple rig

四钛酸钡陶瓷 barium tetratitanate ceramic

四探极法 four-probe method

四探针 four-point; four-point probe

四探针测量 four-point probe measurement

四探针电阻率测试系统 four-point resistivity test system

四探针法 four-probe method

四探针装置 four-probe arrangement

四套马 four-horse team

四体 tetrasome

四体浮座 <浮式钻台的> quadramaran

四体问题 four body problem

四体性 tetrasomy

四条道路交会于一点的 quadrivial

四条(或以上)马路的交叉路口 carfax

四铁芯 four-limbed

四烃基硅 silicane

四通 cross-nailed material; cross joint(ing); double lee; four-start; pipe cross

四通长度 length of 4 way pipe

四通道乘法器 four-channel multiplier

四通道多光谱带照相机 four channel multiband camera

四通道燃烧器 four channel burner

四通道设计 <锅炉的> four-pass design

四通的 four way

四通电磁阀 four-way solenoid valve

四通阀 cross cock; cross valve; four-channel valve; four-port pump; four-way port; four-way tap; four-way valve

四通管 cross(bar); cross branch; cross pipe; cross tube; four-way branch; four-way pipe; four-way tube

四通管接 four-way junction

四通管接头 cross connection; cross joint(ing); cross pipe; four-way junction; four-way union

四通管头 four-way piece

四通回动阀 reversing four way valve

四通活管接 cross union

四通接头 four-channel union; four-way box; four-way connection; four-way coupling; four-way joint; four-way tee; four-way union

四通节头 double tee

四通开关 four-way cock; four-way switch; four-way tap

四通连接 cross connection; four-way connection

四通联管节 four-way union

四通龙头 four-way cock

四通路 four way

四通路循环器 four-terminal circulator

四通轮缘扳手 four-way rim wrench

四通驱动与操纵的分级器 four-wheel drive and steer grader

四通塞阀 four-way plug valve

四通塞门 four-way cock

四通栓塞 four-way cock

四通旋塞 four-way cock

四通支管 cross branch

四通制模板 four-entry pallet; four-way pallet

四筒绞车 four-drum hoist

四头带 four-tailed bandage

四头灯 quadruple burner

四头翻边机 four-head pressure

四头螺旋 four-start spiral; quadruple screw

四透镜 four-element; four-lens

四透镜变焦系统 four-lens zoom system

四透镜物镜 four-lens objective; quadruplet lens

四凸耳凸轮 four-lipped cam

四弯矩定理 theorem of four moments

四弯曲柄轴 four-throw crank shaft

四烷基季铵盐 tetraalkyl ammonium salt

四桅船 four-master

四桅帆船 four-masted bark

四维标准正交标架 orthonormal tetrad

四维的 four-dimensional

四维基本形 four-dimensional fundamental form

四维几何 four-dimensional geometry

四维空间 four-dimensional space

四维流形理论 <几何及拓扑学>【数】four-manifold theory

四维密度 four-density

四维全息记录 four-dimensional holographic recording

四维矢量 four-dimensional vector

四维资料同化 four-dimensional data assimilation

四位二进制 tetrad

四位码 four-figure code

四位片电路 four-bit slice circuit

四位片微处理机 four-bit slice microprocessor

四位片系统 four-bit slice system

四位式信号机【铁】four-position signal

四位数 four figures

四位数代码 four-figure code

四位数字显示 four digit display; four-figure display

四位制 four digit system

四位置透镜转头 four-position lens turret

四位字节 nibble; quartet(te)

四位组 nibble; tetrad

四纹螺旋 quadruple screw

四屋式磨碎机 four-compartment mill

四系弹簧悬挂式转向架 bogie with quadruple suspension

四系悬挂的转向架 bogie with quadruple suspension

四弦索多边形 four-chord polygon

四显示三闭塞区段系统 four-aspect three-block system

四显示三闭塞区段制 four-aspect three-block system

四显示色灯信号 four-aspect colo(u)r light signal

四显示色灯信号机【铁】four-aspect colo(u)r light signal

四显示信号法 four-aspect signal(l)ing

四显示信号机【铁】four-position signal

四显示自动闭塞【铁】four-aspect automatic block

四显示自动闭塞系统 four-aspect automatic block system

四显示自动闭塞制 four-aspect automatic block system

四显性组合 quadruplex

四线八芯电缆 quad pair cable

四线触发器 quadded flip-flop

四线的 four-wire; quadded; quadruple

四线电缆 quadded cable

四线电路 four-wire circuit

四线电路制 four-wire circuit system

四线端接装置 four-wire terminating set

四线对绞 quad pairing

四线对绞电缆 quad pairing cable

四线二线变换设备 four-wire terminating set

四线发送 four-wire transmission

四线横担 four-wire arm

四线交换 four-wire switching

四线接收 four-wire reception

四线馈线线夹 four feeder wire clamp

四线连接 four-wire connector

四线逻辑 quadded logic

四线螺纹 quadruple thread

四线螺旋 quadruple screw

四线木担 four-pin wooden cross arm

四线期 four-strand stage

四线区段 four-track section; quadruple-track section

四线塞孔 four-way jack

四线三相制 four-wire three-phase system

四线实线电路 four-wire side circuit

四线式电路 four-wire type circuit

四线式轧机 four-strand mill

四线双交换 four-strand double crossing over

四线铁路 four-track line; four-track railway; quadruple line; quadruple railway; quadruple track

四线通道 four-wire channel

四线蜗杆 four-start worm

四线系统 four-wire system

四线线路 four-wire line

四线形的 quadrilateral

四线性形式 quadrilateral form

四线预加应力千斤顶 four-wire prestressed jack

四线载波制 four-wire carrier system

四线增音机 four-wire repeater

四线制 four-wire system

四线制传输线 four-wire line

四线制的转发器 four-wire translator

四线制电路 four-wire channel; four-wire circuit

四线制多路复用设备 four-wire multiplex facility

四线制交换 four-wire switching

四线制交换中心 four-wire switching center[centre]

四线制增音器 four-wire repeater
四线制中继线 four-way trunk
四线制终端设备 four-wire terminating set
四线中继器 four-wire repeater
四线终端电路 four-wire line termination circuit
四线纵横交换制 four-wire cross bar switching system
四线组对绞(电缆) quad pairing
四箱砂矿跳汰机 four cell placer jig
四镶板门 four-panel door; four panel(l)ed door
四镶门板 four panel(led) door
四向刀架 four-way tool block; four-way tool post
四向的 four way
四向阀 four-way valve
四向翻倾混凝土运载车 circular-tipping concrete skip
四向钢筋 four-way reinforcement
四向拱 four-way arch
四向管接 four-way union
四向开关 four-way switch
四向锚泊法 all fours moorings; four arm mooring
四向配筋 four-way reinforcement; four-way system of reinforcement
四向配筋的 four-way reinforcing
四向平板 four-way flat slab
四向调节 four-way control
四向停车 four-way stop
四向停车标志 <用于同级公路交叉口处> four-way stop sign
四向系泊 four arm mooring
四向(小)龙头 four-way cock
四向装卸叉车 four direction side loader
四项表 quadruple table
四项的 quadrinomial; tetrachoric
四项分布 quadrinomial distribution
四项式 quadrinomial
四项相关 fourfold correlation
四相步进电动机 four-phase stepper motor
四相差分相移键控 quadrature differential phase shift keying
四相单级液力变矩器 four-phase single stage torque converter
四相点 quadruple point
四相动态逻辑电路 four-phase dynamic logic circuit
四相方波发生器 four-phase square-wave generator
四相平衡 quaternary phase equilibria
四相曲线 quadruple curve
四相双级液力变矩器 four-phase two stage torque converter
四相调制 four-phase modulation
四相图 quaternary diagram
四相相移键控 quadrature phase shift keying; quarternary phase shift keying
四相循环 four-phase cycle
四相制 four-phase system; quadraphase; quadriphase system
四象限 quadra-phase
四象限乘法器 four-quadrant multiplier
四象限固态传感器 four-quadrant solid-state sensor
四象限结构 four-quadrant construction
四象限特性图 four quadrant characteristic diagram
四象限图 four-quadrant diagram
四象限运算 four-quadrant operation
四象限整流器 four-quadrant rectifier
四硝基甲烷 tetranitromethane
四小时快干漆 four hour varnish
四小时轮班制 four-hour shift
四效 quadruple effect
四效蒸发器 four-effect evapo(u)rator; quadruple effect evapo(u)rator
四楔结合 four wedge joint
四心插塞 four-pin plug
四心都德式拱 four-centered[centred] Tudor arch
四心拱 depressed arch; Tudor arch
四心尖顶拱 four-centered[centred] pointer arch
四心螺线 four-center[centre] spiral
四心桃尖拱 four-centered[centred] arch
四心外心桃尖拱 tented arch
四心直线尖顶拱 four-centered[centred] Tudor arch
四芯导线 quad
四芯的 quadded
四芯电缆 four-core cable; four-wired cable; quad; quadded cable; quadraplex cable
四芯电缆系统变换 systematic transposition of cable quad
四芯扭绞 spiral four
四芯扭绞电缆 quad cable
四芯线 four-core wire
四芯线组 quad
四芯型电缆 quadplex type cable
四芯柱 four-limbed
四芯组电缆 quadded cable
四信道转换开关 four-channel switch
四信路制 twinplex
四行程的 four-stroke
四行程发动机 four-stroke engine
四行程循环 four-stroke cycle
四型钾霞石 tetrakalsilite
四性 quadruplex
四溴(代)邻苯二甲酸酐 tetrabromophthalic anhydride
四溴代乙烯 perbromo-ethylene
四溴双酚 tetrabromobisphenol
四溴酞酸酐 tetrabromophthalic anhydride
四溴乙烷 tetrabromoethane
四溴荧光素 tetrabromofluore scein
四溴萤光素 eosine
四溴萤光素颜料 nacarat
四牙轮钻头 four cutter bit; four-roller bit
四亚甲基二胺 tetramethylene diamine
四亚乙基五胺 tetraethylene pentamine
四盐基铬酸锌 zinc tetroxy chromate
四眼插座 four-pin plug
四眼联结板 four eyed joining piece
四氧化锇 osmium tetr(a)oxide
四氧化二氮 nitrogen tetr(a)oxide
四氧化二钒 vanadium tetr(a)oxide
四氧化铬 chromium tetr(a)oxide
四氧化钌 ruthenium tetr(a)oxide
四氧化镍 nickel superoxide
四氧化三锰 mangano-manganic oxide
四氧化三镍 nickelous-nickelic oxide
四氧化三铅 lead orthoplumbate; lead oxide red; lead tetr(a)oxide; minium; red lead oxide
四氧化三铁 ferriferous oxide; iron oxide black; magnetic iron ore; magnetite black
四氧化三铁锈层 magnetite
四氧化物 quadroxide; tetroxide
四摇臂式酸洗机 four-arm pickling machine
四叶窗花格 <哥特式的> tracery with four-leaf-shaped curves
四叶的 quadrifoil; quadrifoliate; tetraphyllous
四叶电容器 four-bladed capacitor
四叶风扇 four-bladed fan; four-bladed ventilator
四叶花雕饰 <盛饰时期的建筑特征之一> four-leaved flower
四叶花(装)饰 dogtooth; nail-head-(ed) mo(u)lding; four-leaved flower
四叶螺旋桨 four-blade propeller; quadruple screw
四叶片刀头 four-bladed bit
四叶片快门 four-flanged disc shutter; four-flanged shutter
四叶片圆抓斗 four-segment circular grab
四叶片钻头 four-bladed bit
四叶式 quaterfoil
四叶式布置 cloverleaf layout
四叶式道路交叉 quaterfoil crossing
四叶式花窗棂 quatrefoil tracery
四叶式环行道 cloverleaf loop
四叶式环形坡道 cloverleaf loop ramp
四叶式交叉 cloverleaf
四叶式交叉布置【道】 cloverleaf layout
四叶式交通布置 cloverleaf crossing
四叶式立体交叉 cloverleaf crossing; cloverleaf grade separation; cloverleaf interchange; cloverleaf intersection; cloverleaf junction
四叶饰 dogtooth; quarterfoil[quartrefoil]
四叶形 quatrefoil
四叶形窗花格 four-leaved tracery
四叶形立体交叉 cloverleaf; cloverleaf interchange; cloverleaf intersection; grade separation
四页破斜纹 broken crow twill
四页斜纹 four-harness twill; four-leaf twill
四乙铵 tetraethylammonium
四乙基 tetraethyl
四乙基放射铅 tetraethyl radio lead
四乙铅 tetraethyl-lead
四乙铅污染 tetraethyl lead pollution
四乙铅中毒 tetraethyl lead poisoning
四乙氧基钛 titanium tetraethoxide
四乙氧甲替苯代三聚氰二胺 tetrakis ethoxymethyl benzoguanamine
四翼飞机 quadriplane[quadruplane]
四翼刮刀钻头 four-way bit
四翼机 quadruplane
四翼扩孔钻头 four-wing reaming bit
四翼式转门 four-compartment revolving door
四翼圆盘式扩孔钻头 four-disc[disk] reaming bit
四翼缘刀具 four-wing bit
四翼缘钻头 four-wing bit
四翼钻头 four-edged bit; four-point bit; four-wing drilling bit; four-wing rotary bit
四因次的 four-dimensional
四英寸大钉 twopenny nail
四英寸厚混凝土块体 half height block
四用扳手 four-way wrench
四用斗铲 four in one bucket
四用螺钉起子 four-screw driver
四用自记计 quadruple recorder
四游仪 component of the four displacements; sighting-tube ring
四元 quaternionic
四元操作符 quarternary operator
四元醇 tetrabasic alcohol
四元的 quarternary
四元电流 four current
四元电流密度 four-current density
四元电位 four potential
四元法 quaternion
四元酚 tetra-atomic phenol; tetrahydric phenol
四元附件 quadruple unit attachment
四元共晶合金 quaternary eutectic alloy
四元合金 four-component alloy; four-part alloy; quad alloy; quarternary alloy
四元合金钢 quarternary steel; quinery steel
四元化合物 quaternary compound
四元环 four-membered ring
四元混合物 quaternary mixture
四元基 tetrad
四元加成物 quaternary adduct
四元加速度 four acceleration
四元加速器 four accelerator
四元件检测器 four-element detector
四元聚合物 quadripolymer
四元力 four-force
四元配料 quaternary mix
四元齐式 quarternary form
四元群 quaternion group
四元射影空间 quaternionic projective space
四元矢量 four vector; Lorentz four-vector
四元矢量势 four-vector potential
四元数 hypercomplex number; quaternion; quaternion number
四元数代数 quaternion algebra
四元数格拉斯曼流形 quaternion Grassmann manifold
四元数函数 quaternion function
四元数环 quaternion ring; ring of quaternions
四元数群 quaternion group
四元数双曲空间 quaternion hyperbolic(al) space
四元数体 quaternion field
四元数椭圆空间 quaternion elliptic space
四元数向量丛 quaternion vector bundle
四元数域 quaternion field
四元素钢 quarternary steel
四元素合金 quaternary alloy
四元素合金钢 quaternary alloy steel
四元素说 four-element theory
四元酸 quadribasic acid; tetrabasic acid; tetrahydric acid
四元(体)系 four-component system; quaternary system
四元通信[讯] quaternary signal(l)ing
四元网络 four-element grid
四元相图 quaternary system phase diagram
四元向量 four vector
四元形式 quarternary form
四元型 quarternary form
四元液体系统 quaternary liquid system
四元易熔合金 quarternary eutectic alloy
四元制 quaternary system
四元组 quadruple
四元组表示 quadruple notation
四元组类型 quadruple type
四元组算符 quadruple operator
四元组形式 quadruple form
四原色 four basic colo(u)rs
四原色法 basic four colo(u)r system
四原子氧 oxozone
四圆单晶衍射仪 quaternary monocrystalline diffractometer
四圆单晶衍射仪法 quaternary monocrystalline diffractometer method
四圆心近似法 <画椭圆用> four-center[centre] approximate method
四圆柱式庙宇 tetrastyle temple
四圆坐标 tetracyclic coordinate
四月牙形的 tetraselenodont
四凿岩机钻车 four-drill rig

四则 four species
四则计算机 arithmometer
四则运算 arithmetic; arithmetic(al) operation; four arithmetic(al) operation; four fundamental rules; four operations
四则运算机 arithmometer
四则运算器 arithmometer
四胀式蒸汽机 quadruple expansion engine
四爪 four-jaw
四爪单动车床卡盘 four-jaw independent lathe chuck
四爪分动卡盘 four-jaw independent chuck
四爪夹盘 four-jaw chuck; four armed spider; four-jaw independent chuck; four-jaw plate; independent chuck
四爪链钩 four paws
四爪锚 four-fluked anchor; grapple
四爪平面卡盘 branch chuck
四爪同心卡盘 four-jaw concentric chuck
四爪拖扫沉船 locate a wreck with towing grapnel
四爪小锚 grapnel; grapnel anchor; grapple; grappling hook; grappling iron
四折叠板 fourfold block
四折门扇之一 quadrivalve
四折木尺 fourfold rule; fourfold wood rule
四正丁氧基钛 tetra-n-butoxy titanium
四支点的 four-bearing
四支脚(开挖)深度 four-foot depth
四支脚挖沟试验 <反铲挖土机的> four-foot trenching test
四止尺寸 <土地的> metes and bounds
四中心反应 four-center [centre] (type) reaction
四中心过渡状态 four-center[centre] transition state
四种主要形式 four principal form
四周 all-around; circumference; entourage; environment; periphery
四周凹圆的格子平顶 coved checker ceiling
四周凹圆的细格平顶 coved lattice ceiling
四周不靠墙的浴盆 free-standing(bath) tub
四周大气 surrounding atmosphere
四周挡土墙 <地下室> area wall
四周的 ambient; circumferential
四周环水 compassed by the sea
四周加荷 all-around loading
四周界柱式(建筑) peripteral; periptery
四周视野 ail-round vision
四周双列柱廊式建筑 dipteral; dipteros
四周搜索 round-looking scan
四周无边的(工作平台) open sides and ends
四周镶玻璃的区域 glazed area
四周镶嵌玻璃的地区 glassed-in area
四周压力 <三轴仪的> cell pressure
四周用板桩围住的基坑 building pit closed by sheet pile
四周有墙或建筑物围绕的庭院 enclosed court
四周有柱的房屋 peritral
四周有柱的建筑 peritery
四轴 four-axle
四轴半自动 four-spindle semiautomatic
四轴车 four-axle car
四轴车床 four-spindle lathe
四轴承的 four-bearing
四轴承曲轴 four-bearing crankshaft
四轴法 four-axis method
四轴光点颤动 four-axis spot wobble
四轴货车 four-axled wagon
四轴铰接车 four-axle articulated vehicle
四轴卷纸机 four drum revolving reel
四轴客车 four-axled coach
四轴螺钉机 quadruple bolt cutter
四轴牵引车 four axle tractor
四轴式底卸载货车 four axles bottom-dump
四轴数字控制车床 four-spindle numerically controlled lathe
四轴系统 four-axis system
四轴载货汽车 four-axle truck
四轴载重汽车 four-axle truck
四轴针 tetraxon
四轴装置 four-axis mounting
四轴自动 four-spindle automatic
四柱插销 four-pin plug
四柱床架 four-poster bedstead
四柱大床 four poster
四柱大木 four poster
四柱导承 four-bar guide
四柱径式 araeostyle
四柱立面式 tetrastyle
四柱汽车举升器 four-post car lift
四柱塞操舵装置 four-ram steering gear
四柱式的(建筑) tetrastyle
四柱式门廊 tetrastyle; tetrastylic portico
四柱式水压机 four-post hydraulic press
四柱式液压机 four-column hydraulic press
四柱式柱廊 tetrastyle colonnade
四柱水压机 four-column hydraulic press
四柱厅 four-column hall
四柱装药 quadruple charge
四砖层的 four-bed
四桩腿平台 four-leg platform
四锥孔螺母 four-pin driven nut
四锥孔圆缘螺母 four-pin driven collar nut
四足鼎 quadripod
四足动物 tetrapod
四组并排椭圆弹簧 quadruplet; quadruplet elliptic spring; quartet(te) elliptic spring
四组分系统 four-component system
四组元红外物镜 four element infrared objective
四组元物镜 four-component objective
四嘴包装机 four spout packing machine
四座车身 four-passenger body
四座乘员分离舱 four seat crew capsule
四座篷顶小客车 foursome cabriolet
四座式(汽车) four seater
四座式小客车 four seater

寺庙 house of God; temple

寺庙后的厕所 reredorter
寺庙建筑 temple architecture; hypaethral <古典无屋顶的>
寺庙境域的入口 propylaeum[复 propylaea]
寺院 temple; abbey; cloister; joss house; monastery; monkery; religious house; sanctuary
寺院藏经楼 monastic library
寺院藏书楼 monastic library
寺院唱诗班 monastic choir; monastic quire; monk's choir
寺院厨房 monastic kitchen
寺院大厅 monastic hall
寺院的 cloistral
寺院的卧室部分 dorter[dortour]
寺院房屋 monastic building
寺院歌唱班席位 monastic choir; monastic quire; monk's choir
寺院会堂 monastic hall
寺院建筑 conventional architecture; monastic architecture; temple architecture
寺院居住场所 monastic house
寺院墓地 monastery cemetery; monastic cemetery
寺院前庭 parvis
寺院前院 parvis
寺院群落 monastic community
寺院社区 monastic community
寺院食堂 frater house; fratry
寺院宿舍 dorter[dortour]
寺院庭园 monastery garden
寺院围墙 peribolos; temenos
寺院文娱室 monastic recreation room
寺院小室 monastic cell
寺院休息室 monastic recreation room
寺院娱乐室 monastic recreation room
寺院园林 monastery garden
寺院中可接待客人的房间 parlatory
寺院住宅 monastic house

汜溢 debacle

伺服 servo

伺服安全阀 servo-relief valve
伺服泵 service pump; servo-pump
伺服笔绘记录仪 servo-scribe recorder
伺服补偿机 servo-tab
伺服部件 servo component
伺服操纵 servo-control
伺服操纵阀 servo-valve
伺服操纵杆 servo-assisted lever
伺服乘法器 servo-multiplier
伺服程序设计 servo-programming
伺服触头 servo contact
伺服传动 servo-drive
伺服传动电机 servo-drive motor
伺服传动机构 servo-operated mechanism
伺服传动器 servo-driver
伺服传动系统 servo-mechanism
伺服传动装置 servo-driver; servo-link; servo-system drive
伺服磁场 servofield
伺服磁场调节器 servofield regulator
伺服磁鼓 servo-head wheel
伺服磁盘 servodisk
伺服磁头 servo-head
伺服单位 servo-unit
伺服单元 servo-unit
伺服的 servo-actuated
伺服电磁铁 servo-magnet
伺服电动机 actuating motor; pilot motor; servomotor
伺服电动机操纵阀 power-operated valve
伺服电动机容量 servomotor capacity
伺服电机 servo-actuator
伺服电势计 servo-potentiometer
伺服电位计 servo-potentiometer
伺服定位 servo-positioning
伺服断路开关盘 servo-cutout switch panel
伺服发电机 servo-generator
伺服阀 pilot-controlled valve; servo-motor valve; servo-operated back pressure valve; servo-valve
伺服阀缓冲作用 servomotor cushion
伺服反馈 servofeedback
伺服反馈系统 servofeedback system
伺服方位发送 servobearing transmission
伺服仿真机 servo-simulator
伺服放大器 servo-amp(lifier)
伺服分解器 servo-resolver
伺服分析器 servo-analyser[analyzer]
伺服分压器 servo-potentiometer
伺服干扰 servo-noise
伺服缸 servocylinder
伺服功率 service power
伺服函数发生器 servo-function generator
伺服换挡 power shift
伺服回路 servo-loop; servomotor loop
伺服回路模型 servo-loop model
伺服回路系统 servo-loop system
伺服回授系统 servofeedback system
伺服绘图器 servo-curve plotter
伺服活塞 servo-piston
伺服机 relay; servo
伺服机缸 relay cylinder
伺服机构 operator; servo; servo-actuator; servo-control mechanism; servo-gear; servo-mechanism; servo-system; servo-unit
伺服机构动力学 servo-dynamics
伺服机构类型 servo-mechanism type
伺服机构理论 theory of servomechanism
伺服机构能源 servomotor
伺服机构试验台 servoboard
伺服机件 servo-mechanism
伺服机械 servo-unit
伺服机械手 servo-manipulator
伺服机制 servo-mechanism
伺服唧筒 servocylinder
伺服积分器 servo-integrator; velocity servo; velodyne
伺服及自动增益控制检波器 servo and automatic gain control detector
伺服记录器 servo-recorder
伺服加力转向 power-assisted steering
伺服接触 servo contact
伺服接收机 servo-control receiver
伺服解算器 servo-resolver
伺服控制 servo-control; servo-operated control
伺服控制传动 servo-controlled drive
伺服控制传送装置 servo-controlled transport device; servo-controlled transport equipment; servo-controlled transport unit
伺服控制电路 servo-controlled circuit
伺服控制阀 pilot valve
伺服控制光圈调节 servo-controlled aperture setting
伺服控制机构 servo-control mechanism
伺服控制机器人 servo-controlled robot
伺服控制镜 servo-control mirror
伺服控制论 servo-controlled theory; theory of servomechanism
伺服控制马达 servo-controlled motor; servomotor
伺服控制器 servo-controller
伺服控制扫描干涉仪 servo-controlled scan(ning) interferometer
伺服控制设备 servo-control equipment
伺服控制系统 servo-control system
伺服控制线路 servo-control circuit
伺服控制压力 pressure servo-control
伺服离合器 servoclutch

S

伺服理论 servo-theory
伺服连接 servo-connection
伺服马达 slave unit
伺服马达部件 oil power cylinder block
伺服马达阀 servomotor valve
伺服模拟机 servo-simulator
伺服模拟计算机 servo-analog(ue) computer
伺服模拟装置 servo-analog(ue) device;servo-simulator
伺服盘面编码 servosurface encoding
伺服汽缸 servocylinder
伺服千斤顶 servojack
伺服驱动绘图机 servo-driven plotting table
伺服驱动(装置) servo-drive; slave drive
伺服扫描 servo-scanning;servo-scribe
伺服扫描记录器 servo-scribe recorder
伺服刹车 servobrake
伺服设备 servo-system
伺服摄影机 servocamera; slave camera
伺服式传感器 servo-sensor; servo-transducer
伺服式的 servo-type
伺服式加速度计 servoaccelerometer
伺服调速器 servo-governor
伺服调谐 servo-tuning
伺服调整片 control servo-tab; servo-tab
伺服调制 servo-modulation
伺服调制器 servo-modulator
伺服通道 servochannel
伺服拖动 servo-drive
伺服拖动的 servo-driven
伺服拖动机构 servo-actuator
伺服拖动装置 servo-actuator
伺服稳定性 servo-stabilization
伺服系统 following-up mechanism; follow-up system; servo-drive system; servo-link; servo-loop; servo-system;follow-up
伺服系统参数 servo-parameter
伺服系统测试仪 servo-mechanism tester
伺服系统的动力传动(装置) servo-dyne
伺服系统电势计 follow-up potentiometer
伺服系统电位器 follow-up potentiometer
伺服系统驱动 servo-driven
伺服系统噪声 servo-noise
伺服响应率 servo-response rate
伺服信号电路 servochannel
伺服压力 servo-pressure
伺服压力计 servo-manometer
伺服液力系统 servohydraulic system
伺服液压激振器 servohydraulic actuator
伺服液压马达 servomotor
伺服液压系统 servohydraulics
伺服油缸 power cylinder;servocylinder;slave cylinder
伺服元件 actuating unit;servoelement
伺服运算电路 servo-operation circuit
伺服增量 servo-increment
伺服执行机构 servo-actuator
伺服制动(器) servobrake
伺服助力转向 servo-assisted steering
伺服转向 <履带拖拉机的> power turn
伺服装置 actuating device;servo-mechanism;slave unit
伺服作用 servoaction
伺服作用背压阀 servo-operated back pressure valve
伺服作用器 servo-actuator

似

似斑状花岗岩 porphyroid granite
似斑状结构【地】 porphyritic-like texture
似板法 quasi-slab method
似碧玉化 jasperoidization
似玻璃状断口 vitreous fracture
似槽痕 setulf
似层状 likelihood bedded;parabedded
似层状构造 stratoid structure
似层状矿体 stratoid ore body
似长石 feldspathoid
似长石的玄武岩 feldspathoidal basalt
似长石矿物 feldspathoid mineral
似长石类 lenad
似长石岩 foidite
似臭氧的 ozonic;ozonous
似粗面结构 trachytoid texture
似粗面状 trachytoid
似大地水准面 quasi-geoid
似大地水准面高程 quasi-geoidal height
似弹头的 bullet-nosed
似地球面 telluroid
似地震事件 seismic like event
似动力高 quasi-dynamic(al) height
似动力学的 pseudo-dynamic
似二度量板法 graticule method for quasi two-dimensional bodies
似芳族化合物 quasi-aromatic compound
似非而可能是的说法 paradox
似霏细状【地】 felsitoid
似肥皂的 saponaceous
似沸腾 pseudo-boiling
似概差 quasi-probable error
似橄榄石 olivinoid
似钢的 steely
似根 rooty
似宫殿的 palatial
似共轭效应 hyperconjugation; quasi-conjugation
似钩状构造 hook-like structure
似光学性 quasi-optic(al)
似核形石 oncoid
似乎合理的 plausible
似乎真实的 plausible
似糊的 pappy
似互穿聚合物网络 quasi-interpenetrating polymer network
似花岗石的 allotriomorphic granular
似花岗石状的 granitoid
似花岗岩状 granitoid
似花岗岩状结构 granitoid texture; xenomorphic granular texture
似滑石的 talcose
似化学方法 quasi-chemical method
似黄铜 brassy
似黄锡矿 stannoidite
似灰 ashy
似灰的 cinereous
似灰物质 ashy substance
似辉石 pyroxenoids
似活塞作用 <钻头包泥后的> piston-like action
似或然误差 quasi-probable error
似尖塔 spiry
似剪切 pseudo-shear
似剪应力 apparent shear stress
似碱玄岩 tephritoid
似箭的 arrow
似焦油的 tarry
似角砾岩的 breccioid
似金刚石 diamantin(e)
似金属 metalloid
似金属的 metallike
似金属相 metalloid phase
似近色 advancing colo(u)r
似晶的 crystalloid
似晶格 quasi-lattice
似晶结构 quasi-crystalline structure
似晶结晶 mimetic crystallization
似晶石 phenacite
似晶石的 sparry
似晶石铍矿石 phenakite ore
似晶质 crystalloid
似静态变化 quasi-static change
似静态试验 quasi-static test
似绝对伏特 semi-absolute volt
似抗剪强度 apparent shear-strength
似抗剪强度角 apparent angle of shearing resistance
似壳的 shell like
似可畏码 formidable-looking code
似块状结构 blocklike structure
似矿物 gel mineral;mineraloid
似蜡的 ceraceous;wax-like
似蓝闪石 naurodite
似棱镜媒质 prism-like medium
似棱体改正 prismoidal correction
似棱体公式 prismoidal formula
似棱体校正 prismoidal correction
似棱形的 prismoidal
似棱形改正 prismoidal correction
似棱柱体公式 prismoidal formula
似棱柱状结构 prism-like structure
似沥青的 asphaltic
似粒子模型 quasi-particle model
似连续变化 quasi-continuous variation
似邻接面 vicinaloid
似鳞的 squamose
似流体动力学的润滑 quasi-hydrodynamic(al) lubrication
似六边形 quasi-hexagon
似六角形 quasi-hexagon
似卵形 ovaloid
似煤的 coaly
似模板模式 template-like pattern
似木的 xyloid
似难码 formidable-looking code
似能 plausibility
似黏[粘]聚力 apparent cohesion
似凝灰岩的 tuffaceous
似偶极子 pseudo-dipole
似泡沫的 spumescent
似片麻岩状 gneissoid
似片岩的 schistoid
似片状的 plate-like
似平底晶洞状构造 stromatactoid
似平流层 stratosphere
似铅的 leady;plumbean;plumbeous
似前期固结压力 pseudo-preconsolidation pressure; quasi-reconsolidation pressure
似球粒 pelletoid
似然比 likelihood ratio
似然比泛函 likelihood ratio functional
似然比检验 likelihood ratio test
似然比值测验 likelihood ratio test
似然比值检定 likelihood ratio test
似然比值检验 likelihood ratio test
似然法 likelihood method
似然估计 likelihood estimation
似然估计量 likelihood estimate
似然估值 likelihood estimate
似然率 likelihood
似然率函数 likelihood function
似然性准则 criterion of likelihood
似然准则 likelihood criterion
似沙蚕迹 <遗迹化石> nereites
似沙漠的 xeric
似砂石的 calculous
似舌形沙坝 linguloid bar
似蛇的 ophidian;serpentine
似蛇纹石 miskeyite;pseudo-phite
似石的 petrosal
似石灰 limy
似石榴子石 garnetoid
似石棉 asbestiform
似石棉的 asbestoid
似树状的 arborescent
似衰减光度 quasi-attenuance
似速度 apparent velocity
似塑度 quasi-plasticity
似塑料橡皮沥青填缝板 para-plastic rubberized asphalt sealing strip
似塑性 quasi-plasticity
似酸的 <土壤等> acidoid
似缩色 contractive colo(u)r
似塌建筑 ramshackle
似苔的 mosslike
似弹簧的 springy
似弹性散射 quasi-elastic scattering
似碳的 carbonous;charry
似碳物 <沥青组分> carboids
似陶瓷的 ceramic-like
似条纹的 streaky
似铁 chalybeate
似同步观测 subsynchronous observation
似团粒 pelletoid
似推覆构造 nappe-like structure
似瓦片排列的 tegular
似弯管 siphonal
似网状的 reticulate
似伟晶岩的 pegmatoid
似文象结构【地】 graphic(al) like texture
似稳电流 quasi-stationary current
似稳定的 quasi-stable;quasi-stationary
似稳定流 quasi-steady flow
似稳定状态解 quasi-steady state solution
似稳过程 quasi-stationary process
似稳态 quasi-stability;quasi-stationary state
似稳态的 metastable; quasi-static; quasi-stationary
似稳状态 quasi-stable state; quasi-steady state
似锡的 tinny
似先期固结压力 pseudo-preconsolidation pressure; quasi-reconsolidation pressure
似线 liny
似线性化 quasi-linearization
似象牙色的 eburnean
似橡胶的 rubbery
似橡胶特性 rubber-like behaviour
似橡胶液体 rubber-like liquid
似型 homotype
似雄黄 dimorphite
似玄武岩的 basal-like
似牙物 tusks
似氧的 oxygenic
似曜石 obsidianite
似曜岩斑状体 marekanite
似耀斑现象 flare-like phenomenon
似耀斑增亮 flare-like brightening
似液体 quasi-liquid
似银 silvern
似鹰 aquiline
似永久变形 quasi-permanent deformation
似油的 oily
似油质的 oily
似有理 plausibility
似鱼的 piscina
似远色 receding colo(u)r
似月 lunar
似云母黏[粘]土 micaceous clay
似噪声信号 noise-like signal
似胀色 expansive colo(u)r
似真 plausibility
似真排序 plausibility ordering
似真强度 probable strength
似真推理 abduction
似真误差 actual error
似蒸气的 vapo(u)rish;vapo(u)rous

似整合【地】 paraconformity
似直线应力图 quasi-rectilinear stress diagram
似中线 symmedian
似柱 analogous column
似柱的 prismoidal
似柱法 column analogy method

泗尔贝格地震烈度表 Sirberge scale

饲草收割机 forage harvester

饲料 chaff; feed (stuff); fodder; forage
饲料拌和分送小车 feed mixer-and-elevator combination wagon
饲料拌和机 feed mixer
饲料比率 forage ratio
饲料仓 feed bin
饲料槽 bunk; feed bunk; feeding trough; manger
饲料储藏 feeder reservoir
饲料传送机 feed conveyer[conveyor]
饲料吹送器 feed blower
饲料地 food patch
饲料分配器 feed distributor; fodder distributor
饲料分送车 feed cart
饲料分送机 feeding machine
饲料分送器 feed dispenser
饲料分选小车 bunk feed wagon
饲料粉碎机 feed grinder; feed mill; fodder grinder
饲料干燥机 forage drier
饲料混合机 feed mixing machine
饲料集存器 feed collector
饲料计量 feed metering
饲料计量分送器 meter feeder
饲料加工厂 provender mill
饲料加工机 feed processor
饲料加工间 feed processing building
饲料架 fodder rack; cratch <英国方言>
饲料搅拌机 fodder mixing machine
饲料搅拌器 feed agitator
饲料进给机 feeding machine
饲料库 feed storage; feed store
饲料框架 feed frame
饲料抛送机 forage blower
饲料切割机 feed cutter
饲料切碎机 fodder cutter; forage chopper
饲料切碎机具 forage chopping equipment
饲料切碎装置 forage-chopper cutter-head
饲料清选机 feed cleaner
饲料筛 feeding grid
饲料收割机 forage harvester
饲料收获和青贮设备 forage harvesting and silage equipment
饲料收获机 forage harvester
饲料收集器 feed collector
饲料输送管道 feed line
饲料塔 fodder tower
饲料提升机 feed elevator
饲料调制设备 forage equipment
饲料调制室 forage kitchen
饲料通道 feed alley
饲料箱 feed box
饲料制粒机 feeding stuff cuber
饲料装卸分送设备 feed handling equipment
饲料作物 feed(ing) crop; forage crop
饲煤机 coal feeder
饲喂槽 feed bunk
饲喂场 feed barn
饲喂过道 feed alley; feedway
饲喂间 feeding floor
饲喂区 feeding area
饲畜舍 feeding house
饲养 feeding; husbandry; raising
饲养场 dry lot; farm; feedlot; live box
饲养场废水 rear farm wastewater; wastewater from poultry farm
饲养场废物 waste of stockfarming
饲养家畜建筑物 livestock building
饲养圈 feedlot
饲养设备 feeding facility
饲养塘 rearing pond
饲养业 livestock raising; livestock rearing
饲养员 breeder
饲用石灰 feed lime

松柏醇 coniferol; coniferyl alcohol; lubanol

松柏甙 coniferin(e)
松柏科的 coniferous
松柏科木材 coniferous wood
松柏类 coniferals
松柏类植物 acicular-leaved tree; conifer; coniferophyte
松柏林 needle-leaved forest
松柏园 conifer garden
松板 yellow deal
松包 bales slack
松包机 bale breaker
松爆破 loose blasting
松比容 bulk specific volume
松边 loose edge; loose selvedge; slack side <皮带或传动链的>
松扁吉丁 <拉> Anthaxia proteus
松冰山 sugar berg
松冰团 frazil ice; lolly ice; lolly sludge lump; rubber ice
松饼 loosened cake
松布混凝土 loosely spread concrete
松衬套 loose hub
松弛 flabbiness; looseness; relax; slack(ening); slackness; slake; unbend
松弛边 slack side
松弛变量 relaxation variable; slack variable
松弛表 loose list; relaxation table; thin list
松弛参数 relaxation parameter
松弛测量器 relaxometer
松弛长度 relaxation length
松弛常数 relaxation constant
松弛带范围 range of relaxed zone
松弛单缆系泊 slack single line mooring
松弛的 flagging; untensioned
松弛的履带 loose track
松弛的绳索 slackline
松弛法 method of relaxation; relaxation method; relaxation technique
松弛方案 relaxation procedure
松弛复制调控 relaxed replication control
松弛格式 relaxation pattern
松弛光谱 relaxation spectrum
松弛过梁 loose lintel
松弛函数 relaxation function
松弛合成 relaxed synthesis
松弛荷载 load of loosen bedrock; load of loosen ground
松弛回火 letting down
松弛回缩 relaxation shrinkage
松弛机理 relaxation mechanism
松弛积分定律 relaxation integral law
松弛极化 relaxation polarization
松弛角 angle of dilatancy
松弛结点 slack busbar
松弛结合 loosely bound
松弛介质 relaxing medium
松弛缆道 slackline; slackline cableway
松弛缆索 slackline cableway
松弛棱镜 relieving prism
松弛理论 theory of relaxation
松弛疗法 relaxation therapy
松弛路线 slack path
松弛模量 relaxation modulus
松弛模数 relaxation modulus
松弛模楔 releasing mo(u)ld wedge
松弛耦合 low binding
松弛平衡 relaxation balance
松弛强度 loosening strength
松弛曲线 relaxation curve
松弛绳索 slackline cableway
松弛时间 relaxation time; slacking time; time of relaxation
松弛时间效应 effect of relaxation time
松弛式集材机 slaker
松弛式架索道 slacking sky line
松弛试剂 relaxation reagent
松弛试验 relaxation test
松弛试验机 relaxation tester; relaxation testing machine
松弛速度 relaxation velocity
松弛速率 release rate
松弛损失 <预应力筋> relaxation of prestress tendons
松弛弹性模量 elastic modulus of relaxation; relaxed modulus of elasticity
松弛土压 loosening pressure
松弛现象 relaxation phenomenon
松弛相 relaxation phase
松弛效应 relaxation effect
松弛泄漏 slacken leak
松弛型黏[粘]度计 relaxation viscometer
松弛型色散 relaxation type dispersion
松弛型吸收 relaxation type absorption
松弛岩体 relaxed rock
松弛因素 relaxing factor
松弛因子 relaxation factor
松弛应变 relaxation strain
松弛应力 relaxed stress
松弛约束 loose constraint
松弛在轴上 loose running
松弛振荡 relaxation oscillation
松弛指数 relaxation index
松弛作用 relaxation; tendon release
松出绳索 veer rope
松垂 sagging; swag
松脆 shortness
松脆物 crisp
松村氏钢球硬度试验 Matsumura indentation hardness test
松村氏硬度计 Matsumura hardness meter
松带 loose wheel
松丹斯统 Sundance series
松单板 loose veneer
松的 lax
松的炭渣 loosened carbon
松的约束 loose constraint
松底层 loose base
松底土器 pan breaker
松地 loose ground
松吊速度 lowering speed
松掉绞盘上绳索 surge the capstan
松叠轧制 loose pack rolling
松动 amount of looseness; back lash; break loose; rap <模型的>
松动爆破 blasting for loosening rock; concussion blasting; loose(ning) blasting; mudcapping; standing shot
松动爆破漏斗 loose blasting crater
松动爆破药包 loose blasting charge
松动爆炸 induced shot firing
松动部分 loose part
松动车轮检测器 loose wheel detector
松动触点 loose contact
松动的轨枕 loose tie
松动工作 loosening work
松动活塞 loose-lifting piston
松动活塞环 slap ring
松动角 angle of cutting; angle of disintegration; angle of loosening <开挖的>
松动结合 loose coupling
松动金刚石 loose-stone
松动连接 loose connection
松动量 amount of slack; shake allowance
松动零件 loose fitting
松动轮规 loose to ga(u)ge
松动螺柱 loose bolt
松动模型 rapping
松动配合 clearance fit; extra slack running fit; free running fit; loose fit; running fit; working fit
松动区 loose(ned) zone
松动圈 loosening zone
松动圈厚度 thickness of relaxation ring
松动调整螺钉 give more screw
松动围岩压力 relaxation pressure of surroundings
松动楔块 easing wedge
松动性放炮 back-off shooting
松动压力 loosening pressure
松动压力设计 loosening pressure design
松动岩石 loosened rock
松动药包 loose charge
松动硬土 break-up hard ground
松动枕木 pumping sleeper
松动制动盘 loose brake backing plate
松度 looseness
松端支承 free end bearing
松短角幽天牛 <拉> spondylis buprestoides
松堆密度 bulk-dry specific gravity
松堆容重 loose packing unit weight; loose volume weight
松发引信 pressure-release action fuze; release fuze
松方 loose measure; loose volume; loose cubic meter
松方计量 loose measure
松方量 <土的> loose yard
松方体积 loose measure volume
松放杆 <犁安全器的> trip arm
松放滑车 round down
松放拉杆 engaging tie rod
松放缆索 slack out wire
松放链轮 snubber sprocket
松放锚缆 slack off cable
松放曲线 release curve
松放圈 engaging lever
松放圈套头 engaging lever head
松放圈头 head engaging lever
松放踏板 release pedal
松放弹簧 release spring
松放销 trip pin
松粉 loose powder
松粉机 scroll mill
松粉烧结 sintering of loose powder
松腐朽菌 stem-rot fungus
松杆率 loosening leverage
松杆密度 loose dry density
松根层孔菌 pine root fungus
松钩 off-hook
松管接头器 loose tube splicer
松光 light wood
松果形装饰品 pineapple
松横坑切梢小蠹 <拉> Blastophagus minor
松厚度 bulk; loose thickness
松黄星象甲 <拉> Pissodes nitidus
松基层 loose base; loss base

S

松级配 open grading
松级配的 open-graded
松级配混合料 loose grade mix;open-graded mix
松级配结合层 open binder; open binder course
松江鲈鱼 Trachidermus fasciatus
松浆油 fin oil;fluid resin;liquid rosin;sulfate resin;sulfate rosin;swedish olein;swedish pine oil;swedish rosin oil;sylvic oil;tall oil;tallol
松浆油醇酸树脂 tall alkyd
松浆油的初馏分 tall oil heads
松浆油脚 sulfate pitch
松浆油沥青 tall oil pitch
松浆油清漆 tall varnish
松浆油松香 tall oil rosin
松浆油酸 talloleic acid
松浆油酸钙 calcium tallate
松浆油酸铁 iron tallate
松浆油酸酯 tallate
松浆油酸钴 cobalt tallate
松浆油皂 tall oil soap
松浆油脂肪酸 tall oil fatty acid
松焦油 country tar; pine tar; softwood tar;Stockholm tar;tar oil
松焦油沥青 archangel pitch; pine pitch; pine tar pitch; softwood tar pitch;wood tar pitch
松接触 loose contact
松接式管道 loose-coupled pipe;loose-coupled type pipe
松接式混凝土管道 unjointed concrete pipe
松接式压力钢管 loose penstock
松接头 loose joint;open joint
松节头 loose joint
松节瓦管 turf drain
松节油 gum spirit;gum spirit turpentine;gum turpentine;oil of turpentine;pine oil;spirit of turpentine;terebenthene;terebinthina;turpentine;turpentine oil;wood turpentine(oil)
松节油沉积试验 turpentine residue test
松节油澄清槽 turpentine separator
松节油代用品 subturps;turpentine substitute
松节油灰泥 gum-spirit cement
松节油精 terebene
松节油硫酸试验 turpentine sulphuric acid test
松节油喷漆 turpentine lacquer
松节油清漆 turpentine varnish
松节油溶剂 turpentine medium;turpentine vehicle
松节油溶解试验 turpentine solvency test
松节油瑕疵 turpentine stain
松节油纤维 turpentine fiber[fibre]
松节油与松油精之混合物 <用作防腐剂,涂料等> terebene
松节油脂着色 turpentine stain
松节油中毒 terebinthinism
松结合的 loosely bound
松结合体系结构 loosely coupled architecture
松解 unbuttoning;unlock
松解按钮 trip push button
松解层 pressure equalizing layer
松解工作 loosening work
松解机构 tripping mechanism
松解缆 release line
松解手柄 release handle
松紧撑 brake toggle
松紧带 elastic;webbing
松紧度 tightness
松紧度调整轮 tension idler
松紧环 loose collar
松紧接头 turn buckle
松紧口装置 sphincter
松紧扣 adjustable nut
松紧螺钉 parbuckle screw
松紧螺扣 strainer
松紧螺丝 bottle screw
松紧螺丝机 turnbuckle
松紧螺丝接头 turnbuckle
松紧螺丝扣 turnbuckle
松紧螺旋机 twisting stick
松紧螺旋扣 adjusting screw;rigging screw; stainer; tightening screw; turnbuckle screw
松紧绳 bungee
松紧调节的拉杆 stay tightener
松紧调整器 slack adjuster
松紧线 elastic cord
松紧楔 folding wedge;striking wedge <调整高度用的一对木楔>
松紧针钩 draw-bit
松紧指示器 up-and-down indicator
松紧装置 take-up
松劲 looseness;slacken
松卷 loose winding;paying-off
松卷机 coil opening machine;delivery reel
松卷螺旋 loosely wound spiral
松卷退火 open coil annealing
松卷退火炉 open coil annealing furnace
松卷纸卷 loosely wound roll
松卷装置 reel-off gear
松开 loosen(ing);unbolt;uncage;uncaging;unclamp;unclasp;unfasten;unfix;unlink;unlock;untie;unclamp <夹子等>
松开的 wide
松开的打捞筒 releasing overshot
松开管接头 loosening of coupling
松开机构 release mechanism;tripping gear
松开驾驶杆 <指飞机、机车等> stick-free
松开控制杆 trip paddle
松开离合器 declutch
松开联轴节 loose coupling
松开轮箍 loose type
松开螺钉 let out screw
松开弹簧 release spring
松开位置 release position
松开信号 unlock signal
松开游动滑车 unstring the block
松开制动器 brake release;releasing of brake
松开制动踏板 loosen the brake;release the brake
松科【植】Pinaceae
松孔 porosity
松孔镀铬 spongy chromium plating
松扣打桩机 trip pile driver
松扣链条 spinning chain
松旷动作 <车钩> slack action
松捆 bundles slack loose
松类树 coniferous tree
松离 release
松离离合器 clutch release
松砾石 loose gravel
松联结处理机 loosely coupled(inter) processor
松联轴节 loose coupling
松量 <按松容重计量> loose measure
松料 aerate;deals;loose metal
松料空气 aeration air
松裂穹 fracture dome
松裂推土铲 rip bulldozer
松林 pine forest;pinery;pinetum[复pineta]
松林泥炭地 pine barren
松林石 dendrite
松溜油 tar oil
松螺钉 unscrew
松螺旋 loose spiral end
松毛虫 bordered white moth
松毛虫属 <拉> Dendrolimus
松铆钉 loose rivet
松铆(接) leaky riveting;loose riveting
松密度 apparent density;bulk density
松面 loose grain;loose side
松模 rapping
松模棒 rapping bar;rapping pin
松模杆 rapping iron
松模工具 rapper
松模剂 mo(u)ld release agent
松模楔 releasing key
松木 common pine; dealwood; fat wood;pine-wood
松木板 deal;deal board;pine board
松木层孔菌 phellinus pini;ring scale fungus
松木地板 deal floor(ing);pine-wood flooring
松木电杆 piney pole
松木夹板 pine plywood
松木胶合板 pine plywood
松木焦油(沥青) pine tar;pine-wood tar
松木焦油脂 pine tar pitch
松木节 loose knot
松木毛地板 deal sub-floor
松木木材 deals
松木墙面板 pine shingles
松木深度 ripping depth
松木推土机的液压缸 cylinder of rip-bulldozer
松木瓦 pine shingles
松木屋顶板 pine shingles
松木屋架 deal frame
松木硬沥青 pine tar pitch
松木硬脂 piney tallow
松木油 pine oil;pine-wood oil;turpentine;wood turpentine
松木圆锥体 pine cone
松木枕木 pine sleeper
松纳镜头 Sonnar lens
松拈机 untwisting machine
松捻 loose twist
松捻双股细绒线刺绣装饰 crewel
松耦合 loose coupling
松耦合电路 loose coupled circuit
松耦合系统 loosely coupled system
松泡货 high cube cargo;light cargo
松配(合) free fit; loose fit; running fit;movable fit
松配合活塞 loose fitting piston
松皮天牛 <拉> Stenocorus inquisitor japonicas
松飘的束帆索 deadman;Irish pennant
松铺 loose placement
松铺地砖 loose laid tile
松铺厚度 laydown thickness; loose depth; loose laying depth; loose thickness
松铺梢褥 loose brush mattress
松铺屋面 loose laid roofing
松铺系数 coefficient of loose laying
松铺压顶屋面 loose laid and ballasted roofing
松潜 pine tar
松墙 pine(-tree) wall
松球 pine nut
松球菱镁片岩 pinolite
松圈 loose ring
松壤土 mellow loam
松软 mellow;spongy
松软表土层 mollic epipedon
松软冰 cream ice;sludge ice;slush ice
松软冰块 sludge cake
松软沉积(物) soft sediment;unconsolidated deposit; unconsolidated sediment
松软大冰块 sludge floe
松软道床 soft roadbed
松软道路 mushy road
松软的 flabby;floppy;fluffy
松软的帆 slab
松软的钢索头 porcupine
松软的轨道 soft track
松软地层 broken ground;loose ground
松软地层隧道施工 soft ground tunnel(l)ing
松软地点 soft spot
松软地基 hover ground;yielding ground
松软地面 loose ground
松软顶板 loose roof
松软度 mellowness
松软而富有弹性的 spongy
松软方解石 rock milk
松软肥沃的土壤 unctuous soil
松软骨料 soft aggregate
松软硅质岩 tripoli
松软灰岩 clunch
松软集料 soft aggregate
松软节 unsound knot
松软结持 mellow consistency;mellow consisting
松软类 soft type
松软路基 soft spot;weak subgrade
松软煤 yolk coal
松软黏[粘]土 soft clay;weak bind
松软壤土 fibrous loam soil
松软土地 yielding ground
松软土(壤) mellow soil; mollisol; spongy soil;yielding ground;yielding soil;soft soil
松软物质 loose material
松软性 mellowness
松软性土 soft soil
松软岩层 incompetent bed
松软岩石 crumbling rock;loose rock;scall
松软岩石类 type of soft rocks
松散 loosening; decompacting; dilatancy; fretting; incoherence; ravel(ling);unravel;untwisting
松散饱和沙 loose saturated sand
松散保护层 loose protection course
松散保温材料 loose insulation
松散臂销 engaging arm pin
松散变质岩 soft and altered rock
松散冰盖 unconsolidated ice cover
松散冰碛物 unconsolidated glacial deposits
松散剥落 <路面的> unraveling
松散材料 bulk(material); cohesionless material; discrete material; divided material;free flowing material; incoherent material; material in bulk;non-cohesive material;unconsolidated material;loose material
松散材料搬运机 bulk material loading plant
松散材料搬运设备 bulk material loader
松散材料铲斗 loose material bucket
松散材料等级 bulk material scale
松散材料流动 flow of bulk material
松散材料撒布机 bulk distributor;bulk spreader
松散材料输送机 bulk handling machine
松散材料卸货器 bulk material unloader
松散材料压缩机 bulkload compressor
松散材料运输 bulk transport
松散材料抓斗 bulk material grab;loose material grab;loose material bucket
松散材料抓扬机 bulk material grab
松散材料转运设备 bulk material transfer equipment

松散材料装载机 bulk material loader
松散层 unconsolidated formation
松散产品 bulk product
松散沉积 flusch
松散沉积层 incoherent alluvium
松散沉积物 unconsolidated deposit; unconsolidated sediment
松散充填带 green pack
松散充填料 loose fill
松散冲积层 loose alluvium
松散处 ravel(l)led spot
松散单粒结构 loose texture of single particle
松散的 bulky; fluffy; friable; incompact; loose (ned); non-coherent; non-cohesive; non-preformed; open-textured; unconsolidated
松散的烧结粉 loose powder sintering
松散的珍珠岩 loose pe(a)rlite
松散的主斜杆 loose main diagonal
松散地 hover ground
松散地层 friable formation; unconsolidated strata
松散地基 loose foundation; loose ground
松散地面 hover ground; loose foundation; loose ground
松散冻土 loose frozen soil
松散度 degree of looseness
松散堆石 loose rock-fill
松散飞灰 loose ashes
松散废石充填料 loose rock-fill
松散覆盖层 loose overburden; unconsolidated coverage; unconsolidated overburden
松散隔热填充料 loose fill insulation material
松散骨料 discrete aggregate; loose aggregate
松散固体 bulk solid
松散固体流量计 bulk solid flowmeter
松散固体流量喂料机 bulk solid flow feeder
松散含水层 loose deposit aquifer
松散厚度 loose depth
松散互连 loose interconnection
松散护坦 loose apron
松散滑塌 incoherent slump; slump incoherent
松散灰 light ash
松散回填 loose fill
松散混合料成分 bulk composition
松散货物 bulk goods; loose goods
松散机 hog shred
松散积雪 cohesionless snow
松散集料 discrete aggregate; loose aggregate
松散夹芯 <混凝土的水泥浆流失及骨料离散> loose core
松散角 angle of disintegration
松散结构 arenaceous texture; loose structure
松散结构的 free-open-textured
松散介体 discrete medium[复 media]
松散介质 discrete medium[复 media]
松散介质极限平衡理论 theory of ultimate equilibrium of a loose medium
松散绝热材料 loose fill type insulant
松散颗粒 discrete particle
松散孔隙率 bulk porosity
松散块体 discrete block
松散垃圾 bulky refuse
松散立方码 bulk cubic (al) yard [bcy]; loose cubic(al) yard
松散立方米 loose cubic (al) meter; bulk cubic meter <爆破岩石>
松散沥青混合料 loose bituminous mixture
松散砾石 loose gravel
松散粒状绝热材料 granular loose-fill thermal insulation
松散联合(体) loose combination
松散料保温 loose fill insulation
松散料铲斗 chip bucket
松散料单位重量 loose unit weight
松散料堆 loose stockpile
松散料隔热层 loose insulation
松散料绝热层 fill insulation
松散料容量 loose yards
松散料重量 loose weight
松散垆坶 mellow loam
松散路面 loose pavement; ravel (l) -ing of pavement
松散煤 crumble coal
松散密度 bulk density; loose density; loose thickness
松散面层 loose surface
松散面层的道路 loose-surface road
松散泥煤 crumble peat; moor coal
松散泥沙 unconsolidated sediment
松散泥炭 crumble peat
松散耦合堆芯 loosely coupled core
松散抛填混凝土块体 loose concrete block
松散泡沫轻质填料绝缘 loose fill insulation
松散容积 loose volume
松散容重 bulk density; bulk specific gravity; loose bulk density
松散砂 free-open-textured sand; open sand; loose sand
松散砂压实 loose sand compaction
松散砂振密 loose sand compaction by vibration
松散梢褥 loose brush mattress
松散深度 loosening depth
松散石灰喷洒器 bulk lime spreader
松散石料护板 loose apron
松散石料护坦 loose apron
松散水磨扁砾岩 shingle
松散水泥 loose cement
松散体积 bulk volume; loose measure; loose volume; loose yards
松散体积配合比 mix proportion by loose volume
松散体积配料(称量)法 loose volume batching
松散填充 loose fill
松散填充隔热层 loose fill insulation
松散填充绝热材料 loose fill insulation
松散填料 random packing
松散透水路面 loose and open surface
松散透水面层 loose and open surface
松散土层 loose formation
松散土方 loose measure volume
松散土方量 loose yards
松散土(壤) free soil; friable soil; loose ground; loose soil; unsonsolidated soil; yielding ground; loose earth
松散土压实 loose soil compaction
松散土质 detached soil material
松散纹理 loosen(ed) grain
松散污染 loose contamination
松散物料 fragment; loose material
松散物质 loose material
松散系数 coefficient of volumetric expansion; loosening coefficient; bulk factor
松散纤维污染 loose fiber[fibre] pollution
松散线圈型 loose-loop model
松散楔体 loose wedge
松散性 bulking property; fluffiness; friability; incoherence; looseness
松散锈蚀 <钢筋由于保护材料剥离而产生的> fretting corrosion; loose rust
松散雪 loose snow
松散雪崩 loose snow avalanche
松散压力 <松散材料的重量引起的压力> loosening pressure
松散岩层 loose stuff; ravelly ground
松散岩层支护 running-ground timbering
松散岩石 friable rock; loosely grained rock; loose rock; uncemented rock
松散岩体 loose ground; loose (ned) rock mass; unconsolidated rock
松散岩土 non-cohesive soil
松散岩芯 loose core
松散油灰 loose putty
松散有机质砂 loose organic sand
松散状况 friable state
松散状矿石 loose ore
松散状态 friable state; loose condition; loose state
松散状态下测定的 loose measure
松散状态下测量 loose measure
松砂 aeration of mo(u)lding sand; blending; light sand; open sand; running sand
松砂工作 sand cutting
松砂机 aerator; blender; fluffer; sand blender
松山反向极性带【地】 Matuyama reversed polarity zone
松山反向极性时 Matuyama reversed polarity chron
松山反向极性时间带 Matuyama reversed polarity chronzone
松山反向期 Malsuyama
松山负极性期 Matuyama reversed polarity epoch
松杉暗孔菌 Phaeolus schweinitzii; velvet top fungus
松杉木材 larch fir
松烧结块 soft sinter
松生拟层孔菌 <拉> Fomitopsis pinicola
松生拟原孔菌 red belt fungus
松绳 slack
松绳保护装置【机】 safe device against slack rope
松绳补偿器 rope creep compensator
松绳钩 releasing hook
松绳开关 slack-rope switch
松绳塔式挖土机 slackline scraper
松石 loose (ned) rock; loose-stone; rammel; soft rock
松石机 <推土机> rock rake
松石开挖 loose rock excavation; soft-rock excavation
松石器 rock ripper
松石坍落 loose rock fall
松实比 compacting ratio
松式法兰 loose type flange
松式环状洗涤机 slack loop washer
松式绳状洗布机 slack washing machine
松式索道 slackline
松释 <内聚力的> disbonding
松树 pine tree
松树皮 pine bark
松树石 dendrite
松树式天线阵 pine-tree array
松树外状叶 needle-like leaves of pine
松树形天线 pine-tree antenna
松树形天线阵 pine-tree array
松树油 pine(-tree) oil
松树园 pinetum[复 pineta]
松树脂 abietic resin; pine-tree resin
松树脂醇 lariciresinol
松水土壤 weak soil
松碎 ravel
松碎沉积 <一般为异重流所形成> flusch
松碎带 loosened zone
松碎机 chipper
松索缆道 <供挖运用的> slackline cableway
松锁 uncage
松坍 slacktip
松套边盘 loose beam heads
松套法兰(盘) loose flange
松套法兰盘接头 loose flange joint
松套光纤 loose tube optic (al) fiber [fibre]
松套连接 loose joint
松套凸缘 loose flange
松蹄 secondary shoe; trailing shoe
松体积 loose volume
松天止 <拉> Monochamus alternatus
松填材料法隔热 fill-type insulation
松填方 bulk fill; loose fill
松填管道隔热套 pipe lagging loose fill
松填土 bulk fill; loose fill
松填蛭石 loosely filled vermiculite
松填重量 bulk weight
松桶板料 slack shook
松透性土 mellow soil
松土 dirt; loose; loose earth; loose ground; loose soil; movable material; rip; rotovate; scarification; scarify(ing)
松土拌和 pulvimix(ing)
松土拌和机 pulverizing mixer; pulvimixer; rotary hoe
松土部位 ripping lip
松土铲 break shovel
松土铲运装置 scarifier-scraper
松土齿 ripper tooth; ripping tooth; scarifying tooth; tine
松土除草机 cultivator
松土单位成本 unit ripping cost
松土翻土机 ripper-scarifier
松土覆盖镇压机 roller mulcher tiller
松土工具 getter; loosening equipment
松土工(作) loosening earthwork
松土机 loosener; loosening machine; loosing equipment; pneumatic pick; ripper; rooter; rotovator; scarifier; soil pulverizer; soil ripper; road ripper【道】
松土机齿 ripper tip
松土机齿杆 scarifier tine[tyne]
松土机串联 tandem ripper
松土机刀杆 ripper shank
松土机耙齿 scarifier tooth
松土机手柄 scarifier shank
松土机械 loosening equipment; ripper equipment
松土角 ripping angle
松土搅拌机 pulverizing mixer; pulvimixer; rotary hoe
松土搅拌转子 pulvimixer rotor
松土掘根机 ripper-rooter
松土宽度 scarifying width
松土犁 pulverator plough; scarifier plough; scarifier plow
松土路线 ripper line
松土耙 rake attachment (for graders); scarifier
松土耙土器 ripper-scarifier
松土皮 scarification
松土平地机 scarifier-scraper
松土平整机 scarifier-scraper
松土器 cultivator tooth; ridger; ripper linkage; ripper pin
松土器拔出 ripper kick-out
松土器撑架 ripper bracket
松土器齿柄 ripper shank
松土器横架 ripper beam
松土器护板销 shank pin
松土器耙齿 ripper tooth
松土器耙齿支架 scarifier shank holder

S

松土器液压缸 ripper cylinder
松土器支臂 scarifier arm
松土器支撑杆 ripper carriage bar
松土设备 earth-loosening equipment; scarifier attachment; scarifying attachment; ripper equipment
松土深度 loosening depth
松土时间 rip time
松土型推土铲 rip bulldozer
松土型推土机 rip bulldozer
松土照明灯 ripper lighting
松土筑路机 rip rooter
松土装置 rake attachment (for graders); ripper attachment; scarifier device
松土作业 loosening earthwork; ripping
松土作业工况 ripping condition
松团作用 deagglomeration
松推配合 easy push fit
松脱 fetch(a)way; get loose
松脱安全离合器 release clutch
松脱履带 loosing a track
松脱振动 ratchetting
松脱纸 release paper
松纬 loose pick
松纬起绒 loose pile
松物料单位体积重量 bulk weight
松下 lower away
松线夹 loose wire gripper
松线式集材机 slacker
松线式索道 slackline
松香 abietyl; colophany; colophonium; colophony; glyptal resin; liquid rosin; pine resin; rosin; white resin; wood rosin
松香胺 abietic amine
松香标样 rosin standard
松香醇 abietinol; abietyl alcohol
松香的颜色等级 grades of rosin
松香定性试验 rosin qualitative test
松香粉 powder of resin
松香改性醇酸树脂 rosin modified alkyd resin; rosin modified glyptal resin
松香改性酚醛树脂 rosin modified phenolic resin
松香改性马来树脂 rosin modified maleic resin
松香改性失水苹果酸酸树脂 rosin modified maleic resin
松香改性顺丁烯二酸树脂 rosin modified maleic resin
松香钙皂 limed rosin
松香钙脂清漆 chian varnish
松香甘油酯 ester gum
松香含量 rosin content
松香焊膏 colophony soldering paste
松香焊锡丝 colophony soldering wire
松香基树脂 rosin resin
松香加成物 rosin adduct
松香胶 rosin size
松香精 rosin essence; rosin spirit; spirit of rosin
松香聚合物 vinsol resin
松香蜡膏 basilicon
松香沥青 rosin pitch
松香清漆 rosin varnish
松香热聚物 vinsol resin
松香乳液 rosin milk
松香水 mineral spirit; solvent naphtha; volatile mineral spirit; white spirit
松香-顺丁烯二酸酐缩合物 rosin-maleic anhydride condensation compound
松香酸 abietic acid; colophonic acid; rosin acid
松香酸苄酯 benzyl abietate; benzyl resinate

松香酸钙 calcium resinate
松香酸酐 abietic anhydride
松香酸基 abietyl
松香酸甲酯 methyl abietate
松香酸铝 alumin(i)um rosinate
松香酸锌 zinc resinate
松香酸盐 abietate
松香酸乙酯 ethyl abietate
松香酸酯胶 ester gum
松香烃 pinoline
松香烯 abietene
松香心焊锡条 resin-cored solder
松香心焊锡线 solder wire with rosin core
松香型试样 rosin type sample
松香烟 rosin smoke
松香衍生物 vinsol
松香乙二醇酯 rosin ethylene glycol ester
松香油 abies oil; abietic acid oil; abietic oil; cod oil; resin oil; retinol; rosin oil
松香皂 resin soap; rosinate soap
松香皂泡沫剂 rosinate soap foamer
松香皂热塑料树脂 saponified vinsol resin
松香皂树脂 vinsol resin
松香酯 rosin ester
松象田 <拉> Hylobius abietis haroldi
松销机构 <测斜仪> uncaging section
松些 ease off
松懈 slacking
松懈方法 loosening
松心 shrinkage porosity
松型 easing
松型纹孔 pinoid pit
松压盖 loose gland
松压器 discompressor
松烟 lampblack pigment; turpentine soot
松烟黑 vegetable black
松岩 scall
松岩机 rock ripper
松叶蛾【动】eggar
松叶油 pine leaf oil; pine needle oil
松隐皮菌 stem-rot fungus; Cryptoderma <拉>
松幽天牛 <拉> Asemum amurense
松油 pine oil
松油精苎烯 dipentene
松油萜醇 terpineol
松油烯 terpinene
松余量 amount of slack
松鱼油 bonito oil
松豫时谱 relaxation spectrum
松闸 declutch; release of brakes
松胀 dilatation
松折 loose fold
松枝晶 pine-tree crystal
松枝状的 pine tree
松脂 abietic resin; colophonium; oleoresin; pine resin; rosin; suglgum; turpentine
松脂斑岩 pitchstone porphyry
松脂醇 pinolin; rosin spirit; spirit of rosin
松脂光泽 resinous luster[lustre]
松脂合剂 rosin wash
松脂价格废水 pine gum processing effluent; rosin processing wastewater
松脂孔 pitch pocket
松脂蜡膏 basilicon
松脂类原料 naval stores
松脂裂纹 pitch seam; pitch streak(ing)
松脂林 turpentine-orchard
松脂木 terebinth
松脂石 fluolite; pitchstone
松脂酸盐 rosinate
松脂酸衍生物 rosin acid derivative

松脂酸酯处理的颜料 rosinated pigment
松脂酸酯制清漆 rosinate varnish
松脂条纹 pitch streak(ing)
松脂芯焊料 rosin-core solder
松脂芯焊条 resin-cored solder wire
松脂芯软钎料 resin-cored solder
松脂岩石 pitchstone
松脂眼 pitch pocket
松脂油 raw pine oil; rosin oil
松脂油膏 rosin grease
松脂油酸 sylvic acid
松脂皂 rosin soap
松脂制品 naval stores
松质土 light soil
松质瓦 porous tile
松重量 loose weight
松皱结 baggy wrinkle
松属 pine; Pinus <拉>
松爪 loose jaw
松转配合 coarse clearance fit; loose running fit
松装比容 apparent volume
松装密度 apparent density; loose density; loose packed density
松装配 loose fit
松装烧结 loose sintering
松装体积 apparent volume; bulk volume
松装重量 loading weight
松子 pine nut
松子油 pine-seed oil
松纵坑切梢小蠹 <拉> Blastophagus piniperda
松组织 open texture

菘

菘蓝 woad
菘蓝染料 pastel

耸

耸立 spire
耸起 haunch up; upraise

楤

楤木 angelica-tree

讼

讼费保证金 security for cost

宋

宋 <响度单位, 1000 赫的纯音声压级在闻阈上 40 分贝时的响度> sone
宋体字 Song typeface

送

送包舌板 delivery tongue
送标书 submission of tenders
送别换乘 kiss-and-ride
送波机 transmitter
送波器 sender; transmitting oscillator
送菜窗口 butler's window; pass-through; service hatch
送菜吊车井 dumbwaiter shaft
送菜升降机 dumb waiter
送菜升降机井 dumbwaiter shaft
送菜提升器 service lift
送车线 feed track
送出量 sendout
送达 service
送达的背书 indorsement of service
送达签条 delivery docket
送达通知书 service of notice
送达证明书 certificate of service
送带机构 tape feed
送带键 tape load key

送袋机 bag conveyer[conveyor]
送到时间 time of delivery
送递快件的信差 courier
送电 energize; power-on; power transmission
送电变电站 sending(sub)station
送电端 feed end; sending end
送电端电压 feed end voltage; sending-end voltage
送电端电压表 pilot voltmeter
送电端阻抗 sending-end impedance
送电发电站 sending substation
送电线 feed cable
送电线路 power transmission sequence
送钉定位板 nailing marker
送锭车 ingot bogie; ingot buggy
送锭车的锭座 ingot chair; ingot pot
送饭到客车座席 meals service at seat
送粉器 powder feeder
送粉速率 powder feeding speed
送风 air blast; air blow; air supply; blast; blowing in; delivery air; draft blast; force-in air; on blast; supply air; supply of blast
送风道 air supply duct; delivery conduit; inlet duct; supply duct
送风段 air supply section
送风阀 blow valve
送风方法 air supply method
送风方式 air supply mode
送风分配 distribution of blast
送风风扇 plenum fan
送风格栅 supple grille
送风格子 plenum grid
送风管 air pipe; air supply pipe; ajutage; blast tube; pneumatic hose; supply air pipe; supply air tube; wind pipe
送风管道 air supply duct; plenum duct; supply air duct
送风管道系统 supply air duct system
送风管连接器 supply spigot
送风机 air feeder; air-handling fan; air supply fan; air ventilator; blast engine; blower; fan blower; forced draft fan; forced draught fan; fresh-air fan; supply fan; ventilator
送风机节流挡板 supercharger blast gate
送风机进口 fresh-air inlet
送风机内装式空气净化设备 self-contained air cleaner
送风机室 air supply fan room; supply fan room
送风机组 make-up air unit
送风加热 plenum heating
送风加热器 blast heater
送风胶管 air supply hose
送风井 blowing shaft
送风孔板 air supply panel
送风控制板 blast gate
送风口 air(discharge) outlet; air supply grille; air supply outlet; blowing opening; blow(ing) vent; diffuser; supply air outlet; vent port
送风口格栅 ejector grille
送风口面积 vent area
送风口散流板 plaque
送风口消声器 outlet sound absorber; outlet sound attenuator
送风立管 air duct riser
送风连接管 air connection
送风量 amount of blast; supply air rate
送风能力 draft capacity
送风喷口 jet diffuser
送风喷嘴 blast nozzle
送风期 on air
送风栅 supply air grille
送风软管 air supply hose

送风设备 air-circulating installation; air supply system; blowing plant; ventilating installation
送风式冷却塔 forced draft cooling tower
送风式凉水塔 forced draught cooling tower
送风式暖气设备 all-blast heating system
送风式通风系统 plenum ventilation system
送风竖井 blowing in shaft
送风损失 draft loss
送风调节器 air governor; air regulator; furnace pressure controller; transfer register
送风温差 effective temperature difference; supply air temperature difference
送风温度 draft temperature; supply air temperature
送风吸音吊顶 air-distributing acoustic(al)ceiling
送风系统 air supply system; supply (air)system
送风巷道 blowing duct
送风小室 plenum chamber
送风小室清扫口 plenum cleanout
送风噪声 discharge noise
送风支管 blow pipe
送风总管 header pipe
送弧 flash
送话电平 transmitting level
送话机吊架 mike boom
送话器 microphone(transmitter); telephone transmitter; transmitter; mike
送话器支架 mike stand
送话筒 telephone transmitter
送回 bringing back; put back
送回风调节阀 supply return damper
送回空气的风机 return air fan
送话器放大器 transmitter amplifier
送话器口承 microphone lip
送话器中炭精末结块 packing of microphone
送货簿 delivery book
送货车 delivery car; package delivery truck
送货车至岔线 setting-in of wagons toward sidings
送货带 supply conveyer[conveyor]
送货单 consignment invoice; delivery form; delivery note
送货到户 delivery to domicile
送货到户单 delivery note; delivery sheet
送货到户费 delivery charges; delivery fee
送货到户费已付 delivery free; paid home
送货到门车 home delivery van
送货到门费已付 paid home
送货到用户的费用已付 delivered home
送货道 supply road
送货电梯 service lift
送货吊机 service lift
送货工 trucker
送货回单 delivery receipt
送货机 supply conveyer[conveyor]
送货篷车 delivery van
送货凭证 goods delivery receipt
送货汽车 delivery van
送货人 deliveryman
送货上门 deliver goods to the customers; delivery at door; store door delivery
送货上门运输距离 door-to-door transport distance
送货竖井 supply shaft
送货用汽车 door-to-door car
送交 deliver; order of
送进 feed-in; feed-through
送进搭边 feed bridge
送进孔型 forward journey; forward pass
送进量 feed
送进速度 <焊丝的> feed rate
送卷装置 coil handling apparatus
送卡 card feed
送卡穿孔机 feed punch
送卡箱 card hopper; hopper
送客换乘火车 <小汽车在站旁短暂停留> kiss-and-ride
送款簿 deposit(e)receipt
送款单 deposit(e)slip
送缆 line transfer
送缆船 line boat
送链齿 link feed tooth
送料 feed; payoff
送料板 delivery sheet
送料泵 charging pump; feed(ing) pump
送料槽 feeding tank
送料叉 feeding finger
送料车 delivery car; delivery wagon; stack pallet
送料搭边 feed bridge
送料道 feed(ing)channel
送料斗 charging feeder; feed hopper; feeding funnel; discharge hopper
送料阀 charging valve; feed valve
送料管 feed pipe
送料管道 charging line
送料辊 feed roll
送料计量器 batcher
送料节距 feeding pitch
送料孔型 swabbing pass
送料溜槽 feed chute; magazine chute
送料轮 feed wheel
送料磨 feed mill
送料盘 feed table; feed tray
送料喷嘴 delivery nozzle
送料皮带 feed belt
送料平台 charging platform
送料器 feeder
送料设备 feeder equipment
送料射流连接管 delivery jet pipe
送料输送机 feed(er) conveyer[conveyor]
送料速率 rate of feed
送料调节管 adjustable pipe for feed flow
送料筒 feed drum
送料斜槽 feed chute
送料支架 stock support
送料装置 feeder; feeder apparatus; feed gear; feeding attachment; pushing device
送墨辊 form roller
送片暗盒 supply film cassette
送气 air-feed(ing); air-in; air-on; air supply; plenum[复 plenums/plana]
送气泵 delivery pump
送气斗 plenum hopper
送气风道 plenum duct
送气风扇 plenum fan
送气格栅 plenum grid; transfer grille
送气格子 plenum grid
送气管 air line; air transmission pipe; air tube; induction pipe; pneumatic hose; snorkel
送气管道 air duct
送气管阀门 air line valve
送气机 air feeder
送气孔 delivery orifice
送气框格 plenum grid
送气期 change on gas
送气室 plenum chamber
送气(通)道 plenum duct
送气通风 plenum ventilation
送气系统 plenum system
送汽 initial steam admission
送钎工 tool nipper
送请试验 submit to test
送去拍卖 being to the hammer
送人工具 running tool
送入 enter; load
送入轨道 orbital injection
送入数据计数器 load data counter
送入压缩空气 air-on
送入桩 displacement pile
送砂螺旋机 sand screw
送膳窗口 service hatch
送审本 approval copy
送审货物 goods on approval
送收话器 electrophone; receiver and transmitter; receiver-transmitter
送受分开电路 remote end feeding circuit
送受话器 electrophone; hand microtelephone; monophone; receiver and transmitter; receiver-transmitter
送受话器叉簧 cradle
送受话器叉托开关 cradle switch
送水暗渠 filling culvert
送水管 delivery pipe; feed pipe; flow pipe; induction pipe; pumping line
送水管路 delivery pipe line
送水管线 delivery pipe line
送水管压力 flowline pressure
送水落差 feeder drop
送水器 douche
送水软管 discharge hose
送水嘴 delivery nozzle
送丝 electrode feed; wire feed
送丝机构 wire feeder; wire feeding device
送丝轮 wire feed rolls
送丝器 wire feeder
送丝速度 rate of feed; wire feed rate
送丝系统 wire feed system
送丝装置 wire drive unit; wire feeder
送酸器 acid feeder
送妥回单 goods delivery receipt
送尾风 following wind
送尾流 following sea
送文函 letter of transmittal
送吸两用风机 supply-return diffuser
送线器 wire feeder
送线装置 wire feeder
送像装置 iconoscope
送芯小车 core truck
送信人 bearer; poster
送压泵 inflator
送烟试验 smoke test
送油泵 fuel circulating pump; oil-line pump
送油槽 feeding bank
送油管堵塞 choke for oil delivery pipe
送油急止阀 delivery check valve
送油节止阀 delivery check valve
送柱 follower; long dolly
送柱器 chaser
送桩 <打桩时用的> sett; false pile; follower(pile); tube-pile following; long dolly
送桩锤 beetle head
送桩机 chaser; follow block; pile follower
送桩木 follow block
送桩器 <打桩时用的> long dolly; chaser
送桩装置 pile follower
送钻头工 tool carrier

颂

颂经台 ambo(n)

搜

搜查 ransack; rummage; searching
搜查令 warrant
搜查证 search warrant
搜出的物件 rummage
搜根 dike-root caving
搜根刨 bottom plane
搜购 coemption
搜集 collection; gathering; compile
搜集标本 collect specimen
搜集剂 gathering agent
搜集圃 collective nursery
搜集数据 collection of data
搜集已有的资料 search for existing information
搜集者 gleaner
搜集资料 collection of data; search for information
搜救 search and rescue
搜救结束 termination of SAR[search and rescue] operation
搜救区 search and rescue region
搜救任务协调员 search and rescue mission coordinator
搜救卫星辅助跟踪 search and rescue satellite-aided tracking
搜救业务 search and rescue service
搜索 beat about; bird dogging; hunting; reconnoiter [reconnoitre]; scan; scout; search
搜索波法 searching wave method
搜索菜单 search menu
搜索策略 search strategy
搜索长度 length of search
搜索地区 sector of search
搜索动作 hunting action
搜索度盘 search dial
搜索队 search party
搜索法 search method
搜索反射器 search reflector
搜索范围 hunting zone; search band; search coverage
搜索方法 search technique
搜索方向 direction of search
搜索飞机 search plane
搜索飞行 search mission
搜索跟踪 search track
搜索跟踪雷达 search and track radar; track-while-scan radar
搜索(工作)方式 search mode
搜索光点 exploring spot
搜索过程 search process
搜索航线 scouting course
搜索航线中心 search line center[centre]
搜索和救援信标设备 search and rescue beacon equipment
搜索和营救 search and rescue
搜索和自动跟踪雷达 search-lock-on radar
搜索激光器 acquisition laser
搜索技术 search technique
搜索监视 search surveillance
搜索角 angle of aspect
搜索阶段 search phase
搜索接收机 search receiver
搜索接收器 search receiver
搜索救援航天器 search and rescue spacecraft
搜索救援区 search and rescue area
搜索救援网 search and rescue net
搜索救援与自动导引无线电信标 search and rescue and homing
搜索距离 range of detection; scouting distance; search range
搜索控制 searching control
搜索雷达 acquisition radar; radar search unit; search radar(set); spotter

搜索雷达显示器 search scope
搜索雷达站 search radar installation
搜索雷达终端 search radar terminal
搜索理论 search theory
搜索脉冲 search pulse
搜索目标 scanning
搜索目标望远镜 scanning telescope
搜索能力 search capability
搜索频率发生器 search frequency generator
搜索器 hunter;searcher
搜索切断 search disconnect
搜索情报 search information
搜索情况标图板 search plotter
搜索区(域) region of search;field of search; searching area; searching sector
搜索圈 search turn
搜索扫描 search sweep
搜索删除法 search and kill method
搜索扇形区 search sector
搜索设备 search equipment
搜索声呐 search sonar
搜索式干扰机 search jammer
搜索守所 scanning watch
搜索树 search tree
搜索数据 search data
搜索速度 scouting speed
搜索速率 search rate
搜索天空 search sky
搜索天线 search(lighting) antenna
搜索调谐 search tuning
搜索停止机构 search-stopping mechanism
搜索艇 search craft;searcher
搜索图 search graph;search diagram
搜索系统 search system
搜索显示器 search indicator
搜索线 scouting line
搜索选通脉 search gate
搜索巡逻 search patrol
搜索抑制器 search stopper
搜索营救费用 search and rescue expenses
搜索营救用海图 search and rescue chart
搜索用的潜望镜 search periscope
搜索与测高雷达 search and height finder radar
搜索与测绘雷达 search and mapping radar
搜索与测距雷达 search and ranging radar
搜索与导航雷达 search and navigation radar
搜索与火控雷达 search and fire control radar
搜索与救难 search and rescue
搜索与救援 search and rescue aid
搜索与救援电话 search and rescue telephone
搜索与救援区 search and rescue region
搜索与救援演习 search and rescue exercise
搜索与救助 search and rescue
搜索与救助船 search and rescue vessel
搜索与救助费 search and rescue expenses
搜索与救助控制中心 search and rescue control center[centre]
搜索与救助艇 search and rescue boat;search and rescue craft
搜索与救助通信[讯] search and rescue communication
搜索与救助卫星 search and rescue satellite
搜索与救助行动 search and rescue operation
搜索与救助演习 search and rescue exercise
搜索与救助用位置图 search and rescue chart
搜索与救助中心 search and rescue center[centre]
搜索与目标捕捉雷达 search and target acquisition radar
搜索与扫荡 search and clear
搜索圆圈区 search circle
搜索援救飞行图 search and rescue chart
搜索直升机 search helicopter
搜索周期 hunting period;search cycle
搜索装置 locator;searcher
搜寻 frisking;searching;seeking
搜寻边缘线 line retirement
搜寻范围 search area
搜寻方式 search pattern
搜寻基点 search datum
搜寻技术 search technique
搜寻起始线 line of departure
搜寻器 hunter
搜寻失踪船只 search for missing boat
搜寻速度 search speed
搜寻线 track
搜寻线间距 track spacing
搜寻线路 search track
搜寻线圈 search coil
搜寻星历表 searching ephemeris
搜寻与救助 search and rescue
搜寻与救助船 search and rescue vessel
搜寻与救助费 search and rescue expenses
搜寻与救助控制中心 search and rescue control center[centre]
搜寻与救助艇 search and rescue boat;search and rescue craft
搜寻与救助通信[讯] search and rescue communication
搜寻与救助卫星 search and rescue satellite
搜寻与救助行动 search and rescue operation
搜寻与救助演习 search and rescue exercise
搜寻与救助用位置图 search and rescue chart
搜寻与救助中心 search and rescue center[centre]
搜寻与营救 search-and-find
搜寻遇难船舶空间形态 space system for search of distress vessels
搜寻援救飞行图 search and rescue chart
搜眼 towing eyelet

飕飕声<风等的声音> sough

艘/时 vessels per hour

嗽口器 dental lavatory

苏氨酸 threonine

苏必利尔湖 Superior Lake
苏长辉长岩 noritegabbro
苏长岩 hypersthenfels;norite;hypersthenite
苏打 salt soda;soda(salt);sodium carbonate;yellow soda ash
苏打电池 soda cell
苏打法 soda process
苏打粉 soda ash
苏打粉煅烧机 soda ash roaster
苏打湖 natron lake;soda lake
苏打灰 anhydrous sodium carbonate;barilla;calcined soda;soda ash
苏打灰除非硅酸盐硬度法 removal of non-carbonate hardness by soda-ash
苏打焦油 soda tar
苏打结晶 soda crystals
苏打卤 soda brine
苏打明 soda mint
苏打溶液 soda solution
苏打熔融物 soda smelt
苏打软化 soda process;soda softening
苏打软化法 soda softening method;soda softening process
苏打软水法 soda process;soda softening
苏打石 nahcolite
苏打石膏 soda gypsum
苏打石灰 soda lime
苏打石灰处理 soda-lime process
苏打石灰混合物 soda-lime mix
苏打石灰法 soda-lime process
苏打石灰软化法 soda-lime softening process
苏打石棉 ascarite;soda-asbestos
苏打水 carbonated water;soda water
苏打酸灭火器 soda-acid(fire) extinguisher
苏打土 soda soil;sodic soil
苏打洗涤 soda wash
苏打循环泵 soda circulating pump
苏打盐土 soda-saline soil;soda-solonchak;soda solonetz
苏打淤渣 soda sludge
苏打纸浆 soda pulp
苏打纸浆废水 soda pulp mill wastewater
苏丹阿拉伯树胶 Sudan gum arabic
苏丹红 Sudan red
苏丹黄 Sudan yellow
苏丹可乐果 cola tree;common cola;gorra;kola tree;Sudan colanut
苏尔泽二冲程发动机 Sulzer two cycle engine
苏尔泽填料 Sulzer packing
苏尔泽锌基轴承合金 Sulzer alloy
苏方木 bukkum wood;log wood
苏方树 log wood
苏复 resurgence
苏格兰的房屋立面 Saxon facade
苏格兰东岸渔帆船 nabby
苏格兰胶带 Scotch tape
苏格兰冷杉 Scots fir
苏格兰礼拜堂 kirk
苏格兰砌合 Scotch bond
苏格兰曲柄 Scotch crank;Scotch yoke
苏格兰石 Scotch stone;scotlandite;snakestone
苏格兰式船用锅炉 scotch marine boiler
苏格兰式锅炉 Scotch boiler
苏格兰式屋顶天沟 rone
苏格兰式檐口托座 Scotch bracketing
苏格兰式檐口隅撑 Scotch bracketing
苏格兰松 Scots pine
苏格兰早期的帆船 gabbard
苏格兰支架<一种粉刷用支架> Scotch bracketing
苏格兰棕颜料 Caledonian brown
苏硅矾钡石 suzukiite
苏合香木材 hazel pine
苏柯特式变压器接线法<把三相电变为二相电的接线法> Scott connected transformer
苏库匹拉硬木<一种产于中美洲光亮深棕色硬木> Sucupira
苏莱尔调焦狭缝系统 focusing Soller-slit system
苏黎世分类 Zurich classification
苏黎世数 Zurich number
苏里南<拉丁美洲> Surinam
苏里南红木 Surinam mahogany
苏里锌白铜 Suhler-white copper
苏利文角形压气机 Sullivan angle compressor
苏联<前苏维埃社会主义共和国联盟的简称> Soviet Union
苏美尔建筑 Sumerian architecture
苏门答腊风 Sumatra
苏门答腊萝芙木 Rauvolfia sumatrana
苏门答腊松 Tenasserium pine
苏门答腊铁木 Borneo ironwood
苏模鞣料 shumac;sumac(h)
苏木 Brazil wood; bukkum wood;campeachy wood;hematoxylon;log wood
苏木浸膏 hemartine extract;hematin extract
苏木精 hematine;logwood extract
苏木精锶染剂 hemastrontium
苏木提取物 logwood extract
苏帕柚楠<一种黄铜色~深棕色圆木,产于印度-马来西亚> Supa
苏塞克斯花园墙砌合 Sussex garden-wall bond
苏塞克斯砌合<一种三顺一丁砌合法> Sussex bond
苏式彩画 Suzhou-style decorative painting;Suzhou-style pattern
苏斯平索尔<一种多彩石墨防水漆> Suspensol
苏台德运动【地】Sudetian orogeny
苏钽铝钾石 sosedkoite
苏特罗式量水堰 Sutro measuring weir
苏特罗式堰 Sutro weir
苏铁纲<古植物> Cycadales
苏通方程式 Sutton's equation
苏通扩散公式 Sutton's formula
苏通石 Sutton stone
苏瓦<斐济首府> Suva
苏维埃社会主义共和国联盟<即原苏联> Union of Soviet Socialist Republics
苏醒 resurgence;resurrection;revive
苏伊士地堑 Suez graben
苏伊士运河 Suez Canal;the Suez Canal
苏伊士运河表 Suez canal form
苏伊士运河航线 via Suez Canal shipping line
苏油 perilla oil
苏云石英闪长岩 opdalite
苏州园林 Suzhou traditional garden

酥碎岩石 crumbling rock

酥性土<易碎的土> friable soil
酥炸岩 friable rock

俗丽城<好莱坞的俚称> Tinsel City

俗名 popular name;trivial name;usual name

诉讼案件 contentious case

诉讼当事人 contesting party;litigant;party in a lawsuit
诉讼费加损害赔偿费 costs and damages
诉讼费清单 bill of legal cost
诉讼费(用) expenses in litigation;costs of litigation;court fee;legal cost;legal expenses
诉讼各方 contesting parties
诉讼和解 settlement of action
诉诸仲裁 appeal to arbitration
诉诸仲裁人 peal to arbitration

素

素布 untreated fabric
素材 fodder;raw data;source material;unsawn timber
素餐厅 beanery
素除子 prime divisor
素垂饰 plain pendant
素瓷 biscuit porcelain;bisque
素瓷焙烧 bisque firing
素瓷膜 porous membrane
素磁 biscuit porcelain
素点 prime spot
素雕 bisque sculpture
素吊架 plain pendant
素多项式 prime polynomial
素封面 plain cover
素环 prime ring
素灰 neat;neat plaster
素灰石膏条板 plain lath
素混凝土 non-reinforced concrete; plain concrete; plane concrete; unreinforced concrete
素混凝土单个基础 plain concrete single base
素混凝土垫层 plain concrete bedding
素混凝土墩 plain concrete pier
素混凝土墩台基础 plain concrete foundation pier
素混凝土结构 plain concrete structure
素混凝土块 plain concrete block
素混凝土路面 non-reinforced pavement;plain concrete pavement;unreinforced concrete pavement
素混凝土路面板 non-reinforced slab
素混凝土面层 plain concrete surface; unreinforced surface
素混凝土柱 plain concrete column
素净色 quiet shade
素矩阵 prime matrix;primitive matrix
素理想 prime ideal
素炼机 masticator
素炉钧釉 single oven Jun glaze
素描 adumbration; limn; outline; sketch
素描薄 sketch book
素描地质剖面图 sketched geologic(al) profile
素描剖面图 sketch profile
素描图 pictorial diagram;sketch map
素描纸 sketching paper
素模 prime modulus
素木 plain wood
素喷法 gunite
素坯 biscuit
素坯彩涂 biscuit painting
素坯瓷 biscuit porcelain
素坯粉制成的模具 pitcher mo(u)ld
素坯模具 biscuit mo(u)ld
素坯印花 biscuit printing
素平移 primitive translation
素奇异立方 prime singular cube
素三彩 plain colo(u)r glaze; plain tricolo(u)r
素色 plain colo(u)r
素色幻灯片 tinter
素色人造石 plain reconstructed stone
素色熔合宝石 plain reconstructed stone
素色亚麻油地毡 plain linoleum
素色轧花 plain embossed design
素砂浆 plain mortar
素商 prime quotient
素烧 biscuit fire[firing];biscuiting
素烧板 porous tile
素烧瓷 unglazed porcelain
素烧瓷板 hard baked slab; porous plate;porous tile
素烧瓷的 porous
素烧瓷筒 porous cell
素烧瓷砖 unglazed tile
素烧的 unglazed
素烧坩埚 unglazed crucible
素烧管 unglazed pipe
素烧滤筒 porous filter cylinder
素烧耐土材料 keramzite
素烧黏[粘]土填料 keramzite
素烧坯 biscuit
素烧坯滚磨抛光 scouring
素烧坯黏[粘]附的垫砂 white dirt
素烧瓶 porous cell
素烧器皿 biscuit ware
素烧陶瓷 bisque; hard-burnt clayware
素烧陶器 hard-burnt stoneware; unglazed earthenware
素烧瓦 red roof tile;unglazed roofing tile
素烧窑 biscuit furnace;biscuit kiln
素石 flat band
素数 prime integer;prime number
素数对 prime pair
素数分解 prime decomposition
素数偶 prime couple
素数生成 prime number generation
素水泥浆 neat cement grout
素水泥浆砂浆结合层 neat cement binding course
素胎 plain white body
素陶瓦 biscuit tile;bisque tile
素填土 plain fill;pure made land
素铁 plain sheet
素铁罐 plain can
素图 blank chart; blank map; blank sheet; grey base; griblet; outline drawing;sketch map
素土 plain earth;plain soil
素土地面 earth surface
素土垫层 plain soil cushion
素土夯实 packed soil;rammed earth
素圬工 plain masonry
素系 prime system
素线 element line
素项 prime implicant
素信息 prime information
素压 blind
素压凸印 antique
素压印 blind blocking;blind embossing
素因 predisposition
素因数 prime factor
素因性的 predisposing
素因子 prime divisor;prime factor
素域 prime field
素元 primitive element
素元素【数】prime element
素蕴含 prime implicant
素枕 untreated timber sleeper
素阵 prime matrix;primitive matrix
素质 diathesis;disposition;predisposition;stuff
素砖石砌体 plain masonry
素装饰线条 band(e)let

速

速办 despatch[dispatch]
速饱和变压器 quick saturation converter
速爆炸药 fast powder
速比 gear ratio; progressive ratio; speed ratio; transmission ratio; velocity ratio
速比范围 ratio coverage
速比范围选择杆 range selector lever
速比范围选择器 range selector
速闭 quick close stop
速闭阀 quick-closing valve; rapid-closing valve
速闭截止阀 quick-closing stop valve
速闭门 quick-closing door
速变 quick changing
速变齿轮箱 quick change gearbox
速变振荡器 velocity-variation oscillator
速波放大 magnification for rapid waves
速测 hasty survey;rapid test;tach(e)ometry
速测地震仪 tachyseismic instrument
速测断面图 hasty profile
速测法 quick test;rapid method;tach(e)ometric(al) method; tachymetry
速测仪 tach(e)ometer;tachymeter
速查表 zoom table
速差测量装置 speed difference measuring device
速差沉淀 differential settling
速差传感器 differential speed sensor
速差式信号法 speed signal(l)ing
速差式信号系统 speed signal(l)ing system
速差式信号制 speed signal(l)ing system
速差制信号 speed signal(l)ing
速差转向装载机 skid steering loader
速长比 speed length ratio; Taylor's quotient;velocity ratio
速沉(土)颗粒 quick-settling particles
速沉质点 quick-setting particle
速成班 accelerated course; crash course;streamlined course
速成测量 hasty survey
速成地图 hasty map
速成堆肥法 composting method
速成费用 crash cost
速成计划 crash program(me);crash project
速成胶质 accelerated gum
速成绿化 quick planting works
速成图 hasty map
速成训练班 intensive training course
速定胶粘剂 rapid cure adhesive
速动 immediate action;quick-action; snap;snap action
速动安全阀 guard valve
速动保险丝 quick-action fuse
速动报警 immediate action alarm
速动比率 quick circulating percentage;quick ratio
速动齿轮 fast motion gear
速动锤 rapid action hammer
速动的 quick acting;quick-operating
速动电压调整器 quick-acting voltage regulator
速动阀 fast-acting valve;guard valve; quick-action valve; quick-release valve; rapid action valve; snap valve
速动阀簧 quick-action valve spring
速动负债 quick liability
速动合闸 quick-acting switching
速动继电器 instantaneous relay; quick-acting relay;speed relay
速动夹具 quick-action clamp
速动开关 quick-action switch;quick-break switch; quick-made-and-break switch;snap switch
速动螺旋 quick-motion screw
速动逆电流断路器 quick-acting reverse-current circuit-breaker
速动起动器 quick-acting starter
速动熔断器 quick-action fuse
速动三通阀 quick triple valve
速动事故阀 guard valve
速动弹簧开关 quick-acting spring switch
速动调节器 quick-acting regulator
速动阴极 quick start cathode
速动永磁铁 snap magnet
速动闸门 guard valve;stop gate
速动折叠百叶门 rapid acting folding shutter door
速动转接 quick-acting switching-over
速动资产 quick assets
速动资产流转表 statement of net quick assets flows
速动资金 quick money
速冻 quick freeze;quick freezing
速冻产品 quick-frozen product
速冻货物 quickly frozen goods
速冻间 quick-freezing room; sharp-freezing room
速冻冷库 blast freezer;deep freezer; quick freezer
速冻器 quick freezer
速冻溶液 quick-frozen solution
速冻水果 quick-frozen fresh fruit
速冻装置 deep-freezing plant;quick-freezing plant
速度 speed;velocity
速度摆 velocity sensitive pendulum
速度保持 speed hold
速度保持信号灯 speed lock light
速度比 ratio of speeds; speed ratio; velocity ratio
速度比较指示器 velocity comparison indicator
速度比例 velocity scale
速度比率 velocity rate
速度变动 perturbation of velocity; speed fluctuation;speed variation
速度变化 variation in speed;variation in velocity;velocity change; velocity fluctuation;velocity variation
速度变化率 percentage speed variation;rate of speed
速度变换 speed changing; speed transformation
速度变换测验计 <测流水、血液等速度,如流速计、血行计等> tach(e)ometer
速度变换阶段 speed transition zone
速度变换器 velocity transducer
速度变率 velocity gradient
速度变慢 inching
速度变速箱 pick-off gearbox; quick gearbox
速度标杆 velocity rod
速度标志 speed marker
速度表 autometer; odometer; speed counter;speed ga(u)ge;speed meter; speedometer; velograph; speed chart <指图表>
速度波动 velocity perturbation; velocity pulsation
速度波动系数 speed fluctuation coefficient
速度补偿 velocity compensation
速度补偿器 velocity compensator
速度不均匀性 wow and flutter
速度不快 underspeed
速度不连续 velocity jump
速度不连续性 velocity discontinuity
速度不灵敏度 speed insensibility
速度不稳定性 speed instability
速度不足 underspeed
速度参考值 speed reference
速度测定 velocity determination
速度测定基线 speed course
速度测井 velocity logging; velocity shooting
速度测量 speed measurement;tach(e)ometric(al) survey; velocity measurement;velocity survey
速度测量范围 speed measuring range
速度测量器 velocity meter
速度测量数据 velocity survey data
速度测量误差 data noise

S

速度测量系统 velocity measuring system
速度测量学 velocimetry
速度测量装置 velocity measuring device
速度测针 velocity probe
速度层 velocity layer
速度差 speed difference
速度差反馈 counting rate difference feedback
速度差值 velocity deficiency
速度常数 rate constant;velocity constant
速度场 velocity field;velocity pattern;velocity profile
速度场测定 Pitot investigation
速度场计算 velocity distribution calculation
速度场矢量 velocity field vector
速度场有向量 velocity field vector
速度冲击 velocity shock
速度冲击硬化 velocity impact hardening
速度储备 speed margin
速度传感 speed sensing
速度传感可变的定时装置 speed sensing variable timing unit
速度传感器 speed pick-up;velocity pick-up;speed sensing device;speed sensor;speed transducer;speed transmitter;velocity sensitive transducer;velocity sensor;velocity transducer
速度存储器 velocity memory
速度挡数 gear speed;number of speeds
速度挡位 speed stage;speed threshold
速度导线 velocity traverse
速度倒数区 reciprocal velocity region
速度的合成 composition of velocities;resultant of velocities
速度的平行四边形 parallelogram of velocities
速度等值线剖面 velocity contour section
速度地裂检波器 velocity detector
速度地震计 velocity seismograph;velocity seismometer
速度地震仪 velocity seismograph
速度递减法 velocity reduction method
速度递减法风道计算 velocity reduction method duct sizing
速度递减求径法 velocity method duct sizing
速度电势 speed voltage
速度电压 velocity voltage
速度调查器 speed control device
速度动力感觉 speed power feel;Q-feel
速度动力系数 speed power coefficient
速度读出 speed sensing
速度读出可变的定时装置 speed sensing variable timing unit
速度读出器 speed sensing device
速度读出装置 speed sensing device
速度多边形 polygon of velocities
速度法 tach(e)ometric(al)method;velocity method
速度反常 velocity inversion
速度反共振 velocity anti-resonance
速度反馈 rate feedback;velocity feedback
速度反馈回路 velocity feedback loop
速度反馈控制机构 feedback speed control mechanism
速度反馈系统 rate feedback system
速度反向 velocity reversal
速度反演 velocity inversion
速度反演方法 velocity inversion procedure
速度反应 speed responsing;velocity response
速度反应谱 velocity response spectrum
速度反应谱包线 velocity response envelope spectrum
速度反应谱曲线 velocity response envelope spectrum
速度反应器 speed responser;governor <液力传动的>
速度范围 speed range;velocity band;velocity interval
速度范围数据 speed range data
速度方位显示器 velocity-azimuth display
速度放大 speed amplification
速度放大器 velocity-variation amplifier
速度分布 velocity distribution;velocity spread
速度分布测定 velocity exploration
速度分布定律 velocity distribution law
速度分布函数 velocity distribution function
速度分布律 law of distribution of velocities
速度分布曲线 speed distribution curve
速度分布图 speed profile;velocity contour;velocity(distribution)diagram;velocity profile
速度分布线 speed distribution line
速度分布型 velocity distribution pattern
速度分布噪声 velocity noise
速度分级 velocity stage
速度分解 resolution of velocity;velocity resolution
速度分类 velocity sorting
速度分离器 velocity separator
速度分量 component of velocity;velocity component
速度分散 velocity dispersion
速度分析 velocity analysis
速度分析器 velocity analyser[analyzer]
速度分选 velocity sorting
速度分选原理 velocity-sorting principle
速度幅值 amplitude of velocity
速度负反馈 velocity negative feedback
速度干扰 velocity disturbance;velocity jamming
速度感 speed sense
速度感传器 velocity sensor
速度感觉 velocity-feel
速度感应 velocity response
速度高度比(值)velocity/height ratio
速度高度配平位置 speed-altitude trim position
速度高度曲线图 velocity versus altitude graph
速度各向异性 velocity anisotropy
速度跟踪 velocity tracking
速度跟踪系统 rate servo system
速度公式 velocity formula
速度功率谱密度 velocity power spectral density
速度共振 velocity resonance
速度构造 velocity structure
速度关联控制 speed concatenation control
速度管制 speed regulation
速度惯性导航系统 velocity inertial navigation system
速度惯性系统 velocity inertial system
速度规 speed transducer
速度规律性变化 cresceleration
速度过程理论 rate process theory
速度过调量 speed overshoot
速度过渡区 speed transition zone
速度函数 velocity function
速度航程发送器 <电磁计程仪> speed distance transmitter
速度航向纬度误差 speed-course latitude error
速度合成 velocity complex
速度合成定律 composition of velocity law
速度和打滑电机 speed and slide generator
速度和高度优势 speed and altitude supremacy
速度和牵引重量关系曲线 speed and hauled weight curve
速度核心区 potential core
速度荷重曲线 speed-load curve
速度横向分布 velocity traverse
速度滑脱 velocity slip
速度环 <水轮机的> speed ring
速度环境 <大多数驾驶人能接受的速度的具体环境> speed environment
速度环量 circulation;velocity circulation
速度缓慢的船 slug
速度换能器 velocity transducer
速度换算 speed conversion
速度恢复期 velocity recovery phase
速度回声 velocity resonance
速度机动 speed maneuver
速度积分器 velocity integrator
速度畸变 speed distortion;velocity distortion
速度及阻滞调查【交】speed and delay study
速度级 speed stage;velocity level
速度级别 speed step
速度级冲动式汽轮机 impulse turbine with velocity stages;velocity-stage impulse turbine
速度级次变化 velocity staging
速度极限 speed barrier;speed limit;velocity limit
速度计 dromometer;rate indicator;ratemeter;speed ga(u)ge;speed indicator;speed meter;speedometer;tach(e)ometer;vegraph;velo(ci)meter;velocity ga(u)ge;velocity indicator;velograph
速度计齿轮 speedometer gear
速度计传动轴 speedometer take-off shaft
速度计传感器 speedometer transducer
速度计从动齿轮 speedometer driven gear
速度计接线套 speedometer cable casing
速度计刻度误差表 speedometer scale error card
速度计挠性轴 speedometer flexible shaft
速度计双速接头 speedometer two-speed adapter
速度计算 velocity determination
速度计算器 speed calculator
速度计主动齿轮 speedometer drive gear
速度计主轴 speedometer main shaft
速度计座 speedometer base
速度记录 velocity photogram
速度记录带 speed recording strip;speed recording tape
速度记录计 tachograph
速度记录器 recording speed indicator;speed recorder;tach(e)ograph
速度记录曲线 speed recording graph
速度记录图 speed recording graph;tach(e)ogram;tach(e)ograph
速度记录仪 speed recorder;vegraph;velograph
速度记录指示器 speed recording indicator
速度继电器 quick action relay;speed relay
速度加快 shifting up;speed rise
速度监督 speed supervision
速度监控 velocity monitor(ing)
速度监视 speed monitoring
速度监视站 <汽车> speed trap
速度减慢 speed drop
速度检波器 velocity geophone
速度检测器 speed detector;speed pick-up;velocity detector
速度检查器 speed checker
速度降(低)speed drop;speed reduction
速度校对器 speed checker
速度校验 speed calibration;speed checking
速度校正 velocity correction
速度节点 velocity node
速度结构 velocity structure
速度界面 velocity interface
速度界限 speed envelope
速度矩 moment of velocity
速度距离关系曲线 velocity-distance curve
速度距离关系图 speed-distance diagram
速度距离曲线 speed-distance curve;speed versus distance curve;velocity-distance curve
速度距离曲线图 speed-distance diagram
速度距离图 speed-distance chart
速度聚焦 velocity focusing
速度聚焦质谱仪 velocity-focusing mass spectrograph
速度空间 velocity space
速度空间不稳定性 velocity space instability
速度控制 rate-controlling;speed control;velocity control
速度控制标 speed control sign
速度控制程序器 velocity control programmer
速度控制程序设计器 velocity control programmer
速度控制的开关 speed-sensitive switch
速度控制电路 speed control circuit
速度控制电压 velocity control voltage
速度控制阀 speed control valve
速度控制方式 speed control system
速度控制器 accelerator-decelerator;speed controller;speed governor;velocity controller
速度控制伺服机构 velocity control servo
速度控制伺服系统 speed control servo system
速度控制系统 speed control system
速度控制信号 speed control signal
速度控制信号法 speed control signal(l)ing
速度控制旋钮 velocity control knob
速度控制圆盘 speed control disc
速度控制装置 speed control device
速度扩散 velocity dispersion
速度廓线 velocity profile
速度累积曲线 speed accumulation curve
速度灵敏触点 speed-sensitive contact
速度灵敏触头 speed-sensitive contact

速度灵敏度 rate-sensitivity; velocity sensitivity
速度灵敏释放装置 speed-sensitive release
速度灵敏脱扣器 speed-sensitive release
速度流量计 velocity meter
速度流量曲线 speed-flow curve
速度滤波 velocity filtering
速度滤波器 velocity filter
速度轮廓 velocity profile
速度脉动 velocity fluctuation; velocity pulsation
速度慢的 slow-footed
速度弥散度 dispersion of velocity; velocity dispersion
速度密度曲线 speed-density curve
速度面 velocity surface
速度面积法 velocity-area method
速度敏感开关控制 speed sensitive switch control
速度敏感特性 speed sensitivity characteristics
速度模型 velocity model
速度能变换的 variable speed
速度能高 velocity head
速度能高图 velocity head graph
速度扭矩曲线 speed-torque curve
速度匹配 speed match(ing); velocity match(ing)
速度偏差 hunting; jitter; velocity misalignment
速度偏差系数 velocity misalignment coefficient
速度偏流指示器 speed-and-drift indicator
速度频散 velocity dispersion
速度剖面图 velocity profile
速度谱曲线 velocity spectrum curve
速度谱仪 velocity spectrograph
速度起伏 eddy velocity; velocity fluctuation
速度牵引力曲线 speed-tractive effort curve
速度牵引重量关系曲线 speed-hauled weight curve
速度切变 velocity shear
速度切变界限 velocity shear boundary
速度曲线 rate curve; velocity curve; velocity graph
速度曲线表 speed chart
速度曲线斜度 velocity slope
速度圈 speed circle
速度三角形 speed triangle; velocity triangle
速度三角形机械计算器 avigraph
速度扫描 velocity scan(ning)
速度色散 velocity dispersion
速度上升 speed rise
速度上升特性 rising-speed characteristic
速度上下范围 range of speed control
速度深度分布 velocity-depth distribution
速度失配 velocity misalignment
速度失配系数 velocity misalignment coefficient
速度时间计算 speed timing calculation
速度-时间-距离关系 speed-time-distance relationship
速度时间曲线 speed-time curve
速度拾震器 velocity pick-up
速度矢端曲线变形 hodograph transformation
速度矢端线法 velocity hodograph method
速度矢量 velocity vector
速度矢量(曲线)图 hodograph
速度矢量图解 velocity vector diagram
速度式触发器 velocity sensitive trigger
速度式地震计 velocity-type seismometer
速度式地震检波器 seismic detector of the velocity type
速度式流量计 current-type flowmeter; rotating meter; velocity-type flowmeter
速度式制冷机 velocity refrigerator
速度势 velocity potential
速度势函数 velocity potential function
速度试验 speed test; speed trial
速度收集器 speed trap
速度数据 speed data
速度水表 velocity meter
速度水头 dynamic(al) head; velocity head
速度水柱 velocity head
速度瞬心 instantaneous centre of velocity
速度伺服(机构) speed servo; velocity servo
速度伺服系统 rate servosystem; velocity of servosystem; velocity servo
速度随动系统 <拖缆机的> speed-holding servo
速度损失 loss of speed; speed loss
速度探针 velocity probe
速度特性 speed characteristic
速度特性曲线 speed characteristic curve
速度梯度 gradient of velocity; velocity gradient
速度梯度输入 input velocity gradient
速度梯度张量 velocity gradient tensor
速度提高 speed rise; velocity increase
速度体 body of velocity
速度调节 speed adjustment; speed control; speed governing; speed regulation; speed setting
速度调节杆 speed lever
速度调节感应离合器 speed governed induction clutch
速度调节器 rate fixer; speed control device; speed control governor; speed governor; speed regulator; speed timer
速度调节手柄 speed controller
速度调节系统 velocity-controlled system
速度调整 speed adjustment; speed regulation; speed set; velocity adjustment; velocity regulation
速度调制 speed modulation; velocity modulation
速度调制度 depth of velocity modulation
速度调制管 klystron(tube); velocity-modulated tube; velocity-variation tube
速度调制管放大器 klystron repeater
速度调制束 velocity-modulated beam
速度调制系数 velocity modulation coefficient
速度调制效应 velocity modulating effect
速度调制型聚束器 velocity modulation type buncher
速度调制振荡器 velocity-modulated oscillator
速度跳跃 velocity jump
速度同步器 speed synchronizer; speed timer
速度头 impact pressure; kinetic head; velocity pressure
速度头损失 loss of velocity head
速度头压缩 ram compression
速度图 diagram of velocities; velocity photograph; velocity plan
速度图变换 hodograph transformation
速度图法 hodograph method
速度图平面 hodograph plane
速度陀螺仪 rate gyro(scope)
速度椭球 velocity ellipsoid
速度椭球分布 ellipsoidal distribution of velocities
速度椭圆 adiabatic ellipse; velocity ellipse
速度位 velocity potential
速度稳定的 speed stable
速度稳定器 speed stabilizer
速度稳定性 speed stability
速度误差 speed error; velocity error
速度误差补偿器 velocity error compensator
速度误差常数 velocity error constant
速度误差改正器 speed corrector; speed error corrector
速度误差校正 velocity error correction
速度误差系数 velocity error coefficient
速度误差修正表 speed correction table
速度系数 coefficient of velocity; speed coefficient; speed factor; velocity coefficient; velocity factor
速度显示 speed display; speed sensing
速度显示可变的定时装置 speed sensing variable timing unit
速度显示装置 speed sensing device
速度线 speed line
速度限度 margin of speed; speed limit
速度限制 limitation of speed; rate limitation; speed limit(ation); speed restriction
速度限制伺服机构 velocity servomechanism; velocity limiting servo
速度限制伺服系统 velocity servosystem
速度限制器 speed limit device
速度限制线 speed limit line
速度限制信号 restricted speed signal; restricting speed signal
速度箱 quick change gear
速度响应 velocity response
速度响应谱 velocity response spectrum
速度向量 velocity vector
速度象差 velocity aberration
速度效应 speed effect; velocity effect
速度谐振 velocity resonance
速度信号 rate signal; speed signal
速度信号发生器 speed signal generator
速度信号输入 speed input
速度信息 velocity information
速度形成的压力 velocity formed pressure
速度型调节器 velocity-type governor
速度性能 speed ability
速度修正 velocity correction
速度选配齿轮箱 speed-matching gear box
速度选通 speed gate
速度选择开关 speed selector valve
速度选择控制 speed selection control
速度选择控制系统 speed selection control system
速度选择控制 speed selection control system
速度选择器 velocity selector
速度压力 velocity pressure
速度压力控制 speed governor control
速度沿旋翼直径的分布 velocity along rotor diameter
速度液力 circulation
速度仪 speed indicator; speedometer; speed ga(u)ge
速度异常 velocity anomaly
速度异常带宽度 width of velocity anomaly band
速度因数 speed factor
速度因子值 values of velocity factor
速度应变 velocity strain
速度优势 speed advantage
速度优先模式 shutter-priority mode
速度与密度分解 separation of density and velocity
速度与温度的相关性 temperature dependence of velocity
速度预测尺 prediction scale
速度圆 speed circle
速度跃变 discontinuity; velocity jump
速度增长 speed increment
速度增减 fluctuation of speed
速度增量 increment of speed; velocity gain; velocity increment
速度增益 speed gain
速度振幅 velocity amplitude
速度振谱 velocity vibrating spectrum
速度支数补偿拉丝机 speed compensated winder
速度指示标 speed indicator sign
速度指示牌 speed board
速度指示器 motion indicator; rate indicator; speed indicator; velocity indicator
速度指示器的发送机 transmitter for speed indication
速度指示仪 speed indicator
速度指数 velocity index
速度滞后 velocity lag
速度滞后误差 velocity lag error
速度重量曲线 speed-load curve
速度周期性变化 periodic(al) velocity disturbance
速度助流区 velocity inlet region; velocity intake region
速度准确控制 speed sensing control
速度自动控制 automatic speed control
速度自动控制器 automatic speed controller
速度自动控制系统 automatic speed control system
速度自动调节 automatic speed regulation
速度自动调节器 automatic speed governor
速度自动调整 automatic speed regulation
速度自动选择系统 automatic speed selecting system
速端平面 hodograph plane
速端曲线【物】 hodograph
速端曲线法 hodograph method
速端图 hodograph
速端图法 hodograph method
速断 quick break
速断保险丝 quick-break fuse; rapid acting fuse
速断开关 instantaneous breaker; perking switch; quick-break switch
速断馈电保险丝 quick-break feeder fuse
速断泡沫 quick-breaking foam
速断熔断器 quick-break fuse
速断闸刀开关 quick-break knife switch
速断装置 quick-release fitting
速发气锅炉 flash boiler
速封耦合 quick seal coupling
速缝机 folder
速干 rapid-curing cutback

S

速干剂 drier[dryer] sol;dry solution
速干印油 heat set ink
速干油漆 fast paint;Marb-I-cote <干后呈大理石面状>
速高比 speed-height ratio
速化石灰 rapid slaking lime
速换 quick change
速换发动机部件 quick change engine unit
速换夹头 quick change chuck
速换销 quick change pin
速回握柄 quick-return lever
速记 stenograph
速记测距仪 stenometer
速记员 steno(grapher)
速降 dowse;prompt drop
速降阀 quick drop valve
速接齿轮装置 knuckle gear(ing)
速接联轴节 quick coupling
速接联轴器 rapid quick coupling
速进键 forward wind key
速距关系 velocity-distance relation
速开阀 quick-opening valve
速开式降落伞 quick-release parachute
速开手柄 quick-release handle
速开锁销 quick-release pin
速开逃脱舱口 quick-opening escape hatch
速开信用证 rush letter of credit
速开装置 quick-opening device;quick-release device
速控调阀 governor valve
速控压阀 governor valve
速冷(处理)deep freezing
速力 turn of speed
速力计 veeder counter
速力型 speed type
速立脚手架 <商品名> Rap-rig
速裂乳胶体 quick-breaking emulsion
速流动力学 rapid flow kinetics
速流技术 rapid flow technique
速流曲线 flow curve
速滤剂 accelerator
速率 velocity rate;rate of speed;rate-plus-displacement control;speed rate
速率比控制 speed ratio control
速率比控制器 speed ratio controller
速率变化 speed change;speed variation
速率变化指示器 rate of change indicator
速率变换区 speed transition zone
速率标准 speed standard
速率表 speed indicator
速率波动 speed fluctuation
速率测量 speed measurement
速率测量装备 rate instrumentation
速率测试 rate test
速率差 speed difference
速率常数 rate constant;speed constant
速率传感器信号 rate sensor signal
速率导数 rate of rate
速率的透明性 rate transparency
速率电动势 speed electromotive force
速率动作控制器 rate action controller
速率发射机 rate transmitter
速率发生器 rate generator
速率反馈 rate feedback
速率范围 speed range
速率方程 rate equation
速率分布函数 speed distribution function
速率公式 rate equation
速率管制 speed regulation
速率过程 rate process
速率过程理论 rate process theory
速率过程曲线 rate process curve
速率过渡区 speed transition zone
速率环 speed ring
速率积分陀螺 rate integrating gyro(scope)
速率极限 speed limit
速率计 ratemeter;speed ga(u)ge;speed meter;speedometer;tach(e)ometer
速率计数器 speed counter
速率计主动齿轮 speedometer drive gear
速率加速剂 rate accelerating material
速率降低 speed reduction
速率接收机 rate receiver
速率决定阶段 rate-determining step
速率控制 rate-control(ling);speed control
速率控制步骤 determining rate-controlling step;rate-determining step
速率控制开关 speed control switch
速率控制器 rate controller;speed controller
速率控制器作用 rate controller action
速率控制系统 speed control system
速率控制型 rate control pattern
速率浓度数据 rate-versus-concentration information
速率披盖 cocooning
速率偏差信号 off-speed signal
速率曲线 rate curve
速率取样脉冲 speed sampling pulse
速率上升 relative increase of speed
速率失真 rate distortion
速率失真编码 rate-distortion coding
速率时间曲线 speed-time curve
速率式传声器 velocity microphone
速率式话筒 velocity microphone
速率试验 speed trial
速率适应能力 speed capacity
速率送话器 velocity microphone
速率随动部件 rate follow-up unit
速率特性曲线 speed characteristic curve
速率梯度 velocity gradient
速率梯形波 speed trapezoidal wave
速率调节 speed regulation
速率调节器 rate regulator
速率调整 speed adjustment
速率陀螺(仪)rate gyro(scope)
速率微调控制 vernier-rate control
速率稳定法 rate stabilization
速率误差 rate error
速率系数 speed coefficient
速率限度 speed limit
速率限制 rate limitation
速率限制过程 rate-limiting process
速率限制阶段 rate-controlling step
速率限制区段 speed zone
速率限制区段预告标志 speed zone ahead sign
速率响应 rate response
速率效应 rate effect
速率信号接收器 rate receiver
速率选择开关 speed selector switch
速率学说 rate theory
速率因素 speed factor
速率因子 speed factor
速率增长 increase of speed;rate increase
速率增量 incremental speed
速率指令系统 rate-command system
速率指示计 speed indicator
速率指示器 rate indicator
速率转矩曲线 speed-torque curve
速率准确度 rate accuracy
速率阻尼 rate damping
速率作用 rate action
速率作用时间 rate action time
速率作用因数 rate action factor
速密耦合 quick seal coupling
速凝 flash set;grab set;quick hardening;quick set;rapid-curing cutback;rapid hardening;accelerated set
速凝玻璃 fast-setting glass
速凝材料 quick-setting material;rapid-curing material
速凝掺和剂 quick-setting admix(ture)
速凝掺和料 acceleration admixture;quick-setting admix(ture)
速凝灌浆 quick-setting grout
速凝灰浆 ga(u)ged mortar;quick-setting mortar;rapid-curing mortar
速凝灰泥 adamant plaster
速凝混合料 quick-setting mixture
速凝混凝土 early strength concrete;quick-setting concrete
速凝剂 accelerating admixture;accelerating agent;accelerator;flash-setting admixture;flash-setting agent;quick-setting additive;quick-setting agent;rapid setting admixture;set accelerator;set-accelerating admixture
速凝减水剂 set-accelerating water-reducing admixture;water reducing set-accelerating admixture
速凝胶泥 quick-setting mastic
速凝聚合物 instant-set polymer
速凝沥青 quick-curing bitumen;rapid-curing(cutback)asphalt
速凝喷射层 flash coat
速凝乳化沥青 quick asphaltic emulsion;quick bitumen emulsion
速凝砂浆 ga(u)ge mortar
速凝石膏粉刷 adamant plaster;quick-hardening gypsum plaster
速凝石膏料 adamant plaster
速凝石膏炮泥 hardstem
速凝石膏砂浆 ga(u)ged mortar
速凝水泥 accelerated cement;early setting cement;fast coagulating cement;flash-setting cement;one-hour cement;quick(-setting)cement;quick-taking cement;rapid setting cement;very quick-setting cement
速凝添加剂 quick-setting additive
速凝型减水剂 accelerating water reducer
速凝性 quick-jellying property
速漂爆发 fast drift burst
速启阀 quick-opening valve
速启阀门 quick-opening gate valve
速启杠杆 quick-opening level
速启闸阀 quick-opening gate valve
速潜 quick dive
速潜水舱 quick-diving tank
速遣费 despatch[dispatch] money
速遣费仅在卸货时计付 despatch discharging only
速遣费仅在装货时计付 despatch loading only
速遣费率 rate of despatch money
速遣日数 dispatch days
速燃 deflagration;quick firing
速燃层 accelerant coating
速燃成分 fast-burning composition;quick burning composition
速燃导火索 fast-burning fuse;quick burning fuse;quick match
速燃导火线 fast-burning fuse;quick burning fuse;quick match
速燃的 conflagrant;free-burning;quick burning
速燃点火药 first fire mixture
速燃法 explosion method
速燃火药 fast-burning powder
速燃期 rapid combustion period
速燃物 flash fuel
速燃引信 quick burning fuse
速燃引信头 quick match
速热式电子管 quick heater tube
速溶的 instant;quick dissolving
速溶化 instantizing
速闪 speed flash
速渗(透)rapid permeability
速渗性 rapid permeability
速升凸轮 quick-lift cam
速生的 high-growing
速生林 fast-growing trees
速生人工林 quick-growing plantation
速生树种 fast-growing species of trees;fast-growing tree species;quick-growing species of trees
速生植物 fast grower;fast-growing plant
速矢 velocity pressure
速矢端迹 hodograph;polar plot
速矢端线 hodograph
速矢端线变换 hodograph transformation
速矢端线法 hodograph method
速矢端线方程 hodograph equation
速矢描迹系统 hodoscope system
速矢曲线方程 hodograph equation
速示 dead beat
速示测试 deadbeat measurement
速示电流计 deadbeat galvanometer
速示电压表 aperiodic(al)voltmeter
速示罗盘 aperiodic(al)compass;deadbeat compass
速示器 tachystoscope
速示响应 deadbeat response
速示仪表 deadbeat meter
速势函数 velocity potential function
速视记录 quick-look record(ing)
速视系统 quick-look system
速释 quick release
速释继电器 quick-releasing relay
速释卡栓 quick-release grip
速算 rapid calculation
速调测微仪 quick-adjusting micrometer cal(l)ipers
速调放大管 amplifying klystron
速调管 klystron(tube);Shepard tube;single-beam transit tube;transit time tube;velocity-variation tube
速调管混频器 klystron mixer
速调管三倍倍频器 klystron tripler
速调管振荡器 klystron generator;klystron oscillator
速调管座 klystron mount
速调式振荡器 velocity modulating generator
速调水平仪 quick-setting level
速调水平装置 quick level(l)ing head
速调水准仪 quickset level;quick-setting level
速调系统 velocity modulation
速调振荡器 velocity-modulated oscillator
速通开关 quick-make switch
速头 velocity head
速头测杆 velocity head rod
速头测流计 velocity head current meter;velocity head flowmeter
速头测速计 velocity head current meter;velocity head flowmeter
速头改正因数 velocity head correction factor
速头海流计 velocity head current meter;velocity head flowmeter
速头流量计 velocity head current meter;velocity head flowmeter
速头系数 velocity head coefficient
速头转速计 hydraulic tachometer;velocity head tacheometer
速退 fast rewind

速脱钩 hold-back hook; pelican hook; slip hook
速脱钩式半拖车 quick-detachable semitrailer
速位差 kinetic head; velocity head
速熄火花 quenched spark
速效 quick result
速效保险丝 quick-acting fuse
速效的 quick effective
速效肥料 active fertilizer; readily available fertilizer
速效混凝土沉淀装置 high-speed coagulative precipitation unit
速效金属 available metal
速效磷 rapidly available phosphorus
速效凝结沉淀装置 high-speed coagulative precipitation unit
速效杀虫剂 fast-acting insecticide
速效水 readily available water
速效药物 active remedy
速写工具 sketch tool
速写图 sketch map
速卸沉积 dumped deposit
速卸接合 quick-disconnecting arrangement; quick-disconnecting coupling
速卸接头 quick-release joint
速卸联轴器 quick-disconnecting coupling
速卸式轮辋 quick-detachable rim
速压 kinetic pressure; ram compression
速压头 dynamic(al) pressure; pressure differential; ram effect; ram pressure
速应时间常数 promptitude time constant
速硬 quick hardening
速硬剂 hardening accelerating admixture; hardening accelerator (admixture); rapid hardener
速震动放大 magnification for rapid ground movements
速止 short stopping
速止泵 shortstop pump
速止剂 shortstopped; short stopping agent
速装 quick-mounting
速钻器 drill speeder

宿

宿处 berth

宿根植物 perennial; perennial plant
宿命论 determinism
宿舍 dormitory; dormitory house; dorsum; habitacle; hostel; lodging house
宿舍单元 dormitory unit
宿舍房屋 bedroom building
宿舍结构 housing construction
宿舍楼 bedroom unit; dormitory block; dormitory building
宿舍区 dormitory area; living quarter
宿舍区面积 dormitory quarter area
宿务港 <菲律宾> Port Cebu
宿营 cantonment
宿营车 boarding car; bunk car; camp car; dormitory car; dormitory van; living van; lodging van; outfit car
宿营地 accommodation; bull pen
宿营设备 tentage
宿主地层 host formation
宿主钉螺 host snail
宿主寄生物 xenoparasite
宿主寄生物间相互作用 host-parasite interaction
宿主节点 host node
宿主颗粒 host grain
宿主数据库 host data base
宿主数据语言 host data language
宿主选择 host choice

粟

粟 Chinese chestnut; Turkestan millet

粟粒状砂岩 millet-seed sandstone

塑

塑变 flow; galling; plastic deformation

塑变变质方式 plastic deformation metamorphic way
塑变柔皱构造 crystalloplastic corrugated structure
塑变双晶 mechanical twin
塑变值 plastic yield; yield point value; yield value
塑玻透镜 plastyle lens
塑成式电弧隔板 mo(u)lded arc chute
塑成式熄弧沟 mo(u)lded arc chute
塑度计 plasticorder; plastograph; plastometer
塑度计常数 plastometer constant
塑法 mo(u)lding
塑焊 plastic welding
塑化 plasticisation [plasticization]; plasticise [plasticize]; plasticising [plasticizing]; plastify
塑化波特兰水泥 plasticized Portland cement
塑化材 plasticizing wood
塑化的 plasticised[plasticized]
塑化粉剂 powder plasticizer
塑化硅酸盐水泥 plasticized Portland cement
塑化烘箱 fusing oven
塑化灰浆 plasticised[plasticized] mortar
塑化剂 curing agent; elasticizer; plasticising[plasticizing] agent; plasticity agent; plasticizer; plastifier; plastifying agent; water-reducing admixture; workability agent; dubbin; dubbing
塑化聚氯乙烯 plasticized polyvinyl chloride
塑化木(材) plastic wood; plasticizing wood
塑化熔体纺丝 plasticized melt-spinning
塑化砂浆 plasticised[plasticized] mortar
塑化树脂 plasticising[plasticizing] resin
塑化水泥 plastic cement; plasticized cement
塑化相 plastic phase
塑化效果 plasticizing effect
塑化(硬)沥青 plasticized pitch
塑化浴 stretch bath
塑化纸板 papreg
塑化作用 plasti(fi)cation; plasticizing action
塑胶 plastic(cement); plastic glue; fibestos <一种醋酸纤维素>
塑胶板 plastic board
塑胶玻璃 perspex
塑胶材料 plastic material
塑胶袋 plastic bag
塑胶的 plastic
塑胶地板 plastic flooring
塑胶粉 mo(u)lding powder
塑胶覆盖围篱 plastic-coated chain link fencing
塑胶攻击艇 plastic assault boat
塑胶夹层玻璃 laminated glass
塑胶铰链 plastic hinge
塑胶浸透处理 plastic impregnation
塑胶绝缘料 <一种用纸屑做的> lignin(e)
塑胶冷上釉 plastic cement cold glaze
塑胶楼面 plastic floor
塑胶铺面 plastic floor
塑胶体 plastic body
塑胶涂层 plastic coating
塑胶箱 plastic box
塑胶炸药 slurry explosive
塑解剂 peptizer; peptizing agent
塑孔器 extruder
塑口互搭接头 mo(u)lded chime lap joint
塑炼 plasticate; plastication; plasticization; plasticize; plastify
塑炼机 plasticator
塑炼橡胶 softened rubber
塑料 plastics; material plastics; plastic compound; plastic mass; plastic material; Amianite <一种用石棉作填料的>
塑料板 plastic board; plastic plate; plastic sheet; sheet plastic
塑料板壁 plastic siding
塑料板防水层 plastic impervious membrane
塑料板排水 plastic drain
塑料板排水预压法 plastic plate drain preloading method
塑料板坯 hide
塑料板条 plastic strip
塑料板预制模型 panel mo(u)ld
塑料棒 plastic rod
塑料包层光纤 plastic clad optic(al) fiber
塑料包层石英光纤 plastic-clad silica-(optic) fiber[fibre]
塑料包封线圈 plastic encapsulated coil
塑料包埋 plastic embedding
塑料薄板 plastic sheet
塑料薄壳 plastic shell
塑料薄膜 plastic film; plastic foil; plastic membrane; plastic sheeting
塑料薄膜包线 plastic film covered wire; plastic film covered wire
塑料薄膜包装 plastic film package
塑料薄膜覆盖 plastic film mulching
塑料薄膜覆盖机 plastic-laying machine
塑料薄膜护罩 shrink film
塑料薄膜基底 plastic foil base
塑料薄膜揭除机 plastic remover
塑料薄膜屋顶 plastic film for roofing
塑料薄膜养护 plastic film curing
塑料薄膜粘贴机 laminating machine
塑料薄片 plastic flake; plastic foil; plastic tab
塑料保持架 plastic cage
塑料保护涂层 plastic protective coating
塑料被覆管 plastic-coated tube
塑料被盖 cocoon
塑料泵 plastic pump
塑料边框板 plastic framed plaque
塑料编织袋 plastic woven bag
塑料变性记录仪 plastograph
塑料表面薄膜 plastic surface film
塑料波纹板 corrugated plastic sheeting
塑料波纹瓦 plastic corrugated tile
塑料波形瓦 plastic corrugated tile
塑料玻璃 plastic glazing; pollopas; resin glass
塑料玻璃窗 plastic window light
塑料玻璃纤维管 plastic and glass fiber pipes
塑料玻璃纤维加劲材 plastic-glass fiber reinforcement
塑料钵 polypots
塑料布 plastic cloth; plastic textile
塑料布青贮器 plastic silo
塑料部件 plastic component; plastic section
塑料采光设备 plastic light fixture
塑料采光圆屋顶 plastic(ceiling) light cupola
塑料采光装置 plastic light fitting; plastic luminaire(fixture)
塑料槽 plastic channel(section)
塑料槽车 plastic tanker
塑料草绳 plastic grass
塑料测定法 plastometry
塑料测斜管 plastic tube for incline measuring
塑料层压板 laminated plastics; plastic laminate
塑料层压避热层 plastic-laminated heat shield
塑料层压钢 plastic-laminating of steel
塑料层压胶合板 plastic-laminated plywood
塑料层压热防护层 plastic-laminated heat shield
塑料层压纤维面板 plastic-laminated surfaced hardboard
塑料插板式侧面风口 plastic lateral air opening with damper
塑料插入管件 plastic insert fitting
塑料掺和剂 plastic admix(ture)
塑料厂 plastic factory
塑料厂废水 plastic factory wastewater
塑料厂废物 plastic factory wastes
塑料车顶 plastic car roof
塑料车身 plastic body
塑料车身小客车 plastic-bodied car
塑料车厢 plastic body
塑料衬 plastic lining
塑料衬垫 plastic gasket
塑料衬金属管 plastic-lined metal pipe
塑料衬金属管件 plastic-lined metal fitting
塑料衬里 plastic liner; plastic lining
塑料衬里钢管 plastic-lined steel pipe
塑料衬片接头 plastic gasket joint
塑料衬砌 plastic lining
塑料成型 shaping of plastics
塑料齿轮 plastic gear; silent gear
塑料船 plastic boat; synthetic(al) boat
塑料窗 plastic framed window; plastic window
塑料窗衬垫 plastic window gasket
塑料窗框 plastic window frame
塑料窗剖面图 plastic window section
塑料窗纱 plastic window yarn
塑料窗外形 plastic window section
塑料锤 plastic hammer
塑料(瓷)砖 plastic tile
塑料磁盘 diskette
塑料淬火剂淬火 plastic quench
塑料打印色带 plastic printing ribbon
塑料带 plastic tape
塑料带排水井 drainage wick
塑料带竖向排水 prefabricated vertical drain; wick drain
塑料袋 plastic bag; plastic sack
塑料袋青贮 plastic bag silage
塑料袋橡胶 poly bag rubber
塑料袋育苗 polythene bagged plant
塑料单体 plastic monomer
塑料单宅房屋 plastic house
塑料挡水板 plastic flashing piece
塑料导线管 plastic conduit
塑料导向器 plastic guide
塑料道钉垫片 plastic screw spike washer
塑料道路条纹 plastic roadline
塑料的 plastic

S

塑料的固性极限 plastic yield-point
塑料的抗磨能力 abrasion resistance of plastics
塑料的抗磨性 abrasion resistance of plastics
塑料的老化 ag(e)ing of plastic material
塑料的模塑涂层 mo(u)ld coating of plastics
塑料的应力-应变曲线 flow curve
塑料灯檐 plastic visor
塑料底座镜 plastic-base mirror
塑料地板 plastic floor
塑料地板材料 flexible flooring
塑料地板铺设 plastic floor covering; plastic flooring
塑料地面 plastic floor(ing)
塑料地图 plastic map
塑料地形模型 plastic embossed model
塑料电镀 electroplating on plastics; plating on plastics
塑料电缆夹 plastic cable clip
塑料电缆终端盒 plastic cable end box
塑料电木盒 mo(u)lded bakelite case
塑料电容器 plastic capacitor
塑料电位计 plastic potentiometer
塑料电线 vinyl insulated wire
塑料垫 plastic cushion
塑料垫片 plastic gasket
塑料垫圈 plastic gasket; plastic washer
塑料垫条 <混凝土路面纵缝拉杆下的> plastic cushion(of tie bar)
塑料叠合板 plastic laminate
塑料碟形穹隆 plastic ceiling saucer dome
塑料碟状穹隆 plastic saucer dome
塑料丁字尺 plastic T-square
塑料顶棚 plastic ceiling
塑料顶棚粉饰 plastic ceiling plaster
塑料定位器 <钢筋用> plastic spacer
塑料堵漏 grouting with plastics
塑料渡槽 plastic flume
塑料断面模数 plastic section modulus
塑料发白 clouding
塑料阀 plastic valve
塑料方形风阀 square plastic damper
塑料防潮材料 amazu
塑料防尘堵头 plastic dust-proof plug
塑料防护软管 plastic sock
塑料防水 plastics(-coated) waterproofing
塑料防水膜 plastic waterproof membrane
塑料防水片材 plastic waterproofing sheet
塑料飞机 plastic airplane
塑料废物 plastic waste
塑料分隔板条 plastic divider strip
塑料分散体 plastisol
塑料焚烧炉 incinerator for plastics; plastic incinerator
塑料粉 mo(u)lding powder; synthetic(al) resin mo(u)lding compound
塑料粉末磨塑成型 powder plastics mo(u)lding
塑料粉末喷涂 plastispray
塑料粉球 plastic microballs
塑料风管及附件 plastic duct and fittings
塑料风扇 plastic fan
塑料封闭混合剂 plastic sealing compound
塑料封闭剂 plastic sealing compound
塑料封壳 plastic capsule
塑料封装 plastic capsulation; plastic casing; plastic packaging
塑料敷层 plastic coating
塑料敷面 rubber-based coating
塑料扶手 plastic handrail
塑料浮标 plastic buoy
塑料浮子 plastic float
塑料符号 plastic numeral
塑料辐射防护措施 plastic radiation shielding wall
塑料辐射隔离器 plastic biological shielding wall
塑料复合薄钢板 plastic clad composite steel sheet
塑料复合材料 plastic composition
塑料复型 plastics replica
塑料复制版浇铸机 plastic plate mo(u)lding machine
塑料覆盖薄膜铺放机 plastic layer
塑料覆盖层 plastic cladding; plastic cover(ing)
塑料覆面 plastic overlay
塑料覆面带材 plastic-coated strip
塑料改性焦油油膏 plastic modified coal tar ca(u)lking compound
塑料盖 plastic cover
塑料盖层 plastics coating
塑料钢层板 plastic-steel laminate
塑料钢层压板 plastic-laminating of steel
塑料钢组合 plastic-steel combination
塑料格片顶棚 plastic eggcrate ceiling
塑料隔墙 plastic partition(wall)
塑料隔热板 plastic heat shield
塑料隔声材料 plastic insulation
塑料工厂 plastic factory
塑料工业 plastic industry
塑料工业废水 plastic industry wastes
塑料工作帽 plastic helmet
塑料构件 plastic element
塑料骨料 plastic aggregate
塑料固化 plastic solidification
塑料管 plastic pipe [piping]; plastic tube[tubing]; silent stock tube[tubing]
塑料管材 plastics pipe; plastic tubing
塑料管道系统 plastic piping system
塑料管件 plastic pipe fittings
塑料管接 plastic pipe joint
塑料管接头 tube couples
塑料管壳 plastic case
塑料管帽 plastic end cap; plastic pipe cap
塑料管蠕变 plastic pipe creep
塑料管子 pipe plastics; plastic pipe
塑料罐 plastic tank
塑料光栅 plastic grating
塑料光纤 plastic optic(al) fiber[fibre]
塑料光学 plastic optics
塑料光学纤维 plastic optic(al) fiber [fibre]
塑料轨道 plastic rail
塑料轨条 plastic rail
塑料轨枕垫板 plastic sleeper bearing plate; plastic sleeper shim; plastic tie-plate
塑料过滤器 plastic filter
塑料过滤器介质 plastic filter medium
塑料海绵底层 plastic sponge underlay
塑料焊接 plastic welding
塑料焊枪 welding pistol
塑料合缝钉 plastic dowel
塑料合金 plastic alloy
塑料和聚合物废物 plastic and polymer waste
塑料和树脂废水 plastic and resin waste(water)
塑料盒 mo(u)lded bakelite case; plastic case; plastic casing
塑料糊 plastic paste
塑料护轨 plastic guard rail
塑料护栏 plastic guard rail
塑料护面 plastic coating
塑料护皮 plastic sheath
塑料护墙板 plastic panel; sheeting plastics
塑料护套电缆 plastic-sheathed cable
塑料护套光缆 optic(al) fiber[fibre] cable with plastic jacket
塑料缓冲器 plastic bumper
塑料换向器 plastic commutator
塑料混凝土 plasto-concrete
塑料混凝土掺和料 elastomer
塑料火焰抑制剂 flame-retardant of plastics
塑料火焰预处理 flame-pretreatment
塑料机壳 cabinet
塑料机身 plastic fuselage
塑料机械 plastics machinery
塑料机翼 plastic wing
塑料基板 plastic base
塑料基质 plastic matrix
塑料级滑石 plastic-grade talc
塑料集料 plastic aggregate
塑料集装箱 plastic container
塑料挤出机 plastic extruder
塑料挤瓶 <挤压时能排出所装之物> squeeze bottle
塑料挤压机 plastic squeezer
塑料记性 plastic memory
塑料剂量计 plastic dosimeter
塑料继电器室 plastic relay house
塑料加工设备 plastic processing equipment
塑料加固的导爆索 plastic reinforced primacord
塑料夹 <钢筋定位用> plasclip
塑料夹层结构 plastic sandwich structure
塑料夹层系统 plastic sandwich system
塑料夹子 plastic grip
塑料家具 plastic furniture
塑料间隙规测定法 plastic filler method
塑料建筑 plastic construction
塑料胶 plastic cement
塑料胶带 adhesive plastic strip
塑料胶合的 plastic-laminated
塑料胶片 plastic film
塑料铰接法 plastic hinge method
塑料铰链 plastic hinge
塑料接头垫片 plastic preformed gasket joint
塑料接头垫圈 plastic preformed gasket joint
塑料结构 plastic construction; plastic structure
塑料结构系统 plastic structural system
塑料介质 plastic media
塑料金属粉末制品 plastic-metal powder product
塑料金属喷镀 plastic plating
塑料金属陶瓷复合材料 plastic-metal-ceramic composite
塑料金属轴承合金 plastic-metal bearing alloys
塑料金字塔 plastic pyramid
塑料浸涂层 plastic dip coating
塑料径迹探测器 plastic track detector
塑料救生筏 plastic liferaft
塑料救生浮 synthetic(al) lifefloat
塑料救生圈 synthetic(al) lifebuoy
塑料救生艇 plastic lifeboat
塑料救生衣 synthetic(al) lifejacket
塑料矩形空气分布器 rectangular plastic register
塑料聚焦光纤 plastic focusing fibre
塑料卷材地面 plastic sheet flooring
塑料卷帘百叶窗 plastic roller shutter
塑料卷升式升降门 plastic up-and-over door of the roll-up type
塑料绝热材料 plastic insulation
塑料绝缘 plastic insulation
塑料绝缘材料 plastic insulating material; plastic insulation; plastic insulation material
塑料绝缘电缆 plastic insulated cable
塑料绝缘电线 plastic insulated wire cord; thermoplastic-covered wire
塑料绝缘管 plastic insulating tube
塑料绝缘胶带 plastic insulating tape
塑料绝缘线 plastic insulated wire
塑料铠装电缆 plastic sheath cable
塑料抗蚀剂 plastic resist
塑料壳体 plastic casing
塑料空心地板填料 plastic hollow floor filler
塑料块地面 plastic tile flooring
塑料框 plastic frame
塑料框窗 plastic framed window
塑料垃圾袋 plastic refuse bag; plastics waste sack
塑料拉链式蝶阀 plastic butterfly valve with chain
塑料拉手柄 <门窗上用的> plastic pull handle
塑料蓝图 plastic print
塑料老化 ag(e)ing of plastics
塑料篱笆 plastic fence
塑料立体地图 embossed plastic relief map; plastic relief map
塑料立体模型 plastic relief model
塑料立体图 plastic relief map
塑料立柱 plastic column
塑料立锥体 plastic pyramid
塑料连接 plastic bonding
塑料连接垫片 plastic structural gasket joint
塑料连接垫圈 plastic structural gasket joint
塑料磷光体 plastic phosphor
塑料零件 plastic parts
塑料零配件 plastic fittings
塑料流点 plastic yield-point
塑料流动性 flowability of plastic
塑料楼梯扶手 plastic stairrail
塑料楼梯栏杆 plastic stairrail
塑料滤波器 plastic filter
塑料滤布 plastic filter cloth
塑料滤层布 plastic filter cloth
塑料滤料 plastic filter medium; plastic media
塑料滤料滤池 plastic media filter
塑料滤色镜 plastic filter
塑料滤水器 plastic filter
塑料轮胎的 plastic-tired[tyred]
塑料螺钉锚座 plastic screw anchor
塑料螺塞 plastic screw plug
塑料慢化堆 plastic-moderated reactor
塑料漫射光顶棚 plastic light-diffusing ceiling
塑料门 plastic door
塑料门窗横挡 plastic rail
塑料门窗冒头 plastic rail
塑料门挡 plastic stopper
塑料门拉手柄 plastic door pull handle
塑料门扇 plastic door leaf
塑料密封 plastic seal
塑料密封材料 plastic sealant; plastic sealing material
塑料密封垫片 plastic sealing gasket
塑料密封垫圈 plastic sealing gasket
塑料密封膏 plastic sealant
塑料密封环 plastic sealing ring
塑料密封混合剂 plastic sealing compound
塑料密封剂 plastic sealing compound
塑料密封胶 plastic sealant
塑料密封圈 plastic packing ring
塑料面层 plastic finish

塑料面扶手 plastics covered handrail
塑料面胶合板 plastic-faced plywood
塑料面模具 plastic-faced die
塑料面砖 plastic;plastic furring tile;plastic tile
塑料模 plastic mo(u)ld
塑料模板 plastic formwork;plastic mo(u)ld;plastic shuttering
塑料模壳 plastic shuttering
塑料模塑涂装法 in-mo(u)ld coating of plastics
塑料模型 plastic model;plastic mo(u)ld;plastic pattern;plastic shape
塑料模压封装 plastic mold package
塑料模制冷却系统 plastic mo(u)ld cooling system
塑料膜 plastic coating;plastic film;plastic foil
塑料膜电容器 plastic film capacitor;plastic film condenser
塑料膜敷层 plastic film coating
塑料尼龙水栅 plastic-nylon boom
塑料黏[粘]结剂 plastic binder;plastic cement
塑料黏[粘]结料 plastic cement
塑料黏[粘]结炸药 plastic-bonded explosive
塑料镊子 plastic tweezer
塑料排泥管线 plastic delivery pipeline
塑料排水板 plastic drain;plastic drain board
塑料排水板处理 foundation area improved by plastic drain
塑料排水带 plastic drain
塑料排水管 plastic drainage pipe;wick drain
塑料排水管敷置机 plastic drain layer
塑料盘 vinyl disc[disk]
塑料抛光 plastic polishing
塑料跑道 plastic track
塑料泡沫 foamed plastics;plastic foam
塑料配件 plastic fittings
塑料喷水池 plastic fountain basin
塑料喷涂 cocooning;plastics spray coating
塑料喷涂机 spray plastic machine
塑料喷嘴 plastic nozzle
塑料披盖 cocooning
塑料皮电缆 plastic cable;plastic sheath cable
塑料片 plastic plate;plastic sheet
塑料片叠层地形模型 plastic-mounted lamination
塑料片基 plastic base
塑料片基抗划力 plastic jamming pressure
塑料片刻图 plastic engraving;plastic scribing;scribing on plastic
塑料片排水 <代替砂井处理软土地基> band drain
塑料铺面 plastic facing
塑料漆 plastic paint
塑料器具 plastic ware
塑料铅字 plastic letter
塑料浅穹隆 <天窗采光用> plastic saucer dome
塑料嵌入式护墙板 plastic lay-in panel
塑料墙板 plastic siding;plastic wall panel
塑料墙面 plastics wall finish
塑料墙面涂料 plastic wall cover(ing)
塑料墙纸 plastic wall paper;wall covering
塑料青贮器 plastic silo
塑料穹隆 plastic dome
塑料穹隆采光 plastic ceiling dome light
塑料穹隆顶天窗 plastic domed roof-light
塑料圈装订本 plastic binding
塑料燃烧法 plastic combustion method
塑料热敏电阻器 plastic thermistor
塑料热释光探测器 plastic thermoluminescence detector
塑料容器 plastic container
塑料溶胶 plastisol
塑料蠕变 creep of plastics
塑料乳胶 plastic emulsion
塑料乳胶地面 plastic emulsion flooring
塑料乳胶漆 plastic emulsion paint
塑料软管 plastic hose
塑料塞子 plastic stopper
塑料伞形风帽 plastic cowl
塑料散光罩 plastic diffuser
塑料色层光纤 plastic clad fiber[fibre]
塑料色料起霜 colo(u)r chalking of plastics
塑料砂 plastic sand
塑料砂袋 sand-filled plastic bag
塑料砂浆 plastic plaster
塑料闪光块 plastic flashing piece
塑料闪烁计数器 plastic scintillation counter
塑料闪烁器 plastic scintillator
塑料闪烁体 plastic scintillant;plastic scintillator
塑料闪烁体探测器 plastic scintillator detector
塑料商品 plastic goods
塑料设备 plastic fixture
塑料渗透膜 plastic filter membrane
塑料饰面 plastic facing;plastic veneer
塑料饰面板 decorative plastics laminate;plastic decorative board;plastic veneer
塑料室 plastic chamber
塑料收缩裂缝 plastic shrinkage crack(ing)
塑料手锤 plastic hammer
塑料手套 plastic glove
塑料输水管道 plastic water pipeline
塑料树脂 plastic resin
塑料双层支管块 plastic dual lateral block
塑料双面模板 plastic match-plate
塑料水粉涂料 plastic water paint
塑料水管 plastic water pipe
塑料水龙带 plastic hose
塑料水溶性涂料 plastic cold water paint
塑料水溶性颜料 plastic water paint
塑料水性漆 plastic cold water paint
塑料丝 plastic wire;plastic filament
塑料四面体 plastic pyramid
塑料太阳蒸馏锅 plastic solar still
塑料弹簧锁 plastic latch
塑料套 plastic coating;plastic hood;plastic cover
塑料套管 plastic bushing;plastic sheath;plastic sleeve
塑料天窗 plastic roof-light;plastic skylight
塑料天沟 plastic gutter
塑料天蓬 plastic awning
塑料添加剂 plastic additive
塑料填料 plastic material;plastic stuffing
塑料填密 plastic packing
塑料填密环 plastic packing ring
塑料条板卷帘 plastic slatted roller blind
塑料调和漆 plastic distemper paint
塑料贴接面 plastics meeting face
塑料贴面 paper face overlay;plastic overlay;plastic-(sur)faced
塑料贴面板 board backed laminated plastic;laminated plastics;plastic veneer;decorative laminate
塑料贴面板透明保护层 cover sheet of plastics
塑料贴面的 plastics-coated
塑料贴面胶合板 plastic-faced plywood
塑料贴面犁 plastic plow
塑料贴面纤维板 plastic-faced hardboard
塑料贴面硬纸板 plastic-faced hardboard
塑料艇 plastic boat
塑料砼 <不用水泥而用高分子材料作为胶凝材料的> cementless concrete
塑料桶 plastic pail
塑料筒装置 plastic pit setter
塑料透镜 plastic lens
塑料透镜系统 plastic lens system
塑料凸缘 plastic lip(ping)
塑料涂层 plastic coat(ing);plastic cover;plastic membrane;plastic-surfaced
塑料涂层的 plastic-coated
塑料涂层纤维 plastic-coated fiber
塑料涂敷金属 metallization of plastics;plastic-coated metal
塑料涂覆的钢材 plastic-coated steel
塑料涂盖 cocooning
塑料涂料 plastic coating;plastic paint
塑料涂面纤维 plastic-coated fabric
塑料涂面织物 plastics coated fabric
塑料涂面纸 plastics coated paper
塑料瓦(片) plastic(made) tile
塑料外挂板 plastic structural cladding
塑料外壳 plastic sheath
塑料外皮电缆 plastic sheath cable
塑料外形 plastic profile
塑料网 plastic net
塑料网眼 screening
塑料围护结构 plastic structural cladding
塑料围栏 plastic rail
塑料温室 plastic greenhouse
塑料稳定剂 plastic stabilizer
塑料污染 plastic pollution
塑料屋顶板 plastic roof panel
塑料屋顶薄膜 plastic roof(ing) foil
塑料屋顶雨水槽 plastic roof shutter
塑料屋面薄板 plastic sheeting for roofing
塑料屋面薄片 plastic sheeting for roofing
塑料屋面片材 plastic roofing sheet
塑料洗涤器 plastic washer
塑料系统 plastic system
塑料细管 plastic straw
塑料纤管 mo(u)lded plastic pirn
塑料纤维 plastic fiber[fibre]
塑料纤维板 plastic-coated hardboard
塑料纤维布 plastic fiber[fibre] sheet
塑料纤维光学 plastic fiber[fibre] optics
塑料纤维面板 plastic-surfaced hardboard
塑料纤维网 plastic fiber[fibre] mesh
塑料线 plastic cord;plastic wire
塑料线间隙规 <测曲轴轴承游隙用的> plastiga(u)ge
塑料线烫订机 thread sealing machine
塑料镶板 plastic panel
塑料镶边 plastic lining;plastic lip(ping)
塑料橡胶混合物 polyblend
塑料楔形空气分布器 wedge-type plastic register
塑料泄水管 plastic drainage pipe
塑料芯 plastic core
塑料芯板 quilted plastic panel
塑料芯滤清器 plastic filter
塑料信号透镜 plastic signal lens
塑料型材 plastic section;plastic shape
塑料修饰 plastic trim
塑料蓄水池 plastic cistern
塑料悬浮 plastic suspension
塑料悬挂 plastic suspension
塑料压花机 plastic embossing machine
塑料压敏带 plastic pressure-sensitive tape
塑料檐槽 plastic gutter
塑料叶片 plastic fan
塑料异型材 plastic profile
塑料翼缘 plastic lip(ping)
塑料印刷 plastic printing
塑料印刷用油墨 plastic printing ink
塑料应变 plastic strain
塑料硬质高压板 plastic-laminated hardboard
塑料硬质纤维板 plastic-laminated hardboard
塑料用涂料 paint for plastics
塑料油膏 plastic caulk;plastic modified coal tar ca(u)lking compound
塑料油灰 plastic putty
塑料油箱 flexible bag;plastic fuel tank
塑料游泳池 plastic swimming pool
塑料有机黏[粘]土 plastic organic clay
塑料鱼尾板 plastic fishplate
塑料雨水槽 plastic rainwater gutter
塑料雨水斗 plastic rainwater outlet
塑料雨水附件 plastic rainwater articles
塑料雨水管件 plastic rainwater goods
塑料雨水屋檐槽 plastic rainwater gutter
塑料浴盆 plastic bath(tub)
塑料预制垫片 plastic preformed gasket
塑料预制垫圈 plastic preformed gasket
塑料圆花窗 plastic rose
塑料圆屋顶 plastic cupola
塑料圆锥体 plastic cone
塑料在园艺上的应用 plastics for horticultural use
塑料再生木材 plastic lumber
塑料闸瓦 composite brake shoe
塑料栅栏 plastic fence
塑料帐篷 plastic tent
塑料罩 plastic housing;plastic jacket
塑料整体件 plastic monoblock
塑料正面 plastic facade
塑料支座 plastic bearing
塑料织物 plastic fabric
塑料止水带 plastic water stop(ping strip)
塑料止水条 plastic water bar;plastic water stop(ping strip)
塑料纸 plastic paper
塑料指板 plastic finger plate
塑料制模 cut foundry pattern
塑料制品 plastic product;plastics
塑料制品成型机 plastic mo(u)lding press
塑料制造工 plastic manufacturer
塑料中的气泡 pocket air of plastics
塑料终饰 plastic finish
塑料种植法 plastoponic
塑料轴承 plastic bearing
塑料注射模型孔 plastic injection mo(u)lding machine
塑料贮水器 plastic cistern
塑料铸模 plastic mo(u)lding
塑料铸造齿轮 plastic mo(u)lded

gear
塑料桩 plastic pile
塑料装甲板 plastic armour plate
塑料装配槽 plastic mounting channel
塑料装配线 plastic mounting channel
塑料装饰板 decorative plastic board
塑料装饰层压板 decorative plastic laminate
塑料装饰嵌线(条) plastic mo(u)lding
塑料装置 plastic fixture
塑料阻燃剂 flame-retardant of plastics
塑料阻止器 plastic stopper
塑料座<钢筋定位用> plaschair
塑流 creepage;creeping;plastic flow;plastic yield;yielding flow
塑流比 creep ratio
塑流冰碛(土)【地】flow till
塑流点 plastic yield-point
塑流定律 law of plastic flow
塑流极限 limit of plastic flow
塑流黏[粘]度 plastic viscosity
塑流破坏 failure by plastic flow
塑流区 zone of plastic flow
塑流试验 plastic yield test
塑流损失 plastic flow loss
塑流条件 plastic flow condition
塑流现象 flow phenomenon
塑流线 line of creep
塑流型滑坡 flow slide
塑流学 rheology
塑流应力 flow stress
塑流褶曲【地】flow fold
塑流值 plastic yield value
塑面线条<模板上的> form liner
塑模 die;mo(u)ld;wooden mo(u)ld
塑模承套 die block
塑模型槽 die land
塑模旋转装置 die turning gear
塑泥 plasticine
塑黏[粘]弹性体 plasto-visco-elastic body
塑黏[粘]性 plastic viscosity
塑坯模制法 preform mo(u)lding
塑坯预塑 preform
塑溶胶 plastisol
塑饰面料 plastic veree
塑态 plastic state
塑态焊接 plastic welding
塑态沥青材料 plastic asphaltic material
塑态水泥 plastic cement
塑弹平衡状态 state of plastoelastic equilibrium
塑弹形变 plastoelastic deformation
塑弹性 plastoelasticity
塑弹性变形 plastoelastic deformation
塑弹性材料 plastoelastic composition
塑弹性的 plasto-elastic
塑弹性合成物 plastoelastic composition
塑弹性密封层 plastoelastic sealant
塑弹性物 plastelast
塑弹性物质 plastoelastic composition
塑体 plastomer
塑体的隐匿气孔 boil
塑物脱膜 knockout
塑限 limit of plasticity;plastic limit
塑限荷载 plastic limit load
塑限设计 plastic limit design
塑限试验 plastic limit test
塑限图 plasticity chart
塑限下限<土壤含水量指标> lower plastic limit
塑限状态<土的> plastic limit state
塑像 statue
塑像台座 statue pedestal
塑像用青铜 statuary bronze
塑形光 accent light
塑型 mo(u)lding;plastotype
塑型模 mo(u)lding-die
塑型泥 model(l)er's clay
塑型云母板 mo(u)lding micanite
塑型照明 accent lighting
塑型者 mo(u)lder
塑型铸造 full-mo(u)ld(cavityless) casting
塑性 ductility;plasticity;plastic nature;plastic property
塑性安全炸药 security explosive of plasticity
塑性白漆 white plastic
塑性保持率 plasticity retention percentage
塑性保持指数 plasticity retention index
塑性比 plasticity ratio
塑性边界 plastic bound
塑性变位 plastic deflection
塑性变形 after flow;creep;irrecoverable deformation;irreversible deformation;plastic deflection;plastic deformation;plastic flow;plastic strain;plastic yield(ing);plastometric(al) set;scuffing;yielding
塑性变形的极限表面 limiting surface of yielding
塑性变形点 set point
塑性变形记录仪 plastograph
塑性变形区 plastically deforming area
塑性变形区范围 plastic deformation area
塑性变形区域 plastic deformation zone
塑性变形区最大深度 maximum depth of plastic deformation area
塑性变形曲线描记仪 plasticorder;plastograph
塑性变形时间图 yield-time diagram
塑性变形图描记器 plastograph
塑性变形线 yield line
塑性变形应力 flow stress;plastic deformation stress
塑性变质 plastic metamorphism
塑性波 plastic wave
塑性玻屑 plastic-vitreous fragment
塑性不稳定性 plastic instability
塑性材料 plaster material;plastic material
塑性测定法 plastometry
塑性层 ductile bed
塑性常数 plastometer constant
塑性沉淀裂缝 plastic settlement crack
塑性沉降 plastic set
塑性沉陷裂缝 plastic settlement crack
塑性成型法 plastic making
塑性承重结构 plastic bearing structure;plastic supporting structure;plastic weight-carrying structure
塑性稠度 plastic consistence[consistency];wet consistency[consistence]
塑性处理 ductile treatment
塑性锉 plastic file
塑性带 plastic zone
塑性导线管 plastic conduit
塑性的 plastic
塑性底座的 plastic-based
塑性地层 plastic formation
塑性地基 plastic ground
塑性地面 plastic ground
塑性地压 plastic earth pressure;plastic ground pressure
塑性垫板 plastic pad
塑性垫片材料 plastic shim material
塑性定理 plastic theorem
塑性定律 plasticity law
塑性定形 plastic setting
塑性冻土 plastic frozen soil
塑性度 degree of plasticity
塑性断口 plastic fracture
塑性断口百分率 shear fracture percentage
塑性断裂 plastic fracture
塑性断裂理论 plastic-fracturing theory
塑性断面 plastic section
塑性断面模量 plastic section modulus
塑性断面系数 plastic section coefficient
塑性范围 plasticity range;plastic range;plastic region;plastic zone
塑性范围内应力 plastic range of stress
塑性范围试验 plastic range test
塑性防潮材料 plastic barrier material
塑性非弹性 plastic inelasticity
塑性分散系 plastic disperse system
塑性分析 plastic analysis;plastic calculation
塑性分析法 analysis by plasticity;plastic method
塑性粉质黏[粘]土 plastic silty clay
塑性盖层 plastic coating
塑性干扰 plastic intrusion
塑性干缩裂缝 plastic shrinkage crack
塑性杆件 plastic bar;plastic member
塑性刚度 plastic stiffness
塑性刚度矩阵 plastic stiffness matrix
塑性高分子物质 plastomer
塑性高岭土 plastic kaolin
塑性隔绝材料 plastic barrier material
塑性各向异性 plastic anisotropy
塑性铬矿 plastic chrome ore
塑性功 plastic work done
塑性共振 plastic resonance
塑性构件 plastic component
塑性构思的 plastically conceived
塑性骨料 plastic aggregate
塑性固体<这种物体在它的屈服点发生之后才破裂> plastic solid
塑性管 plastic tube
塑性灌缝水泥浆 plastiment
塑性含水量<土的> plastic limit
塑性焊<如锻接压接等> plastic welding
塑性焊接 plastic welding
塑性合金 plastic metal
塑性河床护面 plastic riverbed lining
塑性荷载的承重结构 plastic load-bearing structure;plastic load-carrying structure
塑性荷载法 plastic-load approach
塑性后效 plastic after effect
塑性护面钢板 plastic-coated sheet steel
塑性花砖地面 polyvino
塑性滑动 plastic slip
塑性滑移<晶粒的> plastic flow;plastic slip
塑性化 fluxion;plastification;plastifying
塑性化纺织物 plasticised fabric
塑性化油 plastifying oil
塑性划线材料<道路> plastic making material
塑性灰浆 plastic mortar
塑性灰土护岸 plastic soil cement lining
塑性恢复 plastic recovery
塑性恢复值 plasticity-recovery number
塑性回弹 plastic recoil;plastic resilience
塑性混合料 plastic mix
塑性混合物 plastic compound;Seelastic<一种嵌填框架和连接建筑材料的>
塑性混凝土<即坍落度大的混凝土> plastic concrete;quaking concrete;buttery concrete;cement bentonite concrete;plasticised[plasticized] concrete;pumpable concrete;wet concrete;workable concrete
塑性混凝土防渗墙 plastic concrete cut-off wall
塑性混凝土混合料 plastic mixture;wet concrete mix
塑性混凝土配筋定位件 plastic reinforcement spacer
塑性混凝土配筋间隔块 plastic reinforcement distance piece
塑性混凝土配筋间隔物 plastic reinforcement spacer
塑性混凝土柱 quaking concrete column
塑性极限 limit of plasticity;plastic limit
塑性极限荷载 plastic limit load
塑性极限设计 plastic limit design
塑性极限设计理论 plastic theory of limit design
塑性集料 plastic aggregate
塑性挤出法 soft extrusion
塑性计 phastograph;plasticimeter;plastometer
塑性计算 plastic calculation
塑性记录媒质 plastic recoding medium
塑性记忆 plastic memory
塑性加工 plasticity processing;plastic working
塑性夹 plastic clip
塑性剪切 plastic shear
塑性剪切模量 plastic shear modulus
塑性建筑材料 plastic building material;plastic construction material
塑性建筑构件 plastic building component
塑性胶 plastic glue
塑性胶粘剂 plastic bonding adhesive;plastic cementing agent
塑性胶砂 plastic mortar
塑性铰 plastic hinge
塑性铰长度 plastic hinge length
塑性铰机构 hinge mechanism
塑性铰区 plastic hinge zone
塑性铰位置 plastic hinge position
塑性铰线 plastic hinge line
塑性铰转动 hinge rotation
塑性阶段 plasticity range;plastic stage
塑性阶段的承重结构 plastic loaded structure
塑性阶段破坏 plastic stage collapse
塑性阶段设计法 plastic method
塑性阶段弯矩 plastic moment
塑性结持度 plastic consistency
塑性结构 plastic structure
塑性截面模量 plastic section modulus
塑性界限 plastic bound;plastic limit
塑性矩心<钢筋混凝土设计的> plastic centroid
塑性绝缘体 plastic insulator
塑性开裂 plastic cracking
塑性抗矩 plastic reactance moment
塑性抗力 plastic reactance
塑性抗弯强度 plastic bending strength
塑性苦土 plastic magnesia
塑性框架 plastic frame
塑性扩散现象 plastic dispersion
塑性拉伸 plastic yield
塑性冷黏[粘]合剂 plastic cold bonding agent
塑性理论 plastic theory;theory of plasticity
塑性理论设计 plastic theory design;collapse design
塑性力矩 plastic moment
塑性力矩分配法 plastic moment dis-

tribution method
塑性力学 mechanics of plasticity; plastic mechanics
塑性沥青膏 plastic asphalt cement
塑性沥青胶泥 plastic asphalt cement
塑性沥青结合料 <一种专卖的用于防水面层的> Creetex
塑性粒料 plastic aggregate
塑性梁弯曲 plastic beam bending
塑性裂纹 <指混凝土> plastic cracking
塑性流 plastic flow; plastic yield
塑性流变 plastic flow
塑性流动 plastic flow; plastic yield; yield flow
塑性流动变形 plastic flow deformation
塑性流动传质机理 material transfer by plastic flow
塑性流动带 zone of plastic flow
塑性流动方程 plastic flow equation
塑性流动极限 flow limit
塑性流动曲线 plastic flow curve
塑性流动值 flow limit
塑性流度 plastic fluidity
塑性流幅 plastic flow range
塑性流理论 theory of plastic flow
塑性流模型 plastic flow model
塑性流体 plastic fluid
塑性流体动压润滑 plasto-hydrodynamic(al) lubrication
塑性路面 plastic surface
塑性帽分量 plastic cap component
塑性煤 plastic coal
塑性猛烈炸药 high-explosive plastic; plastic explosive
塑性密封层 plastic sealant
塑性模量 modulus of plasticity; plastic(ity) modulus
塑性摩擦 plastic friction
塑性木粉膏 plastic wood
塑性木粉浆 plastic wood
塑性耐火材料 fireclay plastic refractory; plastic refractory
塑性耐火黏[粘]土 bond fireclay; plastic bond fireclay; plastic fireclay; plastic refractory clay
塑性挠度 plastic deflexion
塑性挠曲 creep deflection; creep deflexion; inelastic bending
塑性能 plastic energy
塑性能量条件 yield condition
塑性泥团 plastic paste
塑性黏[粘]度 plastic stickness; plastic viscosity
塑性黏[粘]度系数 coefficient of plastic viscosity
塑性黏[粘]合料 plastic bonding adhesive
塑性黏[粘]接料 plastic cement
塑性黏[粘]结剂 plastic adhesive; plastic cementing agent
塑性黏[粘]结料 plastic cement
塑性黏[粘]结乳剂 plastic bonding emulsion
塑性黏[粘]磐土 plastic clay pan soil
塑性黏[粘]土 plastic clay
塑性黏[粘]土碾磨机 plastic clay grinding mill
塑性黏[粘]性土 plastic cohesive soil
塑性黏[粘]滞流动 <即土的次固结> plasto-viscous flow; plastic-viscous flow
塑性黏[粘]滞流动理论 theory of plastic-viscous flow
塑性凝胶 plastigel[plastogel]
塑性凝胶模 plastigel mo(u)ld
塑性凝结 plastic set
塑性扭力 plastic torsion
塑性扭曲 inelastic torsion
塑性泡沫混凝土 plastic foam concrete
塑性泡沫塑料 plastic foam
塑性盆地理论 plastic though theory
塑性疲劳 plastic fatigue
塑性疲劳破坏 plastic fatigue rupture
塑性平衡 plastic equilibrium
塑性平衡主动状态 active state of plastic equilibrium
塑性平衡状态 state of plastic equilibrium
塑性平台 plastic platform
塑性破断 ductile fracture; plastic fracture
塑性破坏 plastic collapse; plastic failure
塑性破坏荷载 plastic collapse load
塑性破坏理论 plastic theory of failure
塑性破裂 plastic fracture; plastic rupture
塑性漆 plastic paint
塑性铅 plastic lead
塑性铅青铜 plastic bronze
塑性强度 plastic strength
塑性亲液溶胶 plastic lyophilic sol
塑性青铜 plastic bronze
塑性区 plastic range; plastic region; plastic zone
塑性区尺寸 dimension of plastic zone
塑性区法 plastic zone method
塑性区扩大 extension of plastic zone
塑性区最大深度 maximum depth of plastic zone
塑性屈服 plastic yield(ing)
塑性屈服点 plastic yield-point
塑性屈服试验 plastic yield test
塑性屈服应力 plastic yield stress
塑性屈服值 plastic yield value
塑性屈曲 non-elastic buckling
塑性全量理论 total theory of plasticity
塑性容器 plastic container
塑性溶胶 plastic sol; plastisol
塑性溶胶涂布 plastogel coating
塑性溶液 plastic solution
塑性乳剂 plastic emulsion
塑性软磁盘机 plastic flexible disc
塑性润滑剂 semi-solid lubricant
塑性砂浆 plastic mortar; wet mortar
塑性上限 upper limit of plasticity; upper plastic limit; upper yield point; upper yield point of plasticity
塑性上限顶点 upper plastic limit
塑性上限含水量 upper plastic moisture content limit
塑性设计 limit state design
塑性设计法 plastic design method
塑性伸长 plastic elongation
塑性失稳 plastic buckling; plastic instability
塑性失效准则 plastic failure criterion
塑性湿度 mo(u)lding moisture content; mo(u)lding water content
塑性石板 plastic slate
塑性石灰砂浆 wet lime mortar
塑性石蜡 plastic wax
塑性势 plastic potential
塑性势面 plastic potential surface
塑性试验机 plastic testing machine
塑性试针 plasticity needle
塑性收缩 plastic shrinkage
塑性收缩裂缝 hairline crack(ing); plastic shrinkage crack(ing)
塑性树脂基铺成的无缝地面 plastic resin-based jointless cover(ing)
塑性树脂胶 plastic-resin glue
塑性水 plasticity water; water of plasticity
塑性水泥 plastic cement
塑性水泥浆 plastic cement grout
塑性水泥冷上釉 plastic cement cold glaze
塑性水泥涂釉面 plastic vitreous surfacing
塑性水泥土 plastic soil cement
塑性水泥土混合衬砌 plastic soil cement lining
塑性碎屑 plasticlast
塑性损失 plastic loss
塑性缩裂 plastic shrinkage crack(ing)
塑性态 plastic state
塑性弹性的 plasto-elastic
塑性弹性平衡状态 state plastic-elastic equilibrium
塑性淌度 plastic mobility
塑性套管 plastic sheathing
塑性特征 plastic behavio(u)r; plastic nature; plastic property
塑性体 plastic body; plastomer
塑性体改性沥青 plastic modified asphalt; plastomer modified asphalt
塑性体改性沥青油毡 plastic modified asphalt membrane; plastomer modified asphalt membrane
塑性体积应变 plastic volumetric(al) strain
塑性体静力学 plasticostatics
塑性填充物 <用于填塞木缝孔洞的> plastic filler
塑性填料 plastic filler; plastic wood
塑性条件 condition of plasticity; plasticity condition
塑性调整 <路面长期荷载产生变形后的> plastic adjustment
塑性图 diagram of forgeability behavio(u)r; plasticity chart
塑性涂层 plastic coating
塑性涂料 plastic coating; plastic painting
塑性土 plastic clay; plastic soil
塑性土力学 soil plasticity
塑性团块 gob
塑性推进剂 plastic propellant
塑性弯沉 plastic deformation
塑性弯矩 plastic bending moment; plastic moment
塑性弯曲 plastic bend(ing)
塑性弯曲顶板 plastically flexible roof
塑性弯曲量 plastic compliance
塑性围岩 plastic ground
塑性位能 plastic potential
塑性温度范围 plastic temperature range
塑性稳定理论 plastic stability theory
塑性稳定性 plastic(ity) stability
塑性物质 plastic mass; plastic material
塑性系数 coefficient of plasticity; plastic(ity) coefficient
塑性细粉 clay-like fines; plastic fines
塑性细颗粒 plastic fines
塑性下弯 plastic downbuckling
塑性下限 lower plastic limit
塑性下限含水量 lower plastic moisture content limit
塑性限度 plastic limit; plastic limit of soil
塑性限度试验 plastic limit test
塑性想象的 plastically conceived
塑性效应 plastic effect
塑性胁变 plastic strain
塑性行为 plastic behavio(u)r
塑性形变记录仪 plastograph
塑性形变理论 plastic deformation theory
塑性形心 plastic centroid
塑性修整 plastic finish
塑性徐变 plastic creep
塑性悬浮体 plastic suspension
塑性压曲 inelastic buckling; plastic buckling
塑性压制 plastic process
塑性延缓 plastic lag
塑性延伸 plastic elongation
塑性岩 incompetent rock
塑性岩层 flowing rock formation; plastic formation; plastic stratum
塑性岩石 plastic rock
塑性岩屑 plastic-detritus
塑性颜料 plastic pigment
塑性氧化镁 plastic magnesia
塑性液 plastic fluid
塑性液体 Bingham liquid; plastic fluid
塑性仪 plastometer
塑性应变 anelastic strain; plastic strain
塑性应变增量 increment of plastic strain
塑性应力 plastic stress
塑性应力重分布 plastic stress redistribution
塑性应力范围 plastic stress range
塑性应力分布 plastic stress distribution
塑性应力阶段 plastic stress range
塑性应力区域 plastic stress range
塑性应力应变关系 plastic stress-strain relation
塑性应力应变矩阵 plastic stress-strain matrix
塑性应力应变图 plastic stress-strain diagram
塑性硬黏[粘]土 plastic clay pan soil
塑性油驳 flexible oil barge
塑性有机黏[粘]土 plastic organic clay
塑性淤泥 plastic silt
塑性约束 plastic constraint
塑性杂土水泥 plastic soil cement
塑性造型 plastic form
塑性增厚 plastic thickening
塑性增量理论 incremental theory of plasticity
塑性炸药 plastic type explosive
塑性振动 plastic vibration
塑性支承 <指支柱反力均匀分布在板柱的交面上> plastic support
塑性值 plasticity index
塑性指标 plasticity index
塑性指数 index of plasticity; plasticity index; plasticity number
塑性制品 plastic article
塑性质心 plastic centroid
塑性滞后 plastic hysteresis
塑性中心 plastic centroid
塑性重心 plastic centroid
塑性砖坯 plastic process
塑性转变温度 ductility transition temperature
塑性转动 plastic rotation
塑性装饰 plastic decoration
塑性状态 condition of plasticity; inelastic behavio(u)r; plastic behavio(u)r; plasticity condition; plastic stage; plastic state; state of plasticity
塑性状态焊接 plastic welding
塑性状态时楔体 wedge in plastic state
塑性锥分量 plastic cone component
塑性阻力 plastic resistance
塑压 plastic compression
塑压成型 plastic pressing
塑压法 ram pressing
塑液阳联合试验 plasticity and liquidity test
塑造 figured; mo(u)lding
塑造的 plastic
塑造术 plastic arts
塑造者 model(l)er
塑罩 plastics cover
塑制纹理 texturing

S

溯河(产卵)的 anadromous

溯河洄游 anadromous migration; upstream migrant; upstream migration
溯河性鱼类 anadromous fishes
溯河鱼 anadromous fish
溯流(而上的) upstream
溯流回游 concrete saddle
溯上 run up
溯源 affiliation
溯源冲刷 backward erosion; headcutting; headwater erosion; regressive erosion; retrogressive erosion
溯源的 headward; retrogressive
溯源侵蚀 backward erosion; headward erosion; headwater erosion; retrogressive erosion
溯源性 traceability
溯源淤积 retrogressive deposition

酸败 rancidity; spoilage

酸败气味 rancid flavor
酸斑 acid stain
酸泵 acid pump
酸比 acid ratio
酸比重计 acidometer
酸不溶木质素 acid-insoluble lignin
酸不溶物 insoluble residue in acid
酸不溶物含量 insoluble substance content
酸残渣 acid residue
酸槽 acid bath; acid tank
酸测定 acidity test
酸沉降 acid deposition; acid precipitation
酸沉降槽 acid settler; acid settling tank
酸沉降池 acid settler
酸沉降监测 acid deposition monitoring
酸沉降评估模型 acid deposition assessment model
酸沉降器 acid settler
酸沉降体系 acid deposition system; acid precipitation system
酸橙 bitter orange; lime
酸橙绿 <暗黄绿色> lime green
酸橙油 lime oil
酸池 acid bath
酸处理 acidation; acidization; acidize; acidizing; acid treatment
酸处理的 acid-processed
酸处理法 acid polishing; acid treated method
酸处理粉饰 acid-treated plaster
酸处理剂 acidizer
酸处理抛光 acid polishing
酸处理松木 pickled pine
酸醇 acid alcohol
酸催化剂 acid catalyst
酸脆 acid brittleness
酸脆性 acid brittleness; acid embrittlement; hydrogen embrittlement
酸萃取 acid extraction
酸当量 acid equivalent
酸的 acid; sour
酸的质子理论 proton theory of acid
酸滴 acid droplets
酸滴定法 acidimetry
酸电解液 acid electrolyte
酸淀池 acid settling tank
酸定 acid cut
酸定量法 acidimetry[acidometry]
酸定量器 acid estimation apparatus
酸毒症 acidosis
酸度 acid degree; acidic content; acidity; acidness; degree of acidity; sourness
酸度变化 changes of acidity
酸度测定 acidity test
酸度常数 acidity constant
酸度的标准测定法 standard acid test
酸度的标准试验 standard acid test
酸度计 acetimeter; acidimeter; acidometer; pH meter
酸度检定 acidity test
酸度试验 acidity test
酸度误差 acidity error
酸度系数 acid factor; acidity coefficient; coefficient of acidity; oxygen ratio
酸度指标 acidity index
酸度指数 acidity index; index of acidity
酸对混凝土的影响 acid effect on concrete
酸发生消化 acidogenic digestion
酸法 acid system
酸法制浆 acid polishing; sulphite process
酸反应供热 acid heat
酸反应热 acid heat
酸防腐法 acid preservative method
酸防护 acid protection
酸废水石灰处理法 limestone treatment for acid wastewater
酸分解 decomposition with acid
酸分解法 acid decomposition method; acid splitting
酸分析仪 acid analyser[analyzer]
酸封漆 acid seal paint
酸浮秤 acidimeter
酸腐 rancidity
酸腐蚀 bite
酸酐 acid anhydride; anhydride
酸酐类固化剂 anhydrite curing agent
酸根 acid group; acid radical
酸固化 acid cure; acid curing
酸固化丙烯酸树脂系统 acid curing acrylic resin system
酸固化环氧树脂 acid-cured epoxy resin
酸固化清漆 acid-cured varnish
酸固化树脂 acid-cured resin
酸管 acid tube
酸灌丛樱 sourbush cherry
酸罐车 acid tank truck
酸过多 hyperacidity
酸含量 acid content; degree of acidity
酸耗量 acid consumption
酸虹吸管 acid siphon
酸化 acidate; acidation; acidising[acidizing]; acidize; acidulate
酸化八叠球菌 Sarcina acidificans
酸化层段 interval acidized
酸化池 stabilization pond
酸化处理 acidizing; acid treatment
酸化萃取法 acidification-extraction process
酸化法冷却水处理 acidic cooling water treatment
酸化反应动力学 acidifying reaction kinetics
酸化废水 acidized wastewater; pickling waste(water)
酸化粉刷处理 acid-washed plaster
酸化工艺 acidizing technology
酸化骨 acidulated bone
酸化硅酸钠 acid-treated sodium silicate
酸化过程 acidization
酸化剂 acid former; acidifier; acidizer; acidulant; acidulating agent
酸化井 sour well
酸化黏[粘]土 acid-treated clay
酸化器 acid former; acidifier
酸化软水 acidified soft water
酸化树脂 resin ester
酸化水 acidified water; acidulated water
酸化水解 acidification hydrolysis
酸化污染物输移与沉淀模型 transport and deposition of acidifying pollutant model
酸化物质 acidifying substance
酸化效应 acidification effect
酸化油 acidifying oil
酸化作用 acidification; acidization; acidulation
酸辉绿结构的 oxybasiophitic
酸回收 acid recovery; acid recyling
酸回收设备 acid reclaiming plant; acid-recovery plant
酸回收装置 acid reclaiming plant
酸活化 acid activation
酸或碱中毒 acid or alkali poisoning
酸基 acidic group; acid radical; acidyl
酸价 acid number; acid value
酸减量 acid loss
酸碱泵 acid-alkaline pump
酸碱测定法 acid-base determination
酸碱处理 acid and alkali treatment; acid-base treatment
酸碱催化 acid-base catalysis
酸碱代谢 acid-base metabolism
酸碱当量 equivalent weight of acid and base
酸碱的 acid-base
酸碱滴定法 acid-alkali titration; acid-base titration; neutralization titration
酸碱滴定检测器 acid-base titration detector
酸碱电池 acid-alkali cell
酸碱度 pH value; potential of hydrogen
酸碱度试验 acidity and alkalinity test
酸碱对 acid-base pair
酸碱法 pH-stat method
酸碱反应 acid-base reaction
酸碱废水 acid and alkaline wastewater; acid-base wastewater
酸碱分析仪 acid-base analyser[analyzer]
酸碱共轭偶 acid-base conjugate pair
酸碱关系 acid-base relationship
酸碱基 acid-base group
酸碱计 pH meter
酸碱络合反应 acid-base complexation reaction
酸碱灭火机 alkali acid extinguisher
酸碱灭火器 soda-acid extinguisher; soda-acid fire extinguisher
酸碱喷洒除臭 odo(u)r treatment by acid and alkali spray
酸碱平衡 acid-base balance; acid-base equilibrium
酸碱平衡反应 acid-base balance reaction; acid-base equilibrium reaction
酸碱污染 acid(and) base pollution
酸碱物质 acid and alkaline matter
酸碱洗液 caustic and acidic cleaning solution
酸碱相对平衡 relative balance between acid and alkali
酸碱性系数 acid-base ratio
酸碱液计量箱 acid-alkaline measurement device
酸碱值测定 pH value determination
酸碱指示剂 acid-base indicator; hydrogen ion indicator; neutralization indicator; pH-indicator
酸浆法 acid slurring
酸胶基 acidoid
酸焦油 acid tar
酸结合力的 oxidetic
酸解产物 acid hydrolysate
酸解固相物 digestion cake
酸解器 digester[digestor]
酸解作用 acidolysis; acyl exchange
酸浸 acid etching; acid pickling
酸浸槽 pickling tank
酸浸出 acid leach
酸浸法 acid leaching method; pickling process
酸浸废水 acid pickling wastewater; pickle wastewater; pickling waste(water)
酸浸废液 spent pickling solution; wastewater pickling
酸浸钢丝 pickled steel wire
酸浸剂 pickling agent
酸浸滤 acid leaching
酸浸时滞性试验 pickle lag
酸浸蚀 acid attack; acid etch
酸浸蚀的 acid-etched
酸浸酸 pickling acid
酸浸提液 acid extract
酸浸析 acid embossing
酸浸样品 acid digested sample
酸浸液 pickle liquor
酸浸渍 acid dipping; acid maceration
酸精制油 acid refined oil
酸净化系统 acid purification system
酸刻 acid mark
酸冷凝器 acid condenser
酸冷却器 acid cooler
酸离解常数 acid dissociation constant
酸离解反应 acid dissociation reaction
酸离气浮 acid separation air-floated
酸沥滤玻璃纤维 leached glass fiber[fibre]
酸量 amount of acid
酸量测定 acidimetric estimation
酸量滴定的 acidimetric
酸量滴定法 acidimetric method; acidimetry
酸量滴定分析 acidimetric analysis
酸裂化处理 acid cracking treatment
酸硫镍 nickelous sulfate
酸露点 acid dewpoint
酸露点腐蚀 acid dewpoint corrosion
酸滤 acid leach
酸模 Rumex acetosa; sour dock; sour grass; sour leek
酸磨法 acid milling
酸凝固法 acid coagulation
酸凝集 acid agglutination
酸凝清漆 acid-cured varnish
酸凝树脂 acid-cured resin
酸浓度 acid concentration
酸浓缩器 acid concentrator
酸抛光 acid polishing
酸泡沫 acid foaming
酸漂妥尔油 acid refined tall oil
酸漂亚麻仁油 acid-refined linseed oil
酸漂油 acid refined oil
酸瓶 carboy; acid bottle <测定钻孔偏差测量用的>
酸瓶车 carboy wagon; jar wagon
酸气 acid gas; sour gas <含多量硫化氢的天然气>
酸气吸收 acid gas absorption
酸气蒸化机 acid ager
酸强度 acid strength; strength of acid
酸侵蚀 acid attack
酸青贮料 acid silage
酸清洗 acid treatment
酸缺乏 anacidity
酸热试验 acid heat test
酸容器 acid container
酸溶传像束 leached image guide
酸溶的 acid-soluble
酸溶法 acid pasting
酸溶率 acid-soluble rate
酸溶木质素 acid-soluble lignin
酸溶液 acidic solution

酸溶油 acid-soluble oil
酸乳剂 acid emulsion; cation emulsion
酸涩的 tart
酸蚀 acid embossing; acid frosting; aciding; acid pickling; erosion; obscuring; pickle
酸蚀脆性 acid brittleness; pickle brittleness
酸蚀打印 acid badging
酸蚀法 acid etch method; aciding; acidizing
酸蚀法磨光 acid polishing
酸蚀法抛光 acid polishing
酸蚀粉饰 acid-etched plaster; acid-washed plaster
酸蚀粉刷 acid-washed plaster
酸蚀痕 acid mark; erosion mark
酸蚀后退火 acid annealing
酸蚀混凝土 etched concrete
酸蚀刻 acid etching; acid treatment
酸蚀刻的 acid-etched
酸蚀刻浮雕 acid embossing
酸蚀孔道 etched channel
酸蚀毛玻璃 acid-ground glass; etched glass
酸蚀商标 acid stamping
酸蚀试验 etch test
酸蚀条纹 <在玻璃表面进行的> acid streaking
酸蚀铁钉 acid-etched nail
酸蚀印记 acid badging; etching stamp; stamping
酸式 acidic
酸式草酸盐 acid oxalate
酸式醋酸钾 acid potassium acetate
酸式氟化钠 sodium acid fluoride
酸式焦磷酸钠 sodium acid pyrophosphate
酸式磷酸钠 sodium acid phosphate
酸式硫化钠 sodium acid sulfide
酸式硫酸盐 acid phosphate; bisulfate [bisulphate]
酸式砷酸钠 sodium acid arsenate
酸式碳酸铵 acid ammonium carbonate
酸式碳酸钾 potassium acid carbonate
酸式碳酸钠 acid sodium carbonate
酸式碳酸盐 acid carbonate; bicarbonate
酸式碳酸盐硬度 bicarbonate hardness
酸式亚硫酸钾 potassium bisulfite
酸式亚硫酸盐 acid sulfite [sulphite]; bisulfite[bisulphite]
酸式盐 acid salt; bisalt
酸数 acid number
酸水 acidulous water; acid water
酸水解 acid hydrolysis
酸塔 acid tower
酸坛 acid carboy; carboy; demijohn
酸提出物 acid extract
酸提取 acid extraction
酸土 acid soil; sour soil
酸土植物 acidophytes; acid plant; oxylophyte
酸味 sour odour; sour taste
酸污染 acid pollution
酸雾 acid fog; acid fume; acid mist
酸雾捕集器 acid mist eliminator
酸雾涤气器 acid-fume scrubber
酸雾洗涤器 acid-fume scrubber
酸雾液 acid fog liquid
酸吸收法脱氮 control of NO_x by absorption with acid
酸吸收器 acid absorber
酸洗 acid cleaning; acid dip; acid etch; acid pickling; acid wash (ing); dipping; etching; pickling
酸洗斑点 pickle patch; unpickled spot
酸洗薄板 pickled sheet; white finished sheet
酸洗薄钢板 acid pickling steel sheet; pickled sheet steel; sheet-metal free from oxides; spinkled steel sheet
酸洗不够 underpickling
酸洗残渣 smut
酸洗槽 descaling bath; dipping tank; pickling bath; pickling vat
酸洗车间 pickling department; pickling plant
酸洗池 pickling bath; pickling cell; pickling tub
酸洗冲洗水 pickling rinse water
酸洗脆性 corroding brittleness; pickling brittleness
酸洗底层 acid wash primer
酸洗法 pickle; pickling process
酸洗废水 acid cleaning waste(water); pickle liquor; pickle wastewater
酸洗废液 drag-out; pickling waste liquor; spent pickle liquor; spent pickling solution; waste pickle; acid washing liquor
酸洗粉饰 acid-washed plaster
酸洗粉刷 acid-washed plaster
酸洗缸 pickling vat
酸洗钢板 acid pickling steel plate; pickled plate
酸洗钢丝 pickled steel wire
酸洗工段 pickling bay; pickling zone
酸洗柜 acid bath
酸洗过的金属线 pickled wire
酸洗过度 overpickling
酸洗黑砂 acid-washed darkened sand
酸洗痕 pickle house scratch
酸洗缓蚀剂 restrainer
酸洗活性炭 acid-washed active carbon
酸洗机 pickling machine
酸洗机组 pickling line
酸洗剂 etch reagent; mordant; pickling agent
酸洗间 etching room
酸洗检查 pickling test
酸洗(浸)废水 pickling waste (water)
酸洗井 acidizing of wells; well cleaning with acid
酸洗筐 dipping basket
酸洗篮 dipping basket
酸洗泡 pickling blister
酸性熔岩 acid lava
酸洗色斑 pickle stain
酸洗设备 pickler
酸洗试验 acid washing test; corroding proof
酸洗损失 loss on acid washing
酸洗桶 pickling vat
酸洗污斑 pickle stain
酸洗消色试验 acid wash colo(u)r test
酸洗液 pickle(liquor); pickler; pickle solution
酸洗抑制剂 pickling inhibitor
酸洗用酸 pickling acid
酸洗油 acid-treated oil
酸洗装置 acid dip pickler; pickler; pickling unit
酸洗作业线 pickling line
酸显影 acid development
酸相 acid-phase
酸相升流厌氧污泥层系统 acid-phase upflow anaerobic sludge blanket system
酸相厌氧消化 acid-phase anaerobic digestion
酸形成 acid formation
酸性 acidic property; acidity; acidness
酸性铵矾 letovicite
酸性白土 acid clay; Fuller's earth
酸性不溶残渣 acid-insoluble residue
酸性材料 acid material
酸性残余物 acid residue
酸性残渣 acid residue
酸性测验比率 acid test ratio
酸性差色试验 acid wash colo(u)r test
酸性常数 acidic constant
酸性尘雾 acid fume
酸性沉积 acidic deposit(e)
酸性沉积物 acidic deposit(ion)
酸性沉降物 acid fallout; acid precipitation
酸性衬里 acid lining
酸性橙 acid orange
酸性池 sour oil
酸性冲天炉 acid lined cupola
酸性处理 acid cure; acid treatment
酸性处理整面法 <混凝土的> acid-treated finish
酸性脆裂 acid brittleness
酸性萃取剂 acidic extractant
酸性萃取物 acid extract
酸性淡红 <偶氮染料之一> acid carmoisine
酸性的 acidic
酸性滴定剂 acidic titrant
酸性地沥青 acidic asphalt
酸性地下水 acidic groundwater
酸性地下水羽 acidic groundwater plume
酸性电极 acidic electrode
酸性电解 acidic electrolysis
酸性电解液 acidic electrolysis bath
酸性电解质 acidic electrolyte bath
酸性电炉 acid electric(al) furnace
酸性电炉钢 acid electric(al) steel
酸性定影液 acid fixer; acid fixing bath; acid hypo
酸性蒽醌蓝 acid anthraquinone blue
酸性蒽棕 acid anthracene brown
酸性发酵 acid fermentation
酸性发酵成熟期 acid-ripening stage
酸性法 acidic method
酸性矾泉 acid alum spring
酸性反应 acid(ic) reaction; acidity reaction
酸性反应供暖 acid heat
酸性废水 acid drain; acid effluent; acid(ic) waste (water); pickling waste(water)
酸性废水的同化作用 acid waste assimilation
酸性废水污染 acid water pollution
酸性废污泥 acid waste sludge
酸性废物 acid(ic) waste
酸性废液 acid effluent; acid waste liquid; acid waste liquor
酸性废渣 acid sludge
酸性分 <罗斯特勒的沥青煤馏分分类中的一种> acidaffins
酸性分解 acid decomposition
酸性腐泥 moder
酸性腐蚀 acid attack; sour corrosion; acid corrosion
酸性腐殖化作用 acid humidification
酸性腐殖质 acid humus
酸性腐殖质化 acid humidification
酸性干气 sour dry gas
酸性钢 acid steel; Bessemer iron; Bessemer pig
酸性高锰酸钾氧化法 acid potassium permanganate oxidation method
酸性高锰酸钾指数法 acid potassium permanganate index method
酸性高锰酸盐氧化法 acid permanganate oxidation method
酸性高频感应电炉钢 acid high-frequency induction furnace steel
酸性铬黑 acid chrome black
酸性铬兰 acid chrome blue
酸性铬媒染料 acid chrome dye
酸性铬酸铜 <铜铬防腐剂> acid copper chromate
酸性铬盐 acid chrome salt
酸性骨料 acidic aggregate
酸性官能团 acidic functional group
酸性官能团含量 acidic functional group content
酸性硅石 acid silica
酸性含量 acidic content
酸性焊剂 acid flux(material)
酸性焊条 acid electrode
酸性河 acid river
酸性黑 acid black
酸性红 acid red
酸性红色淀 scarlet lake
酸性红色染料 ponceau
酸性湖泊 acid lake
酸性湖泊再酸化 acid lake reacidification
酸性湖泊再酸化模型 acid lake reacidification model
酸性花岗岩含稀有元素建造【地】acid granite rare element-bearing formation
酸性化合能力 acid combining capacity
酸性化合物 acidic compound
酸性环境 acid environment
酸性黄 Indian yellow
酸性挥发硫化物 acid volatile sulfide
酸性活度 acidic activity
酸性火成岩 acidic igneous rock
酸性甲酚红 acid cresol red
酸性坚膜定影剂 acid hardening fixer
酸性间胺黄 metanil yellow
酸性降解 acid degradation
酸性降水 acidic precipitation
酸性交换 acid exchange
酸性胶体 acidoid
酸性焦炭 acid coke
酸性焦油沥青油渣 acid tar
酸性介质 acid medium
酸性浸出 acid leaching
酸性浸出设备 acid leach plant
酸性酒精 acid alcohol
酸性橘红 acid orange
酸性菌分解 acid digestion
酸性矿井水 acid mine drain
酸性矿坑水 acid pit water
酸性矿坑污水 acid mine drainage wastewater
酸性矿泉 acidulous spring
酸性矿山废水污染 acid mine pollution
酸性矿山废物 acid mine drainage waste
酸性矿山排水 acid mine drainage
酸性矿石 acid drift; acid ore
酸性矿水 acid mine water
酸性矿物排放 acid mine drainage
酸性矿渣 acid(ic) slag
酸性喹啉黄 quinoline yellow
酸性蓝 acid blue
酸性冷凝物 sour condensate
酸性离子 acid ion
酸性沥青乳液密封剂 cationic emulsion slurry seal
酸性炼钢法 acid process
酸性亮蓝 acid brilliant blue
酸性裂化法 acid cracking process
酸性磷酸盐 acid phosphate; superphosphate
酸性流出物 acid effluent
酸性硫酸盐—氯化物泉 acid sulfate-chloride spring
酸性硫酸盐氯化物型热水 acid sulphate chloride thermal water
酸性硫酸盐泉 acid sulfate spring

S

酸性硫酸盐土 acid sulfate soil
酸性硫酸盐型热水 acid sulphate thermal water
酸性馏分 sour distillate
酸性炉衬 acid lining
酸性炉钢 acid steel
酸性炉渣 acidic slag;acid slag
酸性绿 acid green;Guignet's green
酸性氯化物泉 acid-chloride spring
酸性氯酸钠 acidic sodium chlorate
酸性络合染料 acid complex dye
酸性马丁炉钢 acid open-hearth steel
酸性玫瑰红 acidic rose red
酸性玫瑰红印染废水 acidity rose red dyeing wastewater
酸性媒介红 acid mordant red
酸性媒染黑 solochrome black
酸性媒染染料 acid mordant dyes
酸性煤尘 acid smut
酸性煤烟 acid smut
酸性面 acidic surface
酸性磨黑 acid milling black
酸性磨绿 acid milling green
酸性木材 <化学加工用材> acid wood
酸性内衬 acid lining
酸性耐火材料 acid(ic) refractory
酸性耐火制品 acid refractory product
酸性耐火砖 acid(fire) brick;acid refractory
酸性嫩黄 acid light yellow
酸性泥炭 acid peat
酸性泥炭土 acid peat soil
酸性黏[粘]合剂 acid binding agent
酸性黏[粘]土 acid clay
酸性凝固 acid coagulation
酸性偶氮黄 azoflavine
酸性偶氮染料 acidazo-colo(u)r azo-dye
酸性排水 acid drainage;acid effluent
酸性抛光 acid polishing
酸性喷气孔 acidic jet
酸性片状炭黑 acid smut
酸性漂白 acidic bleaching
酸性漂白液 acidic bleaching solution
酸性贫营养湖 acidic dystrophic lake
酸性品红 acid fuchsin
酸性平炉 acid open-hearth furnace
酸性平炉法 acid open-hearth process
酸性平炉钢 acid open-hearth steel
酸性平炉炼钢法 acid open-hearth process
酸性漆 acid-cured varnish
酸性气体 acid gas
酸性汽油 sour gasoline
酸性茜素黑 acid alizarine black
酸性茜素蓝 acid alizarine blue
酸性桥接羟基 acidic bridged hydroxyl
酸性侵蚀指标 acid erosion index
酸性去污剂 acid detergent
酸性泉 acid spring
酸性染剂 acid dye(stuff)
酸性染料 acid(ic) dye(stuff)
酸性染料色淀 acid dye lake
酸性染色剂 acid stain
酸性热水塘 acid hot pool
酸性溶液 acid(ic) solution
酸性熔池 acid bath
酸性熔剂 acid flux(material)
酸性熔岩 acidic lava
酸性熔岩流 acidic lava flow
酸性乳剂 cationic emulsion
酸性乳浊液 acid emulsion
酸性软化法 acid bating
酸性生铁 acid pig
酸性蚀斑 <在玻璃表面进行的> acid streaking
酸性蚀变 acid alteration
酸性试验 acid test
酸性树皮 sour bark
酸性树脂 acid(eluted) resin; acidic resin
酸性水 acidic water; acidulous water;sour water
酸性水剥离剂 sour water stripper
酸性水腐蚀 acid water corrosion
酸性水磷铵镁石 dittmarite
酸性水泥浆 acid slurry
酸性水排放 acid water discharge; acid water drainage
酸性水热淋滤 acid hydrothermal leaching
酸性水溶青 China blue
酸性水洗提剂 sour water stripper
酸性饲料 acidic feed
酸性炭黑 acid smut
酸性碳酸岩 bicarbonate
酸性碳酸盐 acid carbonate;hydrocarbonate
酸性铁废物 acid iron waste
酸性停影液 acid stop bath
酸性铜电解溶液 acid copper
酸性土的中和 acid soil neutralization
酸性土(壤) acidic ground; acid(ic) soil;acidity soil;sour soil
酸性土指示植物 oxylophyte
酸性微粒散发 acidic particle emission
酸性污泥 acid sludge
酸性污水 sour water
酸性污物 acid smut
酸性物 acid
酸性物质 acidoid;acid substance
酸性稀浆 acid slurry;cationic slurry
酸性洗煤浸沥液 acidic coal-cleaning leachate
酸性系数 acidity coefficient
酸性显影液 acid developer
酸性橡浆 positex
酸性消退阶段 acid regression stage
酸性斜长石 acid plagioclase
酸性絮凝 acid coagulation
酸性蓄电池 acid accumulator; acid battery
酸性蓄电池组 acid storage battery; lead battery
酸性悬液 acid suspension
酸性压花 <采用氟氢酸在玻璃表面形成闪光装饰花纹> acid embossing
酸性亚硫酸盐半化学纸浆 acid sulfite semichemical pulp
酸性烟尘 acid smut
酸性烟灰 acid soot
酸性岩 acidic rock;acidite
酸性岩浆元素 elements of acidic magma
酸性岩浆作用 acid magmatism
酸性岩类 acid rocks
酸性岩石 <以石英为主要矿物的火成岩> acid(ic) rock
酸性盐 acid(ic) salt
酸性颜料 acid colo(u)r; acid pigment
酸性阳离子型乳浊液 acid emulsion
酸性氧化剂 acidic oxidizing agent
酸性氧化物 acid(ic) oxide
酸性窑衬 acid lining
酸性萤石 acid spar
酸性油 acid oil;acid stage oil;sour oil
酸性油分 acidic oil
酸性有机质 mor
酸性淤渣 acid sludge
酸性原油 sour crude oil
酸性月岩 lunarite
酸性枣红 bordeaux
酸性枣红色淀 bordeaux lake
酸性渣 acid slag
酸性沾污 acid stain
酸性障 acidic barrier
酸性沼(泽) acid bog
酸性沼泽地 acid moor
酸性蒸气 acid vapo(u)r
酸性植物 acid(soil) plant;oxylophyte
酸性中和剂 acid acceptor
酸性砖 acid brick
酸性转炉 acid(-lined) converter;Bessemer converter
酸性转炉吹炼法 bessemerizing
酸性转炉低碳钢 Bessemer low carbon steel
酸性转炉法 acid process
酸性转炉钢 acid Bessemer steel;Bessemer(steel);Bessemer iron
酸性转炉钢渣 acid Bessemer furnace slag
酸性转炉炼钢法 acid Bessemer converter;acid Bessemer process;acid converter process; Bessemer process
酸性转炉炉料 Bessemer charge
酸性转炉熔炼 Bessemer heat
酸性转炉生铁 acid Bessemer pig;acid converter pig iron
酸性转炉铁 acid Bessemer iron;Bessemer iron
酸性转炉铁矿 Bessemer ore
酸性转炉冶炼 acid Bessemerizing
酸性转炉铸铁 acid Bessemer cast iron
酸性转炉砖 acid Bessemer brick
酸性状况 acidic condition
酸性着色剂 acid stain
酸性紫 acid violet
酸性棕色土 acid brown soil
酸性组分 acidic component; acidic constituent
酸蓄电池 acid accumulator
酸雪 acidic snow
酸驯化接种污泥 acid acclimated seed sludge
酸循环泵 acid circulating pump
酸烟垢 acid soot
酸烟(雾) acid fume
酸衍生物 acid derivative
酸样收集器 drip pan;dripping cup
酸液 acid liquor;pickling liquor
酸液泵 acid-handling pump; acid pump
酸液比重计 acidometer[acidimeter]
酸液定量法 acidometry
酸液配方 composition of acidizing fluid
酸液输送泵 acid-handling pump
酸液稀释 acid dilution
酸液总量 amount of acidizing fluid
酸以前的碱性 pre-acidification alkalinity
酸抑制剂 acid inhibitor
酸阴离子总和 sum of acid anions
酸硬化清漆 acid-cured varnish
酸雨 acid deposit;acid precipitation; acid rain
酸雨监测 acid rain monitoring
酸雨监测网 acid rain (monitoring) network
酸雨量 acid precipitation
酸雨危害 acid rain damage
酸雨影响 effect of acid rain
酸浴 acid bath;sour bath
酸浴冲洗 acidulated rinsing
酸浴镀锌 acid bath
酸再生的 acid-regenerated
酸增强 acid strengthening
酸增强处理 acid fortification
酸渣 acid residue;acid sludge;goudron
酸渣柏油 acid tar
酸渣柏油脂 acid-sludge pitch
酸渣地沥青 acid(ic)-sludge asphalt; sludge asphalt
酸渣回收 recycle(d) sludge
酸渣焦油(沥青) acid tar
酸渣焦油脂 acid-sludge pitch
酸渣煤焦沥青 acid-sludge pitch
酸渣煤沥青 acid tar
酸渣石油沥青 acid-sludge asphalt
酸渣硬(煤)沥青 acid-sludge pitch
酸胀法 acid swelling
酸沼 bog; moor; peat bed; peat bog; peat moor
酸沼草原 moor grass
酸沼水域 bog waters
酸蒸煮 acid cooking
酸值 acid number;acid value
酸酯交换 acidolysis
酸中毒 acidosis;acid poisoning
酸中和能力 acid neutralizing capacity
酸中和容器 acid knock-out drum
酸中和装置 acid neutralizing unit
酸煮分析 acid digestion analysis
酸紫树 sour tupelo
酸自由基 acid free radical
酸渍 acid cure; acid pickling; acid soaking

蒜头桩 pile with bulk shaped base

算表程序 table simulator

算出 figure out(at);work(ing)-out
算错 miscalculate; miscalculation; miscount;wrongly calculated
算法 algorithm;algorithmic method
算法逼近 algorithmic approach
算法比较 algorithm comparison
算法程序 algorithm(ic) routine
算法的 algorithmic
算法的收敛 convergence of algorithm
算法调度 algorithmic dispatching
算法翻译 algorithmic translation
算法分析 algorithmic analysis
算法复杂性 algorithm complexity
算法级 algorithmic level
算法理论 theory of algorithm
算法联系 algorithmic connection
算法流程图 algorithmic flow chart
算法模拟 algorithmic simulation
算法模式 algorithmic model
算法排队控制 algorithmic queue control
算法图 algorithmic pattern
算法推敲 algorithmic elaboration
算法详细描述 algorithmic elaboration
算法形式 algorithmic form
算法语句 algorithmic statement; arithmetic(al) statement
算法语言 algorithmic language; programming language
算法转接 algorithmic dispatching
算法子程序 algorithmic subroutine
算法字组 algorithmic block
算符 functor;nabla;operator
算符表 operator table
算符部分 operator part
算符定义性出现 operator-defined occurrence
算符方程【数】 operator equation
算符分类 operator class
算符寄存器 operator register
算符校验表 operator verification table
算符文法 operator grammar
算符应用性出现 operator-applied occurrence
算符优先法 operator precedence method
算符优先方式 operator precedence system
算符优先分析程序 operator preced-

ence parser
算符优先矩阵 operator precedence matrix
算符优先情况 operator precedence case
算符优先(数) operator precedence
算符优先顺序矩阵 operator priority order matrix
算符优先文法 operator precedence grammar
算后编辑 post edit(ing)
算后编辑程序 post-edit program-(me)
算后编排 post edit(ing)
算后检查 postmortem
算后检查程序 postmortem program-(me);postmortem routine
算后检查法 postmortem method
算后增量 post increment
算后诊断程序 postmortem program-(me)
算后转储 postmortem dump
算例 arithmetic(al) example
算流域水账 basin accounting
算盘 abacus [复 abaci/abacuses]; counting frame
算盘式输送机 abacus conveyer[conveyor]
算前编排 preediting
算前减量 predecrement
算入 count in
算术 arithmetic
算术 IF 语句【计】 arithmetic(al) IF statement
算术比 arithmetic(al) ratio
算术标识符 arithmetic(al) identifier
算术表达式 arithmetic(al) expression
算术不变量 arithmetic(al) invariant
算术不变数 arithmetic(al) invariant
算术操作 arithmetic(al) operation
算术差 arithmetic(al) difference
算术常数 arithmetic(al) constant
算术乘积 arithmetic(al) product
算术初等项 arithmetic(al) primary
算术错误 arithmetic(al) error
算术递增滤色镜 arithmetic(al) filter
算术动词 arithmetic(al) verb
算术分位数差 arithmetic(al) quartile deviation
算术分位数峭度 arithmetic(al) quartile kurtosis
算术俘获 arithmetic(al) trap
算术俘获屏蔽 arithmetic(al) trap mask
算术赋值 arithmetic(al) assignment
算术赋值语句 arithmetic(al) assignment statement
算术概率 arithmetic(al) probability
算术函数 arithmetic(al) function
算术和 arithmetic(al) sum
算术和逻辑单元 arithmetic(al) and logic(al) unit
算术化 arithmetization
算术及逻辑寄存器栈 arithmetic(al) logic register stack
算术级数 arithmetic(al) progression; arithmetic (al) range; arithmetic-(al) series
算术几何级数 arithmetic(al)-geometric series
算术几何平均 arithmetic-geometric-(al) mean
算术计算 arithmetic(al) computation
算术计算机 arithmetic(al) computer
算术加法 arithmetic(al) addition
算术校验 arithmetic(al) check
算术句子 arithmetic(al) sentence
算术距离 arithmetic(al) distance
算术例外 arithmetic(al) exception
算术例行子程序 arithmetic(al) subroutine
算术连续统 arithmetic(al) continuum
算术逻辑单元 arithmetic(al) logic unit
算术码 arithmetic(al) code
算术描述 arithmetic(al) statement
算术平均 arithmetic(al) average
算术平均法 arithmetic (al) mean method; center line average method; center[centre] line average
算术平均厚度 arithmetic(al) average thickness
算术平均粒径 arithmetic(al) mean diameter
算术平均品位 arithmetic(al) average grade
算术平均声压级 arithmetic(al) mean sound pressure level
算术平均数 arithmetic(al) average (value); arithmetic(al) mean (value)
算术平均体重 arithmetic(al) average volume weight
算术平均温度差 arithmetic(al) mean temperature difference
算术平均误差 arithmetic(al) mean error
算术平均压力法 arithmetic(al) average pressure method
算术平均直径 arithmetic(al) mean diameter
算术平均值 arithmetic(al) average; center[centre] line average; arithmetic(al) mean
算术平均指数 arithmetic(al) average index; index number of arithmetic-(al) average
算术权 arithmetic(al) weight
算术扫描 arithmetic(al) scan
算术上的 arithmetic
算术式 arithmetic(al) expression
算术数组 arithmetic(al) array
算术四分位(偏)差 arithmetic(al) quartile deviation
算术算子 arithmetic(al) operator
算术条件语句 arithmetic(al) IF statement
算术图表 arithmetic(al) chart
算术图像数据 arithmetic(al) picture data
算术无功因数 arithmetic(al) reactive factor
算术误差码 arithmetic(al) error code
算术系统 arithmetic(al) system
算术项 arithmetic(al) term
算术型内部函数 arithmetic(al) built-in function
算术移位 arithmetic(al) shift
算术异常 arithmetic(al) exception
算术溢出 arithmetic(al) overflow
算术右移 arithmetic(al) shift right
算术语句 arithmetic(al) sentence; arithmetic(al) statement
算术元素 arithmetic(al) element
算术运算 arithmetic(al) operation
算术运算单元 arithmetic(al) unit
算术运算符 arithmetic(al) operator; mathematic(al) operator
算术运算符优先权 operator hierarchy
算术运算四元组 arithmetic(al) quadruple
算术运算溢出 arithmetic(al) overflow
算术增长 arithmetic(al) growth
算术直线增长 arithmetic(al) growth
算术中项 arithmetic(al) mean
算术属性 arithmetic(al) attribute
算术转换 arithmetic(al) conversion
算术子程序 arithmetic(al) subroutine
算术自陷 arithmetic(al) trap
算术自陷赋能 arithmetic(al) trap enable
算术自陷陷阱 arithmetic(al) trap mask
算术左移 arithmetic(al) shift left
算术坐标(方格)纸 arithmetic(al) plotting paper
算态 computing mode; problem mode; problem status
算图 abac; alignment chart; alignment diagram; alignment nomogram; nomogram; nomograph(chart)
算图法 nomographic(al) method
算账 cast account; reckoning
算子 operator
算子波函数 operator wave function
算子代数 operator algebras
算子方程 operator equation
算子分裂 operator compact implicit
算子分裂法 operator-splitting method
算子符号 operator notation
算子广义函数 operator-valued distribution
算子函数 operator(-valued) function
算子环 operator ring; ring of operators
算子矩阵 operator matrix
算子理论 operator theory
算子谱论 spectral theory of operators
算子群 operator group
算子三角形 del
算子同构 operator isomorphism
算子同态 operator homomorphism
算子伪单调的 operator pseudomonotone
算子位势 operator potential
算子序列 sequence of operators
算子演算 operational calculus
算子优势 operator dominance
算子域 operator domain
算子自同态 operator endomorphism

随

随被动轴 driven shaft
随边 lagging edge; trailing edge
随变 covariant
随波 flowing wave
随场地变化的 site-dependent
随场地更换 to be changed depending on the site
随潮流的冰川 tide water glacier
随潮流的冰山 tidal glacier
随潮漂流物 tide walker
随潮水涨落出没的礁石 tide rock
随车测速法 moving vehicle method
随车吊 truck crane attachment; truck loading crane; vehicle mounted attachment
随车调查 on vehicle survey
随车工具 basic hand tool; driver's tool
随车工具及附件全套 driver's tool and accessories
随车供给的涂料 factory pack
随车观测法 moving observer method
随车观察 <车速及行车情况等> riding check
随车控制 on-board control
随车流观测法 floating vehicle method
随车流现测法 moving car method
随车起重装置 truck crane attachment; vehicle mounted attachment
随车式吊 lorry loading crane; lorry-mounted crane; trailer crane; truck crane
随车式起重机 lorry loading crane; lorry-mounted crane; trailer crane; truck crane
随车液压起重臂 escort hydraulic arm
随车邮局职工 travel(l)ing postal staff
随车职工 staff provided to accompany trains
随车制动工 car rider
随车装卸机械 escort handling machinery
随车装卸器 materials handling augmenter
随乘制【铁】 caboose working system
随船架 launching cradle
随船理货员 master of the hold
随船送达邮件 captain's mail
随船提单 captain's bill of lading; ship's bill of lading
随船提单副本 captain's copy
随从室 valet room
随打随抹 placing and trowel(l)ing
随打随抹光 trolled as soon as placed
随到随批计划 plan checked and approved as its arriving
随到随装 prompt
随到随装船 prompt ship
随电压变化的电阻器 voltage-variable resistor
随订单付款 cash with order
随订单支付现金 cash with order
随动 following; follow-up
随动板 follower plate
随动棒 follower
随动泵 jackhead pump
随动变压器 follow-up transformer
随动部分 phantom element; phantom part
随动部分轴颈 phantom stem
随动部件 follow-up unit
随动操纵 pilot-operated; servo-control
随动操纵装置 servo-control unit
随动乘法器 servo-multiplier
随动齿轮 follower gear; following gear; idler gear; phantom gear
随动齿轮轴 driven gear shaft
随动传动 servo-drive
随动传动力 auxiliary force
随动磁头 tracer head
随动刀架 follower rest; movable support
随动的 servo-actuated
随动电动机 follower motor; follow-up motor; slave motor
随动多谐振荡器 driven multivibrator
随动阀 follower valve; follow-up valve; servo-valve; slave valve
随动放大器 follow-up amplifier
随动缸筒 slave cylinder
随动跟踪装置 mobile tracking mount
随动滚筒 roller follower
随动滚柱 roller follower
随动滑轮 idler sheave
随动环 follower ring; phantom ring
随动簧 spring follower
随动回路油 slave circuit oil
随动活塞 piston follower
随动机构 detector; follower; follow-up mechanism; slaving ring
随动机件 follower
随动激光探测器 slave laser detector
随动继电器 slave relay
随动加压 follow-up pressure
随动件 follower
随动件拖动 follower drive
随动开关 follow-up switch
随动开关装置 bang-bang device
随动控制 follow-up control; servo-control
随动控制扫描干涉仪 servo-controlled scan(ning) interferometer
随动控制系统 following control sys-

tem; self-aligning control system; follow-up control system
随动连杆 follow-up linkage
随动流动 random current
随动轮 driven wheel; follower wheel; follow-up pulley; idle pulley; idler gear; supporting roll(er); track idler
随动木塞 follower plug
随动强化模型 kinematic(al) hardening model
随动桥轴 dead axle
随动球 outer sphere
随动圈 follower ring
随动设备 follow-up device
随动摄影 panning
随动双桥组 tandem trailing axle
随动速度 follow-up speed
随动弹簧 follower spring
随动套筒 follow-up sleeve
随动天线 slave antenna
随动调节 follow-up control
随动调节系统 servo-actuated regulating system
随动托轮 carrying idler
随动万向架 unlock gimbal
随动污泥锥 slave sludge cone
随动误差 following error
随动系统 auto-follower; follow(ing)-up mechanism; follow-up system; servo-system; synchrosystem
随动系统参数 servo-parameter
随动系统测试仪 servo-mechanism tester
随动系统电位计 follow-up potentiometer
随动系统分析器 servo-analyser[analyzer]
随动系统理论 follow-up theory
随动系统灵敏度 sensitivity of follow-up system
随动系统调谐操纵台 servo-mechanism tester
随动系统元件 servoelement
随动线圈 follow-up coil
随动研究 follow-up study
随动硬化 kinematic(al) hardening
随动油缸 power cylinder; servocylinder; slave cylinder
随动圆片 flat follower
随动增压 follow-up pressure
随动闸 servant brake
随动振动 random vibration
随动指针 follow-up point
随动中心架 movable support
随动轴 idler shaft
随动装置 follower; follow-up device; follow-up gear; servo; slave unit
随动自动同步机系统 servo-selsyn system
随动组件 follow-on subassembly
随堆随烧 progressive burning
随帆化 randomize
随方法变化的 method-dependent
随访研究 follow-up study
随附 annex(e); appendix[复 appendices/appendixes]; attach
随附标记 trailer label
随附记录 trailer record
随附信息组 trailer block
随附债务 contiguous obligation
随附属件 subsidiary attribute
随钻测井 follow-up log
随观察角度变化的颜色 colo(u)r-flop
随航作业 towing operation
随后的 subsequent; succeeding
随后的冷却 cooling after
随后电告 telegram to follow
随后发生的事情 subsequence
随后热处理 post-heat treatment
随后位置 lag position
随后注浆法 follow-up grouting
随货运列车发料 freight train shipment
随机 stochasticism
随机按序存储器 random sequential memory
随机包长度 random packet length
随机逼近 stochastic approximation
随机边试边改试验 random trial-and-error test
随机编码 hash coding; hat
随机编码限 random-coding bound
随机编制 random organization
随机编制文摘 stochastic abstracting
随机变参数系统 random variation parameter system
随机变动 random fluctuation
随机变化 random change; random variation; stochastic variation
随机变换 stochastic transformation
随机变量 chance variable; extraneous variable; random variable; stochastic variable; stray parameter
随机变量的独立性 independence of random variables
随机变量的分布 distribution of a random variable
随机变量的概率分布 probability distribution of random variable
随机变量的离差 dispersion of a random variable
随机变量的众数 mode of random variables
随机变量分布 probability distribution
随机变量函数 function of random variable
随机变量序列 sequence of random variables
随机变数 chance variable; random variable; stochastic variable
随机变异 random variation
随机遍历定理 random ergodic theorem
随机标记 random labelling
随机表面形状 random surface profile
随机波 random sea; random wave
随机波动 random fluctuation
随机波形 random wave shape
随机波造波机 random wave generator
随机不确定度 random uncertainty
随机不贴合性结构 random lack-of-fit structure
随机不稳定性 stochastic instability
随机不相关 statistic(al) independence; stochastic independence
随机不匀率 random irregularity
随机部分 random component
随机擦除 selective erasing
随机采试样 grab sample
随机采样 chance sampling; random sampling; stochastic sampling
随机参数 random parameter; stochastic parameter
随机参数系统最优化 optimization of systems with random parameters
随机测度 random measure
随机测量误差 random measurement error
随机测试 random test
随机测试产生法 random test generation
随机策略 random device; randomized policy; random strategy
随机查找 random searcher; stochastic searching
随机差错 random error
随机差分方程 random difference equation; stochastic difference equation
随机差(数) random difference
随机长度 random length
随机场 random field
随机车辆组成的列车 random cars
随机成对法 random pairs method
随机成对技术 random pairs technique
随机成分 random component; random element; stochastic element
随机成群的 random clumped
随机程序 random process
随机程序试验 randomized program(me) test
随机冲击 random shock
随机冲量 random shock
随机重发间隔 random retransmission interval
随机重复 randomized replication
随机重合【计】 random coincidence
随机抽查 random check
随机抽检 inspection at random; random inspection
随机抽样 chance sampling; random sample; random sampling; stochastic sampling
随机抽样法 method of random sampling
随机抽样分布 random sampling distribution
随机抽样检查 random inspection; random test
随机抽样检验 random inspection; random test
随机抽样试验 random sampling test
随机抽样误差 random sampling error
随机出料 random ejection
随机处理 random processing
随机次级样品 random subsample
随机存取 direct access; random access
随机存取 X-Y 定位装置 random access X-Y positioning unit
随机存取程序设计 random access programming
随机存取程序设计与检查设备 random access programming and checkout equipment
随机存取存储计算机 random access memory computer
随机存取存储器 random access memory; random access storage; uniformly accessible store
随机存取存储器数模转换器 random access memory digital/analog(ue) converter
随机存取存贮器 random access memory
随机存取电子束定位 random access beam positioning
随机存取电子束定位系统 random access beam positioning system
随机存取电子束偏转装置 random access beam deflection system
随机存取分类 random access sort
随机存取分类程序 random access sorter
随机存取辅助存储器 random access auxiliary storage
随机存取功能块 random access module
随机存取光存储器 random access optic(al) memory
随机存取机 random access machine
随机存取计算机 random access computer
随机存取计算机设备 random access computer equipment
随机存取卡片装置 random access card equipment
随机存取控制器 random access controller; random access control unit
随机存取库 direct access library
随机存取器 random access device
随机存取软件 random access software
随机存取时间 random access time
随机存取输入输出 random access input-output
随机存取输入输出程序 random access input/output routine
随机存取文件 random access file
随机存取消息存储器 random access message storage
随机存取寻址 random access addressing
随机存取遥测 random access telemetry
随机存取装置 random access device
随机错误 random error
随机单调性 stochastic monotonicity
随机的 accidental; arbitrary; probabilistic; stochastic
随机地震反应分析 stochastic seismic response analysis
随机地震分析 stochastic seismic analysis
随机地震激发 stochastic seismic excitation
随机地震运动 random earthquake motion
随机点过程 random point process
随机点体视图 random dot stereogram
随机电位计 servobalance potentiometer
随机电子流 random electron current
随机调查 random search(ing)
随机叠加 random superposition
随机叠加峰 random sum peak
随机定律 law of chance
随机定向 random orientation
随机定向(放置)纤维 randomly oriented fiber[fibre]
随机动荷载 random dynamic(al) loading
随机动力学 stochastic dynamics
随机动力预测 stochastic dynamic prediction
随机动态规划 stochastic dynamic programming
随机动态预报 stochastic dynamic prediction
随机读数误差 random reading error
随机读写存储器 random access memory
随机独立 probability independence; stochastic independence
随机独立分量 stochastically independent component
随机断层模型 stochastic fault model
随机断层作用 stochastic faulting
随机断裂力学 stochastic fracture mechanics
随机对策 stochastic game
随机多级过程 random multi-level process
随机多路存取 random multiple access
随机多值过程 random multi-level process
随机法 randomized process
随机反射率 random reflectivity
随机反演 stochastic inverse
随机反应 random response
随机反应特性 stochastic response characteristic
随机反应特征 stochastic response characteristic
随机反应性函数 random reactivity

function
随机方案 randomizing scheme; stochastic scheme
随机方差分析 random analysis of variance
随机方程式 stochastic equation
随机方法 stochastic method
随机方式 random fashion
随机方向 random direction
随机仿真 stochastic simulation
随机访问 direct access; random access
随机访问分类 random access sort
随机访问文件 random access file
随机放置防波堤块体 pell-mell placement of blocks
随机非线性系统 stochastic non-linear system
随机分布 chance distribution; random distribution; stochastic distribution
随机分段文件 arbitrarily sectioned file
随机分离法 random isolation method
随机分量 random component
随机分配 random allocation; random assignment; random assortment; random distribution; stochastic assignment
随机分配多址 random assignment multiple access
随机分散策略 random diversification strategies
随机分酸 random assortment
随机分析 random analysis; stochastic analysis
随机分组 blind assignment
随机分组法 randomized block
随机服务 random service; service in random order
随机服务系统 random service system; stochastic service system
随机服务系统理论 random service system theory; stochastic service system theory; theory of stochastic service system
随机符合 random coincidence
随机附标产生式 stochastic indexed production
随机干扰 random disturbance; random jamming; stochastic disturbance
随机干涉效应 random interference
随机工程师 flight engineer
随机工具 attachment tools; tools attachment
随机共振 accidental resonance
随机故障 random failure
随机故障间隔期 random failure period
随机故障率 random failure rate
随机关系 stochastic relation
随机观测 casual observation
随机观察 <在工作抽样中所作的随机观察> random observation
随机规划法 stochastic programming
随机轨道 random orbit
随机轨道卫星系统 random-orbit satellite system
随机过程 stochastic process; random process(ing)
随机过程包线 envelop of stochastic process
随机过程的概率描述 probability description of random processes
随机过程的截尾 tail of a stochastic process
随机过程的微分 differentiation of random process
随机函数 random function; stochastic function
随机函数发生器 randomizer
随机函数方程 stochastic functional equation
随机核 stochastic kernel
随机荷载 random load(ing)
随机横向力 random lateral force
随机后验概率 random posterior probability
随机化 randomisation[randomization]
随机化表 randomized table
随机化抽样 randomized sampling
随机化方案 randomizing scheme
随机化方法 randomization method
随机化估计值 randomized estimator
随机化检定 randomized test
随机化判决函数 randomized decision function
随机化区组 randomized block
随机化区组设计 randomized block design
随机化试验 randomized test
随机化完全区组设计 randomized complete-block design
随机化行动 act of randomization
随机环境 random environment
随机绘图误差 random plotting error
随机积分 stochastic integral
随机积分方程 random integral equation; stochastic integral equation
随机基底运动 random base motion
随机畸变 fortuitous distortion
随机畸变模型 stochastic deformation model
随机激励 arbitrary excitation
随机激振 random excitation
随机激振反应 response to random excitation
随机级数 random series
随机几何学 random geometry
随机计算 stochastic calculus
随机技术 randomizing technique; stochastic technique
随机加热 stochastic heating
随机加速度 casual acceleration; random acceleration
随机加速器 stochastic accelerator
随机加载 stochastic assignment
随机加载(交通)算法 stochastic assignment algorithm
随机间隔 random interval
随机检索 random search(ing); stochastic retrieval
随机检验 randomized test
随机简并度 accidental degeneracy
随机交通分配算法 stochastic assignment algorithm
随机校核 random check
随机结构体系 stochastic structural system
随机介质 random medium
随机近似法 stochastic approximation
随机精度 accidental accuracy
随机救生船 air-borne lifeboat
随机矩阵 random matrix; stochastic matrix
随机决策树分析 stochastic decision tree analysis
随机决定过程 stochastic decision process
随机开关函数 random switching function
随机考虑 stochastic consideration
随机可观测性 stochastic observability
随机可观察性 stochastic observability
随机可微性 stochastic differentiability
随机克立格法 random Kriging
随机客梯 airsteps; built-in air stair(case); self-contained air stair(case)
随机控制 random control; stochastic control
随机控制理论 stochastic control theory
随机扩展转移网络 stochastic augmented transition network
随机累积 random accumulation
随机离差 random deviation
随机理论 random theory
随机利润 chance of a profit
随机连续性 stochastic continuity
随机链规则 stochastic chain rule
随机链模型 random chain model
随机量 random entity; stochastic quantity
随机量线能量 random quantity stochastic quantity linear energy
随机流域系统 stochastic watershed system
随机路径时间【交】random link time
随机路径选择 random routing
随机路线 random path
随机率 law of chance
随机逻辑 random logic
随机逻辑部件 random logic device
随机逻辑测试 random logic testing
随机逻辑设备 random logic device
随机逻辑设计 random logic design
随机码 hatted code; random code
随机脉冲 random pulse
随机脉冲串 random pulse train
随机脉冲发生器 random pulse generator
随机脉冲过程 random pulse train process
随机脉冲调制 random pulsing
随机脉动 random fluctuation
随机模拟 stochastic analog; stochastic model(l)ing; stochastic simulation
随机模拟法 random simulation method
随机模拟模型 stochastic simulation model
随机模式 random pattern
随机模型 probabilistic model; probability model; stochastic model; stochastic sampling
随机母体 random population
随机能量 random energy
随机耦合 random coupling
随机拍特性 random beat characteristic
随机排队 stochastic queue
随机排队模式 stochastic queuing systems
随机排列 random arrangement; random order; random permutation
随机判定过程 stochastic decision process
随机判定准则 random criterion
随机判据 random criterion
随机配位法 random-coordinate method
随机疲劳 random fatigue
随机偏差 random deviate; random deviation
随机偏心 random eccentricity
随机偏振 random polarization
随机偏置全息图 random-bias hologram
随机漂变 random drift
随机漂移 random drift
随机漂移率 random drift rate
随机拼合 random matching
随机破坏 chance failure; stochastic failure
随机起动 random start
随机起伏 chance fluctuation; random fluctuation
随机起伏数据 randomly fluctuating data
随机起伏系统 random fluctuating system; random fluctuation system
随机器材包 flyaway kit
随机求积公式 random quadrature formula
随机区间 random interval
随机取数 direct access
随机取样 chance sampling; random grab; stochastic sampling; random sample
随机取样法 random sampling
随机取样伏特计 random sampling voltmeter
随机取样示波器 random sampling oscilloscope
随机群体 random population
随机扰动 random disturbance; stochastic disturbance
随机扰动函数 random forcing function
随机扰动最优化 random-perturbation optimization
随机热应力 stochastic thermal stress
随机溶解氧模型 stochastic dissolved oxygen model
随机入射吸声系数 random-incidence sound absorption coefficient
随机散射 random scatter(ing)
随机扫描 random interlace; random scan(ning)
随机设计地震 stochastic design earthquake
随机生成 random generation
随机失配 random mismatch
随机失效 random failure
随机失真 fortuitous distortion
随机时变通道 random time-varying channel
随机时变线性系统 randomly time-varying linear system
随机时变信道 random time-varying channel
随机时间间隔 random time spacing
随机时间序列模型 stochastic time-series model
随机时序 random sequence
随机时序存储器 random sequential memory
随机实验 random experiment
随机式计划 randomizing scheme
随机事件 chance event; random event; random occurrence; stochastic event
随机事件仿真 simulation of random events
随机事件模拟 simulation of random events
随机试件 random sample
随机试验 random experiment; random test
随机试验法 random trial method
随机试样 random sample
随机收敛 convergence in probability; random convergence; stochastic convergence
随机收缩算子 random contraction operator
随机输入 stochastic input
随机输入法 random entry method
随机束 stochastic beam
随机数 random number
随机数表 table of random digits
随机数产生程序 Monte-Carlo generator
随机数的产生 generation of random numbers
随机数发生 random number generation
随机数发生器 random number generator
随机数据 random data
随机数据处理 random data processing

S

随机数据生成系统 stochastic data generating system
随机数生成 generation of random numbers
随机数生成程序 random number generator; random number program(me)
随机数位 random digit
随机数序列 random number sequence;random number series
随机数字 random digit
随机数(字)表 table of random numbers
随机水力学 random hydraulics; stochastic hydraulics
随机水文学 statistic(al) hydrology; stochastic hydrology; synthetic(al) hydrology
随机水质系统 stochastic water quality system
随机顺序 random order; random sequence
随机顺序存储 random sequential access
随机顺序存取 random sequential access
随机伺服系统 stochastic service system
随机搜索 random search(ing)
随机搜索法 Monic method; Monte-Carlo analysis; random search method
随机速度 random velocity
随机算子 random operator; stochastic operator
随机损失 chance of a loss
随机榫 law of chance
随机索得率 browsability
随机探测 random search(ing)
随机探寻 random search(ing)
随机特征 stochastic characteristic
随机填料 random packing
随机调制 stochastic modulation
随机调制载波 randomly modulated carrier
随机跳动 randomized jitter; random jump
随机跳位 random skip position
随机通信[讯]卫星 random communication satellite
随机通信[讯]卫星系统 random communication satellite system
随机突变 chance mutation; random mutation
随机图【数】random graph
随机图表 randomizing scheme
随机图像跳动 randomized image jitter
随机网络 random network; stochastic network
随机网络模拟技术 random network simulation technique
随机微分 stochastic differential
随机微分方程 stochastic differential equation
随机微商 stochastic derivative
随机卫星系统 random satellite system
随机位 random order
随机位移 random displacement; stochastic movement
随机文件 attachment paper; random file
随机问题 random problem; stochastic problem
随机误差 accidental error; chance error; random error; stochastic error
随机误差防护 random error protection
随机误差校正卷积码 random-error-correcting convolutional code
随机误差校正能力 random-error correcting ability
随机误差项 random error term
随机误差修正 stochastic error-correction
随机系集 random assembly
随机系统 probabilistic system; servo-system; stochastic system; random system
随机下推自动机 stochastic pushdown automata
随机显示 selective-access display
随机现象 random phenomenon
随机线性规划 stochastic linear planning; stochastic linear programming
随机线性离散系统 stochastic linear discrete system
随机线性系统 stochastic linear system
随机相变动 random phase shift
随机相关 stochastic dependence
随机相角 random phase angle
随机相位 random phase
随机相位近似 random phase approximation
随机相位偏心 randomly phased eccentricity
随机相位误差 random phase error
随机相依 random dependence; stochastic dependence
随机相移 random phase shift
随机响应 random response
随机向量 random vector
随机向量生成程序 random-vector generator
随机项 random entry
随机效果 random effect; stochastic effect
随机效应 random effect; stochastic effect
随机效用最大理论 random utility maximization theory
随机信号 random signal; stochastic signal
随机行走过程 random walk process
随机型 random type; stochastic pattern
随机型变化 variation of random type
随机型加速 stochastic type of acceleration
随机型马赛克 random mosaic
随机型式 random fashion
随机性 probabilistic nature; randomness; stochasticism
随机性成分 randomness component
随机性存储 random inventory; stochastic inventory
随机性地球化学模型 random geochemical model
随机性地质地球物理模型 random geologic(al) geophysical model
随机性地质模型 random geologic(al) model
随机性故障 random fault
随机性检验 randomization test; randomness test
随机性模型 random model; stochastic model
随机性试验 randomization test; randomness test
随机性污染 random pollution
随机性研究 stochastic study
随机性自动机 random automation
随机需要 probabilistic demand
随机序列 random sequence; random series; stochastic sequence; stochastic series
随机序列模型 stochastic sequential model
随机选通信[讯]号 random gate signal
随机选样 sample taken at random
随机选择 randomize; random selection; stochastic selection
随机选择算法 stochastic selection algorithm
随机选择原理 random selection principle
随机延误 random delay
随机样本 chance sample; probability sample; random sample
随机样品 chance sample; probability sample; random sample
随机移动法 random walk method
随机因素 enchancement factor
随机引出 stochastic ejection; stochastic extraction
随机引出系统 stochastic extraction system
随机引晶 random seeding
随机应变 according to circumstances; random strain; random variable; stochastic strain
随机应力 random stress; stochastic stress
随机映射 stochastic mapping
随机用户均衡【交】stochastic user's equilibrium
随机游动 random walk
随机游动法 random walk method
随机游走扩散模拟 random walk simulation
随机有限元 stochastic finite element
随机有限状态 stochastic finite state
随机有限状态自动机 stochastic finite state automata
随机有序样本 random ordered sample
随机预报 random forecast; stochastic forecast
随机预测 random prediction; stochastic prediction
随机元件计算机 probabilistic machine
随机元素 random element; stochastic element
随机原点间隔 random origin interval
随机原样法 random sampling method
随机源 random source; stochastic source
随机运动 random motion; stochastic motion
随机再散列法 random rehash method
随机再散列技术 random rehash technique
随机噪声 random noise; stochastic noise
随机噪声电压 random noise voltage
随机噪声发生器 random noise generator
随机噪声干扰 random noise interference
随机噪声功率 random noise power
随机噪声级 random noise level
随机噪声加权网络 random-noise-weighting network
随机噪声相关 correlation of random noise
随机增量 random increment
随机张量场 random tensor field
随机涨落 random fluctuation
随机阵 random array; stochastic array; stochastic matrix
随机振动 random vibration; stochastic vibration
随机振动分析 random vibration analysis
随机振动环境 random vibration environment
随机振幅 random amplitude
随机振幅分布 random amplitude distribution
随机震动 random shock
随机震群模型 stochastically seismic clustering model
随机整数 random integer
随机正射坐标 randomized orthographic(al) coordinate
随机正弦波 random sine wave
随机支承加速度 random support acceleration
随机值变化 random value variance
随机指标 random index
随机制宜 contingency approach
随机质配 perittogamy
随机种群 random population
随机转变函数 stochastic transition function
随机转移网络 stochastic transition network
随机装载 random loading
随机状态 stochastic regime
随机状态转移矩阵 stochastic state transition matrix
随机自变量 random argument
随机自动机 probabilistic automaton [复 automata]
随机自动机模型 stochastic automata model
随机自动机推断 stochastic automata inference
随机自同态 random endomorphism
随机走动 random walk
随机走动理论 random walk theory
随机走动问题 problem of random walk; random walk problem
随机走向 random walk
随机组构 random fabric
随机最佳控制 stochastic optimal control
随机最佳器 random optimalizer
随即不贴合性结构 random lack-of-fit structure
随角度变化模式 angularly dependent mode
随角异色 colo(u)r-flop; flop
随角异色涂层 colo(u)r-flop coating
随角异色效应 flop effect
随角异色效应涂层 effect coating
随角异色效应颜料 effect pigment
随角异色效应中间涂层 effect basecoat
随角异色性能 down-flop property
随叫通道 on-call channel
随叫信道 on-call channel
随进钻头 follower
随景物而变化的 scene-dependent
随军医院 rolling hospital
随开存单 tap certificate of deposit
随客运列车发货 passenger train shipment
随控布局 control coordination overall arrangement
随矿物落下的顶板岩石 following stone
随里程递远递减的运价 tapering mileage scale rate
随炉冷却 furnace cooling
随路信令 channel associated signal(l)ing; in-channel signal(l)ing
随码轨道继电器 code following track relay
随模成型性 shapeability
随模浇口 set gate
随年龄变化的自然死亡率 change in natural mortality with age
随铺轨道 <坑道掘进后> following the track
随气候更换 to be changed depending on the weather
随气排出水分量 primage
随砌层 brought to course
随砌随手勾缝 struck joint; struck

joint work
随身单开小刀 pen knife
随身电话装置 bellboy
随身工具 paraphernalia
随身携带的行李 accompanied baggage
随身行李 accompanied luggage; personal luggage
随身用具 paraphernalia
随深度变化 variation in depth
随深度变化系数 coefficient of variation with depth
随深度取的砂样 depth-integrating sediment sample
随深度取砂器 depth-integrating sediment sampler
随时 at all time;at any time
随时变化的电容 time-varying capacitance
随时操纵 immediate manoeuvre
随时间变动的振幅 time-varying amplitude
随时间变化 time dependence; time history;time-variation
随时间变化的 time-dependent; time-varying
随时间变化的变形 time-dependent deformation
随时间变化的力 time-varying force
随时间变化的灵敏度 time-varied sensitivity
随时间变化的挠度 time-dependent deflection
随时间变化的曲线图 time-history plot
随时间变化的数据 time-variable data
随时间变化的梯度 time-varying gradient
随时间变化的信号测量 time-varying signal measurement
随时间变化的信号程序 time-dependent signal program(me)
随时间变化的增益 time-varied gain
随时间变化的正弦曲线梯度 time-varying sinusoidal gradient
随时间变化过程 time-dependent process;time-varying process
随时间的减少 time-dependent reduction
随时间而变的变形 time deformation
随时间而变化的关系 time history
随时间而发生的变形 time deformation
随时间改变的 time-varying
随时谨慎右转 right turn at anytime with care
随时谨慎左转 <相当于我国的右转> left turn at any time with care
随时可使用的 ready to operate
随时可收回的贷款 money on call
随时可收回的借款 money at call
随时可以安装 ready to mount
随时可用 ready-to-use
随时可用的材料 ready-to-use material
随时可用的胶 ready-to-use paste
随时适应系统 on-going system
随时修缮制度 maintenance system of random repair
随时右转 free right turn
随时支取账户 on call account
随手地质剖面 freehand geologic(al) profile
随手画草图 freehand sketching
随手可得的代用品 drop in substitute
随手移动 freehand motion
随水灌溉 application with irrigation water
随所规则 anywhere rule
随特派团出差 mission assignment
随挖随衬法 concrete-as-you-go lining method
随挖随盖 cut-and-cover;cut-and-fill
随挖随填 cut(-and)-fill; cut-and-fill excavation;cut-and-cover
随挖随填的隧道 cut-and-cover tunnel
随挖随填(施工)法 cut-and-cover method;cut-and-fill method
随挖随支护 immediate installation of tunnel support
随温度变化的 temperature-dependent
随温度而变的 temperature-dependent
随纹 tracking
随纹失真 tracking distortion
随纹误差 tracking error
随项目不同的 project-dependent
随行刀架 follower rest;travel(l)ing stay;travel(l)ing steady
随行电缆 travel(l)ing cable
随行扶架 following rest
随行工作台 pallet
随行夹具 follower fixture; pallet; travel(l)ing fixture
随行夹具交换装置 pallet changer
随行夹具式组合机床自动线 palletized transfer line
随行夹具梭动系统 pallet shuttle system
随行夹具主文件 pallet master file
随行客车 accompanied vehicle
随行人员 party
随行托板 pallet
随行托板交换装置 pallet changer
随行中心架 follower rest
随行装药 Langweiler charge; travel(l)ing charge
随压力变化的 pressure-dependent
随要随付 payable on demand
随意 at will
随意并联【电】permissive paralleling
随意布局 random placement
随意布线法 discretionary wiring (method)
随意抽查 random check
随意抽样 accidental sampling; random sample
随意存取 arbitrary access
随意的 arbitrary;permissive;voluntary
随意过账 random posting
随意划分 random work
随意活动用能 energy of voluntary activity
随意可溯图 randomly traceable graph
随意量 arbitrary quantity; quantum liberty;quantum placet
随意浓度 arbitrary concentration
随意排列 random work
随意区位工业 foot-loose industry
随意取材 random access
随意取样 grab sample
随意试样 random specimen
随意停车 optional stop
随意停车排列 haphazard parking arrangement
随意停机 request stop
随意停机指令 optional stop instruction
随意停生 optional halt
随意消除法 optional suppression
随意消费 optional consumption
随意行为 voluntary behavio(u)r
随意选取 cherry picking
随意选择 option
随意选择资料 optional information
随意运动 voluntary movement
随意涨价 wilful inflation of prices
随意整方的毛石 square random stone
随意支配的收入 discretionary income
随意中断 voluntary interrupt
随意驻留例行程序 optional resident routine
随意组合方石板路 flagstone path paved at random
随应力而变的 stress-dependent
随用灯 occasional light
随遇面波 ubiquitous surface wave
随遇平衡 indifferent equilibrium;neutral balance;neutral equilibrium
随遇生物 tychocoen; ubiquitous organism
随遇稳定 neutral stability
随遇植物 ubiquist
随员 attendant
随运乘客自用汽车的旅客列车 autotrain
随运动而变的阻力 motion-dependent resistance force
随运随拌混凝土 transit-mix(ed) concrete
随转顶尖 loose centre
随转阀 puppet valve
随转尾架 poppet
随转尾座 <车床的> puppet (head tail stock);poppethead
随转尾座顶尖 <车床的> puppet head center[centre]
随转尾座套筒 barrel of puppet head; sleeve of puppet head
随着季节的推移 as the season progresses
随着里程递远递减的运价 mileage scale rate;tapering rate
随着时间的推移 in-process of time
随钻测量 inclination measurement with drilling
随钻测斜仪 measurement with drilling inclinometer
随钻定向法 orientation while drilling

髓斑 <木材> medullary spot;pithray fleck

髓部不直(的木材)wandering heart
髓节 pith knot
髓裂 pith shake
髓射线 medullary ray;pith ray
髓石 pulpstone
髓线 medullary ray;pith ray
髓心 heart center[centre];pith <树木>
髓心板 heart board
髓心部 heart
髓心材 boxed heart
髓心的 medullary
髓心干裂 heart check
髓心厚板 heart plank
髓心偏离 wandering heart
髓心线 ray
髓质 medulla
髓状斑(纹)pith fleck

岁变 annual variation;yearly variation

岁差 precession of equinoxes
岁差变化率 rate of precession of equinox
岁差常数 constant of precession;precessional constant
岁差圈 precession circle
岁差系数 precession coefficient
岁差与章动改正量 precession and nutation correction
岁差运动 precessional motion
岁差周期 period of precession;period of precession of equinox; precessional period
岁差锥 precession cone
岁出 annual expenditures
岁出科目 account titles for annual expenditures
岁出预算 budget for annual expenditures;budget for annual expenses
岁入 annual income;annual revenue; revenue;yearly income
岁入财源 way and mean
岁入法案 revenue act
岁入分配 revenue haring
岁入分享 <美> revenue haring
岁入概算书 estimate of revenue
岁入关税 <为财政收入而征收的关税> revenue tariff
岁入税 revenue tax
岁入弹性 revenue elasticity
岁入下降(下滑)revenue slippage
岁入余缺 annual surplus or deficit
岁入预算 budget for annual receipts; estimate revenue
岁入债券 revenue bond
岁收 revenue;tunnel
岁首 beginning of year
岁首月龄 epact
岁星纪年 Jupiter cycle
岁修 annual maintenance; annual overhaul;annual repair;yearly maintenance;yearly overhaul
岁修养护 routine maintenance

遂安石 suanite

遂步退磁 progressive demagnetization
遂点爆炸 roll-along shooting

碎白云石 crushed dolomite

碎斑沉积 mortar bed
碎斑构造【地】mortar structure;porphyroclastic structure
碎斑结构 mortar texture
碎斑晶 clastic phenocryst
碎斑熔岩 mortar lava
碎斑岩 crushed porphyry;porphyroclasite
碎斑岩混凝土 crushed porphyry concrete
碎斑状 porphyroclastic
碎板 slashings
碎边机 chopper
碎边剪 scrap cutter
碎边剪切 cropping cut
碎边剪切机 scrap chopper
碎边结构 crush borer texture
碎冰 crushed ice;ice sludge;trash ice <混着水的>
碎冰堆 screw ice
碎冰机 icebreaking machine; ice crusher
碎冰块 brash ice;broken ice
碎冰楼 icing tower
碎冰片 bit;chipped ice;chipping ice
碎冰群 brash;brash ice;mush
碎冰制造机 chopped ice machine
碎冰装置 deicer;deicing
碎冰锥 ice pick
碎饼机 cake mill
碎波 breaker;breaking wave;broken water; broken wave; lipper; lop; wind lop;surf
碎波波高 wave height on breaking
碎波波高指数 breaking height index
碎波仓 cullet bin
碎波带 breaker zone;surf zone
碎波堤 wave breaker
碎波点 breaking point;wave breaking point
碎波点水深 breaker depth;breaking

S

depth
碎波级 scale of surf
碎波距离 breaker distance
碎波浪 breaker
碎波临界波高 breaker height
碎波临界水深 breaker depth; breaking depth
碎波区 breaker zone; breaking zone; surf zone
碎波区内向岸的海流 inshore current
碎波深度 breaker depth
碎波水花 white water
碎波水深 breaker depth; breaking depth; depth of breaking
碎波水深指数 breaking depth index
碎波特征 feature of breakers
碎波线 breaker line; breaking line; surf line
碎波相似性参数 surf similarity parameter
碎波行程 breaker travel; breaking travel
碎波在近岸区域的不规则振荡 surf beat
碎波指数 breaker index; breaking index
碎玻璃 cullet; scrap glass
碎玻璃比 cullet ratio
碎玻璃仓 cullet bunker; cullet silo
碎玻璃承接盘 cullet pan; cullet tray
碎玻璃承接器 cullet catcher
碎玻璃传送机 cullet conveyer [conveyor]
碎玻璃传送链 cullet conveyor chain
碎玻璃堆场 cullet yard
碎玻璃返回量 returned quantity of broken glass
碎玻璃划伤 cullet cut
碎玻璃沥青路面 glassphalt
碎玻璃料拣选机 cullet picker
碎玻璃料破碎机 cullet crusher
碎玻璃料压痕 cullet impression
碎玻璃溜槽 cullet chute
碎玻璃配合料 raw cullet
碎玻璃片 chunk(glass); cullet
碎玻璃清洗机 cullet washer
碎玻璃熔成的玻璃 glass melted from cullet
碎玻璃熔制玻璃 glass melted from cullet
碎玻璃桶 cullet catcher
碎玻璃箱 cullet catcher; edge catcher
碎玻璃屑 splinters
碎布 clout; macerated fabric; rag; rag cuttings
碎布打浆机 rag engine
碎布滚花涂装 rag-rolled finish
碎布胶料 rag mix
碎布胶料压延机 rags calender
碎布开松机 rag devil
碎布敲打机 rag thrasher
碎步点记载手薄 detail point description
碎步走 <避免桥梁共振> broken step
碎部测量 detail(ed) survey; detailing (survey); surveying in parts; survey of details; survey situation
碎部测量记录 record for surveying in parts
碎部测图 detailed survey; survey of details; survey situation
碎部地形 detailed topography
碎部点 detail point; place mark; stadia point
碎部描绘 rendering of detail
碎部清晰度 sharpness of detail
碎部图 detailed plan
碎部遗漏 detail missing; missing detail
碎层云 fracto-stratus
碎碴隔拦 debris barrier
碎成角砾 brecciate
碎瓷粉 bit stone
碎搓板形地面 washboarded surface
碎搓板形路面 washboarded surface
碎大理石 broken marble; crushed marble
碎的熔岩 crushed lava
碎锭机 ingot breaker
碎断 cataclasm
碎废玻璃 cullet
碎废钢 shred scrap
碎干草 chopped hay; ground hay
碎钢 scrap steel
碎钢粉 crushed steel
碎高层云 altostratus fractus
碎高炉泡沫矿渣 crushed foamed blast-furnace slag
碎高炉渣 crushed ballast-furnace slag
碎稿筛 short-straw sieve
碎谷 boussir
碎谷机 grain crusher
碎骨机 bone cutter; bone grinder
碎果壳 <堵漏用> ground nutshell
碎核 fragmentation nucleus
碎花 shivering
碎花岗岩 broken granite; crushed granite
碎花岗岩混凝土铺面 granolith
碎花铁片 break iron
碎化 fragmentation
碎黄铜 scrap brass
碎混凝土 rubble
碎混凝土骨料 aggregate of broken concrete
碎混凝土块 broken concrete
碎混凝土块丁坝 broken concrete groin; broken concrete groyne
碎火山灰岩 crushed slag
碎积云 fracto-cumulus
碎集煤 attritus
碎集石 clusterite
碎甲 scabbing
碎减式破碎机 reduction crusher
碎减式旋回破碎机 reduction gyratory crusher
碎件 aster
碎浆机 pulper; pulping engine; pulp kneader
碎浆机槽 kneading trough
碎胶机 waste grinding machine
碎胶器 glue grinder
碎焦火钩 pricker; slash bar
碎焦机 coke breaker
碎焦炭 crushed coke
碎解 size degradation
碎金属 scrap metal
碎茎打麻机 hemp-broken and scutcher; ribboner
碎茎机 breaker; breaking machine
碎晶凝灰岩 peperino
碎块 broke; chip(ping)s from masonry ruins; chop; dice; fragment; mammock; quarry waste; shiver; shives; bat <黏[粘]土等的>
碎块的 fragmental; fragmentary
碎块构造 lumpy structure; rubbish structure
碎块混凝土丁坝 broken concrete groin
碎块混凝土块 hard core
碎块结构 lumpy texture
碎块熔岩 block lava
碎块散体结构 clastic loosen texture
碎块(体)力学 clastic mechanics
碎块体模型 clastic model
碎块土 lumpy soil
碎块形状 fracture pattern
碎块岩石 rock fragments
碎块状 clastic
碎块状构造 cloddy pulverescent structure; clod structure
碎块状结构 cloddy pulverescent texture; cloddy structure; clod texture
碎矿板 anvil; bucking iron plate
碎矿仓 crushing pocket
碎矿车 crusher car
碎矿车间 ore-breaking plant
碎矿锤 bucker
碎矿工 bucker
碎矿机 bruising mill; bucker; bucking hammer; lump breaker; ore breaker; ore crusher
碎矿机表皮效果 alligator skin effect
碎矿机颊板 cheek plate
碎矿石 crushed ore
碎矿石给矿机 crushed ore feeder
碎矿石给料机 crushed ore feeder
碎矿石喂料机 crushed ore feeder
碎矿用捣棒 dolly
碎矿渣 crushed slag; slag sand
碎矿渣块 <铺路用> slag ballast
碎浪 breaker; broken sea; broken wave; comber; surge wave
碎浪带 surf zone
碎浪海面 short sea
碎浪器 wave breaker; wave splitter
碎浪区 surf zone
碎浪深度 depth of breaking
碎浪特征 feature of breakers
碎浪艇 surf boat
碎浪艇船员 surf boat man; surf man
碎浪线 line of breakers; surf line
碎砾沉积(物) rudaceous sediment
碎砾石 broken gravel; crushed gravel
碎砾石骨料 crushed gravel aggregate
碎砾石集料 crushed gravel aggregate
碎砾石路 crushed-gravel road
碎砾石砂 broken gravel sand; crushed gravel sand
碎砾石土 debris gravelly soil
碎砾岩 psephite [psephyte]; rudite [rudyte]
碎粒 crushed shot; dice; particle
碎粒包尔特金刚石 crushed bort
碎粒长石 crushed feldspar
碎粒构造 granulitic texture
碎粒化方式 cataclastic granulation way
碎粒金刚石 industrial diamond
碎粒料 crushed material
碎粒泥煤 crumble peat
碎粒石英石 crushed quartzite
碎粒饲料 granulated feed
碎粒岩 micro-cataclasite
碎料 crushed aggregate; outthrow; sliver
碎料板 flake board; particle board <用合成树脂将碎木粒黏[粘]合制成>
碎料板芯 chipcore
碎料板芯胶合板 particle board core plywood
碎料板压机 chip board press
碎料的 scrappy
碎料分类 conditioning of scrap
碎料工 stocker
碎料胶合板实心门 particle board solid core door
碎料筛 bull screen; sliver screen
碎裂 crash; blasting; cleavage fracture; crumble away; crumple; disintegration; disruption; egg-shelling; fire crack; fractionation; fragmentation; shatter; shivering; spall away; spalling; splintering; breaking up <冰的>
碎裂斑状(结构)的 clastoporphyritic
碎裂爆破 splitting blasting; splitting shot
碎裂变质【地】 catachosis
碎裂变质带 katamorphic zone
碎裂变质作用 catamorphism; cataclastic metamorphism
碎裂带 shattered belt; shattered zone
碎裂的 cataclastic; spalt
碎裂法 fragmentation method
碎裂反应 fragmentation reaction
碎裂缝 <水泥混凝土路面的> spalled joint
碎裂构造【地】 cataclastic structure
碎裂花纹 crackle; crackling
碎裂机 fragmentiser
碎裂角砾岩 cataclastic breccia
碎裂结构【地】 cataclastic texture; clastic texture; kataklastic texture
碎裂矿脉 rubbly reef
碎裂力 spalling
碎裂砾石 cataclastic conglomerate
碎裂流 cataclastic flow
碎裂流动 cataclastic flow
碎裂煤 cataclasitic coal; crumble coal; shatter coal
碎裂模型 fragmentation model
碎裂片 shatter
碎裂器 cracker; disintegrator
碎裂体系 fragmentary system
碎裂纹 chip crack; shatter crack
碎裂物 sliver
碎裂屑堆【地】 shatter cone
碎裂岩 cataclastic rock; kataklastic rock; cataclasite
碎裂岩系 cataclastic series; cataclasite series
碎裂应力 spalling stress
碎裂褶皱【地】 disjunction fold; disjunctive fold
碎裂作用 clastation
碎裂作用方式 cataclasis way
碎鳞 discale
碎鳞机 scale breaker
碎鳞轧辊 discaling roll
碎流板 stream breaker
碎炉渣 crushed lump slag
碎铝 scrap alumin(i)um
碎铝片 alumin(i)um foil
碎卵石 crushed gravel
碎乱天空 amorphous sky
碎落 chip; slough
碎落台 berm at the foot of catting slope; stage for heaping soil and broken rock; stage for soil and broken rock
碎煤 culm; drossy coal; fine coal; fracture of coal; rice coal; slack; small coal
碎煤锤 coal hammer
碎煤堆积场 culm bank
碎煤机 coal breaker; coal cracker; coal crusher; coal pulverizer
碎煤射流 <水力碎煤的> breakdown agent
碎煤装载机 Anderton's shearer-loader
碎煤装置 coal pulverizing installation
碎米 cracked grain
碎磨 disintegration
碎木 matchwood; slashings; wood refuse
碎木板 flake board; particle board; wood chipboard
碎木板心 core board
碎木材料 wood particle material
碎木电板 wood chipboard
碎木堵塞 spackling
碎木混凝土 chipped wood concrete; wood waste concrete; wood cement particle board
碎木机 bucker; defibrator; edging grinder; grinder; hogger; wood

S

grinder
碎木胶合板 chipboard; particle board; shredded wood fibre board; wood(en) chipboard
碎木焦油 wood splitting tar
碎木料 particles
碎木片 cutstuff
碎木片吸音天花板 wood chip absorbent ceiling
碎木丝胶合板 shredded wood fibre board
碎呢除尘机 rag shaker
碎呢地毯 rag rug
碎呢机 rag grinder
碎呢开松机 knot breaker
碎泥刀 mud breaker; mud crusher; mud cutter
碎黏[粘]土 shattered clay
碎黏[粘]土砖 clay brick hardcore
碎膨胀矿渣 crushed expanded cinder
碎膨胀炉渣 crushed expanded cinder
碎坯 pitchers
碎皮 scrap leather
碎片 brash; broken stone; chat; chippings; crumb; fraction; fragment; fragmental grain; junk; mammock; patch; rasura; rive; scrap; shard; shives; shred; smithereens; smithers; splinter
碎片层 layer of chippings
碎片挡板装置 debris guard
碎片的 chippy; cragged; fractional; fractionary; fragmental; fragmentary
碎片度 chipping degree
碎片堆 heap of debris
碎片防护板 debris guard
碎片防护装置 debris guard
碎片飞散 separation of fragments
碎片废料 filament waste
碎片峰值 fragment peak
碎片护罩装置 debris guard
碎片垃圾 debris
碎片离子 fragment ion
碎片泥层 argille scagliose
碎片谱法 fragmentography
碎片石 spawl
碎片体煤岩 micrinite
碎片新月形结构地 bogenstruktur
碎片压痕 <木材> chip marks
碎片硬合金镶嵌物 scrap bit inserts
碎片云 pannus
碎片榨干离心机 chip wringer
碎片榨干器 chip wringer
碎片状结构 fragmental structure
碎片状金刚石 chip diamond
碎片状破裂 splintering fracture
碎片状态试验 fragmentation test
碎片钻头 chip bit
碎漂石 crushed boulder
碎拼大理石地面 broken-marble patterned flooring
碎平板玻璃 sheet cullet
碎铅 scrap lead
碎青石 blue metal
碎青铜 scrap bronze
碎熔渣 crushed slag
碎散 perish
碎砂 waste
碎砂机 sand cutter; sand grinding mill
碎砂砾堆积 gruss
碎湿 wet comminution
碎石 angular cobble; brash; bray stone; break stone; broken rock; broken stone; channery; chat; cobble; crushed aggregate; crushed rock; crushed stone; crushing; detritus; fragmental stone; galet; reduced stone; rock destruction; spall; stone ballast; stone fragment; stones
碎石坝 debris dam; rock debris dam
碎石柏油路面 macasphalt type pavement
碎石板 <锤式碎石机> breaker plate
碎石保持能力 stone retention
碎石崩落 debris avalanche
碎石边坡 stone slope
碎石编框 <安放在水工建筑物下游防止冲刷，系印度北部习用的名词> tarungar
碎石仓 crushing pocket; rock storage bin
碎石层 broken stone course; crushed-stone course; crushed-stone stratum; layer of broken stone; metal(l)ing; tie-bed <铺铁路的>
碎石叉 stone fork
碎石厂 crushed-stone plant; crushing mill; crushing plant; rock-crushing plant; stone mill
碎石场 crushing plant
碎石充填 rock-filling
碎石充填滴滤池 rock-filled trickling filter
碎石除藻池 rock filter for algae removal
碎石船 rock breaking vessel; rock-crushing vessel
碎石锤 ballast hammer; boss hammer; bucker; chipping hammer; chisel breaker; flail; granulated hammer; granulating hammer; gravel hammer; knapper; knapping hammer; mickle hammer; spall(ing) hammer
碎石锤重力冲击球 <从不同高度丢下，利用重力冲击，以破碎超尺寸的岩块和孤石> headache ball
碎石打底 broken stone base
碎石大锤 boss hammer
碎石带 <土内> stone line
碎石道 Tolford-foundation
碎石道床 ballast bed
碎石道砟 broken stone metaling; crushed rock ballast; crushed-stone ballast
碎石的 calculifragous; lithoclastic
碎石底 rubble bed
碎石底层 crushed-stone base course; macadam base
碎石底料 crushed rock base material
碎石底座 equalizing bed
碎石地基 macadam base
碎石垫层 broken stone course; crushed-stone base; gravel underlayer
碎石堆 pile of rubble; rubble drift; scree; rock debris; debris <积在山底等处的>
碎石堆场 crushed-stone stockpile; gravel yard
碎石堆积层 tipped stone rubble
碎石墩 stone pillar
碎石防挡装置 debris guard
碎石覆盖层 crushed stone blanket
碎石镐 rock picker
碎石工 knapper; moellon <填塞墙缝>
碎石工厂 stone breaking plant
碎石工程 stone packing
碎石工具 breaker tool
碎石工人 spalder
碎石供料机 crusher feeder
碎石沟 dribble
碎石谷坊 loose rock check dam
碎石骨架 stone skeleton
碎石骨料 ballast aggregate; crushed (rock) aggregate; crushed-stone aggregate; macadam aggregate; stone aggregate
碎石骨料的混合路面 macadam aggregate type road mix surfacing
碎石骨料混合料 macadam aggregate mix
碎石骨料混凝土 chip(ping)s concrete
碎石过程 stoning
碎石和筛选移动作业车 mobile stone crushing and screening plant
碎石护岸 stoning
碎石滑动 debris slide
碎石灰石 broken limestone; crushed limestone
碎石混合骨料 macadam aggregate mix
碎石混合料 <拌有沥青或其他黏[粘]结料的> macadmix
碎石混合料工厂 macadam mixing plant
碎石混合料路 mixed macadam
碎石混凝土 ballast concrete; concrete made with crushed stone aggregate; crushed-stone concrete
碎石机 ballast crusher; ballastic crusher; breaker; breaker block; crusher; crushing machine; crushing mill; grainer; granulator; grinder; gritting machine; knapper; knapping machine; lump crusher; reduction crusher; rock breaker; rock crusher; rock-crushing machine; rock cutter; rock smasher; stone breaker; stone crusher; stone crushing machine; stone mill; stoner
碎石机的工作指数 work index of crusher
碎石机的加料漏斗 breaker feed hopper
碎石机的加料器 breaker feeder
碎石机的润滑油 rock crusher lubricating grease
碎石机颚板 alligator jaw; jaw crusher plate; jaw plate
碎石机粉尘 breaker dust
碎石机钢带接头 alligator steel belt lacing
碎石机轧辊 crushing roll
碎石机轧石的石屑 crusher screenings
碎石机轧碎石块比例 reduction ratio
碎石机罩 alligator bonnet
碎石基层 aggregate base course; broken stone packing; crushed-stone base; crushed-stone bed; macadam base; macadam foundation
碎石基础 broken stone base; crushed-stone base; macadam foundation; stone base
碎石基床 ballast bed; broken stone bed; crushed-stone bed; macadam base
碎石基底 macadam base; stone base
碎石级配 crushed-stone grading
碎石集料 ballast aggregate; crashed stone; crushed rock aggregate; crushed-stone aggregate; macadam aggregate; stone aggregate
碎石集料混合料 macadam aggregate mix
碎石集料混凝土 chip(ping)s concrete
碎石剂 lithontriptic
碎石夹层 broken stone interlayer
碎石搅拌机 macadam mixing plant
碎石块 channery; crushed hard rock; gal(l)et
碎石立方体系数 cubicity factor of crushed stone
碎石沥青磨耗层 stone seal
碎石料 <铺路用的> metal(ling)
碎石料溜槽 material debris chute
碎石流 debris flow; rock fragment flow; rubble flow
碎石榴石 crushed garnet
碎石滤器 stone filter
碎石路 ballast road; ballast road bed; broken rock road; crushed-stone macadam; crushed-stone road; gravel path; macadam(ized) road; metalled road; rubble bed; stone road; stoneway; macadam
碎石路基 broken stone bed; hard core; macadam bed; macadam substructure
碎石路面 broken rock pavement; cement-bound surface; cement-bound surfacing; crushed-stone pavement; gravel pavement; macadam aggregate type; macadamized road surface; macadam pavement; macadam surface; macadam surfacing; metal surface; stone road; Telford pavement; broken stone road; metal road; premixed macadam; macadam <冷铺焦油沥青的>
碎石路面层 macadam surface
碎石路面道路 macadam road
碎石路面的夹层铺筑法 sandwich process macadam
碎石路面施工法 Macadam's construction
碎石路面筑路法 Macadam's construction
碎石路碾 macadam roller
碎石路施工方法 macadam road construction method
碎石路修筑法 macadamization (of road)
碎石盲沟 spall drain(age)
碎石冒落 dribbling
碎石面层 broken stone pavement; crushed rock surfacing
碎石面饰 <灰泥未干时，压碎石入内的一种墙壁面饰法> depreter[depeter]
碎石碾压路 macadamized road
碎石耙 ballast rake
碎石排水层 broken stone drainage layer; crushed-stone drainage layer
碎石排水沟 spall drain(age); stone-filled drain(age)
碎石跑道 macadamized runway
碎石片 chat; chip of stone; chipping; chip stone; fragments of stone; gal(l)et(ing) quarry chip(ping)s; spall; splinter of stone; stone chip(ping)s; stone fragment; stone splitter
碎石片嵌灰缝 galleting
碎石铺道法 macadamizing
碎石铺底 bottoming; pitching
碎石铺路 macadamized road
碎石铺路的 metalled
碎石铺路法 macadamization(of road)
碎石铺路机 macadam spreader
碎石铺面道路 metalled road
碎石铺面的 aggregate-surfaced
碎石铺砌层 broken rock pavement; crushed-stone pavement
碎石砌谷坊 rubble masonry check dam
碎石砌筑 stone rubble masonry(work)
碎石器 lithoclast; lithoconion; lithotriptor; lithotrite
碎石钳 lithotrite
碎石清除术 lithocenosis
碎石清晰性 clarity of detail
碎石区 crushed zone
碎石取样 crushed-stone sampling

S

碎石撒布机 chip(ping)s spreader; crushed aggregate spreader; spreader of crushed stone; stone spreader; macadam spreader
碎石砂 broken stone sand; crushed sand; stone sand
碎石筛 rock screen
碎石筛分机 rock screening machine
碎石筛石厂 crushing and screening plant
碎石设备 crushing equipment; crushing plant; detritus equipment; rock-crushing plant; stone breaking plant
碎石生物滤池 rock biologic(al) filter
碎石石工 boss hammer
碎石饰面 pebbledash; roughcast
碎石术 lithoclasty; lithoplaxy; lithothrypty; lithotripsia; lithotripsy; lithotrity
碎石摊铺机 aggregate spreader; chip spreader; macadam spreader; spreader; stone spreader
碎石填充 hand rubble fill
碎石填缝屑 <细粒> chippings
碎石填料 flood coat for chip(ping)s
碎石填料系统 rock-media system
碎石填平层 crushed-stone level(l)ing course
碎石填筑的 debris-filled
碎石土 debris soil; soil aggregate; soil-aggregate mixture; gallet
碎石土面层 granular soil stabilization; soil-aggregate surface
碎石圬工 scrabbled rubble; stone rubble masonry(work)
碎石洗出术 litholapaxy
碎石下部结构 macadam sub-structure
碎石小路 metal(led) path
碎石楔 spalling wedge
碎石选分设备 chip chisel; chip rejector
碎石压路机 macadam roller
碎石样 gravel sample
碎石英石 crushed quartz
碎石錾 chipping chisel
碎石凿 chipping chisel
碎石凿船 chisel breaker vessel
碎石渣 chip ballast; fragments of rock; knocking; gallet; broken stone chip(ping)s
碎石找平层 broken stone level(l)ing course
碎石整平层 broken stone level(l)ing course
碎石柱 gravel pile; stone column
碎石筑路法 macadamization(of road)
碎石(砖)垫层 hard core
碎石桩 broken stone pile; gravel pile; stone column
碎石状的 rubbly
碎石锥 scree cone
碎树脂 broken resin
碎燧石 crushed flint
碎陶片 crock
碎条 shred
碎条形贴面板 broken-strip(e) veneer
碎条(状花)纹 broken strip(e)
碎铁 bushel iron; scarp iron
碎铁堆 scrap heap
碎铁机 pig breaker
碎铁片 jacks; iron waste
碎铁用起重机 stamp work's crane
碎铜 copper cuttings; scrap copper
碎土 crumble; hack(ing)
碎土车 break barrow
碎土机 break barrow; clay crusher; clay cutter; clay mill; clay shredder; cultivator; grubber; soil pulverizer; soil shredder
碎土角 <工作部件的> approach angle
碎土块状的 cloddy
碎土耙 brake
碎土器 pulverizer; soil agitator; soil shredder
碎土松土压土器 cultimulcher
碎土镇压器 clod breaking roller; cultipacker
碎土整地刀板耢 soil conditioner
碎土转筒 cutting rotor
碎瓦片铺面 testaceum
碎纹 crackle
碎纹石路 crazy pavement
碎纹式玻璃器皿 crackled glassware
碎纹饰玻璃 crackled glass
碎纹涂料 crackle coating
碎纹釉 crackle glaze
碎污机 garbage disposer
碎无烟煤 flaxseed coal
碎雾 fog patches
碎匣钵熟料 saggar grog
碎屑 chippings; clast; crushed chip(ping)s; dregs; dross(coal); fines; litter; odd-come-short; offal; patch; riddings; rubbish; scum; smithereens; smithers; thrum; cultch <美>
碎屑白云岩 detrital dolomite
碎屑比 clastic ratio; detrital ratio
碎屑变形 clastic deformation
碎屑滨岸带体系 clastic shoe-zone system
碎屑槽 chip-breaker
碎屑产物 fragmentary product
碎屑场 spoil dump; spoil tip
碎屑车 spoil wagon
碎屑沉积层 detrital sediment
碎屑沉积(物) clastic deposit; detrital deposit; detrital sediment; clastic sediment; fragmental deposit; mechanical sediment
碎屑沉积岩 clastic sedimentary rock
碎屑成分 clastic constituents
碎屑储存器 spoil bin
碎屑储油层 detrital reservoir
碎屑带 detritus zone
碎屑的 chippy; clastic; drossy; fragmental; fragmentary; petroclastic
碎屑斗 spoil bucket
碎屑度指数 clasticity index
碎屑堆 spoil dump; spoil tip; talus
碎屑堆积线 debris line
碎屑惰性体 inertodetrimite
碎屑分散 clastic dispersion
碎屑腐殖体 humodetrinite
碎屑腐质 detritus
碎屑构造【地】fragmental structure; crumb structure
碎屑滑动 detritus slide
碎屑滑移 detritus slide
碎屑灰岩 calclithite
碎屑火山岩 fragmental volcanic rock
碎屑胶合板 particle board
碎屑角砾岩 clastic breccia
碎屑结构 clastic texture; fragmental texture
碎屑结晶质 clasto-crystalline
碎屑镜质体 vitrodetrinite
碎屑颗粒 clastic grain
碎屑颗粒结构 texture of clastic grains
碎屑壳质体 liptodetrinite
碎屑坑 spoil dump; spoil tip
碎屑矿床 detrital ore deposit
碎屑矿物 detrital mineral
碎屑矿物质 detrital mineral mater
碎屑垃圾 detritus rubbish
碎屑流 debris flow
碎屑泥灰岩 clastic marl
碎屑凝灰岩 detrital tuff
碎屑凝胶体 detrogelinite
碎屑坡 detrital slope
碎屑器 chiprupter
碎屑侵蚀变形 clastomorphic deformation
碎屑熔岩结构 pyroclastic lava texture
碎屑砂 detrital sand
碎屑剩余磁化强度 detrital remanent magnetization
碎屑石 clasolite; debris
碎屑石膏 gypsum sand
碎屑石灰凝灰岩 detrial lime tuff
碎屑石灰岩 clastic limestone
碎屑丝质体 fusinite splitter
碎屑钛铁矿 menaccanite
碎屑微粒体 detritomicrinite
碎屑稳定体 liptodetrinite
碎屑物 clastics
碎屑物的变化 change of fragments
碎屑物质 fragmentary material
碎屑楔形层 clastic wedge
碎屑楔(状)体 clastic wedge
碎屑压痕 chip marks
碎屑压制板 particle board
碎屑岩 clastic rock; conglomerate; detrital rock; fragmental rock; fragmentary rock; petroclastic rock
碎屑岩储集层 fragmental reservoir; detrital reservoir
碎屑岩胶结物 cements of clastic rocks
碎屑岩类 clastic rocks
碎屑岩类型 clastic rock type
碎屑岩脉 clasolite; clastic dike[dyke]
碎屑岩潜山油气藏趋向带 burial hill pool trend of clastic rock
碎屑岩墙 clasolite; clastic dike[dyke]
碎屑岩石 ravelly ground
碎屑岩筒 clastic pipe
碎屑岩楔 clastic wedge
碎屑岩杂基 matrix of clastic rocks
碎屑异常 clastic anomaly
碎屑油储 clastic reservoir
碎屑织构 clastic texture
碎屑质的【地】fragmental
碎屑皱片 scrap crepe
碎屑状的 clastic; detrital; fragmental
碎屑状废金刚石 scrap diamonds
碎屑状土(壤) crumbling soil
碎屑锥 cinder cone; detrital cone; pyroclastic cone
碎屑组分 attritus
碎屑组构 clastic fabric
碎屑组合 clastic association
碎锌 scrap zinc
碎性 fragility
碎性系数 crushability factor
碎修 jobbing
碎锈铁片 mill scale
碎岩 rock debris
碎岩堆 pile of rubble
碎岩滑动 creep slide; detritus slide
碎岩机 rock breaker; rock cutter
碎岩筛分机 rock screen
碎岩石 crushed rock
碎岩屑 detritus(rubbish); fragmental debris; fragmentary debris
碎岩屑扇形地 detrital fan
碎研机 breakdown mill
碎盐 rubbish salt
碎阳极残头 crushed scrap anode butt
碎样机 sampling mill
碎硬石 crushed hard rock
碎雨云 fracto-nimbus; fractus nimbus; scud
碎云 detached cloud; fractus
碎云玻璃 sideromelane
碎云母 mica scrap
碎云雨 fracto-nimbus
碎渣 crushed slag; disintegrating slag
碎渣机 ash breaker; ash crusher; cinder mill; clinker crusher; slag breaker; slag crusher
碎渣填料 flood coat for chip(ping)s
碎胀系数 breaking and expanding coefficient
碎枝杈 slashings
碎纸机 kneader; kneading machine
碎纸机刀片 kneader bar; kneader blade
碎纸机搅拌臂 kneader arm
碎纸胶合板 chipboard
碎纸片 tatter
碎纸塑料 paper-based laminate
碎制矿渣 crushed cinder sand
碎制炉渣 crushed cinder
碎砖 brick bat; brick dust; brick rubble; broken brick; chip(ping)s from brick ruins; crushed brick; ground brick
碎砖垫层 broken brick hardcore
碎砖堆 pile of rubble
碎砖骨料 aggregate of broken bricks; brick aggregate; brick hardcore; broken aggregate
碎砖混凝土 brick concrete; broken brick concrete; crushed brick concrete
碎砖混凝土衬块 broken brick concrete filler brick
碎砖混凝土墙板 broken concrete wall slab; crushed brick concrete wall slab
碎砖混凝土填充块 crushed brick concrete filler slab
碎砖混凝土填充块体 broken brick concrete filler brick
碎砖混凝土填块 broken brick concrete filler brick
碎砖机 brick breaker
碎砖基层 broken brick base
碎砖集料 brick aggregate; brick hardcore; broken aggregate
碎砖块 hard core
碎砖盲沟 brickbat drain
碎砖排水盲沟 brick bat drain
碎砖片 brick bat
碎砖器 brick breaker; brick crusher
碎砖三合土 lime earth-broken brick concrete
碎砖石 chip(ping)s from masonry ruins
碎砖石垫层 hard core
碎砖头 brick bat
碎砖屑 broken chip(ping)s
碎砖渣土 brick rubbish

隧道 shaft alley; tunnel; syrinx <古埃及石墓中的>; black hole <俚语>

隧道帮柱 jamb
隧道报警器 tunnel alarm
隧道爆破 tunnel blasting
隧道爆破最小抵抗线 burden
隧道被覆 tunnel lining
隧道被坍流土体充塞 fill-up with flow-out earth
隧道崩坍 collapse of tunnel
隧道避车洞 side pocket
隧道边墙 tunnel(side) wall
隧道边墙底座混凝土 curb concrete
隧道标 tunnel post; tunnel sign
隧道标高 tunnel level
隧道标志与报警 tunnel indicators and alarms
隧道病害 tunnel deterioration
隧道薄壳衬砌 thin-shell lining

隧道补挖 chisel(l)ing of tunnel surface
隧道擦除 tunnel erase
隧道裁弯 tunnel cut-off
隧道裁弯取直 tunnel cut-off
隧道采空区入口堵墙 cross-off
隧道操作人员 tunnel operator
隧道槽 tunnel slot
隧道测量 tunnel survey(ing)
隧道测量范围 scope of survey for tunnel
隧道插板 forepole;poling plate;spile
隧道插板法 forepoling
隧道插板掘进法 poling board method of tunnel(l)ing
隧道长度 tunnel length
隧道超前纵梁 horsehead
隧道超挖 overbreak
隧道车道管理信号 lane-control signal
隧道衬垫 tunnel pack(ing)
隧道衬砌 lining of tunnel;tunnel liner;tunnel lining
隧道衬砌防排水图 drainage and waterproofing system of tunnel lining
隧道衬砌钢板 tunnel liner(steel) plate
隧道衬砌厚度 thickness of tunnel lining
隧道衬砌环 ring of tunnel lining;tunnel ring
隧道衬砌机 tunnel lining machine
隧道衬砌几何形状 tunnel lining geometry
隧道衬砌内轮廓 inner contour of tunnel lining
隧道衬砌填料 sealing of tunnel lining
隧道衬砌用缸砖 tunnel clinker
隧道撑板掘进法 poling board method of tunnel(l)ing
隧道初期支护 initial support;preliminary support
隧道初试衬砌 primary tunnel liner
隧道打眼车 tunnel drilling car
隧道单位阻力 specific tunnel resistance
隧道导电 tunnel conduction
隧道导洞 pioneer bore;pioneer hole;pioneer tunnel;tunnel heading;pilot tunnel
隧道导洞掘进法 pilot-tunnel method
隧道导坑 tunnel heading
隧道导坑法 drift method
隧道导水管 tunnel(led)aqueduct
隧道导向控制 direction control of tunnel
隧道倒拱 tunnel invert
隧道的顶部 top heading
隧道的覆盖率 rate of covering of tunnel
隧道的人行道 footway in tunnel
隧道的中央导洞 center heading
隧道灯具 tunnel luminaire
隧道底板 tunnel floor;tunnel invert
隧道底板宽度 tunnel floor width
隧道底部掏槽 holing of slab tile
隧道底撑 invert strut
隧道底高程 tunnel invert elevation
隧道底沟 bottom channel
隧道底混凝土 invert concrete
隧道底基 sole of tunnel
隧道底面积 tunnel floor area
隧道地表沉陷 ground surface subsidence over tunnel;surface settlement
隧道电流 tunnel current
隧道电压 tunnel voltage
隧道电阻器 tunnel resistor
隧道吊管上的支撑与定位 support and registration from drop tube
隧道调查<地形、地质、有无断层以及旋工条件等> investigation of tunnel
隧道顶板 ceiling slab;roof of tunnel;tunnel roof
隧道顶部合拢 top closing
隧道顶盾 top shield
隧道顶拱衬砌 arch lining of tunnel
隧道顶管施工法 tunnel(l)ing by pipe jacking
隧道顶截面 crown-section of tunnel
隧道顶钎 roof bolting
隧道定位 layout of tunnel;tunnel location
隧道定线 alignment of tunnel;alignment of tunnel;fixing the tunnel alignment;tunnel location
隧道冻结法施工 freezing method of tunnel(l)ing
隧道冻结施工法 freezing method of tunnel(l)ing
隧道洞顶 tunnel roof
隧道洞顶锚栓 roof bolt
隧道洞顶坍落 roof fall
隧道洞孔 tunnel opening
隧道洞口 tunnel access;tunnel portal
隧道洞口棚架 head-house
隧道洞口伪装 tunnel entrance pretension
隧道洞门 portal;tunnel face;tunnel portal
隧道洞门室 portal chamber
隧道洞门限界 portal clearance
隧道洞身 tunnel body;tunnel trunk
隧道洞围岩 tunnel surrounding rock
隧道堵塞段 tunnel plug
隧道堵头 tunnel plug
隧道断面轮廓 tunnel contour
隧道断面收敛 tunnel cross section convergence
隧道对角变形 tunnel diagonal deformation
隧道盾构 jumbo;tunnel(ling)shield
隧道盾构的液压导向千斤顶 hydraulic steering jack of tunnel shield
隧道盾构法 shield method;shield tunnel(l)ing method
隧道躲避所 manhole in tunnel
隧道二极管探测器 tunnel-diode detector
隧道二极管振荡器 oscitron;tunnel-diode oscillator
隧道发射 tunnel emission
隧道法施工 tunnel(l)ing;tunnel(l)ing construction method
隧道防护 protection of tunnel
隧道防护板支架 tunnel shield
隧道防护信号 tunnel signal
隧道防护信号机【铁】tunnel signal
隧道防护支架 tunnel(ling)shield
隧道防水 tunnel waterproofing
隧道防水(密闭)层 tunnel sealing
隧道防烟问题 smoke problem in tunnel
隧道防灾设施 disaster prevention facility of tunnel
隧道放样 setting-out of tunnel
隧道飞拱 flying arch
隧道非破损探查 non-destructive investigation of tunnel
隧道分岔段 tunnel fork
隧道风 tunnel wind
隧道风风口 air vent of tunnel
隧道风机控制箱 control box for tunnel fan
隧道风流 tunnel airflow
隧道附加阻力 additional resistance due to tunnel;additional resistance for tunnel
隧道附加阻力换算坡度 equivalent gradient of additional resistance on tunnel
隧道覆盖率 rating of covering of tunnel
隧道改建 reconstruction of tunnel
隧道改建工程 tunnel reconstruction work
隧道干燥 tunnel drying
隧道干燥器 canal drier[dryer]
隧道干燥窑 canal drier[dryer]
隧道钢拱架 tunnel steel arch truss
隧道钢模板 tunnel steel form(work);tunnel steel moulding plate
隧道钢支护 steel tunnel support
隧道高程 tunnel level
隧道工 sand hog;tunneler
隧道工程 tunnel engineering;tunnel construction;tunnel driving;tunnel(l)ing;tunnel works
隧道工程地质勘察 engineering geologic(al)exploration of tunnel
隧道工程队 tunnel(l)ing gang
隧道工程工业 tunnel(l)ing industry
隧道工程学 tunnel engineering
隧道工人 drifter;ground-hog;sand hog;tunnel(l)er
隧道工人安全帽 Greathead shield
隧道工作面 tunnel face;tunnel front;tunnel heading
隧道拱 tunnel arch
隧道拱顶 tunnel vault;underpitch vault
隧道拱顶梁 crown bar of tunnel
隧道拱顶下沉 tunnel roof settlement
隧道拱肩 arch shoulder of tunnel
隧道拱圈 tunnel arch;tunnel ring
隧道管 tunneltron
隧道管段 tunnel tube section
隧道管节 tunnel ring
隧道管片膨胀衬砌 expanded liner of tunnel pipe pieces
隧道贯穿 tunnel(l)ing
隧道贯通误差 tunnel through error
隧道荷载 load on tunnel;loads for tunnel
隧道横断面 tunnel(cross-)section
隧道横断面测量 tunnel cross profiling;tunnel cross-section survey
隧道洪水防治 flood protection of tunnel
隧道护板 umbrella arch
隧道护盾尾部 tail of shield
隧道护拱 umbrella arch
隧道环形开挖法 ring cut
隧道换算坡度 converted gradient of tunnel
隧道回填 packing
隧道混凝土工程 tunnel concrete work
隧道混凝土浇灌车 tunnel concreting train
隧道火焰试验 tunnel flame test
隧道火灾 tunnel fire hazard
隧道火灾监测 tunnel fire monitoring
隧道机 boring machine
隧道基础 tunnel foundation
隧道激光导向 tunnel alignment by laser;tunnel laser
隧道激光系统 tunnel laser system
隧道级别 tunnel class
隧道几何形状 tunnel geometry
隧道计价线 payment line
隧道计算机监控系统 computerized tunnel supervisory system
隧道加强长度 strengthened section for tunnel
隧道监控量测 tunnel monitoring measurement
隧道监控设备 monitoring equipment in tunnel
隧道监控系统 tunnel supervisory system
隧道建造 tunnel operation
隧道建筑 tunnel construction;tunnel(l)ing
隧道建筑界线 structural approach limit of tunnel;tunnel clearance
隧道建筑界限 structural approach limit of tunnel;tunnel clearance;tunnel structure ga(u)ge
隧道建筑限界 construction clearance of tunnel;tunnel construction clearance;tunnel construction ga(u)ge;clearance of tunnel
隧道渐变段 tunnel transition
隧道结 tunnel junction
隧道结构 tunnel structure
隧道结构钢筋监测端子 testing terminal for tunnel structural reinforcement
隧道结构钢筋连接端子 linking terminal for tunnel structural reinforcement
隧道捷径 tunnel cut-off
隧道截面 tunnel cross-section
隧道介质 tunnel media
隧道进尺丈量 footage measurement of tunnel
隧道井 tunnel shaft
隧道警告牌 tunnel warning board
隧道警卫人员 tunnel guard
隧道净长 length of closed section;length of tunnel proper
隧道净断面 inside cross-section of tunnel;tunnel clearance
隧道净高 tunnel headroom
隧道净空 tunnel clearance
隧道掘进 construction of tunnels;gallery driving;piercing a tunnel;piercing of a tunnel;tunnel drilling;tunnel drivage;tunnel driving;tunnel driving method;tunnel(l)ing;tunnel(ling)operation;tunnel piercing
隧道掘进定位 setting-out for tunnel(l)ing
隧道掘进队 tunnel(l)ing gang
隧道掘进方法 method of tunnel(l)ing
隧道掘进工 mud hog
隧道掘进工作 tunnel works
隧道掘进机 mechanical modulus;mole;rock tunnel(ling)machine;tunnel borer;tunnel boring machine;tunnel(l)er;tunnel(ling)machine
隧道掘进机机头 header
隧道掘进机起重臂 erector arm of tunnel(l)ing
隧道掘进机械 tunnel machinery
隧道掘进记录 tunnel(l)ing record
隧道掘进技术 tunnel(l)ing technique
隧道掘进开挖 tunnel(l)ing operation
隧道掘进设备 tunnel(l)ing equipment
隧道掘进速度 tunnel(l)ing speed
隧道掘进速率 tunnel(l)ing rate
隧道掘进用钻车 tunnel jumbo
隧道掘进中无支撑可承受时间 bridging time
隧道掘进装置 tunnel(l)ing plant
隧道掘进钻车 tunnel drill;tunnel(ling)jumbo
隧道掘进作业 tunnel(l)ing operation
隧道掘通 tunnel(l)ing through
隧道开掘车 tunnel carriage
隧道开挖 tunnel excavation;tunnel(l)ing;tunnel(l)ing operation
隧道开挖的弃土 tunnel spoil
隧道开挖法<有中央导坑式、上半部断面开挖、大断面开挖等方法> tunnel driving method
隧道开挖法则 tunnel(l)ing law
隧道开挖方式 tunnel excavation meth-

od
隧道开挖机 tunnel(l)ing machine
隧道开挖面 facing;tunnel face
隧道开挖平行钻孔 <中孔不装药> burn cut
隧道开挖器 tunnel driver
隧道开挖日进尺 daily tunnel footage
隧道开凿机 tunnel borer;tunnel boring machine
隧道铠框 jumbo;tunnel(ling) shield
隧道勘测 tunnel reconnaissance
隧道勘察 engineering geologic(al) investigation of tunnel
隧道靠河侧 riverside of tunnel
隧道坑道木支柱 pitprop
隧道坑内设备 tunnel equipment
隧道坑外设备 surface equipment
隧道空气 tunnel atmosphere
隧道孔 tunnel opening
隧道孔机 jumbo
隧道控制区 tunnel control area
隧道控制室 tunnel control room
隧道口 tunnel face;tunnel front;tunnel opening;tunnel portal
隧道口部 tunnel portal
隧道口部推出的盾构 tunnels driven from portal
隧道口净空 portal clearance; portal clearance of tunnel
隧道口凌空面 <洞口三面为岩石,一面即爆破面凌空> free face
隧道快速掘进 rapid tunnel driving
隧道快速掘进法 high-speed tunnel(l)ing
隧道快速施工 rapid tunnel(l)ing
隧道矿山法 mine tunnel(l)ing method
隧道扩大改造 tunnel remodelling
隧道扩径式支撑 expanding support
隧道冷子管 tunnel cryotron
隧道力学 tunnel mechanics
隧道连接点 junction of tunnel
隧道连续打插板护顶 driving and timbering;forepoling
隧道连续打插板桩 spiling
隧道连续施工机械 continuous tunnel(l)ing machine
隧道联合掘进机 continuous heading machine;tunnel boring machine
隧道两端挖方 approach cutting
隧道料车 tunnel skip
隧道临时支撑 horsehead;primary lining
隧道临时支护 temporary tunnel support
隧道炉 continuous tunnel furnace
隧道轮廓 profile of excavation;tunnel profile
隧道螺旋线 tunnel spiral
隧道落底 cutting down of tunnel bed
隧道埋深 depth of tunnel
隧道埋置深度 depth of tunnel; embedment depth
隧道门 portal
隧道门入口 entrance portal
隧道面 tunnel face
隧道面修整台架 scaling rig
隧道面支承板 face board of tunnel face
隧道名牌 tunnel name plate
隧道明挖法 cut-and-cover method
隧道明挖改建 daylighting
隧道模板 tunnel(ling) form(work)
隧道内部排水 drainage within tunnel
隧道内部照明水平 luminance level in (the) tunnel interior
隧道内超挖 surplus lining overbreak of tunnel
隧道内底标高 tunnel invert elevation
隧道内棘轮装置 tension wheel in tunnel
隧道内亮度水平 luminance level in (the) tunnel interior
隧道内轮廓位移 displacement of inner contour of tunnel
隧道内灭火皮管连接 fire hose connection in the tunnel
隧道内曲线段 curved track tunnel
隧道内用柴油机发动的自行式车辆 tunnel diesel locomotive
隧道内直线段 tangent track tunnel
隧道内装饰 tunnel finish
隧道内装修 interior finishing in tunnel;tunnel finish
隧道泥沙分流设施 tunnel-type sediment diverter
隧道排水 drainage of tunnel; tunnel drainage
隧道排水槽 tunnel spillway
隧道排水设备 tunnel drainage facility
隧道棚子 tunnel set
隧道偏差 drift of tunnel
隧道平峒掘进机 tunnel(l)ing machine
隧道平巷 tunnel heading
隧道平巷支撑板 astel
隧道坡道折减 compensation of gradient in tunnel
隧道坡度标定 setting-out of tunnel
隧道坡度折减 compensation of gradient in tunnel
隧道起拱处缺口 notch at arch springing
隧道起拱处上挑梁 collar stretcher at tunnel springing
隧道气流 tunnel airflow
隧道弃土 tunnel spoil
隧道弃渣 tunnel muck
隧道砌块 tunnel segment
隧道欠挖 underbreak;underbreak in tunnel excavation
隧道墙体位移 tunnel wall displacement
隧道墙座 tunnel abutment
隧道倾斜度 grade in tunnel
隧道清扫 tunnel cleaning
隧道清洗操作 tunnel washing operation
隧道清洗车 tunnel cleaning vehicle
隧道穹顶 tunnel vault
隧道曲线 curvature of tunnel
隧道渠 tunnel aqueduct
隧道取直 tunnel cut-off
隧道圈 tunnel ring
隧道圈构件 tunnel ring
隧道全长 overall length of tunnel
隧道全断面掘进法 full face tunnel(l)ing
隧道全断面掘进机 mechanical mole; moling machine
隧道全面开挖 full face digging; full face tunnel(l)ing
隧道绕组 tunnel windings
隧道日光利用 daylight use in tunnel
隧道入口 portal entrance; portal of tunnel;tunnel access
隧道入口段 entrance zone;threshold zone of tunnel
隧道入口加强照明 entrance reinforcement(of tunnel)
隧道三角网 tunnel triangular network;tunnel triangulation network
隧道上导洞 top heading
隧道上方覆盖层 burden
隧道设备 tunnel equipment
隧道设计规范 tunnel design specifications
隧道射极放大器 tunnel emitter amplifier
隧道施工 tunnel construction;tunnel(l)ing;construction of tunnels
隧道施工比利时法 Belgian method of tunnel(l)ing
隧道施工的奥地利法 Austrian method of tunnel(l)ing
隧道施工防尘 tunnel construction dust controlling
隧道施工机械化 mechanization of tunnelling
隧道施工控制 tunnel construction control
隧道施工历史 history of tunnel(l)ing
隧道施工速度 speed of tunnel(l)ing
隧道施工所引起的地表影响 surface effects of tunnel construction
隧道施工坍方 collapse in tunnel construction
隧道矢板 forepoling
隧道使用分类 tunnel service classification
隧道式冻结间 freezing tunnel
隧道式干燥机 canal drier[dryer];tunnel drier[dryer]; tunnel drier for bobbins
隧道式干燥炉 tunnel drier[dryer]
隧道式干燥器 tunnel drier[dryer]
隧道式干燥室 tunnel chamber for drying
隧道式干燥窑 tunnel drier[dryer]; tunnel drying oven
隧道式高压引水管 tunnel-type penstock
隧道式鼓风冻结机 air blast freezer tunnel
隧道式烘炉 tunnel furnace
隧道式混凝土蒸汽养护窑 tunnel curing kiln
隧道式进水口 tunnel intake
隧道式冷却 ducted fan
隧道式冷却器 tunnel cooler
隧道式冷却装置 cooling tunnel
隧道式泥沙分流设施 tunnel diverter; tunnel-type sediment diverter
隧道式排出物运载工具 tunnel-type discharge carrier
隧道式皮带输送机 tunnel belt conveyer[conveyor]
隧道式曲柄箱 tunnel-type crankcase
隧道式输送机 tunnel belt conveyer [conveyor]
隧道式搪烧炉 tunnel enamelling furnace
隧道式退火炉 tunnel annealing furnace;tunnel furnace;tunnel lehr
隧道式泄水道 tunnel-type tail race
隧道式熏白机 tunnel stoving machine
隧道式养护窑 tunnel curing chamber
隧道式窑 car tunnel kiln
隧道试验 tunnel test
隧道饰层 tunnel finish
隧道输送机 tunnel conveyer[conveyor]
隧道输送量 capacity of transport tunnel
隧道竖井 sollar;tunnel shaft
隧道水平变形 tunnel horizontal deformation
隧道随压力的变化 tunnel(l)ing variation with pressure
隧道碎部测量 detail survey of tunnel
隧道碎渣搬运 tunnel haulage
隧道踏勘 tunnel mapping
隧道坍流土体 flow out earth
隧道坍塌 collapse of tunnel
隧道套线 tunnel loop
隧道通风 tunnel ventilation
隧道通风风道 air duct of tunnel
隧道通风机 tunnel fan;tunnel ventilation fan;tunnel ventilator
隧道通风施工 construction ventilation of tunnel
隧道通风系统 tunnel ventilation system
隧道通信[讯] tunnel communication
隧道通行限界 clearance for traffic of tunnel
隧道通知设备 tunnel annunciating device
隧道推进率 tunnel advance speed; tunnel excavation speed
隧道托柱 underpinning post
隧道挖出的泥石 tunnel spoil
隧道挖出的弃渣 tunnel muck
隧道挖出土方 tunnel muck
隧道挖出土石料 tunnel muck
隧道挖进机起重臂 erector arm
隧道挖掘机 tunnel excavator
隧道挖运机械 muck loader
隧道挖凿机 tunnel borer;tunnel boring machine
隧道腕臂调整底座 swivel cantilever bracket in tunnel
隧道围岩 tunnel surrounding rock
隧道围岩质量指标 tunnel(l)ing quality index
隧道位置 tunnel location
隧道位置选定 tunnel site selection
隧道无轨掘进 trackless tunnel(l)ing
隧道无线增音机 tunnel radio repeater
隧道吸声设施 acoustic(al) absorbing device of tunnel
隧道纤维 tunnel-fiber[fibre]
隧道现象 tunnel(l)ing
隧道线形 tunnel alignment
隧道限界 clearance of traffic in tunnel;tunnel clearance;tunnel structure ga(u)ge
隧道效应 barrier effect; quantum leakage;tunnel(l)ing;tunnel effect <电子通过金属边界势垒的现象>
隧道效应电流 tunnel(l)ing current
隧道效应漏模 tunnel(l)ing leaky mode
隧道效应系数 channel(l)ing effect factor
隧道效应元件 tunnel effect element
隧道信号 tunnel alarm;tunnel signal
隧道信号机【铁】tunnel signal
隧道形状 tunnel shape
隧道修复 tunnel of tunnel
隧道选址 tunnel site selection
隧道巡逻车 tunnel patrol car
隧道压气掘进法 plenum process of tunnel(l)ing
隧道压缩空气开挖法 compressed-air method of tunnel(l)ing
隧道仰拱 tunnel invert
隧道养护 tunnel maintenance;tunnel maintenance and protection
隧道养路工班 length gang
隧道窑 continuous pusher-type furnace; tunnel furnace; tunnel kiln; tunnel oven
隧道窑工厂 tunnel kiln plant
隧道窑海绵铁生产工艺 tunnel kiln sponge iron process
隧道窑运料车 tunnel kiln car
隧道引道 approach of tunnel;tunnel approach
隧道营运和维修 tunnel operation and maintenance
隧道涌水 tunnel water; water flow into tunnel
隧道用电瓶车 tunnel battery locomotive
隧道用激光器 tunnel laser
隧道用门式台架 tunnel(l)ing gantry
隧道用内燃机车 <具备排气、冷却、消防等设备> tunnel diesel locomotive
隧道用凿岩机 tunnel drill
隧道与车站接头 junction between

tunnel and station
隧道员工 tunnel personnel
隧道圆环 tunnel ring
隧道运河 tunnel canal
隧道运营通风 transportation ventilation of tunnel
隧道运渣车 jumbo
隧道凿挖 piercing
隧道凿岩机 tunnel drill
隧道造价 tunnel cost
隧道掌子面 tunnel heading
隧道照明 tunnel lighting
隧道照明设计 tunnel lighting design
隧道遮断信号 tunnel monoindication obstruction signal
隧道整体吊弦 dropper in tunnel
隧道正洞 main tunnel
隧道支撑 tunnel support
隧道支撑臂与定位 tunnel arm support and registration
隧道支撑的横挡木 ground bar
隧道支护 tunnel support
隧道支护顶撑 bracing in tunnel support
隧道支护间距 pitch of tunnel support;support spacing
隧道支护型钢 steel-shape for tunnel support
隧道支模车 jumbo
隧道支托拱顶板条 crown lagging
隧道枝杈型柱支撑 branch shaped timbering
隧道直径 tunnel diameter
隧道中同时平行掘进的面 mixed face
隧道中推运泥车的小车 bullfrog
隧道中线标定 setting-out of center line of tunnel
隧道中(心)线 center[centre] line of tunnel;tunnel center line;alignment of tunnel;tunnel axis
隧道中中纵向隔板 sollar
隧道终端照明 threshold lighting
隧道周边 tunnel perimeter
隧道轴线 tunnel axis
隧道轴线之标定 laying-out the tunnel axis
隧道主轴线 tunnel center line
隧道装修 tunnel finishing
隧道装修剖面 cross-section for tunnel finish
隧道装载机 tunnel loader
隧道纵撑 collar bracing
隧道纵断面 tunnel profile
隧道纵轴 axis of tunnel
隧道纵轴线 tunnel axis
隧道阻力 tunnel resistance
隧道钻车 jumbo;tunnel drilling rig
隧道钻车臂 jumbo boom
隧道钻进 tunnel drilling
隧道钻进机 tunnel borer;tunnel boring machine;tunnel drill
隧道钻孔 tunnel(bore)hole;tunnel drill hole
隧道钻孔车 tunnel drilling car
隧道钻孔台车 drilling carriage;tunnelling rig
隧道钻探 boring of tunnel
隧道作业 tunnel(l)ing operation
隧底排水管 tunnel floor drain
隧洞 tunnel tube
隧洞爆破 tunnel blasting
隧洞测量 tunnel survey(ing)
隧洞超前纵梁 horsehead
隧洞超挖 overbreak
隧洞衬砌 tunnel lining
隧洞衬砌环 tunnel ring
隧洞出渣车 muck car
隧洞出渣机 mucking machine
隧洞粗糙度 tube roughness
隧洞打通 breakthrough
隧洞导洞 pioneer bore
隧洞导坑 tunnel heading
隧洞导流 tunnel diversion
隧洞底 tunnel invert
隧洞底部 tunnel floor
隧洞底拱 tunnel invert
隧洞地质测绘 geology mapping of tunnel
隧洞掉顶 roof fall
隧洞顶板 tunnel roof
隧洞顶板开裂 roof break
隧洞顶部防护板 top shield
隧洞顶拱 umbrella arch
隧洞顶面 tunnel soffit
隧洞顶钎 roof bolting
隧洞顶栓 roof bolting
隧洞洞顶防塌护板 poling board
隧洞洞顶锚杆支护 roof-bolting support
隧洞洞顶锚栓 roof bolt
隧洞堵塞段 tunnel plug
隧洞堵头 tunnel plug
隧洞盾构机 tunnel shield
隧洞防护支架 tunnel shield
隧洞废渣 tunnel spoil
隧洞分流 tunnel diversion
隧洞风摩擦 tunnel wind friction
隧洞钢模板 tunnel steel form
隧洞拱顶 tunnel soffit
隧洞和隧洞工程 <英国月刊> Tunnels & Tunnel(l)ing
隧洞横截面面积 tube cross-section area
隧洞机具 tunneling rig
隧洞渐变段 tunnel transition section
隧洞交叉(口) tunnel intersection
隧洞进洞支撑构架 headframe
隧洞掘进 drive;tunnel driving;tunnel(l)ing
隧洞掘进法 tunnel driving method
隧洞掘进机 tunnel(l)er;tunnel(l)ing machine
隧洞掘进记录 tunnel(l)ing record
隧洞掘进速率 tunnel(l)ing rate;tunnel(l)ing speed
隧洞掘进装置 tunnel(ling) plant
隧洞掘进作业班 tunnel(l)ing crew
隧洞开挖 tunnel excavation
隧洞开挖钻爆法 drilling and blasting method of tunnel(l)ing
隧洞铠框 jumbo
隧洞口 tunnel face
隧洞快速掘进法 high-speed tunnel(l)ing;rapid tunnel(l)ing
隧洞快速开挖 rapid tunnel(l)ing
隧洞扩大掘进法 tunnel reaming method
隧洞内底 tunnel invert
隧洞内拱底 tunnel invert
隧洞排水 tunnel drainage
隧洞平巷顶支撑板 astel
隧洞弃土 mullock;tunnel spoil
隧洞全断面掘进机 full face(excavation) machine
隧洞施工 tunnel(l)ing
隧洞施工循环 tunnel cycle
隧洞矢板 poling board
隧洞式出水口 drilled outfall
隧洞式炉 continuous pusher-type furnace
隧洞式泥沙分流设施 tunnel-type sediment diverter
隧洞式排污口 tunnel(l)ed outfall
隧洞式污水排海口 tunnel(l)ed oceanic outfall
隧洞竖井 sollar
隧洞挖掘机 rocker shovel
隧洞仰拱底 tunnel invert
隧洞用机铲 tunnel shovel
隧洞运渣车 jumbo
隧洞支撑 tunnel support
隧洞支护 tunnel support
隧洞支护结构 tunnel support
隧洞周围固结灌浆 envelope grouting
隧洞钻车 jumbo
隧盾 <开挖隧道的盾构> tunnel(ling) shield
隧拱变形 deformation of vault
隧拱下沉 sagging of vault
隧拱修复 renew of vault
隧管冷却 ducted cooling
隧射线 anode ray;canal ray;positive ion rays

燧

燧石 flint stone; amausite; arrow stone; chert(y)(flint); fire stone; hornstone; silex

燧石板岩 flinty slate;siliceous rock;lydite <又称试金石>
燧石玻璃 flint glass
燧石的 cherty
燧石骨料 chert aggregate;flint aggregate
燧石管磨机 flint stones tube mill
燧石化作用 chertification
燧石灰色 flint gray
燧石灰岩 chert limestone;flint-containing limestone
燧石集料 chert aggregate
燧石角砾岩 chert breccia
燧石结核 chert nodule
燧石块 flint rubble
燧石砾岩 chat
燧石面层 chert surface
燧石黏[粘]土 clay with flints;flints clay
燧石球 flint ball
燧石球磨 flint mill
燧石砂屑岩 chert arenite
燧石石灰岩 cherty limestone
燧石似的 flinty
燧石陶器 flint faience
燧石圬工墙 <一种组合圬工墙> flint walling
燧石屑锯面 chat-sawn finish
燧石岩 silexite
燧石页岩 chert shale
燧石渣 flint rubble
燧石质的 flinty
燧石质砂岩 <铺路用> malmstone
燧石质石灰岩 flinty limestone
燧石质土 flint clay;flinty ground
燧石砖 flint brick
燧石状(黏[粘])土 flint clay
燧石子 flint pebble
燧石钻头 chert bit
燧烁石 chert gravel
燧土 flint clay

穗

穗花杉 amentotaxus

穗状的 bunchy
穗状胶束 fringed micelle
穗状排列的 spiciform
穗子 tassel

孙

孙子剩余定理 Chinese remainder theorem

损

损管器材柜 damage control locker

损害 damage; damnify; deface; detriment; disadvantage; disserve; disservice; harm; impair; lepton; mutilate; nuisance; prejudice
损害补偿 compensation for damage
损害不赔 free of damage
损害程度 degree of damage
损害程度的测定 measure of damages
损害臭氧的排放物 ozone-damaging emission
损害的发育 impaired development
损害范围 damaging range
损害防止费用 sue and labo(u)r charges
损害防止条款 <海上保险船方的救护条款> sue and labo(u)r clause
损害费 damage
损害风险标准 damage risk criterion
损害风险准则 damage risk criterion
损害管制 damage control
损害函数 damage function
损害环境 damage to the environment
损害率 <汽车对路面的> damage coefficient;damage factor
损害赔偿 compensation for damage; indemnity
损害赔偿的分担 apportionment of damages
损害赔偿估量 measure of damages
损害赔偿金 liquidated damage
损害赔偿诉讼 action for indemnity; action for the recovery of damage
损害赔偿条款 damage done clause
损害赔偿责任 liability for damages
损害频率曲线 damage-frequency curve
损害索赔 damage claim
损害索赔权 right to claim damages
损害索赔要求 claim
损害通知书 notice of damage
损害威信 damage to prestige
损害系数 coefficient of injury
损害信用的 discreditable
损害行动 action for damages
损害阈值 nuisance threshold
损害原因 cause of damage
损害庄稼 damage crops
损耗 attrition; depletion; loss; shrinkage; tear and wear; waste; weak-out;wear and tear
损耗百分率 percentage loss
损耗倍增器 loss multiplier
损耗比 loss ratio
损耗波 evanescent wave
损耗补偿额 breakage
损耗补偿器 loss compensator
损耗补贴 tret
损耗测度表 loss meter
损耗测量 loss measurement
损耗测量仪 loss measuring instrument
损耗的材料 lost material
损耗电导 loss conductance
损耗电度表 loss meter
损耗电流 loss current
损耗电阻 loss resistance
损耗费用 cost of wear and tear; depletion expenses;wastage expenses
损耗分离 separation of losses
损耗概率 loss probability
损耗功率 wasted power
损耗估计 loss estimate
损耗估算 loss evaluation
损耗故障 wear-out failure
损耗函数 loss function
损耗机理 loss mechanism
损耗件 wear and tear part
损耗角 loss angle
损耗角技术 loss angle technique
损耗角正切 loss tangent
损耗角正切试验 loss tangent test
损耗距离 loss distance
损耗控制 loss control
损耗量 abraded quantity; breakage; wastage
损耗率 attendance rate;attrition rate; percentage of loss; rate of loss;

wastage rate
损耗密度 loss density
损耗模量 loss modulus
损耗求和 summation of losses
损耗曲线 damage curve
损耗柔量 loss compliance
损耗失效期 loss failure period
损耗时间 loss time
损耗试验 loss test
损耗衰减 dissipative alternation
损耗水 wastewater
损耗特性 loss characteristic
损耗调制 loss modulation
损耗调制锁模 mode-locking by interactive loss modulation
损耗土壤水分 loss of soil moisture
损耗瓦数 wattage dissipation
损耗维护部件 loss maintenance component
损耗系数 coefficient of losses; loss coefficient; loss factor; loss modulus
损耗限制 loss limit
损耗相加法 loss-summation method
损耗效应 loss effect
损耗性衰减 dissipative attenuation
损耗因数 dissipation factor; loss factor
损耗因素 dissipation factor; loss factor
损耗因子 dissipation factor; loss factor
损耗元件 losser
损耗折扣 allowance for damage
损耗指数 loss index
损耗转矩 loss torque
损耗资产 waste assets
损坏 blemish; breakdown; bug; cripple; derangement of service; destroying; deterioration; detract; disablement; distress; dysfunction; eating away; failure; get out of order; impairment; mishap; out of condition; out-of-order; out-of-repair; tear and wear
损坏报告书 damage report
损坏变形 strain at failure
损坏标卡 <红牌> defect card
损坏不保 free from damage
损坏不能修的 damage beyond repair
损坏不赔 <船体> free of damage absolutely
损坏材料 spoiled material
损坏财产 damage to property
损坏查定 damage assessment
损坏车 bad order car
损坏车辆收集站 recovery collecting point
损坏车停留线 bad order track; track for damaged wagon
损坏程度 damaged condition
损坏处 breakage
损坏处测定 location of faults
损坏的 bad order; defective; disabled; out-of-service; spoiled; damage
损坏的程度 level of damage
损坏的东西 spoilage
损坏的刮路机 fault drag
损坏的井 injured well
损坏的路刮 fault drag
损坏的违章建筑 damaged non-conforming building
损坏调查 damage survey
损坏段落 failure section
损坏阀 damage threshold
损坏范围 degree of damage; extent of damage
损坏估价 damage assessment; evaluation of damage
损坏和遗失 damage and loss
损坏货物 damaged goods; distress cargo
损坏检查 damage survey
损坏检验 survey for damage
损坏检验设备 fault finder
损坏截面 failure section
损坏零件箱 unfit parts box
损坏路面 defacement
损坏率 damage ratio
损坏面 failure surface
损坏面层的修复 replacement of lost surfacing
损坏面积 impaired area
损坏难修的 damaged beyond repair
损坏赔偿 indemnity for damages
损坏赔偿金 damage awards
损坏赔偿诉讼 damage suit
损坏赔偿要求 damage claim
损坏情况简图 damage plan
损坏情况快速判定 rapid damage assessment
损坏曲线 damage curve
损坏时间 time-to-failure
损坏事故 damage accident
损坏速度 rate of decay
损坏索赔 damage claim
损坏通报 advice of damage
损坏通知单 damage note
损坏外形 defeature
损坏危险 risk of damage
损坏威信 damage to prestige
损坏维护 breakdown maintenance
损坏物 spoilage
损坏系数 damage coefficient
损坏线 damage line
损坏项目 damage item
损坏修理 damage repair
损坏一览表 damage catalogue
损坏因素 deadline factor
损坏应变 strain at failure
损坏阈 damage threshold
损坏原因 source of damage; source of trouble
损坏征象 evidence of damage
损坏指数 damage index
损坏周期平均值 mean cycles between failures
损坏状态 destructed condition; distress condition
损毁 broken down
损毁材料 spoiled material
损毁的房屋 damaged premise
损毁费用 spoilage expenses
损毁货物 spoiled goods
损毁修理费清单 schedule of dilapidations
损毁作业汇总表 summary of spoiled work
损流河 losing stream
损漏 breakage
损缺三角形 missing triangle
损伤 blemish; blight; damage; damnify; impairment; injury; lepton; mar; shatter; trauma[复 traumas/traumata]
损伤层 affected layer
损伤长度 damage length
损伤点 impaired loci
损伤钢轨 damaged rail
损伤控制 damage control
损伤力学 damage mechanics
损伤率 damage rate; damage ratio
损伤螺纹 stripping
损伤容限 damage tolerance
损伤外貌 disfigure
损伤外形 disfigure
损伤线试验 damage line test
损伤效应 damage effect
损伤应力 damaging stress
损伤与断裂 damage and rupture
损伤阈值 damage threshold
损伤指示器 fault indicator
损失 damage (and loss); detriment; disadvantage; loss; toll
损失保险 damage insurance
损失报告书 loss report
损失补偿 allowance for the loss; compensation for damage
损失补偿保险 indemnity insurance
损失补偿合同 contract of loss and compensation
损失补贴 allowance for damage
损失产品会计处理 accounting for lost unit
损失长度 lost length
损失超额赔款 excess-of-loss
损失程度 degree of damage; extent of damage
损失的百分(比)率 percentage of damage
损失的能量 wasted energy
损失调查 damage survey
损失额 amount of loss
损失额评定 loss assessment
损失费用 failure cost
损失分数 fraction of losses
损失分摊 loss apportionment
损失分摊条款 contribution clause
损失浮力法 loose of buoyancy method; lost buoyancy method
损失功率 wasted power
损失估计 loss assessment
损失估价师 <保险索赔的> loss assessor
损失估算 assessment of loss
损失函数 loss function
损失回收 loss recovery
损失汇总法 loss-summation method
损失或损坏证明书 certificate of loss or damage
损失集 loss ensemble
损失角 angle of loss; loss angle
损失角正切值 loss tangent
损失理算 adjustment of loss
损失率 damage rate; loss factor; loss rate; percentage loss
损失率计算方法 computational method of ore losses ratio
损失螺纹 strip
损失面积 lost area
损失明细表 loss bordereaux
损失模量 loss modulus
损失能量 off-energy; waste energy
损失排队系统 queue system with loss
损失赔偿 compensation for damage
损失赔偿保证书 letter of indemnity
损失赔偿的预防 protection against damage claims
损失赔偿估量 measure of damages
损失赔偿要求报告 unsatisfactory report
损失赔偿责任 liability for damages
损失频率曲线 damage-frequency curve
损失清单 claim statement
损失情况 damaged condition
损失区分 separation of losses
损失曲线 damage curve
损失热量 lost heat
损失上限 loss ceiling
损失时间 lost time
损失数额估计 estimated amount of damage
损失数据 lost data
损失水量 water loss
损失所负责任 liability for loss
损失索赔 claim for loss and damage
损失索赔种类 class of loss and damage claims
损失调整 adjustment of loss
损失通知单 notice of loss
损失通知书 loss advice; notice of loss
损失物 loser
损失系数 coefficient of losses; loss coefficient; sputter loss coefficient <焊接>
损失险 risk of loss
损失线 loss contour
损失性费用 loss cost
损失因数 loss factor
损失由公司负责 company's risk
损失与开支索赔 claim for loss and expenses
损失与损伤 loss and damage
损失源 loss source
损失者 loser
损失证明 proof of loss
损失证明书 certificate of damage; certificate of loss
损失制排队系统 loss queue system
损失锥 loss cone
损失锥不稳定性 loss-cone instability
损失锥角 loss-cone angle
损失总值 total damage
损蚀 deterioration
损头 lost head
损形 disfigurement
损益 gain and loss; operation statement; profit and loss
损益比率 income sheet ratio; profit and loss ratio
损益表 account of business; income sheet; income statement; operating statement; profit and loss statement; statement of loss and gain
损益表分析 income statement analysis
损益表及利润分配表 income statement and profit appropriation statement
损益法 profit and loss method
损益分配 distribution of profit and loss; division of net gains or loss
损益分配率 profit and loss sharing ratio
损益分配账户 profit and loss appropriation account
损益份额 profit and loss contribution
损益核算制度 loss and profit accounting system
损益汇总表 consolidated profit and loss statement; profit and loss summary account
损益汇总账户 income summary account
损益计算表 income sheet
损益计算书 earnings statement; income account; income statement; statement of profit and loss
损益计算书审计 income account audit
损益计算书原则 income account principle
损益进款表 profit and loss and income statement
损益矩阵分析法 cost-benefit matrix analysis
损益两平点 breakdown point; break-even point
损益两平点范围 <不盈不亏点的> range of break-even points
损益平衡成本 break-even cost
损益平衡点 breakdown point; break-even point
损益平衡分析 break-even analysis
损益平衡贴现率 breakdown discount rate; break-even discount rate
损益平衡图 break-even chart
损益清单 profit and loss statement
损益收费法 charged profit and loss method
损益说明书 statement of loss and gain
损益调整账户 profit and loss adjustment account
损益相抵 gains offset the losses

损益账 loss and gains account
损益账户 nominal account;profit and loss account
损益转让 profit and loss transfer
损益转账事项 profit and loss transaction

笋 革菌属 <拉> Lloydella

笋状沉积物 stalagmite

榫 gib;jib;joggle;pallet【机】

榫板接合法 tenon-bar splice
榫槽 grooving;keyway;mortise;open mortise;slip mortise;slot mortise;tongue and groove
榫槽不合 mismatching
榫槽夹板结合 tabled fishplate joint
榫槽接合 groove and tongue connection;groove and tongue joint;groove connection;jointing by mortise and tenon;tenon-and-slot mortise;tongue and groove joint
榫槽接合板材 tongue and groove material
榫槽接合墙板 tongue and groove siding
榫槽连接 ploughed-and-tongued joint
榫槽密封面 groovy fit seal face;tongue and groove seal contact face
榫槽面 tongue and groove face;tongued and grooved surface
榫槽面法兰 tongued and grooved flange
榫槽刨 bevel(l)ing plane;shoulder plane
榫槽刨机 rabbeting machine
榫槽式 tongue and groove
榫槽铣刀 tonguing cutter
榫齿 tenon tooth
榫齿接合 cogged joint;mortise joint;tongue joint;cogging joint
榫钉 dowel
榫钉缝 dowel(led) joint
榫钉接合 dowel bar;mortise dowel joint
榫钉支持强度 dowel bearing strength
榫沟 open mortise
榫构合 framed
榫规 counter ga(u)ge;mortice[mortise] ga(u)ge
榫焊接 bar past
榫和榫眼 tenon and mortise
榫肩 relish;shoulder <榫根部的横断面>
榫接 joggle(d) joint;lock joint;mortice[mortise] joint;mortise-and-tenon joint;scarf;socketing;tenon and mortise joint;tenon jointing;tongue joint
榫接凹槽 tabled scarf
榫接边 badger
榫接部件 joggle piece
榫接的 joggled
榫接凳 joint stool
榫接缝 joggle joint
榫接高踢脚板 double skirting
榫接合 cogged joint;joggle;feather joint;joint mortised;mortice[mortise];tenon joint
榫接横挡 rabbet ledge
榫接机 joggling machine
榫接机操作工 joggler
榫接件 joinery timber;joinery unit;joggle piece
榫接角缝 <木工> laminated joint
榫接接合 comb joint
榫接接头 joggle joint
榫接口 gain joint
榫接栏 bar post
榫接肋骨 joggle frame;joggle timber
榫接梁 joggle(d) beam;keyed beam
榫接木工作业 finger-jointed
榫接墙板 tight sheathing
榫接踏板楼梯梁 mitred[mitered] stringer
榫接头 carpenter's joint;coak;finger joint;mortise[mortice] joint
榫接外墙板 rabbeted siding
榫接柱 bar post;joggle post
榫接桩 bar post
榫进 tenoned into
榫锯 tenon saw
榫孔 mortice[mortise];tenon hole
榫孔斧 mortice[mortise] axe
榫孔机 square chisel
榫孔四周表面 abutment cheek
榫孔凿 mortise chisel;firmer chisel
榫孔周围与榫肩对抵的颊面 abutment cheek
榫孔钻 slotting auger
榫口 joinery component
榫连接 feather joint;ploughed-and-tongued joint
榫卯 tenon and mortise work
榫刨 dovetailing plane
榫砌砖缝颜料 join colo(u)r
榫舌 feather;feather piece;joint tongue;key protrusion;male dovetail;single-tenon;tenon;tongue
榫舌刨 double plane;tonguing plane
榫条 dowel bar
榫条并接 tenon-bar splice
榫条的内边线 bearding;bearding line;stepping point
榫头 cog;rabbet;tenon
榫头垫块 sprocket
榫头放宽根部 <木工用语> root
榫头划线 dimensioning of joint(s)
榫头加腋宽度 relish
榫头减少 rebate
榫头锯 dovetail saw;miter[mitre] saw;tenon saw
榫头连接 coggea
榫头榫眼接合 mortise and tenon
榫尾芯头 tail core print
榫销 draw(bore) pin;mortise pin
榫销孔(洞) drawbore
榫眼 dapt;housing;mortice[mortise](hole);socket <燕尾榫的>
榫眼板 dovetail plate
榫眼槽 chase mortice[mortise]
榫眼侧 cheek
榫眼接合 joining by mortice[mortise] and tenon
榫眼开凿 mortice[mortise] preparation
榫眼去屑凿 ripping chisel
榫眼圈线 morticed[mortised] astragal
榫眼锁 mortice[mortise] lock
榫眼凿 mortice[mortise] chisel;mortising bit
榫眼钻床 boring and mortising machine
榫眼钻头 mortising bit
榫腋脚 haunch;hauncheon
榫凿 heading chisel;mortice[mortise] chisel;socket chisel;tenon cutter

唆 使 suborn

娑 罗双树 sakhu

娑罗双树属 <拉> Shorea
娑罗双属 mangasinoro

梭 车 shuttle car;shuttle tram

梭车转载 transfer by shuttle tram
梭床 shuttle race
梭挡转子式水泵 shuttle block pump
梭挡转子水泵 shuttle block pump
梭道 race;shuttle path
梭动 push-pull
梭动阀 shuttle valve
梭动机操作间 shuttle drive house
梭动夹头 shuttle chuck
梭动驾驶间 shuttle drive house
梭动式进料 shuttle-type feed
梭动式脉冲 shuttle pulse
梭动式悬臂 shuttle boom
梭动随行工作台 shuttle pallet
梭动运输机 shuttle conveyer[conveyor]
梭阀 shuffle valve
梭夹型连续挤出机 shuttle-clamp type continuous extrusion machine
梭尖 shuttle point;tip
梭口形式 shed forms
梭库 shuttle magazine
梭式 shuttle-type
梭式板形运输机 shuttle apron conveyer[conveyor]
梭式浮标 bobbing buoy
梭式给料机 shuttle feeder
梭式刮板输送器 shuttle-action scraper
梭式矿车 shuttle mine car
梭式矿车集矿 shuttle car gathering
梭式矿车驾驶员 shuttle car driver
梭式皮带运输机 shuttle belt conveyer[conveyor]
梭式输送机 shuttle-type conveyer[conveyor]
梭式送料器 shuttle loader
梭式送料装置 shuttle feeder
梭式窑 shuttle kiln
梭式运输机 shuttle conveyer[conveyor]
梭式自卸车 shuttle dump truck
梭饰 vesica
梭梭林 saxoul wood land
梭梭属 saxoul;Haloxylon <拉>
梭挑邑港 <泰国> Sattahip Port
梭箱 shuttle box
梭箱板 swell
梭箱扬起背板 wood fly back
梭箱油渍 shuttle marks
梭芯 shuttle-cock;shuttle peg
梭芯舌头 shuttle tongue
梭行 shuttle
梭行矿车 shuttle car
梭行拖拉机 reversible tractor
梭行运动 shuttling movement
梭行作业 shuttle working
梭形车场 shuttle yard
梭形电枢 H-armature;shuttle armature
梭形电枢型号 shuttle armature type
梭形桁架 lens truss
梭形滑阀 shuttle valve
梭形滞变曲线 spindle-shaped hysteresis curve
梭眼 shuttle eye
梭眼剪 shuttle-eye cutter
梭织 tat
梭柱 shuttle-shaped column
梭状的 fusiform;spindle-shaped
梭子 shuttle
梭子校正器 shuttle-adjusting machine
梭子稳定装置 shuttle stabilizer
梭子制动 shuttle checking

羧 化物【化】carboxylate

羧化作用 carbonation;carboxylation
羧基 carbonyl group
羧基法粉末坯块 carbonyl compact
羧基化丁苯橡胶 carboxylated styrene butadiene rubber
羧基树脂 carboxylic resin
羧基铁 carbonyl iron
羧基酰胺 carboxamide
羧基阳离子交换剂 carboxyl cation exchanger
羧甲基醚 carboxymethyl ether
羧甲基羟乙基纤维素 <泥浆添加剂> carboxy methyl hydroxyethyl cellulose
羧甲基纤维素 <泥浆添加剂> carboxy methyl cellulose
羧甲基纤维素钠 sodium carboxy methyl cellulose
羧酸 carboxylic acid
羧酸聚合物 carboxylic acid polymer
羧酸树脂 carboxylic(acid) resin
羧酸盐 carboxylate
羧酸阳离子树脂 carboxylic cation resin
羧酸酯 carboxylate

蓑 蛾 bagmoth;bagworm moth;case moth

蓑状线 hatchure

缩 氨基硫脲 thiosemicarbazone

缩氨酸 peptide
缩凹 puncture
缩拌 preshrunk
缩拌混凝土 <使拌和后总体积缩小> shrink-mixed concrete;partially mixed concrete;preshrunk concrete
缩倍 demagnification
缩倍器 demagnifier
缩倍望远镜 demagnifying telescope
缩苯胺 anil
缩比两脚规 reduction compasses
缩边 narrowing ribbon;crawling;cissing[sissing] <一种漆病>
缩边露角 <一种漆病> corner defect
缩并 contraction
缩并的曲率张量 contracted Riemann-Christoffel tensor;Ricci tensor
缩波 condensation(al) wave
缩波系数 shorted factor of wave length
缩补 truncate
缩插接 Irish splice
缩差率 <弯曲木材内外曲面相差百分率> upset
缩沉 shrink mark;sink
缩沉量 sinking volume
缩尺 diminished scale;diminishing scale;pattern-maker's rule;reduced scale;reduction scale;representative fraction;scale;scale of reduction;shrinkage scale;shrink scale
缩尺比 scaling factor
缩尺比例 scale factor
缩尺变态 scale distortion
缩尺表示 scaled down version
缩尺定床模型 solid bed scale model with fixed
缩尺定律 scaling law
缩尺法 scaling method
缩尺反应 condensation reaction
缩尺关系 scale relation
缩尺模型 reduced scale;scale model
缩尺模型试验 scale(d) model test
缩尺模型水轮机 scale model turbine
缩尺模型研究 scale model investigation;scale model study
缩尺试片 reduced specimen

缩尺水工模型 hydraulic scale model
缩尺条件 scale condition
缩尺图 scale drawing
缩尺图形 scale picture
缩尺误差 error of scale
缩尺效应 scale effect
缩尺因数 scale factor;scaling factor
缩尺真型试验 reduced scale prototype model test
缩带离合器 contracting band clutch
缩到最低程度 minimize
缩到最小 minimize
缩叠式连接 concertina connection
缩丁醛 butyral
缩丁醛树脂 butyral resin
缩顶 < 钢锭缺陷 > top hat
缩度计 compressometer
缩短 abbreviation;abridge;abridg(e)ment; boil down; contraction; curtail(ing);cut-short;cutting down; decrease; negative elongation; reduction; shorten (ing); shrink(age);telescoping
缩短差距 narrow the gap
缩短长的闭塞区间 shorten the long blocks
缩短的工艺流程 shortened process
缩短的瞭望距离 reduced visibility
缩短的显示距离 reduced visibility
缩短动作 under reach
缩短渡线 curtailed crossover; shortened crossover
缩短工期 reduction of construction time;reduction of erection time
缩短管筒透镜 reducing lens
缩短轨 curtailed rail;fabricated short rail;standard shortened rail
缩短航道 channel shortening
缩短节距 shortening of pitch
缩短里程效益 benefit from distance shortening
缩短率 < 试件的 > contraction
缩短锚链 heave a cable short; shortening-in
缩短弥合差距 close the gap
缩短器 shortener
缩短时间的可能性 potential reduction of time
缩短使用寿命 reduction of service life
缩短式股道终端连接 shortened connection at ends of tracks
缩短受载期 narrow laydays
缩短套管 reentrant bushing
缩短梯线 curtailed ladders;shortened ladders
缩短投影 foreshortening
缩短系数 shortening coefficient
缩短项吊钩 shortening hook
缩短行距 reducing the row-spacing
缩短型麦氏真空计 shortened McLeod ga(u)ge
缩短型喷管 shortened nozzle
缩短冶炼时间 shorten the heat
缩短引入线 reentrant bushing
缩短硬化 acceleration of set(ting)
缩锻 upset(ting)
缩多酸 polyacid
缩额单位 skeleton unit
缩二苯乙酮 dypnone
缩二胍 biguanide
缩二脲 allophanamide;biuret
缩二脲键合 biuret linkage
缩帆 reef a sail;reefing;shorten sail; take in sail
缩帆带 bag reep;gab rope
缩帆绞辘 reef tackle
缩帆结 bow knot;reef knot
缩帆索 knittle;reef line
缩放 convergent-divergent;reproduce by pantograph; scaling; zoom; zooming
缩放百分率 percent of enlargement/reduction
缩放比 pantograph ratio
缩放比例系数 zoom scale factor
缩放杠杆 mechanical proportional
缩放绘图仪 eidograph
缩放机 reducing and enlarging machine
缩放刻度仪 cutting pantograph
缩放刻膜机 router
缩放刻图仪 cutting pantograph; pantograver
缩放喷管 De Laval nozzle
缩放器 eidograph;pantograph
缩放晒印 radio print
缩放图 zoom graph
缩放图法 pantography;planigraphy
缩放图器 planigraph
缩放相片 ratio print
缩放型流道 convergent-divergent channel
缩放仪 pantograph; redactor; reductor
缩放仪复制 reproduce by pantograph
缩放仪刻模铣床 pantagraph engraving machine
缩放仪式仿形磨床 pantograph copying grinder
缩放仪式拉门 pantograph gate
缩放仪台 pantograph base
缩放因子 zoom factor
缩放轴 zoom axis
缩分 cutting down
缩分次数 reducing number
缩分方法 reducing method
缩分公式 reduction formula
缩分系数 reduction coefficient
缩缝 contraction joint; shrinkage crack(ing);dummy joint
缩根法 diminution of roots
缩管 pipe cavity;pipe reduction
缩管接头 reducing joint;union reducer
缩管偏析 pipe segregation
缩规 contraction ga(u)ge
缩号 sucking
缩合 condensation
缩合产品 condensation product
缩合促进剂 condensation accelerator
缩合催化剂 condensation catalyst
缩合反应 condensation reaction
缩合废水 condensed wastewater
缩合管道 collector
缩合化合物 condensation compound
缩合剂 condensating agent; condensing agent
缩合加速剂 condensation accelerator
缩合聚合物 condensation polymer
缩合偶氮颜料 condensed azo-pigment
缩合式容器 shrink-fit vessel
缩合树脂 condensation resin
缩合温度 condensation temperature
缩合物 condensation compound
缩合型 condensed type
缩合型多环颜料 condensed polycyclic(al) pigment
缩合型双偶氮颜料 disazo condensation pigment
缩痕 shrink mark;sink mark
缩喉管【给】Venturi
缩后的入口 recessed portal
缩花 frizzing
缩回 retract(ion)
缩绘 contracted drawing; squaring down
缩甲醛 formal
缩减 curtailment;cutback;cut down; decrement; detruncate; diminution; down size;dwindle;lessening;minify; negative growth; recede; reduction; retrenching; run down; short out
缩减……的规模 downscale
缩减比(例) reduction ratio
缩减比率 scalage
缩减闭段 diminished block
缩减长度 foreshortened length
缩减成本系数 reduced cost coefficients
缩减抽样检查 reduced sampling inspection
缩减次数 reduced degree
缩减的受荷载线 reduced line of loading;reduced loading line
缩减的文件长度 reduced file size
缩减的褶皱 lessening fold
缩减的主应力 reduced principal stress
缩减点 reduced point
缩减法 economization;flop-out method < 按束强度缩减观察共振的方法 >
缩减哥特式 reduction Gothic(style)
缩减雇员 diminution of employment
缩减迹 reduced trace
缩减集 reduced set
缩减进口 import curtailment
缩减开支 curtail expenditures; expenses deducted
缩减力矩 reduced moment
缩减量 reduced quantity
缩减率 economy;percent reduction
缩减曲线 reduction curve
缩减圈矩阵 reduced cycle matrix
缩减时间 < 水文研究的 > condensed time
缩减系数 reduction factor
缩减映射圆柱 reduced mapping cylinder
缩减余因子 reduced cofactor
缩减装置 reducing fittings
缩焦器 focal reducer
缩角 diminished angle
缩接 reducer coupling;reducer jointing;union reducer
缩接(分)支管 reducing lateral
缩接管 bushing
缩接连接管 reducing nipple
缩节 < 即套接管 > casing coupling; faucet pipe;socket pipe;spigot and socket pipe
缩节管 reducer
缩结 sheep shank
缩截线 cut-off line
缩金属 receding metal
缩紧 scrunch
缩紧环 shrink ring
缩紧螺旋扣 turnbuckle
缩进 backsetting;retraction;setback; shrink
缩进部分 < 堤或岸坡上的 > scarcement
缩进规定 setback ordinance
缩进扭弯内缘 neck-in
缩颈 bottling; choke; constriction; gapping; necking; necking down; necking waist;waist < 桩工等的 >
缩颈处 bottleneck
缩颈的最小直径 minimum thickness of neck
缩颈接头 reducer coupling
缩颈联轴节 reducing pipe coupling
缩颈泥芯 riser pad
缩颈现象 necking;pinch phenomenon [复 phenomena]
缩颈旋压 necking in;spin down
缩景 abbreviated scenery; miniature scenery
缩径 crimping
缩径侧流三通 side outlet tee[T] reducing
缩径衬套 reducing sleeve
缩径承窝 reducing socket
缩径丁字管节 reducing tee[T]
缩径短三通 short body reducing tee[T]
缩径管 diminished pipe; diminishing pipe;reducing pipe
缩径管接 reducer
缩径管接头 reducer [reducing] coupling
缩径接管 reducer coupling
缩径接头 reducing joint
缩径卡钻 hole shrinking sticking
缩径孔数 restricted area
缩径联轴节 reducing coupling
缩径量 cut-off
缩径模 reducing die
缩径内外丝管接头 < 俗称补心 > pipe bushing
缩径三通(管) reduced tee; reducing tee[T]
缩径套节 reducing socket
缩径弯头 reducing bend
缩径支管 reducing lateral
缩径轴套 reducing bushing
缩径钻孔 tight(bore) hole;tight drillhole
缩聚 fasculation
缩聚合作用 condensation polymerization
缩聚沥青 condensed asphalt
缩聚树脂 condensation resin
缩聚体 condensation polymer
缩聚物 condensed polymer;polycondensate
缩聚作用 polycondensation; condensation polymerization
缩卷 crinkle
缩孔 blister;cissing;contraction cavity;font cavity;funnel;orifice;pipe cavity;shrinkage hole;shrink hole; sink hole; sunk spot; throttle; vug; shrinkage cavity < 一种漆病 >
缩孔测流法 contracted opening discharge measurement
缩孔垫圈 reducing bushing
缩孔法 contracted opening method
缩孔法测流量 contracted-opening discharge measurement
缩孔防止剂 anti-piping compound
缩孔分层 pipe lamination
缩孔金属 piped metal
缩孔控制 crater control
缩孔裂纹 crater crack
缩孔流量测定法 contracted opening discharge measurement
缩孔喷嘴 restriction jet
缩孔偏析 pipe segregation
缩孔起层 pipe seam
缩孔钳 pipe dog
缩孔试验 orifice test
缩口 necking; necking down; tube sinking
缩口部 choke
缩口测流计 constriction meter
缩口短管 converging tube
缩口法兰 reducing flange
缩口管 throat
缩口接头 constricted end joint; constricted end point; reducing coupling;reducing jointing
缩口喷漆头 constricted paint sprayer
缩口试验 reducing test
缩口水表 constriction water meter
缩口损失 contraction loss
缩口系数 necking coefficient
缩口堰 contracted weir
缩口用压力机 reducing press
缩粒机 hog
缩裂 shrinkage crack(ing); casting

crack <铸件>
缩流堰 weir with end contraction
缩流装置 side gatherer
缩拢 purse
缩略 breviary
缩略语词典 acronym dictionary
缩略字 literation
缩脉 <射流的最小断面点> vena contracta[复 contractae]
缩灭场 collapse field
缩模 miniature
缩模率 shrinkage ratio
缩膜包装 shrink-film package
缩呢槽 fulling trough
缩呢辊 fulling roller
缩呢机 fulling machine
缩呢折痕 mill rigs
缩片机 reduction printer
缩圈 shrunk-on ring
缩醛 ethylidene ether
缩醛共聚物 acetal copolymer
缩醛均聚物 acetal homopolymer
缩醛类 acetal
缩醛树脂 acetal resin
缩醛塑料 acetal plastics
缩绒工 fuller
缩绒黏[粘]土 fulling clay
缩绒织物 board of felt(ed fabric); felted fabric
缩绒织物布匹 web of felted fabric
缩入 intussusception
缩入管式炉墙 recessed tube wall
缩入式照明设备 regressed luminaire
缩入天线 suppressed antenna
缩三甘油 epihydric alcohol; glycide; glycidol; triglycerol
缩摄一半 reduction to half-frame area
缩水比例 reduction ratio
缩水甘油 glycidol
缩水甘油苯醚 glycidyl phenyl ether
缩水甘油基 glycidyl
缩水甘油醚树脂 glycidyl ether novolac; glycidyl ether resin
缩水甘油醚双酚 A 环氧树脂 glycidyl ether bisphenol-A epoxy
缩水甘油烯丙醚 glycidyl allyl ether
缩水甘油乙醚 glycidyl ethyl ether
缩松 shrinkage porosity
缩酮 ketal; ketone
缩图 contracted drawing; epitome; mini(ature); minimizing chart; model
缩图器 omnigraph; planimegraph
缩图器连杆 pantograph link
缩图仪 reduction printer
缩洼 shrinkage depression
缩微 demagnify; micro
缩微版 microtext
缩微版本 microform
缩微本 microcopy
缩微长条 microstrip
缩微成象 microimaging
缩微出版(物) micropublication
缩微传真 microfacsimile
缩微存储 microimage storage
缩微档案 microfile
缩微地图传输装置 micromap transport device
缩微地图绘图机 micromap generator
缩微地图记录 micromap recording
缩微地图起止控制 micromap start and stop control
缩微地图原片 micromap master
缩微范围 reduction range
缩微放大照片 microfilm; microphotograph
缩微幅度 reduction range
缩微复制 microcopy; microproduction
缩微复制片 microprint
缩微复制品 microform
缩微复制图 microcopy
缩微副本 microcopy
缩微稿 microcopy
缩微光圈 stop down
缩微过程 microform
缩微化 micromation
缩微技术 micrographics
缩微胶卷 microfilm; microphotograph; microstrip
缩微胶卷原件 original edition in microfilm
缩微胶卷正片 positive microfilm
缩微胶片 microfiche
缩微胶片电子检索法 filmorex system
缩微胶片放大 microfilm enlargement
缩微胶片(复制用)母片 master fiche
缩微胶片记录 film record
缩微胶片记录器【计】 microfilm recorder
缩微胶片记录阅读器 film recorder scanner
缩微胶片卷 roll microfilm
缩微胶片扫描装置 film scanner
缩微胶片系统 minicard
缩微胶片阅读复印器 microform reader-printer
缩微胶片阅读器 film reader; reader for microfilm
缩微卡(片) microcard; microfiche; micro-opaque
缩微卡片观察器 microfiche viewer
缩微率 minification rate; rate of reduction; reduction ratio; reduction scale; scale of reduction
缩微模型 reduced model
缩微目录 microbibliography
缩微片 microphotograph
缩微片复制 microreproduction
缩微片缩放仪 micropantograph
缩微片终端 microfilm terminal
缩微平版印刷术 microlithography
缩微平片 microfiche; sheet microfilm; superfische
缩微摄影 micro(photo)graphy
缩微索引卡片 film card
缩微图书 filmbook
缩微图书胶片 microbook fiche
缩微图像 microimage
缩微图像数据 microimage data
缩微图形学 micrographics
缩微文本 microtext
缩微文件 microfile
缩微文献 microdocumentation
缩微像传真 microfacsimile
缩微像片 reduced print; reduction print; reduced photograph
缩微形式 <文件等的> microform
缩微仪 multiplex reduction printer; reduction printer
缩微印刷品 microform; microprint
缩微印刷品阅读复印机 microform reader copier
缩微印刷品阅读印刷机 microform reader-printer
缩微影片 microfilm
缩微影片条 film strip
缩微影像 microimage
缩微阅读器 film card reader; film card view
缩微照片 microcopy; microdot; microfilm; micro(photo)graph; microprint; photomicrograph
缩微照相复印本 photomicrocopy
缩微照相机 microcamera; photoreducer
缩微照相卡片 film card
缩微照相术 microphotography; photomicrography
缩微制品 microfiche; microfilm
缩微资料 microform; microtext
缩位拨号 abbreviated dial(l)ing; speed dialing
缩位拨号方式 abbreviated dial system
缩位拨号系统 abbreviated dial system
缩位号码呼叫 speed calling
缩误【物】 shrinkage
缩狭断面 contracted cross-section
缩狭工程 contraction works
缩线 diminishing line
缩限【岩】 Atterberg limit; contraction limit
缩限含水量 shrinkage limit
缩限仪 shrinkage limit apparatus
缩陷 <漆病> sinking
缩相 condensed phase
缩像数据 microimage data
缩像仪 reducing printer
缩小 coarctation; contraction; deflate; deflation; diminish; dwindle; lessen; narrowing; narrow shrink; reduction; zooming in; minification
缩小百分数 percentage reduction
缩小比例 descaling; ratio of reduction; reduce in scale; drawdown ratio; reduction ratio
缩小比例尺 contraction scale; reduced scale; scale(of) reduction
缩小比例尺度过程 scaling-down process
缩小比例的 scaled down
缩小比例混凝土 microconcrete
缩小差价 narrow price gap
缩小尺寸 minification; minify
缩小尺寸的 scale down
缩小尺寸的配件 undersize part
缩小到规定比例尺 reduce to scale
缩小的 miniature; reductive
缩小的比例(尺) diminishing scale; reduced scale
缩小的螺栓杆身 relieved shank
缩小幅度 reduction range
缩小复制 reduction print
缩小复制图 reduction factor
缩小管接头 reducing piece
缩小管径衬套 pipe reducing bushing
缩小规模 reduced scale
缩小接管 reducing coupling; reducing piece
缩小镜 minifier
缩小孔径 reduced bore
缩小口径成颈状 necking
缩小率 ratio of reduction; reduction factor
缩小模型 reduced model
缩小平面 reduced plan
缩小迁移率 immobilization
缩小绕组图 reduced winding diagram
缩小摄像机 reduction camera
缩小摄影 microfilming
缩小摄影法 microfilm
缩小套管直径 reduce casing
缩小套节 diminishing socket
缩小体积 reduced the size; reduced volume
缩小透镜 reducing glass
缩小图 scale drawing
缩小图样 reduce a plan
缩小系数 reduction factor
缩小像 diminished image; reduced image
缩小一半的 half-size
缩小仪 photoreducer; reducer; reduction printer
缩小仪镜头 reduction printer lens
缩小因数 reduction factor
缩小影印图 reduced print
缩小振幅 amplitude contraction
缩小钻孔直径 decrease the hole diameter
缩写词表 index of abbreviation
缩写词汇 abbreviated words; contracted words
缩写呼叫 abbreviated call letters
缩型 mini
缩性极限 shrinkage limit
缩性界限 shrinkage limit
缩性试验器 shrinkage apparatus
缩压 contractive pressure
缩样 sample reducing; sampling reduction
缩腰炉膛 constricted furnace
缩印仪 reducing printer
缩影 epitome; mini(ature)
缩影版的放大照片 printback
缩影复制 microcopying
缩影胶片卡 microcard
缩影镜头 reduction lens
缩影片 spot film
缩影仪 reducing apparatus
缩影印刷品 microcopy
缩釉 curling; rolling
缩余釉上皮 reduced enamel epithelium
缩语 abbreviated expression
缩窄 constriction; contraction; narrowing; intake <管子等的>
缩窄带 intake belt course
缩窄堤 contracting dike[dyke]
缩窄段 contracted section; contracting reach
缩窄工程 contraction works
缩窄管 diminished pipe; diminishing pipe
缩窄建筑物 contraction works
缩窄损失 contraction loss
缩窄系数 coefficient of contraction
缩胀势 shrink-swell potential
缩制模型 scale model
缩皱 crenation
缩注 blink

所

所包括的范围 <指区域、数量等> coverage
所测的函数 function measured
所持有的有价证券 securities portfolio
所出现的构造层 occurred structure layer
所得 earning; income
所得贷款 loan proceed
所得额来源资料 information at source
所得分析 income analysis
所得及利润税 tax on income and profit
所得扣除项目 income deduction
所得收入法 income received method
所得税 income duty; income tax; tax on income
所得税的累进度 progressiveness of income tax
所得税等级 income tax bracket
所得税额 income tax liability
所得税额减少 decrease of income tax
所得税法 income tax law
所得税费用 income tax expenses
所得税分配 income tax allocation
所得税负债 income tax liability
所得税附加额 income tax surcharge
所得税杠杆作用 income tax leverage
所得税合并申报 consolidated income tax return
所得税缓冲额 income tax cushion
所得税减免 income tax credit
所得税减免额 deduction from income tax
所得税减免界限 marginal relief of income tax
所得税扣款 income tax deduction
所得税率 income tax rate

S

所得税申报表 income tax return
所得税税率表 income tax rates table
所得税以外其他捐税 taxes other than income
所得税预扣表 income tax withholding table
所得税预扣法 pay-as-you-earn
所得税准备 income tax reserve
所得效果 income effect
所得政策 income policy
所得值 income value
所得资本比 income-capital ratio
所发生的构造阶段 occurred tectonic stage
所附单据 attached documents; document attached
所附证件 attached documents; document attached
所规定的比价 rate established
所积累的资料 background of information
所及范围 reach range
所加荷载方向 direction of the applied load
所间区间 section between block posts
所经时期 elapsed time; time elapsed
所罗门的五角星饰 pentacle of Solomon
所罗门海 Solomon Sea
所罗门海盆 Solomon basin
所罗门柱 Solomonic column
所涉经费问题 financial implication
所施荷载 applied load
所施力矩 applied moment
所属地理区类型 included geographic(al) region type
所属构造旋回【地】 included tectonic cycle
所属区 precinct
所属上缴利润 subordinate unit paying profit
所谓 alleged; so-called
所需材料 material requested
所需场强 required field intensity
所需的信号 desired signal
所需地面点 desired ground point
所需经费 necessary fund; required fund
所需空格符 required space character
所需库容 required reservoir capacity
所需劳动力 labo(u)r requirement; required labor
所需面积 required area
所需强度 desired strength
所需预算经费 budgetary requirement
所需值 desired value
所用变压器 house transformer
所用舱容 space occupied
所有保险 anyone risks
所有船舶 anyone vessel
所有的 all-inclusive; proprietary
所有的处理 all treatments
所有各车场的顺序配置(纵列式) consecutive arrangement of all yards
所有国 host country
所有合同的总金额 contract volume
所有阶段勘察 all stages of the investigation
所有局 <如机车车辆> owning administration
所有可能回归法 all possible regression
所有楼层高的墙 wall rising through all the stories
所有权 holding; legal title; ownership; possession; proprietary; proprietorship; situs of ownership; tenure; title
所有权保留 retention of ownership
所有权保险 title insurance
所有权的保护 property protection
所有权的取得 acquisition of ownership
所有权的让渡 livery of seisin
所有权的让与 transmutation
所有权的丧失 loss of ownership
所有权的移交 transfer of title
所有权的转移 passage of title; transmutation
所有权调查 title search
所有权独有条款 sole and unconditional ownership clause
所有权法则 ownership rule
所有权及风险 title and risk
所有权记号 ownership mark
所有权理论州 title theory state
所有权名牌 ownership plate
所有权凭证 certificate of title; documents of title; title deed
所有权契据 title deed for land
所有权清理 clearing title
所有权权利 proprietary rights
所有权尚未归属的房产 beneficial estate
所有权审查 title search
所有权文据 documents of title
所有权益 ownership equity
所有权益总额 owner's equity
所有者权益及负债 owner's equity and liability
所有权优势 ownership advantages
所有权原理 title theory
所有权证 certificate of title
所有权证件 document of title
所有权证据 evidences of title
所有权证明 evidence title
所有权证(明)书 certificate of ownership; document of title
所有权注册 title registration
所有权转让证书 transfer certificate of title
所有权转移 transfer of property
所有人 owner; possessor; proprietary; proprietor
所有入口 <人口普查区内的> enumerated population
所有事件 anyone event
所有损失 anyone loss
所有物 belongings; holding; property; seisin[seizin]
所有者 owner; proprietary; proprietor
所有者权益 ownership interest
所有者权益核算 owner's equity accounting
所有者权益审计 owner's equity audit
所有者责任 owner's responsibility
所有者责任保险 owner's liability insurance
所有这类生物体 all organisms of this type
所有证件 documents of title
所有制 ownership; ownership system; proprietorship; system of ownership
所有制结构 pattern of ownership
所有制形式 pattern of ownership
所有组成部分 repertoire
所在的构造单元 located tectonic unit
所在地 locality; location; locus; site
所在地法律 lex loci rei sitae; lex situs
所增建的小套间 in-law suite
所占舱容 space occupied
所长 director; superintendent
所走距离 distance covered; distance run
所做的功 workdone

索

索鞍半径 saddle radius
索鞍(铲) cable saddle
索鞍摩阻板 saddle friction-plate
索拜珠光体 curly pearlite; sorbitic pearlite
索辫 English sennit
索槽 cable trough
索铲 cable drag scraper; dragline machine; dragline scraper; pull shovel
索铲铲斗 dragline bucket; dragline scoop
索铲挖掘机 dragline; dragline excavator
索铲挖掘机的放松索 slackline cableway
索铲挖土机 dragline conveyor [conveyer]
索偿陈述 statement of claim
索偿的背书 indorsement of claim
索车 cable car
索带 strap
索带的 funicular
索带滑车 stropping block
索带筛 rope-band screen
索单元 cable element
索道 cable(rail) road; cable track; cableway; road railway; rope railway; ropeway; wire ropeway; wire tramway; blondin <料场运石用的>
索道测流缆车 ganging car on cableway
索道车 shuttle car
索道车辆 cable car
索道电车 cable tramway
索道吊车 ropeway car
索道吊斗 cableway bucket
索道吊罐 cableway bucket
索道吊篮 ropeway gondola
索道钢索 ropeway cable
索道高压线 high line
索道滑轨 siding
索道集材 cable skidding
索道集材机 cable skidder
索道解脱器 uncoupling device
索道立柱 tramway mast
索道起重机 cableway transporter
索道起重设备 cable crane
索道式起重机 blondin; cable crane
索道式输送机 ropeway conveyer [conveyor]
索道输送机 cable(belt) conveyer [conveyor]; cable hoist conveyer [conveyor]; telpher conveyer [conveyor]
索道输送系统 telpherage
索道塔 cableway tower
索道线路 cable run
索道小车 ropeway car
索道运输机 cableway transporter
索得率 browsability
索德奈尔坐标 Soldner coordinate
索第位移定律 Soddy's displacement law
索吊桥 cable suspension bridge
索定支座 guyed pole
索斗铲 dragline crane shovel; dragline excavator
索斗式挖掘机 boom dragline; dragline excavator; dragline-type shovel
索斗式挖土机 boom dragline; dragline excavator; dragline-type shovel
索端 bitt(er)-end; rope end
索端固定装置 rigging end fitting
索端结扣眼圈 eye splice
索端配件 cable terminal fittings
索端之结节 wall
索多边形 equilibrium polygon; funicular polygon; line polygon; link polygon; rope polygon; string polygon
索多边形方程 funicular polygon equation; string polygon equation
索多角形 funicular polygon
索尔公式 <用于测定混凝土成熟度> Saul's formula
索尔克滤光器 Solc filter
索尔维法废液 <含有氯化钙> Solvay liquor
索尔维循环制冷机 Solvay cycle refrigerator
索尔兹伯里 <罗得西亚(津巴布韦)首府> Salisbury
索耳 eyelet grommet; rope grommet
索饵场保护 protection of nursery ground
索饵洄游 feeding migration
索饵机制 feeding mechanism
索饵习性 feeding habit
索非亚 <保加利亚首都> Sofia
索缚浮标 captive float
索高价的 dear
索高价贿赂法 accommodation payment
索格利特抽取器【化】 Soxhlet's extractor
索格利特提取法 Soxhlet's extraction method
索格利特提取器 Soxhlet's extractor
索箍 cable band
索股股数 lay
索股纹路 lay
索桁架 cable truss
索环 cord grommet; eye splice; garland; grommet; grommet ring; sling; string wreath
索环碰垫 fisherman's fender
索环套头端 biglet
索回多收运费的要求 overcharge claim
索汇 claim reimbursement
索加劲 cable bracing
索夹 cable(band) clamp; cord clip; rope clamp; rope clip; rope-grip
索价 charge
索价比行情低 undercharge
索结 hitch
索结板 hitch plate
索具 cordage(cord); patch cord; rig(ger); rigging; tackle
索具车间 rigging shop
索具传动装置 rigging loft
索具工场 rigging shop
索具工人 erector; rigger; ship rigger
索具工油盒 rigger's horn
索具卡环 rigging shackle
索具库 cordage room
索具配置 rigging
索具绳索 rigging line
索具套环 bight
索具装配(员)工 rigger
索距 cable spacing
索卡 cable clamp
索控机械 cable control unit
索控式铲运机 cable-controlled scraper
索控式起重小车 rope trolley
索控式装置 rope control unit
索控小车门式起重机 gantry crane with rope trolley
索控装置 cable control unit
索快速振动 cable vibration
索拉铲土机 walker
索拉渡船 rope ferry
索拉结构 cable-supported structure
索拉耙铲机 track cable scraper
索拉桥 cable-supported bridge
索拉油 <石油蒸馏时提取煤油后的馏份> solar oil
索缆 cable
索勒尔胶结料 magnesium oxychloride cement; oxychloride cement
索勒米尼铝焊剂 Soluminium
索勒坦化学灌浆法 Sotetanche system
索勒坦奇体系 Soletanche system
索雷尔电炉炼铁法 Sorel's process

索雷尔胶结料 Sorel's cement
索雷尔水泥 Sorel's cement
索雷尔锌合金 Sorel's alloy
索雷尔组成 Sorel's composition
索雷特棱镜 Soret prism
索累补偿器 Soleil compensator
索里蓝 <一种无臭木材防腐剂> Solignum
索力 staying force
索力测定计 cable tension measurement device
索力测量 cable force measurement
索力控制 cable force control
索利特高度光亮镜合金 Sollit's alloy
索链渡船 chain ferry
索硫锑铅矿 sorbyite
索仑棱镜单色仪 Thollon prism monochromator
索伦石 suolunite
索轮 cable pulley;rope sheave
索轮式提升机 rope-pulley hoist
索罗铸轧法 Soro process
索马里 <非洲> Somalia
索马里板块 Somalia plate
索马里海盆 Somali basin
索马里深海平原 Somalia abyssal plain
索米威乐地板 <一种耐火地板,由空心加混凝土管组成> Somerville floor
索末菲辐射条件 Sommerfeld radiation condition
索末深海平原 Somme abyssal plain
索诺管 Sonotube
索诺拉雷暴 sonora
索诺马造山运动 Sonoma orogeny
索盘 cable spool;offering wanted
索赔 claim;claim for compensation; claim indemnity;demand compensation;lodge a claim
索赔保险单 claims-made policy
索赔报告 claim report;statement of claim
索赔补偿费 claim for compensation
索赔部门 claims board
索赔程序 procedure for claims
索赔代理人 claim settling agent;settling agent
索赔单据 claim documents
索赔的有效性 validity of a claim
索赔的证明 proof of claims
索赔法院 court of claims
索赔费用 claim expenses
索赔函 claim letter
索赔合同 contract of indemnity
索赔金额 claim
索赔金支付 payment for claims
索赔局 claiming administration
索赔科 claim section
索赔期限 time limit for filing claims; time limit for payment of claim
索赔清单 claim statement;statement of claims;written statement of claims
索赔清理 liquidation of claim
索赔权 rights of claim;rights of redress;rights of appeal
索赔人 claimant;claimer
索赔涉及第三方 third party claim
索赔审理委员会 claims commission
索赔时数 time limit for filing claims
索赔时限 time limit for filing claims
索赔时效 indemnity period
索赔事项咨询委员会 Advisory Board on Compensation Claims
索赔书 claim letter
索赔损害赔偿费 claim for damages
索赔损失 claim for damages
索赔条款 claim clause
索赔通知 advice of charges
索赔通知书 claim notice;notice of claims
索赔通知条款 notice of claim clause
索赔委员会 claim board
索赔文件 claim documents
索赔要求 damage claim
索赔意向 intension for claims
索赔有效期 time of validity of a claim
索赔有效性 validity of a claim
索赔圆满解决 satisfactory settlement of a claim
索赔(债权)人 claimant
索赔债务人 claimee
索赔证件 claims documents;document for claims
索赔证据 substantiation of claims
索赔证明 substantiation of claims
索赔准备金 claim reserve
索赔总代理人 general claim agent
索平面 cable plane
索起重设备 cable crane
索钳 <架空索道> rope clamp
索桥 cable bridge;rope bridge;cable suspension bridge
索取 subpoena
索取补偿 claim compensation
索取承包商估价书 calling for proposals
索取号 accession number
索取价格 price appeal
索取利息 claim for interest
索取特权 demand special privileges
索取应得价款 claim for proceeds
索圈 cringle
索升式倾开桥 cable lift bascule bridge
索石 flat band
索氏体 sorbite
索氏体钢轨 sorbite rail
索式测距仪 wire distance measuring machine
索式冲击钻进法 cable tool drilling
索式冲击(钻探)法 cable tool method
索式结构 cable structure
索式卷扬机 capstan
索式耙斗 cable-hauled bucket
索式拖运 cable towing traction
索式斜拉桥 bundle type cable-stayed bridge
索式卸料车 cable-dump truck
索式运土机铲斗 rope-hauled bucket
索式抓斗 cable clamshell
索式自卸卡车 cable-dump truck
索思韦尔松弛法 Southwell's relaxation method
索塔 cable bent tower;cable support tower;head mast <挖掘机的>
索套 cable sleeve
索梯 jack ladder;Jacob's ladder
索筒 load barrel
索头联结器 end connector
索托风 vento di sotto
索挖式铲运机 cable excavator;cable scraper
索挖式起重小车 rope trolley
索网 cable network
索网体系 cable net system
索桅式打桩架 false leader
索桅式动臂起重机 guyed derrick; guyed derrick crane
索尾插结 back splice;back splicing
索系 cable system
索系门 corded door
索系桩 guy stake
索线 bands;cords;funicular line
索线多边形 string polygon
索线力 funicular force
索线曲线 string curve
索线头 slips
索芯 heart;hemp cord;hemp core
索星卡 star finder;star finder and identifier;star identifier
索星图 star identification map
索型壳 funicular shell
索绪 groped ends
索绪槽 groping end though
索绪锅 groping end basin
索绪机 groping end machine
索绪器 cocoon agitator
索绪效率 groping end efficiency
索绪帚 groping brush
索绪装置 cocoon beater
索悬挂屋盖 cable-suspended cantilever roof
索靴 strand shoe
索压力线法 funicular pressure line method
索压力线拱 funicular pressure line arch
索压力线拱顶 funicular pressure line vault
索眼 bight;cringle;grommet;wiring grommet
索眼插接 eye splice
索眼钩 <系在绳索末端的> toggle hook
索眼心环 thimble
索样 ask for samples
索要虚价 overcharge price
索曳铲运机 cable drag scraper
索曳刮斗 scraper bucket
索引编号 index number
索引编制人员 indexer
索引标记 index marker
索引表 concordance list;indexed list
索引簿 directory;index book
索引程序 concordance program(me); indexing sequence
索引抽屉 index drawer
索引杆挂钩孔 drawbar eye
索引光泽纸板 index bristol
索引号(码) index number
索引化合物 index compound
索引机 index machine
索引目录 index catalogue
索引平面 index plan;key plan
索引平面图(幅) index plan sheet
索引术语 index term
索引图 base map;chart index;index chart;index diagram;index drawing;index map;index plan;key diagram;key drawing;key map;key plan;location index;location map
索引维护 index maintenance
索引文件 index(ed) file;inverted file
索引文献 indexed document
索闸 cable brake;cable-control brake
索债文件 recourse note
索张力 cable tension
索张力线 funicular tension line
索状 funicular;strand
索状多边形 funicular polygon;link curve;string polygon
索状拱 funicular arch
索状浇注 cord-pour
索状曲线 funicular curve
索状图 funicular diagram
索钻 boring with line;cable drilling; jump drilling
索钻法 rope drilling
索钻工作 cable drilling
索钻机 cable rig;rope drill(er)
索钻架 cable rig
索钻器 rope drill(er)
索钻设备 cable drilling outfit
索钻钻具 cable drilling tool;cable tool
索座 strand shoe

琐

琐碎的 trivial;unimportant

锁

锁坝 closed dike[dyke];closure dam;closure dike[dyke]

锁板 jam plate;lock plate
锁板齿轴 locking-plate pinion
锁比法 method of chain relatives
锁闭 blind;bolting;locking;lockup
锁闭板 locking slide
锁闭表示 lock(ing) indication
锁闭表示灯 light lock indicator;lock lamp;lock light
锁闭齿 locking tooth
锁闭齿轮 locking gear
锁闭磁铁 locking magnet
锁闭的 blocked
锁闭的闭塞机 closed block
锁闭点 block point
锁闭电磁铁 locking magnet
锁闭电路 locking circuit
锁闭动程 locking movement
锁闭法 locking
锁闭方法 locking medium
锁闭风压 locking air pressure
锁闭盖 locking lip
锁闭杆 locking bar;lock(ing) rod; locking stud
锁闭杆操纵握柄架 lock bar operating lever stand
锁闭杆齿 locking bar tooth
锁闭鼓 locking drum
锁闭轨道电路 locking track circuit
锁闭活塞 lock piston
锁闭机构 locking gear;lock(ing) mechanism
锁闭棘轮 locking ratchet
锁闭继电器 lock(ing) relay
锁闭夹 locking clamp
锁闭夹板 locking clamp
锁闭检查接点 lock proving contact
锁闭检查器 detector
锁闭检查器复示器 detector indicator
锁闭接点 locking contact
锁闭开关 locking cock;lock-up cock
锁闭孔 locking hole
锁闭块 locking block;locking piece
锁闭块导框 locking block guide
锁闭力 locking force
锁闭了的进路 locked(-up) route
锁闭马镫铁 locking stirrup
锁闭面 locking face
锁闭片 locking piece;locking segment
锁闭片架 segment support
锁闭器箱 casing of lock
锁闭区段 locked section
锁闭缺口 locking notch
锁闭设备 locking arrangement
锁闭深度 locking depth
锁闭式汽车库 lock-up garage
锁闭式信号铃系统 locked-bell system
锁闭手柄 locking handle;locking lever
锁闭手段 locking medium
锁闭栓 lock(ing) bolt
锁闭条 locking bar;locking slide
锁闭条件 locking requisition
锁闭筒 locking drum
锁闭图 locking scheme
锁闭位置 locked position
锁闭握柄 locking lever
锁闭系统 locking system
锁闭箱 bolt-lock
锁闭元件 lock(ing) unit
锁闭圆弧 locking arc
锁闭圆盘 locking disc
锁闭钥匙 locking key
锁闭值 locking value
锁闭爪 locking pawl
锁闭装置 latch-out device;locking arrangement

锁闭子 locking piece
锁臂 locking arm
锁边 lockrand; selvage edge; serging <地毯>; decorating seaming <缝帆法>
锁槽 key seat; locked groove
锁插销 lock(ing) bolt
锁车轮链条 lock chain
锁匙眼形长孔 keyhole slot
锁尺 lock(ing) rod; locking stud; locking tappet; lock plunger; plunger; sword iron; tappet; tappet rod
锁尺齿 locking bar tooth
锁尺传动器 locking bar driver
锁尺电路控制器 tappet circuit controller
锁尺回路管制器 tappet circuit controller
锁尺锁闭 bar locking; tappet locking
锁齿式排种轮 studded roller
锁床 locking bed
锁床式电气联锁 all-electric(al) interlocking
锁存器 latch
锁挡 backset; stopper
锁导销 lock guide pin
锁的机构 lockwork
锁的销栓 broach
锁底对接接头 lock butt joint
锁钉 locking pin; lock spike
锁定 anchor(age); anchoring; blockage; dog; latchdown; latch-up; lockage-on; lock(ing) in; locking pin snap ring; lock-on; lockout; lockup
锁定按钮 cocking button; lock(ing) (press) button; locking push-button; press button; push-button; stay-put button
锁定板 lock plate
锁定边界 lock-on boundary
锁定插头 locking-type plug
锁定插销 dead bolt
锁定充填 locking fill
锁定磁铁 lock(ing) magnet
锁定带 locking strip
锁定挡板 key plate
锁定点 keyed end
锁定电路 lock-on circuit; lockout circuit
锁定端 keyed end
锁定二极管 locking diode
锁定阀 lock valve
锁定范围 lock(ing)-in range(on synchronization)
锁定方法 <闸门> method of dogging
锁定放大器 lock-in amplifier
锁定附件 stabiliser[stabilizer]
锁定杆 catch arm; gripping lever; locking-up lever; securing rod; stay rod; stop lever
锁定光圈 click stop
锁定轨温 fastening-down temperature of rail; stress-free temperature
锁定过程 locking process
锁定环 lockloop; snap ring
锁定换码 locking escape
锁定混频器 lock-in mixer
锁定机构 caging mechanism; lock mechanism
锁定继电器 lock(ing) relay; lock-up relay
锁定寄存器 locked register
锁定键 locking key
锁定绞车 hold-back winch
锁定开关 locking key
锁定控制 locking control
锁定块 locking segment
锁定联杆 locking link; locking lever
锁定链 locked chain; sway chain
锁定螺钉 lock screw
锁定螺母 block nut; clamp nut; jam nut; locking nut; retaining nut; safety nut
锁定螺栓 lock(ing) bolt
锁定脉冲 locking pulse; lockout pulse
锁定模式 locked mode
锁定钮 lock knob
锁定片 locking piece
锁定频率 locking frequency
锁定器 lock
锁定器壳 locking device shell
锁定切换触点 locking transfer contact
锁定区 lock room
锁定曲柄 locking crank
锁定圈 locked coil
锁定时间 locking time; lock-up time
锁定特征 lock-in feature
锁定同步 lock-in synchronism
锁定凸轮 locking cam
锁定位置 latched position
锁定文件 locked file
锁定线 locked-in line
锁定线圈 holding-on coil; holding out coil
锁定相(位) locking phase
锁定销定位座 lock pin positioning plate
锁定销壳体 lock pin housing
锁定楔 locking wedge
锁定斜杆 <悬挂装置的> sway bar
锁定芯轴 locking mandrel
锁定信号 locking signal; lock-on signal
锁定选择 lock option
锁定移位符号 locking shift character
锁定振荡器 locked oscillator
锁定振荡器检波器 locked-oscillator detector
锁定爪 lock(ing) dog; lock(ing) pawl
锁定转换控制 cage switch control
锁定转义 locking escape
锁定转子 locked rotor
锁定转子频率 lock rotor frequency
锁定装置 checking device; fixing device; locking device; locking gear; lock-up device; retaining device
锁定组件 locking assembly
锁端 lock front
锁端斜面 lock front bevel
锁分解 lock resolution
锁风装置 air lock
锁封 lock seal
锁封房顶钢线钉 lock seal steel wire roofing nail
锁缝 crimping; eyelet; whipping
锁缝编织 lock knitting
锁缝操作 lock seaming operation
锁缝分度凸轮 lockstitch indexing cam
锁缝机 chain stitch machine; overlock machine
锁缝线圈横列 lockstitch course
锁盖 cap of lock
锁盖板 nose plate
锁盖开关 asylum switch; lockedcover switch; locking switch; secret switch
锁杆 lock(ing) bar; locking lever; lock(ing) rod; plunger; sword iron
锁杆轮座 <集装箱> cam lock bracket
锁杆凸块 lock lever projection
锁杆凸轮 <集装箱> lock door cam
锁杆托架 <集装箱> door lock rod bracket
锁工 locksmith; lock smith's work
锁钩 latch hook
锁箍 locking clip
锁骨 clavicle
锁固地板系统 lock-tile floor system
锁固体系 locking system
锁光圈 iris diaphragm
锁光(圈)环 iris ring
锁国主义 seclusionism
锁合 closing component
锁合凹座 locking recess
锁合横杆 locking cross bar
锁合装置 lock unit
锁环 check ring; circlet; clamping nut; clamping ring; locking loop; lockloop; lock ring; retainer ring; soffit cusp
锁环槽 lock ring groove
锁环轮辋 locking rim
锁簧 bolt; die; locking dog; lock(ing) spring; locking tappet; tappet
锁簧槽 locking box channel; locking box trough; locking tray
锁簧传动器 dog driver
锁簧床 interlocking compartment; lock box; locking bed; locking box trough; lock(ing) bracket; locking room
锁簧床底座 locking box bracket
锁簧杆 locking slide; tappet rod
锁簧杆传动器 locking bar driver
锁簧联锁保护法 dog lock protection
锁簧配列图 dog chart; dog sheet
锁簧锁闭 dog locking; tappet locking
锁簧图 dog chart; dog sheet
锁加闩 lock and bolt
锁夹 lock clamp; locker
锁件 latch fitting
锁键 arresting device; arresting stop; catch; holding-down latch
锁键换装器 lock key replacer
锁键弹簧 pawl spring
锁键调节式 snap up
锁匠 locksmith
锁匠的锤 lock smith's hammer
锁匠工作 lock smith's work
锁交引线 lock-fit lead
锁角 interlocking angle; locking angle
锁接 halving
锁接缝 lock seam
锁接箍 coupling with wrench flat
锁接砌层 key course
锁接头 detent adaptor; flush coupling; lock joint
锁接头量规 joint template
锁接头突肩 shoulder of the box
锁结 interlock(ing)
锁结角 lock corner
锁结颗粒 interlocked grain
锁结良好的结构 well-keyed structure
锁结式骨料 <指相似尺的粗集料，压实后空隙很大> macadam aggregate
锁结式集料 macadam aggregate
锁结式建筑板 interlocking building panel
锁结性 interlocking properties; interlocking quality
锁结阻力 interlocking resistance
锁结作用 keying action
锁紧 crimping; lock in; lock nut
锁紧把手 locking handle
锁紧半径 lock radius
锁紧柄 locking handle
锁紧操作杆 lock operating lever
锁紧带 lock-strip
锁紧带槽 lock-strip cavity
锁紧的 tightlock
锁紧垫圈 limpet washer; lock(ing) washer; lock nut washer; nut lock washer; retaining washer; safety washer; split-load washer; stop washer
锁紧舵栓 lock pintle; pintle; rudder lock(ing)
锁紧阀 locking valve; lock(-up) valve
锁紧杆 check lock lever; locking bar; lock lever yoke
锁紧杆桩 lock lever stud
锁紧杆座 lock support
锁紧缸 locking cylinder
锁紧钢球 lock ball
锁紧钢丝 locking steel wire
锁紧钩 locking bail
锁紧横列 locking course
锁紧环 binding ring; check ring; circlip; clamping ring; end ring; retaining ring; locking nut; locking ring
锁紧簧 lock spring
锁紧机构 caging section; catch gear; locking device; lock(ing) mechanism; retaining mechanism
锁紧件 retaining member
锁紧键 tightening key; locking key
锁紧角 lock corner
锁紧接合 lock joint
锁紧接头 lock-on connection
锁紧缆 locked cable
锁紧离合器 locking clutch
锁紧力 locking force; tight locking force
锁紧链套 locking link sleeve
锁紧螺钉 gib screw; lock(ing) screw
锁紧螺杆 lock bolt
锁紧螺帽 back nut; blocking nut; check nut; cheek-nut; jamb nut; jaw nut; latch nut; lock(ing) nut; retaining nut; keeper
锁紧螺帽座 lock nut bearing
锁紧螺母 block(ing) nut; check nut; clamp nut; jam nut; keeper; latch nut; lock(ing) nut; nut lock; retaining nut; safety nut
锁紧螺母垫圈 lock nut washer
锁紧螺母箍 lock nut snap ring
锁紧螺栓 lock(ing) bolt; dead bolt <没有弹簧>
锁紧螺丝 lock(ing) screw
锁紧螺旋 gib screw; lock(ing) screw
锁紧面 locking face
锁紧木片 keeper plate; locking plate
锁紧木栓 keeper plate; locking plate
锁紧捏手 lock knob
锁紧钮 lock knob
锁紧盘式主轴端部 bayonet spindle nose
锁紧片 stay
锁紧嵌接 locked scarf
锁紧球 lock ball
锁紧圈 clamp ring; ring clamp
锁紧式防尘罩 retaining shield
锁紧式软管接头 lock type hose coupling
锁紧手柄 set lever
锁紧栓 lock plunger
锁紧弹簧 latch spring; locking spring
锁紧弹簧环 lock spring ring
锁紧套 lock(ing) sleeve
锁紧填隙片 lock-strip gasket
锁紧条 lock-strip; zipper strip
锁紧条空腔 lock-strip cavity
锁紧凸轮 locking cam
锁紧系统 caging system
锁紧销 adjusting pin; clamp pin; fitting pin; lock(ing) pin; prisoner; steady pin
锁紧楔 locking wedge
锁紧压片 lock cover
锁紧叶片 gate blade; locking blade
锁紧应力 locked-up stress
锁紧用钢丝 lock(ing) washer; lock(ing) wire
锁紧制动偏心销 lock banking eccentric
锁紧中心销 locking center pin

锁紧装置 catch gear;locking apparatus; locking arrangement; locking clamp; locking device; locking gear;locking latch
锁紧锥 locking cone
锁紧座 <火车车架上的> locking bed
锁具 locker;lockset
锁壳 case of lock
锁孔 locking hole
锁孔板 selvage[selvedge]
锁孔锉 warding file
锁孔盖 escutcheon;key escutcheon
锁孔式槽口 keyhole type notch
锁口 fore shaft; locking notch; lock joint;preliminary shaft
锁口板桩 tongued and grooved sheet pile
锁口处断续焊 intermittent weld at interlock
锁口防渗 interlocking seepage prevention
锁口分离 out of interlock
锁口风 lock-jaw
锁口缝 interlocking joint;lock seam
锁口管 interlocking pipe; joint tube; stop casing tube
锁口极限强度 ultimate interlock strength
锁口接缝 lock seaming
锁口截面 interlocking section
锁口拉力 interlock tension
锁口锚链 die-lock chain
锁口模 counter locked die
锁口盘 collar set
锁口破坏 interlock failure
锁口张力 interlock tension(force)
锁口阻力 interlock resistance
锁口钻孔 collar hole
锁扣 lock catch; snap-close; turning bolt
锁扣材料 locking material
锁扣环 locker ring
锁扣绝缘子 shackle insulator
锁扣门闩 locking bar
锁扣弹簧 latch spring
锁扣支柱 lock pillar
锁扣装置 locker;locking device
锁块 closing piece;gate piece
锁牢的 lockfast
锁链 chain;curb chain
锁链状矿脉 chain like veins
锁量 hold depth
锁轮滑行检测器 locked wheel skid tester
锁轮滑行检验器 locked wheel skid tester
锁螺帽 safety nut
锁螺片 threaded plate
锁帽 cap of lock
锁门 detent catch
锁门闩 slip bolt
锁面 lock face
锁面至钥匙孔中心的距离 backset
锁模 locked mode; mode locked; mode locking
锁模倍频器 mode-locking frequency doubler
锁模倍频运转 mode-locked frequency-doubled operation
锁模激光器 mode-locked laser
锁模技术 mode-locking technique
锁模力 mo(u)ld locking force
锁模列 mode-locked train
锁模器 modelocker
锁模双偏振运转 mode-locked dual polarization operation
锁模态 mode-locked state
锁模效应 mode-locking effect
锁母 nosing
锁内圆形凸挡 ward
锁钮 cage knob;lock knob
锁片 nut retainer
锁片离合器 lock-up plate clutch
锁气阀 blocking valve
锁气器 clapper; rotary seal; rotating seal
锁气器式挡板 swing gate
锁气室 air lock
锁气装置 air lock device; gas lock system
锁嵌 keyed scarf
锁圈 locking collar; locking ring; oil seal snap ring;safety ring;snap-ring internal
锁上 keying
锁上行动 lockup
锁梢 bayonet catch
锁舌 bolt;spring bolt;tumbler
锁舌板 lock plate
锁舌导板 striker
锁舌盒 lock keeper; lock plate; lock strike
锁舌孔板 lip strike
锁舌片 striker; strike plate; striking plate
锁舌片加固件 strike reinforcement
锁舌锁住钮 trigger bolt
锁舌弯碰板 curved lip strike
锁舌斜面 lock bevel
锁舌眼凿子 bolting iron
锁舌罩 box strike
锁声廊 sound lock
锁石 keystone
锁式 loctal
锁式暗缝 lockstitch blind stitch
锁式插头 locking-type plug
锁式管 local tube;lock-in tube
锁式管座 loctal socket;loktal base
锁闩 catch; detent catch; latching; plunger latch
锁闩的 latched
锁闩控制钮把手 turn piece
锁闩型机构 latch-type mechanism
锁栓 latch bolt;lock plunger;set bar
锁栓强度 <车门的> latching strength
锁栓眼板 lock strike
锁栓制动的凹槽 gating
锁栓轴 latch pin
锁丝钢丝绳 locked coil rope;locked-wire rope
锁丝结构 <钢丝绳的> locked coil construction
锁套 lock tube
锁条 hatch cleat; locking strip; locking tappet; lock plunger; plunger; sword iron;tappet(rod)
锁条管 lock-bar pipe
锁条式密封垫 lock-strip gasket
锁条锁闭 bar locking
锁铁 kicker;thrower
锁托 <车窗> latch bracket
锁网名 name of chain(net)
锁网平均边长 average side length of chains
锁位压烫机 locked press
锁匣 <金属门上的> lock reinforcing unit
锁线订书机 book sewing machine
锁线机 thread sewing machine
锁线装订机 sewing-press
锁箱 casing of lock;lock box
锁相 phase-locked detection; phase lock(ing)
锁相的 phase-locked
锁相环(路)phase-locked loop
锁相环路检波器 phase-locked loop detector
锁相回路 phase-locked loop
锁相激光器 phase-locked laser
锁相技术 phase-locked technique
锁相解调器 phase-locked demodulator
锁相伺服系统 lock-servosystem
锁相振荡器 locked oscillator; phase-locked oscillator
锁像 picture lock
锁消解 lock resolution
锁销 cotterel; link locking pin; lock-(ing) pin; pin in groove
锁销导板 locking pin guiding plate
锁销垫圈 locking pin washer
锁销防松螺母 anchor pin locknut
锁销机构 latch gear
锁销接触器 latched-in contact
锁销弹簧 lock pin spring
锁销提臂 lock lifter
锁销与锁销提臂 lock and lock lift
锁心后退量 lock backset
锁心柱 lock barrel
锁眼 eyelet hole;key hole;loop
锁眼盖 escutcheon
锁眼工 buttonhole maker
锁眼机 lockstitch button holder
锁眼式刻槽 <用于冲击试件> keyhole notch
锁以挂锁 padlock
锁执手摆线作用 cycloid knob action
锁止 deadlock
锁止点 locked point
锁止断层 locked fault
锁止杆 lock(ing) rod
锁止件 lock piece
锁止离合器 lock-up clutch
锁止片 lock banking stop
锁止式差速器 controlled differential
锁止位置 lock position
锁中心到门边距离 lock backset
锁轴 locking shaft;stud shaft
锁轴柄 latch pin handle
锁轴杆栓 locking shaft lever pin
锁轴杠杆 locking shaft lever
锁轴锁紧 lock clamp
锁轴压板 locking pin locating plate
锁住 enchain; latch(ing); lock-on; lockup;shut
锁住侧面 locking profile
锁住电路 latch(ing) circuit
锁住回转平台 turntable lock
锁爪 locking claw; locking pawl; locking ratchet
锁砖 closing brick; keystone; neck brick
锁转臂 tumbler
锁着 keyed
锁座 lock(ing) bracket
锁座托 lock bracket support

T

他

他变量 dependent variable
他变数 dependent variable
他变质作用 allometamorphism
他感作用物质 allelopathic substance
他国飞地 diplomatic enclave
他化变质 allochromatic metamorphism
他化学变质作用 allochemical metamorphism
他激 independent excitation
他激差绕直流电动机 differential-field motor;split-field motor
他激电 separate excitation
他激电焊发电机 arc welding generator with independent excitation
他激电机 separately excited machine
他激电路 separately driven circuit
他激发电机 separate excited generator
他激式发射机 driven transmitter
他激式间歇振荡器 separate-excited block oscillator
他激双稳态 slave bistable
他激外差法 separate heterodyne
他激振荡 sustained oscillation
他激振荡器 independent drive oscillator;separately excited oscillator;slave oscillator
他冷式电机 separately cooled machine
他力通风式电机 separately ventilated machine
他励 separate excitation
他励电动机 separately excited motor
他励发电机 separately excited dynamo;separately excited generator
他励励磁机 separate exciter
他人的实际反应 objective group ranking
他人物件的扣押 detainer
他色 allochromatic colo(u)r
他色性 allochromatism
他伤 trauma by other people
他生的 allogenic;allothigenic
他生化学作用 allochemical
他生矿物 allogenic mineral
他生物【地】allogene;allothigene
他生演变 allogenic change
他维亚 <一种用于筑路的沥青焦油> Tarvia
他形变晶【地】xenoblast;allotrioblast
他形变晶结构 xenoblastic texture
他形的【地】xenomorphic;allotriomorphic;anhedral
他形花岗变晶结构 granulitic texture
他形晶 anhedron;xenomorphic crystal
他形晶粒 xenomorphic grain
他形晶粒状的 xenomorphic-granular
他形晶粒状结构 allotriomorphic granular texture;xenomorphic granular texture
他形矿物 allotriomorphic mineral
他形粒状 allotriomorphic granular
他形体 allothimorph
他形组构 xenotopic fabric
他型 allotype
他养型湖 allotrophic lake
他治性 heteronomy

铊

铊玻璃 thallium glass
铊臭氧试纸 thallium ozone test paper
铊化合物 thallium compound
铊黄 thallium yellow
铊矿 thallium ore
铊矿床 thallium deposit
铊束流钟 thallium-beam clock
铊探测器 thallium detector
铊污染 thallium pollution
铊氧硫光电管 thalofide cell
铊中毒 thallium poisoning;thallotoxicosis

塌

塌岸 bank caving;bank failure;eroding bank
塌崩 breakdown
塌壁 sunk panel
塌边 edge slump(ing);slip-off edge
塌车 hand barrow;cart <双轮轻便运货车>
塌的 cavernous
塌顶 breakdown of roof
塌洞 funnel(l)ing inside the kiln
塌方 <地层结构上的> earth slide;earth slip;land fall;landslide;mountain creep;overbreak;rock slide;slumping;debris avalanche;ground loss;landslip;rock fall;cave-in <修筑上的>;collapse <修筑上的>
塌方处理 collapse treatment
塌方防御板 avalanche shed
塌方防御建筑物 avalanche baffle
塌方防御栅 avalanche defense[defence]
塌方湖 landslide lake
塌方区 landslip area
塌滑 slumping
塌滑构造 slump structure
塌积沉积 colluvial deposit
塌积的 colluvial
塌积盖层石 colluvial mantle rock
塌积黏[粘]土 colluvial clay
塌积碎屑 collapse rubbish
塌积土 colluvial soil
塌积物 colluvial deposit;colluvium
塌积物异常 colluvium anomaly
塌孔 snakes in hole
塌砾 talus
塌砾石流 tulus glacier
塌料 dropping of the charge
塌落 caving;caving-in;crumble;falling;slumping;tearing away
塌落的顶板 shet
塌落岩石堵塞钻孔 squeezed borehole
塌坡 landslip
塌奇主义 Tachism
塌缩隔离器 collapsing liner;projectile
塌陷 cave-in;collapse;downthrow;fall-in;sink;slump settlement;subsidence;wash-in;excessive penetration <焊接时>
塌陷岸 caving bank
塌陷地震 earthquake due to collapse
塌陷构造 collapse structure;gravity collapse structure
塌陷灰岩坑 collapsed doline
塌陷火山口 collapse crater
塌陷角 angle of draw
塌陷坑 collapse dolina
塌陷裂缝 collapse fissure
塌陷落水洞 collapsed doline;collapse sink;collapse sinkhole
塌陷破火口 collapse caldera
塌陷砌块 slump block
塌陷区 area of subsidence;subsidence area
塌陷溶陷坑 collapsed doline
塌陷洼地大小 size of collapse depression
塌陷洼地分布规律 distributed law of collapse
塌陷洼地形状 shape of collapse depression
塌陷型坍落度 collapsed slump
塌陷性 caving nature
塌陷性态 collapse behavio(u)r
塌箱 crush
塌芯 sag
塌窑 material collapse inside the kiln

塔

塔巴斯地块 Tabus massif
塔板 derrick panel <起重机>;fourble board <在四根钻杆组成立根的高处的>;column plate;column tray;tower tray;tray【化】
塔板操作弹性 operating flexibility of tray
塔板定距管杆长度 length of tray pitch spacer
塔板高度 plate height
塔板和泡罩 plates and bubble caps
塔板理论 plate theory
塔板式蒸馏塔 plate column
塔板数 number of plates
塔板水平度 tray flatness
塔板外径 outside diameter of tray;tray outside diameter
塔板效率 plate efficiency;tray efficiency
塔板效率因子 plate efficiency factor
塔板支座间距 distance of tray support to support
塔板至塔板支座的距离 distance of tray to tray support
塔崩 tabun
塔壁料 tower wall material
塔臂 tower boom
塔臂配置 mast tower attachment
塔标 tower beacon
塔伯拉伯造山运动【地】Tabberbbran orogeny
塔侧抽出物 side run-off
塔层形台 tier table
塔车 tower wagon
塔池 tower basin
塔尺【测】extension rod;level(l)ing pole;level(l)ing rod;level(l)ing staff;sliding staff;Sopwith staff;telescopic(al)staff;adjusting latch;box staff;telescopic(al)rod
塔尺垫 foot pan
塔齿轮 stepped gear;stepped wheel
塔的分馏效率 fractionation efficiency of tower
塔的汽提段【化】exhausting section of column
塔的生产能力 output of column
塔底 tower base;tower bottom
塔底泵 column bottoms pump
塔底残油 bottoms;tower bottoms
塔底产品 <炼油产生的> bottom product
塔底产物 tower bottoms
塔底基础 tower base foundation
塔底集液管 bottom collector
塔底框 base square
塔底流出物 bottom stream
塔底压力 tower bottom pressure
塔底油 bottom oil
塔底座 tower base
塔吊 monotower crane;tower crane;tower gantry;tower hoist;whirly
塔吊配件 tower attachment
塔吊行走机臂 saddle jib
塔吊助手 bank(s)man
塔顶 overhead;tower roof;tower top;tower head
塔顶拔杆【岩】line burner
塔顶产品 top product
塔顶产物冷凝器 overhead condenser
塔顶吊杆 top davit
塔顶工人 derrick man;derrick monkey;derrick skinner
塔顶回流 top reflux;trim the top of column
塔顶回转式塔式起重机 tower crane with slewing cat head
塔顶空气冷却器 overhead air cooler
塔顶馏出物 overheads
塔顶排出的蒸汽 over-the-top vapour
塔顶台板栏杆 pigpen
塔顶调节 top tower control
塔顶温度 tower top temperature
塔顶循环回流 top reflux
塔顶压力 tower top pressure
塔尔博特带 Talbot bands
塔尔博特平炉炼钢法 Talbot process
塔尔博特双联炼钢法 Talbot duplex process
塔尔博特液心钢锭轧制法 Talbot ingot process
塔(尔博特)<光能单位> talbot
塔菲石 taaffeite
塔费尔斜率 Tafel slope
塔夫绸 taffeta
塔夫-斯塔夫铝青铜 Tuf-Stuf alumin(i)um bronze
塔釜残液 bottom product
塔釜液 tower bottoms
塔高 tower height
塔沟瓦 tower gutter tile
塔硅锰铁钠石 taneyamalite
塔基 column foot;stupa base;tower base;tower footing
塔吉克矿 tadzhikite
塔吉克盆地 Tadzhik basin
塔吉克斯坦 <亚洲> Tadzhikistan
塔架 derrick;gantry scaffold;gin pole;mast frame;pylon bent;tower pier;head tower <架空索道或缆索起重机的>;pylon <架线缆用的>;tower <输电线路的>
塔架大梁 tower girder
塔架底坑 <混凝土> tower pit
塔架顶 tower head
塔架格床 tower grillage
塔架基础 tower foundation
塔架基础桩 tower foundation post
塔架间距 tower span
塔架间跨度 tower space;tower span
塔架可倾式缆道 luffing cableway
塔架可倾式索道 luffing cableway
塔架拉索体系 system of pylons and stays
塔架缆索挖土机 tower cable excavation
塔架门式起重机 tower gantry crane
塔架内提升斗 tower bucket
塔架平台 tower platform
塔架式桥墩 tower pier
塔架式挖掘机 tower machine
塔架提升机 tower hoist
塔架托梁系统 system of binders and joists for tower
塔架围梁 tower girt
塔架下部结构 tower substructure
塔架支枢 tower pivot
塔架支柱 tower leg
塔架支座 tower support
塔尖 epi;spire
塔尖放电 corposant
塔尖用瓦 steeple tile
塔间距离定比 intercolumniation

塔礁 pinnacle reef
塔脚 leg of tower
塔浸 tower leaching
塔径 tower diameter
塔卡斯铜镍合金 Toucas
塔康导航天线 Tacan antenna
塔科马港<美> Port Tacoma
塔科马海峡桥<美> Tocoma Narrows Bridge
塔克拉玛干沙漠 Takla Makan
塔拉沃德<一种可受钉砌块> Terrawode
塔兰<一种装配房屋> Tarran
塔冷却 tower cooler
塔利酸 tariric acid
塔梁墩 pylon-girder pier
塔梁铰接 pylon to girder hinged connection
塔淋浸出 tower leaching
塔楼 bicoca; tower building; tower dwelling; turret; rood tower<教堂平面十字架交叉处的>
塔楼的 turriculated
塔楼地下室 tower cellar
塔楼尖顶 spire
塔楼脚手架 tower scaffold(ing)
塔楼墙上十字形孔 balistraria
塔楼踏步 turret step
塔楼梯级 turret step
塔楼屋顶 turreted roof
塔楼状的 turret-like; turriculated
塔轮 cone pulley(with steps); cone wheel; driving cone; pulley cone; stepped cone; stepped gear; stepped pulley; stepped wheel
塔轮车床 cone pulley lathe
塔轮齿轮 cone gear
塔轮传动 cone pulley drive
塔轮护罩 cone wheel guard
塔轮转动 cone drive
塔轮转动装置 cone gear
塔轮装置 stepped wheel gear
塔罗斯型 Tarros type
塔马努里黏[粘]合剂<商品名> Tamanori
塔曼温度 Tammann temperature
塔帽的安装<附着式塔式起重机的> mounting of the crane head
塔门 pyller; pylon
塔锰石 takanelite
塔庙 stupa shrine; ziggurat
塔姆铁钛合金 Tam alloy
塔姆锡青铜 Tamtam
塔内件 tower internals
塔内竖放钻杆容量 stacking capacity of derrick
塔内台板<下管时用> belly board
塔内污泥 tower sludge
塔内淤渣 tower sludge
塔内作业 inside work
塔那那利佛<马达加斯加首都> Tananarive
塔纳利斯防腐剂<一种氟铬砷酚木材防腐剂> tanalith
塔能固件<一种专利的镀锌墙板紧固件> Talon fixing
塔逆流 tower counter-flow
塔帕纤维布 tapa cloth
塔盘 column tray; tower tray; tray floor; tray of column
塔盘板 tray deck
塔盘环 tray ring
塔盘间距 tray spacing
塔盘内件材料 tray internal materials
塔盘泡罩 tray cap
塔盘设计 tray design
塔盘升气管 tray riser
塔盘水力学 tray hydraulics
塔盘梯度 tray gradient
塔盘下流管 tray down-spout
塔盘压力降 tray gradient
塔盘堰 tray weir
塔盘振荡 tray oscillation
塔盘蒸汽上升口 tray riser
塔盘支撑环 tray support ring
塔盘支承 tray stiffer; tray support
塔泡罩 cap of column
塔硼锰镁矿 takeuchiite
塔-珀二氏塑性指数 Talwalker-Parmelee plasticity index
塔器 column
塔墙 tower wall(ing)
塔裙 tower skirt
塔上爆炸 tower burst
塔上工人<钻探时> attic hand; elevator latcher; sky-hooker
塔上架设的 tower-mounted
塔上起下钻工作平台 tubing board
塔上无人 without top man
塔上信号 semaphore signal
塔设备 tower equipment
塔身 tower body; tower shaft; tower section
塔身截面 mast section
塔身伸缩 central telescoping
塔身透光窗孔 dream hole
塔什蒂普极性超带 Tashtyp polarity superzone
塔什蒂普极性超时 Tashtyp polarity superchron
塔什蒂普极性超时间带 Tashtyp polarity superchronzone
塔式拌料箱 column mixer
塔式泵 turret type pump
塔式比率 pyramid ratio
塔式篦子 step grate
塔式标尺 box staff; extension rod; sliding rod
塔式仓 tower silo
塔式称量装置 tower type batch plant
塔式齿轮 cluster gear; nest
塔式单斗装载机 tower loader
塔式灯桅 structural mast; tower mast
塔式滴滤池 tower trickling filter; trickling tower
塔式滴滤池废水 trickling towers effluent
塔式电动挖掘机 cableway power scraper
塔式吊车 crane tower
塔式对流干燥机 counter-flow tower drier[dryer]
塔式发酵罐 tower-type fermentor
塔式发射架 tower launcher; vertical tower launcher
塔式发射台 tower-type launch pad
塔式法 tower process
塔式法硫酸 tower acid
塔式反应器 tower reactor
塔式房屋正面 tower block facade
塔式分批干燥机 tower-type batch drier[dryer]
塔式粉碎机 tower mill
塔式干燥器 tower drier[dryer]
塔式钢板烟囱 steel tower supported stack
塔式高层建筑<单独耸立的> point tower block
塔式高层住房 point-block housing
塔式高效滴池 tower high rate trickling filter
塔式戈培提升机 tower-type Koepe winder
塔式拱 tower arch
塔式谷物干燥机 tower-type grain drier[dryer]
塔式锅炉 updraft boiler
塔式回收系统 tower reclaiming system
塔式回转起重机 tower slewing crane
塔式架间距 tower span
塔式建筑 point building; tower block; tower building
塔式(建筑的)上部建筑 tower superstructure
塔式建筑群 point block
塔式建筑物 tower structure
塔式建筑压升水管 tower block rising line
塔式脚手架 cross bracing; scaffold tower; tower scaffold
塔式脚手架卷扬机 scaffold tower hoist
塔式教堂 tower-shaped church
塔式结构 tower structure
塔式截煤机 turret coal cutter
塔式进水口 tower intake
塔式井架 derrick; tower-type headframe
塔式净化器 tower purifier
塔式卷扬机 tower hoist
塔式拉铲挖掘 dragline tower excavation
塔式拉索挖土机 slackline cableway excavator; slackline cableway scraper
塔式缆索挖掘机 cable excavator; tower cable excavation
塔式缆索挖土机 tower cable excavation
塔式冷却器 tower cooler
塔式立窑 tower shaft kiln
塔式龙门架 tower gantry
塔式龙门起重机 tower gantry crane
塔式炉 tower furnace
塔式门道 towered gateway
塔式门与龙式缆索起重机 cable tower/gantry crane
塔式磨粉机 tower abrasion mill
塔式木笼 tower crib
塔式浓缩器 tower concentrator
塔式配合料仓 batch tower
塔式漂白装置 bleaching tower system
塔式平台 tower platform
塔式破碎机 beating tower
塔式起重机 tower(gantry) crane; tower hoist; tower derrick; column crane; monotower crane; pillar crane; stacker crane; hammerhead-boom<美>
塔式起重机的主塔 king tower
塔式起重机爬升法 tower crane climbing system
塔式起重机上部操纵台 tower head
塔式起重机悬臂 hammerhead jib
塔式起重机旋臂 hammerhead jib
塔式起重机摇臂 hammerhead jib
塔式起重机支柱 hammerhead pier
塔式起重架 tower derrick
塔式起重台架 derrick tower gantry
塔式起重桅杆 tower derrick
塔式起重装配设备 tower crane erection equipment
塔式气体冷却器 tower gas cooler
塔式汽车库<具备升降机的> autosilo
塔式汽船 turreted steamer
塔式汽蒸机 tower ager
塔式桥 tower bridge
塔式取水口 tower intake
塔式筛 screen tower
塔式升降机 tower hoist
塔式生物反应器 tower bioreactor
塔式生物滤池 biologic(al)tower filter; biotower; biowater; column type biofilter; tower biologic(al) filter; tower high rate trickling filter
塔式时钟 tower clock
塔式双翼风向标 tower bivane
塔式水准尺 extension rod
塔式酸洗机 tower pickle
塔式碎解机 beating tower
塔式索铲 dragline tower excavator
塔式索斗挖掘机 tower excavator
塔式提升 tower winding
塔式提升机 headframe hoist; tower hoist
塔式提升机料斗 tower hoist bucket
塔式天窗 lantern
塔式天线 tower-type antenna
塔式挺杆起重机 tower jib crane
塔式筒仓 tower silo
塔式拖铲(挖掘机) dragline tower excavator
塔式拖铲挖土机 dragline tower excavator
塔式脱硫 tower purification
塔式挖铲 dragline tower excavator
塔式挖掘机 cable scraper; cableway power scraper
塔式挖掘机开采法 tower system
塔式挖土机 tower excavator
塔式望远镜 tower telescope
塔式桅杆起重机 monotower derrick
塔式吸收系统 tower absorption system
塔式洗涤 tower washing
塔式洗涤器 tower scrubber; tower washer
塔式洗矿机 tower washer
塔式卸船机 unloading tower
塔式卸料篦子 cone-type discharge grate
塔式卸料机 tower unloader; unloading rig
塔式悬臂吊车 hammerhead crane
塔式悬臂起重机 column jib crane; hammerhead crane; tower slewing crane
塔式旋臂起重机 column jib crane
塔式旋动起重机 rotary-tower crane
塔式氧化 continuous air blowing
塔式蒸发 tower evapo(u)ration
塔式蒸发器 tower evapo(u)rator
塔式蒸馏器 tower still
塔式支座 tower support
塔式住房 housing tower
塔式住宅 residence tower; residential tower; tower dwelling
塔式筑坝起重机 whirly
塔式转摆式卸料机 conic(al) swing discharger
塔式转动起重机 rotary-tower crane
塔式装船机 shiploading tower; loading tower
塔式装载机 tower loader
塔式钻具 tapered drill string
塔式钻头 pilot type bit
塔斯康柱头 capital of Tuscan column
塔斯马尼高大乔木<材色淡黄,有波浪纹,用于雕刻和造船> Huon pine
塔斯曼地槽 Tasmania geosyncline
塔斯曼海 Tasman Sea
塔斯曼海槽 Tarim marine trough
塔斯曼海路 Tasmania seaway
塔斯曼煤 Mersey yellow coal
塔斯曼油页岩 combustible shale; Mersey yellow coal; tasmanite; white coal; yellow coal
塔斯曼纸浆和纸公司<新西兰> Tasman Pulp and Paper Co. Ltd.
塔酸 tower acid
塔酸回收系统 tower acid system; tower reclaiming system
塔台<机场的> control tower; pylon
塔台防护设施 tower shield facility
塔台飞行指挥员 tower officer
塔台控制中心 tower control center [centre]
塔台频率 tower frequency
塔台通信[讯] tower communication
塔台望远镜 tower telescope
塔台指挥人员 tower man

塔特尔纹 Tuttle lamellae
塔体 tower body
塔填充物 tower packing
塔填料 tower packing
塔腿 tower leg
塔腿构件 leg members
塔腿加固 leg reinforcing
塔外壳 shell
塔沃克-珀米利塑性指数 Talwalker-Parmelee plasticity index
塔下部引出物 side bottom
塔形 sugar-loaf fashion;turriform
塔形冰块 serac
塔形电子管 megatron
塔形堆 pyramidal pile
塔形辐射器 tower radiator
塔形干燥器 tower drier[dryer]
塔形拱 tower arch
塔形罐 pagoda shaped jar
塔形轮系 castle-wheel system
塔形门 pylon
塔形墓 high tomb
塔形平衡缠绕 <钢丝绳往滚筒上的> counter-balance spooling
塔形桥墩 tower pier
塔形桥楼 tower bridge
塔形砂轮座 cone wheel adapter
塔形双翼风向标 tower bivane
塔形轴 step shaft
塔形钻头 crowned bit
塔型结构 tower-type structure
塔中部液体入口 intermediate feed inlet
塔中浓缩段 enriching section
塔中形成沟流 channel(l)ing in column
塔中蒸气通过速度 throughput of column
塔钟 tower clock;turret clock
塔钟打点轮系 turret clock trains
塔钟打点凸轮 turret clock striking cams
塔重 tower weight
塔柱 pylon
塔柱起重机 post crane
塔柱座 stupa stylobate
塔状矗立的 spiring
塔状地貌 tower landform
塔状积云 towering cumulus
塔状喀斯特 tower karst
塔状抹面 <印度庙宇建筑中> sikhara
塔状上部结构 sikhara
塔状石英 Babel quartz
塔状室顶 <印度庙宇建筑中> sikhara
塔状树木 <如槐、榕树等> pagoda-tree
塔状屋顶 turreted roof
塔状物 pylon;turricula
塔兹冰阶【地】Taz stage
塔兹冰期【地】Taz glacial stage
塔组 tower system

榻 couch

榻榻米 <日本> tatami

踏板 balance brow;floor sheet;foot board;foot hold;foot path;foot plate;foot step;foot treadle;pad;paddle(board);planer;running board;side step;step(tread);striker;toe bead;toe board;tread(board);dome platform <罐车>;pedal(lever) <脚踏车、缝纫机、卫生洁具配件等的>;treadle <一端与地板铰接的>

踏板安全装置 pedal guard
踏板臂 pedal arm
踏板材料 tread material
踏板操纵的阀门 treadle valve
踏板操作阀 treadle operated valve
踏板操作系统 foot operated system
踏板长度 tread length
踏板齿轮 pedal gear
踏板锤 treadle hammer
踏板的 pedal
踏板电路 pedal circuit
踏板吊架 foot hanger;foot pedal hanger
踏板端翼 tread return
踏板阀 pedal valve
踏板杆 pedal-rod
踏板杆护环 pedal-rod grommet
踏板杆回动弹簧 pedal-rod return spring
踏板杆头 pedal push rod knob
踏板杆托架 pedal-rod bracket
踏板横轴 pedal cross shaft
踏板横轴杆 pedal cross shaft level
踏板护面铁(护边铁) step nosing
踏板护圈 pedal retaining collar
踏板回动弹簧 pedal return spring
踏板间隙 pedal clearance
踏板阶高 step height
踏板接点 treadle contact
踏板控制 pedal control
踏板宽度 tread width
踏板拉杆 pedal pull rod;pedal-rod
踏板拉后弹簧 pedal pull back spring
踏板路 tread plank road
踏板曲柄 pedal crank
踏板驱动 treadle drive
踏板声防止 footstep sound attenuation
踏板声强度 footstep sound intensity
踏板式泵 foot pump
踏板式电动走道 pallet-type moving ramp
踏板式杠杆 pedal lever
踏板式楼梯 ladder stair(case)
踏板式扭杆 pedal lever
踏板式起落机构 foot lift
踏板试验 walkway test
踏板套 pedal bushing
踏板调整 pedal adjustment
踏板调整环 pedal adjusting collar
踏板调整机 pedal adjustment machine
踏板调整连杆 pedal adjusting link
踏板停止器 pedal stop
踏板推杆 pedal push rod
踏板推杆导承 pedal push rod guide
踏板托 dome platform bracket;step tread board support
踏板托架 pedal bracket
踏板限位块 pedal stop
踏板橡皮面 step tread board rubber cover
踏板销 pedal pin
踏板延侧面 tread return
踏板验车器 vehicle treadle
踏板照明灯 courtesy light
踏板制动 pedal brake;pedal braking
踏板轴 pedal shaft
踏板轴衬 pedal bushing
踏板转向 pedal steer(ing)
踏板座 pedal base
踏步 gha(u)t;step;tread
踏步板 flier;pace-board;step board
踏步板槽口 pien(d) check
踏步板的圆形凸缘 rounded step
踏步边栏杆柱 bracket baluster
踏步步高 tread rise
踏步侧边栏杆 bracket baluster
踏步长度 tread length
踏步单元 step unit
踏步的踏板与踢板 step treads and risers
踏步底板 subtread
踏步地砖 tread tile
踏步垫 step pad
踏步顶盖 step cover
踏步定距标杆 going rod
踏步对搭接头 joint stepped
踏步泛水 step flashing
踏步覆盖物 tread cover(ing)
踏步高 tread rise
踏步高度 height of step;rise
踏步高计算表 <梯级> rise table
踏步级高宽比 <楼梯> rise and run ratio
踏步级高与踏步宽之比 rise and run ratio
踏步级宽 tread run
踏步检验 crippled leapfrog test
踏步进深 going of tread
踏步宽(度) going;run;tread run;width of step;tread width
踏步栏杆柱 bracket baluster
踏步面 <楼梯> tread surface
踏步木 stepping
踏步木板 step-plank
踏步平台 half pace
踏步前缘 nosing;stair nosing
踏步石 stepping stone
踏步式挡土墙 retaining wall with stepped back
踏步式拱楔块 stepped voussoirs
踏步式夯实 <蛙式打夯机的> straddling composition
踏步式护岸 contour bank
踏步式护堤 contour bank
踏步式基础 offset footing;offset foundation
踏步式街道 step street
踏步式楼板 slab-on-grade
踏步式楼梯梁 stepped string
踏步式坡道 stepped ramp
踏步式山墙 corbel gable
踏步式止推轴承 stepped thrust-bearing
踏步式座位 stepped seating
踏步首步 bottom step
踏步竖板 riser board
踏步(竖板)高度 riser height
踏步凸边 tread nosing
踏步凸面竖板 commode step
踏步突缘 nosing of tread
踏步涡饰 teram
踏步斜长 pitch dimension
踏步形的 crowfooted
踏步形泛水 stepped flashing
踏步形基础 benched foundation;stepped foundation
踏步形山墙 catstep;corbie gable;crow gable
踏步压型机 <楼梯> step press
踏步振动压型机 <楼梯> step vibrating press
踏步转角延伸的小突缘 return nosing
踏步桌 step table
踏车功能试验器 bicycle ergometer
踏成道路 tracking
踏成的小路 trail
踏乘自动车 autoped
踏出的小径 creteway
踏锤 treadle hammer
踏道 steps
踏蹬 treadle disc[disk]
踏凳 footstool
踏垛 banquette
踏垛坡 banquette slope
踏杆 foot bar;footstock lever;pedal lever
踏杆式织机 treadle loom
踏杆转子 treadle bowl
踏花被 quilted fabric;quilting
踏脚 cripple;step tread;tread
踏脚板 run(ning) board;foot board <车辆的>;foot step <楼梯>;wood clapper <木的>;buckboard <在软土面上作业用>
踏脚板灯 running board lamp
踏脚板护罩 running board shield
踏脚处 scarcement
踏脚凳 footstool;stool
踏脚格子 foot grating
踏脚孔 foot hole
踏脚拉杆 pedal pull rod
踏脚石 <浅水中的> stepping stone;stepped stone
踏脚石法 <一种线性规划方法> stepped stone method;stepping-stone method
踏脚索 foot rope
踏脚台 <上下马用> stepped stone;stepping stone
踏脚铁条 foot iron;step iron
踏阶吊铁 step hanger
踏阶支架 step hanger
踏勘 field inspection;perambulation;preliminary investigation;preliminary prospecting;reconnaissance survey;reconnoiter[reconnoitre];running survey;site reconnaissance;walkover survey
踏勘报告 reconnaissance report
踏勘测量 exploration survey;exploratory survey;reconnaissance survey
踏勘的目的和范围 purpose and scope of reconnaissance
踏勘地图 reconnaissance map
踏勘调查 investigation;reconnaissance
踏勘队 reconnaissance party
踏勘分类 classes of reconnaissance survey
踏勘检查 reconnaissance examination
踏勘阶段 reconnaissance phase
踏勘阶段的勘察 reconnaissance stage exploration
踏勘解译标志建立 reconnaissance/establishment of interpretation
踏勘目的 purpose of reconnaissance survey
踏勘者 reconnoiterer;reeonnoitrer
踏勘组 reconnaissance party
踏勘作业 reconnaissance work
踏口线 <楼梯的> nosing line
踏面 step board;tread
踏面剥离 flanking on wheel tread;tread shell-out
踏面擦伤 scotching of tread;scotching of wheel tread;slid flat spot on tread
踏面基点 taping point
踏面磨平 slid flat;wheel flat
踏面突沿 tread projection
踏面斜度 tread taper
踏面制动 tread brake[braking]
踏面制动传动装置 tread brake actuator
踏面制动单元 tread brake unit
踏钮 foot bolt
踏盘织机 cam loom;tappet loom
踏坪 tread
踏上高跷的高度 stilted height
踏石小径 stepway
踏踏板 pedal
踏梯 step ladder
踏旋器 tread mill
踏腰线板 foot plate;frame board;tread board;treadle pedal
踏综杆栅子 treadle grate
踏钻 foot drill

胎边 bead;tire[tyre] bead

胎垫 boot;tire[tyre] cushion;tire[tyre] chain
胎儿期的 prenatal
胎钢圈断裂 rim cut
胎环 tire[tyre] rim
胎火山 embryo volcano
胎肩 tire[tyre] shoulder
胎具 positioner

胎壳 tire[tyre] carcass
胎肋 tire[tyre] shoulder
胎里 inner liner
胎链 tire[tyre] chain
胎链备修链节 tire[tyre]-chain repair link
胎链钳 tire[tyre]-chain pliers
胎链调整器 tire[tyre]-chain adjuster
胎链修理钳 tire[tyre]-chain repairing pliers
胎裂救急套 boot
胎面 tread
胎面半径 tread radius
胎面边缘磨损 heel-and-toe wear
胎面剥离 tread detachment
胎面剥落 separation of tread
胎面底层 tread base;tread cushioning layer
胎面翻新 retread; tire[tyre] soles process
胎面翻新的轮胎 recap;recapped tire [tyre]
胎面翻新硫化模 recapping mould
胎面冠部 tread cap
胎面横向花纹 tread rig
胎面厚度 tread thickness
胎面厚度规 tread ga(u)ge
胎面弧宽 tread arc width
胎面花凹纹 tread pattern groove
胎面花纹 tread contour;tire[tyre] engraving
胎面花纹层 tread ply
胎面花纹翻新 regrooving of worn tire[tyre]
胎面花纹高度 skid
胎面花纹沟 tread groove;tread void
胎面花纹加强筋条 tread reinforcement fillet
胎面花纹接地面积 tread bar contact area
胎面花纹磨损 tread wear
胎面花纹设计 tread design
胎面花纹深度 tread depth
胎面机 profiling calender
胎面开槽机 regroover
胎面宽度 tread width
胎面磨损寿命 tread-wear life
胎面突起花纹 tread lug
胎面脱壳 tread separation
胎面外层 tread configuration
胎面弦宽 tread chord width
胎面压扁 tread concavity
胎面样模 tread contour plate
胎模 mo(u)lding bed
胎膜成型 socket form mo(u)lding
胎内衬带 tire[tyre] flap
胎内衬片 tire[tyre] boot
胎内的 prenatal
胎片 tire[tyre] shoe
胎圈 building ring;tire[tyre] bead
胎圈包布 chafer
胎锁 tire[tyre] lock
胎体 base;carrier;matrix body【岩】
胎体滴状突出式金刚石钻头 teardrop set bit
胎体开卷机 raw felt unroller
胎体帘布层 casing ply
胎体种类 <金刚石> kind of matrix
胎行高架起动机 travel(l)ing overhead tire[tyre]-mounted crane
胎压表 tire[tyre] pressure ga(u)ge
胎压记号 tire[tyre] deflection mark
胎缘加强层 bead reinforcement
胎缘填充心 bead core

台 balcony;dais;pedestal

台班 machine-team;shift
台班进尺 meterage per rig-shift
台班使用费 working day cost
台班数 amount of rig-shift
台班效率【岩】 drilling rate per rig-shift;meterage per rig-shift
台板 bed piece;bedpiece;bed plate;derrick board;platen;rig timber
台背 quarry bench
台背斜 anteclise[anteklise]
台泵抽水的群井 gang of wells
台编单位 cooperative compiled unit
台布 table cloth
台车 balanced full trailer;bogie[bog(e)y];carriage;dolly;jack plane;jumbo;loop wheel machine;skeleton log car;supply car;trailer converter dolly;troll(e)y
台车底盘 wagon drill carriage
台车盾构 truck jumbo
台车回转装置 bogie swing mechanism
台车架 bogie bracket;bogie frame
台车起重机 platform crane
台车式烘炉 car-type oven
台车式加热炉 car furnace
台车式炉 car-bottom hearth furnace;car type furnace
台车式泡包 buggy ladle
台车式热处理炉 car-type heat treating furnace
台车式下水 truck launching, craddle launching
台车式窑 platform car kiln;truck kiln
台车式装岩机 jumbo loader
台车式装载机 jumbo loader
台车窑 bogie kiln;car-bottom kiln;shuttle kiln;truck chamber kiln
台车钻架 carriage mounting
台城墙 Tai city wall
台秤 bench scale;platform balance;platform scale;platform weigher;platform weighing machine;platform weighing scale;platform weight machine;table balance;weigh(ing) bridge;weighing machine
台船 pontoon
台锤 bench hammer
台唇 apron;forestage;proscenium[复 proscenia]
台从同步 general locking
台从同步设备 genlock equipment
台导承 table guide
台灯 desk lamp;desklighting;reading lamp;table lamp;table light
台灯插头盖 bayonet cap
台灯柱环 bobeche
台地【地】 rising ground;bed terrace;bench terrace;chapada;elevated plain;land-berm;mesa;mesa terrace;plateau[复 plateaus/plateaux];platform;tableland;terrace;terrace land;upland plain
台地边缘沉积 platform-marginal slope deposit
台地边缘浅滩相 platform-marginal shoal facies
台地边缘生物礁相 platform-marginal organic reef facies
台地边缘斜坡相 platform-marginal slope facies
台地成因的【地】 kratogenic
台地城镇 <指美籍意大利建筑师索莱里所建议的方案> town on a table mountain
台地工程 simple terracing works
台地盥洗盆 pedestal lavatory
台地脊线 terrace ridge
台地峭壁 terrace cliff
台地泉 terrace spring
台地松 table mountain pine
台地碳酸盐岩组合 platform carbonate rock association
台地相 platform facies
台地园 terrace garden
台垫 cushion;mattress;pad
台度 dado
台度框架 dado frame
台段 quarry bench
台段采矿法 benching
台对 pair of stations;station pair
台份 component parts for assembly of a machine
台风 typhoon;hurricane wind
台风暴潮 typhoon storm surge
台风暴雨 typhoon rainstorm
台风避风处 typhoon shelter
台风边缘 edge zone of typhoon;marginal zone of typhoon
台风飑 typhoon squall
台风波涛 typhoon surge
台风搬运小组 typhoon evacuation team
台风地区 typhoon region
台风动向 typhoon movement
台风风荷载 typhoon wind load
台风轨迹 track of typhoon
台风紧急警报 typhoon imminent warning;typhoon urgent warning
台风警报 blue-alert;typhoon warning
台风警戒线 typhoon detective line
台风狂涌 hurricane surge;hurricane tide;hurricane wave
台风路径 path of typhoon;track of typhoon;typhoon path;typhoon track
台风路线转折点 cod;vertex[复 vertices/vertexes]
台风区 typhoon zone
台风扇 table fan
台风损害 typhoon damage
台风消息 typhoon information
台风信号 typhoon signal
台风行程 typhoon course
台风眼 eye of typhoon;typhoon eye
台风涌浪 typhoon surge
台风雨 typhoon rain
台风预报 typhoon forecasting
台风源地 typhoon source
台风中心 center[centre] of typhoon;eye of typhoon
台风中心气压指数 central pressure index
台风转向 typhoon recurvature
台佛尔磨耗试验机 Dever abrasion tester
台浮 flo(a)tation-tabling;table flo(a)tation
台盖梁 capping
台拱【地】 anteklise
台辊 bench roller
台虎钳 anvil vice;bench screw;bench vice[vise];clamp-on vice;leg vice;stock vice[vise];table vice
台基 boule;platform;stereobate;stylobate
台基技术 pedestal technique
台基上部线脚 dado mo(u)lding
台基上部装饰线脚 surbase
台级色谱法 step chromatography
台级形乱石堆 terraced riprap
台级形乱石堆砌 terraced riprap
台架 bench;desk frame;fiddle;gantree;gantry;jack horse;jumbo;rack;saddle piece;skid bed;staging;stand;trestle
台架充电 bench charge
台架级试验 bench-scale test(ing)
台架加速试验 accelerated bench test
台架架梁(法) staging erection
台架脚手架 ladder jack scaffold;ladder scaffold(ing)
台架模拟试验 bench simulation test
台架起重机 pedestal crane
台架时间 bench time
台架式烘干小车 rack dryer car
台架式集装箱 platform based container
台架式起重机 abutment crane
台架式水槽 bench flume
台架试验 bench-scale test(ing);bench test;block test;captive test;indoor test;rig test(ing);stand test
台架试验结果 rig result
台架试验时间 bench testing time
台架试验寿命 dynamometer life
台间联络线 interposition trunk
台间碳酸盐岩组合 interplatform carbonate rock association
台肩 circular bead;shoulder
台肩厚度 shoulders thickness
台肩棱边 land corner
台肩磨床 shoulder grinding machine
台肩式凸模 shoulder punch
台剪机 bench shears;bench snip;stock shears
台礁 table reef
台礁圈闭 table reef trap
台脚 plinth wall;table foot
台脚线 pedestal mo(u)lding
台铰 abutment hinge;table-hinge
台阶 bench;bench terrace;pedestal sit;plateau[复 plateaus/plateaux];steps;stoop;terrace slope;box steps <带台阶立板>;altar <干船坞壁的>
台阶板 foot step
台阶爆破 bench blasting
台阶爆破循环 benching round
台阶布置 step arrangement
台阶采矿法 bank method of attack
台阶齿轮 stepped gear wheel
台阶穿孔 bench drilling
台阶大爆破 large scale bench blasting
台阶的平均漂移值 average shift value of step
台阶垫 step pad
台阶法 <隧道开挖> bench cut method;benching tunnel(l)ing method
台阶高度 bench height;shoulder height;step height
台阶工作面 bench face
台阶工作面爆破(法) heading and bench blasting
台阶夯实机 bench rammer
台阶结构 step structure
台阶结构面 step structural plane
台阶掘进法 heating and bench method
台阶开挖法 bench cut method;benching
台阶力 step force
台阶立板 step riser;string board
台阶流 step flow
台阶螺钉 stepped screw
台阶码头岸壁 stepped wharf wall
台阶坡面角 bench angle
台阶熔岩 bench lava;bench magma
台阶色谱法 step chromatography
台阶式 setback;setback type
台阶式爆破 bench blasting
台阶式边坡 bench slope
台阶式波形发生器 staircase generator
台阶式插进工作面的炮眼组 bench round
台阶式成本 stepped cost
台阶式底脚 benched footing
台阶式工资增加 stepped increase
台阶式工作面 buttock face;stepped face
台阶式花园住宅 terrace garden housing

T

台阶式回采 benching cut
台阶式基础 benched foundation;crepidoma;stepped footing
台阶式基脚 stepped footing
台阶式建筑 setback building
台阶式掘进施工法 bench cut method
台阶式开采 bench mining;bench stoping
台阶式开挖 bench(ing) cut(ting);bench excavation;stepping
台阶式开挖法 bench method
台阶式看台 <跑道周围的> stadium [复 stadia]
台阶式牛眼灯玻璃 stepped bull-eye lamp glass
台阶式坡面 bench scale
台阶式墙基础 stepped wall foundation
台阶式渠底 stepped bed
台阶式生长 step growth
台阶式掏槽 bench cut
台阶式通道 <剧院等的> terraced aisle
台阶式挖掘 bench excavation
台阶式挖土 benching cut;step cutting
台阶式挖土法 bench cutting;bench excavation;bench method;benching
台阶式挖土路 bench cutting
台阶式运率 step-like rate
台阶式住宅 stepped house;terraced house
台阶式作业 benching mode-of-operation
台阶试验 step test(ing)
台阶掏槽 benching cut
台阶挖土法 benching(cut)
台阶形波发生器 staircase generator
台阶形垫铁 step block
台阶形基础 benched foundation
台阶形开挖 step excavation
台阶形量规 step ga(u)ge
台阶形砌体 corbelling
台阶形塞规 stepped plug ga(u)ge
台阶形线 step wire
台阶修整机 bench trimmer
台阶旋轮 <旋压用> stepped roller
台阶延迟时间 amplitude step time
台阶支板 bench board
台阶装载 bank loading
台阶状的 benched
台阶状断层【地】 staircase fault
台阶状接合 step joint
台阶状结构面 stepped discontinuity
台阶状外延生长 step-like epitaxial growth
台阶钻进 bench drilling
台阶钻孔 bench drilling
台景的框架 proscenium arch
台锯 bench saw;table saw
台锯机 breast bench
台卡 bench hook;Decca;side hook <使工件不移动的设施>
台卡导航 Decca navigation
台卡导航覆盖区 Decca coverage
台卡导航航线 Decca lane
台卡导航网 Decca chain
台卡导航系统 Decca navigation system;Decca(system)
台卡导航仪 Decca navigator
台卡导向定位 Decca homing fixing
台卡导向技术 Decca homing technique
台卡电台链 Decca chain;star chain
台卡电子定位系统 Decca electronic positioning system
台卡定位 Decca fixing;Decca positioning
台卡定位系统 Decca positioning system
台卡格网 Decca grid
台卡跟踪测距导航系统 Decca tracing ranging navigation system;trunk route Decca
台卡跟踪测距系统 Dectra
台卡海图 Decca chart
台卡航迹记录器 Decca track plotter
台卡活页资料 Decca data sheet;Decca sheet
台卡基波频率 fundamental frequency of Decca
台卡计 decometer
台卡接收机 Decca receiver
台卡链 Decca chain
台卡领航系统 Decca navigator system
台卡区 Decca zone
台卡双曲线导航图 Decca chart
台卡图网海图 Decca lattice chart
台卡位置线图 Decca chart
台卡无线电导航系统 Decca radio system
台卡系统选台字母 Decca suffix letter
台卡巷识别 Decca lane identification
台卡仪 Dacca
台卡远程覆盖区 Decca long range area coverage
台卡指示器 Decca indicator
台坎 banquette;berm(e)
台坎边坡截水沟 berm(e) ditch
台科拉克隆 <一种电控制动机防滑器> Decelakron
台口 proscenium arch;proscenium opening
台口侧幕 tormenter[tormentor]
台口高度 height of proscenium
台口坡度 slope of a cornice
台口墙 <舞台的> stage wall
台口式断面 benched section
台口式剧场布置 proscenium stage type
台口式路基 benched subgrade
台口线 cornice of pedestal;finish casing;mo(u)lded projecting course;stor(e)y band
台口檐幕 teaser
台链 chain of stations;station chain
台灵架 portable derrick crane
台曼干涉仪 Twyman interferometer
台帽 abutment cap(ping);coping(of abutment)
台面 bottom bed;floor;mesa;tabletop
台面板 deck plate
台面材 bench lumber
台面二极管 mesa diode
台面腐蚀 mesa etch
台面工艺 mesa technology
台面厚木板 floor board
台面检验 desk check
台面距 <压机的> daylight opening
台面宽度 land width
台面呢 baize
台面式晶体管 meca
台面型器件 mesa device
台面掩模 mesa mask
台面引导 land riding
台幕索具设备 curtain set
台年进尺【岩】 meterage per rig-year
台年效率 drilling rate per-year
台刨 bench plane
台刨把手 toat
台钳 bench clamp;bench vice[vise];carpenter's screw clamp;pliers;table vice[vise];vice[vise]
台钳夹(具) viscous clamp;vice[vise] clamp
台球桌 billiard table
台扇 desk fan;table fan
台上车床 bench lathe
台上计算机 desk-top computer
台上试验 <汽车发动机的> block test
台上手工作业 bench work
台上旋回起重机 titan crane
台上指示表 dial bench ga(u)ge
台上柱形灯 table standard
台身 abutment body;body of abutment
台施预应力 bench prestressing
台石 socle
台时 equipment-hour;machine-hour
台时利用(程度)指标 indicator of utilization of machine-hour
台时效率【岩】 meterage per rig-hour
台式 table model
台式冰箱 desk refrigerator
台式测微计 stage micrometer
台式插接板 desk-type control panel
台式车床 bench lathe
台式称量机 platform weighing machine
台式传声器 desk microphone
台式吹芯机 bench blower;bench-type core blower
台式锉锯机 bench filing and sawing machine
台式的 tabletop
台式低中速离心机 table type low and medium speed centrifuge
台式点焊机 bench type spot-welder
台式电动砂轮机 electric(al) bench grinder
台式电话机 desk telephone set;telephone set
台式电联锁器 electric(al) table interlocker
台式电子计算机 electronic desk computer
台式电子计算器 electronic desk calculator;electronic table calculator
台式电钻 electric(al) bench drill
台式多用木工机床 table multi-purpose wood working machine
台式风磨机 bench pneumatic grinder
台式风扇 bracker fan;desk fan
台式高频对讲机 mounted high frequency
台式攻丝机 bench tapping machine
台式光度计 bench photometer
台式弧焊机 bench arc welding machine
台式虎钳 stock vice[vise];table vice [vise]
台式混凝土切割机 table model concrete saw
台式机器 desk machine
台式激光器 mesa surface laser
台式计算机 desk calculating machine;desk computer;desk(-top) machine;desk calculator
台式计算器 desk calculator
台式计算装置 desk-top computer
台式计算机 desk-top computer
台式交换机 desk switchboard
台式浇筑 table casting
台式校验 desk check
台式结构 mesa structure
台式晶体管 mesa transistor
台式烤箱灶 table cooker
台式控制板 desk-type control panel
台式控制屏 desk-type control panel
台式老虎钳 bench vice[vise];leg vice [vise]
台式离心机 desk centrifuge
台式料堆 <集料堆方的> bermed pile
台式炉灶 table-mounted cooker
台式磨床 bench grinder
台式磨光机 bench sander
台式拧螺丝机 stand mat
台式牛床挤奶厅 elevated milking parlor
台式抛光机 table type buffing machine
台式刨床 bench shaper;bench shaping machine
台式配电板 desk switchboard
台式配电盘 bench board;desk switchboard
台式普通车床 bench engine lathe
台式齐边锯 table edger
台式器械 desk mounting apparatus;desk-top apparatus
台式刃磨机 bench grinding machine
台式三脚架 table(-top) tripod
台式砂轮机 floor stand grinder;bench sander
台式砂桩 bench rammer
台式实验用拉挤机 bench-type laboratory pultruder
台式收音机 table set
台式手摇砂轮 bench grinder
台式镗孔机 bench boring machine
台式天线 table antenna
台式投影仪 table viewer
台式弯板机 bench brake
台式屋顶 deck roof
台式洗面器 countertop basin;pedestal washbasin
台式显示器 console display
台式修边机 bench trimmer
台式修整机 bench trimmer
台式压床 bench press
台式压机 bench press
台式压力机 table press
台式仪器 table-top unit
台式音频视频通信[讯]机 desk-top audio video communicator
台式硬度计 bench-mounted hardness tester
台式油墨仪 bench inkometer
台式凿岩机 platform drill
台式振荡器 table shake
台式振动 plate vibration
台式振动器 deck shock machine;plate vibrator;platform vibrator;table vibrator
台式制动机 bench brake
台式自动计算机 automatic desk computer
台式钻床 bench drill;bench drilling machine
台式钻机 platform drill
台式钻架 bench drill stand
台数控制 sequence control
台数控制元件 sequence control kit
台田 platform field;raised field
台湾果松 master pine
台湾海峡 Taiwan Strait
台湾漆 <大漆> semecarpus vernicifera
台湾杉 Taiwan cypress;taiwania
台湾杉属 taiwania
台网 network of stations
台网探测能力 network detection capability
台维斯阶 <晚侏罗世>【地】 Divesian
台位利用系数【铁】 utility factor of the position
台下的 <指显微镜> substage
台下聚光器 substage condenser
台线坡度 slope of a cornice
台向斜 syneclise
台榭 high terraced
台雅鱼属 daya
台轧头 bench dog
台用插头 table tap
台用锤 bench hammer;fitter's hammer
台用虎钳 anvil vice;bench vice[vise]
台用千分尺 bench micrometer
台用水准仪 bench level
台用小车 turn bench

台月【岩】rig-month
台月进尺 meterage per rig-month
台月利用率 time availability of rig-month
台月数 amount of rig-month
台月效率【岩】meterage per rig-month;monthly efficiency
台站 station
台站传输矩阵节点 station transfer matrix
台站方位角 azimuth of station
台站改正 station correction
台站海拔高度 station elevation
台站号 station index(number);station number
台站校正 station correction
台站经纬度 longitude and latitude of station
台站距 station-to-station distance
台站连续图 station continuity chart
台站密度 station density
台站日志 station operation log
台站数据处理机 station processor
台站索引 station index
台站网 network of stations
台站误差 station error
台站线路设施 station-line facility
台站选址 station siting
台站异常 station anomaly
台站震中距 station-epicenter
台站自动管理 automatic station keeping
台站坐标 station coordinates
台账 machine account
台褶带 platform(al) folded belt
台砧 bench anvil
台阵地震学 array seismology
台阵反应 array response
台阵谱 array spectrum
台阵响应 array response
台阵站 array station
台震观测 <地震> array observation
台址 site of station;station location
台主 troll(e)y-table
台柱(子) main stay;pillar
台状冰山 table iceberg
台组 chain
台钻 bench drill(ing machine)
台座 bay;bed;bed box;bench;cabinet base;cabinet leg;fiddle;foot stall;pedestal
台座舱 stool tank
台座底脚 pedestal foot(ing)
台座雕带 pedestal frieze
台座发生器电路 pedestal generator circuit
台座法 stand method
台座基础 pedestal foot(ing);pedestal foundation
台座基脚 pedestal foot(ing)
台座夹钳 bench holdfast
台座脉冲 pedestal pulse
台座起重机 abutment crane
台座式 pedestal type
台座式盥洗盆 pedestal lavatory
台座式千斤顶 stand jack;table type jack
台座式小便池 pedestal urinal
台座式烟灰缸 pedestal ashtray
台座式叶轮搅拌器 pedestal-type impeller agitator
台座式饮水器 pedestal-type drinking fountain
台座体 solidium
台座线脚 pedestal mo(u)lding

抬

抬板机 pallet loader
抬包 shank;two-man ladle
抬包架 hand shank;ladle shank
抬车垫 jacking pad
抬车螺丝千斤顶 <救援起重机> bolster screw jack
抬船设备 lifting equipment
抬刀 cutter back-off;cutter lifting;cutter relieving
抬刀滑块 clapper block
抬刀机构 lifter
抬刀装置 clapper
抬刀座 clapper box
抬斗 hand frame
抬高 drive up;elevate;jack;uphold
抬高坝面 run-up on dam face
抬高的 raised
抬高的讲坛 elevated pulpit
抬高的圈梁 raised girt(h)
抬高的人行道 raised walkway
抬高价 mark-up
抬高价格 huff;jack up the price
抬高利润 profiteering
抬高路面板 slab jacking
抬高时价 raise the current price
抬高市价 jacking up price;jack-screwing
抬高水库水位 raising of pool
抬高水位 heading up;raising of water level
抬轨梁 track-carrying beam
抬轨器 rail lifter
抬机 jumping
抬价 force up commodity price;jacking up;jack up the price;raise;raise a price
抬价出卖 by-bidding
抬价者 raiser
抬举感 elevating sensation
抬缆方趸 anchor line scow
抬梁式构架 post-and-lintel construction
抬起 raise;raising;rear up;upheaval;upheave;uphold;uplift;uprear
抬起杆 raising wire
抬起轮胎的起重量 <汽车起重机的> free-on-wheels capacity
抬起器 raiser
抬前轮速度 rotation speed
抬圈钩 pull-out hook
抬升海岸 elevated shore face terrace
抬升礁海岸 raised reef coast
抬升量 ascendant quantity
抬升凝结高度 isentropic condensation level;lifting condensation level
抬升雾 lifting fog
抬头 name of buyer or payee
抬头人 payee
抬头支票 check to order;order check [cheque]
抬箱铸件 strained casting
抬芯 core flo(a)tation;core lift(ing);core raise
抬闸止杆 rack tail

苔

苔草沼泽 carex swamp;sedge moor
苔淀粉 moss starch
苔钙铁钛矿 cafetite
苔纲 hepaticae
苔类 <苔藓植物> hepaticae
苔芦泥炭 moss sedge peat
苔绿色 mossy green
苔绒绉 moss crepe
苔色素 orchil;orseille
苔色酸 orsellinic acid
苔似的 mossy
苔纹泉华 moss sinter
苔纹铜 moss copper
苔藓 lichen;merceya latifolia;moss
苔藓层 moss layer
苔藓虫动物地理区 bryozoan faunal province
苔藓虫纲 Bryozoa
苔藓虫灰岩 bryozoan limestone
苔藓地 mossery
苔藓动物 moss animal
苔藓风化 green efflorescence
苔藓泥炭 eriophorum peat;sphagnum moor
苔藓泥炭沼 eriophorum peat moss
苔藓色 mosstone
苔藓湿地 sphagnum wetland
苔藓学 muscology
苔藓沼泽 acid moor;moss bog;moss moor;sphagniopratum;sphagnum bog
苔藓植物 bryoflora;bryophyte;bryopyta
苔藓状 mossy
苔星体 asty
苔原 cold desert;tundra
苔原景观 tundra landscape
苔原气候 climate of tundra;tundra climate
苔沼土 <加拿大> muskeg
苔状的 mossy
苔状金 moss gold
苔状银 moss silver

薹

薹属 sedge

太

太 <1 太 = 10^{12}> tera

太尔各特测定纬度法 Talcott method of latitude determination
太尔各特法纬度 latitude by Talcott's method
太尔各特方法 Talcott method
太尔各特水准 Talcott level
太古【地】Grey Past;Archaeoid
太古代【地】Archaean(era);Archaeozoic era
太古代的【地】primeval
太古代基底【地】Archean basement
太古代陆核【地】Archean nucleus
太古的 archaean;pristine
太古及老元古巨旋回 archaean and eon-algonkian megacycle
太古界【地】Archaeozoic erathem;Archaean group
太古生代【地】Archeozoic
太古岩【地】Archaean rock
太古宇【地】Archaean Eonothem
太古宙【地】Archaean Eon
太赫 terahertz
太湖石 eroded limestone;Taihu Lake stone;water model(l)ed stone
太康造山运动 <晚奥陶世> Taconic orogeny
太靠风 all in the wind
太空 firmament;interplanetary space;outer space;space
太空城市 cosmograd;space city
太空船 satelloid
太空(发射)建筑 space building
太空飞船 space ship
太空飞行器 deep space vehicle;space craft
太空跟踪网 deep space net;deep space network
太空科学 space science
太空垃圾 space junk
太空模拟器 space simulator
太空企业 space enterprise
太空时代 space age
太空实验室 skylab;sky laboratory
太空试验室 skylab;sky laboratory
太空探测器 deep space probe;space probe
太空通信[讯] deep space communication
太空卫星 deep space satellite
太空运载器 space vehicle
太空站 cosmodom;space station
太梅尔地槽【地】Tajmyr geosyncline
太庙 Imperial Ancestral Temple
太欧姆表 teraohmmeter
太平岸洋红松 Pacific red cedar
太平舱口 escape hatch;escape hole
太平窗(户) escape window;fire escape window
太平岛-金盾暗沙拗陷地带【地】Taipingdao-Jindun'ansha depression region
太平灯 exit lamp
太平舵机 auxiliary steering gear
太平间 dead house;lich-house;morgue;mortuary
太平龙头 fire hydrant;fire plug
太平楼梯 escape stair(case);fire escape stair(case)
太平楼梯井 fire escape staircase well
太平门 port of exit;emergency door;emergency exit;escape;escape door;escape hatch;exit door;exit opening;fire escape;fire exit;free-escape;safe exit;sally port
太平门机械装置 panic exit mechanism
太平门门把手 emergency door handle
太平门栓 fire exit bolt;panic bolt;panic exit device;panic hardware
太平门锁止装置 exit lock
太平门五金配件 fire exit hardware
太平门照明 exit illumination
太平门指示灯 exit lighting
太平盛世 <未来的> the Y2K
太平竖井 escape shaft
太平隧道 escape tunnel
太平梯 emergency stair(case);escape ladder;escape stairway;fire escape ladder;refuge stairs;safety ladder
太平梯井口 escape staircase well
太平通道 fire escape corridor
太平桶 deck bucket
太平洋 the Pacific Ocean
太平洋岸柏树 Pacific coast cypress
太平洋岸黄雪松 Pacific coast yellow cedar
太平洋岸银杉 Pacific silver fir
太平洋岸云杉 Pacific coast spruce
太平洋岸紫杉 Pacific yew
太平洋板块【地】Pacific plate
太平洋北赤道洋流 Pacific North Equatorial Current
太平洋赤道反向洋流 Pacific Equatorial Countercurrent
太平洋北极风 Pacific arctic front;Pacific polar front
太平洋标准时间 <美国太平洋沿岸地区采用的西八区时间> Pacific Standard Time
太平洋带状地区 Pacific Rim
太平洋的 Pacific
太平洋地槽 Pacific geosyncline
太平洋地区经济合作委员会 Pacific Basin Economic Co-operation Committee
太平洋地区经济理事会 Pacific Basin Economic Council
太平洋地区旅游协会 Pacific Area Travel Association
太平洋地区通信[讯]系统 Pacific area communication system
太平洋动物区 Pacific faunal region
太平洋反气旋 Pacific anticyclone

T

太平洋-非洲地壳波系 Pacific-Africa crustal-wave system
太平洋风系 Pacific wind system
太平洋杆栎 pacific post oak
太平洋高气压 Pacific anticyclone
太平洋高压 Pacific high
太平洋鲑 Pacific salmon
太平洋国际航线 Pacific international line
太平洋海岸 Pacific coast
太平洋海岭地震构造带 Pacific ridge seismotectonic zone
太平洋海流 Pacific Ocean current
太平洋海啸警报系统 tsunami warning system in the Pacific
太平洋海啸警报中心 < 美 > Pacific Tsunami Warning Centre[Center]
太平洋海啸警报组 international coordination group for tsunami warning system in the Pacific
太平洋海洋观测 oceanographic(al) observation of the Pacific
太平洋航路指南 Pacific Ocean route instructions
太平洋航线 trans Pacific trade
太平洋花楸 S. americana tree; western mountain ash
太平洋活火山带 Pacific ring of fire
太平洋火山-珊瑚岛链 Pacific coral-volcano chain
太平洋金融共同体法郎 Communaute Financiere du Pacifique franc
太平洋经济合作会议 Pacific Economic Cooperation Conference
太平洋经济圈设想 Idea of Pacific Economic Sphere
太平洋壳块 Pacific crustal cupola
太平洋冷杉 Pacific silver fir
太平洋煤气电力公司 < 美 > Pacific Gas, Electric Co.
太平洋美国汽船协会 Pacific American Steamship Association
太平洋美国油轮协会 Pacific American Tankship Association
太平洋南赤道洋流 Pacific South Equatorial current
太平洋漆树 poison oak
太平洋气象观测网 Pacific meteorological network
太平洋区 Pacific Ocean area; Pacific region
太平洋区域通信[讯]卫星 Pacific satellite
太平洋时间 Pacific Time
太平洋树花 Ramalina pacifica
太平洋套 anapeirean; circum-Pacific province
太平洋铁木 ipil(e)
太平洋危险区 danger area in the Pacific
太平洋温带动物区 Pacific temperate faunal region
太平洋西北集装箱班轮公司 Pacific North West Container Line
太平洋西海岸航运联盟 Pacific coast West Bound Conference
太平洋夏季时间 Pacific daylight time
太平洋型大陆边缘 Pacific-type continental margin
太平洋型海岸线 coastline of Pacific type
太平洋型造山运动 Pacific-type orogeny
太平洋学会 < 美 > Institute of Pacific Relations
太平洋岩套 Pacific suite
太平洋沿岸标准时 Pacific Coast Standard Time
太平洋沿岸天然气协会 Pacific coast Gas Association
太平洋邮船公司 Pacific Mail Steamship Company
太平洋运动 Pacific orogeny
太平洋中隆 mid-Pacific rise
太平洋紫杉 Oregon yew
太平闸 emergency lock; escape lock
太沙基(承载力)公式 Terzaghi's solution
太沙基承载力理论 Terzaghi's bearing capacity theory
太沙基固结理论 Terzaghi's consolidation theory
太沙基水冲式土壤贯入仪 Terzaghi wash-point soil penetrometer
太沙基土体拱线 Terzaghi's ground arch
太沙基岩石荷载 Terzaghi's rock load
太熟 overripe
太弯曲的河流 misfit stream
太位 terabit
太位存储器 terabit memory
太阳 X 射线 solar X-ray
太阳 X 射线辐照度 solar X-ray irradiance
太阳 X 射线天文学 solar X-ray astronomy
太阳 X 射线耀斑 solar X-ray flare
太阳暗条 solar filament
太阳斑 solar flare
太阳半径 solar radius
太阳半日潮 lunar semidiurnal tide; solar semidiurnal tide
太阳半日分潮 solar semidiurnal component tide
太阳半日周期分潮 solar semidiumal tide
太阳半日主潮 main solar semidiurnal tide; principal solar semidiurnal tide
太阳保护装置 sun control
太阳暴 solar burst; solar eruption; solar storm
太阳暴风 solar gale
太阳爆发 solar eruption; solar outburst
太阳背点 solar antapex
太阳奔向点 solar apex
太阳扁率 solar oblateness
太阳标志 sun strobe
太阳波 solar wave
太阳测微计 heliomicrometer
太阳常数 solar constant
太阳潮 solar tide; solstitial tide
太阳成分 solar component
太阳齿轮 sun gear; central gear
太阳齿轮驱动 output sun gear
太阳赤经的增量 increment of sun's right ascension
太阳赤纬 solar declination
太阳传感器 solar sensor; sun sensor
太阳磁场 solar magnetic field
太阳磁学 solar magnetism
太阳丛 solar plexus
太阳大爆发 solar outburst
太阳大气 solar atmosphere; sun atmosphere
太阳大气潮 solar atmospheric(al) tide
太阳大气光学质量 solar air mass
太阳大气模型 model solar atmosphere
太阳单色光电影仪 spectroheliocinematograph
太阳单色光观测镜 spectrohelioscope
太阳单色光谱摄影 spectroheliograph
太阳单色光照片 spectroheliogram
太阳单色光照相术 spectroheliography
太阳单色光照相仪 spectroheliograph
太阳单色像 filtergram; monochromatic image; spectroheliogram
太阳导航镜 solar periscope
太阳的运动 sun's motion
太阳灯 artificial daylight; cold quartz mercury vapo(u)r lamp; mercury vapo(u)r lamp; sun burner; sun lamp; sun light; sunlight lamp; sunspot
太阳灯丝 helion filament
太阳等离子体 solar plasma
太阳等离子体流 solar plasma stream
太阳地磁潮 solar geomagnetic tide
太阳地形说 solar-topographic(al) theory
太阳电池 solaode; solar battery; solar cell
太阳电池板 solar blade; solar panel
太阳电池帆板 solar paddle
太阳电池叶片 paddle; solar paddle
太阳电池翼板 solar paddle
太阳电池阵列 solar cell array
太阳电池组件 solar battery modules; solar cell modules; solar modules
太阳电磁辐射 solar electromagnetic radiation
太阳电子事件 solar electron event
太阳订正 reduction to the sun
太阳定向飞行器 sun-oriented vehicle
太阳定向器 sun seeker
太阳定向仪 sun finder; sun-seeking device
太阳动力卫星 solar-powered satellite
太阳发电机 solar generator
太阳发射谱线 solar emission line
太阳法 < 天文观测 > solar method
太阳帆 photon sail; solar sail
太阳反光玻璃 solar reflecting glass
太阳反辉区 sun glint
太阳反射辐射 reflected solar radiation
太阳反射仪 heliotrope
太阳方位 solar position
太阳方位角 azimuth; azimuths of the sun; declination of sun; solar azimuth; sun's azimuth
太阳方位角表 red azimuth table
太阳房 solar building; solar house
太阳分潮 solar component tide; solar tidal component
太阳分光热量计 spectropyrheliometer
太阳分光仪 spectrohelioscope
太阳风 solar wind; sun wind
太阳风边界 solar wind boundary
太阳风层 heliosphere
太阳风层顶 heliopause
太阳风地磁场边界 solar wind geomagnetic field boundary
太阳服务 solar service
太阳辐射 solar radiation; sun's radiation
太阳辐射爆发辐射计 solar burst radio meter
太阳辐射标尺 solar radiation scale
太阳辐射波谱 solar radiation spectrum
太阳辐射测量卫星 solar radiation measuring satellite
太阳辐射测量仪 solar radiation measuring set
太阳辐射观测 solar radiation observation
太阳辐射计 actinometer
太阳辐射量 quantity of solar radiation
太阳辐射流 solar flux
太阳辐射模拟器 solar radiation simulator
太阳辐射能 solar radiation energy
太阳辐射能吸收器 solar absorber
太阳辐射平衡 balance of solar radiation
太阳辐射谱 solar radiation spectrum
太阳辐射强度 intensity of solar radiation; solar radiation intensity
太阳辐射强度计 solarimeter
太阳辐射热 solar heat; solar radiant heat
太阳辐射热吸收系数 < 表面吸收的太阳辐射热与其所接受到的太阳辐射热之比 > absorptance for solar radiation; absorbance for solar radiation
太阳辐射事件 solar radiation event
太阳辐射输出 solar output; sun's output
太阳辐射卫星 solar radiation satellite
太阳辐射误差 solar radiation error
太阳辐射吸收率 absorption factor of solar radiation
太阳辐射压 solar radiation pressure
太阳辐射仪 actinometer
太阳附件 solar attachment
太阳伽马射电暴 solar γ-ray burst
太阳干扰 sun interference
太阳高度 solar altitude; sun elevation
太阳高度表 solar altitude table
太阳高度光渗差改正量 irradiation correction
太阳高度角 solar altitude angle; solar elevation angle; sun's altitude angle
太阳高度指示器 gnomon
太阳高度总正量 total correction of sun's altitude
太阳跟随器 sun follower; sun seeker
太阳跟踪 sun tracking
太阳跟踪仪 sun tracker
太阳观测 solar observation; sun observation
太阳观测附件 solar attachment
太阳观测目镜 helioscope eyepiece
太阳观测与预报网 solar observing and forecasting network
太阳观察镜 helioscope
太阳光 sun light; sunshine
太阳光斑 solar facula; solar flare
太阳光抽运 sun pumping
太阳光抽运激光器 solar-excited laser; sun-pumped laser
太阳光催化 solar photocatalysis
太阳光催化降解 solar photocatalytic degradation
太阳光催化矿化 solar photocatalytic mineralization
太阳光催化脱色 solar photocatalytic decolo(u)rization
太阳光催化氧化反应器 solar photocatalytic oxid reactor
太阳光灯 sun burner
太阳光电磁象仪 photoelectric(al) solar magnetograph
太阳光电过程 helioelectric process
太阳光度 solar luminosity
太阳光度计 heliograph
太阳光度仪 heliograph
太阳光化过程 heliochemical process
太阳光谱 solar spectrum
太阳光谱光通量 solar spectral flux
太阳光谱强度 solar spectral intensity
太阳光谱线 Fraunhofer lines
太阳光线 sunray; sun streams; sunbeam
太阳光压摄动 sunlight pressure perturbation
太阳光子 solar photons
太阳轨道 solar orbit
太阳轨道图 sun path diagram
太阳过渡区 solar transition region
太阳黑子 macula[复 maculae]; sun spots

太阳黑子带 sunspot zone
太阳黑子分量 sunspot component
太阳黑子辐射 sunspot radiation
太阳黑子光谱 sunspot spectrum
太阳黑子活动 sunspot activity
太阳黑子极大期 sunspot maximum
太阳黑子极小期 sunspot minimum
太阳黑子极性 sunspot polarity
太阳黑子理论 sunspot theory
太阳黑子群 sunspot group
太阳黑子日珥 sunspot prominence
太阳黑子数 sunspot number
太阳黑子相对数 relative sunspot number; sunspot relative number
太阳黑子耀斑 sunspot flare
太阳黑子指数 solar spot index
太阳黑子周期 sunspot cycle; sunspot periodicity
太阳红外辐射 infrared solar radiation; solar infrared radiation
太阳红外线 solar infrared
太阳后向散射紫外线仪臭氧探测仪 solar backscattered ultraviolet
太阳弧光 sun arc
太阳换热器推进 solar heat exchanger drive
太阳活动 solar activity
太阳活动峰年 solar maximum year
太阳活动区 solar activity region
太阳活动效应 solar activity effect
太阳活动预报 prediction of solar activity; solar activity prediction
太阳活动预报中心 solar forecast center[centre]
太阳活动周 solar activity cycle; solar cycle
太阳火箭 solar rocket
太阳集热器效率 collector efficiency
太阳继电器 sun relay
太阳焦点 solar focus
太阳角 solar angle; sun angle
太阳角传感器 sun-vector sensor
太阳角模拟器 heliodon
太阳金字塔 <位于墨西哥特奥蒂瓦坎> Pyramid of the Sun
太阳经度 solar longitude
太阳经纬仪 <装有太阳观测附件的经纬仪> solar transit
太阳静日 solar quiet day
太阳静日变化 solar quiet daily variation
太阳静日变化场 solar quiet daily variation field
太阳镜 helioscope; solar telescope; sunglasses; sun visor
太阳局部射电源 solar local radio source
太阳聚光灯 solar spotlight
太阳空间 solar space
太阳空气温度 solar air temperature
太阳雷达 solar radar
太阳理论 solar theory
太阳历 solar calendar
太阳历书 solar ephemeris[复 ephemeredes]
太阳粒子 solar particle
太阳粒子警戒网 solar particle alert network
太阳粒子事件 solar particle events
太阳连续谱辐射 solar continuum emission
太阳流 solar stream
太阳流量单位 solar flux unit
太阳滤光片 solar filter; sun filter
太阳路径 sun's way
太阳轮 sun gear; sun wheel
太阳罗盘 solar compass; sun compass
太阳罗盘定向 sun compass orientation
太阳漫射辐射 diffuse solar radiation
太阳庙 Temple of the Sun
太阳目镜 solar eyepiece
太阳(目视)观测镜 helioscope
太阳能 insolation energy; solar energy; solar energy source; solar power; sun energy; sun power
太阳(能)板 <指装有太阳电池的平翼或板> solar panel
太阳能表 solar watch
太阳能冰箱 solar energy refrigerator
太阳能材料 solar energy materials
太阳能采暖 solar heating
太阳能测量计 solarimeter
太阳能常数 solar constant
太阳能储存介质 solar storage
太阳能储存器 solar storage
太阳能储存系统 active system
太阳能储热器 solar storage
太阳能穿透率 solar energy transmission
太阳能船 solar energy ship
太阳能得热系数 solar heat gain factor
太阳能电池 cell battery; solaode; solar battery; solar cell
太阳能电池板 solar panel
太阳能电池极 solar energy panel
太阳能电池屋盖材料 solar cell roofing material
太阳能电池叶片 solar cell paddle
太阳能电池阵 solar array
太阳能电池阵激励组件 solar array drive assembly
太阳能电源 sun-generated electric(al) power
太阳能电站 solar power station
太阳能动力 solar power
太阳能动力的 solar-powered
太阳能动力航天器 solar-powered space vehicle
太阳能发电 solar electric(al) energy generation; solar electric(al) power generation
太阳能发电厂 solar power plant
太阳能发电机 solar generator
太阳能发电屋顶 solar electric(al) roof
太阳能发电屋面 solar electric(al) roof
太阳能发动机 solar energy prime mover; solar engine
太阳能发生器 solar generator
太阳能房 solar building
太阳能辐射 solar irradiation
太阳能辐射系数 solar heat coefficient
太阳能辐射照度 solar irradiance
太阳能辅助热泵 solar-assisted heat pump
太阳能辅助热泵系统 solar-assisted heat pump system
太阳能辅助吸收制冷系统 solar-assisted absorption cooling system
太阳能干燥 solar drying; solar energy drying
太阳能干燥器 solar drier[dryer]
太阳能耕作 solar farm
太阳能工程(学) solar(energy) engineering
太阳能供电无线电设备 solar-powered radio; sun-powered radio
太阳能供暖 solar heating
太阳能供暖和热水系统 solar heating and hot-water system
太阳能供热系统 solar heating system
太阳能固定 fixation of solar energy
太阳能光电池板 solar panel
太阳能光伏电池 solar photovolt cell
太阳能海水蒸馏器 solar sea water still
太阳能获得量 solar gain
太阳能集热板 solar panel
太阳能集热器 solar collector; solar energy collector; solar heat collector
太阳能集中器 solar concentrator
太阳能技术 heliotechnics; solar energy technology
太阳能加热 solar heating
太阳能加热器 solar radiation collector
太阳能加热式沥青贮仓 solar energy heating asphalt storage
太阳能加热系统 solar heating system
太阳能建筑(物) solar building; solar construction
太阳能建筑物负荷比 solar building load ratio
太阳能交流发电机 solar-powered alternator unit
太阳能聚集器 solar energy collector
太阳能开发 development of solar energy
太阳能空间加热器 solar space heater
太阳能空气加热器 solar air heater
太阳能空气调节 solar air-conditioning
太阳能空气调节系数 absorption air conditioning
太阳能空调系统 solar air-conditioning system
太阳能冷却负荷 solar cooling load
太阳能冷却系统 absorption refrigeration
太阳能利用 solar energy utilization; solar utilization; utilization of solar energy
太阳能利用系统 solar energy system
太阳能利用协会 <美> Association for Applied Solar Energy
太阳能利用装置 helioplant
太阳能沥青熔化装置 asphalt solar energy melter
太阳(能)炉 solar furnace
太阳能汽车 solar energy car
太阳能区域精炼炉 solar zone refiner
太阳能取暖 solar heating
太阳能热泵 solar heat pump
太阳能热交换器 solar heat exchanger
太阳能热水器 solar water heater
太阳能热搪瓷收集器 solar heat enamel collector
太阳能热温控性能 solar heat controlling property
太阳能赛车 solar power racing car
太阳能收集面积 solar collection area
太阳能收集器 solar collector; solar energy collector
太阳能水泵 solar water pump
太阳能通道 solar chimney
太阳能推进 solar propulsion
太阳能温室 solar greenhouse
太阳能涡轮发电装置 solar turboelectric(al) power unit
太阳能无线电设备 solar-powered radio
太阳能吸热片 absorber plate
太阳能吸收器 flat plate collector; solar collector
太阳能吸收涂层 solar energy absorbing coating
太阳能吸收应用系统 solar absorbing system
太阳能吸收制冷机 solar absorption refrigerator
太阳能系统最佳经济成本 economic optimal solar system
太阳能蓄水池 solar pond
太阳能学会 <美> Solar Energy Society
太阳能养护 sun-energy curing
太阳能再生器 solar regenerator
太阳能蒸发池 solar evapo(u)ration pond
太阳能蒸发器 solar energy evapo(u)rator; solar still
太阳能蒸馏 solar distillation
太阳能蒸馏池 solar distillation basin
太阳能蒸馏法 solar distillation process; solar still process
太阳能蒸馏器 solar distiller
太阳能蒸煮 solar cooking
太阳能指数 solar fraction
太阳能制冷 solar cooling; solar refrigeration
太阳能制冷系统 solar cooling system
太阳能助航系统 solar-powered navigation
太阳能转换器 flat plate collector; solar energy converter; solar regenerator
太阳能转换系统 solar power thermal conversion system
太阳能装置 solar energy equipment; solar heating system
太阳年 <365日5时48分46秒> astronomic(al) year; equinoctial year; natural year; solar year; tropic(al) year
太阳年周潮 solar annual tide
太阳平均辐射率 solar constant
太阳谱 solar spectrum
太阳谱斑 solar faculae; solar flocculus[复 flocculi]; solar plage
太阳气候关系 solar-weather relationship
太阳潜望镜 <航测用> solar periscope
太阳强度 solar intensity
太阳桥 solar bridge
太阳青铜合金 sun bronze alloy
太阳倾斜角 declination of sun
太阳全光谱传感器 whole-sun sensor
太阳全日潮 solar diurnal tide
太阳全色照片 photoheliogram
太阳全色照相仪 photoheliograph
太阳染料 <酸性偶氮染料> solar colo(u)rs
太阳扰动 solar disturbance
太阳扰日变化 solar disturbed daily variation
太阳热 heliothermal; solar heat; sun heat
太阳热储存 solar heat storage
太阳热发射板 solar reflection surface
太阳热反射面 solar reflection surface
太阳热辐射 solar heat radiation
太阳热负荷 solar heat load
太阳热力 solar heat power
太阳热力交换器 solar heat exchanger
太阳热量测量学 pyrheliometry
太阳热量计 pyrheliometer
太阳热蒸馏法 solar distillation process
太阳日 natural day; solar day; tropic(al) day
太阳日变化 solar daily variation; solar diurnal variation
太阳入射角 angle of solar incidence
太阳软X射线暴 solar soft X-ray burst
太阳三角测量 solar triangulation
太阳伞 umbrella
太阳散射辐射 diffuse solar radiation
太阳色球 solar chromosphere
太阳色球网络 solar network
太阳上侧正切弧 supralateral tangent arcs
太阳射电 solar radio radiation
太阳射电爆发 solar radio burst
太阳射电波 solar radio waves
太阳射电发射 solar radio emission
太阳射电辐射 solar radio emission; solar radio waves
太阳射电干扰 solar noise

T

太阳射电天文学 solar radio astronomy
太阳射电运动频谱仪 solar radio dynamic(al) spectrograph
太阳射电噪暴 solar radio noise storm
太阳射电噪声 solar noise
太阳射电噪声干扰 radio noise from the sun
太阳射线电子 solar-ray electron
太阳射线收集器 solar radiation collector
太阳摄动 solar perturbation
太阳摄谱仪 solar spectrograph; spectroheliograph
太阳摄影机 heliograph; solar camera; stellar camera
太阳摄影仪 photoheliograph
太阳神 <希腊神话> Apollo; Phoebus
太阳神神庙 <古埃及> Temple of Horus
太阳石 aventurin(e) feldspar; sunstone
太阳时 solar time; sun time
太阳时角 sun hour angle
太阳时钟 mean time clock; sun time clock
太阳视差 solar parallax
太阳视宁度监测仪 solar seeing monitor
太阳室 solar house
太阳输入 solar input; sun's input
太阳塔 solar tower; tower telescope
太阳探测器 solar probe; sunblazer
太阳探测宇宙飞船基地 base for solar probe spacecraft
太阳特大高度求船位 fixing by very high altitude sun sights
太阳天顶角 solar zenith angle; sun's zenith angle
太阳-天气关系 sun-weather relationship
太阳天文台 solar observatory
太阳通量 solar flux
太阳通信[讯] solar communication
太阳同步轨道 sun-synchronous orbit
太阳同步极地卫星 sun-synchronous near polar orbit satellite
太阳同步近极轨道 near polar sun-synchronous orbit; sun-synchronous nearly polar orbit
太阳同步卫星 sun-synchronous satellite
太阳同步准极轨道 quasi-sun-synchronous orbit
太阳图 solar atlas
太阳椭圆分潮 solar elliptic(al) component; solar elliptic(al) constituent
太阳望远镜 helioscope; solar eyepiece; solar telescope
太阳微波暴 solar microwave burst
太阳微粒 solar corpuscle
太阳微粒爆发 solar corpuscular burst
太阳微粒发射 solar corpuscular emission
太阳微粒辐射 solar corpuscular radiation
太阳微粒流 solar corpuscular flow; solar corpuscular stream
太阳微粒束 solar corpuscular beam
太阳卫星 solar satellite; sun satellite
太阳位置图 sun path diagram
太阳位置线 sun line
太阳温度 solar temperature
太阳温度计 heliothermometer; solar thermometer
太阳屋 solar house
太阳物理学 heliophysics; solar physics
太阳吸收率 solar absorptance
太阳吸收指数 solar absorption index
太阳吸水 sun drawing water
太阳系 solar system; sun system
太阳系齿轮 sun-and-planet gear
太阳系仪 orrery; planetarium
太阳现象 solar phenomenon
太阳向点 solar apex
太阳小齿轮 sun pinion
太阳小时 solar hour
太阳星云 solar nebula
太阳行星标介质 solar interplanetary medium
太阳行星际磁圈 solar interplanetary magnetic loop
太阳行星角距 angle of commutation
太阳行星运动 sun-and-planet motion
太阳型反应 solar-type reaction
太阳型恒星 solar-type star
太阳学 heliology
太阳巡视 solar patrol
太阳研究航天器 solar space vehicle
太阳盐 solar salt
太阳眼镜 preserve; sunglasses
太阳耀斑 solar flare
太阳仪 solar attachment
太阳移线定位 sun-run-sun fixing
太阳引力潮 gravitational tidal force of the sun; tide-generating force of the sun
太阳油 solar oil
太阳鱼 sunfish
太阳宇宙粒子 solar cosmic particle
太阳宇宙射线 solar cosmic ray; space cosmic ray
太阳浴层面 sun deck
太阳圆面 solar disc[disk]
太阳远紫外线暴 solar extreme ultraviolet burst
太阳月 <30 日 10 时 29 分 3.8 秒> solar month
太阳运动 solar motion
太阳灶 solar cooker; solar energy stove; solar furnace
太阳噪声 solar noise
太阳照片 photoheliogram
太阳照射的气团 solar air mass
太阳照射卫星 solar illuminated satellite
太阳照相磁像仪 photographic(al) solar magnetograph
太阳照相机 heliograph
太阳照相图 heliogram
太阳照相仪 heliograph; photoheliograph
太阳针状物 solar spicule
太阳蒸馏法 solar distillation process
太阳蒸馏锅 solar still
太阳蒸馏器 solar still
太阳直接辐射 direct solar radiation
太阳直射 vertical incident solar rays
太阳直射辐射 direct solar radiation
太阳直射光线 direct rays of the sun
太阳质量 solar mass
太阳质子 solar proton
太阳质子监视仪 solar proton monitor
太阳质子流 solar proton flux
太阳质子事件 solar proton event
太阳质子效应 solar proton event
太阳质子耀斑 solar proton flare
太阳中天高度 sun's meridian altitude
太阳中天观测 noon sight
太阳中微子 solar neutrino
太阳中微子单位 solar neutrino unit
太阳中微子流量 solar neutrino flux
太阳中心说 heliocentricism
太阳周 cycle of the sun; heliacal cycle; solar cycle
太阳周变化 solar cycle variation
太阳周年视运动 annual solar motion
太阳周期 solar period
太阳周期潮 solar annual tide
太阳周日变化 solar daily variation
太阳周效应 solar cycle effect
太阳主要半日分潮 major solar semidiurnal constituent; principal solar semidiurnal component; principal solar semidiurnal constituent
太阳主要全日分潮 major solar diurnal constituent
太阳主要日分潮 principal solar diurnal component; principal solar diurnal constituent
太阳状破裂 <锆合金的一种腐蚀破裂形式> sun burst
太阳状物 sun
太阳紫外成像望远镜 solar ultraviolet imaging telescope
太阳紫外辐射 solar ultraviolet radiation
太阳紫外线 solar ultraviolet
太阳自转 solar rotation
太阳综合图 synoptic(al) map of the sun
太阳总辐射 total solar radiation
太阳总改正增订量 additional sun's correction
太阳最大辐射强度 maximum solar radiation intensity
太阳最大高度求纬度 latitude by sun's maximum altitude
太阳最高高度 maximum solar altitude
太阳作用气温 sol-air temperature
太阴 moon
太阴半日变化 lunar semi-diurnal variation
太阴半日潮 lunar semidiurnal tide
太阴变化 lunar variation
太阴潮 lunar tide
太阴潮变化 lunar tide variation
太阴潮的 lunitidal
太阴赤纬分潮 lunar declinational diurnal constituent
太阴出差分潮 lunar evectional component; lunar evectional constituent
太阴大气潮 lunar atmospheric tide
太阴太阳分潮 lunisolar tide
太阴的 lunar
太阴分潮 lunar component tide; moon tidal component
太阴轨道 moon's orbit
太阴间隙 lunar tide interval
太阴历 lunar ephemeris
太阴年 <354 日> lunar year
太阴浅水六分之一日分潮 shallow water lunar 1/6 diurnal tidal component
太阴浅水四分之一日分潮 shallow water lunar 1/4 diurnal tidal component
太阴全日潮 lunar diurnal tide
太阴全日分潮 lunar diurnal component
太阴日 <24 时 50 分，相当 24.84 太阳时> lunar day; tidal day
太阴日变化 lunar daily variation; lunar diurnal variation
太阴日变化场 lunar daily variation field
太阴日周期分潮 lunar diumal tide
太阴时 lunar hour; lunar time
太阴视差 lunar parallax
太阴视差半日分潮 lunar parallactic semidiurnal constituent
太阴视差不均衡 lunar parallax inequality
太阴双周潮 lunar fortnightly tide
太阴太阳半日潮 lunisolar semidiurnal tide
太阴太阳半日分潮 lunisolar semidiurnal component; lunisolar semidiurnal constituent
太阴太阳赤纬半日分潮 lunisolar declinational semidiurnal constituent
太阴太阳赤纬全日分潮 lunisolar declinational diurnal constituent
太阴太阳分潮 lunisolar component; lunisolar constituent
太阴太阳(合成)全日(分)潮 lunisolar diurnal component
太阴太阳浅水四分之一 shallow water luni-solar 1/4 diurnal tidal component
太阴太阳日潮不等 lunar and solar diurnal inequalities
太阴太阳日分潮 lunisolar diurnal component; lunisolar diurnal constituent
太阴太阳日周潮 lunisolar diurnal tide
太阴通信[讯] lunar communication
太阴椭率主要半日分潮 major lunar elliptic(al) semidiurnal constituent
太阴引潮力 tide-forming force of the moon; tide-generating force of the moon
太阴月 lunar month; lunation; month of the phases; synodic(al) month
太阴月分潮 lunar monthly component; lunar monthly constituent
太阴正午 lunar noon
太阴中天到下一高潮的平均间隔时间 mean establishment
太阴周期 lunar cycle
太阴逐日影响 lunar diurnal influence
太阴主要半日分潮 major lunar semidiurnal constituent; principal lunar semidiurnal component; principal lunar semidiurnal constituent
太阴主要全日分潮 major lunar diurnal constituent
太阴主要日分潮 principal lunar diurnal component; principal lunar diurnal constituent
太拥挤 over-crowding
太子港 <海地首都> Port an Prince

汰石子 granitic plaster

汰石子粉刷 granitic plaster; granitoid
汰石子面 washed finish

态的纯度 purity of state

态的合并 uniting of states
态度群集 attitude cluster
态函数 state function
态空间法【数】 state space method
态密度 density of states
态势图 situation map; tactical setting; tactical situation
态-态反应 state-to-state reaction

肽【化】 peptide

钛白 titanium dioxide; titan white

钛白粉 titanium pigment; titanium white
钛白刻图膜 titanium white coat(ing)
钛白涂层 titanium white coat(ing)
钛(白)颜料 titanium pigment
钛板 titanium sheet
钛钡白 tatan-barium white; titanox
钛钡铭石 lindslengite
钛钡颜料 titanium-barium pigment
钛泵 titanium getter pump; vacion pump

钛病 titanosis
钛材板片 titanium plate
钛材换热器 titanium heat exchanger
钛材平板式换热器 titanium plate heat exchanger
钛尘肺 titanosis
钛磁铁矿 titanic magnetism; titanomagnetite
钛磁铁矿含量 titanomagnetite content
钛磁铁矿矿石 titanomagnetite ore
钛的 titanic
钛电极 titanium electrode
钛电解槽 titanium electrolytic cell
钛钒矿 berdesinskiite
钛粉 titanium powder
钛覆盖层 titanium coating
钛钙白 rutile-calcium composite; titan-calcium white
钛钙白颜料 calcium base titanox; calcium extended titanium dioxide
钛钙型 titanium-calcium type
钛钙型焊条 limetitania type electrode
钛钙颜料 titanium-calcium pigment
钛酐 titanic anhydride
钛橄榄石 titanhydroclinohumite
钛锆钍矿 zerkelite[zirkelite]
钛铬黄 chrome rutile yellow
钛铬铁矿 titano-chromite
钛工具钢 Titanor metal
钛共晶石 ulvoespinel; ulvtite
钛硅合金 titanium silicon
钛硅镁钙石 rhonite
钛硅钠泥矿 pyrhite
钛硅铁 ferro-silicon-titanium
钛硅钇铈矿 perrierite
钛合金 titanium alloy
钛黑云母 titanbiotite
钛化 titanize; titanizing
钛化锡 tin titanium
钛黄 titan(ium) yellow
钛辉方钠正长岩 assyntite
钛辉沸煌岩 fourchite
钛辉石 titanaugite
钛辉石岩 titanopyroxenite
钛辉无球粒陨石 angrite
钛基涂层电极 Ti-based oxides coating anode
钛钾铬石 mathiasite
钛尖晶石 titanspinel; ulvospinel
钛角闪石 kaersutite; titanian hornblende
钛金红石 teshirogitlite
钛聚合物 titanium polymer
钛矿石 titanium ore
钛矿物 titanium mineral
钛榴石 ferrocyanide titanate; ivaarite; schorlomite; titangarnet
钛铝锆合金 Mallory Sharton alloy
钛铝合金 titanium-aluminium alloy
钛铝钼合金 titanium-aluminium-molybdenum alloy
钛铝锡合金 titanium-aluminium-tin alloy
钛绿 titanium green
钛镁尖晶石 baikorite
钛镁铁矿 kennedyite
钛镁颜料 titanium-magnesium pigment
钛母合金 titanium master alloy; titanium mother alloy
钛铌锰石 gerasimovskite
钛镍黄 nickel rutile yellow; nickel titanate yellow; titan(ium) yellow
钛铅钡白 titone
钛铅石 macedonite
钛闪石 kaersutite; titanhornblende
钛升华泵 titanium-sublimation pump
钛石膏 titanium gypsum
钛试剂 tiron
钛铈钙矿 loveringite
钛水蛭石 zonolite
钛酸 metatitanic acid; titanic acid; titanic hydroxide
钛酸钡 < 用于振动探测仪探头 > barium titanate
钛酸钡瓷器 barium titanate porcelain
钛酸钡热敏电阻器 barium titanate thermistor
钛酸钡陶瓷 barium titanate ceramic
钛酸钡陶瓷传感器 ceramic pickup
钛酸钡压电陶瓷 barium titanate piezoelectric(al) ceramics
钛酸铋 bismuth titanate
钛酸铋陶瓷 bismuth titanate ceramics
钛酸丁酯 butyl titanate
钛酸钙 calcium titanate
钛酸钙瓷 calcium titanate porcelain
钛酸锆 zirconia titanate; zirconium titanate
钛酸锆酸镧铅 lanthanum doped lead zirconate-lead titanate; lead lanthanum zirconate titanate
钛酸根 titanate radical
钛酸铪 hafnium titanate
钛酸钾 potassium titanate
钛酸镧陶瓷 lanthanum titanate ceramics
钛酸锂 lithium titanate
钛酸铝 < 一种低膨胀陶瓷 > alumin(i)um titanate
钛酸铝瓷器 alumin(i)um titanate porcelain
钛酸铝陶瓷 alumina titanate ceramics; alumin(i)um titanate ceramics
钛酸铝制品 alumin(i)um titanate product
钛酸镁 magnesium titanate
钛酸镁瓷 magnesium titanate porcelain
钛酸镁陶瓷 magnesium titanate ceramics
钛酸锰 manganese titanate
钛酸镍黄 nickel titanate yellow
钛酸镍黄色颜料 nickel titanate yellow pigment
钛酸铅 lead titanate
钛酸铅陶瓷 lead titanate ceramics
钛酸铅压电陶瓷 lead titanate piezoelectric(al) ceramics
钛酸锶 strontium titanate
钛酸铜 copper titanate
钛酸烷基酯 titanium alkoxide
钛酸辛二醇酯 octylene glycol titanate
钛酸锌 zinc titanate
钛酸锌陶瓷 zinc titanate ceramics
钛酸亚铁 ferrous titanate; ilmenite black
钛酸盐 titanate; titanate salt
钛酸盐电容器 titanate capacitor
钛酸盐陶瓷 titanate ceramics
钛酸乙酯 titanium ethoxide; titanium tetraethoxide
钛酸异丙酯 titanium isopropoxide
钛酸酯 titanate
钛钽铁矿 strueverite
钛铁 ferrotitanium
钛铁长橄岩 cumberlandite
钛铁磁铁矿 ilmenomagnetite
钛铁合金 ferrotitanium
钛铁黑 ilmenite black
钛铁混凝土 ilmenite loaded concrete
钛铁金红石 ilmenorutile
钛铁晶石 ulvospinel
钛铁矿 gregorite; ilmenite; menaccanite; serite; titanic iron ore; washingtonite
钛铁矿骨料 ilmenite aggregate
钛铁矿含量 ilmenite content
钛铁矿集料 ilmenite aggregate
钛铁矿矿石 ilmenite ore
钛铁矿型 ilmenite type
钛铁矿型(电)焊条 ilmenite type electrode
钛铁矿型花岗岩 ilmentite type granite
钛铁磷灰岩 nelsonite
钛铁砂 menakanite
钛铁双阳极电混凝法 electrocoagulation with titanium-iron double anodes
钛铁霞辉岩 jacupirangite
钛铁岩 ilmenitite
钛铁铀钇矿 delorenzite
钛铜 titanium copper
钛铜合金 titanium copper alloy
钛透辉石 titandiopside
钛涂层 titanium paint
钛涂层增强 tatanising; titaisation
钛钍矿 thorutite
钛网 titanium mesh
钛稳定的 titanium stabilized
钛钨金属陶瓷 Titait
钛钨硬质合金 dia-titanit
钛稀金矿 kobeite
钛锡合金 titanium-tin alloy
钛系珠光颜料 titanium dioxide coated mica
钛斜硅镁石 titanhydroclinohumite; titanclinohumite
钛斜钠锆石 titan-elpidite
钛锌钠矿 murataite
钛型 titanium type
钛型焊条 rutile rod; rutile type electrode; titania type electrode
钛盐 titanium salt
钛盐滴定(法) titanometric titration; titanometry
钛阳极 titanium anode
钛氧陶瓷 titania ceramics
钛钇钍矿 yttrocrasite
钛铀矿 brannerite
钛云母 odenite; titabiotite; titanmica; wodanite
钛制设备 titanium equipment
钛质白色陶瓷 titania whiteware
钛质瓷 titanate porcelain; titania porcelain
钛质陶瓷 titanate ceramics
钛铸铁 titanium cast iron
钛棕 titanium brown

泰

泰伯尔法耐磨试验 Taber abrasion test
泰伯尔法中的循环次数 wear cycles in Taber test
泰伯尔耐磨性测定仪 Taber abraser
泰伯尔耐磨性试验仪 Taber abraser
泰尔贝克煤气燃烧器 Terbeck gas burner
泰尔福式基层 Telford base; Telford foundation; Telford macadam; Telford base
泰尔福式路面 Telford pavement
泰尔福式碎石路 Telford macadam
泰尔福装配式房屋 Telford
泰尔红紫 Tyrian purple
泰尔科铁镍钴合金 Telcoseal
泰尔斯方法 < 地下水力学的 > Theis method
泰尔铜锰镍合金 Telcuman
泰革敞杯法闪(火)点 flash point Tag open-cup
泰格·鲁宾逊色度 Tag-Robinson colo(u)r
泰格·鲁宾逊比色计 Tag-Robinson colo(u)rimeter
泰格板 < 一种石膏板 > Tiger board
泰格闭杯点试验器 Tag closed cup tester
泰格闭杯闪点试验 Tag flash point test
泰格敞杯 Tagliabue open-cup
泰格敞杯法闪(火)点 < 试验重质油类及沥青材料用 > flash point of Tag open cup
泰格敞杯试验 Tagliabue open-cup test
泰格敞杯试验器 Tag open-cup tester
泰格法闪点 flame point Tag
泰格开杯闪点试验器 Tag open-cup tester
泰格密闭闪点试验器 Tag closed cup tester
泰格闪(火)点试验 Tagliabue flash point test
泰国 < 亚洲 > Thailand
泰国电力管理局 Electricity Generating Authority of Thailand
泰国皇宫 < 建筑 > Thai Imperial Palace
泰国建筑 Thai architecture
泰国式 Thai style
泰国湾 Gulf of Thailand
泰国银行 Bank of Thailand
泰姬玛哈陵 < 印度 > Taj mahal
泰加林气候 climate of Taiga; subarctic climate; taiga climate
泰科斯自钻孔螺钉 < 商品名 > Teks
泰勒比值 Taylor's quotient
泰勒标准筛 Tyler screen; Tyler standard sieve; Tyler standard screen
泰勒标准筛分标度 Tyler standard screen scale
泰勒标准筛号 Tyler mesh
泰勒标准筛网 Tyler standard mesh
泰勒标准筛制 Tyler standard screen scale
泰勒波 Taylor's wave
泰勒不稳定性 Taylor's instability
泰勒差别计件工资制 Taylor's differential piece-rate system
泰勒袋滤器 Taylor's bag filter
泰勒的职能工长制 Taylor's system of functional foremanship
泰勒的职能组织 Taylor's functional organization
泰勒定理 Taylor's theorem
泰勒方程式 Taylor's equation
泰勒公式 Taylor's formula
泰勒怀特炼钢法 Taylor-White process
泰勒级数 Taylor's series
泰勒级数展开 Taylor's series expansion
泰勒架空索系 tyler
泰勒接线法 Taylor's connection
泰勒近似 Taylor's approximation
泰勒扩散 Taylor's diffusion
泰勒拉丝法 < 封入玻璃管拉丝法 > Taylor's method
泰勒粒级标准 Tyler standard grade scale
泰勒论件计划 Taylor's Piecework Plan
泰勒模型池 < 美国的一个水力实验所 > Taylor Model Basin
泰勒筛 Tyler mesh; Tyler screen; Tyler sieve
泰勒筛号尺寸 Tyler scale
泰勒筛孔尺度 Taylor scale
泰勒式筛号尺寸 Tyler's scale
泰勒数 Taylor's number
泰勒斯科特自动调压器 Taylor-Scotson type automatic voltage regulator
泰勒四滚筒磨 Taylor's four-roll mill
泰勒算子 Taylor's operator
泰勒弹力擒纵机构 Taylor's resilient
泰勒图解法 Taylor's graphic(al) construction

泰勒土坡稳定图 Taylor stability chart
泰勒土坡稳定系数 Taylor's stability numbers for earth slope
泰勒土坡稳定(因数)图解 Taylor's stability chart
泰勒稳定数 Taylor's stability number
泰勒稳定图 Taylor's stability chart
泰勒涡流 Taylor's vortex flow
泰勒系列 Taylor's series
泰勒线法 Taylor's wire method
泰勒线性化 Taylor's linearization
泰勒效应 Taylor's effect
泰勒旋涡 Taylor's vortex
泰勒雪夫氧化物测量 Tallysurf oxide measurement
泰勒亚麻油地毡 Taylor's lino(leum)
泰勒原理 Taylor's principle
泰勒圆 Taylor's circle
泰勒展开式 Taylor's expansion
泰勒制标准筛序 Tyler standard series
泰勒制筛序 Tyler series
泰勒主义 <科学管理> Taylorism
泰勒柱 Taylor's column
泰勒组织理论 organization method put forward by Taylor
泰雷铝硅合金 Telectal alloy
泰利 <一种压敏非线性电阻材料,其电阻值随所加电压大小而改变> Thyrite
泰利避雷器 thyrite arrester[arrestor]
泰罗 <计算机用的一种电子管> thyrode
泰罗(科学管理)制 Taylor system
泰罗斯卫星 television and infrared observation satellite;Tiros satellite
泰姆氏双管式螺钻取土器 Tams double-tube auger sampler
泰姆氏双管式心管取土器 Tams double-tube core barrel soil sampler
泰纳里夫山脉 Montes teneriffe
泰森多边形 Thiessen polygon
泰森多边形网格 Thiessen network
泰森洗涤机 Thiessen disintegrator; Thiessen gas cleaner; Thiessen gas scrubber
泰山群【地】 Taishan group
泰山系【地】 Taishan system
泰斯公式 Theis formula
泰斯井函数 Theis well function
泰伍德无声打桩机 Taywood pile master
泰晤士河 Thames River
泰晤士河国际水公司 <英> Thames Water International
泰西封宫殿 <伊拉克> Palace at Ctesiphon
泰越缅古陆 Thailand-Vietnam-Burma old land
泰泽利饰用铸锌合金 Tyseley alloy

酞化青绿 phthalocyanine green

酞化青染料 phthalocyanine dyestuff
酞菁 phthalocyanine
酞菁蓝 copper phthalocyanine;phthalocyanine blue
酞菁蓝颜料 cyanin(e) blue
酞菁绿 chlorinated copper phthalocyanine;phthalocyanine green
酞菁绿颜料 cyanine green
酞菁染料 phthalocyanine dye
酞菁素 phthalocyanine
酞菁(系)颜料 phthalocyanine pigment
酞类 phthalein
酞酸 phathalic acid
酞酸二丁酯炸药 dibutyl phthalate
酞酸二戊酯 diamyl phthalate
酞酸二烯丙酯 diallyl phthalate
酞酸甘油酯 glyceryl phthalate
酞酸盐 phthalate
酞酰亚胺石 kladnoite

坍岸 bank collapse; bank caving; bank failure; bank sloughing; bluff failure;collapsed portion of the bank;sloughing bank;shoreline regression <水库等的>; streambank failure

坍岸调查 bank collapse investigation
坍岸量 collapse volume of bank
坍岸速度 collapse velocity of bank
坍崩防护林 avalanche preventing forest
坍崩阻滞工程 avalanche protection works
坍倒 crash down
坍到其自然休止角 collapse to take up their natural angle of repose
坍堤 levee slide
坍顶 roof fall;collapse of roof <隧洞的>
坍动边坡 slipped embankment
坍动面积 slipping area
坍方 avalanche;cave-in;creep(ing); downfall;earth creep;fall(ing)-in; landslide;landslip;mountain creep; rock slide; slide; sloughing; soil slip;soil slump
坍方防护工程 landslide protection works
坍方防护墙 avalanche baffle wall
坍方防护设备 avalanche protection
坍方防御板 avalanche shed
坍方防御建筑物 avalanche baffle
坍方防御廊 avalanche gallery
坍方防御棚 avalanche shed
坍方防御墙 avalanche defense[defence];avalanche wall
坍方防御体 avalanche baffle
坍方防御栅 avalanche defense[defence];avalanche fence
坍方防止林 avalanche preventing forest
坍方检测器 slide detector
坍方检测栅 slide detector fence
坍方控制 landslide control
坍方线 line of slide[sliding]
坍滚破波 collapsing breaker wave
坍滑路径 slide path
坍滑面 plane of sliding
坍坏房屋 hurley house
坍坏油池 <地下储油箱> collapsed storage tank
坍毁弧 <即滑动弧> failure arc
坍积土 collapse soil
坍流土体 flow out earth
坍落 cave-in;caving;fall;slump
坍落(稠)度实验 slump consistency test
坍落的岩石碎块 <孔内> cavings
坍落度 <测定混凝土的流动性> slump(ing)(constant);slump test value
坍落度测定仪 slump meter
坍落度常量 slump constant
坍落度常数 slump constant
坍落度计 slump meter
坍落度试验 <混凝土> gravimetric-(al) yield test; slump consistency test;slump test
坍落度(试验)筒 slump cone
坍落度试验用圆锥体 slump cone
坍落度试样 slump specimen
坍落度损失 slump loss
坍落度筒 slump cone
坍落度限值 slump limitation
坍落度圆锥筒 <混凝土> slump cone
坍落堆 <混凝土> slump block
坍落防御设施 avalanche defense[defence]
坍落拱 collapse arch
坍落拱高度 height of collapse arch
坍落流 slump flow
坍落前落石 pickings
坍落试验堆 slump block
坍落体 caved material
坍落危险 cave-in risk
坍落物 slough
坍落线 caving line
坍落(圆锥)常数 slump constant
坍坡 failure slope; land creep; land fall; landslide; landslip; mountain creep;slide
坍坡分类 landslide classification
坍坡面 slip surface
坍坡探测器 slide detector
坍缩 collapse;pinch
坍塌 cave-in; caving; collapse; dilapidation; eboulement; give way; sloughing;slump(ing)
坍塌带 zone of loose rock
坍塌的土壤 caving soil
坍塌堤岸 caving bank;sloughing bank
坍塌卡钻 collapse sticking
坍塌破坏 slump failure
坍塌区 caving zone
坍塌山体 disintegrated earth
坍塌试验 slump consistency test; slump test
坍塌条件 collapse condition
坍塌土体 disintegrated earth
坍塌形坍落度 collapse slump
坍塌型 collapse type
坍塌岩层 caving formation
坍塌岩石堵塞的井眼 cave-obstructed borehole
坍塌岩石堵塞的孔眼 cave-obstructed borehole
坍陷【地】 downthrow; cave-in; caving;collapse;sloughing;slump failure;subsidence
坍陷变形 collapse deformation
坍陷侧 downcast side
坍陷地 caving ground;collapse land
坍陷地面 caved ground
坍陷洞穴 collapse sink
坍陷构造 collapse structure;gravity-collapse structure
坍陷机理 collapse mechanism
坍陷角砾岩 collapse breccia
坍陷裂缝 collapse fissure
坍陷区范围 collapse area extent
坍陷深度 collapse layer
坍陷现象 collapse phenomenon

摊布料杆支柱 boom post

摊草机 hay tedder;tedder
摊偿债款 amortization
摊店 booth
摊额 quota
摊额比例 scale of assessment
摊额基准 basis of assessment
摊贩 gutter man; stall holder; stall keeper
摊贩业税 tax on stalls
摊费乘数 <劳务的定价法> overhead multiplier
摊付分保现金赔款 cash call
摊还 amortization;dividend
摊还本金 principal amortization
摊还表 amortization schedule;amortization table
摊还股本 capital returned to stockholders in dividends
摊还借款 amortization loan
摊还率 amortization factor
摊还期 period of amortization
摊胶机 adhesive spreader
摊蜡器 paraffin(e) spreading unit
摊派 apportion;contribute;proration; prorate <美>
摊派成本 apportioned charges;apportionment cost
摊派的计划费用 assessed program(me) cost
摊派的间接费 allocated indirect expenses
摊派的税捐估价 assessment valuation
摊派法 amortization method
摊派费用 apportioned charges; assessed cost
摊派捐税基金 assessment fund
摊派捐税收入 assessment receipt
摊派粮款 demand grain and money from production teams
摊派税(捐) apportioned tax; assessment;apportionment tax
摊派税捐基金 assessment fund
摊派税捐收入 assessment receipt
摊派税款 apportioned tax;assessment
摊派税款清册 assessment rolls
摊派提款 amortization payment
摊派预算 assessed budget
摊配成本法 absorption costing
摊平 flattening;stretching
摊平玻璃的铁器 flattening iron
摊平工具 flattening tool
摊平机 trimmer
摊平炉 <平板玻璃> flattening kiln; flattening oven;smoothing kiln
摊平锹 flattening tool;hoe
摊平台 flattening table
摊平铁器 flattening iron
摊平窑 flattening kiln
摊铺 paving spread; spraying; spread-(ing)
摊铺板 spreading beam;bowl spreader <铲土机的>
摊铺材料压实试验 paved material compaction test
摊铺层 paving course
摊铺场地 spreading site
摊铺底子 spreading ground
摊铺地沥青 melted asphalt; poured asphalt;guss asphalt <德式>;floated asphalt
摊铺地沥青混合料 poured asphalt mix(ture)
摊铺地沥青(混凝土)面层 guss-asphalt surfacing
摊铺地镘整沥青 floated asphalt
摊铺斗 distributing hopper;operating bucket <水泥混凝土摊铺机的>
摊铺工人 spreader
摊铺工作速度 paving speed
摊铺刮板 distributing blade
摊铺厚度 paving thickness;spreading depth; spreading thickness; lift thickness <按滚压要求施工的每层摊铺厚度>
摊铺混合料 mixture placing; spreading mixture
摊铺混凝土 guss concrete;spreading of concrete
摊铺混凝土铲 come-along
摊铺混凝土锄 come-along
摊铺混凝土路面 concrete paving
摊铺机 <摊铺沥青混凝土用> laydown machine; mechanical spreader; mechanical spreading paver;pav(i)er; paving machine;paving plant;paving spreader;spreader;spreading machine
摊铺机的机架 paver frame
摊铺机桁架 <混凝土> paver boom
摊铺机料斗 spreader hopper

摊铺机上的轻型水平测量系统 eze-ski
摊铺机卸料斗 paver bucket
摊铺加拌合[和]的填土法 fill-spreading and blending
摊铺宽度 paving width; spreading width
摊铺沥青路面 asphalt paving
摊铺砾石 gravel(l)ing
摊铺料拌和机 paving mixer
摊铺料斗 spreading hopper
摊铺路面平整度试验 paved surface evenness test
摊铺螺旋布料器 spreading screw
摊铺煤渣 cindering work
摊铺面积 spreading area
摊铺器 spreader
摊铺砂料 spreading of sand
摊铺设备 <装在汽车后面的道路材料> paving plant; spreader device; spreading device
摊铺石渣 spreading ballast
摊铺式面层 floating cover
摊铺速度 spreading rate
摊铺温度 <沥青混合料的> laying temperature; placing temperature
摊铺系数 spreading coefficient
摊铺现场 <筑路工程> placing point
摊铺行程 spreading run
摊铺修整机 laying and finishing machine
摊铺修整两用机 spreader-finisher
摊铺一次成型的机械 one-pass machine
摊铺找平层 spreading of screed
摊铺-整修机 <路面的> paver-finisher
摊铺作业试验 paving operation test
摊晒 ted
摊晒机 tedder
摊石膏器 plaster spreader
摊提 amortization; written off
摊提表 amortization schedule; amortization table
摊提成本 amortized cost
摊提固定资产 amortization of fixed assets
摊提净值 net of amortization
摊提款 amortization payment
摊提期间 amortization period
摊提前递延费用 deferred charges before amortization
摊提贴水 amortization of discount; premium and bond issuance expenses
摊提准备 reserve for amortization
摊提租金 amortization rent
摊涂量 spreading rate
摊位 stallage
摊销 amortization; charge off
摊销备抵 allowance for amortization
摊销表 amortization schedule; amortization table
摊销的开办费 amortization
摊销法 amortization method
摊销费(用) amortization charges; distribution expenses
摊销费用净额 net amortization charges
摊销明细表 schedule of amortization
摊销期 amortization period
摊销期限 period of amortization
摊销准备 reserve for amortization
摊药器 spreader
摊余成本 amortized cost
摊余债券折价 unextinguished discount on funded debt
摊渣槽 <土路基中> ballast boxing
摊子 booth
摊租 stallage

滩岸 beach strand; frontage

滩岸侵蚀 beach strand erosion
滩壁 berm(e) scarp
滩边阶地 beach berm(e); berm(e)
滩槽 riffle and pool; swale
滩槽高差 elevation difference between riffle and pool; riffle-pool difference
滩槽间距 riffle-pool spacing
滩槽序列 pool-and-riffle sequence; riffle-and-pool sequence
滩沉积 beach deposit
滩池 beach pool
滩底 beach bottom
滩地 beach land; bottomland; coastal plain; flood plain; foreland; holm(e); overbank; washland; bottom land <洪水时受淹地>; landwash <涨水时淹没的地>
滩地槽蓄量 overbank storage
滩地冲刷 flood plain scour
滩地过流流量 flood plain discharge
滩地横剖面 beach profile
滩地后退 beach retreat
滩地后延 beach extension
滩地砾石 beach gravel
滩地林 bosquet
滩地侵蚀防治 beach erosion control
滩地调蓄 flood plain storage
滩地外延 beach extension
滩地(物)质 beach material
滩地淤积 flood plain deposit
滩段 step
滩埂 beach ridge
滩海 beach-shallow sea
滩积内陆湖 etang
滩积岩 beachrock
滩脊 beach fulls; beach ridge; fulls; ridge of shoal
滩尖 beach cusp
滩尖嘴 cusp
滩肩 beach berm(e); berm(e)
滩肩顶 berm(e) crest
滩肩海侧边缘 berm(e) crest
滩肩脊(顶) crest of berm(e); berm(e) crest
滩肩坎 berm(e) scarp
滩肩外缘 berm(e) edge
滩肩崖 beach berm scarp; berm(e) scarp
滩礁 bank reef
滩角 beach cusp; cusp
滩坎 beach scarp
滩口 rapids throat
滩宽 beach width
滩砾 pit-run gravel
滩砾石 beach gravel
滩流 shoal flow
滩面 beach face
滩面槽蓄量 overbank storage
滩浅影响 shoaling effect
滩丘 backshore terrace
滩群 serial rapids; serial rapids of shoals; serial shoals; shoal group
滩砂 bank sand; beach sand
滩砂圈闭 beach sand trap
滩上浮标 bar buoy
滩舌 rapids tongue
滩头 <急滩的> rapids head; beachhead
滩头浮标 middle-ground buoy
滩头系留 beaching
滩外沙洲 offshore barrier
滩险 rapids and shoals; shoals and rapids; traffic hazard
滩险整治 rapids regulation
滩线 beach line; shore line
滩线漂沙 beach drift
滩崖 beach scarp
滩岩 beach rock; beach scarp
滩沿 beach approach
滩与滩的间距 riffle-to-riffle spacing
滩缘礁 bank-inset reef
滩锥 cusp
滩嘴 beach cusp; cusp

瘫痪 paralysis

坛 altar; demijohn; jar

坛甲板 turret deck
坛甲板船 turret deck vessel
坛囊 ampiclla
坛前踏步 predella
坛式石棺 altar-tomb
坛形的 ampullaceal; ampullaceous
坛形感器 sensillum ampullaceum
坛盏 tan cup
坛子 carboy; crock

谈话室 conversation room

谈判 bargaining; conference; negotiation; palaver; transact
谈判代理人合格证 certification of bargaining agent
谈判单位 bargaining unit
谈判合同条款 negotiation of contract terms
谈判机构 negotiating machinery
谈判决定工资论 bargain theory
谈判权 bargaining right
谈判人 negotiator
谈判失败 fail in negotiations
谈判室 negotiation room
谈判投标 negotiated tender
谈判招标 negotiate(d) bidding
谈判者 negotiator
谈判者角色 negotiator role
谈判中的僵局 tie-up in negotiations
谈判阻力 negotiating resistance

弹簧 collet chuck; elastic spring; mechanical spring; spring; prongs <岩芯采取器的>

弹簧安全阀 spring-loaded relief valve; spring-loaded safety valve; spring reducing valve; spring relief valve; spring-return valve; spring safety valve; spring valve
弹簧安全钩 clevis; clivvy
弹簧安全联结装置 spring safety hitch
弹簧安全器 spring release
弹簧安全系数计算公式 spring formula factor of safety
弹簧安全销 spring-loaded pin; spring-loaded stop
弹簧安装(法) spring mounting
弹簧鞍座 spring saddle; stirrup
弹簧板 buckboard; latch plate; leaf mechanical spring; spring board; spring lamination
弹簧板牙 spring die
弹簧保安器 grasshopper fuse
弹簧保持压力 spring holding pressure
弹簧保险圈 spring clip retainer
弹簧保险丝 spring fuse
弹簧鼻吊挂装置 spring nose suspension installation
弹簧比率 spring constant; spring rate
弹簧闭锁开关 spring-loaded cock
弹簧避障器 spring trip
弹簧避振垫 spring bumper
弹簧避振器 spring bumper
弹簧变位 spring deflection
弹簧变形限度 range of spring
弹簧柄 snapholder; spring holder
弹簧薄片 spring tab
弹簧补偿器 spring tensioner
弹簧布置 spring arrangement
弹簧材料 spring material
弹簧采血针 spring puncture needle
弹簧操纵浮阀 spring-controlled float valve
弹簧测力计 spring dynamometer; spring-loaded thrust meter; weight beam
弹簧叉 fork; spring fork
弹簧插销 cabinet lock; kickout latch; latch head; night bolt; spring bolt; spring cotter; spring latch; spring stop
弹簧插座 cushion socket; spring socket
弹簧铲柄 spring shank
弹簧颤动 spring surge
弹簧颤动频率 spring surge frequency
弹簧常量 spring constant
弹簧常数 <弹性支承每单位弯沉所产生的力,黏[粘]弹性模型中弹簧单元的常数> spring constant; spring rate
弹簧车刀 heal bite; spring tool
弹簧车钩 elastic draw gear; spring draw gear
弹簧车架缓冲块座 spring chassis bumper seat
弹簧车间 spring smithy
弹簧掣子 spring catch
弹簧撤叉【铁】 spring points
弹簧衬套 spring bushing
弹簧称 balance spring; spring balance
弹簧撑杆 spring spacer bar
弹簧撑条 spring stay
弹簧成型淬火机 bending and quenching machine
弹簧成型机 spring forming machine
弹簧承杯传动 cup drive
弹簧承力波纹管 spring-opposed bellows
弹簧承力活塞 spring-opposed piston
弹簧承力空气阀 spring-opposed air valve
弹簧承力膜片 spring-loaded membrane
弹簧承力调节器 spring-loaded regulator
弹簧承梁 spring bolster
弹簧承受的重量 spring weight
弹簧秤 spring balance; spring scale; steelyard; weight beam
弹簧持钎器 spring-steel holder
弹簧齿除草耙 spring-tine weeder
弹簧齿耙 <附于平地机后端的> spring-tine harrow; spring-tooth harrow
弹簧齿耙路机 spring-tooth harrow
弹簧齿片 spring toothed disc
弹簧齿中耕机 spring-tine cultivator
弹簧冲击扳手 spring wrench
弹簧初张力 initial tension of spring
弹簧触点 spring contact
弹簧传动 spring drive
弹簧传动装置 spring gear
弹簧窗扇 spring sash; swing sash
弹簧床 bedspring; spring bed
弹簧床垫 spring washer
弹簧床垫形天线 bedspring
弹簧垂度 spring sag
弹簧捶布机 spring beetling machine
弹簧槌布机 spring hammer beetle
弹簧锤 spring hammer
弹簧唇形密封 flexible-lip seal; spring-loaded lip seal

弹簧从板托 <缓冲装置> check plate bridle iron
弹簧从板座 check plate; draft lug
弹簧促动式推种爪 spring-actuated knockout pawl
弹簧搭扣 snap fastener; snap lock; spring buckle
弹簧带式空气加热器 spring strip air heater
弹簧挡板 spring stop
弹簧挡圈 check ring; clip ring; spring collar
弹簧挡装置 spring catch device
弹簧刀 gooseneck tool
弹簧刀杆 split holder
弹簧刀夹 spring tool holder
弹簧导承 spring guide
弹簧导承组合 spring and guide assembly
弹簧导杆 spring guide; spring rod
弹簧导杆耳轴 spring guide trunnion
弹簧导管 spring guide
弹簧导环 spring guide collar
弹簧导架 spring horn
弹簧导头 spring guide
弹簧导销 spring guide pin
弹簧道岔 spring points; spring switch
弹簧道钉 elastic rail spike; spring rail spike
弹簧的 resilient
弹簧的防尘套 spring dust cover
弹簧的可靠指标 reliability index of spring
弹簧的拉伸 tension of a spring
弹簧的屈服点 yield of spring
弹簧的松弛 relaxing of spring
弹簧的调压螺钉 spring tensioner
弹簧的限程 spring range
弹簧等级 spring rate
弹簧底板 level (l) ing spring plate; spring plate
弹簧底座 spring base
弹簧底座销 spring retainer lock
弹簧地板 spring floor (ing); sprung floor cover (ing)
弹簧地震计 spring seismometer
弹簧点圆规 drop-bow compasses; pump-bow compasses
弹簧电缆绕线盘 spring cable winding drum
弹簧电缆绕线筒 spring cable winding drum
弹簧垫 spring pad
弹簧垫层 spring cushion
弹簧垫固定螺帽 spring-pad clip nut
弹簧垫夹 spring-pad clip
弹簧垫架 spring mounting
弹簧垫块 spring cushion
弹簧垫耐磨板 spring-pad wear plate
弹簧垫片 expansion washer; lock washer; split lock washer; spring shim; spring washer
弹簧垫圈 elastic ring; expansion washer; grower washer; lock (ing) washer; retaining washer; split lock washer; spring washer; washing screw
弹簧垫子 spring cushion; spring mattress
弹簧吊 suspension link
弹簧吊耳 spring shackle
弹簧吊杆 spring hanger link
弹簧吊钩保险 spring shackle bolt lock
弹簧吊架 spring carrier arm; spring hanger; spring shackle; suspension arm
弹簧吊架扁栓 spring hanger gib
弹簧吊架杆 spring hanger bar
弹簧吊架托 spring hanger bracket
弹簧吊架座 spring hanger seat
弹簧吊销 suspension link pin
弹簧碟阀 spring-disk valve
弹簧钉(座椅) spring stop
弹簧顶杆 spring spindle
弹簧顶尖 spring center[centre]
弹簧顶件器 spring knockout
弹簧定位器 spring space
弹簧定位圈 spring collar
弹簧定位销 spring holder; spring plunger; spring sheet-holder
弹簧定中(心) spring centered; spring centering
弹簧动力计 spring dynamometer
弹簧动挠度 dynamic (al) deflection of spring
弹簧端部 spring end
弹簧端垫 spring end block
弹簧端圈 bearing coil
弹簧端头 spring tip
弹簧断裂 spring breakage; spring fracture
弹簧锻工车间 spring forge
弹簧锻模 spring tup
弹簧对中的 spring centered
弹簧对中的方向控制阀 spring-centered direction control valve
弹簧对中心 spring centered
弹簧对准中心的盖 spring centering cap
弹簧对准中心的帽 spring centering cap
弹簧对准中心的套 spring centering cap
弹簧舵柄弧 elastic quadrant; spring quadrant; spring tiller
弹簧额定负载量 rated spring capacity
弹簧阀(门) spring-loaded valve; spring valve
弹簧反作用 spring reaction
弹簧返回式卸粮管 spring-backed discharge spout
弹簧防爬器 resilient anchor; resilient anti-creeper; spring anti-creeper
弹簧防水圈 spring water seal
弹簧防松螺母锁紧件 spring stop-nut locking fastener
弹簧防振器 spring-coupler damper
弹簧分度规 hair compasses; spring hair compasses
弹簧分规 bow divider; spring division
弹簧分离机构 spring disconnecting mechanism
弹簧浮动模 spring-floated die
弹簧浮动式捡拾器 spring-cushioned pick-up
弹簧复位的换向阀 spring-return directional valve
弹簧复位开关 spring-return switch
弹簧复位控制器 spring-return controller
弹簧复位装置 spring reverse motion
弹簧副缓冲块 spring additional buffer
弹簧盖 spring cap; spring cover; spring cup
弹簧盖板 spring bumper cover
弹簧杆 spring bar; spring beam; spring lever; spring pole; spring rod
弹簧杆衬套 spring rod bushing
弹簧杆导套 push rod guide
弹簧杆接头 <边车的> spring bar connection
弹簧杆润滑器 snap lever oiler
弹簧杆水阀 push rod
弹簧刚度 spring constant; spring rate; spring stiffness
弹簧刚性系数 spring rate
弹簧钢 spring steel; steel spring
弹簧钢板 spring plate; steel spring plate
弹簧钢板隔片 spring separator
弹簧钢板套 spring puttee
弹簧钢带 spring-steel band
弹簧钢活门 spring-steel shutter
弹簧钢卷帘门 spring-steel shutter
弹簧钢丝 spring-steel wire; spring wire
弹簧钢丝沟管清通器 spring-cable sewer cleaner
弹簧钢丝筛 spring-steel wire screen
弹簧钢椅 resilient chair
弹簧隔板 spring insert
弹簧隔离件 spring separator
弹簧隔振器 metal spring isolator
弹簧隔震器 spring shock absorber
弹簧工作的 spring-actuated
弹簧弓 spring bow
弹簧弓卡环 spring arch clasp
弹簧功 work of spring
弹簧拱 spring arch
弹簧共振 spring resonance
弹簧钩 carbine; clasp; latch hook; snap hook; spring dog; spring hook
弹簧钩杆 spring shackle bar
弹簧钩环 spring shackle
弹簧钩环托架 spring shackle bracket
弹簧钩环销 spring shackle pin
弹簧钩簧隔片 spring clip bar spacer
弹簧箍 spring band; spring buckle
弹簧箍圈 spring coupling plate; spring rebound clip
弹簧固定 spring attachment
弹簧固定端 spring fixed end
弹簧固定架 spring anchorage bracket
弹簧固定塞 spring anchor plug
弹簧挂钩 spring carrier arm; spring hanger; spring shackle
弹簧挂钩衬套 spring shackle bushing
弹簧挂钩衬套换装器 spring hanger bushing replacer
弹簧挂钩销 spring hanger pin
弹簧挂闸 <起作用的> spring applied
弹簧管 Bourdon spring
弹簧管式压力计 spring-tube manometer; spring-tube pressure ga(u)ge
弹簧管式压力温度计 pressure spring thermometer
弹簧管压力表 Bourdon tube ga(u)ge
弹簧管压力计 Bourdon manometer; Bourdon tube pressure ga(u)ge; spring pressure ga(u)ge
弹簧管压力式温度计 pressure spring thermometer
弹簧规 spring cal(l)ipers
弹簧辊 spring roller
弹簧辊磨机 spring roll mill
弹簧辊式破碎机 spring roll crusher
弹簧辊碎机 spring roll
弹簧滚柱 <轴承的> flexible roller; spring roller
弹簧合页 spring butt (hinge); spring hinge
弹簧和黏[粘]壶模型 spring and dashpot model
弹簧荷载 spring compression; spring load (ing)
弹簧荷载试验 spring load test
弹簧盒 spring box; spring case; spring housing
弹簧盒挡板 spring box guide plate
弹簧盒导板 spring box guide plate
弹簧盒底 end plate
弹簧盒垫板 spring case patch plate
弹簧盒盖 end cap
弹簧横轴 spring cross shaft
弹簧护板 spring fender; spring guard
弹簧护套 spring protecting sleeve
弹簧护舷(设备) spring buffer; spring bumper pad; spring fender
弹簧花键 spring spline
弹簧滑块模型 spring ride model
弹簧滑脂杯 spring-loaded grease cup
弹簧环 ring spring; spring collar
弹簧缓冲 spring reaction fendering
弹簧缓冲块 spring block; spring cushion
弹簧缓冲器 die cushion; spring buffer; spring bumper; spring cushion
弹簧缓冲器帽 spring buffer cap
弹簧缓冲设施 spring fender
弹簧缓冲式捡拾器 spring-cushioned pick-up
弹簧缓冲室 spring buffer chamber
弹簧缓冲压缩行程 spring buffer stroke
弹簧缓冲装置 spring bumper
弹簧黄铜 spring brass
弹簧回程 spring return
弹簧回程活塞 spring-return piston
弹簧回火 spring temper
弹簧回弹片 spring rebound plate
弹簧活顶尖 spring-loaded live centre [center]
弹簧活门 spring valve
弹簧活塞环 snap piston ring
弹簧活页夹 spring-binder
弹簧机 coiling machine
弹簧机构 spring mechanism
弹簧激发液压缸 spring activated cylinder
弹簧及活塞环弹性检查器 spring and piston ring elasticity tester
弹簧几何挠度 geometric (al) free camber of spring
弹簧基础 <机器的> sprung foundation
弹簧脊封面 spring back cover
弹簧计 blenometer
弹簧加荷的 spring-loaded
弹簧加荷的活塞 spring-loaded piston
弹簧加力的操纵杆 spring-loaded stick
弹簧加力开沟器 spring-loaded drill colter
弹簧加料器 spring feeder
弹簧加强卷耳 spring reinforced eye
弹簧加强器 spring tensioning tool
弹簧加压 spring-weighted
弹簧加压怠速阀 spring-loaded idling valve
弹簧加压杆 spring pressure rod
弹簧加压密封 spring-loaded seal
弹簧加压喷嘴 spring-loaded nozzle
弹簧加压皮带轮 spring-loaded sheave
弹簧加压破碎机 spring-loaded crusher
弹簧加油杯 spring grease cup
弹簧加载插入式热电偶 spring-loaded bayonet thermocouple
弹簧加载齿轮 <消除啮合间隙的一种齿轮结构> spring-loaded gear
弹簧加载齿式中耕机 spring-loaded tine cultivator
弹簧加载的 spring-loaded
弹簧加载的挡块 spring-loaded stop
弹簧加载的阀 spring-biased valve; spring-opposed valve
弹簧加载的螺栓 spring-loaded bolt
弹簧加载的压力泄放阀 spring-loaded pressure relief valve
弹簧加载剪式齿轮 spring-loaded scissor gear
弹簧加载径向密封片 spring-loaded apex seal
弹簧加载式土刮器 spring-loaded scraper
弹簧加载式针阀 spring-loaded needle valve
弹簧加载式制动器 spring-loaded brake
弹簧加载式中耕机 spring-loaded cul-

T

tivator
弹簧加载调压器 spring-loaded regulator
弹簧夹 alligator clip; Mohr's clamp; pinch clamp; pinchcock; pinch cock clamp; snap terminal; spring clip; spring finger; spring pinch cock
弹簧夹板 clip board; spring clamping plate; spring clipboard; spring clip plate
弹簧夹持器 split holder
弹簧夹定位架 spring clip spacer
弹簧夹紧螺栓 spring clamp bolt
弹簧夹具 spring clamp; spring perch
弹簧夹块 spring block; spring clip
弹簧夹盘 split chuck; spring chuck
弹簧夹钎器 spring-steel holder
弹簧夹套 spring collet
弹簧夹条 spring clip bar
弹簧夹头 collet chuck; collet clamping head; draw-in chuck; expansion chuck; spring carrier; spring collet
弹簧夹头闭合器 collet closing tool
弹簧夹头钳 collet closing and broaching tool
弹簧架 spring support; spring-supported mount
弹簧架吊件 suspension arrangement
弹簧尖轨 spring switch blade; spring tongue
弹簧尖轨转辙器 spring tongue switch
弹簧尖叫声 spring squeak
弹簧减声器 spring anti-squeak
弹簧减压阀 spring-loaded pressure reducing valve; spring reducing valve
弹簧减振垫 spring cushion
弹簧减振垫托架 spring bumper bracket
弹簧减振器 rebound check; spring bumper
弹簧减震发动机 spring-mounted engine
弹簧减震器 spring damper; spring shock absorber; spring snubber
弹簧减震筛 spring-supported screen
弹簧减震系统 spring damper system
弹簧减震柱 spring shock absorber strut
弹簧剪钳 cut-nippers with spring
弹簧检验 spring tester
弹簧键 spring key
弹簧矫准的盖 spring centering cap
弹簧矫准中心的帽 spring centering cap
弹簧矫准中心的套 spring centering cap
弹簧铰刀 spring reamer
弹簧铰链 spring butt < 门窗的 >; spring hinge < 门窗的 >; spring pivot < 装在地板支承摆门的 >
弹簧接触 spring contact
弹簧接触插头 spring contact plug
弹簧接触器 spring contactor
弹簧接点 spring contact
弹簧接合 spring joint
弹簧接合的 spring-engaged
弹簧接合卡钳 spring-joint caliper
弹簧接受器 spring receiver
弹簧接头 flexible connector
弹簧节流夹 pinch cock clamp; spring pinch cock
弹簧介质 spring medium
弹簧金 spring gold
弹簧紧固垫圈 spring lock washer
弹簧进给 spring feed
弹簧静挠度 static deflection of spring
弹簧卷布装置 spring cloth wind-up motion
弹簧卷尺 spring-return rule; spring tape measure; spring wind tape
弹簧卷耳 spring eye
弹簧卷筒 spring spool
弹簧卡 spring clamp
弹簧卡尺 spring cal(l)ipers
弹簧卡打捞器 dog-leg jar
弹簧卡盘 cutter chuck; split chuck; spring split chuck
弹簧卡钳 spring cal(l)ipers; spring clamp
弹簧卡头 split chuck; spring chuck
弹簧卡子 spring clip; spring-loaded catch
弹簧开关 clip-spring switch; jack; snap switch; spring points; spring switch; spring valve
弹簧开口环 spring snap ring
弹簧开口销 spring cotter; spring key
弹簧抗压强度 spring compression strength
弹簧控制 spring control
弹簧控制的 spring-actuated
弹簧控制阀 spring-controlled valve
弹簧控制器 spring controller; spring governor
弹簧扣 latch catch; snapback hook; snap hook; spring buckle; spring fastener
弹簧扣盖 snap-on cover
弹簧框架 spring frame
弹簧扩张器 spring expander; spring spreader
弹簧拉钩 snap clasp
弹簧篮式盛土器 spring basket retainer
弹簧离合器 spring clutch; spring-loaded clutch
弹簧力 spring rate
弹簧连杆 spring link
弹簧连接板 spring coupling plate
弹簧连接夹 spring rebound clip
弹簧联结器 spring hitch
弹簧两脚规 bow divider; spring dividers
弹簧量规 spring dividers
弹簧龙头 spring faucet
弹簧笼 < 压机的弹簧压力机构 > spring cage
弹簧炉顶 sprung arch
弹簧路耙 spring harrow
弹簧履带 spring track
弹簧螺钉 spring screw
弹簧螺距 turn-to-turn distance
弹簧螺帽 spring nut
弹簧螺栓 spring bolt
弹簧帽 spring cap; spring compressor
弹簧门 double-acting door; spring-loaded door; swing(ing) door
弹簧门插销 friction(al) catch; friction(al) latch
弹簧门碰锁 bullet catch
弹簧门碰头 bale's catch; ball catch
弹簧门闩 spring latch; spring stop
弹簧门锁 bescot; catch bolt; night latch; spring door latch
弹簧迷宫汽封 spring-back labyrinth gland
弹簧模 spring-loaded floating die
弹簧模具 floating die assembly
弹簧模量 spring modulus
弹簧模压机 dual-pressure press; floating die press
弹簧模预压机 dual-pressure preformer
弹簧模座 floating base
弹簧膜 spring diaphragm
弹簧膜盒压力计 spring bellows ga(u)ge
弹簧膜压力计 harmonic membrane
弹簧磨床 spring grinding machine
弹簧内卡钳 spring inside caliper
弹簧挠度 deflection of spring; flexibility of spring; spring deflection
弹簧挠度裕量 surplus deflection of spring
弹簧挠性 flexibility of spring; spring flexibility
弹簧啮合的 spring-engaged
弹簧镊子 spring pincers; spring-type forceps
弹簧盘 spring holder
弹簧配重机构 spring rigging
弹簧喷射法 method of spring spurt
弹簧喷射系统 spring injection system
弹簧碰垫 spring buffer; spring bumper; spring fender
弹簧碰锁 spring latch
弹簧片 coach spring; spring leaf; spring plate
弹簧片刮管器 go-devil
弹簧片悬挂 spring strip suspension
弹簧偏压 bias of spring
弹簧偏置式换向阀 spring offset changeover valve
弹簧平衡安全阀 spring-balance safety valve
弹簧平衡窗框 spring balancing sash
弹簧平衡的吊挂卫板 spring-balanced guard
弹簧平衡阀 spring and lever loaded valve
弹簧平衡杆 springs and compensating beam
弹簧平衡式接轴托架 spring-loaded spindle carrier
弹簧平衡系统 spring balance system
弹簧平衡装置 spring equalizing device
弹簧平头插销 cupboard catch
弹簧骑马螺栓 spring U-bolt
弹簧启动的 spring-actuated
弹簧起动机构 spring starting mechanism
弹簧起动器 < 发动机 > spring starter
弹簧起模钉 spring draw nail
弹簧起始张力 initial spring tension
弹簧牵动器 spring tensioning tool
弹簧钳子 spring pincers
弹簧腔 spring cavity; spring pocket
弹簧强度 spring strength
弹簧擒纵器 spring detent
弹簧清洁器 spring cleaner
弹簧球节 spring ball joint
弹簧曲度 camber of spring
弹簧驱动 spring-actuated
弹簧驱动的 spring driven
弹簧圈 coil spring; garter spring; spring coil; spring ring
弹簧圈机和装置 spring coiling machines and apparatus
弹簧圈式铲柄 spring coil shank
弹簧圈式推动螺旋 coil-type auger
弹簧圈数 number of coils
弹簧全压缩的 solid
弹簧韧带 spring ligament
弹簧容量 spring capacity
弹簧柔度 pliability of spring
弹簧软垫 box spring
弹簧软系统 spring softening system
弹簧润滑器 spring grease cup; spring lubricator
弹簧塞孔 spring jack
弹簧筛 spring screen
弹簧上部质量 sprung mass
弹簧上座 upper site of springs
弹簧舌小钩 lock hook
弹簧深度规 spring depth ga(u)ge
弹簧失效 bottom of springs
弹簧式 spring-loaded
弹簧式安全阀 spring-loaded pressure relief valve; spring-loaded safety valve
弹簧式安全装置 spring-loaded safety system
弹簧式测边定位楔块 spring side ga(u)ge cam
弹簧式导料板 spring stock guide
弹簧式道岔握柄 spring type point lever
弹簧式地震仪 spring seismometer
弹簧式顶出装置 spring-loaded shedder
弹簧式顶件工具 spring knockout
弹簧式顶件器 spring knockout
弹簧式缓行器 spring-type retarder
弹簧式回转盘 < 冲击钻的 > spring swivel
弹簧式间接导正销 spring-backed type indirect pilot
弹簧式减速器 spring loaded type retarder
弹簧式可动卸料板 spring-suspended movable stripper
弹簧式美兹取土器 spring-loaded Mazier sampler
弹簧式坯料定位装置 spring stock ga(u)ge
弹簧式切割器 spring-assisted cutter
弹簧式取芯器 spring core catcher
弹簧式书脊 spring-back
弹簧式调速器 spring-type governor
弹簧式土芯取样器 spring core catcher
弹簧式脱开装置 spring release
弹簧式弯头车刀 gooseneck
弹簧式蓄能器 spring accumulator
弹簧式蓄压器 spring accumulator
弹簧式压力表 spring manometer
弹簧式重力仪 spring gravimeter
弹簧式装置 spring mounting
弹簧试验机 spring testing machine
弹簧试验器 spring tester
弹簧试验仪 spring tester
弹簧室 chamber of spring
弹簧室式停车制动器 spring chamber parking brake
弹簧释放销 spring-loaded release pin
弹簧手柄 spring handle
弹簧枢架 spring pivot bracket
弹簧数 spring constant
弹簧闩 latch bolt; live bolt; spring catch
弹簧松脱式联结装置 spring release hitch
弹簧锁 cabinet lock; cylinder latch; cylinder lock; friction(al) latch; latch; latch lock; night latch; rim latch; rim lock; rim night latch; snap lock; snap terminal; spring latch lock; spring lock; Yale lock; bored latch < 装在门孔中的 >
弹簧锁的锁舌匣 staple
弹簧锁定装置 spring barrel
弹簧锁环 snap ring
弹簧锁紧垫圈 spring lock washer
弹簧锁紧装置 spring cocking device; spring lock device
弹簧锁开关 snap-lock switch
弹簧锁式销 spring locking pin
弹簧锁销 spring cotter; spring plunger
弹簧锁钥匙 latch key; night key
弹簧弹力 spring force
弹簧弹力测量仪 blenometer
弹簧弹力试验器 spring tension tester
弹簧弹射器 < 撞车试验用的 > spring catapult
弹簧套 spring housing; spring pocket; spring sleeve
弹簧套筒夹头 open collet; spring col-

let
弹簧特性 spring performance
弹簧特征 spring characteristic
弹簧天线 spring antenna;spring-loaded antenna
弹簧调节 spring adjustment
弹簧调节的 spring-actuated
弹簧调节螺塞 <喷嘴的> spring cap nut
弹簧调节器 spring governor;spring-loaded fly governor;spring-loaded governor
弹簧调速器 spring governor;spring-loaded governor;spring regulator
弹簧调整 spring adjustment;spring control
弹簧调整环 spring adjusting collar
弹簧调整螺钉 spring adjusting screw
弹簧调整螺丝 spring adjusting screw
弹簧调整片 spring tab;spring tab assembly
弹簧调整器 spring adjuster
弹簧筒夹 collet chuck;contracting chuck;expanding chuck
弹簧筒夹架 spring collet holder
弹簧头 spring head
弹簧土 spongy soil;springy fill
弹簧推车器 cushion feeder
弹簧推动式击针 spring-activated firing pin
弹簧推力 spring thrust
弹簧退让式导销 spring pilot
弹簧托 latch bracket;spring abutment;spring retainer
弹簧托板 swing spring plank;spring plank
弹簧托板安全吊 spring plank safety hanger
弹簧托板座 spring plank seat
弹簧托辊 spring-loaded idler
弹簧托架 spring bracket;spring carrier
弹簧托架衬套 spring bracket bushing
弹簧托座 spring suspension
弹簧托座螺栓 suspension bolt
弹簧拖拉装置 spring draw gear
弹簧脱开装置 spring release
弹簧外壳 spring housing
弹簧弯度 spring camber;spring sag
弹簧弯管 lyre-type expansion piece
弹簧弯管器 spring-tube bender
弹簧弯曲 spring bend(ing)
弹簧弯曲器 spring bender;spring bending tool
弹簧微震震实 shockless jolt
弹簧握 spring holder
弹簧无效圈 end coil
弹簧系紧螺栓 spring tie bolt
弹簧现象 <路面等> springing
弹簧线夹 crocodile clip;grip clip
弹簧限动板 spring retaining plate
弹簧限动器 spring stop
弹簧限位器 spring retainer
弹簧限制式加速度计 spring-restrained accelerometer
弹簧陷型模 spring swage
弹簧箱 spring box
弹簧箱铸型 spring box mo(u)ld
弹簧橡胶衬套 spring rubber bushing
弹簧消声器 spring silencer
弹簧销 spring bolt;spring catch;spring cotter;spring pin;spring pivot
弹簧销拖挂装置 spring-trip hitch
弹簧销油嘴 spring bolt pressure lubricator
弹簧销支架 spring pivot bracket
弹簧销装置 spring catch device
弹簧小圆规 bow spring compasses;spring bow compasses
弹簧楔片 spring wedge plate
弹簧楔形闸板阀 spring wedge valve
弹簧卸料板 spring-operated stripper;spring stripper
弹簧心阀 valve with spring core
弹簧心杆 spring spindle
弹簧心杆套 spring spindle bush
弹簧心杆托螺丝 spring spindle bracket screw
弹簧心夹 spring center[centre] clamp
弹簧心离合器板 spring center clutch plate
弹簧心瘤 spring center[centre] hump
弹簧行程 spring travel
弹簧型软线 coil cord
弹簧型重力仪 spring-type gravimeter
弹簧蓄能制动 spring-loaded brake
弹簧悬吊 spring-suspended;spring suspension
弹簧悬挂 spring suspension
弹簧悬挂杆 spring suspension rod
弹簧悬挂连接销 spring suspension link-pin
弹簧悬挂有眼螺栓 spring suspension eyebolt
弹簧悬挂装置 spring suspension damper;spring suspension device
弹簧悬架 spring carrier arm;spring rigging
弹簧悬架车轮 spring wheel
弹簧悬架的 spring-mounted
弹簧悬架座椅 sprung seat
弹簧悬置(法)spring suspension
弹簧旋塞 spring action cock;spring-loaded cock
弹簧压板 spring bearer plate;spring pad;spring pressure plate
弹簧压紧板 spring holddown
弹簧压紧皮碗 spring-loaded leather seal
弹簧压紧式离合器 spring-type clutch
弹簧压紧调整装置 spring compression adjustment
弹簧压紧装置 spring attachment
弹簧压力 spring compression;spring load(ing);spring pressure
弹簧压力标 spring scale
弹簧压力表 spring manometer;spring pressure ga(u)ge
弹簧压力计 spring manometer;spring tension ga(u)ge
弹簧压力空气阀 spring-opposed air valve
弹簧压力磨 spring pressure mill
弹簧压力探测仪 spring pressure sounding apparatus
弹簧压力调节螺钉 spring tension adjusting screw
弹簧压模 floating die
弹簧压缩高度 buffer height;compressed height of spring
弹簧压缩试验 closure test
弹簧压条 band spring
弹簧压罩 spring cup
弹簧压制膜盒 spring-opposed bellows
弹簧延动离心式离合器 spring centrifugal clutch
弹簧眼螺栓 spring eye bolt
弹簧移位的阀 spring offset valve
弹簧溢流阀 spring-loaded overflow valve
弹簧翼轨 spring wing rail
弹簧应变率 spring rate
弹簧油封 spring-loaded oil seal
弹簧油环 spring-loaded oil control ring
弹簧油箱 hydraulic pillow;spring pillow;torsional membrane
弹簧游间 play of spring
弹簧游筘装置 spring reed release motion
弹簧预紧力 initial tension of spring
弹簧预紧装置 spring preloading device
弹簧元件 spring element
弹簧圆板 cushion disc[disk]
弹簧圆规 bow compasses;bow instrument;screw compasses;spring bow;spring bow compasses;spring compasses
弹簧圆眼衬套 spring eye bushing
弹簧圆柱滚子轴承 Hyatt roller bearing
弹簧增力器 spring booster
弹簧增能器 spring energizer
弹簧闸 spring brake;spring-operated brake
弹簧张紧钩环 tension shackle
弹簧张紧器 spring tensioning
弹簧张紧装置 spring take-up
弹簧张力 spring tension
弹簧胀圈油环 conformable oil ring;spring-backed oil ring;spring expander oil ring
弹簧爪 spring detent;spring holder;spring pawl;spring tension pawl
弹簧爪式定位装置 spring finger
弹簧找中心 spring centered[centred]
弹簧罩 spring housing;spring shield
弹簧折合刀 spring blade knife
弹簧折弯机 spring bender
弹簧辙叉【铁】spring rail frog
弹簧辙尖 spring points
弹簧针 spring beard needle;spring needle
弹簧针针织机 spring needle machine
弹簧振动间隙 bump clearance
弹簧支撑 spring support;spring brace
弹簧支撑的 spring-supported
弹簧(支撑式)推力轴承 <水轮发电机的> spring thrust bearing
弹簧支承 spring hanger;spring support
弹簧支承的 spring-backed;spring-loaded;spring-supported
弹簧支承的挡板 spring-loaded apron
弹簧支承的螺旋 spring-loaded screw;spring-supported screw
弹簧支承的振动输送器 spring-supported vibrating conveyer[conveyor]
弹簧支承滚轴运输机 spring-mounted roller conveyer[conveyor]
弹簧支承圈 bearing coil
弹簧支承式推力轴承 spring-supported thrust bearing
弹簧支承座 spring seat
弹簧支架 spring bracket;spring holder;spring shoe;spring support
弹簧支架筛箱 spring-supported screen box
弹簧支枢 spring pivot
弹簧支枢座 spring pivot seat
弹簧支柱 <独立悬挂的> spring leg;spring strut
弹簧支座 spring abutment;spring bearing;spring bracket;spring fastening;spring support
弹簧直浇口 collapsible sprue
弹簧直浇口模棒 spring gate stick
弹簧直径比 spring index
弹簧止挡 spring catch
弹簧止回阀 spring-loaded check valve
弹簧止水夹 spring water stopper
弹簧止水条 spring water seal
弹簧指式捡拾器 spring-finger pickup
弹簧指数 spring index
弹簧制动杆 spring-loaded clamp lever
弹簧制动气动松开的制动腔 spring set and air released chamber
弹簧制动器 spring brake;spring-loaded brake
弹簧制动器室 spring brake chamber
弹簧制动装置 spring catch device
弹簧制销 spring cotter
弹簧质量系统 spring-mass system
弹簧质量阻尼系统 spring-mass-damping system
弹簧致动液压缸 spring activated cylinder
弹簧中心螺栓 spring center[centre] bolt
弹簧钟形压力计 spring-balanced bell ga(u)ge
弹簧轴 spring shaft
弹簧轴承 spring bearing
弹簧主片拱度 spring main leaf camber
弹簧柱螺栓 spring stud
弹簧柱塞 spring-loaded plunger;spring plunger
弹簧柱塞式卸料板 spring plunger stripper
弹簧抓取装置 spring catch device
弹簧转辙器 spring switch
弹簧桩帽 spring cap for pile driving;spring cap for piling driving
弹簧装置 spring device;spring gear;spring installation;spring rigging
弹簧装置横均衡梁 transverse equalizer
弹簧装置角 spring seat angle
弹簧锥式贯入硬度计 spring cone penetrometer
弹簧自动闭门器 adjunct spring;adjust spring
弹簧自动还原按钮 spring-return button
弹簧自动起动器 spring-type self-starter
弹簧自紧油封 spring-loaded seal;spring-tensioned oil seal
弹簧自由端 spring free end
弹簧自振阻尼器 spring damper
弹簧总成 main assemblage of spring
弹簧阻车器 spring stop block
弹簧阻尼系统 spring-dashpot system
弹簧组 group spring;spring group;spring nest;spring pack;spring set;spring assembly
弹簧最大压缩高度 solid height of spring
弹簧作用 spring action
弹簧作用的 spring operated
弹簧作用的活塞 spring-actuated piston
弹簧作用和气动释放的制动器 spring applied-air released brake
弹簧座 base of spring;nest of springs;spring abutment;spring base;spring block;spring cup;spring holder;spring perch;spring seat(ing)
弹簧座板 spring plate
弹簧座承杯 spring cup
弹簧座挡圈 spring seating collar
弹簧座环 spring seating collar
弹簧座架 spring frame of seat
弹簧座摩擦板 spring seat chafing iron;spring seat wearing plate
弹簧座盘 spring cap
弹簧座腔 spring cavity
弹簧座圈 spring(-leaf) retainer;spring retaining in position
弹簧座锁夹 spring retainer lock
弹簧座锁销 spring seat lock pin
弹簧座闸阀 resilient seated gate valve
弹簧座中心线 spring seat center[centre] line
弹簧座轴承 spring seat bearing

弹簧座轴承盖 spring seat bearing cap
弹簧座轴承调整螺母 spring seat bearing adjusting nut
弹回 bounce; rebound; rebound action; recoil; reflect; repercussion; resilience; reverb; reverberate; reverberation; spring; spring-back
弹回的 resilient
弹回率 elastance
弹回线圈 flip coil
弹回性 elastance; rebound elasticity
弹回原位 kick back
弹回运动 snappy return
弹回装置 repeller
弹力 elastica; elasticity; resilience; spring
弹力测定法 elastometry
弹力持骨螺钉镊 self-holding screwdriver
弹力感 elastic sensation
弹力辊磨机 spring roll mill
弹力焊 spring(contact arc)welding
弹力护腿 elastic shinguard
弹力恢复 spring return
弹力恢复系数 restitution coefficient
弹力回程 spring return
弹力回综 <提花机的> elastic return
弹力计 blenometer; elastomeric meter; elastometer
弹力盘 puck
弹力曲线 spring curve
弹力式焊接 spring-type welding
弹力式继电器 non-safety relay; spring-type relay
弹力特性 spring-loaded characteristic
弹力天线 body-type antenna
弹力网 elasticated net
弹力纤维 elastic fiber[fibre]
弹力型继电器 spring returned relay
弹力叶片 spring leaf
弹力圆锥 conus elasticus; elastic cone
弹沥青 balkacchite
弹料 elastomer
弹率比 modular ratio
弹毛机 picker
弹棉机 picker
弹模试验数 number of elastic modulus test
弹能下降 elastic energy degradation
弹黏[粘]度 elastic-viscosity
弹黏[粘]计 elasto-viscometer
弹黏[粘]塑性 elastic-viscous plasticity
弹黏[粘]塑性体 elastic viscoplastic body
弹黏[粘]塑性性态 elasto-visco-plastic behavio(u)r
弹黏[粘]体系 elasto-viscous system
弹黏[粘]性 elasticoviscosity; elasto-viscosity
弹黏[粘]性层状体理论 elasto-viscous layered theory
弹黏[粘]性层状体系 viscoelastic layered system
弹黏[粘]性场 elastic-viscous field
弹黏[粘]性的 elasticoviscous; elasto-viscous
弹黏[粘]性固体 elasticoviscous solid; elasto-viscous solid; Kelvin body
弹黏[粘]性流 elasto-viscous flow
弹黏[粘]性流体 elasto-viscous fluid
弹黏[粘]性体系 elasto-viscous system
弹黏[粘]性液体 elasticoviscous liquid; elasto-viscous liquid
弹黏[粘]性状态 elasto-viscous behavio(u)r
弹起 bounce; eject
弹起和降落轮对 take-off/landing wheel sets
弹起键 eject key
弹起轮 spring-loaded roller
弹起驱动 eject drive
弹起装置 pop-up
弹韧油毡 stretchy asphalt roofing felt
弹射 catapult; eject; launch; lift-off
弹射波 bullet wave
弹射钉 spit nail
弹射发射装置 ejection firing device
弹射轨道 ejection orbit
弹射机构 ejection mechanism
弹射机制 catapult mechanism
弹射架 launching cradle
弹射紧固工具 powder-actuated fastening tool
弹射力 ejection force
弹射路径 trajectory path
弹射能力 ejectability
弹射器 catapult; ejector; jettison device; knockout attachment
弹射器发射 catapult launching
弹射前准备 preejection
弹射式发射装置 jettison launcher
弹射式喷洒器 ball drive sprinkler
弹射位置 ejection seat
弹射硬化 peen hardening
弹射装置 ejection device; jettison device
弹射座舱 bail-out capsule; ejection capsule
弹射座椅 ejection seat; ejector seat
弹升式喷洒头 pop-up head
弹式测热器 bomb calorimeter
弹式量热法 bomb calorimetry
弹式量热计 bomb calorimeter
弹式量热器 bomb calorimeter
弹式量热试验 bomb calorimeter test
弹塑界面 elastic-plastic interface
弹塑平衡状态 state of plastoelastic equilibrium
弹塑态 elasto-plastic state
弹塑体 elastics
弹塑性 elastic-plasticity; elasto; elasto-plasticity
弹塑性板 elasto-plastic plate
弹塑性包络谱 elasto-plastic envelop spectrum
弹塑性边界 elasto-plastic boundary
弹塑性变形 elastic-plastic deformation; elasto-plastic deformation
弹塑性变形理论 theory of elasto-plastic deformation
弹塑性材料 elastic-plastic material; elasto-plastic material
弹塑性的 elasto-plastic
弹塑性动力反应分析 elastic-plastic dynamic(al)analysis
弹塑性断裂 elastic-plastic fracture; elasto-plastic fracture
弹塑性断裂力学 elastic-plastic fracture mechanics; elasto-plastic fracture mechanics
弹塑性反应包络谱 elasto-plastic response envelop spectrum
弹塑性反应分析 elasto-plastic response analysis
弹塑性反应谱 elasto-plastic response spectrum
弹塑性范围 elastic-plastic range; elasto-plastic range
弹塑性分析 elastic-plastic analysis; elasto-plastic analysis
弹塑性刚度 elasto-plastic stiffness
弹塑性刚架 elastic-plastic frame
弹塑性固体 elasto-plastic solid
弹塑性混合体 elasto-plastic body
弹塑性阶段 elastic-plastic stage
弹塑性阶段破坏 elastic-plastic stage collapse
弹塑性结构 elasto-plastic structure
弹塑性介质 elasto-plastic medium
弹塑性界面 elasto-plastic interface
弹塑性矩阵 elasto-plastic matrix
弹塑性矩阵位移分析 elasto-plastic matrix displacement analysis
弹塑性聚合物 elasto-plastic polymer
弹塑性理论 elasto-plasticity theory; plastoelasticity
弹塑性力学 elasto-plastic mechanics; mechanics of elasto-plasticity; plastoelasticity
弹塑性梁 elasto-plastic beam
弹塑性流 elastic-plastic flow
弹塑性密封层 elasto-plastic sealant
弹塑性模量 elasto-plastic modulus
弹塑性模型 elastic-plastic model; elasto-plastic model
弹塑性耦合 elasto-plastic coupling
弹塑性平衡 elasto-plastic balance
弹塑性区 elasto-plastic area
弹塑性区域 elasto-plastic region
弹塑性驱动 elastic-plastic drive; elasto-plastic drive
弹塑性屈曲 elasto-plastic buckling
弹塑性蠕变分析 elastic-plastic-creep analysis
弹塑性失效准则 elasto-plastic failure criterion
弹塑性体 elastic-plastic body; elastic-plastic material; elastic-plastic solid; elasto-plastic solid; plastoelastic body
弹塑性体系 elasto-plastic system; plastoelastic system
弹塑性土 elasto-plastic soil
弹塑性弯矩 elastic-plastic moment
弹塑性弯矩-曲率关系 elastic-plastic moment-curvature relationship
弹塑性弯曲 elastic-plastic bending; elasto-bending; elasto-plastic bending
弹塑性位移 elastic-plastic displacement
弹塑性问题 elasto-plastic problem
弹塑性物体 elasto-plastic body; plastoelastic body
弹塑性系统 elasto-plastic system
弹塑性现象 elastic-plastic behavio(u)r; elasto-plastic behavio(u)r
弹塑性行为 elastic-plastic behavio(u)r; elasto-plastic behavio(u)r
弹塑性形变 elasto-plastic deformation
弹塑性型 elasto-plastic type
弹塑性性质 elasto-plastic behavio(u)r
弹塑性性状 elasto-plastic behavio(u)r
弹塑性应变硬化破坏 elastic plastic strain hardening fracture
弹塑性应力应变关系 elastic-plastic stress-strain relation
弹塑性有限元法 elasto-plastic finite element method
弹塑性振动 elasto-plastic vibration
弹塑性状态 elastic-plastic behavio(u)r; elasto-plastic behavio(u)r; elasto-plastic state
弹塑性阻力 elasto-plastic resistance
弹缩性变形带 elastic deformation zone
弹锁机构 snap over mechanism
弹踢器 kicker
弹条式扣件 elastic rod rail fastening
弹跳 bounce; bound; dap; recoil; spring(ing)
弹跳杆 bouncing pin
弹跳痕迹 bounce mark
弹跳频率 bounce frequency
弹跳铸型 bounce cast
弹跳装置 bouncer; pop-up
弹涂机 catapult
弹涂效率 catapult productivity
弹性 elastane; elastica; elastomeric property; stretch; elastic characteristic
弹性鞍形垫圈 elastic saddle shaped washer
弹性摆 elastic pendulum
弹性摆轮 resilient escapement
弹性板 elastic plate; elastic slab; elastomeric sheet; resilient board
弹性板法 elastic plate method
弹性板模拟 elastic sheet analog
弹性板线路 resilient plate track
弹性版印刷 aniline printing
弹性半空间 elastic half-space
弹性半空间弹簧刚度 elastic half-space spring stiffness
弹性半空间理论 elastic half-space theory
弹性半无限地基 elastic semi-infinite foundation
弹性半无限体 elastic half-space; elastic semi-infinite body
弹性绷带 elastic bandage; elasto-plast
弹性比 elastic(ity)ratio
弹性比功 elastic strain energy
弹性比例极限 elastic proportional limit; proportional elastic limit; proportion elastic limit
弹性比值 <物体弹性界限和其拉力的比值> elastic ratio
弹性边界 elastic boundary
弹性边界条件 elastic boundary condition
弹性边墙 elastic side wall
弹性边缘座椅 spring edge seat
弹性变定当量 equivalent elastic set
弹性变换 elastic translation
弹性变流动力润滑剂 elastorheodynamic(al)lubricant
弹性变位 elastic deflection
弹性变形 elastic deformation; elastic distortion; elastic expansion; elastic give; non-permanent deformation; non-residual deformation; recoverable deformation; resilience; resilient deformation; temporary set
弹性变形断裂 elastic deformation breakdown
弹性变形范围 regime(n)of elastic deformation
弹性变形功 recoverable strain work; resilience work of deformation
弹性变形恢复 resilient-elasticity recovery
弹性变形极限 elastic deformation limit
弹性变形阶段 elastic deformation stage; stage of elastic deformation
弹性变形力矩 compliance torque; elastic torque
弹性变形面积 elastic deflection area
弹性变形内力 internal resilience
弹性变形曲线 elastic deformation curve
弹性标准 flexible standard
弹性标准预算 flexible standard budget
弹性表面 elastic surface
弹性表面波 elastic surface wave
弹性波 elastic wave
弹性波测探 elastic wave exploration
弹性波层析成像 elastic wave computerized tomography
弹性波传播 elastic wave propagation
弹性波传播速度 rate of elastic wave propagation
弹性波发生器 elastic wave generator
弹性波法 elastic wave method
弹性波理论 elastic wave theory
弹性波偏移理论 elastic migration theory
弹性波全息摄影术 elastic wave ho-

lography
弹性波速度 elastic wave velocity; seismic velocity
弹性波探查 test by elastic wave
弹性波型垫圈 elastic wave shaped washer
弹性玻璃丝垫 vibramat
弹性薄板 elastic sheet; resilient sheet
弹性薄壳方程 elastic shell equation
弹性薄膜模拟法 elastic membrane analogy; membrane analog(y)
弹性不变合金 Elinvar alloy
弹性不连续性 elastic discontinuity
弹性不完整 elastic imperfection
弹性不稳定(度) elastic instability
弹性不稳定性 elastic instability
弹性不足 logu
弹性材料 elastomer; elastomer material; resilient material; elastic material
弹性材料零件 elastomeric parts
弹性参数 elastic parameter
弹性残存变形 elastic drift
弹性残留变形 elastic drift
弹性残余变形 elastic drift
弹性槽楔 elastic wedge
弹性槽形夹 resilient channel
弹性槽形龙骨 resilient furring channel
弹性侧向弯曲 elastic lateral buckling
弹性测定 elasticity measurement
弹性测定法 elastometry
弹性测定器 compliance test apparatus; elastomeric tester
弹性测力计 elastic dynamometer
弹性测量仪 resilience meter
弹性层 elastic layer
弹性层状路面体系 elastic multilayered pavement system
弹性层状体理论 elastic layered theory
弹性层状体系理论 elastic multilayer theory; theory of elastic laminated system
弹性常量 elasticity constant
弹性常数 elastic constant; elasticity constant
弹性场 elastic field
弹性车顶 flexible car roof
弹性车轮 resilient wheel
弹性沉降 elastic settlement
弹性沉降当量 equivalent elastic set
弹性沉陷 elastic depression
弹性衬垫 cushion shim
弹性衬垫承座 elastomeric pad bearing
弹性衬垫支座 rubber bridge bearing
弹性衬套 resilient bushing
弹性成层的 elastically bedded
弹性弛豫时间 elastic relaxation time
弹性迟滞性 elastic hysteresis
弹性尺 elastic scaler
弹性齿轮 resilient gear; resilient gear wheel
弹性齿圈 resilient gear rim; resilient gear ring
弹性冲击 elastic impact; elastic shock
弹性储存量 elastic storage
弹性传导体 elastic medium
弹性传动 elastic drive; elastic transmission; resilient drive; resilient transmission
弹性传动片 flexible drive plate
弹性垂度 elastic deflection
弹性锤击 elastic impact
弹性唇片 elastic lip
弹性脆性材料 elastic brittle material
弹性存储器 elastic store
弹性带 elastic zone
弹性挡圈 circlip; elastic collar; snap ring
弹性导正销 spring-backed pilot
弹性道钉 elastic (rail) spike; spring (rail) spike
弹性道耙 spring-tooth rake
弹性的 buoyant; elastic; resilient
弹性的连续价值 elastic continuum
弹性的束梁原理 theory of elastically restrained beams
弹性的贴胶 elastic coating
弹性的涂层 elastic coating
弹性的相容 elastic compatibility
弹性地 elastically
弹性地板 resilient floating floor; resilient floor
弹性地板材料 resilient flooring
弹性地板面 resilient floor covering
弹性地基 elastic bed; elastic foundation; elastic subgrade
弹性地基板 slab on elastic subsoil
弹性地基梁 beam on elastic foundation
弹性地基梁比拟法 analogy method for beam on elastic foundation; elastic foundation beam analogy method
弹性地基梁法 elastic foundation supported beam method
弹性地基上梁 beam on elastic subgrade
弹性地蜡 helenite
弹性地沥青 elaterite
弹性地面 elastic ground
弹性地面材料 elastic material; resilient flooring
弹性地面铺料 resilient floor covering
弹性地震能量 elastic seismic energy
弹性电阻 elastoresistance
弹性电阻系数 elastoresistance coefficient
弹性垫 cushion
弹性垫板 resilient sleeper bearing; resilient tie pad
弹性垫层 cushioning; elastic cushion
弹性垫料 cushioning material; elastic packing
弹性垫圈 elastic ring; elastic spacer; elastic washer; elastomeric gasket; flexible washer; spring washer
弹性垫条 elastic spacer
弹性垫座 bumper block; resilient chair
弹性吊挂 resilient hanger
弹性吊架 elastic suspension; flexible suspension
弹性吊卡 resilient hanger clip
弹性顶尖 movable center[centre]
弹性定价法 flexible markup pricing
弹性定律 law of elasticity
弹性定位装置 elastic positioning device
弹性锭子 elastic spindle; elastic stator
弹性动力辐射 elastodynamic(al) radiation
弹性动力模型 elastodynamic(al) model
弹性动力位移场 elastodynamic(al) field
弹性动力学 elastodynamics; elastokinetics
弹性动力学的 elastodynamic(al)
弹性动力应力强度因数 elastodynamic(al) stress-intensity factor
弹性动态挤出机 elastodynamic(al) extruder
弹性端部约束 elastic end-restraint
弹性断裂 elastic failure
弹性断裂转变温度 fracture transition of elastic
弹性多层体系理论 elastic multiple layered theory
弹性舵扇 elastic quadrant; spring tiller
弹性阀 cushion valve; feather valve
弹性反冲 elastic recoil
弹性反冲分析 elastic recoil analysis
弹性反馈 elastic feedback
弹性反馈控制器 elastic feedback controller
弹性反力 elastic resistance
弹性反应 elastic reaction; resilient response
弹性反应分析 elastic response analysis
弹性反应截面 elastic cross-section
弹性反作用力 elastic reactance
弹性范围 elastic range; elastic zone
弹性方案 elastic scheme
弹性方程式 elastic equation; elasticity equation
弹性防水 elastomeric waterproofing
弹性防水层 <基础的> elastomeric waterproofing
弹性防松螺母 elastic stop nut
弹性防撞装置 elastic fender; elastic fendering device; resilient fender; spring fender
弹性非均匀性 elastic heterogeneity
弹性非均匀压缩系数 coefficient of elastic non-uniform compression
弹性分布 elastic distribution
弹性分量 component of elasticity
弹性分区制 elastic zoning
弹性分析 analysis of elasticity; elastic analysis
弹性封闭层 elastic seal(ing)
弹性封口 resilient seal
弹性峰值 elastic peak
弹性敷层 elastomeric coating
弹性浮筒 spring buoy
弹性辐操向轮 spring-spoke steering wheel
弹性辐条轮 spring-spoked wheel
弹性辐条转向盘 spring-spoked steering wheel
弹性负载方法 elastic load method
弹性复数模量 complex elastic modulus
弹性复原 elastic recovery
弹性复原试验 back test
弹性覆层 resilient coating
弹性覆盖层楼板 resilient floor covering
弹性干扰物 resilient chaff
弹性杆支撑 elastic rod support
弹性感 elastic sensation
弹性感应线 elastic response line
弹性刚度 elastic rigidity
弹性刚度常量 elastic stiffness constant
弹性刚度常数 elastic stiffness constant
弹性刚度系数 elastic stiffness coefficient
弹性钢板 elastic sheet
弹性钢表带 link expansion band
弹性钢底梁 spring-steel floor member
弹性钢顶梁 spring-steel roof member
弹性钢轨扣件 flexible rail fastening
弹性钢丝应变计 elastic wire strain meter
弹性高分子物质 elastomer
弹性高聚物 elastopolymer
弹性搁置的 elastically bedded
弹性隔块 elastic space
弹性隔片 resilient spacer
弹性各向同性 elastic isotropy
弹性各向同性材料 elastically isotropic material; elastic isotropic material
弹性各向异性 elastic anisotropy
弹性工时 flexible hours; flexible working hours
弹性工时制 elastic work schedule system; flexible work hours system
弹性工作时间 flexible time; flexible work hour; flextime
弹性工作时间表 flexible schedule
弹性工作时间制度 elastic work schedule system; flexitime system
弹性工作制度 flexible job system; flex-job system
弹性公式 elastic formula
弹性供应 elastic supply
弹性拱 elastic arch
弹性拱法 elastic arch method
弹性拱理论 elastic arch theory
弹性拱座 elastic abutment
弹性构件 elastic component; elastic member
弹性估计 estimation of elasticity
弹性固定 flexible fastening
弹性固定的 elastically built-in; elastically fixed
弹性固接 elastic fixing
弹性固体 elastic solid
弹性关节 elastic joint
弹性管 elastic pipe; elastic tube; flexible pipe
弹性管式压力计 spring-tube manometer
弹性惯性 elastic lag
弹性灌浆料 elastomer
弹性光生 elastic photoproduction
弹性光学 elastooptics
弹性广义复变函数方法 complex variable method in theory of elasticity
弹性规划 flexible plan
弹性合成物 elastic compound; elastic synthetic
弹性合成橡胶材料 elastomer material
弹性合成橡胶支座 elastomeric bearing pad
弹性和塑性流 elastic and plastic flow
弹性和塑性体 elastic and plastic body
弹性核 elastic kernel
弹性核心 elastic core
弹性荷载 elastic load; elastic weight
弹性荷载法 angle weights method; method of elastic weights <计算静定梁与桁架的挠度的方法>; elastic weight method <求结构位移的一种方法>
弹性荷重 elastic load; elastic weight
弹性横断面 elastic section
弹性横向弯曲 elastic lateral buckling
弹性后效 delayed elasticity; elastic after effect; elastic after-working; elastic drift; elasticity after effect; elastic reaction; elastic spring back; post-elastic behavio(u)r; residual elasticity; spring-back
弹性后效变形 elastic after deformation
弹性后效应 delayed elasticity
弹性后张力 elastic backpull
弹性护舷 elastic fender(ing)
弹性滑移 elastic slip
弹性化 elastification
弹性化合物 elastomeric compound
弹性环 elastic ring; resilient ring; snap ring
弹性环功率计 elastic loop dynamometer
弹性环销 spring ring dowel
弹性缓冲器 elastic buffer; spring bumper; stop buffer
弹性缓和 elastic moderation
弹性恢复 elastic rediscovery; elastic restitution; recovery of elasticity; elastic rebound
弹性恢复力 elastic restoring force
弹性恢复率 elastic recovery

弹性恢复系数 coefficient of elastic recovery
弹性回复 compensation return;elastic comeback;elastic recovery;spring-back
弹性回复力 elastic restoring force
弹性回复能力 elastic resilience
弹性回火 spring temper
弹性回能 elastic resilience
弹性回缩 elastic recoil
弹性回弹 elastic rebound;elastic recoil;elastic resilience
弹性回弹假说 elastic rebound hypothesis
弹性回弹理论 elastic rebound mechanism;elastic rebound theory
弹性回弹系数 elastic rebound factor;rebound coefficient of elasticity
弹性回跳 elastic rebound;spring-back
弹性回跳假说 elastic rebound hypothesis
弹性回跳理论 elastic rebound theory
弹性回跳学说 elastic rebound theory
弹性回跃 elastic rebound
弹性汇率 flexible exchange rate
弹性火棉胶 collodion elasticity
弹性火焰管 expanding flue
弹性货币供应 elastic money supply
弹性机座 elastic frame
弹性基层反力 elastic subgrade reaction
弹性基础 elastic foundation;flexible foundation
弹性基础连续梁 beam on elastic foundation
弹性基础上安装钢板 elastically bedded plate;elastically embedded plate
弹性基础上定长梁 finite beam on elastic foundation
弹性基础(上)梁 beam on elastic subgrade
弹性基床反力 elastic subgrade reaction
弹性畸变 elastic distortion
弹性极限 elastic(ity) limit;k-point;limit of elasticity
弹性极限的抗压强度 crushing strength at elastic limit
弹性极限荷载 <不引起残余变形的最大荷载> proof load
弹性极限强度 proof strength
弹性极限应变 elastic limit strain
弹性极限应力 elastic proof stress;proof stress
弹性极限值 elastic limit value
弹性急变 immediate strain
弹性挤曲 elastic buckling
弹性挤压作用 elastic compaction
弹性计 elasticity gage;elastometer;resilience meter;resiliometer
弹性计划 elastic plan;flexible plan
弹性计量单位 elastic modulus
弹性记忆 elastic memory
弹性记忆效应 elastic memory effect
弹性剂 elasticator;elasticizer
弹性夹 collet;pinch clamp;resilient clip
弹性夹紧联结杆 snap tie
弹性夹扣板 resilient clamp
弹性夹头 collet;drawback collet
弹性价格政策 flexible-price policy
弹性间接生产费预算 flexible factory overhead budget
弹性减缩 elastic shortening
弹性减震板 resilient damping board
弹性减震垫 resilient cushion member
弹性减震悬挂装置 resilient shock absorbing suspension
弹性剪切 elastic shear
弹性剪切模量 shearing modulus of elasticity
弹性剪切系数 coefficient of elastic shear
弹性剪切形变 elastic shear deformation
弹性简单悬挂 stitched simple catenary suspension;stitched tramway type equipment
弹性建筑 elastic construction
弹性键 captive key
弹性胶 elastic glue;elastic gum
弹性胶布 tensoplast
弹性胶垫 elastomeric pad
弹性胶结磨轮 elastic bond wheel
弹性阶段 elastic stage
弹性阶段设计法 elastic (design) method
弹性接触 elastic contact
弹性接缝 elastic joint;resilient joint
弹性接缝料 resilient joint material
弹性接合 elastic joint;resilient joint
弹性接头 elastic connector;elastic joint;flexible connector;resilience connector;resilient connector
弹性接头密封料 elastic joint seal(ing) compound
弹性结构 elastic structure
弹性结构拱 elastic structural arch
弹性结构体系 elastic construction system
弹性结构系统 elastic structural system
弹性截面 elastic cross-section
弹性介质 elastic medium
弹性介质生物接触氧化塘 biologic(al) contact oxidation pond of elastic medium
弹性界限 limit of elasticity;elastic region
弹性金属接头 swing joint
弹性金属曲管式压力计 Bourdon ga(u)ge
弹性金属胀圈 elastically metallic expansion ring;elastic metallic expansion ring
弹性紧固装置 elastic fastening device
弹性锦纶 elastic nylon
弹性劲度 elastic stiffness
弹性劲度常量 elastic stiffness constant
弹性劲度常数 elastic stiffness constant
弹性静力学 elastostatics
弹性矩 elastic moment
弹性矩阵 <应力和应变关系的矩阵> elastic(ity) matrix
弹性聚合物 elastomeric polymer
弹性聚合物液体 elastic polymer liquid
弹性绝缘填料 expansion strip
弹性均一程度 degree of elastic homogeneity
弹性均匀压缩系数 coefficient of elastic uniform compression
弹性均质材料 elastic isotropic material
弹性卡环 circlip
弹性卡环装卸器 snap remover;snap-ring expander
弹性卡子 resilient clip
弹性抗矩 elastic reactance moment
弹性抗力 elastic resistance
弹性抗震设计 elastic seismic design
弹性靠船墩 elastic dolphin
弹性壳 elastic container
弹性壳体 elastic shell
弹性可变含水量 elastic deformable aquifer
弹性可弯式尖轨转辙器 switch rail with flexible heel joint
弹性孔 spring hole
弹性控制 elastic control
弹性扣板 elastic pinch plate;resilient clip;spring clip;flexible rail clip <混凝土轨枕的>
弹性扣件 elastic fastener;elastic rail fastener;flexible fastening;resilient rail fastenings;spring clip;spring fastening
弹性筘 spring reed
弹性矿质硬沥青 elastic mineral pitch
弹性框架 elastic frame
弹性拉钩 elastic draw hook
弹性拉伸 elastic stretch(ing)
弹性拉伸应变 elastic tensile strain
弹性离合器 elastic clutch;resilient clutch
弹性理论 elasticity theory;elastic theory;theory of elasticity
弹性理论计算法 elastic theory computational method
弹性力 elastic force
弹性力矩 moment of elasticity
弹性力学 elasticity;elastic mechanics;elastomechanics;mechanics of elasticity
弹性力学反应分析 elastic-plastic dynamic(al) response analysis
弹性力学平面问题 plane problems in theory of elasticity
弹性利率 flexible interest rate
弹性利率抵押贷款 flexible rate mortgage
弹性沥青 dopplerite;elastic bitumen;mineral caoutchouc;elaterite <一种焦油沥青>
弹性连接 elastic coupling;elastic fixing;flexible coupling;flexible fixing;resilient coupling;resilient fixing
弹性连接器 elastic connector
弹性连续介质 elastic continuum
弹性连续体 elastic continuum
弹性连续支承体 continuous elastic support;continuous supporting elastic-media
弹性联轴节 coupling with resilient;elastic clutch;elastic joint;elastomeric coupling;flexible coupling;flexible joint;sleeve flexible joint;elastic coupling
弹性联轴器 elastic clutch;elastic coupling;elastic joint;flexible coupling;flexible joint;resilient coupling;sleeve flexible joint;spring coupling
弹性链形悬挂 stitched catenary suspension;stitched catenary type equipment
弹性梁 elastic beam;spring beam
弹性梁支承法 elastic supported beam method
弹性裂纹增长断裂 elastic-crack-growth fracture
弹性临界负荷 elastic critical load
弹性临界荷载 elastic critical load
弹性临界剪应力 elastic critical shearing stress
弹性流动 elastic flow
弹性流动蠕变 elastic flow creep
弹性流量 elastic flow
弹性流体 elastic fluid
弹性流体动力润滑 elastohydrodynamic(al) lubrication
弹性流体动力润滑理论 elastohydrodynamic(al) lubrication theory
弹性流体动力学 elastohydrodynamics
弹性硫 elastic sulfur;plastic sulphur
弹性硫塑料 elastopolymer;elastothiomer;thioplast
弹性隆起 elastic heave
弹性路基 elastic subgrade
弹性履带系统 resilient track system
弹性氯丁橡胶条 flexible neoprene seal
弹性轮 <研磨的> elastic wheel
弹性轮胎 elastic tire[tyre];spring tire[tyre]
弹性轮辋车轮 elastic-rim wheel
弹性螺母 elastic nut;spring nut
弹性螺旋圈 elastic helical coil
弹性玛琋脂 resilient mastic
弹性埋没的 elastically embedded
弹性埋置的 elastically embedded
弹性埋置的圆形隧道 elastically embedded circular tunnel
弹性锚链系泊 resilient mooring
弹性门 <采用橡胶等材料做门扇> flexible door
弹性门窗插销 spring snib
弹性门窗钩 spring snib
弹性密封 elastic packing;elastomeric seal;elastomer seal(ing);flexible seal;resilient seal;wiper seal
弹性密封材料 elastic seal(ing)
弹性密封层 elastic sealant
弹性密封剂 elastomeric sealant
弹性密封料 elastic sealing compound;elastomeric sealant;resilient sealant
弹性面砖 resilient tile
弹性敏感元件 elastic sensing element
弹性模量 chord modulus;dynamic(al) modulus of elasticity;elasticity coefficient;elasticity modulus;elastic modulus;module of elasticity;modulus of elasticity;resilient modulus;Young's modulus
弹性模量比 elastic modular ratio;modular ratio;modular ratio of elasticity;ratio of modulus elasticity
弹性模拟 elastic analogy
弹性模片 elastic sheet;flexible sheet;resilient sheet;snap
弹性模数 chord modulus;elasticity modulus;elastic modulus;module of elasticity;modulus of elasticity;resilient modulus;Young's modulus
弹性模型 elasticity model;elastic model
弹性模约束力 binding force of elasticity film
弹性膜 elastic membrane
弹性膜校正曲线 correction curve of elastic film
弹性膜模型 elasticity membrane model
弹性磨具 elastic grindstone
弹性磨轮 buffing wheel
弹性挠度 elastic deflection;elastic deformation;rebound deflection
弹性挠曲 elastic bending;elastic deflection;elastic deformation
弹性挠曲变形 elastic springing
弹性挠曲变形指数 elastic deflection exponent
弹性挠曲线 elastic line
弹性能量 elastic energy
弹性能量降级 elastic energy degradation
弹性尼龙 stretchy nylon
弹性泥炭 peaty pitch coal
弹性腻子 <装玻璃用> elastic glazing compound
弹性黏[粘]度 elasticoviscosity;elastoviscosity
弹性黏[粘]度计 elasto-viscometer
弹性黏[粘]度特性 elasto-viscous character

弹性黏[粘]膏 elasto-plast
弹性黏[粘]合剂 elastomer adhesive
弹性黏[粘]结材料 elastomeric adhesive
弹性黏[粘]性场 elasto-viscous field
弹性凝胶 elastic gel;elastogel
弹性扭力 elastic torsion
弹性扭曲 elastic distortion
弹性耦合 elasto-plastic coupling
弹性耦连 resilient coupling
弹性旁承 spring-load side bearer
弹性泡沫保温套 flexible foam lagging
弹性膨胀 elastic expansion;elastic extension
弹性膨胀接头 elastic expansion joint
弹性碰垫 elastic fendering;resilient fender;spring fender
弹性碰撞 billiard-ball collision;bumping collision;elastic collision;elastic impact
弹性碰撞模型 elastic collision model
弹性碰撞引起的能量损失 elastic energy degradation
弹性疲劳 elastic failure;elastic fatigue
弹性疲劳的橡胶 strained rubber
弹性片 elastic trip;flexure strip;resilient trip
弹性片调速器 vibrating-reed escapement device
弹性平衡 elastic equilibrium
弹性平衡系统 spring balance system
弹性平衡状态 elastic state of equilibrium;state of elastic equilibrium
弹性屏蔽材料 elastomeric shield material
弹性破坏 elastic breakdown;elastic failure
弹性破裂应变 elastic cracking strain
弹性铺设的钢轨 elastically mounted rail
弹性谱 elastic spectrum
弹性谱动能 elastic spectral kinetic energy
弹性气室 elastic air chamber
弹性气压表 elastic barometer
弹性气压车轮 pneumatic elastic wheel
弹性迁移 elastic removal
弹性牵引 elastic traction
弹性前效 elastic fore-effect
弹性强度 elastic strength;elastic stress;resilient strength
弹性墙 elastic wall
弹性桥台 elastic abutment
弹性切变波 elastic transverse wave
弹性切力极限测定计 pachimeter
弹性球体 elastic sphere
弹性区域 elastic domain;elastic range;elastic region;elastic zone
弹性曲面 elastic surface
弹性曲线 <在屈服点内梁或柱受力时弯曲的线形> elastic curve
弹性驱动 elastic drive
弹性屈服 elastic buckling;elastic yielding
弹性屈曲 elastic buckling
弹性屈曲负载 elastic buckling load
弹性圈 elastic packing ring
弹性权限 elastic limit
弹性扰动 elastic turbulence
弹性热应力 elastic thermal stress
弹性容积模数 volume modulus of elasticity
弹性容量 elasticity volume
弹性溶胶 elastic sol
弹性溶液 elastic solution
弹性柔顺常量 elastic compliance constant
弹性蠕动 elastic creep
弹性蠕动复原 elastic creep recovery
弹性软横跨 head span
弹性散射 elastic scattering
弹性散射波 elastic scattering wave
弹性散射电子 elastic scattering electrons
弹性散射共振 elastic scattering resonance
弹性散射截面 elastic scattering cross-section
弹性散射碰撞 elastic scattering collision
弹性砂轮 elastic grinding wheel;shellac-bond wheel
弹性熵 elastic entropy
弹性上班时间 flexitime
弹性上限 upper elastic limit
弹性设计 allowable stress design;elastic design
弹性设计方法 elastic design method
弹性伸长 elastic elongation;elastic extension;elastic stretch(ing)
弹性伸缩纱框 elastic reel
弹性伸缩筒管 elastic expandable bobbin
弹性声(阻)抗 elastic acoustic(al)reactance
弹性失稳 elastic instability
弹性失稳破坏 elastic instability failure
弹性失效 elastic breakdown;elastic failure
弹性失效压力 elastic breakdown pressure
弹性失效准则 elastic failure criterion
弹性石英 elastic quartz
弹性实体 solid elastic body
弹性势 elastic potential
弹性势理论 elastic potential theory
弹性势能 elastic potential energy
弹性试验 elastic(ity)test;resilience test
弹性室 elastic chamber
弹性释放 elastic release
弹性收缩 <预应力混凝土> elastic shortening;elastic shrinkage
弹性受托发行制 elastic fiduciary issue system
弹性束 bungee
弹性树胶 gum elastic
弹性树脂 elastic resin;resilient resin
弹性数 elasticity number
弹性衰减 elastic extenuation
弹性水驱 elastic water drive
弹性顺从 elastic compliance
弹性顺度 elastic compliance
弹性顺度常量 elastic compliance constant
弹性顺度系数 elastic compliance coefficient
弹性松弛时间 elastic relaxation time
弹性塑料 elastomer;elasto-plast;elasto-plastic;elastoplasts
弹性塑性工作条件 elasto-plastic working condition
弹性塑性力学 elasto-plasticity
弹性随动 spring following
弹性随动件 spring follower
弹性隧道效应 elastic tunnel(l)ing effect
弹性损耗因子 elastic loss factor
弹性损失 elastic loss
弹性缩短 elastic shortening
弹性索 elastic cable
弹性锁紧环 circlip
弹性锁紧螺母 elastic stop nut
弹性锁销 latch lock
弹性套管 spring-backed quill
弹性套环扣件 resilient stirrup
弹性套筒 elastic sleeve;resilient sleeve
弹性套筒轴承 elastic sleeve bearing
弹性特性 elastic behavio(u)r;elastic characteristic;elastic property
弹性特征 elastic characteristic
弹性体 elastic body;elastic mass;elastic solid;spring mass
弹性体部件 elastomeric unit
弹性体材料 elastomer
弹性体的 elastomeric
弹性体的线性振动 linear vibrations of elastic bodies
弹性体动力学 elastodynamics
弹性体防水 elastomeric waterproofing
弹性体改性热塑性塑料 elastomer modified thermoplastic
弹性体合金 elastomer alloy
弹性体护舷部件 elastomeric fender unit
弹性体加成物 elastomer adduct
弹性体片材 elastomer sheet
弹性体塑料 elastomer plastic
弹性体系 elastic system
弹性体支承 elastomeric bearing
弹性体止水材料 elastomeric sealant
弹性体贮罐模塑成型 elastic reservoir mo(u)lding
弹性填层 cushioning
弹性填缝材料 resilient joint material
弹性填缝料 expansion joint material
弹性填料 elastic filler;elastic packing;resilient packing
弹性填密环 elastic packing ring
弹性条件 elasticity condition
弹性通货 elastic currency
弹性图形 elasticity figure
弹性涂层 resilient coating
弹性土堤 elastic embankment
弹性土楔 elastic soil wedge
弹性土压力 elastic earth pressure
弹性湍流 elastic turbulence
弹性推针导管 tickler guide
弹性托板 spring bed
弹性拖油容器 dracone
弹性椭圆 ellipse of elasticity
弹性外套 elastomer covering
弹性弯沉 elastic deflection
弹性弯曲 <屈服点范围内> elastic bending
弹性弯曲的形状 <指梁而言> elastica
弹性万向节 fabric joint
弹性万向节叉 cushion disc[disk] spider
弹性网 elastic network
弹性网络 elastomeric network
弹性网眼 elastic net
弹性位错 elastic dislocation
弹性位错理论 elastic dislocation theory
弹性位能 elastic potential energy
弹性位移 elastic displacement
弹性稳定 elastic stabile
弹性稳定度 elastic stability
弹性稳定分析 elastic stability analysis
弹性稳定性 elastic stability
弹性稳定性的本征值问题 eigenvalue problem of elastic stability
弹性问题 elasticity problem
弹性物 elastic mass;elastic matter;elastomer
弹性物的 elastomeric
弹性物质 elasticizer;elastic material
弹性物质平衡法 elastic material balance method
弹性吸收 elastic absorption
弹性系船柱 elastic dolphin
弹性系缆桩 resilient bollard
弹性系数 coefficient of elasticity;elastic coefficient;elasticity coefficient;module of elasticity;modulus of elasticity
弹性系统 elastic system
弹性纤维 elastane fiber[fibre];elastic fiber[fibre];Spandex;spandex fiber[fibre]
弹性纤维板 resilient fiberboard
弹性纤维组织 elastic tissue
弹性弦线 elastic string
弹性现象 elastic behavio(u)r
弹性线 elastic line
弹性线法 elastic line method
弹性线接触密封 elastic linear contact seal
弹性限度 elastic limit;elastic strength;limit of elasticity
弹性限制法 elastic limit method;elastic limit system
弹性相互作用 elastic interaction
弹性相容性 elastic compatibility
弹性相似定律 Cauchy's law of elasticity;elastic similarity law
弹性相似性 elastic similarity
弹性箱 expellant bag
弹性箱底接缝 elastomeric tank base joint
弹性响应 elastic response
弹性橡胶 elastic caoutchouc;elastic rubber;India rubber
弹性橡胶新切口法 new northway method
弹性橡皮 caoutchouc;indian rubber
弹性消能器 elastomeric energy absorber
弹性销 captive key;spring pin
弹性效应 buoyancy effect;elastic effect
弹性楔接头 elastic wedge joint
弹性芯子填密物 elastic core packing
弹性信托 flexible trust
弹性行动 elastic behavio(u)r
弹性行为 elastic behavio(u)r
弹性形变 elastic deformation;elastic deforming
弹性形变极限 elastic deformation limit
弹性形状因素 elasticity form factor
弹性性能 elastic property
弹性性质 elastic property
弹性修正 elasticity correction
弹性需求 elasticity of demand
弹性需求法则 law of elastic demand
弹性需求量 elastic demand
弹性徐变总回复应变 total elastic-plus-creep recovery strain
弹性悬吊架 elastic support
弹性悬挂 elastic suspension;flexible mounting;resilient suspension
弹性悬挂吊架 spring suspension
弹性悬挂法 resilient mounting
弹性悬挂架 elastic support
弹性悬挂装置 resilient hanger
弹性悬架【机】elastic suspension
弹性悬架式车架 spring frame
弹性学 elasticity
弹性压密 cushioning
弹性压曲 elastic buckling
弹性压屈 elastic buckling
弹性压屈极限荷载 critical elastic buckling load
弹性压缩 elastic compression
弹性压缩度 elastic compressibility
弹性压缩量 elastic compression;elastic shortening
弹性压缩模量 modulus of elastic compression
弹性压头 compensating squeeze head
弹性延伸 elastic extension;elastic stretch(ing)
弹性样物质 elastoid
弹性液体 elastic liquid
弹性移位 elastic displacement

弹性抑制 elastic restraint
弹性因素 elasticity factor
弹性应变 elastic strain; non-permanent strain; non-residual strain; rebound strain
弹性应变恢复 elastic strain recovery
弹性应变力 work of elastic strain
弹性应变能(量) elastic strain energy
弹性应变仪 elastomeric strain gate
弹性应变增量 incremental elastic strain
弹性应力 elastic stress
弹性应力分布 elastic stress distribution
弹性应力损失 elastic stress loss
弹性硬蛋白 elastin
弹性硬度 elastic hardness
弹性硬度计 elastodurometer
弹性油灰 bouncing putty
弹性油膜 elastic film
弹性油泥 bouncing putty
弹性余震 elastic aftershock
弹性预算 flexible budget; sliding budget; variable budget
弹性预压缩 elastic precompression
弹性元件 elastic cell; elastic element
弹性原理 elastic(ity) theory
弹性原则 flexibility principle; law of elasticity; principle of flexibility
弹性圆盘 cushion disc[disk]
弹性圆形夹合梁 elastic circular sandwich beam
弹性圆柱 elastic cylinder
弹性圆柱销联轴节 kick-latch
弹性圆锥 conus elasticus; elastic cone
弹性约束 elastic restraint
弹性约束边 elastically restrained edge
弹性约束的 elastically restrained
弹性约束系数 elastic restraint coefficient
弹性匀质的 elastic-homogeneous
弹性运动 elastic movement
弹性载荷 elastic load
弹性藻沥青 coorongite; hangelite
弹性增进剂 elasticizer
弹性增值 elastic gain
弹性毡 resilient felt; resilient quilt
弹性张力 elastic tensor
弹性张量 elasticity tensor; elastic tensor
弹性胀圈 spring expander
弹性折合模数 reduced modulus of elasticity
弹性辙叉 spring frog
弹性褶裥带 elastic ruche tape
弹性振荡 elastic oscillation
弹性振动 elastic shock; elastic vibration
弹性振动元件 elastic vibrating element
弹性震动 elastic vibration
弹性整理稀薄平布 elastic finish sheeting
弹性正割模量 secant modulus of elasticity
弹性正应力系数 elastic normal stress coefficient
弹性支撑 elastic support
弹性支承 continuous beam on elastic supports; elastic bearing; elastic mounting; elastic support; floating mounting; resilient support; elastomeric bearing
弹性支承板 elastically supported plate; resilient support plate
弹性支承边(缘) elastically supported edge
弹性支承大梁 elastically supported girder
弹性支承刚度 elastic supporting stiffness
弹性支承轨道 resilient-supported track
弹性支承连续梁 continuous girder on elastic supports
弹性支承连续梁法 elastically supported continuous girder method; method of continuous beams on elastic supports
弹性支承梁 beam on elastic support; elastically supported beam; elastically supported girder
弹性支承悬臂梁 elastic supported cantilever
弹性支承压力分布 elastic bearing pressure distribution
弹性支承轴承 elastically yielding bearing
弹性支承装置 resilient supporting unit
弹性支点法 elastic fulcrum method
弹性支架 spring support
弹性支座 elastic bearing; elastic mounting; elastic support; elastomeric bearing; flexible mounting; resilient mounting; resilient supporting member; yielding seat
弹性支座梁 beam on elastic support
弹性织物 elasticized fabric; elastic webbing
弹性执行 flexibility of execution
弹性直浇口 collapsible sprue
弹性值 elasticity number; elastic number
弹性制动爪 elastic stop
弹性制造系统 flexible manufacturing system
弹性滞后 elastic lag; elastic hysteresis
弹性滞后闭合回线 elastic hysteresis loop
弹性滞后现象 elastic hysteresis
弹性滞回模量 hysteresis modulus of elasticity
弹性中间垫层 elastic intermediate layer
弹性中心 elastic center[centre]
弹性中心法 <解超静定结构的一种方法> method of elastic center[centre]; elastic center[centre] method
弹性中子 elastic neutron
弹性终端抑制 elastic end-restraint
弹性终端约束 elastic end-restraint
弹性终止压力 elastic breakdown pressure
弹性重力刚度 elastic gravitational stiffness
弹性重力耦联振动 elastic gravitational coupled vibration
弹性重量 elastic weight
弹性轴 axis of elasticity; elastic axis; expanding shaft
弹性轴承 elastically supported bearing; elastic bearing
弹性注油器 squirt can
弹性贮能 elastic stored energy
弹性驻环 spring collar
弹性柱体 elastic cylinder
弹性铸铁 modular iron
弹性转向盘 spring steering wheel
弹性转移 elastic translation
弹性转子 elastic rotor
弹性桩 elastic pile
弹性装置【机】 springing
弹性状态 elastic behavio(u)r; elastic condition; elastic stage; elastic state
弹性自由度 elastic degree of freedom
弹性阻抗 elastic impedance; elastic reactance
弹性阻力 elastic resistance; elastic resisting force
弹性阻尼 elastic damping
弹性组合物 elastic composition
弹性组件 elastic modulus
弹性作用 elastic reaction
弹性坐垫 seat cushion
弹性座位架 cushion frame
弹性座 spring support
弹絮机 batting machine
弹压齿式离合器 snap clutch
弹压导板 spring guide plate
弹压式脱粒机 spring sheller

痰

痰盂 spittoon; cuspidor <美>

潭

潭 pot-hole; deep pool; pond
潭泉 pool spring

檀

檀香(木) santal; sandalwood; Santalum album
檀香(木)油 sandalwood oil
檀香属 sandalwood

坦

坦白 self-disclosure
坦泊土 dambo soil
坦查洛依铝锌铸造合金 Tenzaloy
坦登串联式烘干机粉磨设备 Tandem drying grinding plant
坦度比 flatness ratio
坦拱 flat arch; shallow arch
坦拱桥 flat arch bridge; shallow arch bridge
坦卡维尔桥 <法国> Tancarville Bridge
坦克车 tanker
坦克登陆艇 landing craft; landing tank
坦克履带 tank track
坦克推土机 tank dozer
坦克用起重器 tank jack
坦克用千斤顶 tank jack
坦克运输挂车 tank trailer
坦姆扑拉克卷材 Temperac quilt
坦纳格拉无釉陶塑人像 Tanagra figure
坦能贝格圬工砌筑法 Tannenberg masonry bond
坦皮科湾盆地 Tampico embayment
坦皮赖克示温器 tempilac
坦坡风 flat trough
坦普拉斯特 <几种建筑材料的商品名称> Templaster
坦普莱特铅锡铋镉易熔合金 Temperite alloy
坦桑尼亚 <非洲> Tanzania
坦桑尼亚陆核 Tanzania nucleus
坦斯提尔 <一种密实高熔点合金钢> Tannsteel
坦索维克 <一种浸透树脂的加强木材> Tensovic
坦圆拱 depressed arch
坦赞铁路管理局 Tanzania-Zambia Railway Authority

钽

钽板 tantalum sheet
钽铋矿 bismutotantalite
钽测辐射热计 tantalum bolometer
钽衬 tantalum lining
钽电容器 tantalum capacitor
钽覆盖层 tantalum coating
钽钙矿 rynersonite
钽管 tantalum tube
钽黑稀金矿 tanteuxenite
钽夹钳 tantalum clips applying forceps
钽介质电容器 tantalum capacitor
钽金红石 strueverite; tantalorutile; tantalum rutile
钽矿 tantalum ore
钽铝青铜 tantalum bronze
钽铝石 simpsonite
钽锰矿 manganotantalite
钽模 <热压用> tantalum die
钽膜电阻器 tantalum film resistor
钽钠石 rankamaite
钽铌铁钇矿 khlopinite
钽镍铬合金工具钢 Tantal tool alloy
钽烧绿石 microlite; tantalpyroclore
钽石 tantite
钽铈钇矿 loranskite
钽丝灯 tantalum lamp
钽酸 tantalic acid
钽酸钾 potassium tantalate
钽酸铅 lead tantanate
钽酸盐 tantalate
钽钛离子泵 tantalum-titanium ion pump
钽锑矿 stibiotantalite
钽铁 ferro-tantalum
钽铁矿 ferrotantalite; tantalite
钽铁矿含量 ferrotantalite
钽铁矿矿石 tantalite ore
钽锡矿 thoreaulite
钽锡石 tantalian-cassiterite
钽钇矿 yttrotantalite
钽钇易解石 tantalaeschynite-(Y)
钽铀烧绿石 tantalo hatchettolite
钽整流器 tantalum rectifier

毯

毯层 carpet veneer
毯式投料 blanket feed
毯式投料机 blanket batch charger; blanket feeder
毯式压气浮选机 blanket-type pneumatic machine
毯型隔水层 drainage blanketing
毯状 blanketlike
毯状绝热材料 blanket-type insulant
毯状绝缘材料 blanket-type insulant
毯子 rug

炭

炭 charcoal; vegetable charcoal
炭斑 <黑心> black core; black heart
炭板 carbon slab
炭棒 carbon rod; crayon; kryptol
炭棒传声器 carbon-stick microphone
炭棒灯 carbon arc lamp
炭棒电阻炉 Kryptol stove
炭棒熔融法 graphite-rod melting method
炭笔 charcoal crayon; charcoal pencil; scorch-pencil; score-pencil
炭笔画 charcoal drawing
炭材 charcoal wood; raw material of wood-charcoal
炭尘 carbon dust
炭尘沉降 ash fall
炭尘肺 anthracosis; melanedema
炭沉积 carbon laydown
炭的百分含量 content of carbon
炭的摩擦损失 abrasion loss of carbon
炭电极 charcoal electrode
炭电阻温度计 carbon resistance thermometer
炭额度 carbon credit
炭肺病 anthracosis
炭粉 carbon powder; pounce
炭粉电阻器 cracked-carbon resistor
炭腐病 charcoal rot
炭复制器 carbon replica
炭管膜曝气生物反应器 carbon tube membrane-aerated biofilm reactor
炭过滤器 anthrafilter

炭黑 black pigment; carbon black; carbon ink; coom; gas black; smoke black; soot carbon
炭黑厂 carbon black factory
炭黑尘肺 carbon black pneumoconiosis; silicate of carbon <炭黑的别名>
炭黑法底色 carbon black undertone
炭黑防光晕层 rem-jet
炭黑分散体 carbon black dispersion
炭黑工业 carbon black industry; off-colo(u)r industry
炭黑混合物 black stock
炭黑计数器 black carbon counter; Libby counter
炭黑浆状液 black slurry
炭黑凝胶复合体 carbon black gel complex
炭黑生产过程 carbon black process
炭黑水分散体 black slurry
炭黑填充量 black loading
炭黑涂料 charcoal blacking
炭黑橡胶 black rubber
炭黑颜料 colo(u)r black
炭弧 carbon arc
炭弧(切)割 carbon arc cutting
炭化 carbonization; char(k); charry; gas carburizing
炭化斑 charred spot; charring spot
炭化层 charring layer
炭化防腐 charring
炭化钙石灰 by-product lime
炭化骨 charred bones
炭化红壤 podzolized red earth
炭化炉煤气 retort gas
炭化喷油处理 furnos treatment
炭化烧蚀材料 charring ablative material
炭化烧蚀体 charring ablator
炭化速率 charring rate
炭化钨 tungsten carbide tool
炭画 charcoal drawing
炭环 carbon ring
炭环压盖 carbon ring gland
炭火蓝 kangri
炭基金 carbon fund
炭极 baked carbon
炭极电池 carbon cell
炭极电弧焊 electronic tornado welding
炭极弧割 carbon arc cutting
炭极弧焊 carbon arc weld(ing); electronic tornado welding
炭角菌属 <拉> Xylaria
炭精 black lead; graphite; plumbago
炭精棒 carbon electrode; carbon stick
炭精避雷器 carbon arrester[arrestor]
炭精电弧 carbon arc
炭精电极 carbon electrode
炭精电极炉 carbide furnace
炭精盒 carbon capsule; carbon chamber
炭精夹 carbon clip
炭精块 carbon block
炭精膜(片) carbon diaphragm
炭精气封圈 carbon ring gland
炭精铅笔 charcoal pencil
炭精式换能器 carbon transducer
炭精振动膜片 carbon diaphragm
炭阱 charcoal trap
炭疽病 anthrax
炭疽杆菌 bacillus anthracis
炭沥青质 anthracoxene
炭粒 carbon granule; granule of carbon
炭粒电阻炉 kryptol furnace
炭粒炉 kryptol furnace
炭粒凝集试验 charcoal agglutination test
炭粒送话器电阻的周期性小变化 breathing of microphone
炭炉 carbon furnace
炭滤池 carbon filter; charcoal filter
炭滤器 charcoal filter
炭密封 carbon-type seal
炭膜 carbon diaphragm; carbon film
炭膜电阻 carbon resistance film
炭膜电阻器 deposited carbon resistor
炭膜曝气生物膜反应器 carbon membrane-aerated biofilm reactor
炭末 carbon powder
炭末沉着病 anthracosis
炭末沉着性变形 anthracotic deformity
炭末电阻 carbon dust resistor
炭末石末沉着病 anthracosilicosis
炭末炸药 amidpulver
炭墨 carbon ink; soot
炭凝试验 charcoal agglutination
炭片 carbon plate
炭屏 carbon screen
炭铅铀矿 widenmannite
炭青质 carboids
炭砂滤池 carbon-sand filter
炭刷 carbon brush
炭刷握 carbon holder
炭刷组 brush assembly
炭丝灯 carbon lamp
炭素厂 charcoal ink factory
炭素电极 carbon electrode
炭素普通工具钢 carbon ordinary tool steel
炭素印相法 carbon pigment printing
炭素印相纸 carbon paper
炭素制品 carbon product
炭酸盐碱度 carbonate alkalinity
炭团菌属 <拉> Hypoxylon
炭吸附测定 charcoal test
炭吸附测定法 charcoal test method
炭吸附法 char adsorption; charcoal adsorption process
炭吸附剂 carbon sorbent
炭吸附作用 char absorption
炭屑 carbon dust; coke breeze
炭屑检波器 carbon coherer
炭窑 charcoal kiln
炭铀矿 joliotite
炭黝铜矿 binnite
炭毡 carbon-felt
炭纸 <传真电报收录用> teledeltos paper
炭质 carboneous
炭质避雷器 carbon arrester[arrestor]
炭质变阻器 carbon rheostat
炭质残渣 oil-carbon sludge
炭质电阻噪声 carbon noise
炭质堆积的 cumulose
炭质放电器 carbon arrester[arrestor]
炭质滚轮 carbon roller
炭质褐煤 coaly lignite
炭质活塞环 carbon piston ring
炭质接点 carbon contact
炭质开关接点 carbon switch contact
炭质沥青 carbene
炭质耐火材料 carbon refractory
炭质泥岩包裹体 carbonaceous mudstone inclusion
炭质页岩 bone; carbonaceous shale; criggling; danby; macker; rattle jack
炭质组 anthrinoid group
炭砖 brick fuel; coke briquette
炭砖炉衬 carbon line

探棒 probe; sonde; sounding pole; test rod

探边井 extension well; outpost well
探波针 demodulator probe
探采对比图 comparative map of exploration and exploitation data; correlation map between exploration and exploitation data
探采工作失调 incoordination from of prospecting by mining
探采工作协调 coordination from of prospecting by mining
探采结合勘探 drilling-mining combine exploration
探采结合孔 exploration and developing (bore) hole; exploration and developing drill-hole
探采资料对比法 contrast method of between exploration and mining data
探槽 exploration trench; exploratory trench; prospecting trench; prospect pit; test trench; trial trench
探槽编录 channel documentation
探槽采样 trench sampling
探槽地质编录 geologic(al) documentation of exploratory trenches
探槽工作面 face of the channel
探槽剖面图 profile of exploratory
探槽数 number of exploratory trench
探槽素描图 sketch map of exploratory trench; sketch of trench
探槽原始地质记录 trenching initial geologic(al) logging
探测 detection(survey); electromagnetic subsurface probing; exploration survey; exploratory survey; finding; localization; position finding; pricking; probing (survey); sounding; acquisition
探测标杆 pricker staff; probing staff
探测表面 searching surface
探测不到的废气 non-detectable emissions
探测部件 exploring block
探测层 detecting layer
探测程序 acquisition processing; locater[locator]
探测传感器 acquisition sensor
探测船 recording ship; recording vessel; sounding vessel; exploration ship
探测锤 sounding bob
探测磁棒 magnetic field balance
探测带宽 detective bandwidth
探测的 exploratory
探测地球物理学 exploration geophysics
探测电极 sounding electrode
探测电流 probe current; searching current
探测电路 detection circuit
探测断面 sounding profile
探测范围 acquisition range; detection limit; detection range
探测范围图 operational coverage diagram
探测方式 acquisition mode
探测方位 direction bearing
探测杆 dowsing rod; feeler; ga(u)ge rod; proof stick
探测工具 instrumentation tool
探测管 exploring tube; sound pipe
探测和识别系统 intelligence system
探测火箭 probe rocket; probe vehicle; sounding rocket
探测机 explorer
探测极限 detection limit
探测技术 probing technique
探测剂 probe
探测角控制 search-angle control
探测阶段 detection phase; probing step
探测进程 acquisition processing
探测井 prospecting borehole; sounding well; wildcatting <石油>
探测距离 detection range; probing distance; range of detection; range of detector
探测孔 exploration hole; try hole
探测缆索器 cable detector; cable locator
探测灵敏度 detection sensitivity; detectivity
探测率 detectivity
探测轮 calibration wheel
探测脉冲 direct impulse; direct pulse; main pulse; sounding impulse
探测模块 monitor zone adapter module
探测目标最小通量密度 minimum acquisition flux density
探测能力 detectability; detectivity
探测片 foil; foil detector
探测片计数 foil count
探测片支架 foil holder; foil support
探测片支架杆 foil holding rod
探测气球 registering balloon; sounding balloon
探测气体 probe gas
探测器 detecting device; detecting element; detector; detector cell; detector tester; explorer; finder; localizator; localizer; locator; probe(r); prover; searcher; sensor; sonde; sounder; survey meter
探测器材料 detector material
探测器窗 window of a detector
探测器的灵敏体积 sensitive volume of a detector
探测器分布 <用于地震研究中> detector spread
探测器覆盖率 detector coverage
探测器类型 detector type
探测器灵敏度 sensitivity of detector
探测器灵敏面积 detector area
探测器扫描全息图 scanned detector hologram
探测器试场 detector field
探测器透镜 finder lens
探测器外壳 probe body; probe housing
探测器线圈 probe coil
探测器效率 detector efficiency
探测器阈值灵敏度 detector threshold sensitivity
探测器噪声滤波器 detector noise discrimination
探测器阵(列) detection array; detector array
探测器止回阀 detector check valve
探测器止回阀的水表 detector check valve meter
探测器状况监视 detector condition monitoring
探测区域 search coverage
探测曲线 detection curve
探测缺陷光电装置 aniseikon
探测人 prober
探测设备 detecting device; detection equipment; detector; exploration equipment; prospecting apparatus; senser [sensor]; sounding equipment; diving device <探查地下水用的>
探测设备舱 probe compartment
探测深度 exploration depth
探测时间 detection time; exploration time
探测时间常数 detective time constant
探测速记图 sounding tachygraph
探测隧道 exploratory tunnel
探测索 log line
探测天线 exploring antenna; search antenna
探测头 detecting head; detector head
探测图 exploration map
探测望远镜 telescopic(al) finder
探测卫星 acquisition satellite
探测物质 detecting material
探测系统 detection system

探测系统传感器 detecting instrument
探测线圈 director coil;explorer;exploring coil;magnetic test coil
探测线圈测试法 search coil test
探测线圈传感器 search coil transducer
探测效率 detection efficiency
探测信号 detected signal
探测性技术预测 exploratory forecast
探测性研究 exploratory study
探测压力 detection pressure
探测研究 research findings
探测仪(器) detecting instrument;detector;fathometer;prospecting instrument;survey meter
探测用磁秤 magnetic field balance
探测与报警系统 detection and alarm system
探测与测距声呐 detecting-range sonar;sonar detecting-ranging set
探测元件 detecting element;detector;detector element;exploring element
探测员 explorer
探测者 explorer
探测针 probe
探测指数 detection equipment;detection index
探测装置 detection device;detecting device;detector;search unit
探测钻孔 exploration boring;prospecting borehole
探查 ascertain;exploration;probe;probing;prospect(ing);search
探查抽样 exploratory sampling
探查的 prospecting
探查灯 probe lamp;trouble lamp
探查点 exploration point
探查电极 exploring electrode
探查法 equipotential method
探查分类 exploration-classification
探查坑 prospect pit
探查坑道 exploration gallery;prospecting tunnel;survey tunnel
探查孔 exploring drill;feeder hole;trial boring
探查试验 exploratory test
探查术 exploratory operation
探查竖井 exploration shaft
探查性监测计划 detection monitoring program(me)
探查仪 searcher
探查指令 look-up instruction
探察电流 searching current
探察计划 prospecting schematization
探察井 prospecting shaft
探尺 stock rod;trial rod
探出 smell
探触器 feeler wire
探锤 prospecting hammer
探磁圈 magnetic probe
探地雷达 ground penetrating radar
探地雷达法 ground probing radar method;georadar
探洞 exploratory drift;exploratory heading;exploratory tunnel;investigation hole;trial heading;trial tunnel
探洞开挖 exploratory heading excavation;exploratory tunnel excavation;trial heading excavation;trial tunnel excavation
探杆 feeler lever;probe rod
探杆测 rod test
探钩 feeler;finger pin
探管 searching tube;sonde
探管传声器 probe microphone
探管仪 pipe finder;pipe locator
探管锥套式 probe-drogue
探管锥套系统 probe-drogue system
探海钩 creeper
探海锚 admiralty creeper
探海球 bathysphere
探海艇 bathyscaph(e);creeping boat
探火器 fire detector
探火装置 fire detecting arrangement
探获率 detection rate
探及深度 accessible depth
探计式读出 pin sensing
探价 enquiry[inquiry]
探井 bore pit;conduit pit;discovery bore;discovery boring;exploratory pit;exploratory shaft;exploratory well;inspection pit;inspection shaft;inspection well;investigation pit;investigation shaft;manhole;open test pit;prospect hole;prospecting shaft;prospect pit;test pit;test shaft;trial hole;trial pit;trial shaft;ascertaining well <探查井深与井径的工序>【岩】
探井壁 manhole wall
探井编录 prospect pit documentation
探井费用 exploration well cost
探井盖 manhole cover;manway cover
探井原始地质编录 initial geologic(al) logging in exploratory well
探究 exploration
探究反射 investigatory reflex
探究反应 investigatory reaction
探究行为 exploratory behavio(u)r
探勘图 exploration map
探勘者 explorer
探坑 costeaning pit;exploration trench;exploratory drift;exploratory pit;exploratory shaft;inspecting pit;inspection pit;inspection shaft;investigation pit;open test pit;pit;prospect pit;test boring;test pile;test pit(ting);trial boring;trial hole;trial holing;trial pit
探坑取样 pit sampling
探坑取样器 pit sampler
探坑展示图 test pit unfolding graph
探空 sounding;air sounding
探空火箭 probe;rocket sounder;sounding rocket
探空气球 sounding balloon
探空气球测风 rabal
探空气象仪 sounding meteorograph
探空仪 sonde
探空仪观测 radio sonde observation;raob
探空仪气球 radio sonde balloon
探空装置 sounder;sounding unit
探孔 discovery bore;discovery boring;exploratory hole;handhole;inspecting hole;investigation hole;probe hole;prospect(ing) hole;proving hole;trail bore-hole;trial hole
探孔垫片 handhole cover gasket
探孔门 handhole door
探矿 mineral exploration;mineralize;mineral prospecting;mining prospecting;ore-search;prospecting
探矿工程 exploration engineering;prospecting work
探矿工程平面图 prospecting engineering planimetric(al) map
探矿工程资料 data of mineral exploration engineering
探矿机 explorer
探矿技术低 exploration lower technology
探矿技术高 exploration higher technology
探矿技术研究 explorative technologic(al) research
探矿坑道 exploration drift;exploration tunnel
探矿坑道断面 cross-section of exploration tunnel
探矿坑道类型 type of exploration tunnel
探矿平洞 exploration adit
探矿平巷 prospect tunnel
探矿器 douser;mine detector
探矿权 mineral exploration right
探矿闪烁辐射计 scintillator prospecting radiation meter
探矿设备 divining device
探矿天井 exploratory raise
探矿条 twig
探矿巷道 prospecting drift
探矿辛迪加 search syndicate
探矿仪 mine locator;prospecting meter
探矿用音响指示器 prospecting audio-indicator
探矿杖 divining rod
探矿者 mineralizer
探矿准备金 mine prospecting reserves
探矿钻机 drill for quarrying
探雷器 mine detector
探雷声呐 mine detecting sonar
探料尺 ga(u)ge rod;ga(u)ging rod
探料尺装置 stock line indicator
探漏器 detectaphone;leak detector;leakiness detector;leak localizer
探漏试验 immersion test
探漏仪 leakage detector;leak detector;leakiness detector
探路灯 spotlight
探路航行 feeling walk;feel the way
探锚 anchor drag;grapnel;sweeping anchor
探明 ascertain;ascertainment;glean
探明储量 demonstrated reserves;explored reserves;known reserves;proven reserves
探明的填料储量 proven fill reserves
探明面积 proved area
探明资源 demonstrated commodities
探摸检查 diving inspection before salvage operation
探摸丈量 measurement of wreck dimension
探木钻 increment borer
探潜器 anti-submersible detection indicator;asdic
探亲假 home leave
探亲假旅费 travel on home leave
探求 disquisition;look-up
探求顾客需要 detecting customer's needs
探区雷达 earth-detective radar
探热器 heat probe
探砂面 detecting sand level
探伤 defect detection;detection of defects;fault detect(ion);flawmetering
探伤标准 detection standard
探伤部位图 drawing of position to be detected
探伤地点 place of examination
探伤定位 flaw location
探伤法 crack detection;defectoscopy;flaw detection;detectoscope
探伤范围 inspected area
探伤方法 method of detection;test method
探伤回波 defect echo
探伤记录 detection record
探伤技术 faulty technology
探伤检查 defect detecting test
探伤结果 result of detection
探伤镜 inspectoscope
探伤灵敏度 flaw detection sensitivity
探伤能力 flaw detectability
探伤器 crack detector;detector;detectoscope;fault detector;fault finder;fault indicator;faulty indicator;flaw detector;scanner
探伤器波形图 reflectogram
探伤设备 detecting equipment
探伤试验 test for fault
探伤图 defectogram;reflectogram
探伤仪 crack meter;defect detector;deflectoscope;detector of deflects;detectoscope;fault detector;fault finder;flaw detector;flawmeter
探伤仪灵敏度 sensitivity of tester
探伤转轮机构 detecting turning wheel mechanism
探深杆 feeler
探深管 <探测压浆混凝土中灰浆高度用> sounding well
探深线 line of sounding
探声棒 sound probe
探声管 acoustic(al) probe;aural probe;sound probe
探声器 sonoprobe;sound probe;sound probe sonoprobe
探声头 sound probe
探示器接口 prober interface
探视检查 invasive test(ing)
探视孔 spyhole
探视器 probe;view finder
探视取景器 contour finder
探试 probe;test
探试程序 heuristic program(me)
探试的 heuristic
探试法 heuristic method;heuristics;heuristic technique
探试非抢先算法 heuristic non-preemptive algorithm
探试函数 heuristic function
探试技术 heuristic technique
探试器测试 probe test
探试搜索 heuristic search
探试算法 heuristic algorithm
探试修剪 heuristic pruning
探试者 feeler
探水侧孔 flank hole
探水杆 watching stick;water diving rod
探水警戒线 warning line of water probing
探水设备 divining device
探水树杈 divining rod
探水温器 bathythermograph
探水线 line of water probing
探水者 water finder
探水钻孔 borehole of water probing
探梭器 shuttle detector
探梭指 shuttle feeler
探索 burrow;exploration;finding;probing;research;searching
探索臂 <可伸入棚车内装卸货物> searcher jib
探索布局 search plan
探索长度 search length
探索程序 heuristic program(me);heuristic routine
探索次序 search sequence
探索调查 exploratory survey
探索法 exploratory method;heuristic method
探索工作 search work
探索光点 exploring spot
探索规划 heuristic programming
探索计划 prospecting scheme;search plan
探索角 search angle
探索脉冲 main bang
探索气体 search gas
探索器 explorer;survey meter
探索试验 exploratory experiment
探索问题 <优选法> search problem
探索线圈 exploring coil

T

探索线圈磁强计 search coil magnetometer
探索性的 heuristic;pathbreaking
探索性发展 exploratory development
探索性分析 exploratory analysis
探索性开发 exploratory development
探索性模型研究 pilot model study
探索性试验 investigative test; pilot test;scouting test;test by trial;trial test
探索性试验分析 trial-test investigation
探索性数据分析 exploratory data analysis
探索性研究 exploration investigation; pilot study; exploratory research; exploratory work
探索性研制 exploratory development
探索性预测 exploratory forecasting
探索性预测法 exploratory forecasting technique
探索研究 exploratory development; pilot study
探索优先法 search plan
探索预测技术 exploring forecasting technique
探索者 explorer;pathfinder
探讨 exploratory investigation; investigation;research
探讨设计方案 trial design
探讨性磋商 exploratory consultation
探讨性会谈 exploratory talk
探讨性协商 exploratory consultation
探条 bougie; dip rod; doodlebug; rod divining
探听器 listening device
探通术 sounding
探头 contact probe; detector; feeler; measuring element; measuring head;probe;searching unit;sonde
探头半径 probe radius
探头标定系数 coefficient of probe demarcation
探头测斜仪 probe inclinometer
探头插孔 probe insertion
探头窗口 probe window
探头定位器 probe positioner
探头法 probe method
探头计数器 probe counter
探头灵敏度 sensor sensitivity
探头屏蔽 probe shield
探头踏步板 tread return
探头线圈 probe coil
探头型振动传感器 sonde-type vibropickup
探头移位 probe-shift
探头与取样面间距 space of detector and sampling section
探头种类 type of probe
探头锥底面积 area of probe bottom
探途元素 pathfinder element
探途指标 pathfinder index
探土钻 sounding borer
探微技术 microprobe technique
探纬板 feeler plate
探纬运动 feeler motion
探(纬)针 feeler
探纬针杆 feeler lever
探温计 temperature detector
探险 expedition;exploration
探险车 roving vehicle
探险船 discovery ship;exploration ship
探险队 expedition
探险游乐场 adventure ground
探险者 explorer
探险者航天器 prospector
探险者卫星 Explorer Satellite
探向 direction finding
探向方位 direction bearing
探向接收机 loop receiver
探向罗盘 bearing compass
探向器 direction finder;sense finder
探向线圈 Helmholtz coil
探销式塞规 flash pin ga(u)ge
探穴 test boring;trial boring
探寻 nose;seek
探寻器 seeker
探寻装置 search unit
探询 inquisition;polling
探询表 polling list
探询电路 polling circuit
探询反应测验 inquiry test
探鱼顶 fish locating
探云雷达 cloud-detection radar
探照灯 beacon light;flood light;projector;search(ing) lamp; search(ing) light
探照灯安装座 searchlight flat mounting base
探照灯标 searchlight beacon
探照灯部队 searchlight unit
探照灯车 searchlight car; searchlight vehicle
探照灯的协同 searchlight co-operation
探照灯灯管 searchlight lamp
探照灯灯光保障 searchlight support
探照灯灯泡 projector lamp
探照灯发电机 searchlight generator
探照灯反射镜 searchlight mirror
探照灯反射器 searchlight reflector
探照灯防区 searchlight sector
探照灯分级 luminaire classification
探照灯固定叉架 searchlight mounting
探照灯光 projector light
探照灯光交叉 searchlight intersection
探照灯光束 searchlight beam
探照灯控制设备 searchlight control device;lady
探照灯雷达 searchlight radar
探照灯滤光镜 searchlight screen
探照灯排 searchlight platoon
探照灯汽车 searchlight truck
探照灯式声呐 searchlight sonar
探照灯式天线 searchlight antenna
探照灯式信号机【铁】searchlight signal
探照灯双轴挂车 searchlight tandem axle trailer
探照灯搜索 searchlighting
探照灯塔 beacon light tower
探照灯透镜 searchlight lens
探照灯信号 searchlight signal
探照灯掩蔽部 searchlight shelter
探照灯与声波定位器 searchlight and sound locator
探照灯照射的天空 searchlight canopy
探照灯照射区 searchlight belt
探照灯阵地 searchlight site
探照反射器 intensive reflector
探照式色灯信号 colo(u)r searchlight signal
探照式色灯信号机【铁】colo(u)r searchlight signal; searchlight signal;semaphore
探照式信号 searchlight type signal
探照式信号机【铁】searchlight type signal
探照式信号机构 searchlight signal head; searchlight signal mechanism;searchlight signal unit
探照头灯 far-reaching headlamp
探针 detecting probe; detector head; dipstick;explorer;feeler pin;feeling pin; needle; pecker; probe; probe handle; probe unit; proof stick; sonde;sounder;sounding pole;stylet;survey probe;wire probe
探针板 probe card
探针采样 probe sampling
探针测量 probe measurement
探针测试法 probe test method
探针插入 needle implant
探针触探强度 <混凝土> probe-strength
探针电导 probe conduction
探针电流 probe current
探针技术 probe technique
探针架 probe carriage
探针间距 probe spacing
探针检验 feeler inspection
探针耦合 probe coupling
探针膨胀仪 feeler dilatometer
探针器 prober
探针取样 probe sampling
探针绕射 probe diffraction
探针式电子管电压表 probe-type vacuum tube voltmeter
探针式读出 pin sensing
探针式高温计 probe-type pyrometer
探针式换纬装置 feeler filling changing device
探针式料位指示器 probe-type level indicator
探针式液位计 probe-type liquid level meter
探针台 probe station
探针探查 probing
探针天线 probe antenna
探针显微分析仪 probe-microanalyser [microanalyzer]
探针线圈 probe coil
探针形成系统 probe-forming system
探针型检漏器 probe-type leak detector
探针型探头 needle probe
探针衍射 probe diffraction
探针仪 probe machine
探针直径 probe diameter
探知 ascertain;ascertainment
探子 probe;sound
探钻 prospecting bore bit;prospecting drill;prospecting hit
探钻机 prospecting drilling machine

碳13 核磁共振 carbon-13 magnetic resonance

碳 14 carbon-14
碳 14 年代测定法 carbon-14 dating
碳 14 同位素年代测定法 carbon-14 dating
碳铵石 teschemacherite
碳板火花室 carbon-plates chamber
碳板架 carbon plate shelf
碳棒 carbon piece; crayon <弧光灯的>
碳棒灯 carbon lamp
碳棒电炉 carbon bar furnace
碳棒式话筒 carbon rod microphone
碳棒原子化器 carbon rod atomiser [atomizer]
碳背送话器 carbon back transmitter; carbon disc[disk] transmitter
碳钡矿 witherite
碳钡钠石 khanneshite
碳比 carbon ratio
碳比理论 carbon ratio theory
碳铋钙矿 beyerite
碳变阻器 carbon rheostat
碳丙铁 austenite
碳丙铁圈 <超声发生器的> austenite (generator) ring
碳材料 carbon material
碳草酸钙石 mourolite
碳测井 carbon log
碳测井曲线 carbon log curve
碳尘 carbon dust
碳尘消除剂 carbon remover
碳沉积 carbon deposit;carbon laydown
碳触点 carbon contact; carbon(electric(al)) contact
碳触点开关 carbon break switch
碳醇萃取物 carbon alcohol extract
碳大理石 lucullite
碳胆碱 carbocholine
碳氮比(率) carbon-nitrogen ratio; carbon to nitrogen ratio
碳氮共渗 carbonitriding; nicarbing; nitrocementation
碳氮共渗法 tufftride method
碳氮共渗钢 carbonitrided steel
碳氮化钚 plutonium carbonitride
碳氮化钢 carbonitrided steel
碳氮化(合)物 carbonitride
碳氮化合物夹杂 carbonitride segregation
碳氮化钛 titanium carbonitride
碳氮化钛涂层 titanium carbonitride coating
碳氮化物涂层 carbonitride coating
碳氮键 carbon nitrogen bond
碳氮磷比(率) carbon-nitrogen-phosphorus ratio
碳氮循环 carbon-nitrogen cycle
碳氮氧基 carbon-nitrogen-oxygen group
碳氮指数 carbon-nitrogen index
碳当量 carbon equivalent
碳当量仪 carbon equivalent meter
碳的 carbolic; carbonaceous; carbonic; carbonous
碳的百分数 percentage of carbon
碳的补充部分 make-up carbon
碳的载体气体 carbon carrier gas
碳碲钙石 mroseite
碳电池 carbon battery
碳电极 baked carbon; carbon electrode; carbon resistance rod; carbon resistor rod;carbon rod
碳电刷 carbon brush;starter brush
碳电阻 carbon resistance
碳电阻温度计 carbon resistance thermometer
碳定年 carbon dating
碳断路器 carbon breaker
碳堆 carbon pile
碳堆变压器 carbon pile voltage transformer
碳堆电压调整器 carbon pile volt(age) regulator
碳堆调压器 carbon pile regulator
碳堆稳压管 carbon pile regulator
碳堆稳压器 carbon pile regulator
碳堆压力传感器 carbon pile pressure transducer
碳沸腾期 carbon boil
碳分支 carbon branch
碳分子筛 carbon molecular sieve
碳粉 powdered carbon
碳粉传感器 carbon transducer
碳粉电气淬火炉 fire-cloud furnace
碳粉电阻 carbon dust resistance
碳粉含量 powdered carbon content
碳氟化合物 fluorocarbon
碳氟化物 carbon-fluoride
碳氟树脂 fluorocarbon resin
碳复型 carbon replica
碳复型法 carbon replica method
碳钙铋矿 beyerite
碳钙镁石 huntite
碳钙镁铀矿 swartzite
碳钙铈矿 calcioancylite
碳钙铀矿 zellerite
碳刚玉 carbo-corundum
碳钢 carburizing steel;case-hardened steel;common carbon steel;simple steel

碳钢钢筋 carbon steel reinforcement
碳钢管 carbon steel pipe
碳钢焊条 carbon steel covered electrode
碳钢及低合金钢容器 carbon and low-alloy steel vessels
碳钢局部索氏体化处理 Sandberg process
碳钢幕墙 carbon steel curtain wall
碳钢钻杆 carbon drill steel
碳钢钻钢 carbon drill steel
碳钢钻头 metal-carbide rock drill
碳膏 carbon paste
碳垢 carbonated deposit; carbon deposit
碳管电阻炉 graphite-tube resistance furnace
碳管炉 carbon shirt-circuiting furnace; carbon tube furnace
碳硅棒 Globar
碳硅电阻材料 Silit
碳硅钙石 scawtite
碳硅钙钇石 caysichite
碳硅碱钙石 caretonite
碳硅铝铅石 surite
碳硅砂 carborundum
碳硅石 moissanite
碳硅铈钙石 kainosite
碳硅钛钕钠石 tundrite-(Nd)
碳硅钛铈钠石 tundrite
碳硅砖 carbide brick
碳过滤器 carbon filter
碳含量 carbon content
碳含量高的 high-carbon
碳耗率 carbon consumption rate
碳和木炭 carbon and charcoal
碳核磁共振 carbon magnetic resonance
碳盒送话器 carbon back transmitter; carbon disc[disk] transmitter
碳黑 black carbon; carbon black; gas carbon
碳黑沥青 carbon black-asphalt
碳黑片 sootflake
碳黑填料 carbon black filler
碳黑涂料 charcoal blacking
碳黑颜料 carbon black pigment
碳弧 carbon arc
碳弧保护焊 shielded carbon arc welding
碳弧灯 carbon arc lamp
碳弧电焊 carbon arc welding
碳弧电极 carbon arc electrode
碳弧光灯 carbon lamp
碳弧焊焊把 carbon arc welding electrode holder
碳弧焊(接) carbon arc weld(ing)
碳弧聚光灯 arc spotlight; carbon arc spotlight; carbon arc lamp
碳弧气刨 carbon arc air gouging
碳弧切割 carbon arc cutting
碳弧切割金属 metal carbon arc cutting
碳弧硬钎焊 carbon arc brazing
碳糊 carbon paste
碳滑板切割打磨装置 cutting and grinding device for sliding carbon block
碳滑条 carbon strip
碳化 acieration; caking; carbonate; carbonify; carbonize; carbonizing; carbon pick-up; carburization; carburize; carburizing; charring
碳化布 carbonized cloth
碳化层 carburization zone
碳化叉齿 carbide-tipped tine
碳化厂 carbonization plant
碳化沉淀物 carbonated deposit
碳化程度 degree of carbonation
碳化出水 carbonization effluent
碳化处理 artificial carbonation; charring treatment
碳化二钽 tantalium carbide
碳化二钨 ditungsten carbide
碳化二亚胺 carbodiimide
碳化法 carbonization method
碳化钒 vanadium carbide
碳化反应 carburizing reagent
碳化腐蚀 graphitic corrosion
碳化钙 calcium acetylide; calcium carbide; carbide; carbide of calcium
碳化钙测型砂水分法 carbide method
碳化钙反射率 calcium carbide reflectivity
碳化钙石灰膏 by-lime putty
碳化钙石灰(油灰) carbide lime
碳化铬 chrome carbide; chromic carbide; chromium carbide
碳化铬陶瓷 chromium carbide ceramics
碳化工 carbonizer
碳化骨 carbonized bones
碳化硅 carbide of silicon; carbofrax; carbolon; carborundum; moissanite; samite; silicate of carbon; silicon carbide; silicide of carbon
碳化硅板 carbide of silicon slab; carbon silicide slab; silicon carbide slab
碳化硅棒 silicon carbide rod
碳化硅变阻器 silicon carbide varister
碳化硅刀片 silicon carbide cutting blade
碳化硅电炉 silicon carbide furnace
碳化硅电阻器 silicon-carbide resistor
碳化硅发光二极管 transitron
碳化硅反射率 silicon carbide reflectivity
碳化硅非线性电阻器 thyrite resistor
碳化硅粉 carborundum powder
碳化硅高级耐火材料 Silfrax
碳化硅管 carborundum tube
碳化硅合成物 silicon carbide composite
碳化硅灰浆 carbide of silicon mortar; carbon silicide mortar
碳化硅检波器 carborundum detector
碳化硅晶须 silicon carbide whisker
碳化硅晶须增强氧化铝切削工具 Al_2O_3-SiC whisker cutting tool
碳化硅颗粒增强铝合金复合材料 SiC particle reinforced alumin(i)um alloy composite
碳化硅可变电阻 SiC varistor
碳化硅磨料 silicon carbide abrasive
碳化硅耐火材料 carbon silicide refractory material; carborundum refractory; refrax; silicon carbide refractory
碳化硅耐火材料制品 silicon carbide refractory product
碳化硅耐水砂纸 silicon carbide waterproof paper
碳化硅切削片 silicon carbide cutting blade
碳化硅砂浆 silicon carbide mortar
碳化硅砂锯 carborundum saw
碳化硅砂轮 carbonundum wheel; silicon carbide grinding wheel
碳化硅砂纸 carborundum paper; silicone-carbide paper
碳化硅陶瓷 silicon carbide ceramics
碳化硅陶瓷材料 thyrite
碳化硅纤维 silicon carbide fiber[fibre]
碳化硅载体 carboround
碳化硅砖 carbide of silicon brick; carbon silicide brick; carborundum brick; silicon carbide brick; rubbing brick
碳化过程 carbonization
碳化核燃料 carbide nuclear fuel
碳化褐煤 lignite char
碳化黑 carbide black
碳化辉石 allagite
碳化火焰 carburizing flame; soft flame
碳化剂 carburetant; carburizer
碳化裂缝 carbonation crazing
碳化炉 carbide furnace; retort
碳化炉室 retort house
碳化铝 alumin(i)um carbide
碳化锰 manganese carbide
碳化钼 molybdenum carbide
碳化钼陶瓷 molybdenum carbide ceramics
碳化钠 sodium carbide
碳化铌陶瓷 niobium carbide ceramics
碳化浓度 depth of carbonation
碳化泡沫石灰混凝土 carbonated lime foam concrete
碳化喷涂处理 <木材防腐> charring-and-spraying treatment
碳化硼 boron carbide; norbide
碳化硼铝 boral
碳化期 carburizing cycle
碳化气 carbon gas
碳化器 carbonizer
碳化燃料 carbonized fuel
碳化热 heat of carbonization
碳化软木 carbonized cork
碳化三铁 triferrous carbide
碳化砂 carbonized sand; carbosand
碳化烧蚀材料 charring ablative material
碳化深度 carbonized depth; carbonation depth by phenolphthalein test <酚酞试验>
碳化石膏板 carbonized gypsum slab
碳化石灰混凝土 carbonated lime concrete
碳化石灰空心板 carbonated lime hollow-core slab
碳化石灰砂砖 carbonated lime sand brick
碳化石屑砖 carbonated rock rubbish brick
碳化试验 carbonized test
碳化室 carbonating chamber; carbonization chamber; retort
碳化铈 cerium carbide
碳化收缩 carbonize shrinkage; carbonation shrinkage
碳化塔 carbonizer
碳化钛 titanic carbide[TiC]; titanium carbide[TiC]
碳化钛烧结合金 titanium carbide sintering alloy; WZ alloy
碳化钛碳化钨复式碳化物 titanium carbide-tungsten carbide complex carbide
碳化钛-碳化钨-钴硬质合金 titanium carbide-tungsten carbide-cobalt hard alloy
碳化钛涂层 titanium carbide coating
碳化钛涂层刀片 TiC coated tip
碳化钛硬质合金 Himet
碳化钽 ramet; tantalous carbide; tantalum carbide
碳化钽合金 ramet
碳化钽陶瓷 tantalum carbide ceramics
碳化钽硬质合金 Ramet alloy
碳化铁 carburet(t)ed iron; ferric carbide; ferrous-carbide; iron carbide
碳化铁体 cementite
碳化钍 thorium carbide
碳化钍钨阴极 carbonized thoriated tungsten cathode
碳化温度 carburizing temperature
碳化污泥吸收剂 carbonization sludge adsorbent
碳化污水 carbonization effluent
碳化钨 <一种硬质合金> tungsten carbide; carbide tungsten; tungsten car
碳化钨冲击式凿岩 carbide percussion drilling
碳化钨冲击式钻眼 carbide percussion drilling
碳化钨弹心 tungsten-carbide core
碳化钨刀片 tungsten-carbide blade
碳化钨堆焊的(切削刃) fused tungsten carbide tipping
碳化钨粉末 tungsten-carbide powder
碳化钨敷层 tungsten-carbide coating
碳化钨复合材料 tungsten-carbide composite
碳化钨工具 tungsten carbide tool
碳化钨钴硬质合金 tungsten-carbide-cobalt hard-alloy
碳化钨合金 tungsten-carbide alloy
碳化钨合金补强的可拆卸式钻头 detachable tungsten carbide insert bit
碳化钨合金镶嵌块 tungsten-carbide tipped inserts
碳化钨合金钻头 tungsten-carbide alloy bit
碳化钨球 tungsten-carbide ball
碳化钨球齿钎头 tungsten-carbide bit
碳化钨陶瓷 tungsten-carbide ceramics
碳化钨镶片钎头 tungsten-carbide insert bit; tungsten-carbide tipped bit
碳化钨镶嵌镀片 tungsten-carbide insert
碳化钨镶嵌物 tungsten-carbide insert
碳化钨牙钻 tungsten-carbide bur
碳化钨硬质合金 carballoy; carboloy (metal); cemented tungsten carbide; diamondite
碳化钨硬质合金刀片 tungsten-carbide insert
碳化钨硬质合金钢钎 tungsten-carbide drill steel
碳化钨硬质合金截齿 tungsten-carbide tipped cutting teeth
碳化钨硬质合金一字形钻头 tungsten-carbide chisel
碳化钨硬质合金轧辊 tungsten-carbide roll
碳化钨硬质合金珠齿钻头 tungsten-carbide button bit
碳化钨硬质合金钻头 tungsten-carbide boring bit
碳化钨制品 tungsten-carbide composition
碳化钨柱齿钎头 tungsten-carbide button bit
碳化钨钻头 button bit; tungsten-carbide chisel
碳化钨钻头凿岩 tungsten-carbide drilling
碳化钨钻头钻眼 tungsten-carbide drilling
碳化物 carbide; carbonide; carburet
碳化物不均匀性级别 grade of inhomogeneity of carbide
碳化物促进剂 carbide promoter
碳化物带状组织 carbide lamellartiy
碳化物刀具 carbide cutter
碳化物刀头 carbide bit
碳化物合金 carbide alloy
碳化物灰 carbide ash
碳化物基金属陶瓷 carbide base(d) cermet
碳化物金属陶瓷 carbide cermet
碳化物类 carbon compounds
碳化物耐火材料 carbide refractory
碳化物耐火制品 carbide refractory product
碳化物耐磨镶块 carbide insert

碳化物偏析 carbide segregation
碳化物热还原法 carbide reduction process
碳化物陶瓷 carbide ceramics
碳化物退火 carbide annealing
碳化物稳定剂 carbide stabilizer
碳化物型金属陶瓷 carbide type cement
碳化物牙钻 carbide bur
碳化物阴极 carbonized cathode
碳化物硬质合金 cutanit
碳化物渣 carbide slag
碳化物铸造厂 carbide casting plant
碳化纤维 carbonized fibre[fiber]
碳化纤维素 carbonized cellulose
碳化锌 carbon zinc
碳化絮凝法 carbofloc process
碳化焰 carbonizing flame
碳化窑 carbonizing chamber
碳化钇 yttrium carbide
碳化铀 uranium carbide; uranium monocarbide
碳化周期 carbonizing period
碳化砖 carbide brick;carbonated lime brick
碳化装置 carbonization plant;carbonizing plant;carburizing apparatus
碳化作用 carbon(iz)ation;carbonification
碳化作用处理 carbonation treatment
碳还原 carbon reduction
碳还原法 carbon reducing method
碳环 carboatomic ring; carbocycle; homocycle
碳环的 carbocyclic
碳环化合物 carbocyclic(al)compound; isocyclic compound
碳环压盖 carbon gland
碳缓冲 graphite buffer
碳迹<玻璃制品缺陷> carbon mark
碳积 carbon deposit
碳基 carbon matrix
碳及其他物质 carbon along with other materials
碳极棒 carbon-point rod
碳极电池 carbon cell
碳极电弧 carbon arc
碳极断路器 carbon breaker
碳极弧割 carbon arc cutting
碳极弧焊 carbon arc weld(ing)
碳钾钙石 fairchildite
碳钾钠矾 hanksite
碳钾铀矿 grimselite
碳交换率 carbon exchange rate
碳胶<脱水剂> carbon cement;carbon paste;karbogel
碳接点 carbon contact
碳结构钢丝 carbon structural steel wire
碳结合耐火材料 carbon bonded refractory
碳金属键 carbon metal bond
碳精 carbon;plumbagine
碳精按钮 carbon button
碳精棒 carbon stick;carbon rod
碳精避雷器 carbon protector
碳精电极 baked carbon;carbon;carbon electrode
碳精粉 carbon powder
碳精盒 capsule;inset transmitter
碳精弧光灯 carbon arc lamp
碳精环 carbon ring
碳精块 carbon block;carbon segment
碳精粒 granular carbon
碳精砂 granular carbon
碳精式话筒"button" microphone
碳精刷 carbon brush
碳精送话器 carbon(granule)transmitter;carbon powder transmitter;granular carbon transmitter
碳精填密涵盖 carbon-packed gland
碳精轴封 carbon packing
碳聚乙烯充氧阴极 carbon-polytetrafluoroethylene O_2-fed cathode
碳块 carbon block
碳块避雷器 carbon block protector
碳块炉底 carbon bottom
碳蜡 carbowax
碳离子聚合作用 carbonium ion polymerization
碳沥青<含固定碳 97.2%> anthraxolite
碳沥青质 anthracoxene
碳粒 carbon grain;carbon granule
碳粒沉积室 carbon settling chamber
碳粒传感器 carbon transducer
碳粒传声器倾斜角特性 positional response of carbon microphone
碳粒发热管<石墨管> Tammann tube
碳粒发热体电炉 Arsem furnace
碳粒接触拾音器 carbon contact pickup
碳粒凝合检波器 granular coherer
碳粒凝集反应 carbon granular agglutination
碳粒清除试验 carbon clearance test
碳粒拾音器 carbon contact pick-up
碳粒式传声器 carbon grain microphone; granular microphone; granule microphone; powder microphone
碳粒式话筒 carbon grain microphone; granular microphone; granule microphone; powder microphone
碳粒送话器 carbon(granule)transmitter;carbon powder transmitter;granular carbon transmitter
碳粒噪声 carbon noise
碳粒转移记录 carbon transfer recording
碳粒子 carbon particle
碳链 carbochain;carbon chain
碳链分解作用 desmolysis
碳链异构 carbon chain isomer
碳磷灰石 carbonate-apatite
碳磷锰钠石 sidorenkite
碳鳞 kish carbon
碳卤树脂 halocarbon resin
碳卤塑料 halocarbon plastic
碳滤池 carbon filter;char filter
碳滤器 char filter
碳氯仿萃取(法) carbon chloroform extraction
碳氯仿萃取液 carbon chloroform extract
碳镁钙石 borvarite
碳镁铬矿 stichtite
碳镁铀矿 bayleyite
碳锰合金钢<低锰合金钢> carbon-manganese steel
碳模制 carbon replication
碳膜电位器 carbon-film potentiometer
碳膜电阻 carbon;carbon film resistance
碳膜电阻器 carbon-film resistor
碳末 carbon powder
碳末颗粒 carbon particle
碳末润滑剂 aquadag
碳钠矾 burkeite
碳钠钙铝石 tunisite
碳钠钙铀矿 andersonite
碳钠铝石 dawsonite
碳钠镁石 eitelite
碳黏[粘]泥 carbon cement
碳镍合金 nickel-carbon alloy
碳镍化钛合金陶瓷 titanium carbide-nickel cermet
碳钕石 lanthanite-(Nb)
碳排 carbon discharge
碳盘传声器 carbon disk[disc] microphone
碳盘电极 carbon disk[disc] electrode
碳硼电阻器 borocarbon resister[resistor]
碳硼钙镁石 sakhaite
碳硼硅镁钙石 harkerite
碳硼锰钙石 gaudefroyite
碳硼烷 carborane
碳硼烷甲基硅酮 carborane methyl silicone
碳片避雷器 carbon plate lightning protector
碳平衡 carbon balance
碳气化期 carbon blow
碳迁移模式 model of carbon migration
碳羟磷灰石 carbonate-hydroxylaptite
碳青铜 carbon bronze
碳氢比 carbon-hydrogen ratio
碳氢氮元素分析仪 carbon-hydrogen-nitrogen analyser[analyzer]
碳氢黑 hydrocarbon black
碳氢化 hydrocarbonize; Nytron <硫酸钠清洁剂>
碳氢化合物 carbon hydrogen compound; hydrocarbon; hydrocarbon compound
碳氢化合物测量方法 hydrocarbon survey
碳氢化合物产品 hydrocarbon production
碳氢化合物的高温转化 pyrolytic conversion of hydrocarbons
碳氢化合物废物泥浆燃烧处理法 slurry-burning process for hydrocarbon waste
碳氢化合物分析仪 hydrocarbon analyser[analyzer]
碳氢化合物改质处理燃料电池 hydrocarbon reforming cell
碳氢化合物类似物 hydrocarbon analog(ue)
碳氢化合物硫酸钠清洁剂 Nytron
碳氢化合物燃料电池<烃燃料电池> hydrocarbon fuel cell
碳氢化合物(烃)溶剂 hydrocarbon solvent
碳氢化合物污染 hydrocarbon pollution
碳氢化合物制冷剂 hydrocarbon refrigerant
碳氢化合物中毒 hydrocarbonism
碳氢活化剂泡沫 hydrocarbon surfactant type foam
碳氢基层 hydrocarbon base course
碳氢基团分析 hydrocarbon group analysis
碳氢结合料 hydrocarbon binder; hydrocarbon binding material
碳氢聚合物 hydrocarbon polymer
碳氢可溶性染料 hydrocarbon soluble dyestuff
碳氢镁石 nesquehinite
碳氢钠石 wegscheiderite
碳氢黏[粘]合层 hydrocarbon binder film
碳氢黏[粘]结剂泵 hydrocarbon binder pump
碳氢黏[粘]结剂喷雾机 hydrocarbon binder spraying machine
碳氢黏[粘]结剂融化壶 hydrocarbon binder melting kettle
碳氢黏[粘]结剂蒸煮器 hydrocarbon binder cooker
碳氢气体 hydrocarbon gas
碳氢酸 hydrochloric acid
碳氢氧比例 carbon-hydrogen-oxygen ratio;C-H-O ratio
碳氢氧相图 carbon-hydrogen-oxygen diagram;C-H-O phase diagram
碳氢有害排出物监测器 hydrocarbons monitor
碳氢原子比 atomic ratio of carbon to hydrogen
碳圈密封 carbon ring
碳燃烧 carbon burning
碳热法炼铝 carbothermic smelting of aluminum
碳热还原 carbon thermal reduction
碳热还原法 carbothermic method; carbothermic process
碳绒铜矾 carbonate-cyanotrichite
碳闪 carbon flash
碳石墨纤维 carbon graphite fibre[fiber]
碳石英岩 carbon quartz
碳势 carbon potential
碳势计 carbon potential meter
碳铈钙钡石 ewaldite
碳铈镁石 sahamalite
碳铈钠石 carbocernaite
碳树脂电阻器 carbon-resin resistor
碳刷 brush; carbon brush; carbon shoe;starter brush
碳刷柄 brush holder;carbon holder
碳刷盒 brush box
碳刷架 brush holder;carbon holder
碳刷罩【电】carbon brush cover
碳刷组 brush assembly
碳水化合物 carbohydrate; carbonaceous compound;carbon hydrates
碳水化合物供给量 carbohydrate allowance
碳丝 carbon filament
碳丝灯 carbon filament lamp
碳丝原子化器 carbon filament atomizer
碳锶矿 strontianite
碳素材料 carbonaceous material;carbons
碳素测定年代【地】carbon-date
碳素尘肺 carbonaceous pneumoconiosis
碳素衬垫 carbon packing
碳素电极 carbon pole
碳素电阻体 carbon resistor block
碳素垫料 carbon packing
碳素发热材料 carbon heating element
碳素钢 carbon steel;plain carbon steel; plain steel
碳素钢板(和型材) carbon steel plate;shapes and bars
碳素钢管 carbon steel tubing
碳素钢轨 carbon steel rail
碳素钢条 plain steel bar
碳素钢钻头 carbon steel bit
碳素工具钢 carbon tool steel; plain carbon tool steel
碳素护板 carbon backplate
碳素机械零件 carbon component for mechanical engineering
碳素结构钢 carbon construction(al) steel;carbon structural steel
碳素结构圆钢 carbon construction(al)steel round
碳素晶格 carbon lattice
碳素耐火材料 carbonaceous refractory;carbon refractory
碳素耐火砖 carbon brick;carbon refractory block
碳素黏[粘]结剂 carbon cement
碳素砂 carbon sand
碳素弹簧钢 carbon spring steel
碳素填料 carbon packing
碳素填料环 carbon filler ring
碳素铁 carbon iron
碳素同化作用 carbon assimilation
碳素涂底 carbon lining

碳素涂料 black wash;wet blacking
碳素物 carbonaceous material;carbonaceous matter
碳素像纸 carbon tissue;industrial tissue
碳素循环 carbon cycle
碳素印刷 carbonizing
碳素印相纸 carbon tissue
碳素营养 carbon nutrition
碳素纸 carbon tissue
碳素纸过版机 carbon tissue transfer machine
碳素纸式记录器 carbon paper type recorder
碳素质堆积层 cumulose deposit
碳酸 acidum carbolicum; carbonic acid
碳酸铵 ammonium carbonate;sal volatile
碳酸铵液 ammonium carbonate solution
碳酸胺 amine carbonate
碳酸包裹体 carbonic acid inclusion
碳酸饱充用气 carbonation gas
碳酸饱和 carbonation
碳酸饱和处理 carbonation treatment
碳酸饱和作用 carbonatization
碳酸钡 barium carbonate
碳酸钡钙矿 bromlite
碳酸钡矿 witherite
碳酸钡中毒 barium carbonate poisoning
碳酸丙二酯 propylene carbonate
碳酸丙烯 propylene carbonate
碳酸丙烯酯 propylene carbonate
碳酸(测定)计 calcimeter
碳酸测定仪 calcimeter
碳酸次乙酯 ethylene carbonate
碳酸定量 alkalimetry
碳酸定量法 carbometry;carbonometry
碳酸定量计 alkalimeter
碳酸定量器 kalimeter
碳酸二苯胍 diphenyl guanidine carbonate
碳酸二苯酯 diphenyl carbonate
碳酸二甲酯 dimethyl carbonate
碳酸二乙酯 diethyl carbonate; ethyl carbonate
碳酸二正丁酯 di-n-butyl carbonate
碳酸法 carbonic acid method
碳酸复红 carbolfuchsin
碳酸钙 aragonite;calcium carbonate; carbonate of lime;chemical chalk; clef-non; lime carbonate; lime rock;limestone
碳酸钙饱和度 calcite saturation(level);calcium carbonate saturation
碳酸钙保护层 calcium carbonate protective scale
碳酸钙钡矿 alstonite
碳酸钙补偿深度 calcium carbonate compensation depth
碳酸钙层 calcium carbonate layer
碳酸钙沉淀 precipitation of calcium carbonate
碳酸钙沉淀的 lime-depositing; lime-precipitating;lime-secreting
碳酸钙沉淀势 calcium carbonate precipitation potential
碳酸钙沉淀物 lime deposit
碳酸钙沉积物 lime deposit
碳酸钙当量 calcium carbonate equivalent
碳酸钙的半水化合物 calcium sulphate hemihydrate
碳酸钙粉 Paris white
碳酸钙粉末 calcium carbide powder
碳酸钙胶结物 calcium carbonate cement
碳酸钙结晶 calcium carbonate crystal
碳酸钙累积层 Ca-horizon
碳酸钙氯化铵烧结 sintering with mixture of $CaCO_3$ and NH_4Cl
碳酸钙镁 calcium magnesium carbonate
碳酸钙镁矿 huntite
碳酸钙镁石 huntite
碳酸钙水泥 calcium sulfate cement
碳酸钙填料 pearl filler
碳酸钙小结核 race
碳酸钙絮凝物 calcium carbonate floc
碳酸钙盐 polcard
碳酸钙油性填孔料 mineral white primer
碳酸钙制品 lime product
碳酸酐 carbonic anhydride
碳酸根 carbonate radical
碳酸化 carbonating
碳酸化器 carbonator;carbonizer
碳酸化装置 carbonator
碳酸化作用 carbonation;carbonatization
碳酸计 calcimeter; carbacidometer; carbonometer
碳酸甲酯 methyl carbonate
碳酸钾 carbonate of potash; kalium carbonate; potash; potassium carbonate;sal tartari;salt of tartar
碳酸钾钙石 fairchildite
碳酸钾氯化钾混合物 kalinor
碳酸钾熔融 fusion with K_2CO_3
碳酸矿泉水 apollinaris water
碳酸矿水 acidulae
碳酸镭 radium carbonate
碳酸冷冻机 carbonic acid refrigerating machine
碳酸磷灰石 carbonate-apatite; dahllite;podolite
碳酸芒硝 hanksite
碳酸镁 carbonate magnesia; magnesite;magnesium carbonate
碳酸镁薄板 magnesite sheet
碳酸镁底层地板 magnesite subfloor
碳酸镁地板砖 magnesite flooring tile
碳酸镁钙 magnesium calcium carbonate
碳酸镁建筑板 magnesite building board
碳酸镁岩石 magnesite rock
碳酸锰 manganese carbonate
碳酸灭火器 soda-acid fire extinguisher
碳酸钠 carbonate of soda; natron(ite);salt soda;soda(salt);sodium carbonate; washing soda; yellow soda ash
碳酸钠的存在 sodium carbonates' presence
碳酸钠矾 burkeite
碳酸钠钙石 shortite
碳酸钠过氧化钠熔融 fusion with mixture of Na_2CO_3 and Na_2O_2
碳酸钠镁矿 eitelite
碳酸钠灭火器 soda-acid extinguisher;soda-acid fire extinguisher
碳酸钠浓度 sodium carbonate concentration
碳酸钠去垢(法)sodation
碳酸钠熔融 fusion with Na_2CO_3
碳酸钠熔融法 sodium carbonate fusion
碳酸钠烧结 sintering with Na_2CO_3
碳酸钠石 trona
碳酸霓辉岩 carbonate aegirine-pyroxenite
碳酸霓霞岩 carbonate ijolite
碳酸镍 nickel carbonate; nickelous carbonate
碳酸喷气孔【地】mof(f)ette
碳酸气 blackdamp; carbon dioxide; choke damp
碳酸气量计 carbonic acid gas meter
碳酸气灭火器 asphyxiator
碳酸气体 carbonic acid gas
碳酸铅 carbonate of lead;ceruse;lead carbonate;plumbic carbonate
碳酸铅白 basic carbonate of white lead;ceruse;flake white
碳酸羟铅 lead hydroxyl carbonate
碳酸羟锌 zinc hydroxyl carbonate
碳酸侵蚀 carbonic acid attack
碳酸侵蚀指标 carbonic acid erosion index
碳酸氢铵 ammonium bicarbonate
碳酸氢钙 calcium bicarbonate; calcium hydrogen carbonate
碳酸氢根 bicarbonate radical
碳酸氢钾 potassium acid carbonate; potassium bicarbonate
碳酸氢钠 baking soda; bicarb; dicarbonate;sodium bicarbonate;sodium hydrogen carbonate
碳酸氢钠浓度 sodium bicarbonate concentration
碳酸氢三钠 sodium sesquicarbonate
碳酸氢亚铁 ferrous bicarbonate
碳酸氢盐 bicarbonate[dicarbonate]; hydrocarbonate
碳酸氢盐缓冲液 bicarbonate buffer
碳酸氢盐碱度 bicarbonate alkalinity
碳酸氢盐离子 bicarbonate ion
碳酸氢盐碳酸盐系统 bicarbonate carbonate system
碳酸氢银 silver bicarbonate
碳酸泉 apollinaris spring; calcareous spring; carbonated spring; carbon dioxide spring; carbonide spring; carburet(t)ed spring
碳酸热水 carbonated thermal water
碳酸水 aerated water;carbonate water;carbonic acid water
碳酸水淋浴器 ombrophore
碳酸锶 strontium carbonate
碳酸锶铈矿 ancylite
碳酸铽 terbium carbonate
碳酸铜 copper carbonate
碳酸铜矿 bergblau
碳酸铜绿 mineral green
碳酸锌 carbonate of zinc;zinc carbonate
碳酸亚铊 thallous carbonate
碳酸亚铁 ferrous carbonate
碳酸岩补偿深度 carbonate compensation depth
碳酸岩的分类图 classification of carbonatites
碳酸岩矿床 carbonatite deposit
碳酸岩熔岩 carbonatite lava
碳酸岩熔岩类 carbonatite lava group
碳酸盐 carbonate
碳酸盐饱和指数 calcite saturation index
碳酸盐泵 carbonate pump
碳酸盐补偿界面 carbonate compensation level
碳酸盐补偿深度 carbonate compensation depth
碳酸盐潮坪 carbonate tidal flat
碳酸盐沉淀法 calcite sedimentation method
碳酸盐沉积物 carbonate sediment
碳酸盐沉积周期 carbonate deposit period
碳酸盐储层沉积环境 depositional environment of carbonate reservoir
碳酸盐储集层 carbonate rock reservoir
碳酸盐地层 carbonated rock
碳酸盐二氧化碳 carbonate carbon dioxide
碳酸盐法 carbonate method;carbonation
碳酸盐法处理 carbonation treatment
碳酸盐分布 carbonate distribution
碳酸盐分析测井图 carbonate analysis logging map
碳酸盐风化壳 carbonate weathering crust
碳酸盐骨料 carbonate aggregate
碳酸盐黑土 black calcareous soil
碳酸盐湖 carbonate lake
碳酸盐化 carbonation
碳酸盐化作用 carbonatization
碳酸盐还原带 carbonate reduction zone
碳酸盐缓冲系 carbonate buffer system
碳酸盐活度系数 carbonate activity coefficient
碳酸盐基 carbonate group
碳酸盐集料 carbonate aggregate
碳酸盐夹层 carbonate interbed
碳酸盐碱度 carbonate alkalinity
碳酸盐建造 carbonate formation
碳酸盐胶结物 carbonate cement
碳酸盐礁圈闭 carbonate reef trap
碳酸盐结核 carbonate concretion
碳酸盐结石 carbonate calculus
碳酸盐介质 carbonate medium
碳酸盐浸出 carbonate leach
碳酸盐矿泉水 carbonate mineral spring water
碳酸盐矿石 carbonate ore
碳酸盐矿物 carbonate mineral
碳酸盐硫酸盐盐土 carbonate-sulfate salinized soil
碳酸盐陆棚 carbonate continental shelf
碳酸盐锰矿石 Mn carbonate ore
碳酸盐黏[粘]土 effervescing clay
碳酸盐浓度 carbonate concentration
碳酸盐盆地 carbonate basin
碳酸盐平衡 carbonate equilibrium
碳酸盐浅滩 carbonate shallow shoal
碳酸盐侵蚀 carbonate attack
碳酸盐丘圈闭 carbonate mound trap
碳酸盐融烧 molten carbonate
碳酸盐酸/碱基 carbonate acid/base group
碳酸盐台地 carbonate platform
碳酸盐团块状 carbonate nodular forms
碳酸盐系 carbonate system
碳酸盐相 carbonate facies
碳酸盐斜坡 carbonate slope
碳酸盐斜坡沉积模式 sedimentation model carbonate slope
碳酸盐型风化壳 residuum of carbonate type
碳酸盐循环 carbonate cycle
碳酸盐岩 carbonatite
碳酸盐岩残余结构 relic texture of carbonate rocks
碳酸盐岩层控铅锌矿床 strata-bound lead-zinc deposit in carbonate rocks
碳酸盐岩层中层状汞矿床 stratiform mercury deposit in carbonate rocks
碳酸盐岩层中层状锑矿床 stratiform antimony deposit in carbonate rocks
碳酸盐岩二氧化碳水体系 carbonatite-carbon dioxide-water system
碳酸盐岩基质和胶结物 matrix and cements of carbonate rocks
碳酸盐岩及砂页中硫铁矿床 pyrite deposit in carbonate rock and sandstones hale
碳酸盐岩建造 carbonate rock formation
碳酸盐岩交代铅锌矿床 metasomatic lead-zinc deposit carbonate rocks
碳酸盐岩晶粒结构 crystalline granular texture of carbonate rocks
碳酸盐岩颗粒结构 allochem grain texture carbonate rocks

碳酸盐岩隆圈闭 carbonate build-up trap
碳酸盐岩隆相 carbonate buildup facies
碳酸盐岩潜山油气藏趋向带 burial hill pool trend of basement rock
碳酸盐岩圈闭 carbonate trap
碳酸盐岩溶解动力学 kinetics of carbonate rock dissolution
碳酸盐岩石 carbonate rock
碳酸盐岩石矿床 carbonate rock deposit
碳酸盐岩滩相 carbonate bank facies
碳酸盐岩异常 anomaly of carbonatite
碳酸盐岩油储 carbonate reservoir
碳酸盐盐渍土 carbonate salinized soil
碳酸盐乙烯 ethylene carbonate
碳酸盐乙酯 carbonic ether;ethyl carbonate
碳酸盐异化粒 carbonate allochems
碳酸盐硬度 alkaline hardness; carbonate hardness; temporary hardness
碳酸盐障 carbonate barrier
碳酸盐质石膏 calcareous gypsum
碳酸盐质石膏-硬石膏矿石 calcareous gypsum-anhydrite ore
碳酸盐质硬石膏矿石 calcareous anhydrite ore
碳酸盐质硬石膏石膏矿石 calcareous anhydrite-gypsum ore
碳酸盐渍土 carbonate-saline soil
碳酸盐总量 total carbonates
碳酸乙丁酯 ethyl butyl carbonate
碳酸银 carbonate of silver;silver carbonate
碳酸脂 carbonate
碳酸酯 carbonic ester
碳损 carbon loss
碳索 fusain
碳钛锆钠石 sabinaite
碳碳复合材料 carbon-carbon composite
碳碳合金 carbon-carbon alloy
碳碳键合 carbon-to-carbon linkage
碳陶瓷 carbon ceramics
碳陶质耐火材料 plumbago
碳铁矿 haxonite
碳铁钠矾 ferrotychite
碳同位素 carbon isotope;radiocarbon
碳同位素比 carbon isotope ratio
碳同位素地热温标 carbon isotopic geothermometer
碳同位素地质温度计 carbon isotope geothermometer
碳同位素年代测定 radiocarbon dating
碳同位素年龄测定 radiocarbon dating
碳铜钙铀矿 voglite
碳铜氯铅矾 wherryite
碳铜铅钙石 schuilingite
碳涂层 carbon coating
碳脱氧的金属 carbon-deoxidized metal
碳稳定同位素标准 standard of carbon stable isotope
碳稳定同位素值间距 separation of carbon stable isotope value
碳吸附 carbon adsorption
碳吸附法 carbon absorption method
碳吸附柱 carbon absorption column
碳吸收 carbon absorption
碳吸收剂 absorbent carbon
碳烯 carbene
碳烯化学 carbene chemistry
碳纤维 carbon fiber[fibre]
碳纤维布 carbon cloth
碳纤维带 carbon fiber[fibre] tape
碳纤维复合材料 carbon fiber[fibre] composite

碳纤维合成索 carbon fiber[fibre] composite cable
碳纤维混凝土 carbon fiber[fibre] concrete
碳纤维加筋聚酯 <一种高强轻质材料> carbon fiber[fibre] reinforced polyester
碳纤维加筋(强)塑料 carbon fiber[fibre] reinforced plastic
碳纤维加筋塑料板 <一种结构加强用新材料> carbon fiber-reinforced plastic plate
碳纤维盘根 carbon fiber[fibre] packing
碳纤维陶瓷 carbon fiber[fibre] ceramics
碳纤维网预浸渍法 preimpregnated carbon fiber[fibre] web method
碳纤维增强复合材料 carbon fiber[fibre] reinforced composite
碳纤维增强环氧复合材料 carbon epoxy composite
碳纤维增强混凝土 carbon fiber[fibre] reinforced concrete
碳纤维增强金属 carbon fiber[fibre] reinforced metal
碳纤维增强聚合物 carbon fiber[fibre] reinforced polymer
碳纤维增强水泥 carbon fiber[fibre] reinforced cement
碳纤维增强塑料 carbon fiber[fibre] reinforced plastics
碳纤维毡 carbonized felt
碳纤维锥 carbon fiber[fibre] cone
碳酰胺树脂 carbamide resin
碳酰二胺 carbamide
碳酰氟 carbonyl fluoride
碳酰硫 carbonyl sulfide
碳酰氯 carbonyl chloride;phosgene
碳酰亚胺 carbimide
碳线 carbon line;carbon-point curve;grey line
碳屑检波器 carbon coherer
碳屑凝合检波器 granular coherer
碳锌电池 carbon-zinc battery
碳锌钙石 minrecordite
碳锌锰矿 loseyite
碳修饰剂 carbon modifier
碳、溴或卤的气体化合物 halon
碳序 carbon sequence
碳循环 cycle of carbon;carbon cycle
碳压记录法 carbon pressure recording
碳烟灰 carbon black
碳岩 chalkstone
碳阳离子 carbocation
碳氧比测井 carbon/oxygen log
碳氧比测井曲线 carbon/oxygen log curve
碳氧化物污染 pollution by carbon oxides
碳氧同位素相关性 carbon-oxygen isotope correlation
碳氧血红蛋白 carboxyhemoglobin
碳钇钡石 mckelveyite
碳钇锶石 donnayite
碳优势指数 carbon preference index
碳铀钙石 liebigite
碳原子网 carbon atom net
碳原子网间距 spacing of carbon atom net
碳源 carbon source
碳渣 carbon dust;carbon residue
碳渣(值)试验 carbon residue test
碳正极 carbon anode
碳正离子 carbenium ion
碳值 carbon value
碳质板岩 carbonaceous slate
碳质变阻器 carbon rheostat
碳质残渣 carbonaceous residue

碳质测辐射热计 carbon bolometer
碳质沉积(物) carbonaceous deposit; carbonaceous sediment
碳质打结炉底 carbon hearth bottom
碳质的 carbonaceous
碳质灯丝 carbon filament
碳质电阻 carbon resistance
碳质电阻器 carbon resistor
碳质堆积土 cumulose deposit
碳质废弃物 carbonaceous refuse
碳质沸石 carbonaceous zeolite
碳质化合物 carbonaceous compound
碳质还原剂 carbonaceous reducing agent
碳质黄铁矿 coal pyrite
碳质集电弓条 carbon collector strip
碳质接点 carbon contact
碳质矿物 carbonate mineral
碳质沥青 carbene
碳质模型 carbon mo(u)ld
碳质膜 carbon membrane
碳质内衬 carbon lining
碳质耐火材料 anthracite coal base refractory;carbon refractory
碳质耐火堵泥 carbonaceous brasque
碳质耐火砖 carbon brick
碳质泥岩 carbonaceous mudstone
碳质黏[粘]土 bast;carbonaceous clay
碳质气氛 carbonaceous atmosphere
碳质球粒陨石 carbonaceous chondrite
碳质燃料 carbonaceous fuel
碳质砂岩 carbonic sandstone;carboniferous sandstone
碳质烧舟 carbon boat
碳质石灰岩 carbonic limestone; carbonaceous limestone
碳质填料 carbon packing
碳质铁石 carbonaceous ironstone
碳质涂料 carbon-base coating;carbon paint
碳质涂料碾磨机 blacking mill
碳质土 cumulose soil
碳质脱膜剂 carbonaceous parting
碳质岩 carbonaceous rock;carbonolite;carbon rock
碳质氧化作用 carbonaceous oxidation
碳质页岩 carbonaceous shale;carbon shale;culm;jerry
碳质页岩废物 carbonaceous shale waste
碳质页岩与煤线互层 batt
碳质陨石 carbonaceous meteorite
碳质炸药 carbonite
碳质中间相小球体 carbon mesophase microbead
碳质砖 carbon brick
碳珠混凝土 carbon beads concrete
碳柱 carbon column
碳柱调压器 carbonstat voltage regulator
碳砖 brick fuel
碳转化率 carbon transformation
碳族化合物 carbon family
碳族元素 carbon group elements
碳阻棒 carbon resistor rod
碳阻测辐射热计 carbon resistor bolometer
碳阻电炉 carbon resistance furnace

汤格阶 <早渐新世>【地】Tongrian

汤硅钇石 tombarthite
汤河原石 yugawaralite
汤加 <大洋洲> Tonga
汤加板块 Tonga plate
汤加海沟 Tonga trench
汤加开发银行 Tonga Development Bank
汤加银行 Bank of Tonga
汤浸釉 glazing by dipping
汤姆森方式 Thomson's type
汤姆森三角量水堰 Thomson V-notch weir
汤姆逊保护环 Thomson's guard-ring
汤姆逊电流计 Thomson's galvanometer
汤姆逊法 Thomson's method
汤姆逊公式 Thomson's formula
汤姆逊灰岩 Thomson's limestone
汤姆逊静电计 Thomson's electrometer
汤姆逊量水堰 Thomson measuring weir
汤姆逊抛物线 Thomson's parabolas
汤姆逊抛物线装置 Thomson's parabola apparatus
汤姆逊平衡 Thomson's balance
汤姆逊热 Thomson's heat
汤姆逊热电效应 Thomson's effect
汤姆逊散射 Thomson's scatter(ing)
汤姆逊散射公式 Thomson's scattering formula
汤姆逊散射截面 Thomson's cross section
汤姆逊式 Thomson's type
汤姆逊式仪表 Thomson's meter
汤姆逊系数 Thomson's coefficient
汤姆逊效应 Thomson's effect
汤姆逊斜圈式仪表 Thomson's inclined coil instrument
汤盘 soup plate
汤普逊电弧焊接法 Thompson process
汤普逊滤波器 Thompson filter
汤普逊摩擦试验机 Thompson machine
汤森德电流 Townsend current
汤森德特性曲线 Townsend characteristic
汤森德系数 Townsend coefficient
汤山古溶洞 Tangshan grottoes of ancient stalagmite
汤氏桁架 Town(lattice)truss
汤霜晶石 thomsenolite
汤似的 soupy
汤团 dumpling
汤釉 glazing by rinsing
汤蘸釉 glazing by immersion

羰化作用 carbonylation

羰基 carbonyl;carboxide;carboxy(l)
羰基粉末 carbonyl powder
羰基合成 oxo-process
羰基化合物 carbonyl compound
羰基化作用 carbonylation
羰基镍 nickel carbonyl;nickel tetracarbonyl
羰基铁 iron carbonyl
羰基铁粉 carbonyl iron powder
羰基铁粉磁芯 carbonyl iron dust core
羰基铁芯 carbonyl core
羰基铁压粉 sirufer(core)
羰基钨 tungsten hexacarbonyl
羰基物 carbonyl
羰基亚铂氯 carbonyl platinous chloride
羰基值 carbonyl value
羰络金属 metal carbonyl

镗车两用机床 boring and turning machine; boring and turning mill;boring lathe

镗床 borer;boring lathe;boring ma-

chine;boring mill
镗床工作台 boring table
镗床主轴箱 boring head
镗床柱 upright of a boring machine
镗刀 boring cutter; boring tool; cornish bit
镗刀杆可伸缩定中心爪 cat's-paw
镗刀盘 boring head
镗刀头 cathead;cutter head
镗杆 boring bar; boring rod; boring spindle;cutter spindle;sabot
镗杆刀具 boring bar tool
镗缸机 boring bar; cylinder borer; cylinder(re)boring machine
镗工 borer;boring machine operator
镗阶梯孔 counterbore;counterboring
镗锯法 boring cut
镗孔 bore(hole); boring; counterbore; hole boring; borizing <金刚石的>
镗孔车刀 boring(turning)tool
镗孔车面机床 boring and facing machine
镗孔车削机床 boring and turning machine
镗孔刀 hole boring cutter;borer
镗孔刀具 boring apparatus
镗孔法 method of boring
镗孔工具 uprighting tool
镗孔工作 boring work
镗孔光车刀 boring finishing turning tool
镗孔和成面两用车床 boring and surfacing lathe
镗孔锪端面加工机床 bore face machine
镗孔机 boring mill
镗孔机械 boring machinery
镗孔夹具 boring fixture;boring jig
镗孔检查仪 <气缸镜面或通道的> bore searcher
镗孔深度 boring depth
镗孔头 head for boring bar
镗孔样板 template for boring
镗孔凿榫机 boring and mortising machine
镗孔支柱 boring stay
镗孔轴线 axis of bore
镗孔锥度 boring taper
镗孔钻具 borer
镗孔钻头 bore bit
镗口直径 bore diameter
镗磨 honing
镗磨头 honing tool
镗磨行程 honing stroke
镗汽缸内孔 cylindric(al)boring
镗同心孔 align boring
镗头 bore;boring head
镗屑 bore chip(ping)s;borings
镗圆筒孔 cylindric(al)boring
镗钻两用机(床) boring and drilling machine

唐钧窑 Tang Jun kiln

唐纳德式整流环 Townend ring
唐南薄膜理论 <离子交换> Donnan membrane theory
唐人街 Chinatown
唐三彩 Tang sancai
唐太宗青花瓶 <瓷器名> in white and blue; vase of Emperor Tangtai Zhong
唐窑 Tang kiln;Tang ware

堂皇的教堂 stately church

堂皇的门道 stately gateway
堂皇的入口 stately portal
堂前内柱廊 prodomus

塘处理 lagooning

塘堤岸 dike[dyke];embankment
塘地 wetland
塘泥 pond silt;pond sludge
塘式间歇泉 pool geyser
塘堰 small reservoir;stank
塘预处理 lagoon pretreatment

搪玻璃 enamel(ling); glass coating; glassed steel; glass enamel; glass lining

搪玻璃的 glass-lined
搪玻璃钢件 <耐化学侵蚀的钢件> glass-lined steel
搪玻璃换热器 enamel heat exchanger
搪玻璃容器 enamel(l)ed vessel
搪玻璃微月坑 glass-lined microcrater
搪衬 patching
搪瓷 enamel(ling); glass lining; porcelain enamel; vitreous enamel; vitrified enamel
搪瓷泵 enamel lined pump
搪瓷臂板 enamel(l)ed blade;enamel(l)ed semaphore arm
搪瓷边釉 beading enamel
搪瓷表面的不平缺陷 orange peel
搪瓷冰箱 enamel lined refrigerator
搪瓷玻璃 enamel(led)glass
搪瓷材料 enamel(led)material
搪瓷层 enamel layer
搪瓷的 enamel(l)ed; enamel lined; glass-lined
搪瓷灯罩 enamel lampshade
搪瓷底层 grip coat
搪瓷底料 undercoat
搪瓷底釉 base enamel
搪瓷法 vitreous enamel coating
搪瓷珐琅 porcelain enamel; vitreous enamel
搪瓷反射罩 enamel(l)ed reflector
搪瓷反应罐 enamel reactor
搪瓷封接 enamel seal
搪瓷敷层管 enamel back tubing
搪瓷釜 enamel still
搪瓷钢 enamel(l)ed steel; glassed steel
搪瓷钢板 enamel(l)ed pressed steel; enamel(l)ing iron;porcelain enamel steel panel
搪瓷高压釜 enamel high pressure still
搪瓷工匠 enameller
搪瓷管 enamel(l)ed tube; enamel pipe;glass lining pipe
搪瓷锅 enamel pan
搪瓷厚度仪 Brenner ga(u)ge
搪瓷画 enamel(l)ed paint(ing)
搪瓷画家 enamel painter
搪瓷计量罐 enamel gauging tank
搪瓷计量器 enamel counter
搪瓷建筑板材 porcelain enamel(l)ed building panel
搪瓷建筑物 enamel building
搪瓷聚合釜 enamel polymerization still
搪瓷烤花窑 enamel kiln
搪瓷冷却器 enamel cooler
搪瓷脸盆 enamel wash basin
搪瓷粮仓 enamel(l)ed silo
搪瓷面层 porcelain enamel finish
搪瓷面冷轧钢板 yodowall
搪瓷面釉气泡 boiling
搪瓷漆 enamel(led)paint;lacquer enamel
搪瓷器(皿) enamel ware; porcelain enamel ware
搪瓷墙板 porcelain enamel panel;vitreous enamelled steel wall panel
搪瓷青 enamel blue
搪瓷球形物 enamel(l)ed bulb
搪瓷缺陷 enamel defect
搪瓷容器 enamel(l)ed vessel
搪瓷熔块 enamel frit; porcelain enamel frit
搪瓷伞形罩灯 umbrella-shoe enamelled metal reflector lamp
搪瓷上釉涂层 enamel glazed coating
搪瓷烧器 enamel(l)ed cooking utensil
搪瓷饰面 porcelain finish
搪瓷铁器 enamel(l)ed ironware
搪瓷铁浴盆 enamel(l)ed cast-iron (bath)tub
搪瓷涂层 enamel coating; porcelain enamel finish;vitreous coating
搪瓷涂敷 enamel painting
搪瓷土 enamel clay
搪瓷卫生洁具 porcelain enamel sanitary wares
搪瓷颜料 enamel colo(u)r; enamel dye
搪瓷颜料彩釉 overglaze colo(u)r
搪瓷颜色 enamel colo(u)r
搪瓷氧化物 enamel oxide
搪瓷窑 enamel furnace;glaze kiln
搪瓷硬板 enamel(l)ed hardboard
搪瓷用玻璃料 frit
搪瓷用钢板 enamel sheet;enamel(l)ing sheet steel;porcelain enameling sheet
搪瓷用铸铁 cast-iron for enameling
搪瓷釉 enamel glaze;porcelain enamel;vitreous enamel
搪瓷釉商标 enamel badge;enamel label;vitreous enamel label
搪瓷灶 enamel(l)ed hotplate
搪瓷罩 enamel reflector
搪瓷蒸发器 enamel evapo(u)rator
搪瓷制品 enamel;enamel(l)ed product; enamel ware; vitreous enamelled product
搪瓷珠 enamel bead
搪瓷贮罐 enamel storage tank
搪瓷铸铁 enamel(l)ed cast iron
搪瓷铸铁浴盆 enamel(l)ed cast-iron (bath)tub
搪缸机 boring machine
搪灰泥 <在烟道或炉膛中涂抹灰泥> parg(et)ing
搪孔直径 bore diameter
搪料 daubing mud
搪炉料 ramming mix(ture)
搪炉砖 stove tile
搪磨 hone(d finishing);honing
搪磨后的表面(光洁度) honed finish
搪磨机 honer;honing machine
搪磨头 honing head
搪塞 dodge
搪烧 enamel(ling)firing
搪烧炉 enamel firing furnace
搪烧支架 buck
搪釉熔炉 enamel melting furnace

膛 bore

膛内保险销 bore riding pin
膛内瞄准 boresighting
膛内偏转角 yaw in bore
膛内压力 breech pressure
膛式炉 hearth furnace;muffle furnace
膛线 groove;rifle;rifling
膛线缠度 twist of rifling
膛线拉刀 rifling head
膛线炮 rifled cannon;rifled gun
膛线炮管 rifled barrel;rifled bore
膛线起点 origin of rifling
膛线周径 lead
膛压 gun pressure
膛压计 pressure ga(u)ge
膛炸 bore premature

糖桉 sap gum;sugar gum

糖仓 sugar bin
糖厂 sugar house;sugar refinery
糖淀粉 amylose
糖分含量 <以水泥重量的百分率计，糖分是一种水泥缓凝剂> sugar content
糖缸 sugar basin
糖果 confection
糖果车 <俚语> candy wagon
糖果点心店 confectionary[confectionery]
糖果店 candy store;sweet shop
糖果吊桶 <俚语> candy bucket
糖果加工厂废水 sweet refinery wastewater
糖【化】 saccharide
糖桦 black birch;cherry birch;sweet birch
糖浆 molasses;sirup;syrup
糖浆废液 molasses slop
糖浆运输船 molasses tanker
糖胶树 red gum
糖精 benzosulfimide;saccharin
糖类 saccharide;carbohydrate
糖粒状 sugar-granular
糖粒状结构 sucrosic texture
糖料作物 sugar-producing crop
糖蜜废液 molasses slop
糖蜜外掺剂 sugar-sludgy admixture
糖蜜运输船 molasses tanker
糖朴 sugarberry
糖槭 hard maple; rock maple; sugar (hardwood)maple
糖溶解度试验法 sugar-solubility test
糖松 sugar pine
糖衣 ice
糖皂 sugar soap
糖蔗车 sugarcane car
糖汁 sirup[syrup]
糖状石英 sugary quartz

螳螂 mantis[复 mantises/mantes]

淌度 mobility

淌度计 mobilometer
淌航 carrying way with engine stopped
淌凌 drift ice;ice-drift(ing);ice floe;ice gang;ice run;run of ice
淌漆 curtaining
淌漆纹 tear
淌油 dribbling
淌釉 crawling of glazes;smear glaze

躺板 <修理用> creeper

躺板极堆焊 surfacing by fire cracker welding process
躺倒不干 lying down
躺滴法 sessile drop method
躺焊 fire cracker welding
躺铺 reclining berth
躺筛 horizontal screen
躺下 slumber
躺椅 bed chair; day bed; deck chair; dormette; lounge chair; reclining chair;sling chair;triclinium <古罗马圆餐桌椅的>
躺椅客车 coach with couchetts

躺着的 recumbent
躺座 reclining seat

烫画 poker work;pyrograph;pyrogravure

烫金 gild;gilt;gold blocking;gold stamping
烫金边 gilt edge
烫金箔 adhesive foil
烫金垫 gold cushion
烫金工具 gold tooling
烫金机 bronzing machine; gilding press;press gilder;stamping press
烫金精印封面 gilt stamped cover
烫开 seal(ing)-off
烫沥青路面机 asphalt road burner
烫模铸件 dummy casting
烫平 ironing
烫平机 heater planer;ironer
烫伤 blister;scald
烫锡 tinning
烫下焊头 tip-off
烫衣板 ironing board
烫衣人 ironer
烫衣室 ironing room
烫印 hot stamping
烫印的标题 titling
烫印油墨 alumin(i)um foil ink
烫字 stamping

趟车 coasting

绦虫【生】cestode

掏爆破 springing blast

掏爆破炮眼 springing blast hole
掏槽 burn cut;cut;kerf[复 kerve];sump;undercut;undermine
掏槽爆破 breaking(-in) shot;cut blasting;snubbing shot;sump shot
掏槽爆破孔 easer
掏槽法 <装药集中在孔底> burn cut
掏槽放炮 sump shooting
掏槽孔 breaking-in hole;snubber hole
掏槽孔布置形式 cut pattern
掏槽落煤法 holing;thirling
掏槽炮孔 cut hole
掏槽炮眼 breaking shot;cut hole;cut shot;key cut hole;snubber hole
掏槽平巷掘进法 cut-hole drifting
掏槽深度 bearing-in;depth of kerf
掏槽位置 cutting position
掏槽眼 cut hole;gouging shot;snubber
掏槽凿 cross chisel
掏槽钻臂 cut-boom
掏底 floor cut
掏底槽 bottom cutting
掏堵 access eye;clean-out
掏粪工 <一般在夜间工作> nightman
掏壶 blast-hole springing; borehole springing
掏壶爆破 springing blast
掏壶爆破眼 springing blast hole
掏灰门 cleaning port
掏金摇动槽 rocker
掏空 hollow out
掏料机 pulp digger
掏泥斗 bagger
掏泥机 bagger
掏泥筒 bailer;mud barrel
掏蚀岸 undercut bank
掏蚀作用 undercutting
掏刷 undercut
掏碎玻璃孔 opening for taking out cullet
掏箱 unstuffing
掏心 <地应力测量用的> overcore
掏心法 <装药集中在孔底> burn cut
掏心钻进 overcore
掏岩粉 bite
掏渣 de-drossing
掏渣池 skim pocket
掏渣勺 sludge ladle
掏渣箱 dross box

逃避 back-out(of);dodge;evasion

逃避承诺 welsh
逃避城市枯燥生活的处所 urban escape hatch
逃避纳税者 <口语> tax dodger
逃避实际 flight from reality
逃避税收 evade tax;evasion of tax
逃避条款 escape clause
逃避行为 escape behavio(u)r
逃避义务 shift off obligations
逃避隐蔽植被 escape covert
逃避资本 refugee capital
逃避资金 refuge capital
逃汇 evade foreign exchange;evasion of exchange control
逃离塔 escape tower
逃跑 break free
逃入病中 flight into disease
逃生安全系数 escape margin
逃生舱 escape compartment
逃生出口 escape hatch
逃生孔 escape trunk
逃生路线 escape route
逃生闸 escape lock
逃生者 survivor
逃税 escape of taxation;evade duty;evade paying taxes;evasion of taxation;tax avoidance;tax evasion
逃税乐园 tax haven
逃税人 tax dodger
逃税手段 tax shelter
逃税途径 tax escape valves
逃税(与)避税 evasion and avoidance
逃税侦查员 tax ferret
逃脱 escape
逃脱共振几率 resonance escape probability
逃脱共振因子 resonance escape factor
逃脱几率 escape probability
逃脱因子 escape factor
逃亡工厂 runaway shop
逃亡者 runaway
逃向幻想世界 flight into fantasy
逃向另一现实 flight into reality
逃向生理疾病 <藉生理上某种机能障碍以避免面对困难> flight into disease
逃逸峰 <探测器的> escape peak
逃逸构造 escaping structure
逃逸迹 escape trace
逃逸临界高度 critical level of escape
逃逸率系数 escape rate coefficient
逃逸能级 escape level
逃逸能力 escape capability
逃逸趋势 escaping tendency
逃逸试验 escape test
逃逸速度 escape rate;escape speed;escape velocity;speed of escape;velocity of escape
逃逸因子 escape factor
逃逸锥(面) escape cone;loss cone
逃走条件作用 escape conditioning

桃铲 buttering trowel

桃红色 peach blossom;peach(red);pink;pink-colo(u)red
桃红色釉 <中国瓷器上的> peach blow
桃红釉 pink glaze
桃花心黑 peach-black
桃花心木 acajou; mahogany; sipo (mahogany);citywood <产于圣多明各最佳质量的>;gedunoha <红褐色硬木,产于西非和东非>
桃花心木中短的断断续续的条纹 roe
桃花心木棕 mahogany brown
桃尖拱 lancet arch
桃尖拱拱尖 tierce point
桃尖线脚 keel mo(u)lding
桃金娘【植】downy rosemyrtle; hill gooseberry; Ribes hill gooseberry; myrtle;myrtus
桃金娘科 <拉> Myrtaceae
桃金娘蜡 myrtle wax
桃轮 heart cam;heart piece
桃轮轴【机】camshaft
桃色釉 peach blow
桃树 peach(tree);Prunus persica
桃树蛀虫 peach-tree borer
桃汛 snow flood; snowmelt flood; snow water flood;spring flood
桃汛期 freshet period
桃叶栎 peach leaved oak
桃叶珊瑚 aukuba
桃子夹头 driving dog;heart carrier
桃子轮 cam
桃子轴 tappet shaft
桃棕色 peach tan

陶板 stoneware tile

陶贝尔试验 Tauber test
陶槽 kirve
陶厂废水 potter wastewater; pottery waste
陶瓷 ceram;earthenware;pottery and porcelain
陶瓷把手 ceramic knob
陶瓷板 ceramic plate
陶瓷板镶面 ceramic veneer
陶瓷板印刷电路 ceramic print circuit
陶瓷棒火焰喷涂法 ceramic rod flame spraying
陶瓷包装纸 pottery tissue
陶瓷杯 ceramic cup
陶瓷泵 ceramic pump;stoneware pump
陶瓷避雷器 ceramic arrester[arrestor]
陶瓷扁平磁头 ceramic flat head
陶瓷变压器 ceramic transformer
陶瓷变阻器 ceramic varistor
陶瓷表层 ceramic coat(ing)
陶瓷表面起泡 blowing
陶瓷玻璃釉 ceramic glass enamel
陶瓷薄膜 ceramic film; ceramic thin film
陶瓷材料 ceramic material;ceramics
陶瓷材料反应堆 ceramic reactor
陶瓷材料分子工程 ceramic material molecular engineering
陶瓷彩(色)釉(面) ceramic colo(u)r glaze
陶瓷测头 ceramic tip
陶瓷产品 ceramic product
陶瓷厂 ceramic factory; clay mill; pottery
陶瓷厂废气和废水 porcelain manufacture waste gas and water
陶瓷厂废水 pottery wastewater
陶瓷超导体 ceramic superconductor
陶瓷超塑性 ceramic superplasticity
陶瓷车刀 Sintex
陶瓷衬底 ceramic substrate
陶瓷衬底釉 ceramic substrate glaze
陶瓷衬垫推力室 ceramic-lined thrust chamber
陶瓷衬里的燃烧室 ceramic-lined chamber
陶瓷衬料 ceramic lining; terra-cotta lining
陶瓷衬套 ceramic chamber lining
陶瓷成分 ceramic constituent
陶瓷成型 ceramic forming
陶瓷成型机 jolley;jollie
陶瓷传感器 ceramic sensor; ceramic sensor element
陶瓷传声器 ceramic microphone
陶瓷词汇 ceramic glossary
陶瓷磁体 ceramagnet;ferro-magnetic ceramic;ceramic magnet
陶瓷次口 <挑炼后的> lump
陶瓷搭板 ceramic column plate
陶瓷带加热器 ceramic strip heater
陶瓷刀 <美国商品> Stupalox
陶瓷刀具 ceramic cutting tools; cutting ceramics;tool tip;Degussit <主要成分为三氧化二铝>
陶瓷刀头 ceramic bit
陶瓷的 ceramic; porcelainous; vitrified
陶瓷等离子体 ceramic plasma
陶瓷地面 ceramic floor
陶瓷地面覆盖层 ceramic floor cover(ing)
陶瓷地面砖 ceramic flooring tile
陶瓷地砖 ceramic mosaic; pottery mosaic
陶瓷点火器 ceramic lighter
陶瓷点火装置 ceramic ignitor
陶瓷电解质 ceramic electrolyte
陶瓷电容器 ceramic capacitor;porcelain capacitor
陶瓷电阻器 ceramic resistor
陶瓷垫片 ceramic packing
陶瓷垫圈 ceramic packing
陶瓷雕塑 ceramic sculpture
陶瓷顶棚面砖 ceramic ceiling tile
陶瓷动叶片 ceramic rotor blade
陶瓷断裂功 ceramic fracture power
陶瓷断裂韧性 ceramic fracture toughness
陶瓷多孔板 ceramic hollow slab
陶瓷多孔管 ceramic porous tube
陶瓷多孔砖 ceramic cellular block
陶瓷发动机 ceramic engine
陶瓷法 china clay method
陶瓷反应器 ceramic reactor
陶瓷防滑铺地砖 <一种专利的有突起点的> Durogrip
陶瓷防热瓦 ceramic thermal protection system tile
陶瓷放大器 ceramic amplifier
陶瓷废料 shraff
陶瓷废品 stuck ware
陶瓷分隔墙 terra-cotta partition(wall)
陶瓷粉体的造粒 ceramic powder granulation
陶瓷封接 ceramic seal;crunch seal
陶瓷封料 ceramic encapsulation; encapsulation
陶瓷封装 ceramic package
陶瓷蜂窝结构 ceramic honeycomb
陶瓷敷层 ceramic coat(ing)
陶瓷腐蚀浮雕 white acid embossing
陶瓷复合材料 ceramic composite material
陶瓷盖 ceramic cap
陶瓷坩埚 ceramic crucible;clay pot;porcelain crucible
陶瓷缸套 <水泵的> porcelain liner
陶瓷工厂 ceramic plant; pot bank <英国旧用语>
陶瓷工尘肺 porcelain worker's pneumoconiosis

陶瓷工具 chemical tool
陶瓷工人 ceramist;potter
陶瓷工业 ceramic industry; pottery industry;potty industry
陶瓷工业用润滑油 brockle oil
陶瓷工业用石膏粉 potter's plaster
陶瓷工艺 ceramics
陶瓷工艺学 ceramic technology
陶瓷工作者 ceramist
陶瓷构件 ceramic component
陶瓷骨料 ceramic aggregate
陶瓷固化 ceramic solidification
陶瓷管 ceramic tube;hall of pottery and porcelain;porcelain form
陶瓷管道 earthenware duct
陶瓷盥洗盆 ceramic wash basin
陶瓷盥洗室盆 ceramic lavatory basin
陶瓷过滤池 ceramic filter
陶瓷过滤除尘器 ceramic filter dust separator
陶瓷过滤介质 ceramic filter media
陶瓷过滤器 ceramic filter;Pasteur filter;porcelain filter
陶瓷过滤芯 candle
陶瓷合金 ceramal; cermet; cermetallic
陶瓷核燃料 ceramic nuclear fuel
陶瓷红外线发热器 ceramic infrared emitter
陶瓷红外线加热器 ceramic infrared heater
陶瓷厚膜 ceramic thick film
陶瓷滑模油 press oil
陶瓷化工泵 ceramic process pump
陶瓷画屏 plaque
陶瓷环 ceramic ring
陶瓷环状绝缘器 porcelain loop insulator
陶瓷换能器 ceramic transducer
陶瓷换能头 ceramic cartridge
陶瓷换热器 ceramic heat exchanger; ceramic recuperator
陶瓷基复合材料 ceramic matrix composite
陶瓷基黏[粘]结剂 ceramic adhesive
陶瓷基片 ceramic substrate; ceramic wafer
陶瓷及耐火工业 ceramic and refractory industry
陶瓷级滑石 ceramic grade talc
陶瓷集料 ceramic aggregate
陶瓷加工废水 pottery wastewater
陶瓷夹板 ceramic cleat;cleat
陶瓷建筑 terra-cotta architecture
陶瓷建筑材料 ceramic building material
陶瓷建筑单元 ceramic building unit
陶瓷建筑构件 ceramic building member
陶瓷鉴频器 ceramic discriminator
陶瓷交叉流微滤 ceramic crossflow microfiltration
陶瓷胶粘剂 ceramic adhesive
陶瓷接收放大管 stacktron
陶瓷结构材料 ceramic structural material
陶瓷结构构件 ceramic structural member
陶瓷结构学 ceramography
陶瓷结合 ceramic bonding
陶瓷结合剂 ceramic bond; vitrified bond
陶瓷结合砂轮 vitrified abrasive;vitrified wheel
陶瓷介电材料 ceramic dielectric(al) material
陶瓷介质 ceramic dielectric
陶瓷金属 ceramal; ceramet; ceramic metal
陶瓷金属封接 ceramic to metal seal
陶瓷金属管 ceramic-metal tube
陶瓷金属化 ceramic metallization;ceramic metallizing
陶瓷金属系统 ceramic-metal system
陶瓷金属撞入封接 ram seal
陶瓷锦砖 ceramic mosaic tile;mosaic
陶瓷浸渍 cerdip
陶瓷晶须 ceramic whisker
陶瓷晶须复合材料 ceramic whisker composite
陶瓷绝热材料 ceramic insulation material;crolite
陶瓷绝缘 ceramic insulation
陶瓷绝缘材料 ceramic insulation material; ceramic isolating material; crolite
陶瓷绝缘带式加热器 ceramic-insulated band heater
陶瓷绝缘的 ceramic-insulated
陶瓷绝缘器 ceramic insulator;porcelain insulator
陶瓷绝缘线圈 ceramic-insulated coil
陶瓷绝缘子 ceramic insulator;porcelain insulator
陶瓷铠甲 ceramic armo(u)r
陶瓷颗粒增强剂 ceramic particulate reinforcing agent
陶瓷壳型 ceramic shell mo(u)ld
陶瓷空心砖 hollow ga(u)ged brick
陶瓷块 vitrified block
陶瓷快速烧成 rapid ceramic firing
陶瓷蜡笔 ceramic crayon
陶瓷粒料 ceramic granules
陶瓷粒状饰面材料 ceramic granular facing material
陶瓷脸盆 ceramic wash basin;ceramic washbowl
陶瓷裂纹 crayon
陶瓷淋浴(浅)盆 ceramic shower tray;clay shower tray
陶瓷滤波器 ceramic filter; porcelain filter
陶瓷滤波器振荡器 oscillator with ceramic filter
陶瓷滤头 ceramic filter;porcelain filter
陶瓷马赛克 ceramic mosaic
陶瓷帽 ceramic cap
陶瓷美术史 art history of ceramics
陶瓷门把手 ceramic door knob
陶瓷门执手 ceramic door knob
陶瓷密封 ceramic seal
陶瓷面板 terra-cotta panel
陶瓷面玻璃 ceramic-faced glass
陶瓷面层 terra-cotta facing
陶瓷面砖 ceramic tile; ceramic tile panel;ceramic veneer
陶瓷模 ceramic die
陶瓷模制层压板 ceramic mo(u)lding plyboard
陶瓷模注法 ceramic mo(u)ld casting
陶瓷膜 ceramic membrane
陶瓷膜超滤 ceramic membrane ultrafiltration
陶瓷膜处理 ceramic membrane treatment
陶瓷膜过滤 ceramic membrane filtration
陶瓷膜生化工艺 ceramic membrane-biochemical process
陶瓷膜生物反应器 ceramic membrane-bioreactor
陶瓷膜微滤 ceramic membrane microfiltration
陶瓷磨球 ceramic grinding ball
陶瓷墨水 <含陶瓷色料的墨水> ceramic ink
陶瓷耐火黏[粘]土 glass-pot clay
陶瓷耐火土托座 <窑具> saddle
陶瓷耐蚀耐高温镍基合金 ceramic hastelloy
陶瓷黏[粘]合剂 ceramic binder; ceramic bond;porcelain cement
陶瓷黏[粘]结 ceramic bond
陶瓷黏[粘]结剂 ceramic bond;porcelain cement;vitrified bond
陶瓷黏[粘]土矿石 pottery clay ore
陶瓷排水管 ceramic discharge pipe; ceramic drain; ceramic draining pipe
陶瓷泡沫 ceramic foam
陶瓷配料 ceramic body composition
陶瓷喷敷层 ceramic spray coating
陶瓷喷淋浴盆 clay spray shower tray
陶瓷喷淋浴浅盆 ceramic spray shower tray
陶瓷喷嘴 ceramic nozzle; porcelain nozzle
陶瓷盆景 ceramic landscape
陶瓷坯的吸水试验 absorption test
陶瓷坯刻花车床 engine turning lathe
陶瓷坯料的制备 preparation of ceramic body
陶瓷坯泥 ceramic paste
陶瓷坯体 ceramic body
陶瓷坯体着色剂 pottery-body stain
陶瓷坯体组分 ceramic body composition
陶瓷片 ceramic chip; ceramic plate; ceramic wafer
陶瓷瓶 mason jar
陶瓷瓶式窑 Hob mouthed oven
陶瓷铺地砖 ceramic flooring tile
陶瓷铺面(材料) terra-cotta(sur)facing;ceramic surfacing
陶瓷汽缸 ceramic cylinder
陶瓷砌块 ceramic block; terra-cotta block
陶瓷器 pottery
陶瓷器表面的光泽 reflet
陶瓷器皿黏[粘]连 kiss
陶瓷器皿摄影术 ceramic photography
陶瓷器铸坯法 pottery casting
陶瓷嵌板 ceramic lay-in panel
陶瓷墙面砖 ceramic wall tile
陶瓷墙砖 terra-cotta wall tile
陶瓷燃料 ceramic fuel
陶瓷燃料堆 ceramic reactor
陶瓷燃料元件 ceramic fuel element
陶瓷燃烧器喷嘴 ceramic burner
陶瓷燃烧室 ceramic-lined chamber
陶瓷热交换器 ceramic heat exchanger
陶瓷容器 ceramic vessel; earthenware porcelain receptacle
陶瓷散热器 porcelain radiator
陶瓷色材 ceramic colo(u)rant
陶瓷色剂 ceramic stain
陶瓷色料 ceramic colo(u)r
陶瓷色素 ceramic stain
陶瓷色釉 ceramic colo(u)r;ceramic pigment
陶瓷刹车片 ceramic brake
陶瓷砂轮 vitrified abrasive
陶瓷烧结 ceramic post sintering
陶瓷烧结术 ceramic sintering technology
陶瓷烧蚀材料 ablator ceramics
陶瓷生物材料 ceramic biomaterial
陶瓷声控砖 ceramic sound-control brick
陶瓷拾声器 ceramic pickup
陶瓷拾音器 ceramic pickup
陶瓷拾音器芯座 ceramic cartridge
陶瓷拾音头 ceramic cartridge
陶瓷史 ceramic history
陶瓷饰边机 machine for decorating rim of ceramic ware
陶瓷饰面 ceramic facing;ceramic veneer
陶瓷饰面板 ceramic wafer
陶瓷受话器 ceramic receiver
陶瓷衰减器元件 ceramic attenuator element
陶瓷双列直插式封装 ceramic dual in-line package
陶瓷素坯 bisque
陶瓷塑料 ceramoplastic
陶瓷塑像 terra-cotta statue
陶瓷碎坯 pitcher
陶瓷碎片 pot sherd;sherd
陶瓷套管 porcelain bushing(shell)
陶瓷体 ceramic body
陶瓷填充圈 ceramic filling ring;Pall ring
陶瓷填料 ceramic filler;ceramic packing
陶瓷贴花 ceramic transfer picture
陶瓷贴花纸 decal paper for ceramic ware
陶瓷涂层 ceramic coat(ing)
陶瓷涂层金属切削工具 ceramic coated metal cutting tool
陶瓷涂层切削工具 ceramic coated cutting tool
陶瓷涂釉 ceramic glazing
陶瓷土 ceramic clay
陶瓷土湿治 <以改善匀度与塑性> souring
陶瓷团粒 ceramic aggregate
陶瓷脱模油 repress oil
陶瓷瓦砖 vitrified-clay tile
陶瓷外壳 ceramic package
陶瓷微滤膜 ceramic microfiltration membrane
陶瓷卫生洁具 china sanitary wares
陶瓷涡轮机 ceramic turbine
陶瓷屋瓦 terra-cotta roof(ing) tile
陶瓷洗面器 ceramic wash basin
陶瓷洗手盆 ceramic lavatory
陶瓷下水管 stoneware drain pipe
陶瓷纤维 ceramic fiber[fibre];nextel
陶瓷纤维板 ceramic fiber board
陶瓷纤维编织物 ceramic fiber mat
陶瓷纤维带 ceramic fiber strip; ceramic fiber tape
陶瓷纤维浇注料 ceramic fiber castable
陶瓷纤维绝热材料 ceramic fiber insulation material
陶瓷纤维绝热层 ceramic fiber insulation
陶瓷纤维绳 ceramic fiber rope
陶瓷纤维毯 ceramic fiber blanket
陶瓷纤维异型制品 shaped ceramic fiber product
陶瓷纤维增强剂 ceramic reinforcing agent
陶瓷纤维毡 ceramic fiber felt
陶瓷纤维纸 ceramic fiber paper
陶瓷显微结构 ceramic microstructure
陶瓷相 ceramic phase
陶瓷镶板 clay lay-in panel
陶瓷镶面 ceramic veneer
陶瓷谐振器件 ceramic resonant device
陶瓷芯 ceramic core
陶瓷型法 Shaw process;unicast process
陶瓷型铸造 Shaw process;slurry mo(u)lding process
陶瓷学 ceramics
陶瓷压电元件 ceraminator;cerap
陶瓷压挤机 ceramic extrusion machine
陶瓷延迟线 ceramic delay line
陶瓷岩 porcellanite
陶瓷研磨体 ceramic grinding media
陶瓷颜料 ceramic colo(u)r;ceramic paint
陶瓷扬料板 ceramic lifter
陶瓷扬声器 ceramic loudspeaker;ce-

ramic microphone
陶瓷窑 pottery kiln
陶瓷窑炉 ceramic kiln
陶瓷业 ceramics
陶瓷业用润滑油 brockle oil
陶瓷叶片 ceramic blade
陶瓷艺术 art of pottery
陶瓷引擎 ceramic engine
陶瓷印花 ceramic printing
陶瓷印像法 photoceramics
陶瓷用黏[粘]土 ceramic clay
陶瓷用熟石膏 ceramic pottery plaster; gypsum pottery plaster
陶瓷用油 ceramic oil
陶瓷釉 ceramic colo(u)r glaze; ceramic glaze
陶瓷釉刀痕 metal marking
陶瓷釉面 ceramic glazed coat(ing)
陶瓷元件 ceramic unit
陶瓷元件反应堆 ceramic unit reactor
陶瓷原料 ceramic material; ceramic raw material
陶瓷原料矿产 ceramic raw material commodities
陶瓷增添美化 terra-cotta enrichment
陶瓷增添装饰 terra-cotta enrichment
陶瓷照相法 photoceramic process; photoceramics
陶瓷针型栅格阵列 ceramic pin grid array
陶瓷真空管 ceramic vacuum tube
陶瓷蒸发计 clay atmometer
陶瓷执手 ceramic knob
陶瓷制备科学 ceramic preparation science
陶瓷制品 ceramic article; ceramic product; ceramic ware; clay article; saltern ware; ceramics
陶瓷制造废水 potter wastewater
陶瓷制造工艺 ceramic process
陶瓷制造者 ceramist
陶瓷轴承 ceramic bearing
陶瓷烛式过滤器 ceramic candle filter
陶瓷铸型 ceramic mo(u)ld
陶瓷砖 ceramic screen brick
陶瓷砖板 biscuit tile
陶瓷砖胶结剂 ceramic tile adhesive
陶瓷砖块 ceramic tile block
陶瓷砖墙 ceramic wall
陶瓷转盘 jolley; jollie
陶瓷装潢 terra-cotta decoration
陶瓷装潢面貌 terra-cotta decorative feature
陶瓷装潢特色 terra-cotta decorative feature
陶瓷装潢终饰 terra-cotta decorative finish; terra-cotta ornamental finish
陶瓷装潢装置 terra-cotta decorative fixture
陶瓷装饰品 terra-cotta ornament
陶瓷装饰特色 terra-cotta ornamental feature
陶瓷着色剂 ceramic colo(u)rant
陶瓷组织结构 ceramic constitutional structure
陶瓷组织学 ceramography
陶瓷钻头 ceramic bit
陶鬲 cooking vessel
陶工 potter
陶工修坯用牛角刀 potter's horn
陶工旋轮式积分器 potter's wheel integrator
陶工旋盘 potter's wheel
陶管 ceramics; earth conduit; earthenware pipe; pottery pipe; stoneware pipe; terra-cotta pipe; vitrified-clay pipe; vitrified pipe; clay pipe
陶管电阻器 ceramic resistor
陶管喇叭头 male end of a pipe
陶管连接配件 oddment
陶管排水 tile drain
陶管砂床试验法 sand-bearing method for testing clay pipes
陶管小端头 female end of a pipe
陶罐 clay pot; gallipot; pottery jar
陶规鬲 < 中国古代陶质炊具 > pottery gui
陶壶 pottery pot
陶化 vitrification
陶化的 vitrified
陶记 notes on ceramics
陶结块 keramzite
陶钧 potter's wheel
陶凯公式计算法 Turc formula method
陶块 brick ware
陶类滤器 earth-type filter
陶立克波纹线脚 Doric cyma(tium)
陶立克雕带 Doric frieze
陶立克风格 Doric style
陶立克结构 Doric structure
陶立克门廊 Doric portico
陶立克式 < 希腊式建筑 > dorian
陶立克式波状花边的 cymatium recta
陶立克式的 doric
陶立克式建筑 Doric structure
陶立克式楣梁 Doric architrave; Doric epistyle
陶立克式神庙 Doric temple
陶立克式凸圆线脚 Doric echinus
陶立克式檐壁 Doric frieze
陶立克式柱座 Doric base
陶立克体 doric
陶立克柱 Doric column
陶立克柱顶过梁 Doric architrave
陶立克柱顶盘 Doric entablature
陶立克柱基础 Doric base
陶立克柱廊 Doric colonnade
陶立克柱上飞檐雕带 Doric frieze
陶立克柱上飞檐过梁 Doric frieze
陶立克柱式 Doric order
陶立克柱式的原型 proto-Doric order
陶立克柱头 capital of Doric column; Doric capital
陶立克柱头顶板下四分之一圆饰 Doric echinus
陶立克柱型 Doric order
陶立克柱檐口 Doric cornice
陶立克柱檐座 Doric entablature
陶粒 aglite; ceramsite; clay aggregate; haydite; porcelain granule; sintered aggregate
陶粒骨料 haydite type aggregate; sintered aggregate
陶粒骨料混凝土 haydite aggregated concrete
陶粒过滤 ceramic particle filtration
陶粒混凝土 all-haydite concrete; ceramic concrete; ceramisite concrete; haydite concrete; sand-haydite concrete; sintered aggregate concrete; sintered clay concrete
陶粒混凝土墙板 ceramsite concrete wall panel; haydite concrete wall panel
陶粒集料 haydite type aggregate; sintered aggregate
陶粒集料混凝土 haydite aggregated concrete
陶粒介质 ceramic particle medium
陶粒路面 ceramsite pavement
陶粒-锰砂粒滤池 ceramic particles-manganese sand particles filter
陶粒黏[粘]土矿床 earthenware deposit
陶粒轻骨料 lightweight expanded clay aggregate
陶粒轻集料 lightweight expanded clay aggregate
陶粒砂混凝土 sand-haydite concrete
陶粒生物滤池 bio-ceramic filter
陶粒页岩 ceramisite shale
陶粒页岩矿床 earthenware shale deposit
陶粒用黏[粘]土 ceramisite clay
陶路管 earthenware duct
陶轮 potter's wheel
陶轮的顶部表面 wheel-head
陶面釉 earthenware glazed finish
陶面砖 veneered brick
陶面砖铺地 earthenware tile pavement
陶奈黄铜 Tournay brass
陶泥釉 Albany slip
陶坯 green ware
陶坯体 earthenware body
陶片砖铺面【建】 earthenware tile pavement
陶器 ceram; crockery; earthenware; figuline; keramics; potter's work; pottery(ware); vitreous china; vitrified china
陶器板 earthenware slab
陶器的 ceramic; fictile; keramic
陶器的图案描绘者 figurer
陶器干燥泛白 drier[dryer] white
陶器挂釉底料 all clay body
陶器罐 earthenware jars
陶器烘炉 biscuit oven
陶器烘箱 biscuit oven
陶器烤炉 biscuit oven
陶器裂纹 craze
陶器面图画 china painting
陶器面砖 faience tile
陶器模 clay mould
陶器模制 pottery mo(u)lding
陶器黏[粘]土 earthenware clay
陶器坯 unfired ceramic body
陶器器皿 pottery ware
陶器墙砖 majolica wall tile
陶器上的支撑痕 pluck(ing)
陶器上釉层 majolica glazed coat(ing)
陶器生产方法 china process
陶器素坯 biscuit; bisque
陶器素坯焙烧 biscuit firing
陶器素坯窑 biscuit kiln
陶器瓦 < 石灰质的 > majolica tile
陶器下水道 sewer stoneware
陶器修理所 China-hospital
陶器旋床 pottery lathe
陶器旋坯 pottery turning
陶器窑 pottery kiln
陶器用褐色釉 brown salt glaze on pottery
陶器釉 earthenware glaze; pottery glaze
陶器制造 potting
陶器制造术 pottery
陶器制造所 pottery
陶器铸坯 pottery casting
陶器转盘 Jolly
陶砂 ceramsite; fine ceramisite
陶石 pottery stone
陶塑工程 terra-cotta work
陶塑用黏[粘]土 potter's clay; potter's earth; pottery clay; terra-cotta clay
陶胎琉璃珠 beads with pottery body
陶土 argil(la); baked clay; bolus; carclazyte; earthenware clay; figuline; glass-pot clay; porcelain clay; pot clay; potter('s) clay; pottery clay; soft clay; vitrified clay; ball clay < 高塑性的细土 >
陶土板岩 argillaceous slate
陶土标准溶液 china clay standard solution
陶土槽形板楼盖 Fawcett's floor
陶土厂 clay mill
陶土瓷砖 clay tile
陶土导管 clay conduit
陶土电缆覆盖 earthenware cable cover
陶土粉 kaolin powder
陶土坩埚法 clay crucible process
陶土坩埚拉丝炉 fireclay bushing furnace
陶土构件 clay member; clay unit
陶土构件的企口 lip joint
陶土管 earth ware pipe; pottery pipe; stoneware pipe; terra-cotta conduit; terra-cotta pipe; terra-cotta tube; tile pipe; vitrified-clay pipe; vitrified pipe; vitrified tile; ware pipe; ceramic pipe; clay pipe; earthen pipe
陶土管道 clay conduit
陶土管修理班 tile repair screw
陶土化作用 kaolinization[kaolinisation]
陶土锦砖 earthenware mosaic
陶土精制 earth refining
陶土空心砖加筋的密肋楼板 tile lintel floor
陶土滤器 earth filter
陶土滤水管 earthenware filter pipe
陶土马赛克 earthenware mosaic; pottery mosaic
陶土排水管 clay pipe drainage; tile drain
陶土排水管道 clay pipe line
陶土片岩 argillaceous schist
陶土墙面砖 earthenware wall tile
陶土套管 tile envelope; tile jacket
陶土通风管 clay vent pipe
陶土瓦 vitrified tile; clay tile
陶土瓦管管道 clay tile line
陶土污水管配件 clay sewer pipe and fittings
陶土下水管 clay sewer pipe
陶土岩 ceramicite
陶土用转盘 potter's wheel
陶土釉面修饰 earthenware glazed finish
陶土制品 clay product; clayware; salt-glazed ware; terra-cotta
陶土制品工厂 clay plant
陶瓦 clay shingle; stoneware tile; terra-cotta tile
陶瓦半管 stoneware half-pipe
陶瓦槽 stoneware tank
陶瓦产品 stoneware product
陶瓦厂 stoneware plant
陶瓦衬里 stoneware lining
陶瓦地砖 stoneware floor cover(ing) tile; stoneware tile floor cover(ing)
陶瓦废水管 stoneware water pipe
陶瓦工厂 stoneware factory
陶瓦沟槽 stoneware half-pipe
陶瓦沟管 terra-cotta sewage conduit
陶瓦管 terra-cotta pipe; vitrified tile
陶瓦管接头水泥 stoneware pipe joint cement
陶瓦管零配件 stoneware pipe fittings
陶瓦接合 stoneware junction
陶瓦块 flashing block
陶瓦滤器 stoneware filter
陶瓦明沟 stoneware gutter
陶瓦排管 stoneware drain
陶瓦排水沟 stoneware discharge gutter; stoneware drainage gutter
陶瓦排水管 stoneware discharge pipe; stoneware drainage pipe
陶瓦排水瓦筒 stoneware discharge pipe
陶瓦配件 stoneware fittings
陶瓦平顶砖 stoneware ceiling tile
陶瓦器 stoneware
陶瓦器件 stoneware article
陶瓦饰面 stoneware facing

陶瓦水管接头组合件 stoneware joint(ing) compound
陶瓦水箱 stoneware tank
陶瓦天沟 stoneware roof gutter
陶瓦贴面 stoneware(sur) facing
陶瓦桶 stoneware tank
陶瓦污水管 stoneware for sewer pipes
陶瓦污水管接头组合物 stoneware sewer joint(ing) compound
陶瓦屋谷 clay tile valley
陶瓦屋面 clay tile roofing
陶瓦下水管 terra-cotta sewage pipe
陶瓦修饰 stoneware(sur) facing
陶瓦油箱 stoneware tank
陶瓦制品 stoneware goods
陶形曲面 mo(u)lding surface
陶雅 book on ceramics
陶衣 a thin coating on ceramics
陶俑 pottery figurine
陶釉 ceramic glaze;Bristol glaze
陶釉搪瓷 majolica enamel
陶釉污水管 ceramic glazed sewer pipe
陶渣 grog
陶枕 pottery pillow
陶制泵 stoneware pump
陶制大缸 ark
陶制的 earthen;earth-type;figuline
陶制电缆护盖 earthenware cable cover
陶制缸 craggan
陶制管 stoneware pipe; terra-cotta pipe;vitrified pipe
陶制管道 stoneware conduit
陶制过滤器 earth-type filter
陶制壶 olla
陶制滤管 earthenware filter pipe
陶制马赛克 earthenware mosaic
陶制面砖 earthenware tile
陶制品 fictile;brown ware <褐色的>
陶制墙面砖 earthenware wall tile
陶制容器 earthen container;earthenware container;earthenware vessel
陶制塔 earthenware tower
陶制下水管 terra-cotta sewage pipe
陶制小块马赛克 earthenware small-sized mosaic
陶制蒸馏塔 earthen tower
陶制贮罐 earthen storage tank
陶质板<印刷电路用> ceramic wafer
陶质材料 ceramic material
陶质的多孔滤筒 bougie
陶质顶棚面砖 clay ceiling tile
陶质反应堆 ceramic reactor
陶质焊剂 baked flux;bond(ed) flux;ceramic flux
陶质换能器 ceramic transducer
陶质换热器 brick recuperator;ceramic recuperator
陶质建筑材料 clay building material;clay-pottery building material
陶质建筑构件 clay building member;clay building unit
陶质结构材料 clay structural material
陶质砌块 clay block
陶质酸坛或玻璃瓶 earthenware or glass carboy
陶质涂层 ceramic coat(ing)
陶质污水管 ceramic waste pipe
陶质烟囱顶管 clay chimney pot
陶砖 brick ware;terra-cotta;vitrified brick
陶砖板 terra-cotta slab
陶砖护岸 terra-cotta block protection
陶砖铺面 earthen tile pavement;earthware tile pavement
陶砖饰面 terra-cotta block finish

淘

淘岸 bank caving
淘簸筛 jig(ger)
淘成洞穴 caving bank
淘出的黏[粘]土 washed clay
淘涤 sluice
淘底冲刷 undermine;undermining
淘分机 elutriator
淘分试验 elutriation test
淘结块 keramite
淘金 alluvial mining;placer-mining
淘金木桶 batea
淘金盘 prospecting disk[disc]
淘金热 gold rush
淘空 hollowing;caving <河岸的>
淘空的堤岸 caving bank
淘矿(法) vanning
淘矿机 Frue vanner;vanner;vanning machine
淘泥机 clay slurry preparator;wash mill
淘盘 pan
淘盘洗选法 panning
淘盘选 panning
淘砂盘<又称淘沙盘> batea;rocker;panning plate
淘砂盘洗选 rocking
淘蚀 scour;undercut
淘蚀岸 river cliff
淘蚀岸坡 undercut slope
淘蚀保护 scour protection
淘蚀深度 scour depth
淘刷 downcutting;incaving;scouring;undercutting; undermine; underwash(ing)
淘水戽 bailing bucket
淘汰 boult; filtering; roguing; sift; washout
淘汰标准 culling level
淘汰策略 replacement policy
淘汰的 selective
淘汰法 method of elimination;screening method
淘汰或改造燃煤锅炉 eliminate or convert coal-burning boiler
淘汰机 cradle
淘汰技术 sweeping technique
淘汰旧机器 phasing out of obsolete machinery
淘汰矩阵 sweeping matrix
淘汰落后产品 eliminating of outmoded products
淘汰落后工艺、技术和设备 retirement of outdated processes, technologies and equipment
淘汰盘 buddle
淘汰盘选矿 buddle-work;tabling
淘汰型号 superseded model
淘挖 scoop
淘析 elutriate;elutriating;elutriation;elutriation wet analysis
淘析法 elutriation analysis;elutriation method
淘析环 elutriation ring
淘析精矿 elutriated product
淘析离心机 elutriator-centrifuge
淘析漏斗 elutriating funnel
淘析瓶 elutriating flask
淘析气体色层分离法 elution gas chromatography
淘析器 elutriating apparatus;elutriator
淘析试验 elutriation test
淘析装置 elutriation apparatus
淘洗 elutriate;elutriation;washing
淘洗的黏[粘]土 washed clay
淘洗的污泥 elutriated sludge
淘洗法 elutriation method
淘洗分选器 elutriation separator
淘洗高岭土 washed kaolin
淘洗过程 elutriation process
淘洗机 wash mill
淘洗金刚石用淘盘 diamond pan
淘洗离心机 elutriator-centrifuge
淘洗率 washing rate
淘洗盘 washing pan
淘洗器 elutriator
淘洗设备 elutriation apparatus
淘洗试验 elutriation test
淘洗水 elutriation water
淘洗污泥 elutriation sludge
淘洗重矿物工艺 panning
淘选 buddle-work; buddling; elutriation;levigate;levigation;vanning
淘选铲 van
淘选带 vanner
淘选法 elutriation method
淘选试验 elutriate test
淘渣 chamot(te)

讨

讨价 asking price;make a price
讨价还价 argybargy;bargain;bargain over the price;chaffer;drive a bargain;haggle;higgle;higgling;make a bid;price bargain
讨价还价的筹码 bargaining counter
讨价还价的地位 bargaining position
讨价还价的能力 ability to bargain;bargaining power
讨价还价买卖策略 bargaining strategy
讨价还价者 bargainer
讨论会 seminar; colloquium; panel discussion; symposium; workshop;forum[复 forums/fora]
讨论会论文集 symposium
讨论会主持人 panelist
讨论教室 seminar
讨论式决算表 discussion statement
讨论终结 closure
讨厌的工作 stinkers
讨厌的天气 putrid weather
讨厌的味道 troublesome taste
讨厌的味觉 unpleasant taste
讨厌的植物 undesired plant
讨厌气味 unpleasant smell
讨债人 dun

套

套扳手 spanner wrench
套班运行 package run
套板 cleading;lag(ging);strap;toggle plate
套版 registering
套包 collar for a horse
套包伤 collar injury
套杯 retainer cup
套壁 jacket wall
套柄铁锤 fuller;hand fuller
套柄凿 socket chisel
套餐巾用的小环 napkin ring
套槽 grooving
套层 jacket layer;mantle layer
套层的 jacketed
套层阀 jacket valve
套层结晶器 jacketed crystallizer
套层蒸发器 jacketed evapo(u)rator
套层蒸馏器 jacketed still
套层注射器 jacketed syringe
套车 harness an animal to a cart
套车的马 cart-horse
套撑 socketed stanchion
套抽样 nested sampling
套床 trundle bed
套锤锻前 fuller
套带螺钉 jacket band screw
套袋机 bag placer
套堤 ring levee;setback levee
套垫 sleeve gasket
套迭 intussusception;invagination
套叠 telescoping
套叠的 telescopic(al)
套叠式的 telescopiform
套叠形状 telescoped shape
套叠钻塔 telescopic(al) derrick
套洞扳手 tommy
套阀 jacket valve;sleeve valve;sleeving valve
套阀发动机 knight engine
套阀灌浆管 sleeved grout pipe
套阀孔 sleeve-port
套方亭 linked square kiosk
套房 adult suite;flat;suite(of rooms)
套房楼 block of flats
套分解算法 nested decomposition algorithm
套封包装机 sleeve wrapper
套覆玻璃 cased glass
套盖箱 case-lid box
套杆 loop bar
套格子花纹 overcheck
套钩 hitch;rocker
套钩臂 toggle lever
套钩角铁 hitch angle
套购 arbitrage
套购人 arbitrage dealer
套购业务 arbitrage business
套购账 arbitrage account
套箍 cuff
套挂断错 hitch
套挂断错故障 hitch
套管 casing(pipe);casing tube;adapting pipe;barrel core;barrel fitting;barrel of pipe; bolster; boot; bor(d) casing; bored tube; borehole lining; borehole tube; boring casing; boring tube; branch piece; branch pipe; bushing; case pipe; cat's head;cathead;collar;conductor string; connecting; connector; coupling;double pipe;double tube;drive shoe; ferrule; guide tube; hoisting pipe;housing pipe;jacket; jacket pipe;joint box;junction box; lining pipe; lining tube; annular tube;mantle(d) pipe;mouthpiece; nestable pipe; pipe casing; pipe duct;pipe liner;pipe sleeve;shackle;sheath;sleeve;sleeve(d) pipe; sleeve(d) tube;sleeve piece;sleeving;socket (ed) pipe; socketed tube;spigot;thimble;drive pipe <防钻孔坍壁的>;casing spear <深孔技术>;cannula[复 cannulae/cannulas]
套管安装钻孔机 casing-mounted drill rig
套管扳钳 casing tongs
套管扳手 tubler key;tubular key
套管壁厚 casing wall thickness
套管标记 casing mark
套管部件 casing part
套管测井 casing log
套管测漏器 casing(leakage) tester
套管测试器 casing tester
套管插入术 cannulation
套管长度 length of casing
套管车 bull wagon;casing wagon
套管成本 casing cost
套管承托环 casing seat
套管尺寸 casing size
套管齿轮 quill gear
套管冲孔器 casing perforator;tubing perforator
套管冲孔枪 casing gun
套管冲头托架 thimble punch holder
套管冲洗铣鞋 rotary washover shoe
套管传动 quill drive
套管锤 drive block
套管打捞 fishing for casing
套管打捞公锥 casing tap

套管打捞矛 bulldog casing spear; casing dog; tubing dog; tubing spear; casing spear
套管打捞器 tubing socket
套管打捞筒 casing bowl; trip casing spear; tubing socket
套管打捞抓 casing grab
套管打入头 casing drive head; pipe drive head
套管打入靴 casing drive shoe
套管打头 drive pipe head
套管大小头 casing sub(stitute)
套管单位重量 casing unit weight
套管导向帽 casing guide
套管导正装置 casing guide
套管的 telescopic(al)
套管的接箍端 collar end of the casing
套管的开端 open-end of tubing
套管的连顶接箍 landing collar
套管的压坏 collapse of casing
套管底开式井 open-bottomed well
套管吊钩 casing hook
套管吊机 casing elevator
套管吊卡 casing elevator; casing hanger
套管吊钳 casing tongs
套管顶端 bell nipple
套管定中器 casing centralizer
套管堵塞器 casing anchor packer
套管短截线 sleeve stub
套管短柱 sleeve stub
套管段长 length of casing section
套管段重 weight of casing section
套管法灌浆 sleeve pipe grouting
套管法兰(盘) casing flange
套管反应器 double tube reactor
套管费 casing cost
套管封隔器 casing packer
套管敷设 casing installation
套管扶正器 casing centralizer
套管浮阀 casing float; casing valve
套管浮箍 casing float collar
套管浮靴 casing float shoe
套管浮重 casing buoyed weight
套管干重 dry weight of the casing
套管钢材最小极限强度 casing steel minimum limit strength
套管钢级 casing grades
套管钢级系数 casing steel grade coefficient
套管钢丝绳 casing line
套管钢丝绳滚筒 casing spool
套管钢丝绳滑轮 casing line pulley
套管钢靴 casing shoe
套管割刀 case ripper; casing cutter; casing knife
套管割刀加重杆 casing cutter sinker
套管割刀楔 casing cutter wedge
套管割刀震击器 casing cutter jar
套管隔离 casing off
套管跟进装置 casing advancement system
套管工具 casing tool
套管公称直径 casing nominal diameter
套管公接头 casing nipple
套管钩 thimble hook
套管固井 wall off
套管固孔 casing off hole
套管刮刀 rotary casing scraper
套管挂 bell-weevil hanger; casing hanger
套管管靴 casing shoe; drive shoe; driving shoe
套管灌浆 casing grouting; sleeve grouting
套管锅炉 thimble-tube boiler
套管焊接 socket weld
套管护壁法 casing pipe shield method; sleeve wall protection method
套管护壁钻孔法 casing hole-boring method
套管护箍 casing protector
套管护孔 casing off hole
套管护圈 casing protector; casing screw protector <公扣>; casing screw head <母扣>
套管滑阀式液压自封接头 sleeve valve type hydraulic self-sealing coupling
套管环 drive pipe ring; driving pipe ring
套管环状间隙 casing clearance
套管换热器 double pipe heat interchanger
套管混凝土桩 cased concrete pile; shelled concrete pile
套管或击入管 casing or driving pile
套管机 jacketing machine
套管夹 casing grip; clamp sleeve
套管夹板 casing clamp
套管夹具 casing clamp
套管间隙 shell clearance
套管减震柱 telescopic(al) shock strut
套管角度 angle of bend
套管矫直器 mandrel socket
套管接箍 case adapter; casing adapter; casing collar; casing connections; casing coupling; casing joint
套管接箍定位器 casing collar locator
套管接合 bell and spigot; bell joint; casing joint; sleeve joint; spigot joint; telescopic(al) joint; thimble joint(ing)
套管接口 bell joint; casing joint; collar joint; spigot joint; thimble joint(ing)
套管接头 casing head; bell-and-plain end joint; casing joint; muff joint; pipe coupling; sleeve tubing connector; spigot joint; swagelok coupling; telescoped joint; thimble joint(ing)
套管结合 thimble joint(ing)
套管进水回转接头 casing water swivel
套管进水转座连接器 casing water swivel
套管井段 casing well section
套管径厚比 casing diameter-thickness ratio
套管就地灌筑桩 driven cast-in-place pile
套管绝缘子 bushing insulator; pothead insulator
套管卡 casing clamp
套管卡盘 casing spider
套管卡瓦 casing slips
套管开孔钻头 starting casing barrel
套管抗挤下入深度 casing collapse depth
套管抗拉强度 tensile strength of casing
套管控制头 control casing head
套管扣型 thread type of casing
套管缆绳滑轮 casing line sheave
套管类型 casing form
套管累计长 total casing length
套管累计重量 total casing weight
套管冷凝器 tube-in-tube condenser
套管冷却 jacket cooling
套管连箍 casing coupling
套管连接 box connection; box coupling; casing joint; muff coupling; muff joint; sleeve coupling; sleeve joint; telescope joint; tubular splice
套管连接件 bushing fitting
套管连接器 casing collar; thimble connector
套管连接装置 telescoped joint device
套管联结 box coupling
套管联轴节 thimble coupling
套管联轴器 thimble coupling
套管漏泄 casing leak
套管螺母 sleeve nut
套管螺丝起子 box driver
套管螺纹 casing threads
套管螺纹规 casing threads ga(u)ge
套管螺旋钻钻孔 shell and auger boring
套管帽 drive cap; drive collar; drive head; driving cap; sleeve cap
套管密封 bobbin seal; seal by sleeve
套管内壁腐蚀电子检查仪 electronic casing caliper
套管内径 casing inside diameter
套管内径规 casing drift ga(u)ge
套管内液体水平 liquid level in casing
套管内钻孔桩 drilled-in-tube pile
套管偶极子 sleeve dipole
套管盘根盒 tubing packer
套管配件 casing fittings
套管劈裂器 casing splitter
套管平台 casing platform
套管破坏分类 casing breakdown type
套管起拔工具 spring dart
套管起拔机 pulling machine
套管千斤顶 casing jack
套管强度 casing strength
套管屈服破坏 casing yield breakdown
套管热交换器 double pipe exchanger
套管设计 casing design
套管射孔器 casing perforator
套管射孔枪 casing gun
套管伸缩喷枪 telescopic(al) extension lance
套管伸缩轴 extension shaft
套管深度 casing depth; depth of socket; tubing depth
套管深井 cased well
套管失稳破坏 casing unstable breakdown
套管式 double pipe; extension-type; telescoping
套管式伴随加热法 double pipe trace heating; double type trace heating
套管式变流器 bushing current transformer
套管式的 telescopic(al)
套管式电流互感器 bushing current transformer
套管式电容器 bushing-type condenser
套管式换热器 double pipe heat exchanger
套管式接合 muff coupling
套管式冷凝器 double pipe condenser; tube-in-tube condenser
套管式毛毯热风干燥辊 pocket ventilation roll
套管式热交换器 double pipe cooler; double pipe heat exchanger; double tube type heat exchanger; tube-in-tube heat exchanger
套管式柔性接头 sleeve flexible joint
套管式三脚架 telescopic(al) tripod
套管式伸缩接头 sleeve flexible joint
套管式伸缩器 sleeve-type compensator; slip-type expansion joint
套管式提升机 tubing elevator
套管式天线杆 extension mast; mast extension
套管式调整器 sleeve-type compensator
套管式油加热器 double pipe oil heater
套管式锗电阻温度计 capsule-type germanium resistance thermometer
套管式装配 telescopic(al) mount(ing)
套管事故 casing trouble
套管试验抽头 bushing test tap
套管试验器 casing tester
套管受力分类 casing force type
套管数据 casing data
套管双耳 clevis end holder
套管水龙头 casing swivel; casing water swivel
套管水泥塞 casing cement plug
套管丝扣强度 casing thread strength
套管丝扣数据 casing thread data
套管送入工具 transportation tool
套管送下程序 casing drill-in
套管塑性失稳破坏 casing plastic unstable breakdown
套管弹性不稳定破坏 casing elastic unstable breakdown
套管弹性过渡区不稳定破坏 casing elastic transition zone unstable breakdown
套管提引器 casing elevator
套管天线杆 telescopic(al) mast; telescopic(al) tower; tubular mast
套管筒 casing barrel
套管头 case head; casing housing; casing shoe; control casing head
套管头储罐 casing-head tank
套管头气体 casing-head gas; well-head gas
套管头上端 casing-head top
套管头填料函 casing-head stuffing box
套管突缘 casing flange
套管外(部)封隔器 external casing packer
套管外部腐蚀 external casing corrosion
套管外径 casing outside diameter
套管弯曲临界长度 casing buckling critical length
套管握持器 bushing holder
套管铣刀 casing section mill
套管下放到孔内台阶上 landing casing
套管下入层数 layer number of setting casing
套管下入深度 setting casing depth
套管下入直径 setting casing diameter
套管下至孔底 land the casing
套管楔 socket wedge
套管型电离箱 thimble chamber
套管修整器 casing roller; casing swage; roller swage
套管修正器 casing roller; casing swage; roller swage
套管悬挂卡瓦 casing hanger slips
套管悬挂器 casing suspender
套管靴 drive pipe shoe; footpiece; casing shoe
套管靴深度 casing shoe depth
套管靴下扩孔 under-ream(ing)
套管靴下扩孔器 underreamer
套管靴下扩孔器的推出式切刃 underreamer cutter; underreamer lug
套管靴下扩孔钻头 undereaming bit
套管靴钻头 casing shoe bit; set casing shoe
套管压机 socket press
套管压力 casing pressure
套管压入机 bushing press
套管压入头 casing snubber
套管压缩距 amount of casing compression
套管液压吊钳 casing hydraulic tongs
套管异径接头 casing adapter; casing sub(stitute)
套管引靴 casing guide
套管凿 bolster chisel
套管炸药包 casing squib
套管找中器 casing centralizer
套管蒸汽 jacket steam
套管整形器 casing mandrel
套管支撑 jack shore
套管支柱 jack shore

套管止水 water sealing with casing
套管重量 casing weight
套管轴 quill; quill shaft; sleeve; telescopic(al) shaft
套管注浆 casing injection
套管注水泥法 casing method of cementing
套管注水泥作业 casing cementing job
套管柱 borehole line; casing column; casing string; string of casing
套管柱拧紧度 spring of the casing
套管柱与井壁间的间隙 hole clearance
套管转接器 casing sub(stitute)
套管桩 belaying pile; cased pile
套管自重伸长 casing weight elongation
套管总重 total casing weight
套管(纵向)割刀 casing ripper; casing splitter
套管组件 thimble assembly
套管钻机 sleeve drill
套管钻进 casing drilling; drilling with casing; drill with casing tubes; drive pipe drilling
套管钻孔 cased bore hole; cased hole; cased hole completion
套管钻孔平台 casing boring platform
套管钻孔试验 cased-hole-test
套管钻孔用钻头 casing bit
套管钻孔桩 socketed pile
套管钻孔桩基突堤码头 socketed column pier
套管钻孔桩在基岩内的灌浆销子 grouted pile pin
套管钻头 pipe bit
套管座 casing seat; thimble seat
套规 socket ga(u)ge
套轨线路 interlaced line; overlapped line
套锅 jacketed kettle
套合标志 corner tick
套合差 register difference
套合定位孔 register hole punch; register punch
套合定位栓 registration bar
套合规矩线 hairline register
套合级数 nested level
套合精度 registration accuracy
套合式容器 shrink-fit vessel
套合试验 register trail
套合栓 register pin
套合误差 register error; registration error
套合线 register tick
套合样图 register proof
套红 red printing
套红料 flash ruby
套环 collar; eye ring; lantern ring; shrink ring; shroud ring; sleeve; socket ring; staple bolt; toggle
套环换向器 shrink-ring commutator
套环接头 collar joint
套环联结 collar joint
套环螺栓 box closure; clevis bolt; draw bolt; toggle bolt
套环瓶 <转心瓶> revolving vase
套环式测渗仪 concentric(al) ring infiltrometer; ring infiltrometer
套环制动器 toggle brake
套环钻 annular borer
套汇 arbitrage; currency arbitrage; exchange arbitrage
套汇股票 arbitrage bond
套汇汇率 cross rate
套汇价 cross rate
套汇率 cross rate
套汇人 arbitrage dealer; arbitrager; arbitrageur
套汇商 arbitrageur
套汇账 arbitrage account
套几 nested tables; nest of tables
套加印 overprint
套夹 cartridge clip
套价 cross rate
套假设 nested hypothesis
套间 closet; compartment; inner room; stanza; suite; zotheca; flatlet <英国只有一个房间而附有厨房、浴室的>
套间房 suite room
套间客房 suite
套间门 communicating door
套间门锁 communicating door lock; connecting door lock
套间式病房 suite room for patient
套简 chuck
套简销 socket pin
套简轴承 bush bearer
套件 external member; sleeve piece
套接 bell socket; cup joint; dowel joint; housed; house joint; muff coupling; socket joint; thimble joint(ing); muff joint <两个管筒之内用另一管筒套入连接起来>
套接棒 socketed bar; socketed rod
套接的 belled
套接短管 flue pipe brick
套接法兰 sleeve flange
套接管 <俗称缩节> casing coupling; socket pipe; spigot and socket pipe; socketed tube
套接管插入端 spigot end of pipe
套接管子 faucet pipe; socket pipe
套接接件 socket connector
套接接头 socket connector; telescopic(al) lap splice
套接口 faucet pipe
套接圈 mating collar
套接式减速齿轮 rested reduction gear
套接榫 tusk tenon
套接套管 <套管一端套入另一套管的一端后加焊> slip-joint casing
套接头 spigot and socket pipe
套接瓦筒 flue pipe brick
套接(支)柱 socketed stanchion
套接字对 socket pair
套节 connecting sleeve; hub; socket
套节管 socket pipe
套节焊接配件 socket-weld fitting
套节及套管 hub and spigot
套节熔融接合 socket-fused joint
套节压机 socket press
套结瓦管 bell and spigot tile
套(井壁)管钻管 casing pipe
套颈 hub
套具 harness
套壳式轮 shrouded wheel
套孔 trepan; trepan boring; trepanning
套孔刀 trepanning tool
套孔法 over-coring method
套孔机 trepan borer; trepan boring machine
套孔锥 trepanning drill
套口 linking; mouthing; mouthpiece; muzzle bell; pocket
套口缝 linking seam; looping seam
套口横列 linking course; looper course
套口机 linking machine; looper; looping machine
套口机剪线装置 looper clip
套口架 linking jig
套口控制器 looper control
套口设备 <钻杆> stock and dies
套口针 linking machine needle; looping needle
套口针片 looper point
套拉绳 parbuckle
套肋 jacket rib
套利 arbitrage
套利公司 arbitrage house; arbitrated house
套料 casing jacking; jacking【机】; overlaying <玻璃>
套料玻璃 overlay glass
套料玻璃器皿 cased hollow ware
套料乳白玻璃 flashed opal; flashed opal glass
套料乳白平板玻璃 flashed opal plat glass
套料铣刀 hollow fraise
套料钻 core drill; cylindric(al) drill; trepanning drill
套料钻探法 core drill method
套炉 muffle furnace
套罗口 running-on
套买 hedge purchase
套买保值 hedge buying; hedge selling
套卖 hedge; hedge sale
套帽 drive head
套模 cover die
套膜沟 mantle groove; pallial groove
套膜节 mantle region; mantle segment
套膜褶 mantle fold(ing)
套片 nest plate; plate-fin
套片式翅片管 casing fin tube
套瓶帽机 capsuler
套期保值 hedging
套期保值条款 hedge clause
套腔 mantle cavity
套墙 jacketed wall
套切细纹规格 casing thread size
套球式温度计 globe thermometer
套取脱落岩芯 running over core
套取岩芯 overcoring
套圈 bring forward; collar; cover ring; spigot ring; thimble; ferrule【机】
套圈定心保持架 ring-centered cage
套圈加热器 ring heater
套圈轮 pushing-in wheel
套圈选配装置 ring gauging device; ring matching device
套染 resist-dye
套入 telescope; telescoping
套入接头 housed joint
套入式法兰 slip-on flange
套入式接头 shouldered housed joint
套塞接头 hub and spigot joint
套色 casing; colo(u)r registration; overlaying <玻璃>
套色版 chromatic printing; chromatogram; chromatograph; colo(u)r printing; multicolo(u)r printing
套色玻璃 cased glass; casing glass; flashed glass
套色打样 progressive sheet
套色法 chromatography
套色夹层玻璃 tinted laminated glass
套色刻花玻璃 cased glass cutting
套色木刻 colo(u)red woodcut
套色拼隔版 dissecting
套色漆 scumble
套色区 area mask
套色石印术 chromolithography
套色石印图 chromotype
套色印刷 chromaticity printing
套色印样 combined sheet
套色装置 colo(u)r register
套筛 bushing screen; nest of sieves; set of screens; set of sieves
套晒制版 double shooting
套上 place on; slip on; strap
套上轮缘 rimming
套上游梁 <冲击钻动作> hitch to the beam
套绳 running knot
套式焙烧炉 muffler roaster
套式插齿刀 shell slotting cutter
套式粗齿面铣刀 shell coarse tooth face(milling) cutter
套式地表水监测网 nested surface water monitoring network
套式轨道 gantlet(ted) track
套式弧形键槽铣刀 shell radial keyway cutter
套式加热 jacketing heating
套式加温 jacketing heating
套式铰刀 shell reamer
套式校对规 quill master
套式接合 slip-fit connection
套式扩孔钻 shell drill
套式拉刀 shell broach
套式冷却 jacketing; jacketing cool
套式立铣刀 shell end mill
套式炉 blind roaster
套式面铣刀 facing-type cutter
套式喷雾器 sleeve atomizer
套式手钻 shell gimlet
套式铣刀 arbor-type cutter; shell cutter
套式细齿面铣刀 shell fine tooth face milling cutter
套式桩帽 hood pile cap
套数 number of sets
套数齿轮刀具 number cutter
套栓式窗撑 peg stay
套栓式风撑 peg stay
套丝 mantle fiber[fibre]; threading
套丝板 threaded plate
套丝机 threading machine
套塑光纤 plastic-jacketed optic(al) fiber
套算汇率 cross rate of exchange
套算汇率差距 break-in cross rates
套索 lasso; noose
套索钉 toggle
套索柱 belaying pin
套索桩 belaying pin; toggle
套弹簧 sleeve spring
套筒 bolt sleeve; breast; bushing; cartridge; draw tube; jacket(ing); liner; lining; muff; rosette; sleeve; sleeve barrel; thimble; sleeve piece <管间接头>
套筒扳手 box key; box socket set; box spanner; box wrench; cap key; casing tongs; casing wrench; female spanner; impact wrench; socket key; socket spanner; socket wrench; tommy wrench; volume box; wrench socket
套筒扳手扳杆 socket wrench crossbar
套筒扳手柄 nut spinner
套筒扳手加长杆 socket wrench extension bar
套筒扳手接头 socket wrench adapter
套筒扳手头 socket head
套筒扳手组 socket wrench set
套筒扳子 casing tongs
套筒闭塞螺母 sleeve blocking nut
套筒臂 telescopic(al) arm
套筒插座 sleeve socket
套筒拆卸器 bushing puller
套筒铲 socket spade
套筒成桩法 <一种桩的施工方法> casing off
套筒尺寸 sleeve dimension
套筒齿轮 sleeve gear
套筒储气柜 telescopic(al) gas-holder
套筒的 telescopic(al)
套筒底板 thimble bat
套筒电扳手 electric(al) socket wrench
套筒阀 cage valve; sleeve valve; telescopic(al) valve; telescoping valve
套筒法兰 sleeve flange
套筒工作台 telescopic(al) working platform

套筒钩 thimble hook
套筒管 telescope pipe;telescope tube
套筒灌浆 sleeve grouting
套筒灌浆法 sleeve grouting method
套筒滚子传动链 sleeve(d) roller transmission chain
套筒滚子链 bush roller chain;Gall's chain;rotary chain
套筒滚子牵引链 sleeved roller traction chain
套筒焊接 sleeve weld
套筒滑板 casing guide
套筒混凝土 muff concrete
套筒活塞泵 sleeve pump
套筒棘轮扳手 ratchet wrench;socket ratchet wrench
套筒继电器 sleeve relay
套筒夹 collet head
套筒夹头 collet chuck
套筒减速升程 reduced lift of sleeve
套筒接管 bell and spigot pipe
套筒接合 bell and spigot joint;box coupling;butt and collar joint;female joint;fit joint;muff coupling;sleeve(d) joint;socket and spigot joint(ing);socket joint;spigot and faucet joint;spigot and socket connection;spigot and socket joint;spigot joint;telescope joint;telescopic(al) joint
套筒接合器 female connector
套筒接口 butt and collar joint
套筒接头 adapter;barrel fitting;faucet joint;hub and spigot joint;muff joint;sleeve coupling;sleeve(d) joint;sleeve(flexible) joint;socket adapter;spigot joint;telescope joint
套筒结构 tube-in-tube structure
套筒绝缘 hub insulation
套筒卡盘 muff chuck;muff clamp chuck
套筒控制交换机 sleeve control switchboard
套筒口 sleeve-port
套筒扩孔器 bushing reamer
套筒拉出器 sleeve puller
套筒拉链舵机 end-on barrel steering gear
套筒离合器 sleeve clutch
套筒连接 butt and collar joint;fit joint;muff joint;sleeve joint;socket joint;spigot;thimble joint(ing)
套筒连接器 sleeve coupler
套筒联结 muff coupling
套筒联轴节 box coupling;butt coupling;clamp coupling;couping muff;coupling sleeve;muff coupling;sleeve coupling
套筒联轴器 muff coupling
套筒链 bush(ing) chain;sleeve-type chain
套筒溜槽 telescopic(al) chute
套筒炉 sleeve burner
套筒螺母 sleeve nut
套筒螺旋 telescopic(al) screw
套筒锚 <美国 Roebling 等公司创制，用于钢绞线及钢丝索> trulock
套筒密封 sleeve seal
套筒密封垫 sleeve packing
套筒偶极天线 skirt dipole antenna
套筒偶极子 skirt dipole
套筒排泥阀 telescopic(al) valve
套筒炮身 jacket tube
套筒膨胀螺栓 bolt socket
套筒千斤顶 telescoping jack
套筒钳 casing tongs
套筒倾卸槽 telescopic(al) chute
套筒燃烧器 sleeve burner
套筒伸缩器 sleeve expansion joint;slide type expansion joint;sliding expansion joint
套筒深度 depth of socket
套筒式 shingle construction;telescope-feed
套筒式臂杆 telescopic(al) boom
套筒式补偿节 sleeve expansion joint
套筒式补偿器 slip-type expansion joint
套筒式测力计 hub dynamometer
套筒式沉陷计 telescopic(al) settlement ga(u)ge
套筒式沉箱 telescopic(al) caisson
套筒式灯杆 telescopic(al) light pole
套筒式电流互感器 bushing-type current transformer
套筒式电枢 sleeve armature
套筒式吊杆 telescoping boom
套筒式端头 bushing terminal
套筒式发动机 sleeve-type engine
套筒式钢梁 telescopic(al) metal joist
套筒式钢模板 telescopic(al) steel shuttering
套筒式滑阀 sleeve valve
套筒式换热器 double shell recuperator
套筒式活接头 sleeve flexible joint
套筒式减震器 direct acting shock absorber;telescopic(al) shock absorber
套筒式接合 faucet joint
套筒式接头 barrel fitting;sleeve-type coupling;telescopic(al) joint
套筒式进料 telescope-feed
套筒式颈轴承 sleeve-type journal bearing
套筒式喇叭口 telescopic(al) bellmouth
套筒式离合器 sleeve-type clutch
套筒式联轴节 sleeve-type coupling
套筒式联轴器 adapter coupling;case butt coupling
套筒式溜槽 telescopic(al) chute
套筒式溜子 telescopic(al) chute
套筒式漏斗口 telescopic(al) bellmouth
套筒式螺旋桨 sleeve mounted propeller
套筒式模板 telescopic(al) form(work);telescoping form(work);telescoping shuttering
套筒式模具 <多用于隧道衬砌> telescopic(al) form
套筒式挠性接头 sleeve flexible joint
套筒式(普通)轴承 sleeve-type journal bearing
套筒式气门 sleeve valve
套筒式千斤顶 telescope jack;telescopic(al) jack
套筒式倾卸槽 telescopic(al) chute;telescoping chute
套筒式伸缩接头 sleeve expansion joint;sleeve flexible joint
套筒式史腿 double box outrigger
套筒式水下灌注用混凝土导管 telescoping-type tremie
套筒式调速器 sleeve governor
套筒式凸模 sleeved type punch
套筒式下料口 telescopic(al) spout
套筒式消声器 concentric(al) cylinder muffler
套筒式卸料槽 telescopic(al) chute;telescoping chute
套筒式烟囱 telescope chimney;telescopic(al) chimney
套筒式液压顶杆 telescopic(al) hydraulic ram
套筒式凿岩机 stopper drill
套筒式涨缩接合 sleeve expansion joint
套筒式涨缩器 slip-type expansion joint
套筒式轴承 sleeve-type bearing
套筒式装配 telescopic(al) mounting
套筒松紧螺旋扣 sleeve turnbuckle
套筒套 jacket casing
套筒天线 sleeve antenna
套筒网 sleeve net
套筒铣刀 arbor-type milling cutter
套筒销 sleeve pin
套筒楔块式钢筋连接器 sleeve and wedge coupler
套筒形铣刀 hollow fraise
套筒型夹紧夹头 socket chuck
套筒性升降机 collar type elevator
套筒靴 drive shoe
套筒压力 sleeve force due to centrifugal action
套筒烟囱 sleeve chimney
套筒眼 thimble-eye
套筒支架 telescopic(al) support
套筒支柱 telescopic(al) strut
套筒指轴 telescopic(al) spindle
套筒置换设备 bushing replacer
套筒中心 hollow center[centre]
套筒轴 quill shaft;sleeve;telescopic(al) axle
套筒轴承 bearing bush(ing);bushed bearing;sleeve bearing;spigot bearing
套筒轴承槽 grooving for sleeve bearing
套筒主轴 quill spindle
套筒砖 sleeve brick
套筒转位夹具 collet index fixture
套筒装拆器 bushing driver
套筒装卸工具 bushing tool
套头 cuff
套头交易 hedge;hedging
套头交易条款 hedge clause
套图塑料膜 overlay
套铣长度 length of milling
套铣次数 times of milling
套铣时间 time for milling
套铣筒 milling tap
套线【铁】 gauntletted track;overlapping line;detour
套线道岔 mixed ga(u)ge turnout
套线汇合的一段 gauntlet
套线继电器 sleeve relay
套线隧道 loop tunnel
套箱 pouring jacket;slip jacket;snap flask band
套箱围堰 precast-boxed cofferdam
套(橡胶)管法 <木材防腐> tire-tube method
套鞋 overshoe
套芯铰链 lift-off hinge
套形活塞杆 hollow piston rod
套形铰刀 shell reamer
套袖 oversleeve
套碹 recessed arch
套靴 <长钢管桩脚的> over boot
套靴式冲洗钻头 washover shoe
套压的 pressed-on
套印 overprint;register work
套印规矩线 laying corner
套印孔 register hole
套印图样 complete proof map
套印性能不良 <油墨> bad trapping
套印资料 overprint information;overprinting data
套用 nesting
套用设计 duplicate design
套用设计图 duplicate design drawing
套闸 simple lock;simplified basin lock
套罩 housing;shroud
套针 narrowing point;trocar
套制 arbitrage
套种 interplanting of another crop;tender-crop sowing;interplant
套轴 sleeve;sleeve shaft;sleeve spindle
套轴管接 spigot and socket joint
套爪【机】 collet
套爪夹 collet holder
套爪夹头 collet chuck;grip chuck;spring for grip chuck
套爪卡盘 collet chuck;grip chuck
套砖 sleeve brick
套桩 lag(ged) pile;sheath pile
套装驳 nested barge
套装测试引线 casing test leading line
套装的 nested
套装电线的非金属软管 gilflex
套装扩孔钻 shell drill
套装立铣刀 shell end mill;shell end milling cutter
套装软体 software package
套装润滑系统 self-contained lubrication system
套装式喷油器 injection nozzle
套装物品 nested article
套装叶轮 shrunk-on disc
套装转子 rotor bushing
套状铰刀轴 arbor for shell reamer
套状沙丘 loop bar
套状沙洲 loop bar
套撞 telescoping
套撞列车 telescope train
套准 matching;register
套准定位装置 register lock-up
套准调节辊 register rollers
套准装置 register device
套桌 nested tables
套子 encasement;encasing
套钻 overcore;overcoring
套嘴 nozzle bushing
套做买卖订单 spread order

特

特昂价格 fancy price

特白 extra white
特比德烧结耐热合金 Turbide
特比斯通高强度黄铜 Turbiston
特便用途线路 special type of track
特别安全的 ultra-safe
特别包厢 proscenium box;stage box;state room
特别保密措施 special security measure
特别报酬 separate compensation
特别报告 special report
特别报告方格坐标 special reporting grid
特别报价 special offer
特别背书 endorsement in full;full endorsement;special endorsement
特别标注货 label(l)ed cargo
特别拨款 special allocation
特别补充规定 special additional regulation
特别部门资料源 special sectoral source
特别裁定 special verdict
特别仓储 <危险品> special storage
特别常化法 brunorizing
特别超载 exceptional overload
特别成本 abnormal cost
特别承兑 special acceptance
特别程序研究 special process study
特别出口商声明书 special exporter's declaration
特别储备基金 special reserve fund
特别储藏费率 <对贵重货物> special storage rate
特别处理过的 prepared
特别传输制 special transmission system
特别存款 special deposit

特别存款银行 special depositary
特别大减价 giving great bargain
特别代表 special envoy
特别代理人 special agent
特别贷方专栏 special credit columns
特别的 especial; extraordinary; king-sized; special
特别地区 special area
特别调查 special investigation
特别定座(客运) ad hoc reservation
特别动力作用 specific dynamic(al) action
特别对照 extra check
特别吨位税 special tonnage tax
特别多栏式 special columnar journal
特别发行市场 specific issue market
特别法人 special juristic person
特别放大比例 exaggerated scale
特别费用 extraordinary cost; extraordinary payment; particular charges; special charges; special expenses; special fee
特别分档 special bracket
特别福利基金 special welfare fund
特别附加税 superimposed tax; super-tax
特别附加险 special extraneous risks
特别富配合砂浆 ceiling mortar
特别工班 extra gang; special gang
特别工业业务 special industrial service
特别工作队 extra gang
特别工作组 task force; task group
特别公积金账户 special surplus account
特别攻击 special strike
特别观察员 ad hoc observer
特别光纸 patent coated paper
特别规定 special provision; special regulation
特别规则 ad hoc rule
特别过程 singular process
特别海关发票 special customs invoice
特别海损 particular average
特别合伙 special partnership
特别合伙人 special partner
特别红利 extra dividend; special bonus
特别呼叫 special call
特别滑动构件 extra sliding member
特别画线支票 check[cheque] crossed specially; specially crossed check
特别黄色炸药 extra dynamite
特别会议 ad hoc meeting
特别火险 extra hazard
特别货物运费率 specific commodity rate
特别积分 particular integral
特别基金 special fund
特别记账单位 special unit of account
特别加急业务 special urgent service
特别价格 special price
特别减低的价目 special reduced rate
特别减价运费 exception rate
特别减税 special tax cut; special tax reduction
特别检查报告 special case report
特别检验 special survey
特别奖金 special bonus
特别降低的车架 <车辆重心的> kick-up frame
特别借方专栏 special debit columns
特别借土 special borrow
特别津贴 extra allowance
特别紧迫的 extra-urgent
特别进口 special import
特别进口措施法 <加拿大> Special Import Measures Act
特别进口许可证 special import licence[license]
特别经理 special manager
特别捐税 special assessment
特别决议 special resolution
特别开发区 special development area
特别开支 special expenses
特别空中活动 special air operation
特别快车 de luxe train; express; varnished car; limited <美>
特别快车机车 express engine
特别快车加价票 supplement for express train
特别快凝水泥 extra-rapid-hardening cement; jet cement
特别利益 exceptional advantage
特别列车 train de luxe
特别留置权 particular lien; special lien
特别旅客快车 express train
特别买方 special buyer
特别密码 special cipher
特别纳税扣除 special taxes deduction
特别派款 special assessment
特别评价 special assessment
特别评价法 special valuation method
特别普查 special census
特别轻载 extra-light loaded
特别清算终结委员会 Special Settlement Finality Committee
特别全权代表 ad hoc plenipotentiary
特别任务 ad hoc task
特别任务完成情况图 special job cover map
特别容易受污染威胁的时候 pollution episode
特别融资 special finance
特别商品 specialties; special(ty) goods
特别射向 special sheaf
特别申请 special requisition
特别审计 special audit
特别市 special municipality under direct central control
特别收入分成 special revenue sharing
特别收益 abnormal gain
特别收(筑路受益)费 special assessment
特别税 extra duty; extraordinary tax; special assessment; special tax
特别说明 special explanation; special version
特别说明书 special direction; special prescription
特别松软岩石 special competent bed
特别搜索 special search
特别损失 abnormal loss; extraordinary loss; special loss
特别所得税 special income tax
特别索夹 special cable clamp
特别提款权 special drawing rights; special rights of drawings
特别提款权本位 special drawing rights standard
特别提款权存款单 special drawing rights certificate of deposit
特别提款权的分配额 allocations of special drawing rights
特别提款权的逐日汇率 daily special drawing rights rate
特别提款权估值 valuation of special drawing rights
特别提款权累计分配净额 net cumulative allocation of special drawing rights
特别提款权平价 special drawing rights parity
特别提款权证明书 special drawing rights certificate
特别提款权证券账户 special drawing rights certificate account
特别条件 special provision
特别条款 special clause; special provision; special term; stamp clause
特别调停 special mediation
特别调整 special tuning
特别通信[讯] special communications
特别通信[讯]中心 special communications center[centre]
特别通知信用证 specially advised letter of credit
特别投递 special delivery
特别往来存款 special current account
特别危险品 extremely dangerous material; extremely hazardous substance
特别维护条款 entrenched provision
特别委员会 ad hoc committee
特别险 extra risks
特别销售税 selective sales taxes
特别小组 ad hoc group; project group
特别协定 specific agreement
特别卸货泊位 special unloading berth
特别信贷风险 special credit risk
特别信贷基金 special credit fund
特别信息音 special information tone
特别信用证 special credit; special letter of credit
特别修理费 extraordinary repairs
特别许可证 special licence[license]; special permit
特别押汇信用证 special documentary credit
特别养护 extraordinary maintenance
特别要求 ad hoc request
特别业务 special service
特别意外准备金 special contingency reserves
特别应急车辆 special emergency vehicle
特别用途线【铁】 special using siding
特别用途线延展长度【铁】 extended length of special using siding
特别优惠待遇 special preferential treatment
特别优先权 special priority
特别优先通行权 own right-of-way
特别邮递 pneumatique
特别预算 special budget
特别运费率 special rate
特别运价 exceptional rate
特别运价表 exceptional tariff
特别增压 extra pressurization
特别账户 special account
特别折旧 abnormal depreciation; extraordinary depreciation
特别折扣 channel discount; special discount
特别折让 special allowance
特别征税 special assessment
特别支出 special outlay
特别支付代理人 special disburser
特别职位津贴 special post allowance
特别指定结汇银行的信用证 specially negotiating bank letter of credit
特别致密 particularly compact
特别中间检验 special intermediate survey
特别仲裁 ad hoc arbitration
特别仲裁人 special referee
特别仲裁条款 special compromissory clause
特别重物 heavy weight
特别注意悬移质的泄放 special attention paid to the release of suspended solids
特别专家小组 ad hoc expert group; ad hoc panel; ad hoe group
特别咨询电报 special advisory message
特别咨询工程师 special advisory engineer
特别资本 specific capital
特别资金 specialized funds
特别资源 ad hoc resources
特别总价目表 special through rate
特薄壁套管 extra thin-walled casing
特薄壁钻头 extra thin-walled bit; extremely thin-walled bit
特薄玻璃板 extra thin sheet glass
特薄层 sharp layer
特薄片 sharp layer
特薄细牙六角小螺母 extra thin fine thread small hexagonal nut
特餐厨房 special kitchen
特产 special(i)ty; special(local) product
特产出口商 special exporter
特产商店 specialty shop
特产水生植物 specialized aquatic flora
特长 exceptional length
特长货物 lengthy cargo
特长平车 extra-long flat car
特长射孔孔道 extra-long perforation tunnel
特长型斗杆 <挖掘机> extra-long stick
特超长虚线 extra-super long dash
特超声 Debye wave; hypersonic sound; hypersound
特超声频(率) hypersonic frequency
特超声速 hypersonic speed; hypervelocity
特超声速风洞 hypervelocity wind tunnel
特超声速空气动力学 hypersonic aerodynamics
特超声速流状态 hypersonic flow condition
特超声速模型 hypersonic model
特超声速稳定性 hypersonic stability
特超声速相似参数 hypersonic similarity parameter
特超声速相似定律 hypersonic similarity law
特超声速相似律 hypersonic similitude
特超声速研究 hypersonic study
特超声速研究装置 hypersonic research device
特超声尾流 hypersonic wake
特超声学 hypersonics; praetersonics
特超缩微胶片 ultra-microfiche
特超音速的 hypersonic
特超音速飞机 <飞行速度五倍于音速以上的飞机> hypersonic transport
特超蒸发器 hypervapotron
特称命题 particular proposition; singular proposition
特此 here-by
特粗钢丝 extra thick steel wire
特粗砂 mo(u)lding gravel
特大 outsize(d)
特大暴雨 catastrophic cloudburst; exceptional shower; extraordinary storm
特大暴雨量 outstanding rainfall
特大波 extraordinary wave
特大材料 over-sized material
特大超级油轮 <300000 ~1000000DWT> giant tanker
特大潮 equinoctial spring tide
特大城市 mega(lo)polis; super-city
特大城市的 mega(lo)politan
特大城市的居民 mega(lo)politan
特大冲刷 extraordinary scour
特大床 king-size bed
特大的 extra(large); imperial; jumbo; king-sized; super-giant; ultra-large
特大的自然灾害 extraordinarily seri-

ous natural calamities
特大都市 megapolis
特大断面 very large cross-section
特大反差的 extra contrasty
特大房屋抵押货款 balloon mortgages
特大飞行器 aerodreadnaught
特大废气燃烧器 jubo burner
特大丰收 an exceptional bumper harvest
特大高潮 extra-high tide; extraordinary high tide; extraordinary spring tide
特大高度定位法 very high altitude method
特大工程 superproject
特大公园 megapark
特大功率 super-power
特大功率的 extra heavy
特大功率电台 superstation
特大功率电网 <275 ~ 38 × 10^4 伏> super-grid; super-power net(work)
特大功率管 super-power tube
特大功率锅炉 super-power boiler
特大功率系统 super-power system
特大规模 imperial scale
特大规模集成电路 ultra large scale integration
特大号 extra large; king size; outsize(d)
特大号包装 giant size; jumbo size
特大号的 over-size
特大荷载 very high load
特大洪水 cataclysm; catastrophic flood; eventual flood; exceptional flood; exceptionally high flood; extraordinary flood; extreme flood; record flood; severe flood; superflood; unusual flood
特大洪水位 exceptional flood level
特大滑车 <用来架设临时吊重货专用吊杆的大型滑车> wrecking block
特大绘图纸 imperial paper
特大货物 exceptionally bulky goods
特大降水 extraordinary rainfall
特大降雨量 outstanding rainfall
特大街坊规划 superblock plan
特大开口率摄影物镜 wide-aperture photo objective
特大块煤 huge coal
特大缆绳 wrecking cable
特大浪 extraordinary sea
特大流量 extreme discharge
特大喷口喷水灭火系统 extra-large orifice sprinkler system
特大企业 megacorporation
特大强度 super-strength
特大桥 extra-long bridge; grand bridge; super-bridge; super-major bridge
特大商场 hypermarket
特大水位 exceptional water level
特大碎屑 <沉积岩中的> distant admixture
特大突水 exceptionally big bursting water
特大尾数抵押货款 balloon mortgages
特大型 huge size
特大型船 largest vessel
特大型的 super-huge
特大型发电厂 superstation
特大型花体大写字母 factotum
特大型滑坡 very large landslide
特大型矿床 huge-size ore deposit
特大型矿山 extraordinary-tonnage mine
特大型汽轮发电机 super-huge turbogenerator
特大型水源地 super-huge water source
特大雨 very heavy rain
特大雨量 excessive rain (fall of long duration)
特大值 exceptional value
特大质量旋体 supermassive rotator
特大重件 exceptionally heavy goods
特等白铅调合漆 special liquid white lead
特等包厢 dress circle; state room
特等舱 cabin de luxe; state room; suite cabin; suite room
特等的 super-duty
特等房间 state room
特等钢丝绳 special grade steel wire rope
特等硅酸盐砖 super-duty silica brick
特等红铅调合漆 special liquid red lead
特等客舱套间 cabin de suite; cabin en suite
特等客车 parlo(u)r car
特等客车座椅 parlo(u)r car chair
特等列车 de luxe train
特等轮船灰色 special marine grey
特等品 imperial
特等卧车 drawing room car; roomette car; drawing room <客车列车的>
特等站 superclass station
特低场 very low-strength field
特低灰煤 special low-ash coal
特低磷煤 special low-phosphorus coal
特低硫煤 special low-sulfur coal
特低频 ultra-low frequency
特低强度煤 special low-strength coal
特低压照明 extra-low voltage lighting
特低折射率光学玻璃 extra-low refractive index optic(al) glass
特蒂锡铅焊料 Tertiarium
特蒂锡铅焊条 Tertiarium
特点 characteristic; distinguishing feature; hallmark; specific character; train; unique feature; style device【建】; style feature【建】
特定保单 specific policy
特定保险 specific insurance
特定编码 specific coding
特定标目 specific entry
特定补贴 selective grants
特定部门失业 specific unemployment
特定财产保险 specific covered property insurance
特定产品 specific products
特定长度 specific length
特定偿债基金 specific sinking fund
特定场地 given site
特定成本 special cost; specific cost
特定承包合同 negotiated contract
特定程序 ad hoc program(me)
特定程序错误 program(me) sensitive error; program(me) sensitive malfunction; program-sensitive error; program-sensitive fault
特定程序故障 program-sensitive fault
特定程序计算机 target computer
特定尺寸 cut size
特定处理方法 specific treatment method
特定代理商 special agent
特定单元 discrete cell
特定的 restrictive
特定的流水号 special series
特定的流水号数 special running series of numbers
特定的水流 specific flow
特定的弹性弯沉 <分段量出的> specific elastic deflection
特定的危险保险单 single policy
特定的运费率 <铁路> differentiate rate
特定的沾污物 specific contaminant
特定抵押 special mortgage; specific mortgage
特定地区 designated area
特定订单 specific indent
特定订单生产 specific order production
特定毒物 specific poisonous substance
特定多数 qualified majority
特定发行 contingent issue of securities
特定范畴 specific category
特定范围 specified range
特定范围放射线检测器 area monitor
特定方法 ad hoc approach
特定方式 ad hoc fashion
特定沸点的醇类 special boiling point spirits
特定沸点的酒精 special boiling point spirit
特定费用 specific cost
特定分布 identified distribution
特定概念 specific concept
特定工程用砖 engineered brick
特定规则 ad hoc rule
特定函数 particular function; specific function
特定航程 particular leg
特定荷载 specific load; superload
特定呼叫业务 custom calling service
特定画线 specific crossing
特定环境 specific environment
特定活期存款 specific current deposit; specified current deposit
特定活载 superload
特定货物 specific goods
特定货物运费率 particular commodity rate; specific commodity rate
特定级别 specific rating
特定价 central special price; given price
特定价格或更优价格 at or better
特定价值 characteristic value; specified value
特定检验水平 specific level of test
特定解 particular solution
特定就业税 selective employment tax
特定卷宗 specific volume
特定决策规则 ad hoc decision rule
特定离子电极 specific ion electrode
特定历史模拟法 specific historical analogy method
特定列车 dedicated train
特定零件 specific component
特定留置权 particular lien
特定脉冲重复率 specific pulse recurrence rate
特定密度 specific density
特定命题 particular proposition
特定模式故障 pattern-sensitive fault
特定目的财务报表 special-purpose financial statement
特定排污标准 special emission standard
特定(判断)标准 specific criterion
特定频率 spot frequency
特定频率干扰 spot jamming
特定气候费用 climate-specific cost
特定区域 location
特定日温差 degree-day
特定商品 particular commodity
特定设计法 ad hoc approach
特定深度 specific depth
特定生产指数 specific productive index
特定试剂 specific tester
特定试验 specific test
特定数量的社会需要 quantitatively definite social need
特定双折射 specific birefringence
特定水流 unique flow
特定水体 specific body of water
特定水头 specific head
特定水位 exceptional water level; specific level
特定水域 specific bodies of water
特定速度 specific speed
特定添加剂 intentional additives
特定条件 conditions for particular applications
特定条件的使用 conditional use; special exception use
特定推力 specific thrust
特定危险性 particular risk
特定微功能 microspecific function
特定文本 definitive text
特定污染源 specific source
特定相 specific phase
特定相关参照 specific cross reference
特定项目技术标准 project specification
特定信用证 special credit
特定循环 specific cycle
特定样本调查 specific sample survey
特定应力循环 specified stress cycle
特定用途 specific end use
特定预算分配 specific allotment
特定元素法 specific-element method
特定原油 selective crude
特定运价 special tariff
特定运输工具模型【交】independent mode-specific model
特定账产留置权 <对借契上指明的财产> specific lien
特定值 designated value; designed value; specific value
特定重复率 specific repetition rate
特定主义 ad-holism
特定资源 ad hoc resources
特定组配 specific combination
特定作业用端机 job-oriented terminal
特定坐标(系) preferred coordinates
特定座位测验 specific locus test
特陡边 infinitely sharp edge
特尔费纳式铁轨 Telfener rack
特尔尼克耐蚀青铜 Telnic bronze
特尔史密斯脉动筛 Telsmith pulsator
特尔史密斯破碎机 Telsmith breaker; Telsmith crusher
特尔史密斯型旋球式破碎机 Telsmith gyrasphere crusher
特发性反应 idiosyncratic reaction
特发性水污染 sudden water pollution
特发性污染 sudden pollution
特发浊流 spasmodic turbidity current
特夫棱镜 Dove prism
特夫南等效发生器 Thevenin's generator
特夫南定理 Thevenin's theorem
特氟隆 <聚四氟乙烯> Teflon
特氟隆板支座 Teflon plate bearing
特氟隆薄膜 Teflon film
特氟隆导向条 Teflon guide strip
特氟隆管 Teflon pipe
特氟隆滑板支座 Teflon plate-coated sliding bearing
特氟隆夹入物 Teflon insert
特氟隆石墨 graphite filled Teflon
特氟隆涂层 <混凝土钢模接触面> Teflon-coating
特氟隆橡胶支座 Teflon-neoprene bearing
特氟隆柱塞密封 Teflon plunger
特富金属星 super-metal-rich star
特富矿体 specimen ore
特赋留置权 special assessment liens
特赋清册 special assessment roll
特干混合料 extremely dried mix
特干硬混凝土拌合料 very stiff mixture
特干硬混凝土混合料 very stiff mixture
特高大楼 skyscraper
特高电压 extreme high voltage; extra-

T

high tension
特高分辨率乳胶 special maximum resolution emulsion
特高高耸建筑 super-skyscraper
特高功率晶体管 very high-power transistor
特高海浪 sea bore
特高亮度 extra-bright
特高邻差值【测】 value of extreme high contiguous deviation
特高灵敏度 extra-sensitivity;ultra-sensitivity
特高浓度 extra heavy concentration
特高频 super-frequency; super-high frequency;ultra-high frequency
特高频变频器 ultra-high frequency converter
特高频波 <频率 300 ~ 3000 兆赫兹,波长 1 ~ 0.1 米> ultra-high frequency wave
特高频差转机 ultra-high frequency translator
特高频带 ultra-high frequency band
特高频带通滤波器 ultra-high frequency bandpass filter
特高频发射机 super-high frequency transmitter; ultra-high frequency transmitter
特高频功率放大管 amplitron
特高频功率放大器 amplitrans
特高频接收机 ultra-high frequency receiver
特高频频道 ultra-high frequency channel
特高频前置放大级 ultra-high frequency preamplifier stage
特高频天线 ultra-high frequency aerial;ultra-high frequency antenna
特高频调谐机构 ultra-high frequency tuning mechanism
特高频调谐器 ultra-high frequency tuner
特高频调谐装置 ultra-high frequency tuning device
特高频图像发射机 ultra-high frequency vision transmitter
特高频外差波长计 ultra-high frequency heterodyne wave meter
特高频无线电话设备 super-high frequency radio telephone equipment; ultra-high frequency radio telephone equipment
特高频载波 ultra-high frequency carrier
特高频转播机 ultra-high frequency translator
特高品价样品 sample of erratic high-grade
特高品位 erratic high-grade; extra-high grade;storm grade
特高品位样品处理 treating of erratic high-grade sample
特高强度钢 extra-high tensile steel; mar-ag(e)ing steel
特高强度钢丝 extra high strength steel wire
特高强管 double extra strong pipe
特高强混凝土 very high strength concrete
特高速 ultra-high speed; very-high-speed
特高压 extra-high voltage
特高压部件 extra-high tension unit
特高压电缆 extra-high tension cable
特高压发生器 extra-high voltage generator
特高压力 extreme pressure
特高压输电线路 ultra-high frequency transmission line
特高压整流器 extra-high tension rectifier
特高烟囱 skyscraper
特格德碳化硅耐火材料 Tercod
特供住房 <公司专供退职员工的> tied cottage;tied house
特古西加尔巴 <洪都拉斯首都> Tegucigalpa
特光滑饰面 type C finish
特光洁铸造表面 hardware finish
特广角航空摄影 super-wide-angle photography
特号钢丝绳 plough steel wire rope
特号斜纹布 imperial
特号制钢丝 <经冷拉的索氏体钢丝> plough steel wire
特厚壁钢管 extra-strong pipe
特厚玻璃板 extra heavy sheet;extra-thick sheet glass
特厚管 extra heavy pipe
特化 specialization
特化内尔氏自动吸管 Trenner's automatic pipet
特环曼干涉仪 Twyman interferometer
特惠 special preference
特惠酬金 preference premium
特惠待遇 preferential treatment
特惠贷款 concessional loan
特惠的 preferential
特惠关税 preference duty; preferential duty;preferential tariff
特惠关税制 preferential tariff system
特惠汇率 preferential exchange rate
特惠减税 preferential tariff cut
特惠贸易 preferential trade
特惠贸易协定 preferential trade agreement; preferential trade arrangement
特惠税 preferential duty
特惠税率 cheap tariff; concessionary rate of tax;reduced tariff
特惠条件 concessional term
特惠条款 preference clause;preferential clause
特惠折扣 special rate
特惠制度 preferential system
特惠状态 privileged state
特级 extra grade;special class;special grade;super-fine
特级大港 specially designated major port
特级大厦 megastructure
特级导线测量 zero order traversing
特级硅氧材料 high-duty silica
特级硅质耐火材料 super-duty silica refractory
特级浑圆形 special round
特级精度配合 extra-fine fit
特级耐火材料 super-duty refractory
特级耐火砖 super-duty fireclay brick
特级黏[粘]土耐火砖 super-duty fireclay brick
特级黏[粘]土砖 super-duty clay brick
特级品 extra
特级浅色 extra-pale
特级商品 first line
特级市场 hypermarket
特急 <电报时限等级> most immediate
特急电报 flash traffic
特急呼叫 flash call
特急闪光 scintillating light
特急时间 crash(ed) time
特急通报 flash traffic
特急通话 flash call
特辑 reference issue;special issue
特技 special techniques;tricky
特技布景 trick scene
特技插盘 effect disk[disc]
特技场影 trick scene
特技车间 stunt photography department
特技的 acrobatic
特技合成桌 animation desk
特技机器 effect machine
特技镜头 trick shot
特技慢转换 animation dissolve
特技模式 trick mode
特技品 trick work
特技设备 animation equipment; special effect <电视>
特技摄像机 animation camera
特技摄影 trick photography
特技声 effect sound
特技台 animation stand
特技同步 animation timing
特技匣 stunt box
特技效果 special effect;trick effect
特技信号发生器 special effect generator
特技用玻璃 effect glass
特技用混合放大器 gizmo montage amplifier
特技照明 effect lighting
特技装置 effect machine
特加重管 double extra heavy pipe; double extra strong pipe
特佳能见度 exceptional visibility
特价 bargain price; differential rate; exceptional price;special price
特价表 special tariff; tariffs of exceptions
特价出售 on sales
特价率 special rate
特价品 leader;leading article
特价商店 leader
特价提供 special offer
特价线路 differential route
特奖 grand prize
特解 particular integral;particular solution;special solution
特近运输 exceptionally short-distance-traffic
特精密加工 extra fine
特景图 close up view
特净 ultra-clean
特净煤 ultra-clean coal
特刊 special
特克卢燃烧器 Teclu burner
特克萨斯栎 Texas oak
特克萨斯热 red water fever
特克斯特隆公司 <美> Textron Inc.
特克头轧辊 Turk's head rolls
特快班轮 express liner
特快车 express coach; express railway coach
特快的 ultra-rapid
特快断路器 super-chopper
特快服务 <运输> railway express and fast freight
特快货物列车 express freight train
特快货运业务 express goods service
特快机车 express locomotive
特快加价票 express extra
特快客票 express ticket
特快客运列车 express passenger train
特快空运 air express
特快列车 limited express; red ball <俚语>
特快旅客列车 express passenger train
特快情报 express information
特快通话 lightning call
特快投送 express delivery
特快信息服务 flash information service
特快型法兰西斯式水轮机 express type Francis turbine
特快型混流式水轮机 express type Francis turbine
特快硬波特兰水泥 extra rapid hardening Portland cement
特快硬高铝水泥 super-aluminous cement
特快硬硅酸盐水泥 extra rapid hardening Portland cement
特快硬水泥 extra rapid hardening cement; jet cement; rapid hardening cement;special rapid hardening cement;ultra-rapid hardening cement
特快运列车 express train
特快载重汽车 express freight car
特快斩波器 super-chopper
特宽角 super-wide-angle
特宽角多倍仪 super-wide-angle multiplex
特宽角航空摄影机 super-wide-angle aerial camera
特宽角航摄机 super-wide-angle aerial camera
特宽角镜头 <视场角大于 100 度> super-wide-angle lens; ultra-wide-angle lens
特宽角摄影 super-wide-angle photography
特宽角摄影机 super-wide-angle camera;ultra-wide-angle camera
特宽角投影器 super-wide-angle projector
特宽角透镜 super-wide-angle lens
特宽角相片 super-wide-angle photograph
特宽轮胎 extra-wide base tire
特拉布克锡镍合金 Trabuk alloy
特拉德水泥着色剂 Tellard
特拉尔风 Terral
特拉福德牌石棉水泥瓦 <一种波纹石棉水泥瓦> Trafford tiles
特拉福利特纸 <一种墙纸> Trafolyte
特拉华水渠 Delaware aqueduet
特拉科巴式体系建筑 Tracoba
特拉科巴式体系建筑法 Tracaba method
特拉科巴预制构造系统 Tracoba system
特拉蒙塔那风 Tramontana[Tramontano]
特拉拍实式震动辗 <土方压实用> Terrapac vibratory roller
特拉特克斯板 <一种绝缘板> Treetex
特拉托尔 <一种玛琋脂、水泥快凝剂等建筑材料> Tretol
特拉维斯池 Travis tank
特拉扬尼广场 <古罗马帝王的> Forum of Trajan
特莱 <一种粉煤灰陶粒沥青骨料> Terlite
特兰科尔合金 Trancor
特兰斯泰纳 transtainer
特朗秤 trone
特朗衡器 trone weight
特劳伍德线材电流加热法 Trauwood process
特劳泽(铅柱试验)法 Trauzl method
特劳泽试验 Trauzl test
特勒花环填料 Teller Rosette
特勒花环型填料 Tellerette packing
特勒花型填料 Teller Rosette packing
特勒环填料 Tellerette packing
特勒填料 Tellerette
特勒维索软瓷 <意大利> Treviso porcelain
特勒维索陶器 <意大利> Treviso faience
特雷斯卡加屈服面 Tresca yield surface
特雷斯卡屈服函数 Tresca yield criterion
特雷斯卡屈服准则 Tresca yield criterion
特雷斯卡准则 <强度理论的> Tresca

T

criterion
特雷沃附近的伊格尔塔形墓碑 Igel Monument near Treves
特里奥利特防腐剂 <一种木材防腐剂> Triolith
特里芬计划 Triffin plan
特里洛牌振动捣实器 Trillor vibrator
特里奈斯库沥青液 <一种修补地面和屋面的沥青液> Trinasco
特立库塞尔 <一种水泥防水硬化液> Tricosal
特立尼达地沥青 Trinidad asphalt
特立尼达和多巴哥 <拉丁美洲> Trinidad and Tobago
特立尼达和多巴哥石油公司 Trinidad and Tobago Oil Co.
特立尼达和多巴哥中央银行 Central Bank of Trinidad and Tobago
特立尼达(湖)地沥青 <美洲特立尼达岛所产天然地沥青> Trinidad lake asphalt
特立尼达湖沥青胶泥 Trinidad lake asphalt cement
特立尼达精制沥青 Trinidad refined asphalt
特立尼达沥青湖 Trinidad Pitch Lake
特立尼达天然沥青 Trinidad asphalt; Trinidad pitch
特立中央胎座式 free-central placentation
特利板 <一种不滑的铁路面> Tripedal
特利夫湿磨矿渣水泥法 Trief process
特例 extreme case; special case; special instance
特亮抛光 planish extra-bright
特林尼特阶【地】Trinitian(stage)
特留树 special tree
特硫锑铅矿 twinnite
特隆布墙 Trombe wall
特鲁贝对比 Trube's correlation
特鲁顿比值 Trouton's ratio
特鲁顿常数 Trouton's constant
特鲁顿定律 Trouton's law
特鲁顿法则 Trouton's rule
特鲁顿黏[粘]度 Trouton's viscosity
特鲁顿黏[粘]性引力系数 Trouton's coefficient of viscous traction
特鲁顿伸长黏[粘]度 Trouton's elongational viscosity
特伦顿阶 <美国中奥陶世>【地】Trentonian(stage)
特伦间冰阶【地】Treene interstage
特伦普分选点 Tromp cut-point
特伦普井筒注水通风法 Tromp blast
特伦普面积曲线图 Tromp area diagram
特伦普重介质分选法 Tromp process
特伦特生物指数 Trente biotic index
特罗贝雪夫尺 coordinate grid scale
特罗尔液位调节器 Level-Trol
特罗菲尔 <一种聚乙烯纤维> Trofil
特罗哈顿电炉 Trollhatten furnace
特罗黄铜 Sterro metal
特罗莫赖特烧结磁铁 Tromolite
特罗彭纳斯侧吹转炉 Tropenas converter
特罗水磷铝石 trolleite
特罗韦牌檐口玻璃条 <每节 3 英尺长,内装条形灯> Trower cornice
特洛坡图 <对流层内的高空气象图>【气】tropogram
特马多克板岩 Tremadoc slate
特马多克阶 <英国早奥陶世>【地】Tremadoc stage
特米特铅基轴承合金 Termite
特密封修理管箍和接头 super-seal repair clamp and coupling
特敏极化继电器 high sensitive polarized relay

特命全权大使 ambassador extraordinary and plenipotentiary
特命全权公使 envoy extraordinary and minister plenipotentiary
特命使节 envoy extraordinary
特纳德蓝 Thenard's blue pigment
特纳尔牌 <为多种石棉水泥制品> Turnall
特纳法 Turner method
特纳黄 Turner's yellow
特纳莱牌水泥 <一种快硬水泥> Tunnelite
特纳水泥 <一种快速凝固水泥> Tunelite
特耐火的 super-refractory
特黏[粘]的 extremely viscous
特纽阿尔高强度铜铝合金 Tenual
特派领事馆信使 ad hoc consular courier
特派新闻记者 accredited journalist
特派员 commissioner; delegate; special
特平板 dead-flat sheet
特强 extra heavy
特强钢管 extra-strong steel pipe
特强管 extra heavy pipe; extra-strong pipe
特强厚壁管 double extra strong pipe
特强混凝土管 extra heavy strength concrete pipe
特强接点 extra heavy contact
特强石灰炉渣 extra-limy slag
特轻材料 super-light material
特轻火石玻璃 extra-light flint glass
特轻加感 extra-light loading
特轻加载 extra-light loading
特轻加载线路 extra-light-loaded circuit
特轻型起重机 extra-light duty crane
特轻型抓斗 extra-light duty crane
特轻质填方 ultra-light weight fill
特区 special administrative region; special district; special economic zone; special zone
特区文化 culture in the special economic zones
特屈儿 <一种烈性炸药> tetralite
特屈儿混合炸药 tetrytol
特屈儿炸药 tetryl [trinitrophenylmethylnitramine]
特屈儿-重氮混合物引爆剂 tetryl-azide detonator
特权 liberty; privilege; franchise <美>; concession <铺设权、使用权等的>
特权的 prerogative
特权的控制 control of a privilege
特权独占 special privilege monopoly
特权费 royalty expenses
特权阶层 privileged stratum
特权用户【计】power user
特权者使用的房屋 secos[sekos]
特权指令 privileged instruction
特权指令异常 privileged operation exception
特权状态 privileged mode
特软钢 extra mild steel; extra soft steel; special soft steel
特软回火薄钢板 <布氏 5 号硬度> dead-soft temper sheet
特三裁 extra thirds
特色 characteristic; distinctiveness; feature; motif; outstanding feature; particular; salient feature; speciality; unique feature
特色饭店 specialty restaurant
特色化学物种 distinct chemical species
特色酸 distinct acid
特沙尔透镜 Tesaar lens

特上品质 super-fine quality
特设的 special
特设法庭 ad hoc tribunal
特设机构 special entity
特设桥(梁) accommodation bridge
特设试验 ad hoc test
特设委员会 ad hoc committee
特设仲裁 ad hoc arbitration
特设专家小组 ad hoc expert group
特设专家咨询委员会 Ad Hoc Advisory Committee of Experts
特深埋 very deep
特省油汽车 quadricycle
特使 special envoy
特饰建筑 barock
特殊 VSP 反褶积 special VSP deconvolution
特殊爱好 predilection
特殊安排住所 special placement residence
特殊安全克罩 special safety housing
特殊安全轴套 special safety housing
特殊安全住宅 special safety housing
特殊扳手 special key
特殊包装 extrapacking; particular package
特殊保险 special insurance
特殊保险单 special policy
特殊暴露潜水 exceptional exposure
特殊暴露潜水方案 exceptional exposure dive schedule
特殊暴露巡潜时间 exceptional exposure excursion time
特殊备用金 special reserve fund
特殊比例尺 particular; special scale
特殊编码选择器 special code selector
特殊编址 specific addressing
特殊变换 special transformation
特殊变星 peculiar variable
特殊变形 idio-morphosis
特殊表壳 special case
特殊表示 special representation
特殊兵种 corps
特殊波纹纸板 specifically corrugated board
特殊波形信号发生器 special waveform signal generator
特殊薄钢板 Artz press sheet
特殊薄灰泥 special grout
特殊补给品 special supplies
特殊补救方法 exceptional remedy
特殊(部门)收据 extraordinary receipts
特殊部位 privileged sites
特殊部位炉衬 zoned lining
特殊部位砖衬 zoned lining
特殊材料 exotic material; special material
特殊材料制的模板 special form(work)
特殊采购 special purchasing
特殊彩色摄影 non-conventional colo(u)r photography
特殊参照 special cross-reference entry
特殊餐厅 specialty restaurant
特殊操作 particular operation
特殊操作异常 privileged operation exception
特殊插入编辑 special insertion editing
特殊产品 outlay; special product
特殊产品订货 special product order
特殊产品规范说明 closed specification
特殊铲斗活口开度 bucket opening
特殊厂 specialist plant
特殊厂区 particular territory
特殊超重 exceptional overload
特殊车间 specialist plant
特殊车牌 <钉在装有爆炸品及其他危险物品的车上> special placard

特殊车轴 special vehicle axle
特殊成本调查 special cost study
特殊程序 special program(me)
特殊处理 special treatment
特殊处理钢 special treatment steel
特殊处理剖面 particular processing section
特殊处理指令 special handling instruction
特殊处理资料解释 special processing data interpretation
特殊传动 special drive
特殊待遇 perquisite; special treatment
特殊担保 specific guarantee
特殊的 especial; individual; odd; particular; special; specific
特殊的不景气或紧急补助金 special depression or emergency aid
特殊的臭味 specific odo(u)r and taste
特殊的等价形式 particular equivalent form
特殊的防渗构造 special protective feature
特殊的和非标准的项目 peculiar and non-standard items
特殊的建筑用具 special builder's furniture
特殊的品种 particular strain
特殊的商品经济 special kind of commodity economy
特殊的直馏煤焦油硬沥青 special straight run coal tar
特殊的柱截面 special column section
特殊等价物 special equivalent
特殊等效点系 special equivalent point system
特殊底盘起重机 special mounted crane
特殊地质 special geology
特殊地质体平面图 plan diagram of special geologic(al) body
特殊地质条件 special geologic(al) condition
特殊点高程 special point elevation
特殊点焊 gap weld
特殊点集系 special system of point-group
特殊点坐标 coordinates of special point
特殊电动机 special type motor
特殊电机 special machine
特殊电力 non-firm power
特殊吊车 special crane
特殊订货 special order
特殊订货加工车间 jobbing shop
特殊定向钻头 special for directional drilling bit
特殊动力作用 specific dynamic(al) action
特殊断面的钢丝 special cross-section steel wire
特殊(断面)钢材 <钢板桩、轨枕、连接件等特殊型钢> special rolling-mill section
特殊断面牵出线 lead track of special section
特殊堆焊 special build-up welding
特殊法兰 special flange
特殊方式 particular form
特殊方向 exceptional direction
特殊防渗结构 special protective feature
特殊废水管 special waste pipe
特殊废物 special waste
特殊沸程汽油 special boiling point gasoline
特殊费用 extraordinary expenditures; privileged expenses; special expenses
特殊粉刷 special plaster
特殊风险 abnormal risk; special risk

特殊风险保险基金 special risk insurance fund
特殊符号 additional character;special character; special numeric; special symbol
特殊附件 ancillary attachment; special accessory;special attachment
特殊覆盖层 special manhole
特殊感觉的 organoleptic
特殊感觉能力说 law of specific energy of sensation
特殊钢 fine steel;special steel
特殊钢材 special rolling-mill section
特殊钢厂 special steel plant
特殊高潮 high water exceptional spring tide
特殊高能燃料 exotic composition
特殊高强度铸铁 Promal
特殊哥特式建筑风格 particularistic Gothic
特殊工程 special construction;special project;special works
特殊工程设计程序 special project program(me) order
特殊工作条件 special operating condition;specific operating condition
特殊功能 special function
特殊功能玻璃 specific function glass
特殊估价地区 special assessment district
特殊观测 special observation
特殊观察用扁形显像管 mejatron
特殊管塞 special pipe plug
特殊灌浆(法)special grouting
特殊光谱 peculiar spectrum
特殊规格 special requirement
特殊规格门窗型材的制造及供应工厂 open-planning millwork
特殊规划项目 unique project
特殊规律研究法 idiographic
特殊规则 special rules
特殊规则区 special rule zone
特殊贵重物品 article of exceptional value
特殊国际法 particular international law
特殊函数 special function; specific function
特殊航道 special channel
特殊合金钢 special alloy steel
特殊合同 special appointment contract;specific contract
特殊合同条件 special conditions of contract
特殊和偶然发生的项目 unusual and non-recurring item
特殊荷载 extreme load;special load;extraordinary load
特殊荷载条件 extreme loading condition
特殊虎钳 special vise[vice]
特殊化 particularize; specialization; specialize
特殊化学品 speciality chemicals
特殊环境 envirium[复 enviria];particular surroundings; specific environment
特殊环境保护法 law of special environmental protection
特殊灰浆 special grout
特殊混凝土搅拌机 special concrete mixer;mud slinger <俚语>
特殊火灾危险区 special fire hazard area
特殊货签 special docket
特殊货物 special cargo;special goods
特殊货物搬运 handling of special cargo
特殊积分 special integral
特殊集装箱 specific container
特殊记录仪 special recording device
特殊技术标准 particular specification
特殊技术用语 technicals
特殊加工面金属 specially surfaced metal
特殊加强筋 extra reinforcing
特殊夹头 special carrier
特殊家庭补助 special family subsidy
特殊监测 special monitoring
特殊监察设备 <检查窃听或拆线> special observation post
特殊检查 special inspection
特殊建设项目 unique project
特殊建造的工厂 specially constructed plant
特殊建造的码头 specially constructed quay; specially constructed wharf
特殊建筑物 special building
特殊降压闸 special decompression chamber
特殊校正电路 special correcting circuit
特殊教育 special education
特殊街坊 superblock
特殊街坊规划 superblock plan
特殊结构 special construction;special structure
特殊结构混凝土 special structural concrete
特殊紧固装置 specific restraint device
特殊紧急事态 exceptional urgency
特殊井 exceptional well
特殊举证责任 special burden
特殊矩阵法 special matrix method
特殊掘进法 special excavating method;special drivage method <采矿>
特殊卡爪 special jaw
特殊可溶物质 specific soluble substance
特殊扣除 particular deduction
特殊扣除项目 particular deduction
特殊矿物 distinctive mineral
特殊矿渣水泥 special slag cement
特殊垃圾 <法律规定需要特殊处理的有害垃圾> special refuse;special waste
特殊类型 special type
特殊类型的设备 special type of equipment
特殊类型矿石 ore of special type
特殊利益保护计划 advocacy planning
特殊料斗 special bucket
特殊零件 peculiar part
特殊滤波器 rectangular window
特殊路面研究 specific pavement study
特殊铝青铜 Arms bronze
特殊脉冲重复频率 specific pulse recurrence rate
特殊脉冲重复周期 specific pulse repetition interval
特殊锚 special anchor
特殊锚卸扣 special anchor shackle
特殊贸易 special commerce
特殊灭火器 special extinguisher
特殊名 special name
特殊模拟系统 special analog(ue) system
特殊模板 special form(work); special shuttering
特殊目的 special-purpose
特殊耐磨钢轨 special wear-resisting rail
特殊耐磨铝青铜 dynamobronze
特殊能源 non-conventional energy
特殊腻子 special putty
特殊年洪水量 specific year flood
特殊排放 special emission
特殊配合力 specific combining ability
特殊配制的油 specially formulated oil
特殊配置 peculiar setup
特殊喷水灭火系统 special sprinkler system
特殊票价 special fare
特殊频率 distinct frequency
特殊品 specialty goods
特殊品质的选择 breeding for special qualities
特殊起重机 special crane
特殊气候应用计划 special climate application program(me)
特殊墙板 specialty siding
特殊桥梁 special bridge
特殊侵蚀作用 special erosion
特殊青铜 special bronze
特殊情况 abnormal condition;an exceptional case;exceptional circumstance;extreme scenario;particular case; special case; special circumstance; special condition; unique feature
特殊情况下 in particular cases
特殊请求 specific request
特殊权数 specific weight
特殊染色 specific stain
特殊染色法 special staining
特殊热容量 specific heat capacity
特殊任务 special-purpose
特殊日期 technical date
特殊溶性物质 specific soluble substance
特殊溶液 singular solution;special solution
特殊乳胶体 special emulsion
特殊乳状液 special emulsion
特殊软管 special hose
特殊润滑油 special lubeoil
特殊三轴试验仪 special triaxial test cell;special triaxial test device;special triaxial test equipment
特殊三轴仪 special triaxial device; special triaxial equipment
特殊三轴仪试验盒 special triaxial cell
特殊色指数 special colo(u)r-index
特殊筛布 special screen cloth
特殊商品 particular kind of commodity;specialty goods
特殊商品出口港 specific commodity port
特殊商品交换 special form of commodity exchange
特殊蛇管 special hose
特殊设备 extra equipment;special equipment; special installation; specialist plant;special provision;specific installation
特殊设备检验规范 inspection code for special equipment
特殊设计 special designing; special planning
特殊设计标准 special design criterion
特殊设计项目 unique project
特殊射影线性群 projective special linear group
特殊摄动 special perturbation
特殊摄影 non-conventional photography
特殊深度 <危险物顶部的最浅水深> detached sounding
特殊生产方法的技术标准 closed specification
特殊生产领域 particular sphere of production
特殊十字管接头 special pipe cross
特殊石面 special finish
特殊石油产品 petroleum specialties
特殊识别法 special identification method
特殊使用流量 special discharge
特殊使用条件 particular service requirement
特殊事件 special event;specific event
特殊试验 special test;specific test
特殊试验费 special examination fee
特殊试验机 special testing machine
特殊试验条件 abnormal test condition
特殊试(样)件 specialized coupon
特殊(饰)面 special finish
特殊适应 idioadaptation
特殊适应性演化 idioadaptive evolution
特殊鼠笼式感应电动机 special squirrel cage induction motor
特殊数据组合故障 pattern-sensitive fault
特殊水化白云石石灰 special-hydrated dolomitic lime
特殊水龙带 special hose
特殊水深 special sounding
特殊水组分 specific water constituent
特殊隧道 special tunnel
特殊损害 special damage
特殊台阵 special array
特殊陶瓷 special ceramics
特殊特征 special feature
特殊天气报告 special weather report
特殊条件 special condition; special provision
特殊条件下的路基 roadbed in especial condition; subgrade under the special condition
特殊条款 special clause; special provision;specific item
特殊贴面 special overlay
特殊图纸 special map
特殊涂层 individual coat;special coat
特殊土地基 special soil foundation
特殊土路基 subgrade of special soil
特殊土壤 special soil; problematic(al)soil
特殊挖泥法 special dredging method
特殊外形 distinctive appearance
特殊网纱 special screen cloth
特殊危险 catastrophe risk;extraneous risks
特殊问题 special issue;special problem
特殊污染物 specific pollutant
特殊污染源 specific source
特殊细则 special instruction
特殊显色指数 special colo(u)r rendering index
特殊线性群 special linear group
特殊相互作用 specific interaction
特殊详图 special detail drawing
特殊响应 specific response
特殊响应计划 specific response plan
特殊项目 extraordinary item; special item
特殊效果 special effect
特殊效率 special vehicle
特殊信道 special channel
特殊信号浮标 special-purpose buoy
特殊形的种类 kind of special form
特殊形式 special shape
特殊型 special type
特殊型钢 special section
特殊性 particularity; specific characteristics
特殊性能 special performance
特殊性土 regional soil
特殊性岩土 special rock and soil
特殊性质 special property
特殊许可证 special licence[license]
特殊训练 special training
特殊养护浸透树脂 special nutritive contact resin

特殊要求 specific requirements
特殊叶轮 special-purpose impeller
特殊仪表用钢 special steel for making instruments
特殊异谱同色指数 special metameric index
特殊易损件 special wear item
特殊应用 particular application
特殊用地 <城市规划> special use area
特殊用法 exceptional use; special usage; special use
特殊用钢 steel for special purposes
特殊用户 individual consumer
特殊用途 special application; special-purpose; special service; special usage; special use
特殊用途车辆 purpose-built vehicle
特殊用途船 special-purpose vessel
特殊用途的料斗 special application bucket
特殊用途的模子 special-purpose mold
特殊用途的批准 special use permit
特殊用途地产 special-purpose property
特殊用途焚化炉 special-purpose incinerator
特殊用途浮标 special-purpose buoy
特殊用途混凝土 special-purpose concrete
特殊用途免税物品 non-taxable-goods for special use
特殊用途石砖 special-purpose tile
特殊用途水泥 special-purpose cement
特殊用途梯 special-purpose ladder
特殊用途运输机 special-purpose transport
特殊优惠 special preference
特殊预报 special forecast
特殊预防措施 extraordinary precaution
特殊预算 extraordinary budget; special budget
特殊预制构件 special prefabricated element
特殊预制孔空心块体 special prefabricated hollow block
特殊遇难信号 special distress signal
特殊原因的收益 but-for income
特殊原子 weird atom
特殊运动迟缓 specific motor retardation
特殊运价 special tariff
特殊运价表 special tariff; tariffs of exceptions
特殊运输车 special transporter
特殊运输器 special transporter
特殊灾害保险 special hazards insurance
特殊载重 exceptional load
特殊凿井法 special methods of shaft sinking; special shaft-sinking method; special sinking method
特殊账户 special account
特殊照顾的收容所 special care home
特殊照明 special lighting
特殊折扣 channel discount
特殊正交群 special orthogonal group
特殊支出 extraordinary expenditures
特殊指令 special instruction
特殊制备 special preparation
特殊质量砌块 special quality block
特殊种植法 special planting
特殊重复频率 specific repetition frequency
特殊周期 specific cycle
特殊轴承 special bearing
特殊铸铁 special cast iron
特殊专业计划 special career program(me)
特殊砖 special brick
特殊转运 special transport
特殊桩 special pile
特殊装潢 individual fabrication
特殊装配 individual mounting
特殊装卸 special handling; special loading and unloading
特殊装载 <贵重品的> special stowage
特殊装置 individual mounting
特殊状况 special status
特殊状态 authorised[authorized] state
特殊字符 special character
特殊字符组 special character set
特殊钻进情况 particular drilling condition
特殊钻头 special drill
特殊作业时间 special work time
特殊作用 to play a special role
特殊坐标 special coordinates
特殊座位 special seat
特殊座席 special seat
特殊座椅 special seat
特水硅钙石 truscottite
特斯拉 <磁通量密度单位> Tesla
特斯拉变压器 Tesla transformer
特斯拉-达松伐耳安培计 Tesla-D'arsonval ammeter
特斯拉-达松伐耳电流 Tesla-D'arsonval current
特斯拉电流 Tesla current
特斯拉空心变压器 Tesla coil; Tesla transformer
特斯拉耦合 Tesla coupling
特松动配合 extra slack running fit
特提斯斑岩铜矿成矿带 metallogenic belt of Tathyan porphyry copper deposits
特提斯北缘大区 north Tethyan region
特提斯构造活动带 Tethys tectonic active zone
特提斯海 <古地中海> Tethys
特提斯海路 Tethys seaway
特提斯海洋 Tethys ocean
特提斯南缘大区 south Tethyan region
特提斯区系 Tethyan realm
特提斯-喜马拉雅构造域 Tethys-Himalayan tectonic domain
特瓦德尔比重计 Twaddell's hydrometer
特瓦德尔度 <液体比重表示法之一> Twaddell's degree
特威切耳试剂 Twitchell reagent
特纬纱形成的花纹 extra weft figuring
特沃德尔比重标 Twaddle scale
特沃德尔比重标度 Twaddle gravity scale
特沃德尔比重计 Twaddle's hydrometer
特沃德尔度 <比重单位> degree of Twaddell
特污带 supersaprobic zone
特戊酸盐 pivalate
特细粉 ultra fines
特细粉末 ultra-fine dust
特细粉碎机 ultra-fine grinder
特细钢丝 extra thin steel wire
特细骨料混凝土 finest concrete
特细过滤器 ultra-fine fibre[fiber]
特细号 extra fine
特细集料混凝土 finest concrete
特细煤烟 ultra-fine soot
特细漆包线 ultra-fine enameled wire
特细牙螺纹 extra-fine screw
特细研粉机 ultra-fine pulverizer
特狭面型 hyperleptoprosopic
特狭上面型 hyperleptene
特项成本 cost memo
特项推销 special sales promotion
特销渠道 channel of promotion
特小断面 mini cross-section
特小间隙滑动配合 close sliding fit
特小型版 miniature edition
特小型轴承 extra small bearing
特小值 exceptional value
特效 specially good effect
特效除莠剂 special weed chemicals
特效反应 specific reaction
特效解毒剂 specific antidote
特效离子电极 specific ion electrode
特效黏[粘]合剂 specific binding agent
特效黏[粘]结剂 specific binding agent
特效试剂 specific reagent
特效试验 specific test
特效药 sovereign remedy; specific medicine; specific remedy
特效预防 specific prophylaxis
特写镜头 close up; close up shot; close up view
特写摄影 close up photography
特写图 close up view
特形边 featured edge
特形光束 shaped beam
特形焊接 contour weld
特形喷管 shaped nozzle
特形铁芯电缆 shaped-conductor cable
特形铣刀 special formed milling cutter
特型管道 contoured duct
特型集装箱 contoured container
特型接头 sub
特型轴 profile shaft
特型铸造 shaped casting
特型砖 purpose-made brick
特性 behavio(u)r; characterization; prerogative; special character; specific character; specific property; train; unique feature
特性 X 辐射 characteristic X-radiation
特性 X 射线 characteristic X-ray
特性半径 characteristic radius
特性丙烯酸 special acrylic acid
特性不均匀的传输线 heterogeneous line
特性参数 characteristic parameter; natural parameter
特性残余流 characteristic residual current
特性测定 characteristic test
特性测井 signature log
特性测试 characteristic test
特性长度 characteristic length
特性尺寸 characteristic dimension
特性抽样(法) attribute sampling
特性导纳 natural admittance; surge admittance
特性导线电缆 shaped-conductor cable
特性的 characteristic
特性点 characterisation point; characteristic point; point of singularity
特性恶臭 characteristic odo(u)r
特性方程 characteristic equation
特性非点源污染 characterizing nonpoint pollution
特性废物 special waste
特性分类 property sort
特性分歧 characteristic divergence
特性辐射 characteristic radiation
特性附着 specific adhesion
特性格 property lattice
特性关系曲线 characteristic relation
特性归化 performance reduction
特性函数 characteristic function; eigenfunction
特性化 characterization
特性环境 special environment; specific environment
特性畸变 characteristic distortion
特性计算 performance calculation; performance computation
特性记述 characterization
特性简化预计法 simplified performance prediction
特性鉴定 characterization
特性解 characteristic solution
特性经营权 franchise
特性颗粒污泥 specific granule sludge
特性离子电极 specific ion electrode
特性列举法 attribute listing method
特性灵敏检测器 property selective detector
特性流量 characteristic discharge
特性流速 characteristic velocity
特性伦琴辐射 characteristic X-radiation
特性论 ethology
特性描述 characterization
特性黏[粘]度 inherent viscosity; intrinsic(al) viscosity; limiting viscosity
特性黏[粘]度数 inherent viscosity number; intrinsic(al) viscosity number; limiting viscosity number
特性黏[粘]附 specific adhesion
特性黏[粘]合(力) specific adhesion; inherent adhesion
特性黏[粘]结 specific adhesion
特性排气速度 characteristic exhaust velocity
特性频率 characteristic frequency
特性强度 characteristic strength
特性曲线 characterisation curve; characteristic curve; performance(characteristic) diagram; performance chart; performance curve
特性曲线陡度 characteristic curve slope
特性曲线法 characteristic method; tie-line-bias <能源供应>
特性曲线拐点 knee of characteristic
特性曲线图 characteristic diagram
特性曲线弯点 knee
特性曲线网 characteristic net
特性曲线族 characteristic family; family of characteristic curves; family of characteristics
特性曲线最大弯曲处 knee of characteristic
特性溶质物质 specific soluble substance
特性设计 characteristic design
特性失真 characteristic distortion
特性时间 characteristic time
特性时间反应 characteristic time response
特性试剂 specific reagent
特性试验 attribute test(ing); characteristic test; performance test; testing of characteristic
特性数 characteristic quantity; performance number
特性数据 performance data
特性水深 characteristic depth of water
特性水体 specific bodies of water; specific water body
特性水文曲线 characteristic hydrograph
特性水域 specific bodies of water; specific water body
特性说明 characterization
特性速度 characteristic speed; intrinsic(al) speed
特性损失 performance loss
特性态 eigenstate
特性图 character diagram; performance plot

特性图解 cause of effect diagram; characteristic diagram
特性温度 characteristic temperature; representative temperature
特性污染物 characteristic contaminant; particular pollutant; specific contaminant; specific pollutant; typical pollutant
特性污染物浓度 concentration of particular pollutant; particular pollutant concentration
特性误差 characterisation error; characteristic error
特性吸附 specific adsorption
特性吸附的吸附质 specifically adsorbed adsorbate
特性吸附离子电荷 charge of specifically adsorbed ions
特性吸附容量 specific adsorption capacity
特性吸附速率 specific adsorption rate
特性吸附物种 specifically adsorbed species
特性吸收峰 characteristic absorption peak
特性系数 character coefficient; characterization factor; characterizing factor; coefficient of performance
特性下降(曲线) false characteristic
特性线 characteristic line
特性线法 method of characteristics
特性型号 characteristic type
特性阳极电压 characteristic anode voltage
特性氧化还原环境 specific redox environment
特性要素 characteristic element
特性要图 characteristic diagram
特性要因分析图 cause and effect analysis chart
特性因数 characteristic factor; characterization factor; characterizing factor
特性因素 characteristic factor; characterization factor
特性有机污染物 typical organic pollutant
特性值 characterisation value【数】; characteristic value
特性指标 index of characteristics; index property
特性指数 index of specialty; performance index; quality number; characteristic exponent【数】
特性质量 characteristic quality
特性滞后时间 characteristic delay time
特性周波带 formalin
特性转速 characteristic speed
特性资料 characteristic data
特性自适应控制 characteristic adaptive control
特性阻抗 characteristic impedance; characteristic wave impedance; intrinsic(al) impedance; natural impedance; surge impedance
特性阻抗终端负载 characteristic impedance termination
特性阻抗终接 characteristic impedance termination
特性组分 characteristic component
特需供应 supply in accordance with special needs
特许 charter; franchising; privilege
特许安排 concession arrangement; franchising arrangement
特许保留地 charter-land
特许被授予人 franchisee
特许捕猎证 special hunting license[licence]
特许材料 licensed material
特许操作 privileged operation
特许测量员 chartered surveyor
特许成本及例外事项 cost favor and exceptions
特许成立公司 charter
特许承包公司 concessionary and contracting company
特许程序 licence program(me)
特许出口 special permission export
特许出口商品 goods exported under special license
特许出口证 special permission export
特许存储 authorized access
特许存储器操作 privileged memory operation
特许存取 authorized access; privileged access
特许代理人 chartered agent
特许代码 authorization code
特许的 chartered; concessional; concessionary
特许的运输学会 <英> Chartered Institution of Transport
特许房屋勘测员 chartered building surveyor
特许工程师 chartered engineer; graduated engineer
特许工程施工交通管制 encroachment permits
特许工会 certified union
特许公司 chartered company
特许规范 privileged norm
特许护板 patent protective
特许价格 price concession
特许建筑 concession building
特许建筑服务设施协会 Chartered Institution of Building Service
特许交易 franchise deal
特许进口证明书 certificate of import licence[license]
特许经销代理人 franchisee
特许经营方式 franchising
特许会计师 certified public accountant; chartered accountant
特许会计师学会会员 <英> Fellow of Chartered Accountants
特许连锁 franchise chain
特许零售商 franchised dealer
特许码头 sufferance quay; sufferance wharf; legal quay <英>
特许锚地 restricted anchorage
特许期限 duration of franchise
特许权 charter; chartered concession; chartered right; patent right; right of patent; right of special permission; royalty; concession; franchise <美>
特许权保证书 franchise bond
特许权持有人 concessionaire
特许权合同 contract of concession
特许权获得者 concessionaire; concessioner
特许权使用费 loyalities
特许权税 franchise tax
特许权所有人 concessionaire; concessionaire granter; concessioner
特许权所有者 concessionaire; concessioner
特许权协议 licensing agreement
特许商店 franchised shop; franchised store
特许商人 authorized dealer
特许生产 license production
特许施工证 variance
特许使用费受益所有人 beneficial owner of the royalties
特许使用金 royalty payment
特许收储机构 licensed deposit takers
特许收集 franchise collection
特许授予人 franchiser[franchisor]
特许输入 privileged input
特许数据 authorization data; authorized data
特许体系 franchise system
特许通行标志 priority permit sign
特许销售 franchise distribution
特许销售网 franchise distribution network
特许协议 concession agreement
特许意向书 concession agreement
特许银行 chartered bank
特许用途 conditional use
特许邮票 franchise stamp
特许运输学会 chartered institute of transport
特许在车站经营的零售商 concessionaire
特许证(书) letter of patent; letter patent <指专利>; charter(ed concession); charter of concession; diploma; license[licence]; special permit
特许止漏器 patent protector
特许指令 privileged instruction
特许制 franchise system
特许转载 courtesy
特许状 charter
特许状持有人 franchisee
特许状态 privileged state
特选 preference
特压 extreme pressure
特压润滑剂的振动试验 shock test of extreme-pressure lubricants
特压添加剂 extreme pressure additive
特压皂 extreme pressure soap
特邀报告 guest lecture; invited lecture
特邀投标者 invited bidder
特异的耗竭性 specific exhaustibility
特异毒害作用 specific toxic effect
特异反应性 atony; idiocrasy; idiosyncrasy
特异景观风景区 specific natural scenes area
特异离心 differential centrifugation
特异凝集 specific agglutination
特异凝聚 special agglutination
特异品质 idio(syn)crasy
特异危险性 attributable risk
特异型耐火砖 special-shaped fireclay brick
特异型制品装窑法 box in
特异型砖 special shape brick
特异性 specificity
特异性颗粒 specific granule
特异性损害 special damage
特异性转移 transspecific
特异序数 singular ordinal
特异选择 differential selection
特异作用 specific action
特异作用带 zone of specific effect
特异作用阀 threshold of specific effect
特硬 extremely hard
特硬的 ultra-hard
特硬的铺面砖 <其中的一种> pavio(u)r
特硬方砖 clink paving brick
特硬钢 dead-hard steel; extra hard steel; glass-hard steel
特硬花岗岩 dead granite
特硬木板 super-hardboard
特硬耐磨合金 Akrit
特用存储区 dedicated memory
特用电动机 special-duty motor
特用林 special-purpose forest
特用作物 special crop
特优等级 special elected quality
特优跑道 <机场> preferential runway
特优饮用水 extra drinking water
特优质量 extra quality
特优种 predominant species
特有财产 peculiar
特有的 endemic; particular; specific
特有的风格 idiosyncrasy
特有的化学关系 specific chemical relation
特有分布(现象) endemism
特有经济规律 laws governing special economy
特有树 proper tree
特有物权 special property
特有物种 endemic species
特有现象 endemism
特有种属 endemic genus
特远运输 exceptionally long-distance traffic
特约保险单 specified policy
特约比价 restricted tender
特约编辑 contributing editor
特约不赔 warranted free
特约订货 custom millwork
特约检修服务站 authorized service shop
特约经销处 special sales agency
特约经销商 franchised dealer
特约商店 appointed store; tied shop
特约数量 specified amount
特约条款 warranties
特约通话 subscription call
特约维修店 special repair shop
特约险 extra risks
特约窑干 custom kiln drying
特造的 purpose-built
特窄平齐头锉 extra narrow pillar file
特征 cachet; characteristic; characterization; distinguishing feature; earmark; feature; lineament; physiognomy; salient feature; signature; texture; token; train; tag【计】
特征 n 维 properly n-dimensional
特征 X 辐射 characteristic X-radiation
特征 X 射线点扫描像 characteristic X-ray point scanning image
特征 X 射线面扫描像 characteristic X-ray area scanning image
特征 X 射线强度 characteristic X-ray intensity
特征 X 射线图像 characteristic X-ray image
特征 X 射线吸收系数 characteristic X-ray absorption coefficient
特征 X 射线线系 characteristic X-ray systems
特征白色 characteristic white
特征比 characteristic ratio
特征边界条件 characteristic boundary condition
特征编码 characteristic coding feature coding
特征变换 eigentransformation
特征变量 characteristic variable
特征辨认 character recognition
特征标的正交性 orthogonality of character
特征标记存储器 tag storage
特征标量 characteristic scalar
特征标群 character group
特征标志 characteristic mark
特征表 mark sheet
特征表面 figuratrix
特征并矢 characteristic dyadic
特征波高 characteristic wave height
特征波前 eigen wavefront
特征波阻抗 characteristic wave impedance
特征参数 characteristic parameter; characterizing parameter; latent parameter; characterisation parameter【数】
特征测验图 profile
特征产品 characteristic product

特征长度 characteristic length
特征常数 characteristic constant
特征场地周期 characteristic site period
特征超平面 characteristic hyperplane
特征尺寸 characteristic dimension
特征抽出程序 characteristic extraction program(me); feature extraction program(me)
特征抽取 feature extraction
特征初值问题 characteristic initial value problem
特征穿孔机 tag marker
特征词 feature word
特征单色谱 characteristic isochromatic spectroscopy
特征的 eigen
特征的正规形式 normal form of the characteristic
特征点 characterisation point; characteristic point; silent point
特征点类型 special point type
特征电阻 characteristic resistance
特征度量 measure of characteristics
特征多项式 characteristic polynomial; eigen polynomial; proper polynomial
特征二次型 characteristic quadratic form
特征发生器 character line generator
特征法 characteristic method; signature method
特征方程 characteristic equation; proper equation; secular equation; characterisation equation【数】
特征方向 characteristic direction
特征分类 feature selection; tag sort
特征分析 characteristic analysis
特征分析法 characteristic analysis method
特征分析技术 characteristic analysis technique; signature analysis technique
特征分析软件 signature analysis software
特征峰 characteristic peak
特征符按钮 flag button
特征伽马射线常数 specific gamma ray constant
特征伽马射线照射率 specific gamma-ray emission
特征概率 characteristic probability
特征高峰小时 characterisation rush hour
特征根【数】 characterisation root; characteristic root; latent root
特征根检验 characteristic root test
特征光谱 characteristic spectrum
特征广义函数 eigendistribution
特征过程线 characteristic hydrograph
特征函数【数】 fundamental function; characterisation function; characteristic function; eigenfunction; proper function
特征函数的重零点 repeated zeroes of a characteristic function
特征函数展开式 eigenfunction expansion
特征行列式 characteristic determinant
特征和连续荧光修正 characteristic and continuous fluorescence correction
特征河段 characteristic river reach
特征荷载 characteristic load
特征红外基团频率 characteristic infrared group frequency
特征化 characterization
特征化石 characteristic fossil; diagnostic fossil
特征计数检验 mark counting check
特征记号 characteristic mark; tag mark(ing)
特征记录 <展现声波波列的> character log
特征监测 feature detection; feature monitoring
特征简介 diagnosis
特征交叉点 characterisation crossing point
特征解 characteristic solution
特征矩阵 characteristic matrix; eigenmatrix
特征矩阵法 eigenmatrix method
特征卡(片) aspect card; tag card; token card
特征卡阅读器 tag card reader
特征空间 eigenspace; feature space
特征控制符号 feature control symbol
特征宽度 characteristic width
特征矿物 characteristic mineral; diagnostic mineral
特征离子 characteristic ion
特征理论 characteristic theory
特征力法 eigen force method
特征量 characteristic quantity
特征量度 measure of characteristics
特征流量 characteristic discharge
特征流速 characteristic velocity
特征流形 characteristic manifold
特征码 characteristic code; condition code; feature attribute
特征码表 characteristic code list; feature code list
特征码定义卡 characteristic definition card
特征码清单 feature codes menu; menu card
特征码清单法 menu technique
特征码索引 object code header
特征美 characteristic beauty
特征描述制 profiling systems
特征明显的变种 distinct variety
特征模型 eigenmode
特征能量 characteristic energy
特征年龄 characteristic age
特征年流量过程线 discharge hydrograph for characteristic year; discharge time curve for characteristic year
特征年水位过程线 stage hydrograph for characteristic year; stage-time curve for characteristic year
特征浓度 characteristic concentration
特征浓度曲线 characteristic concentration curve
特征排气速度 characteristic exhaust velocity
特征频率 characteristic frequency; eigenfrequency
特征频谱 characteristic frequency spectrum
特征谱带 characteristic band; key band
特征强度 characterisation strength; characteristic strength
特征曲线 characteristic curve; indicatrix; intrinsic(al) curve; eigencurve
特征曲线窗口法监督分类 character curve-window based supervised classification
特征曲线法 characteristic curve method
特征曲线类型 characteristic class
特征曲线族 family of characteristic curves; family of characteristics
特征群 syndrome
特征三角形法 characteristic triangle method
特征扫描 mark scan(ning)
特征上溢【计】 characteristic overflow
特征射线 characteristic ray
特征时间 characteristic time
特征时间尺度 characteristic time scale
特征识别 characteristic identification; trick recognition
特征矢量 characteristic vector; eigen vector; eigenvector; proper vector
特征数 characteristic number; performance number
特征数据 attribute data; characteristic data
特征数列 characteristic series
特征衰变率 character(istic) rate of decay
特征双峰 characteristic doublet peak
特征水力参数 characteristic hydraulic parameter; significant hydraulic parameter
特征水力性能 characteristic hydraulic feature
特征水深 characteristic depth
特征水头 characteristic head
特征水位 characteristic stage; characteristic water level
特征瞬间 characteristic instant
特征松弛 key relaxation
特征速度 characteristic speed; characteristic velocity
特征速率 characteristic rate
特征损失能谱学 characteristic loss spectroscopy
特征淌度 eigen-mobility
特征提取【测】 feature extraction
特征条件 characteristic condition; representative condition
特征图 characteristic pattern
特征椭圆 eigenellipse
特征位 flag bit; stencil bit
特征位标记 flag
特征位操作数 flag operand
特征位触发器 flag flipflop
特征位错 characteristic dislocation
特征位移法 eigen displacement method
特征温度 characteristic temperature
特征问题 eigenproblem
特征污染图 characteristic pollution graph
特征误差 characteristic error
特征吸收峰 characteristic absorption peak
特征系数 characteristic coefficient
特征系统 tag system
特征线 characteristic line; typical line
特征线差分混合法 characteristic difference hybrid method
特征线法 characteristic line method; method of characteristic curves; method of characteristics
特征线方法 method of characteristics
特征线四点法 four-point method of characteristics
特征线图 characteristic diagram
特征线系 set of characteristics
特征相量 proper phasor
特征向量【数】 eigen vector; characteristic proper; characteristic vector; eigenvector; latent vector
特征信号 characteristic signal
特征信息 attribute information; characteristic information
特征形式 characteristic form; characteristic shape
特征形状 characterisation shape; characteristic shape
特征形状的砖 purpose-made brick
特征型号 characteristic type
特征性质 index property
特征选择 feature selection
特征因数 characterization factor
特征应力 characteristic stress
特征元素 characteristic element; eigenelement
特征圆锥曲线 characteristic bowl-shaped curve
特征源标记 aspect source flags
特征阅读器 tagging reader
特征振动 characteristic vibration; eigenvibration
特征振幅 <结构振动> characterization amplitude
特征直径 characteristic diameter
特征值 characterisation value; characteristic constant; characteristic number; characteristic root; characteristic value; eigenvalue; eigenwert; proper value
特征值的渐近分布 asymptotic(al) distribution of eigenvalue
特征值的数值计算 numeric(al) computation of eigenvalues
特征值法 eigenvalue method; method of characteristics
特征值问题 characteristic value problem; eigenvalue problem
特征值析取固有振动频率 eigenvalue extraction
特征值域问题 eigenvalue field problem
特征植被 characteristic vegetation
特征指标 characterisation index; characteristic index; index property
特征指数 character index; characteristic exponent; characteristic index
特征质量 characteristic mass
特征种(类) characteristic species
特征周期 characteristic period; eigenperiod
特征状态 eigenstate
特征锥 characteristic cone; characteristic conoid
特征锥面 eigencone
特征字 characteristic letter; tag(ged) word
特征阻抗 characteristic impedance; intrinsic(al) impedance; natural impedance
特征组分 characteristic component
特征坐标 characteristic coordinates
特值 particular value
特指的 designated
特指值 designated value
特制 made-to-measure; made-to-order; special make
特制爆炸胶 gelatin(e) extra
特制闭合板桩 <非标准型的> closer
特制铲斗 purpose-made bucket
特制的 purpose-built; tailored; tailor-made; purpose(ly) made
特制的防护薄片 purpose-made protecting foil
特制的混合物 purpose-made composition
特制的建筑工人用具 special builder's fitting
特制的结构型材 purpose-made structural profile
特制的金属线 purpose-made wire
特制的黏[粘]合剂 purpose-made (bonding) adhesive
特制的配筋钢 purpose-made reinforcement steel
特制的升降机 purpose-made lift
特制的施工用的小五金 purpose-made builders hardware
特制的施工用器具 purpose-made builders fitting
特制的箱形天沟 purpose-made box (roof) gutter
特制的斜坡清洗射流箱 purpose-built

slope cleansing jet box; purpose-made slope cleansing jet box
特制的型材 purpose-made section
特制的装饰材料 purpose-made ornamental block
特制底地毯 patent-back carpet
特制垫带 specially made gasket
特制斗 purpose-made bucket
特制方向架 specially made tool for orientation
特制防护(金属)薄片 special protecting foil
特制放气阀 special air cock
特制粉饰 patent plaster
特制缝带 feature strip
特制缝条 feature strip
特制高硅铝合金活塞 specialloid piston
特制供暖供水或回水三通 special supply or return tee
特制骨料 tailor-made aggregate
特制光纸 patent coated paper
特制盒子 purpose-built box
特制灰泥 patent plaster
特制集料 tailor-made aggregate
特制架子 purpose-built rack
特制建筑板材 patent board
特制胶合板 weldwood
特制接头 special fit
特制磨切纤维 extra-milled fiber[fibre]
特制黏[粘]结剂 purpose-made bonding agent
特制片 featurization
特制品 special(i)ty
特制品订单 special production order
特制气动铺料机 job-designed air-actuated spreader
特制气动摊铺机 job-designed air-actuated spreader
特制砌块 special-quality block
特制清坡射流箱 purpose-built slope cleansing jet box; purpose-made slope cleansing jet box
特制烧透砖 special burnt brick
特制升降机 special lift
特制石膏灰浆 <整体混凝土涂料> bond(ing) plaster
特制饰面混合料 face mix
特制熟石膏 patent plaster
特制水磨石(地面) special matrix terrazzo
特制陶瓦 <美> clay facing tile
特制天沟瓦 tile valley
特制细菌 tailor-made microbe
特制型材 special profile
特制硬煤沥青 special pitch
特制油 tailor-made oil
特制纸片 special-carte
特制砖 purpose-made brick; specially made brick; special-quality brick; squint brick
特质 aptitude; idiosyncrasy; particularity
特质淬火油 super-quench oil
特质交易 idiosyncratic exchange
特种扳手 special spanner; special wrench
特种板材 specialty sheet
特种泵 special pump
特种编码 specific coding
特种变压器 special transformer
特种表面处置 special surface treatment
特种病防治中心 control center[centre] of specific disease
特种玻璃 special glass
特种玻璃纤维 specialty glass fibre[fiber]
特种材料 special material
特种车 car of particular class; particular class wagon
特种车辆 car of particular class; special vehicle; specific vehicle
特种车辆及装置 special vehicles and devices
特种车辆优先信号系统 very important person system
特种车身 special body
特种乘车证 special service pass
特种船舶 <巡逻船工作船等> special boat
特种船货 special cargo
特种船检(验)合格证 special vessel inspection certificate
特种粗插刀 roughing special slotting tool
特种存款 special deposit; special loan
特种弹药补给区 special ammunition supply area
特种弹药仓库 special ammunition depot
特种弹药供应站 special ammunition supply point
特种的 special
特种底漆 special primer
特种地图 special-purpose map; thematic map
特种电码 special code
特种电阻 special resistance
特种订货 <按特种设计要求制造的> specially built
特种订货成本制度 special order cost system
特种订货单 special order for goods
特种动力作用 specific dynamic(al) action
特种断面尖轨 full-web-section switch rail; special heavy section switch rail; tongue rail made of special section rail
特种断面尖轨转辙器 switch rail of special section
特种多样性指数 species diversity index
特种方法 special process
特种防火石膏板 type X gypsum wallboard
特种防火石膏板条 type X lath
特种非金属矿产 special non-metallic commodities
特种费用 special charges
特种分类账 special ledger
特种粉刷 special plaster
特种氟硅酸盐 special fluorosilicate
特种福利基金 special welfare fund
特种赋税 special assessment
特种覆盖金属箔片 special covering foil
特种钢 special steel
特种钢板 special steel plate
特种钢管 special steel pipe; special steel tube
特种钢筋 alloy reinforcing steel; special reinforcement (steel); special reinforcing steel
特种钢丝 special wire
特种钢丝玻璃 special wire(d) glass
特种钢丝绳 special wire rope
特种高程曲线 hypsocline
特种高级锌 special high-grade zinc
特种工程结构 special engineering structure
特种工程塑料 special engineering plastic
特种工具钢 special tool steel
特种工具与器材 special tool and equipment
特种工业建筑物 special industrial building
特种工业溶剂 special industrial solvent
特种工艺 handicraft; special arts and crafts; special handicraft products
特种公司 special corporation
特种公园 special park
特种构件 special unit
特种关税减让 special tariff concession
特种管接头 special coupling
特种管理信托 specialized trust management
特种光纤 special optic(al) fibre[fiber]
特种规格 special format
特种航空图 special aeronautical chart
特种合金 special alloy
特种合金钢 special alloy steel
特种核材料 special nuclear material
特种花纹轮胎 extra tread tire[tyre]
特种化合物冷却阀 special chemical composition cooling valve
特种化油器 special carburettor
特种灰泥 special plaster
特种绘图纸 plotting paper
特种混合料 special composition
特种混合物 special compound
特种混凝土 special(ized) concrete
特种混凝土制品 special concrete product
特种活载 special live load
特种火山灰水泥 special pozzolan(a) cement
特种火石玻璃 extra dense flint
特种货车 wagon for special purposes
特种货车挂运计划 cars of particular class dispatching plan
特种货品费率 specific commodity rate
特种货物 special cargo; special goods
特种货物固定编组列车 special commodity unit train
特种货物集装箱 special cargo container
特种货物清单 special cargo list
特种货物托运 exceptional consignment
特种货物托运通知 advice of special consignment
特种货物装载 special load
特种货运调度员 goods of particular class controller
特种机动车 special motor vehicle
特种基金 special fund
特种基金拨款账户 special fund appropriation account
特种基金存款 special fund deposit
特种基金收入账户 special fund receipt accounts
特种集装箱 special container
特种技术报告 special technical report
特种加工 non-conventional machining; non-traditional machining
特种夹头 special carrier
特种减压仓 special decompression chamber
特种建筑材料 special building material; special construction(al) material
特种建筑方法 special building method
特种建筑构件 special construction unit
特种建筑物设计 one-off design
特种建筑型材 special construction profile; special construction shape
特种建筑装饰件 special construction trim
特种浇灌混合料 special pouring compound
特种浇注 special casting
特种胶 special gelatin
特种胶合板 special(ty) plywood
特种胶凝剂 special cementing composition; special cementing compound
特种胶乳 special-purpose latex
特种胶粘剂 special bonding adhesive; special bonding agent; special bonding medium
特种胶质炸药药卷 special gelatine cartridge
特种校正电路 special correcting circuit
特种接头 special joint; transition fittings
特种结构材料 special structural material
特种结构钢 special structural steel
特种结构构件 special structural unit
特种结构型材 special structural profile; special structural section; special structural shape
特种结构用混凝土 special structural concrete; special structure concrete
特种结构装饰件 special structural trim
特种金属垫 special metallic gasket
特种经纪人 special broker; specialist
特种空砌块 special hollow block
特种空中钩取 special air pick-up
特种矿产 special commodities
特种垃圾 special refuse
特种类型建筑物 special type of construction
特种类型施工 special type of construction
特种沥青 special asphalt
特种沥青混合料 special asphalt mixture
特种沥青屋面卷材 special asphalt roofing felt
特种练习 special service practice
特种路用沥青 special road asphalt; special road oil
特种路用沥青拌砂 special road asphalt sand
特种轮胎 special tyre[tire]
特种螺栓 special bolt
特种码 absolute code; specific code
特种码头 special terminal
特种贸易 special commerce
特种煤焦油沥青 special coal tar pitch
特种煤沥青屋面油毡 special tarred roof(ing) felt
特种煤沥青预制屋面 special tarred ready roofing
特种敏感性 special sensibility
特种模型定律 special model law
特种抹灰 special plaster
特种目的 special-purpose
特种目的监测 special monitoring
特种目的税 special-purpose tax
特种耐火材料 special refractory
特种耐火隔热砖 special insulating brick
特种耐火砖 special-quality fire-brick
特种黏[粘]结剂 special adhesive
特种黏[粘]结剂混合物 special adhesive composition; special adhesive compound
特种配件 special casting; special fittings
特种铺用煤沥青混合物屋顶 special tarred composition roofing
特种漆 special varnish
特种汽车 special car
特种砌合 special bond
特种器材 special device
特种铅笔 chinagraph pencil; china-marking pencil; glass pencil; special pencil
特种墙纸 specific wallpaper

特种青铜 special bronze
特种清漆 special varnish
特种日记账 special journal
特种容器 special container
特种溶剂 exotic solvent
特种熔线 <滞后断开电流的> fuseron
特种软铁 special soft iron
特种瑞典油灰 special Swedish putty
特种润滑脂 special greases
特种三相变压器 special three-phase transformer
特种砂浆 special mortar
特种商店 specialty store
特种商号 special shop
特种商品 special cargo
特种商品运价表 special commodity tariff
特种设备 special equipment
特种审计 special audit
特种生物显著区 area of special biological significance
特种声阻抗 specific acoustic(al) impedance
特种失业补助 special unemployment assistance
特种石油产品 petroleum specialties
特种食品店 specialty food store
特种试验与操纵装备 special test and handling equipment
特种手势信号 special arm and hand signal
特种梳状滤波器 special comb filter
特种数量表 special table of allowance
特种刷 rigger
特种水管 special pipe
特种水泥 special cement
特种水泥管 special cement pipe
特种水泥砂浆 special cement mortar
特种锁闭 special locking
特种台阵 special array
特种弹簧安全阀 special spring safety valve
特种陶瓷 special ceramics
特种陶瓷纤维 special ceramic fiber [fibre]
特种特赋债券 special-special assessment bond
特种铁 special iron
特种同步电机 special synchronous machine
特种铜 complex brass
特种图 special chart; specific chart
特种涂层织物 specialty coated fabric
特种涂料 special coating
特种土 property soil
特种外加剂 special additive
特种文献 specialized information
特种屋顶排水沟槽 special valley gutter
特种屋面玻璃工 patent roofing glazier
特种无机纤维 special inorganic fiber [fibre]
特种无机纤维增强金属 special inorganic fiber[fibre] reinforced metal
特种无机纤维增强塑料 special inorganic fiber[fibre] reinforced plastics
特种无机纤维增强陶瓷 special inorganic fiber[fibre] reinforced ceramics
特种武器储存地 special weapons storage site
特种武器弹药补给站 special weapons ammunitions supply point
特种武器库 special weapons depot
特种武器中心 special weapons center [centre]
特种武器作战中心 special weapons operation center[centre]
特种稀砂浆 special slurry
特种线路 special type of track
特种橡胶 special rubber; specialty elastomer; specialty rubber
特种橡胶筛网 special rubber screen deck
特种消费品 specialty goods
特种消费行为税 excise tax; tax on special consumption behavio(u)r
特种销 special pin
特种楔形密封 closure using special wedge seal; seal by special wedge-like gasket
特种信贷 special loan
特种信号灯 optiphone
特种信托 special trust
特种行动措施 operational plan(ning)
特种形式集装箱 special type container
特种型材 special shape
特种修理车 special repair truck
特种循环 special cycle
特种训练器材 special training device
特种岩土工程 special geotechnic(al) engineering
特种厌氧滤床 special anaerobic filter bed
特种养护 special curing
特种业务话务员 special service operator
特种业务中继器 special service repeater; special service trunk circuit
特种液体泵 pumps for special liquids
特种液压油 skydrol
特种银行账户 special bank account
特种营业税 special tax
特种用户 special subscriber
特种用途船舶 special-purpose ship
特种用途的砖 purpose-made brick
特种用途焚烧炉 specific purpose incinerator
特种用途雷达 special-purpose radar set
特种用途林 forest for special purpose; forest for special use
特种油井水泥 special oil-well cement
特种油品 specialty oil
特种釉 extra-duty glaze
特种釉面砖 extra-duty glazed tile
特种运输 special traffic
特种运输方式 specialized transport mode
特种运输费率 special rate
特种灾害保险 special hazards insurance
特种轧制构件 special rolled unit
特种轧制泡沫高炉熔渣 special crushed foamed iron blast-furnace slag
特种轧制膨胀性熔渣 special crushed expanded cinder
特种轧制型材 special rolled profile; special rolled section; special rolled shape
特种政策型资产保险 property-special policy insurance
特种支援装备 special support equipment
特种执照 special licence[license]
特种直流机 special direct current machine
特种直馏硬煤沥青 special straight run coal tarpitch
特种纸张 specific paper
特种终接器 special service final selector
特种注射和分配装置 special injection and distribution device
特种铸件 special casting
特种铸铁 cast-iron alloy; Ni-tensilorin; special cast iron
特种砖 special(-quality) brick
特种转动换流机 special rotary converter
特种转换纸 special transfer paper
特种桩 special pile
特种装饰构件 special rolled trim
特种装饰件 special trim
特种装饰砌块 special decorative block; special ornamental block
特种装饰砖 special decorative tile; special ornamental tile
特种准备 special reserves
特种准备基金 special reserve fund
特种字符 special character; special sorts
特种字状 special character
特种自动真空焊接室 special automatic vacuum welding chamber
特种自卸汽车 specific purpose dump truck
特种钻进工艺 special drilling practices
特种钻头 special bit
特种作业费 special job cost
特重钡冕玻璃 extra-dense barium crown glass
特重车架 extra heavy frame
特重冲洗管 extra heavy wash rod
特重捣棒 giant weight tamper
特重阀和附件 extra heavy valves and fittings
特重管(子) extra heavy pipe
特重夯 giant weight tamper
特重荷载 extra heavy loading
特重厚玻璃板 extra heavy sheet
特重混凝土 heavy weight concrete
特重火石 extra-dense barium crown glass
特重火石玻璃 extra dense flint; extra dense flint glass
特重货件 very high load
特重货物 heavy lift; heavy cargo; exceptional weight
特重级 extra heavy
特重交通 ultra-heavy traffic; very heavy traffic
特重冕玻璃 extra heavy crown glass
特重特长运费率 heavy and lengthy rates
特重硪 giant weight tamper
特重物件 heavy weight volumetric(al) component
特重型 extra heavy duty type
特重型锤式破碎机 special heavy duty hammer breaker; special heavy duty hammer crusher; special heavy duty hammer mill
特重型绞车 extra heavy duty winch
特重型起重机 extra heavy duty crane
特重型弹簧 extra heavy spring
特重型抓斗 extra heavy type grab; extra heavy type grab bucket
特重载 extra heavy duty
特重载轮胎 extra heavy duty tire [tyre]
特重载重车 extra heavy duty truck
特重钻杆 extra heavy drill rod
特装本 edition binding; publisher's binding
特准储备物资 charactered material reserve; spee
特准储备资金 special approved stocking funds
特准的 under special authorization
特准极限 authorized limit
特准运照 navicert

铽

铽矿 terbium ores

疼

疼痛的强声 painfully loud sound

腾

腾冲矿 tengchongite

腾出 evacuate; vacate
腾出资金 save and divert funds to; transfer funds to
腾格林瓦合金 Tungelinvar
腾贵的物价 inflated price
腾空单元 vacated cell
腾空机 lift engine
腾空艇 air cushion boat; air cushion vehicle; hovercraft
腾涌 slugging
腾涌流试验 lug-flow test

誊

誊入总账 post(ing)

誊写板 papyrograph
誊写稿 fair copy
誊写者 copier

藤

藤本 voluble shrub

藤本的 lianoid
藤本橡胶 vine rubber
藤本型 lianoid form
藤本植物 climbing shrub; liana; vindicareine; vine
藤编细工 cane work
藤丛 canebrake
藤垫 cane mattress
藤顶凉亭 pergola
藤壶 acorn barnacle; acorn shell; barnacle; sea acorn
藤壶成虫 adult barnacle
藤壶属 Balanus
藤黄 cambogia; clusia rosea; gamboge
藤黄树 gamboge tree
藤黄油 garcinia butter
藤架 arbor; pergola; trellis
藤胶 liana rubber
藤茎分离筛 vine rack
藤科 canes
藤科编织物 cane mat
藤孔状的 caney
藤料 cane; rattan
藤料防冲器 cane fender; rattan fender
藤萝花架 wistaria trellis
藤蔓 bind
藤蔓攀附植物 climbing shrub
藤蔓植物 liano
藤器 rattan articles
藤青 enamel blue
藤色釉白坯陶瓷 cane-and-white ware
藤索 rattan rope
藤条 ratran(bar)
藤条碰垫 cane fender; fender rattan; rattan fender
藤网桥 rattan net bridge
藤席 cane mat
藤悬木 chinar
藤椅 cane chair; rattan chair
藤荫小径 pergola
藤制工艺品 cane work
藤制品 rattan product; rattan work
藤子 vine

剔

剔茬结合 slip joint

剔出 cull
剔出期间 elimination period
剔出器 ejector
剔出枕木 cull-tie
剔除 abate; charge off; disallowance; pick out; rejection
剔除病木 cutting out disease wood
剔除法 scalping method
剔除器 flush trimmer

剔除之物 cull
剔雕 cut-out carving
剔缝 raked joint
剔骨机 bone trimmer
剔红 <朱红雕漆> carved red lacquer;cut decoration
剔口 broking
剔清灰缝 rake-out(of joint)
剔清接缝 raking out of joint
剔野值 mute value

梯 ladder;catwalk

梯壁 altar
梯臂式装车机 hayrack boom loader
梯步木块 rough brackets
梯槽 ladder niche
梯层状构造 terracing
梯层状轮廓 terracing
梯车 tower wagon
梯承脚手架 ladder jack scaffold;ladder scaffold(ing)
梯尺 step wedge
梯齿轮离合器 trapezoidal tooth clutch
梯次打钢板桩法 echelon driving method
梯次配置 echelon(ment)
梯道 ladder access
梯的横木 rung
梯的竖梃 ladder stile
梯的一级 stair
梯蹬 ladder step
梯凳 step ladder
梯磴 stile
梯底背面 fluing soffit
梯地 terrace land
梯跌坡水 stepped energy destroying spillway
梯顶围栏 ladder ring
梯斗式起重绞车 ladder winch
梯斗式挖沟机 bucket ladder
梯度 declivity;gradient;gradient ratio;slope
梯度本征向量法 gradient eigenvector method
梯度比试验 gradient ratio test
梯度变化 change of gradient
梯度变化曲线 gradient
梯度波导纤维 graded-index waveguide fiber[fibre];gradient waveguide fiber[fibre]
梯度薄层色谱(法) gradient thin-layer chromatography
梯度材料 functionally gradient material
梯度操作 gradient operation;gradient run
梯度测量 gradient survey
梯度测量参数 parameter of gradient survey
梯度层 gradient layer
梯度常数 gradient constant
梯度磁强计 gradiometer
梯度电极系 lateral sonde;normal resistivity device
梯度发展 stratified development
梯度法 gradient method
梯度范围 gradient range
梯度方程 gradient equation
梯度分布 graded distribution;gradient distribution
梯度分离 gradient separation
梯度分离技术 gradient separation process;gradient separation technique
梯度分离因素 gradient separation factor
梯度分析 gradient analysis
梯度风 gradient wind
梯度风高度 gradient level
梯度风公式 gradient wind equation
梯度风气流 gradient flow
梯度风速 gradient velocity
梯度功能材料 functionally graded materials
梯度和拉普拉斯算子 gradient and Laplacian operator
梯度混合器 gradient mixer
梯度混合装置 gradient mixing device
梯度混合装置系统 gradient mixing device system
梯度机能陶瓷 functionally gradient ceramics
梯度计 gradient meter[metre];gradiometer
梯度校正器 gradient corrector
梯度结光电晶体管 graded-junction phototransistor
梯度矩阵 gradient matrix
梯度冷凝 gradient freezing
梯度离心(法) gradient centrifugation
梯度理论 gradient theory
梯度淋洗 gradient elution
梯度淋洗色谱法 gradient elution chromatography
梯度流 gradient current;slope current
梯度炉 gradient furnace
梯度模型 gradient former;gradient model
梯度凝固技术 gradient freeze technique
梯度凝固生长装置 gradient freeze growth apparatus
梯度浓度培养皿 gradient plate
梯度耦合 gradient coupling
梯度匹配滤波器 gradient matched filter
梯度漂移电流 gradient drift current
梯度坡度 gradient slope
梯度起点 gradient start
梯度倾斜度 gradient steepness
梯度曲线 gradient curve
梯度曲线剖面平面图 profiling-plan figure of gradient curve
梯度曲线剖面图 profiling figure of gradient curve
梯度确定 gradient determination
梯度溶剂 gradient solvent
梯度色调底片 gradient tint negative
梯度筛选调查 gradient screen survey
梯度时间 gradient time
梯度矢量 gradient vector
梯度适应 gradient preference
梯度输送理论 gradient transfer theory
梯度搜索(法)【数】 gradient search
梯度速度 gradient speed;gradient velocity
梯度算子 gradient operator
梯度体积 gradient volume
梯度投影 gradient projection
梯度投影法 gradient projection method
梯度投影算法 gradient projection algorithm
梯度洗提 gradient elution
梯度洗提吸附剂系统 gradient elution adsorbent system
梯度洗脱【化】 gradient elution
梯度洗脱泵 pump for gradient elution
梯度洗脱分离 gradient elution separation
梯度洗脱分配色谱(法) gradient elution partition chromatography
梯度洗脱分析 gradient elution analysis;solvent programming analysis
梯度洗脱谱带宽度 gradient elution bandwidth
梯度洗脱谱带压滤 gradient elution band press filter
梯度洗脱器 gradient elution device
梯度洗脱吸附剂系统 gradient elution adsorbent system
梯度洗脱中谱带宽度 bandwidth in gradient elution
梯度下降 gradient descent
梯度线 gradient line
梯度相关法 gradient related method
梯度向量 gradient vector
梯度协变矢量 gradient covariant vector
梯度形测热器 gradient calorimeter
梯度形状 gradient shape
梯度形式 gradient type
梯度掩模 gradient mask
梯度仪 gradiometer
梯度仪灵敏度 gradiometer sensitivity
梯度异常曲线 gradient anomaly curve
梯度映射 gradient mapping
梯度运算符 gradient operator
梯度折射率光纤 gradient index fiber[fibre];parabolic(al) index fiber[fibre]
梯度折射率光学纤维 gradient index optic(al) fibre[fiber]
梯度指数 gradient index
梯度装置 gradient device
梯端 head of ladder
梯段 flight;flight of stair(case);flight of step;going;stepway
梯段板 flight slab;waist slab
梯段爆破 blasting benches;heading blast(ing)
梯段标杆 going rod;pitch dimension;rake dimension
梯段布置 step arrangement
梯段步高 flight rise
梯段步距 flight run
梯段长度 flight length;go
梯段(超前)掘进爆破(法) heading and bench blasting
梯段尺寸 rise;rise of flight;stair rise
梯段的 stepwise
梯段底部炮眼 toe hole
梯段顶 stairhead
梯段法开石 stope
梯段高度 rise;rise of a flight;stair rise
梯段横梁 flight header
梯段回采 bench stoping
梯段净高 headroom of flight
梯段掘进爆破 heading and bending blasting
梯段开采法 bank method of attack
梯段开挖法 benching
梯段开挖掌子面 <隧洞工作面> forestope
梯段宽度 flight width
梯段裂缝 <露天开挖> angular crack
梯段坡度线 <各级踏步凸缘或前缘的连线> nose line;nosing line
梯段式长壁工作面 hitch and step
梯段式开采 benching
梯段式露天矿 terrace-type pit
梯段水平距离 going of the flight;run;stair run
梯段形开采 bench stoping
梯段形开采面 stope
梯段凿岩 bench drilling
梯段支架 spring tree
梯段钻眼 bench drilling
梯队 echelon
梯队构造【地】 en echelon structures
梯队铺沥青作业法 echelon paving
梯恩梯 <一种烈性炸药> TNT
梯恩梯当量 TNT equivalent
梯恩梯药块 Triton block
梯恩梯炸药 trinitrotoluene[TNT]
梯恩梯炸药块 Triton block
梯缝 ladder stitch
梯格林温暖期 Tiglian warm epoch
梯工 ladder work
梯钩 ladder hook
梯基础 ladder foundation
梯基加固材 rough carriage
梯级 stair step;step component;step member;step unit;cascade;flier;flight;flyer;ladder stair(case);rime;rounds of ladder;rundle;rung;spoke;spoke tread;stave;step;apples and pears <俚语>;steyre <中世纪教堂的>
梯级坝 cascade dams;step dams
梯级坝群 barrage steps
梯级波形 step wave form
梯级布置 layout of steps
梯级长度 step length
梯级成本 stair-step cost;step cost
梯级船闸 series of locks;chain of locks;flight of locks;staged-lift lock
梯级单元 step component
梯级地理倾差 graded topocline
梯级电站 chain of power plants;series of power stations
梯级跌流 cascading flow
梯级跌水 cascade drop;cascade fall;fall drop in series;flow cascade;ladder of cascades;steps;velocity reducing steps
梯级跌水流 cascading flow
梯级跌水式冷凝器 cascade condenser
梯级跌水式水加热器 cascade heater
梯级分配法 stepladder method
梯级高度 rise of a flight;riser height
梯级高进比 rise/run ratio
梯级高链节 angled stud
梯级搁板 <木楼梯的斜帮,承载踏步> bridge board;notch board
梯级工程 ladder of cascades
梯级构件 step member
梯级光楔 discontinuous wedge;step wedge
梯级规划 flight plan
梯级棍 rung;spoke;stave
梯级河槽 stepped channel
梯级河道 river steps
梯级横木 slat;stave;stave ladder
梯级厚度 thickness of step
梯级化 <河道的> continuous canalisation[canalization]
梯级化河道 canalization river;canalized river
梯级教室 lecture theatre[theater]
梯级接头 graded joint
梯级开发 cascade development;stepped development
梯级开发工程 canalization project
梯级理论 ladder theory
梯级炉格 stepped grate
梯级频率图 histogram
梯级平原 scarped plain
梯级曝气法 cascade aeration
梯级起步 stair raiser;stair riser
梯级筛子 series of sieves
梯级山墙顶 crown-step gable
梯级栅 gradient grid
梯级上防滑条 stair reeding
梯级式步道 terrace walk
梯级式船闸 chain of locks;flight of locks;lock chain;lock flights;lock steps;train of locks
梯级式大方脚 step footing
梯级式挡土墙 terrace wall
梯级式基脚 step footing
梯级式架桥 cable ladder
梯级式模板装置 stepping formwork equipment
梯级式水池 stepped pool
梯级式座位 gradin(e)
梯级试射 range ladder
梯级竖板 riser board

梯级水电站 cascade hydroelectric(al) power plant; cascade hydroelectric(al) station; power station in cascade; step hydroelectric(al) station; power stations in cascade
梯级水库 cascade reservoirs; serially linked reservoirs; step reservoirs
梯级水库调节 regulation of cascade reservoirs; regulation of reservoir cascade; regulation of step reservoirs
梯级水头 water head of step
梯级踏板 step risers
梯级台地 flight of terraces
梯级通气器 cascade aerator
梯级凸沿 stair nosing
梯级突边【建】 nosing
梯级拖影测试卡 graduated streaking chart
梯级圬工工程 step masonry work
梯级下水道 flight sewer
梯级先导 stepped leader
梯级信号 stair-step signal
梯级形齿轨 ladder rack
梯级堰 step weir
梯级样板 height board
梯级异径螺旋 stepped diameter auger
梯级鱼道 stepped fish pass
梯级预算法 step budget method
梯级园 terrace garden
梯级运河 canal ramps; step-shaped canal
梯级制造机 step making machine
梯级状 step-like
梯级状抬高 <混凝土路面因冻胀而形成的> high joint
梯架 ladder frame; scarcement
梯架卡车 ladder truck
梯架立柱 standard ladder
梯架汽车 automobile tower wagon; ladder truck
梯架式钻进 ladder drilling
梯架式钻井法 ladder drilling method
梯架式钻孔 ladder borehole; ladder drill hole
梯架式钻孔法 ladder drilling method
梯架钻孔 ladder drilling
梯降法 escalator method
梯阶 ladder rung; step
梯阶波形信号发生器 stair-step generator
梯阶式房屋 terraced house
梯阶式房屋建筑 spirit level building; spirit level construction
梯阶竖板 riser
梯矩阵 echelon matrix
梯科镍锰铜钢 Tico
梯口 ladder way
梯口梁 pitching piece
梯口门 gangway port
梯栏杆 ladder rail
梯利尔燃烧器 Tirrill burner
梯链式挖沟机 ladder ditcher
梯链式挖泥机 elevator-ladder dredge(r)
梯列 echelon
梯列拱点 apses in echelon
梯流 cascading
梯流粉碎机 cascade pulverizer
梯流润滑 cascade oiling
梯流式重介质旋流器 cascade dense-medium cyclone
梯流水洗 cascade washing
梯螺 ladder shell
梯牧草 timothy
梯坡田宽度 terrace width
梯坡消能减速结构 stepped energy destroying spillway
梯墙 <支承楼梯踏步的> stringer wall
梯瑞尔(电压)调整器 Tirrill regulator
梯塞绕组 teaser winding
梯塞式变压器 teaser transformer
梯塞线圈 teaser coil
梯式变动成本 step-variable cost
梯式变动能量成本 step-variable capacity cost
梯式成本 step cost
梯式分摊法 stepladder method
梯式钢筋网 ladder tie
梯式沟渠 ridge terrace
梯式基础 benched foundation
梯式加筋锚定墙 ladder wall
梯式加筋锚碇墙 multianchored wall
梯式脚手架 ladder scaffold(ing)
梯式控制 ladder control
梯式拉条 ladder bracing
梯式滤波器 ladder-type filter
梯式起重机 ladder hoist
梯式挖沟机 ladder-type ditcher; ladder-type trenching machine
梯式挖沟机偏离机心作业 offset digging
梯式挖掘机 ladder excavator
梯式钻孔(法) ladder drilling
梯式钻眼法 ladder drilling
梯斯克拉姆铬锰钢 Tiscrom
梯算符 ladder operator
梯索 ratlin(e)
梯塔 graduation tower
梯踏步 ladder rung
梯台 footpace; half pace; half-space; landing; pace <楼梯转弯处的平台>; quarter pace; quarter space <直角转弯处>; ladder landing [platform] <钻塔的>
梯台架 ladder scaffold(ing)
梯台模型 half-space model
梯台式炉 terrace furnace
梯台式墙型重整炉 terrace wall type reforming furnace
梯田 bed terrace; bench terrace; contour check; contoured field; field terrace; ladder farm; terrace; terraced field; terraced land; terrace farmland; terrace-field
梯田边缘 terrace edge
梯田出水口 terrace outlet
梯田地 terraced land
梯田地坎 bank of terraced field
梯田耕作(法) terracing; terrace cultivation
梯田灌溉 bench border irrigation
梯田化 terracing
梯田化土地 terraced land
梯田间距 terrace interval; terrace spacing
梯田阶地 bench terrace
梯田宽度 terrace width
梯田农业 terrace agriculture
梯田排灌水渠 terrace water channel
梯田排水沟 terrace outlet channel
梯田排水口 terrace outlet
梯田坡度 terrace grade
梯田台地沟壑 channel pipe
梯田土埂 ridge terrace
梯田系统 <坡地上的> terrace system
梯田修筑机 terracing machine
梯田崖 linehet[lynchet]
梯田种植 terrace cropping
梯田专用整平机 special terracing grader
梯温法 method; temperature gradient
梯温炉 gradient furnace; temperature gradient furnace; thermal gradient furnace
梯纹 scalariform marking
梯纹导管 scalariform duct; scalariform vessel
梯纹加厚 scalariform pitting
梯下安全网 ladder screen
梯线【铁】 connecting track; gathering track; ladder track; parent track; gathering line
梯线和平行股道的夹角 angle between ladder track and body tracks
梯线扩距 spread of ladder track
梯厢 car; elevator car; lift carriage
梯厢尺寸 car dimensions; elevator car dimensions
梯厢减振器 car buffer; elevator car buffer
梯箱形断面 trapezoidal box section
梯形 echelon form; ladder type; trapezoid <美>; trapeze <英>; trapezium[复 trapopia/trapeziums] <英>
梯形凹口 trapezia notch
梯形凹面 stepped concave face
梯形坝 trapezoidal buttress dam
梯形边沟 trapezoidal ditch
梯形波发生器 trapezoidal generator
梯形波形 keystone wave; trapezoidal waveform
梯形剥蚀面 step-like surface of denudation
梯形槽 dovetailed groove; trapezoid slot
梯形槽口 trapezoidal notch
梯形槽口板 notch board
梯形槽口量水堰 trapezoidal notch weir; trapezoidal weir
梯形槽刨 dovetail plane
梯形槽式密封面对焊法兰 ring joint welding neck flange
梯形槽式密封面法兰 ring joint flange
梯形侧断面 trapezoidal profile
梯形测流槽 trapezoidal flume
梯形铲斗 trapezoidal bucket
梯形车场 trapezium yard
梯形车架 ladder-type frame
梯形沉淀池 settlement reservoir; trapezoidal grit chamber
梯形沉箱式防波堤 trapezoidal caisson breakwater
梯形冲击脉冲 trapezoidal shock pulse
梯形除杂机 step cleaner; super-cleaner
梯形穿孔板 latticed perforation plate
梯形锉 swaged file
梯形大梁 trapezoidal girder
梯形带 <百叶窗的> ladder web
梯形导标 trapezoidal leading mark
梯形的 cuneiform; trapezoidal <美>
梯形的箱形(断面)钢梁 <利用桥面作为上翼缘> trapezoidal steel box girder
梯形电路 ladder circuit; ladder-type circuit
梯形电压 trapezoidal voltage
梯形垫块 ladder gasket
梯形跌水 cascade fall
梯形丁坝 trapezoidal groin; trapezoidal groyne
梯形定律 <应力分布的> trapezoidal law
梯形陡槽 trapezoidal chute
梯形度 taper
梯形渡槽 trapezoidal flume
梯形段 trapezoidal piece
梯形段采面 stope
梯形断面 trapezoidal profile; trapezoidal section
梯形断面活塞环 keystone piston ring
梯形堆垛法 stepped stacking
梯形法 trapezoidal method
梯形法则 trapezoidal rule
梯形分布 trapezoidal distribution
梯形分割绘制法 trapezoidal divisor
梯形分配法 graded distribution method; ladder-shaped distribution method
梯形复合板 trapezoidal sandwich panel
梯形刚构 trapezoidal frame
梯形刚架 trapezoidal rigid frame
梯形刚架桥 trapezoidal frame bridge
梯形钢筋混凝土槽 trapezoidal reinforced concrete channel
梯形钢丝 trapezoidal steel wire
梯形钢箱梁桥 trapezoidal steel box girder bridge
梯形隔距 step ga(u)ge
梯形公式 trapezoidal formula
梯形拱 trapezoidal arch
梯形拱顶开间 trapezoidal vault bay
梯形拱顶跨度 trapezoidal vault bay
梯形沟 flat-bottom ditch
梯形构架 trapezoidal frame
梯形构件 trapezoidal piece; trapezoidal unit
梯形管 laddertron
梯形规则 trapezoidal rule
梯形过程线 trapezoidal hydrograph
梯形河槽 trapezoidal channel
梯形荷载 trapezoidal load
梯形桁架 parallel-chord truss; trapezoidal truss
梯形桁架式构架 trapezoidal truss frame
梯形桁条 trapezoidal purlin(e)
梯形横断面 trapezoidal cross-section; trapezoidal-type section
梯形横截面 trapezoidal cross-section
梯形猴耳环 ladder apes-ear ring
梯形厚板 trapezoidal slab
梯形环 keystone ring
梯形灰板条 <抹灰用> trapezoidal wood lath(ing)
梯形混合机 stepblender
梯形混凝土箱梁桥 trapezoidal concrete box girder bridge
梯形活塞环 keystone-type piston ring
梯形积分 trapezoidal integration
梯形积分法 trapezoidal integration
梯形畸变 keystone distortion; trapezoidal distortion
梯形畸变校正 keystone correction
梯形畸变效应 keystone effect
梯形夹芯板 trapezoidal sandwich panel
梯形假山 stepped artificial hill
梯形件 trapezoidal piece
梯形阶地 bench terrace; trapezoidal terrace
梯形接合 scalariform conjugation
梯形结构 trapezoidal frame
梯形结合 scalariform conjugation
梯形截面 trapezoid cross-section
梯形截面槽 trapezoidal channel
梯形截面铝材 alumin(i)um trapezoidal section
梯形截面受钉木条 bevel(l)ed nailing strip
梯形截面运河 trapezoidal canal
梯形金属板 trapezoidal metal plate
梯形金属薄板 trapezoidal metal sheet
梯形聚合物【化】 ladder polymer
梯形开间 trapezoidal bay
梯形开口缝 bar fagoting
梯形开棉机 inclined cleaner; step cleaner
梯形靠背椅 ladder-back chair
梯形孔型 trapezoidal pass
梯形扣套管 buttress thread casing
梯形块 trapezoidal piece
梯形框架 trapezoidal frame
梯形拉条 ladder bracing
梯形喇叭 scalar horn
梯形肋 trapezoidal rib
梯形棱镜 Dove prism
梯形犁铧 slab share; trapezoidal share
梯形立轧孔型 squabbing pass

梯形链斗式挖泥机 ladder dredge(r)
梯形链条 ladder chain
梯形(量水)堰 Cippolletti weir
梯形料斗 trapezoidal bucket
梯形裂缝 ladder
梯形裂缝凹槽 ladder well
梯形檩条 trapezoidal purlin(e)
梯形零件 trapezoidal piece
梯形溜槽 trapezoidal chute
梯形滤波器 ladder-type filter
梯形螺 <英制> Acme(screw)thread
梯形螺纹 Acme thread;leaning thread;trapezoidal thread
梯形螺纹规 Acme thread ga(u)ge;trapezoidal thread ga(u)ge
梯形螺纹丝锥 tap for trapezoidal thread
梯形脉冲 trapezoidal pulse
梯形脉冲发生器 trapezoidal generator
梯形慢波线 taped slow-wave line
梯形门窗贴脸 trapezoidal trim
梯形门洞 atticurge
梯形门框 pylon
梯形磨瓦 trapezoid abrasive tile
梯形目标 mire
梯形牛腿(托架) trapezoidal corbel
梯形排架 sloping-leg bent;trapezoidal bent
梯形排架间距 trapezoidal bay
梯形排列拱点 apses in echelon
梯形排水沟 trapezoidal gutter
梯形棚子 ladder-type support
梯形劈理 trapezoidal cleavages
梯形皮带 trapezoidal belt;trapezoidal-shaped belt
梯形片 trapezoidal piece
梯形偏转 trapezoidal deflection
梯形剖面 trapezoidal profile
梯形铺砌 paving in echelon
梯形起重臂 trapezoidal boom
梯形桥台 trapezoidal abutment
梯形曲线 step curve
梯形渠道 trapezoidal channel
梯形缺口 trapezia notch
梯形绕带 echelon strapping
梯形三层胶合板 trapezoidal three-layer(ed) panel
梯形扫描 keystone scanning
梯形山墙 corbie gable;corbie step gable;crow step gable;step gable
梯形山墙顶 catstep;corbie step;crow step
梯形失真 <电视光栅> keystone distortion;keystoning;trapezoidal distortion
梯形失真校正 keystone correction;keystoning correction
梯形失真效应 keystone effect
梯形石香肠 trapezoidal boudin
梯形蚀痕 ladder marking;washboard erosion
梯形试样法 trapezoid method
梯形试样撕破强力 trapezoid tear strength
梯形饰条 trapezoidal trim
梯形衰减器 ladder attenuator
梯形水槽 trapezoidal flume
梯形丝扣 buttress thread;trapezoidal thread
梯形丝扣接头 step-jointed coupling
梯形丝锥 Acme thread tap;trapezoidal thread tap
梯形撕裂 trapezoidal tear
梯形撕裂试验 trapezoidal tear test
梯形速度-时间曲线 trapezoidal speed-time curve
梯形掏槽 step cut;trapezoidal cut
梯形淘汰机 trapezoidal jig
梯形套管接箍丝扣抗拉强度 buttress casing coupling thread tensile strength
梯形套管丝扣抗拉强度 buttress casing thread tensile strength
梯形天线 echelon antenna
梯形调制 trapezoidal modulation
梯形淘汰机 trapezoidal jig
梯形桶 trapezoidal bucket
梯形投影 trapezoidal projection
梯形透镜 stepped lens
梯形透镜式天线 echelon lens antenna
梯形凸面 stepped convex face
梯形图 ladder chart;ladder diagram;trapezoidal pattern
梯形图案数据缓冲存储器 trapezoid pattern data buffer
梯形图案数据寄存器 trapezoid pattern data register
梯形图幅 quadrangle;quadrangle map
梯形图幅投影 trapezoidal projection
梯形挖沟机 boom type trenching machine;ladder ditcher
梯形挖掘机 trapezoidal excavator
梯形网络 ladder network;ladder-type filter;ladder-type network;periodic(al) line
梯形蜗杆 trapezoidal worm
梯形屋架 trapezoidal roof truss
梯形系杆 bevel tie
梯形线 ladder track
梯形箱梁 trapezoidal box girder
梯形箱式大梁 trapezoidal box girder
梯形效应 ladder effect
梯形斜面受钉木条 bevel(l)ed nailing strip
梯形檐槽 trapezoidal gutter
梯形檐沟 trapezoidal eaves gutter
梯形堰 Cippolletti weir;trapezia notch;trapezoidal weir
梯形溢流管 trapezoid overflow pipe
梯形造材台 bucking ladder
梯形枕木 trapezoidal sleeper
梯形柱头 Byzantine capital
梯形砖 trapezoidal brick
梯形状 en echelon
梯形组织 ladder weave
梯形钻头 ripper step bit;tapered bit
梯形座位群 <希腊剧场阶梯过道间的> cercis
梯型去耦滤波器 ladder-type decoupling filter
梯型线路 ladder circuit
梯阵 echelon
梯阵式构造 en echelon structures
梯支撑 ladder bracing
梯柱 cleat stanchion;ladder post;notch stanchion;step stanchion
梯状波纹 ladders
梯状薄壁组织 scalariform parenchyma
梯状穿孔 scalariform perforation
梯状穿孔板 scalariform perforation plate
梯状穿孔导管 scalariform vessel
梯状次极直线悬浮电机 ladder secondary linear floating motor
梯状的 scalar(iform)
梯状地形 riser
梯状阶地 step-like terrace;step terrace
梯状矿脉 ladder lodes;ladder veins
梯状链 ladder chain
梯状配置 echelonment
梯状平原 klimakotopedion;stepped plain
梯状纹孔式 scalariform pitting
梯状物 ladder
梯状云 echelon clouds
梯状织物 ladder web
梯子 ladder;foot path;lean-to ladder;pair of steps;staircase;staircase way
梯子板 ladder board
梯子侧板 ladder string
梯子插口 ladder bracket
梯子道 footway;ladder road;ladder way
梯子迭代 treppeniteration
梯子和护罩 ladder and cage
梯子横档 rounds of ladder;rung;stave
梯子横木 ladder rung;rounds of ladder
梯子架 ladder stay
梯子间 footway;ladder roadway;ladder-way compartment
梯子拉出长度 extension length of ladder
梯子平台 ladder landing[platform];ladder sollar
梯子形轨道 ladder track
梯子形拉条 ladder bracing
梯子之间平台 sollar
梯子支承的轻型脚手架 ladder-jack scaffold
梯子支架 ladder bracket
梯子状的 scalariform
梯字型 trapezium-type
梯座 altar;ladder base;ladder cradle <配合空气腿用>

锑钯矿 stibiopalldinite

锑白 antimony oxide;antimony white;Timonex <一种氧化锑>
锑白颜料 antimony pigment
锑斑 antimony spot
锑贝塔石 stibiobetafite
锑焙砂 antimony calcine
锑饼 antimony cake
锑波酚 stibophen
锑玻璃 antimonial glass;antimony glass
锑铂矿 geversite
锑尘肺 antimony pneumoconiosis
锑橙 antimony orange
锑雌黄 wakabayashilite
锑电极 antimony-electrode
锑锭 antimony slab
锑粉 iron black
锑钙矾 peretaite
锑钙镁非石 welshite
锑钙石 romeite
锑汞矿 shakhovite
锑合金 antimony alloy
锑黑 antimony black;iron black
锑红 antimony red
锑红玻璃 antimony ruby glass
锑红颜料 crimson antimony
锑华 antimony bloom;valentinite;white antimony
锑化锆 zirconium antimonide
锑化镉 cadmium antimonide
锑化铝 alumin(i)um antimonide
锑化镍 nickel antimonide
锑化铷光电阴极 rubidium antimonide photocathode
锑化(三)氢 stibine
锑化铜 copper antimonide
锑化物 antimonide;stibnide
锑化锌 zinc antimonide
锑化铟 indium antimonide
锑化铀 uranium antimonide
锑黄 antimony yellow
锑火石玻璃 antimony flint glass
锑基合金 antimony containing alloy
锑检波器 antimony detector
锑桔黄 antimony orange
锑块 antimony regulus
锑矿石 antimony ore
锑矿异常 anomaly of antimony ore
锑酪 antimony butter
锑疗法 stibiation
锑硫镍矿 ullmannite
锑硫砷铜银矿 antimonpearceite
锑镁矿 byströmite
锑钠铍矿 swedenborgite
锑铌矿 stibiocolumbite
锑镍矿 antimonial nickel
锑铅 antimonial lead;antimony lead
锑铅薄板 hard lead sheet
锑铅合金 antimonial lead;hard lead;lead regulus;regulus[复 reguli/reguluses];regulus lead;regulus metal
锑铅合金管 antimonial lead pipe
锑铅矿石 Sb-bearing lead ore
锑铅压力管 hard lead pressure pipe
锑青铜 antimony bronze
锑熔块 regulus of antimony
锑砷锰矿 manganostibite
锑酸钙 calcium antimonate
锑酸钠 antimony sodiate;sodium antimonate
锑酸铅 antimoniate of lead;lead antimonite;Naples yellow
锑酸盐 antimonate
锑铊铜矿 cuprostibite
锑钛钙矿 lewisite
锑钛铁矿 derbylite
锑铁合金 <其中的一种> Reaumur alloy
锑铁矿 tripuhyite
锑铁钛矿 derbylite
锑铜矿 horsfordite
锑铜锌轴承合金 Karmash alloy
锑污染 pollution by antimony
锑钨烧绿石 scheteligite
锑锡焊条 antimonial tin solder
锑锡矿 stistaite
锑细晶石 stibiomicrolite
锑线石 holtite
锑星 antimony star
锑伊碲镍矿 Sb-imgreite
锑银矿 dyscrasite
锑藏红 crocus of antimony
锑渣 regulus antimony
锑赭石 cervantite
锑中毒 antimony poisoning;stibialism
锑朱 antimony vermilion
锑朱颜料 crimson antimony

踢板 kick plate;mop plate

踢板凹进处 kick recess
踢板和滴水槽 kick plate and drip
踢管 kickpipe
踢花 cut decoration
踢回 kick back
踢脚 kick strip;skirting
踢脚凹(进)处 kicking recess
踢脚凹空 kicking recess
踢脚板 adjacent plank;base block;baseboard;bottom panel;foot plate;kicking plate;kicking strip;mopboard;mop plate;quadra;scrub board;skirting;skirt(ing) board;subbase;toe board;wall base;wash board;dado <踢脚板俗称>
踢脚板标高 skirting level
踢脚板部件 sanitary cove member;washboard member;washboard unit
踢脚板采暖 sanitary cove heating;scrub board heating
踢脚板采暖器 scrub board heater
踢脚板成件 mopboard member
踢脚板处的暖气管 base plate heater
踢脚板处供暖气 base heating

踢脚板处供暖器 base heater
踢脚板处暖气 base plate heating
踢脚板处暖气片 base plate radiator; base radiator
踢脚板单元 base plate member; base plate unit; sanitary cove unit; scrub board unit
踢脚板底缝压条 base shoe mo(u)lding; floor mo(u)lding; shoe mo(u)lding; carpet strip < 与地板之间的线脚 >
踢脚板放热器 skirting heater
踢脚板高程 skirting level
踢脚板供暖器 baseboard heater
踢脚板供暖装置 baseboard heating unit; skirting board heating device
踢脚板构件 baseboard member; baseboard unit; mopboard component; mopboard member; mopboard unit; scrub board component; scrub board member; skirting component; skirting member; skirting unit; washboard member; washboard unit
踢脚板固定块 foot block
踢脚板后面的砌块 skirting block
踢脚板加热 washboard heating
踢脚板加热器 washboard heater
踢脚板角压条 baseboard corner moulding
踢脚板进气口 skirting air inlet
踢脚板木龙骨 rough ground for nailing skirting
踢脚板内管道 shirting duct
踢脚板内暖气管 baseboard radiator; shirting heater
踢脚板内调温装置 baseboard register [registor]
踢脚板暖气管 mopboard radiator
踢脚板暖气片 mopboard radiator; sanitary cove radiator
踢脚板企口 washboard groove
踢脚板散热器 baseboard convector; baseboard heater; baseboard radiator; baseboard radiator unit; baseboard unit; mopboard radiator; scrub board radiator; washboard radiator
踢脚板上缘线脚 base cap; base mo(u)lding; base table
踢脚板式暖气片 baseboard radiator
踢脚板式暖气设备 baseboard radiator unit
踢脚板式器件单元 baseboard type unit
踢脚板式取暖器 baseboard heater
踢脚板送风口 baseboard diffuser
踢脚板线脚(样板) base shoe
踢脚板压顶条 base cap; base mo(u)lding
踢脚板压顶线脚 surbase
踢脚板压条 base shoe
踢脚板元件 sanitary cove component
踢脚板砖 skirting brick; skirting tile
踢脚边沿管道 skirting duct
踢脚挡板 kicker
踢脚垫 foots scraper
踢脚墩 < 堵头 > base block; plinth block; skirting block
踢脚护(条)板 toeplate
踢脚花格板调温装置 baseboard register[registor]
踢脚金属板 kicking strap
踢脚栏 kick rail
踢脚木 < 即护墙木 > lower wainscot rail
踢脚线 plinth
踢脚线顶部线脚 mo(u)lding base
踢脚压顶条 base table
踢脚压条 base shoe; carpet strip; floor mo(u)lding; shoe mo(u)lding
踢脚砖 shirting tile
踢马刺 spur
踢面 raking riser; riser
踢木器 log kicker

啼

啼锰锌石 spiroffite

啼状双晶 knee shaped twin

提

提 < 表面亮度单位 > stilb

提案 proposal
提案人 sponsor
提案制 proposal system; suggestion system
提把 bale handle; carrying handle
提板式加料机 lifting-gate feeder
提棒 rod withdrawal
提包 catch-all
提柄 lift handle
提倡国货 encourage native products; promote home products
提倡者 advocate; entrepreneur; proponent
提成 amortization; deduct a percentage from a sum of money
提成费 royalty
提成合同 percentage contract
提成率 percentage allocation; royalty rate
提成支付 royalty
提秤 Roman balance; Roman beam; Roman steelyard; steelyard
提秤湿度计 steelyard moisture meter
提抽泥筒作业 wet job
提出保留 formula of reservation
提出报告 inform
提出被海关扣存的货物 < 完税后 > take the goods out of bond
提出被卡钻具 backing-off stuck drilling tools
提出抽油杆 < 修理深井泵时 > stripping job
提出初步试验方案 put forward a preliminary plan for test
提出低级维修要求 making low maintenance demands
提出海事声明 note of protest
提出计划 propound
提出建议 make a suggestion; propose; suggest commend
提出结论和建议 present the conclusion and recommendations
提出解散申请书 file an application for dissolution
提出惊人的意见 explode a bombshell
提出控告 sue
提出履行合同 tender performance
提出轮廓草图 block out
提出赔偿要求 filing a claim; lodge claim; submit a claim; submit a claim for compensation
提出人 exhibitor; introducer
提出上诉 entry of appeal
提出申请 apply
提出式艇架 mechanical boat davit
提出诉讼 litigate; sue
提出索赔 institute a claim
提出索赔的时效 time limit for submission of loan
提出讨论 put forward for discussion
提出问题 pose
提出物 educt; extractive; extraction
提出物储器 extract storage
提出要求人 claimant; claimer
提出要求者 claimant; demandant
提出议案(英) tabling of a bill
提出异议 interposition
提出优缺点 pros-and-cons
提出预算 open the budget
提出预算案 introduce the budget; open the budget
提出债权作为抵销 assert a set-off
提出者 raiser; starter; tenderer
提出证明文件 documentation
提出制造资料 offering information on manufacture
提出专业性意见 expertise[expertize]
提出追偿 filing a claim for recovery
提出钻杆作业 stripping the pipe; stripping the rod
提出钻孔 pulling out the hole
提窗把手 window lift
提纯 cleansing; clean-up; depurate; depuration; plaining; purify(ing); refinement; refining
提纯泵 sublimation pump
提纯比 purification ratio
提纯段 rectifying section
提纯方法 method of purification
提纯剂 plaining agent; purifying agent
提纯目估法 purification ocular estimate method
提纯器 clean(s)er; purifier
提纯塔 purification tower
提纯信号 purified signal
提纯氧化锆 zircon alba
提纯仪器 purifying apparatus
提纯油 good oil
提纯柱 decontaminating column
提纯作用 purification
提存 drawing
提存基金的准备 covered reserve
提存利息 covered interest
提存权 drawing rights
提存账户 drawing account
提存准备金 covered reserve
提大纲 block out
提单 bill of lading [B/L]; cartier's note; warehouse book
提单背书申请书 application for bill of lading endorsement
提单抄本 duplicate bill of lading; memorandum copy of bill of lading
提单持有人 holder of the bill of lading
提单单位运价 bill of lading minimum charges
提单的船长用抄本 captain's copy of a bill of lading
提单吨 bill of lading ton
提单副本 duplicate bill of lading; memorandum bill of lading; non-negotiable bills of lading
提单批注(的免责事项) exceptions noted on the bill of lading
提单日期 bill of lading date
提单上的重量 billed weight
提单条款 bill of lading clause
提单限额超装许可证 freight release; release
提单信 consignee's covers
提单运费 bill of lading freight
提单正本 original bills of lading
提单转让 negotiability of bills of lading
提单总运费 aggregate B/L freight
提刀 griffe; lifting knife
提灯 hand lantern; lantern; lantern light
提灯腔 lantern coelom
提吊柱 hoisting mast
提顶臂 top lifter
提动阀 poppet valve
提动式溢流阀 poppet relief valve
提斗 bucket grab
提斗扣胶带输送机的联合卸车机 belt conveyer
提斗提升 < 指加速 > bucket lift
提断环 ring lifter
提断器外壳 < 岩芯 > ring lifter case
提放工具 handing tool
提风 < 一种防风雨窗 > Ti-foon
提杆 lifting stem
提纲 compendium[复 compendiums/compendia]; conspectus; outline; outline plan; outline program(me); syllabus
提纲式询问 skeletal query
提高 aggrandizement; elevate; elevation; enhancement; heighten(ing); jack; promote; rev; sublimate; sublime; upgrading; uplift; upswing; upturn; upgrade < 指等级质量的 >
提高比表面积 increasing specific surface area
提高标度计数器 advance range counter
提高标价 make-up; mark-up; over bid
提高标价的方程 price markup equation
提高补给土壤的储水量 more replenishing soil moisture reserves
提高采收率的措施 enhanced recovery factor technique
提高采收率法采油 enhanced oil recovery
提高产量 boost output; yield improvement
提高产品质量 improve the quality of products
提高出力 uprating
提高出力的潜力 uprating capacity
提高出力的设计 uprating project
提高代替能力 increasing exchange power
提高贷款利息 marking-up loans
提高单位面积产量 raise the per unit yield
提高到同一水平 level up
提高的 elevated
提高的发射速度 increased rate of fire
提高的工资 raised wage
提高等级 upgrade; upgrading
提高等级更新产品 upgrade
提高定额 uprating
提高定额处理厂 uprating treatment plant
提高多种生产的能力 raising capabilities for multi-production
提高复种指数 raise the multiple cropping index
提高工资 advance of wages; wage kikes
提高功率 increase of power; uprate; uprated
提高灌溉效率 increasing irrigation efficiency
提高含金量 raise gold content
提高货币价值 raise the value of money
提高计划 upgrading program(me)
提高剂量效应 increasing dosage effect
提高金客支票 raised check
提高经济效应 increase economic of efficiency
提高竞争能力 boost competitive power; boost the competitiveness
提高(矿物的)品位 upgrade
提高劳动生产率 raising labour productivity
提高劳动者素质 create a more highly skilled workforce
提高利用率 increase operation rate
提高灵敏度装置 activator
提高票面价值 increase the par value
提高频率 increase of frequency
提高品位 upgrading; upgrade
提高品种质量 improve the quality of

varieties
提高平价 increase the par value
提高人力资源的质量 upgrading of human resources
提高商品质量 improvement of commodity quality
提高设备的效能 improve the utility of equipment
提高生产率 boost productivity;increase productivity;raise productivity
提高生产资金利用效果 production funds put to a better use
提高时价 raise the current price
提高水位 raising of water level
提高水质 increasing water quality
提高饲料转化率 improving food conversion ratio
提高图像对比度 image intensification
提高(图像中物体)轮廓的明显性 contour accentuation
提高土壤肥力 increasing of soil fertility
提高土壤渗透性能 increasing soil permeability
提高土壤水分含量 increasing soil moisture content
提高土壤酸含量 increase the acid content of soil
提高土壤吸收性能 increasing absorbing power
提高物价 raise price
提高硝酸盐水平 raising the nitrate level
提高效率 increase of efficiency; souped-up
提高效益 improve economic performance
提高效应 enhancement effect
提高蓄水量 raising water storage
提高蓄水位 raising the level of impoundage
提高岩芯采取率 increase core recovery
提高业务水平 heighten one's vocational level
提高艺术的本能 instinct of workmanship
提高因素 improvement factor
提高应力装置 stress raiser
提高增长速度 growthmanship
提高账面价值 write up
提高真空度 gas clean-up
提高质量 upgrading
提高质量装置 upgrader
提高转数 revved up
提高准确度 enhanced accuracy
提高资源质量 resource enhancement
提高自我 ego-enhancement
提高总产量 increasing the total yield
提供 afford;furnish;offer(ing);provide
提供庇护所 sheltering
提供标价 escaladed bid
提供玻璃纸袋 supply glassing bag
提供仓库容量 provision of warehouse accommodation
提供成套服务 providing a complete set of services; to provide a complete set of services
提供程序 distribution program(me)
提供程序库 distribution library
提供充分服务的批发商 full-service wholesaler
提供出售 offer for sale
提供出售项目 offer item
提供贷款 borrow offer
提供贷款的新方法 innovative financing technique
提供担保 offering for security
提供的成果 feedback
提供的文件 documentation
提供的证书 documentation
提供的座席数 seats provided
提供抵押 collateralize
提供地下水 supply subsurface water; supply underground water
提供电传 offering telex
提供多孔管喷灌装置 supply perforated sprinkler
提供服务 render service; service delivery
提供服务的批发商 service wholesaler
提供服务的设计 design of service offered
提供辅助地质图件 auxiliary geologic(al)map
提供灌溉龙头 supply irrigation hydrant
提供灌溉系统图 supply irrigation layout plan
提供广告版面 space selling
提供国际投入 delivery of international inputs
提供活动喷灌装置 supply potable sprinkler
提供或随附的单据、证件 supporting documentor
提供基金者 foundationer
提供技术情报协议 supply of technical information agreement
提供技术资料 supply technical data
提供进度计划 program(me) to be furnished
提供救援列车 provision of relief train
提供就业 employment creation; employment generation
提供就业机会 job creation
提供卷宗 distributed volume
提供可以征信的企业名称 give trade reference
提供可以征信的银行名称 give bank reference
提供劳务 providing labor service; provision of labo(u)r;rendering of services;utility service
提供劳务合同 contract for supply of labo(u)r
提供牢固的基础知识 to provide a solid foundation
提供理论基础 providing fundamental basis for
提供农田用水的水库 service reservoir for farming;supply farm reservoir
提供排水计划 supply water management scheme
提供情报者 informant
提供曲线 offer curve
提供全面服务的银行业务 multiple banking
提供容量 <以客位/小时计的线路容量,或线路运送乘客的理论最大能力> offered capacity
提供商业资金 commercial financing
提供设计 design procurement
提供社会劳动日 contribute the social working day
提供生活服务的小型设施 retail service
提供输水管 supply main pipe
提供说明的责任 burden of persuasion
提供贴现 discount offered
提供土地规划 providing land planning
提供土壤调查情况 providing soil investigation conditions
提供土壤耕性 providing tilth to a soil
提供土壤详图 providing detailed soil map
提供土样 supply soil sample
提供维修服务的租赁 service lease
提供物 tender
提供线路 implementation
提供项目 tender item
提供消息者 informant
提供信贷 allowance of credit
提供选择权的人 optionor
提供银行信贷 granting bank credit
提供用户的最后成果 user-oriented end product
提供原始材料 supply parent material
提供原种 supply mother seed; supply pedigree seed; supply stock seed
提供援助 delivery of assistance
提供援助的机构 aid-giving agency
提供援助国 aid-giving nation
提供远期资金 forward funding
提供折扣 discount offered
提供者 provider;supplier;tenderer
提供证据 production of evidences
提供证据的 probative
提供证明 production of proof
提供证明的责任 burden of proof
提供证明文件 documentation
提供职业 provision of employment
提供职业岗位 job creation
提供住宿的 accommodating
提供住所 sheltering
提供专用泊位 appropriated berth
提供装货详单 present a ship's manifest
提供咨询服务 offering consultation service
提供资本 capital financing; funding; funds provided
提供资金 financing
提供资金的办法 financing arrangement
提供资金的银行 financing bank
提供资料 submit data
提供资料的 informative;informatory
提供资源的国家 resources-providing country
提钩 cleaner; drilling hook; lancet; lifter;lift(ing)lever <车辆>
提钩销 <自动车钩> lock lifter
提管器 pipe elevator
提管绳 tubing line
提花带 jacquard ribbon
提花地毯 jacquard weave carpet
提花垫纬凸纹布 trapunto
提花吊线 jacquard embroidering
提花杆 jacquard lever
提花滚筒 jacquard drum;trick drum
提花横机 jacquard flat knitting machine
提花机 jacquard;needle board
提花棱柱撑动装置 jacquard prism racking mechanism
提花轮 pattern wheel;trick wheel
提花轮片 trick wheel bit
提花三角 accordion cam
提花三角装置 jacquard lock
提花凸纹织物 figured pique
提花图案 figured pattern
提花织物 broche; figured cloth; figured fabric; jacquard fabric; jacquard weaves
提花装置 jacquard attachment; jacquard device;jacquard mechanism
提花组织 jacquard weaves
提环 elevator bail
提灰机 ash lift
提货 picking;pick-up goods;take delivery of goods
提货报关代理人 delivery and customs agent
提货不着 non-delivery
提货不着险 risk of non-delivery
提货单 bill of lading; carrier's note; delivery order; general bill of lading;ship's release;waybill
提货单处理(转让) disposition of bill of lading
提货单的转让 transfer of bill of lading
提货单汇票 documentary draft
提货单位 delivery unit
提货即付款协议 take-and-pay agreement
提货凭单 bill of lading
提货权 drawing rights;drawn rights
提货手续 delivery formality
提货通知(单) cargo delivery notice
提货与付款合同 take-and-pay contract
提货证 carrier's note
提级 upgrade
提价 add to the prices; advance(in) price; boost price; mark-up; price advance;raise price
提价百分率 mark-up percentage
提价率 mark-up rate
提架 hand barrow
提浆抹光 floating
提交 submission; presentation; submittal
提交公断 submission
提交汇票 delivery of the bill of exchange
提交建议书预备会 pre-proposal conference
提交日期 submission date
提交试验 submit to test
提交投标价 tendering of bid
提交投标时间 time for bid submission
提交图纸 submit drawings
提交仲裁 refer to arbitration; submission for arbitration; submission to arbitration
提交仲裁裁决 deposit(e) of the award
提交状态 submit state
提紧套管柱 pull tension on casing
提净 defecate;defecation;edulcorate; sublimation
提举辊子 lifting of roll
提开门 lever gear door
提款 drawing;draw money;withdrawal; withdraw deposit; withdraw money
提款单 withdrawal slip; withdrawal ticket
提款卡 debit card
提款权 drawing rights;drawn rights
提款申请 application for withdrawal
提款收据 withdrawal receipt
提款通知 advice of drawing;drawing advice
提款账(户) drawing account
提款账目 drawing account
提矿箕斗 ore skip
提拉窗 double hung window; vertical sliding window
提拉窗导轨 guide bead;inner bead
提拉窗吊链 sash chain; sash cord; sash line
提拉窗吊索 sash cord;sash line
提拉窗横档 meeting rail
提拉窗滑车槽 hanging stile; pulley stile;sash run
提拉窗滑轮 axle pulley;sash pulley
提拉窗扣手 sash lift;window lift
提拉窗平衡重 sash counterweight; sash weight
提拉窗扇 vertically sliding sash
提拉窗上口槽 pulley head
提拉法 pulling method
提拉方向 draw direction; pull direction
提拉悬浮法 draw-flo(a)tation method

提篮 rod elevator

提篮式 basket type

提捞 bail down; bailing; bail-out; bali

提捞次数 bailing times

提捞倘容积 volume of bailer

提捞筒长度 length of bailer

提捞筒外径 outside diameter of bailer

提捞筒下入深度 bailing depth

提勒耳氏钩 iridectomy hook; Tyrrell's hook

提离孔底 <钻局> pull-off the bottom

提炼 abstraction; extract; extraction; extraction and purification; refine; refinement; winning

提炼产品产量 refinery output

提炼厂 refinery

提炼程序 refinery procedure

提炼方法 extraction procedure

提炼工序 extraction procedure

提炼过程 refining process; extractive process

提炼过的沥青 extracted asphalt

提炼剂 extraction agent

提炼器 extractor

提炼物 extractive material; extractive matter

提料斗 chain bucket

提料翼板 <干燥滚筒中的> lifting flight

提列斯蒂利昂大会堂 <公元前希腊> Telesterion

提留 profit deduction and reserving

提馏段 stripping section

提罗尔防水砂浆 <防雨漏用> Tyrolean

提罗尔防水砂浆罩面 Tyrolean finish

提罗尔抹灰 <一种防水抹灰> Tyrolean

提罗喷漆枪 <一种手持的喷涂彩色水泥砂浆的喷涂器> Tyrol

提落离合器 lift clutch

提蔓 lifting up of vines

提煤机 coal hoist

提名 nominate; nomination

提名受保人 named insured

提名委员会 nomination committee

提名者 nominator

提浓 concentration

提浓精馏段 enriching section

提浓物 concentrate

提起 lifting; pulling up

提起的 elevatory

提起公诉 initiation of public prosecution

提起式下水口 pop-up sewer opening; pop-up waste

提起诉讼 initiation of proceedings

提起钻具 short trip

提前 advance; antedate; in advance; preact

提前报警 advance warning

提前报知 ample warning

提前背书人 antecedent party

提前拨款 advance appropriation

提前偿付 prepayment; voluntary payment

提前偿付特权 prepayment privilege

提前偿付条款 acceleration clause; call provision

提前偿付债款 loan calling

提前偿还 advanced redemption; voluntary prepayment

提前偿还的承兑汇票 anticipated acceptance

提前偿还权 call privilege

提前偿还日期 call date

提前偿还债款 loan calling

提前偿还债券价格 call price

提前偿还债券贴水 call premium on bonds

提前成本 anticipated cost

提前出现 anticipation

提前处理 bring forward

提前处理的资料 anticipatory data

提前处理种子 pretreated seed

提前打顶 pretopping

提前到期日 accelerate the maturity

提前的 advanced

提前点 predicted point

提前点火 advance (d) ignition; early ignition; prefiring; preignition; sparking advance

提前点火环 advance ring

提前点火机构 ignition advance mechanism

提前点火阶段 preignition phase

提前点火器 advanced igniter

提前调度法 forward scheduling

提前定时 advance timing

提前兑付 anticipated redemption

提前兑换债券 anticipated redemption of bonds

提前兑回条款 call provision

提前兑回债券的条款 call provision on hand

提前风向角 predicted wind angle

提前付款 anticipated payment; anticipation; make payment beforehand; prematurity payment; prepayment; rebated acceptance

提前付款利率 anticipation rate

提前付款信用证 anticipatory letter of credit; packing letter of credit

提前告警 early warning

提前供油 fuel lead

提前灌水 advance irrigation; prewatering

提前归还未到期借款 break a loan

提前还船 underlap

提前还款罚金 prepayment penalty

提前还款权 right of anticipation

提前还款条款 prepayment clause

提前还债 hono(u)r a debt in advance

提前还债期 accelerated debt maturity

提前火花 early spark

提前交付的票据 bill for premature delivery

提前交货 early delivery of goods; premature delivery

提前角 advance angle; angle of advance; angular advance

提前解锁 premature release

提前进口 anticipated import

提前进气 admission lead; advance of admission; preadmission

提前警告装置 advance warning device

提前竣工 accelerated completion; earlier completion

提前竣工奖金 bonus for early completion of works

提前控制 advanced control; anticipating control; control in advance

提前量 aiming off; initial lead; lead

提前量表 lead table

提前量测定时间 predicted dead time

提前量测定准备 predicting dead time

提前量分划盘 set-forward scale

提前量计算尺 lead computer; set-forward rule

提前量计算器 set-forward device

提前量调节 predictor control

提前量图表 lead chart

提前量预测 lead prediction

提前排气 advance of release; exhaust lead; prerelease

提前喷射 early injection

提前喷药 prespray

提前期 lead time; period in advance; time in advance

提前期累计编号法 accumulative numbering method in advance

提前起爆 preinitiation

提前切断(绿灯信号) early cut-off

提前切断信号 early cut-off signal

提前全部付清外债 payoff foreign debt ahead of schedule

提前日期提单 predate bill of lading

提前上 move-up

提前上死点 top center[centre] heading

提前时间 predicted time; prediction period; preset time

提前释放 premature release

提前收割 premature harvesting

提前赎回押品的权利 prime redemption privilege

提前赎回证券 call

提前赎债条款 call provision

提前赎债溢额 call premium

提前随动系统 predictor servomechanism

提前退休养老金 early retirement benefit

提前完成 fulfil ahead of schedule

提前完成计划 ahead of schedule; ahead of time

提前完工 early completion; accelerated completion; constructive acceleration

提前完工奖 bonus for early completion

提前位置 set-forward position

提前位置距离 range at future position

提前艏装 advance fitting-out

提前(修正)量 preact

提前一期的预测 one-period-ahead forecast

提前引燃防止剂 ignition controller

提前支付 advance payment; payment in advance

提前支付的承兑汇票 anticipated acceptance

提前支付条款 acceleration clause

提前支取未到期存款 break a deposit

提前终止合同 terminate contract before the data of expiration

提前仲裁 submit a dispute for arbitration

提前周期 lead time

提前诸元 prediction data

提前转期 advance refunding

提前转期偿还 <美国公债> advance refunding

提前装运 anticipate shipment

提前准备 advance preparation

提前坐标 predicted coordinate

提琴背板纹 fiddleback figure

提琴背纹(木纹) fiddleback grain

提琴式背靠椅 fiddleback chair

提琴式滑车 fiddle block; long tackle block; thick and thin block

提琴形的 tiddle-shaped

提琴形滑车 fiddle block

提琴叶栎 overcup oak; swamp post oak

提请批准 submit for approval

提请仲裁的资格 capacity to submit to arbitration

提取 extraction; abstraction; drawing; extracting; withdrawal

提取层 extract layer

提取常数 extraction constant

提取储蓄存款 dissaving

提取存款 withdraw deposit

提取的 extractive

提取的沥青 extracted bitumen

提取滴定(法) extractive titration

提取段 extraction-step

提取法 extraction method; separation

提取反应 abstraction reaction

提取分离(法) extraction separation

提取酚 pull phenol

提取工艺 extraction process

提取管 extraction tube

提取光度法 extraction-photometric method

提取货物 take delivery of merchandise or goods

提取积累 accumulation fund draw by

提取积累金 draw the accumulation fund

提取基金 appropriation of the funds

提取技术 extraction technique; extractive technique

提取剂 extractant; extracting agent; extraction solvent; extractive agent

提取加热炉 extract furnace

提取加压蒸发塔 extract pressure flash tower

提取壳筒 extraction thimble

提取款项 make a draft of money

提取率 extraction percentage; extraction rate; extraction ratio; extraction yield; rate of extraction; recovery ratio

提取奶油 cream

提取逆转汇票 redraw

提取瓶 extraction flask

提取器 extraction apparatus; extractor; taker

提取器接头 extractor sub

提取器筐 extractor basket

提取热交换器 extract exchanger

提取色谱(法) extraction chromatography

提取设备 extractor

提取式浅海海底地震仪 draw shallow sea bottom seismograph

提取室 extraction chamber

提取数据 extraction data

提取水 extraction water

提取塔 extracting column; extraction column; extract tower

提取筒壳 extraction shell

提取土壤溶液 extraction of soil solution

提取物 abstract; extractive; extract

提取物质 extraction matter; extraction substance; extractive substance

提取-洗涤柱 extraction-scrubbing column

提取系数 distribution ratio; extraction coefficient

提取系统 extraction system

提取现金 cash drawing

提取相 extract phase

提取效率 extraction efficiency

提取行李 baggage claim

提取行李处 luggage claim

提取性 extractability

提取序列 abstraction sequence

提取岩芯 take a core

提取冶金学 extraction metallurgy; extractive metallurgy

提取液 extract; extracting solution

提取因数 extraction factor

提取荧光分析 extraction fluorometric analysis

提取油 extract oil

提取债券 drawing of bonds

提取者 extractor

提取蒸馏 extractive distillation

提取蒸汽 bleeding steam

提取指令 extract instruction

提取指示剂 extraction indicator

提取制造程序 extractive process

提取柱 extraction column

提取资产 withdrawal of assets

提取资金 drawdown

提人罐笼 man cage
提溶极谱分析 stripping analysis
提砂斗 sand bailer
提砂筒 <打井用> sand bailer
提升 advancement; dolly; elevate; hoisting; lifting; preferment; promote; sublimate; sublimation; winding; windlass
提升靶 flip-up target
提升百叶门 lift away shutter door
提升板 liftout plate; riser plate
提升泵 elevator pump; lift(ing) pump
提升泵站 lift pumping station
提升臂 lift arm
提升臂延伸 lift arm extension
提升臂支撑销钉 lift arm support pin
提升柄 lift handle
提升部件 lifting piece
提升操作手柄 lift lever
提升槽钢 lifting channel
提升叉 lifting fork
提升叉车 forklift truck
提升铲 elevating scraper
提升超速保护装置 hoist overspeed device
提升车 lift truck
提升齿轮 elevating gear
提升齿轮制动器 hoisting gear brake
提升齿条 lifting rack
提升传动装置 lifting gear
提升窗口 climbing aperture
提升窗扇的吊窗拉手 sash lift knob
提升锤 lift hammer
提升大吊桶 bowk; kibble; sinking bucket; hoppit <采矿用>
提升带 elevator belt(ing)
提升的不可预见费 escalation contingency
提升的施工技术 lifting construction technique
提升电磁 electric(al) lifting magnet
提升电动机 winding motor
提升电压 booster tension
提升垫板 jacking plate
提升吊杆 elevating boom
提升吊架 lift yoke
提升吊桶 hoist bucket
提升动臂 raise boom
提升动力消耗 power consumption of hoisting
提升动作 hoisting motion
提升斗 lift pot
提升斗式挖沟机 trenching machine of the bucket elevator type
提升短节 elevator plug
提升阀 level(l)ing valve; lift valve; mushroom valve; poppet; poppet(type) valve
提升阀分配器 poppet-valve distribution
提升阀装置 lifting valve gear; poppet-valve gear
提升法 lift method
提升方钻杆并放入鼠洞装置 bozo line
提升分级机 elevator classifier
提升分线管 lifting pipe
提升负荷 hoist load
提升附加装置 lifting attachment
提升复式滑轮 hoisting pulley-block
提升干管 rising main
提升杆 hoisting mast; lift(ing) arm; lift(ing) bar; lift(ing) lever; lift(ing) rod; lift(ing) link; link arm; raising lever
提升杆调节器 lift rod adjuster
提升杆调整螺杆 lift link screw
提升杆销 lift rod pin
提升钢丝绳 hoisting cable; holding line; lift rope; raising sling; winding cable; winding rope
提升钢丝绳安全系数 hoisting rope safety factor
提升钢丝绳速度 rope speed
提升钢丝绳系数 rope factor
提升钢索 steel hoist wire rope
提升高度 height of lift(ing); hoisting depth; hoisting height; lift above ground; lift(ing) height; winding depth
提升高度限位器 hoisting limiter
提升格间 hoisting compartment
提升工时消耗 man hours consumption of hoisting
提升工资 raise wages
提升工作量 amount of hoisting
提升工作时间 time of hoisting
提升钩 hoisting hook; lifting bail; lifting hook
提升鼓 lifting drum
提升鼓制动器 lifting drum brake
提升鼓轴 lifting drum shaft
提升管 lift(ing) pipe; riser tube
提升管反应器 reactor riser
提升管路 lift line
提升管线 lifting piping line; lift line
提升管子 removal of pipe
提升罐 lift pot
提升罐笼 hoisting cage; lift car; winding cage
提升罐笼门 lift-car door
提升辊 liftout roller; take-out roller
提升滚筒 haulage drum; hoisting drum; lifting cylinder
提升过紧 over-travel
提升和调动 promotion and transfer
提升荷载 hoist(ing) load
提升滑车 hoisting tackle; lifting tackle; purchase tackle
提升滑车组 hoisting pulley-block
提升滑窗 lifting sliding window
提升滑动门 lifting sliding door
提升滑轮 hoisting tackle; lifting tackle; purchase tackle
提升滑轮组 lifting block; load block
提升环 lifting dog; retrieving ring
提升环总成 <在打捞筒端的> lifting dog assembly(in overshot head)
提升活塞 jigger lifting piston
提升机 elevator; gig; gin; hoisting engine; hoisting machine; hoist(ing) unit; inverted lever; lifter; mechanical elevator; vertical transporter; winding engine; winding machine
提升机臂 <挖掘机> lifter arm
提升机电动机 winder motor
提升机斗的固定钢轨 rail for fixing elevator buckets
提升机房 hoist-engine house
提升机杠杆 hoist lever
提升机构 elevating gear; elevating mechanism; hoisting appliance; hoisting gear; hoisting mechanism; lifting gear(ing); lifting mechanism; raising gear; raising mechanism
提升机构杆系 lift linkage
提升机构曲柄 raising crank
提升机构蜗杆 raising worm
提升机刮板 elevator flight
提升机活塞旋转耳轴 lift cylinder trunnion
提升机驾驶员 hoistman; lift attendant
提升机进料箱 elevator boot
提升机井 shaft for risers
提升机卷筒 lifting drum
提升机缆索 hoist cable
提升机缆索卷筒 hoist cable socket
提升机链条 elevator chain; hoist chain
提升机链托 idler of elevator chain
提升机料仓 elevator hopper
提升机料斗 elevator hopper; elevator scoop
提升机容量 lift capacity
提升机刹车 hoist brake
提升机勺斗 elevator bucket
提升机深度指示器 winding engine indicator
提升机司机 hoistman
提升机外壳 elevator casing
提升机橡胶吊桶 rubber elevator bucket
提升机橡胶戽斗 rubber elevator bucket
提升机械 lifting gear; raising machinery; lifting plant
提升机制 elevating mechanism; lifting mechanism
提升机制动杆 hoisting gear brake
提升机装卸设备 landing equipment
提升机最大拉力 maximum tensile force of hoisting machine
提升极限开关 lifting limit switch
提升集电弓的信号 signal to raise pantograph
提升技术 lift technique
提升技术经济指标 technical-economic index of hoisting
提升夹钳 lifting tongs
提升架 hoisting frame; pulling yoke; raising legs
提升间 hoist hole; hoistway
提升间歇时间 winding interval
提升绞车 hoisting gear; hoisting machine; hoisting winch; lifting hoist; lifting winch; reversing winch
提升绞车平衡重 dolly
提升绞筒 winding barrel; winding drum
提升搅拌器 lift agitator
提升阶段 lifting stage
提升进料铲 lifting skip
提升井 hoisting shaft; hoistway; pulley shaft; raising shaft; riser shaft
提升井搭叠式闸门 hoistway telescoping gate
提升井道 hoisting well
提升井对开拉门 hoistway biparting door
提升井检修通道入口开关 hoistway access switch
提升井门联锁装置 hoistway door interlock; hoistway unit system
提升井门锁定装置 hoistway door locking device
提升井入口 hoistway entrance
提升井伸缩式闸门 hoistway telescoping gate
提升井筒 lift tower
提升井围栏 hoistway enclosure
提升井闸门 hoistway gate
提升距离 lift above ground
提升卷筒 hauling drum; lifting drum; winding drum
提升卷筒制动器 hoisting drum brake; lifting drum brake
提升卷筒轴 lifting drum shaft
提升掘进 raise driving
提升卡钳弹簧 lifting dog spring
提升控制杆 lift control
提升控制装置 lift control
提升跨度 lift
提升框架 lift frame
提升拉杆 raising link; raising rod
提升缆索 hoist line
提升力 elevating power; feed pull; lifting force; raising force
提升力值 lifting value
提升连杆 lifting link
提升链 elevator chain; lift chains
提升链条 elevating chain
提升量 hoisting rating; lifting capacity; lifting power
提升料斗 hoisting bucket
提升螺栓 elevator bolt; lift bolt
提升门 lever gear door; lift away shutter door; lift-up door
提升密封 lift-off seal
提升模板 climbing shutter(ing); jumpforming; lifting form(work)
提升能力 elevating capacity; hoisting duty; lift(ing) capacity; lifting power
提升喷射法 <用电动泵将钻井中的泥砂喷出> jet lifter method
提升平板车 elevating platform truck
提升平板闸门 flat vertical gate; lifting flat gate; Stony gate
提升平车 dukey
提升平台 lifting platform
提升铺丝绳 steel hoist wire rope
提升(起重)笼 hoisting cage
提升气体 lift gas
提升气体返回口 lift gas return
提升器 erector; lifter; lifting dog; raiser
提升器臂 lifter arm
提升器断绳保险 parachute
提升牵引角 hoisting angle
提升钳 lifting clamps
提升桥 lifting bridge
提升倾卸式煤炭翻车装船机 coal chute crane
提升曲柄 lift crank; raising crank
提升曲线 lifting curve
提升设备 hoisting appliance; hoisting equipment; hoisting gear; hoisting installation; hoisting plant; hoisting unit; hoisting winch; lifting apparatus; lifting appliance; lifting device; lifting equipment; winding gear; pulling equipment <升降钻具>
提升设备拆旧摊销及大修费 hoisting plant depreciation apportion and overhaul charges
提升设施传动 transmission for lifting gear
提升深度 hoisting depth
提升生产定额 hoisting production quota
提升绳 hoisting rope
提升绳道 wireway
提升绳索 winding rope
提升绳索臂端滑轮 hoist line boom point sheave
提升时丢失岩芯 lost cores during lifting
提升时间 bucket raise time; lift time
提升时损失岩芯 lost cores during lifting
提升时脱离岩芯 lost cores during lifting
提升式百叶窗 lifting shutter
提升式百叶(门) lifting shutter
提升式舱口盖扳手 lift lock arm
提升式铲运机 elevating scraper
提升式船闸门 lift gate
提升式单级船闸 lift inland lock; lift navigation lock
提升式底阀 poppet foot valve
提升式阀 lift(ing) valve
提升式翻车机 lifting car dumper
提升式过道 lifting walkway
提升式开闭器 lift(ing) shutter
提升式开合桥 elevator bridge
提升式门 lift gate; lift(ing) door
提升式模板 climbing form(work); jumping shuttering; leaping shuttering; leaping formwork
提升式平板闸门 Stoney gate
提升式平台 raising platform
提升式启闭器 lifting shutter

提升式气动搅拌机 airlift agitator
提升式气流干燥机 airlift drier[dryer]
提升式（汽车）翻车机 car dumping crane
提升式桥孔 lift(ing) span
提升式桥跨 lift span
提升式升船机 ship lifter of lifting type
提升式升降桥 vertical-lift bridge
提升式通道 lifting walkway
提升式通气阀 poppet air valve
提升式挖掘机 lifting shovel
提升式挖泥机 elevating dredger
提升式坞门 vertical lift gate
提升式卸船机 <扬卸机> elevator
提升式运砖机 brick elevator
提升式闸门 caterpillar gate; drop gate; lifting gate; lifting hook-type gate; lifting lock gate; vehicle lift gate; lift gate
提升式闸门堰 draw door weir
提升式装料机 elevating loader; elevator-type loader
提升式装载机 elevator-type loader
提升式装载挖土机 elevator loader-excavator
提升手把 lifting handle
提升手柄 lifting handle
提升手杆 hoist hand lever
提升输煤栈桥 elevated coal track
提升输送带 elevating conveyer[conveyor]
提升输送机 convelater[convelator]; elevating conveyer[conveyor]
提升栓 hoisting plug
提升水的水压损失 elevation loss
提升水头 delivery head
提升速度 ascending velocity; hoisting speed; hoisting velocity; lift(ing) speed; lift(ing) velocity; raising speed; speed lifting
提升速度与起重量 hoisting speed and loads
提升速率 lifting speed; lift rate
提升索 hoisting rope; lift cable; lifting rope
提升塔（架）hoisting tower; take-off tower
提升台 lifter board; lift platform
提升套管 pull casing
提升天轮 hoisting pulley
提升头 poppet
提升推土机 C-frame
提升腿 lift leg
提升系统 hoisting system; lift system
提升限位开关 lift limit switch
提升限位器 lift stop
提升限制 lifting barrier
提升限制器 lift stop
提升小车 lifting truck
提升小绞车 lifting puffer; puffer
提升效率 lifting efficiency
提升信号 booster signal
提升行程 lift stroke; power stroke
提升性 building-up property
提升旋转液压缸 lift-turn cylinder
提升循环 winding cycle
提升循环时间 hoisting cycle; hoisting period
提升压紧装置 lifting press
提升液压缸 lifting cylinder; lift ram
提升液压缸活塞 lift piston
提升液压系统 lifting hydraulic system
提升移送装置 lift and carry transfer
提升用气体 lift gas
提升用液压缸 lift hydraulic cylinder
提升用真空吸盘 vacuum lifter pad
提升用桩箍 pitching ferrule
提升油缸 lift(ing) cylinder; lift(ing) ram
提升油缸臂 lift cylinder arm
提升油缸活塞 lift piston
提升与操纵键 lifting and operating key
提升与压力泵 life and force pump
提升运动 lifting movement; lifter motion
提升运动限制开关 limit switch for hoist motion
提升运输带 elevating belt
提升运载桥 lifter transporter
提升闸 lift(ing) lock
提升闸板式加料机 lifting-gate feeder
提升闸门式加料机 lifting-gate feeder
提升闸门堰 lift gate weir
提升闸墙 lift(ing) wall
提升站 lift station
提升止回阀 lift check valve
提升重量 hoisting weight; lifting weight
提升周期 hoisting period; raising cycle
提升轴 lift shaft
提升轴臂 lift shaft arm
提升柱塞 liftout plunger
提升抓 lifting dog
提升转向机构 lifting and traversing mechanism
提升装载机 elevating loader
提升装置 elevating appliance; elevating device; elevating gear; elevating unit; hauling-up device; index unit; lifter; lifting appliance; lifting device; lifting equipment; lifting gear; lifting linkage; lifting piece; lifting rig; lifting unit; riser; take-up; bont <提升用的钢丝及其附件>
提升装置接合器 lift engager
提升总全高变 full lift
提升钻杆 breakout; removal of pipe; strip the drill pipe; strip the drill rod
提升钻进 raise boring; raise drilling
提生产率 hoisting productivity
提示 cue; hint; prompting; reminder; tip-off
提示程序 attention program(me)
提示符 prompt
提示光标【计】prompt
提示交单 documents against presentation
提示卡 cue card
提示时 on presentation
提示提单 order bill of lading
提示席 prompt box; prompter's box
提示销售法 suggestive selling
提示性文摘 annotative abstract
提示银行 presenting bank
提示罩 prompt hood
提式打桩机 oil engine
提收率 extraction rate
提手 eye ring; hanging ring; lifting yoke
提手柄 bar lift
提闩 lift latch
提水 lifting of water
提水板 pallet
提水吊杆 water spar
提水工具 water lift; tabboot <埃及>
提水灌溉 irrigation by pumping; lift irrigation; pumped irrigation; pumping irrigation
提水灌溉面积 lift irrigated area
提水灌溉区（域）lift irrigated area; lift irrigation area
提水虹吸式配件 fittings for single trap siphon
提水机 water elevator; water lifting machine; water raiser; water-raising engine
提水机械 lifting machine; water lifting machinery
提水井 draw well; open bucket well
提水辘轳 <印度用语> rati
提水器 water lifts
提水设备 water lifting device
提水桶 pail; water barrel
提送所得税申报表 file income tax return
提送印鉴 filing of a seal-impression
提速 speed-up
提锁 lift latch
提桶 bowk; lifting barrel; pail
提桶取样 sampling with a pail
提旺诺风 tivano
提问档 profile
提问方式 question formulation
提问格式 question format
提问逻辑表 question logic table
提问失败 question failure
提问式 question profile
提问输入阶段 question input phase
提问术语 question term
提问术语表 question term table
提问向量 question vector
提问有效性表 question validity table
提物手柄 bar lift
提吸速率 rate of aspiration
提下钻杆 pulling and running the drill pipe
提现挪用 kiting
提携手柄 carrying handle
提醒注意 warn
提要 brief outline; digest; prospectus; synopsis
提液机 extracting machine
提议人 mover
提议者 proponent; proposer
提银 desilver
提银炉 cupellation furnace
提引工具【岩】lifting tool
提引钩 hoisting hook; tackle hook
提引钩提升速度 hook speed
提引环 eye ring
提引器 elevator; lifting bail; puller; pulling dog
提引塞 <钻杆的> elevator plug; swivel hoisting plug
提引栓 hoisting plug
提引水接头 water swivel with bail
提引水龙头 heave hook; hoisting type water swivel; shackle swivel hook
提余液 good oil; raffinate
提早断裂期 early stage of failure
提早灌水 early watering
提早结束 curtail
提早排水 predrainage
提早入场门 <戏院> early door
提早势 advanced potential
提早退休 early retirement
提早预灌 advance irrigation
提早装船 early shipment
提闸门式加料机 lifting-gate feeder
提折旧费 amortise[amortize]
提支 called forward
提租的可能额度 up rent potential
提钻 pull out
提钻能力 auger lifting capacity
提钻作业 hoisting operation

题

题材 subject matter

题解目录 <附有简介的目录> annotated catalogue
题铭 legend
题目输入带 problem input tape
题目数据 problem data
题目说明 problem definition; problem description
题目文件 problem file
题目语言 problem language
题外关键词索引 keyword out-of-context index

蹄

蹄式车轮制动器 shoe wheel brake

蹄式闸 shoe brake
蹄式制动器 block brake; double shoe brake; horseshoe type brake; shoe (-type) brake
蹄式制动器紧蹄 primary brake-shoe
蹄酸 ungulic acid
蹄铁 horseshoe
蹄系状的 ansate
蹄形磁体 horseshoe shape magnet(iron)
蹄形垫圈 horseshoe washer
蹄形航道 horseshoe channel
蹄形家具腿 hoof foot
蹄形卡规 cal(l)iper ga(u)ge
蹄形松土铲 hoof shovel
蹄型 style of the shoe
蹄型垫圈 horseshoe washer
蹄印 hoof print
蹄支 bars of foot
蹄状体 ungula

体

体比重 apparent specific gravity

体变模量 modulus of volume change; volume change modulus
体变潜量 potential volume change
体变形 body deformation
体表 body surface
体表面积 body surface area
体表面联合扩散理论 Gilmer Ghez Cabrera theory
体波 body wave
体波辐射 body wave radiation
体波滤波器 bulk wave filters
体波谱 body wave spectrum
体波震级 body wave magnitude
体操房 gymnastic room
体侧平直的 flat-bodied
体掺杂红外探测器 bulk doped infrared detector
体长 body length; length of body; trunk length
体尺 body measurement
体尺比例 body proportion
体磁导率 bulk permeability
体存根 body stub
体大而笨拙的东西 jumbo
体电导率 bulk conductivity; volume conductivity
体电荷 body charge
体电荷密度 volume charge density
体电荷转移器件 bulk charge transfer device
体电离 volume ionization
体电离密度 volume ionization density
体电流 bulk current
体电势 bulk potential
体电致发光 bulk electroluminescence
体电阻 bulk resistance; bulk resistor; cubic(al) resistance
体电阻率 bulk resistivity
体定量分析法 gasometry
体对角线 body diagonal
体发射率 volume emissivity
体分布力 force per unit volume
体辐 antimere
体负阻效应 bulk effect
体复合 bulk recombination
体复合速率 volume recombination rate
体高 height at withers; withers height
体格 physique
体格大的 large-framed
体格大小 body size
体格分类等级 physical profile serial

体格分类等级代码 physical profile serial code
体格检查 health checkup; medical examination; physical checkup; physical examination; physical inspection; check up <美>
体格适应能力检查 physical aptitude examination
体光电效应 volume photoelectric(al) effect
体含量 body burden
体击穿电压 bulk breakdown voltage
体积 volume; cubic; cubic (al) content; cubic (al) measure (ment); cubic (al) yardage; measure volume; solid measure
体积安定性 stability of volume
体积安定性试验 autoclave expansion test
体积百分比 percentage by volume; percent by volume; volume percent(age)
体积百分含量 volume percentage content
体积百分率 percentage by volume; volume fraction; volume percent(age)
体积百分浓度 concentration expressed in percentage by volume
体积百分数 part per volume; percent(age) by volume; volume in volume; volume percent (age); volumetric(al) part
体积泵 volume pump
体积比储水系数 specific storativity
体积比例 part by volume
体积比(率) volume ratio; volumetric(al) proportion; bulk factor; proportioning by volume; ratio by volume; ratio of volume; volume composition; volumetric(al) ratio
体积比配合法 proportioning by volume; volume mix; volumetric (al) method; volume method
体积比配料 volume batching
体积比配料法 proportioning by volume; volume method; volume mix; volumetric (al) method; arbitrary proportion method
体积比配料器 volume batcher
体积比配料箱 volumetric (al) batch box; volumetric(al) batcher
体积比热 volumetric(al) specific heat
体积比吸湿率 volumetric(al) absorption
体积比吸收量 volumetric(al) absorption
体积比重 bulk specific gravity
体积变更 volume change
体积变化 volume change; volumetric(al) change
体积变化模量 modulus of volume change
体积变化能力 potential volume change
体积变化系数 coefficient of volume change
体积变化(引起的)应力 stress from volume change
体积变化自动记录器 auxograph
体积变形 bulk strain; cubic (al) deformation; cubic strain; volumetric(al) deformation
体积变形模量 modulus of volume deformation
体积表面积比 mass-surface ratio; volume-to-surface-area ratio
体积补偿器 pressurizer
体积不变而形状变化时的应力状态 deviatoric state of stress distortion
体积不变性 constancy of volume; volume constancy; volume constant
体积采出量 volume withdrawal
体积测定(法) volumetric (al) measurement; volumetric(al) determination
体积测量 cubing
体积产量 volumetric(al) production
体积产率 volume yield
体积长度平均径 volume length mean diameter
体积沉积速率法 depositional rate method of sedimentary rock volume
体积成本 volume cost
体积成型 bulk forming
体积尺码 cubic(al) dimension
体积充满度 voluminal space filling factor
体积磁化率 volume susceptibility
体积磁化系数 volume magnetizing coefficient
体积磁致伸缩 volume magnetostriction
体积磁致伸缩效应 volume magnetostrictive effect
体积存储密度 volumetric(al) packing density
体积大的 bulky; voluminous
体积大的货物 volume cargo
体积大的饲料 bulked feed
体积单位 elementary volume; unit of volume; volume element; volume unit
体积单元 elementary volume; volume unit
体积导电 volume conduction
体积倒数 inverse volume
体积得率 volume yield
体积的 volumetric (al); voluminal; voluminous
体积等同定律 law of equal volumes
体积电导率 bulk conductivity; volume conductance
体积电荷 volume charge
体积电离度 volume ionization
体积电量计 volume voltameter
体积电致伸缩效应 volume electrostrictive effect
体积电阻 volume resistance
体积电阻率 dissipation resistivity; mass resistivity; specific volume resistance; volume resistivity
体积电阻系数 specific insulation resistance; volume resistivity
体积动态 volumetric(al) performance
体积度 specific volume
体积吨 freight ton (nage); measurement ton(nage); scale ton; shipping ton; volumetric(al) ton
体积发射 volume emission
体积发射率 volume emission rate
体积法 volume basis; volumetric (al) method
体积法测流 volumetric (al) measurement of discharge
体积反向散射强度 volume backscattering strength
体积分 part by volume
体积分布 volume distribution
体积分法 volume integration method
体积分量 volumetric(al) component
体积分率 volume fraction
体积分批箱 volumetric (al) batch box; volumetric(al) batcher
体积分数 volume fraction
体积分析 volume analysis; volumetric(al) analysis
体积分析法 volume analysis method; volumetric(al) analysis method
体积份量 part by volume
体积丰度 abundance by volume
体积复合 volume recombination
体积改变 stereomutation
体积估计法 volumetric(al) estimation method
体积估计方差 estimation variance of volumes
体积估计误差 estimation error of volumes
体积估价法 cubing; cubitage; volume method
体积估算法 <一种用体积毛估建筑造价的方法> volume method of estimating cost; volume method
体积固定性 volume constancy
体积固体 volumetric(al) solids
体积固体分 solids by volume; volume solid
体积光电导性 volume photoconductivity
体积光电吸收指数 volumetric(al) photoelectric(al) absorption index
体积光电效应 volume photoeffect
体积含量 content by volume; volume content; volumetric(al) content
体积含量率 moisture content by volume
体积含沙量比例 volumetric(al) sediment concentration scale
体积含水量 volumetric(al) moisture content; volumetric(al) water content
体积含水率 volumetric(al) moisture content; volumetric(al) water content
体积恒定性 resistance to expansion and contraction
体积后向散射 volume backscattering return
体积缓冲瓶 volume bottles
体积换算 cubic(al) conversion
体积换算系数 volume conversion factor
体积回收率 volume recovery
体积混合 volumetric(al) mixing
体积混响 volume reverberation
体积火焰扩散 volumetric (al) flame spread
体积或重量总损耗 wastage in bulk or weight
体积货物 measurement cargo; measurement goods
体积积分 volume integral
体积基准比表面 specific surface on volume basis
体积级 volume level
体积几何 volumentary geometry
体积计 content ga(u)ge; stereometer; volume meter; volumenometer; volumescope; volumeter; volumometer
体积计量 cubing; volume measurement; cubic measurement
体积计算 cubage
体积剂量 volume dose
体积加权的 weighed volumetrically
体积加权平均灰分产率 weighted mean ash production rate by volume
体积减缩 reduction of bulk
体积减小 volume reduction; volumetric (al) reduction; loss of volume; reduction in bulk
体积减小量 volume-decrease potential
体积检查 volume inspection
体积界(限) volume bound
体积精压 bulk coining
体积矩 moment of volume
体积绝热 bulk insulation
体积克式浓度 volume formality
体积空位扩散 volume vacancy diffusion
体积孔隙率 bulk porosity
体积控制 volume control
体积扩散 bulk diffusion; volume diffusion
体积扩胀 volume expansion
体积理论 volume theory
体积力 body force; body stress; mass force; bulk strength <爆炸物质的>
体积力势能 body force potential
体积量测(法) volumetric (al) measurement; measurement of volume
体积裂隙率 volume fissure ratio
体积流动 volumetric(al) flow
体积流动速率 rate of volume flow
体积流量 volume flow; volume rate of flow; volumetric(al) flow rate
体积流量计 volumetric (al) displacement meter; volumetric (al) water meter
体积流率 volume flow rate
体积流速 rate of volume flow
体积氯度 chlorosity
体积密度 bulk density; bulk specific gravity; volume concentration; volume density
体积面积平均径 volume surface mean diameter
体积描绘仪 plethysmograph
体积描记法 plethysmography
体积描记器 plethysmograph
体积敏感度 volume susceptibility
体积模量 bulk modulus; modulus of volume; volume modulus
体积摩尔浓度 molar concentration; molarity; molar solution; volumetric(al) molar concentration
体积摩尔浓度的 molar
体积目标 volume target
体积能 volume energy
体积黏[粘]度 bulk viscosity; volume viscosity
体积黏[粘]弹性 bulk viscoelasticity; volume viscoelasticity
体积黏[粘]性 bulk viscosity
体积浓度 bulk concentration; volume concentration; volumetric (al) concentration
体积排除效应 excluded volume effect
体积排放流量 volume rate of discharge
体积排量 swept volume; volumetric(al) displacement
体积排水法 volumetric (al) dewater method
体积排阻色谱法 size exclusion chromatography
体积配合 volumetric(al) mixing
体积配合比 nominal mix proportioning; nominal mix proportions; proportioning by volume
体积配合法 batching by volume; volume batching; volumetric(al) method
体积配料 volumetric(al) batching
体积配料斗 volumetric(al) batch box; volumetric(al) batcher
体积配料法 batching by weight; volume batching; volumetric(al) method of batching
体积膨胀 bulking; cubic (al) dilatation; cubic (al) expansion; inconstancy of volume; measure expansion; volume dilatation; volume expansion; volume increase; volumetric (al) expansion; volumetric (al) strain; volumetric(al) swell(ing)
体积膨胀比 volumetric(al) expansion

T

ratio
体积膨胀回弹系数 volume expansion coefficient of resilience
体积膨胀计 volume dilatometer; volume expansion meter
体积膨胀率 swelling rate of volume
体积膨胀模量 modulus of volume expansion
体积膨胀模数 modulus of volume expansion
体积膨胀碎石 cubic(al) expansion crushed stone
体积膨胀系数 coefficient of bulk increase; coefficient of cubic(al) thermal expansion; coefficient of cubic(al) expansion; coefficient of volumetric expansion; volume expansion coefficient; volumetric(al) expansion coefficient; cubic(al) expansion coefficient
体积平衡方程式 equation of volumetric(al) balance
体积平均径 volume mean diameter
体积平均通量 volume-averaged flux
体积气孔率 bulk porosity
体积全息图 volume hologram
体积全息图存储器 volume hologram memory
体积全息元件 volume holographic(al) element
体积缺陷 volume defect
体积热膨胀率 volumetric(al) thermal expansibility
体积热容量 volumetric(al) heat capacity
体积溶胀 volume swelling
体积柔量 bulk compliance
体积散射 volume scattering
体积散射函数 volume scattering function
体积散射强度 volume scattering strength
体积散射系数 volume scattering coefficient
体积色谱法 volumetric(al) chromatography
体积设计 volumetric(al) design
体积式水表 volumetric(al) water meter
体积势能 body force potential
体积收缩 volume contraction; volume shrinkage; volumetric(al) contraction; volumetric(al) shrinkage
体积收缩极限 volumetric(al) shrinkage limit
体积收缩量 volume shrinkage mass
体积收缩率 shrinkage of volume; volume shrinkage ratio
体积收缩试验 volumetric(al) shrinkage test
体积寿命 volume lifetime
体积衰减系数 volume attenuation coefficient
体积松散系数【疏】 bulking factor
体积速度 volume velocity
体积塑性流动 bulk plastic flow
体积随变生物 volume regulator
体积损失 volume loss
体积缩减 reduction in bulk; volume decrease
体积缩减系数 coefficient of volume decrease
体积缩小 volume decrease
体积弹性 bulk elasticity; elasticity of bulk; volume elasticity
体积弹性模量 bulk elastic modulus; bulk modulus of elasticity; elasticity of volume; modulus of bulk elasticity; modulus of elasticity of volume; modulus of volume; modulus of volume elasticity; modulus of volumetric(al) elasticity; volume modulus of elasticity; volumetric(al) modulus of elasticity
体积弹性模数 bulk modulus; bulk modulus of elasticity; elasticity of volume; modulus of bulk elasticity; modulus of volume elasticity; modulus of volumetric(al) elasticity; volume modulus of elasticity; volumetric(al) modulus of elasticity
体积弹性系数 coefficient of cubic(al) elasticity
体积-体积浓缩系数 volume-volume concentration factor
体积调节 volume control
体积调节器 volume regulator
体积通量 volume flux
体积投配设备 volumetric(al) batching equipment
体积投配箱 volumetric(al) batch box; volumetric(al) batcher
体积图 volume graph
体积温度系数 volume temperature coefficient
体积稳定成分 sound ingredient
体积稳定性 constancy of volume; stability in bulk; volume stability; volume stabilization; stability of volume
体积稳定组分 <混凝土> sound ingredient
体积吸收 volume absorption
体积吸收系数 volume absorption coefficient
体积吸收项 volume absorption term
体积吸水 volume water absorption
体积吸水率 volumetric(al) coefficient of water absorption
体积系数 bulk coefficient; bulk factor; volume coefficient; volume factor; volumetric(al) factor; volumetric(al) coefficient
体积消毒 volume sterilization
体积效率 volume efficiency; volumetric(al) efficiency
体积效应 bulk effect; volume effect; volumetric(al) effect
体积胁变 cubic(al) strain
体积形变 volume deformation
体积形状系数 volume shape factor
体积形状因子 volume shape factor
体积性质 bulk property
体积学说 volumetric(al) theory
体积压实比 die fill ratio
体积压实机 volume compressor
体积压缩 volume compression
体积压缩变化 polytropic(al) change
体积压缩模量 bulk modulus of compressibility
体积压缩试验 volumetric(al) compression test
体积压缩系数 coefficient of volume compressibility; volume compressibility; coefficient of volume decrease
体积压缩性 bulk compressibility; volumetric(al) compressibility; voluminal compressibility
体积压缩指数 polytropic(al) index
体积岩溶率 volume karst factor
体积颜料浓度 volumetric(al) pigment concentration
体积仪 content ga(u)ge
体积因素 bulk factor; volume factor
体积因子 bulk factor; volume factor
体积应变 bulk strain; cubic(al) dilatation; cubic(al) strain; volume strain; volumetric(al) strain
体积应变计 volume strain ga(u)ge
体积应变能 volumetric(al) strain energy
体积应力 body stress; volumetric(al) stress
体积应力状态 <物体受力形状不变，而只体积变化时的应力状态> volumetric(al) state of stress
体积与表面比 volume surface ratio
体积元 element of volume; volume element
体积元内溶质量的纯变化率 pure change rate of solute in volume element
体积元素 volume element
体积原理 volume theory
体积运输 volume transport
体积增加 volume gain
体积增量 dilatation
体积张力 cubic(al) strain
体积胀缩振动 volume dilatational vibration
体积折合的重量 measurement weight
体积植入 <放射源> volume implant
体积指标 volume index
体积指示器 volume indicator
体积指数 bulk index; volume index
体积中心 center[centre] of figure; center[centre] of volume; centroid center of volume; volume center [centre]
体积重 gravity of volume
体积重量比 volume weight ratio
体积状态 volumetric(al) behavio(u)r
体积自由能 volume free energy
体积阻力 volume resistance
体积组成 bulk composition; volume composition
体极化 volume polarization
体极化系数 coefficient of volume polarization
体减系数 coefficient of volume decrease
体接触 physical contact; stereoscopic(al) contact
体节板 segmental plate; somatic plate
体结构 volume structure
体结构缺陷 bulk structural defect
体结合 volume junction
体晶 bulk crystal
体绝缘漏泄 volume insulation leakage
体壳 sleeve
体孔 body opening
体宽 body width
体离子密度 volume ion density
体力 body force; muscle; physical effort; physical energy; physical strength
体力不支的 unfortified
体力放大器 man amplifier
体力工人 blue-collar worker[employee]; manual worker
体力工作 blue-collar jobs; muscular work
体力活动分析 manual activity analysis
体力劳动 blue collar; manual labo(u)r; manual work; physical exertion; physical labo(u)r
体力劳动的工业 blue-collar industry
体力劳动和脑力劳动的对立 antithesis between manual and mental labor
体力劳动者 blue-collarite; blue-collar worker[employee]; manual worker
体力疲劳 physical fatigue
体力衰竭 adynamia
体力重量锻炼房 weight training room
体量 size
体量法规 bulk regulation
体量管理条例 bulk regulation
体量水泥系数 content of cement in a cubic(al) meter
体密度 bulk density; density by volume
体模材料 phantom material; tissue equivalent material
体内无线电探头 endoradiosonde
体能 physical energy
体黏[粘]滞性 mass viscosity
体浓度 bulk concentration
体盘 body disc[disk]
体膨胀 cubic(al) dilatation; cubic(al) expansion
体膨胀系数 coefficient of dila(ta)tion; coefficient of volume expansion; efficiency of dila(ta)tion; volume dilatation coefficient
体腔 body cavity
体缺陷 body defect; bulk defect
体容度 bulkiness
体容重 bulk unit weight
体散度 volume divergence
体散射反射率 reflectivity of volume scattering
体色(彩) body colo(u)r; bulky colo(u)r
体声波换能器 bulk acoustic(al) wave transducer
体声波延迟线 bulk acoustic(al) wave delay line
体式动臂 monoblock boom
体视 stereo; stereoscopic(al) vision; stereovision
体视半径 radius of stereoscopic vision
体视比较镜 stereocomparator
体视比较仪 stereocomparagraph; stereocomparator
体视测距仪 stereo range finder; stereoscopic(al) height finder
体视测量仪 stereometer
体视测图仪 stereocartograph
体视测微计 stereomicrometer
体视的 stereoscopic(al)
体视法 stereography; stereoscopy
体视观测 stereoscopic(al) observation
体视光学 stereooptics
体视化仪表 stereoscopic(al) instrument
体视绘图仪 stereoautograph
体视技术 stereoscopic(al) technology
体视经纬仪 stereotheodolite
体视镜 stereoscope; stereoscopic(al)
体视量角 stereogoniometer
体视率 stereo power; stereoscopic(al) power; total relief
体视锐度 stereoscopic(al) acuity
体视术 stereography; stereoscopy
体视图 stereogram; stereoscopic(al) graph
体视望远镜 relief telescope; stereotelescope; telestereoscope
体视显示 perspective three-dimensional display
体视显微镜 stereomicroscope; stereoscopic(al) microscope
体视显微术 stereomicrography
体视像片 stereoscopic(al) picture
体视效应 stereoscopic(al) effect
体视学 stereology; stereoscopy
体视照片 stereograph
体视照片摄制术 stereography
体视照相 stereoscopic(al) photography
体缩 shrinkage of volume
体缩率 volumetric(al) shrinkage rate; volumetric(al) shrinkage

体态 body style
体特性 bulk property
体外层 volume surrounding
体外预(加)应力 external prestressing
体外预应力索 external prestressing tendon
体围 body girth
体位移 land monument
体温 body temperature
体温表 clinical thermometer; rafraichometer
体温计 clinical thermometer;thermometer
体温曲线 temperature curve
体系 body system; edifice; regime(n); set-up;syntax;system
体系不稳定性 system instability
体系差异 architectural difference
体系定义 architectural definition
体系刚度矩阵 system stiffness matrices
体系工程 system construction
体系函数 system function
体系行列式 system determinant
体系化 systematization;systematize
体系间比较 intersystem comparison
体系建筑 system building
体系结构 architecture; system architecture
体系结构功能 architecture function
体系结构模拟 architecture simulation
体系结构评价 architecture evaluation
体系结构设计 architecture design
体系矩阵 system matrix
体系论 architectonics
体系模型 system modal
体系设计 system design
体系特性 architectural characteristic
体系特征 architectural feature
体系限制 architectural limit
体系相图 system phase diagram
体系转换 system transformation
体系转换法 system transform method
体系阻尼 systematic damping
体现 embodiment; embody; externalization
体现原则 principle of stereoscopic vision
体像 body image
体效率 bulk effect
体效应 block effect;bulk effect
体效应功率放大器 Gunn effect power amplifier
体效应技术 bulk technology
体效应器件 bulk effect device
体心 body-center[centre]
体心的 body-centered[centred]
体心点阵 body-centered lattice
体心格子 body-centered lattice
体心晶格 body-centered lattice
体心立方 cubic(al) body-centered; cubic(al) face-centered
体心立方的 body-centered cubic(al)
体心立方点阵 body-centered cubic(al) lattice
体心立方堆积 body-centered cubic(al) packing
体心立方结构 body-centered cubic(al) structure
体心立方晶格 body-centered cubic(al) lattice
体心立方晶体 body-centered cubic(al) crystal
体心立方晶系 body-centered cubic(al) system
体心立方体 body-centered cube
体心四方的 body-centered tetragonal
体心正方点阵 body-centered tetragonal lattice

体型 body type;form;set-up
体型标准 type standard
体型等级 type grade
体型规划 physical planning
体型鉴定 type classification
体型交联分子 three-dimensional crosslinked molecule
体型结构 physical conformation; three-dimensional structure
体型结构紧凑 compact conformation
体型聚合物 network polymer
体型美观的 clean-cut
体型(躺)椅 <如宇航用> contour couch
体型特征 physical characteristic
体型图 body-type graph;profile graph
体型网络结构 three-dimensional network structure
体型性状 conformation trait
体型氧化锌压敏电阻器 bulk type zinc oxide baristor
体型一致 uniformity of type
体型椅 contour chair
体型阻力 shape resistance
体循环 general circulation; systemic circulation
体验 taste;undergo
体液相 bulk fluid phase
体应力 body stress
体育 physical education
体育场 athletic ground; sports field; sports ground; stadium[复 stadia]; pal(a)estra[复 palestrae] <古希腊>
体育场的大看台 stadium grandstand
体育场高塔照明 tower lighting
体育场馆经营型管理 operating management of stadium and gymnasium
体育场建筑 stadium construction
体育场上的看台座位 stadium seat
体育场设施 stadium facility
体育场施工 stadium construction
体育的多元化功能 diversified functions of sports
体育锻炼 physical training
体育发展战略 sports development strategy
体育宫 sports palace
体育馆 boys gymnasium; games hall; gym; gymnasium[复 gymnasia]; indoor arena; indoor stadium; sport arena;sports building;sports hall
体育馆的大看台 stadium grandstand
体育馆建筑 stadium construction
体育馆设备 gymnasium equipment
体育馆设施 stadium facility
体育广告 sports advertisement
体育活动 physical exercise
体育建筑 sports architecture; sports building;sports structure
体育俱乐部 sports club
体育旅游 sports tourism
体育设施 athletic facility;sports facility
体育社会化 socialization of sports
体育统计 physical education statistics;sports statistics
体育卫生 sports hygiene
体育文化 sports culture
体育型体质 mesomorphy
体育用品 sport goods; sport requisites
体育用品壁橱 sports equipment closet
体育用品室 sports equipment closet
体育运动 athletics
体育中心 sports center[centre];sports forum
体噪声 bulk noise
体胀 bulking
体胀率 volume expansion

体胀模量 modulus of volume expansion
体胀系数 coefficient of dila(ta)tion
体砧 body stock
体制 frame; frame work; institutional framework;regime(n);system
体制的障碍 institutional barrier
体制改革 system reform
体制工程 systems engineering
体制函数 institutional function
体制基础 institutional setting
体制结构 institutional structure
体制下放 transfer administrative functions to lower levels
体制因素 institutional factor
体制障碍 institutional obstacle
体制转换 replacement of the old structure with the new;structure conversion
体质 organization;physique
体质论 craseology[crasiology]
体质颜料 body pigment; extender; extender pigment; loading pigment; non-opaque pigment; pigment extender
体重 body weight
体重比 weight ratio
体重变化系数 variation coefficient of unit weight
体重超过一千磅 weigh more than one thousand pounds
体重秤 scales;weighing machine standard
体重辅助 weight aid
体重管 volume weight tube
体重计 weighing machine
体重样品 sample for unit weight of ore
体重增加 gain weight;put on weight
体轴 axon
体轴形成 axiation
体状 body form
体锥面 body cone;polhode cone

剃

剃齿 gear shaving;shaving
剃齿刀 gear shaver;gear shaving cutter;shaving cutter
剃齿机 gear shaving machine; gear tooth shaving machine;shaving machine
剃齿留量 shaving stock
剃齿屑 shavings
剃齿心轴 shaving arbor
剃刀 razor
剃刀钢 razor steel
剃刀磨皮挂钩 shaving strop hook
剃鼓形齿 crown shaving
剃锯 razor saw
剃前 preshaving
剃前插齿刀 preshaving pinion cutter
剃前刀具 <齿轮> preshaving cutter
剃前滚刀 preshaving hob

替

替班 relay
替班乘务组 relief crew
替班工人 spare hand;swing man
替班职工 relief staff
替班职务 relief duty
替打(垫块) cap block;cushion block; dolly;piling dolly;hammer cushion
替打垫桩 follower;follower pile
替打动力 replacing power
替代设计 alternative design
替代 interchangeability; replace; replacement;stead;substitute for;supersede

替代变量 substitute variable
替代标准 alternate standard
替代材料 alternative material; equivalent material
替代参数 surrogate parameter
替代成本 alternative cost; opportunity cost
替代乘数 substitution multipliers
替代船 substitute vessel
替代担保 substituted security
替代担保附加条款 superseded suretyship riders
替代的 alternate;fall back
替代电路 replacement circuit
替代电站 alternative power plant; competitor
替代电阻 substitutional resistance
替代调换 substitition swap
替代定理 substitution theorem
替代法 alternative method; method of substitution;substitution method
替代方案 alternate design; alternative project; alternative scenario; alternative scheme; alternative solution; additive alternate; alternative design;alternative offer <承包商在投标中提出的>
替代方案的决定 determining alternative course;determining alternative project
替代方案分析 alternative analysis
替代方案评价 evaluating alterative course
替代方案选择 alternative choice
替代费用 substituted expenses
替代分析 substitution analysis
替代根 replacement root
替代固定资产 substitute fixed assets
替代固溶体 substitutional solid solution
替代关系 substitutional relation
替代灌浆 displacement grouting
替代规律 substitution law
替代汇率 stand-in rate
替代货币 substitute money
替代货物 substitute cargo
替代价值会计 replacement value accounting
替代进口 substitutes for imported goods
替代径路 alternate route; alternative route
替代纠纷调解法 alternative dispute resolution method
替代可能性 fallback possibility
替代可能性递减 diminishing substitution possibility
替代库容 replacement storage
替代扩散 substitutional diffusion
替代劳工 displacement of labo(u)r
替代零件 replacement of parts
替代率 substitution rate
替代能源 alternative energy sources; energy alternative
替代农业 alternative agriculture
替代品 succedaneum
替代区 alternate area
替代曲线 substitution curve
替代群落 substitute community
替代群落型 substitute type
替代商品 substitute goods; substitute products
替代设备 alternate device;back-up
替代设计 alternative design
替代生产函数 substitutional production function
替代式受主 substitutional acceptor
替代输入库 alternate input library
替代送达 substituted service
替代弹性 elasticity of substitution

T

替代条款 superseding clause
替代物 substitute;succedaneum
替代系统常驻设备 alternate system residence device
替代效果 substitution effect
替代效能 substitution effect
替代效应 effect of substitution; substitution effect
替代信道 alternate channel
替代行为 substitutive behavio(u)r
替代学说 alternation theory
替代原理 principle of substitution; substitution principle; substitution theorem
替代杂质 substitutional impurity
替代债务人 expromissor
替代账户 substitution account
替代者 substituent
替代阵列 replacement array
替代职权 alternative competence
替代中心 substitute
替代种 substitute species
替代仲裁员 substitute arbitrator
替灯 follower;long dolly;puncheon
替氟烷 telfurane
替工 accommodator;fill in;temporary substitute; temporary worker; turnover labo(u)r
替换 changing; replace; replacing; substitute for
替换泵 relay pump
替换表层土 replacing top soil
替换部件 replacement component;replacement unit
替换材料 alternate material;replacement material; substitutional material
替换测验 alternative test; duplicate test
替换成本 alternative cost
替换措施表列法 alternative check list method
替换单元 replacement unit
替换的荷载传递路线 alternative load path
替换电缆 <修理故障时用> interruption cable;breakdown cable
替换定理 replacement theorem
替换发动机 alternative engine
替换发票的转让 substitution of invoice transfer
替换法 <桁架分析用> interchange method;method of interchange
替换方式 substitute mode
替换分析 replacement analysis
替换杆件 replacing member; substitute member
替换公式 replacement formula
替换功率 replacing power
替换活动 displacement activity
替换货物 cargo alternative
替换技术 alternative technology
替换件 replacement piece
替换框架 substitute frame
替换零件 renewal parts;replacement (of) parts
替换路签 alternate staff
替换率 rate of substitution; replacement rate
替换论 replacement theory
替换(模板)支撑 reposting
替换期 replacement period
替换溶液 alternative solution
替换入口 alias
替换设备 spare attachment
替换式 alternate form
替换试验 alternative test
替换数组 replacement array
替换弹性 elasticity of substitution
替换图 replacement chart
替换问题 replacement problem
替换/无应答选择 alternate/no answer option
替换物 alternative
替换系数 replacement coefficient
替换线路 alternative routing
替换性 replaceability
替换性假设测试 alternative hypothesis test
替换需求 replacement demand
替换选择技术 replacement selection technique
替换油槽式运油车 relay tank truck
替换债券 refunding bonds
替换账户 substitution account
替换中心 substitute center[centre]
替换轴承 replacement bearing
替换装置 alternative device
替换钻头 replacement bit
替换作用 substitution
替泥浆量 amount of mud displacement
替身机器人 augmentor
替手 substitutor
替位 displacement
替位固溶体 substitutional solid solution
替位活动 displacement activity
替续泵 relay pump
替续常数 transfer constant
替续管 relay tube
替续器 chopper;commutator;relay
替用材料 alternate material
替用灯船 relief lightship
替用燃料 alternate fuel
替罪羊 goat

天安门 Cloud pillar

天板式溶解器 slab dissolver
天边 skyline
天波 atmoshperic wave; ionospheric wave; sky wave; space wave; spatial wave
天波传输延迟 skywave transmission delay
天波电台误差 skywave station error
天波辐射图 skywave pattern
天波改正量 skywave correction
天波干扰 skywave interference; skywave trouble
天波后向散射 skywave backscatter
天波校正 skywave correction
天波时延 skywave-delay(curves)
天波使用范围 skywave range
天波同步劳兰 skywave synchronized loran
天波同步远程导航 skywave synchronized long range navigation
天波误差 skywave error
天波误差图 skywave accuracy pattern
天波效应 skywave effect
天波修正量 skywave correction
天波作用距离 skywave operating distance
天测船位 astronomic(al) fix
天测罗经差 compass error by celo-observation
天测双星 astrometric binary
天测位置 astronomic(al) position
天测位置线 astronomic(al) position line
天车 crown block; head block; head sheave; overhead crane; overhead travel(l)ing crane; shop traveler; telfer; telpher; top-running crane; travel(l)ing overhead crane
天车大梁 runway girder
天车导轨 crane rail
天车工 crane man
天车横梁 hoist girder
天车滑道 crane runway
天车滑轮 crown pulley;crown wheel; crown sheave
天车梁 water-table beams
天车轮 crowned pulley; derrick pulley
天车木架 crown beam
天车桥架 bridge of overhead travel(l)ing crane
天车台 crown block floor; crowns nest
天车台检视口 water-table opening
天车运行接触导线 crane trolley wire
天车轴承外壳 pulley box
天车最大负荷 maximum load crown block
天成井 <石灰岩溶蚀塌陷形成的> cenote
天池风景区 crater lake scenic spot
天赤道 equinoctial
天赤道平面图 diagram on the plane of the celestial equator; time diagram
天窗 abatjour; barrel light; clear stor(e)y;dormant window;dormer;dormer window; dream hole; fanlight transom (e) window; hypaethron; karst window; lantern light; louver [louvre]; lying light; lying window; overstor(e)y; roof glazing; roof light; scuttle; shed dormer; shutter; skylight(opening); skylight window; top light;trap door;curb plate <复折屋顶两斜面相交处>;open gap between humped cuts <俗语>
天窗板 louver board
天窗玻璃 glass curved in two planes; skylight glass
天窗玻璃护栅 skylight grating
天窗玻璃压边框 glass check
天窗补给 recharge through skylight
天窗采光 clerestor(e)y lighting; illumination from skylight; lighting by skylight; skylighting; skylight lighting
天窗采光照明 louver lighting
天窗窗洞 scuttle hole
天窗窗扇 monk
天窗(窗扇)芯子 roof glazing bar
天窗的侧翼 wing of a dormer; wing of a lucarne
天窗的透明或半透明塑料波纹板 rooflight sheet
天窗顶板 clerestor(e)y roof
天窗顶盖 skylight cap
天窗泛水 roof-light flashing
天窗防水压条 <瓦楞石棉板屋面> bottom glazing flashing
天窗盖 skylight cover
天窗阁楼 clerestor(e)y
天窗刮水器 roof window wiper
天窗或入孔围坎 <防水用的> coaming
天窗架 skylight frame
天窗开关弧条 skylight quadrant
天窗开关装置 skylight gear
天窗拉杆 roof glazing bar
天窗檩条 skylight purlin(e)
天窗滤光片 skylight filter
天窗气楼屋顶 clerestor(e)y roof
天窗人孔围坎 coaming
天窗通风控制 scuttle ventilation control
天窗网格 skylight net
天窗围板 skylight coaming
天窗云母 skylight mica
天窗照度 illuminance of skylight;skylight illumination
天窗照明 illumination from skylight
天的 celestial
天敌 controlling parasite;native predator;natural enemy
天底点【测】 vertical point;map plumb point;nadir; nadir point; photograph nadir point; photograph plumb point;plate nadir point;plate plumb point;plumb point;V-point
天底点法 nadir-point method
天底点辐射线 nadir radial line
天底点解析三角测量 analytic(al) nadir point triangulation
天底点三角测量 nadir-point triangulation
天底读数 nadir reading
天底观测 <子午仪校正视轴与横轴正交时用> nadir observation
天底角 nadir angle
天底距 nadir distance
天地波识别 identification of ground and sky-waves
天地销 double door bolt
天电 atmospheric electricity; atmospherics;natural wave;sferics
天电的 static
天电定位 sferics fix
天电放电 atmospheric discharge
天电放电线圈 atmospheric discharge coil
天电干扰 atmospheric disturbance;atmospheric interference; atmospheric noise;static disturbance;static noise; statics
天电干扰电平 static level
天电干扰哼声 static hum
天电干扰声 grinders
天电干扰限制器 static eliminator; static limiter
天电干扰消除器 static eliminator
天电干扰抑制器 atmospheric suppressor;static suppressor
天电衡消器 static balancer
天电极 static level
天电记录器 lightning recorder
天电扰乱 atmospherics
天电突增 sudden enhancement of atmospherics
天电学 sferics[spherics]
天电噪声 atmospheric noise;sky noise;static noise
天顶 ceiling【建】;vertex【复 vertices/vertexes】;zenith【天】
天顶泵 zenith pump
天顶赤纬 declination of zenith
天顶大气吸收 zenith absorption
天顶大气折射 zenithal refraction
天顶的 zenithal
天顶等积投影 Lambert's projection; zenithal equal-area projection
天顶等距投影 zenithal equidistant projection
天顶点 zenith point
天顶读数 zenith reading
天顶季雨 zenithal rain
天顶间断 zenith discontinuity
天顶角 angle at zenith;zenith(al)angle
天顶角变化 <宇宙射线强度的> zenith-angle variation
天顶角分布 zenith-angle distribution
天顶角图 zenithal angle diagram
天顶角效应 zenith-angle effect
天顶距 zenith angle;zenith distance
天顶距测量 zenith distance measurement
天顶距观测 zenith measurement
天顶蓝 <淡紫光蓝色> zenith blue
天顶棱镜 zenith prism

天顶流星出现率 zenithal hourly rate
天顶每时出现率 zenith hourly rate
天顶目镜 zenith eyepiece
天顶扇形仪 zenith sector
天顶摄影 zenith photography
天顶摄影机 zenith camera
天顶水准器 vertical index level
天顶太阳 zenith sun
天顶筒 zenith tube
天顶投影(法) azimuthal projection; zenithal projection
天顶投影海图 azimuthal chart; zenithal chart
天顶望远镜 zenith telescope
天顶位置 zenith position
天顶吸引 zenith attraction
天顶星 zenith star
天顶星等 zenith magnitude
天顶仪 zenith instrument; zenith telescope; zenith tube
天顶引力 zenith attraction
天顶照相仪 zenith astrograph
天鹅绒 velour; velvet
天鹅绒地毯 velvet carpet
天鹅绒短绒纸 velvety flock paper
天鹅绒挂毯 velvet tapestry carpet
天鹅绒似的 velvety
天鹅绒织物 brocade
天鹅式管座 swan socket
天蛾 sphinx moth
天蛾科 <拉> Sphingidae
天蛾属 <拉> Sphinx
天帆横桁式船 skysail-yarder
天赋自然资源 natural resource endowment
天盖 awning; canopy
天盖式车顶 canopy top
天干 celestial stem; heavenly stem
天宫图 horoscope
天沟 caves gutter; crown ditch; drain(age) gutter; eaves gutter; gutter; intercepting channel; intercepting ditch; overhead ditch; rain gutter; roof gutter; roof valley; launder <俗称>; laced valley <瓦片犬牙交错铺砌的>
天沟板 rainwater gutter board; valley slab
天沟边角瓦 gutter corner tile
天沟薄片 rainwater gutter sheet
天沟槽 valley gutter
天沟侧板 cant board; gutter plate
天沟椽 valley rafter
天沟椽子 sleeper
天沟次椽木 valley jack
天沟次椽条 valley jack rafter
天沟挡水板 gutter bed
天沟底板 valley board
天沟底面 valley soffit
天沟端瓦 valley cut
天沟泛水 valley flashing; valley flashing piece
天沟防雨板 valley flashing
天沟分水线 high point of gutter; ridge of gutter
天沟钢吊 gutter steel hanger
天沟钢托 gutter steel bearer
天沟钩钉 rainwater gutter hook
天沟梁 valley girder
天沟两侧挂瓦条 valley side batten
天沟两侧瓦下的间隙 clearance under valley tiles
天沟螺栓 gutter bolt
天沟排水口 gutter outlet
天沟披水 valley flashing
天沟平口泛水 valley soaker
天沟坡度 valley slope
天沟水斗 cistern head
天沟水密接头 rust joint
天沟托板 gutter board; layer board; lear board; leat board
天沟托架 gutter bearer; gutter bracket
天沟托木 gutter bearer
天沟瓦 gutter tile; valley clay roof(ing) tile; valley tile
天沟网罩 wire gutter top
天沟屋面板 valley shingle
天沟下托梁板 gutter plate
天沟檐槽 valley channel
天沟支承 gutter support
天沟支脚 gutter hanger
天沟支柱 valley post
天光 daylight; skylight
天光光谱 sky spectrum
天光滤光镜 skylight filter
天光滤光片 skylight filter
天河【天】 Milky Way; ceiling river; galaxy
天河石 amazonite; amazon stone
天河石化 amazonitization
天候龟裂 exposure cracking
天候信号器 aeroclinoscope
天弧拱 coved and flat ceiling
天花 smallpox; variola
天花板 ceiling; boarded ceiling; ceiling board(ing); ceiling panel; overhead; planceer; lacunaria <庙宇内殿外走廊的>; camp ceiling <斜的或凸圆的>; overhead
天花板安装 ceiling installation
天花板凹槽 troffer
天花板边缘的金属支托 metal tray
天花板材料 ceiling material
天花板出风口 ceiling outlet
天花板出入口 hatchway
天花板的平线圈 plain loop for ceilings
天花板的镶板 laquear
天花板灯线盒 ceiling rosa
天花板电线匣 rosette
天花板吊杆 ceiling hanger
天花板吊扇 ceiling fan
天花板防护条 ceiling guard
天花板辐射供暖 radiant ceiling heating
天花板格栅 ceiling joist
天花板格 travis
天花板格栅梁 <古罗马> trabes
天花板格子 ceiling lattice work
天花板管道 ceiling duct
天花板回风口 ceiling return grille
天花板空气散流器 ceiling air diffuser
天花板离楼地面高度 ceiling height
天花板梁 ceiling beam
天花板漫射灯具 ceiling diffuser
天花板漫射照明系统 diffusing ceiling system
天花板木块 wood ceiling blocks
天花板上的彩画 plafond
天花板上的方形通风孔 compluvium
天花板上的方形通气孔 compluvium
天花板上空调散发器 ceiling diffuser
天花板上水平窗 lay light
天花板上信道 trapdoor
天花板透气孔 air diffuser for troffer
天花板图案 ceiling design
天花板镶板 coffer; lacunaria
天花板小梁 ceiling joist
天花板效应 ceiling effect
天花板悬挂物 ceiling hanger
天花板用装饰面砖 ornamental ceiling tile
天花板照明 ceiling lighting
天花板装饰 ornamental ceiling
天花彩画 ceiling pattern
天花反射板 ceiling reflector
天花梁 ceiling beam
天花下垂 sagging of ceiling
天花镶面板 ceiling panel(l)ing
天灰(色) sky gray; sky grey
天鸡壶 <瓷器名> chicken pot; combscomb ewer
天极高度 altitude of the pole
天极距 celestial polar distance
天极仪 polar telescope
天际线 skyline
天经加载 antenna loading
天井 atrium[复 atria/ triums]; courtyard; hypaethron; impluvium; light court; parvis; patio; raise(d shaft); small yard
天井的 hypaethral
天井的进深 depth of court
天井高度 court height; height of court
天井工作面 raise face
天井井口 raise opening
天井掘进 raise driving
天井掘进工作台 raise lift
天井溜口 raise chute
天井爬罐 raise climber
天井深孔掘进法 deep hole driving
天井式住宅 patio housing
天井支架 raise set
天井钻机 raise borer; raise borer machine; raise boring machine
天井钻进 raise boring; raise drilling
天井钻孔 raise-bore hole
天井钻孔进尺 raise-bore advance
天居 <日本建筑师菊竹清训 1958 年设计的住宅> Sky House
天空 vault; vault of heaven
天空背景 sky background
天空背景噪声 sky background noise
天空布景 <舞台上的> soffit
天空部分 sky area; sky portion
天空的 celestial; skyey
天空电波 ionosphere wave; sky wave
天空电波传输延迟 skywave transmission delay
天空风景 skyscape
天空辐射制冷系统 radiant sky cooling system
天空覆盖 sky cover
天空光 skylight
天空光偏振 polarization of the sky light
天空回散照射屏蔽 sky shine shield
天空基准 sky reference
天空景 skyscape
天空扩散辐射 diffused sky radiation
天空喇叭开关 sky-horn switching
天空蓝度 blue of the sky
天空亮度 brightness of sky; sky brightness
天空亮度温度 sky brightness temperature
天空滤光片 sky filter
天空罗盘 sky compass
天空漫(幅)射 diffuse sky radiation
天空漫射光 diffuse skylight
天空强光 sky-glare
天空清晰度 clearness number of sky
天空情况 sky condition
天空区间 skyspace
天空(散射)辐射 sky radiation
天空扫描光度计 sky scanning photometer
天空射电温度 radio sky temperature
天空实验室 skylab; sky laboratory
天空实验室对地摄影机 skylab earth terrain camera
天空实验室摄影 skylab photography
天空视见线 no-sky line
天空试验室 sky laboratory
天空投射系数 sky factor
天空文字 sky-writing
天空因数 sky factor
天空噪声温度 sky noise temperature
天空照度 skylight illumination
天空照亮 sky glow
天空状况 sky condition; state of the sky
天蓝符山石 fresno
天蓝计 cyanometer
天蓝计量 cyanometry
天蓝(色) azure; blue colo(u)r of sky; cerulean blue; navy blue; Paris blue; sky-blue; blue; cerulean; lapis lazuli
天蓝色玻璃 skylight window
天蓝色的 cerulean; sky-blue
天蓝石 azure spar; azure stone; berkyite; blue spar; false lapis; lazulite; mollite; tyrolite
天蓝釉 sky blue glaze
天蓝釉盘 <瓷器名> dish with sky blue glaze
天绿色 sky green
天轮 head pulley; head sheave; hoisting sheave; pull wheel
天轮架 pulley frame
天轮平台 sheave wheel platform
天轮驱动 head pulley drive
天轮水平 sheave wheel horizon
天轮销 head pulley pin
天罗盘 sky compass
天落水 meteoric water
天(门)冬氨酸 aspartic acid
天门冬氨酰胺 asparagin(e)
天幕 awning; backcloth; back drop; canopy; cyclorama; marquise; velarium
天幕边绳 leech rope
天幕边索 awning jackstay; bending jackstay
天幕叉口 sharks' mouth
天幕撑架 awning brace
天幕灯 strip light lamp
天幕灯槽 horizon light
天幕吊板 eufroe; euphroe; uphroe
天幕桁 awning boom
天幕横木 awning spar
天幕甲板 awning deck; hurricane deck; hurricane roof; promenad deck
天幕角孔 dog ear
天幕角眼 bull earing
天幕结 magnus hitch; Roband hitch
天幕捆绳 awning stops
天幕毛虫 <拉> malacosoma neustria testacea
天幕通桅孔 sharks' mouth
天幕下斜的压绳 housing line
天幕缘材 awning stretcher
天幕柱 awning stanchion
天幕纵梁 ridge pole; ridge spar
天幕纵梁支柱 ridge support
天幕纵木 awning rafter; awning ridge
天南极 south celestial pole
天南星 rhizoma arisaematis
天牛 longhorn beetle; longicorn; stem borer
天牛科 Cerambycid(beetle); Cerambycidae <拉>
天牛科昆虫 Capricorn-beetle
天棚 canopy; kitchen hood; sun shade
天棚按钮 ceiling button
天棚板 ceiling slab
天棚的平线圈 plain loop for ceilings
天棚灯 ceiling fitting; ceiling lamp; dome lamp; dome light
天棚电加热板 electric(al) heating ceiling panel
天棚电气采暖装置 electric(al) ceiling heating device
天棚防火层 ceiling protection
天棚开关 canopy switch
天棚拉线开关 ceiling switch
天棚梁 ceiling rafter
天棚通风机 ceiling ventilator
天棚照明 dome illumination; skylight
天棚照明灯具 dome lamp fixture

天棚照明配件 ceiling lamp fixture
天棚照明设备 ceiling fitting
天篷 awning;canopy;canopy top;velarium
天篷边绳 bolt rope
天篷侧檐 awning curtain
天篷顶框 sperver
天篷桁 awning boom
天篷脊梁 awning rafter
天篷甲板 awning deck
天篷甲板船 awning deck vessel
天篷肋骨 canopy frame
天篷支柱 awning stanchion
天篷装饰法 <能产生天篷低矮幻觉的> drop ceiling
天平 balance;beam balance;lever balance;measuring balance;scale
天平臂 balance arm;balance bar;balancing arm
天平测量 balance measurement
天平秤 balance type scale
天平秤杆 scale beam;scale arm
天平动 libration
天平动理论 libration theory
天平动偏异 libration deviation
天平动椭圆 libration ellipse
天平动效应 libration effect
天平动轴 librational axis
天平读数 balance reading
天平法 <测密度> balance(d) method
天平法测量 balance measurement
天平砝码 balance weight; balancing weight
天平杆 balance arm;rocker-bar;weigh-beam
天平杆给棉调节器 pedal evener
天平杆给棉装置 pedal feed motion
天平杆横轨 pedal rail
天平杆自动均匀装置 self-acting piano link regulator
天平架 balance column
天平校准 balance calibration
天平梁 balance beam;equalizer;scale beam
天平零点 balance zero
天平零件 balance parts
天平罗拉 pedal roller
天平盘 pan of balance;scale pan
天平盘托 scale pan arrester
天平示数 balance reading
天平式摆动 balance-like oscillations
天平式松包机 pedal bale breaker
天平室 balance room
天平调节装置 pedal regulating motion
天平托盘 balance pan arrest
天平系统 balance system
天平游码 balance rider;rider
天平预报 weather forecast(ing)
天平预报器 weather detector
天平支点 balance pivot
天平指针 balance indicator
天启瓷 Tianqi porcelain
天气 weather
天气报告 synoptic(al) report;weather actual;weather report
天气变动 fluctuation of climate
天气变化 weather change; weather variation
天气变坏 break-up
天气标志 weather sign
天气不正常 rough weather
天气参数 weather parameters
天气测定法 sferics
天气尺度 synoptic(al)-scale
天气尺度风 synoptic(al)-scale wind
天气尺度过程 synoptic(al)-scale process
天气尺度环流系统 synoptic(al)-scale circulation system
天气尺度流 synoptic(al)-scale current;synoptic(al)-scale flow
天气尺度预报 synoptic(al)-scale forecast
天气尺度运动 synoptic(al)-scale motion
天气船 weather ship
天气的 synoptic
天气电码 synoptic(al) code;weather code
天气发展 weather development
天气反常 weather anomaly
天气分 synoptic
天气分类 weather type[typing]
天气分析 synoptic(al) analysis; weather analysis
天气分析表 synoptic(al) table
天气分析电码 weather analysis in abbreviation form
天气分析和气候监测 weather analysis and climate monitoring
天气服务 weathering service
天气符号 weather notation; weather sign
天气概况图 synoptic(al) chart
天气干旱 atmospheric drought
天气公报 forecast bulletin; weather bulletin
天气观测 synoptic(al) observation; weather observation
天气观测站 synoptic(al) station; weather-observing station
天气观测站网 synoptic(al) network
天气广播 weather broadcast(ing)
天气过程 synoptic(al) process;weather process
天气和技术趋势 weather and technology trend
天气和气候监测 weather and climate monitoring
天气和气候小组 panel on weather and climate
天气滑尺 sliding scale
天气及气候监测 weather and climate monitoring
天气急变 snap
天气集合 collective;sequence
天气记录 weather recording
天气监测 weather monitoring
天气监视雷达 weather surveillance radar
天气警报 weather warning
天气控制 weather control; weather modification
天气控制咨询委员会 Advisory Committee on Weather Control
天气雷达 weather radar
天气雷达资料处理分析器 weather radar data processor and analyser [analyzer]
天气类型 weather ultimate
天气良好工作日 weather working days
天气模式 synoptic(al) model;weather model;weather pattern
天气模式变化 change in weather pattern
天气偏航调整 weather yaw adjustment
天气气候学 synoptic(al) climatology
天气气象图 synoptic(al) weather chart
天气情报 weather information
天气情报接收网 weather information network and display
天气情报网与显示 meteorologic(al) information network and display; weather information network and display
天气情报委员会 Commission for Synoptic(al) Weather Information
天气情况 meteorologic(al) situation; synoptic(al) situation; weather condition;weather situation
天气晴朗 clear weather
天气区划分 weather divide
天气区界限 weather parting
天气趋势 weather outlook; weather prospect;weather trend
天气扰动 synoptic(al) disturbance; weather disturbance
天气时件 weather event
天气试验 synoptic(al) experiment
天气数据通信[讯]网 weather data communication network
天气数值预报 numeric(al) weather forecasting
天气条件 weather condition
天气条件下可能工作天数 weather permitting days
天气调整装置 weather adjustment
天气图 meteorologic(al) chart; synoptic(al) chart <大范围的>;synoptic(al) map; weather chart; weather map
天气图报告 synoptic(al) report
天气图的 synoptic
天气图电码 synoptic(al) code
天气图方法 analysis of weather map; synoptic(al) method for weather forecasting
天气图分析 synoptic(al) analysis; weather chart analysis; weather map analysis
天气图类型 weather map type
天气图模型 synoptic(al) model
天气图气候学 synoptic(al) climatology
天气图统计分析 statistic(al) analysis of synoptic(al) chart
天气图预报 synoptic(al) forecast
天气卫星 weather satellite
天气稳定晴朗 set fair
天气系统 weather system
天气现象 weather phenomenon
天气信号 weather signal
天气形势 general weather situation; meteorologic(al) situation; synoptic(al) situation; weather prospect; weather situation
天气形势图 synoptic(al)(weather) chart
天气形势预报 forecasting of synoptic(al) position; general inference; weather prognostics
天气型 synoptic(al) type; weather-map type;weather type
天气型的变化 change in weather pattern
天气学 synoptic(al) meteorology; synoptics
天气学委员会 Commission for Synoptic(al) Meteorology
天气循环 weather cycle
天气严酷的 inclement
天气演变 weather development;weather modification
天气演变序列 weather sequence
天气谚语 weather lore;weather maxim;weather proverb
天气要素 weathering element
天气异常 weather anomaly
天气因素 weather factor
天气阴 dull
天气阴沉 loury
天气阴湿的 soupy
天气影响 weather effect
天气预报 meteorologic(al) report; weather advisory warning;weather cast;weather forecast(ing);weather prediction; weather prognosis; weather prognostics; weather report
天气预报局 Weather Service Offices
天气预报器 weather detector;weather prophet
天气预报所 weather forecasting center[centre]
天气预报图 prebaratic chart; prognostic chart
天气预测 aeromancy;weather prediction
天气预告 aeromancy;weather prediction;weather prognosis
天气预兆 weather prospect; weather sign
天气允许 weather permitting
天气灾害 weather hazard
天气(造成的)效果 weathered effect
天气展望 weather outlook; weather prospect
天气站 weather station
天气侦察飞机 weather plane
天气侦察飞行 weather reconnaissance flight
天气之岸 weather shore
天气中心 weather center[centre]
天气重现 weather recurrences
天气周期 spell of weather
天气骤变 sudden change of weather
天气状况 weather condition;weather regime
天气资料 weather data; weather information
天气资料加工 weather data processing
天气资料整编 weather data processing
天气(字码)符号 weather symbol
天气自然作用试验 natural outdoor weathering test
天气阻挠 weather-bound;weather fast
天堑 natural chasm
天桥 catwalk; cross-bridge; flying bridge;foot bridge;footpath bridge; monkey bridge; overbridge; overhead floor bridge;overline bridge;overpass;passenger foot-bridge;platform bridge;transition bridge;walkway;fly door <舞台的>;fly gallery <舞台的>;fly stair(case) <舞台的>
天桥爬梯 fly ladder
天青钴矿 julienite
天青黑云岩 girekenite
天青蓝 azure blue;azurite blue;copper blue
天青蓝颜料 celestial blue pigment
天青色 azure
天青色料 celeste blue
天青石 celestine; celestite; lazurite; schatzite;zolestin
天青石含量 celestine
天青石胶结物 celestite cement
天青石结核 celestite concretion
天青石矿石 celestite ore
天青石蓝 lapis lazuli blue
天青石色 lapis lazuli
天青釉 celeste glaze;Tianqing glaze
天穹 celestial concave; heavenly sphere
天球 celestial sphere;coelosphere
天球北极 north celestial pole
天球赤道 celestial equator; equator; equinoctial
天球赤道坐标系(统) celestial equator system of coordinates
天球地平(圈) astronomic(al) horizon;celestial horizon;rational horizon
天球极 celestial pole
天球历书极 celestial ephemeris pole

天球两极 poles of the heaven
天球南极 south celestial pole
天球平面图 planisphere
天球瓶 vault-of-heaven vase
天球切面 plane of the sky
天球三角形 celestial triangle
天球投影图 projection of the celestial sphere
天球纬圈 astronomic(al) parallel;celestial parallel
天球仪 celestial globe;cosmosphere;star globe
天球子午圈 celestial meridian;principal vertical circle
天球子午线 astronomic(al) meridian;celestial meridian
天球坐标 celestial coordinates;coordinates of celestial sphere
天球坐标系 celestial coordinate system
天然安息角 natural angle of repose
天然岸坡 natural bank;raw bank
天然凹陷 natural depression
天然坝 avalanche dam;crude dam;natural dam
天然白垩 natural chalk
天然白铅 native white lead
天然白土 natural clay
天然白云母 natural muscovite
天然棒 natural rod
天然宝石 natural gemstone
天然保护层 < 如水、土 > natural containment
天然保护区 landscape reservation
天然本底 natural background
天然本底放射性当量活度 natural background radiation equivalent activity
天然本底辐射 natural background radiation
天然本色毛纱 natural gray yarn
天然本色织物 natural-tinted fabric
天然比尺 natural scale
天然比尺模型 natural scale model
天然比例 natural scale
天然比例模型 natural scale model
天然铋 native bismuth
天然避风港 natural haven
天然边坡 natural slope
天然变化 natural change
天然变色 mineral streak
天然标石 natural monument
天然表面 self-faced
天然表面层色 natural finish
天然表面石材 self-faced stone
天然表土面层 topsoil surfacing
天然冰 natural ice
天然冰晶石 ice spar;ice stone
天然波纹 natural moire
天然波状土 self-wallowing soil
天然玻璃 natural glass
天然剥理【地】crude foliation
天然播种再生 natural regeneration by seeds
天然铂 native platinum;platina
天然铂钯矿矿石 native Pt-Pd ore
天然薄层砂岩 natural flag(stone)
天然补给量 natural recharge
天然材料 natural material
天然财富 natural resources;natural wealth
天然裁弯(段)natural cutoff
天然采光 daylight illumination;daylighting;natural illumination;natural lighting
天然采光的 naturally lighted
天然采光计算 daylight calculation
天然采光率 daylight ratio
天然采光系数 daylight factor;natural illumination factor
天然彩色摄影 natural colo(u)r photography
天然彩色陶瓷 natural colo(u)red earthenware
天然(彩)色照片 heliochrome
天然糙率 natural roughness
天然草地 native meadow;native pasture;natural grassland;natural meadow
天然草皮 natural sod
天然侧石 natural curbstone
天然层 natural bed;natural layer;quarry bed
天然层理 natural seam
天然产地的建筑用砂 building sand from natural sources
天然产卵场 natural spawning place
天然产品 native products;natural product
天然产物 natural product
天然产状 natural occurrence
天然场法 natural field method
天然(沉积)坡度 depositional gradient
天然沉积物 natural deposit;natural sediment
天然辰砂 native vermillion
天然衬底 native substitution;native substrate
天然成双石像 natural stone mate figure
天然澄清 natural clarification
天然尺寸 natural scale;natural size;physical dimension;physical size
天然充水 natural recharge
天然充水通道 passage of natural water filling
天然冲积堤 levee
天然稠度 natural consistency
天然稠度试验 natural consistence [consistency] test
天然臭氧 natural ozone
天然出(水)口 natural outlet
天然储藏量 natural reserves
天然储库 natural reservoir
天然储量 nature reserves
天然纯清漆 natural clear varnish
天然磁石 loadstone;lodestone;natural lodestone
天然磁铁 loadstone;lodestone;native magnet;natural magnet
天然磁性 natural magnetism;spontaneous magnetism
天然粗骨料 natural coarse aggregate
天然粗集料 natural coarse aggregate
天然粗砂 natural coarse sand
天然醋酸胺 acetamide
天然簇叶 natural foliage
天然催化剂 natural catalyst
天然脆沥青 grahamite
天然大分子絮凝剂 natural macromolecular flocculant
天然大块石 natural stone block
天然大理石 natural marble
天然大理石荒料 natural marble block
天然大砾石 native boulder
天然大气 clean atmosphere;natural atmosphere
天然大气分散胶体 natural atmospheric dispersion;natural atmospheric dispersoid
天然淡水水体污染 contamination of natural freshwater bodies
天然的 crude;inartificial;native;natural;naturalistic
天然的防冲刷面层 erosion pavement
天然的热贮藏 natural geothermal reservoirs
天然的水凝砂浆 natural hydraulic mortar
天然的透明石英 rock crystal
天然的物质 < 尤指石油、石棉等 > crude
天然等离子体 natural plasma
天然堤 levee;natural dike[dyke];natural levee
天然堤沉积 levee deposit
天然堤后沉积 backland deposit
天然堤后泛滥地 backland
天然堤后湖 lateral levee lake
天然堤相 levee facies
天然底盘 natural bed
天然抵抗线 natural line of resistance
天然砥石 gritstone
天然地基 natural base;natural foundation;natural ground;natural subgrade;natural subsoil;untreated foundation
天然地基壤 subsoil
天然地基允许荷载 subsoil permissible load
天然地沥青 bitusol;native asphalt;pit asphalt;rock asphalt;natural asphalt;original asphalt
天然地沥青质焦沥青 native asphaltic pyrobitumen
天然地面 native ground;natural ground(surface);open ground;original ground
天然地面标高 natural ground level
天然地面高程 natural ground level;natural surface level;original ground level
天然地面线 natural ground line;natural surface line
天然地平 natural grade
天然地平线 ground line;natural grade line
天然地热井 natural geothermal well
天然地热电站 natural generation of electric(al) power
天然地热系统 natural geothermal system
天然地热蒸汽 direct geothermal steam
天然地势 natural feature
天然地物 natural feature;physical feature
天然地物点 natural point
天然地下灌溉 natural subirrigation
天然地下渗灌 natural subirrigation
天然地下水 native groundwater
天然地下水面 natural water table
天然地下水位 natural groundwater level;natural water table
天然地形 natural feature;physical feature
天然地震 natural earthquake;nature shock
天然地震学 earthquake seismology
天然地质环境 geologic(al) norm
天然地质力 natural geologic(al) force
天然碲化银 hessite
天然靛青 natural indigo
天然叠标 natural range
天然动物园 natural zoological garden;paradise
天然冻结 weather freezing
天然冻结法 natural freezing method
天然洞室 natural cavern
天然洞穴 cavity;natural cavity
天然陡坡 bluff
天然断层 natural fault
天然堆石堤 natural rock-fill dike [dyke]
天然凡士林 natural vaseline
天然矾土水泥 natural bauxite cement
天然反式聚异戊二烯 natural trans-polyisoprene
天然方解石粉 marble white
天然防护层 < 如水、土 > natural containment
天然防护功能 natural preventive function
天然防御工事 natural fortification
天然防治 natural prevention and treatment
天然放射系 natural radioactive series
天然放射系母体 natural parent
天然放射性 natural activity;natural radioactivity
天然放射性背景值 natural radioactive background value
天然放射性本底 natural radioactivity background
天然放射性测量 natural radioactive survey
天然放射性核 natural radioactive nucleus
天然放射性环境 natural radiation environment
天然放射性气体 natural radioactive gas
天然放射性同位素 natural radioactive isotope
天然放射性物质 naturally occurring radioactive substance
天然放射性元素 natural radioactive element
天然肥料 natural fertilizer
天然沸石 natural zeolite;permufit
天然沸石介质 natural zeolite medium
天然分隔 natural split
天然分解 < 岩石的 > natural quarrying
天然分开 natural split
天然分离 natural split
天然分裂 natural split
天然分水岭 natural water shed
天然分水线 natural water shed
天然粉笔 natural chalk
天然粉尘 natural dust
天然丰度 natural abundance
天然风 natural wind
天然风洞 gloup[gloap]
天然风化砾石 uncrushed gravel
天然风景 natural landscape;physical landscape
天然风景区 area of outstanding natural beauty
天然风桥 natural air crossing
天然浮石 natural pumice
天然浮石灰 pumicite
天然浮石混凝土 natural pumice concrete
天然辐射带 natural radiation belt
天然辐射环境 natural radiation environment
天然腐蚀试验 natural condition test
天然复壮 self-restoration
天然覆盖物 natural cover
天然伽马射线背景 natural γ-ray background
天然干容重 natural dry unit weight
天然干燥 air seasoned;air seasoning
天然干燥材 natural-seasoned wood
天然杆 natural rod
天然刚度 natural stiffness
天然刚玉 emery;Naxas emery
天然钢 natural steel
天然港(口)natural harbo(u)r;natural port
天然港湾 landlocked harbo(u)r;natural harbo(u)r;natural port;bay harbo(u)r;mutual harbo(u)r
天然海湾内的港口 port in natural bay
天然高分子 natural polymer
天然高分子絮凝剂 natural polymeric flocculant

天然高聚物 natural high polymer
天然高铝水泥 natural bauxite cement
天然更新 natural regeneration；nature regeneration；self-restoration
天然公差 natural tolerance
天然公园 landscape park；natural park；wild park
天然汞 native mercury
天然汞齐 native amalgam
天然拱 natural arch；natural self-supporting arch
天然谷 nature valley
天然骨料 crude aggregate；natural aggregate
天然骨料混凝土 natural aggregate concrete
天然固体绝缘体 natural solid insulator
天然固体燃料 natural solid fuel
天然光 daylight；natural light
天然光利用时间 serviced period of daylight
天然光源 natural light source
天然光照明 natural illumination
天然硅酸钙 natural calcium silicate
天然硅酸钙镁 tremolite
天然硅酸矿 willemite
天然硅酸铝无机物 natural aluminosilicate mineral
天然贵金属块 nugget
天然过程 natural process
天然过滤器井 natural strainer well
天然海水 natural seawater
天然海滩 natural beach
天然含沙量 natural load
天然含水当量 field moisture equivalent
天然含水量 natural moisture content；field capacity；field moisture capacity
天然含水量随深度变化图 variation of natural moisture content with depth
天然含水率 natural moisture content；natural water content
天然含油树脂 oleoresin
天然航道 natural channel；natural navigable waterway
天然合金 native alloy；natural alloy
天然合金铁 natural alloy iron
天然河岸 raw bank
天然河槽 natural channel；natural stream channel
天然河槽的槽蓄演算 storage routing in natural channel
天然河槽卡口段 natural channel control
天然河槽控制断面 natural channel control section
天然河槽中洪水储蓄路程推算 storage routing in natural channel
天然河槽中洪水储蓄路程追踪 storage routing in natural channel
天然河床 natural bed；natural stream channel；unprotected river bed；unregulated bed
天然河道 leat；natural channel；natural course of river；natural water course
天然河道沉积（物）natural river deposit
天然河堤 natural levee
天然河段 natural reach
天然河流 natural river；natural stream
天然核反应 spontaneous nuclear reaction
天然荷载 natural load
天然红丹 native minium
天然红的铅氧化物 native red oxide of lead
天然红铅 native minium
天然红土 Vandyke red
天然洪水还原 re-establishment of natural flood
天然湖滩 natural beach
天然护面石 natural face stone
天然花岗石荒料 natural granite block
天然花园 natural garden
天然化合物 native compound
天然化学性质 physio-chemical property
天然环境 natural environment
天然缓冲 natural buffer
天然灰色 natural gray
天然灰质黏[粘]土 natural adobe clay
天然回补 <地表水或雨水渗入含水层> natural recharge
天然汇水 natural catchment
天然混合骨料 all-in aggregate
天然混合集料 all-in aggregate
天然混合水泥 <波特兰水泥和火山灰的混合物> natural pozz(u)olan(a)
天然混合物 physical mix(ture)
天然混凝剂 natural coagulant
天然混凝土比热 natural concrete heat
天然混凝土骨料 natural concrete aggregate
天然混凝土集料 natural concrete aggregate
天然混凝土释放热 natural concrete heat
天然火山灰 natural pozz(u)olan(a)；natural trass；Santorin <其中的一种>
天然火山灰材料 natural pozz(u)olanic material
天然积雪 natural snow cover
天然基层 natural bed
天然基础 natural foundation
天然级配 natural grading；prototype gradation
天然级配道砟 open-graded ballast
天然级配骨料 pit-run aggregate；as-dug aggregate
天然级配集料 open graded aggregate
天然级配砾石 pit-run gravel
天然级配砂 pit sand
天然集料 natural aggregate
天然集料混凝土 natural aggregate concrete
天然纪念物 natural monument
天然技艺 natural arts
天然夹缝 <岩石中的> natural seam
天然坚硬性 natural stiffness
天然检波器 natural rectifier
天然碱 natron；natrum；natural base；trona；urao
天然碱矿床 trona deposit
天然碱流程 Trona process
天然建材勘察级别 grade of prospecting of natural building material
天然建筑材料 natural construction material
天然建筑材料产地分布图 distribution plan of a quarry and borrow areas for natural construction
天然建筑材料勘察 engineering geologic(al) investigation of natural building materials
天然建筑材料种类 type of natural building materials
天然建筑石料 natural building stone
天然建筑用砂 natural building sand
天然浇浴水标准 natural bathing water standards
天然胶 natural glue；natural gum
天然胶合剂 natural adhesive
天然胶结剂 natural cementing agent
天然胶乳 natural latex
天然胶乳橡胶 natural latex rubber
天然胶树胶 natural gum
天然胶体 eucolloid；natural colloid
天然胶粘合剂 natural rubber adhesive
天然胶粘剂 natural adhesive；natural glue
天然胶质 lac
天然焦 carbonite；cokeite；natural coke；mineral coke
天然焦炭 burnt coal；carbonification；blind coal；cinder coal；dandered coal；natural coal；smudge coal
天然阶沿石 natural curbstone
天然界碑 natural monument
天然界限 natural margin；terrain line
天然金刚石 natural diamond
天然金刚石钻头 natural diamond bit
天然金块厚度 nugget thickness
天然金属 native metal
天然进水量 natural inflow
天然浸渍 natural retting
天然晶体 mineral crystal；natural crystal
天然井 cenote；karst well；pozo
天然景观 natural landscape；natural scenery
天然景色 natural landscape；natural scenery
天然净化过程 natural purification process
天然径流 natural flow；natural runoff；nature flow；virgin flow
天然径流场 natural runoff field
天然径流还原 re-establishment of natural flood
天然聚合体 natural aggregate
天然聚合物 natural polymer
天然聚集砾石 crushed conglomerate
天然均衡 natural balance
天然开裂面的石板 slate in natural cleft surface
天然抗病性 natural resistance
天然抗氧剂 natural oxidation inhibitor
天然柯巴脂 raw copal
天然可浮性 inherent floatability；native floatability
天然可制造水泥的岩石 cement rock
天然空气污染 natural air pollution
天然空气污染源 natural sources of air pollution
天然孔雀石绿 mountain green
天然孔隙比 natural void ratio
天然块度 native blockness
天然块金 nugget
天然快干树脂 natural copal
天然矿蜡 native mineral wax
天然矿石 natural crystal
天然矿物材料 natural mineral aggregate
天然矿物骨料 natural mineral aggregate
天然矿物集料 natural mineral aggregate
天然矿物油 natural oil
天然（矿物质）颜料 earth pigment
天然蜡 native paraffin；natural wax
天然来水量 natural inflow
天然来源的放射性 naturally occurring radio activity
天然蓝细菌水华 natural cyanobacterial bloom
天然老化 natural weathering；weather ag(e)ing
天然离散电磁波 atmospherics
天然沥青 asphaltic bitumen；bitusol；crude asphalt；dopplerite；lake asphalt；mineral caoutchouc；native asphalt；native bitumen；native pitch；natural bitumen；natural tar；pit asphalt；rock asphalt；asphaltic sand <含大量矿物质的>；courtzilite <其中的一种>；Imposite <一种不溶于松节油的>
天然沥青混合物 natural mixture of asphalt
天然沥青基底封层 native asphalt-base sealant
天然沥青色的 natural asphalt colo(u)r
天然沥青砂岩 natural bituminized sandstone
天然沥青瓦片 natural asphalt tile
天然沥青脂 crude petrolene
天然砾料 natural gravel material
天然砾石 gravel by nature；natural gravel
天然砾石和砂的混合物 natural gravel-sand mix(ture)
天然砾石和砂混合料 bank gravel-sand mix(ture)
天然砾石料 pit-run gravel
天然粒面 natural grain
天然粮仓 natural granary
天然料坑泥木 pit-run clay gravel
天然料坑砂砾石 natural pit-run gravel
天然裂缝 natural cleft；natural split
天然裂化催化剂 natural cracking catalyst
天然裂口 natural cleft；natural split
天然林 natural forest；wild wood
天然林保护工程 natural forest protection project
天然流场 natural flow field
天然流理论 natural flow theory
天然流量 natural discharge；non-regulated discharge
天然流速 ambient velocity
天然流域 natural water shed
天然硫 native sulfur；native sulphur；natural sulphur
天然硫化镉 orient yellow
天然硫矿床 native sulphur deposit
天然硫矿石 native sulfur ore
天然硫酸钡 baryte；native barium sulphate；native sulfate of barium；natural barium sulfate
天然硫酸钡骨料 barytes aggregate
天然硫酸钡集料 barytes aggregate
天然硫酸钡矿石 barytes ore
天然硫酸铜 blue stone
天然垄断性的行业 industries that are monopolistic in nature
天然露头【地】natural exposure；natural outcrop
天然露头观测点 observation point of natural outcrop
天然卤水 natural brine
天然滤井 natural strainer well
天然滤水井 natural strained well；natural strainer well
天然路 natural road
天然路边石 natural curbstone
天然路基 natural roadbed；raw subgrade
天然路面道路 natural surfaced road
天然路缘石 natural curbstone
天然绿 native green
天然氯化钠 halite
天然落差 natural fall
天然煤 blind coal；coke coal
天然煤气 natural gas；rock gas
天然煤气的储藏 natural gas storage
天然煤气的处理方法 natural gas processing
天然煤气燃烧器 natural gas burner
天然锰水化物 native manganic hy-

drate
天然面石料 rock-faced stone
天然明矾 native alum
天然膜 natural membrane
天然磨料 natural abradant; natural abrasive
天然磨石 grindstone; gritstone; natural grindstone
天然磨蚀剂 natural abradant
天然木色 natural wood colo(u)r
天然木素 native lignin; protolignin
天然木炭 mineral charcoal; motherham; mother of coal; natural charcoal
天然牧草 native grass
天然牧场 native pasture; natural feed land; natural grassland; natural grazing ground; natural pasture; wild pasture
天然内摩擦角 angle of true internal friction
天然内在增长率 intrinsic(al) rate of natural increase
天然耐火材料 natural refractory
天然耐火建筑材料 naturally refractory construction material
天然能量 energy of nature; nature energy
天然泥浆 native mud; natural mud
天然泥面线 natural mud surface
天然黏[粘]合剂 natural adhesive; natural cement; natural glue
天然黏[粘]结剂 natural adhesive; natural cementing agent
天然黏[粘]结砂 natural bond sand
天然黏[粘]土 natural clay
天然黏[粘]土砂 naturally bonded sand
天然年产量 natural turnover
天然凝聚过程 <地表空隙水的> hydrogenesis
天然凝聚力 original cohesion
天然凝析油 natural condensated oil
天然农圃 native estate
天然浓黄土 natural sienna
天然排水 free drainage; free draining; natural drainage
天然排泄 natural discharge
天然配对石像 natural stone mate figure
天然喷油井 well kicked off natural
天然硼砂 borax; sassolite; tincal; tincalconite
天然硼酸 sassolite
天然膨润土 natural bentonite
天然皮脂 natural sebum
天然劈开的石料 cleft stone
天然漂白 natural bleaching
天然频率 natural frequency
天然平衡拱 natural arch
天然平衡坡度 equilibrium slope
天然屏蔽港 natural sheltered harbo(u)r; natural sheltered port
天然屏障 natural barrier
天然坡度 natural slope
天然坡度角 natural angle of repose
天然坡角 <松散体的> angle of repose; angle of rest; natural angle of slope; angle of natural repose
天然坡面休止角 angle of repose of natural slope
天然剖面 natural profile
天然铺盖 natural blanket
天然铺路板石 natural paving flag (stone)
天然铺路砂岩薄板 natural paving flag(stone)
天然铺路石板 natural flag(stone)
天然铺路小方石块 natural paving sett
天然铺石 natural sett
天然曝气(池) natural aeration
天然曝晒 natural exposure; weathered exposure
天然曝晒试验 weather exposure test
天然漆 Chinese lacquer; japan; Rhus lacquer
天然气 casing-head gas; commercial rock gas; earth gas
天然气按成分分类 classification of natural gas according to its composition
天然气比重计 gas specific gravity meter
天然气采集系统 gathering system of natural gas
天然气槽黑 natural gas-based channel black
天然气产量 natural gas production rate
天然气产区 gas field
天然气产状 natural gas occurrence
天然气成熟度图 maturation diagram of natural gas
天然气除湿 dehumidification of gas
天然气储备 reservoir of natural gas
天然气储存 natural gas storage
天然气储量 reserves of natural gas
天然气处理法 natural gas processing
天然气处理系统 gas treating system
天然气处理者协会 <美> Natural Gas Processors Association
天然气串流 gas by-passing
天然气的含氢指数 hydrogen index of gas
天然气的开采 gas extraction
天然气的密度 density of gas
天然气的收集 natural gas gathering
天然气电力驱动 gas electric(al) drive
天然气动力站 natural gas power station
天然气发电驱动钻井装置 gas electric(al) rig
天然气发动机 natural gas engine
天然气返回 gas return
天然气分析 gas analysis
天然气干管 gas main
天然气工程与管理 gas engineering and management
天然气工业 natural gas industry
天然气管 nature gas pipe
天然气管网 natural gas grid
天然气管线 natural gas line; gas pipeline
天然气管线压气机 gas line compressor
天然气和石油相图 phase map of oil and gas
天然气滑升 gas slippage
天然气化工 gas chemical industry; natural gas chemical industry
天然气化工厂 natural gas chemical plant
天然气化学成分分析 analysis of chemical composition of natural gas
天然气回采 gas withdrawal
天然气回采量 deliverability of gas
天然气回注 gas cycling; gas re-injection; repressuring
天然气回注管线 reinjection line of gas; repressing line of gas
天然气回注井 gas injection well
天然气火炬 flaring
天然气机 natural gas engine
天然气机车 natural gas locomotive
天然气积聚 gas accumulation
天然气加工 natural gas processing
天然气加工厂凝析油 plant condensate
天然(气)井 <火井> gas well; gasser; natural gas well
天然气聚集 natural gas accumulation
天然气开采 natural gas extraction
天然气勘探 gas prospection; natural gas exploration
天然气裂解 natural gas pyrolysis
天然气露头 gas show
天然气苗 gas show; natural gas seepage
天然气凝析液 natural gas condensate
天然气凝析油 natural gas condensate
天然气泡 gas bubble
天然气膨胀液化工艺流程 natural gas expander cycle liquefaction process
天然气偏差系数 deviation coefficient; gas deviation factor
天然气(气)田 natural gas field
天然气汽车 gas-fueled vehicle
天然气汽油厂 casing-head plant
天然气泉 natural gas spring
天然气燃烧器 natural gas burner; neat gas burner
天然气溶于石油 solubility of gas in oil
天然气砂(岩) gas sand; natural gas sand
天然气渗透率 gas permeability
天然气生产井 producing gas well
天然气输送 natural gas transportation
天然气水化合物 gas hydrate
天然气水溶液 gas in water solution
天然气弹性压缩能 elastic compression energy of gas
天然气炭黑 gas-produced black
天然气提纯 natural gas refining
天然气体回注 gas injection
天然气体积换算系数 volume conversion factor of natural gas
天然气体积热膨胀系数 thermal expansivity of gas volume
天然气体汽油 natural gas gasoline
天然气田 gas field
天然气田开发 gas field exploitation
天然气同位素分析 analysis of stable isotope of natural gas
天然气推动力说 propulsive force theory of gas
天然气脱水 removal of water in gas
天然气脱水装置 gas dewatering device
天然气物理参数的测量 measurement of physical parameter of natural gas
天然气物理性质 physical property of natural gas
天然气系统 natural gas system
天然气相对渗透率 gas relative permeability
天然气消耗量 gas consumption
天然气销售合同 gas sale agreement
天然气压气站 gas compressor station
天然气压缩机 natural gas compressor
天然气样瓶 gas sampling bottle
天然气液化 natural gas liquefaction
天然气液化厂 natural gas liquefaction plant
天然气液化汽油 gas gasoline
天然气液料 natural gas liquid
天然气液态产物 natural gas liquid
天然气运移 gas migration
天然气蕴藏量 gas reserve
天然气增味剂 natural gas odorant
天然气政策法 natural gas policy act
天然气终端 natural gas terminal
天然气主要物理性质参数 physical parameter of natural gas
天然气贮藏 natural gas deposit
天然气资源 natural gas resource
天然气组分分析设备 gas composition analysis equipment
天然气钻进 gas drilling
天然汽油 casing-head gasoline; natural gasoline
天然汽油厂 natural gasoline plant
天然汽油稳定塔 natural gasoline stabilizer
天然铅丹 native minium; native red oxide of lead
天然潜水面 natural water table
天然潜洲 natural bar
天然浅滩 natural bar
天然桥 natural bridge
天然青蓝 natural ultramarine
天然轻(浮)石 natural pumice
天然轻骨料 natural lightweight aggregate
天然轻集料 natural lightweight aggregate
天然倾斜 original dip
天然倾斜角 natural angle of slope
天然清漆树脂 natural varnish resin
天然曲材 grown; natural crook timber
天然渠道 natural channel
天然群青 genuine ultramarine; lapis lazuli
天然燃料 natural fuel
天然染料 natural dyestuff
天然热储流体 natural reservoir fluid
天然热绝缘软木板 natural cork insulating board
天然热流 natural heat flow
天然热流量 natural heat output
天然热释光法 nature heat releasing light method
天然热资源 natural heat resource
天然容许度 natural tolerance
天然容蓄 <湖泊沼泽洼地等的> natural storage
天然容重 natural unit weight
天然溶质 natural solute
天然乳化液 natural emulsion
天然乳剂 natural emulsion
天然乳胶 crude emulsion; natural emulsion
天然乳液 crude emulsion; natural emulsion
天然入流量 natural inflow
天然软瓷 natural soft porcelain
天然软沥青 pissasphalt(um)
天然软木 natural cork
天然软土层 natural soft deposit
天然三相土 realistic two-phase soil
天然色 autochromatic colo(u)r; natural colo(u)r; physical colo(u)r; real colo(u)r; technicolo(u)r; trichromatic
天然色的 natural colo(u)red
天然色地形表示 natural colo(u)r-relief presentation
天然色接收 natural colo(u)r reception
天然色料 natural colo(u)ring matter
天然色素 natural colo(u)ring matter; natural pigment
天然色釉面砖 natural finish tile
天然色照片 autochrome
天然色照相 autochrome
天然色照相法 natural colo(u)r photography
天然色照相干板 autochrome
天然色照相术 heliochromy
天然森林 natural forest; natural woods
天然杀虫剂 natural insecticide
天然沙坝 natural bar
天然沙滩 natural bar; natural beach
天然沙土 top soil
天然沙洲 natural bar
天然砂 natural sand; pit-run sand; pit

sand
天然砂料 quarry-run sand
天然砂土 top soil
天然砂土路 topsoil road
天然砂土面层 topsoil surfacing
天然砂细骨料 natural sand fine aggregate
天然砂细集料 natural sand fine aggregate
天然砂洲 natural bar
天然山水旅游区 recreation resources
天然砷 native arsenic
天然深度 natural depth
天然渗透率 natural permeability
天然生产业 <矿业、农业、渔业等> extractive industries
天然生物处理 natural biologic(al) treatment
天然声呐 natural sonar
天然剩余磁化强度 natural remanent magnetization
天然湿地 natural wetland
天然湿地处理系统 natural wetland treatment system
天然湿度 field moisture capacity;natural moisture content;natural wetness
天然石板 natural flag(stone)
天然石板祭桌 natural stone table altar
天然石板瓦 natural slate tile
天然石板装饰件 natural slate decorating unit;Excelate <商品名>
天然石壁缘 natural stone frieze
天然石材 natural stone
天然石材铺砌工 natural stone layer
天然石层 natural bed;natural stone course;natural stone layer;quarry bed
天然石撑 natural stone corbel
天然石城堡 natural stone keep
天然石凳 natural stone bench
天然石垫层 natural stone matting
天然石垫块 natural stone matting
天然石垫子 natural stone matting
天然石雕带 natural stone frieze
天然石雕刻品 natural stone sculpture
天然石雕塑品 natural stone sculpture
天然石顶 natural stone coping
天然石方工程 natural stonework
天然石粉状沥青胶泥 powdered natural rock asphalt mastic
天然石膏 gypsum rock;massive gypsum;natural gypsum;plaster rock;rock gypsum;untreated calcium sulphate
天然石拱 natural stone arch
天然石含水量 quarry sap;sap
天然石灰 natural lime
天然石灰硅砂水泥 natural lime-silica cement
天然石灰石 natural limestone
天然石基础 natural stone foundation
天然石槛 natural stone threshold
天然石阶 natural stone bench
天然石块铺面 natural stone-block paving
天然石块铺砌 natural stone-block paving
天然石块铺砌的路 road with paving of natural setts
天然石蜡 native paraffin;natural paraffin wax
天然石料 natural rock;natural stone
天然石料擦光机 natural stone polishing machine
天然石料裂开器 natural stone splitter
天然石料抛光机 natural stone polishing machine
天然石料琢面机 natural stone dressing machine
天然石棉 crude asbestos
天然石面抛光机 natural stone polishing machine
天然石磨机 natural stone rubbing machine
天然石墨 native graphite;natural graphite
天然石墨电极 natural electrode
天然石墨电刷 natural graphite brush
天然石墨粉 natural graphite powder
天然石头覆盖物 natural stone capping
天然石头祭坛 natural stone altar
天然石托 natural stone corbel
天然石瓦 natural stone tile
天然石纹理 natural stone texture
天然石英 natural quartz
天然石油 natural oil
天然石油沥青 native asphalt;original asphalt;pit asphalt;rock asphalt
天然石油沥青混合物 natural mixture of asphalt
天然石油沥青路面 rock asphalt pavement
天然石油裂化气 crude oil gas
天然石油气 commercial rock gas
天然石中含水量使表面硬化 caseharden(ing)
天然石柱 natural stone column
天然石柱型 <尤指古典建筑的柱型> natural stone order
天然石装修 dressed fair face
天然时效 natural ag(e)ing;seasoning;weathering
天然时效裂纹 season crack
天然蚀像 natural etched figure
天然食物 natural food
天然示踪剂 environmental tracer;natural tracer
天然示踪物 natural tracer
天然饰面材料 natural finish
天然饰面石 natural face stone
天然收成 fructus naturales
天然输沙量 <稳定河流的> natural load
天然树胶 natural gum
天然树脂 elemi;natural copal;natural resin;natural wood resin;wood resin
天然树脂改性酚醛树脂 natural resin modified phenol resin
天然树脂基的砂胶 plant resin-based mastic
天然树脂基混凝土养护剂 plant resin-based concrete curing compound
天然树脂基玛琋脂 natural resin-based mastic;plant resin-based mastic
天然树脂基腻子 natural resin-based mastic
天然树脂基养护剂 natural resin-based concrete curing agent
天然树脂胶合剂 natural resin adhesive
天然树脂胶水 natural resin glue
天然树脂胶粘水泥 natural resin mastic
天然树脂胶粘物 natural resin glue
天然树脂黏[粘]合 natural resin bonding
天然树脂黏[粘]合剂 natural resin adhesive
天然树脂黏[粘]结剂 natural resin adhesive
天然树脂漆 natural resin paint
天然树脂清漆 natural resin varnish
天然竖坑 gloap
天然衰变 natural disintegration
天然水 native water;natural water;conjunction water;connate water;crude water;raw water
天然水变质 metamorphism of natural water
天然水成因类型 genetic(al) classification of nature water
天然水成因系数 genetic(al) factor of natural water
天然水道 leat;natural water course;natural waterway
天然水含量 natural water content
天然水化学特征 chemical characteristics of natural water
天然水晶 mineral crystal
天然水井 natural well
天然水库 natural reservoir;natural storage
天然水类型 natural water categories
天然水利资源 natural water resources
天然水流 natural flow
天然水面线 natural water line
天然水泥 <用一种天然泥土制成的水泥,类似水硬性石灰> natural cement;eminently hydraulic lime;Parker's cement;Roman cement
天然水泥产地 cement deposit
天然水泥灰岩 natural cement rock
天然水平基床 natural bedding
天然水深 <当水流流量、比降、糙糙系数和横断面都保持不变时的水深> normal depth of flow;natural depth
天然水生腐殖质 natural aquatic humic substance
天然水生生态系统 natural aquatic system
天然水生植物 native aquatic plant
天然水水质 natural water quality
天然水体缓冲能力 buffering capacity in natural water bodies
天然水体酸化 acidification of natural waters
天然水体系 natural aquatic system
天然水体氧化还原作用 oxidation and reduction in natural water bodies
天然水头 natural head
天然水位 natural level of water;natural water table
天然水位的平均降速 average velocity of nature water level drawdown
天然水污染 pollution of natural water
天然水系 natural water way
天然水异常 water anomaly
天然水域 free waters;natural waters
天然水域生态系统 ecosystem of natural water bodies
天然水源 natural water source
天然水质量 quality of the natural water
天然丝 natural silk
天然饲料地 natural forage basis
天然苏打 sal soda;trona
天然塑料 natural plastic
天然塑料性材料 natural plastic material
天然酸 natural acid
天然态 native state
天然滩肩 natural berm
天然炭 natural carbon
天然碳氢化合物 crude hydrocarbon;native hydrocarbons;natural hydrocarbon
天然碳酸钙 chalk;ground oyster shell;natural whiting
天然碳酸镁 magnesite
天然碳酸铅 cerus(s)ite
天然淘汰 natural selection
天然特权 natural privilege
天然体系 natural system
天然体质颜料 natural extender
天然填石堤 natural rock-fill dike [dyke]
天然条件 in-situ condition;natural condition;natural endowment
天然调节 natural regulation
天然调节河(流) naturally regulated river
天然铁 native iron;natural iron
天然铁黄 Mars yellow
天然铁镍合金 taenite
天然铁氧体 natural ferrite
天然铁棕颜料 Caledonian brown
天然通道 natural lane
天然通风 natural draft ventilation
天然通航河道 natural navigable waterway
天然通航水道 natural navigable waterway
天然同位素 natural isotope
天然同位素丰度 natural isotopic abundance
天然同位素组成 natural isotopic composition
天然铜 native copper;natural copper;virgin copper
天然涂料 natural finish
天然土 natural earth;natural soil
天然土表面 natural earth surface
天然土层 natural soil stratum
天然土机场 natural ground airfield
天然土路 dirt road;earthen road;natural road;natural soil road;primitive road
天然土路面 natural earth surface
天然土面 natural soil surface
天然土面层 topsoil surfacing
天然土铺面 topsoil surfacing
天然土壤 natural soil
天然土釉 loam glaze
天然团块 natural aggregate
天然团粒 natural granule
天然蜕变 radioactive disintegration
天然脱水石膏 natural anhydrite
天然脱枝 self-pruning
天然洼地 natural depression
天然弯曲的肋骨木材 natural frame
天然弯曲的肘材 natural knee
天然微晶均密石英粉 novacite;novaculite
天然尾水 natural tail water
天然卫星 natural satellite
天然未筛选的混合骨料 bank gravel;run-of-bank gravel
天然温井 hot well
天然稳定极限 natural stability limit
天然污染 natural pollution
天然污染物 natural pollutant
天然污染源 natural pollution source
天然无机物质 natural minerals
天然无机颜料 natural inorganic pigment
天然无水芒硝 thenardite
天然无水石膏 natural anhydrite
天然物质 naturally occurring substance;natural matter
天然误差 natural error
天然吸附剂 natural adsorbing agent;original adsorbent
天然吸水石 water-absorbing natural stone
天然系 natural series
天然细骨料 natural fine aggregate
天然细集料 natural fine aggregate
天然细晶粒 natural fine grain
天然细颗粒 natural fine grain
天然下种 wild seeding
天然下种造林法 seed shedding
天然纤维 native fiber[fibre];natural fiber[fibre]

天然纤维素 native cellulose; natural cellulose
天然咸水 natural brine
天然香料货 essential oil
天然橡胶 caoutchouc; gum rubber; gun rubber; India rubber; native rubber; nat-rubber; natural rubber; par-rubber; plantation rubber; plant rubber; polyisoprene; wild rubber
天然橡胶粉末 natural rubber powder
天然橡胶胶粘剂 natural rubber adhesive
天然橡胶乳化涂料 natural rubber emulsion paint
天然橡胶乳化液 natural rubber emulsion
天然橡胶乳化油漆 natural rubber emulsion paint
天然橡胶乳剂 natural rubber emulsion
天然橡胶生产国协会 Association of Natural Rubber Producing Countries
天然橡胶烃 natural rubber hydrocarbon
天然小道 natural trail
天然小方石(块) natural sett; natural stone paving sett
天然小方石铺路 paving of natural setts
天然小径 <树林中的> natural trail
天然斜锆石矿 zircite[zirkite]
天然斜坡 natural slope
天然卸载拱 natural load-transmitting arch
天然形成的盆地 barrier basin
天然形成的水井 naturally developed well
天然形态 physical form
天然形状原貌 natural form
天然型砂 naturally bonded mo(u)lding sand; natural mo(u)lding sand
天然性 naturalness
天然休养林 natural refreshment forest
天然休止角 natural angle of repose; natural angle of slope; angle of natural repose
天然休止角试验 test natural angle of repose
天然选择 natural selection
天然雪盖 natural snow cover
天然岩层 natural bed; undisturbed rock
天然岩床 natural bed
天然岩沥青 natural rock asphalt
天然岩沥青(乳)液 natural rock asphalt mortar
天然岩沥青砂浆 natural rock asphalt mortar
天然岩石 natural rock
天然岩石表面 natural bed
天然岩石骨料 natural rock aggregate
天然岩石混凝土 natural rock concrete
天然岩石集料 natural rock aggregate
天然岩石群 rock works
天然岩石碎块 <用作路面或基础下的底基层的> hard core
天然研磨料 <如金刚砂> natural abradant; natural abrasive
天然盐 natural salt
天然盐水 bittern; mire-drum; natural brine
天然颜料 natural colo(u)r; natural dyestuff; natural pigment
天然掩蔽 natural cover
天然演替 natural succession
天然羊毛脂 degras
天然氧化铝 natural alumina
天然氧化铁 Indian ochre; Indian red ochre; natural iron oxide; natural oxide of iron; terra disiena
天然氧化铁黄 yellow limonite
天然氧化铀 uraninite
天然叶板 natural acanthus leaf
天然叶理【地】 crude foliation
天然叶饰 natural foliage
天然液体燃料 natural liquid fuel
天然铱锇合金 nevyanskite
天然遗传性 natural heritage
天然抑制 natural suppression
天然易蚀性 inherent erodibility
天然逸水口 natural escape
天然溢洪道 <小水库的> natural escape
天然银 native silver
天然银朱 natural vermillion
天然应力 native stress; virginal stress
天然荧光 natural fluorescence
天然硬度 natural hardness; natural stiffness
天然硬度钢 natural steel
天然硬钢 naturally hard steel
天然硬骨料混凝土瓦 natural hard aggregate concrete tile
天然硬集料混凝土瓦 natural hard aggregate concrete tile
天然硬沥青 uinta(h)ite
天然硬石膏 natural anhydrite
天然硬树脂(胶) natural copal
天然硬水 naturally hard water
天然油 natural oil
天然油石 nature oilstone
天然油脂色素 pigment in natural fats
天然铀 native uranium; naturally occurring uranium; natural uranium; normal uranium
天然铀堆 natural-uranium reactor
天然铀反应堆 natural-uranium reactor
天然铀裂变室 natural-uranium fission chamber
天然铀原子反应堆 natural-uranium reactor
天然游泳场 swimming place
天然有机螯合剂 natural organic chelating agent
天然有机肥 natural organic fertilizer
天然有机高分子絮凝剂 natural organic polymeric flocculant
天然有机化合物 natural organic compound
天然有机化学 natural organic chemistry
天然有机染料 natural organic dye (stuff)
天然有机酸 natural organic acid
天然有机污染物 natural organic pollutant
天然有机物 natural organic matter; natural organism
天然有机颜料 natural organic pigment
天然幼树 seeding growth
天然鱼类生境 native fish habitat
天然隅石块 natural stone (angle) quoin
天然元素 native element; natural element
天然原料配成的釉 natural glaze
天然圆条 natural rod
天然云母 natural mica
天然蕴藏量 nature reserves
天然杂质 natural impurity
天然再生植被 natural revegetation
天然皂石 natural steatite
天然造成的效果 weathered effect
天然造林法 natural afforestation
天然噪声 natural noise
天然增量剂 natural extender
天然增殖 natural increase
天然障碍 natural barrier; natural obstacle; natural terrain obstacles; physical obstacle
天然沼泽区 natural everglades area
天然照度系数曲线 distribution curve of daylight factor
天然照明 natural illumination; natural lighting
天然照明率 daylight ratio
天然遮障 natural screen
天然赭土 <一种矿物颜料> natural sienna
天然(震)源 natural source
天然蒸汽 natural steam
天然蒸汽田 natural steam field
天然整枝 pruning
天然支护采矿法 self-supported opening
天然脂 <重质原油> natural grease
天然脂肪 natural fat
天然植被 native vegetation; natural vegetation
天然植被类型 type of natural vegetation
天然中等粒度砂 natural medium sand
天然中子 natural neutron
天然重度 natural density
天然朱 natural vermillion
天然朱砂 native cinnabar; natural cinnabar
天然转曲 natural convolution; natural twist
天然装饰 natural finish
天然状态 native state; natural state
天然资源 gratuitous goods; natural resources
天然资源储藏所 naturally occurring repository
天然自承拱 natural self-supporting arch
天然自净能力 natural self-purification capacity
天然鬃刷 natural bristle brushes
天然纵剖面 equilibrium profile
天色测量 cyanometry
天色灰暗 leaden
天山石 tianshanite
天上的 superterranean
天社蛾科 <拉> Notodontidae
天生持久性 natural durability
天生的 native; natural-born; congenital <疾病、缺陷等>
天生的经理人员 born manager
天生反射 inborn reflex
天生桥 karst bridge; natural bridge
天生桥拱高 arch height of a natural bridge
天生桥拱跨 arch width of a natural bridge
天生石拱桥 natural stone arch bridge
天使雕饰梁 angel beam
天使梁 <在梁的尾雕刻人形> angel beam
天使像 portrait of angel
天书 sealed book
天坛 Temple of Heaven
天坛回音壁 Echo Wall
天坛九龙壁 Nine-Dragon Screen
天坛祈年殿 Hall of Prayer for Good Harvests
天坛圜丘坛 circular mound altar
天堂 paradise
天体 celestial body; celestial object; celestial sphere; heavenly body; orb
天体测定学 astrometry
天体测光学 astronomic(al) photometry; astrophotometry
天体测量 astrogeodesy; celestial observation
天体测量赤经 astrometric right ascension
天体测量赤纬 astrometric declination
天体测量法 astrometric method
天体测量方位 astrometric aspect
天体测量轨道 astrometric orbit
天体测量基线 astrometric baseline; astronomic(al) base line
天体测量经纬仪 <同时测量经度和方位角的> altazimuth
天体测量期望 astrometric expectation
天体测量卫星 astrometry satellite
天体测量学 astrometry; uranometry
天体测量仪 astrometer
天体测温计 astronomic(al) pyrometer
天体成因 celestial origin
天体赤道 celestial equator
天体赤经 right ascension of a celestial body
天体出没磁方位角 magnetic amplitude
天体出没幅角 amplitude of a heavenly body; bearing amplitude
天体磁学 astromagnetism
天体大地测量学 celestial geodesy
天体大气 stellar atmosphere
天体大气学 aeronomy
天体导航 astronavigation; astronomic(al) navigation; celestial navigation; navigation by space reference; star navigation
天体的 celestial; celestial body's; spheric(al); stellar
天体的出没方位角 amplitude
天体地理位置 geographic(al) position of celestial body
天体地理学 astrogeography
天体地质学 astrogeology; astronomic(al) geology; cosmogeology
天体顶距 co-altitude
天体定位 celestial fix
天体定向 celestial orientation
天体方位 azimuth of heavenly body; celestial body azimuth
天体方位角 azimuths of celestial bodies
天体分光度测量 astronomic(al) spectrophotometry
天体分光计 astrospectrometer
天体干扰 star statics
天体高度 altitude of the heavenly body
天体高度测量仪 prism astrolabe
天体高度定位线法 method of position line by altitude
天体高度方位计算表 tables of computed altitude and azimuth
天体高度曲线 altitude curve
天体观测 astronomic(al) observation; celestial observation
天体观测群 round of sights
天体观测卫星 orbiting astronomic(al) observatory
天体观测仪 nocturnal
天体惯性导航稳定平台 stellar-inertial stable platform
天体光度测量 astronomic(al) photometry; astrophotometry
天体光度计 astrophotometer
天体光度学 astronomic(al) photometry; astronomic(al) spectroscopy; astrophotometry; celestial photometry
天体光谱学 astrospectroscopy
天体轨道式磨砂机 orbital sander
天体化学 astrochemistry; cosmochemistry

T

天体黄纬 latitude of a heavenly body
天体基准 space reference
天体极距 celestial polar distance; codeclination; pole distance
天体经纬仪 altazimuth
天体镜 celescope
天体距角 elongation
天体力学 celestial mechanics; gravitational astronomy
天体立体照片 astrostereogram
天体罗盘 celestial compass
天体碰撞坑 astrobleme
天体偏振计 astropolarimeter
天体气象学 astrometeorology
天体上边缘 upper limb
天体摄谱仪 astronomic(al) spectrograph; astrospectrograph
天体摄影 antenna pick
天体摄影位置 astrographic(al) position; astrometric position
天体摄影物镜 astrographic(al) objective
天体摄影学 astrography
天体摄影仪 astrograph
天体生物物理学 bioastrophysics
天体生物学 astrobiology
天体时角 hour angle of heavenly body
天体视向速度仪 astrovelocimeter
天体视增大 augmentation
天体图 astronomic(al) chart; astronomic(al) map; celestial chart; celestial map
天体图集 celestial atlas
天体望远镜 astronomic(al) telescope
天体位置 position of heavenly body
天体物理学 astrophysics
天体学论文 uranology
天体演化 cosmogony
天体仪 celestial globe; globe; sphere
天体照相 astrophotograph
天体照相机 astrocamera; astrophotocamera
天体照相平时 astrograph mean time
天体照相术 astrophotography
天体照相双合透镜 astrographic(al) doublet
天体照相位置 astrographic(al) position
天体照相物镜 astrographic(al) objective
天体照相学 astrography; astronomic(al) photography; astrophotography; celestial photography
天体照相仪 astrograph
天体真方位 true azimuth
天体真运动 celestial body true motion
天体制图学 celestial cartography
天体钟 celestial clock
天体周日视运动 celestial body diurnal apparent motion
天体最大高度 maximum altitude
天体坐标 celestial coordinates
天体坐标量测仪 celestial coordinate comparator; stellar comparator
天铁矿水泥 titan cement
天铁斯特绝缘纤维板 Tentest
天王星 Uranus
天文爱好者 astrophile
天文比对表 chronometer watch
天文表 sidereal watch
天文测定 astronomic(al) determination
天文测量 astronomic(al) measurement; astronomic(al) surveying
天文测量基线 astronomic(al) base line
天文测量学 astrometry
天文测量用导线 astronomic(al) traverse
天文常数 astronomic(al) constant
天文常数系统 astronomic(al) constant system; system of astronomic(al) constants
天文潮 astronomic(al) tide; equilibrium tide; fundamental tide
天文潮波 astronomic(al) tidal wave
天文潮位 astronomic(al) tide level; equilibrium tide level; fundamental tide level
天文晨光始 beginning of astronomic(al) morning twilight
天文晨昏蒙影 astronomic(al) twilight
天文赤道 astronomic(al) equator
天文赤道坐标系 equatorial astronomic(al) coordinate system
天文船位 celestial fix
天文船位三角形 astronomic(al) triangle
天文船位线 astronomic(al) position line
天文船位圆 astronomic(al) position circle
天文垂线 astronomic(al) vertical
天文大地测量方法 astrogeodetic method
天文大地测量偏差 astrogeodetic deflection
天文大地测量数据 astrogeodetic data
天文大地测量系统 astrogeodetic system
天文大地测量学 astronomic(al) geodesy; celestial geodesy
天文大地垂线偏差 astrogeodetic deflection; astrogeodetic deflection of the vertical
天文大地点 astrogeodetic point
天文大地基准 astrogeodetic datum; astronomic(al) datum
天文大地基准定向 astrogeodetic datum orientation
天文大地偏差 astrogeodetic deflection; relative deflection
天文大地水准测量 astrogeodetic level(ling)
天文大地(水准面)起伏 astrogeodetic undulation
天文大地网 astrogeodetic net(work); astronomic(al) network
天文大地网平差 adjustment of astrogeodetic network; astrogeodetic net adjustment
天文单位 <长度单位, 1 天文单位 = 149,599,000 千米> astronomic(al) unit
天文单位距离 unit distance
天文弹道学 astroballistics
天文导航 astrogation; astronavigation; astronomic(al) navigation; celestial guidance; celestial navigation; celonavigation; coelonavigation
天文导航系统 astronavigation system; astronomic(al) navigation system; celestial system
天文导航装置 astronavigation system
天文导引 stellar guidance
天文的 astronomic(al)
天文底片 astronegative; astronomic(al) plate; astrophotogram
天文地理 astronomic(al) geography
天文地平(线) astronomic(al) horizon; celestial horizon; rational horizon; sensible horizon; true horizon
天文地球动力学 astrogeodynamics
天文地球物理学 astrogeophysics
天文地质学 astrogeology; astronomic(al) geology
天文点 astronomic(al) point; astronomic(al) position; astronomic(al) station
天文点标石 astronomic(al) (observation) monument
天文点连测导线 astronomic(al) traverse
天文电子学 astrionics; astronomic(al) electronics
天文定点网 astrofix(ation) network; astronomic(al) fixation network
天文定位 astrofix; astronomic(al) fix; astronomic(al) fixation; astronomic(al) orientation; astronomic(al) position determination; astronomic(al) position finding; celestial fix; celestial fixing position by astrocalculation
天文定位器 astrograph
天文定位线 astronomic(al) line of position
天文定向 astronomic(al) fix; astronomic(al) orientation; celestial orientation
天文动力学 astrodynamics
天文反射望远镜 astrometric reflector; astronomic(al) reflector
天文方位角 astroaz; astronomic(al) azimuth; astronomic(al) bearing
天文方位角点 astronomic(al) azimuth point; astronomic(al) azimuth station
天文方位角观测 astronomic(al) azimuth observation
天文方向点 astronomic(al) azimuthal station
天文分潮 astronomic(al) tidal constituent
天文分量 astronomic(al) constituent
天文符号 astronomic(al) sign
天文高度图 astrograph
天文观测 astronomic(al) measurement; astronomic(al) observation; astronomic(al) sight; celestial observation
天文观测舱 astral dome; astrodome; navigation dome
天文观测窗 astral dome; astrodome; navigation dome
天文观测计数器 counter for astronomic(al) observation
天文观察 astronomic(al) observation
天文观察窗 <在飞机顶部> astrodome
天文馆 astronomic(al) observatory; planetarium
天文惯性导航 automatic celestial navigation; celestial inertial guidance; stellar-inertial guidance
天文惯性导航设备 astro-inertial navigation equipment; astronomic(al) inertial navigation aids; celestial inertial navigation aids
天文惯性导航系统 stellar-inertial guidance system
天文光度计 astrophotometer; telescopic(al) photometer
天文光行差 astronomic(al) aberration
天文光学 astronomic(al) optics
天文归化纬度 reduced astronomic latitude
天文航法 astronavigation; astronomic(al) navigation; celestial navigation; celonavigation
天文航海表 astronomic(al) navigation tables
天文航海法 cele-navigation
天文航海学 astronavigation; astronomic(al) navigation; celestial navigation; celonavigation
天文航海仪表 celestial navigation instrument
天文怀表 pocket chronometer
天文回照目镜 helioscopic ocular
天文昏影终 end of astronomic(al) evening twilight
天文计算 astronomic(al) calculation
天文监测 astrosurveillance
天文检验 astronomic(al) test
天文经度 astronomic(al) longitude; celestial longitude
天文经济周期理论 astronomic theory of business cycles
天文经纬仪 astronomic(al) theodolite; meridian circle; transit circle
天文距离 astronomic(al) distance
天文考古学 astroarchaeology
天文控制 <用天文方法测施的地面控制> astronomic(al) (ground) control
天文历 almanac; astronomic(al) calendar; astronomic(al) ephemeris; ephemeris[复 ephemerides]
天文历表 ephemeris[复 ephemerides]
天文历书 astronomic(al) chronicle
天文领航 astronavigation; celestial navigation
天文罗经 astrocompass; astronomic(al) compass
天文罗盘 astrocompass; astronomic(al) compass; celestial compass
天文蒙气差 astronomic(al) refraction; celestial refraction
天文命名 astronomic(al) nomenclature
天文年 astronomic(al) year; equinoctial year; natural year; solar year; tropic(al) year
天文年代学 astrochronology
天文年历 astronomic(al) almanac; astronomic(al) year book
天文气候 astroclimate; astronomic(al) climate; solar climate
天文气候学 astroclimatology
天文日 astronomic(al) day
天文日期 astronomic(al) date
天文三角形 astronomic(al) triangle; celestial triangle; navigation triangle; parallactic triangle
天文闪烁 astronomic(al) scintillation; stellar scintillation
天文摄动技术 astronomic(al) perturbation technique
天文摄影机 astronomic(al) (photographic) camera
天文摄影术 astronomic(al) photography
天文摄影学 astronomic(al) photography; astrophotography
天文摄影仪 astrophotocamera
天文时 sidereal time
天文时计 astronomic(al) clock
天文时间 astronomic(al) time
天文时间标度 astronomic(al) time scale
天文时期 astronomic(al) time
天文曙暮光 astronomic(al) twilight
天文数字 astronomic(al) figures
天文水准(测量) astronomic(al) level(l)ing
天文台 astronomic(al) observatory; astronomic(al) station; observatory; planetarium
天文台圆顶 observatory dome
天文天顶 astronomic(al) zenith
天文透镜 astronomic(al) lens
天文图 celestial map
天文望远镜 astronomic(al) telescope; inverted image telescope; inverting image telescope; telescope

of inverted image
天文望远镜玻璃毛坯 glass blank for astronomic(al) telescope
天文纬度 astronomic(al) latitude;astronomic(al) parallel
天文纬圈 astronomic(al) parallel
天文卫星 astronomic(al) satellite
天文位置 astronomic(al) position
天文位置线 astronomic(al) line of position;astronomic(al) position line;astro-position line;celestial line of position
天文物镜 astronomic(al) objective
天文物理学 astrophysics
天文象限角 astronomic(al) bearing
天文学 astronomy;uranology
天文学方法 astronomic(al) method
天文学家 astronomer
天文学(上)的 astronomic(al)
天文学史 history of astronomy
天文研究中心 astronomic(al) research centre[center]
天文仪 astroscope
天文仪器 astronomic(al) instrument
天文圆顶 astrodome
天文站 astronomic(al) station
天文章动 astronomic(al) nutation
天文照相机 astronomic(al) camera
天文折光差 astronomic(al) refraction
天文钟 astronomic(al) clock;board chronometer;chronometer
天文钟比差 chronometric difference
天文钟比对 comparison of chronometers
天文钟差 chronometer correction
天文钟擒纵机构 detent escapement;Earnshaw escapement
天文钟室 chronometer house;chronometer room
天文钟误差率 chronometer rate;rate of astronomic(al) clock
天文重力测量 astrogravimetry
天文重力测量的 astrogravimetric
天文重力测量点 astrogravimetric point
天文重力水准 astronomic(al) gravity level
天文重力水准测量 astrogravimetric level(l)ing
天文子午面 astronomic(al) meridian(al) plane
天文子午圈 astronomic(al) meridian
天文子午线 astronomic(al) meridian;terrestrial meridian
天文自动定位系统 automated astronomic(al) positioning system
天文自动定位装置 automated astronomic(al) positioning device
天文坐标 astronomic(al) coordinates
天文坐标测量仪 astronomic(al) coordinate measuring instrument;Ascorecord <商品名>
天文坐标系 astronomic(al) coordinate system
天雾 sky fog
天险 nature hazard
天线 aerial;aerial wire;air wire;high line;skyline;antenna[复 antennas/antennae] <美>
天线安培计 antenna ammeter
天线安装 aerial installation
天线安装机械 antenna mount
天线棒 aerial rod;antenna rod
天线背对背隔离度 back to back directivity separation of antenna
天线变频器 antennaverter
天线波束方向 antenna beam direction
天线波束宽度 beam width of antenna
天线侧对侧隔离度 side-to-side directivity separation of antenna
天线插孔 antenna jack
天线插口 connection socket for aerial
天线插座 aerial plug-in point;aerial socket;antenna jack;antenna socket
天线长度 antenna length
天线场 antenna field
天线场强增益 antenna field gain
天线串话 antenna crosstalk
天线串音 antenna crosstalk
天线磁心 antenna core
天线导线 antenna conductor
天线的方向性 antenna directivity
天线的方向性系数 antenna directivity
天线等效电阻 phantom antenna resistance
天线地网 antenna counterpoise
天线电报机 aerograph
天线电抗 antenna reactance
天线电缆 aerial cable;antenna cable
天线电流 aerial current;antenna current;radiation current
天线电路 aerial circuit;line circuit;radiating circuit;radiative circuit
天线电路断路器 antenna circuit breaker
天线电器中产生的起伏电压 antenna pick-up
天线电容 aerial capacitance;aerial capacity;antenna capacitance
天线电容器 antenna condenser
天线电设备 radio set;wireless set
天线(电)收发转换开关 duplexer
天线电阻 aerial resistance;antenna resistance
天线顶部 aerial head
天线顶帽 top hat
天线定位控制装置 antenna positioning control unit
天线定向作用系数 directive gain
天线动态特性 antenna dynamics
天线短路开关 antenna blocking up switch
天线断路器 aerial circuit breaker;antenna circuit breaker
天线对称装置 aerial choke;antenna choke
天线扼流圈 aerial choke;antenna choke
天线发射机 antennamitter
天线反射器 antenna reflector;parasite
天线反射阵 mattress reflector
天线方位 antenna;antenna bearing
天线方向 antenna bearing
天线方向图 antenna radiation pattern
天线方向图分析器 antenna pattern analyser[analyzer]
天线方向图宽度 beam width
天线方向性调整器 goniometer
天线方向性图 antenna directivity diagram;antenna pattern
天线方向性转接装置 beam switching device
天线放大器 antennafier
天线分离滤波器 diplexer
天线风挡玻璃 antenna windshield
天线辐射 aerial radiation;antenna radiation
天线辐射电阻 aerial radiation resistance;antenna radiation resistance
天线辐射能量 antenna energy
天线辐射图 antenna radiation pattern;lobe
天线俯角 antenna depression angle
天线俯仰电动机 elevation drive motor
天线负载 antenna load
天线复用器 antenna duplexer
天线杆 aerial mast;aerial pole;antenna mast;antenna pole;antenna spike;antenna support;antenna tower;mask;radio mast
天线杆拉线 guy;mast rope
天线高度 antenna height
天线隔电子 aerial insulator;antenna insulator
天线跟踪系统 antenna tracking system
天线功率 antenna energy;antenna power
天线功率增益 antenna power gain
天线共用器 antenna diplexer;antenna splitting device;diplexer;duplexer;notch diplexer <锐截式>
天线共用装置 community set of antenna
天线共振频率 antenna resonance frequency
天线合股线 antenna strand
天线互换器 polyplexer
天线基座 antenna mounting
天线激励 antenna excitation
天线架 antenna frame
天线间隔 aerial spacing
天线交换器 antenna switching device
天线角开度 angular aperture
天线绞车 antenna winch
天线绞合线 antenna strand
天线绞盘 antenna winch
天线接触电刷 aerial wiper
天线接地开关 aerial earth switch
天线接收图 reception diagram
天线接头 aerial adapter;antenna terminal
天线接线 antenna connection
天线接线端 antenna terminal
天线结构 aerial structure;antenna structure
天线介质耦合损耗 aperture-to-medium coupling loss
天线卷盘 antenna reel
天线绝缘导管 antenna trunk
天线绝缘子 aerial insulator;antenna insulator
天线孔径 antenna aperture
天线控制板 antenna control table
天线控制系统 antenna control system
天线宽度 bandwidth of an antenna
天线馈点阻抗 aerial feed-point impedance
天线馈电 antenna feed
天线馈电线 antenna feeder
天线馈线 aerial feeder;antenna feeder;down lead
天线馈线端子 antenna feeder connector
天线馈线连接管 antenna connector
天线扩展 mast extension
天线拉线 guy
天线廊道 <广播塔> antenna gallery;aerial gallery
天线面积 antenna area
天线模型 antenna simulator
天线幕 antenna curtain
天线耦合 aerial coupling;antenna coupling
天线耦合电容器 antenna coupling capacitor
天线耦合器 antenna coupler
天线旁瓣 antenna side lobe
天线旁瓣转换 sidelobe switching
天线配电板 antenna control table
天线匹配 antenna matching
天线匹配变量器 aerial matching transformer
天线匹配网络 antenna matching network
天线匹配装置 antenna matching unit
天线平台 aerial gallery;antenna gallery
天线屏蔽器 radom(e)
天线屏幕 antenna curtain
天线区 antenna field
天线群方向特性 group-directional characteristic
天线扫掠 lobing
天线扫描中心 antenna scanning center[centre]
天线设备 aerial installation;antenna installation
天线射束 antenna beam
天线射束的控制 lobing
天线升高部分和引下线的连接线束 rat tail
天线升降索 aerial ha(u)lyard
天线升压器 antenna booster
天线收发转换开关 polyplexer
天线收集面 antenna capture area
天线输出 aerial output;antenna output
天线输入电路 antenna input circuit
天线输入功率 aerial input;antenna input power
天线输入阻抗 antenna input impedance
天线衰减量 antenna decrement
天线双工器 antenna diplexer
天线伺服系统 antenna servo system
天线损耗 antenna loss
天线塔 aerial mast;antenna mast;antenna tower;radio tower
天线塔架 hi-line tower
天线特性阻抗 antenna characteristic impedance
天线特征 lineament
天线调谐 aerial balance;aerial tuning;antenna tuning
天线调谐电感 aerial(tuning) inductance;antenna tuning inductance
天线调谐电容器 aerial tuning capacitor;antenna tuning capacitor
天线调谐器 antenna tuner
天线调谐线圈 aerial tuning coil;antenna tuning inductor
天线调谐用电动机 antenna tuning motor
天线调谐指示器 antenna tuning indicator
天线铁塔 antenna iron tower
天线头 aerial head
天线网 aerial network
天线位置控制 antenna position control
天线位置调节 antenna positioner
天线位置指示器 antenna position indicator
天线吸收表面 action radius
天线系统 aerial system;antenna system
天线下引线 down lead
天线销 antenna spike
天线效率 antenna efficiency
天线斜度转换开关 antenna tilt switch
天线旋转机构 aerial rotating equipment;antenna turning unit
天线旋转开关 antenna rotation switch
天线旋转器 antenna rotator;antenna spinner
天线旋转位置图盘 antenna repeat dial
天线仰角 antenna tilt;tilt
天线仰角机构 antenna tilt mechanism
天线仰角阻尼 elevation antihunt
天线抑制器 antenna eliminator
天线引入绝缘子 antenna lead-in insulator
天线引入线 antenna lead(-in)
天线引下线 antenna down-lead
天线引线 down lead
天线引线管 aerial fairlead

T

天线引向器 sender
天线用的合股线 radio wire
天线用金属线 antenna wire
天线用线 antenna wire
天线有效高度 effective antenna height
天线有效面 absorption plane
天线有效面积 antenna area; antenna effective area; capture area of antenna
天线噪声 antenna pick-up
天线增益 antenna gain
天线张角 angular aperture
天线罩 antenna cap; antenna dome; antenna housing; blister; aerial radome; radom(e) <雷达>
天线阵 aerial array; aerial system; antenna array; antenna panel; array antenna; multilateral system; trap
天线阵辐射元 elementary antenna
天线阵帘 array curtain
天线支架 aerial structure
天线指向角 antenna directional angle
天线中心架 central gallows for antenna
天线重合 antenna coincidence
天线主瓣 antenna main lobe
天线主背瓣比 front-to-back ratio of antenna
天线主侧瓣比 front-to-side ratio of antenna
天线驻波比 antenna standing wave
天线柱 aerial mast
天线转换开关 <接收机保护设备> reprod[receiver protective device]; aerial switch; antenna change-over switch; duplexer; transmitter blocker switch
天线转换装置 antenna switch
天线转接 aerial changeover
天线转数 antenna rotation number
天线装置 aerial installation; aerial system; antenna installation
天线阻抗 antenna impedance
天线阻尼 antenna damping
天象 astronomical phenomenon; celestial phenomenon
天象馆 hall of heavenly phenomenon
天象勤务通信[讯] weather communication
天象厅 hall of heavenly phenomenon
天象图 sky diagram
天象仪 planetarium
天行 the nature's behavio(u)r
天演 natural selection
天灾 <如洪水、风暴、地震等> act of God; act of nature; natural calamity; natural disaster; natural hazard; plague
天灾及海难除外条款 acts of God clause
天职 vocation
天职的 vocational
天植物学 astrobotany
天轴 celestial axis; connection shaft; countershaft; line shaft(ing); main shaft
天轴传动 line shaft transmission; overhead driving gear
天竹子 heavenly bamboo
天竺桂(籽)油 yabunikkei seed oil
天竺牡丹 dahlia
天主教的十字架 papal cross
天主教卡尔特修道院 Carthusian monastery

标 subscript; suffix

添播 interseeding intersowing
添定 repeat order
添附证件 attached documents
添改 interpolation
添购自用资产通知单 betterment order
添画 touch in
添火 poke
添加 affix(at)ion; intercalation; superimpose; super-induction
添加插件板 add-on board
添加成分 component of additive
添加存储器 add-on memory; add-on storage
添加的 supplementary
添加法 additive process
添加固定设备 addition of fixtures
添加合金 alloying addition
添加合金元素 alloying agent
添加化学保藏剂的冰 chemical ice
添加剂 addition agent; additive; adjunct; admixture; affix; chemical admixture; extender; furnace addition; intrusion agent; accelerating additive <混凝土>; densifying agent <水泥的>
添加剂的消耗 depletion to additive
添加剂耗减 additive depletion
添加剂耗损 additive depletion
添加剂和隔离液 additives and spacer fluid
添加剂进料装置 additive feeder
添加剂浓度 additive concentration
添加剂迁移 additive migration
添加剂损失 additive loss
添加剂污染 additive pollution; adjunct pollution
添加剂箱 add-mix tank
添加卷尺 adding tape
添加量 refill capacity
添加料 additive
添加钼化合物 adding molybdenum compounds
添加泥浆 ada mud
添加燃料停车 refuel stop
添加润滑油 oil addition
添加受热面 extended surface
添加塑料 plastic addition
添加炭黑 black loading
添加同位素 spike isotope
添加物 accretion; additional matter; affixture; applied addition; superaddition; supplement; addition
添加系统 add-on system
添加硬件 add-on hardware
添加元素 additional element
添加增塑剂 additive plasticizer
添加折叠式椅子 <在教堂长凳子旁> pew chair
添建的 built-on
添量 tret
添埋构件 building in
添煤 firing; stoke
添煤器 shaking grate stoker; stoker
添燃料 fuel(l)ing
添燃油 fuel(l)ing
添入 intercalate
添纱织物 plated fabrics
添上 tail
添塔料 tower filling
添味剂 odo(u)rant
添印 overprint
添印上去的东西 overprint
添油 topping up
添油孔 oil feeding hole
添油液量 service refill capacity
添置财产拨款 appropriation for addition of property

田边 boundary of field

田边沟 contour ditch; contour furrow
田边植物 field border plantings
田塍 ridge
田地 cropland; farmland; field
田地交易注册经纪人 accredited farm and land broker
田地经纪人 farm broker
田地掮客 farm broker
田地清除 ground clearing
田赋 land tax
田埂 ba(u)lk; border; border dike [dyke]; low bank of earth between field; rand; ridge
田埂坎 linch(et) [lynch(et)]
田沟 border ditch; field ditch
田间 farm; field
田间保水量 field capacity; field moisture capacity
田间保雪量 snow retention
田间测定 field test(ing)
田间测图 field mapping
田间持水量 field capacity; field moisture capacity
田间持水率 normal field capacity
田间持水能力 field carrying capacity
田间抽水试验 field pumping test
田间储藏 field storage
田间处理 <污水的> land treatment
田间的 on-farm
田间地块 field plot
田间调查 field investigation
田间斗门 farm head gate
田间放水需要量 farm delivery requirement
田间废水 field waste
田间供水程序 field supply schedule
田间供水方案 field supply schedule
田间沟渠 field ditch
田间沟渠效率 field ditch efficiency
田间沟网 field ditch system
田间观察 field observation
田间管理 care of field; farm management; field care; field practice; handling of plants
田间灌溉渠 <英> feeder canal
田间灌溉废水 field waste
田间灌溉供水规格 field supply schedule
田间灌溉沟 field ditch
田间灌溉机具 irrigation implement
田间灌溉率 field duty of water
田间灌溉水渠 sub-minor canal
田间灌溉小沟 field ditch
田间灌溉效率 efficiency of farm irrigation
田间灌溉需水量 field irrigation requirement
田间灌后的废水 field waste
田间灌水定额 farm consumptive use; farm duty(of water); net irrigation requirement
田间灌水沟 farm ditch
田间灌水管理 water application management
田间灌水率 farm duty(of water)
田间灌水渠 feeder canal; field ditch; quarternary canal
田间灌水效率 field application efficiency; irrigation application efficiency
田间含水当量 field moisture equivalent
田间含水量 field moisture capacity; field capacity
田间含水量差值 <实际含水量与饱和含水量之差以水深计> field moisture deficiency
田间含水能力 field carrying capacity
田间耗水量 consumptive use of crop; farm consumption use; water consumption in the field
田间检查 field examination; field inspection
田间流失 capacity waste; field waste
田间路 farm road
田间毛细管含水量 field capillary moisture capacity
田间描述 field description
田间内部配水 internal field distribution
田间排水 field drain
田间排水沟 drainage field ditch
田间排水梁 field ditch
田间排水系统 field drainage
田间排水支沟 field drain(age) branch
田间配水 internal field distribution
田间配水渠 head ditch
田间配水系统 farm distribution system
田间喷灌系统 field sprinkler system
田间禽舍 rustic bird house
田间渠道分水闸 farm turnout
田间渠系 field ditch system
田间热 field heat
田间入渗 field infiltration
田间洒水装置渗透仪 field sprinkler infiltrometer
田间沙地 sand break
田间设备 field equipment
田间渗漏损失 farm percolation loss
田间渗入 field infiltration
田间实验小区 field plot
田间使用 field operation
田间试验 field experiment; field test(ing)
田间试验计划 field plan
田间试验室 field laboratory
田间水分 field moisture
田间水分不足 moisture field deficiency
田间水分当量 field moisture equivalent
田间水量损失 field loss
田间水塘 farm pond
田间通道行树 boundary vista
田间土壤 field soil
田间土壤持水量 field moisture capacity
田间土壤渗透系数 field coefficient of permeability
田间脱粒 field shelling
田间脱粒机 field thresher
田间下渗 field infiltration
田间小区试验 field plot experiment; field plot trial
田间需水量 farm water requirement; irrigation water demand; irrigation water requirement; net irrigation requirement
田间蓄水 on-farm storage
田间蓄水池 listing basin
田间蓄水坑塘 farm pond
田间用水 field application
田间用水管理 on-farm water management; water application management
田间用水量 field duty of water
田间用水率 farm duty(of water); field duty of water
田间用水效率 field application efficiency
田间蒸发 evaporation in the field
田间支渠 farm lateral; field lateral
田间最大持水量 maximum field capacity; maximum field carrying capacity
田间最大耗水率 peak use rate
田间作业机械 field machinery
田间作业面积计数器 acreage meter
田芥菜 charlock
田界 boundary strip
田径运动 <英> athletics
田径运动员 trackman

田埂 field ridge;ridge
田亩分派 acreage allotments
田纳西 <美国州名> Tennessee
田纳西河 Tennessee River
田纳西流域管理局 <美> Tennessee valley Authority
田纳西州大理石 <美> Tennessee marble
田纳西州软木材 <美> Tennessee softwood
田权 land ownership rights
田鼠 field mouse
田鼠丘 mole-hill
田树 elite tree
田涛 cultivated field
田调和函数 tesseral harmonics
田头菇属 <拉> Agrocybe
田谐函数 tesseral harmonics
田形调和函数 tesseral harmonics
田熊式锅炉 Takuma's boiler
田旋花 bindweed;field bindweed
田野 dol
田野草皮 field sod
田野的 campestral
田野发掘 field excavation
田野间或树丛中光秃不毛的地方 gall
田野间砂地 sand break
田野木舍 rustic home
田野木屋 rustic home
田野千里光 Senecio oryzetorum
田野燃烧 field burning
田野小舍 rustic home
田园 fields and gardens
田园城市 garden city
田园城市理论 garden city theory
田园地带 agricultural belt; country belt;rural belt
田园风光 rurality
田园风光陶器 rustic ware
田园郊区 garden suburb
田园市郊 garden suburb
田中新煤贫 Neocapnodium tanakea
田庄 grange
田庄出入道路 land-access road
田庄出入受益 <道路筑成后由于田庄土地增值所受的利益> land access benefit
田庄通道受益 <道路筑成后田庄的得益> land access benefit

甜白 <一种白釉品> sweet white

甜白釉 lovely white glaze
甜菜 beet cancer;sugar beet
甜菜地下贮藏室 beet silo
甜菜顶切碎器 beet top chopping machine;top pulverizer
甜菜顶青贮 beet top silage
甜菜顶清除器 top cleaner
甜菜顶收集器 top gatherer
甜菜堆藏机 beet piler
甜菜废丝提升器 pulp elevator
甜菜废丝压榨水 pulp press water
甜菜捡拾装载机 sugar beet pickup loader
甜菜浆 beet pulp
甜菜块根 beet root
甜菜切片机 beet slicer
甜菜切碎机 beet cutter
甜菜清理装载机 beet cleaner loader
甜菜清洗机 beet washing machine
甜菜收购站 beet collecting station
甜菜松土器 beet loosener
甜菜糖厂 beet sugar factory
甜菜糖厂废水 beet sugar process wastewater
甜菜条播装置 sugar beet drill unit
甜菜挖堆机 beet digger; sugar beet lifter
甜菜渣 bagasse;megass(e);beet pulp <堵漏材料>
甜菜制糖废弃物 sugar beet waste
甜菜制糖废水 beet sugar process wastewater;beet sugar waste;sugar beet waste;sugar beet wastewater
甜菜制糖废物生长厌氧污泥 sugar beet waste-grown anaerobic sludge
甜菜制糖废液 beet sugar waste liquor;Steffen's waste
甜菜制糖废渣 beet sugar waste
甜橙 Chinese evergreen chinquapin; sweet orange;Washington naval orange
甜栗木 sweet chestnut
甜眠型床垫 beauty rest mattress
甜气 <含少量硫化氢的天然气> sweet gas
甜水 sweet water
甜松 shade pine;sugar pine
甜土植物 glycophyte
甜玉米 sweet corn
甜槠栲 Chinese evergreen chinquapin

填白 white glaze

填板 fille(d) plate;filler panel;filler sheet;filler slab;packing plate
填报乘客名单 bill passengers
填报工程进度 make a progress report on a project
填报者 informant
填背砂 <铸造用语> backup
填背圬工 backup masonry(work)
填表机 billing machine
填冰孔隙 ice-filling pore space
填补 blind up; filing chock; fill-up; spot priming;supply
填补材 filing piece
填补差距的策略 gap-filling strategy
填补焊 dot weld(ing)
填补空白 fill gap
填补空隙 plugging the gap
填补料 stopping
填补片 plug and filler
填补砂浆 patching mortar;repair mortar
填补摊铺(法) fill spreading
填补物 expletive;fill in
填补用混凝土 backfill concrete
填补用耐火水泥 refractory patching cement
填补桩 filling pile
填材 loading
填彩搪瓷 cell enamel
填舱货 berth cargo;fittage
填舱物料 dunnage
填层 chocked layer; filling course; packing course
填层绝缘 fill-type insulation
填沉 fill settlement
填衬材料 backup material
填成的土地 reclaimed land
填成地 made(-up) ground;made land
填迟日期 postdate
填迟日期的 post dated
填迟日期票据 post-dated bill
填迟日期支票 post-dated check [cheque]
填充 fill(ing)-in;filling up;packaging; padding;stow;stuffing;packing
填充板 infill panel;infill slab
填充棒 filler rod
填充保温 packed heat insulation
填充比 packing fraction;packing ratio
填充变料的砂浆 matrix[复 matrixes/matrices]
填充补强 filler reinforcement
填充材料 backup material; blinding material;filled composite;filler material;fill(ing) material; furniture; packing material; slurry <由自云石、沥青组成的>
填充材料坡度 slope of stowed material
填充操作 padding operation
填充测试法的 cloze
填充层 bed of packing;filler course; filling layer; packed bed; packed layer;packing course;packing layer
填充层过滤器 packed-bed filter
填充抽提塔 extraction packed column
填充床 packed bed
填充床脱氮作用 packed-bed denitrification
填充床洗涤器 packed-bed scrubber
填充萃取塔 packed extractor
填充带 filled band
填充道路面层孔的细石屑 blinding
填充的 filled;packed
填充的混凝土 staunching piece
填充的破裂口 fill fracture
填充的墙 infill(ed) wall(ing)
填充的热固性塑料 filled thermoset
填充的热塑性塑料 filled thermoplastic
填充的圬工工作 infill masonry work
填充的砖石砌体 nogging
填充地基 filling ground
填充地面 filling ground
填充地图 outline map for filling
填充点 charging point;filling point
填充垫环 filler ring
填充垫片 joint sheet
填充垫圈 joint packing
填充度 compactedness;compactness; degree of fill
填充反应塔 packed reaction tower
填充反应柱 packed reaction column
填充方式 fill mode
填充废巷道 bash
填充缝 filled joint
填充符 filler;character fill
填充符号 filling symbol
填充钢丝 <钢丝绳用> filler wire
填充高度 packed height;packing height
填充隔热 fill insulation
填充工程 infilling work
填充公式 charging formula
填充骨料 filler aggregate
填充管片 filler piece
填充管柱 filled column
填充过程 filling process
填充焊丝 filler wire
填充糊 filling-in paste
填充环 ring-type filling
填充灰浆 mortar fill(ing)
填充灰砂砖 backing calcium silicate brick
填充会切位形 stuff cusp configuration
填充混凝土 blinding concrete; filler concrete; filling concrete; infilling concrete
填充混凝土的双层板桩围堰 double wall cofferdam with concrete fill
填充混凝土地面板洞的沥青 under-sealing asphalt
填充混凝土式管柱 filled pipe column
填充剂 bulk additive; bulking agent; extender;filler;filling agent;loading material;stuffing;weighting agent
填充剂配合量 quantity of filler present
填充剂效果 effect of bulking agent
填充剂增强 filler reinforcement
填充夹板 plugged plywood
填充建筑材料 infiller material; infilling material
填充胶乳 compounded latex
填充搅拌 blowdown
填充结构 interstitial texture
填充介质 packing medium
填充金属 added metal; filler metal; metal filler
填充紧密度 closeness of packing
填充经 wadding warp
填充精馏塔 rectification packed column
填充聚四氟乙烯的青铜 bronze filled polytetrafluoro ethylene
填充空间 packing space
填充孔 filler opening;filling opening; fill orifice
填充孔隙 pore filling
填充孔隙的细石屑 blinding
填充块 filler brick
填充块体 infill block
填充矿床 filled deposit
填充理论 voids filling theory
填充沥青 filled bitumen
填充粒状隔音隔热材料 granular fill insulation
填充量 filled grade
填充料 aggregate; bulking agent; filler; filler material; packing filler; plugging compound;stuff(ing)
填充料称量器 stone batcher
填充料储存装置 filler storage unit
填充料供料定量器 stone batcher
填充料混凝土 backing concrete
填充料计量箱 stone batcher
填充料抗弯试验机 stopper bending tester;stopping bending tester
填充料收缩 pack compression
填充料用量 filler loading
填充料中木条 battens in the filling
填充裂隙 filled opening
填充淋塔 packed spray tower
填充楼板 filler floor
填充率 filling ratio;specific charge
填充脉冲间歇系统 fill-in-the-blank system
填充毛细管柱 packed capillary column
填充密度 density of packing; filling density; packed density; packing density
填充密封 stopping seal(ing)
填充模壳 filler form
填充模式 fill mode
填充木屑板 filled particleboard
填充能力 filling power
填充喷淋塔 packed spray tower
填充片 filler piece
填充平瓦 slip tile
填充气 blanket gas
填充气垫 blanketing gas
填充气相色谱法 packing gas chromatography
填充砌块 infill block; masonry filler unit
填充砌体 backfill(ing)
填充墙 filled wall; filler wall; beam filling <格栅间的>
填充墙框架 frame with infill wall
填充区表示法 fill area representation
填充区内部类型 fill area interior style
填充区设备 fill area facility
填充区索引 fill area index
填充区域 fill area;filled band
填充区域颜色索引 fill area colo(u)r index
填充圈 Raschig ring;Raschig tube
填充热绝缘 packed heat insulation
填充容积 packed space; packed volume;packing volume
填充润滑 packing lubrication
填充润滑脂 repack with grease
填充三角测量 supplementary triangulation

T

填充色料 filling colo(u)r
填充色谱(法) packing chromatography
填充色谱柱 packing of a chromatography column
填充纱线 stuffer yard
填充砂 backing sand;heap sand
填充上限 top filling
填充设施 filling-up device
填充生长 intussusception growth
填充石膏 filler gypsum
填充石灰 filling lime
填充石料 filled stone;filling stone
填充时间 filling time
填充式萃取塔 packed extraction column;packed extraction tower
填充式反应器 packed-bed reactor
填充式管柱 filled cylinder
填充式混凝土砌块 fill-up concrete block
填充式吸收器 packed absorber
填充式吸收塔 packed absorption;packed absorption column;packed absorption tower
填充试验 test filling
填充手续 packing procedure
填充树脂 additive resin
填充数 occupation number;padding;filler【计】
填充塑料 filled plastics
填充碎石作为铺砌基础 packed broken rock soling
填充塔 packed column;packed tower
填充提取塔 packed extractor
填充体 obturator
填充体系 infill system
填充填料 placing the fill(ing)
填充图 blank map;outline drawing;outline edition;silent map
填充网 filling-in net;fill(ing) net-(work)
填充纬 wadding picks;wadding weft
填充位 filler;padding bit
填充圬工 infilling masonry
填充物 adjutage;filler;filling material;gap filler;infilling;loading;pack-hardening;packing;packing filling;packing material;padding;weighting material;filling
填充物不均匀因数 packing non-uniforming factor
填充物尺寸 packing size
填充物的紧密化 packing tightening
填充物干燥器 packing drier[dryer]
填充物结构 packing structure
填充物支撑 packing support
填充物支架 packing support
填充物质 filling compound
填充吸收器 packed absorber
填充系数 block coefficient;charge coefficient;coefficient of charge;packing factor;packing fraction;solidity ratio;volumetric(al) coefficient;volumetric(al) factor
填充细胞 complementary cell
填充细矿料的地沥青 <膏体地沥青中掺 10% ~50% 细矿粉> mineral-filled asphalt
填充细矿料沥青 mineral-filled asphalt
填充限度 filling limit
填充型绝缘 fill-type insulation
填充性骨料 void-filling aggregate
填充性水泥混合材料 cement extender
填充性颜料 bulking pigment;extended pigment;extender pigment
填充穴 filled cavity
填充颜料 filler pigment
填充因数 fill factor;packing factor;stacking factor
填充因素 packed factor;packing factor
填充用块石 filling rock;packing rock
填充用纤维 fiber[fibre] fill
填充用原料 extender pigment;filler
填充油 packeted oil
填充圆 filled circle
填充蒸馏塔 packed distillation column
填充植物 filling plant
填充纸板 cardboard space former
填充柱 packed column
填充砖 filler brick;infill tile
填充砖块 infill brick
填充砖墙 brick nog(ging);fill-in brickworks
填充状态 occupied state
填充字符【计】 fill character;padding character
填出堤 banquette of levee
填出高地 filled-up ground
填粗土 coarse fill
填大陆海 aggrading continental sea
填到设计标高 fill-up to grade;fill-up to the design level
填道砟 boxing up
填堤材料 material composing the embankment
填堤运河 canal on embankment
填地 landfill;made-up ground
填地基硬材料 hard core
填垫扁铁 gasket iron
填垫作用 shimming
填堵进水通道 blocking and stuffing intake passage
填发票据 preparation of documents
填发运输票据的权力 authority to issue carriage documents
填方 backfill;earth fill;filling;landfill;made ground
填方材料 filling material
填方沉降 settlement of fill
填方沉陷 fill settlement
填方锤 backfill tamper
填方挡土墙 embankment
填方地面 filled-up ground
填方断面 fill section
填方堆放机 stacker for building up fills
填方高度 depth of fill(ing)
填方工程量 volume of filling
填方工程完成时的地面线 ground line
填方夯 backfill tamper
填方和挖方模拟 simulation of embankment and excavation
填方滑坡 fill slide
填方基础 embankment foundation
填方计量【疏】 measurement of fill
填方加宽 widening of fill
填方开挖 fill excavation
填方量 amount of fill;bank measure
填方溜失检测器 fill slip detector
填方路基 filled-up ground
填方坡脚 fill toe
填方铺摊推土铲 fill-spreading bulldozer
填方铺摊推土装置 fill-spreading bulldozer
填方渠道 canal on embankment
填方输送机 stacker for building up fills
填方数量 bank measure;filling volume;fill quantity;fill yards
填方摊铺(法) fill spreading
填方填充材料 deposit(e) fill material
填方下炸泥 mud blasting under fill
填方形成的地面 made ground
填方压实机 landfill compactor
填方压实机本体 basic landfill compactor
填方压实机主机 basic landfill compactor
填方用土 fill earth
填方用推土板 landfill bulldozer
填方用推土机 landfill bulldozer
填非均匀质土 heterogeneous fill
填非黏[粘]性土 granular fill
填粉煤灰 fly-ash fill
填封 blinding;tamping
填封度 <装炸药> degree of confinement
填封土 sealed earth
填缝 ca(u)lking;ca(u)lk joint;crack filling;crack sealing;joining;joint closure;jointing;rejointing;scratch in;sealing;sealing of cracks
填缝板 joint filler;joint fillet;joint plate
填缝板接缝 joint with sealing plate
填缝边 ca(u)lked edge
填缝材料 ca(u)lking material;crack filler;joint filling material;fill material;joint filler;jointing material
填缝槽 ca(u)lking groove
填缝锤 calking mallet;ca(u)lking hammer
填缝刀 filling knife
填缝对接 ca(u)lking butt
填缝法 joint filling
填缝工具 ca(u)lker;ca(u)lking set;ca(u)lking tool
填缝工作 ca(u)lking work;sealing work
填缝骨料 chippings;choker aggregate;key aggregate
填缝灌浆 joint grouting
填缝焊 ca(u)lking weld
填缝混合料 ca(u)lking compound
填缝混合物 joint filling composition
填缝混凝土 filler concrete;filling concrete;sealed concrete
填缝机 joint applicator
填缝集料 choker aggregate
填缝剂 sealant compound
填缝浆膏 plastic wood
填缝浇口 joint runner
填缝胶 gap-filling glue;joint sealant
填缝胶结料 joint cement
填缝胶泥 gap-filling cement
填缝胶粘剂 gap-filling adhesive
填缝接合 filled joint
填缝接头 ca(u)lked joint;gap filled joint
填缝冷黏[粘]合剂 gap-filling no-heat adhesive
填缝沥青板条 asphalt board strip
填缝沥青拌合[和]料 blended asphalt joint filler
填缝料 aggregate filling;building mastic;chinking;choker;filler;fissure feeder;gap filler;joint filler;joint filling compound;jointing material;joint sealer;joint sealing material;putty;sealant;sealer
填缝料加热炉 kettle for heating joint filler
填缝料熔化炉 melting furnace for joint sealer
填缝料熔化器 melter for joint sealer
填缝料熔炉 joint compound cooker;joint compound melting furnace;kettle for heating joint filler
填缝料溢流 sealant overflowing
填缝路渣 blinding
填缝麻屑 ca(u)lking hards;ca(u)lking tow
填缝麻絮 oakum
填缝玛琋脂 joint sealing mastic
填缝密封剂 joint sealing compound;gap-filling sealant;gap-filling adhesive
填缝密封胶 joint sealing compound
填缝黏[粘]结剂 gap-filling adhesive
填缝膨胀条 filling expansion strip
填缝片 expansion shim;sealing strip
填缝器 jointer;joint(ing) tool;sett feeder;sett jointer <用于石块路>
填缝铅 ca(u)lking lead
填缝枪 ca(u)lking gun
填缝乳剂 latex pitching compound
填缝软木板 cork board
填缝砂浆 sealing mortar
填缝勺 paying shell
填缝绳 rope sealing
填缝绳索 seal rope
填缝石 choke stone;filler stone;keystone
填缝石棉绳 <承口处> asbestos (joint) runner
填缝石屑 blinding
填缝(石油)沥青掺和料 blended asphalt joint;blended asphalt joint filler
填缝水泥 cement for joint filling;cement for joints;gap-filling cement
填缝索 joint runner
填缝套圈 ca(u)lking ferrule
填缝条 joint runner;sealing strip
填缝铁条 dumb iron
填缝涂料 joint finish
填缝土 <铺石地面> top dressing
填缝小砖 glut
填缝油膏 ca(u)lking compound
填缝油灰 <木材的> wood filling
填缝錾 ca(u)lking iron
填缝凿 ca(u)lking chisel;ca(u)lking iron;ca(u)lking tool;fuller
填缝止水材料 joint sealant (compound)
填覆 bury
填高 banking;depth of fill(ing);fill height;fill-up;raise;raising
填高场地用砂 sand for raising sites
填高地 filled-up ground
填高沟渠 elevated ditch
填高条 stemming piece
填沟机 backfiller;trench filler
填沟机平铲 backfiller blade;blade backfiller
填沟耐火材料 runner refractory
填沟造地 gull(e)y reclamation
填谷造地 gull(e)y reclamation
填灌混合料 grouting compound
填海 marine fill
填海单价 unit price marine fill;unit rate of marine fill
填海扩地 sea reclamation
填海用取砂区 sand borrow areas for reclamation
填海造地 sea reclamation;seashore new land;land reclamation
填海造地工程 reclamation works
填海造陆 sea reclamation
填海造陆用的填筑料 fill material for reclamation
填函连接 packed slip joint
填函料 packing gland
填焊 seal(ing) weld(ing);wed up
填好投标单价的工程量清单 priced bill of quantities
填黑的 black-filled
填弧坑 crater filler;crater filling
填混凝土的 concrete filled
填积层 <焊接> padding
填积沉积【地】 aggradational deposit
填积堤 bed groin
填积基准面 aggradation base level
填积料 fill
填积平原 aggradated plain;aggradation plain

填积型 aggrading pattern
填积型生长层序 aggrading growth sequence
填积作用 aggradation
填集 packing
填集料镐 packing pick
填集趋近度 packing proximity
填集指数 packing index
填间结构 interspatial texture
填肩拱 spandrel-filled arch
填肩式拱桥 spandrel-filled arch bridge
填焦过滤器 coke filter
填焦洗涤器 coke scrubber
填角 fillet
填角焊 bead welding;fillet weld(ing)
填角焊分缝 parallel fillet weld
填角焊缝 back bead;filler weld;fillet welded joint
填角焊缝尺寸 fillet weld size
填角焊缝喉长 throat of fillet weld
填角焊缝轮廓 profile of fillet weld
填角块 corner fillet
填角密封接缝 fillet sealant joint
填金 filled gold
填卷 snake;snaking;worming
填开纳税凭证 issue payment receipt
填坑 backfill;dump pit;landfill
填空白 space filling
填空部件 infilling
填空材料 infiller material;infilling material
填空度 compactedness
填空短管 spacer nipple
填空混凝土墙板 infiller concrete panel;infilling concrete panel
填空料 infilling
填空砌块 infiller block;infilling block
填空铅 quotation
填空铅条 quad
填空嵌条 quadrat
填空墙 infiller wall;infilling wall
填空石 expletive stone
填空小石 sneck
填空性建设 infill development
填空性建筑 infill development
填空性住房 infill housing;vest-pocket housing
填空砖 infiller brick;infilling brick
填孔 pore filling
填孔导纱器 filler point
填孔及擦平 stopping and smoothing down
填孔剂 packing compound
填孔浆 porous filling paste
填孔木楔 filling-in piece
填孔腻子 beaumontage;filling putty
填孔漆 block filler
填孔涂层 filler coat
填块 caul;cover block;fill;fill block;filler block;wad
填块格栅 <防火地板中用的> filler joist
填块石 rock fill
填沥青伸缝 bituminous expansion joint
填砾 gravel pack(ing)
填砾材料 material of gravel stuffing
填砾层厚度 thickness of gravel stuffing
填砾的给水度 specific yield of gravel stuffing
填砾段起止深度 depth of gravel stuffing from top to bottom
填砾高度 height of gravel stuffing
填砾过滤器 gravel stuffing filter
填砾井 gravel wall well
填砾料井 gravel wall well
填砾石 gravel fill
填砾石衬筒 prepacked liner
填砾石井 gravel envelope well
填砾石排水沟 pebble-filled trench
填砾体积 volume of gravel stuffing
填砾直径 diameter of gravel stuffing
填粒状土 granular fill
填料 filler material;filling(up);aggregate;brasq(ue);ca(u)lking;dumped packing;extender;filter;gasket material;gaskin;gland packing;liner;loaded stock;loading;loading material;packing;packing block;packing filter;packing material;padding;placed material;plums;sparkling;stuff(ing);wad(ding);backup material <接缝用>
填料斑(瑕) filler specks
填料板 filler plate
填料保持 filler retention
填料补强 filler reinforcement
填料槽 filler trough;packing chamber;packing groove;stuffing box
填料层 filter pack;packing layer
填料层收尘器 gravel dust filter
填料衬环 throat ring
填料称量斗 filler weigh hopper
填料秤 filler scale
填料储器 filler reservoir
填料床 packed bed
填料床法 packed-bed process
填料床反应器 packed-bed reactor
填料床洗涤除尘器 packed-bed dedusting scrubber
填料斗式升运机 filler bucket elevator
填料反应器 filling reactor
填料粉 filler powder
填料盖 gland cap;gland cover;packing gland;stuffing box gland
填料高度 packing height
填料割刀 packing knife
填料隔圈 packing ring
填料工 filler;fuller
填料钩 packing hook;packing worm
填料管 <弯管时管内填料以防变形> loading pipe
填料过滤器 wadding filter
填料含量 filler content;filler loading
填料函 gland body;gland box;packing box;packing chamber;stuffing box;stuffing case
填料函盖 follower;stuffing box
填料函螺母 gland nut;stuffing box nut
填料函热补偿器 stuffing box thermal compensator
填料函式换热器 stuffing box heat exchanger
填料函压盖 follower plate;stuffing (box) gland;packing gland;junk ring
填料函压盖螺帽 packing gland nut
填料和装潢用工业石膏 industrial plaster for fillers and decoration
填料盒 packing box;packing chamber;stuffing box
填料盒螺栓 stuffing box bolt
填料盒压盖 stuffing box cover;stuffing box gland;stuffing gland
填料盒压环 junk ring
填料护盖 bull ring
填料环 lantern gland;packing ring;ring of packing
填料换装 filler replacement
填料回收 filler reclamation
填料混炼胶 compounded rubber;pigmented compound
填料混入 filler absorption
填料级滑石 filler-grade talc
填料加强效应 filler reinforcement effect
填料介质 packing medium
填料金属 filler metal
填料紧螺栓 packing bolt
填料可混入量 filler acceptance
填料空隙度 packing void fraction
填料孔 fill opening
填料-沥青比 filler bitumen ratio
填料硫化胶 filled vulcanizate
填料螺环 packing stuff screw ring
填料螺旋起子 packing screw;packing worm
填料密度 packing density
填料密封 packing seal
填料模壳 filler form
填料黏[粘]土 filler clay
填料喷注 filling spout
填料膨胀 expansion of packing medium
填料曝气塔 packed tower aeration
填料器 sizer
填料圈 insertion ring
填料容量比 matric capacity ratio
填料润滑 pad lubrication
填料式润滑装置 pad lubricator
填料塔 filled column;filled tower;packed column;packed tower;packing column;packing tower
填料塔曝气 packed tower aeration
填料塔式过滤器 packed tower filter
填料塔式滤池 packed tower filter
填料弹簧 packing spring
填料提升机 filler elevator
填料添加 filler adding
填料填缝用设备 application equipment
填料筒 loading bin
填料筒仓 filler silo
填料头 packing head
填料瓮 sizing vat
填料物 packing material
填料纤维 filler fibre[fiber]
填料箱 gland box;packing box;packing case;stuffing box
填料箱补偿器 packing box compensator;stuffing box compensator
填料箱衬套 stuffing box bushing;stuffing box insert
填料箱刮油环 scraper ring for stuffing box
填料箱涵 stuffing box
填料箱密封环 sealing ring for stuffing box
填料箱密封填料 packing of stuffing box
填料箱内套 neckbush in stuffing box
填料箱式伸缩器 gland type joint
填料箱水冷夹套 water-cooled jacket for stuffing box
填料箱型号 stuffing box model
填料箱压盖 stuffing gland
填料箱压缩环 pressure ring for stuffing box
填料橡胶 loaded rubber;packing rubber
填料压板 hold-down grid;packing restrainer
填料压盖 junk ring;packing gland;stuffing box gland;stuffing gland;stuffing ring
填料压盖板 gland plate
填料压盖衬套 packing bush
填料压盖法兰 follower flange
填料压盖颈 gland neck
填料压盖螺母 packing nut
填料压盖室 stuffing chamber
填料压盖凸缘 follower flange
填料压盖轴承 stuffing box bearing
填料压盖轴颈衬套 gland-neck bush
填料压机 bale press;baling press;packing press
填料压力 filling pressure
填料压圈 packing ring
填料压缩器 packing compressor
填料因子 packing factor
填料硬化 packing hardening
填料油 gland oil
填料增强效应 filler reinforcement effect
填料胀圈 packing flange
填料蒸馏塔 packed distillation column
填料支承架 packing support
填料纸 loaded sheet
填料贮存装置 filler storage unit
填料柱 filled column
填零 zero fill;zeroize
填路堤边坡 fill bank
填卵石排水沟 gravel-filled drain trench
填埋 landfill
填埋处理 landfill disposal;packing disposal;packing treatment
填埋处置 packing disposal
填埋法 burying method
填埋管 buried pipeline
填埋滤液特性 characteristics of landfill leachates
填埋气体 landfill gas
填埋污泥 bury sludge
填满 chock;filling-in;fill-up;padding
填满板缝 filled joint
填满到坡度线 fill-up to grade
填满概率 occupation probability
填满砂的 sand-filled
填满喂料 chock feed
填满细胞法 <木材防腐的一种处理方法> Bethell's process;full-cell process
填密 ca(u)lk;filleting;fuller;pack(ing);stuffing
填密材料 padding material
填密垫片 joint sheet
填密垫圈 jointing material
填密缝 filler seam
填密函盖 packing gland;packing retainer
填密函盖螺母 packing gland nut
填密函盖弹簧 packing gland lock spring;packing retainer spring
填密盒 packing case
填密环 dummy ring;packing ring
填密簧 packing spring
填密接头 filling connection
填密空间 packing space
填密螺母 packing nut
填密螺母扳手 packing nut wrench
填密螺栓 packing bolt
填密片 ca(u)lking segment;ca(u)lk piece;gasket;packing piece;packing sheet;packing strip;sheet gasket;sheet packing
填密片板式连接 gasket mounting
填密圈 packer;packing ring
填密砂 close sand
填密石棉绳 asbestos rope packing
填密凸缘 packing flange
填密橡胶圈 rubber-ring packing
填密毡 felt packing
填密胀圈 expansion gland
填密状态 state of packing
填没 blinding
填木 dunnage;wooden filler
填木孔料 <树脂胶泥> beaumontage
填泥 filler;spackling
填泥料 spackle
填腻子 filling;putty filling
填腻子用刮刀 filling knife
填黏[粘]性土 clay fill
填排水砂垫层 filling drained sand cushion
填炮泥 stemming
填炮眼工 tamperer
填配料嵌缝 spackling compound seal-

(ing)
填配料涂层 spackling compound coat
填片厚度 hinge
填片密封镶嵌 gasket glazing
填票机 billing machine
填平 dubbing out
填平补齐 transition patching
填平补齐用楔 level-up wedge
填平地坪 level(l)ing work
填平绳纹 worm
填坡 fill slope
填铅 leaded
填铅承插口 socket for lead joint
填铅接缝 lead joint
填铅接合缝 lead joint;plumb joint
填嵌粉料 stopper powder
填嵌胶泥 badigeon
填嵌料 filling compound
填嵌石膏 stopper gypsum
填嵌物密封 stopper seal(ing)
填墙木砖 nogging;nogging piece
填圈 gaskin
填入 fill(ing)-in;offering
填入料 loading material
填入土 buried dump
填塞 choke;choking;closure;clutter;cramp;dubbing out;feeding;filling(up);pack(ing);pad(ding);pinning-in;plug;stop;stop-off;stuff(ing);tamp(ing);tucker;wad
填塞材料 blinding material;stemming material;tamping material;choke material
填塞槽 ca(u)lking pocket
填塞袋 tamping bag
填塞刀 filling knife
填塞的 blind;choke;expletive;expletory
填塞的拱 blind arch
填塞垫 filler wad
填塞钉孔 backnailing
填塞洞眼 ca(u)lking up hole
填塞堆积阶地 fill-in fill terrace
填塞法 plugging
填塞缝 filled joint
填塞缝隙 pay a seam;garret <用小石块>
填塞干硬性混凝土 dry pack(ing)
填塞杆 stemming rod;tamping bar
填塞工程 packwork
填塞工具 filler tool;tamper
填塞工作 filling operation;filling process;packwork
填塞骨料 choke aggregate
填塞管 loading pipe
填塞过滤器 wadding filter
填塞灰板条缝的草泥灰 cat
填塞集料 choke aggregate
填塞坚实的 impregnable
填塞件 packing piece
填塞接缝 packed joint
填塞接头 tamped joint
填塞进料 chock feed
填塞绝热 fill-type insulation
填塞绝缘 fill insulation
填塞绝缘材料 fill insulation material
填塞空洞 block openings
填塞孔 fill-up hole
填塞孔洞 block openings
填塞块 fill block;filling-in piece;packing block;packing piece
填塞矿物 stuffed mineral
填塞料 blinding material;choke material;packing;packing material;stuffing;tamping;filling element;filling material;stopping
填塞泥土隔墙 pugging
填塞炮孔 <用炮泥> tamped well
填塞炮泥 tamp
填塞炮眼 tamp a hole
填塞盆地 stuffed basin
填塞片 filling piece
填塞气旋 filling cyclone
填塞深度 depth of packing
填塞绳 packing cord
填塞石 expletive
填塞石缝 galleting;garnet(ing)
填塞石块 stone packing
填塞式绝缘 fill-type insulation
填塞膛 ca(u)lking pocket
填塞条 wadding strip
填塞土(或砂石)blinding
填塞物 bolus;bysma;crammer;padding;plugger;tamping;tamp(i)on;stemming;wadding
填塞穴 ca(u)lking pocket
填塞硬化 pack-hardening
填塞凿 chinsing iron
填塞炸药杆 stemming rod
填塞者 tamper
填塞注浆 confined injection;packed injection;packer system
填塞砖 closer;closure brick
填塞桩 chock pile;fill pile
填塞作业 packwork
填沙护滩 beach fill;beach nourishment
填砂 <又称填沙> backing sand;backup sand;blinding sand;fill sand;floor sand;sand feed;sand fill(ing);sand(ing)up
填砂垫层 sand-filled cushion
填砂护滩 artificial nourishment;artificial renourishment;beach fill;beach nourishment;beach renourishment;beach nourishing
填砂排水井 sand-filled drainage well
填砂渗水井 sand-filled drainage well
填砂升高施工法 sand jacking method
填砂石料基层 sand-filled stone base
填砂土工袋 sand-filled bag
填砂围垦 sand fill reclamation
填砂支柱 sandow
填上 fill-up
填上部分 fill section
填梢工 packwork
填梢护岸 kidding
填梢石 packwork
填石 drop fill rock;enrockment;placed rockfill;rock-fill(ing);rock riprap;stone fill(ing);stone packing
填石暗沟 stone-filled trench
填石暗沟排水 rock underdrain
填石坝 backfill dam;rock-fill dam
填石板桩格笼 rock-filled sheet pile cell
填石板桩格体 rock-filled sheet pile cell
填石材料 rock-fill material
填石沉排 stone mattress
填石的 stone-filled
填石地下排水管 blind subdrain
填石丁坝 rock-fill groin;rock groyne
填石垛 rock-filled cribbing
填石防波堤 rock-filled breakwater;rock-filled jetty
填石钢丝笼 wire bolster
填石和填砂 rock and sand fill
填石护岸 packed stone revetment;stone-fill revetment
填石护坡 packed stone revetment
填石集水沟 rubble catchwater channel
填石块 rocking
填石沥青板 <粗骨料达 25% 的沥青板> stone-filled sheet asphalt
填石笼 stone cage
填石路堤 rock embankment
填石路面填筑石料 stone packing
填石盲沟 French drain;rock drain;rock-fill drain;rubble drain;stone drain;stone-filled drain;stone-filled trench
填石木垛 filled crib;filled timber crib;rock-fill crib
填石木垛坝 rock-filled timber crib dam
填石木框坝 stone-filled crib dam
填石木笼 cribwork filled with stone;rock-fill(ed)crib;rock-filled timber crib;stone-filled crib;timber crib;timber rock-fill crib
填石木笼坝 crib dam filled with stone;rock-filled crib dam
填石木笼挡土墙 crib retaining wall
填石木笼防波堤 timber crib breakwater
填石排水 stone drainage
填石排水沟 blind drainage conduit;gravel fill;hard core drain;rock drain;rock-fill drain;rubble drain;stone-filled trench
填石排桩护岸 stone-filled pile revetment;pile with stone fill revetment
填石片地沥青(混合料)stone-filled sheet asphalt
填石铅丝笼 lead wire stone cage
填石梢笼 brush rolled filled stones
填石铁丝笼 bolster;gabion;rock-filled wire gabion
填石突堤 filled jetty;rock-filled jetty
填石围堰 rock-fill cofferdam
填石细粒式沥青混凝土 <德国> stone-filled fine asphalt
填石竹笼 bamboo gabion;bamboo stone cage;stone-filled bamboo gabion
填石竹笼围堰 bamboo gabion cofferdam;rock-fill bamboo basket cofferdam
填实 compaction;pack(ing);solid filling;tamp
填实部分 solid filled portion
填实层 packing course
填实底座 packed bottom
填实度 degree of packing;packing degree
填实缝 ca(u)lk seam;packed joint;sealed joint;spline joint
填实管道接口用 O-ring
填实管柱 <用混凝土> filled(pipe)column
填实管桩 filled pipe column;filled pipe pile
填实接缝 gasketed joint
填实空间 partition coverings
填实空隙 infilling
填实孔 tamped well
填实框架 infill frame
填实料 packing material
填实密度 packing density
填实凿紧 ca(u)lk
填丝 filler rod
填碎石的排水沟槽 stone-filled drain trench
填碎石的排水沟渠 stone-filled drain trench
填碎石排水沟 stone-filled drain
填碎石排水管 stone-filled drain
填碎砖排水(盲)沟 brick bat drain;brickbat drain
填塘混凝土 dental concrete
填图 charting;map exercise;mapping;map plotting;map spotting;plotting
填图单位 mapping unit
填图单位界线控制点 control point of boundary between different mapping units
填图地质单位 unit of geologic(al) mapping
填图符号 plotting symbol
填图格式 station model
填图孔 mapping drill hole
填图员 chart plotter
填土 backfill(soil);banket(te);banking;banquette;earth embankment;embankment fill;filled soil;fill(ing);ground fill;landfill;made land;made-up ground;soil embankment;soil filling
填土坝 earth-fill dam;fill dam
填土包 earth-filled bag
填土边坡 fill slope
填土边坡角 <路堤> angle of fill slope
填土标桩 fill stake;gradient peg;gradient stake
填土不多的 lightly covered
填土部分 fill section
填土层 fill layer;reclamation soil layer
填土长度 fill length
填土场 buried dump
填土沉降 fill settlement;settlement of embankment
填土沉陷 fill settlement;fill subsidence;settlement of fill
填土处理 earth filling
填土袋 earth-filled bag
填土的底 base of fill
填土的岩洞 soil-filled bag
填土地 made ground
填土地基 filled(-in)ground;fill(ed)-up ground
填土顶部 fill crest
填土动物穴 crotovina
填土堆 bunker
填土方 earth fill
填土高度 depth of fill(ing);depth of foundation;fill height;height of fill
填土高渠 elevated ditch
填土工 backfiller
填土工程 fill construction;filling-in work;fill(ing)work;reclamation works
填土工程面积 fill construction area
填土工作 filling-in work;filling operation;filling process;filling work
填土过程 filling process
填土含水量 placement moisture content;placement water content
填土夯实机 backfilling tamper;soil compactor
填土厚度 height of fill
填土机 backfiller;pushfiller
填土基层 fill base
填土结构物 banking structure
填土截面 fill section
填土宽度 fill width
填土路 made road
填土木坝 earth-fill timber dam
填土抛高 extra-banking;extra fill
填土坡脚 fill toe;toe of fill
填土深度 depth of fill(ing)
填土升运机 fill lift
填土施工 fill construction
填土石方法 pay formation
填土实腹石拱桥 earth-filled stone arch bridge
填土围垦 land reclamation by filling
填土围堰 earthfill cofferdam
填土位置 filling position
填土下沉 slumping of fill;subsidence in fill
填土压力 filling pressure
填土压实 compacting of fill;compaction of fill;compaction of landfill;fill compaction
填土压实机 landfill compactor;soil compactor

填土掩埋 embankment fill
填土用铲刀 landfill blade
填土用的推土板 landfill blade
填土用水枪 slushing giant
填土造地 land reclamation by filling; made ground; made-up ground; reclamation of land by filling
填土栈桥 filling trestle
填土振动压实器 vibrating backfill compactor
填土筑地 made land
填挖标 grade stake
填挖方高程标准桩 grade stake
填挖方平衡 cut-fill transition
填挖方设计曲线 cut-fill design curve; mass-haul curve
填挖高度 cut-fill height
填挖计划 cut-and-fill program(me)
填挖渐变段 fill-cut transition
填挖平衡 equalization of embankments and cuttings
填挖平衡的坡度 balanced grade; balanced grading
填挖平衡的土方工程 balanced earthwork
填挖平衡的纵断面 balanced profile
填挖平衡线 supported line
填洼水量 depression storage
填丸加固 shot backing
填无字的报表"nil" return
填隙 ca(u)lk(ing); filling of crack; skimming
填隙板 filler plate; shim(ming) plate
填隙比 void-filling ratio
填隙材料 interstitial material
填隙锤 ca(u)lking hammer; ca(u)lking mallet
填隙的 intersertal; interstitial
填隙垫衬 shim liner
填隙垫圈 shim washer
填隙钉 ca(u)lking nail
填隙工具 ca(u)lking hammer; ca(u)lking tool
填隙骨料 choker aggregate; void aggregate; void-filling aggregate
填隙固熔体 interstitial solid solution
填隙合金 ca(u)lking metal; metal filler
填隙化合物 interstitial compound
填隙环 packing spool
填隙货 beam filling; short stowage; small stowage
填隙货物 filler cargo; short stowage cargo
填隙集料 choker aggregate; void-filling aggregate
填隙结构 intersertal texture
填隙结核 intercretion concretion
填隙金属 filler metal; ca(u)lking metal
填隙空缺偶 interstitial vacancy pair
填隙空位偶 interstitial vacancy pair
填隙离子 interstitial ion
填隙量 void-filling capacity; void-filling content
填隙料 calking material; ca(u)lker (material); gap filler; gasket; filler
填隙率 interstitial rate; void-filling ratio
填隙能力 void-filling capacity
填隙黏[粘]合剂 gap-filling adhesive
填隙捻缝凿 yarning iron
填隙凝胶量 <水泥净浆颗粒间的空隙> void-filling gel
填隙片 dutchman; filler piece; filling piece; shim; shim piece; shim plate
填隙器 ca(u)lking iron; ca(u)lking tool; clincher iron
填隙嵌缝能力 <胶粘剂的> gap-filling capacity
填隙圈 junk ring
填隙石 pinning
填隙式 interstitial
填隙式合金 interstitial alloy
填隙式结构 interstitial structure
填隙式扩散 interstitial diffusion
填隙数据 gap filler data
填隙替位模型 interstitial-substitutional model
填隙物 chinker
填隙物质 interstitial material; interstitial matter
填隙压板 fuller board
填隙阳离子 interstitial cation
填隙因数 full factor
填隙阴离子 interstitial anion
填隙用细石屑 blinding
填隙原子 interstitial atom
填隙凿 ca(u)lking chisel; ca(u)lking hammer; ca(u)lking iron; ca(u)lking tool
填隙子 interstitial
填隙作用 infilling; skim action
填闲作物 alternate crop; catch crop
填写 fill in; fill-up; interpolation
填写单据 billing
填写领料单 writing requisition
填写路程单 preparation of waybill
填心板 core panel
填心混凝土 hearting concrete
填心石块 hearting
填絮锤 reeving beetle
填压 pressurization
填鸭式施教 cram
填药棒 stemming rod
填以黏[粘]土的袋穴 <防止气压盾构在砂砾土中逸气而在工作面挖成的> clay-filled pocket
填有内容的登记项 filled entry
填有碎石的 debris-filled
填载运费 distress freight
填早日期 antedate; backdate; foredate; predate
填早日期票据 foredated bill
填早日期支票 antedated cheque; foredated check
填渣 slag fill
填炸药棒 stemming rod
填脂孔 arming hole
填制器 controller
填质【地】 matrix[复 matrixes/matrices]
填注料价 pricing
填柱物 column filling
填柱液 column hold-up
填筑 land accretion; land reclamation; reclaim; reclamation (by filling)
填筑坝 embankment dam; fill-type dam; reclaimed dam; reclamation dam
填筑标准 <土石方的> placement criterion
填筑材料 embankment material; fill material
填筑层厚 <土石料的> placement lift
填筑层厚度 lift thickness
填筑长度 fill length
填筑成堤 bank against
填筑成坡 bank against
填筑成滩 bank against
填筑到设计线 bring to grade
填筑的 filled
填筑的填料 fill material for reclamation
填筑地 filled land; fill-up ground; made ground; made land; make-up ground; reclaimed ground; reclaimed land
填筑地点 point of placement
填筑地区 reclaimed area
填筑点 point of placement
填筑高程 reclamation level
填筑工程 reclamation engineering; reclamation project; reclamation works
填筑含水量 placement moisture
填筑含水率 placement water content
填筑好的基础 prepared foundation
填筑好的路基 prepared foundation; prepared subgrade
填筑和挖槽模拟 simulation of embankment and excavation
填筑混凝土 fill(er) concrete
填筑机 reclaimer
填筑机械 positive appliance
填筑块石 stone packing
填筑立方码 fill yards
填筑露天矿 strip mine reclamation
填筑率 rate of filling
填筑轮廓点测量 setting-out of footing foundation peripheral points
填筑密度 density of fill
填筑密实度 placement density
填筑面 esplanade; surface of filling
填筑面积 reclaimed area
填筑耙 reclaiming barrow
填筑强度 fill intensity
填筑区 area under reclamation; built-up land; reclamation area; reclamation district; reclamation site; reclaimed area
填筑湿度 placement moisture
填筑石料 stone packing
填筑式堤坝 fill-type dam
填筑碎石 debris-filled
填筑土坝 fill dam
填筑土的改良 improvement of reclaimed soil
填筑土地 land accretion; land reclamation
填筑土地材料 fill material
填筑土路 reclamation land
填筑心墙 hearting
填筑用砂 sand for raising sites
填筑用杂石 gob
填筑柱箍 casing of pile
填砖 bricking-up
填砖密肋楼板 ribbed(clay)brick floor
填装 packing; padding
填装器 precompressor; wad box
填装炸药棍 loading stick

舔液辊 lick roller

挑板 bridging piece; cantilever slab; launching pad

挑鼻坎 bucket curve of spillway
挑拨 provoke
挑出 corbel out; destuffing; undercut; sail-over <构件端头>; jetting out <如托臂等由墙突出>
挑出臂 cantilever arm
挑出层 projection course
挑出窗 bay window
挑出的 corbelled out
挑出的长方形踏步 rectangular cantilever(ed) step
挑出的扬声器 projecting diffuser
挑出的承重砌体 bragger
挑出的格栅尾端 tail joist
挑出的脚手架 flying scaffold
挑出的金属旗杆插座 braciale
挑出的扩散器 projecting diffuser
挑出的梁 outrigger beam
挑出的楼梯 cantilever stair(case)
挑出的楼梯纵梁 bracket stringer
挑出的平板 cantilever slab
挑出的腿 projecting leg
挑出的线脚 corbel mo(u)ld
挑出的压顶 projecting coping
挑出的雨罩 <一般指戏院门口的> marquee
挑出的圆顶 cantilevering dome
挑出的圆盘 cantilevering disk[disc]
挑出的圆锥体 cantilevering conoid
挑出的招牌 projecting sign
挑出的整平板 <混凝土> flying screed
挑出的砖石层 oversailing course
挑出地板 oversailing floor
挑出焊 projection welding
挑出脚手架 projecting scaffold; outrigger scaffold
挑出跨 cantilever span
挑出块 corbel piece
挑出栏杆 bracket baluster
挑出梁 outrigger beam
挑出楼层 overhang; overhanging stor(e)y
挑出楼梯边梁 bracket stringer
挑出轮 outrigger wheel
挑出面层 <由一片牛腿支承的> round-arched corbel-table; corbel table
挑出牌 facia
挑出式脚手架 bracket scaffold(ing); bracket staging; cradle scaffold; flying scaffold; outrigger scaffold(ing)
挑出式楼梯边梁 bracket stringer
挑出式压顶 projecting coping
挑出踏步 corbel steps; crow step; bracket-step
挑出铁件 tail iron
挑出托拱 corbel arch
挑出圬工 corbelling
挑出屋顶 overhang; overhanging roof
挑出屋面 overhanging roof
挑出物下的线脚 thickness mo(u)lding
挑出线脚 corbel mo(u)ld
挑出斜坡 counter slope
挑出摇摆式招牌 swinging projecting sign
挑出砖 brick corbel; corbel piece
挑出砖层 sailing course
挑出砖牙 brickwork corbel
挑出砖牙压顶山墙 corbel step gable; corbie step gable
挑椽接木 sprocket
挑窗 canopy window; projected window
挑窗台 moucharaby
挑错测验 absurdity test
挑带柱 tape lifter
挑担石 cornice stone
挑顶 back brushing; blowdown; brushing of tunnel top; datalling; overbreakage; overcut; second ripping; top canch; top breaking【矿】; top picking <隧道工程>
挑顶爆破 ripping blasting; top shooting
挑顶边界 ripping edge
挑顶的 ripping
挑顶工作面支护 ripping face support
挑顶机 scaler
挑顶面工作端 ripping lip
挑顶刷帮炮眼 brushing shot
挑顶卧底 canch; kanch
挑顶岩石 ripping dirt; ripping stone
挑顶钻车 sealing jumbo
挑杆 ram
挑杆摆动机构 rod oscillation mechanism
挑杆回转机构 rod tipping mechanism
挑杆起重机 arm crane
挑杆式装卸机 ram truck
挑杆锁紧机构 rod locking mechanism

T

挑焊运条法 whipping method
挑弧焊 whipping method
挑花窗帘 lace curtain
挑架 builder's jack
挑尖梁 main aisle exposed tiebeam
挑拣 pick
挑脚手架 projecting scaffold
挑坎 ramp;rim
挑口板 <檐槽的> fascia board;gutter board
挑口过梁 boot lintel
挑口梁 fascia beam
挑口饰 fascia[复 fa(s)ciae/fa(s)cias]
挑扣刀 threading tool
挑框游丝 Breguet hair spring;Breguet spring;overcoil
挑廊式入口 balcony access
挑梁 cantilever beam;outrigger;overhanging beam;projecting beam
挑梁板 bridge beam;bridging piece
挑梁式基础 cantilever foundation
挑料棒 gathering iron
挑料杆 <挑玻璃液用> gathering iron;dolly
挑料工 bit gatherer;gatherer;post gatherer;post holder
挑料环 boat;gathering ring
挑料(加入的)气泡 gathering bubble
挑料口 gathering hole;ring hole;working hole
挑料温度 gathering temperature
挑流 deflected stream;ski jump
挑流坝 spur dike[dyke];spur jetty;transverse dike[dyke]
挑流板 deflector hood
挑流鼻坝 bucket
挑流鼻槛 deflecting bucket
挑流鼻坎 bucket;bucket lip;deflecting bucket;deflector bucket;trajectory bucket;flip bucket
挑流鼻坎曲线 bucket curve
挑流齿槛 <溢洪道的> dentated bucket lip
挑流冲刷 jet scour
挑流唇 deflector lip;flip bucket
挑流堤 spur levee
挑流反口弧 trajectory bucket
挑流反力 deflecting force
挑流构物筑 current deflector
挑流弧坎 deflecting bucket;deflector bucket
挑流戽斗 deflecting bucket;deflector bucket
挑流离岸 divert current away from the bank
挑流式消力戽 trajectory bucket type energy dissipater[dissipator]
挑流式消能工 flip trajectory bucket dissipater;ski-jump energy dissipater[dissipator]
挑流式消能设备 flip trajectory bucket dissipater;ski-jump energy dissipater[dissipator]
挑流式溢洪道 ski-jump spillway
挑流物 baffle
挑流消能 energy dissipation of ski jump type
挑流消能坎 flip bucket
挑楼 sail-over
挑棚 canopy;marquee;marquise
挑棚式人行道 canopy sidewalk
挑篷 umbrella roof
挑起 shoulder
挑砌腰线 projecting belt course
挑砂坝 bed-load deflecting apron
挑砂底槛 bed-load deflecting sill
挑砂护坦 bed-load deflection apron
挑射路径 trajectory path
挑式便桶 corbel WC
挑水坝 bankhead;groin-dam;groin(e);groyne;repelling groin;spur dike[dyke];wing dam
挑水倾斜面板 canting strip
挑台 bragger;cantilever platform;sail-over
挑台窗 meshrebeeyeh;moucharaby
挑剔 pick on
挑剔索赔 picking claim
挑头 corbel out
挑头接合 corbel joint
挑选传送器 sorting conveyer[conveyor]
挑选的 alternative;selective
挑选的材料 selected material
挑选的填充物 selected filling
挑选购买品 shopping goods
挑选临时职工 shape up
挑选试验 percentage test(ing)
挑选输送器 picking conveyer[conveyor]
挑选投标者的竞争投标 selective tendering
挑选投标者的投标 closed bidding;closed competitive selection
挑选物 option;picking
挑选型抽样检验 sampling inspection with screening
挑选最佳方案 select the most favorable alternatives
挑檐 corbel table;cornice;creasing;overhanging eaves;projecting eaves;raking cornice;roof overhang;roof overhung door
挑檐板 gutter board
挑檐杵 pole plate
挑檐处带形花边饰 <陶立克建筑的> tenia
挑檐的转向 cornice return
挑檐滴水板 corona
挑檐滴水板下的线脚 thickness mo(u)lding
挑檐底板 cornice soffit;planceer piece;plancier;plancier piece;soffit board
挑檐底面 soffit
挑檐枋 eaves purlin(e)
挑檐封檐板 verge board
挑檐拱顶 corbel vault
挑檐桁 pole plate system
挑檐就位加工 revale
挑檐梁 eaves beam
挑檐抹灰样板 horse mo(u)ld
挑檐平顶 deflecting bucket
挑檐平顶板 eaves soffit
挑檐墙以上的楼层 attic stor(e)y
挑檐穹隆 corbel vault
挑檐上部窗 luthern
挑檐石 cantilever stone on eave
挑檐托块的间距 intermutule
挑檐托梁 bearer of a gutter;cornice bracket
挑檐托座 cornice bracket
挑檐瓦 water-rib tile
挑檐外包饰面 cornice casing;cornice facing
挑檐形线脚 <圆形窗头的> round pediment
挑檐照明 cornice lighting
挑檐支架 lookout
挑檐装饰 lacunaria
挑檐装饰抹面 <古罗马> coronarium
挑阳台 balcony;cantilevered balcony
挑战书 challenge
挑战者 challenger
挑战者破裂带 Challenger fracture zone
挑针器 narrowing picker
挑砖 corbelling;corbel piece;oversailing bricks
挑砖层 oversailing brick course;oversailing course;projecting brick course
挑砖饰线 mo(u)lded projecting course

条

条案 long narrow table

条斑 cord
条斑窃蠹 furniture beetle
条斑状褐腐 brown mottled rot
条板 board;lath;louver[louvre];louver board;luffer board;plank;ribbon;ribbon board;riffler;slat;stripe;sparring
条板包装 crating
条板边马车 slat-side stock car
条板边牲畜车 slat-side stock car
条板锯 slat saw
条板靠背椅 slat-back chair
条板轮拖拉机 cleated-wheel tractor
条板抹灰隔墙 lath and plaster partition
条板蕊细木工板 batten board
条板式输送机 slat conveyer[conveyor]
条板箱 crate;skeleton case;slat crate
条板芯细木工板 lumber-core board;batten board
条孢牛肝菌属 <拉> Boletellus
条壁犁体 slat bottom
条冰 ice in sheet
条播 line seeding;row cropping;sowing in drills;sowing in furrows
条播法 drilling method
条播沟 drill furrow
条播机 driller;seed drill;sowing drill
条播开沟机 drill colter
条播开沟器 drill coulter;drill shoe
条播犁 seed drill plough
条播器 drill planter
条播装置总成 drill assembly
条播作物 row crop
条材 bar stock
条材钢料 steel billet
条材进给 stock feed
条尺 narrow scale
条虫状的 tapeworm-shaped
条虫状气孔 worm hole
条船柱 bitt
条带 banding;braid;image strip;railing
条带测深系统 swath(e) sounding system
条带法 bands method
条带耕作 strip cropping
条带构造 streaky structure
条带结构 ribbon texture
条带聚合体 ribbon polymer
条带录音机 strip-chart recorder
条带绿板岩 desmosite
条带片麻岩 banded gneiss
条带式裁剪机 tape cutting machine
条带式型钢 strip steel section
条带式照明设备 strip-line lighting fixture
条带式照明装置 strip-line lighting fixture
条带涂层法 strip coating method
条带纹的 scored
条带形灯 tubular discharge lamp
条带形扩孔器 <镶焊金刚石的> slug-type reaming shell;strip type (reaming) shell
条带形绕组 strip winding
条带形衰减器 strip attenuater[attenuator]
条带形诱缝设置 crack inducer strip
条带状 ribbon;stripped
条带状白云岩 zebra dolomite
条带状层面 stripped bedding plane
条带状大理石 oriental alabaster
条带状大理岩 banded marble;oriental alabaster
条带状辐射体 strip shaped radiator
条带状构造 banded structure;ribbon structure
条带状构造面 stripped structural surface
条带状含铁建造 banded iron-bearing formation
条带状混合片麻岩 ribbon amphogneiss
条带状混合岩 striped migmatite
条带状假流纹构造 streaked pseudofluidal structure
条带状结构 banded structure
条带状矿石 bended ore
条带状煤 banded coal
条带状泥炭 banded peat
条带状黏[粘]土 ribbon clay
条带状片麻岩 banded gneiss
条带状铅锌矿石 bended Pb-Zn ore
条带状石灰岩 streaked limestone
条带状铁建造 banded iron formation
条带状铁矿 banded iron ore
条带状影纹 banded texture
条地 strip
条锭 wire bar
条锭铜 wirebar copper
条段光谱 fluted spectrum
条沸石 yugawaralite
条分法 <用以验算土坡稳定> method of slices;slice method;finite slice method;Swedish method
条缝抽风罩 slot exhaust hood
条缝出风口 slotted outlet
条缝地板 slotted floor
条缝回风口 return air slots
条缝滤水管 slotted pipe
条缝式散流器 slot-type air diffuser
条缝式散热器 slit type diffuser
条缝送风口 breeze-line diffuser;slotted outlet
条缝形风口 slot diffuser;slot outlet
条缝形散流器 continuous diffuser;slot diffuser
条缝形送风系统 ejector type grille
条缝型送风口 slit type outlet
条幅 kakemono
条幅式摄像机 strip-film camera
条幅式摄影机 strip-film camera
条幅式照相机 strip camera
条幅式侦察照片 flight strip;reconnaissance strip
条盖松开 carton flap open
条杆带式升运器 open-web elevator
条杆节 bar grate
条杆强力试验 strip test
条杆筛 bar grate
条钢 band steel;bar iron;bar steel;billet bar;billet rod;billet steel;billet steel reinforcing bar;billet steel reinforcing rod;commercial steel;flat bar;flat bar iron;flat iron;merchant steel;ribbon iron;rolled bar;steel bar
条钢粗轧机 stranding roll
条钢架 bar steel rack
条钢热剪机 hot bar shears
条钢轧机 merchant bar mill;merchant mill
条钢折叠 pinchers
条钢折叠缺陷 pinchers
条格平布 gingham
条格式结构 grate bar structure
条根比率 shoot-root ratio
条工 roustabout
条箍巾 band
条管辐射加热器 strip tube radiator

条管加热器 strip tube heater
条焊 fillet welding
条盒 barrel
条盒反冲棘爪 recoiling ratchet going barrel
条盒心轴 barrel arbor
条痕 doctor line; doctor mark; hash mark; knife mark; streak flaw; streak(ing); striation; surface waviness; tramline
条痕板 streak plate
条痕的 striate; striated
条痕状冲刷 rill erosion
条痕状构造 streaky structure
条痕状混合片麻岩 streaky amphogneiss
条痕状漆膜 streaky coat
条痕状涂层 streaky coat
条脊 <沿屋脊砌墙> roof comb; roof crest(ing)
条件 condition; qualification; term; terms and condition
条件保守法 conditional conservation
条件背书 conditional endorsement
条件编译 conditional compilation
条件变更条款 change condition
条件变化 conditions change
条件变量 conditional variable
条件标记 conditional flag
条件标准误差 conditional standard error
条件表 condition list
条件表达式 conditional expression
条件不等式 conditional inequality
条件不明的契约 open contract
条件不明合同 open contract
条件不稳定性 conditional instability
条件采样 conditional sample
条件测试 condition(al) test
条件差的国家 less privileged country
条件承兑 <票据> qualified acceptance
条件抽样 conditional sample
条件刺激 conditioned stimulus
条件存活函数 conditional survivor function
条件等色 metamer; metameric colo(u)r; metameric match; metamerism
条件等色光 metamer
条件等色物体 metameric objects
条件等色指数 metamerism index
条件等式 conditional equality
条件电源 conditional power source; conditional power supply
条件电源屏 conditional power supply panel
条件调度 conditional scheduling
条件定理 conditional theorem
条件断点 conditional breakpoint
条件对数似然函数 conditional log likelihood function
条件反射 condition(ed) reflex; conditioning
条件反射传递 transfer of conditioned reflex
条件反射反应 conditioned response
条件反射分化 differentiation of conditioned reflex
条件反射疗法 conditioned reflexes therapy
条件反射系统 conditional reflex system
条件反射性分泌 conditional secretion
条件返回 conditional return
条件方差 conditional variance
条件方程 condition equation
条件方程式 condition(al) equation; equation of condition; equation of state
条件方程式常数项 constant term of conditional equation
条件方框 conditional box
条件分布 conditional distribution
条件分支 conditional branch
条件风险 conditional risk
条件改变 changed condition
条件改善 improvement of terms
条件概率 conditional probability
条件概率的树形图 tree diagram of conditional probability
条件概率分布【数】conditional probability distribution
条件概率函数【数】conditional probability function
条件概率计算机 conditional probability computer
条件概念 conditional concept
条件供水收费 charges for conditional water service
条件故障率 conditional failure rate
条件关系符 conditional relator
条件观测 conditional observation; conditioned observation
条件观测平差 adjustment of condition observations
条件规划 conditional plan
条件和加法器 conditional-sum adder
条件和逻辑 conditional-sum logic
条件荷载 specified rated load
条件宏表达式 conditional macroexpression
条件宏扩展 conditional macroexpansion
条件宏指令 conditional macroinstruction
条件回归 conditional regression
条件汇编 conditional assembly
条件汇编表达式 conditional assembly expression
条件汇编功能 conditional assemble function
条件汇编指令 conditional assembly instruction
条件级数 condition series
条件极值 conditional extreme value; conditional extremum
条件价值 conditional value
条件间接观测方程 indirect observation with conditional equation
条件检验 conditional test
条件交付信用证 escrow letter of credit
条件交付账户 escrow account
条件结构 construction of condition
条件紧对策 conditionally compact game
条件紧集 conditionally compact set; relatively compact set
条件紧空间 conditionally compact space
条件劲度 conditioned stiffness
条件进入 condition entry
条件矩 conditional moment
条件均值 conditional mean
条件苛刻的贷款 interest-hard loan; hard loan
条件苛刻的条款 hard loan
条件可靠性 conditional reliability
条件空间 condition space
条件控制系统 controlled condition system
条件控制转移指令 conditional control transfer instruction
条件利润 conditional profit
条件利润表 conditional profit table
条件联锁 conditional interlocking
条件联系 conditioned connection
条件流限 condition flow limit
条件码 condition code
条件码标记 bar code label
条件码寄存器 condition code register
条件蒙特卡罗法 conditional Monte Carlo method
条件密度函数 conditional density function
条件名 condition(al) name
条件名描述项 condition name description entry
条件名条件 condition name condition
条件模量 offset modulus
条件模拟 conditional simulation
条件模拟变差函数 variogram of conditional simulation
条件模拟变程 range of conditional simulation
条件模拟方差 variance of conditional simulation
条件模拟基台 sill of conditional simulation
条件模拟均值 mean of conditional simulation
条件模拟块金常数 nugget of conditional simulation
条件模拟漂移值 value of drift of conditional simulation
条件模拟剩余值 residual value of conditional simulation
条件模拟套合级数 nested level of conditional simulation
条件模拟值 value of conditional simulation
条件内部函数 condition built-in function
条件配色 metameric colo(u)r match; metamerism
条件配色指数 index of metamerism
条件频率 conditional frequency
条件平差 adjustment by condition; adjustment of condition; adjustment of condition equations; condition adjustment
条件平衡常数 conditional equilibrium constant
条件期望【数】conditional expectation
条件期望值 conditional expected value
条件气候学 conditional climatology
条件前缀 conditional prefix
条件前缀域 conditional prefix scope
条件屈服点 conditional yield point; constrained yield point
条件屈服强度 offset yield strength
条件屈服限 constrained yield stress
条件容许闭塞 permissive conditional block
条件熵 conditional entropy
条件收敛 conditional convergence
条件收敛级数 conditionally convergent series
条件收入弹性 conditional income elasticity
条件输入 condition entry
条件数 <矩阵病态的> condition(al) number; number of conditions
条件说明 condition stub
条件斯鲁茨基弹性 conditional Slutsky elasticity
条件似然 conditional likelihood
条件损失 condition(al) loss
条件锁闭 conditional locking; special locking
条件锁簧 adjustable dog; swing dog
条件提取常数 conditional extraction constant
条件条款 condition clause
条件跳变 conditional branch; conditional jump
条件跳跃指令 conditional jump instruction
条件停车 conditional halt; conditional stop
条件停机 optional stop
条件停机指令 conditional stop instruction; conditional stop order; optional half instruction; optional stop instruction
条件统计量 conditional statistic
条件图 information drawing
条件图表 bar chart
条件伪变量 condition pseudo-variable
条件位 condition bit
条件稳定常数 conditional stability constant
条件稳定的 conditionally stable
条件稳定电路 conditionally stable circuit
条件稳定度 conditional stability
条件稳定性 conditional stability; limited stability
条件无偏估计量 conditionally unbias(s)ed estimator
条件误差 conditional error
条件吸附平衡常数 conditional adsorption equilibrium constant
条件系数 <以含水率小于4%的统货坑采材料为1> condition factor; conditional coefficient
条件显性 conditional dominance
条件线 <空气调节> condition line
条件线应变 conventional linear strain
条件限制 conditionality
条件销售 conditional sale
条件协方差矩阵 conditional covariance matrix
条件信息 conditional information
条件信息量 conditional information content
条件信息量总平均值 equivocation
条件形成常数 conditional formation constant
条件性 conditionality
条件性不稳定的 conditionally unstable
条件性不稳定度 conditional instability
条件性不稳定性 conditional unstability
条件性概率密度 conditional probability density
条件性可接受的每日摄入量 conditionally acceptable daily intake
条件性平衡 conditional equilibrium
条件性情绪反应 conditioned emotional reaction
条件性容度积 conditional solubility product
条件性抑制 conditional inhibition
条件性指示常数 conditional indicator constant
条件性指示物 conditional indicator
条件性致死诱变体 conditional lethal mutant
条件选择【交】conditional choice
条件颜色匹配 conditional match
条件优惠的信贷 credit of favorable conditions
条件优越的坝址 attractive dam site
条件语句【计】IF statement; conditional statement; control statement
条件预测 conditional forecast
条件域 condition field
条件约束 condition restraint
条件运算 conditional operation
条件指令 conditional instruction; conditional order; conditional statement
条件指令组 conditional order unit
条件致死突变体 conditional lethal mutation
条件中断点 conditional breakpoint

T

条件中断指令 conditional breakpoint instruction
条件周期运动 conditionary periodic motion
条件转储 conditional dump
条件转移 conditional branch; conditional jump; conditional transfer; decision instruction; discrimination; jump if not
条件转移命令 conditional transfer command
条件转移权力 conditional branch capability
条件转移指令 conditional branch instruction; conditional jump instruction; conditional transfer command
条件子句 condition(al) clause
条件字段 condition field
条件最佳化 constrained optimization
条件最优方案 conditionally optimal plan
条件最优配对 conditionally optimal pair
条件作用 conditioning
条胶压延机 gum strip calender; strip calender
条金 bullion
条锯机 tie mill
条卷 sliver lap
条卷机 sliver lap machine
条卷机导条板 sliver-doubling plate
条孔长度 length of strip hole
条孔垂向净距 net spacing between strip hole
条孔骨架 strip hole framework
条孔管测压计 slotted-pipe piezometer
条孔间距 interval of strip hole
条孔宽度 width of strip hole
条块砌筑 sliced blockwork
条宽 strip width
条款 article; clause; item; point; provision; stipulation; subsection; term; terms and condition
条款编号 numbering of clause
条款草案 draft provision
条款的议定 arrangement of clause
条款解释 interpretation
条款解说 definition of term
条例 act; ordinance; regulation; rule; rules and regulations
条例生效日期 effective date of regulations
条例手册 rulebook
条料 billot
条令 assize
条龙骨 bar keel
条垄耕作 strip cultivation
条炉 bar oven
条码 <商业用的> bar code
条码阅读器 bar code reader
条面 side face
条面斜砖 end skew on edge
条木 barwood
条木地板 strip flooring; wood strip flooring
条木管 stave pipe
条木胶合板 compo board
条木楼板 wood strip flooring
条目检索号 coded items
条黏[粘] channel mopping; strip mopping
条片 sliver
条片填充 slat packing
条屏沉降器 rod-curtain precipitator
条屏显像管 stripe screen tube
条铺沥青法 strip mopping
条砌 pavement in rows; stretcher bond
条筛 bar grit
条栅 bar screen

条石 chipped ashlar; block stone; boulder strip; rag; sized slate; slabstone; stone band
条石堤 rag dike[dyke]
条石护面 rag armo(u)ring; rag pitching
条石连续底脚 continuous stone footing
条石梁 stone lintel
条石料 regular stone
条石路面 ashlar pavement; sett paving; stone block pavement
条石磨孔机 honing machine
条石砌合 ashlar bond
条石砌面中填毛石圬工桥墩身 bridge pier body of ashlar stone face filled with rubble masonry
条石砌筑 ashlar masonry work
条石圬工 ashlar masonry
条式电加热器 electric(al) strip heater
条式格栅 bar type grating
条式花坛 ribbon bed
条式输送带 cleated belt conveyer [conveyor]
条式输送机 cleated conveyer[conveyor]
条式双凸透镜 ribbon lenticular
条饰花格窗 bar tracery
条条极板蓄电池 block accumulator
条条框框 rules and regulations
条铁 bar iron; flat iron; strap plate
条铁感应器 strap iron inductor
条铜 flat copper; strip copper
条筒 sliver can
条筒揿压器 can packer
条头沟 row ditch
条图 bar chart; bar diagram; going map
条图心中线 bisector of the strip map sheet
条涂 channel mopping; strip mopping
条文 citation
条文草案 draft articles
条文范例 standard clause
条纹 ribbing; ribbon; riveling; streak(ing); striation; stripe; striped figure; stripping; surface waviness; stria [复 striae] 【地】; striature 【地】; fringe【物】
条纹斑杂岩 eutaxite
条纹斑状的 eutaxitic
条纹斑状构造 eutaxitic structure
条纹板 channel(l)ed plate
条纹包络 fringe envelope
条纹倍增 fringe multiplication
条纹碧玉 ribband jasper
条纹表面 striated surface
条纹布 chambray; stripe
条纹长石 perthite
条纹长石结构 perthitic texture
条纹长岩正长岩 perthite syenite
条纹衬度 fringe contrast
条纹次生壁 striated secondary wall
条纹粗制毡 ribbed rag felt
条纹大理石 onyx marble; striated marble
条纹大理岩 Mexican onyx; onyx marble; oriental alabaster; striated marble
条纹单板 striped veneer
条纹的明暗配合 fringe value
条纹地板 strip floor(finish)
条纹地毯 stria carpet
条纹定位 fringe location
条纹定位法 fringe localization technique
条纹读出器 fringe follower
条纹法摄影 schlieren
条纹反差 contrast of fringes; fringe contrast

条纹方砖 striped tile
条纹仿古纸 laid antique
条纹幅度 fringe amplitude
条纹钢板 ribbed steel
条纹构造【地】 streaky structure; ribbon structure
条纹级数 <等色线的> order of fringes
条纹计数法 <光弹仪的> fringe counting technique
条纹计数器 fringe counter
条纹计数式干涉仪 fringe count micrometer
条纹记录 strip record
条纹间隔 fringe spacing
条纹间距 fringe separation; fringe spacing
条纹结构【地】 perthitic texture; ribbon texture
条纹晶体 striated crystal
条纹可见度 fringe visibility
条纹可见度谱 fringe-visibility spectrum
条纹空化 streak cavitation
条纹空间 fringe spacing
条纹宽度 width of fringe
条纹滤色片 fringed filter
条纹率 fringe rate
条纹煤 banded coal
条纹密码 bar code
条纹模 striation cast
条纹能见度 fringe visibility
条纹片麻岩 ribbon gneiss
条纹槭 striped dogwood
条纹器皿 combed ware
条纹强度 fringe intensity
条纹锐度 fringe sharpness
条纹砂岩 ribband stone; ribbon stone
条纹摄像机 streak camera
条纹摄影 schlieren photograph
条纹式样 stripped-down style
条纹饰面 striated finish
条纹调制 fringe modulation
条纹调制度 modulation of fringes
条纹图案 candy stripe; strip figure
条纹图形 <等色线> fringe pattern; striated pattern
条纹图样 <等色线> fringe pattern
条纹褪绿病 streak chlorosis
条纹稳定性 fringe stability
条纹霞正长岩 kakor tokite
条纹线 streak line
条纹相位 fringe phase
条纹响应 fringe response
条纹橡胶地面 ribbed rubber flooring
条纹信号 stripe signal
条纹序数 fringe order
条纹亚麻 rigid flax
条纹岩 ribbon rock
条纹羊毛毡 ribbed wool felt
条纹仪 schlieren apparatus; stria projector; zebra instrument
条纹釉 streaked glaze
条纹照片 streak photograph
条纹照相 schlieren; streak photograph
条纹照相法 schlieren method
条纹证认 fringe identification
条纹值 <材料的> fringe value
条纹纸 stripping paper
条纹驻留 fringe stopping
条纹驻留器 fringe stopper
条纹驻留系统 fringe stopping system
条纹驻留中心 fringe stopping center [centre]
条纹装饰 laid finish; streak decoration
条纹装饰玻璃 combed glass
条纹状 striation; striature
条纹状构造 striped structure
条纹状混合片麻岩 striped amphogneiss

条纹状混合岩 striped migmatite
条纹状锈蚀 striated rust pattern
条纹组织 striation
条线 strip line
条线表 <统计用> bar chart
条线代码扫描器 bar code scanner
条线图 bar chart; bar diagram; bar graph
条响应 bar response
条项 clause
条橡胶 slat gum
条信号 bar
条信号边沿 bar edge
条形板 ribbon strip
条形保险丝 band fuse
条形草地 meadow strip; turf strip
条形城市 strip city
条形翅片 strip fin
条形磁铁 axial magnet; bar magnet
条形导线 strip conductor
条形底脚 long strip footing
条形地板 parquet strip flooring; strip flooring
条形地毯 carpet strip
条形电磁铁 bare electromagnet; bar electromagnet
条形吊杆 strap hanger
条形镀锌板 zinc strip
条形发展 strip development
条形风口 linear grille
条形浮雕 anaglyphic strip
条形钢 bar steel
条形灌溉 strip irrigation
条形光源 strip source
条形含水层 strip aquifer
条形荷载 strip load
条形横挡 ribbon rail
条形花边 banded lace
条形基础 band-like foundation; connecting footing; continuous footing; long strip footing; ribbon foundation; strap footing; strip footing; strip foundation
条形基脚 strip footing
条形极板蓄电池 block accumulator
条形建筑 ribbon building; strip building
条形节 splay knot
条形介质光波导 slab dielectric(al) optical waveguide
条形金属 bullion
条形料堆 windrow
条形码 bar code
条形码标记 bar code label
条形码读出器 bar code reader
条形码扫描器 bar code scanner
条形码阅读机 bar code reader
条形模板 strip forms
条形耐火砖 soaps
条形屏 line-screen
条形畦田 border strip
条形砌窑砖 lath brick
条形绕组 bar winding
条形散流器 linear diffuser; slot diffuser; strip diffuser
条形色谱法 paper strip chromatography
条形商业区 strip development
条形视镜 strip sight glass
条形水草 strip aquatic mat
条形送风口 slot outlet
条形天线 stick antenna
条形铁 bar iron
条形图 bar chart; bar graph; columnar graph
条形图表 bar chart
条形图存储器 bar graph memory
条形图记录仪 strip-chart recorder
条形图样 bar pattern

T

条形污水槽 bar sink
条形污水沟 bar sink
条形屋面板 strip shingle
条形细锉 ridged-back file
条形线 strip line
条形橡皮垫 bonded rubber cushion(ing)
条形一次绕组 bar primary coil
条形应变片 strip ga(u)ge
条形纸 paper strip
条形砖 <5 厘米厚,标准长和宽> soap;lath brick
条样板 <桥梁号料用> band strip
条样逼近 spline approximation
条音针 recording stylus
条银 bullion
条元 bar element
条约 convention;pact;treaty
条约保障 treaty protection
条约本 text of treaty;treaty text
条约的废弃 denunciation of treaty
条约的废止 denunciation of treaty
条约的解释 interpretation of treaties
条约的批准 treaty ratification
条约的条文 article of a treaty
条约的终止 termination of treaties
条约法 law of treaties
条约附件 treaty annex
条约汇编 treaty series
条约集 treaty series
条约批准 treaty ratification
条约签字 sign treaty
条约全文 text of treaty;treaty text
条约生效 treaty valid
条约生效日期 effective date of regulations
条约实施情况 operation of treaty
条约文本 text of treaty;treaty text
条约义务 treaty obligation
条约展期 renew of treaties
条约正本 original of the treaty
条植 plant in rows;strip cultivation
条植法 drill planting method
条纸记录器 strip-chart recorder
条轴套夹 pendant collet
条砖 stretcher
条状包覆物 strip coating
条状波形干扰 Venetian blind interference
条状剥皮 remove the bark in strips;strip peel
条状窗 strip window
条状导线 strip conductor
条状的 banded appearance;banding
条状垫板 backing strap;backing strip
条状阀 beam valve;feather valve
条状废料 scrap ribbon
条状封接 strip sealing
条状覆板 strap cover
条状覆盖 strap cover
条状盖板 strap cover
条状耕作 strip farming
条状管 strip tube
条状焊接 strip sealing
条状荷载 strip load
条状化 banding
条状加热 strip heating
条状胶片 strip film
条状节 <沿木节长度方向切开时露出的节疤> splay knot
条状结构 list structure
条状结核 rod nodule
条状裂痕 streak flaw
条状隆起部 wale
条状滤色镜 banded colo(u)r filter
条状螺旋线 tape helix
条状锚固钢板 strap anchor
条状模型加固片 tie piece
条状木瓦 strip shingle
条状排水沟 strip drain
条状刨花 sliver
条状嵌缝 strip sealing
条状切碎机 barminutor
条状筛 bar screen
条状商业地区 strip commercial area
条状碎片 sliver
条状缩微胶卷 <其中的一种> dekafilm
条状填缝 strip sealing
条状铁 bar iron
条状铁素体 ferrite band(ing)
条状图 band chart;bar chart
条状纹理 ribbon stripe
条状纤维 feltlike fiber[fibre]
条状信号 flag pole
条状照明装置 strip light
条桌 side table
条子 sliver
条子不匀率 sliver unevenness
条子混合 sliver blending;sliver mixing
条子集合器 sliver condenser
条子条纹 stripe
条子展幅装置 sliver spreading device

调

调成灰色 turn grey
调成腻子 running to putty
调成水平 level off
调成一直线 line up
调成油灰 running to putty
调程序 debug(ging)
调稠剂 set-controlling admixture
调稠器 absorptiometer
调出河流 donor stream
调出流域 donor basin
调出区 donor region
调除 <调节收音机、电视机等以去除干扰> tune out
调磁合金 magnetic shunt alloy
调淡 lightening
调挡 gear changing;gear shift(ing);shift;shifting gear
调挡齿轮 shift gear
调挡装置 shift gear
调刀 spatula [spatule];tool - setting
调刀混合法 rub - out method
调刀千分表 tool setting ga(u)ge
调刀轴环 adjustable cutter bush
调到零位 resetting of zero
调到零(值) adjust to zero;set to zero
调低 downward adjustment
调低挡 down shift
调低速挡 shift down
调定 setting;setting up
调定的泄压压力 relief set pressure
调定点 set point
调定电路 set-up circuitry
调定时间 setting-up time
调定误差 set-up error
调定旋钮 setting knob
调定压力 set pressure
调定压力公差 tolerance of set pressure
调定仪表 set-up instrument
调风门 air shutter;butterfly damper
调风器 air register;register
调风闸 flue damper
调峰 peak regulation
调峰电厂 peaking plant;variable load plant
调峰电站 peak power station
调峰工程 peak power project
调峰管理 peak power management
调峰机组 peak load unit
调峰气 peak load gas;peak-shaving gas;stand-by gas
调峰型天然气液化装置 peak-shaving LNG facility
调幅 amplitude;amplitude modulation;modulation;overhoist
调幅半径 <起重机> handing radius (of crane)
调幅伴音发射机 amplitude-modulated sound transmitter
调幅编码 amplitude modulation coding
调幅变压器 amplitude modulation transformer
调幅标志 amplitude modulation signature
调幅波 amplitude-modulated wave;modulated wave
调幅测量 amplitude modulation measurement
调幅带通滤波器输出 modulator band filter out
调幅度 amplitude modulation factor;modulation factor
调幅扼流圈 modulating choke
调幅扼流图 amplitude modulation choke
调幅发射机 amplitude-modulated transmitter
调幅阀 modulating valve
调幅范围 range of regulation
调幅分度线 amplitude modulation reticle
调幅幅度 range of regulation
调幅杆 radius bar
调幅钢索 <起重机> derricking rope
调幅广播 amplitude modulation broadcast(ing)
调幅接收机 amplitude modulation receiver
调幅控制 modulating control
调幅脉冲 amplitude-modulated pulse
调幅器 amplitude modulator;modulator
调幅器带通滤波器 modulator band filter
调幅器带通滤波器输出 modulator band electric(al) system out
调幅器带通滤波器输入 modulator band filter in
调幅器盘 modulator panel
调幅锁模 amplitude modulation mode locking
调幅-调频 amplitude modulation frequency modulation[AM/FM]
调幅-调频变换器 AM/FM converter
调幅调相转换 amplitude modulation phase modulation conversion
调幅调制盘 amplitude modulation retic(u)le
调幅通信[讯]系统 amplitude modulation communication system
调幅无线电设备 amplitude modulation radio
调幅系统 amplitude modulation system
调幅信号 amplitude-modulated signal
调幅抑制 amplitude modulation rejection
调幅载波 amplitude-modulated carrier
调幅噪声 amplitude modulation noise
调幅指示器 amplitude-modulated indicator;deflection-modulated indicator;intensity-modulated indicator
调感变压器 slide transformer
调高衬垫 height adjustment pad
调高档 up-shift
调高为整数 round up
调高物价水平 adjust prices upward;upward adjustments of prices
调功率 power setting
调管黄杨木 <管工调直铅皮或铅管用> dresser
调光 adjustment of light level
调光电阻 dimming resistor
调光孔径 curtain aperture
调光器 dimmer;light dimmer;light modulator
调光器开关 dimmer switch
调光室 dimmer room
调光台 illumination desk;lighting console
调光制 intensity scale
调好的 no-float
调好的漆 prepared paint
调合 blending;chime;due proportion;syncretize;temper
调合板 mixing-plate
调合剂 blending agent
调合漆 diluted paint;prepared paint;ready-mixed paint
调合器 dispenser;mixing machine
调合油 tempered oil
调和 accordance;chimb(e);chime;consonance;harmonization;harmony;proportion(ing);reconcile;tempering
调和比(例)【数】 harmonic proportion;harmonic ratio
调和边界 harmonic boundary
调和变化 harmonic variation
调和波 harmonic wave
调和波分析器 harmonic wave analyser[analyzer]
调和测度 harmonic measure
调和常数 harmonic constant
调和潮候推算 harmonic tide prediction
调和的比例卡尺 proportional
调和等能性 harmonious equapotentiality
调和点 harmonic point
调和点列 harmonic range of points
调和对策 compromise strategy
调和法 harmonic method
调和方程 harmonic equation;Laplace's equation
调和分布 harmonic distribution
调和分潮 harmonic constituent
调和分割 harmonic division;harmonic section
调和分隔 harmonic separation
调和分量 harmonic component;tidal component
调和分析 Fourier analysis;harmonic analysis;harmonic reduction
调和分析法 system of harmonic analysis
调和分析机 harmonic analyser[analyzer]
调和分析仪 harmonic analyser[analyzer];periodometer
调和共轭 harmonic conjugate
调和共轭点 harmonically conjugate points
调和共轭射线 harmonic conjugate rays
调和共轭直线 harmonic conjugate line
调和构形 harmonic configuration
调和关税 compromise tariff
调和关税制度 harmonized system
调和函数 ellipsoidal harmonics;harmonic function;harmonics
调和花瓶饰 harmonical vase
调和积分 harmonic integral
调和级数 harmonic series;harmonic progression
调和均数 harmonic mean
调和开拓 harmonic continuation
调和空气 tempered air
调和矿浆 conditioning pulp;pulp climate
调和流 harmonic flow
调和绿化 harmony planting

T

调和面束 harmonic pencil of planes
调和平衡器 harmonic balancer
调和平均(数) harmonic mean; harmonic average
调和平均直径 harmonic mean diameter
调和平均值 harmonic average; harmonic mean
调和平均指数 harmonic average index
调和漆 distemper paint; mixed paint; mixing varnish; prepared paint; ready-mixed paint; tempered paint
调和强函数 harmonic majorant
调和清漆 mixing varnish
调和曲线 harmonic curve
调和曲线图 harmonic diagram
调和色 compound colo(u)r; harmonizing colo(u)r; secondary colo(u)r
调和式 harmonic expression
调和束 harmonic pencil
调和数列 harmonic progression; harmonic series
调和水 tempering water
调和水系 accordant drainage
调和算子 Laplace's operator; Laplacian; Laplacian operator
调和涂料 mixed pigment; prepared paint
调和推算 < 潮汐的 > harmonic prediction(of tide)
调和退算机 harmonic synthesizer
调和微分 harmonic differential
调和微分形式 harmonic differential form
调和维数 harmonic dimension
调和温度系数 harmonic temperature coefficient
调和物 admix
调和系数 harmonic coefficient
调和线 harmonic line
调和线丛 harmonic complex
调和线束 harmonic pencil of lines
调和线向 harmony of alignment
调和相关 harmonic correlation
调和相角 harmonic phase angle
调和相位 harmonic phase
调和向量场 harmonic vector field
调和项 harmonious term
调和性 harmonicity
调和序列 harmonic sequence
调和颜料 filling colo(u)r
调和颜料的溶剂 gumpnon
调和颜色 colo(u)r mixing
调和音阶 harmonic scale
调和映射 harmonic mapping
调和油漆 mixed pigment; ready-mixed paint
调和预报 harmonic prediction
调和预报方法 < 潮汐的 > harmonic prediction
调和元素 harmonic element
调和展开 harmonic expansion
调和张量场 harmonic tensor field
调和振幅函数 harmonic amplitude function
调和指数 harmonic index number
调和滞后 harmonic lag
调和中数 harmonic mean
调和中项 harmonic mean
调和中心 harmonic center[centre]
调和子流 harmonic subflow
调和坐标 harmonic coordinates
调洪 flood mitigation; flood regulation; mitigation of flood
调洪河槽 flood-control channel
调洪计算 storage routing
调洪建筑物 flood-relating structure
调洪库容 flood storage; flood storage capacity
调洪能力 flood storage capacity
调洪水库 flood-control basin; flood-control reservoir
调洪水位 flood regulating level
调洪洼地 flood-control basin
调糊机 paste mixer
调绘像片 annotated photograph; identified photograph
调绘像片数量 number of annotated photographs
调绘(照)片【测】 annotated photograph; identified photograph; surveying picture
调剂 dispense
调剂电路 dispensable circuit
调剂会议 meeting to adjust shortages and surpluses
调剂价 adjusting price
调剂劳动力 adjust the use of labor force
调剂液 dispersing liquid
调剂余缺 regulate supply; regulate supply and demand
调价 price adjustment
调价浮动 range of price changes
调价公式 price adjustment formula
调价条款 escalation clause
调浆槽 surge tank
调浆机 paste mixer
调浆桶 mixing vessel
调胶机 dough mill; glue mixer
调焦 eyesight adjustment; focus setting; focusing action; focusing adjustment; focusing movement; focusing
调焦拔杆 focusing lever
调焦保留时间 adjusted retention time
调焦杯 focusing cup
调焦标 focusing mark
调焦玻璃 focus glass; punt glass
调焦差误 error of focusing
调焦传动齿轮 focusing gear
调焦范围 adjustment range; focal length range; focal range
调焦放大镜 focusing magnifier
调焦幅度 range of focus setting
调焦滚花螺旋 focusing knurl
调焦滑筒 focusing slide
调焦环 focusing ring
调焦机理 focusing mechanism
调焦精度 accuracy of focusing
调焦镜筒 focusing barrel
调焦距离 focusing distance
调焦控制系统 focus control system
调焦螺母 focusing nut
调焦螺旋 focusing screw
调焦目镜 focusing eyepiece
调焦平面 focusing plane
调焦屏 focusing glass
调焦清晰区域 zone of sharp focus
调焦圈 focusing ring
调焦设备 adjusting appliance
调焦筒 focusing tube
调焦投影器 focusing projector
调焦透镜 focusing lens
调焦误差 error in focusing
调焦显微镜 focusing microscope
调焦旋钮 focus knob
调焦装置 focus control; focusing device; focusing mechanism
调焦作用 focusing action
调角 angle modulation; sinusoidal angular modulation
调角电容 book capacitor
调角装置 < 平地机刮刀的 > circle reverse
调校 timing
调节 accommodation; adjustment; conditioning; control; governing; modulate; modulating; modulation; piloting; regulate; regulating; temper; tune; tune-up; tuning
调节安全风门 modulating relief damper
调节凹口 notch
调节坝 regulating barrage; regulating dam
调节板 adjusting plate; control damper; damper; regulating plate; shutter
调节板振动 shutter vibration
调节瓣 pallet
调节瓣阀 regulating flap valve
调节棒 regulating control rod; regulating rod; regulator rod; rod follow-up
调节保证计算 calculation of regulation guarantee; guaranteed calculation for regulation
调节备料 blank stock
调节泵 governor pump gear; metering pump
调节泵齿轮轴 governor pump gear shaft
调节比 turn-down ratio
调节臂 pitch arm; regulating arm
调节变量 controlled variable
调节变压器 regulating transformer
调节变阻器 regulating rheostat; standardizing rheostat
调节标度 degree scale
调节标记 alignment mark
调节表 reconciliation statement
调节波 modulated wave
调节不当 maladjustment
调节不当的 maladjusted
调节不均匀度 cyclic(al) irregularity factor
调节不善 maladjustment
调节步骤 regulating step; shunting step < 货车自到达线至出发线之间一系列位置移动 >
调节部件 regulating unit
调节部位 regulatory site
调节参数 controlling parameter
调节参数的偏差 control deviation
调节仓 balancing bin; gate road bunker; surge bin; surge bunker
调节仓库 buffer storage; surge storage
调节舱 trimming chamber; trimming tank
调节槽 conditioning tank; regulating tank; retention basin; surge bin
调节层 regulating course
调节常数 regulating constant; regulation constant
调节池 balancing reservoir; conditioning tank; equalization basin; equalization pond; equalization reservoir; equalization tank; equalizing tank; pondage reservoir; regulating pondage; regulating tank; regulation tank; retention basin; surge tank
调节池容积 volume of equalization basin
调节池式发电站 pondage type power plant
调节齿轮 adjusting gear; regulating gear
调节储存 regulating storage
调节储量 regulating reserves
调节处理(方法) conditioning process
调节触点 regulating contact
调节锤 governor weight
调节存货 reconciling inventory
调节代谢物 regulatory metabolite
调节单元 correcting member
调节挡板 regulating damper; regulation damper; damper regulator
调节的 adjust(ing); controlled; regulative
调节的发散过程 undamped control
调节点 control point; place of control
调节电池 regulating cell
调节(电)磁铁 regulating magnet
调节电动机 modulating motor; regulating motor
调节电流 regulating current
调节电路 regulating circuit
调节电位器 regulator potentiometer
调节电压 regulation voltage; regulated voltage
调节电阻 adjusting resistance; regulating resistance; regulation resistance
调节电阻器 regulating resistor; varistor[varister]
调节垫 spacer
调节垫层 regulating underlay
调节垫圈 adjusting disc[disk]; adjusting washer
调节度 < 径流的 > degree of regulation
调节对象 controlled plant
调节对象放大系数 amplification factor of controlled plant
调节对象飞升曲线 response curve of controlled plant
调节对象时间常数 time constant of controlled plant
调节对象响应曲线 response curve of controlled plant
调节对象滞后 lag of controlled plant; plant lag
调节对象自平衡 inherent regulation of controlled plant
调节发电厂 regulating station
调节发电机 regulator generator
调节阀 adjustable valve; adjust(ing) valve; control(ling) valve; damper regulator; damper valve; expansion valve; governing valve; governor; governor valve; graduating valve; hydraulic valve; manoeuvering valve; modulating valve; modulation valve; range of adjustment; register valve; regulating valve; regulator; regulator valve; spill valve
调节阀防松螺母 governor valve lock nut
调节阀杆 regulating valve stem
调节阀开度计 governor valve position indicator
调节阀流量特性 flow characteristic of control valve; flow characteristics of regulating valve
调节阀流通能力 flow capacity of control valve
调节阀螺母 regulating valve cap nut
调节阀帽 regulating valve bonnet
调节阀门 regulating gate
调节阀塞 metering pin
调节阀行程记录仪 governor valve position recorder
调节阀行程指示器 governor valve position indicator
调节阀压盖 regulating valve gland
调节阀座 regulating valve seat
调节阀座衬套 regulating valve seat bush
调节法兰盘 spacer spool
调节反射 accommodation reflex; accommodatory reflex
调节范围 adjustable range; adjustment range; range of regulation; regulating range; regulation range; throttling range; turn-down range; turn-down region; turn-down zone; variable range; modulating range < 自动控制的 >

调节方法 adjusting mode; control method; control mode; regulating system

调节放大器 regulating amplifier; resonance amplifier

调节风道 conditioning duct

调节风门 adjusting damper; air damper; control damper; damper; modulating damper; regulating damper; tempering air damper; trap door

调节缝 control joint

调节浮选剂 disperser

调节幅度 amplitude of accommodation; range of accommodation; range of adjustment

调节负载 regulating load

调节杆 adjustable lever; adjusting rod; advance lever; doctor bar; dolly bar; pitch arm; pitch rod; regulating arm; regulating stem; regulator handle; standard lever

调节杆止动螺钉 regulating stem set screw

调节高峰负荷 peak shaving

调节工程 regulating works

调节工具 adjusting appliance

调节工作 control work

调节功 regulatory works

调节供气门 supply register

调节拱 adjustable arch

调节拐肘 accommodating crank; accommodation crank

调节管 adjutage; control valve; overflow pipe

调节管网 regulating network

调节罐 compensator tank

调节柜 regulator cubicle

调节辊 dancer roll

调节过程 conditioning process; controlled process; controlling process; regulating processing

调节函数 adjustment function

调节河流 regulated stream

调节河流流量 regulated stream flow

调节恒温器 regulation thermostat

调节后的流量 regulated flow

调节滑阀 regulating slide valve

调节滑块 adjusting slider

调节滑轮 control block

调节环 adjusting ring; landing ring; regulating ring; setting ring

调节环节 adjusting link; governing loop

调节簧螺钉 regulator spring screw

调节回路 governing loop; regulating loop

调节活门 governor valve

调节活门流动特性 control valve flow performance

调节活塞 control piston; control piston follower; regulating piston

调节机构 controlling element; correcting element; governing mechanism; governor motion; operating unit; regulating gear; regulating mechanism; setting mechanism

调节机制 regulating mechanism; regulatory mechanism

调节机组 regulating set

调节级 governing stage; regulating stage

调节极限 limit of accommodation

调节剂 conditioner; conditioning agent; corrective; modifier; regulating additive; regulating agent; regulator

调节继电器 regulating relay

调节间隔 control interval; spacing

调节间距 distance control; distance setting

调节建筑物 regulation structure

调节降水记录 adjusting precipitation records; regulating precipitation records

调节交通 accommodate the traffic

调节结构 regulator

调节经济 regulating economy

调节经济活动 regulating economic activities

调节精密度 sharpness of regulation

调节井 regulating well

调节径流 regulated flow

调节聚合反应 telomerization

调节开关 by-pass cock; regulating switch

调节空气 artificial atmosphere

调节孔 regulating port; regulation hole; regulator opening

调节控制盘 dial

调节控制器 adjustment controller

调节口 gate

调节库存 buffer stock; regulating stock

调节库存量 buffer storage; regulating storage

调节库容 balance storage; balancing storage; pondage capacity; regulating pond; regulating pondage; regulating storage; regulation storage; usable storage

调节拉杆 adjusting yoke; regulating pull rod

调节冷热 mitigate

调节力 adjusting force

调节联动装置 regulating linkage

调节量 controlled variable; regulated quantity

调节料仓 balancing bin; surge pocket; booster bin; buffer bin; surge bin

调节料堆 buffer stockpile; surge pile

调节灵敏度 sensitivity of regulation

调节流阀 flow regulating valve

调节流量 regulated discharge; regulated flow; regulated stream flow

调节漏斗 surge hopper

调节轮 idle pulley; regulating wheel

调节螺钉 adjusting screw; control screw; governing screw; quarter screw; set screw; temper screw; tuner screw; tuning screw

调节螺杆 adjusting screw rod; temper screw

调节螺帽 adjusting nut; setting nut

调节螺母 adjusting nut; rating nut; regulating nut

调节螺栓 adjustable bolt; adjusting bolt

调节螺丝 temper screw

调节螺旋 adjustable screw; adjusting screw; adjustment screw; rectifying screw; regulating screw; turnbuckle screw

调节马达 governor motor

调节脉冲 regulating impulse

调节门 adjustable gate; adjusting shutter

调节命令 regulating command

调节能力 modulability; surge capacity

调节能量 regulating energy

调节凝固混凝土 regulated-set concrete

调节盘 setting dial

调节配件 doctor bar; regulating fitting

调节膨胀 variable expansion

调节膨胀阀 variable expansion valve

调节片 adjustment sheet

调节频率 regulating frequency

调节坡 accommodating ramp; accommodation ramp

调节期 period of regulation

调节气温 climate comfort; climate conditioning

调节汽缸 adjusting cylinder

调节器 accommodator; actuating apparatus; actuator [adjuster]; adjusting device; adjusting instrument; buffer; butterfly damper; conditioner; control apparatus; controller; controls; dampener; damper; moderator; modifier; modulator; regulating apparatus; regulator; setter

调节器板 adjuster board

调节器衬套 governor bushing

调节器传动装置 governor actuator

调节器的摆动 hunting of governor

调节器的作用 controller action

调节器电枢 regulator terminal

调节器轭 regulator yoke

调节器阀盒 governor valve box

调节器阀簧 governor valve spring

调节器阀门 adjuster valve

调节器阀内弹簧 regulator valve inner spring

调节器阀外弹簧 regulator valve outer spring

调节器防松螺母 regulator check nut

调节器封 governor seal

调节器盖 governor cap; regulator cover

调节器盖衬 regulator cover gasket

调节器杆 governor lever

调节器杆滚轮销 governor link roller pin

调节器高速限制弹簧 governor high limit spring

调节器滑阀 governor slide; governor slide valve

调节器静止时间 dead time of governor

调节器壳体 governor body

调节器空转弹簧 governor idling spring

调节器控制的 governor-controlled

调节器控制图 governor control diagram

调节器连杆螺钉 governor link screw

调节器连接杆 regulator link

调节器连接管螺母 governor union nut

调节器连接管柱螺栓 governor union stud

调节器帽 dome of regulator

调节器旁通开关 regulator bypass switch

调节器歧管 governor manifold

调节器曲柄 regulator crank

调节器扇形齿轮 regulator quadrant

调节器试验装置 regulator tester

调节器手柄 regulator handle

调节器弹簧 governor spring; spring for governor

调节器套节 governor socket

调节器套筒 governor sleeve

调节器套筒衬套 governor sleeve bushing

调节器体 body of regulator

调节器调整螺钉 governor adjusting screw

调节器线圈 regulator coil

调节器限制单元 regulator limiter

调节器压力 regulator pressure

调节器压力调节范围 reduced pressure range

调节器延迟 controller lag

调节器一次滑阀 relay piston

调节器增益 controller gain

调节器轴 regulator shaft

调节器座 regulator base

调节千斤顶 adjustment jack

调节球阀 regulating globe valve

调节区 conditioning zone

调节曲线 regulating curve; regulation curve

调节渠 regulator canal

调节燃料 fuel metering

调节绕组 regulating winding

调节人员 conditioner

调节容量 pondage; regulating capacity; regulation storage; regulator storage; volume of pondage

调节塞 regulating plug; regulator plug

调节散流器 control diffuser

调节晒片时间测时计 print-meter

调节商品交换 regulate the exchange of commodities

调节设备 adjusting appliance; conditioning equipment; governing device; regulating apparatus; regulating device; regulating equipment; setting appliance; setting device

调节渗透的 osmoregulatory

调节时间 control period; control time

调节式导口【船】adjustable precentering device

调节式仪表 regulation meter

调节室 chamber regulator; conditioning chamber; regulating chamber; regulation chamber

调节手柄 regulating handle; regulating shaft

调节手段 leverage

调节手轮 adjusting handle; telegraph wheel < 司钻台上的 >

调节数量 control rate

调节衰减器 regulated attenuator

调节栓 regulating cock; regulation cock

调节水 conditioning water

调节水池 balancing tank; equalizing pond; make-up reservoir; regulating reservoir; regulating storage; storage regulator; store tank; equalizing reservoir

调节水库 balancing reservoir; compensating reservoir; detention reservoir; equalising[equalizing] reservoir; flood-control reservoir; make-up reservoir; pondage reservoir; regulating pond; regulating reservoir; reservoir for low-flow augmentation

调节水流 regulated flow; regulated stream flow

调节水位 regulated level; regulated water stage

调节水箱 make-up tank; regulating tank

调节税 adjustable tax; adjustment tax; regulatory business tax; regulatory tax

调节松紧 adjust tension

调节松紧的螺栓 draw bolt

调节速度 regulation speed

调节损失 regulation loss; regulations waste

调节锁紧螺母 adjusting lock nut

调节塔 surge tank

调节弹簧 regulating spring

调节套(筒) regulating sleeve

调节特性 control characteristic; regulating characteristic; regulating property; regulation characteristic; governing characteristic

调节特性曲线 regulating characteristic curve

调节梯度 regulating gradient

调节体 control agent; setting cylinder

调节条件 controlled condition

调节通风门 ga(u)ge door

调节通路 conditioner channel

调节头 control head

调节图 regulating diagram; regulation

diagram
调节网络 regulating network
调节委员会 Board of Adjustment
调节温度 attemper;attemperation;attemperment;regulating temperature
调节温室温度 regulate the green house temperature
调节稳定性 stability of regulation
调节蜗杆 adjusting worm; setting worm
调节误差 governing error; regulating error
调节系杆 pitch brace
调节系数 accommodation coefficient; modulation factor; pondage factor; storage ratio < 水库 >
调节系统 conditioning system; governing system; regulating system; regulator system; regulatory system
调节系统图 governor control diagram
调节线圈 regulating coil; regulating winding
调节箱 regulating box; regulating chest
调节响应 governing response
调节楔块 adjusting wedge
调节楔座 adjusting block
调节泄水 pondage
调节泄水工程 regulated outlet work
调节信号 conditioning signal; control signal; regulatory signal
调节信号相时 timing offset
调节信息 control information
调节型料斗 < 可容纳多余物料的 > surge bin
调节型水源地 regulating water source
调节性交易 accommodating transactions
调节性库存储备 buffer stock
调节性库存储备的资金供应 buffer stock financing
调节性能 quality of regulation; variability
调节性抑制 regulatory inhibition
调节性援助 adjustment assistance
调节蓄水 pondage
调节蓄水池 regulating reservoir
调节蓄水量 regulating pondage
调节旋钮 adjusting knob
调节旋塞 adjusting cock; regulating cock; stop cock
调节亚单位 regulating subunit; regulator subunit
调节延迟 control lag
调节堰 regulating weir
调节样板 Maco template
调节液 regulator solution
调节液流浓度 flow-adjusted concentration
调节仪 accommodometer
调节仪表 control instrumentation; monitoring instrument
调节仪器 adjusting appliance
调节因数 control factor; pondage factor < 水库 >
调节因子 regulatory factor
调节银行存款账户 reconciling bank account
调节应力 accommodatory ability
调节用变阻器 governor
调节用的导频 regulating pilot
调节用水分 tempering moisture
调节油 pilot oil
调节油箱 make-up tank
调节余缺 regulate supply and demand
调节元件 final controlling element; regulating element
调节闸 regulator
调节闸门 register gate; regulating gate; regulator gate
调节账户 accommodating accounts; reconciliation account
调节者 accommodator; adjuster; modifier
调节针 metering needle; regulating needle
调节针阀 adjusting needle valve; metering pin valve
调节支柱 adjustable fixing
调节值 regulated value
调节指令 regulating command
调节制动器 regulating brake
调节质量 quality of regulation
调节滞后 control lag
调节轴 regulating shaft; regulating spindle
调节轴爪 regulator shaft jaw
调节主令 regulating command
调节贮仓 regulating storage
调节柱塞 regulating plunger
调节砖 gate; stopper
调节装置 accommodator; actuating device; actuating mechanism; adjuster; adjusting device; adjusting gear; adjusting unit; conditioner; control apparatus; controlling device; control set-up; control valve gear; governing device; governing gear; regulating apparatus; regulating device; regulating unit; setting device; compensating unit < 自动安平水准仪的 >
调节状态 regulated condition
调节锥 cone governor
调节准确度 control accuracy
调节资金 buffer fund
调节资金供求 adjust supply and demand for funds
调节子 regulon
调节子配件 control subassembly
调节阻气门 registration choke
调节作用 equalizing effect; pondage action; regulating action; regulating effect; retention effect; reservoir action < 水库 >
调解 conciliation; mediacy; mediation; reconcilement
调解处 mediation abode
调解法 paramodulation
调解法庭 conciliation court; court of conciliation
调解及仲裁规则 Rule of Conciliation and Arbitration
调解纠纷 mediation of disputes
调解人 accommodator; bridge builder; conciliator; intercessor; intermediator; mediator; middleman; moderator; defuser < 紧急局面的 >
调解式 paramodulant
调解书 amicable settlement; mediation decision; reconciliation agreement
调解委员会 board of conciliation; commission of conciliation; conciliation committee
调解者 mediator
调金漆料 bronzing fluid; bronzing medium
调金色法 gold toning
调晶剂 habit modifier
调距隔片 spacer shim
调距螺母 adjusting screw nut
调距螺栓 setting-up screw
调距螺旋 setting-up screw
调距螺旋桨 adjustable pitch airscrew; controllable pitch propeller
调聚反应 telomeric reaction; telomerization reaction
调聚剂 telogen
调聚体 telogen
调聚物 telomer
调均 equalization
调控 adjust and control; regulation and control
调控职能 adjust and control function
调宽扼流圈 width choke
调宽线圈 width coil
调理 conditioning
调理池 conditioner
调理剂 amendment; conditioner; set-controller; set-controlling agent; set-modifying agent; set regulator
调理指数 opsonic index
调理作用 opsonic action; opsonization
调梁设备 car beam straightening equipment
调梁线【铁】 car beam straightening siding
调量表 adjusting quantity meter
调量的幅度变化范围 rangeability
调料 mixtion
调料槽 conditioner
调料锅 crutcher
调料库 condiment storage
调零 < 仪器等 > adjust to zero; zero adjust(ing); zero alignment; zero set(ting); zeroing
调零电路 zeroing circuit
调零电位计 zero potentiometer; zero trimmer
调零法 null method
调零位 zero point adjustment
调零旋钮 zero adjusting screw
调零装置 balancing control; null setting
调铝粉漆用漆料 alumin(i)um mixing varnish
调螺距油泵 pitch control oil pump
调面机 dough mill
调墨 abrasive ink
调墨板 table-inking
调墨刀 palette knife
调墨料 printer's varnish
调墨台 ink slab
调墨用熟油 linseed oil stand oil; stand oil
调墨油 ink compounds; stand linseed oil; thickened oil
调凝剂 set-controller; set-controlling admixture; set-controlling agent; set-modifying agent; set regulator
调凝石膏浆 gypsum ga(u)ging plaster
调凝水泥 regulated-set cement
调凝水泥和喷射水泥 regulated-set cement and jet cement
调扭矩 torque setting
调牌通报器 drop annunciator
调配 allotment; deployment; rearrangement; recombined; redeploy
调配槽 surge tank
调配场 marshalling yard
调配电路 dispensable circuit
调配劳动力 deployment of labo(u)r force
调配沥青 blended asphalt
调配膜片 matching window
调配牛油 < 固体润滑剂 > tallow compound
调配色 matched colo(u)r
调配系统 distributing system; distribution system
调配原材料 allocation of raw materials
调频 frequency modulation[FM]
调频备用机组 frequency modulation reserve
调频波 frequency modulation wave
调频波段分配 frequency modulation allocation
调频电报 frequency-modulated telegraphy
调频电厂 frequency regulating plant
调频度 frequency modulation index
调频发射机 FM transmitter
调频发射机综合测试仪 panalyser [panalyzer]
调频法 frequency modulation method [FM-method]
调频干扰 frequency modulation jamming
调频管 frequency-modulated tube; phasitron
调频广播 frequency modulation broadcast(ing)
调频广播波段 frequency-modulated broadcast band
调频广播发射机 frequency modulation broadcast transmitter
调频广播频道 frequency modulation broadcast channel
调频轨道电路 frequency-modulated track circuit
调频晃动阻尼器 tuning liquid damper; tuned liquid damper; tuned sloshing damper
调频回旋加速器 frequency-modulated cyclotron
调频畸变 frequency modulation distortion
调频极限 modulation frequency limit(ation)
调频鉴别器 frequency modulation discriminator
调频接收机 FM receiver; fremodyne
调频接收机门限电平 threshold level of FM receiver
调频接收器 fremodyne
调频立体声 frequency modulation stereo
调频立体声波段 frequency-modulated stereoband
调频录音 frequency modulation recording
调频脉冲压缩 frequency modulation pulse compression
调频门限 frequency modulation threshold[FM-threshold]
调频扭力仪 frequency-modulated torque regulator
调频频偏 frequency modulation deviation
调频器 frequency modulator; frequency regulator; frequency shift keying; frequency variator
调频深度 warble rate
调频声 frequency-modulated sound
调频声呐 frequency-modulated sonar
调频声载波 frequency modulation sound carrier
调频失真 frequency modulation distortion
调频式测高计 frequency modulation type altimeter
调频式电容脉波计 capacitoplethysmograph
调频式载波电话 frequency-modulated carrier current telephony
调频收音机 frequency modulation receiver; receiver of frequency modulation
调频特性 frequency modulation characteristic
调频-调幅变换器 FM/AM [frequency modulation/amplitude modulation] converter
调频调谐器 frequency modulation tuner
调频通信[讯]制 frequency modulation

telecommunication system
调频同步加速器 frequency-modulated synchrotron
调频图 frequency modulation chart
调频无线电测高计 frequency modulation radio altimeter
调频无线电台 frequency modulation radio station
调频无线电通信[讯]制 frequency modulation telecommunication system
调频系数 coefficient of frequency modulation
调频系统 frequency-modulated system;frequency modulation system
调频限制器 cycle-by-cycle device
调频信号 frequency-modulated signal
调频信号发生器 frequency-modulated signal generator; frequency modulation generator
调频信号检波器 frequency modulation detector; frequency-sensitive detector
调频遥测发射机 frequency modulation telemetry transmitter
调频遥控系统发生器 frequency-modulated telecontrol system generator
调频叶片 frequency-modulating blade
调频液体阻尼器 tuning liquid damper;tuned liquid damper;tuned sloshing damper
调频抑制 frequency modulation rejection
调频音调 frequency-modulated audio tone
调频音频电报 modulation frequency-voice frequency telegraph
调频引信 frequency-modulated fuze
调频阈值 frequency modulation threshold[FM-threshold]
调频运行 speed governor operation
调频运行方式 holding frequency
调频杂波电平 frequency modulation noise level
调频载波 frequency-modulated carrier
调频噪声 frequency modulation noise
调频振荡器 frequency-modulated generator; frequency-modulated oscillator;frequency-modulated signal generator;frequency modulation oscillator
调频指数 frequency modulation index
调频制 frequency modulation system
调频质量阻尼器 tuning modulation damper;tuner mass damper
调频中心频率 average frequency
调频终端设备 frequency modulation terminal equipment
调频转速控制 AC[automatic control] and frequency speed control
调频装置 frequency regulator
调频最大频偏 frequency swing
调平 equalization;level(l)ing
调平垫片 level(l)ing pad
调平阀 level(l)ing valve
调平范围 level(l)ing range
调平机 level(l)ing machine
调平机构 level(l)ing gear;level(l)ing mechanism
调平(机构)油缸 level(l)ing cylinder
调平镜 level(l)ing mirror
调平连通管 level(l)ing pipe
调平螺钉 level(l)ing screw <测量仪器的>;level(l)ing bolt
调平螺杆 linkage level(l)ing screw
调平螺丝 level(l)ing screw
调平螺旋 <测量仪器的> level(l)ing screw
调平器 level(l)er
调平塞 level plug
调平栓 level plug
调平用千斤顶 level(l)ing jack
调平油缸 level(l)ing ram
调平支脚 level(l)ing shoe
调平装置 level(l)ing device;level(l)ing adjustment
调屏的 tuned-anode
调屏调栅电路 tuned-plate-tuned-grid circuit
调屏调栅振荡器 Armstrong oscillator; tuned-plate-tuned-grid oscillator
调屏振荡器 tuned-plate oscillator
调坡 adjusting gradient
调漆 breaking up; let down; paint mixing
调漆刀 palette knife;spatula[spatule]
调漆罐 paint conditioner tank
调漆机 colo(u)r matching system; paint mill;paint mixer
调漆料 vehicle
调漆器 paint conditioner
调漆刷 distemper brush
调漆油 painter's naphtha;paint naphtha;paint oil
调气 conditioning
调气阀 choker valve
调气装置 air register
调热杆 heat regulating lever
调热器 heat regulator;thermoregulator
调色 adjusting colo(u)r;colo(u)r bleeding;colo(u)r blend;colo(u)r matching;colo(u)r mixing;colo(u)r modulation;colo(u)r working;tinting;toning
调色板 palette;palette board;pallet
调色杯 plettecup
调色比 pigment binder ratio
调色程序 colo(u)r matching program(me)
调色刀 painting knife;palette knife
调色碟 colo(u)r mixing tray
调色工 stainer
调色机 colo(u)r mixer
调色基料 colo(u)r base
调色激光器 tunable-dye laser
调色剂 tinting colo(u)r;toner
调色浆 stainer; tinter; tinting paste; toner
调色蓝 toning blue
调色料 printer's varnish;tinting vehicle
调色盘 palette
调色漆 colo(u)r in oil;tinting paste
调色染料 dope dye
调色桶 colo(u)r pan
调色油 colo(u)r oil;Victory <俗称维力油>
调色照明 colo(u)rama lighting
调砂及筛砂机 sand cutting and screening machinery
调栅的 tuned grid
调栅电路 tuned grid circuit
调梢装置 tape set
调伸长度 overhang; overhanging in length
调声 voicing
调湿 conditioning
调湿的 moisture-conditioned
调湿机 damping machine
调湿器 humidifier
调湿溶液 fountain solution
调湿设备 humidifying equipment
调湿箱 humidistat
调湿装置 humidifier
调时 timing
调时标灯 timing lamp
调时齿轮 timing gear
调时垫片 timing washer
调时阀 tarry valve;timing valve
调时螺钉 mean time screw
调时螺丝 timing screw
调时器 time clock
调时制 time scale
调时装置 tarry device
调视 accommodation
调试 commissioning; debugging; setting; set-up; shakedown test; trouble shoot
调试部件 debugging unit
调试车间 adjusting shop; tune-up shop
调试程序 debugged program(me); debugged routine;debugger;debugging program(me);debugging routine
调试程序包 debugging packet
调试冲模用压力机 die spotter
调试方法 adjustment method
调试方式 debugging mode
调试辅助程序 debug aids;debugging-aid program(me); reference debugging aids
调试工具 debug(ging) aids;debugging tool
调试功能 debug function; debugging facility
调试宏指令 debug macroinstruction
调试活动 debugging activity
调试机 debug machine
调试机器设备 alignment
调试监督程序 debug monitor
调试检查 bug check
调试阶段 debug(ging) phase
调试开关 debug switch
调试命令 debug command
调试模型 debugging model
调试屏幕 aiming screen
调试时间 development time
调试实用程序 debugging utility
调试系统 debug(ging) system
调试指令 debug command
调束系统 beam-control(led) system
调水活门 hydraulic valve
调水平 level(l)ing adjustment
调速 governing; speed control; speed governing; speed regulation; speed timing;velocity modulation
调速臂 speed-regulating arm
调速变阻器 speed adjusting rheostat; speed-regulating rheostat
调速秤 speed-regulated scale
调速齿轮 speed control gear
调速传动 regulator drive
调速带 speed belt
调速电动机 adjustable speed motor; regulating speed motor; variable-speed motor; speed-regulating motor
调速电动气动阀 speed regulation electro-pneumatic valve
调速电极 buncher
调速电空阀 speed regulating electro-pneumatic valve
调速电气阀 speed regulation electro-pneumatic valve
调速电子管 prionotron
调速惰轮 governor idle gear
调速发电机 velodyne
调速阀 compensated flow control valve; governor valve; speed control valve;speed-regulating valve
调速范围 regulating range
调速方式 mode of speed regulation
调速杆 gear change lever;speed control rod;speeder rod
调速杠杆 speed control lever
调速功 servomotor capacity
调速管 klystron(tube); prionotron; velocity-modulated valve; velocity modulation tube
调速环 adjustable ring;gate operating ring; regulating ring; speed ring; guide ring <水轮机的>
调速环节 governing loop
调速机 governor
调速机构 speed adjusting gear;speed control mechanism;speed gear
调速继电器 speed-regulating relay
调速接力器 speed-regulating servo-motor
调速聚束 velocity modulation bunching
调速控制器 actuator governor;speed setting controller
调速皮带轮 adjustable speed wheel
调速汽门 throttle
调速汽门前汽温 throttle steam temperature
调速汽门前压力 throttle pressure
调速器 governor of velocity; rate governor; regulator; speed controller; speed governor; speed regulator;velometer;actuator
调速器把手 governor handle
调速器泵 governor pump
调速器闭油路 governor trapped oil
调速器拨叉 governor fork
调速器不等率 governor regulation
调速器操纵杆 governor lever
调速器操纵箱 governor control box
调速器(操作)柜 governor cabinet; regulator cubicle
调速器测速装置 governor head
调速器超速控制 governor overriding control
调速器齿轮 governor gear
调速器传动 governor motion
调速器传动装置 governor gearing; governor movement
调速器磁铁阀 governor magnet valve
调速器的不动作时间 dead time of governor
调速器的传动装置 governor actuator
调速器的控制装置 governor control
调速器的灵敏度 sensitivity of governor
调速器电动机 governor motor
调速器发电机 governor generator
调速器阀 governor valve
调速器反应率 governor response rate
调速器副弹簧 governor assist spring
调速器杆 governor arm
调速器杆螺钉 governor arm screw
调速器杠杆的叉头 governor fork
调速器高压油 governor pressure oil
调速器工作能力 governor work capacity;governor work output
调速器关闭时间 closing time of governor
调速器滚柱轴承 governor roller bearing
调速器滑阀 rotating pilot valve
调速器环 governor ring
调速器机构 governor mechanism
调速器间断油压油 governor intermittent oil
调速器开关 governor switch
调速器壳体 governor housing
调速器空气信号压力 governing air signal pressure
调速器控制 governor control
调速器控制安全阀 governor control safety valve
调速器控制的 governor-controlled
调速器控制电缆 governor control cable
调速器控制盘 regulator cubicle
调速器控制皮带轮 governor-controlled sheave

T

调速器控制轴 governor control shaft
调速器拉杆 governor rod
调速器廊道 governor gallery
调速器离心锤 governor weight
调速器连杆 governor link(age)
调速器联动装置 governor linkage
调速器灵敏度 sensitiveness of governor;sensitivity of governor
调速器螺管 governor solenoid
调速器螺管阀 governor solenoid valve
调速器马达 governor motor
调速器皮带轮 governor pulley
调速器偏转 governor deflection
调速器球杆 governor ball arm
调速器驱动机构 governor drive
调速器扇形齿轮 speed setting segment
调速器试验 governor test
调速器死区 governor dead time
调速器速度变换器 governor speed changer
调速器踏板 governor control pedal
调速器弹簧 governor spring
调速器特性 governor characteristic
调速器调整螺钉帽 governor adjusting screw cap
调速器停车手柄 governor stop arm
调速器停止螺管 governor stop solenoid
调速器外壳 housing of governor
调速器稳定性 stability of governor
调速器下降特性 governor droop
调速器限定速度 governed speed
调速器箱 governor box
调速器销 regulator pin
调速器引导阀 actuator governor
调速器用电动机 governor motor
调速器油槽油 <无压油> governor sump oil
调速器油系统 governor oil system
调速器油压 governor oil pressure
调速器振荡 regulator oscillation
调速器执行机构 governor actuator
调速器止块 governor stop block
调速器中的摩擦 governor friction
调速器中间油压油 governor intermediate oil
调速器重锤 <离心式> governor weight
调速器周期性振动 governor hunting
调速器轴 governor shaft; regulator spindle
调速设备 speed control device
调速伸缩率 flexibility ratio
调速时间 governing time
调速手柄 speed governor handle; speed shank
调速手轮 speed-regulating hand wheel
调速伺服电动机 speed control servomotor;speed-regulating servomotor
调速随动系统 speed follow-up system;speed servo system
调速弹簧 speed control spring
调速特性曲线 speed regulation characteristic
调速系统 speed governing system; speed-regulating system
调速箱 governing box
调速型液力耦合器 variable-speed fluid coupling
调速选粉机 speed controlled separator
调速叶轮泵 governor impeller
调速制动 speed control braking
调速助力器 governor booster
调速助力器活塞杆 governor booster piston rod
调速装置 speeder
调索 cable adjustment

调停 arbitrament;conciliation;harmonize; intervention; mediation; reconcile
调停的 intervenient
调停工资纠纷 wage arbitration
调停劳资纠纷 arbitration in disputes between labor and capital
调停器 arbiter
调停人 accommodator;adjuster;arbitrator;intermediator
调停协议 conciliation agreement
调停者 intervenient;intervenor
调位 positioning;position modulated
调位装置 rack work
调味 seasoning
调味剂 condiment;seasoning
调味品小瓶 <餐桌上的> cruet
调味瓶架 castor
调温 attemper(ment);conditioning
调温表面涂层 thermal control surface coating
调温层 temperature control coating
调温风 tempering air
调温柜 thermotank
调温机 tempering machine
调温空气 tempering air
调温冷却器 trim cooler
调温盘管 tempering coil
调温膨胀阀 thermal expansion valve
调温器 attemperator; damping machine;temperature controller; temperature regulator; thermoregulator;thermosistor;thermostatic regulator
调温蛇管 tempering coil
调温室 controlled temperature cabinet
调温调湿箱 conditioning cabinet
调温调湿箱试验 humidity cabinet test
调温涂料 thermal control material
调温箱 controlled temperature cabinet
调温旋管 tempering coil
调温装置 register
调温组件 temperature-controlled package
调线 realignment
调相 phase modulation
调相包线 phase modulation envelope
调相波 phase-modulated wave;phase modulation wave
调相插塞 phasing plug
调相常数 phase modulation constant
调相叠加 phased stack
调相短截线 phasing stub
调相工况 condenser operation
调相管 phase-tuned tube;phasitron
调相光束 phase-modulated beam
调相机 compensator; condenser; phase modifier;phase regulator;rotary condenser
调相记录(法) phase modulation recording
调相技术 phasing technique
调相检波器 phase modulation detector
调相接收机 phase modulation receiver
调相聚焦 phase focus(s)ing
调相开关 phasing switch
调相脉冲 phase-modulated pulse
调相器 compensator; condenser; dynamo condenser; phase modifier; phase modulator; rotary condenser; rotatable phase-adjusting transformer
调相容量 capacity of synchronous condenser
调相式载波电报 phase modulation carrier telegraph

调相通信[讯]系统 phase modulation communication system
调相同步机 phasing synchro
调相信号 phase-modulated signal
调相用真空管 phasitron
调相运行 condenser operation
调相载波 phase-modulated carrier
调相制 phase modulation system
调谐 attune; debug; dialing; harmonization;tune(in);tune-up;tuning
调谐摆 tuned pendulum
调谐变压器 tuned transformer
调谐槽路 tuned circuit;tuning circuit
调谐单元 tuner unit;tuning unit
调谐的 syntonic
调谐电动机 tuning motor
调谐电路 tuning circuit
调谐电路响应 response of tuned circuit
调谐电容器 tuning capacitor
调谐电眼 tunoscope
调谐度盘 tuning dial
调谐短线 stub line
调谐钝度 broadness of tuning
调谐范围 tuning range
调谐范围扩展 bandspread
调谐放大器 resonance amplifier;tuned amplifier
调谐非调谐交替放大电路 tuned-aperiodic-tuned circuit
调谐公式 <计算静载斜缆拉力的> harmonic formula
调谐回路 tuned circuit;tuning circuit
调谐激荡减振器 tuned sloshing damper
调谐减振器 tuned vibration absorber
调谐界限误差 threshold tuning error
调谐精确度 degree of balance
调谐开关 change-tune switch
调谐刻度盘 tuning dial
调谐控制 tuning control
调谐滤波器 tuned filter
调谐膜片 tuned window
调谐扭矩减震器 tuned torsional vibration damper
调谐偶极子 tuned dipole
调谐耦合电路 tuned coupled circuit
调谐偏差 tuning deviation
调谐器 tuner(unit);tuning unit
调谐曲线 tuning curve
调谐锐度 sharpness;sharpness of tuning
调谐三极管 tuning triode
调谐舌簧继电器 tuned-reed relay
调谐时间 tuning period
调谐时间常数 tuning time constant
调谐天线 tuned array
调谐推潮法 harmonic method
调谐卫星轨道 resonant satellite orbit
调谐无源线圈 tuned inert coil
调谐误差 turning error
调谐匣 tuner unit;tuning unit
调谐线圈 syntonizing coil;tuning coil
调谐(线圈)电感 tuning inductance
调谐箱 dog house
调谐销 tuned post
调谐信号 harmonic ringing
调谐旋钮 selector;tuning knob;turning control
调谐用短线 stub tuner
调谐指示灯 tuning lamp
调谐指示器 cathode-ray tuning indicator;tuning indicator;tuning meter;tunoscope
调谐质量阻尼(减振)器 tuned mass damper
调谐转换系统 tuned transducer system
调谐装置 tuner unit;tuning arrangement;tuning device;tuning unit

调谐阻抗联结器 tuned impedance bond
调谐组件 tuning block
调心球面 aligning seat
调心式托辊 self-aligning belt idler
调心套筒 aligning sleeve
调心托辊 self-centering idler;training idler
调心销钉 aligning bar;aligning pin;aligning stud
调心止推轴承 tilting pad thrust bearing
调心轴承 self-aligning bearing; tip bearing
调心座圈 self-aligning seat washer
调休 holiday for working an extra shift
调休制【铁】 lodging system
调修 reconditioning
调蓄池 regulation tank
调蓄量 pondage
调蓄水库 equalized reservoir
调蓄作用 pondage action; reservoir action
调压 pressure regulation;voltage regulation
调压板 orifice plate
调压泵 surge pump
调压变压器 regulating transformer; variable transformer
调压变阻器【电】 rheostat voltage adjust rheostat
调压薄膜 compensation diaphragm
调压舱 surge chamber
调压槽 surge bin
调压差值 override pressure
调压池 equalizing reservoir;surge basin;surge tank
调压电位计 voltage-regulating potentiometer
调压阀 air vent valve; dual relief valve;dump valve;modulating ring pressure valve;pressure modulation valve; pressure regulating valve; regulating valve
调压阀手柄 valve handle
调压管 equalizing pipe; surge pipe; voltage regulator tube
调压罐 compensator
调压活塞 relief piston
调压机 booster
调压机构 pressure regulating device
调压继电器 voltage relay
调压进出式潜水器 pressure regulating lock-in and lock-out submersible
调压井 stilling well;surge shaft;surge tank
调压井的扩大室 expansion chamber
调压井的稳定性 stability of surge tank
调压井功能 action of surge chamber
调压井下室 lower surge chamber
调压井涌浪 tank surge
调压器 compressor governer[governor]; constant voltage generator; extraction pressure governor; governor; pressure governor; pressure regulator;regulator;voltage adjuster;volt(age)regulator【电】
调压器活塞 governor piston
调压器绝缘连接管 governor pipe insulating joint
调压器连接管 governor union
调压器同步系统 governor synchronizing system
调压式多辊矫直机 pressure regulating roller level(l)er
调压式振荡器 voltage-tuning oscillator
调压室 adjusting chamber; distributing chamber;surge chamber;surge

tank
调压室功能 action of surge chamber
调压竖管 standpipe;surge pipe
调压水槽 surge tank
调压水池 equalizing tank
调压水库 equalized reservoir;equalizing tank;equalizing reservoir
调压水塔 equalized tank;equalizing tank;surge tank
调压水箱 equalized tank;equalizing reservoir;equalizing tank;surge tank
调压水柱 surge shaft
调压塔 surge tower
调压凸轮 compression relief cam
调压箱 pressure regulating box
调压溢流阀 pressure adjustment relief valve
调压站 governor house;governor station;regulator station
调压转速控制 ac adjustable voltage speed control
调压装置 pressure regulating device
调压自耦变压器 voltage-regulating autotransformer
调药槽 reagent tank
调页【计】paging
调液厚器 absorpti(o)meter;absorption meter
调音 attune;tone control;tone tuning;tuning;voicing
调音管 pitch pipe
调音室 audio control;sound regulation room
调音台 audio mixer;sound console;sound mixing console
调音台对讲装置 talk back of sound console
调音者 tuner
调优运算 evolutionary operation
调油墨 blending of ink
调油器 oil feeder
调鱼机 gag press
调匀砂 temper sand
调整 adjust(ing);adjustment;alignment;balancing;checkout;compensation;conditioning;debug;demodulation;ga(u)ging;governing;harmonization;justify;level(l)ing;line up;modulate;modulating;ordering;processing;readjust;readjustment;rearrange;rearrangement;rectification;rectify;redress;regularization;regulating;re-structuralization;restructuring;rigging;take-up;trade off;trueing;truing;tune-up;tuning;updating;adequation
调整按钮 control button;regulating knob
调整凹口 regulating notch
调整摆 regulator pendulum
调整板 adjustment panel;adjustment plate;wearing strip
调整保留时间 adjusted retention time
调整保留体积 adjusted retention volume
调整比价 adjust the price ratios
调整比例尺 adjust scale
调整变数 adjusting variables
调整标高 trued for level
调整标记 adjusting mark;adjustment mark
调整柄 adjusting handle
调整波形 demodulation
调整不当 misalignment
调整不良 maladjusted
调整不善 maladjustment
调整步骤 set-up procedure
调整槽口 adjustment notch
调整测量 balancing survey
调整测量仪器透镜中的视准线 collimation adjustment
调整层 regulating course
调整叉 adjusting fork
调整产品方向 reorienting production
调整产品结构 adjustment productive texture
调整产业结构 readjusting industrial structure
调整长期趋势后的时间数列 trend adjusted time series
调整车间 adjusting shop
调整衬条 adjusting strip
调整成本 adjusted base cost
调整成本制 adjusted cost basis
调整程度 degree of regulation
调整程序 debugging program(me);set-up procedure;tune-up procedure
调整齿架 adjuster rack
调整齿轮 adjuster pinion;adjusting gear
调整齿轮压板 adjuster pinion holder
调整齿轮压板螺丝 adjuster pinion holder screw
调整处理 conditioning treatment
调整触点 set feeler
调整触片 adjustable contact plate
调整船舶吃水 adjust a ship's draft
调整磁铁 compensating magnet
调整存货价格 inventory valuation adjustment
调整措施 redeployment
调整大地水准面 adjusted geoid;cogeoid;regulated geoid
调整带 speed belt
调整单位 unit of adjustment
调整挡铁 adjustable stopper
调整档 regulating notch
调整导板 adjustable slide
调整到零 adjust to zero;zeroing
调整到正确尺寸 set to exact size
调整的 regulative
调整的测定系数 adjusted coefficient of determination
调整的复相系数 adjusted multiple correlation coefficient
调整的基本造价 adjusted base cost
调整的频率表 adjusted frequency table
调整的实际库存量 adjusted base
调整的指数 rectified index number
调整点 adjust point;setting point
调整电动机 control motor
调整电容 control capacitance
调整电压 regulation voltage
调整垫板 adjusting plate
调整垫块 adjusting block;setting block
调整垫片 adjusting shim;adjustment plate;dutchman;ring riser;spacer shim
调整垫圈 adjusting spacer;adjusting washer;adjustment washer
调整垫铁 taper parallel
调整对象 controlled marker
调整耳 adjusting ear
调整发动机 make true an engine
调整阀 adjustable valve;adjusting valve;regulating valve;regulator valve;trim valve
调整阀隙 adjustment of valve clearance
调整阀座 regulating valve seat
调整阀座套 regulating valve seat bush
调整范围 adjusting range;adjustment range;governor deflection;range of adjustment;setting range
调整方案 readjustment plan
调整方法 adjustment method;regulation means
调整方式 adjusting mode
调整放大器 control amplifier
调整费率 adjustment of rate
调整费用 adjustment cost;adjustment work expenses;set-up cost
调整分贝 adjusted decibel;decibel adjusted
调整分录 adjusted entry;adjusting entry;adjustment entry
调整风险后贴现率 risk-adjusted discount rate
调整缝 compensation joint
调整幅度 adjustment range
调整杆 adjustable stem;adjuster bar;adjusting rod;doctor bar;dolly bar;regulating crank;regulating lever;regulation rod
调整杠杆 adjusting lever;regulating lever
调整高程 trued for level
调整工 adjuster;fettler;finisher;setter
调整工件位置 adjust work for position
调整工具 adjusting appliance;adjustment tool
调整工商业 readjusting industry and commerce
调整工资率 adjustment of wage rates
调整工作 stoppage of work
调整工作结构 textural of geologic(al)working adjustment
调整供求关系 adjust supply to a current demand;readjust supply and demand;regulate supply and demand
调整拱模位置 swinging arch
调整管 compensating pipe
调整管理机构 adjustment administrative setup
调整管子工具 bending pin
调整光圈 orifice;set diaphragm
调整轨缝 adjustment of gaps;evenly distribution joint gaps
调整轨距工作 regauging work
调整轨距器 tie spacer
调整轨枕间距 respacing of cross ties
调整辊 adjusting roller
调整滚子 adjusting roller
调整过程 adjustment process;process of adaptation;process of setting
调整函数 adjustment function
调整焊缝 control seam;pilot seam
调整合同总价 adjusted total price of contract
调整和修补工程 adjusting and repairing work
调整河流 adjusted river;adjusted stream
调整荷载 factored load
调整后采购成本 adjusted acquisition cost
调整后成本基础 adjusted cost basis
调整后的抵押 adjusted mortgage
调整后的基本价格 adjusted base cost
调整后的加权平均数 weighted average of post adjustments
调整后的净填方 adjusted net fill
调整后的售价 adjusted sales price
调整后的死亡率 adjusted mortality(rate)
调整后的总收益 <房地产> adjusted gross income
调整后价格 adjusted price
调整后净利润 adjusted net profit
调整后历史成本 adjusted historical cost
调整后毛收入所得额 adjusted gross income
调整后年金 adjusted pension
调整后实际库存量 adjusted actual inventory
调整后试算表 adjusted trial balance
调整后收益 adjusted income
调整后数字 adjusted figures
调整后损益计算表 adjusted income statement
调整后所得 adjusted income
调整后银行存款(余)额 adjusted bank balance of cash
调整后银行余额 adjusted bank balance
调整后债务 restructured debt
调整后账面现金额 adjusted book balance of cash
调整后总所得 adjusted gross income
调整后最早到期日收益率 yield to adjusted minimum maturity
调整滑块 adjusting slider
调整环 adjustable collar;adjusting ring;binding ring;Fink-ring;holding ring;regulating ring;shifting ring
调整缓行器 <在峰顶调整前后溜放钩车的间隔> adjusting retarder
调整换位 coordinated transposition
调整汇价 adjustment of exchange rate;exchange rate adjustment
调整汇率 adjustment of exchange rate;exchange rate adjustment
调整活动百叶窗 adjustable louvers
调整活塞 set piston
调整机 adjusting machine;reconditioner
调整机构 adjusting gear;adjusting mechanism;regulating organ;setting device
调整机器 set-up a machine
调整机器设备的工人 set-up man
调整(机用)电动机 governor motor
调整极 regulating pole
调整计算 adjustment computation
调整(计算机)密码 debugging code
调整记录 adjusting entries
调整技术结构 adjustment technologic(al)texture
调整技术经济结构 textural adjustment of technologic(al) and economics
调整技术经济指标 adjustment indicator of techno-economics
调整剂 conditioning agent;modifier;modifying agent;regulating agent
调整季节变动后的数据 seasonally adjusted data
调整季节变动后的数字 seasonally adjusted figure
调整价格 adjust price;modify price;price adjustment;price change;readjust price
调整(价格伸缩)条款 <工资等的> escalator clause
调整间隙 control gap
调整间隙垫板 H-block
调整检查 tune-up inspection
调整键 adjusted key;adjusting key
调整搅拌器 correction agitator
调整接点 regulating contact
调整进度表 revise a schedule
调整经济关系 reforming economic relations
调整经济结构 adjustment economic texture
调整精度 adjusting accuracy;degree of regulation;setting accuracy
调整精整机 set-up the finishing mill
调整井 adjusting well
调整就位 reshifting
调整空间 setting space
调整孔 access;access port
调整控制数字 justification control

digits
调整块 equalizer disk [disc]; push wedge; set piece; setting block
调整快慢针的螺钉座 screw support for adjusting regulator
调整矿区形状 swinging a claim
调整拉杆 regulating pull rod
调整力矩 trim moment
调整利率 interest rate adjustment; re-adjustment of interest rate
调整连接杆 adjusting link
调整连续条件 adjustment continuation condition; adjustment continuous condition
调整联杆 regulating link
调整量规 adjusting ga(u)ge
调整列线图 alignment chart
调整裂 regulative cleavage
调整零点 zero adjustment; zeroing
调整流量 regulated stream flow
调整流量脉冲 flow rate impulse
调整流速 modified velocity
调整流向 adjustment in flow direction
调整滤波器 compensation filter
调整路拱 adjusting crown; crown adjustment
调整率 adjustment rate; regulation factor; relative regulation
调整轮 adjusting wheel; control wheel; dancing pulley; dirigible wheel; ga(u)ge wheel; regulating wheel
调整罗盘 adjustment compass
调整螺钉 adjust(ing) screw; adjusting spring; plus-minus screw; regulating screw; sett(ing) screw; tuning screw
调整螺钉的固定钢丝 securing wire of adjustment screw
调整螺杆 adjusting screw; adjusting spring; plus-minus screw; stretching screw; tensioning device adjusting screw
调整螺帽 adjusted nut; adjusting nut; set nut
调整螺母 adjusted nut; adjusting nut; regulating nut; set(ting) nut
调整螺栓 adjustable bolt; adjusting bolt; adjustment bolt; adjustment stud; jack(ing) bolt; stud
调整螺栓以增加钢丝绳的给进长度 fan-out screw
调整螺丝 adjusting screw; adjusting spring; regulating screw; stretching screw; timing bolt; turnbuckle
调整螺旋 adjusting screw; adjustment screw; rectifying screw; regulating screw; turnbuckle screw
调整面 adjusting surface
调整模型比例尺 scaling model
调整捏手 regulating knob
调整盘 adjusting plate
调整盘存 adjusting inventory
调整配方 batch blending; blending batch
调整片 adjusting strip; regulation chip; trimmer; trim plate; trim tab; chock plate <下作用水阀端部>
调整片钮 trim knob
调整片随动机构 trim-tab servo
调整偏差 offset
调整凭单 adjustment memo; adjustment memorandum
调整曝光 modulated exposure
调整期 adjustment phase
调整期间 adjustment period; period of adjustment
调整器 adjuster; adjusting device; back-pressure regulator; compensator; conditioner; control apparatus; controller; controls; corrector; governor; modulator; regulator; setting block; trimmer
调整器板 adjuster board
调整器测量机构 primary measuring element
调整器阀门 adjuster valve
调整器架 compensator frame; compensator stand
调整器控制活门 controller pilot
调整器扇形齿轮 regulator quadrant
调整器座 compensator frame
调整钎子 setting chisel
调整牵引定额 adjusted tonnage rating
调整前试算表 preadjustment trial balance; trial balance before adjustment; unadjusted trial balance; unadjusted trial statement
调整清晰度 adjustment for definition
调整区间 control interval
调整区组 adjustment group
调整曲线 adjustment curve; regulation curve
调整圈 adjusting ring; Fink-ring; regulating ring; shifting ring
调整绕组 regulating winding
调整溶液 adjusted solution
调整色光 shading
调整砂垫层 regulating carpet of sand
调整设备 adjusting appliance; adjusting device
调整生产计划 revise production plans
调整时间 setting time
调整时期 adjust the time; governing time; period of adjustment
调整时滞 adjustment lags
调整实际库存量 adjust actual inventory
调整式阶梯规 adjustable stair ga(u)ge
调整式卡规 adjustable snap ga(u)ge
调整试验 tune-up test
调整室 control room
调整手把 adjusting handle
调整手柄 regulating handle
调整手续 alignment procedure
调整数字 justifying digit
调整水舱 adjusting tank; compensation tank
调整水柜 adjusting tank
调整水平 adjustment of cross level
调整顺序 setting-up procedure; setting-up sequence
调整速度 adjusting speed; regulate the speed; regulate the velocity
调整索赔金额 adjusted claim
调整锁紧螺母 adjusting lock nut
调整弹簧 adjusted spring; adjusting spring; check spring; compensation spring; regulating spring; set-up spring
调整填料函 compensating stuffing box
调整填片 adjustment plate
调整条缝形状 pattern control adjustment
调整条款 escalation clause
调整通信[讯]线路 setting-up the link
调整通知 advice of adjustment
调整投资结构 better distribution of investment; changing the distribution of investment
调整投资使用方向 redirecting the use of investment
调整图 adjusting diagram; adjustment diagram
调整土地使用权 regularization of land tenure
调整外汇汇率 readjustment of foreign exchange rate
调整网 level(l)ing network
调整位置 trued for position
调整物价 readjust price; regulate price
调整物料 reconciliation of inventory
调整误差 alignment error
调整系数 adjustment coefficient; adjustment factor; regulation factor
调整系统 control system; regulating system
调整现值法 adjustment value method; adjust present value method
调整相位 phase
调整镶条 adjustable gib
调整项目 reconciliation item
调整销 adjusting pin; spacer pin
调整效应 corrective action
调整楔(块) adjusted wedge; adjusting wedge; pull wedge; push wedge; stay wedge; taper gib; taper wedge; wedge adjuster
调整楔用螺钉 gib screw
调整斜铁 pull wedge
调整(信号灯)控制系统 closed loop control system
调整行程长度装置 adjustment for length of stroke
调整型抽样检验 sampling inspection with adjustment
调整性变幅 non-operating luffing
调整性降价 corrective price decline
调整性能 adjustability
调整性信贷 adjustment credit
调整需要量 backing-out; requirements
调整旋钮 adjusting knob; setting knob
调整旋塞 by-pass cock; control cock; regulating cock
调整循环 modification loop
调整压力 <安全阀的> regulating pressure
调整压下装置 roll gap set
调整延迟 control lag
调整氧化物 modifying oxide
调整液 correction liquor
调整仪器 adjusting appliance; regulating instrument
调整翼片 tab
调整因数 adjustment factor; regulation factor
调整因素 adjustment factor; regulation factor
调整印刷电路板 regulating printed circuit board
调整盈余额 adjustment of surplus
调整用磁盘 alignment disc[disk]
调整用惰轮 adjusting idler wheel
调整用工具 setting tool
调整用螺钉 plus-minus screw
调整用双头螺栓 adjusting stud
调整余量 fitting allowance; margin for adjustment
调整裕量 alignment reserve
调整元件 coordinating element; pass element
调整员 regulator
调整原价 adjusted original cost
调整载波接收 reconditional-carrier reception; reconditioner-carrier reception
调整轧辊位置 roll setup
调整债券 adjusted bond; adjustment bond; escalator bond
调整张力 tensioning
调整账户 adjusting account; adjustment account
调整辙杆夹 transit clip
调整者 regulator
调整针塞 exhaust cone
调整值 adjusted value; setting value; trimmed value; value of set(ting)
调整职工工资 give wage increases to workers and staff members
调整指数 adjusted indexes
调整指针刻度盘 set-hands dial
调整至零点 adjust to zero
调整中心 center[centre] adjustment
调整轴 adjusting spindle
调整轴承间隙 take-up bearing
调整轴器 axometer
调整装车 diverse loading
调整装置 adjuster; adjusting apparatus; adjusting device; adjusting gear; correcting member; level(l)ing device; regulating apparatus; regulating device; setting device
调整准确度 accuracy of adjustment
调整准线 adjustment notch
调整资本 recapitalization
调整资本结构 recapitalize
调整资本移动 accommodating capital movement
调整作用 correct action; corrective action
调正 correction; redress; regulate
调正方位 realignment
调正精度 adjustment accuracy
调正器 corrector
调帧器 framer
调直 aligning; alignment; straightening
调直处理 <木材的> reconditioning
调直钢筋 straighten steel bar
调直工具 <管子工用> long dummy
调直管器 pipe straightener
调直管子工具 pipe straightener
调直辊 straightener roll
调直机 bar straightener; straightener
调直器 aligner
调直速度 straightening speed
调直筒 straightening drum
调直望远镜 alignment telescope
调值 value call
调值计算 adjustment calculation
调值优化 adjustment optimization
调值总价合同 escalation lumpsum contract; total and adjustable price contract
调职补助金 benefit for the change of job
调至市价 mark-to-market
调制 chop; confection; delta modulation; modulate; modulating; modulation
调制百分比 modulation percentage; percentage modulation; percent modulation
调制倍频器 modulator multiplier
调制本领 modulability
调制比 modulation ratio
调制变换器 modulation converter
调制变换转发器 modulation type switched transponder
调制变压器 modulation transformer
调制标准 modulation standard
调制表 modulometer
调制波 modulated wave; modulating wave; modulation wave
调制(波)包络线 modulation envelope
调制波形 modulation waveform
调制不足 undermodulation
调制不足的 undermodulated
调制参数 modulation parameter
调制产物 modulation product
调制成浆状 form into a paste
调制程序包 debugging package
调制传递函数 modulation transfer function
调制传递系数 modulation transfer factor

调制带宽 modulation bandwidth
调制的 modular;modulatory
调制的理想瞬间 ideal instants of a modulation
调制的有效状态 significant conditions of modulation
调制点 modulation point
调制电极 modulation electrode;signal grid
调制电流 modulated current
调制电路 modulated circuit;modulator circuit;tone circuit
调制电平 modulation level
调制电压 modulating voltage
调制定理 modulation theorem
调制定律 modulation law
调制度 degree of modulation;index of modulation; modulation degree; modulation depth; modulation factor; modulation index; percentage modulation
调制度百分数 amount of modulation
调制度测定器 modulation meter;percentage modulation meter
调制度测量 modulation measurement
调制度测量仪 modulation meter
调制度测试器 modulation meter
调制度器 modulation meter
调制度增高 modulation rise
调制段 modem section
调制扼流圈 modulating choke;modulation reactor
调制发光管 modulator glow tube
调制发生器 modulation generator
调制法 modulation method
调制反响 <测量吸声能力的一种方法> moderate reverberation
调制范围 modulation range
调制方案 modulation scheme
调制方法 method of modulation; modulator approach
调制方式 modulation mode;modulation system
调制放大器 modulation amplifier; modulator amplifier
调制分集 diversity of modulation
调制分量 modulation component; modulation product
调制峰值 modulation crest
调制辐射 chopped radiation
调制副载波法 modulated subcarrier technique
调制功率 modulation power
调制管 modulating tube; modulating valve;modulation tube;modulation valve; modulator tube; modulator valve
调制光 modulated light
调制光谱 modulated spectrum;modulation spectrum
调制光谱学 modulation spectroscopy
调制光束 modulated light beam
调制光源 modulated light source; modulation light source
调制函数 modulating function;modulation function
调制哼声 modulation hum
调制红外线光束 modulated infrared light beam
调制机 preparer
调制积 modulation product
调制畸变 modulation distortion
调制级 modem stage
调制极 Wehnelt cylinder;Wehnelt electrode
调制极谱法 modulation polarography
调制计 modulation meter;modulometer
调制技术 modulation technique
调制加宽 modulation broadening
调制间隙 modulation gap
调制监视器 modulation monitor
调制接收机 modulated receiver
调制解调法 modulation-demodulation method
调制解调方式 modulation-demodulation system
调制解调器 data set;modem;modem modulator-demodulator;modulator-demodulator(unit)
调制解调器多路转换器诊断 modem multiplexer diagnostics
调制解调器共享装置 modem sharing unit
调制解调器接口 modem interface
调制解调器连接访问协议 link access procedure for modem
调制解调(器)芯片 modem chip
调制解调器硬件 modem hardware
调制解调器转换功能 modem conversion function
调制解调设备 modem equipment
调制晶体 modulation crystal
调制警告 modulation alarm
调制控制器 modulation controller
调制控制系统 modulating control system
调制控制信号 modulator control signal
调制宽度 modulation width
调制类型 modulation type
调制量 modulated quantity
调制率 amount of modulation;modularity
调制码 modulation code
调制脉冲 modulating pulse
调制脉冲干扰 pulse modulated jamming
调制能力 modulability; modulation capability;modulation capacity
调制能量分布 spread of the modulation energy
调制盘 chopper wheel;retic(u)le
调制盘结构 reticle configuration
调制盘调制 reticle chop
调制盘轴 reticle axis
调制频带 modulation band
调制频率 modulation frequency
调制频率反馈 modulation frequency feedback
调制频率响应 modulation frequency response
调制频率与载频之比 modulation-frequency ratio
调制平衡 modulation balance
调制器 keyer; modulator; modulator driver;signal grid
调制器电容器 modulator condenser
调制器电压 modulator voltage
调制器-发射机-接收机 modulator transmitter receiver
调制器分频器 modulator divider
调制器激励器 modulator driver
调制器晶体 modulator crystal
调制器灵敏度 modulator sensitivity
调制器启动 modulator firing
调制器输出电压 modulator output voltage
调制器线性 modulator linearity
调制器箱 modulator cabinet
调制器元件 modulator element
调制器载频 modulator carrier
调制器振荡器 modulator-oscillator
调制器作用 modulator function
调制腔 buncher
调制曲线 adjustment curve;easement curve;junction curve
调制驱动电路 bootstrap driver
调制散焦 modulation defocus(s)ing
调制栅 inner grid; modulating grid; modulator grid
调制栅极 intensity grid
调制设备 modulating equipment;modulation device
调制深度 depth of modulation;modulation depth;percentage modulation
调制深度差 difference in depth modulation
调制声压 audio modulating voltage
调制声因数 modulation-tone factor
调制失真 modulation distortion
调制失真特性 modulation distortion characteristic
调制时号 modulated time signal
调制式乘法器 modulation type multiplier
调制式恒温器 modulating thermostat
调制(速)率 modulation rate
调制损耗 modulation loss
调制特性 drive characteristic;modulating characteristic
调制图形 modulation pattern
调制位移 modulation deviation
调制误差 modulation error
调制系数 index of modulation;modulation factor;percentage modulation
调制系数测试计 modulation factor meter
调制线圈 modulation coil
调制相 chopping phase
调制消除器 modulation eliminator
调制效率 modulation efficiency
调制效率系数 modulation efficiency factor
调制效应 modulation effect
调制信号 modulating signal;modulation signal
调制型补偿 modulation type compensation
调制型乘法器 modulation type multiplier
调制型干涉仪 modulation interferometer
调制性 modularity
调制要图 diagram mapping
调制抑制 modulation suppression
调制因数 modulation factor
调制因素 modulation factor
调制音频 modulating audio frequency
调制用电流【电】modulating current
调制元件 modulation element
调制载波 modulated carrier
调制载波信道 modulated carrier channel
调制载频衰落 scintillation fading
调制噪声 modulation noise;noise behind the signal;zoop
调制噪声的改进 modulation-noise improvement
调制振荡器 modulating oscillator; modulation generator
调制振幅 modulated amplitude
调制指示器 modulation indicator
调制指数 index of modulation;modulation index;ratio deviation
调制周期 modulation period
调制转换函数 modulation transfer function
调制状态 modulation condition
调制准直器 modulation collimator
调制作用 chopping action;modulating action
调治构筑物 regulating structure
调治建筑物 regulating structure
调治结构物 regulating structure
调质 hardening and tempering; refine;slack quench;tempering
调质处理 modified treatment;quenching and tempering;thermal refining; water conditioning【给】
调质处理结构 thermal refined structure
调质钢 converted steel;hardened and tempered steel;homogeneous steel; quenched and tempered steel
调质钢板 quenched and tempered steel plate
调质钢丝 final-hardened and tempered steel wire
调质高强度钢板 quenched and tempered high tensile strength plate
调质过的 quenched and tempered
调质平衡 balancing
调质砂 tempering sand
调质水 temper water
调质温度 refining temperature
调质轧制 skin pass
调质硬质纤维板 tempered hardboard
调置螺旋 antagonizing screw; clip screw
调轴机构 shaft position mechanism
调轴器 axometer
调柱机 pole-and-chain
调转方向 reversal of direction
调桩机 pole-and-chain
调准 adjust(ing); adjustment; aligning; alignment; harmonize; true-up;tune in;tune-up
调准板 aligning strip
调准处理 alignment treatment
调准度 degree of regulation
调准工具 adjusting tool
调准过程 alignment procedure
调准簧盒 adjusting spring case
调准校正 alignment adjustment
调准螺丝 adjusting screw; adjusting spring
调准螺旋 adjusting screw
调准误差 alignment error
调准延迟 corrective lag
调准用示波器 alignment scope
调准余量 alignment reserve
调准针 adjusting needle
调准装置 adjuster; alignment scope; co-alignment

眺

眺台 balcony; gazebo [复 gazebo(e)s]

眺台前沿 balcony front
眺台入口 balcony access
眺台式窗栏 Balconet(te)
眺台式桁架 balcony truss
眺望 outlook;through view
眺望长廊 prospecting gallery
眺望窗 picture window
眺望台 prospect deck
眺望用角塔 mirador
眺望游廊 prospecting gallery

跳

跳板 access board; apron; bridge ramp; brow; catwalk; foot plank; gang board; gangplank; gangway; pier; ramp; rampway; removable bridge; running plank; scaffold; shore approach;shore board;stage plank; transfer bridge; walk board;gang plank <码头用>; diving board; spring board <体育用>; treadway <军用桥、应急桥临时铺跳板,便利行车>

跳板结 stage lashing
跳板平台 brow platform
跳板受力面 bearing surface of a ramp
跳板台 <船舷上的木格板> brow landing

跳板原则 gangplank principle
跳背法 leapfrog method
跳背式 leapfrogged
跳背试验【计】leapfrog test
跳变 jump
跳变地址 jump address
跳变电压 leaping voltage
跳变概率 jumping probability
跳变关系 jump relation
跳变频率 jump frequency
跳变器 jump set
跳变延拓 jump continuation
跳波 spray
跳步测试 leapfrog test
跳步电阻 step resistance
跳步法 leapfrog;leapfrog method
跳步夯 leapfrog
跳步记录 cycle skip(ping);skip logging
跳步检验 leapfrog test
跳步键控 skip-keying
跳步开关 leapfrog switching
跳步码 skip code
跳步模 progressive die
跳步前进 leapfrog
跳步图 galloping pattern
跳仓施工(法)<建筑混凝土路面的> alternate bay construction
跳槽 job hopping
跳层公寓单元<占有两层的公寓单元> duplex apartment
跳层(楼面)skip floor
跳厂换职业 job hopping
跳场制式 field skip system
跳车 injury accident of passenger jumping from train
跳冲动 beating
跳出 trip-out
跳穿 skip draft;skip draw
跳打 staggered piling
跳打桩<桩工的> staggered piling
跳带指令 tape skip
跳挡 trip-over stop
跳挡装置 trip dog
跳刀 chattering
跳岛式安放 jump process method
跳点法 leapfrog
跳点格式 stagger scheme
跳点光栅 polka-dot raster
跳点扫描 dot interlacing;picture-dot interlacing
跳丁砖砌法 flying bond; Yorkshire bond
跳丁砖砌合 flying bond; Yorkshire bond
跳动 bounce; jigging; jitter; jumping; rippling; shimmy; ship motion; springing; wabble; bouncing<汽车前后钢板>
跳动床离子交换 jigged-bed ion-exchange
跳动床树脂矿浆吸附法 jigged-bed resin-in-pulp process
跳动锤 tilt hammer
跳动顶杆 vibration rod
跳动对应 jumping correspondence
跳动阀门 jump valve
跳动负荷 saltation load
跳动杆 tripping bar
跳动公差 run-out tolerance
跳动刮板 jump scrape
跳动光电导 hopping photoconduction
跳动辊 dancer roll
跳动过程 hopping process
跳动继电器 jumping relay
跳动距离 jump distance
跳动块 jumping block
跳动能 bound energy
跳动频率 jump frequency
跳动燃烧 pulsating combustion; pulsation combustion
跳动筛 jigging screen
跳动式滚轮轴 vibration roller
跳动式起动机 kickstarter
跳动式小开关 toggle switch
跳动特性 bouncing characteristic
跳动凸轮 tripping cam
跳动现象 jumping phenomenon
跳动响应 bounce response
跳动型造波机 plunger type wave generator
跳动运送机 jigging conveyer[conveyor]
跳动运转 run untrue
跳动沼 quaking bog
跳动重影 galloping ghost
跳动注油泵 jerk pump
跳动装置 trip gear;trip mechanism
跳动钻 jumper(drill);jumping drill
跳段<爆破时> skipping a period
跳返继电器 reset relay;step-back relay
跳飞 ricochet
跳幅 skip frame
跳杆 jumper;jumping bar
跳杆定位 jumper locate
跳杆簧螺钉 screw for jumper spring
跳格符号 figure shift
跳弓 shock
跳轨 jump on rail
跳过 hurdle;jump over;skip over
跳过标记 skip flag
跳过任选程序段 optional block skip
跳过头 overshoot
跳焊 skip sequence welding;skip welding;stitch bonding;stitch welding
跳焊法 wandering block sequence; wandering welding sequence
跳夯<俚语> jumping jack
跳行 line skipping
跳合联轴节 jump coupling
跳痕 skip mark
跳簧 bungee;fly spring
跳回 rebound;recoil;resile
跳火 arcing;arc over;flashover;jump spark;spark over
跳火电刷 arcing brush
跳火电压 arcing voltage; flashover voltage; sparking voltage; spark-over voltage
跳火花 overlap
跳火距离 disruptive distance; explosive distance; flashover distance; sparking distance
跳火信号灯 flashover sign
跳击夯 rampactor
跳击夯锤 rampactor
跳级小区试验 promotion plot trial
跳甲 flea beetle
跳间隔 jump space
跳阶级联 leapfrog cascade
跳接 jumper connection
跳接线 jumper wire
跳进 plunge
跳进的 plunging
跳进式脉冲计数器 jump counter
跳开 rebound;trip(ping) off;tripping out
跳开电路 tripping circuit
跳开式熔断器 dropout fuse;flip-open cutout fuse
跳开装置 tripper
跳浪 spray
跳雷 bounding mine
跳离起飞 jump-off
跳立 rampant
跳流鼻坎 flip bucket;deflecting bucket
跳楼货<亏本出售的商品> distress merchandise
跳轮定位杆 jumping wheel jumper
跳秒<每隔一秒钟跳动一次的秒针> jump second
跳秒杆星轮 second jumper whip star
跳秒轮 jumping second whip; second jumper wheel
跳秒轮驱动器 second jumper wheel driver
跳秒轮用压板 jumper bridge for jumping second wheel
跳秒抬杆 jumper lift
跳秒针叉瓦 jumping second pallet jewel
跳秒针传动轮 jumping second driving wheel
跳秒针拉档 jumping second detent
跳模 double moding;mode jump;skip cast【地】
跳频 frequency-hopping
跳频多址(连接) frequency-hopping multiple access
跳频扩展频谱 frequency-hopping spread spectrum
跳桥 swing span
跳入 diving;plunge
跳入者 plunger
跳伞塔 free tower;parachute tower
跳纱 skipped thread
跳闪 bounce flash
跳升式熔断器 dropout fuse
跳时拨针 jumping hours finger
跳时定位杆 jumping hours jumper cover
跳时定位杆簧铆钉 jumping hours jumper spring rivet
跳时定位杆铆钉 jumping hours jumper rivet
跳时定位簧 jumping hours jumper spring
跳时杆压板 jumping hours jumper maintaining plate
跳时星轮座 jumping hours star support
跳时指示盘 jumping hours indicator
跳式三辊机座 jumping three-high stand
跳水坝 transverse dike[dyke]
跳水板潜水箱 diving board
跳水池 diving pool
跳水凸缘<溢洪道的> priming nose
跳塔 diving tower
跳台 diving platform; diving stand; diving tower
跳汰床 jigged bed
跳汰床层 jig bed;mobile bed
跳汰床吸附塔 jigged-bed adsorption column
跳汰法 jigging method
跳汰分室 jigging cell
跳汰机 hurley;jig(ger);jigging box; vanning jig;wash box
跳汰机产物 hotching
跳汰机冲水 transport water
跳汰机床层 ragging
跳汰机床层厚度 height of jig bed
跳汰机给料拦板 washbox feed sill
跳汰机工 jigger
跳汰机滑风阀 washbox piston valve; washbox slide valve
跳汰机活塞 jigging plunger
跳汰机空气进出周期 washbox air cycle
跳汰机排渣器 jig plunger
跳汰机筛板 jigging screen; jigging sieve;washbox screen plate
跳汰机筛上分室 washbox compartment
跳汰机筛下产品 hotchwork
跳汰机筛下气室 washbox cell
跳汰机筛下室 hut(ch)
跳汰机洗箱 jigging box
跳汰机选槽 jig tank
跳汰机选矿 hotching
跳汰机溢流拦板 washbox discharge sill
跳汰机溢流堰板 tailboard
跳汰机中部拦板<在中部排渣室上> washbox center[centre] sill
跳汰精矿 jig concentrate
跳汰力 jig force
跳汰面积 jigging area
跳汰室 jigging chamber
跳汰洗煤机 jig washer
跳汰洗选机 jig washer
跳汰箱 jig tank
跳汰选 jigging
跳汰选槽 jigging box
跳汰选矿(法)jigging;skimping
跳汰摇床 jig table
跳汰周期曲线 jigging cycle curve
跳停器 troll(e)y pole catcher
跳脱试验 trip test
跳蛙式 leapfrogging
跳蛙式打夯机 leaping frog
跳蛙式夯实机 leap lean frogging
跳蛙式模板 leapfrog formwork
跳蛙式排水<指基坑逐步深挖时,泵像跳蛙般移动的排水方式> leapfrog fashion
跳蛙战术 leapfrogging tactics
跳舞厅 dancery;salo(o)n dancing
跳线 bonding jumper; bridle wire; jumper; jumper wire; jumping wire;wire jumper;office wire<配线架上的>
跳线插座 jumper receptacle
跳线电路 jumping circuit
跳线端子 jumper terminal
跳线盒 jumper holder
跳线接头 jumper terminal
跳线连接器 jumper coupler
跳线路 jumping circuit
跳线耦合器 jumper coupler
跳线设置 strap setting
跳线线夹 jumper clamp
跳陷电路 trap circuit
跳相<感应式信号机的一种功能> skip-phase
跳像 skip image
跳页【计】skip page
跳移 slew
跳移高度 height of jump-movement
跳移输沙量 silt discharge of jump-movement
跳远沙坑 long-jump pit
跳跃 bounce; hop; jump(ing); saliance[saliency];skipping;vault
跳跃搬运 saltation transport
跳跃变化 jumping variation
跳跃标记 skip flag
跳跃不连续点 jump discontinuity
跳跃不连续性 jump discontinuity
跳跃传播 hop propagation
跳跃传导 saltatory conduction
跳跃单位 skip
跳跃的 skipped
跳跃的高度或距离 leap
跳跃点 saltation point
跳跃电火花 jump spark
跳跃阀 mushroom valve;poppet valve
跳跃方法 skip philosophy
跳跃符 skip symbol
跳跃负载 saltation load
跳跃杆 skip bar;skip stop insert bar
跳跃高度 jump height
跳跃格式项 skip format item
跳跃拱顶<古波兰西利西亚的地方格式> jumping vault
跳跃轨道 jump over;skip path
跳跃轨迹 skip trajectory
跳跃函数 jump function
跳跃痕 skip mark
跳跃火花 jump spark
跳跃火花点火 jump spark ignition

跳跃火花机 jump spark
跳跃火花制 jump spark system
跳跃级配 jump grading
跳跃计数字段 hop count field
跳跃间断点 jump discontinuity
跳跃进给 jump feed;skip feed
跳跃进位 carry skip
跳跃进位法 carry skip method
跳跃矩阵 jumping matrix
跳跃距离 skip distance
跳跃块焊接 skip block welding
跳跃连接环 skip-linked ring
跳跃连接链 skip-linked chain
跳跃连续不规则 zigzag continual non-regular variation
跳跃连续有规则 zigzag continual regular variation
跳跃链路 skip link
跳跃码 skip code
跳跃模型 hopping model
跳跃膜片 flick diaphragm
跳跃器 leaping organ
跳跃前进 saltation
跳跃区 skip zone;zone of silence
跳跃任选 skip option
跳跃式安全阀 pop safety valve
跳跃式程序 skip program(me)
跳跃式打夯机 frog rammer;jumping frog
跳跃式发展 development by leaps and bounds
跳跃式夯实机 frog compactor
跳跃式接触 bouncing against
跳跃式连续锤击法 <地震勘探> skip continuous thumper method
跳跃式脉冲计数器 jump counter
跳跃式潜水 bounce dive
跳跃衰落 skip-fading
跳跃顺序存取 skip sequential;skip sequential access
跳跃速度 skip speed
跳跃算子 jump operator
跳跃弹簧 inshot spring
跳跃特性 jump characteristic
跳跃条件 jump condition
跳跃统计量 jump statistic
跳跃现象 chattering;jumping phenomenon;kipp phenomenon
跳跃谐振 jump resonance
跳跃演化 saltatory
跳跃溢出法 skip spill method
跳跃原子时 stepped atomic time
跳跃运动 jumper motion;leaping motion
跳跃运输 skipping transport
跳跃运行 <车辆的> hunting movement
跳跃噪声 popcorn noise
跳跃者 jumper;leaper;lipper;skipper
跳跃指令 jump instruction
跳跃铸型 skip cast
跳跃总线输入 skip bus input
跳跃足 leaping legs
跳越坝 jump dam
跳越标记 skip flag
跳越分度 jump index
跳越级配 jump grading
跳越继电器 jumping relay;skip relay
跳越检索链 skip-searched chain
跳越进位 carry skip
跳越距离 jump-over distance;skip distance
跳越码 skip code
跳越区(域)skip area;skip zone
跳越式航天器 skip vehicle
跳越式回绕 skip back
跳越顺序运行 skip sequential run
跳越速度 skip speed
跳越效应 skip effect
跳越信息组 skip field
跳越堰 leaping weir;leap-weir
跳越堰溢流 leaping-weir overflow
跳越指令 jump instruction;jump over
跳蚤 flea
跳蚤市场 flea fair;flea market
跳闸 breakaway;opening;trip gear;trip-out;tripping;tripping operation
跳闸报警 trip alarm
跳闸电流 tripping current
跳闸电压 tripping voltage
跳闸断路 trip-out
跳闸杆 tripping bar
跳闸回路 tripping circuit
跳闸机构 tripping mechanism
跳闸继电器 trip(ping) relay
跳闸开关 trip wire switch
跳闸力矩 trip-out torque
跳闸率 trip-out rate
跳闸母丝 trip bus
跳闸频率 tripping frequency
跳闸时间 trip time
跳闸试验 trip test
跳闸线圈 breaking coil;trip(ping) coil
跳闸值 trip value
跳闸装置 trip;trip dog;trip gear
跳针 plunger pin;skipping stitch;bouncing pin <仪表的>
跳针仪 <用以确定发动机爆燃特性> bouncing-pin apparatus
跳针指示仪 bouncing-pin apparatus
跳值排列 leapfrog
跳指标 jump set
跳指令 immediate skip
跳周 cycle skip(ping)
跳转 skip
跳转程序 jump routine
跳转链 skip chain
跳转指令 jump instruction;skip instruction
跳桌 flow table;shaking sieve;shock table
跳桌流动试验 flow table test
跳字计数器 cyclom;cyclometer counter
跳字钟表盘 jumping dial
跳字转数表 cyclometer

贴

贴岸冰 border ice;collar ice;fast ice;ice ledge;land floe;shore floe;shore ice

贴岸礁 fringing reef
贴岸码头 quay
贴岸吸力 bank suction
贴岸吸力作用 bank suction
贴板 boarding;flitch;panel;pasting
贴板边接法 butt-strap(ped) joint
贴板梁 flitch-plate girder
贴本 loss money in a business
贴壁式垃圾箱 wall-mounted waste receptacle
贴壁紊流 wall turbulence
贴壁纸 paperhanging
贴边 seam;welted edge;welted nosing;welt(ing)
贴边界坐标 boundary-fitted coordinates
贴边开口缝 seam placket
贴边木条 <平板门的> lipping
贴边弯下的铁皮条 strip welting
贴边线 hemline
贴标签 affix a label;label(l)ing
贴标签部位 label band
贴标签存车 disc[dick] parking
贴标签法 <调查车流流量流向的> tagged or sticker method
贴标签机 label(l)er;label(l)ing machine
贴标签停车 disc[dick] parking
贴柄机 handle stickling machine
贴玻璃线 applied thread;laid-on thread
贴补 <集装箱等> patching
贴彩色标签 applied colo(u)r label
贴传单的人 poster
贴瓷砖 tiling
贴底泥沙流 tractive current
贴地层 <墙基、柱基、台基等的> earth table;ground table;ground layer
贴地大气层 bottom layer;lowest atmospheric layer
贴地面砖工程 engineering brick floor(ing)
贴地石座 earth stone;earth table
贴雕 applique carving
贴顶天花 contact ceiling
贴动系数 <计算预应力筋在管道内或弯折处由于贴动而引起的阻力用> wobble coefficient
贴动效应 <预应力钢索在管道内由于贴动而引起的阻力> wobbling effect
贴风航行 ahold
贴风行驶 luff
贴封条机 banderoling machine
贴缝 welt
贴缝玻璃纤维带 joint glassfiber tape
贴缝胶纸带 joint tape
贴缝带 joint masking tape
贴附的 planted
贴附砾岩 plaster conglomerate
贴附穹顶 counter-vault
贴附射流 wall-attachment jet
贴附式门挡 planted stop
贴附饰 applied mo(u)lding;planted mo(u)lding;plated mo(u)lding
贴附水舌 adherence of nappe;adhering nappe;clinging nappe;depressed nappe;weir nappe
贴附温度计 surface thermometer
贴附线脚 applied mo(u)lding;laid-on mo(u)lding;planted mo(u)lding
贴附效应 surface effect
贴附于墙的墩子 attached pier
贴膏药 plaster(ing)
贴海报 bill-posting
贴合 laminating;reconcilement
贴合布 bonded fabric
贴合辊 doubling roller
贴合机 combining machine;make-up machine
贴合胶粘剂 close-contact adhesive
贴合面 abutted surface;binding face;faying surface;seating surface
贴合黏[粘]结剂 close-contact adhesive
贴合墙板夹 application wall clips
贴合墙支架 application wall clips
贴合桩靴 fitted pile shoe
贴花 applied colo(u)r label;applied design;transfer decoration
贴花薄膜 transfer film
贴花的 applied
贴花法 <陶瓷装饰法> decal;decal comania method
贴花膜 decal comania
贴花镶饰 applique decoration
贴花印花法 decal
贴花印刷 transfer printing
贴花印刷法 decal comania process
贴花织物 applique
贴花纸 decal tissue;paper transfer
贴花纸转印 decal comania transfer
贴花转印法 decal comania transferring
贴汇水 discount on exchange
贴笺纸 paster
贴胶 rubberizing;sticking patch
贴胶的 rubberized
贴胶合板面层 veneering
贴胶压延机 coating calender
贴角(搭)焊 fillet welding
贴角焊缝 corner fillet;fillet weld(ing seam)
贴角焊接 angle weld(ing);fillet weld(ing)
贴角条 fillet mo(u)lding
贴角周边焊 periphery fillet welding
贴接 amplexiform;approach grafting
贴接面 meeting face
贴金 gilding;gold foil painting;gold leafing;gold-overlaid;putting leaf
贴金箔 leaf-gilding
贴金箔的 gilded
贴金挂钟 <宝塔尖顶的> tee
贴金技术 gilding technique
贴金漆 gold silver jeweleries;gold size;sizing varnish
贴金叶装饰 gold leafing
贴金属箔 putting leaf
贴近度 applicability
贴近继电器 proximity relay
贴近开关 proximity switch
贴近上风向 hard up
贴近下风向 hard down
贴进 exchange premium
贴旧换新 trade away;trade-in
贴旧换新价值 trade-in value
贴旧换新交易 trade-in transaction
贴旧换新折价 trade-in allowance
贴靠接合 abutting joint
贴靠他船 get aboard
贴扣现金 discounted cash
贴扣现金流动 discounted cash flow
贴丽 cladding
贴砾过滤器 gravel sticking filter
贴脸 architrave;door casing;finish casing;preen;trim
贴脸板 architrave;back band;lambrequin <门窗的>
贴脸墩 architrave block;plinth block;skirting block
贴了印花的正式文件 clause stamped
贴邻边墙 counter wall
贴邻地 skin to skin
贴邻房地产的业主 abutter
贴码 signature letter;signature mark;signature symbol
贴码线 signature line
贴毛毡的圆筒 felted fabric cylinder
贴面 facing;line post;overlaid;overlay;veneering;meeting faces【船】
贴面板 face board;insert;laminated board;overlay sheet;surfacing board;veneer;veneer board;veneer wood
贴面板衬背 backing of veneer;veneer backing
贴面板门 veneer door
贴面板拼花 veneer panel matching
贴面玻璃棉制品 faced glass wool product
贴面薄板 overlaid veneer;weather shingling
贴面薄石板 bastard ashlar
贴面材料 facing material
贴面锤 veneer hammer
贴面瓷板 thick china
贴面瓷砖 furring tile;structural clay tile
贴面单板 overlaid veneer
贴面的 veneered
贴面法 method of facing
贴面钢材 coated steel
贴面工 face work
贴面工程 tiler work
贴面混凝土砌块 face block
贴面胶合板 face board;faced ply-

T

wood; hardboard-faced plywood; overlaid plywood; veneered plywood
贴面接合 enfacial junction
贴面结构 veneered construction
贴面结合胶结物结构 enfacial junction cement texture
贴面卷材 faced building mat; faced building quilt
贴面绝热材料 faced insulation (material)
贴面片材 overlay sheet
贴面墙 veneer of wall; faced wall
贴面石 facing stone
贴面式肥皂粉配出器 surface-mounted powdered soap dispenser
贴面式卫生巾配出器 surface-mounted sanitary napkin dispenser
贴面式纸巾配出器 surface-mounted paper towel dispenser
贴面塑料 formica
贴面碎木板 veneered flake board; veneered particle board
贴面修饰 exposed finish
贴面硬质酚醛泡沫塑料保温板 faced rigid cellular phenolic thermal insulation
贴面用胶粘剂 furring compound
贴面纸 face paper; paper overlay
贴面纸面石膏板 laminated plaster board
贴面砖 face tile; facing brick; facing clay brick; facing tile; furring brick; masonry veneer; tile; tiling; veneered with brick
贴面砖层 brick veneer
贴面砖工 tiler
贴面砖工程 engineering brick facing
贴面砖墙 tile hung wall
贴面砖石墙结构 masonry veneer wall construction
贴面砖作业 tiler work
贴模边 tool side
贴膜法 film applicator coating
贴膜机 laminator
贴铺胶合板 plywood sheathing
贴墙布 fabric wall covering; wall-covering fabric
贴墙衬材 backing for wall covering
贴墙光源照射墙面 wash light
贴墙立木 wall piece
贴墙楼梯小梁 wall string(er)
贴墙面 ashlering; veneer of wall
贴墙面用石板 wall tile stone
贴墙面砖 wall tile
贴墙面砖工作 hung tilework
贴墙托梁 tail trimmer
贴墙织物 fabric wall covering
贴墙纸 paperhanging; wall papering; wall fabric
贴墙纸的 wall-papered
贴墙纸工作 wall-papering work
贴墙纸糨糊 hanging paste; wall-paper hanging paste
贴墙柱 column engaged to the wall; engaged column
贴切的 appurtenant
贴切的资料 pertinent data
贴色玻璃 flashed glass
贴沙砾泥球 armo(u)red mud ball; pudding ball
贴纱布机 gauzing machine
贴砂砾泥球 mud ball
贴商标 label(l)ing
贴上 affixture; stick-on
贴身包装 wrapping and package
贴身传声器 <美> body mike
贴生 adhesion; adnation【生】
贴水 at discount; discount; premium <差价补贴>; agio <升水>; disagio <证券面额>
贴水率 contango rate; discount rate
贴水日 marking-up day
贴水账户 agio account
贴水账目 agio account
贴陶瓷面砖 ceramic vertical tiling
贴体包装 skin packaging; skin packing
贴体坐标 body-fitted coordinate
贴条 furring
贴条吊顶 furred ceiling
贴铜 copper-surfaced
贴图 chartlet
贴图改正 chartlet for correction
贴图纸胶条 Scotch tape
贴现 discount for cash; discount(ing); time discount
贴现表 discount table
贴现差额 discount difference
贴现窗口 discount window
贴现的 discounted
贴现的现金流 discounted cash flow
贴现额度 line of discount
贴现法 discounting methods; discount method
贴现费用 discount charges; discount cost; discount expenses
贴现分类 discount category
贴现公司 discount house
贴现行情 discount quotation
贴现行市 discount quotation
贴现后的现值 discounted present value
贴现回收率 discount return rate
贴现回收期 discounted payable period
贴现汇票 bill discount
贴现基础 discount basis
贴现及买汇 bills discounted and remittance bought
贴现技术 discounting technique
贴现价值 discounted value
贴现净得 net avails
贴现利率 discount rate; rate of discount
贴现利息 discount interest
贴现流量 discounted flow
贴现率 bank rate; discounted rate; discounting rate; discount rate of interest
贴现率差 discount margin
贴现率差数 discount difference
贴现率的波动 fluctuation of discount rate
贴现率政策 discount rate policy
贴现票据 bill discounted; discounted bill; discounted note; discount paper; notes on discount
贴现票据及汇票 note and bill discounted
贴现票据经纪人 discount broker
贴现票据日记账 bills discounted daily list
贴现票据账簿 discount cash book
贴现期 acceptance period; term of discount; discount period
贴现期票 bill discounted
贴现期限 term of discount
贴现曲线 discount curve
贴现人 applicant for a discount; discount clerk; discounter; discount teller
贴现商行 discount house
贴现市场 acceptance market; discount market
贴现收入 discount earned; discount income
贴现水酌减 discount
贴现投资回收期 discounted payback period
贴现投资收益率 discounted rate of return
贴现外汇 discount on exchange
贴现系数 discount coefficient; discount factor
贴现现金簿 discount cash book
贴现现金流量表 discounted cash flow(sheet)
贴现现值 discount present value; present discount value
贴现信贷 discounted credit
贴现业者 bill discounter; discounter
贴现银行 bank of discount; discount(ing) bank
贴现银行债券 discount bank debenture
贴现债券 discount debenture
贴现政策 discount policy
贴现支票 discount on check
贴现值系列 discounted cash flow
贴线法 wool tuft technique
贴压时间 shuffling time
贴印法 reflex print(ing)
贴印花税 put on stamp duty
贴用印花 affix revenue stamp
贴邮票 affix stamps; stamp
贴有浮雕图样的器皿 sprigged ware
贴有危险品标签的货物 label(l)ed cargo
贴于表面 applicate
贴毡木板 felted fabric board
贴招贴的人 sticker
贴纸图案 papier colle
贴砖 patch block; tile; tile feed; tiling
贴砖工程 tiling work
贴砖工作 tiling work
贴砖护墙板 tile-faced panel
贴砖镶板 tile-faced panel
贴砖毡 tile lino(leum)
贴琢石(墙面) ashlaring

萜

萜二烯 terpinene
萜类化合物【化】terpenoids
萜品 terpine
萜品醇 terpineol
萜品烯 terpinene
萜品烯-顺(丁烯二酸)酐加成物 maleic anhydride terpinene adduct; terpinene maleic anhydride adduct
萜品油烯 terpinolene
萜(烃) terpene
萜烷 terpane
萜烷的分子参数 molecular parameter of terpanes
萜烷含量 terpanes content
萜烷/甾烷 terpances/steranes
萜烯 terpene
萜烯酚醛树脂 terpene phenol resin
萜烯(类)的 terpenic
萜烯树脂 terpene[terpine] resin
萜烯树脂体 terpene resinite
萜烯-顺(丁烯酸)酐缩合物 terpene maleic anhydride condensation product
萜烯系 terpenic series

铁

铁(Ⅱ)滴定 ferrometry
铁T角 steel tee-plate; three-way strap; T-plate
铁铵矾 <硫酸铵铁> ferric alum; iron alum
铁暗壁 iron dowel
铁扒钉 cramp iron; dog iron
铁扒锯 iron dog
铁白云母 ferrimuscovite
铁白云石 ankerite; brown spar; ferrodolomite
铁白云石岩 ankerite rock
铁白云石质煤 carbankerite
铁拜来石 ferribeidellite
铁斑 iron mo(u)ld; iron speck; iron spot; iron stain
铁斑点 iron spot
铁板 Armenia plate; armo(u)r plate; armo(u) red plate; clout; iron plate; iron sheet; metal plate; plate iron; sheet iron; sheet-iron plate; slab; slab iron
铁板电阻器 cast-iron resistor
铁板护栏 metal plate guard rail
铁板铰链 doorbrand
铁板开槽机 lock former
铁板梁 plate girder
铁板料 iron brick
铁板溜槽 sheet-iron chute
铁板铺面 iron pavement; iron sheet paving
铁板起重机 plate crane; sheet-iron crane
铁板铅被覆 spot-homogen
铁板热焊 hot tapping
铁板砂 slab iron sand
铁板砂垫层 slab iron sand cushion
铁板输送机 plate apron conveyer [conveyor]
铁板钛矿 pseudo-brookite
铁板条 iron lath(ing)
铁板灶 hot plate
铁板桩 iron sheet pile
铁版照相(术) ferrotype
铁棒 boast; gavelock; iron bar; wrecking bar
铁包皮门 kalamein door
铁包石棉垫片 metal jacket gasket
铁贝得石 iron beidellite
铁钡氧化物烧结而得的永磁材料 ferroxdure
铁钡永磁合金 magnadur
铁笔 hand spike; spike; steel marline spike; steel pointed marline spike; stencil pen; stylus; stylus pen
铁篦子 grate; gridiron; grill(e); iron grating
铁扁担门铰链 strap hinge
铁表面渗铝法 altierfen
铁柄 hand spike
铁柄地锚 raw iron plug bolt anchor
铁柄螺丝刀 iron handle screw driver
铁饼 discus
铁饼型蝴蝶阀 lenticular butterfly valve
铁波特兰水泥 iron Portland cement
铁波纹管涵洞 corrugated culvert
铁玻璃 iron glass
铁铂矿 tetraferroplatinum
铁箔 iron foil
铁材腐蚀 iron sick
铁槽形油断路器 tank-type oil circuit breaker
铁草酸盐络合物 iron-oxalate complex
铁叉 broach
铁插销 iron sliding bolt
铁柴架壁炉 iron firedog
铁尘肺 arc-welder's disease; siderosis
铁沉积物 iron deposit
铁撑螺栓 iron stay bolt
铁成球 balling up
铁承窝 iron socket
铁齿 iron teeth
铁齿板 <防滑用> creeper
铁窗挡 iron window bar
铁窗格条 iron munting
铁窗樘 iron glazing bar
铁窗条 iron window bar
铁锤 aboutsledge; iron hammer; tupid
铁磁饱和稳压器 saturation core regulator

铁磁参量放大器 ferro-magnetic parametric amplifier
铁磁场 ferrite-magnet field
铁磁存储器 ferro-magnetic store
铁磁带 ferro-magnetic tape
铁磁的 ferro-magnetic
铁磁电动式仪表 ferrodynamic(al)instrument
铁磁电动陶瓷 ferrodynamic(al)ceramics
铁磁电动系仪表 ferrodynamic(al)instrument
铁磁电抗器 ferristor
铁磁电枢 ferro-magnetic armature
铁磁放大器 ferro-magnetic amplifier
铁磁粉 ferro-magnetic powder
铁磁杆 ferrod
铁磁感应 ferric induction; intrinsic(al)induction
铁磁各向异性 ferro-magnetic anisotropy
铁磁共振 ferro-magnetic resonance; ferroresonance
铁磁共振触发器 ferro-resonant flip-flop
铁磁共振计算 ferro-resonant computing
铁磁共振线宽 ferro-magnetic resonance line width
铁磁核共振 ferro-magnetic nuclear resonance
铁磁回路 ferromagnetic circuit
铁磁计 ferrometer
铁磁继电器 ferro-magnetic relay
铁磁尖晶石 ferrospinel
铁磁晶体 ferro-magnetic crystal; polar crystal
铁磁矩 ferro-magnetic moment
铁磁励磁 ferro-magnetic excitation
铁磁流体 ferrofluid
铁磁路 iron circuit
铁磁屏蔽 ferro-magnetic shield
铁磁谱仪 iron magnetic spectrometer
铁磁示波器 ferro-graph
铁磁式功率计 ferrodynamometer
铁磁探测器 ferroprobe
铁磁探伤器 ferro-magnetic crack detector
铁磁陶瓷 ceramic magnet; ferro-magnetic ceramic
铁磁体 ferro-magnet; ferro-magnetic body; ferro-magnetics
铁磁体的 ferro-magnetic
铁磁体移位寄存器 ferro-magnetic shift register
铁磁调制器 ferro-magnetic modulator
铁磁物 ferro-magnetic
铁磁物质 ferro-magnetic substance
铁磁吸着剂 ferro-magnetic sorbent
铁磁谐振 ferroresonance
铁磁芯 ferro-magnetic core
铁磁性 ferro-magnetic; ferro-magnetism; iron magnetic property
铁磁性棒 ferro-magnetic bar
铁磁性薄膜 ferro-magnetic thin film
铁磁性材料 ferro-magnetic material
铁磁性合金 ferro-magnetic alloy
铁磁性记录法 ferro-magnetography
铁磁性胶质 ferro-magnetic colloid
铁磁性居里点 ferro-magnetic Curie point
铁磁性矿物 ferro-magnetism mineral
铁磁性微晶玻璃 ferro-magnetic glass-ceramics
铁磁性物质 ferro-magnetic material
铁磁性悬浮液 ferro-magnetic suspension
铁磁学 ferro-magnetics; ferro-magnetism
铁磁氧化矿物 ferro-magnetic oxide mineral
铁磁振子 ferro-magnon
铁磁质 ferro-magnetics
铁磁转变 ferro-magnetic transition
铁次透辉石 ferrosalite
铁催干剂 iron drier[dryer]
铁催化剂 iron catalyst
铁催化剂反应 iron catalyst reaction
铁存水弯 ferrous trap; iron trap
铁锉屑 iron filings
铁搭 cramp
铁带 ferric tape; sheet-iron strip
铁带滑车 ironbound block; iron strapped block
铁丹 bengala; calcothar[colcothar]; red oxide; roude; rubigo
铁挡边 iron slips
铁挡块 break iron; dressing iron
铁刀木 bomb(ay)black wood; Indian rose chestnut; Siamese senna
铁导管 iron conduit
铁导线 iron wire
铁道 railroad; trackage <铁道总称>
铁道兵 railway corps
铁道部 Ministry of Railways
铁道部运输指挥中心 Ministry of Railways Traffic Management Center[Centre]
铁道财务会计信息库 railway finance and accounting information library
铁道岔道 track on turnout side
铁道岔尖 railroad point
铁道岔尖轨 tongue rail
铁道车 railway carriage
铁道车辆 railroad car; railroad vehicle; rail truck; rolling stock
铁道车辆运输 haulage by rail-mounted vehicles
铁道出口提单 railway export bill of lading
铁道电气化 railway electrification
铁道钉 track spike
铁道干船坞 railway dry dock
铁道干坞 railway dry dock
铁道干线 trunk railway
铁道工兵部队 railway engineer troops
铁道工程 railway construction; railway engineering
铁道工程测量学 railroad engineering surveying
铁道工程运料车 push car
铁道轨距 rail ga(u)ge; railroad ga(u)ge; railway ga(u)ge; track ga(u)ge
铁道航空测量 railway aerosurveying
铁道护道员工 truckman
铁道护路员 trackwalker
铁道机车车辆 rolling stock
铁道机动车 rail motor car; railroad motor car
铁道计时器 railroad chronometer
铁道建筑 railway construction
铁道建筑学 railroad architecture
铁道交叉 railway crossing
铁道交通噪声 railway transportation noise
铁道界限 railroad clearance; railway clearance
铁道经济 railway economics
铁道警告标志 railroad warning sign
铁道聚居地 <靠港口> railway colony
铁道口 railroad opening
铁道跨距 track span
铁道路堤 railway embankment
铁道路线 track system; railway
铁道路线长度 trackage
铁道螺钉电扳手 electric(al)wrench for railway bolt
铁道起重机 rail-mounted crane
铁道桥 railroad bridge
铁道曲线板 railway curve
铁道入口孔 track opening
铁道设备 railway equipment
铁道弯尺 French curve; railway curve
铁道系统 railway system
铁道斜坡道 railroad ramp
铁道学院 railway institute
铁道巡查车 railway inspection trolley
铁道巡查手摇车 railway inspection trolley
铁道运材 railroad logging; railway logging
铁道运费 railage; rail rate; railway fare; railway freight
铁道运输 railroad transport(ion); rail transport(ation); railway transportation
铁道运输部 Ministry of Railway Transport
铁道运输系统 rail haulage system
铁道杂工 spiker
铁道辙尖 railroad point
铁道支线 railroad branch; railroad siding; railway siding
铁的 ferric; ferrous; ferruginous; siderous
铁的不足 iron deficiency
铁的工资规律 iron law of wages
铁的净化 iron purification
铁的利用效率 iron utilization
铁的密度 iron density
铁的同素异形 allotropic forms of iron
铁的正常含量 normal lead content
铁镫 pole climbers; hand iron
铁底阀 iron foot valve
铁点 iron spot
铁点污染 specking
铁电 ferro-electricity
铁电变换器 ferro-electric(al)converter
铁电变压器 ferro-electric(al)transformer
铁电玻璃陶瓷组合体 ferro-electric(al)glass-ceramic composition
铁电材料 ferro-electric(al)material
铁电参量器 ferro-electric(al)parametron
铁电池 iron cell
铁电畴 ferro-electric(al)domain
铁电存储矩阵 ferro-electric(al)memory element; ferro-electric(al)memory matrix
铁电存储器 ferro-electric(al)memory; ferro-electric(al)storage
铁电的 ferro-electric
铁电电子性 ferro-electronic
铁电光导器件 ferro-electric(al)photo-conductor device
铁电光闸 ferro-electric(al)shutter
铁电化合物 ferro-electric(al)compound
铁电机 iron machine
铁电介质阻抗 transpolarizer
铁电晶体 ferro-electric(al)crystal
铁电居里点 ferro-electric(al)Curie point
铁电膜 ferro-electric(al)film
铁电气石 iron tourmaline
铁电式电容器 ferro-electric(al)condenser
铁电式放大器 ferro-electric(al)amplifier
铁电态 ferro-electric(al)state
铁电陶瓷 ferro-electric(al)ceramics
铁电陶瓷显示 ferro-electric(al)ceramic display
铁电特性 ferro-electric(al)property
铁电体 ferro-electrics; seignette-electrics
铁电体存储器 ferro-electric(al)memory
铁电体感应相变 ferro-electric(al)induced phase transition
铁电体移位寄存器 ferro-electric(al)shift register
铁电调制器 ferro-electric(al)modulator
铁电铁氧体 ferro-electric(al)ferrite
铁电效应 ferrite electric(al)effect
铁电性 ferro-electricity
铁电性的 ferro-electric
铁电影响 ferro-electric(al)effect
铁电有序化 ferro-electric(al)ordering
铁电元件 ferro-electric(al)cell
铁电载流子 ferro-electric(al)carrier
铁电振荡器 ferractor
铁电质 ferro-electrics
铁电轴 ferro-electric(al)axis
铁电阻 iron resistance
铁垫 metal-bearing
铁垫板 cast baseplate; iron chair
铁垫圈 iron washer
铁钉 iron nail; wire tach
铁钉脱落 nails off
铁钉污染 nail staining
铁钉箱 nail box
铁钉与胶结的屋顶桁架 nail-glued roof truss
铁碇 iron ballast
铁锭 ingot(iron)
铁动态 ferrikinetics
铁冻蓝闪石 ferrobarroisite
铁豆 cold-shot; drops; shot metal; splashings
铁耳 iron lug
铁二铝酸六钙 hexacalcium dialuminoferrite
铁矾 iron vitriol; siderotil
铁矾土 aluminoferric; bauxite; laterite; laterite soil; red bauxite
铁矾土砾石 laterite gravel; lateritic gravel
铁矾土卵石 lateritic gravel
铁矾土碎屑 laterite chip(ping)s
铁钒矿 nolanite
铁方钴矿 iron-skutterudite
铁方解石 ferrocalcite
铁方硫镍矿 Fe-vaesite
铁方镁石 ferropericlase
铁方硼石 eisenstassfurtite; ericaite; iron boracite
铁肺 siderosis
铁粉 ferrous powder; ferrum pulveratum; iron dust; iron-powder; irons; powdered iron
铁粉磁芯用镍铁合金 Gecalloy
铁粉防水材料 metallic waterproofing
铁粉焊条 iron-powder type electrode
铁粉记录术 ferro-graphy
铁粉记录图 ferro-graph
铁粉检波器 filings coherer
铁粉结合剂 steel bond
铁粉末 iron dust
铁粉漆磨坊 iron paint mill
铁粉切割法 iron-powder process
铁粉水泥 <防水用> iron-powder cement
铁粉填料环氧树脂模 iron filled epoxy pattern
铁粉涂料 iron-powder coating
铁粉涂料焊条 iron-powder electrode
铁粉线圈 iron-dust coil
铁粉芯 ferrocart core; iron-dust core; dust core; powdered-iron core
铁粉芯线圈 compressed-iron-core coil; iron-dust core coil; dust-core coil
铁粉氧炔切割法 iron-rich powder process

铁粉氧焰精整 powder cutting process
铁粉助熔表面清理 powder washing
铁风 iron winds
铁封玻璃 iron-sealing glass
铁峰 iron peak
铁扶手 grab iron; handrail iron; iron railing
铁扶梯 iron stair(case)
铁氟硅酸盐 iron fluosilicate
铁氟化钾 potassium ferric fluoride
铁釜 iron still
铁腐蚀 iron corrosion
铁副矾石 ferriparaluminite
铁钙橄榄石 iron-monticellite
铁钙钠石 imandrite
铁钙闪石 ferrotschermakite
铁盖 iron covering
铁盖板 <下水道用> stop iron
铁干料 iron drier[dryer]; iron soap
铁杆 gavelock; horse; iron bar; iron pole; iron rod; puntee; punty; tommy bar
铁杆清扫 rodding
铁杆扫床【疏】bar sweep
铁杆钻进 rod boring
铁橄榄石 fayalite; ferrocyanide orthosilicate
铁橄苏辉岩 anabohitsite
铁淦氧棒形天线 ferrite-rod antenna
铁淦氧磁芯可调滤波器 ferrite-tunable filter
铁淦氧磁芯线圈 ferrite-cored coil
铁淦氧回转器 ferrite gyrator
铁高岭石 canbyite; faratsihite; hisingerite
铁镐 iron pike
铁格架 gridiron
铁格筛 grizzly
铁格条 iron sheeting
铁格栅 bar strainer
铁格子 careening grid; gridiron
铁格子板 grill(e) flooring; steel grating
铁隔断 iron interceptor
铁铬二元图 iron-chromium binary diagram
铁铬合金 ferro-chrome(iron)
铁铬尖晶石 ferropicotite
铁铬铝电阻合金 radiohm alloy
铁铬铝电阻丝合金 ferropyr
铁铬铝合金 Aludirome
铁铬铝耐蚀耐热合金 Alcres
铁铬铝系电炉丝 Alchrome; Aludirome
铁铬绿 green vermilion; mineral green
铁铬系不锈钢 Mischrome
铁工 hammer man; ironsmith; iron worker; smith; bucker <铆工的助手>
铁工厂 iron mill; iron works
铁工场 blacksmith shop; iron worker's shop
铁工炉 smithy chimney
铁工砧 smith anvil
铁弓门闩 strong back
铁汞齐 iron amalgam
铁拱 iron arch
铁拱桥 iron arch bridge
铁拱主梁 iron arch(ed) girder
铁钩 cramp iron; dog; dog iron; grappling iron; hasp iron; iron hook
铁钩和钩环 hasp and staple
铁钩环 iron shackle
铁构架 iron framework; iron framing
铁构建筑 iron architecture
铁箍 agraf(f)e; attachment clip; band iron; cramp; eyelet; ferrule; hoop iron; iron band; iron hoop; metal strap; strip iron
铁箍方格子构造 iron hoop lacing
铁箍条 iron strap
铁箍枕木 steel reinforced wooden sleeper
铁箍桩帽 driving band
铁骨 gagger
铁骨板 gagger board
铁骨构架 skeleton frame
铁骨构架配置设计 grid formation
铁钴比色计 iron cobalt liquid colo(u)r scale
铁钴钒合金 supermendur
铁钴镍合金 Kovar
铁固定 ferropexy
铁挂锁 iron pad lock
铁管 ferrocyanide pipe; iron pipe; iron tube
铁管标志 <用铁管做的点位标石> iron pipe mark
铁管尺寸 iron pipe size
铁管道 iron pipe conduit
铁管涵洞 black culvert; iron pipe culvert
铁管架温室 pipe frame house
铁管接合油泥 rust cement
铁管内径 iron pipe size
铁管隧道 iron tube tunnel
铁管线 iron pipeline
铁龟头橡胶长筒靴 steel toe rubber boots
铁硅白云母 ferriphengite
铁硅尘肺 siderosilicosis
铁硅合金 ferro-silicium; iron-silicon alloy
铁硅灰石 ferrobustamite; ferrowollastonite
铁硅铝磁合金 Sendust
铁硅铝合金 Alsifer
铁硅酸钙 calcium iron silicate
铁硅酸盐 ferro-silicate; silicate of iron
铁硅酸盐水泥 iron Portland cement
铁轨 rail; railroad rail; railroad track; railway rail; railway track; track rail
铁轨道钉 track spike
铁轨顶面 upper surface of rail
铁轨端部 rail end
铁轨分岔 track pitch
铁轨附件 rail attachment
铁轨混凝土衬砌 rail-cement lining
铁轨接座 joint chair
铁轨连接器 track rail bond
铁轨面 rail level
铁轨膨胀 expansion of rail
铁轨上蓄电池操作的小车 battery railcar
铁轨式发射台 railway launching platform
铁轨与道路两用车 combined rail and road vehicle
铁轨辙印 tracking ruts
铁轨枕 iron sleeper
铁轨枕木 wooden railroad tie
铁轨状况良好 tracks in surface
铁柜条款 iron safe clause
铁棍 pinch bar; wrecking bar
铁滚筒 iron roller
铁滚轧平 iron out
铁棍 crow; crowbar; granny bar; iron rod
铁锅 iron pan
铁锅熔化炉 iron-pot furnace
铁含量 iron content
铁涵洞 black iron culvert; iron dust
铁涵管 black iron culvert
铁焊料 iron solder
铁夯 iron ramp; ramming iron; tamping iron
铁耗 core loss
铁耗分量 core-loss component
铁耗试验 core-loss test
铁耗系数 iron loss factor
铁合金 alloy iron; ferroalloy; iron alloy
铁合金厂 ferroalloy work
铁合金的 ferro
铁合金法 ferroalloy process
铁合金粉末 ferroalloy powder; plastalloy
铁合金管 iron-alloy pipe
铁合金金属 ferroalloy metal
铁合金炉 ferroalloy furnace
铁褐色 iron brown
铁黑 black rouge; iron black; iron oxide black; tri-iron tetroxide
铁红 allite red; gulf red; imperial red; iron oxide red; iron red; Persian gulf red; Pompeian red; Pompey red; Prague red; red ferric oxide; Spanish red; Turkey red; Venetian red
铁红锰铁矿 ferro-friedelite
铁红色料 Tuscan red
铁红锌矿 ferrozincite
铁红颜料 iron red pigment
铁红釉 iron red glaze
铁弧 iron-arc
铁弧灯 iron arc lamp
铁弧拱 iron-arc
铁花 iron openwork; ornamental work of iron
铁花格 grill(e); iron grill(e); ornamental grill(e)
铁花饰 open iron work
铁花栅门 iron grill(e) door
铁华 flos ferri; flower of iron
铁华绿泥石 aphrosiderite
铁滑车 iron block; metal block; steel block
铁滑轮 steel sheave
铁滑石 liparite; minnesotaite
铁化合物 ferric compound; iron compound
铁画 iron picture
铁环 eyelet; iron eye; iron ring
铁黄 iron oxide yellow; nankin yellow; yellow ferric oxide; yellow iron oxide; yellow oxide
铁黄碲矿 ferrotellurite
铁黄颜料 iron yellow (pigment); yellow ocher[ochre]
铁簧继电器 ferreed(relay)
铁簧接线器 ferreed switch
铁簧开关 ferreed switch
铁灰色 gunmetal gray[grey]; iron gray [grey]; steel gray[grey]
铁辉长岩 ferrogabbro
铁辉钴矿 ferrocobaltite
铁辉石 ferrosilite
铁回丝 metal wool
铁惠丝 <一团细钢线> steel wool
铁混凝剂 iron coagulant
铁混凝土 iron aggregate concrete; iron-loaded concrete
铁混浊 iron turbidity
铁活 iron works
铁基材料 iron-based material
铁基超强度合金 iron-base superstrength alloy
铁基粉末冶金轴承 iron-base bearing
铁基粉末冶金轴套 iron-base bushing
铁基高温合金 iron-base superalloy
铁基合金 ferroalloy; ferrous alloy
铁基合金粉末 iron-base powder
铁基轴承 iron-base bearing
铁畸变性反铁电体 ferrodistortive antiferro-electrics
铁畸变性铁电体 ferrodistortive ferro-electrics
铁及钢板桩法凿井 iron and steel sheet piling
铁极蓄电池 iron storage battery
铁剂 chalybeate
铁加工 ironworking
铁夹 clamp iron; grip iron; iron clamp; iron clip; iron dog; shim liner
铁夹环 iron clamp ring
铁夹头 break iron
铁夹子 agraf(f)e
铁甲 iron clad; steel-clad
铁甲合同 iron-clad contract
铁甲协定 <绝对不能违反的协定> iron-clad agreement
铁钾矾 potassium ferric sulfate
铁架 brandreth; iron stand; steel tower
铁架帆布床 bunk
铁架高架铁路 elevated railway on iron structure
铁架构架工 iron worker
铁间空隙 <电机的> entrefer
铁肩垫圈 iron washer
铁剪 iron shears
铁件 iron works; rig irons
铁件加工锤 fuller
铁件制造者 iron worker
铁件装卸机 iron carrier
铁件自动装卸机 automatic iron carrier
铁建造 iron architecture; iron formation
铁匠 blacksmith; hammer man; hammer smith; ironsmith; smith
铁匠锤 blacksmith's hammer
铁匠的 vulcanic
铁匠的方柄凿 blacksmith's hardy
铁匠活儿 smithery
铁匠锄头 blacksmith's hammer
铁匠铺 forge shop; smithery
铁胶合剂 iron cement
铁胶合料 iron rust cement
铁胶结料 iron cement
铁胶粘土 gluing
铁焦 iron coke
铁焦比 iron-coke ratio; melting ratio
铁角砾岩 canga
铁角闪石 ferrohornblende
铁角砧 beakiron
铁脚手架 iron scaffold
铁接线盒 iron connector; iron electric(al) connector
铁结构 iron construction
铁结构工厂 constructional iron works
铁结合力 iron-binding capacity
铁结砾岩 ferricrete
铁结硬质合金 iron-cemented carbide
铁金红石 nigrine
铁金星玻璃 iron-aventuring glass
铁金云母 ferriphlogopite
铁金属 iron metal
铁金属保护特性 ferrous metal protection characteristics
铁金属须 iron whiskers
铁金属铸件 ferrous metal casting
铁筋 iron reinforcement
铁堇青石 sekaninaite
铁精矿 beneficiated iron core; iron-ore concentrate
铁精砂 beneficiated iron core; iron-ore concentrate
铁精制 ferrofining
铁韭闪石 ferropargasite
铁锔 <用于连接固定木结构构件的铁制品> cramp iron
铁绢云母 ferri-sericite
铁菌 iron bacteria
铁卡爪 soft jaw
铁卡爪卡盘 soft jaw chuck
铁开尾销 iron split pin
铁康铜 iron-constantan
铁康铜热电偶 iron-constantan cou-

ple;iron-constantan thermocouple
铁康铜温差电偶 iron-constantan thermocouple
铁壳 iron case;iron clad
铁壳摆动式防逆阀 iron body swing check valve
铁壳板 metal lining
铁壳变压器 iron-clad transformer; shell-type transformer
铁壳的 metal-clad;metal-enclosed
铁壳的橡胶挡块 pillow block
铁壳电池 iron-clad battery
铁壳电动机 iron-clad motor
铁壳电流计 iron-clad galvanometer
铁壳断流器 iron-clad cutout
铁壳汞弧整流器 steel-tank mercury-arc rectifier
铁壳开关 enclosed knife switch; iron box switch; iron-clad switch; iron cover switch
铁壳开关装置 iron-clad switchgear
铁壳配电装置 iron-clad distribution equipment
铁壳式 shell type
铁壳式变压器 shell-type transformer
铁壳线圈 iron-clad coil
铁壳压力表 pressure ga(u)ge with iron case
铁壳整流器 iron-container rectifier; tank rectifier
铁壳砖 metal-cased brick; metal-encased brick
铁空气电池 iron-air cell
铁口 brown edge
铁口渣 <搅烧炉渣> tap cinder
铁扣 hasp;shackle
铁块 bloom;iron block;slab
铁块路面 iron block pavement
铁块破碎机 pig breaker
铁块铺面 iron pavement
铁筐 <手工成型时盛放吹制玻璃瓶> casher box
铁矿 iron mine;iron ore
铁矿处理 iron-ore processing
铁矿床 iron-ore deposit
铁矿分选机 iron-ore separator
铁矿粉 iron-ore dust
铁矿浮选(法) iron-ore flo(a)tation
铁矿搅炼 ore puddling
铁矿结核 cathead
铁矿颗粒 iron-ore pellet
铁矿露天采矿场 iron-ore stripe mine
铁矿炉渣粒料 iron slag aggregate
铁矿砂 iron sand
铁矿砂出口国协会 Association of Iron-Ore Export(ing) Countries
铁矿砂浆泵 iron-ore slurry pump
铁矿山 iron-ore mine
铁矿石 ferrolite;iron oxide;iron ore; ironstone
铁矿石骨料 iron-ore aggregate
铁矿石还原 reduction of iron ores
铁矿石混凝土骨料 iron-ore concrete aggregate
铁矿石集料 iron-ore aggregate
铁矿石路面 iron-ore road surface
铁矿石码头 iron-ore terminal; iron-ore wharf
铁矿石球团 iron-ore pellet
铁矿石水泥 iron-ore cement
铁矿石尾矿 iron-ore tailing
铁矿水泥 Ferrair cement;hard-setting cement;iron-ore cement
铁矿物 iron mineral
铁矿岩 chromitite;ferrolite
铁矿研磨 iron-ore grinding
铁矿运载车 iron-ore carrier
铁矿运载工具 iron-ore carrier
铁矿渣 iron-ore dust
铁矿渣水泥 ferro-cement; iron-ore cement
铁矿指示植物 indicator plant of iron
铁框格 gridiron
铁框格系统 gridiron system
铁框刮路器 gridiron drag
铁框架 iron framework;hearse <重要人物墓上的>
铁框架构造 iron carrying framing construction
铁框架结构 iron framing construction; iron skeleton framing construction
铁框路刮 gridiron drag
铁拉杆 iron stay
铁拉条 iron stay
铁栏杆 handrail iron; iron railing; metal rail;railing
铁蓝 Erlanger blue; ferriferous cyanide;iron blue; Prussian blue; gas blue <旧称>
铁蓝色 electric(al)blue;iron blue
铁蓝闪石 ferroglaucophane
铁蓝透闪石 ferrowinchite
铁类金属 ferrous metal
铁类金属压模铸造 ferrous die casting
铁楞窗(花)格【建】bar tracery
铁梨木 lignum vitae
铁梨木嵌环 lignumvitae strip
铁梨木条 lignumvitae strip
铁梨木轴衬 lignumvitae bush
铁梨木轴承 lignumvitae bearing
铁离子 iron ion
铁篱笆 fang
铁锂闪石 ferroholmquistite
铁锂云母 zinnwaldite;zinnwald mica
铁利蛇纹石 ferrolizardite
铁链平轮 horizontal chain wheel
铁链桥 iron chain bridge
铁梁 iron beam
铁梁桥 iron beam bridge;iron girder bridge
铁磷铝钙石 ferrian millisite;pallite
铁磷锰矿 magniotriplite
铁鳞 dross;iron scale;oxidation scale
铁鳞铁粉 Linz powder
铁菱镁矿 breunnerite;brown spar
铁菱锌矿 ferroan smithsonite;monheimite
铁硫砷钴矿 glaucodot(e)
铁楼梯 iron stair(case)
铁炉 iron stove
铁路 form of transport; line of rail; rail;railroad;railway(track);tread road
铁路爱好者 <美> railfans
铁路安全侧线 safety siding
铁路安全运营 safe railway operation
铁路案例 railroad case
铁路包裹车 parcel van
铁路保价运输 railway keep-price transport
铁路保温车 insulated wagon
铁路保养场 railway roundhouse;railway shop
铁路背驮式运输 rail piggyback
铁路闭塞工程 blockwork
铁路边界 railway boundary
铁路编组场 railway classification yard
铁路编组站 marshalling station;marshalling yard;switch yard
铁路变电所 railway substation
铁路标准及设计规范 standards and specifications for railway design
铁路/驳船水道转运点 rail barge river transfer point
铁路博物馆 railway museum
铁路不同业务部门间的协调 coordination of the rail business sectors
铁路部门 railway interest
铁路簿记 railway bookkeeping
铁路材料 railway material
铁路财产 railway property
铁路财务 railway finance
铁路财务分析 railway financial analysis
铁路财务工作 railroad financing;railway financing
铁路财务管理体制 railway financial management system
铁路财务决策 railway financial decision
铁路财务控制 railway financial control
铁路财务会计管理信息系统 railway finance and accounting management information system
铁路财务预算 railway financial budget
铁路槽车 railway tank car;tank wagon
铁路侧线 lay-aside; railroad siding; rail siding; railway siding; sidetrack;switch
铁路测量 railroad[railway] survey(ing)
铁路岔道 passing place;railroad siding;railroad spur;railway spur
铁路岔线 railroad[railway] siding
铁路长期计划 railway long plan;railway long-term program(me)
铁路长途电话网 railway long distance telephone network
铁路长途字冠 prefix number for railway toll call
铁路敞车 railroad gondola;railway flatcar;rail-truck <带拖车的卡车>
铁路敞车交货价 free on train
铁路敞篷车辆 gondola flat
铁路车辆 railroad car; rail vehicle; railway car; railway rolling stock; railway vehicle; rolling stock <包括机车>
铁路车辆秤 railway track scale
铁路车辆动力学 rail vehicle dynamics
铁路车辆段废水 wastewater from railway vehicle depot
铁路车辆服务 car service
铁路车辆检修车间 railroad repair shop;railway repair shop
铁路车辆轮渡 car ferry
铁路车辆统计 cars statistics of railway
铁路车辆卸货机 wagon unloading machine
铁路车辆在港停留时间 port time of rail car
铁路车辆制造 railway coach building
铁路车辆装卸作业时间 loading-unloading time of rail car
铁路车轮 rail wheel
铁路车皮 rail car;railway truck
铁路车厢 rail car
铁路车站 railroad depot;railway depot;railway station
铁路车站候车大厅 railway station hall
铁路车站及枢纽设计规范 railway yard and hub designing standards
铁路车站通过能力 carrying capacity of station
铁路车站线路长度 track length of railway station
铁路车站中间广场 nave
铁路承运人 rail carrier
铁路乘车证 railway staff service pass
铁路乘务员 train staff
铁路乘务组 crew
铁路抽样调查 railway survey in selected items from totality at random
铁路筹资 fund raising for railway
铁路打桩机 railway pile driver
铁路贷款 loan to railroads
铁路到港口的终点 railroad connection
铁路道班 railway maintenance gang
铁路道班工人 gandy dancer
铁路道岔 railroad switch; railroad turnout
铁路道岔转辙器 railway point of frogs
铁路道床 rail bed
铁路道床结构 ballast bed structure
铁路道钉 railroad spike; railway spike;track spike
铁路道口 grade crossing;level crossing; railroad crossing; railroad grade crossing;railway grade crossing
铁路道口标志 railroad crossing sign
铁路道口交叉标志 railroad crossing sign
铁路道口交叉角 railroad crossing angle
铁路道口交叉信号 railroad crossing signal
铁路道口交叉栅门 railroad crossing gate
铁路道口警告标志 railroad crossbuck sign
铁路道口预警标志 railroad advance warning sign
铁路道口自动管理 automated railroad-grade crossings
铁路道砟 railroad ballast
铁路的 rail-borne
铁路的废弃 abandonment of railway [railroad]
铁路的连带责任 joint responsibility of railway
铁路的上部建筑 permanent way
铁路的辙叉角 crossing angle
铁路等级 classification of rail(way); class of railway;grade of railway; line classification; railway classification
铁路地产 railway estate;railway territory
铁路地区 railway territory
铁路地图集 railroad atlas;railway atlas
铁路电报 railway telegraph
铁路电话 railway telephone
铁路电流 railway current
铁路电气化 railroad electrification; railway electrification
铁路电气化和运输控制集中化系统 centralized railroad electrification and traffic control system
铁路电信 railway telecommunication
铁路电影图书馆 railway film library
铁路吊臂起重机 railroad boom crane
铁路调车场 railroad yard; railway marshalling yard; railway yard; switch yard
铁路调车机车 shunting engine
铁路调车驼峰 railway shunting hump
铁路调车线 railroad siding; railroad spur
铁路调度长 yardmaster
铁路调度电话 railroad dispatching telephone;traffic control telephone
铁路调度集中 railway despatch concentration
铁路调度室 traffic controller's office on railway
铁路调度员 train dispatcher
铁路定期旅客乘车证 rail pass
铁路定线 railway location

铁路定线测量 railroad location survey
铁路动车传动（装置）railcar transmission
铁路渡线 cross-over
铁路端点 railhead
铁路端式装车斜坡台 railway end-loading ramp
铁路兑换价率 <国际联运> railway rate of exchange
铁路多元经营 railway diversified economy
铁路发信号装置 tripper
铁路发展 railway development
铁路发展规划 railway development planning
铁路发展联合会 Federation of Railway Progress
铁路发展协会 <美国设在弗吉尼亚州的亚历山大市> Railway Progress Institution
铁路法 railway act；railway law
铁路法规 railroad legging
铁路反限制（监督）专题讨论会 <美> Rail Deregulation Symposium
铁路防护林 railway protection forest
铁路防雪棚 railroad snowshed；railway showshed
铁路房产 railway premise
铁路房屋 railroad building
铁路非经营性国有资产 railway non-operating state-owned capital
铁路费率 rail rate
铁路费用 railway charges
铁路费用已付 free on rail
铁路分局 railway branch administration；railway sub-bureau
铁路分局经济效益审计指标体系 indices system of economic efficiency audit of railway subbureau
铁路分线终点 dead end
铁路封闭方案 rail closure program(me)
铁路服务 railway service
铁路服务条件规程 regulations governing conditions of service
铁路辅助干线 subarterial railway
铁路付款交货 free on rail
铁路负责 at railway risk
铁路附属的汽车运输单位 railroad (affiliated) motor carrier
铁路附属的汽车运输公司 railroad (affiliated) motor carrier
铁路复线里程 kilometrage of double-line railway
铁路复线率 percentage of double tracks to total route kilometers[kilometres]
铁路改线 railway realignment；railway relocation；route change of railway
铁路改组 railroad reorganization
铁路干船坞 railway dry dock
铁路干线 arterial railway；backbone road；main railway line；main stem；main track；through line；trunk railway
铁路干线运输 rail trunk haulage；trunk rail haulage
铁路岗位测评 railway post appraisal
铁路钢 rail iron
铁路高架装车机 decker
铁路高速运输 railway high speed traffic
铁路各种统计调查中的计算机应用 computerized application of railway statistic(al) surveys
铁路更新 railroad rehabilitation
铁路更新改造项目 railway renewing，transforming project
铁路工厂 railroad workshop；railway workshop
铁路工程 railway engineering；railroading；railway works
铁路工程测量 railway engineering survey
铁路工程测量学 railway engineering surveying
铁路工程技术规范 code for technology of railway engineering
铁路工程图 railway engineering drawing
铁路工程学 railroad engineering；railway engineering
铁路工程质量检查 railway project quality check
铁路工段道工 section hand
铁路工会 rail labor union
铁路工人失业保险条例 <美> Railroad Labor Unemployment Insurance Act
铁路工务设备统计 statistics of m-of-w facility
铁路工务设备统计报表 statistic(al) report on m-of-w facility
铁路工业 railroad industry
铁路工作班制 shifting system of railway enterprise
铁路公共（群众）关系协会 <美> Railroad Public Relation Association
铁路公路车辆 <不同运输方式联运用> rail-highway vehicle
铁路公路交叉 railroad intersection
铁路公路交叉口 rail highway xing
铁路公路交叉口拦路杆 railway crossing bar
铁路公路交替运输 rail-road combined transport
铁路公路联合枢纽 railroad terminal
铁路公路联合运输 combined rail and road transport；rail and road traffic
铁路公路联合运输公司 <西欧> Rail-Road Company
铁路公路联运 combined rail and road transport；rail-road combined transport
铁路公路联运的 rail motor
铁路公路联运业务 railroad service
铁路公路两用半挂车 railroad semi-trailer
铁路公路两用车 combination rail-highway vehicle；combined rail and road vehicle
铁路公路两用挂车 railroad trailer
铁路公路两用桥 combined bridge；double deck(ed) bridge；railway and highway combined bridge
铁路公路平交叉道（口）railroad grade crossing；rail-highway crossing
铁路公路双层桥 double decked bridge for railway and highway
铁路公路水平交叉 railway grade crossing
铁路公路转换器 rail-to-road converter
铁路公司 rail carrier；railway company；railway corporation
铁路公司化 railway private
铁路公务运输 traffic conveyed on railway service
铁路公用事业 railway utilities
铁路供电网 railway power supply network
铁路供应工业 railway supply industry
铁路估定（价）值 railway valuation
铁路估值 railroad valuation
铁路股票 rails
铁路固定资产管理 railway fixed assets management
铁路顾问委员会 <邀请工、商、农、运输各界人员参加> Railway Advisory Commission
铁路雇员 railroad employee；railway employee
铁路挂车 trailer car
铁路挂车平车 trailer on flat car
铁路挂车现有数 <客车和货车> trailer stock
铁路挂车运输 rail-trailer shipment
铁路管界 railway precinct
铁路管理 railway administration
铁路管理机构 railway authority
铁路管理局 railroad administration；railway administration
铁路管辖的 railway-controlled
铁路罐车 railcar tanker
铁路规定的车端装车 railway-specified end loading
铁路规章 railway regulation
铁路轨道 rail(road) track；rail(way) track
铁路轨道车 handcar
铁路轨道地磅 railway track scale
铁路轨道检查车 inspection car of railway track
铁路轨道内燃起重机 diesel crane on railway track
铁路轨道平面图 track plan
铁路轨道中线 axis of railway track
铁路轨道轴线 axis of railway track
铁路轨距 ga(u)ge；ga(u)ge of railway track；railroad ga(u)ge；railway ga(u)ge
铁路轨距样板 alignment ga(u)ge
铁路轨枕 railroad sleeper；railroad tie；railway sleeper
铁路轨座 rail base；rail chair；rail foot
铁路锅炉工 iron skull
铁路国有化 nationalization of railways；railway nationalization
铁路国有资产 state-owned assets of railway enterprise
铁路国有资产管理 state-owned assets of railway enterprise management
铁路国有资产收益 revenue of railway state-owned assets
铁路国有资产统计 statistics of railway state-owned capital
铁路过车磅 railway track scale
铁路过车秤 railway track scale
铁路海路联运运价率 through rail-ocean rate
铁路涵洞 culvert for railway；railway culvert
铁路行业统计 railway industrial statistics
铁路航空测量 railway aerosurveying
铁路航空快速联运业务 train-to-plane service
铁路合并 railway amalgamation
铁路合并法庭 <英> Railways Amalgamation Tribunal
铁路合理化 rail rationalization
铁路合作组织 <设在华沙> Organization for Collaboration of Railways
铁路和公路交叉 railway crossing
铁路和公路交叉角 railway crossing angle
铁路和公路平交 railway grade crossing
铁路和水道联运 raft and water traffic
铁路荷载 train load(ing)
铁路盒线 railroad siding；railway siding
铁路宏观（大）系统 railway macrosystem
铁路护轨 side rail
铁路环线 loop track
铁路环行线 railcar loop
铁路换装站场 railroad intermodal yard；railway intermodal yard
铁路货车 rail-truck；rail(way) wagon；wagon
铁路货车底架 truck
铁路货车调车场 drill yard
铁路货车共用制 <西欧> Europe(-Wagon) Pool
铁路货车每辆积载量 carload
铁路货车轴承润滑脂 railway journal (box) grease
铁路货物分等 railroad freight classification
铁路货物列车编组计划 plan for the formation of freight trains
铁路货物通知书 railroad advice
铁路货物运价 railway freight price
铁路货物运价规则 rules relating to railway goods tariff
铁路货物运价率表 list of railway goods tariff rate
铁路货物运价体系 railway freight transport tariff system
铁路货物运输 railway goods transport
铁路货物运输产品 product of railway goods transport
铁路货物运输公司 <英> Railfreight
铁路货物运输统计 railway freight transport statistics
铁路货物运输组织 <英> Railfreight Organization
铁路货物周转量 goods moved by rail
铁路货运场 railroad freight yard
铁路货运车 railroad freight car
铁路货运费 railroad freight charges
铁路货运积载量 car loadings
铁路货运趋势 rail freight traffic trend
铁路货运提单 railroad bill of lading
铁路货运业务 <英> rail freight business
铁路机车及车辆 rolling stock
铁路机车司机司炉联合会 <澳大利亚> Federated Engine Drivers and Firemens Association
铁路机车运用 locomotive operation
铁路机车运用统计 statistics of locomotive operation
铁路机动车 rail motor car
铁路机务段 railroad depot
铁路机务段废水 wastewater from railway machine depot
铁路基本建设 railway basic construction；railway infrastructural construction
铁路基本建设程序 procedure of capital construction
铁路基本建设管理 railway basic construction management
铁路基本建设管理机构 railway basic construction management agency
铁路基本建设管理内容 railway basic construction management content
铁路基本建设管理制度 railway basic construction management system
铁路基本业务会计科目 account titles for railway fundamental services
铁路绩效 railway performance
铁路及卡车运输 rail and truck
铁路及空运 rail and air
铁路集装箱平车 container on flat car
铁路集装箱运价 tariff of railway container
铁路集装箱站 railway container terminal
铁路给水 water supplies for railroad
铁路给水厂 water plant of railway
铁路计划 railway plan(ning)

铁路技术 railway technics
铁路技术改造 betterment and improvement of railway; technical reform of railway; technical renovation of railway
铁路技术(工艺)应用研究 railroad technology applied research
铁路技术管理规程 railway technical regulation; regulations of railway technical operation; rules for railway technology management
铁路技术经济特征 technologic(al) and economic characteristics of railway
铁路间账目清算 clearance of railway accounts
铁路间中转运输 rail-to-rail
铁路建设 railroading; railway construction
铁路建设基金 funds of railway construction
铁路建设监理 railway construction supervision
铁路建设前期工作 prework of railway construction
铁路建设项目 railway construction project
铁路建设项目可行性研究 feasibility study of railway construction
铁路建设项目总投资 railway project investment
铁路建造 railroad building; railroad construction; railway construction
铁路建筑 railroad construction; railway construction
铁路建筑公司 railway construction company
铁路建筑机械 railway construction machine
铁路建筑界限 railroad construction clearance
铁路建筑物 railway structure
铁路建筑限界 clearance for railway; railroad clearance
铁路交叉 railway crossing
铁路交叉安全装置 safety appliance of railroad crossing
铁路交叉处的轨枕配列 tie layout at railroad crossing
铁路交叉点(道口) railroad crossing; railway crossing; railway junction
铁路交叉检查条 crossing bar
铁路交叉口 railroad crossing; railroad grade crossing; railway crossing
铁路交货 ex rail; free on rail
铁路交货价(格) free on rail
铁路交通噪声 railroad noise
铁路接轨 railway junction
铁路界标 railway boundary
铁路尽头站 railhead
铁路进入车站或枢纽方式 rail approach
铁路经济核算 railway economic accounting
铁路经济效益审计 railway economic benefit audit
铁路经济效益审计程序 railway economic efficiency audit procedure
铁路经济效益审计的评价标准 evaluation standard of railway economic efficiency audit
铁路经济效益审计方法 railway economic efficiency audit method
铁路经济效益审计依据 grounds of railway economic efficiency audit
铁路经营 railroading
铁路经营性国有资产 railway operating state-owned capital
铁路警察 railway police
铁路警告标志 railroad warning sign; railway warning sign
铁路净空 railroad clearance; railway clearance
铁路竞争 railroad competition
铁路局的运输营业收入 transport operating earnings of railway bureau
铁路局对固定资产管理的权限 fixed assets management of rights of railway bureau
铁路局对固定资产管理的责任 fixed assets management responsibility of railway bureau
铁路局经济效益审计指标体系 indices system of economic efficiency audit of railway bureau
铁路局经营权 operating rights of railway bureau
铁路局局长 director of railways
铁路局旅客发送量 volume of sending passengers by railway bureau
铁路局线长途通信[讯]网 railway administration toll communication network
铁路局资产经营责任制 responsibility system of assets management of railway bureau
铁路局资金周转额 railway bureau turnover
铁路举重器 railroad jack
铁路距离 rail distance
铁路开发 railway development
铁路开挖 railroad excavation
铁路开展公司(社) <英> Railway Development Society
铁路勘测 railroad survey(ing); railway reconnaissance; railway survey(ing)
铁路勘测设计 railway survey and design
铁路客车 carriage; coach; passenger railway stock; railway carriage
铁路客货车辆运用统计 statistics of railway wagon and coach operation
铁路客货营销战略 marketing strategy of railway passenger and freight transport
铁路客货运容量 railway accommodation
铁路客运 railway transport of passengers
铁路客运调度工作 railway passenger traffic control
铁路客运公司 railway passenger company
铁路客运合同 railway passenger contract
铁路客运目标市场选择 market section of passenger traffic
铁路客运运价里程表 railway passenger tariff kilometrage table
铁路客运走廊 <美> alley way; corridor
铁路控制论 railroad cybernetics; railway cybernetics
铁路跨线桥 railroad overcrossing; railroad underbridge; railway overbridge; railway underbridge; turnpike; underbridge; underpass; railroad overbridge <公路在铁路下面>
铁路会计 railway accounting
铁路会计报表 railway accounting statement
铁路会计核算 railroad accounting; railway accounting
铁路会计核算方式 railway accounting pattern
铁路会计制度 railway accounting system
铁路快运包裹代理业 railway fast parcel transport agency
铁路劳动定额 railway labo(u)r quota
铁路劳动统计 labo(u)r statistics of railway enterprise
铁路劳动组织 railway labo(u)r organization
铁路冷藏车 railway refrigerated wagon
铁路连接 railroad connection
铁路连接线 rail link; railway connection
铁路连续立交 continuous grade separation of railroad crossing
铁路联轨站 railway junction
铁路联络线 adjacent strip; rail connecting lines; rail link; railway connection
铁路联锁设备 railway integration mechanism
铁路联营 railroad pool
铁路联运 railroad interline traffic; railroad through transport
铁路联运货物 interline freight
铁路联运客票 interline ticket
铁路列车 railroad train
铁路列车编组站 train assembly station
铁路列车调度设备 railway dispatching system
铁路列车调度系统 rail train dispatch system; railway dispatching system
铁路列车无线电通信[讯] railroad mobile radio; road train radio
铁路列车运行统计 statistics of train operation
铁路列车运行图 train schedule chart
铁路临时营业长度 length of temporarily operating railway
铁路零担运价 tariff of railway partload
铁路流动资产管理 railway current assets management
铁路留置权 railway's right of lien
铁路漏斗车 railroad hopper
铁路路标 track sign post
铁路路标支柱 track sign post
铁路路堤 railroad embankment; railway fill
铁路路基 body of railroad; ground work; railroad bed; railroad embankment; railway bed; railway embankment; railway subgrade
铁路路基宽度 width of railway subgrade formation
铁路路基面 railway roadbed
铁路路基坡度 railway slope
铁路路肩 railway shoulder
铁路路签 train staff
铁路路堑 railroad cut; railway cut(ting); railway in a cut
铁路(路)网规划 railway network planning
铁路(路)网密度 railway network density
铁路路线 roadway
铁路路线图 railroad map; railway map
铁路路线网 railway network
铁路旅客运价 railway passenger fare
铁路旅客运输 railway passenger traffic
铁路旅客运输计划 railway passenger transportation plan
铁路旅客运输价格 railway passenger traffic tariff; railway passenger transport tariff
铁路旅客运输统计 railway passenger traffic statistics
铁路旅客运输组织 organization of railway passenger traffic
铁路旅行 rail travel
铁路旅行的中途下车 break in continuity of rail journey
铁路旅行指南 railway guide
铁路铝电阻丝 Kanthal (resistance) wire
铁路绿化 railroad greening; railway planting
铁路绿色通道 railway "green passage"
铁路轮渡 railway ferry; railway train ferry
铁路轮渡码头 railway ferry terminal
铁路轮渡线路 rail-ferry route; railway ferry route; train-ferry track
铁路罗盘仪 railway compass
铁路码头 railroad wharf; railway wharf
铁路民营化 railway private
铁路名称 title of railroad
铁路模型爱好者 model-railway enthusiast
铁路内部控制度评审 estimation of railway internal control system
铁路内部劳动力市场 railway internal labo(u)r market
铁路内部审计 internal audit of internal control system
铁路内燃机车 bee-liner
铁路能力 railway capacity
铁路年度报告审计 railway annual report audit
铁路年度计划 railway annual plan; railway yearly plan
铁路旁交货(价格) ex rail
铁路配电系统 railway dispatching system
铁路棚车 railway box wagon
铁路癖 railroad hobbing
铁路票价 railroad fare
铁路拼合文字标记 railway monogram
铁路平板车 railway flat-car
铁路平板货车 rail platform car
铁路平车运输 piggyback service
铁路(平车)运送公路重载卡车和挂车 highway on wheels
铁路平交道(口) railroad crossing; railway level-crossing; railroad level-crossing; rail(way) crossing
铁路平均货运量 average line capacity
铁路平面交叉 railroad grade crossing
铁路铺设权 charter; railway charter
铁路普查 railway general investigation
铁路普通电报 railway telegraph ordinary
铁路企业多元化经营战略 diversified operating strategy of railway enterprise
铁路企业岗位技能工资制 wage system based on post technique in railway enterprise
铁路企业劳动组织 the organization of labo(u)r of railway enterprise
铁路企业内部控制制度 inside-controlled system of railway enterprise
铁路企业人工成本 labo(u)r cost in railway enterprise
铁路起重机 locomotive crane; rail-mounted crane; railway crane; railway-wheeled crane
铁路汽车联运运价率 joint rail motor rate
铁路牵引 railway traction
铁路牵引动力 railway motive power; railway traction power
铁路牵引力 railway haulage capacity
铁路潜在运输能力 railway potential capability
铁路桥渡 bridge crossing of railway
铁路桥(梁) railroad bridge; railway bridge

铁路桥梁荷载 railway bridge load
铁路桥面 railway deck
铁路轻油车 rail car
铁路区段 railway district; railway division
铁路区段电话 railway division communication system
铁路区段技术情报室 cabinets of technical information of railway divisions
铁路区段通信[讯]系统 railway division communication system
铁路区段运输成本 transport cost of railway line section
铁路区截信号 block signal
铁路曲线(板) curve in the track; railroad curve; railway curve
铁路曲线半径 radius of railway curve
铁路曲线表 railway curve table
铁路曲线测量 railroad curve survey
铁路曲线尺 railway curve rule
铁路全程的 all-rail
铁路散装水泥车 railroad container
铁路散装水泥罐车 railcar tank
铁路上部建筑物统计 statistics of railway superstructure
铁路上交货 free on rail
铁路上用以称过热物的轴箱 hot box
铁路设备 railway equipment
铁路设备登记簿 < 指车辆 > railway equipment register
铁路设备制造 railway equipment manufacturing
铁路设计标准 railway design criterion
铁路设计准则 railway design criterion
铁路设施 railroad infrastructure; railway infrastructure; railway facility
铁路设施能力 railway infrastructure capacity
铁路审计 railway auditing
铁路施工 railway construction
铁路施工单位审计 railway construction enterprise audit
铁路施工规范 specifications of railway construction
铁路施工企业 railway construction enterprise
铁路施工企业标准 railway construction enterprise standard
铁路施工企业的技术管理 technic management of railway construction enterprise
铁路施工企业的经营计划 operation plan of railway construction enterprise
铁路施工企业管理 railway construction enterprise management
铁路施工企业机械设备管理 machinery equipment management of railway construction enterprise
铁路施工企业技术管理制度 technical management system of railway construction enterprise
铁路施工企业技术素质 technic quality of railway construction enterprise
铁路施工企业会计 accounting of railway construction enterprise
铁路施工企业年度经营计划 annual operation plan of railway construction enterprise
铁路施工企业素质 railway construction enterprise quality
铁路施工企业中长期经营计划 medium and long-term plan of railway construction enterprise
铁路施工组织设计 railway construction organization design
铁路实际运输能力 railway actual capability
铁路使用费 trackage
铁路使用权 < 一个铁路企业对于另一个铁路企业的 > trackage
铁路市郊定期客运业务 rail commuter service
铁路事故 railroad accident; railroad disaster; railway accident
铁路事业 railroading; railway undertaking
铁路事业单位预算 budget of railway utilities unit
铁路视察团 railway inspectorate
铁路试验车(辆) railway test vehicle
铁路收签机 staff catcher
铁路收入资金 railway income capital
铁路收入综合分析 comprehensive analysis of railway revenue
铁路枢纽(站) railroad terminal; railway terminal; railway junction terminal; railway junction; multiple junction; railway center [centre]; railway hub
铁路输送 rail delivery
铁路输送垃圾 transportation of refuse by rail road
铁路输送能力 rail carrying capacity
铁路数据系统 railroad data system
铁路双线率 percentage of double tracks to total route kilometers[kilometres]
铁路水道联合运输 rail and water traffic
铁路水道联运 combined rail-water traffic
铁路水道枢纽 rail-water terminal
铁路水路联运码头 rail-water terminal
铁路水路联运设备 rail-water facility
铁路水路联运业务 rail-water service
铁路水路联运运价 rail-water rate
铁路水路铁路联合运输 rail-water-rail movement
铁路水路中转设备 rail-water facility
铁路水泥罐车 railroad car tank
铁路水运联运码头 rail and water terminal
铁路税 railway duty
铁路死岔 cul-de-sac(street)
铁路四线化 quadrupling of line
铁路隧道 railroad tunnel; railway tunnel; train tunnel
铁路隧道工程 railway tunnel works
铁路损益管理 railway income management
铁路所属的 railway-owned
铁路所属的港口 railway-owned port
铁路提货单 railway bill of lading
铁路体系控制 railway system control
铁路天桥 railroad overcrossing; railway bridge crossing; railway overcrossing
铁路填土 railway fill
铁路条例 railway code
铁路通车长度 length of line in operation
铁路通车里程 rail mileage open to traffic
铁路通过能力 carrying capacity of railway; rail capacity; railway carrying capacity
铁路通勤 railway commutation
铁路通信[讯] railway communication
铁路通学 railway commutation
铁路统计 railway statistics
铁路统计编码 coding of railway statistics
铁路统计标准体系 system of railway statistic(al) standard
铁路统计调查 railway statistic(al) survey
铁路统计调查队 investigation team in railway statistics
铁路统计调查方法体系 investigation methodical system in railway statistics
铁路统计法制建设 legal construction of railway statistics
铁路统计分析 railway statistic(al) analysis
铁路统计工厂 railway statistic(al) factory
铁路统计监察 railway statistics supervision
铁路统计监督 surveillance railway statistics
铁路统计教育培训 railway statistic(al) educational training
铁路统计科学研究 railway statistic(al) scientific research
铁路统计手段现代化 modernization of railway statistic(al) means
铁路统计信息管理系统 railway statistic(al) management information system
铁路统计整理 railway statistics arrange
铁路统计指标体系 index system of railway statistics
铁路统计专项调查 special investigation of railway statistics
铁路统计咨询 railway statistic(al) consultation
铁路统计组织结构 organization structure of railway statistics
铁路统一技术联盟 < 西欧 > Railway Technical Unity
铁路投资经济评价 economic evaluation of railway investment
铁路投资体制改革 railway investment system reform
铁路图 railway diagram
铁路土地 railway territory
铁路土地使用权 eminent domain
铁路土方工程队 construction gang
铁路土方工人 construction labo(u)rer
铁路退休协会 < 美 > Railroad Retirement Association
铁路托运单 railway consignment note
铁路拖车 railway trail-car; trail car
铁路驮背式系统 piggyback system
铁路驮背式装卸车 piggy packer
铁路驼峰 hump
铁路挖方 railroad cutting
铁路挖方顶部 railway cutting crest
铁路外的 extra-railway
铁路外贸提单 railroad export bill of lading
铁路弯道 railroad curve
铁路网 grid rail; network of railroad; network of railway; railroad network; railroad system; railway system
铁路网结构 composition of the railway system
铁路网联轨点 node of rail network
铁路网密度 density of railway net
铁路网通过能力 railway network capacity
铁路维护人员 railroad maintainer
铁路维修工 < 美 > lineman
铁路文件的编制 railway documentation
铁路文物 railroad hobbing
铁路文献的编集 railway documentation
铁路卧车 sleeping car; sleeping carriage
铁路无盖货车 gondola
铁路物资供销财务管理的原则 financing management rule of railway material supply and marketing
铁路物资供销企业财务管理 financing management of railway material supply and marketing enterprise
铁路物资供应目录 catalog(ue) of railway material supply
铁路物资合同制供应管理 railway material contracting supply management
铁路物资库存量 railway material stocks
铁路物资流通企业 intranet; railway material circulation enterprise
铁路物资收入量 railway material incomes
铁路物资消耗量 railway material consumption
铁路物资总公司 head office of railway material
铁路系统 railroad system; railway system
铁路系统控制 railway system control
铁路下穿桥 railway underbridge
铁路现代化 modernization of railway
铁路线 railway line; railway route; track(age); trainway; railway track
铁路线打桩(放线) pegging out of railway line
铁路线等级 classification of railway lines
铁路线概定(放线) pegging out of railway line
铁路线联合使用区段 joint section of line
铁路线路 rail(road) line; railway line; rail track; railway track
铁路线路长度 length of railway line; route length
铁路线路长度统计 statistics of railway line length
铁路线路公里里程 railway route kilometrage
铁路线路公里数 railway route kilometrage
铁路线路设计规范 code for design of railway line
铁路线路通过能力 carrying capacity of line; track capacity
铁路线路图 railway route map
铁路线路折合长度 equivalent length of tracks; equivalent track kilometers[kilometres]
铁路线全长 trackage
铁路线输送能力 carrying capacity of line
铁路线通过能力 carrying capacity of line
铁路线图 route map
铁路线终点 railhead
铁路限界 railroad clearance; railway clearance
铁路箱车 railway box wagon
铁路小说 < 美 > rail-fiction
铁路协调 railroad coordination
铁路协会铁路经济(研究)局 < 美 > Bureau of Railway Economics
铁路协作组织 < 欧洲 > Railway Co-operation Organization
铁路卸车点 rail clearance point; rail unloading point
铁路新线建设 newly-built railway construction
铁路信号 railroad signal(ling); railway signal(ling)
铁路信号工程师学会 < 英 > Institution of Railway Signal Engineers
铁路信号规则 railway signal(l)ing regulations; railway signal(l)ing

rules
铁路信号设备 railway signal(ling); railway signals equipment
铁路信号室 <英> cab(in)
铁路信号维护 railway signal service
铁路信号学 railway signal(ling)
铁路信息化工程 railway information project
铁路行包运输合同 railway luggage and traffic contract
铁路行包运输统计 railway luggage and parcel transport statistics; railway parcel transportation statistics
铁路行车指挥自动化 train operation and control automation
铁路行车自动化 train control automation
铁路行车组织 organization of train operation
铁路行李车 blind car; van
铁路修复 railroad rehabilitation
铁路修建营 railway construction battalion
铁路修理车间 railroad repair shop; railway repair shop
铁路修配厂 railway repair shop
铁路修筑 railroading
铁路虚拟长度 suppositional length of railway line
铁路悬臂起重机 railway jib crane
铁路选线 location of railway route selection; railway location
铁路选线工程师 railway locating engineer
铁路选线设计 design of railway location; railway location design
铁路延长线 extension
铁路延展长度 extended length of railway; railway track line length
铁路延展里程 extended length of railway
铁路沿线电缆 lineside cable
铁路沿线以外地区 out-of-the-way region
铁路验道车 railway inspection car
铁路养护班 track maintenance gang
铁路养护车 railway maintenance car
铁路养护队 railway maintenance gang
铁路养路工人 line man
铁路业务 railroading; railway interest; railway service
铁路引入站 adjoining railway
铁路英里里程 track mileage; railroad mil(e)age; rail mileage
铁路营业部分 rail market segment
铁路营业里程 length of railroad lines in service; railroad lines opened to traffic; railway operating kilometrage
铁路营业招揽 railroad marketing
铁路营业资金 railway working capital
铁路拥有的港口 railway-owned port
铁路用变流机 railway convertor
铁路用粗润滑剂 railway freight car oil
铁路用地 railroad right-of-way; railway land; railway right-of-way; railway territory; right of way
铁路用地标牌 land monument
铁路用地界 right-of-way boundary
铁路用电动机 railway motor
铁路用电负荷 railway load
铁路用房 railroad building
铁路用经纬仪 railway transit
铁路用曲线板 railroad curve; railway curve
铁路用润滑脂 railway grease
铁路用语 railwayese
铁路用制动器 railway brake
铁路邮局 railway post office
铁路邮政 railway mail service
铁路邮政分局 railway post suboffice
铁路油槽车 petroleum tank wagon; tank railway car; rail tank car; rail tanker
铁路与道路交叉口的栅门 crossing gate
铁路与公路交叉 rail-highway crossing; railroad intersection
铁路与公路平交道 level crossing
铁路与公路平面交叉口 level railway crossing
铁路与轨道平交道口 railroad grade crossing; railway grade crossing
铁路与铁路交叉 track crossing
铁路员工退休制度 railroad retirement system
铁路远程自动化信息网 <美国铁路协会总部监督全国货车的运用情况> Tele-Rail Automated Information Network
铁路运材 railroad logging
铁路运程 average distance of freight transportation
铁路运单 railway receipt
铁路运单副本 duplicate of way bill
铁路运费 railage; railway freight
铁路运费表 railroad rate schedule
铁路运费率 railroad freight rates; railway tariff
铁路运货棚车 railroad boxcar
铁路运价 railway rate
铁路运价表 railway tariff
铁路运价分类表 label of railway tariff
铁路运价结构 railway rate structure
铁路运输 railage; rail freight; rail service; rail traffic; railway traffic; railway transport; railway transportation; shipping by rail(way); transportation by railway; transportation railroad
铁路运输安全 safety of railway traffic
铁路运输补贴政策 rail transport subsidy policy
铁路运输成本 railroad transportation cost; railway transportation cost
铁路运输成本报表 statement of railway transportation cost
铁路运输成本费用管理 management of railway transportation cost and expenditures
铁路运输成本费用计划 railway transportation cost and expense calculation
铁路运输成本费用决策 railway transportation cost and expense decision
铁路运输成本费用控制 railway transportation cost and expense control
铁路运输成本费用预测 railway transportation cost and expense forecast
铁路运输成本分析 railway transportation cost analysis
铁路运输成本计划 railway transportation cost planning
铁路运输成本计算 railway transportation cost calculation
铁路运输成本计算方法 railway transportation cost calculating
铁路运输成本运营统计 operating cost of statistics of railway transport
铁路运输单位资产 assets of railway transport unit
铁路运输地图 railroad transportation map
铁路运输工作技术计划 technical work plan in railway transportation
铁路运输公司 railway transport company
铁路运输管理体制改革 railway transport management system reform
铁路运输规程 railway transportation regulation; regulations concerning carriage by rail
铁路运输换算周转量 converted turnover of railway transportation
铁路运输活底车 railroad hopper car
铁路运输货车 railroad boxcar
铁路运输及海运 rail and ocean
铁路运输及水运 rail and water
铁路运输价格 railway transport rate
铁路运输价格指数 railway transport rate index
铁路运输结构 structure of railway transportation
铁路运输进款管理 railway transportation revenue management
铁路运输进款科目 railway transport revenue account
铁路运输进款会计 railway transport revenue accounting
铁路运输进款会计报表 railway transport revenue financial statement
铁路运输进款会计核算 railway transport revenue accounting
铁路运输进款凭证 railway transport revenue receipts
铁路运输经济调查 investigation of railway transportation economics
铁路运输开拓 development of railroad transport
铁路运输会计账户 accounts of railroad transportation; accounts of railway transportation
铁路运输劳动生产率 railway transport labo(u)r productivity
铁路运输量 total volume of railway freight
铁路运输量预测 railway transportation volume forecasting
铁路运输零担(货) less-than-carload
铁路运输露天矿 rail pit
铁路运输密度 density of railway traffic; railway transport density
铁路运输模拟市场分区定价制度 simulating system of railway price based on region
铁路运输木材 railway transport timber
铁路运输企业财务管理体制 financial management system of railway transport enterprise
铁路运输企业点到点成本核算 cost calculation for point to point of railway transport enterprise
铁路运输企业改革 railway transport enterprise reform
铁路运输企业工效挂钩 wage in railway transport enterprise based on efficiency
铁路运输企业固定资产 fixed assets of railway transport enterprise
铁路运输企业会计 accounting of railway transport enterprise
铁路运输企业劳动定员标准 labo(u)r personnel quota standard in railway transport enterprise
铁路运输企业目标成本管理 target cost management of railway transport enterprise
铁路运输企业市场营销 marketing of railway transport enterprise
铁路运输企业责任 railway traffic enterprise responsibility; railway transport enterprise responsibility
铁路运输设备统计 railway transport equipment statistics
铁路运输市场份额 market share of railway transportation
铁路运输收入 railway traffic revenue; railway transport(ation) revenue
铁路运输收入稽查 railway transport action receipts check
铁路运输收入会计 accounting of railway transport income
铁路运输收入审计 income audit of railway transportation
铁路运输收入与支出审计 railway transportation revenue and expense audit
铁路运输收支审计 income and cost audit of railway transportation
铁路运输数据处理 railway data-processing
铁路运输数据处理中心 railway data-processing center[centre]
铁路运输速度 speed of railway transportation
铁路运输统计 railway transport statistics
铁路运输统计报表 statistics statements of railway transport
铁路运输统计分组 subgroup of railway transport statistics
铁路运输统计加工整理 processing and arranging of railway transport statistics
铁路运输统计理论与方法 railway transport statistics theories and methods
铁路运输统计数据质量控制 quality control of railway transportation statistical data
铁路运输统计现代化 modernization of railway transport statistics
铁路运输统计信息网络 railway transportation statistics information network
铁路运输图 railway transportation map
铁路运输业的净值 net output value of railway transport enterprise
铁路运输业的增加值 added value of railway transportation enterprise
铁路运输业中的间投入 intermediate input in railway transport industry
铁路运输业总产量 total product of railway transportation
铁路运输业总产值 gross output of railway transport industry
铁路运输营销 railway transport marketing
铁路运输营业收入管理 railway transport operation income management
铁路运输增加值 value-added of railway transport
铁路运输支出科目 railway transport expense entry
铁路运输中间的投入价格指数 rate index of intermediate input in railway transport
铁路运输周转量 railway traffic turnover
铁路运输自动化管理 automatic railway management
铁路运输自动化管理中心 automatic railway management center[centre]
铁路运输自动化控制 automatic railway control
铁路运输自动化控制中心 automatic railway control center[centre]
铁路运输综合作业方案 integrated program(me) for rail transport
铁路运输总产值 total value of pro-

T

duction of railway transportation
铁路运输组织 organization of rail transport
铁路运送 rail delivery;ship by rail
铁路运营 operation of railroad;railroad operation;railway operation;railway working
铁路运营费用 railing operating expenses
铁路运营固定资产 railing operating fixed assets
铁路运营固定资金 railing operating fixed capital
铁路运营管理理论 railway operating management theory
铁路运营管理信息系统 railway management information system
铁路运营计划 railway operating plan
铁路运营进款 railway operating income
铁路运营里程 railway operating length of railway
铁路运营流动资金的运用 operation of railway operating current capital
铁路运营流动资金计划 plan of railway operating current capital
铁路运营收入 railway operating revenue;railway traffic revenue
铁路运营线路全长度 railway trackage operated
铁路运营限额分配模型 railway operation quota assignment model
铁路运转部门基层工人 <美俚> monkey
铁路载客车辆 <包括客车、电力动车和内燃动车等> passenger accommodation carrying vehicle;passenger accommodation service vehicle;passenger vehicle
铁路在上交叉 elevated track crossing
铁路在下交叉 bridge overgrade;crossing below;flyover bridge;railroad underbridge;railway underbridge;undergrade crossing;underline bridge;underpass
铁路噪声 railway noise
铁路炸药 railroad powder
铁路债券 railroad bond;railway bond
铁路债务 railroad debt
铁路站场 railway yard
铁路站场设计 railway yard designing
铁路站场作业班组 drill crew
铁路站界入口 access to railway premises
铁路站内电话 railway station communication system;railway station telephone
铁路站内通信[讯]系统 railway station communication system
铁路站台 railroad platform
铁路照明器附件 railway lamp fitting
铁路照明设备 railway lighting equipment
铁路辙尖 railroad point;railway point
铁路辙枕 railway chair
铁路枕木 railroad sleeper;railroad tie;railway sleeper;wooden railroad cross-tie;wooden railroad tie
铁路蒸汽起重机 steam railway crane
铁路整车运价 tariff of complete railway vehicle
铁路正线长度 railway main line length
铁路支线 feeder;feeder line;feeder railway;local railway;railroad feeder;railroad spur;rail siding;rail spur;railway feeder;railway siding;secondary railroad;secondary railway;side-track;siding;spur(line)
铁路支线高速联络列车 speedlink feeder train
铁路支线转辙器 turn branch switch
铁路直达运输 rail transit
铁路职工 railman;railroader;railway man
铁路职工联合理事会 <英> Railway Staff Joint Council
铁路职工全国法庭 <英> Railway Staff National Tribunal
铁路职工全国理事会 <英> Railway Staff National Council
铁路指南 railway directory
铁路智能运输系统 railway intelligent transportation system
铁路中期计划 railway mid-term program(me)
铁路中线 railway axis
铁路中线间距 distance between centers of tracks
铁路中心 railway center[centre]
铁路中心线 center[centre] line of track;railway centerline[centreline]
铁路中转运输 rail transit
铁路终点储备的物品 railhead reserves
铁路终点站 dead-ended railroad station;terminal railway station;end of railway;end of track;railroad terminal;rail terminal;railway terminal
铁路重新调查 railroad readjustment
铁路重载运输 heavy haul railway transport;railway heavy haul traffic
铁路重组 railway reorganization
铁路周围环境 railway surroundings
铁路轴线 railway axis;track axis
铁路轴箱 railroad journal box;railway axle box
铁路主要技术标准 main technical standard of railways
铁路主要技术要求 main technical requirements of railways
铁路筑路机械 track building machinery
铁路筑路权 railroad right-of-way
铁路专用频率 located frequency for railway use
铁路专用无线电 railroad radio;railway radio
铁路专用线 access railway;bay-line;rail spur;railway siding;railway special line;special rail;special spur line;special spur track
铁路专用线长度 length of railway special line
铁路专用支线 bay-line;jerkwater
铁路转盘 traverse table
铁路转水路运输 rail-to-water
铁路转运点 railhead
铁路装备 railway facility
铁路装车点 railway filling point;railway loading point;special spur railway
铁路装车设施 railway loading facility
铁路装货台 staith(e)
铁路装卸长度 length of handling;length of loading/unloading siding
铁路装卸线通过能力 throughput capacity of rail siding
铁路装卸作业成本 operating cost of railway handling
铁路装运 railing
铁路装运卡车的装卸设备 rail-truck loading and unloading facility
铁路装运卡车、挂车、集装箱、小型棚车等的枢纽 raft-van terminal
铁路资本结构 railway capital structure
铁路资产 railway property
铁路资产负债审计 audit for railway assets and liabilities
铁路资产经营 railway assets operating
铁路自动车管理协会 <美> Railway Automotive Management Association
铁路自动化信息 railway automated information
铁路自动信息中心 automated railway information center[centre]
铁路自用机车 service locomotive
铁路自有的汽车线路 railroad-owned truck line
铁路综合运价 freight of all kinds
铁路总厂 central railroad shop
铁路总段管理局 railway grand division
铁路总经理 president of railway
铁路纵轨枕 track stringer
铁路组合 railroad combination
铁路作业计划 railway operating plan
铁铝包层 iron-aluminum coating
铁铝比(率)iron-alumina ratio
铁铝的 ferrallitic
铁铝矾 halotrichite
铁铝富化作用 ferrallitization
铁铝铬电阻丝合金 Megapyr
铁铝合金 Fe-Al alloy;ferro-alumin(i)um;iron-aluminum alloy
铁铝红土 ferrallite
铁铝尖晶石 hercynite
铁铝榴石 almandine[almandite]
铁铝榴石白云母片岩 almandine-muscovite-schist
铁铝榴石黑云母片岩 almandine-biotite schist
铁铝率 iron-alumina ratio
铁铝锰钢 ferro-alumin(i)um-manganese steel
铁铝钠闪石 ferro-eckermannite
铁铝蛇纹石 berthierine
铁铝酸钙 calcium aluminoferrite
铁铝酸钙水泥 calcium aluminoferrite cement
铁铝酸矿物 aluminoferrite mineral
铁铝酸四钡 tetrabarium aluminoferrite
铁铝酸四钙 <水泥中矿物成分> tetracalcium aluminoferrite;tetracalcium alumina ferrite
铁铝酸四锶 tetrastrotium aluminoferrite
铁铝土 feral(l)ite;ferralitic soil;ferralsol;pedalfer;red bauxite
铁铝性黏[粘]土 sesquioxidic clay
铁铝岩 feral(l)ite
铁铝盐 feral(l)ite
铁铝氧耐火材料 diamantin(e)
铁铝氧石 bauxite
铁铝直闪石 ferrogedrite
铁率 iron modulus <水级的,即铝率=%三氧化二铝/%三氧化二铁>;iron modulus of cement <水泥的>
铁绿矾 iron vitriol
铁绿泥石 aphrosiderite;chamosite;daphnite;ripidolite
铁绿松石 chalcosiderite
铁绿纤石 ferropumpellyite
铁轮 iron tire[tyre]
铁轮车辆 metal-tired vehicle;metal-tyred vehicle
铁螺栓 iron bolt
铁络合剂 iron-retention agent
铁马 car body rest frames after lifting jack released;iron horse;sweeping mo(u)lder's horse
铁马凳 bolster;chair;beam bolster <支梁的>
铁镘板 smoothing iron
铁毛矾石 tektite
铁毛毡的金属箔 felted fabric foil
铁铆钉 iron rivet
铁冒型铁矿床 Gossan-type iron deposit
铁帽 blossom;capping;gossan[gozzan];iron cap;iron gossan;iron hat;siderosphere
铁帽矿物 gossan mineral
铁媒染剂 iron mordant
铁楣 iron lintel
铁镁催化剂 ferro-magnesium catalyst
铁镁黄长石 ferroakermanite
铁镁辉石 violaite
铁镁尖晶石 ceylonte
铁镁矿物 dark mineral;ferro-magnesium mineral;melane;sideromelane
铁镁绿泥石 brunsvigite
铁镁氟铝石 zirklerite
铁镁镍矾 ferroan-magnesian retgersite;ferro-magnesian retgersite
铁镁坡缕石 mountain leather
铁镁齐 ferro-magnesium
铁镁质 femic;ferro-magnesium material
铁镁质组分 femic constituent
铁门 iron gate
铁门偏置铰链 offset pivot
铁门销子 door catch
铁蒙脱石 ferrimontmorillonite;stolpenite
铁锰 ferrimanganic
铁锰沉淀(物)ferro-manganese precipitate
铁锰钙辉石 ferro-johannsenite;iron schefferite
铁锰橄榄石 ferrotephroite
铁锰共沉淀物 coprecipitate of iron and manganese hydroxide
铁锰合金 ferro-manganese;spiegel(eisen)
铁锰黄铜 iron manganese brass
铁锰尖晶石 ferrojacobsite;galaxite
铁锰结核 iron-manganese concretion
铁锰鳞绿泥石 ferro-pennantite
铁锰绿铁矿 rockbridgeite
铁锰钠闪石 kozulite
铁锰齐 ferrimanganese
铁锰氢氧化物覆膜 Fe-Mn hydroxide coating
铁锰氢氧化物结核 Fe-Mn hydroxide concretion
铁锰铜合金 ferro-manganin
铁锰质结核 ferrous-manganese nodule
铁锰质条纹 ferrous-manganese stripe
铁锰质土 umber
铁锰着色玻璃 iron-manganese colo(u)red glass
铁密高岭土 iron lithomarge
铁棉 iron shavings;iron wool
铁面 iron surface
铁明矾 butter rock;feather alum;ferric alum;halotrichine[halotrichite];iron alum;mountain butter;redingtonite;rock butter
铁模型 iron pattern
铁模 ingot mo(u)ld;iron mo(u)ld;swage;swedge
铁模板 iron forms;iron shuttering;iron formwork
铁摩辛柯梁 Timoshenko beam
铁磨 iron mill
铁磨光器 polishing iron
铁磨砖石 float stone
铁抹赶光 smoothing with trowel;trowel(l)ing
铁抹子 plaster's trowel;trowel
铁末 iron dust
铁姆肯滚柱轴承 Timken roller bearing

铁木 hop hornbeam; iron tree; quebracho
铁木混合结构车身 composite body
铁木结构 ironwood structure
铁木结构船 composite vessel
铁钼华 ferrimolybdite
铁钠 ferrisodium
铁钠沸石 <钠沸石与绿泥石的混合物> iron natrolite
铁钠钾硅石 fenaksite
铁钠磷锰矿 ferri-alluaudite
铁钠透闪石 ferrorichterite; waldheimite
铁南针 iron-compass
铁泥刀 finishing trowel
铁腻子 iron filler
铁黏[粘]土 clunch; iron clay
铁碾 iron roller
铁镍铂矿 ferro-nickel platinum
铁镍薄板 Deltamax
铁镍磁软合金 Sinimax
铁镍磁性合金 Anhyster; hipernik
铁镍低膨胀系数合金 guillaume alloy
铁镍电池 Hawkins cell
铁镍矾 honessite
铁镍铬合金 iron-nickel-chromium alloy
铁镍钴铬合金 Fernichrome
铁镍钴合金 Elcolloy; Fernico; Kovar; teleoseal; Sealvar <其中的一种>
铁镍合金 dilvar ferronickel; ferronickel; ferro-nickel iron; iron-nickel; iron-nickel alloy; Mutemp; Permenorm; radio metal; thermopermalloy; thermopermalloy orthonik alloy
铁镍合金钢 invar steel
铁镍尖晶石 nickel ferrite
铁镍交联改性膨润土 Fe-Ni crosslinked modified bentonite
铁镍矿 awaruite
铁镍铝磁铁 iron-nickel-aluminum magnet
铁镍铝合金 Calite
铁镍铝系磁铁合金 Alnic
铁镍耐热耐蚀合金 thermalloy
铁镍热磁合金 thermalloy
铁镍铜高电阻和高导磁率的合金 mu-metal
铁镍铜钼合金 iron-nickel-copper-molybdenum alloy
铁镍透磁合金 audiolloy; audiolloy Hypernic
铁镍蓄电池 Edison-Junger accumulator; Edison storage battery; iron-nickel accumulator; iron-nickel storage battery; Jungner battery
铁镍整磁合金 climax alloy
铁爬梯 access hook; foot iron; step iron
铁耙 iron rake
铁盘 griddle; iron pan
铁磐 gut hammer; hard pan; iron pan
铁刨子 spokeshave
铁炮 cone drums
铁炮式粗砂机 cone reducer; reduction cone
铁硼合金 ferroboron
铁坯 balled iron
铁皮 iron sheet; metal sheet; sheet iron; sheet-iron panel; algam <威尔士语>
铁皮桉 grey gum; iron bark; ironbark gum
铁皮绑条 strapping
铁皮包角 armo(u)red corner
铁皮带 strap
铁皮泛水 flexible metal flashing (piece)
铁皮泛水圆形卷边 beaded drip; headed drip
铁皮风管 plate air conduct
铁皮封闭式电池 steel-seal type cell
铁皮工 tinsmith
铁皮管 sheet-iron tube
铁皮痕 scale pattern
铁皮黄杨 Ironbox
铁皮货车 <铁路> blind baggage
铁皮剪(刀) iron sheet shears; tinsmith shears; snips; tin snips
铁皮剪子 iron sheet shears; tinsmith shears
铁皮坑 scale pit
铁皮轮 iron tire[tyre]
铁皮轮车 iron-tired vehicle; iron-tired wheel
铁皮轮车交通 iron-tired cart traffic
铁皮门 armo(u)red door
铁皮木 iron bark
铁皮松开 iron strap loosened
铁皮条 iron strap; strap
铁皮条打包 iron strapping
铁皮屋顶 flexible metal sheet roof cladding; flexible metal sheet roofing; flexible metal sheet roof sheathing; iron roofing; iron sheet roofing
铁皮屋顶覆盖层 iron sheet roof cover(ing)
铁皮屋顶盖板 iron roof cladding
铁皮屋顶夹衬板 iron sheet roof sheathing
铁皮屋面 iron roofing
铁皮屋面板 iron roof cover(ing); iron roof sheathing
铁皮屋面工 iron roofer; sheeter
铁皮镶板 sheet-iron encasing
铁皮镶套耐火材料 metal-cased refractory
铁皮芯 core shield
铁皮折页 strap hinge
铁皮制品 sheet metal work
铁皮砖 metal-cased brick; steel-clad brick
铁片 iron sheet; sheet iron
铁片式安培表 iron vane type ammeter
铁片式伏特计 iron vane type voltmeter
铁片式仪表 iron vane instrument
铁器 hardware; iron ware; sheet iron
铁器工人 iron worker
铁器商 ironmonger
铁器时代 Iron Age; iron period
铁器业 ironmongery
铁器制造商 iron master
铁钎 crooked chisel; gavelock
铁牵条 bar rigging; rod rigging
铁钳 hawkbill; iron dog; pinchers
铁浅黄 iron buff
铁锹 iron shovel; shovel; spade
铁锹捣实 spading
铁锹工作 spade-work
铁桥 iron bridge
铁撬 crow(bar)
铁撬棍 grab iron; grapple iron; jimmy
铁切割机 iron cutter
铁切机 chipper
铁青色 iron blue
铁氰化钾 potassium ferricyanide
铁氰化钾溶液 potassium ferricyanide solution
铁氰化钠 sodium ferricyanide
铁氰化物 ferricyanide
铁球 iron ball
铁圈 eyelet; iron ring; quoit
铁圈支架 iron-ring support
铁泉 chalybeate; iron spring
铁泉水 chalybeate water
铁溶出值 iron solution value
铁熔渣 iron slag
铁三角 L-plate; steel L-plate; corner reinforcement <加固门框上角的>
铁三脚架 iron-triangle
铁纱 iron gauze; wire fabric
铁纱片 wire fabric sheet
铁纱网 gauze screen; wire gauze
铁砂 ferro-sand; iron sand; iron shot; fine chilled iron sand <锯解石料时掺用的>
铁砂布 steel wire cloth
铁砂出口国联盟 Association of Iron-Ore Export(ing) Countries
铁砂喷射处理 abrasive steel shot
铁砂石 iron sandstone
铁砂石层 carstone
铁砂箱 iron flask
铁砂岩 iron sandstone
铁砂釉 tessha glaze
铁砂钻进 shot drilling
铁砂钻岩法 shot drill
铁筛 bar screen
铁杉 Chinese hemlock; Chinese hemlock-spruce
铁杉丹宁 hemlock tannin
铁杉木材 hemlock fir
铁杉属 Hemlock fir; Hemlock spruce; Tsuga <拉>
铁杉油 hemlock needle oil
铁杉脂 Canada pitch
铁闪石 grunerite
铁闪石片岩 grunerite schist
铁闪锌矿 christophite; ferroan sphalerite; marmatite; new boldite
铁舌 iron lug; lug
铁蛇纹石 greenalite
铁蛇纹岩 greenalite rock
铁砷硅锰矿 ferroschallerite
铁砷石 karibibite
铁砷铀云母 kahlerite
铁渗碳体平衡图 iron-cementite diagram
铁施特伦茨石 ferrostrunzite
铁十字律 iron cross law
铁石 ironstone
铁石结核 dogger
铁石笼 wire bolster
铁石棉 amosa asbestos; amosite; mountain wood
铁石墨平衡图 iron-graphite diagram
铁石英 eisenkiesel; ferruginous quartz; sinopole
铁石陨石 aerosiderolite; stony-iron meteorite
铁石渣 iron ballast
铁栓 iron bar; iron bolt
铁双头螺栓 iron stud
铁水 hot iron; hot metal; iron melt; liquid iron; melted iron
铁水包 foundry ladle; ladle; ladle pot
铁水包衬砖 ladle brick
铁水包吊车 ladle crane
铁水包吊运车 ladle crane trolley
铁水包倾注装置 ladle tilter
铁水包运输车 ladle car
铁水车 hot-metal car
铁水穿漏 hot-metal break-out; run-out
铁水沟 iron runner; sow
铁水罐 hot-metal bottle; pig-iron ladle
铁水罐车 hot-metal ladle and carriage; iron ladle car; transfer ladle
铁水混合包 reservoir ladle
铁水混合桶 reservoir ladle
铁水货物联运规则 rules relating to rail-water through goods
铁水货物联运计划 rail-water through goods transport plan
铁水静压力 ferrostatic pressure
铁水口堵眼机 Mutegun
铁水粒化 granulation of pig iron
铁水磷锰矿 baldanfite
铁水铝英石 sinopole
铁水镁石 eisenbrucite; iron brucite
铁水面龟纹 tortoise shell figure
铁水泥 Erz cement; iron cement
铁水上的漂浮石墨 kish
铁水蛇纹石 eisengymnite; iron gymnite
铁水脱氧转炉炼钢法 killed Bessemer process
铁水液面花纹 play film
铁水预处理 pretreatment of hot metal
铁水支沟 lateral channel
铁丝 ferrocyanide wire; iron wire
铁丝布 iron wire cloth
铁丝(粗)筛 iron wire screen
铁丝打捆装置 wire-tying device
铁丝钉 wire brad; wire tack
铁丝钉套订 stabbing
铁丝防护罩 wire guard
铁丝格栅 wire grating
铁丝格子 grill(e); iron grill(e)
铁丝盒 wire box
铁丝嫁接 wire grafting
铁丝剪 snips; wire cutter
铁丝卷 bundle iron; iron wire coil
铁丝铠装 iron wire armo(u)ring
铁丝筐 wire basket
铁丝捆绑方木排 mat packs
铁丝捆扎式捡拾压捆机 wire-tying baler
铁丝捆扎装置 wire-tying mechanism
铁丝拉紧器 plain wire strainer
铁丝拦污栅 wire fence groin; wire fence groyne
铁丝篮 wire basket; wire box
铁丝笼 gabion; wire basket; wire box
铁丝木杆沉排 pole and wire mattress
铁丝扭结器 wire knotter; wire spinner
铁丝耙 wire rake
铁丝盘 roll of wire
铁丝钳 hand snips
铁丝纱 iron wire gauze; screen wire cloth; wire cloth
铁丝石笼 steel gabion
铁丝石笼 wire cage packed with stone
铁丝刷 iron wire brush
铁丝弹簧圈 spring ring; wire spring ring
铁丝网 cattle guard; chicken wire; entanglement; festoon; iron gauze; net of iron wire; wire fence; wire gauze; wire lath; wire mesh; wire netting; wire screen; woven wire; chain link
铁丝网玻璃 wired glass
铁丝网柴(褥沉)排 brush and wire envelop mattress
铁丝网窗护栏 woven-wire window guard
铁丝网底笼 wire-floored coop
铁丝网地面鸡圈 wire screen floor pen
铁丝网地面养鸡法 wire-floored house system
铁丝网防护栏 chain link fence
铁丝网隔堵 wire-mesh bulkhead
铁丝网护栏 woven wire guard
铁丝网络柴排 brush and wire envelop mattress
铁丝网门 screen door
铁丝网石笼 crushed rock wrapped in wire mesh
铁丝网水幕 wire-mesh water screen
铁丝网围栏 woven-wire fence
铁丝网院门 woven-wire gate
铁丝网栅 wire-mesh hurdle; wire net fencing
铁丝网栅栏 wire net fencing

铁丝网栅栏桩 wire fence picket
铁丝网障碍物 wire entanglement
铁丝围栏 wire fence
铁丝(细)筛 iron wire sieve
铁丝下水口箅子 wire basket strainer
铁丝芯骨 core wire
铁丝烟筒刷子 wire flue brush
铁丝扎捆机 wire binders
铁丝栅栏 wire fence
铁丝织网 chicken wire
铁丝制的无头钉 wire brad
铁似 irony
铁素体 ferrite
铁素体不锈钢 ferritic stainless steel
铁素体钢 ferrite steel;ferritic steel
铁素体合金 Alfer
铁素体化 ferritising[ferritizing]
铁素体化退火 ferritizing annealing
铁素体界 ferrite region
铁素体晶粒 ferrite grain
铁素体楼面硬化剂 ferrocyanide floor hardener
铁素体中的机械孪晶 Neumann band
铁素质 ferrite
铁酸 ferrous acid
铁酸二钙 dicalcium ferrite
铁酸钙 calcium ferrite
铁酸镁 magnesium ferrite
铁酸锌颜料 zinc ferrite
铁酸盐 ferrate[ferrite]
铁酸盐矿物 ferrite mineral
铁酸盐水泥 ferrite cement
铁酸盐相 ferrite phase
铁燧石 taconite
铁燧岩 taconite
铁损分量 cores-loss component
铁损(失) iron loss;core loss
铁损试验 cores-loss test; Epstein test;iron loss test
铁损系数 iron loss factor
铁索道 mountain lift
铁索吊车 funicular railway
铁索护栏 cable guard rail(ing)
铁索桥 iron chain bridge;wire bridge
铁索斜拉桥 iron rope stayed bridge
铁索眼环接头 capel
铁锁 iron lock
铁塔 iron tower;pylon;steel tower
铁塔菲石 pehrmanite
铁塔负载 tower loading
铁塔广播天线 broadcast-tower antenna
铁塔桁架 tower truss
铁塔式天线 mast antenna;pylon antenna;tower antenna
铁踏步 step iron
铁胎搪瓷 enamel(l)ed cast-iron (bath) tub
铁钛闪石 ferrokaersutite
铁弹体 ferro-elastics
铁弹效应 ferro-elastic effect
铁弹性材料 ferro-elastic material
铁弹性的 ferro-elastic
铁弹性晶体 ferro-elastic crystal
铁弹性效应 ferro-elastic effect
铁炭化微电解分析 ferreous charry micro-electroanalysis
铁炭曝气微电解 iron-carbon aeration micro-electrolysis
铁炭水解反应器 iron-carbon hydrolysis reactor
铁炭微电解(法) iron-carbon micro-electrolysis
铁炭微电解混凝沉淀法 ferric-carbon microelectrolysis-coagulation sedimentation process
铁炭微电解生化法 Fe-C micro-electrolysis-biochemical process
铁探头 ferroprobe
铁碳比 iron-carbon ratio
铁碳合金 iron-carbon alloy;pearlite
铁碳化合物 iron-carbon compound
铁碳化铁平衡图 iron-iron carbide equilibrium diagram
铁碳化铁系 iron-iron carbide system
铁碳化物 ferrous-carbide
铁碳磷母合金 pacteron
铁碳平衡 carbon-iron balance
铁碳平衡图 iron-carbon diagram; iron-carbon equilibrium diagram
铁碳相图 iron-carbon phase diagram
铁糖 sugar of iron
铁套管 metal sleeve
铁锑钙石 schneebergite
铁锑化合物 iron-antimony compound
铁锑酸钙石 schneebergite
铁蹄型河谷 cigar-shaped valley
铁体阀 iron body valve
铁天蓝石 iron-lazulite;lipscombite
铁条 bar iron;iron bar;strap iron
铁条格 bar screen
铁条格栅 bar grate
铁条拉门 folding trellis gate
铁条片 sheet bar
铁条起重机 bar iron crane
铁贴体(块) <一种防波堤护面异形块体> akmon
铁挺 crow(bar)
铁铜催化剂 iron-copper catalyst
铁铜合金 iron-copper;iron-copper alloy
铁铜混合粉末 Sintropac
铁铜矿石 Fe-Cu ore
铁铜蓝 idaite
铁桶 iron drum; iron pail; metal drum;steel drum
铁头登山杖 alpenstock
铁头木棍 quarter staff
铁透花釉 rust colo(u)red glaze
铁透闪石 ferrotremolite
铁钍石 ferrithorite
铁托 iron bracket
铁托板 shell feed plate
铁弯刨 spokeshave
铁丸 iron shot;shot iron
铁丸和铁砂 shot and grit
铁丸混凝土 iron shot concrete
铁丸砂分离器 shot separator
铁丸通道 shot gate
铁顽火辉石 protobastite
铁网包坝 wire wrapped dam
铁网笼 cage screen
铁微粒 iron granules
铁微生物 iron organism
铁纹石 alpha nickel-iron;kamacite
铁纹石藻沥青 balkhashit
铁硪 tamping iron
铁污染 pollution by iron
铁屋脊 iron ridging
铁屋面 iron roof
铁钨合金 ferrotungsten
铁钨华 ferritungstite
铁钨硬质合金 iron-cemented tungsten carbide
铁矽尘肺 siderosilicosis; silicosiderosis
铁矽末沉着 siderosilicosis
铁矽末沉着病 siderosilicosis
铁硒合金 ferro-selenium
铁硒铜矿 eskebornite
铁锡石 ferrian-cassiterite
铁系杆 iron tie;iron tie rod
铁系混凝剂 Fe-chain coagulant
铁细菌 <使铁沉积管道的一种丝状细菌> Crenothric;ferribacteria;ferric bacteria;iron bacteria
铁狭条 iron strap
铁线材轧机废水 iron wire mill rinse water
铁箱 <储存泥浆的> steel pits
铁销 gudgeon
铁斜硅镁石 ferrohumite
铁斜拉桥 iron cable-stayed bridge; iron rope stayed bridge
铁携带 iron carryover
铁鞋 brake block;metal shoe;skate
铁鞋缓行器 skate retarder
铁鞋减速器 skate brake
铁鞋制动员 brake-slipper operator
铁屑 floor hardener;hammer scale; iron-carrier particle; iron chippings and shavings; iron chip(ping)s; iron dust;iron filings;iron fittings;iron scurf; nill; scrap iron; iron shavings <车、刨床的>
铁屑槽 scale trap
铁屑地 buckshot land
铁屑骨料 iron aggregate
铁屑骨料混凝土 iron aggregate concrete
铁屑灰泥 sal ammoniac
铁屑混凝土 iron aggregate concrete
铁屑活性炭内电解法 iron chipping-activated carbon interior-electrolysis
铁屑集料 <掺入混凝土中的> ferrolite;iron aggregate
铁屑集料混凝土 iron aggregate concrete
铁屑检查器 sideroscope
铁屑流化床预处理-催化氧化-混凝沉淀组合工艺 combined iron shaving fluidized bed pretreatment-catalytic oxidation-coagulating sedimentation process
铁屑内电解法 iron chip internal electrolysis process;iron chipping interior-electrolysis;scrap iron inner electrolysis
铁屑喷射器 chip injector
铁屑砂浆 iron-filing mortar
铁屑收集装置 scrap collection system
铁屑双氧水氧化法 iron chipping-H_2O_2 oxidation process
铁屑水泥 iron cement;rust cement
铁屑水泥密接头 iron aggregate joint
铁屑团块 iron briquette
铁屑微电解法 iron chip micro-electrolysis
铁屑预处理 scrap iron pretreatment
铁芯 core; iron core; magnetic steel core;mandrel;slug;yoke
铁芯半径 radius of the core
铁芯饱和 core saturation
铁芯变压器 closed core transformer; iron-core transformer
铁芯薄片 core lamination
铁芯材料 iron-core material
铁芯长定子线性电动机 iron-core long-stator linear motor
铁芯磁密 core induction
铁芯电感(线圈) iron core inductance
铁芯叠片 core-lamination stack
铁芯叠片绝缘纸 core disc[disk] paper
铁芯扼流圈 iron-cored reactor
铁芯感应 core induction
铁芯感应线圈 ironcore inductor
铁芯耗 iron core loss
铁芯间隙 interferric space
铁芯开槽 open slot
铁芯绕组 cored winding
铁芯式变压器 core type transformer
铁芯损耗 core loss;iron core loss;iron loss
铁芯损失 core loss;iron loss
铁芯调谐 slug tuning
铁芯涂漆 core plating
铁芯温度跳闸 core temperature trip
铁芯线圈 iron-core coil
铁型梁 preflex beam
铁型箱 iron flask
铁锈 aerugo;corrosion;iron scale;rubigo;rust(ing)
铁锈斑 iron mottling
铁锈粉 iron rust
铁锈红色 rust red
铁锈红涂料 iron red plastering
铁锈花釉 light brown glaze;rust colo(u)red glaze
铁锈黄色 rust yellow
铁锈迹 iron mo(u)ld
铁锈结节 tubercule
铁锈皮 sinter
铁锈色 rust colo(u)r
铁锈色的 ferruginous
铁锈色砂 ferruginous sand
铁锈色釉 iron rust glaze
铁锈试验 ferroxyl test
铁锈水 red water
铁锈水污浊 red water trouble
铁锈污染 iron stain
铁锈指示剂 ferroxyl indicator
铁蓄电池 iron storage battery
铁玄武岩 iron-basalt
铁悬索桥 iron rope suspension bridge; iron suspension bridge
铁靴 stirrup strap;wall hanger
铁循环 ferrikinetics[ferrokinetics]
铁压头 kentledge
铁烟囱 funnel pipe;iron plate chimney; iron smoke tube; iron stove pipe
铁岩 ironstone
铁研式抗裂试验 Tekken type cracking test
铁盐 ferric salt;molysite
铁盐减薄液 ferric salt reducer
铁盐矿泉 chalybeate spring
铁盐类 iron salts
铁盐晒图 ferric cyanide blueprint
铁盐渗透的 chalybeate
铁眼杆悬索桥 iron eyebar suspension bridge
铁阳起石 ferroactinolite
铁氧磁材料 ferrimagnet;ferrimagnetic material
铁氧存储器 magnetic-ferrite memory
铁氧电池 iron-oxygen battery
铁氧化剂 ferrooxidant;iron-oxidizer
铁氧化皮 iron scale;mill scale
铁氧化物 iron oxide
铁氧化物的清除作用 scavenging action of iron hydroxide
铁氧化细菌 iron-oxidizing bacteria
铁氧体 ferrite;oxyferrite
铁氧体板存储器 ferrite-plate memory;ferrite-plate storage
铁氧体棒 ferrite bar
铁氧体棒形天线 ferrite bar antenna; ferrite-rod antenna; ferrod; loop-stick antenna
铁氧体保通片 ferrite keeper
铁氧体波导管 ferrite filled waveguide
铁氧体薄膜 ferrite film
铁氧体薄片 ferrite lamina
铁氧体不锈钢 ferrite stainless steel
铁氧体参量倍频器 ferrite harmonic generator
铁氧体参量放大器 ferrite parameter amplifier
铁氧体处理 ferrite treatment
铁氧体磁棒 ferrite bar;ferrite rod
铁氧体磁轭 ferrite yoke
铁氧体磁放大器 ferractor
铁氧体磁粉 ferro-magnetic oxide powder
铁氧体磁杆存储器 ferrite-rod memory
铁氧体磁环 ferrite bead

铁氧体磁芯 ferrite core
铁氧体磁芯存储器 ferrite-core memory
铁氧体磁性 ferrimagnetism
铁氧体磁致伸缩振动子 vibro(c)s
铁氧体存储器 ferrite memory;ferrite storage
铁氧体存储铁芯 ferrite memory core
铁氧体法 ferrite process;ferrite technique
铁氧体检波器 ferrite detector
铁氧体开关 ferrite switch
铁氧体器件 ferrite device
铁氧体陶瓷 ferrite ceramics
铁氧体调谐元件 ferrite-tuning element
铁氧体微波器件 ferrite microwave device
铁氧体吸收材料 ferrite absorbent material
铁氧体(小)珠存储器【计】 ferrite-bead memory
铁氧体移相器 ferrite phase shifter
铁氧系 iron-oxygen
铁叶腊石 ferripyrophyllite
铁叶绿泥石 delessite
铁叶蛇纹石 ferroantigorite;iron antigorite;jenkinsite
铁叶式安培计 iron vane type ammeter
铁叶式测量仪表 iron vane instrument
铁叶式伏特表 iron vane type voltmeter
铁叶式仪表 iron vane instrument
铁叶云母 eastonite;siderophyllite
铁一碳化铁合金 iron and iron carbide alloy
铁阴极 iron cathode
铁银制品 iron-silver composition
铁英岩 <含铁高达 64% > itabirite;itabaryte;taconite
铁硬铬尖晶石 ferrichromspinel
铁油灰 iron putty
铁油酸酯 iron oleate
铁铀合金 ferrouranium
铁铀云母 bassetite;iron uranite
铁云母 annite;iron mica
铁云母片岩 iron mica schist
铁云母涂料 iron mica paint
铁云片岩 iron mica schist
铁云霞霓辉岩 salitrite-annite jacupirangite
铁陨石 aerosiderite;iron meteorite;meteoric iron;palasite;sider(aer)olite
铁杂质 iron tramp
铁载体 siderophore
铁皂 iron drier[dryer];iron soap
铁渣 crust of iron;dross;dross coal;dry dross;iron cake
铁渣水泥 iron Portland cement
铁扎线 iron binding wire
铁栅 clathri;fence bar;grill
铁栅篦 bar screen
铁栅栏 iron bars
铁栅栏门 folding gate;openwork iron gate
铁栅筛 bar grizzly;bar screen;bar strainer;grizzly;grizzly screen
铁栅(筛栅)条 grizzly bar
铁栅推拉门 grilled sliding door
铁栅网 bar sieve
铁赭色 iron ocher[ochre]
铁赭石 iron ocher[ochre];paint rock
铁赭土 paint rock;paint soil
铁针石 ferrithorite[ferrothorite]
铁砧 anvil;beckern;bick-iron;hammer anvil;stithy
铁砧插模孔 hardy hole
铁砧底座 anvil bed
铁砧垫 anvil cushion
铁砧虎钳 anvil vise[vice]
铁砧角 horn of the anvil beak
铁砧台 anvil block
铁砧状块体 akmon block
铁砧嘴 anvil beak
铁砧座 anvil stand
铁枕 girder block
铁支承 iron-bearing
铁直闪石 ferro-anthophyllite
铁指南针 iron-compass
铁指数 <水泥的> iron modulus
铁制边缘 edge iron
铁制薄板桩 iron sheet pile
铁制部分 iron works
铁制窗 iron window
铁制墩木 girder block
铁制隔断拦截器 iron disconnecting (air) trap
铁制空气拦截器 iron intercepting air trap
铁制垃圾臭气阻闭器 ferrocyanide trap
铁制拦截器 iron interceptor;iron trap
铁制连接件 iron connector;iron timber connector <木结构的>
铁制零件 iron works
铁制楼梯 iron stair(case)
铁制抹灰底 iron lath(ing)
铁制木螺钉 ironwood screw
铁制品 ferrocyanide product;iron article;iron product;iron ware;iron works
铁制器 iron ware
铁制嵌玻璃条 iron glazing bar
铁制嵌轮 thimble
铁制熔锅 iron-melting pot
铁制润滑油注射器 iron oil syringe
铁制型材 iron section
铁制油桶 iron oil drum
铁制阻截空气用存水弯 iron intercepting air trap
铁质 ferruginous
铁质斑彩 iron shot
铁质比率 the percentage of iron-rich matter
铁质薄膜 Iron film
铁质超镁铁岩 ferro-ultramafic rock
铁质沉积物 ferruginous sediment
铁质沉凝灰岩 ferruginous tuffite
铁质沉着 siderosis
铁质沉着病 siderosis
铁质沉着性真菌病 sideromycosis
铁质地层 iron pan
铁质垫圈 iron washer
铁质分离 iron separation
铁质粉砂岩 ferruginous siltstone
铁质硅铝土 ferric siallite;ferric-siallitic soil
铁质红土 ferruginous laterite
铁质环境 ferruginous environment
铁质灰壤 iron podzol
铁质火山灰水泥 ferric-pozzolan cement
铁质夹杂物磁力分离器 tramp-iron magnetic separator
铁质夹杂物排除器 tramp-iron rejector;tramp-iron remover
铁质建造 iron formation
铁质胶合剂 iron cement
铁质胶结 ferruginous cement
铁质胶结层 ferruginous bonding layer
铁质胶结砂 ferruginous bonding sand
铁质胶结物 ferric cement;ferruginous cement
铁质胶料 ferruginous cement
铁质角砾岩 ferruginous breccia
铁质结核 ferruginous concretion;iron concretion
铁质结壳 ferricrust
铁质结皮作用 ferruginous incrustation
铁质壳 ferricrete
铁质矿泉水 ferric mineral spring water
铁质砾岩 ferricrete;ferruginous conglomerate
铁质密致材料 ferrous dense material
铁质泥岩 ferruginous mudstone
铁质黏[粘]土 ferruginous clay
铁质片麻岩 ferriferous gneiss
铁质泉 chalybeate spring;ferruginous spring
铁质缺乏 iron deficiency;sideropenia
铁质溶洞 ferriferous fluxing hole
铁质砂 ferriferous sand;Hastings sand
铁质砂岩【地】 ferruginous sandstone
铁质石灰石 ferruginous limestone
铁质石英砂岩 ferruginous quartz sandstone
铁质水 chalybeate water
铁质水泥 ferritic cement;ferruginous cement
铁质燧石 ferruginous chert
铁质填料 iron filler
铁质土 ferruginous soil
铁质系数 ferruginous coefficient
铁质细菌 <污水中各种铁细菌的总称> iron bacteria
铁质橡胶 iron rubber
铁质岩 ferruginous rock;paint rock
铁质岩相 ferruginous rock facies
铁质页岩 ferruginous shale
铁质硬磐 iron hardpan
铁质砖红壤 ferruginous laterite
铁质砖红壤性土 ferruginous lateritic soil
铁轴棒 axle bar
铁珠 iron shot
铁柱 iron pole;iron prop
铁柱绿泥石 strigovite
铁柱石 ferrocarpholite
铁铸件 iron casting
铁爪 iron claw
铁砖 electric(al) cast mullite brick;grey iron block
铁砖铺砌 iron brick paving
铁转门 metal revolving door
铁桩 iron pile
铁撞柱 tamping iron
铁坠陀 iron balance weight
铁着色玻璃 iron-colo(u)red glass
铁籽 myrtle;myrtus
铁紫 iron violet
铁紫苏辉石 ferrohypersthene
铁棕(色) dark brown;iron oxide brown;ulmin brown;Vandyke brown
铁棕颜料 iron brown pigment
铁足 <青瓷的底脚> iron foot
铁族 iron group
铁族元素 iron-family element;iron group elements
铁阻气 iron-air trap
铁钻杆 driving iron
铁钻矿 wairauite
铁座 iron seat

厅

厅式大楼 hall-type building

厅式公寓 hall system apartment
厅式建筑 hall-type building
厅式平面(各室人口开向大厅的)布置方式 hall access type
厅堂的声学模型分析 acoustic(al) model(l)ing
厅堂钢门 hall steel door
厅堂建筑 hall building
厅堂式房屋 hall-type block;hall-type building
厅堂屋顶 hall roof
厅堂屋(顶桁)架 hall roof truss
厅堂翼部 hall transept
厅堂噪声 hall noise
厅堂座椅排列 auditorium seating
厅长 director general
厅座 baignoire

汀

汀太特 <一种专利的蒸汽活塞阀> Twintite

听

听测声呐 listening sonar

听到的 auditory
听到的声波频率 audio frequency
听得到的信号 audible call
听地器 geophone
听度表 audibility meter;audiometer[audiometre]
听度测量 shunt telephone measurement
听度级 level of audibility
听度计 audibility meter;audiometer[audiometre]
听度器 audiometer[audiometre]
听感因数 acoustic(al) condition factor
听话线 listening-in line
听讲堂 audience hall
听觉保护 hearing conservation
听觉不良 hard of hearing
听觉测向 auditory direction-finding
听觉范围 audible range;auditory area;auditory sensation area;range of audibility;range of hearing;sensation area
听觉防护器 hearing protector
听觉警报 audible alarm
听觉警告信号 acoustic(al) warning signal
听觉距离 ear reach;earshot
听觉力 auditory sense modality
听觉临界 threshold of audibility;threshold of hearing
听觉零点 aural null
听觉阈图 audiogram
听觉阈值 sensation level
听力 acuity of hearing;audition;auditory acuity;hearing ability
听力保护 hearing conservation;hearing protection
听力保护器 hearing protector
听力测定 audibility test
听力测定法 audiometry
听力测验器 acoumeter
听力范围 earshot
听力防护器 ear protector;hearing protector
听力计 acoumeter;acousimeter;audibility meter;sonometer
听力器 audiometer[audiometre]
听力损失 deafness;hearing loss
听力损失程度 deadness
听力损失率 deafness percent
听力图 audiogram;threshold audiogram
听力域 hearing threshold
听漏器 aquaphone
听录机 dictaphone[dictophone]
听录音打字 audiotyping
听敏度 acuity of hearing;auditory acuity
听取汇报 debriefing
听取请愿人申诉 hearing of petitioners
听任处理 placing at disposal

听任使用的房屋 disposable house
听声法 <检查地下管道漏水> listening method
听声检查 audible inspection
听声检验 sonic inspection
听筒 earphone; ear piece; eartrumpet; hand receiver; head set; receiver; stethoscope; telephone receiver; watch receiver
听筒壳 receiver housing
听筒软垫 ear pad
听筒软线 phone cord
听筒塞孔 phone jack
听筒塞子 phone plug
听筒托架 receiver hook; telephone hook
听野 auditory field
听音辨光器 optophone
听音器 audiphone; detectagraph; listening apparatus; listening device; listening gear
听音器插头 listening plug
听阈 audibility threshold; auditory threshold; threshold detectability; threshold of audibility; threshold of detectability; threshold of hearing; zone of audibility
听阈变化 auditory threshold shift
听阈频率范围 audio-frequency range
听者 auditor
听证会 evidentiary hearing
听知觉 aural perception
听众 audience; auditory
听众席 auditorium[复 auditoria]; nave
听字饰 swastika

烃 hydrocarbon

烃饱和度 hydrocarbon saturation
烃合成干性油 hydrocarbon drying oil
烃黑 hydrocarbon black
烃化 alkylation
烃化合物 hydrocarbon compound
烃化剂 alkylating agent
烃化理论 hydroxylation theory
烃换硫酸 sulphovinic acid
烃混溶驱动 hydrocarbon miscible flooding
烃基 alkyl(group); alkyl radical; paraffin(e) base
烃基氟硅烷 organofluorosilane
烃基化合物 alkyl compound
烃基硫酸 sulfovinic acid
烃基烷氧基硅烷 organoaroxy silane
烃降解菌 hydrocarbon degradation bacteria
烃类 hydrocarbon family
烃类爆炸极限 explosive limits of hydrocarbon
烃类的含氢指数 hydrogen index of hydrocarbon
烃类的视密度 apparent density of hydrocarbon
烃类分解细菌 hydrocarbon utilizing bacteria
烃类分析 hydrocarbon analysis
烃类分析自动记录仪 hydrocarbon automatic recorder
烃类丰度 abundance of hydrocarbon
烃类合成 hydrocarbon synthesis
烃类检测技术 hydrocarbon indication technique
烃类检测剖面 hydrocarbon indicator section
烃类结合料 hydrocarbon binder
烃类燃料 hydrocarbon fuel
烃类溶剂 hydrocarbon solvent
烃类树脂 hydrocarbon resin
烃类塑料 hydrocarbon plastics
烃类中的氢含量 hydrogen content of hydrocarbon
烃类组成 composition of hydrocarbon
烃利用微生物 hydrocarbon-utilizing microorganism
烃裂化 hydrocarbon cracking; cracking of hydrocarbons
烃裂解 hydrocarbon cracking
烃硫金属 thiolate
烃齐聚物 hydrocarbon oligomer
烃气 hydrocarbon gas
烃气成因类型 genetic(al) classification of hydrocarbon gas
烃燃料 hydrocarbon fuel
烃属酸 acetylene acids
烃污染 hydrocarbon contamination
烃系 hydrocarbon series
烃氧化微生物 hydrocarbon-oxidizing microorganism
烃氧基金属 alcoholate; alkoxide
烃氧基乙酸 carboxymethyl ether
烃油 hydrocarbon oil
烃源成因联系 genetic(al) relation of hydrocarbon and source
烃源岩的密度 density of source rock
烃源岩的体积 volume of source rock
烃源岩的岩石类型 lithologic(al) classification of source rock
烃源岩地理分布 geographic(al) distribution source rock
烃源岩分析方法 analytic(al) method of source rock
烃源岩及其评价 hydrocarbon source rock and its evolution
烃源岩经历的最高温度 exposure maximum temperature of source rock
烃源岩类型 type of hydrocarbon source rock
烃源岩评价 evaluation of source rock
烃源岩时代和层位分布 distribution of source rock in geologic(al) age and strata
烃源岩相分布 facies controlled source rock
烃源岩一般特征 general characteristic of source rock
烃源岩中有机质 organic matter of source rock
烃蒸气转化炉 steam hydrocarbon reformer
烃质谱 hydrocarbon spectrum
烃柱高度 height of hydrocarbon column
烃转化 hydrocarbon conversion
烃转化率 transformation ratio of hydrocarbon
烃族组分分析 hydrocarbon type analysis

廷布 <不丹首都> Thimbu

廷德尔冰花 Tyndall flowers
廷德尔测尘计 Tyndall meter
廷德尔法 tyndallimetry
廷德尔光 Tyndall light
廷德尔计 tyndallimeter; Tyndall meter
廷德尔亮锥 Tyndall cone
廷德尔灭菌法 tyndallization
廷德尔散射光 Tyndall scattering light
廷德尔现象 Tyndall phenomenon
廷德尔效应 Tyndall effect
廷德尔悬浮体测定法 Tyndallometry
廷德尔悬浮体浓度计 Tyndallometer; Tyndalloscope
廷斧石 tinzenite
廷磷钾铝石 tiptopite
廷挖带运的 cut-and-carry

亭 pavilion; t'ing <中国>

亭暗箱 booth
亭可马里暗红色硬木 trincomalee
亭桥 pavilion bridge
亭式屋顶 pavilion roof
亭子 alcove; kiosk; pavilion
亭子般的 pavilion-like

庭外和解 settle-out of the court

庭荫树 shade enduring plant; shade tolerant tree; shade tree
庭园 curtilage; flower garden; garden; garden court; garth; house garden; outdoor garden; pleasance
庭园布置 landscaping
庭园长椅 garden bench
庭园灯 garden lamp
庭园凳子 garden bench
庭园地 garden plot
庭园雕刻装饰 ornamental vessel
庭园雕像 garden statuary
庭园废物 yard rubbish; yard waste
庭园风景工程 landscape engineering
庭园工程 landscape work
庭园家具 garden furniture; outdoor furniture
庭园建筑 ornament architecture
庭园建筑单元 garden building unit
庭园建筑物 garden building
庭园建筑学 garden architecture
庭园垃圾 yard rubbish; yard waste
庭园露台 garden village
庭园喷头 garden sprinkler
庭园青草地 grass plot
庭园洒水器 garden sprinkler
庭园设计 garden design
庭园设计师 landscaper; landscapist
庭园式喷灌机 garden sprinkler
庭园术 garden craft
庭园陶瓷制品 garden ceramics
庭园舞台 garden terrace
庭园小径 alley
庭园小品 garden furniture
庭园小品点缀 garden ornament
庭园用的小型抽水机 garden engine
庭园住宅 garden apartment
庭院 court; courtyard; front stead; garth; house garden; patio; yard; apodyterium <古罗马、古希腊的>
庭院便门 garden wicket
庭院带格栅下水道进口 yard gull(e)y
庭院道路 court pavement
庭院废物 yard waste
庭院花园 courtyard garden; courtyard house; home garden
庭院集水井 yard catch basin
庭院建筑 court architect
庭院截流井 yard catch basin
庭院经济 courtyard economy; truck farming
庭院刻物 garden sculpture
庭院空地 yard space
庭院宽度 width of court
庭院垃圾 yard rubbish
庭院门 patio door
庭院门零件 patio door fitting
庭院门配件 patio door furniture
庭院门上小五金件 patio door hardware item
庭院门装配 patio door fitting
庭院木门 wood patio door
庭院排水 yard drainage
庭院排水沟 yard drain; yard gull(e)y
庭院排水口截污设备 yard trap
庭院砌块 patio block
庭院入口 court entrance
庭院生态系统 ecosystem of courtyard
庭院式窗 patio door
庭院式建筑 court building
庭院式住宅 courtyard house
庭院术 garden craft
庭院圬工墙 patio masonry wall
庭院雨水井 yard catch basin
庭院雨水口 yard gull(e)y
庭院遮篷 patio awning
庭院遮阳伞 patio umbrella
庭院住宅 house with court(yard)

停班 idle shift

停板订单 limit order
停板价 limit up or down
停办企业 relinquishing of business
停闭 close out; lockout
停闭电路 shut-off circuit
停闭信号 stoping signal
停避机场 refuge aerodrome; refuge airport
停臂 detent plate stop
停变期 standstill
停表 second stop watch; stopclock; stop-wat; stop watch
停播时间 off-air time
停泊 berthing <指靠码头作业>; anchor(age); call at a port; laid up <指锚泊作业>; lie at anchor in a harbor; mooring <指系泊作业>; tied-up
停泊场 berth space; moorage; sea berth
停泊池 anchorage basin
停泊处 anchorage; barge berge; berth; dock; harbo(u)rage; harbo(u)r ago; moorage; moorings; tie-up; wharf
停泊处水位 quay level
停泊船 idle ship
停泊待命 tie-up; lay-by
停泊灯 anchor light; moor light; riding light; stem light
停泊地 anchorage; bay; berth; lying-up basin; roadstead
停泊地海图 anchorage chart
停泊地平面图 berthing plan
停泊点 mooring point
停泊吨位天 tonnage-day in port
停泊发电机 port duty generator
停泊发电机组 port duty generating set
停泊发动机组 harbo(u)r generating set
停泊费 berthage; berth charges; dockage; docking charges; docking dues; groundage; harbo(u)r dues; keelage; moorage; tie-up basin; dock charges; dock dues
停泊负载 port load
停泊港 port of anchorage; port of call
停泊管理员 parking tender
停泊器具 <锚、锚链等总称> ground tackle
停泊区 anchorage; berthage space; berthing area; berthing room; berthing space; mooring basin
停泊权 rights of mooring; shore rights
停泊日记 port log
停泊日期 lay days
停泊日志 port log
停泊时机舱值班员 donkeyman
停泊时间 berth time; lay days; laytime

停泊时值班水手 deck watchman
停泊税 anchorage dues; mooring charges; keelage <英>
停泊所 anchorage; moorings; quay; roadstead
停泊提示 berthing note
停泊条款 berthing clause; in regular turn clause; turn berth clause
停泊退费 laid-up return; layup return
停泊销 harbo(u)r pin
停泊延期 days of demurrage
停泊用泵 harbo(u)r pump
停泊在港 laid-up in port
停泊值班 harbo(u)r watch
停泊周旋余地 berthing room; berthing space
停泊装置 mooring device
停泊撞击力 berthing impact
停埠船运费率 berth cargo rate
停采矿区 stopping mining area
停测声呐 listening sonar
停产 closedown; off production; phase-out production; production halt; shut-down; stop production; suspend production
停产成本 shut-down cost
停产井 shut well
停产期 idling period
停产时间 downtime; idling period; no-productive time
停产整顿 stop operation and undergo shake-up
停潮 standing water; stand of tide; still tide; tidal stand; water stand; slack water
停车 bring to rest; engine cut-out; final cut-off; halt; hold-up; motor stoppage; park; power-off; shut-down; stabling; stall; stop; depriming <电动机的>
停车按钮 stop button
停车饱和度 degree of parking saturation
停车保险 waiting insurance
停车比率 <商业建筑的> parking ratio
停车臂板 stop arm
停车标 stop sign
停车标签 <一种贴于车窗上写明车辆到达及应开走时间的圆形标签纸> parking disc
停车标志 <指示停车方式、时间、车型等或禁止停车> parking sign; buffer stop indicator; halt sign; stop(ping) sign; stop signal; train stop mark【铁】
停车标志街 <车辆进入主干道必须先停车的支路> stop sign street; parking sign street
停车标准 parking standard
停车表示灯 stop light
停车波 stopping wave
停车不足 parking deficiency
停车布置 parking allocation; parking arrangement; parking configuration
停车层 parking floor; parking tier[tyer]
停车场 park(ing block); parking ground; parking lot; parking place; parking space; car dump; car park; hard standing; hold yard; motor court; motor pool; official parking area; standing; stopping place; vehicle depot; vehicular parking (area); wagon yard; car parking; parking; holding yard【铁】; park-and-ride lot <停好自备汽车后换乘火车的>
停车场标志 parking area sign
停车场标准 parking standard
停车场存车累计数 parking accumulation
停车场地 parking area; parking apron; parking compound; parking point; parking space; standing area
停车场调查 parking interview
停车场调查法 parking interviews method
停车场机械设施 parking structure
停车场技术档案 parking inventory
停车场技术档案调查 parking inventory study
停车场建筑 parking structure
停车场进出口 parking access
停车场清册 parking inventory
停车场区 parking area
停车场上划分好的汽车停放位置 car shall
停车场位置 location of parking area
停车场现状 parking inventory
停车场现状调查 parking inventory study
停车场选址 location of parking facility
停车场询问法 parking interviews method
停车场周转率 parking turnover rate
停车场周转数 parking turnover
停车车道 <路上的> pull-off lane; parking lane
停车车位 car place; parking set; parking space; parking stall
停车车位布置 parking layout
停车车位尺位 parking dimension
停车车位需求量 parking need
停车车影 <车辆前的一段距离,等于其在某已知制动速率下的停车距离> stopping shadow
停车持续时间 parking duration
停车处 car parking site; car stop; parking (apron); parking place; parking point; pull-up; standing; vehicle depot; waiting place
停车处的让车岔道 parking turnout
停车窗 stopping window
停车次数计数器 stop counter
停车带 <路上的> pull-off strip; parking bay; parking strip
停车待发 stopping for departure
停车待接 stopping for being received
停车挡板 stop bar
停车岛 <供公共车辆上下乘客用的> loading island; stop island
停车道 parking apron; parking lane; stopway; pull-off strip <路上的>
停车道坡度 parking grade
停车的 unoperated
停车的价值当量 cost of stops
停车灯 parking lamp; parking light; stop lamp; stop light
停车灯和牌照灯线 stop-and-license plate light cable
停车灯及车尾灯结合 stop-and-tail lamp combination
停车等待信号 stop-and-stay signal
停车地带 parking strip
停车地点 parking lot; parking point
停车地段 parking lot; parking point
停车点 break point; halting point; stopping point; target point <溜放车辆的>
停车点冲挂 target shoot
停车点制动 target point braking
停车电控阀 shut-down electro-pneumatic valve
停车调查 parking census; parking study; parking survey
停车定额 parking standard
停车动作 stop motion
停车段车道 parking lot lane
停车罚款 parking fine
停车阀 engine stop valve; shut-down valve; shut-off valve
停车方式 parking module
停车防冲装置 stop bumper
停车防滑器 <坡道上汽车> hill holder
停车房 car park building
停车费 parking fee
停车服务 parking supply
停车负荷 parking load
停车杆 shut-off rod; stop rod; throw-out lever
停车缸 shut-off cylinder
停车港 parking bay
停车杠杆 engine stop lever
停车供应量 parking supply
停车股道 occupied track
停车固轮器 rim holder
停车管理 stop control
停车管理程序 parking management program(me)
停车管制 stop control
停车管制区 parking control area
停车广场 parking square
停车规则 parking ordinance; parking regulation
停车轨道 hold track
停车和牌照灯的组合灯 stop-and-license plate light
停车和尾灯的组合灯 stop-and-tail light
停车横条 stopping line
停车红灯信号 stop-on red signal
停车后穿越空 stopped gap-acceptance mode
停车后再进信号 stop(-and)-go signal
停车后再前进灯光 stop-and-go light; stop-and-proceed light
停车后再前进示像 stop-and-go aspect; stop-and-proceed aspect
停车后再前进显示 stop-and-proceed indication
停车后再前进信号 stop-and-go signal; stop-and-proceed signal
停车后再行信号 stop(-and)-go signal
停车滑行距离 stopping distance
停车滑行时间 stopping time
停车缓冲器 stop buffer
停车换乘 <一种交通乘行方式,乘客驾车到公交站,将其汽车存放在车站的停车场并换乘其他公交车辆> park-and-ride system
停车机构 stop mechanism
停车计费表 parking meter
停车计划 parking plan
停车计时器 parking meter; parking time meter
停车计数法 <观测交叉口延误及停车率的一种方法>【交】 stopped vehicle counts
停车继电器 plugging relay; shut-down relay; stop train relay
停车加宽段 parking bay
停车价值 cost of stops
停车架 jiffy stand
停车间隔 clearance at parking; cut-off interval; stall
停车间隔时间 stop interval
停车间距 parking space
停车检修日 parking maintenance day
停车交通口 stop street
停车角度 parking angle
停车接触器 shut-down contactor
停车街道 parking street
停车结构 parking structure; structure parking
停车界限标 stopping limit mark
停车禁令 parking prohibition
停车精度 stopping accuracy
停车净空限界 parking space limit
停车距离 stopping distance; vehicle stopping distance
停车距离试验 stopping distance test
停车开关 shut-down switch; shut-off cock; shut-out cock; stop motion switch
停车可能性 possibility of parking
停车空地 parking space
停车控制区 parking control area
停车库 automobile parking structure; parking garage; stabling shed
停车库出空时间 dump time
停车廊 <入口处> carriage porch
停车累计量 parking accumulation
停车楼 parking structure; parking tower
停车楼层 parking stor(e)y
停车楼面 parking deck
停车露天电影场 drive-in theater[theatre]
停车率 stop rate
停车门廊 carriage porch; porte cohere
停车密度 cut-off concentration
停车模式 parking module
停车排列方式 parking arrangement; parking configuration
停车牌 red board; stop board
停车盘点总数 parking inventory
停车棚 car parking roof; carport; car shed; parking shed; traffic shed
停车票证自动发放机 automatic parking ticket issuing machine
停车平台 landing; parking deck; parking terrace
停车凭证 parking voucher
停车坪 parking deck; stopping pad
停车起步试验 stop-start test
停车-起动频率 stop-start frequency
停车器 boundary member; scotch; stopping device; wagon arrester
停车器轭 Scotch yoke
停车区 parking area; parking block; parking lock; parking lot; stabling zone; zone of parking
停车区段 stop section
停车(区间)限界标示 parking space limit mark(ing)
停车区域 <交叉口的> stop-light area
停车容量 parking capacity
停车肉眼距离 sight distance for stopping
停车设备 parking facility
停车设施 parking facility
停车升降机 parking lift
停车剩余 parking surplus
停车时间 cut-off interval; idle period; parking duration; parking time; run-out time; stop(ping) time; stopped time <旅程中由于其他交通关系而停车的时间>; parking period <征收基本停放费的>
停车时限 parking time limit
停车实数 parking demand
停车示像 stop aspect
停车视距 non-passing sight distance; sight stopping distance; stopping sight distance
停车试验 cut-off test
停车收费表 <短时> waiting meter
停车收费计 parking meter; waiting meter
停车收据 parking voucher
停车手柄 shut-down handle
停车数 <因交通信号的停车> number of starts
停车伺服马达 shut-down servomotor
停车速度 cut-off velocity
停车损失 stop penalty
停车塔 parking tower

停车塔楼 parking tower
停车通道 parking aisle
停车位 parking bay; parking location; parking set; parking stall; vehicular stall
停车位清单 parking inventory
停车位置 on position; parking space; stop(ping) position
停车位置计算 stopping position calculation
停车位置控制 stopping position control
停车位置牌 stop location board
停车位置指示装置 stopping position indicator
停车屋顶 car parking roof; parking roof
停车吸引量 parking attraction volume
停车系统 parking system; shut-down system
停车显示 stop indication
停车显示点 stop indication point
停车显像 stop indication
停车线 car stop; holding track; parking line; park-lane; stabling siding; standing siding; stop line
停车线到达图 stop-line arrival profile
停车线排队 stop-line queue
停车线驶离图 stop-line departure profile
停车线延误 stop-line delay
停车限界标志 parking space limit mark(ing)
停车限制 parking restriction
停车限制点信号 stabling limit signal
停车小院(子) parking court
停车效率 stopping distance
停车斜面 parking ramp
停车信号 cut-off signal; parking signal; shut-down signal; stop(ping) signal
停车信号灯 stop lamp; stop light
停车信号杆 halt signal baton
停车信号机前方的轨道电路段 berth section
停车信号圆牌 stop signal disc
停车信号圆盘 stop signal disc
停车性能 stopping performance
停车需求量 parking demand
停车需要量 parking demand
停车循环 stopping cycle
停车延误 idling delay; stop delay; stopped delay; stopped-time delay
停车延续时间 parking duration; parking load
停车液压操纵阀 shut-down hydraulic pilot valve
停车以保持闭塞区间隔 stop to maintain block distance
停车余隙 clearance at parking
停车逾时 penalty period
停车预订【交】 park booking
停车院子 parking compound
停车运动 disengaging movement
停车闸 deadman control; parking brake
停车闸操纵杆 parking brake control
停车闸凸轮 parking brake cam
停车闸瓦 parking brake shoe
停车站 parking station; stop; stopping place; major stop < 乘客很多的 >
停车站和停车点的位置 station and stopping locations
停车站台 car stop
停车站台设备 queue arrangement
停车站线 holding yard
停车振动带 rumble strip
停车指数 parking index
停车指针 stop finger
停车制动 braking to a stop; stop(ping) braking
停车制动杆 brake parking lever
停车制动器 parking brake
停车周转率 parking turnover rate
停车轴 knock-off shaft; shut-off spindle
停车柱塞 shut-down plunger
停车转步行 < 停车结合步行的一种方式 > park and walk
停车装卸台 landing
停车装载标杆 spot log
停车装置 arresting gear; parking facility; shutting-down device; stop(ping) device; stop(ping) motion
停车着陆 stopped engine landing
停车自理 self parking
停车阻力 stopping resistance
停传 closedown
停船场 lay-bay
停船池 anchorage basin
停船处 ship berth
停船段 lay-bay
停船港 boat harbo(u)r
停船港池 laying-up basin; laying-up berth
停船滑行距离 stopping distance
停船甲板 dock floor; pontoon deck
停船码头 berthing dock
停船试验 stopping test
停船坞 berthing dock
停船效率 stopping distance
停吹 blow-off; blowout
停吹气压 blow-off pressure
停打阻力 home
停待时间 stopping and waiting time
停点挡块 end stop
停电 blackout; current absence; current failure; current interruption; cut-off; electric(al) power outage; interruption; lack of current; outage; power cut; power failure; service interrupt(ion); switch out
停电报警信号 current failure alarm
停电表示灯 power-failure indicator
停电表示器 power-off indicator
停电继电器 power-failure relay; power-off relay
停电警报 current failure alarm
停电警告信号 current failure alarm signal
停电率 outage rate
停电时间 power-off time; outage time
停电事故 power outage
停电线路 deadline
停电信息不丢失随机存储器 non-volatile random memory
停电用(电源转换)继电器 power-off relay
停电振打 power-off rapping
停动连接 knock-off joint
停读时表 hack chronometer
停堆 shut-down
停堆按钮 shut-down switch
停堆棒 shut-down rod
停堆操作 shut-down work
停堆程序 shut-down procedure
停堆反应性 shut-down reactivity
停堆放大器 shut-down amplifier; trip amplifier
停堆功率 shut-down power
停堆控制元件 shut-off member
停堆冷却泵 shut-down cooling pump
停堆冷却器 shut-down cooler
停堆日程 shut-down schedule
停堆通道 shut-down channel
停堆系统 reactor shut-off system
停堆信号 shut-down signal; shut-off signal
停堆余热 shut-down heat
停堆裕度 shut-down margin
停顿 aberrance [aberrancy]; breakdown; break off; deadlock; halt; hold-up; paralysis; quiescing; stalemate; standstill; stop; tie-up
停顿控制 pause control
停顿时间 dead time
停顿政策 standstill policy
停放 park
停放场所 < 汽车 > parking stall
停放车辆起讫点调查 parking origin-destination survey
停放车起讫调查 parking origin-destination survey
停放的车辆 parker
停放的汽车 parked vehicle
停放分析 analysis of materials' placement
停放好自备汽车后换乘火车 park-and-ride
停放机构 stop-and-release mechanism
停放汽车信号灯 parking light
停放时间 storage period
停放拖车的可调节支腿 parking stand
停放(无人照管的)车辆地点 parking lot
停放支架 parking leg
停放制动 parking brake
停风 blowing down; blowing-out; off-blast; shut-down; blowout < 鼓风炉 >
停风期 off-blast period
停付 respite; stoppage
停付利息 stoppage of interest
停付支票 stop a check; stop payment on a check
停付资金的恢复支付 resumption of payment of suspended funds
停港时间 lay days; port time; time-in port; terminal time
停港时供气或者供电 port service
停港用发电机 port duty generator
停港用给水泵 port feed pump
停港作业泵 port pump
停给 cut-off
停工 aid-off work; cease work; industrial action; job action; laid-off work; laying off; out-of-work; quit work; run-out of work; shut-down; stand-down; stand-off; standstill; stoppage; stoppage of work; stop work; suspend work; suspension of works; work stoppage
停工成本 inactivity cost; shut-down cost
停工待料 work being held up for lack of materials
停工的工厂 standing factory
停工的机器 standing machine
停工费用 idle time cost; shut-down expenses
停工工时 man-hour in idleness
停工减产 suspending operation or curtail production; suspending operation or slashing production
停工检查 shut-down inspection
停工检修 shut-down breakdown; shut-down maintenance
停工检修时间 down period; stream-to-stream time < 上一次停工与下次开工之间的时间 >
停工津贴 allowance for work stoppage
停工率 downtime ratio
停工命令 stop work order
停工期 downtime
停工期工资 idle time pay
停工期工作计划 shut-down schedule
停工期间 lay-off
停工日 downtime day; lay days; no-productive time day; shut-down day; stoppage day
停工时间 bad time; breakdown period; breakdown time; broken time; dead time; dwell time; idle time; knock off; lie time; lost time; outage time; shut-down period; standing time; time-out; downtime < 特指机器设备等 >
停工时间成本 idle time cost
停工时间的会计处理 accounting for idle time
停工时期 shut-down period; stoppage period
停工时期工资 idle time pay
停工事故 lose-time accident
停工损失 idle cost; loss in the suspension of work; loss on idle time; stand-by loss; work stoppage loss
停工维修 shut-down maintenance
停工小时 stood off hour
停工指令 stop work order
停工指示 quitting instruction of work
停工装置 off-stream unit
停棺处 mortuary block
停航 immobilize; lay by(e); navigation pause; suspension of shipping/air service
停航保险 port risks insurance
停航船舶 laid-up tonnage; unemployed ship
停航期 air/shipping service suspending period; non-shipping period
停航退费 lay up refund
停航闲置船 idle ship; laid-up ship
停号桥 signal gantry
停火线 armistice line; cease-fire line
停机 closedown; closing down; complete shut-down; engine cut-off; halt; machine halt; machine halting idle; machine hard stopping; parking; shut-down; stop; stoppage; trip-out; stop calculation【计】
停机按扣 engine stop push-button
停机按钮 engine stop push-button; stop button
停机场 hard stand
停机程序 pause program(me); shut-down procedure
停机道 stopway
停机地址 stopper
停机阀 shut-down valve
停机方式 halt mode
停机费用 idleness expenses; idle time cost
停机分析【计】 downtime analysis
停机钩 arrester hook
停机号 halt number
停机后输出【计】 postmortem dump
停机滑动长度 sliding length after stop
停机回路 stop loop
停机键 stop key
停机接触器 shut-down contactor
停机开关 halt switch; shut-down switch; stop(ping) switch
停机控制程序 stop-controlled sequence
停机拦截网 safety barrier
停机率 outage rate
停机码 stop code
停机码头 aeroquay
停机面最大挖掘半径 maximum cutting radius at ground level; maximum reach at ground level
停机命令 halt command
停机坪 aircraft parking apron; aircraft parking(area); apron; apron area; hard stand(ing); park; parking apron; parking ramp; standing apron; standing area
停机坪标志 apron mark(ing)

停机坪泛光照明 apron flood lighting
停机坪闪光 berge
停机坪设备 ramp equipment
停机坪系统 aircraft parking system
停机期 inoperative period
停机区 landing zone;parking apron
停机日 no-productive time day;stoppage day
停机时间 downtime;floor time;idle hours;idle time;machine downtime;off-time;pause period;stop time;unused time;outage time
停机时间比率 downtime ratio
停机时间控制 rest time control
停机时间率 downtime rate
停机时期 outage;shut-down period
停机台时 machine idle hour
停机凸轮 knock-off cam
停机问题 halt(ing) problem
停机握手 stop-engine button
停机系统 shut-down system
停机线 alignment bar;holding line;parking ramp
停机橡胶 cut-off rubber
停机橡皮 cut-off rubber;cutting buffer;cutting strip
停机信号 silence signal;stopping signal
停机站台 aeroquay
停机指令 halt instruction;stop instruction
停机指示器 halt indicator;stop indicator
停机重量 ramp weight
停机装置 arrester [arrestor];engine shutdown device;shut-down device
停机状态 halt mode;outage;stopped state;stopped status
停机总计 down total
停建缓建项目 ceased and deferred projects;deferred project;projects were suspended or deferred;suspended or concealed projects
停建矿区 stopping building area
停建项目 canceled project;ceased project
停井 well-off
停柩门道 lich gate
停开的机器 standing machine
停开关 guard's van valve
停开循环 on-off cycling
停刊的 defunct
停靠 alongside;call at;call on;lay alongside;touch at
停靠表 parking meter
停靠港 <沿途> port of call
停靠列车 parking a train
停靠码头费 wharfage
停靠时间 berthing time
停靠拖船 berthing tug(boat)
停靠拖轮 berthing tug(boat)
停靠信号 berthing signal
停靠在码头的船只 berthed vessel;docked vessel
停靠站 parking station;bus bay <公共车辆>
停拉 up behind
停流点 dead point
停流法 stopped-flow method
停留 continuance;detention;standing;stay;stoppage;tarry
停留按钮 locking button;press button;push-button
停留半衰期 residence half-time
停留车 standing car
停留车辆 waiting vehicle
停留池 detention tank
停留费 charges for detention at station
停留机 looper
停留机车 standing locomotive
停留期 period of retention
停留期间 stationary period
停留时间 detention period;detention time;dwell time;hold-up time;residence time;resident time;resistance time;retention period;retention time;standing time;stay(ing) time;stop time
停留时间分布 residence time distribution
停留时间显示器 residence time indicator
停留时期 detention period;retention period
停留损失 stand-by loss
停留天数 <不能使用或未动用的天数> idle days
停留条款 touch and stay clause
停留调时阀 tarry valve
停留线 pause line;stabling siding
停留线路 inactive line
停留周期 retention period
停漏 leak detection by listening
停炉 blowing-out;blow-off;blowout;blowout the furnace;closedown;furnace shut-down
停炉期 shut-down period
停煤燃烧 dead bank
停门器 door stay
停秒 second-hand shop;stop-second
停秒杠杆 blocking lever
停炮场 gun park
停启程序 terminator/initiator
停气阀 stop(ping) valve
停气调节 cut-off governing
停汽 occlusion
停汽车车位 car place
停(汽)车处 car park
停汽点 point of exhaust closure
停汽阀 closing valve;cut-off value;steam cut-off valve;steam stop valve;stop(ping) valve
停塞阻抗 blocked impedance
停闪频率 flicker-fusion frequency;fusion frequency
停尸房 dead house;lich-house;mortuary
停驶 stagnation of movement
停驶车辆 <统计车辆总数时用> stalled vehicle
停驶车日 non-operative vehicle-days
停水 service interrupt(ion);stop of water supply
停水事故 water failure
停水位 standing level;standing water level;stationary level;still level
停松 avast veering
停停开开(运行) stop-and-go operation
停停走走 stop-go
停停走走运行 stop-and-start operation
停停走走状态 <指车流> stop-go condition
停筒 head set
停拖 avast hauling
停息 sputter
停息段 suspension period
停息痕 resting mark
停息迹 resting trace
停显液 shortstop;stop bath
停歇 closedown;closure;lay-off
停歇机器时间 idle machine time
停歇时间 dead time;downtime;off-time;pause period;standing time;stopping time;unemployed time
停薪留职 leave without pay;remaining at post without wage
停休时间 inaction period
停修工期 downtime
停修时间 maintenance downtime
停蓄期 period of retention
停悬 hover
停延时间 dwell time
停窑 furnace shut-down;kiln shutdown;shut-down kiln
停业 business closed;business suspended;cessation of business;close a business;closedown;closure;fold up(wards);out of business;suspension of business;termination of business;tie-up;winding up;wind-up of suspense
停业保险 use and occupancy insurance
停业的 defunct
停业公司 defunct company
停业清理 go into liquidation
停业清理(大)拍卖 liquidation sale
停业日 closing day
停业时期 shut-down period
停业损失 loss from suspension;stoploss
停业削价出售 close business sale
停业整改 stop routine work and rectify
停业状况 close-down case
停用 block up;deactivate;non-use;out-of-commission;out-of-service;turn-off
停用标志 not-in-use sign
停用程序 dead program(me)
停用触点 inoperating contact
停用的 off-stream;unused
停用的工作面 inactive face
停用的跑道 inactive runway
停用管线 off-stream(pipe)line
停用卡片 dead card
停用设备 idle unit;off-stream unit
停用时间 outage time
停用外存储器 inactive file
停用文件 dead file
停用一条管路 block a line
停淤场 <泥石流> deposit(e) site of sludge;mud avalanche retarding field
停运 off-stream;off-the-line
停运的 non-operating;out of action;out-of-operation
停运列车 train withdrawn from schedule
停运期 shut-down period;shut-down phase;shut-down stage;shut-down time
停运权 stoppage in transit
停运时间 idle period;idling period;off period
停运油管线 inactive filled line
停运装置 idle off-stream unit
停运状态 down state;shut-down condition
停在道岔限界外 stop clear of the switch
停在港外 lay-off
停在进站信号机近前 stop short of home signal
停在零位 zero dead stop
停战 truce
停站距离 distance between stops
停站区 station stop zone
停站时分 dwell time
停站时间 station dwell(ing) time;station stop time;terminal time
停针计时表 stop watch
停振 failure of oscillations
停止 blow over;break-up;caging;cease;cessation;close up;come to a stand-still;die out;discontinuance;dwell(ing);halt;interruption;lay-off;phase down;quiescing;run down;shut;standstill;stand stop;stoppage;stop(ping);suspension;trip-out;waive
停止按钮 stop button
停止办公 stop working
停止标志 stop signal
停止标柱 stop post
停止不用的 out-of-work
停止操作 shut-down operation;stopping the service
停止操作指令 no-operation;no-op(s)
停止操作中断作业 cut off
停止偿付 transfer moratorium
停止承运货物 stoppage on acceptance of goods
停止抽水时间 water-pumping stop time
停止出售证券 suspension of the sale of securities
停止出租通知 notice to quit
停止磁铁 stop magnet
停止代码 stop code
停止单元 stop element
停止导通 stop conducing
停止道 stopway
停止的 inactive;suspended
停止的地方 stopping place
停止点 break point;hold point;stagnation point;stopping point;arrests;arrest point <加热或冷却的>
停止电路 halt circuit;shut-down circuit
停止调车示像(显示) shunt-away aspect
停止订货 stop an order
停止动作 stall
停止兑现 moratorium
停止发电 generation outage
停止发行 stoppage of publication
停止阀 cut-off value;cut-out valve;relief valve;shut-down valve;stopgate valve
停止阀球 stop valve ball
停止反应 stopped reaction
停止服务的通知 notice of closing of service
停止符 stop element
停止符号 <串联传输中的结束符号> stop element
停止付款 stop payment;suspension of payment
停止付款通知书 notice dishonour
停止杆 stop bar
停止高度 shut height
停止给燃料 failure of fuel
停止跟踪 cease tracking
停止工作 knock off;place out of service;quit work;shut-down;stoppage of work;suspension
停止工作点 cut-out point
停止工作时间 closing time;idle time
停止供电 cut-off supply;out-of supply
停止供气 cut-off supply
停止供水 cut-off supply;stop of water supply;suspension of water supply
停止供应订货 kill order
停止广播 off the air
停止规则 stopping rule
停止过程 stopped process
停止号码 halt number
停止和通行信号 stop-and-go signal
停止痕 arrest mark;dwell mark
停止虹吸 stop siphoning;un-priming
停止活动期 lay-off
停止机构 locking gear;shut-down mechanism;stop device;stop motion
停止棘轮 stop gear
停止继电器 slew relay

停止降价 stop-out price
停止交付情况通知书 advice of circumstances preventing delivery
停止交通 stop traffic
停止交易 close an account with; withhold business
停止交易的企业 company suspending transactions
停止绞缆 stop heaving
停止绞锚 stop heaving
停止接地 off-ground
停止进行诉讼 stay of proceedings
停止警报试验 stop alarm test
停止距离 stop distance;stopping length
停止开关 halt switch; limit switch; shut-down switch;stop switch
停止块 stop piece
停止累积生物量 accumulating biomass
停止力矩 stalling torque
停止令 order of suppression
停止码元 stop bit
停止脉冲 stop pulse
停止排气(阀门)【机】exhaust close
停止盘 stop disc
停止期 withholding period
停止期间 stopping period
停止起动开关 stop-start switch
停止器 hold-back;stopper;stop piece
停止前进(航行)deaden the way
停止侵害 cessation of infringement
停止权 sign-off power
停止权利 suspension of right
停止燃烧 burn-out
停止上市 delist(ing)
停止设备 arrestment;safety gear
停止生产点 shut-down point
停止生效 cease to have effect
停止时间 closing period; off-time; quiescent time; stopped time; stop(ping)time
停止时效 suspension of prescription
停止使用的船队 laid-up fleet
停止收货 close for cargo
停止手续 stay proceedings
停止售货 withhold from the market
停止输送推进剂 propellant cut-off
停止诉讼申请 caveat
停止损失指令 stop-loss order; stop order
停止锁 lock stop
停止套筒 stop sleeve
停止条件交易 conditional precedent transaction
停止调节 stop adjustment
停止通行标志 stop sign
停止通行信号 stop sign
停止投资 disinvestment; negative investment
停止托运 stoppage of consignment
停止委托 stop order
停止位 stop bit
停止位置 closed position; idle position;stop position
停止线 stop line
停止线圈 stop coil
停止信号 break alarm; stop(ping) sign(al)
停止信号(灯用)开关 stop switch
停止行车 stoppage of traffic; stop traffic;train suspension
停止旋塞 stop cock
停止旋转 despin(ning)
停止循环 stop loop
停止循环的 off-cycle
停止摇摆装置 sway-stop device
停止营业 close of business;shut up a business; stoppage of business; suspension of business
停止营业时间 closing hour; closing time
停止营业线路 line closed to traffic
停止语句 stop statement
停止运动 stop motion
停止运货权 stoppage en route;stoppage in transit
停止运送情况通知书 advice of circumstances preventing carriage
停止运行 bring to rest; out-of-service;stopping the service
停止运转 shut-down; throw out of motion
停止运转周期 off-cycle
停止运转周期融霜 off-cycle defrosting
停止再行的驾驶 <按交通规则> stop-and-go driving
停止再行的信号 stop-and-go signal
停止账目 sleeping account
停止照明 lighting-off
停止振荡 failure of oscillations
停止支付 cease payment; non-payment; stop payment; suspension of payment
停止支付通知 caveat; stop-payment notice
停止支付支票 stop payment on a check; stopped check; stopping payment of check
停止执行 stay of execution
停止指令 halt instruction; stop instruction;sign-off
停止注水 stop siphoning
停止柱 stop post
停止柱塞 stop plunger
停止装车 embargo on wagon loading
停止装置 arresting stop; shut-down device; shut-down feature; stop gear; stopping arrangement; stopping device
停止状态 halted state;shut-down condition;state of rest;stop position
停止准则 stopping criterion
停止走纸 <打印机> space suppression
停止阻力 stopping resistance
停止作业 cease operation;deadlock
停滞 deadlock; detention; hold-up; sluggishness;stagnancy;stagnation; standstill;stasis
停滞边缘 stagnating margin
停滞冰川 dead glacier;stagnant glacier
停滞不前 at fault
停滞层玻璃 <池窑底部> stagnant glass
停滞潮 stand of tide
停滞带 stagnant zone
停滞的 dead;sluggish;stagnant;stale; standing
停滞的部门 stagnating sectors
停滞的经济 dormant economy
停滞的市场 dull market
停滞底水 stagnant bottom water
停滞点 point of stagnation; stagnant point; stagnating point; stagnation point
停滞过剩人口 stagnated overpopulation
停滞阶段 lag phase
停滞进化 stagnant evolution
停滞空气 dacker;dead air;stagnant air
停滞空气团 stagnated air mass
停滞膜 stagnant film
停滞膜模型 stagnant-film model
停滞逆温 stagnant inversion
停滞膨胀 stagflation
停滞期 lag period;lag phase
停滞气体 stagnant gas
停滞气团 stagnated air mass
停滞区 stagnant area;stagnant wake; stagnation zone;zone of stagnation
停滞区腐蚀 stagnant area corrosion
停滞时段 period of stagnation
停滞时间 dead time;dwell time;shut-off period;shut-off time;stagnation period;insensitive time
停滞时期【地】plateau[复 plateaus/plateaux]
停滞试验 <检查罗经灵敏度> swing test
停滞水 dead water; stagnant water; standing water
停滞温度 stagnation temperature
停滞型 stationary form
停滞性 stagnation
停滞性通货膨胀 stagflation
停滞压力 stagnation pressure
停滞蒸汽 stagnant steam
停滞政策 standstill policy
停滞状态 backwater;dead state;stagnation condition; stagnation state; state of stagnancy
停滞作用 stagnation
停住 anchor;hold-up;stall;tacky
停住冷结 teeming arrest
停住脉冲 stop impulse
停住引擎 seize up
停驻 park
停驻车辆 parked vehicle; parker; parking vehicle
停驻的车辆 stationary vehicle
停驻水跃 stationary jump
停转 stall(ing)
停转电流 stalling current
停转负荷 stalling load
停转力矩 breakdown torque
停转扭矩 breakdown torque
停转式凸版平台印刷机 stop-cylinder press
停转转差率 breakdown slip
停转转矩 breakdown torque
停转转速 breakdown torque speed
停走相地工作 stop-and-go operation
停租 off-hire
停租船 suspension of hire
停租条款 breakdown clause;off-hire clause
停租证书 off-hire certificate

蜓

蜓绝灭 <一种古生物> fusulinid extinction

挺

挺臂式装载机 arm loader

挺穿孔器 lever brace
挺度 stiffness
挺度计【物】deflectometer
挺杆 crossbar;gib;jib(boom);lifter; lifting bar; push rod; tail arm; tappet(rod); transverse member; top ladder
挺杆扳手 tappet wrench
挺杆臂 tappet arm
挺杆冲锤式凿岩机 tappet machine
挺杆导槽 tappet-guiding groove
挺杆导承 tappet guide
挺杆导管 lifter guide; tappet rod guide
挺杆叠置焊道 string beading
挺杆滚轮 lifter roller;tappet roller
挺杆滑车 <起重机> jig sheave
挺杆架 jib mounting
挺杆间隙 tappet clearance
挺杆间隙调节螺钉 tappet screw
挺杆磨损面 tappet wear surface
挺杆起重机 gib crane;jibcrane
挺杆润滑油道 tappet oil gallery
挺杆式起重机 arm crane;gib crane; jib crane
挺杆锁 tappet lock
挺杆弹簧 lifter spring
挺杆套 tappet sleeve
挺杆调节装置 tappet adjusting device
挺杆调整 tappet adjustment
挺杆头(部)tappet head
挺杆推出式安全装置 tappet pull-out guard
挺杆行程 tappet motion
挺杆余隙 tappet clearance
挺杆运动 jib motion
挺杆柱塞 tappet plunger
挺杆装置 tappet gear
挺杆组合 tappet assembly
挺进叠置焊道 string bead
挺冷 pretty cold; quite cold; rather cold
挺器 jim crow
挺水植物 emergent aquatic plant
挺托 <一种水泥喷涂材料> Tintocrete
挺托巴尔 <一种彩色半透明玻璃> Tintopal
挺相弹簧 tappet spring
挺针片 needle jack;pushing rod jack
挺柱 lifter
挺柱弹簧 lifter spring

艇

艇 boat;watercraft

艇鞍架 boat saddle
艇背部 dossal
艇侧纵条 boat rising
艇的铭牌 boat label plate
艇底塞 boat plug
艇底泄水孔 boat drain
艇垫架 boat nest
艇队 flotilla
艇筏 raft
艇筏配员 manning of life boat
艇钩联动脱钩装置 releasing gear of hooks
艇护舷 boat fender
艇机 lifeboat capstan
艇甲板 boat deck
艇架 davit
艇架底座 keel block
艇库 boathouse
艇缆 boat fast
艇锚 boat anchor
艇铆钉 boat rivet
艇牵索 boat guy
艇式起重机 boat derrick
艇首缆 boat painter; bow painter; painter;sea painter
艇首缆系环 bow strap;bow strop
艇首碰垫 pudding fender
艇首座板 headsheets
艇外马达 outboard motor
艇外推进器 kicker
艇尾舱 cockpit
艇舾装 boat fitting
艇腰挂锚索 belly sling;belly strap
艇用信号锣 boat gong
艇长 coxswain
艇罩 boat cover
艇罩结 boat cover hitch
艇支柱 boat stand
艇座 keel block
艇座架 block

通

通报 announce; bulletin; circular; report;telegraphy

通报船 advice ship
通报电流 signal(l)ing current
通报呼叫 report call

通报频率 traffic frequency
通报时间 holding time
通报速度 telegraph speed
通报系统 reporting chain
通报用语 service signal
通报终了 end of message
通波器 acceptor
通播发射台 collective broadcast sending station
通播接收台 collective broadcast receiving station
通仓浇筑 pouring without longitudinal joint
通草纸 <米纸> rice paper
通长 continuous length
通长窗 continuous-vision window
通长垫板 continuous plate
通长钢筋 continuity tendon
通长镜子 full-length mirror
通长小便池 slab urinal
通常 according to rule
通常包装 conventional packing
通常采用的 commonly adopted
通常车辆 conventional stock
通常齿横割锯 plain tooth cross-cut saw
通常的 customary;natural
通常的关闭时间 normal closing hours
通常的利润 usual profit
通常的运输途径 ordinary course of transit
通常短缺量 ordinary shortage
通常反应 conventional reaction
通常方法 conventional way
通常费用 ordinary disbursement
通常工作时间 ordinary working hours
通常公认的会计原则 generally accepted accounting principles
通常公认的审计标准 generally accepted auditing standards
通常供应 current supply
通常开支 overhead charges
通常利率 conventional rate
通常漏损 ordinary leakage
通常密度 current density
通常破损 ordinary breakage
通常认为安全的 generally recognized as safe
通常闪光 ordinary flash
通常设计 conventional design
通常收入 ordinary income
通常损耗 permissive waste
通常损失 ordinary loss
通常条件 usual condition
通常习惯 ordinary practice
通常险别 usual risks
通常运输线 ordinary course of transit
通常租税 customary rent
通畅水区航行 clear water circulation
通潮闸坞 closed dock; enclosed dock;impounded dock;tidal dock
通车 open to traffic
通车便道 drive-through access
通车车道 moving traffic lane;traffic lane
通车道路 traffic road
通车地带 traffic area
通车典礼 inauguration
通车空间 traffic space
通车口 <桥隧的> traffic opening
通车里程 railway track in use
通车量 traffic discharge;traffic volume
通车年份 opening year
通称 generic term
通称的 nominal
通称关键字 generic key
通称函数 generic function
通称描述符表 generic descriptor list
通称内径 nominal inside diameter
通称直径 nominal diameter
通称值 nominal value
通称族 generic family
通船桥孔 navigation opening
通达地面的平巷 day drift
通达距离 distance range
通达卡 access card
通大气管路 atmospheric pipe
通大气管线 atmospheric pipe
通代符 unifier
通代算法 unification algorithm
通带 band pass;transmission band
通带边缘振荡 band-edge oscillation
通带波动 passband ripple
通带范围 <滤波器> free transmission range
通带共用制 band sharing; passband sharing
通带宽度 bandwidth;pass bandwidth
通带滤波器 passband filter
通带滤光片 passband filter
通带频率特性 bandpass shape
通带频率响应 hand-pass response
通带上限 upper cut-off frequency
通带调节 bandwidth control
通带限制电路 bandwidth-limited circuit
通导角 angle of flow
通导截面 admittance area
通导区 bottomed region
通导时间 conduction time
通导特性 turn-on characteristic
通导延迟 turn-on delay
通到港口的河道 harbo(u)r reach
通到工地的道路 construction site access road
通道 ai(s)le;allure;ambulatory aisle;catwalk; channel way; debouche;duct;enterclose;gang(way);ingress;modality;oil passage <油路>;passage(way);pass road;path of travel;pathway;service road;thoroughfare;walkway;gateway【电】;electric(al) duct <安装电缆的>;aisleway;alley way;couloir <堆货用>;tablinum <古罗马建筑正厅>;balteus <古罗马竞技场上的或两讲堂之间的>;areamgy <建筑物之间的>;fauces <罗马建筑中的一种>;drong <英国方言>;tresaunce <中世纪建筑>;admittance
通道安全报警系统 pass alert safety system
通道板 channel plate
通道编码 channel coding
通道标志 gap marker
通道表 channel table
通道波 channel wave
通道布线算法 channel routing algorithm
通道布置 access arrangement
通道层网络 path layer network
通道长度 length of conduit
通道车站 thoroughfare station
通道程序 channel program(me)
通道程序转换 channel program(me) translation
通道处理机 channel processor
通道传输能力 channel capacity
通道窗 flanking window
通道带宽 channel bandwidth
通道倒换环 path-switched ring
通道地板 gangway floor plate
通道地毯 carpet runner
通道地址 port address
通道地址字 channel address word
通道垫 walkway pad
通道调度程序 channel schedule
通道陡度 aisle gradient
通道多路复用器 channel multiplexor
通道多路调制器 channel multiplexor
通道多路转换器 channel multiplexor
通道盖 duct cover
通道拱顶 flue bridge
通道管理 control of access
通道号 channel number
通道横截面形态 cross-section shape of karst conduit
通道环境研究 corridor study
通道缓冲存储器 channel buffer storage
通道换乘 passageway transfer
通道建筑物 approach structure
通道交叉电路 channel cross circuit
通道交通 corridor traffic
通道交通分配 corridor traffic assignment
通道接口 channel interface
通道接续器 channel-to-channel adapter
通道开销 path overhead
通道孔 access opening;hatchway
通道控制 corridor control;control of access
通道控制器 channel controller
通道控制字 channel control word
通道口 access hole;opening of gangway;pass-through opening
通道门 access door
通道面积 area of passage;port area
通道命令 channel command
通道命令字 channel command word
通道模型 corridor model
通道频带宽度 channel capacity
通道频率 channel frequency
通道坡度 <剧场的> aisle gradient
通道权 access right
通道人孔 access manhole
通道容量 channel capacity
通道设计 channel design
通道设施 passage appliance
通道失谐 detuning of channel
通道式驾驶室 walk-through cab
通道适配器 channel adapter
通道收缩 laning
通道竖井 access shaft
通道数 number of channels;port number
通道梯 access ladder
通道天桥 communicating bridge
通道跳变 channel trap
通道同步器 channel synchronizer
通道吞吐量 channel throughout
通道挖掘 access excavation
通道围阱 trunk to a space
通道稳定性 channel stability
通道衔接器 channel adapter
通道相互衔接器 channel-to-channel adapter
通道折棚 gangway bellows;gangway diaphragm
通道折棚盖 gangway bellows cover
通道直径 diameter of conduit
通道中断 chain break
通道终结点 path termination point
通道轴线 axis line
通道转移 channel trap
通道状态 channel status
通道状态字 channel status word
通道阻力 resistance of passage
通道作用 admittance function
通地 earthing
通地的 earthed;grounded
通地漏电 earth leakage
通地漏电测试装置 earth leakage testing device
通地漏电检测器 earth-leakage detector
通地漏电指示器 earth-leakage detector
通地线 earth return system
通地指示器 earth detector; ground detector
通电 circular telegram; electromotion;open telegram;power-on
通电部件 energized part
通电磁化法 end contact method
通电的 energized
通电电线 live wire
通电继电器 energized relay
通电流 energise[energize];galvanism;switch
通电流的 electrically active
通电路 closed circuit
通电去极化 electric(al) depolarization
通电时间(电焊) weld time
通电试验 power-on test
通电铁丝网 live wire
通电位置 energized position
通电周期 energized period
通电状态 on position;on-state
通斗墙 all-rowlock wall
通断 make-and-break;on-off
通断比 break-make ratio
通断比值 on-off ratio
通断操作 make-break operation
通断操作回合 make-break operation cycle
通断触头 make-and-break contact
通断动作 on-off action
通断工作状态 on-off operating mode
通断机构 make-and-break mechanism
通断开关 on(-and)-off switch
通断控制 on-off control;stqp-go control
通断控制动作 on-off control action
通断控制器 open and shut controller
通断器 contact maker
通断线路输送拨号脉冲法 loop-disconnect impulsing
通断型信息 on-off information
通断作用 on-off action;open and shut action
通多种语言的人 multilingual
通防篱 party fence
通分(母) reduction of fractions to a common denominator;reduce fractions to a common
通风 aerate; aeration; air-condition; air-conditioning; air draft; air draught;airiness;airing;air removal;atmospheric ventilation;breath;deaerate; draftiness; drafting; draught;draughtiness;extract ventilation; fanning; perflation; reaeration;ventilate;vent(ing)
通风安全旋塞 ventilating relief cock
通风百叶窗 ventilating shutter;ventilation louver [louvre]; ventilation shutter
通风板 air-handling panel; air plate; ventilation sheet
通风板条 <贮存混凝土、矿砂石等用物的> venting stave
通风瓣 ventilating flap
通风保温层 ventilated insulation
通风篦子 air grate;air grill(e)
通风标准 ventilation standard
通风表 draft ga(u)ge
通风玻璃块 air glass block;ventilating glass block
通风玻璃砖 air glass block;air glass brick;air-handling glass brick;glass ventilating brick; ventilating glass block;ventilating glass brick
通风不良 improper ventilation
通风不良的 ill-ventilated;poorly ventilated;stuffy
通风不足 deficiency in draft
通风布置 ventilating arrangement
通风部件 ventilating piece
通风采光权 light and air easement
通风舱 cool compartment;ventilated

cabin;ventilated compartment
通风舱口 air hatch
通风槽 ventilating air slot;ventilating slot;ventilation slot
通风槽片 vent segment;vent spacer
通风侧板 air stave
通风测量 ventilation survey
通风测量站 air-handling station;ventilating station;ventilation station
通风差动继电器 ventilation differential relay
通风车 <运水果、蔬菜等> ventilator car; ventilated vehicle; ventilation car
通风橱 exhaust fume hood;fume cupboard; fume hood; fuming cupboard;stinks cupboard;strike cupboard;ventilating cabinet
通风橱入口损失 hood entry loss
通风窗 air-handling window;air window;deck sash;fan window;night vent;ventilating shutter;ventilating window;ventilation casement;ventilation louver [louvre]; ventilation window; ventilator (sash); venting window;ventlight;vent louver[louvre];window ventilator
通风窗扉 ventilation casement
通风窗拉杆 deck sash opener; pull hook;ventilation port level;ventilator opener;ventilator staff
通风窗手柄 <车厢> ventilator handle
通风窗枢销 ventilator pivot
通风窗铁丝网 ventilator netting
通风窗用的磷铜合金密封带 atomic draught strip
通风吹管 foul air duct
通风次层 perforated under layer
通风次数 ventilation number
通风大门 bearing door
通风挡板 ventilation flap
通风导流栅 ejector grille
通风道 air channel;air chimney;air chute; air conduit; air drain; air duct; air-handling cavity; air passage; air stack; airway; fan drift; trunking; vent duct; vent gutter; ventiduct; ventilating flue; ventilating shaft;ventilation chimney;ventilation duct;ventilation stack;ventilation trunk;windway
通风道壁板 air course board
通风道隔离片 vent finger
通风道竖井 duct shaft
通风的 aerated;air dry;airy;drafty;ventilated;ventilating
通风的储桶 vented tank
通风的房屋正面 facade with air circulation
通风的建筑物立面 ventilated facade
通风的平屋顶 flat roof with air circulation;ventilated flat roof
通风的镶装顶棚 ventilating lay-in ceiling
通风底层 perforated under layer
通风地板 airflow floor
通风地点 ventilating position
通风电枢 armature with ventilation
通风吊杆柱 ventilator derrick
通风顶 monitor;monitor roof
通风顶管 chimney cowl
通风顶棚 air ceiling;air-conditioned ceiling;air-handling ceiling;air lay-in ceiling; ventilated ceiling; ventilating ceiling
通风顶棚板 ventilating ceiling board
通风顶棚薄板 ventilating ceiling sheet
通风顶棚方式 air-handling ceiling system
通风顶棚系统 air-handling ceiling system;ventilating ceiling system
通风洞 air opening; air passage; air tunnel;ventilation tunnel;vent tunnel
通风洞平面 plan of ventilation adit
通风斗 air funnel;vent cowl;ventilator scoop
通风斗罩 cowl cover
通风段 ventilating section
通风阀 airing valve; air valve; air vent valve; blow valve; chimney valve;draft damper;draught damper;pressure vent valve;ventilation valve;vent(ing)valve
通风阀导承 vent valve guide
通风阀簧 ventilator valve spring;vent valve spring
通风阀簧座 ventilator valve spring seat
通风阀活塞 vent valve piston
通风阀座 vent valve seat
通风法 ventilation method
通风翻窗 hopper ventilator window
通风范围 air-handling area;air-handling arrange;ventilation area
通风方案 ventilation scheme
通风方法 ventilating method;ventilation method
通风方式 draft type;form of ventilation; method of vent; ventilation system
通风防水板 vent flashing
通风房屋 ventilation building
通风飞行服 ventilated flight clothing
通风风流 ventilating air; ventilation air
通风风扇 air-circulating fan;draught fan;ventilator fan;draft fan;ventilating fan
通风服 ventilated suit; ventilating suit;vent suit
通风负荷 ventilation load
通风盖 vent flap;cowl <引擎盖后面的>
通风盖托臂 <冷藏车> ventilator bar
通风干道 ventilating trunk
通风干管 main vent;arterial vent
通风干路 ventilating trunk
通风干湿表 aspirated psychrometer; aspiration psychrometer; ventilated psychrometer
通风干湿计 aspiration psychrometer; sling psychrometer
通风干湿球湿度表 aspirated dry-(and-)wet bulb thermometer;aspirated dry-wet bulb hygrometer;aspiration psychrometer
通风干湿球湿度计 aspirated dry-(and-)wet bulb thermometer;aspirated dry-wet bulb hygrometer;aspiration psychrometer
通风干燥(法) air seasoning;aeration drying;pneumatic drying
通风干燥机 through-circulation dryer [drier];ventilation drying unit
通风干燥窑 ventilating dry kiln
通风干燥装置 ventilation drying unit
通风阁楼 ventilated loft
通风格 night vent
通风格窗 ventilating grill(e)
通风格栅 air grating;air grate;ventilating grill(e);ejector grin
通风格子 vent grating
通风格子板 air grill(e)
通风格子窗 air-handling grille;ventilating grill(e);ventilation grill(e);venting grill(e)
通风隔墙 abat(t)is;air barrage
通风工 gasman
通风工程 ventilation engineering; ventilation work;ventilator works
通风功能 ventilation function
通风沟隔片 ventilating duct spacer
通风构筑物 ventilating structure; ventilation building
通风谷物干贮塔 ventilated silo grain drier[dryer]
通风管 air chimney; air drain; air ducting;air flue;air-handling pipe; air-handling tube; air pipe; air tube;air vent pipe;blow(er)pipe; draft pipe;draft tube;draught pipe; draught tube;night vent;pipe vent; stack; stack pipe; trunking; upcast shaft; ventilating flue; ventilation chimney;ventilation conduit;ventilation piping;ventilation tube;ventilation tubing; ventilator; venting duct;venting pipe; vent line; vent pipe;vent tube;schnorkel <水下作业用>
通风管布置图 ventilation plan
通风管瓷砖 vent pipe tile
通风管道 air conduit;air(-handling) duct;air-handling line; air intake; pipe for conducting air;ventilating ducting;ventilating line;ventilation duct;ventilation gallery;ventilation line; ventiduct; ventilating pipe; ventilation pipe; ventilation stack; ventilating duct; venting tube; local vent <在存水弯前的>
通风管道配件 ventilating piece
通风管道系统配件 air-handling piece
通风管道支架 air-handling pipe tray
通风管道中的防火阀 fire damper in vent duct
通风管的防水板 vent pipe flashing
通风管盖 breather pipe cap
通风管护罩 <防海浪溅入> Liverpool head
通风管坑道 draught tube tunnel
通风管联箱 vent header
通风管路 air canal
通风管帽 ventilating jack
通风管摩擦系数 fanning friction factor
通风管内衬 vent lining
通风管配件 air-handling piece
通风管渠式设障沟 draft tube channel barrier ditch
通风管塞子 vent stopper
通风管式感烟探测器 duct type smoke detector
通风管通道 vent duct
通风管托架 air pipe tray
通风管网 ventilation pipe network
通风管系统 vent system
通风管线 ventilation line
通风管闸门 draft tube gate; draught tube gate;stack door
通风管罩 ventilating jack; ventilator pipe hood
通风管支架 air-handling tray;air pipe tray;air tray
通风管柱 ventilation column
通风管座 ventilating pipe tray; venting tray
通风罐顶 breather cap
通风柜 draft cupboard;draught hood; fume chamber;fume cupboard;fume hood;fuming cupboard;hood;stink cupboard; ventilated case; ventilated chamber;ventilation cabinet
通风过滤器 ventilation filter
通风耗热量 ventilation heat loss
通风恒压室 plenum chamber
通风横管 vent header
通风横巷 opening
通风烘箱 ventilated drying oven
通风花窗 air grill(e)
通风花格 air grate
通风滑板 ventilater sliding panel
通风滑动窗 venting sliding window
通风换气 exhaust ventilation
通风换气次数 air change rate;number of ventilating times per unit; ventilation rate
通风换气流量 <按小时计算> air change
通风回路 ventilation circuit
通风回纹饰 ventilating fret
通风活盖 ventilating flap
通风货柜 ventilated container
通风货物 ventilated cargo
通风机 aerator; air blast; air machine;blast fan; draft engine; draft fan;draught fan;fan(ner);forced draught fan; impeller; ventilating fan;ventilating machine;ventilating set; ventilation fan; ventilator (fan);wind blower
通风机传动装置 fan drive
通风机导管 ventilator duct
通风机的电动机防雨罩 weatherproof covering for electric motor of fan
通风机的鼓风 fan blast
通风机法则 fan law
通风机房 air-handling station; fan house; ventilating station; ventilation building; ventilation machine room;ventilation plant room
通风机供风量 air supply volume of ventilation
通风机供暖 fan air heating
通风机间 fan room;ventilator room
通风机进口段 inducer
通风机进气边 suction side of fan
通风机进气口 fan inlet
通风机开关 ventilator blower switch
通风机壳 ventilated enclosure
通风机空气入口 ventilator air inlet
通风机控制机构 ventilator control
通风机排风口 ventilator dash drain
通风机平台 fan platform
通风机扇 ventilating blower
通风机扇叶 ventilator blade
通风机式加热器 fan heater
通风机室 fan house;fan room
通风机送热集中供暖 fan-assisted warm-air central heating
通风机速率 fan speed
通风机调节板 fan shutter
通风机调节风门 blower
通风机调节器 blower
通风机通风 fan-assisted ventilation
通风机外壳 fan casing
通风机涡轮 draft turbine
通风机效率 fan efficiency
通风机械 ventilating machinery
通风机性能 blower performance
通风机压头 fan head
通风机叶片 air vane;fan blade;ventilating vane;ventilator blade
通风机噪声 air-blower noise
通风机闸门 fan shutter
通风机罩 fan casing;ventilator cover;ventilator hood
通风机转速 fan speed
通风机组 blower set;unit ventilator
通风及调温系统 ventilation and temperature control system
通风集装箱 hide container
通风脊瓦 ventilating ridge tile;venting ridge tile
通风计 draft fan;draft ga(u)ge;draft indicator;draught ga(u)ge;draught indicator
通风计算计算室外温度 calculating

outdoor temperature for ventilation
通风技术员 ventilation foreman
通风间 ventilation cell
通风减速板 draft retarder;draught retarder
通风建筑物 intake structure
通风降温 aeration cooling;cooling (by) ventilation
通风筋 ventilated rib
通风进气管 inlet ventilation duct
通风井 air chimney;air pit;air shaft;air trunk;air well;discharge air shaft;downcast;funnel;shaft;upcast shaft;ventilating shaft;ventilation shaft;vent shaft;wellhole
通风井道 ventilation shaft
通风绝热 ventilated insulation
通风开关 draft switch
通风壳 plenum casing
通风坑 air pit;upcast
通风坑道 air heading;airway;ventilation road
通风空气 vent air
通风空气过滤系统 vent air filter system
通风空气试样 ventilation sample
通风空腔 air-handling cavity;vent cavity
通风空调电控室 electric(al) control room for ventilation and air-conditION
通风空调机房 ventilation and air-conditioner room
通风空调设备 vent and air-conditioning equipment;ventilation and air-conditioning equipment
通风空心砌块 air-handling block
通风空心砖 air brick;ventilating brick;ventilation brick
通风孔 air breaker;air exhaust;air-handling cavity;air-handling opening;air hole;air opening;air vent;bleeder hole;draft hole;draught hole;tewel;vent;ventage;vent hole;vent holing;ventilating opening;ventilating pit;ventilation hole;ventilation opening;ventilation slot;ventilator;blow-hole <隧道>
通风孔盖 vent capping;ventilation vent cover;ventilator cap
通风孔盖板 ventilating panel;vent panel
通风孔塞 vent peg
通风孔式板状炉排 liveplate
通风孔砖 vent brick
通风控制 control of ventilation;ventilation control
通风控制盘 venting panel
通风控制器 draft controller
通风口 air end;air hatch;air opening;air slot;air vent;blow-off;blow vent;breathing place;chimney breast;gutter;night vent;spiracle;ventage;ventilating opening;ventilation aperture;ventilation opening;ventilation orifice;ventilator(scoop);vent opening
通风口盖 ventilator cap
通风口格栅 air grill(e);air-handling grille
通风口值班工人 <采矿> trapper
通风栏栅 hopper
通风冷凝器 vent condenser
通风冷却 aeration cooling;air blast cooling
通风冷却塔 ventilator cooling tower
通风立管 ventilation pipe;vent stack
通风连接管 air-handling connection pipe
通风联络孔 air hole
通风联络小巷 breakthrough;room crosscut
通风联络斜巷 monkey drift
通风良好的 well-ventilated;draughty
通风良好的房间 drafty room
通风量 air rate;air ventilation;amount of ventilation;ventilation;ventilation quantity;ventilation rate;volume of ventilation air
通风量分配 air distribution
通风量控制阀 ventilator valve
通风流 ventilation current
通风漏斗 plenum hopper
通风炉 air furnace;draft furnace;draught-furnace;ventilating furnace;wind furnace
通风率 ventilation rate
通风螺钉 ventilating screw
通风帽 abat-vent;air capping;cowl;cowl head ventilator;cowling;cowl ventilator;ventilating capping;ventilating cowl;ventilation cap;ventilation cowl;ventilation hood
通风门廊 ventilated lobby
通风门用的磷铜合金密封带 atomic draught strip
通风面积 air-handling area;air-handling arrange;draught area;vent area;ventilated area;ventilating area;ventilation area
通风面具 <矿工用压缩空气面具> aerophore
通风模拟设备 pneumatic analog(ue) simulator
通风木垛 boule
通风能力 delivery volume;draft power;draught capacity
通风凝结器 vent condenser
通风排气 vented exhaust
通风排气管路 ventilation exhaust line
通风排气系统 exhaust system of ventilation
通风棚车 ventilated box car
通风片 ventilating disk[disc]
通风平衡系统 balanced system of ventilation
通风平面图 ventilation layout
通风平巷 air head;ventilating entry;ventilation gallery;ventilation lateral;ventilation level;vent level
通风歧管 vent(ed) manifold
通风起阀器 valve spring lifter
通风气洞 ventilating cavity
通风气流 draft air;draught air
通风气流压力 draft head
通风气体 ventilated gas
通风气象计 aspiration meteorograph
通风砌块 ventilating block
通风器 aerarium;air ventilator;fan;fanning machine;vacuum breaker;ventilator
通风器出口 outlet ventilator
通风器出气管 ventilator outlet pipe
通风器出气管滤气器 ventilator outlet pipe air cleaner
通风器导风板 ventilator deflector
通风器的鼓风机 ventilator blower
通风器的调风器 ventilator register
通风器管 ventilator pipe
通风器管道 ventilator duct
通风器夹滑板 ventilator clamp slide
通风器夹紧翼形螺钉 ventilator clamp thumb screw
通风器夹托架 ventilator clamp bracket
通风器夹销 ventilator clamp pin
通风器联杆 ventilator link
通风器联节托销 ventilator link bracket pin
通风器螺栓紧固器 fan bolt tightener
通风器螺栓调整滑轮支架 fan bolt adjusting pulley bracket
通风器帽 ventilator cap
通风器门 ventilator door
通风器罩 ventilator hood;ventilator valve
通风强度 draft intensity;intensity of draft;intensity of draught;strength of draft
通风墙 shaft wall
通风墙系统 shaft wall system
通风区段 ventilation section
通风容积 ventilation volume
通风软管 air hose;flexible ventilation ducting;vent hose
通风筛 vent screen
通风栅格 air grill(e)
通风扇 blowdown fan;circulating fan;draught fan;vent fan;ventilating fan;ventilation fan;ventilator fan
通风设备 aeration equipment;air equipment;air-handling installation;air installation;draft apparatus;draft equipment;ventilate;ventilating appliance;ventilating device;ventilating equipment;ventilating facility;ventilating installation;ventilating plant;ventilation appliance;ventilation device;ventilation equipment;ventilation facility;ventilation installation;ventilation plant;venting equipment;venting installation
通风设备室 ventilating plant room;venting plant room
通风设备折旧摊销及大修费 ventilation equipment depreciation apportion and overhaul charges
通风设计 draft design;ventilation design;ventilation planning
通风湿度计 sling psychrometer
通风石门 ventilation cross-cut
通风式电动机 ventilated motor
通风式防毒掩蔽部 ventilated shelter
通风式干燥机 ventilating drier[dryer]
通风式集装箱 ventilated container
通风式磨煤机 air-swept pulverizer
通风式喷雾机 low sprayer
通风式潜水 ventilative diving
通风式潜水工具 ventilative diving equipment
通风式潜水器 ventilation diver
通风式潜水医务保障 medical security of ventilative diving
通风式人井盖板 ventilated manhole cover
通风式湿度计 ventilation psychrometer
通风式箱式货车 ventilated box car
通风式整流器 ventilated commutator
通风式装饰气灯 vent decorative appliance
通风试验 draft test
通风饰 ventilating fret
通风室 air compartment;draft chamber;draught chamber;plenum chamber;ventilating chamber;ventilation chamber
通风竖管 discharge air shaft;stack vent;upcast shaft;ventilating stack;ventilator stack;venting shaft;vent(ing) stack;main stack <建筑物中从下水道直通屋顶的>
通风竖井 air intake shaft;downcast;light well;vent(ing) shaft;air-handling shaft;air inlet shaft;air(relief) shaft;air well;ventilating shaft;ventilation shaft
通风竖坑 downcast
通风水平 ventilation level
通风水平仪 air level
通风速度 ventilation rate
通风损耗 breathing loss;windage loss;wind resistance loss <电机的>
通风损耗量 draft loss;draught loss;ventilation loss
通风损失量 draft loss;stack loss;ventilation loss
通风塔 aeration tower;ventilating tower;ventilation tower
通风天窗 aeration skylight;ventilating skylight
通风天花板 air ceiling;air-conditioned ceiling;air-handling ceiling;air lay-in ceiling;ventilated ceiling;ventilating ceiling
通风天花板方式 air-handling ceiling system
通风天花板系统 air-handling ceiling system;ventilating ceiling system
通风天井 vent court
通风调节 draft control;draft regulation
通风调节阀 blast regulation valve;blower regulation valve
通风调节门 ventilation damper
通风调节器 draft regulator;draught regulator
通风调节装置 ventilation conditioner
通风筒 air chimney;air funnel;air shaft;air trunk;chimney;ventilating trunk;ventilation tubing;ventilator;ventilator duct;vent sleeve
通风筒顶罩 ventilator hood
通风筒口开关 louver slide
通风筒围板 ventilator coaming
通风图 ventilation map
通风推拉窗 air sliding window
通风瓦 ventilating tile
通风网 ventilation network
通风网路 ventilating network
通风网路模拟装置 ventilation network simulator
通风位置 ventilating position
通风温度 aspiration temperature;draft temperature
通风温度表 aspiration temperature meter;aspiration thermometer
通风温度计 aspiration temperature meter;aspiration thermometer;ventilated thermometer
通风温度记录器 aspiration thermograph
通风温湿度计 sling psychrometer
通风稳定器 draft stabilizer
通风屋顶 ventilated attics;ventilated roof
通风屋脊 air-handling ridge;ventilating ridge;ventilation ridge;venting tile
通风吸顶灯 <兼有通风和照明功能> airlight troffer
通风吸入管路 ventilation supply line
通风吸声顶棚 ventilating acoustic (al) ceiling
通风吸声天花板 ventilating acoustic (al) ceiling
通风系 ventage
通风系统 air system;plenum system;quenching system;supply system;ventilating system;ventilation system;venting system;vent line
通风系统出口 outlet of ventilating system
通风系统的压力调节阀 vent pressure vacuum valve
通风系统进口 inlet of ventilating system
通风系统配件 air piece;air plant
通风系统示意图 ventilation system diagram

T

通风系统图 ventilation plan; ventilation system map
通风卜层 perforated under layer
通风舷窗 airside scuttle
通风线 breathing line
通风线路 ventilation circuit
通风箱 ventilated box; ventilator box
通风箱式谷物干燥机 ventilated bin grain drier[dryer]
通风镶板 air-handling panel
通风巷道 air head(ing); back heading; draughty workings; exhaust drift; monkeyway; aircourse
通风小室 ventilating room
通风小巷 breakthrough
通风效率 efficiency of draught
通风性能曲线 characteristic curve of fan
通风休息室 ventilated lobby
通风旋塞 ventilating cock
通风压差 ventilating column
通风压力 ventilating pressure; ventilation pressure
通风压力测量 pressure survey
通风压头 draft head; draught head
通风烟囱 air chimney; ventilating chimney; venting stack
通风眼 air vent
通风叶轮 fan wheel
通风仪表盘 ventilating panel; venting panel
通风用百叶窗 louver[louvre]
通风用电动机 ventilating motor
通风裕量 draught margin
通风原理 ventilation principle
通风运输 ventilated transport
通风再循环 recirculation of air
通风噪声 ventilation noise
通风增压系统 ventilation and pressurization system
通风闸 opening damper; ventilating damper
通风栅(栏) stack bond; vent bar; ventilation fence
通风栅门 grilled ventilation door
通风站 air-handling station; ventilation station
通风照明配件 ventilating light fittings
通风照明设备 ventilated light fitting
通风照明装置 air-handling lighting fixture; air luminaire(fixture); luminaire; ventilating light(ing) fixture; ventilation light fitting; ventilation light(ing) fixture; ventilation luminaire(fixture)
通风罩 air bleeder; air bleeder cap; airhood; air scoop; cowling; draft hood; draught hood; fume hood; hood; ventilated casing; ventilating capping; ventilating scoop; ventilation hood; ventilation scoop; ventilation shield; ventilator cowling; ventilator hood
通风折翼 vent flap
通风支管 branch duct; branch vent; branch ventilating pipe
通风织物 vent fabric
通风止回阀 check damper
通风指示器 draft indicator
通风指数 ventilation index
通风周期 ventilation cycle
通风主道 ventilating trunk
通风砖 air-handling block; air-handling brick; ventilating brick; ventilation tile
通风转向器 draft diverter
通风装置 air breaker; air-breather; air fixture; air-handling installation; air installation; chimney breast; fan unit; register; ventilating device; ventilating fixture; ventilating installation; ventilating plant; ventilating unit; ventilation device; ventilation ducting; ventilation installation; ventilation system; ventilation unit; ventilator; ventilator unit; venting device; venting installation
通风装置隔板 draft curtain
通风总管 air truck; air trunk; main duct; trunk duct
通风阻力 resistance of ventilation; ventilation resistance
通风阻流器 draft retarder; draught retarder
通风作业 ventilating
通缝 straight joint
通缝砌法 stack(ed) bond
通杆 through rod
通高坛耳堂 chancel aisle
通告 announcement; announcing; circular; notice; notification; public notice
通告的 encyclic(al)
通告废除(条约、协定等) denounce
通告分红 declare a dividend
通告话筒 announcement transmitter
通告价格 publicly posted price
通告扣押债权者 garinsher
通告牌 noticeboard; notice board
通告器 announcer; annunciator
通告送话器 announcement transmitter
通告信号 announcing signal; right of way signal
通告员 announcer
通告照会 circular note
通告者 annunciator
通告装置 announcing device
通拱 through arch
通沟 thorough cut
通沟木牛 sewer pill
通沟木球 sewer pill
通沟牛 bucket
通沟室盖 cleaning cover
通沟条 drained rod; sewer rod
通沟污泥 sewer sludge
通古斯海盆 Tungus sea basin
通关 clearance
通关卡 carnet
通关手续 customs clearance procedures
通管孔(盖) rodding eye
通管器 pipe cleaner; snake
通管丝 stylet
通过 clearance; enact; gate through; go through; pass(age); through; transit
通过 00 号筛的细料 minus #00 material
通过按钮 through button
通过按钮电路 through button circuit
通过百分率 percentage passing
通过臂板 passing arm; through blade
通过波段 throttling band
通过补偿系试验 by means of compensation lines
通过不停 by-passing
通过不通过控制器 < 即检音器 > go-no-go monitor
通过侧线 passing through the siding
通过测试 pass test
通过车 transit car
通过车场【铁】 by-pass(ing) yard; through yard; transit yard
通过车站允许速度 permissive speed of train passing through station
通过抽查核对 verification by test and scanning
通过穿孔套管注水泥 cementing through
通过磋商成交的合同 negotiated contract
通过大气扰流区 rough air penetration
通过带速度 band speed
通过的 non-stop
通过的曲线半径 radius of curvature negotiated
通过的议案 resolution passed
通过地役权 servitude of passage
通过第三国汇兑 cross exchange
通过第三国汇付 cross exchange
通过电话交易的市场 over-the-telephone market
通过端 go side
通过端极限 go limit
通过对向道岔 run over the facing points
通过吨数 passing tonnage; tonnage passed by
通过吨位 passing tonnage; tonnage passed by
通过法律手段的支配 control through a legal device
通过方向 direction of passage
通过费 navigation toll
通过分类法 pass sorting
通过腐烂有机物 by causing organic material to rot
通过港 junction port
通过公开投标 by public tender
通过功率测量仪 feed-through power meter
通过管 through pipe
通过管道的速度 duct velocity
通过轨道总重 million gross tonnes
通过国 transit country
通过过境货物 freight in transit
通过海关后交货条件 duty-paid term
通过焊接连接处的开孔 openings through welded joints
通过横向交通 passing cross-traffic
通过或靠近焊缝处的开孔 openings through or near welded joints
通过货物 transit goods
通过极地 transpolar
通过检验 pass test
通过交叉道口的时间 clearance time at crossing
通过交换托收的银行票据 bank exchanges
通过交通 through traffic; transit traffic
通过交通分配 corridor traffic assignment
通过介质起作用 action through(the) medium
通过金额 amount of vote
通过进路 through route
通过进路锁闭 through route locking
通过经纪人事务所而不通过交易所 < 买卖证券 > over-the-counter
通过局 transit administration
通过客流 through passenger flow
通过空间 through space
通过空气或水的传染 infection by air or water
通过孔 clearing hole
通过拦江沙信号 bar signal
通过梁 ring beam
通过量 quantity passed; throughput; throughput capacity; thruput
通过量规 go-ga(u)ge
通过量因数 throat opening factor
通过列车 non-stop train; run-through train
通过零点的 zeroaxial
通过楼层的干管 rising main
通过路线 pass-course
通过率 percentage passing; percent of pass; through rate
通过贸易 transit trade
通过秘密(非法)途径的 back door
通过能力 carrying capacity; discharge capacity; passing ability; trafficability; traffic capacity; transmissivity; throughput capacity; capacity of track; carrying capacity of track【铁】
通过能力估计 capacity estimate of passage; traffic capacity estimate
通过能力曲线 traffic capacity curve
通过能力限制区间 restriction section of carrying capacity
通过能力研究 capacity study of passage; traffic capacity study
通过旁通输沙道输沙 bypassing
通过频带 throttling band
通过墙的泛水件 through-wall flashing piece
通过墙身的顶砖 thorough
通过桥梁 passing bridge
通过区 key-in region; key-on region
通过曲线 curve negotiation; curve passage; negotiating curve; passing through curve; running through curve; through curve
通过曲线的能力 negotiability
通过权 right of way
通过容积 through volume
通过溶液的水化 through-solution hydration
通过三通管的冷却液流 tee coolant flow
通过色辉 tint of passage
通过时间 passage time; time of passage; transition time; transit time; passage period < 从检车感应器的位置起,车辆以平均速度驶至交叉口的最近点所需的时间 >
通过实验作更多的调查 more investigation by means of experiment
通过式车站 through-type station
通过式风管 through air duct
通过式货场 through-type goods yard
通过式货运站 through-type goods station
通过式挤奶台 walk-through stall
通过式均化 throughput homogenizing
通过式客运站 through-type passenger station
通过式热水器 push-through geyser
通过式枢纽站 through terminal
通过式通风管 passage ventilation duct
通过式总站 through terminal
通过试验 pass test
通过水道 pass waterway
通过税 passage duty; transit dues; transit duty
通过顺向道岔 run over the trailing points
通过司法裁定确定水量分配 judicial allocation
通过速度 transit velocity
通过速率 through(put) rate
通过台 gangway connections between coaches; vestibule
通过台安全链 platform chain; vestibule chain
通过台侧梁 vestibule side sill
通过台出入口 vestibule entrance
通过台窗 vestibule end window
通过台的敞开车顶 canopy
通过台灯 vestibule lamp
通过台地板 vestibule floor; vestibule floor plank
通过台顶 vestibule hood
通过台顶棚纵向梁 end deck rail
通过台端板 vestibule end sheathing
通过台端部顶弯梁 vestibule end carline
通过台端梁 vestibule end sill
通过台端门 tail gate; vestibule gate

通过台端柱 vestibule end post
通过台护杆 vestibule guard rail
通过台缓冲板 vestibule buffer plate
通过台角立柱 vestibule body corner post
通过台脚蹬 vestibule step
通过台门 vestibule door
通过台门楣 vestibule door lintel
通过台三杆缓冲装置 Buhoup platform; three stem platform equipment
通过台上操纵的提杆柄 platform lever
通过台上操纵的提杆柄销 platform lever pin
通过台外伸车顶 platform hood
通过台栅栏 <客车的> platform railing
通过特性 through characteristic
通过梯阶 negotiating stair(case)
通过提单 through bill of lading
通过铁路 transfer railway; transit railway
通过铁路运输 by rail
通过投标报价 offer by tender
通过投标发盘 offer by tender
通过无人机 unattended repeater with ground temperature compensation and power-passing
通过效率 throughput efficiency
通过协商 through consultation
通过新技术和培训来改善管理效能 improve the operational efficiency through modern technology and training
通过信号 through signal; automatic block signal; passing signal; block section signal; intermediate automatic block signal
通过信号机【铁】automatic block signal; block signal
通过信号继电器 through signal relay
通过型检测器 passage detector
通过性 trafficability characteristic
通过性能 flotation
通过窑炉的时间 time of passage through kiln
通过仪器目镜调整 diopter adjustment
通过议程 adoption of agenda
通过预付款降低贷款初期利率 <1~5年内的> buydown
通过远距信标台的高度 outer marker crossing height
通过运河吨位 canal tonnage
通过运输 bridge traffic; transit traffic
通过站 through station
通过招标方式购买 buying by tender; purchase by tender
通过折棚 communication bellows
通过整个种子表面 through the entire surface of the seed
通过整流器充电 rectifier charger
通过整流器和扼流线圈的直流充电 direct-current choke rectifier charger
通过正线 passing through the main line
通过质疑进行核查 verification by challenge
通过资格预审的投标者 qualification bidder
通过自交或回交 by self-crossing or backcrossing
通过阻碍能力 obstacle overcome ability
通过最小曲线半径 minimum track radius negotiated
通过坐标原点的 zeroaxial
通海船闸 sea lock; maritime lock; maritime lock
通海地点 sea gate
通海阀 hull valve; hull ventilator; Kingston valve; sea inlet; sea (suction) valve
通海阀舱 scuttling chamber
通海航道 sea access; seaward channel
通海航道入口浮标 sea buoy
通海接头 sea connection
通海轮船闸 sea lock
通海轮的深水运河 maritime canal
通海深水航道 sea canal
通海水道 sea approach; seaway; ship pass
通海水系 external drainage
通海吸入阀 sea inlet cock
通海吸水箱 sea chest; suction box
通海小河 creek; inlet
通海旋塞 sea plug
通海运河 maritime canal; sea(way) canal; open(sea) canal <无闸的>
通海闸门 sea gate; sea lock
通海直水道 sea reach
通海主吸入阀 main injection valve
通函 circular; circular letter
通函征询 circularization
通航 navigation
通航坝 navigation dam
通航保证率 navigable stage frequency
通航标准 navigation standard
通航冰况标准 operational ice criterion
通航汊道 active arm
通航尺度 navigable dimension
通航船闸 canalization lock
通航道 route way
通航灯浮标 navigation flame float
通航渡槽 navigable aqueduct; navigable canal bridge; navigable flume
通航发电渠道 navigable power canal
通航费用 navigation toll
通航分道 traffic lane
通航分隔带 traffic separation zone
通航分隔线 traffic separation line
通航高度 headway
通航工程 navigation works
通航拱 fairway arch
通航拱跨 fairway arch span
通航号志 passage signal
通航河槽 navigable channel
通航河道 canalization river; canalized river; navigable river; navigable waterway; navigation channel; open river; navigable channel; ship channel; open river waterway
通航河道船闸 sasse
通航河道深度 depth of the navigable channel
通航河段 canalized river stretch; navigable reach; navigable stretch
通航河段起点 beginning of floating; beginning of navigation; beginning of navigation shipping; beginning of shipping
通航河流 navigable river; navigable stream
通航护照 passport
通航活动坝 navigable movable dam
通航建筑物 navigation construction; navigable structure; navigation structure
通航净高 navigable clear height; navigation net height
通航净空 clearance for navigation; navigable clearance; navigational clearance; overhead clearance for navigation
通航净跨 navigable net span
通航净宽 navigable clear with; navigable net width
通航孔 fairway arch; shipping channel
通航口 <桥的> traffic opening
通航口门 navigation opening
通航跨度 <桥的> traffic span
通航宽度 navigable spalling; navigable span; navigational width
通航里程 air route in use; mileage open to traffic
通航流量 navigation discharge
通航流速 navigable current velocity
通航期 navigable period; navigation period; open river period
通航起讫点 beginning and end of water way
通航桥孔 navigable bridge opening; navigation opening; navigation span; navigable space; navigable span
通航桥跨 navigable span
通航情况 navigable condition
通航渠道 communicating canal; navigable canal; navigable channel; navigable waterway; navigation channel; navigation pass; ship pass
通航权 right of way
通航设施 navigation facility
通航深度 navigable depth
通航深度疏导 training for depth
通航水道 canalized waterway; channel pass(age); navigable channel; navigable passage; navigable water; navigable watercourse; navigable waterway; navigation channel; navigation pass; ship channel; ship pass
通航水道空间 navigation channel space
通航水道宽度 width of navigable passage
通航水库 navigation reservoir
通航水流条件 flow condition for navigation
通航水路 navigable watercourse; navigable waterway
通航水深 depth of navigable channel; navigable channel depth; navigable depth; nautical depth
通航水位 navigable stage; navigable water level; navigation(al) water level
通航水位保证率 guaranteed rate of navigable water level; guaranteed rate of navigable water stage
通航水域 navigable waters
通航隧道 navigation tunnel
通航隧洞 navigation tunnel
通航梯级 navigation steps
通航条件 navigation condition
通航条约 navigational treaty
通航峡口 navigable pass; navigation pass
通航巷道 traffic lane
通航效益 navigation benefit
通航要求 navigation(al) requirement
通航影响 navigation effect; navigation impact
通航用坝 <用坝将河道分段> canalization dam
通航用闸 <用闸将河道分段> canalization lock
通航运河 artificial navigation canal; navigable canal; navigation canal; ship canal
通航运河的供水 water supply to navigation canals
通航终点 head of navigation
通航状况 navigation condition
通号车间 telecommunication and signal(l)ing workshop
通号综合楼 telecommunication and signal(l)ing building
通候机楼的滑行道 terminal taxiway
通弧 arc-through
通(花)花瓶 openwork vase
通花制品 cage work
通话管 communicating pipe; speaking tube; vice pipe
通话盒 <内部通信[讯]用扬声器> squawker
通话室 booth; cabinet; cope
通话位置 speaking position
通话线路 talk line
通话箱 intercom box
通汇契约 agency agreement; agency arrangement; contract agreement; correspondent agreement
通汇银行 correspondent; correspondent bank
通汇银行信用证 correspondent letter of credit
通火 fire poking
通火车的码头泊位 railway berth
通货 circulating medium; currency; currency coin; currency money; current coin; current money; medium of circulation
通货保证险 risk of currency depreciation
通货贬值 currency devaluation
通货贷款 currency liability; currency loan
通货和银行钞票法例 Currency and Bank Notes Act
通货互换协定 reciprocal currency agreement; swap agreement
通货汇率 currency rate
通货集团 currency bloc
通货检查员 controller of currency
通货紧缩 currency deflation; deflation; disinflation
通货紧缩效应 deflationary effect
通货紧缩政策 deflationary policy; deflation policy; disinflation policy
通货流动性 currency liquidity
通货膨胀 currency inflation; inflate; inflation; monetary inflation
通货膨胀压力 inflationary pressure
通货条款 currency clause
通机油散热器的舱口盖 oil cooler access cover
通积分 general integral
通角 current flow angle
通街道出口 access to street
通解【数】general integral; general solution
通井机 wire-line workover unit
通径 latus rectum
通径规 drift diameter ga(u)ge
通径井试验 full hole testing
通径钻孔 full ga(u)ge(bore) hole
通克条 <书柜内承受搁板的可调镀锌铁条> Tonk's strip
通孔 body size hole; interlock pore; open-end hole; through hole
通孔镀敷 through hole coating; through hole plating
通孔喷吹法 hole blowing process
通孔套管 continuous boss
通孔直径 drift diameter
通栏标题 streamer
通廊 breezeway; connected traverse; dogtrot; traverse <与对面建筑物相通的>
通廊式壁龛 gallery niche
通廊式窗 gallery window
通廊式大厦 access corridor block
通廊式地窖 gallery crypt
通廊式洞龛 gallery niche
通廊式公寓 gallery apartment house
通廊式拱顶 gallery vault
通廊式喷火口 gallery burner; gallery port

通廊式平面布置 corridor access type
通廊式燃烧器 gallery burner
通廊式屋顶 gallery roof
通力合作 pool effort;pull together
通连板 through plate
通连列板 passing strake
通连性阶 order of connectivity
通梁 continuous beam;through beam
通量 flux
通量比 flux ratio
通量变化周期 flux period
通量波 flux wave
通量补偿 flux flattening
通量不变量 flux invariant
通量采样器 flux sampler
通量测绘 flux mapping
通量测绘孔 flux plot hole
通量测绘系统 flux mapping system
通量测量 flux measurement
通量测量道 flux-measuring channel
通量穿透 flux penetration
通量穿透测量 flux-traverse measurement
通量的单位管 unit tube of flux
通量额定值 flux rating
通量法 flux method
通量反照率 flux albedo
通量分布 flux distribution
通量分布测量 flux-traverse measurement
通量分布测量管道 flux scanning channel
通量分布图 flux pattern
通量分步法 flux step method
通量分析 flux distribution
通量峰 flux peak
通量辐散 flux divergence
通量管 flux tube
通量函数 flux function
通量恢复 flux recovery
通量畸变 flux distortion
通量级 flux level
通量集中 flux concentration
通量集中器 flux concentrator
通量计 fluxmeter
通量计数器 flux counter
通量记录仪 fluxgraph
通量监测 flux monitoring
通量监测箔 flux-monitoring foil
通量监测器 flux monitor
通量减弱修正 flux-depression modification
通量界限 flux boundary
通量阱 flux trap(ping)
通量阱(反应)堆 flux trap reactor
通量阱区 flux-trap region
通量坑 flux dip
通量力冷凝洗涤器 flux force-condensation scrubber
通量率 flux rate
通量每秒 flux-second
通量门 flux gate
通量密度 flux density
通量评定 flux evaluation
通量强度 flux level
通量散度 flux divergence
通量衰减 flux depression
通量水平 flux level
通量探测器 flux detector
通量图 flux map; flux pattern; flux plot
通量线 flux line;line of flux;streamline
通量压低校正 flux-depression correction
通量阻隔系数 orificing factor
通量作图 flux plotting
通流 through-flow
通流阀 passage valve;vent valve
通流加热器 through-flow heater
通流室 flowing-through chamber
通楼天桥 <带桥梁的上层建筑> bridgework
通路 accessment; road access; passage; clearway; closed path; completed circuit; corridor; duct; entry way; fairway; gangway; gateway; going; highway; intercommunication; land-access road; lead; path (of travel);route;turn-on【电】
通路表达式 path expression
通路操作 closed circuit operation
通路测试 path testing
通路测试振荡器 channel test oscillator
通路长度 path length
通路传输能力 channel capacity
通路串杂音防卫度 channel signal crosstalk noise level difference;channel signal to crosstalk and noise ratio
通路带通滤波器 channel bandpass filter
通路地址 port address
通路段 forehearth section;forehearth side rail
通路非线性失真系数 channel distortion coefficient; channel non-linear distortion coefficient
通路分类 way sort
通路分配 channel allocation;channel assignment
通路分支 forehearth branch;forehearth limb
通路隔断 barred access
通路固有杂音 background noise of channel;channel basic noise
通路固有噪声 channel idle noise
通路管 eircuitron
通路解调器 channel demodulator
通路净衰耗 channel net loss
通路净衰耗持恒度 channel net loss stability
通路净损失持恒度 channel net loss stability
通路净增益 channel net gain
通路矩阵 path matrix
通路开关 channel selector
通路孔 access port;via hole
通路控制 path control
通路口 access port
通路宽度 duct width;path width
通路连接砖 forehearth connection block
通路流槽砖 flow block
通路密度 roading
通路面积 area of passage;passage area
通路敏化 path sensitization
通路敏化法 path sensitizing
通路耦合器 access counter
通路器 circuit closer
通路器与断路器 circuit closer and breaker
通路权 easement of way
通路容量 channel capacity
通路入口砖 forehearth entrance block
通路衰耗频率特性 channel loss frequency characteristic
通路特性 characteristics of channel
通路条件 path condition
通路调制器 channel modulator
通路位置 close position; connected position
通路稳定度 channel stability
通路下游端 downstream end of forehearth
通路线圈 closed coil
通路相关性 path dependency
通路效应 channel(ling) effect
通路信息容量 information-handling capacity
通路形成 path generation
通路形成法 path generation method
通路旋塞 passage cock
通路碹 canal arch;forehearth arch
通路压缩 path compression
通路引线 via pin
通路有效传输频带 channel effective transmission bandwidth; effective transmission band of channel
通路增益系数 path-gain factor
通路振幅特性 channel amplitude characteristic
通路振铃边际 channel ringing margin
通路争用 contention
通路指示 routing
通路指示字 port pointer
通路砖 channel block
通路追迹法 path tracing method
通路走廊 passageway
通路组合 place mix
通马狭桥 <不通车的> bridle bridge
通门 open gate
通磨 continuous grinding
通盘处理法 comprehensive approach
通盘管理 vertical control
通盘计算 take together
通配符 wild card;wildcard character
通票 through ticket
通频带 band pass;passband
通频带宽度 pass bandwidth
通频带宽度特性 bandwidth characteristic
通频带利用系数 passband utilization factor
通频带上限 upper cut-off frequency
通频带调谐 bandpass tuning
通频带下限 lower cut-off frequency
通气 aerate; airing; air vent; breathing;gassing-up;ventilating
通气板条 venting stave
通气棒 venting pin
通气槽 aeration channel;aeration slot; air channel;vent groove;vent(ing) channel;leakage groove <闸缸>
通气槽刀片 vent knife
通气厕所 ventilated pit closet
通气层 aeration zone;deaerating layer;unsaturated zone;vadose zone; zone of aeration;zone of suspended water;weathering zone <地球岩石圈的>
通气层的水 kremastic water
通气层水 argic water;kremastic water;wandering water
通气池 aeration basin
通气冲击 ventilated shock
通气橱 stink cupboard
通气储量百分比 ventilatory reserve percentage
通气窗 air vent window;louver[louvre];vent sash;window ventilator
通气窗框 ventilator frame
通气次数 aeration frequency
通气带 aeration zone; zone of aeration
通气当量 ventilation equivalent
通气道 air duct;air flue;air gate;air passage;foul air flue;gas vent
通气的 airy
通气洞 bleed hole
通气多叉管 air vent manifold
通气阀 air relief valve; air vent valve; breather valve; breathing valve;vent valve
通气阀塞头 breather adapter
通气法处理活性污泥 ditch oxidation
通气法处理生物垃圾 contact aerator
通气房 air chamber
通气缝 gas slot
通气腐蚀 aerating corrosion
通气干管 main vent;vent main
通气格窗 night vent
通气格子板 air vane
通气根 aerating root
通气功能 ventilatory function
通气沟 air drain
通气鼓风机 aeration blower
通气管 aeration pipe;aerator pipe;air chimney;air drain;air duct;air inlet pipe;air pipe;air vent;air vent pipe; back vent; breather pipe; breathing pipe;drainage tube;flowing-through pipe; snorkel; vent flue;ventilating pipe;venting duct; vent(ing) pipe;vent shaft;air draft
通气管道 aerated conduit; aeration conduit; air duct; ventilating line; breather line
通气管的滤气管 breather pipe air cleaner
通气管孔 breather
通气管帽 vent cap
通气管气封 breather pipe packing
通气管头 air pipe head
通气管系统 vent system
通气管罩 ventilator cowl
通气管嘴 aeration nozzle
通气管座 ventilating pipe tray
通气环 air ring
通气机 ventilator
通气检视器 visible output
通气接管 breather adapter
通气井 airway;breathing well;ventilating shaft;vent shaft
通气坑 air pit
通气孔 air breaker;air-breather;air drain;air funnel;air hole;air vent; bleeder hole; breather; breather hole;breathing hole;burn-out pipe; gas vent;pick-up hole;snore-hole; snore piece; spiracle; vent hole; ventilating opening; ventilation vent;vent opening;whistler
通气孔阀 air vent valve
通气孔盖 vent cap
通气孔格栅 air-hole grate
通气孔连接装置 vent connector
通气孔塞 vent cover; vent peg; vent plug
通气孔隙 ventilating slit
通气孔隙度 <土壤的> aeration porosity;aerated porosity
通气控制阀 breather valve; breather vane
通气口 air port;air vent;blow vent; breather; breather vent; gas vent; vent(age);ventilating opening
通气口纯度控制 vent purity control
通气口盖 breather cap
通气蜡 vent wax
通气锂电池 vented lithium cell
通气立管 main vent;vent stack(er)
通气列板 air strake
通气流 ventilated flow
通气炉 air furnace
通气帽 abat-vent;cow
通气门 air door;loophole door;trap door
通气模量 ventilation modulus
通气模板 vented form
通气喷嘴 aeration jet
通气起始 ventilation inception
通气器 aerator blower;breather;ventilation breather
通气区 aeration zone
通气渠 aeration channel;aeration ditch
通气人孔盖 ventilated manhole cover
通气软管 air hose;steam hose
通气塞 air plug;breather plug;ventilation plug

通气栅 air grill(e)
通气设备 aeration equipment;breather;breathing apparatus;ventilating device
通气式便桶 privy vault
通气式风洞 ventilated wind tunnel
通气室 air box
通气竖管 air shaft;vent stack
通气竖坑 downcast
通气水舌 free nappe
通气隧洞 air tunnel
通气挑坎 aeration ramp
通气筒 air funnel;air hole;breather;ventilator
通气网 vent network
通气洗涤器 vent scrubber
通气系统 aerating system
通气箱 air box
通气小孔 air bleeder;spile-hole <桶的>
通气芯 atmospheric core;pencil core;puncture core;William's core
通气性 aeration;permeability
通气性能 breathing ability
通气性试验 air-permeability test
通气需要 aeration need
通气选择性沉淀 selective gas precipitation
通气液体氮化 aerated bath nitriding
通气因素 aeration factor
通气造型坑 occasional mo(u)lding pit
通气罩 vent hood
通气针 air vent needle;pricker;venting pin;vent(ing) wire;vent rod
通气支管 branch vent;vent branch
通气指数 ventilation index
通气珠状花缘 ventilating bead
通气主管 main vent;vent stack
通气砖 extract ventilation tile
通气装置 aerator;air-breather;breather;breathing apparatus
通气总管 air trunk;common air chamber
通气组织 aerating tissue;ventilating tissue
通汽软管 steam hose
通墙 parts wall;party fence;party wall
通切刀 bull-nose tool
通勤 commuting
通勤出行 commuter trip;commuting trip
通勤交通 commuter traffic;commuting traffic
通勤交通距离 commuting distance
通勤交通率 commuting ratio
通勤交通区 commuter zone
通勤交通圈 commuting sphere
通勤客流 commuter movement
通勤口 service entrance
通勤列车 service train
通勤旅客区 commuter zone
通勤圈 commutable area
通勤(上下班)往返 commute
通勤铁路 commuter transit
通勤行程 commuter journey
通勤者 commuter
通勤者共汽车 commuter
通球率 through ball rate
通球试验 through ball test
通渠孔 rodding eye
通融 financing;refinancing
通融背书 accommodation endorsement
通融背书人 accommodation endorser
通融承兑(汇票) accommodation acceptance
通融承兑人 accommodation acceptor
通融船席 accommodation berth
通融贷款 accommodation
通融贷款人 accommodator
通融付款申请 ex-gratia claim;ex-gratia application
通融汇票 accommodation bill;accommodation draft;finance bill
通融接受远期信用证 accommodate by accepting time letter of credit
通融赔款 ex gratia payment
通融票据 accommodation bill;accommodation kite;accommodation note;accommodation paper;kite
通融票据出票人 accommodation maker
通融票据关系人 accommodation party
通融申请付款 ex gratia payment
通融(受保)业务 accommodation line
通融索赔 ex gratia claim
通融提单 accommodation bill of lading
通融信贷 concessional credit
通融性 flexibility
通融性的 accommodating
通融性金融交易 accommodating financial transaction
通融有关方面 accommodation party
通融支票 accommodation check;accommodation cheque;kiting cheque
通融资金 accommodation;allow temporary credit;finance;financier;make an accommodation
通融资金背书人 endorser for accommodation
通入 admission;inlet
通入检验器 run a rabbit
通入蒸汽 steaming
通商 commercial intercourse;have trade relations;trade
通商保护法 trade safeguarding;trade safe-guarding act
通商待遇公平 equity treatment of commerce
通商港口 open harbo(u)r;open port
通商航海条约 treaty of commerce and navigation
通商局 Bureau of Commercial Affairs
通商口岸 open port;trading port;treaty port <条约规定的>
通商路线 trade route
通商破坏舰 commerce-destroyer
通商条款 commerce clause
通商条约 commercial treaty;trade treaty
通商通航条约 treaty of commerce and navigation
通商协定 agreement on commerce;trade agreement
通商许可证 license to trade
通商自由 freedom of commerce
通商自由港 open port
通式 general expression;general form;general formula;general normal equation
通式取样 pitcher barrel sampler
通视 exist visibility;exit visibility;intervisibility
通视不良 poor visibility
通视范围 observation range;visual range;visual sector
通视浑浊度仪 all-through turbidimeter
通视角 angle of visibility
通视距离 observation distance;observation range;visibility distance;visibility range
通视能见度检验仪 all-through visibility tester
通视区 <即在交叉口、转角处或曲线处的展宽带> visible zone
通视试验 intervisibility test
通视条件 sighting condition;visibility condition
通视条件图 intervisibility map
通视性 intervisibility
通视障碍 visibility restriction
通水管路 water passage
通水接头 flush head;water coupling
通水孔 limber
通水平巷 ventilation lateral
通水试验 water test
通俗的 exoteric;vulgar;trivial <指名称>
通俗文化 pop culture
通俗象征主义 popular symbolism
通俗艺术 pop art
通索孔 <滑车等> swallow
通体尺寸 overall dimension
通天插销 Cremo(r)ne bolt;double door bolt;espagnolette bolt;level bolt
通天长插销开关 espagnolette
通天通气管 continuous vent
通天用天窗 femerell
通条 cleaning rod;gavelock;poker;ramrod
通铁棒 tap-out bar
通往地下古墓或古庙的通道 dromos
通往工地入口 access to a site
通往海洋 access to the sea
通往客室门 door to passenger compartment
通往码头泊位的栈桥 access trestle
通往桥的道路 approach of bridge
通往下水道 to drain
通往乡村住宅的林荫小路 <英> avenue
通往自然景区的小径 nature trail
通洗管 rodding pipe
通洗口 rodding opening
通洗装置 rodding fitting
通向 lead up
通向地面的人行道 boutgate
通向房屋的车道 vanway
通向建筑物的连接管 building connection
通向牧场道路 ranch access road
通向农场道路 ranch access road
通向室外的门窗口 outdoor opening
通向室外的烟道 uptake
通向用具的管道 branch line
通向中央舞台的进出口 arena vomitory
通项 general term
通宵 overnight
通宵存车 overnight parking
通宵服务 all-night service;overnight service
通宵检修 all-night service;overnight service
通宵停车 overnight parking
通晓 acquaint oneself with
通晓的 deep read
通晓数门知识者 generalist
通信[讯] bonding;communicate;communication;correspondence;signal(l)ing;traffic
通信[讯]安全 communication security
通信[讯]保密 communication security
通信[讯]报告 letter report
通信[讯]兵补充人员训练中心 signal replacement training center[centre]
通信[讯]波段 communication band
通信[讯]不清 miscible
通信[讯]车 communication car
通信[讯]部件 communication element
通信[讯]操作台 communication operation station
通信[讯]车天线 over-car antenna
通信[讯]成员国委员会 Committee of Communication Member Countries
通信[讯]程序 communication programme);signal procedure
通信[讯]程序包 communication package
通信[讯]冲突 communication conflict
通信[讯]处 communication department;communication department office;signal office
通信[讯]处理机 communication processor
通信[讯]处理器 communication processor
通信[讯]处理系统 communication processing system
通信[讯]处理中心 communications processing center[centre]
通信[讯]船 advice boat;aviso boat;communication boat;communication ship;despatch boat;dispatch boat;dispatch vessel
通信[讯]存款 mail deposit
通信[讯]大队 communication group
通信[讯]导航设施 communication and navigation facility
通信[讯]地址区 channel address field
通信[讯]电缆 communication cable;sound cable
通信[讯]电路 communication circuit
通信[讯]电路载频 channel carrier frequency
通信[讯]电源闸 power source room for communication
通信[讯]电子学 communication electronics
通信[讯]调查 mail inquiry;mail survey
通信[讯]端局 communication terminal station
通信[讯]端站 terminal office for communication
通信[讯]段 communication section
通信[讯]断开时间 communication clearing time
通信[讯]对抗 communication countermeasure
通信[讯]多路传输机 communication multiplexer
通信[讯]多路复用器 communication multiplexer
通信[讯]多路转接器 communication multiplexer
通信[讯]发射机 communication transmitter
通信[讯]方案 communication scheme
通信[讯]方式 signal(l)ing;signal(l)ing method
通信[讯]方式控制 communication mode control
通信[讯]仿真系统 communication emulator
通信[讯]费(用) communication expenses
通信[讯]服务器 communication server
通信[讯]辅助设备 communication auxiliary equipment
通信[讯]干扰器 communication jammer
通信[讯]工厂 telecommunication factory
通信[讯]工程 communication engineering
通信[讯]工具 communication tools;means of communication;telecommunication facility
通信[讯]工作人员 signalman
通信[讯]工作站 communication workstation
通信[讯]公司 common carrier
通信[讯]观察法 investigation from correspondent method
通信[讯]官 signal officer

通信[讯]管理 call management; telecommunication management
通信[讯]光缆 communication optic(al) cable
通信[讯]规约 communication protocol
通信[讯]号 communicator
通信[讯]和数据组合系统 communications and data system integration
通信[讯]呼号 call letter; call sign; ship's number; signal letters
通信[讯]缓冲器 communication buffer
通信[讯]缓冲区 communication buffer
通信[讯]会员 corresponding member
通信[讯]机 telegraph
通信[讯]机偏压用电池 Acorn cell
通信[讯]机试验器 set analyser[analyzer]
通信[讯]机试验仪 set analyser[analyzer]
通信[讯]计时器 communication timer
通信[讯]计算机 communication computer
通信[讯]记录 communication log
通信[讯]记者 correspondent
通信[讯]技术 communications art; communication technique; communication technology; mechanics of communication
通信[讯]技术卫星 communication technical satellite
通信[讯]寄存装置 communication register unit
通信[讯]架设中队 communication construction squadron
通信[讯]监督系统 traffic monitoring system
通信[讯]建筑 communication building
通信[讯]舰 aviso
通信[讯]接口 communication interface
通信[讯]接收机 communication receiver
通信[讯]矩阵 communication matrix
通信[讯]距离 communication distance
通信[讯]距离延长台 extender
通信[讯]科 communication section
通信[讯]可靠度 communication reliability
通信[讯]控制程序包 communication control package
通信[讯]控制处理机 communication control processor
通信[讯]控制处理装置 communication control processor
通信[讯]控制单元 communication control unit
通信[讯]控制点 communication controller node
通信[讯]控制符号 communication control character
通信[讯]控制规程 communication control procedure
通信[讯]控制器 communication controller; communication control unit
通信[讯]控制设备 communication control unit
通信[讯]控制台 communication console
通信[讯]控制系统 communication control system
通信[讯]控制中心 communication control center[centre]
通信[讯]控制字符 communication control character
通信[讯]口 communication port
通信[讯]困难 difficult communication
通信[讯]理论 <研究所传递的信息性质的数学理论> communication theory; information theory
通信[讯]连接控制器 communication linkage controller
通信[讯]连接设备 communication linkage
通信[讯]连接装置 communication link; tie line
通信[讯]联络 communication link; liaison; signal communication
通信[讯]联络的 communicative
通信[讯]联络中断 loss of contact
通信[讯]链(路) communication link
通信[讯]量 communication traffic; traffic; traffic load
通信[讯]量分析 traffic analysis
通信[讯]量交换局 traffic exchange office
通信[讯]量数据 traffic data
通信[讯]量速率 traffic rate
通信[讯]量拥挤 traffic congestion
通信[讯]流量 communication flows
通信[讯]录 address
通信[讯]滤波器 telefilter
通信[讯]路径 communication path
通信[讯]论 communication theory
通信[讯]码流保密 traffic flow security
通信[讯]媒介 communication medium [复 media]
通信[讯]密度 intensity of traffic
通信[讯]密码 communication code; signal code
通信[讯]模块 communication module
通信[讯]模型 traffic model
通信[讯]能力 ability to communicate; communication capacity
通信[讯]频带 communication band
通信[讯]频率干扰 radio frequency interference
通信[讯]器 communicator
通信[讯]情报 signal intelligence
通信[讯]情报报告 signal intelligence report
通信[讯]情报处 signal intelligence service
通信[讯]情报科 signal intelligence section
通信[讯]区 communication zone
通信[讯]区域 communication area
通信[讯]区域指示器 communication zone indicator
通信[讯]全套设备 communication complex
通信[讯]任务 communication task
通信[讯]日记簿 signal record book
通信[讯]容量 communication capacity; message capacity
通信[讯]软件 communication software
通信[讯]闪光灯 light flashing signal; signal light; signal(l)ing lamp
通信[讯]设备 communication apparatus; communication equipment; communication facility; communicator; signal equipment; telecommunication facility
通信[讯]设备室 communication equipment room
通信[讯]设施 communication utility
通信[讯]申请 apply by letter
通信[讯]实体 communication entity
通信[讯]适配器 communication adapter
通信[讯]枢纽 communication center [centre]
通信[讯]数据处理机 communication data processor
通信[讯]数据处理器 communication data processor
通信[讯]数据处理装置 communication data processing
通信[讯]数据系统 communication data system
通信[讯]台 called station
通讯[信]体系 communication system
通信[讯]铜线 communication copper wire
通信[讯]筒 streamer
通信[讯]投标 correspondence bidding
通信[讯]图 communication scheme; traffic diagram
通信[讯]网管理电台 net control station
通信[讯]网呼号 net call sign
通信[讯]网控制程序 communication network control program(me)
通信[讯]网(络) communication network; network of communication
通信[讯]网络处理机 communication network processor
通信[讯]网损耗 net loss
通信[讯]维修功能 communication serviceability facility
通信[讯]委员 corresponding member
通信[讯]卫星 communication satellite; comsat; geostationary satellite; telesat [telecommunications satellite]; telstar
通信[讯]卫星操作控制中心 operational control center [centre] of COMSAT
通信[讯]卫星测控站 communication satellite tracking telemetry and command station
通信[讯]卫星地面站 communication-satellite earth station
通信[讯]卫星覆盖范围 communication satellite coverage
通信[讯]卫星覆盖区 coverage area of communication satellite
通信[讯]卫星公司 <美> Communications Satellite Corporation
通信[讯]卫星轨道 communication satellite orbit
通信[讯]卫星空间站 telecommunication satellite space station
通信[讯]卫星上转发器 transponder
通信[讯]卫星太空电台 communication-satellite space station
通信[讯]卫星转发器 communication satellite transponder
通信[讯]文字处理机 communicating word process; communication word processor
通信[讯]系统 communication; communication system; reporting system; transducer
通信[讯]系统包 package of communication system
通信[讯]系统的模型 model of communication system
通信[讯]系统控制中心 signal systems control center[centre]
通信[讯]系统设计 communication design
通信[讯]线 order wire
通信[讯]线路 communication circuit; communication link; interchange circuit; telecommunication line; tie line; communication line
通信[讯]线路保密 link encryption
通信[讯]线路处理机 communication link handler
通信[讯]线路的复用 multiplexing of communication line
通信[讯]线路绝缘子 communication line insulator; telephone insulator
通信[讯]线路控制信息 link control message
通信[讯]线路连接器 communication adapter
通信[讯]线路折旧费 communication line depreciation expenses
通信[讯]线路终端 communication line terminal
通信[讯]向量表 communication vector table
通信[讯]小组 communication group
通信[讯]协议 communication protocol
通信[讯]信道 communication channel
通信[讯]信号 signal of communication
通信[讯]信号处 signal(l)ing and communication department
通信[讯]信号机【铁】block signal
通信[讯]信号设备折旧费 communication and signal(l)ing equipment depreciation expenses
通信[讯]信息转移通路 communication bus
通信[讯]学 telecommunications
通信[讯]询问 traffic inquiry
通信[讯]延迟 communication delay
通信[讯]业务 communication service
通信[讯]业务的流向 traffic route
通信[讯]业务量 communication traffic volume; message volume
通信[讯]业务量密度 traffic density
通信[讯]应用程序接口 messaging application programming interface
通信[讯]营 signal battalion
通信[讯]用单元 communication cell
通信[讯]用户程序 communication user program(me)
通信[讯]用激光器 communication laser
通信[讯]用声呐 communication sonar
通信[讯]油基 congestion of traffic
通信[讯]有误 miscible
通信[讯]与跟踪系统 communication and tacking system
通信[讯]与跟踪子系统 communication and tracking subsystem
通信[讯]与数据传输系统 communications and data-link system
通信[讯]语言 communication language
通信[讯]预处理机 communication preprocessor
通信[讯]员 communicator; correspondent; messenger; orderly
通信[讯]载重汽车 signal truck
通信[讯]站 communication site; communication station
通信[讯]征求意见(法) post card questionnaire
通信[讯]证明 correspondence proving
通信[讯]执行程序 communication executive
通信[讯]值班台 patrol desk
通信[讯]指挥 traffic guidance
通信[讯]中断率 communication interruption rate
通信[讯]中继船 communication relay ship
通信[讯]中继线 communication trunk
通信[讯]中继站 communication relay (station)
通信[讯]中心 communication center [centre]; message center[centre]
通信[讯]中心站 communication report center[centre]; communication tie-station; report center[centre]
通信[讯]终端 communication terminal
通信[讯]终端设备 communication terminal equipment
通信[讯]周期 communication cycle
通信[讯]转接 communication switching
通信[讯]转接部件 communication switching unit
通信[讯]转接器 communication adapter; communication switching unit

通信[讯]装置 communicator
通信[讯]资源 communication resources
通信[讯]子系统 communication subsystem
通信[讯]自动监听器 automatic communication monitor
通信[讯]总线 communication bus
通信[讯]作业 communication task
通信[讯]作业区标志 communication operating zone identification
通信[讯]作业室 signal operation room
通信[讯]作业须知 signal operation instruction
通信[讯]作业营 signal operation battalion
通信[讯]作业中心 signal operations center[centre]
通行 thoroughfare
通行坝 navigation dam
通行标记 transit marker
通行冰况标准 operational ice criterion
通行车辆 traffick
通行程度等级 degree of pass through
通行程度图 trafficability map
通行船闸 navigable lock;navigation lock
通行道 passageway;wayleave
通行定位系统 transit position fixing system
通行费 passage fare; passage money; toll;toll charges;toll dues;wayleave
通行服务设施 passing service facility
通行管沟 passable trench
通行轨道 < 指铁路 > running track
通行过路费 road toll
通行汇率 going rate;prevailing rate
通行货币 current money
通行价格 observed price; prevailing price
通行价格时价 going price
通行检查站 gate house
通行卡 visa card
通行跨度 navigable span;traffic span < 桥隧的 >
通行利息 current interest
通行路面 traffic surface
通行路线 pass-course
通行路线指示牌 < 公共汽车等 > destination board
通行美元 current dollar
通行门 pass door
通行面积 passage area
通行能力 capacity; discharge capacity; trafficability; traffic capacity; traffic-carrying capacity; transportation capacity
通行能力调查 capacity study
通行能力降低 capacity reduction
通行能力-交通需求分析(法) capacity-demand analysis
通行能力控制模式 capacity mode
通行能力限制 capacity restraint
通行能力限制交通分配【交】 capacity restrained assignment
通行能力研究 capacity study
通行能力指数【交】 capacity index
通行球节 through ball joint
通行权 easement(of way); passage; right of ingress; right of way; wayleave; right of access
通行权规则 right-of-way rule
通行设施 access facility
通行试验 performance test
通行水道 navigable pass; navigable water
通行税 pike; road tax; road toll; toll; toll charges; toll on transit; toll tax; transit dues; transit duty; transit tax
通行税收费处 toll house
通行税征收所 toll house;turnpike
通行税征税关卡 toll bar;toll gate
通行税征税所 toll house
通行税征税员 toll keeper;toll man
通行屋顶 access roof
通行无阻 passage clear
通行线 < 指铁路 > open track; running line; running track; service track;through track
通行限界 traffic ga(u)ge
通行信号 passage signal; right of way signal
通行信号标 traffic control signal mark
通行信号时间 traffic movement phase
通行信号台 traffic control signal station
通行信号显示 traffic movement phase
通行许可 < 交通 > safe conduct
通行运河 navigable canal; navigation canal
通行证 laissez-passer; permit; protection; safe conduct; pass-check < 车辆 >; traffic permit < 车辆 >
通行字地址 password address
通行阻力 mobilizable resistance
通行阻滞 < 交通 > stream friction
通性 general characteristic
通讯报导 news dispatch
通讯编辑 editorial correspondent
通讯社 news agency; news service; press
通讯员 orderly
通衢 thoroughfare
通烟囱 chimney sweeping
通烟道 smoke uptake
通烟试验 < 检查管道方法 > smoke test
通洋运河 interoceanic canal; transoceanic canal
通夜照明灯 night light(ing)
通阴沟 unstop a drain
通阴沟工人 < 俚语 > sewer hog
通阴沟条 sewer rod
通用 all-function; common usage; current general; general usage; general use; multiapplication
通用按钮开关 general-use snap switch
通用按钮箱 universal button box
通用拔桩机 universal pulling machine
通用扳手 all wrench; general utility wrench
通用板材 ga(u)ge plate
通用包络代数 universal enveloping algebra
通用泵 convectional pumping unit; general purpose pump; universal pump
通用比长仪 universal comparator
通用比较器 general comparator
通用比例尺 general scale
通用比重计 panhydrometer; universal hydrometer
通用编译程序器 general compiler
通用变数【数】 generic variable
通用变压器 general purpose transformer
通用标志 conventional sign
通用标志和符号 general sign and symbol
通用标准 acceptable standard; accepted standard; generally accepted standard; general standard; universal standard; working standard; general criterion
通用表 univariate meter
通用表面处理机 universal finisher
通用表面修整机 universal finisher
通用表示法 conventional representation
通用并行语言 general parallel language
通用部分 generic element
通用部件 common component
通用材料模型 universal material model
通用材料试验机 universal material testing machine
通用财务报表 all-purpose financial statements
通用参照(条目) general cross reference
通用操纵器 general purpose manipulator
通用操作程序 general purpose operating program(me)
通用操作系统 general purpose operating system
通用槽车 general purpose tank car
通用测量仪表 universal instrument; universal measuring instrument
通用测量仪器 multimeter; universal instrument
通用测试器 general purpose tester
通用测试信号发生器 general purpose test-signal generator
通用测试仪器 general testing meter
通用插件 universal card
通用插件板 universal board
通用插座 blanket socket
通用查询 general polling
通用产品 general goods; universal products
通用产品代码 universal product code
通用铲叉 utility fork
通用铲斗 bucket for general purpose; multipurpose bucket; utility bucket
通用铲运机 carryall scraper
通用常数 general constant; universal constant
通用车 conventional car
通用车载无线电设备 universal vehicular radio
通用程序 generalized program(me); generalized routine; general (purpose) program(me); general routine; universal program(me)
通用程序设计 generalized programming; general-purpose programming
通用程序设计语言 common programming language; generalized programming language
通用尺寸的薄钢板 utility sheet
通用齿轮润滑剂 regular-type gear lubricant
通用齿轮润滑油 regular-type gear lubricant
通用齿轮润滑脂 universal gear lubricant
通用齿轮油 universal gear lubricant
通用出入港许可 blanket clearance
通用处理程序 general processor program(me); universal processor
通用处理过程控制器 universal process controller
通用处理函数 universal function
通用处理机 general processor; general purpose processor
通用触发器 general purpose flip-flop
通用船 general purpose ship; universal type ship
通用船舶建造规范 general specifications for the building of ship
通用串行总线 universal serial bus
通用辞典 general dictionary
通用丛 universal bundle
通用存储方程 general storage equation
通用存储器 general purpose memory; general purpose storage
通用打桩机 universal pile driver; universal pile driving plant; universal piling plant
通用打桩设备 universal pile driving plant
通用大口径铜管 heavy ga(u)ge copper tube for general purposes
通用刀 all-purpose knife
通用刀板 universal blade
通用刀架 universal tool post
通用刀具 universal cutter
通用刀头 all-purpose bit
通用导航计算机 general navigation computer; universal navigation computer
通用导航信标 general navigation beacon; universal navigation beacon
通用的 all-around; all-purpose; commercial; common; general duty; general purpose; general utility; interchangeable; multipurpose; universal; varsal; versatile; general service
通用的弛豫时间 universal relaxation time
通用的模拟系统程序 general purpose simulation system
通用的组织设计理论 universal design theory
通用灯 utility lamp
通用登陆舰 utility landing ship
通用登陆艇 utility landing craft
通用滴滤池 conventional trickling filter
通用底盘 universal bed
通用底漆 all-purpose primer; universal primer
通用抵押契约 fictitious mortgage
通用地毯 anywhere carpet
通用地震仪 universal seismograph
通用地址阅读装置 general address reading device
通用点 universe point
通用电表 universal electric(al) meter; universal meter
通用电动机 general motor; general purpose machine; general purpose motor; universal electric (al) motor; universal motor
通用电话电子公司 < 美 > General Telephone Electronics Corporation
通用电话(交换)网 general switched telephone network
通用电缆 general purpose cable
通用电流计 universal galvanometer
通用电气公司 < 美 > General Electric Co. Limited
通用电器公司 General Electric(al) Company
通用电桥 universal bridge
通用电位计 universal potentiomerer
通用电像系统 general electric(al) image system
通用电源 multiple power source
通用电源接收机 all-electric(al) receiver; all-electric(al) set; all-mains set
通用电源收音机 all-electric(al) set; all-mains set
通用电子管 general vacuum tube
通用吊桶 general purpose bucket
通用定标器 multiscaler
通用定额资料 universal standard data
通用定位纸 universal plotting sheet
通用动力公司 < 美 > General Dynamics Corp.
通用动态计量仪 generally used dynamic(al) quantifier
通用对准仪 univariate aligner
通用多极 universal multipole
通用多支电路 general purpose branch circuit

T

通用耳塞 universal ear insert
通用阀 universal valve;valve for general use
通用翻斗 general purpose tipper
通用反应堆 general purpose reactor
通用方法 universal method
通用防锈剂 multipurpose inhibitor
通用防锈漆 anti-corrosive paint
通用仿真程序 general purpose simulation program(me)
通用放大 universal amplification
通用放大管 general purpose amplifier tube
通用放大器 general purpose amplifier;universal amplifier
通用放大曲线 universal amplification curve
通用飞机 general purpose aeroplane;general purpose aircraft;general purpose plane;utility aircraft;utility aviation
通用飞机公司 <美> General Aircraft Corporation
通用费用 utility cost
通用分辨力 general resolution
通用分带序列 universal zoning
通用分类程序 generalized sort program(me)
通用分流器 general shunt;universal shunt
通用分组无线电业务 general packet radio service
通用弗兰德利吸附等温线 conventional Freundlich isotherm
通用浮标(装置) universal buoyage
通用符号 conventional sign;conventional symbol;general symbol;universal signal
通用符号集【计】universal character set
通用辅助操作 general utility function
通用辅助设备 general purpose accessory
通用附件 general purpose accessory
通用复式炸弹架 universal multiple bomb rack
通用覆盖 universal covering
通用覆盖空间 universal covering space
通用覆盖流形 universal covering manifold
通用覆盖面 universal covering surface
通用覆盖群 universal covering group
通用干燥机 all-purpose drying unit
通用干燥剂 multipurpose dryer[drier]
通用钢 primary steel
通用钢板 universal steel plate
通用钢拱架 universal steel centering
通用港(口) general purpose harbo(u)r;general purpose port
通用高铜合金 common high brass
通用格式 general format
通用格式提单 common form bill of lading
通用隔声材料 all-purpose insulation
通用个人通信[讯] universal personal telecommunication
通用跟踪功能 generalized trace facility
通用工兵牵引车 universal engineer tractor
通用工程 general engineering
通用工程工具 general engineering tool
通用工程系统 general engineering system
通用工具 general purpose tool;general utility tool;multipurpose instrument
通用工字钢梁 universal I-beam
通用工作灯 utility light
通用公式 general equation;general formula
通用拱架 common centering
通用钩 conventional hook
通用构件 general member;standard construction element;standard member
通用构件系统 general member system
通用估算方法 general estimate method
通用鼓风机 universal blower
通用管箍 universal clamp
通用光度计 universal photometer
通用光栅系统 uniras
通用规则 blanket rule;rule-of-thumb
通用辊压机 universal roller
通用滚压机 universal roller
通用国际浮标系统 Universal International Buoyage System
通用国际量规 universal international ga(u)ge
通用海关用语 common tariff nomenclature
通用海图图式 conventional signs and abbreviations
通用函数 general purpose function
通用函数发生器 arbitrary function generator;general purpose function generator;universal function generator
通用函数发生器部件 universal function generating unit
通用行间中耕机 multipurpose inter-row cultivator
通用航空 general aviation;utility aviation
通用合同条款 general conditions of contract
通用盒子构件 universal box
通用横向磁迹 general transverse tracks
通用横轴墨卡托投影 universal transverse Mercator projection
通用横轴墨卡托投影坐标 universal transverse Mercator coordinate
通用宏功能生成程序 general purpose macrogenerator
通用后处理程序 universal postprocessor
通用互见条目 general cross reference
通用滑车 patent block
通用化 commonality;generalized;unification
通用化常数 generalized constant
通用化的 unitized
通用化设计 unitized design
通用环节 univariate link;universal link
通用环式破碎机 universal ring-type crusher
通用换算 multiscale
通用换算线路 multiscaler
通用黄油 multipurpose-type grease
通用混凝土骨料 all-purpose concrete aggregate
通用活度系数 conventional activity coefficient
通用活性污泥法 conventional activated sludge process
通用货币 current money
通用货币单位 current money unit
通用货车 general wagon
通用货船 all-purpose cargo ship;combination carrier;combined carrier
通用货单代码 universal product code
通用货轮 general purpose cargo vessel;multipurpose cargo carrier
通用机车 general purpose locomotive;general purpose machine
通用机床 universal machine tool
通用机架 implement carrier;implement porter;tool bar;toolbar carrier;tool frame
通用机架式中耕机 toolbar cultivator
通用机架总成 toolbar assembly
通用机具架 tool bar
通用机器 general purpose machine;universal machine
通用机器语言 common machine language;general purpose language
通用机械 all-purpose machinery;universal machine
通用机械厂 universal machine works
通用机械式推土机 toolbar bulldozer
通用机械手 general purpose manipulator;robot;Unimate;universal manipulator
通用机械制造 sectionalized machine manufacture
通用机油 all-purpose engine oil;multiservice oil
通用基本数据 general purpose data base
通用基础 universal bed
通用基础地质数据库 general basic geologic(al) data base
通用极球面投影格网 universal polar stereographic(al) grid
通用极球面投影坐标网 universal polar stereographic(al) grid
通用集 universal set
通用集装箱 dry cargo container;general purpose container;international pallet;pool pallet;universal container;universal utility container
通用计数器 universal counter
通用计数式计算机 verdan;verdant [versatile digital analyzer]
通用计算机 all-purpose computer;general computer;general purpose computer;multipurpose computer;universal computer
通用计算机程序 generalized computer program(me)
通用计算机语言设备寄存器 universal machine language equipment register
通用计算器 all-purpose calculator;general purpose calculator;universal calculator
通用计算图式 versatile model
通用计算系统 general purpose computing system
通用技术保养站 superservice station
通用技术标准 reference standard;general specification;standard specification
通用技术条件 general technical specifications
通用技术卫星 general technology satellite
通用继电器 general purpose relay
通用寄存器 general purpose register
通用寄存器单元 general register unit
通用寄存器地址 general address;general register address
通用寄存器文件 general purpose register file
通用寄存器组 general purpose register file
通用价值尺度 common utility scale
通用价值指标 common utility scale
通用架 univariate stand;universal stand
通用监察器 general purpose monitor
通用监控 general surveillance monitoring
通用监视装置 general monitor unit
通用剪切机 multipurpose shearing machine
通用检测器 all-purpose detector
通用建筑 utility architecture
通用建筑体系 open system
通用键 universal key
通用键控器 general purpose manipulator
通用键盘 universal keyboard
通用键组 universal keyset
通用交换台 <磁石、共电两用电话交换台> universal switchboard
通用(交直流两用)电动机 universal motor
通用胶管 general service hose
通用胶凝剂 general purpose cement;general purpose cementing agent
通用胶粘剂 general purpose bonding medium;universal cementing agent
通用焦虑量表 general anxiety scale
通用绞车 universal hoist
通用搅拌机 universal mixer
通用接口 general purpose interface
通用接口适配器 general purpose interface adapter
通用接口总线 general purpose interface bus
通用接收机 general purpose receiver;universal receiver;universal set
通用结构电机 convenient construction machine
通用结算表 general purpose statement
通用解释程序 general interpretative program(me)
通用金属箔 general purpose foils
通用经纬仪 universal theodolite;universal transit
通用晶体管指示仪器 versatron
通用剧场 universal theater[theatre]
通用决算表 all-purpose financial statements;general purpose statement
通用绝热材料 all-purpose insulation
通用绝热隔声材料 all-purpose insulation
通用卡盘 universal chuck
通用卡钳 leg cal(l)ipers
通用开表壳器 universal case opener
通用开关 universal switch;general-use switch
通用客车 all-purpose coach
通用空白海图 universal plotting sheet
通用空间 universal space
通用控制阀 universal control equipment
通用控制器 general purpose controller;universal controller
通用控制台 general service desk
通用库【计】universal library
通用矿物质混合料 all-purpose mineral mix
通用扩散曝气池 conventional diffused aeration tank
通用拉出器 general utility puller
通用拉软机 universal staking machine
通用栏 unclassified column
通用雷达 general purpose radar
通用肋剂 common agent
通用冷藏船 general refrigerated ship;general refrigerated vessel
通用犁 general purpose plow
通用犁铧 general share;regular-cut share
通用犁体 general purpose body;general purpose bottom
通用(沥青)针入度仪 universal penetrometer
通用例程 general purpose routine
通用连杆对准器 universal connecting rod aligner
通用连接器 combination connector;regular connector
通用连接装置 universal interconnec-

ting; universal interconnecting device
通用联轴节 universal coupler; universal coupling
通用链 universal chain
通用链节 univariate link
通用梁 universal beam
通用量测仪 multimeter
通用量测仪表 universal instrument
通用量规 go-ga(u)ge
通用量热器 universal calorimeter
通用列车 utility train
通用零备件 parts common
通用炉 versatible furnace
通用炉黑 all-purpose furnace black; general purpose furnace black
通用滤波器 universal filter
通用逻辑模块 universal logic module
通用逻辑元件 universal logic element
通用逻辑组件 universal logic module
通用螺纹切头 plurality dies
通用马达 universal motor
通用码头 general purpose terminal
通用脉冲形成器 versatile pulse shaper
通用毛细管黏[粘]度计 all-purpose capillary viscosimeter
通用门式起重机 gantry crane for general use
通用密封件 common seal
通用面粉公司<美> General Mills Inc.
通用面罩 universal facepiece
通用灭火器 all-purpose extinguisher
通用名称 common name; versatile nomenclature
通用名词 generic term
通用模架 universal die set
通用模拟计算机 general purpose analog(ue) computer
通用模拟系统 general purpose simulation system
通用模式 versatile model
通用磨床 universal grinding machine
通用木材加工机 general joiner
通用木工机床 universal woodworker
通用内燃机油 general service oil
通用黏[粘]结剂 general purpose adhesive
通用黏[粘]土砖 general purpose burnt(clay) brick; general purpose fired(clay) brick
通用捏练机 universal kneading machine
通用排放标准 general purpose discharge standard
通用排水泵 general service pump
通用判断元件 universal decision element
通用培养基 collective medium
通用配合比 standard mix
通用配件 interchangeable parts; universal spare parts
通用喷燃器 universal burner
通用棚车 all-purpose box car; general purpose box car
通用篷车 general purpose van
通用疲劳试验机 univariate fatigue testing machine; universal fatigue machine
通用片蚀方程 universal sheet-erosion equation
通用票据 passable bill
通用频率 universal frequency
通用平衡方程 general balance equation
通用平面设计图 standard plan
通用铺料厂 multipurpose coating plant
通用起阀器 universal valve lifter
通用起重机 all-terrain crane; convertible crane; univariate crane; universal crane
通用气体常数 universal gas constant
通用气体定律 general gas law
通用气象色谱仪 universal gas chromatograph
通用汽车公司 General Motors; General Motor Corporation<美>
通用汽车公司研究所 General Motor Research
通用汽油 all-purpose gasoline
通用契据 fictitious instrument
通用器件 general purpose device
通用钳 engineer's pliers; universal forceps; universal pliers
通用潜水器 versatile underwater submersible; versatile underwater vehicle
通用潜艇模拟器 universal submarine simulator
通用嵌缝膏 general purpose sealant
通用桥式起重机 general purpose overhead crane
通用切碎碾磨机 all-purpose chopper mill
通用清漆 general purpose varnish
通用曲线 master curve; universal curve
通用取景器 universal viewfinder
通用缺省值 universal default
通用燃料箱组 universal fuel tank kit
通用燃气炉 all gas burner
通用燃烧器 all gas burner; universal burner; universal combustion burner
通用热容校正 generalized heat capacity correction
通用软件 common software; general purpose software; universal software
通用润滑剂 all-purpose grease; general purpose grease; multiservice oil; universal grease; universal lubricating grease
通用润滑脂 universal grease
通用润剂 multipurpose-type lubricant
通用润脂 multipurpose-type grease
通用塞绳电路 universal cord circuit
通用塞氏黏[粘]度计 universal Saybolt visco(si)meter
通用散(装)货船 general purpose bulk ship; universal bulk carrier; universal bulk ship
通用扫描算法 general scanning algorithm
通用色浆 universal tinter
通用纱线支数计 universal thread counter
通用筛号 scalping number
通用上浆干燥机 universal starching and drying machine
通用设备 flexible unit; general equipment; general purpose equipment; universal equipment
通用设备包 general equipment package
通用设计 conventional design; general design; interchangeable design; stable design
通用设计方法 conventional design method
通用摄影机 all-purpose camera; general purpose camera; universal camera
通用审计方案 general purpose audit program(me)
通用施工吊车 universal building crane
通用十进制分类法 universal decimal classification
通用十进制系统 Brussels system
通用时间 universal time
通用时间坐标 universal time coordinate
通用实验安培计 univariate test ammeter
通用食品公司<美> General Foods Corp.
通用矢量 vector general
通用示波器 general purpose oscilloscope
通用式 general purpose type; universal type
通用式锯 all-purpose saw
通用式络筒机 universal winder
通用式耐磨试验仪 universal wear tester
通用式捻线机 universal twisting frame
通用式纱支测定天平 universal yarn numbering balance
通用式拖拉机 universal tractor
通用式挖掘机 all-purpose excavator
通用式挖土机 all-purpose excavator
通用式线圈 universal coil
通用式针阀 convectional type needle valve
通用试验 performance test
通用试验安培计 universal test ammeter
通用试验机 universal test(ing) machine
通用试验台 universal test bench
通用试纸 universal indicator paper
通用手钳 general purpose pliers
通用输出变压器 universal output transformer
通用输入方式 general purpose entry method
通用输入或输出 universal input or output
通用术语 common terminology; generic term
通用树脂 general-use resin
通用数据 conventional data
通用数据处理 general data processing
通用数据管理系统 general purpose data management system
通用数据话音多路复用 universal data voice multiplexing
通用数据信道 universal data channel
通用数字地图数据库 universal digital cartographic(al) data
通用数字读出机 common digitizer
通用数字计算机 all application digital computer; general digital computer; general purpose digital computer; universal digital computer; versatile digital computer
通用数字转换器 common digitizer
通用数字自动驾驶仪 universal digital autopilot
通用双壁开沟犁体 general purpose middle breaker bottom
通用水泥砂浆 common cement mortar; standard paste
通用水下移动式机器人 universal underwater mobile robot
通用水准器 universal level
通用税则 general tariff
通用撕断纸带系统 universal torn tape system
通用速度分布图 universal velocity profile
通用塑料 general purpose plastics
通用算法 universal algorithm; very general algorithm
通用锁 convectional lock
通用台(架) universal stand; univariate stand
通用条 universal bar
通用条播机 all crop drill
通用条分法 generalized slice method
通用条款 interchangeable term
通用铁路信号<美> General Railway signal
通用铁路信号公司<美> General Railway Signal Company
通用停车架<停自行车用> universal stand
通用艇 all-purpose launch
通用通道存储器 unitized channel storage
通用通风机 utility fan
通用通信[讯]适配器 common communication adapter
通用通信[讯]系统 all-purpose communication system
通用通信[讯]转换装置 universal communications switching device
通用同步齿轮箱 all-synchromesh gearbox
通用同步示波器 universal synchroscope
通用投影 multipurpose projection
通用透射函数 universal transmission function
通用图 conventional drawing; currency drawing; standard drawing; universal drawing
通用图灵机 universal Turing machine
通用土壤流失公式 universal soil loss equation
通用湍流常数 universal turbulence constant
通用推土机 universal blade
通用拖拉机 all-purpose tractor; common tractor; general purpose tractor; utility tractor
通用挖斗 multipurpose bucket
通用挖掘机 all-purpose shovel-crane; convertible excavator; convertible shovel; universal excavator
通用挖土车 universal mobile excavator
通用挖土机 all-purpose shovel-crane; convertible excavator; convertible shovel; universal excavator; universal mobile excavator
通用外特性 universal external characteristic
通用外围控制器 universal peripheral controller
通用网络 universal network
通用网络模拟器 generalized network simulator
通用微程序设计 general purpose microprogramming
通用微探针分析器 universal microprobe analyser[analyzer]
通用微型计算机 minicomputer
通用文件 fictitious document; general file
通用文件处理系统 generalized file processing
通用文件翻译程序 general file translator
通用紊流常数 universal turbulence constant
通用问题解算机 general problem solver
通用问题求解程序 general problem solver
通用无线电接收机 general purpose radio receiver
通用熄弧电抗器 universal arc suppressing reactor
通用铣床 universal miller; universal milling machine
通用系统 general purpose system
通用系统仿真语言 general purpose systems simulation
通用系统模拟程序 general purpose system simulator
通用系统模拟(语言) general purpose

system simulation
通用细部设计图 standard detail
通用显示逻辑 common display logic
通用显示系统 general purpose display system
通用线路 universal link
通用限界(架) universal ga(u)ge
通用详图 standard details
通用橡胶 general purpose rubber
通用削匀机 universal shaving machine
通用消弧电抗器 universal arc suppressing reactor
通用小艇 utility boat
通用楔形管片环 universal tapered ring
通用协调时间 universal coordinated time
通用斜角规 combination bevel; universal bevel
通用谐振曲线 universal resonance curve
通用信号细则 standing signal instructions
通用信托契据 fictitious deed of trust
通用信息处理机 versatile information processor
通用信息处理系统 general information processing system; generalized information system
通用信息系统 generalized information system
通用型 universal
通用型板 master plate
通用型铲斗 general purpose bucket; utility bucket
通用型反铲挖掘装置 <装在拖拉机上的> utility backhoe
通用型辊式张力装置 universal roller tension unit
通用型混合机 universal mixer
通用型货板 utility pallet
通用型聚酯树脂 general purpose polyester resin
通用型犁壁 general purpose mo(u)-ld board
通用型履带拖拉机 utility crawler
通用型捏和机 universal kneading machine
通用型铁 commercial iron
通用型推土铲 universal bulldozer
通用型推土机 universal bulldozer
通用型推土装置 universal bulldozer
通用型装载机 all-purpose loader; multipurpose loader
通用性 commonality; generality; general purpose; interchangeability; versatility; utility; relatability <评定交通标志设计质量指标之一>
通用性车辆 general purpose vehicle
通用性港口 general purpose harbo(u)r; general purpose port
通用性工业用轮胎 general purpose and industrial tread tire[tyre]
通用性质 universal property
通用修理工程车 general purpose shop truck
通用蓄水池 <水力发电、航行、灌溉、给水等> multiple-purpose reservoir
通用旋转台 universal stage
通用压力 universal pressure
通用压力锅炉 universal pressure boiler
通用压缩机拖拉机 universal compressor tractor
通用压缩因数 generalized compressibility factor
通用压延机 general purpose calender; universal calender
通用研磨机 universal grinding machine
通用要求条款 common requirement clause
通用液体比重计 panhydrometer
通用仪器 all-purpose instrument; universal apparatus
通用仪器接口标准 GB-IB
通用移动通信[讯]系统 universal mobile telecommunication system
通用异步接收发送器 universal asynchronous receiver and transmitter
通用异步收发两用机 universal asynchronous receiver and transmitter
通用异步收发器 universal asynchronous receiver and transmitter
通用引物 universal primer
通用映射性 universal mapping property
通用油产品 universal oil product
通用油压打包机 universal oil pressure bale press
通用语法分析程序 general parsing program(me)
通用语言 all-purpose language; general purpose language
通用语言定义 universal language definition
通用预应力张拉台 universal beam bed
通用预制构件 unified precast element
通用圆锯台 universal saw bench
通用钥匙 master key
通用钥匙锁 master-keyed lock
通用运输 general purpose hauling
通用运输车辆 universal transport vehicles
通用运输工具 universal carrier
通用运输公司 General Transport Co.
通用载体 universal support
通用凿岩钻车 universal drilling jumbo
通用炸药 all-purpose explosive; orthodox explosive
通用遮板 slick sheet
通用砧 utility dolly
通用振动器 universal vibrator
通用整流电源 universal mains unit
通用正反铲斗 common shovel
通用支架 universal mount
通用支数制 universal numbering system
通用值 current value
通用纸张 general-use paper
通用指令 general purpose instruction; universal command
通用指令组 universal instruction set
通用指示剂 univariate indicator; universal indicator
通用指示植物 universal indicator plant
通用指数 general purpose indexes
通用制模机 universal mo(u)lding machine
通用制图程序 universal plotting program(me)
通用中继线 general purpose trunk
通用终端 general terminal
通用终接机 combination connector
通用终接器 toll and local combination connector
通用主处理机 universal host processor
通用主题索引 neutral index
通用柱 universal column
通用装备 interchangeable equipment
通用装订机 general binding machine
通用装甲车 utility armored car
通用装配式冷却器 step-in utility cooler
通用装卸货物码头 multiple-purpose terminal
通用装载机 all-purpose loader
通用装置 flexible unit; general arrangement
通用着色剂 universal stains
通用着色器 universal tinter
通用资产负债表 all-purpose balance sheet; balance sheet for all purposes
通用资源指示程序 uniform resource indicator
通用子程序 common subroutine; general subroutine
通用子例行程序 common subroutine
通用字符法 universal character method
通用字母 conventional alphabet
通用自动编码器【计】 general compiler
通用自动电子(数字)计算机 universal automatic computer
通用自动计算机 universal automatic computer
通用自动控制与测试设备 automatic control and test equipment; universal automatic control and test equipment
通用自动平行光管 universal autocollimator
通用钻床 full universal drill
通用钻杆夹持器 universal rod holder
通用钻机 universal drill
通用钻探机 universal boring and drilling machine
通用钻头 all-purpose bit
通用钻凿机 univariate boring and drilling machine
通用坐标 conventional coordinates; world coordinates
通用坐标系 common coordinate; world coordinate system
通用座架 universal mount
通油孔 oil through
通有电流的 alive
通鱼路 <拦河坝上的> fish ladder
通约 commensurability
通约轨道 commensurate orbit
通约性 commensurability
通约运动 commensurable motion
通约秩 rank of commensurability
通则 common guideline; general guideline; generality; generalization; general principle; general regulation; general rules
通栈 transit shed
通站台的地道 <车站> subway leading to platforms
通针 scratch awl
通蒸汽期 steam period
通知 advertise[advertize]; advertisement; advice notice; circular letter; impart; inform; information
通知保险 declaration insurance
通知(并)付款 advice and pay
通知拆放 callable loan; call loan; call money; money at call; money on call
通知偿还价格 call price
通知偿还债券 callable bond; called bond
通知储蓄存款 savings at call
通知存款 call deposit; deposit(e) at call; deposit(e) at notice; notice deposits
通知存款利率 rate of call
通知存款资金 notice money
通知贷款 callable loan; call loan
通知贷款额度 advised line of credit
通知贷款经纪人 call broker
通知贷款利率 call rate
通知单 advice note; job order; letter of advice; letter of notices; notification
通知到期日 notice of due date
通知登记簿 advice book
通知抵押放款 call secured loan
通知地址 notify address
通知放款 call loan; call money; day-to-day loan
通知放款利率 call loan rate
通知放款市场 call market
通知废除(条约、协定)等 denunciation
通知费用 advising charges; notification charges
通知分配器 <长途试验> toll offering distributor
通知付款 notice of payment
通知工资 call pay
通知公债券还本 call a bond
通知行 advising bank; notifying bank
通知即付资本 callable capital
通知即缴的股本 callable capital
通知交货 delivery on call
通知交款 advice and pay
通知借款经纪人 call broker
通知借款余额 call return met; call transaction made
通知借款周转余额 call turnover made
通知卡 information card
通知栏 noticeboard
通知利率 rate of call
通知铃 announcing bell; annunciator bell
通知码 attention code
通知牌 notice board
通知期 notice day
通知期限 notice period
通知期限的第一天 first notice day
通知器 annunciator
通知人 notifier
通知日 declaration date; reporting day
通知日期 date of notification
通知收益率 yield to call
通知手续费 advising charges; advising commission
通知书 advice note; circuit letter; covering letter; letter of advice[advise]; notification
通知双方 both advice
通知送货制 call-off system
通知损失的义务 obligation of giving claim notice
通知台 interception desk
通知条款 notify clause
通知退出 withdrawal by notice
通知退约 withdrawal by notice
通知文句 notification clause
通知线分配器 order wire distributor
通知信 letter of instruction
通知信号 announcing signal
通知信托债券 callable trust bond
通知信息 broadcast message
通知性电传打字电报 radio note
通知音 warning tone
通知银行 advising bank; notifying bank; transmitting bank
通知者 notifier
通知装运日期 advise shipping date
通直阀 straight-way(line) valve
通轴式变速器 straight-through transmission
通柱 through post

同

同……定约 enter into engagement with

同伴 cohort; companion
同傍内角 interior angles of the same side
同胞 countryman
同辈 contemporary

同比律 law of similitude
同变晶状的 homeoblastic
同标记 isolabel(l)ing
同标准的设计路段 road section design by same standard
同表型配合 isophenogamy
同波道 co-channel
同波道分离 co-channel separation
同波道干扰 co-channel interference; common channel interference; Venetian blind interference
同波道干扰比 co-channel interference ratio
同波道干扰信号 co-channel interfering signal
同步 bring into step; holding; in synchronism; interlocks; isochronism; lock(ing) in; synchronize; syntonization
同步按钮 lockage button
同步百分比 percentage synchronization
同步保持范围 following range of synchronization; hold in range
同步保护的 synchronous protective
同步爆破 simultaneous blast(ing)
同步备用 synchronized reserves
同步比较器 synchrous comparator
同步边限 synchronous margin
同步边缘静噪 sync edge muting
同步变换 cogredient transformation
同步变流机 synchronous converter
同步变速齿轮 synchromesh change gear
同步变速箱 synchronized transmission
同步变速装置 synchromesh(type) transmission
同步变形体系 synchrotexturing system
同步变压器 synchrotrans
同步标记 synchronous mark; sync mark; sync pip
同步表 synchronoscope
同步波 synchronous wave
同步波动 synchronous oscillation
同步波形 sync waveform
同步补偿 synchroballistic
同步补偿器 dynamic(al) condenser; synchronous compensator
同步捕获 synchronization acquisition
同步部分 sync section
同步采样器 synchronized sampler; synchronous sampler
同步参考器 reference synchronizer
同步操作 synchronizing; synchronous operation; synchronous working
同步测量法 synchronous surveying
同步测试器 synchroscope
同步测试仪 synchroscope
同步插孔 lockout jack
同步潮位观测 observation of synchronous tide stage
同步沉积(物) synchronous deposit; synchronous sediment
同步成圈 synchronized timing
同步齿轮 sync-gear; synchronizer gear; synchronizing gear; synchronous gear
同步齿轮离合器 synchro-clutch; synchromesh clutch gear
同步齿轮系 sycromesh; synchromesh
同步齿形带 timing belt
同步赤道轨道 synchronous equatorial orbit
同步出口程序 synchronous exit routine
同步储存电路 synchrolock
同步处理 synchronous process(ing); sync process
同步触发 syncburst
同步触发脉冲发生器 synchronized trigger generator
同步触发器 synchronizer trigger
同步穿孔 synchronous punching
同步传动 synchrodrive; synchronous drive
同步传动装置 synchronized driving device; synchronizing gear; synchronous gear
同步传动装置的受信仪箱 receptacle box
同步传感器 synchropickoff; synchropickup; synchrotransmitter
同步传声 sound-in-syncs
同步传输 synchotransmission; synchronized transmission; synchronous transmission
同步传输装置 synchronous transmission device
同步传送 synchro; synchronous driving; synchronous transfer; synchronous transmission
同步传送的概略射角 quadrant elevation coarse synchrodata
同步传送的精确射角 quadrant elevation fine synchrodata
同步传送模块 synchronous transfer module
同步传送模式 synchronous transfer mode
同步串行数据适配器 synchronous serial data adapter
同步串联系统 synchronous serial system
同步磁阻电动机 synchronous reluctance motor
同步从动装置 synchronizing slave unit
同步存储方法 synchronous storage method
同步错误 timing error
同步带 locking band; lock(ing)-in range(on synchronization)
同步单电动机 synchronous monomotor
同步单极电动机 sychronous homopolar motor
同步单元 l lock unit; synchronized unit
同步导频 synchronizing pilot
同步倒相 sync inversion
同步的 cogradient; isochronal; isochronous; isohronal; synchro; synchronal; timed; in-step; synchronizing
同步的基本命令 synchronizing primitive command
同步灯 synchronous light
同步地球观察卫星 synchronous earth observation satellite
同步电动发电机 synchronous motor-generator
同步电动机 synchromotor; synchronous motor
同步电动机机车 synchronous motor locomotive
同步电动机控制器 < 电动机与电源频率同步 > synchronous motor controller
同步电动机振荡 synchronous motor hunting
同步电动势 synchronous electromotive force
同步电机 synchromotor; synchronous dynamo; synchronous machine; synchronous motor
同步电抗 synchronous reactance
同步电抗器 synchronizing reactor
同步电流 synchronizing current
同步电流转换器 synchronous inverter
同步电路 lock(ing)(on) circuit; sync-circuit; synchrodyne circuit; synchronizing circuit; synchronous circuit
同步电平 sync level
同步电容器 synchronous capacitor; synchronous condenser
同步电压 synchronizing voltage; synchronous voltage; timing voltage
同步电压表 synchronizing voltmeter
同步电钟 synchroclock; synchronized clock; synchronous electric(al) clock; synchronous motor clock
同步电子开关 synchronous electric(al) sampler; synchronous electronic sampler; synchronous electronic switch
同步调相机 synchronous compensator; synchronous condenser
同步调相机容量 capacity of synchronous condenser
同步调相器 synchronous compensator; synchronous condenser
同步调相运行 synchronous condenser operation
同步顶升 synchronisation of jacking up; synchronization of jacking up
同步定时比较器 synchronous timing comparator
同步定时器 synchrotimer
同步定时振荡器 synchronized-timing oscillator
同步定相 synchrophasing
同步动量 synchronous momentum
同步动作 synchronization action
同步动作图解 simo chart
同步动作周期图 simultaneous motion cycle chart
同步读出器 synchroreader
同步断路器 synchronized breaker
同步断续器 synchronous contactor
同步对照管理 management through synchronization
同步多路转接器 synchronous multiplexer
同步多址通信[讯]系统 synchronized multiple access communication system
同步发电机 alternator; synchrogenerator; synchronizing generator; synchronous generator
同步发送机 control synchro
同步发送器 control synchro; synchrotransmitter
同步发展 sympathetic(al) development
同步发展论 synchronous developmentalism
同步阀 synchronous valve
同步范围 capture range; holding range; lock(ing)-in range(on synchronization); lock(ing)(in) range; pull-in range; synchronizing range; synchronous range
同步方式 synchronizing system
同步方向 sync direction
同步放大器 sync amplifier; synchronizing amplifier
同步放电 synchronous discharge
同步分隔 sync split
同步分离 synchronizing separation; synchronous separation
同步分离电路 synchronizing separator circuit; sync separating circuit
同步分离管 synchronizing separator tube
同步分裂 synchronized division
同步分配放大器 sync distribution amplifier
同步分频 synchronization frequency division
同步分时多路换接器 synchronous time division multiplexer
同步峰值 sync peaks
同步伏特计 synchronizing voltmeter
同步辐射衰减 synchrotron radiation decay
同步复用器 synchronous multiplex
同步干扰 synchronous jamming
同步干扰器 synchronous jammer
同步感应电动机 simplex motor; synchronous induction motor
同步杠杆 synchronous lever
同步高度 synchronous altitude
同步高度通信[讯]卫星 synchronous altitude communications satellite
同步高度重力梯度实验 synchronous altitude gravity gradient experiment
同步高度自旋稳定实验 synchronous altitude spin stabilized experiment
同步跟踪 locking; synchronous tracking
同步跟踪计算机 synchronous tracking computer
同步跟踪扫描装置 lock-and-follow scanner
同步跟踪天线 synchronous tracking antenna
同步工作 synchronous operation; synchronous working
同步功率 synchronizing power; synchronous power; synchronous watt
同步功能 synchronizing function
同步共振 synchroresonance
同步共振波 synchronous resonant wave
同步观测 simultaneous observation; synchronous observation
同步管理 management by synchronization
同步惯性轮 synchronizing wheel
同步光谱分析器 synchronized spectrum analyzer[analyser]
同步光网络 synchronous optic(al) network
同步广播 synchronized broadcast(ing)
同步轨道 geostationary orbit; stationary orbit; synchronous orbit; timing track
同步过程 synchronizing process; synchronous process(ing)
同步航天站 synchronous station
同步化 synchronization; synchronize
同步化的异步电动机 synchronized asynchronous motor
同步化运转齿轮 synchronized transport gear
同步环境应用卫星 geostationary operational environment(al) satellite
同步环线 synchronous loop
同步缓冲 sync buffering
同步换挡 synchronized shift
同步换流机 rotary converter[convertor]; synchronous converter
同步换流器 rotary converter[convertor]; synchroconverter; synchronous converter
同步换向器 synchronous commutator
同步恢复 synchronization recovery
同步回波显示 fruit
同步回路 synchronizing circuit
同步回填注浆 simultaneous back filling injection
同步回旋加速器 synchro-cyclotron
同步混频器 lock-in mixer; synchrodyne mixer
同步火花 synchronized sparks; synchronous spark
同步火花隙 synchronous spark gap

同步获得 synchronization gain
同步机 magslip; self-synchronous device; self-synchronous repeater; synchro; synchrodevice; synchrodyne; synchrogenerator; synchronizer; synchronous machine
同步机构 lazy tongs; synchronization mechanism; synchronizing linkage; synchronizing mechanism
同步机架 synchronization bay
同步机馈线 synchrotic
同步机数字转换器 synchroto digital converter
同步基本命令 synchronizing primitive command
同步基底电平 sync pedestal
同步基座 sync pedestal
同步计 synchronometer
同步计量器 lock-on counter
同步计时器 synchronous timer; synchrotimer
同步计数 synchronous counting
同步计数器 coincidence counter; lock-on counter; synchronous counter
同步计算机 synchronous computer
同步计圆盘 synchronoscope disc; synchroscope disc
同步技术 simultaneous technique
同步继电器 lock-on relay; synchronizing relay
同步加速 synchronous acceleration
同步加速不稳定性 synchrotron instability
同步加速辐射 synchrotron radiation
同步加速辐射源 synchrotron radiation source; synchrotron source
同步加速谱 synchrotron spectrum
同步加速器 synchroaccelerator; synchronous accelerator; synchrotron; synchrotron accelerator
同步加速器电压 synchrotron voltage
同步加速器辐射 synchrotron light; synchrotron radiation
同步加速阻尼速率 synchrotron damping rate
同步间隙脉冲发射器 half-time emitter
同步监控装 synchronous supervision mechanism
同步监控装置 synchronous supervision mechanism
同步检波器 commutator detector; sampling detector; synchronous demodulator; synchronous detector
同步检测器 synchronous detector
同步检查 sync check
同步检查字 sync check word
同步检定能力 synchronization detection capability
同步检定器 synchronous detector
同步检验 sync check; synchronization check
同步建设 synchronous construction
同步鉴别器 sync discriminator; syncriminator
同步降低 <指上下游水位同时降低> sympathetic(al) retrogression
同步交叉式双旋翼直升机 synchropter
同步交流发电机 synchronous alternator
同步交通信号 synchronous traffic signal
同步角 synchroangle; synchronizing angle
同步角速度 synchronous angular velocity
同步绞车 synchronized winch
同步校验 synchronization check
同步校验继电器 synchronism check relay
同步校验字 synchronization check word
同步校正 synchronous correction
同步接触器 synchronous contactor
同步接口 synchronous interface
同步接收电动机 synchronous receiving motor
同步接收机 receiving synchro; synchronous receiver; synchroreceiver
同步接收器 synchronous receiver
同步解算器 synchroresolver
同步解调器 synchronous demodulator
同步进度表 synchronized time schedule
同步进相机 synchronous advancer; synchronous phase advancer
同步经济指示数字 coincident economic indicators
同步均衡 sync equalizing
同步开关 synchronized switch; synchronizer switch; synchronous switch; syncrho-switch
同步空气制动操纵 pneumatic synchronization of brake control
同步空载 synchronous idle
同步空转符号 synchronous idle character
同步控制 hold(ing) control; locking control; sync control; synchrocontrol; synchronization control; synchronized control; syncrho-control
同步控制变压器 synchrocontrol transformer; synchrotransformer
同步控制电压 locking control voltage
同步控制发射机 synchrocontrol transmitter
同步控制机构 synchronizing controls
同步控制接收机 synchrocontrol receiver
同步控制接收器 synchrocontrol receiver
同步控制器 isochronous controller
同步快门 synchronized shutter; synchro-shutter
同步拉伸 synchro-draw
同步离合器 synchronizing clutch; timing clutch
同步离散地址信标系统 synchronous disperse address beacon system
同步离子 synchronous ion
同步力矩 synchronising [synchronizing] torque; synchronous torque
同步连杆 synchronising(connecting) rod
同步连接 synchrotic
同步联动装置 synchronizing linkage; synchronous link
同步链传动箱 timing-chain case
同步灵敏度 synchronization sensitivity; synchronizing sensitivity
同步零差 synchronous homodyne
同步流速 simultaneous velocity
同步流向仪 synchro-current direction detector; synchronometer
同步轮 synchronizing wheel
同步马达 sync motor
同步马达调速器 synchronizing motor governor
同步脉冲 clock pulse; lockout pulse; synchronization pulse; synchronizing(im) pulse; synchronous clock; synchronous pulse; synchropulse; sync-pulse
同步脉冲道 clock track
同步脉冲定时误差 synchronizing pulse-timing error
同步脉冲发生器 synchronized pulse generator
同步脉冲发生设备 synchronizing pulse generating equipment
同步脉冲方向 sync direction
同步脉冲分离电路 sync separator circuit
同步脉冲干扰 synchronous pulse jamming
同步脉冲"扩展"电路 sync stretch circuit
同步脉冲压缩 synchronization compression
同步脉冲再生器 synchronizing pulse regenerator; sync pulse regenerator
同步脉冲振荡器 synchronized pulse generator
同步脉冲重复频率 clock rate
同步脉冲周期 clock cycle
同步门 synchronous gate
同步秒表 synchronous timer
同步摩擦锥轮 engaging friction cone
同步母线 synchronizing bus-bar
同步能力 synchronizing capacity; synchronous capacity
同步能量 synchronous energy
同步能量增量 synchronous increment of energy
同步能量增益 synchronous energy gain
同步逆变器 synchronous inverter
同步啮合 sycromesh; synchromesh
同步啮合变速 synchromesh change
同步啮合变速器 synchromesh(type) transmission
同步啮合变速箱 synchromesh gearbox
同步啮合齿轮 synchromesh gear
同步啮合齿轮副变速箱 synchromesh gearbox
同步啮合齿轮式变速箱 synchromesh (type) transmission
同步啮合齿轮箱 synchromesh gearbox
同步啮合副轴式变速器 synchronized countershaft transmission
同步啮合机构 synchromesh gear
同步扭矩差动式发送器 synchro torque differential transmitter
同步扭矩差动式接收器 synchro torque differential receiver
同步扭矩接收器. synchro torque receiver
同步纽锁系统 simultaneous twist lock system
同步耦合 synchro-coupling; synchronous tie; synchrotic
同步耦合器 genlock
同步旁通阀 synchronous bypass valve
同步培养 synchronized culture; synchronous culture
同步配电板 synchronization panel
同步配合 sycromesh; synchromesh
同步配合变速器 synchromesh(type) transmission
同步配合齿轮机构 synchromesh gear
同步配合穿孔机 synchro-mating punch
同步皮带 synchronous belt; timing belt
同步皮带轮 timing belt pulley
同步偏差 synchronism deviation
同步频带 synchronizing band
同步频率 synchronization frequency; synchronizing frequency; synchronous frequency
同步频率放大器 synchronous frequency booster
同步平面 plane of synchronization; synchronous plane
同步破坏 desynchronize
同步曝光 simultaneous exposure; synchronous exposure
同步起爆雷管 instantaneous cap
同步起动 synchronous initiation; synchronous start(ing)
同步起动电路 trigger circuit
同步起升方式 synchrolift system
同步气象卫星 geostationary meteorologic(al) satellite; synchronous meteorological satellite
同步气象卫星系统 geostationary meteorological satellite system
同步器 lock(ing) unit; speed changer; synchro(nizer); synchronizing device
同步器闭锁环 synchronizer ring
同步器变速箱 non-clashing gear set
同步器电动机 speed-changer motor
同步器毂 synchronizer hub
同步器摩擦锥轮 synchronizer cone
同步器摩擦锥轮的锥形工作面 synchronizing cone
同步器啮合套 synchronizer sleeve
同步牵入转矩 synchronous pull-in torque
同步箝位 synchronized clamping; synchronous clamping
同步切变 synchro-shear
同步倾斜航摄照片 synchronous oblique air photographs
同步区(域) retaining zone; retention range; synchronization zone; synchronizing band; zone of synchronization
同步驱动机 synchrodrive
同步取样 synchronous sampling
同步取样器 synchronous sampler
同步绕转 synchronous rotation
同步热反应堆 synchrotherm reactor
同步热反应器 synchrotherm reactor
同步扫描 isochronous scanning; synchronized sweep; synchronous scanning
同步扫描激光电视系统 synchronously scanned laser television system
同步闪光灯 synchroflash
同步设备 synchronizer equipment; synchronizing apparatus; synchronizing arrangement; synchronous apparatus
同步设备定时源 synchronous equipment timing source
同步设备管理功能 synchronous equipment management function
同步设备物理定时物理接口 synchronous equipment timing physical interface
同步摄影机 synchronous camera
同步伸缩 synchronous telescope
同步升船机系统 synchrolift system
同步升船台 synchrolift
同步升船装置 synchrolift
同步升降系统 synchrolift system
同步升压变流机 synchronous booster converter
同步升压器 synchronous booster
同步失效 sync fail
同步失真 synchronous distortion
同步时差 simultaneous offset
同步时分多路复用器 synchronous time division multiplexer
同步时计 synchroclock
同步时间 lock-in time; synchronization time; synchronizing time
同步时间分配多路转换器 synchronous time division multiplexer
同步时间继电器 synchronous timer
同步时间选择器 synchronous time selector
同步时序系统 synchronous sequential system
同步时钟 synchronizing clock; synchronous clock
同步示波器 oscillosynchroscope; synchrooscilloscope; synchroscope
同步示波仪 oscillosynchroscope; synchroscope

同步式 synchronous system
同步式比较器 synchronous mode comparator
同步式系统 synchronous system
同步式振动器 synchronous type vibrator
同步式自动机 synchronous automaton
同步收发机 synchronous transmitter-receiver
同步输出 sync-output
同步输出分析 synchronous output analysis
同步输出脉冲 sync-out pulse
同步输入 lock input;synchronous inputs;sync input
同步输入脉冲 sync-in pulse
同步数据 synchrodata
同步数据传输 synchronous data transmission
同步数据传送 synchronous data transmission
同步数据发送 synchronous data transmission
同步数据链控制 synchronous data link control
同步数据链控制规程 synchronous data link control regulations
同步数据网(络) synchronous data network
同步数字分级系统 synchronous digital hierarchy
同步数字键路控制 synchronous data link control
同步数字交叉跨接设备 synchronous digital cross connect equipment
同步数字体系 synchronous digital hierarchy
同步数字体系管理网 synchronous digital hierarchy network
同步数字体系管理子网 synchronous digital hierarchy management subnetwork
同步数字体系物理接口 synchronous digital hierarchy physical interface
同步衰减 synchronous fading
同步水位 simultaneous water level
同步速度 synchronous speed; synchronous velocity;synchro-speed
同步速率 no-load speed
同步随动系统 synchrofollow-up system
同步损失 locking out
同步锁 synchrolock
同步锁订单元 sync-lock unit
同步锁定设备 sync-lock equipment
同步锁相 general locking;genlocking
同步探测 synchronizing detection; synchronous detection
同步探测器 synchronous detector
同步套 < 齿形离合器 > synchronizing sleeve
同步特性 synchronizing characteristics
同步天文罗盘 synchronous astrocompass
同步天文钟 astronomic(al) synchronome
同步条件 in-step condition; synchronous condition
同步条款 pari passu provision
同步调节 synchronization regulation
同步调节器 isochronous governor; synchronous governor; synchronous regulator; synchrostat regulator
同步调速器 isochronism speed governor; isochronous governor; synchronism governor; synchronism regulator;synchronization regulator
同步调谐 synchronous tuning
同步调压逆变器 synchronous booster-inverter
同步调整 hold(ing) control;synchrocontrol; synchronization adjustment;synchronized hold control
同步调整装置 synchronization control;synchronizing controls
同步调制 hold control; isochronous modulation; synchronizing modulation;synchronous modulation
同步调制解调器 synchronous modem
同步通道 synchronizing channel
同步通路 synchronizing channel
同步通信[讯] synchronous communication
同步通信[讯]电气动态系统 synchronous communication electrodynamic(al) system
同步通信[讯]接口 synchronous communication interface
同步通信[讯]卫星 geostationary satellite; synchronous communication satellite;syncom
同步通信[讯]系统 synchronous communication system
同步头 synchronous head
同步图形 synchronizing pattern
同步退潮 sympathetic(al) retrogression
同步瓦 synchronous watt
同步网(络) synchronized network; synchronizing network; synchronous network
同步微处理机 synchronizing microprocessor
同步卫星 geostational satellite; motionless satellite; stationary satellite;synchronous satellite;synchrosatellite
同步卫星轨道 synchronous satellite orbit
同步卫星通信[讯]系统 synchronous satellite communication system
同步卫星系统 synchronous satellite system
同步位 sync bit
同步位置控制 synchro-position control
同步喂料机 feed synchronization
同步稳定性 stability of synchronization;synchronous stability
同步稳相加速器 cosmotron;synchrophasotron
同步无线电通路 synchronous radio telegraph channel
同步无线电信道 synchronous radio telegraph channel
同步误差 synchronization error;synchronizing error; synchronous error;timing error
同步系数 synchronization factor
同步系统 synchronizing system;synchronous system; synchrosystem; simultaneous system
同步系统发送装置 synchrotransmitter
同步系统接收装置 receiving selsyn
同步显示器 synchronization indicator
同步现金流转 synchronizing cash flows
同步现象 synchronia
同步线路 synchronous link
同步线性电动机 synchronous linear motor
同步线性控制 synchronous line control
同步限制 synchronization constraint
同步相位 locking phase
同步相位变换器 synchronous phase converter
同步相位补偿器 synchronous phase modifier
同步相位调制器 synchronized phase modulator
同步相位计 synchrophasemeter
同步相位加速器 synchro-fazotron
同步销售 synchro-marketing
同步效应 pulling effect
同步斜齿轮变速箱 synchro-spiral gearbox
同步信道 synchronizing channel
同步信号 locking signal;simultaneous signal; synchronization signal; synchronized signal;synchronizing signal;synchronous signal;sync signal
同步信号波形 synchronization waveform;synchronizing waveform
同步信号插入 sync insertion; sync signal insertion
同步信号电平 synchronizing level; synchronizing signal level;sync signal level
同步信号对齐 sync line-up
同步信号发生器 sync generator;synchronized-signal generator; synchronizing generator;synchronizing signal generator;sync signal generator
同步信号法 synchronous signalling
同步信号放大器 synchronizing amplifier
同步信号非线性 sync non-linearity
同步信号分离 synchronizing signal separation
同步信号分离电路 sync separator circuit
同步信号分离器 synchronizing signal separator;sync separator
同步信号幅度 synchronizing signal amplitude
同步信号后消隐间隔 post-sync field-blanking interval
同步信号极性 sync direction
同步信号控制 synchronized-signal control
同步信号滤清器 sync signal purifier
同步信号频率 synchronous signal frequency
同步信号限幅器 sync limiter;sync signal limiter
同步信号压缩 sync compression;synchronizing signal compression;sync signal compression
同步信号源 synchronizing signal source
同步信号振荡器 synclator
同步信号振幅 sync amplitude;sync signal amplitude
同步信号重插入 sync signal re-insertion
同步信号注入电路 injector circuit
同步信息 simultaneous information; synchronizing information
同步行程 synchronized stroke
同步行销 synchro-marketing
同步性 isochromism;synchronism
同步性破坏 loss of synchronism
同步性起伏 fluctuating synchronism
同步修理法 synchronous repair approach
同步序列 synchronizing sequence
同步旋转 synchronous revolution; synchronous rotation
同步选通脉冲 lock-following strobe
同步选择开关 sync select switch
同步选择脉冲 synchronous gate
同步选择器 synchronizing selector; sync selector
同步选择制 synchronous selector system
同步循环 synchronized cycle
同步信号电平 synchronizing potential
同步验潮 tidal synobservation
同步摇臂 synchronous lever
同步摇臂轴 synchronous lever shaft
同步遥控 synchronous remote control;synchrostep remote control
同步业务卫星 geostationary operational environment
同步移位器 synchro-shifter
同步移相器 synchrophase shifter
同步异步电动机 synchronous asynchronous motor
同步因数 synchronization factor
同步引导器 synchronous leader
同步引入 pull-in
同步引线 synchronizing leader
同步印刷器 synchroprinter
同步圆 circle of simultaneity;circle of spontaneity
同步源 sync source
同步远距控制 synchronous remote control
同步运动 synchronized motion; synchronized operation; synchronous motion
同步运行 synchronized operation; synchronous operation; synchronous running
同步运行系数 factor of synchronous operation
同步运转 run-in synchronism; synchronized operation; synchronous running
同步载波 sync carrier
同步载波系统 synchronous carrier system
同步再生 synchronizing regeneration
同步再现波形 synchronous recurrent waveform
同步增压变流机 synchronous booster converter
同步增压机 synchronous booster
同步增长 grow in pace with;grow in phase with;grow in step with;pari passu
同步闸门 synchronous gate
同步闸门电路 synchronous gate
同步展宽 synchronization stretching
同步振荡 isochronous oscillation; synchronized oscillation; synchronous oscillation
同步振荡器 lock(ed)-in oscillator; synchronization generator;synchronous generator; synchronous oscillator
同步振动 synchronous vibration
同步振子 synchronous vibrator
同步整流电路 synchronizing rectifier circuit
同步整流器 < 变交流为直流 > synchronous rectifier
同步整相机 synchronous phase modifier
同步执行周期 synchronous executive cycle
同步指标 coincidental indicator;coincident indicator; coinciding indicator
同步指令【计】 synchronization tag
同步指示灯 lamp synchronizer;lamp synchroscope;synch light;synchrolight; synchronization light; synchronizing lamp; synchronizing light
同步指示计 synchrometer
同步指示器 synchro-indicator; synchronism indicator; synchronized indicator;synchronometer;synchro(no)scope;synchronizing indicator
同步制 synchronous system
同步质量 synchronizing quality

T

同步质谱仪 mass synchrometer
同步中断 simultaneous interruption
同步中继系统 synchronous relay system
同步终端 synchronous terminal
同步周期 synchronizing cycle; synchronous period
同步轴 synchronizing shaft
同步逐稿轮 synchronizing beater
同步注浆系统 synchronized grouting system
同步转动 turn in synchronism
同步转动开关 synchronous rotary switch
同步转鼓 synchronizing drum
同步转换 synchrotrans
同步转换机构 synchro-shifter
同步转换接点 synchronized changeover contact
同步转接 synchronous switching
同步转矩 synchronizing torque
同步转速 synchronous speed; synchronous speed of rotation; synchro-speed
同步转移轨道 synchronous transfer orbit
同步装配线 synchroassembling line
同步装置 synchronizer
同步状态 in-step condition; synchronous regime
同步字符 sync character
同步字头 synchronization character
同步自动电报机 synchronograph
同步自脱离合器 synchro-self-shifting clutch
同步阻抗 synchronous impedance
同步作用 sync; synchronization; synchronous effect
同步作用制动机 synchronized brake
同步作用制动器 synchronized brake
同操作 equivalence operation
同侧对向道岔 two-sets of turnouts facing each other and laid on the same side the original track
同侧顺向道岔 two-sets of turnouts trailing each other and laid on the same side the original track
同层 no host; peer-to-peer; same layer; sharing
同层等价 equivalence of same layer
同层型 costrototype
同潮的 cotidal
同潮流时线 cocurrent line; current line
同潮时 cotidal and coranged; cotidal hour
同潮(时)图 cotidal chart
同潮(时)线 cotidal line
同车人 fellow passenger
同沉积 cosedimentation; syndeposition
同沉积凹陷 syndepositional depression
同沉积背斜 synsedimentary anticline
同沉积变形作用 synsedimentary deformational process
同沉积断裂 syndepositional fracture
同沉积构造 contemporaneous structure; syndepositional structure
同沉积构造类型和性质【地】type and property of synsedimentary structure
同沉积隆起 syndepositional uplift
同沉积向斜 synsedimentary syncline
同沉积褶皱 post-depositional fold; syndepositional fold; synsedimentary fold
同沉积作用 cosedimentation; syndeposit
同成分 congruent
同成分熔点 congruent melting point
同成控制 conjugate control
同成矿床 idiogenite
同成因的 idiogenous; isogenetic
同城的 intra-city
同城结算 city-wide settlement of accounts
同城票据交换 local clearing
同城托收 city collection
同城往来 intra-city
同程式布置 reversed return scheme
同程式回水系统 reverse-return system
同程式系统 reversed return system
同尺寸骨料 one-size(d) aggregate; single-size(d) aggregate
同酬 parity income
同船舶所有人条款 sister ship clause
同船船员 ship mate
同船人 fellow passenger
同次 homogeneous
同次方程式 homogeneous equation
同次性 homogeneity
同簇矿脉 domestic vein
同大小 equidimension
同代人 contemporary
同带可以互换 exchangeable for same zone
同带信号传输 inband; inband signal(l)ing
同单体组分聚物 azeotropic copolymer
同单位的 commensurate
同等 coordination; par; parity
同等保证条款 equal coverage
同等边际生产率的法则 law of equimarginal productivity
同等层 peer layer
同等层通信[讯] peer layer communication
同等大地网 homogeneous network
同等待遇 parity of treatment
同等的 coequal
同等的劳动 equal performance of labo(u)r
同等的人 equal
同等地位人员团体 cousins group
同等关系 coordination; identity relation
同等机会 equal chance
同等几率原则 insufficient reason rule
同等级大地网 homogeneous network
同等级网 homogeneous network
同等级债权 debts of equal degree
同等就业机会 equal employment opportunity
同等就业机会委员会 <美> Equal Employment Opportunity Commission
同等利润曲线 isoprofit curve
同等利润线 isoprofit line
同等利益 equal advantage; unity of interest
同等连续的 equally continuous
同等连续性 equicontinuity
同等排列 identical permutation
同等权力 concurrent authority
同等收敛 equiconvergent
同等数量 equal proportions
同等物 coordinator; equivalent
同等效力 equal authenticity; equal effect; equal validity
同等学历 equivalent education; have the same educational level
同等者 coordinate; coordinator
同等状态 same condition
同等组 congruence class
同点共极 mutual pole of alike point
同电车轨道 tramway
同电位的 idiostatic
同调 ganged tuning; ganging; homologue; homology
同调边缘 homology boundary
同调边缘同态 homology boundary homomorphism
同调变换 homologous transformation; homology transformation
同调代数 homological algebra
同调的覆盖面 homology covering surface
同调点 homologous point
同调函子 homological functor; homology functor
同调环 homology ring
同调基 homology base
同调角 homologous angles
同调结构 homology structure
同调类 homology class
同调棱 homologous edges
同调连通空间 homologically connected space
同调流形 homology manifold
同调论 homology theory
同调面 homologous faces
同调模 homology module
同调谱序列 homology spectral sequence
同调球 homology sphere
同调群 homology group
同调维 homology dimension
同调维数 homological dimension
同调系 homology system
同调线 homologous lines
同调项 homologous terms
同调型 homology type
同调性质 homological property
同调映射 homological mapping
同调有限滤子 homologically finite filtration
同调于零 homologous to zero
同调元素 homologous element
同调正合序列 homology exact sequence
同度分布 homograde distribution
同度量 homometric
同度量变换 isometric(al) transformation
同度系数 coefficient of commensurability
同多钼酸盐 isopolymolybdate
同多钨酸盐 isopolytungstate
同多形现象 isopolymorphism
同多形性 isopolymorphism
同二晶现象 isodimorphism
同方 party
同方差表 homoscedastic table
同方差性 homoscedasticity
同方差性检验 test for homoscedasticity
同方位线 loxodrome
同方向不同比例 same direction but in different proportions
同方向的 synclastic
同方向列车连 time interval for two trains dispatching in succession in the same direction
同方向同比例 same direction and proportion
同方向线 isogon
同分布 same distribution
同分异构的【化】allo
同分异构体 isomer; isomeride
同分异构现象 isomerism
同分异构作用 isomerism
同风向线 isogon
同附贴样品严格一致 strictly same as per attached sample
同概率转换 probability-preserving transformation
同杆架设 joint use
同感性光反射 consensual light reflex
同高程面积曲线 area-elevation curve
同高度 co-altitude
同高架空线(路) ungraded pole line
同格 apposition
同工同酬 equal pay for equal work
同工种竞赛 emulation among workers of the same type of work
同工作有关的管理活动 work-related activity
同功结构 analogous structure
同共轭效应 homoconjugation effect
同构变晶 topotaxy
同构的 isomorphous
同构集 isomorphic sets
同构空间 isomorphic space
同构群 isomorphism group
同构体 isolog(ue)
同构图 isomorphic graph
同构图形 isomorphic image
同构问题 isomorphism problem
同构型多处理机 homogeneous multiprocessing
同构异量质凝胶 isogel
同构异素的 isologous
同构异素体 isolog(ue)
同构映射 isomorphic mapping
同构造 isostructure
同构造的 synkinematic; syntectonic
同构造等深线 syntectonic pluton
同构造交叉褶皱 simultaneous cross-folds
同构造结晶纤维 syntectonic crystal fiber[fibre]
同构造期花岗岩 synkinematic granite
同构造深成岩体 syntectonic pluton
同构造生长脉 syntectonic growth vein
同构造重结晶 syntectonic recrystallization
同广阔的 coextensive
同国人 countryman
同行 craft brother; fellow trader; profession
同行价格 trade price
同行间应付账款 account payable interline
同行评定 peer review
同行审查 peer review
同行业单位之间应付账款 account payable interline
同行业务标准 standards of professional practice
同行折扣 trade discount
同好曲线 indifference curve
同号【电】jack per line
同号标本 isotype
同号电 like electricity
同号电荷 homocharge
同厚度、不同种类硬木材货垛 stack pile
同厚度宽度的木材垛 tier[tyer] pile
同化 assimilate
同化层 assimilation layer
同化产物 assimilation products
同化代谢 assimilation; assimilatory metabolism
同化反应 assimilative reaction
同化呼吸 assimilation respiration
同化能力 assimilation ability; assimilation capacity; assimilative capacity; assimilatory power
同化能量 assimilation capacity
同化容量 assimilative capacity
同化商【化】assimilation quotient; assimilatory quotient
同化烧结现象 extra-sintering phenomenon
同化衰竭 assimilatory depletion
同化物 assimilation quotient
同化系数 assimilatory coefficient; coefficient of assimilation
同化系统 assimilation system
同化效率 assimilatory efficiency

同化效应 assimilation effect
同化序列 homotopy sequence
同化学性质指示 isochemical indication
同化于新工作环境 assimilate into new working environment
同化组织 assimilatory tissue
同化作用 assimilation
同化作用类型 assimilation type
同机种方式 homogeneous way
同机种网络 homogeneous network
同机种系统 homogeneous system
同基双柱 twin columns
同级 in-step;sister
同级官员 opposite number
同级节点 brother node
同级列车 train of one class
同级任务表 brother task table
同级(速度)列车 trains of one class
同级迂回 alternative route of the same stage
同级制订目标 peer goal-setting
同极场磁铁 homopolar field magnet
同极磁铁 homopolar magnet
同极的 homopolar
同极发电机 homopolar dynamo;homopolar generator
同极发电机式电磁泵 homopolar generator electromagnetic pump
同极高压直流输电系统 homopolar high-voltage direct current system
同极键 homopolar bond
同极键联 homopolar binding
同极晶彩 homopolar crystal
同极晶体 homopolar crystal
同极聚合物 homopolar polymer
同极型轴对称 homopolar type of axial symmetry
同极性 like polarity
同极性胶体 homopolar colloid
同价买卖 straddle
同价品 equivalent
同架 unit frame
同架电杆 joint pole
同浆岩浆 comagmatic rocks
同阶 same order
同阶有偏估计量 same order bias estimators
同结构 isostructure
同结构性 isostructuralism
同结晶作用 syncrystallization
同结线 cotectic line
同晶化合物 isomorphous compounds
同晶体 allomeric
同晶替换 isomorphous replacement
同晶系 isomorphous system
同晶形取代 isomorphous substitution
同晶形现象 homomorphism
同晶型的 isomorphic;isomorphous
同晶型混合物 isomorphous mixture
同晶型聚合物 isomorphous polymer
同晶型体 isomorph
同晶型系 isomorphous series
同晶型现象 isomorphism
同晶型置换 isomorphous replacement
同晶型组与化学成分的关系 parallelosterism
同晶置换 isomorphous substitution
同径侧流三通 straight size side outlet tee
同径侧肘管 straight size side elbow
同径阀 full-way valve
同径井身 tube construction with same diameter
同径三通 straight size tee
同径四十五度分支管 straight size 45 degree lateral
同径四通 straight size cross
同径止水 water sealing in same diameter
同径肘管 straight size elbow
同久远的 coextensive
同距选排叠加 common-range-gather stack
同聚物 homopolymer
同均 isomorphism
同开同关的联窗 continuous lights
同颗粒组成的土<即大小颗粒相近的土> closely graded product;closely graded soil
同空间 isospace
同空气混合的 aerated
同口径三通 straight tee
同类财产 like in kind property
同类沉积 isopical deposit
同类丛结构 homogeneous plex structure
同类大宗货物 homogeneous bulk cargo
同类的 intraclass;related
同类房间成组布置的平面 battery plan
同类工作 job family
同类构件 homogeneous member
同类规则 ejusdem generis rule
同类货 analogous articles
同类结构 homogeneous structure
同类列表 column-homogeneous table
同类散货压舱包装 homogeneous cargo bagged for stowage purpose
同类石料 allied rocks
同类试件 companion specimen
同类树 homogeneous tree
同类数 like numbers
同类土壤 undifferentiated soil
同类物质 allied substances
同类线型地图 isopleth map
同类相残 cannibalism
同类相关 intraclass correlation
同类相关器 intraclass correlator
同类相食 cannibalism
同类项【数】 similar terms;like terms
同类型地图 homogeneous series
同类样本 companion specimen
同类异性物 heterotype
同类意识 consciousness of kind
同类组 homogeneous group
同离子 coion
同离子 homo-ion
同离子溶液 homoionic solution
同离子土 homoionic soil
同离子效应 common ion effect
同粒度 one-size
同粒度混合料 one-size mixture;single-size mixture
同粒度集料 all-in aggregate
同粒径材料 single-size material
同粒径的 uniformly graded
同粒径的沉积物 nongraded sediment
同粒径的土 nongraded soil
同粒径骨料 single-size(d)aggregate
同粒径集料 one-size(d)aggregate;single-size(d)aggregate
同粒径稳定砂 uniform-sized stabilised sand
同量 commensurability
同量的 commensurate
同量加权 equal weighting
同量异位的 isobaric
同量异位多重态 isobaric multiplet
同量异位核 isobaric nucleus
同量异位模型 isobar model
同量异位三重态 isobaric triad;isobaric triplet
同量异位素 isobar;isobaric element;isobaric heterotope
同量异位素的丰度 isobaric abundance
同量异位素空间 isobaric space
同量异位素模型 isobaric model
同量异位素相似态 isobaric analogue
同量异位转化 isobaric transformation
同量异序素 isobar;heterotope
同量异序元素 heterotope;isobar
同量异序原子 isobaric atom
同列式铆接 chain-riveted joint
同龄的 coetaneous;coeval;even-aged
同龄级木 age class
同龄林 even-aged forest;uniform crop
同龄林分 even-aged stand
同龄人 age grade;contemporary
同龄树 same-age tree
同流 cocurrent flow
同流换热 recuperative heat exchange
同流换热的 recuperative
同流换热法 recuperation of heat;recuperative system
同流换热空气预热器 recuperative air heater
同流换热炉 recuperative burner;recuperative furnace; recuperative oven
同流换热器 recuperative heater;recuperative heat exchanger;recuperator
同流换热式空气预热器 recuperative air preheater
同流换热式燃气轮机 recuperative gas turbine
同流换热系统 recuperative heat exchange system;recuperative system
同流混合器 concurrent flow mixer
同流节热法 recuperative system
同流浸出 concurrent leaching
同流态取样 isokinetic sampling
同路干扰 common channel interference
同律分节 homonomous segmentation
同律性 homonomy
同伦 homotopy
同伦边缘同态 homotopy boundary homomorphism
同伦边缘运算 homotopy boundary operation
同伦不变量 homotopy invariant
同伦不变性定理 homotopy invariance theorem
同伦测地线 homotopic geodesic line
同伦道路 homotopic paths
同伦的障碍 obstruction to a homotopy
同伦地 homotopically
同伦定理 homotopy theorem
同伦分类 homotopy classification
同伦分类问题 homotopy classification problem
同伦公理 homotopy axiom
同伦关系 relation of homotopy
同伦集 homotopy set
同伦加法 homotopy addition
同伦加法定理 homotopy addition theorem
同伦扩张定理 homotopy extension theorem
同伦扩张性 homotopy extension property
同伦类 homotopy class
同伦理论【数】 homotopy theory
同伦链映射 homotopic chain-mapping
同伦临界点 homotopic critical point
同伦球面 homotopy sphere
同伦群 homotopy groups
同伦上边缘运算 homotopy coboundary operation
同伦上链 homotopy cochain
同伦系 homotopy system
同伦型不变量 homotopy type invariant
同伦映射 homotopic mapping
同伦于零射 homotopic to zero morphism
同伦运算 homotopy operations
同伦障碍 homotopy obstruction
同伦正合序列 homotopy exact sequence
同媒沉积 isomesical deposit
同门【计】 equivalence gate;exclusive NOR gate; equivalence element; equivalent-to-element;equality gate; identity gate; identity unit; match gate
同盟 alliance;confederation;link-up;union
同盟罢工 joint strike
同盟船公司 participating carrier
同盟定期船 conference line vessel
同盟费率 conference rate
同盟费率表 conference tariff
同盟会员 conference member
同面的 coplanar
同面弯曲 uniplanar bending;uniplane bend
同面型盒式磁带 coplanar tape
同面性 coplanarity
同面应变 coplanar strain
同名 isonym
同名地物 identical object
同名点 homogeneous point; identical point;matching point
同名光线 conjugate image ray;conjugate ray;corresponding image ray
同名核线 corresponding epipolar ray
同名极 like pole
同名数 like numbers
同名投影 same-name projection
同名投影光线 corresponding projection ray
同名物 homonym
同名像点 corresponding image points;homogeneous;image point
同模气体区 homologous gaseous sphere
同模式标本 homeotype
同模输入 common-mode input
同模增益 common-mode gain
同谋 aid and abet
同能量 co-energy
同胚 homeomorphism;homeomorphy
同胚不可约图形 homomorphically irreducible graph
同皮丁顺砖交错砌筑 single Flemish bond
同偏 parallel avertence
同频带通信[讯]方式 inband signal(l)ing
同频带信号传输 inband
同频道广播电台 shared frequency station
同频的 in-line
同频广播 common-frequency broadcast(ing)
同频校正 coincidence correction
同频调谐 in-line tuning
同频信道 shared channel
同频增音机 same frequency repeater
同品质样品 same quality sample
同平面 coplanar
同平面单元 equal planar unit
同平面的 isoplanar;uniplanar
同平面减速器 single-plane gear
同平面内的力 coplanar force
同平面应变 coplanar strain
同平面应力 coplanar stress
同谱线 co-spectrum
同谱异色 isomeric colo(u)r
同栖共生 calobiosis
同期 corresponding period;hold;syntonization;synchronism
同期比较法 contemporary comparison

同期不相关变量 contemporaneously uncorrelated variables
同期沉积 codeposition; synchronous deposit
同期成的 isogenetic
同期的 simultaneous; synchronizing
同期调相机 dynamic(al) condenser
同期风化 contemporaneous weathering
同期花岗岩 synchronous granite
同期纪录 contemporary records
同期降水量差数 precipitation deficiency
同期界线 synchronous boundary
同期开关 synchronizing switch
同期平均法 corresponding period average method
同期砌筑的承重墙 contemporary bearing wall
同期侵蚀作用 contemporaneous erosion
同期深成岩体 synchronous pluton
同期误差 contemporaneous errors
同期显微共生结构 implication texture
同期相关 contemporaneous correlation
同期相依 contemporaneous dependence
同期协方差 contemporaneous covariance
同奇偶性 like parity
同气候带 isoclimate zone
同气候地区 homoclime
同强震声 isacoustic
同情罢工 strike in sympathy; sympathetic(al) strike; sympathy strike
同情罢工损失 sympathetic(al) damage
同情的 sympathetic(al)
同区沉积 isotopic deposit
同区相邻地址 regional address
同日资金 same day funds
同容积 co-content
同容量 co-content
同熔花岗岩 syntactic(al) granite
同熔岩浆 syntectic magma
同熔作用 syntectic; syntexis
同熔作用方式 syntexis way
同三晶形 isotrimorphism
同色 concolo(u)r; homochromy
同色测光 homochromatic photometry
同色的 concolo(u)rous; isochromatic
同色光 <不同光谱能量分布的> metamer
同色光度测量 homochromatic photometry
同色光度学 homochromatic photometry
同色浓淡的 duotone
同色条纹法 isochromatic fringe method
同色同谱色 isomeric colo(u)r
同色性 homochromatism
同色异构 homochromo-isomerism
同色异构体 homochromic; homochromo-isomer
同色异谱程度 degree of metamerism
同色异谱刺激 metameric colo(u)r stimuli
同深度面积曲线 area-depth curve; area-depth distribution curve
同生 apposition
同生白云岩 syngenetic dolostone
同生变形 contemporaneous deformation; penecontemporaneous deformation
同生层控矿床 syngenetic strata bound deposit
同生成矿说 syngenetic ore-forming theory
同生成矿作用 syngenetic ore-forming process
同生成岩期 syndiagenetic stage
同生成岩作用 syndiagenesis
同生的 connate; syngenetic; syngenic
同生断层 contemporaneous fault; growth fault; synsedimentary fault
同生断层-逆牵引背斜聚集带 accumulation zone of growth fault-roll-over anticline
同生分散 syngenetic dispersion
同生分散模式 syngenetic dispersion pattern
同生构造 syngenetic structure
同生角砾岩 autobreccia; contemporaneous breccia
同生结核 syngenetic concretion
同生喀斯特 syngenetic karst
同生矿床 syngenetic deposit
同生矿物 syngenetic mineral
同生砾岩 contemporaneous conglomerate
同生泥炭沼泽 contemporaneous peat swamp
同生侵蚀【地】contemporaneous erosion
同生区 station
同生群 consortive group of flora
同生水 native water
同生异常 syngenetic anomaly
同生异常包裹体 syngenetic anomalous inclusion
同生褶皱 contemporaneous fold
同生组分 syngenetic component
同生组构 apposition fabric; primary fabric
同生作用 metabiosis; syngenesis
同声传译 simultaneous interpretation
同声传译设备 facility for simultaneous interpretation
同时爆发 simultaneous firing
同时爆破 simultaneous blast(ing); simultaneous firing
同时逼近 simultaneous approximation
同时比较法 simultaneous comparison methods
同时变形 contemporaneous deformation
同时采集 simultaneous acquisition
同时操作 concurrent operation; concurrent working; simultaneous operation
同时操作计算机 simultaneous computer
同时测定 simultaneous determination; simultaneous measurement
同时测量 simultaneous measurement
同时潮 cotidal hour
同时潮流线 cocurrent line
同时沉淀 synchronous precipitation
同时沉积 codeposition; synchronous deposit
同时沉积物 synchronous deposit
同时沉积作用 cosedimentation
同时乘法器 simultaneous multiplier
同时充填 simultaneous filling
同时出错 simultaneous error
同时出现线 isopipteses
同时除磷脱氮 simultaneous phosphorus and nitrogen removal
同时处理 concurrent processing; simultaneous processing
同时处理方式 simultaneous mode
同时触发 simultaneous triggering
同时传递 simultaneous transmission
同时传染 concurrent infection
同时传输 parallel transmission; simultaneous transmission
同时传送 concurrently transmission; concurrent transmission; simultaneous transmission
同时萃取金属 simultaneously extracted metal
同时存取 simultaneous access
同时存在 coexist; coexistence
同时代 contemporaneity
同时代的 coetaneous; coeval; contemporary; contemporaneous
同时代的人 contemporary
同时代性 contemporaneity principle
同时单音制 simultaneous tone system
同时氮氧分析程序 simultaneous nitrogen and oxygen analysis program(me)
同时的 simultaneous; synchronal
同时地平纬度 simultaneous altitude
同时点火 simultaneous firing; simultaneous ignition
同时迭代 simultaneous iteration
同时迭代方法 simultaneous iteration method
同时动作 coaction; simultaneous motion; simultaneous operation
同时动作图解 simultaneous motion cycle chart
同时短程硝化反硝化 simultaneous shortcut nitrification and denitrification
同时对比 simultaneous contrast
同时多片解析改正 simultaneous multiframe analytic(al) calibration
同时多元分析 simultaneous multielement analysis
同时而起的 coincidental
同时发起的冲击 simultaneous assault
同时发生 concurrence; conjuncture; contemporize; simultaneity; sync; synchronization; synchronize
同时发生的 cocurrent; coinstantaneous; simultaneous; contemporaneous
同时发生的地震 simultaneous earthquake
同时发生性 contemporaneity principle
同时翻译 simultaneous translation
同时反馈误差分量 simultaneous feedback error component
同时反硝化 simultaneous denitrification
同时反应 coincident reaction; simultaneous reaction
同时放炮 instantaneous blast
同时分离 coseparation
同时幅频调制 simultaneous frequency and amplitude modulation
同时付清条款 simultaneous settlements clause
同时复合制动 simultaneous combined braking
同时高度 simultaneous altitude
同时工作 simultaneous work(ing)
同时工作方式 simultaneous mode of working
同时工作系数 coincidence factor
同时估计法 simultaneous estimation method
同时故障检测 concurrent fault detection
同时观测 cocurrent observation; simultaneous observation
同时观测的天体高度 simultaneous altitude
同时观测法 cocurrent observation method; simultaneous observation method
同时观察 cocurrent observation; simultaneous observation
同时广播 simultaneous broadcast(ing)
同时行迭代 simultaneous row iteration
同时好氧缺氧生物反应器 simultaneous oxic and anoxic bioreactor
同时烘干和粉磨 combined drying and grinding; simultaneous drying and grinding
同时呼叫 simultaneous call(ing)
同时呼叫次数 traffic density
同时滑移理论 simultaneous slip theory
同时还原 coreduction
同时回程注浆 simultaneous grouting
同时挤进的 co-extruded
同时给水百分率 percentage of simultaneous water demand
同时加热 concurrent heating
同时加热感应淬火 single-shot induction hardening
同时间阻滞 synchronizing delay
同时兼两个职业 moonlight
同时检测 simultaneous detection
同时检测携带式探测计 portaprobe simul-tester
同时检测携带式探测针 portaprobe simul-tester
同时检索 simultaneous scanning
同时降水 simultaneous precipitation
同时交易 simultaneous settlement
同时接地故障 simultaneous ground fault
同时接收 simultaneous reception
同时解冻线 isotac
同时解扣 simultaneous tripping
同时进位 simultaneous carry
同时进行电报电话制 simultaneous telegraph and telephony
同时进行受话和送话 simultaneous talking
同时具有两种特性的 dimorphic
同时开花线 isophene[isophane]
同时开启的双拉门 biparting door
同时控制 simultaneous control
同时扩散 simultaneous diffusion
同时拉伸变形机 simultaneous draw texturing machine
同时连接 simultaneous connection
同时溜放【铁】double humping(of two trains); dual humping(of two trains); simultaneous humping of two trains
同时流量 concurrent discharge
同时流溜放 simultaneous humping
同时流速 simultaneous velocity
同时率 coincident factor; simultaneity factor
同时轮 simultaneous wheels
同时模拟 concurrent simulation
同时排水百分率 percentage of simultaneous drainage
同时喷出 simultaneous ejection
同时碰撞 simultaneous collision
同时平衡 simultaneous equilibration
同时平行运行 simultaneous parallel movement
同时普查 simultaneous census
同时期性 contemporaneity
同时期性原理 contemporaneity principle
同时起爆 simultaneous initiation; simultaneous shotfiring
同时起动 <多发动机> simultaneous firing
同时起作用的横向载荷 simultaneous transverse load
同时侵蚀 contemporaneous erosion
同时热解重量差热分析仪 simultaneous thermogravimetric-differential thermal analyser[analyzer]
同时扫掠制 simultaneous lobing system
同时扫描 simultaneous scan

同时审议 concurrent consideration
同时生产热电系统 cogeneration system
同时生物除磷脱氮 simultaneous biological nitrogen and phosphorus removal
同时使用系数 coincidence factor;simultaneous;usage factor
同时适应 simultaneous adaptation
同时适用的进路 compatible route;simultaneously possible route
同时收发 simultaneous transmission and reception
同时收发双工制 duplex system
同时输入脉冲 simultaneous input pulse
同时输入输出 concurrent input-output;simultaneous input-output
同时双向操作 full-duplex operation;simultaneous two-way operation
同时双向工作 simultaneous two-way operation
同时水坡线 isochron(e)
同时水位 simultaneous stage;simultaneous water level
同时替换法 simultaneous displacement method
同时条件反射 simultaneous conditioned reflex
同时调谐 gang tuning
同时通话 simultaneous talking
同时通信[讯] simultaneous communication
同时统计学推论 simultaneous statistic(al) inference
同时投料法 <混凝土搅拌时> ribbon loading
同时突变 simultaneous mutation
同时吞吐量 simultaneous throughput
同时脱氮除磷 simultaneous denitrification-dephosphorization
同时外部操作 concurrent peripheral operation
同时外围处理 concurrent peripheral processing
同时完工 concurrent completion
同时位移 simultaneous displacement
同时吸附 simultaneous adsorption
同时系数 simultaneity factor;simultaneous factor
同时系统 simultaneous system
同时显示 simultaneous display
同时线 isochron(e)
同时线法 isochron(e) method
同时相 parvafacies
同时相关 simultaneous dependence
同时硝化反硝化 synchronous nitrification-denitrification
同时硝化反硝化脱氮技术 synchronous nitrification-denitrification denitrogen technique
同时效应 simultaneous effect
同时协调的 in phase
同时信号控制法 simultaneous traffic signal controlling
同时型 simultaneous type
同时性 isochroneity;simultaneity;synchronism;synchronization
同时性叠加 simultaneous summation
同时性观察研究 concurrent study
同时需气量 coincident demand
同时选择 concurrent selection;simultaneous choice
同时氧化 simultaneous oxidation
同时异相沉积 contemporaneous heterotopic facies;heteropic(al) deposit
同时异相的 heteropic(al)
同时诱导 simultaneous induction
同时原则 contemporaneity principle
同时运算计算机 simultaneous computer
同时占用计数器 traffic recorder
同时张拉 simultaneous tension
同时整理法 simultaneous finished process
同时指令系统 simultaneous instruction system
同时制 simultaneous system
同时制彩色电视 field simultaneous colo(u)r television;simultaneous colo(u)r television
同时制无线电信标 simultaneous type radio range
同时中断 concurrent interruption
同时装罐 simultaneous decking
同时着陆 simultaneous landing
同时总线操作 simultaneous bus operation
同时最大需用功率 coincident demand power
同时最大需用效率 coincident demand efficiency
同时作用 synchronization
同实体大小的 full-sized
同实体大小满刻度 full-scale
同实物一样大小 life size
同事关系 fellowship
同势差的 idiostatic
同势差连接法 idiostatic method
同势面 equipotential surface
同属的 congeneric;congenerous
同素还化合物 homocyclic compound
同素环的 bomocyclic
同素异形的 allotropic
同素异形体 allotrope;allotropic modification
同素异形(现象) allotropism;allotropy
同素异形性 allotropy
同素异形转换 allotropic transformation
同素异重体 allobar
同速 synchronize
同速联动 electric(al) shaft
同速生长 isogony
同速转动 corotate
同塔双回线 double circuit lines on the same tower
同态 homostasis;multiple isomorphism;homomorphism
同态变换 isomorphic transformation
同态的 isomorphic
同态的核 kernel of a homomorphism
同态反滤波 homomorphic inverse filtering
同态反褶积 homomorphic deconvolution
同态函数 homomorphic function
同态环 homomorphism ring
同态滤波 homomorphic filtering
同态滤波器 homomorphic filter
同态模 homomorphism module;module of homomorphisms
同态内插定理 homomorphism interpolation theorem
同态群 homomorphism group
同态调节器 homeostat;homestat
同态像 homomorph;homomorphic image
同态映射 homomorphic mapping
同态语音处理 homomorphic speech processing
同碳耦合 isocarbon couple
同体的 coessential
同体关系 same body relation
同体链 same body chain
同体式电弧焊接变压器 built-in current regulator type electric arc transformer
同体式焊机 one-body welding machine;one-body welding set
同外延 homepitaxy
同纬度航法 parallel sailing
同纬圈图形 images of identical parallel
同纬映象 suspension
同纬映象序列 suspension sequence
同位 apposition;coordination;parity;parity bit
同位边 homologous sides
同位标脉冲 cylinder pulse
同位层 equivalent layer
同位差连接法 idiostatic method
同位穿孔 batten;cordonnier;peek-a-boo
同位穿孔(检索)系统 Batten system;peek-a-boo system
同位穿孔检验 batten check
同位穿孔校验法 batten check;sight check
同位穿孔卡(片) peek-a-boo card
同位的 corresponding
同位地层【地】 equivalent bed
同位多重态 isomultiplet
同位多重性 isomultiple
同位角 corresponding angle;exterior-interior angle
同位孔 coordinating holes
同位孔系统 cordonnier system
同位孔隐字 peephole mask
同位面 equipotential surface
同位数 isotopic number
同位素【化】 isotope
同位素靶 isotopic target
同位素比 isotopic ratio
同位素比放射性 isotopic specific activity
同位素比分析 isotope ratio analysis
同位素比活度 isotope specific activity
同位素比示踪法 isotope ratio tracer
同位素比示踪剂法 isotope ratio tracer method
同位素比示踪物法 isotope ratio tracer method
同位素比值 isotope ratio
同位素比值记录质谱仪 isotope-ratio-recording mass spectrometer
同位素比值质谱计 isotope ratio mass spectrometer
同位素比质谱计 isotope ratio mass spectrometer
同位素变数 isotopic variable
同位素辨别法 isotope discrimination
同位素标记 isotope-label(l)ing;isotopic tag(ging)
同位素标记的 isotopically tagged
同位素标记法 isotope-label(l)ing method
同位素标志 isotope tag
同位素标准 isotopes standard;isotopic standard
同位素标准样品 isotopic standard samples
同位素表 isotope chart;table of isotopes
同位素箔 isotopic foil
同位素不均-性 isotope heterogeneity
同位素操作计算器 isotope handling calculator
同位素测定(法) isotopic assay;isotopic measurement
同位素测定年龄 isotopic dating
同位素测定水样 water sample of isotope measurement
同位素测厚仪 isotropic(al) thickness ga(u)ge
同位素测量 isotopic determination
同位素测量参数 parameter of isotope determination
同位素测流法 isotope-velocity method
同位素测沙仪 isotope sediment concentration ga(u)ge
同位素测温法 isotope thermometry
同位素差异 isotopic differentiation
同位素成分 isotope composition;isotopic composition;isotopic constituency
同位素处理 isotope handling
同位素纯 isotopically pure
同位素纯度 isotopic purity;isotopic quality
同位素代换的磷酸盐 isotopically exchangeable phosphate
同位素代替物 isotope substitute
同位素的 isotopic
同位素的同质异能素 isotopic isomer
同位素等值线 isotope contour
同位素等值线图 isotope isogram
同位素地球化学 isotope geochemistry
同位素地球化学参数 isotopic geochemical parameter
同位素地热温标 isotopic geothermometer
同位素地质年代表 isotopic geochronologic(al) scale
同位素地质年代学 isotopic geochronology
同位素地质年龄 isotope geologic(al) age
同位素地质温度计 isotope geothermometer
同位素地质学 isotope geology
同位素电源 isotope generator
同位素电中和器 isotope neutralizer
同位素动力 isotopic power
同位素毒物 isotopic poison
同位素多重性 isotopic multiplicity
同位素发射密度仪 back-scattered densitometer
同位素发生器 radionuclide generator
同位素法 method of isotope
同位素法测定的年代 isotopic age
同位素法测定的年龄 isotopic age
同位素反应性系数 isotopic reactivity coefficient
同位素方法 isotope method;isotopic method
同位素分布 isotope distribution;isotopic distribution
同位素分丰度 isotopic abundance
同位素分丰度测量 isotopic abundance measure
同位素分离 isotope fractionation;isotope separation;isotopic separation;separation of isotopes
同位素分离方法 isotope separation method
同位素分离工厂 isotope separation plant
同位素分离器 isotope separation apparatus;isotope separator;isotron;mass separator
同位素分离因数 isotope separation factor
同位素分馏 isotope fractionation
同位素分馏方程 isotope fractionation equation
同位素分馏机理 isotope fractionation mechanism
同位素分馏系数 isotope fractionation factor
同位素分馏效应 isotope fractionation effect
同位素分析 isotope analysis;isotopic analysis
同位素分析方法 isotope analysis method
同位素分析方法及仪器设备 the methods and instruments for isotope anal-

T

ysis
同位素分析器 isotron
同位素分析设备 some equipment of isotope analysis
同位素分析样品 sample for isotopic analysis
同位素分析质谱仪 the mass spectrometers for isotope analysis
同位素分子 isotopic molecule
同位素丰度 isotope abundance
同位素丰度形式 the pattern of isotopic abundance
同位素峰 isotope peak
同位素伽马常数 gamma constant of isotope
同位素含量 isotope content; isotopic content
同位素含砂量计 isotopic turbidity meter
同位素含水量密度测定仪 moisture-density nuclear ga(u)ge
同位素含水量探测仪 moisture probe; nuclear soil moisture meter
同位素含水量仪 nuclear soil moisture densometer
同位素化探 isotopic geochemical method
同位素化学 isotope chemistry; isotopic chemistry
同位素混合 isotope contamination
同位素混合交换模式 model of mixed-exchanged isotopes
同位素混合-交换平衡模式 model of isotopes mixed-exchanged equilibriums
同位素混合-交换总模式 general model of mixed-exchanged isotopes
同位素混合模式 model of mixing isotopes
同位素混合物 isotopic mixture
同位素混合作用 isotope mixing
同位素计算尺 isotope handling calculator
同位素加入法 isotope addition method
同位素检测器 isotope locator
同位素检查器 isotope locator
同位素交换 isotope exchange
同位素交换法 isotope exchange method
同位素交换反应 isotope exchange reaction
同位素交换平衡 isotope exchange equilibrium
同位素交换速率 isotope exchange rate
同位素校正 isotopic correction
同位素均质化 isotope homogenization
同位素矿物学 isotope mineralogy
同位素类型 isotopes types
同位素类型曲线 isotope pattern curve
同位素类型曲线图 isotopic type curve diagram
同位素灵敏度 abundance sensitivity; isotopic sensitivity
同位素密度计 nuclear densometer
同位素密度探测仪 density probe; isotope density probe; nuclear densometer
同位素内部测温法 isotope internal thermometry
同位素年代测定 isotopic age determination
同位素年代年龄测定 isotopic dating
同位素年龄 isotopic age
同位素年龄测定样品 sample for isotopic age determination
同位素年龄梯度 isotopic age gradient
同位素浓缩 isotopic enrichment
同位素配分函数 isotope distribution function
同位素配分系数 isotope partition coefficient
同位素频率曲线 isotope frequency curve
同位素平衡 isotope equilibrium
同位素平衡常数 isotope equilibrium constant
同位素平衡温度 isotope equilibrium temperature
同位素群 pleiad; pleyade
同位素扫描器 isotope scanner
同位素扫描仪 isotope scanner; radioisotope scanner
同位素射线种类 type of isotope ray
同位素深海海流分析仪 deep-water isotopic current analyser[analyzer]
同位素湿度密度探测仪 nuclear moisture-density apparatus
同位素实验室 isotope laboratory
同位素示踪 isotope tracing; isotopic tag(ging)
同位素示踪的 isotopically tagged
同位素示踪法 tracer method
同位素示踪技术 isotope tracer technique; isotropic (al) tracer technique
同位素示踪剂 isotope tracer; isotropic(al) tracer; radioactive tracer
同位素示踪器 isotope tracer
同位素示踪试验 tracer test(ing)
同位素示踪物 isotopic tracer
同位素试验 radioactive testing
同位素输出函数 isotopic output function
同位素输入函数 isotopic input function
同位素熟化曲线图 isotopic maturation curve diagram
同位素衰变法 isotopic decay method
同位素水位计 isotope water level ga(u)ge
同位素水文地质 isotopic hydrogeology
同位素水文方案 isotope hydrology program(me)
同位素水文学 isotope hydrology
同位素探测仪 density probe; nuclear soil moisture meter
同位素探伤仪 isoscope
同位素土壤含水量探测仪 nuclear soil moisture meter
同位素外部测温法 isotope external thermometry
同位素温标 isotope temperature scale
同位素温度 isotopic temperature
同位素温度方程 isotope thermometer equation
同位素温度计 isotopic thermometer
同位素污染 isotope contamination; isotope pollution; isotopic contamination
同位素污染物 isotope contaminant; isotope pollutant
同位素稀释法 isotope dilution method
同位素稀释分析 isotope dilution analysis
同位素稀释火花源质谱法 isotope dilution spark source mass spectrometry
同位素稀释质谱法 isotope dilution mass spectrometry
同位素相关图 isotope correlogram
同位素相关性 isotope correlation
同位素效应 isotope effect; isotopic effect
同位素学 isotopy
同位素衍生法 isotope derivation method
同位素演化 isotope evolution
同位素氧交换 isotopic oxygen exchange
同位素移动 isotope shift
同位素运输罐 isotope transport container
同位素载体 isotope carrier
同位素指示剂 isotopic tracer; tracer element
同位素质量比 isotope mass ratio
同位素质量测量 isotopic mass measuring
同位素质谱计 isotope mass spectrometer
同位素重量 isotopic weight
同位素组分 isotope composition; isotopic composition
同位素组分分布图 distribution graph of isotope composition
同位素组分频率分布图 frequency distribution of isotope composition
同位系数 corresponding coefficient
同位线 corresponding lines
同位相似比 homothetic ratio
同位相似变换 homothetic transformation
同位相似道路 homothetic paths
同位相似对应 homothetic correspondence
同位相似二次曲线 homothetic conics
同位相似加工硬化 homothetic work-hardening
同位相似解 homothetic solutions
同位相似曲线 homothetic curves
同位相似三角形 homothetic triangle
同位相似图形 homothetic figures
同位相似中心 homothetic center[centre]
同位相似轴 homothetic axis
同位小数 similar decimals
同位型 isotype
同位(元)素 isotopic element
同温变 isothermal change
同温层 homothermy; isothermal layer; stratosphere
同温层堡垒 stratofortress
同温层电视 stratovision
同温层飞机 stratocruiser; stratoplane
同温层观测镜 stratoscope
同温层货运机 stratofreighter
同温层加油机 stratotanker
同温层客机 stratoliner
同温层喷气机 stratojet
同温层气球 stratospheric(al) balloon
同温层运输机 stratoliner
同温床 sirafo bed
同温动物 homoeothermic animal
同温度的 synthermal
同温条件 homothermal condition
同文 identical text
同文电报 same text
同文电报接线设备 multiaddress telegraph connector
同纹理砾石 like-grained gravel
同物价指数相联系的 index linking
同物异量质溶胶 isosol
同物异名 synonym
同物质 commaterial
同系 homologize
同系波 homologous waves
同系场 homologous field
同系化合物 homologous compound
同系机械 homologous machines
同系聚合物 polymer-homologous series; polymer-homologue; polymeric homologue
同系列 homologous series; homogeneous series
同系列的 uniserial; homologous
同系列的水轮机 homologous turbine
同系配合 endogamy
同系色谱法 homochromatography
同系射电暴 homologous radio burst
同系统的 cognate
同系物【化】 homolog(ue)(compound)
同系现象 homology
同系线对 homologous pair
同系星 homologous star
同系移植物 isograft
同系移植系统 isotransplant system
同系株 ramet
同线 colinear; multiparty line
同线的 collinear
同线电话 bridging telephone; multiparty telephone; party line; party-line telephone
同线电话线 single-party line
同线电话选择振铃 party-line ringing
同线电话用户间互叫机键 reverting call switch
同线电话振铃电键 party-line ringing key
同线互叫终接器 reversing call connector
同线群 syntenic group
同线性 colinearity
同线用户间的呼叫(电话) reverting call
同线用户终接器 reversing call connector
同相 cophase; homogeneous phase; homophase; in-phase state; in-step; phase coincidence; phase lock; syniphase
同相波 cophase wave
同相波痕纹层 ripple laminae in phase
同相部分 in-phase component
同相彩色信号 in-phase chrominance signal
同相垂直天线 cophased vertical antenna; same-phase vertical antenna
同相的 cophasal; cophased; equiphase; isopic; in phase
同相电流 in-phase current
同相电位计 in-phase potentiometer
同相电压 in-phase voltage
同相垫纱 in-phase overlap
同相端子 in-phase terminal
同相多普勒跟踪系统 phase-locked Doppler tracking system
同相反馈 in-phase feedback
同相放大器 non-inverting amplifier
同相分量 cophase component; in-phase component
同相工作 push-push operation
同相供电 cophase supply
同相回输 in-phase feedback
同相混频器 in-phase mixer
同相激励 cophase excitation; excitation in phase; excited in phase; syniphase excitation
同相检波器 phase-locked detector
同相交叉相关 in-phase cross correlation
同相角 in-phase angle
同相接触联结 phase contact bond
同相接法 noninverting connection
同相粒子 in-phase particles
同相滤波器 in-phase filter
同相母线 isophase bus
同相能量谱 in-phase energy spectrum
同相偶极天线阵 phased array of dipoles
同相谱 coincidence spectrum; cospectrum
同相沙纹交错纹理构造 ripple cross-lamellar structure
同相纱 in-phase yarn
同相输入 common-mode input; noninverting input
同相束 in-phase beam
同相水平天线 cophased horizontal antenna; same-phase horizontal antenna

同相水平天线阵 cophased horizontal antenna array; in-phase horizontal antenna array
同相损失 in-phase loss
同相天线 broadside antenna
同相天线阵 cophased array
同相条件 in-step condition
同相调制 cophasal modulation
同相位 same phase
同相位的 in-step
同相线 isopen
同相信道 in-phase channel
同相信号 in-phase signal
同相性接触 phase contact
同相压缩比 in-phase compression ratio
同相抑制 in-phase rejection
同相抑制比 common-mode rejection ratio
同相运动 simultaneous movements
同相运行 in-phase operation
同相运用 in-phase operation
同相振荡器 phase synchronized oscillator
同相振动 in-phase vibration
同相终端 in-phase terminal
同相轴分辨率 event resolution
同向 equidirectional
同向波簇 wave train
同向差 difference of readings in the same orientation
同向车道 concurrent flow lane
同向车道境界线 lane line in same direction
同向断层 synthetic(al) fault
同向断曲线 broken-back curve
同向分布 isodirectional distribution
同向辊涂法 direct roll coating
同向滚削 climb hobbing
同向流 parallel flow
同向流沉淀池 cocurrent tank
同向流式隔油池 isolating-oil pool of same-direction flow
同向流斜板式沉淀池 inclined plank settling tank of same-direction flow
同向磨削 climb-cut grinding
同向捻(法) Albert lay; lang lay
同向捻钢丝绳 Albert lay wire rope; lang-lay rope
同向凝结作用 orthokinetic coagulation
同向凝聚 shear flow-induce aggregation
同向偏斜 conjugate deviation
同向曲率 synclastic curvature
同向曲面 synclastic surface
同向曲面薄壳 synclastic shell
同向曲线 adjacent curve in one direction; curve of same sense; same adjacent curve; same direction adjacent curve; same-sense curve
同向三开道岔 similar flexure turnout; three-throw turnout of similar flexures
同向生长型结晶纤维 syntaxial growth type
同向双工电报 diplex telegraphy
同向双工电路 diplex circuit
同向双工发送器 diplex sender
同向双工接收 diplex reception; duplex reception
同向双工器 diplexer
同向双工(制) diplex
同向双讯器 diplexer
同向弯曲 similar flexure
同向弯曲式 homotropy
同向纹理层压 parallel lamination
同向铣切 down cut milling
同向铣削 climb cutting; climb milling
同向线 isogonic line
同向行车 train movement with the current of traffic
同向絮凝作用 orthokinetic flocculation
同向旋转 homodromy
同向移动 orthokinetic
同向右捻 right-hand lang-lay
同向褶皱 accordant fold
同向左捻 left-hand lang-lay
同项归并统计 statistics of groping the same item
同消色线 isogyre
同效对 homometric(al) pair
同效抑制剂 isoinhibitor
同斜层 homocline
同斜屋谷 isoclinal valley
同斜褶皱 congruent fold
同心 U 形钻头 concentric(al) U bit
同心泵 overcenter pump
同心变径管 concentric(al) reducer; concentric(al) taper
同心波 circular wave
同心剥离 concentric(al) exfoliation
同心槽 locked groove
同心层结核 rattle stone concretion
同心大小头 concentric(al) reducer; concentric(al) taper
同心带 belt
同心导线电缆 coaxial line
同心的 concentric(al); homocentric; stigmatic
同心电极弧光灯 concentric(al) candle
同心电缆 concentric(al) lay cable
同心定位 concentric(al) locating
同心度 concentricity; trueness
同心度检查装置 concentricity checking fixture
同心度量规 bearing ga(u)ge
同心断层 concentric(al) faults
同心二次曲面 concentric(al) quadrics
同心二次曲线 concentric(al) conics
同心阀 concentric(al) valve
同心风化 concentric(al) weathering
同心缝隙喷雾器 tubular atomizer
同心复式转筒 concentric(al) compound trommel
同心钢筋束 concentric(al) tendon
同心钢丝索 concentric(al) tendon
同心拱 concentric(al) arch
同心拱坝 constant radius arch dam
同心管 concentric(al) tube
同心管换热器 concentric(al) tube exchanger
同心管(精密蒸馏)柱 concentric(al) tube column
同心管热交换器 concentric(al) tube heat exchanger
同心光束 concentric(al) pencil; homocentric pencil concentric (al) beam; homocentric pencil of rays
同心虹吸 concentric(al) siphon
同心弧 concentric(al) arc
同心环 concentric(al) rings
同心环测渗仪 concentric(al) ring infiltrometer; ring infiltrometer
同心环带模型 concentric(al) zone model
同心环和十字刷 ring-and-brush
同心环渗透计 concentric(al) ring infiltrometer
同心环形空气散流器 concentric(al) ring air diffuser
同心环形山构造 concentric(al) ring structures
同心环形障板 shell-type baffle
同心环状的 concentric(al) girdle
同心活塞环 concentric (al) piston ring
同心夹盘 concentric(al) chuck
同心渐扩踏步式卷 arch with order
同心绞线 concentric(al) lay conductor; concentric(al) strand(ed wire)
同心节理 concentric(al) joint
同心卡盘 concentric(al) chuck
同心刻度 concentric(al) scale
同心扩压段 concentric(al) diffuser
同心梁 concentric(al) beam
同心裂隙 concentric fracture
同心笼式粉磨碎机 concentric(al) squirrel cage mill
同心模 concentric(al) die
同心膜 concentric(al) coat
同心尼龙管 concentric (al) nylon bush
同心劈理 concentric(al) cleavage
同心片状 concentric(al) lamellar
同心屏蔽筒 concentric(al) terminal shield
同心球 concentric(al) sphere; homocentric sphere
同心球粒 orbicule
同心区理论 concentric(al) zone theory
同心圈 concentric(al) rings
同心绕法 concentrated winding
同心绕丝管换热器 concentric(al) tube exchanger with wire spacer
同心绕组 concentrated winding; concentric(al) winding
同心锐孔隔板 concentric(al) orifice plate
同心食 concentric(al) eclipse
同心式 concentric(al) pattern; concentric(al) type
同心式钢丝绳 concentric(al) wire rope
同心式汽化器 concentrically built carburet(t)er
同心式装配 concentric(al) mounting
同心树 concentric(al) tree
同心双转子电动机 duocentric motor
同心套轴的两个对转螺旋桨 contra-turning propeller
同心筒式黏[粘]度计 concentric(al) cylinder visco(si)meter
同心透镜 concentric(al) lens
同心椭圆 concentration ellipse
同心维管束 concentric(al) vascular bundle
同心线 concentric(al); concentric(al) line
同心线布线 concentric(al) wiring
同心线圈 concentric(al) coil
同心线束 concentric(al) pencil
同心橡胶管 rubber line
同心协力 pull together
同心性 concentricity
同心药皮 concentric(al) covering
同心异跨联拱 compound arch
同心预应力钢索 concentric tendon
同心圆 concentric(al) circle
同心圆城市 concentric(al) city
同心圆发展理论 concentric(al) zone theory
同心圆发展说 concentrate zone theory
同心圆拱 rainbow arch
同心圆理论 concentric(al) circles theory
同心圆论 concentric(al) zone concept
同心圆模型 concentric(al) model
同心圆排列式样 <金刚石在钻头唇部的> concentric(al) pattern
同心圆喷嘴 concentric(al) nozzle
同心圆平面 concentric(al) plan
同心圆绕射板 zone plate
同心圆说 concentric(al) zone theory
同心圆筒 concentric(al) cylinder
同心圆筒式黏[粘]度计 concentric(al) cylinder visco(si)meter
同心圆筒式消声器 concentric(al) cylinder muffler
同心圆筒旋转式测黏[粘]度法 concentric(al) cylinder rotation visco(si)metry
同心圆图 concentric(al) circle diagram
同心圆柱体 concentric(al) cylinder
同心圆状 concentric(al)
同心(圆状)构造【地】 concentric(al) structure
同心晕 concentric(al) halo
同心褶曲 concentric(al) fold; parallel fold
同心褶皱 concentric(al) fold; parallel fold
同心转炉 concentric(al) converter
同心状断裂 concentric(al) fractures
同心锥形管 concentric(al) reducer; concentric(al) taper; concentric(al) taper pipe
同心锥形件装配 concentric(al) taper fitting
同信道 co-channel
同信道干扰 co-channel interference; in-channel interference
同形 homoemorphy; homomorphism
同形变态 homomorpha
同形层 isomorphous layer
同形代替 isomorphous substitution
同形的 isomorphic; isomorphous
同形度 conformity
同形花现象 homogamy
同形晶体 isomorphous crystal
同形偶 isomorphous pair
同形取代 isomorphic substitution
同形射线 homogeneous ray
同形射线组织 homogeneous ray tissue
同形体 homomorphs; isomorph
同形网状 homobrochate
同形现象 isomorphism
同形性 homomorphism; homomorphy; isomorphism
同形映射 homomorphic mapping
同形置换 isomorphous replacement
同形组织 homogeneous tissue
同型 allomerism; homoplasy; homotype; same type
同型变异 isotypic variation
同型齿 homodont
同型的 homoplastic; isotypic
同型二聚体 homodimer
同型分析 common-size analysis
同型合子 homozygote
同型联合 homosyndesis
同型联会 homeosynapsis
同型水轮机 homologous turbine
同型损益表 common-size income statement
同型特异性 isotypic specificity
同型同境群落 synusia
同型物 analog(ue); isoreagent; isotype substance
同型性 homotypy; isotypism
同型原理 homotyposis
同型装配 cannibalization; cannibalize [cannibalise]
同型作用 homotypic(al) effect
同性磁极 like pole
同性低聚物 pleionomer
同性电 like electricity
同性极 like pole
同性接插件 <接插部分完全相同> hermaphroditic connector
同性相 magnafacies
同性质的 cognate; congeneric; congenerous; congenial; connatural
同血缘 consanguinity
同压吹管 balanced-pressure blow pipe

同样办法 same manner
同样大小(颗粒)的 equigranular
同样的 similar;uniform
同样的全副活字 font
同样货物 identical goods
同样品大致相同 about equal to sample
同样品相似 similar to sample
同样品质的酒 liquor of the same tap
同样作用 same purpose
同业 profession
同业拆款 interbank loan
同业存款 deposit(e) due to other banks;due from banks
同业工会 craft union;horizontal labor union;horizontal union;trade association;trade council;trade union
同业公会 business guild trade organization;guild(-ship);trade association;trade association guild;trade council;trade guild;trading association
同业公会会长<英> warden
同业合并 horizontal combination;horizontal integration;horizontal merger
同业汇率 interbank exchange rate
同业会所 guildhall
同业价格 fellow-trader price
同业间 within the trade
同业竞争 horizontal competition;horizontal trade competition
同业联合 horizontal combination;horizontal integration
同业协议 trade agreement
同业银行 bank correspondent;correspondent bank
同业佣金 fellow-trader commission
同业折扣 fellow-trader discount;trade discount
同业组合 guild
同业组织 business guild;business league;trade organization
同一标准的 uniform
同一产品 identical product
同一车轮偏心距 wheel spinning reading for run-out
同一城市内各铁路枢纽间的调车 interterminal switching
同一尺寸配合料 uniformly graded
同一单价开挖 unclassified excavation
同一单元 identity unit
同一的 identical
同一的造价 identifiable cost
同一地层标高 the same stratigraphic level
同一定价的一批货物 price line
同一度盘位置的测回中数 zero mean
同一方向的 parallel
同一费率 freight-all-kinds rate
同一符号 prosign
同一个资本 self-same capital
同一规格的大宗货物 homogeneous cargo
同一国家海岸与海岸间的贸易 intercoastal commerce
同一国家海岸与海岸间的运输 intercoastal traffic;intercoastal transport
同一海域和地区 same sea and country
同一海域和国家 same sea and country
同一海域和海岸 same sea and coast
同一号码电话 jack per line
同一级配的土壤 uniformly graded soil
同一建筑物内的 inside plant
同一接收 identity reception
同一刻度 identical graduation
同一律 identity law;law of identity
同一年龄的人们 age grade;age-group
同一年同一路上平均每公里出事车辆或车祸死伤率 accident involvement rate
同一配方运转 operate according to a same formula
同一配位体 identical ligand
同一平面的 coplanar
同一日内透支 intra-day overdraft
同一日期的 of even date
同一色的 whole-colo(u)red
同一时期 contemporaneity
同一水平面的 horizontal coplanar
同一线圈 identity coil
同一型号车辆组成的列车 unit train
同一性 homogeneity;identity;uniformity
同一性定理 identity theorem
同一性假设 identity hypothesis
同一性系数 coefficient of homogeneity
同一性原则 identificability principle
同一元件 identity element
同一元素 identity element
同一运费率 freight-all-kinds rate
同一责任制<统一由签发联运提单的承运人对货主负全程运输责任> uniform liability system;uniform liability principle
同一种材料 commaterial
同一装置 identity unit
同意按钮 approval button
同意按钮盘 agreement button panel
同意按照合同开工建设 release for construction
同意代用品 approved equal;approved equal substitution
同意换用合格 approved equal
同意加入合同 contract of adhesion
同意建造 consent to build
同意接车【铁】accept a train
同意接车的闭塞机 co-acting block;cooperating block
同意赔偿破损 award a damage
同意票 concurring vote
同意旗 affirmative flag
同意条件用地区划 consent zoning
同意协助书 letter of comfort
同意选举 consent election
同意样品 approved sample
同意债券 assented bond
同音异字编码 homophonic enciphering
同硬橡皮一样的绝缘材料 vulcalose
同余 congruence[congruency];coresidual
同余点集 coresidual point group
同余关系 congruence relations
同余类 congruence class
同余模 modulus of congruence
同余式 congruent expression
同余式的模 modulus of a congruence
同余数 congruent number
同余数生成程序 congruent(ial) generator
同余整数 congruence integer
同域的 sympatric
同域物种 sympatric species
同源 consanguinity;homologisation[homologization];isogeny
同源包体 autolith;cognate inclusion;enclave homogene;homologue
同源变异 homologous variation
同源捕虏体 cognate xenolith
同源的 cognate;congeneric;congenerous;connate;consanguineous;homologous;isogenetic
同源的工业问题 congeneric industrial problem
同源发生 isogenesis
同源花岗岩 congeneric granite
同源空气污染质 primary air pollution
同源联会 autosyndesis
同源流行 common sources epidemic
同源论 homologous theory
同源配对 homogenetic association;homologous pairing
同源喷出物 cognate ejecta
同源融合 homomixis
同源图形 homological figures
同源系列群落 homologous series
同源现象 homology
同源学说 homologous theory
同源岩浆的【地】comagmatic
同源岩浆区 comagmatic region
同源岩浆岩组 comagmatic assemblage
同源岩石 rock association
同源异构包体 allomorph;enclave allomorphe
同源异形 homoeosis
同源组合 consanguineous association
同在(一起)all together
同造山期 synorogenic period
同造山期的 synorogenic
同造山期煤化作用 synorogenic coalification
同造山期盆地 synorogenic basin
同造山运动 synorogenesis
同站干扰 interference in the co-located station
同站台换乘 same platform transfer
同震带 coseismic zone
同震的 coseismal;coseismic;homoseismal
同震区 coseismal area;coseismic area
同震曲线 coseismal;coseismic curve
同震圈 coseismic circle
同震位移 coseismic displacement
同震线 coseismal line;coseismic line;homoseis;homoseismal line;isochronal line
同支索平行 astay
同质 homogenesis;unity
同质产品 homogeneous product
同质衬底 native substrate
同质成核 homogeneous nucleation
同质大宗货物 homogeneous cargo
同质的 coessential
同质的劳动 qualitatively identical labor
同质多晶(现象)pleomorphism;polymorphism;polytropism
同质多晶型 polytropism
同质多晶型物 polymorphic substance
同质多象 polymorph
同质多象变体 polymorphic modification
同质多象转变类型 type of polymorphic transition
同质多象转变温度 polymorphic transition temperature
同质多形现象 isomeromorphism;pleomorphism
同质二形 isodimorphism
同质二形的 isodimorphous
同质二形体 dimorphic
同质房顶 homogeneous roof
同质分类 homogeneous classification
同质化 homogenization
同质假象 allomorphism;paramorphism
同质胶体 isocolloid
同质接合子 homozygote
同质结 homojunction
同质结构 uniform structure
同质结激光器 homojunction laser
同质垃圾 homogeneous waste
同质量位态 isobaric state
同质楼板 homogeneous floor
同质期 homogeneous period
同质热电效应 Thomson's effect
同质熔化 congruent melting
同质三形 isotrimorphism
同质社会集团 homogeneous social group
同质衰变 metastasis[复 metastases]
同质双链 homoduplex
同质弹性梁 isotropic(al) elastic beam
同质套毛 homogeneous fleece
同质体 homoplasmon
同质同晶 allomerism
同质土壤 homogeneous soil
同质蜕变 metastasis[复 metastases]
同质外延 homoepitaxy;iso-epitaxy
同质外延金刚石 homoepitaxial diamond
同质外延生长 isoepitaxial growth
同质屋顶 homogeneous roof
同质性 homogeneity
同质性分类 grouping for homogeneity
同质性活动 homogeneous activity
同质性检验 homogeneity test
同质性试验 nick and break test
同质性系数 coefficient of homogeneity
同质异构性 isomerism
同质异晶 allomorph
同质异晶的 allomorphous
同质异晶假象 allomorphism
同质异晶体 paramorph
同质异晶现象 allomorphism;paramorphism
同质异晶性 paramorphism;allomorphism
同质异能的 isomeric
同质异能裂变 isomeric fission
同质异能位移 isomeric shift
同质异能性 isomerism
同质异能跃迁 isomeric transition
同质异位素 isobar
同质异位素衰变 isobar decay
同质异相胶【化】allocoiloid
同质异象 heteromorphism
同质异象(变)体 polymorph;allomorph;dimorph
同质异形(晶)体 paramorph
同质异形(现象)pleomorphism
同质异形性 paramorphism
同质异性胶 isocolloid
同质异重 genetic(al)polymerism
同质硬木拼花地板 parquet solid hardwood floor
同中心的 homocentric
同中子异荷素 isotone
同种车辆的列车编组 homogeneous train formation
同种的 congeneric;congenerous
同种类货 homogeneous cargo
同种凝集 iso-agglutination
同种凝集素 isoagglutinin
同种期权 class of options
同种特异性 homospecificity
同周期 synchronous period
同周期波 synchronous wave
同周期振动 synchronous vibration
同轴波长计 coaxial wavemeter
同轴波导变换器 coaxial-waveguide transition
同轴波导型输出器 coaxial-waveguide output device
同轴玻璃纤维 coaxial glass fibre[fiber]
同轴操作 gang-operated
同轴测辐射热计 coaxial bolometer
同轴测辐射热组合 coaxial bolometer unit
同轴插孔 coaxial jack
同轴插头 coaxial plug
同轴齿轮 in-line gears

同轴抽运激光器 coaxial pump laser
同轴传输线 coaxial transmission line
同轴磁控管 circular electric (al) mode magnetron; coaxial magnetron
同轴磁偏角 codeclination
同轴的 centred; coaxial; on-axis; uniaxial
同轴等离子体发动机 coaxial plasma engine
同轴电极间火花隙 coaxial rod gap; rod gap
同轴电缆 coax; coaxial cable; coaxial feeder; coaxial transmission line; concentric (al) cable; concentric(al) lay cable; concentric (al) line; concentric(al) transmission line
同轴电缆插座 coaxial socket
同轴电缆段 coaxial section
同轴电缆功率吸收器 coaxial dry load
同轴电缆接头 coaxial fitting
同轴电缆均衡器 coaxial cable equalizer
同轴电缆连接器 coaxial cable connector
同轴电缆脉冲测试仪 pulse echo tester for coaxial cable
同轴电缆去耦装置 coaxial line isolator
同轴电缆心线 coaxial inner conductor
同轴电缆心线的绝缘垫圈 axial wire bead
同轴电缆信息系统 coaxial cable information system
同轴电缆载波电话 coaxial cable carrier telephone
同轴电缆载波电话系统 coaxial cable carrier telephone system
同轴电缆载波通信[讯]系统 coaxial cable carrier communication system
同轴电力电缆供电方式 coaxial power cable feeding system
同轴电路 coaxial circuit; concentric(al) cylinder circuit
同轴(电路转换)开关 coaxswitch
同轴电容器 gang capacitor; ganged condenser
同轴电位器 gang potentiometer
同轴电子开关 coaxial trochotron
同轴动力转向装置 coaxial power gear
同轴度 coaxiality; concentricity; proper alignment; trueness
同轴度试验 alignment test
同轴短截线 coaxial stub
同轴短线 coaxial stub
同轴对转螺旋桨 coaxial contrary rotating propeller
同轴多孔镗杆 line boring bar
同轴反转式螺旋桨 contraprop
同轴分离器 coaxial separator
同轴分支 coaxial bifurcation
同轴峰值脉冲功率计 coaxial peak pulse powermeter
同轴辐射热计 coaxial bolometer
同轴隔离器 coaxial isolator
同轴功率吸收器 coaxial dry load
同轴共振器 concentric(al) resonator
同轴固定定向耦合器 coaxial fixed direction-coupler
同轴固定衰减器 coaxial fixed attenuator
同轴关系 coaxial relation
同轴管 coaxitron
同轴管内置偶极振子 sleeve dipole element
同轴管天线 sleeve antenna
同轴管振荡器 concentric(al) line oscillator
同轴过渡器 coaxial adapter
同轴环形电容器 coaxial-torus condenser
同轴回转黏[粘]度计 coaxial rotary viscosimeter
同轴混合器 coaxial hybrid
同轴激光抽运 coaxial laser pumping
同轴集光器 coaxial concentrator
同轴挤出 coaxial extrusion
同轴挤出料粒 coaxial extruded pellet
同轴继电器 coaxial relay
同轴溅射 on-axis sputtering
同轴结构 coaxial configuration
同轴开关 coaxial switch; deck switch; dynaform; gang switch
同轴可变电容器 gang capacitor; gang condenser
同轴空腔 coaxial cavity
同轴控制 gang control
同轴馈(电)线 coaxial feeder; coaxial supply line; concentric (al) feeder; coaxial transmission line
同轴连接器 coaxial connector
同轴联动控制 ganged control
同轴联动谐振电路 ganging circuit
同轴流束喷射器 coaxial streams injector
同轴偶极天线 isometric (al) dipole antenna; sleeve antenna
同轴偶极子 sleeve dipole
同轴球面系统 system of centered spheric(al) surfaces
同轴全息摄影 in-line holography
同轴全息图 in-line hologram
同轴软线 coaxial cord
同轴射流 coaxial jets
同轴射频量热计 coaxial of calorimeter
同轴式滤波器 coaxial filter
同轴输电线 coaxial transmission line
同轴衰减插头 coaxial pad insert
同轴衰减器 coaxial attenuator; coaxial pad
同轴双筒库 coaxial silo
同轴水泵水轮机 isogyre pump turbine
同轴水流 coaxial stream
同轴探头 coaxial probe
同轴天线 coaxial antenna
同轴调节 gang control
同轴调谐 gang (ed) tuning; one-spot tuning; unicontrol; uni-tuning
同轴调谐电容器 gang tuning capacitor; tracking capacitor
同轴调谐器 coaxial tuner
同轴调整 gang adjustment; gang control
同轴调整电容器 dovetail capacitor; dovetail condenser
同轴调制盘叶片 on-axis chopper blade
同轴筒下落式黏[粘]度计 falling coaxial cylinder viscometer
同轴线 concentric(al) line
同轴线安装 center[centre] line mount
同轴线波导管过渡 coaxial-to-waveguide transition
同轴线波导管耦合变压器 coaxial-to-waveguide transformer
同轴线波导管匹配换能器 coaxial-to-waveguide transducer
同轴线波导转换器 coaxial-waveguide transducer
同轴线传输 coaxial transmission
同轴线对 coaxial pair
同轴线功率分配器 coaxial line power divider
同轴线共振器 coaxial line resonator
同轴线接插件 coaxial connector
同轴线接头 coaxial connection
同轴线空腔谐振器 coaxial cavity resonator
同轴线馈电的直线天线阵 coaxial fed linear array
同轴线路 coaxial line
同轴线路终接 coaxial line terminal; coaxial line termination
同轴线去耦装置 coaxial line isolator
同轴线圈 coaxial coil
同轴线式管 coaxial line tube
同轴线收发转换装置 duplexer of coaxial line system
同轴线输出器 coaxial output circuit
同轴线调谐器 coaxial line tuner
同轴线镗孔 line boring
同轴线谐振器 concentric(al) line resonator
同轴线心线 coaxial inner conductor
同轴线旋转接头 coaxial rotary joint
同轴线振荡器 coaxial line oscillator; concentric(al) line oscillator
同轴线终端负载 coaxial line terminal; coaxial line termination
同轴相关性 coaxial correlation
同轴向磨削 up-cut grinding
同轴谐振器 coaxial resonator
同轴谐振腔 coaxial resonant cavity
同轴谐振腔磁控管 coaxial cavity magnetron
同轴型多角经营 concentric (al) diversification
同轴型组件 coaxial package
同轴性 coaxality
同轴性检验规 concentricity ga(u)ge
同轴旋转 circumvolute; circumvolution
同轴选择器 gang selector
同轴扬声器 coaxial speaker
同轴液态涂覆(法) coaxial liquid coating
同轴液相镀技术 coaxial liquid coating
同轴移相器 coaxial phase shifter
同轴应变计 triaxial strain ga(u)ge
同轴圆 coaxial circles
同轴圆筒 coaxial cylinder
同轴圆柱体 coaxial cylinder
同轴褶曲作用【地】homoaxial folding
同轴褶皱 homoaxial folds
同轴振荡 coaxial hunting
同轴振子 sleeve stub
同轴正反转螺旋桨 contra-rotating propellers
同轴驻波检测器 coaxial standing-wave detector
同轴转接器 coaxial adapter
同宗 consanguinity
同族 consanguinity
同族的 cognate[connate]; connatural
同族凝集 coagglutination
同组试件 companion specimen
同组试样 companion specimen
同组元素 group element

茼

茼麻 piemarker

茼麻属 <拉> abutilon

桐

桐柏矿 tongbaiite

桐棉 portia tree
桐漆 <一种仿石油漆> Tungcrete
桐树 Paulownia; tung oil tree
桐酸 el(a)eomargaric acid; el(a)eostearic acid
桐酸精 elaeostearin
桐酸酯 eleostearate
桐油 China wood oil; Chinese oil; Chinese wood oil; tung oil; wood oil
桐油的胶化时间 gel time of tung oil
桐油底漆料 primer oil
桐油活性树脂 wood oil reactive resin
桐油胶化试验 Browne heat test
桐油清漆 super-varnish; tung oil varnish; wood oil varnish
桐油清漆加碱后的增稠度试验 alkali increase test for tung oil varnishes
桐油树 tung oil tree; tung tree; wood oil tree
桐油污渍 tung oil stain
桐油衍生的金属皂 tungate
桐油制成的催干剂 tungate
桐油中毒 Moentjang tina
桐油籽 tung seed

砼

砼贯入阻力测定仪 concrete penetration resistance tester

砼结构 concrete structure

铜

铜埃洛石 cuprohalloysite

铜氨液油分离器 copper liquor oil separator
铜铵 cuprammonium
铜铵合剂 copper ammoniacal
铜铵(人造)丝 cuprammonium rayon
铜铵纸 Willesden paper
铜凹板 copperplate
铜钯接点 copper-palladium contact
铜板 copper panel; copperplate; copper sheet; copper sheet plate; copper slate; sheet copper; sheet plate
铜板电量计 copper coulometer
铜板火箱 copper fire box
铜板送经器 brasses
铜板瓦 copper shingle
铜板屋面 copper roofing
铜板形裂缝 penny-shaped crack
铜板移印法 intaglio printing
铜版 <印刷用> copperplate
铜版画 copperplate; copperplate engraving; copperplate etching
铜版卡 chromo board
铜版印刷 copperplate printing
铜版纸 chrome paper; coated paper; copperplate paper; enamel(led) paper; half-tone paper; art paper
铜棒 bar copper; copper bar; copper rod
铜包 copper-clad; copper-coated
铜包钢 copper covered steel
铜包钢缆 copper-clad steel cable; copper-clad steel conductor; copper-clad steel wire; copper covered steel conductor
铜包钢线 copper covered steel conductor; weld wire
铜包铝板 copper-clad alumin(i)um sheet
铜包铝(导)线 copper-clad alumin(i)um conductor
铜包镍铁线 Dumet wire
铜包皮 copper sheath
铜包石棉 copper-asbestos
铜包石棉衬垫 copper-asbestos packing
铜包石棉垫片 copper asbestos gasket
铜包石棉环 copper-asbestos ring
铜币 brass coins; bronze coins; copper
铜币合金 ounce metal
铜铋合金 Guillaume's metal
铜边 <宋朝定窑瓷盘瓷碗铜质包边> copper bound
铜标牌 bronze table

铜玻璃封接 copper-to-glass
铜玻璃密封 copper-glass seal
铜铂矿 cuproplatinum
铜箔 clutch gold; copper foil; raised copper
铜箔叠层板 copper foil laminate
铜材 copper product
铜衬 brass bush(ing);pillow
铜衬扁销 brass cotter pin
铜衬料 copperlining
铜撑栓 copper staybolt
铜赤色 cupreous
铜翅 copper fin
铜冲头 brass drift;drift
铜穿孔器 brass drift
铜吹炼 copper converting
铜吹炉 copper converter
铜锤 copper hammer
铜磁黄铁矿 chalcopyrrhotite
铜催干剂 copper drier[dryer]
铜催化乙酸盐雾试验 copper accelerated acetic acid salt spray test
铜脆 copper brittleness
铜锉 brass file
铜带 copper strip; ribbon copper; strip copper
铜带绕的电枢 strip-wound armature
铜带绕组 copper-tape winding
铜带试验 copper strip test
铜带屋面 copper strip roofing
铜带线圈 strap coil
铜单管 copper one-pipe
铜单管环系统 copper one-pipe ring system
铜弹 copper shot
铜导体 copper conductor
铜导线 copper conductor
铜的 coppery;cupreous
铜的多因复成矿床 polygenetic compound copper deposit
铜的氟硅酸盐 copper fluosilicate
铜的可溶性 copper solubility
铜的溶解度 copper solubility
铜的脂肪盐 sebaceous salt of copper
铜碲合金 Kuttern
铜电镀废水 copper electroplating wastewater
铜电极 copper electrode
铜电解精炼厂 copper electrolytic refinery
铜(电解式)电量计 copper volt(a)meter
铜电缆 copper cable
铜电线 copper wire
铜电阻器 copper resistor
铜电阻温度计 copper resistance thermometer
铜垫 copper packing
铜垫片 copper backing
铜靛矿 chalcocyanite
铜靛石 chalcocyanite;hydrocyanite
铜雕刻 copper-engraving
铜雕饰 incised decoration in bronze
铜钉 brass tacks;copper nail
铜鼎 bronze tripod
铜锭 copper ingot;ingot copper
铜端子 copper tip
铜钝化剂 <其中的一种> Kuplex
铜矾 lapis divinus
铜矾石 chalcoalumite
铜钒铅锌矿 ramirite;schalfnerite
铜防潮层 copper dampproof(ing) course
铜废料 copper junk
铜粉 bronze tablet;copper powder
铜粉浆 bronze paste;bronzing liquid
铜粉漆 bronzing fluid
铜覆盖层 copper coating
铜覆钢线 copper-clad steel wire;copperweld
铜盖顶 capping
铜干料 copper drier[dryer]
铜杆 copper bar;copper rod
铜坩埚 copper crucible; copper mo(u)ld
铜坩埚壁 copper mo(u)ld wall
铜钢 copper-bearing steel; copper steel
铜镉黄锡矿 cernyite
铜镉线 copper cadmium conductor
铜铬复合黑色颜料 copper-chrome complex black
铜铬矿 macconnellite
铜铬木材防腐剂 acid cupric chromate; celcure; celcure acid; cupric chromate
铜工 brass smith
铜工车间 brass smith shop; coppersmith shop
铜汞合金 copper amalgam
铜箍 copper clamp
铜骨无纹汝釉 copper-boned Ruglaze without crazing
铜骨鱼子纹汝釉 copper-boned Rugalze with fish-roe crazing
铜钴铝合金 sun bronze
铜钴锰土 lubeckite;rhabdionite
铜鼓 <釜状> kettledrum
铜官窑器 Tongguan ware
铜管 copper pipe[piping];copper tube [tubing]
铜管封接 pinch-off seal
铜管管接头 copper pipe coupling
铜管和配件 copper pipe and fittings
铜管接环 for copper pipe ring
铜管接头 copper coupling; copper pipe union
铜管连接器 copper pipe coupling
铜管联合会 copper pipe union
铜管涂膜 coating scale of copper pipe
铜管弯头 copper pipe bend
铜光泽(彩) copper lustre
铜光泽彩露花 <陶瓷装饰方法> copper resist
铜硅焊条 copper-silicon welding rod
铜硅合金 copper-silicon alloy; cupro silicon
铜硅铝明 copper silumin
铜硅锰合金 Everdur;Jacob alloy
铜含量 copper content
铜焊 brass brazing; braze; braze welding; brazing; brazing solder; copper brazing; copper welding; hard solder
铜焊插座 brazing socket
铜焊挡板 brazing flap
铜焊灯 brazing lamp
铜焊粉 brazing powder
铜焊缝 brazed seam
铜焊工具 brazing apparatus
铜焊焊料 brass brazing alloy
铜焊焊枪 brazing burner
铜焊剂 cubond
铜焊件 brazed fixture
铜焊接 braze
铜焊接的 brazed
铜焊接灯 brazing lamp
铜焊接合 brazed joint;copper bond
铜焊接双层钢管 bundyweld tube
铜焊接头 brazed joint; copper bit joint
铜焊炬 brazing torch
铜焊连接 brazed joint
铜焊料 brazing brass;brazing solder; spelter <用于焊接铜、铁等>
铜焊零件 braze-on fitting
铜焊炉 brazing furnace
铜焊喷嘴 brazing burner
铜焊钳 brazing clamp; brazing pincers;brazing tongs
铜焊熔剂 brazing flux
铜焊套筒 brazing socket
铜焊填充金属 brazing filler metal
铜焊条 brazing rod;copper arc welding electrode; copper electrode; copper welding rod
铜焊头 copper(ing)bit
铜焊线 brazing wire
铜耗 copper loss;ohmic loss
铜合金 aldary; Ambraloy; berylco alloy; copper alloy; cupro alloy; pinkus metal
铜合金存水弯 copper-alloy trap
铜合金废料 copper-alloy scrap
铜合金钢 copper-alloy steel
铜合金管和配件 copper-alloy pipe and fittings
铜合金焊条 copper-alloy arc welding electrode
铜合金凝气筒 copper-alloy trap
铜合金线 copper-alloy wire
铜合金铸件 copper-alloy casting
铜红 copper red
铜红玻璃 copper ruby glass
铜红扩散着色 copper staining; red staining
铜红色 copper
铜红铊铅矿 wallisite
铜红釉 copper red glaze
铜护环 copper thimble
铜华 copper bloom
铜滑块 copper shoe
铜化合物 copper compound; cupric compound
铜化合物法 copper compound process
铜化加铬砷酸锌 <木材防腐剂> copperized Boliden salt;copperized chromated zinc arsenate
铜环 copper collar;copper links <测梁连的标环>
铜环钼合金 copper-molybdenum
铜换向片 copper commutator segment
铜黄色的 brass yellow;copper yellow
铜辉铅铋矿 retzbanyite
铜混汞板 copper amalgamating plate
铜活工 brass smith;coppersmith
铜基合金 acid bronze alloy; basis brasses; copper base alloy; copper base metal;keen alloy
铜基合金基体 copper base matrix
铜基合金摩擦材料 copper base friction material
铜基锰铝磁性合金 Heusler's alloy; Heusler's magnetic alloy
铜基体 copper matrix
铜基中间合金 copper master alloy
铜吉丁 <拉> chalcophora japonica
铜加劲条 copper cleat
铜夹钳 copper clamp
铜夹丝外露 zips
铜钾钙矾 leightonite
铜尖晶石 cuprospinel
铜减活剂 copper deactivator
铜匠 brazier;coppersmith
铜匠锤 coppersmith's hammer
铜浇注 copper casting
铜铰链 brass hinge
铜接头 copper fitting
铜接线端子 copper tip
铜解伏安计 copper volt(a)meter
铜金矿 auricupride;cuproauride
铜金镍合金 nioro
铜金星玻璃 copper aventurine glass
铜浸焊 dip-braze
铜精矿 copper concentrates; copper core concentrates
铜镜 bronze mirror
铜卷尺 bronze tape
铜康 <一种铜基合金> Tungum
铜康铜补偿导线 copper-constantan compensating conductor
铜康铜热电偶 copper-constantan thermocouple
铜壳救生圈 Franklin lifebuoy
铜扣钉 copper clamp
铜块 copper billet;copper block
铜矿化探 geochemical exploration for copper
铜矿石 copper ore
铜矿尾矿 copper mining tailing
铜矿渣 copper slag
铜矿渣砖 copper slag brick
铜矿指示植物 indicator plant of copper
铜拉杆 copper tie
铜蓝 azurite;blue verdigris;blue verditer; Bremen blue; copper blue; covellite; cupric sulphide; green verditer; indigo copper; lime blue; verditer blue
铜蓝色 bronze blue
铜缆分布式数据接口 copper distributed data interface
铜雷管 copper detonator
铜冷却板 copper cooling plate
铜离子 copper ion;cupric ion
铜锂合金 cupralith
铜砾 copper granule
铜砾岩 copper conglomerate
铜粒 copper shot;shot copper
铜连接器 copper coupling
铜菱铀矿 voglite
铜领 <俚语,指铁路职工> brass collar
铜硫 white metal
铜硫层 blue metal
铜硫合金 copper-sulphur alloy
铜硫矿石 copper-bearing sulfur ore
铜锍 copper matte regulus
铜铝硅合金 copper silumin
铜铝焊条 copper-aluminium welding rod
铜铝合金 albronze;copper-alumin(i)um alloy;X alloy
铜铝金装饰合金 Nuremberg gold
铜铝金装饰用合金 <纽伦堡合金> Nuremberg gold
铜铝锰土 rabdionite
铜铝镍合金 Batterium(alloy)
铜铝铁镍耐蚀合金 Alcumite
铜铝系耐蚀合金 Ambraloy
铜铝轴承合金 X alloy
铜绿 aerugo; basic copper acetate; Bremen green;copper green;greenish patina;green verdigris;Olympic green;verdigris;verditer
铜绿矾 cuproferrite;pisanite
铜绿色的 aeruginous
铜绿污斑 copper stain
铜绿锈 patina
铜绿釉 copper green glaze
铜氯矾 connellite;footeite
铜轮 copper wheel
铜毛 capillary copper
铜锚 copper anchor
铜镁矾 kellerite;pulszkyite
铜蒙脱石 <硅孔雀石和云母的混合矿> medmontite
铜锰硅合金 Argofil alloy
铜锰合金 cupromanganese
铜锰铝标准电阻合金 Kumanal(alloy)
铜锰铝合金 Durcilium
铜锰铝合金电阻丝 Halman
铜锰镍电阻合金 contamin
铜锰镍合金 copper-manganese-nickel alloy;cuniman;Manic
铜锰土 lampadite
铜密封条 copper sealing strip
铜面饰 copper facing
铜皿胶质 copper dish gum

T

铜明矾 chalcoalumite
铜模 copper mo(u)ld;matrix[复 matrixes/matrices]
铜模底 copper mo(u)ld base
铜模雕刻机 matrix cutting machine
铜母线 copper busbar
铜钼矿石 Cu-bearing Mo ore
铜钠云母 green mica
铜泥 copper sludge
铜镍超级耐蚀合金 super-nickel
铜镍低合金高强度钢 Yoloy
铜镍钴合金 cunico
铜镍钴永磁合金 cunico;cunico alloy
铜镍硅高强度合金 cunisil
铜镍硅合金 C-alloy;Corson alloy
铜镍焊条 copper-nickel welding rod
铜镍焊条合金 soft-weld
铜镍合金 Ambrac;colomony;copper-nickel alloy;cupron;cupro-nickel;cupro-nickel constantan;Hecnum;Ideal;lecnum;midohm alloy;Monel metal;nickel-copper(alloy);nicolite;teleconst;Jae metal < 含铜 30%、镍 70% >
铜镍合金钎料 German silver solder
铜镍合金丝 Contra wire;Eureka wire
铜镍基合金 steamalloy
铜镍锍 copper-nickel matte;nickel-copper matte
铜镍锍转炉 copper-nicked converter
铜镍铝弹簧合金 cunial
铜镍锰合金 cuniman;Manic;nickeline;ohmal
铜镍耐蚀合金 Silverine;silvore;super-nickel
铜镍硼合金 colomony
铜镍锑铅合金 copper-nickel-antimony-lead bronze
铜镍铁合金 calmalloy;cunife;Kunifer alloy
铜镍铁锍 copper-nickel-iron matte
铜镍铁永磁合金 cunife;cunife magnet alloy
铜镍锡硅合金 barberite
铜镍锡合金 Adonic
铜镍系合金 Oda metal
铜镍锌电阻合金 platinoid
铜镍锌焊料 platinoid solder
铜镍锌合金 Ambrac;copper-nickel-zinc alloys;German silver;Maillechort;neusilber;nickel brass;Nickelin;nickel silver;oldsmoloy
铜镍锌合金电阻丝 platinoid
铜镍锌合金丝 platinoid wire
铜镍锌合金线 German silver wire
铜镍锌耐蚀合金 Mercoloy
铜镍装饰合金 Silveroid
铜牛油杯 brass grease cup
铜排 copper bar
铜排水管 copper discharge pipe;copper drainage pipe
铜抛光轮 copper burnishing wheel
铜泡石 tyrolite;kupaphrite < 天蓝石 >
铜配件 copper fitting
铜盆 brass basin
铜盆试验 copper dish test
铜坯珐琅 < 铜胎搪瓷 > copper enamel
铜铍合金 beryllium-copper alloy;copper beryllium alloy;Trodaloy
铜铍中间合金 Bealloy
铜皮 copper sheet;sheet copper
铜皮层面板 copper roofing sheet
铜皮泛水 copper flashing(piece);copper slate
铜皮管脚泛水 copper slate
铜皮屋面 copper roofing;copper sheet roofing
铜皮屋面板 copper roofing sheet
铜片 copper sheet;sheet copper;copper form < 镶焊钻头用的 >
铜片腐蚀试验 copper strip test
铜平头钉 copper tack
铜屏蔽 copper shield
铜屏蔽线圈 copper-jacketed coil
铜起爆雷管 copper blasting cap
铜起爆器 copper blasting cap
铜气泡 gassing of copper
铜器 brass;bronze;bronze ware;copper ware
铜器时代 Bronze Age;Copper Age
铜器制造人 coppersmith
铜钎焊 brass solder;copper brazing
铜钎接双层钢管 bundyweld tube
铜铅玻璃纤维复合板 copper-lead fiber[fibre] glass
铜铅薄板 copper-lead sheet
铜铅矾 elyite
铜铅合金 copper-lead alloy;copper-lead-bronze alloys;pot metal
铜铅合金钉 lead nail
铜铅合金料 cuprolead
铜铅矿石 Cu-bearing lead ore
铜铅锍 copper-lead matte
铜铅铝矾 osarizawaite
铜铅铁矾 beaverite
铜铅锡合金 copper-lead-tin
铜铅霰石 schuilingite
铜铅织物板 copper-lead fabric sheet
铜铅轴承 copper-lead bearing
铜铅轴承合金 copper-lead(bearing) alloy;Kelmet
铜钱 copper;copper cash
铜锹钉 copper tack
铜青铜薄片 copper-bronze
铜青铜漆 copper-bronze paint
铜青釉 copper blue glaze
铜球 copper shot
铜溶金属 copper-soluble metal
铜溶金属粉末 copper-soluble metal powder
铜熔炼炉 copper smelter
铜软电缆 flexible copper cord
铜散热器 copper fin(ned) pipe
铜色斑病 copper spot
铜色斑点 copper spot
铜色的 copper colo(u)red;coppery;cupreous
铜色黄条跳甲 Phyllotreta cupreata
铜色口沿 brown edge
铜色疫病 copper blight
铜杀菌剂 copper fungicide
铜纱 gauze
铜闪绿矿 rahtite
铜闪速熔炼 copper flash smelting
铜砷铬防腐剂 Ascu
铜砷铀云母 zeunerite
铜十字(接头) copper cross
铜石墨电刷 copper-graphite brush
铜石墨合金 copper-graphite alloy
铜石墨制品 copper-graphite composition
铜实心型材 copper solid section
铜蚀斑 copper stain
铜蚀污染 copper staining
铜试剂 cuprisone;cupron
铜刷 copper brush
铜水绿矾 boothite
铜水钴矿 mindingite
铜丝 bronze wire;copper rod;copper wire
铜丝布 copper wire cloth;copper wire gauze;wire fabric
铜丝布电刷 copper brush;gauze brush
铜丝排线器 brass wire traverse
铜丝绳 copper rope
铜丝刷 brass wire brush
铜丝网 copper mesh;copper netting;copper rod gauze;copper wire gauze;copper wire net;net of copper wire
铜丝网过滤器 copper wool filter
铜丝网滤水器 copper wire strainer
铜四通(管) copper cross;copper pipe cross
铜酸钙 calcium cuprate
铜损(耗) copper loss
铜榫 copper dowel
铜胎画珐琅 drawing enamel with copper body
铜胎掐丝珐琅 filigree enamel with copper body
铜胎搪瓷 copper enamel
铜酞菁 copper phthalocyanine
铜碳电刷 compound brush
铜碳酸盐 mineral green
铜套 bronze bush(ing);copper jacket;copper sheathing
铜套管 copper sleeve
铜套止螺丝 bushing screw
铜填料 copper gasket
铜条 bar copper;copper bar;copper billet;copper rod;copper strip;copper strip bar copper
铜条剥落腐蚀 copper strip corrosion
铜条导接线 copper strip bond
铜条试验 copper strip test
铜条镶嵌玻璃 copper glazing;copper light glazing;electrocopper glazing
铜条整流子 bar commutator
铜条装配玻璃 copper(lite) glazing
铜条装配窗玻璃 copper light glazing
铜条装配法 < 一种防火玻璃装配方法 > copperlite
铜调整垫 brass shim
铜贴面 copper revetment;copper (sur)facing
铜铁铂矿 tulameenite
铜铁矾 ransomite
铜铁尖晶石 cuprospinel
铜铁矿 delafossite
铜铁矿石 copper-bearing iron ore
铜铁灵 copperon;cupferron
铜铁试剂 cupferron
铜铁双金属 copper-iron bimetal
铜铜镍补偿导线 copper-copper-nickel compensating conductor
铜筒管 brass bobbins
铜头 copper head
铜瓦 brasses;copper tile;journal brass
铜瓦片 copper shingle
铜弯管 copper bend;copper elbow;copper pipe elbow
铜弯头 copper bend;copper elbow
铜弯嘴旋塞 brass cock angle
铜网校正辊 wire guide
铜网接头机 wire lying machine
铜网屏蔽套管 copper wire gauze shielded bushing
铜网刷 copper gauze brush
铜污染 copper pollution;pollution by copper
铜污染物 copper pollutant
铜屋顶衬板 copper roof sheathing
铜屋顶覆盖(材料) copper roof cladding
铜屋顶覆盖物 copper roof cover(ing)
铜屋顶实施 copper roofing practice
铜屋顶檐沟 copper roof gutter
铜钨合金 copper-tungsten
铜钨华 cuprotyngstite
铜钨铅矿 chillagite
铜吸热剂 heat-sink copper
铜硒铁矿 eskebornite
铜硒铜矿 marthozite
铜锡钯 copper stannopalladinite
铜锡存水弯 copper-tin trap
铜锡单元 copper-tin unit
铜锡垫圈 copper-tin washer
铜锡锻炼合金 copper-tin wrought alloy
铜锡废水管 copper-tin waste pipe
铜锡管平顶辐射供暖板 copper-tin tube ceiling radiant heating panel
铜锡焊条 copper-tin welding rod
铜锡合金 copper-tin alloy;gun metal;Rhinemetal;signal bronze;speculum metal
铜锡合金汞齐 copper-tin amalgam
铜锡贴面 copper-tin trim
铜锡屋顶斜沟 copper-tin valley gutter
铜锡系合金 Phono-bronze
铜锡线 copper wire
铜锡镶边 copper-tin trim
铜锡楔 copper-tin wedge
铜锡锌合金 admiralty metal;Albaloy;G metal;Jackson alloy
铜锡止水器 copper-tin water stop
铜锡制品 copper-tin unit
铜锡阻水片 copper-tin water stop
铜洗塔 cuprammonium washing tower for removing CO
铜系缆墩 copper cleat
铜线 brass wire;copper busbar;copper conductor;copper wire
铜线锭 < 水平浇铸的轧制线材用的铜锭 > copper wire bar;copper wire bar ingot
铜线接头夹 copper terminal clamp
铜线密垫 copper wire gasket
铜线占空系数 copper space factor
铜像 bronze statue
铜硝石 gerhardtite
铜销钉 copper dowel
铜楔子 copper cleat
铜屑 copper cuttings;copper scale
铜屑沉着病 copper dust disease
铜芯碳 copper cored carbon
铜芯(线) copper core
铜芯橡胶绝缘电线 rubber-insulated copper wire
铜芯轧制法 copper core process
铜锌氨酚防腐剂 aczol
铜锌焊条 copper-zinc welding rod
铜锌焊药 copper-zinc brazing mixture
铜锌合金 common brass;complex brass;copper-zinc alloy;Delta alloy;Kinghoren metal;Macht's metal;mosaic gold;yellow brass;gilding metal
铜锌合金焊 braze
铜锌合金屋面板钉 composition nail of copper-zinc alloy
铜锌基锡镍钴合金 Lemarquand
铜锌矿石 Cu-bearing zinc ore
铜锌镍合金 Alfenide;Frick alloy;nickel silver;Olsmoloy;plumber white;white copper
铜锌钎料 copper-zinc solder
铜锌铅合金 architectural bronze
铜锌铅装饰合金 chrysochalk alloy
铜锌铁合金 Sterro alloy;Sterro metal
铜锌锡合金 French gold;Jackson alloy;ormolu;Oroide;Potin
铜锌锡青铜 Tobin brass;Tobin bronze
铜锌蓄电池 copper-zinc accumulator
铜型材 copper profile;copper section;copper shape
铜型离子 ions of the chalcophile type
铜锈 aerugo;copper rust;patina;verdigris
铜锈绿色 patina green
铜悬吊滑轮 copper suspension pulley
铜盐 copper salt;cupric salt;nantokite
铜盐加厚法 copper intensification
铜盐颜料 verditer
铜(檐)槽(沟) copper gutter
铜阳极 copper anode

铜叶绿矾 cuprocopiapite
铜页岩 kupferschiefer
铜银汞膏 cuproarquerite
铜银共晶合金 leval alloy
铜银合金 Kufil alloy
铜铀矾 johannite
铜铀矿 roubaultite
铜铀云母 chalcolite; copper uranite; kupferphosphoruranit; torbernite; uranphyllite
铜釉 copperlite glaze; copper (lite) glazing
铜与玻璃封焊 copper-glass seal
铜雨水槽 copper rainwater gutter
铜雨水制品 copper rainwater goods
铜皂 copper soap
铜渣 copper ashes; copper scale
铜渣块 copper slag block
铜渣瓦 copper slag tile
铜渣小方块 copper slag sett
铜扎线 copper binding-wire
铜织物 copper fabric
铜直嘴旋塞 brass cock straight
铜值 copper number
铜止水带 copper waterstop
铜止水片 copper seal
铜止水(条) copper waterstop
铜纸 half-tone news; half-tone paper
铜纸复合板 <铜与防水皱纹牛皮纸的> copper paper
铜制的 coppery
铜制工人 coppersmith
铜制品 brass work; copper manufactures; copper product
铜制鳍管 copper fin(ned) pipe
铜制容器 copper vessel
铜制屋顶配件 copper rainwater articles
铜制轴衬 axial brass
铜质的 copperish
铜质结合垫圈 copper joint washer
铜质漆 copper paint
铜质型片 copper matrix
铜质贮水器 copper cistern
铜中毒 copper poisoning
铜轴衬 axle brass
铜轴承 copper-bearing
铜轴套 copper bush
铜轴瓦 bearing brass; brass
铜肘形管 copper pipe elbow
铜珠 copper shot
铜柱测压器 copper crusher
铜铸铁 copper cast iron
铜阻电压降 copper drop
铜阻尼器 copper damper

童车 pram

童车储藏室 pram storage room
童工 child labo(u)r(er); worker under age
童工保护 protection for the worker under age
童工保护法 child labor law
童工的保护 protection of child labo(u)r
童帽 bonnet
童期的 juvenile
童山 bald mountain
童子鸡工厂 broiler plant
童子军露营 camporee

酮胺 keto-amine

酮败 ketone rancidity
酮苯脱蜡法 ketone-benzol-dewaxing process
酮醇 keto-alcohol; ketol; ketone alcohol
酮化 ketonization; ketonize
酮基 carboxide; carboxy(l); ketone group; ketonic group
酮(类) ketone
酮类甲醛树脂 ketone-formaldehyde resin
酮(类)树脂 ketone resin
酮醚 ketone ether; ketonic ether
酮醛 ketone-aldehyde
酮式 keto form; ketone form; ketonic form; ketonic type
酮酸 keto-carboxylic acid; ketone acid; ketonic acid
酮酸皂基润滑脂 keto-acid soap grease
酮酸中毒 ketoacidosis
酮缩醇 ketal; ketone acetal
酮缩醛树脂 ketone resin
酮烷基化 ketoalkylation
酮肟 ketoxime
酮烯醇系 keto-enol system
酮亚胺 ket(o)imine
酮氧化 ketone oxidation
酮乙烯化作用 ketovinylation
酮异己酸 ketoisocaproate; ketoisocaproic acid
酮异戊酸 ketoisovalerate; ketoisovaleric acid
酮硬脂酸 ketostearate; ketostearic acid
酮油 ketone oil
酮脂酰 ketoacyl
酮酯 ketone ester

瞳孔 pupil

瞳孔角放大率 angular magnification of pupils
瞳孔距离 interpupillary distance

统【地】series

统包工业 turnkey industry
统包合同 all-in contract
统包价格 all-in price; lump-sum price
统包统配的劳动制度 centralized labor allocation system
统保 blanket insurance
统保单 blanket policy; block policy
统驳分类账 controlling ledger
统材 commons
统舱 deck passage; steerage; steerage passage
统舱餐厅 steerage mess
统舱乘客 <客轮的> steerage passenger
统舱旅客 deck passenger; steerage passage; unberthed passenger
统长窗 ribbon window
统长窗采光 continuous lighting
统长扶手 continuous handrail
统扯收益 flat yield
统称 general term
统筹 plan as a whole; pooling; unified planning
统筹安排 give an overall consideration; overall arrangement; unified planning and overall arrangement
统筹保险基金 overall financing for insurance
统筹财政补贴 general grant
统筹法 network planning; program(me) evaluation and review technique
统筹法会计 critical path accounting
统筹法施工 construction by overall planning
统筹方法 critical path diagram; program(me) evaluation and review technique; critical path method
统筹分析 critical path analysis
统筹规划 make overall plans; overall planning
统筹规划管理法 complete project management
统筹计划 integrated plan(ning)
统筹兼顾 give overall consideration; overall planning and all-round consideration
统筹兼顾适当安排 overall planning and proper arrangement
统筹全局 take the whole situation into account
统筹人 coordinator
统筹社会保险基金 overall raising of social insurance funds
统筹学 critical path methodology
统调 aligning; gang adjustment; padding; tracking; unicontrol
统调电路 ganged circuit; ganging circuit
统调电容器 tracking capacitor
统调控制 tracking control
统负盈亏 be responsible for losses and profits
统购统销 planned purchase and supply
统合 <多国性企业> synergism
统灰 unsorted ash
统货材料 pit-run material
统货道砟 <未经筛分的> all-in ballast
统货荒漠砾石 reg
统货集料 <采石场或轧石场中未经筛分的集料> all-in aggregate; pit-run aggregate; as-dug aggregate
统货石料 quarry-run stone; run-of-quarry stone
统货碎石 all-in ballast; crushed-run macadam; crushed-run rock; crusher-run
统货碎石路 crusher-run macadam
统级 irrespective of size
统计 count; numeric(al) statement
统计保证特性值 statistically guaranteed characteristic value
统计报表 numeric(al) statement; statistic(al) returns; statistic(al) statement
统计报表制度 statistic(al) returns system; system of statistic(al) report
统计报单 statistic(al) copy
统计报告 statistic(al) report
统计报告日 reporting date of statistics
统计报告制度 accounting system
统计比较 statistic(al) comparison
统计比转速 statistic(al) specific speed
统计边界 statistic(al) boundary
统计变宽 statistic(al) broadening
统计变量估计 statistic(al) variable estimation
统计变式 variant
统计变数 statistic(al) variable
统计变异 statistic(al) variation or dispersion
统计遍历定理 statistic(al) ergodic theorem
统计辨别力 statistic(al) discrimination
统计标准 statistic(al) standard
统计表 statistic(al) chart; statistic(al) table; summary sheet; statistics
统计表达 statistic(al) present
统计表式 model of statistic(al) table
统计波动 statistic(al) fluctuation
统计博弈 statistic(al) game
统计补偿 statistic(al) compensation
统计不变性 statistic(al) invariance
统计不可逆方法 statistic(al) irreversible procedure
统计不确定性 statistic(al) uncertainty
统计不准 statistic(al) uncertainty
统计步骤 statistic(al) procedure
统计部门 statistic(al) department
统计簿 statistic(al) book
统计采样 statistic(al) sampling
统计参数 statistic(al) parameter
统计测定 statistic(al) determination; statistic(al) survey
统计测度 statistic(al) measurement
统计测验 statistic(al) test
统计差异 statistic(al) discrepancy
统计常数 statistic(al) constant
统计成本分析 statistic(al) cost analysis
统计程序 statistic(al) procedure; statistic(al) program(me)
统计尺度 statistic(al) yardstick
统计抽样 statistic(al) sampling
统计抽样调查法 statistic(al) sampling investigation method
统计抽样模型 statistic(al) sampling model
统计处理 statistic(al) disposition; statistic(al) treatment
统计处理系统 statistic(al) processing system
统计处理与分析 statistic(al) processing and analysis
统计单位 statistic(al) unit
统计单元 statistic(al) cell
统计当局 statistic(al) authority
统计的变差或离差 statistic(al) variation or dispersion
统计的独立性 statistic(al) independence
统计的观点 statistic(al) viewpoint
统计的均匀性 statistic(al) uniformity
统计的可靠性 statistic(al) certainty
统计的图解分析 statistic(al) graphic-(al) analysis
统计的质量控制法 statistic(al) quality control system
统计的准确性 statistic(al) accuracy
统计的总热点因子 statistic(al) overall hot spot factor
统计地理学 statistic(al) geography
统计地图 mapgraph; statistic(al) map
统计地图程序 statistic(al) map program(me)
统计地图图形 statistic(al) cartogram
统计地震反应 statistic(al) seismic response
统计地震活动性 statistic(al) seismicity
统计地震学 statistic(al) seismology
统计地质学 geostatistics; statistic(al) geology
统计点估计 statistic(al) point estimation
统计点数 statistic(al) point number
统计电平分布 statistic(al) level distribution
统计电平分析 statistic(al) level analysis; statistic(al) level distribution
统计调查 statistic(al) inquiry; statistic(al) investigation; statistic(al) questionnaire; statistic(al) survey
统计调查法 method of statistic(al) survey
统计调查设计 design of statistic(al) inquiry
统计定额 statistic(al) norm
统计定轨理论 statistic(al) orbit determination theory

统计动差 statistic(al) moment
统计动力学 statistic(al) dynamics
统计动力预报 statistic(al) dynamic-(al) prediction
统计独立 statistic(al) independence [independency]; statistically dependent
统计独立变量 statistically independent variable
统计度 statistic(al) range
统计对策 statistic(al) game
统计对象 statistic(al) object
统计法 law of statistics; statistic(al) law; statistic(al) method; statistic-(al) technique; statistics act; statistics law
统计法则 statistic(al) law
统计法质量管理 statistic(al) quality control
统计法质量控制 statistic(al) quality control
统计范围 scope of statistics; statistic-(al) range
统计方案 statistic(al) project
统计方差 statistic(al) variance
统计方程 statistic(al) equation
统计方法 method of average; statistic-(al) approach; statistic(al) method
统计方法论 statistic(al) methodology
统计分布 statistic(al) distribution
统计分布律 statistic(al) distribution law
统计分解 statistic(al) decomposition
统计分类法 statistic(al) classification
统计分配定律 statistic(al) distribution theory
统计分析 statistic(al) analysis; statistic(al) breakdown
统计分析参量 statistic(al) analysis parameter
统计分析程序库 statistic(al) analytic-(al) program(me) library
统计分析方法 method of statistic(al) analysis; statistic(al) (analysis) method
统计分析技术 statistic(al) analysis technique
统计分析监视器 statistic(al) analysis monitor
统计分析图件 statistic(al) diagram
统计分组 statistic(al) grouping; statistics by group
统计风险 statistic(al) risk
统计复用 statistic(al) multiplexing
统计赋权变量 statistic(al) weighting variable
统计概率 statistic(al) probability
统计概率论 statistic(al) probability theory
统计概念 aggregate concept; statistic-(al) concept
统计工具 statistic(al) tool
统计工作 statistic(al) service; statistic(al) work
统计工作人员 statistic(al) staff
统计工作者 statistician
统计公报 statistic(al) bulletin
统计估测方法 statistic(al) approach to estimation
统计估计 statistic(al) estimate; statistic(al) estimation
统计估计量 statistic(al) estimate
统计估计值 statistic(al) estimate; statistic(al) estimated value
统计估算 statistic(al) estimate
统计关联性 statistic(al) association
统计管理 statistic(al) control; statistic(al) management
统计管制图 statistic(al) control chart
统计归纳 statistic(al) induction
统计归纳法 statistic(al) inductive method
统计规范 statistic(al) specification
统计规律 statistic(al) law
统计规律性 statistic(al) regularity
统计规则性 statistic(al) regularity
统计过程 statistic(al) processes
统计过程控制 statistic(al) process control
统计海洋学 statistic(al) oceanography
统计函数 statistic(al) function
统计号 registration number
统计合流分析 statistic(al) confluence analysis
统计核算 statistic(al) accounting; statistic(al) calculation
统计荷载 statistic(al) load
统计恒性法则 law of statistic(al) constancy
统计宏观经济计量模型 statistic(al) macroeconometric model
统计后报 statistic(al) hindcast
统计环境 statistic(al) environment
统计回归分析 statistic(al) regression analysis
统计回归模型 statistic(al) regression model
统计回旋加速器过程 statistic(al) betatron process
统计汇总 statistic(al) summary
统计机 statistic(al) machine
统计机构 statistic(al) agency; statistic-(al) body
统计级数 statistic(al) series
统计计算 statistic(al) calculation; statistic(al) evaluation
统计计算机 statistic(al) computer
统计记录 statistic(al) record
统计记录功能 statistics recording feature
统计技术 statistic(al) technique
统计加工 statistic(al) treatment
统计加速度仪 statistic(al) accelerometer
统计假定 statistic(al) assumption
统计假设 statistic(al) hypothesis
统计假设检验 assumed statistic(al) inspection; statistic(al) hypothesis testing
统计假设序贯检验 sequential test of statistic(al) hypotheses
统计监督 supervision by statistic(al) means
统计检波器 statistic(al) detector
统计检查 check-up through statistic-(al) means
统计检验 statistic(al) test
统计检验法 statistic(al) testing method
统计检验理论 theory of statistic(al) test
统计接收 statistic(al) reception
统计结构 statistic(al) framework
统计结果 statistic(al) result
统计解释 statistic(al) interpretation
统计解释语言 statistic(al) interpretive language
统计界 statistic(al) circle
统计经济学 statistic(al) economics
统计精度 statistic(al) accuracy
统计精度区间 statistic(al) precision interval
统计精确度 statistic(al) accuracy
统计局 Statistic(al) Bureau; statistics department
统计矩 statistic(al) moment
统计矩阵 statistic(al) matrix
统计决策 statistic(al) decision-making
统计决策法 statistic(al) decision method
统计决策过程 statistic(al) decision procedure
统计决策函数 statistic(al) decision function
统计决策理论 statistic(al) decision theory; statistic(al) experiment theory
统计决策论 statistic(al) decision-making theory
统计决策原则 statistic(al) decision rule
统计均衡 statistic(al) equilibrium
统计均匀一性 statistical uniformity
统计科目 statistic(al) account
统计可靠性 reliability of statistics; statistic(al) reliability
统计可识别性 statistic(al) identifiability
统计控制 statistic(al) control
统计控制图 statistic(al) control graph
统计控制图和差异分析 statistic(al) control chart and variance analysis
统计离差 statistic(al) dispersion
统计离散 statistic(al) straggling
统计理论 statistic(al) theory; stochastic theory
统计理论初步 elements of statistic-(al) theory
统计力学 statistic(al) mechanics; statistic(al) thermodynamics
统计立柱图 histogram
统计链 statistic(al) chain
统计链段 statistic(al) segment
统计量 statistic(al) magnitude; statistic(al) quantity; statistics
统计量的抽样分布 sampling distribution of the statistic
统计流体力学 statistic(al) fluid mechanics
统计论据 statistic(al) evidence
统计逻辑 statistic(al) logic
统计面 statistic(al) surface
统计模拟 statistic(al) modelling; statistic(al) simulation
统计模式 statistic(al) mode
统计模式识别 statistic(al) pattern recognition
统计模型 statistic(al) model
统计模型评价 statistic(al) models evaluation
统计内的火灾 statistic(al) fire
统计能量分析 statistic(al) energy analysis
统计拟合 statistic(al) fit
统计年报 annual bulletin
统计年鉴 statistic(al) annual; statistic-(al) yearbook
统计排列 statistic(al) arrangement
统计判别方法 statistic(al) discriminant technique
统计判别函数 statistic(al) discriminant function
统计判定 statistic(al) decision
统计判定法 statistic(al) decision method
统计判定理论 statistic(al) decision theory
统计判断 statistic(al) judgment
统计判决过程 statistic(al) decision procedure
统计判决函数 statistic(al) decision function
统计判决理论 statistic(al) decision theory
统计判决问题 statistic(al) decision problem
统计判优法 statistic(al) arbitration
统计偏差 statistic(al) bias
统计频率 statistic(al) frequency
统计品质管理 statistic(al) quality control
统计平衡 statistic(al) equilibrium
统计平衡表法 statistic(al) balance method
统计平均法 statistic(al) averaging method
统计平均(值) statistic(al) average; assembly average; ensemble average; statistic(al) mean
统计评定 statistic(al) evaluation
统计齐性 statistic(al) homogeneity
统计起伏 statistic(al) fluctuation
统计区 statistic(al) area
统计曲线 statistic(al) curve
统计权数 statistic(al) weight
统计权(重) statistic(al) weight
统计权重定理 statistic(al) weight theorem
统计权重因数 statistic(al) weight factor
统计缺口 statistic(al) gaps
统计群体 statistic(al) population
统计热点因子 statistic(al) hot spot factor
统计热力学 statistic(al) thermodynamics
统计热力学方法 statistic(al) thermodynamic(al) method
统计热力学分析 statistico-thermodynamic(al) analysis
统计人员 statistician
统计容许极限 statistic(al) tolerance limit
统计容许域 statistic(al) tolerance region
统计软件 statistic(al) software
统计软件包 statistic(al) package
统计散布 statistic(al) scattering
统计上不显著的 statistically non-significant
统计(上)的 statistical
统计上的独立性 statistic(al) independence
统计上的相关性 statistic(al) dependence
统计上显著的 statistically evident; statistically significant
统计设计 statistic(al) design
统计设计值 statistic(al) design value
统计生产函数 statistic(al) production function
统计生态学 statistic(al) ecology
统计生物气候学 statistic(al) bioclimatology
统计声功率吸收系数 statistic(al) sound power absorption coefficient
统计声级 statistic(al) sound level
统计师 statist; statistician
统计实验 statistic(al) experiment; statistic(al) test
统计实验法 statistic(al) experiment method
统计实验方案 planning of statistic-(al) experiments
统计实验分析 statistic(al) analysis of experiments
统计实验设计 statistic(al) design of experiments
统计式 statistic(al) equation
统计事件 statistic(al) phenomenon
统计事实 statistic(al) fact
统计势差 statistic(al) moment
统计视差 statistic(al) parallax
统计试验 statistic(al) test
统计试验法 <即蒙特卡罗法> Monte-Carlo method
统计试验方法 statistic(al) experiment method; statistic(al) test method

统计试验理论 statistic(al)test theory
统计试验设计 statistic(al)design of experiments
统计室 statistic(al)room
统计手册 statistic(al)handbook
统计数 statistic;statistic(al)number
统计数据 statistic(al)data
统计数据采样控制系统 statistic(al) sampled-data control system
统计数据处理 statistic(al)data processing
统计数据分析 analysis of statistic(al)data
统计数据记录器 statistic(al)data recorder
统计数据库 statistic(al)database
统计数量 statistic(al)magnitude
统计数列 statistic(al)series
统计数学 statistic(al)mathematics
统计数字 statistic(al)figure;statistics
统计水文学 statistic(al)hydrology;stochastic hydrology;synthetic(al) hydrology
统计说明 statistic(al)description
统计算法 statistic(al)algorithm
统计台账 statistic(al)ledger;statistic(al)record
统计特征 statistic(al)character;statistic(al)property
统计特征参数 statistic(al)characteristic parameter
统计特征数 statistic(al)characteristics
统计特征值 statistic(al)characteristics
统计体重 statistic(al)weight
统计天文学 statistic(al)astronomy
统计通信[讯](理)论 statistic(al) communication theory
统计同质性 statistic(al)homogeneity
统计突发误差校正 statistic(al)burst
统计图 cartogram;isotype;statistic(al)chart;statistic(al)graph;statistic(al)map
统计图表 graphic(al)statistics;pictograph;statistic(al)diagram;statistic(al)pictograph
统计图法 statistic(al)graphic(al) method
统计图符号 statistic(al)graphic(al) symbols
统计图解分析 statistic(al)graphic(al)analysis
统计图示 statistic(al)graphic(al) representation
统计图示法 statistic(al)graphic(al) presentation method
统计土力学 statistic(al)soil mechanics
统计团体 statistic(al)body
统计推断 statistic(al)inference
统计推断法 statistic(al)inference method
统计推理 statistic(al)reasoning
统计推论的梯度相关法 gradient-related method of statistic(al)inference
统计推论的图解法 graphic(al)method of statistic(al)inference
统计推论法研究设计 statistic(al)inference research design
统计外推 statistic(al)extrapolation
统计外延 statistic(al)extrapolation
统计委员会 Statistic(al)Commission
统计文件 statistics files
统计文献 statistic(al)literature
统计文字说明 notes to statistics
统计问题 statistic(al)problem
统计物理学 statistic(al)physics
统计误差 statistic(al)discrepancy;statistic(al)error;statistic(al)uncertainty
统计误差估计值 statistic(al)estimate of error
统计吸收系数 statistic(al)absorption coefficient
统计系列 statistic(al)series
统计系统 statistic(al)system
统计系综 statistic(al)ensemble
统计先验信息 statistic(al)prior information
统计显示 statistic(al)present
统计显著性 statistic(al)significance
统计显著性检验 statistic(al)significance test
统计线性化 statistic(al)linearization
统计线性模型 statistic(al)linear model
统计相对频率 statistic(al)relative frequency
统计相关 statistic(al)dependence
统计相关的 statistically dependent
统计相关法 statistic(al)correlation method
统计相关性 statistic(al)correlation
统计相联 statistic(al)association
统计项目 statistic(al)item
统计像 statistic(al)image
统计心理学 statistic(al)psychology
统计信息 statistic(al)information
统计信息系统 statistic(al)information system
统计性的控制 statistic(al)control
统计性分类 statistic(al)breakdown
统计性估计 statistic(al)estimate;statistic(al)estimation
统计性估计值 statistic(al)estimated value
统计性能 statistic(al)property
统计性试验 statistic(al)experiment
统计性沾污 statistic(al)contamination
统计性涨落 statistic(al)fluctuation
统计性质 statistic(al)property
统计需求分析 statistic(al)demand analysis
统计需求函数 statistic(al)demand function
统计序列 statistic(al)series
统计学 statics;statistics
统计学的 statistical
统计学的卵形曲线 statistic(al)ogive
统计学会 statistic(al)institution
统计学家 statist;statistician
统计学理论 general theory of statistics
统计学评价 statistic(al)evaluation
统计学群体 statistic(al)population
统计学上的分布 statistic(al)distribution
统计延时 statistic(al)time lag
统计岩相 statistic(al)lithofacies
统计研究 statistic(al)investigation;statistic(al)research
统计样本 statistic(al)sample
统计样本矩 statistic(al)sample moment
统计样本数 statistical sampling
统计要求 statistic(al)requirements
统计一体化 statistic(al)integration
统计有效变数 statistically significa-tion variables
统计有效性 statistic(al)efficiency
统计与经济分析 statistic(al)and economic analysis
统计语言 statistic(al)language
统计预报 statistic(al)forecast(ing);statistic(al)prediction
统计预测 statistic(al)forecast;statistic(al)prediction
统计预测模型 statistic(al)prediction model
统计员 statistic(al)clerk;actuary;statician;statist;statistician;statistics controller
统计允许极限 statistic(al)tolerance limit
统计噪声 statistic(al)noise
统计摘要 digest of statistics;statistic(al)abstract;statistic(al)summary
统计展延 statistic(al)extrapolation
统计涨落 statistic(al)fluctuation;statistic(al)variation
统计账户 statistic(al)account
统计真实性 statistic(al)validity
统计整理 statistic(al)treatment
统计直径 statistic(al)diameter
统计指标 statistic(al)indicator
统计指标体系 statistic(al)indicator system;system of statistic(al)indicators
统计指数 statistic(al)index
统计制度 statistic(al)system;system of statistics
统计制图法 statistic(al)cartography;statistic(al)graphics
统计质量管理 statistic(al)quality management
统计质量管理学 statistic(al)quality control
统计质量控制 statistic(al)quality control
统计主体 statistic(al)subject
统计准确度 statistic(al)accuracy
统计准则检验 statistic(al)criterion test
统计咨询 statistic(al)consultation
统计资料 statistic(al)data;statistic(al)information;statistic(al)material
统计资料的加工整理 processing of statistic(al)material
统计资料的散发 dissemination of statistic(al)information
统计总体 statistic(al)ensemble;statistic(al)mass;statistic(al)population;statistics as a whole;statistics for entire group
统计组织 statistic(al)organization
统计最小平方拟合 statistic(al)least square fit
统间式办公室 loft plan office;open plan office
统间式建筑 loft building
统间式平面 open plan
统间式设计 open planning
统建 unified construction;unified development
统建方式 mode of unified construction of houses
统建住房 unified housing development
统建的住宅区 housing estate
统觉 apperception
统括保险 blanket insurance
统括的 omnibus
统括价格 blanket price
统括条款 omnibus clause
统括预算调整法 omnibus budget reconciliation act
统楼层 <工厂的> mo(u)ld loft
统率能力 leadership
统配包销 state monopoly of material supply and marketing
统配部管物资 state-controlled goods
统配煤 coal production under unified central planning
统配煤产量 output of coal under unified central planning
统配物资 materials earmarked for unified distribution;materials placed under unified distribution;producer goods under unified control;unified distributed material
统收统支 keep a tight control on revenue and expenditures;monopoly control over its income and expenditures;total control over revenue and expenditures;unified state control over income and expenditures
统售价 flat rate
统售价格 flat price
统税 general rate
统一 unification;uniformity;unify
统一安排 unified arrangement
统一版 collected edition
统一保单条件 uniform policy condition
统一保险费率 blanket rate
统一报价 uniform price
统一报价协定 open price agreement
统一币制 unification of the currency
统一编码 drilling unify coding
统一变额折旧法 depreciation method of uniformity varying amounts;depreciation-uniformity varying amounts method
统一变数折旧法 depreciation method of uniformity varying amounts
统一标度 unified scale
统一标准 standardization;unified standard;uniform standard
统一标准超细牙螺纹 unified extra-fine thread
统一标准粗牙螺纹 unified coarse thread
统一标准螺纹 unified screw thread;uniform thread;unity thread
统一标准细牙螺纹 UNF thread;unified fine thread
统一标准装置 unified equipment
统一拨款法案 consolidated appropriation bill
统一补充规定 additional uniform regulation
统一偿债基金 consolidated sinking fund
统一场论 unified field theory
统一车辆法规 uniform vehicle code
统一车辆规则 uniform vehicle code
统一成本核算法 uniform costing
统一成本计算制度 uniform costing system
统一成本控制 uniform cost control
统一成本会计 uniform cost-accounting
统一成本会计标准 uniform cost accounting standards
统一成本制度 uniform cost system
统一程序设计系统 unified programming system
统一尺寸 uniform in size
统一尺寸包装 uniform-sized package
统一赤字 unified deficit
统一抽样比 overall uniform sampling fraction
统一打包货品特别费率 special rates for unitized consignment
统一大地基准(面)Universal Geodetic Datum
统一大市场 unified large market
统一刀盘 unitool
统一的 unified;uniform;unitized
统一的财政信用体系 unified finance and credit system
统一的方法 unified method;uniform method
统一的工资底数 unified base wages
统一的累进税 consolidated progressive tax

统一的社会平均盈利率 unified average social rate of profit
统一的社会主义市场 single socialist market
统一的信贷政策 unified credit policy
统一的预制构件 unified precast element
统一的重型线路 uniform heavy type of track
统一的准备金率 unified ratio of reserves
统一地层学【地】 universal stratigraphy
统一电力系统 unified power system; power pool system <美国采用的>
统一调度 centralized dispatching; unified transfer
统一定期租船合同 uniform time charter
统一对外的原则 unified approach in our external dealings
统一发货包括运费定价法 uniform freight-allowed pricing
统一发货定价法 uniform delivered pricing
统一发票 uniform invoice
统一法 unified law
统一法理学 integrative jurisprudence
统一方法 unified approach
统一房地产交易风险法案 uniform vendor and purchaser risk act
统一仿真程序 integrated emulator
统一费率 flat rate
统一费率减少 flat rate reduction
统一分等表 uniform classification
统一分类法 unified approach
统一分配 distribution in a unified way; unified distribution; uniform invoice
统一浮标 universal buoyage
统一浮标系统 uniform buoyage(system)
统一工商税 consolidated industrial and commercial tax
统一工资标准 unified wage scale
统一工资分 uniform wage points
统一工作制度 uniform practice
统一公债 consolidated annuities; consolidated bond; consolidated stock; rentes
统一公债基金收支 consolidated fund service
统一构件 unified member; unified structure element
统一管理 regimentation; unified management; unified regulation
统一管理机构 centralized authority
统一管理下的男女修道院 double monastery
统一惯例 uniform customs
统一规定 standardization
统一规范 unified regulation; uniform code
统一规格的预制构件 unified precast element
统一规划 unified planning
统一规则 uniform rule
统一化 unification; uniformization; unitize
统一化厨房单元 unitized kitchen unit
统一化单元 unitized unit
统一化电机 unitized machine
统一化电机理论 unified electric(al) machine theory
统一化房屋单元建造法 unitized unit method
统一化木构件 wooden unitized unit
统一化浴室(连厕所)单元 unitized bathroom unit
统一换算单位 effective unit; equivalent unit
统一回扣卡特尔 aggregated rebate cartels
统一汇率 flat rate
统一货物分等 consolidated freight classification
统一货物税 consolidated tax
统一货物销售条例 <美> Uniform Sales Act
统一基金 consolidated fund
统一计划 unified planning
统一计划分级管理 unified plan and management at different levels
统一计时制 central timing system
统一价(格) flat price; flat rate; flat yield; price on a uniform basis; uniform price
统一价格表 flat list prices
统一价格代码 uniform pricing code
统一价格拍卖 uniform price auction
统一建设 unified construction
统一建筑法规 uniform building code
统一建筑规范 uniform building code
统一建筑指标 uniform construction index
统一交通法规 uniform traffic regulations
统一交通管理(控制)设施规程 <美> Manual on Uniform Traffic Control Devices
统一交通管理系统 universal traffic management system
统一交通管理协会 Universal Traffic Management Society
统一接头 uniform coupling
统一结算 global settlement
统一解释 unified interpretation
统一经济计划 unified economic plan
统一经营 unified management
统一抗震性能指数 unified aseismic performance index
统一控制 consolidated control; statistic(al) control; unified control
统一会计 uniform accounting
统一会计制度 uniform accounting system
统一蜡克板 <一种表面涂清漆的硬质板> Unilac
统一累进所得税 consolidated progressive income tax
统一里程 through chainage
统一理论 general theory
统一利率 flat interest rate
统一利润率 uniform rate of profit
统一连带经济法律关系 unified joint economic legal relation
统一领导 unified leadership
统一领导分级管理 unified planning and management at different levels
统一路面 one-sole course
统一螺纹 unified thread
统一马桶 <一种用化学液体的马桶> Unisan
统一命名 standard nomenclature; unified nomenclature
统一命名约定 universal naming convention
统一模型 unified model
统一排放标准 uniform effluent standard
统一排污标准 uniform emission standard
统一牌号 unified brand
统一跑道 one-sole course
统一赔偿自然限额规则 uniform monetary limitation of liability rule
统一品牌 family brand; uniform brand
统一平均值 statistic(al) average
统一汽车登记 uniform motor vehicle registration
统一汽车规则 uniform motor vehicle code
统一清算单价 uniform clearing unit price
统一曲线系 homaloidal curves
统一塞科 <一种装配式住房牌号> Uniseco
统一砂 all-purpose sand; identical sand; unit sand
统一商法 Uniform Commercial Code
统一设计标准和方法 consolidated design criterion and methodology
统一施工 unified construction
统一时间 unidied time
统一实体 consolidated entity
统一世界大地基准 unified world geodetic datum
统一市场 uniform market
统一市场价格 uniform market prices
统一试验 uniform tests
统一试验程序 uniform test procedure
统一试验法 uniform test procedure
统一收费率 flat rate
统一收费系统 unified toll system
统一收购统一调拨 state purchase and allocation on a monopoly basis
统一收款法 uniform receipt method
统一手续费 flat commission
统一书号 standard book number
统一输入信号 total input
统一术语系统 uniterm system
统一数据处理 integrated data processing
统一数据库语言 unified database language
统一税 consolidated tax; single tax
统一税率 flat rate
统一所得税 unified income tax
统一特斯克板 <一种硬质板牌号> Unitex
统一提单 uniform bill of lading
统一提单法 <美> Uniform Bills of Lading Act
统一体 integral unit
统一天然气公司 <美> Consolidated Natural Gas
统一条例 uniform regulation
统一调剂资金 overall regulation of funds
统一调配 unified allocation of resources
统一图案 unity of plan
统一土分类法 unified classification system
统一土壤分类体系 unifies soil classification system
统一土壤分类体制 unifies soil classification system
统一土壤分类系统 unifies soil classification system
统一土壤分类制 unified soil classification system
统一外汇汇率 unified foreign exchange rate
统一网络 unified network
统一危险性谱 uniform risk spectrum
统一习惯和做法 uniform customs and practice
统一系统 integrated system; uniform system
统一现金预算 consolidated cash budget
统一行车规则 Standard Code of Operating Rules
统一性 unitarity; unity
统一盈余留存表 consolidated retained earnings statement
统一佣金 flat commission
统一有限合伙条例 uniform limit partnership act
统一预算 single budget; unified budget; uniform budget; unitary budget
统一原则 principles of integrity; uniform principles
统一运费分类表 consolidated freight classification
统一运费率 uniform rate
统一运价表 standardization of tariffs; standard tariff
统一运价率和票价 standardization of rates and fares
统一运输系统 unified transport system
统一运送订价法 uniform delivered pricing
统一运用专款 consolidated working fund
统一责任标准 uniform level of liability
统一责任制 uniform liability system
统一债权 equalization claim
统一债券 unified bonds
统一债务 consolidated debt
统一战线 united front
统一震级 unified earthquake magnitude; unified magnitude
统一整体 uniform entity
统一制度 uniform system
统一种类的货物 homogeneous cargo; straight cargo
统一仲裁法 <美> Uniform Arbitration Act
统一资源定位器 uniform resource locator
统一资源定义符 uniform resource identifier
统一总线 unified bus
统一作业计划 unified work program(me)
统驭分类账 controlling ledger
统驭记录 controlling record
统驭账户 collective account; controlling account
统制 control
统制对外贸易 controlled foreign trade
统制分类账 controlling ledger
统制股权 controlling interest
统制价格 controlled price
统制经济 controlled economy
统制经济学 economics of control
统装材料 bulk material

捅 poke

捅杆 ramrod; tapping pin
捅料棍 poke rod
捅箱机 punch-out equipment
捅窑内物料 poking in the kiln

桶 basquet; cask; hopper; ladle; pail; tub; bac; keg <钉的重量单位,=100 磅>; barrel bulk <松散物料体积单位,1 桶 = 松方 0.142 立方米>; barrel <液体度量单位,英国 = 163.65 升,美国 = 119 升,重量度量单位,随所装物质而异,约 89 公斤>

桶板 clapboard; stap
桶板材 stave bolt
桶板短圆材 stave bolt
桶板缝渗漏 cask leaking through seam
桶板管 wood-stave pipe
桶板锯 bilge saw; plug saw
桶板料 shook
桶板破裂 cash stave broken
桶板条 stave
桶板脱落 cask stave off; stave off
桶包装 drum can package

T

桶孢锈菌属 < 拉 > crossopsora
桶泵 barrel pump
桶部 < 桶等的 > bilge
桶材 Cooper's wood; hoop wood
桶侧孔 bunghole
桶厕 bucket-type privy; pail closet
桶车 ladle car
桶衬 barrel liner; ladle lining
桶穿洞并渗漏 barrel punctured and leaking
桶唇 ladle lip
桶的通气孔 spile; spile-hole
桶底 barrel head
桶底板 heading
桶底虹吸池 ladle well
桶底凸缘 chime
桶吊具 barrel sling
桶镀 barrel plating
桶端 drum end
桶堆楔块 cantick quoin; cating coin; quoin
桶浮标 can buoy; keg float
桶腹胀大 blow the bilges
桶盖 bung
桶盖生锈 cover rusty
桶盖松 barrel lid loose; drum lib loose
桶盖桶底材 heading
桶盖脱落 cap top, head off; drum lid missing
桶杆 pike pole
桶工斧 cooper's axe
桶工工作台面 cooper's bench plane
桶工用槽刨 cooper's rabbet
桶工用长刨 cooper's bench plane
桶钩 bali; cask hook off; cask sling
桶箍 barrel hoop
桶箍脱落 hoops off bands off
桶箍锈 hoops rusty
桶号 barrel number
桶混机 tumble blender
桶货堆排机 barrel pitching machine
桶货加箍费 cooperage
桶间凹隙 cantline
桶壳 ladle bowl; ladle casing
桶壳式水轮机 barrel cased turbine
桶孔 bung(hole)
桶口 bung(hole)
桶口熟炼油 bunghole boiled oil
桶筐 tub basket
桶哩运费 barrel mile
桶梁结 matthew walker knot; walker knot
桶漏 drums leaky leaking
桶内混合 ladle mixing
桶内加料 ladle addition
桶内结壳 ladle scull
桶破裂 drum rupture
桶破漏 drum broken and leaking
桶容量 ladle capacity
桶塞 stopper
桶塞板条 bung stave
桶塞朝上 bung up and bilge free
桶塞处漏 barrel leaking at plugs
桶塞脱落 barrel lung off; bung off
桶身 ladle body
桶渗漏 barrel oozing; drum oozing
桶升降机 barrel elevator; barrel hoist
桶式浮标 barrel buoy
桶式混汞法 barrel amalgamation
桶式接缝 coopered joint
桶式喷雾泵 barrel pump
桶式热水器 storage-type geyser
桶式软垫椅 tub chair
桶式研磨机 barrel mill
桶式钻头 bucket auger
桶数 barrelage
桶形 barrel shape; drum
桶形拌和机 barrel mixer
桶形齿 crowning
桶形的 barrel-shaped
桶形电极 bucket
桶形反射器 spun-barrel reflector
桶形浮标 barrel buoy; cash-buoy; drum buoy; drum can buoy; key-buoy
桶形滚柱 crowned roller
桶形畸变 barrel distortion
桶形搅拌机 barrel mixer
桶形接触活塞 < 波导管中的 > bucket piston
桶形锯 bilge saw; cylinder saw
桶形开关 barrel switch
桶形绕法 barrel winding
桶形绕组 barrel winding
桶形失真 barrel-shaped distortion
桶形失真校正 anti-barreling
桶形试验 ladle test
桶形掏槽 box cut; box type cut; cylinder cut
桶形天线 barrel antenna
桶形系船浮筒 drum mooring buoy; barrel mooring buoy
桶形轧辊 barrel-shaped roll
桶性畸变 barrel distortion
桶样 ladle sample
桶样成分 ladle chemistry
桶样分析 ladle analysis
桶样试验 ladle test
桶腰 bilge; bulge
桶腰箍 bilge hoop
桶腰架空法 bilge free
桶业制造者 cooper
桶栽果园 tub orchard
桶栽园艺 tub-gardening
桶栽植物 tub plant
桶中精练 ladle refining
桶重焊 barrel resoldered; barrel rewelded
桶重修理 cask-re-coopered
桶砖 ladle brick
桶装 barrel; can filling; in bulk
桶装货 barrel cargo; barrel goods; cargo in cask; cargo in drum; barrels and drums
桶装沥青 barreled asphalt; barreled bitumen
桶装沥青熔化装置 asphalt barreled melter; asphalt heater
桶装啤酒店 cask beer store
桶装水泥 barrel(ed) cement; cement in barrels; barrel of cement < 美国标准桶装水泥每桶 376 磅,合 170.5 公斤 >
桶装物 barreled goods; goods in barrels
桶装亚麻子油 bunghole boiled linseed oil
桶装液体漏耗量 ullage
桶装油 barrel oil; case oil; dump oil
桶装重量法 barrel gravity method
桶状货柜 tank container
桶状青贮塔 stave silo
桶状绕组 barrel winding
桶状体 barrel
桶状物 pipe barrel; lumber core < 木条板拼装成的 >; stave core < 木条板拼装成的 >
桶嘴 bucket nose

筒

筒背椅 barrel chair
筒表变薄旋压 tube spinning
筒仓 cylindric(al) silo; silo; storage silo
筒仓出口 silo outlet
筒仓出料 silo discharge
筒仓出料槽 silo drawing channel
筒仓出料器 bin discharger
筒仓储藏 silo storage
筒仓存储能力 silo storage capacity
筒仓底 silo bottom
筒仓顶部传送带 oversilo conveyer
筒仓顶部卸料器 oversilo tripper
筒仓分隔间 silo compartment
筒仓供暖 silo heating
筒仓建筑 silo construction
筒仓进料吊斗提升机 feeder skip hoist for silos
筒仓库存指示器 silo level detector; storage level detector
筒仓料位探测器 silo level detector
筒仓料位指示器 silometer
筒仓漏斗 silo bunker
筒仓模板 silo formwork
筒仓墙 silo wall
筒仓群 group of bins; silo block
筒仓设备 silo storage
筒仓渗漏 silo seepage
筒仓式焚化炉 silo-type incinerator
筒仓涂布塑料 silo coating plastic
筒仓效应 silo effect
筒仓卸料门 silo gate
筒仓压力 silo pressure
筒仓液面探测器 silo level detector
筒仓砖 silo tile
筒仓装料斗 silo hopper
筒仓装置 silo installation
筒仓组 group of bins
筒灯 flush light panel
筒镀 barrel plating
筒篙 crown daisy
筒拱 barrel arch; coving; cylindric(al) vault
筒拱顶隔块 severy
筒拱壳体 barrel vault shell
筒拱模板 barrel vault form; barrel vault formwork
筒拱矢高 rise of barrel vault
筒拱术【建】vaulting
筒拱圬工 vaulting masonry
筒拱弦宽 chord width of barrel vault
筒拱柱 vaulting pillar; vaulting shaft
筒钩 barrel hook
筒管 bobbin
筒管出水口 barrel outlet
筒管纺纱机 bobbin spinning machine
筒管分径机 bobbin sorter
筒管架 spool stand
筒管(硫铵)炸药 bobbinite
筒管式涡轮喷气发动机 spool turbojet
筒管形线圈 solenoid coil
筒环形燃烧室 cannular burner; cannular combustion chamber
筒夹 collet; collet chuck; contracting chuck; shell ring
筒夹控制凸轮 collet cam
筒夹装置 collet attachment
筒接 muff joint; sleeve joint; thimble joint(ing)
筒接头 box coupling; muff coupling; sleeve joint
筒节 cylindric(al) shell section; shell ring; vessel course
筒节和封头 shell sections and heads
筒紧炮 built-up gun
筒井从下加固基础 underpinning with cylinders
筒壳 barrel roof; barrel shell; barrel shell roof; barrel vault shell; cylindric(al) shell
筒壳屋顶 arched barrel roof; barrel shell roof; barrel vault roof; cylindric(al) shell roof
筒口 nozzle
筒檩条 barrel purlin(e)
筒铝过渡板 copper to aluminum adapter bar
筒磨机 tube mill
筒频率响应 microphone response
筒瓶 elephant-leg shaped vase
筒裙式活塞 trunk piston
筒筛 shaft screen; trunnion screen
筒式拌和机 drum mixer
筒式拌和设备 drum plant
筒式焙烧炉 cylinder roaster
筒式泵 barrel-type pump
筒式剥皮机 barking drum
筒式插装式阀 cartridge inserted valve
筒式柴油机打桩锤 tubular diesel pile hammer
筒式成球机 drum nodulizer
筒式吹风转炉 barrel converter
筒式磁选机 drum magnetic separator; drum-type electromagnetic separator
筒式存仓 silo bin
筒式打磨机 drum sander
筒式电镀 barrel plating
筒式阀 sleeving valve
筒式分离机 drum separator
筒式干燥 cylinder drying
筒式干燥机 cylinder drier[dryer]
筒式干燥器 cylinder drier[dryer]; drum drier[dryer]
筒式给煤机 rotary feeder
筒式拱桥 barrel arch
筒式谷仓 grain silo
筒式管连接头 sleeve connector
筒式锅炉 shell boiler
筒式过滤机 drum filter
筒式过滤(结构) tubular filtration
筒式过滤器 cartridge filter
筒式过滤筛 drum filter
筒式海洋检波器 cylindric(al) hydrophone
筒式烘干机 cylindric(al) drier[dryer]
筒式回转冷却器 tubular rotary cooler
筒式混凝土拌和机 drum type concrete mixer
筒式活塞 pot-type piston
筒式给水泵 barrel casing feed pump
筒式加热器 cartridge heater
筒式减震器 telescopic(al) shock absorber
筒式搅拌器 drum mixer
筒式接头 barrel fitting
筒式结构 core structure
筒式卷取机 drum reel; drum-type coiler; mandrel coiler
筒式卷绕机 drum reel; drum-type coiler
筒式卷扬机 drum hoist
筒式粮仓 grain(storage) silo
筒式炉 cylindric(al) furnace
筒式卵石过滤器 cylindric(al) gravel filter
筒式煤炭干燥箱 coal cylindrical drier [dryer]
筒式磨碎机 kibbling mill
筒式喷丸机 barrel-type shot blasting machine
筒式千斤顶 telescope jack
筒式曲轴箱 barrel crankcase; barrel-type crankcase
筒式去皮机 barking drum
筒式热水采暖系统 cylinder system
筒式筛分机 drum separator
筒式筛砂机 rotating screen
筒式湿磨机 wet tube grinding mill
筒式送料机 feed drum
筒式碎石机 roller crusher; roll mill
筒式筒仓 horizontal silo
筒式喂料机 feed drum
筒式吸尘器 cylinder vacuum cleaner
筒式旋转粉碎机 rotary drum pulverizer
筒式旋转氧气炼钢炉 Graef rotor
筒式选分机 drum separator
筒式选分机滚轮静线压力测定 drum separator

筒式选矿机 drum separator
筒式烟雾发生器 can-type generators
筒式窑 drum-type kiln
筒式运煤气车 gas reservoir truck
筒式闸阀 cylinder-type of lock valve
筒式种子精选机 cylinder grader
筒式钻头 shell bit
筒室 silocell
筒体 barrel;bowl;cylinder;main body;shell;tube
筒体变形 deformed kiln shell
筒体衬板 shell liner
筒体结构 bull core structure;tube structure
筒体紧固轴 shaft for cover close
筒体紧固轴轴承 bearing for cover close shaft
筒体热损失 loss of heat through wall
筒体扫描仪 shell scanner
筒体椭圆度测量仪 shell test apparatus
筒体弯形 elliptic(al) distortion
筒瓦 arc-shaped tile;astragal tile;cylindric(al) tile;mission roofing tile;mission tile;old Roman tile;round tile;semi-cylindric(al) tile;Spanish roofing tile;Spanish tile;tile
筒螅 tubularia
筒箱形檐沟 copper box roof gutter
筒形 trunk
筒形把手 cylinder knob
筒形板 arched plate
筒形拌和器 barrel mixer
筒形壁灯 barrel wall lamp
筒形玻璃 cylinder glass
筒形薄壳 barrel shell;cylindric(al) shell
筒形薄壳结构 thin-shelled cylindrical structure
筒形薄壳屋顶 barrel roof;barrel shell roof;thin-shell barrel roof
筒形插销 barrel bolt
筒形插装式阀 cartridge valve
筒形长拱 wagon-headed vault
筒形沉井 cylinder open caisson
筒形打磨器 cylinder sander
筒形的 barrel;cored;tubular
筒形顶 wagon top
筒形顶棚 barrel ceiling;cylindric(al) ceiling;wagon ceiling
筒形顶篷 cylindric(al) ceiling
筒形断面 tubular section
筒形断面车辆 tubular car
筒形发动机 truck engine
筒形阀 cylindric(al) valve
筒形分离机 drum separator
筒形风帽 cylindric(al) cowl;cylindric(al) ventilation
筒形风帽保温蝶形阀 insulated butterfly valve of cylindrical ventilation
筒形风帽滴水盘和滴水槽 drip disk [disc] and drip groove of cylindrical ventilator
筒形扶手 barrel handrail
筒形浮标 barrel buoy;can buoy;trunk buoy
筒形干燥机 cylindric(al) drier[dryer]
筒形钢板 wagon plate
筒形格筛 Burch grizzly
筒形拱 barrel arch
筒形拱坝 barrel arch dam
筒形拱顶 annular vault;barrel vault(ing);cradle vault;half-round tunnel vault;half-round wagon vault;tunnel vault;tunnel-vault(ed) roof;wagon-headed vault;wagon vault
筒形拱顶屋顶 barrel vaulted roof
筒形拱桥 tubular arch bridge
筒形拱式大梁 barrel arched girder
筒形拱屋顶 wagon-vaulted roof
筒形沟 barrel drain
筒形构架 barrel frame
筒形构件 tubular member
筒形骨架 barrel skeleton
筒形管 cylinder-type pipe
筒形辊 barrel roll
筒形锅炉 barrel boiler;cylindric(al) boiler;shell boiler
筒形过滤器 drum filter;rotary filter
筒形涵洞 barrel culvert
筒形烘炉 Salamander
筒形混合机 cylinder mixer
筒形混凝土薄壳(面) cylindric(al) concrete shell
筒形混凝土薄壳屋顶 cylindric(al) concrete shell roof
筒形混凝土壳 cylindric(al) concrete shell
筒形混凝土壳体 cylindric(al) concrete shell
筒形活塞 ring piston;trunk piston
筒形火管锅炉 trunk boiler
筒形记录器 drum recorder
筒形件挤压 can extrusion
筒形件卷边接合偏心冲床 horning press
筒形交叉穹顶 barrel vault with intersecting;cylindric(al) intersecting vault
筒形铰刀 shell reamer
筒形节流阀 barrel throttle
筒形结构 tube structure;tubular construction
筒形截面 cored section
筒形锯 barrel saw;crown saw;cylinder saw;drum saw;hole saw;tubular saw
筒形绝缘套 cylindric(al) insulator
筒形开关 barrel switch;switch cylinder
筒形开挖断面 cylinder excavation section
筒形控制器 cylindric(al) controller
筒形扩孔钻 shell core drill
筒形拉刀拉床 pot broaching machine
筒形栏杆 barrel railing
筒形老虎窗 wagon-headed dormer
筒形冷凝器 cylindric(al) condenser
筒形联轴节 cylindric(al) coupling
筒形联轴器 branch sleeve
筒形梁 arched beam;tubular girder
筒形隆起屋顶坡面 tarus
筒形轮 cylindric(al) wheel
筒形螺纹接套 barrel nipple
筒形螺线管 cylindric(al) solenoid
筒形模 cylindric(al) mould
筒形膜 core
筒形摩擦轮 spur friction wheel
筒形排气泵 barrel pump
筒形排水 barrel drain
筒形排水泵 barrel pump
筒形排水管 barrel drain pipe
筒形排水渠 barrel drain canal
筒形配合 cylindric(al) fit
筒形屏蔽罩 cannon tube shield
筒形铺砂机 cylinder sanding machine
筒形气泵 barrel pump
筒形戗脊 hip roll
筒形戗脊构件 ridge roll
筒形桥墩 barrel pier;cylindric(al) pier
筒形擒纵机构 cylinder escapement
筒形清洗设备 barrel washing device
筒形穹的 wagon-headed
筒形穹顶 barrel shell roof;barrel vault(ing);barrel vault roof;cradle vault;tunnel vault;wagon-headed vault;wagon vault
筒形穹隆 barrel vault(ing);cradle vault;cylindric(al) vault
筒形曲轴箱 barrel-type crankcase
筒形燃烧室 can combustor;can-type combustor
筒形热水锅炉 cylindric(al) hot water boiler
筒形熔断器 cartridge fuse
筒形砂轮 cylinder grinding wheel;cylindric(al) grinding wheel
筒形筛分机 drum separator
筒形筛面 curved surface
筒形水泵 barrel pump
筒形丝锥 shell tap
筒形掏槽 cylinder cut
筒形天窗 barrel light
筒形天花板 wagon ceiling;wagon-headed ceiling
筒形桶 barrel
筒形凸轮 barrel cam;cylinder cam
筒形涂沥青机 barrel pitching machine
筒形瓦 spool tile
筒形外圆铣刀 hollow fraise;hollow mill
筒形万向联轴器 muff coupling joint
筒形桅杆 barrel mast
筒形蜗轮 cylindric(al) worm gear
筒形屋顶 arched barrel roof;barrel roof;barrel shell;barrel shell roof;cradle roof;cradle vault;tunnel vault;wagon-headed vault;wagon roof;wagon vault
筒形屋顶的镶板装饰 <教堂十字架或圣坛上> celure
筒形屋顶肋 barrel roof rib
筒形屋脊 hip roll
筒形屋脊包层 hip roll
筒形物 barrel;cylinder body;gaiter
筒形系船浮筒 barrel mooring buoy;drum mooring buoy
筒形销 cylindric(al) pin
筒形岩粉积聚管 cylindric(al) rock-powder collector
筒形堰 cylinder weir
筒形遮光器 barrel shutter
筒形支墩 barrel pier
筒形轴 sleeve shaft
筒形砖 chimney block;spool tile
筒形桩 cylindric(al) pile
筒形钻 cylindric(al) auger;cylindric(al) drill;shell bit;shell drill
筒型汽缸 barrel-type casing
筒枕 bearer
筒制法 core method
筒制机 core-building machine
筒制外胎 core-built tire[tyre]
筒中筒 tube-in-tube
筒中筒结构 core-in-core structure;framed tube-core structure;tube-in-tube structure
筒中筒结构体系 tube-in-tube structural system;tube-in-tube system
筒柱 barrel mast;cylindric(al) column
筒柱式码头 cylinder wharf;cylindric(al) wharf
筒砖 tubing brick;tubular brick
筒桩 cylindric(al) pile
筒装嵌缝胶 ca(u)lking cartridge
筒装炸药 <爆破工程用> stick powder;stick of explosive
筒状保险装置 cartridge tube fuse out
筒状玻璃量器 cylinder glass
筒状薄壳用三夹板 cylindric(al) shell sandwich panel
筒状储仓 silo storage
筒状盾构 cylindric(al) shield
筒状阀 spool valve;thimble valve
筒状隔膜 tubular diaphragm
筒状滚动支承 barrel-shaped roller bearing
筒状滚动轴承 barrel-shaped roller bearing
筒状花 tubular flower
筒状活塞 trunk piston
筒状活塞发动机 trunk engine
筒状火管锅炉 trunk boiler
筒状铰链式提引器 ideal type elevator
筒状快门 barrel shutter
筒状容器栽植 tube planting
筒状三夹板 cylindric(al) three-layers panel
筒状瓦 cored tile
筒状弯曲 cylindric(al) bending
筒状旋风分离器 tubular cyclones
筒子 bobbin
筒子板 door dressing;ingo plate;lining;reveal lining;siding <门窗的>
筒子插钉 bobbin peg
筒子车 spooler
筒子架 creel
筒子纱 cheese
筒子纱染色机 yarn package dyeing machine
筒子瓦 imbrex
筒座 cylinder base

痛

痛阀 <声学术语> threshold of pain
痛觉缺失 analgesia
痛觉阈限 pain threshold
痛苦 discomfort
痛痛病 itai-itai disease
痛阈 pain threshold

偷

偷乘者 stowaway
偷倒人 midnight dumper
偷盗保险 burglary
偷盗、提货不着保险 insurance against TPND[theft, pilfer-age, non-delivery]
偷渡封锁线 blockade running
偷渡通路 sneak path
偷工 scamp work;underwork
偷工减料 cheat in work and cut down material;do shoddy work and use inferior material;jerry-build;scamp work and stint material;shoddy work and inferior material;short-changing on work and materials;ghosting <油漆作业>
偷工减料包工头 <俚语> jerry-builder
偷工减料的 jerry
偷工减料的建筑 jerry-building;jerry-building construction
偷工减料的营造商 jerry-builder
偷工减料的制造商 jerry-builder
偷工减料建筑 jerry-built construction
偷进尺 steal hole
偷漏 evasion
偷漏税 evasion of taxation;tax fraud;evasive
偷入领海的走私船只 hovering vessel
偷水 water-pirating
偷税 dodge a tax;evade a tax;tax dodging
偷税漏税 evade paying taxes;evasion and avoidance;tax evasion;evade tax
偷税漏税行为 ways of going about tax evasion
偷税漏税者 tax dodger;tax-evader
偷税人 tax dodger
偷偷地做 sneak
偷越国境的船 runner
偷运 contraband

头

头 blaenau[复 blaen];caput
头班 first night watch;first watch

头遍筛 primary screen
头标 first bid;header
头波 bow wave;head wave;leading wave
头部 head end;header
头部保护装置 head restraint
头部槽 head groove
头部带凸缘的螺钉 collar screw
头部的 cephalic
头部镦粗螺栓 upset bolt
头部镦粗铆钉 upset rivet
头部镦锻工具 heading tool
头部发动机 head motor
头部防护 head protection
头部附件 head attachment
头部桁条 batten
头部激波 bow wave;forward shock
头部尖拱 ogive
头部减速器 main retarder;primary retarder;master retarder【铁】
头部减震针 nose spike
头部交锁砖 head-lock tile
头部可调进气口 controllable nose intake
头部连接(法)head attachment;head construction
头部卵形部分 nose ogive
头部驱动轮 driving wheel on head
头部热处理轨 head-hardened rail
头部水槽 head flume
头部填密 head packing
头部为 L 型的顶撑 L-shore
头部为喇叭形圆锥破碎机 flared-head gyratory
头部压力 head pressure;nose pressure
头部整流罩 transparency nose;transparent nose
头部支肋 hose batten;nose batten
头部制动位置 primary retarder position
头部锥角 nose angle
头部阻力 head resistance
头部阻力系数 head-drag coefficient
头部最佳形状 optimum nose curve
头舱甲板 salo(o)n deck
头测量法 cephalometry
头测量器 cephalometer
头长 head length
头场霜 the first front
头车 first attendance;leading vehicle;lead-vehicle;A unit <带司机室及操纵控制的机车单元>
头池 head vat
头尺寸中数 cephalic module
头冲 primary drying
头寸 liquidity
头寸短绌 tight position
头寸宽裕 easy position
头寸松 loose money
头带 <头戴受话器的> headband
头带表示灯按钮 headlighted button
头带式护面罩 helmet
头戴耳机 earphone unit
头戴式耳机 bi-telephone;ear piece;ear receiver;head band receiver
头戴式耳机和送话器 head set
头戴式收音器 headphone unit
头戴式受话机 headphone
头戴式受话器 ear receiver;head band receiver
头戴式双耳机 double headphone
头戴式送受话器 telephone headset assembly
头戴式送受话器电路 head circuit
头戴受话机 earphone;headphone
头戴受话器 bi-telephone;headgear;headphone;head piece;head set;head-telephone;telephone headphone
头戴听筒 headgear
头挡 bottom gear
头挡齿轮 first gear
头挡速率 first gear rate
头道并条机 breaker drawing frame;first drawing frame
头道操作 work first
头道锉纹 first course
头道底漆 base coat
头道灰 first coat;first coat of plaster;rendering coat
头道灰浆内涂层 first plaster undercoat
头道灰泥 rough coat
头道沥青 asphaltic-bitumen primer;bitumen prime coat
头道抹灰 first coat;prime coat;render coat
头道漆 first coat;priming coat;priming paint;undercoating lacquer;undercoating paint
头道筛 primary screen;scalper screen;scalping screen
头道涂层 prime coat(ing);primary coat;first coat
头的转动角度 head-turned angle
头的转动中心 head pivot point
头灯 cap lamp;front light;heading light;head lamp;headlight
头灯玻璃 head lamp lens
头灯大光灯泡 head lamp driving bulb
头灯灯泡 head lamp bulb;headlight bulb
头灯灯泡护圈 headlight bulb retainer ring
头灯灯泡座 headlight bulb mounting seat
头灯电线 headlight cable
头灯反射镜 head lamp reflector;headlight reflector
头灯反射镜封垫 headlight reflector seal
头灯防眩罩 head light visor;headlight visor
头灯光度指示器 headlight beam indicator
头灯护架 headlight guard
头灯回光罩 head lamp deflector
头灯活节螺栓 head lamp swivel
头灯级 headlight step
头灯架 head lamp holder;head lamp support;headlight bracket
头灯接线板 headlight cable terminal block
头灯聚焦 focusing of headlamps
头灯开关 headlight switch
头灯开关钮 headlight switch knob
头灯壳 head lamp casing;head lamp housing;headlight body;headlight casing;headlight shell
头灯壳座 headlight casing stand
头灯控制距离 headlight control
头灯框 head lamp door;headlight door;headlight mo(u)lding
头灯框闭锁弹簧 head lamp door spring
头灯框闩 head lamp door latch
头灯框座 head lamp door rocket
头灯视距 headlight sight distance
头灯试验器 head lamp tester
头灯托架 headlight bracket
头灯小光灯泡 <汽车> head lamp parking bulb
头灯眩光 <汽车的> headlight dazzle;headlight glare
头灯罩 head lamp case
头灯座 headlight socket
头等舱 first-class cabin;salo(o)n cabin
头等车【铁】first-class car
头等的 first-class;first rate;blue ribbon;first grade;first line;tiptop
头等风险 primary risk
头等服务 front-rank service
头等货 clinker;first-class quality;first-rater
头等客车 salo(o)n car
头等利率 prime rate
头等列车 salo(o)n car
头等木材 A-grade wood;grade A wood
头等票据 first-class paper
头等试件 research class
头等橡胶 first latex rubber
头等样品 research class
头等证券 blue chip shares;first-class paper
头等质量 first-class quality;first-rate quality
头垫 <汽车座椅的> head rest
头顶防水板 head flashing
头顶花篮女神像 canephora
头顶净空 head clearance
头顶空隙 head clearance
头顶上保护 overhead protection
头顶有槽沉头平尖螺钉 slotted countersunk head flat point screw
头顶有槽圆柱头平尖螺钉 slotted cylindrical head flat point screw
头兜 head helmet
头端 head end
头端单元 headend unit
头对头安放 placing head to head
头对头接合 butt and butt
头对头结构 head to head structure
头对尾键合 head to tail linkage
头(对)尾结构 head to tail structure
头二道混合底漆 primer surfacer
头、二、三、四档速度 first, second, third, fourth speed
头阀 head valve
头阀扫气 valve-in-head scavenging
头发湿度计 hair hygrometer
头杆 head rod
头高 head height
头攻丝锥 first tap
头沟 header
头冠纹章 <散射光芒的> antique crown
头光油 headlight oil
头滚筒 head roller
头号牌子 brand leader
头环 head band
头昏 dizziness
头昏眼花的 dizzy
头架 headstock;spindle stock
头尖舱 forepeak tank
头桨手 bowman;bowoar
头接头 head to head
头锯 head saw;log saw
头靠 head rest;head roll
头壳 scalp
头宽 head breadth
头款 down stroke;original equity <购置房地产的>
头盔 headgear;helm;helmet;skull guard
头盔安全锁 helmet safety lock
头盔垫子 helmet cushion
头盔附属物 helmet attachment
头盔供气 helmet air supply
头盔潜水 helmet diving
头盔软管式潜水 helmet-hose diving
头盔上所附的活动脸罩 beaver
头盔式潜水器 helmet diving apparatus
头盔式潜水装具 helmet-type diving apparatus
头盔式双目望远镜 helmet mounted binoculars
头盔无线电话机 helmet radiophone
头盔与领盘 helmet and breast plate
头缆 bow fast;bowline;head fast;headline;head rope;head wire
头缆抬缆方趸 head wire pontoon;head wire scow
头馏分 head fraction;heads;overhead distillate
头轮 head pulley;head sheave
头轮驱动皮带运输机 head-pulley-drive conveyor[conveyer]
头锚 fore anchor
头帽 head cap
头描记器 cephalograph
头模 headform
头木 pollard
头木作业法 pollard system
头脑产业 brain industry
头排钉 head nail
头批生产的产品 early production
头皮 scalp
头侵蚀 head erosion
头上的 overhead
头上滑轮 <皮带运输机的> head pulley
头上滑轮驱动 head pulley drive
头上开裂 end splitter
头上空间 headroom
头绳 headline;head rope
头饰 headgear
头饰带 head band
头数 number of starts
头痛 headache
头头 boss
头头相接的 head to head
头尾加成 head to tail addition
头尾结构 head-tail structure
头尾绳运输 main-and-tail haulage
头尾绳运输系统 main-and-tail system
头尾系统连接 head to tail connection
头尾相接的 head to tail
头形装饰 headworks
头晕 dizziness
头载式单耳机 monaural receiver
头载式双耳机 double head receiver
头渣 first-run slag
头枕 head rest
头正对着 end-on
头柱 head pin
头状的 capitate
头锥 nose cone
头子 honcho
头足纲【动】Cephalopoda
头足类灰岩 Cephalopoda limestone
头座 headstock
头座顶尖 headstock center[centre]
头座随转顶尖 loose headstock center[centre]

投 throw

投棒短节 bar dropper
投保 apply for insurance policies;effect an insurance;effect a policy;insure;proposal for insurance
投保保险 cover insurance;effect insurance
投保单 application for insurance;insurance slip;proposal form
投保方 applicant;insurance policy holder;insurant;policy holder
投保工程险 insure works
投保海上保险申请书 application for marine insurance
投保价值 insured value
投保价值条款 insured value clause
投保金额 amount insured;face amount;insurance amount;insured amount
投保期限 validity of tenders
投保全险 against all risks
投保人 applicant;applicant for insur-

ance; assured; insurant; person insuring; policy holder
投保人身意外险 secure oneself against accidents
投保人失踪 lost policy holder
投保人资格预审 prequalification of (prospective) bidders
投保声言 representation
投保书 proposal of insurance
投保条款 coverage
投保通知 insurance instruction
投保"一切险" cover of all risks
投币操作时间 cash operating time
投币电话站 coin telephone station
投币电话机 coin(box) telephone set
投币阀 action valve; autovalve; prepayment valve
投币公用电话 prepayment coin box
投币加(气)油站 gasateria
投币孔 coin slot
投币口 coin slot
投币式电度表 slot meter
投币式煤气表 prepayment meter; slot meter
投币箱 coin box
投币制公用自动电话站 pay station
投标 bid bond; bidding; bid on; bid upon; competitive bid; enter a bid; enter a bid for; enter bid; make a bid; make a bid for; make tender; submission of bids; submission of competitive tenders; submission of tenders; submission tender; submit a tender; submit a bid; submitting the bid; tender; sealed proposal; proposal <美>
投标保单 bid bond; tender bond
投标保函 bank's guarantee for bid bond; bid bond; bid guarantee; tender bond; tender guarantee
投标保证 bid guarantee
投标保证金 bid bond; bid deposit; bid security; guaranty bond; guaranty money of tender; initial guarantee; tender bond
投标保证书 bid guarantee; bid security; letter of guarantee for bid; tender guarantee; tender security
投标报价 bid price; tender offer
投标报价表 schedule of bidding price
投标报价单 bid price quotation
投标编写 bid preparation
投标表格 bid form; form of tender; proposal form
投标表格格式 <按表格填写> bid form
投标步骤 bidding process
投标裁定 award of bid
投标裁决 bidding procedure
投标厂商国籍 bidder's home country
投标承包 bidding contract; bid on; competitive contract; make a bid for; make a tender for; put in a tender for
投标承包部分项目 partial bid
投标承包方式 form of tender
投标承包人 competitive contractor
投标承包商 competitive contractor
投标承包者 competitive contractor
投标承建 tender for construction; tender for the construction of
投标程序 bidding procedure; competitive tender(ing) action; hid procedure; tender procedure; tender system
投标澄清会议 bid clarification meeting
投标出价 tender offer
投标出价的比较 comparison of bid
投标代理商 agent for bidding
投标单 bid(ding) sheet
投标单价 priced bill of quantities
投标担保 bid guarantee; bid security; tender bond; bid bond
投标得标人 successful bidder
投标的初步查询 preliminary tender enquiry
投标的根本性反应 bid's substantial responsiveness
投标的密封和标志 sealing and marking of bids
投标底价 base bid price
投标定价法 sealed-bid pricing
投标费用 bid cost; cost of tendering
投标格式 bid form; tender form
投标购货 buying tender
投标购买企业 take-over bids
投标估计 bid evaluation
投标估价 bid evaluation; tender evaluation
投标估算 bid estimate
投标广告 tender advertisement
投标规则 bidding rules
投标过程 tender process
投标过低 underbid
投标合同 bidding contract
投标和报价 bid and quotation
投标和估价 bid and quotation
投标和授标 bidding and award
投标核查 bid examination
投标后的协商 post-tender negotiation
投标候选人 candidate for tendering
投标汇总表 bid abstract
投标或议标阶段 bidding or negotiation phase
投标货币 bid currencies
投标基本方案的替代方案 deductive alternate
投标基本要求 basic requirements for tendering
投标集团 bidding group; tendering ring
投标计划 tender design
投标价单一览表 abstract of bids
投标价(格) competitive price; price tendered; tender(ed) price; bid price; contractor's price; bid rate
投标价格表 bid schedule of price
投标建议(书) bid proposal; tender proposal
投标阶段 bidding period; bidding phase; bidding stage; tender stage
投标接受书 letter of acceptance
投标结果 bidding result; tender result
投标结果展示 <所有投标者的> bid result
投标截止日期 closing date; final date; bid(ding) deadline; deadline for submission of bids; deadline for submission of tenders; deadline of bid
投标截止时间 deadline for submission of tenders
投标金额 tender sum
投标津贴 tender allowance
投标经费 cost of tendering
投标竞争 competitive bid(ding)
投标开标纪要 bid opening minutes
投标开价 tender offer
投标理论 bidding theory; theory of bidding
投标联合体 bidding combination
投标履约情况 tender performance
投标履约信用证 bid bond and performance bond letter of credit
投标目录 tender list
投标内容的变化 variation in bidding conditions
投标评定 voter assessment
投标评估 evaluation of bids; evaluation of tender
投标评价 bid evaluation
投标期(间) tender period; tender stage; bidding period
投标期间汇期权 tender to contract option
投标期限 tendering period; bidding period <美>
投标企业 bid business
投标契约 bidding contract
投标人 bidder; tenderer
投标人分项标价综合单 bid summary
投标人各项标价一览表 bid tabulation
投标人国籍 bidder's home country
投标人名单 bidders sheet; bid sheet; tenderers sheet
投标人投标日期 date for latest delivery of bids by bidder
投标人须知 instructions to bidders; instructions to tenderers; notice to bidders
投标人选择名单 selected list of bidders
投标人资格预审 bidder's prequalification; prequalification of tenderer
投标日 tender date
投标日期 bid date; tendering date
投标商 bidder; tenderer
投标商的无欺诈宣誓书 noncollusion affidavit
投标商分项标价综合单 bid summary
投标商附加标价 additive alternate
投标商各项标价一览表 bid tabulation
投标商国籍 bidder's home country
投标商合约保证金 bid security
投标商名单 bidders sheet; tenderers sheet
投标商须知 instruction to bidders; instruction to tenderers; notice to bidders
投标商业务收购人 bid shopper
投标设计 tender design
投标审查 analysis of tenders; review of tenders
投标审查报告 report on tenders
投标审核 examination of bids; screening of bids
投标审计 inviting bids audit
投标时间 bid time
投标实质内容 bid substance
投标手续 bidding procedure; tendering procedure adjudication; tender procedure
投标书 application for tenders; bid; bidding book; book of tender; contractor's proposal; tender; tender book
投标书承诺 tender offer
投标书的接受 tender acceptance
投标书的拒绝 rejection of tender
投标书的完备性 sufficiency of tender
投标书的谢绝 tender rejection
投标书分析 analysis of tenders
投标书附件 appendixes to the tender
投标书附录 appendixes to bid
投标书格式 form of tender
投标书有效期 bid validity; validity of tenders
投标书中的工程量 bid quantity
投标书中问题的澄清 bid clarifications
投标说明 instructions to tendering
投标说明书 instruction to bidders
投标说明书文件 bidding document
投标诉讼 bidding procedure
投标提交的最后期限 deadline for submission of bids
投标条件 bidding condition; conditions of bid; tender(ing) condition
投标条款 bid item; proposal item; tender clause
投标图样 tender drawing
投标图纸文件押金 document deposit
投标委员会 tender committee
投标文件 bid(ding) documents; tender(ing) documents
投标文件包 bid package
投标文件修改 amendment of bidding documents; amendment of tendering documents; modification of bids; modification of tenders
投标文件押金 deposit(e) for bidding documents; plan deposit
投标限期 bidding period; bid time
投标相互竞争 bid against each other
投标项目 bid item; item in the tender; proposal item; tender item
投标项目与数量清单 list of bidding items and quantities
投标小组 bidding group
投标形式 bid format; by-tender
投标须知 bidding requirements; general instruction to tenders; instruction for bidding
投标须知事项 conditions of bid
投标押金 bid bond
投标邀请 tender invitation
投标邀请函 bid invitation letter
投标邀请书 invitation for bidding; invitation for tender(ing); invitation to bid
投标要求 bidding condition; bidding requirements
投标要求表 schedule of requirements
投标业务 bid shopper
投标一览表 tender list
投标有效期 validity of bids
投标有效性 validity of tenders
投标语言 bid language
投标预备会 pre-bid meeting
投标预备会议 prebid conference
投标摘要 abstract of bids
投标者 bidder; tenderer
投标者的限价单 restricted list of bidders
投标者的资格预审 qualification of bidder
投标者国籍 bidder's home country
投标者建议的担保 proposal guaranty
投标者建议形式 proposal form
投标者名单 bidder's list
投标者须知 instructions to bidders; instructions to tenderers
投标指示 bidding requirement; conditions of bid; instructions to tenders
投标者资格预审 prequalification of (prospective) bidders
投标争取 bid for
投标指南 instructions to bidders; instructions to tendering
投标制 bidding system; competitive bidding system
投标制度 tender system
投标中标人 successful bidder
投标准备 bid preparation; preparation of bids
投标总报价 bid price
投标总价 tender sum
投标总价的细分 breakdown of bid price; breakdown of lump sum bid price
投标最后期限 deadline for acceptance of tenders
投产 commission(ing); operating; place in operation; put into commission; put into operation; put into service; put on production; starting-up to put into production; start-up

投产出产分析 input-output analysis
投产的机器 production machine
投产方式 pattern of starting production
投产后的援助 post-operational assistance
投产期 running-in period; start-up time; commissioning date
投产前需时 lead time
投产日期 commissioning date; in-service date
投产时间 date of mining begins
投产试验 commissioning test
投产外加费 associated cost
投产原型 production prototype
投产运转中的 on-stream
投产周期 lead time
投产追加费 associated cost
投产准备阶段 lead time
投弹 bomb
投弹角 lead angle
投低标 underestimation
投低标价 under-bidding
投低价标 under-bidding
投递 deliver; deliverance
投点 shaft plumbing
投点误差 projection point error
投吊式声呐 dipping sonar; dunking sonar
投放 put…into circulation; release by
投放电路 release circuit
投放高度 release altitude
投放溶液器 solution feeder
投放市场 commercialization; release
投放式电导海水温度计 expendable conductivity temperature depth sonde
投放式海水温度法 expendable bathythermography
投放式海水温度计 expendable bathythermograph
投放式仪器 drop type instrument
投放试验 drop test
投放顺序 release sequence
投放通货 money supply
投放现金 currency issued
投放重量 launching weight
投高标 over estimating
投稿人 contributor
投稿者 contributor
投工量 labo(u)r input
投光灯 flood light; projector; spotlight
投光灯具 projecting illuminator
投光器 light projector
投光照明 floodlighting
投海处理 ocean disposal
投海处置 sea disposal
投机 jobbing; play the market; speculation; venture; flyer <俚语>; adventure
投机倒把 engage in speculation; engage in speculation and profiteering; profiteering; speculation
投机的 speculative
投机动机 speculative motive
投机购买 buying long
投机过度行为 speculative excesses
投机活动 aleatory operation; profiteering
投机交易 speculation; speculative transaction
投机交易所 bucket shop
投机买卖 scalping; speculation; speculative trade
投机买卖账户 venture account
投机能力 leverage
投机企业 venture business
投机曲线 speculation curve
投机商人 gambler; profiteer
投机商业 speculative business
投机市场 speculation market; speculative market
投机事业 adventure; speculation business
投机套利 speculation arbitrage
投机投资 speculative investment
投机物 venture
投机效率假说 speculative efficiency hypothesis
投机心理 take a profiteering attitude
投机信用 speculative credit
投机性采购 speculative purchasing
投机性成分 speculative component
投机性风险 speculative risk
投机性高涨 speculative boom
投机性股票 cats and dogs
投机性合同 aleatory contract
投机性失业 speculative unemployment
投机性土地投资 speculative land investment
投机性需求 speculative demand
投机性营造商 speculative builder
投机需求 speculation demand
投机诈骗 speculation and swindling
投机者 speculator; stock jobber
投机者的需求 speculator's demand
投机证券商 risk arbitrager
投机指数 index of speculation
投机资本 risk capital; venture capital
投加率 application rate
投建可行性评议【岩】 feasibility assessment of(mine) building
投进 drop in
投距 throw
投考者 candidate
投料 batch charging; batch filling; charging; feeding
投料点 feeding point
投料斗 charging bucket
投料端 batch charging end; batch feeding end; feeding end; feeding side
投料端墙 backwall; end wall; feed end wall; filling end wall
投料机 batch charger; batch feeder
投料计量 dosing
投料计量室 dosing chamber
投料孔 access door
投料控制 dosing cock
投料口 dog house; feeder nose; filling pocket; inlet
投料口盖砖 dog house mantle block
投料口拐角砖 dog house corner block
投料口旋拱 dog house arch
投料量 dose rate; inventory rating
投料料位 charge level
投料平台 feeding platform
投料前生产试验 preproduction test
投料深度 batch depth
投料试生产 commissioning test run; commission test run
投料试运转 commissioning test run
投料速率 rate of feeding
投氯泵 chlorine feed pump
投落作用 accretion
投铆钉手 rivet catcher
投煤机 mechanical stoker
投抛 sling
投配 batching; dosing
投配泵 dosing pump
投配比 dosing ratio
投配槽 dosing tank
投配池 <污水处理> dosing tank; dosing basin
投配单元 dosing unit
投配罐 dosing tank
投配虹吸 dosing siphon[syphon]
投配间 dosing room
投配间隔 dosing interval
投配开头 dosing cock
投配量 calculated dose
投配龙头 dosing cock; dosing control
投配频率 frequency of dosing
投配器 batch box; batch(er) bin; batching bin; dosing apparatus; dosing device; dosing tank; dosing unit; measuring hopper; proportioning plant
投配设备 dosing equipment
投配室 dosing chamber
投配箱 dosing box; dosing tank
投配周期 dosing cycle; periodicity of dosing
投配装置 dosing apparatus
投票 drawing of ballot; poll; vote; voting
投票表决 polling
投票反对 black ball
投票记录 poll
投票结果 results of a poll
投票决定 determine by votes
投票权 ballot; voting power; voting right
投票人 voter
投票信托 voting table; voting trust
投票总数 ballot
投弃货物 <船舶遇险时投弃的货物> jetsam; jettison
投弃减速发动机 retro pack jettison
投弃式温深计 expendable bathythermograph
投弃式温深计观测断面 expendable bathythermograph section
投弃式温深计观测资料 expendable bathythermograph data
投弃式仪器 expendable instrument
投弃装置 jettison device; jettison gear
投溶液器 solution feeder
投入 input; phase in; plunge; pull-in; throw-in
投入表 input table
投入产出 input-output
投入产出比 input-output ratio
投入产出比率 input-output ratio
投入产出变量 input-output variable
投入产出表 diagrammatic representation of input-output; input-output table
投入产出的理论基础 theoretic(al) basis of input-output
投入产出动态分析 dynamic(al) input-output analysis
投入产出法 input-output approach
投入产出分析 input-output analysis
投入产出分析法 input-output analysis method
投入产出关系 input-output relationship
投入产出过程 input-output process
投入产出核算 input-output accounting
投入产出技术 input-output technique; input-output technology
投入产出结构 input-output framework
投入产出结果 cost-yield result
投入产出解 input-output solution
投入产出经济学 input-output economics
投入产出矩阵 import-outport matrix; input-output matrix
投入产出会计 input-output accounting
投入产出模型 cast-in yield-out model; input-output model
投入产出矢量 input-output vector
投入产出数学模型 input-output mathematical model
投入产出系数 input-output coefficient
投入产出系统 input-output system
投入产出优化模型 input-output optimization model
投入产出预测模型 input output forecast model
投入产业 invested in plant
投入成本 input cost
投入单位 input unit
投入导向 input orientation
投入的现金抵押 cash collateral invested
投入的转移 transfer of inputs
投入的资本 contributed capital; vested capital; venture capital
投入分布 input distribution
投入分析法 input analysis
投入沟内 toe ditch
投入海内的货物 <船只遇难时> lagan; lagend
投入海中 toe ditch
投入河中 toe ditch
投入记录 input recording
投入间隔期 interval between inputs
投入结构 input structure
投入开采 placing on production
投入量 input amount
投入(品)需求函数 input demand function
投入期 input time
投入切除(网络)卡 on-off card
投入生产 bring into operation; bring into production; bring on stream; go into operation; go into stream; place in operation; placing on production; put into operation; put into production; put on stream
投入使用 commissioning; place in service
投入式乙炔发生器 carbide to water generator
投入式钻头 wire-line bit
投入输出成本 input-output price
投入顺序 ordering in launching
投入替代弹性 elasticity of input substitution
投入调整 input adjustment
投入系数 input coefficient
投入需求量 input requirement
投入一断开方式 on-off service
投入营运 put into service
投入运行 bringing into service; bring into operation; commissioning; on-the-line; place in operation; place in service; put into operation; put into service; putting into operation
投入运行日期 commissioning date
投入运营 put into service
投入运营线路 line open to traffic
投入运用 place in operation
投入运转 put(ting) into operation
投入正常工作状态 putting in working order
投入重复期 input repeat time
投入资本 capital incorporation; capital input; equity capital; invested capital; paid-in capital; vested proprietorship
投入资本的商号 invested firm
投入资本合法性审计 called-up capital legality audit
投入资本利润率 rate of returns on invested capital
投入资本审计 called-up capital audit
投入资本收益率 return in invested capital
投入资本所得率 rate of return on invested capital
投入资本盈利 earnings on invested capital

投入资本真实性审计 called-up capital authenticity audit
投入资本总额 all amount invested in capital;total invested capital
投入资产 invested assets
投入资金 invested funds
投砂量 shot feeding per run
投砂器 sand slinger;sand slinging machine
投射 dart;launching;projection
投射标度天平 shadowgraph scale
投射表示器显示 projection indicator display
投射测验 projective test
投射成型法 <耐火材料> slinging process
投射到街道 projection into a street
投射的 projectile;projecting
投射灯 projection lamp; projection light;projector lamp;reflectoscope
投射法 projection method;projective method
投射辐射 incident radiation
投射管 projection tube
投射光 projection light
投射光束 projecting beam
投射光束式烟感火灾探测器 projected linear foam smoke detector
投射光线 throw light
投射光学 projection optics
投射光源 projection source
投射技术 projection technique;projective technique
投射角 angle of incidence;angle of projection;coign(e);projecting angle;projection angle
投射角分布 pitch angle distribution
投射截止点 projected cut-off
投射镜 projecting mirror
投射距离 projection distance
投射料 slinger mix
投射面积 projected area
投射面积比 projected area ratio
投射剖面 projected profile
投射器 missile; projector; projectoscope
投射区 projection area
投射示波器 projection oscillograph
投射式 projection type
投射式白炽灯 projector type filament lamp
投射式单位散热器 projector type unit heater
投射式放映机 opaque projector
投射式进路表示器 projector type route indicator
投射式进路指示器 projector type route indicator
投射式显像管 projection cathode-ray tube
投射式阴极射线管 projection cathode-ray tube
投射体 projectile
投射筒 projecting cylinder
投射武器的进入 weapon delivery run
投射物屏障 missile barrier
投射纤维 projection fiber[fibre]
投射线 projecting line
投射效率 projection efficiency
投射仪 overhead project
投射阴影 cast shadow
投石的距离 stone cast
投石器 ballista
投视式罗经 projection compass;projector compass
投水流送 watering
投送 deliver
投送等级 priority of delivery
投送通信[讯] communication of delivery
投送状态 delivery status
投诉期 <公众向当地政府投诉税收不当的> grievance period
投诉与索赔 complaints and claims
投梭装置 picking motion
投纬 wefting
投物点 payload-release point
投物伞吊带 cargo parachute harness
投下 downthrow;hurl
投向误差 projection direction error
投像平面 plane of delineation
投像器 camera obscure
投信口 letter drop;mail slot
投信口滑槽 letter drop chute
投信口遮板 letter drop plate
投药 chemical dosing;chemical feeding
投药法 medication
投药机 chemical feed machine
投药剂量监测测量图 dosage monitoring survey design
投药剂量控制系统 dosing control system
投药间 chemical feed room
投药漏斗 feed hopper
投药瓶 dispensing bottle
投药器 balling gun
投药设备 dosing apparatus
投以强烈的光 in the highlight
投以阴影 adumbrate
投影 cast shadow;deep shadow;deep shallow; plane; projecting; projection; projection of image; projecture
投影比长 projectometer
投影比长计 protect meter
投影比较器 projection comparator
投影比较仪 optimeter; projection comparator
投影比例尺 projected scale
投影比例系数 projection scale factor
投影编图仪 map projector
投影变换 projection change; projection transformation
投影变形 distortion of projection
投影标尺 projected scale
投影标尺天平 projected-scale balance
投影标度尺仪表 projected-scale instrument
投影表 projection table
投影薄膜 transparency[transparence]
投影测图仪 projecting plotter;projection plotter; Balplex plotter <商品名>
投影测温术 projection thermography
投影差 deformation due to relief; height displacement;relief displacement;relief distortion
投影差改正 correction for relief;correction of relief displacement
投影差校正 altitude correction; correction for relief
投影长度 projected length
投影垂直面 plane perpendicular to projection plane
投影带 projection zone
投影带号 projection zone number
投影导杆 guide rod
投影的 projected;projective
投影灯 projecting lamp
投影底片 autochrome
投影地 projectively
投影地图显示器 projected map display
投影地图显示系统 projected map display system
投影点 projection point; projective point;subpoint <天体或飞行器在地面上或其他星球上的投影点>; vanishing point <透视图的>
投影电路 project circuit
投影电视 projection television
投影电视接收机 projection television receiver
投影电位示波器 iatron
投影电子显微术 shadow casting
投影定理 projection theorem
投影度带 projection zone
投影断链 projective broken chain
投影对比度系数 projected contrast ratio
投影多度 projective abundance
投影法 method of projection; projection method; projective method; sciagraphy
投影翻印 projection printing
投影方程式 projection equation
投影方式 projection mode
投影放大器 balopticon; episcotister; episcope <不透明文件的>; projection amplifier;stereopticon
投影盖度 projective cover degree
投影干涉仪 projection interferometer
投影杆 shadow bar
投影格网 graticule; projection grid; projective grid
投影格网延伸短线 projection tick
投影管 light guide tube; projection cathode-ray tube;projection tube
投影光 projecting light
投影光束 projected light beam; projecting beam
投影光学 projection optics
投影光学比较仪 projection optimeter
投影光学系统 projection optic(al) system
投影光源 projection light source
投影光泽 gloss distinctness of image
投影函数 projection function
投影画 shade and shadow; shadowgraph
投影环 projection nucleus
投影幻灯机 overhead projector
投影换算程序 projection conversion program(me); projection conversion routine
投影火花室 projection chamber;projection spark chamber
投影机 projecting camera;projection machine;projector
投影机灯泡 projector lamp
投影机构 projector mechanism
投影机效率 projector efficiency
投影基线 baseline of projection;model base(line);projected baseline
投影几何码 projective geometry code
投影几何学 descriptive geometry;geometry of projection; perspective geometry;projective geometry
投影计算 projection calculation;projection computation
投影技术 shadow casting technique
投影检验器 projection detector;shadow projector
投影焦深 projection focus depth
投影角 angle of projection;angle of refraction;projected angle;projecting corner;projection angle;refraction angle
投影教学材料 projectual
投影经线 projected meridian
投影径 projected diameter
投影纠正 rectification of projection
投影矩阵 projection matrix
投影距离 projection distance;projector distance;scaled distance
投影刻度仪 projected-scale instrument
投影控制器 projector controller
投影立体测图仪 projection stereoplotter
投影密度 projected density
投影面 picture plane;plane of projection; projecting plane; projection plane;projection surface;projective plane
投影面垂直线 line perpendicular to projection plane
投影面积 area of contour; projected area;projection area;projective area;shadow area
投影面基准线 alignment of the projection surface
投影面位置 altitude of projection surface
投影模 projective module
投影目标 projected object
投影目镜 projection eyepiece
投影片 screen sheet
投影平均长度 projected mean length
投影平面图 projected planform
投影屏 projection screen; projector screen;wall screen
投影器 camera of project;camera of projection; projecting instrument; projector
投影器方向节 camera cardan
投影器光度头 optic(al) head
投影器位置 projector position
投影器主距 principal distance of projector;projector principal distance
投影器座架 projector frame
投影曲率 projecting curvature;projection curvature
投影曲线 projected curve
投影取样 projection sampling;projective sampling
投影全景显示器 projection panoramic display
投影晒像 projection print
投影晒印 projection printing;projection ratio-printing;ratio print
投影晒印器 projection printer;projection ratio-printer
投影设备 projector equipment
投影射线 projecting line
投影示波器 projection oscillograph
投影式电视接收器 projection receiver
投影式坐标量测仪 projection comparator
投影算符 projection operator
投影算子 projection operator
投影缩减 foreshortening
投影缩减效应 foreshortening effect
投影天平 projection balance
投影通信[讯]系统 projection communication system
投影透镜 projecting lens; projection lens;projector lens
投影透镜套 projection lens jacket
投影图 perspective plane;projection; projection drawing; projection figure;reflected plan; sciagraph[sciograph];skiagram;skiagraph
投影图法 axiometric;axonometry
投影图绘制者 projectionist
投影图像 projected image
投影图像尺寸 projection picture size
投影图像显示器 projection-view-display;projection viewer
投影图形 projecting figure
投影网 projective net
投影位置 air position;projection position
投影物镜 projection objective
投影系统 optic(al) projection system;optic(al) projector;projection system;projector
投影显示(器) projected display;projection display

T

投影显示仪 projector scope
投影显微镜 projection microscope
投影显微射线照相术 projection microradio graphy
投影显微照片 projected micrograph
投影显像管 projection kinescope
投影线 line of projection; projection line
投影相片 projection print
投影效应 projection effect
投影型 projection type
投影压力计 projection manometer
投影仪 optic(al) projector; overhead project; projecting apparatus; projecting camera; projection apparatus; projection camera; projector
投影影像 projected image
投影圆棒 projecting bar
投影中心 air position; center[centre] of projection; projection center [centre]
投影轴 axis of projection
投影轴线 axis of perspectivity; perspective axis
投影转换 projection transformation; projective transformation
投影转换参数 projection transformation parameter; projective transformation parameter
投影转绘 sketch projection
投影转绘仪 compilation camera; mapograph; projector; sketch projector
投影装置 make-up projector; projecting apparatus; projecting unit; projection device
投影追踪分析 projection pursuit analysis
投影字行 projected line
投影总盖度 projective total cover degree
投影坐标系(统) project coordinate system
投硬币启动的干燥机 coin-operated dryer[drier]
投硬币售货机 slot-machine
投硬币售票机 slot-machine
投源点号 number of reagent dropping point
投源孔到接收孔距离 distance between reagent dropping well and receiving well
投掷 cast(ing); jettison; pitch(ing); shoot; sling; throw(ing)(off)
投掷爆管或爆竹 squib
投掷过筛 throw screening
投掷角 angle of departure
投掷开关 throw-over switch
投掷器 sling(er); thrower
投掷矢石台 assommoir
投掷试验 dart test
投掷物 <尤指武器> missile
投掷者 slinger
投资 bring in; capital cost; capital formation; capital invested; capitalization; capitalize; capital spending; investment; investment cost; money invested; payout; sinking
投资包干 fixed investment responsibility; investment lump-sum contracting
投资包干责任制 investing responsibility system; system of investment responsibility
投资保护 investment protection
投资保护协定 agreement on protection of investment; investment protection agreement
投资保护协议 investment protection agreement
投资保险 investment insurance; investment political risks insurance
投资保证 investment guarantee
投资保证方案 investment guaranty program(me)
投资报酬 return of investment
投资报酬率 accounting rate of returns; returns of capital investment
投资报酬率定价法 rate-of returns pricing
投资比较方案 alternative investment
投资比例 ratio between investments
投资边际效率 marginal efficiency of investment
投资边际效率表 marginal efficiency of investment schedule
投资边际效益 marginal efficiency of investment
投资变换收益 gain on conversion of investment
投资表 investment schedule
投资并且建造房屋出售的建筑商 speculative builder
投资拨款 capital appropriation; investment grant
投资补贴 investment allowance; investment grant; subsidy for capital expenditures
投资补助金 investment grant
投资不足 undercapitalize; under investment
投资不足的 undercapitalized
投资财 investment goods
投资参考事务 investment reference service
投资残值 residual value in capital investment
投资策略 investment strategy
投资产出比 capital-output ratio
投资产出率 ratio of investment to output
投资产业 investment property
投资偿还 cost reimbursement contract
投资偿还年限 payback period of investment
投资场所 cost of the investment; outlet for investment
投资成本 capitalized cost; cost of investment; investment cost
投资承诺 investment commitment
投资乘数 investment multiplier
投资乘数作用 investment multiplier effect
投资纯收入率 investment-net income ratio
投资刺激 investment incentive
投资搭配 portfolio
投资搭配理论 portfolio theory
投资搭配效果 protfolio effect
投资搭配选择 portfolio selection
投资贷款 investment loan
投资单位持有者 unit-holder
投资的多样化规划 diversified program(me) of investment
投资的减税率 investment tax credit
投资的内部收益率 internal rate of returns on investment
投资的时间构成 time structure of investment
投资的时间延滞 investment lag
投资的双重效益 dual effect of investment
投资的双重性 dual nature of investment
投资的需求 investor's demand
投资的总现金收益 total dollar return
投资低风险 tertiary risks
投资抵减 investment allowance
投资、抵押、担保权 the right of investment, mortgage and guarantee
投资动态 investment behavio(u)r
投资额 amount invested; amount of capital invested; amount of investment; capital cost; capital intensity; investment cost; investment dollars
投资法 investment act; investment law; law of investment
投资繁荣 investment boom
投资返本年限法 payout period; payout time
投资返回率 rate of return
投资方案 investment project; investment proposal
投资方案排队格式 ranking form
投资方案排队问题 ranking problem of capital investment
投资方案选择 capital budgeting
投资方向 investment along proper lines; investment direction; investment orientation
投资方向调节税 accommodation taxes for the investing direction
投资费用 capital cost; cost of investment; expenditure of capital; investment cost; investment expenditures; initial cost
投资费用的节省 saving in investment cost
投资分段评估法 band of investment method
投资分级 investment rating
投资分类 classification of investment
投资分类账 investment ledger
投资分配 allocation of investment; budget allocation; cost allocation
投资分散 investment diversification
投资分摊 allocation of investment; budget allocation; investment allocation
投资分析 investment analysis
投资分析工作表 investment-analysis worksheet
投资分析师 investment analyst
投资份额收益 dividend yield
投资风险 investment risk; risk in investment
投资风险分析 investment risk analysis
投资风险率 risk rate
投资服务 investment service
投资赋税优惠 investment tax credit
投资概算 investment estimate
投资高涨 investment boom
投资高指标 bigger investments
投资公式 investing formula
投资公司 invested firm; investment company; investment corporation; investment firm; investment house; investment trust
投资公司法案<美> Investment Company Act
投资公司股票 investment company shares
投资公司税收抵免 regulated investment company credit
投资构成 composition of investment
投资购买 investment buying
投资估计 investment estimate; investment estimation
投资估价 investment appraisal; investment evaluation
投资估价基础 rate base
投资估算 construction cost estimate; investment estimate; investment estimating; investment estimation
投资估算分项金额 estimate breakdown
投资股份 investment shares; investment stocks
投资顾问 investment adviser[advisor]; investment counselor
投资顾问法 investment adviser[advisor] act
投资管理 investment management
投资归属流动资产 investments classifiable as current assets
投资规划 investment programming
投资过度 overcapitalization; overcapitalize; overinvestment; top-heavy
投资过多 overcapitalization
投资过多危机论 crisis theory of overinvestment
投资过热 high demand for investment
投资过剩 overcapitalization; overinvestment
投资函数 investment function
投资合伙 cooperating investment
投资合同 investment contract
投资后续行动 investment follow-up
投资还本 capital recovery; investment recovery
投资还原成本 capital recovery cost
投资环境 investment climate; investment environment
投资回报 return of investment
投资回收 capital recovery; capital pay-off; investment recovery; return of investment; turn on investment
投资回收率 payback; rate of return on investment; rate-of-turn on investment
投资回收年数法 payoff method for capital investment
投资回收年限 years in return of capital investment
投资回收期 capital pay-off time; investment recovery period; payback period of investment; payoff period; payout time; period for recovery of investment; period of returns (from investment); recovery period of investment; repayment period of investment; return period of investment; time period for recovery of investment
投资回收期限 rate of return
投资回收系数 coefficient of investment recovery; recovery coefficient of investment; capital recovery factor
投资回收因数 capital recovery factor
投资回收周期 capital recovery period; period of return investment
投资回收总额 payoff gross investment
投资汇票 investment bill
投资活动 investment activity
投资货币 investment currency
投资机构 funding agency; investment agency; investment institution
投资机构购入 buying by institutional investors
投资机会 investment opportunity
投资机会表 investment opportunity schedule
投资机会函数 investment-opportunities function
投资机会研究 opportunity study; study of investment opportunities
投资基础 investment base
投资基金 capital fund; funded reserve; investment funds
投资基年 base year
投资基准 investment base
投资计划 capital planning; investment plan; investment program(me); investment project
投资计划完成率 planned performance rate of investments

投资加速因素 investment accelerator
投资价值 investment value; investment worth; value of each contribution
投资价值模型 investment value model
投资减税 investment allowance; investment credit
投资减税额 investor tax credit
投资建议 capital proposal
投资奖励 investment incentive
投资结构 composition of investment; investment makeup; investment structure; pattern of investment; structure of investment
投资结构调整 readjustment of investment structure
投资截止点 cut-off point
投资紧缩 undercapitalize
投资经纪人 investment broker
投资经营者 investment manager
投资景气 investment boom
投资净额 net investment
投资净收入税 tax on net investment income
投资净收入现值 net present value of income from investment
投资净收益 net investment income
投资净数 net investment
投资净值 net value of capital invested
投资矩阵 investment matrix
投资决策 capital investment decision; investment decision
投资决策的最小风险原则 minimum principle in investment decision
投资决策分析 analysis on investment decision
投资控股 share control takeover
投资控制 control of investment
投资宽减税额 investment broad tax credit
投资扩张 expanding investment
投资理论 theory of investment
投资利得率 rate of return on investment; return on investment
投资利润 investment return; return on investment; returns of investment
投资利润率 earning power of real assets; profit ratio of investment
投资利税率 profit-duty rate of investment; profit plus taxes ratio of investment; ratio of profit payments and tax turnover
投资利息 interest on investment
投资量 flow of investment
投资率 investment rate; rate of investment
投资毛收入 gross investment income
投资媒介 investment medium
投资模型 investment model
投资目标 investment objective
投资能力 ability to invest; investment ability
投资膨胀 investment inflation; swollen investment
投资平衡表 balance sheet of investment
投资评估 investment appraisal
投资评级 investment rating
投资评价 investment appraisal; investment evaluation
投资期间 investment period
投资企业公司 investment and enterprise corporation
投资气候 investment climate
投资契约计划 investment contractual plan
投资前的研究 pre-investment study
投资前方案 pre-investment program(me)
投资前活动 pre-investment activity
投资前基金 pre-investment fund
投资前时期(阶段) pre-investment phase
投资前项目 pre-investment project
投资前研究费用 cost of preinvestment studies
投资前援助 pre-investment assistance
投资潜力 investment potential
投资倾向 investment propensity; propensity to invest
投资权 capital authority
投资权益 investment interest
投资人不用支付佣金的互相信托投资基金 no-load fund
投资人权益 partner's equity of a joint venture; partner's investment interests
投资上的一般风险 secondary risks
投资设备 investment goods
投资升水 investment premium
投资失败 investment failure; take a bath
投资失当 investment dislocation
投资时滞 investment lag
投资实得率 yield rate
投资实物构成 physical composition of investment
投资使用年限 service time of investment
投资市场 investment market
投资收回 capital pay-off; capital recovery
投资收回期 capital pay-off; return on investment
投资收入 gain returns on investment; income from investment; income on investment; investment income; investment revenue
投资收入报酬率 rate of return on investment
投资收入比率 investment income ratio; return on investment ratio
投资收入的现值 present value of income investment
投资收入附加税 investment income surcharge; investment surcharge
投资收入净额 net investment income
投资收入净现值 net present value of income from investment
投资收入率 capitalization rate; investment yield; rate of return; return on investment
投资收入现值 present value of income from investment
投资收缩的最低限度 maximum disinvestment floor
投资收益 earning on investment; income from investment; income on investment; investment income
投资收益比 cost-benefit ratio; investment-earnings ratio
投资收益的免税 exemption for reinvested earnings
投资收益率 capital recovery factor; rate of income from investment; rate of return on investment
投资收益率法 rate of return on invested capital method
投资收益上交率 ratio of turned over income from investment
投资收益值 investment yield
投资手册 investment manual
投资水平变动 investment level movement
投资税 investment tax
投资税减免 investment tax credit
投资税收抵免 investment tax credit
投资税收减让 tax incentives for investment
投资损失 investment loss; capital loss
投资所得 investment yield
投资特性 investment feature
投资条件 condition of investment
投资统计 investment statistics
投资投入产出模型 model of input-output in investment
投资托拉斯<即合股投资公司> investment trust
投资物量指数 volume index of capital investment
投资误差 investment error
投资喜好 investment appetite
投资系数 investment coefficient
投资现值 present value of investment; time-adjusted investment
投资限额 capital allowance; control budget; investment norm
投资项目 investment project
投资项目基金 capital project funds
投资消耗费 investment consumption cost
投资效果 cost efficiency; effect of investment; efficient of investment; investment effect; investment effectiveness
投资效果系数 coefficient of investment yields
投资效果系数与投资回收期的审计 audit of investment result coefficient and payback period
投资效益 investment benefit; investment result; investment return
投资效益比 cost-benefit ratio
投资效益法 cost-benefit approach
投资效益率 investment and effect rate; rate of return
投资效益设计 cost-effective design
投资效应 investment benefit
投资协会 investment association
投资新观点 new view of investment
投资信贷 investment credit
投资信托 investment credit; investment trust
投资信托公司 investment trust company
投资信托基金未偿本金 outstanding principal of investment trust fund
投资信托业 trust investment
投资信息库 investment information bank
投资信用<指长期信用> investment credit
投资信用等级 investment credit rating
投资性成本 capital cost
投资性债券 investment bond
投资性证券 investment security
投资需求 investment demand
投资需求表 investment demand schedule
投资需求配套 investment fits the needs of funds
投资循环 investment cycle
投资研究 investment research
投资业务 investment portfolio
投资意向 investment intention
投资意愿 investment willingness
投资溢价 investment premium
投资银行 bond house; investment bank
投资盈余率 rate of return
投资优待 investment incentive
投资优惠扣除 capital allowance
投资诱导论 theory of inducement to invest
投资诱因 inducement of investment; inducement to invest
投资于股票 invest in stock
投资于机械或厂房的决策 capital investment in machine or plant decision
投资于某企业 embark money in an enterprise
投资于企业 invest in an enterprise
投资于设备或厂房的决策 capital investment in machine or plant decision
投资于新企业的资本 equity capital
投资预算 capital budget; investment budget
投资预算约束 investment budget constraint
投资预算支出 below-the-line expenditures
投资酝酿阶段 brewing period of investment; gestation period of investment
投资增长速度 investment growth rate
投资增加额 increment investment
投资增益乘数 investment multiplier
投资账 investment account
投资折价 investment discount
投资者 investor; investor-sponsor; moneyman; placer
投资者存放 lodgment for investor
投资者的利益 investor's yield
投资者方法 investor's method
投资者进行的建设项目 investor project
投资者进行的项目 investor project
投资者支持的协作 investor-sponsored cooperation
投资证明 investment certificate
投资证券 investment securities
投资证券的市值 market value of investment securities
投资证券买卖 investment securities traded
投资证券组合 investment portfolio
投资证书 investment certificate
投资政策 investment policy
投资支持活动 investment support activity
投资支出 capital expenditures; investment expenditures; investment outlay; investment spending; capital investment
投资支出函数 investment outlay coefficient
投资指数 investment index number; national significant number
投资中间人 investment middleman
投资中心 investment center[centre]
投资重点 investment priority; key investment project; priority of investment
投资周期 investment cycle
投资周转额 investment turnover; turnover of investment
投资周转率 investment turnover; investment turnover rate; rate of capital turnover
投资主导型经济繁荣 business boom led by investment
投资主体 subject of investment
投资助销 investment performance requirements
投资专家 investment expert
投资准备金 investment reserve
投资准则 investment criterion
投资咨询服务中心 investment consultancy service center[centre]
投资咨询公司 investment consultancy corporation; investment consultant corporation
投资资本 investment capital; venture capital
投资资金筹措 investment financing
投资资金的流量 flow of investment

fund
投资资金交付使用率 rate of use of invested funds
投资资源 investment resources
投资总额 aggregate investment; capitalized cost; gross assets; total amount of investment; volume of total investment
投资总额周转率 equity capital turnover
投资总规模 volume of total investment
投资总金额 gross investment
投资总量 aggregate investment
投资组合 investment portfolio
投资组合风险 investment portfolio risk
投资组合管理 investment portfolio management

透背 bleed-through
透层 prime coat(ing); primary coat
透层材料 prime material; priming material
透层处理 <沥青> priming treatment
透层油 primer; priming oil
透长花岗岩 kaukasite
透长辉煌岩 eustratite
透长凝灰岩 krablite
透长石 glassy feldspar; ice spar; kalifeldspath; rhyacolite; sanidine; sanidine feldspar
透长石粗面岩 sanidine trachyte
透长霞岩 sanidine nephelinite
透长岩 sanidinite
透长岩相 sanidinite facies
透潮的 poromeric
透潮性 moisture permeability
透彻的 diaphanous; transparent
透彻度 diaphaneity
透彻煅烧 dead burn
透彻了解 digest
透尘试验 dust test
透出 grinning through
透穿率 penetration rate
透穿数 permeability number
透穿照射 transillumination
透磁棒 permeable rod
透磁合金 <一种强磁性铁镍合金> permalloy; permeability alloy
透磁率 magnetic susceptibility
透磁性 magnetic permeability
透淬 through-hardening
透淡闪石 kievite
透蛋白石 magic stone
透底 bleeding; telegraphing
透底珐琅 plaque-a-jour enamel
透雕 openwork carving
透雕细工 fretwork; openwork
透雕型粉刷 sgraffi(a)to
透雕圆花饰 openwork rosette
透顶涵洞 <公路涵洞顶部留孔供路面排水> open-top culvert
透度计 dutch penetrometer; penetrameter[penetrometer]
透度计灵敏度 penetrameter sensitivity; wire sensitivity
透度试验 penetration test
透反射两用幻灯机 epidiascope
透放射线 radio transparent
透风 draught penetration; wind penetration
透风机 aerator
透风帽 cowl
透辐射的玻璃 radiation transmitting glass
透辐射能力 radiabiliy
透钙磷石 brushite
透橄斑岩 garewaite
透橄无球粒陨石 nakhlite
透光 shining through; sunlight penetration; transmission
透光百分率 percent transmission; percent transmittance
透光板 light-admitting board; light-passing board
透光玻璃砖 light-directing block
透光薄板 light-admitting sheet
透光不均匀的 cruddy
透光层 euphotic layer; photic layer; photic region; photic zone
透光层的 euphotic
透光层浮游生物 phaoplankton
透光长度 transparent length
透光尺寸 <玻璃装配后的> daylight size; sight size
透光窗 light inlet window
透光带 euphotic zone; photosynthetic-(al) layer
透光的 euphotic; light-admitting; non-light tight; photic
透光地板 naked floor
透光雕刻 ajour
透光度 light transmittance; luminous transmittance; transmissivity; transmittance [transmittancy]; transmittance of light(ing)
透光度计 light-transmittance meter
透光伐 released thinning
透光珐琅 jour-a-jour enamel
透光反射镜 transparency mirror
透光范围 passband
透光高层云 altostratus translucidus
透光高积云 alto cumulus translucidus
透光格栅 light-admitting grill(e)
透光格子窗 light-admitting grill(e)
透光拱廊 <教堂侧廊上的> transparent triforium
透光海区 photic region
透光混凝土 glass tile for glass concrete
透光计 penetrameter[penetrometer]
透光刻图 backlight scribing
透光孔 loophole
透光口 light opening
透光蜡石 specular alabaster
透光量 light transmission
透光率 light-transmission value; light transmittance; luminousness; rate of transmission; throughput; transmittance[transmittancy]
透光率计 opacimeter; sight meter; VI-meter
透光面积 fenestration
透光膜 transmitting film
透光穹顶 light-cupola
透光区 photic zone
透光式液面计 transparent liquid level ga(u)ge
透光塑料 light-admitting plastic material; light-passing plastic
透光塑料薄板 light-admitting plastic sheet (ing); light-passing plastic sheet(ing)
透光塑料片 light-passing sheet(ing) plastic
透光塑性材料 light-passing plastic material
透光缩小 transmitting reduction
透光陶瓷 light-transmittance ceramics; light-transmitting ceramics
透光系数 light-transmission coefficient
透光性 light transmission; light-admitting quality; light permeability; light transmittance
透光云 translucidus
透光云量 transparency sky cover; transparent sky cover
透光轴 translucent enamel
透光总长度 total transparent length
透过 penetrance; penetrate; penetration; permeance; permeation; reachthrough; transmission
透过法 <超声波探伤的> penetrant method
透过缝隙的一束可见光 chink
透过辐射热的 diathermanous
透过光 transmit light
透过光强度 transmission light intensity
透过光速 transmitted beam
透过距离 penetration distance
透过率 degree of turbidity; permeability; transmissibility; transmittance[transmittancy]
透过能力 infiltration capacity
透过浓度 penetrating concentration
透过片吸收限 absorption limit of penetrated flat
透过式测厚计 absorption type ga(u)ge; transmission ga(u)ge
透过式屏幕 transmissive viewing screen
透过性 perviousness
透过转换波垂直时距曲线 vertical hodograph of transmitted convert wave
透焊缝 through weld
透红外(线)玻璃 heat transmitting glass; infrared ray transmitting glass; infrared transmitting glass; infrared transparent glass
透红外(线)材料 infrared transmitting material
透红外(线)陶瓷 infrared transmitting ceramics
透红外(线)纤维 infrared transmitting fiber[fibre]
透花 pieced carving
透花铁门 openwork iron gate
透化瓷 vitreous china
透环流 through shake
透辉橄无球粒陨石 nakhlite
透辉角闪斜长麻粒岩 diopside amphibole plagioclase granulite
透辉角闪岩 diopside amphibolite
透辉闪大理岩 diopside-tremolite marble
透辉石 diopside; malacolite; mussite; vermiculite diopside
透辉石玻璃 diopside glass
透辉石大理岩 diopside marble
透辉石橄榄岩 diopside peridotite
透辉石角岩 diopside hornfels
透辉石苦橄岩 diopside picrite
透辉石榴岩 griquaite
透辉石岩 diopsidite
透辉石质瓷 diopside porcelain
透辉石紫苏辉石角岩 diopside hypersthene hornfels
透辉岩 bistagite
透胶 bleeding; bleed-through; strike-through
透胶的污迹 <胶合板面> strike-through
透角闪石 grammatite
透进材料玻璃管 sight feed glass
透景线 perspective line
透镜 lens; optic(al) lens; perspective
透镜板 lens board; lens plate
透镜边缘 lens periphery; lens rim
透镜表 lens measure ga(u)ge
透镜薄膜 lens blooming
透镜部件 lens component; lenticular unit
透镜测试表 lens testing chart
透镜层 lens jacket
透镜层理 linsen bedding
透镜磁罗盘 lensatic compass
透镜的 lenticular
透镜的几何中心 geometric(al) centre [center] of lens
透镜的焦点 focus of lens
透镜的中性焦强 neutralizing power of lens
透镜垫 German-style metal lens ring; grooved metallic gasket; lens ring washer; lens washer
透镜垫密封 seal with metal lens ring
透镜垫片 lens gasket; lens shim
透镜镀膜 lens coating
透镜对屏间距 lens-to-screen distance
透镜反射光斑 lens flare
透镜方程 lens equation
透镜放大倍数 diameter
透镜放大率仪 auxiometer
透镜分辨率 lens efficiency
透镜分辨(能)力 lens efficiency; lens factor
透镜分辨周数 lens factor
透镜分光计 lens spectrometer
透镜盖 lens cap; lens cover
透镜高温计 lens pyrometer
透镜公式 lens formula
透镜共焦线 confocal lens line
透镜鼓 lens drum
透镜光 lens light
透镜光导管 lens light guide; lens wavebeam guide
透镜光焦度 lens strength
透镜光阑 lens stop
透镜光阑的伺服控制 servo control of lens iris
透镜光阑孔 lens diaphragm opening
透镜光圈 lens stop
透镜光栅 lenticulation; transmission grating
透镜光栅膜制造方法 lenticulation
透镜光学能力 lens power
透镜光闸 lens shutter
透镜光轴 lens axis
透镜柜 lens board; lens carrier; lens holder
透镜厚度 lens thickness
透镜护圈簧 lens retainer spring
透镜环 lens ring
透镜回转头 lens turret
透镜会聚 lens convergence
透镜会聚面 lens convergence plane
透镜畸变 lens distortion
透镜畸变误差 error of lens distortion
透镜加工 lens processing
透镜架 lens carrier; lens holder
透镜检查仪 lens meter
透镜检验 lens test
透镜检验器 lens tester
透镜检验仪 lens tester
透镜胶合 cementation of lenses
透镜焦点 lens focus
透镜焦度 power of lens
透镜焦距 focal length of lens; lens focal length
透镜焦强 power of lens
透镜角 lens angle
透镜校正角 lens-corrected horn
透镜接头 lens mount adapter
透镜径向畸变 radial distortion of lens
透镜径向失真 lens radial distortion
透镜镜筒 lens barrel
透镜矩阵 lens matrix
透镜聚焦 lens focus
透镜可变光阑 lens iris
透镜可随时调节的 pancreatic
透镜孔 lens opening
透镜孔道 lens channel
透镜孔径 lens aperture
透镜孔径比 ratio of lens aperture
透镜孔径数 lens aperture number
透镜控制 lens control

透镜块 block of lenses
透镜框 lens attachment
透镜框架 lens mount
透镜棱栅 grens
透镜率 speed of lens
透镜轮扫掠器 lens drum scanner
透镜罗盘 lens compass
透镜毛坯 lens blank
透镜面 lens face;lens plane;lens surface
透镜磨光 lens grinding
透镜偶极子 lens doublet
透镜耦合 lens coupling
透镜耦合瞄准器 lens-coupled viewfinder
透镜盘 lens disc[disk]
透镜配曲调整 lens bending
透镜片 lens plate
透镜剖面 lens profile
透镜切向畸变 tangential distortion of lens
透镜曲度 meniscus curvature
透镜曲率 lens curvature
透镜圈 lens ring
透镜扫描盘 lens scanning disk[disc]
透镜深度规 lens watch
透镜式立体镜 lens stereoscope
透镜式梁桥 lenticular girder bridge
透镜式色灯信号机【铁】multilens colo(u)r light signal;multiple light signal
透镜式信号(机) lens-type signal;multilens signal
透镜视场 lens field
透镜视场照明 lens field illumination
透镜视角 lens coverage
透镜适配圈 lens-adapter ring
透镜水平角 horizontal lens angle
透镜速度 lens speed;lens velocity
透镜速率 lens rate;lens speed
透镜损耗 lens loss
透镜锁光圈 lens iris
透镜体 lenticle;lens【地】
透镜体层 lenticular bed
透镜体油藏 oil lens
透镜天线 antenna lens;lens antenna
透镜筒 lens barrel;lens drum;lens mount
透镜透光率 lens transmission
透镜透光因数 lens transmission factor
透镜涂层 lens coating
透镜望远镜 lens telescope;meniscus telescope
透镜系统 lens combination;lens system
透镜响应 lens response
透镜像差 lens aberration
透镜像场 lens field
透镜形摆锤 lenticular bob
透镜形的 lens shaped
透镜形构造 phacoidal structure
透镜形零件 oyster
透镜形纹孔口 lenticular pit aperture
透镜型圈闭 lens-type trap
透镜旋转台 lens turret
透镜选择 lens selection
透镜押板 lens stopper
透镜因数 <透镜光学分辨能力> lens factor
透镜有效孔径 effective lens aperture
透镜域 lenticular domain
透镜元件 lens element
透镜圆片 lens panel
透镜罩 lens cap;lens shade
透镜遮光片 gobo;lens screen
透镜遮光罩 lens hood;zip pouch
透镜折射率计 phacometer
透镜阵列成像 lens array imaging
透镜直径 lens diameter
透镜中继系统 relay-lens system
透镜中心 lens center[centre]
透镜中心面 central lens plane
透镜轴 axis of lens
透镜状 lensoid
透镜状层理 lenticular bedding
透镜状层理构造 lenticular bedding structure
透镜状沉积 lenticular deposit
透镜状沉积砂 lenticular deposit sand;shoestring sand
透镜状成层矿床 lens-shaped stratified deposit
透镜状的 lens shaped;lenticular;lentoid;lentiform
透镜状地层 lensing
透镜状地震岩相单元 lens seismic facies unit
透镜状构造 lenticular structure
透镜状河床 lenticular bed
透镜状假流纹构造 lenticular pseudofluidal structure
透镜状交错层理构造 lenticular cross-bedding structure
透镜状结构 lentoid
透镜状空隙 lenticular void
透镜状矿体 lenticular ore body;lentiform ore body
透镜状马氏体 lenticular martensite
透镜状丘 lenticular hill
透镜组 battery of lens;lens assembly;set of lenses
透镜组合 combination of lens
透镜组聚光能力测定计 phakometer;phokoscope
透空花格 brandishing;brattishing;bretisement
透空花格窗 openwork tracery
透空花格装饰 bratticing
透空花山头 openwork gablet
透空花纹的顶部装饰 <哥特式的> brandishing;bretisement
透空结构 open construction;open structure;openwork
透空楼梯 <仅有踏板的> skeleton steps
透空墙 pierced wall;perforated wall;screen wall <有规则孔洞的>
透空山墙 openwork gable
透空式地板 open deck;suspended floor
透空式防波堤 curtain breakwater;open-type breakwater;permeable breakwater
透空式防波栅 perforated wave screen
透空式混凝土管柱码头 open-type wharf on concrete cylinder
透空式结构 cancelled structure;open structure;openwork;suspended structure
透空式楼板 suspended floor;open floor
透空式码头 open piling wharf;open jetty;open quay;open (-type) wharf;dock of open construction
透空式码头建筑物建筑界限线 pierhead line
透空式码头面板 suspended deck
透空式面板结构 suspended deck structure
透空式木地板 open timbered floor
透空式木结构 open timbering
透空式桥面板 suspended deck
透空式水工建筑物 open hydraulic structure
透空式顺岸码头 false quay;open wharf
透空式顺岸栈桥码头 open-type wharf
透空式突堤码头 open(type) pier;open jetty
透空式栈桥码头 open jetty
透空式桩基结构 open-piled structure
透空式桩基突堤 open pile jetty
透空式桩基突堤码头 open-pile pier
透空小山墙 openwork gablet
透空圆花窗 openwork rosette
透空桩基 open piling
透空桩基式 open-piled type
透空桩基式码头 open-piled type quay
透空桩靠船建筑物 open-type berthing structure
透空桩排 open piling
透空桩水工建筑物界线 pierhead line
透空桩突堤 openwork jetty
透孔 through hole
透孔法 method of penetration
透孔塔顶 lace-like spire
透孔性 conductivity of an aperture
透孔织物 openworking
透孔制品 openworking
透孔装饰 openwork
透锂长石 castorite;petalite
透锂铝石 bikitaite
透磷钙石 brushite
透露 reveal
透炉 poke
透铝英石 hyaloallophane
透绿帘石 oisanite
透绿泥石 sheridanite
透明 lucency
透明板 dispositive plate;surveying panel;translucent panel
透明保护敷料 impermephane
透明保护涂料 impermephane
透明杯脂 transparency cup grease;transparent cup grease
透明背心 hi-visibility vest
透明比率 transparency ratio
透明边墙 transparent side wall
透明变性 hyaline degeneration;hyalinosis
透明冰 clear ice;glaze ice;transparency ice;transparent ice
透明丙烯酸板 acrylic lens
透明玻璃 clear glass;transparency glass;transparent glass
透明玻璃板 transparency glass plate
透明玻璃(灯)泡 clear bulb
透明玻璃块 glass-crete
透明玻璃料 transparent frit
透明玻璃门 transparent glass door
透明玻璃陶瓷 transparent glass ceramics
透明玻璃纤维增强塑料 transparent glass fibre[fiber] reinforced plastics
透明玻璃纸面 cellophane cover
透明箔 astrafoil
透明薄冰 black ice
透明薄玻璃 clear sheet glass
透明薄膜 clear film;transparency film;transparent film;transparent membrane;transparent overlay
透明薄膜背面印刷 bottom printing
透明不透明矿物 transopaque mineral
透明材料 apparent material;craftone;glassy material;transparent material;vitreous material;craftone <印注记符号的>
透明舱 transparency cabin
透明舱盖 transparency canopy
透明层 diaphanotheca;glassy layer;hyaline;hyaline layer;stratum lucidum;transparency layer;transparent layer
透明场致发光电池 transparency electro-luminescent cell
透明衬板 see-through panel;vision panel
透明传导电极 transparency conducting electrode
透明传输 transparent transmission
透明窗 transparency window;transparent window
透明窗玻璃 clear window glass
透明瓷器 vitreous ware
透明瓷釉 transparent enamel
透明瓷砖 glass tile
透明磁性玻璃 transparent magnetic glass
透明磁釉 transparent enamel
透明带 transparency zone;transparent tape
透明带溶素 zonallysin
透明蛋白 hyaline
透明导电(薄)膜 nesa coating;transparency conductive film
透明导电层 transparency conducting layer
透明导电电极 transparency conductive electrode
透明导电涂料 transparent conductive coating
透明的 clear;crystalline;crystalloid;diaphanous;glass clear;glassy;hyaline;hyaloid;lucid;see through;translucent;transparent;vitreous
透明的窗 see-through window
透明的建筑物 see-through building
透明的照相正片 diapositive
透明灯(泡) clear lamp;transparent lamp
透明底片 dianegative;transparency negative;transparent negative
透明地面密封剂 clear floor sealer
透明点状压花 clear stipple embossed
透明电极 transparency electrode
透明电介质层 transparency dielectric(al) layer
透明电文方式 transparent text mode
透明冻胶 jelly
透明度 clarify;clarity;coefficient of transparency;degree of transparency;diaphaneity;limpidity;pellucidity;pellucidness;see-through clarity;transmittance [transmittancy];transparency clarity;transparency [transparence](degree)
透明度板 Secchi's disc
透明度表 potometer
透明度测定法 diaphanometry
透明度测定器 transparency meter
透明度测定仪 opacimeter;transparency meter
透明度级别 degree of diaphaneity
透明度极限 Secchi's depth
透明度计 diaphanometer;potometer;transparency meter
透明度盘 transparent scale
透明度仪 diaphanometer;potometer
透明多晶陶瓷 transparent polycrystalline ceramics
透明多晶氧化铝陶瓷 transparent polycrystalline alumina ceramics
透明反光镜 transparency reflector
透明反应器 transparency reactor;transparent reactor
透明方案 transparency scheme
透明方解石 Iceland spar
透明防雨聚合物涂层 rain-repellent polymer coatings
透明分度盘 transparency scale
透明分解腐殖物质 translucent humic matter
透明浮法玻璃 clear float glass
透明负片 negative transparency;transparent negative
透明复写纸 transparency manifold
透明盖 transparency cover
透明革 transparency leather
透明隔断 transparent partition
透明拱顶防弹罩 bubbletop

T

透明观察孔线 transparent window
透明冠 hyaline cap
透明管 hyaline tube;hyaloid canal
透明管型 hyaline cast
透明光电阴极 transparency photocathode;transparent photocathode
透明光度计 vitreousness photometer
透明光滑的 hyaline smooth
透明焊料玻璃 vitreous solder glass
透明核模型 transparency-nucleus model
透明厚玻璃 transparent plate glass
透明琥珀 clear amber
透明花纹 <瓷器中的> grain of rice
透明化 hyalinize; transparentizing; vitrification;vitrify
透明化作用 hyalinization
透明黄色 <亮黄色> glassy yellow
透明灰度梯度 transparency-grey-scale
透明绘图纸 template
透明基板 transparency carrier;transparent carrier
透明集料 glassy aggregate
透明剂 clarifier
透明胶 Scotch glue
透明胶表面涂层 gelatin(e) supercoat
透明胶带 cellulose tape;Scotch tape
透明胶片 transparent film
透明胶性漆 transparent adhesive varnish
透明胶纸 sticker
透明介质 transparent medium
透明介质层 transparent dielectric(al) layer
透明晶体 transparent crystal
透明晶质 hyalocrystalline
透明镜 transparent mirror
透明聚甲基丙烯酸甲酯 transpex
透明绝缘敷层 transparency insulating coating
透明刻度盘 transparent scale
透明亮漆 clear lacquer; transparency lacquer;transparent lacquer
透明量图器 roamer
透明滤光片 clear glass filter; transparent filter
透明媒质 transparency medium
透明镁铝尖晶石陶瓷 transparent magnesium alumin(i)um spinel ceramics
透明蒙皮漆 clear dope
透明面层 glaze coat
透明面漆 clear finish
透明描图纸 transparent tracing paper
透明膜 hyaline membrane; transparent coating;transparent film
透明内视图 phantom view
透明黏[粘]合剂 clear binder
透明黏[粘]合媒质 clear medium
透明喷漆 transparent lacquer
透明膨胀 transparency swelling
透明片 clear foil;overlay
透明平纹薄织物 mousseline
透明屏 transparency screen
透明漆 clear lacquer; transparency varnish; transparent lacquer; zapon lacquer
透明漆膜 clear felling;clear film
透明嵌金属钢玻璃 clear wire glass
透明嵌丝玻璃 clear wire glass
透明切削油 transparency cutting oil
透明/清 transparent/clear
透明清漆 clear varnish; transparent varnish
透明清漆介质 clear varnish medium; clear varnish vehicle
透明清漆体系 clear varnish system
透明清漆罩面 clear varnish finish
透明区 transparency area
透明全息图 transparent hologram
透明染料 transparency dye
透明熔块 clear frit
透明柔性弹簧门 transparent flexible swing door
透明乳液 transparency emulsion
透明朊的 hyaline
透明软管 transparent hose
透明软片 transparency[transparence]
透明赛璐珞 transparency celluloid; transparent celluloid
透明色料 transparency colo(u)r; transparent colo(u)r
透明色漆 colo(u)red varnish
透明色釉 transparency enamel
透明蛇皮管 transparent hose
透明生物 transparent organism
透明石膏 fraueneis;selenite
透明石膏水泥 <生石灰加5%石膏> Scott's cement
透明石膏质水泥 selenitic cement
透明石英 vitreous silica
透明石英玻璃 transparency silica glass; transparency vitreous silica; transparent fused silica; transparent vitreous silica;vitreous fused silica
透明树胶 copal gum
透明水晶 transparent crystal
透明水印 water mark
透明顺磁材料 transparency paramagnetic substance; transparent paramagnetic substance
透明丝绒 transparency velvet
透明素 hyaline;hyaline hyalin
透明速干漆 transparent quick(ly)-dried varnish
透明塑胶 perspex
透明塑料 lucite; perspex (sheet); transparent plastics
透明塑料板 clear plastic sheet; transparent plastic sheet
透明塑料管 lucite pipe; transparent plastic pipe
透明塑料建筑构件 transparent plastic building unit
透明塑料模型 transparency plastic model;transparent plastic model
透明塑料膜 clear plastic sheet
透明塑料片 laminated plastic sheet; transparent plastics; transparent plastic sheet
透明塑料瓦 transparent plastic tile
透明塑料屋面覆盖层 transparent plastic roof cover(ing)
透明酸蚀刻 bright etching;clear etching
透明酸蚀刻槽 clear etching bath
透明碎屑 transparent attritus
透明搪瓷 vitreous enamel
透明陶瓷 crystalline ceramics; light-transmitting ceramics; transparency ceramics;transparent ceramics
透明陶瓷光釉面 clear ceramic glaze
透明体 transparent body; vitreous body
透明铁电陶 transparent ferro-electric(al) ceramics
透明铁电陶瓷 transparency ferro-electric(al) ceramics
透明通道 transparent passage(way)
透明图样纸 template;templet
透明涂层 clear coat(ing); transparent coating
透明涂料 clear dope; transparent coating
透明涂料层 optic(al) coating
透明涂面【建】clear coating; transparent finish
透明涂膜 clear coating
透明瓦 vitreous tile
透明微晶玻璃 transparent glass ceramics
透明温室玻璃 transparent horticultural glass
透明物 hyaline
透明物体 transparency[transparence]
透明物质 transparency material; transparent material
透明系数 coefficient of transparency
透明纤维素 transparency cellulose; transparent cellulose
透明显示窗 transparency display window
透明相片正片 diapositive
透明消失 devitrification
透明小容器 cuvette
透明型放射同位素测量计 transmission-type radio isotope ga(u)ge
透明性 diaphaneity; pellucidity; pellucidness; transmissivity; transmittance[transmittancy]; transparency [transparence]
透明颜料 bright dye; transparent colo(u)r; transparent dye; transparent pigment
透明羊皮纸 transparent parchment
透明氧化铝陶瓷 transparent alumina ceramics
透明氧化铍陶瓷 transparent beryllia ceramics
透明氧化钍陶瓷 transparent thoria ceramics
透明氧化物陶瓷 transparent oxide ceramics
透明氧化钇陶瓷 transparent yttria ceramics
透明液体 transparency liquid
透明液体比色计 absorptiometer
透明油 bright stock;clean oil
透明油墨 diaphanous ink; transparent ink
透明油漆 celluloid paint
透明有光塑料涂料 gloss clear plastic coating
透明釉 clear ceramic glaze; clear glaze; glass enamel; transparency glaze;transparent glaze
透明釉面砖 clear ceramic glaze tile
透明雨伞 bubbletop
透明原稿 transparent original
透明运行方式 transparency mode of operation
透明憎水涂料 clear water repellent coating
透明遮阳棚 transparent shade
透明真空箱 vitreous evacuated container
透明正片 diapositive; diapositive film; transparent positive
透明正片晒像机 diapositive printer
透明纸 cellophane; cellophane paper; ice paper; light-tracing paper; tracing cloth; tracing paper; transparent paper;transparent sheet
透明纸法 tracing paper method
透明纸后方交会法 resection by tracing paper
透明指针片 cursor
透明质 enchylema; hyalomitome; hyalotome;paramitome
透明质浆 hyaloplasm
透明质酸 hyaluronidase
透明质酸盐 hyalurate;hyaluronate
透明注记 lettering foil; slice-lettering;type-stick
透明注记剪贴字 trans-adhesive map type
透明状态 vitreous state
透明桌 light table
透明坐标纸 screen tone
透膜 permeable membrane
透墨 strike-through
透硼砂 ezcurrite
透平 turbine;turbo
透平泵 roturbo; turbine pump; turbopump
透平出气壳 gas outlet casing
透平电动推进 turbo-electric(al) propellant
透平动力 turbine power;turbo-power
透平端 turbine side
透平发电机 turbine generator; turbo-set
透平工作轮 turbine wheel
透平功率输出 power output of turbine
透平后轴承 rear turbine bearing
透平机 turbine;turbine engine
透平机房 turbine hall
透平机桨叶 turbine blade
透平机外壳 turbine casing
透平机械试验台 turbomachine test stand
透平交流发电机 turbo-alternator
透平交流发电机组 turbo-alternator set
透平进气壳 gas inlet casing
透平静止部件 stationary turbine components
透平壳 turbine casing; turbine cylinder
透平喷射器 gasifier nozzle diaphragm
透平膨胀机 expansion turbine; turbo-expander
透平汽车 turbine-powered automobile
透平汽缸 turbine casing
透平式风机 turbine fan;turbo-blower
透平式搅拌器 turbine-type agitator
透平式模具用砂轮机 turbine-powered die grinder
透平式膨胀机 turboexpander
透平式气马达 turbine air motor
透平式清管器 turbine go-devil; turbine pig;turbine scraper
透平式选粉机 turbo-classifier; turbo-separator
透平室 turbine pit
透平旋转部件 running turbine components
透平循环压缩机 circulating turbo-compressor
透平叶 turbine blade
透平叶轮 turbo-wheel
透平叶片 segment of blading; turbine blade
透平油 turbine oil
透平增压 turbocharge
透平增压柴油机 turbocharged diesel engine
透平增压器 turbocharger
透平增压装置 charging turbine
透平制冷机 turbo-refrigerator
透平制造厂 turbine builder
透气 aeration; air vent; pervious to air;ventilation;venting
透气百叶板 louvered air outlet
透气百叶窗 louvered air outlet
透气板 aerating panel
透气玻璃 gas permeable glass
透气薄膜 breather film; breather membrane
透气材料 air pervious material
透气舱 non-pressurized interior
透气的 air permeable; poromeric; vapo(u)r-permeable
透气度 air permeability; degree of aeration;unfilled porosity
透气度测定仪 densometer; permometer
透气度试验仪 air-permeability apparatus

透气额定值 permeability rating of air
透气阀 vent trap
透气法 airflow method
透气法比表面积测定仪 air-permeability specific surface apparatus;permeability apparatus
透气法比表面积值 air-permeability value
透气法细度试验仪 air-permeability fineness tester
透气方式 method of vent
透气防潮纸 breather paper
透气盖 ventilating cover
透气管 permeability cell;vapo(u)r pipe;vent pipe
透气井 breathing well
透气孔 breather hole;loophole;riser vent
透气亮子 night vent;ventlight;vent sash
透气率 air permeability;air permeance;gas permeability;permeability rating of air;venting quality
透气率测定仪 permeability apparatus
透气面层 breathable coating;permeable coating
透气膜 breather membrane;gas-permeable membrane
透气模板 vented form
透气漆膜 breather film
透气式燃料元件 vented fuel element
透气室 permeability cell
透气竖管 vent stack
透气屋顶 breather roof
透气系数 coefficient of air permeability;coefficient of transparency
透气系统 gas venting system;venting system
透气细度仪控制磨机 mill control by permeability fineness meter
透气性 air conductivity;air permeability;breathing property;gas permeability;permeability;permeability for gas;permeability to air;permeability to gas;perviousness to air;vapo(u)r permeability;venting quality
透气性材料 gas pervious material
透气性测定仪 permeability meter;permeater;permometer;respirometer
透气性测量 permeability measurement
透气性法 < 测量比表面法 > air-permeability method
透气性化纤织物 poromeric material
透气性试验 air-permeability test
透气性试验法 permeability test method
透气性试样室 permeability cell
透气性涂层 air permeable coating;breathing coating
透气仪 air-permeability apparatus
透气圆木 boule
透气主井 main shaft
透气砖 air brick
透气装置 air-breather
透热包壳 diathermous envelope
透热玻璃 heat transmitting glass
透热的 diabatic;diathermal;diathermanous;diathermic
透热电凝法 diathermic coagulation;diathermocoagulation
透热法 diathermy;transthermy
透热辐射的 diathermanous
透热辐射性 diathermancy;diathermaneity
透热机 diathermy machine;dynatherm
透热器 diathermy apparatus
透热深度 heat penetration
透热体 diathermanous body
透热性 breathability;diathermancy;diathermaneity;heat permeability;permeability of heat;permeability to heat
透日光 photic
透入 imbibe;imbibition
透入度试验 penetration test
透入剪切 penetrative shear
透入络合物 penetration complex
透入深度 depth of penetration;penetration depth;penetration of current;skin depth
透入湿气 moisture entrance
透入式封面料 < 混凝土防冻用的 > penetrating type sealant
透入性 penetration
透入性线理 penetrative lineation
透入性要素 penetrative element
透入性组构 penetrative fabric
透入值 penetration number
透沙丁坝 permeable groyne;permeable spur dike[dyke]
透闪蛇纹片岩 tremolite serpentine schist
透闪石 hopfnerite;raphilite;sebesite;tremolite
透闪石大理岩 tremolite marble
透闪石角岩 tremolite hornfels
透闪石棉 Italian asbestos
透闪石片岩 tremolite schist
透闪石石棉 tremolite asbestos
透闪石石英岩 tremolite quartzite
透烧砖 body brick
透蛇纹石 tangiwaite
透射 homology;penetrance;transmission;transmitted intensity
透射靶 transmission target
透射倍增极 transmission dynode
透射比(率) transmission;transmittance[transmittancy];transmittivity
透射边界 transmitting boundary
透射表 transmittance meter
透射波 transmission wave;transmissive wave;transmitted wave
透射波法 transmission wave exploration method
透射测定法 transmission measurement
透射测厚仪 penetron
透射测量法 transmissometry
透射传输 through transmission
透射单位 transmission unit
透射的伦琴射线 radiolucent
透射电镜法 transmission electron microscope method
透射电子技术 transmission electron microscopy
透射电子图像 transmission electron image
透射电子显微镜 scanning electron microscope;transmission electron microscope
透射电子显微镜检查法 transmission electron microscopy
透射电子显微术 transmission electron microscopy
透射电子衍射 transmission electron diffraction
透射度 transparency[transparence]
透射对流 penetrative convection
透射法 transmission-beam method;transmittance method
透射高能电子衍射 transmission high energy electron diffraction
透射光 transmission light;transmitted light
透射光电阴极 transmission photocathode
透射光干涉显微镜 transmission light interference microscope
透射光观察工作 transmitted work
透射光密度计 transmission densitometer
透射光盘 transmissive optic(al) disc [disk]
透射光谱 transmitted spectrum
透射光谱法 transmission spectrometry
透射光强度 transmitted light intensity
透射光栅 transmission grating
透射光通量 penetrating luminous flux
透射机理 transmission mechanism
透射级 transmission level
透射计 < 利用透光程度测定水浊度的计量器 > transmissometer
透射截面 transmission cross section
透射矩阵法 transmission matrix method
透射孔 beam orifice
透射滤波器 transmission filter
透射率 light transmittance;transmissivity;transmittance[transmittancy]
透射率检验仪 transmissivity tester
透射密度 transmitting density
透射膜 transmission film
透射能 transmitted energy
透射平面 plane of homology
透射谱 transmission spectrum
透射强度 transmitted intensity
透射区 region of transmission;transmission region
透射曲线 transmission curve
透射色 transmission colo(u)r
透射深度 penetration;transmission depth
透射声 transmitted sound
透射实验 transmission experiment
透射式电子显微镜 transmission type electron microscope
透射式放射同位素测量仪 transmission type radio isotope ga(u)ge
透射式光弹性仪 transmission polariscope
透射式密度探测器 transmission density probe
透射式偏光镜 transmission polariscope
透射束 transmitted beam
透射损失 transmission loss
透射梯 transmission echelon
透射调节器 osmoregulator
透射投影仪 stereopticon
透射系数 transmission coefficient;transmission factor;transmissivity;transmittance[transmittancy]
透射线 radio parent
透射形貌学 transmission topography
透射衍射法 transmission diffraction
透射因素【物】transmission factor
透射因子 transmission factor
透射荧光镜 transmission fluoroscope
透射映画器 diascope
透射照明 transmitted lighting
透射照明法 transparant illumination
透射指数 transmissive exponent
透射中心 center[centre] of homology
透射轴 axis of homology
透射纵波垂直时距曲线 vertical hodograph of transmitted longitudinal wave
透砷铅矿 schultenite
透砷铅石 schultenite
透声 entrant sound;sound transparent
透声壁 soft wall
透声界面 sound-soft termination
透声率 acoustic(al) permeability
透声穹室 kuppe
透声系数 acoustic(al) transmission coefficient;acoustic(al) transmission factor;acoustic(al) transmissivity;sound transmission coefficient
透声压容器 acoustically transparent vessel
透湿 drench;pervious to moisture
透湿度 penetration dampness;penetration of dampness;water vapo(u)r permeability
透湿量 vapo(u)r transfer rate
透湿气性 green permeability;moisture vapo(u)r permeability
透湿试验 moisture test(ing)
透湿系数 coefficient of permeance;moisture permeance;vapo(u)r permeance
透湿现象 moisture penetration
透湿性 moisture permeability;vapo(u)r permeability;vapo(u)r transmission;water vapo(u)r permeability
透湿阻力 vapo(u)r resistance
透石膏 < 无色透明结晶石膏 > selenite;gypsum spar;specular gypsum
透石膏作促凝剂的水泥 selenitic cement
透视 fluoroscopy;penetrate;perspective view(ing);perspectivity;radiographic(al) inspection;show-through
透视暗物图像 perspective dark field imaging
透视变换 perspective collineation;perspective transformation
透视表示法 perspective representation
透视草图 perspective sketch
透视差 collimation error;error of collimation
透视窗板 vision panel
透视错觉 perspective illusion
透视的 perspective
透视的建筑图 perspective construction
透视灯 tankoscope
透视地平面 ground plane
透视点 perspective center[centre];perspective point
透视断块图 fence diagram
透视对应 perspective correspondence
透视法 perspective representation
透视方位投影 perspective azimuthal projection
透视符号 perspective symbol;stick-up symbol
透视格网 perspective grid
透视格网法 perspective grid method
透视光栅 perspective grating;transmission grating
透视航空照片 perspective picture
透视画 < 包含透视中心的任何平面 > perspective plane;diorama;perspective picture;quadratura
透视画法 perspective;scenography
透视画仪器 perspectograph;perspectometer
透视绘画器 perspectograph
透视绘图仪 perspective drawing instrument
透视畸变 perspective distortion
透视几何学 perspective geometry
透视近点角 perspective anomaly
透视镜 photoscope
透视镜荧光屏 photoscope
透视纠正仪 perspectograph
透视孔 sight vane
透视立体符号 perspective stereo-symbol;perspective volume symbol
透视面 perspective surface
透视模型 perspective model
透视平面 perspective plane;plane of perspectivity
透视摄影 metric(al) photograph

T

透视石 dioptase; dioptasite; emeraudine
透视收缩 foreshortening
透视速写 sketch in perspective
透视缩图 foreshortening
透视投影 perspective projection
透视投影地图 perspective chart
透视投影格网 perspective grid
透视投影器 panoramic projector
透视图 artist's impression; broken-open view; diorama; expanded view; panorama sketch; panoramic picture; panoramic sketch; perspective chart; perspective drawing; perspective plan(e); perspective view(ing); phantom view; rendering; scenograph; skeleton view; vista
透视图变换 perspective viewing transformation
透视图法 perspective method; scenography
透视图设计 scenographic design
透视图像 perspective picture
透视图形 perspective figure
透视网 perspective net
透视网格 perspective grid
透视位置 perspective position
透视线 line of collimation; perspective ray
透视相片 metric(al) photograph
透视性 perspectivity
透视旋转定律 axiom of perspective rotation; rotation axiom of perspective
透视渲染 perspective rendering
透视映射 perspective mapping
透视制图 perspectograph
透视中心 center[centre] of perspective; center[centre] of vision; center[centre] of perspectivity; perspective center[centre]
透视轴 axis of homology; axis of perspective; axis of perspectivity; map parallel; perspective axis
透视主点 center[centre] of perspective; center[centre] of vision
透水 water penetration; water percolation
透水坝 filter dam; permeable dam; permeable dyke[dike]; porous dam
透水坝壳 <土石坝的> pervious shell; permeable shell
透水板 permeable plate; permeable slab; pervious plate; pervious slab; porous disc[disk]; porous plate; porous slab
透水材料 blinding material; permeable material; pervious material; seepage material; seepy material; water pervious materials
透水层 filter layer; permeable bed; permeable course; permeable foundation; permeable ground; permeable layer; pervious bed; pervious course; pervious stratum; pervious to water; water-bed; drainage course
透水常数 permeable constant; pervious constant
透水床 pervious bed
透水带 pervious zone
透水导治建筑物 permeable training structure; pervious training structure
透水的 non-watertight; permeable; permeable to water; pervial; pervious; seepy
透水堤(坝) permeable dyke[dike]; pervious dyke[dike]; porous dyke[dike]
透水底层 pervious bed
透水底土层 permeable subsoil; pervious subsoil
透水地层 permeable stratum; pervious stratum[复 strata]
透水地带 pervious zone
透水地基 pervious foundation
透水地基上的土坝 earth dam on pervious foundation
透水地面 filter floor; pervious surface; phreatic surface
透水垫层 pervious blanket; pervious cushion
透水丁坝 permeable groin; permeable groyne; permeable spur; pervious groin; pervious spur; pile and waling groin; pile and waling groynes
透水度 perviousness
透水度试验 perviousness test
透水防波堤 permeable groyne; pervious breakwater; porous breakwater
透水封闭层 permeable confining bed
透水封水层 permeable confining bed
透水缝 infiltration slit; permeable joint
透水覆盖层 permeable overburden; pervious overburden
透水盖层 blanket course
透水格笼堤 dike[dyke]; permeable crib dyke[dike]; pervious crib; pervious crib dyke[dike]
透水格笼堤透空堤 permeable dyke[dike]
透水隔层 permeable confining bed
透水工程 permeable works
透水构造 permeable structure
透水管测压计 porous tube piezometer
透水河床 permeable riverbed
透水护岸 pervious revetment
透水护坦 pervious blanket; seepage apron
透水混凝土 pervious concrete
透水基层 permeable base
透水基床 permeable bed; pervious bed; porous bed
透水基底 permeable base
透水基土 pervious subsoil
透水建筑物 permeable structure; permeable works; pervious structure; pervious works; porous structure
透水结构 open structure; permeable structure; pervious structure; porous structure
透水井 filter well; penetrating well
透水壳层 permeable shell; pervious shell
透水壳体 permeable shell; pervious shell
透水栏栅 permeable stake; pervious fence
透水沥青碎石 pervious macadam
透水沥青纤维砂胶沉排 fibrous open stone asphalt mattress
透水路堤 permeable embankment
透水路面 <道路的> porous pavement
透水面层 porous pavement
透水磨耗层 pervious wearing course
透水模板 vented form
透水木笼堤 dike[dyke]; permeable crib dyke[dike]; permeable dyke[dike]; pervious crib; pervious crib dyke[dike]
透水排桩堤 pile hurdle dyke[dike]
透水排桩丁坝 pile hurdle dyke[dike]
透水平面 perspective plane
透水铺盖 pervious blanket
透水铺盖层 permeable blanket; pervious blanket
透水汽性 permeability to water vapo(u)r
透水墙 pervious wall
透水区 pervious zone
透水砂砾石层 pervious sand gravel
透水砂岩 filter sandstone
透水石 perforated stone; porous disc[disk]; porous stone
透水石渠 rubble drain
透水式导流堤 permeable dyke[dike]
透水式防波堤 permeable breakwater
透水式护岸 permeable revetment; pervious revetment
透水式沥青碎石路面 pervious bitumen macadam surface
透水式马克当路面 pervious bitumen macadam surface
透水试验机 permeability testing machine
透水陶管 land tile
透水填充体 porous backfill(ing)
透水填料 <排水暗管周围的> blinding material
透水挑水坝 permeable dyke[dike]
透水通道 permeable passage
透水土 pervious soil
透水土层 permeable ground; pervious ground
透水土壤 permeable ground; permeable soil; pervious ground; pervious soil
透水外堤 intermittent external wall
透水围岩 permeable ground; pervious ground
透水污水池 pervious cesspool
透水物料 seepy material
透水系数 coefficient of water permeability; permeability to water; permeation coefficient
透水下层土 pervious subsoil
透水线 line of percolation; phreatic line
透水消波器 permeable wave absorber
透水消波设备 permeable wave absorber
透水性 hydraulic conductivity; hydraulic permeability; permeability to water; perviousness; water permeability
透水性测定仪 infiltrometer
透水性防波堤 porous breakwater
透水性黑色碎石(路) pervious coated macadam
透水性回填材料 pervious backfill material
透水性沥青(路面) pervious asphalt
透水性试验 perviousness test; water permeability test
透水性土壤 pervious soil; permeable soil
透水性微弱的含水层 aquiclude
透水性系数 coefficient of permeability to water
透水性岩层 permeable rock; pervious rock
透水性岩石 permeable rock; pervious rock
透水性要求 permeability requirement
透水岩层 permeable stratum; pervious stratum
透水岩床 permeable bed; porous bed
透水岩类 permeable rocks; pervious rocks; porous rocks
透水岩石 permeable rock; pervious rock; porous rock
透水岩芯 weeping core
透水圆板 porous disc[disk]
透水圆板透水板 <试验土质的> porous disc[disk]
透水栅(栏) open fence; permeable stake; pervious fence
透水折流坝 permeable groyne
透水桩坝 permeable pile dyke[dike]; pile hurdle dyke[dike]
透水桩堤 permeable pile dyke[dike]; pile hurdle dyke[dike]
透水桩基丁坝 permeable pile dike[dyke]
透榫 through tenon
透析【化】dialyse; dialysis
透析结晶作用 percrystallization
透析器 dialyzer
透析液 dialysate
透析蒸馏作用 perdistillation
透写 tracing
透写台 glass tracing table; light box; light table; tracing stand
透写图 plate sketch; selection overlap
透写桌 tracing table
透性膜 permeable membrane
透性漆 <潮气等可以透过的油漆> porous paint
透性障 permeability barrier
透眼 through hole
透移率 transport ratio
透音系数 acoustic(al) transmission coefficient
透印 offsetting; show-through; telegraphing
透印蹭脏 offset
透影 shadow-mark
透油计 <测量纸的> vancometer
透照法 transillumination
透照镜 diaphanoscope
透照术 diaphanoscopy
透蒸气材料 vapo(u)r pervious material
透蒸汽的 permeable to steam
透支 advance by overdraft; call money; make an overdraft; overcheck; overdraft; overdraught; overdraw(ing); take-out an overdraft
透支贷款 call loan
透支户 overdraft account; overdrawing account
透支利息 overdraft interest
透支限额 credit line; limit of overdrawn account
透支账户 on call account; overdraft account
透支制 overdraft system
透紫外线玻璃 apollo glass; Calowlex glass; glass for ultraviolet rays; Lindemann's glass; quartzite glass; quartzolite glass; Sunalux glass; ultraviolet ray transmitting glass; uviol glass; vitaglass
透字 show-through
透字签名 perforated signature

凸岸 convex bank; convex shore; inner bank

凸岸边滩 accretionary meander point; convex bar; point bar; meander bar
凸岸边滩坡度 point bar slope
凸岸边滩斜坡 point bar slope
凸岸边滩淤积 point bar sedimentation
凸岸沙洲 convex bar
凸岸沙嘴 point bar
凸岸沙嘴沉积 point bar deposit
凸岸水流 convex flow
凸岸一侧 convex side
凸凹变形阻力 buckling resistance
凸凹的 convexo-concave
凸凹地形 knob-and-basin topography; knob-and-kettle topography

凸凹分型面 matched parting
凸凹接头 match joint(ing)
凸凹面 male and female face
凸凹模 punch die
凸凹砌筑砖工 skintled brickwork
凸凹榫 groove and tongue
凸凹榫接 rabbet
凸凹榫接合 tenon-and-mortise joint; tenon-and-slot mortise; tongue-and-groove joint
凸凹榫接头 mortise-and-tenon joint
凸凹套接管 rebated pipe
凸凹透镜 convexo-concave lens; meniscus[复 menisci/meniscuses]
凸凹铣刀 convex and concave milling cutter
凸凹形 convex-concave
凸凹形的 convexo-concave
凸凹形断面 concavo-convex profile
凸凹纸 embossed paper
凸板 embossed plate; scab
凸板打印机 embossed plate print
凸板发电机 salien-pole generator
凸板交流发电机 salient pole alternator
凸版 relief block; relief printing plate
凸版的 anastatic
凸版胶印 dry-relief offset; letter-set
凸版轮转印刷机 rotary letterpress
凸版平台印刷机 cylinder flat bed machine
凸版印刷 anastatic printing; cameo printing; letter press printing; letter press typographic(al) printing; relief printing; surface printing; typographic printing
凸版印刷法 anastatic printing process; gelatin(e)-pad printing
凸版印刷机 relief printing machine; letter press <总称>
凸版印刷纸 letter press paper
凸版油墨 letter press ink
凸版油墨用调墨油 varnish for typographic(al) ink
凸版纸 Indian paper
凸版转印法 anastatic transfer
凸半圆成型铣刀 convex cutter
凸半圆铣刀 convex milling cutter
凸瓣 pawl
凸包 convex closure; convex hull
凸背构架 hog frame
凸壁 spur
凸边 convex edge; convex shore; flange; nosing; unset rim
凸边垫圈 tongued washer
凸边格形模板 depressed panel form
凸边箍 chime hoop
凸边加强件 chime reinforcement
凸边螺帽 collar nut
凸边刨 rabbet plane
凸扁嵌缝 tuck and pat pointing
凸部 banquette
凸部爪 prong
凸槽上轧辊 <组成闭口孔型的> tongue roll
凸侧 convex shore
凸齿 double wedge
凸出 bilge; bulge out; embossment; outshoot; projecture; prominence [prominency]; protrusion; stick-out
凸出比 projection ratio
凸出壁外的窗 oriel window
凸出边缘 <木作的> cock(ed) bead
凸出部 bulging; salience [saliency]; lug; proud; sally; tongue; lip
凸出部分 banquette; bulge; feather; outshot; projecting portion; protruding portion; shoulder; teat; bulge; projection
凸出长度 projecting length
凸出处 crest
凸出带层 projecting belt course
凸出的 abutting; overhung; projecting; prominent; protuberant; salient; convex
凸出的雕带 swelled frieze
凸出的角 salient junction
凸出的立柱 outstanding leg
凸出的石块 popping rock
凸出的形状 belling
凸出的锥曲面 oversailing conoid
凸出顶层 starling coping
凸出顶盖 starling coping
凸出风斗 projecting scoop
凸出高度 convexity
凸出沟槽 projecting groove
凸出钩 projecting hook
凸出焊(接) projection welding
凸出弧 bulge arc
凸(出)角 salient corner
凸出角石 rusticated quoin
凸出来的 proud
凸出连接 salient junction
凸出瞭望窗 oriel window
凸出平面的线脚 bolection; bilection; bolexion mo(u)lding
凸出嵌线 bilection[bolection]
凸出嵌线饰 bolection mo(u)ld(ing); bolexion mo(u)lding
凸出墙 scrotch wall
凸出墙面标志 projecting sign
凸出(墙面的)砖 projecting brick
凸出墙面式标示牌 projection wall sign
凸出墙外的凸窗 window bay
凸出容差 protrusion allowance
凸出伸臂 overhanging
凸出四分之一圆(饰) ovolo
凸出舞台 thrust stage
凸出物 ancon(e); projecture
凸出线脚 bilection mo(u)lding
凸出岩架 ledge rock
凸出叶片 bulged blade
凸出圆饰 bowtel(l)
凸出圆形转角 bullnose corner
凸出折边 male flange
凸处 hump; summit
凸窗 bay window; bow window; compass window; jut window; oriel; projected window; architectural projected window <一种较高级的>
凸窗处座位 bay stall
凸窗高台 half pace
凸窗前空间座位 <礼拜堂中的> bay stall
凸窗前座位 <礼拜堂中> carol(le)
凸窗座 bay stall; caroll; carol(le)
凸锉 bellied file
凸导尺 nib guide
凸堤码头 quay pier
凸底 <玻璃缺陷> convex bottom
凸底锅炉 dished end plate boiler
凸底箱形孔型 convex box pass
凸点 salient point
凸点电阻焊机 projection weld machine
凸点钢板 button plate
凸点焊 mesh weld
凸雕 embossment
凸顶形活塞 deep-domed piston
凸肚窗 bow window; compass window; oriel (bay window); oriel window; mirador
凸肚墙 bulging wall; wall entasis
凸肚线型板 entasis reverse
凸肚型橱柜 breakfront
凸肚型书柜 breakfront
凸肚状 entasis[复 entases]
凸度 camber; convexity; crown; degree of convexity; protuberance
凸端门槛 lug sill
凸断口 convex fracture
凸多边形 convex polygon
凸多面体 convex polyhedron
凸多面锥体 convex polyhedral cone
凸耳 convex lug; dummy club; ear(ing); ledge; lug; projecting lug
凸耳卡环 lugged shackle
凸耳螺栓 lug bolt
凸耳手柄盖桶 lug cover pail
凸法兰 boss flange; male flange
凸方格图案 raised chequer pattern
凸方和凹线圆环线脚 square and rabbet
凸方线脚 <中世纪建筑> roll-and-fillet mo(u)lding
凸峰态 leptokurtosis
凸峰态分布 leptokurtic distribution
凸缝 bulk joint; bulky joint; convex joint; high joint; tongue
凸腹凸轮 convex flank cam
凸腹形颚板破碎机 bellied jaw crusher
凸勾缝 bastard pointing; bastard tuck pointing; high-joint pointing
凸箍桶 flange drum
凸鼓变形 bulging and deflection
凸鼓形锯 bilge saw
凸辊环 outer collar; positive collar
凸海岸 convex coast
凸海岸国家 convex state
凸函数 convex function
凸焊 beading weld(ing); point welding; projection weld(ing); relief welding
凸焊(焊)缝 convex weld; projection weld
凸焊机 projection welder; projection welding machine
凸焊接头 projection welded joint
凸焊帽 projection welded cap
凸焊填角 reinforcing fillet
凸河岸 convex stream bank
凸弧形坡屋顶 rainbow roof; whaleback roof
凸花厚缎 brocatel(le)
凸花细工 fretwork
凸极 magnet limb; projecting pole; salient pole
凸极电机 radial rotor machine; salient pole machine
凸极电枢 pole armature; radial armature; salient pole armature
凸极发电机 salient pole generator
凸极交流发电机 salient pole alternator
凸极式 salient pole type
凸极式发电机 salient-pole generator; salient pole machine
凸极式同步发电机 salient pole synchronous generator
凸极式同步感应电动机 salient pole synchronous-induction motor
凸极式转子 salient pole type rotor
凸极转子 field spider; projecting pole rotor; rotor with salient poles
凸集 <线性规划等用> convex set
凸肩 relish; rib; shoulder
凸肩支承座 rib seat
凸件 male member
凸件起动开关 tumbler switch
凸角 arris; convex angle; convex corner; lobe; quoin; salient; salient angle
凸角堡 redan
凸角石 rusticated ashlar
凸角式剃前插齿刀 protuberance type shaper cutter
凸角式剃前滚刀 protuberance type hob
凸界面 convex interface
凸进 bulge in
凸晶 convex crystal
凸镜 convex glass
凸镜状结构 lenticular structure
凸口 bulged finish
凸块 projection
凸块滚轮 padfoot drum
凸块碾 padfoot roller; pad roller
凸块式压路机 padfoot drum roller
凸棱 fin
凸棱衬板 lining plate with rib
凸棱式衬板 ribbed liner
凸棱窑衬 refractory lifter
凸莲瓣茶壶 <瓷器名> ewer with raised lotus petals
凸梁 cam bearing
凸露的 bold
凸率 convexity
凸轮 actuating cam; cam (wheel); jaw; latch; wabbler[wobbler]
凸轮板 lobe plate
凸轮(包)角 cam angle
凸轮泵 lobe pump
凸轮表面 cam face
凸轮操纵的 cam-operated
凸轮操纵阀 cam actuated valve
凸轮(操纵)开关 cam switch
凸轮操纵式气动制动器 cam-operated air brake
凸轮操纵式气闸 cam-operated air brake
凸轮操作的 cam-operated
凸轮操作的联锁装置 cam-operated interlock
凸轮槽 cam path; cam slot
凸轮槽盘 grooved cam
凸轮车床 cam lathe
凸轮冲击筛 Leahy screen
凸轮传动 cam drive; cam-driven
凸轮传动阀 tappet valve
凸轮传动器 cam driver
凸轮传动装置 cam driving gear
凸轮从动杆 cam follower lever
凸轮从动滚轮销 cam follower pin
凸轮从动件 cam follower; tappet
凸轮从动轮 cam follower
凸轮从动球 cam follower ball
凸轮从动销 cam follower pin
凸轮带式制动器 cam band brake
凸轮挡 cam catch
凸轮挡块 cam dog
凸轮的凸角 cam lobe
凸轮垫圈 cam ring
凸轮顶出装置 cam ejector
凸轮定时阀 tappet valve
凸轮动程 cam stroke; cam throw
凸轮动作的限流继电器 cam-operated current limit relay
凸轮仿形机床 cam forming and profiling machine
凸轮分度角 cam angle
凸轮分配机构 tappet gear
凸轮杆 cam lever
凸轮杆轴 cam lever shaft
凸轮杠杆机构 cam and lever mechanism
凸轮跟随器 cam follower
凸轮工作角 cam angle
凸轮鼓 barrel cam; cam drum; tappet drum
凸轮滚轮 cam roller
凸轮滚子 cam bowl; cam roller
凸轮夯实机 cam-ram machine
凸轮和棘轮传动装置 cam and ratchet drive
凸轮盒 cam case
凸轮后侧边 back of cam
凸轮花键盘 serrated camshaft
凸轮滑块 cam slide

凸轮环 cam ring; contour ring; track ring
凸轮回位弹簧 cam return spring
凸轮回行机构 eccentric-motion reversing gear
凸轮回转角度 dwell angle
凸轮活塞式水泵 cam and piston pump
凸轮机 cam press
凸轮机床 cam cutter
凸轮机构 cam gear; cam mechanism
凸轮夹紧装置 cam clamp
凸轮架 cam carrier; cam holder; cam spider
凸轮尖 cam nose; nose of cone
凸轮间隙 cam clearance
凸轮减速装置 cam reduction gear
凸轮件 cam member
凸轮接触器 cam contactor
凸轮节流阀 cam throttle
凸轮卡盘 cam chuck
凸轮开关 cam-operated switch
凸轮控制 cam control
凸轮控制机构 cam control gear
凸轮控制装置 cam control gear
凸轮块 cam bit
凸轮廓 cam contour
凸轮连接器 cam adapter
凸轮联轴节 cam sleeve
凸轮联轴器 cam adapter; shifting sleeve
凸轮轮廓 cam contour; cam profile
凸轮面 cam surface
凸轮磨床 cam grinder
凸轮磨削 cam grinding
凸轮磨削的 cam ground
凸轮盘 cam drum; cam plate; circular cam; edge cam; sheave
凸轮配磨机 cam ground
凸轮启动装置 cam breakout
凸轮曲线的同心部分 dwell
凸轮驱动 cam drive
凸轮驱动出坯 cam-driven knockout
凸轮驱动的 cam-actuated
凸轮驱动旋转喷洒头 cam drive rotary head
凸轮升程 cam-lift; cam rise; lift of cam
凸轮升度 cam-lift
凸轮升降器 cam-lift
凸轮式冲击研磨机 pin beater mill
凸轮式搅拌器 cam agitator
凸轮式遮板 cam type screen
凸轮塑性计 cam plastometer
凸轮随动件 cam follower
凸轮随动件导承 cam follower guide
凸轮随动轮 cam follower
凸轮锁 cam-lock
凸轮锁紧 cam-lock
凸轮锁紧机构 cam lock holding mechanism
凸轮锁紧轴端 cam lock spindle nose
凸轮调节阀 cam throttle
凸轮调速器 cam governor
凸轮调整装置 cam adjusting gear
凸轮同速器 cam governor
凸轮凸尖 cam lobe
凸轮图 cam diagram
凸轮推动的 cam-operated
凸轮推杆 cam carrier
凸轮推杆滚柱 cam follower roller
凸轮推杆随动件 shedding tappet
凸轮推杆锁定 locking tappet
凸轮铣床 cam milling machine
凸轮铣切装置 cam milling attachment
凸轮系统 camming
凸轮箱 cam box
凸轮镶片 toe piece
凸轮样板 cam template
凸轮(圆)盘 cam disc[disk]
凸轮运动 cam motion
凸轮运转组 cam-operated group
凸轮支承的空隙 cam bearing clearance
凸轮止推回转轴承 kingpin bearing; slewing journal
凸轮制动器 cam brake
凸轮轴 axial cam; camshaft; cam spindle; countershaft; drum cam; tappet shaft; tumbling shaft; wobbler shaft
凸轮轴臂 <气闸制动腔与制动凸轮连接的> slack adjuster
凸轮轴车床 camshaft turning lathe
凸轮轴衬套 cam bush; camshaft bushing
凸轮轴齿轮 cam gear
凸轮轴传动 camshaft drive
凸轮轴的主动轴 drive shaft of the camshaft
凸轮轴定时齿轮 camshaft time gear; camshaft timing gear
凸轮轴定时齿轮壳 camshaft timing gear hub
凸轮轴盖 camshaft cover
凸轮轴夯实机 cam-operated (tamping) block machine
凸轮轴夹布胶木齿轮 camshaft textolite gear
凸轮轴接触器 camshaft contactor
凸轮轴孔校正铰刀 reamer for camshaft aligning
凸轮轴控制 camshaft control
凸轮轴控制器 camshaft controller
凸轮轴链轮 camshaft sprocket
凸轮轴磨削机 camshaft grinding machine
凸轮轴(上)齿轮 camshaft gear
凸轮轴上控制盘 camshaft phasing gear
凸轮轴填密压盖 camshaft packing gland
凸轮轴凸轮 cam of camshaft; camshaft cam
凸轮轴楔 camshaft wedge
凸轮轴油泵齿轮 camshaft oil pump gear
凸轮轴止推塞 camshaft thrust plunger
凸轮轴止推轴承 camshaft thrust bearing
凸轮轴止退板 camshaft check plate
凸轮轴中间齿轮 camshaft idler gear
凸轮轴轴承 camshaft bearing
凸轮轴轴套 camshaft bushing
凸轮轴转角度 camshaft degree
凸轮爪 cam pawl
凸轮转角 cam angle
凸轮转子泵 lobe rotor pump
凸轮转子液压马达 lobed rotor motor
凸轮装置 cam mechanism; cam gear
凸轮组 cam group
凸轮组式 cam-group type
凸螺纹 male screw
凸埋式管道 projecting conduit; projecting pipe; projecting tube
凸面 convex; convexity; convex shore; convex surface; crown; gibbosity; raised face; bulb <抹灰的>; crown face <皮带轮、轴承承载鞍等>
凸面边界 convex boundary
凸面玻璃 convex glass
凸面刀具 raised face tool
凸面对焊法兰 raised face welding neck flange
凸面法兰 convex flange; raised face flange
凸面焊 convex weld
凸面滑轮 crowned pulley
凸面加工装置 cambering gear; crowning set
凸面角焊缝 convex fillet weld(ing)
凸面接头 male adapter[adaptor]
凸面镜 burning glass; convex lens; convex mirror
凸面卷尺 convex rule
凸面壳体 convex shell
凸面皮带轮 crowned pulley; crown(ing) pulley
凸面踏脚板踏步 commode step
凸面体 convexity
凸面外视镜 convex exterior mirror
凸面圆形淘汰盘 convex table
凸面轧辊 cambered roll
凸面状 gibbosity
凸模 force plug; male; male die; male mo(u)ld; male punch; mo(u)ld plunger force plug; terrace die; top mo(u)ld half
凸模背靠块 punch heel
凸模端部圆角 radiused punch nose
凸模固定板 punch-holder; punch plate; punch retainer
凸模固定凸缘 punch flange
凸模接头 punch adapter; punch carrier; punch plate
凸模拉延 punchless drawing
凸模面刃口宽度 punch land
凸模坯料 punch block
凸模围栅 punch inclosure
凸模位移 punch displacement
凸模行程 punch travel
凸模压板 punch pad
凸模圆角 punch-nose angle; punch-nose radius
凸模圆角半径 punch profile radius
凸模最大长度 maximum punch length
凸磨光 relief polishing
凸钮揉包机 button breaking machine
凸盘秤 cam scale
凸片 bulge; fin; lug
凸片钢板接合 table steel plate connection
凸片钢板结合 tabled steel plate connection
凸平面线脚 bolection
凸平形的 convexo-plane
凸平形磨瓦 flat and raised abrasive tile
凸坡 convex slope; protruding slope; summit grading
凸坡发育(上升大于剥蚀) waxing development
凸起 bulge; bulging; convex; convexity; crowning; doming-up of pavement; embossment; hump; lobe; projection; prominence [prominency]; relieve; rilievo; saddle-backing; salience[saliency]; swell
凸起边 lug
凸起表面 convex area; raised face
凸起部 hub
凸起部分 boss(ing); rising part; lug boss
凸起处 <路面> high-spot; chatter bump
凸起的 bellied; bossed; bossy; embossed; protrudent; salient
凸起的(木)纤维 raised fiber[fibre]
凸起的条纹 wale
凸起点 <冷轧板表面缺陷> high-spot
凸起分针 varnished skeleton minute hand
凸起高度 height of projection
凸起高位沼泽 raised umbrageous bog
凸起管 bulged tube
凸起甲板 trunk deck
凸起甲板横梁 trunk deck beam; turtle-back beam
凸起空心分针 varnished skeleton cannoned minute hand
凸起空心时针 varnished skeleton cannoned hour hand
凸起路面 doming-up of pavement
凸起轮辋 ridge rim
凸起翘曲 convex bow
凸起时针 varnished skeleton hour hand
凸起天窗 coaming
凸起图案 loaf
凸起形 protuberance
凸起直径 projection diameter
凸嵌板 raised panel
凸嵌缝 tuck-pointed joint
凸嵌勾缝 tuck and pat pointing; tuck (joint) pointing
凸嵌灰缝 tuck joint; tuck-pointing joint
凸嵌线 bolection mo(u)ld(ing)
凸嵌线脚背 <椅子> reedback
凸墙处座位 bay stall
凸墙内角 internal angle
凸墙外角 external angle; salient angle
凸球壳属 <拉> Nitschkia
凸曲 camber
凸曲力矩 hogging moment
凸曲率 convex curvature
凸曲面 convex camber
凸曲线 crest curve; summit curve
凸曲线半径 convex radius
凸圈 spigot ring
凸刃装岩铲斗 V-edge rock bucket; V-nose rock bucket
凸舌轧辊 tongue roll
凸式码头 pier
凸式线脚 bolection mo(u)ld(ing)
凸饰 boss; bossed ornament
凸榫 cog; cut joint; tenon
凸榫接合 square cogging
凸榫吞肩 relish
凸台 boss; paraskenion <古希腊剧场后台向前伸出两翼的>
凸梯度 convex gradient
凸体 convex
凸挑饰件 <充当拱顶石的> key console
凸条 batten; raised line; rib(bing)
凸条纹 stria[复 striae]
凸条线脚 rail bead
凸条砖 <侧边有凸纹的铺路砖> vertical-fiber brick; cal-fibre brick
凸筒瓦 convex tile
凸桶 drums bulged
凸头 raised head
凸头钢板 button plate
凸头钢线钉 convex head steel wire nail
凸头活塞 convex head piston
凸头螺钉 raised head screw
凸头螺栓 raised head bolt; snug bolt
凸头铆钉 raised head rivet
凸头式 bonnet type; cab behind engine type
凸头式载重汽车 bonnet type cab
凸头形电极 convex-end electrode
凸透镜 bulls eye; convex lens; positive lens
凸凸 convex-convex
凸图版腐蚀墨 process engraving ink
凸瓦 <俯瓦> convex tile
凸弯度 camber
凸弯管 convex bend; male bend
凸弯形 convexity
凸弯形的 pulvinated
凸尾带钢 full strip
凸纹 blister design; burr; lug; relief
凸纹锤制 repousse
凸纹电闸 cord switch
凸纹雕刻 relief engraving
凸纹粉刷 pargetry; parget-work
凸纹钢板 nipple plate
凸纹辊筒印花 kiss printing
凸纹辊筒印花机 surface printing ma-

chine
凸纹厚浆涂料 texture coating
凸纹花边 repousse lace
凸纹花样瓦 relief pattern tile
凸纹漆 textured paint; texture-finished paint
凸纹墙纸 cameoid relief;embossed wall paper;lincrusta;cordelova relief
凸纹图案<立体图案> embossed pattern;relief pattern
凸纹涂料 relief coating;texture coating;texture pebble finish
凸纹线脚 convex mo(u)lding
凸纹效应 raised effect
凸纹轧辊<钢板防滑> island roller
凸纹织物 blister
凸纹制作 repousse
凸纹砖 vertical-fiber brick
凸纹转筒印花 cameo printing
凸线 convex;facet(te)【建】
凸形 convexity;gibbous
凸形变坡点 point of gradient change on convex section
凸形导向器 knurled deflector
凸形的 convex;male;protrudent;umbonate
凸形的底 bumped head
凸形断面 convex (cross-) section; crowned section
凸形多孔板 convex perforated plate
凸形腭板 bellied jaw
凸形颚板 bellied jaw
凸形封头 convex head
凸形盖人孔 manhole with convex cover
凸形钢坯 convex billet
凸形拱 convex arch
凸形勾缝 rusticated joint
凸形构件 camber piece
凸形规划(法) convex programming
凸形规划问题 convex programming problem
凸形焊缝 convex weld
凸形河漫滩 convex flood plain
凸形角焊缝 convex fillet weld(ing)
凸形梁 upstand beam
凸形梁板 upstand T-beam slab
凸形轮 male rotor
凸形螺旋转子 male screw rotor
凸形密封环 male adapter[adaptor]
凸形瓶底 pushed punt;push(ed)-up bottom
凸形坡 convex slope
凸形气门头 convex head valve
凸形曲线 convex curve;crest curve
凸形竖曲线 convex vertical curve; crest vertical curve; summit vertical curve;vertical curve at summit
凸形特征 crown feature
凸形外壳 convex hull
凸形铣刀 convex cutter
凸形线脚 risen mo(u)lding
凸形圆饰 botel
凸形整体铣刀 convex solid cutter
凸形纵断面 summit profile
凸形钻头<回转钻进中金刚石不取芯钻头,用于不同硬度的岩层> pilot bit
凸性 convexity
凸性公理 axiom of convexity
凸序列 convex sequence
凸压模 ejector die
凸檐盾构 bulkhead shield
凸檐墙 bulkhead wall
凸眼球 eyeballs out
凸阳台 balcony;exterior balcony
凸印版 printing block; relief printing plate
凸隅角石 canton
凸域 convex domain
凸圆 gibbosity
凸圆带形线脚 scroll mo(u)lding
凸圆的 convex
凸圆勾缝 bead pointing
凸圆弧金刚石钻头<圆弧直径等于壁厚> double round nose bit
凸圆接缝 convex joint;convex tooled joint
凸圆面 circular face
凸圆曲线 bellied
凸圆饰 boltel;boulevard boultin(e); boultin(e)
凸圆体 convex
凸圆条线脚 nosing strip
凸圆头 cheese head
凸圆线脚 bo(u)ltel;boultine;bowtel(l);convex round mo(u)lding;ovolo;reel carriage;torus[复 tori]
凸圆线脚刨 bead plane
凸圆线刨 forkstaff plane
凸圆线饰 beading fillet
凸圆形装饰线条 ovolo mo(u)lding
凸缘 bump;cock bead;collar;crimp; crown edge;flaring;ledge;lid;lip; lug;pad;rand;reeding;rim;set-off; shaft collar;teat;fla(u)nch(ing) <烟囱防水>
凸缘安装 flange mount
凸缘扳手 flange wrench
凸缘板 flange plate;lip block
凸缘边 flanged lip
凸缘边接合 flanged edge joint
凸缘补强 reinforcement by flange
凸缘叉 flange yoke
凸缘叉臂 flanged yoke
凸缘车工 flanger
凸缘衬垫 flange(d)gasket
凸缘衬圈 flange back
凸缘衬套 flanged bushing;flange liner
凸缘承窝 flanged pipe socket
凸缘冲模 flanging die
凸缘穿墙管 flanged wall piece
凸缘窗台板 lug sill
凸缘磁电机 flange-mounted magneto
凸缘的主轴瓦 flange engine bearing
凸缘底 flanged bottom
凸缘底板 bottom flange plate; pan-shaped base
凸缘底座电动机 flange motor
凸缘电动机 flanged motor
凸缘垫环 flange grommet
凸缘垫圈 flange gasket;flange washer
凸缘顶砖 flanged ridge tile
凸缘断面 flange section
凸缘断面图 profile flange
凸缘法兰 male flange
凸缘分接头 flange tap
凸缘盖桶 lug cover pail
凸缘钢 flange steel
凸缘沟 flange way
凸缘构件 flange member
凸缘管 flange(d)pipe;flange(d)tube
凸缘管接 flanged union
凸缘管散热器 flanged-tube radiator
凸缘规 flange ga(u)ge
凸缘滚辗 cam roll
凸缘滚形 roll flanging
凸缘后盖表壳 flange back case
凸缘花盘 flange chuck
凸缘滑轮 flange pulley
凸缘环 angle staple;flange(d)ring
凸缘回转弯头 flanged return bend
凸缘活接头 lip union
凸缘活塞 flanged piston
凸缘机 flanger
凸缘基础螺栓 foundation bolt with nose
凸缘脊瓦 flanged ridge tile
凸缘加劲 flange stiffening
凸缘加强 flange stiffening
凸缘夹紧板 flanged clamping plate
凸缘架 flange mount
凸缘架的 flange-mounted
凸缘角钢 flange angle;flange angle steel
凸缘角铁 flange angle
凸缘接缝 flanged seam
凸缘接缝铆钉 flanged seam riveting
凸缘接合 flange connection;flange(d) joint;flange(d)seam;lipped joint
凸缘接头 flange(d)joint
凸缘结合 flange coupling;joint flange
凸缘紧固的 flange-mounted
凸缘卷曲法 flange crimping
凸缘卡盘 flange chuck
凸缘孔型 flange pass
凸缘肋 flange rib
凸缘连接 flange(d)connection;flange joint
凸缘连接的 flange connected
凸缘连接器 flange connector
凸缘连接用颈圈 collar for flange connection
凸缘联结 flanged coupling
凸缘联轴节 flanged coupling;flange(d) joint;flange(d)(type)shaft coupling
凸缘联轴器 flange coupling
凸缘链轮 grip disk[disc]
凸缘梁 flanged beam
凸缘轮 flanged wheel;flanged wheel flange pulley; flanged rail wheel <火车的>
凸缘螺帽 collar nut
凸缘螺母 flange(d)nut
凸缘螺栓 flange(d)bolt
凸缘铆钉 flange rivet
凸缘密封 flange seal
凸缘密封垫 flanged packing
凸缘面积 flange area
凸缘盘 flange
凸缘刨 nosing plane
凸缘喷嘴 flange(d)nozzle;flange spout
凸缘皮带轮 flange(d)pulley
凸缘片式散热器 flanged radiator
凸缘剖面 flange section
凸缘汽缸 flanged cylinder
凸缘软木塞封口 flange cork closure
凸缘散热器 flanged radiator
凸缘式安装件 flange mounting
凸缘式衬筒 flanged-type liner
凸缘式连续放泄阀 flange type everlasting blow-off valve
凸缘式联轴器 flanged joint
凸缘式排气管 flange exhaust
凸缘试验 flanged test
凸缘胎 lug tire
凸缘弹簧弯头 flanged spool piece
凸缘套 flange sleeve
凸缘套管 flanged pipe spigot
凸缘体 body flange;flange body
凸缘条 nosing strip
凸缘铜阀 flanged brass valve
凸缘线脚 lip mo(u)ld;treacle mo(u)lding
凸缘销钉 lug dowel
凸缘小齿轮 shrouded pinion;shrouded wheel
凸缘斜坡 flange slope
凸缘型电动机 flange motor
凸缘型钢 flanged section
凸缘压力管 flanged pressure pipe
凸缘元件 flange element
凸缘制造机 flanger
凸缘装配 flanging arrangement
凸月 gibbous moon
凸止口 male half coupling
凸珠 bur
凸砖层 string course
凸装型仪表 salient instrument
凸状 convex
凸状壁 embattlement
凸状壁的 embattlemented
凸状部分 male member
凸状雕饰带 cushioned frieze
凸状联锁铣刀 convex interlocking cutter
凸锥 convex cone
凸字 anastatic
凸字的 embossed
凸组合 convex combination

秃

秃斑 bald spot
秃的 bald
秃顶背斜 bald-headed anticline
秃积雨云 cumulonimbus calvus cloud
秃壳贫属<拉> Irenina
秃山 bald mountain;bare mountain
秃山顶 bald
秃土 bare soil
秃状(云) calvus

突

突阿斯<法国长度单位 toise
突岸 convex bank
突爆 detonation
突边【建】 nib;flange
突边尺 nib guide
突边导板 nib guide
突边瓦 nibbed tile
突边桌 projecting table
突变 abrupt change;mutation;revulsion;saltation
突变表型 mutant phenotype
突变波 abrupt wave
突变不平整度 abrupt irregularity
突变层<水中的> saltation layer
突变沉积 sudden change
突变成分 mutagenic components
突变单位 muton
突变的 abrupt;discontinuous
突变地形<如冰川形成的地形> accident(al)form;abrupt relief
突变点 abrupt junction; catastrophe point;point of discontinuity
突变段 sharp transition;sudden transition
突变发生 mutagenesis
突变发育 saltative evolution
突变反应 jump reaction
突变负荷 mutational load
突变固定 mutation fixation
突变过渡 sharp transition
突变荷载 discontinuous load
突变剂 mutagen
突变加强 mutafacience
突变结 abrupt junction;step junction
突变结型二极管 abrupt junction diode
突变界面模型 suddenly changing interface
突变理论 catastrophe theory;mutation theory
突变率 mutation rate;rate of mutation
突变论 catastrophe theory
突变面 surface of discontinuity
突变频率 mutation frequency
突变平衡 mutational equilibrium
突变区 saltation zone
突变曲线 abrupt curve
突变色 accidental colo(u)ring
突变势 abrupt potential
突变试验 mutation test
突变体 mutant
突变体污染 sudden pollution
突变体细菌 mutant bacteria
突变体性状 mutant character
突变位点 mutant site

T

突变无花果 caprifig
突变现象 jumping
突变形成 mutagenesis
突变型 mutant;saltant
突变性 discontinuity;mutability
突变性的负载 discontinuous load
突变性能 mutability
突变性危害 mutagenic hazard
突变性状 mutant character
突变学说 theory of mutation
突变压力 mutation pressure
突变应力 mutation stress
突变振动 catastrophic vibration
突变指数 mutation index
突变株 mutant
突变子 mutation;muton
突层(砌体)offset course
突齿 lobe
突出 beedle; booming(out); highlight; jut out; overhang; oversail; predominance; project(ion); projecture; protract(ion); protrude; protrusion;sail-over;stick-out
突出边缘 eaves
突出部 projection; bump; jut; lug; nose; peak; salient; sally; set-off; shoulder;teat;ledge;protuberance
突出部分 high-spot;ledge;margin;outshot; peak; projecting part; protrusion; rising; overshot <建筑物>; cusp <月牙形沙滩海岸线的>
突出部分防雨板 head flashing
突出部分建筑 corbel;pannier
突出长度 projecting length
突出窗 jut window
突出的 emergent; eminent; overhanging; pointed; predominant; projective; prominent; protrusive; salient; projecting
突出的扁平岩石 shelf
突出的法兰 projecting flange
突出的前端 end prow;prow
突出的翼缘 projecting flange
突出堤头 jetty head
突出底部 projecting foot
突出地面的管道 projected pipe
突出地面截水墙 negative cutoff
突出雕饰 knop
突出陡岸 projecting bluff
突出度 <分布曲线中的高峰程度> kurtosis
突出短杆 stub column
突出覆盖层 projecting cover
突出钢铁生产 play up steel production
突出公差 projected tolerance
突出公差带 projected tolerance zone
突出共振 prominent resonance
突出海面的浪蚀岩柱 rauk[复 rauker]
突出横饰线 fillet
突出后崩 afterburst
突出换向条 high bar
突出建筑 advancing longwall system
突出轮缘 collar rim
突出木 emergent
突出嵌线的装饰线条 bolection mo(u)lding
突出墙面的雕带 corbel-table frieze
突出墙面的砖或石块 diatonous
突出墙面石支托 stone corbel
突出墙面砖 projecting brick
突出桥台 projecting abutment
突出三角洲 projected delta; protruding delta
突出石梁 projecting ledge rock
突出式结构 projection type construction
突出式桥端 projecting end
突出式桥台 projecting abutment
突出式上旋窗 projecting top-hung window
突出式舞台 thrust stage
突出树 prominent trees
突出竖线条的建筑 <哥特式> vertical accent
突出体 emergence;excrescence
突出物 joggle; overhang; projection; projecture;protrusion;spur
突出斜隅角 <建筑物的> squint quoin
突出性 highlighting
突出翼缘 outstanding flange
突出隅石 projecting quoins
突出圆顶 projection dome
突出照明 accent lighting
突出肢 outstanding leg
突出重工业 setting the pace for heavy industry
突出主体网点版 cut-out halftone
突出砖层 oversailing course
突出自己 self-assertion
突窗 oriel
突堤 croy; groyne; jetty; jetty pier; mole; pier; projecting pier; shore-connected breakwater
突堤堤头 groin head;pier head
突堤端部码头 head wharf
突堤端尖形拔水 cutwater
突堤港 jetty harbo(u)r
突堤建筑线 pierhead line
突堤码头 jetty; jetty pier; marine pier;open pier;right-angle pier;slip wharf;finger pier
突堤码头长度 pier length
突堤码头端部界线 pierhead line
突堤码头构件 pier component
突堤码头间港池 pier slip
突堤码头间水域 slip
突堤码头宽度 pier width
突堤码头面高程 pier elevation
突堤码头排架 pier bent
突堤码头前方仓库 pier shed
突堤码头前沿 pier apron
突堤码头前沿泊位 berthing head;jetty-head berth;pierhead wharf
突堤码头前沿龙门式集装箱起重机 pierside container gantry
突堤码头之间的港池净宽 clear width of slip
突堤前端 pier head
突堤前沿 jetty head
突堤式防波堤头 mole head
突堤式码头 finger pier; fork-shaped wharf; jetty wharf; jib wharf; quay pier; finger wharf; landing pier; landing stage
突堤式码头的弹性缓冲设施 spring fender for jetty
突堤式码头系统 jetty system;pier system
突堤式煤码头 coal pier
突堤式木结构码头 timber jetty
突堤式石码头 stone jetty
突堤头 jetty head
突堤外端 mole head
突堤栈桥 jetty
突雕刻花饰 embossing
突堞 machicolation
突动触点 snap action contact
突端门槛 lug sill
突额 corbel
突额圬工工程 corbel masonry work
突额圆顶 corbel cupola
突额砖 corbel brick
突耳 lug;nib
突发 bursting;burst out;gust;outburst
突发波 erupting wave
突发擦除 erasure burst
突发差错 burst error;error burst
突发长度 burst length
突发传输 burst transmission
突发错误 burst error
突发错误校正能力 burst-correcting ability
突发方式 burst mode
突发故障 sudden failure
突发纠正 burst-correction
突发删除 erasure burst
突发误差 sporadic fault
突发误差校正 burst-error-correction
突发性环境污染事故 emergency environment pollution accident
突发性客流 outburst passenger flow
突发性排放 accidental discharge
突发性喷发 paroxysmal eruption
突发性破坏故障 sudden death failure
突发性水环境风险 accident water environmental risk
突发性污染 accidental contamination; accidental pollution;sudden pollution
突发性污染事件 emergent pollution event
突发性污染源 accidental pollution source
突发压力 bursting pressure
突发噪声 burst noise
突发阵风 sharp-edged gust
突放热量临界温度 recalescence point
突风风洞 gust tunnel
突风梯度距离 gust-gradient distance
突拱 <突出于立面的一种假拱> corbel(led) arch; Maya arch; offset arch
突拱大梁 corbel arched girder
突硅钠锆石 terskite
突击 onrush
突击拨号 attack dialing
突击部 tit
突击动作 shock action
突击队 shock brigade
突击队员 commando
突击工程 rush work
突击工作 rush work;sharp work
突击工作班 shock brigade
突击工作者 shock worker
突击计划 taut planning
突击检查 snap check
突击检查站 surprise checkpoint
突击进度安排 crash schedule
突击力 shock power
突击施工成本 crash cost
突击式测验 pop test
突击手 shock worker
突击销售队 commando sales team
突击销售法 shotgun approach
突击运输机 assault aircraft
突击作业班 shock brigade
突加 impact
突加负载 impact load(ing); pop-in load;shock load(ing)
突加荷载 impulsive load(ing);shock load(ing);suddenly applied load
突加应力 suddenly applied stress
突加应力的 abruptly stressed
突加载荷 shock load(ing)
突肩【建】crossette
突降 dive;sudden drawdown
突降法 anti-climax
突降阵雨 sudden shower
突角 pien(d)
突角堡 redan
突角补偿 horn balance
突角拱 <墙角支承上层结构的角拱> squinch arch
突角构件 quoining
突角露头石 quoin header
突角支柱 quoin post
突进 dart
突进阀 dart valve
突开动作 pop action
突开阀 pop-off valve;pop valve
突开式安全阀 pop safety valve;pop valve
突开压力 popping pressure
突口 <道路的> hub
突口吸接 hub and spigot joint
突肋 finned
突肋瓦 water-rib tile
突肋形接缝 <金属薄板屋顶咬口的> conic(al) roll
突棱 lug;nib <瓦端的>
突利夫水泥 Trier cement
突梁 jib
突码头 finger pier; jetty [jutty]; landing stage;mole;pier
突码头前端 pier head
突面法兰 raised face flange; raised flange
突面镶板 fielded panel; raised and fielded panel;raised panel
突尼斯海峡 Strait of Tunis
突尼斯禾草 Tunis grass
突尼斯棱纹钩编花边 Tunis crochet
突尼斯中央银行 Banque Centrale de Tunisia
突破 breakthrough
突破定额 over-fulfil a quota
突破贯穿 breakthrough
突破口 sally port
突破量 breakthrough capacity
突破顺序 breakthrough sequence
突破体积 breakthrough volume
突起 blow-up; prominence [prominency]; protrusion; protuberance; relief;salience[saliency];tubercle
突起变形 bowing
突起部 jut
突起的 mammilar; prominent; protuberant
突起的齿条 <防车辆进入路面> raised bars
突起的埋头螺钉 raised countersunk head screw
突起的平行脊 projected parallel ridges
突起的小泡 <陶瓷缺陷> bleb
突起的圆头螺钉 raised cheese head screw
突起地 salient
突起拱 raised arch
突起构件 haunched member
突起路缘 raised curb
突起面 raised face
突起式分隔带 raised separator
突起纹理 raised grain
突起物 outrigger
突起线 flash line
突墙墙角塔 tourelle
突球 knuckling
突然爆开 pop
突然爆裂 popouts
突然崩坍 inrush
突然变化 abrupt change; saltation; sudden change;sudden variation
突然变音 judder
突然操纵 abrupt maneuver
突然插入 interject
突然沉降 sudden settlement
突然沉陷 sudden settlement; sudden subsidence
突然冲击 jab
突然抽回 revulsion
突然出现 burst(up)on
突然出现的裂缝 pop-out cracks
突然大量降落 <雨、雪等> downfall
突然的大变动 catastrophe
突然的停止 fetch up
突然的弯曲 quirk
突然地质大变动 sudden catastrophe
突然吊起 snatch-lift
突然丢弃 sudden thrown

突然短路 suddenly applied short circuit
突然发出 burst into
突然发出火焰 flare-up light
突然发亮 snap
突然发生 crop cut;crop up;pop
突然放水 sudden water release
突然放松绳线 surge
突然故障 catastrophic failure
突然关闭 sudden closure
突然荷载 sudden load(ing)
突然回动 < 卸钻杆时 > snapping of drill pipe
突然回转 slue
突然加大油门 < 美 > peel tire
突然加载 immediate loading
突然降压 explosive decompression
突然截止 sharp cut-off
突然解锁 sudden release
突然开大油门 goose
突然开始 onset
突然扩大 abrupt expansion; sudden enlargement
突然扩大系数 coefficient of sudden expansion
突然来到 blow in
突然蔓延 outbreak
突然拿去 whip off
突然排出 slug discharge
突然偏转 < 声波测井中,由于周波跳跃或仪器的停顿而产生的曲线突然偏转 > tent poling
突然频率偏差 sudden frequency deviation
突然频移 sudden frequency deviation
突然破裂 sudden rupture
突然起燃 flare-up
突然倾斜 lurch;sudden lurch
突然燃烧 deflagrate
突然燃烧爆炸 deflagration
突然刹车 abrupt stop
突然射出 pop-shoot(ing)
突然升起 pop-up
突然失掉 popouts
突然失效 catastrophic failure;sudden failure
突然失压 rapid decompression
突然施荷载 instantaneous applied load
突然事故 < 大的 > sudden catastrophe
突然释放 sudden release
突然收缩 abrupt contraction; sudden contraction
突然甩负荷 sudden load rejection
突然缩小 abrupt contraction
突然缩窄 abrupt contraction
突然坍塌 sudden collapse;cave in
突然跳动 jerk
突然停车 abrupt stop;dead stop;sudden stop
突然停机 abrupt halt;drop-dead halt;hard shut-down;sudden halt;unexpected halt
突然停止 stop short;to conk(out);unexpected halt;abrupt halt
突然停住 crash stop
突然退化 catastrophic degradation
突然脱出力 breakout force
突然弯曲 abrupt bend
突然位移 sudden displacement
突然无风 fall calm
突然袭击 raid;surprise attack
突然泄放 abrupt discharge
突然泄降 sudden drawdown
突然泄水 sudden water release
突然移动 hike
突然涌出力 breakout force
突然涌水 breaking through of water;sudden flooding
突然灾变 sudden catastrophe
突然灾难 shock hazard
突然折断 snap
突然蒸发 flash distillation
突然制动 sharp braking
突然制止 snub
突然注入溶液法 < 测流方法之一 > sudden injection method
突然转变 flop;pitch out
突然转弯 pitch out
突然转弯的 hairpin
突然转向 jog
突燃 deflagration
突燃器 deflagrator
突入水中的码头 jetty
突山顶 scalp
突伸式舞台 arena type stage
突升 skyrocket
突升海水深度 overfall
突式键槽 projecting key
突水 gushing water;water bursting
突水点高程 altitude of water bursting point
突水点位置 location of bursting water
突水点坐标 coordinate of water bursting point
突水分类 classification of bursting water
突水规模 scale of bursting water
突水可能性预测 prediction of water bursting possibility
突水量 quantity of water bursting
突水时间 time of water bursting
突水水源 source of bursting water
突水突泥 water-mud bursting
突水系数 water bursting coefficient
突水系数等值线图 contour map of bursting water coefficient
突水延续时间 continues period of water bursting
突水预测图 map of bursting water prediction
突榫锯 miter[mitre] saw;tenon saw
突跳 kick
突跳电路 kick circuit
突跳电压 step voltage
突跳响应 ramp-forced response
突头螺栓 headed bolt
突头试件 headed test specimen
突透容量 breakthrough capacity
突围 sally
突纹理 raised grain
突屋角 angle quoin
突线条 bolection mo(u)ld(ing);raised mo(u)lding
突型脊【地】 bounce cast
突岩 ledger rock;tor
突沿 butt
突沿的轮廓线 profiled nosing
突檐 projecting eaves
突眼光学玻璃 < 低色散和低折射的聚焦元件 > crown optic(al) glass
突腰线 corbel course
突涌 soil burst;heave piping
突隅石 diatoni;projecting quoins
突圆边饰 treacle mo(u)lding
突圆线刨 forkstaff plane
突圆转角的 return-cocked
突缘 flange;fla(u)nching;lip;lug;outstanding flange;profiled nosing
突缘匙形钻 nose bit
突缘的 nosed
突缘角钢 flange angle
突缘接合 flange joint
突缘接头 flange joint
突缘连接 flange connection
突缘木条 nosing piece
突缘饰 nosing
突缘饰线 nosing line
突缘线刨 forkstaff plane
突缘销钉 lug dowel
突缘圆饰 nosing bead
突跃 discontinuity;jump;sudden change
突跃变换 abrupt transformation
突跃弯头 sharp bend
突增 overshoot
突砖 projecting brick
突阻水跌 check jump
突嘴 projecting point;protruding point
突嘴型急滩 rapids of protruding point pattern; rapids of protruding point type

图 drawing;figure;map

图案 design;dessin;draft;draught;figuration;figure;graphic(al) form;graphic(al) pattern;pattern;stencil
图案矮篱 pattern dwarf hedge
图案表现法 figuration
图案玻璃 configurated glass;figured glass;figured rolled glass;figured sheet glass;muranese;patterned glass;plate glass
图案不清晰 < 压花玻璃缺陷 > dim design;dim letter
图案发生器 pattern generator
图案方毯 art square
图案花坛 design bedding;mosaic flower bed
图案花纹重叠 pulled pattern
图案花纹重复 pulled pattern
图案激励 raster excitation
图案记录器 chart recorder
图案结构 patterning
图案模糊 dim pattern
图案平板玻璃 figured plate glass
图案砌合 decorative bond;pattern bond
图案砌筑 ornamental brickwork;pattern
图案散光罩 patterned diffuser
图案色 pattern colo(u)r
图案设计 pattern design;pattern layout
图案识别 pictorial pattern recognition
图案速度 pattern velocity
图案塑膜 pattern mo(u)ld
图案填充 pattern-filling
图案镶板 figured veneer
图案形成 patterning
图案形纹理 patterned grain
图案颜料 trick paint
图案艺术家 graph artist
图案纸 Bristol paper
图案重叠控制 composite layering
图案装饰 patterning
图案装饰法 figuration
图案走样 < 陶瓷表面彩饰缺陷 > floating
图板 chart board;chart desk;drawing base;manuscript sheet;map board;plotting sheet;trestle board
图板套 map pocket
图板藻煤 torbanite
图版纸 enamel(l)ed paper
图边 sheet margin
图边补充说明 supplemental note
图边地物 marginal object
图边注记 marginal name;marginal note;supplemental note
图边资料 title block
图标 icon;title;name block < 图纸右下角的 >
图标板 legend plate
图标菜单 icon menu
图标符号 < 插图说明 > legend
图标式菜单 icon menu
图表 bar diagram;chart;diagrammatic map;diagrammatic view;graph;graphic(al) chart;graph-table;pictograph;plan;plot(ting table);scheme;tabular representation
图表板 graphic(al) panel
图表比例尺 chart scale
图表边线 neat line
图表编号 numbering of figure;sheet number
图表表示法 graphic(al) presentation
图表测度法 graphic(al) rating scale
图表阐释 graphic(al) interpretation
图表处理法 < 设计制图的 > graphic(al) processing
图表的 diagrammatic
图表法 diagram method;graphics;schedule method
图表分解 diagrammatic decomposition
图表分解法 diagrammatic decomposition method
图表符号 legend;schematic symbol
图表管理 management through figures
图表记录 chart recording;recording card
图表记录器 chart recorder
图表记录式仪表 chart-recording instrument
图表记录速度 chart speed
图表技术 table technique
图表零点 datum zero
图表模型 schematic model
图表盘 graph disc
图表曲线 diagrammatic curve
图表设计 graphic(al) design
图表设计表 graphic(al) design table
图表设计法 graphic(al) design(method)
图表式指示器 card type indicator
图表室 chart room
图表说明 illustration statement
图表索引 chart index
图表台 chart table
图表完备程度 efficiency of diagram relative
图表文字符号 graphic(al) language
图表显示 graphic(al) display
图表显像 graphic(al) display
图表项目 graph-table entry
图表信息检索语言 graph information retrieval language
图表形式 diagrammatic form
图表研究 charting
图表艺术 graphic(al) arts
图表语言 diagram language
图表阅读者 chart reader
图表纸 chart paper
图表中没有的线 non-diagram line
图表注释编号 numbering of note to table and figure
图表资料 graph data
图表自动记录 automatic autographic(al) record;automatic graphic(al) record
图册 atlas
图带 image strip
图袋 chart board
图档室 drawings and documents room
图的半径 radius of graph
图的边界点 boundary point of graph
图的邻接顶点 adjacent vertices of a graph
图的起点 source of graph
图的说明 legend
图的右侧 right side of the figure
图的终点 sink of graph
图的左侧 left side of the figure
图底 intermediate guide key

图钉 drawing pin; tack (nail); thumb pin; thumb tack
图段变换 segment transformation
图段优先级 segment priority
图厄系统 Thue system
图尔-马歇尔弯曲试验 Tour-Marshall test
图尔奈大理石 <比利时> Tournai marble
图尔尼黄铜 Tournay metal
图分析 map analysis
图幅 chart sheet; dimension of chart; drawing size; mappable unit; map sheet; plate dimension; sheet
图幅边缘 map margin
图幅边缘匹配 edge matching
图幅编号 map number; nomenclature; sheet designation; sheet number
图幅编号字母 map number; sheet letter
图幅编辑计划 sheet editing instruction
图幅尺寸 map dimension; map face; sheet dimension
图幅大小 printing dimension
图幅代号 number of divisional map
图幅范围 sheet coverage
图幅分幅 sheet division
图幅分幅略图 chart relationship; interchart relationship; intermap relationship; map relationship
图幅高程基准面 map plane
图幅号 call number
图幅接边 constitution of details; continuation of details; continuity of details; joint the map; map adjustment; map join
图幅接合表 assemblage index; chart index; location index; map relation(ship); sheet assembly
图幅接合索引表 assemblage index; neighbo(u)rhood index
图幅面积 <内图廓线内的面积> map face; sheet area; sheet coverage
图幅名称 name of divisional map; quadrangle name; sheet name
图幅内航线数 number of strips in the sheet
图幅内相片数 number of photos in the sheet
图幅拼接索引 chart index
图幅拼贴 sheet assemble
图幅数量 amount of divisional map
图幅说明书编制要求 requirement for map introduction
图幅天底点 map nadir
图幅验收 reception of maps
图幅验收单位 accepting unit of geologic(al) map
图幅移动 map-sheet motion
图幅右下角 bottom right-hand corner; lower right-hand corner
图幅中心 center[centre] of sheet
图幅中心相片 quadrangle-centered photograph
图幅中心照片 quadrangle-centered photograph
图幅左下角 bottom left-hand corner; lower left-hand corner
图格 lattice; map grid
图根测量 mapping control survey
图根导线测量 mapping traversing
图根点 mapping base point; mapping control point; supplemental control; supplementary control point; traverse point
图根高程测量 mapping control vertical survey; mapping height survey
图根解析补点 complementary control point of analysis mapping
图根控制 mapping control; topographic(al) control
图根控制点 mapping control point
图根平面位置测量 mapping control horizontal survey
图根平面位置测量方法 method for horizontal mapping control survey
图根三角测量 detail triangulation; mapping triangulation; topographic(al) triangulation
图根水准测量 graphic(al) level(l)ing; mapping control level(l)ing
图根网 survey network of lower name
图功率放大器 image power amplifier
图号 drawing number; graphic(al) symbol(ization); sheet number
图合并 union of graph
图盒 map case
图横线 map parallel
图划分 graphic(al) partition(ing)
图画 drawing; painting; pictogram
图画的 pictorial
图画的构成 pictorial composition
图画明暗法 chiaro(o)scuro
图画模型 pictorial model
图画清漆 picture varnish
图画示意图 pictorial diagram; picture diagram
图画书 picture book
图画缩放仪 eidograph
图画影光 chiaro(o)scuro
图画杂志 pictorial
图画纸 cartridge paper; drawing paper
图画专藏 picture collection
图画资料档 picture file
图基间隙检验 Tukey's gap test
图基快速检验 Tukey's pocket test; Tukey's quick test
图基统计量 Tukey's statistic
图集 atlas; collected drawings; collection of drawings; collective drawings; graphics set
图集位置指示格网 atlas grid
图记样方 chart quadrat
图夹 chart board; plan file
图架 chart holder; chart rack; map holder
图件 map sheet
图件比例尺 map scale
图件编号 map number
图件编辑 map editing
图件编制 map compilation
图件规格 map dimension
图件绘制 map development
图件类别 class of map sheet
图件密级 security classification of map
图件名称 name of graphs
图件说明 direction of map
图件统一编号 set of maps
图件性质 property of graphs
图件褶集 map convolution
图件正常规格 map norm dimension
图件种类 map type
图角点 sheet corner
图角点坐标 corner geographic coordinates
图角坐标 corner coordinates; sheet corner coordinates
图解 scheme; graphic(al) calculate representation; graphic(al) expression; graphic(al) symbol(ization); graphics solution; iconography; illustrated diagram; illustration; pictogram; schematization; cartogram; delineation
图解比较 graphic(al) comparison
图解比例尺 bar scale; divided scale; graphic(al) scale; scale line
图解变换 graphic(al) transfer
图解表 graphic(al) chart
图解表示(法) diagrammatic representation; graphic(al)(re)presentation
图解测定(法) graphic(al) determination; diagrammatic determination
图解测井(记录) graphic(al) log
图解测图法 graphic(al) plotting
图解觇标 graphic(al) target
图解处理 graphic(al) manipulation; graphic(al) treatment
图解词典 illustrated dictionary; picture dictionary
图解代数学 graphic(al) algebra
图解导线测量 graphic(al) traverse [traversing]
图解的 diagrammatic; graphic(al); pictorial; schematic
图解点 geometric(al) point; graphic(al) dot; graphic(al) point
图解定位 graphic(al) position finding; graphic(al) position fixing
图解定位法 graphic(al) location (method)
图解法 descriptive geometry solution; diagrammatic solution; diagram method; graphic(al) illustration; graphic(al) measurement; graphic(al) technique; graphic(al) treatment; graphology; schematic way; graphic(al) analysis; graphic(al) method; graphic(al) representation; graphic(al) solution
图解法步骤 graphic(al) procedure
图解法反应器设计 graphic(al) reactor design
图解法级配集料 graphic(al) proportioning of aggregate
图解法计算 graphic(al) computation
图解法经验公式 empiric(al) formula by plotting
图解法配料 graphic(al) method for proportioning
图解放大 graphic(al) enlargement
图解分解法 diagrammatic decomposition method
图解分类 graphic(al) classification
图解分析 graphic(al) analysis
图解分析的 graphic(al) analytic(al)
图解分析法 graph(ic) analytic(al) method; method of graphic(al) analysis; selected ordinate method
图解符 graphic(al) mark
图解符号 diagrammatic symbol; graphic(al) symbol(ization)
图解辐射三角测量 graphic(al) radical triangulation
图解拱的分析 graphic(al) arch analysis
图解估计 graphic(al) estimation
图解估计量 graphic(al) estimator
图解后方交会(法) graphic(al) resection
图解机 graphic(al) type machine
图解积分法 graphic(al) illustration; graphic(al) integraph; graphic(al)(method of) integration
图解计算(法) graphic(al) calculation (method); graphic(al) evaluation; graphics
图解计算机终端装置 graphic(al) computer terminal
图解计算系统 graphic(al) computing system
图解记录 graphic(al) record
图解记录器 graphic(al) recorder
图解记录仪 graphic(al) meter
图解技术 graphic(al) technique
图解加法 graphic(al) addition
图解加密 graphic(al) extension
图解加密控制 graphic(al) extension control
图解交会 graphic(al) intersection
图解交会法【测】 graphic(al) intersection method; alidade method
图解校正 graphic(al) correction
图解阶段 diagram stage
图解结构 graphic(al) texture
图解结构分析 graphic(al) structural analysis
图解解析法 graphic(al) analytic(al) method
图解近似法 graphic(al) approximation method
图解精度 graphic(al) accuracy
图解静力学 graphic(al) statics
图解纠正 <航空照片> graphic(al) transformation; graphic(al) rectification; graphic(al) restitution
图解控制 graphic(al) control
图解理论 graph theory
图解理论模型(法) graph theory model
图解力学 graphic(al) mechanics
图解立体测图 graphic(al) stereocompilation; graphic(al) stereoplotting
图解立体测图仪 graphic(al) stereometer
图解例证 graphic(al) illustration
图解量表 graphic(al) scale
图解量测 graphic(al) measurement
图解模型 graphic(al) model; graph model; schematic model
图解内插法 graphic(al) interpolation
图解抛物线法 graphic(al) parabola method
图解平差法 graphic(al) adjustment
图解平均值 graphic(al) mean
图解平面 diagrammatic plan
图解评价 graphic(al) evaluation
图解评价法 graphic(al) appraisal method
图解评价量表法 method of graphic(al) rating scale
图解评审技术 graph evaluation and review technique
图解剖面 diagrammatic profile; diagrammatic section; graphic(al) profile
图解求积分法 integration by graphical method
图解曲线 diagrammatic curve
图解软件系统 graphics software system
图解三角测量 graphic(al) triangulation; plane-table triangulation
图解设计 graphic(al) design
图解实体 graphic(al) entity
图解式 graphic(al) formula
图解式测距仪 diagram tacheometer; graphic(al) tach(e) ometer
图解式计算机 pictorial computer
图解式记录装置 graphic(al) recording unit
图解式面板 graphic(al) panel
图解视距仪 diagram tacheometer
图解视图 diagrammatic view
图解视准仪 diagram tacheometer
图解数据 graph data; graphic(al) data
图解数据处理 graphic(al) data processing
图解顺序 graphic(al) sequencing
图解说明 graphic(al) extension; graphic(al) illustration
图解算法 graph algorithm
图解缩小转绘 graphic(al) reduction
图解特性 graphic(al) performance
图解统计分析 graphic(al) statistic(al) analysis

图解透视网格 graphic(al) perspective
图解图 graphic(al) drawing
图解图表 graphic(al) diagram
图解图根点 graphic(al) mapping control point
图解图例 graphic(al) symbol(ization)
图解图样 nomogram;nomograph
图解推导(法) graphic(al) derivation
图解推求法 graphic(al) derivation; graphic(al) deviation of results from data
图解推算船位法 graphic(al) dead reckoning
图解推算船只法 graphic(al) dead reckoning
图解外推 graphic(al) extrapolation
图解外推法 graphic(al)-extrapolation method
图解网络评核法 graphic(al) evaluation network method
图解微分法 graphic(al) differentiation
图解误差 graphic(al) error
图解显示器 graphoscope
图解限差 graphic(al) tolerance
图解相关 graphic(al) correlation
图解像片三角测量 graphic(al) phototriangulation
图解形式 graphic(al) aspect;graphic(al) form;graphic(al) model
图解学 graphics
图解研究 graphic(al) investigation
图解验证 graphic(al) check
图解样方 chart quadrat
图解值 graphic(al) value
图解转换 graphic(al) transformation
图解资料 graphic(al) material
图解子程序 graphic(al) subroutine
图解最优化 graphic(al) optimization
图解作法 graphic(al) construction
图距单位 map unit
图距扩散 map expansion
图距(离) map distance
图卡 graph card
图康试验器 Tukon tester
图康显微硬度 Tukon hardness
图康显微硬度计 Tukon tester
图库 chart gallery;map depot;map library;map storage
图块 block of graph;segmenting
图块属性 segment attribute
图框 frame;map boundary
图框滤波 picture frame filter
图廓 edge of the format; limit of map; limit of sheet; map border; map edge;map frame;map margin; sheet border; sheet edge; sheet margin
图廓比例尺 border scale
图廓标注说明 border information
图廓尺寸与理论尺寸较差 deviation between map size and nominal size
图廓点 marginal point;sheet corner
图廓点横坐标 sheet corner abscissa
图廓点经度 sheet corner longitude
图廓点纬度 sheet corner latitude
图廓点位移值 position error of sheet corner
图廓点直角坐标表 plane coordinate intersection tables
图廓点纵坐标 sheet corner ordinate
图廓点坐标 corner coordinates; corner geographic coordinates; sheet corner coordinates
图廓点坐标值 sheet corner value;values at sheet corners
图廓花边 cartouch(e)
图廓角 marginal angle
图廓经线 bounding meridian
图廓蒙片 edge mask
图廓内部分 map body
图廓数字注记 border figure
图廓外 legendary
图廓外测图说明 survey note
图廓外说明 explanatory;explanatory notes;marginal note
图廓外整饰 marginal representation; title block
图廓外资料 edge data;legendary data
图廓纬线 bounding parallel
图廓下边 bottom border
图廓线 border line; boundary line; limit of map; limit of sheet; map boundary;perimeter of figure;sheet line
图廓整饰 map decoration
图廓整饰样图 letterhead; provide in the margin;sheet layout
图廓注记 border information; border note; letterhead; marginal information
图廓资料 border data;mapped data
图拉真纪功柱 <古罗马> Trajan's column
图拉真纪念柱 <古罗马> column of Trajan
图勒群 Thule group
图类 type of map
图历表 compilation history;history of cartographic(al) work; map history;sheet history
图历薄 file of a map;mapping recorded file;quadrangle report
图立电传机 graphtyper
图例 legend(symbol); characteristic sheet; coding legend; conventional legend; conventional sign; conventional signs legend; conventional symbol; cut line; descriptive data; descriptive statement; explanatory notes;explanatory pamphlet;graphic(al) example; graphics symbol; list of signs;list of symbols;manual of symbols; map legend; map symbol;symbol table;table of conventional signs
图例板 legend plate
图例表 explanatory label
图例符号 graphic(al) symbol(ization);legend of symbols
图例说明 key to legend;key to symbol;marginal data;explanatory legend
图灵机 <计算自动化的一种数学理想化机器>【计】 Turing machine
图灵计算机 Turing machine
图噜喀尔特构造段 Turugart tectonic segment
图滤波技术 map filtering technic
图论 theory of graph;graph theory
图面 face of map; map body; map face;map surface;picture plane
图面布置 laying drawing
图面点 map point
图面定位 map reference
图面格网 map reference
图面积 area of pictorial surface
图面排字 lettering of map
图面配置 map face position;map layout
图面设计 map face design;preliminary layout
图面天底点 map nadir
图面位置 map place
图面注记 lettering on map surface
图面注记机 map surface lettering machine
图面自动片字 automated map lettering
图名 drawing name;map title;quadrangle name; sheet designation; sheet title
图谋 contrive
图囊 map case;map pocket;satchel
图盘式监视【计】 dial supervision
图片 photograph;wafer;picture plane
图片边缘 image border
图片裁切 cropping
图片处理技术实现 picture processing implementation
图片的 pictorial
图片发射 photo transmission
图片放映机 opaque telop
图片复制 picture reproduction
图片烘干上光机 print glazer
图片剪贴艺术 papier colle
图片摄影机 magazine camera
图片转绘仪 sketch master
图片自动机 picture automata
图谱集 atlas
图签 caption of a drawing; drawing logo;name block;title block
图切剖面图 transverse cutting profile
图上标指示器 map target indicator
图上表示隐藏结构的虚线 hidden lines
图上测距仪 map measurer
图上定位 paper location
图上定位点 map point
图上定位装置 chart position indicating apparatus
图上定线 paper location
图上加印的空中领航数据 aerooverprint
图上加印的领航资料 aeronautical information overprint
图上距离 map distance;map range
图上快速定向能力 eye for map
图上量标 map measurement
图上量表 map measurement
图上量测 scaling on map
图上量算 cartometry;map measure
图上路线 map course;traverse course
图上没有(标明)的 uncharted
图上目测 map eye for map
图上设计的导线网 proposed traverse net
图上体积量测(法) measurement of volume on map
图上位置 charted position
图上选点 map reconnaissance
图上选址 paper location
图上战术作业 tactical scheme
图上作业 map man(o)euver; mapping problem
图上作业法 graphic(al) dispatching method;graphic(al) method;operation method expressed in graphs; paper method
图示 diagrammatic presentation;graphic(al) expression; graphic(al) presentation;graphics;schematic representation
图示比例尺 graphic(al) scale
图示测速法 tachygraphometry
图示测微计 graphotest
图示处理机分机 <装设在各车站、机务段、列车段输送实时信息> agent set of graphic display processor
图示地图 pictorial map
图示电平记录器 graphic(al) level recorder
图示二相土 schematic two-phase soil
图示法 charting; chart representation;graphic(al) instrument method; graphic(al) method; schematic manner; diagrammatic representation;graphic(al) representation
图示符号 graphic(al) mark; graphic(al) symbol(ization)
图示工程进度表 graphic(al) progress chart
图示航行计算机 pictorial computer with courser
图示记录 graphic(al) record;graphic(al) recording
图示记录仪 graphic(al) recorder
图示解法 graphic(al) solution
图示控制盘 graphic(al) panel
图示控制台 graphic(al) console
图示录井图 graphic(al) log
图示面板 <测量系统的> graphic(al) panel
图示目录 illustrated catalogue
图示盘 graphic(al) panel
图示配电盘 graphic(al) panel
图示评审技术 evaluation and review technique;graph evaluation and review technology
图示剖面 diagrammatic section
图示器 graphic(al) instrument;graphic(al) recording instrument
图示曲线 diagrammatic curve
图示设计 graphic(al) design
图示深度记录 graphic(al) depth record
图示湿度计 hygrodeik
图示式计数器 graphed counter
图示数据处理 graphic(al) data processing
图示水位表 graphic(al) water-stage register
图示土类 <土壤调查的> mapping unit
图示位置显示器 graphic(al) position indicator;pictorial position indicator
图示行为法 charting
图示型录 illustrated catalogue
图示性 pictorialness
图示仪(器) graphic(al) instrument; grapher
图示字符 graphic(al) character
图式 characteristic sheet; symbol table;table of conventional signs
图式比例尺 graphic(al) scale
图式布置 diagrammatic arrangement
图式的 diagrammatic
图式符号 cartographic(al) symbol(ism); characteristic sheet; list of signs;manual of symbols;published symbol(ism); symbolism; symbolization;symbol list
图式符号标准化 symbol standardization
图式符号系统 set of conventional signs
图式化 schematization
图式曲线 diagrammatic curve
图式显示 graphic(al) display
图释 graphic(al) interpretation
图书 credential
图书编号法 number for books system
图书传送带 book conveyer[conveyor]
图书发行对象 readers interested
图书分馆 branch library
图书馆 athen(a)eum;library
图书馆藏书 bibliotheca
图书馆书籍的提要说明 <注明页数、插图等> collation
图书馆数据处理 library data processing
图书馆数据库 library data base
图书馆装帧 library binding
图书馆自动化系统 library automation system
图书馆租书处 lending library
图书管理员 librarian
图书架 bookrest
图书入藏目录 acquisition list
图书升降机 booklift
图书缩微胶片 bibliofilm

T

图书提要说明 collation
图书显微软片 bibliofilm
图书阅览室 browsing room
图书资料 books and reference materials
图书资料外借 home lending
图搜索策略 graph search strategy
图搜索控制 graph search control
图素 graphic(al)element;pixel
图算 graphic(al)calculation
图算法 nomograph alignment chart;nomography
图塔尼阿锡锑铜合金 Tutania metal
图腾 totem
图腾柱 <刻有图腾像的> totem post;totem-pole
图腾柱放大器 totem-pole amplifier
图贴 block
图外整饰 marginal decoration
图网 grid;lattice;map grid
图网对准 map align
图网坐标 grid reference
图文电视 teletext
图文法 graph grammar
图文检索 videotext
图文缺陷 defect of images
图文转印材料 transfer lettering system
图误 out of drawing
图显示器 graphic(al)alphanumeric display
图箱 chart gallery
图像 icon;iconography;image;image picture;pattern;pictogram;picture;presentation;scheme;video;vision frequency;crucifix <耶稣钉在十字架上的>
图像暗区 dark picture area
图像暗影 shading error
图像白色 image white;picture white
图像百分率 picture-percent
图像斑点调整 shading correction
图像板 image plate
图像半色调跃迁 picture halftone transition
图像保留 image retention;picture sticking
图像保留时间 image retention time
图像背景 image background;picture background
图像倍增恢复法 multiplicative reconstruction method
图像逼真度 image fidelity
图像比例尺 contact scale;image scale;picture scale
图像边界 image boundary
图像边界点 sharp point
图像边框 framing mask
图像边缘 image border;margin of image;picture border;picture edge
图像边缘分块 edge segmentation
图像边缘分析 edge analysis
图像边缘幅度 edge amplitude
图像边缘模糊 soft edge
图像编辑 picture editing
图像编辑软件【计】image editor
图像编码 image coding;image encoding;picture coding
图像编码技术 image coding technique
图像编码理论 theory of image coding
图像编码器件 image coding devices
图像变坏 image deterioration
图像变换 image conversion;image transformation
图像变换器 image transformer;image translator
图像变换摄像机 image converter camera
图像变劣 picture degradation
图像变位 pattern displacement
图像变形 anamorphose;anamorphosis;deformation of image
图像标准转换设备 picture transfer converter
图像表示 graphic(al)representation;image representation;pictorial representation
图像表示法 iconic representation;iconography
图像表式 graphic(al)present
图像波段 image band
图像波段号码 band number of image
图像波函数 image wave function
图像波纹横条 hum bar
图像补充 picture replenishment
图像不同性 image dissimilarity
图像不稳定 judder;jitterbug
图像不稳定性 picture instability
图像部分 image section
图像彩色质量 picture colo(u)r quality
图像参数 image parameter
图像残留 picture retention
图像测量装置 image measuring apparatus
图像插行 image interpolation
图像差 image difference
图像场修正信号 image flattening lens
图像持久性 image persistance
图像持续时间 image persistance
图像持续性 perpetual of pattern
图像尺寸 dimension of picture;image dimension;image size;picture dimension;picture size
图像重叠 patterning
图像重复 image repetition
图像重复循环 pattern repeat cycle
图像重合 image registration;picture registration;picture superimposition
图像重合失调 misregistry
图像重合失调校正 misregistration correction
图像重建 image reconstruction
图像重显 image reconstruction;image reproduction;reconstruction of image;reproduction of image
图像重显精细度 reproduced image fineness
图像重显装置 image reconstructor;image reproducer;picture reproducer
图像重现 image repetition
图像重影 streaking
图像出处 image source
图像处理 graphic(al)processing;image manipulation;image process(ing);image treatment;picture processing
图像处理产品 image processing products
图像处理程序包 image processing package
图像处理功能 frame process
图像处理机 image processor
图像处理激光扫描机 laser image processing
图像处理技术 image processing technique;image processing technology
图像处理解译方法 image processing interpretation method
图像处理设备 image processing equipment
图像处理数字系统 image processing digital system
图像处理系统 image processing system
图像处理语言 image manipulation language
图像传感 image sensing
图像传感器 imaging sensor;picture sensor
图像传输 image transfer;image transmission;graphic(al)transmission;picture transmission
图像传输部分 image stage
图像传输扫描 image transmission scanning
图像传输系统 image transfer system;image transmission system;picture transmission system
图像传输与处理 image transmission and processing
图像传送 picture transmission period
图像传真 image transmission
图像传真机 picture facsimile apparatus
图像传真台 facsimile broadcast station
图像传真信号 facsimile signal
图像串扰 picture crosstalk
图像存储 image storage
图像存储变换和再生 image storage translation and reproduction
图像存储管 image storing tube;picture storage tube
图像存储器 picture memory;video memory
图像存储器件 image storage device
图像存储阵列 image storage array
图像存取法 graphic(al)access method
图像存取空间 image storage space
图像打印 image printing
图像大小 <帧的宽与高之比> picture shape
图像代数修复 algebraic(al)approach restoration
图像带 picture strip
图像带宽压缩 image bandwidth compression;picture bandwidth compression
图像单元 elemental area;elementary area;picture unit
图像档案 image file
图像的 iconic
图像的变换编码【电】transform coding of image
图像的多重性 multiplicity of image
图像递降 image degradation
图像点处理 point processing
图像电传机 graphtyper
图像电荷 image charge;picture charge
图像电话 face-to-face picture phone;picture-phone
图像电流 picture current
图像电流密度 image current density
图像电路 video circuit;vision circuit
图像电平 picture level
图像电平控制 picture level control
图像电位分布图 picture charge pattern
图像电子放大 zoom
图像叠化 video dissolve
图像定位 framing
图像定心 picture centring[centering]
图像定心调整机构 picture centring adjustment mechanisms
图像动乱 busy picture
图像抖动 floating
图像对比度 image contrast;picture contrast
图像对角线 diagonal of picture;image diagonal;picture diagonal
图像多边检测器 polyhedral edge detector
图像多余度 image redundancy
图像惰性 picture lag
图像发射机 image transmitter;picture transmitter;vision transmitter
图像发射机测试柜 vision transmitter monitoring equipment
图像发射机功率 image transmitter power;vision transmitter power
图像发射机功能 picture transmitter power
图像发射机输出 vision transmitter output
图像发射机输出功率 output power of visual transmitter;vision transmitter output power
图像发送 facsimile transmission
图像发送装置 picture transmitter device
图像发晕 blooming
图像法 image method
图像反差 image contrast
图像反转 image inversion;picture inversion
图像放大 image multiplication
图像放大灯 image enlarger lamp
图像放大光电摄像管 image amplifier iconoscope
图像放大器 image amplifier;image multiplier;picture amplifier;vision amplifier
图像放大器自动增益控制 vision automatic gain control
图像分辨力 image resolution;picture resolution
图像分辨率 image resolution ratio
图像分段 image segmentation
图像分割 picture cup apart
图像分割法 image segmentation
图像分解 image dissection;scansion
图像分解力 picture resolution
图像分解器 dissector tube;image dissector
图像分解扫描 scanning in transmission
图像分解摄影 image dissection photography
图像分类 image classification
图像分量 picture content
图像分排 pictorial section
图像分析 graphic(al)analysis;image analysis;picture analysis
图像分析法 image analysis method
图像分析器 picture analyser[analyzer];mage analyser[analyzer]
图像分析仪 picture analyser[analyzer];mage analyser[analyzer]
图像分析照相机系统 image dissector camera system
图像份数 number of copies
图像浮散 bloom
图像符号化 image symbolization
图像幅度 picture amplitude
图像幅面 image form
图像负拖尾 negative streaking
图像复合 image overlaying
图像复原 image restoration
图像覆盖面积 image area coverage
图像改正器 image corrector
图像干扰 image interference;picture breakdown;picture interference;second channel interference;vision interference
图像干扰比 image disturbance ratio
图像干扰限制器 vision interference limiter
图像感应系统 image sensing system
图像感知系统 image sensing system
图像高度 field height;image height;picture altitude
图像高度控制 picture altitude control
图像高度调整 height control
图像格式项 picture format item
图像跟踪 picture-chasing;video track
图像构成 image construction
图像估价 image evaluation
图像管 image tube;converter tube
图像光电变换管 image converter;image converter tube;image-viewing

tube
图像光电管 image photocell
图像光电摄像管 image iconoscope
图像光栅 image raster;picture raster
图像光心 optic(al) picture center[centre]
图像光学恢复 image optic(al) reconstruction
图像滚动 image roll;image scrolling;picture roll
图像行畸变 line bend
图像合成 image composition;image synthesis;pictorial composition;picture composition;picture synthesis
图像合成扫描(显像) scanning in reception
图像和伴音 picture-and-sound
图像和伴音同步 sight-sound synchronization
图像和伴音载波的幅度比 picture-to-sound ratio
图像和波形监视器 image and waveform monitor;picture and waveform monitor
图像和消隐信号 picture and blanking signal
图像黑白对比 picture contrast
图像黑白反转 picture inversion
图像黑斑补偿信号 picture-shading signal
图像黑部分 picture black
图像黑色信号 picture black signal
图像哼扰 picture hum
图像红外线摄影术 pictorial infrared photography
图像滑动 picture slip
图像缓冲器 frame buffer
图像灰度校正 gamma correction
图像灰度控制 gamma control
图像恢复 image restoration
图像恢复技术 image recovery technique
图像混合 image mixing
图像混合操作员 vision mixture operator
图像混合控制台 vision mixer control panel
图像混合器 video mixer;vision mixer
图像混合系统 vision mixing system
图像机构 picture mechanism
图像基础 image basis;image fundamental
图像畸变 deformation of image;frame distortion;image distortion;pattern distortion;picture distortion
图像激励级 vision driver stage
图像激励器 vision driver
图像几何畸变 picture geometry fault
图像几何校正 geometry correction of imagery
图像几何误差 image geometric(al) error
图像几何形状 picture geometry
图像几何(性质) image geometry
图像记录 image record(ing);picture record
图像记录程序 picture recorder
图像记录媒体 image recording medium
图像记录器 picture recorder;scanner recorder
图像记录仪 graphic(al) recording instrument
图像加工 image process(ing)
图像加速电极 image accelerator
图像加速极电压 image accelerator voltage
图像监控器 image monitor
图像监视接收机 picture monitoring receiver
图像监视器 image monitor;picture monitor
图像监视设备 vision monitoring equipment
图像剪辑放大器 montage amplifier
图像检波 image detection
图像检波器 detection unit picture;vision detector;vision signal detector
图像检测 image detection
图像检测板 image sensing panel
图像渐显 fade in
图像渐隐 fade out
图像交流干扰 picture hum
图像角 angle of image;image angle
图像角清晰度 corner detail
图像洁化 map cleaning
图像结构 image texture;picture structure
图像结束 end of image
图像截止 blackout;visual cut-off;visual cut-out
图像截止电压 picture cut-off voltage
图像解码 picture decoding
图像解调器 visual demodulator
图像解析器 image dissector
图像解译标志 image interpretation criterion
图像解译方法 image interpretation method
图像解译设备 image interpretation equipment
图像解译样片法 method of image interpretation keys
图像景色 visual scene
图像聚焦 image focus(ing)
图像聚焦检查镜 focusing glass
图像聚焦检验器 focusing eyeglass
图像卷起 image scrolling
图像卷曲 image warping
图像均衡器 graphic(al) equaliser[equalizer]
图像卡片 image card
图像开关 presentation switch
图像颗粒结构 picture grain
图像可测性 image measurability
图像可见度 image visibility
图像客观保真度标准 objective fidelity criterion
图像空间 image space;picture space
图像空间坐标系(统) image space coordinate system
图像孔径 picture aperture
图像控制 image control;image manipulation;vision control
图像控制技术人员 vision supervisor
图像控制台 picture control desk
图像控制线圈 picture control coil
图像库 image library
图像宽度 image width;picture traverse;picture width
图像宽度调整 width control
图像宽高比 aspect ratio;image aspect ratio;picture aspect ratio;picture ratio
图像类型 image type
图像累积跟踪器 image integrating tracker
图像理解系统 image understanding system
图像亮度 brightness of screen picture;image brightness;image intensity;image luminance;picture brightness
图像亮度放大管 image intensifier tube
图像亮度校正器 gamma corrector
图像亮化器 image intensifier
图像量化 image quantization
图像劣化模型 image degradation model
图像录制 image record(ing);image transcription;video recording
图像轮廓 picture contour
图像轮廓加重电路 crispening circuit
图像面 pictorial surface
图像面积 image area;picture area
图像面积测量 image area test
图像描述 image description;picture description
图像描述文法 picture description grammar
图像描述语言 picture description language
图像命令中心 graphic(al) command centre[center] for maps
图像模糊 image blurring;image diffusion;picture blurring
图像模糊警铃 fog bell
图像模糊现象 blooming
图像模糊消除 image deblurring
图像模拟 image simulation
图像模式 iconic model
图像模式识别 image pattern recognition
图像模型 iconic model;image model;pictorial model
图像内插 image interpolation
图像内容 picture material
图像能 image energy
图像弄脏 picture smear
图像判读 image interpretation;photoreading
图像判读者 photointerpreter
图像判读专家系统 expert system of image interpretation
图像配准 image register;image registration;picture registration;picture superimposition
图像匹配 image matching
图像匹配法 image matching technique
图像匹配监视器 picture matching monitor
图像匹配系统 image matching system
图像偏移 image shift;picture shift
图像频率 image frequency;picture frequency
图像频率范围 vision frequency range
图像频谱分析 image frequency spectral analysis
图像品质 image quality
图像平滑化 image smoothing
图像平均 image averaging
图像平均电平 mean picture level
图像平均亮度 mean picture brightness
图像平面 image plane;plane of delineation
图像屏幕 picture screen
图像起始 picture start
图像强度 graphic(al) strength;image intensity;strength of figure
图像强化 image intensification
图像切换 image switching;picture switching
图像切换矩阵 vision switching matrix
图像切换器 vision switcher
图像切换误差 picture switching error
图像清除器 piclear unit
图像清晰度 definition of image;image definition;image detail;image resolution;image sharpness;picture definition;picture sharpness
图像清晰化 image sharpening
图像清晰化处理 sharp processing of imagery
图像区分表 picture specification table
图像取均值 image averaging
图像取向 image orientation
图像取样 image sampling
图像圈 image circle
图像全息术 pictorial holography
图像缺陷 defect of images;image defect;image fault;picture defect;picture fault;picture impairment
图像人机联系处理 interactive image processing
图像容积形状变换器 image volume
图像锐度 picture sharpness
图像锐化 image sharpening
图像扫描 picture scanning;picture sweep unit
图像扫描电路 frame time base
图像扫描电压 picture sweep voltage
图像扫描频率 image sweep frequency
图像扫描器 cartographic(al) scanner;image analyser[analyzer];image scanner;picture scanner
图像扫描输出仪 image scanner;magic scanner
图像扫描线 image line;picture line
图像扫描装置 image-scanning device
图像色 pattern colo(u)r
图像色调 picture tone
图像色调质量 picture tonal quality
图像闪变 flicker of image
图像闪烁 image flicker
图像上下摆动 vertical hunting
图像上小条<颤噪效应引起的> microphonic bar
图像烧伤 image burn
图像设备 vision facility
图像射频信号 vision radio frequency signal
图像伸长 pulling
图像深淡程度 gradations of image
图像深度 picture depth
图像深浅等级 gradations of image
图像生成 image generation
图像声频干扰 sound on vision
图像声音模糊 blur
图像失落 picture drop-out
图像失锁 loss of picture lock
图像失真 frame distortion;image distortion;pattern distortion;picture smear
图像识别 image(ry) recognition;pattern recognition;picture identification;picture perception
图像式符号表 set of conventional signs;set of symbols
图像视频检波器 picture video detector
图像适配器 video graphic adapter
图像收缩法 shrinking-raster method
图像输出 image output;image printout
图像输出变压器 picture output transformer
图像输出功率 vision output power
图像输出信号 picture output signal
图像输入 image input;video input
图像输入变压器 picture input transformer
图像数据 graphic(al) data;image data;picture data
图像数据处理 image data processing
图像数据传真电话机 viewdata phone
图像数据发送-接收系统 viewdata transmit-receive system
图像数据格式 image data format
图像数据工作 viewdata operation
图像数据接收器 viewdata receiver
图像数据结构 image data structure
图像数据库 graphic(al) database;image data base;picture database
图像数据库系统 image database system
图像数据索引页面 viewdata index page

T

图像数据调制解调器 viewdata modem
图像数据图形 viewdata graphics
图像数据网 image data network; viewdata network
图像数据系统 viewdata system
图像数据信号 viewdata signal
图像数据压缩 image data compression
图像数据业务 viewdata service
图像数据页面 image data page; viewdata page
图像数值化 image digitization
图像数字化 image digitization; image digitizing; imagery digit
图像数字仪 photodigitizing system
图像数字语言 figurative language
图像数字转换器 image digital converter; image digitizer; picture quantizer
图像衰减 fade down; image attenuation; vision fading
图像衰减常数 image attenuation constant
图像衰减系数 image attenuation coefficient
图像水平摆动 < 行频不稳 > horizontal hunting
图像水平倒置 lateral inversion
图像水平偏移失真 slipping
图像水平位移 horizontal image shift
图像说明 picture specification
图像说明字符 picture specification character
图像撕裂 < 因同步不准造成的 > picture break-up; picture tearing
图像撕碎 tearing
图像搜索 picture search; picture seeking
图像素材 picture material
图像速调管 vision klystron
图像速度传感器 image velocity sensor
图像损缺 image burn
图像缩放转绘仪 zoom transferscope
图像索引 image index
图像锁定 pixlock
图像锁定技术 picture locking technique
图像锁定重放方式 pixlock playback mode
图像特性 picture characteristics; picture feature; picture property
图像特征抽取 image feature extraction
图像特征提取 image characteristics extraction
图像条(带) image strip
图像调整 image control; picture control
图像调整线圈 image control coil; picture coil
图像调制 picture modulation; vision modulation
图像调制摆幅 picture modulation swing
图像调制度 picture modulation percentage
图像调制器 vision modulator
图像跳动 bouncing; bouncing motion; image jitter; jitter (bug); jumping of picture; picture bounce; picture disturbance; picture jitter; roll-over
图像通道 image channel; picture channel
图像通道串扰 vision crosstalk
图像通道的同步灵敏度 image channel synchronizing sensitivity
图像通道的有限杂波灵敏度 image channel noise limited sensitivity
图像通道最大灵敏度 image channel gain limited sensitivity
图像通过量 image throughput
图像通信[讯] picture communication; visual communication
图像同步 image lock; picture lock; picture synchronization; synchronizing of images
图像同步传输系统 picture synchronization transmission system
图像同步信号 picture synchronization signal; picture synchronizing signal
图像统计图表 pictograph
图像投影 pictorial projection
图像投影法 < 超声波检测的 > image-projected method
图像投影方式 image projection form
图像透镜 image lens
图像图案 picture pattern
图像退化 image deterioration
图像退化模型 image degradation model
图像拖尾 hand-over; streaking
图像拖尾测试卡 streaking chart
图像外形失真 picture outline distortion
图像微音效应 microphonics
图像位移 image degradation; image shift; picture displacement
图像位置 picture position
图像位置调整信号 phasing signal
图像位置调整帧偏转 picture deflection
图像位置显示 pictorial position indicator
图像文件 image file
图像纹理分析处理 image texture analysis technique
图像稳定 image stabilization
图像稳定性 picture steadiness
图像系统 image system
图像细部 image detail
图像细节 image detail; pictorial detail; picture detail
图像细节区 image detail region; picture detail region
图像细节因数 image detail factor; picture detail factor
图像显示 graphic(al) presentation; map display; pictorial display; picture showing
图像显示板 video display board
图像显示定义方式 image display definition mode; pictorial display definition mode; picture display definition mode
图像显示接口 graphic(al) display interface
图像显示控制屏 graphic(al) panel
图像显示屏 cartoscope
图像显示器 image display; picture display; graphic(al) display
图像显示终端 graphic(al) display terminal
图像线路放大器输出 image line-amplifier output
图像相关 correlation of image(ry)
图像相关器 image correlator
图像相位变化系统 image phase-change coefficient
图像相位稳定器 video phase stabilizer
图像镶边 fringe
图像详细解译 image interpretation in detail
图像消失 fade down; picture drop-out; vision fading
图像消隐 picture blanking
图像信道 picture channel; video channel; vision channel
图像信号 image signal; picture intelligence; picture signal; visual signal
图像信号存储 picture signal store
图像信号带宽 vision bandwidth
图像信号对伴音的干扰 vision on sound
图像信号发射机 visual signal transmitter
图像信号发生器 image signal generator; picture signal generator
图像信号放大器 image signal amplifier; picture signal amplifier
图像信号分配放大器 image signal distribution amplifier; picture signal distribution amplifier
图像信号幅度 image signal amplitude
图像信号极性 image signal polarity; picture signal polarity
图像信号监视器 picture monitor; picture signal monitor
图像信号检波器 picture demodulator; picture video detector
图像信号检测 visual detection
图像信号接收机 picture receiver
图像信号接收器 picture sensor
图像信号频带 picture signal band
图像信号失落 image drop-out
图像信号输出 picture output
图像信号输出变压器 image output transformer
图像信号特性 vision signal characteristic
图像信号通路 vision channel
图像信号性质 picture signal property
图像信号载波 image carrier; luminance carrier; picture carrier
图像信号增益 picture signal gain
图像信号振幅 picture signal amplitude
图像信息 graphic(al) information; image information; pictorial information; picture information
图像信息程序包 video package
图像信息处理 graphic(al) information processing; image information processing
图像信息处理系统 image information processing system
图像信息电话 visual telephony
图像信息几何中心 geometric(al) center[centre] of picture information
图像形成 image formation
图像形成装置 image processing system
图像形式 graphic(al) form
图像形状 picture shape
图像旋转 picture orientation
图像旋转控制 picture rotate control
图像旋转设备 image orbiting facility; picture orbiting facility
图像学 iconography; iconology
图像压缩 image compression; image press; packing < 非线性扫描引起的几何失真 >; picture compression
图像移动 image motion; picture-area shift; picture traverse
图像移动补偿 image motion compensation
图像印录器 picture recorder
图像印刷机 image printer; video printer
图像映射 image map
图像映象 picture image
图像与伴音的偏量电路 picture/sound offset unit
图像与伴音载频的间隔 sound-to-picture separate; sound-to-picture separation
图像与地图叠合 chart matching
图像语言 graphic(al) language; picture language
图像原稿 original copy for illustrative matter
图像原样 picture original
图像源 eikonogen; image source; picture source
图像运算 image operation
图像载波 image carrier; picture carrier; video vision carrier; vision carrier; visual carrier
图像载波间隔 vision carrier spacing
图像载波陷波器 picture carrier trap
图像载波抑制 image carrier suppression; image rejection
图像载频 picture carrier; pix carrier; vision carrier; vision carrier frequency
图像载频抑制 image rejection
图像再现 image reconstruction; image reproduction; reconstruction of image
图像再现设备 picture reproducer
图像再现装置 picture reproducer
图像噪声 pattern noise
图像增亮 fade up
图像增强 image enhancement; image intensification; photographic(al) enhancement; picture enhancement
图像增强器 image-amplifying device; image intensifier
图像增强设备 image-enhancing equipment
图像照度 image illumination
图像照相刻蚀法 graphic(al) blast
图像照相制版法 graphic(al) blast
图像阵列 image array
图像整饰处理 image cosmetic processing
图像正拖尾 positive streaking
图像正析像管 image orthiconoscope
图像正析像管摄像机 image-orthicon camera
图像正析像管装置 image-orthicon assembly
图像直方图 image histogram
图像直接转换胶印法 direct-image offset system
图像值 image value
图像质量 image quality; picture quality
图像质量标准 image quality criterion; standard of picture quality
图像质量降级 image degradation
图像质量评价 image quality evaluation
图像中频 vision intermediate frequency
图像中频调制器 vision intermediate frequency-modulator
图像中心 picture center[centre]; picture centring[centering]
图像中最亮处 highlight
图像终端设备 image termination
图像终了信号 end-of-copy signal
图像逐点照明 progressive illumination
图像主监视器 master picture monitor
图像主要参量 principal image components
图像主要分量 principal picture components
图像转换 image conversion; image inversion; inversion of the image
图像转换比 image transfer ratio
图像转换法 conversion system
图像转换技术 image transfer technique
图像转换控制 turn-picture control
图像转换器 image converter
图像转换设备 image transfer converter

图像转换指数 image transfer exponent
图像转移加速器 image acceleration
图像转移特性 image transfer characteristics
图像自动处理设备 automatic image processing equipment
图像自动传输 automatic picture transmission
图像自动发射 automatic picture transmission
图像自动扫描器 automatic image scanner
图像自动扫描仪 automatic image scanner
图像综合 image synthesis;picture synthesis
图像综合天线阵 image synthesis array
图像纵横(尺寸)比 image ratio;aspect ratio;dimension of picture;picture ratio
图像最白部 picture white
图像最亮部分 highlight;hi-lite
图像左右倒置 lateral inversion
图心 centroid
图形 artwork;drawing;figure;graph;pattern(ing);logo <抽象的>
图形板 graphic(al)board;graphic(al)panel
图形报表 pictorial statement
图形报告生成程序 graphic(al)report generator
图形背景辨别 figure-ground discrimination
图形背景清晰度 figure ground articulation
图形比较 graphic(al)comparison
图形编辑程序 graphic(al)editor
图形变比 zooming
图形变换 graphic(al)transformation
图形变换功能 graphic(al)manipulation function
图形变量 graphic(al)variable
图形变示器 graph follower
图形标记 pictorial symbolization
图形标志 diagrammatic sign;symbol sign
图形标准 graphics standard
图形表示(法)graph notation;pattern representation;graphics(re)presentation
图形捕获 frame grab
图形部分 graphic(al)parts;picture parts
图形裁剪 graphic(al)clipping
图形参考点 pattern reference point
图形插入 drawing insertion;graphics insertion
图形程序 graphics program(me)
图形程序包 graphic(al)package
图形程序编辑 graphic(al)programming
图形程序设计 graphic(al)programming
图形尺寸分档程序 pattern size grading program(me)
图形抽象程序设计语言 graphic(al)abstract programming language
图形处理 graphic(al)manipulation;image manipulation
图形处理程序 graphic(al)processor
图形处理功能 graphic(al)manipulation function
图形处理机 graphic(al)processor
图形处理系统 graphic(al)processing system
图形处理语言 graphic(al)processing language
图形处理中心 graphic(al)processing center[centre];processor
图形传输 graphics transmission
图形窗口 graphic(al)window
图形存取法 graphic(al)access method
图形打印机 graphic(al)printer
图形大小 pattern size
图形单元 graphic(al)primitive
图形点阵打印机 graphic(al)matrix printer
图形叠加 graphic(al)overlay
图形叠置法 <环境影响评价的> overlay approach of environmental impact assessment
图形对象 graphic(al)object
图形发生 artwork generation
图形发生器 graphics generator;pattern generator
图形法 graphic(al)arts technique;graphic(al)mean
图形方程 graphic(al)equation
图形方式 graphics mode
图形分解 graph decomposition
图形符号 graphic(al)symbol(ization);graphics symbol;graph notation
图形符合 chart matching
图形复制 graphic(al)reproduction
图形复制器 graph follower
图形改正 pattern correction
图形概括 simplification of figure
图形干扰 picture disturbance
图形跟踪器 graph follower;graphic(al)follow;graphic(al)tracker
图形工作站 graphics workstation
图形管理功能 graphic(al)manipulation function
图形光标 graphic(al)cursor
图形很好三角形 well-conditioned triangle
图形积分器 chart integrator
图形基本要素 graphic(al)entity
图形基线 baseline of diagram
图形畸变 picture distortion
图形集 graphic(al)aggregation;graphic(al)atlas
图形集合 graphics set
图形记录仪器 graphic(al)recording instrument
图形加速端口 accelerated graphics port
图形交换能力 drawing interchange capability
图形校正 figure adjustment
图形接口 graphic(al)interface
图形结构 graphic(al)entity
图形结构输入 graphics structure input
图形结构算法 graph structure algorithm
图形解释程序 graphic(al)interpreter
图形静力学 graphics statics
图形决算表 graphic(al)statement;pictorial statement
图形孔眼 patterned hole
图形库 graphic(al)bank;graphic(al)base;graphic(al)repertoire;shape library
图形库设计 graphic(al)base design
图形块 graph block
图形类别 pattern class
图形联机编辑系统 on-line graphic(al)editing system
图形码扩展 graphic(al)code extension
图形面积 area of diagram
图形描述 pattern description
图形描述指令 picture description instruction
图形名称 pattern name
图形命令 graph command;graphic(al)command
图形模拟程序 graphic(al)simulator
图形目标 graphic(al)object
图形匹配 chart matching;graphic(al)matching;pattern matching
图形片段 picture segment
图形平差 adjustment of figure;figure adjustment;graphic(al)adjustment
图形评估技术 graphic(al)evaluation and review technique
图形强度 graphic(al)strength;image intensity;strength of figure
图形强度系数 coefficient of equal triangle
图形清晰度 pattern definition
图形曲线拟合程序 graphics curve fittings program(me)
图形软件 graphics software
图形软件包 graphic(al)package
图形设备 pattern facility
图形设备接口 graphics device interface
图形设计 graphic(al)design
图形设计协会国际理事会 International Council of Graphic Design Association
图形设计站 compuscope
图形失真 graphic(al)distortion
图形识别 figure identification;identification of drawing;pattern recognization[recognition]
图形识别对比法 image identification correlation method
图形识别对比法曲线元素 curve element in image identification correlation method
图形识别算法 graphic(al)recognition;pattern recognition algorithm
图形适合性试验 graphic(al)suitability test
图形输出 graphic(al)output;pattern printout;video output
图形输出板 plotter board;plotting board
图形输出设备 graphic(al)output unit
图形输入 graphic(al)input
图形输入板【计】graphic(al)tablet;graphics tablet;plotting tablet;tablet
图形输入交互式技术 graphic(al)input interaction technique
图形输入模式 graphic(al)input mode
图形输入设备 graphic(al)input unit
图形输入系统 graphic(al)input system;graphics input system
图形输入语言 graphic(al)input language
图形输入终端 graphic(al)input terminal
图形输入装置 graphic(al)input device;tablet
图形数据 graph data;graphic(al)data;graphics
图形数据表示 pictorial data representation
图形数据处理 digital graphics processing;graphic(al)data processing
图形数据分析 graphic(al)data analysis
图形数据结构 graphic(al)data structure
图形数据库 graphic(al)data bank
图形数据库管理系统 graphic(al)database management system
图形数据输入装置 graphic(al)data input device
图形数据系统 graphic(al)data system
图形数据整理 graphic(al)data reduction
图形数字化器 drawing digitizer;graphic(al)digitiser
图形缩放 pantography
图形索引表 list of pattern indices
图形条件 figure condition
图形同步并接 graphics hold
图形投影屏 chart screen
图形图 pie chart
图形土 patterned ground
图形外围设备 graphics peripheral
图形网络软件 graphic(al)network software
图形文本组合 graphics text composition
图形文档资料 graphic(al)documentation
图形文件 graphic(al)file;picture file
图形文件化 graphic(al)documentation
图形文件维护 graphic(al)file maintenance
图形系统 graphics system
图形显示程序 graphic(al)display program(me)
图形显示定义方式 graphic(al)display definition mode;image display definition mode
图形显示分辨率 graphic(al)display resolution
图形显示技术 graphic(al)display technics
图形显示控制 graphic(al)display control
图形显示器 graphic(al)alphanumeric display;graphic(al)display;graphic(al)display unit;pattern display;plotter;image display;picture display
图形显示软件 graphic(al)display software
图形显示设备 graphic(al)display unit
图形显示台 output table;plotting table
图形显示图 plotting table
图形显示系统 graphic(al)display system
图形显示终端 graphic(al)display terminal
图形显示装置 graphic(al)display unit
图形显示字符 graphic(al)display character
图形显示字符变换 graphic(al)character conversion
图形效果 graphic(al)effect
图形信息 figure information;graphic(al)information
图形学 graphics
图形要素 graphic(al)element
图形硬件 graphics hardware
图形用户接口 graphics user interface
图形用户界面 graphics user interface
图形语句 graphics statement
图形语言 graphic(al)language
图形域 graphics field
图形元 pattern primitive
图形元素 graphic(al)element
图形原语 graphic(al)primitive
图形阵列 graphic(al)array
图形中的回路 cycle in graph
图形中心 center[centre] of figure
图形终端 graphics terminal
图形重合仪 chart matching device
图形轴 axis of figure
图形转换 graphic(al)transformation
图形子程序包 graphic(al)subroutine package
图形子程序系统 graphics subroutine system
图形子空间 graph subspace
图形字段 graphics field
图形字符修改 graphic(al)character modification
图形字母数字发生器 graphic(al)alphanumeric generator

图形字母数字显示器 graphic(al) alphanumeric display
图形自动识别技术 computer pattern recognition technique
图形作业处理 graphic(al) job processing
图形作业处理程序 graphic(al) job processor
图形坐标变换 figure coordinate conversion
图型 graphic(al) form; graphic(al) pattern
图序列 graphic(al) sequence
图样 design; draft; drawing; pattern
图样爆破 pattern shooting
图样规范 graphic(al) standards
图样控制 model control
图样识别 pattern recognition
图页 chart sheet; map sheet
图移速率 chart speed
图应用 graph application
图元 pel; pixel
图元阵列 pixel array
图章 chop; seal; signet; stamp
图阵 block graphics; graphic(al) array
图镇 chart weight; paper weight
图纸 drawing(sheet); working drawing; blueprint <指蓝图>; drawing paper
图纸比例 scale of a drawing
图纸比例尺 echelle
图纸编号 drawing number
图纸编目 drawing list
图纸的保管 custody of drawings
图纸的标题页 title sheet
图纸的正面 face of drawing
图纸扉页 title sheet
图纸复制 duplication of drawing; reproduction of drawing
图纸更改通知 drawing change notice
图纸更改一览 drawing change summary
图纸管理 management of blueprint
图纸号数 blueprint number
图纸会审制度 drawing joint examination system
图纸夹 paper clip
图纸校正 revision of map
图纸空间 paper space
图纸类别 type of drawings
图纸名称 designation of drawing; title of drawing
图纸目录 drawing list; list of contents; list of drawings; table of drawings
图纸清单 drawing list; list of drawings
图纸上的文字书写 lettering on drawing
图纸上着重注意点的标注 callout
图纸设计 blueprint; drawing design; layout design
图纸说明 explanatory notes of drawings
图纸误期 delays of drawings
图纸一览 drawing summary
图纸中线 sheet line
图纸资料 information and drawing
图中标尺寸的箭头 crowfoot[复 crowfeet]
图注 cut line; explanatory text; notes on drawings
图注尺寸 nominal dimension
图注灯标射程 charted visibility of light
图注海岸线 charted coast
图注水深 charted depth
图注位置 charted position
图注资料 descriptive data
图组 family of maps; set of diagrams; sheet series

徒

徒步出行 pedestrian trip
徒步处 ford
徒步购物区 pedestrian mall
徒步旅行 tramp
徒步旅行者 wayfarer; way rarer
徒步涉水 ford
徒步上下旅客 walk on/walk off passenger
徒步游历 through migration
徒动速度 migration velocity
徒工 apprentice; roustabout
徒工制 apprentice system
徒里特 tolite
徒然 all for naught
徒然的 unproductive
徒涉处 ford
徒手搬运 manual weight carrying
徒手操纵 hand-operating
徒手法 method with hand
徒手画的 free hand
徒手画的草图 freehand drawing; freehand sketch
徒手画的线 freehand line
徒手绘曲线 freehand curve
徒手切片 free-hand section
徒手素描 freehand sketch
徒手图 freehand sketch
徒手圆滑 smoothing by hand
徒刑执行令 mittimus

涂

涂暗色 scumble
涂暗色技术 scumbling technique
涂白 lime wash; whitewash
涂白灰 whitening
涂白色颜料的 white pigmented
涂柏油 asphalt; tar down; tarring
涂柏油的 tarry
涂柏油的粗糙碎片 tarred coarse chip(ping)s
涂柏油的密封材料 tarred rope sealing
涂柏油的石灰石 tarred limestone
涂柏油的拖绳 black tarred tow; tarred tow
涂柏油的屋顶卷材油毛毡 tarred rolled-strip roofing felt
涂柏油的屋顶油毛毡 tarred felt
涂柏油的油毛毡织物 tarred felted fabric
涂柏油的预制薄板屋顶材料纸 tarred prepared sheet roofing paper
涂柏油的预制屋顶材料 tarred prepared roofing; tarred ready roofing
涂柏油管 tar-coated pipe; tar-coated tube
涂柏油混合物屋顶材料 tarred composition roofing
涂柏油麻绳 tarred rope
涂柏油碎片 tarred chip(ping)s
涂版 painting out
涂保护层 stopping off
涂保护色 baffle painting
涂边 beading
涂冰 ice glazed
涂玻璃粉 glassing
涂箔 foliate; foliation
涂薄胶泥 grouting
涂不明色 scumble
涂布 coat(ing); empire cloth; spread(ing); superimposition
涂布边缘控制 spreading edge control
涂布边缘准直器 spreading edge-collimator
涂布标本 smear preparation
涂布薄膜 coating film
涂布材料 coated material
涂布操作方式 coating action
涂布层 coat; covert
涂布刀 coating knife; spreading knife
涂布丁苯涂料 styrene-butadiene coating
涂布法 spreader process; spreading process
涂布方法 coating method
涂布钢材 coated steel
涂布工 coater
涂布辊 coating roll; doctor roll
涂布辊压机 coating calender
涂布厚度 application thickness; coating thickness
涂布机 coater; coating applicator; coating machine; laminator; spreader
涂布机边挡板 edge-dam of spreader
涂布机导边传感器 edge sensor of spreader
涂布机导边器 edge guide of spreader
涂布机集料器 catch pan of spreader
涂布机气刮刀 air doctor of spreader
涂布机涂料盘 coating pan of spreader
涂布基层 coating substrate
涂布级瓷土 coating grade clay
涂布量 coating content; coating weight
涂布量控制辊 doctor roll
涂布料 coating medium
涂布率 spreading rate
涂布面 coated side; coating surface
涂布面积 coated area; spreading area; spreading rate
涂布能力 spreading capacity; spreading power
涂布黏[粘]合剂 dope adhesive
涂布培养法 spread plate method
涂布喷气刀 air jet
涂布漆 dope
涂布器 spreader
涂布热量 coating heat
涂布设备 coating equipment
涂布温度 coating temperature
涂布物 coated material
涂布系数 spreading coefficient
涂布性能 coatability
涂布压光机 coating calender
涂布液 coating solution
涂布用树脂 coating resin
涂布原纸 coating body stock; coating stock
涂布织物 coated fabric
涂布纸 coated paper; enamel paper
涂布助剂 coating aids
涂布装置 coating device
涂擦 dab; inunction
涂擦处 erasure
涂擦剂 embrocation; inuncta; inunctum
涂层 coating; cover(age); covering(flux); finishing; garment; layer; lining; overcoat; overlay; paint-coat(ing); paint skin
涂层板 clad plate
涂层保护 coating protection
涂层边缘接合 picking up
涂层表面 coat surface
涂层表面发裂 checking
涂层玻纤织物 fiberglass coated fabric
涂层剥落 film detachment
涂层薄板 coated sheet
涂层薄膜 coat skin
涂层不连续性 coating discontinuity
涂层材料 coating material
涂层测厚仪 coating thickness ga(u)ge; coating thickness testing instrument
涂层打底 spackling
涂层刀片 coated chip; coated tips
涂层道数 number of coats
涂层的 coated
涂层的连续性 continuity of coating
涂层的裂缝模型 smeared cracking model
涂层的硫化物锈蚀 sulfide staining
涂层的随角异色效应 down-flop of coats
涂层的最大施工厚度 full coat
涂层发白 white spotted finish
涂层发黏[粘]期 tacky dry
涂层法 coating method
涂层帆布 enamel(l)ed duck
涂层翻新 knitting
涂层方法 coating process
涂层防水 waterproof painting film
涂层粉化 chalking
涂层粉末 coated powder
涂层附着力 coating bond
涂层复合粉 coated composite powder
涂层干硬 <油漆> hard dry; dry hard of coats
涂层坩埚 lined crucible
涂层钢板 coated steel sheet
涂层钢材 coated steel
涂层钢管 coated pipe
涂层钢筋 coating bar
涂层工具 coated tool
涂层光纤 coating optic(al) fibre[fiber]
涂层和衬层 coating and lining
涂层和易剂 <粉光抹平掺和剂> flow promoter[promotor]
涂层厚度 coat(ing)(layer) thickness
涂层基布 coating fabric
涂层技术 coating technology
涂层剂 coating agent
涂层加厚 coat thickening
涂层间附着力 intercoat adhesion
涂层间污染 intercoat contamination
涂层结合强度 coating bond
涂层金属电极 coated metallic electrode
涂层镜片 coated optic
涂层抗震性能 thermal shock resistance of coating
涂层颗粒 coated particle
涂层面 coated face; coated side
涂层明度 coating brilliance
涂层黏[粘]着力破坏 film bond(ing) failure
涂层片 coated foil; paint-coated plate
涂层起泡 gaul
涂层气泡 coating blister
涂层器 coater
涂层强度 coating strength
涂层缺陷 coat defect
涂层熔烧工艺 fused slurry coating technique
涂层色斑 coating colo(u)r spot
涂层色谱(法) spread-layer chromatography
涂层纱 coated yarn; coating yarn
涂层石棉水泥 enamel(l)ed asbestos-cement
涂层石棉水泥管 coated asbestos cement pipe
涂层石屑 lacquered chip(ping)s
涂层试验 coating test
涂层损坏 coating damage
涂层探伤仪 coating inspector
涂层条痕 coating streak
涂层脱落 film dislodgement; film displacement
涂层外的保护罩层 protective cover on a coat
涂层网状织物 coated mesh fabric
涂层雾罩 coating hood
涂层系统 coat system

涂层纤维 clad fiber[fibre];coated fiber[fibre]
涂层相对密度 coating density ratio; coating relative density
涂层消除机 stripper
涂层形式 coating type
涂层性能 coating performance
涂层液 coating solution
涂层硬质合金 coated carbide
涂层载体开关柜 support-coated open tubular column
涂层织物 coated fabric;coating fabric
涂层织物的轨状压花 railroad tracks of coating textiles
涂层铸铁管 coated cast-iron pipe
涂层装置 plater
涂层渍浸 coating immersion
涂衬炉床 fettle
涂衬炉床材料 fettle
涂成"斑马"纹(人行)横道 "zebra" crossing
涂成木纹的 wood-grained
涂成霜面 frosting
涂充漆 shellac(k)
涂除 obliteration
涂瓷 porcelainization
涂瓷漆 enamel(l)ing
涂瓷漆的 enamelized;enamel(l)ed
涂瓷漆反射器 enamel(l)ed reflector
涂瓷釉 glass-lined
涂磁粉带 coated tape;magnetic powder-coated tape
涂磁胶片 magnetic film
涂大漆 urushi work
涂底 bottom sizing;pad
涂底层 primary coat(ing); prime coat(ing);priming operation
涂底剂 primer coat(ing)
涂底漆 coat;paint primer;prepainting work;prime(r) coat(ing);priming (application);undercoating
涂底漆的 back primed
涂底漆前填塞木料的孔隙 primer-sealer
涂底漆设备 priming device
涂底色 blot
涂底刷 ground brush
涂底用熟炼油 primer oil
涂地板清漆 floor varnish
涂掉 dash out
涂镀设备 coater
涂镀装置 coater;plater
涂二道漆 two-coat
涂珐琅的 enamelized;enamel(l)ed
涂珐琅钢带 enamel(l)ed strip
涂泛水用沥青胶结材料 asphalt flashing cement
涂防护漆 slurrying
涂防霉剂的 fungusized
涂防水胶的 waterproof-glued
涂防水物料 waterproof
涂防锈油 cosmoline
涂粉 dusting
涂粉磁带 coated tape;magnetic powder-coated tape
涂粉式上釉 glazing by dusting
涂封闭底漆 sealing
涂缝镘 <泥工用的> jointer
涂敷 coat(ing);mortar dab;overlay;smear;spreading
涂敷电极 coated electrode
涂敷粉粒 coated particle
涂敷光学 coated optics
涂敷机 coating machine
涂敷浆膏 <木材防腐> plastering
涂敷金属层 metallizing
涂敷沥青 dressing compound
涂敷量 spread
涂敷螺旋布料器 spreading screw
涂敷面积 coated area
涂敷磨料 coated abrasive
涂敷抹刀 spatula
涂敷衰减器 coating attenuator
涂敷脂 creaming
涂附玻璃 glassivation
涂复耐火材料的 refractory faced
涂覆 cladding;coating;wash
涂覆电极 coated electrode
涂覆光纤 coated fiber[fibre]
涂覆滚子 paint roller
涂覆机械 coating machine
涂覆技术 coating practice
涂覆金属 plate
涂覆力 throwing power
涂覆量 coated weight;coating weight
涂覆磷光体印刷法 phosphor printing
涂覆牛皮纸 coated kraft paper
涂覆清漆 coat with varnish
涂覆色漆 coat with paint
涂覆树脂 coating resin
涂覆性能 coating performance; coating property;paintability
涂覆业务 coating practice
涂改 alter(ation)
涂改的客票 altered ticket
涂改的支票 altered check; check alteration
涂改痕迹 sign of erasure
涂改支票 raised check
涂改支票提高票面值 <俚语> hike
涂盖 buttering
涂盖料 coating material
涂钢丝的油 wire rope grease
涂膏 bonding paste coating
涂膏腐蚀试验 corrod(o)kote test
涂膏密室耐蚀试验 corrod(o)kote test
涂膏耐蚀试验 corrod(o)kote test
涂隔离剂 separant coating
涂镉 cadmium plating
涂铬 chrome
涂汞 amalgamate
涂刮 paste coating
涂刮法 coating process
涂刮辊 coating roll
涂管壁(用的)沥青乳剂 pipe enamel
涂管机 pipe coater
涂光 burnish;burnish gilding
涂光性 brushing property
涂光泽瓷漆 glossy glazing
涂过白漆的 white-painted
涂过的管道 coated pipe
涂过的管子 coated tube
涂焊剂 sweeting
涂焊药的金属焊条 covered metallic weld rod
涂黑 blacken
涂黑的 smudged
涂黑法 blackening
涂黑接收器 blackened receiver
涂黑漆 japanning
涂厚浆式绳索 mastic cord
涂糊阴极 paste cathode
涂灰浆 buttering;parge
涂灰浆刀 <泥工的> finishing knife
涂灰镘板 butterflying trowel
涂灰镘刀 <砌砖用的> butterflying trowel
涂灰泥 parget(ing);plaster(ing)
涂灰泥拱形顶棚 coom ceiling
涂灰泥天花板 coon ceiling
涂灰泥阻止传音 pug
涂混合漆 combination painting
涂火酒漆 bodying-up
涂剂 pigmentum;vernix
涂剂焊条 coated electrode; coverage electrode;covered electrode
涂假漆 shellac(k)
涂浆 wash
涂浆机 slurry spreader
涂浆式极板 pasted plate
涂浆式极板铅蓄电池 pasted plate accumulator
涂浆型极板 grid plate;faure-type plate; pasted plate
涂胶 adhesive coating; glue spreading;gumming;paste;paste spread; rubber coating; rubber cover; rubber covering;rubberizing;top with gum
涂胶标签 gummed label
涂胶并相互摩擦后的接缝 rubbed joint
涂胶带 rubber fabric
涂胶刀 doctor bar;doctor blade;doctor scraper; knife bar; spreading knife
涂胶的 gummed;rubberized
涂胶钉 cement-coated nail
涂胶封条 adhesive band
涂胶辊 doctor roll;glue spreader;roll coater
涂胶滚子 glue spreading roller
涂胶机 adhesive spreading machine; glue spreader;gummer;paster;rubber spreading machine;spreader
涂胶帘布层 rubberized cord plies
涂胶量 glue spread
涂胶铝箔 tenaplate
涂胶泥砖防水 <水压力较大时施工用> brick in mastic water-proofing
涂胶器 gluer;glue spreader;glue gun
涂胶软管 Indian rubber hose
涂胶水 sizing
涂胶丝 gelatinized silk
涂胶体 adherend
涂胶外层 gel coat
涂胶压延机 spreading calender
涂胶砑光机 calender coater
涂胶用喷枪 coating pistol
涂胶纸 paster;rubber-coated paper; sized paper
涂焦炭粉 coke blacking
涂焦油的 tarred;tarry
涂焦油等的防水布 tarpaulin
涂金 gilding;gold plating
涂金的 gilt;gold-coated
涂金胶 gold silver
涂金膜玻璃 gold-film glass
涂金色的 gilded
涂金油 gold silver jeweleries
涂金属 metallize
涂金属层板 plymetal
涂金属卷板 coil-coated metal
涂金属膜电阻 metallized film resistor
涂聚氯乙烯钢板 Vynitop
涂矿物质的底层 mineral substrate
涂矿物质的基层 mineral substrate
涂蜡 waxing;cere
涂蜡的 cerated
涂蜡机 waxer
涂蜡漆布 <铺地板用> wax cloth
涂蜡纱包线 waxed cotton covered wire
涂蜡装饰法 wax resist painting
涂蓝 blu(e)ing
涂蓝薄层 blue wash
涂蓝色 blue wash
涂沥青 bitumenizing;tar
涂沥青薄钢板 pitch-on metal
涂沥青材料 bitumen-coated material
涂沥青的 asphalt-coated; asphaltic-bitumen coated; bitumen-coated; bitumen-lined;pitchy
涂沥青的波状铁管 pavement invert
涂沥青的材料 bitumen-coated material
涂沥青的底层毡 coated base felt
涂沥青的骨料 asphalt-coated aggregate;bituminized aggregate
涂沥青的矿物骨料 bituminized mineral aggregate
涂沥青的铺路片石 asphalt-coated chip(ping)s
涂沥青的铺路石屑 asphalt-coated chip(ping)s
涂沥青的绳子 bituminized cord;bituminized rope
涂沥青的石子 bitumen gravel
涂沥青的屋面 asphalt coating roof(ing)
涂沥青的纸 bituminized paper
涂沥青钢管 bitumen-dipped steel tube
涂沥青骨料 bituminized aggregate
涂沥青管(子) asphalted pipe;asphalted tube;bitumen-coated pipe;tar-coated pipe
涂沥青集料 bituminized aggregate
涂沥青金属管 asphalt-coated metal pipe
涂沥青矿物骨料 bituminized mineral aggregate
涂沥青矿物集料 bituminized mineral aggregate
涂沥青砾石 asphalt-coated gravel; bitumen-coated gravel
涂沥青楼面覆盖材料 bituminized floor covering material
涂沥青片石 bitumen-coated chip(ping) s;bituminized chip(ping)s
涂沥青片石铺面层 asphalt-coated chip(ping)s carpet
涂沥青片石毡层 bitumen-coated chip(ping)s carpet
涂沥青砂 asphalt-coated sand
涂沥青绳 bituminized cord
涂沥青石屑 lacquered chip(ping)s
涂沥青石屑磨耗层 asphaltic-bitumen-coated chip(ping)s carpet
涂沥青水管 bitumen-coated pipe
涂沥青松散骨料 bituminized discrete aggregate
涂沥青碎石 <筑路用> asphaltic-bitumen-coated(road) metal
涂沥青碎石磨耗层 asphaltic-bitumen-coated chip(ping)s carpet
涂沥青硬纸板 asphalt-coated pasteboard
涂沥青用的布 granny rag
涂沥青油毡 coated roofing felt
涂料 coating (compound); coating medium; coat of paint; covering material; daub (ing); dopant material;dope;doping material;inhibitory coating;paint(ing);painting material; pek; plastering; protective cover; restrictive coating; beaded paint;stuff
涂料暴露试验 paint exposure test
涂料泵 paint pump
涂料泵送压力 paint pumping pressure
涂料比较试验板 paint patch panel
涂料表被 coating
涂料薄膜 film of paint
涂料仓库 paint store
涂料层 brush coat;dope(d) coat(ing); coat
涂料层数 number of layers
涂料厂 coating factory
涂料车间 paint workshop
涂料沉淀物 paint residue
涂料成分 coating composition;paint-ingredient
涂料稠度 body;paint consistency
涂料触变性 paint thixotropy
涂料催干剂 paint drier[dryer]
涂料刀 coating knife
涂料的反光 reflectorizing of paint
涂料的腐败 putrefaction of paints
涂料的浸涂 dipping of paint
涂料的磨蚀 abrasion of paint
涂料的黏[粘]附 adhesion of paint

涂料的起霜(现象) frosting of paint
涂料的渗透性 permeability of paint
涂料的作业性 working property of paint
涂料底层 primer coat(ing)
涂料底度 inner primer
涂料方法 coating system
涂料防护性试验层 isolating test coating
涂料分配辊 paint distributing roller
涂料分散性试验机 multi paint shaker
涂料粉化 powdering of paints
涂料干性油 paint oil
涂料工业 coating industry; paint industry; surface coating industry
涂料工业废水 coating industry wastewater; pigment industry wastewater
涂料工业废水处理 coating industry wastewater treatment; pigment industry wastewater treatment
涂料工艺 paint technology
涂料工艺学会联合会 <美> Federation of Societies for Coatings Technology
涂料勾缝 <砂浆缝加涂饰> pencil(l)ing
涂料刮刀 cleaning doctor; paint scraper
涂料刮涂试验 drawdown test
涂料罐 paint tank
涂料规范书 paint specification
涂料焊条 coated electrode; coated rod; coated stick electrode; coated wire electrode; covered electrode
涂料化学家 paint chemist
涂料混合后的安定性 stability after mixing
涂料混合机 paint mixer
涂料混合器 paint mixer
涂料混合室 paint mixing room
涂料基本名称代号 basic paint coding
涂料级滑石 paint grade talc
涂料技师 paint technician
涂料技术 coating technology
涂料加压罐 paint pressure tank
涂料搅拌机 mo(u)ld wash mixer
涂料颗粒 paint particle
涂料块 coating lump
涂料扩展试验 paint spreading test
涂料拉起 coating pulling up
涂料类型 coating type
涂料量 coating content
涂料密封层 sealing coat of paint
涂料膜 paint film; paint frame; paint membrane
涂料磨机 paint roller mill
涂料黏[粘]附 clagging
涂料碾磨机 paint mill
涂料碾盘 paint mill
涂料凝硬试验 paint livering test
涂料排出物 coating effluent
涂料配方 coating formula(e); coating formulation; paint formulation
涂料配套系统 paint system
涂料喷枪 blackwash sprayer; coating pistol; paint sprayer; paint-spray(ing) gun[pistol]
涂料喷枪导管 paint guide
涂料喷枪导管盖 paint guide cover
涂料喷枪喷嘴 painting gun nozzle; paint-spray gun nozzle
涂料喷枪针形阀 material needle valve
涂料喷射机 paint sprayer
涂料喷涂泵 paint spray pump
涂料喷涂室 paint spray room
涂料喷嘴 paint nozzle; paint spray nozzle
涂料平光剂 flattening agent
涂料破坏剂 paint-destroying agency
涂料企业管理 enterprise management of paints industry
涂料气孔 blacking hole
涂料器 coater; coating applicator
涂料轻度痕迹 ropey
涂料清除剂 paint remover
涂料染色 pigment dyeing
涂料溶剂 surface coating medium
涂料溶液 coating solution
涂料乳胶色浆 pigment resin emulsion colo(u)r
涂料软管 paint hose
涂料色调试验 paint shade test
涂料色浆 pigment printing paste
涂料生产设备 paint production plant
涂料失效 paint failure
涂料施工 coating application
涂料施工程序 painting procedure
涂料实用试验 paint service test
涂料蚀洗处理 wash primer process
涂料收缩 traction
涂料受挤压作用 painting squeezing action
涂料刷 brush; distemper; distemper brush; flat brush; knot brush; paint brush
涂料刷柄 paint brush handle
涂料刷箍 paint brush ferrule
涂料刷毛 paint brush bristle
涂料淌流 creeping
涂料弹涂机 paint catapult
涂料添加剂 coating adhesive
涂料涂覆量 coatings pick-up
涂料涂刷试验 paint brushing test
涂料调和剂 paint liquid
涂料污泥 paint sludge
涂料稀释剂 paint thinner
涂料细度试验 paint fineness test
涂料消除剂 paint remover; paint scrubber
涂料研磨机 paint grinder(mill)
涂料液体 coating fluid
涂料印花 pigment printing; pigment resin printing; resin-bonded pigment printing; resin-fixed pigment printing
涂料用二氧化钛 paint grade titanium dioxide
涂料用毛刷 blacking brush
涂料用树脂 coating resin
涂料用钛白 paint grade titanium dioxide
涂料用亚麻籽油 linseed oil for paints
涂料油 paint oil
涂料载色体 paint base
涂料展色剂 coating vehicle
涂料遮盖力计 cryptome(te)r
涂料蒸发试验 paint evaporation test
涂料纸 coated paper
涂料制造厂 paint manufacturer
涂料制造工业 paint manufacturing industry
涂料助剂 coating additive
涂料转移 pigment transfer
涂料着色 pigment colo(u)ration
涂料组成 coating composition
涂料组分 coating composition
涂磷管 phosphor-coated tube
涂磷光体的底面玻璃 face glass
涂炉材料 fettling
涂炉床材料 fettle
涂铝 aluminizing
涂铝薄钢板 alumin(i)um-coated sheet steel
涂铝钢 alumetized steel; aluminized steel; calorized steel
涂铝(钢丝)网 alumin(i)um-coated fabric
涂氯乙烯钢板 vinyl-coated steel plate; vinyl covered steel plate
涂满黏[粘]合剂的 adhesive backed
涂没 obliteration
涂面斑点 mottling
涂面变色 bleaching; fading
涂面钉 coated nail
涂面法 <研究边界层流态的> method of surface coating
涂面方法 facing method
涂面粉末 coating powder
涂面钢材 coated steel
涂面钢丝 coated wire
涂面工 face work
涂面混合料 facing mix(ture)
涂面料 facing material
涂面率 spreading rate
涂面漆 finishing coat; top coat; top-coating
涂面树脂 coating resin
涂面塑料 coating plastics
涂面质量 coating quality
涂面桩 coated pile
涂模材料 adhering mo(u)lding material
涂膜 coated film; coating film; film; film coating; paint film
涂膜材料 adhering mo(u)lding material; emulsion-coated material
涂膜搭接 coating film lap
涂膜鳄纹 alligatoring
涂膜疙瘩 crawling
涂膜厚度 film thickness
涂膜机 applicator; coater
涂膜积垢 dirt settling
涂膜面 coated surface
涂膜明晰度 depth of finish
涂膜片基 coating base
涂膜片落 chipping
涂膜片落等级标准 chipping rating standards
涂膜破坏 paint film failure
涂膜器 applicator; film coater
涂膜浅裂 checking
涂膜上凸起的小块或颗粒 peppery
涂膜塑料 coated plastics
涂膜塑料片 painted plastic; painted plastic piece
涂膜碎落 chipping
涂膜涂布器 drawdown bar; drawdown blade
涂膜蜕变破裂 coated film disintegration
涂膜修补剂 coat film repair agent
涂膜硬度 film hardness
涂膜皱皮 alligatoring
涂膜总厚度 total thickness of coat film
涂抹 brush application; butter; daub(ing); efface; lay; paint; smear
涂抹层 trowel-applied coat
涂抹稠度 trowel(l)ing consistency
涂抹底层灰 rendered finish
涂抹法 smear technique
涂抹符号 delete character
涂抹工具 dauber
涂抹厚度 application thickness; distribution thickness
涂抹技术 paint on technique
涂抹剂 liniment
涂抹节疤 killing; knotting
涂抹力 obliterating power
涂抹派 <一种抽象画> Tacher
涂抹器 obliterator
涂抹效应 <砂井排水的> smear-effect(on drainage)
涂抹油 liniment
涂抹者 dauber
涂抹作用 smear
涂末道漆 finish coating; upper coating
涂墨 blocking out
涂墨辊 brayer roll
涂泥焙烧 roasting in stalls
涂泥浆 slurrying
涂泥浆层 slip coating
涂泥炭的拱形顶棚 coom ceiling
涂泥釉 slip coating
涂泥釉机 slip coating machine
涂硼的 boron-coated; boron loaded
涂片 smear
涂铺的 coated
涂铺机 coating machine
涂铺器 spreader; spreading device
涂铺设备 coating plant
涂漆 coating; doping; japanning; lacker; lacquering japanning; lacquer work; paint coating; painting; stop-off lacquer; varnishing
涂漆板 lacquered board
涂漆包皮 doped coating
涂漆保护 protection painting(work)
涂漆标线 painted road stripe; paint line; paint marking
涂漆标志 paint mark
涂漆玻璃织物 varnished glass fabric
涂漆薄钢板 lacquer-coated steel sheet
涂漆层 coat of paint; coat of tar; japanning; paint coating
涂漆场 paint plant
涂漆带钢 enamel(l)ed strip
涂漆道数 number of passes
涂漆的 enamel(l)ed; japanned; painted
涂漆底 paint primer
涂漆底层 base coat
涂漆垫层 paint coat cushion
涂漆镀锡薄钢板 lacquered plate
涂漆钢板 painted steel
涂漆钢材 lacquered steel
涂漆工艺 painting technology
涂漆辊 doctor roll; ductor; paint roller
涂漆机 coater; coating machine; dope machine; painting machine
涂漆间 coating room
涂漆绝缘纸 varnish paper
涂漆量 coating weight
涂漆料纸带 varnished bias tape
涂漆炉 japanning oven
涂漆马口铁皮 tole
涂漆蒙布 doped fabric
涂漆蒙布面 doped fabric covering
涂漆面 coating surface; painted surface
涂漆面积 painting area
涂漆膜 paint film
涂漆色 varnish colo(u)r
涂漆手套 painter's mitt; paint mitt
涂漆刷 dabber
涂漆套管 varnished cotton tube
涂漆锡器 tole
涂漆用炉 japanning oven
涂漆用橡胶片 padder
涂铅的 lead-coated
涂铅的金属 lead-coated metal
涂铅管 lead-coated pipe
涂铅锑锡合金的钢材 ternecoated steel
涂铅铜板 lead-coated copper sheet
涂铅铜带 lead-coated copper strip
涂黔 mud
涂墙石灰乳 lime wash
涂墙稀灰泥 thin plaster
涂墙用的灰泥 wall stuff
涂青铜粉 bronzing
涂清漆 varnish coating; lacquering
涂清漆的硬质木纤维板 lacquered hardboard
涂清漆工 varnisher
涂清漆机 varnish coater
涂清漆纱布 varnished cambric
涂清漆刷 <圆端头> dabber
涂去 blotting; dropout; obliterate; obliteration
涂染成胭脂红色 carmin(e)
涂熔剂焊条 flux-coated electrode
涂乳化沥青甘蔗板 bitumen emulsion

cane fiber[fibre] board
涂润滑脂 coating with grease
涂色 coat with paint; face up; overtone; paint(ing); tintage
涂色斑点试验 smear test
涂色玻璃 colo(u)r coating glass; painted glass
涂色层 pigmented coating
涂色的混凝土骨料 colo(u)r coated concrete aggregate
涂色的混凝土集料 colo(u)r coated concrete aggregate
涂色钢材 colo(u)r-coated steel
涂色规定 colo(u)r code
涂色剂 marking compound
涂色检查 smear test
涂色试验 <齿轮接触斑痕检试> smear test
涂色作标记 colo(u)r code
涂砂浆镘刀 buttering trowel
涂上聚乙烯的 poly-coated
涂上橡胶的带 proofed tape
涂上橡胶液的帆布 rubberized canvas
涂上终饰 thrown-on finish
涂渗 coating penetration
涂石灰 lime coating; liming
涂石墨 graphitization; graphitizing
涂石墨机 blackleading machine
涂石墨剂的 graphited
涂石墙粉 stone plaster
涂石墙粉材料 stone plaster stuff
涂石墙粉混合物 stone plaster mix(ture)
涂饰 finish; illumination; overcoat(ing); parget
涂饰板 coated board
涂饰剂 coating agent
涂饰夹具 finishing fixture
涂饰墙板的灰泥 aegrit
涂饰墙板的墙粉 aegrit
涂饰深度 depth of finish
涂饰时间 overcoating time
涂饰饰面水泥 cement paint
涂树脂的缆索 coated resin cable
涂树脂的钻杆接头 rosined joints
涂刷 brush application; brushing; brush-on; brush work; overcoat(ing); paint(ing); shelling-out; slap; stencil; wash
涂刷不当的油漆涂层 <俚语> piss coat
涂刷操作 painting practice
涂刷处理(法) <木材防腐> brush treatment
涂刷次序 painting system
涂刷的瓷漆 painted enamel
涂刷的屋面薄膜 brushed roof(ing) membrane
涂刷的增饰 painted enrichment
涂刷的装饰面层 painted ornamental finish
涂刷(方)法 application method; brush application <油漆技术中的>; brush (treatment) method <木材防腐>
涂刷防腐 brush treatment
涂刷防腐剂 brush treatment
涂刷工作 painting work
涂刷灰膏涂层 fining-off
涂刷或喷洒(防水)层 <在地面下基础外表> brush or spray coating
涂刷计划 painting scheme
涂刷剂 daub material
涂刷面层防水法 surface coating method for waterproofing
涂刷黏[粘]度试验 working viscosity test; draining test of brush
涂刷器 coater; squeegee
涂刷清漆 varnish work; varnishing
涂刷热分解氧化法 paint thermal decomposition oxidation
涂刷设备 painting device
涂刷石灰水 lime wash
涂刷性 brushability
涂刷乙烯基的 vinyl-coated
涂刷易脱性彩粉浆 colo(u)r wash
涂刷用腻子 stopper for brush application
涂刷用瑞典油灰 Swedish putty for brush application
涂刷用油灰 stopper for brush application
涂刷质量 brushing quality
涂刷装置 squeegee assembly
涂水泥的钉 cement coated nail
涂水泥法 gunite process
涂松节油 turpentine
涂松香 rosin
涂塑玻璃纤维窗纱 plastic-coated glass screen
涂塑布 plastic-coated fabric
涂塑布机组 plastic-coated fabric manufacturing aggregate
涂塑窗纱 vinyl-coated screening
涂塑窗纱机组 plastic impregnation aggregate for insect screening
涂塑钢片 enamel strip
涂塑料的棉手套 plastic-coated cotton glove
涂塑料钢片 enamel strip
涂塑料硬纸板 plastic-coated hardboard
涂塑墙纸 plastic-bonded wallpaper; plastic-coated wallpaper
涂塑纱 polymer-coated yarn
涂塑相纸 resin-coated paper
涂碳层 carbon coating
涂碳粉 blackening
涂搪 application of enamel; enamel(l)ing
涂搪瓷的 coated with baked enamel
涂搪瓷用黏[粘]土 enamel(ling) clay
涂铜 copper coating
涂铜板用玻璃织物 glass fabric for copper coated laminate
涂铜纸 copper-coated paper
涂透明漆 transparent painting
涂钍放射体 thoriated emitter
涂钍钨电极 thoriated tungsten electrode
涂污 blur; spotting
涂污函数 smearing function
涂污效应 smearing effect
涂屋面的沥青 coating asphalt
涂无光漆 mat(te) finish(ing)
涂吸附剂的玻璃条 adsorbent coated glass strip
涂硒鼓 selenium-coated drum
涂硒圆筒 selenium-coated drum
涂显剂 streak reagent
涂线序列 wire train
涂橡胶 coat with rubber
涂橡胶的 rubberized
涂橡胶的油地毡 rubberized linoleum
涂橡胶垫层 rubberized cushion
涂橡胶织物 rubberized fabric
涂消除剂 stripper
涂硝化纤维板清漆 coating with zapon lacquer
涂销 deface(ment)
涂写污染 graffito pollution
涂芯型浆 core wash
涂锌 spelter coating; zinc coating
涂型芯浆 core wash
涂颜料 pigment finish
涂氧化铁砂 iron oxide-coated sand
涂氧化铜的 aeruginous
涂氧化物的 oxide-coated
涂药 <焊条的> coating; covering
涂药电焊条 coated electrode
涂药焊条 coated electrode; covered electrode; flux-coated electrode; fluxed electrode
涂药厚度 thickness of covering
涂药器 applicator
涂液 masking liquid
涂以松香胶的木纤维片 rosin sized sheathing
涂以塑料 coating with plastics
涂银 silver coating; silver pastebrushing
涂银膜 silver coating film
涂硬质清漆的铁器 japanned
涂油 brush application; dub; fat liquoring; grease liquor; greasing; lubricate; oiling; primer; priming; unction
涂油白钢皮 oil finished terne plate
涂油捕收 smear-collection
涂油槽 coating mass container
涂油池 coating tank
涂油的 oil-coated; oiled
涂油垫层 paint coat cushion
涂油封边 <混凝土胶合板模板的> oiled and edge sealed
涂油干料 dry form paint material
涂油膏 inunction
涂油革 oiled leather
涂油工作 oiling job
涂油辊 oiling roll; sizing roller
涂油过多 overstuff
涂油滑道【船】 greased ship-building berth; shipbuilding berth with greased launching way
涂油灰 slushing
涂油机 oiler
涂油麻填料 hemp tallowed packing
涂油蒙布 doped fabric
涂油墨 inking
涂油漆 overcoat(ing); painting; lay on
涂油漆层 coating-in
涂油漆法 paint-and-batten process
涂油器 greaser
涂油脂 grease
涂油者 oiler
涂油织物 oil-treated fabric
涂油脂的套管 greased sleeve
涂油装置 oiling station
涂有……的 coated
涂有保护层的钢筋 coated bar
涂有玻璃釉的 glass-glazed
涂有糊状放射物质的阴极 pasted cathode
涂有灰浆的 mortar-bound
涂有灰泥的 plastered
涂有沥青的 pitchy
涂有熔剂的焊条 flux-coated electrode
涂有涂料的表面 surface to be painted
涂有硝棉的防水布 keratol
涂有硬膜的 hard coated
涂有油漆的表面 surface to be painted
涂釉 glaze
涂釉层 vitreous coating
涂釉电阻 ceramet resistance; glazed resistance
涂釉电阻器 glaze resistor; vitrified resistor
涂釉坩埚 glazed pot
涂釉水管 coating pipe
涂釉陶瓷 glazed ceramics
涂釉土坯 enamel(l)ing clay
涂脏 blot
涂罩光漆 bodying in; bodying-up
涂脂 grease coating
涂脂抽芯管 greased bar
涂脂样物 butter
涂助焊剂 prefluxing
涂装 application; coating; finishing; painting
涂装板的自动卸件装置 automatic unloader for coating plate
涂装范畴 coating facet
涂装方法 coating method
涂装工艺 painting technology
涂装供料辊 furnishing roll
涂装规范 painting schedule
涂装环境 painting environment
涂装机 decorating machine
涂装间隔 interval between coating
涂装浸没导辊 immersed guide roll for coating
涂装室 coating room
涂装塑化炉 coating fusing oven
涂装体系 paint system
涂装用混料 sheathing compound
涂渍操作 coating operation
涂渍层 coating layer
涂渍溶液 coating solution
涂渍效率 coating efficiency

途

途程衰减 range attenuation

途次 en route
途耗 ullage
途经 curriculum[复 curricula/curriculums]
途经交通 non-access traffic
途经时间 travel time
途径 gateway; path
途径目标领导理论 path-goal theory of leadership
途径终点 terminus of walk
途栈桥 pier bridge
途中拌和 concrete mixing 'en route'; concrete mixing on route; mixed-in-transit; mix en route
途中报关进口 immediate transportation entry
途中编组场 intermediate marshalling yard
途中不搅拌混凝土运送车 non-agitating truck
途中货物 floating cargo
途中搅拌 concrete mixing on route; mixed-in-transit; mix(ing) on route
途中旅馆 rest house
途中修理 road repair
途中舀水设备 water scoop
途中运缓 out-of-course
途中运输 time-haul
途中运行时间 <自起运站至到达站> time in transit; transit time
途中装卸 wagon-load goods loaded or unloaded not at goods station
途中自然减重 natural loss of weight during transit; normal loss

屠

屠刀 butcher's knife; whittle

屠杀鱼群 fish kill
屠宰场 abattoir; butchery; killing work; lethal chamber; shambles; slaughter establishment; slaughter hall; slaughterhouse
屠宰场废弃物 abattoir waste
屠宰场废水 abattoir waste; abattoir wastewater; slaughterhouse wastewater
屠宰场废物 slaughterhouse wastes
屠宰废弃物 cutting offal
屠宰废水 slaughter wastewater
屠宰加工厂 meat packing plant; packing house; packing plant

土

土 soil

土岸 soil bank
土芭树脂 tubain
土坝 earth dam; earth embankment

dam; earth-fill(ed) dam; earthwork dam
土坝坝身 fill dam body
土坝边坡压重 weighting of slope
土坝底土 fill dam subsoil
土坝固结 consolidation of earth dam
土坝夯土坡 tamped slope of earth dam
土坝护面 earth dam paving; earth-fill dam paving
土坝浸润线观测 observation of saturation line in soil dam
土坝宽度 fill dam width
土坝老化 earth dam ag(e)ing
土坝铺面 earth dam paving
土坝施工 fill dam construction
土坝天然地基 fill dam subsoil
土坝位置 fill dam site
土坝芯墙 core of dam; core of earth dam
土坝芯墙基槽 core trench of dam
土坝压实 compaction of earth dam; consolidation of earth dam; earth dam compaction
土坝涌毁 fountain failure
土坝中央排水道 drain-zone chimney
土办法 indigenous method; local method
土包 earth bag
土包防冲谷坊 sack erosion check dam
土饱和密度 density of saturated soil
土保持 soil conservation
土堡 revetment
土被 earthing cover(ing); ground layer; ground over; mantle of soil; mantle rock; regolith; soil cover(ing); soil mantle; soil mulch
土崩 avalanche; earth creep; earth fall; earth flow; earth slide; earth slip; land fall; landslide; shear slide; soil avalanche; soil fall; soil slip; soil slump
土崩安全系数 safe factor for shear sliding
土崩防护林 forest for earth fall prevention
土崩坍 soil slump
土壁计算方法 bank measure
土壁支撑 earth wall bracing
土变形 earth deformation
土表层 soil surface
土表层泥流 solifluction mantle
土表面潜在的升高 <由于毛细管作用或含水量增加引起的> potential vertical rise
土鳖虫 ground beetle
土饼(块) soil pat
土拨鼠 ground-hog
土不均一性 soil heterogeneity
土布 native cloth
土布量 spread
土布散 thiophahate
土槽 soil box
土槽模拟 model(l)ing with soil tank
土槽模拟法 soil tank modelling
土槽试验 soil bin test
土厕(所) <无水冲厕所> earth closet
土层 horizon; horizon of soil; layer of soil; layer of earth; soil formation; soil horizon; soil layer; soil stratum [复 strata]; solum; solum of soil; top soil; true soil
土层被覆 mantle of soil
土层标高 elevation of soil formation; elevation of soil layer
土层垂直剖面图 soft profile
土层的地质学问题 geologic(al) aspects of soil formation
土层的厚度 the thickness of the layer
土层地板 subfloor
土层断面 section of soil
土层风化 soil weathering; solum
土层厚度 depth of soil; soil depth; soil stratum thickness; thickness of soil layer
土层滑动 earth slide; earth slip
土层加固 soil consolidation; soil improvement; soil stabilization
土层架空 soil piping
土层界限 well-defined horizons
土层螺旋钻(具) mud auger
土层锚杆 earth anchor; ground anchor; soil anchor
土层锚固 earth anchor
土层描述 description of materials; soil formation description
土层名称 name of soil formation; name of soil layer
土层剖面 soil section
土层剖面特性 soil-profile characteristics
土层剖面图 soil profile
土层蠕动 creep of soil; soil creep
土层深度 depth of soil
土层隧道 earth tunnel
土层探测 soil sounding
土层藻类 subterranean algae
土层钻探 spudding through soil
土层钻探记录 soil boring log
土层钻探取样 soil boring
土层钻头 mud bit
土产 aboriginal; aborigines
土产的 home-made; indigenous
土产地质阶段 step of production geology
土产品 domestic products; local product; native goods; native products
土产商店 native product shop
土产税 excise duty
土颤动 earth trembler
土常数 soil constant
土场地 soil site
土车 ash car; navvy barrow
土尘 ground dust
土沉淀池 earthen settling tank
土成分 soil constituent
土承压力 bearing pressure on foundation; soil pressure
土承载比 soil bearing ratio
土承载力 soil bearing capacity
土承载量 soil bearing capacity
土承载(能力)试验 soil bearing test
土池 earthen basin
土赤水蓝色 dusty aqua(blue)
土臭物质 earthy smelling substance
土锄 grafting tool
土触探仪 soil sounding device
土触探装置 soil sounding device
土传病害 soil-borne disease
土槌 claying bar
土搓条 soil ribbon; soil thread
土搓线 soil thread
土带 soil zone
土袋 earth baffling; earth bag; earth-fill bag
土袋临时挡水坝 bag dam
土袋埝坝 bag dam
土单元 soil element
土刀 spatula
土捣密工作 soil puddling
土捣实 soil puddling
土道砟 earth ballast
土的阿太堡极限 Atterberg limits of soil
土的饱和度 degree of saturation of soil
土的饱和度分级 degree of saturation grade of soil
土的饱和强度 saturated strength of soil
土的饱和容重 saturated unit weight of soil
土的保水能力 retention capacity of soil
土的爆炸压密 blasting compaction of soils
土的被动压力 passive thrust of earth
土的本构关系 constitutive relation of soil
土的本构律 constitutive law of soil
土的比热 specific heat of soil
土的比重 specific gravity of soil
土的变形 soil deformation
土的变形参数试验 soil deformation parameters test
土的变形模量 modulus of soil deformation
土的表层 mantle of soil; waste mantle
土的不规则结构 erratic soil structure
土的不均一性 soil heterogeneity
土的不排水剪切强度 undrained shear strength of soil
土的不透水性 soil impermeability
土的侧推力 soil lateral thrust
土的层理 soil bedding; soil stratification
土的长期三轴压缩试验 long-term triaxial compression test
土的超单元 soil superelement
土的沉陷 settlement of soil
土的成因类型 genetic(al) type of soil
土的承载力 bearing capacity of soil; bearing power of soil; soil bearing capacity
土的承载力试验 soil bearing test
土的承载试验 bearing test of soil
土的持水性 soil water retention
土的稠度 consistency of soil
土的稠度界限 consistency limit of soil
土的稠度试验 soil consistency test
土的稠度限界 consistency limit of soil
土的初始吸水量 soil priming
土的初始压缩破坏 primary compression failure of soil
土的触变性 soil thixotropy
土的垂直摩擦力 vertical soil friction force
土的单粒结构 single-grained structure
土的单位潜容重 submerged unit weight of soil
土的单位体积质量 bulk density of soil
土的导热系数 coefficient of thermal conductivity of soil
土的导水性 soil hydraulic conductivity
土的导温系数 coefficient of thermometric conductivity of soil
土的捣固试验 compaction test
土的电法加固 electric(al) stabilization
土的电渗加固法 electric(al)-osmotic stabilization
土的电渗稳定(法) electroosmotic stabilization of soil; electric(al)-osmotic stabilization of soil
土的电阻 electric(al) resistance of soil
土的电阻率 electric(al) resistivity of soil
土的电阻系数 resistance coefficient of soil
土的调查报告 soil survey report
土的动力试验 dynamic(al) test of soil
土的动三轴试验 dynamic(al) triaxial test of soil
土的动弹性模量 modulus of soil dynamic(al) elasticity
土的冻胀 uplift
土的翻掘改良 soil conditioning
土的放大效应 soil amplification
土的非湿胀 non-expansive condition of soil
土的非线性动力学特性 non-linear dynamic(al) property of soil
土的非线性性质 non-linear behavio(u)r
土的分层 soil stratification
土的分级 soil rating
土的分类 soil classification; soil grade
土的分类试验 soil classification test
土的分类体系 soil classification system
土的粉碎机 soil pulverizer
土的风力移动 blowout of soil
土的浮容重 soil buoyancy
土的负孔隙压力 soil suction
土的附着力 soil adhesion
土的概念 earth concept
土的干二相容重 dry two-phase unit weight of soil
土的干密度 dry density of soil
土的干容重 unit dry weight of soil
土的干重 dry weight of soil
土的各种成分密度 density of soil constituents
土的工程地质分类 engineering geologic(al) classification of soil
土的工程分类 engineering classification of soil
土的工程性质 engineering property of soil
土的工地含水当量 field moisture content
土的构成 earth formation
土的构造 soil structure
土的骨架 soil skeleton
土的骨架净高 reduced height of soil
土的固化 soil solidification
土的固结 soil consolidation
土的固结理论 consolidation theory of soil
土的固结曲线 consolidation curve of soil
土的固结状态 consolidation state of soil
土的固结作用 soil consolidation
土的固体成分 solid constituent of soil
土的管涌 soil piping
土的含水量 soil moisture content; soil water content; water content of soil
土的含水量指数 moisture index
土的含水量状态评价试验 moisture condition value test
土的含水密实度测定仪 hydrodensimeter
土的核状结构 nutty structure of soil
土的横向保护物 earth traverse
土的烘干 oven drying of soil
土的化学加固 chemical consolidation of soil; chemical soil stabilization
土的缓冲作用 buffer action of soil
土的回填压力 backfill pressure of soil
土的活性 <指土的塑性指数与黏[粘]粒成分之比> activity of soil
土的活性指数 activity number of soil
土的击实试验 soil compaction test
土的击实性 compactibility of soil
土的机械加固法 mechanical soil stabilization[stabilisation]
土的机械稳定法 mechanical soil stabilization[stabilisation]
土的基本测试 basic test on soil
土的基本试验 basic test on soil
土的级配 grading of soil; grain-size distribution of soil
土的级配和球度系数 gradation sphericity factor

土的极化曲线试验 polarization curve test of soil
土的加固 soil improvement;soil stabilization;soil strengthening
土的加固剂 soil stabilizer
土的加筋(法) soil reinforcement; earth reinforcement
土的剪切破坏 soil shear failure
土的剪切试验 shear test of soil;soil shear test
土的简易分类法 quick soil classification;rapid soil classification
土的鉴定 soil identification
土的胶结物 soil binder
土的结构 soil fabric;soil structure; soil texture;structure of soil
土的结构特征 structural type of rock mass
土的界限含水量试验 Atterberg test
土的静弹性模量 modulus of soil static elasticity
土的抗剪角度 angle of soil shear resistance
土的抗剪强度 shear strength of soil; soil shear strength
土的抗剪强度记录器 soil sheargraph
土的抗剪试验 shearing test of soil
土的抗剪性 shear property of soil
土的抗力 resistance of soil
土的抗渗强度 seepage resistance strength of soil
土的颗粒成分分析 grain composition analysis of soil
土的颗粒大小分布测定 determining particle-size distribution of soil
土的颗粒组成 granulometric composition of soil
土的可侵蚀性 soil erodibility
土的可塑性 plasticity of soil
土的可塑性分级 plasticity grade of soil
土的可塑值 plastic number of soil
土的空隙体积 non-soil volume
土的孔隙 soil pore
土的孔隙比 soil porosity ratio;void ratio of soil;soil void ratio
土的孔隙度 porosity of soil;soil porosity
土的孔隙率 porosity of soil;soil porosity
土的孔隙面 interstitial surface of soil
土的矿物成分 mineral composition of soil
土的矿物种类 mineral kinds of soil
土的类别 soil types
土的离心含水当量 centrifugal moisture equivalent;centrifuge moisture content;centrifuge moisture equivalent
土的离心湿度当量 centrifugal moisture equivalent;centrifuge moisture equivalent
土的力学性质 mechanical property of soil;soil mechanical characteristic
土的沥青加固 asphalt soil stabilization
土的沥青稳定法 asphalt soil stabilization
土的粒度成分 granulometric composition of soil
土的粒度分级取样 fractional sampling
土的粒径 soil size
土的粒径分类 size classification of soil particles
土的粒径分析 particle size analysis of soil
土的粒组 soil fraction
土的临界孔隙比 critical void ratio of soil
土的灵敏度 sensitivity of soil
土的流动曲线 <试验土的含水比和落锤次数间的关系曲线> flow curve
土的流动指数 flow index
土的流失 soil loss
土的隆起 flow of ground;uplift
土的氯盐稳定法 chloride stabilization of soil
土的慢剪试验 slow test
土的密度 density of soil;soil density
土的密实度与含量关系 density moisture relation of soil
土的密实化 soil densification
土的敏感性 <一不扰动的土样无侧限抗压强度/重塑的土样无侧限抗压强度> sensitivity of soil
土的模量 soil modulus
土的内部冲蚀 internal erosion
土的内部侵蚀 internal erosion
土的内聚力 cohesion of soil;soil cohesion
土的内摩擦角 angle of soil internal friction
土的内摩擦力 internal friction of soil
土的内摩擦系数 coefficient of internal friction of soil
土的黏[粘]结键 soil binder
土的黏[粘]结力 soil cohesion
土的黏[粘]结性 earth cementation; soil cohesion
土的黏[粘]结作用 earth cementation
土的黏[粘]聚性 cohesive property of soil
土的黏[粘]限 soil sticky limit
土的黏[粘]性 viscidity of soil
土的凝固作用 earth solidification
土的凝聚强度 soil cohesive strength
土的排水剪力试验 drained shear test
土的排水三轴试验 drained triaxial test
土的膨胀 soil bulging;swelling of soil
土的膨胀量 degree of swelling
土的膨胀能力 swelling capacity of soil
土的膨胀试验 swelling test of soil
土的膨胀性 dilatancy of soil
土的平均稠度 average consistency of soil
土的破坏 failure of the soil
土的前期固结比 rate of soil preconsolidation
土的前期固结系数 coefficient of soil preconsolidation
土的强度 soil strength; strength of soil
土的强度包线 failure envelope
土的球度系数 soil sphericity factor
土的人工冻结法 artificial freezing of soil
土的容许压力 allowable soil pressure
土的容重 bulk density of soil;soil unit weight;volume weight of soil; unit weight of soil
土的肉眼分类 visual soil classification
土的蠕变 soil creep
土的蠕动 soil creep
土的撒布 spreading of soil
土的三角坐标分类法 triangle classification of soils;triangular soil classification system
土的三角坐标图 <分类用> soil triangle
土的三相草图 three-phase diagram of soil
土的上层 top soil
土的烧失量 ignition loss of soil
土的烧灼 soil burning
土的渗透固结类型 permeation consolidation type of soil
土的渗透系数 coefficient of soil permeability
土的渗透性 soil permeability;soil hydraulic conductivity
土的渗透性测定 permeability determination of soil
土的生成 soil formation
土的湿度 ground humidity;soil moisture
土的湿度密(实)度仪 soil moisture-density meter
土的湿化时间 time of slaking
土的时间固结关系曲线 time-consolidation curve
土的时间固结曲线 time-consolidation curve of soil
土的识别 identification of soils
土的收缩限界 shrinkage limit
土的收缩限界试验 shrinkage limit test
土的收缩指数 shrinkage index of soil
土的水分速测仪 soil moisture teller
土的水分应力 soil moisture stress
土的水分运动 soil water movement
土的水泥加固法 cement stabilization of soil; soil cementation; soil-cement processing
土的水平摩擦力 horizontal soil friction force
土的水上容重 drained bulk weight density of soil
土的水下容重 submerged bulk weight density of soil
土的水压力 soil water tension
土的塑限 plastic limit of soil
土的塑限试验 plastic limit test of soil
土的塑性 soil plasticity
土的塑性变形 flow of ground
土的塑性分类 plasticity classification of soil
土的塑性流动区 zone of plastic flow of soil
土的塑性指数 plasticity index of soil
土的酸碱度试验 soil pH test
土的酸煮分析 acid digestion analysis of soil
土的摊铺 spreading of soil
土的弹塑性 elastic-plasticity of soil; elastic-plastic of soil
土的弹性 elasticity of soil
土的弹性均压模量 modulus of reaction of soil
土的弹性模量 modulus of soil elasticity
土的弹性模数 resilient modulus of soil
土的特性 soil characteristic
土的特性试验 index property test
土的体变率 <挖方或碾压> bulking factor of soil
土的体积热 volumetric(al) heat of soil
土的体积压缩系数 coefficient of volume of compressibility of soil
土的体缩 soil volume of shrinkage
土的体胀 soil volume of expansion; volumetric(al) expansion of soil
土的天然稠度 natural consistence of soil;natural consistency of soil
土的天然含水当量 field moisture equivalent
土的天然容重 natural unit weight of soil
土的条件 soil condition
土的统一分类(法) unified soil classification system
土的统一分类简要流程图 classification brief flowchart for soil
土的统一分类塑性图 a plastic chart for use in the Unified Soil Classification system
土的透水性 soil permeability
土的图例 soil legend;soil symbol
土的团粒 soil aggregate
土的稳定 soil stabilization
土的稳定技术 soil stabilization technique
土的稳定剂 soil stabilizer
土的稳定性 soil stability
土的污染 soil contamination
土的无湿胀状态 non-expansive condition of soil;no-swell condition of soil
土的无收缩状态 no-shrink condition of soil
土的物理力学性质 physical-mechanical property of soil
土的物理性质 physical property of soil
土的吸力 soil suction
土的吸湿系数 hygroscopic coefficient of soil
土的吸收能力 soil absorption capacity
土的吸收系统 soil absorption system
土的吸水系数 hygroscopic coefficient of soil
土的细粒部分 soil fines
土的现场鉴别法 field identification procedure of soil
土的相对含水量 relative moisture content of soil
土的形成 earth formation;occurrence of soils
土的形成因素 soil-forming factor
土的性质 nature of soil;soil property
土的虚质量 virtual mass of soil
土的徐变 soil creep
土的絮凝化比 flocculation ratio
土的絮凝结构 flocculated structure
土的悬浮液 soil suspension
土的压密 soil compaction
土的压密性 firmness of soil
土的压实 soil compaction
土的压缩固结试验 compression consolidation test of soil
土的压缩模量 constrained modulus of soil;modulus of soil compression
土的压缩试验 compression test of soil;soil compression test
土的压缩系数 compression coefficient of soil
土的压缩性 compressibility of soil;soil compaction;soil compressibility
土的压缩性分级 compressibility rank of soil
土的压缩指数 compression index of soil;soil compression index
土的氧化还原电位试验 soil redox test
土的野外鉴定法 field identification procedure of soil
土的野外压实 field compaction of soil
土的液化 liquefaction of soil;soil liquefaction
土的液限 liquid limit of soil
土的应力-应变-时间特性 stress-strain time characteristics of soil
土的应力应变状态 stress-strain state of soil
土的硬化 soil solidification
土的有效粒径 effective grain size of soil;effective soil size
土的有效容重 effective bulk weight density of soil
土的有效重度 effective weight of soil
土的原位密度测定 field density determination of soil
土的原状 in-place conditions
土的支承力 bearing value of soil
土的质地分类 textural classification of soil

土的肿胀 swelling
土的种类 type of soil
土的重度 unit weight of soil
土的柱状结构 columnar soil structure
土的自然坡度 earth natural slope
土的总应力 total stress of soil
土的阻力 resistance of soil
土的组成 soil composition
土的组分比重 constituent density of soil
土的组构 soil fabric
土的组织 soil texture
土的钻凿开挖 soil boring cut(ting)
土堤 bank of soil; berm(e); earth dike[dyke]; earth embankment; earth embankment dam; earth fill; earth levee; mound; mound of earth; soil bank; soil embankment; clay dike
土堤岸 earth bank
土堤岸护面 chemise
土堤岸护墙 chemise
土堤坝填筑材料 bank material
土堤保护政策 land conservation policy
土堤堆土模 ground mo(u)ld
土堤敷设管线 earth embankment laying pipeline
土堤护岸 chemise
土堤决口 break of an earth bank
土堤破坏 embankment failure
土堤斜坡样板 batter ga(u)ge
土底冰丘 ground ice mound; ice mound
土底座 earth table
土地 holding; land; soil; terra; territory; tract
土地保存 land conservation; land construction
土地保护 land conservation; protection of land; soil protection
土地保护法 land protection law
土地保护和发展委员会 Land Conservation and Development Commission
土地保护性契约 protective covenant; restrictive covenant
土地报酬 payment based on land shares
土地报酬递减律 law of diminishing returns
土地崩塌 devolution
土地边界(线) metes and bounds; lot line
土地编号 parcel number
土地变干 exsiccation
土地补偿费 compensation fee for land; compensation for land; land compensating fee; land compensation fee
土地补偿和安置费 land compensation and rehabilitation
土地布置 laying-out land
土地裁判庭 Lands Tribunal
土地册 cadastre
土地测链 <每链 66 英尺> land chain
土地测量 cadastral survey; earth survey; land measure(ment); land survey(ing)
土地测量地块 land survey section; quarter-quarter section
土地测量联测点 witness corner
土地测量师 land surveyor
土地测量学 chorometry
土地测量员 land surveyer[surveyor]
土地产权 land property right
土地沉陷 land subsidence
土地承受能力 land carrying capacity
土地承载容量 land carrying capacity
土地持有费 carrying charges
土地重划 land readjustment work
土地重划分 subdivision of land
土地重新规划 reallocation of land
土地重新配置 reallocation of land
土地(重新)调整工作 <改造城市的> land readjustment work
土地出让 land leasing
土地处分权 right to dispose of land
土地处理 land disposal; land application; land treatment
土地处理法 land treatment method
土地处理负荷量 loading capacity of land treatment
土地处理权 right to dispose land
土地处理设施 land treatment facility
土地处理系统 land treatment system
土地处置 land application; land disposal
土地处置系统 land disposal system
土地处置需求 <废物的> land disposal needs
土地从属租契 subordinated ground lease
土地粗平整 rough grading
土地贷款 land loan
土地单元 land unit
土地档案 land archives
土地的 agrarian
土地的分配 land appropriation
土地的附属用途 ancillary land use
土地的公共所有 community of land
土地的获得 land access
土地的(季节)状态 condition of ground
土地的价值 land value
土地的居住用部分 residential portion
土地的开垦 reclaim
土地的死守保有 mortmain
土地的占用 appropriation of land
土地的征用 expropriation of land
土地的租用 land tenure
土地的最终用途 end-use of land
土地登记 land register; land registration; rememberment
土地登记测量 cadastral survey; land registry survey; property survey
土地登记处 land registry office
土地登记法例 Land Registration Act
土地登记费 land-register fee
土地电 telluric electricity
土地调查 land survey
土地冻结(政策) <指政府对土地的出卖或转让等所作出的限制> land freeze
土地断面 section of soil; soil section
土地发展 site development
土地发展工作 land development; land development
土地发展规划 site design
土地发展平面图 site development plan
土地法(规) agrarian law; land law; land legislation; law of land
土地肥力 land fertility
土地肥力递减规律 law of diminishing land fertility
土地肥力下降 exhaustion of soil
土地费用 cost of land; land cost
土地分层耕作 land bedding
土地分等 land classification
土地分红 dividend on land shares; payment on land shares
土地分类 classification of land; land classification; soil taxonomy
土地分类图 land-classification map
土地分配 hideland; land allocation; parceling-out
土地分配图 allocation map; allotment plan
土地分配制度 land allotment system
土地分区计划 land zoning regulations
土地分区原则 land zoning regulations
土地负担 land charge
土地负荷能力 land carrying capacity
土地复归 escheat
土地复垦 reclaimed land; reclamation of land
土地复垦规定 regulations on reclamation of land; the statute of reclamation of land
土地复原 land reconversion
土地改革 agrarian reform; land reform
土地改革与土地占有 land reform and tenure
土地改良 agrarian reform; betterment of land; bonification; land accretion; land betterment; land forming; land improvement; land melioration; land reclamation; land reform; land treatment
土地改良贷款 land improvement loan
土地改良计划 reclamation project
土地改良区域 land improvement district
土地改善工作 land improvement
土地改造 land reclamation
土地高度利用 intensive use of land
土地革命 agrarian revolution
土地耕作 land farming
土地公有制 common ownership of land; public land ownership; public ownership of land
土地共有 commonage
土地共有权 commonage
土地购置 land purchase; purchase of land
土地估价 evaluation of land; land appraisal
土地固结 land consolidation
土地管理 land management
土地管理部门 land administration department
土地管理处 land agency
土地管理法 law of land management
土地管理局 Bureau of Land Management; land office
土地管理人 estate agent; land agent
土地管理所 land office
土地管制 land control
土地管制措施 land control measures
土地灌溉(法) land irrigation
土地规划 area planning; land planning; land-use program(me); plan of land utilization; reallocation of land; territory planification
土地规划测量 land planning survey
土地规划管理 zoning
土地规划师 land planner
土地规划图 field layout; land capability map; landscape project
土地国有化 land nationalization; nationalization of land
土地国有制 ownership of land by the state
土地过滤 land filtration
土地合并 land assembly
土地和无形资产的净购买额 net purchase of land and intangible assets
土地核配 land allotment
土地划拨 land assignment
土地划分 division of land; land subdivision; partition of land
土地划分条例 subdivision regulation
土地划分图 land parcelling plan
土地环境 geoenvironment
土地恢复 land restoration
土地回填 land reclamation
土地或资金分配 apportionment
土地获得 acquisition of land
土地获得权 land acquisition right
土地及建筑物 land and building
土地级差收益 differential earnings from land; differential incomes of land
土地集约 land intensive
土地集约工业 land intensive industry
土地集约化 land intensification
土地集约经营 land intensification
土地集中 concentration of landholding
土地计量 land measure
土地记录索引 tract index
土地加固区 land consolidation area
土地价格 ground price; land price; price of land
土地价值 value of land
土地兼并 land annexation
土地鉴定 land appraisal
土地交易冻结 land freeze
土地结构 agrarian structure; land structure
土地界标 land mark
土地界线 borderline of land
土地借款 land loan
土地浸润 land filtration
土地浸润灌溉 land filtration irrigation
土地经管人 <苏格兰> factor
土地经纪人 land agent
土地经济容力 economic capacity of land
土地经济学 land economics
土地经营 soil management
土地经营的垄断 land management monopoly
土地经营管理 soil management
土地经营权 land management right; right to manage land
土地局 land board
土地开发 development of land; development of soils; estate development; handling of land; land development
土地开发贷款 land development loan
土地开发公司 redevelopment company
土地开发管理 management of land development
土地开发强度 intensity of development
土地开发业者 land developer
土地开垦 land development; land reclamation; reclamation of land
土地开垦机械 land reclamation machinery
土地勘测 property survey; site investigation
土地可用性地图 land capability map
土地垦拓 land reclamation; reclamation of land; soil reclamation
土地垦殖工程 land settlement project
土地垦殖规划 land settlement project
土地垦殖计划 land settlement project
土地垦殖与改良 land reclamation and improvement
土地库 land bank
土地浪费 land wastage; land abuse
土地类别 land classification
土地类别分区 <城市区域规划> land use
土地类型 land type
土地类型图 land-type map
土地立法 land legislation
土地利用 land usage; land utilization; land use
土地利用测量 land-use survey
土地利用的主要根据 the functional basis for land utilization
土地利用调查 land-use survey
土地利用方式 kind of land use; land utilization type
土地利用费 land-use cost
土地利用分类 land-use classes
土地利用分区 land-use zoning
土地利用分析 land-use analysis

土地利用规划 land-use planning
土地利用规划图 land-use plan
土地利用及交通最优选择 land-use and transport optimization
土地利用级别 land-use classes
土地利用计划 land-use planning
土地利用结构 land-use structure
土地利用经济学 land-use economics
土地利用类型 land-use type;land utilization type
土地利用率 land-use capability;land-use rate;land-use ratio;land utilization rate;ratio of land utilization
土地利用密度 land-use intensity
土地利用强度 land utilization intensity
土地利用情况 land-use status
土地利用调整 land-use adjustment
土地利用统计 land-use statistics
土地利用图 land capability map;land-use map
土地利用现状 present land-use
土地利用现状图 existing land-use map
土地利用形式 land-use pattern
土地利用选择方案 land-use alternative
土地利用研究 land-use study
土地利用预测模型 land-forecast model;land-use forecast model
土地利用诊断标准 diagnostic criterion
土地利用种类 kind of land use
土地利用专题图 thematic land-use map
土地利用状况调查 land-use survey
土地沥青配合比公式 job mix formula
土地撂荒 leave the land uncultivated
土地流动 flow of ground
土地隆起 upheaval of land
土地隆起监测器 heave ga(u)ge
土地买卖代办所 <美> land agency
土地买卖合约 agreement for deed
土地没收 land confiscation
土地密集 land intensive
土地面积 acre(age);area of land;land area
土地面积的计算单位 a unit of land measurement
土地排水 land drainage
土地排水道 land drain
土地排水法 land drainage act
土地排水沟 land drain
土地排水网 land drainage network
土地批租 leasehold of land;leasing land in batches
土地贫乏 land poor
土地平整 bedding of land;grading;land forming;land grading;land level(l)ing;land smoothing;preparation of land
土地平整测量 survey for land smoothing;surveying for land level(l)ing
土地平整的包络线 envelope of grading
土地平整费 cost of grading
土地平整工程 land level(l)ing project
土地平整工作 grading operation
土地评价 bonitation;land evaluation;land judging;land valuation
土地坡度 land slope
土地普查 census tract
土地期望价 capital value
土地期望期 soil expectation value
土地契约 land contract
土地潜力 capacity of the land;land potential;potential capacity of land
土地侵蚀 land erosion
土地侵蚀治理 land erosion control
土地倾斜度 land pitch
土地清理 land clearing
土地区划整理 land readjustment
土地取得 land requisition(ing)
土地权 land rights
土地权属 allocation of landownership and land-use right
土地容量 land capability
土地蠕变 land creep
土地入股 pooling of land
土地入股分红 receive payment on the basis of their shares of the land
土地沙漠化 desertification;desertization;land desertification
土地上房屋拆除 land clearance
土地渗滤处理 land treatment
土地生产力 fertility;land capability;productivity of land
土地生产能力 capability of land;land capacity;land capability
土地施加污泥 land application of sludge
土地湿润 ground moistening
土地使用 land tenure;land use;tenure
土地使用标示 land use designation
土地使用调查 land-use survey
土地使用法规 land-use act
土地使用发展规划 land use planning
土地使用法规 land use regulation
土地使用费 fee for land use;land royalty;land-use fee
土地使用分类 land-use classes;land-use classification
土地使用功能 land service function
土地使用规划 land-use planning
土地使用规则 land-use regulation
土地使用规章 land-use regulation
土地使用活动率 land-use-activity ratio
土地使用控制 land-use control
土地使用能力分类 land-use capability class
土地使用期 land tenure
土地使用强度 land-use intensity
土地使用权 easement;land tenure;land-use right;right of land usage;right to use land;right to use site;wayleave
土地使用权范围 easement boundary
土地使用权转让 transfer of land-use rights
土地使用适宜性评价 land-use suitability evaluation
土地使用图 graphic(al) method of land use;land use map
土地使用现状 existing land-use pattern
土地使用现状图 existing land-use map
土地使用与交通运输规划 land-use and transportation plan
土地使用预测技术【道】 land-use forecasting techniques of transportation planning
土地使用远景图 future land-use pattern
土地使用证 land warrant
土地使用证书 land-use certificate
土地使用资料要目 land-use inventories
土地适用性分类 <指宜农、宜林、宜牧、宜耕等而言> land suitability classification
土地适用性(能)land suitability
土地收购 land purchase
土地收回诉讼 ejectment
土地收入 land revenue
土地收益递减律 law of diminishing returns of land
土地收益估值区 assessment district
土地收益权 right to derive benefit from land
土地收用 eminent domain
土地税 land tax
土地私有化 privatizing ownership of the land
土地私有制 private ownership of land
土地所有 land ownership
土地所有权 landed property;land ownership rights;quarter's rights;right to ownership of land;squatter's rights;title to land
土地所有权的分配 division of land property
土地所有权和使用权 ownership and use rights of land
土地所有权简介 abstract of title
土地所有权证书 certificate of possession of land;certificate of title
土地所有权转移(过户)land transfer
土地所有人 landholder;landlord;land owner;property owner
土地所有人的矿业使用费 landowner's royalty
土地所有者 landed proprietor;landholder;land owner
土地所有者的工作者 in the land owner's employ
土地所有制 land ownership
土地台账 land files
土地探测 acquisition of land;land prospecting
土地特性 land characteristic;land quality
土地体系 land system
土地填埋 landfill
土地填筑 landfill;reclamation of land
土地条例 land regulation;land tenure
土地调整 land readjustment
土地调整规划 land readjustment project
土地投机 land speculation;speculation in land
土地投机商 land jobber
土地投资 investment in land
土地图 land map;land plat
土地退化 land deterioration;land retirement
土地围垦 land accretion
土地污染 land pollution
土地污染控制 land pollution control
土地无偿使用制度 system of free use of land;system of uncompensated land use
土地无限使用权 equitable servitude
土地系统 land system
土地细分 land subdivision
土地细整 land smoothing
土地下沉 land subsidence
土地闲置税 land holding tax;unused land tax;vacant land tax
土地消耗 land consumption
土地协议文书 compound settlement
土地信托 land trust
土地信托证 land trust certificate
土地信息系统 land information system
土地兴业公司 proprietary company
土地形态 conformation of ground
土地休耕 fallow soil;land resting;land retirement
土地许用证 land grant
土地盐渍化 soil salinization
土地银行 land bank
土地用途分区制 land-use zoning
土地用途区 land-use zone
土地有偿使用管理制度 management system for the paid use of land
土地再植被 revegetation of land
土地增价税 land increment value duty
土地增值 appreciation of land value;land increment
土地增值税 canal advantage rate;inclusion fee;increment in land;increment tax on land value;land increment value duty;land value added tax;land value increment tax
土地债券 land bond
土地占用权 land tenure
土地占有 demesne
土地占有权 right to hold land
土地占有热、领土扩张热 land hunger
土地占有税 land holdings tax
土地占有制度 land tenure
土地丈量 land survey;property survey
土地丈量单位 land measure
土地丈量员 measurer
土地沼泽化 land swamping
土地征购价 land price
土地征购价格 land purchasing price
土地征用 compulsory land acquisition;expropriation;land acquirement;land expropriation;land requisition(ing)
土地征用标准 land requisition criterion
土地征用补偿 condemnation award
土地征用贷款 land acquisition loan
土地征用法 land expropriation law
土地征用费 land acquisition cost;site-acquisition cost
土地征用后价值 value after the taking
土地征用前价值 value before the taking
土地征用权 eminent domain;land expropriation right;rights of eminent domain
土地征用线 land requisition line
土地整备 land consolidation
土地整理 land arrangement;land consolidation;land preparation;land readjustment;reallotment
土地整理略图 simplified land consolidation scheme
土地整理区 land consolidation area
土地整理图 land capability map
土地整平机 land plane
土地整形 land shaping
土地整治 land consolidation;land reclamation
土地整治计划 physical planning
土地整治小区 parcel
土地证 land deed;title deed for land
土地证抵押 deposit(e) of title-deeds
土地证书 land certificate
土地政策 land policy
土地执照 land license
土地植物 ground cover;ground vegetation
土地制度 land system
土地质量 land characteristic;land quality
土地质量特性鉴别基准 diagnostic criterion
土地终身占有人 life tenant
土地主管部门 agency in charge of land
土地专利 land patent
土地转让 land transfer
土地转让证 land patent
土地转位 <法语,因洪水冲裂而致土地突然转入他人地产内> avulsion
土地状况自动分析系统 land status automated system
土地资本 land capital;soil capital;terre-capital
土地资产 real estate property
土地资金利息 interest occurring from land funds
土地资金收益 land fund benefit
土地资金折旧费 depreciation fees from land funds
土地资源 capacity of the land;land resources;land sources
土地资源保持 land conservation

土地资源保护 land resources conservation; protection of land resources; conservation of land
土地资源调查 land resources survey
土地资源经济学 economics of land resources; land resource economics
土地资源区域 land resource area; resource area of land
土地自然增价 unearned increment
土地综合开发 multiple-purpose land development
土地租佃 tenancy
土地租费 land rental
土地租赁 land tenancy
土地租用权 leasehold
土地租约 ground lease
土地阻力 soil resistance
土垫层 bedding layer of soil; soil cushion
土调查者 soil investigator
土钉 brad; round lost head nail; soil nail(ing)
土钉坝 earth spur dike[dyke]
土钉挡土结构 soil nailed retaining structure
土钉法 <地基处理> soil nailing
土钉墙 nailing wall; soil nailed wall; soil nailing wall
土钉墙技术 soil nailing wall technology
土动力特性 soil dynamic(al) property
土动力学 soil dynamics
土动力学与地震工程 <英国季刊> Soil Dynamics and Earthquake Engineering
土洞 cave-in soil; karstic earth cave; soil cavity; soil hole; earth cavity
土洞埋深 burying depth of soil cave
土洞塌陷 collapsed earth cavity
土洞体积 volume of soil cave
土洞钻掘机 bore tunnel(l)ing machine
土斗 dipper
土斗车 muck car; spoil car
土斗列车 muck car
土豆 potato
土豆废水 potato waste(water)
土豆泥机 potato masher
土堆 bunker; burrow; earth deposit; earth hummock; earth mass; earth mound; heap; hill(ock); mound; mound of earth; prism; soil stack; agger <尤指古罗马军营外围所筑的土堤>
土(对板桩的)自由支承 free-earth support
土墩 earth mound; earth pillar; hill(ock); knoll; mound(of earth); carn; mott(e) <城堡的>
土墩和外廊 motte-and-bailey
土尔戈式水轮机 Turgo impulse turbine; Turgo turbine
土尔戈式透平 Turgo turbine
土尔戈式涡轮机 Turgo turbine
土耳其 <亚洲> Turkey
土耳其-爱琴海板块【地】Turkish-Aegean plate
土耳其板块 Turkish plate
土耳其厕所 Turkish closet
土耳其船级社 Turk Loydu
土耳其地毯 Turkish carpet
土耳其红 Turkey red
土耳其红油 Turkey red oil; monopole oil <别名>
土耳其建筑 Turkish architecture
土耳其蓝 Turkey blue
土耳其栎木 <包括红栎及平滑栎木材> Turkey oak
土耳其没食子丹宁 Turkish gallotannin
土耳其砂浆 <1/3 砖粉加 2/3 石灰粉拌和的砂浆> Turkish mortar
土耳其式尖塔 <伊斯兰教寺院的> Turkish minaret
土耳其式建筑 Ottoman architecture
土耳其式清真寺 Turkish type mosque
土耳其式浴室 hamman; Turkish bath
土耳其玉 turquoise
土耳其赭土 Turkey umber
土耳其-中伊朗-冈底斯中间板块 Turkish-Central Iran-Gangdise plate
土耳其棕土 Turkey umber
土法 native experience
土法炼焦 heap coking
土法上马 adoption of indigenous method
土法熟铁吹炼炉 bloomary
土法炭化 pile charring
土翻拌机 soil tiller
土反力模数 modulus of soil reaction
土方 cubic(al) meter of earth; yardage
土方搬运 earthmoving; muck shifting; soil removal
土方搬运工程 earthmoving works
土方搬运设计曲线 mass-haul curve
土方崩塌 earth slide
土方边坡 earthwork side slope
土方表 quantity sheet; table of earthwork; yardage table <英制>
土方铲斗 earth bucket; mud bucket
土方铲运机 earth(moving) scraper
土方超运距费 cost of overhaul
土方承包班组 earthwork contract section
土方粗平 rough grading
土方的 earthwork
土方的水力机械化 jetting
土方的最大免费运距 limit of free haul
土方地形整理图 grading map
土方调配 earthwork adjustment; groundwork management
土方调配计划 cut-fill transition program(me)
土方叠积图 mass diagram
土方动力学 mechanics of landslide
土方断面 section of earthwork
土方分布数量 mass yardage distribution; yardage distribution
土方分配 yardage distribution; yardage table
土方工 mucker
土方工程 earthmoving works; earthwork (engineering); earth handling; earthmoving; earthmoving job; excavation works; grading; ground work; handling of earth; mucking; navvy work; subgrade construction; trench works; agger
土方工程车辆 earthmoving vehicle; earthwork vehicle
土方工程承包地段 earthworks contract section
土方工程工序 earthwork operation
土方工程机械 earthmoving construction machine; earthmoving machine; earthwork machinery
土方工程机械设备 earthworking machinery
土方工程计划 earthwork plan
土方工程界线 earthwork outline
土方工程进度表 schedule of earthworks
土方工程控制统计 earthwork control statistics
土方工程量 earthwork quantity; quantities of earthwork; volume of earthwork
土方工程量测定 earth quantity determination
土方工程量平衡 balanced earthwork
土方工程设备 earthwork plant
土方工程设计 earthwork design
土方工程师 earthmoving engineer
土方工程施工 earthwork construction
土方工程图 earthwork drawing
土方工程现场 earthwork site
土方工具 hack iron
土方工人 <俚语> humper
土方工艺 earthmoving process
土方工作 ground work
土方过程 earthmoving process
土方行业 earthmoving; earthmoving industry
土方机具 earthwork's tool; ground engaging tool
土方机械 dirt mover; earth handling equipment; earth mover; earthmover equipment; earthmoving; earthmoving equipment; earthmoving machine; earthmoving machinery; earthmoving plant
土方机械化 mechanization of earthmoving
土方机械化施工 earthwork mechanical construction
土方机械驾驶室 environmental cab
土方机械设备 earthmoving plant
土方机械设备驾驶员 earthmoving plant operator
土方机械施工 mechanical earthwork
土方机械用轮胎 earthmover tire [tyre]
土方积累图 mass diagram
土方计 <英制的> yardage meter
土方计划 earthwork plan
土方计算 calculation of earth volume; cut-and-fill estimate; mass calculation for earth works
土方技术 earthmoving
土方截面 section of earthwork
土方开挖 earth excavation
土方控制 earthwork control
土方累积曲线 mass curve; summation curve
土方累积体积 mass volume
土方累积图 mass diagram
土方累计体积 mass volume
土方量 earth quantity; earth volume; mass profile; quantity sheet; volume of earthwork
土方量测定 earth quantity determination
土方量估计 quantity survey(ing)
土方量估算 quantity survey(ing)
土方量计算 calculation of cutting and filling; mass calculation; computing quantities of earthwork
土方平衡 balanced grading; balance of cut and fill; cut-and-fill balance; earthwork balance; equalization of earthwork
土方平衡表 earthwork balance sheet; earth work balancing chart; earthwork balancing chart; table of earthwork balance
土方平衡计算步骤 balancing procedure
土方平衡系数 earthwork balance factor
土方平整 grading; grading operation; grading work
土方平整设备 grading outfit
土方剖面 section of earthwork
土方生产量 earthmoving production
土方数 <英制> yardage
土方调配经济运距 economic hauling distance of soil
土方调配图 cut-fill transition diagram
土方图 quantity diagram
土方挖掘 earth excavation; soil excavation
土方(挖填)调配 cut-fill transition
土方挖填计算 calculation of cutting and filling
土方挖填计算图 mass diagram
土方外运 soil transportation
土方修整 grading
土方修筑费 cost of grading
土方压实机 landfill compactor
土方压实机械 soil compactor
土方用砂 mason sand
土方运距 earthwork haul distance
土方运距按英里里程分区 mileage zones of haulage
土方运距费 cost of haulage
土方运距曲线 mass-haul curve
土方运输工具 earthwork vehicle
土方运输机械 earth hauler; hauling plant
土方运输设备 load-and-carry equipment
土方运算(工程) earthwork operation
土方站 yard station
土方筑坝 earthwork dam
土方纵剖面图 mass profile
土方作业 earth handling; earthwork; handling of earth; earth-moving operation
土房 earth house; keekwilee-house
土沸现象 boiling of soil
土分布图 engineering soil map; soil distribution diagram; soil map
土分级 soil rating
土分类法 soil classification system
土分类简要流程图 brief flowchart of soil classification
土分类三角图 triangular chart for soil classification
土分类试验 soil classification test
土分类体系 soil classification system
土分类完整流程图 complete flowchart of soil classification
土分析 soil analysis
土分析误差 soil analysis error
土分选机 soil separator
土风建筑 vernacular architecture
土氟磷铁矿 richellite
土覆盖房屋 <地下式和半地下式的> earth-sheltered housing
土盖 earth mulch
土岗 earth hummock
土高炉 indigenous blast furnaces; native style blast furnace
土镐 clay pick; navy pick
土埂 border check; earth banking
土工 banker; hog; soil engineering; subgrade construction
土工(边)坡度 earthwork slope
土工(薄)膜 geomembrane; plastic foil
土工布 civil engineering fabric; geofabric; geotechnic(al) fabric; geotextile(fabric); non-woven geotextile; petromat; road rug; supac; synthetic(al) fabrics
土工布沉排 fabric mattress
土工布挡土墙 fabric retaining wall
土工布反滤层 filter fabric mat
土工布反滤层试验 filter fabric soil retention test
土工布灌浆沉排 grouted fabric mattress
土工布加固层 geotextile reinforcement
土工布加强 polymer grid reinforcement
土工布滤层 geotextile filter
土工测试仪器 soil testing instrument
土工产品 geoproduct
土工处理法 geotechnic(al) process
土工袋 geobag
土工的 geotechnical; earthwork

土工垫 geocushion; geomat; geospacer
土工复合材料 geocomposite
土工格室 geo-grating
土工格栅 geogrid; geo-grille
土工隔膜 geomembrane
土工工程 soil engineering
土工工程师 geotechnic(al) engineer; soil engineer
土工工程学 geotechnic(al) engineering; soil engineering
土工工具 earthworker's tool
土工构造 earth structure
土工构造物 earthen structure
土工构筑物 earthen structure
土工合成材料 geosynthetics; geotextile; geomaterial
土工合成物 geosynthetics; geotextile
土工技术 geotechnics; geotechnique; soil technics; soil technology
土工技术参数 geotechnic(al) parameter
土工技术的 geotechnic(al)
土工技术方法 geotechnic(al) method; geotechnic(al) process
土工技术工程 geotechnic(al) engineering
土工加筋带 geostrip
土工建筑(物) earth(en) structure
土工建筑物的维护 maintenance of earthworks
土工工程破坏 geotechnical failure
土工管 geotube
土工过滤层 geotextile filter
土工合成材料配黏[粘]土衬砌 geosynthetic clay liner
土工盒 geocell
土工技术失败 geotechnical failure
土工加强斜坡 geotextile-reinforced slope
土工建筑物 earth structure
土工结构物 earth structure
土工聚合物 geopolymer; geosynthetics
土工离心试验机 geotechnic(al) centrifuge
土工历史 geotechnic(al) history
土工帘 geocurtain
土工模袋 geotechnic(al) mo(u)lded bag; geofabriform
土工模型试验 geotechnic(al) model test
土工膜 geomembrane
土工膜基衬垫 geomembrane-based liner
土工膜网铺法 fabric sheet reinforced earth
土工排 geomat
土工排水板 geodrain
土工平整工作 grading job; grading operation
土工平整机 grading machine
土工平整设备 grading equipment
土工评价 geotechnic(al) evaluation
土工器具 intrenching tools
土工刃具 hack iron
土工实验室 soil lab; soil laboratory; soil mechanics laboratory
土工试验 engineering geotechnic(al) test; geotechnic(al) test; soil test(ing)
土工试验规范 soil test regulation; soil test specification
土工试验室 soil lab; soil laboratory; soil mechanics laboratory
土工试验手册 <美国垦务局的> earth manual
土工手册 earth manual
土工图 geotechnic(al) map
土工网 geonet
土工网垫 geotechnic(al) net mat
土工网格加固 geogrid reinforced
土工网排水 geonet drainage
土工网状结构 geotech-mesh
土工围堰 dike-type cofferdam
土工席垫 geonet
土工系统 geosystem
土工纤维材料 geotextile
土工纤维织物 geosynthetics; geotextile
土工箱 geocontainer
土工修样器 soil lathe
土工修整 grading
土工修筑费 cost of grading
土工学 geotechnics; soil technology; geotechnique
土工学的 geotechnic(al); geotechnological
土工仪器 soil mechanic instrument
土工艺术品 <以泥沙石块制成> earthwork
土工整平工作 grading work
土工整平设备 grading outfit
土工织物 geotextile(fabric); civil engineering fabric; geofabric; geotechnic(al) fabric; petromat; supac; synthetic(al) fabrics
土工织物的顶破强度 breaking force of geotextile in a puncture test
土工织物垫层 geotextile cushion
土工织物加固层 geotextile reinforcement
土工织物加固地基 geosynthetic reinforcement
土工织物夹层 geotextile interlayer
土工织物排 geotextile mattress
土工作台 earth bench
土拱架 earthen centring
土拱作用 soil arching action
土沟 earth ditch
土构成 soil formation
土构造分类图 textolite classification chart
土骨架 soil skeleton
土骨架重 dry weight
土骨料混合物 soil-aggregate mixture
土圭 template
土过筛 dirt screening
土海堤 earth sea dike[dyke]
土含水量 soil moisture
土夯 earth rammer
土和结构的相互作用 soil-structure interaction
土和剩余水荷载 soil and differential water load
土和水的关系 soil and water relationship
土和水泥表面层 soil-cement surface course
土荷载 load of earth pressure; overburden load; soil load(ing)
土盒 moisture can
土褐煤 earth coal; wax coal
土褐色 peat brown
土褐色的 drab
土黑铜矿 melaconite
土红 princess mineral
土红钴矿 remingtonite
土红色的 latericeous; lateritious
土滑(坡) earth slide; earth slip; soil creep; soil slide; earth creep; soil slip
土化学全分析试验 total chemical analysis of soil
土黄褐色 siennese drab
土黄色 khaki; sienna
土黄色面砖 brownish yellow(facing) brick
土黄色颜料 ochre
土黄(颜料) ocher
土黄棕色 sienna brown
土灰橙色 dusty orange
土灰粉红色 dusty pink
土灰橄榄绿色 dusty olive
土灰浆 earth mortar
土灰蓝色 dust blue
土灰绿色 dusty green
土灰绿松石色 dusty turquois
土灰玫瑰红色 dusty rose
土灰沙三合土 clay-lime-sand mixture
土灰水绿色 dusty aqua green
土灰桃红色 dusty peach
土灰玉绿色 dusty jade green
土灰紫红色 dusty mauve
土火箭 autochthonal rocket
土积曲线 grading curve; mass curve
土积图 quantity diagram
土基 earth base; earth foundation; earth pad; ground base; soil ground; soil matrix; subbase; subgrade
土基长杆贯入仪 penetration test equipment
土基承载能力 subsoil bearing capacity
土基础 soil foundation
土基础系统 soil-foundation system
土基的板承值 <一种指示土基承重能力的指标> plate bearing value of subgrade; plate hearing value of subgrade
土基干湿类 subsoil moistness classification
土基干湿类型 type of dry and damp soil base
土基或路面材料抗力(值) resistance value; R-value
土基接触面 soil-footing contact
土基浸渗 subgrade intrusion
土基浸透 subgrade intrusion
土基湿度稳定(法) moisture stabilization
土基压实 rolling of soil
土基支承值 soil support value
土基锥承试验 cone bearing test of subgrade
土及岩性的记录 lithologic log
土级 soil class
土级配系数 soil gradation factor
土级数 <美国农业部的> soil series
土集料混合物 soil-aggregate mixture
土脊 bank; earth ridge
土技术 indigenous technology
土加固 soil fixation
土加固剂 soil stabilizer
土加权剪切模量 weighted shear modulus of soil
土建承包商 civil contractor; civil engineering contractor
土建费用 civil cost; civil engineering cost
土建工程 building project; civil construction; civil engineering; civil works
土建工程投资 investment of civil work
土建和服务技术规范 specifications for civil works and service
土建机械 building machinery
土建设计 civil design
土建设施 civil feature
土建施工说明书 general specification of building construction
土建室 civil works office
土鉴定图 soil identification chart
土浆 soil paste
土胶体 soil colloid
土胶质 soil colloid
土窖 cob house; crypt
土结的 soil-bound
土结构 earth structure; soil structure
土结构的基本单元 texture fundamental element of soil
土结构等级 texture grade of soil
土结构分类 soil textural classification; textolite soil classification
土结构(交)界面 soil-structure interface
土结构类型 texture type of soil
土结构破坏 soil failure
土结构相互作用 soil-structure interaction
土结合料 soil binder
土结皮 soil crust
土界 pedosphere
土金属 earth metal
土井 uncased well
土静力的 geostatic
土居动物群落 soil animal population; soil fauna
土居生物 edaphon(e)
土坎填筑层 lift
土抗力 earth resistance; passive earth pressure; passive resistance; passive soil pressure; resistant earth pressure
土颗粒 soil particle
土颗粒表面积 surface area of soil
土颗粒间力 intergranular force
土颗粒间吸力 interparticle attraction
土颗粒压力 intergranular pressure
土壳 soil crusting
土坑 bury; earth pit; lysimeter pit <土壤蒸发器的>
土孔隙 soil pore
土孔隙体积 non-soil volume
土库曼斯坦 <亚洲> Turkmenistan
土库曼装饰织物 Turkoman
土块 clay ball; clay lump; clod; clump; earth mass; soil block; soil cake; soil clod; soil lump; soil mass; soil pat; lysimeter monolith <土壤蒸发器的或地中渗透仪的>
土块比重 specific gravity of soil
土块分离机 soil extractor
土块分离器 clod eliminator
土块粉碎器 clod disintegrator
土块击碎器 cold breaker
土块类 soil great group
土块排除法 cold sweep
土块排除器 clod sweeper
土块破碎机 clod buster; clod smasher
土块清除器 clod clearer; clod separator
土块实验 clod test
土块试验法 <测土壤密度> chunk method
土块细碎板 crusher board
土块压碎器 clod breaker
土筐 corf
土况 soil condition
土矿物学 soil mineralogy
土拉姆法 <定源式双线框交流电法> Turam method
土篮 corf
土牢 dungeon; oubliette
土类 great group of soils; great soil group; soil class; soil grade; soil group; soil type
土类分级法 system of land classification
土类树脂 earth-type resin
土力侵蚀 soil corrosion
土力学 geotechnics; geotechnique; soil mechanics
土力学触探仪 soil penetrometer
土力学词汇 soil mechanics vocabulary
土力学工程学 geotechnic(al) engineering
土力学实验室 soil mechanics laboratory
土力学特性 soil mechanic property
土力学与基础工程 <美期刊名>

T

Journal of the Soil Mechanics and Foundations Division
土力学与基础工程学 soil mechanics and foundation engineering
土力学原理 soil mechanics principle
土力学中强度概念 strength concept in soil mechanics
土沥青 earth pitch;land asphalt;maltha;malthite;mineral tar;oil coal;stellarite
土粒 grog;soil grain;soil particle
土粒比重 specific gravity of soil particles
土粒表面结合水 bound water on mineral surface
土粒大小 soil particle size
土粒大小尺度 grade scale
土粒复合体 grain complex
土粒干重 dry weight of soil solids
土粒间的法向力 normal force between particles
土粒接种法 soil particle inoculating method
土粒径累积曲线 soil grain size accumulation curve
土粒聚集 aggregate of soil particles;aggregation of soil particles
土粒孔隙体系的稠度 consistency of soil particle and cementing substances
土粒累积曲线 soil grain size accumulation curve
土粒密度 density of soil grain;density of soil particle
土粒排列 arrangement of soil particles
土粒培养法 soil particle culture method
土粒容重 unit weight of soil grains;unit weight of solid particles
土粒实容重 unit weight of soil constituents;unit weight of solid constituents
土粒体积 soil solid volume;volume of solids
土粒团 aggregate of soil particles
土粒稳定性 particle stability
土粒与水的作用方式 the mode of action between grain and water
土粒运动 movement of soil particles
土链 catena[复 catenae/catenas];soil catena
土链丝菌素【医】terramycin
土梁 earth beam
土梁试验 earth beam test;soil beam test
土料 earth material
土料铲运机 earthmoving scraper
土料场 borrow area;local borrow
土料场地质平面图 geological plan of soil borrow area
土料场地质剖面图 geological section of soil borrow area
土料场工程地质剖面图 engineering geological profile of soil material borrow area
土料的压实参数 earth compaction factor
土裂 soil cracking
土裂缝 fissure in ground
土裂隙 fissure in ground
土磷灰石 osteolite
土磷铁矿 picite
土磷锌铝矿 kehoeite
土溜 solifluxion
土溜阶地 solifluction terrace
土溜坡 solifluction slope
土溜舌 solifluction lobe;solifluction tongue
土流 earth flow;mudflow;quick soil;soil flow;solifluxion
土流堤 toe of earthflow
土流舌 toe of earthflow
土硫铀矿 uraconite
土垄 earth ridge
土垅 lence
土垅大白蚁 <拉> Macrotermes annandulei
土滤池 earth filter
土路 <无路面的> dustroad;earth road;natural surfaced road;packway;soil path;dirt path;dirt road;dirt track;earthen road;earth path;soil road;unsurfaced road;unmetallized road <无硬质路面的>;drove <英>
土路基 basement soil(-subgrade);basement subgrade;bedding-in soil;earth grade;unsurfaced subgrade
土路基面 earth roadbed
土路肩 earth shoulder;unprotected shoulder
土路面 earth surface;soil surface;soil surface(d) road
土路面桥 earth-paved bridge
土路喷油稳定处理 suboiling of soil road
土路平整器 road hone
土路渗油稳定 suboiling of soil road
土路整平器 road hone
土绿磷铝石 planerite
土仑阶 <晚白垩世>【地】Turonian
土仑-赛诺期海浸 Turonian-Senonian transgression
土仑统【地】Turonian series
土螺钻 earth auger
土麻黄 beef wood
土埋耐腐试验 graveyard test
土埋试验 <纺织品耐腐性试验> burial trial;soil burial test
土锚 earth bolt;ground anchor
土锚钉 soil nail
土锚法 soil nailing;soil nailing method
土锚杆 earth anchor;tieback anchor
土锚钻机 soil anchor driller
土煤 earthy coal;smut;sooty coal
土霉素 terramycin
土幂 soil mulch
土面粗糙 roughness of soil surface
土面路 soil surface(d) road
土面施肥 top dressing
土面学 saturnigraphy
土面有光泽 luster of soil surface
土面蒸发量 evaporation discharge from soil surface
土面整修 soil-shaping measures
土面坐标 saturnigraphic(al) coordinates
土名 soil name;trivial name
土模成型混凝土 soil-mo(u)lded concrete
土木材料 engineering material
土木工程 civil engineering
土木工程承包人 civil engineering contractor
土木工程承包商 civil engineering contractor
土木工程承包商联合会 <英> Federation of Civil Engineering Contractors
土木工程程序 civil engineering procedure
土木工程处 civil engineer's department
土木工程定额测算方法 civil engineering standard method of measurement
土木工程费用 civil engineering cost;cost of civil engineering works
土木工程公司 civil engineering firm
土木工程构筑物 civil engineering structure
土木工程合同 construction contract
土木工程绘图员 civil engineering draughtsman
土木工程机械 civil engineering machinery
土木工程计算机辅助设计 civil CAD
土木工程技术员 civil engineering technician
土木工程检查员 civil engineering inspector
土木工程建设项目 civil engineering project
土木工程建筑 civil engineering work
土木工程勘测 civil works investigation
土木工程设计 civil design
土木工程设施 civil works
土木工程师 civil engineer
土木工程师项目 civil engineering project
土木工程师学会 <英> Institute of Civil Engineers;Institution of Civil Engineers
土木工程师学会学报 <英国双月刊> Proceedings of Institution of Civil Engineers
土木工程施工 civil engineering works
土木工程施工场地 civil engineering site
土木工程施工合同条件 conditions of contract for(works of) civil engineering construction
土木工程系 civil engineering department
土木工程学 civil engineering
土木工程学士 bachelor of civil engineering
土木工程研究协会 Civil Engineering Research Association
土木工程与公共建筑综论 <英国期刊名> Civil Engineering and Public Works Review
土木工程与结构工程师评论 <期刊> Civil and Structure Engineer Review
土木工程助理人员 civil engineering assistant
土木基础设施系统 civil infrastructure system
土木技术 geotechnique
土木技术员 civil technician
土木建筑 civil architecture;civil construction
土木建筑材料 civil engineering material
土木建筑工程 civil engineering and building construction
土木建筑设施 civil construction facility
土木主任工程师 chief civil engineer
土木总工程师 chief civil engineer
土内冰 ground ice
土内等孔隙水压线 soil lines of equal pore pressure
土内含气 soil air
土内孔隙水压力等压线 soil lines of equal pore pressure
土内气体 soil gas
土内水流 <降雨入渗的> interflow
土内碎石带 stone line
土囊 pockets of soil
土黏[粘]结 soil cement
土牛拱架 earthen centering
土牛拱胎 arch formed on filled earth
土爬 soil creep;surficial creep
土盘 pan
土跑道 earth runway;natural surfaced runway;unimproved runway
土跑道上的网眼钢板 surface mat
土跑道上起飞 rough-strip takeoff;rough-field take-off
土培 soil culture
土膨胀 soil expansion
土坯 adobe;sun-dried mud brick
土坯房 adobe building
土坯房屋 adobe house
土坯构造 adobe(clay) construction
土坯建筑 adobe building;adobe (clay) construction
土坯结构 adobe(clay) construction
土坯块 adobe block
土坯女儿墙 pretil
土坯砌体 adobe masonry
土坯墙 adobe(brick) wall;cob wall;cob walling;earth wall
土坯圬工 adobe masonry
土坯砖 adobe;adobe brick;clay body brick;cob;pisay
土坯作用 ground cylinder effect
土皮层钻进 overburden drilling
土皮植草法 topsoil planting
土平台 earth platform
土坡 bank of slope;bank of soil;earth slope;hillock
土坡安全稳定措施 safety measures of soil slope
土坡挡土墙 tallus wall
土坡道 earth ramp;soil ramp
土坡的极限高度 ultimate height of soil slope
土坡的极限坡角 ultimate angle of soil slope
土坡滑动弧 failure arc of earth slope
土坡滑坍 failure of earth slope
土坡基底破坏 base failure of slope
土坡极限分析法 slope limit analysis method
土坡临界高度 critical height of slope
土坡临界圆 critical circle of slope
土坡深度系数 depth factor of slope
土坡坍毁 failure of earth slope
土坡坍塌 failure of earth slope
土坡稳定破坏类型 type of slope failure
土坡稳定系数 stability factor of setting
土坡稳定性 earth slope stability;stability of setting;stability of slope;stability of soil slope
土坡稳定性分析 earth slope stability analysis;stability analysis of soil slope
土坡稳定性计算结果 computation result of soil slope stability
土坡稳定性评价方法 evaluation method of soil slope stability
土坡稳定性因素分析 factor analysis of soil slope stability
土坡稳定性与土压力 soil slope stability and earth pressure
土剖面 soil profile;soil section
土铺路面 earth surface road
土栖白蚁 ground termite;subterranean termite
土漆膜【地】agricere
土潜动 soil creep
土潜滑 soil creep
土强度试验 soil strength test
土强度歇后增长 <沉桩后桩周> soil set-up
土墙 adobe;clay wall;cob wall;dirt wall;earth wall;loam wall(ing);mud wall;peasey hut
土墙房屋 adobe(clay) construction
土墙茅舍 peasey hut
土桥 earth-paved bridge;natural bridge;highway embankment built in loess plateau <黄土高原路堤>

土丘 earth(en) mound; earth hummock; mound; rideau
土丘系数 terrain factor
土球包扎 <植物移植> balled and burlapped
土球包扎植物 ball-plant
土球包扎装置 ball and burlap planting
土球移植 ball transplanting
土渠 earth canal; earth channel
土取样器 soil sampler
土圈 pedosphere
土圈采样 soil sampling
土圈充水状态 replete state of soil
土圈结构 structure of pedosphere
土壤 soil; soil culture; earth; ground; ranker <在硅质岩或沉积上与 AC 层一起发育的>
土壤 pH 值图 soil pH map
土壤斑驳层 mottling zone
土壤板结 hardening of soil
土壤拌和场机 soil mixing plant
土壤拌和机 mix-in-place machine
土壤饱和 soil saturation
土壤保持 earth conservancy; soil conservancy; soil conservation
土壤保持措施 soil-conservation measure; soil conserving practice
土壤保持调查 soil conservation survey
土壤保持法 Act of Soil Conservation
土壤保持方法 soil-conservation measure
土壤保持覆盖 soil-protective cover
土壤保持技术 soil conserving practice
土壤保持局 <美> Soil Conservation Service
土壤保持区 soil conservation district
土壤保持推广 soil conservation extension
土壤保护 soil protection
土壤保护和分配法案 Soil Conservation and Allotment Act
土壤保护作物 soil conserving crop
土壤保洁 soil sanitation
土壤保水化学剂 water retention chemicals
土壤保水能力 retention capacity of soil
土壤保水特性 water-holding characteristic
土壤保养 soil saviour
土壤暴露试验 soil exposure test
土壤背景值 background value of soil; soil background value
土壤背景值图 chart of background value of soil; chart of soil background value
土壤被动压力 passive earth pressure
土壤本构 soil constitution
土壤崩塌 soil slump
土壤比表面 specific surface of soil
土壤比热 specific heat capacity
土壤比重 soil density; specific gravity of soil
土壤比重计 hydrometer
土壤变形 soil deformation
土壤变异 soil variation
土壤变质 soil deterioration
土壤变种 soil variety; variety of soil
土壤标本 boring test
土壤标准物质 soil standard substance
土壤表层 mantle of soil; solum; upper soil layer
土壤表层夯实 soil packing
土壤表面 soil surface
土壤表面板结 seal(ing) of soil
土壤表面碟状沉陷 dishing of surface
土壤冰冻 soil freezing
土壤病害 soil disease
土壤病症 soil disease
土壤剥蚀剖面 truncated soil profile
土壤剥蚀作用 soil denudation
土壤薄膜水 soil moisture film
土壤不均匀性 soil heterogeneity
土壤不透水层 water-tight stratum
土壤不易受侵蚀部分 non-erodible fractions of soil
土壤部分 soil fraction
土壤材料试验室 earth materials laboratory
土壤采样 soil sampling
土壤采样器 soil sampler
土壤采样位置 soil sampling locality
土壤采样钻 soil auger
土壤参数 soil parameter
土壤残留 pedo relict
土壤残留处理 soil residue treatment
土壤残留性农药 soil persistent pesticide
土壤操作 soil operation
土壤草测 reconnaissance soil survey
土壤测量 soil surveying
土壤测渗计 soil lysimeter
土壤测试概要 soil test summaries
土壤层次 horizon of soil
土壤层的 edafic; edaphic
土壤层理 soil bedding; soil stratification
土壤层内侵蚀 internal erosion
土壤层(位) soil horizon
土壤层下灌溉 subsurface irrigation
土壤差异性 soil heterogeneity
土壤长期潮湿 long periods of wet soil
土壤长湿状 udic
土壤常数 soil constant
土壤超载 soil surcharge
土壤潮湿采样 soil moisture sampling
土壤沉积 soil deposit
土壤沉降 soil displacement; soil settlement
土壤沉陷 settlement of soil
土壤成分 composition of soil; constituents of soil; soil constituent; soil fraction; soil separate
土壤成分单位重量 unit weight of soil constituents
土壤成型混凝土 soil-mo(u)lded concrete
土壤成因学 soil genesis
土壤承压力 bearing force of soil
土壤承载比 soil bearing ratio
土壤承载力试验 bearing test of soil
土壤承载(能)力 bearing capacity of soil; bearing value of soil; soil bearing capacity
土壤持水化学剂 water retention chemicals
土壤持水量 field carrying capacity; moisture-holding capacity; soil retention; water capacity
土壤持水能力 field capacity
土壤持水特性 water-holding characteristic
土壤充水 soil priming
土壤充水状态 replete state of soil
土壤冲蚀 soil erosion; soil wash(ing)
土壤冲蚀度 abrasivity of ground
土壤冲刷 soil erosion; soil wash(ing)
土壤冲刷控制 soil erosion control
土壤稠度 soil consistency[consistence]
土壤稠性 soil consistency[consistence]
土壤初始吸水量 <产生径流前的> soil priming
土壤处理 soil treatment
土壤触探仪 soil sounding device
土壤传播的 soil-borne
土壤传染 soil infection
土壤吹失 blowing of soil; soil blow-off; wind soil erosion
土壤垂直钻孔机 continuous flight auger
土壤次生盐碱化 secondary salinization of soil
土壤次生盐渍化 secondary salinization of soil
土壤大肠杆菌群值 soil colititre
土壤大动物区系 soil macrofauna
土壤大生物 soil macroorganism
土壤大团聚体 macroaggregate
土壤代谢 soil metabolism
土壤带 soil belt
土壤单位 pedologic(al) unit
土壤氮的有机类型 organic forms of soil nitrogen
土壤氮素 soil nitrogen
土壤氮循环 soil nitrogen cycle
土壤导电热率 soil conductivity
土壤导热率 soil conductivity
土壤的 earthy; edafic
土壤的不可渗透性 soil impermeability
土壤的铲装性能 loadability of soil
土壤的陈化 ag(e)ing of clay
土壤的成分 the content of soil
土壤的地质分类 geologic(al) classification of soil
土壤的叠层形成 stratification of the ground
土壤的防潮处理 Callendar's system
土壤的放射性净化 radioactive decontamination of soils
土壤的放射性污染 radioactive contamination of soils
土壤的分层作用 stratification of the ground
土壤的附着水 attached ground water
土壤的化学改良 chemical improvement of soil
土壤的缓冲作用 soil buffering
土壤的恢复 restoration of soil
土壤的间接改良 indirect improvement of soil
土壤的可渗透性 permeability of soil
土壤的离心含水量 centrifugal moisture of soil
土壤的密实性 compactibility of soil
土壤的热加固法 thermal stabilization
土壤的生成 formation of soil
土壤的生物改良 biologic(al) improvement of soil
土壤的水压力 hydraulic of soils
土壤的体积变化 volume changes in the soils
土壤的通过特性 soil-mobility characteristic
土壤的脱氮作用 denitrification in soils
土壤的微型结构 micromorphology
土壤的物理适耕性 soil physical maturity
土壤的形成 forming of soil
土壤的休耕 fallowing of soil
土壤的旋转压实 gyratory compaction of soil
土壤的压实 soil packing
土壤的有机质含量 organic content of soils
土壤的允许承载压力 allowable bearing force of soil; allowable soil pressure
土壤的允许耐力 allowable soil bearing strength
土壤的震动夯实 solidifying of earth by vibration
土壤登记 land register
土壤等温线图 soil isotherm
土壤地层单位 soil-stratification unit; soil-stratigraphic(al) unit
土壤地带 soil zone
土壤地带性 soil zonality
土壤地理学 pedogeography; soil geography
土壤地力图 soil and capacity map
土壤地沥青混合物 soil-asphalt mix(ture)
土壤地球化学 pedogeochemistry
土壤地球化学测量 geochemical soil survey
土壤地球化学勘探 pedogeochemical prospecting
土壤地图 pedologic(al) map
土壤地下水浸润 ground moistening
土壤地下水排出 soil discharge of ground water
土壤地下水渗流网 soil flow net
土壤地形 soil relief
土壤地质勘测 geologic(al)-soil investigation
土壤地质图 agrogeological map
土壤地质学 soil geology
土壤电参数 electric(al) parameter of soil
土壤电化学 soil electrochemistry
土壤电介质 soil electrolyte
土壤电流 earth current
土壤电阻 soil resistance
土壤电阻比率 soil electric(al) resistivity
土壤电阻率 soil resistivity
土壤垫层 bedding layer of soil
土壤淀积层 illuvial horizon
土壤调查 soil exploration; soil investigation; soil survey
土壤调查手册 soil survey manual
土壤调查图 reconnaissance soil map
土壤调查钻探 drilling for soil investigation
土壤顶极 edaphic climax; pedoclimax
土壤动力学 soil dynamics
土壤动物 soil animal
土壤动物区系 soil fauna
土壤冻结 freezing of the soil; soil freezing
土壤冻结法 ground freezing method
土壤冻结法下沉基础 freezing method of sinking foundation
土壤冻结深度 depth of soil-freezing
土壤冻裂搅动作用 cryoturbation
土壤断面 profile; soil profile
土壤堆放 soil placement
土壤发生 genesis of soil; pedogenesis; pedogenetic relation; soil formation
土壤发生层 soil genetic horizon; soil genetic layer
土壤发生过程 pedogenic process
土壤发生学 soil genesis
土壤发生演替 edaphogenic succession
土壤发育 development of soils; soil development
土壤翻拌机 soil tiller
土壤反力模量 modulus of reaction of soil
土壤反力系数 coefficient of soil reaction
土壤反压力 earth back pressure
土壤反应 soil reaction
土壤返盐作用 salt return of soil
土壤防护 soil defence
土壤防水 soil waterproofing
土壤放大作用 ground amplification
土壤放线菌 soil actimomycetes
土壤肥度 soil fertility
土壤肥力 fertility; soil fertility
土壤肥力保持 preservation of soil fertility
土壤肥力补救措施 soil fertility remedial measures
土壤肥力等级 degree of soil fertility

土壤肥力递减定律 law of diminishing fertility of soil
土壤肥力递减值 value of diminishing of fertility
土壤分布 distribution of soil
土壤分布图 map of soil; pedologic(al) map; soil(strip) map
土壤分层图 boring log
土壤分级 soil rating
土壤分类单元 category of soil classification; soil taxonomic unit
土壤分类(法) classification of soils; land classification; soil classification; soil grade
土壤分类级别 soil taxonomic classes
土壤分类三角图 soil classification triangle; triangular soil classification chart
土壤分类三角坐标图 triangular classification chart
土壤分类试验 soil classification test
土壤分类系统 soil classification system
土壤分类学 soil taxonomy
土壤分散 soil dispersion
土壤分析 soil analysis
土壤分析方法 methods of soil analysis
土壤分析样品 sample for soil analysis
土壤分析用研磨机 soil grinder
土壤分选工 grader man
土壤分选机 soil separator
土壤风化层 soil mantle
土壤风蚀 soil drifting; wind erosion of soil; wind soil erosion
土壤封闭层 soil sealant
土壤封面 earth cover
土壤腐蚀 soil corrosion
土壤腐蚀性 soil corrosivity
土壤腐殖化 soil humification
土壤腐殖酸 soil humic acid
土壤腐殖物质 earth humus; soil humic substance; soil humus
土壤腐殖质 soil humus
土壤腐殖质含量 the humus content of the soil
土壤附着力 soil adhesion
土壤附着水 soil moisture film
土壤复式组合 soil parabination
土壤复域 soil complex
土壤复原 regradation of soil
土壤复壮 soil renewal
土壤覆盖 earth cover; rail mulch; soil mulch
土壤覆盖层 mantle of soil; soil cover(ing); soil mantle
土壤覆盖机 mulcher; mulch layer
土壤改良 amelioration(of soil); improvement of agrarian structure; improvement of soil; land amelioration; land betterment; land improvement; melioration; reclamation; soil amelioration; soil conditioner; soil improvement; soil modification; soil reclamation; soil amendment
土壤改良方法 improvement method of soil
土壤改良剂 amendment; soil ameliorant; soil improvement agent
土壤改良水文地质 soil improvement hydrogeology
土壤改良水文地质调查 hydrogeologic(al) survey of soil improvement
土壤改良水文地质图 hydrogeologic(al) map of reclamation
土壤改善 soil amendment
土壤盖层影响 soil overburden influence
土壤概测 reconnaissance soil survey
土壤概图 reconnaissance soil map
土壤干密度 dry density of soil
土壤感染 soil infection
土壤钢材相互作用 soil-steel interaction
土壤根际微生物群 rhizospheric microflera of soil
土壤耕地 soil tillage
土壤耕性 soil tilth
土壤耕性好 a nice tilth
土壤耕种层 top soil
土壤耕作 soil cultivation; soil tillage; soil working; work of cultivating soil
土壤耕作层 plough horizon; top soil
土壤耕作方法 methods of soil tillage
土壤耕作工作 the work of cultivating soil
土壤耕作机具 ground-working equipment
土壤耕作机械 soil cultivating machine
土壤工程师 soil engineer
土壤工程学 soil engineering
土壤构成 soil formation
土壤构成过程 soil building process
土壤构造 soil constitution; soil structure
土壤骨骼 soil skeleton
土壤骨架 soil skeleton
土壤固定 soil fixing
土壤固化 earth cementation; earth solidification; ground solidification; soil compaction
土壤固化剂 soil solidifier
土壤固结作用 earth consolidation; soil consolidation
土壤固相 solid-phase of soil
土壤管理 soil management; soil regulation
土壤管理不良 bad soil management
土壤管理制度 soil management system
土壤管涌 soil piping
土壤管状物 pedotubule
土壤贯入度仪 soil penetrometer; sounding apparatus
土壤贯入器 soil penetrometer
土壤贯入仪 penetrameter[penetrometer]
土壤灌溉 soil irrigation
土壤灌浆 ground cementation; ground grouting; soil injection
土壤灌注 soil perfusion
土壤过载 soil surcharge
土壤害虫 soil pest
土壤含氮污染物 soil nitrogenous pollutant
土壤含石量 stoniness
土壤含水 soil-laden water
土壤含水百分比 water-percentage of soil
土壤含水带 soil water belt
土壤含水量 moisture of the soil; soil moisture; soil moisture content; soil water content
土壤含水量测定 soil moisture measurement
土壤含水量常数 soil moisture constant
土壤含水量指数 moisture index
土壤含水率 percentage of soil moisture content; soil moisture content; water-percentage of soil
土壤含水区 soil moisture zone
土壤含水曲线 moisture content profile
土壤含水指数 soil moisture index
土壤含盐量 soil salinity
土壤夯实 earth compaction; earth densification; soil compaction
土壤夯实机 compactor; earth compacting machine; earth compactor; soil compacting machine
土壤耗竭 exhaustion of soil; soil depletion; soil exhaustion
土壤荷载试验 soil bearing test
土壤呼吸 soil respiration
土壤呼吸仪 soil respirator
土壤滑动 soil creep
土壤滑溜状态 soil slipperiness
土壤滑坡 landslide
土壤滑泻 creepwash
土壤滑移长度 soil sliding length
土壤滑移线 soil sliding path
土壤化学 soil chemistry
土壤化学成分 chemical composition of soil
土壤化学处理 chemical treatment of soil
土壤化学分析 chemical analysis of soil
土壤化学稳定法 chemical soil stabilization
土壤化学性质 soil chemical property
土壤化学循环 geochemical cycle
土壤环境 edatope; soil environment
土壤环境容量 environmental capacity of soil; soil environmental capacity
土壤环境容量区域分异 division of soil environmental capacity
土壤环境影响评价 impact assessment of soil environment
土壤环境质量标准 quality standard of soil environment; soil environmental quality standard
土壤环境质量评价 soil environmental quality assessment
土壤环境质量图 soil environmental quality map
土壤环境质量指数 soil environmental quality index
土壤缓冲能力 buffering power of soil
土壤缓冲作用 soil buffer action
土壤灰化作用 podzolization of soils
土壤混合机 soil mixing machine
土壤混合(物) soil mix(ture)
土壤或岩石的地层传震波 ground wave
土壤击实 earth compaction
土壤击实试验 soil compaction test
土壤机械化耕作 mechanical soil cultivation
土壤机械适应率 soil workability
土壤积毒 soil poisoning
土壤积热器 thermointegrator
土壤积热仪 thermointegrator
土壤基础工程 soil and foundation engineering
土壤基础系统 soil-foundation system
土壤基质 soil matrix
土壤级配 grading of soil
土壤极限承载力 ultimate bearing capacity of soil
土壤集合体 soil aggregate
土壤集料混合物 soil-aggregate mixture
土壤挤密加固法 soil compaction
土壤技术 soil technology
土壤寄居菌 soil invader
土壤加肥 enrichment of soil
土壤加富法 soil enrichment method
土壤加固(法) soil stabilization; soil fixation; soil solidification; strengthening of soil; soil improvement
土壤钾素 soil potassium
土壤监测 soil monitoring
土壤剪(切)力 soil shear
土壤剪切试验 shear tests for soil
土壤检验 soil inspection
土壤碱度 soil alkalinity
土壤碱性提取物 alkaline extracts of soils
土壤鉴别 soil identification
土壤鉴定 soil identification
土壤交换量 exchange capacity of soil
土壤交换使用 alternative usage of soil
土壤胶态颗粒 soil colloidial particle
土壤胶体 soil colloid
土壤胶体颗粒 soil colloidial particle
土壤矫正计划 soil remediation program(me)
土壤搅拌机 soil mixer
土壤校正剂 soil correctives
土壤接触面 soil contact
土壤接种 <对土壤注射造氮菌> inoculation of the soil
土壤结持度 soil consistency[consistence]
土壤结构 soil constitution; soil structure; soil texture; structure of soil; texture of soil
土壤结构常数 soil consistency constant
土壤结构分类 textural classification of soil; textural soil classification
土壤结构改良剂 soil conditioner
土壤结构剖面 texture profile
土壤结构特性 soil morphology
土壤结构相互作用分析系统 system for analysis of soil-structure interaction
土壤结构组成 mechanical component of soil
土壤结合 soil association
土壤结壳 soil crust
土壤结皮 soil crust
土壤结特性 soil consistency[consistence]
土壤界线 soil boundary
土壤金属量测量 soil metallometric survey
土壤紧实指标 soil compaction index
土壤浸出琼脂 soil extract agar
土壤浸出液 soil extract
土壤浸染 land retirement; soil erosion; soil infection
土壤浸湿器 soil soaker
土壤浸蚀率 soil erosion index
土壤浸蚀图 soil erosion map
土壤浸提的化学组分 soil extract's chemical composition
土壤浸提液 soil extract
土壤经营 soil management
土壤净化 soil decontamination; soil disinfection; soil hygiene; soil purification; soil sanitation
土壤净交换量 net exchange capacity of soil
土壤静压力 earth pressure at-rest
土壤聚合体大小 soil-aggregate size
土壤绝对年龄 absolute age of soil
土壤开辟利用 soil development
土壤开发 soil exploitation
土壤抗冲(刷)能力 anti-eroding ability of soil
土壤抗剪强度 shearing strength of soil; soil shear strength
土壤抗剪强度环形测定仪 shear vane
土壤抗剪试验 shear test of soil
土壤抗力 passive earth pressure; resistance of soil
土壤抗蚀性 erosion durability of soil
土壤抗污性 soil resistance
土壤科学 soil science
土壤科学家 soil scientist
土壤科学杂志 <半年刊> Journal of Soil Science
土壤颗粒 soil grain; soil particle
土壤颗粒分级 class of soil particles
土壤颗粒分组 soil separate

土壤颗粒密度 soil bulk density
土壤颗粒排列 arrangement of soil particles
土壤颗粒实有体积 volume of the solid substance
土壤可供水量 available soil water supply
土壤可溶盐分析 dissolvable salt analysis of soil
土壤可溶盐淋滤实验 leaching experiment of dissolvable salt of soil
土壤可塑性 soil plasticity
土壤可用水分 chresard
土壤垦殖 soil reclamation;soil exploitation
土壤空气 soil air;soil atmosphere
土壤空气热泵 earth-air heat pump
土壤空隙 interstices of soil;soil void
土壤空隙度 soil porosity
土壤孔隙 soil pore;soil pore space;void of soil
土壤孔隙比 soil porosity ratio
土壤孔隙度 soil porosity
土壤孔隙和结构 the porosity and structure of soil
土壤孔隙内静水压力 neutral pressure
土壤孔隙性 porosity of soil
土壤块状结构 blocky soil structure
土壤快速检验法 quick tests for soils
土壤快速试验 quick tests of soils
土壤矿物 soil mineral
土壤矿物学 soil mineralogy
土壤矿物质 soil mineral matter
土壤昆虫 soil insect
土壤类别 soil class;soil group;types of soil
土壤类型 soil category;soil type;type of soil
土壤里的石灰 lime in the soil
土壤里含有有机物 soil contains organic materials
土壤里加碱 alkali added to the soil
土壤力学 soil mechanics
土壤力学家 soils mechanician
土壤力学实验室 soil mechanics laboratory
土壤力学研究 soil mechanics investigation
土壤立管 soil stack
土壤利用 soil exploitation;soil utilization
土壤利用分类 soil use classification
土壤利用图 soil utilization map
土壤沥青测量 soil bitumen survey
土壤砾石接触层 paralichic contact
土壤粒度 grading of soil particle;soil grade
土壤粒度分级取样 fractional sampling
土壤粒级 soil fraction;soil grade;soil separate
土壤粒径 soil size
土壤粒径分布试验 soil grain distribution test
土壤粒径累计曲线 soil grain size accumulation curve
土壤粒组 soil separate
土壤裂缝 interstices of soil;soil crack
土壤裂隙 interstices of soil;soil crack
土壤淋溶作用 soil leaching
土壤磷 soil phosphorus
土壤磷的平衡 soil equilibrated phosphorus
土壤灵敏度 sensitivity of soil;soil sensitivity
土壤零温度层 zero curtain
土壤流失 depletion of soil;losses in soil;losses of soil;soil loss
土壤流失强度 intensity of soil erosion
土壤流失容许限度 soil loss tolerance
土壤流失容许值 soil loss tolerance
土壤流失总量 gross soil loss
土壤流蚀 soil corrosion
土壤隆起 upheaval
土壤隆起容许量 allowance for uplift
土壤路线图 route map of soil
土壤螺旋钻 soil auger
土壤慢剪试验 drained shear test
土壤毛 soil capillarity
土壤毛管水压 soil moisture suction
土壤毛(细)管水上升高度 capillary lift of soil
土壤毛细管现象 soil capillarity
土壤毛细管作用 soil capillarity
土壤毛细含水量 field capacity
土壤毛细水 capillary soil moisture;capillary soil water
土壤密度 compactness of soil;soil density
土壤密度检验 soil density test
土壤密度探测器 density probe
土壤密实 earth densification
土壤密实度 soil compactness
土壤密实度测定针 penetration needle
土壤密实化 soil densification
土壤幂 soil mulch
土壤面层 soil surface
土壤描述性分类法 descriptive classification of soil
土壤描述学 pedography
土壤灭菌 soil defence;soil disinfection;soil sterilization
土壤灭菌剂 soil sterilant
土壤名称 soil name
土壤模制混凝土 soil-mo(u)lded concrete
土壤磨石 gretstone
土壤母质 parent soil material
土壤木块试验 <木林防腐的> soil block test
土壤内部排水 internal soil drainage
土壤内聚力 soil cohesion
土壤内摩擦力 internal friction of soil
土壤内渗流 soil seepage
土壤内渗透 soil seepage
土壤内蒸发 inner evapo(u)ration of soil
土壤耐压力 bearing force of soil;soil capacity
土壤能力 soil capability;soil capacity
土壤泥炭混合物 soil-peat mixture
土壤年龄 soil age
土壤年内湿度变差 difference of moisture yearly variance
土壤黏[粘]附 soil adhesion
土壤黏[粘]合性 cohesion of soil;soil cohesion
土壤黏[粘]结力 cohesion of soil;soil cohesion
土壤黏[粘]聚力 cohesion of soil;soil cohesion
土壤黏[粘]聚临界垂直高度 cohesion height
土壤黏[粘]性 cohesion of soil;soil stickiness
土壤凝结剂 soil coagulant
土壤凝聚力 soil cohesion
土壤凝聚性 cohesion of soil
土壤农化图 agrochemical soil map
土壤农药污染 pesticide contamination of soil
土壤农药相互作用 soil pesticide interaction
土壤排水 soil discharge;soil drainage
土壤排水能力 drainability of soil
土壤排水图 soil drainage map
土壤排水性 soil drainability
土壤培养 soil culture
土壤喷油处理 <道路> suboiling
土壤膨润土混合物 soil-bentonite mix(ture)
土壤膨胀 abrupt expansion;soil expansion;soil swelling;swelling of earth;swelling of soil
土壤疲乏 fatigue of soil
土壤疲劳 soil exhaustion;soil sickness
土壤贫瘠化 soil degeneration;soil depletion
土壤平板法 soil plate method
土壤平衡 soil balance;soil equilibrium
土壤平衡含水量 equilibrium moisture content of soil
土壤平整 land level(l)ing;land smoothing
土壤评价 soil assessment
土壤破坏 soil deterioration;soil fail(ure)
土壤破坏作用 destruction of soil;soil deterioration
土壤破碎机 soil shredder
土壤剖面 profile of soil;soil profile
土壤剖面特征 soil-profile characteristics
土壤普查 soil census;soil survey
土壤普查和土壤规划 general survey of soil and land planning
土壤气测量 soil gas survey
土壤气候 soil climate
土壤气候关系 soil climate relation
土壤气候相关性 soil climate relation;soil-climate relationship
土壤气候学 soil climatology
土壤气取样法 sampling method of soil gas
土壤气体 soil gas
土壤气相 gaseous phase of soil
土壤前期含水量 antecedent soil moisture;antecedent soil water
土壤潜动 soil creep
土壤潜力 soil capability
土壤潜水面 soil water table
土壤潜育层 gley
土壤潜在的冻胀能力或冻胀高度 potential vertical rise
土壤强度 resistance of soil;soil strength
土壤切片制备 soil split preparation
土壤切削能力 earth cutting ability
土壤侵蚀 erosion of soil;land retirement;soil corrosion;soil erosion
土壤侵蚀对其他地方的影响 off-site impacts of soil erosion
土壤侵蚀过程 soil erosion process
土壤侵蚀基准面 soil erosion datum
土壤侵蚀模数 soil erosion modulus
土壤侵蚀性 soil aggressivity
土壤侵蚀预报 soil erosion prediction
土壤侵蚀指标 soil erosion index
土壤侵蚀治理措施 control measures of soil erosion
土壤侵蚀作用 soil denudation
土壤清毒 soil sterilization
土壤情况 soil regime
土壤区分 soil division
土壤区划 soil delimitation;soil regionalization
土壤区域背景 background value of soil region
土壤取样 soil sample;soil sampling
土壤取样法 sampling method of soil
土壤取样管 soil pencil;soil sampler
土壤取样器 soil pencil;soil sampler
土壤取样筒 soil sampler
土壤圈 pedosphere
土壤(圈)地球化学异常 pedogeochemical anomaly
土壤全部体积 the total volume of a soil
土壤缺氮 nitrogen starved plot
土壤缺乏孔隙 soil lake porosity
土壤缺水度 soil moisture deficit
土壤缺水量 soil moisture deficit
土壤群 soil group
土壤群落分布区 edapho-phytocoenotic area
土壤群体 polypedon
土壤群系 edaphic formation
土壤染色制片法 stained soil suspension preparation
土壤热导率 soil thermal conductivity
土壤热量平衡 heat balance of soil
土壤人工冻结法 artificial freezing of soil
土壤韧度 tenacity of soil
土壤容积 soil volume
土壤容水性 soil water retention
土壤容许压力 permissible soil pressure
土壤容重 volume weight of soil
土壤溶蚀作用 soil erosion
土壤溶液 soil solution
土壤溶液的酸碱度 acidity-alkalinity reaction of soil solution
土壤溶液内氧化状况 oxygen status of soil solution
土壤溶液浓度 concentration of soil solution
土壤蠕变 soil creep
土壤蠕动 soil creep
土壤入渗 soil infiltration
土壤撒布密实机 bulk spreader
土壤三角形坐标分类(法) triangular classification of soil
土壤三角坐标分类法 triangular soil classification system
土壤三相 three-phase of soil
土壤杀虫剂 soil insecticide
土壤墒情 soil moisture content
土壤上层 top soil
土壤设备 soil study device
土壤射气浓度值 emanation concentration value in soil
土壤深处 depth of soils
土壤深度 depth of soils
土壤深度与含水量的关系 moisture content profile
土壤深色制片术 stain soil suspension preparation
土壤渗流网 soil flow net
土壤渗漏 soil filtration;soil percolation
土壤渗漏仪 lysimeter
土壤渗滤 soil filtration;soil percolation
土壤渗滤处理污水 land treatment
土壤渗入速率 soil infiltration rate
土壤渗水场 soil absorption field
土壤渗水计 lisimeter[lysimeter]
土壤渗水仪 lisimeter[lysimeter]
土壤渗透 soil percolate
土壤渗透性 soil permeability
土壤渗透仪 lisimeter[lysimeter]
土壤渗液 soil percolate
土壤生产力 producibility of soil;soil productivity
土壤生成群落 edaphic community
土壤生理学 soil physiology
土壤生态系(统) soil ecosystem;soil ecological system
土壤生态型 edaphic ecotype;soil ecotype
土壤生态学 edaphology;soil ecology
土壤生物 geobiont;soil organism
土壤生物动力学 soil biodynamics
土壤生物工程 soil bioengineering
土壤生物化学 soil biochemistry
土壤生物群落 edaphon(e)

土壤生物污染 biologic(al) pollution of soil
土壤生物学 soil biology
土壤施加石灰 liming of a soil
土壤湿度 soil humidity; soil moisture; ground moisture
土壤湿度不足 soil moisture deficiency
土壤湿度测定 soil moisture measurement
土壤湿度测定仪 soil moisture meter
土壤湿度等级 soil moisture grade
土壤湿度动态观测井 observation well for soil moisture regime
土壤湿度含水差 soil moisture deficit
土壤湿度计 irrometer; soil moisture meter; tens(i)ometer
土壤湿度水分差 soil moisture deficit
土壤湿度塑限 plastic limit of soil
土壤湿度梯度 soil moisture gradient
土壤湿度张力表 soil moisture tension meter
土壤湿度中子测定仪 soil moisture neutron probes
土壤湿容重 unit humid weight of soil
土壤湿润 ground moistening
土壤十字板剪力试验 vane shear test of soil
土壤石灰施用图 soil liming map
土壤识别 identification of soils
土壤实方 bank density; bank state
土壤实验室 earth materials laboratory
土壤试验 ground test; soil test(ing)
土壤试验取样器 soil test probe
土壤试验设备 soil tester
土壤试验探测器 soil test probe
土壤试样 test soil
土壤适耕性 soil tilth; workability of soil
土壤收缩限 shrinkage limit of soil
土壤手册 earth manual; soil manual
土壤熟化 cultivation of soil; maturation of soil; soil maturation
土壤熟化过程 soil maturation process
土壤数量的确定 soil quantity determination
土壤衰竭 exhaustion of soil; soil sickness
土壤衰颓 soil sickness
土壤水 holard; rhizic water; soil water; vadose water
土壤水百分率 soil water percentage
土壤水带 belt of soil moisture; belt of soil water; discrete film zone; soil water belt; soil water zone; zone of soil water
土壤水的负压力 water tension
土壤水的滞后(现象) hysteresis of soil water
土壤水分 soil moisture
土壤水分百分率 soil water percentage
土壤水分半连续带 zone of semicontinuous soil moisture
土壤水分饱和度 soil percent saturation
土壤水分不足 shortage of soil moisture
土壤水分测定 soil moisture measurement
土壤水分测定仪 soil moisture meter
土壤水分抽汲 soil moisture suction
土壤水分储量 soil moisture storage
土壤水分带 belt of soil water
土壤水分导电性 hydraulic conductivity of soil
土壤水分电测仪 electric(al) soil moisture meter
土壤水分电导仪 electric(al) soil moisture meter
土壤水分供应 soil moisture supply
土壤水分偏干(状态) ustic
土壤水分平衡 soil water balance
土壤水分实验 soil water experiment
土壤水分特征 soil water characteristic
土壤水分特征线 soil moisture characteristic curve
土壤水分蓄水量 soil moisture storage
土壤水分应力 soil moisture stress; soil water of stress; stress of soil moisture
土壤水分有效性 availability of soil water
土壤水分预报 forecast(ing) of soil moisture; soil moisture forecast
土壤水分运动 soil water movement
土壤水分张力 soil moisture tension
土壤水分蒸发(蒸腾)的损失总量 evapotranspiration
土壤水分贮藏量 soil water storage
土壤水分总应力 total soil-moisture stress
土壤水负压 soil moisture suction
土壤水扩散度 soil water diffusivity
土壤水扩散系数 soil water diffusivity
土壤水量平衡 soil moisture budget
土壤水生生物 edaphonekton
土壤水蚀 soil erosion
土壤水势(能) soil water potential
土壤水试验 soil moisture experiment; soil water experiment
土壤水收支平衡 soil water budget
土壤水文学 pedohydrology
土壤水吸力 soil moisture suction
土壤水张力 soil water tension
土壤水中栖居生物 edaphonekton
土壤水总潜力 total soil water potential
土壤松散 looseness of soil
土壤速测法 quick tests for soils; rapid test for soil
土壤塑限 plastic limit of soil
土壤酸度 soil acidity
土壤酸度探测器 pH value computer
土壤酸含量 the acid content of soil
土壤酸碱变试剂 soiltex
土壤酸碱度 soil acidity or alkalinity
土壤酸解分析 acid digestion analysis
土壤酸性 soil acidity; soil sourness
土壤碎块 clod
土壤碎屑 soil crumb
土壤-隧道动力相互作用 dynamic(al) soil-tunnel interaction
土壤损失 soil loss
土壤缩限 shrinkage limit; shrinkage limit of soil
土壤弹性 elasticity of soil
土壤弹性均压系数 modulus of reaction of soil
土壤探测 rod sounding
土壤探测杆 soil probe
土壤探测钎 soil probe
土壤探查 soil exploration
土壤探查系统 soil exploration system
土壤碳酸盐测量方法 soil carbonate survey method
土壤特性 characteristics of soil; nature of soil; soil characteristic; soil feature; soil property
土壤特质的碎裂 breakdown of soil materials
土壤提取 soil extract
土壤体积 soil volume
土壤体系 soil system
土壤天然排水 natural soil drainage
土壤田间持水量 field capacity of soil
土壤填埋处理 landfill disposal
土壤填筑 soil placement
土壤条件 edaphic condition; soil condition
土壤调节剂 soil conditioner
土壤调治剂 soil conditioner
土壤通气不良 defective aeration
土壤通气层 zone of suspended water
土壤通气器 soil aerator
土壤通气性 aeration of soil; soil aeration
土壤透度仪 sounding apparatus
土壤透气性 soil venting quality
土壤透水性 soil permeability
土壤图 edaphic map; pedologic(al) chart; soil map
土壤图表 edaphic scale
土壤钍射气浓度值 thorium emanation concentration value in soil
土壤团块 aggregate of soil
土壤团粒 soil granule
土壤团粒化 granulation of soil
土壤团粒结构 soil aggregate texture
土壤团粒(形成)作用 granulation of soil; soil granulation
土壤团粒总体积 total volume of the soil aggregate
土壤推移 soil bulldozing
土壤退化 land retirement; soil degeneration; soil degradation
土壤退化作用 degradation of soil; soil degradation
土壤脱盐作用 salt exclusion of soil
土壤挖掘 earth excavation; soil cutting; soil digging; soil excavation
土壤挖掘工作 soil digging work
土壤外部摩擦系数 soil coefficient of external friction
土壤微结构 soil fabric
土壤微生物 soil microorganism; terrestrial life
土壤微生物区系 soil microflora
土壤微生物群 edaphon(e)
土壤微生物群系 soil microbial population
土壤微生物学 soil microbiology
土壤微生植物群落 phytoedaphon
土壤微团粒 soil granulation
土壤微植物区系 soil microflora
土壤卫生 soil hygiene; soil sanitation
土壤卫生值 sanitary value of soil
土壤位移 earth displacement; soil displacement
土壤温度 soil temperature
土壤温度过程线 soil thermograph
土壤温度计 soil thermometer
土壤稳定(法) soil solidification; soil stabilization; stabilization of soil; electric(al) stabilization
土壤稳定机械 soil stabilizing machine
土壤稳定剂 earth stabilizer; soil stabilizer; Plasmofalt <一种废糖浆和燃料油混合物>
土壤稳定时的沥青覆盖 primer
土壤稳定外加剂 additive to stabilized soil
土壤稳定性 soil stability
土壤稳定用机械 soil stabilizing machine
土壤稳定作用 soil solidification; soil stabilization
土壤稳固法 soil stabilization
土壤污染 soil contamination; soil pollution
土壤污染调查 soil pollution investigation
土壤污染防治 prevention and control of soil pollution
土壤污染分析 soil contamination analysis; soil pollution analysis
土壤污染化学 soil pollution chemistry
土壤污染监测 control of soil pollution
土壤污染控制 control of soil pollution; soil pollution control
土壤污染图 soil pollution map
土壤污染物 soil pollutant
土壤污染预测 soil pollution prediction
土壤污染源 soil pollution source
土壤污染综合指数 comprehensive index of soil pollution
土壤无机胶体 inorganic colloid in soil
土壤无脊椎动物 soil invertebrate
土壤物理化学 soil physicochemistry
土壤物理学 soil physics
土壤物理学研究 soil physics investigation
土壤物质 soil material
土壤物质的移动 removal of soil material
土壤吸附除锈剂 soil adsorption of herbicides
土壤吸附系统 soil adsorption system
土壤吸力 soil suction
土壤吸收 soil absorption
土壤吸收降雨量能力 infiltration capacity
土壤吸收能力 absorptive capacity of soil; soil absorption capacity
土壤吸收系统 soil absorption system
土壤吸收性复合体 absorbing complex of soil
土壤吸水负压 soil moisture suction
土壤吸水量 water-absorbing capacity
土壤吸水能力 <通过毛细管作用> sorptivity
土壤吸水速率 intake rate
土壤吸水性试验 soil absorption test
土壤稀释平板技术 soil dilution plating technique
土壤习居菌 soil inhabitant
土壤习居者 soil inhabitant
土壤系列 soil series
土壤系统 soil system
土壤细菌 soil bacteria
土壤细菌学 soil bacteriology
土壤细粒结构 fine-textured soil
土壤细粒物质 soil plasma
土壤隙缝侵蚀 internal erosion
土壤下层 substratum[复 substrata]
土壤下沉 settlement of soil; soil shrinkage
土壤下沉基础冻结法 <用于流沙地基等> freezing method of sinking foundation
土壤下渗 soil infiltration
土壤现场试验仪 geogauge
土壤线虫 soil nematode
土壤相对含酸量 the relative acid content of the soil
土壤相对年龄 relative age of soil
土壤相对湿度 relative moisture of soil
土壤箱 soil box; soil tank
土壤详测 detailed soil survey
土壤详图 detailed soil map
土壤消毒 ground disinfection; soil sterilization
土壤消毒剂 soil disinfectant; soil sterilant
土壤消毒作用 soil disinfection
土壤消耗 soil depletion; soil exhaustion
土壤消化作用 nitrification in soil
土壤小片 soil crumb
土壤小区划分 plot assignment
土壤新生体 new growth of soil
土壤形成 pedogenesis; soil development; soil formation
土壤形成的几个因素 some major factors in soil formation
土壤形成过程 soil-forming process
土壤形成及类似过程 soil formation

and similar processes
土壤形成物 pedologic(al) feature
土壤形成因素 soil-formation factor; soil-forming factor
土壤形成作用 soil formation
土壤形态学 soil morphology
土壤性能 soil quality
土壤性质 nature of soil; property of soil; soil characteristic; soil property
土壤性质特征 soil coefficient; soil constant
土壤休止角 angle of repose of the earth
土壤需水量 soil requirement
土壤需氧孢子体 aerobic spore populations of soil
土壤悬液 soil suspension
土壤学 agrology; edaphology; pedology; soil science
土壤学的 pedologic(al)
土壤学的土分类法 pedologic(al) system of soil classification
土壤学的准备工作 preliminary pedological work
土壤学分类 pedologic(al) classification
土壤学分类系统 pedologic(al) classification system
土壤学家 pedologist; soil scientist
土壤学者 soil scientist
土壤熏蒸 soil fumigation
土壤熏蒸剂 soil fumigant
土壤熏蒸器 soil fumigator
土壤熏蒸仪 soil fumigator
土壤循环流动疲劳模型 fatigue model of soil cyclic(al) mobility
土壤压出液分析 liquid chemistry analysis from compressive soil
土壤压力 bearing pressure on foundation; earth pressure; ground pressure; soil pressure
土壤压力状态 soil pressure condition; soil pressure phenomenon
土壤压密 compaction of soil
土壤压密性 firmness of soil
土壤压实 earth compaction; earth densification; soil compaction; soil compression; soil densification
土壤压实程度 degree of compaction
土壤压实方法 soil compaction method
土壤压实滚筒 soil compaction roller
土壤压实机 compactor; earth compacting machine; earth compactor; soil compacting machine; soil compactor
土壤压实机械 soil compaction equipment
土壤压实排出的水 water compaction
土壤压实性 firmness of soil
土壤压实阻力 soil compaction resistance
土壤压缩性 compressibility of soil; soil compaction
土壤亚纲 soil suborder
土壤亚类 subtype of soil
土壤亚系统 soil subsystem
土壤岩 soilstone
土壤岩面接触层 lithic contact
土壤研究 soil investigation; soil research; soil study
土壤研究设备 soil investigation device
土壤研磨机 soil grinder
土壤盐测量 soil salt survey
土壤盐度 soil salinity
土壤盐碱化 salinization of soil; salting of soil; soil salination; soil salinization
土壤盐渍度 soil salinity
土壤盐渍度分级 grading of soil salinity
土壤演替顶极 edaphic climax
土壤演替顶极群丛 edaphic climax association
土壤演替顶极群落 edaphic climax community
土壤养分 soil nutrient
土壤氧化还原作用 oxidation-reduction in soil; redox of soil
土壤样品 soil sample; pedotheque
土壤样品类型 type of soil sample
土壤样品粒径组分 soil sample composition
土壤样品特征 soil sample characteristic
土壤液度计 soil lysimeter
土壤液化潜势 soil liquefaction potential
土壤液化塌垮 liquefaction failure; soil liquefaction failure
土壤液限 liquid limit of soil
土壤液相 liquid phase of soil
土壤仪器 soil study device
土壤异常 soil anomaly
土壤异常类型 type of soil anomaly
土壤抑菌作用 soil fungistasis
土壤抑菌作用和线虫的真菌 soil fungistasis and nematophagus fungi
土壤因素 edaphic factor; soil factor
土壤银行计划 soil bank program(me)
土壤应变比特性 soil strain-rate behaviour
土壤应力 soil stress
土壤应力表 soil stress ga(u)ge
土壤应力仪 soil stress ga(u)ge
土壤硬表层 soil crust
土壤硬度计 stratometer
土壤用作农业耕地 soils in use for cropland
土壤有机胶体 organic colloid in soil
土壤有机污染 soil pollution by organic matter
土壤有机物 soil organic matter
土壤有机质 gein; organism matter of soil; soil organic matter
土壤有机质部分 soil organic matter fractions
土壤有机质层 organic soil horizons
土壤有效持水量 available water-holding capacity
土壤有效供水量 available soil water supply
土壤有效含水量 available moisture capacity of a soil; available water capacity
土壤有效浓度 available moisture capacity of a soil
土壤有效湿度 available soil moisture
土壤有效水分 chresard
土壤有效压力 active thrust of earth
土壤有效养分 available nutrient of soil
土壤有效应力 <垂直于单位面积的颗粒间平均压力> soil effective stress
土壤圆锥贯入指数 cone index
土壤约测 reconnaissance soil survey
土壤运动 landslide
土壤运输 soil transport
土壤再生 soil renewal
土壤藻类 soil algae
土壤沼泽化 swamping of soil
土壤真菌 soil fungi
土壤振动夯实法 vibratory soil compaction
土壤振动器 soil vibrator
土壤振动压实 vibration compaction of soil
土壤振动压实器 vibro-soil compactor
土壤振实器 soil vibrator
土壤蒸发 evaporation from soil; land evapo(u)ration; soil evapo(u)ration
土壤蒸发测定 lysimetry
土壤蒸发计 potential evapo(u)rimeter; soil evapo(u)rimeter
土壤蒸发皿 soil evapo(u)ration pan; soil pan
土壤蒸发器 lysimeter; soil evapo(u)ration pan; soil evapo(u)rimeter; soil pan; soil tank
土壤蒸发箱 <测量蒸发和蒸发散发联合作用的> soil tank
土壤蒸气处理 steam treatment of soil; treatment of soil steam
土壤整段标本 monolith; soil monolith
土壤整治 land forming
土壤支承结构 soil supported structure
土壤支承力值 soil support value
土壤值采样 sampling of soil values
土壤植被 soil cover(ing)
土壤-植物-大气连统体 soil-plant-atmosphere sowing continuum
土壤-植物-大气连续体 soil-plant-atmosphere continuum
土壤-植物-大气系统 soil-plant-atmosphere system
土壤植物区系 soil flora
土壤植物群落分布区 edapho-coenotic series; edapho-phytocoenotic area
土壤植物群落系列 edapho-coenotic series
土壤植物系统 soil-plant system
土壤植物系统的净化 purification of soil-plant system
土壤植物相互关系 soil-plant relationship
土壤指示剂 edaphic indicator
土壤指示特征 soil index property
土壤指示植物 edaphic indicator; indicator plant of soil
土壤制备 <试验时> soil processing
土壤制图 soil mapping
土壤制图文件 soil names file
土壤制图学 cartography of soil; soil cartography
土壤质地 soil texture; texture of soil
土壤质地分类 soil texture classification
土壤质地图 soil texture map
土壤质量 soil quality
土壤质量标准 soil quality standard
土壤质量分类 soil quality classification
土壤质量评价 soil quality assessment
土壤质量指数 soil quality index
土壤滞留蓄水量 detention storage
土壤滞水量 soil detention storage; soil moisture retention
土壤中铵和氨 ammonium and ammonia in soils
土壤中残留性农药 residue-prone agricultural chemical in soil
土壤中氮循环模拟 model(l)ing of the soil nitrogen cycle
土壤中得到的富里酸 soil-derived fulvic acid
土壤中的冰冻标高 freezing level in soil
土壤中的空气孔隙 air void
土壤中的空气孔隙百分比 percentage of voids
土壤中的炮眼 foot hole
土壤中的水 soil water
土壤中毒 soil poisoning
土壤中发酵性微生物区系 zymogeneous flora in soil
土壤中腐烂 soil rot
土壤中间粒径 <与平均粒径有区别> median diameter
土壤中空气 soil air
土壤中水传递 transfer of water in soil system
土壤中水相对活动性 relative activity of soil water
土壤中水相对流动性 relative activity of soil water
土壤中游离铁 free iron in soil
土壤种类 type of soil
土壤重金属污染 soil pollution by heavy metal
土壤主剪切面 <与机具前进方向平行的> soil primary shear surface
土壤主要类型 major soil types
土壤注射 soil injection
土壤状况 soil condition; soil regime
土壤资料 soil information
土壤资源 soil resource
土壤资源丰富 soil abundance
土壤自记温度计 soil thermograph
土壤自净率实验 self-cleaning experiment of soil
土壤自净作用 soil self-purification
土壤自然结构体 ped
土壤自然排水 natural soil drainage
土壤自身结构 structure of the soil itself
土壤总含水量 holard
土壤总碱度 total alkalinity of soil
土壤总酸度 total acidity of soil
土壤纵断面 pedologic(al) profile
土壤族 edaphic race
土壤阻力 resistance of soil; soil resistance
土壤阻力计 dynamometer
土壤阻尼 soil resistance
土壤组成 soil composition; soil constitution; texture of soil
土壤组分 soil constitution
土壤组合 association of soil; soil association
土壤组织 soil texture
土壤钻杆 soil drill
土壤钻探深度测量 soil boring yardage
土壤钻探原始资料 soil boring source
土壤最大强度 soil peak strength
土壤最佳含水量 optimum soil moisture
土壤最佳湿度 optimum soil moisture
土壤最适含水量 optimum soil moisture
土壤最优含水量 optimum soil moisture
土热量交换【气】thermal economy of soil
土塞 soil plug
土塞效应 plugging effect
土塞作用 plug effect
土色 ground colo(u)r; soil tone
土色的 terreous
土砂沉积 silted deposition
土砂和碎石混合料 tabby
土砂流量 sediment discharge
土砂流失防护林 forest for erosion control
土砂流送 sediment transport
土砂石 soil aggregate
土砂石路面 soil-aggregate surface
土砂芯堤 rubble-mound with a loam core
土筛分 dirt screening
土山 artificial mound; earth piled hill; mound
土勺贯入仪 spoon penetration test
土设备 indigenous equipment
土砷铁矾 pitticite; pittizite
土渗流网 soil flow net
土渗透性 soil permeability
土生的 endemic; home-grown
土生动植物 aborigines

土湿度常数 soil moisture constant
土石坝 earth rock(fill)dam;embankment(composite)dam;riprap dam;fill dam
土石坝加固 embankment dam stabilization
土石坝填筑 embankment fill
土石层 <矿脉中的> jamb
土石方 cubic(al)meter of earth and stone;earth and stone work
土石方爆破 earth rock blasting;earthwork blasting
土石方爆破法 earthwork blasting procedure
土石方的散方数 loose yards
土石方调配 cut-fill adjustment;cut-fill balance
土石方分配 distribution of earth quantities
土石方工程 earth and stone work;earthwork;earth-rock works;groundworks
土石方工程材料 bank material
土石方工程费 cost of grading
土石方工程工长 grade-staff
土石方工程量 earth quantity;quantity of earthwork
土石方工程施工 earthmoving operation
土石方工程作业 earthwork operation
土石方横断面图 earthwork cross-section
土石方横向围堰 earth rock transversal cofferdam
土石方机械 earthmoving equipment;mucker
土石方计算 earth mass calculation;mass calculation;mass calculation for earth works
土石方开挖 earth rock excavation
土石方开挖出来后的松散体积 loose cubic meter
土石方挖出来后的松方数量 loose cubic meter;loose cubic yard
土石方量 excavated volume;bank measure <原地的>
土石方量比例尺 scale of earth quantities
土石方量测定 bank measure
土石方量图 section of earthwork
土石方零点 null point of earthwork
土石方露明面 <新开挖土石方的> high wall
土石方设备 mucking unit
土石方数量 earth quantity;quantity of earthwork
土石方挖填平衡 balance of cut and fill
土石方纵向围堰 earth rock longitudinal cofferdam
土石膏 clay gypsum;gypsite
土石混合坝 composite earth-rockfill dam;earth and rockfill dam
土石混合料 rock-earth mixture
土石礁 moraine
土石筐垒成的堤 gabionade
土石料 soils and rocks
土石流 earth flow
土石填方 earth rockfill;earth and rock fill
土石围堰 earth rock cofferdam
土石支撑 ground support
土蚀 soil erosion
土试样质量等级 quality classification of soil samples;quality grade of soil samples
土收缩 soil shrinkage
土受侵蚀 soil erosion
土水池 earth reservoir;earth tank
土水关系 clay-water relationship;soil water relationship
土水库 earth reservoir;earth tank
土水渠 earth supply ditch
土水体系 soil water system
土塌 earth fall;land fall
土台 earth pad;stereobate
土坍 soil fall
土弹簧常数 soil spring constant
土探测 rod test
土塘 <海工的> clay wall
土套 soil suite
土特产 local speciality
土特产品 native products
土体 clay body;earth body;earth mass;soil block;soil body;soil mass;solum
土体边坡 soil slope
土体常数 soil constant
土体沉降 soil settlement
土体的沉降及变形计算 settlement and deformation calculation of soil mass
土体的工程性质 engineering property of soil mass
土体的摩擦圆分析法 soil friction circle method
土体电阻测定计【电】 earth Megger tester
土体冻胀 soil blister
土体浮容重 submerged unit weight of soil mass
土体工程地质 engineering geologic(al)of soil mass
土体拱线 ground arch
土体灌浆剂 GV5 <硅酸钠或水玻璃类> earth firm GV5
土体滑动 earth creep;earth slide;earth slip;land monument;soil mass slide;land slide
土体滑坍 soil creep
土体加固 soil stabilization
土体结构 soil structure
土体结构类型 structural type of soil mass
土体抗碎强度 body crushing strength;clay body crushing strength
土体抗震刚度 seismic rigidity of ground
土体可压缩性 compressibility of soil
土体量 soil mass
土体流变学 soil rheology
土体隆起 bulging of soil mass
土体剖面略图 soil-profile sketch
土体强度检查 soil strength inspection;strength inspection of soil
土体侵蚀 soil erosion
土体蠕动 earth creep;soil creep
土体渗透变形 seepage deformation of soil mass
土体收缩 shrinkage of soil mass
土体特性 soil characteristic;soil mass property
土体通过汽车能力 soil trafficability
土体位移 displacement in earth mass;land monument;shifting of earth;shifting of soil
土体稳定 soil stabilization
土体稳定问题 problem of stability of soil mass
土体性质 soil characteristic;soil mass property
土体休止角 angle of repose of the earth
土体压缩变形阶段 compressive deformation stage of soil mass
土体移动 land movement;movement of earth;movement of earth mass
土体原位测试 in-situ soil mass test
土体胀缩 soil mass swelling
土体震动 soil mass shock
土体重 weight of soil mass
土体状况 soil condition
土挑水坝 earth dike[dyke]
土条带 soil ribbon;soil stripe
土亭 tutin
土团 clay ball;clay lump;mass of soil;ped <采掘或其他扰动时出现的一种独特的天然土壤>
土团粒 aggregation of soil particles;soil aggregate
土团粒作用 soil granulation
土推环 thrust ring
土推力 active earth pressure;earth thrust;soil thrust;thrust of earth;thrust of soil
土挖机 clay digger
土挖掘器 clay digger
土围墙 earth barrier
土围堰 earth(-fill)cofferdam;earthwork cofferdam
土卫布聚合物 geotextile polymer
土温 soil temperature
土温超热状况 hyperthermic
土温的年变化 annual variation in soil temperature
土温的日变化 daily fluctuation in soil temperature
土温计 soil thermometer
土稳定(法) soil stabilization
土稳定剂 soil stabilizer
土屋 earth house
土物理学 soil physics
土吸力势 soil suction potential
土吸收作用 soil suction
土吸(水能)力 soil suction
土系 soil series
土下沉 soil subsidence
土相 soil phase
土香胶(脂) earth balsam
土楔分析法 wedge analysis
土楔理论 soil wedge theory;wedge theory
土楔体 soil wedge
土楔体的重量 weight of soil wedge
土芯 soil core;lysimeter monolith <土壤蒸发器的>
土芯墙 soil core
土芯试样 soil core sample
土芯样 core of soil
土芯钻取筒 soil coring tube
土星 Saturn
土腥味 argillaceous odo(u)r;earthy taste
土型 soil pattern;soil type
土性参数 soil parameter
土性植物 land-flora
土性指标 soil indicator
土蓄水池 earth-dammed reservoir
土悬(浮)液 clay suspension;soil suspension
土悬液单位重 unit weight of soil suspension
土穴 soil pit
土学家 soil investigator;soil scientist
土压测定盒 soil pressure cell
土压的 geostatic
土压等分布曲线 pressure bulb
土压拱 geostatic arch
土压或地压 ground pressure;surrounding rock pressure
土压力 earth pressure;soil pressure;geostatic pressure
土压力测定 earth pressure measurement;soil pressure measurement
土压力测定盒 soil pressure cell
土压力测试 earth pressure test
土压力传感器 earth pressure transducer
土压力的 earth pressure;geostatic
土压力的静水压分布 hydrostatic distribution of the earth pressure
土压力等分布曲线 pressure bulb
土压力分布 earth pressure distribution
土压力荷载 earth pressure loading
土压力盒 earth pressure cell;geocell;soil pressure cell
土压力计 earth pressure cell;earth pressure ga(u)ge;earth pressure probe;soil pressure cell
土压力计算 earth pressure computation
土压力建筑物 <如挡土墙、护岸、堤岸> bulkhead
土压力库尔曼图解法 Culmann construction of soil pressure
土压力理论 earth pressure theory
土压力量测 soil pressure measurement;earth pressure measurement
土压力量测压力盒 earth pressure measuring cell
土压力泡 bulb of pressure
土压力平衡 earth pressure balance
土压力平衡盾构 earth pressure balance shield
土压力平衡盾构机 earth pressure balanced shield machine
土压力平衡盾构施工 earth pressure balance shield tunneling
土压力平衡盾构隧道挖掘机 earth pressure balanced tunneling machine
土压力图 soil pressure diagram
土压力系数 coefficient of earth pressure;coefficient of soil pressure;earth pressure coefficient
土压力线 earth pressure line;soil pressure line
土压力楔体 wedge of earth
土压力楔(形体) earth pressure wedge
土压力仪 earth-pressure ga(u)ge
土压力元件 earth pressure cell
土压力作用 earth-pressure behavio(u)r
土压平衡盾构 earth(pressure)balance(d)shield
土压平衡式盾构 shield with balanced earth pressure
土压实 soil compaction
土压实控制仪 soil compaction control kit
土岩组合地基 soil-rock composite subgrade
土洋结合 combine indigenous and foreign method
土样 bore plug;bore specimen;representative sample;sample;soil core;soil pattern;soil sample;soil specimen
土样标记 sampling label(l)ing
土样标签 sample label
土样采取率 percent of sample recovery
土样测量 soil sample measure
土样长度 length of soil sample
土样尺寸 dimension of sample
土样锤击数 blown count
土样袋 sample bag
土样底面 bottom of soil sample
土样顶面 top of soil sample
土样法 soil sample measure
土样分析 analysis of samples
土样国际标准 draft international standard
土样盒 sample can
土样横截面积 cross-sectional area of soil sample
土样密封剂 soil sealant
土样面积 area of soil sample
土样目测鉴定 soil sample identification with naked eye
土样切割器 core cutter;core lifter
土样取样工具 soil sampling tool
土样扰动 sample disturbance

土样深度 sample depth of soil
土样试块 ground test pieces
土样试验 soil sample test
土样数 number of soil sample
土样水理性质测定 determination of hydrophysics property of soil sample
土样筒 sample barrel;soil can
土样筒中的水位深度 water-level depth in soil sample cylinder
土样制备 preparation of soil sample
土样总体积 total sample volume;total volume of soil sample
土腰【地】isthmus
土窑 earth(en) kiln
土窑砖 clamp-burned brick; clamp-burnt brick;dome brick
土窑砖垛 clamp
土液 soil solution
土液化机制 mechanism of soil liquefaction
土移 boil mud
土应变 soil strain
土应力 soil stress
土涌 boiling soil
土油池 earth storage
土釉 clay glaze
土与废料相互作用 soil-waste interaction
土与建筑物间的相互作用 soil-structure interaction
土与结构物的相互作用 soil-structure interaction
土与容器相互作用 soil-container interaction
土与水的相互作用 interaction of soil and water
土与桩之间的附着力 soil-pile adhesion
土与桩之间的相互作用 soil-pile interaction
土语 patois
土圆锥仪 soil cone penetrator
土源疾病 soil-borne disease
土匝道 earth ramp
土在自然状况下可通车能力 soil trafficability
土葬 inhumation
土胀 soil blister
土障 earth barrier
土蒸发计 soil evapo(u)rimeter
土支撑 earth buttress
土支承结构(物) earth-supported structure; earth-supporting structure;soil supported structure
土制的 earthen
土制浮标 debiteuse
土制艺术品 earthwork
土质 nature of soil; quality of soil; soil characteristic;soil quality
土质不分类开挖 unclassified excavation
土质查勘 soil exploration;soil reconnaissance
土质处理 soil improvement;soil treatment
土质的 earthy
土质的天然角 angle of repose
土质地基 earth foundation;soil foundation
土质调查 geotechnic(al) investigation; geotechnic(al) survey; soil exploration;soil investigation;soil survey
土质调查表 soil survey chart
土质调查图 soil survey map
土质防渗体分区坝 zoned earth-rock-fill dam with impervious soil core
土质分类 soil classification;soil type
土质分析 soil analysis
土质改良 ground improvement; soil amelioration; soil conditioning; soil improvement
土质改良学 reclamation of geotechnique
土质工程 geotechnic(al) engineering
土质钴 earthy cobalt
土质混凝土 tabby
土质技术 geotechnics; geotechnique; soil technology
土质鉴定 soil identification;soil interpretation;terrain interpretation
土质勘测 soil investigation
土质勘察 soil investigation
土质勘探 soil exploration; soil prospecting
土质矿石 earthy ore
土质类别 soil classification
土质类型 type of soil
土质力学法 geodynamic(al) method
土质泥浆池 earthen pit
土质石膏 earthy gypsum
土质石灰 earthy lime
土质试验 soil test(ing)
土质水库 earth reservoir
土质隧道 earth tunnel;tunnel in earth
土质踏勘 soil reconnaissance
土质探察 soil exploration
土质条件 soil condition;subsoil condition
土质调整器 soil conditioner
土质图 soil map
土质稳定 soil stabilization
土质稳定剂 soil stabilizer
土质芯墙 earth core
土质芯墙堆石坝 earth core rockfill dam
土质学 geotechnique;science of ground soil;soil technology
土质研究 soil investigation;soil study
土质颜料 earth pigment
土质资料 soil data;soil information
土中垫层 bedding-in soil
土中动态水 soil water in motion
土中夹石 traprock
土中静态水 static water in soil
土中孔隙类型 pore type in soil
土中孔隙体积 volume of void in soil
土中零星冻土 sporadic permafrost
土中流网 soil flow net
土中毛吸水作用带 boundary zone of capillary in soil
土中锚碇装置 soil anchorage
土中气隙 air-space of soil
土中气隙比 air-void ratio
土中热传导 heat transmission in soil
土中渗流 soil seepage
土中水 soil moisture;soil water
土中水带 soil water belt
土中水的类型 type of soil water
土中水的体积 volume of water in soil
土中水分渗出(压力) exudation pressure
土中水分移动 movement of water in soil
土中水分张力 soil moisture tension
土中心架支套 cathead
土中悬着水 zone of suspended water
土中应力 stress in earth mass
土中应力分布 stress distribution in earth mass
土中住宅 earth contact house
土中阻尼 damping in soil
土重 earth load
土主曲线 navvy curve
土贮水池 earth-dammed reservoir
土柱 earth column;earth pillar;earth pyramid; hoodoos; pillar; soil column;temoin <标记挖方深度用>
土柱截面面积 cross-section area of soil column
土柱样品 soil core sample
土柱直径 diameter of soil column
土著 autochthon(e) [复 autochthon(e)s];native
土著的 aboriginal
土著居民 aboriginal;aborigines;indigenous inhabitant;native inhabitant
土著语 native language
土著种 autochthon(e) [复 autochthon(e)s];endemic species
土筑防洪堤 earthen flood bank
土筑防御工事 earthwork
土筑房 earth house
土筑结构 earth-fill structure
土筑墙 adobe
土筑土路 earth path
土筑住宅 earth-sheltered home
土抓斗 <挖掘机> grab bucket
土专家 indigenous expert; local expert;self-taught expert
土砖 bauxite brick;cob brick
土砖圬工 adobe masonry
土桩 earth pile;soil pile
土桩挤密 compacted earth pile
土桩结构相互作用 soil-pile-structure interaction
土桩与灰土桩 soil-pile and lime-soil pile
土状 earthy
土状赤铁矿 argillaceous hematite;red bole;reddle
土状风化岩 lithorelics
土状褐煤 bog coal;brown earthy coal
土状铝土矿 earthy bauxite
土状氯铋矿 daubreeite
土状煤 druss
土状石膏 gypseous marl; gypsite; gypsum earth
土状石墨 amorphous graphite;earthy graphite
土棕色 adobe brown
土纵断面 soil profile;soil section
土族元素 earth family element;earthy element
土阻力 passive earth resistance
土组 local soil type;soil type
土组成 soil constitution
土组织 soil texture;texture of soil
土组织分类 textolite soil classification;textural soil classification
土钻 earth auger; earth borer; earth boring auger; earth drill; earth screw;ground auger;ground drill; mud auger; soil auger; soil boring auger;well auger
土钻孔 hand drill(ing machine)
土钻孔数 number of land drilling
土钻总进尺 total drilling footage of land drilling

吐

吐出 disgorge;regorge

吐出口 spout
吐酒石 tartar emetic
吐量 export cargoes
吐露 effuse
吐鲁胶树 tolu tree
吐纳特 <一种烈性炸药> tonite
吐舌垫圈 tongued washer
吐氏酸 Tobias acid
吐水作用【给】guttation
吐焰 blaze
吐渣 scum;kiln scum <表面析出白色浮渣>

钍

钍230镤231过剩法 230Th-231Po excess method

钍230镤231亏损法 230Th-231Po deficiency method
钍232的热产率 heat productivity of Th-232
钍232含量 content of Th232
钍道窗宽 thorium channel window width
钍道灵敏度 sensitivity of Th-channel
钍的铀当量 uranium equivalence of thorium
钍含量等值图 contour map of thorium content
钍含量平剖图 profile on plane of thorium content
钍含量异常 thorium content anomaly
钍后元素 tranthorium
钍化物 thoride
钍钾比值 ratio of Th/K
钍钾比值平剖图 profile on plane of Th/K ratio
钍钾等值图 contour map of Th/K ratio
钍钾异常 thorium/potassium anomaly
钍矿 thorium ore
钍铅208等时线 232Th-208Pb isochron
钍铅年代测定法 thorium-lead age method
钍射气 thorium emanation;thoron
钍射气浓度 thorium emanation consistence
钍石 thoride
钍石含量 thoride
钍钛铀矿 absite
钍土 thoria
钍蜕变系 thorium series
钍钨极 thoriated tungsten electrode
钍钨矿 thorotungstite
钍钨酸盐 thoriated tungstate
钍线储量外检误差 external examining errors of thorium linear reserves
钍铀法 thorium-uranium method
钍铀矿 thor-uraninite
钍铀增殖堆 valubreeder
钍脂铅铀矿 thorogummite

兔

兔耳式龙头 rabbit ear faucet

兔耳形水龙头 rabbit ear faucet

湍

湍动 isotropic(al) turbulence; turbulence;turbulent motion

湍动波 turbulent wave
湍动潮流 turbulent tidal current
湍动磁场 turbulent magnetic field
湍动等离子体 turbulent plasma
湍动地幔对流 turbulent mantle convection
湍动法 turbulence method; turbulent method
湍动风结构 turbulent wind structure
湍动混合室 turbulent mixing chamber
湍动激发 excitation of turbulence
湍动加热 turbulence heating
湍动空气 turbulent air
湍动雷诺应力 turbulent Reynolds stress
湍动刘易斯数 turbulent Lewis number
湍动流化床 turbulent fluidized bed
湍动能 turbulence energy; turbulent energy
湍动谱 spectrum of turbulence; turbulence spectrum
湍动气流 turbulent airflow
湍动施密数 turbulent Schmidt number
湍动尾流 turbulent wake
湍激海面 chopping sea;choppy sea
湍急河流 angry river; flashy stream; rapid river;rapid stream;tachydrom-

ile;torrential river;torrential stream
湍急急流 swift race
湍急水流 flashy flow;impetuous torrent
湍急系数 coefficient of hydraulic flow;coefficient of turbulence(flow)
湍降冰川 cascading glacier
湍流 flashy flow;hasty flow;hydraulic flow;rapid flow;sinuous flow;stirring motion;super-critical flow;swift;torrential flood;torrential flow;turbulence current;turbulent current;turbulent flow;turbulent fluid;turbulent movement;turbulent stream;whirling current
湍流边界层 turbulent boundary layer
湍流边界层分离 turbulent separation
湍流标度 scale of turbulence;turbulence scale;turbulent scale
湍流表面摩擦系数 turbulent skin-friction coefficient
湍流层 turbosphere;turbulent layer
湍流层顶 homopause;turbopause
湍流场 field of turbulent flow;turbulent field
湍流尺度 scale of turbulence;turbulence scale;turbulent scale
湍流促进器 turbulent promoter
湍流的 turbulent
湍流动能 turbulent kinetic energy
湍流动能耗散率 turbulent kinetic energy dissipation rate
湍流动能平衡方程 turbulent kinetic energy equation
湍流度 degree of turbulence;turbulence level;turbulent level;turbulent number;turbulivity
湍流对流 turbulent convection
湍流发电机 turbulent dynamo
湍流法 turbulence method;turbulent method
湍流分离(现象) turbulence separation;turbulent separation
湍流分量 turbulent component
湍流风 turbulent wind
湍流附面层噪声 turbulent boundary layer noise
湍流过程 turbulent process
湍流河段 roil
湍流核 turbulent flow core
湍流核心阻力 core resistance
湍流混合 turbulent mixing
湍流混合区 turbulent mixing zone
湍流积分尺度 turbulent integral scale
湍流级 turbulence level;turbulent level
湍流加热 turbulent heating
湍流夹卷 turbulent entrainment
湍流剪力 turbulent shear
湍流剪切力 turbulent shear force
湍流交换 turbulent exchange
湍流交换系数 austausch coefficient;turbulent exchange coefficient
湍流搅动 trubulation
湍流接触吸收器 turbulent flow contact absorber
湍流结构 turbulent structure
湍流聚结法 turbulent coagulation
湍流扩散 turbulent diffusion
湍流扩散模式 turbulent diffusion
湍流扩散系数 coefficient of eddy diffusion;turbulent diffusion coefficient;turbulent diffusivity
湍流理论 turbulence theory
湍流力学 mechanics of turbulence
湍流流动 turbulent flow
湍流脉动 turbulent fluctuation
湍流模型 turbulence model
湍流磨坊 cascade grinding mill
湍流能(量) energy of turbulence;turbulence energy;turbulent energy
湍流能谱 turbulent energy spectrum
湍流逆流 turbulence inversion;turbulent inversion
湍流逆温 turbulence inversion;turbulent inversion
湍流黏[粘]滞性 turbulent viscosity
湍流谱 spectrum of turbulence
湍流谱极大尺度 peak wavelength in turbulent spectrum;wavelength corresponding to peak in turbulent level
湍流谱线宽度 turbulent line width
湍流器 turbulator
湍流强度 intensity of turbulence;turbulence level;turbulent intensity;turbulent level
湍流切应力 Reynold's stress;turbulent shear stress
湍流区 turbulent region;turbulent zone
湍流渠 canal rapids
湍流圈 turbulent annulus
湍流燃烧器 turbulent burner;turbulent flow burner
湍流扰动 turbulent perturbation
湍流散射 turbulent scattering
湍流声学 aerothermoacoustics
湍流式燃烧室 turbulence chamber
湍流输送 turbulent transfer
湍流衰减 decay of turbulence;turbulent decay
湍流水 roil water
湍流水流 impetuous torrent
湍流速度 turbulent velocity
湍流速度分布 turbulent velocity distribution
湍流通量 turbulent flux
湍流尾迹 turbulent trail
湍流尾流 turbulent wake
湍流涡 turbulent eddy
湍流系数 coefficient of turbulence(flow);turbulent coefficient;turbulivity
湍流相关系数 correlation coefficient of turbulence
湍流型泥石流 turbulent type mud flow
湍流悬浮分速 turbulent suspended composite velocity
湍流漩涡 turbulent eddy;vortex
湍流焰 turbulent flame
湍流应力 turbulent stress
湍流元 turbulence cell;turbulence element
湍流运动 turbulent motion;turbulent movement
湍流噪声 turbulence noise;turbulent noise
湍流增进器 turbulence promoter
湍流指数 index of turbulence
湍流致宽 line broadening by turbulence;turbulence broadening
湍流转变 turbulence transition
湍流状态 turbulent condition
湍流阻力 turbulent resistance
湍谱 turbulence spectrum;turbulent spectrum
湍球除尘 turbulent ball tower dedusting
湍球塔 turbulent ball tower;turbulent contact tower
湍速 turbulent velocity
湍滩 rapids
湍滩急流 cataract
湍珠洗涤 turbulent ball tower scrubber

团斑 blob

团花<地毯图案形式> medallion
团花桉 gray Stringybark eucalyptus
团集 agglomeration
团集化湿度 aggregation moisture
团集结构 aggregate structure
团集素 conglutinin
团集体分析 aggregate analysis
团集体形成 aggregate formation
团集体状态 aggregation state
团集体组织 fabric of aggregate
团集作用 aggregation
团结 rally
团结的 united
团结剂 agglomerator
团结一致 solidarity;unite as one
团聚 agglomerate;birdnesting;conglobate;conglobation;glomeration
团聚的 conglobate;coacervate
团聚度 degree of aggregation
团聚构造 aggregated structure
团聚结构 aggregated texture;flocculent texture
团聚颗粒 agglomerated particle
团聚力 agglomerating force
团聚体 aggregate
团聚体构造 aggregate structure
团聚性土壤 aggregated soil
团聚作用 agglomeration;coacervation
团块 aggregate;block mass;bolus;cake;conglomeration;glomerate;lumped mass;ped;briquet(te)【地】
团块腐殖体 corpohuminite
团块焦 coke briquette
团块结构 lumpiness
团块镜质体 corpocollinite
团块菌质体 corposclerotinite
团块磷块岩 lump phosphorite
团块炉料 charge of briquettes;power-compacted charge
团块铝质岩 lump aluminous rock
团块耐火黏[粘]土 nodular fireclay
团块泥状铝质岩 lump pelitomorphic aluminous rock
团块压制盒 wad box
团块制造机 briquet(te) making machine
团块状 blocky shape;crumby
团块状构造 cloddy structure;crumb structure
团块状结构 cloddy texture;crumb texture
团块状结核 massif nodule
团块状土壤 crumby soil
团矿 agglomerate;agglomeration;briquetting;coalette;nodulizing
团矿机 briquetting press
团粒 aggregate;crumb;cumulate sharolith;granule;lump;pellet
团粒构造 aggregate(d) structure;cluster structure;crumble structure
团粒化 aggregation
团粒灰岩 lump limestone
团粒结构 cluster structure;cluster texture;crumb-(le) texture;crumb-(le) structure;duster structure;granular texture
团粒结构土壤 aggregated soil
团粒理论<土结构的> domain theory
团粒台浮 agglomerate tabling
团粒土 granulated soil
团粒微晶灰岩 lump-micritic limestone
团粒状结构 crumble texture
团粒状土壤 crumbling soil
团粒组构<土的结构> domain fabric
团粒作用 granulation
团球状耐火土 nodular fireclay
团球状黏[粘]土 nodular fireclay
团绒机 ball winder
团伞花 flower cluster
团伞花序 flomerule
团伞花序的 glomerulose
团绳 pouch line
团石灰岩 lump limestone
团体 community;fellowship;gesellschaft;group;organization;squadron
团体保险 group insurance
团体保险费 group insurance premiums
团体标准 group standard
团体财产 community property
团体残废保险 group disability insurance
团体成员小组 membership groups
团体定期人寿保险 group term life insurance
团体动力学理论 group dynamics theory
团体动力训练 group dynamics training
团体动态学 group dynamics
团体奖工计划 group incentive plan
团体奖金 group bonus
团体奖励(制度) unit-wide incentives;group-incentives
团体聚合室 chapter-room
团体力学<社会心理学> group dynamics
团体旅客 group passenger
团体旅行 collective tour
团体旅游 group tourism;group travel
团体弥补免税法 group relief
团体目标 group goal
团体内聚力 group cohesiveness
团体年金 group annuity
团体凝聚力 group cohesiveness
团体票 collective ticket
团体气氛 group atmosphere
团体签证 group visa
团体群聚 adoption society
团体人寿险 group life insurance;life insurance for groups
团体生态学 group ecology
团体特征 group property
团体投标 group bidding
团体投资人 institutional investors
团体洗手池 water fountain
团体相互影响过程 group processes
团体行为 group behavio(u)r
团体压力 group pressure
团体预订 group booking
团体预约 group reservation
团体折扣 group discount
团体中的角色 roles in groups
团体作用学<研究团体成员间相互关系及如何最大限度地发挥该团体作用等的科学> group dynamics
团体作者 corporate author
团压机 briquetting press
团中心 cluster centre[center]
团状空化 patch cavitation
团状流动 slug flow
团状模塑料 dough mo(u)lding compound
团状石墨 nodular graphite
团子 dumpling

推拔床 push bench

推拔钢管机 push bench
推拔器 push puller
推板 hand plate;knockout plate;pusher;push pedal;push plate;push bumper<装置在机械或车辆前面的>
推板式平路机 blade drag;push-type motor grader

推板式切坯机 side cutter
推板式窑 push bat kiln
推板窑 pushed-bat kiln; pushed slab kiln; pusher kiln; slab kiln; sliding-bat kiln
推棒 push rod
推臂 push arm
推臂斜撑 <推土机> push-arm brace
推波 push wave
推驳搭配 pusher barge pair
推槽除雪机 channel snow plough
推测 calculate; conjecture; extrapolation; guess (work); presume; presumption; suppose; supposition
推测变动 conjectural variation
推测变化 conjectural variation
推测变量 inferred variable
推测储量 expected reserves; hypothetical reserve; inferred reserves; speculative reserves
推测的 presumptive; speculative; tentative
推测的(地基)承载力 presumed bearing value
推测的航位 dead position
推测断层 supposed fault
推测航行法 dead reckoning
推测井 image well
推测矿量 inferred ore; possible ore; prospective ore
推测使用期 expected life
推测数据 tentation data; tentative data
推测行为 conjectural behavio(u)r
推测值 guess value
推测资源 inferred resources
推测资源量 speculative resources
推铲卸车机 shovel car unloader
推铲支撑架 bulldozer stabilizer
推铲支撑桩 bulldozer stabilizer
推车 hand trolley; trammer; trip feeding
推车工 bandsman; hand trammer; harrier; headsman; trammer; wheelbarrow man
推车工助手 helper-up
推车轨 troll(e)y guide
推车机 car puller; car pusher; pusher; pusher gear; trip feeder
推车机系统 ejector mechanism
推车起动 push starting
推车器 car pusher
推车式喷粉器 push-type duster
推车式手压喷雾器 hand-operated barrow mounted sprayer
推车装车工 car pincher
推车装置 pusher gear
推撑 push brace
推撑器 extruding device
推成包络 envelope glissette
推成曲线 glissette
推承凹座 thrust recess
推迟 hold off; phase back; put over; retardation; retard(ing)
推迟标量位 retarded scalar potential
推迟偿债期 reschedule the debt
推迟场 retarded field
推迟处理 deferred processing
推迟电位 retarded potential
推迟反应 retarded reaction
推迟方程 retarded type equations
推迟符合 delayed coincidence
推迟付款时间 external of time for payment
推迟购买力 deferred purchasing power
推迟基线 retarded baseline
推迟交货 deferment delivery
推迟距离 retarded distance
推迟控制 retarded control
推迟量 retardation
推迟流 retarded flow
推迟膜片 secondary diaphragm
推迟偏置 retarded offset
推迟谱 retardation spectrum
推迟时间 retardation time
推迟势 retarded potential
推迟弹性 retarded elasticity
推迟弹性函数 retarded elasticity function
推迟弹性形变 retarded elastic deformation
推迟网络 retardation network
推迟线圈 retarder
推迟相位 retarding phase
推迟效应 retarded effect; retarding effect
推迟寻址 deferred address
推迟运行 putting back into operation
推迟自发回复 retarded spontaneous recovery
推迟作用 retardation function; retarded action
推斥 repulse; repulsion
推斥感应电动机 repulsion-induction motor
推斥感应式电动机 repulsion and induction type motor
推斥感应型单相电动机 repulsion and induction type single phase motor
推斥感应型电动机 repulsion-induction motor
推斥激励 repulse excitation
推斥力 repelling force; repulsion force; repulsive force
推斥起动感应电动机 repulsion-start induction motor
推斥式电动机 repulsion motor
推斥态 repulsive state
推斥型电动机 repulsion type motor
推斥型仪表 repulsion instrument; repulsion type meter
推冲器 side thrust; thruster
推出 ejection; kick; protrusion; pulsion; push off; push-out
推出安全装置 <冲床用> pull-out guard
推出板 head board; push board
推出的 protrusive
推出杆 push-off pin
推出机 kick-off mechanism; ram
推出机构 ejecting mechanism
推出舰首水平舵 rig out bow planes
推出平面调车 kicking; kick-off
推出器 expeller; pusher(bar); push off; pushover; throw-out
推出式安全装置 sweep guard
推出式挡风玻璃 pop-out windscreen; push-out windshield
推出试验 push out test
推出速度 kick-off speed
推出系数 pushing figure
推出新产品 launch new products
推出新技术 launch new technology
推出装置 liftout attachment
推船 push boat
推窗杆 sash pole
推窗杆钩 sash pole hook
推床齿条 manipulator rack
推床的导板 manipulator slide beam
推床痕 <轧材缺陷> manipulator marks
推袋小车 bag cart
推刀 pusher blade; push-type broach
推导 derivation; development
推导比例尺 derived scale
推导关键词 derived key
推导数据 derived data
推导图 derivation graph
推倒 detrude; pulling down; topple
推倒重建派 buldozerite
推钉 ejector pin
推顶臂 ejector arm
推顶杆 ejecting plug; ejector pin; penetrator ram
推顶杆板 ejector plate
推顶构造 jacking frame
推顶管系统 pipe-forcing system
推顶活塞 ejector ram
推顶键 ejector key
推顶力 ejection force
推顶器 ejector
推顶套 ejector sleeve
推顶拖拉机 push tractor
推顶柱 ejector rod
推定 estimation; presumption
推定储存矿量 probable ore
推定储量 inferred ore; inferred reserves; possible reserve
推定的 constructive; presumptive
推定的损失总额 constructive total loss
推定的证据 presumptive evidence
推定恶臭浓度 counted odo(u)r concentration
推定过失 constructive fault
推定(换算)距离 <适用于高低不同运价区之间> constructive distance
推定(换算)里程 <适用于高低不同运价区之间> constructive-mileage
推定极限 derived limit
推定价格 computed price; constructed price
推定价值 constructed value; constructive value
推定交货 constructive delivery; symbolic delivery
推定结账 constructive closing
推定解雇 constructive discharge
推定拒付 constructive dishonour
推定空气浓度 derived air concentration
推定趋势 determining tendency
推定全损 constructive total loss; presumed total loss; technical total loss
推定全损理算 adjustment of constructive total loss
推定全损条款 constructive total loss clause
推定试验 presumptive test(ing)
推定收入 constructive receipts; presumptive income
推定数值 extrapolated value
推定损害 <保险> sentimental damage
推定条款 constructive clause
推定意思 presumptive intention
推定运费率 constructed rate
推定占有 constructive possession
推定证据 presumptive evidence
推定资源 indicated resources
推锭机 ingot pusher
推动 impetus; motivation; propel; propulsion; pushing; shove; shoving
推动篦式冷却器 reciprocating grate cooler; stoker type grate cooler
推动臂 actuating arm
推动的 impellent
推动(发展)……市场 make a market of
推动杆 catch bar; pusher arm
推动杠杆 <手摇车> propelling lever
推动滑轮 propelling sheave
推动控制杆 push lever
推动力 expulsive force; impellent; impelling force; mover
推动轮 propel wheel
推动螺杆 lead screw
推动其他机器的机械 prime mover
推动器 driver unit
推动曲柄 throw crank
推动式滑坡 pushing landslide; shored landslide
推动式试验车 push cart
推动鳌尖 thrust pad; thrust pallet
推动水 push water
推动凸轮 actuating cam
推动芯头 clearance print
推动业务 active business
推动爪 driving pawl
推动装置 thrust unit
推断 assertion; deduction; draw a conclusion; extrapolation; presume; presumption
推断承载力 presumptive bearing pressure
推断储(备)量 inferred reserves; prospective reserves
推断的矿石 inferred ore
推断断层 inferred fault
推断规则 rule of inference
推断角 conclude angle
推断接触带 inferred contact
推断解释图 interpretation map
推断力 extrapolability
推断统计学 inferential statistics
推垛机 handler
推阀杆【机】 valve push rod
推翻 demolition; explode; overthrow; overturn; pulldown; repudiation
推翻的 overturned
推翻计划 disconcert
推峰 humping
推峰部分 humping part
推峰程序 humping program(me)
推峰机车 push-up engine
推峰速度 approach speed to the hump; humping speed; push-up speed
推峰线【铁】 track leading from receiving yard to hump
推峰信号 humping signal
推峰运动 humping movement
推峰作业【铁】 pushing-up operation at hump; humping operation
推覆构造【地】 nappe structure; nappe tectonics
推覆构造带【地】 nappe structure belt
推覆构造掩盖型 nappe structure covered type
推覆体【地】 nappe; decken
推覆体底基逆掩断层 detachment fault
推覆体根带位置 location of nappe root zone
推覆体运移方向 movement direction of nappe
推覆体运移距离 distance of nappe transporting
推覆体展布面积 area of nappe
推覆作用 napping
推杆 pusher; pusher bar; pushing bar; push pole; push rod; armature buffer; armature pusher; ejector pin; follower; kilhig; peel; push arm; push brace; push bracing; ramrod; tappet; tappet rod; peel bar; travel(l)ing bar <加热炉推料机的>; crash bar <太平门五金>
推杆操纵的阀 stem-operated valve
推杆操纵系统 push rod system
推杆插口 push pole pocket
推杆铲车 free-lift mast forklift truck
推杆衬套 push rod bushing
推杆传动装置 push rod actuator
推杆传送器 push-bar conveyer[conveyor]
推杆导承 push rod guide
推杆阀快速接头 stem valve coupling
推杆阀自封接头 stem valve self-sealing coupling
推杆滚柱 tappet roller
推杆机构 pushrod mechanism; transfer bar mechanism

推杆间隙 push rod clearance
推杆绞盘 bar capstan
推杆密封环 rod packing
推杆驱动 radius rod drive
推杆式继电器 pusher relay
推杆式料盘 pusher tray
推杆式炉 pusher furnace
推杆式送料 pusher feed
推杆式送料导向装置 pusher-type stock guide
推杆输送机 throw transporter
推杆弹簧 push rod spring
推杆套管 push rod tube
推杆套筒 push rod housing
推杆提升机 pusher bar booster
推杆调整螺钉 tappet screw
推杆头 push rod cup
推杆凸轮 push-cam
推杆托架 push rod holder
推杆托销 push rod holder pin
推杆销 push rod pin
推杆型雨淋阀 push rod deluge valve
推杆窑 pushed-bat kiln
推杆运输机 bar conveyer[conveyor]; pusher bar conveyer[conveyor]
推杆运输链 power-free conveyer[conveyor]
推杆闸缸 push rod brake cylinder
推杆执行机构 push rod actuator
推杆爪簧 pusher-click spring
推钢杆 pusher bar
推钢机 ejector; pusher; pushing device; ram
推钢机的推杆 pusher ram
推杠 hand spike
推梗 push and guard bar
推挂式筑路机械 tilt dozer
推关窗扇 side-hung casement
推管装置 pipe pusher; thrust boring machine
推光 ashing; brightening
推广 circularize; generalization; popularization; popularize
推广费用 promotion expenses
推广津贴 promotional allowance
推广贸易及商务 extend trade and commerce
推广媒介 promotion media
推广模型 spread model
推广先进经验 spread advance experience
推广小册子 sales brochure
推广研究评论 review of extension research
推广站 extension services
推广组合 promotion mix
推辊压路机 push roller
推滚 push roller
推过去 push through
推合座 leakproof fit; push fit
推环 throw-out collar; thrust ring
推换 push off; pushover wipe
推换分布 push-pull distribution
推换装料系统 push-through loading system
推回 push-back
推回火灾烟雾 backing of smoke
推机焦侧 pusher end
推积土滑坡 drift soil landslide
推集机 pushrake
推挤 pushing; shoving
推挤成型法 pushing process
推挤裂缝 crescent cracking
推挤裂纹 crescent cracking
推挤式离心机 push-type centrifuge
推挤现象的 pushed
推挤型 extruded shape
推件盘 ejector pad; knockout pad
推荐 commend; nominate; nomination; propose; recommendation
推荐标准 recommended limit; recommended practice; recommended standard
推荐参数 recommended parameter
推荐车速标示 advisory speed indication
推荐车速标志 advisory speed sign
推荐尺寸 preferred size
推荐代用器 recommend substitute
推荐的备件单 recommended spare parts list
推荐的储量分级术语 pushed grade term of reserves
推荐的热处理温度范围 recommended temperature ranges for heat treatment
推荐的照度标准 recommended level of illumination
推荐地点 proposed site
推荐方案 proposal; proposed scheme; recommended alternative; recommended plan; recommended project; selected project alternative; suggested design
推荐方法 recommended practice
推荐工程 recommended alternative; recommended project; selected project alternative
推荐规范 recommended tolerance
推荐航路 recommendation route; recommended route; recommended track
推荐航线 recommendation route; recommended route; recommended track
推荐航向 course recommended
推荐厚度 preferred thickness
推荐截面 preferred section
推荐零售价格 recommended retail price
推荐流程 recommended flowsheet
推荐品种 recommended variety
推荐平面布置方案 proposed layout
推荐坡度 proposed grade
推荐书 letter of recommendation; reference letter
推荐速度 advisory speed; recommended velocity
推荐天然采光系数 recommended daylight factor
推荐项目 recommended alternative; recommended project; selected project alternative
推荐信 letter of recommendation
推荐性标准 recommended standard
推荐性简介 recommendatory annotation
推荐业务条例 code of recommended practice
推荐照度 recommended illumination
推荐者 nominator; presenter
推荐值 recommended value
推荐专用工具 recommend special tools
推荐装载方法 stowage recommendation
推荐自行车道路线 advisory cycle route
推荐自行车路线 advisory cycle route
推键式工作 push-to-type operation
推桨 breast stroke
推焦车 pusher machine; pushing machine
推焦程序 carbonizing program(me)
推焦杆 pushing ram
推焦杆头 coke pusher shoe
推焦机 coke pusher
推焦机侧 pusher end
推焦面 pusher side
推焦顺序 pushing schedule
推接端 push-joint end
推紧螺套 sheeting jack
推进 advance; beat in; carry forward; drive-in; facilitation; feed control; impel(ling); propel; propulsion; push-on
推进泵 boost pump
推进比 impelling ratio
推进滨线 prograding shoreline
推进冰碛 push moraine
推进波 advanced wave; progressive wave; propagating wave; translation(al) wave; translatory wave; travel(l)ing wave; wave of translation
推进补机 helper
推进参数 propulsive parameter
推进舱 rudder-propeller; rudder-screw
推进操纵 propulsive control
推进操纵手把 feed lever
推进长度 feed length; length of run; pushed length
推进沉淀作用 displacive precipitation
推进冲程 feed stroke
推进传动 propulsion drive; propulsion transmission
推进的 projectile; propulsive
推进的矿井 pushing pit
推进的沙丘 dune on the march
推进电动机 propulsion motor
推进调车 backing movement; backing shunting; backup movement; backward movement
推进动力类型 propulsion type
推进动力数据显示板 propulsion data display panel
推进陡波 abrupt translatory wave
推进发动机 propelling motor; propulsion engine; propulsive engine; propulsion generator
推进法 end-on system
推进分量 propulsion component
推进分系统 propulsion subsystem
推进辅机控制设备 propulsion auxiliary machinery control equipment
推进杆 propel arm
推进杆托架 pusher bracket
推进缸筒 thrust cylinder
推进钢丝绳 traction rope
推进工 driver
推进功率 propeller power; propelling power; propulsion power; propulsive output; propulsive power; thrust power
推进和电气操纵系统 propulsion and electric(al) operating system
推进活塞 feed piston
推进机 mover; pusher
推进机车 banking locomotive
推进机风洞 propulsion wind tunnel
推进机构 propelling mechanism
推进机械 propelling machinery; propelling plant; propulsion machinery
推进剂 propellant[propellent]; propelling agent; propergol
推进剂爆炸 propellant explosion
推进剂残余物 propellant residual
推进剂仓库 propellant terminal
推进剂舱 propellant module
推进剂成分 propellant constituent
推进剂储存区 propellant storage area
推进剂的使用保管 propellant handling
推进剂的输送 propellant supply
推进剂点火 propellant fire
推进剂断路开关 propellant shutoff valve
推进剂供给系统 propellant feed system
推进剂化学 propellant chemistry
推进剂化学原料 propellant chemical
推进剂加注操作 propellant-loading operation
推进剂加注区 propellant-servicing area
推进剂结构 propellant structure
推进剂井 propellant silo
推进剂冷却的 propellant-cooled
推进剂黏[粘]合剂 propellant binder
推进剂牌号 model of propellant
推进剂喷头 propellant injector
推进剂燃烧 propellant combustion
推进剂燃烧速度 propellant burning velocity
推进剂容积 propellant volume
推进剂输送设备 propellant transfer equipment
推进剂添加物 propellant additive
推进剂雾化 propellant atomization
推进剂箱 propellant container
推进剂消耗 propellant expenditures
推进剂消耗量 propellant consumption; propellant waste
推进剂性能 performance of propellant; propellant performance
推进剂性质 propellant nature
推进剂止送装置 propellant cut-off mechanism
推进剂质量比 propellant mass fraction; propellant mass ratio
推进剂重量比 propellant weight fraction
推进剂组分比 reactant ratio
推进加料机 spreader screw
推进阶段 propulsion phase
推进井 push well
推进空气 actuating air
推进栏 crowd chain
推进力 ahead power; boosting power; drive power; driving power; feed force; impelling force; propelling effort; propelling power; propulsion; propulsive force; propulsive power
推进力系数 propulsive coefficient
推进链 crowd chain
推进轮 feed wheel; propel wheel
推进螺管 outer feed screw
推进螺栓 push bolt
推进马达 feed motor
推进马力 propulsive horsepower; thrust horsepower
推进脉冲 channel pulse; advance pulse
推进面 pressure surface
推进能力 propulsion capability
推进扭矩 propelling torque
推进喷管 propelling nozzle; propulsion nozzle
推进汽轮机 ahead turbine
推进器 feed attachment; feed swing-jaw; impellent; impeller; mover; propeller; propulsion; propulsor; pusher; thrust device; thruster
推进器伴流 propeller wake
推进器车叶背面 back
推进器导流槽 fairing cap
推进器的受力面 thrust surface
推进器的推进力 feed thrust
推进器毂 propeller boss; propeller hub
推进器后叶缘 following edge
推进器护材 screw guard
推进器护杆 propeller boom
推进器护栏 propeller guard
推进器滑距 propeller slip
推进器滑失率 propeller slip
推进器滑脱 slip of propeller
推进器架 <双推进器船的> A-bracket
推进器紧锁螺帽 propeller lock nut
推进器框架 propeller frame
推进器框穴 propeller aperture; pro-

peller port; propeller well; screw aperture
推进器流 race of screw
推进器螺杆 feed screw
推进器螺距 pitch of propeller; propeller pitch
推进器马力 propeller horsepower; screw horsepower
推进器帽 propeller cap; propeller cone
推进器射流 propeller jet
推进器伸缩液压缸 <凿岩机> feed extension cylinder
推进器式流量计 propeller meter
推进器式流速计 propeller current meter
推进器推力 propeller thrust
推进器尾流 propeller current; propeller race; slip stream; propeller jet
推进器尾柱 screw post
推进器效率 propeller efficiency
推进器行程 feed travel
推进器仰俯液压缸 <凿岩机> feed dump cylinder
推进器叶 propeller blade
推进器叶端 blade tip; tip of propeller
推进器叶片 blade
推进器造成的冲刷 propeller wash
推进器噪声 propulsion noise
推进器轴 screw shaft
推进器轴孔 propeller shaft hole
推进器轴离合器 tail clutch
推进器轴轴承 propeller shaft bearing
推进器柱 heel post; propeller post
推进器座 feed holder
推进千斤顶 pushing jack; segment pushing jack; thrust jack; main jack; propelling jack <盾构工程用>
推进取样 drive sampling
推进三角洲 prograded delta
推进上浮取样管 free fall rocket core sampler
推进射流 propulsive jet
推进式的 push type
推进式发动机 pusher engine
推进式飞机 pusher; pusher airplane; pusher-type airplane
推进式风洞 propulsion wind tunnel
推进式干燥窑 progressive type kiln
推进式管头 driving shoe
推进式架设 erection by launching
推进式交通控制信号联动系统 progressive system of traffic control
推进式量角器 propeller protractor
推进式螺旋桨 pusher airscrew; pusher propeller; pusher screw
推进式螺旋桨飞行器 pusher vehicle
推进式输送机 push conveyer[conveyor]; pushing conveyer[conveyor]
推进式旋翼机 pusher autogiro
推进式运输机 push conveyer[conveyor]
推进式运送机 pushing conveyer[conveyor]
推进式运行 progressive movement
推进式凿岩机 push-feed drill
推进数据 propulsion data
推进送料 push feed
推进速度 advance speed; forward speed
推进特征 propulsion characteristic
推进体 pushing body
推进透平 ahead turbine
推进腿 feed leg; pusher leg
推进涡轮机 ahead turbine
推进物 propellant[propellent]
推进系数 coefficient of propulsion; propulsive coefficient
推进系统 propulsion system
推进线 axis of thrust
推进效率 propulsion efficiency; propulsive efficiency
推进性能 propulsive performance
推进压力 feed pressure
推进叶轮 impulse impeller
推进用堆 propulsion reactor
推进用反应堆 propulsion reactor
推进运行 backing movement; backup movement; backward movement; propelling movement; reverse running; running-in reverse
推进者 propellant[propellent]; propeller[propellor]
推进支架控制阀 feed control valve
推进轴 cardan shaft
推进转矩 propelling torque
推进装置 feeding device; feed equipment <凿岩机的>; propelling unit; propulsion system; propulsion unit; propulsive arrangement; propulsive device; propulsive machinery; propulsive unit; thrust system; thrust unit
推进装置舱 propulsion bay
推进装置模拟器 propulsion simulator
推进装置试验设备 propulsion test facility
推进阻力 resistance to propulsion
推进钻臂 feed boom
推进钻车 tunnel jumbo
推进作用 propellant action; pushing action
推举 recommendation
推卷机 coiler kickoff
推开 shove; sweep
推开窗扇 side-hung casement
推拉 push and pull
推拉把撑 push and pull brace
推拉把手 <门上的> push and pull brace; push and pull bar; push and pull handle
推拉百叶窗 sliding shutter
推拉板 drawback plate
推拉泵送功率 push and pull pumping power
推拉边 <门窗> edge pull
推拉玻璃墙 sliding glass wall
推拉舱盖 sliding deck
推拉操纵缆 push-pull control cable
推拉撑 <桁架的> push and pull brace
推拉锄 scuffle hoe
推拉窗 austral window; double sash window; slidable window; sliding sash; sliding sash window; sliding window
推拉窗扇 sash
推拉窗扇中梃 parting bead; parting strip
推拉窗锁位器 thumb screw
推拉窗之平衡锤箱 space for balance
推拉的 push-pull
推拉吊钩 sliding hanger
推拉调车法 push-pull switching; tail drilling; tail switching
推拉动车组 push-pull set
推拉杆 connecting rod; pull-push rod; push and pull brace; push and pull rod; push-pull rod; side arm <气窗两侧的>
推拉隔断 sliding partition(wall)
推拉工作 push-pull working
推拉挂钩装置 <铲土机> push puller
推拉滑轨 sliding rail; sliding track
推拉键 sliding key
推拉开关 push(-and)-pull switch
推拉力 push-pull effort
推拉连接机构液压缸 <铲土机> push-pull cylinder
推拉联杆 push and pull brace
推拉列车运行 push-pull service
推拉螺杆 push-pull screw
推拉门 draw gate; horizontal sliding door; rolling door; slide gate; slide panel; sliding door
推拉门板 push plate
推拉门吊钩 sliding door hanger
推拉门吊轨 barn-door hanger
推拉门吊架 sliding door hanger; barn-door hanger <库房的>
推拉门吊架滑轮 <库房的> barn-door stay
推拉门吊件 sliding door hanger
推拉门滑道 sliding door track
推拉门滑轨 sliding door rail; sliding door track
推拉门键 sliding door key
推拉门门挡 sliding door stop
推拉门门架 door pocket
推拉门驱动装置 sliding door gear drive
推拉门锁 sliding door catch; sliding door lock
推拉门箱 door pocket; sliding door pocket
推拉门压力驱动 pneumatic sliding door drive
推拉内窗 sliding inner window
推拉钮 push-pull knob
推拉千斤顶 push-pull jack
推拉切片机 sliding microtome
推拉设备 push-pull equipment
推拉摄影 dolly shot
推拉式铲运机 push-pull scraper
推拉式顶进装置 push-pull jacking rig
推拉式列车 <前后两端各有一台机车 pull(-and)-push train
推拉式铝门 alumin(i)um sliding door
推拉式门 sliding gate
推拉式双层市郊旅客列车 double deck push-pull commuter train
推拉式送进装置 push-pull feed arrangement
推拉式弯管 push-draw mode
推拉式握柄 push-pull lever; push-slide lever
推拉式坞门 rolling caisson; sliding caisson; sliding pontoon; traversing caisson
推拉式闸门 <船闸的> traversing caisson; sliding caisson
推拉式折叠窗 sliding folding window
推拉手把 push-pull knob
推拉输出电路 totem-pole
推拉锁 sliding lock
推拉台 scenery wagon
推拉台面 draw top
推拉台面桌 draw top table
推拉推土机 grade builder
推拉镶板 slide panel
推拉型 <铲土机> push-pull
推拉业务 push-pull service
推拉用小把手 button stem
推拉运行 push-pull operation; push-pull running
推拉折叠百叶门窗 sliding folding louvred shutter
推拉折叠隔断 sliding folding partition (wall)
推拉折页窗 sliding folding window
推拉折页门 sliding folding door
推拉装置 push-pull arrangement
推拉作业式铲运机 push-pull scraper
推拉作用 push-pull process
推理 ratiocinate; ratiocination; reasoning; theorizing
推理的 theoretic(al)
推理法 inference method; method of induction; organon; rational method
推理分析 rational analysis
推理公式 rational formula
推理估测 inferential measurement
推理规则 inference rule
推理机 inference engine; inference machine
推理机构 inference mechanism
推理机制 <专家系统的> inference mechanism
推理计算机 inference computer
推理径流公式 <小流域用的> rational runoff formula
推理能力 inferential capability
推理网络 inference network
推理网络法 inference network method
推理危险性 speculative risk
推力 counter force; impellent; propulsive force; push(ing) force; push thrust; thruput; thrust(ing force); thrust push
推力板 thrust plate; toggle plate
推力磅 thrust-pound
推力泵 bull pump
推力布置 thrust arrangement
推力测定计 thrust jack
推力测量装置 thrust-measuring device; thrust-measuring equipment; thrust-measuring installation
推力车 push car
推力衬套 thrust bush(ing)
推力持续时间 thrust duration
推力储备 thrust reserves
推力传感器 thrust pickup
推力挡板 set thrust plate
推力点 thrust point
推力垫圈 thrust washer
推力定额 thrust rating
推力定向操纵 control of thrust orientation
推力轭 back end; thrust yoke
推力伐树机 tree pusher
推力反向器 reverse thrust device; thrust-reversal device
推力范围 thrust range
推力方向 thrust direction
推力方向调节 thrust-direction control
推力方向调整 thrust-line adjustment
推力方向性 thrust orientation
推力分布曲线 thrust grading curve
推力分量 thrust component
推力风机 impulse fan; thrust ventilator
推力风机控制箱 control box for impulse fan
推力风速表 thrust-anemometer
推力辅助抛锚 thruster assisted anchor(ing)
推力负荷 thrust load(ing)
推力杆 distance rod; propelling rod; push and pull brace; thrust arm
推力拱 arch with thrusting force
推力滚动轴承 thrust ball bearing
推力滚针轴承 needle roller thrust bearing
推力滚珠轴承 ball-thrust bearing
推力荷载 thrust load
推力环 thrust collar; thrust ring
推力环轴承 thrust collar bearing
推力缓变 smooth thrust variation
推力缓冲拱系统 arched system for absorption of thrust; arcual system for absorption of thrust
推力换向式飞机 convertiplane
推力恢复 thrust recovery
推力级发动机额定推力 thrust level
推力计 thrust meter
推力减额 thrust deduction; thrust reduction
推力减额系数 thrust deduction coefficient; thrust deduction factor; thrust deduction fraction
推力减少 thrust decay

推力角 thrust angle
推力校正装置 propellant-actuated device
推力截止阶段 thrust-cutoff phase
推力可调的发动机 adjustable thrust engine
推力控制 control of impulse
推力块 thrust block;thrust pad
推力力矩 thrust moment
推力螺钉 thrust screw
推力螺母 thrust nut
推力螺旋 thrust screw
推力马力 thrust horsepower
推力锚 propellant anchor
推力面 displaced surface;thrust face;thrust surface;driving face <桨叶的>;thrust side
推力面积 area of thrust surface
推力逆转机构 thrust-reversing mechanism
推力盘 thrust disc
推力喷管 propelling nozzle;thrust nozzle
推力偏差 thrust deflexion;thrust deviation
推力偏心矩 eccentricity of thrust
推力偏心率 thrust misalignment
推力平衡装置 thrust balancing device
推力器 thrustor
推力器辅助下锚系统 thruster assisted mooring system
推力千斤顶 pusher jack;shoving jack;thrust jack
推力球 thrust ball
推力球面滚子轴承 spheric(al) roller thrust bearing
推力球轴承 axial contact ball bearing;thrust ball bearing
推力曲线 thrust curve
推力扰流器 thrust spoiler
推力升力喷管 thrust-lift nozzle
推力矢量控制 thrust vector control;vector steering
推力式液压缸 displacement type ram
推力试车台 thrust bed
推力试验 thrust test
推力试验室 test thrust chamber
推力试验台 thrust stand
推力室 thrust barrel;thrust chamber;thrust cylinder;thrust section
推力室阀门开关 thrust chamber valve switch
推力室压力 thrust chamber pressure
推力受力元件 thrust web
推力输出量 thrust output
推力衰减拱系统 arched system for absorption of thrust;arcual system for absorption of thrust
推力衰减期 thrust-decay period
推力套筒 thrust bush(ing);thrust sleeve
推力套筒轴承 Jordan bearing
推力体系 system with thrusting force;thrust system
推力调节 thrust control;thrust level control
推力调节器 thrust controller
推力头 collar of thrust bearing;rotating thrust collar;thrust bearing runner;collar <水轮发电机组的>;bearing collar <推力轴承的>
推力图线 thrust graph
推力瓦块 thrust pad;thrust segment;thrust shoe
推力误差 thrust error
推力吸收拱系统 arched system for absorption of thrust;arcual system for absorption of thrust
推力系杆 push brace
推力系数 coefficient of thrust;propulsive coefficient;thrust coefficient
推力下降 thrust decay;thrust drop
推力下降信号器 thrust-loss indicator
推力线 axis of thrust;line of thrust;thrust line
推力线拱 thrust line arch
推力线拱顶 thrust line vault
推力限制器 thrust limiter
推力向量线性位移 linear thrust misalignment
推力向心球轴承 annular contact thrust ball bearing
推力修正 thrust correction
推力影响线 thrust influence line
推力油缸 push action cylinder
推力余量 thrust margin
推力载荷 thrust load
推力增大 thrust augmentation;thrust buildup
推力增大的 thrust-augmented
推力增加控制 thrust-augmentation control
推力增量 thrust increment
推力增强器 thrust augmenter
推力炸药 impulse charge
推力止推轴承 thrust bearing
推力指示 thrust indication
推力指示器 thrust indicator
推力中心 center[centre] of thrust;thrust center[centre]
推力-重量 thrust-weight
推力-重量比 thrust(-to)-weight ratio
推力轴 thrust axis;thrust shaft
推力轴承 anti-thrust bearing;axial contact bearing;axial lid;block bearing;journal bearing;plummer block;thrust bearing;thrust block
推力轴承板 thrust bearing disc
推力轴承的挡圈 thrust bearing race;thrust race
推力轴承垫座 thrust bearing pad
推力轴承负荷 thrust bearing load
推力轴承盖 thrust block keep
推力轴承滑道 thrust runner
推力轴承穴 thrust bearing recess;thrust recess
推力轴承止推板 thrust bearing resisting plate
推力轴承轴瓦 thrust bearing pad
推力轴承座 thrust metal
推力轴瓦 thrust bearing liner;thrust bearing shoe
推力轴瓦衬 thrust bearing liner
推力轴瓦扇形块 thrust bearing segment
推力转向 thrust deflexion
推力转向器 thrust deflector;thrust diverter
推力转向叶栅 thrust-reverser cascade
推力装置 thrust device
推力阻流片 thrust spoiler
推力钻 impulse pallet
推力作用线 thrust line
推力座 journal bearing;thrust bearing;thrust block
推梁 <推土机的> push beam
推梁C形架宽度 <推土机的> C-frame of push beams width
推料 stoking
推料车 lorry
推料杆 charging ram;push-off arm
推料机 discharger;pusher;shedder;stock pusher
推料机的推杆 travel(l)ing bar
推料机构 pusher
推料机构装置 pusher mechanism
推料机加煤机 stoker
推料机小车 pusher carriage
推料孔 showing hole
推料耙 shoving rake
推料盘 showing pan
推料器 stowing tool
推料式加热炉 pusher-type furnace
推料叶片 pusher blade
推流 plug flow
推流式反应器 plug-flow reactor
推流式活性污泥法 plug-flow activated sludge process
推流式曝气 plug-flow aeration
推流式曝气池 plug-flow aeration pond
推轮【船】 push boat;pusher;pusher tug;pushing vessel
推论 deduction;consequence;illation;ratiocinate;ratiocination;corollary
推论储量 indicated reserve
推论的 deductive
推论(方)法 deductive approach;deductive method;method of deduction
推论技术 inference technique
推论技艺 inference technique
推论统计学 inferential statistics
推落式挤切修边模 push-through pinch-trim die
推落式拉杆 push-through tie
推落式连杆 push-through tie
推落式模 push-through die
推落式深拉延模 push-through deep drawing die
推毛板 pushing-up board
推煤机 bulldozer;coal dozer;coal transporter
推煤器 coal pusher
推门板 push plate
推门横杆 push bar
推门横条 push bar
推门五金 push hardware
推摩法 stroke
推拿室 massage room
推泥机 sludge moving mechanism
推耙 push frame
推耙机 pushing-collecting machine
推排式并条机 push-bar drawing frame
推盘式炉 pusher tray furnace
推频 frequency pushing
推平 flattening out;raze to the ground
推碛 shoved moraine
推敲 bat;deliberate;elaboration;refine;thrash over;weigh
推切式牛头刨 push-cut shaper
推入 push-and-guide
推入密封套 push-on gland
推入配合 push fit;sucking fit;tuning fit;tunking fit
推入式PVC管 push-on PVC pipe
推入式扁平窑 push-type slab kiln
推入式插座 push socket
推入式接头 push-on joint
推入式配件 push-in fitting
推入硬度 push-in hardness
推扫器 calf-dozer
推扫式传感器 push-broom sensor
推扫式扫描 push-broom scanning
推上 run-up
推上驼峰 run-up to hump
推石机 rock blade
推式插削 pusher-type shaping
推式铲运机 push loaded scraper;push-loading scraper
推式锄 scuffle hoe
推式粪沟清理器 pushing-type dung channel cleaner
推式焊枪 push gun
推式机具 push-type machine
推式联合收获机 pusher-type combine
推式路帚 push broom
推式平地机 push-type grader;push-type motor grader
推式平刨 pushing-type flat plane
推式输送机 push-type conveyer[conveyor]
推式送丝 push-type wire feed
推式送丝机构 push-type wire feeder
推式镗孔 push boring
推式拖拉机 push dozer;push(er) tractor;push-loading tractor
推式研钵 buckmortar
推式肘板夹紧装置 push toggle clamp
推式柱塞压捆机 push plunger baler;push plunger press
推式装载 push-loading
推式自动平地机 push-type motor grader
推手 push handle
推手板 finger plate
推手柄 pushing handle
推手车工 wheeler
推树机 tree cutter;tree dozer;tree pusher;tree stinger
推树器 tree pusher
推闩 push bolt
推送 propelling movement;pushing
推送补机 assisting locomotive for pushing
推送部分【铁】 humping section;pushing section
推送调车(法) push-pull shunting
推送辊 push rolls
推送机车 pusher;pusher engine;pusher locomotive;pushing engine;push-up engine
推送连接车 push car
推送列车上峰 delivery of train to hump
推送溜出 backkick
推送坡 pushing gradient
推送坡度 assisting grade
推送式离心机 push-type centrifuge
推送式炉 pusher-type furnace
推送式隧道窑 pusher-type kiln
推送速度 pushing speed;push-up speed
推送调至 pushing off the wagons or cars
推送喂料分离机 push-feed separator
推送喂料分选机 push-feed separator
推送线 hump lead;pushing track;push-up track
推送小车 mule
推送信号 start humping signal
推算 calculate reckoning;calculation;computation;prediction;reckoning
推算边 computation line
推算波浪 hindcasted wave
推算波浪资料 hindcast wave data
推算潮流 predicted tidal flow
推算潮汐 predicted tide
推算成本 imputed cost
推算程序 <在有限单元法中采用电子计算机前把零散程序排成一致程序> master driving program(me)
推算出的生长率 calculated growth
推算船位 dead reckoning;dead-reckoning position;position by dead reckoning;reckoning;estimated position
推算的潮汐曲线 predicted tide curve
推算的价值量 imputation
推算定位 dead reckoning
推算概率 prediction probability
推算公式 prediction equation;prediction formula
推算价格 computed price
推算价值 imputed value
推算利息 imputed interest
推算流量 computed discharge

推算路线 computation line
推算模型 prediction model
推算时间 dead-reckoning time
推算收入 imputed income
推算温度 inferred temperature
推算者 predictor
推算资料 hindcast data
推算子 estimator
推算租金 imputed rent
推算坐标 calculated coordinates
推填机理 interstitialcy mechanism
推条 pusher bar
推土 <用推土机> bulldozing;bulldoze
推土板 blade; bulldozer(blade); dozer(blade); dozing blade; mo(u)ld board
推土板到端角处的长度 <推土机> length over end bits
推土板的侧板 sidewall
推土板高度 blade height
推土板宽度 blade width
推土板切土角度的调整 adjuestment of blade cutting angle
推土板切削角 blade cutting angle
推土板提升高度 blade lifting height
推土板提升时间 blade lifting time
推土板下降深度 blade lowering depth
推土板下降时间 blade lowering time
推土板翼板部分 wing section
推土板中段 center[centre] section of blade
推土铲 blade; bulldozer; dozer; dozer blade; dozer shovel; dozing blade; push plate
推土铲成套零件 bulldozer kit
推土铲切土深度 blade depth of cut
推土铲稳定装置 dozer stabilizer
推土铲运机 scraper-dozer
推土铲运视 scraper-bulldozer
推土铲装置 bulldozer kit
推土刀 bulldozer blade; pusher blade
推土工具 dozing tool
推土回填机 pushfiller
推土机 blade machine; blader; bulldozer; dirt mover; dozer; earth mover; pull dozer; push-type scraper; scrape dozer; skimmer; soil shifter; tractor dozer; tractor equipped with bulldozer; tractor; tractor with bulldozer; push dozer <后部作压实用>
推土机班 dozer company
推土机铲 bulldozer loader; bulldozer shovel
推土机铲板 bulldozer blade
推土机铲刀 bulldozer blade; dozer blade
推土机铲刀角片 bulldozer end bit
推土机铲挖 dozer shovel(l)ing
推土机挡板 dozer apron
推土机刀片 bulldozer blade; dozer blade
推土机的旱地作业 tractor earth blading
推土机的推板 back sloper
推土机的作业 dozer operation
推土机附件 bulldozer attachment
推土机刮铲 straight pusher-blade
推土机刮刀型刀片 dozer type blade
推土机刮土铲 bulldozer blade
推土机回填机 blade backfiller
推土机驾驶员室 environmental cab
推土机开沟槽(法) troughing
推土机领班 dozer boss
推土机轮胎 earthmover tire[tyre]
推土机平行作业法 blade to blade dozing
推土机前括铲 front pusher blade
推土机清除产量 dozing output
推土机清除铲刀 dozing blade
推土机上犁板 moldboard
推土机升举汽缸 dozer lift cylinder
推土机手 grade man
推土机松土器 bulldozer ripper
推土机提升油缸 bulldozer lift cylinder
推土机推土 plaster shooting
推土机推(土)铲 bulldozer blade
推土机稳定器 bulldozer stabilizer; dozer stabilizer
推土机撞锤 dozer ram
推土距离 dozing distance
推土犁 draw plough
推土器 soil pusher
推土设备 earthmoving equipment; dozer equipment
推土生产 production dozer
推土生产量 dozing production
推土拖拉机 tractor dozer; tractor shovel
推土挖掘机 dozer shovel
推土装载机 caterpillar loader
推土装载两用机 dozer-loader
推土作业 dozing
推推放大器 push-push amplifier
推托 dodge; evasion
推拖两用船 pusher tug; pushtow boat
推拖两用港作船 push-pull port tug
推挽变流器 push-pull inverter
推挽变压器 push-pull transformer
推挽叉式装卸车 push-pull forklift
推挽的 push-pull
推挽电路 push-pull arrangement; push-pull circuit; two-cycle scheme
推挽定心 push-pull centring
推挽放大 push-pull amplification
推挽放大方式 push-pull amplification system
推挽放大器 balanced valve amplifier; push-pull amplifier
推挽功率放大 push-pull power amplification
推挽激励 push-pull drive
推挽激励器 push-pull driver
推挽级 push-pull cascade; push-pull stage
推挽级联 push-pull cascade
推挽检波 push-pull detection
推挽检波器 push-pull detector
推挽接法 push-pull connection; push-pull connection method
推挽锯齿波 push-pull sawtooth wave
推挽力 pushing and pulling forces; traction thrust
推挽连接 push-pull connection
推挽偏转 push-pull deflection
推挽坡的经济 economy of pusher-grade
推挽坡度 assisting grade; helper grade; pusher grade
推挽扫描 push-pull sweep
推挽设备 push-pull equipment
推挽射频放大器 push-pull radio frequency amplifier
推挽声迹 push-pull sound-track
推挽式 push-pull mode; push-pull type
推挽式倍频器 push-pull multiplier
推挽式变压器 push-pull transformer
推挽式铲运机 push-pull scraper
推挽式传声器 push-pull microphone
推挽式堆 push-pull reactor
推挽式放大器 differential amplifier; push-pull amplifier
推挽式光电管 push-pull photocell
推挽式话筒 push-pull type microphone
推挽式锯齿波形 push-pull sawtooth wave form
推挽式临界实验 push-pull-critical experiment
推挽式螺旋桨搅拌器 push-pull propeller stirrer
推挽式驱动器 push-pull driver
推挽式三倍倍频器 push-pull tripler
推挽式市郊客车 push-pull commuter car
推挽式输出放大器 push-pull output amplifier
推挽式送话器 push-pull type microphone
推挽式运算放大器 push-pull computer amplifier
推挽式中和 cross neutralization
推挽输出放大器 push-pull output amplifier
推挽输出级 push-pull output stage
推挽调制 push-pull modulation
推挽误差 push-pull error
推挽系统 push-pull system
推挽信号 push-pull signal
推挽振荡 push-pull oscillation
推挽振荡器 push-pull oscillator
推挽制 push-pull system
推挽中和法 push-pull neutrodyne method
推挽驻极体换能器 push-pull electret transducer
推挽装置 push-pull arrangement
推网 push net
推下 depress; push-down
推下破玻璃 pushing down the cullet
推削 broaching; push-cut
推销 boost sales; brainwash; canvassing; promote sale; promotion; push-piece; sales promotion; tout
推销部 market development-division
推销产品 promote the sale of products
推销成本 allocated cost; sales promotion cost; selling cost
推销成功 crack a prospect
推销的 promotive
推销地区 market
推销定价 promotional pricing
推销费用 distribution cost; marketing cost; marketing expenses; promotion expenses; sales promotion expenses; selling cost; selling expenses
推销工程师 sales engineer
推销鼓励佣金 push money
推销机构 distribution mechanism
推销集团 selling group; selling syndicate
推销计划 sales promotion planning
推销刊物 sale literature
推销力量 sales force
推销联营组合 selling syndicate
推销人员意见综合法 sales-force-composite method
推销商 promoter
推销商品的信件 call letter
推销术 salesmanship; sell
推销锁 push bolt
推销铁 repelling lug
推销途径 channel of promotion
推销网 sales promotion network
推销系统 system of sales promotion
推销线索 referral leads
推销辛迪加 selling syndicate
推销学 marketing
推销研究 promotional study
推销样品 selling sample
推销业务 business of distributor; salesmanship
推销员 bagman; canvasser; dealer aids; dealer helps; detailer; detail man; merchandising salesman; missionary; promotion worker; roundsman; salesman; salesperson
推销员差旅及交际费 salesmen's traveling and entertainment expenses
推销员许可证 pocket card
推销员佣金 push money
推销战略 sales strategy
推销政策 promotion policy
推卸器 rejector
推卸器导轨 <铲土机> guide rail of ejector
推卸器导轨轮 <铲土机> ejector guide roller
推卸器防溢栅 <铲土机> overflow guard of ejector
推卸器托轮 <铲土机> carrier roller of ejector
推卸器液压缸 <铲土机> ejector cylinder
推卸器引导轮 <铲土机的> guide roller of ejector
推卸器支重滚轮 carrier roller of ejector
推卸式铲斗 ejector bucket
推卸式堆垛机 push-off stack
推卸责任 buck-passing; shift off
推卸责任者 buck-passer
推心置腹 heart to heart
推选标商名单 restricted list of bidders
推雪机 snow grader
推压 bulldoze
推压操纵杆 crowd lever
推压动作 racking; thrust motion
推压方法 <挖掘机> crowding
推压盖 thrust cap
推压工作 <挖土机> racking
推压管系统 pipe-forcing system
推压荷载 racking
推压活塞 crowd ram
推压精加工 press finish(ing)
推压卷筒制动器 <挖掘机> drag brake
推压力 thrust pressure; crowding force <挖掘机的>
推压力承受面 abutment cheek
推压连接的管道 push-on joint pipe
推压链轮 crowd sprocket
推压能力 crowd capacity
推压千斤顶 push-jack
推压式铲运机 push-type scraper
推压式滑脂枪 push-type grease gun
推压式水阀 push valve
推压腿架 <凿岩机的> raising pusher leg
推压挖掘 crowd
推压弯曲 racking moment
推压液压缸 <挖掘机> crowd cylinder
推压运动 <挖土机斗柄的> thrust motion
推压作用 thrusting action
推延 shunt
推延寻址 deferred addressing
推样器 sample extruder
推液式条形泡罩塔板 Thorman tray
推移 shoving; traction; wag
推移波 translation wave; wave of translation
推移齿轮 removing gear
推移荷载 tractional load
推移机构 thrust gear
推移力 dragging power; tractional force; tractive force
推移流 <推移河床上土砂> traction
推移能力 traction capacity; tractive capacity
推移式滑坡 slumping slide; translational slide
推移式输送机 push-type conveyer[conveyor]
推移式水磨石机 plate terrazzo grinder
推移速度 tractional velocity

T

推移梯 troll(e)y ladder
推移质 bed material load; bed sediment; bottom bed load; contact load; rolling bed loading; sediment load; solid load; tractional load; traction transport; bed load <河床上的粗沙和石砾>
推移质搬运 bed-load transport
推移质比率 specific bed load
推移质采样器 bed-load sampler; bottom load sampler
推移质测定 bed-load measurement
推移质测验 bed-load measurement
推移质底沙 debris bed load; debris bed sand
推移质动床模型试验 movable-bed model with bed load
推移质公式 bed-load formula
推移质轨迹 bed-load trajectory
推移质函数 bed-load function
推移质河工模型 river engineering model with bed load
推移质集沙槽 bed-load trap
推移质拦截井 bed trap
推移质泥沙 bed-load sediment
推移质泥沙量 silt flux of bed load discharge
推移质泥沙模型试验 bed-load model test
推移质泥沙形成 bed production
推移质取样器 bed-load sampler
推移质沙波法测验 dune tracking
推移质输沙量 bed-load discharge; bed material discharge; bed sediment discharge; transport capacity of bed load
推移质输沙率 bed-load discharge; bed-load rate; rate of bed load discharge; rate of transportation of bed load
推移质输沙率测验 bed-load discharge measurement
推移质输沙强度 bed load intensity; intensity of bed-load transport
推移质输沙试验 bed-load transport test
推移质输送 bed-load transport
推移质输送比(率) specific bed load transport
推移质输送量 bed-load discharge; bed material discharge
推移质输送量测定仪 bed load transport meter
推移质输移 bed-load transportation; transportation of bed load
推移质输移量 bed-load discharge; bed sediment discharge; discharge of solids; transport capacity of bed load
推移质输移强度 bed load intensity; bed-load transport intensity; intensity of bed-load transport
推移质输移试验 bed-load transport test
推移质移动带 bed-load moving strip
推移质运动 bed-load motion; bed-load movement
推移质运动层 bed layer
推移作用 traction
推运螺旋 screw auger
推运螺旋与壳间间隙 screw housing clearance
推运螺旋轴 auger shaft
推爪 <与棘轮相配合的> dog; pawl; push pawl
推爪驱动轮 pusher-click driver
推爪止挡 pusher-click banking stop
推者 pusher
推针 push pin
推针排式针梳机 push-bar gill
推针弹簧 tickler spring
推针装置 needle pushing-up device
推枕 thrust block
推阵排序 heap sort
推助公共车辆乘客上车的人 meat packer
推装的铲斗 heaped bucket
推装式铲运机 push-loading scraper
推装式拖拉机 push-loading tractor
推阻力 thrust drag
推座 push block

颓

颓废派 decadent school
颓废派型 decadence style
颓废派艺术 decadent art
颓废派艺术形式 decadent style
颓积土 collapse soil

腿

腿的劳动保护垫 leg protector
腿靠 calf rest; leg rest
腿圈杆 holding-down rod
腿饰 leglet

退

退拔扩口管 taper
退保金额 cash surrender value; surrender value
退保通知 notice of cancellation
退保注销 cancellation
退币电键 refund key
退币机构 <投币式电话机> refund mechanism
退币口 coin refund
退避 back-off
退变重结晶作用 degrading recrystallization
退步 regression; retrogression; setback; relapse
退步茬 racking; racking back
退步的 retrogressive
退藏式阴极 recessed cathode
退槽让航 to give way to navigation; to move away from navigation channel
退层 backstep
退插(值) backward interpolation (formula)
退差(值) backward difference (method); receding difference
退潮 ebb (tide); falling tide; fall of tide; go-out; outgoing tide; reflowing; refluence; refluent tide; reflux; tidal fall; tide ebb
退潮波 subsidence wave
退潮持续时间 duration of fall
退潮的 refluent
退潮航道 ebb channel; ebb tide channel
退潮回流比 ebb ratio
退潮历时 duration of ebb (current); duration of fall
退潮流 ebb current; ebb tide current; outgoing ebb; rip current; sea push; ebb flow
退潮流间隔时间 ebb interval
退潮末 last of ebb
退潮平均流量 mean flow at outgoing flow
退潮期间 period of decreasing tide
退潮强度 ebb strength
退潮曲线 recession curve
退潮三角洲 ebb tidal delta
退潮三角洲沉积 ebb tidal deltaic deposit
退潮时船搁浅 sewed up
退潮时的静止水位 slack-water on the ebb
退潮时间 retard of tide
退潮时间间隙 ebb interval
退潮时露出的海滩 dry beach
退潮水道 ebb channel
退潮水量 volume of ebb; volume of water discharging on the ebb tide
退潮涌浪 ebb surge
退潮闸门 ebb gate; ebb tide gate
退出 back-out(of); resign; voidance; withdrawal; withdrawing
退出侧 receding side
退出处理程序 exit handle
退出穿孔机 <铆钉> back-out punch
退出队列 dequeue
退出俯冲 pull out
退出呼声模型 exist-voice model
退出螺旋 recovery
退出钎杆 withdrawal
退出曲线 ejection curve
退出扰动区 recovery from a disturbance
退出时间 post-set time
退出使用 out-of-service
退出示像 pull aspect
退出速度 rate of withdraw
退出效率 ejection efficiency
退出运行 decommissioning
退出终端 logoff
退磁 degaussing; de-magnetize; de-magnetization
退磁场衰减速率 decay rate of demagnetizing field
退磁磁场 coercive field
退磁电阻 de-magnetizing resistance
退磁扼流圈 degausser
退磁法冷却 magnetic cooling
退磁方法 de-magnetization methods
退磁力 de-magnetization force; demagnetizing force
退磁设备 degaussing apparatus; demagnetizing apparatus
退磁系数 de-magnetizing factor
退磁线圈 degaussing coil; demagnetizing coil
退磁因数 de-magnetizing factor
退磁装置 de-magnetizer; depolarizer [depolorizer]
退磁作用 demagnetising effect
退弹器 rammer
退刀 retract
退刀槽 escape; recess; relief; tool escape; tool recess
退刀动作 tool backlash movement
退刀伤痕 tool withdrawal mark
退刀纹 run-out; vanishing of thread
退佃 cancel a tenancy
退钉冲孔器 backing-out punch
退钉器 backing-out punch
退订货 countermand
退动 deactuate
退动凸轮 relief cam
退镀 deplate
退废会计 retirement accounting
退废损失 loss on retirement
退废政策 <指折旧> retirement policy
退费手续 procedure for refund
退覆 offlap; regressive overlap
退覆层序 offlap sequence
退覆式分岔 offlapping splitting
退耕还林 return farmland to forests; returning land for farming to forestry
退股 withdrawal shares
退关 shut out
退关单 shut-out memo
退关的 short-shipped
退关货物 shoutouts; shut-out cargo; shut-out goods
退关货物清单 shut-out memo
退光剂 <涂料的> flattening agent; flatting agent
退焊 backhand welding
退焊法 back welding
退洪 flood decline; flood fall; flood recession
退后 back-out(of); fall back
退后堤防 setback levee
退化 atrophy; catagenesis; degeneracy; degradation; degrade; degrading; deteriorate; regress; retrograde; retrogression
退化变态 regressive metamorphosis
退化变质 deterioration; regressive metamorphism
退化变质岩 diaphthorite
退化变质作用【地】 diaphthoresis; retrogressive metamorphism
退化草地 deteriorated grassland
退化程度 deterioration level
退化次数 degree of degeneracy; order of degeneracy
退化单元 singular element
退化的 degenerated; degenerative; degraded; obsolete; regressive; retrograde; retrogressive; rudimentary; vestigial
退化点集 set of degeneracy
退化度 degree of degeneracy
退化二次曲面 degenerate quadric
退化二次曲线 degenerate conic; improper conic
退化二次型 degenerate quadratic form
退化分布 degenerate distribution
退化分布函数 degenerate distribution function
退化根 reduced root
退化故障 degradation failure
退化过程 degenerate process
退化函数 degenerate function
退化河流 defeated stream
退化黑钙土 degraded chernozem
退化恢复力 deteriorating restoring force
退化回归 degenerate regression
退化机理 deterioration mechanism
退化简并 accidental degeneracy
退化结晶作用 annealing crystallization
退化解 degenerate solution
退化矩阵 degenerate matrix; singular matrix
退化可行解 degenerate feasible solution
退化控制 degeneration control
退化类型 degenerated form
退化临界点 degenerate critical point
退化率 rate of degradation
退化判别式 criterion of degeneracy
退化情况 case of degeneracy
退化曲线 decay curve; degenerated curve
退化事故 degradation failure
退化试验 regression test
退化适应 regressive adaptability
退化水流 degenerate flow
退化速率 deterioration rate
退化算子 degeneracy operator
退化碳氢化合物 degrading hydrocarbons
退化特性 degradation characteristic
退化特征 degenerative character; degradation characteristic
退化突变 retrogressive mutation
退化土壤 degenerated soil; degraded soil; degrading soil
退化位形 degenerate configuration
退化系数 degeneration factor; degree of degeneration
退化现象 retrogressive phenomenon
退化线性系统 degenerate linear sys-

tem
退化型 degenerated type; involution form
退化型冻土 degenerative type frozen soil
退化型失效 degradation failure
退化性状 degenerative character
退化序列 degenerate series
退化旋转潮波系统 degenerate rotary tidal waves system
退化伊利石 degraded illite; stripped illite
退化因数 degeneration factor
退化振荡模 degeneration mode
退化正态分布 degenerate normal distribution
退化滞变结构 deteriorating hysteresis [hysteretic] structure
退化种 regression species
退化状态 degenerate state
退化组织 degenerate tissue
退化作用 degeneration; deterioration; regression
退还 kick back; refund
退还保险费 cancelling returns; return premium
退还保证金 deposit(e) releases
退还材料 store returned
退还抽样 sampling with replacements
退还的押金 deposit(e) released
退还缴纳的部分所得税 refund of a part of the income taxes already paid
退还金额 amount retroceded
退还全部票价 refund of fare
退还停泊保险费 laying-up returns
退还已缴税款 drawback for duties paid
退还以前年度税款 refund retroactive to the past taxable years
退还溢缴税款 refund of overpayment of tax
退还佣金 return commission
退还准备金 deposit(e) released
退辉 decalescence
退回 back; put back; recede; rejection; retraction; withdraw; release
退回保险费 return premium
退回保险费条款 return clause
退回材 culls
退回材料 released material; returned material
退回材料厂 return to store
退回材料的处理 handling of released materials
退回超额进口税证书 over-entry certificate
退回贷款 return loan
退回担保 backbond
退回的拒付票据 return item
退回股利 back dividend; rescission of dividends
退回股票 stocks returned
退回关税 <对进口后又出口的材料> drawback
退回货物 returned cargo; returned goods
退回计时法 flyback timing
退回库存物资 stores returned
退回款 refund
退回命令 backspace command
退回文件 backspace file
退回线 line or retirement
退回信件 back letter
退回已付税金 drawback for duties paid
退回预付款 restitution of advance payment
退回原物诉讼 action for restitution
退回证券 backbond
退回支票 check returned; returned check
退汇汇票 reexchange bill; return bill
退汇(要求) reexchange
退火 bakeout; blow-off; drawing; softening; temper(ing); annealing
退火玻璃 annealed glass
退火薄钢板 annealed sheet steel
退火不良的 badly annealed
退火不完全 annealing slack
退火材料 softening material
退火敞炉 annealing hearth
退火车间 annealing room
退火程度 annealing grade
退火程序 annealing schedule
退火处理 annealing treatment
退火的 annealed
退火的铜 annealed copper
退火的应力释放 stress-relief annealed
退火点 annealing point; annealing temperature
退火辅助工 lehr assistant; lehr end serviceman
退火钢 annealed steel; tempered steel
退火钢丝 annealed wire; stone dead wire
退火钢线 annealed steel wire
退火工 lehr attendant; lehr man; lehr minder; lehr operator
退火工段 annealing room
退火工序 annealing operation
退火罐 annealing pot
退火硅钢线 annealed silicon steel wire
退火过程 annealing process
退火焊波 annealing welding wave
退火焊道 normalizing pass
退火焊条 annealing welds
退火后的抗拉强度 annealed tensile strength
退火后的拉力 annealed tensile
退火后的拉力强度 annealed tensile strength
退火后无应力 stress-free annealed
退火浇铸 annealed casting
退火角砾岩 annealing breccia
退火绞铜电缆 stranded annealed copper cable
退火结晶 annealing crystallization
退火金属 annealed metal
退火金属线 annealed wire
退火裂(纹) annealing crack; fire cracking
退火炉 annealer; annealing furnace; annealing kiln; annealing oven; drawing furnace; leer; lehr; lier; stress-relief furnace; stress-relieving furnace
退火炉出炉工 emptier
退火炉跨 annealing bay
退火炉送料装置 rack stacker
退火铝 annealed alumin(i)um
退火铝线 annealed alumin(i)um wire
退火孪晶 annealing twin
退火黏[粘]结板 annealing sticker
退火盘 leer pan
退火破裂 lehr crack
退火破损量 lehr breakage
退火曲线 annealing curve
退火热处理 annealing heat treatment
退火(软)铜 annealed copper
退火色 annealing colo(u)r
退火石墨 graphite aggregate
退火收缩 firing shrinkage
退火双晶 annealing twin
退火丝 annealing wire
退火酸洗作业线 anneal pickle line
退火隧道 annealing tunnel
退火铜线 annealed copper wire
退火温度 annealing point; annealing temperature
退火温度范围 annealing range; annealing region; annealing temperature range
退火温度上限 upper annealing temperature
退火线镀锌钢丝 stone wire
退火箱 annealing box; annealing can; annealing container; seggar
退火效应 annealing effect
退火窑 annealing furnace; annealing kiln; leer; lehr
退火窑输送网带 lehr belt; lehr mat
退火窑送料装置 lehr loader; rack stacker; stacker
退火油 annealing oil
退火制度 annealing schedule
退火重结晶作用 annealing recrystallization
退火周期 annealing cycle
退火铸件 annealed casting
退火铸铁 annealed cast-iron
退火状态 annealed condition
退火组织 annealed structure
退火作业线 annealing line
退伙 retirement of a partner; withdraw from partnership
退伙人 retiring partner
退伙人股份清偿 settlement with retiring partner
退货 cancel the order; goods rejected; goods returned; rejection of goods; return of goods; sales returns; take back goods; vendor charges backs
退货比率 sales return ratio
退货成本汇总表 summary of cost of returned goods
退货冲减 credit for returned goods
退货单 credit note
退货单据 returned purchase invoice
退货的销售折扣 discount on returned sales
退货发运单 return shipping order
退货费用 back goods freight
退货及销货折让日记账 sales return and allowance journal
退货及折让 return and allowance
退货及折让簿 return and allowance book
退货记录簿 sales returns book
退货清单 merchandise credit slip
退货特约条款 rejection clause
退货运费 back freight; back goods freight; return cargo freight
退积层序 retrograding sequence
退积金 retirement allowance
退积型 retrograding pattern
退积型生长层序 retrograding growth sequence
退极化 depolarize; depolarization
退极化场 depolarization field
退极化剂 depolarizer[depolorizer]
退极化谱带 depolarization band
退极化因子 depolarization factor
退极性 depolarization
退减离解 retrograde dissociation
退减作用 retrogradation
退浆 desize; desizing
退降温度计 retreater
退降效应 effect of degrading
退缴税 back tax
退进 setback
退进房屋 setback building
退卷 unreeling; unwind(ing); wind(ing) off
退卷的 unrolled
退卷架 reel-off stand
退卷装置 uncoil-stand; unwinding installation
退壳 ejection
退壳槽 extractor groove
退壳孔 ejection port
退壳器 case discharger; ejector; extractor
退壳器支框 extractor pivot
退壳器柱塞 extractor plunger
退壳器爪 extractor claw
退库 cancelling stocks
退款 drawback; money returned; refund(ing)
退款保证 money-back guarantee
退款单 credit note
退款和回扣 refunds and rebates
退款票据 refund check
退浪 backrush; undertow
退浪冲刷 backwash
退离 recession
退料 material returned; return of material; stores returned
退料报告 material return report
退料单 material credit slip; material returning slip; store credit; stores returned note
退料汇总表 summary of materials returned
退料器 stripping attachment
退料日记账 material returned journal
退料入库 returning materials to storeroom
退料账 return materials journal
退流 undertow; undertow current
退流波痕 regressive ripple
退落 ebb
退落地下水位 receding water table
退落潜水面 receding water table
退落泉 ebbing spring
退落水流 degenerate flow
退黏[粘]剂 reducer
退耦 decouple; decoupling
退抛双锚 dropping moor
退赔 kick back; pay compensation; restitute; return one has unlawfully taken
退赔费 fee for refund
退赔品 returned goods
退票 dishono(u)red bill; return a ticket
退票通知 note of dishono(u)r
退期票据 bill at sight
退铅 de-lead(ing)
退钱 drawback
退圈轮 push back wheel; push-down wheel
退圈三角 needle clearing cam; push-back cam
退却 fall back; retreat; retrocede
退却半周期阶段 <冰川> receding hemicycle
退却速度 <冰川的> retreat velocity
退却套 <滚动轴承的> taper clamping sleeve
退让性 collapsibility
退绕 backing-off; reeled off; unreel; wind(ing) off
退绕长度 unwinding length; winding-off length
退绕机 unwinder
退绕装置 unwinding installation; winding-off installation
退溶 de-solvation
退入界线 setback line
退水 back-out(of); recession; streamflow depletion; water recession
退水道【给】 wasteway
退水段 falling curve; falling limb; falling segment; wasteway section
退水方程 equation of recession; recession equation
退水过程线 depletion hydrograph; re-

cession hydrograph
退水开始期 beginning of fall
退水浪花 backwash
退水浪花痕 backwash marks
退水流速 recession velocity;velocity of retreat <水工建筑物下游的>
退水期 falling(flood) stage;recession period
退水曲线 depletion curve; falling curve;flow recession curve;recession curve; recession hydrograph; recession limb; regression curve; lowering limb
退水曲线方程 equation of regression line
退水渠 escape canal;tail channel;waste canal;waste channel
退水时间 recession time;time of fall; time of recession
退水速度 recession velocity; retreat velocity
退水土地的所有权 reliction
退水线 line of regression; regression line
退水堰 waste weir
退水闸 escape
退水闸门 gate escape;waste sluice gate
退税 back tax; drawback; drawback for duties paid;duty drawback;refund of duty;refund of tax;tax rebate;tax refund;tax reimbursement
退税补偿 refund offset
退税计划 negative tax plan
退税凭单 debenture
退速 back speed
退缩 blench;hold-back;recoil;retroaction;setting back
退缩波 retrogressive wave
退缩的 fall back
退缩分枝 receding branch
退缩固溶度 retrograde solid solubility
退缩海岸 retrograding coast
退缩可能性 fallback possibility
退缩碛 recessional moraine
退缩时间 recedence time
退缩式分岔 shrinking splitting
退缩线 <建筑物的> receding line
退缩行为 withdrawal behavio(u)r
退缩性 retractility
退缩性溶解度 retrograde solubility
退缩檐口 receding cornice
退缩运动 narrowing movement
退缩照明装置 <装在顶棚线之上的照明装置> regressed luminaire
退缩砖砌阶 brick offset
退缩装置 retraction device
退滩 from a rapid;retrocession
退滩锚 hauling-off anchor
退滩装置 pago stick
退弹簧 extractor
退套楔 drift
退位 abdicate;back space
退伍金 gratuity
退伍军人财产免税权 veterans ' s tax exemption
退伍军人管理局 Veterans Administration
退伍军人管理局住房贷款 veterans administration loans
退伍军人管理局住房抵押贷款 veterans administration mortgage
退吸热 heat of desorption
退吸作用 desorption
退息 interest rebate
退席 walk out
退下 back-off
退向变质 retrogressive metamorphism
退向变质作用【地】 retrometamorphism
退楔 drift bolt
退卸工具 withdrawal device
退卸套 adapter sleeve; withdrawal sleeve
退心接头 core plunger
退行 backing movement;back-up;back-up movement;backward movement; recession;regression
退行波 retrograde wave;retrogressive wave
退行示像 backup aspect
退行速度 velocity of recession
退行信号 backing signal;backup signal
退休 retire(ment)
退休表 retirement table
退休的 emeritus
退休方针 retirement policy
退休工人 retired worker
退休雇员 retired employee
退休后工作 sunlighting
退休基金 superannuation fund
退休家庭 retirement home
退休金 pension; retired pay; retirement allowance; retirement benefit;retirement pay;retirement pension; superannuation; superannuation benefit; superannuation payment;retirement annuity
退休金基金 retirement fund;superannuation fund
退休金计划 pension plan;retirement pension plan
退休金计算基础 pension basis
退休金授予 pension vesting
退休金提成 pension contribution
退休金专款 pension fund
退休金准备 reserve for retirement allowance; retirement allowance reserves
退休年龄 age at retirement; age at withdrawal;age limit;age of retirement;pensionable age;retirement age
退休年龄极限 limit of age for retirement
退休年龄前丧失劳动力 disablement before retirement
退休期 date of retirement
退休收入 retirement income
退休收入保险 retirement income insurance
退休优惠 retirement privilege
退休者 retiree
退休者居住区 retirement communities
退休职工 retired staff
退休制度 retirement system
退休住处 retirement home
退休总津贴 total retirement benefits
退押 return a deposit;return deposits to tenants in the land reform;returning security money
退岩芯装置 sample extruder
退曳线 trail line
退移 retire
退役 mothball;retire from service
退役船 ship out of commission
退役的 ex-service;on the shelf
退役废物 decommissioned waste
退役旅馆 retirement hotel
退役旅社 retirement hotel
退约 denunciation
退约条款 denunciation clause
退约行为 act of denunciation
退褶合 deconvolution
退职 quit office;resignation;superannuation;unseating
退职的 emeritus
退职金 gratuity; retirement benefit; severance pay;superannuation
退职者 retiree
退装 <航运> shut out
退装货物 back loading;shut-out cargo
退子钩 extractor
退租 eviction;throw a lease
退租检验 off-hire survey;redelivery survey

蜕变 spall(ing);splitting;transmutation;transmute

蜕变常数 destruction constant
蜕变电压 disintegration voltage
蜕变电子 decay electron
蜕变湖 molt lake
蜕化 degeneracy
蜕晶质 metamict
蜕皮 sloughing;acdysis <昆虫>

褪光 deluster(ing);mat(te);matting <使玻璃、金属等表面无光泽>;mattness <油漆的>

褪光剂 delusterant
褪光油 flatting oil
褪绿 chlorosis
褪色 colo(u)r degradation;colo(u)r deterioration;decolo(u)rize;decolo(u)rment; fade; fading; decolo(u)ration; discolo(u)r(ation); discolo(u)rization;off-colo(u)r; receding colo(u)r
褪色带 bleached zone
褪色的 bleaching-out;washed-out <照片等>;fugitive
褪色的颜色 faded colo(u)r
褪色或变黄 lose their colo(u)r or turn yellow
褪色计 fad(e)ometer
褪色剂 decolo(u)rant;decolo(u)rizing agent
褪色灵 eradicator
褪色毛 cloudy wool
褪色能量灵敏度 bleaching energy sensitivity
褪色黏[粘]土 discolo(u)red clay
褪色染料 fugitive dye
褪色试验 discolo(u)ration test
褪色熟料 discolo(u)red clinker
褪色效应 bleaching effect
褪色颜料 fugitive pigment
褪色作用 decolo(u)rization;fading; bleaching
褪彰 desaturation

吞并 absorb;merger

吞加管 tungar;tungar tube
吞加管整流器 tungar rectifier
吞加整流管 tungar bulb
吞肩 relish
吞喀姆硅黄铜 Tungum
吞没 engulf
吞耐特 <乙酸乙酯纤维素塑料> Tenite
吞食的 phagotrophic
吞噬体(微生物) phagome
吞吐 turnover
吞吐量 cargo-handling capacity;handling capacity; loading and unloading capacity;throughput;throughput capacity;turn volume;volume of freight handled; volume of incoming and outcoming freights
吞吐量不平衡系数 unbalance coefficient of cargo handled at the port
吞吐量大的 large volume
吞吐量大港口 large volume harbo(u)r
吞吐量等级 through put class
吞吐率 throughput rate
吞吐能力 handling capacity;throughput;throughput capacity
吞下 swallow

饨钻模【机】 drilling

豚草 hog weed

豚草病 ambrosia
豚脊 hogback

臀部 breech;buttock;hip

臀的 sciatic

托 <真空压强单位> torr

托铵云母 tobelite
托把 <一种冷加工钢筋> Torbar
托拜厄斯酸 Tobias acid
托板 bearer plate; bearing plate; blade; brace; bracket; capping plane;cap plate;carrier plate;cope plate;corbel plate;cradle;fascia[复fa(s)ciae/fa(s)cias];follower; holding plate;lay board;pallet;saddle strap;stool(ing);tie bar;drop panel <无梁楼盖的柱顶>
托板搬运车 pallet
托板叉车 pallet fork
托板化 palletisation
托板化货物 palletized cargo
托板化机械 palletizing plant
托板化运送 pallet handling
托板架 carrier
托板连接 table steel plate connection
托板式干燥器 pallet drier[dryer]
托板式集装运输 ferry system;pallet system
托板式集装运输系统 pallet system
托板提升机 pallet elevator
托板装卸系统 pallet system
托板装载的袋装水泥 palletized bag; palletized sack
托苞 palea
托杯形电动机 drag-cup motor
托贝莫莱特胶体 <混凝土中起胶凝作用的胶体> Tobemorite gel
托彼卡 <一种粒径在1/2英寸以下的细粒沥青混凝土> Topeka
托彼卡(沥青混凝土)混合料 Topeka mixture
托臂 bracket; bracket arm; corbel; lifter;support;supporting arm
托臂滴水石 label-corbel table
托臂拱顶 corbel vault
托臂荷载 bracket load
托臂梁 hammer beam; haunched beam
托臂梁屋顶 hammer-beam roof
托臂模板 pannier
托臂起拱大梁 corbel arched girder
托臂穹隆 corbel vault
托臂式支柱 bracket-like column
托臂托座【建】 bracket support
托臂弯曲大梁 corbel curved girder
托臂圆顶 corbel cupola
托臂圆屋顶 corbel dome
托臂支承门楣 shouldered arch
托臂支柱 bracket pole;bracket post; bracket support
托臂支座 bracket support
托臂(座)载荷 bracket load
托宾黄铜 Tobin brass
托宾青铜 Tobin bronze
托宾通风管 Tobin's tube
托柄扁担 bracket arm

托柄尾部 tailstock
托病旷工 sickout
托波尔间冰期 Tobol interglacial stage
托钵僧团教堂 mendicant order church
托持压力 backing pressure
托窗梁 breast summer
托词 plead
托带 send through others
托带轮 carrier wheel; return roller; top idler; track-carrier roller; upper track wheel
托带轮导轮 jockey wheel
托刀口角度 blade angle
托端梁 beam-supported at both ends
托儿室 nursery room
托儿所 baby farm; child care center [centre]; child care home; child care institution; children's nursery; creche; day nursery; nursery; nursery school; nursing home; public nursery; school nursery
托儿中心 child care center[centre]
托尔顿阶 <晚中新世>【地】Tortonian
托尔钢 <一种螺纹圆钢筋> Torsteel
托尔克建筑 Toltec architecture
托尔锡釉精陶 Toul faience
托法尼 imipramine hydrochloride; tofranil
托范风暴 tofan
托非特镍铬电阻合金 tophet alloy
托福轮板 <为一种绝缘板> Torfoleum
托付 apply for remittance; authority to pay; commend; commit to the hands of; entrust; give in charges
托付单 payment order
托付人 applicant for payment; applicant for remittance
托付信用证 domiciled credit; escrow credit; escrow letter of credit
托付易货贸易 escrow barter
托杆 die-pin; push-off pin
托杠棘轮系统 pallet system
托购清单 indent invoice
托管 deposit(e); entrust; trust; trusteeship
托管材小车 pipe buggy
托管港 trust port
托管(领)地 trust(ee) territory
托管领土 trust(ee) territory
托管人的职责 trusteeship
托管制度 trusteeship
托轨 <钢轨伸缩接头的> receiving rail
托轨梁 rail bearer; track joist
托辊 carrier roller; carrying idler; guide roller; idler; idler roller; pulley; roller; supporting roll(er)
托辊支承 idler stand
托辊支架 roller support
托辊座 roller seat
托滚 idler pulley
托滚架 roller carriage
托滚座 idle carrier
托环 bearing ring
托环压板 <暖气联结器> bearing ring holder
托换 underpinning
托换工程 underpinning; underpinning engineering
托换基层 <从下面加固基层> underpin
托换基础 <从下面加固基础> pin-up; underfoot; underpin(ning foundation)
托换基础法 underpinning
托换基础工程 underpinning work
托换基础用的支柱 springing needle
托换基础支柱 springing needle
托换基础桩 jacked pile; underpinning pile
托换技术 underpinning technology
托换(支)柱 <隧道坑道> underpinning post
托换柱沉降 settlement of underpinning pile
托换柱基 underpinning to column foundation
托换桩 underpinned pile
托换座墩 underpinned pier; underpinning
托簧 bearing spring
托灰板 <泥工的> hawk; fat board; floating scum-board; mortar-board
托汇 apply for remittance
托基 underpinning
托基支柱 springing needle
托基桩 underpinned pile
托架 bracket(ing); bracket bearing; bracket block; bracket mount; bracket support; ally arm; back stop; bearer; bragger; cantilever; carr; carriage; carrier; carrier frame; chair; chariot; console; corbel brick; cradle; cross arm; gudgeon; jacking frame; lug support; mounting base; pedestal; retriever; saddle strap; seat; slab back; stool(ing); supporting block; supporting bracket; supporting truss; poppet <车床等的>; feed holder; guide shell <凿岩机的>
托架板 bracket plaque
托架臂 bracket arm
托架导板 carriage guideway
托架导轨 carriage rail
托架灯 bracket lamp
托架垫块 bracket block
托架定位 carriage positioning
托架返回符号 carriage return character
托架风扇 bracket arm fan
托架回车符号 carriage return character
托架及滚轴 cradle and roller
托架夹 bracket clamp
托架角钢 shelf angle
托架角铁 bracket angle
托架绞车 bracket winch
托架脚 bracket foot
托架接合 bracket connection
托架绝缘器 bracket insulator
托架空推键 carriage space key
托架控制带 carriage control tape
托架控制字符 carriage control character
托架瞄准具 pedestal sight
托架上盖 yoke cap
托架式导堤 bracket leading jetty
托架式脚手架 bracket scaffold(ing)
托架式空调器 console air conditioner
托架提升机 tray elevator
托架铁件 bridle iron
托架喂料机 cradle feeder
托架系统 rack system
托架销 bracket pin
托架小拱 hance
托架信号 bracket signal
托架信号机【铁】bracket signal
托架悬挂式脸盆 bracket-hung wash basin; bracket-hung wash bowl
托架压盖 frame gland
托架照明装置 bracket lighting fixture
托架支承 bracket bearing; mechanical positioner
托架支座 bearer carrier; bracket support
托架轴承 bracket bearing; pedestal bearing; pedestal-type bearing
托架轴承合金 bracket metal
托架轴承式 pedestal bearing type
托架柱 bracket column; bracket mast; bracket post; cantilever mast; cantilever post; bung of setters <窑具>
托架装置 bracket system
托架组件 bracket assembly
托架钻床 bracket-drilling machine
托架座 bearer carrier
托肩 corbel
托脚 diagonal strut
托脚隔电子 base insulator
托脚绝缘子 base insulator; stand-off insulator
托阶式山墙 gable with corbel steps
托卷辊 cradle roll
托卡 holderbat
托坎廷斯河 Tocantins River
托扛运输系统 pallet system
托克思螺丝刀的旋凿套筒 screwdriver socket for Torx screw
托块 pillow
托拉斯 trust
托拉斯港 trust port
托莱都 <一种可折叠钢架> Toledo
托兰 <一种近程无线电定位系统> Toran
托兰法 Toran method
托勒玫纪 ptoleruian
托勒玫庙宇 Ptolemaic temple
托雷斯海峡 Torres Strait
托里东砂岩【地】Torridonian sandstone
托里切利定理 Torricellian theorem; Torricelli's theorem
托里切利定律 Torricellian law; Torricelli's law
托里切利公式 <流体从小孔流出,受其黏[粘]性的影响,并按 $2gh$ 算出,g 为重力加速度,h 为水深> Torricelli's formula
托里切利管 Torricellian tube
托里切利气压计 Torricellian barometer
托里切利真空 Torricellian vacuum
托链轮 top tread roller; upper roller; chain carrier roller
托梁 backbar; bearer(bar); boot; bracket; bridle; carrier bar; corbel beam; crosshead; downstand beam; jack beam; joist; spandrel beam; spandrel girder; transfer girder; trim(med) joist; trimmer(beam); trimmer joist; trimming joist
托梁承座 joist bearing
托梁垫板构件 trimmer plank unit
托梁拱 trimmer arch
托梁横木 ledger board
托梁脚手架 needle beam scaffold
托梁接头 <与小梁相接处> trimmer joint
托梁连接器 joist connector
托梁漏板 beam-supported bushing
托梁木屋架 <晚期哥特式的一种> hammer-beam roof truss
托梁挑檐 corbel table
托梁系统 system of binders and joists
托梁支承 joist bearing
托梁支承的格栅 tail joist
托炉 burner with stand
托氯铜石 tolbachite
托轮 carrying roll(er); roller carrier; supporting roll(er); support roller; carrier roller
托轮架 roller carrier bracket
托轮式离心成型机 roller-driven centrifugal mo(u)lding machine; rolling-impelled spinning machine
托轮支承装置 roller supporting mechanism
托轮轴 drum supporting roller shaft; roller carrier shaft; roller shaft
托罗斯-安纳托里亚山字型构造体系 Toros-Anatolia epsilon structural system
托马氏钢 Thomas steel
托马数 Thoma number; Thonm number
托马斯碱性炉渣粉 Thomas meal
托马斯生铁 Thomas iron
托马斯铸铁 Thomas pig-iron
托马斯转炉 Thomas converter
托模板 backing plate; flask board; stamping board
托木 bolster; pillow
托泥板 wooden board for taking mortar
托帕石 topaz
托帕特朗射频质谱仪 topartron
托盘 pallet; pan arrest; salver; saucer; tote pan; tray salver; underpan; guide shell <凿岩机的>
托盘搬运车 pallet truck
托盘包装 pallet packing
托盘叉车道 pallet fork lift track
托盘成组货物 palletized cargo
托盘船 pallet conveyance ship; pallet conveyance vessel; pallet ship
托盘船装卸方式 pallet ship system
托盘吊索 pallet sling
托盘化 palletize; palletisation[palletization]
托盘化储藏 palletized storage
托盘化货物 palletised load; palletized cargo
托盘化运输 palletized traffic; palletized transport
托盘化组合装运 palletized unit load
托盘货 pallet load
托盘货船 pallet ship
托盘货物 pallet cargo
托盘货装船机 pallet loader
托盘集装箱 pallet container
托盘架 pallet frame
托盘降送机 tray lowerer
托盘联营组织 pallet pool
托盘码垛 palletisation[palletization]
托盘码垛机 palletizer; pallet stacker
托盘热收缩包装 pallet shrink package
托盘升送机 tray elevator
托盘式叉齿 pallet fork
托盘式冻结法 tray freezing method
托盘式架桥 cable tray
托盘天平 counter-balance; table balance
托盘托架 bearer
托盘拖车 pallet trailer
托盘下管 pipe sinking by boiling
托盘箱 pallet box
托盘运输 pallet traffic; pallet transport
托盘运输船 pallet ship
托盘装运 palletizing; pallet shipment
托盘装载波 pallet stowage
托盘组织装机 palletiser[palletizer]
托皮卡式级配 Topeka grading
托皮卡型沥青混凝土 Topeka type asphaltic concrete
托片 buffer; buffer strip
托片间隙 <继电器的> buffer gap
托普列茨矩阵 Toeplitz matrix
托钎器 drill rod holder; drill steel retainer; drill steel support
托墙梁 breast beam; breast summer; bressummer; summer beam
托圈 backing ring
托什干构造结【地】Toxkan Dao tectonic knot
托饰 modillion
托收 apply for collection; collection;

collection of a bill;encashment
托收安排 collection arrangement
托收报告书 collection report
托收背书 endorsement for collection
托收不承付 collection without acceptance
托收成本 collection cost
托收承付 apply for collection and remittance;collection and acceptance
托收代理 agent for collection
托收单 collection memo;collection order
托收垫款 advance against collection
托收费用 collection expenses
托收服务 collection service
托收付款方式 payment by collection
托收负债 collected liability
托收跟单汇票 collection of documentary bill
托收光票 clean bill for collection; clean collection
托收汇兑款项 exchange for collection
托收汇票 bill for collection;collection bill;draft for collection
托收价值 value in collection
托收款项 bills sent for collection;due form bank;items sent for collection
托收票据 bill for collection;collect a draft;collection bill
托收票据费 collection fee
托收凭单 bill of collection;collection voucher
托收期 collection period
托收期票 bill for collection;collect a draft
托收人 applicant for collection
托收手续费 collecting[collection] commission;collection charges
托收通知书 advice for collection
托收统一规则 uniform rules for collection
托收委托费 advice for collection
托收委托书 advice for collection;collection letter;collection order
托收无承付 collection without acceptance
托收现款 collection of cash
托收项目 collection items;items sent for collection
托收信贷行 factoring
托收银行 remitting bank
托收债款 collection of debt
托收账 collection ledger
托收证书 collection permit
托收中款项 cash items in process of collection
托斯卡红 Tuscan red
托斯卡式【建】Tuscan style
托斯卡式拱 < 古罗马 > Tuscan arch
托斯卡式建筑 < 古罗马 > Tuscan architecture
托斯卡式楼板 < 古罗马 > Tuscan floor
托斯卡式柱 < 古罗马 > Tuscan column
托斯卡式柱基 < 古罗马 > Tuscan column base
托斯卡式柱帽 < 古罗马 > Tuscan capital
托斯卡式柱式 < 古罗马 > Tuscan order
托斯卡式柱型 < 古罗马 > Tuscan order
托斯卡柱 Tuscan column
托斯卡柱建筑 Tuscan order
托斯卡柱型 Tuscan style
托斯科水泥 < 一种防潮层水泥 > Toxement
托索轮 rope support sheave
托弹簧 magazine spring
托臂【建】corbel
托瓦 < 推力轴瓦的 > pillow
托碗砖 bowl brick
托销 consignment; delivery on consignment
托销成品 finished goods on consignment
托销存货 consignment stock
托销货主 consigner[consignor]
托销人 consignor
托销商品预收款 advance consignment-out
托芯 < 用托板 > stooling
托檐石 mutule
托运 consign for shipment; consign(ment); shipment; tender(ing) for conveyance
托运办公室 shipping office
托运代理人 consignment agent; shipper's aggregate
托运单 booking note;consignment bill;consignment note;forward order
托运的 consigned
托运的货物 freight consignment
托运方 consignor;shipper
托运方式 dragmode
托运号码 registration number
托运荷重 load hauled
托运局 registering administration
托运人 consigner[consignor];shipper
托运人的行为 act of shipper
托运人或货主的过失 act of default of shipper or owner
托运人(客商)联合会 Shippers' Association
托运人责任 act of shipper;shipper's liability
托运人转让 shipper alienation
托运申请单 consignment application; shipping application
托运申请书 consignment application; shipping application
托运收据 consignment note
托运行包受托人 consigned luggage
托运行李 baggage check-in; registered luggage
托轴架 pedestal
托住套管 pick-up the casing
托住头的 cephalophorous
托砖 bracket tray;curtain block;rider brick;skewback;springer
托子 < 窑具 > ancient tray support
托钻 cup-jewel; end stone; jewelled end-plate
托座 angle table; bearer supporting bracket; bracket; bracket bearing; bracket support;chair;fixing bracket;gibbet;saddle;prothyride < 砖石砌体上突挑的 >
托座飞檐 bracketed cornice
托座工作台 bracket table
托座荷载 bracket load
托座连接 seated connection
托座联条 bracket bracing
托座式挡土墙 bracket-type retaining wall
托座式支柱 bracket-like column
托座弯头 duckfoot bend;rest bend
托座支撑 bracket bracing
托座支柱 bracket post
托座砖 corbel brick;false brick

拖

drag; pull; tow(ed) target; tugging

拖把 mop;swab;swabber
拖把柄 mopstick
拖把槽 mop sink
拖把涂刷 mop rendering
拖靶 towed target
拖靶机 tow(ed)-target aircraft;trailer aircraft
拖靶装置 tow-target system
拖白 trailing white
拖白边 following white
拖板 carriage;planker;tool carriage; trailing bar
拖板鞍架 saddle of carriage
拖板计程仪 chip log
拖板式耕耘机 sled cultivator
拖板手轮 carriage hand wheel
拖板锁紧螺钉 carriage lock screw
拖板箱车床 apron lathe
拖半挂车的牵引车 tractor-truck for semitrailer
拖杯式测速发电机 drag-cup tachometer
拖杯式电动机 drag-cup motor
拖杯式发电机 drag-cup generator
拖杯形感应电机 drag-cup induction machine
拖杯形转子测速发电机 drag-cup tachogenerator
拖驳 barge in tow;towboat;towed boat
拖驳船队 barge train;barge-tug train; pull-tows; tug and barge fleet; tug and barge-train;tug and lighter fleet; tug-barge combination
拖驳船队系统 barge train connecting ropes
拖驳船队运输 barge train transportation
拖驳队 barge train unit
拖驳运输 towboat transport;tugboat transport
拖捕 trolling
拖布 floor cloth;mop;swab
拖布池 mop sink
拖擦清洁管道内部 swabbing
拖材结 timber and half hitch;timber hitch and half hitch
拖铲 drag scraper;dredge(r) scraper; pulling scraper
拖铲铲斗 drag scraper bucket
拖铲刮土机 slackline scraper
拖铲刮运机 slackline cable dragscraper
拖铲绞车 drag scraper hoist
拖铲拉索 inhaul
拖铲牵引式铲运机 drag scraper
拖铲索 inhaul
拖铲挖掘机 drag shovel;slackline cableway;slack-line excavator;tower excavator;tower machine
拖铲挖土机 dragline excavator;dragline scraper; drag shovel; slackline cableway excavator; slackline cableway scraper;slackline scraper
拖长 draw out
拖长的 longdrawn(-out);protracted
拖长的缆索 slack rope
拖车 articulated lorry;articulated vehicle; bull clam; cart; car trailer; farm tractor; full trailer; hauling truck; hind carriage; pony truck; stake body trailer;tow;towing carriage; towing troll(e)y; tractor wagon; tractor; trailer; trail(er) car;trailer wagon;trailing box;trail transfer car; wheeled trailer; wheeled trolley; articulated trailer < 用铰链连接的 >
拖车部分 trailer portion
拖车超速闸 trailer-type overrun brake
拖车车架 trailer chassis
拖车车盘 trailer chassis
拖车车体 trailer body
拖车车厢 trailer body
拖车底盘 trailer chassis
拖车底盘部分 wagon gear
拖车地带 < 在不准停车地带乱停车即被拖走 > tow-away zone
拖车垫板 riser board
拖车房屋 trailer house
拖车改换架 < 将半拖车改成全拖车的 > trailer converter dolly
拖车构造 trailer construction
拖车固定器 stanchion
拖车挂车列车 tractor trailer combination
拖车挂钩 draw tongue; trailer connector
拖车挂钩孔 trailer pintle eye
拖车及渡船驶进驶出运输 roll on/roll off traffic by trailer and ferry
拖车减震弹簧 cushion hitch spring
拖车交接 trailer interchange
拖车绞盘 car puller
拖车接轮 trailer fifth wheel
拖车接头 trailer connector
拖车居住营地 trailer camp
拖车开上敞车的联运方式 < 铁路 > trailer on flat car
拖车联结器 stanchion; trailer coupling
拖车链 tow chain
拖车轮胎 trailer tire[tyre]
拖车平台 trailer platform
拖车起落架 trailer landing gear
拖车牵引环 lunette
拖车桥轴 trailer axle
拖车倾翻装置 trailer tipping device
拖车区 < 不准停车的 > tow-away zone
拖车上平板车 trailer on flat car
拖车上平板车和集装箱上平板车运输 trailer on flat car/container flat car service
拖车上铁路平板车运输系统 trailer on flat car
拖车升降架 erector
拖车式 trailer type
拖车式拌和机 trailer mixer
拖车式颠簸累积仪 BI trailer
拖车式发射台 trailer launching platform
拖车式发射装置 trailer launcher
拖车式房屋 < 美 > trailer
拖车式混凝土搅拌机 trailer-type concrete mixer
拖车式活动住房 trailer coach home
拖车式活动住房营地 trailer camp; trailer coach home park; trailer coach space; trailer court; trailer mobile home park; trailer park; travel trailer park
拖车式活动住房驻地污水管道 trailer park sewer
拖车式空(气)压(缩)机 mobile air compressor;trailer air compressor
拖车式灭火机 trailer-type extinguisher
拖车式起重机 trailer crane
拖车式撒布机 spreader-trailer
拖车式撒肥机 wagon-type spreader
拖车式移动试验车 trailer-typed mobile laboratory vehicle
拖车式住宅 mobile home
拖车式自动电焊机 tractor automatic arc welding machine
拖车式钻机 trailer-rig
拖车试验装置 trailer test equipment
拖车停车场 trailer parking area
拖车停放场 trailer court;trailer park
拖车头 tractor-truck
拖车拖挂器 trailer coupling
拖车系紧装置 trailer hitch
拖车型拌和车 trailer mixer
拖车营地 trailer camp;trailer court; trailer park
拖车运输系统 trailer system
拖车载运的集装箱 trailer-mounted container
拖车闸 trailer brake
拖车制动器 trailer brake

拖车轴 trailer axle
拖车主销 trailer kingpin
拖车住房 accommodation trailer;trailer dwelling
拖车住房停放面积 area of house trailer spaces
拖车住户集中地 trailer camp
拖车转盘连接装置 5th wheel hitch
拖车转向架 trailer bogie;trailer truck
拖车转向销 trailer kingpin
拖车转运点 trailer transfer point
拖车装卸用的铰接跳板 flip-up ramp
拖车装载升运器 trailer-loading elevator
拖车装钻机 trailer-mounted drill
拖车总长度 over length of trailer
拖出舱角货 dragging out
拖船 dumb barge; towboat; towing vessel;tug boat;tugs and tow boat
拖船安定绳 chest rope; gift rope; guess rope;guest warp
拖船船长 tugmaster
拖船道 trackman;track road
拖船灯 tug light
拖船费留置权 towage lien
拖船费(用) towing charges;towage
拖船服务 tug service
拖船工作 towage
拖船路 <人或马沿河岸拖船时所行的路> tow path
拖船票 towing warp
拖船上岸 hauling ship ashore
拖船索 cordelle;towline;warp
拖船租金 tug hire
拖垂天线 trailing-wire aerial
拖带 take in tow;towage;tow(ing);trail;tuggage
拖带船队 fleet by tow;pull-tows;towing unit;tug-barge combination;tug-barge towing fleet
拖带的平车 trailer dolly
拖带灯 towing lantern;towing light
拖带费 towage;towage dues;tuggage
拖带费用 towage;towing charges
拖带杆 tow bar
拖带挂车 balanced full trailer;dolly;trailer converter dolly
拖带挂车的汽车驾驶员 trailerist
拖带行业 towage service
拖带航程 towing voyage
拖带航速 towing speed
拖带合同 towage contract
拖带和公务交通艇 towing and general service launch
拖带荷载 towed load
拖带机构 chart drive mechanism
拖带绞车 towing winch
拖带力 <河沙的> tractive force
拖带联结 towing connection
拖带链(条) tow(ing) chain
拖带马力 towing horsepower
拖带契约 towage contract
拖带桥楼 <带拖钩的> towing bridge
拖带设备 towing apparatus; towing appliance; towing arrangement; towing rig
拖带式串联铺路机 hitched tandem paver
拖带式串联摊铺机 hitched tandem paver
拖带式空气压缩机 track air compressor
拖带式联合铺面机组 paving train
拖带式输送机 trailing conveyer[conveyor]
拖带式鱼型密度计 towed fish densimeter
拖带式振动压路机 vibrating roller trailer;trailer type vibrating roller
拖带试验 towing trial

拖带条款 towage clause
拖带系统 towing system
拖带小艇的绳或小链条 towing painter
拖带效率 towing efficiency
拖带信号 towing signal
拖带眼板 towing pad eyes
拖带运输 towage transportation
拖带责任条款 tower's liability clause
拖带装置 towing device
拖带阻力 towing resistance;tow rope resistance
拖底大围网 seine
拖地扫测 bottom sweeping with towing rope
拖地扫海 aground sweeping
拖钓 troll
拖钓船 long liner;troller
拖钓绳 trolling line
拖钓渔船 line trawler
拖钓作业 trolling
拖动 dragging
拖动电动机 drive motor;driving motor
拖动力 driving power
拖动能力 luggability
拖动试验 motoring ring test; towing test
拖动性能 luggability
拖动装置 driving unit
拖动阻力测定 towing resistance measurement
拖斗 drag bucket;trail car
拖斗桥轴 trailer axle
拖斗清沟机 dragline ditch cleaner
拖斗挖泥船 trailer dredge(r)
拖斗尾绳 scoop tail rope
拖斗转向装置 scooter deflector
拖渡两用船 combination ferryboat-trailer ship
拖发动 tow starting
拖罚 <违例停车> towaway
拖放 drag and drop
拖杆 draw bar;draw tongue;pull lever; towbar; towing bar; towing handle
拖杆轮衬套 hauling rod wheel bushing
拖杆牵引杆 draw bar
拖杆式扫测 wire and bar sweeping
拖杆万向节衬套 hauling rod universal joint bushing
拖竿 picaroon
拖钩 clevis joint;drawbar hook;pull-devil; pull hook; towbar; tow-hook; towing hitch; towing hook; towing shackle
拖钩处高度 towing height
拖钩浮标 trawl buoy
拖钩宽度 fork width
拖钩离地高度 drawbar height
拖钩牵引的拖车 drawbar trailer
拖钩牵引杆 draw bar
拖钩牵引力 drawbar pull;raw pull
拖钩牵引马力 raw bar horsepower
拖钩试验 drawbar test
拖钩台(架) towing hook platform
拖钩销 draw pin
拖刮处理 drag treatment
拖刮(样)板 drag screed
拖挂拌和机 trailer mixer
拖挂泵车 trailer pump
拖挂表示 dragging indication
拖挂车 towed vehicle
拖挂车车队 tractor-trailer train
拖挂车负荷 train load(ing)
拖挂车滑溜试验 towboat-vehicle skid test;towed-vehicle skid test
拖挂车临时停车场 temporary trailer park
拖挂车停车处 trailer park
拖挂车拖车 tractor-truck
拖挂车运输 articulated traffic

拖挂的活动住房营地 trailer site
拖挂方法 trailer-type method;trailing system
拖挂杆 tow(ing) pole
拖挂夯实机 trailer compactor
拖挂荷载 towed load
拖挂化的 trailerized
拖挂混凝土泵 trailer concrete pump
拖挂锯车 trailer saw
拖挂空气压缩机 trailer compressor
拖挂连接装置 towing gear
拖挂列车 combination vehicle; tractor-drawn train
拖挂轮胎压路机 towed pneumatic-tired roller
拖挂模型 towing model
拖挂铺砂机 towed sand spreader; trailer grit spreader;trailer gritter; trailer gritting machine
拖挂式汽车 combination vehicle
拖挂牵引试验 towing dynamometer test
拖挂扫路车 trailer brush
拖挂扫路机 broom drag
拖挂设备检测环线 dragging detector loop
拖挂设备检测继电器 dragging equipment detector relay
拖挂设备检测器 dragging equipment detector
拖挂示像 dragging aspect
拖挂式 pull(ing) type
拖挂式拌和机 towed-type mixer
拖挂式槽车 tank trailer
拖挂式铲斗平地机 trailing scoop grader
拖挂式车身架 towed-type chassis
拖挂式除雪机 draw plough
拖挂式大型客车 coupled bus
拖挂式底盘 towed-type chassis
拖挂式分布机 trailer distributor
拖挂式刮土机 trailer scraper
拖挂式活动模型 mobile model trailer type
拖挂式活动住房卫生站 trailer sanitation station
拖挂式活动住房营地 tourist camp
拖挂式活动钻机 portable trailer-mounted drill rig
拖挂式机械 trailer-type machine
拖挂式沥青车 asphaltic-bitumen trailer
拖挂式料斗式铺砂机 towed hopper type gritter
拖挂式喷布机 distributor-trailer
拖挂式平地机 towed-type grit spreader
拖挂式铺砂机 towed-type gritter
拖挂式起重机 trailer crane
拖挂式汽车 combination vehicle
拖挂式撒布车 trailer-spreader
拖挂式撒砂机 towed-type sand spreader
拖挂式洒布机 trailer distributor
拖挂式扫路机 towed-type road brush; trailer road-sweeping machine
拖挂式扫雪机 draw plough
拖挂式压路机 pull-type roller
拖挂式油罐车 trailerized tank
拖挂式载重(汽)车 coupled truck
拖挂式筑路机(械) trail builder;tilldozer; bulldozer with angling blade;tilt dozer
拖挂死角 jackknife
拖挂挖土机 trailer excavator
拖挂显示 dragging indication
拖挂型 pull(ing) type
拖挂压路机 towed roller
拖挂运输 tractor-trailer transport; trailer pick-up transport
拖挂振动式压路机 trailed vibrating roller
拖挂重量 trailing load
拖挂装置 augmenter;hitch bar;towbar;trailer coupling
拖挂组成的营区 trailer court
拖挂作业 towing operation
拖关 <装卸货物> towing a sling
拖管驳 pulling barge
拖管索 trailing line
拖管头 pull(ing) head
拖罐车 tank wagon
拖光 <混凝土路面> belting
拖航 towage;towing
拖航合同 towage contract
拖航机 tow(ing) aircraft;towing airplane
拖航检验 survey of tow
拖航卡环 towing shackle
拖航三角板 towing fish-plate
拖航式拖船 pusher-propelled tug
拖航水标志 towing waterline mark
拖航条款 towage clause
拖航直升机 towing helicopter
拖痕 drag mark
拖后指标 lagging indicator
拖戽式取样器 drag-bucket sampler
拖环 towing jaw;towing shackle;tow loop
拖集的木材 trail
拖集木 drag
拖肩 haunch, fillet
拖件前进的连续模 cut-and-carry die
拖角 towing angle
拖绞出搁浅船 refloat a stranded vessel with drag
拖鲸艇 whale-towing craft
拖救 salvage towing
拖救条款 towing and salving clause
拖开 tow out
拖开费 streaming
拖开设施 haul-away equipment
拖拉 drag; lug; tote; tow(ing); traction;tug
拖拉铲土机 drag scraper;drag shovel;tractor shovel
拖拉铲运机 drag scraper;drag shovel; tractor-carryall unit; tractor scraper;tractor shovel
拖拉道 dragging track; skid(ding) trail;skid(ding) road;skidway
拖拉道砟机 ballast drag
拖拉法 erection by launching;towing method
拖拉钢丝绳 drag rope
拖拉刮土机 pulling scraper
拖拉后卸式搬运机 turnarocker
拖拉滑轨 sliding rail
拖拉环设计 two ring layout
拖拉机 hauler;prime mover;traction engine;tractor;donkey <美>
拖拉机安全司机室 Saf-T-Cab
拖拉机半挂车 tractor-semitrailer
拖拉机本体 basic tractor
拖拉机操纵挖沟机 tractor-operated trench hoe
拖拉机铲 tractor shovel
拖拉机铲土车 tractor shovel
拖拉机铲土机 tractor scraper
拖拉机厂 tractor plant
拖拉机带铲土机 tractor-scraper
拖拉机带装载机 tractor loader
拖拉机的行驶速度 tractor ground speed
拖拉机底盘的转臂式装载机 jib-type tractor loader
拖拉机发动机 tractor engine
拖拉机附属设备 tractor attachment
拖拉机构 drag mechanism
拖拉机挂车 tractor trailer; tractor wagon

T

拖拉机挂接装置 hitch
拖拉机挂引的 V 形铲刀 tractor-mounted V blade
拖拉机后部装的平地装置 rear-mounted grader attachment for tractors
拖拉机后悬挂混凝土搅拌机 rear-mounted concrete mixer
拖拉机化 tractorization
拖拉机驾驶室 cockpit
拖拉机驾驶员 tractor operator
拖拉机检验 tractor testing
拖拉机绞盘 tractor rope winch
拖拉机可调式座位 adjustable tractor seat
拖拉机拉杆 < 悬挂装置的 > tractor link
拖拉机离合器 tractor clutch
拖拉机犁 tractor plough
拖拉机履带 tractor tread
拖拉机履带板 tractor shoe
拖拉机履带片 tractor shoe
拖拉机履带蹄齿 tractor grip lug
拖拉机轮距 tractor tread
拖拉机轮胎 tractor tire
拖拉机煤油 tractor kerosene
拖拉机耙 gang harrow
拖拉机皮带盘孔 tractor pulley opening
拖拉机平地机 pull-grader
拖拉机起重机 tractor crane
拖拉机起重绞车 tractor hoist
拖拉机牵引 tractor traction
拖拉机牵引板 tractor-draw-plate
拖拉机牵引的 tractor-dragged; tractor-drawn; tractor-hauled
拖拉机牵引的充气轮胎压路机 pneumatic-tyred tractor-drawn roller
拖拉机牵引的二轮铲运机 tractor-pulled carrying scraper
拖拉机牵引的货车 tractor wagon
拖拉机牵引的羊足碾 tractor-pulled sheep-foot roller
拖拉机牵引加重网格碾 tractor-pulled ballasted grid roller
拖拉机牵引设备 tractor-drawn equipment
拖拉机牵引系数 tractor factor
拖拉机前置装载机后置挖掘机 tractor-loader backhoe
拖拉机驱动泵 tractor-driven pump
拖拉机驱动的 tractor-propelled
拖拉机驱动的发电机 tractor-driven generator
拖拉机燃油 power kerosene
拖拉机润滑油 tractor luboil; tractor oil
拖拉机式叉车 tractor forklift
拖拉机式铲土机 tractor shovel
拖拉机式牵引铲运机 tractor-drawn scraper
拖拉机式台架 tractor jumbo
拖拉机式推土机 tractor dozer; tractor shovel
拖拉机式压路机 roller tractor
拖拉机式装卸机具 front and loader
拖拉机式装载工具 frontal and loader
拖拉机式装载机 tractor loader
拖拉机试验 tractor testing
拖拉机手 tractor operator
拖拉机推动的铲土机 tractor pulled scraper
拖拉机拖车 tractor trailer
拖拉机拖带的 tractor-hauled
拖拉机拖曳铲土斗 tractor scraper
拖拉机拖曳铲运机 tractor scraper
拖拉机拖运 tractor haulage
拖拉机械 towing machinery
拖拉机悬挂的叉式升运器 tractor lift
拖拉机悬挂的叉式装载机 tractor fork-lift loader
拖拉机悬架式驾驶室 tractor suspension cab
拖拉机旋转起重机 tractor revolving crane
拖拉机曳引多轮胎压路机 tractor-drawn multi-rubber-tyre roller
拖拉机曳引压路机 tractor-drawn roller
拖拉机液压系统 tractor hydraulics; tractor hydraulic system
拖拉机用粗汽油 tractor ligroin
拖拉机用绞盘 tractor winch
拖拉机用燃料 tractor fuel
拖拉机用油 tractor oil
拖拉机用直接传动式绞车 direct drive tractor winch
拖拉机油 tractor oil
拖拉机油门调节 tractor throttle setting
拖拉机运输 tractor haulage
拖拉机载泵 tractor-mounted pump
拖拉机载扫路机 tractor-mounted sweeper
拖拉机载挖沟机 tractor-mounted trench excavator
拖拉机载挖土机 tractor-mounted excavator
拖拉机载运的箱式撒布机 tractor-mounted box spreader
拖拉机载重车 coupled truck
拖拉机站 tractor station
拖拉机主机 basic tractor
拖拉机装具铲刀 tractor-mounted blade
拖拉机装载机 tractor loader
拖拉机装钻机 crawler-mounted drill
拖拉机钻机 tractor drill
拖拉架桥法 nosing method; launching erection
拖拉缆索 haulage rope
拖拉平板车的牵引车 tractor towing platform truck
拖拉起重机 tractor crane
拖拉器 tugger
拖拉牵引的 tractor-drawing
拖拉牵引式铲运机 tractor-drawn scraper
拖拉桥 pullback draw bridge
拖拉设备 towing equipment; trailed equipment; winding gear
拖拉绳索 hauling rope
拖拉施工 dragging construction; erection by dragging
拖拉石滚式磨碎机 drag-stone mill
拖拉式铲运机 turn-pull
拖拉式卷扬机 tugger hoist
拖拉式平地机 pull-grader
拖拉式推土机 tractor dozer
拖拉式闸门 tractor gate
拖拉式钻机 traction drill; traction drill(ing) machine
拖拉推土机 cable control unit
拖拉挖土机 drag shovel
拖拉帷幕轨 curtain track
拖拉雪犁 tractor snow plough
拖拉装料机 tractor loader
拖拉装载机 tractor shovel
拖拉装置 trailed equipment
拖缆 cordelle; drag falls; haulage rope; streamer; towing hawser; towing line; towing rope; towline
拖缆长度 towed cable length
拖缆承梁 towing arch; towing bar; towing beam; towing gallows; towing rail; tug arch
拖缆垂度 sag of towing rope
拖缆的递送 delivery of towline
拖缆的跨距 span of tow-rope
拖缆端短索 towing pendant
拖缆短链 chain pendant
拖缆辅助绳 guest rope
拖缆搁架 towing block
拖缆拱架 tug arch
拖缆钩掣 towing tripper
拖缆滚子组 haulage rope roller battery
拖缆缓冲器 towing accumulator
拖缆绞车 towing engine; towing machinery; towing winch
拖缆绞盘 towing cable winch; towing capstan
拖缆卷扬机 towing cable winch
拖缆孔 towing port
拖缆勒头 towing bridle
拖缆力 tow rope power
拖缆马力 tow rope horsepower
拖缆木桩 timber head
拖缆扫海法 drag; drag sweep
拖缆索 tow pendant
拖缆托板 hawser board
拖缆系船柱 towing bitt
拖缆限位器 stop device for towing line
拖缆销 towline toggle
拖缆卸扣 towing shackle
拖缆有效长度 effective length of towing rope
拖缆柱 towing post
拖缆桩 nigger head; towing bitt; towing bollards; towing head; towing post; towing timber
拖捞船 trawboat; trawler
拖离设施 haul-away equipment
拖离速度 velocity of escape
拖离用拖缆 < 油轮 > towing off line
拖力 hauling power; pulling; towing force; towing tension
拖力图 drag diagram
拖链 drag chain; drag link; hauling chain; snaking chain; tow chain
拖链槽 dee
拖链除灌 chaining
拖链孔 drag hole
拖流 tractive current
拖轮 barge tug; launch tug; motor tug; steam tug(boat); towboat; towing launch; towing tug; tug; tug boat
拖轮驳船队 tug-barge combination
拖轮船员 tugman
拖轮船长 tugmaster
拖轮打横 girded; girt(h)
拖轮费 charges for tug's service; towage
拖轮护舷 rubber fender for tug
拖轮及驳船动态显示系统 towboat and barge status system
拖轮兼驳渡船 tug-tender
拖轮式船尾 tugboat stern
拖轮协助 tug assistance
拖轮协助操纵 maneuvering with the aid of tug; maneuver with tugs
拖轮助航 unassisted navigation
拖轮总拖力 total towing force of a tug
拖锚 clubbing; dragging anchor; towing anchor at short stay
拖锚掉头 dredging turning
拖锚漂流 clubbing
拖木材的线路 log track
拖耙 drag harrow
拖盘式钻机 strip bore drill
拖期 behind schedule
拖期完工罚款 liquidated damages for delay
拖漆刷的 gummy
拖牵力 drag force
拖欠 arrearage; arrears; be behind in payment; be in arrears; default; in arrears; remaining arrears
拖欠贷款率 delinquent-loan ratio
拖欠的 owing
拖欠的保险费 premium in arrears
拖欠抵押贷款 delinquent mortgage
拖欠额 amount of arrears
拖欠费用 towing charges
拖欠分期付款 delinquent instalment
拖欠分期款项 back money; payments in arrears
拖欠付款 default of payment
拖欠工资 arrears of wages
拖欠红利 arrears of dividends
拖欠金额 amount in arrear
拖欠款 arrearage; back money; payment in arrear
拖欠利息 arrears of interest
拖欠年金 annuity in arrears
拖欠(清偿的)债券 defaulted bonds
拖欠人 defaulter
拖欠上交利润 being in arrears in handing over profits
拖欠税款 be in arrears with tax payment; default of tax payments; delinquent tax
拖欠应收款项 delinquent receivables
拖欠债款 default; delinquency
拖欠账款 delinquent account
拖欠者 defaulter
拖欠最后期限 delinquency date
拖去车辆 haul-away vehicle
拖圈 towing loop
拖入井架内 tail into the derrick
拖梢护树 pulling tops
拖绳 drag rope; towing rope; towline; trail rope
拖绳绞盘 towing rope winch
拖绳轮 tug wheel
拖绳应急舵 hawser rudder
拖式 V 形刮铲 V-drag
拖式铲运机 drag scraper; drag shovel; land lever; pull-type scraper; towed scraper; tractor-towed scraper; turnapull scraper; slusher
拖式单辊压路机 tow type single drum roller
拖式的 drawn
拖式翻土机 towed scarifier
拖式割草机 towed mowing machine
拖式刮扳 drag screed
拖式刮土机 hauling scraper
拖式光面压路机 plain towed roller
拖式光碾压路机 towed smooth drum roller
拖式光碾振动压路机 towboat smooth drum vibratory roller; towed smooth drum vibratory roller
拖式机械 drawn machinery
拖式沥青混合料摊铺机 towed asphalt paver
拖式沥青延布机 towed asphalt sprayer
拖式裂土机 towed ripper
拖式路碾 pull-type roller; towed roller; tractor articulated roller
拖式埋缆刀板 pull blade
拖式碾压机 pull-type roller; towed roller
拖式平地机 drawn grader; land lever; pull-grader; towboat bowed blade grader; towboat grader; towed blade grader; tow(ed) grader
拖式平路机 tractor grader
拖式倾卸铲运机 tilting-type drag scraper
拖式扫路机 trailer road-sweeping machine
拖式升降机 towed hoist
拖式石屑撒布机 towed chip spreader
拖式双犁扫雪机 tractor snow plough
拖式双联路碾 tractor-drawn two unit articulated roller
拖式松土机 towed ripper
拖式碎石机 traction crusher
拖式碎石摊铺机 towed aggregate

spreader
拖式稳定土搅拌机 towed pulvi-mixer
拖式雪犁 tractor snow plough
拖式压路机 pull tire roller; pull-type roller; tow behind roller; towed roller; tow type roller
拖式压路碾 towboat roller
拖式羊足碾 towed sheep-foot roller
拖式圆板路耙 offset disc[disk] harrow
拖式圆盘路耙 offset disc[disk] harrow
拖式圆盘耙 offset disc[disk] harrow
拖式振荡压路机 towed vibro-roller
拖式振动压路机 towed vibratory roller
拖式柱塞压捆机 pull plunger press
拖刷 brush drag; mopping
拖速仪 tow speed log
拖损 damaged owing to towing
拖索 hauling wire; messenger; tail rope; tow; towing bridle; towing wire; towline; warping hawser
拖索除灌 cabling
拖索防擦段 <一段铁链> chafing chain
拖索扫测 wire-drag survey; wire-drag sweeping
拖索中垂阻力 sag resistance
拖筒式取样器 drag-bucket sampler
拖头 towing the bow; tow the head
拖头掉头 turning by pulling the bow
拖头顶尾掉头 turning by pulling the bow and pushing the stern
拖头和拖车相连 double combination
拖头牵引的起重运输机 lift transporter
拖头丝 pulling-off silk end
拖网 drag net; dredger; ground net; sled; trail net; trawl net
拖网板 trawl board
拖网捕虾船 shrimp trawler
拖网采样 trawl sampling
拖网承梁 trawl beam
拖网船 otter trawler; trawler
拖网灯 trawling light
拖网电缆 electric(al) trawl cable
拖网帆船 sail trawler
拖网拱梁 tow beam; towing arch; tow rail
拖网加工渔船 factory-trawler
拖网架 trawl head
拖网兼漂网渔船 trawler-drifter
拖网绞车 dandy winch; trawl winch
拖网结 herring knot
拖网木滚 bobbin
拖网漂网渔船 trawler-drifter
拖网取样 sampling by dredge(r)
拖网围网渔船 trawler-seiner
拖网下部 belly
拖网渔船 trawboat; trawler; drag boat; dragger; trawl(er) boat; trawler ship
拖网渔船船队 trawler fleet
拖网渔工 dragman
拖网中段 bosom
拖网装置 trawling gear
拖网作业 trawling
拖围网 drag seine
拖尾 hang-over; hangover; towing the stern; trail(ing); trailing smear
拖尾电缆 <电动机械的> trail cable
拖尾掉头 turning by pulling the stern
拖尾峰 non-Gaussian peak; tailed peak; tailing peak
拖尾区域 region of streaking
拖尾图像 streak image
拖尾效应 smearing effect
拖尾因子 tailing factor
拖尾影踪 tailing hangover
拖尾重影 smear ghost
拖物代舵 drag steering
拖下 tear down
拖线输送机 <在货棚内修建的椭圆形的> towing conveyer; tow-line conveyer
拖行速度 drawing speed; towing speed
拖行小车 tractor-driven carrier
拖延 arrear; delay; demurrage; detainment; detention; hang-up; linger; protract; put over; run-around
拖延费 demurrage
拖延付款 slow pay
拖延工作 stretch-out
拖延起动 delayed starting
拖延施工项目 delayed project
拖延时间 demurrage
拖延债务的罚款 late charges
拖腰 towing the waist; tow the midship
拖曳 drag(ging); entrain; haulage; pull; skidding; tow(age); trail(ing); tug
拖曳搬运 traction; traction transport; transportation by traction
拖曳柄 pull handle
拖曳铲斗 drag dipper
拖曳车 hauling vehicle; towed vehicle; tractive machine
拖曳单元 towing unit
拖曳的 tractive
拖曳的机器 towed machine
拖曳点 towing point
拖曳电缆 trailing cable
拖曳断层 drag fault
拖曳耳孔 pull eye
拖曳方法 traction method
拖曳钢缆 towing wire cable
拖曳刮土机 hauling scraper
拖曳环索 <空中索道用,使吊车移动> traction rope
拖曳机具 towing unit
拖曳交通 tug-tender
拖曳角 drag angle; towing angle
拖曳绞车 pulling winch
拖曳绞盘 towing winch
拖曳距离 trail distance
拖曳缆 trailing cable
拖曳力 drag(ging) force; grabbability; gradeability; tractive force; towing force <船舶>; current drag force; entrainment force
拖曳轮式括板的履带式车 crawler for towing wheel scrapers
拖曳埋置锚 drag embedment anchor
拖曳能力 towing capacity
拖曳偏位 towed offset
拖曳桥 retractile(draw-) bridge
拖曳倾卸桩【疏】 trailing tilting spud
拖曳设备 towing gear
拖曳绳 travel(l)ing rope
拖曳式采样器 dredge; towed sampler
拖曳式浮标天线装置 trailing buoy antenna system
拖曳式航速计 patent log; taffrail log
拖曳式基阵监视系统 towed array surveillance system
拖曳式计程仪 <船尾舷> patent log; tafferel log; taffrail log
拖曳式计程仪记录器 taffrail recorder
拖曳式卡车 trailing truck
拖曳式铺路骨料撒布机 towed paver type aggregate spreader
拖曳式吸泥船 trailing suction dredge(r)
拖曳式雪犁 draw plough
拖曳式阵列监视系统 towed array surveillance system
拖曳试验槽 <航模> towing channel
拖曳试验池 <船模的> towing basin
拖曳输送机 drag-in conveyer[conveyor]
拖曳水槽 towing tank
拖曳水池 towing tank
拖曳水流 tractive current
拖曳速度 towed speed
拖曳索 tow rope; traction rope
拖曳坦克 <集装箱> tirtank
拖曳体系 towing system
拖曳天线 antenna trailer; trailing aerial; trailing antenna; trailing-wire antenna
拖曳条件 conditions of tow
拖曳推力 traction thrust
拖曳物 hauler; trail
拖曳现象 traction phenomenon
拖曳线 pull-in line
拖曳效率 trawling efficiency
拖曳效应 drag effect
拖曳性能 towing performance
拖曳用孔眼 eye for towing
拖曳运货车 trailing truck
拖曳载人潜水器 towed manned submersible
拖曳噪音 traction noise
拖曳者 drawer; haul(i)er; trailer
拖曳褶曲【地】 drag fold
拖曳褶皱 drag fold
拖曳之人(或物) trailer
拖曳转矩 drag torque
拖曳装置 towing device; towing gear
拖曳走向 drag strike
拖曳阻力 drag resistance; tow rope resistance
拖曳作用 effect of dragging; entraining action
拖移电缆 <电动起重机的> trailing cable
拖引叉 towing fork; trail hitch
拖引车 towing vehicle
拖引船 towing vessel
拖引列车 tractor train
拖引汽车 towing vehicle
拖印 tire skid
拖影 streaking; trailing smear
拖淤 agitation dredging
拖运 haul(ing); roading; skid; towage
拖运车 tractor trailer
拖运道路 haul(age) road
拖运费 haulage
拖运工具 haulage appliance
拖运货物 trucking cargo
拖运机 chain-and-ducking dog mechanism; drag; drag-over unit; hauler; hauling machine; pull-over gear
拖运机部件 drag-over assembly
拖运机构 transfer mechanism
拖运架装卸的货轮 trailer ship
拖运角 haulage angle
拖运绞车 haulage winch
拖运距离 haulage distance
拖运卡车 haul truck
拖运缆索 haulage cable
拖运力 haulage capacity
拖运链 <木材拖拉机的> bull chain
拖运料车 tractor wagon
拖运路线 towing route
拖运率 truck towing rate
拖运能力 hauling capacity; towability; traction capacity
拖运器 pulling sled; sled
拖运设备 hauling equipment; towing appliance
拖运式线盘传送带 drag-type coil conveyer
拖运速度 hauling speed
拖运小车 drag carriage
拖运性能 towability
拖运应力 hauling stress
拖运重量 shipper's weight
拖运装置 <横移轧件用的> transfer arrangement
拖运钻塔 skid the derrick
拖运作业 tow(ing) operation
拖载运的集装箱 trailer-mounted container
拖在船模后的尾流栅 trailgate
拖枕 bearer
拖纸轮 tape puller
拖滞 dragging
拖着走 trail
拖走 towaway
拖阻力 drag resistance

脱

脱氨 ammonia removal
脱氨法 ammonia stripping process
脱氨基作用 deaminating; deamination; deaminization; deaminizing
脱靶距离 miss-distance
脱靶量指示器 miss-distance indicator
脱白云石化作用 dedolomitisation [dedolomitization]
脱斑 bate pits; bate stains
脱版 misplacing of stencil
脱饱和 desaturation
脱苯 benzol removal; debenzolization
脱边 edge peeling; loose edge; off-clip; shelling border
脱苄基作用 debenzylation
脱变作用 disintegration
脱冰机 can dump
脱丙烷塔 depropanizator; depropanizer; depropanizing column
脱玻安武岩 apoandesite
脱玻包裹体 devitrification inclusion
脱玻雏晶结构 devitrified crystalline texture
脱玻霏细结构 devitrified felsitic texture
脱玻构造 devitrified structure
脱玻化 devitrification
脱玻束状结构 devitrified bunchy texture
脱玻结构 devitrified texture
脱玻晶腺构造 devitrified drusy structure
脱玻璃化作用 devitrification
脱玻球颗结构 devitrified variooitic texture
脱玻球粒结构 devitrified spherolitic texture
脱玻梳状构造 devitrified comb structure
脱玻微嵌晶结构 devitrified micropoikilitic texture
脱玻玄武岩 apobasalt
脱玻隐晶结构 devitrified cryptocrystalline texture
脱玻作用 devitrification; devitrify
脱层 delaminate; delamination; pull(ing) up <旧涂层的软化>
脱产培训 off-the-job training; training off job
脱尘 dedust
脱尘机 deduster
脱臭 deodo(u)ration; deodo(u)rizing; off odo(u)r; air sweetening <使油中有臭味的硫醇经空气氧化成为无臭二硫化物>
脱臭机 deodo(u)rizer
脱臭剂 deodo(u)rant; deodo(u)rizer
脱臭煤油 deodo(u)rized kerosene; refined kerosene
脱臭器 deodo(u)rizer
脱出 break loose; emersion; pull out
脱出杆 eject lever
脱出水 water of dehydration
脱出同步 hold off
脱除 deprivation
脱除粗苯的粗馏分(炼焦) defronting

T

脱除锅垢 descaling
脱除旧漆 stripping of old paint
脱除石墨 degraphitization
脱除污迹 de-ink
脱除性 detachability
脱除油脂 deoil
脱除(杂质)过程 subtractive process
脱醇 dealcoholizing
脱氮 denitriding; denitrify; nitrogen loss; denitrogenation
脱氮法 denitrogenation method
脱氮过程 denitrification process
脱氮化层 denitridation
脱氮剂 denitrifying agent
脱氮菌 denitrifier
脱氮硫杆菌 thiobacillus denitrificans
脱氮滤池 denitrifying filter
脱氮曲线 nitrogen desorption curve
脱氮作用【化】denitrification; denitriding; denitrogenation; nitrogen removal
脱挡 trip dog
脱挡滑行 coasting in neutral; ride the clutch
脱挡销 disengaging latch
脱底 decollement; bottom off
脱底褶皱 decollement fold
脱掉 pull-off
脱丁基作用 debutylizing
脱顶断层 detachment fault
脱顶构造 decollement
脱顶石油 <蒸去轻质油后的石油> topped petroleum
脱顶原油 <蒸去轻质油后的石油原油> topped crude oil
脱顶褶皱 decollement fold
脱顶重力滑动 detachment gravity slide
脱锭吊车 ingot drawing crane
脱锭机 ingot drawing machine; stripper
脱锭机构 stripper mechanism
脱锭跨 stripping bay
脱锭力 stripping strength
脱锭起重机 stripping crane
脱锭钳 stripping tongs
脱二氧化碳反应 decarbonation reaction
脱二氧化碳作用 decarbonation
脱方 <轧件缺陷> off-square
脱方轧件 diagonal stock
脱芳构化 dearomatization
脱酚剂 dephenolizer
脱酚煤气(废)液 dephenolated gas liquor
脱酚作用 dephenolization
脱蜂器 bee-escape
脱缝 <深钻技术中的> joint failure
脱氟瓷石 defluorinated stone
脱氟磷肥 defluorinated phosphate
脱氟作用 defluorination
脱附 desorption
脱附等温线 desorption isotherm
脱附剂 desorption agent
脱附速率 desorption rate
脱钙 decalcify
脱钙棕色土 terra fusca
脱钙作用 decalcification
脱盖 top off
脱汞盐水 mercury-depleted brine
脱钩 breakaway; detaching hook; disconnecting hook; off-hook; release catch; running away; shut out; unhitch; unhook; unlocking
脱钩操纵杆 uncoupling lever
脱钩点 hook separation point; separation point
脱钩杆 release lever; tripping bar
脱钩机构 release mechanism; tripping mechanism; uncoupling rigging
脱钩架 tripper
脱钩拉索 release lanyard
脱钩链段 senhouse slip
脱钩器杆 knock-off post
脱钩式联结器 trip hitch
脱钩索 releasing line
脱钩装置 coupler release rigging; knock-off joint; release device; releasing gear; tripping gear
脱钩撞车 breakaway collision
脱垢剂 detergent
脱垢溶剂 degreasing solvent
脱谷板棚 threshing barn
脱谷机 hulling machine; threshing machine
脱谷装备 threshing outfit
脱管机 extraction device for pipe
脱硅吹炼 silicon blow
脱硅(过程) desilication
脱硅作用 desilication; desilicification; desiliconization
脱轨 deorbit; derail(ment); disorbition; jumping of wheel-flange on rail; jump the track; runoff rails; run the track; on the floor <指车辆>
脱轨安全综合试验 off-track safety test
脱轨保护 derailment protection
脱轨标志 derail indicator
脱轨车起重器 swing jack
脱轨道岔 derailing point; derailing switch; throw-off points
脱轨公式 criterion for wheel climbing; Nadal's formula
脱轨器 derailer; derailing block; derailing stop; wedge block
脱轨器标志 derailer target
脱轨器表示器 derailer indicator
脱轨器不在遮断位置 derailer open
脱轨器操纵机构 derailer operating mechanism
脱轨器握柄 derailer lever
脱轨器在开放位置 derailer open
脱轨器在遮断位置 derailer closed
脱轨事故 derail accident
脱轨试验 derailment test
脱轨位置 derailing position
脱轨系数 coefficient of derailment
脱轨辙叉 derail frog
脱轨转辙器 switch point derail
脱轨装置 derailer
脱辊 roll release
脱焊 loose weld; sealing-off; thrown solder; tip-off; unsoldering
脱焊的 sealed-off
脱黑 deinking
脱黑过程 deinking process
脱环 decyclization
脱换 cast off
脱灰 ash removal; bate pits; de-ash; delime; deliming
脱灰槽 drench pit
脱灰剂 bating material; deliming agent
脱灰燃料 de-ashed fuel; de-ashing fuel
脱灰液 bate
脱灰作用 de-ashing
脱挥发反应 devolatilization reaction
脱挥发分 devolatilization
脱挥发分作用 devolatilization
脱挥发器 devolatilizer
脱挥发作用 devolatilization
脱混 demixing
脱机【计】off-lining
脱机编辑 off-line editing
脱机操作 off-line operation
脱机测试 off-line test(ing)
脱机成批处理系统 off-line batch processing system
脱机处理 off-line process(ing); off-line working
脱机存储器 off-line storage; off-line memory
脱机打印机 off-line printer
脱机的 off-line
脱机方式 off-line mode
脱机工作 off-line working
脱机故障检测 off-line fault detection
脱机绘图 off-line drawing; off-line plot
脱机绘图机 off-line plotter
脱机计算 off line computation
脱机计算机 off-line computer
脱机检索 off-line retrieval
脱机卷取动作 off-loom take-up motion
脱机控制 off-line control
脱机描绘器 off-line plotter
脱机模拟 off-line simulation
脱机设备 off-line equipment
脱机输出 indirect output; off-line output
脱机输出设备 off-line output device
脱机输出装置 off-line output device
脱机输入设备 off-line input device
脱机输入装置 off-line input device
脱机数据处理 off-line data processing
脱机数字化 off-line digitizing
脱机系统 off-line system
脱机印刷装置 off-line printer
脱机支持系统 off-line support system
脱机状态 off-line state
脱机字符 character outline
脱机作业 off-line operation; off-line job
脱机作业控制 off-line job control
脱基乳胶 stripped emulsion
脱己烷塔 dehexanizer; dehexanizing column
脱甲基作用 demethylating
脱碱 dealkalize
脱碱化碱土 solodized solonetz
脱碱化土壤 solodic soil
脱碱土 solod; solod(ic) soil
脱碱作用 dealkalization; solodization <土壤的>
脱浆 desize; desizing; misfurnish
脱浆废水 desizing wastewater
脱浆/染色废水 desizing/dyeing wastewater
脱浆轧布机 desizing mangle
脱胶 adhesion loss; debond(ing); degelatinizing; deglue; de-gumming; delamination; delaminate <夹层玻璃的>; disbond; stick-free
脱胶骨 deglued bone
脱胶罐 <骨灰瓷生产用> glue kettle
脱胶剂 degelling agent
脱胶结作用 de-cementation
脱胶强度 delamination strength
脱胶脂 deresinsate
脱胶助剂 boil-off assistant
脱焦 decoking
脱焦炭 decarbidize; decarbonize
脱焦油 detar; tar removal
脱焦油设备 detarrer
脱焦油设施 detarring plant
脱节 broken baseline; disjunction; dislocation; divorcement
脱节异常 disrupted anomaly
脱节晕 disrupted halo
脱节褶皱 disjunction fold; disjunctive fold
脱结合水 water of dehydration
脱净率 threshing performance
脱静电剂 destaticizer
脱臼 disjoint; dislocation
脱开 declutch; disengage; relieve; uncouple; uncoupling; unlock; unpin
脱开安全器 safety release
脱开齿轮 out-of-gear
脱开刀闸开关 disconnecting knife-switch
脱开的 disconnected; disjointed; out-of-gear
脱开管道 pipe away
脱开后桥 rear axle disconnecting
脱开离合器 declutching
脱开力 breakout effort; release load; tearaway force; tearaway load
脱开联轴节 uncoupling
脱开式离合器 disengaging clutch
脱开式套管打捞器 trip casing spear
脱开凸轮 tripping cam
脱壳 decorticate; decrustation; exuviate; hulk; husk; peel-back; pulled out of its binding; scour; shelling
脱壳谷物 hulled grain
脱壳滚筒 hulling cylinder; hulling drum
脱壳机 decorticator; huller; hulling machine; hulling mill; hulling separator; scouring mill
脱壳肋条 huller rib
脱壳磨机 nut mill
脱壳盘 hulling disk[disc]
脱扣 decoupling; dropout; release; releasing; thread losing; tripping; ungear; dis【计】
脱扣秤 trip scale
脱扣电流 drop-away current; tripping current
脱扣电路 trip circuit
脱扣杆 release lever
脱扣杆叉 release lever yoke
脱扣钩 releasing hook; trip dog
脱扣机构 trip gear; tripping mechanism
脱扣继电器 drop relay; trip(ping) relay
脱扣桨片 trip paddle
脱扣开关 trip switch
脱扣控制杆 trip paddle
脱扣脉冲 tripping pulse
脱扣母线 trip bus
脱扣手柄 free hand
脱扣速度 trip speed
脱扣踏板 trip paddle
脱扣凸轮 deflecting cam
脱扣蜗杆 trip worm
脱扣线 trip line
脱扣线圈 trip(ping) coil
脱扣销 trip pin
脱扣信号 trip signal
脱扣闸板 trip paddle
脱扣整定值 trip setting
脱扣值 drop-away value; dropout value
脱扣装置 releasing arrangement; tripper; tripping device
脱矿泥筛 desliming screen
脱矿物质作用 demineralization
脱矿质 demineralize
脱矿质剂 demineralizer
脱矿质净化器 purification demineralizer
脱矿质器 demineralizer
脱矿质水 demineralized water
脱矿质装置 demineralizing plant
脱矿质作用 demineralization
脱框起模 draw on a frame
脱框式芯盒 loose frame type core box
脱蜡 dewax(ing)
脱蜡车间 dewaxing plant
脱蜡虫胶 dewaxed shellac
脱蜡达玛树脂 dewaxed dam(m)ar
脱蜡工厂 dewaxing plant
脱蜡过程 dewaxing process
脱蜡后的油料 blue oil
脱蜡剂 dewaxing agent
脱蜡漂白虫胶 refined shellac; wax-free white shellac
脱蜡漂白紫胶 refined bleached lac
脱蜡石蜡油 neutral oil
脱蜡油 dewaxed oil

脱蜡装置 dewaxing plant
脱缆钩 disengaging gear; disengaging hook; release mooring hook; releasing hook
脱离 abjunction; abscission; break free; depart; detachment; disengage(ment); divorce from; fall from the plant
脱离板 disengaging plate
脱离层 abscission layer
脱离常规 out-of-the-way
脱离齿轮 releasing gear
脱离点 breakaway point; breaking down point
脱离电线指示器 dewirement indicator
脱离法 spin-off method
脱离关系 disown
脱离轨道 deorbit
脱离架空电线指示器 dewirement indicator
脱离节 abscission joint
脱离锯线的部位 breakaway force; variation in sawing
脱离开的 out-of-gear mesh
脱离控制的市场 runaway market
脱离离合器 disengaging coupling
脱离能级 escape level
脱离啮合 demeshing; ungear
脱离区 abscission zone
脱离群众的经理 separated manager
脱离设备 separation instrumentation
脱离实际的 unbodied
脱离试验设备 separation test facility
脱离速度 escape velocity; velocity of escape
脱离太阳引力区 solar escape
脱离现实 divorced from reality
脱离镶嵌法 breakaway method
脱离协联的 off cam
脱离协联时的飞逸转速 off-cam runaway speed
脱离行星引力区 planet escape
脱离中心 off-center[centre]
脱离状态 disengaged position
脱离子剂 de-ionizer
脱离子水 de-ionized water
脱离子作用 de-ionization
脱力衬套 power takeoff bushing
脱力移动轴 power takeoff shifter shaft
脱沥青 deasphalt(ing); debituminization; diasphaltene
脱沥青剂 deasphalting agent
脱沥青溶剂 deasphalting solvent
脱沥青油 deasphalted oil
脱沥青质 deasplaltene
脱粒 seed extraction; shelling seed husking; thresh grain; threshing
脱粒方法 methods of threshing
脱粒干净 through threshing
脱粒滚筒 beater; shelled cylinder
脱粒滚筒调速器 cylinder governor
脱粒滚筒右向槽纹的纹杆 right-hand cylinder bar
脱粒滚筒轴 cylinder shaft
脱粒滚筒阻稿板 cylinder cutoff
脱粒机 separator; sheller; thresher; threshing machine; threshing mill
脱粒机构 separating mechanism
脱粒机组 threshing rig; threshing set
脱粒率 percentage of thresh
脱粒装置 seed husking establishment; sheller unit; threshing mechanism
脱粒装置离合器 separator clutch
脱粒装置离合器操纵手柄 separator clutch control
脱磷 dephosphorize; dephosphorizing; phosphor removal; phosphorous removal
脱磷法 phosphorous removal method
脱磷生铁 dephosphorized pig iron
脱磷酸 dephosphorylation
脱磷作用 dephosphorization; dephosphorylation
脱鳞锈 descale rust
脱流 breakaway; stall
脱硫 desulfuration[desulphuration]; desulfurize; sulfur removal; sulphur elimination; sulphur removable; sweetening
脱硫焙烧 desulphurizing roasting
脱硫槽 desulfurizer
脱硫车间 sulfur removal plant
脱硫处理 sweetening treatment
脱硫单元 desulfurization unit
脱硫的 desulfurated; desulfurating; devulcanized; sweet
脱硫法 desulfurization method; doctor treatment
脱硫斧 devulcanizing pan
脱硫过程 sweetening process
脱硫化 devulcanization
脱硫化物 sulfide removal
脱硫剂 desulfurater[desulphurater]; desulfurating agent; desulfurizer; desulfurizing agent; reclaimer
脱硫精炼厂 desulfurization refinery
脱硫净化 desulfurization purification
脱硫菌 desulphurization bacteria
脱硫馏分 sweet distillate
脱硫炉 desulfurizing furnace
脱硫气 sweet gas
脱硫气体 processed gas
脱硫器 desulfurizer
脱硫燃料 desulfurized fuel
脱硫设备 sweetener
脱硫石膏 desulfogypsum
脱硫速度 rate of sulphur expulsion
脱硫酸盐作用 desulfation; desulfidation[desulphidation]; desulfurization[desulphurization]; desulfuridation
脱硫塔 desulfurizer; desulfurizing tower; thionizer
脱硫天然气 sweet natural gas
脱硫液 doctor solution
脱硫用石膏 desulfurization gypsum
脱硫油 sweet oil
脱硫重整 Diesulforming
脱硫装置 desulfuring installation; desulfurizer; desulfurizing unit
脱硫作用 desulfation; desulfidation[desulphurization]; desulfurization[desulphidation]
脱漏部分 lacuna[复 lacunae]
脱卤反应 dehalogenation reaction
脱铝作用 dealumination
脱氯 anti-chlorination
脱氯化氢作用 dehydrochlorination
脱氯剂 anti-chlor; dechlorinating agent
脱氯作用 dechlorination
脱落 break off; crumble away; drop-off; dropout; fall off; fallout; pull out; scaling off; shed; shelling; slough(ing); off-grain <指金刚石从钻头脱落>
脱落步骤【计】drop procedures
脱落部分 fallaway section
脱落层 stratum disjunctum
脱落插头 pull-off plug
脱落插座 separation connector
脱落的 caducous; deciduous; shedding
脱落的泥皮 <孔壁> sluff
脱落腐蚀 <氧化膜> runaway corrosion
脱落节 hollow knot; loose knot
脱落连接 pull-off connection
脱落区 obscission zone
脱落生物膜 slough film
脱落式芯盒 knockout core box
脱落素 abscisin
脱落酸 abscisic acid; abscisin
脱落蜗杆 out-of-gear worm
脱落现象 obscission
脱落性 caducity
脱落釉层 stripped glaze
脱锚 anchorage slip; clubbing; slip
脱媒 disintermediation
脱镁 demagging
脱镁叶绿母环类 phorbin
脱镁叶绿素 pheophorbide
脱镁叶绿酸 pheophoebin
脱锰 demanganization; demanganizing
脱棉辊 doffing roller
脱棉滚筒 doffing drum
脱棉器 doffing mechanism
脱棉器圆盘 doffer disk[disc]
脱敏 desensitize
脱敏感(现象) desensitization
脱模层 release layer
脱模衬套 stripper bush
脱模冲程 ejector stroke
脱模冲头 stripper punch
脱模吊车 ingot stripping crane; stripper crane; stripping crane
脱模法 method of stripping
脱模方法 stripping method
脱模粉 separating powder
脱模负荷 ejection load
脱模干燥器 white-hard dryer[drier]
脱模膏 form paste; shuttering paste
脱模隔离涂料 parting agent
脱模工 stripper
脱模工具 extractor
脱模工艺 form scabbing
脱模后的收缩 mo(u)ld shrinkage
脱模后收缩 die shrinkage
脱模混凝土 off-formwork concrete; off-shuttering concrete
脱模机 ingot stripper; knockout machine; stripper(crane); strip(per) machine; stripping device; stripping machine
脱模机构 mo(u)ld emptier
脱模剂 anti-sticking agent; demo(u)lding agent; form coating material; form oil; form paste; form release agent; form release compound; form stripping agent; formwork compound; mo(u)ld lubricant; mo(u)ld oil; mo(u)ld parting agent; mo(u)ld release agent; mo(u)ld separator; parting agent; parting compound; parting medium; parting powder; release agent; releasing agent; separating compound; separating material; separator; shuttering agent; shuttering lube; shuttering oil; shuttle cream; stripping agent; stripping compound; mo(u)ld release
脱模角 draft angle
脱模跨 stripping bay
脱模快速模板 quick strip formwork
脱模困难遗留原处的模板 lost formwork
脱模蜡 forms wax; release wax; wax release agent
脱模力 ejector force; knockout press; release force
脱模起重机 stripping crane
脱模器 stripper
脱模前收缩 mo(u)ld shrinkage
脱模强度 <混凝土> release strength; demoulding strength; stripping strength
脱模日程表 stripping schedule
脱模乳剂 <混凝土> release emulsion
脱模(润滑)膏 release paste
脱模润滑剂 release lube
脱模润滑油 form lube; forms oil; formwork oil; release oil
脱模时间 stripping time
脱模时间表 stripping schedule
脱模式喷头 rubber porous membrane sprayhead
脱模涂层 strippable coating
脱模涂料 mo(u)ld wash; release coating
脱模销 stripper pin
脱模销紧销 knockout latch
脱模楔 releasing key
脱模斜度 draft
脱模性质 release property
脱模压力 ejection pressure; stripping pressure
脱模液 liquid parting; pattern spray
脱模用膜材 sheet release agent
脱模油 form oil; mo(u)ld oil; mo(u)ld release oil; shutter oil
脱模油膏 formwork grease
脱模油剂 shutter cream
脱模装置 knockout
脱膜 ejection; parting; release
脱膜后加工 post-mo(u)lding operation
脱膜机 stripper
脱膜剂 remover
脱膜楔 releasing key
脱墨 deinking
脱墨现象 deinking phenomenon
脱萘 naphthalene removal
脱泥 desilt(ing); desliming; slime separation
脱泥槽 sloughing-off tank
脱泥地带 desilting strip
脱泥分级机 desliming classifier
脱泥弧形筛 desliming sieve bend
脱泥机 deduster; desilter[desiltor]; deslimer
脱泥剂 de-sludging agent
脱泥圆锥分级机 desliming cone; sloughing-off cone
脱黏[粘] abhesion; sticky point; debond(ing); disbonding
脱黏[粘]剂 detackifier; tack eliminator
脱泡 deaerate; deaeration
脱泡机 deaerator
脱泡剂 defoamer agent
脱皮 decrustation; dejacket; desquamation; flake off; peeling; peeling off; scale off; scaling; spalling
脱皮腐蚀 breakaway corrosion
脱片 flaking
脱贫 lift-off poverty
脱坡 sloughing
脱漆 coating removal; de-coat(ing); depaint
脱漆剂 paint and varnish remover; paint remover; paint stripper; stripper; varnish remover
脱漆器 stripper
脱漆溶剂 wiping solver
脱气 deaerate; degasity; degas(sing); outgas; deaeration
脱气槽 deaerating tank
脱气方法 method of degassing
脱气钢 degasified steel
脱气罐 degassing tank
脱气混凝土 deaerated concrete
脱气挤泥机 de-airing auger
脱气剂 deaeration agent; degaser; degasifier; degasifying agent; degassing agent
脱气加热器 deaerating heater
脱气器 apparatus for degasification; deaerator; de-airing machine; degasifier; degasser
脱气式喷洒器 degassing sprinkler
脱气式洒水车 degassing sprinkler

脱气室 deaerating chamber
脱气水 deaerated water;de-aired water;dewater
脱气塔 deaerator;degasser;degassing column;degassing tower
脱气原油 degassed crude oil
脱气原油的比重 specific gravity of degassed oil
脱气装置 deaerating plant;degasser;degassing installation; degassing system
脱气作用 degasification
脱铅 de-lead(ing)
脱浅 refloat(ing)
脱浅拉力 hauling-off pull
脱羟基 dehydroxylation
脱羟基表面 dehydroxylated face
脱氢 dehydrogenate;dehydrogenize
脱氢苯 benzyne;dehydrobenzene
脱氢枞胺 dehydroabietylamine
脱氢枞酸 dehydroabietic acid
脱氢芳烃 aryne
脱氢环化作用 dehydrocyclization
脱氢剂 dehydrogenated agent
脱氢松香 dehydrogenated rosin
脱氢松香胺 dehydroabietylamine
脱氢松香酸 dehydroabietic acid
脱氢乙酸 dehydroacetic acid
脱氢乙酸钠 sodium dehydroacetate
脱氢乙酸盐 dehydroacetate
脱氢异构化作用 dehydroisomerization
脱氢油 dehydrogenated oil
脱氢作用 dehydrogenation
脱轻 <蒸去轻质油> topping
脱轻法 topping process
脱轻石油 topped crude;topped petroleum
脱轻原油 topped crude oil
脱氰作用 decyanation
脱去夹带 deentrainment
脱去旧漆 stripping
脱去离子 de-ionize
脱圈 knocking over
脱圈板 knocking-over bar;knocking-over comb
脱圈沉降片 comb;knocking-over sinker
脱圈动作 knocking-over action
脱圈横列 knocking-over row
脱圈轮 knocking-over wheel
脱圈片 knocking-over bit
脱圈三角 press-off cam
脱圈镶条 knocking-over segment
脱圈栅状齿口 knocking-over verge
脱燃素作用 dephlogistication
脱溶 exsolution;precipitation
脱溶剂器-蒸炒缸 desolventizer-toaster
脱溶硬化 precipitation hardening
脱乳化 de-emulsification;emulsion resolving
脱乳化度试验 demulsibility test
脱乳胶剂 emulsion inhibitor
脱乳作用 demulsionfication
脱色 colo(u)r removal;decolo(u)r(ize);decolo(u)ration;decolo(u)rizing;discolo(u)r;removal of colo(u)r;tarnish
脱色虫胶 white lac;white shellac
脱色的 tarnished
脱色动力学 decolo(u)rization kinetics
脱色计 decolo(u)rimeter
脱色剂 colo(u)r stripper;decolo(u)rant;decolo(u)ring agent;decolo(u)riser[decolo(u)rizer];discharging agent;discolo(u)ration agent;discolo(u)ring agent
脱色力 decolo(u)r power
脱色率 decolo(u)rization index;decolo(u)r ratio
脱色黏[粘]土 discolo(u)ring clay
脱色试验干搓 dry rubbing
脱色水 decolo(u)red water
脱色炭 decolo(u)rizing carbon
脱色效率 decolo(u)rization efficiency
脱色絮凝剂 decolo(u)rization flocculant
脱色紫胶 white lac
脱色作用 decolo(u)rization;discolo(u)ration
脱砂 sand screening
脱砂干燥 sand dry
脱砂箱格子 shakeout grid
脱蛇纹石化 deserpentinization
脱砷 arsenic removal;dearsenification
脱湿 moisture desorption;moisture-free
脱湿器 moisture eliminator;moisture separator;moisture trap
脱湿作用 desorption of moisture
脱手 unload
脱树脂 deresinsate
脱水 dehumidification; dehydrating; desiccation; desorption of moisture;dewater(ing);dry out;elimination of water;evaporation;expulsion of water; hydroextracting; moisture-removal;unwater(ing); water deprivation;water removal;withdrawal of water
脱水柏油 dehydrated tar
脱水保藏法 dehydrated preservation
脱水泵 dewatering pump
脱水蓖麻油 dehydrated castor oil
脱水蓖麻油醇酸树脂 dehydrated castor oil alkyd
脱水蓖麻油脂肪酸 dehydrated castor oil fatty
脱水仓 dewatering bin;dewatering box;dewatering bunker
脱水槽 dehydration tank; dewatering channel;dewatering tank;drench pit
脱水车间 dewatering plant
脱水沉淀池 storage pond for dewatering
脱水池 dewatering tank
脱水处理设施 facility for dehydration treatment
脱水传送带 dewatering conveyer[conveyor]
脱水带 dewatering zone
脱水的 dehydrated
脱水冻结 dehydrofreezing
脱水段 dewatering period
脱水法 evaporation
脱水反应 dehydration reaction
脱水峰 dehydration peak
脱水浮游生物 water-free plankton
脱水干燥设备 dehydrogenation drying equipment
脱水工厂 dewatering plant
脱水谷仓 dewatering silo
脱水固结沉降 dehydrate consolidation settlement
脱水管 dewatering conduit
脱水罐 drain sump;water separating tank;dehydration tank
脱水过滤 dewatering filtration
脱水弧形筛 dewatering sieve bend;draining sieve bend
脱水戽斗 dredging bucket
脱水混凝土 desiccated concrete
脱水机 decker;dehydrater[dehydrator];densifier;dewaterer;dewaterizer;hydroextractor;water extractor
脱水剂 dehydrant;dehydrating agent;dehydrolyzing agent;dewatering agent
脱水加固 dewatering stabilization
脱水焦油(沥青)boiled tar;dehydrated tar
脱水阶段 dehydration stage;water smoking
脱水酒精 dehydrated alcohol
脱水冷冻(法)dehydrofreezing
脱水离心机 dewatering centrifuge
脱水滤池 dewatering filter
脱水滤器 dewatering filter
脱水率 dehydration rate
脱水轮 dewatering wheel
脱水能力 water separation capability
脱水排污舱 dewatering blow-down tank
脱水器 dehydrator;densifier;dewaterer;dewatering unit;dewaterizer;drain separator;hydrodehazer;hydroextractor;separator;water extractor;water separator;water trap
脱水曲线 dewater curve
脱水热 dehydration fever;heat of dehydration
脱水溶胶 hydrosol
脱水筛 dewatering screen;drainage screen;draining screen
脱水山梨(糖)醇 anhydro-sorbite;sorbitan
脱水山梨(糖)醇脂肪酸酯 sorbitan fatty acid ester
脱水设备 dewatering equipment
脱水设施 dewatering facility
脱水石膏 anhydrite;calcined gypsum
脱水石油 dewatering oil
脱水试验 dehydration test
脱水收缩裂缝 syneresis crack
脱水收缩作用 syneresis
脱水输入装置 off-line input device
脱水蔬菜 evaporated vegetables
脱水水煤气焦油 dehydrated water-gas tar
脱水塔 dehydrating tower
脱水提升机 dewatering elevator;drainage elevator
脱水筒仓 dewatering silo
脱水温度 dehydrating temperature;dehydration temperature
脱水污泥 dewatered sludge' dried sludge;dewatering sludge
脱水物 anhydride;calcinate;dehydrant;dehydrate
脱水系数 dewatering coefficient
脱水系统 dewatering system;unwatering system
脱水细筛滤网 dewatering milliscreen
脱水箱 dehydration box;dewatering box
脱水像 dehydration figure
脱水性 dewaterability
脱水性能 dehydration property
脱水压力机 dehydrating press
脱水压榨机 dewatering press
脱水盐 desiccant salt;water desalting
脱水摇动筛 dewatering shaker
脱水用的耙式分级机 rake classifier for dewatering
脱水用的耙式分级器 rake classifier for dewatering
脱水油 dehydrated oil
脱水油墨制造法 flushing process
脱水原油 dehydrated crude(oil)
脱水圆锥分级机 dewatering cone
脱水转鼓 dewatering drum
脱水装置 dehumidifier;dewatering bin;dewatering plant;dewatering unit
脱水作用 anhydration;dehydration;dehydrolysis;deaquation;desiccating action;hydraulic extraction;hydroextraction
脱酸 deoxidate;deoxidize;depickling
脱酸装置 deacidification plant
脱酸作用 deacidification
脱羧基作用 decarboxylation
脱胎 bodiless
脱胎瓷 bodiless chinaware;bodiless porcelain;bodyless porcelain;eggshell porcelain
脱胎器 bodiless ware
脱碳 decarbonization;decarbonize
脱碳薄层 bark
脱碳层 decarbonizing layer;decarburized layer;ferrite banding
脱碳层厚度 practical decarburized depth
脱碳沉积 decarbidize
脱碳反应 decarburizing reaction
脱碳沸腾 carbon boil
脱碳钢 decarburized steel
脱碳钢条 <泡钢的一种废品> aired bar
脱碳化作用 decarbonation;decarburization
脱碳剂 decarboniser[decarbonizer];decarburiser[decarburizer];decarburizing agent
脱碳矿渣 decarburisation[decarburization] slag
脱碳炉渣 decarburisation slag
脱碳气氛 decarburizing atmosphere
脱碳深度 decarburized depth
脱碳时间公式 decarbonating time formula
脱碳酸气塔 decarbonater
脱碳酸盐化作用 decarbonation
脱碳酸作用 decarbonization
脱碳退火 decarburizing annealing
脱碳铸铁粒 decarburized cast iron shot
脱碳组织 decarburized structure
脱碳作用 decarbonation;decarburizing;decarburation;decarburization
脱套位置 press-off position
脱套自停装置 press-off detector
脱体波 bow wave
脱体激波 detached shock wave
脱铁 deferrization
脱铁器 skate throw-off device
脱烃基作用 dehydrocarbylation
脱铜 decopper(ing)
脱铜槽 liberator cell
脱铜电解 copper-stripping electrolysis
脱铜高压釜 copper-stripping autoclave
脱涂膜剂 remover
脱烷基化 de-alkylation
脱网呼叫 off-net call
脱尾 taking-out
脱纬 sloughed filling;sloughed-off weft;slough off
脱位 dislocation
脱位旋钮 off-button
脱纹 hiatus
脱污染 depollution
脱锡 de-tin(ning)
脱锡法 stripping method
脱纤维 defiber
脱险 out of danger
脱险舱 escape compartment
脱险舱口 escape hatch
脱险口 escape scuttle
脱线 derailer
脱线处理【计】off-line processing
脱线的 off-line
脱线器 derailing block;wedge block
脱线数据处理 off-line data processing
脱线铁鞋 derailing block;wedge block
脱线转辙器 derailing point
脱相 out-of-phase
脱箱 drawing;slip flask;snap flask
脱箱造型 removable flask mo(u)lding;slip flask mo(u)ld(ing);snap flask mo(u)ld(ing)
脱箱铸型 slip flask mo(u)ld(ing)
脱硝 denitrate;denitration

脱硝剂 denitrifying agent
脱硝作用 denitrification
脱销 drain on supplies; out of stock; drainage on supplies <商品>
脱销费用 stockout cost
脱鞋道岔 skate throw-off switch
脱鞋器 escapement; skate escapement; skate throw-off device
脱卸次序 stripping order
脱卸接头 tipping link
脱屑 desquamation
脱芯机 mandrel stripper
脱锌 dezinc; dezincify[dezinkify]
脱锌作用 dezincification
脱压实数 decompaction number
脱盐 demineralization; desalinate; desalinization
脱盐槽 salting-out tank
脱盐厂 desalination plant
脱盐车间 desalination plant
脱盐的 desalted; salt-removal
脱盐法 desalination process
脱盐过程 process of salt exclusion
脱盐环境 desalting environment
脱盐技术 desalting technology
脱盐剂 desalter; desalting agent
脱盐率 desalinization ratio; desalting ratio
脱盐膜 desalination membrane; salt rejecting membrane
脱盐期 desalting period
脱盐器 demineralizer; desalter
脱盐强度 desalting strength
脱盐设备 desalination apparatus
脱盐水 demineralized water; desalted water
脱盐系统 desalination system
脱盐因素 desalination factor
脱盐装置 demineralizer; desalination plant; desalter; desalting unit
脱盐装置含盐污水 desalter brine
脱盐作用 desalification; desalination; desalinization; desalt(ing)
脱颜料 depigment
脱颜料作用 depigmentation
脱阳离子作用 decationizing
脱氧 deacidize; deaerating; deaeration; de-airing; degasification; deoxidate; deoxidize; oxygen removal; removal of oxygen
脱氧不良钢 wild steel
脱氧的 deacidizing; deoxidized
脱氧的钢 killed steel
脱氧(方)法 deoxidation method; method of deoxidation
脱氧钢 capped steel; dead setting steel; deoxidized steel; killed steel; piping steel; solid steel
脱氧(合)常数 deoxygenation constant
脱氧合作用 deoxygenation
脱氧核糖【生】 deoxyribose
脱氧核糖核酸 deoxyribonucleic acid
脱氧剂 deoxidant; deoxidation reagent; deoxidizer; deoxidizing addition; deoxidizing agent; desoxidant; dioxidant; killer; killing agent; reductive; scavenger
脱氧玫红初卟啉 deoxy-rhodo-etioporphyrin
脱氧器 deaerator; degasser
脱氧铜 deoxidized copper
脱氧系数 deoxygenation coefficient
脱氧性水体 deoxygenated waters
脱氧渣 deoxidizing slag
脱氧作用 deacidification; deoxid(iz)ation; deoxygenation; desoxidation; desoxygenation; disoxidation
脱叶处理 defoliation process
脱叶剂 defoliant; defoliator
脱液剂 fluid loss agent
脱乙基作用 de-ethylation
脱银 deprivation of silver; desilverization
脱印部分 bite
脱荧光 de-blooming
脱油 deoil(ing); oil removing
脱油沥青 deoiled asphalt
脱油炉 desizer
脱釉 exposed body; peeling; scalding
脱渣能力 slag detachability
脱渣性 detachability; removability of slag
脱脂 defat; degreasing; deoiling; removal of grease; unoil
脱脂槽 degreasing bath
脱脂车间 degreasing plant
脱脂的 defatting; fat-extracted
脱脂剂 degreaser; degreasing agent
脱脂棉 absorbent cotton; cotton wool
脱脂溶剂 degreasing solvent
脱脂溶液 degreasing solution
脱脂乳 butter milk
脱脂纱布 absorbent gauze
脱脂设备 degreasing plant
脱脂装置 degreaser
脱字记号 caret
脱座 unseating

驮背式 piggyback

驮背式集装运输 <小车集于一大车上的运载输送方式> piggyback operation
驮背式联运交通 piggyback traffic
驮背式模架 piggyback formwork
驮背式运输方式 piggyback plan
驮背运输 pick-a-back; pigback; piggyback
驮道 bridge road; bridle path; bridle road; bridle track; bridleway; drove road
驮(货运输的牲)畜 beast of burden
驮筐 dosser
驮马 pack horse
驮马道 pack road
驮兽 <骡、马等> sumpter
驮抬打捞法 method of overhanging lifting
驮箱式喷雾机 saddle tank machine
驮畜 pack animal
驮运道 mule track
驮运路 bridle road; mule track
驮载炮 packing gun
驮载式喷粉机 saddle duster
驮载液箱式喷雾机 saddle tank sprayer
驮子 pack

陀螺 gyro; spinning top; whirligig

陀螺摆 gyropendulum; gyroscopic pendulum
陀螺部件 gyrounit
陀螺操舵装置 gyrosteering gear
陀螺操纵 gyrocontrol
陀螺测试仪 gyrometer
陀螺测速仪 gyrometer
陀螺测斜仪 gyroclinometer; gyrolevel; gyroscopic clinometer; gyroscopic inclinometer
陀螺测斜仪法 gyroscopic-clinograph method
陀螺差 gyro-compass error
陀螺成球机 gyro-granulator
陀螺传感器 gyroscope transducer
陀螺船首防摇器 gyroship stabilizer
陀螺船首偏荡记录器 gyrolog
陀螺船首偏荡指示图 gyrograph
陀螺磁罗经 gyromag; gyromagnetic compass
陀螺磁罗盘 gyromag; gyromagnetic compass; gyrostabilized magnetic compass
陀螺的悬挂外框架 outer gimbal suspension of gyro
陀螺地平 gyroscopic horizon
陀螺地平仪 altitude gyroscope; flight indicator; gyrohorizon; gyrohorizon indicator
陀螺地震计 gyroscopic seismometer
陀螺电池引线 connection for gyrobattery
陀螺电动机 gyromotor
陀螺电机 gyromachine
陀螺定向测量 gyroscopic orientation survey
陀螺定向误差 gyroorientation error
陀螺定向销 pin for fixing gyroposition
陀螺定轴性 gyroscopic inertia
陀螺动力学 gyrodynamics
陀螺动作 gyroscopic action
陀螺方位 gyroscopic bearing; gyrocompass bearing
陀螺方位角 gyroazimuth; gyrobearing
陀螺方向仪 directional gyro(scope)
陀螺防摇装置 gyroscopic stabilizer; gyrostabilizer
陀螺房 case for gyro; rotor case
陀螺俯仰力矩 gyroscopic pitching moment
陀螺俯仰力偶 gyroscopic pitching couple
陀螺复示器 gyrorepeater
陀螺感应罗经 gyro flux-gate compass
陀螺感应罗盘 gyrosyn
陀螺感应同步罗盘 gyro flux-gate compass
陀螺惯性 gyroscopic inertia
陀螺光学导航 gyroscopic erected navigation
陀螺光学导航系统 gyroerectional navigation system
陀螺航向 gyro-compass course
陀螺航向指示器 gyroscopic drift indicator
陀螺回转指示计 gyroscopic turn meter
陀螺机构 gyromechanism
陀螺积分部位 integrating gyrounit
陀螺积分环节 integrating gyrounit
陀螺积分加速度计 gyrointegrating accelerometer
陀螺积分器 gyrointegrator; gyroscope integrator; integrating gyro
陀螺积分仪 gyroscopic integrator
陀螺基准系统 gyroreference system; gyroscope reference system
陀螺加速度表 gyroscopic accelerator
陀螺加速计 gyroaccelerometer; gyroscopic accelerometer
陀螺驾驶仪 gyropilot
陀螺驾驶仪操舵 gyropilot steering
陀螺减振器 gyrovibration absorber
陀螺进动性 gyroprecession
陀螺经纬仪 gyroazimuth theodolite; gyroazimuth transit; gyromeridian indicating instrument; gyroscopic theodolite; gyro-theodolite; gyrotransit; survey gyro(scope)
陀螺经纬仪(定向)法 gyro-theodolite method; gyrotransit method
陀螺开关 gyroon-off switch
陀螺控制 gyrocontrol
陀螺力 gyroscopic force
陀螺力矩 gyro-moment; gyroscopic couple; gyroscopic moment
陀螺力偶 gyroscopic couple
陀螺力学 gyroscopics
陀螺六分仪 gyroscopic sextant; gyrosextant
陀螺罗北 gyrocompass north
陀螺罗方位 gyrocompass bearing
陀螺罗航向 gyrocompass course
陀螺罗经 gyrocompass; gyroscopic compass
陀螺罗经操舵仪 gylot
陀螺罗经差 gyroerror
陀螺罗经迟滞 lag of the gyrocompass
陀螺罗经的水银稳定器 mercury ballistic of gyrocompass
陀螺罗经对准 gyrocompass alignment
陀螺罗经方位 bearing per gyrocompass
陀螺罗经复示器 gyrocompass repeater; repeater gyrocompass
陀螺罗经改正量 correction of gyrocompass
陀螺罗经航向记录器 gyrocourse recorder
陀螺罗经校准指北 gyrocompass alignment
陀螺罗经控制箱 gyrocontrol unit
陀螺罗经日志 gyrocompass log
陀螺罗经室 gyrocompass room; gyroscopic compass room
陀螺罗经艏向误差 error of gyrocompass heading; error of heading of gyrocompass
陀螺罗经随动部分 phantom
陀螺罗经误差 error of gyrocompass; gyrocompass error
陀螺罗经主仪 master gyrocompass
陀螺罗经状态 gyrocompassing
陀螺罗经自动舵 gyropilot
陀螺罗盘 gyro; gyrocompass; gyroscopic compass; gyrostatic compass
陀螺罗盘偏转 gyroscopic deflection
陀螺马达 gyromotor
陀螺瞄准具 gyroscopic sight; gyrosight
陀螺瞄准器 gyrosight
陀螺扭矩 gyroscopic torque
陀螺漂移 drift of the gyro; time running of gyroscope
陀螺漂移值 amount of drift of the gyro; amount of gyroscopic drift
陀螺平衡 gyroscopic equilibrium
陀螺平台 gyropanel; gyroplatform; gyrostabilized platform
陀螺平台指北 gyrocompassing
陀螺倾斜计 gyroscopic inclinometer
陀螺倾斜仪 gyroscopic clinometer
陀螺球 gyroball; gyrosphere
陀螺球顶电极 upper current conducting cap
陀螺球高度位置指示器 height indicator of gyrosphere
陀螺球密封环 sealing ring of gyrosphere
陀螺球稳心高度 metacenter height of gyrosphere
陀螺球座 support for gyrosphere
陀螺式倾斜仪 gyroscopic inclinometer
陀螺式摇摆 wabble
陀螺输出轴信号传感器 gyrooutput-axis pick-up
陀螺水平仪 gyrohorizon; gyroscopic horizon
陀螺水准仪 gyrolevel; gyroscopic level
陀螺探管 gyroprobe
陀螺同步罗经 gyrosyn(compass)
陀螺同步罗盘 gyrosyn(compass)
陀螺稳定磁罗经 gyrostabilized magnetic compass
陀螺稳定法 gyroscopic stabilization
陀螺稳定罗盘 gyrostabilised[gyrostabilized] compass

T

陀螺稳定平台 gyrostabilised platform
陀螺稳定器 gyroscopic stabilizer; gyrostabilizer; gyrostat
陀螺稳定太阳卫星 gyrostabilized solar satellite
陀螺稳定天线 gyrostabilized antenna
陀螺稳定系统 gyratory stabilizing system; gyrostabilized system
陀螺稳定效应 gyrostatic effect
陀螺稳定性 gyroscopic stability
陀螺稳定仪表 gyrostabilized instrument
陀螺稳定重力仪 gyrostabilized gravimeter
陀螺稳定装置 gyrostabilization unit; gyro-stabilized mount; gyrostabilized mount; gyrostabilized unit
陀螺系统 gyrosystem
陀螺效应 gyroscopic effect
陀螺形的 pegtop; trochoid
陀螺修正电动机 torque motor
陀螺旋转轴 spin axis of the gyro
陀螺学 gyrostatics
陀螺液压操舵系统 gyrohydraulic steering control
陀螺液压操舵装置 gyrohydraulic steering system
陀螺仪 gyro; gyrocompass; gyroinstrument; gyroscope; gyroscopic apparatus; gyroscopic equipment; gyrostat
陀螺仪表 gyroscopic instrument
陀螺仪的稳定 gyrostabilization
陀螺仪定向 gyroscopic orientation
陀螺仪方位 gyrobearing
陀螺仪附件 gyroattachment
陀螺仪基准组件 gyroreference package
陀螺仪记录簿 gyrolog
陀螺仪减摇震装置 gyroantihunt
陀螺仪进动 gyroscopic precession
陀螺仪控制装置 gyrocontrol installation
陀螺仪瞄准器 gyro gunsight
陀螺仪内框架 inner gimbal suspension of gyro
陀螺仪内框架轴 inner gimbal axis of gyro
陀螺仪漂移 gyroscopic drift
陀螺仪曲线图 gyrograph
陀螺仪水平 automatic horizon
陀螺仪外壳 gyrohousing
陀螺仪外框架轴 outer gimbal axis of gyro
陀螺仪位置传感器 gyroscope position pick-up
陀螺仪误差 gyroerror
陀螺仪学 gyroscopics
陀螺仪照相机 gyrocamera; gyroscope-camera; gyroscopic camera
陀螺仪指示 gyroindication
陀螺仪(主)轴 axis of gyroscope; gyroaxis
陀螺仪转子 gyrowheel
陀螺仪装备 gyroscope equipment
陀螺仪子午线 gyromeridian; gyroscope meridian line
陀螺仪组 gyropackage
陀螺运动 circumgyrate
陀螺运动规律 gyrolaw of precession
陀螺增益 gyrogain
陀螺振子 gyrotron
陀螺振子微波激射器 gyrotron maser
陀螺轴 gyroaxis
陀螺主罗经室 gyroroom
陀螺转换器 gyroconverter
陀螺转子 gyroflywheel; gyro-rotor; gyroscope rotor
陀螺转子端隙测量计 gyrorotor end play micrometer
陀螺转子转动惯量 moment of inertia of gyrorotor
陀螺子午定向仪 gyromeridian indicating instrument
陀螺自动操舵(装置) gyropilot
陀螺自动导航系统 gyroautomatic navigation system
陀螺自动驾驶仪 gyropilot; gyrorudder; gyroscope autopilot
陀螺自动驾驶仪操纵 gyrorudder control
陀螺自动驾驶仪控制 gyrorudder control
陀螺自转矢量轴 gyrovector axis
陀螺自转指示器 gyroscopic turn meter
陀螺自转轴 gyrospin axis
陀螺总漂移值 total gyroscopic drift
陀螺组合件的外壳 gyrocase
陀螺组件 gyrounit
陀螺作用 gyroscopic action; gyroscopic effect
陀螺作用的流量计 gyroscopically balanced flowmeter

沱 small bay

沱 <动力黏[粘]度单位> stoke

驼背 hump

驼背上弦杆 <桁架的> camel-back top chord
驼背式桁架 camel-back truss
驼背形桥 hump-back bridge
驼队 team
驼峰 camel back; cat's back; double incline; hump; hump-shaped pad block【铁】
驼峰办公室【铁】hump office
驼峰编解效率 <每分钟溜放车数> classification rate of cars
驼峰编组场 gravitation yard; gravity yard; hump yard; summit yard
驼峰编组场低速区段 low-speed section of hump yard
驼峰编组场头部 grouping of hump yard; headend of hump yard; head of hump; hump end of yard
驼峰编组场尾部 bowl end; tail of a hump yard
驼峰编组场纵断面最低处 bowl end
驼峰编组场作业能力 capacity of hump yard
驼峰编组站 hump yard; summit yard
驼峰编组站低速区段 low-speed section of hump yard
驼峰编组站作业能力 capacity of hump yard
驼峰编组作业 gravity marshalling operation
驼峰操纵台 hump console
驼峰操作员 hump conductor
驼峰测流槽 flume with hump
驼峰场道岔区坡 switching grade
驼峰初速能高 <推峰速度能高> humping velocity head
驼峰道岔 hump switch
驼峰道路 hump road
驼峰底 bowl end
驼峰电气集中 electric(al) interlocking for hump yard
驼峰电视 hump television
驼峰电压 hump voltage
驼峰调车 gravity shunting; gravity switching; hump shunting; switching by gravitation
驼峰调车场【铁】double incline; gravitational yard; hump classification yard; hump(ed) yard; gravity yard; summit yard
驼峰调车场头部 grouping of hump yard; hump yard classification throat; head of hump
驼峰调车场尾部 tail of hump yard; tail throat of a hump yard
驼峰调车场线路坡度 grade of hump yard tracks
驼峰调车场线群 group of hump yard tracks
驼峰调车场咽喉 hump yard neck
驼峰调车场制动位的坡度【铁】grade through brake position
驼峰调车法【铁】hump switching
驼峰调车轨道 hump track
驼峰调车机车 hump locomotive
驼峰调车区 <自峰顶至编组线计算停车点之间> distribution zone
驼峰调车线 humping track
驼峰调车员 fixed man
驼峰调车长 hump master
驼峰调车作业通知单 humping list; hump switch list
驼峰顶 crown of hump; summit
驼峰顶端 crest end
驼峰顶控制计算机 crest control computer
驼峰分溜放线 hump sub-lead
驼峰峰底 class bowl; classification bowl
驼峰峰底信号楼 bowl tower
驼峰峰顶 apex of hump; crest; crest of hump; hump crest; hump summit; summit of hump
驼峰峰顶股道 bowl track
驼峰峰顶至计算停车点间距离 crest-to-clearance distance
驼峰复示信号机【铁】hump signal repeater
驼峰高度 height of hump; hump height
驼峰轨道 hump track
驼峰轨道衡 hump scale
驼峰桁架 camel-back truss
驼峰缓行器 hump rail brake
驼峰缓行器编组场 hump retarder classification yard; hump retarder yard
驼峰回头线 hump and kickback
驼峰机车 hump engine; hump locomotive
驼峰机车控制需要的推峰速度 hump locomotive control-desired hump speed
驼峰机车遥控 remote control of hump engines
驼峰机车折返取车 run-round of hump locomotive
驼峰及调车设备 hump and marshalling facility
驼峰计算机 hump computer
驼峰技术作业间隔 <即列车推峰作业周期> hump technological interval
驼峰加速坡 accelerating grade of hump; acceleration grade of hump
驼峰解体能力 break-up capacity of hump; humping capacity
驼峰控制机 hump control machine
驼峰控制楼 hump cabin; hump control tower; retarder tower
驼峰控制楼作业员 hump tower operator
驼峰控制盘 hump control panel
驼峰控制台 hump control platform
驼峰连接员室 couper's cabin at hump crest; hump crest coupler's cabin
驼峰溜车方向 rolling direction of hump
驼峰溜放部分 hump distribution zone; rolling section of hump
驼峰溜放部分道岔 <自峰顶至编组线计算停车点间> distribution points
驼峰溜放部分中间坡 intermediate grade
驼峰溜放点 roll-off point
驼峰溜放坡 gravity incline; humping gradient; hump lead; hump track
驼峰溜放线 hump lead; lead track; rolling track of hump
驼峰溜放装置 runoff installation
驼峰路段的加速设备 accelerating device for humps
驼峰平均下坡度 average down grade of hump
驼峰坡度 hump gradient
驼峰区 <在到达场与编组场之间> hump area
驼峰曲线 camel-back curve; hump curve
驼峰式静水池 hump-type stilling basin
驼峰式桥 camel-back bridge; hump-back bridge
驼峰式退火炉 hump-back conveyor furnace
驼峰式消力池 hump-type stilling basin
驼峰式消能池 humping-type stilling basin
驼峰式纵断面 hump profile
驼峰头部 hump end of yard
驼峰头部缓行器 hump end retarder
驼峰头部减速器 hump end retarder
驼峰推峰部分 humping part
驼峰推峰线 track leading from receiving yard to hump
驼峰推送部分 pushing section of hump
驼峰推送坡 humping gradient
驼峰推送坡段 humping gradient section
驼峰推送线 hump track; pushing track of hump
驼峰推送线平均速度 average gradient of track leading to crest of hump
驼峰尾部 trim end
驼峰尾部存车清单 bowl inventory
驼峰尾部缓行器 foot hump retarder
驼峰尾部减速器 foot hump retarder
驼峰下调车 pulldown
驼峰下缓行器 valley brake
驼峰下线路 track under the hump
驼峰下线路群 classification bowl
驼峰下整理车辆作业 trimming operation
驼峰线 hump track
驼峰线路 humping track
驼峰效应 hump effect
驼峰信号 hump signal
驼峰信号机【铁】hump signal
驼峰信号控制 hump signal control
驼峰信号楼 hump signal tower; retarder tower
驼峰形锤式破碎机 camel-back-type crusher
驼峰咽喉 <进入编组线以前的> track lead
驼峰迂回线 hump avoiding line; loop track of hump; roundabout line of hump; thoroughfare track of hump; track around the crest of hump
驼峰预排进路【铁】programmed switching
驼峰值班员 hump master
驼峰制动器 hump rail brake
驼峰主任 hump master
驼峰自动化 humping automation
驼峰自动集中 automatic interlocking for hump yard
驼峰自动集中机 automatic switching machine
驼峰自动集中控制盘 automatic switc-

hing panel
驼峰纵断面模拟器 hump profile simulator
驼峰纵向断面 < 铁路 > longitudinal hump profile
驼峰作业能力 humping capacity
驼峰作业区 < 从向驼峰推送起点至编组线计算停车点为止的地区 > humping zone
驼峰作业员 hump conductor; hump operator
驼毛刷 camel hair
驼色 Beige; camel

柁 girder

柁墩 wooden pier
柁头 girder head

妥尔油 < 蒸煮硫酸盐木浆的副产品 > liquid rosin; tallol; tall oil

妥尔油醇酸树脂 tall alkyd
妥尔油的初馏分 tall oil heads
妥尔油脚 sulfate pitch
妥尔油沥青 tall oil pitch
妥尔油清漆 tall varnish
妥尔油松香 tall oil rosin
妥尔油酸 talloleic acid
妥尔油酸季戊四醇酯 pentaerythritol tallate
妥尔油酸山梨醇酯 sorbitol tallate
妥尔油酸锌 zinc tallate
妥尔油酸酯 tallate
妥尔油皂 tall oil soap
妥尔油脂肪酸 tall oil fatty acid
妥流 tranquil flow
妥善的投送回单 good delivery receipt
妥帖装配 forced fit
妥协方案 half measure; compromise option
妥协条款 compromise clause
妥协条例 compromise act

椭变面镜 ellipsoidal mirror

椭率测量术 ellipsometry
椭率计 ellipsometer
椭球扁率 ellipticity of ellipsoid; flattening of ellipsoid
椭球变星 ellipsoidal variable
椭球参数 ellipsoidal parameter
椭球长半径 major radius of ellipsoid
椭球法线 spheroidal normal
椭球反射镜 ellipsoidal reflector
椭球管板 ellipsoidal tubesheet
椭球函数 ellipsoidal function
椭球汇合处 ellipsoidal junction
椭球极扁率 polar flattening of ellipsoid
椭球界 ellipsoidal bound
椭球粒 axiolite
椭球螺坐标 elliptic(al) coordinates
椭球面 ellipsoid; ellipsoidal surface
椭球面大地测量学 geodesy on the ellipsoid; spheroid geodesy
椭球面的 ellipsoidal
椭球面法线 spheroidal normal; ellipsoidal normal
椭球面方位角 spheroidal azimuth
椭球面高 ellipsoidal height; spheroidal height
椭球面光栅 ellipsoidal grating
椭球面归算 reduction to ellipsoid
椭球面角超 ellipsoidal excess
椭球面镜 ellipsoidal mirror
椭球面距离 ellipsoidal distance; spheroidal distance
椭球面曲率 ellipsoidal curvature
椭球面三角形 ellipsoidal triangle
椭球面纬度 spheroidal latitude
椭球面坐标 ellipsoidal coordinates; spheroidal coordinates
椭球模数 ellipsoidal norm
椭球模型 ellipsoidal model
椭球偏心率 eccentricity of ellipsoid
椭球偏振 ellipsoidal polarization
椭球平均轴 mean axis of an ellipsoid
椭球区分线 spheroidal junction
椭球双星 ellipsoidal binary
椭球体 ellipsoid
椭球体扁率 ellipticity of spheroid
椭球体的 ellipsoidal
椭球体法 < 探伤定位的 > ellipsoid method
椭球体法线 ellipsoidal normal
椭球体泛光灯 ellipsoidal floodlight; scoop
椭球体高程 ellipsoidal height
椭球体聚光灯 ellipsoidal spotlight
椭球体群 spheroidal group
椭球体天顶距 ellipsoidal zenith
椭球天顶 ellipsoidal zenith; geodetic zenith
椭球(调和)函数 ellipsoidal harmonics
椭球投影 elliptic(al) projection
椭球图纬度渐长率 meridian parts for spheroid
椭球弦距 ellipsoidal chord distance
椭球星系 spheroidal galaxy
椭球形 ellipsoidal shape; spheroidicity
椭球形漂移管 ellipsoidal drift tube
椭球旋转轴 rotational axis of the ellipsoid
椭球状的 ellipsoidal; spheroidal
椭球状结核 ellipsoidal nodule
椭球状熔岩 ellipsoidal lava; pillow lava
椭球状水准面 ellipsoidal geoid
椭球状效应 ellipticity effect
椭球子午线 ellipsoidal meridian; spheroidal meridian
椭球坐标 ellipsoidal coordinates
椭形函数 < 解决力学问题的一个数学方法 > elliptic(al) function
椭性的 elliptic(al)
椭性对合 elliptic(al) involution
椭性透射 elliptic(al) homology
椭性线汇 elliptic(al) congruence
椭性圆束 elliptic(al) pencil of circle
椭性直射 elliptic(al) collineation; elliptic(al) projectivity
椭曳冲断层 stretch fault; stretch thrust
椭圆摆轮 ovalizing balance
椭圆摆线 elliptic(al) trochoid
椭圆板 elliptic(al) plate
椭圆半日分潮 elliptic(al) semidiurnal constituent
椭圆半软式波导 elliptic(al) semiflexible waveguide
椭圆半圆孔 ellipsoidal semi-spheric(al) pore
椭圆半轴 semi-axis of an ellipse
椭圆蚌线 elliptic(al) conchoid
椭圆边 elliptic(al) edges
椭圆扁率 ellipticity of ellipse; ellipticity of spheroid
椭圆变换 elliptic(al) transformation
椭圆表 ellipticity tables
椭圆表示法 elliptic(al) representation
椭圆波 elliptic(al) wave
椭圆波导 elliptic(al) guide
椭圆槽 oval groove
椭圆槽密封面 oval groove seal contact face; ovally grooved seal face
椭圆测位制 distance sum measurement
椭圆长短轴比 ratio of major and minor axis of ellipse
椭圆长轴 major axis of an ellipse
椭圆场 elliptic(al) field
椭圆潮 elliptic(al) tide
椭圆车轮 elliptic(al) wheel
椭圆尺 oval scale
椭圆齿轮 elliptic(al) gear; oval gear
椭圆齿轮流量计 oval gear flowmeter; oval gear type water meter
椭圆船尾 elliptic(al) stern
椭圆窗 oeil de boeuf
椭圆次系 elliptic(al) subsystem
椭圆粗轧孔型 oval-shaped roughing pass
椭圆锉 cross file; oval file
椭圆弹孔 key hole
椭圆导电柱 elliptic(al) conducting cylinder
椭圆的 elliptic(al); oval
椭圆的短轴 minor axis of an ellipse
椭圆的曲率 elliptic(al) curvature
椭圆底贮槽 round bottom tank
椭圆点 elliptic(al) point
椭圆垫 oval ring
椭圆定位制 distance sum measurement system
椭圆度 ellipticity; out-of-roundness; ovality; ovalness
椭圆度比 ellipticity ratio
椭圆断面 elliptic(al) section
椭圆对称 ellipsometry
椭圆对称分布 elliptically symmetric distribution
椭圆对合 elliptic(al) involution
椭圆阀 elliptic(al) valve
椭圆阀图 slide valve ellipse
椭圆反光镜 elliptic(al) glass reflector
椭圆反射镜 elliptic(al) reflector
椭圆方程 elliptic(al) equation
椭圆方孔型系统 oval-square passes
椭圆分布 elliptic(al) distribution
椭圆分水尖 elliptic(al) cutwater
椭圆封头 ellipse head; elliptic(al) head
椭圆缝舟藻 Raphoneis elliptica
椭圆复形 elliptic(al) complex
椭圆钢丝 oval steel wire
椭圆钢丝曲头钉 oval-wire brad
椭圆格子砖 Moll checker
椭圆拱 basket handle arch; elliptic(al) arch; oval arch
椭圆拱坝 elliptic(al) arch dam
椭圆拱顶 elliptic(al) vault
椭圆股钢丝绳 oval strand wire rope
椭圆管 elliptic(al) tube
椭圆管板 elliptic(al) tubesheet
椭圆管式翅片管 elliptic(al) fin tube
椭圆光轮 vesica-piscis
椭圆规 ellipsograph; elliptic(al) compasses; elliptic(al) trammel; oval compass; trammel
椭圆规原理补偿法 means of compensation with principle of ellipsograph
椭圆轨道 ellipsoidal orbit; elliptic(al) orbit; elliptic(al) trajectory
椭圆轨道捕获 elliptic(al) orbiting capture
椭圆轨道寿命 elliptic(al) orbit life
椭圆轨道速度 elliptic(al) orbiting velocity
椭圆轨迹 elliptic(al) orbit
椭圆函数 elliptic(al) function
椭圆函数波 conoidal wave
椭圆函数的模 modulus of elliptic(al) function
椭圆函数域 elliptic(al) function field
椭圆和谐运动 elliptic(al) harmonic motion
椭圆横断面 elliptic(al) (cross-) section; oval (cross-) section
椭圆弧 elliptic(al) arc
椭圆弧轨迹 elliptic(al) segment trajectory
椭圆花环形链幕 elliptic(al) curtain in the chain system
椭圆化 ovalization[ovalisation]
椭圆画规 elliptic(al) compasses
椭圆环 ba(s) ton; elliptic(al) ring; torus[复 tori]
椭圆环形构造 elliptic(al) ring structure
椭圆回转曲面 elliptic(al) rotational curved surface
椭圆混凝土管 elliptic(al) concrete pipe
椭圆活塞 oval piston
椭圆积分 elliptic(al) integral
椭圆积分的模 modulus of elliptic(al) integral
椭圆积分模数 modulus of elliptic(al) integral
椭圆基准 ellipsoidal datum
椭圆极化 elliptic(al) polarization
椭圆极化波 elliptically polarized wave
椭圆极化天线 elliptic(al) polarized antenna
椭圆几何 elliptic(al) geometry; Riemannian geometry
椭圆计 ellipsometer
椭圆计法 ellipsometry method
椭圆夹头 oval chuck
椭圆尖头拱 elliptic(al) pointed arch
椭圆交换 elliptic(al) transformation
椭圆角钉 oval brad
椭圆节 oval knot
椭圆经线 elliptic(al) meridian
椭圆精轧孔型 finishing oval pass
椭圆镜 elliptic(al) mirror
椭圆聚光灯 ellipsoidal spotlight
椭圆卡盘 elliptic(al) chuck
椭圆壳体圆屋顶 elliptic(al) shell cupola
椭圆空间 elliptic(al) space
椭圆孔 elliptic(al) aperture; elliptic(al) hole; oblong hole
椭圆孔口 elliptic(al) opening; elliptic(al) orifice; oblong aperture
椭圆孔筛 oblong-hole screen
椭圆孔式 oval orifice
椭圆孔隙 oblong aperture
椭圆孔型用围盘 oval repeater
椭圆粒 ellipsoid particle
椭圆粒体 ellipsoid
椭圆量规 tram trammels
椭圆流量计 oval flowmeter
椭圆楼梯 elliptic(al) stair(case)
椭圆率 elliptic(al) rate; flattening; ellipticity
椭圆率参数 ellipticity parameter
椭圆率条件 ellipticity condition
椭圆轮 elliptic(al) wheel
椭圆轮胎 elliptic(al) tire
椭圆螺栓孔 elliptic(al) bolt hole
椭圆埋头螺钉 oval countersunk screw
椭圆埋头铆钉 oval countersunk head rivet
椭圆门执手 oval door knob
椭圆面方位角 elliptic(al) azimuth
椭圆面积 ellipse area; elliptic(al) area
椭圆面斜边 beaded bevel
椭圆面坐标 ellipsoidal coordinates
椭圆模函数 elliptic(al) modular function
椭圆模函数域 elliptic(al) modular function field
椭圆模群 elliptic(al) modular group

椭圆模型 elliptic(al) model
椭圆盘绕 <管子的> elliptic(al) coiling
椭圆抛物面 elliptic(al) paraboloid; paraboloid
椭圆抛物面壳顶 elliptic(al) paraboloidal roof
椭圆抛物面壳体 elliptic(al) paraboloid shell
椭圆配汽图 <蒸汽机的> ellipse valve diagram
椭圆偏光 elliptic(al) polarized light
椭圆偏心齿轮传动 oval-eccentric gearing
椭圆偏心率 eccentricity of ellipse
椭圆偏振 elliptic(al) polarization
椭圆偏振波 elliptically polarized wave
椭圆偏振辐射 elliptically polarized radiation
椭圆偏振光 elliptically polarized light
椭圆平板 elliptic(al) slab
椭圆平面几何学 elliptic(al) plane geometry
椭圆平头钉 oval brad head nail
椭圆奇点 elliptic(al) singular point
椭圆穹隆 elliptic(al) vault
椭圆球 ellipsoid
椭圆球柄 ellipsoid knob
椭圆球面壳 ellipsoidal shell
椭圆球形体 toroid
椭圆区域 elliptic(al) region
椭圆曲面 elliptic(al) surface; oval calotte
椭圆曲线 elliptic(al) curve
椭圆裙活塞 elliptic(al) skirted piston
椭圆绕线轴 oval reel
椭圆扫描 elliptic(al) scanning; elliptic(al) sweep; elliptic(al) trace
椭圆色度副载波 elliptic(al) subcarrier
椭圆色散 elliptic(al) dispersion
椭圆扇形 elliptic(al) sector
椭圆时差 elliptic(al) equation
椭圆时基(扫描)线 elliptic(al) time base
椭圆双曲线系统 ellipse-hyperbolic system
椭圆算子 elliptic(al) operator
椭圆弹簧 double laminated spring
椭圆体 ellipsoid
椭圆体法 <探伤定位的> ellipsoid method
椭圆庭院 oval court
椭圆筒 cylindroid
椭圆头开口铆钉 oval head split rivet
椭圆头螺钉 oval head screw
椭圆头螺丝 oval head screw
椭圆头铆钉 oval head rivet; oval rivet
椭圆头铜铆钉 oval head brass rivet
椭圆投影 elliptic(al) projection; oval projection
椭圆凸轮 elliptic(al) cam
椭圆凸缘 oval flange
椭圆纬线 elliptic(al) parallel
椭圆屋顶 elliptic(al) dome
椭圆屋顶建筑 elliptic(al) cupola
椭圆无理函数 elliptic(al) irrational function
椭圆误差 elliptic(al) error
椭圆系统 elliptic(al) system
椭圆线圈 elliptic(al) coil
椭圆形 ellipse; elliptic(al) shape; oblong; elliptic(al) type
椭圆形斑 elliptic(al) spot
椭圆形板 elliptic(al) flat-plate; oval plate; oval slab
椭圆形板弹簧 elliptic(al) plate spring
椭圆形板弹簧垫 elliptic(al) plate spring bolster
椭圆形波导管 elliptic(al) waveguide
椭圆形槽 elliptic(al) slot
椭圆形测深(重)锤 elliptic(al) type weight
椭圆形差分方程 elliptic(al) partial differential equation
椭圆形长轴 major axis of ellipse; transverse
椭圆形充气室 oval air chamber
椭圆形船尾 elliptic(al) stern
椭圆形磁轴 elliptic(al) magnetic axis
椭圆形锉 elliptic(al) file
椭圆形大厦 ellipsoid dome
椭圆形刀头 skew cutter
椭圆形的 disciform; ellipsoidal; elliptic(al); oval-shaped
椭圆形的筒形穹隆 elliptic(al) barrel vault
椭圆形等值线 elliptic(al) isoline
椭圆形地板钉 oval lost head nail
椭圆形电线钉 brad
椭圆形叠板弹簧 elliptic(al) spring
椭圆形顶盖 ellipsoidal head
椭圆形断面 egg-shaped cross-section; ellipse section
椭圆形断面钢筋 oval-section steel
椭圆形断面钢丝 oval wire
椭圆形二次超曲面 elliptic(al) quadric hypersurface
椭圆形二次曲面 elliptic(al) quadratic surface
椭圆形方程 elliptic(al) equation
椭圆形非欧几何学 elliptic(al) non-Euclidean geometry
椭圆形封头 ellipsoidal head
椭圆形钢筋 oval wire
椭圆形拱 oval arch
椭圆形拱坝 elliptical arch dam
椭圆形拱顶 ellipsoid vault
椭圆形股钢丝绳 elliptic(al) strand rope
椭圆形股绳 oval strand
椭圆形光圈式波导管 elliptic(al) iris waveguide
椭圆形规 oval scale
椭圆形涵洞 egg-shaped culvert; elliptic(al) conduit; elliptic(al) culvert; oval culvert
椭圆形活塞 oval piston
椭圆形尖拱 elliptic(al) pointed arch
椭圆形绞盘 elliptic(al) winch
椭圆形截面 elliptic(al) (cross-) section; oval (cross-) section
椭圆形截面织针 elliptic(al) wire
椭圆形井筒 elliptic(al) shaft
椭圆形井眼 elliptic(al) hole
椭圆形救生筏 elliptic(al) liferaft
椭圆形卷装 elliptic(al) build package
椭圆形壳 toroidal shell
椭圆形壳顶 ellipsoidal shell roof
椭圆形坑道 elliptic(al) gallery
椭圆形孔 slotted eye
椭圆形喇叭 elliptic(al) horn
椭圆形缆索屋顶 cable elliptic(al) roof
椭圆形老虎窗 oval luthern
椭圆形黎曼曲面 elliptic(al) Riemann surface; elliptic(al) type of Riemann surface
椭圆形立柱 oval-shaped column
椭圆形连铸机 oval bow machine
椭圆形炉栅 elliptic(al) grate
椭圆形埋头钉 oval lost head nail
椭圆形模 elliptic(al) norm
椭圆形喷涂图案 tapered pattern
椭圆形偏微分方程 elliptic(al) partial differential equation
椭圆形偏微分算子 elliptic(al) partial differential operator
椭圆形前大灯 ellipsoidal head lamp
椭圆形切割 almond(shaped) cut
椭圆形穹顶 ellipsoid dome; elliptic(al) dome; oblong dome
椭圆形球顶 ellipsoid dome
椭圆形球面镜 ellipsoidal mirror
椭圆形容器 elliptic(al) vessel
椭圆形栅极 oval grid
椭圆形蛇形管 coiled oval
椭圆形石拱桥 elliptic(al) stone arch bridge
椭圆形穗 elliptic(al) spike
椭圆形弹簧 elliptic(al) spring
椭圆形天线 elliptic(al) antenna
椭圆形铁芯的磁性天线 ellipsoidal core antenna
椭圆形庭院 oval court
椭圆形通道 elliptic(al) conduit
椭圆形头部汽缸 elliptic(al) nosed cylinder
椭圆形凸轮 oval cam
椭圆形围栏 elliptic(al) gallery
椭圆形微分算子 elliptic(al) differential operator
椭圆形伪微分算子 elliptic(al) pseudo-differential operator
椭圆形问题 elliptic(al) problem
椭圆形屋顶 elliptic(al) roof
椭圆形洗面器 oval lavatory
椭圆形限制性问题 elliptic(al) restricted problem
椭圆形消声器 oval muffler
椭圆形销 elliptic(al) pin
椭圆形小湖 bay lake
椭圆形谐振器 ellipsoid resonator
椭圆形星轮 oval-shaped sprocket
椭圆形行程的振动筛 vibratory screen with elliptic(al) movement
椭圆形旋管 coiled oval
椭圆形旋转磁场 elliptic(al) rotating field
椭圆形旋转式空腔谐振器 ellipsoidal cavity resonator
椭圆形旋转式谐振器 ellipsoid resonator
椭圆形压力圈 pressure ellipse
椭圆形扬声器 ellipse speaker; elliptic(al) loudspeaker
椭圆形样板 oval copy
椭圆形窑 oval kiln
椭圆形油漆刷 oval varnishing brush
椭圆形运转的进料器 feeder with elliptic(al) movement
椭圆形振动筛 elliptic(al) vibrating screen
椭圆形执手门的家具 oval knob door furniture
椭圆形重锤 elliptic(al) type weight
椭圆形轴承 elliptic(al) bearing
椭圆形柱 oval column
椭圆性 ellipticalness
椭圆悬链线 elliptic(al) catenary
椭圆旋转场 elliptic(al) rotating field
椭圆循环群 elliptic(al) cyclic group
椭圆轧槽 oval pass
椭圆样板 ellipse template
椭圆叶片 <推进器> elliptic(al) blade
椭圆叶片截面 elliptic(al) blade section
椭圆仪 elliptic(al) trammel
椭圆油刷 oval paint brush
椭圆余摆线波 ellipsoidal trochoidal wave; elliptic(al) trochoidal wave
椭圆余弦波 conoidal wave; elliptic(al) cosine wave
椭圆余弦波理论 conoidal wave theory
椭圆运动 elliptic(al) motion
椭圆运动振动筛 elliptic(al) motion vibrating screen
椭圆运行轨道 <波浪运动中水质点的> elliptic(al) orbit
椭圆振动 elliptic(al) vibration
椭圆正弦函数 elliptic(al) sine function
椭圆中轴【数】 mean axis of an ellipsoid
椭圆轴承 elliptic(al) bearing
椭圆主轴 principal axis of ellipse
椭圆柱 elliptic(al) column; elliptic(al) cylinder
椭圆柱的 cylindroid
椭圆柱函数 elliptic(al) cylinder function
椭圆柱面 elliptic(al) cylindric(al) surface
椭圆柱面波函数 elliptic(al) cylindric(al) wave function
椭圆柱坐标 elliptic(al) cylinder coordinates; elliptic(al) cylindric(al) coordinates
椭圆状的 axiolitic
椭圆状拱顶 ellipsoid vault
椭圆状异常 elliptic(al) anomaly
椭圆锥面【数】 elliptic(al) cone
椭圆锥磨头 oval tapered grinding head
椭圆锥曲面 elliptic(al)-conic(al) curved surface
椭圆锥体 elliptic(al) conoid
椭圆族 family of ellipse
椭圆坐标 elliptic(al) coordinates
椭振 elliptic(al) polarization

拓 fathom

拓本 rubbing edition; rubbing from stone or metal inscription
拓边铲斗 profile bucket
拓荒 site preparation
拓荒道路 pioneer road
拓荒井 pioneer well
拓荒者 pioneer
拓垦 land reclamation
拓垦局 <美> Bureau of Reclamation
拓宽 broaden(ing); frontier; widening
拓宽道路 widening road
拓宽的河槽 widened channel
拓宽的河道 widened channel
拓宽的河段 widened channel
拓宽的渠道 widened channel
拓宽式交叉口【道】 outspread intersection; flared intersection
拓宽线【道】 widening line
拓木 dao
拓扑 analysis situ
拓扑阿贝耳群 topologic(al) Abelian group
拓扑半群 topologic(al) semigroup
拓扑编辑 topologic(al) editing
拓扑变换 topologic(al) transformation
拓扑变换群 topologic(al) transformation group
拓扑不变的 topologically invariant
拓扑不变量 topologic(al) invariant
拓扑不变性 topologic(al) invariance
拓扑不可约的 topologically irreducible
拓扑不可约性 topologic(al) irreducibility
拓扑乘积 topologic(al) product
拓扑代换 topologic(al) substitute
拓扑代数 topologic(al) algebra
拓扑代数系 topologic(al) algebraic systems
拓扑单位 quasi-interior point; topologic(al) unit
拓扑单形 topologic(al) simplex
拓扑的射影系 projective system of topologic(al) group
拓扑的相对化 relativization of a topology
拓扑等价空间 topologically equiva-

lent spaces
拓扑地理编码 topologic(al) geocoding
拓扑叠合(加) topologic(al) overlay
拓扑动力学 topologic(al) dynamics
拓扑度 topologic(al) degree
拓扑对应的 topologically corresponding
拓扑多面体 topologic(al) polyhedron
拓扑方法 topologic(al) method
拓扑分类 topologic(al) classification
拓扑分析 topology analysis
拓扑复合形 topologic(al) complex
拓扑格 topologic(al) lattice
拓扑共轭的 topologically conjugate
拓扑关系 topologic(al) relation-(ship)
拓扑观点 topologic(al) viewpoint
拓扑广群 topologic(al) groupoid
拓扑和 topologic(al) sum
拓扑恒等模型 topologically identical model
拓扑化 topologize
拓扑环 topologic(al) ring
拓扑积 topologic(al) product
拓扑极限 topologic(al) limit
拓扑结构 topologic(al) structure
拓扑截面 topologic(al) cross section
拓扑浸入 topologic(al) immersion
拓扑矩阵 topologic(al) matrix
拓扑可解群 topologically solvable group
拓扑空间 topologic(al) space
拓扑空间的基本群 fundamental group of a topologic(al) space
拓扑扩张 topologic(al) extension
拓扑连通分支 topologically connected components
拓扑临界点 topologic(al) critical point
拓扑流形 topologic(al) manifold
拓扑幂零算子 topologic(al) nilpotent operator
拓扑幂零元 topologically nilpotent element
拓扑描绘子 topologic(al) descriptor
拓扑描术符 topologic(al) description
拓扑偶 topologic(al) pair
拓扑嵌入 topologic(al) embedding; topologic(al) insertion
拓扑球 topologic(al) solid sphere
拓扑球面 topologic(al) sphere
拓扑群 topologic(al) groups
拓扑三角形 topologic(al) triangle
拓扑上限 topologic(al) upper limit
拓扑收敛 topologic(al) convergence
拓扑树 topologic(al) tree
拓扑数据 topologic(al) data
拓扑数据结构 topologic(al) data structure
拓扑图论 topologic(al) graph theory
拓扑现象 topologic(al) phenomenon
拓扑相容性 topologic(al) consistency
拓扑学 topology; analysis situs
拓扑学的 topologic(al)
拓扑要素 topologic(al) feature
拓扑映射 topologic(al) mapping
拓朴域 topologic(al) field

唾

唾沫 spittle

T

W

娃

娃娃(玩具) doll

娃娃鱼 giant salamander

挖

挖板 bulldozing blade

挖补 patching
挖补混凝土 dental concrete
挖槽 chamfering; channel(l)ing; cut; dap; dredge cut; entrenchment; intrench; rut; trenching
挖槽边坡 side slope of dredge-cut
挖槽边线 sideline of dredge-cut
挖槽边线控制装置 dredge-cut sideline control unit
挖槽除雪机 channel snow plough; channel snow plow
挖槽定线 dredge-cut alignment
挖槽放线 dredge-cut setting out
挖槽工具 groover; intrenching tools
挖槽横撑 waling
挖槽(混凝土)灌注基础 <一种条形基础> trench fill foundation
挖槽机 basin forming machine; channel(l)er; groover; sewerage dredge(r)
挖槽具 intrenching tools
挖槽宽度 width of dredge-cut
挖槽宽度控制 controlling of dredge-cut width; width control of dredge-cut
挖槽嵌接 dapping
挖槽区 trenching area
挖槽上端 upper end of dredge-cut
挖槽设计 dredge-cut design
挖槽深度 depth of dredge-cut
挖槽深度控制 controlling of dredge-cut depth; depth control of dredge-cut
挖槽水力计算 hydraulic calculation for dredge-cut
挖槽水平跳板 byatt
挖槽下端 lower end of dredge-cut
挖槽现浇式隧道 dredged trench built-in-place tunnel
挖槽右边线 right sideline of dredge-cut
挖槽质量 quality of dredge-cut
挖槽左边线 left sideline of edge-cut
挖槽作业 grooving
挖层厚度【疏】 thickness of dredged layer; thickness of dredging layer
挖产量 bulldozing output
挖铲装置 <挖掘机的> excavator equipment
挖成的泊位 dredged berth
挖成的底部 <水底> dredged bottom
挖成的海底 dredged bottom
挖成的航道 dredged channel
挖成的河底 dredged bottom
挖成沟的 trenched
挖成区 dredged area
挖成纵断面 cut a profile
挖匙 excavator spoon
挖出 dugout; gouge out; uprooting
挖出的废料 dredged spoil
挖出的腐殖土 dredged muck
挖出的泥沙[砂] dredged material; dredged matter; dredged spoil
挖出的泥炭土 dredged peat
挖出的土 dugout earth; excavated earth
挖出废料 dredging spoil
挖出管内的土 casing off soil
挖出块根 lifted root
挖出料 borrow
挖出区 dredged area; dredging site
挖出土 dugout out earth
挖出土方的运输 transportation of excavated material
挖出土料 excavated material
挖出物 dredged material; dredged spoil; dredging material
挖出物质 excavated material; excavated substance
挖除 excavation
挖除表土 cutting; stripping; topsoil stripping <以备修筑路面>
挖除场地表土 stripping of the site
挖除泥沙 dislodging of sediment
挖除软土 muck out
挖除石块 ripen
挖倒 dig down
挖到规定标高 cut to line
挖到设计标高 cut(-down) to grade
挖底 scour; trimming
挖底机 trimming machine
挖地 break ground
挖地造型 bedding-in
挖洞 burrow; dibber; dibble
挖洞动物 burrowing animal
挖洞机 boring machine
挖洞器 posthole auger; posthole borer; posthole digger
挖斗 bucket; excavator bucket; excavator grab
挖斗不正确的行程 bucket wander
挖斗刀齿 digging prong
挖斗底卸式闸门 bucket bottom gate
挖斗机铲 basket
挖斗链式传动 bucket line drive
挖斗轮 excavating wheel
挖斗前缘 bucket lip
挖斗容量 bucket volume
挖斗式卸货吊索插座 bucket closing cable socket
挖斗梯状支架 <多斗挖掘机的> digging ladder
挖斗运输车 bucket transfer car
挖方 cubage of excavation; cut(ting); excavation
挖方板架支撑 planking and strutting
挖方爆破 cutback blasting
挖方边衬 cut lining
挖方边坡 cutting slope; excavation slope
挖方表 cut sheet
挖方尺寸 size of digging
挖方挡板 lagging of a cut
挖方的边坡 side of an excavation
挖方底部隆胀 bottom heave
挖方底土 base of excavation
挖方断面 cut section
挖方高程 excavation level
挖方工程 cutting work; excavation works
挖方工程量 volume of excavation works
挖方和填方 cut-and-fill; cut-and-fill excavation
挖方和填方计算 calculation of cutting and filling
挖方和填方建筑工程 <敞开式施工> cut-and-cover construction
挖方和运土 excavation and carting away
挖方护坡道处 bench flume
挖方滑坡 sliding in cut
挖方机械 negative appliance
挖方量 amount of cut; amount of excavation; excavation quantity; excavation volume; quantity of excavation; volume of cut(ting); volume of excavation; excavated volume
挖方坡度 cutting slope; excavation slope; slope of cutting
挖方渠道 canal in a cut; canal in cutting
挖方深度 depth of cut; depth of excavation
挖方体积 excavated volume; volume of excavation
挖方填方段 cut-and-cover section
挖方填方分界处 cut-to-fill location
挖方土柱 <标记挖方深度> temoin
挖方与填方 cut-and-cover
挖方与填方量纵断面图 mass profile
挖方与填土连接处 cut-fill contact
挖方运河 canal in cutting
挖方运土工程 excavation and cart-away
挖方支撑 ditching; planking and strutting; sheeting; sheeting for excavation; timbering of a cut; trench excavation; trenching
挖割 parting
挖割的分型面 stepped joint
挖根 grub(bing)
挖根铲 root shovel
挖根锄 grubbing hoe
挖根钩 grub hook
挖根机 root extractor; root plow; stump puller
挖根犁 rooter plow
挖根螺旋钻 stump auger
挖根者 grubber
挖沟 ditch cut(ting); ditch excavation; ditching; entrenchment; toe ditch; trench cutting; trench digging; trench excavation; trenching
挖沟爆破 ditch blasting
挖沟爆破法 trench method of blasting
挖沟臂角度 trenching angle
挖沟槽 plough
挖沟杈 trench excavator
挖沟铲 ditching shovel; trench-forming shovel; trenching hoe
挖沟铲齿 trencher tooth
挖沟铲斗 ditch digging bucket; ditch(ing) bucket; ditch-making bucket; ditch scoop; trenching bucket
挖沟车 lorry-mounted ditcher; truck-mounted ditcher
挖沟的水平挡土木板 timbering
挖沟法 trench method
挖沟杆 trench cutting boom
挖沟工 ditcher; hitch cutter; trencher
挖沟工程 trench(ing) excavation; trench(ing) works
挖沟工具 intrenching tools
挖沟工人 sand hog
挖沟工作 trenching; trench work
挖沟刮板 furrowing sled
挖沟回填 trench backfill(ing)
挖沟机 ditch-and-trench excavator; ditch cutter; ditch digger; ditcher; ditching and trenching machine; ditch(ing) excavator; ditch(ing) machine; ditch(ing) plough; ditchman; ditch-scoop excavator; ditch shovel; drainage trencher; pipeline excavator; sewerage dredge(r); sewer hog; trench-cutting machine; trench digger; trencher; trench excavating machine; trench excavator; trench hoe; trenching machine; trenching plant; banker; channel(l)er; cutting machine; cutter-trencher
挖沟机臂 ditcher boom
挖沟机铲斗 ditcher bucket; trencher bucket
挖沟机斗 trench bucket
挖沟机反铲设备 trench hoe attachment
挖沟机拖球 mole ball
挖沟机械 trenching machinery
挖沟机抓斗 ditching grab
挖沟犁 digger plough; digging plough; ditching plough; ditching plow; mole plough
挖沟铺管机 pipe-laying trencher; trencher type machine; trenching and pipelaying machine
挖沟弃土 ditch spoil; trench spoil
挖沟器 trenching element
挖沟器转子 digging wheel
挖沟锹 trench hoe
挖沟切土链 trenching chain
挖沟土方量 trench volume
挖沟土方丈量 <根据边沟挖方体积量算土方> ditch excavation measurement
挖沟用抓斗 trenching clamshell
挖沟余土 trench surplus
挖沟元件 trenching element
挖沟者 ditcher
挖沟支撑法 trench cut method
挖沟装置 trench-digging attachment; trencher attachment; trenching element
挖沟作业 trench works
挖壕(沟) entrench(ment); intrench
挖壕沟机 trench excavator
挖壕机 ditcher; ditch machine; ditch-scoop excavator; trench-digging machine; trencher
挖壕锹 entrenching shovel
挖壕人 trencher
挖河 river dredging
挖花织机 swivel loom
挖花织造 swivel weaving
挖基 pit excavation
挖基槽的水平挡土木板 lagging
挖基排水工程 unwater(ing)
挖基用木框 case frame
挖结构物下面的地基 undermine
挖进 cut into
挖进面 point of collection
挖井 dig a well; sink a well
挖井工 well driller
挖井机 well driller
挖井机具 <用以取出井内断杆> beche
挖掘 basal sapping; claw; cramp out; cutting; delve; dig; dredge; drivage; excavate; excavating; excavation; grubbing; hoeing; sapping
挖掘摆幅 digging speed
挖掘办法 digging method
挖掘半径 cutting radius; digging range; excavating radius; radius of clean-up; radius of cut; radius of excavation
挖掘臂 digging arm
挖掘标线 lockspit
挖掘剥离循环 excavator pass
挖掘叉 digging fork
挖掘铲 digger blade; digging share; digging shovel; dipper shovel; ditching scoop; ditching shovel; grubbing hoe; lifting attachment; lifting blade
挖掘铲斗 digging bucket
挖掘铲斗柄柱 dipper stick
挖掘场地 digging site
挖掘锤 digging ram
挖掘错动 excavation deformation
挖掘地点 digging point
挖掘地段 digging reach
挖掘点 digging point
挖掘吊盘绞车 platform hoist
挖掘斗 excavating bucket

W

挖掘堆积 excavation heap
挖掘法 dredging
挖掘范围 digging envelope; digging range; excavating range
挖掘方法 digging method; haul-away <一种包括清运土屑的>
挖掘附件 digging attachment
挖掘高度 cutting height; digging in height
挖掘工 shovel(l)er
挖掘工程 excavation works
挖掘工程支柱 plumb post
挖掘工具 digging tool; intrenching tools
挖掘工具附件 excavating attachment
挖掘工人 digger; navvy; excavator
挖掘工作 navvy work
挖掘工作舱 <气压沉箱的> working chamber
挖掘工作循环 digging cycle
挖掘管沟 trench excavation
挖掘回转时间 dig-and-turn time
挖掘机 digger; digging machine; dirt-digger; dirt mover; dredge; earth mover; excavating machine; excavator; grab; ground-hog; hydraulic grab; mechanical digger; mechanical excavator shovel; mechanical quarry shovel; power navvy; raiser; shovel; shovel(l)ing machine; scope <铲斗具有手腕功能的>; patching hoe <修补路面用的>
挖掘机半径 digging radius
挖掘机臂 boom; ditcher boom
挖掘机臂吊索 boom support guy
挖掘机(杓)斗杆 dipper handle
挖掘机铲臂 ditcher stick
挖掘机铲斗 basket; digger bucket; excavator bucket; excavator grab; shovel bucket; shovel dipper; skimmer scoop
挖掘机铲斗撑杆 straddling dipper handle
挖掘机铲斗齿 shovel teeth
挖掘机铲斗刀口 shovel lip
挖掘机铲斗可调臂杆 pitch arm
挖掘机顶杆 shovel ram
挖掘机斗 bowl
挖掘机斗臂 skipper arm
挖掘机斗柄 skipper arm
挖掘机斗杆 skipper arm
挖掘机斗架 digging ladder; ladder excavation
挖掘机附带的活动剪切机 mobile shear
挖掘机附件 skimmer attachment
挖掘机工作半径 shovel access; shovel reach
挖掘机工作尺寸 digging dimension
挖掘机戽斗导架 digging flight; digging ladder
挖掘机戽斗梯状支架 digging flight
挖掘机机身 excavator base machine
挖掘机加油工 shovel oiler
挖掘机加油器 shovel oiler
挖掘机驾驶员 shovelman; shovel runner
挖掘机绞车 skimmer hoist; skimmer winch
挖掘机绞车滚筒 digging drum
挖掘机绞盘 skimmer hoist; skimmer winch
挖掘机具 digging tool
挖掘机(勺)斗 excavator bucket
挖掘机式推土机 <侧板与刀片成斗形> excavator type bulldozer
挖掘机司机 shovelman; shovel runner
挖掘机推压链轮 crowd sprocket
挖掘机拖带的土块破碎器 clod buster
挖掘机挖斗 digger
挖掘机挖掘能力图 capacity chart
挖掘机械 digging machinery; excavating equipment; excavating machinery; excavation equipment; excavation machinery
挖掘机械底架 undercarriage
挖掘机抓斗 excavator grab
挖掘机钻头附件 drill attachment for excavators
挖掘机作业循环时间 cycle time of excavator
挖掘机作用半径 jib head radius
挖掘角度 digging angle
挖掘界限 digging envelope boundary; limit of excavation
挖掘井 digging well; dug well
挖掘开荒犁 grub-breaker plow
挖掘宽度 digging width; width of cut
挖掘矿工 digger
挖掘犁 excavating plough; grubber
挖掘犁体 raising body
挖掘链 digger chain; digging chain
挖掘两铲深度 double digging
挖掘量 excavation volume
挖掘料 excavated material
挖掘轮 excavating wheel; lifter wheel
挖掘面 digging face; point of collection; cutting face <在隧道掘进方向的>
挖掘能力 digging ability; digging power; digging force
挖掘黏[粘]土的风动铲 clay spade
挖掘黏[粘]土用的一种双柄弯刀 draw knife
挖掘平整机 <商品名> Gradall
挖掘企业潜力 tapping the potentialities of enterprises
挖掘起重机 crane-excavator
挖掘起重两用机 convertible shovel-crane; shovel-crane
挖掘器 sapper
挖掘器操纵 digger control
挖掘潜力 tap potentiality
挖掘区 cut area; digging area; excavating area
挖掘容量 struck capacity
挖掘设备 excavating equipment; excavation equipment
挖掘深度 depth of cut; digging depth; digging height; digging in depth; excavating depth; excavation depth
挖掘升运器 primary lifting elevator
挖掘时间 digging time
挖掘式推土机 elevating grader
挖掘试验 test by digging
挖掘速度 digging rate; digging speed; excavating speed
挖掘隧道的人 tunnel(l)er
挖掘隧洞的人 tunnel(l)er
挖掘掏槽 <疏浚河道时> cutting
挖掘体积 excavation volume
挖掘图 digging diagram
挖掘土址 spoil
挖掘位置 digging point; digging position; excavating position
挖掘稳定条件 excavation-stability condition
挖掘物料堆 spoil pile
挖掘系数 digging coefficient; excavating related factor; excavator factor
挖掘现场 digging site
挖掘线 closing line; digging cable
挖掘效率 digging rate
挖掘效应 excavation effect
挖掘效应校正 correction for excavation effect
挖掘型铲斗 digging bucket
挖掘型抓斗 digging grab
挖掘性 dredgeability
挖掘性能 digging function
挖掘性下陷量 <原地打滑引起的> excavation sinkage
挖掘循环 digging cycle
挖掘岩石的绞吸式挖泥船 rock cutter suction dredger
挖掘叶轮 scoop vane
挖掘一铲深度 single digging
挖掘用抓斗机 grab for excavating
挖掘与行走的独立控制 <挖掘机> independent excavating and travel(l)ing control
挖掘圆盘 digging wheel
挖掘运动 digging motion
挖掘运输机械 excavating-hauling unit
挖掘运输机组 excavating-hauling unit
挖掘者 digger; sapper
挖掘支撑 underpinning
挖掘支架 excavation support
挖掘直径 excavated diameter
挖掘属具 shovel attachment
挖掘抓斗 excavating clamshell; excavating grab
挖掘转轮 digging reel
挖掘转子 digging reel
挖掘装袋机 digger-sacker
挖掘装载机 backhoe loader; cut-and-load machine; excavator-loader; loader-digger
挖掘装载两用机 loader-digger; excavator loader; backhoe loader
挖掘装置 digging tool; excavating gear
挖掘状态 digging position
挖掘作业 digging operation
挖掘作用 digging action
挖开 dig up
挖开沟 trench cutting
挖刻 undercut
挖坑 bore; pit digging
挖坑铲 pit blade
挖坑沉基 foundation by pit sinking
挖坑道 undermine; undermining
挖坑机 earth boring machine; earth drill; trench excavator
挖坑填埋垃圾技术 pit technique
挖空 hollow; hollow(ing) out; scrape
挖空的空间 gob
挖孔板 cored slab
挖孔灌注桩 cast-in-place concrete pile; excavated and cast-in-place pile; shelless; bored pile
挖孔灌桩法 installing pile shaft by excavation
挖孔桩 cast-in-situ pile by excavation; digging pile; hand-dug caisson; bored pile; hand-dug shaft
挖孔桩基础 bored foundation
挖梁窝 needling
挖煤铲斗 coal bucket
挖煤机 coal shovel
挖泥 dredge; dredging; mucking; sludging; mud dredging
挖泥泵 dredge pump; excavating pump; dredging pump
挖泥泵马力 dredging pump horsepower
挖泥臂架 dredge(r) ladder; dredging ladder
挖泥标高 dredged level
挖泥泊位 dredged berth
挖泥参数 dredging parameter
挖泥操作 dredging operation
挖泥操作程序自动控制系统 automated control system of dredging process
挖泥槽 dredged trench; dredging trench; ladder well
挖泥铲 dredging shovel
挖泥铲斗 dipper dredge(r); dredge(r) bucket; muck bucket
挖泥超深 depth of extra-dredging; extra depth of dredging; over-depth of dredging
挖泥承包人 dredge(r) contractor; dredging contractor
挖泥承包商 dredge(r) contractor; dredging contractor
挖泥承包者 dredge(r) contractor; dredging contractor
挖泥处置区 disposal area
挖泥船 digging dredge; drag boat; dredge(boat); dredger; dredge(r) ship; dredging barge; dredging craft; dredging engine; dredging machine; floating dredge(r); hydraulic dredge(r); marine dredge(r); mud boat; mud drag; mud dredge(r); mud lighter; hog barge <俚语>; aque motrice; creeper
挖泥船摆动移进 dredge(r) walking
挖泥船编队 fleeting of dredge(r)s
挖泥船撑柱 dredge(r) spuds
挖泥船出泥管(道) dredge(r) pipe line
挖泥船船身 dredge(r) hull
挖泥船船长 dredge(r) master; dredge(r) captain
挖泥船大桩的套箍 spud well
挖泥船的工作船 dredge(r) tender
挖泥船的戽斗 dredge(r) bucket
挖泥船调遣 mobilization of dredge(r)
挖泥船定位桩 dredge(r) spuds
挖泥船定位桩提升柱塞 dredge(r)-spud hoisting ram
挖泥船斗架 bucket ladder
挖泥船断面监测 dredge(r) profile monitor
挖泥船队 dredge(r) fleet; dredging fleet
挖泥船对开泥舱 split hull
挖泥船非生产性停歇 dredge(r)'s unproductive downtime
挖泥船废料 dredge(r) spoil
挖泥船港口 dredge(r) port
挖泥船工人 mucker
挖泥船工作艇 bum boat
挖泥船戽斗 dredge(r) bucket
挖泥船绞刀 dredge(r) cutter head
挖泥船开体 split hull
挖泥船利用率 utility factor of dredge(r)
挖泥船链斗井 dredge(r) well
挖泥船泥舱 dredge(r) hopper
挖泥船排泥管(道) dredge(r) pipe (line); window pipe
挖泥船桥梁架 dredge(r) ladder
挖泥船驱动带 rotating band
挖泥船取样 sampling by dredge(r)
挖泥船设备利用率 utilization rate of equipment and installation of dredge(r)
挖泥船生产率 dredge(r) productivity; dredge(r)'s production rate; output rate of dredge(r); production rate of dredgers
挖泥船生产率折减系数 reduction factor of dredge(r) productivity
挖泥船生产能力 dredge(r)'s production capacity; production capacity of dredge(r) productivity
挖泥船生产性停歇 productive downtime of dredge(r)
挖泥船时间利用率 time utility factor of dredge(r); time utilization rate of dredge(r)
挖泥船首吹泥成的 rainbowed
挖泥船首向岸吹填 rainbowing
挖泥船艄部 dredge(r) stern
挖泥船台 dredged berth
挖泥船头雨虹形吹沙 spraying(rain-

bowing) of sand
挖泥船挖出物 dredged material; dredged spoil
挖泥船挖斗 dredging bucket
挖泥船艉部 dredge(r) bow
挖泥船坞 mud dock
挖泥船选择 selection of dredge(r)s
挖泥船运转时间 operating time of dredge(r); running time of dredge-(r)
挖泥船支队 dredging branch fleet; dredging sub-fleet
挖泥船桩柱 dredge(r) spuds
挖泥刀 dredging knife
挖泥地点 dredging site
挖泥吊桶 mud kibble
挖泥斗 dredging bucket; dredging scoop; mud bucket; scoop; dredge hopper <挖泥船上的>
挖泥斗式提升机 dredge(r) bucket type elevator
挖泥法则 dredging rule
挖泥范围 area to be dredged
挖泥浮标 dredging buoy
挖泥富裕深度 overdredging depth
挖泥高程 dredged level
挖泥工地 dredging site
挖泥工(人) mucker; dredger
挖泥工作 dredge work; dredging operation; dredging works
挖泥公司 dredging firm
挖泥轨迹图 dredging track plot
挖泥海底 dredged bottom
挖泥航道 dredged channel
挖泥厚度 dredging thickness; height of cut
挖泥戽斗 dredge(r) scoop; dredging box
挖泥浑浊 dredge turbidity
挖泥机 bagger; digging dredge; dredger; dredging engine; dredging machine; excavating machine; land dredge(r); mucking machine; mud drag; mud dredge(r) bulb; mud drum; navvy; sewerage dredge(r); sludger
挖泥机出泥管 dredge pipe(line)
挖泥机斗支架 elevator ladder
挖泥机戽斗 dredge(r) bucket
挖泥机绞车 dredge(r) winch
挖泥机捞出物 dredged spoil matter
挖泥机水泵 dredging pump
挖泥机挖斗 dredging bucket
挖泥机下的钢柱 anchorage spud
挖泥计算机 dredging computer
挖泥井 <挖泥船上泥斗或吸泥管通向水下的> dredging well; dredge well
挖泥卷流特性 dredging plume behavio(u)r
挖泥掘土机 dredging and excavating machine
挖泥竣工标高 final dredged level
挖泥竣工高程 final dredged level; actual dredged level
挖泥坑 dredged pit; excavated pit
挖泥控制系统 dredging control system
挖泥宽度 dredging width; width of cut
挖泥链斗槽 dredging well
挖泥量 dredged volume
挖泥能力 dredging capacity
挖泥耙 mud rake; hand dredge(r) <人工的>
挖泥起重机 dredging crane
挖泥器具 <其中的一种> aquamotrice
挖泥情况显示器 dredging situation display
挖泥区 area to be dredged; dredging site
挖泥区海底略图 subsea outline of dredged areas
挖泥区域 dredged area; excavation zone
挖泥取样 dredge sampling; sampling for dredging
挖泥取样器 sampler for dredging
挖泥沙[砂] bailing
挖泥沙[砂]井取样 bailing well sampling
挖泥沙[砂]试验 bailing test
挖泥沙[砂]用滑车 bailing sheave
挖泥沙[砂]用滑轮 bailing pulley
挖泥沙[砂]有绳 bailing rope
挖泥设备 dredging equipment; dredging plant; dredging unit; excavation equipment; excavation plant; excavation unit
挖泥深度指示器 dredging depth indicator
挖泥施工 dredging works
挖泥时间 dredging time
挖泥示意图 dredging diagram
挖泥炭 dug peat
挖泥提升机 elevator dredge(r)
挖泥填筑 dredge-placed fill
挖泥筒 sludger
挖泥土量计 yardage meter
挖泥信号灯 dredging signal light
挖泥遥控装置 dredging remote control system
挖泥周期 dredging cycle; dredging period
挖泥抓斗 bottom grab; excavator grab
挖泥装载吃水指示仪 dredge(r) load draft indicator
挖泥组件 dredging module
挖泥作业 dredging operation
挖黏[粘]泥用栅底铲斗 slat bucket
挖黏[粘]土 dredging clay
挖排水沟 <用挖沟犁> moling
挖排水沟用铲斗 drainage bucket
挖漂石和卵石 dredging boulders and cobbles
挖坡沉基 foundation by pit sinking
挖器 excavator
挖前切顶 pretopping
挖墙脚 undermine
挖渠 ditch
挖渠机 diker; trencher; ditcher <美>
挖取海底骨料 marine aggregate dredging
挖去 cut-out
挖入的 intrenched
挖入式港池 cut dock; dig-in basin; excavated dock basin; excavated-in harbo(u)r basin
挖入式港口 artificially excavated port; excavated-in port
挖入式人工港 excavated artificial harbo(u)r
挖入式钻模 recessed jig
挖砂 <又称挖沙> coping down
挖砂铲斗 sand bucket
挖砂船 grit dredge(r); sand dredge-(r)
挖砂工(人) sand hog
挖砂机 grit dredge(r); sand dredge-(r)
挖砂机抓斗 sand grab(bing bucket)
挖砂用抓斗 sand grab
挖深 deepen(ing); dredging depth
挖深限制装置 depth limitation device
挖石铲斗 rock bucket
挖石船 gravel dredge(r)
挖石斗 rock bucket
挖石(工程) rock excavation
挖石工具 stone drawing tool
挖石机 rock excavator
挖蚀岸 cutbank
挖树根工具 grubber
挖树机 tree excavation; tree mover
挖竖井 shaft sinking
挖松 dig up
挖松的材料 ripped material
挖隧道法 pilot-tunnel method
挖探坑 test pitting; test trenching
挖淘 trench excavation
挖填 fill by digging other part of rock
挖填法 excavation and replacement
挖填方 cut(-and)-fill; excavation and filling
挖填(方)比 <等于挖方体积/填实后体积> ratio of cut to fill
挖填方的分界 boundary of cutting and filling
挖填方调度(平衡) cut-fill transition
挖填方分界线 cut-and-fill location
挖填方平衡 balanced cuts and fills
挖填交点 cut-and-fill location; cut-to-fill location
挖填平衡 balance of cut and fill
挖填平衡的土方工程 balanced earthwork; balanced grading
挖填土方平衡 balance cuts and fills
挖通 cut-through; dig through
挖土 clay digging; coulisse; cullis; cutting; earth cutting; excavation; excavation of earth; mucking; soil excavation; earth excavation
挖土边 cut side
挖土边坡 borrow bank
挖土铲 clay digger; earth shovel
挖土铲斗 digging bucket; dirt bucket; dirt scoop; earth scoop
挖土铲斗定位杆 pitch arm; pitch brace
挖土车 lorry-mounted excavator; truck-mounted excavator
挖土锄 earth hoe
挖土底板 bucket floor
挖土动力索 digging line
挖土斗 dipper
挖土断面 cut section
挖土方 <结合狭路出土的> gulleting; earth excavation
挖土费 cost of excavation
挖土镐 clay pick; heavy pick; navvy pick
挖土工 gopherman; groundman; mucker; shovel(l)er
挖土工程 earth excavation works; excavation works
挖土工具 spader
挖土工人 banker; hog
挖土工作 soil excavation work; spadework
挖土滑运槽 excavating chute
挖土机 crane navvy; digger; digging machine; dirt-digger; dredge; dredging machine; earth excavating machine; earth mover; excavating machine; excavator; ground-hog; land dredge(r); machine shovel; mechanical shovel; navvy; power navvy; power shovel; rooter; shovel(l)ing machine; spader; spoon
挖土机臂 boom; shovel arm; shovel stick
挖土机臂套管 shovel arm sleeve; shovel stick sleeve
挖土机操作的捣实机 excavator-operated stamper
挖土机铲斗 digger; hoe dipper; shovel bucket; shovel dipper; skimmer scoop
挖土机铲斗柄的推压运动 thrust motion
挖土机(铲)斗齿 shovel teeth
挖土机铲斗容量 excavator bucket capacity
挖土机铲刃 shovel lip
挖土机车架 bogie [bogey/bogy]
挖土机齿 shovel teeth
挖土机电动机 excavator motor; shovel motor
挖土机垫物 excavator mat
挖土机吊臂索 boom support guy
挖土机斗 excavator bucket
挖土机斗柄 skipper arm
挖土机斗架 bucket ladder; ladder
挖土机发动机 excavator engine; excavator motor; shovel engine
挖土机附加装置 attachment for excavators
挖土机附件 skimmer attachment
挖土机钢丝绳 shovel cable
挖土机钢丝索 excavator rope
挖土机钢索 shovel cable
挖土机工作 excavator work
挖土机工作半径 shovel access; shovel (out) reach
挖土机鼓 excavator drum
挖土机活动伸臂 live boom of excavator
挖土机驾驶人 excavator driver
挖土机驾驶员 shovel(l)er; shovelman; shovel runner
挖土机驾驶员室 excavator cab(in)
挖土机绞车 skimmer hoist; skimmer winch
挖土机绞盘 skimmer hoist; skimmer winch
挖土机控制站 shovel control station
挖土机缆索 excavator cable
挖土机链条 shovel chain
挖土机轮胎 earthmover tire [tyre]
挖土机磨损部件 excavator wearing part
挖土机平台 excavator deck
挖土机启动注油器 excavator primer
挖土机倾翻器 shovel trip
挖土机曲柄 shovel handle
挖土机曲柄套管 shovel handle sleeve
挖土机润滑脂 excavator grease
挖土机手 boom cat; pitman
挖土机司机 shovelman
挖土机推进 excavator drive
挖土机推压 racking
挖土机挖斗臂 shovel dipper arm; shovel dipper stick
挖土机挖斗臂套管 shovel dipper arm sleeve; shovel dipper stick sleeve
挖土机挖斗齿 shovel dipper tooth
挖土机挖斗顶杆 shovel dipper ram
挖土机挖斗附件 shovel bucket attachment
挖土机挖斗门 shovel dipper door; shovel door
挖土机挖斗倾翻器 shovel dipper strip
挖土机挖斗曲板套管 shovel dipper handle sleeve
挖土机挖斗刃瓣 shovel dipper tooth
挖土机挖斗容量 shovel bucket capacity; shovel dipper capacity
挖土机挖斗挖掘角度 shovel bucket digging angle
挖土机挖斗挖掘宽度 shovel bucket digging width
挖土机挖斗闸板 shovel dipper slide
挖土机挖斗转臂 shovel bucket boom
挖土机械设备 excavating plant
挖土机悬臂 shovel boom
挖土机旋臂 shovel jib
挖土机旋臂曲柄 shovel jib winch
挖土机牙齿 excavator tooth
挖土机引擎 shovel engine
挖土机在地面一下挖掘 excavator digging below ground level

挖土机支承底板 excavator supporting mat
挖土机支承垫 shovel supporting mat
挖土机支垫 excavator mat
挖土机轴杆 shovel axle
挖土机主机 excavator base machine
挖土机抓斗 bucket grab; earth grab-(bing) bucket; grab; grab bucket
挖土机转臂 shovel boom
挖土机转臂曲柄 shovel boom winch
挖土机总量 total volume of excavated soil
挖土及运走 earth removal
挖土井 dug(out) well; pit well
挖土螺钻 mechanical earth auger
挖土面 cut side
挖土能力 earth cutting ability
挖土皮带输送机 excavating hand conveyer [conveyor]
挖土平衡 balanced excavation; balancing of cut and fill
挖土平台 digging platform
挖土起重两用机 crane shovel
挖土器 idiot stick
挖土前打入土内的竖向木桩或宽翼缘H形钢桩 soldier beam
挖土区 area to be dredged
挖土取样 ditch sample
挖土深度 cut depth
挖土深度标桩 temoin
挖土数量 excavation quantity
挖土稳定条件 excavation-stability condition
挖土线标高 dredge level
挖土掩盖 dig in
挖土用架空索道 excavating cableway
挖土支撑 planking and strutting
挖土作业 excavator work
挖污泥斗 muck bucket; muck sinking bucket
挖心 coring
挖穴法 excavated cavity
挖穴机 earth auger; earth borer; earth boring machine
挖穴螺旋机 mechanical earth auger
挖穴螺旋钻 digging auger
挖穴器 dibber; spotting board
挖穴手铲 dibbler
挖穴钻 earth screw
挖岩机 rock excavator
挖岩石管沟 rock ditching; rocking trenching
挖有沟槽的 trenched
挖运工具 cut-and-carry tool
挖运工作 excavation and cart-away
挖运泥土 splitter shield
挖凿 gouge
挖凿机 navvy
挖至设计标高 cut down to grade
挖柱洞器 posthole borer
挖柱孔(螺旋)钻 posthole auger
挖筑运河 canal in a cut; canal in cutting

洼部警告标志 <多指道路横向的地形> dip sign

洼地 basin; bottom land; callow; depressed area; depression; dimple; dish; ground depression; incavation; lacuna [复 lacunae]; laeuna; low ground; lowland; low-lying land; swag; swale
洼地冰川 depression glacier
洼地储积量 depression storage; pocket storage; surface depression storage
洼地等高线 depression contour
洼地电站 depression power station
洼地改造 improvement of polder
洼地汇水 lowland catchment
洼地区 area of depression
洼地泉 depression spring; pool spring
洼地蓄水(量) depression storage; pocket storage; surface depression storage
洼地滞洪 depression detention
洼淀 dambos
洼坑 hollow spot
洼(坑)泉 dimple spring
洼陷作用 downwarping
洼蓄 surface depression storage

蛙板 frog board

蛙暴 frog storm
蛙面 frog-face
蛙目锡石 toad's-eye tin
蛙人 aqualunger; frogman
蛙式打夯机 frog hammer; frog rammer; frog-type jumping rammer; jumping frog; trench compactor
蛙式捣实机 frog tamper
蛙式夯 electric(al) frog rammer; frog rammer; leapfrog
蛙式夯实机 frog compactor
蛙式夯实器 jitterbug
蛙式夯土机 frog tamper; leapfrog
蛙跳格式 leapfrog scheme
蛙跳式路线 leapfrog routine
蛙腿夹 clip for frog legs
蛙腿式 frogleg
蛙腿式绕组 frogleg windings
蛙位 frog position
蛙形腹 frog-belly
蛙眼 frogeye
蛙眼黏[粘]土 frogeye clay
蛙跃法 leapfrog method
蛙状鼻 frog-shaped nose

瓦 building tile; roof tile; tile; watt【电】

瓦板岩矿床 roofing slate deposit
瓦渤格呼吸计 Warburg respirometer
瓦槽玻璃 fluted glass
瓦层面 tile roof(ing)
瓦厂 tilery
瓦出檐 tile creasing
瓦当【建】 tails
瓦挡 antefix(ae); antefix tile; eaves tile; tile-end
瓦刀 bricklayer's cleaver; cleaver; laying trowel; mason's knife; trowel
瓦德赫斯特黏[粘]土 Wadhurst clay
瓦德乌斯棱镜 Wadworth prism
瓦的 tegula [复 tegulae]; tegular
瓦的抗压强度 tile compression strength
瓦的突边 stub
瓦垫 bearing wedge
瓦叠式列板 inner and outer strake
瓦钉 tile nail; tile pin
瓦顶 tile roof cover(ing)
瓦顶饰 antefix(ae)
瓦杜兹 <列支敦士登首都> Vaduz
瓦尔 <英国压力单位,1 瓦尔 = 10 牛/米> val
瓦尔堡模型 Warbury model
瓦尔登大梁 Valton girder
瓦尔登地板 Valton floor
瓦尔登转化 Walden inversion
瓦尔迪维亚海渊 Caldivia deep
瓦尔间冰期【地】 Waalian interglacial stage
瓦尔莱合金 Valray
瓦尔纳线 Wallner line
瓦尔温暖期 Waalian warm epoch
瓦泛水 tile fillet; tile listing
瓦房 tile-roofed house
瓦分 watt-minute
瓦钙镁硼石 wardsmithite
瓦格纳比表面积 Wagner specific surface; Wagner surface area
瓦格纳(浑)浊度计 <测水泥细度> Wagner's turbidimeter
瓦格纳接地线路 Wagner's earth connection
瓦格纳接地装置 <一对阻抗的连接地点> Wagner's earth device
瓦格纳金属型铸造机 Wagner's casting machine
瓦格纳码 Wagner's code
瓦格纳盆 Wagner pot
瓦格纳溶液 Wagner's solution
瓦格纳氏学说 Wagner's theory
瓦格纳试剂 Wagner's reagent; Wagner's solution
瓦格纳锡基合金 Wagner's alloy
瓦格纳细度 Wagner's fineness
瓦工 bricklayer; brick mason; brick worker; slater-and-tiler; tiler; trowel man
瓦工操作平台 bricklayer's square scaffold
瓦工槌 brick ax(e); brick hammer; bricklayer's hammer
瓦工锤 brick ax(e); brick hammer; bricklayer's hammer
瓦工方(台)脚手架 bricklayer's square scaffold
瓦工工长 foreman bricklayer
瓦工脚手架 bricklayer's scaffold-(ing)
瓦工砍刀 scotch
瓦工镘 bricklayer's trowel
瓦工镘刀 mason's float
瓦工配件 tile fitting
瓦工平头凿 mason's flat-ended chisel
瓦工小锤 scotch; scutcher
瓦工用多点可调悬挂式脚手架 mason's adjustable multiple-point suspension scaffold
瓦工用可调悬挂式脚手架 mason's adjustable suspension scaffold
瓦工油毡 slater's felt
瓦沟 tile; tile drain
瓦管 clay drainage tile; clay pipe; clay tile; crock; earth conduit; earthen (ware) pipe; earth pipe; flashing tile; perforated tile; pottery pipe; tile; tile conduit; tile drain; tile pipe; tile tube; ware pipe
瓦管暗沟 tile subdrain; tile underdrain
瓦管道 tile tubing
瓦管地下排水 tile underdrainage
瓦管敷设机 tile draining machine; tile layer; tile laying machine
瓦管管道挖掘机 tile trenching machine
瓦管灌溉系统 tile field
瓦管涵(洞) tile culvert
瓦管基座 tile cradle
瓦管集水系统 tile filter bottom
瓦管接合 tile joint; tile junction
瓦管接头 tile joint
瓦管接头盖 harmus
瓦管滤池底 tile filter bottom
瓦管滤池底瓦管 tile filter bottom
瓦管排水 clay drain(age); tile drainage
瓦管排水场 tile field
瓦管排水沟 tile drain
瓦管排水管 tile drain
瓦管排水系统 clay pipe drainage
瓦管铺设机 tile layer
瓦管切割机 tile cutting machine
瓦管清边 back edging
瓦管天沟 tiled valley
瓦管屋面天沟 tiled roof valley
瓦管阴沟 tile culvert
瓦管座 tile cradle
瓦罐 crock(ery); pottery jar
瓦硅钙钡石 walstromite
瓦红(橙色) tile red
瓦灰色 slate gray [grey]
瓦脊筒 ridge tile
瓦加杜古 <上沃尔特首都> Ouagadougou
瓦夹 tile clip
瓦间嵌灰泥 torching
瓦解 crumble; disintegration; disorganization; disruption; subversion
瓦解时间 disruption time; dissociation time
瓦卡头 stub
瓦克罗斯层【地】 Vaqueros formation
瓦克岩 wacke
瓦口 tile edging
瓦块式制动器 block brake
瓦拉赤造山运动 Wallachian orogeny
瓦拉赫转变 Wallach transformation
瓦拉伊斯风 valais wind
瓦莱塔 <马尔他首都> Valletta
瓦兰水银扩散泵 Waran's pump
瓦蓝 <浅灰蓝色> tile blue
瓦朗斯港 <法国> Port Valence
瓦棱玻璃 corrugated steel glass
瓦棱衬板 ribbed liner
瓦棱网眼钢皮 high rib lath
瓦楞白铁 corrugated sheet iron
瓦楞板 corrugated plate; corrugated sheet
瓦楞板轧机 corrugator
瓦楞玻璃 corrugated glass; corrugated sheet glass; fluted glass; reeded glass; wave glass
瓦楞玻璃屋面 corrugated roof glazing
瓦楞槽 corrugated pipe; corrugated sheet; valley of corrugation
瓦楞成型机 corrugator
瓦楞垫圈 diamond washer
瓦楞复合铝板 Rigidal Mansard
瓦楞钢板 corrugated steel plate; keystone plate
瓦楞辊 fluted roll
瓦楞胶合板 corrugated plywood
瓦楞铝板 corrugated alumin(i)um sheet
瓦楞铝屋面 corrugated alumin(i)um roofing
瓦楞石棉板 corrugated-asbestos board
瓦楞石棉(水泥)瓦 corrugated-asbestos cement sheet
瓦楞石棉水泥屋面 corrugated-asbestos cement roofing; corrugated-asbestos cement sheet
瓦楞石棉屋面防锈层 asbestos protection
瓦楞铁 corrugated iron; corrugated metal
瓦楞铁棚 corrugated iron shed
瓦楞铁皮 corrugated sheet; corrugated sheet iron; elephant; undulated sheet iron
瓦楞铁皮涵管 tin whistle
瓦楞铁皮建筑物 corrugated iron building
瓦楞铁皮屋面 corrugated iron roofing; corrugated sheet iron roofing
瓦楞纹 washboarding
瓦楞屋顶 corrugated roof(ing)
瓦楞屋面 corrugated roof(ing); waved roof
瓦楞屋面钢板 corrugated roof steel
瓦楞形抛光轮 corrugated buff

W

瓦楞原纸 fluting board
瓦楞纸 corrugated paper
瓦楞纸板 corrugated board
瓦楞纸盒 corrugated paper box
瓦楞纸皮 corrugated board
瓦楞纸箱 corrugated case
瓦楞纸印刷油墨 corrugated board ink
瓦楞绉缩 ribbing
瓦里宁石 vaeyrynenite
瓦里斯乘积 Wallis product
瓦里斯公式 Wallis formula
瓦里斯定理 Wallis theorem
瓦利风 ouari
瓦利农(计算力矩和)定理 Varignon's theorem
瓦利特(混凝土)配合比设计法 Vallette method of gap grading
瓦利特间断级配法 <混凝土骨料级配法> Vallette method of gap grading
瓦利通花岗岩 <一种带有黑灰色斑点的暗红色的> Varitone Mahogany
瓦利兹过滤机 Vallez filter
瓦砾 brash
瓦砾堆 debris;demolition spoil;demolition waste
瓦砾状的 rubbly
瓦垄 rows of roof tiles
瓦垄(波)纹钢板 keystone plate
瓦垄薄钢板 steel corrugated sheet
瓦垄钢板 corrugated sheet steel;deck plate
瓦垄水泥屋顶板 corrugated cement roofing sheet
瓦垅板 corrugated sheet metal
瓦垅板轧机 corrugator
瓦垅薄板 curved corrugated sheet
瓦垅薄钢板 corrugated iron;undulated sheet iron
瓦垅钢板 corrugated steel sheet
瓦垅管 corrugated pipe
瓦垅辊 <瓦垅面镇压器> corrugated roller
瓦垅面式冷却器 corrugated cooler
瓦垅铁 corrugated iron;zores bar
瓦垅轧辊 <轧制瓦垅板用> corrugating roll
瓦垅轧机 corrugation rolling mill
瓦垅纸 corrugated board
瓦伦特阶 <中奥陶世>【地】Valentian stage
瓦伦特石 walentaite
瓦伦特统【地】Valentian series
瓦伦西系统 Valensi system
瓦面 roof covering;tile;tiling
瓦模工 tile slabber
瓦内内漆 <一种油质着色剂> Varnene
瓦努阿图(大洋洲)Vanuatu
瓦排水屋面斜沟 tiled roof valley
瓦排水斜沟 tiled valley
瓦坯 green tile
瓦坯干燥槽架 hake
瓦坯干燥器 hake
瓦片 tile
瓦片尺寸 tile size
瓦片搭接 shingle lap
瓦片锯 shingle saw
瓦片切割 Winchester cutting
瓦片切割器 centrix
瓦片式炉壳结构 shingle shell construction
瓦片式压力容器 half shell pressure vessel
瓦片式圆筒 segmentally welded(monolayered)cylinder
瓦片填充塔 tile-packed column
瓦片屋面 covering in scale tiles
瓦片状排列 tegula [复 tegulae]
瓦铺砌承包人 tile contractor
瓦砌封角条 tile fillet
瓦器 crockery;earthenware
瓦切割机 shingle cutter
瓦圈 earthenware ring
瓦石作 masonry
瓦时 watt-hour
瓦时计 active-energy meter;watt-hour meter
瓦时容量 watt-hour capacity
瓦时消耗量 watt-hour consumption
瓦时效率 watt-hour efficiency
瓦氏平衡锥 cone penetrometer for liquid limit test
瓦数 wattage
瓦栓钉 tile peg
瓦水砷锌石 warikahnite
瓦斯 firedamp;gas;marsh gas;methane
瓦斯保护 gas protection
瓦斯保护装置 gaseous shield
瓦斯爆发 gas outburst
瓦斯爆炸 fire explosion;gas explosion
瓦斯边界层 firedamp fringe
瓦斯边界区 firedamp fringe
瓦斯测定 firedamp testing
瓦斯成分及含量 composition and content of gas
瓦斯冲出 gas rush
瓦斯存在形式 form of gas occurrence
瓦斯对出量 absolute outflow of gas
瓦斯分层 layering of firedamp;methane layering
瓦斯分带 zoning of gas
瓦斯浮标 gas buoy
瓦斯管线上加工厂 gas on-line plant
瓦斯焊接机 gas welding tube mill
瓦斯和煤尘混合爆炸 mixed explosion
瓦斯积聚 fire lamp accumulation;methane accumulation
瓦斯积聚层 firedamp layer
瓦斯及烟雾取样仪 gas and mist sampler
瓦斯极限含量 firedamp limit
瓦斯集合管 gas manifold
瓦斯记录器 methane recorder
瓦斯继电器 Buchholz relay;gas relay
瓦斯监测 gas monitoring
瓦斯监控系统 methane monitoring system
瓦斯检测装置 gas detector
瓦斯检查员 gasman
瓦斯检定器 gas detector
瓦斯警报 gas alarm
瓦斯警报器 fire alarm;methanophone
瓦斯矿 gassy
瓦斯螺纹管 gas pipe thread
瓦斯煤 bottle coal
瓦斯煤矿分级 <中国> classification of gassy mines [China]
瓦斯煤矿停电标准 electricity cut-off standard
瓦斯煤样 coal sample for determination of gas
瓦斯浓度 gas concentration
瓦斯排泄道 gas drain
瓦斯喷出 blow by;gas outburst
瓦斯平稳泄出 placid out-flow gas;placid out-flow methane
瓦斯瓶 gas cylinder
瓦斯气体 fiery mine
瓦斯迁移 firedamp migration
瓦斯入口 gas inlet
瓦斯散放 gas emission
瓦斯渗透率 gas permeability
瓦斯炭黑 gas-produced black
瓦斯探针 firedamp probe
瓦斯梯度 gas gradient
瓦斯通过炉子砌体喷出 bosh breakout
瓦斯筒 gas cylinder
瓦斯突出 outburst
瓦斯析出 gas emission
瓦斯相对涌出量 relative outflow of gas
瓦斯泄出 gas blower;gas emission;methane blower
瓦斯压力 gas pressure force
瓦斯压力室排放法 firedamp pressure-chamber method
瓦斯引火烧嘴 gas pilot
瓦斯涌出 emission of gas;gas outburst
瓦斯涌出形式 form of gas outflow
瓦斯油 gas oil
瓦斯指示器 firedamp indicator
瓦斯装置 bobtail plant
瓦斯着火 gas ignition
瓦特 <电功率单位> watt
瓦特表 power meter;watt-meter;electric(al)power meter;electrodynamic wattmeter
瓦特单位 watt
瓦特定律 Watt's law
瓦特分 watt-minute
瓦特怀特牌玻璃 <一种高度漫光压制玻璃> Waterwite
瓦特计 watt-hour meter;watt-meter
瓦特计法 wattmeter method
瓦特计式继电器 wattmetric relay
瓦特继电器 watt relay
瓦特库剂 <一种木材杀虫剂> Watco
瓦特林页岩 Watling shale
瓦特曼纸 <一种分离色层的滤纸或适于水彩画的优质图画纸> Whatman's paper
瓦特/米开 watt/meter-Kelvin
瓦特秒 watt-second
瓦特牌装配式房屋 <商品名> Wate
瓦特/球面度 watt/steradian
瓦特生方程 Watson equation
瓦特生四电极系统法 Watson four electric system method
瓦特生-索末菲变换 Watson-Sommerfeld transformation
瓦特数 wattage
瓦特炭黑 gas black
瓦特调速器 watt governor
瓦特小时 watt-hour
瓦特小时计 <通称电表> active-energy meter;watt-hour meter
瓦特小时容量 watt-hour capacity
瓦特逊变焦距镜头 Wattson zoom lens
瓦特指示表 indicating wattmeter
瓦天沟 tiled valley
瓦填料 tile packing
瓦铜矿 tile ore
瓦筒 pottery pipe;tile pipe;ware pipe
瓦筒铺设斗 <挖掘机> tile shoe
瓦头钉法 head nailing
瓦托吊托 brake hanger bracket
瓦托品 valtropine
瓦瓦 tile laying
瓦味利岩群【地】Waverly group
瓦屋顶 pantiled roof;tile(d)roof(ing);tiling
瓦屋顶覆盖 tile roof sheathing
瓦屋面 tile(d)roof(ing)
瓦西布恩-威廉黏[粘]度计 Washburn-William's viscometer
瓦希塔统【地】Washita series
瓦希塔(油)石 Washita stone
瓦小时 watt-hour
瓦行 tile course
瓦形垫圈 roofing washer
瓦形翘 cupping
瓦形弯 cupping
瓦檐 antefixal tile
瓦檐饰 ant(a)efix(ae);antefix tile;cheneau
瓦窑 tile kiln;tilery
瓦应尼牌玻璃 <一种供隔墙等用的压制玻璃> Wavene
瓦用板岩 roofing slate
瓦渣基础 broken-tile foundation
瓦闸 block brake
瓦状叠覆【建】imbrication
瓦状头泥芯棒 pipe nail
瓦状物 shoe
瓦兹利石 wadsleyite
瓦作 tilework

歪 swaying

歪脖壳属 <拉> Cryptoderis
歪长石 analbite;anorthoclase;anorthose;soda microcline
歪长岩 anorthoclasite;anorthosite
歪齿轮 skew gear
歪倒 overbalance
歪倒的 cranky
歪的 askew;skew
歪度 skewness
歪尔韦德牌钢筋 <商品名> Wireweld
歪分布 skew distribution
歪碱正长岩 larvikite
歪角 angle of distortion
歪角曲尺 bevel;bevel scale
歪晶 distorted crystal
歪颈 bend neck;bent neck
歪口 <玻璃制品缺陷> bent finish;crocked finish
歪扭 crookedness
歪扭波痕 swept ripple mark
歪扭角 skew angle
歪扭矩形 skew
歪扭曲线 skewed curve
歪扭作用 distortion
歪盘菌属 <拉> Phillipsia
歪曲 contort;garble;misrepresentation;swerve;torture;twist;wrench;wrest
歪曲层理 distorted bedding
歪曲拱 askew arch
歪曲作用 distortion
歪伞齿轮 skew bevel gear
歪头扁嘴钳 thin nose bent pliers
歪霞正长岩 laurdalite
歪像 anamorphosis;anamorphoser
歪像透镜 anamorphote lens
歪像转正镜 anamorphoscope
歪斜 angular deformation;obliqueness;out of plumb;out-of-square
歪斜的黏[粘]土砖 cliff brick
歪斜的托梁 club skew
歪斜地 slantways
歪斜度 skewness
歪斜故障 skew failure
歪斜井 crooked hole
歪斜连接 skewed connection
歪斜扭转 distortion
歪斜失真 skew
歪斜树 skewed tree
歪斜系数 skew factor
歪斜字符 skew character
歪斜钻孔投影图 projection map of coal seam and strata
歪斜坐标 warped coordinates
歪形能 distortion energy
歪形能理论 distortion energy theory
歪形尾 heterocercal
歪圆形 bias
歪轴 skew shaft
歪嘴白蚁 <拉> Capritermes nitobei
歪嘴龙头 bib(b)
歪嘴钳 bent nose pliers
歪嘴斜把 distortion of handle and spout

外

外Q值 external Q
外鞍 external saddle
外岸 outer bank
外岸相 offshore facies
外岸纵主梁 <台式码头> outshore longitudinal girder
外八角砖 external octagon brick
外八字脚 toe(d)-out
外靶 external target
外百叶 outside blind; outside slatted unit
外百叶窗 external shutter; external window shutter; outer blind; outer slatted blind; pinoleum blind; persiennes <可调叶板的>
外摆角 <钻车的> outward divergence
外摆线 epicycloid; outer cycloid
外摆线齿 epicycloidal tooth
外摆线齿轮 epicycloidal gear; epicycloidal wheel tooth
外摆线齿轮铣刀 epicycloidal gear cutter
外摆线的 epicyclic(al); epicycloidal
外摆线轮 epicycloidal wheel
外摆线轮系 epicyclic(al) train
外摆线运动 epicyclic(al) motion
外板 outside plate; planking; shell plating; strake
外板窗 external shutter; external window shutter
外板平铺内加横镶条舢板船壳 ribband carvel planking
外板式集装箱 outside skin type container
外板展开图 shell expansion; shell expansion plan
外板装配工 bolter-up
外板纵搭接 shell landings
外半次 exterior semi-degree
外半规管 canalis semicircularis lateralis; lateral semicircular canal
外半规管凸 prominence of lateral semicircular canal
外半径 external radius; outer radius; outside radius
外帮砖 backing brick
外包 exterior cladding; outer pack
外包板列板 raised strake
外包层 cladding; surround
外包尺寸 outside measurement; out-to-out dimension; out-to-out distance; overall dimension; overall size
外包缝合法 outpocketing
外包工 labo(u)rer
外包混凝土 concrete encasement
外包混凝土钢梁 steel beam enclosed in concrete
外包角 corner angle
外包金属板 plating
外包金属的 metal-covered
外包聚四氟乙烯 wrapped in Teflon
外包聚四氟乙烯双金属 seal by bimetallic O-ring with teflon
外包颗粒 coated grain
外包劳务 outsourced labor
外包梁 cased beam
外包络断面 outer enveloping profile
外包络轮廓 outer enveloping profile
外包络面 outer enveloping profile
外包内陷 epibolic invagination
外包皮 sheathing
外包式楼梯 enclosed stair(case)
外包套 outerwrap jacket
外包线 envelope line
外包箱梁 cased beam
外包橡胶的钢肋轮缘 steel-ribbed rim with rubber cover(ing)
外包橡胶电缆 rubber-isolated cable
外包效应 coating effect
外包(支)柱 cased post
外包直径 overall diameter
外包装 end packing; external packing; outside packing; shipping package
外保持场 externally maintained field
外保护层 outer jacket
外保角半径 outer conformal radius
外保温层 outer jacket; outside insulation
外保温饰面系统 exterior insulation and finish system
外堡 barbican; fortalice; fortilage
外堡垒 outer fort
外抱 reveal
外抱斜削门或窗 bonnet headed door or window
外抱闸 external contracting brake
外抱制动器 external brake
外贝加尔海湾 outer Caecilian gulf
外贝加尔山地 outer Baical mountains
外背辐肋 externo-dorsal ray
外泵壳 outer pump housing
外币 foreign currency; foreign exchange
外币持有额 foreign currency holdings
外币存款 deposit(e) in foreign currency; foreign currency deposit
外币存款户 foreign currency deposit account
外币的折算 foreign currency conversion
外币兑换 exchange of foreign currency; foreign currency exchange
外币兑换水单 exchange memo
外币兑换业执照 money exchange's licence
外币兑换指定代理处 authorized agency for foreign currency exchange
外币付款 exchange payment
外币付款票据 foreign currency bills payable
外币公库 pool of foreign exchange
外币头付 foreign currency position
外币现金投资 foreign cash investment
外币支付 payment in foreign currencies
外币支付票据 foreign payment instrument
外币支付凭证 foreign currency payment instruments
外币转换 translation of foreign currency
外壁 extine; outer wall; outside wall; shell; ectotheca【地】
外壁板 exterior siding
外壁开孔防波堤 breakwater structure with a perforated face
外壁内层 endexine; endexinium; exitine
外边 outer edge; outer side; outskirt
外边车道 far side traffic lane
外边界 exterior boundary; outer edge
外边界区 external boundary region
外边界线 external boundary line
外边内天沟 parapet gutter
外边刃金刚石 shoulder stone
外边缘线 <拱顶室的> back of vault
外变质的 exomorphic; exomorphosed
外变质作用【地】 exomorphism; exometamorphism
外标 external standard
外标道比 ratio of external standard
外标法 external reference method
外标记 external indication; external label(l)ing
外标准源 external standard source
外表 apparel; appearance surface; exterior finish; outward appearance; resemblance; superficies; surface appearance
外表比热容 apparent specific heat
外表比重 apparent specific gravity
外表测查 visual inspection
外表层 exoexine
外表沉积物 epigenic sediment
外表粗的砖石建筑 scappling
外表的 apparent; exterior; external; ostensive
外表的个性品质 surface trait
外表镦锻钻管 exterior upset drill pipe
外表改善 cosmetic improvement
外表观察 visual examination
外表和尺寸 views and sizes
外表化 externalization
外表加工 external operation
外表价 apparent valence
外表检查 observational check; outward inspection; superficial inspection; visual examination; visual inspection
外表检视法 ectoscopy
外表接缝 exterior crack
外表结皮的混凝土 built-up concrete
外表结皮的砂浆 built-up mortar
外表良好 in apparent good order and condition
外表良好的混凝土 fair-faced concrete
外表良好的砌砖工程 fair-faced brickwork
外表裂纹 external crack
外表面 case; exteriority; exterior of surface; exterior skin; externality; outer skin; exterior surface; external surface
外表面镀膜反射镜 first-surface mirror
外表面放热系数 outside film coefficient
外表面粉刷过的 externally rendered
外表面积 external surface area
外表面加工 external work
外表面涂层 external coat(ing); outer coat(ing)
外表面涂饰 exterior finishing
外表面油漆涂层 external coat of paint
外表摩擦 skin friction
外表偏振光 epipolarized light
外表铺面 exterior surfacing
外表情况良好 apparent good order and condition
外表缺陷 appearance defect
外表上的 formal
外表特征 garment tag
外表涂料层 outer paint coat
外表涂泥的篱笆墙 wattle and da(u)b
外表涂漆 external painting(work)
外表圬工 exterior masonry work
外表行为 external behavio(u)r
外表油漆层 outer paint coat
外表装饰 outside finish
外表整修面 external surfacing
外表装修(准备) reprofiling
外表状况 apparent order and condition
外滨(带) nearshore(area)
外滨海流 <碎波区以外的> offshore current
外滨流 inshore current
外滨线 outer shoreline
外冰川沉积 extraglacial deposit
外波 external wave
外玻璃窗 outside window glass door
外剥离物 outer stripping material
外补 primary lining
外不相容性 external inconsistency
外布洛牌振捣器 <一种附着式振捣器> Vibroplat; Vibropyl
外部 exterior portion; outer portion; outside; surface
外部安全特征 external safety feature
外部安装 outfit
外部凹陷 external concavity
外部百叶窗 outside shutter
外部保护 outside protection
外部保温 exterior insulation; external insulation
外部保温层 outer insulation
外部保险 external failsafe
外部保险计时器 external failsafe timer
外部暴露 exterior exposure
外部暴露于火 external fire exposure
外部爆破 adobe blasting
外部壁板 exterior panel
外部边界条件 external boundary condition
外部标号 exterior label; external label
外部标记 external label
外部标志 external mark
外部表面 outer surface
外部表示 external representation
外部玻璃墙 exterior glass wall
外部薄壳 exterior shell
外部薄壳构件 exterior shell component; exterior shell member
外部补偿函数 exterior penalty function
外部不经济性 external diseconomy
外部不经济性的价格和产出效应 price and output effect of external diseconomy
外部不经济性内部化 internality of external diseconomy
外部不利条件 external diseconomics
外部不溶胶 exterior bond
外部不整合晶格 external discontinuity lattice
外部布道讲坛 exterior pulpit
外部布线 outside wiring
外部材料升降机 exterior materials lift; external material lift
外部财务 external finance
外部参考 external reference
外部参考符号 reference external symbol
外部参数 external parameter
外部操作 auxiliary operation; peripheral operation
外部操作时间比 external operating ratio
外部测微计 external micrometer
外部查找 external searcher
外部长廊 exterior gallery; external gallery
外部衬垫 external lining
外部衬砌 exterior lining
外部程序设计 external programming
外部程序式计算机 externally programmed computer
外部程序中断 external program(me) interrupt
外部尺寸 external dimension; outside dimension; overall external dimension
外部尺度 external measurement
外部充填料 alien filling
外部筹资 external financing; outside financing
外部传感及控制线 external sense and control line
外部传送 peripheral transfer
外部窗扇 exterior leaf
外部吹风冷却 blow-over cooling

W

外部吹袭襟翼 externally blown flap
外部瓷面砖 exterior tile
外部瓷砖 outside tile
外部磁场 external magnetic field
外部催化剂 external accelerator;external activator;external catalyst
外部催化剂树脂 externally catalyzed resin
外部存储程序 externally stored program(me)
外部存储器 external memory(storage);external store
外部错误 external error
外部打底材料 exterior rendering stuff
外部打底方案 exterior rendering practice
外部打底工艺 exterior rendering technique
外部打底骨料 exterior rendering aggregate
外部打底灰 exterior rendering coat
外部打底集料 exterior rendering aggregate
外部打底技术 exterior rendering technique
外部打底习俗 exterior rendering practice
外部大理石 external marble
外部大梁 exterior girder
外部大气层区 external atmospheric zone
外部的 exterior;external;foreign;outer;outside;outward;superficial
外部的有利条件 external economics
外部的最后粉刷 exterior final rendering
外部的最后粉刷材料 exterior final rendering stuff
外部的最后粉刷混合物 exterior final rendering mix(ture)
外部等离子体 external plasma
外部底漆 exterior primer;external primer;outer primer;outer undercoat(er)
外部底涂层 exterior rendering coat;exterior undercoat(er);external undercoat(er)
外部底涂层灰浆 exterior undercoat plaster
外部地板 exterior floor slab
外部地形 <重力测量的> external topography
外部地址输入 external address in
外部电话分机 outside extension
外部电缆 external cable
外部电流 extraneous current;foreign current
外部电路 exterior circuit
外部电源 external power source;external power supply
外部电源防蚀法 cathodic protection by power-impressed methods
外部电阻 non-essential resistance
外部调用 external call;external reference
外部定货费 outside order expenses
外部定时 external clocking
外部定位 outer orientation
外部定相 <天线> external phasing
外部定向 exterior orientation;outer orientation
外部定义 external definition
外部定义符号 defined external symbol;external definition symbol
外部动机 external motivation
外部动力 outside power
外部动力供应 outside power supply
外部动能 external kinetic energy
外部镦锻 exterior upset
外部发光漆 external gloss paint
外部发光颜料 external paint
外部反馈 external feedback
外部反馈信号 external feedback signal
外部反射 external reflection
外部方差 external variance
外部防波堤 external breakwater
外部防护工程 external protection works
外部防火淋水器 external drencher
外部非线性共振 exterior non-linear resonance
外部费用 external cost
外部分级 external classification
外部分类 external sort
外部粉刷 exterior coat;external rendering
外部粉刷过的 externally plastered
外部敷设层 exterior layer
外部扶手 outer handrail
外部扶手栏杆 exterior handrail
外部符号 external symbol
外部符号引用 external symbolic reference
外部符号字典 external symbol dictionary
外部符合 external consistency
外部辐射 external radiation
外部辐射暴露 external radiation exposure
外部辐射剂量 external radiation dose
外部辅助柱 external auxiliary column
外部腐蚀 external corrosion
外部腐朽 sap rot
外部负荷 external load
外部复位 external reset
外部覆盖面缸砖 brick-size external cladding klinker
外部覆面 exterior cladding
外部覆面瓷砖 external cladding tile
外部干预的停车 unintentional stop
外部感觉 external sensation
外部感受器 external receptor
外部钢筋束 exterior tendon;external tendon
外部钢索 external tendon
外部隔绝 exterior insulation
外部隔热 external insulation
外部隔热材料 insulating back-up material
外部隔音层 outer insulation
外部工程 outer works;external works
外部工作 outer work
外部供能的电力机车 externally-powered electric(al) locomotive
外部供水机 external water-feed machine
外部供压气体润滑 externally pressurized gas lubrication
外部共振 exterior resonance
外部沟 external access
外部沟槽 outer ditch
外部构件 external member
外部构造 external construction;external structure
外部估价 external valuation
外部固定 external stability
外部故障 external fault
外部故障成本 external failure cost
外部观测 external observation
外部观察 external observation;visual observation
外部管道系统 exterior pipe system;external pipe system;outside network
外部管网 exterior pipe system;external pipe system;outside network
外部管柱 external column
外部光电效应 external photoelectric(al) effect
外部光阑 external stop
外部光强度 external light intensity
外部光线 extraneous light
外部光泽涂料 exterior gloss paint
外部光泽油漆 exterior gloss paint
外部规定的符号 externally defined symbol
外部过程 external procedure;out-of-line procedure
外部过滤 bulk filtration
外部含水量 shell moisture content
外部函数 external function
外部函数翻译程序 peripheral function translator
外部合成 external composition
外部荷载 external load;outer loading
外部横墙 outer cross wall
外部呼叫 external call
外部护面板 exterior panel
外部环境 exotic environment;external environment
外部环境情况 ambient condition
外部环境状态 ambient condition
外部换向机构 external reversing mechanism
外部灰 free ash
外部灰板墙柱 exterior stud
外部回流 cold reflux;external reflux
外部回路 external circuit
外部回输 external feedback
外部混合 external mix
外部混合型喷枪 external mix type gun
外部混凝土模板 exterior concrete form
外部活化体系 externally activated system
外部火 exterior fire;external fire
外部火焰 external flame
外部火灾公害 external fire hazard
外部火灾事故 external fire hazard
外部火灾危险 external fire hazard
外部基标 external reference
外部激发 external excitation
外部激励 external drive
外部集气罩 open hood
外部给水管路 external feed line
外部挤压作用 external loading
外部记录程序 outboard recorder
外部记忆装置 external memory(storage)
外部寄存器 external register
外部加粉刷 exterior varnishing;external plastering
外部加强支柱 external reinforced column
外部加热 external heating;indirect heating
外部加热弧 externally heated arc
外部加热介质 external heating medium
外部加热设备 outside heating device
外部加油 external dressing
外部加载 exterior loading;external loading
外部夹板制动器 exterior cheek brake
外部间柱 exterior stud
外部检查 appearance test;exterior inspection;external inspection;observational check;outer inspection;visual examination;visual inspection
外部检验 external examination
外部建筑风格 exterior architecture;external architecture
外部建筑(面)板 exterior building panel;external building panel
外部建筑艺术 exterior architecture
外部交换业务 foreign exchange service
外部胶结粉刷 bond external plaster;bond external rendering;bonding exterior plaster;bonding exterior rendering;bonding rendering
外部胶结抹灰 bonding exterior plaster;bonding exterior rendering
外部脚手架 exterior scaffold(ing);external scaffold(ing)
外部搅拌叶片 exterior mixing blade;external mixing blade
外部校准 external calibration
外部阶梯 exterior stair(case)
外部接触 exocontact
外部接口 external interface
外部接口适配器 peripheral interface adapter
外部接口转接器 peripheral interface adapter
外部接头 external lug;external terminal
外部接线图 external wiring diagram
外部节点 external joint
外部节间 exterior panel
外部结构 exterior structure;external construction;external structure
外部结构系统 exterior structural system
外部介质条件 environmental factor
外部金属板包皮 external sheet-metal covering
外部锦砖精加工 exterior mosaic finish;external mosaic finish
外部锦砖终饰 exterior mosaic finish
外部禁止中断 external inhibit interrupt
外部经济 external economics
外部经济负效果 external diseconomy
外部经济理论 theory of external economy
外部经济效果 external economy effect;externality
外部经济性 external economy
外部精度 external accuracy
外部径向边界 external radial boundary
外部救援 external rescue
外部绝缘 outside insulation
外部绝缘层 outer insulation
外部开裂 external crack
外部铠装 armo(u)r
外部空间 exterior space;outer space
外部空气 extraneous air
外部空气离析器 extraneous air separator
外部空隙分率 external void fraction
外部孔隙率 external porosity
外部控制 external control;extra control
外部控制器 extra controller;peripheral control unit
外部控制台 outer pulpit
外部框缘 outside architrave
外部扩充性 external extendibility
外部拉条 external bracing
外部廊沿 external corridor
外部冷却 external cooling
外部冷却反应堆 external cooling reactor
外部冷却回路 external cooling circuit
外部冷却器 external cooler
外部力矩 external moment
外部立面 outside face
外部利益 external benefit
外部连接钻杆 externally flush-coupled rods
外部链接 external linking
外部裂化 extraneous cracking
外部临界阻尼电阻 external critical damping resistance
外部领海 exterior territorial waters

外部领示控制 external piloting control
外部楼板 outer floor slab
外部楼梯通道 access stairway
外部漏风 surface air leakage
外部漏泄 external leakage
外部(露)多边形(非直线布置)(预应力)筋 external polygonal tendon
外部(露)无黏[粘]结筋 <预应力钢丝束等> external unbonded tendon
外部旅客电梯 exterior passenger elevator
外部逻辑 external logic
外部螺纹 exterior thread; external thread
外部落水管 external downpipe
外部马赛克精加工 exterior mosaic finish;external mosaic finish
外部马赛克饰面 outer mosaic finish
外部马赛克终饰 exterior mosaic finish
外部慢化堆 external moderated reactor
外部锚定装置 external tieback
外部媒体 foreign medium
外部门扇 exterior leaf
外部(门头)装饰 exterior trim
外部密封 outside seal
外部密封垫 exterior scaling
外部面积的 superficial
外部面砖 outside tile
外部名 external name
外部命令 external command
外部模件 external module
外部模式 external schema
外部模型 external model
外部抹灰拌和物 external plastering mix(ture)
外部抹灰材料 exterior plastering stuff; exterior rendering stuff
外部抹灰层 exterior plastering coat; external plastering coat
外部抹灰方案 exterior plastering scheme;exterior rendering practice
外部抹灰工艺 exterior plastering technique;exterior rendering technique
外部抹灰工作 exterior plastering
外部抹灰骨料 exterior rendering aggregate; external rendering aggregate
外部抹灰过的 externally plastered
外部抹灰混合物 exterior plastering mix(ture)
外部抹灰集料 exterior rendering aggregate; external rendering aggregate
外部抹灰技术 exterior plastering technique;exterior rendering technique
外部抹灰泥 external plastering
外部抹灰实践 exterior plastering practice;external plastering practice
外部抹灰示意图 external plastering scheme
外部抹灰习俗 exterior rendering practice
外部抹灰系统 exterior plastering system
外部抹灰作业 parget-work
外部木材污斑 exterior wood stain
外部目标 external object
外部耐磨寿命 external wear life
外部能量输入 external energy input
外部能源 extra power
外部扭转力矩 exterior torsion(al) moment; exterior twist(ing) moment;external torsion(al) moment; external twist(ing) moment
外部排水 external drainage
外部排水管网 external sewer network
外部排水系统 exterior drainage system; exterior sewer system; external sewer system
外部排土场 external dump
外部盘 peripheral disk
外部盘管 outside coil
外部配件 outside appurtenances
外部喷水灭火器 outside sprinkler
外部频率 foreign frequency
外部破坏 external damage
外部破坏压力 external bursting pressure
外部起动 external activate
外部气顶 external gas cap
外部气候 exterior climate
外部嵌板 exterior panel
外部嵌缝 outer sealing; outside sealing
外部墙板 outer panel
外部桥 outer bridge
外部切除 external cut;external mute
外部清管器 external cleaning pipe machine
外部清洗 external cleaning
外部请求 external request
外部穹隆 exterior dome;external dome
外部驱动 external drive
外部权 external weight
外部燃烧 external burning; external firing
外部燃烧系统 external firing system
外部热交换 extraneous heat removal
外部热交换介质 external heat exchange medium
外部热绝缘 external insulation
外部入口点 external entry point
外部润滑 external lubrication
外部色彩 exterior colo(u)r
外部闪光 external flashing
外部上(清)漆 external varnishing
外部烧蚀 external ablation
外部设备 ancillary equipment; auxiliary equipment;external device; external equipment; external plant; external unit;extra face;extra facility; off-line equipment; outside plant;peripheral equipment;peripheral subsystem; peripheral unit; peripheric equipment;peripheric unit
外部设备操作码 external device operation code
外部设备操作数 external device operand
外部设备地址 external device address
外部设备缓冲器 peripheral buffer
外部设备接口通道 peripheral interface channel
外部设备控制 external device control
外部设备控制程序 peripheral control program(me)
外部设备控制装置 peripheral control unit
外部设备码 external device code
外部设备起动 external device start
外部设备数据流 external device data flow
外部设备数据转移 peripheral transfer
外部设备通信[讯] external device communication
外部设备响应 external device response
外部设备指令 external device instruction
外部设备中断 external device interrupt
外部设备状态 external device status
外部审计 outside auditing; external audit
外部十进制数 external decimal digit
外部十进制项目 external decimal item
外部时钟(脉冲) external clock
外部时钟脉冲信号 external clock signal
外部使用 outside use
外部事件处理机模件 external event processor module
外部事件模块 external event module
外部饰面 exterior facing; exterior veneer; external veneer
外部疏散楼梯 external escape stairway
外部输出设备 output peripheral equipment
外部输入 external input
外部输送指令 external transport instruction
外部竖井 outer shaft; outside shaft
外部数 external number
外部数据结构 external data structure
外部数据库 external database
外部水 extraneous water
外部水流 outside flow
外部水幕喷头 external drencher
外部水幕系统 external drencher system
外部水驱 external water drive
外部水下区 external submerged zone
外部说明 external declaration
外部损害 external damage
外部损失 external losses
外部锁闭 external locking; outside locking
外部探测器 outer locator
外部特性 external behavio(u)r; external characteristic
外部特性曲线 external characteristic curve
外部条件 ambient condition; exterior condition; external condition
外部贴脸 exterior trim
外部停机时间 external idle time
外部通道程序 external channel program(me)
外部通信[讯] external communication
外部投入资本 outside venture capital
外部投资 outside financing
外部涂层 exterior coat; outer coating
外部涂料 exterior finish
外部涂刷 external coat(ing)
外部涂刷油漆工程 exterior painting work
外部涂油 external dressing
外部涂釉的 exterior glazed
外部涂装 exterior cladding
外部弯矩 exterior bending moment; external bending moment
外部位能 external potential energy
外部位移 external movement
外部温度传感器 external temperature sensing device
外部稳定度 external stability
外部圬工空心砖砌筑 external masonry wall of double-leaf cavity construction
外部圬工墙 exterior masonry wall
外部污斑 exterior stain
外部污染 external contamination
外部吸气罩 capturing hood
外部洗刷 external washing
外部系统 external system; peripheral subsystem
外部系统设计 external system design
外部先导控制 external piloting
外部纤维层 exterior fibre [fiber]
外部线路 outside plant
外部镶板 exterior panel
外部镶衬 outside lining
外部镶嵌玻璃的 exterior glazed
外部消光 external delustering
外部消火栓系统 exterior fire hydrant system
外部效果 outside effect
外部效益 external benefit
外部心板 external core
外部信号 external signal
外部信号线 external signal line
外部信号源 outside signal source; external signal source
外部信息 extraneous information
外部信号源 outside source; external source
外部形式 external form
外部性 externality
外部性特征 external characteristic
外部修饰工作 external work
外部锈蚀 external corrosion; extraneous rust
外部悬挂 outfit
外部选择 external selection
外部循环系统 external circulating system
外部压力 external pressure
外部压气机 external compressor
外部烟囱 exterior chimney
外部烟囱隔板 outer withe [wythe]
外部延迟 external delay
外部延伸的混凝土柱 exterior stretched concrete column
外部页地址 external page address
外部页扇 external leaf
外部一层 exterior tier; external tier
外部一排 exterior tier; external tier
外部仪表测量系统 external instrumentation system
外部乙烯基漆 exterior vinyl paint; external vinyl paint
外部乙烯基颜料 exterior vinyl paint; external vinyl paint
外部乙烯树脂漆 outside vinyl paint
外部因素 external effect
外部因子 extrinsic factor
外部引线 outside line
外部引线接合 outer lead bonding
外部引用 external reference
外部引用符号 external reference symbol
外部引用记录 external reference record
外部引用项 external reference item
外部应力 external stress
外部应用 exterior application; exterior use
外部影响 external action; externality
外部硬件 external hardware
外部用的底层涂料 outside undercoat(er)
外部用氯气处理的橡胶涂料 exterior chlorinated rubber paint; external chlorinated rubber paint
外部用途 exterior use
外部用砖 external brick
外部油冷却器 external oil cooler
外部有效灭火 effective external fire-fighting
外部预浇砌块 exterior precast block
外部预应力混凝土桥 externally prestressed concrete bridge
外部预应力混凝土柱 exterior prestressed concrete column; external prestressed concrete column
外部预应力(张拉) external stressing
外部预应力柱 exterior prestressed column; external prestressed column
外部预制块 outer precast block; outer precast brick
外部预制砌块 exterior precast block; external precast block

外部预制砖 exterior precast brick
外部圆屋顶 exterior cupola;external cupola
外部援助 external aid;external assistance
外部运动阻力 external motion resistance
外部运输建设 regional transport building
外部运输设备 outside carrier
外部运算 external arithmetic
外部造型 external styling
外部噪声 exterior noise;external noise
外部噪声穿透 external noise penetration
外部噪声电平 external noise level
外部噪声因数 external noise factor
外部增添装饰 external enrichment
外部张力混凝土柱 exterior tensioned concrete column
外部丈量【船】external measurement
外部照明 exterior lighting;external illumination; outdoor lighting; outside lighting
外部照明设备 external lighting device
外部照明式标志 external illumination sign
外部照明系统 exterior lighting system
外部照明装置 exterior lighting unit
外部照射 external irradiation
外部遮帘 exterior blind;external blind
外部遮阳 exterior blind;external blind
外部振捣 <适用于模板小配筋密的构件> external vibration
外部振捣器 <浇混凝土时用在模板上> external vibrator;form vibrator;shutter(ing) vibrator
外部振动 exterior vibration; external vibration
外部振动器 external vibrator; form vibrator;mo(u)ld vibrator
外部整修 exterior trim;external dressing
外部证据 extrinsic evidence
外部支撑 external bracing
外部支承 exterior support
外部指令缓冲器 peripheral order buffer
外部制导 external guidance
外部质量控制 outside quality control
外部中断 external interrupt
外部中断处理器 external interrupt processor
外部中断符号 external interrupt symbol
外部中断禁止 external interrupt inhibit
外部中断状态字 external interrupt status word
外部中心 exterior core
外部中子 external neutron
外部重力场 external gravity field
外部周围温度 exterior ambient temperature;external ambient temperature
外部属性 external attribute
外部住宅装潢 exterior home decoration
外部住宅装修 exterior home decoration
外部注气 crestal injection; external gas injection
外部专用索引 externally specified index
外部砖 exterior brick
外部砖墙 exterior brick wall
外部砖石墩墙 outer masonry wall
外部砖石墩柱 outer masonry column
外部砖石墙 exterior masonry wall; outer masonry wall
外部砖石墙衬砌 outer masonry wall lining
外部砖石墙饰面 outer masonry wall facing
外部砖石圬工 exterior masonry; external masonry
外部转角处 exterior corner
外部转矩 exterior torque; external torque
外部装潢 external decoration
外部装潢面貌 external decorative feature
外部装潢最后装修 external decorative finish
外部装饰 exterior decoration;facade treatment
外部装饰精加工 external ornamental finish
外部装饰面貌 external ornamental feature
外部装饰品 exterior trim part
外部装饰设计 facade density
外部装饰特征 external ornamental feature
外部装饰最后装修 external ornamental finish
外部装修 exterior decoration;outside finish
外部装药爆破(法) adobe shooting; mudcapping;plaster blasting
外部装置 external device
外部资金 outside finance
外部子程序 external subroutine
外部自锁表征 external self-locking flag
外部走廊 external corridor;outer gallery
外部阻力 external drag; external resistance
外部作用 extraneous action
外埠的 out of town
外埠付款票据 domicile bill
外埠同业存款 outport bank deposit
外埠同业透支 overdraft by outport correspondents
外埠支付汇票 domiciled bill of exchange
外埠支付票据 domiciled bill
外材贸易 foreign timber trade
外操纵阀 external operating valve
外操纵杆 top lever
外槽壁 external groove sidewall
外槽轮排种器 external force feed
外侧 flank;outer flank;outer section; outer side
外侧安装 outboard mounting
外侧鼻突 lateral nasal process
外侧壁 <窗框的> outer reveal
外侧补强圈补强 reinforcement by outside panel
外侧部 lateral portion
外侧舱门 outboard door
外侧车道 curb lane; kerb lane; nearside lane; outer lane; flank-traffic path <紧沿路边或人行道的车道>
外侧车道信号 nearside signal
外侧车轮最小回转半径 minimum turning inner radius
外侧尺寸 outside measurement
外侧窗头线 outside architrave
外侧带 out-band
外侧导轨 follower rail
外侧的 outboard;outer
外侧登陆舰区 outer landing ship areas
外侧吊架悬挂 outside swing hanger suspension
外侧段 lateral segment
外侧舵 out-rudder
外侧房间 outside room
外侧分隔带 outer separator
外侧浮筒 outboard float
外侧辐肋 externo-lateral ray
外侧根 lateral root
外侧沟 lateral ditch
外侧股道 marginal track
外侧拐角线脚 outside corner mo(u)lding
外侧轨 main rail
外侧基础线 outside foundation line
外侧加宽 outside widening
外侧间距 distance between out to out
外侧间柱横木板 outside studding plate
外侧交通 <沿道边车道的交通> flank traffic;flank-riding
外侧结节 lateral tubercle
外侧界 lateral border
外侧框缘 outside casing
外侧廊 outer aisle;outer nave aisle
外侧梁 outside girder;string
外侧裂 lateral fissure
外侧楼梯帮 outer string; outside string
外侧楼梯斜梁 outer stringer
外侧履带 outside crawler
外侧轮 <双轮的> off-side wheel
外侧螺旋桨 outboard propeller;outboard screw
外侧螺旋推进器 wing screw propeller
外侧门头线 outside architrave
外侧面 lateral surface; outside wall <钻头的>
外侧墙板 side sheathing;side sheet
外侧墙板拉板 side sheathing bracing plate
外侧切削具 <钻头的> outside cutter
外侧球齿 outer ball tooth
外侧人行道 outer footpath
外侧刃金刚石 <钻头的> outside stone
外侧升降舵 outer elevator
外侧水槽 outside spline
外侧填料盖 follower gland
外侧贴脸 outer lining
外侧贴面 outside casing; outside facing;outside lining
外侧线 <电力线> outer conductor
外侧鞋式磁性制动器 outboard shoe type magnetic brake
外侧鱼尾板 outer fish plate
外侧预应力 outer prestress
外侧制动 outside hung brake
外侧轴 outboard shaft;outer shaft
外侧主梁 exterior girder
外侧转向半径 <叉车> radius of outside steering
外侧装饰 outside finish
外侧纵纹 lateral longitudinal stria; stria longitudinalis lateralis
外侧组 lateral group
外侧最大尺寸 overall external dimension
外测度 exterior measure;outer measure
外测量规 female ga(u)ge
外测量器 pelvimeter for external measurement
外层 external layer; garment; nappe; outer layer; outer leaf; outer tier; outside layer; outside leaf; sheath; finishing coat;outer coating <涂料的>
外层百页 outer window blind
外层板 stripboard
外层板材 outer sheet
外层包覆 external sheath
外层玻璃 outer-pane
外层薄板 outside sheet
外层窗玻璃 outdoor glass of double-glazing units;outdoor pane
外层大理石 outer marble
外层大气 outer atmosphere
外层大气密度 outer atmospheric density
外层单砖墙 outside with;outside wythe
外层的 outer
外层底板 outer bottom
外层电子 outer-shell electron
外层防风暴窗的窗框 storm window frame
外层防潜幕 outer antisubmarine screen
外层钢丝 <钢丝绳> cover wire; crown wire
外层轨道 out orbit
外层空间 deep space;outer space
外层空间生物学 exobiology
外层空间通信[讯] outer space communication
外层氯化橡胶漆 outside chlorinated rubber paint
外层络合物 outer sphere complex
外层门 outer door;weather door
外层黏[粘]附 adherence of nappe
外层砌好后内部添陷 filling-in
外层穹隆 outer dome
外层守卫区 outer ward
外层受力板 stressed-skin panel
外层通路 skin pass
外层涂料 outer coat of paint; outer paint;outside paint coat;topcoating
外层涂刷工作 outer painting work
外层涂刷油漆工 outside painting work
外层纤维 outer fiber [fibre]
外层旋转发电机 outer field generator
外层循环 surrounding loop
外层焰 outer flame
外层乙烯树脂漆 outer vinyl paint
外层硬化 case-harden(ing)
外层油漆 multitle;outer coat of paint; outer oil paint;outside paint coat
外层有光泽漆 outer gloss paint
外层砖石墙 outside masonry(wall)
外层砖石墙衬砌 outside masonry wall lining
外层砖石墙面 outside masonry wall facing
外层砖石墙柱 outside masonry wall column
外叉 end yoke
外插 extrapolate; extrapolating; extrapolation
外插端点 extrapolated end point
外插法 extrapolation method;method of extrapolation
外插曲线 extrapolated curve
外差 heterodyne
外差变频器 heterodyne conversion transducer;heterodyne converter
外差标志叠加器 heterodyne marker adder
外差电路 heterodyne circuit
外差动补偿齿轮 external differential compensating pinion
外差法 heterodyne method; process of heterodyning
外差法测量 heterodyne measurement
外差分析器 heterodyne analyser [analyzer]
外差蜂鸣振荡器 heterodyne buzzer oscillator
外差干扰 heterodyne interference
外差级 heterodyne stage
外差检波 heterodyne detection

外差检波法 heterodyne detection method
外差检波器 heterodyne detector
外差接收法 heterodyne reception
外差滤波器 heterodyne filter
外差频率 heterodyne frequency
外差频率计 heterodyne-type frequency meter
外差式 heterodyne system
外差式波长计 heterodyne wavemeter
外差式波形分析仪 heterodyne wave analyser [analyzer]
外差式等幅振荡 sustained oscillation
外差式电压计 heterodyne voltmeter
外差式分析仪 heterodyne analyser [analyzer]
外差式接收机 beat frequency receiver; beat receiver; heterodyne receiver
外差式频率计 heterodyne frequency meter
外差式声分器 heterodyne sound analyser [analyzer]
外差式谐波分析 heterodyne wave analysis
外差式谐波分析器 heterodyne harmonic analyser [analyzer]
外差式振荡器 heterodyne oscillator
外差伺服滤波器 heterodyne slave filter
外差啸鸣 heterodyne whistling
外差啸声 heterodyne whistle
外差调制 heterodyne modulation
外差谐波分析仪 heterodyne harmonic analyser [analyzer]
外差振荡器 heterodyne
外差值 extrapolated value
外差中继 heterodyne repeating
外差中继方式 heterodyne relay system
外差中继站 heterodyne repeater
外差转发 heterodyne repeating
外差作用 heterodyne action; heterodyning
外掺剂量 admixture dosage
外缠绕层 outer wrap(ping)
外长尺 outside cal(l)ipers
外长分枝 exogenous branching
外长规 outside cal(l)ipers
外长树 exogen(ous) tree
外长植物 exogenous plant
外场 external field; outstation
外场感应电流 field-generated current
外场设备 outstation equipment
外潮三角洲 outer tidal delta
外沉积弧 outer sedimentary arc
外衬板 outside welt
外衬(砌) outside lining; outer lining
外撑条 outer stay
外成包体 exogenetic inclusion
外成变形 exogenetic deformation; exogenous ejecta
外成沉积 exogenetic sediment
外成的 epigene; epigenic; exogenetic; exogenic; exogenous; exokinetic
外成节理 exogenetic joint; exokinetic joint
外成裂缝 exokinetic fissure
外成穹隆 exogenous dome
外成穹丘 exogenous dome
外成岩 exogenetic rock
外成岩类 exogenous rocks
外成因水库诱发地震 exogenous reservoir-induced earthquake
外成作用【地】 exogenetic action; epigene action; exogenetic process; exogenic process; metadiagenesis; epigenesist
外成作用异常类型 type of exogenic process anomaly
外城 outer city
外城门 outer gate
外乘(法) exterior multiplication; outer multiplication
外尺寸 outside dimension; overall measurement; superficial dimension
外齿 external tooth
外齿层 exoperistome
外齿轮 exterior-teeth gear; external-teeth gear; outer gear
外齿轮传动装置 external gearing
外齿轮油泵 external gear oil pump
外齿片 outer toothed disc
外齿型锁紧垫圈 external-tooth washer
外冲【地】 detrusion
外重窗 <防暴风雨的> storm sash
外重门 storm door; weather door
外出的 outbound
外出(权) egress
外初切 exterior ingress
外储存 external storage
外触发脉冲 external trigger pulse
外触发器 external trigger
外触发输入 external trigger input
外传力法预加应力 <无握裹力的预加应力> prestressing without bond
外传噪声 exterior noise
外传噪声级 exterior sound level
外传噪声量 exterior sound level
外传噪声水平 exterior sound level
外传噪声限度 noise emission limit
外传噪声值 noise emission value
外窗 exterior window; external window; outer window; out-opening; outside window
外窗边槽口 exterior window check; exterior window rabbet
外窗槽口 outer window check; outer window rabbet
外窗窗盘 exterior window sill
外窗挡 outside window stop
外窗框 <防暴风雨用> outer window frame; exterior window frame
外窗台 drip cap; exterior window sill; external window sill; outer window sill; outside window sill; window ledge
外窗台木 belt rail cap
外窗台装饰 bird's beak ornamental
外窗樘 outside window frame
外窗涂层 exterior window lining
外唇面 outside lip surface
外磁场 external field; foreign field
外磁铁 external magnet
外磁效应 extraneous magnetic effect
外次大陆架 <水深 50 ~ 200 米> outer sublittoral
外次序 external ordering
外丛状层 external plexiform layer
外猝灭 external quenching
外猝灭计数管 externally quenched counter tube
外存储 external medium; external storage
外存储计算机 external memory computer
外存储器 external storage; peripheral memory; external store; external memory(storage)
外存储器信息处理机 file processor
外存结束 end of file
外错角 alternate exterior angle
外大陆架 outer continental shelf
外大陆架沉积 outer shelf deposit
外大陆土地使用条例 Outer Continental Shelf Lands Act
外大门 anteport; outer gate
外大气层 <离地 480 ~ 1610 千米处的> exosphere; exoatmosphere
外大气层辐射 exoatmospheric radiation
外大气层射线 exoatmospheric rays
外大气层温度 outer atmospheric temperature
外大气层现象 exospheric phenomenon
外大气圈 exoatmosphere; exosphere
外代数 exterior algebra; external algebra
外代数丛 exterior algebra bundle
外带 outer zone
外带滑车 external bound block
外带夹角 angle between foliation and kink plane outside
外带离合器 external band clutch
外带内紧式制动器 external band brake
外带式制动器 contracting band brake; external-band-type brake
外单位 outside firm
外担子菌 <拉> Exobasidium
外挡 outer side
外挡车道 <最近路边的行车道> outer wheel path
外挡过驳 overside delivery; overside transfer; waterside transfer
外挡换装 waterside transfer
外挡锚 offshore anchor
外挡圈 outer thrust collar
外挡卸货【船】 discharge offshore side
外挡装卸 overside discharging and loading
外导轨 outer guide
外导流堤 external diversion dike [dyke]
外导流砂堤 external diversion sand dike [dyke]
外导数 exterior derivative
外导体 <同轴电缆> outer conductor
外倒角活塞环 bevel piston ring
外倒转术 abdominal version; external version
外到外尺寸 out-to-out dimension
外堤 external breakwater; external dike [dyke]; external wall; outer bank
外堤岸 back scarp
外堤岸趾 outside embankment toe
外堤坝趾 outside embankment toe
外底 outer bottom; outsole
外底支 lateral basal branch
外地裁判庭证据法案 Foreign Tribunals Evidence Act
外地槽【地】 foredeep; delta geosyncline
外地槽带 externide
外地槽的 exogeosynclinal
外地磁场 external geomagnetic field
外地付款期票 domiciled note
外地付款支票 domiciled check
外地集装箱回空 overland transit empty
外地节目来源 remote programme source
外地壳层 supracrust
外地幔 exo-mantle; outer mantle
外地支付票据 <指定支付地点的> addressed bill
外地植物 exotic plant
外地装箱回送 overland transit full
外地作物 alien crop
外点 exterior point; external point; outer point
外电动势 external electromotive force
外电抗 external reactance
外电路 external circuit
外电路电流 external current
外电路电阻 external resistance
外电路电阻器 external resistor
外电势 external electric(al) potential; external potential
外电枢 external armature
外电枢式同步发电机 external armature alternator
外电网 external power grid
外电源 external power
外电源插座 external power receptacle; external socket
外电晕 external corona
外电子 exo-electron
外垫圈 outside washer
外顶板 roof sheet
外顶盖 outer head cover; runner end cover; runner end lid
外顶棚窗 outer deck sash
外顶线 exterior crest
外定索引操作 externally specified index operation
外定位 outside fix
外定位条 exterior stop
外定向元素 extra orientation element
外定子 outer stator
外动力地质作用 exogenic geologic(al) process
外动力裂缝 exokinetic fissure
外动力土壤 ectodynamic(al) soil
外动力型的 ectodynamorphic
外毒素 exotoxin
外端 outside end
外端间的(尺寸) out-to-out
外端梁 outside end sill
外端汽缸 outer cylinder
外端支架 outboard support
外对称 external symmetry
外对管器 exterior clamp; exterior lineup clamp; external clamp; external lineup clamp
外对光 external focusing
外对光望远镜【测】 external focus(s)ing telescope; exterior focusing telescope
外多管加热器 external multitubular heater
外俄耳 violle
外二心桃尖拱 acute arch
外罚函数 exterior function; external function
外法兰 outside flange; outward flange
外法线 exterior normal; outer normal
外法向导数 exterior normal derivative
外翻门 outward opening
外翻状的 valgoid
外反馈控制器 external feedback type controller
外反馈式 external feedback type
外反馈式磁放大器 external feedback magnetic amplifier
外反馈式振荡电路 external feedback type oscillator circuit
外反射 external reflection
外反射镜 external mirror
外反应作用 exterior reaction
外方头 external square; male square; outside square
外方位 exterior orientation; outer orientation
外方位元素【测】 elements of exterior orientation; data of outer orientation
外防火保护 outside fire protection
外防水 exterior waterproofing
外防水层 waterproof coating
外飞航线定向法 outbound course orientation
外非正则点 externally irregular point
外分 exterior division; external division
外分比 external ratio
外分点 external point of division

外分隔带 < 限制进入的干道和服务性道路之间的分隔带 > outer separator
外分隔路段 outer separation
外分角线 external bisector
外分子层 outer molecular layer
外粉饰浆 exterior plaster
外粉刷 exterior plaster;external plaster;outside finish
外粉刷层 stucco work
外风扇 external fan
外峰 outer peak
外缝合线 external suture line
外敷层 outer coating;overcoating
外敷时间 overcoating time
外浮坞门座 outer seat
外辐射带 outer radiation belt
外辐照 external irradiation
外腐蚀 exterior corrosion
外腐殖质 ectohumus
外付报费 outpayment
外负载 external load
外负载电路 external load circuit
外附的 extraneous
外附力 stick force
外附式振捣器 surface type vibrator
外附体 epimorph
外附同态体 epimorph
外附物 extraneous
外附因素 extraneous factor
外赋传导率 extrinsic conductivity
外赋的 extrinsic
外赋区 extrinsic region
外覆盖层 external cladding;external covering;external skin
外覆盖层材料 exterior material;external material
外覆盖层建筑材料 external building material;outer material
外盖 end cover
外盖垫片 outer cover gasket
外干涉 external interference
外干涉效应 external interference effect
外缸 outer casing
外港 < 离开主要海关或商业地区的海港 > outport;fore harbo(u)r;front harbo(u)r;outer harbo(u)r;outer port
外港池 avant port;open basin;outer basin;outer dock
外港单浮筒系泊 exposed location single buoy mooring
外港界线 outer harbo(u)r line
外港区 outer harbo(u)r area
外杠杆 outer lever
外高加索山地 outer Caucasus mountains
外割 external secant
外割刀 external cutter
外格 external grid
外隔板 outer lining;outside lining
外隔圈 outer ring spacer
外隔热导流管 guide pipe with outer insulation
外隔声层 outside insulation
外公切面 external common tangent plane
外公切线 external common tangent
外功 external work
外功函数 outerwork function
外功率 outside power;rate of external work
外攻 exterior attack
外供的【计】 off-line
外供气 distributed gas
外供水 exterior supply water
外拱圈 extrados;extrados of arch
外拱圈半径 radius of extrados
外拱圈起拱线 extrados springing line
外共振 external resonance
外沟 outer fissure
外构造带【地】 externide
外购补给品 outside supplier
外购产品成本 order-getting cost
外购废料 bought scrap
外购件 bought in components
外购商品 commodity purchased
外骨骼 exoskeleton
外鼓 bulging outward;external drum;outer drum
外鼓式过滤机 outside-drum filter
外鼓式过滤器 outside-drum filter
外固定螺钉 outer set screw
外固定术 exopexy
外固数 external stability number
外挂板 cladding panel;siding shingle
外挂的 out-band
外挂架 pylon
外挂脚手架 outrigger scaffold
外挂墙板 outbond wall board
外挂石板瓦 slate hanging;weather slating
外关闭压力 external confining pressure
外观 appearance;exterior;extrinsic feature;facade;facies;general view;outward appearance;physical form;profile;resemblance;superficies
外观保持性 appearance retention
外观标准 appearance standard
外观尺寸检查 appearance and dimension check;visual and dimensional check
外观疵点 visual defect
外观大小 apparent size
外观等级 appearance grade
外观电势 appearance potential
外观反应 overt response
外观分级 visual grading
外观附属建筑物 apparent easement;apparent servitude
外观驾 external view
外观检查 appearance test;exterior inspection;external inspection;observational check;outer inspection;surface check;visual check;visual examination;visual inspection
外观检验 appearance test;outer inspection;visual inspection
外观类型 physiognomic(al) type
外观评定 visual assessment
外观评估 visual evaluation
外观评价 ocular estimate;visual evaluation
外观迁移率 external mobility
外观缺陷 apparent defect;appearance defect;macroscopic irregularity;open defect
外观容积 appearance volume
外观上 seemingly
外观图 outside drawing;outside view;outside view drawing
外观温缩系数 coefficient of apparent thermal shrinkage
外观吸引力 eye appeal
外观形状 visual form
外观应力 apparent stress
外观油表 sight oil ga(u)ge;sight oil glass
外观质量 apparent quality;appearance quality;presentation quality
外观质量评级 appearance rating
外观注油计 oil sight-feed ga(u)ge
外观装饰 decorative appearance
外观状况评定法 visual condition rating
外管 < 双层岩芯管的 > outer barrel;outer pipe;outer tube;outside barrel;outside pipe;outside tube;offside piping
外管理界限 outer control limits
外管路 < 滤池系统的 > face piping
外管螺纹 external pipe thread
外管网 outside network
外管直径 outside tube diameter
外管钻头 < 双层岩心钻的 > outer cutting bit
外光电效应 external photoeffect;outer photoeffect
外光电效应光电管 photoelement with external photoelectric(al) effect
外光环 outer reticle circle
外光阑 external diaphragm
外光路 external light path
外光密度 external optic(al) density
外光谱 external spectrum
外光圈打开 circle-in
外光圈关闭 circle-out
外广延度【数】 exterior extent
外轨 outside rail
外轨超高 outer rail super-elevation;super-elevation of outer rail
外轨超高不足 cant deficiency
外轨道 outer rail
外轨垫高 super-elevation in track
外轨抬高 out track build-up
外轨抬高递增距离 increment distance of outer track build-up
外轨线 outer line of the rail
外国产品 exotic product
外国产业 alien property
外国车 foreign wagon
外国贷款 foreign loan
外国的 alien;foreign
外国电波干扰 interference of foreign radio station
外国电台干扰 interference of foreign radio station
外国法律 foreign law
外国法人 foreign juridical person
外国工人 foreign labo(u)rer
外国公库支票 treasury cheque
外国公司 alien corporation;foreign company;foreign corporation
外国股份 foreign equity holding
外国合营(者) foreign joint venture
外国和国际债券市场 foreign and international bond markets
外国化 foreignize
外国汇票 foreign bills of exchange
外国货物进口报单 application for import of foreign goods
外国借款 foreign borrowing
外国借款人 foreign borrower
外国矿石 foreign ore
外国劳动力 expatriate labo(u)r
外国劳工 foreign worker
外国判决互惠执行法案 Foreign Judgment Reciprocal Enforcement Act
外国企业 foreign firm
外国(汽)车 foreign car
外国侨民 foreign nationals
外国人 alien;foreigner;uintalander < 南非用语 >
外国人的雇用 alien employment
外国人的入境许可 admission of alien
外国商人 alien merchant
外国商行 foreign firm
外国石油 foreign oil
外国税赋 foreign tax
外国税收 foreign revenue
外国税收抵免 foreign tax credit
外国铁路 overseas railway
外国投资 foreign investment
外国投资净额 net foreign investment
外国一般代理商 foreign general agent
外国银行 alien bank
外国银行票据 foreign bank bill
外国债券 foreign bond
外国债券市场 foreign bond markets
外国证券 foreign securities
外国政府贷款 foreign state loans
外国政府公债 foreign government bond
外国支票 foreign check
外国种的 exotic
外国注册 foreign patent
外国专利 foreign patent
外国专利文摘 foreign patent abstract
外国资本 foreign capital
外国资本流量 foreign capital flows
外国资产 foreign assets
外国资产的往来 transactions in foreign assets
外国子公司 foreign affiliates
外海 high sea;main sea;marine sea;midsea;off-lying sea;open ocean;open sea;outer shore
外海暴露条件 marine exposure condition
外海波高 offshore wave height
外海波况 offshore wave condition
外海波浪资料 offshore wave data
外海波谱密度函数 offshore spectral density function
外海泊位 open sea berth;open sea terminal;sea berth
外海驳船运输系紧作业 sea fastening
外海测深学 offshore bathymetry
外海单点系泊码头 offshore monobuoy terminal
外海单浮筒码头 offshore monobuoy terminal
外海岛屿 offshore island
外海的 off-sea
外海底部固定设施 offshore bottom-fixed marine facilities
外海防波堤 offshore breakwater;outer breakwater
外海(防潮)堤 offshore sea wall
外海浮式钻井平台 floating platform
外海干散货码头 offshore dry bulk terminal
外海港 offshore port;open sea port
外海港线 outer harbo(u)r line
外海工程 offshore engineering
外海工程结构 offshore structure
外海管道 offshore pipeline
外海航行 open sea navigation
外海环境 open sea environment
外海混凝土结构 offshore concrete structure
外海建筑物 offshore structure
外海建筑物内部的水下部分 internal submerged zone
外海建筑物内部浪溅区以上部分 internal atmospheric zone
外海结构物 offshore structure
外海勘察 offshore exploration
外海勘探 offshore exploration
外海码头 < 常指水深大于 20 米的外海码头 > open sea terminal;sea berth;offshore terminal
外海锚泊 offshore mooring
外海平台 offshore platform
外海平台的主支柱 main leg
外海平台围绕主支柱 pile sleeve
外海区域 offshore region
外海人工岛 man-made offshore island
外海施工 offshore construction
外海施工驳船 offshore construction barge
外海施工技术 offshore construction technology
外海施工设备 offshore construction equipment
外海石油工业 offshore petroleum industry

外海滩 outer beach
外海突堤码头 open sea pier
外海系泊 offshore mooring
外海系泊系统 offshore mooring system
外海卸煤码头 offshore coal unloading terminal
外海液压自升行走式结构 jack-up structure;jack-up structure
外海油田 offshore oil field
外海渔业 midwater fishery; off-sea fishery
外海装船码头 open sea loading terminal;offshore loading terminal
外海装卸设施 open-sea loading facility
外海装油码头 offshore loading terminal
外海综合企业 offshore complex
外海钻机 offshore rig
外海钻井平台 drilling platform
外海钻塔 offshore rig
外海作业导向架 offshore lead
外函数名 external function name
外函数引用 external function reference
外函数子程序 external function subroutine
外焊 cover weld
外焊缝 outside weld(ing)
外行 inexperience; jack leg; layman; out of one's line;outsider;unadept
外航 outward voyage
外航船 outward bounder; outward-bound ship
外航道 exterior channel
外核 outer core
外核质量 mass of outer core
外荷载 external applied load; outside loading
外横梁 <三轴转向架> outside transom
外后盖 outer rear cover
外弧 positive camber
外弧断层 fault of extrados
外弧脊【地】outer-arc ridge
外弧口凿 outside gouge
外弧面 extrados
外弧曲率 extrados curvature
外弧张节理扇 tensional joint fan of extrados
外护板 external cladding;external covering;outer casing
外护堤 outer banquette; outer berm(e)
外护套 outer sheath
外花键 external spline;male spline
外华夏古陆海槽 outer Cathaysian marine trough
外滑脚 outer shoe
外滑块停歇时间 outer slide dwell
外画廊 outer border
外环 external ring; outer race; outer ring;outer belt【道】
外环滚道 outer-race ball track
外环路 outer belt;outer circumferential highway;outer hoop;outer ring road
外换热器 external heat exchanger
外灰分 free ash
外回路 outer loop
外回旋 external rotation
外汇 foreign currency; foreign exchange;hard currency
外汇管理局 exchange control department
外汇管理中心 control center[centre] of foreign exchange
外混合型多组分喷嘴 multiple component external spray nozzle
外混式雾化喷嘴 external mixing type atomizer
外混式油喷燃器 external mix oil burner
外混式油燃烧器 external mix type gun
外活 outside work
外活套结 outside clinch
外火山脊岭 outer volcanic ridge
外火箱 outer fire box
外火箱板 outside firebox sheet
外火箱顶板 firebox wrapper plate
外迹 exichnia
外积 cross product;exterior product; outer product
外积代数 exterior product algebra
外积分电容器 external integrating capacitor
外基底段 lateral basal segment
外激电子 exo-electron
外极位置 exterior pole position; pole outside figure
外籍工人 expatriate labo(u)r; migrant labo(u)r
外籍劳工 expatriate labo(u)r; migrant labo(u)r
外籍引航员 alien pilot
外挤 outsqueezing
外挤力 collapse of casing
外剂量 external dose
外寄生物 ectozoon;epizoon;ectoparasite
外加 superaddition
外加变形 imposed deformation
外加常数 additional constant
外加成本 on cost
外加冲击 applied shock
外加存储器 extra memory
外加点 extra point
外加电动势 applied electromotive force; impressed electromotive force
外加电荷密度 impressed-charge density
外加电流 <阴极防蚀用> applied current; impressed (electric) current
外加电流保护系统 impressed current protection system
外加电流阳极 impressed current anode
外加电流阴极保护 impressed current cathodic protection
外加电压 <电气线路两端所加的电位差> applied voltage;external voltage; impressed voltage; applied pressure
外加负荷 applied load
外加负载 applied load
外加副翼 external aileron
外加覆盖 oversubmergence
外加工程命令 emergency work order; emergency field order; extra work order;emergency work order
外加工程通知单 emergency work order; emergency field order; extra work order;emergency work order
外加工费用 expense arising from outside manufacture
外加工作队 extra gang
外加功率 applied power
外加荷载 applied load;externally applied load; imposed load; superimposed load(ing);external load
外加恒载 imposed dead load
外加厚 external upset
外加厚端 external upset end
外加厚管 <钻杆端部的> external upset tubing
外加厚管子 outside upset pipe
外加厚套管 external upset casing;extreme-line casing
外加厚油管 external upset tubing; tubing with external upset ends
外加厚钻杆 external upset drill pipe; external upset drill rod
外加厚钻管 external upset drill pipe
外加厚钻钎 external upset drill pipe; external upset drill rod
外加缓凝剂 set-retarding admixture
外加货 berth cargo
外加剂 addition agent; additive; admixture;intrusion agent;aruhuesiru <预填集料灌浆混凝土的>
外加剂搅拌器 admixture agitator
外加剂效果 admixture effect; admixture result
外加晶种 outside nuclei
外加晶种法 induced nucleation method
外加静载 imposed dead load
外加冷却空气 extra cooling air
外加力 applied force
外加力法 externally prestressed
外加力筋 extra tendon
外加力矩 applied moment
外加螺丝 captive screw
外加慢车道 extra slow lane
外加捻度 extra twist
外加频率 impressed frequency
外加钎料的钎焊 face-fed brazing
外加热高压釜 externally heated pressure vessel
外加热器 external heater;guard heater
外加熔岩球 accretionary lava ball
外加失火危险 additional fire hazard
外加套 jacketing
外加套管的 cased
外加填土 extra-banking;extra fill
外加贴面 applied trim
外加伪指令 extra pseudo order
外加限止器 extra limiter
外加限制的 extra limit
外加压力 applied pressure; impressed pressure
外加异常 superimposed anomaly
外加应力 applied stress; impressed stress
外加预应力 external prestress
外加圆柱 applied column
外加噪声 alien tones
外加增塑剂 additive plasticizer
外加蒸气压力 impressed vapo(u)r pressure
外加指标点 extra marker
外加指令 extra order
外加重量 added weight; additional weight
外加驻车车位 additional parking space
外加转弯车道 <交叉口处的> added turning lane
外夹层 outer jacket
外夹角 external angle
外架 outer tower;scaffold <觇标的>
外间 outer room
外间距 between outside;olo
外间距比 ratio of outside interval; outer clearance ratio <取土器的>
外间隙 <钻具与孔壁之间的间隙> outside clearance
外间隙比 <取样器> outside clearance ratio
外检测器 external detector
外检分析 external control analysis
外检实验室 laboratory for external examination
外检样品化学分析 chemical analysis for external examination
外检样品数 number of samples for external examination
外交 diplomacy
外交庇护 diplomatic asylum
外交部 foreign office;Ministry of Foreign Affairs
外交部部长 foreign minister;minister for foreign affairs
外交叉 external X-cross
外交辞令 diplomatic language; diplomatic parlance
外交大臣 foreign minister
外交典礼 diplomatic protocol
外交飞地 diplomatic enclave
外交分 <尖轨位于两端普通辙叉以外的交分线路> outside slip
外交关系 diplomatic relations
外交关系公约 convention on diplomatic relations
外交官衔 diplomatic rank
外交惯例 diplomatic practice; diplomatic usage
外交豁免 diplomatic exemption; diplomatic immunity
外交机关 diplomatic institution
外交来往 diplomatic intercourse
外交礼节 diplomatic protocol; protocol
外交签证 diplomatic visa
外交人员 diplomatic personnel
外交人员行李 diplomatic baggage
外交人员租用房屋条款 diplomatic clause
外交使节 minister
外交事务 foreign affairs
外交谈判 diplomatic negotiation
外交特权 diplomatic prerogative; diplomatic privilege
外交途径 diplomatic channel
外交文书 diplomatic document
外交信使 courier
外交邮袋 diplomatic baffling
外交照会 diplomatic note
外交政策 foreign policy
外交专线 diplomatic channel
外礁弧 outer reef-arc
外角 alternate exterior angle; arris; canton;coign(e);outer corner; salient junction; exterior angle; angle of declination <导线的>; deflection angle <导线的>
外角边条 outside corner mo(u)lding
外角不足 cant deficiency
外角构件 quoining
外角焊缝 outside fillet(weld)
外角砌分墙 angle bond
外角线脚 outside corner mo(u)lding
外角用的短木材 half-timber
外角柱 quoin post
外角砖 coign brick;corner brick
外脚手架 external scaffold(ing)
外接 circumscription;circumscribe
外接插座 external socket
外接触 exterior contact;external contact
外接触变质带 exomorphic zone
外接地 external ground
外接电流 external impressed current
外接电源 external power source
外接电源插座 external electric(al) supply socket
外接动力轴 <拖拉机的> power takeoff
外接蜂鸣器 external buzzer
外接符 outconnector
外接副载波 external subcarrier
外接镜管 extension tube
外接口 external tapping
外接平面位置显示器 plan position indicator repeater
外接球(面)circumscribed sphere

外接圈 described circle
外接色同步 external burst flag
外接式坡道 outer connection ramp
外接式匝道 < 立体交叉的 > outer connection ramp
外接收器 external receiver
外接输出端 < 液压系统 > external delivery point
外接数据调制 external data modem
外接头 coupling; sleeve; external connection
外接行交替 external line alternation
外接仪表 extension instrument
外接元器件 outward element
外接圆 circumcircle; circumscribed circle
外接圆半径 circumradius
外接圆心 circumcenter [circumcentre]
外接蒸汽 steam out
外节点 exterior node; external nodal point
外结点 external nodal point
外解绘图桌 external tracing table
外解接头 sleeve joint
外界 condition; exterior boundary; outer edge; surrounding
外界层 external limiting layer
外界场 extraneous field
外界尺寸 overall dimension
外界传染病 environmental epidemic
外界刺激 visual stimuli
外界的 ambient; exoteric
外界董事 < 非股东董事 > outside director
外界干扰 external disturbance; external interference
外界感应环境 induced environment
外界感应因素 induced environment
外界构筑物 lien structure
外界光 external light
外界环境 external environment
外界环境条件 external environment(al) condition
外界环境温度 ambient temperature
外界空气 ambient air; exterior air
外界膜 external limiting membrane
外界品质因素 external quality factor
外界审计 external audit
外界体验 xenopathic experience
外界条件 environmental condition; external condition; physical condition
外界(条件)因素 environmental factor
外界条件总体 complex of external condition
外界微量影响物 ectocrine
外界温度 ambient temperature; exterior temperature; external temperature; open air temperature; outside temperature
外界温度补偿 ambient temperature compensation
外界温度影响 externally temperature influence; extraneous thermal effect
外界信誉 external credibility
外界压力 ambient pressure
外界因素 environment factor; exogenous factor; external factor; extraneous factor
外界应力龟裂 environmental stress cracking
外界影响 external influence
外界影响的成本 external cost
外界元件 external element
外界杂音 external noise
外界噪声 ambient noise; community noise
外界障碍 physical obstruction
外紧带式制动器 contracting band brake; contracting brake
外进气阀 outside admission valve
外进站信号 outer home signal
外进站信号机【铁】outer home signal
外景 outdoor location
外景窗 picture window
外径 external diameter; major diameter; outer diameter; outside diameter
外径 15/8 寸钻杆 A-drill rod
外径边刃 < 指钻头 > outer diameter kickers
外径标准尺寸 outside diameter ga(u)ge
外径测径器 outside cal(l)ipers
外径测量仪 external measuring instrument
外径测微规 outside micrometer
外径测微计 outside micrometer with counter
外径超高(的坡度) cant
外径尺寸 outside dimension
外径代号 outside diameter character
外径公差 outside diameter tolerance
外径规 external ga(u)ge; female ga(u)ge; outside cal(l)ipers; outside ga(u)ge; snap ga(u)ge
外径弧长 length of outer diameter arc
外径极限规 external limit ga(u)ge
外径精测仪 passameter
外径磨损 outside wear
外径内径组合卡钳 Fay cal(l)ipers
外径配合 outer diameter fitting; outside diameter fitting
外径千分表 outside micrometer
外径千分尺 outside micrometer
外径掏槽刀 outer diameter kickers
外径圆柱规 external cylinder ga(u)ge
外径指示规 passameter; passatest
外静定体系 externally determinate system
外静水压力 external hydrostatic(al) pressure
外居住面积 < 等于公寓外面的面积加户外公用面积 > exterior resident(ial) area
外矩 external moment
外距 external distance; external spur
外距比 outside clearance ratio
外聚焦 exterior focusing
外卷层 wrapping
外卷帘 pinoleum blind
外卷装置 wrapper
外绝缘 external insulation; out insulation
外卡 outside cal(l)ipers
外卡扳手 spanner wrench
外卡尺 outside cal(l)ipers
外卡规 external calliper ga(u)ge; outside cal(l)iper ga(u)ge
外卡钳 outside cal(l)ipers
外开百叶窗 outside shutter
外开槽 external recessing
外开窗 out-opening window; outward opening window; outward-swinging side hung(window) casement; outward-swinging window; swinging-out casement window
外开窗扉 casement opening out; outswinging casement; outwardly opened casement
外开窗扇 casement opening out; outswinging casement; outwardly opened casement
外开门 out-opening door; outwardly opened door; outward opening door; reversed door
外开扇 out-swinging casement; outwardly opened casement
外开上悬窗 top-hung window opening outwards
外开式 outward opening
外开通风窗 out-swing ventilator
外开褶 kick pleat
外坎板 outer sill
外科 surgery; surgical department
外科病房 surgical ward
外科洗手室 scrub-up room
外科医院 surgical hospital
外壳 outer shell; outer skin; outer casing; case(ment); case shell; casing; bell housing; body case; boiler plate; clothing; container; cover; cowl; crust; encase(ment); encasing; encloser; enclosure; envelope; housing; jacket; lantern; mantle; outer cover; outer leaf; outermost shell; outside shell; outside skin; outside tier; package; rind; sheath; shell; surface casing; end bell < 仪器的 >; crustal < 尤指地球或月球的外壳 >
外壳板 outside casing plate; shell plate
外壳壁 shell wall
外壳材料 sheathing material
外壳层 exterior skin
外壳衬板 outside casing liner
外壳承载的结构 exostructure
外壳承载式结构 geodetic construction
外壳顶盖 closure head
外壳盖 housing cover
外壳干燥后 after the shell dries
外壳构件 outside shell member
外壳混凝土桩 shell concrete pile
外壳击穿 shell puncture
外壳基座 package base
外壳加强列板 fender strake
外壳件 casing element
外壳接地 earth(ing) of casing
外壳冷凝温度 screen condensation temperature
外壳连接器 cage adapter [adaptor]
外壳漏电 body leakage
外壳漏泄 casing leak
外壳铆接机 shell riveter
外壳拼合的泵 split-casing pump
外壳前缘 cowl lip
外壳切割 shell cutter
外壳天线 skin antenna
外壳温度 skin temperature
外壳效应 shell effect
外壳锈蚀 rust through
外壳账户 shell account
外壳罩 outer casing
外壳整形变模器 sheath-reshaping converter
外壳支架 casing support
外壳铸型 biscuit
外壳阻力 hood drag
外空间 external space
外空腔调速管 external cavity klystron
外孔 external opening
外孔瓦管 uncemented tile
外控点【测】external control point
外控式三极汞气整流管 cathetron [kathetron]
外口 collar extension
外口司水流 outward flow
外扣扳机控制扳挽 outside trigger
外跨 exterior span; outer bay
外矿送选的煤 foreign coal
外窥孔 outer spyhole; judas(window) < 门、窗上墙上的 >
外奎尔切线法 Vacquir tangential method
外扩光斑 flare
外扩散 external diffusion; outdiffusion
外扩散器 free air diffuser
外扩散效应 outdiffusion effect
外廓 gabarite
外廓尺寸 gabarite; outside dimension; out-to-out; overall dimension
外廓尺寸图 dimension's chart
外廓截面 outline cross section
外廓设施 contour facility
外廓样板 exterior template
外拉 external bracing
外拉撑飞机 externally braced aeroplane
外拉刀 broach for external broaching; external broach
外拉浮筒 hold-off buoy
外拉杆 outside link
外拉格朗日点 outer Lagrangian point
外拉伸 outdraw
外拉索 external tieback
外拉削 surface broaching
外来白磷 foreign white phosphorus
外来包体【地】accidental inclusion; exogenetic inclusion; foreign inclusion
外来病 ecdemic disease; exotic disease
外来材料 foreign material
外来成分 foreign component
外来的 adventive; alien; allochthonous; ecdemic; ektogenic; exogenous; exotic; extraneous; extrinsic; foreign; adventitious
外来的冰碛覆盖层 nappe
外来的不溶物 extrinsic insoluble
外来的不完整性 foreign imperfection
外来的固体杂质 foreign solids
外来的移民 immigration
外来的杂物 forehand solids
外来地幔岩质 exotic mantle materials
外来地栖动物 geoxene
外来地体 allochthonous terrane
外来地下水 allochthonic groundwater
外来电流 extraneous electricity
外来非金属夹杂物 exogenous nonmetallic inclusion
外来废料 external scrap
外来废石 extraneous waste
外来废铁 bought scrap
外来峰 extraneous peak
外来负荷 extraneous load
外来干扰 extraneous interference
外来惯用名 exonym
外来光(线) ambient light; extraneous light
外来含氮化合物 foreign nitrogenous compound
外来河流 inflowing river
外来化合物 xenobiotics
外来灰分 extraneous ash
外来火险 external hazard
外来货币 exotic
外来夹杂物 foreign impurity; foreign inclusion
外来金属夹杂物 exogenous metallic inclusion
外来居民 immigration
外来客户 correspondent
外来矿物质 extraneous mineral mater
外来劳动力 foreign labo(u)rer
外来流湖 outflow lake
外来煤 foreign coal
外来品种 exotic breed
外来破坏 external damage
外来气 extraneous gas
外来热除去 extraneous heat removal
外来人村落 squatter village
外来溶胶 extrinsic sol
外来入库物质 allochthonous

外来商品 commercial goods imported;foreign goods
外来事故 external cause
外来熟料 foreign grog
外来树种 exotic tree species
外来水 foreign water;secondary water
外来水产量 extraneous water production
外来碎屑 extraclast
外来体【地】allochthon(e)
外来填隙 foreign interstitial
外来投入 external input
外来投资 outside financing
外来投资条例 regulation on foreign investment
外来土 borrow soil
外来物 foreign material;foreign particle
外来物的混合溶蚀 mixed corrosion exotic material
外来物体 foreign body
外来物质 allochthonous material;foreign component;foreign matter;foreign substance
外来物种 exotic species
外来夕卡岩 alloskarn
外来系统地层 strata of allochthone
外来线接头 field terminal
外来响应 extraneous response
外来形的 xenomorphic
外来形式 imported style
外来性 foreignness
外来压力 extraneous pressure
外来氩 extraneous argon
外来岩块 allochthon(e);detached block;detached mass;exotic block
外来因素 foreign element
外来鱼 exotic fish
外来语 exoti(ci)sm;foreignism;loan(word)
外来语辞典 dictionary of foreign adopted words;dictionary of loan words
外来元素 extraneous element
外来源数据 foreign source data
外来杂质 exogenous impurity;extraneous dirt;extraneous material;foreign inclusion;foreign matter
外来噪声 extraneous noise
外来振动 external vibration
外来植物 adventitious plant;adventive plant;foreign plant;introduced plant
外来主机 foreign host
外来专家 outside specialist
外来资金 external finance;external fund
外拦江砂(沙)outer bar
外拦门砂(沙)external sand barrier
外廊 gallery;lanai;outer gallery;side corridor;veranda(h)
外廊采灯 outline lighting
外廊尺寸 contour dimension;out-of-out
外廊道 promenade deck
外廊建筑 block with external access galleries
外廊宽度 clearance width
外廊式 balcony-type;gallery corridor type;gallery-type
外廊式公寓 exterior corridor apartment;gallery apartment house
外冷 external cooling
外冷凝法 external condensation process
外冷凝器 external condenser;outer condenser
外冷式电机 separately cooled machine
外冷铁 surface chill
外力 exogenetic force;exogenic;exogenic force;exogenous force;exterior force;external force;externally applied force;extraneous force;out(er)force;outside power
外力变质 epigenesist
外力沉积作用 epiclastic sedimentation
外力大小 magnitude of imposed force
外力的 epigene
外力地质作用 exdogenic force of geologic(al)function
外力地质作用方式 mode of exogenic process
外力地质作用类型 type of exogenic process
外力对流 forced convection
外力工程 <合同以外的> extra work
外力函数 forcing function
外力火山碎屑 epiclastic volcanic fragment
外力控制 power-assisted control
外力偶 outside torque
外力漂移 external force drift
外力式缓行器 external force retarder
外力势 potential of external force
外力碎屑的 epiclastic
外力碎屑砾岩 epiclastic conglomerate
外力碎屑岩 epiclastic rock
外力碎屑岩相 epiclastic fragmental facies
外力所做的功 external work
外力通风式 separately ventilated type
外力图面积 area of force
外力消弧 separately extinction of arc
外力阻抗 external resistance
外力作用【地】epigene action;exogenetic action;epigenetic action;exogenic action;exogenic process;exogenous process
外力作用持续时间 acting duration of force
外力作用方式 acting type of imposed force
外力作用方向 direction of imposed force
外力作用面 acting surface by force
外力作用下不稳定的块体 externally unstable block
外力作用下稳定的块体 externally stable block
外连接器 <插入内连接器的> male connector
外联结线盒 extension terminal box
外联网 extranet
外联箱 external header
外僚机 outer wingman
外列板 outer plank;outer plating;outer strake;outside strake;raised strake
外列板水沟 outer waterway
外磷负荷 external phosphorus loading
外流 drain;external flow;flow out;outflow;spillage
外流管 efflux tube
外流湖 exorheic lake;outflow lake
外流加热器 outflow heater
外流量 external flux
外流流域 exorheic basin;external basin
外流盆地 exorheic basin;external basin
外流气垫艇 external flow vehicle
外流区 exorheic region
外流人才 brain drainer
外流人口 spillover
外流式过滤器 inside-out filter
外流式水轮机 outward flow turbine
外流式涡轮机 outward flow turbine
外流水系 exorheic;exorheic drainage;exorheism;exterior drainage;external drainage
外流速度 efflux velocity
外流涡轮 outward flow turbine
外流阻力 outflow resistance
外龙骨 outer keel
外楼梯 exterior stairway
外楼梯扶手 outer stair rail
外楼梯围栏 exterior stair rail
外楼梯斜梁 exterior string
外漏 leak away
外漏等级 external leakage classification
外露 exposing;open-to-air
外露保险丝 open fuse
外露背斜层 exposed anticline
外露表面 exposed surface
外露布线 exposed wiring;open wiring;surface wiring
外露部分 <屋面板> to the weather
外露部位 exposed portion
外露层 butte(temoin);zeugenberg;outlier【地】
外露齿轮 open gear
外露齿轮润滑油 open gear lubricant
外露椽 show rafter
外露带电部件 exposed live part
外露导线 exposed electric(al)wire
外露的 exposed
外露的丁头石 inband
外露灯光 naked light
外露底基 stereobate
外露电线 exposed electric(al)wire
外露钉 exposed nailing
外露钉钉法 exposed nailing
外露钢索 external tendon
外露骨料 <多指冲洗表面砂浆后的防滑面层> exposing aggregate;exposed aggregate
外露骨料混凝土 exposed aggregate concrete;visual concrete
外露骨料墙板 aggregate panel;exposed aggregate panel
外露骨料饰面 exposed aggregate finish;exposed aggregate texture;scrubbed finish
外露骨料饰面墙板 aggregate facing panel
外露厚度 <石板或瓦片的> margin
外露混凝土 exposed concrete
外露集料墙板 aggregate panel
外露接缝 untreated joint
外露结构 visible structure
外露开口 exposed opening
外露棱边 exposed edge
外露楼梯基 face string
外露密封剂 exposed exterior sealer
外露密封条 vision strip
外露面 exposed edge;exposed face;fair face
外露面混合料 facing mix(ture)
外露面积 exposed area
外露耐久性 exterior durability
外露切削刃 exposed cutting surface
外露擒纵叉 exposed pallets
外露区域 exposed area
外露缺陷 open discontinuity
外露热电偶 bare thermocouple
外露使用 exposed application
外露式插塞 surface plug
外露式钢结构 exposed steel construction
外露式木结构 open-timbered structure;open timbering
外露式蜗壳 exposed spiral case
外露条件 <混凝土等的> field exposure condition
外露(围)墙 exposed wall
外露圬工 exposed masonry
外露悬挂法 exposed suspension system
外露岩石 day stone
外露油灰 face putty
外露预制(砌)块 exposed facing block;facing block
外露张拉钢索 <预应力混凝土的> out cable
外露柱廊 pteron
外露装饰瓷砖 exposed finish tile
外露装修瓷砖 exposed finish tile
外陆盆地 extracontinental basin
外陆缘高地 outer high
外滤式过滤机 outside-drum filter
外滤式过滤器 outside-drum filter
外路车 foreign car;foreign line car;foreign wagon
外路干线 outer main
外路高频 extraneous high frequency
外路机车 foreign locomotive
外路运输 foreign traffic;foreign transportation
外律性 heteronomy
外轮供应公司 ocean shipping supply corporation
外轮毂 outer(wheel)hub
外轮管理规则 regulations governing foreign vessels
外轮轨迹带 outer wheel path
外轮迹带 outer wheel path
外轮迹线 outer wheel path
外轮廓锯法 external contour saw
外轮山【地】somma
外轮缘挤压钢轨 outer flange is crowd the rail
外轮轴承 outer wheel bearing
外螺杆和手轮阀 outside screw and yoke valve
外螺丝 male screw;male thread;running nipple
外螺纹 exterior thread;external screw thread;male screw;male screw thread;male thread;outside spin;pin thread
外螺纹板牙 screwing die
外螺纹半锁接箍 pin coupling
外螺纹半锁接头 pin joint
外螺纹车刀 external screw cutting tool;male screw cutting tool
外螺纹车削 external threading
外螺纹导程仪 outside lead ga(u)ge
外螺纹端 male end
外螺纹阀 external thread valve
外螺纹管端 male end of a pipe
外螺纹管接头 male connector
外螺纹过渡管接头 male adapter[adaptor]
外螺纹过渡接头密封环 male adapter ring
外螺纹(检查)规 external screw ga(u)ge
外螺纹胶管接头 male rubber hose nipple
外螺纹(接)管 male nipple
外螺纹接管套 male coupling
外螺纹接套 male thread nipple
外螺纹接头 male connection
外螺纹接头管 male nipple
外螺纹卡扎里密封 casale closure with external screw
外螺纹联管节 male union;nipple union
外螺纹联结管 male union
外螺纹磨床 external thread grinder
外螺纹配件 male fittings
外螺纹瓶口 external screw thread finish
外螺纹切削 external thread cutting
外螺纹三通 male branch tee

W

外螺纹三通管 male tee [T]
外螺纹三通管接 male branch tee [T]
外螺纹十字接头 male cross
外螺纹梳刀 outside chaser
外螺纹弯头 male bend; male elbow; nipple elbow joint; street elbow
外螺纹异径管接头 male adapter ring
外螺纹钻杆 external threaded drill rod
外螺旋 external spiral
外螺旋线 outer vortex
外落式下水道检查井 outside drop of sewer manhole; outside drop type of sewer manhole
外落式下水道人孔 outside drop of sewer manhole
外落水 exterior drainage
外落水管 outside leader
外卖餐馆 take-out restaurant
外卖食物餐馆 carryout; carry-out restaurant
外卖饮食店 take-out service
外脉冲同步系统 externally pulsed system
外毛管水 outer capillary water
外锚块 block-out(anchor)
外锚式斜拉桥 fully anchored cable-stayed bridge
外冒地槽带【地】external miogeosynclinal zone
外贸港(口) foreign trade harbo(u)r; foreign trade port; port of entry
外贸局 Board of Foreign Trade
外贸区 foreign trade zone
外贸中心 foreign trade center [centre]
外貌 appearance; externality; external view; general view; lineament; physiognomy
外貌均匀度 physioghomic(al) homogeneity
外貌损伤 disfigurement
外貌特征 gross features
外门 anteport; entrance door; exterior door; external door; outer door; out-opening; outside door
外门板拉条 back head brace
外门窗口的侧墙面 outside reveal
外门窗框 outside casing; outside easing
外门道的吊挂 anteport
外门防风雨板 splash board
外门附加率 additional factor for exterior door
外门架 <叉车> outer upright mounting; outer mast
外门架伸出后的最大高度 maximum open height
外门槛 anteport
外门框 outside door frame; outside frame
外门廊 anteporch; open air terrace
外门厅 exonarthex
外门围栏 outdoor rail
外密封 outer gland sealing
外密封冲洗 external seal flush
外幂 exterior power
外幂层 exterior power sheaf
外面 exterior face; superficies; outer face <磁带、纸带的>
外面百叶窗 outside shutter
外面操作开关 externally operable switch
外面的 exoteric; external; outer; outside; outward
外面的反射部分 externally reflected component
外面覆盖层 exterior bedding
外面供给润滑油 external oil supply
外面过梁 front lintel
外面加热 externally heating
外面磨削机 external grinder
外面偏差 outside deviation
外面墙面 exterior wall surface
外面庭院 exterior yard
外面整平的绞合绳 flattened strand rope
外庙门 outer gate
外模 external mo(u)ld; top form
外模板 exterior form; exterior sheathing
外模振荡器 outside vibrator
外模振动法 outer mo(u)ld vibration method
外摩擦角 angle of external friction; angle of wall friction
外摩擦力 external friction
外摩擦橡皮 <水轮发电机的> external rubber
外内比 extreme and mean ratio
外南极洲地槽 outer Antarctica geosyncline
外能除霜 external defrosting
外啮合 external gearing; external toothing
外啮合齿轮 external gear
外啮合齿轮泵 external gear rotary pump
外啮合齿轮沥青泵 outside gear asphalt pump
外啮合齿轮式油泵 external gear-type oil pump
外啮合传动装置 external gearing
外啮合凸轮转子泵 abutment pump
外啮双排行星齿轮机构 external planetary arrangement
外扭矩 outside torsion(al) moment; outside twist(ing) moment
外排废气 atmospheric exhaust
外排流 effluent stream
外排水 outer drainage
外排水道 <外渠的> outer dike [dyke]
外排水管 outside gutter
外排水系统 external drainage system; outside storm system
外排序 external sort
外派津贴 assignment allowance
外盘簧 outer helical spring
外炮台 outer fort
外配合 fit on
外配流径向柱塞泵 radial piston pump with exterior admission
外配位层 second coordination sphere
外喷溢 external eruption
外喷嘴 outer nozzle
外皮 cortex [复 cortices/cortexes]; crust; husk; incrustation sheath; integument; outer bark; outer coating; outside skin; sheathing; skin; wrapper
外皮材料 <如叠合板多孔材料等的外罩薄板> skin material
外皮潮湿包(指货物) cover damp bags
外皮到外皮尺寸 out-to-out dimension
外皮覆盖物 sheath
外皮节子 face knot
外皮结构 skin construction
外皮失落 <指货物> cover missing
外皮涡流 sheath eddies
外皮效应 sheath effects
外皮质 exinite
外皮质煤 periblain
外皮重 taring
外皮砖 outer brick; outside brick
外飘 flare
外飘船首 flare-out bow
外飘船首型 flared bow
外飘角 flare angle
外飘舷 flaring side
外聘审计员 external auditor
外平板龙骨 outer flat keel; outer keel
外平衡管 external equalizer
外平衡环 outer gimbal(ring)
外平衡架 outer gimbal
外平衡装置 external equalizer
外平接头 external flush-jointed coupling
外平连接的(钻杆) flush coupled
外平连接套管 flush-coupled casing
外平连接钻杆 flush-joint drill pipe
外平面的 outerplanar
外平面图 outerplane graph
外平内弯拱 camber arch
外平套管 flush-joint casing
外平钻杆 outside-flush drill-pipe
外屏蔽 external shield
外屏蔽防火塑料 external shield plastics
外屏蔽耐火塑料 external shield plastics
外屏极阻抗 external plate impedance
外坡 outer slope; outside slope; riverside slope; waterside bank; water slope
外坡道 out ramp; outer loop <立体交叉的>
外坡准地槽【地】exogeosyncline
外气套 outside lining
外气相沉积法 outside vapo(u)r deposition process
外汽缸 outer casing; outer sleeve; outside cylinder
外汽缸式(蒸汽)机车 outside-cylinder locomotive
外砌的 out-band; outbond
外砌墙 outbond wall
外砌式燃烧炉 external firing system
外砌物 outerstack
外迁(移) external migration
外牵索 external tieback; outboard stay
外前后径 external conjugate diameter
外前坡 front slope
外潜伏期 extrinsic incubation period
外潜热 external latent heat
外浅海区 outer neritic zone
外嵌凹引线条 <装饰用> laid-on thread
外嵌/下嵌接线 outer rabbet line
外戗道 outer banquette; outer berm(e)
外戗堤 waterside banquette
外墙 exposed wall; exterior wall; external wall; frame work; outer wall; outside wall; periphery wall; wall enclosure; antemural <城堡的>
外墙板 exterior wall slab; external wall panel; facade panel; outer wall slab; outside panel; outside wall panel; sheathing; side fascia; siding; weather-board(ing)
外墙板上部压条 fascia [复 fa(s)ciae/fa(s)cias]
外墙包层 drop siding
外墙采光塑料板 plastic panel in a wall
外墙衬 exterior wall lining
外墙衬砌 enclosing wall lining; external walling lining; outer wall lining; outside wall lining
外墙橱窗下采光窗 stallboard light
外墙窗洞组合 fenestration
外墙垂吊披叠壁板 drop siding
外墙瓷面砖 exterior wall tile
外墙单元 outer wall unit; outside wall unit
外墙底板 outside base
外墙叠板 exterior siding
外墙防水隔墙 retention wall
外墙粉刷 external rendering; stucco
外墙覆盖板 siding shingle
外墙覆面层 vertical siding
外墙构件 exterior wall member; exterior wall unit; outer wall component; outside wall member; outside wall unit
外墙构造 outer structural system; outside system; outside wall construction
外墙挂板 exterior wall cladding; cladding
外墙护角 exterior corner reinforcement; external corner reinforcement
外墙灰泥层 exterior plaster; external plaster
外墙灰泥骨料 external plaster aggregate
外墙灰泥集料 external plaster aggregate
外墙加劲板 lining slab
外墙建筑 exterior wall construction
外墙建筑单元 enclosing wall building unit
外墙建筑构件 enclosing wall building member; exterior wall building component
外墙建筑施工 enclosing wall construction
外墙建筑组成部分 enclosing wall building component
外墙角 outside corner
外墙结构 external wall construction; outer wall construction
外墙结构系统 outer structural system
外墙筋 outer stud; outside stud
外墙空腔 furring
外墙空心砖 Holbric
外墙拉毛粉刷 exterior stucco
外墙梁 periphery beam
外墙门侧 door reveal
外墙面 exterior surface; outside wall surface; surface of external wall
外墙面层 outer wall(sur)facing; outside wall facing
外墙面抹灰 plaster externally
外墙面上浮雕装饰 stucco-embossed surface
外墙面线 ashlar line
外墙面砖 facing tile; outer wall tile; outside wall tile; veneer tile
外墙抹灰基底 exterior plaster base; external plaster base
外墙木板 <木架房屋的> wall siding
外墙内的空间 firring
外墙耐候护板 weather-board(ing)
外墙砌块 exterior wall block; outer wall block; outside wall block
外墙砌筑 external walling; outer wall construction
外墙嵌板 exterior wall panel
外墙嵌饰 exterior masonry wall lining
外墙墙面保护板 siding shingle
外墙热绝缘 out insulation
外墙饰面 enclosing wall facing; exterior cladding; exterior facing; exterior wall facing; exterior wall finish; outer wall finish; outside wall finish
外墙饰面材料 exterior wall(sur)facing material
外墙竖龙骨顶板 outside studding plate
外墙条板 batten
外墙贴釉砖 exterior glazing
外墙涂料 exterior wall lining
外墙退后距离 setback distance of

outer wall
外墙托架 beam for exterior wall; spandrel beam
外墙托梁 beam for exterior wall; spandrel beam
外墙无突出烟囱的壁炉 rear dorse
外墙镶板 exterior wall panel
外墙镶面板 cladding sheet
外墙修饰 enclosing wall finish
外墙腰线 weather mo(u)lding
外墙用披叠壁板 drop siding
外墙用砖 face brick
外墙雨水污纹 festoon staining
外墙预制块 outside precast brick
外墙支柱 exterior masonry wall column
外墙终饰 exterior wall finish
外墙柱 exterior wall column; external walling column
外墙砖 external tile; veneer tile
外墙转角 external corner
外墙装饰 exterior masonry wall lining
外侨 alien; uintalander
外侨身份 alienism
外桥砌块 outside bridge wall block
外桥砖 outside bridge wall
外撬力 prying force
外切 circumscribe; circumscription; exterior contact; external contact; externally tangent
外切刀齿 outside blade
外切断 external cutting off
外切多边形 circumscribed polygon
外切多面体 circumscribed polyhedron
外切棱柱 circumscribed prism
外切棱锥体 circumscribed pyramid
外切三角形 circumscribed triangle
外切四边形 circumscribed about circle
外切图形 circumscribed figure
外切圆 circumscribed circle; excircle; externally tangent circle
外切圆柱 circumscribed cylinder
外切圆锥 circumscribed cone
外切晕 circumscribed halo
外勤 outwork
外勤工作 field operation
外勤工作简报 field bullet
外勤管库员 travel(l)ing storekeeper
外勤人员 outworker; patrolman
外勤任务 field assignment
外勤职工 outdoor staff
外倾 extroversion; fallout; flare; flaring
外倾船首 flare-out bow
外倾船首型 flared bow
外倾角 camber angle; flare angle
外倾角调整 camber angle adjustment
外倾舷 flaring side
外倾舷侧 flared ship sides
外倾(斜)斜度 outward batter
外清管机 external pipe cleaning machine; external tube cleaning machine
外清管器 external pipe cleaning machine; external tube cleaning machine
外球 ectosphere
外区 exterior zone; outer zone
外区段 foreign section
外曲 outcurve
外曲导轨 outside curved lead rail
外曲面 positive camber
外曲物 outcurve
外曲线 outer curve
外圈 < 滚动轴承 > outer race; outer ring; housing washer
外圈大齿轮 < 搅拌机 > circumferential ring gear
外圈框石 out-band
外燃 external combustion; external fire
外燃冲压发动机 external-Bruning ram
外燃发动机 external-burning engine; external combustion engine
外燃锅炉 externally fired boiler
外燃火管锅炉 externally fired fire tube boiler
外燃式过热器 externally fired superheater
外燃式气体涡轮机 external combustion gas turbine; external fired gas turbine
外燃式透平 external combustion turbine
外燃室 external combustion chamber
外燃室热风炉 cowper with internal combustion
外燃水管锅炉【机】externally fired water tube boiler
外燃循环 external combustion cycle
外热 external heat
外热电偶 external thermocouple
外热动物 exothermic animal
外热式烤箱 externally heated oven; indirectly heated oven; indirect oven
外韧致辐射 external bremsstrahlung
外日冕 outer corona
外容度 exterior content
外容量 exterior capacity
外容器 outer container
外色 exogenic colo(u)r
外沙坝 outer bar
外刹车 external brake
外刹车带式制动器 external cheek brake
外刹车毂 outer brake hub
外栅 bailey; zwinger
外扇沉积 outer fan deposit
外伤 trauma [复 traumas/traumata]
外伤后的 post-traumatic
外伤药 traumatic
外商 alien merchant; foreign merchant
外商独资企业 solely foreign funded enterprise
外商投资政策 foreign investment policy
外烧火管锅炉【机】externally fired fire tube boiler
外蛇形管 external coil
外设安全措施 engineered safety feature
外设安全系统 engineered safety system
外设部件互联总线 peripheral component interconnect bus
外伸 lituate; overhang(ing); corbel out
外伸臂 overhanging arm
外伸部分 outshot
外伸层 corbel course
外伸长度 extended length; extension; stick-out
外伸的 overhung
外伸的排水管 stove pipe
外伸二层台 < 钻塔的 > outside monkey board
外伸距 < 起重机 > outreach
外伸梁 extensional beam; overhanging beam; outrigger beam
外伸梁临时支撑 outrigger shore
外伸轮 overhang wheel
外伸曲柄 overhang crank; overhung crank
外伸三角洲 protruding delta
外伸砂轮 overhang grinding wheel
外伸施工法 end-on system; over-end system
外伸式格筛 cantilever grizzly
外伸式脚手架 outrigger scaffolding
外伸弹簧 overhanging spring
外伸托架 outboard support
外伸屋顶 overhanging roof
外伸小齿轮 overhung pinion
外伸悬垂式按钮 overhanging pendant switch
外伸引火箱 end projecting firebox
外伸缘 outshot
外伸支承 outbound bearing
外伸支架 outrigger
外伸支腿拆卸系统 outrigger removal system
外伸支腿液压销子 outrigger pin system
外伸支腿最大压力 maximum outrigger load
外伸轴承 outboard bearing
外渗 extravasation; outward seepage
外渗测定器 exosmometer
外渗透 exosmosis
外渗现象 exosmose; exosmosis
外生变量 exogenous variable
外生变数 exogenous variable
外生层 ectal layer
外生长凝固 exogenous solidification
外生成矿建造 exogenetic metallogenic formation
外生成矿作用 exogenetic metallization
外生成岩作用 exodiagenesis
外生的 exogenetic; exogenic; exogenous
外生洞穴 exogenous cave
外生多源包体 exopolygene
外生鲕状的 exoolitic
外生环 exogenous cycle
外生环境因素 exogenous environmental factor
外生角砾溶岩 exogenic breccia lava
外生巨角砾熔岩 exogenic boulder lava
外生矿床 exogeneous ore deposit; exogenetic ore deposit; exogenic ores
外生裂隙 exokinematic fissure
外生砂屑熔岩 exogenic arenitic lava
外生素 exogenous toxin
外生污染物 exogenous pollutant
外生物相 exobiophase
外生现象【植】ectogeny
外生效应【植】ectogony
外生形成 exogenous formation
外生性植物 exogens
外生铀矿 exogenous uranium ore
外生源 exogenous origin
外生运动 exogenetic movement
外生作用 epigene action; epigensis; exogenous action
外施冲击 applied shock
外施的功 applied work
外施荷载 applied load
外施力矩 applied moment
外施凝固剂 external coagulant
外施推力 applied thrust
外施脱模剂 external mo(u)ld lubricant
外施应力 applied stress
外施张力 applied tension
外施震动 applied shock
外十字区 external cross-region
外时钟分频器 external clock divider
外蚀 external corrosion
外矢距 external distance; external secant
外视场 apparent field
外视海面 offing
外视图 exterior view; external view
外视油表 sight oil ga(u)ge
外视油量计 oil sight ga(u)ge
外视振幅 apparent amplitude
外视注油器 sight feed oiler
外饰面 exterior facing; exterior finish
外饰面板 outer veneer; outside veneer
外饰面挂墙板 < 非承重的 > panel curtain wall
外饰面墙板 < 非承重的 > panel curtain wall
外饰面砖 exposed finish tile
外饰线材 exterior trim
外受热面 external heating surface
外束 external beam
外束流 external beam current
外束路线 external beam path
外树皮 external bark; outer bark
外双峰 out double peak
外水分 free moisture
外水分离器 external water separator
外水分循环 external water circulation
外水合层 outer hydration sphere
外水落管 outside leader
外水压力 external hydraulic pressure; external water pressure; outer water pressure
外水源 exterior supply water
外丝接头 nipple; running nipple
外丝扣 external thread; male thread
外丝扣管 male nipple
外丝钻杆 male rod
外斯晶带定律 zone law of Weiss
外斯理论 molecular field theory
外死点 bottom dead-center [centre]; outer dead-center [centre]; outer dead point
外送餐馆 drive-in restaurant
外送运输机 run-out contouring; run-out conveyer [conveyor]
外隧道 outer tunnel
外缩 external contracting
外缩成环 exocondensation
外缩刹车 external contracting brake
外缩式制动器 external contracting brake
外缩制动器 external cheek brake
外缩(制动)闸 external contracting brake
外缩作用 exocondensation
外锁闭 outside locking device
外锁信号 external lock signal
外胎 cover tire [tyre]; outer casing; outer tube; outer tyre [tire]; tire [tyre] casing; tire [tyre] cover; tire [tyre] shoe
外胎边壁 tire side wall
外胎唇 cover bead
外胎花纹 tread
外胎扩展器 tire spreader; casing spreader
外胎面 tread
外胎胎面 protector
外台阶 external stair(case)
外态模式 behavio(u)r model
外滩 outer bar; shoreface
外滩带 inshore zone
外滩地段 inshore zone
外弹簧 outer spring
外碳源投加量 external carbon dosage
外逃工厂 runaway shop
外套 casing; housing; jacket(ing); lagging; outer coating; outer housing; outer jacket; overcoat; sheath; shell; shroud; top coat; outer liner < 发动机的 >
外套板 lag
外套泵 jacketed pump
外套管 jacket; outer shell; outer sleeve; outer tube
外套管的设置深度 setting depth of

W

surface casing
外套夹子 jacket clamp
外套螺帽 cap nut
外套螺母 box nut
外套筒 outer sleeve;outer thimble
外套筒式空气枪 external sleeve gun
外套折边 jacket edging
外特性曲线 external characteristic curve
外梯玻璃 outer-stepped glass; outside-stepped glass
外梯级 external stair(case)
外梯梁 <不与墙连接的楼梯斜梁> outside string(er)
外梯透镜 outer step(ped) lens
外蹄片 external shoe
外蹄式制动器 external shoe brake
外体暴露 external exposure
外挑 kicking out
外挑脚手架 projecting scaffold
外挑平台 cantilevered platform
外挑阳台 cantilevered balcony
外挑支撑 outrigger shore
外调和 external harmonics
外调焦 external focusing
外调焦望远镜 exterior focusing telescope
外调倾斜度的机械式撑杆 <推土机> mechanical thrust open tilt
外调制 external modulation
外调制器 external modulator
外贴地下防水层 externally applied tanking
外贴式防水层 outer surfacing waterproofing layer
外铁式强制油冷变压器 shell-type forced-oil-cooled transformer
外庭院 outer court
外停泊区 outer berthing area
外艇体 outer hull
外通风 separate ventilation
外通风式电机 separately ventilated machine
外通系统 open system
外通镶板 openwork panel
外同步 external synchronization
外同步复位驱动器 reset driver for external synchronization
外同步脉冲装置 externally pulsed system
外同步器 external synchronizer
外筒 outer sleeve
外投影法 external projection
外透镜 outer lens
外透视中心 exterior perspective center [centre]; external perspective center [centre]
外凸面离合器 outer cam clutch
外凸通风口 project-out vent
外凸字体 convexity of character
外突的桥端 projecting end
外突式进水口 projecting inlet
外突式砖 projecting brick
外突云母 high mica
外图廓 exterior border; exterior frame;exterior margin;external border;external margin;outer border
外图廓蒙片 exterior mask
外图廓线 outermost line
外涂层 outside coating;top coat
外涂磨料 coated abrasive
外涂漆 exterior painting;surface coat
外涂有光漆 outside gloss paint
外推 extrapolate
外推保险数据 extrapolated insurance data
外推边界 extrapolated boundary
外推长度 augmentation distance; extrapolated length
外推程序 extrapolation procedure
外推电离程 extrapolated ionization range
外推电离室 extrapolation ionization chamber
外推电离箱 extrapolation chamber
外推法 extrapolated method;extrapolation(method);method of extrapolation
外推法设计 extrapolation design
外推法卫星制导系统 extrapolation type satellite's guidance system
外推高度 extrapolated height
外推公式 extrapolation formula
外推过程 extrapolation process
外推函数 extrapolation function
外推夹套 push-out chuck; push-out collet
外推交点对应时间 extrapolated intersection time
外推截距 extrapolated intercept
外推截止电压 extrapolated cutoff
外推居里温度 extrapolated Curie temperature
外推距离 extrapolation distance
外推力 outward thrust
外推面 extrapolated side
外推黏[粘]度 extrapolation viscosity
外推平均 extrapolated mean
外推起始点 extrapolated onset point
外推起始温度 extrapolated onset temperature
外推曲线 extrapolated curve
外推上悬窗 hopper window;hospital window <医院>
外推上旋窗 hopper ventilator window
外推上旋气窗 hopper light
外推射程 extrapolated range
外推式表壳 push-out case
外推式窗扇 hopper light
外推数 extrapolation number
外推算子 extrapolation operator
外推弹簧夹头 push-out(collet)chuck
外推误差 extrapolation error
外推系数 extrapolation coefficient
外推辛烷值 extrapolated octane number
外推压力 extrapolated pressure
外推压力恢复值 extrapolated building-up pressure
外推值 extrapolated value;extrapolation value
外推制 extrapolating system
外推状态 extrapolated state
外退解 external unwinding
外托板 outer shoe
外脱模剂 external release agent
外弯 bend outward;excurvature
外弯割口缝合针 reverse cutting needle
外弯矩 outer bending moment
外弯肋骨 convex bend frame
外湾 open bay
外湾生境 outer bay biotope
外网状层 outer plexiform layer
外微分 exterior differential
外微分形式 exterior differential form
外微调电容 external trimmer
外围 off-lying; outer ring; outskirt; periphery;purlieu
外围仓库制度 terminal receiving system
外围操作 peripheral operation
外围层 perisphere
外围程序 peripheral routine
外围处理 peripheral handling
外围处理机 peripheral processing unit;peripheral processor
外围传送 peripheral transfer
外围存车场 perimeter car park; peripheral car park
外围挡板 cockle;cottle
外围挡物 external enclosure
外围岛(屿) off-lying island; out(lying)island
外围的 outlying;peripheral
外围等离子体 environmental plasma
外围地点 peripheral locality
外围地区 external zone;fringe area; outlying zone
外围电路 peripheral circuit
外围电子 peripheral electron
外围防空 outer air defense;peripheral air defense
外围防御 outer defense
外围分系统 peripheral subsystem
外围功能转换器 peripheral function translator
外围管 peripheral tube
外围含水层 peripheral aquifer
外围花岗岩体 circumscribed massif granite
外围缓冲器 peripheral buffer
外围绘图设备 peripheral graphic(al) device
外围减震孔 rib snubber
外围建筑 outbuilding
外围接口 peripheral interface
外围接口适配器 peripheral interface adapter
外围接口系列 peripheral interface family
外围接口转接器 peripheral interface adapter
外围接口组件 peripheral interface module
外围控制 peripheral control
外围控制线路 peripheral control line
外围雷达 peripheral radar
外围流体 ambient fluid
外围内围线交通调查 external-internal-cordon survey
外围墙 enclosing wall;periphery wall
外围切土 enlarging;ripping
外围热冲压压力 peripheral hot stamping press
外围商业发展区 outlying business district
外围设备 ancillary equipment; auxiliary equipment; external equipment;external unit;peripheral;peripheral apparatus; peripheral device;peripheral equipment;peripheral instrumentation;peripheral unit
外围设备程序 peripheral routine
外围设备间传输 peripheral transfer
外围设备接口 peripheral interface
外围设备控制 peripheral control
外围设备控制程序 peripheral control program(me)
外围设备控制符 device control character
外围设备控制器 peripheral controller
外围设备控制装置 peripheral control unit
外围设备调整机 peripheral coordinator
外围设备译码 peripheral decoding
外围设施费用 off-site cost
外围视野 peripheral field
外围输出设备 peripheral output device
外围数据寄存器 peripheral data register
外围数据总线 peripheral data bus
外围掏槽眼 <爆破时的> rib snubber
外围停车场 perimeter car park; peripheral car park
外围头砖 bull-nose
外围图像存储器 peripheral image memory
外围线 object line;external cordon <城市的>
外围线(交通)调查 external-cordon survey
外围牙轮 side cutter
外围业务 out-out business
外围站 outstation
外围障碍物 off-lying danger
外围支撑导桩的木板桩围堰 timber sheet piling cofferdam supported with guard timber piles at the outside perimeter
外围支援计算机 peripheral support computer
外围砖(石)砌墙 enclosing masonry wall
外围资料档 satellite file
外围子系统 peripheral subsystem
外围子系统通道 peripheral subsystem channel
外围总线 peripheral bus
外涡旋 main vortex
外屋 outhouse;outbuilding <指车库、谷仓等>
外物 foreign object
外物杂质 acquired impurity
外吸渗 exsorption
外吸收法 external absorbent method
外矽卡岩 outsharn
外弦杆 exterior chord
外舷部 topside
外现成本 explicit cost
外线 external line; line wire; outerline;outer wire;outward line
外线堡 outer fort
外线拨号自动识别 automatic identification of outward dialing
外线操作 off-line operation
外线查修线务员 outside trouble man
外线电话 external telephone
外线电路 line circuit
外线端钮 mechanical joint
外线热放射器 infrared heater
外线输入 outline input
外限 outer limit
外相 <即连续相>【化】 external phase
外相干 external coherence system
外相干的 external coherent
外相流体 external fluid phase
外镶边石 out-band
外镶玻璃 outside glazing
外向 extraversion;outbound;outward orientation;toe-out
外向窗框口窗 window with sashes opening outwards
外向电流 outward current
外向构型 exoconfiguration
外向化 exteriorization;exteriorize
外向交通 outbound traffic
外向径向流 radical outward flow
外向湿润 dewetting
外向式 outboard type
外向水流 outward flow
外向顺式加成(反应) exo-cis-addition
外向投射 extrajection
外向推力 outward thrust
外向型经济 export-oriented economy;foreign-oriented economy
外向性 exotrophy;extroversion
外向油封 oil seal with lip facing outward
外项 extreme(term);outer term
外消旋固体溶液 racemic solid solution
外消旋混合物 racemic mixture
外消旋体 raceme

外消旋物 raceme
外销 export sale;foreign selling
外销初始期 infant marketing
外销货 export goods
外销机构 export organization
外小纵梁 jack stringer
外斜角接合 external mitre [miter]
外斜肋骨 sloping outward frame
外斜梁 <楼梯井弯曲处> wreath string(er);wreath piece
外心 <外接圆的中心> circumcenter [circumcentre];excenter [excentre]
外心投影 external projection
外信号线 external signal line
外信号中断 external signal interrupt
外星人 extraterrestrial
外行程 outer stroke
外行的 unprofessional
外行买主 good faith purchaser
外行人 jack leg; laity; layman; non-professional
外行星 superior planet
外行星齿轮 outer planetary gear
外形 appearance; configuration; contour(ing); external form; externality; facies; figuration; outline; outline of figure; physical form; planform; profile; resemblance
外形不良的 unshapely
外形布置 physical layout
外形参差不齐的 ragged
外形尺寸 boundary dimension; boxed dimension; contour dimension; external dimension; gabarite; outer dimension; outline dimensions; outside dimension; outside measurement; overall dimension; overall size; physical size
外形尺寸宽度 overall width
外形尺寸图 outline dimensional drawing
外形尺寸有效系数 coefficient of gabarite efficiency
外形淬火 contour quenching
外形的真实性 trueness of shape
外形电视 atobit
外形分析 contour analysis
外形腐蚀(加工) contour-etching
外形公差 contour tolerance
外形后宽 rear span
外形毁损 disfigurement
外形及安装尺寸 outline overall and installing dimensions
外形加工 contouring
外形检查 appearance inspection test; appearance test; configuration inspection; observation check
外形检验样板 receiver ga(u)ge
外形精度 accuracy of configuration
外形精美的 well-shaped
外形控制系统 contour control system
外形拉削 contour broach
外形量规 figure dimension
外形流线化导管 external faired pipe
外形轮廓断面图 clearance diagram
外形轮廓像 outline image
外形描述附注 physical description notes
外形模 epimorph
外形前宽 front span
外形切削 contour cutting
外形设计 styling
外形深度 profile depth
外形十分圆 quite circular in outline
外形式 exterior form
外形竖铰链窗 out-swinging casement window
外形说明 physical presentation statement
外形特征 external physical characteristic; resemblance
外形调整 configuration control
外形凸点 height of contour
外形图 drawing of external shape; external view; figuration drawing; outline drawing; outline map; outside drawing; outside view
外形椭圆 oval in outline
外形稳定性 dimensional stability
外形线 contour line; object line; visible line; outline
外形芯 outside core
外形修复 contouring
外形学 topology
外形凿刀 profiled chisel
外形直径 form diameter
外形制作者 profile maker
外形装配图 outline assembly drawing
外形阻力 shape resistance
外型 <石膏模> case
外休息厅 <客船的> foyer
外悬式厕所 overhung latrine
外悬式离心压缩机 overhung-type centrifugal compressor
外旋 outer side of a vertical surfaces; outward rotation
外旋轮线 epitrochoid
外旋轮线旋转发动机 epitrochoidal engine
外旋式 <渐曲线> evolute
外旋式双推进器 outboard turning screws; outward turning screws
外旋位 external rotatory position
外旋压 outside spinning
外旋转 external rotation
外旋转的 out-swinging
外循环 extrinsic cycle; output loop; outside loop
外循环三相流化床 external loop three phase fluidized bed
外循环三相流化床反应器 external loop three phase fluidized bed reactor
外循环升流式厌氧污泥床反应器 exterior circulation upflow anaerobic sludge bed reactor
外压法兰 external pressure flange
外压封头和壳体 external pressure heads and shells
外压管 external pressure tube
外压花 external knurling
外压力 collapsing pressure; external pressure
外压力约束 external pressure confinement
外压强 collapsing pressure; external pressure
外压强度 external pressure strength
外压球壳 external pressure sphere
外压区 external pressure region
外压容器 external pressure vessel
外压容器的失稳 vessel buckling under external pressure
外压容器开孔补强 opening reinforcement of external pressure vessel
外压式发动机 outward compression engine
外压试验 external pressure test
外压试验器 external pressure tester
外压条 outside casing; exterior stop <玻璃>
外压圆筒 external pressure cylinder
外压纸格 outer tympan
外烟囱 trunk of funnel
外延 breadth; denotation; extension
外延X射线能量损失精细结构 extended X-ray energy loss fine structure
外延X射线能量吸收精细结构 extended X-ray energy absorption fine structure
外延层 epitaxial film; epitaxial layer
外延长度 perimetric length; perimetric pattern
外延沉积 epitaxial deposition
外延迟时间 external delay time
外延抽象方法 method of extensional abstraction
外延储量 extension ore
外延的 epitaxial
外延电流 extrinsic current
外延淀积 epitaxial deposition
外延淀积硅 epitaxial deposited silicon
外延钝化集成电路 epitaxial passivated integrated circuit
外延多晶膜 epitaxial polycrystalline film
外延法 epitaxial method; extrapolation method; method of extrapolation; epitaxy
外延隔离法 epitaxial isolation method
外延硅 epitaxial silicon
外延硅靶摄像管 epicon
外延过程 epitaxial process
外延合金器件 epitaxial alloyed device
外延后填 epitaxial backfill
外延化学气相沉积生长 epitaxial chemical vapo(u)r deposition growth
外延结 epitaxial junction
外延界面 epitaxial interface
外延晶体管 epitaxial transistor
外延扩散 epitaxial diffuse
外延扩散法 epitaxial diffused method
外延扩散型光电二极管 epitaxial diffused phototransistor
外延量 extensive magnitude; extensive quantity
外延榴石膜 epitaxial garnet film
外延炉 epitaxial furnace
外延膜 epitaxial film
外延片 epitaxial wafer
外延平均 extrapolated mean
外延平面集 epi-planar integrated circuit
外延平面技术 epitaxial planar technique
外延平面晶体管 epitaxial planar transistor
外延气相生长 epitaxial vapor growth
外延区 epitaxial region
外延生长 epitaxial growth; epitaxis
外延生长衬底 epitaxial substrate
外延生长硅晶体 epitaxially grown silicon crystal
外延生长作用 epitaxial effect
外延台阶 epitaxial ledge
外延台式晶体管 epitaxial mesa transistor
外延网 extranet
外延误差 extrapolation error
外延辛烷值 extrapolated octane number
外延性 epitaxy; extensionality
外延意义 denotative meaning
外延用衬托器 epitaxial susceptor
外延元件 epitaxial cell
外延自动掺杂 epitaxial autodoping
外岩浆 epimagma
外岩浆的 extramagmatic
外岩浆热液矿床 apomagmatic hydrothermal mineral deposit
外岩芯筒长度 outer core barrel length
外岩芯筒尺寸 outer core barrel size
外沿间距离 distance out to out; out-to-out distance
外檐斗拱 outer-eaves corbel bracket
外檐天沟 cornice gutter
外檐装修 exterior finish work
外衍射 external diffraction
外阳台 drip cap; exterior balcony; open air terrace; outporch
外阳台板 external balcony slab
外洋 broad ocean; mid-ocean; midsea
外洋航路 ocean lane
外洋航线 ocean lane
外洋流域 exorheic; exorheic drainage; exorheism; exterior drainage
外洋区 broad ocean area
外洋区域 exorheism
外业 field operation; field procedure; field survey; field work
外业标准 field standard
外业草图 field sketch; ground sketch
外业测量记录 survey record
外业测量图幅 field surveyor's plot
外业测量资料 survey record
外业测图板 field board
外业测图图幅 survey plat
外业成果 field result
外业的 outdoor
外业队 field party
外业高程 field elevation
外业规范 field manual
外业基地 field survey depot; field survey station
外业计算 field calculation; field computation
外业计算的点位 face position of telescope; field position
外业记录 field note; field record(ing); survey field notes
外业记录本 field book
外业记录簿 field book
外业记录手簿 field record book
外业记事 field description
外业技术改造 field development
外业检测 field completion; field inspection
外业检查 field check; field examination
外业检核 field check
外业校核 field check
外业控制 field control; ground control
外业判读 field identification; identify(ing) on the ground
外业设备 <勘测等工作用的> camp equipment; field equipment
外业收集 field collection
外业手簿 field book; field document; survey field notes
外业手册 field book; field guide; field manual
外业数据 field data
外业修测 revise in the field; revise on the ground
外业仪器 field instrument
外业因素 foreign element
外业用纠正仪 field rectifier
外业用立体镜 field stereoscope
外业原图 basic plate; compilation sheet; composite sheet; field document; field map; field sheet; field survey sheet; master sheet; planetable map; survey plat; survey sheet
外业值 field value
外业指南 field manual
外业装备 camp equipment; field apparatus; field equipment; field outfit
外业资料 field data; field document; field record(ing); survey field notes
外叶轮 outer impeller
外页岩百叶窗 exterior slatted blind
外衣 coat; garb; outerwear; slop
外移桩【测】 offset stake
外逸层 exosphere; outer sphere
外逸电子剂量计 exo-electron dosimeter
外溢 egress; overspill

外因 ectogene; external cause; external factor; transient cause
外因半混合层 exogenetic mesomixis
外因沉降 exogenic subsidence; exogenous subsidence
外因故障 exogenous failure; influenced failure
外因特网【计】extranet
外因误差 external error
外因性半导体 extrinsic semiconductor
外因性中毒 exogenic toxicosis
外因延迟 external delay
外因造成的停工时间 external idle time
外因作用【地】exogenous process
外阴极磁控管 external cathode inverted magnetron
外阴极计数管 external cathode counter
外应力 exterior stress
外应力腐蚀裂纹 external stress corrosion cracking
外营力【地】exogenic; exogenetic force; exogenic force; exogenous force; external agent
外营力类型 type of exogenic process
外营力能量来源 energy source of exogenic process
外映射半径 outer mapping radius
外用瓷漆 external enamel
外用大理石 exterior marble
外用的 exterior
外用的承包商吊车 exterior contractor's hoist; external contractor's hoist
外用的承包商绞车 exterior contractor's hoist; external contractor's hoist
外用底层抹灰 undercoater for external use
外用吊斗提升机 external skip-hoist
外用钉 exterior nailing
外用黑色有光漆 exterior black gloss paint
外用胶合板 exterior plywood
外用胶乳涂料 exterior latex paint
外用抗静电剂 external antistatic agent; external antistatics
外用路标漆 exterior traffic paint
外用抹灰石膏 exterior gypsum plaster
外用耐久性 exterior durability
外用腻子 face putty
外用漆 exterior paint; outer paint
外用气雾剂 aerosol for external use
外用清漆 body varnish; exterior varnish; external varnish; outdoor varnish
外用乳胶漆 exterior emulsion paint; outdoor emulsion paint
外用涂料 exterior coating; external paint
外用油漆 exterior paint; outdoor paint
外用增塑剂 external plasticizer
外用罩面漆 exterior finish
外用装饰漆 exterior trim paint
外用着色剂 exterior stain
外油箱 externally mounted fuel tank
外釉质上皮 external enamel epithelium
外余摆线型转子发动机 rotary piston engine of epitrochoid design
外余面 outside lap
外隅角 external mitre [miter]
外雨量测定法 external pluviometry
外预告信号(机) outer distant signal
外域 external area
外圆 excircle
外圆边角侧砖 round edge external reveal
外圆车削 cylindric(al) turning; external cutting
外圆滚线 epicycloid
外圆珩床 external honing machine
外圆角【建】bull-nose; bull's nose
外圆角半径 bullnose radius
外圆角刨 bull-nose(d) plane
外圆角秋叶 fluted bead
外圆角形混凝土砌块 bull-nose block
外圆角砖 bull-nose brick; thumb
外圆磨床 circular grinder; circular grinding machine; cylindric(al) grinder; cylindric(al) grinding machine; external grinder; plain grinding machine
外圆磨削 cylindric(al) grinding; external grinding
外圆刨 convex round plane
外圆跳动量 outside diameter runout
外圆无心磨 external centerless grinding
外圆无心磨床 cylindric(al) centerless-type grinder
外圆修正 external circle trimming
外圆研磨 cylindric(al) lapping
外圆中心磨床 cylindric(al) center-type grinder
外圆锥 male cone
外援 foreign aid; outside help
外援计划 foreign aid program(me)
外缘 exterior margin; outer edge; outer limit; outer margin
外缘板 curtain plate
外缘的 peripheral
外缘加高的拐弯 raised curve
外缘距离 distance out to out
外缘离地间隙 curb clearance
外缘山麓侵蚀面 peripedinent
外缘推挤带 frontal compression belt
外缘纤维应力 external fiber stress; extreme fiber stress; stress in extreme fibre
外缘应力 external stress; extreme stress
外缘预应力 outer prestress
外缘整修 external shaving
外缘直径 external profile diameter
外缘转弯半径 clearance radius; outer clearance radius; outer radius
外缘转弯圆周 turning circle
外缘最小离地间隙 curb height
外源 extraneous sources
外源暗示 heterosuggestion
外源包体 accidental inclusion; exogenic inclusion; exogenous inclusion
外源捕掳体 accidental xenolith
外源捕掳岩 accidental xenolith
外源场 field of the external source
外源传染 exogenous infection
外源代谢 exogenous metabolism
外源的 allogenic; exogenous; exotic
外源河 allochthonous river; allochthonous stream; allogenic river; allogenic stream; exotic river; exotic stream
外源化学沉积 allochem
外源混合岩化方式 exomigmatization way
外源凝集素 lectin
外源喷出物 accidental ejecta
外源喷块 accident block
外源起电 exogenous electrification
外源生物群落 allobiocoenosium
外源同步因素 exogenous zeitgeber
外源微生物 exogenous microorganism
外源污染物 exogenous pollutant
外源物 allogene; allothigene
外源氧化硅胶结物 extrastratal silica cement
外源因素 extrinsic factor
外源有机物 exogenous organic matter
外源准地槽【地】exogeosyncline
外苑 outer garden
外院 outer court
外院式平面布置 open outside court type
外运材料 outward material
外运的 outward
外运货(物) outbound cargo; outbound freight; outward cargo; outward freight
外运加工 outward processing
外运砾石 shipped-on gravel
外运砂屑 allodapic
外运输区 outer transport area
外运土 borrow soil
外运运费 outward freight
外匝道 outer loop
外杂基 epimatrix
外在磁场 external magnetic field
外在的酬赏 extrinsic rewards
外在副作用 externality
外在活动 external activity
外在节约因素 external economy
外在曲率 extrinsic curvature
外在水分 extraneous moisture
外在问题 extrinsic problem
外在效率 external efficiency
外在性 outness
外在因素 externality
外在原因 external cause; transient cause
外在折旧 external depreciation
外在作用误差 external error
外增塑作用 external plasticization
外增殖比 external breeding ratio
外闸带 external contracting brake band
外闸门 flood tide gate; outer gate
外闸首 < 复式闸门的 > outer head
外闸瓦制动器 external block brake
外债 external; external debt; external loan; foreign debt; foreign loan; loan from foreign powers; overseas debt
外债偿付 external debt servicing
外债偿还率 debt-service ratio
外债清偿 external debt servicing
外展 < 船首 > flam; flare
外展法 abduction
外展过度 hyperabduction; superabduction
外展海岸 accretion coast
外展夹板 abduction splint
外展神经 abducent nerve
外展滩脊 accretionary ridge
外展纹孔口 extended pit aperture
外展作用 accretion; aggradation
外张侧墙 taper side
外张单翼机 externally braced monoplane
外张的 flanning
外张端墙 taper end
外张舷 flaring side
外涨 bulge
外胀式制动器 expanding brake
外照射 external exposure; external irradiation exposure
外照射防护 external irradiation protection; protection of outer radiation
外照射剂量 external exposure dose; external irradiation dose
外照射危害 external irradiation hazard
外罩 bell housing; case; cladding; encloser; end bell; jacket; lagging; mantle; outer cap; outer casing housing; overlay
外罩带 jacket band
外罩钩扣 hood fastener
外罩箍 jacket band
外罩炉膛锅炉 externally fired boiler
外罩清漆 outer varnish(ing); outside varnishing
外遮阳 pinoleum blind
外褶皱带【地】externide
外整修 outside finish
外正齿轮 exterior-teeth spur gearing; external-teeth spur gearing
外正割 exsecant
外正立面 < 房屋 > stylar facade
外支 external branch
外支架 outer support
外支块 outer shoe
外支墙 buttressed wall
外支式挡土墙 buttressed type retain wall
外枝副地槽【地】exogeosyncline
外枝准地槽【地】exogeosyncline; delta geosyncline
外直和 external direct sum
外直径 external conjugate diameter; outer diameter; outside diameter
外直线车削 external straight turning
外止口 male half coupling
外指点标 outer marker
外指示剂 external indicator
外指示器 external indicator
外制动臂 outer brake arm
外制动带 brake band; outer brake band
外制动毂 outer brake hub
外制动器 outer brake
外制动闸 external brake
外质 ectoplasm
外质混合岩化作用 exomigmatization
外置 outlay
外置发动机 external engine
外置机械密封 outboard mechanical seal
外置炉膛 external furnace
外置马达 outboard motor
外置平面位置显示器 remote plan (position) indicator
外置器件 external device
外置球轴承 outboard ball bearing
外置燃烧室 external furnace
外置散热器 remote radiator
外置刹车缸 outer cylinder
外置式过热器 external superheater
外置式冷却水加热箱 external tank-type heater
外置式省煤器 external economizer
外置式旋风筒 external cyclone
外置式液压油缸操纵杆 remote operation cylinder lever
外置式油缸 external ram; portable cylinder; portable ram; pull-behind cylinder
外置术 exteriorization
外置油缸控制阀 external ram control valve
外置轴承 outboard bearing
外中比 extreme and mean ratio
外中层 sarcocyte
外中间加热器 external interheater
外中心缩孔 open centerline [centreline] shrinkage
外终丝 external terminal thread
外周 periphery
外周室 peripheral compartment
外周水压 external hydraulic pressure
外周系统 peripheral system
外周阻力 peripheral resistance
外轴 outer axis; outer shaft; outer spindle
外轴承 outer bearing
外轴承盖 outer bearing cover
外轴承油封 outer oil seal

外轴肩车削 external shoulder turning
外轴颈 outside-journal axle
外轴式相位机构 <单纯旋转发动机的> external phasing gear
外轴套 outer bushing
外住者 absentee
外注入 external injection
外注图 order drawing
外柱 column jacket;exterior column;external column;outer column;outside column
外柱身 exterior shaft
外柱式集装箱 exterior post type container;outside post type container
外砖墙 external brick wall
外转 abversion;exodeviation
外转车道 outer connection
外转换 external conversion
外转换比 external conversion ratio
外转换系数 external conversion coefficient
外转结 outside clinch;outside rolling hitch
外转内商品 export reject
外转设备 ancillary equipment;peripheral equipment
外转式 outboard turning
外转子 outer rotor
外转子式电动机 external-rotor motor
外装玻璃(法) exterior glazing;outside glazing
外装窗玻璃 outside glazing
外装电动机 outboard motor
外装发动机 outboard engine
外装防雨帽 external rain cap
外装风扇干燥窑 external fan kiln
外装马达 outboard;outboard motor
外装配玻璃型的钢窗油灰 steel sash putty of the exterior glazing type
外装品 exterior trim part
外装式 fitted outside
外装式怠速浓度调程限制器 external-type idle limiter
外装式机械密封 external mechanical seal
外装式空气喷射总管 external air injection manifold
外装式密封 external seal
外装饰物用漆 exterior trim paint
外装饰用薄板 exterior sheet
外装弹簧锁 stock lock
外装修 exterior finish
外装修木材 wood exterior
外装修用胶合板 exterior type plywood
外装叶片式气动马达 external vane-motor
外装增压器 external supercharger
外装轴承 outboard bearing
外装轴承座盖 outboard bearing seat cover
外锥 external cone
外锥度 external taper
外锥度车削 external taper turning
外锥面 male cone
外锥体层 external pyramidal layer
外锥形折射 external conic(al)refraction
外资 foreign capital;foreign fund
外资产业 enterprises owned by foreign capitalists
外资法 foreign investment law;law of foreign capital
外资公司 corporation with foreign capital
外资流入 foreign capital inflow
外资逆差 unfavo(u)rable balance of trade
外资企业 business operating with foreign capital;foreign-funded enterprise;foreign-invested enterprise;full foreign-owned enterprises;overseas-funded enterprise
外资商行 foreign-funded firm;overseas-funded firm
外资投向 foreign capital investment target
外资投资 foreign capital investment
外总线指令 external bus instruction
外纵梁 exterior beam;exterior stringer;outside stringer
外走廊 aisle gallery;exterior corridor;outside corridor;outside gallery;promenade deck
外阻抗 external impedance
外作业 external work
外作用闸 externally acting brake
外座盖垫片 outer seat cover gasket
外座圈 outer race;outer race ring;outside race
外座下销 bearing casing dowel pin

弯

弯把手摇钻 angular bit stock
弯把执手 lever handle
弯板 angle block;bending plate;bent plate
弯板锤 beading hammer;seam hammer
弯板工作台 bending trestle
弯板机 angle bender;bending machine;bending roll(er);bending rolls;plate bender;plate bending machine;plate bending roll;plate roll;press brake;squeezer
弯板炉 bending furnace
弯板模 bending mo(u)ld
弯板内面 belly
弯板室 bending chamber
弯板首柱 shaped plate stem
弯板台 plate bending stand
弯板图式 bent pattern
弯板退火炉 bending lehr
弯棒机 rod bending machine
弯孢壳属 <拉> Eutypa
弯背梁 camber beam
弯背手锯 skewback
弯臂式启闭机 operating machinery with curved arm
弯臂形船首柱 soft nose stem
弯臂形船首柱钢板 soft nose plate
弯臂轴 bent axle
弯边 beading;crimp(ing);flange;flanged edge
弯边半径 radius of flange
弯边搭焊缝 sidelap weld(ing)
弯边端接接头 flanged edge joint
弯边对接焊 flanged butt weld
弯边对接接头 flanged butt joint
弯边高度 height of flange
弯边机 flanger;press brake
弯边接头 coach joint
弯边模 seaming die
弯边压力机 flanging press
弯边圆钢板 dish plate
弯边装置 beader
弯柄单头扳手 bent handle single head wrench
弯柄铰刀 shell reamer
弯柄螺母丝锥 bent tap
弯柄漆刷 angle brush;angle paint brush
弯柄式 open handle
弯柄刷 dog's leg brush
弯柄丝锥 bent tap
弯柄套筒扳手 bent socket wrench
弯波导 waveguide bend
弯脖扳手 angle wrench
弯脖撬棍 gooseneck claw bar
弯脖式 gooseneck type
弯脖套筒扳手 offset pattern socket wrench
弯脖压板 gooseneck clamp
弯脖油刷 wryneck paint brush
弯材平台 bending block;bending floor;bending slab
弯插 bend cutting
弯插头 angle plug
弯车头 contra-angle handpiece
弯沉 flexure;deflection【道】
弯沉杯 deflection bowl;deflection cup
弯沉比 settlement deflection ratio
弯沉对照图 <用弯沉值对照说明路面性状> interpretation graph
弯沉法 deflection procedure
弯沉峰值 <一年四季中的> peak deflection
弯沉计算(公)式 deflection equation
弯沉量测 deflection measurement
弯沉盆 deflection basin;deflection bowl
弯沉曲线 deflection curve
弯沉试验【道】 deflection test
弯沉图 deflectogram
弯沉系数 coefficient of deflection;deflection coefficient
弯沉仪 <用于路面弯沉测量> Benkelman beam;deflection beam;deflectograph;deflectometer
弯沉值 deflection value
弯成拱 inflected arch
弯成钩 hook bending
弯成钩形 hook
弯成弧形 embow(ment)
弯成曲柄状 crank
弯成曲柄状的螺丝起子 cranked screwdriver
弯成圆形 circumflexion
弯成肘形的 elbowed
弯齿耙 curve-tined harrow
弯椽 curved rafter
弯垂石松碱 lycocernuine
弯垂下来的 incumbent
弯带闸 knee brake
弯刀 bent blade;hook knife
弯倒温度 end point
弯道 bend;bendway;bent approach;curve;curve bend;curved conduit;road bend;track curve;bend of road <道路曲线段>
弯道安全速度 <汽车的> safe curve speed
弯道半径 bend radius;radius at bend;radius of curve;track curve radius
弯道标 curve post
弯道标志 curve sign
弯道长度 length of curves
弯道超高结构 super-elevated curve structure
弯道道岔 curved turnout
弯道的机械影响 mechanical effects of curvature
弯道的经济影响 economic effect of curvature
弯道顶点 apex of bend;summit of bend
弯道反光镜 traffic mirror
弯道改善 bend improvement
弯道公式 curve
弯道横净距 lateral clear distance of curve
弯道环流 circulation current in river bend
弯道计算修正值 curvature correction
弯道加宽 bend widening;curve widening;widening on curve
弯道加宽递增距离 increment distance of widening on curve
弯道减速标志(牌) curve-speed plaque
弯道宽度 bend width
弯道流量计 bend meter
弯道模型试验 river bend model test
弯道浅滩 bend shoal;shoal at river bend;shoal in a bend
弯道曲率 curved channel curvature;bend curvature;curvature of bend
弯道曲率半径 bend radius;curvature radius of bend
弯道曲率总和 total curvature
弯道上的单坡路拱 banked crown on curve
弯道视距 horizontal sight distance inside of curve
弯道水流 bend flow;spiral flow;water flow in river bend
弯道损失 bend loss;bent loss
弯道土方计算修正值 curvature correction
弯道外缘的超高 canting
弯道站 <索道的> angle station
弯道整治 bend improvement;bend regulation
弯道治理 bend improvement;bend regulation
弯道阻力 curve resistance
弯的 kneed
弯底肘管 shoe elbow
弯丁坝 hooked groin;hooked groyne
弯钉钉合 clench nailing
弯钉钉合用钉 clench nail
弯顶 apex of bend;curve top;summit of bend
弯定位器 curved steady arm
弯斗 curved scoop
弯度 amount of deflection;bending deflection;bilge;camber;circumflexion;flexure;sinuosity
弯度计 deflectometer
弯度线 bending line
弯端 bent-up end
弯段 bendway;curve bend
弯段加宽 curve widening;widening at turns
弯段曲率 bend curvature
弯段水头损失 head loss in bend
弯段损失 bend loss
弯断作用 flexing action
弯防波堤 arched mole
弯缝导板 flexible guide
弯杆 bent lever;curved rod
弯杆法黏[粘]度测定 beam bending viscometry
弯钢板 <闭式桥面用> hang plate
弯钢管扳头 bending iron
弯钢管机 pipe bend
弯钢轨器 jim crow
弯钢化 bending temper(ing)
弯钢化玻璃 bent tempered glass
弯钢机 bulldozer
弯钢筋 bar bending;hickey bar
弯钢筋扳头 bending iron
弯钢筋表 bending list;bending schedule
弯钢筋钢管的条凳 bending bench
弯钢筋工 rod bender
弯钢筋工场 reinforcement bending yard;steel bending yard
弯钢筋工具 bending iron;steel bender
弯钢筋工作台 bending table;bar bending bench
弯钢筋机 bar bender;bar bending machine;bending apparatus;reinforcing steel bender;steel bender;steel bending machine;bender
弯钢筋明细表 bending schedule
弯钢筋台 bar bend table
弯弓 pan bow
弯拱 camber(ed)arch;hog

W

弯拱桥 curved arch bridge
弯拱因数 arching factor
弯拱因素 arching factor
弯拱作用 arching action;arching effect
弯钩<钢筋末端的> anchorage bend; anchor bar;crotch;hook
弯钩朝上 hookup
弯钩朝下 hook down
弯钩的开脚扳手 hooked spanner; hooked wrench
弯钩地脚螺栓 hooked foundation bolt
弯钩钢筋 hooked bar
弯钩螺栓 hooked bolt;J-bolt
弯钩螺丝 round bend screw hook
弯钩锚固铁件 twisted wire anchor
弯钩状 edge bend
弯箍机 stirrup bender; stirrup bending machine
弯骨钻 curved bone drill
弯关螺栓 bolt with one end bent back
弯管 angle branch;angle pipe;barron bend; bend pipe; bend tube; bent pipe; bent tube; connector bend; corner;curved pipe;curved spout; elbow; elbow bend; elbow pipe; knee;knee pipe; offset bend; pipe bend; pipe elbow; winding pipe; bent housing【岩】; forty-five degree swan-neck <如楼梯扶手、管子等>
弯管半径 bending radius;radius of elbow
弯管部分 bent-tube section
弯管车间 pipe bending shop
弯管当量长度<相等水头损失的> elbow equivalent
弯管的内曲度 bass
弯管的弯出距 offset of U-bend
弯管钉 bending pin
弯管顶 pipe crown
弯管段水头损失 bend loss;bent loss
弯管给料器 elbow feeder
弯管工具 bending tool;pipe bending tool
弯管工作台 bending trestle
弯管辊 pipe bending roll
弯管锅炉 bent-tube boiler;drum-type boiler
弯管合金 bend alloy;bend metal
弯管活接头 bend union
弯管机 angle bender;bender;bending machine;pipe bender;pipe bending machine; pipe bending press; tube bender;tube bending machine;tubing bender
弯管夹具 bending fixture
弯管架 creasing stake
弯管角度 angle of bend
弯管接头 angular pipe union;angular pipe unit;bend piece;corner joint; elbow joint; elbow union; ell fitting;knee bend;street elbow;union bend
弯管开盖器 elbow cover opener
弯管开关 bib valve
弯管连接 corner joint
弯管联管节 angular pipe union
弯管流量计 bend flow meter; bend meter; bend pipe meter; elbow (flow) meter
弯管流速仪 elbow meter
弯管流体压力计 sympiesometer
弯管龙头 bib valve;Fuller's faucet
弯管内曲度 Hass
弯管气压计 siphon barometer
弯管器 bender;kick(e)y;pipe bender;pipe bending instrument
弯管钳(子) pipe bending pliers;bending pliers;bending tongs
弯管曲率 elbow curvature
弯管式锅炉 bent-tube boiler
弯管式伸缩接头 bend type expansion joint
弯管式水管锅炉 water-tube boiler with bent tubes
弯管式压力计 U-tube manometer
弯管式蒸发器 bent-tube evapo(u)rator
弯管水表 bend pipe meter
弯管水封 gas trap
弯管水管锅炉 bent-tube boiler
弯管水头损失 head loss in bend;knee head loss
弯管损失系数 curvature factor
弯管胎具 bending shoe;pipe bending shoe
弯管弹簧 pipe bending spring
弯管通风筒 elbow ventilator;gooseneck ventilator
弯管头 knee bend
弯管小队 bending crew; bending gang;bending party;bending team
弯管芯 bending mandrel
弯管芯棒 pipe bending mandrel
弯管芯子 pipe bending mandrel
弯管旋塞 angle cock;bibcock
弯管压头损失 loss of head due to bend
弯管液压拉力 bending force
弯管用的芯杆 mandrels for pipe bending
弯管用芯棒 mandrel for bending
弯管用易熔合金 bend alloy
弯管直径 diameter range of pipe bent
弯管轴流泵 angle type axial flow pump
弯管铸造 corner casting
弯管最大壁厚 maximum wall thickness
弯轨 bending rail; bent rail; curved rail
弯轨机 cambering machine;jack rail bender; rail bender; rail bending machine
弯轨交(道)叉 frog crossing
弯轨器 jack rail bender;jim crow;rail bender
弯辊 bowed roller
弯航道信号 bend whistle
弯合 joggle;joggled joint
弯合试验<厚板> book test
弯河湾 barachois
弯喉深度 depth of throat
弯滑刀式开沟器 curved runner
弯滑褶皱 flexure ship flooding
弯环机 ring bending machine
弯簧机 spring bending machine
弯机头 contra-angle handpiece
弯夹板 bent joint bar
弯剪 curved scissors
弯剪刀 bent blade snips
弯剪相互作用 flexure-shear interaction
弯角 bend(ing) angle;turning
弯角连接板 knee bracket plate
弯角面砖 kerb tile
弯角式工具 angle tool
弯角式螺母扳机 angle nutrunner;angle nutsetter
弯角式气砂轮 angle grinder
弯角式气钻 air angle drill;right-angle drill
弯角式砂轮机 angle-headed grinder
弯角式瞎光机 angle sander
弯角旋转式撒砂器 right-angle rotary sander
弯角旋转式研磨机 right-angle rotary sander
弯角针形阀 angle needle valve
弯角钻 angle drill
弯脚 swan-neck bracket
弯脚钉 crotch;nail clinch
弯脚钉合 clench [clinch] nailing
弯脚钩 clasp hook
弯脚卡钳机 compass cal(l)iper
弯脚式楼梯 staircase of dog-legged type
弯脚羊蹄压路机 club-foot sheep's foot roller
弯脚羊足碾 club-foot roller
弯脚羊足压路机 club-foot roller; club-foot sheep's foot roller
弯脚圆规 caliber compasses; caliper compasses
弯接头 bent sub;connector bend;elbow connection;union elbow
弯结带 kink band
弯结面 plane of kinking
弯筋 bent bar; bent steel; reinforcement bending
弯筋工作台 bending trestle
弯筋机 bending machine; bent apparatus; crimper; reinforcement bar-bender; reinforcement bar-bending machine;reinforcing rod bender
弯筋锚固 bent bar anchorage
弯筋一览表 bending list
弯晶分光计 curved crystal spectrometer
弯晶摄谱仪 curved crystal spectrograph
弯晶体光谱仪 bent crystal spectrometer
弯颈 neck
弯颈龙头 swivel valve
弯颈轴承 neck bearing
弯矩 bending moment; bending of flexure;moment flexure;moment of bending;moment of deflection;moment of flexure
弯矩包络图 bending moment envelope
弯矩包(络)线 bending moment envelope;moment envelope
弯矩比(率) bending-moment ratio
弯矩长度的裕度 allowance for length of moment
弯矩重分配 moment redistribution; redistribution of moment
弯矩传感器 moment sensor
弯矩叠加 superposition of moments
弯矩方程式 moment equation
弯矩分布 moment distribution
弯矩分布法 method of moment distribution; moment distribution method
弯矩分配 distribution of bending moment;moment distribution
弯矩分配法 method of moment distribution; moment distribution method; moment of inertia method; Cross method<主要用于设计连续梁和刚架>
弯矩分配系数 moment distribution factor
弯矩荷载 moment load
弯矩荷载图 moment bending chart; moment load chart
弯矩剪力比 moment shear ratio
弯矩角变位特性曲线 moment-curvature characteristic
弯矩理论 bending-moment theory; bending theory
弯矩零点 point of zero moment
弯矩面积 area of moment;moment area
弯矩面积法 area moment method; moment area method
弯矩挠度曲线 moment-deflection curve
弯矩平衡 moment equilibrium
弯矩曲率定律 moment-curvature law
弯矩曲率关系 moment-curvature relationship
弯矩曲率特性曲线 moment-curvature characteristic
弯矩曲率图 moment-curvature diagram
弯矩曲线 moment curve
弯矩试验 moment test
弯矩塑性截面模量 plastic torque modulus of section
弯矩塑性重分布 plastic redistribution of moment
弯矩塑性重分配 plastic redistribution of moment
弯矩梯度 moment gradient
弯矩图 bending diagram;bending moment diagram; moment curve; moment-diagram
弯矩图面积 bending moment area
弯矩图式 moment pattern
弯矩危险截面 critical section for moment
弯矩系数 moment coefficient
弯矩影响系数 moment-influence factor
弯矩影响线 influence line of moments
弯矩再分配 moment redistribution; redistribution of moments
弯矩增大系数 moment magnifier factor
弯矩中心 moment center [centre]
弯矩重分布 moment redistribution; redistribution of moments
弯矩重分配 moment redistribution; redistribution of moments
弯矩轴 moment axis
弯矩转角关系 moment-rotation relationship
弯矩转角滞后回线<用于梁柱设计> moment-rotation-hysteresis loop
弯距 curve distance
弯靠<中途在一港口暂时停靠> touch at
弯靠某港 touch at a port
弯靠与停留 touch and stay
弯口铲 graft
弯口剪 hawkbill snips
弯口钳 hawkbill snips
弯口(沙)坝 baymouth
弯拉回弹模量 tensile resilient modulus
弯拉模量 flexural-tensile modulus
弯拉强度 bending(-tensile) strength; bending tension strength; flexural tensile;flexural tensile strength
弯拉试验梁 bending-tensile test beam
弯拉应变 flexural tensile strain
弯拉应力 flexural tensile stress
弯肋骨 curved frame
弯力 bending force
弯联管 swing joint
弯梁 cam bearing
弯梁的净空 cam bearing clearance
弯梁机 beam bender; beam bending machine;beam bending press;beaming machine
弯梁压机 beam bend press
弯梁压力机 beam bending press;girder bending press
弯料<坯料或板材的> cobble
弯裂 flex crack(ing)
弯裂模量 modulus of rupture in bending
弯流褶皱 flexure-flow fold
弯路 crooked road;tortuous path
弯路上的视距【道】 visibility on curves
弯螺脚 crotch
弯铆钉 nail clinch
弯面锉 half-round file

弯面盖瓦 bonnet hip tile;bonnet tile
弯面机 cambering machine
弯木机 wood bending machine
弯木料 compass(ing) timber;crooked timber
弯挠度 bending deflection
弯挠位移 flexural movement
弯挠振动 flexural vibration
弯扭 blank tear;crankle
弯扭副翼颤振 flexural-aileron flutter
弯扭接头 twist joint
弯扭耦合 bending twisting coupling
弯扭耦合颤振 classic(al) flutter
弯扭屈曲 flexural-torsional buckling; torsional-flexural buckling
弯扭失稳 flexural-torsional buckling
弯扭相互作用 flexure-torsion interaction
弯盘 kidney basin
弯刨 curved plane
弯棒式搅拌器 bent rod type stirrer
弯起部分 bent-up portion
弯起点 bent-up point
弯起端 bent-up end
弯起钢筋 bend(ing) up reinforcement;bend up bar;bent-up bar;diagonal reinforcement;diagram reinforcement;inclined shear bar;slant bar <美>
弯起钢筋的有效面积 effective area of reinforcement in diagonal bends
弯起钢筋分布 spacing bent-up bars
弯起钢筋间距 spacing bent-up bars
弯起钢缆 <预应力混凝土中的> deflecting of strands
弯起钢丝束 <偏离构件重心轴配置的> harped tendons
弯起缆索 raising of strands
弯起(折)预应力筋 draped tendon
弯铅皮用的硬木板 setting-in stick
弯铅皮用的硬木棒 setting-in
弯钳 curved forceps
弯强化 bending quenching; bending temper(ing)
弯桥 bridge of circular(plan) form; curved bridge
弯桥曲线中心 center [centre] of curve of curved bridge
弯桥中心角 central angle of curved bridge
弯翘 curved projection; warpage; warp(ing)
弯翘缝 warping joint
弯翘面 warped surface
弯翘翼墙 warped wing wall
弯翘应力 curling stress; warping stress
弯翘作用 warping effect
弯切两用机 <钢筋> bender and cutter
弯曲 anacampsis; anfractuosity; arcuation;bend;bending flexure;bent; bight;buckling;circumflexion;crankle; crookedness; crooking; curl; curve;curving;dogleg;flex(ure); hogging; incurvation; incurve; paring hammer;sag;sinuosity;tortuosity;torture;warpage
弯曲摆动 flexural oscillation
弯曲摆动试验 flexural vibration test
弯曲板 snyed plate;twisted plate
弯曲板桥 skew slab bridge
弯曲半径 bending radius;bevel(l)ing radius; curvature radius; curve radius; radius at bend; radius of bend;radius of curvature
弯曲半径极小的弯道 knuckle bend
弯曲包铁皮的运输带 snaking conveyer [conveyor]
弯曲比 bending(-moment) ratio;meander ratio;sinuosity ratio;tortuosity ratio
弯曲臂(扶手) bending arm
弯曲扁铁 cranked flat iron
弯曲变位 bending deflection
弯曲变形 bending deformation;buckling; deformation due to bending; flexional elasticity;flexural deformation
弯曲柄 side crank
弯曲波 bending wave;flexural wave
弯曲波导管 bend wave guide
弯曲波痕 sinuous ripple
弯曲波状路线 sinuous route
弯曲部 edge fold;flexion;flection
弯曲部分 curvature; flexion; curve end post
弯曲材 bend wood;bent wood;crook
弯曲操作 bending operation
弯曲测量 flexural measurement
弯曲层积 laminated bending
弯曲插头 angle plug
弯曲颤振 flexural-torsional flutter
弯曲长度 bending length
弯曲常数 bending constant
弯曲成型 buckling
弯曲成型机 bending and forming machine
弯曲程度 sinuosity;degree of curvature
弯曲持久性 flexural endurance
弯曲持久性极限 flexural endurance limit
弯曲尺度 bending dimension
弯曲冲模 air-bend die
弯曲处 knee;sinuosity;throat
弯曲椽 compass rafter
弯曲脆点 bending brittle point
弯曲搭扣 cranked dog
弯曲大梁 curved girder
弯曲大梁起重臂 boom of an arched girder
弯曲导沙墙 silt-vane-cum-curved wing
弯曲道路 tortuous passage; winding road
弯曲的 aquiline; bent up; buckled; cambered; campylotropal; cranky; curved; embowed; flexuose [flexuous]; flexural; fornicate; incurvate; meandering;sinuous;winding
弯曲的板 skew slab
弯曲的波纹铝片 curved corrugated alumin(i)um
弯曲的构件 buckled frame member
弯曲的河流 anfractuosity
弯曲的河流深槽 meandering thalweg channel
弯曲的桥 skewed bridge
弯曲的桥板 skew bridge slab
弯曲的通路 anfractuosity
弯曲的凿刀 skew chisel
弯曲的窄木条 stave
弯曲底层平面图 curved ground plan
弯曲地堑 flexure graben
弯曲点 bend(ing) point; point of inflection
弯曲顶点 apex of bend
弯曲动作 curving performance;flexure operation
弯曲度 angularity; circumflexion; degree of curve; warpage; bending; bendiness <以每公里转折度数计>;tortuosity <河流>
弯曲度量 amount of deflection;bending amount
弯曲段 sinuous section;curved reach <河流的>
弯曲断层 curved fault;flexure fault
弯曲断裂 bending failure; bending fracture;bending rupture
弯曲断裂模量 modulus of rupture in bending
弯曲断裂模数 modulus of rupture in bending
弯曲断裂曲线 bending failure curve; bending rupture curve
弯曲断裂应力 bending failure stress
弯曲断面 crimped section
弯曲法矫形 straightening by bending
弯曲方程 flexural equation
弯曲放电 wriggling discharge
弯曲分布 bending distribution;distribution of bending
弯曲分析 bending analysis
弯曲负荷 bend loading
弯曲覆盖片 cranked sheet
弯曲杆件 flexural member; arch bar <窗扇中>
弯曲刚度 bending rigidity; bending stiffener; bending stiffness; flexural rigidity
弯曲刚性 flexural rigidity
弯曲钢 bend bar
弯曲钢管用弹簧 bending spring
弯曲钢筋 bend bar; bending reinforcement;bent bar;bent element; bent reinforcement bar; bent rod; bent steel;bent-up bar;bent-up reinforcement; diagonal reinforcement;truss bar
弯曲钢筋机 bending machine
弯曲钢筋束 deflected tendon
弯曲钢筋束的工艺 deflected cable technique; deflected strand technique
弯曲杠杆 bending lever
弯曲工具 bending tool
弯曲工作面 bent face
弯曲工作台 bending table
弯曲公差 bending allowance
弯曲公式 bending formula; flexure formula
弯曲共振 flexural resonance
弯曲构件 bending member;curved element; curved member; flexural member;ribbed arch
弯曲刮刀推土机 bull clam
弯曲拐折摩阻力损失值 <预应力混凝土后张法钢筋> curvature friction
弯曲管 curved pipe;gooseneck;swanneck;half-normal bend <弯角135°>
弯曲管件 sweep fitting
弯曲惯性 bending inertia
弯曲轨道 curved rail
弯曲辊道 snaky track
弯曲滚板机 plate bending roll
弯曲过程 bending process
弯曲海岸 curving beach
弯曲海岸线 curved coastline;winding coastline
弯曲航道 curved channel;meandering channel;tortuous channel
弯曲和凹陷条款 <对桶装货的> bending and denting clause
弯曲和回转 ins and outs
弯曲和扭转组合的应力 flexural and torsional combined stresses
弯曲和轴向组合荷载 combined bending and axial loading
弯曲河 meandering stream
弯曲河槽 curved channel; meander(ing) channel;tortuous channel
弯曲河道 curved channel; sinuous river;tortuous channel;meandering channel
弯曲河段 curved reach;curve on river; meandering reach; sinuous section;tortuous river section;winding section
弯曲河段宽度 channel width in bend
弯曲河段潜堰 bendway weir
弯曲河环 meander loop
弯曲河流 meandering river;meandering stream; sinuous river; snaking river;snaking stream;winding river;winding stream
弯曲河流急弯处裁湾 neck cutoff
弯曲荷载 bending load(ing); crippling load; flection load; flexing load; flexion load; flexural load; transverse load
弯曲滑动 bending plane slip;flexural glide;flexural slip
弯曲滑动褶皱作用【地】 flexural slip folding
弯曲化 meandering
弯曲机 bender;bending machine;crimper
弯曲基本轨 bent stock rail
弯曲激波 curved shock
弯曲极限 bending limit;flection limit; flexing limit; flexural limit; flexure limit
弯曲极限强度 modulus of rupture;ultimate strength
弯曲加工 sweep finish;swirl finish
弯曲剪力裂缝 flexure shear crack
弯曲剪切裂缝 flexural shear crack
弯曲剪切裂纹 flexural shear crack
弯曲剪切破坏 flexural shear failure
弯曲建筑 curved block;curved building
弯曲桨叶型 cambered blade section
弯曲胶合板 bent plywood
弯曲角 angle curvature; angle of bend;angularity
弯曲角度 bending angle
弯曲矫直两用压力机 bending and straightening press
弯曲接管 nipple bent pipe
弯曲接头 bending connection; bending joint; crooked joint; kinked joint
弯曲街道 wind street
弯曲结合力 flexural bond
弯曲截断两用机 bender and cutter
弯曲金属(钢)板 bent sheet metal
弯曲晶体 flexure crystal
弯曲井 curved well
弯曲井段 curved portion
弯曲矩 moment of flexion
弯曲抗拉强度 tensile strength in bending;flexural tensile strength
弯曲抗压强度 bending compression strength; flexural compressive strength
弯曲抗振强度 bending oscillation strength;bending vibration strength
弯曲空间 curved space
弯曲孔 curved hole
弯曲孔段 curved portion
弯曲宽度 bending width
弯曲拉力 bending tension
弯曲拉力破坏 bending tension failure
弯曲拉伸破坏 flexural tensile failure
弯曲拉伸试验 flexural tensile test
弯曲拉应力 flexural tensile stress
弯曲来福线锉 bent riffler
弯曲理论 flexure theory; theory of bending;theory of flexure
弯曲力 bending force;buckling force
弯曲力矩 bending moment; flexure moment;moment of bending
弯曲力矩图 bending moment diagram
弯曲力矩影响线 bending moment influence line
弯曲力距 moment of deflection
弯曲连杆 <支承杆间的> reticuline bar
弯曲连接 bending connection; ben-

W

ding joint
弯曲连接件 gooseneck
弯曲梁 bent beam; flexural beam
弯曲梁流变仪 bending beam rheometer
弯曲梁疲劳试验 flexural beam fatigue test
弯曲梁试验 bending beam test
弯曲量 bending amount; compliance
弯曲裂缝 bending crack; curved crevasses; flexural crack(ing)
弯曲裂缝宽度 flexural crack width
弯曲裂纹 bending crack; curved crevasses; flexural crack(ing)
弯曲流槽 curved-trough flow device
弯曲楼梯 geometric(al) stair(case); navel of winding staircase; wreathed stair(case)
弯曲路 winding road
弯曲路缘石 curved curb
弯曲率 curvature; flexuosity; rate of curving; sinuosity ratio; tortuosity ratio; sinuosity <河道弧线长度与直线长度之比>
弯曲轮辐为圆锥形 <在压力机上整轧车轮> coning
弯曲面 bending plane; bent face; curved face; flexure plane; plane of bending
弯曲模 bending die; snaker
弯曲模具 bending mo(u)ld
弯曲模量 bending modulus; flexural modulus; modulus of bending
弯曲模数 bending modulus; flexural modulus
弯曲模膛 bender impression; edger
弯曲木 bending; crook
弯曲木材 bending wood; bent wood; crooked timber; bent timber; compass timber; curved timber; knee
弯曲木家具 bentwood furniture
弯曲木甲板 sprung planking; swept planking
弯曲内裂 internal bending crack
弯曲耐久力 flexural endurance
弯曲耐久性 bending endurance
弯曲能力 bending capability; bending power
弯曲能量 bending energy
弯曲黏[粘]附性试验 flexural bond test
弯曲黏[粘]结 flexural bond
弯曲黏[粘]结力破坏 bending bond failure
弯曲扭断 flexural-torsional buckling
弯曲扭折 flexural-torsional buckling
弯曲扭转 flexural torsion
弯曲扭转耦合 bending twisting coupling
弯曲疲劳 bending fatigue; fatigue in bending; fatigue under flexing; flexural endurance; flexural fatigue
弯曲疲劳抵抗力 flexural fatigue resistance
弯曲疲劳极限 flexural fatigue limit
弯曲疲劳耐久性 flexural fatigue endurance
弯曲疲劳耐久性极限 flexural fatigue endurance limit
弯曲疲劳破坏 fatigue bending failure
弯曲疲劳强度 bending fatigue strength; flexural fatigue strength
弯曲疲劳试验 bending fatigue test; endurance bending test; fatigue bending test; flexile fatigue test; repeated bend(ing) test; reverse bend test
弯曲疲劳试验机 fatigue bending machine
弯曲疲劳寿命 flexible life; flexing life
弯曲频谱 curved spectrum
弯曲平面 bending plane; plane of bending
弯曲平面图 <房屋的> curved plan
弯曲平瓦 bent plain tile
弯曲破坏 bending failure; bending rupture; fail in bending; flexural failure
弯曲破坏理论 theory on failure by bending
弯曲破坏模量 modulus of rupture in bending
弯曲破坏曲线 bending failure curve; bending rupture curve
弯曲破坏系数 modulus of rupture in bending
弯曲破坏应力 bending failure stress
弯曲破损 bending failure
弯曲起点 point of curvature
弯曲器 bending apparatus
弯曲千斤顶 curved jack
弯曲铅管用弹簧 bending spring
弯曲强度 bending strength; flexural strength; strength of flexure; transverse strength
弯曲强度极限 transverse modulus of rupture
弯曲切片 curved cutting blade
弯曲倾度 curved batter
弯曲区 buckled zone
弯曲曲率 crookedness
弯曲曲线 bending curve; flection curve
弯曲屈服 flexural yielding
弯曲屈曲 flexural buckling
弯曲燃烧室 swirl combustion chamber
弯曲扰动 bending disturbance
弯曲韧性 flexural ductility
弯曲容度 bending allowance
弯曲蠕变 bending creep; flexural creep
弯曲蠕变劲度 flexural creep stiffness
弯曲蠕变试验 bending creep test
弯曲软管 curved spout
弯曲三通管 sweep tee
弯曲沙嘴 curved spit; hook(ed spit); recurved spit
弯曲设备 bending apparatus
弯曲伸出艏 clipper stern
弯曲伸缩率 upset
弯曲时不分层主云母 hard mica
弯曲时空 curved space-time
弯曲矢高 curved height
弯曲式预应力筋 harped tendons
弯曲试验 bend(ing) test; angular test; deflection test; flexural test; flexure test; folding test; normal bending test; pliability test; reversed bend test
弯曲试验的抗拉强度 tensile strength on bending test
弯曲试验机 bend(ing) tester; bending test machine; flex tester; flexure test machine
弯曲试验梁 bending test beam; flexure test beam
弯曲试验试件 bending test specimen
弯曲试验试件强度 bending test specimen strength
弯曲试验试样 bend test specimen
弯曲试样 bend specimen
弯曲寿命 flex life
弯曲受拉 combined tension and bending; tension by flexure
弯曲受压 combined compression and bending
弯曲水道 curved channel; meander channel; sinuous channel; tortuous channel
弯曲水流 curved flow; curvilinear flow; sinuous flow; tortuous flow; tortuous stream
弯曲损耗 bend loss
弯曲损坏 fail in bending; failure caused by bending; rupture in bending
弯曲损失 bend loss
弯曲胎模 former for pipe bending
弯曲弹簧 flexural spring
弯曲弹性 bending resistance; elasticity of bending; elasticity of flexure; flexional elasticity; flexural elasticity; flexural resilience
弯曲弹性模量 modulus of elasticity in bending; modulus of elasticity in static bending
弯曲调直两用机 bending and straightening machine
弯曲图 curved diagram
弯曲瓦管 curved earthenware pipe
弯曲瓦面 bent tile
弯曲尾水管 bent draft tube
弯曲位移 bending displacement; flexural displacement
弯曲误差 flexure error
弯曲系数 bending coefficient; bugle factor; coefficient of sinuosity; coefficient of skew; curved coefficient; meander coefficient; meander ratio; tortuosity factor
弯曲纤维 bending fiber[fibre]; curved fiber [fibre]; flexible fiber [fibre]
弯曲纤维木材 wood with crooked fiber [fibre]
弯曲线 bend(ing) line; flexing curve; flexion curve; flexure curve; meander line
弯曲线条 curvature line
弯曲线形 curved alignment; sinuous alignment
弯曲箱形梁 curved box beam
弯曲小径 swept path
弯曲斜度 bending slope; ding slope
弯曲斜接 <楼梯扶手> miter knee
弯曲芯 snake core
弯曲芯柱 bending mandrel
弯曲形电阻器 zigzag resistance unit
弯曲形构件 curved member
弯曲形老虎窗 barrel light
弯曲形态 meandering pattern
弯曲形线脚 curved mo(u)lding; sprung mo(u)lding
弯曲形折(叠)门 serpentine folding door
弯曲形振荡器 flexural oscillator
弯曲形 flexure type; meandering type
弯曲型河段 meandering reach
弯曲型砌块 bending block
弯曲型扰动 wriggling perturbation
弯曲型式 falling mo(u)ld
弯曲型振动 flexural mode vibration
弯曲性 flexing property; toughness; twistiness
弯曲性能 curving performance; flexural behavio(u)r; bending behavio(u)r
弯曲徐变 flexural bond
弯曲压缩 compression with bending
弯曲延性 bend ductility; curvature ductility
弯曲延性系数 curvature ductiling factor
弯曲延展性 flexural ductility
弯曲岩层遮挡 curved barrier of rock
弯曲叶片 cambered blade; curved blade
弯曲移动 meandering movement
弯曲翼梁 bent spar
弯曲翼墙 warped wing wall
弯曲引起的拉应力 tensile bending stress
弯曲应变 bending strain; buckling strain; flexural strain; strain of flexure; transverse strain
弯曲应变能 bending strain energy; flexural strain energy
弯曲应力 bending power; bending pressure; bending stress; buckling stress; flection stress; transverse stress; flexural stress; flexure stress
弯曲应力波 bending stress wave
弯曲应力测定仪 fleximeter
弯曲应力分布 bending stress distribution; flexural stress distribution
弯曲应力公式 bending stress formula; flexural stress formula
弯曲应力疲劳极限 bending stress fatigue limit
弯曲应力疲劳试验 repeated bending stress test
弯曲应力区 bending compression zone
弯曲硬度 flexural rigidity
弯曲迂回的街道系统 curvilinear street system
弯曲余量 bend allowance
弯曲原理 theory of buckling
弯曲圆棒 bending mandrel
弯曲载荷 transverse load
弯曲造型 bending mo(u)ld
弯曲张力区 zone of bending tension
弯曲张应力 bending tension stress
弯曲褶曲 buckle fold
弯曲褶皱【地】 buckle fold; flexural fold; flexure(d) fold; buckling fold
弯曲褶皱作用 buckle folding; flexure fold(ing)
弯曲振荡 bending oscillation; flexural oscillation
弯曲振荡试验 flexural vibration test
弯曲振动 bending vibration; flexural vibration
弯曲振动棒 flexure bar
弯曲振动断裂 bending oscillation failure
弯曲振动方式 flextensional mode; flexural mode
弯曲振动破坏 bending oscillation failure; bending vibration failure; flexural oscillation failure; flexural vibration failure
弯曲振动强度 flexural vibration strength
弯曲振动试验 bending oscillation test; bending vibration test; flexural oscillation test; flexural vibration test
弯曲振型 flexural mode of vibration
弯曲整直两用机 bending and straightening machine
弯曲指数 flexural index; flexural number; index of meandering
弯曲中心 center [centre] of flexure; flexural center [centre]; shear center [centre]
弯曲中心线 elastic line
弯曲中心线方法 elastic line method
弯曲轴 bending axis
弯曲主平面 principal plane of bending
弯曲状 curved
弯曲状的纵向裂缝 longitudinal meandering crack
弯曲状木甲板 laid deck
弯曲状态 case of bending
弯曲阻力系数 curve resistance coefficient
弯曲阻尼 flexural damping
弯曲钻杆 bent rod
弯曲钻孔 knee hole
弯曲钻孔校直钻头 straight hole bit
弯曲作用 bending action; flex action

弯屈 buckle
弯屈强度 buckling strength
弯渠 curved beam
弯刃剪 curved blades snips
弯入 embayment
弯上钢筋 bent-up bar; bent-up reinforcement
弯身 stoop
弯石(砌)块 curve block
弯矢 sagitta
弯式木工车床 bentwood turning machine
弯受力筋 draped tendon
弯水管 gooseneck
弯水头损失 bend loss
弯胎机 tire [tyre] bender
弯探子 ankylomele; curved probe
弯套管机 eye bender
弯体式轴向活塞泵 inclined axial piston pump
弯条机 bar bender; bar bending machine
弯条试验 bent strip test
弯铁 bending iron
弯铁工作台 bending table
弯铁机 bending machine
弯铁台 bending table
弯挺 curved elevator
弯头 angle head; bend; bent; bent pipe; bent tube; connector bend; corner; elbow; elbow bend; elbow joint; elbow pipe; hog; knee bend; knee pipe; pipe angle; pipe elbow; sweep fitting; T-bend
弯头扳手 bent handle wrench; bent spanner; bent wrench; crescent wrench
弯头插销 necked bolt
弯头车刀 angular tool; bent tool
弯头持针钳 curved needle holders
弯头尺寸 corner dimension
弯头当量 elbow equivalent
弯头刀架 knee tool
弯头道钉 brob
弯头的外半径 heel radius
弯头钉 clench [clinch] nail
弯头顶点 apex of bend
弯头钢筋 hooked rod
弯头刮刀 angle scraper; gooseneck scraper
弯头管 bending; knee; swan-neck
弯头活管接 union elbow
弯头活接管 elbow union
弯头活接头 elbow union
弯头活络扳手 adjustable spanner angle; angle adjustable spanner
弯头键 gib-head(ed) key
弯头角 <刀具的> shank angle
弯头角度 angle of bend
弯头接合 elbow joint; knee joint; toggle coupling; toggle joint(ing); toggle linkage; toggle mechanism
弯头接头 union elbow
弯头连接 corner joint; toggle connection; toggle joint
弯头联管节 elbow union
弯头螺母旋转器 flex head nut spinner
弯头螺栓 bolt with one end bent back; clench bolt; clinch bolt; toggle bolt
弯头螺丝刀 offset screwdriver
弯头捻缝凿 bent iron; crooked iron
弯头配件 angle fitting; bend fitting
弯头钳 hawkbill pliers
弯头撬杆 crown bar
弯头上带风标的回转通风器 gyrator and duct vanes for elbow
弯头手柄 knee lever
弯头水表 bend meter
弯头水头 head in bend
弯头水头损失 head loss in bend; knee head loss; loss on head in bends; knee loss
弯头损失 bend(ing) loss
弯头锁 sweep lock
弯头锁紧螺母 elbow jam nut
弯头套管 elbow union
弯头套环 elbow union
弯头套筒扳手 offset socket wrench; socket offset wrench
弯头脱水器 elbow separator
弯头舷板钉 rove clinch nail
弯头芯盒 elbow core box
弯头旋塞 angle cock
弯头闸阀 angle gate valve
弯头配件 elbow piece
弯头转角 angle of bend
弯腿 bandy knees; crooked leg; cabriole leg <S 形曲线的家具腿>
弯托梁挑檐 arched corbel table
弯瓦 compass tile
弯弯曲曲的 flexuose [flexuous]; cranky <道路、河流等>
弯弯曲曲的道路 winding road
弯弯曲曲的栅栏 worm fence
弯尾夹头 bent-tail carrier
弯线 crooked line
弯形玻璃 bent glass; curved glass
弯形薄壳屋顶 curved thin-shell roof
弯形浮冰 ice bay
弯形杆 curved profiled bar
弯形钢化玻璃 curved tempered glass
弯形辊 reduction roll
弯形航道 horseshoe channel
弯形胶合板 ply plastics
弯形卡环 curved clasp
弯形箝齿 bent dent
弯形块 bending block
弯形切口 hockey stick incision; Meyer's hockey stick incision
弯压 bending compression
弯压机 bending press
弯压破坏 flexural compressive failure; moment compressing failure
弯压强度 bending compression strength
弯压区 bending compression zone
弯压铁 clamp upset
弯叶涡轮式搅拌器 curved turbine type agitator
弯液面 meniscus [复 menisci/meniscuses]; meniscus of liquid
弯液面校正 meniscus correction
弯液面校正表 <氟氢酸测斜> test correction chart
弯翼 curve flank
弯硬石膏 tripestone
弯油的气水界面 curved gas-water interface
弯油的油水界面 curved oil-water interface
弯油管缸 pipe bending cylinder
弯月板强制冷却 forced cooling of meniscus
弯月镜施密特望远镜 meniscus-Schmidt telescope
弯月面 meniscus [复 menisci/meniscuses]
弯月面的 meniscoid
弯月面校正 meniscus correction
弯月面曲度 meniscus curvature
弯月面形 meniscus shape
弯月形 falcate
弯月形改正镜 meniscus corrector
弯月形进料口 <旋回破碎机的> moon-shaped feed opening
弯月形镜头 Topogon lens
弯月形零件 meniscus [复 menisci/meniscuses]
弯月形摄影机 meniscus camera
弯月形饰 meniscus [复 menisci/meniscuses]
弯月形太阳照相仪 meniscus photoheliograph
弯月形天体摄影仪 meniscus astrograph
弯月形透镜 meniscus lens; meniscus shaped lens
弯月形透镜太阳照相机 meniscus heliograph
弯月形透镜太阳照相仪 meniscus heliograph
弯月形透镜望远镜 meniscus telescope
弯月形透镜系统 meniscus shaped lens system
弯月形透镜中星仪 meniscus transit
弯月形望远镜 meniscus photographic(al) telescope
弯月形中星仪 meniscus transit
弯凿 bent chisel
弯扎钢筋棚 reinforcement bending shed
弯轧槽 snaker
弯轧钢筋车间 bending and reinforcement assembly shop
弯窄航道 narrow crooked fairway
弯张换能器 flextensional transducer
弯折 buckling; fluting; kinking
弯折常数 buckling constant
弯折钢筋箍 folding stirrup
弯折共振试验法 flexural resonance method
弯折辊 flexing roll
弯折机 bar folder; bending machine; crimper
弯折模量 flexural modulus; modulus of rupture
弯折疲劳 endurance in bending
弯折破坏 bucking failure
弯折试验 bend-over test; folding test; root-bend test; radius test <油毛毡的>
弯折应力 bucking stress
弯折原理 theory of buckling
弯折作用 flexing action
弯针 bolt rope needle; curved needle; short spur needle
弯直两用轧机 bending and straightening rolls
弯指孢属 <拉> Curvidigitus
弯制钢筋 bent-up bar
弯钟乳石 anemolite
弯轴 bent axle; cambered axle
弯肘杆 elbow lever
弯肘连接梁 elbow stay
弯皱 crimp
弯皱器 crimper
弯转机 bending machine
弯状的 periclinal
弯子 <楼梯扶手的弯曲部分> curved ramp
弯钻杆 crooked drill pipe
弯嘴阀 bib valve
弯嘴开关 bid cock
弯嘴龙头 bib(b); bibcock; bib tap; bid cock; nose cock; sink bib
弯嘴钳 angle jaw tongs; bent nose pliers; curved nose pliers
弯嘴水龙头 bib nozzle; bib tap
弯嘴旋塞 bib(b); bibcock; bib tap; nose cock

湾 embayment; gulf; sinus

湾奥寇梯木 gulf ocotea
湾边海滩 bayside beach
湾冰 bay ice
湾侧滩 bayside beach
湾顶 head of bay
湾顶坝 bayhead bar
湾段 reach
湾港 bay harbo(u)r
湾积土 bay deposit
湾积物 bay deposit
湾尖(海)滩 bayhead beach
湾尖沙洲 bayhead bar
湾口沙坝 bay bar(rier); baymouth bar(rier)
湾口沙洲 bay bar(rier); baymouth bar(rier)
湾口沙嘴 bay bar(rier); baymouth bar(rier)
湾口堰洲 baymouth barrier
湾口洲 bay bar(rier); baymouth bar(rier)
湾流 gulf current; gulf flow; gulf stream
湾流锋 gulf stream front
湾流环 gulf stream eddy
湾流蛇曲 gulf stream meander
湾流系统 gulf stream system
湾内海岸 embayed shore
湾内沙坝 bay(mouth) bar
湾内沙洲 bay(mouth) bar
湾滩桥 beach bridge
湾头 bay head
湾头坝 bayhead bar
湾头海滩 bayhead beach
湾头三角洲 bayhead delta
湾头沙坝 bayhead bar
湾头沙埂 bayhead bar
湾头沙洲 bayhead bar
湾头滩 bayhead beach; pocket beach
湾头小滩 pocket beach
湾头堰洲 bayhead barrier
湾形海岸 embayed coast; estuary coast
湾形物 embayment
湾中坝 mid-bay bar
湾中沙滩 mid-bay bar
湾中沙洲 mid-bay bar
湾中沙嘴 mid-bay spit
湾状湖 bay lake

蜿蜒 meander; serpentine turnings; wriggle

蜿蜒比 meander ratio; sinuosity ratio
蜿蜒波 meander wave
蜿蜒波浪 sinuous ripple
蜿蜒长度 meander length
蜿蜒带 meander belt
蜿蜒带长度 meander belt length
蜿蜒带宽度 meander belt width
蜿蜒的 serpentine; winding; sinuous <特指河流等>; meandering
蜿蜒的冲积河流 meandering alluvial river
蜿蜒度 degree of sinuosity; sinuosity
蜿蜒幅度 amplitude of meander; meander amplitude
蜿蜒公路 serpentine highway
蜿蜒构造 meandering structure
蜿蜒管 serpentine pipe; serpentuator
蜿蜒河道 meandering channel; sinuous channel; twisting channel
蜿蜒河段 meandering reach; meandering river stretch; S-turn; twisting section
蜿蜒河谷 meandering valley
蜿蜒河流 snaking river; snaking stream; winding river; winding stream; meandering river
蜿蜒河湾 meander scroll
蜿蜒化 meandering
蜿蜒宽度 meander width
蜿蜒路 zigzag route

蜿蜒路线 wavy trace
蜿蜒盘管 sinuous coil
蜿蜒期 period of meander
蜿蜒声音 flaring
蜿蜒水道 meandering channel; sinuous channel; twisting channel
蜿蜒水道的演变 evolution of meandering channel
蜿蜒水流 sinuous flow
蜿蜒线 meander line
蜿蜒小路 twisting lane
蜿蜒形航线 sinusoidal (outline) course
蜿蜒形墙 serpentine wall
蜿蜒性河段整治 regulation of meandering reach; regulation of meandering river section
蜿蜒性河流 meandering river; meandering stream; sinking river; wandering river; wandering stream
蜿蜒运动 meandering motion; meandering movement
蜿展因数 flare factor

豌豆 pea

豌豆罐头工厂废水 pea cannery wastewater
豌豆级煤 <美国无烟煤粒级 9/16 ~ 13/16 英寸,英国商用煤粒级 1/2-1/4 英寸,1 英寸 = 0.0254 米> pea
豌豆绿色 pea-green
豌青 cobalt blue
豌青颜料 cobalt blue pigment

丸粒掺和机 pellet blender

丸形膨胀高炉熔渣骨料 pelletized expanded blast-furnace slag aggregate
丸药 bolus
丸状的 pelletized
丸状干燥器 pellet drier [dryer]
丸状软化 pellet softening
丸状炸药 pellet power
丸子饰 pellet mo(u)lding

完备测试集 complete test set

完备存储系统 complete storage system
完备的船 taut ship
完备度量空间 complete metric space
完备函数序列 complete function series
完备河曲 full meander
完备集 exhaustive set; perfect set
完备空间 complete space
完备理论 well-developed theory
完备码 perfect code
完备索引 complete indexing
完备图 complete graph
完备性 completeness
完备有序公理【数】complete-ordering axiom
完成 achievement; clear off; completion; consummate; consummation; effectuate; encompass(ment); finish; fulfil; make good; making good; succeed; top off; carry through <计划等>
完成尺寸 finished dimension; finished size; neat size
完成的 accomplished; finished; ripe
完成的合同金额 contract turnover
完成的井 completed well
完成的路面 finished surface
完成的数量 quantitative performance
完成底层抹灰 brown-out
完成调用 complete call
完成订货 completed order
完成订货时间 pipeline time
完成冻结所需时间 freezing time
完成反应 consummatory response
完成工程担保书 completion bond
完成工程第三方担保书 completion bond; construction bond; contract bond
完成工程价值分析 earned value analysis
完成工程量 completed amount; completed quantity
完成工程有关的行业 related trade
完成工作的规定时间 allowed time
完成工作计划成本 planned cost of work completed
完成工作量 amount of work completed; executed amount
完成工作清算 liquidation of accomplished task
完成管道工程的安装 finish plumbing
完成国家计划 fulfilment of state plan
完成合同 completion of contract
完成合同后的财务结算法 completed-contract method of accounting
完成呼叫 complete call
完成机车车辆运用指标计划 fulfilment of roiling stock utilization index plan
完成计划 fulfilment of plan
完成计划指数 planned index fulfilled
完成阶段 concluding stage
完成结构的框架工程 framing in
完成进度 fulfillment of schedule
完成量 accomplishment; quantity finished
完成旅行 accomplish a journey
完成码 completion code
完成脉冲 final pulse
完成面 finished surface
完成某特定项目的系列设备 equipment train
完成排队 completion queue
完成坡度 finished grade
完成期限 time of completion
完成清理工作 fettle
完成任务的能力 mission capability
完成日期 complete date; completion date; date completed; date of completion; time limit
完成施工保险 completed operation insurance
完成时间 completion time; execution time; finish time
完成数量 quantity performed
完成位 completion bit
完成项目 completed project
完成修缮历史 achieving maintenance history
完成一次旅行 accomplish a journey
完成预算的百分比 percentage of the budgeted figure
完成运输成本计划 fulfilment of transportation cost plan
完成运输计划 fulfilment of transportation plan
完成运输收入计划 fulfilment of transportation revenue plan
完成者 consummator
完成指标 hit the target
完成周期 execution cycle; execution period
完成装配 complete assembly; final assembly
完成状态 completion status
完工 complete a project; completion; completion of works; finish(ing)
完工百分率 percentage of completion
完工百分率法 percentage-of-completion method
完工百分率制 percentage-of-completion basis
完工百分率制的存货估计 percentage-of-completion inventory valuation
完工保证书 construction bond; performance bond
完工报单 work completed report
完工报告 completion report
完工成本 finished cost
完工承包保证书 completion bond
完工程度 stage of completion
完工程度毛利计算法 percentage-of-completion method
完工船 completed ship
完工的地面板 finished floor
完工概率曲线 probability completion curve
完工工程 closed out project
完工工程成本 cost of completed works
完工工艺 workmanship
完工合格证书 certification of completion
完工后测量 post works survey
完工后尺寸 finish size
完工后的服务 post-completion service
完工后附加工程 post-completion service
完工后审计 post completion audit
完工后再插入某种东西 cut in parts after completion
完工计量 final measurement
完工计量时间 period of final measurement
完工检查 final inspection; finish turn inspection
完工件 finished pieces
完工建筑 completed construction
完工进度计算法 degree of completion method
完工卡片 end-of-job card
完工量 quantity finished
完工留量 allowance for finish
完工面 finished surface
完工契约 construction bond
完工前检查清单 inspection list; punch list
完工切削 finishing cut
完工日期 completion date; date of completion; time of completion; finish date
完工(施工)保证书 completion bond
完工时的修整工作 finishing operation
完工时间 throughput time; time of finishing work; period for completion
完工事件 completion event
完工退火 finish anneal
完工温度 finishing temperature
完工物料 finishing material
完工项目清单 completion list
完工者 finisher
完工证书 certificate of completion
完好车辆 vehicle in running order; vehicle in working order
完好的 fully good; in good condition; sound; undamaged
完好的模板 perfect form
完好的质量 sound quality
完好货物 sound cargo; sound goods
完好货物到达时净价 net arrived sound value
完好货物到达时净值 net arrived sound value
完好货物的价格 sound market value
完好交货 delivered sound
完好晶体 well-defined crystal
完好台日数 number of machine-days in good condition
完好卫星 health satellite
完好性检验 <桩工的> integrity test(ing)
完好证明书 safety certificate
完好状态 health state; serviceable condition
完合平衡 complete equilibrium
完建工程 completed project
完结 culminate; wind-up
完结交易 closed trade
完井 well completion
完井测试 testing after completion
完井方法 completion method
完井费 completion cost
完井工艺 well completion technology
完井过程 well completion process
完井技术 well completion technique
完井井深 completion depth
完井类型 classification of well completion
完井日期 completion well date
完井数据 data of well completion
完井系统 well completion system
完井液费 completion fluid cost
完孔日期 complete drilling date
完满的 full-bodied
完满立木度 full stocking
完满涂层 full coat
完满运行保险 completed operation insurance
完美晶体 perfect crystal
完美气体 perfect gas
完美溶液 perfect solution
完模标本 holotype
完全 altogether; completeness; down to the ground; hand and foot; through and through; wholly
完全奥氏体化 complete austenitizing
完全包裹 wraparound
完全饱和 wholly saturation
完全饱和土样 fully saturated sample
完全保护木材 full protection of wood
完全鲍姆巴赫日冕 complete Baumbach corona
完全贝塔函数 complete beta function
完全背书 endorsement in full
完全焙烧 complete roasting
完全闭塞 entirely shut
完全编组 complete grouping
完全变态 complete metamorphosis
完全变态的 holometabolic; holometabolous
完全变态类 holometabola
完全冰冻 complete freezing
完全冰封 complete ice coverage
完全冰结 complete freezing
完全博弈树 complete game tree
完全不变映射 complete invariant map
完全不活动的元素 silent element
完全不可能的事 blank impossibility
完全不连通闭集 totally disconnected closed set
完全不连通的 totally disconnected
完全不连通度量空间 totally disconnected metric space
完全不连通集 totally disconnected set
完全不连通紧群 totally disconnected compact groups
完全不连通空间 totally disconnected spaces
完全不连通群 totally disconnected group
完全不连通图 totally disconnected graph
完全不连续函数 totally discontinuous function
完全布格改正 complete Bouguer re-

duction
完全操作 complete operation
完全拆开 completely knocked down
完全掺气水流 fully aerated flow
完全沉陷 full subsidence
完全成本 general cost
完全成本法 absorption costing; all cost method; full costing
完全成本计算 full costing
完全承保 full coverage
完全程序 complete routine
完全充填 solid filling; solid stowing
完全出料法 perfect discharge
完全除尘 perfect dust collection
完全处理 advanced treatment; complete treatment; total treatment
完全处理系统 complete treatment system
完全处理装置 complete treatment plant
完全粗糙区 wholly rough zone
完全淬火 full hardening
完全代偿间歇 complete compensatory pause
完全贷款 straight loan
完全单调 completely monotonic
完全单位模矩阵 totally unimodular matrix
完全的 down-to-earth; entire; throughing; total; unreserved
完全的国际法主体 full subject of international law; perfect subject of international law
完全滴定曲线 complete titration curve
完全地下截水墙 positive underground cutoff
完全电控制 all-electric(al) operation
完全电离的等离子体 stripped plasma
完全电离气体 fully ionized gases
完全电子化的 all electronic
完全电子化点火系统 all-electronic ignition system
完全定义函数 completely specified function
完全独占 perfect monopoly
完全短路 dead short; dead-short circuit
完全对称 full symmetry
完全对流湖 holomictic lake
完全对偶单调 complete dual monotonic
完全多项式 complete multinomial
完全二叉树 complete binary tree
完全二次型方根法 complete quadric combination method; CQC [complete quadratic cube]
完全二次组合(法) complete quadratic combination
完全发生【地】hologenesis
完全反相关 perfect inverse-correlation
完全反向应力 completely reversed stress
完全反应 complete reaction
完全方向组 complete set of directions
完全防热 full heat protection
完全防渗墙 positive barrier to seepage
完全非弹性碰撞 completely inelastic collision; perfectly inelastic collision
完全肥料 complete fertilizer
完全废品 completely defective product
完全分解 complete decomposition
完全分离 complete separation
完全分离操作 dissociated operation
完全分水 complete diversion
完全分异岩体 completely differentiated intrusive body
完全焚烧 total incineration
完全风化 full exposure to weather
完全风化岩石 completely weathered rock
完全封闭的 fully enclosed; totally enclosed
完全弗罗因德佐剂 Freund's complete adjuvant
完全氟化的烷烃 fully fluorinated paraffin
完全符合标准的数据 full conforming data
完全符合设计要求的路基 true subgrade
完全辐射 perfect radiation
完全辐射体 complete radiator; full radiator; perfect radiator
完全付清 full payoff
完全干燥的 all dry; bone dry
完全刚塑性体 rigid-perfectly plastic body
完全格 complete lattice
完全隔音的房间 dead room
完全公开 full disclosure
完全公开原则 principle of full disclosure
完全固定 full fixity
完全固定的 fully fixed
完全固定的梁端 complete end restraint
完全固化 complete curing
完全固溶体 complete solid solution
完全故障 complete failure
完全关闭 complete closure
完全关税同盟 complete customs union
完全归纳法 complete induction
完全规范化决策 completely specified decision
完全规范化调和函数 fully normalized spheric(al) harmonics
完全国营贸易 full state trading
完全函数序列 complete function series
完全焊透 complete penetration
完全和永久丧失工作能力 total and permanent disability
完全黑体 perfect black body
完全烘干的 thoroughly air dried
完全烘干的木材 oven dry timber
完全弧烧法 total combustion method
完全互换 complete interchangeability
完全化学分析 complete chemical analysis
完全缓解 complete remission
完全回缩 complete retraction
完全混合 complete mixing; perfect mixing
完全混合槽式反应器 completely mixed tank reactor
完全混合池 completely mixed basin
完全混合反应器 complete-mix reactor; continuous flow stirred tank reactor
完全混合(方)法 complete mixing process
完全混合湖 holomictic lake
完全混合活性污泥 completely mixed activated sludge
完全混合活性污泥法 completely mixed activated sludge process; complete mixing activated sludge process
完全混合活性污泥系统 completely mixed activated sludge system; complete mixing activated sludge system
完全混合间歇式反应器 completely mixed batch reactor
完全混合搅拌池式反应器 completely mixed stirred-tank reactor
完全混合空系统 complete-mix air system
完全混合流 completely mixed flow
完全混合曝气 completely mixed aeration; complete mixing aeration
完全混合曝气池 completely mixed aeration basin
完全混合曝气塘 completely mixed aerated lagoon; completely mixed aeration lagoon
完全混合曝气系统 completely mixed aeration system; fully mixed aeration system
完全混合生物膜反应器 completely mixed biofilm reactor
完全混合室 completely mixed cell
完全混合水流 completely mixed flow
完全混合系统 complete mixing system
完全混合性污泥处理系统 completed mixing activated sludge treatment system
完全混合作用 holomixis
完全混流式反应器 completely mixed flow reactor
完全混溶性 complete miscibility
完全活性污泥氧化法 completely activated sludge oxidation process
完全积分【数】complete integral
完全加法族 completely additive family
完全加性集函数 totally additive set function
完全交换 complete exchange
完全胶结的 complete cemented
完全搅拌 complete mixing
完全接长 full splice
完全接地【电】dead earth; dead ground; solid earth
完全结束 finish all over
完全截水墙 complete cutoff; positive cut-off
完全进位【计】complete carry
完全浸渍 full impregnation
完全井 complete penetration well; full-penetrating well
完全净化 complete purification
完全净化水马力 purified hydraulic horsepower
完全净化钻速 purified bit speed
完全竞争 perfect competition; pure competition
完全竞争市场 perfect competition market
完全聚焦质谱计 perfect focusing mass spectrometer
完全聚焦质谱仪 perfect focusing mass spectrometer
完全喀斯特 holokarst
完全开启 complete opening
完全开拓的 fully developed
完全抗磁性 perfect diamagnetism
完全抗弯框架 complete moment-resisting frame
完全抗性 complete resistance
完全可加性的 totally additive
完全可靠的数据 full conforming data
完全可微分的 totally differentiable
完全可微函数 totally differentiable function
完全控制 complete control; positive control
完全控制进入 full control of access
完全矿化 permineralization
完全扩散面 perfectly diffusing plane; perfectly diffusing surface
完全老化 full-ageing
完全类 complete class
完全类质同象 complete isomorphism; perfect isomorphism
完全离解 complete dissociation
完全立方 perfect cube
完全连锁 complete linkage
完全连续的 totally continuous
完全流体 perfect fluid
完全垄断 complete monopoly; perfect monopoly; perfect oligopoly; pure monopoly
完全密封的环境 totally sealed environment
完全模糊变量 perfect fuzzy variable
完全膜生物反应器 full-scale membrane bioreactor
完全耐水胶合板 perfectly water-proofing plywood
完全拟合 perfect fit
完全拟环 complete near ring
完全凝固 thorough consolidation
完全配合 perennial fitting; perfect fitting
完全膨胀 complete expansion
完全膨胀循环 complete expansion cycle
完全平方 perfect square
完全平衡 complete equilibrium; perfect balance
完全平衡的 dead true
完全平面应变 complete plane strain
完全破产 dead broken; strong broke
完全破坏 eventual failure; total runway
完全破坏时强度 total failure strength
完全气化 complete gasification
完全气化过程 complete gasification process
完全气体 perfect gas
完全汽化 complete gasification
完全嵌套 completely nest
完全强化生物除磷工艺 full-scale enhanced biological phosphorus removal process
完全切变裂缝 complete shear crack
完全亲油 entirely oil-wet
完全情报期望值 expected value of perfect information
完全曲线 full curve
完全全缘的 quite entire
完全确定 complete determinate
完全确认 full confirmation
完全群 complete group
完全燃烧 ashing; burn-through; complete burning; complete combustion; economic combustion; perfect combustion; thorough burning; normal combustion <混合物的>
完全燃烧的 all-burnt
完全燃烧极限 complete combustion limit
完全燃烧装置 smoke consumer
完全热交换 complete heat exchange
完全溶合性 complete miscibility
完全溶解的 completely dissolved; fully dissolved
完全熔合 complete fuse
完全熔化(物) complete fusion
完全熔接 complete fusion
完全软化的黏[粘]土强度 fully softened strength of clay
完全润滑 complete lubrication
完全散射 perfect diffusion
完全丧失 bankruptcy
完全丧失劳动力 complete disability; total disability
完全色盲 achromatopsy
完全闪蒸 full flashing
完全烧结 tight burning
完全剩余系 complete residue system

W

完全失败 come to nothing;fall flat
完全失灵 through fault
完全失效 total failure
完全时 perfect time
完全时效硬化 quenching and age-hardening
完全试验 complete test;complete trial
完全适用的水 all suitable water
完全收缩 complete combustion;complete contraction; ultimate shrinkage
完全收缩孔(口) complete contraction orifice;orifice with full contraction
完全寿命表 complete life table
完全受拉 perfect tensioning
完全受拉伸 perfect stretching
完全受约束 perfect restraint
完全疏导 < 汇水面积内水的 > complete diversion
完全数 perfect number
完全双循环 complete double circulation
完全水处理 full-scale water treatment
完全水分析 complete water analysis
完全水化水泥 completely hydrated cement;fully hydrated cement
完全水跃 complete hydraulic jump; perfect hydraulic jump
完全四边形 complete quadrangle
完全四边形的 complete quadrilateral
完全四线形 complete quadrilateral
完全塑性 perfect plasticity
完全塑性的 perfectly plastic
完全塑性屈服 perfect plastic yield
完全酸碱成分 complete acid/base composition
完全随机 completely random
完全随机设计 completely random design;complete randomized design
完全损坏 complete failure
完全锁闭 complete locking;deadlocking;full locking
完全锁闭器 deadlock
完全锁口的 fully interlocking
完全太阴潮 perfect moon tide
完全弹塑性体 elastic-perfectly plastic body;perfectly elasto-plastic body
完全弹塑性体系 perfectly elastic plastic system; perfectly elasto-plastic system
完全弹塑性滞回线 completely elasto-plastic hysteresis
完全弹性 perfect elasticity
完全弹性材料 perfect elastic material
完全弹性冲击 perfect elastic impact
完全弹性的 perfectly elastic
完全弹性固体 ideal elastic solid;perfectly elastic solid
完全弹性体 perfectly elastic body; solid elastic body
完全特性化羧酸 well-characterized carboxylic acid
完全替代 perfect substitute
完全调节 full regulation
完全调节电站 full-time storage plant
完全调制 total modulation
完全停车 complete stop; dead halt; dead stop;drop-dead halt;full stop
完全停顿 at a standstill
完全停机 complete stop; dead halt; drop-dead halt
完全停止 complete stop; dead stop; positive stop
完全通风的 fully vented
完全同步 complete synchronism;perfect synchronism
完全透明体 perfect transmission body; perfect transmitting body
完全透射体 perfect transmission body; perfect transmitting body
完全图 complete graph;perfect graph
完全湍流 complete turbulence
完全退火 complete tempering; dead annealing;full annealing; full dead annealing;soft annealing
完全退火的 dead annealed
完全退缩 complete retraction
完全脱位 complete dislocation
完全脱叶 complete defoliation
完全位错 perfect dislocation
完全紊动 complete turbulence; fully developed turbulence
完全紊流 complete turbulence; fully developed turbulence;rough turbulent flow
完全稳定 complete stabilization;full fixity
完全稳定的 fully stable
完全稳定性 complete stability
完全无法使用的危险 risk of total condemnation
完全无风 dead calm;flat calm
完全无缺的 consummate
完全无水的 bone dry
完全无相关 completely no correlation
完全无序 complete of order
完全吸收 complete absorption
完全吸收体 perfect absorber
完全系 complete set
完全系列 complete series
完全下沉 full subsidence
完全线性群 full linear group
完全限定名 fully qualified name
完全相变退火 total case annealing
完全相关 complete correlation; perfect correlation
完全消耗的 completely consumed
完全消耗系数 complete consumption coefficient
完全消化 catapepsis
完全消失 complete obliteration
完全小区 whole plot
完全卸载 complete discharge
完全信息期望值 expected value of perfect information
完全形蜗壳 full spiral case
完全性 completeness;totality
完全性定理 completeness theorem
完全性断离 complete separation
完全性条件 holomictic condition
完全性约束条件 integrity constraint
完全需要系数 complete demand coefficient
完全压缩 complete compression
完全亚硝化 complete nitridation
完全淹没的 totally buried
完全堰 perfect weir
完全氧化 complete oxidation; total oxidation
完全氧化法 complete oxidation process;total oxidation process
完全液化 complete liquefaction
完全液体 ideal fluid;perfect liquefaction;perfect liquid
完全一致算法 complete unification algorithm
完全异构变化 complete isomeric change
完全抑制点 < 防腐液浓度的 > total inhibition point
完全溢流堰 perfect overfall; perfect overflow
完全硬化 complete curing
完全有界的 totally bounded
完全有界度量空间 totally bounded metric space
完全有界分布 totally bounded distribution
完全有界一致空间 totally bounded uniform space
完全有向图 complete-directed diagram;complete-directed graph
完全有序 complete ordering; perfect order
完全语言 full-language
完全预混式燃烧 pre-aerated combustion
完全预应力 perfect prestressing
完全原函数 complete integral; complete primitive function
完全原理图 complete schematic diagram
完全约束的非晶网 fully constrained non-crystalline network
完全约束的梁端 complete end restraint
完全运算 complete operation
完全再生 full regeneration
完全遭受风化 full exposure to weather
完全责任 full liability
完全债务 perfect obligation
完全展开 fully expand
完全展开流 fully developed flow
完全遮盖 complete hiding
完全真空 perfect vacuum
完全镇静钢料 < 经无气泡处理的 > fully killed steel material
完全蒸发 boil away
完全蒸发时的温度 final boiling point
完全正规空间 completely normal space
完全正交 complete orthogonal
完全正交规范集 closed orthonormal set;complete orthonormal set
完全正交集 complete orthogonal set
完全正交平面 completely perpendicular planes
完全正交系 complete orthogonal system
完全正则变换 completely canonic(al) transformation
完全正则空间 completely regular space;Tychonoff space
完全正则调和函数 fully normalized harmonics
完全指数模型 complete exponential model
完全致癌物 complete carcinogen
完全转导 complete transduction
完全准确 entirely accurate
完全自动的 complete automatic;fully automatic;purely automatic
完全自动化的 supermatic
完全自养脱氮 completely autotrophic nitrogen removal
完全组合 complete combination
完全佐剂 complete Freund's adjuvant
完全坐标 world coordinates
完善 finishing; finish off; sophistication
完善城市基础设施建设 perfect the construction of urban infrastructure
完善程度 degree of integrity
完善的 sophisticated
完善的材料 sophisticated material
完善的道路网 complex network of roads
完善的数学模型 sophisticated mathematical model
完善的塑性理论 perfectly plastic theory
完善的预加应力 perfect prestressing
完善的预应力 perfect prestress
完善度 sophistication
完善化 sophistication
完善计划 perfect one's plan
完善胶片 sound rubber
完善塑性机理 perfectly plastic mechanism
完善信息 perfect information
完善信息期望值 expected value of perfect information
完善性维护 perfective maintenance
完熟期 full-ripe stage
完税 pay tax
完税单 duty-paid certificate
完税后交货(价) delivered after duty paid
完税后买方关栈交货价格 ex buyer's bonded warehouse duty paid
完税货价 duty-paid price;price duty paid
完税价格 dutiable price;dutiable value;duty-paid price;price for tax assessment
完税价值 dutiable value
完税进口报关单 entry for home use
完税凭证 duty-paid proof; tax payment receipt
完税收据 duty receipt
完税证 certificate of tax payment
完税证明 duty-paid proof
完形倾向 law of pregnancy
完整 completion; holonomy; roundness
完整包装 integral packaging
完整背书 endorsement in full;full endorsement
完整补偿筏基础 fully compensated raft foundation
完整操作 complete operation
完整承压井 completely penetrating artesian well
完整程度 degree of integrity;level of integrity
完整的 full-scale;intact;self-contained; unabridged; unbroken; undamaged; undivided
完整的材料 unbroken material; uncrushed material
完整的产品系列 comprehensive range
完整的产权 clear title
完整的东西 integer
完整的钢轨 unbroken rail
完整的(管道)施工队【给】complete spread
完整的计划 complete plan
完整的生态系统 intact ecosystem
完整的体系 rounded system
完整的条缝送风口 integral slot diffuser
完整的资料 complete information
完整的自然保护区 strict nature reserves
完整地平线 round horizon
完整方形波信号 composite square-wave signal
完整浮力 intact buoyancy
完整工业体系 all-round industrial system;comprehensive industrial system
完整构架 perfect frame(work);simple frame(work)
完整光滑面理论 < 晶体生长的 > Kossel theory
完整呼号 full call letters
完整回收 physical recovery
完整货物 clean cargo
完整记录 unit record
完整剪切强度 intact shear strength
完整检验 complete survey
完整晶体 perfect crystal
完整井 completely penetrated well; complete penetration well; fully penetrating well; well of complete penetration
完整空间 holonomic space
完整块裂体结构 complete block fracture texture

完整框架 perfect frame
完整(例行)程序 complete routine
完整砾石格架 intact gravel framework
完整流程图 complete flowchart
完整面 finished face
完整黏[粘]土 intact clay
完整耦合 unity coupling
完整排水渠 drainage canal excavated to aquifer
完整铅封 intact lead seal;unbroken seal;undamaged seal
完整潜水井 completely penetrating gravity well
完整墙 unbraced wall;unbroken wall
完整侵蚀循环 uninterrupted cycle of erosion
完整倾向 pregnance [pregnancy]
完整曲线 unbroken curve
完整群 holonomy group
完整设备 complete system
完整食 complete eclipse
完整水舌 complete nappe
完整条款 entire agreement clause
完整同步信号 composite synchronization signal
完整图 complete graph
完整微分 exact differential; perfect differential
完整文件 self contained document
完整稳性 intact stability
完整无缺的 intact
完整无损的 intact;undamaged
完整无损价值 undamaged value
完整系列 complete series
完整系统 complete system;holonomic system;overall system
完整线路 complete line
完整星表 complete catalogue
完整形蜗壳 full scroll(case);full spiral case
完整性 completeness;degree of integrity; integrality; integrity; solidity; wholeness
完整性法则 law of pregnancy
完整性控制 integrity control
完整性试验 integrity test(ing)
完整性系数 coefficient of sound degree
完整学习 global learning
完整岩层 unbroken formation
完整岩桥 intact rock bridge
完整岩石 block rock; intact rock; sound rock;unaltered rock
完整样品 intact sample
完整叶 intact leave
完整油膜润滑 complete lubrication; viscous lubrication
完整约束 holonomic constraints
完整运算 complete operation
完整运行 complete operation;complete running
完整炸弹 bomb complete round
完整装饰品 self-contained ornament
完整状态 good working condition
完整子波 entire wavelet
完整子系统 integrity subsystem
完整租借 entire tenancy
完整组件 complete package
完整钻样 well-drill(ed) sample
完钻井深 drilling finished well depth
完钻日期 complete dull date

玩

玩忽职守 dereliction;misprision
玩忽职守的 derelict
玩忽职务者 derelict
玩具店 toy shop
玩偶 doll
玩物商店 hobby shop

顽

顽磁 magnetic remanence; remanence
顽磁性 magnetic retentivity;retentivity
顽固 obstinacy
顽辉石 enstatine
顽辉石陨星物质 chladnite
顽火辉石 enstatite;protobastite
顽火辉石橄榄岩 enstatite peridotite
顽火辉石苦橄岩 enstatite picrite
顽火辉石球粒陨石 enstatite chondrite
顽火辉石无球粒陨石 audrite
顽火榴辉岩 newlandite
顽火石 enstatine;enstatite
顽火透辉岩 marchite
顽火无球粒陨石 aubrite;bustite
顽火岩 enstatite
顽劣峰 rouge peak
顽强 tenacity
顽强的 resistant;unyielding
顽石坝 boulder dam

烷

烷芳基胺【化】alkarylamine
烷芳基磺酸盐 alkyl aryl sulfonate
烷化 alkylation
烷化剂 alkylating agent
烷基 alkyl;alkyl group;alkyl radical;amyl
烷基胺 alkylamine
烷基苯 alkyl benzene;dodecyl
烷基苯磺酸 alkyl benzene sulfonic acid
烷基苯磺酸盐 alkyl-benzene sulphonate;ABS alkyl benzene sulfonate <一种工程塑料>
烷基苯磺酸酯<一种合成洗涤剂中产生泡沫的表面活性剂> alkyl-benzene-sulfonate
烷基苯氧基聚氧乙烯醚(乙)醇 alkyl phenoxy poly(ethyleneoxy)ethanol
烷基醇 alkylol
烷基芳基聚醚醇 alkyl arylpolyether alcohol
烷基芳香烃 alkyl aromatics
烷基酚环氧乙烷聚合物 alkyl phenoxy poly(ethyleneoxy)ethanol
烷基酚醛树脂 alkyl phenolic resin;alkylphenol resin
烷基氟氯硅烷 alkyl chlorofluorosilane
烷基汞 alkyl mercury
烷基汞盐 alkyl mercuric salt
烷基硅烷醇 alkylsilanol
烷基化 alkanization;alkylation
烷基化合物 alkyl compound
烷基化焦油 alkylated tar
烷基化物 alkylate
烷基化油蒸馏残液 alkylate bottoms; alkylate polymer
烷基磺酸钠 sodium alkane sulfonate
烷基磺酸酯 alkylsulfonate
烷基季铵盐 alkyl quaternary ammonium salts
烷基磷酸盐 alkylphosphate
烷基膦酸烷基酯 alkyl alkanephosphonate;alkyl alkylphosphonate
烷基硫酸钠 sodium alkylsulfate
烷基硫酸酯 alkyl sulfate
烷基氯硅烷 alkyl chlorosilane
烷基醚 alkyl ether
烷基醚化树脂 alkyl-etherified resin
烷基萘磺酸酯 alkylnaphthalene sulphonate
烷基铅 alkyl lead
烷基取代 alkylation
烷基四氢化菲 alkyltetrahydronaphyhalene
烷基烷氧基硅烷 alkylalkoxy silane
烷基锌 zinc alkyl
烷烃 alkane
烷烃的 paraffinic
烷烃磺酸钠 sodium alkane sulfonate
烷烃基 paraffin(e)base
烷烃溶剂 paraffinic solvent
烷系 methane series
烷氧基硅烷 organoalkoxy silane;organoaroxy silane;silicon alkoxide
烷氧基硅烷涂料 silicon alkoxide coating
烷氧基化 alkoxylate
烷氧基钛 titanium alkoxide
烷氧基酮 ketone ether
烷氧碳酰 alkoxycarbonyl
烷氧羰基 carboalkoxy
烷属 paraffin(e)base
烷属烃 paraffin(e)hydrocarbon
烷族的 paraffinic

挽

挽绑锚链 muzzle anchor
挽车马 draught horse;harness horse
挽钩 adjustable dog hook;boat hook;gaff
挽钩牵引力 drawbar pull
挽钩牵引马力 drawbar(horse)power
挽回 retrieve
挽救 rescue;save
挽救海洋运动 save-our-sea movement
挽缆插栓 belaying pin
挽缆桩 warping bollard
挽牢 make fast
挽牢头/尾缆和倒缆 make fast the head/stern line and the spring
挽力 drafting ability;tractive effort
挽留 detain
挽索插栓架 belaying pin rack;pin rail rack
挽索木挂钩 sling cleat
挽头 short leg
挽在桩上 bitt

晚

晚阿尔卑斯期地槽【地】late Alpine geosyncline
晚奥陶世【地】Late Ordovician epoch
晚奥陶世绝灭【地】Late Ordovician extinction
晚白垩世【地】Late Cretaceous epoch;Upper Cretaceous
晚白垩世海退【地】Late Cretaceous regression
晚白垩世绝灭【地】Late Cretaceous extinction
晚白垩世气候分带 Late Cretaceous climatic zonation
晚班 night shift
晚报 evening paper
晚冰期 late glacial epoch
晚冰期地壳运动 late glacial period crust movement
晚材 autumn wood;late wood;summer wood
晚材层 zone of late wood
晚材带 zone of late wood
晚餐 dinner
晚潮 evening tide
晚成雏的 altricial
晚成的 serotinous
晚成年期 late mature
晚成年期谷 late mature valley
晚成熟期河流 late maturity river
晚春 deep spring
晚得利亚斯冰阶【地】youngest Dryas stage
晚地槽阶段 late-geosyncline stage
晚第三纪【地】Neocene(period)
晚点 behind the schedule;lost time
晚点表示 delay time indication
晚点到达 arrive late;train arrived behind scheduled time【铁】
晚点发车 start late;train left behind booked time; train left behind scheduled time
晚点列车 delayed train
晚二叠世【地】Late Permian epoch
晚二叠世海退【地】Late Permian regression
晚二叠世绝灭【地】Late Permian extinction
晚发矽肺 delayed silicosis
晚发效应 late effect
晚高峰<下班时的高峰交通时间> P.M. peak;evening peak
晚高峰时间 evening peak hours;P.M. hour
晚高峰小时 evening peak hours;evening rush hours;P.M. hour
晚更新世【地】Epipleistocene; Late Pleistocene epoch; Upper Pleistocene
晚构造期花岗岩 late-kinematic granite
晚古生代 Neopaleozoic
晚果松 pond pine
晚海西期地槽 late Hercynian geosyncline
晚寒武纪【地】Neo-Cambrian period
晚寒武纪的 Neo-Cambrian
晚寒武世【地】Late Cambrian epoch
晚寒武世绝灭 Late Cambrian extinction
晚季作物 after-crop
晚加里东期地槽【地】Late Caledonian geosyncline
晚间负荷 night-time load(ed)
晚间照明 night light(ing)
晚密西西比世【地】Upper Mississippian
晚木 autumn timber
晚木材 late wood
晚泥盆世【地】Late Devonian epoch
晚期 advanced stage;late period;late stage
晚期成岩作用 late diagenesis
晚期成岩作用阶段 phyllomorphic stage
晚期法兰克式建筑 late-Frankish architecture
晚期费用 terminal expenses
晚期哥特式教堂 late Gothic hall church
晚期几何形窗花格 late geometric(al)lattice
晚期几何形体式(装饰)late geometric(al)
晚期罗马风格 late Roman
晚期罗马式建筑 late Romanesque
晚期渗滤液 mature leachate
晚期生油说【地】late origin theory of petroleum
晚期效应 late effect
晚期新华夏系【地】Late Neocathaysian system
晚期新艺术风格 late art nouveau
晚期新艺术运动 late art nouveau
晚期修剪 late pruning
晚期岩浆矿床 late magmatic mineral deposit;late magmatic ore deposit
晚期艺术作品 late opus
晚期用火 late burning

晚期支付 late commitment
晚秋 <英> back end
晚秋灌溉 late-fall irrigation
晚秋晴热天 all hallown summer
晚秋作物 late autumn crop; late fall crop
晚三叠世【地】 Late Triassic epoch
晚三叠世绝灭【地】 Late Triassic extinction
晚三叠世气候分带 Late Triassic climatic zonation
晚上捕鱼 owling
晚上新世【地】 Post-Pliocene
晚石碳世【地】 Late Carboniferous epoch
晚熟品种 late variety
晚熟期河流 late maturity river
晚霜 late frost; spring frost
晚霜害 late frost damage
晚维斯康辛【地】 late Wisconsin
晚霞 after glow
晚香玉【植】 tuberose
晚燕山期地槽【地】 late Yanshanian geosyncline
晚燕山亚旋回【地】 Late Yanshanian subcycle
晚元古代冰期【地】 Late Proterozoic glacial stage
晚元古代海浸【地】 Late Proterozoic transgression
晚元古代绝灭【地】 Late Proterozoic extinction
晚造山期的【地】 late-orogenic
晚造山期盆地【地】 late-orogenic basin
晚造山相【地】 late-orogenic phase
晚壮年地形 subdued forms

晥

晥南构造结【地】 Southern Wan tectonic knot

碗

碗厨 dresser; buffet

碗橱开关 cupboard catch
碗橱扣栓 cupboard latch
碗橱门闩 cupboard catch
碗碟柜 chiffonier
碗碟加温器 plate warmer
碗碟架 chine cupboard
碗碟洗涤室 scullery
碗碟洗涤台 scullery table
碗碟洗涤装置 scullery
碗柜 cuddy; cupboard; dish cabinet; kitchen cupboard
碗柜或壁龛 <教堂圣台上的> armarium
碗辊磨 bowl mill
碗模 <焊管用> welding bell
碗磨 bowl mill
碗式离心机 bowl-type centrifuge
碗头挂板 socket-clevis eye
碗形 bowl shape
碗形边型器 <无槽法生产平板玻璃> edge bowl
碗形铲刀的轻土料推土机 light material bowl-dozer
碗形超高 <用于高速弯道的> banked bowl
碗形磁铁 bowl-shaped magnet
碗形灯 bowl lamp
碗形地面 bowled floor
碗形绝缘子 bell-shaped insulator
碗形密封圈 cup seal
碗形磨 bowl-type mill
碗形平行线 <编组场> bowl track
碗形砂轮 flaring cup grinding wheel; tapered cup grinding wheel
碗形深碟 bacile
碗形填密法 cup packing
碗形推土板 bowldozer
碗形柱头 bowl capital
碗形柱头装饰 bowl arrangement; bowl capital
碗状沉降 bowl-shaped settlement
碗状大冰块 corrie ice
碗状构造 bowl arrangement
碗状排列(构造) bowl arrangement
碗状洼地 bowl depression land
碗状物 bowl

万

万 myriad

万测仪 pantometer
万吨采掘比 ratio of development meters to ten thousand tons of mined ore
万吨每年 myriatonne per year
万吨每日 myriatonne per day
万分表 ten thousandth micrometer
万分尺 ten thousandth micrometer
万格盘 checkerboard
万公尺 myriameter [myriametre]
万公尺的 myriametric(al)
万公升 myrialiter [myrialitre]
万拱 long arm
万古 aeon
万国博览会 World's Fair
万国法 law of nations
万国劳动协会 International Labour Association
万国邮政公约 Universal Convention of Post
万国邮政联盟 Universal Postal Union
万花筒 kaleidoscope; stroboscope
万金油 nostrum
万景画 myriorama
万克 myriagram(me)
万里长城 Chinese Wall; the Great Wall
万立方米 myriastere
万立方米每年 myriastere per year
万利板 woodwool board; woodwool slab
万磷铀矿 vanmeersscheite
万米 myriameter [myriametre]
万米波 myriameter wave
万能 multiapplication
万能 V 形弯曲模 universal V-die
万能鞍架 universal saddle
万能拔桩机 universal pulling machine
万能板材 universal plate
万能保护涂层材料 all-purpose protective coating(material)
万能笔 magic ink
万能表 universal electric(al) meter
万能补偿器 universal compensator
万能材料试验机 universal material testing machine
万能操作台 universal operation table
万能测齿仪 universal gear tester
万能测定显微镜 universal measuring microscope
万能测角器 combination set
万能测角仪 pantometer
万能测量显微镜 universal measuring machine
万能测试机 multiple-purpose tester
万能测试器 universal tester
万能测试仪 universal tester
万能叉车 universal fork-lift
万能叉式提升机 universal fork-lift
万能插口 consent
万能拆卸机 universal puller
万能拆装器 universal claw
万能铲齿车床 universal relieving lathe
万能铲床 universal slotting machine
万能车床 universal lathe
万能车刀 universal turning tool
万能尺 isograph
万能齿轮试验机 universal gear testing machine
万能传送装卸机 universal transported loader
万能打桩机 universal pile driving plant
万能刀 universal knife
万能道尺 combined track ga(u)ge and level; universal track ga(u)ge
万能的 all-purpose; all round all; general duty; general service; general utility; universal; varsal; versatile
万能底漆 all-purpose primer; universal primer
万能电表 avometer; multimeter; universal electric(al) meter; volometer; voltimeter; universal meter
万能电动工具 universal electric(al) tool
万能顶锻焊机 universal upsetting welder
万能定标器 multiscaler
万能分度头 univariate dividing head; universal dividing head; universal indexing head
万能分度中心 universal index center [centre]
万能分线规 universal dividers
万能封罐机 universal seamer
万能缝焊机 universal seam welder
万能浮动工具夹具 universal floating tool holder
万能辅助探照灯 universal supplementary search light
万能附件 universal attachment
万能杆件 fabricated universal steel members; universal member; universal rod body
万能钢板 universal mill plate; universal steel plate
万能钢杆件 fabricated universal steel members
万能杠杆 universal lever
万能高速冷冻离心机 universal high speed refrigerated centrifuge
万能工程车 <美军的> universal engineer tractor
万能工具 combination square; multipurpose instrument tool
万能工具机 all-purpose machine
万能工具磨床 universal cutter and tool grinder; universal cutter grinder
万能工具台 universal fixture
万能工具铣床 universal tool miller; universal tool milling machine
万能工具显微镜 universal tool maker's microscope
万能工作机械 universal machine
万能工作台 univariate table; universal table; universal worktable
万能供电的 all-mains
万能刮刀 universal scraper
万能挂钩 universal drawbar
万能管接头 universal(pipe) joint
万能管形钻 universal tube drill
万能轨距尺 universal track ga(u)ge
万能滚焊机 universal seam welder
万能滚节机 universal mill
万能后支架 universal back rest
万能弧形衬砖 universal ladle brick
万能虎钳 universal vise [vice]
万能花盘 universal face plate
万能滑动切片机 universal sliding microtome
万能滑轮 universal pulley
万能绘图机 universal drafting machine; universal plotting instrument
万能绘图仪 universal plotting instrument
万能活动量角规 universal bevel protractor
万能活动量角器 universal bevel protractor
万能活塞压钳 universal piston vice and press
万能机床 combination lathe; universal machine
万能机械手 general purpose manipulator; universal manipulator
万能夹持器 universal clamp
万能夹具(台) universal chuck; universal fixture
万能夹盘 universal chuck
万能夹头 universal chuck
万能剪冲床 universal shearing and punching machine
万能剪钳 universal cutting pliers
万能剪切机 multipurpose shearing machine; universal shears
万能键 universal key
万能胶 all-purpose adhesive; almighty adhesive; universal adhesive; universal glue
万能胶黏[粘]剂 universal cementing agent
万能角尺 bevel protractor
万能角度尺 universal protractor
万能角规 universal angle block
万能接头 spider trunnion; universal contact
万能截煤机 longwall-shortwall coal cutter; universal coal cutter
万能精密切片机 universal precision microtome
万能锯 universal saw
万能锯断机 universal sawing machine
万能锯台 universal saw bench; universal saw table
万能卡 master card
万能卡盘 universal chuck
万能开关 universal switch
万能可逆式粗轧机座 universal reversing roughing mill
万能拉出器 univariate puller; universal puller
万能拉力试验机 universal tensile testing machine
万能犁体 turf bottom
万能立铣装置 universal spindle milling attachment
万能连接杆 universal link
万能连接器 combination connector
万能联合冲剪机 universal combined punching and shearing machine
万能联轴器 universal coupling
万能两脚规 universal compass
万能量规 universal ga(u)ge
万能量角器 universal bevel protractor
万能龙头 universal cock
万能螺帽扳手 come-along
万能螺丝把 universal screw-key
万能螺纹磨床 universal thread grinder; universal thread grinding machine
万能螺旋扳手 monkey spanner; universal screw wrench
万能螺旋钻式回填机 auger backfiller
万能锚碇 all-round anchorage
万能磨床 universal grinder; universal grinding machine
万能磨床附件 universal grinding attachment
万能磨粉机 universal mill
万能磨机 universal grinder
万能磨削 universal grinding
万能木材加工机 general joiner
万能木工车床 universal wood turning

W

lathe
万能木工机 universal woodworking machine
万能内圆磨床 universal internal grinder
万能黏[粘]合剂 all-purpose adhesive
万能黏[粘]结剂 all-purpose adhesive; almighty adhesive
万能牛头刨床 universal shaper; universal shaping machine
万能扭力试验机 universal torsion tester
万能耦合器 universal coupling
万能刨 combination plane; universal plane
万能刨床 universal planer
万能喷嘴 all-purpose nozzle
万能疲劳试验机 universal fatigue testing machine
万能平面规 universal surface ga(u)ge
万能起重机 univariate crane; universal crane
万能牵引车 do-all tractor
万能牵引滑轮 universal extension pulley
万能钳(子) combination cutting and twisting pliers; combination pliers; engineer's pliers; universal tongs; universal pliers
万能强度试验机 universal strength tester; universal strength testing machine
万能强制器 universal compulsator
万能倾斜台 universal tilting table
万能曲尺 universal bevel
万能取景器 multifocus viewfinder; universal viewfinder; varifocal viewfinder
万能润滑油 multipurpose lubricant
万能润滑脂 multipurpose grease
万能塞规 combination standard ga(u)ge
万能三角板 universal set-square
万能摄影操纵仪 universal camera control system
万能升降车 forklifter
万能升降台式铣床 universal head and column miller; universal knee-and-column miller
万能式板坯初轧机 universal slabbing mill
万能式拆卸器 universal puller
万能式粗轧机座 universal roughing mill
万能式干燥机 universal drier [dryer]
万能式钢梁轧机铣床 universal beam mill
万能式轨梁轧机 universal structural mill
万能式精轧机座 universal finishing stand
万能式轧机机座 universal mill stand
万能式装载机 all-purpose loader
万能式钻模 universal drill jig
万能试验机 multiple-purpose tester; universal tester; universal test(ing) machine
万能试验仪 universal tester
万能试验指示器 universal test indicator
万能手柄 universal handle
万能手术台 universal operation table
万能手术台附件 accessories of universal operation table
万能手摇切片机 universal rotary microtome
万能输送机 all-round conveyer [conveyor]
万能数字记录机 universal digital recording machine
万能台 universal stand
万能台式铣床 universal bench mill
万能镗床 universal borer; universal boring machine
万能镗刀盘 universal boring head
万能镗钻床 universal boring and drilling machine
万能镗钻两用机床 universal boring and drilling machine
万能套筒扳手 universal socket wrench
万能提升机 universal lifting gear
万能调整器 universal compressor
万能调整仪 universal setting ga(u)ge
万能推土机 angledozer; side dozer
万能拖拉机 all-purpose tractor; carryall tractor; general purpose tractor; multipurpose tractor; universal tractor; utility tractor
万能挖掘机 universal excavator
万能挖土机 universal excavator
万能外圆磨床 universal grinding machine external
万能弯管机 univariate bender; universal bender
万能维护车 universal handling dolly
万能吸附剂 all-round absorbent
万能铣床 univariate mill; universal mill; universal milling machine
万能铣镗机床 universal milling and boring machines
万能铣削装置 universal milling attachment
万能显微镜 universal microscope
万能显影液 universal developer
万能型灯 porcelain enamel(l)ed standard dome lamp
万能型电子显微镜 universal electron microscope
万能旋转台 universal stage
万能研究显微镜 universal research microscope
万能研磨机 universal grinding machine
万能摇臂钻床 universal radial drilling machine; universal radical drill
万能仪 universal apparatus
万能仪表 all-purpose instrument
万能用的 general purpose
万能圆锯 universal circular saw
万能钥匙 grandmaster key; master key; passkey; pass-partout; skeleton key
万能钥匙锁 master-keyed lock
万能钥匙系列锁 master-keyed lock
万能钥匙型 master-keying
万能轧机 universal mill
万能轧机轧出的工字钢梁 universal mill beam
万能轧制 universal rolling
万能照像显微镜 universal camera microscope
万能真空吸尘器 all-purpose vacuum cleaner
万能支架 universal stand
万能支腿 universal leg
万能指示剂 universal indicator
万能制轮机 universal wheel wright machine
万能制图机 universal drafting machine
万能中心磨床 universal center-type grinding machine
万能终接器 <自动电话> toll and local combination connector; combination connector
万能转换开关 master changeover switch
万能桩架 versatile pile frame
万能装料机 all-purpose loader
万能装卸车 straddle carrier
万能装卸机 cherry picker; universal crane; universal forklift truck; universal loader
万能装岩机 all-purpose loader
万能自动测试设备 versatile automatic test equipment
万能组合式牵引床 universal overhead traction frame assembly
万能钻床 full universal drill
万能钻孔测斜仪 versatile borehole surveying instrument
万尼格型浮选机 Weining flo(a)tation cell
万年冰 perpetual ice
万年积雪 firn cover; perpetual snow
万年历 perpetual calendar
万年青 evergreen
万年雪 firn(snow); neve snow; perpetual snow
万年雪冰 firn ice; ice firn
万年雪带 band of firn
万年雪区 firn area
万年雪线【气】line of firn
万神殿式圆屋顶 Pantheon dome
万神庙 Pantheon
万升 myrialiter [myrialitre]
万圣之圣 Holy of the Holies
万寿菊 French marigold
万泰堪古城 <泰国> the ancient city of Wiang Ta Kan
万瓦 myriawatt
万维网 web; www [world wide web]
万维子网 subinternet
万位 myriabit
万位存储器 myriabit memory; myriabit storage
万向操纵杆 universal lever
万向传动轴 universal drive shaft; universal driving shaft
万向传动轴护罩 universal drive shaft guard
万向导缆器 universal fairleader
万向的 all-around; general purpose; universal
万向吊环 pivot bracket
万向阀 univariate valve; universal valve
万向反射镜 cardanic mirror
万向杆 universal rod
万向挂器 universal hanger
万向关节 cardan link
万向关节托架 cardan bracket
万向(关)节轴 cardan axis
万向虎钳 toolmaker's vise; universal vise [vice]
万向环 cardan ring
万向火箭发动机 gimbaled rocket
万向架 gimbal
万向架位置传感器 gimbal pick-up
万向架支承的发动机 gimbal-mounted engine
万向架支座火箭发动机 gimbaling rocket motor
万向接合器 universal coupling
万向接矢支轴 cardan fulcrum
万向接头 ball coupling; ball joint; cardan; cardanic suspension; cardan joint; free joint; gimbal [gymbal]; gimbal joint; Hooke's coupling; knuckle; swivel(l)ing head; universal connector; universal coupling; universal joint; universal joint body
万向接头叉头 universal yoke
万向接头插销 knuckle pin
万向接头传动 cardan drive; cardan gear
万向接头传动装置 cardan gear
万向接头关节 universal-joint knuckle
万向接头润滑脂 universal-joint lubricant
万向接头伸缩轴 telescopic(al) multiple par shaft
万向接头十字轴 knuckle centre
万向接头套管轴 telescopic(al) multiple par shaft
万向接头轴承 univariate joint bearing; universal-joint bearing; universal-joint cross bearing
万向接头装置 universal-joint assembly
万向节 cardan; cardan joint; cardan link; free joint; gimbal joint; universal joint
万向节叉 universal-joint fork
万向节衬圈 universal-joint knuckle retainer
万向节衬套 universal-joint bushing
万向节衬套盖 universal-joint bushing plate
万向节衬套及球十字架 universal-joint bushing and ring spider
万向节传动 joint drive; universal-joint drive [driving]
万向节传动轴 cardan shaft
万向节传力凸缘 universal-joint transmission flange
万向节轭 universal-joint yoke
万向节耳轴 universal-joint trunnion
万向节耳轴座 universal-joint trunnion block
万向节盖 universal-joint cap
万向节管 universal pipe joint
万向节滚柱 universal-joint roller
万向节护罩 universal-joint boot
万向节花键轴 universal-joint splined shaft
万向节滑叉 universal-joint slip yoke
万向节机构 universal-joint mechanism
万向节加油嘴 universal-joint lubricating nipple
万向节壳 universal-joint casing
万向节壳盖 universal-joint housing cap
万向节壳弹簧 universal-joint casing spring
万向节壳体 universal-joint housing
万向节联接轴 universally jointed axle
万向节流系统 gimbal throttle system
万向节球 universal ball; universal-joint ball; universal joint
万向节十字叉衬环 universal-joint spider bushing ring
万向节十字架 spider; center [centre] piece
万向节十字头 cross connecting piece; univariate spider; universal-joint crossing; universal trunnion crossing; universal joint spider
万向节十字轴 center crossing; spider center [centre]; trunnion; univariate joint cross; universal-joint cross trunnion
万向节十字轴溢油阀 universal-joint cross relief valve
万向节十字轴油嘴 universal-joint cross grease fitting
万向节式测功仪 joint dynamometer
万向节弹簧圈装卸钳 universal-joint snap ring pliers
万向节头 cross pin type joint; gimbal suspension; univariate joint
万向节凸缘 universal-joint flange
万向节凸缘轭 universal-joint flange yoke
万向节系统 gimbal system
万向节销 gudgeon of universal joint; pin of universal joint
万向节压力润滑器 universal-joint pressure lubricator
万向节油封帽环 universal-joint oil

W

seal cap ring
万向节针式轴承 universal-joint needle bearing
万向节支枢 gimbal pivot
万向节中心环 universal-joint center [centre] ring
万向节中心球 universal-joint centering sphere
万向节中心球销 universal-joint center [centre] ball pin
万向节中心销 universal-joint center [centre] pin
万向节轴 universal-joint shaft;cardan shaft
万向节轴承 univariate joint bearing; universal-joint bearing
万向节轴承衬 universal-joint bearing gasket
万向节轴承衬套 universal-joint bearing bushing
万向节爪 universal-joint jaw
万向节座 universal socket
万向连接 universal connection
万向连接杆 universal link
万向连接杆传动 universal link drive
万向连接轴 universal-joint spindle
万向连轴节 Hooke's joint
万向联管节 universal pipe joint
万向联轴节 ball coupling; cardan joint; Hook's joint; Hooke's coupling; Hooke's universal joint; universal coupling
万向联轴节十字头 universal joint spider;universal-joint cross
万向联轴器 ball coupling;cardan joint; universal coupling
万向龙头 universal cock
万向喷头 gimbaled nozzle
万向喷嘴 gimbaled nozzle
万向平衡式吊架组合 gimbaled hanger assembly
万向气钻 universal drill
万向倾斜计 universal inclinometer
万向球度 universal ball joint
万向球关节 universal ball joint
万向球铰接 universal ball joint
万向球铰螺杆 ball jointed screw
万向十字接头 joint cross
万向伺服电动机 gimbal servo motor
万向套节 universal socket
万向弯管机 universal bender
万向吸附剂 all-round absorbent
万向悬挂架 cardanic suspension;cardan mounting; gimbal suspension; gimbal mount
万向悬挂式动力装置 gimbaled power plant
万向悬置 cardanic suspension
万向摇臂钻床 universal rocker arm drilling machine
万向仪 direction instrument
万向支架 gimbal
万向支架连接 gimbaling
万向支架自由度 gimbal freedom
万向轴 cardan; cardan axle; multiple direction shaft;versatile spindle
万向轴传动转盘 shaft-driven rotary
万向轴管 torque tube
万向转镜 articulated mirror
万向转塔铣床 milling machine with universal turret head machine
万向转台 universal angle plate
万向转轴 articulated shaft; universal shaft
万象 <老挝首都> Vientiane
万一的 eventual
万亿 <美国、法国相当于 10^{12},英国相当于 10^{18}> trillion;billion <英>
万亿分之几 parts per trillion
万亿分之一 part per trillion
万用 general service;general utility
万用表 ampere-volt-ohm meter; avometer; circuit tester; multimeter; circuit tester
万用测试器 multiple-purpose tester
万用成型片固定器 universal matrix retainer
万用电表 avometer; circuit tester; multiple meter; universal electric-(al) meter;universal meter
万用电桥 universal bridge
万用附件 universal attachment
万用焊接吹管 universal blowpipe
万用记录器 multiclass sender
万用夹盘 universal chuck
万用键控器 general purpose manipulator
万用控制器 universal controller
万用连接器 thrift mate
万用滤波器 universal filter
万用示波器 multielement oscillograph
万用试纸 universal pH test paper
万用数字计算机 universal digital computer
万用旋转台 Fedorov stage; universal stage;U stage
万用仪表 multipurpose instrument
万用支撑杆 universal support rod
万有 creation
万有斥力 gravitational repulsion
万有覆盖面 universal covering surface
万有集 universal set
万有紧群 universal compact group
万有曲线 universal curve
万有系数定理 universal coefficient theorem
万有引力 gravitational attraction; mass attraction;universal gravitation
万有引力常数 constant of universal gravitation; gravitational constant; universal gravitation constant
万有引力场 gravitational field
万有引力定律 law of universal gravitation;Newton('s) law
万有引力理论 gravitational theories
万有引力理论的证明 evidence for gravitational theories
万有引力学说 gravitational theories
万有域 universal domain
万元产值排水量 water discharged output of ten-thousand-Yuan
万元产值排污量 discharge capacity of output of ten-thousand-Yuan
万元产值取水量 water demand of output of ten-thousand-Yuan
万元定额 ten-thousand-Yuan norm
万字浮雕 fret
万字廊 swastika
万字饰 swastika fylfot
万字细工 fretwork
万字形 fylfot;gammadion;gammatia
万字形饰 swastika

卍字饰 swastika

卐字浮雕 fret

腕臂 cantilever

腕臂底座 cantilever swivel bracket
腕臂底座连接架 connector for cantilever bracket
腕臂活动 cantilever reach
腕臂上底座 upper bracket for cantilever
腕臂下底座 lower bracket for cantilever
腕臂支撑 cantilever support
腕臂柱 cantilever mast
腕尺 <长度单位,1 腕尺 = 45.7 厘米> cubit
腕足动物地理区 brachiopod faunal province
腕足(动物)门 Brachiopoda
腕足类灰岩 brachiopod limestone

汪克尔单纯旋转式发动机 Wankel's simple rotating engine

汪克尔发动机 <一种偏心转子式内燃机> Wankel's engine
汪克尔密封系统 Wankel's sealing grid
汪克尔式压气机 Wankel type compressor
汪克尔旋转压缩机 Wankel rotary compressor
汪克尔转子发动机 Wankel engine; Wankel rotating engine
汪克尔转子发动机排量 Wankel engine's displacement
汪克尔转子活塞 Wankel's rotating piston
汪纳高温计 Wanner's pyrometer
汪纳光测高温计 Wanner optic(al) pyrometer
汪尼埃函数 Wannier function
汪洋大海 expanse of water

亡革菌属 <拉> Thanatephorus

王朝 dynasty

王宫 basilica;imperial palace;palace
王宫建筑师 king's master mason
王冠卫星 <法国大地测量卫星名称> Diademe
王后寝宫 Queen's megaron
王后套房 Queen's suite
王后卧室 Queen's chamber
王家广场 <伊朗伊斯法罕> Royal Square
王莲属植物 victoria
王陵 king's tomb
王权 royalty
王锐分类 Wang Rui classification
王室林 crown forest
王室领地 grown land
王水 chlorazotic acid; chloronitrous acid; nitrohydrochloric acid; nitromuriatic acid; aqua regia <拉>【化】
王水分解 decomposition with aqua regia
王台 royal cell
王座 dais

网板 otter board;raster;screen

网板架【船】 trawl gallows
网板式干燥器 sheeting drier [dryer]
网板拖网 otter trawl
网板注射法 raster injection
网版 half-tone screen
网版行车 printing carriage
网笔石层 dictyonema bed
网边 <金属> selvage edge
网变形 net deformation
网波纹 insertion waves
网玻璃 reticulated glass
网布 textile screen cloth
网布搭接 mesh lapping
网部浆坑 hog-pit
网衬 scrim back
网带 woven-wire belt
网带吊货兜 flour sling
网带式干燥机 <单板用> mesh-belt drier [dryer]
网带式烘燥机 net dryer
网带式输送器 mesh-belt conveyer [conveyor]
网带式隧道退火窑 conveyer belt lehr
网带窑 mesh-belt kiln
网袋 box net;mesh bag
网道运输 network service
网点 grid point; half-tone dot; lattice point; mesh point; network point; screen dot;screen point
网点比例 dot ratio;dot scale
网点尺寸 dot size
网点大小 dot size
网点法 net-point method
网点翻网点 dot for dot
网点腐蚀 dot etching
网点覆盖率 dot area coverage
网点刻蚀 dot engraving;dot etching
网点密度 dot density
网点面积 dot area
网点平滑 grid smoothing
网点清晰度 sharpness of dots
网点剩余值法 grid residual method
网点识别 dots recognition; sharpness of dots
网点线条混合版 combined halftone and line
网点增大 dot enlargement;dot gain
网垫 gauze pad
网顶横杆 top rail to fabric
网兜 box net;net sling;sling net;sling of net;rope sling <起重机用的>; loading net <装卸用的>
网兜渔船 tuck net boat;volyer
网段布点【航测】 allocating point of network
网幅 fabric width
网钢丝 netting wire
网格 frame work; graticule; graticule meridian; graticule mesh; grid; grid module; lattice; mesh; reseau; waffle grid
网格板 grid plate;lattice plate;waffle slab
网格板条 netting lath(ing)
网格曝光框架 grid exposure frame
网格北(向) grid north
网格本初子午线 prime grid meridian
网格比例尺 grid scale
网格筛 mesh screen
网格边界 mesh boundary;net boundary
网格编图 grid square mapping
网格变化 grid variation
网格变形 distortion of the mesh; feather
网格标记 grid mark
网格薄毡 lamella mat
网格布 open weave cloth;scrim cloth; woven scrim
网格步长 mesh spacing
网格参数 mesh parameter
网格测量系统 grid measuring system
网格尺度 mesh scale
网格赤道 grid equator
网格窗口 grid window
网格磁差 grid variation
网格磁方位角校正 grid magnetic azimuth adjustment
网格磁偏角 <格网北和磁北的方向角> grid magnetic angle
网格单元 grid unit
网格导航 grid navigation
网格等斜线 grid rhumb line
网格地图 grid map

网格点 grid point; lattice point; mesh point; net point
网格电路 lattice circuit
网格叠加 grid stacking
网格叠置片 grid overlay
网格顶棚 grid ceiling
网格定点测量法 network-fixed site monitoring method
网格定点能谱测量 location network spectrum survey
网格读数 grid readings
网格发生器 mesh generator
网格法 method of grid; method of lattice
网格法地形测量 grid survey
网格法应变测量 grid method for strain measurement
网格方程 mesh equation
网格方位 grid direction
网格方位角 grid azimuth; grid bearing
网格方向 grid direction
网格分析 netting analysis; network analysis
网格分析程序 grillage analysis program
网格幅度 < 东西方向的 > grid amplitude
网格钢筋 fabricated bar
网格拱顶 diamond vault; net vault
网格构造 boxwork
网格规划图 gridiron plan
网格函数 net function
网格航法 grid navigation
网格航线 grid track
网格航向 grid course
网格恒向线 grid rhumb line
网格化 tessellate
网格化变换 network transformation
网格化数据图 gridded data map; grid value map
网格基础 grid formation; grid foundation
网格基准 grid reference
网格畸变 mesh distortion
网格计算 grid computation
网格加筋板 grid stiffened plate
网格尖塔 latticed steeple
网格间隔 grid spacing
网格间距 grid interval; grid spacing
网格校正法 grid method
网格接法 mesh connection
网格结点 nodal point of mesh
网格结构 cancellation; cellular structure; cellular texture; double cancellated structure; grid structure; lattice structure; lattice work; network; trellis work
网格经度 grid longitude
网格距 mesh scale
网格壳体 latticed shell
网格孔 grid opening
网格宽度 mesh width
网格框架 grid frame(work)
网格雷诺数 mesh Reynolds number
网格理论 netting theory
网格连接线 grid junction
网格梁 grillage beam; grillage girder
网格路面碾 mesh roller
网格门 screen door
网格门芯 mesh core
网格密度 network density
网格模拟 network analog(ue)
网格模拟理论 < 应力分析的 > theory of lattice analogy
网格模拟器 network simulator
网格模式 grid model
网格模型 grid model
网格偏差 grid variation
网格偏角 grid convergence; grid declination
网格平面 gridded plane
网格平面布置 grid plan
网格墙 wire lattice wall
网格穹隆 grid dome
网格球 geodesic cupola; geodetic cupola
网格球顶 geodesic dome
网格区域 mesh region; net region
网格栅 cellular grid
网格识别 grid recognition
网格式爆破网 mesh blasting net
网格式挡土墙 lattice retaining wall
网格式盾构 mesh-type face of the shield
网格式(给排水)系统 < 给排水的 > gridiron system
网格式构架 cellular framing
网格式基础 waffle footing
网格式加热管 strip heating pipe
网格式结构布置 grid formation
网格式拦污栅 lattice screen
网格式楼盖 waffle floor
网格式路碾 grid roller
网格式碾压机 grid roller
网格式配管系统 gridiron system
网格式体系 gridiron system
网格式围篱 screen fence
网格式系统 grid system
网格饰的凸出方形木线脚 lattice mo(u)lding
网格水平航向 grid course
网格缩小转绘 graphic(al) reduction
网格塔楼 lattice tower
网格图 arrow diagram; grid chart; lattice chart
网格图形 grid pattern
网格图型 lattice pattern
网格土工布 geogrid
网格纬度 grid latitude
网格纬线 grid parallel
网格位置 grid position
网格纹板 checkered plate
网格问题 grid problem
网格屋顶 lamella(r) roof
网格屋盖 lamella(r) roof
网格细工 openwork
网格线 grid line; mesh lines
网格线定向 orientation of mesh lines
网格线号数 grid number
网格线圈 honeycomb coil; lattice coil
网格象限角 grid bearing
网格芯 grid core
网格形 net pattern
网格形方孔 lattice square
网格形构造 cellular construction
网格形护岸 geomatrix
网格形土工织物 geogrid
网格形线路 lattice network
网格颜色玻璃 network tinted glass
网格原点 grid origin
网格圆顶 geodesic cupola; geodetic cupola
网格圆屋顶 network cupola
网格闸门 lattice gate
网格值 grid value
网格中心点坐标 coordinate of central point of grid
网格主垂面 grid prime vertical
网格状 cancellate
网格状玻璃 network glass
网格状道路型式 grid road pattern
网格状的 latticed
网格状底脚 grid footing
网格状拱顶 reticulated vault
网格状结构 geodesic construction; network texture
网格状结构系统 grid structural system
网格状图案 waffle-like pattern
网格状细工 openwork
网格状线 grid pattern
网格状线路 lattice network
网格状型式 grid pattern
网格状悬臂底脚 grid cantilever footing; grillage cantilever footing
网格状悬吊顶棚 grid suspension system
网格状影纹 netted texture
网格子午线 grid meridian
网格子栅栏 latticed fence
网格组构 chicken wire fabric
网格组织 grid work; netting fabric
网格坐标 grid coordinates; grid reference; mesh coordinates
网格坐标海图 grid chart; grid coordinates chart
网格坐标航向 grid heading
网格坐标线 grid line
网格坐标指示 grid indication
网构 lattice
网构大梁 lattice girder
网构桁架 lattice truss
网关 gateway
网关路由器 gateway router
网管 network management
网贯交织 interwoven
网硅酸盐 framework silicate; tectosilicate
网号 mesh number
网际电视 web TV
网际金典 < 一种英汉电脑辞典,即指即译 > Roboword
网际协议 < 其作用是将信息从一台计算机传送到另一台计算机 > internet protocol [IP]
网架 screen substructure; space grid; wire frame
网架结构 grid frame(work); spatial grid structure
网架穹顶 geodesic dome
网架式铁塔 lattice mast
网架屋顶结构 lamella roof structure
网架型 cellularity
网架圆顶 net-worked dome
网架组织 trabecula [复 trabeculae]
网间包 internet packet
网间连接计算机 gateway computer
网间连接控制器 gateway controller
网间连接器 gateway
网间连接协议 gateway protocol
网将接线 mesh connexion
网接电路 mesh-connected circuit
网结 anastomosis [复 anastomoses]
网结的 anastomosed
网结河 anastomosed stream
网结河层序 anastomosed stream sequence
网结作用 < 路面上层混合料的 > knitting action
网金红石 sagenite
网具 netting gear
网具浮标 fish net buoy
网具修理场 netting gear mend yard
网具装卸滑车 car block
网卡 network card
网壳 latticed shell; reticulated shell
网孔 loop; mesh opening; mesh screen
网孔长度 mesh length
网孔尺寸 mesh size
网孔电流 mesh current
网孔电路 mesh circuit
网孔分析 mesh analysis
网孔分析法 mesh analysis
网孔结点 mesh node
网孔净径 clear mesh
网孔菌 < 拉 > Dictyopanus
网孔宽度 mesh opening width; mesh width
网孔盆 mesh pot
网孔塔板 Perform tray
网孔形电路 mesh circuit
网孔印刷器 screener
网孔状屏 hole screen
网篮 wire basket
网篮式 fixed-basket type
网篮式格子体 basket ware
网连接 < 用计算方法把许多单独网连接在一起 > net consolidation
网链 network chain
网裂 block crack; check crack; map crack(ing); net-shaped crack
网滤器 net filter
网路 circuit network
网路备用试验 network readiness testing
网路分析 network analysis
网路分析器 network analyser [analyzer]
网路分析小组 network analysis team
网路管理点 network management point
网路管理信号 network management signal
网路规则 network planning
网路判定小组 network assessment group
网路图 network chart
网路位置图 loop-and-trunk layout
网路虚终端 network virtual terminal
网路运输控制站 network control station
网路站 network station
网络 net(work)
网络安全性 network security
网络板 lattice board
网络保护装置 network protector
网络保密 network security
网络倍增器 mesh multiplier
网络比拟法 network-analog(ue) method
网络编号方案 linked numbering scheme
网络变换 network transformation
网络变压器 network transformer
网络拨入 network in-dialing
网络布局 network topology
网络布局图 network topology figure
网络布线 laying of cables
网络参数 network parameter
网络参数矩阵 network parameter matrix
网络操作系统 network operating system
网络操作员命令 network operator command
网络操作中心 network operation center [centre]
网络策略 network strategy
网络层 network layer
网络层内部结构 internal organization of the network layer
网络层转接 network layer relay
网络常数 network constant
网络场 network field
网络程序 network program(me)
网络程序设计 network programming
网络出口局 linked exit office
网络初始条件 network initial condition
网络处理机 network processor
网络处理器 network processing unit
网络传递函数 network transfer function
网络传输 network transmission
网络传输卡 network interface adapter; network interface card; network media card
网络传真 network FAX
网络存取 network access

网络存取处理机 network access processor
网络单元 network element
网络导纳 network admittance
网络导体 network conductor
网络导线 network conductor
网络的对偶性 network duality
网络的实频特性线 network characteristic at real frequency
网络地址 network address
网络地址单位 network address unit
网络点 network point
网络电流 mesh current
网络电线 network cable
网络调度中心 network dispatching center [centre]
网络调度自动化系统 network dispatching automation system
网络定理 network theorem
网络定律(克希荷夫定律) law of electric network
网络定时 network timing
网络定相继电器 network phasing relay
网络独立变量 network independent variable
网络法 mesh method;network metaphor; program(me) evolution and review technique
网络法的解 network solution
网络范围协调实体 network-wide coordination entity
网络方程 network equation
网络仿真 <美国开发的一个城市交通控制系统> Network Simulation
网络仿真程序 network simulation
网络分配器 network distributor
网络分配设备 network distribution equipment
网络分析 circuit analysis;mesh analysis; network analysis; network planning
网络分析程序 network analysis program(me)
网络分析法 network analysis method
网络分析技术 network analysis technique
网络分析理论 network analysis theory
网络分析器 network analyser [analyzer]
网络分析仪 network analyser [analyzer]
网络服务 network service
网络服务程序 network servicer
网络服务接入 network service access point
网络服务器 web server
网络服务协议 network service protocol
网络负载分析 network load analysis
网络改良器 network modifier
网络改性剂 network modifier
网络构造 net structure
网络管理 network management
网络管理部件 network management unit
网络管理员 web master
网络管区 network domain
网络规划(法) network planning;network programming
网络过滤器 web-filter
网络函数 network function
网络函数的极点 pole of network function
网络哼声 network hum
网络互连 network interconnection
网络互连协议 internetwork protocol
网络缓冲器 network buffer
网络换向器 network commutator
网络混合 network mixing
网络几何学 network geometry;geometry of nets
网络计划 network program(me); network planning
网络计划技术 network planning technique
网络计划进度表 network schedule
网络计划逻辑顺序的优先工作项目 predecessor work item
网络计划模型 network planning model
网络(计划)图 network diagram
网络计划中的最后事件发生时间 latest event occurrence time
网络计算工业 network computing industry
网络计算环境 network computing environment
网络计算机 network computer
网络计算机接口 network computer interface
网络计算器 network calculator
网络记发器 network register
网络技术 network technique
网络假说 network hypothesis
网络监视 network monitoring
网络建模 model(l)ing network
网络交换器 network exchange unit
网络交换中心 network switching center[centre]
网络接口 network interface
网络接口程序 network interface program(me)
网络接口单元 network interface unit
网络接口卡 network interface card
网络接线 network connection
网络节 link
网络节点 network node
网络节点接口 network node interface
网络节共振 link resonance
网络结构 network architecture;network configuration;network structure
网络结构单元 structural unit of network
网络结构理论 theory of network structure
网络解法 network solution
网络进度表 network schedule
网络进度计划 construction project schedule network diagram
网络进口局 linked entry office
网络经济 internet economy
网络经济学 cybernomics;internet economics;webnomics
网络局 linked office
网络矩阵 network matrix
网络决策 network decision
网络控制 <即网络计划技术> network control
网络控制程序 network control program(me)
网络控制方式 network control mode
网络控制阶段 network control phase
网络控制模块 network control module
网络控制器 network control unit
网络控制系统 network control system
网络控制站 net control station
网络控制中心 network control center [centre]
网络控制装置 network control centre equipment;network control unit
网络理论 lattice theory;network theory
网络连接 mesh connection
网络连接字符串 network connection string
网络连线 link
网络联机信息检索 on-line retrieval of information over a network
网络联结 network connection
网络灵敏度 network sensitivity
网络浏览器 web browser
网络流 network flow
网络流程程序 network flow routine
网络流量理论 network flow theory
网络流通量 network throughput
网络路线 network path
网络滤波器 network filter
网络码 network code
网络满益法 network flooding technique
网络密度 network density
网络模拟 network analogy;network simulation
网络模拟理论 theory of lattice analogy
网络模型 network model
网络模型模拟 analogy of net model
网络碾 grid roller
网络配置 line configuration;network configuration
网络匹配 network matching
网络频率 network frequency
网络平台 internet service provider
网络评价法 network method of environmental impact assessment
网络剖线 cut-in a network
网络软件 network software
网络商店【计】 on-line-shop
网络设备 network equipment;netdevice;network device
网络设计 network design
网络时间校准 network time adjust
网络识别 network awareness
网络式盾构 shield faced with grid
网络式耐火材料 network refractory (product)
网络数据 network data
网络数据单位 network data unit
网络数据翻译程序 network data translator
网络数据模块 network data model
网络数据模型 network data model
网络数控库 network database
网络死锁 network lock-up
网络算子 network operator
网络特性 network characteristic
网络体外氧化物 network modifying oxide
网络体系结构 network architecture
网络调节剂 network modifier
网络调整体 network modifier
网络调整物 network modifier
网络调整氧化物 network modifier
网络通路 network path
网络通信[讯] internetwork communication;network service
网络通信[讯]线路 network communication circuit
网络通知设备 bridge annunciating device
网络图 arrow diagram; concurrence chart; intersection chart; network chart;network diagram
网络图模型 network model
网络拓扑 network topology;topology of network
网络拓扑结构 network topology structure
网络拓扑学 network topology
网络外离子 extra-network ion
网络外体离子 network modifying ion
网络外体氧化物 network modifier; network modifying oxide
网络维护 network maintenance
网络维护信号 network-maintenance signal
网络位相学 network topology
网络位置 network site
网络文件系统 network file system
网络稳定性 network stabilization
网络物理部件 network physical unit
网络系 system of nets
网络系统 network system;system of nets
网络系统自动程序控制 scheduling control automation by network system
网络纤维 network fiber [fibre]
网络线路 network line
网络相关变量 network related variable
网络协调站 network coordination station
网络协调中心 network coordination center [centre]
网络协议 network protocol
网络协议数据单元 network protocol data unit
网络新闻组 netnews
网络信息联机检索 on-line retrieval of information over a network
网络信息流通 network traffic
网络信息系统 network information system
网络信息中心【计】 network information center [centre]
网络行列式 network determinant
网络形成离子 network forming ion
网络形成熵 entropy of network formation
网络形成物 network former
网络形成氧化物 network forming oxide
网络形成元素 network forming element
网络型计算机 network-type computer
网络型模拟计算机 network-type analog(ue) computer
网络性能目标 network performance objective
网络修改剂 network modifier
网络修饰物 network modifier
网络修整器 network modifier
网络虚拟终端 network virtual terminal
网络寻址 network addressing
网络延迟时间 message transfer time
网络演习 network drill
网络用户 network user
网络优化算法 algorithm for network optimization; arithmetic (al) for network optimization
网络元件 network element
网络原理 network theorem
网络运行者维护通路 network operators maintenance channel
网络运行中心 network operation center [centre]
网络诊断 network diagnosis
网络支路 arm of network;branch of network
网络制 network system
网络质量 network quality
网络智能 network intelligence
网络中继器 network relay
网络终端装置 network terminating unit
网络主机 network host
网络状的中间面 network middle plane
网络状地震测线 net seismic line
网络状二次输电 grid subtransmission;network subtransmission
网络状泉华 network sinter
网络状态 network state
网络状中间夹层 network middle

plane
网络资源 network resource
网络综合 network synthesis
网络阻抗 network impedance
网络组【数】system of nets
网络组成部分 networks component
网络组织 cancellation; net (work) structure
网络作业处理 network job processing
网络作业控制语言 network job control language
网络作业控制中心 network operations control center [centre]
网脉 stockwork
网脉凸疤 mapping
网脉状构造 network structure; reticulated structure
网脉状矿石 network ore
网脉状矿体 stockwork
网茅属 spartina
网密度 density of network
网民 citinet; net citizen; netizen
网膜照度 retinal illumination
网膜字符阅读器 retina character reader
网目 sieve number
网目板 fine screen; mesh ga(u)ge; screen plate
网目半色调 screen halftone
网目变换 screen change
网目测量仪 mesh ga(u)ge
网目尺寸 basic mesh size; mesh size
网目点 screen element
网目点缩小 dot reduction
网目电流法 mesh method
网目分离器 screening disintegrator
网目分量 screen value
网目复制 half-tone reproduction
网目规 mesh ga(u)ge
网目号 mesh ga(u)ge; mesh number
网目角度 screen angle
网目距离 screen distance
网目孔 screen opening
网目片 plastic halftone
网目色调 screen tint
网目摄影 screen photography
网目摄影机 reseau camera
网目铜版 half-tone
网目凸版 half-tone block
网目图形 mezzograph
网目线数 ruling of screen
网目阳图片 half-tone positive
网目印相纸 half-tone paper
网内呼叫 on-net call
网平差 net adjustment
网屏 grill(e); screen
网屏齿轮 screen gear
网屏距离 screen distance
网屏蒸发皿 screened pan
网屏蒸发器 screen pan
网桥 bridge
网球场 <有观众座位的> tennis stadium; racket court; tennis court
网球馆 indoor tennis stadium
网球俱乐部 tennis club
网纱 screen cloth
网筛 cribble; mesh screen; mesh sieve; net sieve
网筛分析 mesh analysis
网上查询器 search engine
网勺 wire screen ladle
网深遥测仪 net-depth telemeter
网深仪 net-depth meter
网式操作运行的网络 mesh-operated network
网式窗(花)格 net tracery
网式床屉 net type bed bottom
网式风口 mesh-type air opening; mesh-type air outlet
网式供热系统 <埋于楼板内的> cable mat heating system
网式过滤器 gauze filter; mesh(y) filter; screen filter; gauze strainer; mesh strainer; well strainer
网式回风口 mesh-type air return opening
网式机油过滤器 gauze oil strainer
网式滤清器 mesh filter; net type filter; screen filter
网式滤水器 gauze filter
网式滤油器 mesh-type oil filter; screen filter
网式碾压机 mesh or grid pattern roller
网式排泄 grapevine drainage
网式收集管线 scree-collector pipeline
网式隧道运碴车 mesh laying jumbo
网式挖泥机 net dredge(r)
网式吸滤器 gauze filter
网式液压油滤清器 screen hydraulic filter
网式制动装置 <飞机着陆> net arrester barrier
网数 netting index; netting number
网刷 gauze brush
网丝规 fabric wire ga(u)ge
网索夹 wire rope clamp
网锁平差 chain adjustment
网筒过滤 filtration of sieve cylinder
网头灯 gauze top burner
网歪斜 out-of-square for wire
网位监视仪 net monitor
网位仪 net-depth telemeter; netsonde
网纹 craze; overlapping curve; texture; webbing
网纹斑 vermiculated mottle
网纹斑杂状结构 harrisitic texture
网纹板 ribbed plate; tread plate
网纹版 half-tone etching
网纹层 plinthitic horizon
网纹导管 reticulate duct; reticulate vessel
网纹分子 reticulated element
网纹干扰 fine-line interference
网纹钢 riffled iron
网纹钢板 checkered plate; checkered sheet(plate); chequered plate; goff(f) ered plate; patterned metal plate; riffled sheet; floor plate
网纹管 reticulated duct
网纹黏[粘]土 lattice clay
网纹喷胶法 cobwebbing
网纹漆 webbing lacquer
网纹铁板 checkered iron
网纹涂覆法 cobwebbing
网纹涂料 cobweb coating; webbing finish paint
网纹涂装(法) webbing
网纹土 patterned ground; polygonal ground
网纹弯曲 curved mesh
网纹形 criss-cross pattern
网纹形混凝土铺面块体 checker block paving
网纹轧辊 knurled roll
网纹砖 textured brick
网纹状 reticulate pattern
网纹状滚胶法 cottoning
网系 network; system of nets
网线 gauze wire; mesh wire; netting twine; network cable
网线板 copying screen; line-screen; ruling film
网线大小 screen size
网线计数 screen-line count
网线胶片 screen
网线角度 angle of ruling
网线金属版 mezzotint
网线片 tint screen
网线普染 straight-line tint
网线摄影机 reticle camera
网线数 screen density; screen equivalent; screen size
网线凸版 half-tone block
网线图 network chart
网线照相机 reticle camera
网箱 live box; net cage
网箱养鱼 cage fish culture
网箱养殖 cage culture
网形 mesh configuration
网形接地器 mesh-form earthing device
网形结构 chain structure; network structure; reticular structure
网形铁 expanded metal
网压 catenary voltage
网眼 mesh; screen; screen mesh
网眼边 mesh segment
网眼玻璃 screened glass
网眼薄钢板 expanded sheet metal
网眼薄钢皮 expanded sheet metal
网眼不正 out-of-square mesh
网眼布 scrim
网眼尺寸 screen size
网眼窗格 openwork tracery
网眼窗花 openwork tracery
网眼大小 mesh size
网眼钢板 expanded metal(sheet); expanded metal fabric reinforcement
网眼钢筋 expanded metal fabric reinforcement
网眼钢皮 expanded metal lath(ing); perforated sheet metal lath(ing)
网眼钢皮抹灰隔墙 expanded metal lath partition; grid metal lath partition
网眼钢皮上抹灰 plaster on metal lath(ing)
网眼钢上抹灰 plaster on metal
网眼号 mesh number
网眼花边机 go-through machine
网眼花格窗 open tracery
网眼花山头 openwork gablet
网眼集装箱 wire-mesh pallet
网眼结 mesh knot; netting knot
网眼结构 mesh texture; meshwork structure
网眼铝栅板 expanded alumin(i)um grating
网眼密度 reticular density
网眼面积 screening area
网眼纱褶裥边饰 quilling
网眼山墙 openwork gable
网眼网板配筋 expanded metal reinforcement
网眼系列 mesh series
网眼针织物 lacework
网眼织物 mesh fabric
网眼装帧 network decoration
网眼装置 mesh facility
网眼状结构 areolation
网页 webpage
网印标记 squeegee mark
网印用金膏 squeegee gold
网用操作系统 network environment operating system
网用金属丝 fabric metal
网运分离 separating the infrastructure from operation
网站 website
网织品 network
网织物 mesh work
网址 web address; website
网重数 number of adjacent chains in a net
网状 reticulation
网状板 lamina reticularis; reticular lamina; reticular plate
网状保护装置 network protector
网状变换 transforming meshwise
网状表面 meshy surface
网状玻璃态碳 reticulated vitreous carbon
网状薄壁组织 reticulate parenchyma
网状薄壳 reticulated shell
网状草皮 netted turf
网状层 lamina reticularis; reticular layer
网状插座 openwork rosette
网状穿孔 reticulate perforation
网状窗花格 reticulated tracery
网状带 reticular zone
网状的 clathrate; meshed; meshy; netshaped; netty; reticular; reticulated; retiform
网状电极电离室 mesh ionization chamber
网状电路 network
网状顶棚 grid ceiling
网状洞穴 anastomotic cave
网状发射极 mesh emitter
网状盖布 mesh tarpaulin
网状钢板 checkered sheet
网状钢筋 mesh reinforcement; reinforced grillage; screen reinforcement; steel fabric reinforcement; steel mesh reinforcement; weldmesh reinforcement; wire-mesh reinforcement
网状钢筋的 mesh-reinforced
网状钢面板 open mesh steel floor
网状高分子 net high-polymer
网状拱 reticulated vault
网状拱顶 net vault
网状拱顶网目 net vault meshes
网状构造【地】mesh structure; net like structure; net-shaped structure; netted structure; network structure
网状管道 mesh duct
网状光阑 network diaphragm
网状光栅 cross hatch pattern
网状滚花 cross knurled
网状过滤 gauze filtration; gauze strain
网状过滤盘 gauze filter tray; gauze strainer stray
网状过滤器 gauze filter; gauze strainer; strainer screen
网状河道 braided channel
网状河段 braided reach
网状河口 braided distributary estuary
网状河流 anastomosed river; anastomosed stream; anastomosing river; anastomosing stream; braided river; braided stream; distributary of river
网状河沙坝 braid bar
网状花饰 plexiform
网状换向器 network commutator
网状混合岩 dictyonite [diktyonite]
网状激活系统 reticular activating system
网状加厚 reticulate thickening
网状加筋 mat reinforcement; mesh reinforcement
网状加强筋 mesh-reinforced
网状加热器 mesh heater
网状交联 networking
网状礁 mesh reef
网状街道 braided street
网状结构 mesh texture; net structure; netted texture; network structure; reticular structure; reticulated structure; reticulated texture; reticulation; screen structure
网状结构工作 mesh work
网状结构树脂 network structure resin

W

网状结合 networking
网状结线 delta connection; mesh connection
网状金红石 needlestone
网状进化 reticulate evolution
网状聚合体 reticular polymer
网状聚合物 net working polymer
网状卷积云 cirrocumulus lacunosus
网状菌丛 netted turf
网状开裂 map crack(ing)
网状矿床 network deposit
网状矿脉 ore stockwork
网状连接 mesh connection
网状联结 mesh connection
网状裂缝 chicken-wire cracking; honeycomb crack; map crack(ing); net-shaped cracking; network of cracks; pattern crack(ing); alligator crack; map cracking
网状裂纹 alligator-hide crack(ing); alligatoring; chicken-wire cracking; cracking; map crack(ing); pattern crack(ing); resillage; shelling
网状裂隙 reticulated cracks
网状滤心 gauze element
网状脉 network vein; reticulated vein; stockwork
网状水道 braid channel
网状名字 network name
网状膜 reticular membrane
网状泥炭 netted turf
网状排水(盲)沟 trammel drain
网状排水系 reticular drainage
网状配筋 mat reinforcement; mesh reinforcement
网状配筋砌体 mesh-reinforced masonry
网状铺砌 reticulated work
网状砌合 reticulated bond
网状墙板 mesh panel
网状穹隆 network dome
网状渠道 braided channel
网状沙坝 reticulated bar
网状沙洲 reticulated bar
网状筛 mesh sieve
网状栅 mesh grid
网状栅结构 meshed gate structure
网状输送器 reticulated conveyer [conveyor]; screen conveyer [conveyor]
网状树根桩 reticulated root piles
网状数据库管理系统 network data base management system
网状双极晶 mesh bipolar transistor
网状水道 braided channel
网状水系 network drainage; reticulate drainage
网状水系模式 braided mode
网状碳化物 carbide network
网状提单 network bill of lading
网状体系 network system
网状填充环 gauze ring
网状填充物 gauze packing
网状填充组织 reticular magma
网状填料 gauze packing
网状通道 network passage
网状图案 net like pattern; reticulated pattern
网状网络 mesh(ed) network
网状文法 web grammar
网状纹石面 reticulated ashlar
网状纹修琢 <石面的> reticulated dressing
网状纹砖石工 reticulated masonry
网状纹琢面 reticulated dressing
网状纹琢石 reticulated ashlar
网状圬工 reticulated masonry
网状物 mesh work; netting; network; reticulated work; reticulation
网状物网络 network
网状物质 reticular substance
网状系统 network; reticular system
网状细裂纹 crackle; crackling
网状下部结构 screen substructure
网状线 cross hatch(ing)
网状线脚 reticulated mo(u)lding
网状形成 reticulation
网状阳极 meshed anode
网状叶 net-veined leaf
网状阴极 mesh cathode
网状油脂 network fat
网状圆屋顶 network cupola
网状云 lacunaris; lacunosus
网状责任制 <签发联运提单的承运人对货主负全程运输责任,遇损失时则按发生损失的运输阶段的责任内容负责> network liability system
网状支架 network support
网状锥体 gauze cone
网状组织 cellular texture; net structure; netted texture; reseau; reticulated texture
网状组织的 reticular
网组 screen pack
网组系统 network system

往

往驳船上卸货 lighterage

往测【测】 direct run; direct measurement; fore measurement
往测净得 net forward gain
往测路线 direct route
往测与返测 direct and reversed observation
往程 <列车或机车的> forward journey
往反距离 round-trip distance
往返擦洗 double rub
往返测较差 difference of forward and backward observation
往返铲刮 <用平地机的> blade back and forth
往返串话 go-to-return crosstalk
往返地一月间的飞船 lunar shuttle
往返法 back-and-forth method; back-and-forward method
往返飞行 round-trip flight
往返观测 back-and-forward observation
往返航程 ply voyage; round trip; round voyage
往返航程船租合同 return trip charter party; round-trip charter party
往返航次 round trip; round voyage
往返航次船租合同 return trip charter party; round-trip charter party
往返航次时间 turnaround time
往返航行 round voyage
往返互换加热器 reciprocating heater
往返货运 cargo shipped to and fro
往返机制 shuttle mechanism
往返距离 round distance
往返旅程 round trip
往返旅行 round trip
往返贸易 countertrade
往返票 round-trip ticket
往返时间 round-trip time; trip time; turnround time; round time
往返式皮带运输机 shuttle conveyer [conveyor]
往返水准测量 bilateral level(l)ing; reciprocal leveling
往返行程 round trip
往返行程环流 out and return travel
往返行程计数器 lift counter
往返行程时间 round-trip cycle
往返行驶 ply
往返循环列车系统 shuttle system
往返延迟 round-trip delay
往返运动 reciprocating motion; shuttling motion
往返运费 freight out and home
往返运输 out and back haul
往返运行【铁】 shuttle service
往返周期 round-trip cycle
往返租赁 round-trip lease
往返租用契约 round strip lease contract
往复 reciprocate
往复摆动式 way of reciprocating motion
往复摆动式溜槽 reciprocating chute
往复摆动运动 oscillating traverse motion
往复摆式给料机 reciprocating feeder
往复摆式输送机 reciprocating conveyer [conveyor]
往复板式给料机 reciprocating plate feeder
往复板式上料器 reciprocating feeder
往复泵 plunger pump; reciprocal pump; reciprocating pump
往复泵的脉动作用 surge of reciprocating pump
往复臂式配水器 reciprocating-arm distributor
往复臂式洗砂器 reciprocating-arm grit washer
往复编织 reciprocal knitting
往复变位 reciprocal deflection
往复变向的单向交通街道 <在每天一定时间内,供某一方向单程交通,另一时间内则供相反方向单程交通> reversible one-way street
往复变向的中央车道 reversible center [centre] path
往复波 oscillating wave; oscillatory wave; wave of oscillation; oscillation wave
往复部件 reciprocating parts
往复槽式输送机 reciprocating trough conveyer
往复潮流 alternating tidal current; rectilinear stream; reversing (tidal) current; reversing tide
往复齿条装置 mangle gearing
往复冲程 double stroke
往复传令车钟 order and reply telegraph
往复次数 reciprocal time
往复的 reciprocal; reciprocating; to-and-fro; up-and-down
往复动作 reciprocal motion; reciprocating action; reversible motion
往复阀 shuttle valve
往复法 to-and-fro method
往复反应 reciprocal reaction
往复杆 reciprocating arm; reciprocating lever
往复杆头块 oscillating rod end block
往复缸 linear actuator
往复工作台式立轴平面磨床 vertical spindle surface grinder with reciprocating table
往复刮板输送机 reciprocating flight conveyer [conveyor]
往复惯性发生器 reciprocating inertia generator
往复滚动 reciprocating rolling
往复滚压法 reciprocating rolling process
往复焊接电极 reciprocating welding electrode
往复航程 out and home
往复和低压汽轮机 reciprocating and low pressure turbine
往复和低压透平 reciprocating and low pressure turbine
往复和低压涡轮机 reciprocating and low pressure turbine
往复荷载 reciprocal load
往复滑动 reciprocating sliding
往复滑块曲柄机构 reciprocating block slider crank mechanism
往复滑轮 propelling sheave
往复滑移 reciprocating sliding
往复回波 round-trip echoes
往复回转混合搅拌器 oscillating mixers agitator
往复回转罗拉 reciprocating and rotating rollers
往复回转式 oscillating type
往复回转式搅拌器 oscillating agitator
往复回转式液体连续萃取塔 oscillating liquid continuous extraction tower
往复活门 shuttle valve
往复活塞式发动机 reciprocating piston engine
往复活塞式内燃机 reciprocating piston internal combustion engine
往复活塞式气体压缩机 reciprocating gas compressor
往复加热器 reciprocating heater
往复架空索道 to-and-fro (aerial) ropeway
往复交通 to-and-fro traffic
往复锯 drag saw; oscillating saw; reciprocal saw; reciprocating blade saw
往复空气泵 reciprocating air pump
往复来回翻斗车 shuttle dumper
往复来回料车 shuttle car
往复来回卸料车 shuttle dumper
往复来回运输 shuttle haul(age)
往复力 reciprocating force
往复流 alternating current; rectilinear current; rectilinear stream
往复炉箅 reciprocal grate
往复炉箅式加煤机 reciprocating grate stoker
往复炉排 reciprocal grate
往复罗拉加捻装置 oscillating roller twister
往复碾压 reciprocating rolling
往复耙式分选机 reciprocating rake classifier
往复喷涂机 reciprocator
往复片式给料机 reciprocating plate feeder
往复偏转反射镜 oscillating mirror
往复曲折 cripple
往复燃烧 reciprocal combustion
往复三角座滑架 reciprocating cam carriage
往复筛 oscillating riddle; reciprocation screen
往复式 ping-pong; reciprocating type
往复式板阀 reciprocating valve
往复式泵 donkey pump; reciprocating pump
往复式布水器 reciprocating-arm distributor
往复式铲掘机 reciprocating spading machine
往复式垂直输送机 reciprocating-motion vertical conveyer [conveyor]
往复式锉床 reciprocating filing machine
往复式单螺杆配料机 reciprocating single-screw compounder
往复式的 double action; shuttle (-type); double-acting
往复式的复动蒸汽打桩机 double-acting steam pile hammer
往复式电锯 reciprocating saw
往复式发动机 reciprocal engine; reciprocal motor; reciprocating engine; reciprocator
往复式阀 reciprocate valve
往复式风钻 reciprocating drill; solid

piston rock drill
往复式割刀 reciprocating knife
往复式隔膜泵 reciprocating diaphragm pump
往复式给料机 oscillating feeder
往复式给料器 oscillating feeder; reciprocating feeder
往复式工作台 reciprocating table
往复式供料机 reciprocating feeder
往复式供料器 reciprocating (-type) feeder
往复式鼓风机 reciprocating blower
往复式刮板输送器 reciprocating scraper
往复式刮板运输机 reciprocating beam conveyer [conveyor]
往复式辊压机 reciprocating roller press
往复式滚压机 reciprocating roller press
往复式换挡装置 shift shuttle
往复式活塞泵 reciprocating piston pump
往复式机器 reciprocating machine
往复式给水泵 reciprocal feed pump
往复式加料泵 reciprocating feeder pump
往复式加料机 reciprocating charger; reciprocating feeder
往复式加料器 reciprocating feeder
往复式架空索道 reversible tramway; shuttle cableway; shuttle ropeway
往复式进料泵 reciprocating feed pump
往复式锯 reciprocating saw
往复式卷扬机 shuttle winding machine
往复式空气压缩泵 reciprocating air pump
往复式空气压缩机 reciprocating (-type) (air) compressor; shuttle air-pressing machine
往复式孔板萃取器 reciprocating plate extractor
往复式冷冻机 reciprocating condensing unit
往复式冷凝器 reciprocating condensing unit
往复式冷凝装置 reciprocating condensing unit
往复式冷油泵 reciprocating cold oil pump
往复式连续推料离心机 reciprocating-conveyor continuous centrifuge
往复式料车 shuttle car
往复式炉箅 reciprocating grate
往复式炉箅片 reciprocating grate bar
往复式炉排 reciprocating grate
往复式路耙 reciprocating rake
往复式螺杆挤出机 reciprocating-screw extrusion machine
往复式磨床 reciprocating grinder
往复式磨平机平台 traverse table
往复式内燃发动机 reciprocating internal combustion engine
往复式内燃机 reciprocating internal combustion engine
往复式耙路机 reciprocating rake
往复式排代泵 reciprocating displacement pump
往复式排锯 deal frame
往复式喷涂机 reciprocating spraying machine
往复式膨胀机 reciprocal expansion engine
往复式皮带输送机 shuttle belt conveyer [conveyor]
往复式平板给料机 reciprocating plate feeder
往复式破碎机 reciprocating crusher; reciprocating cutter
往复式气动发动机 reciprocating air motor
往复式取料法 step back reclaiming method
往复式绕线机 shuttle winding machine
往复式撒布机 reciprocating distributor
往复式砂带磨床 oscillating sander
往复式筛 reciprocating screen; reciprocating sieve
往复式筛分机 sieve travel(l)er
往复式深井泵 deep reciprocation pump; deep-well reciprocating pump
往复式手扶割草机 walking reciprocating mower
往复式手钻 reciprocating drill
往复式输送机 reciprocating conveyer [conveyor]
往复式树篱修剪机 reciprocating hedge cutter
往复式竖锯 jig saw
往复式竖线锯 jigger saw
往复式双缸泵 reciprocating duplex pump
往复式水泵 displacement pump; pump reciprocating type; reciprocating pump
往复式水锤泵 reciprocating ram pump
往复式水轮配水池 reciprocating water wheel distributor
往复式水轮配水器 reciprocating water wheel distributor
往复式索道 to-and-fro (aerial) ropeway
往复式塔 reciprocating column
往复式提升机 reciprocating vertical conveyer [conveyor]
往复式无极绳运输 reversible endless-rope haulage
往复式无极绳运输系统 reversible endless-rope system
往复式铣削夹具 reciprocating-type milling fixture
往复式线锯 jigger saw
往复式卸料篦子 reciprocating discharge grate
往复式循环运输 shuttle loop transit
往复式压板打磨机 straight-line pad sander
往复式压气机 reciprocal compressor; reciprocating blower; reciprocating compressor
往复式压缩机 reciprocal compressor; reciprocating compressor; reciprocating-type compressor
往复式压缩气锤 reciprocating (-type) compressed-air hammer
往复式叶桨分布机 reciprocating blade spreader; reciprocation blade spreader
往复式叶桨铺摊机 reciprocating blade spreader
往复式液体冷却器 reciprocating liquid chiller
往复式移动 shuttle travel
往复式引擎 reciprocating engine; reciprocator
往复式运输机 reciprocating [through] conveyer [conveyor]
往复式凿岩机 quarrymaster piston drill; reciprocating [rock] drill; solid piston rock drill
往复式增压泵 reciprocating intensifier pump
往复式增压器 reciprocating-type supercharger
往复式轧机 reciprocating rolling mill
往复式真空泵 reciprocating vacuum pump
往复式振动筛 ratter; reciprocating screen; reciprocating sieve; reciprocating wedge wire screen <筛面用截面金属丝编成>
往复式蒸汽泵 reciprocating steam pump
往复式蒸汽机 reciprocal steam engine; reciprocating steam engine
往复式整平板 <整压混凝土的> reciprocating screed
往复式整体型冷水机组 reciprocating packaged liquid chiller
往复式制冷机 reciprocating condensing unit; reciprocating refrigerating machine; reciprocating refrigerator
往复式柱塞泵 reciprocating plunger pump
往复式转子发动机 reciprocating rotary piston engine
往复式装载 shuttle loading
往复式自卸车 shuttle dump truck
往复式钻机 reciprocating drill
往复式作动机械 reciprocating machine
往复试验 cycle test
往复水准测量 reciprocal level(l)ing
往复索道 jig back; to-and-fro ropeway
往复推动炉箅炉 reciprocating grate stoker boiler
往复推送器 stroker
往复拖运 towing to and from the site
往复弯曲试验 to-and-fro bending test
往复斜盘式发动机 reciprocating swashplate type engine
往复行车 shuttle service; shuttle traffic; shuttling
往复形装载 shuttle loading
往复旋转 progressive rotation
往复旋转液压油缸 semi-motor
往复循环方式 dual-cycle operation system
往复压电效应 reciprocating piezoelectric(al) effect
往复压挤式破碎机 reciprocating pressure-type crusher
往复压缩泵 reciprocating pump compressor
往复摇动式运输机 chute conveyer [conveyor]; jigger; jigging conveyer [conveyor]
往复移动 reciprocate; traverse <横向>
往复移动式缆索道 shuttle cableway
往复应力 cyclic(al) stress; reciprocal stress
往复运动 alternate motion; alternating motion; alternative motion; back-and-forward motion; in-and-out movement; reciprocal motion; reciprocating motion; reciprocating movement; reversible motion; shuttle; shuttling (movement); to-and-fro motion; to-and-fro movement; reciprocation; oscillatory motion <海洋风生波>
往复运动不平衡 <切割器等> reciprocating unbalance
往复运动电动机 motor with reciprocating movement; pulsating motor; reciprocable motor
往复运动机构 reciprocating mechanism
往复运动机件 reciprocator
往复运动零件用的轴承 reciprocating bearing
往复运动汽缸 actuator
往复运动油缸 actuator
往复运动质量 reciprocating mass
往复运动装置 reciprocating apparatus
往复运货 outward and home freight
往复运输 commodity shunting
往复运输带 shuttle belt
往复直剪试验 reversing shear box test
往复质量 reciprocating mass
往复重量 reciprocating weight
往复轴 reciprocating shaft
往复主运动 reciprocating main motion
往复装载方式 dual loading system
往复装载系统 dual loading system
往复自卸车 shuttle dump truck
往复作用 reciprocal action; reciprocating action
往复作用活塞式工具 reciprocating piston tool
往高处 on high
往高频调谐 tuning upward
往国外的 outbound
往国外汇款 remittance abroad
往海外的 outward; oversea
往后退 keep back
往来 intercourse
往来存款 bank account
往来存款透支账 overdrawn account
往来抵押透支 overdraft on current account secured
往来兜揽生意的出租汽车 <英> crawler
往来赊欠账户 open book credit
往来透支 overdrafts on current account unsecured
往来推算 forward pass
往来银行 bank with credits opened; correspondent bank
往来银行账户 correspondent account
往来账 open account
往来账贷方 current account credit
往来账户 A/C account; account current; book account; current account; reciprocal account; running account; working account
往来账户存款 deposit(e) in current account
往来账户的调节 reconciliation of reciprocal accounts
往来账明细表 statement of current accounts
往来账户明细账 current account ledger
往来账户欠款 due for correspondents account
往来账户赊欠 open book credit
往来账户收支情况 balance of current account
往来账户直接法 direct method of current account
往来账目结算书 final statement of accounts
往来资产 quick assets
往楼下 downstair(case)
往年平均基金总额 hold harmless amount
往年平均基金总额超额拨款 hold harmless grant
往上爬的行为 climbing behavio(u)r
往上通风的竖井 up-shaft
往外去的 outgoing
往往 many a time
往下翻页 page down
往沿海地区 downcountry
往账 <本国银行在国外银行持有的外币账户> nostro

辋板 felloe plate

妄想 illusion

忘忧树 lotus tree

旺季 booming season; busy period; busy season; good season; peak season; peak selling period; rush season

旺盛的 blooming
旺盛生长 luxuriant growth; vigorous growth
旺市 bull market
旺销 brisk sales; sell briskly

望板 bat wing; board sheathing; boxing; over-purlin (e) lining; roof(ing) board (ing); roofer; roof sheathing; sarking(board); sheathing; wing plate

望板坡度 slope of sheathing
望参照执行 for your information and guidance
望后镜 back mirror; back view mirror
望后镜臂 rearview mirror arm
望后镜伸长臂 rear mirror extension arm; rearview mirror bracket
望火塔 fire tower
望火员 tower man
望加锡海峡 Makassar Strait
望加锡乌木 <深褐色纹理装饰硬木> Macassar ebony
望楼 barbican; belvedere; bretesse; lookout (tower); minar; outlook; tower; watch tower
望楼门 barbican entrance
望朔潮 spring tide
望筒 dioptra; sighting tube
望远方位镜 azimuth with telescope
望远镜 binoculars; field glass(es); telescope; telescope arm; visual telescope; aiming telescope <水准仪上的>
望远镜标准位置 normal position of telescope
望远镜测距仪 telescopic(al) range finder
望远镜尺度法 telescope and scale method
望远镜放大倍率 power of telescope
望远镜放大镜组合 telescope magnifier combination
望远镜放大率 enlargement ratio of telescope
望远镜分辨率 telescopic(al) resolution
望远镜辐射点 telescopic(al) radiant
望远镜盖 telescope cap; telescopic(al) cap
望远镜观察窗口 telescope port
望远镜管式伸缩仪 telescopic (al) tube extensometer
望远镜光学系统 telescopic(al) optic(al) system
望远镜盒 binoculars box
望远镜滑动开闭门 telescoping sliding shutter
望远镜回照器 telescopic (al) heliotrope
望远镜架支柱 telescoping shore column
望远镜镜片 telescope disk
望远镜镜筒 telescope tube
望远镜孔径 telescopic(al) aperture
望远镜流星 telescopic(al) meteor
望远镜瞄准镜 telescope
望远镜前天文学 pretelescope astronomy
望远镜清晰度 definition of telescope
望远镜驱动装置 telescope driving system
望远镜色片 telescope shade
望远镜式的 telescopic(al)
望远镜式瞄准具 telescopic(al) sight
望远镜式瞄准器 director telescope; telescopic(al) sight
望远镜视场 range of telescope
望远镜视场角 opening of the telescope
望远镜视野 opening of telescope
望远镜水准管 telescope level tube
望远镜套筒 telescope casing
望远镜天文学 telescopic(al) astronomy
望远镜调整圈 telescope collar
望远镜头帽 telescope cap
望远镜位置 face of theodolite; position of telescope; telescopic(al) position
望远镜系统 telesystem
望远镜效能 efficiency of telescope
望远镜形的 telescopic(al)
望远镜性能 telescopic (al) performance
望远镜学 telescopy
望远镜用火石玻璃 telescope flint glass
望远镜用燧石 telescope flint
望远镜照准仪 telescopic(al) alidade; telescopic(al) sighting alidade
望远镜正像 erect telescope image
望远镜支架 telescope holder; telescopic(al) support; telescope support
望远镜指导镜 finder
望远镜装置 mounting of telescope
望远镜纵转 180° turn one-eighty for telescope
望远物镜 teleobjective; telescope objective
望远显微镜 telemicroscope
望远显微两用镜 panopticon
望月 plenilune; full moon
望柱 baluster column; baluster shaft; newel post
望柱吊饰 newel drop
望柱顶饰 newel cap
望柱接头 newel joint
望柱头 baluster capital
望砖 sheathing brick; sheathing tile

危地马拉 <拉丁美洲> Guatemala

危地马拉城 <危地马拉首都> Guatemala City
危恶地 danger area
危房 dangerous building; deathtrap; derelict building; dilapidated housing
危害 detriment; disserve; disservice; endanger; hazard; menace
危害安全 safe(ty) hazard
危害报告书 hazard report
危害沉降 detrimental settlement
危害沉陷 detrimental subsidence
危害等级 hazard(ous) rating
危害度 density of infection; hazard level
危害度评价 risk evaluation
危害范围 damaging range
危害分级 hazard rating
危害隔绝法 hazard segregation
危害函数 hazard function
危害环境的 environmentally hazardous
危害环境罪 crime against damage to environment
危害可能性 hazard potential
危害率 hazard rate
危害膨胀 detrimental expansion
危害评价 hazard assessment; hazard evaluation
危害潜力 hazard potential
危害人类罪 crime against humanity
危害人民群众生命财产的安全 endanger the safety of people and properties
危害生态 environmental ecological hazard
危害条款 <欧洲货币协定中的一项条款> jeopardy clause
危害物 hazardous material
危害系数 coefficient of injury
危害下限 lower hazard limit
危害性 harmfulness; perniciousness
危害性测量 damage survey
危害性冲击 damaging impact
危害性调查 complaint investigation
危害性分析 hazard analysis
危害性干扰 harmful interference
危害性估计 hazard evaluation
危害性鉴定 hazard identification
危害性评定 hazard evaluation
危害性评估 hazard assessment
危害性识别 hazard identification
危害性屋面材料 hazardous roofing materials
危害性噪声 hazardous noise
危害影响 detrimental effect
危害指数 hazard index
危害准则 damage criterion
危害作物 damage to crops
危机 conjuncture; crunch; precipice
危机对策 crisis game
危机论 theory of crises
危机四伏 beset with crisis
危及 endanger; imperil; jeopardise [jeopardize]
危急 peril
危急安全阀 safety governor
危急程度 criticality
危急反应 critical reaction
危急关头 exigence
危急降落 <飞机> distress landing
危急排水泵 emergency pump
危急情势 position of peril
危急生命的 life-threatening
危急手动脱扣 emergency hand trip
危急信号 emergency signal
危急遮断阀 emergency stop valve
危急遮断器 emergency generator; over-speed trip device
危角 coffin corner
危难 distress
危桥 bridge in danger
危石 fragmented rock; hanging rock; hazardous rock; loose rock
危石判断 rock-fall prediction
危险 dangerousness; hazard; peril; unsafety
危险半圆 dangerous semicircle
危险报警 danger alarming; danger warning
危险报警系统 danger alarm system
危险比 risk ratio
危险标灯 danger beacon; hazard marking light
危险标记 danger label; hazard label; risk label; risk marking
危险标示牌 warning board
危险标志 caution notice; danger arrow; danger mark; danger sign
危险标志牌 hazard marker; warning board
危险标准 risk standard
危险表 danger table
危险泊位 foul berth
危险部位 hazardous location
危险材料 hazardous material
危险场所 deathtrap; hazardous area; hazardous location; hazardous site
危险场所和地点 hazardous site and spot
危险程度 danger level; hazard level; level of risk; risk level
危险持续期限 duration of risk
危险的 critical; dangerous; hazardous; risky; unsafe
危险的道路 red route
危险的地质不连续面 critical geologic(al) discontinuity
危险的堆货 blown up
危险的荷载组合 critical loading combination
危险的化学品 hazardous chemical
危险的货物 hazardous goods
危险的事物 peril
危险的作业 hazardous operation
危险灯标 hazard light beacon; road danger lamp
危险等级 danger class; hazard rating
危险等级标号 hazard code
危险等级界限 hazard level limitation
危险地带 critical zone; danger(ous) zone
危险地点 danger point
危险地段 dangerous section; hazardous location
危险地区 critical zone; dangerous zone; danger zone
危险地震 hazardous earthquake
危险地震协会 Jesuit Seismologic(al) Association
危险点 danger (ous) point; hazard point; peril point
危险电压 dangerous pressure; dangerous voltage
危险顶板 hazardous roof; hazardous top
危险动负载 worst dynamic(al) load
危险度 degree of risk; risk factor
危险断面 critical section; dangerous section; weak section
危险反应 hazardous reaction
危险方位报警指示器 danger bearing alarm indicator
危险房屋 dangerous building; dilapidated building
危险房屋的修复 restoration of unsafe building
危险房屋公告 notice of unsafe buildings
危险废料 hazardous waste
危险废料表示系统 hazardous waste manifest system
危险废(弃)物 hazardous waste
危险废物处置 disposal of hazardous waste
危险废物的出口 export of hazardous waste
危险废物的海域处理 marine disposal of hazardous waste
危险废物的倾弃 dumping of hazardous waste
危险废物的深井灌注 deep-well injection of hazardous waste
危险废物的越界处理 transborder disposal of hazardous waste
危险废物的装运 shipment of hazardous waste
危险废物固定 fixation of hazardous waste
危险废物管理 hazardous waste management

危险废物倾弃于海洋 dumping at sea of hazardous waste
危险废物运输的事先通告 prior notification for hazardous waste transport
危险分布 distribution of risks
危险分极 classification of risk
危险分类 hazard classification
危险分析术 risk analysis technique
危险浮标 hazard buoy
危险符号 danger board;hazard index
危险高度 risk altitude
危险根源 hazard source
危险工程学 <有关于工程中潜伏的事故、破坏和灾害的鉴定和处理> hazard engineering
危险工业 hazardous industry
危险工作 dangerous work; insecure work;peril work
危险工作的额外报酬 danger money
危险(工作)津贴 danger money
危险固体废物 hazardous solid waste
危险固体废物管理 management of hazardous solid waste
危险故障 risk of disturbance
危险海岸 dangerous coast;foul coast
危险海区 foul waters
危险海滩 foul beach;foul coast
危险海域 foul waters
危险化学品 dangerous chemicals
危险化学品标签系统 hazchem
危险化学品防护服 hazchem protective suit
危险化学品数据手册 hazardous chemical data manual
危险化学品作业防护服 protective clothing for hazardous chemical operation
危险货物 dangerous articles; dangerous cargo;hazardous cargo;dangerous goods
危险货物包装标志 indication mark on packages of dangerous goods;labels for packages of dangerous goods;package mark for hazardous goods
危险货物包装性能试验 test for dangerous goods package performance
危险货物标签 danger label;label for dangerous goods
危险货物标志 dangerous mark; instruction mark for dangerous cargo
危险货物法案 Dangerous Cargo Act
危险货物集装箱管理规定 dangerous cargo administrative provision
危险货物集装箱位置 location of dangerous cargo container stowage
危险货物集装箱装载 stowage of dangerous cargo container
危险货物类项 classification and division of dangerous goods
危险货物配装 matched loading of dangerous goods
危险货物配装表 matched loading table of dangerous goods
危险货物品名编号 numbering of name of dangerous goods
危险货物清单 dangerous cargo list
危险货物探测 dangerous cargo detection
危险货物停留线 dangerous goods parking track
危险货物通告单插 placard holder
危险货物运输 carriage of dangerous goods;hazardous material transportation
危险货物运输分委员会 Subcommission on the Carriage of Dangerous Goods
危险货物运输规则 regulations relating to carriage of dangerous goods; transportation of dangerous goods code
危险货物运输委员会 Commission on the Carriage of Dangerous Goods
危险货物装箱单 packing list of dangerous goods
危险极限 danger limit;danger threshold
危险极限值 danger limit value
危险甲板 dangerous deck
危险间接损失 consequential damage
危险建筑 dangerous structure
危险交叉(口) dangerous intersection;dangerous crossing
危险角 danger angle
危险结构 dangerous structure
危险截面 critical section; dangerous (cross-) section; danger section; minimum life section
危险界线 danger line
危险界限 danger limit;limiting danger line
危险界限方位 danger bearing
危险界限水深 danger sounding
危险津贴 hazard pay
危险经理 risk manager
危险警报信号 danger warning signal
危险警告 danger warning
危险警告标志 danger warning sign
危险警告灯 danger warning light; hazard warning light
危险警告牌 danger board;danger warning sign
危险警告信标 hazard beacon
危险警告信号 danger warning signal
危险境遇 deathtrap
危险距离 danger distance; danger range; exposure distance; risk distance
危险考验 <建筑上的> severe test
危险可疑点 <法规海图上可疑点标志> ouvrel loeil
危险控制 hazard control
危险控制程序 risk control program(me)
危险控制区 hazard control zone
危险类别 class of risk
危险量指数 danger number
危险路段 dangerous section; dangerous stretch of road
危险路肩 <不能保证行车安全的路肩> hazardous shoulder
危险路线 hazardous location
危险率 danger scale;risk
危险率估计 risk assessment
危险面 plane of weakness
危险面积 hazardous area
危险浓度极限 hazard concentration limit
危险品 danger;dangerous articles;dangerous cargo; hazardous cargo; hazardous goods;high hazard content;label(l)ed cargo
危险品包装 dangerous articles package
危险品包装标贴 hazardous substance mark
危险品标志 dangerous mark
危险品仓库 dangerous cargo warehouse; dangerous goods warehouse;hazardous cargo house;hazardous material storage
危险品场地 dangerous cargo area
危险品储藏室 hazardous storage
危险品处理工作队通信[讯] hazardous materials team communication
危险品的救援 hazardous material rescue
危险品堆场 dangerous yard
危险品隔离表 dangerous table
危险品工业区 dangerous industrial district
危险品管理 hazardous material management;hazmat management
危险品规则 dangerous goods code; dangerous goods regulations
危险品货场 danger goods team yard
危险品货物 dangerous goods
危险品集装箱 dangerous cargo container
危险品金钢标志 hazard diamond mark
危险品库 storage for dangerous goods
危险品流出量 <溢流、漏出等事故的> discharge of hazardous wastes
危险品码头 dangerous cargo wharf; dangerous cargo terminal; hazardous goods terminal
危险品锚地 dangerous goods anchorage
危险品区 hazardous area
危险品事故 hazardous material incident
危险品事故响应队 hazardous material response team
危险品条款 hazardous cargo clause
危险品箱堆场 hazardous cargo container yard
危险品运输 carriage of dangerous goods;dangerous transport
危险品运输部 Office of Hazardous Materials Transportation
危险品运输船 barge for "hot" cargoes
危险品运输条例 Hazardous Material Transportation Act
危险品制造厂 manufactory of hazardous articles
危险品重箱 dangerous cargo
危险品咨询理事会 Hazardous Materials Advisory Council
危险品资质认定 dangerous cargo certification
危险期 critical stage;dangerous period
危险期前用火 early burning
危险浅滩 vigia
危险浅滩(地)区 critical shoal area
危险情况 hazardous condition;hazard scenario
危险区 danger zone;shoal area
危险区的回避 avoidance of dangerous region
危险区堵墙 cross-off
危险区段 danger sector
危险区间 hazardous area
危险区闪光灯 hazard flasher
危险区域 dangerous area; dangerous space; dangerous zone; deathtrap; endangered area; explosive area; hazard(ous) area;risk area
危险商品 hazardous commodity
危险识别号 hazardous identification number
危险示像 danger aspect
危险事故 accident hazard
危险水道 dangerous channel; dangerous passage;hazardous passage
危险水平 danger level
危险水位 danger level; dangerous stage;flood stage
危险水域 foul waters
危险速率 neckbreaking speed
危险探测 danger sounding
危险特性 hazardous characteristic
危险条件 unsafe condition
危险通报 danger message
危险弯段 awkward bend; dangerous bend;hazardous bend
危险位置 danger position
危险位置指示浮标 danger buoy
危险温度 dangerous temperature
危险污染物 hazardous pollutant
危险物 dangers
危险物料 hazardous material
危险物品 dangerous goods;hazardous material;hazardous matter;hazardous substance
危险物品泄漏 hazardous material release
危险物质 dangerous substance; hazardous matter;hazardous substance
危险系数 dangerous coefficient
危险显示 danger position
危险线 damage risk contours;limiting danger line;danger line
危险线向 dangerous alignment
危险线形 dangerous alignment
危险限度 hazard limit
危险象限 dangerous quadrant
危险信号 damage signal;danger(ous) signal;red lamp;red light;warning signal
危险信号灯 danger light;danger signal lighting
危险信号灯光 danger light; warning light
危险信号浮标 danger signal buoy
危险信号牌 danger board
危险信号闪光 hazard way light
危险行为 hazardous act
危险性 risk
危险性变化 hazard variation
危险性程度 level of risk
危险性处理 risk management
危险性地震 hazardous earthquake
危险性分级 grade of risk
危险性分类 hazardous classification
危险性分类制度 hazard classification system
危险性分析 risk analysis
危险性函数 risk function
危险性化学反应 hazardous chemical reaction
危险性化学药品 hazardous chemical
危险性货物 dangerous goods
危险性模型 risk model
危险性评定 risk assessment
危险性评价 risk assessment
危险性生产材料 hazardous production material
危险性损坏 hazardous failure
危险性特征评定 risk characterization
危险性职业 hazardous occupation
危险性质变更 change in nature of risk
危险性转移 risk transfer
危险性作业 hazardous operation
危险悬帮 heavy ground
危险压力 dangerous pressure
危险药品柜 dangerous drug cupboard
危险药品条例 Dangerous Drugs Act
危险医疗废物 hazardous medical waste
危险抑制剂 dangerous inhibitor
危险溢出物 hazardous spill
危险因素 danger factor;risk factor
危险应变 severe strain
危险应力 danger stress;severe stress
危险影响 dangerous influence
危险由货主负责 owner's risk
危险与有毒物质 hazardous and noxious substance
危险阈(值) hazard threshold;danger threshold
危险圆 danger circle;risk circle
危险圆柱面 dangerous cylinder
危险灾害 hazards
危险载货跟踪 hazardous cargo tracking

危险责任的起期 attachment of risk
危险值 risk value
危险职业 dangerous occupation;hazardous occupation
危险指示标 hazard(ous) marker
危险指数 danger index;hazardous(ness) index
危险状况 danger situation
危险状态 dangerous condition;hazardous behavio(u)r;hazardous condition
危形 <结构的> critical form
危形结构 critical form structure
危崖落石 bluff falling
危岩 dangerous rock;hanging rock;overhanging rock

威

威百亩【化】metham

威布效应 <泥浆解凝时体积变化> Webb effect
威德尔海 Weddell Sea
威德曼-弗郎兹常数 Wiedemann-Franz constant
威德曼-弗朗兹比值 Wiedemann-Franz ratio
威德曼-弗朗兹定律 Wiedemann-Franz law
威德曼加和律 Wiedemann's additivity law
威德曼效应 Wertheim effect;Wiedemann effect
威尔顿管式炉 Wilton still
威尔顿机织绒头地毯 Wilton rug
威尔顿提花地毯 Wilton carpet
威尔豪斯法 <木材防腐,氯化锌与单宁浸注作业> Welhause process
威尔金森蓝 Wilkinson's blue
威尔金逊窑 Wilkinson oven
威尔什石 wilshite
威尔士拱 underpitch groin;Welsh arch
威尔士盆地 Wales basin
威尔士穹顶 Welsh vault
威尔士穹顶肋 Welsh groin
威尔士型充气式浮选机 Welsh cell
威尔士硬砖 Welsh brick
威尔斯登式纸材 Willesden paper
威尔逊-θ(逐步积分)法 Wilson-θ method
威尔逊凹陷 Wilson depression
威尔逊标准相带 standard facies belts proposed by Wilson
威尔逊齿轮 Wilson gear
威尔逊定理 Wilson's theorem
威尔逊密封 Wilson seal
威尔逊射线径迹 Wilson tracks
威尔逊实验 Wilson experiment
威尔逊碳酸盐标准相带沉积模式 sedimentation model of Wilson's carbonate standard facies belts
威尔逊线 Wilson line
威尔逊效应 Wilson effect
威尔逊旋回 Wilson cycle
威尔逊云室【物】Wilson cloud chamber;cloud chamber
威格纳沉降分析管 Weigner sedimentation tube
威格纳法 Weigner method
威吉伍德高温计 Wedgwood pyrometer
威忌州松 Jersey pine
威克板层砂岩群 Wick flagstone group
威克曼黄石 Wakeman buff
威克曼螺纹量规 Wickman ga(u)ge
威克歇姆版楔 Wickersham quoin
威-兰-菲三氏关系式 <液体黏[粘]度与温度关系式> Williams-Landel-Ferry relation
威兰斯线 Willans line
威勒特氏钳 galea forceps;Willett forceps
威里方程 Wieri equation
威利瓦飑 williwaw
威廉管 Williams tube
威廉管存储器 Williams tube storage
威廉生铁杉 Williamson's spruce
威廉斯可塑性恢复值 plasticity-recovery number-Willams
威廉塑度计 William's plastometer
威廉塑性仪 <平板式> William's plastometer
威廉逊合成 Williamson synthesis
威廉逊氏试验 Williamson's test
威廉逊型放大器 Williamson amplifier
威廉逊旋回 Williamson turn
威廉逊窑 Williamson kiln
威廉折射计 William's refractometer
威廉珍尼泵 William Janney pump
威罗机 box willow
威洛变位图 Williot deflection diagram
威洛-摩尔图(解) <桁架结点绝对变形的图解> Williot-Mohr diagram
威洛图 Williot diagram
威洛图解法 Williot graphic(al) method
威洛位移图 Williot displacement diagram
威姆科型浮选机 Wemco flo(a)tation cell
威姆科型旋流器 Wemco cyclone
威尼斯白(色颜料) Venetian white;Venice white
威尼斯百叶窗 Venetian blind;Venetian shutters
威尼斯百叶窗式快门 Venetian blind shutter
威尼斯百叶窗式直接换能器 Venetian blind direct energy converter
威尼斯百叶窗效应 Venetian blind effect
威尼斯百叶窗形倍增器 Venetian blind multiplier
威尼斯(百叶)门 Venetian door
威尼斯-拜占庭风格建筑 Veneto-Byzantine style
威尼斯玻璃器皿 Venice glass
威尼斯玻璃球 Venetian ball
威尼斯淡粉红 Venetian pink
威尼斯岛区 Rialto
威尼斯缎纹织物 Venetian cloth
威尼斯风格 Venetian motif
威尼斯拱 Venetian arch
威尼斯红 <主要成分为氧化铁> Venetian red
威尼斯花边 Venetian lace
威尼斯画派 <文艺复兴时代> Venetian school
威尼斯黄 amber yellow;Venetian yellow
威尼斯尖拱 Venetian pointed arch
威尼斯蓝 Venice blue
威尼斯绿 Venice green
威尼斯马赛克 Venetian mosaic
威尼斯玫瑰红 Venetian rose
威尼斯排齿饰 Venetian dentils
威尼斯排齿饰线脚 Venetian dentil mo(u)lding
威尼斯平底小船 gondola
威尼斯漆树 Venetian sumac
威尼斯嵌镶 terrazzo;Venetian mosaic
威尼斯式窗 picture window;Venetian light;Venetian window
威尼斯式灯具 Venetian light
威尼斯式镜 Venetian mirror
威尼斯式面层 Venetian topping
威尼斯式遮篷 Venetian awning
威尼斯水磨石 Venetian;Venetian terrazzo
威尼斯丝绒 Venetian velvet
威尼斯松节油 Venice turpentine
威尼斯塔牌胶合板 <一种著名高级胶合板> Venesta
威尼斯通风盖 Venetian cover
威尼斯文艺复兴式 Venetian Renaissance style
威尼斯镶面板 Venetian mosaic
威尼斯镶嵌 Venetian;Venetian mosaic
威尼斯小划船 Venetian gondola
威尼斯学派 Venetian school
威尼斯整理 Venetian finish
威尼斯总督宫 Doge's Palace of Venice
威期康星阶【地】Wisconsinian (stage)
威期特发利亚阶【地】Westfalian (stage)
威奇特勒紫金色料 Waechtler's gold purple
威慑能力 deterrent capability
威慑物 deterrence;deterrent
威慑因素 deterrence;deterrent
威式桁架 Howe truss
威斯德桩 West pile
威斯公式 Wiese formula
威斯康辛冰期 Wisconsian glacial stage;Wisconsin glacial epoch
威斯康辛期【地】Wisconsin period
威斯康辛通用测验仪 Wisconsin general testing apparatus
威斯康星 <美国州名> Wisconsin
威斯特福克地体【地】West Fork terrane
威斯特卡德公式 <设计混凝土路面厚度用的> Westergaard's formula
威斯特卡德理论 <设计水泥混凝土路面的> Westergaard's theory
威斯特莱顿层 Westleton beds
威斯特摩兰差动滑轮 Westmorland differential pulley
威斯特摩兰石板(瓦) Westmorland slate
威特 AIO 自动绘图仪 Wild AIO autograph
威特理论 Witt theory
威特色理论 Witt's colo(u)r theory
威特向量 Witt vector
威胁感 threat perception
威兹珀风 wisper wind

微

微 H 圆形无线电导航法 micro H circular radio navigation

微埃 microangstrom
微安 <安培的百万分之一> microammeter;microampere
微安表 microam(pere) meter
微安电路 microampere circuit
微安计 microam(pere) meter
微安灵敏度 microampere sensibility
微安(培)秒 milliampere-second
微暗 duskiness;duskish
微暗的 darkish;dun;dusk
微暗镜煤 durovitrite
微暗亮煤 duroclarite
微暗煤 durite
微暗煤质结构镜煤 durotelovitrite
微暗色 dusk
微凹表玻璃 miconcave
微巴 <旧压强单位> microbar;barie [barye]
微靶 microbarn
微白色的 whitish
微斑板岩 fleckschiefer
微斑橄玄岩 dalmeny basalt
微斑结构 microporphyritic texture
微斑状 microphyric
微斑状的 microporphyritic
微板(块) microplate
微半木质镜煤-丝炭 semi-xylovitrofusite
微半丝煤 semi-fusite
微孢子暗亮煤 sporoduroclarite
微孢子暗煤 sporodurite
微孢子亮暗煤 sporoclarodurite
微孢子亮煤 sporoclarite
微孢子煤 sporite
微胞(团) micell(e) [复 micellae];nannocyte;super-molecule
微爆 microexplosion
微爆凿岩 microblasting drilling
微贝壳灰岩 microcoquina
微比长计 microcomparator
微编码 microcoding
微编码控制器 microcoded controller
微编码指令 microcoded instruction
微变 differential change
微变化 microvariation
微变径规 vernier cal(l)ipers
微变量增益 incremental gain
微变灵敏度 variational sensitivity
微标准 microstandard
微标准学 micromeritics
微表层 microlayer
微冰 slight ice
微冰冻箱 microcryostat
微冰隙 very small fracture
微波 cellular wave;hyperfrequency wave;microray;microwave;riffle;ripple;rippled sea;very smooth sea
微波安装设备 microwave housing facility
微波巴氏灭菌法 microwave pasteurization
微波保护装置 microwave protection unit
微波暴 microwave burst
微波背景 microwave background
微波背景辐射 microwave background radiation
微波本振源 microwave local oscillator
微波标记 microwave label(l)ing
微波表面电阻 microwave surface resistance
微波波长 microwave wavelength
微波波长计 microwave wavemeter
微波波导管 microwave plumbing
微波波段 microwave band;microwave region
微波波谱 microwave spectrum
微波波谱学 microwave spectroscopy
微波测定仪 microwave absorption moisture gage
微波测距 distance measurement;microwave;microwave distance measuring;microwave ranging;radio ranging measurement;tellurometer survey
微波测距导线 microwave ranging traverse;tellurometer traverse
微波测距仪 <用于测量距离的一种微波仪器> cubitape distance meter;microdistancer;microwave distance measuring instrument;microwave distance measuring system;microwave rangefinder;tellurometer
微波测距仪零点差 zero error of microwave distance measuring instrument
微波测量 microwave measurement
微波测湿仪 microwave moisture meter
微波测试接收机 microwave measuring receiver
微波场 microwave field
微波超声 microwave ultrasound

微波超声波 microwave ultrasonic wave
微波超声学 microwave ultrasonics
微波车 microwave vehicle
微波处理机 microprocessing unit; microprocessor; microwave processing unit
微波传播 microwave propagation; propagation of microwave
微波传感器 microwave sensor
微波传输带 microstrip
微波传输电路 microwave transmission circuit
微波传输设备 microwave transmission unit
微波传输塔 microwave transmission tower
微波传输线 microwave transmission line
微波传送 microwave transmission
微波传送带 microstrip; microwave strip
微波窗口 microwave window
微波催化氧化法 catalytic microwave oxidation process
微波萃取 microwave extraction
微波大气遥感 microwave remote sensing of atmosphere
微波带状传输线 microwave strip line
微波导标 microwave course beacon
微波等离子体 microwave plasma
微波等离子体气相沉积 microwave plasma deposition
微波等离子体气相化学沉积 microwave plasma chemical vapo(u)r deposition
微波等离子体枪 microwave plasma gun
微波电话 microwave telephone
微波电路 microwave channel; microwave circuit
微波电子回旋加速度 microwave electron cyclotron
微波定位系统 microwave position fixing system
微波动力等离子体光源 microwave powered plasma light source
微波多普勒效应速度传感器 microwave Doppler speed sensor
微波发射机 microwave transmitter
微波发射检测器 microwave emission detector
微波发射率 microwave emissivity
微波发生器 micro-wave generator
微波发送混频器 microwave transmitting mixer
微波法 microwave method
微波反射计 microwave reflectometer
微波防护眼镜 microwave shielding eyeglass
微波放电检测器 microwave discharge detector
微波分光计 microwave spectrometer
微波分光镜 microwave spectroscope
微波分解 microwave decomposition
微波分量 microwave component
微波分析 microwave analysis
微波风场散射计 microwave wind field scatterometer
微波辐射 microwave radiation
微波辐射测量 microwave radio metry
微波辐射计 microwave radio meter
微波辐射仪 microwave radio meter
微波辅助萃取 microwave assisted extraction
微波辅助氧化法 microwave assisted oxidation process
微波干涉量度法 microwave interferometry
微波干涉仪 microwave interferometer
微波干燥 microwave drying
微波干燥器 microwave drier [dryer]
微波工程学 microwave engineering
微波功率放大器 microwave power amplifier
微波共振 microwave resonance
微波固化 microwave
微波观察仪 microvision
微波管 microwave tube
微波光谱学 microspectroscopy; microwave spectroscopy
微波光学 microwave optics
微波海面 lipper; smooth sea
微波含水率仪 microwave moisture apparatus
微波合成 microwave synthesis
微波烘炉 <快速测定土壤含水率> microwave oven
微波回旋加速器 microwave cyclotron
微波回转器 microwave gyrator
微波混合集成电路 microwave hybrid integrated circuit
微波混频器网络 microwave mixer network
微波活化 microwave activation
微波激射 microwave amplification by stimulated emission of radiation
微波激射弛豫 maser relaxation
微波激射放大器 maser amplifier
微波激射干涉仪 maser interferometer
微波激射器 maser [microwave amplifier by stimulated emission of radiation]
微波激射振荡器 maser oscillator
微波激射作用 maser action
微波集成电路 microwave integrated circuit
微波技术 microwave technique
微波加热 microwave heating
微波加热法 microwave heating method
微波加速器 microwave accelerator
微波检波器 microwave detector
微波检测器 microwave detector
微波检查公路 microwave inspection of highway
微波检验 microwave inspection
微波简谐振荡管 fawshmotron
微波鉴频器 microwave discriminator
微波降解 microwave degradation
微波交指型结构 microwave interdigital structure
微波接力 microwave relay
微波接力分转站 access station
微波接力通信[讯]系统 microwave relay communication system; microwave repeater(communication) system
微波接力通信[讯]站 <简称微波站> terrestrial microwave relay station
微波接力线路 microwave link
微波接收机 hyperfrequency wave receiver; microwave receiver
微波接收器 microwave receiver
微波解冻 microwave thawing
微波开关管 microwave switching tubes
微波烤箱 microwave oven
微波空腔谐振器 microwave resonant cavity
微波雷达 microwave radar
微波联络机 microwave service equipment
微波链路 microwave link
微波亮度温度 microwave brightness temperature
微波量子放大器 maser [microwave amplifier by stimulated emission of radiation]; microwave quantum amplifier
微波漏泄 microwave leakage
微波炉 microwave oven
微波炉垫布 microwave oven gasketting
微波滤波器 microwave filter
微波脉塞 microwave maser
微波能量 microwave energy
微波频率 microwave frequency
微波频谱 microwave frequency spectrum
微波频谱法 microwave spectroscopy
微波频谱学 microwave spectroscopy
微波频谱仪 microwave spectrometer
微波屏蔽门 microwave shield door
微波谱 microwave spectrum
微波谱线 microwave line
微波汽车 microwave car
微波区 microwave region
微波全摄影 microwave holography
微波全息雷达 microwave hologram radar
微波日像仪 microwave heliograph
微波三应答定位仪 microwave trisponder
微波散射计 microwave scatterometer
微波散射仪 microwave scatterometer
微波扫描辐射测量 microwave scanning radiometry
微波扫描辐射仪 microwave scanning radiometer
微波扫描仪 microwave scanner
微波扫频仪 microwave sweep signal generator
微波烧成 microwave firing
微波烧结 microwave sintering
微波设备 micro-wave equipment
微波声学 microwave acoustics
微波视距通信[讯] line-of-sight microwave
微波收发两用机 microwave transmitter-receiver
微波衰减器 microwave attenuator
微波衰减陶瓷 microwave attenuation ceramics
微波水流 ripple flow
微波水平线 microwave horizon
微波损伤 microwave damage; microwave injury
微波锁定 microlock
微波探测器 microwave sounder
微波探测系统 microwave survey system
微波探测装置 microwave sounder unit; microwave survey system
微波探头 microwave probe
微波陶瓷 microwave ceramics
微波天线 microwave aerial; microwave antenna
微波天线屏蔽器 radom(e)
微波调制器 microwave modulator
微波铁氧化体 microwave ferrite
微波铁氧体 microwave ferrite
微波铁氧体材料 microwave ferrite material
微波铁氧体环行器 microwave ferrite circulator
微波通信[讯] microwave communication; short-wave communication
微波通信[讯]设备 microwave communication equipment
微波通信[讯]网 microwave communication network; microwave network
微波通信[讯]系统 microwave communication system
微波通信[讯]站 microwave station
微波透热法 microkymatotherapy; microwave diathermy
微波图像 microwave image
微波脱硫 microwave desulfurization
微波网络 microwave network
微波无线电 microwave radio
微波无线电传送中继系统 microwave radio relay system
微波无线电导航标 microwave beacon
微波无线电导航系统 microwave radio metric(al) navigation system
微波无线电通信[讯]线路 radio communication line of centimetric wave
微波无线电网 microwave radio network
微波无线电中继通信[讯]系统 microwave radio relay communication system
微波无线电中继通信[讯]线路 microwave radio relay link
微波吸收 microwave absorption
微波吸收材料 microwave absorbent material
微波吸收频谱学 microwave absorption spectroscopy
微波系统 microwave system
微波相位均衡器 microwave phase equalizer
微波消解 microwave digestion
微波消声室 microwave anechoic cell
微波信道 microwave channel
微波型 minute wave type
微波旋转光谱学 microwave rotational spectroscopy
微波雪崩二极管 microwave avalanche diode
微波-亚毫米波转换电子谐振器 tornadotron
微波养护 microwave curing
微波遥感 microwave remote sensing
微波遥感技术 microwave remote sensing technology; technology of microwave remote sensing
微波遥感器 microwave remote sensor; microwave sensor
微波移动探测器 microwave motion detector
微波移相器 microwave phase shifter
微波影响 effect of microwave
微波诱导催化剂 microwave induced catalyst
微波诱导氧化工艺 microwave induced oxidation process
微波预测碰撞传感 microanticipatory crash sensor
微波预测碰撞传感器 microwave anticipatory crash sensor
微波预警 microwave early warning
微波预警雷达 microwave early warning radar
微波元件 microwave device
微波噪声 microwave noise
微波噪声系数测试仪 noise-figure measuring set for microwave equipment
微波站 microwave station
微波照射 microwave illumination
微波折射计 microwave refractometer
微波折射率 microwave index of refraction
微波折射系数 microwave index of refraction
微波诊断 microwave diagnostics
微波诊断技术 microwave diagnostic technique
微波诊断仪 microwave diagnostic apparatus
微波振荡器 microwave oscillator
微波振荡源 microwave generating

W

source
微波正常传播距离 microwave horizon
微波中继 microwave relay
微波中继器 microwave link; microwave repeater
微波中继通信[讯] microwave relay communication
微波中继通信[讯]系统 microwave relay communication system
微波中继站 microwave relay station
微波中继装置 microwave relay unit
微波终端设备 microwave terminal equipment
微波终端站 microwave terminal station
微波转换开关 waveguide transfer switch
微波转接机 Interconnection set for micro-wave and wire communication
微波着陆系统 microwave landing system
微波钻机 microwave drill
微泊 micropoise
微薄板 microveneer
微薄层黏[粘]土 book clay
微薄的 microthin; slender
微步 microstep
微操作 micromanipulation; microoperation
微操作码 microcode
微操作设备 micromanipulator
微糙的 roughish
微槽 microflute
微侧向测井 microlaterlog
微侧向测井曲线 microlaterolog curve
微测标 micromark
微测标立体刺点仪 micromark stereo point marking instrument
微测快慢针调整器 micrometer regulator
微测量 micrometering
微测深法 contact log
微测压计 microtonometer
微层 microbedding
微层覆盖工艺 micro sealing process
微层理 microbedding; microstratification
微层填覆工艺 micro sealing process
微层序 microsequence
微差 elemental error; elementary error; hair breadth
微差爆破 fast-delay detonation; millisecond blasting; short delay blasting; split-second blasting
微差测定法 differential methods of measurement
微差测量 differential measurement
微差层 differential stratification
微差迟发电雷管 millisecond delay electric(al) detonator
微差电雷管 split-second delay detonator
微差法 differential method; nuance
微差井温测井图 differential temperature log
微差量测 differential measurement
微差平差法 differential adjustment
微差起爆 millisecond priming
微差水准测量 differential level(l)ing
微差体积描记法 microplethysmography
微差温度表 differential thermometer
微差温度曲线 differential thermal curve
微差隙 differential gap
微差压力计 differential manometer; differential pressure ga(u)ge
微差延发爆炸 short delay blasting
微差延发雷管 fast-delay cap
微差移动 differential movement
微掺和物 microadmixture
微场电机放大器 amplidyne
微场电机放大器激磁机 amplidyne exciter
微场电流放大机 amplidyne generator
微场扩流变换器 metadyne converter
微场扩流发电机 amplidyne generator; metadyne; metadyne generator
微场扩流发电机控制 metadyne control
微超固结黏[粘]土 micro-overconsolidated clay
微尘 dust; fine dust; impalpable powder; mote
微尘成分 composition of dust
微尘学 coniology [koniology]; micrometerits
微程序 microcode; microprogram(me)
微程序化的微微处理机 microprogrammed picoprocessor
微程序计算机 microprogrammed computer
微程序开发 microprogram(me) development
微程序控制 microprogram control; microprogramming
微程序控制的 microprogrammed
微程序控制的处理机 microprogrammable processor
微程序控制器 microprogrammed controller; microprogrammed control unit; microprogramming controller
微程序控制数据处理机 microprogrammed data processor
微程序描述 microprogram description
微程序模块 micromodule
微程序设计 microcoding; microprogram design; microprogramming
微程序设计法 microprogramming approach
微程序设计软件 microprogramming software
微程序设计语言 microprogramming language
微程序设计员 microprogrammer
微程序设计支持软件 microprogramming support
微程序时序器 microprogram sequencer
微程序寻址 microprogram addressing
微程序优化 microprogram optimization
微程序语言 firm ware
微程序支持软件 microprogram support software
微程序只读存储器 microm
微程序转移 microprogram branching
微尺度 microscale
微尺度分布 microscale distribution
微处理机 microprocessing unit; microprocessor
微处理机保护装置 microprocessor protector
微处理机标记 microprocessor mark
微处理机测试系统 microprocessor testing system
微处理机电子学 microprocessor electronics
微处理机检测器 microprocessor detector
微处理机开发系统 microprocessor development system
微处理机控制 microprocessor control
微处理机控制可行性 microprocessor control feasibility
微处理机控制系统 microprocessor control system
微处理机体系结构 microprocessor architecture
微处理机调试程序 microprocessor debugging procedure
微处理机(信号)控制机 microprocessor controller
微处理机信号系统 microprocessor-based signaling system
微处理机掩模 microprocessor mask
微处理机终端 microprocessor terminal
微处理机主从系统 microprocessor master/slave system
微处理机主时钟 microprocessor master clock
微处理机自动错误校正 microprocessor auto error correction
微处理控制的显示终端 microcomputer-controlled display terminal
微处理器 microprocessing unit; microprocessor unit
微处理器编译程序 microprocessor compiler
微处理器超高速缓冲存储器 microprocessor cachememory
微处理器代码汇编程序 microprocessor code assembler
微处理器分析器 microprocessor analyser
微处理器汇编模拟程序 microprocessor assembler simulator
微处理器控制的仪器 microprocessor instrument
微处理器芯片 microprocessor chip
微处理器掩模组 microprocessor mask set
微处理器语言汇编程序 microprocessor language assembler
微处理器终端 microprocessor terminal
微处理随机存取存储器接口 microprocessor ROM interface
微处理随机存取存储器写入程序 microprocessor ROM programmer
微穿刺 micropuncture
微穿孔 microperforate; micropunch
微穿孔板 microperforated panel
微穿孔板吸声结构 microperforated panel acoustic(al) construction
微穿孔板消声器 micropunch plate muffler
微穿孔吸声结构 microperforated absorber
微传感器 microsensor
微疵点 microcracking
微磁测量 micromagnetic survey
微磁磁力仪 micromagnetometer
微丛 microbundle
微粗斑状 magniphyric
微粗糙度 microroughness
微粗糙量 microasperity
微粗粒煤 macroite
微存储器 microstorage
微达因除尘器 microdyne scrubber
微大陆 microcontinent
微大洋 microocean
微代码 microcode
微代码指令系统 microcode instruction set
微代码指令系统成分 microcode instruction set component
微带 microstrip
微带传输线 microstrip transmission line
微带存储器 fine strip memory
微带酸性的泉 acidulous spring
微带酸性的水 acidulous water
微带线 microstrip line
微带线功率分配器 microstrip power divider
微单位 microunit
微倒度 micro-reciprocal degree; mired
微等离子区 microplasma
微等离子体 microplasma
微滴补体结合试验 microtiter complement fixation test
微滴定法 microtitration
微滴乳液共聚合作用 microemulsion copolymerization
微滴乳状液 microemulsion
微滴术 little drop technique
微滴状态 microdroplet status
微地理学 microgeography
微地貌 microrelief; microtopography; minor feature
微地貌学 microgeomorphology
微地形 nanno-relief
微地形学 microtopography
微地形重力测量 gravimetric(al) microrelief
微地震 earth tremor; microseism
微地震测井 microseismogram log
微地震测量仪 tromometer
微地震仪 tremometer
微点 molecule
微点法 dot system
微电池 microbattery
微电池对 microcell
微电机 micromachine; micromotor; print motor; small and special electric machine
微电极 microelectrode
微电极测井 microelectrode log; microlog
微电极测井曲线 microelectrode log curve
微电极测井图 microelectrode log plot
微电极放大器 microelectrode amplifier
微电极系 microsonde
微电极组 multimicroelectrode
微电解 microelectrolysis
微电解反应 microelectrolysis reaction
微电解废水处理 microelectrolysis wastewater treatment
微电解技术 microelectrolysis technology
微电解杀菌 microelectrolysis sterilization
微电解升流厌氧污泥床接触生物氧化 microelectrolysis upflow anaerobic sludge bed contact biooxidation process
微电解消毒 microelectrolysis disinfection
微电解氧化 microelectrolysis oxidation
微电量滴定法 microcoulometric(al) titration
微电量分析法 microcoulometry
微电量检测器 microcoulometric(al) detector
微电量气相色谱法 microcoulometric (al) gas chromatography
微电路隔离 microci isolation
微电路技术 microcircuitry
微电路模板 microci stencil
微电路片 microci chip
微电脑 microcomputer
微电脑操纵的联锁 electronic microcomputer controlled interlocking
微电容器 microcapacitor
微电位测井 micronormal log
微电位测井曲线 micronormal log curve
微电位计 micropot(entiometer); microvolt(o)meter

W

微电位曲线 micronormal
微电泳法 microelectrophoresis
微电泳技术 microelectrophoretic technique
微电子材料 microelectronic material
微电子的 microelectronic
微电子技术 microelectronics; microelectronic technique
微电子控制 microelectronic control
微电子设备 microelectronic device
微电子探测器 <可置于塑囊中吞入体内记录生理数据> endoradiosonde
微电子微型组件装置 microelectronic modular assembly
微电子学 microelectronics; microsystems electronics; microtronics
微电子学雷达相控阵 microelectronic radar array
微电子元件 microelectronic element
微电阻测量 microresistivity survey
微电阻器 microresistor
微定位器 micropositioner
微定值 microposition
微动 fine motion; fine movement; inching motion; jogging; microdrive; microinching; slow motion; weak movement
微动按钮 inching button
微动臂 inching arm
微动操作 inching operation
微动弹簧 hair
微动的 inching
微动电门 microswitch
微动阀 inching valve
微动砝码 jockey weight; sliding weight
微动分解仪 vernier resolver
微动换向作用 light-touch steering action
微动计 microseismograph
微动计时器 microchronometer
微动继电器 microrelay
微动开关 inching switch; microactive switch; microswitch; slow-motion switch
微动控制 inching control; jogging control; slow-motion control; vernier control
微动力清毒 oligodynamic(al) disinfection
微动力学 oligodynamics
微动螺丝 fine motion screw; slow-motion screw; spring-loaded screw; tangent(ial) screw
微动螺旋 fine motion screw; slow-motion screw; spring-loaded screw; tangent(ial) screw
微动脉 arteriole
微动脉造影术 microangiography
微动描记器 micrograph
微动气压计 statoscope
微动气压器 statoscope
微动踏板 inching pedal
微动台 micropositioner
微动态特征 microdynamic feature
微动同步器 microsyn
微动物区系 microfauna
微动物群 microfauna
微动行程 microstroke
微动性钾 slightly mobile potassium
微动液面控制 microlevel(l)ing control
微动制动器 finger-tip brake
微动制动作用 light-touch braking action
微动轴环 microcollar
微动装置 micromotion unit; vernier arrangement
微动作 micromotion
微动作用 oligodynamic(al) action; oligodynamics
微冻结 minute freezing
微度 mired
微度不整合 gentle angular unconformity
微断面 microprofile
微钝锯齿状 crenulation
微惰屑煤 inertodetrite
微惰性煤 inertite
微尔格 microerg
微法计 microfarad meter
微法(拉) microfarad
微放大器 micrograph; micromag
微放射计 minometer
微分【数】 differentiate
微分比容 partial specific volume
微分不变式 differential invariant; reciprocant
微分不等式 differential inequality; reciprocant
微分参数 differential parameter
微分测量 differential measurement
微分测图法 differential methods of photogrammetry
微分差分方程 difference-differential equation
微分差热分析 derivative differential thermal analysis
微分冲淡热 differential heat of dilution
微分磁化率 differential susceptibility
微分导磁率 differential permeability
微分的 differential; differentiated
微分电离室 differential ionization chamber
微分电路 differential circuit; differentiating circuit; differentiating network; differentiator; peaker; sharpening circuit
微分电容 differential capacitance
微分电阻 differential resistance
微分电阻仪 differential resistance type instrumentation
微分动作 differential action
微分对数脉冲偏码调制 differential-logarithmic impulse-code modulation
微分多普勒 differential Doppler
微分多项法 differential polynomial
微分多项式环 ring of differential polynomials
微分发生器 differential generator
微分法 differential calculus; differential method; differentiation; differential calculation
微分法测图 differential methods of photogrammetry; differential photo
微分法全野外布点 differential methods for field point layout
微分法则 differential law
微分反馈 derivative feedback
微分反应 differential response
微分反应速度 differential reaction rate
微分方程 differential equation
微分方程的极限环 limit cycle of a differential equation
微分方程的阶 order of a differential equation
微分方程的(普)通解 general solution of differential equation
微分方程的影响函数核 kernel
微分方程定解条件 conditions of determining solution of differential equation
微分方程计算机 differential equation computer; differential equation solver
微分方程解法 solution of differential equation
微分方程解算机 differential equation solver
微分方程论 theory of differential equations
微分方程模拟计算机 differential equation analog(ue) computer
微分方程式 differential equation
微分方程组 differential system
微分方法 differentiation technique
微分放大器 differential amplifier; differentiator amplifier
微分分光光度测定法 differential spectrophotometry
微分分析 differential analysis
微分分析机 <解微分方程用计算机>【计】 differential analyser [analyzer]
微分分析器 differential analyser [analyzer]
微分分析仪 differential analyser [analyzer]
微分符号 differential sign; differential symbol
微分干涉反衬 differential interference contrast
微分干涉反衬显微镜 differential interference contrast microscope
微分公式 differentiation formula
微分功率表 Johnson power meter
微分号 sign of differentiation
微分和积分方程程序 differential and integral equation program(me)
微分化学反应器 differential chemical reactor
微分回受 derivative feedback
微分积分法 differential integral method
微分几何(学) differential geometry; infinitesimal geometry
微分几何因子 differential geometric factor
微分计算电位器 differential computing potentiometer
微分继电器 rate of change relay
微分节流控制 differential throttle control
微分截面 differential cross-section
微分介质常数 differential permittivity
微分静压控制器 differential static pressure controller
微分纠正 differential rectification
微分纠正仪 differential rectifier
微分矩阵 differential matrix
微分均衡器 derivative equalizer
微分控制 differential control; rate action control; rate control; rate-controlling
微分控制器 differential controller
微分类学 microtaxonomy
微分粒度频率曲线 differential size frequency curve
微分灵敏度 differential sensitivity
微分流动检测器 differential flow detector
微分流形 differentiable manifold
微分马达 differential motor
微分脉冲 differential pulse; differentiating pulse
微分脉冲极谱法 differential pulse polarography
微分脉冲阳极溶出伏安法 differential pulse anode solvation voltammetry
微分面积 elementary area
微分平滑装置 differential smoothing device
微分平行四边形 differential parallelogram
微分谱 differential spectrum
微分谱记录 record of differential spectrum
微分(气压)计 differential ga(u)ge
微分器 differentiator
微分迁移率 differential mobility
微分曲线 differential curve; non-cumulative curve
微分热解重量分析 differential thermal gravimetric analysis; differential thermogravimetric analysis
微分溶解热 differential heat of solution
微分色散 dispersivity quotient
微分筛析 differential screen analysis
微分闪烁图 differential scintigram
微分时间 derivative time; rate time
微分时间常数 derivative time constant; differentiating time constant
微分式 differential expression
微分算符 differential operator; nabla
微分算子的阶 order of a differential operator
微分算子的主部 principal part of a differential operator
微分调节器 derivative controller
微分同胚【数】 diffeomorphism
微分同胚的【数】 diffeomorphic
微分筒 microdrum
微分投影元素 differential projected element
微分图像 differential image
微分陀螺仪 rate gyro(scope)
微分拓扑 differential topology
微分网络 derivative network; differentiating network; lead network
微分温度测量装置 differential temperature measuring device
微分吸附热 differential heat of adsorption
微分吸收率 differential absorption rate
微分吸收热 differential heat of sorption
微分系数 differential coefficient; differential quotient
微分显声器 differential microphone
微分显微镜 micrometric(al) microscope
微分相位 differential phase
微分相位微分增益测试仪 differential phase-differential gain-test equipment
微分相位校正器 differential phase corrector
微分效应 differential effect
微分形式 differential form
微分修正 differential correction
微分学 differential calculus
微分因子 differential divisor
微分音 microtone
微分映射【数】 differentiable maps
微分元件 differentiating element; differentiator
微分运算 differential operation; differentiate
微分运算放大器 differential operational amplifier
微分(运)算子 differential operator
微分增量 differential gain
微分增益 differential gain
微分增益控制 differential gain control
微分增益控制器 differential gain controller
微分增益校正器 differential gain corrector
微分增益失真 differential gain distortion
微分增益特性 differential gain characteristic
微分增益旋钮 differential gain control
微分蒸馏 differential distillation

微分值 differential value
微分重量分布(函数) differential weight distribution
微分装置 derivator
微分子因素 micromolar factor
微粉 micromist;micropowder
微粉出口 fine outlet
微粉二氧化硅 fine particle silica
微粉分离机 micron separator
微粉分粒机 microplex
微粉化 micronizing
微粉机 jiyumill; pulverizer; pulverizing mill
微粉流道 dust chute
微粉密实的 densified with small particles construction material
微粉磨机 micron mill; micronizer [microniser]
微粉喷燃器 pulverised [pulverized] burner
微粉球磨机 microball-mill
微粉碎 micropulverizing
微粉碎机 atomizer;fine grinder
微粉碎黏[粘]土 micromilled clay
微粉振动筛 microvibrating shifter
微风 fair wind; light wind; scant wind; slight breeze; weak wind; whiffle; zephyr; gentle breeze; breeze
微风化 slightly weathering; weak weathering
微风化层 slightly weathered layer
微风化带 slightly weathered zone; slight weathering zone
微风化的 slightly weathered
微风化岩石 slightly weathered rock
微缝 fissure
微缝合线【地】 microstylolite
微缝愈合试验 microcrack healing test
微伏计 microvolter
微伏每米 microvolt per meter
微伏(特) microvolt
微伏(特)计 microvolt(o)meter
微浮沉子 microfloat
微浮选(法) microflo(a)tation
微幅波 small-amplitude wave
微辐射计 microradiometer; radiomicrometer
微腐泥混合煤 sapromixite [sapromyxite]
微腐泥镜煤 saprovitrite
微腐生物 oligosaprobiont
微富丝质暗煤 rich fusodurite
微覆盖 microoverlay
微伽 microgal
微高斯 microgasuss
微功耗锁相环 micropower phase locked loop
微功率 micropower
微构象 microconformation
微构造【地】 microtectonics
微构造地球化学 microtectono-geochemistry
微骨料火山灰质波特兰水泥 fine filler pozzolana Portland cement
微骨料火山灰质硅酸盐水泥 fine filler pozzolana Portland cement
微骨料煤粉波特兰水泥 fine filler fly ash Portland cement
微骨料煤粉硅酸盐水泥 fine filler fly ash Portland cement
微骨料水泥 fine filler cement
微观 microscope;microscopic view
微观包裹体 microscopic inclusion
微观波动 microoscillation
微观不均一性 microheterogeneity
微观不均匀性 microinhomogeneity; microscopic unevenness
微观不稳定性 kinetic instability; microinstability; microscopic instability
微观布朗扩散 micro Brownian diffusion
微观布朗运动 micro Brownian movement
微观常数 microscopic constant
微观场 microscopic field
微观催化反应器 microcatalytic reactor
微观的 microcosmic;microscopic;micrographic
微观地层(分析)仪 microstratigraphical switch
微观地层学 microstratigraphy
微观地层学的 microstratigraphical
微观地理学 microgeography
微观地形学 microtopography
微观电泳 microelectrophoresis
微观反应堆理论 microscopic pile theory
微观方法 microscopic approach; microscopic method
微观分布 microdistribution
微观分析 microanalysis; microscopic analysis
微观各向异性 microanisotropy; microscopic anisotropy
微观共轭 microconjugate
微观构造 microscopic structure; microstructure
微观构造学 microtectonics
微观观察 microscopic observation
微观规整性 microtacticity
微观环境 microenvironment
微观环境政策 microenvironment policy
微观会计 microaccounting
微观机构 micromechanism
微观机制 micromechanism
微观检查 micrographic(al) examination
微观检验 microexamination; microscopic examination
微观交通模型 micro-traffic model
微观结构 microstructure; microtexture
微观结构测量 microtexture measurement
微观结构分析 micro structural analysis
微观进化 microevolution
微观经济 microeconomy
微观经济成本计算 microeconomic cost calculation
微观经济分析 microeconomic analysis
微观经济合理性 microeconomic rationality
微观经济理论 microeconomic theory
微观经济模型 microeconomic model
微观经济评价 microscopic economic evaluation
微观经济收益 microscopic incomes
微观经济学 microeconomics
微观经济政策 microeconomic policy
微观景 microlandscape
微观可逆性 microscopic reversibility;reversibility principle
微观理论 microscopic theory
微观力学 micromechanics
微观粒级 microscopic fraction
微观粒子 microscopic particle
微观连续介质法 microcontinuum approach
微观连续介质模型 microcontinuum model
微观裂缝 microcrack
微观裂纹 microscopic crack
微观流变学 microrheology
微观脉动 microfluctuation
微观毛细管 microscopic capillary
微观弥散 microscopic dispersion
微观模拟(法) microscopic simulation
微观模型 microscopic model
微观磨片 microsection;microspecimen
微观偏析 microsegregation
微观谱学 microspectroscopy
微观起伏 microfluctuation
微观前兆 microscopic precursor
微观强度 microstrength
微观缺陷 microdefect
微观蠕变 microcreep
微观生境 microhabitat
微观世界 microcosm(os)
微观视觉 scotopic vision
微观试验 detail test
微观寿命 microscopic lifetime
微观酸度常数 microscopic acidity constant
微观缩孔 microshrinkage
微观态 microscopic state
微观特征 microcharacteristic
微观图 microgram
微观湍动 microturbulent motion
微观湍流 microturbulence
微观完整性 microperfection
微观纹理 micro-rugosity
微观污染监测 microscosm pollution monitoring
微观污染控制 microscosm pollution control
微观物理学 microphysics
微观显微镜 micrometer microscope
微观现象 microphenomenon [复 microphenomena]
微观效益 microeconomic returns; microeffect
微观效应 microeffect
微观形态连续体 micromorphic continua
微观形态学 micromorphology
微观研究 microexamination
微观因果条件 microcausality condition
微观因果性 microcausality; microscopic causality
微观应力 microstress
微观硬度 microhardness
微观硬度测量 microhardness testing
微观硬度值 microhardness value
微观有效截面 microscopic effective cross section
微观宇宙 microcosm(os); microscopic cosmos
微观照片 micrograph; microphotograph;photomicrograph
微观震中 instrumental epicenter [epicenter];microseismic epicenter
微观滞后效应 microhysteresis effect
微观状态 microstate
微观组构 microfabric
微观组织 microtexture
微观组织检查 microscopic texture examination
微观组织试验 microtexture test
微管 capillary tube;micropipe;microtube
微管采暖系统 microbore heating system;microbore system
微管拆散 microtubule disassembly
微管间桥 intermicrotubular bridge
微管检波器 capillary detector
微管膜 microtubular membrane
微管泡 microtubular vesicle
微管束 microtubule fasolculus
微管系统 microtubule system
微管组织中心 microtubular organizing center [centre]
微管组装 microtubule assembly
微灌注术 microperfusion technique
微光 dimmed light; gleam; glimmer; glimpse;shimmer
微光带 aphotic zone
微光电光度计 microphotoelectric(al) photometer
微光电视 low-light level television
微光度计 microphotometer
微光高岭土 collyrite
微光接纳区 <水体中水面下 100 ~ 600 米深处> dysphotic zone
微光警告信号 low-light indicator
微光密度计 microphoto densitometer
微光摄像电视 night-television
微光视觉 scotopia;scotopic vision
微光束 microbeam
微光束照射 microbeam irradiation
微光夜视六分仪 low-light sextant
微光夜视仪 low-light level night vision device
微光泽测定法 microlustre method
微过滤 microseepage
微过滤器 microfilter
微海绵橡胶 microform rubber
微海啸观测站 microtsunami station
微含粗粒粉质土 silt soil containing lightly
微含粗粒黏[粘]质土 clay soil containing lightly coarse grains
微含粗粒有机质土 organic soil containing lightly coarse grains
微含细粒的砾石 gravel containing lightly fine grains
微含细粒砂土 sand containing lightly fine grains
微焊 microbonding;microwelding
微毫米 micromillimeter [micromillimetre]
微好氧的 microaerobic
微好氧菌 microaerobe; microaerophilic bacteria
微合金 microalloy
微合金工艺 microalloy technology
微合金化 microalloying
微合金扩散型 microalloy diffusion type
微核试验 micronucleus test
微黑 duskiness
微黑的 blackish
微黑色 dusky
微黑色的 dusk
微黑子 microspot
微痕 microscratch
微痕量分析 microtrace analysis
微亨(利) microhenry
微红的 reddish
微红色 reddish colo(u)r
微弧 differential of a arc;elements of an arc;linear element
微弧度 microradian
微弧轨道测定 mini-arc orbit determination
微花斑岩 granulophyre
微花岗结构 hicrogranitic texture
微花岗状 microgranitic
微化分析 microchemical analysis
微化石 microfossil
微画器 micrograph
微环礁 microatoll
微缓磨损 mild wear
微汇编程序 microassembler
微混合暗亮煤 mixoduroclarite
微混合暗煤 mixodurite
微混合亮暗煤 mixoclarodurite
微混凝土 microconcrete
微混浊的 slightly turbid
微活动缝 nonmovement joint
微火焚丝煤 pyrofusites
微火焰电离检测器 microflame ionization detector
微机 personal computer

微机激电仪 induced polarization instrument controlled by microcomputer
微机教学设备 microcomputer teaching equipment
微机开发套件 microcomputer development kit
微机控制降温仪 programmable cooler
微机控制装置说明 microprocessor control description
微机联锁 microprocessor interlocking
微机图像处理系统 microcomputer-based image processing system
微机座 microbase
微积分 calculus of fluxion; differential and integral calculus; fluxion(ary) calculus; infinitesimal analysis; infinitesimal calculus
微积分法 method of fluxions
微积分方程式【数】integral differential equation
微积分学 calculus [复 calculuses/calculi]; differential and integral calculus; infinitesimal calculus; calc <美俗称>
微积分运算 infinitesimal calculus
微级机 micromachine
微级水平仪 microlevel
微级水准仪 microlevel
微级中断 microlevel interrupt
微级中断能力 microlevel interrupt capability
微极 micropolar
微极边界层流 micropolar boundary layer flow
微极化池 microcell
微极距测井 microspaced sonde log
微极性流体 micropolar fluid
微集合体 microaggregate
微计算机 picocomputor
微计算机辅助工程 microcomputer-aided engineering
微计算机功能组织 microcomputer function organization
微计算机检测控制系统 microcomputer detecting control system
微计算机接口 microcomputer interface
微计算机接口套件 microcomputer interfacing kit
微计算机开发成套零件 microcomputer development kit
微计算机开发系统 microcomputer development system
微计算机控制数据采集装置 microcomputer-controlled data collection device
微计算机控制系统 microcomputer control system
微计算机控制钻机 microcomputer control drill
微计算机配套汇编程序 microcomputer kit assembler
微计算机系统 microcomputer system
微计算机样机系统 microcomputer prototyping system
微计算机应用系统设计 microcomputer application design
微计算机执行周期 microcomputer execution cycle
微计算机终端 microcomputer terminal
微剂量测定法 microdosimetry
微剂量学 microdosimetry
微寄生物 microparasite
微尖的 acutate
微间隙 differential gap; microgap
微间隙焊接 microgap welding
微间隙开关 microgap switch
微剪切变形 microshear deformation
微碱化 micro-basification
微碱性 alkalescence; brackishness; slightly alkalinity
微碱性的 alkelescent; subalkaline
微碱性溶液 mild alkaline solution
微件焊接 microwelding
微降速度【机】low load-lowering speed
微降值【机】low-speed down travel
微交错纹层 microcross-bedding
微胶粒 microzyme; plasome; protomere
微胶囊 microcapsule
微角质暗亮煤 cuticoclarodurite
微角质暗煤 cuticodurite
微角质亮暗煤 cuticoclarodurite
微角质亮煤 cuticoclarite
微角质煤 cutite
微校 recalibration
微结构 microarchitecture
微结构变换 microstructure manipulation
微结构镜煤 telite
微结构镜煤质亮煤 teloclarite
微结构丝煤 telofusite
微结构应变 microstructure strain
微截面检测器 microcross section detector
微解冻泥流 cryoturbation
微介电分析仪 dielectric(al) analyzer
微介质 micromedia
微介质水泥磨 mini medium cement mill
微进化 microevolution
微晶 crystallite; micrite; microcrystallite; microlite; microlith; minicrystal
微晶暗色正长岩 micromelasyenite
微晶白云岩 dolomicrite; micritic dolomite
微晶玻璃 devitrified glass; devitroceram; glass ceramics; neoceramic glass; nucleated glass; pyroceram; sitall; sytull
微晶玻璃复合材料 glass ceramic composite
微晶玻璃焊料 glass ceramic solder
微晶玻璃纤维 crystallized glass fiber; devitrified glass fiber
微晶薄膜 microcrystalline film
微晶鲕粒灰岩 oomicrite
微晶二长岩 micromonzonnite
微晶二氧化硅 microcrystalline silica
微晶分散 microcrystalline dispersion
微晶粉末 microcrystalline powder
微晶刚玉 microcrystalline fused alumina
微晶高岭石 montmorillonite
微晶高岭土 montmorillonite
微晶更长花岗岩 microtrotrondhjemite
微晶构造 microcrystalline structure; microlitic structure
微晶花岗闪长岩 microgranodiorite
微晶花岗岩 microgranite
微晶化 controlled microcrystallization; sitallization
微晶化过程 glass ceramic process
微晶化作用 micritilization
微晶灰岩 micrite; micritic limestone
微晶辉长岩 beerbachite; microlitic gabbro
微晶辉绿岩 microdiabase
微晶混凝土 microconcrete
微晶基质 micrite matrix
微晶浆体 microcrystalline paste
微晶角砾岩 microbreccia
微晶结构 cryto-crystalline texture; microlitic texture
微晶结晶 microcrystallization
微晶扩散 minicrystal diffusion
微晶扩散法 minicrystal diffusion method
微晶蜡 ceresin(e); ceresin(e) wax; microcrystal(line) wax; microwax
微晶粒学 micromeritics
微晶粒硬质合金 micrograin hard alloy
微晶粒状 microgranitic; microgranular
微晶粒状的 micromeritic
微晶磷铝石 ceruleolactite
微晶内碎屑灰岩 intramicrite
微晶霓霞钠辉岩 micromelteigite
微晶霓霞岩 microijolite
微晶片 microchip
微晶浅色正长岩 microleucosyenite
微晶球粒灰岩 pelmicrite
微晶软泥 microcrystalline ooze
微晶闪长岩 malachite; microbiorite
微晶砷铜矿 algodonite
微晶石灰岩 microcrystalline limestone
微晶石蜡 microcrystalline paraffin
微晶石英质燧石岩 novaculite
微晶搪瓷 microcrystalline
微晶体 microcrystal
微晶体管 microtransistor
微晶(体)结构 microcrystalline texture
微晶体蜡 microcrystalline wax
微晶团块灰岩 lumpmicrite
微晶歪碱正长岩 microlaikte
微晶霞石正长岩 nepheline microsyenite
微晶纤维素 microcrystalline cellulose
微晶学 microcrystallography; micromeritics
微晶氧化铝 microcrystalline alumina
微晶英碱正长岩 microleucosyenite
微晶云母 microcrystalline mica
微晶质(的) microcrystalline
微晶子假说 crystallite hypothesis
微景观 microlandschaft
微镜暗煤 vitrodurite
微镜惰壳质煤 vitrinertoliptite
微镜惰煤 vitrinertite
微镜亮煤 vitroclarite
微镜煤 microvitrain; vitrite
微镜丝煤 vitrofusi(ni)te
微居里 <放射单位> microcurie
微矩型插头座 microribbon connector
微矩阵 micromatrix
微矩阵法 micromatrix approach
微距 microspur
微距聚焦 macrofocusing
微距离 elementary distance
微距摄影 macroshot
微距摄影测量学 microrange photogrammetry
微距摄影透镜 macroshot lens
微聚焦 microfocus
微聚焦测井 microfocused log
微聚焦测井曲线 microfocused log curve
微聚焦电阻率曲线 microfocused resistivity curve in dip log
微绝热层 subadiabatic layer
微菌类煤 sclerotite
微喀斯特 microkarst
微卡 microcalorie
微开 crack
微开尔文温度 microkelvin temperature
微开着 ajar
微壳灰岩 Indiana limestone
微壳岩屑 spergenite
微壳质煤 liptite
微克 <千分之一毫克> microgram
微克分子 micromolar
微克分子浓度 micromolar concentration
微克立司 <一个氢原子重> microcrith
微克/升 microgram per liter
微克天平 microgram balance
微刻痕 microindentation
微刻(蚀) microetch
微孔 micropuncture; microvoid
微孔薄膜过滤法 millipore membrane filter technique
微孔醋酸纤维素 cellular cellulose acetate
微孔的 finely porous; micropore; millipore; mipor
微孔电导率 pore conductivity
微孔分布 distribution of pores
微孔分气法 atmolysis
微孔构造 microcellular structure; pore structure
微孔管 microporous tube
微孔管过滤 microporous tube filtration
微孔过滤 microporous filtration; microstraining
微孔过滤器 <网孔尺寸以微米计的> micron filter; microporous filter; millipore filter
微孔过滤筛 microseepage
微孔混凝土 cellular concrete
微孔集聚型断裂 microvoid coalescence fracture
微孔几何形术 pore geometry
微孔检测器 pin-hole detector
微孔浆合物 microporous polymer
微孔结构 microcellular texture; pore structure; pore texture
微孔金属过滤器 micrometallic filter
微孔径树脂 microreticular resin
微孔聚合物 microporous polymer
微孔聚偏二氟乙烯膜 microporous polyvinylene difluoride membrane
微孔聚偏氟乙烯膜 microporous polyvinylene fluoride membrane
微孔扩散盒 millipore chamber
微孔滤池法 microporous filter method
微孔滤膜 microporous filtering film; millipore filter
微孔滤网 microscreen; microstrainer
微孔率 microporosity
微孔密集无机膜 microporous and dense inorganic membrane
微孔膜 microporous barrier; microporous film; microporous membrane
微孔膜技术 microporous membrane technique; microporous membrane technology
微孔膜生物反应器 Millipore membrane bioreactor
微孔泡沫胶 microporous rubber; moss rubber
微孔曝气 Millipore aeration
微孔人造革 cellular leather cloth
微孔砂浆 cellular-expanded mortar
微孔筛方法 millipore filtration method
微孔筛(网) micromesh; microscreen
微孔塑料 cellular plastics; microporous plastics
微孔陶瓷 microporous ceramics
微孔陶瓷介质 microporous ceramic medium
微孔陶瓷滤器 microporous ceramic filter
微孔涂层 bubble coating
微孔网 micromesh
微孔微喉型 microthroat with connecting micropore
微孔隙 micropore; microporosity

W

微孔隙度 microporosity
微孔橡胶 microcellular rubber; microporous ebonite; microporous rubber; mipor rubber
微孔橡胶隔膜 mipor schneider
微孔橡皮 microporous rubber
微孔性 microporosity
微孔性的 microporous
微孔氧化铝膜 microporous alumina membrane
微孔硬质胶 microporous ebonite
微孔中空纤维 microporous hollow fibre [fiber]
微孔砖 porous brick
微控制播种机 microcontrolled seeder
微控制器 microcontroller
微控制器接口 microcontroller interface
微控制器结构 microcontroller architecture
微控制器能力 microcontroller capability
微控制器生产系统 microcontroller production system
微控制调制解调器 microcontrolled modem
微控制器外部输入信号 microcontroller external input signal
微控制台 microconsole
微控制终端 microcontrolled terminal
微库(仑) microcoulomb
微库仑滴定分析法 microcoulometry
微库仑法 microcoulometric(al) method
微库仑计 microcoulometer
微库仑检测器 microcoulometric(al) detector
微块 microlith
微亏 bare
微扩散 microdiffusion
微浪 smooth sea; smooth water; smooth wavelet; rippled
微勒克斯 microlux
微类镜煤 provitrite
微冷水 invigorating water
微离子传感器 microionic sensor
微离子学 microionics
微力扳机 hair trigger
微力触发器 hair trigger
微力震动【地】 microtremor
微利 meager profit; small income
微例行程序【计】 microroutine
微粒 atomy; corpusc(u)le; fine grain; fine particle; fines; granule; leptopel; microboulder; microne; microsolid; mote; particle; particulate; pearl; small particle; lutum <直径小于2微米>
微粒爆炸 dust explosion
微粒测定仪 micromerograph
微粒沉淀器 particulate precipitator
微粒沉积分析天平 air sedimentation balance; liquid sedimentation balance
微粒尺寸 particle size
微粒的 impalpable; microdot; micrograined; subsieve
微粒的去除 particulate removal
微粒等级 <一般指1~10微米粒度> microsize grade
微粒度计 microviscosity
微粒二氧化硅 silica fume
微粒发射 corpuscular emission
微粒放射性核素 particulate radionuclide
微粒分散胶体 microdispersoid; particle dispersion
微粒分析器 particle-size analyser [analyzer]
微粒辐射 corpuscular radiation
微粒负荷 particulate loading
微粒干粉 dust
微粒固定相 microparticulate stationary phase
微粒灌浆 particulate grouting
微粒硅土 particulate silica
微粒过滤器 particulate filter
微粒化 micronize
微粒化设备 microniger; micronizer [microniser]
微粒灰岩 micrograined limestone
微粒回降 fallout
微粒回降等强度分布 fallout pattern
微粒回降等强线 fallout contour
微粒混凝土 microconcrete
微粒级 microsize grade
微粒剂 fine granule
微粒夹持 microocclusion
微粒胶片 fine grain film
微粒结构 micrograined texture
微粒结晶 microcrystallization
微粒空气分级器 infrasizer
微粒孔隙率 particle porosity
微粒控制 control of fine particle
微粒流 corpuscular stream
微粒论 emission theory
微粒磨机 microelement grinder
微粒磨砂滚筒机 micromatted roller
微粒磨碎机 micropulverizer
微粒黏[粘]土 micronized clay
微粒凝胶 microgel
微粒凝胶润滑脂 microgel grease
微粒排放 particulate emission; small particle emission
微粒乳剂 fine-grain emulsion
微粒乳液 fine-grain emulsion
微粒散射 particulate scattering
微粒石灰石基层 base bed
微粒食 corpuscular eclipse
微粒收集器 particulate trap
微粒束 corpuscular beam
微粒树脂 finely divided resin
微粒说 emission theory
微粒炭 pepper carbon
微粒特性测定计 hondrometer
微粒体 micrinite; microsome; orbicule
微粒体部分 microsome fraction
微粒体基质 micrinite groundmass
微粒体组 micrinoid group
微粒填充 densified with small particles; microparticulate packing
微粒污染 microparticle pollution; particulate contamination
微粒污染焚烧 fume incineration
微粒污染物 fume
微粒污染物焚化器 fume incinerator
微粒物理学 microphysics
微粒物质 <空气中的> particulate matter
微粒显影 fine grain development
微粒显影剂 fine grain developer
微粒显影液 fine grain developer
微粒相 particulate phase
微粒形应变片 microdot strain ga(u)ge
微粒悬浮酸渣 pepper sludge
微粒学说 corpuscular theory
微粒掩蔽所 false shelter
微粒云 corpuscular cloud
微粒云母 micronized mica
微粒状结构 impalpable texture
微粒状载体 microparticle support
微粒子 finely divided particle; fine particle; particulate
微粒子材质 particulate matter
微粒子论 corpuscular theory
微亮 dimmed light
微亮暗煤 clarodurite
微亮电路 dim light circuit
微亮晶 microspar
微亮晶灰岩 microsparite
微亮晶组构 microspar fabric
微亮煤 clarite
微亮煤质结构镜煤 clarotelite
微亮煤质无结构镜煤 clarocollite
微亮质镜煤 clarovitrite
微亮质丝炭 clarofusite
微量 ace; dregs; drib(b)let; glim; hint; microcontent; microdosage; microdose; microscale; minute quantity; paucity; scruple; shred; thimbleful; touch; trace; trace amount; trace quantity
微量比色测定 microcolorimetric(al) determination
微量比色法 microcolorimetry
微量比色计 microcolorimeter
微量比重瓶 micropicnometer
微量玻璃电极测链 microglass electrode measuring chain
微量薄层色谱法 microthin-layer chromatography
微量采样 microsampling
微量残渣 negligible residue
微量测定 microdetection; microdetermination
微量测定器 microdetector
微量测量 microdetermination; micrometric(al) measurement
微量测压计 micromanometer
微量沉淀 microprecipitation
微量称量 microweighing
微量称样 microsample
微量成分 microconstituent; minor constituent; trace ingredient
微量秤 microbalance
微量代换 microrelief
微量单位 microunit
微量等电聚焦 microisoelectric(al) focusing
微量滴定 microtitration
微量滴定板 microtiter plate
微量滴定法 microtitrimetry
微量滴定管 microburet(te)
微量电解 microelectrolysis
微量电解测定 microelectrolytic determination
微量电流表 microgalvanometer
微量电流计 miniature current meter
微量电渗析 microelectrodialysis
微量电泳法 microelectrophoresis
微量电泳仪 microelectrophoresis apparatus
微量阀 microvalve
微量法 microdetermination; micromethod
微量反应 microreaction
微量反应技术 microreaction technique
微量放射性 trace level activity
微量分光光度计 microspectrophotometer
微量分馏烧管 microfractionating tube
微量分馏柱 microfractionating column
微量分析 microanalysis; trace analysis
微量分析标准 microanalytic(al) standards
微量分析法 microanalytic(al) method
微量分析化学 microanalytic(al) chemistry
微量分析器 microanalyzer
微量分析试剂 microanalytic(al) reagent
微量分析天平 microanalytic(al) balance
微量分析仪 microanalyzer
微量分析质谱计 microanalyzer mass spectrometer
微量分析质谱仪 microanalyzer mass spectrometer
微量浮选 microflo(a)tation
微量浮选试验槽 microflo(a)tation cell
微量浮选试验机 microflo(a)tation cell
微量干燥管 microdrying tube
微量镉 trace cadmium
微量汞中毒 micromercurialism
微量(光)密度计 microdensitometer
微量过程 microprocedure
微量过滤 microfiltration
微量过滤薄膜法 microfiltration membrane process
微量过滤管 microfilter tube
微量过滤器 microfilter
微量含有物 microinclusion
微量和半微量黏[粘]度计 micro and semi-microviscometer
微量恒向线 differential elements thumbline
微量呼吸测定法 microrespirometry
微量呼吸器 microrepiraometer
微量化学 microchemistry
微量化学分析 microchemical analysis
微量化学试验 microchemical test
微量化学天平 microchemical balance
微量化学污染 microchemical pollution
微量化学污染物 microchemical pollutant
微量化学仪器 microchemical apparatus
微量灰化 microincineration
微量活动的 oligodynamic
微量级 trace level
微量级有机物 trace level organics
微量计时表 mierochronometer
微量技术 microtechnique
微量加料器 microfeeder
微量加入物 microaddition
微量检测器 microdetector
微量检流表 microgalvanometer
微量降水 trace of precipitation
微量降雨 <小于0.125毫米的> trace of precipitation; trace of rain
微量搅拌器 microstirrer
微量金属 trace metal
微量金属合剂 mixture of trace metals
微量金属浓度 trace metal concentration
微量金属排入 trace metal input
微量金属污染 pollution by trace metal; trace metal pollution
微量金属污染物 trace metal contaminant
微量进给 microfeed
微量控制 fine inching control
微量库仑滴定 microcoulometric(al) titration
微量库仑气相色谱(法) microcoulometric(al) gas chromatography
微量矿物质添加剂 trace mineral supplement
微量扩散法 microdiffusion
微量扩散分析仪 microdiffusion analyser
微量离心管 microcentrifugal tube
微量磷 trace phosphorus
微量滤器 microstrainer
微量黏[粘]度计 microvisco(si)meter
微量凝集试验 microagglutination test
微量扭力天平 microtorsion balance
微量浓度 trace concentration
微量培养板 microtest plate
微量喷雾润滑 microfog lubrication
微量平底烧杯 microflat bottom beaker

微量平底烧瓶 microflat bottom flask
微量起道 < 找平 > smoothing raiser of track
微量气体 atmospheric trace gas;trace gas
微量气体测定 minor gas measurement
微量气体技术 trace gas technique
微量铅 trace lead
微量切削加工 micromachining
微量氰化物 micro-cyanide
微量区域提纯设备 microscale zone refining apparatus
微量热法 microcalorimetric(al) method;microcalorimetry
微量人口统计学 microdemography
微量容量瓶 microvolumetric(al) flask
微量熔化 microfusion
微量散射浊度计 micronephelometer
微量色谱(法) microchromatography
微量射流 small jet
微量砷 trace arsenic
微量渗析 microdialysis
微量升华 microsublimation
微量试池 microcell
微量试样 microsample
微量塑性计 microplastometer
微量天平 microbalance
微量添加 microadding
微量添加料 minute addition
微量调节器 microadjuster
微量调节注射器 microsyringe
微量调整 microadjustment
微量调整器 microadjuster
微量污染 micropollution; trace contamination;trace pollution
微量污染物 microcontaminant; micropollutant;trace contaminant
微量无机成分 minor inorganics
微量无机物 trace mineral
微量无机元素 trace inorganic element
微量物质 trace material
微量吸收 microabsorption
微量吸(移)管 micropipette
微量循环技术 microcircular technique
微量压痕硬度试验 microindentation hardness testing
微量颜色反应 microcolo(u)r reaction
微量阳离子滴定法 microcationic titration
微量氧化技术 microoxidation technique
微量液-液萃取 micro-liquid-liquid extraction
微量应变 microstraining
微量荧光分析法 microfluorometry
微量荧光光度法 microspectrofluorometry
微量营养(元素) micronutrient;trace nutrient
微量有机污染物 microorganic contaminant; microorganic pollutant; trace organic pollutant
微量有机物 microorganics; trace organics
微量有机元素 trace organic element
微量雨 trace
微量元素 microelement; minor element;trace element
微量元素肥料 microelement fertilizer;trace element fertilizer
微量元素分析 microelement analysis; trace element analysis
微量元素含量 micronutrient level
微量元素浓度 trace element concentration
微量元素污染 trace element pollution
微量圆底烧瓶 microround bottom flask
微量匀浆法 microhomogenisation
微量匀浆器 microhomogenizer
微量杂质 microimpurity;trace impurity;traces of impurities
微量照射 microirradiation
微量蒸馏 microdistillation
微量蒸馏烧管 microdistillation tube
微量蒸馏烧瓶 microdistillation flask
微量指示 microindication
微量指形冷凝管 finger microcondenser
微量中和 microneutralization
微量中和试验 microneutralization test
微量重金属浓度 trace heavy metal concentration
微量纵摆磨削 oscillating grinding
微量组分 microcomponent; minor constituent
微料相 particulate phase
微裂变 microfission
微裂缝 fine fissure; microcrack; microfracture;microflaw < 金属 >
微裂缝地层 creviced formation
微裂缝系统 network of cracks
微裂韧化 microcrack toughening
微裂纹 checking;chip crack;chittering; crazing; hair crack; microcracking; microfissure; microfissuring; microtear; shattercrack; tiny crack
微裂纹陶瓷器底部边缘 check;chittering
微裂系统 microfracture system
微裂隙 microfissure; microfissuring; microscopic fissure
微裂隙压密阶段 microvoid compression stage
微裂相接 joining of microcrack
微裂增韧 microcrack toughening
微邻近触感 microproximity sensing
微流 miniflow
微流结构 microfluxion
微流量 micrometeor; ultratelescopic meteor
微流量电池 microflow cell
微流体 microfluid
微漏 low leakage;slightly leaking
微卢 < 等于 10^{-6} 卢,放射性单位 > microrutherford
微陆桥运输 microland bridge transport
微滤 microfiltration; microscreening; microstraining; microstraining filtration
微滤管式膜 microfiltrate tubular membrane
微滤和生物组合工艺 combined microfiltration and biologic(al) process
微滤机 < 用于污水处理的 > microfilter;microstainer
微滤机分离活性污泥法 microfilter separation-activated sludge method
微滤膜 microfiltration membrane
微滤膜法 microfiltration membrane process
微滤器 microfilter;microstrainer
微滤网 microscreen
微滤污管 microfiltration fouling
微逻辑 micrologic
微逻辑单元 micrologic unit
微逻辑定时电路 micrologic timing circuit
微逻辑加法器 micrologic adder
微逻辑元件 micrologic element
微码 microcode
微码部件 microunit
微脉冲 micropulse
微脉冲发生器 micropulser
微脉动 micropulsation
微脉动场 micropulsation field
微脉动频率 micropulsation frequency
微脉动哨声 micropulsation whistler
微脉动探测设备 micropulsation sensor
微脉动噪声 micropulsation noise
微慢渗透度 moderately rapid permeability
微毛细管 microcapillary
微毛细管孔隙 microcapillary pore
微毛细管作用 microcapillarity
微镁铬铁矿 berezovskite
微米 < 千分之一毫米 > micrometer [micrometre];micron
微米波 micron wave
微米测微表 mu-metron
微米汞柱 micrometer of mercury
微米号数 micron number
微米校验台 microchecker
微米经纬仪 micrometer instrument
微米刻度盘 micrometer dial
微米量级 micron dimension
微米支齿点 < 磨工具用 > micrometer tooth rest
微米轴 micrometer spindle
微密度测定 microdensitometering
微密度测定法 microdensitometry
微密度仪参数 microdensitometer parameter
微密砂岩 compact sandstone
微面 elements of a surface
微秒 < 百万分之一秒 > microsecond
微秒开关 microsecond switch
微秒脉冲 microsecond(im)pulse
微秒脉冲发生器 microsecond pulse generator
微秒延爆雷管 millisecond delay cap
微妙问题 delicate question
微明带 twilight zone
微明区 oligophotic zone;twilight zone
微命令 microcommand;microorder
微模块 micromodule
微模拟系统 < 由模型机、阻抗和其他仪表组成的模拟计算机,在测其基本单位值时,可得到足尺的值的系统 >【计】microsystem
微模塑系列 micromo(u)lding series
微模组件 micromodule
微摩尔 micromole
微摩尔级浓度 micromolar concentration
微磨刀刃 jointing
微磨刃口 jointing
微姆欧 gemmho;micromho
微木栓煤 suberite
微木质镜煤-微镜煤质煤 xulovitro-vitrite
微内核技术 microkernel technology
微泥流 microsolifluction
微黏[粘]度 microriscosity
微黏[粘]弹性流体 slightly viscoelastic fluid
微凝胶体 microgel
微凝聚 microcoacervation
微扭力计 torsion micrometer
微欧米伽 micro-Omega
微欧(姆) < 百万分之一欧姆 > microhm;microohm
微欧(姆)计 microhmmeter
微胖 assignment
微泡 microvesicle
微泡的 vesicular
微泡法 Kalvar process; vesicular process
微泡法胶片 Kalvar film;vesicular film
微泡扩散器 fine bubble diffuser
微泡膜 bubble film
微泡沫凝聚剂 microfoam agglomerating agent
微泡沫(体) microfoam;mini-foam
微泡曝气 fine bubble aeration
微泡涂层 bubble coating
微喷 microjet
微喷发 microflare
微喷灌 micro-sprayer irrigation;minisprinkler irrigation
微喷灌机 microsprinkler
微喷灌喷头 microsprinkler
微膨化 microbulking
微膨胀 microdilatancy
微膨胀计 microdilatometer
微膨胀熔融黏[粘]结 little-expansive fusion caking
微膨胀水泥 mini-expansive cement; slightly expansive cement
微劈石 microlithon
微劈石边界特征 boundary features of microlithon
微劈石结构 texture of microlithon
微片 microchip;speck
微片计算机 microchip computer
微片技术 microchip technology
微平功能 residual function
微平原 microbasin
微破裂 microfracture
微曝气 micro-aeration
微曝气反应器 microaeration reactor
微启动 fine start
微起伏 microrelief;microtopography; nanno-relief;undulating
微起伏地区 normal country
微起伏噪声 fine grain noise
微起伏重力测量 gravimetric(al) microrelief
微起拱模架 < 砌平拱的 > camber slip;trimming piece
微气候测量 microclimatic survey
微气候条件 microclimate condition
微气候影响 microclimate effect
微气化 micro-gasification
微气孔率 microporosity
微气孔群 fine porosity
微气泡纯氧曝气技术 micro-bubble pure oxygen aeration technique
微气象计 micrometeorograph
微气象图 micrometeorogram
微气象学 micrometeorology
微气压表 microbarometer
微气压计 microbarograph; microbarometer;micromanometer
微气压计录器 microbarograph
微气压记录图 microbarogram
微嵌晶结构【地】mikropoikilitic [micropoikilitic] texture
微腔供热系统 mini-bore heating system
微桥 microbridge
微切面检测器 microcross section detector
微倾构造 shallow dipping structure
微倾螺旋 fine tilting screw
微倾水准仪 tilting level
微倾水准仪螺丝 tilting level screw
微倾斜 flat dip;gentle dip
微倾斜的 subaclinal
微倾斜坡道 slightly inclined ramp
微丘区 light hilly area;rolling terrain
微球 microballoon;microballoon sphere
微球粒 framboid
微球粒结构 micropellet texture; microspherulitic texture
微球体 microsphere
微球团 framboid
微球形的 microspheric
微球形聚焦测井 microspheric(al) focused log
微球形聚焦测井曲线 microspherical-

W

ly focused log curve
微区 microregion;microzone
微区尺寸 site size
微区电子探针分析 electron microprobe analysis
微区电子衍射 microzone electron diffraction
微区分布 microdistribution
微区分析 microprobe analysis
微区化学成分分析 microarea chemical analysis
微区划化 microregionalization
微区精炼 microzone refining
微区探查 microprobing
微区衍射 microarea diffraction
微区直径 site diameter
微屈多次反射 peg-peg multiple reflection
微屈多次反射波 peg-leg multiple reflection wave
微屈服强度 microyield strength
微缺陷 microdefect
微扰 infinitesimal disturbance;perturbation
微扰动波 perturbation
微扰方程 perturbation equation
微扰(方)法 perturbation method
微扰理论 perturbation theory
微扰量 perturbation quantity
微扰样品 perturbation sample
微热 eupyrexia;mild fever;subfebrile temperature
微热的 tepid
微热灯丝电子管 dull emitter tube
微热分析 differential thermal analysis;microthermal analysis
微热分析仪 microthermal analyser [analyzer]
微热记录器 microthermograph
微热力学 microthermodynamics
微热量计 microcalorimeter
微热量学 microcalorimetry
微热敏电阻 microthermistor
微容量计数法 microvolumetry
微溶的 slightly soluble;sparingly soluble
微熔鉴定法 microfusion method
微熔颗粒 slightly fused granule
微乳剂 microemulsion
微乳液 microemulsion
微乳液膜 microemulsion membrane
微软【计】Microsoft
微软件 microsoftware
微润滑剂 microlubricant
微弱 faintness;indistinctness
微弱串激绕组 weak series winding
微弱磁场开关 weak-field switch
微弱磁场位置 weak-field position
微弱的 feeble; indistinct; slight; slim;weak
微弱灯光 faint light
微弱地震 <小于里氏 2.5 级的地震> microearthquake;feeble shock
微弱恶臭 faint odo(u)r
微弱胶结的 weakly cemented
微弱声音 faint voice
微弱收敛的 weakly convergent
微弱紊动 microturbulence
微弱运动 slightly movement
微三角洲 microdelta
微扫描 microscanning
微扫描仪器 microscanning instrument
微筛 microtraps
微筛分 microscreening
微筛选 microscreening
微闪长岩 malachite;microdiorite
微闪光 microflare
微闪光泽 glimmering lustre
微伤 scratch;boo-boo <俚语>
微商 differential quotient;derivative;differential coefficient
微商差热分析 derivative differential thermal analysis
微商分布 derivative distribution
微商控制 derivative control
微商热膨胀法 derivative thermodilatometry
微商热谱法 derivatography
微商热重法 derivative thermogravimetry
微商热重曲线 derivative thermogravimetric curve
微商图 derivative map
微商稳定 derivative equalization;lead equalization
微商响应 rate response
微商衍生物 derivative
微商元件 derivative element
微商作用 derivation action;rate response
微商作用时间 derivative action time; rate action time
微商作用因数 derivative action factor;rate action factor
微烧杯 microbeaker
微射角自动摄影术 microautoradiography
微射流 microjet
微射线 microray
微射线自动照相机 microautoradiograph
微摄镜 microlens
微摄影 micrography
微伸 microstretching
微伸长 microstraining
微伸扩幅装置 microstretch expander
微渗(漏) microseepage
微渗滤 microstraining
微渗压计 microosmometer
微渗液 microseepage
微升 microliter [microlitre]
微升速度【机】low load-lifting speed
微升新平原 pastplain
微升值【机】low-speed up travel
微升注射器 microliter syringe
微生态气候 ecidio-climate
微生态系(统) microecosystem
微生态系统试验 microecosystem test
微生态学 microecology
微生物 animalcule; germ; microbe; microbe microscopic organism; microbian;microorganism
微生物比 microorganism ratio
微生物比增长速率 specific growth rate of microorganism
微生物闭锁生态系 microbial closed ecosystem
微生物变体 microbial modification
微生物标准 microbial standard
微生物病 microbiosis
微生物不均衡 microbial imbalance
微生物采水器 bacteriological water sampler
微生物参数 microbial parameter
微生物测定(法) microbioassay; microbiologic(al) measurement; microbiotest; microorganism determination
微生物测量 microbiologic(al) survey
微生物产品 microbiologic(al) products
微生物产物 microbiologic(al) products
微生物成分 microbiologic(al) component
微生物除草剂 microbial herbicide
微生物除磷 microbial dephosphorus
微生物处理 microbiologic(al) treatment;microorganism treatment
微生物处理的污泥 digested sludge
微生物处理污泥罐 digestion tank
微生物处理污泥罐内的热水或蒸汽盘管 digester coil
微生物处理污泥罐下部处理室 digestion chamber
微生物处理污泥系统 digester system
微生物丛 microbiota
微生物代谢活性 microbial metabolism activity
微生物代谢作用 microbial metabolism
微生物的 microbial
微生物的确实性 microbiologic(al) infallibility
微生物的驯化 domestication of microorganism
微生物的转化 microbial transformation
微生物电池 microorganism electric cell
微生物动力学 microbial kinetics
微生物毒素 microbial toxin
微生物对抗作用 microbiologic(al) antagonism
微生物多样性 microbial diversity
微生物繁衍 microbial proliferation
微生物反应 microbial reaction; microbiologic(al) reaction
微生物防治 microbial control
微生物分解工艺 microbial decomposition process
微生物分解系统 microbial decomposition system
微生物分解性(能) biodegradability
微生物分解作用 microbial decomposition;microorganic decomposition
微生物分散剂 microorganism dispersant
微生物腐蚀 microbiologic(al) corrosion
微生物工业 microbiologic(al) industry
微生物工艺 microbiologic(al) process
微生物共代谢 microbial cometabolism
微生物过程 microbial process
微生物合成代谢 microorganism synthetic metabolism
微生物化学电池 biochemical fuel cell
微生物化学反应 microbiochemical reaction
微生物还原 microbial reduction
微生物活力 microbial activity
微生物活性 activity of microorganism;microbial activity
微生物技术 microbial technology
微生物甲基化作用 microbial methylation
微生物鉴定 microbiologic(al) assay; microorganism identification
微生物降解 microbe degradation; microbial degradation; microbiodegradation; microbiologic(al) degradation;microorganism degradation
微生物结合 microbial association
微生物浸出 microbial leaching; microbiologic(al) leaching
微生物菌群 microbial bacterial population
微生物可用磷 microbially available phosphorus
微生物控制 microbiologic(al) control;microorganism control
微生物老化 biodeterioratioration
微生物累积 microorganism accumulation
微生物离异 microbial dissociation
微生物量 microbial biomass
微生物磷 microbiologic(al) phosphorus
微生物密度 microbe density
微生物膜 <有机污泥表面繁殖的> microbial film;microorganism film
微生物内源代谢 microorganism intrinsic metabolism
微生物黏[粘]着碳氢化合物 microbial adhesion to hydrocarbon
微生物农药 microbial pesticide; microbic insecticide
微生物浓度 microbial concentration
微生物培养 microbiologic(al) cultivation
微生物培养法 culture of microorganism
微生物侵蚀 microbial attack
微生物区(系) microbiota;microflora
微生物群 microbiota
微生物群落 microbiologic(al) population;microflora
微生物群落多样性 microbial community diversity
微生物群落结构 microbial community structure
微生物群体 microbial population; micropopulation
微生物群系 microbial population
微生物燃料电池技术 microbial fuel cell technology
微生物杀虫剂 microbial insecticide; microbiologic(al) insecticide
微生物生长 microbial growth
微生物生长曲线 microbial growth curve
微生物生化需氧量试验 microorganism biologic oxygen demand test
微生物生态系(统) microbial ecosystem
微生物生态学 microbial ecology
微生物生物测定 microbial bioassay
微生物生物传感器 microbial biosensor
微生物生物多样性 microbial biodiversity
微生物生物化学 microbial biochemistry
微生物生物物质 microbial biomass
微生物生物物质氮 microbial biomass nitrogen
微生物生物物质磷 microbial biomass phosphorus
微生物生物物质碳 microbial biomass carbon
微生物生长调节剂 microorganism growing adjustment agent
微生物噬菌作用 microbiophagy
微生物数 microbial number
微生物水解 microbial hydrolysis
微生物特性 microbiologic(al) property
微生物特征 microbial characteristic; microbiologic(al) characteristic
微生物条件 microbiologic(al) condition
微生物同化 microbial assimilation
微生物脱氮工艺 microbial nitrogen removal process
微生物脱胶法 bacterial degumming method
微生物脱腊 bacterial dewaxing
微生物脱色 microbial decolo(u)rization
微生物污泥 microbial sludge
微生物污染 microbial contamination; microbial pollution; microbiologic(al) contamination; microbiologic(al) pollution
微生物污染物 microbial contaminant; microbiologic(al) contaminant; microbiologic(al) pollutant; microorganic contaminant
微生物污着 microbiologic(al) fouling
微生物吸附 microorganism adsorption
微生物吸附容量 microbe adsorption

capacity
微生物显微镜 microbioscope
微生物相 microbiota
微生物消化污泥 digested sludge
微生物性原料 microbial material
微生物修复 microbial remediation
微生物絮凝剂 microbial flocculant
微生物学 bacteriology;microbiology
微生物学测定 microbiologic(al) assay
微生物学的 microbiologic(al)
微生物学方法 microbiologic(al) method
微生物学分析 microbiologic(al) analysis
微生物学工业 microbiologic(al) industry
微生物学家 microbiologist
微生物学石油勘探 petroleum prospecting by microbiology
微生物学研究 microbiologic(al) research
微生物学指标 microbiologic(al) index
微生物循环 microbiologic(al) cycle
微生物氧化作用 microbiologic(al) oxidation
微生物冶金学 microbial metallurgy
微生物异常 microbiologic(al) anomaly
微生物引起的变质 microbial deterioration
微生物引起的氧化作用 microbiologic(al) oxidation
微生物引起的转化作用 microbial conversion
微生物属性 microbiologic(al) specification
微生物浊度计 microbiophotometer
微生物资源 microbial resources
微生物作用 action of microorganisms;microbial action
微生藻类 microscopic algae
微生植物 microscopic plant
微声传感器 microphone sensor
微声电子学 acoustoelectronics;pretersonics
微声活动性 <岩体中的> microsonic activity
微声系统 microacoustic(al) system
微湿的 madescent
微石棉 asbestinite
微收缩陶瓷 ceramics with low shrinkage
微束 microbeam
微束 X 射线分析器 microbeam X-ray analyser [analyzer]
微束等离子弧焊 microplasma arc welding
微束照射 microirradiation
微树皮暗煤 bark durite
微树皮煤 barkite
微树脂暗亮煤 resinoduroclarite
微树脂暗煤 resinodurite
微树脂亮暗煤 resinoclarodurite
微树脂亮煤 resinoclarite
微树脂煤【地】 resite
微树脂煤分层【地】 resite layer
微水滴 water droplet
微水位计 microlevel ga(u)ge
微丝 fibril;microfilament
微丝煤 fusite
微丝炭 fusite
微丝炭镜煤 fusovitrite
微丝炭亮煤 fusoclarite
微丝炭木煤 fuso-xylite
微丝炭木煤型暗亮煤 fuso-xyloduroclarite
微丝炭木煤型暗煤 fuso-xyloclarite
微丝炭木煤型亮暗煤 fuso-xyloduroclarite
微丝炭木煤型亮煤 fuso-xylite clarite
微丝炭木煤质煤 fuso-xylain
微丝性 fibrosity
微丝悬挂 fibre suspension
微丝质暗亮煤 fusoducroclarite
微丝质暗煤 fusodurite
微丝质结构镜煤 fusotelite
微丝质亮暗煤 fusoclarodurite
微速 dead slow(speed)
微速后退 dead slow astern
微速前进 dead slow ahead
微速下降机构 precision lowering mechanism
微塑性屈服 microplastic yielding
微酸的 acescent;acidulous
微酸味 acescency
微酸性 subacidity
微酸性的 slightly acidic;subacid
微酸性水 acidulous water
微缩胶片绘图机 computer output microfilm system
微缩图书阅览室 microfilm reading room
微态 microstate
微坍落度 mini-slump
微弹性 microelasticity
微探针 microprobe
微探针度谱术 microprobe spectrometry
微探针法 method of miniature thermal probe
微探针分析 microprobe analysis
微特电机 small and special electric machine
微梯度测井 microlateral log
微梯度测井曲线 microlateral log curve
微提升【机】 low-speed lifting
微体 microbody
微体古生物分析 micropal(a)eontology analysis
微体古生物学 micropal(a)eontology
微体化石 microfossil
微体化石带 microfossil zone
微体生物地层学 microbiostratigraphy
微体生物相【地】 microbiofacies
微调 fine tuning;differential rotation; fine adjustment; fine control; fine inching control; fine set(ting); hairbreadth tuning; inching; inching control; microadjustment; minitrim; minute adjustment; readjust(ing); readjustment; reset; sharp tuning; trimming; vernier adjustment; vernier control; vernier regulation
微调瓷介电容器 ceramic trimmer
微调电感 trimming inductance
微调电容器 aligning capacitor; padder; preset capacitor; trimmer; trimmer capacitor; trimmer condenser; trimming capacitor; trimming condenser; vernier-control capacitor; vernier condenser
微调电位器 trimmer potentiometer
微调电阻器 trimmer resistor
微调定时器 micrometric(al) timing adjuster
微调度盘 vernier dial
微调对中望远镜 microalignment telescope
微调发动机 vernier; vernier engine; vernier rocket
微调阀 reset valve;trim valve
微调分规 hair divider
微调杆 regulating rod
微调卡钳 transfer cal(l)ipers
微调控制 fine control
微调两脚规 hair compass
微调滤波器 trimming filter
微调螺钉 fine adjustment screw; micrometer adjusting screw; micrometer screw; micrometric(al) regulating screw
微调螺杆六分仪 endless tangent screw sextant
微调螺丝 fine adjustment screw; tangent screw
微调螺旋 fine adjustment screw; tangent screw
微调镗刀 microcartridge
微调镗刀头 microboring head
微调线圈 alignment coil
微调小圆规 bow compasses; caliber compasses; caliper compasses
微调旋钮 vernier knob
微调元件 vernier element
微调圆规 hair compass
微调装置 fine setting device; microadjuster; micromatic setting; vernier arrangement; vernier device
微通道 microchannel
微通道板 microchannel plate
微通道变换器 microchannel inverter
微通路 microchannel
微统计学 microstatistics
微透镜 lenticule
微透镜胶片法 lenticular film process
微透明的 subtranslucent
微凸出的断开圆线脚 knulling
微凸曲线 slight convex curve
微土粒 lutum
微土壤学 micropedology
微湍 <流态的一种> microturbulence
微湍流理论 microturbulence theory
微团 micell(e) [复 micellae]
微团聚体 microaggregate
微团粒 microaggregate
微推力 microthrust
微托 <压强单位,1 微托 = 10^{-6}托> microtorr
微洼地 swale
微瓦功率电子学 microwatt electronics
微瓦计 microwattmeter
微瓦(特) microwatt
微弯板 slab with slightly curved bottom
微弯板工字梁组合桥 composite shell-slab and I-beam bridge
微弯板组合梁桥 combination beam bridge with slight curve slab
微弯波导 minor bend
微弯河段 faintly curved reach; slightly curved reach; slightly sinuous reach
微弯曲损耗 microbend loss
微弯渠道 slightly curved canal; slightly sinuous canal
微网格 microgrid
微网栅模型法 microgrid
微微 <10^{-12}> pico;micromicro
微微安(培) micromicroampere; picoampere
微微安培计 micromicroammeter
微微程序 pico program(me)
微微处理机 picoprocessor
微微处理器 picoprocessor
微微法(拉) <电容量单位,常用于无线电> micromicrofarad;picofarad
微微居里 micromicrocurie;picocurie
微微居里/升 micromicrocurie per liter;picocurie liter
微微克 micromicrogram;pico-gram
微微逻辑电路 picologic
微微米 <长度单位,1 微微米 = 10^{-12}> micromicron;picometer
微微秒 picosecond
微微瓦 pico-watt
微微瓦分贝 decibels above picowatt [DBP]
微微下垂的 weeping
微微衍射 micromicrodiffraction
微位 microbit
微位移测量仪 microdisplacement meter
微位移器 micropositioner
微温 lukewarmness; tepefaction; tepefy
微温的 lukewarm;tapid
微温气候 lukewarm climate
微温泉 tepid spring
微温水 tepid water
微文象结构【地】 micrographic(al) texture
微文象岩【地】 micropegmatite
微文象状【地】 micrographic
微纹长石 microperthite
微纹理 microtexture
微稳定煤 liptite
微涡 microvortex
微污染 micropollution
微污染地表水 micro-polluted surface water
微污染富营养化水处理 micro-polluted and eutrophic water treatment
微污染水 micro-polluted water
微污染水体 micro-polluted water body; micro-polluted waters; oligosaprobic waters
微污染水源 micro-polluted water source
微污染水源水 micro-polluted source water; slightly polluted source water
微污染水源水处理 micro-polluted source water treatment
微污染饮用水处理 micro-polluted drinking water treatment
微污染饮用水源 micro-polluted drinking water sources
微污染原水 micro-polluted raw water; slightly polluted raw water
微污着 micro-fouling
微无结构镜煤 collite
微无结构镜质亮煤 colloclarite
微吸附 microadsorption
微吸附柱 microadsorption column
微吸管 micropipette
微熙提 <表面亮度单位> microstilb
微洗刷坡 microwash slope
微系集 microassembly
微系统 microsystem
微系统结构 microsystem organization
微系统设计【电】 micro design
微系震 microafter-shock
微细胞 minicell
微细差别 nuance
微细的区别 nicety
微细二氧化硅 white rouge
微细粉末 attritive powder; finely pulverized powder; fine powder
微细粉碎机 final grinder
微细浮渣 dross fines
微细固体 fine solid
微细管 microcapillary
微细管道 microtubule
微细龟裂 microcrack
微细滑石 micronized talc
微细胶状悬浮剂 finely ground colloidal suspension
微细结构 fine structure
微细孔隙的 finely porous; fine-pored
微细裂缝 microfissure; microfracture
微细裂纹 microcrack
微细路面冷拌铺面产品 <以聚合物改性乳化沥青、碎石、填料、水及添加剂混合而成薄层路面混合物> micro surfacing cold mix paving product
微细砂 ultra-fine sand

微细渗漏 microseepage
微细网眼 microgrid
微细细粉颗粒 fine particle
微细纤维 microfiber [microfibre]
微细屑岩 microclastic rock
微细屑状 microclastic
微细絮凝物 microfloc
微细悬浮液 finely divided suspension
微下降【机】 low-speed lowering
微纤结构 microfibrillar structure
微纤毛 microcilium
微纤丝 microfibril
微纤丝角 microfibrillar angle
微纤维 microfibril; primitive fiber [fibre]
微纤维润滑脂 microfiber grease
微纤维素 dermatosome
微咸的 brackish; mildly brackish
微咸地下水 saltish ground water
微咸湖 brackish water lake
微咸冷却水 brackish cooling water
微咸水 brackish water
微咸水流 brackish flow
微咸水石灰石 brackish water limestone
微咸水域 brackish; brackish waters
微咸水沼泽 brackish marsh
微咸性 brackishness
微限度车轮打滑控制系统 micro-limit wheel slip control system
微相【地】 microfacies
微相理论 microphase theory
微镶嵌 microinsertion
微像数据 microimage data
微小 minification; slenderness; smallness
微小变形 microstrain
微小尺寸 microsize
微小的 miniature; minute; slender; tiny
微小的东西 minikin
微小的数目 small numeral
微小地震 microearthquake
微小动物 animalcule
微小浮游生物 nannoplanktion; nanoplankton
微小改变 minor change
微小功率 micropower; miniwatt
微小故障 minor failure
微小管 microtubule
微小荷载 indivisible load
微小环境 microenvironment
微小剂量 microdosage; microdose
微小角度 minute angle
微小孔隙 minute interstice
微小气候 microclimate
微小气候学 microclimatology
微小驱动 microdrive
微小生态系 microecosystem
微小损伤 microinjury
微小突粒 nib
微小网 micronetwork; minimal lattice
微小位移 infinitesimal displacement; small displacement
微小涡动 microturbulence
微小物 fingerling; minikin; minim
微小误差 light error; slight error
微小系统 microsystem
微小细部 pinpoint detail
微小斜坡屋顶 slightly sloped roof
微小形变 infinitesimal deformation
微小型 microminiature
微小型化电路 microminiaturized circuit
微小延迟 fine delay
微小演替 microsuccession
微小演替系列 microsere
微小应变 small strain
微小应力 microstress
微小圆珠 minute globule
微小真菌 microfungus
微小振动 microvibration
微小振幅波 small amplitude wave
微小终板电位 miniature plate potential
微斜长石 amazonite; amazon stone; kalifeldspath; microcline
微斜长石化 microclinization
微斜长石双晶律 microcline twin law
微斜长石正长岩 microcline syenite
微斜面宽度 joint width
微斜钠长石 rutterite
微斜钠闪正长岩 thuresite
微斜纹长石 microcline perthite
微斜霞石正长岩 itsindrite
微斜正长岩 microclinite
微写器 micrograph
微屑 microclast; mote
微屑结构 microclastic texture
微信道 microchannel
微信息 micromessage
微信息处理器 computer-on a-chip
微形环岛 mini-roundabout
微形结构 micromorphology
微形态分析 micromorphological analysis
微型 micro; miniature; minisize; minitype
微型板块 microplate
微型保险丝 microfuse [microfuze]
微型泵 micropump
微型比例尺 microscale
微型比重瓶 micropycnometer
微型毕托管 micro-Pitot tube
微型变感器 microvariometer
微型步进电动机 microstepping motor
微型测力计 microdynamometer
微型产品 microminiaturization
微型产品目录 minicatolog
微型敞篷马车 go-cart
微型超正析像管 miniature image orthicon
微型城市 micropolitan
微型持针钳 microneedle holder
微型尺 microrule
微型冲击式检尘器 midget impinger
微型稠度计 microconsistometer
微型出租汽车 minicab
微型处理机 microprocessor; miniprocessor
微型处理器 microprocessor
微型触发电路 microflip-flop
微型穿孔 microdrill(ing)
微型传声器 midget microphone
微型催化反应器 microcatalytic reactor
微型打孔 microdrill(ing)
微型打桩机 midget pile driver
微型导管 microguide
微型的 microminiature; midget; mini; miniature; tiny
微型灯 miniature lamp
微型灯泡 midget bulb
微型等离子弧 needle plasma arc
微型狄法尔试验【道】 micro-Deval test
微型狄法尔值 micro Deval value
微型底片 miniature negative
微型地貌 microscale landform
微型地区 microregion
微型地图 micromap
微型地图电子束记录系统 electronic beam recording micromap system
微型地图分色片 micromap colo(u)r separation
微型地图绘制系统 micromap generator system
微型地震仪 microseismograph
微型点焊机 microspot welder
微型电车 minitram
微型电池 minicell
微型电磁离合器 miniature electromagnetic clutch
微型电动机 micromotor; miniature motor; midget motor; subminiature motor
微型电极 microelectrode
微型电控盘 electric(al) mini-panel
微型电离室 miniature ionization chamber
微型电流表 pygmy current meter
微型电路 microcircuit; microminiature circuit
微型电路学 microcircuitry
微型电路元件 miniaturized circuit element
微型电脉池 microelectrophoretic cell
微型电脑 computer-on a-chip; microprocessor
微型电容器 miniature capacitor; walnut capacitor
微型电视 microtelevision; miniature television
微型电视摄像机 miniature TV camera
微型电位计 micropot
微型电站 midget plant
微型电子测距仪 mini ranger/trisponder
微型电子电路 microcircuit; microelectronic circuit; microminiature circuit; microminiaturized circuit
微型电子计算机 microcomputer
微型电子计算机操纵的联锁 electronic microcomputer controlled interlocking
微型电子计算机技术 microcomputer technology
微型电子计算机显示器 minicomputer display
微型电子内诊器 radio capsule
微型电子器件 microelectronic device
微型电子准直仪 miniature electronic autocollimator
微型电阻元件 resistor microelement
微型动物 microzoon
微型断路器 microcircuit breaker; miniature circuit breaker
微型发电机 microgenerator
微型发电站 mini power plant
微型发酵罐 miniature fermenter
微型反应器 microreactor
微型分光光度计 microspectrophotometer
微型封装件 micromodule package
微型浮游动物 microzooplankton
微型浮游生物 nanoplankton
微型浮游生物有机质 nanoplankton organic matter
微型浮游植物 nanoplankton plant
微型钢丝钳 microcutting pliers; microwire cutter
微型高温计 micropyrometer; pyromike
微型割草机 minimower
微型跟踪无线电 minitrack radio
微型跟踪系统 minitrack system
微型工程起重机 midget construction crane
微型公共汽车 microbus; minibus
微型构造 microscopic structure
微型鼓风机 microblower
微型挂车 midget trailer
微型管 microminiature tube; miniature tube
微型惯性导航系统 miniature inertial navigation system
微型滚筒 midget roller
微型海底生物 microbenthos
微型海流计 miniature current meter
微型焊机 microminiature welder
微型焊接 microwelding
微型焊炬 midget; small screw torch
微型珩磨头 microhoner
微型红外测距仪 microranger
微型化 micromation; microminiaturization; miniaturization
微型化系统 miniaturization system
微型环岛【道】 mini roundabout
微型环交【道】 mini-roundabout
微型环芯 microtoroid
微型绘图机 microplotter
微型混合澄清槽 miniature mixer-settler; mini-mixer-settler
微型混合集成电路 moduler circuit
微型混凝土 microconcrete
微型混凝土拌和机 midget concrete mixer
微型混凝土搅拌机 midget concrete mixer
微型机 microcomputer
微型机器人 microrobot
微型机械 micromachinery
微型机械手 microrobot
微型集成电路片 microchip
微型集水沟 micro-catchment
微型计算化(电脑化)系统 microcomputerised system
微型计算机 microcomputer; minicomputer
微型计算机控制自记分光光度计 microcomputer-controlled recording spectrophotometer
微型计算机数字化系统 minicomputer digitizing system
微型计算机显示器 minicomputer display
微型计算器 pocket calculator
微型继电器 microminiature relay; miniature relay
微型继电器联锁 miniature relay interlocking
微型加热器 microheater
微型交错层理【地】 microscale cross-bedding
微型胶片 miniature film
微型搅拌机 midget mixer
微型结构 microstructure
微型经济牵引车 pint-size economy tractor
微型晶体 minicrystal
微型晶体管 minitransistor
微型竞赛汽车 go-kart
微型卡车 midget truck
微型开关 microswitch; miniature switch
微型空穴 microscopic void
微型控制器 microcontroller
微型拉钩 microretractor
微型棱镜 microprism
微型离合器 microclutch
微型裂陷 microaulacogen
微型流速仪 microcurrent meter; midget current meter; miniature current meter; pygmy current meter; pygmy meter <用于低流速测量的>
微型陆块 microcontinental block
微型滤波器 microfilter
微型螺钉 microscrew
微型螺帽套紧器 micronut tightener
微型马达 micromotor
微型煤气灯 micro-gas burner
微型敏感元件 microsensor
微型模型 micro model
微型摩托车 <美> minibike
微型磨料 micro mo(u)ld abrasive
微型浓缩器 microconcentrator
微型培养皿 microculture
微型平板体 platelet
微型汽车 baby car; bubble car; cabin car; midget car; mini; miniature

(motor-) car; minicar; roadlice; road louse
微型器件 miniature device
微型器件制造法 micromation
微型潜水器 microsubmersible
微型潜水器类型 microsubmersible category
微型强力试验仪 microdyn tester
微型区系 microfauna
微型全断面(隧道)掘进机 mini full-facer
微型全息照相 microholograph
微型全息照相术 microholography
微型燃烧器 microburner
微型热电制冷器 miniature thermoelectric refrigerator
微型热量计 microcalorimeter
微型人造卫星跟踪系统 minitrack system
微型赛车 kart
微型色谱板 microchromatoplate
微型栅极干电池 blas
微型设备 micromodule equipment
微型生物群 micropopulation
微型示波器 miniature oscilloscope
微型数字计算机 digital minicomputer
微型双工电台 handie-talkie
微型水电 micro hydropower
微型水电站 micro water power plant
微型水轮机 diminutive turbine; microhydraulic turbine; mini hydro turbine
微型伺服执行机构 microservo actuator
微型塑料细管 mini-straw
微型探针 miniature probe
微型套筒 miniature socket
微型体系结构 microarchitecture
微型填充柱 micropacked column
微型图 microdrawing
微型拖车 midget trailer
微型拖拉机 midget tractor; mini-tractor
微型挖掘机 midget excavator
微型挖掘隧道系统 mini-tunnel system
微型挖泥船 mini-dredger
微型物 midget
微型吸附监测器 microadsorption detector
微型显示装置 microform display device
微型线路 microminiature module
微型线路技术 microci technique
微型小轿车 midget car
微型小客车 midget car
微型小网眼 micromesh
微型信息处理机 microprocessor
微型信息处理机控制 microprocessor control
微型旋桨式流速仪 micro-propeller current meter
微型压路机 midget roller
微型羊蹄滚筒 midget sheep's foot roller
微型羊足压路机 midget sheep's foot roller
微型要素 microfeature
微型移液吸管 micropipette
微型印刷电路 microprinted circuit
微型元件 microcomponent; microelement; micromodule; miniature component; minicomponent
微型元件片 microelement wafer
微型载货汽车 midget truck
微型藻 microscopic algae
微型闸流管 miniature thyratron
微型照片 micrograph
微型照相机 microcamera
微型褶皱【地】 microfold(ing)
微型振荡器 microoscillator
微型振动片 midget vibrating plate
微型振动切入板 midget vibrating plate
微型制冷机 microminiature cryocooler
微型钟 microclock
微型轴承 miniature bearing
微型轴承座圈 miniature bearing race
微型主机 micromainframe
微型柱 microcolumn; mini column
微型桩 micropile; mini-pile; needle pile
微型装载机 miniloader
微型装置 microdevice
微型撞击式检尘器 midget impinger
微型自动化装置 mini-automated device
微型自行式滚筒 midget self-propelled roller
微型自行式压路机 midget self-propelled roller
微型自游泳生物 micronekton
微型组件 chip; microelement; micromodule; micromodule package
微型组件电路 moduler circuit
微型组件电子学 micromodule electronics
微型组件封装 micromodule pack
微型组件技术 micromodule technique
微型组件组 micromodule stable
微型组件组成 modular
微型钻头 microbit
微型钻头凿岩试验 microbit drilling test
微型钻头钻机 microbit drilling rig
微需气细菌 microaerophile; microphilic bacteria
微需氧的 microaeraophilic
微需氧菌 microaerobion
微需氧试验 microaerophilic test
微需氧微生物 microaerophile
微需氧细菌 microaerophilic bacteria
微絮凝 microflocculation
微絮凝法 microfloc process
微絮凝过滤 microflocculation filtration; slight flocculation filtration
微絮凝接触过滤 microflocculation contacting filtration
微絮凝流砂过滤 microflocculation sand flow filtration
微絮凝深床直接过滤 microflocculation deep bed direct filtration
微絮凝物 microfloc
微絮凝纤维 microflocculation fiber [fibre]
微絮凝-直接过滤-生物活性炭工艺 microflocculation-direct filtration-biological activated carbon process
微旋流燃烧室 microturbulence combustion chamber
微循环 microcirculation; microcycle
微循环功能障碍 microcirculation dysfunction
微循环灌流不良 inadequate perfusion of microcirculation
微循环灌流量 microcirculatory perfusion
微循环计算机 microci computer; microcircuit computer
微循环力学 mechanics of microcirculation
微循环系统 microcirculatory system
微循环障碍 microcirculatory disturbance
微压 micropressure
微压表 micromanometer; piezometer
微压计 Chattock ga(u)ge; differential ga(u)ge; differential manometer; inclined manometer; microbarometer; micromanometer; micro-pressure-ga(u)ge; microtasimeter; tosimeter
微压记录计 microbarograph
微压记录图 microbarogram
微压力表 micromanometer
微压力计 micromanometer
微压式供暖系统 differential system
微压式蒸汽系统 steam vapo(u)r system
微氩检测器 micro-argon detector
微岩相指标 micropetrographic(al) index
微岩性学 microlithology
微衍射 microdiffraction
微厌氧性细菌 microanaerobic bacteria
微焰灯 microburner
微氧升流污泥床 microoxygenic upflow sludge bed; microoxygenic upflow sludge blanket
微氧升流污泥床反应器 microoxygenic upflow sludge bed reactor
微摇 jog
微耀斑 microflare
微叶镜煤 phyllovitrite
微液滴 droplet
微液流 microstream
微移动控制系统 micromovement control system
微异地堆积 hypautochthony
微异地生成煤 hypautochthony
微音器 microphone; mike
微音器放大器 transmitter amplifier
微音器架 microphone holder
微音器灵敏度 microphone sensitivity
微音器频率响应 microphone response
微音器碳精粒黏[粘]结 burning of microphone
微音器嘘声 microphone hiss
微音器阵列 microphone array
微音器自鸣 microphone singing
微隐晶质 dubiocrystalline
微应变 microstrain
微应用手册 microapplication manual
微英寸 <长度单位,1 微英寸 = 0.0254 微米> microinch
微盈 full
微影分析器 microimage analyser [analyzer]
微硬 slightly hard
微硬率 microhardness
微硬水 little hard water
微涌 dead swell
微余震 microafter-shock
微雨 dribble; drizzle; drizzling rain; light rain; slight rain; spit; trace of rain
微雨计 ombrometer; trace recorder
微雨量 trace of rainfall
微雨量器 micropluviometer
微语句 microstatement
微语言 microlanguage
微域 microzone
微域地理学 microgeography
微域地形 microrelief
微域区划 microregional plotting
微元 infinitesimal element; minute element
微元段长度参数 parameters of finite segment length
微元反应 elementary reaction
微元分析 infinitesimal analysis
微元件学 micrology
微元平衡方程 infinitesimal balance equation
微云 cloudlet
微陨石 micrometeorite
微藻类亮煤 algo-clarite
微藻类煤 algite
微增压燃烧 pressurized combustion
微涨落 microfluctuation
微针入度 micropenetration
微针入度计 micropenetrometer
微针术 microacupunoture
微针状体 microneedle
微诊断 microdiagnosis
微诊断程序 microdiagnostics
微诊断法 microdiagnostics
微诊断微程序 microdiagnostic microprocessor
微诊断装入器 microdiagnostic loader
微阵雨 sprinkle
微振动计 microvibrograph
微振幅波 wave of infinitesimal amplitude; wave of small amplitude
微振磨损 fretting
微震 earthquake swarm; earthquake tremor; micro(earth) quake; microseism; microtremor; slight shock; very slight shock
微震波勘探 microseismic prospect
微震峰 microseismic peak
微震构造环境 weak shock tectonic environment
微震观测 microearthquake observation; microseismic observation
微震活动 microearthquake activity; microseismic activity
微震活动性 microseismicity
微震计 microseismograph; microseismometer; microvibrograph; tromometer
微震监测网 microseismic monitoring network
微震谱 microseism spectrum; microtremor spectrum
微震区划 microearthquake zoning; microseismic zoning
微震数据 microtremor data
微震台网 microearthquake network
微震台阵 microearthquake array
微震学 microseismology
微震仪 microseismograph; microvibrograph; tremometer
微震预报 microearthquake forecasting; microseismic forecasting
微震预测 microearthquake forecasting; microseismic forecasting
微震运动 microearthquake movement
微震造型机 shockless jolt mo(u)lding machine
微震资料 microearthquake data
微正长石 microsyenite
微正压燃烧 pressurized combustion
微正则的 microcanonical
微植物 microphytes
微植物群 microflora
微指令 microcode; microinstruction
微指令长度 microinstruction length
微指令存储器 microinstruction storage
微指令定序 microinstruction sequencing
微指令段 microfield
微指令寄存器 microinstruction register
微指令排错 microinstruction debug
微指令中断 microinstruction interrupt
微指令周期 microinstruction cycle
微指令字段 microinstruction field
微指示器 microindicator
微中断 microinterrupt
微中子 neutrino
微终板电位 miniature end-plate potential
微终端 microterminal
微重力 microgravity

微周期 microcycle
微周期时间 microcycle time
微珠玻璃 microglass bead
微柱体 microcylinder
微专用功能 microspecific function
微转动 differential rotation
微装置 microdevice
微锥针入度计 microcone penetrometer
微锥针入度试验 microcone penetration test
微浊 less turbid;slightly turbid
微棕色 dusky brown
微阻计 ducter
微组构 microfabric
微组合 microassembly
微座孢属 < 拉 > Microstroma

煨

煨制弯管 hot bending elbow

韦

韦伯 < 磁通单位,1 韦伯 = 10^8 麦克斯韦 > Weber

韦伯比 Weber ratio
韦伯定理 Weber theorem
韦伯定律 Weber law
韦伯分数 weber fraction
韦伯功率计 Weber dynamometer
韦伯光度计 Weber photometer
韦伯海渊 Weber deep
韦伯和白撒勒断裂时间试验 breaking time test of Weber and Bechler
韦伯接合 Weber joint
韦伯空洞 Weber cavity
韦伯炼铁法 Weber process
韦伯数 < 水工建筑物中用以表征流型特征的数量值 > Weber's number
韦伯微分方程 Weber differential equation
韦伯相似准则 Weber similarity criterion
韦伯蒸发器 Weber evapo(u)rator
韦布尔分布 Weibull distribution
韦布尔概率密度函数 Weibel probability density function
韦布尔理论 Weibull's theory
韦布尔模数 Weibull's modulus
韦布尔统计学 Weibull's statistics
韦布尔图 Weibull's plot
韦布尔系数 Weibull coefficient
韦德尔法则 Weddle's rule
韦德曼花纹 Widmanstatten figure;Widmanstatten pattern
韦德曼司特顿结构 Widmanstatten structure
韦德曼组织 < 金相学 > Widmanstatten pattern
韦地亚钻头 Widia bit
韦碲铜矿 weissite
韦都拉特焊接法 < 一种支管焊接法 > weldolet
韦尔德草地 Veld(t)
韦尔登房屋 Wealden house
韦尔登阶 < 早白垩世 >【地】Wealden stage
韦尔登统【地】Wealden series
韦尔登系 < 下白垩纪 >【地】Wealden
韦尔曼法 Wellman's method
韦尔曼-劳德法 Wellman-Lord process
韦尔讷伊(单晶培育)法 Verneuil method
韦尔讷伊式炉 Verneuil furnace
韦尔氏泵 Weir pump
韦尔氏方位图 Weir's azimuth diagram
韦尔统一场论 Weyl unified field theory
韦尔脱蜡过程 Weir process
韦尔稳频器 Weir stabilizer
韦沸石 wairakite
韦康蒂斯石板瓦 Viscountess slate
韦克菲尔德板桩 < 三块错开钉合成榫的板桩 > Wakefield piling;Wakefield sheet(ing) pile
韦利发动机 Walley engine
韦林掺和器 Waring blender
韦林混合器 Waring blender
韦洛克牌木料 < 一种木花树脂做成的木料 > Weyroc
韦茅斯公式 Weymouth formula
韦茅斯理论 Weymouth theory
韦茅斯裂石面 Weymouth seam-face granite
韦茅斯松 Weymouth pine
韦氏比重秤 Westphal balance
韦氏比重天平 Westphal balance
韦氏测温记录仪 Watkin's recorder
韦氏衬套 Wabcolite bushing
韦氏带法兰管接 Wabcogrip fitting;Wabcoseal fitting
韦氏高温计 Watkin's pyroscope
韦氏转向架轻型制动装置 Wabeopac brake assembly
韦斯巴芬祠 Temple of Vespasian
韦斯巴赫三角测量法 < 联系三角形法竖井定向 > Weisbach triangle
韦斯巴契-达西方程 Weisbach-Darcy equation
韦斯顿标准电池 Weston standard cell
韦斯顿差动滑轮 Weston's differential pulley
韦斯顿电池 Weston cell
韦斯特带壳桩系列 West's shell pilling system
韦斯特法比重天平 Westphalt(type) balance
韦斯特法尔及莫尔比重天平 Westphal and Mohr balance
韦斯特法尔亚阶【地】Westphalian (stage)
韦斯特冷铺沥青混合料 < 一种用沥青粉拌和的 > Westphalt
韦斯特(模数化)预制混凝土桩系列 Wests Hardrive
韦斯汀豪斯风闸公司 < 美 > Westinghouse Air Brake Company
韦特斯式建筑 < 英国创造的一种建筑体系 > Wates
韦先阶 < 早石炭世 >【地】Vian
韦宪期海浸 Visean transgression
韦伊斯碘值测定法 Wijs method

圩

圩 empolder

圩堤 polder embankment
圩地 diked land;diked marsh;dykeland;intake < 指排干了的沼泽地区部分 >
圩区 diked area
圩区(围海造陆)开发工程 polder development
圩田 polder
圩垸 enclosing dike [dyke];polder

围

围岸浅滩 barrier beach

围坝 box dam;cofferdam
围坝堵水 dam in
围板 < 围护机械的 > bracttice;enclosure
围板角钢 coaming angle;coaming bar
围蔽舱室 enclosed compartment
围蔽处所 enclosed space
围蔽灌浆 perimeter grouting
围蔽海峡 enclosed fiord
围蔽甲板 enclosed deck
围蔽起居处所 enclosed accommodation space
围蔽散步甲板 enclosed promenade deck
围蔽室 trunk
围壁 casing;trunk
围壁室舱口 trunked hatchway
围壁室甲板 trunk deck
围壁通风筒 trunk ventilator
围标 bidders ring
围舱壁 trunk bulkhead
围测 strapping
围测罐周 tank strapping
围长 girt(h)
围场 amphitheater;enclosure;girdle;inclosure;paddock;yard
围唱诗班座的隔断 choir screen
围唱诗座的栏杆 choir loft;choir rail
围成栅栏 corral
围带 shroud;shroud ring
围道 contour
围道积分 contour integral;contour integration
围道线映射 contour map
围堤 batardeau;border dike [dyke];bordering;circle levee;closing levee;diking;polder dyke [dike];retaining dike [dyke];ring dike [dyke];ring levee; screen dike [dyke];training dike [dyke]
围堤打捞法 salvage with cofferdams
围堤的 diked
围堤面积 diked area;embanked area;impounded area
围堤抛泥 confined placement;diked placement
围堤区域 diked area;impounded area;embanked area
围堤泄水闸 dike drainage lock
围堤蓄水 ponded water
围堤造地 poldering;ponded lake
围地 enceinte;exclosure
围囤木板 bin board
围封舱壁 enclosure bulkhead
围封空间 inclosure of space
围隔唱诗班的栏杆 quire screen
围隔唱诗班的屏障或围栏 < 教堂中 > choir screen
围隔空地 carol(le)
围隔生态系统 enclosure ecosystem
围埂 ridging;containment berm
围沟 perimeter ditch
围谷 zikustal
围裹 muffle
围海围湖 emptying
围海造陆 polder; reclamation; sea reclamation
围海造田 marine reclamation land; reclaim land from the sea
围焊 weld all around
围合板桩 closed sheeting
围湖造田 enclosing lake for land reclamation;reclaim lake bottom land and plant in to crops;reclaim land from the lake
围护 close up;embosom
围护板 apron plate;curb plate;siding;weather-board(ing)
围护侧板 toe board
围护沟 < 防止基础冲刷 > perimeter trench;perimeter ditch
围护构架 enclosure framing
围护火炉的铁丝网 fire guard
围护结构 building enclosure;building envelop(e); enclosing construction; exterior-protected construction; exterior-protected structure; fender structure; perimeter structure
围护结构空腔 envelope cavity
围护结构面积 exposed area
围护结构热损失 heat loss of protection structure
围护结构温差修正系数 temperature difference correction factor of envelope
围护界沟 enclosure border ditches; ha-ha
围护空间 maqsurah
围护墙 cladding wall;curb wall;diaper wall;perimeter wall;enclosure wall
围护墙材料 wall-enclosure material
围护物 fender
围护桩 fender post;guard post;fender pile < 公路安全带上的 >
围级择伐 girth limit cutting
围脊 marginal ridge
围巾 muffler;neckerchief;scarf
围进 fence in
围垦 land reclamation; reclamation; reclamation by enclosure; empolder; emptying; enclose tideland for cultivation
围垦的土地 reclaimed ground
围垦低地 polder
围垦堤 polder dyke [dike]
围垦地 inning; polder; reclaimed land;reclamation land
围垦泛滥地 inning
围垦工程 reclamation project;reclamation scheme;reclamation works
围垦湖泊 reclaimed lake
围垦护堤 reclamation protection bund;reclamation retention bund
围垦面积 reclaimed area
围垦区 embanked area; reclamation district;reclamation site;reclaimed area
围垦滩涂 enclosing beach for land reclamation
围垦土地 inning;reclaimed land
围垦造地 reclamation of land by enclosure
围垦造田 reclamation of land by enclosure
围垦沼泽 inning
围控 contain a fire
围控火势 containing a fire;corralling a fire
围控时间 corral time
围拉网渔船 purse boat;round-haul netter
围栏 embank elongation;enceinte;enclosure;fence;fencing;impalement;perclose;rail(fence);railing;rave;ring fence;corral < 沉箱深水定位的 >;crawl < 浅水中 >
围栏板 ledger board
围栏草地 enclosed pasture
围栏底板 gravel board
围栏框架 enclosure framing
围栏木板 close board
围栏青贮塔 fence silo
围栏上有榫眼的横木 bar post
围栏式乳牛场 corral-type dairy
围栏铁丝 fence wire
围栏铁网 fence wire netting
围栏小路 fencing piste
围栏柱 fence post
围廊 enceinte;peridrome < 古希腊建筑的 >
围篱 boarding;fence;hoarding;pale;paling;palisade [palisado]
围篱顶端木条 riband
围篱钢柱 steel fence post
围篱钢桩 steel fence picket
围篱木条 riband
围梁 gird;girt(h)

围檩 waling stripe
围囹 wale frame; waling strip(e); wale; waling
围笼式拦污栅 basket type trash rack
围垄池 listing basin
围拢 close(up)on
围拢大厅 close-in concourse
围路积分 circulatory integral
围脉 dike-satellite
围密舱壁 enclosure bulkhead
围模料 investment
围内专用交换网络 switched private national network
围埝 spoil dike [dyke]
围埝(堰)打捞法 salvage with cofferdams
围埝炸礁 reef blasting with enclosure
围盘 mechanical repeater
围盘管 looping pipe
围盘轧机 guide mill
围盘轧制 repeat-rolling
围屏隔离幕 parclose
围畦灌溉 check irrigation
围起 fence in
围起的土地 paddock
围起来的场地 enclosed space
围砌 masonry
围碛 peripheral moraine
围墙 barrier wall; boundary wall; bulkhead; closure; coffer wall; enceinte wall; enclosing masonry wall; enclosing wall; enclosure; exterior wall; external wall; fence (wall); fencing(wall); guard fence; hedge; hypaethral; inclosure; inclosure wall; open fence; pale fencing; perimeter wall; property line wall; ring fence; ring wall; wall enclosure; curtain wall; periphery wall
围墙表面 enclosing masonry wall(sur) facing
围墙长度 length of boundary wall; length of fence
围墙衬砌 enclosing masonry wall lining; enclosing wall lining; external walling lining
围墙杆 wall guard
围墙高度 fence height; height of fence
围墙工 enclosing masonry work
围墙花园 wall garden
围墙建筑单元 enclosing wall building unit
围墙建筑构件 enclosing wall building member
围墙建筑施工 enclosing wall construction
围墙建筑组成部分 enclosing wall building component
围墙结构 enclosure framing
围墙框架 curtain-wall frame
围墙密封材料 curtain-wall sealing material
围墙内的房群 compound
围墙砌筑 external walling
围墙前的入口 propylaeum [propylaea]
围墙嵌板 curtain-wall panel
围墙墙面 enclosing masonry wall facing
围墙饰面 enclosing wall facing
围墙修饰 enclosing wall finish
围墙用撑 dead-shore needle
围墙柱 enclosing masonry wall column; exterior wall column; external walling column
围鞘 perisarc
围圈靠背 hoop back
围裙 apron; lap; save-all
围裙气室气垫艇 skirted plenum craft
围裙式输送机 apron conveyer [conveyor]; plate conveyer [conveyor]
围绕 ambient; around; circumvent; close about; encircle; enclosure; encompass(ment); engird(le); enwrap(ping); girt(h); round; surround
围绕并盖于舞台上的天顶 sky-dome
围绕城市环状汽车专用路 orbital motorway
围绕船厂的区域 dock land
围绕道路 beltway
围绕的 circumfluent; circumfluous
围绕的圬工墙 surrounding masonry wall
围绕观众台的矮护墙 parclose
围绕桥墩打的防护桩 starling
围绕天极的 circumpolar
围绕危险品罐的防溢流土埂 storage bund
围绕下水系统 circumferential drainage system
围绕着的 circumjacent
围绕着环境的项目 environment-geared project
围入栅栏的地点 stockade
围山矿 weishanite
围室 peripheral cell
围收漂子 <木材分类后的> pocket boom
围手椅 bergere
围水养护 <混凝土的> curing by ponding; ponding curing; water ponding
围水养护法 ponding method of curing; ponding
围水养生 curing by ponding; water ponding
围田 levee-surrounded field; polder
围填灌浆 backfill grouting
围条 edge strip
围土 build in
围网 purse net; purse seine; seine
围网船 seine boat
围网吊货板 net board
围网浮子 seine float
围网设计 purse seine design
围网渔船 purse boat; purse seiner; seiner
围网渔轮 seine boat
围圩 polder
围涎 bib(b)
围线映像 contour line map
围限压力 confining pressure
围斜层理 periclinal bedding
围斜构造 periclinal structure; pericline
围压 hoop stress; confinement pressure; confining pressure <三轴试验的>
围压率定机 calibrator for transducer-calibrating confining pressures
围压强度 strength under peripheral pressure
围压效应 ambient pressure effect; confined pressure effect; confining pressure effect
围压压缩 hydrostatic compression
围岩 adjacent rock; adjoining rock; country rock; country-tock; deads; dike rock; enclosing rock; enclosing wall; external waste; host rock; mother rock; neighbo(u)ring rock; surrounding material; surrounding rock; wailrock; wall rock
围岩不稳固的 weak-walled
围岩层 enclosing stratum
围岩储存的热量 heat stored in the country rocks
围岩处理 ground treatment
围岩的几何因子 geometric(al) factor of the adjacent bed
围岩的物理力学性质 ground data
围岩的性质 nature of ground
围岩电阻率 resistivity of adjacent formation
围岩调查 surroundings investigation
围岩断裂 fracture around underground opening
围岩二次应力状态 secondary stress state
围岩分级 surrounding rock classification
围岩分类(法) classification of surrounding rock; rock classification
围岩环 wall-rock aureoles
围岩类别 surrounding rock class; surrounding rock type
围岩类型 surrounding rock type
围岩面密度 facial density of surrounding rock
围岩内部变形 deformation of surrounding rock
围岩喷锚支护 lock bolt support with shotcrete
围岩偏压 non-uniform rock pressure; leaning pressure of ambient rock
围岩破坏 failure of surrounding rock
围岩蚀变 country rock alteration; wall-rock alteration
围岩收敛观测 ambient rock convergence monitoring
围岩松弛范围 zone of relaxation of rock
围岩特性 behavio(u)r of surrounding rock; character of surrounding rock
围岩稳定 surrounding rock stability
围岩稳定处理 ground stabilization; soil stabilization
围岩稳定性 stability of surrounding rock
围岩压力 ground pressure; pressure of surrounding rock; rock loading; rock pressure; surrounding rock pressure
围岩压力的施工效应 construction effect of rock pressure
围岩压力的时间效应 time-effect of rock pressure
围岩移动 ground movement
围岩异常 wall-rock anomaly
围岩应力 surrounding rock stress
围岩应力集中系数 factor of stress-concentration in surrounding rock
围岩影响 shoulder bed effect
围岩晕 wall-rock halo
围岩自承能力 self-bearing capacity of surrounding rock
围岩自稳能力 autophasing of wall rock; self-bearing capacity of surrounding rock; self-stabilization capacity of surround rock
围堰 cofferdam; batardeau; box cofferdam; box dam; closing dam; coffer; surrounding dam; embankment; bund wall
围堰板桩 cofferdam piling; pile sheathing
围堰板桩墙 cofferdam sheeting
围堰比较方案 alternative cofferdam scheme
围堰拆除 cofferdam demolition; cofferdam removal
围堰储料场 cofferdam stockpile; cofferdam stockyard
围堰的闸门 cleading
围堰底座 coffered foundation
围堰法 cofferdam method
围堰护档 counterfort
围堰基础 cofferdam foundation; coffered foundation
围堰基础清理 clean-up of cofferdam foundation; cofferdam foundation cleanup
围堰基底 cofferdam foundation
围堰排水 dewatering of cofferdam
围堰抛石护坡 cofferdam riprap protection
围堰戗道 berm(e) of cofferdam
围堰墙 cofferdam; coffer(ed) wall
围堰设计 coffer design
围堰施工 cofferdam construction
围堰施工程序 cofferdam construction sequence
围堰施工方法 cofferdam construction method
围堰石料场 cofferdam quarry
围堰修建 cofferdam construction
围堰迎水面堆石堤 cofferdam outer rockfill dike [dyke]
围堰迎水面堆石体 cofferdam outer rockfill dike [dyke]
围堰支撑 cofferdam bracing; cofferdam support
围堰折流堤 cofferdam deflector
围堰轴线 axis of cofferdam
围堰筑岛 contained sand island
围窑 clamp furnace; stack furnace
围以壕沟 entrench
围以墙 wall
围以石墙 stone wall(ing)
围以水沟 moating
围油浮杆 floating oil barrier
围油栏 oil boom; oil fence
围淤地 warping bank
围垣 fencing
围在墙内 immure
围栅 pale fencing
围植 enclosure planting
围住的板桩 enclosing sheeting
围住的大楼 enclosed block
围住的空间 enclosed space
围住的面积 enclosed area
围住的小教堂 encircling chapel
围住岩脉的岩石 wall rock
围柱殿的柱廊 peripteral colonnade
围柱殿式建筑 peripteral building
围柱殿式庙宇 peripteral temple
围柱式的 peripheral
围柱式殿 peripteral
围柱式房屋 periptery
围柱式建筑(物) peripteral building; peripteros; periptery
围柱式神庙 peripteral temple
围桩 stockade

违背保修条款 breach of warranty
违背担保 breach of warranty
违背(法律等) transgress
违背惯例 out of rule
违背合同 breach of contract; break; break a contract
违背诺言 breach of promise
违背契约 breach of contract; breach of covenant; default
违背事实 be contrary to the fact
违背信托 breach of trust
违法 breach of law; delinquency; infraction; law violation; misfeasance
违法编码模式 code violation pattern
违法超速驾驶者 speedster
违法的 criminal; delinquent; under-the-counter; unlawful
违法复制件 infringing copy
违法合同 illegal contract
违法活动 illegal act
违法火 actionable fire

违法即决 composition of offences
违法建筑 illegal building
违法事件 law violation
违法行为 breach of law;malpractice
违法者 delinquent;lawbreaker
违反 act against; contravention; run counter; transgress(ion); violate; violation
违反保证 breach of warranty
违反闭塞制度 violation of blocking
违反操作程序 failure to follow instruction
违反操作规程 abuse
违反担保 breach of warrant
违反法规 break regulation
违反分期付款的义务 breach of obligation to pay an instalment
违反规程 breach of regulation;infraction of regulations
违反规定的停车时间 penalty period
违反规则 breach of regulation
违反规章 infringement
违反国际法罪行 international crime
违反航行范围的保证 breach of trading warranty
违反合同 breach(of) contract;breach of covenant;break a contract;break of contract;contravention to treaty;infringe contract; infringement of contract; violate a contract; violation of contract
违反合同条件 breach of contract conditions
违反合约 breach of contract;breach of covenant;break a contract;contravention to treaty; infringe contract;infringement of contract
违反和错误操纵信号 signal violation and mishandling
违反计算数据的假设 off-design condition
违反交通规则者 traffic violator
违反交通规章判罪 traffic conviction
违反交通规章事件 traffic offence
违反交通规章者 traffic offender
违反交通信号 violation of signal
违反明确的担保 breach of an express guarantee
违反诺言 breach of promise
违反契约 breach of contract;breach of covenant;break a contract;contravention to treaty; infringe contract;infringement of contract
违反契约的诉讼 covenant
违反事实的 contrafactual
违反特约条款 <水险单上附加的条款> breach of warranty
违反条件 breach of conditions
违反条例 breach of regulation
违反条约 act in violation of the stipulation
违反停车规则(者的)传票 parking ticket
违反协议 break agreement
违反信托 breach of trust
违反行车规则 driving offence
违反许可(证)规定的施工 construction contrary to permit
违反有代理权的保证 breach of warranty of authority
违反有价证券规定的选择权 bad delivery of securities
违反运输契约 breach of the contract of carriage
违反运行规程 mishandling
违反者 violator
违反正弦条件 offense against sine condition
违反自然法则的不可能事 physical impossibility
违犯法律 violate a law
违犯规则 breach of rules
违犯林业规章行为 forest offence
违犯森林规则 rape of forest
违光试验 transillumination
违和 acosmia
违禁捕鱼者 poacher
违禁的 illicit
违禁火 actionable fire
违禁货物 offensive cargo
违禁贸易 contraband of trade; illicit trade
违禁品 contraband;contraband goods;prohibited article;prohibited goods
违禁品检查仪 <射线透视的> inspectoscope
违禁品运输 contraband; illegal traffic;illicit traffic
违禁市场 illicit market
违禁文献 objectionable literature
违禁物品 article of contraband
违例 breach of regulation; breach of rules
违例的航次 illegitimate voyage
违令使用 unauthorized use
违约 breach of contract;break;break an agreement; break of contract; default;failure;in default;indenture defaulting; protracted default; violate a contract
违约偿金 liquidated damages
违约成本 penalty cost
违约当事人 delinquent party
违约的合同当事人 party in breach (default)
违约的投资 default in investment
违约罚金 contractual fines; damages liquidated; default fine; liquidated damages
违约罚金条款 penalty clause
违约罚款 liquidated damages;liquidated damages for delay; penalty for non-performance of contract; penalty of breach of contract; penalty sum
违约方 default party; the breaching party
违约金 forfeit for breach of contract; liquidated damages; penalty for breach of faith
违约利息 penal interest
违约赔偿费 damages for default
违约人 defaulter
违约事件 deviation of the contract
违约事项 event of default
违约损失赔偿准备金 default loss compensation reserve
违约通知 notice of default
违约行为 defaulting behavio(u)r; non-compliance
违约一方 defaulting party;delinquent party
违约债务人 defaulted debtor;defaulting debtor;tardy debtor
违约者 defaulter
违章 breach of regulation;break rules and regulations; peccancy; violation;violation of regulations
违章报告 report on infringement of regulations
违章泊车 tow-away zone
违章操作 operation against regulations;operation against rules
违章操作保险锁 vandalism protection lock
违章操作的成套保护装置 vandalism protection kit
违章处罚 penalty for violation of regulation
违章处理 mishandling
违章处理路签或路牌 mishandling of staff or tablet
违章处理手摇车 mishandling of trolley car
违章罚款 non-compliance penalty
违章建造的 jerry-built
违章建筑 illegal building; squatter house;squatter sectionment;squatter settlement;squatting;unauthorized construction
违章居留地 squatter settlement
违章率 violation rate
违章日期 date of non-conformity
违章使用 non-conforming use
违章通知书 advice of irregularity
违章卸料 indiscriminate dumping
违章行为 act of infringement of regulation
违章占住者 squatter
违章作业 working against regulations;work performed against regulation

桅侧支索牵条 chain plate

桅的鼓出部位 quarter of a mast
桅的下部或下桅甲板下部分 heel
桅的左右支索 shroud
桅灯 headlight;mast lamp;mast light; steaming light;top lantern
桅灯灯座 lamp bracket
桅底座 lutchet
桅钉 sett piling
桅顶 head mast;masthead
桅顶灯 floating light;masthead light; top light
桅顶吊索 gantline;girtline
桅顶放电 corposant
桅顶风标帽 acorn
桅顶高度 height of mast; masthead height
桅顶箍 masthead band
桅顶横桁 cross tree
桅顶滑车 cheek block
桅顶结 collar knot
桅顶瞭望者 masthead lookout;masthead man
桅顶旗索 cap stay;pendant halyard
桅顶前方支索 head stay
桅顶球 mast ball;ox ball
桅顶饰旗 dressed with masthead flags
桅顶天线 masthead aerial
桅顶系绳箍 eye band
桅顶(仰)角 masthead angle
桅顶支索固定板 futtock plate
桅顶支索固定链 futtock chain
桅顶至斜桁间短索 peak line; peak pendant [pendent]
桅顶纵桁 trestle tree
桅斗 bird's nest;crow's nest
桅帆 mast and sail
桅杆 gin pole; guyed mast; kingpost; mast;mast arm;spar
桅杆材 mast timber
桅杆侧支索 shroud
桅杆的加强夹箍 fish
桅杆的甲板下部分 bury;housing
桅杆的竖立 tilting up of the mast
桅杆吊 derrick crane
桅杆吊的顶部滑轮 crown block
桅杆吊的立柱 crane post
桅杆吊中柱 king post
桅杆顶灯 top light
桅杆断面 mast section
桅杆绞辘 mast tackle
桅杆模型 mast mold
桅杆耐水清漆 spar varnish
桅杆漆 mast paint
桅杆起重机 gin pole;guy derrick
桅杆起重机缆风 derrick guy
桅杆起重机双脚架 mast crane
桅杆清漆 spar varnish
桅杆全升起时的高度 full height
桅杆上承座 shaffle
桅杆上的顶球 knob
桅杆上端电火 corposant
桅杆式起重船 derrick crane barge
桅杆式起重吊车 gin-pole truck
桅杆式起重机 derrick(crane); gin pole derrick; mast crane; guyed mast
桅杆式起重机杆 boom derrick
桅杆式天线 mast antenna
桅杆式转臂起重机 derrick crane; light mast crane
桅杆套 smoke cover
桅杆下端 lower end of the mast
桅杆栅栏 life rail
桅杆柱脚 mast table
桅搁架 mitchboard
桅根 mast heel
桅箍 crance [cranse];drift hoop;mast band;mast hoop
桅冠 mast truck
桅冠灯 truck light
桅桁支索 jack stay
桅横杆 yard;yardarm
桅横杆上踏脚挂绳 stirrup
桅横杆上悬挂吊索的铁箍 sling band; sling hoop
桅横杆悬挂吊索部分 slings
桅尖 cap
桅间横牵索 spring stay
桅间索 freshwater stay;jumper stay; signal stay;triatic stay
桅间悬索 spring stay
桅间支索 cap stay
桅肩 cheek plate; hound; hound of a mast;mast hounds;mast shoulder
桅肩箍 hound band
桅肩加强板 bibbs
桅肩下 hounding
桅铰链 mast clamp;mast hasp
桅孔 mast hole;mast opening
桅孔加强板 mast partner
桅缆式动臂起重机 guyed derrick crane
桅缆式转臂起重机 guyed derrick crane
桅楼侧支索 futtock shroud
桅楼扶手 top rail
桅楼瞭望水手 topsman
桅楼升降口 lubber's hole
桅帽 mast truck;truck
桅木 mast timber
桅盘 top
桅盘瞭望水手 topsman
桅盘人孔 lubber's hole
桅盘围 top armo(u)r
桅前缘索 luff rope
桅圈 mast collar
桅裙 boot;petticoat;pittcoat
桅色漆 mast colo(u)r paint
桅上电灯 Saint Elmo's fire
桅上电火 corposant;deed fire
桅上瞭望塔 bird's nest;crow's nest
桅式吊车 mast crane;mast derrick
桅式吊机 mast crane
桅式浮标 pole float
桅式平台 spar platform
桅式起重机 derrick crane; mast derrick
桅室 mast house
桅栓 fid
桅台 mast table
桅套 boot;petticoat
桅梯 mast ladder
桅梯第一级横杆 sheer pole

桅梯横绳 ratlin(e);rattling
桅梯绳 ratlin(e) stuff
桅头滑槽 dead sheave
桅围 dolphin of the mast
桅屋 mast house
桅楔 mast wedging
桅中部支索 belly stay
桅装 <雷达收发机装于雷达天线架上> masthead mounted
桅最上方的前支索 royal stay
桅最上方帆桁 royal yard
桅座 foot step;mast step;saucer;tabernacle
桅座板 mast thwart;mast bench <小艇>
桅座叉柱 crutch
桅座挂滑车铁杆 horse

唯交通论者 traffickist

唯理智论 intellectualism
唯美主义 aestheticism;estheticism
唯实论 realism
唯物辩证法 materialist(ic) dialectic
唯物论 materialism
唯物主义 materialism
唯心论 idealism
唯一遍历的 uniquely ergodic
唯一标识符 unique identifier
唯一存在定理 unique existence theorem
唯一代理 exclusive agency;sole agency
唯一的一致性 unique uniformity
唯一定义类目 uniquely defining class
唯一分解定理 unique decomposition theorem
唯一解(法) unique solution
唯一开拓 unique continuation
唯一开拓定理 unique continuation theorem
唯一可译代码 uniquely decipherable code;uniquely decodable code
唯一可着色图 uniquely colo(u)rable graph
唯一来源设备 sole-source equipment
唯一邻近 unique proximity
唯一数据系统 unique data system
唯一特性 unique trait
唯一析因 unique factorization
唯一析因定理 unique factorization theorem
唯一析因整环 unique factorization domain
唯一性 uniqueness
唯一性条件 uniqueness condition
唯一性原理 uniqueness principle
唯一因子方差 uniqueness factor variance
唯一值 unique value

帷幔 antependium;hanging <窗帘>

帷幕 barrier; curtain; curtain wall; draperies;draping;heavy curtain
帷幕挡土墙 diaphragm retaining wall
帷幕灌浆 curtain grouting;membrane grouting
帷幕灌浆堵水 blocking water with grouting heavy curtain
帷幕灌浆法 curtain grouting process; membrane grouting process
帷幕灌浆孔 curtain grout hole; curtain hole
帷幕截水墙 diaphragm retaining wall
帷幕孔 curtain hole
帷幕宽度 curtain width
帷幕内地下水头 groundwater head in curtain grouting heavy
帷幕内外水头差 difference of water head between the grouting heavy curtain inside and outside
帷幕排水道 curtain drain
帷幕墙 curtain wall;diaphragm wall; kanat
帷幕深度 curtain depth
帷幕式 heavy curtain type
帷幕式防波堤 curtain wall breakwater
帷幕式排渗 curtain drain
帷幕外地下水头 groundwater head outside the grouting heavy curtain
帷幕五金 drapery hardware
帷幕线 curtain line
帷幕注浆法 curtain grouting method
帷墙 curtain wall

维艾尔布拉 <一种铝黄铜> Vialbra

维比稠度 Vebe degree
维比稠度计 Vebe consistency meter; Vebe consistometer
维比稠度计试验 Vebe consistometer test
维比稠度仪 Vebe apparatus; Vebe consistometer
维比尔基-特维尔硬瓷器 verbilki-Tver porcelain
维比时间 Vebe time
维比试件 Vebe specimen
维比试验法 Vebe test method
维比仪 <一种混凝土稠度测定仪> Vebe apparatus
维比振动稠度试验 Vebe test
维比振动台 <用于测定碾压混凝土稠度> Vebe vibratory table
维比值 Vebe value
维勃稠度仪 V-B consistometer
维勃试验仪 <一种测定干硬性混凝土稠度的仪器> Vebe apparatus
维布拉减振材料 <一种减振材料> Vibracork
维持 maintaining; perpetuate; preserving;hold;uphold;upkeep
维持泵 process pump
维持车速 maintained speed
维持电弧 pilot arc
维持电流 current maintenance;maintenance current
维持电压 maintaining voltage
维持费(用) cost of maintenance;running cost; holding cost; maintenance cost;upkeep
维持功率 holding power
维持荷载的试桩法 maintained load test
维持荷载法 <试桩> maintained load method; maintenance load test; maintained load test [ML-test]
维持荷载试验 <试桩> ML-test [maintained load test]
维持荷载法试桩 maintained load pile test
维持机制 support mechanism
维持剂量 maintenance dose
维持价格 vaiorisation
维持价格水平 maintain the price level
维持金 maintenance bond
维持净能 net energy for maintenance
维持库存 maintenance of stock
维持量 maintenance dose
维持能 maintenance energy
维持你方报价 keep your quotation
维持取暖温度所需热量 heating load
维持热 holding heat;maintaining heat
维持生活 support oneself or one's family
维持生活水平 subsistence level
维持时间 hold time
维持天数 carry through
维持土壤肥力 maintenance of soil fertility
维持系数 <照明设备> maintenance factor
维持现状 status(in) quo
维持消耗 maintenance cost
维持行为 maintenance behavio(u)r
维持性价格 support price
维持性托换 maintenance underpinning
维持性挖泥 maintenance dredging
维持性行销 maintenance marketing
维持需要 maintenance requirement
维持压强 take-hold pressure
维持液 maintenance media
维持原状力 do nothing
维持运转 keep on the go
维持账户 maintaining account;servicing account
维持折旧 maintenance depreciation
维持者 sustainer
维持真净能 true net energy for maintenance
维持证券的最低价格 stabilize a security
维持转卖价格 resale price maintenance
维持装置 holdout device
维达尔试验 Widal test
维德尼柯夫数 Vedernikov number
维度 dimension
维多利亚 <加拿大港口城市或塞舌耳首都> Victoria
维多利亚黄 Victoria yellow
维多利亚坚牢紫 <一种紫色酸性染料> Victoria fast violet
维多利亚简型板 <一种防火隔声楼板> Victoria-Simplex
维多利亚蓝 Victoria blue
维多利亚绿 Victoria green
维多利亚女王时代的风格 Victorianism
维多利亚式 Victorian style
维多利亚式建筑 Victorian architecture
维多利亚洲 <澳大利亚> Victoria
维多利亚紫 Victoria violet
维恩常数 Wien constant
维恩定律 Wien's law
维恩分布律 Wien's distribution law
维恩辐射定律 Wien's radiation law
维恩公式 Wien equation
维恩加登文法 Wijngaarden grammar
维恩频率电桥 Wien frequency bridge
维恩图 Venn diagram
维恩位移定律 Wien's displacement law
维恩效应 Wien effect
维尔茨反应 Wurtz reaction
维尔德常数 Verdet constant
维尔弗莱型摇床 Wilfley table
维尔格罗阶 <中三叠世> Virglorian
维尔霍扬斯克-科里亚克地槽 Verchojanskij-Korakoje geosyncline
维尔农页岩 Vernon Shale
维尔斯特拉斯 M 判别法 Weierstrass M test
维尔斯特拉斯逼近定理 Weierstrass approximation theorem
维尔斯特拉斯变换 Weierstrass transform
维尔斯特拉斯标准型 Weierstrass canonical form
维尔斯特拉斯典范积 Weierstrass's canonical product
维尔斯特拉斯点 Weierstrass's point
维尔斯特拉斯函数 Weierstrass function
维尔斯特拉斯曲线 Weierstrass curve
维尔斯特拉斯椭圆函数 Weierstrass's elliptic(al) functions
维尔松管 Wirsung's duct
维尔维兰纤维 vel-velam fibre
维尔烯 versene
维尔型选矿机 Val mineral separator
维格勒柱 Vigreaux column
维格曼屋架 Wiegmann roof truss
维格纳-埃卡特定理 Wigner-Eckart theorem
维格纳超多重态 Wigner supermultiplet
维格纳定理 Wigner theorem
维格纳合理近似 Wigner's rational approximation
维格纳间隙 Wigner gap
维格纳近似 Wigner approximation
维格纳力 Wigner force
维格纳能 Wigner energy
维格纳-赛茨法 Wigner-Seitz method
维格纳-赛茨晶胞 Wigner-Seitz cell
维格纳释放 Wigner release
维格纳同量异位素 Wigner isobar
维格纳效应 Wigner effect
维格纳增长 Wigner growth
维格特效应假彩色编码 Vigoter effect pseudocolo(u)r encoding
维构件 one-dimensional element
维管的 vascular
维管束 vascular bundle
维管形成层 vascular cambium
维管组织 vascular tissue
维管组织的 fibrovascular
维弧 arc maintenance
维弧电流 background current; pilot arc current
维弧电路 keep-alive circuit
维弧电压 keep-alive voltage
维弧阳极 keep-alive electrode
维护 attendance; handling; handling operation; maintain; maintenance; operation activity; service; servicing;upkeep;vindicate;working
维护板 maintenance panel
维护保养 maintaining;tend(ing)
维护保养手册 maintenance manual
维护保养说明书 maintenance instruction
维护保障措施 maintenance safeguard
维护备用时间 maintenance standby time
维护编号卡 maintenance code
维护不良 under-maintenance
维护不足 under-maintained
维护部分 machine assembly department
维护部门 maintenance department
维护操作工作站 workstation for service operation
维护操作控制台 service operation console
维护测试 maintenance test
维护测试法 maintenance test method
维护测试设备 maintenance test equipment
维护策略 maintenance strategy
维护车间 maintenance shop
维护成本 maintenance cost
维护程序 maintenance procedure; maintenance program(me)
维护程序链 maintenance program(me) chain
维护单元 service unit
维护费(用) cost of maintenance;maintenance cost; maintenance fee; sum of maintenance; upkeep cost; service cost; servicing cost; carry-

ing charges;cost of upkeep;maintenance outlay
维护分队 maintenance division
维护分析 maintenance analysis
维护服务 maintenance service
维护辅助程序 maintenance assistance modules
维护工程 maintenance engineering
维护工程师 field engineer;maintenance engineer;service engineer
维护工具 care of instrument;maintenance tool
维护工作 maintenance work
维护工作周期 maintenance task period
维护管理处理机 maintenance processor
维护管理系统 maintenance management system
维护规程 maintenance regulation;service regulation
维护规则 maintenance regulation;operating instruction
维护和外围设备模块 maintenance and peripherals module
维护和修理 operated service and repair
维护荷载 maintenance load
维护及修理 maintenance and repair
维护检查 maintenance test
维护检查程序 test and maintenance program(me)
维护检修 operating repair
维护检修时间 engineering time
维护检修预算 maintenance and repair budget
维护结构 enclosures
维护开支 maintenance expense
维护可靠性 maintenance reliability
维护控制板 maintenance control panel
维护控制数据 maintenance control data
维护控制信号 maintenance control signal
维护控制重算寄存器 maintenance control retry register
维护率 maintenance rate
维护面板 maintenance panel
维护平台 attendant's platform
维护期 defect liability period;maintenance period
维护人员 attendant;attending personnel; maintainer; maintenance personnel;operating personnel;operating staff
维护设备 attention device;maintenance equipment;maintenance facility;service equipment
维护设施 maintenance prevention
维护时间 servicing time
维护时限 maintenance interval
维护试验 maintenance test
维护手册 instructions handbook;maintenance manual; maintenance handbook; maintenance instruction;service manual
维护手段 maintenance service
维护疏浚 maintenance dredging
维护数据 maintenance data; service data
维护数据系统 maintenance data system
维护水平 maintenance level
维护水深 maintenance depth
维护涂料 maintenance finish;maintenance paint
维护挖泥费用 maintenance dredging cost
维护系数 maintenance factor
维护细则 running order
维护性能 maintainability
维护性失常 aids failure due to improper maintenance
维护性疏浚 maintenance dredging
维护性挖泥 maintenance dredging
维护性油漆 maintenance painting
维护业务 maintenance service
维护用起重机 service crane
维护用扫描器 maintenance scanner
维护与操作 maintenance and operation
维护与修理 maintenance and repair
维护质量检查周期 maintenance proof cycle
维护中心 maintenance center [centre];service center [centre]
维护周期 maintenance interval
维护主权 assert
维护状况 state of maintenance
维护组织 maintenance organization
维护钻头 nursing the bit
维卡稠度仪 Vicat consistency apparatus
维卡稠度仪针 Vicat needle
维卡稠度针 Vicat consistency needle
维卡钒钴铁磁性合金 Vicalloy
维卡贯入仪 <测定炸药塑性用> Vicat penetrometer
维卡合金 Vicalloy
维卡计试验 Vicat test
维卡检验计 <测定水泥硬化程度> Vicat needle
维卡软化点 Vicat softening point
维卡透度计 Vicat penetrometer
维卡透度仪 Vicat penetrometer
维卡仪 <用于水泥稠度试验> needle apparatus; Vicat apparatus; Vicat needle apparatus
维卡仪试验 <测定水泥稠度用> Vicat test
维卡针 Vicat needle
维卡针入度试验 <测定水泥凝结时间的> Vicat needle test
维卡针式测试器 Vicat needle apparatus
维卡指数 Vicat hydraulic index
维克玻璃 <一种含硼高硅氧玻璃> vycor glass
维克多铜锌镍合金 Victor metal
维克劳镍铬耐热合金 Vikro
维克托利克型接头 Victoric joint
维克托利克型接头管子 Victaulic pipe
维克硬度 Vicker's hardness
维克硬度计 Vicker's hardness tester
维拉丛风 virazon
维拉德效应 Villard effect
维拉弗朗阶【地】Villafranchian
维拉克斯复合织物 <商品名> Vellux
维拉里倒逆 Villari reversal
维拉里效应 Villari effect
维拉扎诺海峡桥 <美> Verrazano Narrows Bridge
维劳达 <一种转数表传感器> velodyne
维勒方式 Nelle system
维勒米尔制冷机 Vuilleumier refrigerator
维里金公式 Vinikin formula
维里金影响半径公式 influence radius formula Vinikin
维里系数 <液体力学> Virial coefficient
维利迪安颜料 Viridian pigment
维梁 collar(ino)
维量单位 dimensional unit
维量法 method of dimensions
维量分析 dimensional analysis
维量分析器 dimensional analyser [analyzer]
维林特大梁 Vierendeel girder
维林特桁架 Vierendeel truss
维硫铋铅银矿 vikingite
维硫锑铅矿 veenite
维硫锑铊矿 weissbergite
维纶纤维废水 polyvinyl fiber [fibre] wastewater
维罗纳大理石 Verona marble
维罗纳褐色 Verona brown
维罗纳黄 Verona yellow
维罗纳绿 French verones green
维罗纳绿色颜料 Verona green (earth) pigment
维马破裂带 Vema fracture zone
维玛海沟 Vema trench
维玛海渊 Vema deep
维梅特硬质合金 wiemet
维命弧 keep-alive arc
维姆黏[粘]聚力试验 Hveem cohesion test
维姆黏[粘]聚力仪 <测定沥青混合料及稳定土等黏[粘]聚力用> Hveem cohesiometer
维姆柔性路面设计法 Hveem design method of flexible pavement
维姆设计法 Hveem design method
维姆氏(沥青)混合料配合比设计法 California method of mixture design
维姆试验 Hveem test
维姆稳定度 <测定沥青混凝土等稳定性的一种指标> Hveem stability
维姆稳定度试验 Hveem stability test
维姆稳定度仪 <测定沥青混凝土及稳定土等稳定度用> Hveem stabilometer
维姆稳定度仪法 <设计柔性路面厚度的一种方法> Hveem stabilometer method;Hveem method
维纳地图投影 Werner's projection
维纳尔海文石 Vinal Haven granite
维纳反应 Wiener reaction
维纳反褶积 Wiener deconvolution
维纳过程 Wiener process
维纳混合机 Werner mixer
维纳-霍普夫方程 Wiener-Hoof equation
维纳-霍普夫技术 Wiener-Hoof technique
维纳-雷文森算法 Wiener-Levinson algorithm
维纳理论 Werner's theory
维纳滤波 Wiener filtering
维纳滤波器 Wiener filter
维纳频谱 Winer spectrum
维纳谱线 Werner line
维纳-钦辛定理 Wiener-Khinchine theorem
维纳-钦辛关系式 Wiener-Khinchine relation
维纳实验 Wiener experiment
维纳式粉碎机 Werner system crusher
维纳斯雕像 Venus
维纳斯女神 Venus
维纳斯神庙 <罗马> Temple of Venus
维纳土地电阻探测法 Werner System earth resistivity method
维纳效应 Wiener ergodic theorem
维纳信息论 Wiener information theory
维纳最佳化 Wiener optimization
维纳最佳滤光片 Wiener best filter
维纳最佳系统 Wiener's optimum system
维尼昂 <商品名> vinyon
维尼纶 vinylon
维尼纶帆布 vinylon canvas
维尼纶绳 vinylon rope
维铌钙矿 vigezzite
维涅牌玻璃 <一种磨砂玻璃> Verre grave
维宁-曼乃兹海洋三摆仪 Vening-Meinesz sea three pendulum apparatus
维宁-曼乃兹均衡说 Vening-Meinesz isostatic system
维诺格拉多夫石 vinogradovite
维诺沃瓷器 <意大利> vinovo porcelain
维皮计 Vebe apparatus
维羟硼钙石 vimsite
维琴水泥取样装置 Vezin cement sampler
维任第大梁 Vierendeel girder
维塞勒光干涉比长仪 <能测定距离 864 米,精度 10^{-7}> Väcsälä comparator
维生射线 visual ray
维士达牌玻璃 > Vesta
维士法阶 <晚石炭世>【地】Westphalian
维氏稠度计 Vebe meter
维氏法 Vicker's method
维氏金刚石棱锥硬度试验 Vicker's diamond pyramid hardness test
维氏金刚石硬度 diamond penetrator hardness; diamond pyramid hardness;Vicker's diamond hardness
维氏金刚石硬度计 Vicker's diamond hardness tester
维氏金刚石硬度试验 Vicker's diamond hardness test
维氏金刚石硬度值 Vicker's diamond hardness number
维氏刻痕硬度试验 Vickers indentation hardness test
维氏梁桥 Vierendeel girder bridge
维氏密实度 Vede-degree
维氏体 wüstite
维氏椭圆积分 Weierstrass elliptic(al) integral
维氏压痕裂'纹长度 Vicker's indentation crack length
维氏压痕器 Vicker's indenter
维氏硬度 Vicker's hardness
维氏硬度表 Vicker's hardness scale
维氏硬度单位 Vicker's unit
维氏硬度计 Vicker;Vicker's hardness tester
维氏硬度试验 Vicker's harness test
维氏硬度试验机 Vicker's hardness tester; Vicker's pyramid hardness testing machine
维氏硬度值 Vicker's hardness number
维氏钻石压头硬度计 Vicker's diamond pyramid hardness tester
维氏钻石硬度计 Vicker's diamond hardness tester
维数 dimensions; dimensionality; number of dimension
维数的 dimensional
维数后缀 dimension suffix
维数界 dimension bound
维数理论 dimension theory
维数论 dimension theory
维数选择 dimension select
维数语句 dimension statement
维数属性 dimension attribute
维数字 dimension word
维斯马港 <德国> Port Wismar
维苏威火山 Vesuvius Volcano
维苏威式 Vesuvian
维苏威型火山喷发 Vesuvian-type eruption
维他玻璃 <一种能透过紫外线的玻璃> vitaglass
维塔利姆高钴铬钼耐蚀耐热合金 Vitallium
维特克-山农抽样定理 Whittaker-Shannon sampling theorem
维特里页岩 Whittery shale

W

维特鲁威式门窗 vitruvian opening
维特罗维阿斯的埃及会堂 Egyptian Hall of Vitruvius
维特-马古列斯方程 Margules equation;Witte-Margules equation
维脱利牌玻璃 Vitrea glass
维瓦铝基合金 Vival
维谢尔地震仪 Wiechert seismograph
维谢尔法 Wiechert method
维修 inspection and repair;maintain;maintenance and overhaul;maintenance prevention;maintenance service;service action;servicing;upkeep
维修班 maintenance crew;maintenance party;maintenance team
维修班长 maintenance foreman
维修包 service kit
维修保函 maintenance bond
维修保养 maintenance
维修保养班 maintenance gang
维修保养车间 maintenance shop
维修保养费用 cost of upkeep;maintenance cost
维修保养工段 maintenance department;maintenance section
维修保养工作队 maintenance gang
维修保养规程 maintenance and servicing procedures
维修保养机械化 mechanisation [mechanization] of maintenance
维修保养记录 maintenance log
维修保养间 service bay
维修保养时间 maintenance time
维修保养原则 maintenance concept
维修保养站 maintenance point
维修保养组 maintenance team
维修保证金 maintenance bond
维修保证书 maintenance bond
维修备件 awaiting parts;maintenance and repair parts
维修备品和零配件 parts and components required for maintenance
维修比例 maintenance ratio
维修标准 maintenance standard
维修标准规范 maintenance standard code
维修不良 imperfect maintenance;maintenance trouble
维修不足 under-maintenance
维修部费用 maintenance department expense
维修部件 service panel
维修部件清单 maintenance parts list
维修部门 maintenance department
维修材料 maintenance material;servicing material
维修材料储备 service stock
维修策略 maintenance strategy
维修厂棚 maintenance hangar
维修车 maintenance vehicle;maintenance lorry;maintenance truck;tool car;tool wagon
维修车间 maintenance(work) shop;repair and maintenance workshop;repair shop;service maintenance shop;service shop;technical hangar
维修成本 maintenance cost;repair cost
维修程序 maintenance procedure;maintenance routine;program(me) for maintenance
维修程序表 maintenance schedule
维修储备 maintenance reservoir
维修储备量 repair reserve
维修储用器材 service stock
维修次数 frequency of maintenance;maintenance frequency
维修的可靠性 reliability of services
维修点 maintenance point;service point
维修电工 maintenance electrician
维修电缆 recovery cable
维修发动机 service engine
维修方便 maintenance ease
维修飞机库 service hangar
维修费对固定资产比率 ratio of repair to fixed assets
维修费效率 upkeep efficiency
维修费(用) cost of maintenance;cost of upkeep;maintenance expenditures;maintenance outlay;upkeep cost;upkeep expenses;maintenance and repair cost;maintenance charges;maintenance expense;repair cost;attendance cost;maintenance charges;maintenance cost;upkeep
维修费准备金 maintenance reserves
维修服务 maintenance service;maintenance support;service support
维修服务车 maintenance vehicle;service car
维修服务手册 maintenance and service manual
维修工 maintenance man;maintenance operator;maintenance worker;repairman
维修(工)班 maintenance gang
维修工厂 engineering repair works;service plant
维修工场 maintenance and repair shop
维修工程 maintenance engineering;maintenance works;reparation
维修工程部门 service engineering department
维修工程车 mechanics truck;service vehicle
维修工程分析 maintenance engineering analysis
维修工程科 service engineering department
维修工程师 maintenance engineer;service engineer
维修工程用的脚手架 scaffold for maintenance work
维修工段 maintenance shop
维修工具 maintenance tool;service tool
维修工作 maintenance job;maintenance manipulation;maintenance work;operating maintenance
维修工作船 maintenance ship
维修工作队 maintenance crew
维修工作量 maintenance load
维修工作手册 maintenance manual
维修工作台 maintenance platform
维修工作细则 maintenance detailed planning
维修工作周期 cyclic(al) period of maintenance work
维修管理 maintenance management
维修管理报告 maintenance control report
维修规程 maintenance instruction;service manual
维修规则 maintenance regulation;service regulation
维修轨道公里 kilometers of track maintained
维修过剩 over maintenance
维修合同 maintenance contract
维修合同预付款 advance on maintenance contract
维修和更换 maintenance repair and replacements
维修和运营 maintenance and operation
维修后使用期 repair reserve
维修会议 service meeting
维修活动 maintenance activity
维修机构 maintenance agency
维修机库 maintenance hangar
维修机器 maintain machinery
维修机械 servicing machine
维修基地 maintenance center [centre]
维修基地设备 depot equipment
维修及安装后保养工作 maintenance and after installation services
维修及服务时间 repair and servicing time
维修及作业用品 maintenance repair and operating supplies
维修计划 maintenance plan;maintenance project;maintenance schedule
维修计划表 maintenance schedule
维修计划和控制系统 maintenance planning and control system
维修计算机化 computerised maintenance
维修记录 maintenance record;maintenance task log;service action log;service history
维修技术说明书 technical service manual
维修技术训练 maintenance drill
维修技术员 maintenance technician
维修架 maintenance rig
维修间 repair bay;service bay
维修检查 maintenance overhaul;maintenance inspection
维修进度表 maintenance schedule
维修经验 service experience
维修孔 service opening
维修控制 maintenance control
维修库 technical hangar
维修廊道 service gallery
维修类别范畴 lines of maintenance
维修良好 in good repair
维修零件一览表 service action parts list
维修率 maintenance factor;maintenance rate;maintenance ratio
维修轮换 maintenance switching
维修目标 maintenance objective
维修能力 maintainability
维修能量 maintenance energy
维修排 maintenance platoon
维修棚 technical hangar
维修频率 frequency of maintenance
维修平均间隔 mean time between maintenance
维修平均间隔时间 average time between maintenance
维修平台 attendant's platform;maintenance deck;maintenance platform
维修期 defect liability period;maintenance period;period of maintenance;warrant period
维修期间隔 maintenance interval
维修漆 on-line coating;refinishing paint
维修契约 upkeep obligation
维修器材 maintenance package;maintenance supply
维修钳工 maintenance fitter
维修清洁器 maintenance cleaner
维修区 maintenance area
维修人力 maintenance manpower
维修人员 maintenance crew;maintenance man;maintenance personnel;maintenance staff;serviceman
维修人员呼唤按钮 maintainer's call button
维修任务 upkeep undertaking
维修任务书 maintenance task
维修设备 facility for repair;maintenance equipment;maintenance facility;service equipment;service facility;servicing facility;servicing installation
维修设备使用率 rate of use of maintenance equipment
维修设计 maintenance design
维修时的超提升 overlift for maintenance
维修时间 maintenance(down) time;servicing time
维修时期 service period
维修(实施)计划 maintenance schedule
维修手册 maintenance service manual;service manual
维修寿命 service lifetime
维修说明书 instruction for maintenance;service manual
维修台 maintenance station
维修停产时期 maintenance downtime
维修停机坪 service apron
维修停机时间 maintenance downtime
维修统计 maintenance statistics
维修图 service action drawing
维修图表 maintenance chart;maintenance schedule;service chart
维修涂料 maintenance coating
维修涂装 maintenance painting
维修细则 maintenance instruction
维修箱 service kit
维修项目 maintenance clause
维修效率 maintenance efficiency
维修性 maintainability
维修性能不佳 underserviceability
维修性设计原则 maintainability design criterion
维修压力计 service ga(u)ge
维修养护 maintenance and repair
维修养护队 section forces
维修养护分队 maintenance force
维修养护工队 maintenance force
维修养护期 maintenance period
维修要求 maintenance requirement
维修业务工作 service operation
维修义务 maintenance obligation
维修用备件 awaiting parts;repair reserve
维修用便道 maintenance footway;maintenance sidewalk
维修用出入口 access opening for maintenance
维修用阀 service valve
维修用复合膏 maintenance compound
维修用工具 service tool
维修用化合物 maintenance compound
维修用建材 structural repair material
维修用漆 maintenance paint
维修用切断阀 maintenance cutout cock;maintenance isolating valve
维修用套袖 service sleeve
维修油漆工作 maintenance painting work
维修有效性 maintenance effectiveness
维修与保养 maintenance and repair
维修与使用 maintenance and service
维修与调整工具 maintenance and adjustment tool
维修预算 maintenance and repair budget;maintenance budget;maintenance point
维修允许公差 maintenance tolerance
维修站 maintenance depot;maintenance station;servicing center [centre];servicing depot
维修照明 maintenance lighting
维修折旧 maintenance depreciation
维修折旧法 depreciation-maintenance method
维修证书 maintenance certificate
维修帧 maintenance frame
维修政策 maintenance policy
维修支持 maintenance support

维修支援 maintenance support
维修职员 maintenance staff
维修指标值 maintenance target value
维修指南 pathfinder
维修中心 maintenance center [centre]
维修周期 maintenance cycle; maintenance interval; service interval
维修装置 maintenance device; maintenance unit
维修准备时间 maintenance standby time
维修资金 maintenance funds
维修资料登记簿 maintenance register
维修资料文件 maintenance documentation
维修综合场 maintenance complex
维修总费用 total cost of upkeep
维修组 maintenance crew; maintenance gang; maintenance party; maintenance team
维修组织 maintenance organization
维修作业 maintenance operator; maintenance work
维修作业限制 restriction of maintenance work
维也纳 <奥地利首都> Vienna
维也纳定义语言 Vienna Definition Language
维也纳分离派艺术运动 Vienna Secession movement
维也纳宫廷图书馆 Court Library at Vienna
维也纳金属接合剂汞合金 Vienna metallic-Cement amalgam alloy
维也纳蓝 Vienna blue
维也纳石灰 Vienna lime
维也纳式校准器 Vienna regulator
维也纳条约法公约 Vienna Convention on the Law of Treaties
维也纳邮政储蓄银行 Vienna Postal Savings Bank
维易剂 <一种油粉涂层和油漆清洗剂> Veevic
维尤纳夫型计算机组合导航系统 Viewnafa computerized marine navigation system

伟

伟斑岩 pegmatophyre

伟晶构造【地】pegmatitic structure
伟晶花岗岩 giant-grained granite; pegmatitic granite
伟晶矿物 pegmatitic mineral
伟晶蜡石 grave wax; hatchettine; hatchettite; mineral tallow; naphthine; rock tallow
伟晶生铁 very open-grained pig iron
伟晶石膏矿石 pegmatitic gypsum ore
伟晶相 pegmatitic phase
伟晶岩 pegmatite
伟晶岩成矿作用 pegmatitic ore-forming process
伟晶岩化作用 pegmatitization
伟晶岩阶段 pegmatitic stage
伟晶岩矿床 pegmatitic ore deposit
伟晶岩矿田构造 pegmatitic orefield structure
伟晶岩类 pegmatite group
伟晶岩锂矿床 spodumene-lepidolite-pegmatite deposit
伟晶岩相 pegmatoid
伟晶岩型铯矿石 cesium ore of pegmatite type
伟晶岩元素 pegmatitic element
伟晶岩状 pegmatoid
伟晶作用 pegmatitization
伟晶作用地球化学 geochemistry of pegmatitic processes
伟硼镁石 wightmanite
伟人祠 Pantheon
伟震 megaseism

伪

伪凹 pseudo-concave

伪胞 pseudo-cyst
伪币 bogus money; counterfeit coin; false coin; impairing coin; spurious coin
伪边 pseudo-side
伪变换群 pseudo-group of transformations
伪变量 pseudo-variable; pseudo-vector
伪变量法 pseudo-variable method
伪标量 pseudo-scalar(quantity)
伪冰斗 pseudo-cirque
伪彩色 false colo(u)r
伪彩色合成影像 false colo(u)r composite image
伪彩色显示 pseudo-colo(u)r display
伪操作 pseudo-operation
伪操作码 pseudo-operation code
伪长度 pseudo-length
伪场同步脉冲 pseudo-field-sync pulse
伪钞 counter note
伪超常传导 pseudo-supernormal conduction
伪超额需求函数 pseudo excess demand function
伪超椭圆积分 pseudo-hyperelliptic(al) integral
伪程序 pseudo-program(me)
伪冲激波 pseudo-shock
伪传递律 pseudo-transfer law
伪纯量 pseudo-scalar(quantity)
伪纯量场 pseudo-scalar field
伪代码 false code
伪单晶 pseudo-single crystal
伪单眼 pseudo-single eye
伪的 pseudo
伪等色图 pseudo-isochromatic diagram
伪等值线地图 pseudo-isoline map
伪底 phantom bottom
伪地址 dummy address
伪递推 pseudo-recurrence
伪点 ideal point
伪顶 false roof; following dirt; following stone
伪动力试验 pseudo-dynamic(al) test
伪动力学的 pseudo-dynamic
伪度量 pseudo-metric
伪度量空间 pseudo-metric space
伪度量一致性 pseudo-metric uniformity
伪对称 pseudo-symmetry
伪对象语言 pseudo-object-language
伪多边形的 pseudo-polygonal
伪二元图 pseudo-binary diagram
伪法线 pseudo-normal
伪方位投影 pseudo-azimuthal projection
伪峰 spurious peak
伪锋【气】false front; fictitious front; pseudo-front
伪符号 pseudo-symbol
伪辐射 spurious radiation
伪赋值 pseudo-valuation
伪赋值环 pseudo-valuation ring
伪杆 <桁架中的> false member
伪杆状体 rhammite
伪格 pseudo-lattice
伪共晶(的) quasi-eutectic; pseudo-eutectic
伪共析 quasi-eutectoid
伪共振 pseudo-resonance
伪古典主义 pseudo-classicism
伪光标 pseudocursor
伪函数 pseudo-function
伪迹 artefact [artifact]
伪箕舌线 pseudo-witch
伪几何环 pseudo-geometric ring
伪几何网 pseudo-geometric net
伪寄存器 pseudo-register
伪结构 pseudo-structure
伪解析函数 pseudo-analytic function
伪金钟柏 Japanese thuja
伪晶的 pseudo-crystalline
伪晶体 pseudo-cone
伪静力 pseudo-static force
伪静力试验 pseudo-static test
伪静位移 pseudo-static displacement
伪矩阵隔离 pseudo-matrix isolation
伪距测量 pseudo-range measurement
伪距校正 pseudo-range correction
伪距离 pseudo-distance; pseudo-range
伪卷云【气】false cirrus
伪均衡脉冲 pseudo-equalizing pulses
伪科学 pseudo-science
伪空化 pseudo-cavitation
伪扩散 pseudo-diffusion
伪黎曼丛 pseudo-Riemannian bundle
伪黎曼度量 pseudo-Riemannian metric
伪黎曼流形 pseudo-Riemannian manifold
伪黎曼向量丛 pseudo-Riemannian vector bundle
伪力 pseudo-force
伪劣品 fake and poor product
伪临界性质 pseudo-critical property
伪流动构造 flaser bedding structure; flaser structure
伪码 interpreter code; interpretive code; pseudo-code
伪逆矩阵 pseudo-inversion matrix
伪黏[粘]性 pseudo-viscosity
伪黏[粘]滞流 pseudo-viscous flow
伪欧几里得度量 pseudo-Euclidean metric
伪欧几里得空间 pseudo-Euclidean space
伪偶然误差 pseudo-accidental error
伪频率 pseudo-frequency
伪平行面 pseudo-parallel plane
伪平行线 pseudo-parallel lines
伪切面 pseudo-tangent plane
伪切线 pseudo-tangent line
伪球面 pseudo-sphere
伪球形螺旋面 pseudo-spheric(al) helicoid
伪球形曲面 pseudo-spheric(al) surface
伪球状结构 pseudo-globular texture
伪群 pseudo-group
伪群结构 pseudo-group structure
伪三角形 pseudo-triangle
伪三进制 pseudo-ternary
伪三进制信号 pseudo-ternary signal
伪十进制数字 pseudo-decimal digit
伪矢量 pseudo-vector
伪双曲距离 pseudo-hyperbolic distance
伪水文系列 pseudo-hydrologic sequence
伪水文序列 pseudo-hydrologic sequence
伪四维张量 pseudo-four-tensor
伪四位二进制 pseudo-tetrad
伪四元 quasi-quaternary
伪速度测井剖面 pseudo-velocity logging section
伪算术操作 pseudo-arithmetic operation
伪随机传输 pseudo-random transmission
伪随机的 pseudo-random
伪随机二进制序列 pseudo-random binary sequence
伪随机反相 pseudo-random phase reversal
伪随机码 pseudo-noise code; pseudo-random code
伪随机码流 pseudo-random code stream
伪随机脉冲编码震源 pseudo-random pulse coded source
伪随机脉冲响应技术 pseudo-random impulse response technique
伪随机扰动 pseudo-random perturbation
伪随机数 pseudo-random number; quasi-random number
伪随机数据 pseudo-random data
伪随机调制雷达 pseudo-random radar
伪随机数序列 pseudo-random number sequence; pseudo-random sequence of numbers; quasi-random sequence of numbers
伪随机数序列的周期性 periodicity of a pseudo-random number sequence
伪随机数字信号发生器 pseudo-random digital signal generator
伪随机系统 pseudo-random system
伪随机信号 pseudo-random signal
伪随机序列 pseudo-random sequence
伪随机杂波干扰机 pseudo-random noise jammer
伪随机噪声 pseudo-noise
伪随机噪声发生器 pseudo-random noise generator
伪随机噪声信号 pseudo-random noise signal
伪通道 pseudo-channel
伪凸 pseudo-convex
伪凸性 pseudo-convexity
伪脱机工作 pseudo-offline working
伪脱机输入输出 pseudo-off-line input-output
伪椭圆积分 pseudo-elliptic(al) integral
伪弯曲-扭转耦联振动 pseudo-coupled bending-torsion vibration
伪网图 pseudo-net graph
伪微分算子 pseudo-differential operator
伪文件名 pseudo-file name
伪误差 human error
伪相 pseudo-phase
伪相关 spurious correlation
伪向量 pseudo-vector
伪向量场 pseudo-vector field
伪斜对称的 pseudo-skew symmetric
伪心材 false heart
伪行同步脉冲 pseudo-line-sync pulses
伪旋轮类曲线 pseudo-cycloidal curve
伪旋轮线 pseudo-cycloid
伪循环 pseudo-cycle
伪循环码 pseudo-cyclic code
伪曳物线 pseudo-tractrix
伪异步 pseudo-asynchronous
伪引力 mock gravitational force
伪有向点族 pseudo-directed family of points
伪有向集 pseudo-directed set
伪预解式 pseudo-resolvent
伪圆柱体 pseudo-cylinder
伪圆柱投影 pseudo-cylindric(al) projection
伪圆锥投影 pseudo-conic(al) projection
伪运动 pseudo-motion
伪运算 pseudo-operation
伪造 adulterate; adulteration; fabricate; falsification; forge; imitation; reproduction; resemble; sophistica-

tion;forgery < 常指伪造签字 >
伪造背书 forged endorsement
伪造布革 Fabrikoid
伪造的 bogus;false;counterfeit
伪造的单据 forged documents
伪造的客票 falsified ticket
伪造的证件 bogus certificate
伪造记录 false entry
伪造品 counterfeit; fabrication; fakement;forgery
伪造签字 forged signature
伪造文件 false papers;falsification of documents;forged documents
伪造文书 forgery of document
伪造物 bogus;forgery
伪造现场 simulated scene
伪造账目 cook accounts; false accounting;falsify accounts
伪造者 cooker;forger
伪噪声 pseudo-noise
伪噪声序列 pseudo-noise sequence
伪张量 pseudo-tensor
伪张量形式 pseudo-tensorial form
伪折减性质 pseudo-reduced property
伪正则函数 pseudo-regular function
伪值 pseudo-value
伪指令【计】dummy instruction;dummy order;pseudo-instruction;pseudo-order;instructional constant
伪指令形式 pseudo-instruction form;pseudo-order form
伪制机床 copy machine
伪制品 sham
伪周期函数 pseudo-periodic (al) function
伪周期弧 pseudo-periodic(al)arc
伪周期问题 pseudo-periodic (al) problem
伪周期性 pseudo-periodicity
伪珠光体 pseudo-pearlite
伪主机 pseudo-host
伪主解 pseudo-principal solution
伪转置 pseudo-transposition
伪装 cloak;concealment;cryptic mimicry;deception;disguise;garnish;masking;pretend;simulate;simulation
伪装挡板 masking sheeting
伪装道路 dummy road
伪装道路夯实 dummy-road packing
伪装的 deceptive; ostensible; stimulant
伪装的军事工厂 shadow factory
伪装废墟 sham ruin
伪装工事 camouflage work
伪装机场 dummy airfield
伪装建筑物 camouflaged building; dummy
伪装迷彩漆 baffle paint
伪装迷彩涂装 baffle painting
伪装判读 concealment analysis
伪装漆 camouflage paint;dazzle paint(ing)
伪装色 camouflage coat
伪装色彩 camouflage paint; dazzle paint(ing)
伪装商船的军舰 decoy ship
伪装探测胶片 camouflage detection film
伪装涂层 camouflage coat
伪装涂料 baffle paint; camouflage paint
伪装涂漆 camouflage painting
伪装涂色 baffle painting
伪装外罩 camouflage coat
伪装网 camouflage net
伪装文件 masking paper
伪装掩护色用涂料 baffle paint
伪装用涂料 camouflage paint
伪装油漆 camouflage lacquer
伪装侦察 camouflage detection
伪装(着)色 camouflage colo(u)ring
伪自变数 dummy argument
伪自由理想环 pseudo-free ideal ring
伪作业调度 pseudo-job scheduling

尾

尾鞍 caudal saddle
尾巴电缆 stub cable;tail cable
尾摆度 tail swing
尾板 pygidium; tailboard; tail gate; transom flap
尾板嵌灰泥 torching
尾板式铲运机 tailboard scraper
尾板式铺砂机 tailboard gritter
尾板正面宽度 tailboard frontage
尾孢属 < 拉 > Cercospora
尾边 trailing edge
尾鞭病 whiptail
尾标 tail tag
尾波 stern wave; tail wave; wake; wave coda
尾部 aft(er) end; afterpart; aftersection;caudal region;empennage;end part; foot section; stern; tail (piece);tail section;trailer;trailing end
尾部报警 < 飞机 > tail warning
尾部闭锁机构 final lock mechanism
尾部变号电感 tail reversing inductance
尾部标志 trailer label
尾部冰带区 aft ice belt region
尾部补机 bank engine;banker
尾部操纵 tailing
尾部操纵面 rear control surface
尾部长度 shank length
尾部成型床 tail builder
尾部粗毛 say-cast
尾部挡板式石屑撒布器 tail-gate spreader
尾部的 caudal;tail
尾部灯光信号系统 rear-signal system
尾部符号 tail symbol
尾部附属物 tailpiece
尾部工具滑板 end tool slide
尾部航行灯 rear navigation lamp; rear position light
尾部荷载 tail load
尾部护木 stern shield
尾部缓冲器 tail bumper
尾部回旋空间 tail swing
尾部回转半径 rear-end radius
尾部记录 trailer record
尾部减速器 tail gearbox
尾部减缩剂 tailing reducer
尾部绞车 rear-end winch
尾部接收装置 rear receiver
尾部警戒 tail warning
尾部警戒雷达 tail warning radar
尾部卷扬机 rear-end winch
尾部卡片 trailer card
尾部瞭望台 gallery
尾部馏分 tail fraction
尾部螺纹 shank taper;tailing screw flight
尾部螺旋桨 tail rotor
尾部排气管 tail pipe
尾部炮塔 rear turret
尾部起落架 tail undercarriage
尾部千斤顶 tail jack
尾部上货式货机 rear-loading freighter
尾部升降副翼 taileron
尾部式布置 < 地下式水电站 > downstream station arrangement
尾部受热面 tail heating surface
尾部双层底 double bottom afterward
尾部突出的汽车 notchback
尾部吸入式挖掘船 trailing suction dredge(r)
尾部吸入式挖泥船 trailing suction dredge(r)
尾部系留点 stern mooring point
尾部下坐 < 船高速航行时 > squat
尾部卸料车 end dump car
尾部信号 tail signal
尾部信号自动控制 automatic rear signal control
尾部烟道 backpass
尾部异常 rear anomaly
尾部元素 rear element
尾部运输机 rearward conveyer [conveyor]
尾部载荷 tail load
尾部整流锥 tail cone
尾部支点 tail car
尾部重心 tail heaviness
尾部转向架 rear bogie
尾部装货设备 rear-loading facility
尾部装置 rear-mounted installation
尾部阻尼电路 tail damping circuit
尾材 tailings
尾仓 finish department
尾舱 cockpit;stern compartment
尾舱口 after hatch
尾槽 stern notch
尾侧浪 quarter wave
尾车 rear wagon;tail vehicle
尾车倒退 tails-back
尾吃水 after draft;after draught
尾吃水大于首吃水 trim by the stern
尾池 tail vat
尾翅 tail fin
尾冲 tail slide;tail slip
尾处理程序 post-processor
尾传令钟 docking telegraph
尾垂线 after perpendicular
尾到车尾 rear to rear
尾灯 after-light; backlight; overtaking light; rear lamp; rear light; stern light;tail lamp
尾灯插口 tail lamp socket
尾灯插座 tail lamp socket
尾灯导线包皮 taillight wiring conduit
尾灯底座 tail lamp base
尾灯和停车灯 tail and stop light
尾灯架 tail lamp hanger
尾灯开关 taillight switch
尾灯座 taillight socket
尾电池 end cell
尾电流 tail current
尾电阻器 tail resistor
尾顶尖 < 车床的 > back center [centre]
尾堵 < 舢板后段 > box
尾端 back end;caudal end;rear end; stop end;tag end;trailing end
尾端边木 side counter timber;term piece
尾端冲撞 < 车辆尾部与后车相撞 > rear-end collision
尾端处理 tail-end treatment
尾端电池转换开关 end cell switch
尾端废水 backend wastes
尾端固定导叶 baffle vane
尾端挂钩夹具 tail-end hook jig
尾端过程 tail-end process
尾端件 end piece;tailpiece
尾端绞车 tail-end winch
尾端结构砖 end construction tile
尾端进料 end fed
尾端进料刀 end-fed knife
尾端进位数 final carry digit
尾端宽度 aftermost breadth
尾端离地间隙 rear clearance
尾端离地净空 < 汽车车身的 > rear clearance
尾端模板 stop end form
尾端人孔 end manhole
尾端挑梁 < 墙内突出部分 > tail beam
尾端外板 after hood;tuck plate
尾端弯钩 hooked ends;tail-hood
尾端线 tail wire
尾端效应 tail effect
尾端卸车法 end dumping method
尾端窨井 end manhole
尾端运输机 rear conveyer [conveyor]
尾端周边卸料 end peripheral discharge
尾端装载机 backend loader
尾端自动止车器 < 编组线 > end stopper
尾端阻力 end resistance;tail end resistance
尾段 end piece;end section
尾段梁 tail beam
尾墩现象 squat
尾舵 rear control;tail vane
尾轭 end yoke
尾阀 end valve;tail valve
尾(废)气分析器 tail gas analyser [analyzer]
尾粉 tailings
尾峰 end peak
尾浮【船】lift by the stem;lifting;pivoting
尾浮加强区【船】zone of pivoting pressure action
尾盖 tail-hood
尾干 caudal style;tail shaft
尾杆 adapter rod;tail lever;tail rod
尾杆平衡锤 tail lever counterweight
尾杆罩 tail rod catcher
尾隔离舱 after cofferdam
尾钩 tailhook
尾钩弹簧 tailhook spring
尾骨 tail bone
尾管 cartridge container;tail conduit; tail pipe;tail tube;tailpiece < 水温调整阀 >
尾管长度 length of liner
尾管挂入深度 depth of liner hanger
尾管火焰 tailpipe flame
尾管螺母 tail nut
尾管深 liner depth
尾管悬挂器 extension hanger; line hanger
尾管中心件 tailpipe center-piece
尾光纤 pigtail
尾滚筒 tail roller
尾桁 tail boom
尾横骨架 transom floor
尾横缆 aft(er) breast (line); stern breast rope
尾后风 wind after
尾护舷材 after fender
尾花菌属 < 拉 > Anthisrus
尾滑车 end pulley
尾滑轮 tail pulley
尾滑瓦 tail-skid shoe
尾机船 after engine ship;all aft cargo ship
尾机型船 aft-engine boat; aft-engine vessel; aft type ship; stern-engine(d) ship;aft-engine ship
尾迹 back trace;trail;wake
尾迹测量法 wake-survey method
尾迹速度 wake velocity
尾迹移测法 wake traverse method
尾激波 tail(ing) shock wave
尾甲板 quarter deck
尾甲板上止舵楔 rudder brake;rudder chock;rudder deck stop
尾架 deadhead; end bracket; rootstock;tail center [centre];tailstock
尾架顶尖 tailstock center [centre]
尾架顶尖套筒 tailstock-center sleeve;

tailstock quill
尾架顶针套 tail spindle
尾驾驶台天幕 docking bridge awning
尾驾驶台天幕柱 docking bridge awning stanchion
尾驾驶台横梁 docking bridge awning beam
尾驾驶台直梁 docking bridge awning boom
尾尖舱 after peak(water)tank
尾尖舱舱壁 after peak bulkhead; stuffing box bulkhead
尾尖角 angle of run
尾槛 end baffle;end sill
尾桨 auxiliary rotor;tail rotor
尾桨槽 tailings chest
尾角 caudal horn
尾角石块墙砌合 quoin bonding
尾角跳板 quarter rampway
尾接结构 tail-to-tail structure
尾接指令 cue
尾景 terminal feature
尾距 back range
尾卷筒 tail pulley
尾孔 uropore
尾孔楔钉 grappler
尾孔楔形块 grappler
尾跨 tail space
尾款清讫 final payment
尾矿 deads; debris; gangue; mineral tailing; ore tailings; refuse; tails; washery refuse
尾矿坝 mine refuse impoundment; mine tailings dam;tailings dam;tailings fill dam
尾矿坝的建造与治理 construction and improvement of tailings dam
尾矿采样 tailing sampling
尾矿仓 tailings bin
尾矿槽 tail-box;tailing launder
尾矿产品 product from failing
尾矿场 dump pit;refuse yard;tailings area;tailings pile
尾矿沉淀桶 tailings settling tank
尾矿池 impounding dam; tailings dam;tailings pond
尾矿池建设 tailings reservoir building
尾矿充填料 tailings(back)fill
尾矿处理 tailings disposal
尾矿处置 tailings disposal
尾矿带 tailings zone
尾矿吊斗 tailings elevator
尾矿堆 debris dump;spoil stock;tailings heap;tailings pile
尾矿堆祸害 tip disaster
尾矿堆灾难 tip disaster
尾矿法 debris law
尾矿分级机 tailings classifier
尾矿粉 ore tailings
尾矿回收 tailings recycling
尾矿机 ore tailer
尾矿坑 tailings pond
尾矿库 tailings impoundment
尾矿滤饼 tailings cake
尾矿排出端 tailings side
尾矿排出沟 tailings race
尾矿排出口 tailings outlet port
尾矿排放槽 tailings launder
尾矿排放区 tailings disposal area
尾矿排料边 tailings apron
尾矿排卸口 tailings discharge
尾矿品位 refuse ore grade; tailings grade
尾矿砂 mine tailings;tailings
尾矿试金 tailings assay
尾矿试样 tailings sample
尾矿提升机 tailings elevator
尾矿土 tailing soil
尾矿压滤机 refuse filter press; tailings filter press
尾矿溢流堰 tailings overflow weir
尾矿再处理 retreatment of tailings
尾矿渣 mine tailings;mining by-products
尾矿砖 tailing brick;tailings ore brick
尾矿综合利用 waste utilization
尾框 carrier bracket
尾框底部 keel piece;shoe piece;sole piece
尾框顶部 arch piece; bridge arch; bridge piece;propeller arch
尾缆 after line; after rope; pigtail; quarter fast; stern fast; stern line; stern rope
尾缆孔 cathole
尾缆链掣 stern line stopper
尾缆锚定物 tail anchor
尾浪 overtaking wave;sea from aft; stern sea
尾肋板 transom floor
尾肋骨 after frame;transom frame
尾立材 tuck timber
尾链 tail chain
尾梁接铁 beam knee
尾料 quarry rubbish;tailings
尾料分析 tail assay
尾零 tailing zero
尾流 backwash;race of screw;rough air; tailrace; trailing wake; wake; wake current; wake flow; wake stream;wash stream
尾流边界 wake boundary
尾流测量 wake survey
尾流分析与控制 wake analysis and control
尾流结构 wake structure
尾流理论 wake flow theory; wake stream theory
尾流能量 wake energy
尾流强度 wake strength
尾流区 wake zone;zone of wake
尾流湍流 wake turbulence
尾流温度 wake temperature
尾流涡流 wake vortices
尾流系数 wake coefficient;wake factor
尾流效率 wake efficiency
尾流形状 wake shape
尾流引起的振动 wake-induced vibration
尾流阻力 wake drag;wake resistance
尾馏分 end cuts; residual fraction; tailing fraction;tailings
尾楼【船】 poop
尾楼舱壁 poop bulkhead
尾楼舱室 poop cabin
尾楼甲板 poop deck
尾楼甲板倒缆 <靠码头用> poop deck spring
尾楼前端 break of poop
尾楼舷侧缘列板 poop sheer strake
尾轮 rear furrow wheel; tail(ing) wheel;tail pulley
尾轮叉 tail-wheel fork
尾轮刮土板 rear furrow wheel scraper
尾轮减振器 tail-wheel bumper
尾轮螺旋拉紧装置 tail-pulley screw take-up
尾轮起落机构 rear lift linkage
尾轮起落架 tail-wheel landing gear
尾轮起落拉杆 rear-wheel lifting link
尾轮深浅调节器 rear-wheel depth adjustment
尾轮锁 tail-wheel lock
尾轮调节螺钉 rear furrow wheel adjustment screw
尾轮罩 tail-wheel boot
尾轮轴 tail-gear spindle
尾罗经 after compass
尾锚 stern anchor;tail anchor
尾锚吊锚杆 stern davit
尾锚锚泊 anchor by the stern
尾锚锚链 stream cable;stream chain
尾帽 tail cup
尾门 end door; tail gate; stern door【船】;stern port【船】
尾面 tail surface
尾明轮 stern wheel
尾明轮船 stern wheeler;stern wheel steamer
尾磨 tail roll
尾木 tail tree
尾喷管 injection nozzle;jet exit;jet nozzle;tail pipe;tailpipe nozzle
尾喷口 tailing spout
尾碰垫 after fender
尾鳍 dead wood
尾鳍构架 skeg framing
尾气 end gas;off-gas;tail gas
尾气保护 trailing gas shield
尾气处理 tail gas treatment
尾气缓冲罐 tail gas buffer
尾气回收塔 tail gas recovery tower
尾气冷凝器 tail gas condenser
尾气塔 tail gas tower
尾气吸收塔 tail gas absorber
尾碛【地】 terminal moraine;end moraine
尾钎 shank piece
尾橇 tail-skid
尾橇杆 tail-skid bar
尾桥【港】 docking bridge; docking deck;warping bridge
尾桥楼天幕 docking bridge awning
尾桥楼天幕横梁 docking bridge awning beam
尾桥楼天幕直梁 docking bridge awning boom
尾桥楼天幕柱 docking bridge awning stanchion
尾倾 down by stern;stern trim;trim by stern
尾球形舱 aft spheric(al)chamber
尾区 tail region
尾圈 end coil
尾沙丘 wake dune
尾砂 ore tailings;tailings
尾砂坝 tailings dam
尾砂充填法 tailing filling method
尾砂浮选法 sand flo(a)tation process
尾筛 tailings screen;tail sheave
尾声 epilog
尾绳 back guy; balance rope; return rope;tail line;tail rope
尾绳滚筒 tail drum
尾绳滑轮 return sheave
尾绳绞车 tail crab
尾绳轮 tail sheet
尾绳运输法 tail-rope system
尾绳运输系统 tail-rope system
尾室 poop house
尾数 arrears; balance; characteristic mantissa; fixed-point part; floating-point coefficient;mantissa
尾数部分 mantissa part
尾数四舍五入的累加数 round-off accumulating
尾刷 ventral brush
尾栓孔 wedge slot
尾水 <水电站发动机排出的水> tail water
尾水坝 tailrace dam
尾水池 aft-bay;afterbay;lower pool; tail bay;tail pond;tail tank;tailwater pond
尾水池坝 afterbay dam
尾水池墙 tail bay wall
尾水尺修正 stern draft correction
尾水冲蚀 tailwater erosion
尾水冲刷作用 tailwater erosion
尾水电站 tailwater power installation;tailwater power plant
尾水段 tail reach
尾水高程 tailwater elevation;tailwater level
尾水沟 tailrace
尾水管 draft pipe; draft tube; tailpiece;tail pipe;tail tube
尾水管边墙 draft tube wall
尾水管层 draft tube floor
尾水管出口 draft tube exit;draft tube port
尾水管底板 draft tube floor
尾水管顶板 draft tube roof
尾水管盖板 draft tube deck
尾水管拱腹 draft tube soffit
尾水管检查通道 draft tube inspection passage
尾水管交通廊道 draft tube access gallery
尾水管进口 draft tube entrance
尾水管进入井 <水轮机的> access shaft to draft tube
尾水管进入孔 draft tube manhole
尾水管进入廊道 draft tube access gallery
尾水管扩大段 draft tube diffuser
尾水管里衬 draft tube liner
尾水管平台 draft tube deck
尾水管上段 upper draft tube
尾水管损失 draft tube loss
尾水管弯管段 draft tube bend
尾水管效率 draft tube efficiency
尾水管性能 draft tube performance
尾水管压力脉动 draft tube pressure surge
尾水管涌浪 draft pipe surge;draft tube surge
尾水管闸门 draft tube(bulkhead)gate
尾水管闸门槽 draft tube gate slot
尾水管肘管段 draft tube elbow
尾水河段 tail bay
尾水回收灌溉系统 tailwater recovery irrigation system
尾水回收再用 tailwater recovery
尾水坑 tailwater pit
尾水廊道 tailwater gallery
尾水流速 tailwater velocity
尾水面高程 tailwater elevation
尾水平台 draft tube deck; tailrace deck;tailrace platform
尾水渠 tail(race) channel; tailwater canal;tailwater channel;mill tail
尾水渠衬砌 tailrace lining
尾水渠底板混凝土 tailrace slab concrete
尾水渠段 tailrace section
尾水渠开挖线 excavation line of tailwater channel
尾水渠墙 tailrace wall
尾水渠入口 tailrace adit
尾水水库 afterbay reservoir
尾水水深 tailwater depth
尾水水头 tailrace head;tailwater head
尾水水位 level of tail water;tailwater level; tailwater elevation; tailwater stage
尾水水位流量关系曲线 tailwater rating curve
尾水隧道 tailrace tunnel
尾水隧洞 tailrace tunnel
尾水调压池 tailrace surge basin
尾水头 draft head;tail head
尾水位 tailwater level
尾水系统 tailrace system
尾水箱 <水工试验用的> reception tank

W

尾水消退 tail recession
尾水堰 tail-weir
尾水闸门 aftergate;aft gate;tail gate;tail lock;tailrace gate;tailwater gate
尾水锥体 tail cone
尾随 wake
尾随(齿)面 trail side
尾随低压 wake depression
尾随风 tailwind;following wind < 尾随前面的海洋波浪及其方向的风 >
尾随浪 following sea
尾随零 trailing zero
尾随流 following sea
尾随沙(丘)wake dune
尾随台风 wake-following typhoon
尾随行驶 following at a distance
尾随阻力 wake resistance
尾索 < 滑车组上绕过最后滑车的索 > tail rope
尾索套 tail cable sleeve
尾塔 tailing column;tail tower < 缆道起重机的 >
尾滩 back beach;backshore
尾跳板【船】stern rampway
尾推力终止 tail off
尾拖 stern-first tow
尾拖电缆 trailing cable
尾拖电缆槽 trailing cable groove
尾拖网渔船 stern trawler
尾弯【船】tipping
尾尾连接 tail-to-tail linking
尾涡 eddy trail;trailing vortex
尾涡空化 trailing vortex cavitation
尾涡流 wake vortex
尾纤 pigtail cable
尾舷 quarter
尾舷灯 aft sidelight
尾舷吊艇柱 quarter davit
尾舷方向 quartering
尾舷风 quarter(ing) wind
尾舷弧 after sheer
尾舷上风 weather quarter
尾线 buttcock line
尾相 end portion
尾向 caudad
尾向叠标 back range
尾销 end pin;tail pin;terminal pin
尾销扁孔 yoke key slot
尾销止销 draft key retainer
尾斜跳板 mitre stern rampway
尾卸车 end discharge tipper
尾卸车辆 end-tipping vehicle
尾卸的 end-dump(ed)
尾卸法 end dumping method
尾卸卡车 end-dump truck;end-tipping truck
尾卸门 end discharge door
尾卸磨 end discharge mill
尾卸式 end dumping
尾卸式拌和车 end discharge truck mixer
尾卸式车身 end dump body
尾卸式搅拌车 end discharge truck mixer
尾卸式搅拌机 end discharge truck mixer
尾卸式卡车 end dump car;end-tipping lorry
尾卸式汽车 end-dump truck;end-tipping lorry
尾卸式运货车 end-dumped wagon
尾卸式自动倾卸车 end tipper;end-dump haulier
尾卸式自卸(汽)车 end discharge truck;end tipper;rear-dump truck;rear-dump wagon;truck with end dump body
尾卸载重卡车 end-tipping motor lorry
尾卸载重汽车 end-tipping motor lorry
尾旋 tail spin
尾淹 poop down
尾眼 becket
尾焰 wake flame
尾翼 empennage;tail fin;tail vane < 流速仪等的 >
尾翼面 empennage
尾翼液压传动装置 aft foil hydraulic actuator
尾油 tail oil
尾晕 rear halo
尾渣 tailings
尾罩 tail cup
尾震最大波 maximum wave of end portion
尾支承 tail bearing
尾支架【船】after poppet
尾枝渐近线 asymptote of the last part of the curve
尾直跳板 straight stern rampway
尾踵 skeg
尾轴 propeller shaft;screw propeller shaft;stern shaft;tail shaft;tube shaft
尾轴衬 stern bush;stern tube bearing;stern tube bushing
尾轴衬套 propeller shaft lining
尾轴承 tail bearing
尾轴承架 end bracket;tail bracket
尾轴定期检验 periodic(al) propeller shaft survey
尾轴端 tail end
尾轴毂 shaft bossing;shell bossing;stern boss(ing)
尾轴管 screw-shaft tube;shaft tube;stern tube
尾轴管衬套 stern bush
尾轴管堵漏 sealing the leakage of stern tube
尾轴管端板 stern tube end plate
尾轴管螺母 stern tube(ring) nut
尾轴管填料 stern tube packing
尾轴管填料涵 stern tube stuffing box
尾轴管压盖 stern gland
尾轴管止环 stern tube check ring
尾轴管轴承 stern bearing
尾轴颈 tail journal
尾轴孔 < 单推进器船 > shaft hole
尾轴孔部 bossing
尾轴离合器 tail clutch
尾轴螺帽 propeller shaft nut
尾轴密封填料 tail shaft packing
尾轴隧 screw-shaft tunnel;tail shaft passage
尾轴套 screw-shaft tube
尾轴填料涵 propeller shaft;tail shaft packing
尾轴调速器 tail shaft governor
尾轴筒压盖 shaft gland
尾轴推进器轴 propeller shaft;screw shaft
尾轴止环 shroud ring
尾轴轴承 propeller shaft bearing;shaft bearing
尾轴轴承承滑脂 tunnel grease
尾柱 stern frame;stern post
尾柱底骨 skeg
尾柱顶端 top projection
尾柱木 tail spar
尾柱倾角 angle of stern rake
尾柱肘板 heel bracket
尾状管 aeriductus
尾状条纹 gob tail;tail
尾状物 tailing
尾追车 < 溜放时 > catch-up car
尾追冲突 < 列车 > end-on collision;knock-on collision;rear-end collision
尾追撞车 < 列车 > end-on collision;knock-on collision;rear-end collision
尾锥体 tail cone
尾座 footstock;tailstock
尾座导轨 tailstock guide
尾座短横梁 tailstock trimmer

纬背组织织物 filling-backed fabric

纬编 filling knitting;weft knitting
纬编线圈 weft loop
纬差 difference of meridional parts;latitude difference;meridional difference
纬秤动 libration in latitude
纬疵 bias filling
纬档 filling irregularity;weft bar
纬度 degree of latitude;latitude;meridian distance
纬度比例尺 scale of latitude
纬度变化 latitude variation;variation of latitude
纬度补偿器 latitude compensator
纬度采用值 adopted latitude
纬度测定 determination of latitude;latitude determination
纬度测微器 latitude micrometer
纬度差 difference of latitude;tangent latitude error
纬度差订正器 latitude rider
纬度尺度 < 海图上的 > latitude scale
纬度带 latitudinal band;zone of latitude
纬度地带 latitudinal zone
纬度订正 reduction of latitude
纬度度数 degree latitude
纬度法 < 测天球船位 > latitude method
纬度范围 latitude range
纬度分布 latitude distribution;latitudinal distribution
纬度分带 latitudinal zoning
纬度服务 latitude service
纬度改正精度 accuracy of latitude correction
纬度改正量 correction of latitude;latitude correction
纬度改正系数 latitude correction coefficient
纬度改正值 latitude correction value
纬度观测 observation for latitude
纬度归算 reduction of latitude
纬度和船速差订正器 latitude and speed correction mechanism
纬度弧 latitude arc
纬度渐长率 meridianal part
纬度渐长率差 difference of meridianal parts
纬度校正量 correction of latitude;latitude correction
纬度校正器 latitude corrector
纬度截止 latitude cutoff
纬度平衡 latitude poise
纬度气候带 latitude zones of climate
纬度圈 circle of latitude;parallel circle;parallel circle of declination;parallel of latitude;latitude circle
纬度圈曲率半径 transverse radius of curvature
纬度数据计算机 latitude data computer
纬度水准器 < 天顶仪的 > latitude level
纬度误差 error in latitude;latitude error
纬度系数 latitude factor
纬度线 latitude line;parallel of latitude
纬度效应 latitude effect
纬度星对 latitude pair
纬度性海水进退 latitudinal transgressions and regressions
纬度因数 < 天球定位线上经度每变动一分的纬度变化 > latitude factor
纬度因素 latitude factor
纬度影响 latitude effect
纬度圆 latitude circle
纬度站 latitude station
纬度值 latitude value;value of latitude
纬二重织物 weft backed cloth
纬管 quill;weft bobbin
纬管式成型 filling wind
纬滑 filling slippage
纬距 difference of latitude;latitude difference;parallel distance
纬密 pick density
纬密度 pick count
纬面织物 filling-faced fabric;weft faced fabric
纬圈 filling snarl;parallel circle
纬圈弧 arc of parallel
纬圈曲率 curvature of parallel
纬色档 filling band in shade
纬纱 filling yarn
纬丝 < 织物的 > weft wire
纬缩 kink
纬天平动 libration on latitude
纬停装置 weft protector
纬线 filling yarn;parallel;parallel lines;parallel of latitude;weft;woof;weft wire < 织物的 >
纬线尺度 parallel scale
纬线弧线测量 parallel arc measurement
纬线跨度 latitudinal extent
纬线弯曲 curvature of parallel
纬线型 latitudinal type;parallel type
纬线印痕 filling mark
纬向 filling-wise;weft sense;zonal;zonary
纬向动能 zonal kinetic energy
纬向分布 latitudinal distribution
纬向分带性 latitudinal zoning
纬向风 zonal wind
纬向构造体系 latitudinal structural system;latitudinal tectonic system
纬向环流 zonal circulation
纬向距精度 accuracy of latitude distance
纬向流 zonal current;zonal flow
纬向螺旋斜纹组织 weft corkscrew twill
纬向排列错乱 weft disalignment
纬向气流 zonal air flow;zonal circulation
纬向撕裂强度 filling-tear resistance
纬向条痕 filling band;filling bar
纬向条花 barre;filling streak
纬向条纹 weft stripe
纬向条子织物 barre
纬向西风带 zonal esterlies
纬向细线条 weft hairline
纬向斜纹 filling twill weave
纬向指数 zonal index

苇把子沙障 reed sand-break

苇草 bentgrass
苇地 reed bed
苇塘 reed pond
苇席 reed mat;rush mat

委付 abandonment

委付款项 standing instruction
委付权 right of abandonment
委付书 letter of abandonment

委付通知书<海上保险> notice of abandonment
委付移交 tender of abandonment
委付者 abandoner
委陵菜属 cinquefoil
委内瑞拉地沥青 Venezuela asphalt
委内瑞拉桃花心木 Venezuela mahogany
委内瑞拉原油 Venezuelan crude(oil)
委派 accreditation;appointment;delegate;deputation
委派人 assignor
委派者 assigner
委弃的船 abandoned ship
委弃者 abandoner
委任 consign;delegate;entrust;appointment
委任代理 deputation;depute
委任的 mandatory
委任的职务 post(ing)
委任方式 form of proxy
委任国 accrediting state
委任合同 contract of mandate
委任理算人 appointment of an adjuster
委任权 power of appointment
委任使用 authorization to use
委任书 form of proxy;letter of appointment
委任统治 mandate
委任统治地 mandated territory
委任者 mandatory
委任证明 certification of authorization
委任状 certificate of appointment;letter of attorney;power of procuration;warrant of attorney
委托 trust;assignation;assignment;bail(ment);commission(ing);commitment;commit to the hands of;confide;consign;consignation;consignment;contract out;delegation;entrust(ment);leave in trust;mandate;relegate;submission;submit
委托办理 appointment and performance
委托保证金 consignment guarantee money
委托背书 power of attorney endorsement
委托拨款证 letter of authority
委托财产 abandoned property
委托查找 delegated search(ing)
委托船只 abandoned vessel
委托代理 mandate agency
委托代理信 letter of procuration
委托单位 client
委托的 mandatory
委托范围 scope of authority
委托方 clientage;entrusting party
委托费 commitment fee
委托付款证 authority to pay
委托购买书<国外订货单> indent
委托购买证 authority to purchase
委托关系 clientage
委托管理 mandatory administration
委托管理人 receiver
委托函<说明意图的> letter of intent(ion)
委托行 consignment store
委托合同 contract of mandate
委托加工 consigned processing
委托加工材料 consigned materials for processing;outside processing materials
委托加工出口 processing deal for export
委托加工服务 custom processing service
委托加工外销 processing deal for export
委托加工制造 commission manufacture
委托检查员 non-exclusive surveyor
委托校准 custom calibration
委托经纪人 managing operator
委托竞争 proxy fight
委托买卖 consignment business
委托贸易 commission business
委托期 bailable period
委托取款背书 endorsement for collection
委托人 client;clientage;clientele;commissary;consigner [consignor];constituent;depositor;entrusting party;mandator;party giving the mandate;principal;bailer [bailor];charger
委托人的账 entrusting party's account
委托人对受托人行为负责的原则 principle of respondent superior
委托人分账户 client ledger
委托人授予特别权力的文件 special power of attorney
委托人与建造师合同 owner-architect agreement
委托人总分类账 client ledger
委托任务书 assignment terms of reference
委托商 indentor
委托商店 commission house;commission shop
委托商行 commission house
委托试验 commercial test(ing)
委托收购 consignment purchase
委托收款结算 settlement of collecting consignment
委托书 attorney;certificate of entrustment;collection order;form of proxy;letter of attorney;letter of authority;letter of authorization;letter of commitment;letter of contract;letter of intent(ion);letter of proxy;powers of attorney;proxy;proxy statement;trust deed;warrant;warrant of attorney
委托书制度 proxy system
委托条款 abandonment clause
委托托收费 advice for collection
委托文献检索 customized literature searching
委托物 consignment
委托项目 mandated project
委托销售 commission sale;sale on commission;sales of commission
委托销售企业 brokerage business
委托颜色 custom colo(u)r
委托养牛站 custom cow pool
委托银行 authorized bank
委托运送货物 bailment of carriage of goods
委托者 abandoner;assignor;attorney;charger;consigner;shipper
委托证书 certificate of appointment
委托支付证 authority to pay
委为代表 deputize
委用条件 conditions of appointment
委员 commissioner;committeeman;committee member;delegate
委员会 commission;committee;council
委员会办公室 committee room

萎

萎缩 atrophy
萎缩病 shrinking disease
萎缩的下深槽 shrinkage down-stream deep

艉

艉板肋骨 transom frame
艉部构架 stern framing
艉部桥架 stern-mounted ladder
艉部上层建筑 aft superstructure
艉部下沉 stern settlement
艉舱壁 stern bulkhead;afterpeak bulkhead;stuffing box bulkhead
艉垂标 after perpendicular
艉垂线 aft perpendicular
艉导缆孔 stern chock
艉倒缆 after spring(line)
艉灯 stern light
艉定位桩 aft spud;walking spud
艉横缆 after breast(line)
艉护舷 pudding;pudding fender
艉滑道拖网渔船 stern ramp trawler
艉机舱 after engine room
艉机船 aft-engine ship;after engine ship;machinery-aft ship
艉甲板 quarter deck
艉缆 after line;stern fast;stern line;stern rope
艉浪 aftertossing;after-tow
艉楼 poop;poop superstructure
艉楼构架 poop framing
艉楼前舱壁 poop front bulkhead
艉锚 stern anchor
艉锚泊绞车【疏】 stern anchoring winch
艉耙挖泥船 trailing suction dredge(r) with stern well
艉桥楼 after bridge
艉倾 stern trim;trim by the stern
艉水舱 after-peak water tank
艉轴架 A-bracket

鲔

鲔鱼及渔船 tuna and bonito clipper

卫

卫兵哨亭 guerite
卫城<古希腊城市的> acropolis
卫导船位 satellite fix
卫点 satellite point
卫护板端部 blade heel
卫护板加固 blade reinforcement
卫护板开挖角度 blade pitch
卫护板控制 blade control
卫护板升臂 blade lift arm
卫矛【植】 winged spindle tree
卫矛属 spindle tree
卫片地质判译 geologic(al) interpretation of satellite photograph
卫片解译法 analysis method of satellite picture
卫墙杆 wall guard
卫燃带 refractory belt
卫生 health;sanitation
卫生保护区 sanitary protection zone
卫生保健 health care
卫生保健设施 health-care facility;health-care measures
卫生保健事业 health service
卫生保健室<车站内的> public health room
卫生保健员 medical-care worker
卫生泵 sanitary pump
卫生标准 health standard;hygiene standard;hygienic standard;sanitary criterion;sanitary standard
卫生标准化 hygienic standardization
卫生部 Ministry of Health
卫生厕所 sanitary privy
卫生车 vacuum car
卫生城市 hygienic city
卫生处 health department;Health Office
卫生处理 sanitary disposal;sanitary treatment;sanitization;sanitize
卫生处理区 sanitary district
卫生瓷 sanitary china;sanitary porcelain
卫生措施 hygienic measure;sanitary measure
卫生当局 health authorities
卫生的水 sanitary water;wholesome water
卫生调查 sanitary survey
卫生伐 sanitation cutting;sanitation felling
卫生法 health legislation
卫生法规 health regulation;sanitary regulation
卫生方法 sanitation method
卫生防护 health protection;sanitary precaution
卫生防护标准 public health standard
卫生防护带 sanitary protection zone
卫生防护距离 width of sanitary protection zone
卫生防护林 health protection forest
卫生防护区 protective sanitary zone;sanitary protective range;sanitary protective zone
卫生防疫工作 sanitary and anti-epidemic affair
卫生防疫站 quarantine station;sanitation and anti-epidemic station
卫生分析 sanitary analysis
卫生焚化炉 sanitary incinerator
卫生福利设施 welfare facility
卫生工程 sanitary works;sanitary engineering
卫生工程的污水管子 sanitary sewage pipe
卫生工程管道 plumbing piping
卫生工程管道安装 plumbing installation
卫生工程管道铺设 plumbing installation
卫生工程管理 plumbing piping;sanitary plumbing
卫生工程污水管网 sanitary sewer network
卫生工程学 sanitary engineering
卫生管道 plumbing pipe;sanitary line;sanitary plumbing
卫生管道安装 plumbing installation
卫生管道存水弯 plumbing trap
卫生管道工程 plumbing
卫生管道工程系统 plumbing system
卫生管道配件 fitting;plumbing fittings
卫生管道设备 plumbing services
卫生管道装置 plumbing installation;plumbing services
卫生管理 conservancy of sanitation
卫生管理部门 sanitary board
卫生管系 sanitary system
卫生管线 plumbing run
卫生规则 sanitary regulation;sanitary rule
卫生轨道运行数据 satellite orbit data
卫生河流 health stream
卫生化 hygienisation [hygienization]
卫生化学 hygienic chemistry;sanitary chemistry
卫生化学分析 sanitary chemical analysis
卫生化学特征 sanitary chemical feature
卫生机构 health agency;health institution
卫生基础 hygienic basis
卫生间 toilet;washroom
卫生间单元 toilet unit

卫生间隔板 toilet partition
卫生间隔墙 toilet partition wall
卫生间(建筑)模数 sanitary unitized unit; sanitary building block module
卫生间木隔板 duck board
卫生间设备 plumbing unit; toilet equipment
卫生间中的小隔断 toilet enclosure
卫生监测 hygiene monitoring; monitoring of hygiene
卫生监督 health inspection; health supervision; sanitary inspection; sanitary superintendence; sanitary supervision
卫生检查 sanitary inspection
卫生检验书 sanitary inspection certificate
卫生检疫 health quarantine; sanitary quarantine
卫生检疫规定 health and sanitary regulation
卫生检疫所 health quarantine service
卫生检疫站 health quarantine service
卫生检疫中心 health quarantine service
卫生教育电信 health education telecommunication
卫生洁具 plumbing fittings; plumbing fixture; sanitary equipment; sanitary fireclay ware; sanitary ware
卫生洁具配件 plumbing fittings; sanitary fixture
卫生洁具陶瓷 sanitary pottery
卫生(洁具)陶瓷配件 sanitary pottery fittings
卫生局 board of health; public health bureau
卫生距离 sanitary distance
卫生垃圾堆 sanitary landfill
卫生棉 cotton wool; sanitary napkin
卫生排污工程 sanitary drainage; sanitary drainage works
卫生排污系统 sanitary drainage
卫生配件 sanitary fitments
卫生起落终端(站) satellite terminal
卫生气象学 hygienic meteorology; hygieno-meteorology
卫生器具 plumbing fixture; plumbing unit; sanitary apparatus; sanitary device; sanitary fixture; sanitary ware
卫生器具单位 fixture unit
卫生器具当量 fixture unit
卫生器具定额流量 the rated flow of fixture
卫生器具给水管 fixture-supply pipe
卫生器具排水管 fixture drain
卫生器具支管 fixture branch
卫生器皿 sanitary fixture; sanitary ware
卫生器通气管 fixture vent
卫生清洁用具 sanitary ware
卫生球 mothball; naphthalene(ball)
卫生人员 health officer
卫生三通 sanitary tee
卫生(上)的 sanitary
卫生上下水道设备 sanitary plumbing equipment
卫生设备 plumbing equipment; plumbing fittings; plumbing fixture; plumbing system appliance; sanitary accommodation; sanitary appliance; sanitary conveniences; sanitary equipment; sanitary installation; sanitary plumbing; sanitary provision; sanitary unit; sanitation equipment; sanitation facility; sanit sanitation; soil appliance; sanitation <尤指排水设备>
卫生设备安装 sanitation installation
卫生设备单位流率 fixture-unit flow rate
卫生设备废水 sanitary wastewater
卫生设备分支管 fixture branch
卫生设备工程 plumbing work
卫生设备管道 sanitary plumbing
卫生设备管子配件 plumbing fixture
卫生设备给水管 fixture drain; fixture-supply pipe
卫生设备角隅边条压制 sanitary corner pressing
卫生设备排出的废物 sanitary waste
卫生设备排水(管) fixture drain
卫生设备配件 sanitary fittings; sanitary fixture
卫生设备疏水器 fixture trap
卫生设备套件 combination furniture
卫生设备污水 domestic wastewater; household wastewater; wastewater from living
卫生设备系统 plumbing system
卫生设备噪声 plumbing noise
卫生设备支管 fixture branch
卫生设备组合 plumbing unit
卫生设施 health facility; sanitary facility; sanitary fixture; sanitation facility
卫生炻器 sanitary stoneware; stoneware sanitary wares
卫生事业规划 health planning
卫生室 clinic
卫生水泵 sanitary pump
卫生水厕 water closet
卫生水工设备 sanitary plumbing equipment
卫生水管 sanitary pipe
卫生水柜 sanitary tank
卫生四通 sanitary cross
卫生所 aid station; clinic; health center [centre]; health service; sanitary building
卫生搪瓷 enamel sanitary wares
卫生陶瓷 porcelain ware; sanitary porcelain; sanitary pottery; sanitary ware; whiteware
卫生陶瓷制品 china sanitary wares
卫生陶器 sanitary earthenware; sanitary(stone) ware
卫生特性 hygienic feature
卫生特征 hygienic characteristic
卫生填池 sanitary landfill
卫生填埋法 sanitary landfill method
卫生填土 sanitary landfilling
卫生条件 sanitary condition
卫生条例 sanitary regulation
卫生通风 sanitary ventilation
卫生统计学 health statistics
卫生(土地)填埋 <垃圾或废渣的> sanitary landfill
卫生微生物学 sanitary microbiology
卫生污水管(道) sanitary sewer; soil-or-waste pipe
卫生污水立管 soil-or-waste stack
卫生污水排泄系统 sanitary drainage system
卫生物件 sanitary articles
卫生物品 sanitary articles
卫生洗涤剂 sanitizer; sanitizing agent
卫生系统 sanitation system
卫生细菌学 hygienic bacteriology; sanitary bacteriology
卫生消毒剂 sanitizer; sanitizing agent
卫生学 hygiene
卫生学的 hygienic
卫生学家 hygienist; sanitarian
卫生学评价 hygienic evaluation
卫生研究中心 hygiene research center [centre]
卫生掩埋场 sanitary landfill density
卫生掩埋场不透水层 sanitary landfill liner
卫生医疗设备 sanitary and medical equipment
卫生饮水(喷)泉 sanitary drinking fountain; wholesome fountain
卫生饮用水 sanitary drinking water; wholesome water
卫生用具 sanitary fitments
卫生用农药 household pesticide
卫生用品 drug sunchries; sanitary facility
卫生用杀虫液剂 household spray
卫生用水 sanitary use of water; sanitation use of water
卫生员 health guard; sick berth attendant
卫生原则 sanitary principle
卫生院 commune hospital; hospital
卫生整理 hygienic finishing
卫生止门件 sanitary stop
卫生纸 toilet paper
卫生纸废弃场 sanitary napkin disposal
卫生纸分配器 toilet paper dispenser
卫生纸盒 paper-holder; tissue paper holder; toilet paper holder
卫生纸架 paper-holder; toilet paper holder
卫生纸匣 paper-holder; tissue paper holder; toilet paper holder
卫生指标 health indicator
卫生主管部门 health authorities
卫生装置 sanitary fittings; sanitary fixture; sanitary installation
卫生装置零件 <陶瓷> sanitary plumbing fixtures
卫生组合单元 sanitary unitized unit
卫戍部队医院 garrison hospital
卫铁 keeper; magnet keeper
卫线 satellite line
卫星 orbital body; orbiting body; satellite
卫星被淹 occultation of satellites
卫星标准频率和时间信号业务 standard frequency and time signal satellite service
卫星捕捉与收回 satellite capture and retrieval
卫星测得的臭氧资料 satellite ozone data
卫星测风 wind measurement from satellite
卫星测高 satellite altitude determination
卫星测绘雷达 satellite mapping radar
卫星测角测高系统 satellite angle and altitude measuring system
卫星测角导航系统 angle measurement of satellite navigation system
卫星测距差定位 satellite fixing by range difference measurement
卫星测距导航系统 range measurement of satellite navigation system
卫星测距定位 satellite fixing by range measurement
卫星测距仪 satellite rangefinder
卫星测速系统 velocity measurement of satellite system
卫星测量 satellite survey
卫星测站 satellite measurement station
卫星长途中继线 satellite toll trunk
卫星车站 satellite terminal
卫星成像 satellite imagery; satellite imaging
卫星城(市) expanded city; satellite city; satellite town; star town; bed town; new town
卫星城镇 expanded town; satellite band; follow town; satellite city; satellite town
卫星赤道 satellite equator
卫星处理机 satellite processor
卫星传感器 satellite(-borne) sensor
卫星传输业务系统 Sat Stream
卫星传送 satellite transmission
卫星船位 satellite fix
卫星大地测量动力法 dynamic(al) method of satellite geodesy
卫星大地测量轨道法 orbital method of satellite geodesy
卫星大地测量几何法 geometric(al) method of satellite-geodesy
卫星大地测量学 artificial satellite geodesy; satellite geodesy
卫星大地测量洲际联测 intercontinental geodetic connection by satellite observation
卫星大地网 satellite geodetic network
卫星导航 navigation by aid of satellite; satellite-aided navigation
卫星导航地面站 satellite navigation earth station
卫星导航地图 satellite navigation map
卫星导航定位仪 satellite navigator
卫星导航计算机 satellite navigation computer
卫星导航系统 satellite navigation system
卫星导航用户设备 satellite navigation equipment for user
卫星地面接收站 ground station
卫星地面识别 ground identification of satellite
卫星地面站 satellite earth station
卫星地面整体网络 integrated satellite terrestrial network
卫星地面终端站 satellite earth terminal
卫星地球覆盖区 satellite earth coverage
卫星地球站质量因素 figure of merit of satellite earth station
卫星电视 satellite television; stratovision
卫星电视分配业务 satellite television distribution service
卫星电台 satellite radio station; satellite station
卫星电信系统 satellite telecommunication system
卫星电源 satellite power supply
卫星电子 satellite electron
卫星电子对抗系统 satellite electronic countermeasures system
卫星定位 satellite fix; satellite positioning
卫星定位系统 satellite-based positioning system
卫星定向 satellite orientation
卫星定向系统 satellite orientation device; satellite orientation system
卫星动力测地 dynamic(al) satellite geodesy
卫星动力学 satellite dynamics
卫星多普勒测量系统 satellite-Doppler system
卫星多普勒导航系统 satellite Doppler navigation system
卫星多普勒定位(法) satellite Doppler positioning
卫星多普勒声呐组合导航系统 satellite Doppler sonar integrated navigation system
卫星多址技术 satellite multiple-access technique
卫星发射 satellite launching
卫星发射船 satellite launching ship
卫星发射火箭 satellite launching

W

rocket
卫星发射机 satellite transmitter
卫星发射装置 satellite launcher; satellite launching facility
卫星发送 satellite transmission
卫星反射器 satellite retroreflector
卫星方位角 satellite azimuth
卫星飞行 satellite flight
卫星费用 satellite charges
卫星峰 satellite peak
卫星浮标通信[讯] satellite buoy communication
卫星辅助动力装置 satellite auxiliary power unit
卫星辅助系统 satellite auxiliary system
卫星附属系统 satellite auxiliary system
卫星覆盖范围 satellite coverage
卫星覆盖区 satellite coverage area
卫星港 satellite port
卫星高度 satellite altitude
卫星高度角 altitude angle of satellite; angle of satellite altitude
卫星跟踪 moontrack; satellite tracking
卫星跟踪船 satellite tracking ship
卫星跟踪观测 satellite tracking observation
卫星跟踪激光器 satellite tracking laser
卫星跟踪激光系统 satellite tracking laser system
卫星跟踪激光仪 satellite tracking laser
卫星跟踪设备 satellite tracking facility
卫星跟踪设施 satellite tracking installation
卫星跟踪数据 satellite tracking data; spacecraft tracking data
卫星跟踪天线 satellite tracking antenna
卫星跟踪望远镜 satellite tracking telescope
卫星跟踪卫星法 satellite-to-satellite tracking method
卫星跟踪系统 minitrack
卫星跟踪站 satellite tracking station
卫星跟踪中心 satellite tracking center [centre]
卫星工业 satellite industry
卫星攻击警报系统 satellite attack warning system
卫星共振效应 satellite resonance effect
卫星观测 satellite observation
卫星观测船 satellite observation ship
卫星观测镜 moonscope
卫星观测系统 satellite-missile observation system
卫星观测站 satellite observation station
卫星观察者 moonwatcher
卫星惯导组合定位系统 satellite inertial guidance integrated positioning system
卫星光行差 satellite aberration
卫星光学监视系统 satellite optic(al) surveillance system
卫星广播 satellite broadcast(ing)
卫星广播业务 satellite broadcasting service
卫星归算点 reduced point of satellite
卫星轨道 satellite orbit; satellite trajectory
卫星轨道背日时间 satellite orbit shadow time
卫星轨道参数 orbit elements of satellite; orbit parameters of satellite; satellite orbit parameters
卫星轨道长半轴 semi-major axis of satellite orbit
卫星轨道的测定 satellite orbit determination
卫星轨道的地面投影 satellite's ground path; satellite trajectory ground track
卫星轨道动力法 dynamic(al) method of satellite orbit
卫星轨道短弧法 short arc method of satellite orbit
卫星轨道法向误差 normal error of satellite orbit
卫星轨道改进法 improved method of satellite orbit
卫星轨道根数误差 error of satellite orbital elements
卫星轨道跟踪 satellite orbital tracking
卫星轨道几何法 geometric(al) method of satellite orbit
卫星轨道降交点 descending node of satellite orbit
卫星(轨道)接合 satellite coupling
卫星轨道近地点 perigee of satellite orbit
卫星轨道近地点高度 altitude of perigee of satellite orbit
卫星轨道近地点角距 argument of perigee of satellite orbit
卫星轨道径向误差 radial error of satellite orbit
卫星轨道控制 satellite orbit control
卫星轨道偏心率 eccentricity of satellite orbit
卫星轨道切向误差 tangential error of satellite orbit
卫星轨道倾角 inclination of satellite orbit
卫星轨道扰动 orbital perturbation
卫星轨道摄影定位法 positioning method by photographing satellite orbit
卫星轨道升交点 ascending mode of satellite orbit
卫星轨道速度 orbital velocity of satellite
卫星轨道远地点高度 altitude of apogee of satellite orbit
卫星轨道周期 period of satellite orbit; satellite orbital period
卫星轨迹 satellite path; satellite trail
卫星海洋监视鉴定中心 satellite ocean surveillance evaluation center [centre]
卫星海洋监视系统 satellite ocean surveillance system
卫星海洋学 satellite oceanography
卫星航海电子定位系统 Satnav [satellite navigation] electronic positioning system
卫星和空间目标跟踪系统 satellite and space tracking system
卫星红外分光计 satellite infrared spectrometer
卫星红外辐射光谱仪 satellite infrared radiation spectrometer
卫星红外光谱仪 satellite infrared spectrometer
卫星候机楼 satellite building
卫星湖 satellite lake
卫星回合系统 satellite rendezvous system
卫星回收 satellite recovery
卫星回收系统 satellite recovery system
卫星火箭 satellite rocket
卫星机 subhost
卫星机场 satellite airfield
卫星机构 satellization authority
卫星激光测距 satellite laser ranging
卫星激光测距仪 laser ranging apparatus for satellite; satellite laser ranger
卫星及空间探测器测量 satellite and space probe measurements
卫星几何测地 geometric(al) satellite geodesy
卫星计划 Sputnik program(me)
卫星计算机 satellite computer; satellite processor
卫星计算中心 computing centre for processing satellite tracking data
卫星间隔 satellite spacing
卫星间求差 between satellite difference
卫星间数据传输 satellite-to-satellite data transfer
卫星间通信[讯] intersatellite communication
卫星监测 satellite monitoring
卫星监控 satellite monitoring
卫星监视 satellite surveillance
卫星监视器 satellite monitor
卫星检测网 satellite inspection network
卫星教育和新闻电视 satellite educational and informational television
卫星接收机辅助设备 satellite receiver subsystem
卫星接收中心 satellite receive center [centre]
卫星紧急示位无线电信标 satellite emergency position indicating radio beacon
卫星紧急无线电示位电台 emergency satellite position indicating radio beacon station
卫星井采油树 satellite X-mass tree
卫星警报系统 satellite-warning system
卫星局 satellite exchange
卫星距离测量 satellite range measurement
卫星距离差定位 range difference measurement of satellite fixing
卫星可测事件 satellite-sensed event
卫星可见期 visibility of satellite
卫星可用率 satellite availability
卫星控制设备 satellite control facility
卫星控制网(络) satellite control network
卫星控制系统 satellite control system
卫星控制中心 satellite control center [centre]
卫星面质比 area-to-mass ratio of satellite
卫星名称 satellite designation
卫星模拟器 satellite simulator
卫星能见度 satellite visibility; visibility of satellite
卫星逆向反射器 satellite retroreflector
卫星欧米伽组合导航系统 satellite-Omega integrated navigation system
卫星配置 satellite configuration
卫星偏移 satellite drift
卫星(频)带 satellite band
卫星频率误差 satellite frequency error
卫星气球 satellite balloon
卫星气象学 satellite meteorology
卫星情报中心 satellite information center [centre]
卫星全球监测 satellite global monitoring
卫星绕主星运行的平均速度 orbital velocity
卫星三边测量 satellite trilateration
卫星三角测高系统 satellite triangle and altitude measuring system
卫星三角测量 satellite triangulation
卫星三角测量站 satellite triangulation station
卫星扫描与成像系统 satellite scanning and imagery system
卫星上的传感器 satellite-borne sensor
卫星(上的)传感系统 satellite-borne sensing system
卫星摄动 satellite perturbation
卫星摄动运动 satellite perturbation motion
卫星摄影 satellite photography
卫星摄影测量 satellite photogrammetry
卫星摄影及扫描 satellite photography and scanning
卫星摄影(术) satellite photography
卫星升交点赤经 right ascension of satellite ascending node
卫星声学组合定位系统 satellite-acoustics integrated positioning system
卫星识别编号 satellite identification number
卫星式机坪 satellite system apron
卫星式廊道 pier satellite
卫星式实验 <即在重点实验以外,在各处所作的个别小型实验> satellite experiment
卫星试验中心 satellite test center [centre]
卫星收集的气象观测资料 satellite collection of meteorological observation
卫星寿命 lifetime of satellite
卫星数 number of satellites
卫星数据 satellite data
卫星数据处理系统 satellite data processing system
卫星数据记录器 satellite data recorder
卫星数据调制器 satellite data modulator
卫星数据通信[讯]系统 satellite data communication system
卫星数据系统 satellite data system
卫星数字模拟显示 satellite digital and analog(ue) display
卫星双速摄影机 dual-rate satellite camera
卫星水上无线电导航业务 maritime radio navigation satellite service
卫星水上移动业务 maritime mobile satellite service
卫星搜索雷达 satellite searching radar
卫星速率 satellite speed
卫星(台)站 satellite station
卫星探测 satellite sounding
卫星探测线 satellite detection line
卫星体 satelloid
卫星天文学 satellite astronomy
卫星天线 satellite aerial; satellite antenna; satellite dish
卫星天线电测法 method of satellite radio altimetry
卫星天线辐射方向图 satellite antenna radiation pattern
卫星通过 pass of satellite; satellite pass
卫星(通)过近地点时刻 time of satellite perigee passing
卫星通信[讯] satcom; satellite communication
卫星通信[讯]测控站 satellite communication tracking telemetry command station

卫星通信[讯]船 satellite communication ship
卫星通信[讯]地面站 earth-station for satellite communication; satellite ground station
卫星通信[讯]合作委员会 Satellite Telecommunication Coordinating Committee
卫星通信[讯]技术 satellite communication technology
卫星通信[讯]局 satellite communication agency
卫星通信[讯]控制器 satellite communication controller
卫星通信[讯]控制设施 satellite communication control facility
卫星通信[讯]控制室 satellite communication control office
卫星通信[讯]系统 satellite communication system
卫星通信[讯]线路 satellite communication link
卫星通信[讯]用发射天线 reflecting satellite communication antenna
卫星通信[讯]中心 satellite communication center [centre]; satellite communication city
卫星通信[讯]终端(设备) satellite communication terminal
卫星通讯社 satcom
卫星图 satellite imagery
卫星图书馆情报网络 satellite library information network
卫星图像 satellite image
卫星图像解释 interpretation of satellite image
卫星图像解译 interpretation of satellite image
卫星图像判读 interpretation of satellite image
卫星图形系统 satellite graphics system
卫星网络 satellite network
卫星位置 satellite position
卫星位置保持 satellite station-keeping
卫星位置的确定 satellite position determination
卫星位置预测与显示 satellite position prediction and display
卫星污水厂式处理 satellite wastewater plant treatment
卫星污水处理系统 satellite treatment system of wastewater
卫星无线电测量业务 radio determination satellite service
卫星无线电通信[讯] satellite radio communication
卫星无线电线路 satellite radio line
卫星系列 satellite series
卫星线路 satellite circuit; satellite link
卫星线路噪声 satellite circuit noise
卫星相位差 phase error of satellite
卫星像片 satellite image; satellite photograph; satellite picture
卫星像片分析 satellite image analysis
卫星像片判读 interpretation of satellite photograph
卫星像片图 satellite image map; satellite photo map
卫星信道 satellite channel
卫星信息处理机 satellite information processor
卫星信息处理机操作程序 satellite information processor operational program(me)
卫星信息地面接收站 ground station
卫星信息中心 satellite information center [centre]
卫星星历误差 satellite ephemeris error
卫星星下点 subsatellite point; substellar point of satellite
卫星星座 satellite constellation
卫星仰角 satellite elevation
卫星遥测 satellite telemetering
卫星遥测接收机 microlock receiver
卫星遥测数据自动处理系统 satellite telemetry automatic reduction system
卫星遥测系统 microlock
卫星遥测站 satellite telemetry station
卫星遥测自动数据简化系统 satellite telemetry automatic reduction system
卫星遥感 satellite remote sensing
卫星遥感测量 satellite remote sensing
卫星遥感校正站 satellite remote sensing calibration site
卫星遥感系统 satellite remote sensing system; spacecraft remote sensing system
卫星遥感信息处理 information processing for satellite remote sensing
卫星业务 satellite business
卫星移动通信[讯]系统 satellite mobile communication system
卫星影像 satellite imagery
卫星用户电报业务 satellite telex service
卫星预警系统 satellite early warning system
卫星云图 nephanalysis; satellite cloud photograph; satellite cloud picture
卫星运行周期 satellite period of revolution
卫星运载工具 satellite carrier
卫星运载火箭 satellite launcher; satellite-launching vehicle; sputnik rocker
卫星照片 satellite photo(graph)
卫星照片分析 satellite photographic(al) study
卫星照片判读 satellite photographic(al) study
卫星照相术 satellite photography
卫星侦察 satellite reconnaissance
卫星侦察摄影机 spy-in-sky camera
卫星直播电视 direct satellite broadcasting television
卫星制图学 satellite cartography
卫星中继器 satellite relay
卫星中继站 satellite relay station
卫星中天点 point of satellite meridian passage
卫星中心 satellite center [centre]
卫星中心坐标系 satellocentric system
卫星重复周期 repetition cycle of satellite
卫星重力测量 satellite gravimetrical survey; satellite gravimetry
卫星重力梯度法 method of satellite gravity gradient
卫星周期 period of satellite
卫星洲际联测 intercontinental connection by satellite; intercontinental survey by satellite
卫星注入站 station for satellite injection
卫星转播站 relay station satellite
卫星转播资料 satellite relay of data
卫星转发器 satellite repeater; satellite transponder
卫星转换 satellite switching
卫星转频器 satellite transponder
卫星装载的 satellite-borne
卫星装置 satellite asset; satellite equipment
卫星装置传感器 satellite-borne sensor
卫星装置遥测系统 satellite-borne remote sensing system
卫星装置遥感系统 satellite-borne remote sensing system
卫星状态 satellite status; satellitosis
卫星姿态 satellite attitude
卫星资料系统 satellite data system
卫星自动导航 automatic satellite computer aid-to-navigation
卫星自动监控系统 satellite automatic monitoring system
卫星自动控制系统 satellite automation system

为

为安全而驶出海面 safe offing

为安全计 for consideration of safety
为暴风雨所阻的 storm bound
为偿付债务筹集资金 refinancing of debt service payments
为偿付债务开支再筹资 refinancing of debt service payments
为车制螺纹的镦粗 upset for threading
为当时跨度最大的混凝土斜拉桥 Brotonne Bridge
为等候进入戏院而暂时停留的面积 holdout area
为地下室外墙设防潮层 dampproof sheeting
为电影改编 filmize
为防止……提供安全措施 provide insurance against
为粉刷面而作的机械嵌缝 mechanical keying for plastering
为风所吹集的雪堆 snow drift
为付款进行的计量 measurement for payment
为该城市提供劳动力 catchment area
为公共住房建设置地 public housing acquisition
为固定螺钉用的铰孔 reaming holes for screw stays
为害大的杂草 ill weed
为厚层沥青整层翻修而造成的 to be built for full depth reclamation of thick lifts of asphalt
为获利而购进 buying on a yield basis
为检验提供设备 facility for testing
为经济利益而转业 job hopping
为景观目的的土方工程 earthmoving for landscape purposes
为居住者备餐的楼面 service flat
为开挖土方而配备的铁路车辆 ditching car
为了适用陆上/海上作业安装 equipping for land/marine work
为某单位专编的不停站列车 non-stop unit train
为某公司专编的循环直达列车 company train
为期两个月的一段时间 bimester
为其他承包商提供方便 facility for other contractors
为其他承包商提供机会 opportunity for other contractors
为企业开办筹款 flo(a)tation
为热带设计的 tropic(al) designed
为石板屋顶塞泥灰的工人 torcher
为使超挖最小 in order to minimize overdredging
为首班组 heading crew
为损失支付的费用 loss draft
为摊铺道砟而建筑的路槽 <土路基> ballast boxing
为同向运行公共汽车提供的专用车道 with-flow bus lane
为未偿还借款重新筹集的资金 refinancing of outstanding borrowings
为应付意外事件可能性而作准备的状态 open-endedness
为用户服务的附属工作 ancillary customer service
为专门用途的借贷 tributary lending
为专用线服务的铁路线 line serving a siding
为装饰所中断的拱 interrupted arch
为租户服务 tenant services

未

安弹簧的承重量 unsprung weight

未安排的信贷偿还 non-scheduled repayment
未安平 mislevel
未安弹簧的承重量 unsprung weight
未安装的 unerected; unmounted; unset
未安装弹簧的 unsprung
未按规定的 irregular
未拌和的拌容量 unmixed batch capacity
未拌料容量 unmixed batch capacity
未绑扎钢筋 unbonded reinforcement
未包混凝土的 uncased
未包括在内 not included
未包装的 packless; unpack(ag)ed
未包装好的 unsecured
未饱和 undersaturation; unsaturation
未饱和标准电池 unsaturated standard cell
未饱和传导率 unsaturated conductivity
未饱和传导性 unsaturated conductivity
未饱和的 unsaturated
未饱和空气 non-saturated air; unsaturated air
未饱和空隙数量 non-saturated void count; unsaturated void count
未饱和气油藏分布区 distribution area of undersaturated oil pool
未饱和汽 unsaturated vapo(u)r
未饱和铁结合力 unsaturated iron binding force
未饱和土壤对地表自由水的吸收 insoak
未饱和岩 unsaturated rock
未饱和蒸汽 unsaturated steam
未保过险的 uninsured
未保温的管道 uninsulated pipe; unlagged piping
未保温管线 bare pipeline
未保险的 uninsured
未保险的存款 uninsured deposit
未保险货物 uncovered goods
未报 not reported
未报关的 uncustomed
未报关货物 uncustomed goods; undeclared goods
未暴露粗骨料 light scaling
未曝光胶卷 unexposed film reel
未曝光胶片 unexposed film
未爆弹处置 explosive disposal
未爆炮眼 bootleg
未爆破的 unexploded
未爆眼 misfired hole; missed hole; miss shothole
未爆药卷 unexploded cartridge
未爆炸弹 unexploded bomb
未爆炸的 unexploded
未备基金的 unfunded
未背书支票 unendorse check
未被安置的 unplaced
未被饱和的 unsaturated
未被采用的 unadapted
未被超过的 unsurpassed
未被承认的 unadmitted
未被抵偿的残差 uncancelled residual
未被发现的 undiscovered

未被激发的 unavoidable
未被监视的图像分类 unsupervised image classification
未被检出的误差 undetected error
未被冷水掺和的深部热水 original deep thermal water
未被领会的 unapprehended
未被平衡的加速度 non-compensated acceleration; unbalanced acceleration
未被平衡的离心加速度 non-compensated centrifugal acceleration; unbalanced centrifugal acceleration
未被平衡的离心力 non-compensated centrifugal force; unbalanced centrifugal force
未被取代的 unsubstituted
未被认识的 unrecognized
未被市场吸收的 undigested
未被污染 unpollute
未被污染的 uncontaminated; unpolluted
未被污染的淡水湖泊 unpolluted freshwater lake
未被污染的淡水水体 unpolluted freshwaters
未被污染的河流 stream health; uncontaminated stream; unpolluted river
未被污染的冷却水 unpolluted cooling water
未被污染的水 uncontaminated water; unpolluted water
未被污染区 uncontaminated zone; unpolluted zone
未被吸收的成本 unabsorbed cost
未被吸收的气体 unabsorbed gas
未被抑制的 unsubdued
未被占用的 unoccupied
未被占用的建筑空间 <古希腊、罗马剧院舞台后的> choragium
未被证明的 unproved; unvouched
未被遵守的 unobserved
未焙烧的黏[粘]土 unburned clay
未焙烧透石灰 unburnt lime
未闭导管钳 patent duct forceps
未闭合等高线 open contour
未闭合折线 open polyline
未编号的 unnumbered
未编码的 uncoded
未编码字 uncoded word
未编目 uncatalog
未编目文件 uncatalogued file
未贬值的 undepreciated
未贬值的货币 undepreciated currency
未贬值的价值 undepreciated value
未变稠的 unstiffened
未变的 unaltered; unchanged
未变动岩层 undisturbed strata
未变位的接头 undisplaced joint
未变稀的 unattenuated
未变形的 undeformed; unstrained
未变性的 unmodified
未变硬的 unstiffened
未变质的 unmetamorphosed
未变质岩(石) unaltered rock
未变质炸药 undecomposed explosive; undisturbed explosive
未标定界 alignment uncertain boundary; andemarcated boundary; indefinite boundary; undefined boundary; undemarcated boundary; undeterminated boundary
未标记地区 uncharted area
未标记非终结符 unmarked non-terminal
未标记终结符 unmarked terminal
未标价的 unpriced
未标线的 unmarked
未标页码的 unpaged
未标页数 no paging
未标志的 not marked
未标志区 unmarked area
未标注端 unmarked end
未表达的 unexpressed
未并联的 unshunted
未并列的 unparalleled
未拨用盈余 uncommitted surplus; unrestricted surplus
未剥落的 unpeeled
未剥皮材 unbarked wood
未剥皮的 undeformed; unpeeled
未剥皮的原木堆 unpeeled log dump
未剥皮木材 unpeeled wood
未补偿衰减器 uncompensated attenuator
未补偿损耗 uncompensated loss
未补强的接管开孔 unreinforced nozzles openings
未补强接头 unprotected tool joint
未捕海区 unfished ground
未捕群 unfished stock
未布雷航道 not mined channel
未擦亮的 unpolished
未采层 unmined bed
未采掘的煤 living coal
未采矿区 maiden field; virgin field
未采区 unworked country
未采下损失 nofalling losses
未采下损失率 nofalling losses ratio
未采用的 unadopted
未采油层 virgin sand
未采原因 unmined cause
未操纵的 unsteered
未测出的 not detectable
未测定的 undeterminate; unmeasured
未测定损失 unmeasured loss
未测定值 unmeasured value
未测量的 unmeasured; unsurveyed
未测量区 unsurveyed area
未测试的 untested
未测输沙量 unmeasured sediment discharge
未测图的 non-mapping
未测图港区 uncharted port area
未测完深度的 unsounded
未测坐标的 uncoordinated
未测坐标的钻孔 uncoordinated borehole
未查出的误差 undetected error
未拆开的 unopened
未掺和的 unblended
未掺配(过的)地沥青 unblended asphalt
未掺杂的 virgin
未偿贷款 loan outstanding; outstanding item; outstanding loan; outstanding
未偿贷款总额 outstanding loan portfolio
未偿担保信贷 outstanding guarantee credit
未偿股份 outstanding stock
未偿还本金 outstanding principal
未偿还的 unredeemed
未偿还的余额 balance outstanding
未偿还借款 outstanding borrowing
未偿还提款 outstanding drawing
未偿借款余额 outstanding balance of borrowed money
未偿票据 unpaid note
未偿清长期债款的折扣 unextinguished discount on funded debt
未偿外债 outstanding external debt
未偿余额 outstanding balance
未偿债额 outstanding balance
未偿债权 unliquidated claims
未偿债务 debt outstanding; outstanding debt; outstanding obligation; unliquidated debt; unliquidated obligations
未偿资本 outstanding capital
未超载的 unsupercharged
未车光的 non-machined
未车扣管子 unthreaded pipe; unthreaded tube
未撤消前有效 good-till-cancelled
未衬里金属管 unlined metallic pipe; unlined metallic tube
未衬砌的 unlined
未衬砌墙 unlined wall
未衬砌隧道 rough tunnel
未称量的 unweighted
未成部分 fragment
未成年 infancy; minority; under age
未成年工 worker under age
未成品 unfinished product; unfinished work
未成熟沉积物 immature sediments
未成熟虫 adultoid
未成熟的 embryonic; half-baked; immature; rash; rudimentary; sucking; unseasoned; untimely
未成熟(的)土 immature soil
未成熟阶段 immature stage
未成熟节片 immature segment
未成熟木材 immature wood
未成熟体 adultoid
未成熟烃源岩的体积 volume of immature source rock
未成熟纤维 immature fiber [fibre]
未成熟性 immaturity
未成熟源岩 immature source rock
未成型的 unshaped; unfashioned; unformed
未承付的拨款 unobligated appropriation
未充电的 not charged; uncharged
未充电蓄电池 uncharged battery
未充电状态 uncharged state
未充分发展的 unformed
未充分分离纤维的木料 shive
未充分供应(商店等)存货 understock
未充分固结 underconsolidation
未充分固结黏[粘]土 underconsolidated clay
未充分开发的地区 underdeveloped region
未充分理解的 undigested
未充分利用 underutilization
未充分利用的人力 underutilization manpower
未充分利用的土地 underimproved land; underutilized land
未充分养护的 undercured
未充分照射 underexpose
未充满 underfill(ling)
未充满孔型 underfilled pass
未充气的 unaerated; uninflated
未充填 non-filling
未充填垛式支架 unfilled cog
未充填裂隙 unfilled fissure
未冲刷的 unwashed
未冲销 uncovered
未抽出的 unpumped
未抽提完的 unspent
未稠合的多环烷烃 non-condensed polycycloalkane
未稠合多环芳香烃 uncondensed polycyclic aromatic
未出版的 inedited
未出版的报告 unpublished report
未出版的馆藏稿本指南 unpublished guides to manuscript collection
未出版的科研资料 unpublished research materials
未出版的数据资料 unpublished data
未出版的数据资料线索 unpublished data sources
未出版文献 unpublished documents
未出售的 undigested
未除尽锈斑 residual rust
未处理的 raw; unprocessed; unrendered; untreated
未处理的玻璃织物 untreated glass fabric
未处理的木材 untreated timber
未处理的污泥 crude sludge
未处理的污染水 contaminant non-process wastewater
未处理的线圈 untreated coil
未处理的油 untreated oil
未处理的淤渣 raw sludge
未处理的织物 raw cloth; untreated cloth; untreated fabric
未处理地基 unimproved footing; untreated foundation
未处理废气 raw exhaust
未处理废水 crude wastewater; untreated effluent; untreated wastes
未处理废水排放 discharge of untreated effluent
未处理骨料 untreated aggregate
未处理过的负荷指数 untreated loading index
未处理过的河水 untreated river water
未处理过的木材 unseasoned lumber [timber/wood]
未处理过的水 untreated water
未处理过的污泥 untreated sludge
未处理过的污水 untreated effluent; untreated sewage; untreated wastewater
未处理过的污水排放 discharge of untreated effluent
未处理矿石 crude ore
未处理砾石路 <指未经结合料处治> untreated gravel road
未处理木桩 <指未经防腐处理> untreated timber pile
未处理排放物 untreated effluent
未处理汽油 untreated gasoline
未处理数据 unprocessed data
未处理水 untreated water
未处理土 raw soil
未处理土地 unimproved land
未处理污泥 primary sludge; raw sludge
未处理污水 crude sewage
未处理信息 unprocessed information
未处理原油 untreated oil
未处治的 uncured; untreated
未处治底基层 untreated subbase
未处治骨料 untreated aggregate
未处治集料 untreated aggregate
未触动过的 untouched
未穿点 breakaway point; breaking down point
未穿过孔的套管 blank liner
未穿孔的 unperforated; unpunched
未穿透层 <钻探中的> refusal
未穿透的孔 blind hole
未纯化的 unpurified
未催缴股款 uncalled subscription
未淬火的 non-hardened; unhardened; untempered
未存储区 clear area
未存款 undeposit cash
未达标准规定的间隙 wrong clearance
未达到目的 miss the mark
未达到容许应力 distress
未达损益 unremitted profit and loss
未达预定点 undershoot
未达账项 account in transit
未答复的 unresponsive
未打过孔的 non-perforated
未打磨的 unground
未打气的 uninflated

未捣实的<混凝土> loosely spread; unrammed
未捣实混凝土 incomplete compaction concrete; loosely spread concrete; uncompacted concrete; unrammed concrete
未捣实新拌混凝土 uncompacted fresh concrete
未到达 stop short of stop
未到的货 floating cargo
未到货 no delivery
未到价合约 out of the money
未到交叉口的停靠站 near-side stop
未到龄期混凝土 green concrete
未到期保险 unexpired insurance
未到期保险费 unexpired premium
未到期的 unmatured
未到期的长期借款<期限在一年以上> funded debt unmatured
未到期付款 prematurity payment
未到期票据 bill undue; undue bill; undue note
未到期债务 undue debt; unmatured debt
未到商品账户 goods to arrive account
未到支付期的 undue
未得标 lose a bid
未得标人 unsuccessful bidder
未得标商 unsuccessful bidder
未得到安置的 unplaced
未得到的收入 unearned income; unearned revenue
未得到满足的需求 backlog of demand
未得通知的 uninformed
未得支持的 unsupported
未登记的 unlisted; unrecorded; unregistered
未登记的契约 inchoate instrument
未登记借款 off-record loans
未登记资产 unregistered assets
未登录的图书 unrecorded book
未登名上册的 unlisted
未抵到达站前 short of destination
未抵押资产 unmortgaged assets; unpledged assets
未缔合的 unassociated
未点着的 unfired
未雕刻的 uncut
未雕凿石材 self-faced stone
未雕凿石料 self-faced stone
未雕琢的 unkempt
未订合同 not in contract
未定<指所有权> abeyance
未定标脉冲 unscaled pulse
未定长游丝 unvibrated hairspring
未定乘数 undetermined multiplier
未定带 incertitude zone
未定的 problematic(al); undeterminable
未定地位 incertae sedis
未定点<未与控制网连测的点> unfixed point
未定费率 not otherwise rated
未定工程项目 indefinite projects
未定国界 international undefined boundary; undefinite boundary; undetermined boundary
未定级职位 ungraded position
未定价 disputed boundary
未定价的 unpriced
未定角的 uncast
未定界 indefinite boundary; undefined boundary; undemarcated boundary; undeterminated boundary
未定界记录 undefined record
未定界线 alignment uncertain boundary; contested boundary; disputed undefined boundary
未定界限 alignment uncertain boundary; contested boundary; disputed undefined boundary; undeterminate boundary
未定流动资本 working capital suspense
未定日期的 undated
未定误差 indeterminate error
未定系数 indeterminate coefficient; undeterminate coefficient; undetermined coefficient
未定系数法 method of undetermined coefficients
未定向的 unoriented
未定项目 unspecified item
未定形式 indeterminate form
未定型流 unestablished flow
未定义 undefine
未定义符号 undefined symbol
未定义记录 undefined record
未定义数据 undefined data
未定义形式 undefined format
未定义指令 undefined instruction
未定值 undetermined value
未定值保险单 unvalued policy
未动用的期间 period of immobilization
未动用资本 unimpaired capital
未动用资金 unused fund
未冻(结)含水量 non-frosted water content; unfrozen water content
未读出 non-readout
未镀锌薄板 ungalvanized plate
未镀锌的 ungalvanized
未镀锌钢板 ungalvanized steel plate
未镀锌钢铁 ungalvanized iron and steel
未镀锌铁板 black iron sheet
未镀锌线 ungalvanized wire
未堆起来的 unbanked
未对中的 misaligned
未对准 malalignment; misalignment; mismatching; misregistration; misregistry
未兑现支票 outstanding check; outstanding cheque; uncashed check
未恶化路面 non-deteriorated pavement
未发爆 misfired detonation
未发表的 inedited; undelivered
未发表的资料 unpublished materials
未发表过的 unpublished
未发表过的原稿或手稿 unpublished manuscript
未发表资料 unpublished data
未发材料的领料单 requisition unfilled
未发订货单 unfilled order
未发定货单 unfilled order
未发好 underexhaustion
未发货订单 unfilled orders
未发觉的损失 undiscovered loss
未发累积股利 accumulated dividend
未发射(出去)的 unfired
未发现储量 undiscovered reserves
未发现的故障 undetected failure
未发现的损失 undiscovered loss
未发现误差率 undetected error rate
未发现资源量 undiscovery resources
未发行抵押债券 unissued mortgage bonds
未发行股本 unissued capital stock
未发行债券 unissued bond
未发育土层 azonal soil
未发展的 undeveloped
未伐倒的树木 standing timber
未伐商品材 stumpage
未反应的 unreacted
未反应的基团 unreacted radical
未反应状态 unreacted state
未防腐枕木 untreated sleeper
未放大的 unamplified
未放大频率叠加 unamplified frequency overlay
未放牧地区 ungrazed area
未放气的 unvented
未分辨出的峰 unresolved peaks
未分辨的共振 unresolved resonance
未分拨经费 unallotted apportionments
未分大小 unsizing
未分大小的 unsized
未分带的 unzoned
未分的 undivided
未分等级的(锯)<木材> mill run
未分发的经费 unallotted apportionments
未分化 undifferentiation
未分化的 prototaxic; undifferentiated
未分化型 undifferentiated type
未分化型族 undifferentiated race
未分级储量 unclassified reserves
未分级的 un(as)sorted; ungraded; unscreened
未分级开挖 unclassified excavation
未分级矿石 unsized ore
未分解的 uncomposed; undecomposed
未分解的络合物混合物 unresolved complex mixture
未分解的烃类络合物混合物 unresolved complex mixture of hydrocarbon
未分解炸药 undisturbed explosive
未分界的 undemarcated
未分开的片页块云母 booked mica
未分类的 non-sorted; unclassified; unsorted
未分类的砖或瓦 kiln run
未分类河流 unclassified river; unclassified stream
未分类土壤 taxadjunct; unclassified excavation
未分离的 unsegregated
未分利润 undivided profit
未分配成本 unabsorbed cost
未分配的 unappropriated
未分配的利润 undivided profit
未分配费用 expense not allocated; unapplied expenses
未分配利润 retained earning; unappropriated profit; undistributed profit
未分配利润审计 undistributed profits audit
未分配利润税 tax on undistributed profit
未分配收入 unapplied income
未分配盈利 undistributed profit
未分配盈余 undivided surplus
未分配余额 unappropriated balance
未分配预算结余 unappropriated budget surplus
未分品级矿石 undivided ore
未分平纹或竖纹的混合木材 mixed grain lumber
未分区的 unzoned
未分散的小颗粒 speck
未分摊的核定款项 un-assessed authorization
未分选的 non-sorted; unsorted
未分选的石灰 run-of-kiln lime
未分异的 aschistic; undifferentiated
未分异火成岩 undifferentiated igneous rock
未分异岩 aschistite
未分异岩墙 undifferentiated dike rock
未分赢余账 unappropriated surplus account
未分组资料 ungrouped data
未粉刷的 unplastered
未粉刷的顶棚石膏板 unplastered gypsum ceiling plasterboard
未粉刷的石膏板 unplastered plate
未粉刷的石膏墙板 unplastered gypsum wallboard
未风干的 unseasoned
未风化基岩 fresh bedrock
未风化土 unweathered soil
未风化岩石 fresh rock; unweathered rock
未封闭等高线 open contour; unclosed contour
未封冻河道 open river
未封冻河口 open estuary
未封冻水面 open water; unfrozen water surface
未封缝 unsealed joint
未封锁的道路 unblocked road
未封装的片子 naked chip
未封装源 unencapsulated source
未缝合的 unstitched
未敷膜的 unfilmed
未腐败的 uncorrupted
未腐解有机物质 unhumified organic matter
未腐烂的 unputrefied; unrotten
未腐蚀的 unattacked
未腐殖化的 unhumified
未腐殖化的有机物 unhumified organic matter
未付差额 unpaid balance
未付到期负债 matured liabilities unpaid
未付的 owing; unliquidated; unpaid
未付的尾数 payment in arrear
未付的尾数余额 arrears
未付的应付款 outstanding dues
未付费用 outstanding expenses; unpaid expenses
未付工资 unclaimed wage funds; unpaid wage
未付股本 unpaid stock; unpaid up capital
未付股息 unpaid dividend
未付关税的 uncustomed
未付红利 unclaimed dividend
未付款 default of payment; outstandings
未付款补偿 compensation for nonpayment
未付赔款 unpaid claims
未付票据 unpaid bill
未付讫的账款 unpaid account
未付清金额 amount outstanding
未付清赔款准备金 outstanding loss reserves; reserve for outstanding losses
未付税款 unpaid tax
未付特殊项目前的收益 income before extraordinary item
未付尾数 arrears
未付已宣布股利 unpaid declared dividend
未付账款 outstanding account
未付账目 unpaid account
未付支票 unpaid check; unpaid cheque
未付资本 unpaid up capital
未付资本账(目) unpaid up capital account
未付资金报酬 unpaid returns
未复配洗涤剂 unbolt detergent
未富化的 unenriched
未覆盖的 uncovering; unlapped; unmulched
未覆盖工程 open construction
未改变的 unaltered; unchanged
未改建的坝 unmodified dam
未改良路地面 unimproved surface
未改良土地 unclaimed field
未改良型 unimproved type

W

未改善的 unimproved
未改善的土路 primitive road
未改善(的)土地 unimproved land; raw land
未改型飞机 unmodified aircraft
未改性醇酸树脂 straight alkyd resin
未改性的 unmodified
未改性的纤维素 unmodified cellulose
未改性树脂 unmodified resin
未改正的 unrectified
未改正频谱 uncorrected spectrum
未盖戳的 unstamped
未盖章的 unstamped
未干的 tacky dry
未干的湿混凝土 wet concrete
未干的砖砌体 green brick work
未干灰浆 green mortar
未干木材 fresh wood;green lumber
未干扰 0 输出信号 undisturbed-zero output signal
未干色 wet colo(u)r
未干砂浆 green mortar;green plaster
未干透的 unseasoned
未干油漆 fresh paint
未干燥的 undried;unseasoned
未干燥的小片 undried pellets
未干燥(木)材 green lumber; green timber; green wood; unseasoned lumber [timber/wood]
未隔成小间的淋浴室 shower room without cubicle
未隔开的 unisolated
未耕的 untilled
未耕地 uncultivated field; uncultivated land;untilled land
未耕过地区 area bare-cultivated
未耕土地 unploughed land;unplowed land
未耕作的 uncultured
未耕作地 jungle
未公开承认的 unacknowledged
未攻丝管端 plain pipe-end
未共用电子对 unshared electron pairs
未固定的 unstayed
未固定沙 unfixed sand
未固化的树脂 green resin
未固结表层 unconsolidated surface layer
未固结不排水三轴试验 unconsolidated-undrained triaxial test
未固结沉积层 unconsolidated deposit
未固结沉积物 non-consolidated deposit; non-consolidated sediment; unconsolidated deposit; unconsolidated sediment
未固结的 unconsolidated
未固结的后张混凝土板 unconsolidated post-tensioned slab
未固结黏[粘]土 unconsolidated clay
未固结石灰岩 lime rock
未固结土 unconsolidated soil
未固结土料 unconsolidated material
未固结物质 unconsolidated material
未挂号 unregister
未挂机状态(电话)receiver-off-hook condition
未关的 unclosed
未观察到的 unobservable;unobserved
未灌溉的 non-irrigated
未灌溉农田 barani
未灌浆的碎石混合料 open stone mixture
未灌浆固结的后张预应力混凝土板 unbonded post-tensioned slab
未灌浆管道 ungrouted duct
未灌水 un-priming
未光面 unfaced surface
未归类的 unclassed
未归类资料 ungrouped data
未规定比特率 unspecified bit rate
未规定的 not specified;undefined
未过滤的 unfiltered; unscreened; unstrained
未过滤水 unfiltered water
未过期保险费 insurance premium unexpired;unexpired insurance premium
未过筛的 unscreened; unsifted; unsized
未过筛孔石料 by-passed stone; unscreened gravel
未过筛矿石 unsized ore
未过筛砾石 bank gravel; pit(-run) gravel; unscreened gravel; run-of-bank gravel
未过筛料坑砂 pit sand
未过筛砂 pit(-run) sand;unscreened sand
未过时的 non-ag(e)ing
未焊缝 cold lapping
未焊好的 unwelded
未焊接缝 unwelded joint
未焊满 incompletely filling;underfill
未焊满的弧坑 unfilled crater
未焊满的节点 joint bridging
未焊透 faulty fusion;inadequate penetration; incomplete fusion; incomplete penetration; lack of penetration;non-penetration;spilly place
未夯实的 unrammed;untamped
未夯实的回填土 untamped backfill
未夯实混凝土 unrammed concrete
未夯实土吸水能力 field moisture capacity;field moisture equivalent
未耗成本 unexpired cost
未耗用费用 unexpired expenses
未核定的补给品项目 unauthorized item
未核实的产量 uncorrected production
未核准印刷的 unlicensed
未核准资产 non-admitted assets
未烘干砖 unbaked brick
未烘烧的 unbaked
未烘透的 soggy;underbaked
未化合的 uncombined
未划分的 unzoned
未划分区域的 unzoned
未划分土壤 taxadjunct
未还放款 outstanding advance
未还原的 unreduced
未还原芒硝 unreduced salt cake
未还账 account en route
未缓解的 unrelieved
未换位 untransposition
未磺化的 unsulfonated
未恢复的 unrecovered
未回归结合 unbound
未回结淤泥 unconsolidated mud
未回收成本 unrecovered cost
未回收投资差额 unrecorded investment balance
未婚人寓所 bachelor quarters
未混合的 unblended;unmixed;unmixing
未混合区 unmixed zone
未活化的 non-activated;unactivated
未活化的胶料 non-activated stock
未活化态 unactivated state
未获情报的 uninformed
未获专利权的 unpatented
未激活的 unactivated
未极化的 unpolarized
未集合的 unassembled
未计划用地 undersignated area
未计及的损失 unaccounted for loss
未计价燃气量 unbilled gas;unred gas
未计量封闭水 unmetered consumption
未计入损失 uncountable for loss;uncounted for loss;unmeasured loss
未计数 no-count
未计数的 unnumbered
未计水量 uncounted for water
未记号的 unnumbered
未记录 unrecorded
未加保护的 unprotected
未加保护的孔口 unprotected opening
未加保险丝的 unsafetied
未加标签的 untagged
未加防护的火焰灯 open flame light
未加封的 unsealed
未加封土样 unsealed sample
未加工表面 green surface
未加工材料 rough stock;unprocessed material
未加工的 coarse;crude;in the rough; raw; rough; rude; undressed; unfashioned;unprocessed;unworked
未加工的拌和料 raw mix
未加工的窗槛 rough sill
未加工的窗台板 rough sill
未加工的大理石 raw marble
未加工的甘油 raw glycerine
未加工的混合料 raw mix; unprocessed mix
未加工的浇筑玻璃板 rough cast plate;rough plate glass
未加工的门槛 rough sill
未加工的木料 raw timber
未加工的黏[粘]土 raw clay
未加工的软木 raw cork
未加工的石场骨料 pit-run aggregate
未加工的天然气 raw natural gas
未加工的天然汽油 raw natural gasoline
未加工的小山羊皮 undressed kid
未加工的亚麻籽油 raw linseed oil
未加工的圆木 rough log
未加工过石料 raw stone
未加工焊缝 undressed weld
未加工晶体 unworked crystal
未加工精制的气体 crude gas
未加工木材 unwrought timber
未加工木杆 rough lumber;rough spar
未加工木料 unmanufactured wood; unprocessed timber; unwrought timber
未加工木桅 rough mast
未加工石料 crude rock
未加工图书 unprocessed book
未加工纤维 raw yarn
未加工叶板 natural acanthus leaf
未加工原木材 rough timber
未加工状况 raw condition
未加工状态 raw condition
未加固 unreinforcement
未加固的 unreinforced;unstabilized; unstayed
未加固的路肩 earth shoulder
未加固水井 uncased well
未加规定的 unspecified
未加劲边跨板 unstiffened edge slab
未加劲的 unstiffened
未加劲构件 unstiffened member
未加劲式 unstiffened type
未加劲悬索桥 unstiffened suspension bridge
未加括号信号 unbracketing signal
未加拉力的钢筋 untensioned bar reinforcement;untensioned steel(reinforcement);untensioned wire
未加利用资本 dormant capital
未加裂化油的油料 clean oil
未加氯处理的饮用水 unchlorinated drinking water
未加偏压的 unbias(s)ed
未加气混凝土 non-air-entrained concrete
未加铅的 unleaded
未加铅汽油辛烷值 clear octane number
未加强的 unfortified; unreinforced; unstiffened;unstrengthened
未加强的弹性支座 plain pad bearing
未加权的 unweighted
未加权平均数 unweighted mean
未加权信噪比 unweighted signal-to-noise ratio
未加权噪声 unweighted noise
未加燃料的 unfed;unrefueled
未加热的 unheated
未加热的浸灰池 unheated lime
未加热空气干燥 unheated air drying
未加热空气式干燥机 unheated air drier [dryer]
未加热空气式干燥器 unheated air drier [dryer]
未加润滑脂润滑的 ungrease
未加塞的 uncork
未加石膏的水泥 unsulfated cement
未加使用的 unemployed
未加水拌和的商品混凝土干料 dry mix concrete
未加水拌和的商品砂浆干料 dry mix mortar
未加说明的 not elsewhere specified
未加四乙铅的汽油 unleaded gasoline
未加速的 unaccelerated
未加通风的 unventilated
未加稀释油的沥青 unfluxed asphalt
未加星号非终结符 unstarred non-terminal
未加压的 uninflated
未加压的石棉水泥墙板 non-compressed asbestos-cement panel
未加颜色的 uncolo(u)red
未加以注意的 unattended
未加抑制剂的油品 uninhibited oil
未加油漆的 unvarnished
未加约束的 unrestrained
未加整理的 unsorted
未加证明的 unvouched
未加证实的 unvouched
未加支撑的 unbraced;unstayed
未加装饰的 unadorned;unornamented
未加装修的毛窗匣 blind casing
未加籽晶的【地】unseeded
未监测到的污染源 unmonitored polluted sources
未拣过的 unpicked
未减低的 unsubdued
未减(轻)的 unabated
未减少的 unbated
未剪枝的 unpruned
未检波点线终了桩号 end stake number of the last receiver line
未检测出的 not detected
未检查 not examined
未检查过的 unsight
未检出 non-detection; non-readout; undetection
未检出的 not detectable
未检出浓度 no observed effect concentration
未检验出的 not examined
未见过的 unsight
未见矿钻孔 dry hole;negative drill hole
未建成的 unestablished
未建成面积 unbuilt area
未建成区 open district
未建成群落 unestablished coenosium
未建立水位-流量关系的测站 unrated ga(u)ging station
未建面积 unbuilt area
未建区(域)unbuilt area
未建造的 unbuilt
未建筑的 unbuilt
未鉴别的 unidentified
未鉴定的 unidentified
未鉴定的材料 unidentified material

W

未鉴定的产物 unidentified product
未鉴定放射性核素在水中的最高容许浓度 maximum permissible concentration of unidentified radionuclides in water
未降解的表面活性剂 non-degraded surfactant
未降解物质 undegraded material
未交的 uncalled
未交付 non-delivery
未交付的 undelivered
未交付的物品 undelivered article
未交付订货积累 backlog
未交货 failure to deliver the goods; non-delivery
未交货订单 outstanding order; unfilled order
未交货条款 failure to deliver clause
未交清订货 back order
未浇满 poured short
未浇透层 un-priming
未胶结冲积层 incoherent alluvium
未胶结的 unbonded; uncemented
未胶结的粉细砂岩 unconsolidated silt and fine sand type
未胶结的后张预应力混凝土板 unbonded post-tensional prestressing concrete slab
未胶结断层 uncemented fault
未胶结物质 incoherent material
未胶结岩石 loose rock; uncemented rock
未胶牢接头 hungry joint
未胶束化的 unmicellized
未搅拌的 unmixing
未搅动的 undisturbed
未搅动土 undisturbed soil
未搅动土样 undisturbed soil sample
未缴股本 capital stock unpaid; uncalled capital
未叫到的 uncalled
未校核高程 unchecked spot elevation
未校正保留体积 uncorrected retention volume
未校正的 out of adjustment; unregulated
未校正的读数 uncorrected reading
未校正的加速度记录 uncorrected accelerogram
未校正加速度记录 uncorrected acceleration
未校正延迟 uncorrected delay
未校准 misalignment
未校准的 unregulated
未校准分析 uncalibrated analysis
未校准增益控制 uncalibrated gain control
未接触 not in contact
未接地的 off-ground(ed); unearthed; ungrounded
未接合的 unassembled
未接合底盘 <汽车> unassembled chassis
未接通的呼叫 abandoned call; loss call
未接通的呼叫概率 lost-call probability
未接通信号 unobtainable tone
未接用户的自动交换机 hypothetical exchange
未结合补体的 uncomplemented
未结合的 unassembled; unbound
未结合的铬 unfixed chrome
未结合的燃料元件 unbonded fuel element
未结合水 uncombined water
未结合状态 unbound state
未结晶的 uncrystallized
未结平账户 open account
未结清的 unsettled
未结清期货合同 open position
未结清权益 open interest
未结清账户 open account
未结束的 unbalanced; unclosed
未结算成本 unexpired cost
未结算的 unliquidated
未结算账簿 unbalanced book
未结算账目 open account; outstanding account; outstandings
未结硬的混凝土 fresh concrete; green concrete; immature concrete
未结硬的砂浆 green mortar
未结硬的水泥砂浆 fresh cement mortar
未结硬混凝土 wet concrete
未结硬污泥 fresh sludge
未结账销量 unbilled sale
未解除保险 unarming
未解除保险的传爆系统 unarmed explosive train
未解除债务之破产人 undischarged bankrupt
未解决的 unresolved; unsolved
未解决的赔款 outstanding claim; outstanding loss
未解决的问题 open question; outstanding problem
未解决开采条件 unresulted productive condition
未解决冶炼技术 unresulted metallurgic(al) technique
未解离分子 undissociated molecule
未解释的 unsolved
未解释的变量 unexplained deviation
未紧固的 unfixed; unsecured
未尽的 unexpired
未尽寿命 unexpired life
未尽义务 unfulfilled obligations
未进行回填封闭 unstuffing and unsealing
未进行机械加工的 unmachined
未进整的 unrounded
未浸沥青的 untarred
未浸沥青的黄麻细纱 untarred jute spun yarn
未浸染的 unimpregnated
未浸透的 unsoaked
未浸渍的 untreated
未浸渍电杆 untreated pole
未经背书支票 unindorsed check
未经焙烧的 unfired; unroasted
未经编辑的 inedited
未经变质处理的 uninoculated
未经测绘的 uncharted
未经筹划的 unplanned
未经处理的 undressed; untreated
未经处理的废水 raw effluent; raw wastewater; untreated wastewater
未经处理的路面 untreated surface
未经处理的面层 untreated surface
未经处理的木杆 untreated wood pole
未经处理的生活废水 raw sanitary water
未经处理的生活污水 crude sewage
未经处理的水 raw water
未经处理的图像 raw video picture
未经处理的污泥 raw sludge
未经处理的污水 crude wastewater
未经处理的污水污泥 raw sewage sludge
未经处理过的枕木 untreated sleeper
未经处理散发的污染空气 fugitive emission
未经春化处理的 unvernalized
未经定罪的人 unconvicted person
未经煅烧的块材 unburned block
未经煅烧的砖 unburned tile; unfired brick
未经发现资源 undiscovered resources
未经防腐处理的电杆 untreated pole
未经防腐处理的木材 untreated timber
未经防腐处理的木桩 untreated timber pile
未经防腐处理的枕木 untreated sleeper; untreated tie
未经分类的 unsorted
未经分类的开挖 unclassified excavation
未经风化的 unweathered
未经改变的 unadapted
未经改善土地的租赁 net ground lease
未经干扰的 undisturbed
未经过鞣制加工 untanned
未经过挑分的废料 unsorted scrap
未经海关通过的 uncustomed
未经核实的 unchecked
未经核准印刷的 unlicensed
未经机械加工的 non-machined
未经加工的 crude; inartificial; unwrought
未经加工的粗糙锯面 scuffled
未经加工的木材 unwrought timber
未经加工的木料 unwrought log; unwrought timber
未经加固岩石 non-consolidated rock
未经检查的 uninspected
未经检验的 uninspected; unsighted
未经检验的产品 off-test product
未经鉴定的产品 unidentified product
未经校核的 unchecked
未经浸制的电杆 non-treated pole
未经精密设计的造船 shadow building
未经精选的矿石 unbeneficiated ore
未经净化储水池 raw water storage basin
未经净化的水 raw water
未经净化污水 raw sewage
未经开发的 unexploited
未经开发的大片地产 acreage property
未经勘察的 unexplored
未经抗震设计的 non-resistant to earthquake
未经粒度分级的 unscreened
未经率定的 unrated
未经媒染的 unmoradanted
未经(黏[粘]合料)处理的砾石路 untreated gravel road
未经培训的 untrained
未经碰撞的 uncollided; undegraded
未经碰撞辐射 uncollided radiation
未经批准的 off-hand; unauthorized
未经批准的项目 unauthorized project
未经漂白的 unbleached
未经平整的储存区 unimproved area
未经破碎的 uncrushed
未经清理的铸件 raw casting
未经确认的 unconfirmed
未经热处理的 non-heat treated
未经(热)处理铸件 greensand casting
未经任何处理的垃圾 raw refuse
未经任何处理的污水 raw sewage
未经筛分的 ungarbled
未经筛分的集料 all-in aggregate
未经筛洗的天然砾石 as-raised gravel
未经审验许可使用的车辆 unauthorized vehicle
未经时效处理的 non-ag(e)ing
未经实测的 unsurveyed
未经试验的 off-test; untested
未经手工抹光的喷浆面层 gun finish
未经授权的 unwarranted
未经数值黏[粘]结的玻璃棉 unbonded glass wool
未经探测的 unplumbed; unsounded
未经探察的地区 unexplored area
未经特别规定的物品 article not specifically provided for
未经提炼的情报 unrefined information
未经挑选的 unsorted
未经同意的 unapproved
未经挖动的地面 unmade ground
未经挖动的断面 undistorted ground
未经挖动的土地 undistorted ground
未经详细检查的 unsifted
未经协调的 uncoordinated
未经修正的航测拼图 rough mosaic
未经修正的起飞滑跑距离 unfactored takeoff distance
未经修琢的 in the rough
未经许可的 unauthorized
未经宣布的 unannounced; undeclared
未经选分的 unsorted
未经选择的 unselected
未经训练的 raw; untrained
未经压光处理的 uncalendered
未经研磨的材料 unground material
未经阅读的 unread
未经照射的 undosed; unexposed; unirradiated
未经振捣的新浇混凝土 non-vibrated fresh concrete
未经整理的统计数字 crude statistics
未经证实的 unproven; unsupported
未经装饰的 undecorated
未经准直的 uncollimated
未经灼烧的 non-ignited
未经琢磨的 uncut
未精测滨线 coast imperfectly known; shoreline unsurveyed
未精测等高线 approximate contour line
未精测等深线 approximate depth contour
未精测地区界线 limit of unsurveyed area
未精测海岸线 coast imperfectly known; shoreline unsurveyed
未精加工的 unfinished
未精炼的 unrefined
未精整锻件 platter
未精制材料 unrefined material
未精制的 unpurified; unrefined
未精制的天然气 raw natural gas
未精制石蜡 paraffin(e) scale
未精制油 unrefined oil
未精制油料 raw oil; unrefined oil matter
未净化的发生炉煤气 crude producer gas; raw crude producer gas; raw producer gas
未净化的汽车排气 raw auto exhaust
未净化的水 raw water
未净化的污水 raw sewage
未净化的蓄水池 raw water storage basin
未净化废气 raw exhaust
未净化空气 unpurified air
未净化气(体) raw gas
未净化水 raw water
未净化水的隧道 raw water tunnel
未净化水供应 raw water supply
未净化水水网 raw water network
未净化水箱 raw water tank
未净化污泥 raw sludge
未净化污水 non-purified wastewater
未纠正的照片镶嵌图 aerophotographic(al) sketch
未纠正底片 unrectified negative
未纠正镶嵌图 unrectified mosaic
未纠正像片 unrectified photograph
未就业的 unemployed
未就业人口 unemployed population; unoccupied population
未锯开的状况 in the log
未锯木材 unsawn timber; whole timber
未聚合的 unpolymerized

未聚合的单体 unconverted monomer
未聚焦处理 unfocused processing
未聚焦激光器 unfocused laser
未聚束 unbunched
未聚于一点 uncentering
未卷绕的 unwound
未决策 non-decision-making
未决(定)的 suspensive
未决定的设计方案 open plan
未决犯 culprit
未决赔款 losses outstanding;outstanding claim;outstanding loss
未决赔款准备金 reserve for outstanding losses
未决诉讼 pending claim
未决算的账目 open account
未决问题 open question
未决应收账款 suspended accounts receivable
未决账项 unsettled account
未绝热 uninsulation
未绝缘的 uninsulated;unisolated
未均衡河流 ungraded river;ungraded stream
未竣工的孔洞 rough opening
未竣工工程 uncompleted construction
未卡死的 unjammed
未开辟地区 unbeaten track
未开采的 unmined;unquarried
未开采的矿区 unmined field;unquarried field;virgin field;maiden field
未开动的 unactivated
未开发的 sloven;uncivilized;undeveloped;untapped
未开发的并且城市规划也无明确安排的土地 white land
未开发的地区 undeveloped land
未开发的森林 virgin forest
未开发的水能 undeveloped water power
未开发的土地 undeveloped estate;nonuse of land
未开发的渔场 unfished ground
未开发的种群 future expected population
未开发的资源 untapped reserves
未开发地 undeveloped land;unimproved land
未开发国家 undeveloped country
未开发区(域) undeveloped area;unexploited area
未开发区域规划 country planning
未开发水头 undeveloped head
未开发水域 undeveloped waters;unexploited waters
未开发探明储量 proved undeveloped reserves
未开发资源 dormant resource;undeveloped resources;untapped resources
未开放的 unopened
未开放港口 close port
未开封包装 unbroken package
未开封的 unopened
未开化的 uncultivated;wild
未开垦的 fallow;sloven;unbroken;uncultivated;virgin
未开垦(的土)地 raw land;wild land;primitive soil;fallow land;uncultivated land;unreclaimed field;virgin ground;virgin land;virgin soil
未开垦地区 unreclaimed region
未开孔板 solid plate
未开裂路段 uncracked section
未开裂区 uncracked zone
未开裂状态 uncracked state
未开坡口面 unbeveled face
未开(小)口的 unopened
未开账单工程 unbilled works
未开钻的 undrilled
未铠装电缆 inarmo(u)red cable
未勘测岸线 unsurveyed coastline;unsurveyed shoreline
未勘测河流 unsurveyed river;unsurveyed stream
未勘查地区 unexplored area;unexplored region
未勘探的 unproven
未勘探地区 unproven area
未勘探区 unprospected area
未砍伐的木材 stumpage
未砍削的 unhewn
未刻水位 final water level
未刻吸移管 plain pipet(te)
未肯定的危险 risk suspended
未垦丛林地 bush
未垦地 original soil
未垦土壤 virgin soil
未控制区间入流 unga(u)ged local inflow
未扣除折旧的净收益 net income before depreciation
未矿化泡沫 unmineralized froth
未拉紧的 unstrained
未拉伸丝 undrawn yarn
未拉伸丝束 undrawn tow
未来保险的 future sale
未来标记 future label
未来场地使用 future site use
未来成本 future cost
未来城市 future city;visionary city
未来的 futuristic;prospective
未来的道路 future road
未来地址插入码 future address patch
未来发展趋势预测 alternative future
未来费用 future expenses
未来公共陆地移动电话系统 future public land mobile telephone system
未来光锥 future light cone
未来价值 future value
未来(交)通量 future traffic volume
未来进展方向 future thrust
未来(可能)投资者 would-be developer
未来派 <20 世纪建筑的> Futurism
未来派的 futuristic
未来派建筑 Futurist architecture
未来情景设想 future scenarios
未来人口环境系统 environmental system of future population
未来市场 future market
未来事物 futurity
未来投标者 prospective bidder;prospective tender
未来投资者 potential investor
未来学 futuristics;futurology
未来学家 futurist;futurologist
未来学展望 futurological perspectives
未来学者 futurologist
未来展览馆 <1929 年雅各布森与拉尔斯森所设计的展览馆,呈圆形,屋顶可升降直升机> House of the Future
未来值 future worth
未来主义 Futurism
未来租契 reversionary lease
未来作业 future activity
未离解的 undissociated
未离子化氨 unionized ammonia
未利用 non-utilization
未利用的 unimproved;unused
未利用的能力 unused capacity
未利用的(土)地 unused land;unused zone
未利用率 non-availability
未利用热量 unused heat
未利用时间 unused time
未利用水 unused water
未利用水能 undeveloped water power;unused water energy
未利用土地税 unused land tax
未利用资金 dormant capital
未利用资源 idle resources;unused resources
未励磁的 unheated
未联结的 unbound
未晾干的 green
未量化的图像 unquantized picture
未了结的 unsettled
未了事宜 unfinished business
未了责任到期满为止 running off of a portfolio
未了责任转入 entry of portfolio
未了账目 old account
未列等级的道路 unclassified road
未列分项成本的标价 unbalanced bid (quotation)
未列名货物 open cargo
未列入舱单的货物 cargo unmanifested
未列入目录的 uncatalogued
未列入日程的 unscheduled
未列入运价表的 non-tariff
未列入运价表的车站 non-tariff station;station not included in tariff
未列限额的抵押 open mortgage
未列项目 not otherwise provided for
未列账收入 unrecorded income
未裂化残油 uncracked residue
未裂化的 uncracked
未裂化沥青 uncracked asphalt
未裂化烃 uncracked hydrocarbon
未裂开的 uncracked
未裂开的横截面 uncracked cross section
未淋溶木灰 unleached wood ashes
未领工资 unclaimed wages
未领工资卡片 unclaimed wage card
未领股利 unclaimed dividend
未另计数 not otherwise enumerated
未另列出 not otherwise provided/stated
未另列名 not otherwise enumerated;not otherwise provided
未另列明 not otherwise enumerated
未另说明 not otherwise specified
未另特别列出 not specially provided for
未留遗嘱 intestacy
未硫化的 unvulcanized
未露头基岩 concealed bedrock
未滤过的 unfiltered
未履行的 undischarged;unexecuted;unredeemed
未履行的义务 unfulfilled obligations
未履行合同 failure of performance;failure to perform
未履行责任的 on default
未率定测站 unrated ga(u)ging station
未率定的 unrated
未络合金属离子浓度 uncomplexed metal ions concentration
未络合离子 uncomplexed ions
未落实协议 <租船条款> subject open;subject ship being free
未埋石点 unfixed point
未埋石界桩 non-monumented boundary peg
未满轨道 uncompleted orbit
未满晶格结点 vacant lattice site
未满流的 part-full
未满能级 unfilled level
未满期保费 portfolio premium
未满期费用 unexpired expenses
未满期限 unexpired term
未满期业务转出 portfolio withdrawal
未满期业务转移 portfolio transfer
未满期责任 portfolio assumed
未满态 unfilled state
未满载 part cargo
未满载货柜 less than container load
未满状态 vacant state
未满足的 unsatisfied
未满足要求 backlog demand
未慢化的 unmoderated
未慢化束 undegraded beam
未铆紧铆钉 loose rivet
未醚化的 unetherified
未密封 unsealing
未密封的 unencapsulated;unsealed;untight
未灭菌的 unpasteurized
未明确说明的 unstated
未明水量 unaccounted for water
未磨成粉状的 unpulverized
未磨的粗金刚石 brait
未磨光的 unpolished
未磨光的原板玻璃 rough glass
未磨过的 unground
未磨碎树脂 unground resin
未磨损的 unworn
未抹灰板条墙 naked wall
未抹灰的 unplastered;unrendered
未抹灰的顶棚 unplastered ceiling
未抹灰的顶棚板 unplastered ceiling plate
未抹灰的墙 unplastered wall
未纳入成本的支出 unabsorbed cost
未能表达原意的 unexpressive
未能收回的成本 unrecovered cost
未能收回的金额 amounts not recovered
未能探测到海底 no bottom found
未能预见的 unforeseen
未能预见现场条件 unforeseen site condition
未黏[粘]合的 unbonded
未黏[粘]合的玻璃棉 unbonded wool
未黏[粘]合的棉毡 unbonded felt
未黏[粘]结的预应力钢筋 <后张法中> unbonded prestressed bar
未黏[粘]结骨料 unbound aggregate
未黏[粘]结集料 unbound aggregate
未黏[粘]着的 unbonded
未凝固的 not solidified;unset
未凝固的混凝土 as-placed concrete
未凝固稳定性 green stability
未凝结混凝土 immature concrete;unset concrete
未浓集的 unenriched
未浓缩的铀 unenriched uranium
未弄歪的 undistorted
未排出的 undischarged
未排出的岩粉 undischarged cuttings
未排放蓄水层 untapped aquifer
未旁路的 unshunted
未抛光的 unpolished
未抛光的木材 undressed lumber
未刨的木材 rough-sawn;unplaned timber
未刨光(木)材 unwrought timber
未刨木材 <只用钉子钉住的> fir fixed
未刨木料 undressed timber
未刨屋架木材 fir fixed
未配合的电子偶 unshared pair
未配软件的计算机 bare computer
未配弹簧制动器的踏面制动单元 tread brake unit without spring-loaded brake
未膨胀的 uninflated
未批准的 unsanctioned
未匹配 unmatch
未偏转的 undeflected
未偏转束 undeflected beam
未漂白的 ecru
未漂破布浆 unbleached rags

未拼接胶合板 unjoint plywood
未平仓合约 open interest
未平差的 out of adjustment; unadjusted
未平分的 unshared
未平衡摩擦系数 unbalanced friction factor
未评价的 unvalued
未屏蔽部分 unshielded part
未屏蔽的 unscreened
未破坏的 unbroken; undamaged
未破坏岩芯 well-drill(ed) sample
未破浪作用力 force of non-breaking wave
未破裂冰覆盖层的下游界限 downstream margin of unbroken ice cover
未破铅封 unbroken seal; undamaged seal
未破碎冰 unbroken ice
未破碎波 unbroken wave; non breaking wave
未破碎砾石 uncrushed gravel
未破碎砾石骨料 uncrushed gravel aggregate
未铺草皮的 unsodded
未铺草皮的边坡 unsodded slope
未铺草皮地 unsodded land
未铺管道的 unpaved
未铺楼板 naked flooring
未铺路面的 non-surfaced; unmetallized; unpaved; unsurfaced
未铺路面的道路 unsurfaced road
未铺(路)面的土基 unsurfaced soil base
未铺面层 naked flooring
未铺面层的地板 naked flooring
未铺砌的 unpaved
未铺砂石的 unballasted
未铺石渣的 unballasted
未铺石渣的桥面 unballasted deck
未铺筑道面的简易机场 unimproved airstrip; unpaved airstrip
未铺筑道面的跑道 unimproved runway
未铺装的 unpaved
未曝气池 unaerated lagoon
未起动 unstart
未起票行李 unbooked luggage
未气干木材 unseasoned lumber [timber/wood]
未砌合的 unbonded
未钎透 incomplete penetration
未签名本 unsigned copy
未签收的 unreceipted
未签收的票据 unreceipted note
未签字的 unsigned
未遣散的 undischarged
未嵌绕组铁芯 unwound core
未切边 untouched edge
未切边的规格 untrimmed size
未切断长丝 unsevered filament
未切断丝束 endless staple fibre [fiber]
未切口双金属摆轮 uncut bimetallic balance
未切削的 unhewn
未侵染组织 non-infected tissue
未侵蚀纤维 unetched fibre
未清偿的 undischarged; unliquidated
未清偿债券 bond outstanding
未清偿债务 liability outstanding; unliquidated claims; unliquidated debt; outstanding debt
未清股票分割 split
未清借款余额 outstanding balance of borrowed money
未清金额 outstanding account
未清理钢 crude steel
未清理铸件 undressed casting
未清欠账 open debt
未清算的损失 unliquidated damages
未清算账目 outstandings
未清摊款 outstanding contribution
未清洗的 unwashed
未清洗集料 dirty aggregate
未清洗未筛分的回填用砂 dead sand
未清余额 outstanding balance; unpaid balance
未清债务 open debt
未清账 open account
未清账款 outstanding account; unliquidated account
未清账目 outstanding account; unliquidated account
未曲解的 undistorted
未渠化河段 section without canalization
未取代的 unsubstituted
未取代的芳香烃 unsubstituted aromatics
未取得许可证的 unlicensed
未取任何资料的钻进 blind drilling
未取向区域 unoriented region
未取向纤维 unoriented fibre
未取样的 unsampled
未去皮木材 untrimmed timber
未圈的 unenclosed
未确定的 undetermined
未确定的索赔 un-asserted claim
未确定数额的贷款 open-end loan
未确定损害赔偿额 unliquidated damages
未确定装卸货日期 indeterminate lay days
未确认包 unacknowledged packet
未确认的信用证 unconfirmed credit; unconfirmed letter of credit
未燃部分 unburned part
未燃的 unburned; unburnt
未燃固体 unburnt solid
未燃尽的可燃物 unburned combustible
未燃尽可燃物 unburnt combustive
未燃尽气体 unburned gas
未燃垃圾 unburned refuse
未燃气体 unburned gas
未燃燃料 unburned fuel
未燃烧的 unburned; unfired
未燃烧的燃料 unburned fuel; unburnt fuel
未燃烧混合物 unburned mixture
未燃烧气体 unburned gas
未燃碳 unburned carbon
未燃碳氢化合物 unburned hydrocarbon
未燃烃 unburned hydrocarbon
未染色水泥 non-staining cement; white cement
未扰动 0 输出 undisturbed-zero output
未扰动 1 输出 undisturbed-one output
未扰动层理 strata in undisturbed condition
未扰动超固结黏[粘]土 intact overconsolidated clay
未扰动超固结土 intact overconsolidated soil
未扰动沉淀 undisturbed settling
未扰动沉降 undisturbed settling
未扰动的 intact; undisturbed; unperturbed
未扰动的牛顿黏[粘]度 undisturbed Newtonian viscosity
未扰动地基 intact ground; undisturbed foundation
未扰动碟状黏[粘]土样 undisturbed disk-shaped clay sample
未扰动轨道 unperturbed orbit
未扰动剪切强度 intact shear strength
未扰动结构 undisturbed structure
未扰动均质土样 undisturbed homogeneous sample
未扰动流 undisturbed flow
未扰动年龄 undisturbed ages
未扰动黏[粘]土 intact clay; undisturbed clay
未扰动取样 undisturbed sampling
未扰动砂样 undisturbed sand sample
未扰动射流 undisturbed jet
未扰动试件强度 undisturbed strength
未扰动试样 undisturbed sample; undisturbed soil sample
未扰动水 calm water; quiescent water; still water; undisturbed water
未扰动水流 undisturbed stream
未扰动水平面 undisturbed water level
未扰动水位 undisturbed water level
未扰动土(壤) undisturbed soil
未扰动土(石)方量 bank measure; bank yards
未扰动土样 undisturbed sample; undisturbed sample of soil; undisturbed soil sample
未扰动状态 undisturbed state
未扰动状态的强度 strength in undisturbed state
未扰乱流 undisturbed flow
未认购股本 unsubscribed capital stock
未认可的 unsanctioned
未溶解的余渣 insoluble residue
未熔穿 lack of penetration
未熔合 faulty fusion; incomplete fusion; lack of fusion
未熔合点焊接头 stick-out weld
未熔化的 unfused; unmelted
未熔炉料 unmelted charge
未熔透 lack of penetration
未入册的 unlisted
未入账的 unrecorded
未入账存款 unrecorded deposit
未入账费用 unrecorded expenses
未入账负债 unrecorded liabilities
未入账收入 unrecorded revenue
未润滑的 unlubricated
未润色的 unpolished
未杀菌的 unsterilized
未筛的 pit-run
未筛分道砟 all-in ballast
未筛分的 run-of-mine; unsieved; unsized
未筛(分的)机碎骨料 crusher-run aggregate
未筛分的块 unscreened lumps
未筛分的煤 unsized coal
未筛分的天然骨料 all-in aggregate
未筛分的天然砾石 pit run gravel
未筛分的土石料 pit-run
未筛分骨料 all-in aggregate; non-screened aggregate; pit-run aggregate; unsorted aggregate; as-dug aggregate
未筛分机轧碎石 crusher-run stone
未筛分矿石 pit-run ore
未筛分砾石 pit(-run) gravel
未筛分煤灰 unsorted ash
未筛分砂 pit-run sand
未筛分石灰 run-of-kiln lime
未筛分石料 mine run rock; quarry-run stone; run-of-pit stone; run-of-quarry material; run-of-quarry stone; run-of-quarry; unsorted aggregate
未筛分石屑 pit-run fines
未筛分碎石 all-in ballast; crusher-run
未筛分轧骨料 crusher-run stone aggregate
未筛分轧石料 crusher-run rock
未筛过的 unbolted; unscreened
未筛灰泥 unsifted plaster
未筛机碎集料 crusher-run aggregate
未筛集料 pit-run aggregate; as-dug aggregate
未筛砾石 unscreened gravel
未筛砾石道砟 pit-run gravel ballast
未筛料 pit-run material
未筛砂 pit-full sand
未筛试验 latter testing
未筛碎石 crushed-run rock; crusher-run stone
未筛碎石集料 crusher-run aggregate
未筛选采石场石料填筑体 quarry-run rockfill
未筛选的 unscreened
未筛选的材料 pit-run material; random material
未筛选的砾石 pit gravel; pit-run gravel
未筛选(的土石)料 random material
未筛选骨料 pit-run aggregate; as-dug aggregate
未筛选料 pit-run material
未筛选砂料 quarry-run sand
未筛选涂料 pit-run earth; pit-run material
未筛轧碎石料 run-of-crusher stone; crusher-run; crusher-run rock
未删的 full-scale
未删节的 unabridged
未删去的 unpruned
未上发条的 unwound
未上螺栓的 unbolted
未上炮架的 unmounted
未上漆的 unpainted
未上市的 unlisted
未上市股票 unlisted stock
未上市证券 unlisted securities
未上套管钻孔 bare foot
未上釉的 unglazed
未上釉的白瓦载体 Rako
未上釉的结构用砌面块 unglazed structural facing unit
未上釉的陶土(瓦)管 unglazed pipe
未上釉的压制方砖 unglazed pressed tile
未上釉建筑用饰面块 unglazed structural facing unit
未上釉(瓦)管 unglazed pipe
未烧成的黏[粘]土制品 white-hard
未烧的干珐琅层 biscuit
未烧的陶瓷坯 porcelain paste
未烧的砖 unburned brick
未烧过的 unburned; unburnt
未烧过的石灰石 unburned limestone
未烧结的 unsintered
未烧结的压块 green compact
未烧结坯 green pressing
未烧尽的可燃物 unburned combustible
未烧器皿 green ware
未烧透的 insufficiently burnt; unburnt; under-burnt; underfired; soft burned
未烧透的屋面平瓦 unburnt plain roof(ing) tile
未烧透黏[粘]土砖 half-burnt clay brick; half-fired clay brick
未烧透釉面瓦 glazed semiburnt tile
未烧透釉面砖 glazed semiburnt tile
未烧透砖 half-burnt brick; half-fired brick; pale brick; salmon brick; samel brick; soft brick; unburnt brick; unburnt tile; under-burned brick; under-burnt brick; underfired brick; grizzle; place brick; sandal brick
未烧油彩 cold colo(u)rs
未烧砖坯 green block; green brick; unfired brick
未设测站河流 unga(u)ged river; un-

W

ga(u)ged stream
未设测站流域 unga(u)ged watershed
未设防的 unfortified
未设防的城市 undefended city
未设防工事 openwork
未设防护装置的平交道口 unprotected level crossing
未设防建筑物 undefended building
未设信号的道口 unsignalized intersection
未设闸坝的河道 open river waterway
未设闸坝的河流 open river; free flowing stream
未设站点 unoccupied point
未设站流域 unga(u)ged basin
未审核的财务报表 un-audited financial statement
未审核凭单 un-audited voucher
未审核之财务报表 un-audited financial statement
未审批储量 reserve without ratification
未升高的 uninflated
未升起的 unrisen
未生锈的 unrusted
未声明的 unstated
未失时效的 non-ag(e)ing
未失效的 non-ag(e)ing
未施测河流 unga(u)ged river; unga(u)ged stream
未施测流域 unga(u)ged basin; unga(u)ged watershed
未施测区间入流 unga(u)ged local inflow
未施加应力的预应力钢索 unstressed tendon
未时效 non-ag(e)ing
未识别出的目标尖头信号 unidentified pip
未识别的 unidentified
未识别的空中目标 unknown aerial target
未识别的目标 unidentified target
未识别的目标回波 unclassified track; unidentified track; unknown track
未识别的物体 unidentified sight
未实施的 unenforced; unimplemented
未实现固定资产估价增值 unrealized increment by appraisal of fixed assets
未实现利润 unrealized profit
未实现毛利 unrealized gross margin; unrealized gross profit
未实现收入 unrealized income; unrealized revenue
未实现收益 unrealized income
未实现损益 unrealized gains or losses
未实现盈利 unrealized profit
未实现增值 unrealized appreciation
未实现资产增价准备 reserve for unrealized increment in assets
未实现资产重估增值 unrealized appreciation from valuation of assets
未实行的 unexecuted
未实行过的 unpractised
未蚀变岩石 unaltered rock
未蚀变样品 unaltered sample
未使用材料 virgin material
未使用的 unavailable
未使用的起重机 unworked crane
未使用房间 vacant room
未使用固定资产 unused fixed assets
未使用过的润滑脂 unworked grease
未使用率 unavailability
未使用土地 vacant land
未试验 not test
未试验的 untested
未适当冷却的水泥 hot cement
未收保费 uncollected premium
未收到的 unreceived
未收的 uncalled
未收割的干草 standing hay
未收割的作物 standing crop
未收股款 uncalled capital; uncalled shares
未收回成本 unrecovered cost
未收回的 unreclaimed
未收回资本 capital uncalled
未收货款的卖主 unpaid seller
未收口的 unreefed
未收凭单 uncollected voucher
未收凭证 uncollected voucher
未收资本 capital uncalled; uncalled capital
未受潮的 undamped
未受处罚的 unpunished
未受玷污的 unsullied
未受风吹的 unblown
未受腐蚀的 uncorroded
未受干扰的 undisturbed
未受过训练的 untrained
未受警告的 unwarned
未受空气污染的地热流体 air-free geothermal fluid
未受磨损的 unworn
未受培养的 uncultivated
未受破坏的 unspoilt
未受扰的 unperturbed
未受扰动的地热田 undisturbed field
未受扰动等离子体 undisturbed plasma
未受扰林 undisturbed forest
未受(热)影响区 unaffected zone
未受伤害的 unharmed
未受赏识的 unappreciated
未受损船 intact ship; undamaged ship
未受损害的 unimpaired
未受损伤的 undamaged; uninjured
未受损失的 unscathed
未受污染 unpollution
未受污染的 uncontaminated; unpolluted
未受污染的鱼群 good fish population
未受污染河流 uncontaminated river; uncontaminated stream; unpolluted river; unpolluted stream
未受压的纤维板 non-compressed fiberboard
未受抑制的 unchecked; untamed
未受应力的 unstressed
未受影响的 unaffected
未受影响的母材 unaffected base metal
未受支持的 unheeded
未受注意的 unheeded; unmarked; unobserved
未售出的 unplaced
未疏干边坡 undrained slope
未熟的 raw; unbaked; unmatured; unripened
未熟阶段 immaturity
未熟练的 unskilled
未熟黏[粘]胶 unripened viscose
未熟土 immature soil
未衰减的 unabated
未水合离子 unhydrated ion
未水化的 unhydrated
未水化的灰浆 unhydrated plaster
未水化的水泥 neat cement
未水化灰泥 unhydrous plaster
未水化水泥 unhydrated cement; wasted cement
未税货物转口许可单 bill of sufferance
未说明标识符 undeclared identifier
未说明用途的支出 expenditure for unexplained purpose
未撕开的 unriven
未四舍五入的数据 unrounded data
未塑化的聚氯乙烯 unplasticized polyvinyl chloride
未塑炼生胶 unmasticated rubber
未酸洗部分 <钢材上的> black patch
未酸洗的 unpickled
未碎波 unbroken wave
未碎垡片 unbroken furrow
未损伤的 unmarred
未缩短的 full-length
未锁道岔 unlocked switch
未摊还建造期利息 unamortized interest during construction
未摊铺混凝土 heap(ed) concrete
未摊销的 unamortized
未摊销费用 unamortized expenses
未摊销债券贴现 bond discount unamortized
未摊销债券溢价 bond premium unamortized
未摊销债券折价 bond discount unamortized
未探明的热储 undiscovered reservoir
未探明区 unproven area
未提出的 unrendered
未提纯的苏打 sal soda
未提货物 failure to take delivery
未提炼的 unrefined
未提取的汇出汇款 remittance and draft outstanding
未提取货物 undelivered cargo
未提折旧成本 undepreciated cost
未提折旧余额 undepreciated balance
未填补裂缝 yawn
未填充的 unfilled
未填满电子能级 vacant electron site
未填实的 untamped
未填收货人的提单 open bill of lading [open B/L]
未填图区 unmapping area
未调波 unmodulated waves
未调合油料 unblended oil
未调和的 <石灰等> untempered
未调节的 unadjusted; unmodulated; unregulated
未调节的洪水流量 unregulated flood flow
未调节的流量 natural flow; non-regulated discharge; unregulated discharge
未调节电压 unregulated voltage
未调节洪流 unregulated flood flow
未调节流量 non-regulated flow
未调节水流 natural flow
未调纹 blank groove
未调谐的 untuned
未调谐的长反射器 untuned rope
未调谐电路 untuned circuit
未调谐天线 untuned aerial
未调信号 unmodulated signal
未调匀的 unequable
未调整贷项 unadjusted credit
未调整的 unregulated
未调整借项 unadjusted debits
未调整投资回收方法 unadjusted return on investment method
未调制等幅波 continuous waves unmodulated
未调制频率 idle frequency
未调制射束 unmodulated beam
未调制载波 unmodulated carrier
未调制噪声 unmodulated noise
未调制噪音 unmodulated noise
未贴合 unseating
未通过试验 fail test
未同步的扫描 flywheel time base
未同步电路 free running circuit
未透露的 undisclosed
未涂柏油的粗油毡 unsurfaced tar(red) rag felt
未涂层的 uncoated
未涂层钢管 uncoated steel pipe
未涂层石棉水泥管 uncoated asbestos-cement pipe
未涂底漆的 unprimed
未涂底漆的金属底材 unprimed metal substrate
未涂底色的 unprimed
未涂封闭底漆 unsealing
未涂沥青的粗油毡 unsurfaced asphalt rag felt
未涂面层的 uncoated
未涂面的铸铁管 uncoated cast-iron pipe
未涂刷到的 holiday
未涂油的 unoiled
未退火的 unannealed
未退火径迹 unannealed track
未脱钙的 undecalcified
未脱胶丝 unboiled silk; unscoured silk
未脱壳的 unhulled
未脱粒的 unthreshed
未脱硫的 undesulfured
未脱水煤气 moisture-laden gas
未挖掘的潜力 untapped potentiality
未弯起的钢筋 non-bent-up reinforcing bar
未完版 unfinished edition
未完拨号 incomplete dialing
未完成的 crude; half black; immature; unfilled; unfinished
未完成的地基 rough subgrade
未完成的订货 open order
未完成的定货 unfilled order
未完成的工程 construction in progress; uncompleted construction
未完成的工作 arrear; outstanding work
未完成的合同 uncompleted contract
未完成的井 incomplete well
未完成的路基 rough subgrade
未完成的试验 incomplete test
未完成的销售 executory sale
未完成的义务 unfulfilled obligations
未完成的著作 unfinished book
未完成工程 uncompleted project construction
未完成工程占用率的审计 audit of ratio of occupancy of unfinished project
未完成呼叫 incomplete call
未完成会计事项 uncompleted transaction
未完成交易 uncompleted transaction
未完成履行的契约 uncompleted contract
未完成权利 inchoate title
未完成任务记录 incomplete task log
未完工产品资金 unfinished product capital
未完工程 incompleted construction; project under construction; roughing-in; unfinished construction; outstanding work
未完工程价值 value of uncompleted construction
未完工程项目 uncompleted construction project
未完工程资金 capital of unfinished construction work
未完工程资金占用率 rate of funds used by project under construction
未完工的 unfinished
未完工的建筑场所 unfinished building space
未完工建筑物 unfinished building
未完工作 unfinished work
未完航次 unfinished voyage
未完建筑工程 uncompleted construction
未完建筑施工工程 uncompleted building works
未完结的拨号 mutilated selection
未完结交易 open trade

未完全被利用的土地 under-employed land
未完全捣实 incomplete compaction
未完全的 unfulfilled
未完全固结 underconsolidation
未完全劈裂开 dystomic
未完全燃烧气体 hazy atmosphere
未完施工 construction in process;uncompleted construction
未完税的 untaxed
未完税的货物 freight in bond
未完税后交货(价)delivered duty unpaid
未完税价格 inbond price
未完尾工 arrears
未围起的 unenclosed
未稳定的轨道 non-stabilized track
未稳定原油 unstabilized crude
未稳压整流器 unregulated rectifier
未污染带 uncontaminated zone;unpolluted zone
未污染的 uncontaminated;undefiled;unsullied;untainted
未污染的大气 non-contaminated atmosphere
未污染的河流 unpolluted stream;uncontaminated stream
未污染的空气 uncontaminated air
未污染水 uncontaminated water
未污染土地 uncontaminated field
未污染土(壤)uncontaminated soil
未污染岩芯 uncontaminated core;unpolluted core
未污染雨水 clean rain
未吸收成本 unabsorbed cost
未吸收的 unabsorbed
未吸收费用 unabsorbed expenses
未析出的 undecomposed
未稀释的 undiluted;unwatered
未稀释的发动机油 undiluted engine oil
未稀释的环氧树脂 undiluted epoxy resin
未稀释废水 undiluted effluence;undiluted wastewater
未稀释漆料 non-vehicle
未洗矿石 grena
未洗砾石 as-raised gravel
未洗煤 grena;unwashed coal
未系安全带的 unrestrained
未系住的 unstuck
未下套管的井 untubed well
未下套管的井段 bare foot
未下套管井段 blank hole
未下套管钻孔 uncased(bore)hole
未显影的 undeveloped
未限定日期的 undated
未镶面的 unlined
未镶嵌的 unset
未镶嵌牢的金刚石 non-wetted diamond
未镶钻头 steel bit
未向海关申报的 undeclared
未向海关申报的货物 undeclared cargo
未削弱视界 unimpaired visibility
未消除原废水供水 non-sterile raw wastewater feed
未消毒或沸化的 unslaked
未消耗成本 unexpired cost
未消化残余物 undigested residue
未消化的 undigested
未消化石灰 unslaked lime
未消化透的 unsound
未消化污泥 undigested sludge
未消解的 unslaked
未消解的磨细生石灰 unslaked and ground quicklime
未消石灰 calcium lime
未销账款 outstanding account
未楔固支架 unblocked set
未泄露的 undisclosed
未卸 remaining on board
未卸货物 goods afloat
未卸下的 undischarged
未兴建的 unbuilt
未形成的 unformed
未修版像片 unretouched photograph
未修版像片缩小 unretouched photographic(al)reduction
未修的损伤 unrepaired damage
未修改指令 unmodified instruction
未修光滑的曲线 unsmoothed curve
未修剪的 unpruned;unsheared
未修平曲线图 unsmoothed curve chart
未修饰的 unvarnished
未修饰的外底 unfinished bottom
未修匀的 unsmoothed
未修匀曲线 unsmoothed curve
未修整边沟 rough ditch
未修整的 undressed;unfinished
未修整的(石板)表面 self-faced
未修整的斜坡 rough slope
未修整混凝土 rough concrete
未修正场 uncorrected field
未修正的 uncorrected
未修正的指令 unmodified instruction
未修正偏差的无线电方位 observed radio bearing
未修正频谱 uncorrected spectrum
未修琢的 unfinished
未锈蚀的 unattacked
未选裁的板玻璃 uncut sheet glass
未选磁芯 unselected core
未选黏[粘]土 unselected clay
未选用面积 non-designated area
未驯化的活性污泥 unacclimated activated sludge
未压紧 malcompression
未压密土 uncompacted soil
未压实的 uncompacted
未压实堆石(体)uncompacted rockfill
未压实轨道 non-consolidated track;unconsolidated track
未压实混凝土 incomplete compaction concrete;loosely spread concrete;uncompacted concrete
未压实体积 loose measure
未压实土 uncompacted soil
未压实新拌混凝土 uncompacted fresh concrete
未压水 unconfined water
未延迟的 undelayed;unretarded
未研究板材 ungrained plate
未研磨表面 ungrained surface
未掩盖的缺陷 open-defect
未掩护的锚地 exposed anchorage
未邀请投标 unsolicited contract proposal
未应变的 unstrained
未营业的铁路线 line not operated
未硬化的 non-hardened;unhardened;unhardening
未硬化的砌体 green masonry
未硬化混凝土 unhardened concrete
未硬化砂浆 green mortar
未硬结的 undercured
未硬结混凝土 fresh concrete;green concrete;unhardened concrete
未用垂球校准的 unplumbed
未用存储单元 unused storage location
未用的线 disengaged(free)line
未用地 unused land
未用地区 unused zone
未用过带 virgin tape
未用过的 unworked;virgin
未用过的水 unused water
未用过介质 virgin medium
未用过纸 virgin paper
未用坏的 unworn
未用介质 virgin medium
未用金额 amount not taken up
未用尽的 unexhausted
未用空间 vacant space
未用媒体 virgin medium
未用名 unused name
未用完的 unexpended
未用完的拨款 unspent allocation
未用完款项 unspent balance
未用位 unused bit
未用余额 balance outstanding;outstanding balance;unexpended balance;unused balance
未用杂酚油处理的 uncreosoted
未用资源 unused resources
未油漆的 unpainted;unvarnished
未游离的 unionized
未有足够的安全系数的 under designed
未予说明的 unaccounted
未予调整记入贷方的金额 unadjusted credit
未预付估计税金 outstanding advance for estimated tax
未预见水 unaccounted for water
未预见需求 unforseen demand
未预见用水量 unforeseen demand
未预热空气 natural air
未预知的 unforeseen
未约化的 unreduced
未约矩阵 unreduced matrix
未运出货物 unshipped goods
未运行的 unoperated
未运营的铁路线 line not operated
未运营线路 line not operated
未载货的 unladen
未载入史册的 unstoried
未在场物主 absentee owner
未在规定时间内提出(索赔)要求 non-claim
未皂化的 unsaponifiable
未增强的 unfortified
未增强的焊接 unreinforced weld
未增塑的 unplasticized
未增塑聚氯乙烯 unplasticized PVC [polyvinyl chloride]
未增压的 unblown
未轧平头寸 open position
未轧碎的材料 uncrushed material
未轧碎的卵石 uncrushed pebble
未轧完品 unfinished section
未炸的炮弹 blind shell
未沾土的 unsoiled
未展开簇射 undeveloped shower
未占据的 non-occupied
未占领地区 unoccupied area
未占能级 unoccupied level
未占线时 unoccupied time
未占线状态 free condition
未占用的 unappropriated
未占用区段 unoccupied section
未占用区间 unoccupied section
未占用土地 empty land
未遮蔽的 uncurtained
未折旧的 undepreciated
未折旧价值 undepreciated value
未折旧原价余额 undepreciated balance of the cost
未征税的 unrated
未蒸发的 unevaporated
未蒸发溶剂 unevapo(u)rated solvent
未蒸馏的 undistilled
未整理表 unsorted table
未整理的 undigested
未整面的 unfaced
未整面的沥青粗油毡 unfaced asphalt rag felt
未整平的 ungraded;unsmoothed
未整饰的皮革 unfinished leather
未整治的河床 unregulated(river)bed
未整治的河段 unregulated reaches of river
未整治的河流 unregulated river;unregulated stream
未整治的溪流 unregulated stream
未整治河床 unregulated bed
未正确就位的 out-of-position
未证认源 unidentified source
未证实 unverified
未证实的 unconfirmed
未支护顶板 unsupported back
未支护区 unsupported area
未支配款项 uncommitted amount
未支配数额 uncommitted amount
未支配余额 unencumbered balance
未支配预计盈余额 unappropriated estimated surplus
未支用余额 unspent margin
未知参数 unknown parameter
未知参数向量 vector of unknown parameter
未知常数 unknown constant
未知错误指示码 unknown error code
未知的 un(be)known;unsuspected
未知的积累 unknown accumulation
未知的领域 terra incognita
未知地热区 unknown geothermal area
未知点 unknown point
未知方差 unknown variance
未知函数 unknown function
未知晶体 unknown crystal
未知框 black box
未知框测试 black box testing
未知量 unknown(quantity);unknown number
未知脉冲 unknown pulse
未知数 indeterminate;number of unknowns;unknown;unknown number;unknown quantity
未知数数目 number of unknowns
未知素数 unknown prime number
未知损失 unknown loss
未知条件 unknown condition
未知物 unknown
未知向量 unknown vector
未知项【数】unknown term
未知样品 unknown sample
未知液 unknown solution
未知因素 unknown;unknown factor
未知值 unknown value
未知周期分量 unknown periodic component
未知状态 unknown state
未执行的订单 unfilled order
未指拨盈余 free surplus;unappropriated surplus
未指定土地用途区 <城市规划中> zone of undesigned land use
未指定用途的现金 unapplied cash
未酯化的 unesterified
未酯化脂肪酸 unesterified fatty acid
未制动的 unbraked
未制动轴 unbraked axle
未制炼的 raw
未制图的 unmapped
未制造的 unwrought
未置平的 unlevel(l)ed
未中标 lose a bid;losing bid;losing tender
未中毒燃料 unpoisoned fuel
未中和的 unneutralized
未中和的油 sour oil
未终结诉讼责任 liability under pending lawsuits
未终凝的混凝土 immature concrete
未重卷 unrewind
未周转资金 idle money
未煮过的 uncooked
未住人房间 vacant room

未注册的 unregistered
未注明的 not marked
未注明付讫的 unreceipted
未注明货种及卸货港的租船合同 open charter
未注明日期的 undated
未注日期支票 undated cheque
未筑堤的冲积土地 unembanked alluvial land
未铸满 misrun
未转化的 unconverted
未转化石灰 unconverted lime
未装备的 unequipped
未装玻璃的 unglazed
未装废气净化装置的车辆 uncontrolled vehicle
未装工作装置 unequip
未装合底盘 unassembled chassis
未装货时的重量 <即车辆、集装箱自重> unladen weight
未装雷管导火线 uncapped fuse
未装满舱 slack hold
未装满油舱 slack tank
未装满炸药岩孔 partially infilled hole
未装配的 unassembled;unset
未装配的锯 unset saw
未装区别机的支局 <自动电话> full satellite exchange
未装饰的 unenriched
未装弹簧的 unsprung
未装箱的 uncased
未装修的阁楼 blind attic
未装修的毛门窗 subbuck
未装修的毛门窗框 subcasing
未装修的镶板门 framed square
未装药的爆破目标 uncharged demolition target
未装炸药的 uncharged
未装炸药的爆破孔 unloaded hole
未装炸药的炮眼 unloaded hole
未装炸药的岩孔 uncharged bulk hole
未装足 underloading
未装足的 underloaded
未灼烧分析 raw analysis
未灼烧基 raw basis;unignited basis
未着色 black and white
未着色的 non-pigmented;uncolo(u)red; unlayered; unpigmented; unrendered
未着雨地带 rain shadow
未琢磨的金刚石 uncut diamond
未琢磨的托饰 uncut modillion
未琢磨金刚石 uncut stone
未琢凿的 unhewn
未组装的 unassembled
未钻穿层 <钻探> refusal
未钻即可证实的储量 undrilled proved
未钻孔法兰 blank flange
未钻眼钢轨 rail with blind ends
未最后成型的 unwrought
未最后定稿的报告 unsatisfied report
未遵守列车时刻表 nonobservance of time table
未(作)标记的 unlabel(l)ed
未作说明 not specified
未作说明的 unspecified
未做面层的 non-surfaced
未做准备的 unprovided

位 bigit

位比较 bit comparate
位变换 bit map
位变换器 bit changer
位变异构的【化】metameric
位变异构度 degree of metamerism
位变异构态 metameric state; metameride
位变异构体 metamer(ide)
位变异构(现象)metamerism
位变异构性 metamerism
位标 temperature scale
位表 bit mapping;bit table
位并行 bit parallel
位操作 bit manipulation
位差 potential difference;static elevation
位差角 angle of parallax
位差损失 loss in head;loss of head
位差调整 position timing
位场 potential field
位出错率 bit error rate
位处理 bit manipulation
位传递速率 bit transfer rate
位传送率 bit transfer rate
位串 <一串二进位信息> bit string
位串数据 bit string data
位次 locant;rank
位次编排 numbering
位存储密度 bit packing density
位存储组织 bit memory organization
位错 dislocation;line defect
位错半环 dislocation half-loop
位错崩 avalanche of dislocation
位错壁 dislocation wall
位错缠结 dislocation tangle
位错的钉扎 anchoring of dislocation
位错的割切 crossing of dislocation
位错的交互作用 interaction of dislocation
位错的振荡 oscillation of dislocation
位错地震 dislocation earthquake
位错动力学 dislocation dynamics
位错段 dislocation segment
位错堆积 pile-up of dislocation
位错割阶 dislocation jog
位错管线扩散 diffusion along dislocation-pipe
位错惯态 dislocation habit
位错滑移 dislocation glide
位错滑移距离 distance of dislocation glide
位错滑移速度 velocity of dislocation glide
位错环 dislocation loop; dislocation ring
位错角 dislocation angle
位错结 dislocation node
位错结构 dislocation structure
位错界面 dislocation boundary
位错理论 dislocation theory
位错裂缝 dislocation crack
位错露头点 point of emergence of dislocation
位错密度 dislocation density
位错能 energy of dislocation
位错偶极子 dislocation dipole
位错攀移 climb of dislocation; dislocation climb
位错攀移速度 velocity of dislocation climb
位错墙 dislocation wall
位错圈 ring dislocation
位错溶度 dislocation solubility
位错蠕变 dislocation creep
位错时间函数 dislocation-time function
位错锁住 dislocation locking
位错网络 dislocation network
位错线 dislocation line
位错亚结构 dislocation substructure
位错源 dislocation source
位错运动 dislocation motion; dislocation movement
位错中心 dislocation nucleation
位错缀饰法 decoration method of dislocation
位错阻尼 dislocation damping
位单元 bit cell;bit location;bit position
位单元前沿 bit cell leading edge
位单元中心 bit cell center [centre]
位的 n-adic; potential【电】; steric【化】
位的传送 delivery of bits
位的位置 bit position
位灯部件 position light unit
位灯机构 position light unit
位灯信号(机) <用两个以上白色灯表示> position light signal
位灯信号机构 position light signal head
位点 locus
位点的显性 dominance at locus
位点内互作 intralocus interaction
位定址 bit addressing
位读出线 sense-digit line
位对 digit pair
位符号 bit sign
位高 geopotential height
位函数 potential function
位号 item
位缓冲器 digit buffer
位级功能 bit level function
位寄存器 bit register
位降 potential drop
位角 parallactic angle;position angle
位借攀移模型 dislocation climb model
位阱 potential trough;potential well
位距 distance;spacing
位块 bit block
位块传送 bit block transfer
位垒 barrier;potential barrier;potential hill
位理论边值问题 boundary value problem of potential theory
位理论高斯公式 Gauss formula of potential theory
位理论格林公式 Green's formula of potential theory
位力定理 virial theorem
位力定理质量 virial-theorem mass
位力质量 virial mass
位列 bit train
位流 bit stream;potential flow
位流传输 bit stream transmission
位流传输时间 transfer time of byte flow
位论 potential theory
位论反解问题 inverse problem of potential theory
位码 bit-code
位脉冲 digit(al)pulse
位密度 bit density
位面编码 bit plane coding
位面计 level instrument
位面控制 level control
位面控制器 level controller
位面指示器 cut-off level ga(u)ge
位/秒 bits per second [bps]
位模式 bit pattern
位模式发生器 bit pattern generator
位能 energy of position;gravitational energy;latent energy; potential energy;static energy
位能差 difference of potential
位能高 potential energy head
位能失衡 out-of-balance potential
位能损失 elevation loss
位能梯度 potential gradient
位能位差 potential head
位能障碍 potential barrier
位挪用 bit stealing
位偶 digit pair
位片 bit slice
位片处理机 bit-slice processor
位片式系统 bit-slice system
位片体系结构 bit-slice architecture
位片微处理机 bit-slice microprocessor
位平面 bit plane
位平面编码 bit plane coding
位屏蔽 bit mask
位奇偶 bit parity
位起伏 phreatic wave
位扰动 potential disturbance
位热数 geopotential number
位式控制器 positional controller
位势 potential energy
位势不稳定性 potential instability
位势差 potential differential
位势单位 geopotential unit
位势的 geopotential
位势高度 geopotential height
位势高度图 absolute geopotential topography
位势厚度 geopotential thickness
位势控制 position control
位势理论 potential theory
位势力 potential force
位势米 geopotential meter
位势面 geopotential surface; level surface;potential surface
位势能 geopotential energy
位势坡度 geopotential slope
位势数 geopotential number
位势水头 elevation head; position head;potential head
位势调节 position control
位势头 position head;potential head
位势温度 potential temperature
位势涡量 absolute potential vorticity; potential vorticity
位势序列 sequence of potentials
位势英尺 geopotential foot
位数 bit of information;cipher;digit; digit capacity;order of units
位数计算 digit count
位数移调 transplacement
位数组 bit array
位丝测微计 position filar micrometer
位似比 homothetic ratio
位似变换 homothetic transformation
位似生产函数 homothetic production function
位似效用函数 homothetic utility function
位似中心 homothetic center [centre]
位速 potential velocity
位速度 bit speed
位速高 potential velocity head
位速率 bit rate
位损 bit loss
位填充 bit stuffing
位填充算法 bit stuffing algorithm
位挑出 bit destuffing
位调节杆总成 position control lever assembly
位调节凸轮 position control cam
位同步 bit align(e)ment;bit synchronization
位头 elevation head; position head; potential head;static head
位图 bit map
位图表示(法)bit map representation
位图显示 bit map display
位图像 bit image
位温 megadyne temperature;potential temperature
位温梯度 potential temperature gradient
位温效应 potential temperature effect
位线 digit line
位相 argument;phase;tidal stage
位相变易 phase variation
位相差 phasic difference
位相倒置 phase inversion
位相记忆术 topologic(al)mnemonics

位相角 phase angle
位相谱 phase spectrum
位相曲线 phase curve
位相调制密度假彩色编码 pseudo-colo(u)r density encoding through phase modulation
位相突变 phase jump
位相无畸变 phase distortionless
位相系数 phase coefficient
位相显微镜 phase contrast microscope;phase microscope
位相滞后 phase lag
位相状态 phasic state
位向量 bit vector
位向消失 disorientation
位形 configuration
位选择器 digit selector
位压头 geometric(al)head
位移 bias(s)ing; creep; detrusion; dislocation; displacement; displacement in position;judder;offsetting; travel motion
位移边桩 movement pile; side pile displacement
位移变化方程 equation of displacement transformation
位移变量 offset variable
位移波 dislocation wave; displacement wave
位移参数 displacement parameter; shift parameter
位移测定 displacement measurement
位移测量 displacement measurement
位移测量法 displacement-measurement procedure
位移测震仪 displacement seismograph
位移场 displacement field
位移传递系数 displacement transmission coefficient
位移传感器 displacement indicator; displacement pickup; displacement sensor; displacement transducer; movement pickup
位移导纳 displacement mobility
位移的空间变化 spatial variation of displacement
位移地震计 displacement seismometer
位移地震仪 displacement seismograph
位移点 displacement point
位移电流 displacement current
位移电流密度 displacement current density
位移断层 displacement fault;shift fault
位移法 <分析超静定结构的基本方法之一> deflection method;displacement method
位移反向点 inversion of displacement
位移反应谱 displacement response spectrum
位移反应系数 displacement response coefficient
位移反应因数 displacement response factor
位移范围 displacement bound; displacement range
位移分解 resolution of displacement
位移分量 component of displacement;displacement component
位移峰峰值 total excursion
位移峰值 displacement spike
位移幅值 displacement amplitude
位移干涉条纹 shifting interference fringe
位移公式 displacement formula
位移共振 displacement resonance
位移观测 displacement observation
位移轨迹法 displacement path method
位移函数 displacement function
位移合成 composition of displacement
位移荷载 displacement load
位移互等定理 theorem of reciprocal deflection;theory of reciprocity
位移及振动的极限状态 limit state of displacement motion and vibration
位移极限 displacement limit
位移计 displacement ga(u)ge; displacement meter
位移计算 displacement calculation
位移检验 displacement test
位移渐小 gradual reduction of displacement
位移角 angle of displacement; angle of shift; angle of slip; displacement angle
位移矩阵 displacement matrix
位移聚结 dislocation coalescence
位移空位 displaced vacancy
位移控制 displacement control; shifting control
位移力 displacing force
位移连续性 displacement continuity
位移量 amount of movement; displacement value
位移量测 displacement measurement
位移量曲线图 time-displacement chart
位移裂线 rentline of displacement
位移灵敏度 displacement sensitivity
位移脉冲 displacement pulse; shift pulse
位移面 displaced surface
位移能力 movement capability
位移能量 displacement energy
位移频率谱 displacement frequency spectrum
位移谱 displacement spectrum
位移谱密度 displacement spectral density
位移曲线 displacement curve
位移屈服点值 yield point value of displacement
位移韧性系数 displacement ductility factor
位移散射迁移率 dislocation scattering mobility
位移-时间关系 displacement-time relationship
位移-时间关系曲线 displacement-time curve
位移-时间关系图 displacement-time diagram
位移-时间曲线 deformation-time curve;displacement-time diagram
位移矢量 displacement vector
位移式地震检波器 displacement type seismometer; seismic detector of displacement
位移式水表变位计 displacement meter
位移势 displacement potential
位移试剂 shift reagent
位移适应因数 movement accommodation factor
位移守恒 displacement conservation
位移速率 rate of displacement
位移算子 displacement operator
位移调整 shift control
位移调制 displacement modulation
位移通量 displacement flux
位移图 displacement diagram
位移陀螺仪 displacement gyroscope
位移误差 displacement error
位移系数 displacement coefficient
位移线 displacement line
位移相变 displacement phase transition
位移相容条件 displacement compactibility
位移相容性 displacement compatibility
位移向量 displacement vector
位移协调方程式 equation of compatibility of displacement
位移协调条件 displacement compatibility
位移信号发生器 displacement generator
位移形状 displacement shape
位移型地震仪 displacement type seismometer
位移型铁电体 displacement type ferro-electrics
位移型转变 displacive transformation
位移旋钮 displacement knob
位移压差计 displacement manometer
位移延性 displacement ductility
位移延性系数 displacement ductility factor
位移异常 displaced anomaly
位移应力 displacement stress
位移元 displacement element
位移约束 restrain of displacement
位移晕 displaced halo
位移载荷 displacement load
位移噪声 displacement noise
位移增量 incremental displacement
位移张量 displacement tensor
位移振幅 displacement amplitude
位移指示器 displacement indicator; stripping indicator
位移中心 center [centre] of displacement
位移轴 offset axis
位移转变 displacement inversion
位移准则 displacement criterion
位移阻抗 displacement impedance
位移阻力 displacement resistance; resistance to displacement
位移坐标的有限集合 limited set of displacement coordinates
位映象 bit map(ping)
位拥挤 bit crowing
位于边界上的 limitrophe
位于表层下的地层 subcrust
位于低地上的城堡 castle sited on low-lying ground
位于顶点下的 subapical
位于发动机上方的驾驶室 cab over engine
位于高地上的城堡 castle site on high-lying ground
位于焊接接头中 located in welded joint
位于人工湖上的城堡 castle set in artificial lake
位于上部的 superposed
位于斜坡上的 downhill
位于信号桥上的信号楼 bridge signal cabin
位于一线上的设备 in-line equipment
位于中心点以下的 subcentral
位源 potential source
位折射指数 potential refractive index
位置 appointment; locality; localization; location; placing; position; room; seat; site; siting; situs; standing
位置报告 position message; position report
位置报告程序 position reporting procedure
位置编码器 position coder
位置变化 position variation
位置变换器 position transducer
位置变换性能 locomotiveness
位置变量 location variable; position variable
位置变送器 position transmitter
位置标识符 location identifier
位置表示法 positional notation; positional representation
位置表示器 position indicator
位置表象 position representation; Schrodinger representation
位置补偿 position compensation
位置不变性运算 position invariant operation
位置不当的 out-of-position
位置不符 out-of-position
位置不正 misregistration;misregistry
位置参量 locating parameter;position parameter
位置参数 locating parameter;location parameter;position(al)parameter
位置测定 determination of position; localization;position finding
位置测定器 chorograph; position chorograph
位置测定误差 position fixing error
位置测定系统 positioning and location system
位置测量 position measurement
位置测量仪表 position measuring instrument
位置测微器 position micrometer
位置长度 position length
位置常数 location constant
位置触点 position contact
位置传感器 attitude sensor; position sensor;position transducer
位置错乱 dislocation
位置代码 position code
位置的 positional
位置的确定 identification of position
位置灯光 position light
位置电码变换器 position code converter
位置电压 position voltage
位置电压控制 position voltage control
位置电子指示器 electronic position indicator
位置定向 orientation for place
位置度公差带 positional tolerance zone
位置对策 position(al)game
位置对称性 site symmetry
位置反馈 position feedback
位置反馈电桥 position feedback bridge
位置反馈信号 position feedback signal
位置反算 inverse position computation
位置方位角 azimuth at future position
位置浮标 position buoy; marker buoy;station buoy
位置复原图 palinspastic map
位置格式 positional format
位置跟踪器 position tracker
位置跟踪系统 position control servomechanism
位置跟踪装置 positioning control system
位置公差 location tolerance;position(al) tolerance; position of related features
位置固定端 position fixed ends
位置故障 position failure
位置函数 position function
位置函数等值线 isopleth of position function
位置函数梯度 gradient of position function
位置号码 location number; position number
位置号数 placement number
位置回路 position loop
位置基准系 position reference system
位置极限 position limit

位置计数法 notation;positional notation
位置计数器 location counter
位置记录器 position recorder
位置记数法 positional notation;positional representation
位置记忆 position memory
位置记忆术 topologic(al)mnemonics
位置继电器 position relay
位置监测系统 position monitoring system
位置监视浮标 station surveillance buoy
位置检测 location detection
位置检测器 position detector
位置检查 position detection
位置检核 position check
位置交扰 positional cross stalk
位置角 angle of situation;apex angle;position angle;angle of site
位置校正 position correction
位置精度 positional accuracy
位置精度几何因子 position dilution of precision
位置可定性 placeability
位置控制 centering control;position(al)control;positioning;positioning control
位置控制和定向 positioner and orientation
位置控制回路 position control loop
位置控制精确度 control positioning accuracy
位置控制器 position controller;positioner;station controller
位置控制系统 position control system
位置控制信号 position control signal
位置历元 epoch of place
位置灵敏 position sensitive
位置灵敏盖革计数器 position sensitive Geiger counter
位置灵敏探测器 position sensitive detector
位置略图 location sketch;site sketch
位置码 positional code
位置码距离法 position code distance
位置码序列法 position code sequential
位置脉冲 position pulse;p-pulse
位置铆钉 tacking rivet
位置面 position surface; surface of position
位置敏感探测器 position sensitive detector
位置模糊度 position ambiguity
位置模型理论 site model theory
位置拟等位性 position pseudoallelism
位置排列 space occupancy
位置偏差 deviation in position; discrepancy in position;position deviation
位置偏移 offset
位置平衡 position balance
位置平衡制 position balance system
位置强度公式 position strength formula
位置区域 band of position
位置圈 circle of position;position circle;small circle
位置容限 positional tolerance
位置三角形 position triangle
位置色灯信号 position light signal
位置色像差 chromatic aberration of position
位置设计 placement
位置失效 position failure
位置识别 identification of position;position identification
位置矢量 position vector
位置示像 position aspect
位置示意图 adjacent area inset;orientation inset
位置式遥测计 position type telemeter
位置式远测装置 position type telemeter
位置视线 visual line of position
位置数 positional number
位置数据 position data
位置水头 elevation head; potential head;position head
位置伺服机构 position(al)servomechanism
位置算符 position operator
位置算子 position(al)operator
位置索引 location index
位置天文学 astronomy of position
位置调节(器)position control
位置调节手柄 position control lever
位置调整 position adjustment
位置调整器 position regulator
位置调制扫描 position modulation scan
位置头 geometric(al)head; gravity head;potential head
位置图 assembly diagram;drawing of site; locating map; location drawing;location map;location plan;orientation inset; plot plan; site drawing;site plan of site;situation plan of site;space diagram;index map < 建筑物 >
位置陀螺 position gyro
位置稳定性 positional stability
位置无关码 position independence code
位置误差 error of position;position(al)error;site error
位置误差常数 positional error constant
位置误差探测器 positional error detector
位置误差系数 positional error coefficient
位置系数 location coefficient
位置线 line of position;position line
位置线法线方向位移中误差 mean square error of displacement for normal direction of position line
位置线交角 angle between lines of position; angle between position lines
位置线梯度 gradient of position line
位置线移位误差 error of transferring
位置线移线订正 reduction of altitude to another place of observation
位置限制 position limitation
位置向量 position vector;radius vector
位置效应 position effect
位置信号 position signal(ling)
位置信号器 position signal(l)ing apparatus
位置信号指示器 position signal indicator
位置信息 location information;positional information
位置修正 position updating
位置选择 location selection;site selection
位置选择阀 position selector valve
位置遥测计 position telemeter
位置遥测术 position telemetering
位置遥控 remote position control
位置移动指示图 position indicating grid
位置因子 location factor
位置影响 position influence
位置优势 position advantage
位置预测装置 predicted position device
位置预选 preliminary site selection
位置圆 circle of position;position circle
位置正常的 normotopic
位置正反算 direct and inverse position computation
位置指令信号 position command signal
位置指令字 instruction words of location
位置指示 position indication
位置指示符 location pointer
位置指示格网 alphanumeric grid
位置指示开关 position indicating switch
位置指示器 position indicator
位置指示器线圈 position indicator coil
位置指示仪 site marking device
位置指针 position indicator pointer
位置属性 position attribution
位置转换器 position transducer
位置转移 translocation
位置状态 running position
位置自动表示 automatic position indication
位置自动表示系统 automatic position indication system
位置自动调节(器)automatic position control
位置自动显示 automatic position indication
位置自动显示系统 automatic position indication system
位置总图 general location sheet
位置坐标 position data
位总线 bit bus
位阻胺 hindered amine
位阻分离作用 steric separation
位阻酚 hindered phenol; sterically hindered phenol
位阻(现象)steric hindrance
位阻效应 steric effect
位阻因素 steric factor
位阻硬化 steric hardening
位组合 bit combination;bit pattern
位组合格式 bit pattern
位组片 byte slice

味道 taste

味觉 taste
味觉敏度 taste acuity

畏光 photophobia

畏光的 photophobic
畏惧和憎恨 < 对外国人或外国事物的 > xenophobia
畏来风 < 见于澳大利亚西部 > willy-willy

喂槽 spout feeder

喂阀 feed valve
喂料 feeding;material feed(ing)
喂料比例 charge proportion
喂料不足 underfeed(ing)
喂料仓 feed bin;feed boot
喂料槽 chute feeder
喂料池 feed tank
喂料储仓 storage feed bin
喂料斗式提升机 feeding bucket elevator
喂料端 feed end;inlet end
喂料端机罩 air housing
喂料端磨头 feed end head
喂料方法 method of feeding
喂料方向 feeding direction
喂料干管 trunk feeder
喂料杆 charging bar
喂料高度 < 料斗 > hopper feeding height
喂料公差 < 偏离集料和沥青配比值 > delivery tolerance
喂料辊 draw roller; feed(ing) roll(er)
喂料过多 over-feed
喂料喉 feed throat
喂料机 draw roller; feeder; feeding machine
喂料计量器 feed scale
喂料绞刀 feeding screw;screw feeder
喂料截流阀 feed cut-off gate
喂料口 feed port
喂料块度 feed lump size
喂料粒度 feed size
喂料(溜)槽 feed chute
喂料溜子 feed spout
喂料耙 feeding drawer
喂料盘 circular feeder; disk feeder; feed disk [disc];rotary disk feeder;rotary table feeder
喂料皮带秤 weighing feeder; weight belt feeder
喂料皮带机 feeding belt conveyer [conveyor]
喂料平台 charging platform
喂料器 materials feeder
喂料圈 filling ring
喂料热交换器 feed heat exchanger
喂料顺序 order of feed
喂料速度 rate of feed
喂料同步化 feed synchronization
喂料系统 feeding system
喂料箱 feed box
喂料斜槽 feed spout
喂料用斗式提升机 feed bucket elevator;feed elevator
喂料预热器 feed preheater
喂料转盘 dosing rotor
喂料转盘秤 dosing rotor weigher
喂料装置 dosing equipment;feed device;feeding apparatus
喂煤机 stoker
喂煤调节器 coal feed regulator
喂煤系统 coal feeding system
喂模机 template feeding machine
喂气槽 feed groove
喂入材料 feed stock
喂入的料球 nodulized feed
喂入的生料 raw feed
喂入轮 feeder beater
喂入器调节手柄 feeder adjusting crank
喂入输送带 receiving conveyer [conveyor]
喂入物料的喂料比 feed material filling ratio
喂送升运器 feed elevator

蔚蓝色 azure;cerulean

蔚蓝色的 sapphire
蔚欧仿 vioform

慰藉金 solatium

魏茨泽克学说 Weizsacker's theory

魏尔德尼期阶 Wilderness(stage)
魏尔德起重机 Wild elevator
魏尔德栅栏 Wild fence
魏尔温田城市 < 英国伦敦 > Welwyn Garden City
魏格纳大陆漂移说 Wegener hypothesis
魏格特效应 Weigert effect
魏磷石 wicksite

魏纳谱带 Werner band
魏奇塞尔冰期 Weichselian glacial epoch;Weichselian glacial stage
魏森堡法 Weissenberg method
魏森堡图 Weissenberg photograph
魏森堡效应 Weissenberg effect
魏森堡照相机 Weissenberg camera
魏森霍夫居住住宅方案展览会 <1927 年在德国举行> Weissenhof siedlung
魏斯常量 Weiss constant
魏斯常数 Weiss constant
魏斯磁化理论 Weiss theory
魏斯环炉 Weisz ring oven
魏斯式球叉等速万向节 Weiss constant velocity universal joint
魏斯四边形测量法 <竖井联系四边形井下定向法> Weiss quadrilateral
魏特豪尔测试 Wetthauer test
魏因加尔吞曲面 Weingarten surface

温

温拌 warm-mix
温拌混合料 warm-mix
温包 temperature sensing bulb;thermometer bulb
温饱型农业 subsistence agriculture
温变层 thermodine
温变法 temperature variation method
温变周期性 thermoperiodism
温标 scale of thermometer;thermometer scale;thermometric scale
温伯格角 Weinberg angle
温伯格-萨拉姆理论 Weinberg-Salam theory
温桲树 quince
温测深度 thermometric depth
温差 differential temperature;drop in temperature;range of temperature;temperature differential;temperature drop;temperature head;thermal difference;thermal drop
温差比重计 thermohydrometer
温差电 thermoelectricity
温差电池 thermocell;thermoelectric(al) cell;thermoelectric(al) generator;thermopile
温差电池组 thermobattery
温差电的 thermoelectric(al)
温差电定律 thermoelectric(al) law
温差电动势 thermoelectromotive force
温差电(动)势率 thermoelectric(al) power
温差电堆 thermoelectric(al) pile;thermopile
温差电堆供电的无线电设备 thermopile radio
温差电二极管 thermoelectric(al) diode
温差电结 thermojunction
温差电空调系统 thermoelectric(al) air conditioning system
温差电流 thermocurrent;thermoelectric(al) current
温差电流计 thermal galvanometer
温差电偶 pyod;thermal electric(al) couple;thermal element;thermocouple;thermoelectric(al) cell;thermoelectric(al) couple;thermoelement;thermojunction
温差电偶安培计 electrothermal ammeter;thermoammeter;thermocouple ammeter
温差电偶电流计 thermoammeter;thermogalvanometer
温差电偶高温计 thermoelectric(al) pyrometer
温差电偶计 thermocouple meter
温差电偶继电器 thermorelay
温差电偶检流计 thermogalvanometer
温差电偶瓦特计 thermal wattmeter
温差电偶温度计 thermoelectric(al) thermometer
温差电偶仪表 thermocouple meter
温差电偶真空计 thermocouple vacuum ga(u)ge
温差电热泵 thermoelectric(al) heat pump
温差电势 thermoelectric(al) power
温差电势效应 thermoelectric(al) effect
温差电效应 thermoelectric(al) effect
温差电性 thermoelectric(al) property
温差电制冷系统 thermoelectric(al) refrigeration system
温差电致冷 thermoelectric(al) cooling
温差发电器 thermoelectric(al) generator
温差法 temperature difference method
温差范围 temperature range
温差分层取水 thermal selective withdrawal
温差分层作用 thermocline stratification
温差分析 differential thermal analysis
温差风化 temperature differential weathering
温差改正【测】dynamic(al) temperature correction
温差钢筋 temperature reinforcement
温差过大 excessive temperature differentials
温差过滤器 temperature differential filter
温差核电池 thermoelectric(al) nuclear battery
温差虹吸 therm syringe
温差环流冷却 thermocooling
温差环流冷却系统 thermosiphon [thermosyphon] cooling system;thermocooling system
温差环流系统 thermosiphon [thermosyphon]
温差交面 thermal interface
温差交面剪刀 thermal interfacial shear
温差校正 dynamic(al) temperature correction
温差控制 temperature control
温差控制继电器 thermal control relay
温差控制器 differential temperature controller
温差力 thermal force
温差毛细管水运动 thermocapillary movement
温差扭曲 temperature warping;warping due to temperature difference
温差膨胀裂纹 grooving
温差翘曲 temperature warping
温差曲线 difference curve
温差热机 temperature difference heat engine
温差商数 temperature difference quotient
温差式风速表 thermoanemometer
温差式风速仪 thermoanemometer
温差式冷却装置 thermoelectric(al) cooling device
温差式微流速计 thermal small current meter
温差式温度计 manometric(al) thermometer
温差双金属 thermometal
温差探测器 thermodetector
温差温度曲线 differential thermal curve
温差修正系数 correction factor of (mean) temperature difference;modified temperature difference factor
温差修正因子 correction factor of (mean) temperature difference
温差异重流 thermal density current;thermal density flow
温差应变 temperature strain
温差应力 temperature difference stress;thermal stress
温差致冷器 thermoelectric(al) cooler
温差组件 temperature difference module
温差作用 temperature effect
温场照相术 thermography
温床 forcing bed;hot bed;seminary
温床播种 frame seeding
温床轮作 bed crop-rotation
温床栽培 frame culture
温带 extra-tropic(al) belt;moderate belt;moderate zone;temperate belt;temperate zone;warm zone
温带变幅 temperature fluctuation
温带冰川 temperate glacier
温带冰缘 temperate periglacial
温带草原 down land;temperate grassland
温带草原气候 temperate grassy climate
温带臭虫 cimex lectularius
温带大陆性气候 temperate continental climate
温带低压带 temperate low belt
温带冬季气候 <即地中海型气候> temperate winter climate;Mediterranean climate
温带冬雨气候 temperate climate with winter rain
温带多雨气候 temperate rainy climate
温带浮游生物 temperatoplankton
温带干燥气候 temperate arid climate
温带海水碳酸盐岩 temperate water carbonate
温带海洋性气候 temperate marine climate
温带和寒带密灌丛 temperate and cold scrub
温带和寒带稀树草原 temperate and cold savanna(h)
温带湖 temperate lake
温带混交林带 temperate mixed forest region
温带季风气候 temperate monsoon climate
温带季雨气候 temperature rainy climate
温带林 temperature warm zone;warm temperate forest
温带林带 temperate forest region;temperate forest zone
温带落叶林 temperate deciduous forest
温带气候 temperate climate
温带气旋 extra-tropic(al) cyclone
温带群丛 temperature association
温带群落 temperature association
温带森林 temperate forest
温带森林气候 temperate forest climate
温带森林生物层 temperate forest biosphere
温带森林与林地 temperate forest and woodland
温带沙漠气候 temperate desert climate
温带生态系统 temperate ecosystem
温带生态学 temperate ecology
温带湿润气候 temperate rainy climate
温带疏林 park land;temperate woodland;woodland
温带水域 temperate waters
温带西风带 temperate westerlies
温带西风带指数 temperate-westerlies index
温带夏绿林褐土带 temperate zone summer green forest cinnamon soil zone
温带夏雨气候 temperate climate with summer rain
温带亚风林 temperate westerlies
温带岩溶 temperate karst
温带雨林 cloud forest;laurel forest;laurisilva;moss forest;temperate rain forest
温带雨林副热带林 subtropic(al) forest
温带针阔混交林 temperate mixed forest
温带植物区 warm temperate district
温带植物群落 mesothermophytia
温得和克 <纳米比亚首都> Windhoek
温洞 lukewarm cave
温度 climatization;temperature
温度保护 thermal protection
温度饱和 filament saturation;saturation;temperature saturation
温度报警 alarm thermostat
温度报警器 alarm thermometer;temperature alarm;thermostatic alarm
温度报警钟 temperature alarm
温度本底值 temperature noise
温度比 temperature ratio
温度比学 hygrostatics
温度比重换算表 temperature gravity conversion table
温度比重计 thermohydrometer
温度笔 temperature indicating crayon
温度边界层 temperature boundary layer;thermal boundary layer
温度变差 temperature drift
温度变程 march of temperature
温度变动 temperature change;temperature variation curve
温度变动曲线 curve of temperature variation;temperature curve;temperature variation curve
温度变化 hygral change;shift in temperature;temperature change;temperature fluctuation;temperature shift;temperature variation;thermal movement;variation in temperature;thermal change
温度变化产生的变形 deformation due to temperature change;deformation due to thermal change
温度变化冲击 thermal shock
温度变化的影响力 influential force of temperature change
温度变化范围 range of temperature;temperature range limit
温度变化幅度 range of temperature;temperature range limit
温度变化曲线 curve of temperature variation;temperature variation;temperature curve
温度变化上染性能 temperature range property
温度变化适应性 poikilothermy
温度变化速率 rate of temperature change
温度变换器 temperature transducer
温度变位 <膨胀或收缩> temperature movement
温度变形 temperature deformation
温度标 scale of thermometer;thermometric scale;temperature scale
温度标尺 temperature scale
温度标定 temperature calibration

W

温度标准 temperature level
温度表 glass; temperature ga(u)ge; temperature level ga(u)ge; temperature meter; thermometer
温度表百叶箱 thermometer screen
温度表感温球 thermometer bulb
温度表架 thermometer support
温度表现 temperature behaviour
温度波 temperature wave; thermal wave
温度波动 temperature fluctuation; temperature surge; temperature swing; temperature vibration
温度波动范围 temperature fluctuation range
温度波动曲线 temperature wave curve
温度补偿 temperature compensation; thermal compensation
温度补偿摆 temperature compensated pendulum
温度补偿的 temperature compensated
温度补偿电路 temperature compensating circuit
温度补偿电容器 temperature compensated capacitor; thermocompensation capacitor; thermocompensation condenser
温度补偿范围 temperature compensation range
温度补偿基准元件 temperature compensated reference element
温度补偿片 temperature compensated ga(u)ge
温度补偿齐纳二极管 temperature compensated Zener diode
温度补偿器 temperature compensator; temperature equalizer; temperature equilibrator; variator
温度补偿器件 temperature compensation device
温度补偿热敏电阻海流计 temperature compensated thermist current meter
温度补偿闪光断续器 temperature compensated flasher
温度补偿深度 temperature compensation depth
温度补偿(石英)晶体振荡器 temperature compensated crystal oscillator
温度补偿调节器 temperature compensated regulator
温度补偿系统 temperature compensation system
温度补偿线圈 temperature compensating coil
温度补偿应变片 temperature compensated strain ga(u)ge
温度补偿用合金 temperature compensation alloy
温度补偿装置 temperature compensating device
温度不均匀分布 temperature contrast
温度不灵敏的 temperature-resistant
温度不稳定 fluctuation of temperature
温度不稳定性 temperature instability
温度测定 temperature determination; temperature measuring
温度测定学 hygrometry; pyrometry
温度测定仪 temperature indicating instrument
温度测杆 temperature rod
温度测高计 thermobarometer
温度测井(曲线) temperature well log(ging); thermal log(ging); thermolog(ging)
温度测量 measurement of temperature; temperature measurement; temperature survey
温度测量法 thermometry
温度测量方法 temperature measurement method
温度测量器 temperature meter
温度测量系统 temperature measurement system
温度测量仪 temperature measuring set; temperature monitor
温度测量仪表 temperature measuring meter
温度测量仪器罩 temperature instrument shelter
温度测量元件 temperature measuring element
温度测量站 temperature measuring station
温度测深 thermometer sounding; thermometric sounding
温度层结 temperature lamination
温度层析作用 thermal stratification
温度差 difference in temperature; temperature difference; hygrometric deficit; temperature differential; temperature drive; temperature head
温度差别按诊法 thermopalpation
温度差距 temperature span; temperature spread
温度差异 temperature contrast
温度场 field of temperature; temperature field; temperature pattern; thermal field
温度成分图 temperature composition diagram
温度程序 temperature program(me)
温度程序设计 temperature programming
温度赤道 heat equator
温度充足 warm enough
温度冲击 thermal shock
温度传导率 temperature conductivity
温度传导系数 coefficient of temperature conductivity; temperature diffusivity
温度传感的 temperature sensing
温度传感器 heat meter; heat-sensitive sensor; temperature bulb; temperature detector; temperature pick-up; temperature probe; temperature sensing device; temperature sensor; temperature transducer; temperature transmitter
温度垂直带 thermal belt; thermal zone
温度垂直剖面辐射计 vertical temperature profile radiometer
温度垂直梯度 lapse rate; vertical temperature gradient
温度垂直滞后 temperature vertical lag
温度带 temperature band; temperature belt; temperature zone
温度倒布 temperature inversion
温度的波动 fluctuation of temperature
温度的成层作用 stratification of temperature
温度的动力变化 dynamic(al) change of temperature
温度的分布 pattern of temperature
温度的混合溶蚀 temperature mixed corrosion
温度的绝对零度 absolute zero of temperature
温度等级 temperature grade
温度滴定 calorimetric titration; enthalpy titration; thermal titration; thermometric titration
温度递减 lapse of temperature
温度递减率 lapse rate; temperature lapse rate
温度递降率 lapse rate; temperature lapse rate
温度电流曲线 temperature current curve
温度调查 temperature survey
温度调动电路 temperature control circuit
温度订正 temperature correction; temperature reduction
温度动态变化 dynamic(al) temperature change
温度陡度 temperature gradient
温度断面图 temperature section
温度对数压力图 emagram
温度惰性上升 coasting of temperature
温度额定值 temperature rating
温度发送器 temperature transmitter
温度发送装置 temperature sending device
温度阀 temperature valve
温度反馈 temperature feedback
温度范围 heat range; range of temperature; temperature band; temperature interval; temperature level; temperature range
温度防腐作用 temperature antiseptic effect
温度分辨率 temperature resolution
温度分布 temperature distribution; temperature traverse; thermal distribution
温度分布(变化)图 temperature profile
温度分布红外辐射计 temperature distribution infrared radio meter
温度分布曲线 temperature profile
温度分布图 temperature chart; thermogram
温度分布图式 temperature figure
温度分布线 temperature profile
温度分布型式 temperature pattern
温度分层 stratification of temperature; temperature lamination; temperature stratification; thermal stratification
温度分类 temperature classification
温度分派表 temperature schedule
温度分散 temperature dispersion
温度风 temperature wind
温度峰值 thermal peak; thermal spike
温度缝 saw cut; temperature joint
温度浮沉子 temperature float
温度幅度 temperature range
温度辐射 temperature radiation
温度辐射体 temperature radiator
温度改变 temperature change
温度改正 correction for temperature; correction of temperature; temperature correction
温度改正值 temperature correction value
温度感测元件 temperature sensor
温度感生荷载 temperature induced load
温度感受器 temperature receptor; thermoreceptor
温度感受元件 temperature ga(u)ge unit; temperature sensing element
温度钢筋 temperature rebar; temperature reinforcement; temperature steel
温度高度图 temperature altitude chart; temperature height diagram
温度跟踪 temperature tracking
温度公式 hygrometric formula
温度共分散 temperature covariance
温度关系 temperature dependence
温度观测 temperature observation
温度观察 temperature survey
温度过程线 thermogram; thermograph
温度过高 hyperpyrexia
温度过高警报 temperature alarm
温度过高停堆 excess temperature shutdown
温度含盐量及海流观测报告 temperature salinity and current report
温度函数 function of temperature; temperature function
温度和湿度 temperature and humidity
温度和湿度控制 temperature and humidity control
温度荷载 temperature load
温度荷重 temperature load
温度恒定 constancy of temperature; temperature constancy
温度红外辐射计 temperature infrared radio meter
温度环境 temperature environment
温度换算 temperature conversion
温度换算表 temperature conversion table
温度换算系数 temperature conversion factor
温度恢复 temperature recovery
温度恢复曲线 temperature recovery curve
温度恢复系数 temperature recovery factor
温度及收缩钢筋 temperature and shrinkage reinforcement
温度极大值 temperature maximum
温度极限 limits of temperature; temperature extreme; temperature limit; temperature range
温度极小值 temperature minimum
温度极值 temperature extremal
温度急增 thermal shock
温度计 heat ga(u)ge; temperature ga(u)ge; temperature indicator; temperature instrument; temperature meter; temp-stick; thermometer
温度计保护套管接管 nozzle for mounting of thermowell
温度计保护装置 thermometer guard unit
温度计标度 thermometer scale
温度计玻璃 thermometer glass
温度计玻璃陈化 ag(e)ing of thermometer glass
温度计插池 thermometer well
温度计插孔 thermometer well; thermowell
温度计的 thermometric
温度计订正表 thermograph correction card
温度计读数 thermometer reading
温度计读数范围 range of thermometer scale
温度计读数机 thermometer reader
温度计法 thermometer method
温度计干球 dry bulb
温度计干湿球温差 wet-bulb depression
温度计管 temperature ga(u)ge pipe; thermometer tube
温度计换算 thermometer conversion
温度计架 thermometer frame
温度计校正 thermometer calibration; thermometer correction; verification of thermometer
温度计接口 thermometer boss
温度计浸入深度 immersion of thermometer
温度计刻度 temperature scale
温度计鸟 thermometer bird

温度计泡 thermometer bulb
温度计球管 thermometer bulb
温度计式风速表 thermometer anemometer
温度计式热敏电阻 thermometer type thermistor
温度计枢轴 thermometer stem
温度计水银球 thermometer bulb
温度计套管 thermometer boss;thermometer well;thermowell
温度计套圈 thermometer collar
温度计调整 thermometer adjustment
温度计误差 thermometer error
温度计小球 thermometer bulb
温度计延迟 thermometer lag
温度计液柱 steam of thermometer
温度计用电线束 temperature ga(u)ge harness
温度计罩 thermometer shield
温度计组 thermometer set
温度记录 temperature record
温度记录板 temperature recorder panel
温度记录法 thermography
温度记录法检查 thermographic(al) inspection
温度记录和控制 temperature record and control
温度记录计 recording thermometer; temperature recorder; thermograph;thermometrograph
温度记录控制器 temperature recorder controller
温度记录器 recording thermometer; self-recording thermometer; temperature recorder; thermometrograph;thermo(tro)graph
温度记录调节器 temperature recording controller
温度记录图 thermogram
温度记录仪 dummy man; recording thermometer; temperature recorder;thermograph
温度继电器 temperature control relay;thermal relay;thermostat
温度继电器驱动阀 temperature actuated valve
温度假潮 temperature seiche
温度间隔 temperature interval
温度监测装置 <大体积混凝土的> temperature measuring device
温度监视器 temperature monitor
温度减率条件 temperature lapse condition
温度减退 temperature decrement; temperature degeneration
温度检测计 temperature detector
温度检测式过载继电器 temperature overload relay
温度检测装置 temperature detecting apparatus; temperature detecting device;temperature detecting unit
温度降(低) temperature drop; thermal drop
温度降校正 temperature drop correction
温度降率条件 lapse condition
温度降落 drop of temperature
温度校正 correction for temperature; temperature correction
温度校正图表 temperature correction chart
温度校准 temperature correction
温度较差 range of temperature; temperature range
温度结构 temperature structure
温度界面 temperature front
温度筋 <混凝土的> temperature bar;temperature reinforcement
温度警报器 temperature buzzer
温度静振 temperature seiche
温度距平 temperature anomaly; temperature departure
温度绝对年较差 absolute annual range of temperature
温度绝缘 temperature insulation
温度绝缘特性 temperature insulation property
温度觉 thermal sensation
温度均衡 temperature equalization
温度均衡而适中 equable moderate
温度均衡和分配装置 temperature equalization and distribution device
温度均化区 equalizing section
温度均一化 temperature equalization
温度均匀性 temperature homogeneity;temperature uniformity
温度开关 temperature switch
温度开裂 temperature crack(ing)
温度控制 attemperation; heat control; temperature control; thermostatic control
温度控制材料 temperature control material
温度控制传感器 temperature control sensor
温度控制措施 temperature control measure
温度控制的 temperature-controlled
温度控制的阻风间系统 temperature-modulated choke system
温度控制阀 temperature control valve
温度控制放大器 temperature control amplifier
温度控制和深冻货车 temperature-controlled and deep freeze wagon
温度控制回路 temperature control loop
温度控制继电器 temperature relay
温度控制晶体振荡器 temperature-controlled crystal oscillator
温度控制器 attemperater [attemperator]; temperature control equipment; temperature controller; temperature regulator;thermoregulator
温度控制室 temperature-controlled chamber
温度控制系统 temperature control(ling) system
温度控制系统测试设备 temperature control system test set
温度控制箱 temperature-controlled cabinet
温度控制卸压阀 temperature actuated pressure relief valve
温度控制装置 temperature control device; temperature control equipment
温度控制组件 temperature-controlled package
温度亏损 temperature defect
温度扩散 thermal diffusion
温度扩散系数 temperature diffusivity
温度廓线红外(线)辐射计 infrared temperature profile radiometer
温度廓线自记仪 temperature profile recorder
温度类别 temperature type
温度力 temperature force; temperature loading;thermal force
温度量测热电偶 temperature measuring thermocouple
温度量测系统 temperature measuring system
温度裂缝 burning crack; fire check; heat check(ing); heat crack; temperature crack(ing)
温度裂纹 fire check;heat check(ing)
温度灵敏度 temperature sensitivity
温度零度 zero point
温度露点差 depression of the dew point;dew point deficit
温度录井 temperature log
温度落差 heat drop;temperature fall
温度脉冲检测器 temperature pulse transmitter
温度脉动 temperature fluctuation
温度漫散射 temperature diffuse scattering
温度密度 temperature density
温度密度线示图 temperature gravity graph
温度描记器 hygrometrograph
温度敏感的 temperature sensitive
温度敏感度 temperature susceptibility;temperature susceptivity
温度敏感反馈 temperature sensing feed-back
温度敏感期 temperature sensitive period
温度敏感探针 temperature sensing probe
温度敏感系数 temperature sensitivity coefficient;temperature susceptibility factor
温度敏感性 temperature sensitivity; temperature susceptibility; temperature susceptivity
温度敏感元件 temperature sensing device; temperature sensing element; temperature sensitive element;temperature sensor
温度敏感装置 temperature sensing device
温度摩擦应力 <多指混凝土路面板因温度变化而发生的> temperature-frictional stress
温度挠曲 temperature deflection; temperature warping
温度逆变 temperature inversion
温度逆增 inversion of temperature; temperature inversion; thermal inversion
温度逆增的顶点 vertex of temperature inversion;lid
温度年较差 annual range of temperature;annual temperature range
温度黏[粘]度曲线 temperature viscosity curve
温度膨胀 thermal expansion
温度膨胀阀 temperature expansion valve
温度膨胀系数 temperature expansion coefficient
温度膨胀指示器 bench
温度偏离额定值 temperature excursion
温度漂移 temperature drift
温度漂移不稳定性 temperature drift instability
温度平衡 hygral equilibrium; temperature balance; temperature equalization; temperature equilibrium; temperature poise
温度平均值 temperature average; temperature mean
温度平流 temperature advection
温度坡度 temperature slope; thermal gradient
温度剖面 temperature profile
温度剖面记录仪 temperature profile recorder
温度气压表 thermobarometer
温度气压计 thermobarometer
温度桥面连续梁桥 decking bridge continuous for temperature effect; temperature continuous decking bridge
温度翘曲 temperature deflection; thermal curling <路面板的>; temperature warping <由温差引起的翘曲>
温度翘曲应力 temperature warping stress;thermal warping stress
温度情况 temperature condition
温度区 temperature zone
温度区间 temperature interval
温度区域 temperature province
温度曲线 thermal profile
温度驱动阀 temperature actuated valve
温度确定 temperature determination
温度扰动 temperature disturbance; thermal perturbation
温度日际变化 interdiurnal temperature variation
温度日较差 diurnal temperature range
温度日值 degree-day value
温度容限 temperature tolerance
温度散发 temperature drive
温度熵 thermal entropy
温度上升 rise in temperature; temperature pick-up; temperature rise
温度上升率 specific temperature rise
温度上限 limiting high temperature; upper temperature limit
温度伸缩材料 temperature movement material
温度伸缩钢筋 temperature steel
温度深度剖面曲线图 temperature depth profile curve map
温度升高 elevation of temperature; temperature build-up; temperature increment
温度升降 gradient of temperature; temperature fluctuation
温度湿度红外辐射计 temperature humidity infrared radio meter
温度湿度计 thermohygrometer
温度湿度记录器 thermohygrograph
温度湿度曲线图 hygrothermograph
温度湿度指标 temperature humidity index
温度时间等效 temperature-time equivalence
温度时间关系 temperature-time combination; temperature-time relationship
温度时间曲线 temperature-time curve
温度时间因数 temperature-time factor
温度时数 degree hour
温度势位 temperature drive
温度试验 temperature test
温度适应 temperature acclimation; temperature adaptation
温度适应能力 temperature capability
温度适中区 range of thermal neutrality
温度收缩 temperature shrinkage; thermal shrinkage;thermal contraction
温度收缩裂缝 thermal contraction crack(ing)
温度收缩仪 <用以测试试件的收缩开裂> thermal contraction apparatus; temperature contraction apparatus
温度收缩应力 temperature contraction stress
温度受感膜盒 temperature sensing capsule
温度输送 temperature transport
温度衰减 temperature damping
温度双曲线 temperature hyperbola
温度随变生物 thermal conformer
温度随深度增高的 katothermal
温度随时间的变化 temperature history
温度损伤 thermal injury
温度损失 temperature loss
温度探测器 hygrosensor
温度探测设备 temperature sensor
温度探测系统 temperature sensing

system
温度特性 temperature characteristic; temperature property
温度特性曲线 temperature characteristic curve
温度梯度 gradient of temperature; heat drop; temperature gradient; temperature slope; temp grad; thermal drop; thermal gradient
温度梯度不稳定性 temperature gradient instability
温度梯度的区熔提纯 temperature gradient zone refining
温度梯度法 temperature gradient method
温度梯度技术 temperature gradient technique
温度梯度区熔 temperature gradient zone melting
温度梯度色谱法 temperature gradient chromatography
温度梯度型保温箱 temperature gradient type incubator
温度梯度纸色层法 temperature gradient paper chromatography
温度天数法 degree-day method
温度条件 condition of temperature; temperature condition
温度调节 heat control; heat regulator; regulation of temperature; temperature adjustment; temperature conditioning; temperature regulation; thermoregulation
温度调节度盘 temperature setting dial
温度调节阀 temperature control valve; temperature regulating valve
温度调节机理 mechanism of temperature regulation
温度调节节点 joint for temperature adjustment
温度调节器 attemperator; duct thermostat; heating controller; temperature regulator; thermoregulator; thermoswitch
温度调节区 conditioner <平板玻璃>; equalizing section <窑池的>
温度调节设备 temperature control equipment; temperature regulating equipment
温度调节室 temperature conditioning chamber
温度调节箱 temperature-controlled cabinet; temperature regulating chamber
温度调节装置 temperature adjusting device; thermostat
温度调节组件 temperature-controlled package
温度调节作用 attemperation
温度突变 sudden temperature change
温度突变法 temperature jump method
温度突然上升 heat surge
温度突跃 thermal shock
温度图 temperature chart; thermogram
温度弯翘应力 temperature warping stress
温度弯曲 <由温差引起的> thermal bending
温度微差 temperature differential
温度微结构 temperature microstructure
温度位 temperature level
温度位移 <由温差引起的> temperature movement; temperature displacement
温度稳定器系统 temperature stabilizer system
温度稳定系统 temperature stabilizing system
温度稳定性 temperature stability; temperature stabilization
温度无关磁心 temperature independent core
温度误差 temperature error
温度稀释曲线测定 thermodilution curve determination
温度系数 coefficient of temperature; temperature coefficient; temperature factor
温度系数反馈 temperature coefficient feedback
温度系数功率调节 temperature coefficient power regulation
温度隙 expansion gap
温度下降 descent of temperature; drop in temperature; drop of temperature; fall in temperature; lapse; lowering of temperature; temperature drop; temperature fall; temperature reduction
温度显示板 temperature display panel
温度线图 temperature profile
温度限度 temperature margin
温度限制 temperature limitation
温度限制二极管 temperature limited diode
温度限制范围 temperature limited region
温度限制控制器 temperature limit controller
温度限制条件 temperature limited condition
温度限制准则 temperature limiting criterion
温度相关性 temperature dependence
温度相依常数 temperature-dependent constant
温度响应 temperature response
温度效率 temperature efficiency; thermal efficiency
温度效率比 temperature efficiency ratio; thermal efficiency ratio
温度效率指数 temperature efficiency index
温度效应 temperature effect
温度效应连续桥面桥 decking bridge continuous for temperature effect
温度形变曲线 thermomechanical curve
温度性收缩 thermosystaltism
温度修正 temperature correction
温度修正图表 temperature correction chart
温度修正系数 coefficient of temperature correction
温度序 temperature sequence
温度悬差 temperature contrast
温度选择器 temperature selector
温度巡检箱 temperature scanning unit
温度循环 temperature cycle
温度循环变化 thermal cycling
温度循环试验 temperature cycling test
温度压力 temperature press(ing); temperature pressure
温度压力补偿运算器 temperature pressure compensating computer
温度压力曲线 temperature pressure curve
温度压应力 thermal compressive stress
温度延迟系数 temperature coefficient of delay
温度、盐度和流速记录仪 temperature, salinity and current recorder
温度盐分合成对流 thermohaline convection
温度遥测仪 temperature telemeter
温度依赖行为 temperature-dependent behavio(u)r
温度依赖性 temperature dependence
温度异常 temperature anomaly
温度因数 temperature factor
温度因素 temperature factor
温度引起的额定值降低 temperature derating
温度应变 temperature strain; thermal strain
温度应力 stress due to temperature; stress due to temperature change; temperature stress; thermal stress
温度应力变化 variation in temperature stresses
温度应力峰 temperature stress peak
温度应力钢筋 temperature stress rod
温度应力应变图 temperature stress-strain diagram
温度应力折减系数 thermal stress reduction factor
温度影响 influence of temperature; temperature effect; temperature influence
温度影响范围 temperature pattern
温度影响区域 heat-affected zone
温度硬化能谱 temperature hardened spectrum
温度有效指数 temperature efficiency index
温度诱导 temperature induce
温度诱致 temperature induced
温度余值指数 remainder index
温度与露点温度带 depression of the dew point
温度与深度关系图 diagram showing the relation between temperature and depth
温度与湿度关系图 hythergraph
温度与湿度红外辐射计 temperature and humidity infrared radio meter
温度与时间的关系(曲线) temperature history
温度与氧逸度关系图 diagram showing the relationship between temperature and oxygen fugacity
温度雨量曲线 hydrotherm figure
温度雨量图 hytherograph
温度预报 temperature forecast
温度元件 temperature element
温度约束应力 temperature binding stress
温度跃变 temperature jump; temperature step
温度跃(变)层 epilimnion(layer); thermocline layer; transition layer of temperature
温度跃变距离 temperature jump distance
温度跃迁 temperature jump
温度跃升 temperature jump
温度在零度以下的 <尤指华氏度> subzero
温度噪声 temperature noise
温度增高 temperature increase
温度增量 temperature increment
温度胀缩 temperature movement
温度折射 temperature refraction
温度振荡法 temperature oscillation method
温度振动 temperature vibration
温度正常 normothermia
温度正常值 temperature normal
温度直减率 lapse rate; temperature lapse rate
温度指示 temperature indication
温度指示计 indicating thermometer; heat indicator
温度指示记录仪 temperature indicating recorder
温度指示控制器 temperature indicating controller
温度指示控制仪 temperature indicating controller
温度指示屏 thermoindicating panel
温度指示器 dial indicator; temperature indicating device; temperature indicator
温度指示器接头 temperature indicator adapter
温度指示条 temperature indicating strips
温度指数 temperature exponent
温度制度 temperature program(me); temperature schedule
温度致衰 temperature degeneration
温度滞后 temperature lag; thermal hysteresis
温度滞后现象 temperature hysteresis
温度中值 medial temperature
温度周期变化 temperature cycling
温度骤变 temperature shock
温度转化 inversion of temperature; temperature inversion; thermal inversion
温度转换 inversion of temperature; temperature inversion; thermal inversion
温度状况 temperature regime
温度状态 state of temperature
温度状态点 temperature state point
温度准数 temperature number
温度自补偿应变片 self-temperature compensated strain ga(u)ge
温度自动记录器 automatic temperature recorder
温度自动控制 automatic temperature control; thermostatic control
温度自动调节 automatic moisture control
温度自动调节器 automatic temperature controller; thermostat; automatic temperature compressor
温度自记曲线 differential thermal analysis curve; thermogram
温度自记仪 eupatheoscope
温度自控器 automatic heat regulator
温度总和 summation of temperature; temperature sum
温度总和规律 temperature sum rule
温度走时特性 temperature rate characteristic
温度阻限 temperature barrier
温度组成图 temperature composition diagram
温度最适化 temperature optimization
温度作用 temperature action; temperature effect; thermal action; thermal temperature action
温锻 warm forging
温福特红颜料 Winford red
温伽特牌搅拌机 <一种混凝土搅拌机> Winget
温感 thalposis
温感报警器 temperature alarm
温感控制 temperature sensitive control
温感器 temperature detector; temperature receptor
温感探测器 heat detector
温感自动报警器 automatic heat detector
温哥华港 <加拿大> Port Vancouver
温焓表 temperature-enthalpy chart
温焓图 temperature-enthalpy chart; temperature-enthalpy diagram
温和 moderation
温和的 benign; gentle; moderate; sweet; tender; temperate

温和的气候 mild climate
温和的天气 moderate climate
温和地带 temperate belt
温和碱 irenine
温和期 <指间冰期> miothermic period
温和气候 mild climate; moderate climate; temperate climate
温和热度 moderate temperature
温和天气 mild weather
温和条件 mild condition
温和温度 moderate temperature
温和状态数 number of accessible states
温壶 hot-water vase
温惠氏(土)粒径分类表 Wentworth scale
温季植物 warm-season plant
温加工 warm working
温降 drop in temperature; drop of temperature; fall of temperature; heat drop; temperature drop; temperature drop fall; temperature fall; temperature reduction; warm season
温降曲线 temperature decay curve; temperature drop curve
温井 hot well
温卷 warm rolling
温克尔地基 Winkler's foundation
温克尔尔滴定 Winkler's titration
温克尔发生炉 Winkler's generator
温克尔法 Winkler's method
温克尔假设 Winkler's assumption; Winkler's hypothesis
温克尔理论 Winkler's theory
温克尔量管 Winkler's burette
温克尔模型 Winkler's model
温克尔气化法 Winkler's gasification method
温克尔气体分析管 Winkler's burette
温克尔三角形图 Winkler's triangle
温克尔系数 Winkler's coefficient
温克尔值 Winkler's value
温克勒假设 <弹性地基计算的一种假设> Winker's hypothesis
温控继电器 temperature control relay
温控开关 thermal cutout
温控裂纹试验 controlled thermal cracking
温控器 temperator
温控涂层 temperature control coating
温控涂料 thermal control coating material
温控有机涂层 organic thermal control coating
温矿泉 warm mineral spring
温洛克价 <中志留世>【地】Wenloekian
温洛克组 <中志留世>【地】Wenlock formation
温敏阀 temperature responsive valve
温纳(电极)排列 <用于土地电阻测定> Wenner configuration
温纳(电阻测量)法 Wenner method
温纳排列 Wenner arrangement
温暖带 warm-temperature zone
温暖的 calefactory; cosy [cozy]
温暖期 warm epoch
温暖气候 temperate climate
温暖气团 warm-air mass
温暖器 calefactor
温排 <电站> warm water drainage
温排水 thermal discharge
温铺的 <沥青混合料等> warm-laid
温铺沥青混合料 warm bituminous mix
温谱图 thermogram
温切尔特切割 Winchester cutting
温切尔特切割法 Winchester cut method
温切斯特磁盘机 Winchester disk drive
温切斯特(硬盘)技术 Winchester technology
温球 thermometer bulb
温泉 <水温在37℃以上> hot spring; acratotherm(e); hot spa; hot well; spa; spa thermae; thermae; thermal spring; thermopegic; warm spring; watering-place
温泉池 hot pool
温泉的 thermal
温泉的产状分类 occurrence classify of spring
温泉的成因分类 origin classify of spring
温泉的化学成分分类 composition classify of spring
温泉的活动状态分类 activity classify of spring
温泉的流量分类 flow rate classify of spring
温泉的酸碱度分类 acidity classify of spring
温泉的温度分类 temperature classify of spring
温泉的用途分类 usage classify of spring
温泉废水 wastewater of spring
温泉分布图 distribution map of spring
温泉风景区 hot spring scenic spot
温泉湖 hot pool
温泉疗养地 watering-place
温泉疗养院 hot spring sanitarium
温泉区 thermal region
温泉群落 thermium
温泉水 hot spring water; thermal water
温泉水的来源分类 water source classify of spring
温泉余土 hot spring soil residual
温泉浴场 thermae; thermal-bath
温泉浴室 therma room
温泉滓石 plombierite
温热 tepefaction
温热标准 thermal standard
温热处理 chauffage
温热的 tapid; thermal
温热地面 warm ground
温热喷雾 warm spraying
温热区 warm area
温热式风速计 thermoanemometer
温莎堡皇家建筑工匠长 Master of the King's Masons to Windsor Castle
温莎探测器 Windsor probe
温莎椅 Windsor chair
温熵表 temperature-entropy chart
温熵曲线 entropy-temperature curve
温熵图 temperature-entropy chart; temperature-entropy diagram; tephigram
温熵坐标图 temperature-entropy coordinates
温蛇纹石 nephritoid
温深电导记录仪 conductivity temperature depth recorder
温深记录曲线轨迹 bathythermograph trace
温深曲线 temperature depth curve
温深仪 bathythermograph
温深仪绞车 bathythermograph winch
温升 elevated temperature; temperature increase; temperature rise; exotherm <释放化学能引起的>
温升报警信号 heat alarm
温升比 temperature rise ratio
温升的 warm up
温升(过热)报警信号 temperature alarm
温升极限 temperature rise limit
温升曲线 temperature rise curve
温升试验 temperature rise test
温升速率 rate of temperature rise
温升探测器 temperature rise detector
温升系数 temperature rise coefficient
温升效率 temperature rise efficiency
温升周期 temperature rise period
温湿 humit
温湿表 polymeter; thermohydrograph; thermohygrometer
温湿度 humiture
温湿度标准 atmosphere standard
温湿度表 psychrometric table
温湿度的 psychrometric
温湿度电微压计 tasimeter
温湿度过程线 thermohygrograph
温湿度红外辐射计 temperature humidity infrared radio meter
温湿度红外辐射仪 temperature humidity infrared radio meter
温湿度基数 temperature and humidity datum
温湿度记录器 thermohydrograph
温湿度控制 temperature and humidity control
温湿度数字显示器 digital temperature and humidity indicator
温湿度状态线 condition line
温湿计 hygrothemograph; thermohygrograph
温湿气候带 warm wet climate zone
温湿曲线图 temperature humidity graph; thermohygrogram; hythergraph
温湿图 psychrometric chart
温湿图解 temperature moisture diagram
温湿循环 temperature humidity cycling
温湿仪 hygrothermoscope
温湿指数 comfort index; discomfort index; moisture-temperature index; temperature humidity index
温湿自记曲线 thermohygrogram
温湿自记仪 hygrothemograph
温石棉 Canadian asbestos; chrysotile; chrysotile asbestos; chrysotilite; serpentine asbestos
温时曲线 time-temperature curve
温时效 warm hardening
温氏分级表【地】Wentworth scale
温氏分类 Wentworth classification
温氏泥沙粒度分级标准 Wentworth scale
温室 conservatory; forcing house; glasshouse; greenhouse; growing house; horticultural building; hot house; hot spring; stovehouse; warmhouse; warming-house; anthracite stove <英国温室的一种>; palm house <栽培棕榈等用>
温室玻璃 garden glass; greenhouse glass; horticultural cast glass; horticultural glass; transparency horticultural glass
温室的天窗 roof ventilater
温室防潮 moist room dampproofing
温室构造 greenhouse construction
温室管理 greenhouse management
温室加温 greenhouse heating
温室控制器 greenhouse controller
温室模型 greenhouse model
温室木架 greenhouse benches
温室农业 agriculture in greenhouse
温室培养 glasshouse culture
温室喷雾器 greenhouse sprayer
温室气候 glasshouse climate; greenhouse climate
温室气候控制器 greenhouse climate controller
温室气体 greenhouse-effect gas; greenhouse gas
温室气体排放数据库 emission inventory of greenhouse gases
温室设备 greenhouse facility
温室设计 greenhouse planning
温室升温潜能值 greenhouse warming potential
温室试验 greenhouse experiment; greenhouse test
温室蔬菜栽培 greenhouse vegetable growing
温室土壤的灭菌处理 sterilization of greenhouse
温室效果 glasshouse effect; greenhouse effect
温室效应 glasshouse effect; greenhouse effect; hothouse effect; warmhouse effect
温室效应气体计划 program(me) on greenhouse gases
温室效应引起的气候变化 greenhouse-induced variational of climate
温室效应引起的气温升高 greenhouse warming
温室用玻璃 greenhouse cast glass
温室用的镶嵌玻璃 horticultural cathedral glass
温室用涂料 coatings for greenhouse
温室栽培 glasshouse culture; greenhouse culture
温室栽培大豆 greenhouse culture soybean
温室栽培的 hot house
温室栽培小麦 greenhouse culture wheat
温室植物 glasshouse plant; greenhouse plant; hothouse plant; indoor plant; stove plant
温室作用 greenhouse effect; hothouse effect; warmhouse effect
温水 aqua tepida; lukewarm; lukewarm water; mild water; warm water
温水场 thermal field
温水发生器 hot-water generator
温水灌溉 warm water irrigation
温水湖 warm lake
温水器 water heater
温水融化解冻 warm water thawing
温水入口阀 hot-water inlet valve
温水入口止逆阀 check valve for hot water inlet
温水设备 water-warming facility
温水塘 warm pool
温水洗炉 water heating wash boiler
温水循环采暖系统 water circulation heating system
温水浴 lukewarm bath
温水浴池 tepidarium
温水浴室 tepidarium
温顺型 <"朝向他人"的人际关系> compliant type
温斯洛效应 Winslow effect
温缩裂缝 thermal crack(ing)
温缩应力 temperature contraction stress
温塔岩组 Uinta formation
温台显微镜 hot stage microscope
温吞的 lukewarm
温位 potential temperature; temperature level
温相厌氧消化 temperature-phased anaerobic digestion
温性腐殖质 mild humus
温性针叶林 temperate needle-leaf forest
温血动物 warm-blooded animal

温血动物污染 warm-blooded animal pollution
温压 warm-pressing
温压表 thermobarometer
温压波 thermobarometric(al) wave
温压图 pressure-temperature chart
温盐测量仪 thermosalinograph
温盐场 thermohaline field
温盐的 thermohaline
温盐对流 thermohaline convection
温盐海流 thermohaline current
温盐环境 thermohaline environment
温盐环流 thermohaline circulation
温盐结构 thermohaline structure
温盐流报告 temperature salinity and current report
温盐曲线 temperature salinity curve
温盐深度测量装置 salinity temperature depth monitor system; salinity temperature depth recording system; temperature depth salinity system
温盐深度系统 salinity temperature depth system
温盐深度自记仪 salinity temperature depth recorder
温盐图解 temperature salinity diagram
温盐线 thermohaline
温盐洋流 thermohaline current
温硬化 warm hardening
温雨图 hytherograph
温跃层 metalimnion; thermal layer; thermocline
温跃层结构 structure of thermocline
温轧 warm rolling
温胀 swelling
温周期 periodic(al) reaction to temperature
温周期现象 thermoperiodism
温周期性 thermoperiodicity
温阻效应 thermoresistive effect

瘟

瘟疫 pestilence; plague
瘟疫控制 pest control

文

文本字体 text font
文昌鱼 amphioxus; Branchiostoma
文达瓦尔风 vendaval
文带轮 carrying idler
文旦 pomelo
文档 document
文蒂罗克移窗 <一种专利商品> Ventilock
文电处理系统 message handling system; message processing system
文电鉴别 authentication
文电鉴别码 authenticator
文电收据 message acknowledgment
文牍 scribe
文牍主义 officialism; red-tapism
文杜里管 Venturi
文杜里喷嘴 Venturi
文多奔阶 Vindobonian
文法学校 grammar school
文官(职务) civil service
文冠果 shiny-leaved yellow horn
文号 reference number
文化背境 cultural milieu
文化地物【测】 culture
文化堆积物 cultural deposit
文化多元论 cultural pluralism
文化隔阂 cultural barrier
文化公园 cultural park
文化宫 cultural palace; house of culture; lyceum; palace of culture
文化共生状态 cultural symbiosis
文化馆 cultural building
文化馆中心 artistic circle
文化广场 cultural plaza
文化环境 culture environment
文化会堂 cultural hall; culture hall
文化活动 cultural activity
文化活动中心 cultural activity center [centre]
文化机构 cultural organization
文化建筑 cultural building
文化交流 cultural exchange
文化交流中心 cultural exchange center [centre]
文化教育 literacy education
文化教育中心 educational center [centre]
文化界 artistic circle
文化景观 cultural landscape
文化景观保护区 cultural landscape protected area
文化景观论 theory of cultural landscape
文化俱乐部 culture club
文化决定论 cultural determinism
文化连续性 cultural continuity
文化面貌 cultural feature
文化名城 famous cultural city
文化普及 universal access to culture
文化区 cultural area
文化人类学 cultural anthropology
文化商业区 cultural and commercial zone
文化商业中心 metropolis
文化上的 cultural
文化设施 cultural facility
文化社会环境 cultural-social environment
文化生活 cultural life
文化水平 level of education
文化特色区设计 culturally sensitive design
文化特性 cultural identity
文化污染 cultural pollution
文化休憩公园 cultural and recreation park
文化要素 cultural feature
文化遗产 cultural heritage
文化影响 cultural influence
文化运动 cultural movement
文化障碍 cultural barrier
文化珍品 culture valuables
文化指标 cultural indicator
文化中心 cultural center [centre]
文化专员 cultural attaché
文火 low fire; mild fire; smo(u)lder; soft fire
文火燃烧器 simmer burner
文火调节器 simmer control
文集 collected works; omnibus book; omnibus volume
文间注释 <图书资料的> incut note; cut-in notes
文件 document; file; file of information; paper
文件安全控制 file security control
文件保险库房 muniments house
文件保险箱 deed box
文件编号 numbering of documents
文件号 document number; file number
文件夹 file cover; folder; folio; portfolio
文件自动存取与检索 automatic document storage and retrieval
文教车 culture and education car
文教机关 educational institute
文教建筑群 educational block
文教建筑物 educational building
文教区 civil district; cultural district; cultural-educational zone; educational area; education district
文教中心 educational center [centre]
文具 writing materials
文具的 stationery
文具店 stationery shop
文具盒 stationery case
文具库 stationery stock room; stationery store
文具室 stationery room
文具匣 stationery case
文卷 file
文卷柜 file cabinet
文卷结束 end of volume
文科 literary course
文科教育 liberal education
文克勒地基 Winker's foundation; dense liquid subgrade
文克勒假设 Winker's assumption
文库兰牌预制楼板 <商品名> Vinculum
文莱 Brunei
文脉 context
文脉性 contextuality
文脉主义 contextualism
文明城市 culturally advanced city; model city
文明施工 civilized construction
文末字符 end of text character
文凭 diploma
文丘里 <意大利物理学家，又称文氏> Venturi
文丘里比例流量计 Venturi partial meter
文丘里测量管 Venturi-meter
文丘里测流量管 Venturi tube
文丘里测流水槽 Venturi flume
文丘里除尘器 Venturi collector; Venturi dust scrubber; Venturi scrubber
文丘里吹风管 Venturi blower
文丘里涤气器 Venturi gas scrubber; Venturi scrubber
文丘里管 Venturi conduit; Venturi pipe; Venturi tube; Venturi washer
文丘里管混合器 Venturi tube mixer
文丘里管喷嘴 Venturi nozzle
文丘里管式集尘器 Venturi dust trap
文丘里管式流量计 Venturi tube flowmeter
文丘里管温度计 Venturi tube thermometer
文丘里管斜度 Venturi batter
文丘里管形洒布机 Venturi spreader
文丘里管总成 Venturi assembly
文丘里管嘴 Venturi mouthpiece
文丘里管作用 Venturi tube action
文丘里喉管 Venturi throat
文丘里计量管 Venturi-meter
文丘里净气器 Venturi nozzle scrubber
文丘里空气断开 Venturi air break
文丘里扩散喷管 Venturi type expansion nozzle
文丘里扩散器 Venturi diffuser
文丘里量水槽 Venturi flume
文丘里流量计 Venturi-meter; Venturi tube
文丘里流量计变数 Venturi-meter variable
文丘里流量控制阀 Venturi flow control valve
文丘里流量系数 Venturi-meter coefficient
文丘里流速计 Venturi-meter
文丘里耙头 Venturi draghead
文丘里喷射器 Venturi ejector
文丘里气体吸收器 Venturi gas scrubber
文丘里绕闸充水管道 Venturi loop conduit
文丘里式发动机冷却 Venturi engine
文丘里式发酵罐 Venturi type fermentor
文丘里式量水槽 Venturi flume
文丘里式烟囱 Venturi chimney; Venturi stack
文丘里式装药器 Venturi type loader
文丘里竖式向上气流洗涤器 Venturi vertical upward gas flow scrubber
文丘里水表 Venturi flowmeter; Venturi-meter
文丘里速度计 Venturi flowmeter
文丘里调压器 demand-actuated governor
文丘里图 Venn diagram
文丘里喂入管 Venturi feed tube
文丘里雾化洗涤器 Venturi gas scrubber
文丘里吸收器 Venturi absorber
文丘里洗涤器 Venturi scrubber
文丘里洗气法 Venturi scrubbing
文丘里洗气机 Venturi washer
文丘里洗气器 Venturi scrubber
文丘里洗气-旋风式分离装置 Venturi scrubber cyclone separator unit
文丘里细腰式装药管 Venturi loader
文丘里效应 Venturi effect
文丘里型喷射器 Venturi jet
文丘里压差流量计 Venturi flowmeter
文丘利槽 venture flume
文丘利管 venture tube
文丘利流量计 venture(flow)meter
文丘利速度计 venture(flow)meter
文丘利细腰式管 venture loader
文沙剂 Vinsol agent
文砷钯矿 vincentite
文石 aragonite; chimborazite
文石补偿深度 aragonite compensation depth
文石-方解石互层 erzbergite
文石华 flos ferri; flower of iron
文石灰岩 aragonitic limestone
文石律 aragonite law
文氏放射定律 Wein's radiation law
文氏管 Venturi tube
文氏图【数】 Venn diagram
文书 correspondence clerk; diploma
文书的 clerical
文书工作 clerical work
文书检验 verification of documents
文书局 Stationery Office
文泰纳式通风集装箱 ventainer
文特波振子 Van der pol's oscillator
文体识别 character recognition
文体先驱 stylistic forerunner; stylistic predecessor
文尾 file trailer
文伍德牌百叶窗 <一种活动百叶窗> Venwood
文物 art treasure; cultural property; cultural relic; historic(al) relics; relics
文物保护 preservation of cultural relics; protection of historic(al) relics
文物保护区 cultural relic preservation area
文物古迹 cultural heritage
文物古迹保护 conservation of historic(al) landmarks and sites
文物建筑 land mark
文物遗址 point of cultural interest
文献 documentation; literature
文献自动检索装置 documentation automated retrieval equipment
文象斑岩 granophyre; pegmatophyre
文象斑状的 granophyric; pegmatophyric; graniphyric
文象共生 graphic(al) intergrowth
文象构造【地】 pegmatitic structure; graphic(al) structure
文象花岗石 graphic(al) granite
文象花岗岩 graphic(al) granite; He-

W

braic granite;runite;schriftgranite
文象结构 graphic(al)texture
文选 thesaurus
文学 literature
文学硕士 master of arts
文学学士 Bachelor of Arts
文学院 factory of arts
文学作品 literature
文艺复兴 Revival of Learning;the Renaissance
文艺复兴初期 <15 世纪的> Quattrocento
文艺复兴后建筑 post-renaissance architecture
文艺复兴及其派生建筑正面上的线脚和装饰 dressings
文艺复兴建筑 Quattrocento architecture
文艺复兴时期 Quattrocento
文艺复兴时期建筑 Renaissance architecture
文艺复兴式 Renaissance style
文艺复兴式宫殿 Renaissance palace
文艺复兴式教堂 Renaissance church
文艺复兴式穹隆 Renaissance dome
文艺复兴庄园 Renaissance style villa
文艺纪念碑 choragic monument
文艺界 circle of literature and art
文艺领域 realm of literature and art
文艺团体 lyceum
文娱场地 recreational area
文娱场所用水 recreational water
文娱活动室 playroom
文娱建筑 recreation building
文娱设施 recreation(al)facility
文娱室 day room;game room;lounge;playroom;recreation room
文苑 lyceum
文泽尔蓝 Wenzel's blue
文摘 digest;tabloid
文摘服务(处)abstracting service
文摘和索引 abstract and index
文职 white-collar jobs
文职公务人员 civil service
文职雇员 civilian employee
文职官员 magistrate
文职人员 civilian personnel;civil servant;civil service
文中小标题 crosshead
文竹 asparagus fern;emerald feather
文字报告式报表 narrative form of statement
文字比例尺 statement of fraction;verbal scale
文字编辑软件 word processor software
文字编码 literal code
文字标记 <交通标志> word mark(ing)
文字标牌 <道路交通的> symbol word mark(ing)
文字处理软件 word processor software
文字处理系统 words processing system
文字记号 literal notation
文字记录测井仪 digital logging system
文字数据 textual data
文字数字混合标记法 combined numbering and literal symbol method
文字通告 <标志上的> lettered message
文字系数【数】literal coefficient
文字原稿 original copy for type composition

纹

纹板 card;figure sheet;pattern card
纹板冲孔机 card nipper;card puncher
纹板串连孔 lacing hole
纹板架 cradle
纹板链 pattern chain
纹板链传动机构 paper card reading mechanism
纹板岩 varve(d)slate
纹波 ripples
纹波频率 ripple frequency
纹波系数 ripple factor
纹层【地】lamina [复 laminae/laminas]; laminar bedding; laminated bedding
纹层黏[粘]土 laminated clay
纹层岩 laminated rock
纹层状 laminae
纹层状构造 lamellar structure;laminated structure
纹层状互层 interlaminated
纹层组 lamina set
纹长二长岩 mangerite
纹道 track
纹道刻刀 cutter head;cutting head
纹钉保护器 pin guard
纹钉图 lifting plan;peg plan
纹杆 rasp bar
纹杆滚筒 drum with rasp bars
纹杆式滚筒 rasp bar cylinder
纹杆式脱粒滚筒 rasp bar drum
纹沟冲蚀 shoestring washing
纹沟冲刷 shoestring washing
纹脊 wrinkle ridge
纹孔 pit
纹孔次生壁 pitted secondary wall
纹孔导管 pitted vessel
纹孔道 pit canal
纹孔对 pit pair
纹孔环 pit annulus
纹孔口 aperture;pit aperture;pit orifice
纹孔膜 pit membrane
纹孔内口 inner pit aperture
纹孔腔 pit cavity
纹孔塞 pit torus;torus [复 tori]
纹孔式 pitting
纹孔室 pit chamber
纹孔外口 outer pit aperture
纹孔缘 pit border
纹枯利 dimethachlon
纹理 grain;lamel(lation);lamination;streak(ing);vein(ing);venation;texture <木材岩石等的>
纹理板 textured board
纹理不规则的 cross-grained
纹理粗的木材 coarse-textured timber
纹理粗疏的 coarse-grained
纹理方向 grain direction;direction of grain
纹理分析 texture analysis
纹理合成 texture synthesis
纹理黑檀木 veined ebony
纹理灰岩 laminoid-fenestral limestone
纹理交织的 cross-grained
纹理胶合板 textured plywood
纹理角度 grain angle
纹理结构 texture;texture structure
纹理裂缝屋面板 texture split shakes panel
纹理裂隙 grain rupture
纹理木材 veined wood
纹理扭曲的木材 wood with crooked fibre
纹理判别 texture discrimination
纹理漆 texture paint
纹理深度 texture depth
纹理饰面 textured finish
纹理试验 streak test
纹理饰面涂料 texture-finished paint
纹理刷 overgrainer
纹理松的 coarse-textured;open-grained
纹理特征 grain character
纹理梯度 texture gradient
纹理统计 texture statistics
纹理统一法 method of unifying the rock veins
纹理突起 grain raising
纹理图案 textured pattern
纹理图标志 texture map indices
纹理乌木 veined ebony
纹理斜度 slope of grain
纹理斜率 slope of grain
纹理油漆 trick paint
纹理增强 texture enhancement
纹理致密的 <木材> close-grained
纹理组织 veining structure
纹连续图案 swastika
纹裂 broken grain
纹裂砖 chuff
纹流 laminar flow
纹路 mark
纹路交叉的 cross-cut
纹脉结构 ridge configuration
纹泥 <又称季候泥,一层黏[粘]土和一层粉土互叠的泥土,每层很薄,厚度很少大于半英寸> banded clay;laminated clay;varved silt varved clay;varve
纹泥测年法 varved dating method
纹泥沉积 varved deposit
纹泥分析 varve analysis
纹盘钢索卡子 wire rope anchor
纹盘固定式绳头卡子 fixed anchor
纹片釉 crackle glaze
纹石束带层 curstable
纹饰 ornamentation;texturization
纹影 schlieren
纹影法 schlieren
纹影法摄影装置 schlieren set-up
纹影干涉仪 schlieren interferometer
纹影技术 striation technique
纹影屏 schlieren optic(al)screen
纹影设备镜 schlieren mirror
纹影摄影术 schlieren photography
纹影术 schlieren technique
纹影仪 schlieren apparatus;schlieren device
纹影照相 streak photograph
纹影照相术 schlieren photography
纹章装饰 armorial decoration
纹章装饰瓷器 heraldic china
纹状层理 laminar bedding
纹状骨 osteopathia striata
纹状黏[粘]土 laminated clay
纹状区 area striate
纹状体 corpus striatum;striate body;striatum
纹状缘 striated border

闻

闻 smell
闻味探测 <下水管找漏用> scent test
闻阈图 audiogram;audiograph;threshold audiogram

蚊

蚊式飞机 mosquito
蚊帐垂饰 mosquito-net decorative pendant
蚊帐纱 insect cloth
蚊子 mosquito

吻

吻合 anastomosis [复 anastomoses]; close agreement; coincidence;concordance;fit;tally
吻合的 anastomosed
吻合度 curvature tolerance;goodness of fit
吻合断面 concordant cross-section;concordant profile
吻合钢筋腱 concordant tendon
吻合(钢丝)束 concordant tendon
吻合钢索束 concordant tendon
吻合横截面 concordant cross-section
吻合截面 <在预应力混凝土连续梁内将钢索布置成不产生附加支座反力的线形> concordant profile
吻合力筋 concordant tendon
吻合轮廓 concordant profile
吻合面 faying surface
吻合频率 coincidence frequency
吻合剖面 concordant cross-section;concordant profile
吻合索 concordant cable;concordant tendon
吻合涂布 kiss coating
吻合外形 concordant profile
吻合显微镜 concordant microscope;fittings microscope
吻合线 concordant line
吻合线形 concordant profile
吻合线形性 concordant profile
吻合效应 coincidence effect;concordant effect
吻合预加应力 concordant prestressing
吻合预应力钢索 concordant tendon
吻合预应力钢丝束 concordant cable
吻合预应力筋 concordant tendon
吻切点 osculating point
吻切根数 osculating element
吻切轨道 osculating orbit
吻切轨道根数 osculating orbital elements
吻切历元 osculating epoch
吻切平面 osculating plane
吻切椭圆 osculating ellipse
吻兽【建】dragon-head ridge ornament
吻突 muzzle;proboscis
吻涂机 kiss coater

紊

紊动 turbulent fluctuation;turbulent motion
紊动尺度 scale of turbulence
紊动传热 turbulent heat transfer
紊动粗糙度 turbulent roughness
紊动度 degree of turbulence
紊动风荷载 turbulent-wind loading
紊动浮射流 turbulent buoyant jet
紊动光滑的 turbulent smooth
紊动过渡的 turbulent transitional
紊动剪切流 turbulent shear flow
紊动扩散系数 turbulent diffusion coefficient
紊动流体 turbulent fluid
紊动媒质 turbulent medium
紊动明渠流 turbulent open channel flow
紊动能 energy of turbulence
紊动强度 intensity of turbulence;turbulence intensity;turbulent intensity;turbulent strength
紊动射流 turbulent jet
紊动水流 turbulent flow;turbulent stream
紊动尾流 turbulent wake
紊动涡流 eddy flux
紊动性 turbulence
紊动性衰减 turbulence damping;turbulent damping
紊动旋涡上升流动 kolk
紊动异重流 turbulent density current

紊动转换 turbulent exchange
紊剪流 turbulent shear flow
紊流 bumpy flow; burble; eddy flow; erratic flow; non-uniform flow; sinuous flow; swirl; tortuous flow; turbulence; turbulent current; turbulent flow; turbulent fluid; unbalanced flow
紊流边界层 turbulent boundary layer
紊流边界层分离 turbulence boundary separation
紊流标度 scale of turbulence
紊流表面摩擦 turbulent skin friction
紊流层 turbosphere; turbulent layer
紊流层顶 turbopause
紊流场 field of turbulent flow; turbulent field
紊流程度 degree of turbulence
紊流尺度 scale of turbulence
紊流传播 turbulence propagation; turbulent propagation
紊流大气 turbulent atmosphere
紊流的 turbulent
紊流的相似理论 similarity theory of turbulence
紊流动能 turbulent kinetic energy
紊流抖振 turbulent buffeting
紊流度 scale of turbulence; turbulence intensity; turbulence number; turbulence scale; turbulent level; turbulivity
紊流段 turbulent reach; water-break
紊流放大器 turbulent amplifier
紊流分离(现象) turbulence separation; turbulent separation
紊流风模拟技术 turbulent wind simulation technique
紊流腐蚀 turbulent attack
紊流感应噪声 turbulence induced noise
紊流混合 turbulent mixing
紊流火焰 bushy flame; turbulent flame
紊流激发器 turbulence stimulator
紊流计 turbulence indicator; turbulence meter
紊流夹卷 turbulent entrainment
紊流交换 turbulent exchange; turbulent interaction
紊流交换法 <估计蒸发量> eddy correlation method
紊流交换量 austausch
紊流搅拌 turbulent mixing
紊流界限 turbulence level
紊流扩散 eddy diffusion; turbulence diffusion; turbulent diffusion; turbulent exchange
紊流扩散方程 turbulence diffusion equation; turbulent diffusion equation
紊流扩散过程 turbulent diffusion process
紊流扩散率 turbulent flux
紊流扩散系数 eddy diffusivity; turbulent diffusion; turbulent diffusivity
紊流理论 turbulence theory
紊流刘易斯数 turbulent Lewis number
紊流流速分布 turbulent velocity distribution
紊流流态 turbulent condition; turbulent flow regime
紊流率 degree of turbulence; turbulence rate
紊流脉动 turbulent fluctuation
紊流煤粉燃烧器 turbulent burner
紊流密度变动 turbulent density fluctuation
紊流密度波动 turbulent density fluctuation
紊流摩擦 turbulent skin friction
紊流能 turbulence energy
紊流逆温 turbulence inversion
紊流普朗特数 turbulent Prandtl number
紊流谱 turbulence spectrum
紊流器 ebullator; turbulence ring
紊流迁移系数 turbulent coefficient
紊流强度 intensity of turbulence; turbulence intensity; turbulent intensity
紊流区 turbulent range; turbulent region; turbulent zone
紊流燃烧器 turbulent flow burner
紊流燃烧速度 turbulent burning velocity
紊流热传导 eddy heat conduction
紊流热交换 turbulent heat exchange
紊流式翅片管 turbulent fin tube
紊流式燃烧室 baffle type combustor
紊流输送系数 coefficient of turbulent transport
紊流衰减 turbulence decay
紊流衰减定律 law of decay of turbulence; law of density turbulence
紊流水槽 turbulence flume
紊流速度 turbulent speed; turbulent velocity
紊流速度场 turbulent velocity field
紊流速率 turbulent rate
紊流条件 turbulence condition; turbulent flow condition
紊流微尺度 microscale of turbulence
紊流尾迹 turbulent trail
紊流涡旋方程 turbulent vorticity equation
紊流系数 coefficient of turbulence (flow)
紊流型泥石流 turbulent type mud flow
紊流因子 turbulence factor
紊流引起的振动 turbulence-induced vibration
紊流应力 turbulent stress
紊流预燃室 turbulent pre-combustion chamber
紊流运动 stirring motion; turbulence motion; turbulent motion
紊流栅 turbulence screen
紊流指示器 turbulence indicator
紊流指数 index of turbulence
紊流中心 turbulent flow core
紊流状态 turbulence condition; turbulent condition
紊流阻抗 turbulence resistance
紊流阻力 turbulence resistance; turbulent drag; turbulent flow resistance; turbulent resistance
紊乱 derangement; disarrangement; disorder; disturbance; out-of-course; out-of-order; pounding
紊乱的 disorderly; irregular; ragged; unsystematic(al)
紊乱对流 turbulent convection
紊乱情况 race condition
紊乱取样 grab sample
紊乱散布 blown pattern
紊乱水系 aimless drainage; deranged drainage
紊乱水系模式 deranged mode
紊气 rough air
紊气流 rough air

稳变异构 merotropism; merotropy; mesomerism

稳变异构体 allelotrope; desmotrope
稳变异构(体)现象 desmotropism; allelomorphism; desmotropy
稳并励电动机 stabilized shunt-wound motor
稳步增长 steady increase
稳产措施 procedure of stable production
稳产期 stable production stage
稳船块 docking block
稳电路 equalizing network
稳电压器 voltage stabilizer
稳定 level off; settling; stabilization; stabilize; stableness
稳定安全系数 buckling safety factor; stability factor of safety
稳定岸坡 stabilized bank slope
稳定岸滩 stable beach
稳定岸线 shoreline stabilization
稳定白云石 stabilized dolomite
稳定白云石耐火材料 stabilized dolomite refractory
稳定白云石熟料 stabilized dolomite clinker
稳定白云石砖 stabilized dolomite brick
稳定板 hydrofin
稳定板块边界 conservative plate boundary
稳定保护层 stabilized protection course
稳定爆炸 stable detonation
稳定本底 steady background
稳定本机振荡器 stabilized local oscillator
稳定比 stabilization ratio
稳定臂 arm of stability
稳定边际 margin of stability; stability margin
稳定边坡 stabilized slope
稳定边缘海盆 inactive marginal basin
稳定变化的影响力 influential force of temperature change
稳定变速(水)流 permanent varied flow; steady non-uniform flow
稳定变形 steady-state deformation
稳定变形过程 steady deformation process
稳定变压器 stabilizing transformer
稳定表面 surface of stability
稳定表面波 stationary surface wave
稳定并励发电机 stabilized shunt-wound generator
稳定并绕电动机 stabilized shunt motor
稳定波 neutral wave; stable wave; steady wave
稳定玻璃 stabilized glass
稳定补偿器 stabilizer compensator
稳定捕获辐射带 stably trapped radiation zone
稳定捕获量 sustainable catch
稳定不均匀流 steady non-uniform flow
稳定不溶性偶氮染料 stabilized insoluble azo dyes
稳定不锈钢 stabilized stainless steel
稳定部分 steady component
稳定材料 stabilizing material
稳定材料基层 stabilized material base
稳定参数 steadiness parameter
稳定操作 stable operation; steady operation
稳定槽 stabilization tank; sustainable lagoon
稳定层 regular bed; stable layer; stationary layer; stabilized course <多指用结合料稳定的基层、垫层>
稳定层流 stable laminar current; stably stratified flow; steady laminar flow
稳定产量 constant rate of production; firm yield; steady yield; sustainable yield; sustained production; sustained yield
稳定产汽状态 steady steaming condition
稳定场 stabilizing field; steady field
稳定场地 stable court
稳定潮波 stationary tidal wave
稳定车距 steady space
稳定车流 stable flow; steady traffic flow
稳定沉积区 lithotope
稳定成分 stable element
稳定程度 stability degree; degree of stability
稳定池 stability pond; stabilization basin; stabilization lagoon; stabilization pond; stabilization tank; stopping basin
稳定翅 stabilizing fin
稳定充电 stable charger
稳定出水 stable effluent
稳定初等变换 stabilized elementary transformation
稳定储备量 margin of stability
稳定处理 stabilization(treatment)
稳定处理的道路 stabilized road
稳定处理的路肩 stabilized shoulder
稳定处理的土路 stabilized earth road; stable soil road
稳定处理的预应力钢绞线 stabilized strand
稳定处理钢板 stabilized steel sheet
稳定传感器 stability sensor
稳定传热 steady heat transfer
稳定催化剂 rugged catalyst
稳定大气 stationary atmosphere
稳定但可调整的平价 stable but adjustable par value
稳定导热 steady heat conduction
稳定岛 island of stability; stable island
稳定道路 stabilized road
稳定的 braced; indissoluble; non-active; resistant; resisting; stabilizing; stable; stationary; steady; transient-free; unyielding
稳定的贝塔型钛合金 stable beta titanium alloy
稳定的大气条件 stable atmospheric condition
稳定的电弧 stable arc
稳定的合金 dead alloy; quiet alloy
稳定的环节 stable component
稳定的价格 stable price
稳定的结晶区 resistant crystalline region
稳定的抗蠕变钛合金 stable creep-resistant titanium alloy
稳定的裂变产物 stable fission products
稳定的路堑边坡 safe cut
稳定的人口 stationary population
稳定的沙堆 stabilized dune
稳定的生态系统 stable ecosystem
稳定的竖向准直 stable vertical
稳定的无捻回弹力丝 stabilized non-torque yarn
稳定的正面 stabilized front
稳定的种群 balanced population
稳定的重氮化合物 stabilized diazo compound
稳定灯光 steady light
稳定灯光表示 steady light indication
稳定灯光显示 steady light indication
稳定等价的 stable equivalent
稳定等价向量丛 stable equivalent vector bundles
稳定等离子体 stable plasma
稳定等速流(动) stable uniform flow; steady uniform flow; steady flow; uniform flow
稳定等速水流 steady uniform flow

稳定等速湍流 steady uniform turbulent flow
稳定地层 competent formation
稳定地带 stability zone; zone of stability
稳定地盾 stable shield
稳定地基 firm ground
稳定地基土的机械 soil stabilizer
稳定地块 craton; resistant block; stable block; stationary block
稳定地块构造 stable landmass tectonics
稳定地区 stable area
稳定地台 stable platform
稳定点 stable point
稳定点火 stable ignition
稳定碘同位素 stable iodine
稳定电流 standing current; steady current
稳定电流电抗器 current stabilizing reactor
稳定电流式轨道电路 steady current rail circuit
稳定电流系数 current stabilizing factor
稳定电路 stabilizing circuit; stabilizing network
稳定电能 firm energy
稳定电容器 stabilising condenser
稳定电势 stabilizing potential; steady potential
稳定电位 stable potential
稳定电压 burning voltage; flat voltage; regulated voltage; stabilized voltage; steady voltage
稳定电源 regulated power supply; stabilised voltage supply; stabilized power source
稳定电源电压 stabilized supply voltage
稳定电源设备 stabilised power unit
稳定电子轨道 stable electron orbit
稳定电阻 steady resistance
稳定度 stability (constancy); steadiness; steadiness grade
稳定度测量器 regulation meter
稳定度分类 stability category; stability classification
稳定度分析 stability analysis
稳定度极限 stability limit
稳定度警报 stability alarm
稳定度/流值比 stability/flow ratio
稳定度模拟 stability simulation
稳定度试验 incumbator test; stability test(ing)
稳定度调整 stabilizing adjustment
稳定度图 stability chart
稳定度系数 stability factor
稳定度仪 < 常指 Hveem 氏稳定度仪 > stabilometer
稳定度仪设计法 stabilometer design method
稳定度影响 effect of stability
稳定度值 stability value
稳定度指数 stability index
稳定对 stable pair
稳定多晶型物 stable polymorphs
稳定二阶上同调运算 stable secondary cohomology operation
稳定二氧化锆 stabilized zirconia
稳定二氧化氯 stabilized chlorine dioxide
稳定发展 steady progression
稳定法 stabilization; stabilizing process
稳定帆 steadying sails
稳定反馈 stabilized feedback; stabilizing feedback
稳定反馈放大器 stabilized feedback amplifier
稳定反应 stopping reaction
稳定范围 stability range; stabilization range; stable range
稳定方法 anti-hunt means; stabilization
稳定房屋 stable building
稳定放大器 regulating amplifier; stabilizing amplifier
稳定放矿 steady draw
稳定非均匀流 steady non-uniform flow
稳定分布 stable distribution; stationary distribution
稳定分布函数 stable distribution function
稳定分路绕组 stabilized shunt
稳定分析 stabile analysis; stability analysis
稳定风 constant wind; stable wind; steady wind
稳定峰值功率输出 firm peak output
稳定浮力平衡 stable buoyancy equilibrium
稳定浮筒 stabilizing float
稳定俯冲 steady dive
稳定负反馈放大器 stabilized negative feedback amplifier
稳定负荷 firm demand; steady load
稳定负荷试验 steady load test
稳定负载 steady load
稳定附面层 steady-state boundary layer
稳定复制品 stabilized prints
稳定概率 probability of stability
稳定杆 stabilizer rod; sway bar
稳定高炉矿渣 stabilized blast-furnace slag
稳定工况 stabilized condition; stable condition; steady condition; steady working condition
稳定工况试验 steady-state test
稳定工作 steady operation
稳定工作平台 stable working platform
稳定工作状况 steady running condition
稳定功率 firm power; stabilization power
稳定功率极限 stability power limit
稳定功能 firm power energy
稳定供水能力 stabilizing the capacity of water supply
稳定构架 stable framework
稳定构型分析 stable configuration analysis
稳定箍缩 stabilized pinch
稳定谷 stable valley
稳定固定点 stable fixed point
稳定灌浆 stabilizing grout
稳定光 fixed light
稳定光栅 stationary raster
稳定轨道 stable orbit
稳定轨迹 stable trajectory
稳定滚动条件 steady rolling condition
稳定过程 stabilization process; stationary process
稳定海岸 coast of stable region; stable coast
稳定海岸工程 shore stabilization
稳定海岸结构 shore stabilizing structure; shore stabilization structure
稳定海岸线 mature shoreline
稳定海底 bottom stabilization
稳定海流 permanent current
稳定海盆 inactive basin
稳定海滩 stable beach
稳定海滩建筑物 beach stabilization structure
稳定焊料玻璃 stable solder glass
稳定航道尺度 regime (n) channel dimension
稳定航行 steady steaming
稳定航行水路结构 waterway stabilization structure
稳定合金 stabilized alloy
稳定河岸 stable bank
稳定河槽 fixed channel; regime (n) channel; stabilized channel; stable channel
稳定河槽的水力几何形状 hydraulic geometry of stable channel
稳定河槽工程 channel stabilization
稳定河槽几何形状 stable channel geometry
稳定河槽形态 channel stabilization
稳定河床 flow resistance bed; permanent bed; stable channel; stable (river) bed
稳定河床建筑物 stabilization structure of riverbed
稳定河道 poised stream; regime (n) channel
稳定河底 bottom stabilization
稳定河底纵坡 stabilized grade
稳定河段 quiet reach of river; stable waterway section
稳定河谷 mature valley
稳定河流 poised river; poised stream; regime (n) river; regime (n) stream; regulated stream; steady river; steady stream
稳定河弯 stable bend
稳定核素 stable nuclide
稳定荷载 steady load
稳定恒星 stable star; stationary star
稳定横向振荡 stable transverse oscillations
稳定湖 stability lagoon; stability lake; stabilization lagoon
稳定滑动 stable sliding
稳定化 stabilization; stabilizing
稳定化处理 stabilizing treatment
稳定化处理的预应力钢丝 stabilized wire
稳定化钢 stabilized steel
稳定化钢板 stabilized steel sheet
稳定化合物 stable compound
稳定化时效 stabilizing aging
稳定化选择 stabilizing selection
稳定化学品 stable chemical
稳定化影响 stabilizing influence
稳定环形箍缩 stabilized toroidal pinch
稳定灰渣 < 石灰或水泥 > stabilized ash
稳定恢复水位 stably recovering water level
稳定辉光 steady glow
稳定回路 stable loop
稳定回授 stabilized feedback; stabilizing feedback
稳定回填土 stabilized back-fill
稳定回转速度 steady turning speed
稳定回转运动 steady turning motion
稳定汇率 stable exchange rate
稳定混合物 stabilized mixture
稳定火焰 balanced condition of flame; steady flame
稳定货币 stable currency
稳定积 overall stability constant; stability product
稳定基础 stable foundation
稳定基面 stationary datum
稳定基线 steady baseline
稳定级数 stable series
稳定极限 limit of stability; stability limit
稳定(极限)入渗量 ultimate infiltration capacity
稳定极限循环 stable limit cycle
稳定极限振幅 stability limit amplitude
稳定挤压 steady extrusion
稳定计 stabilimeter
稳定计算 calculation of stability
稳定计算图解法 stable computation graphic(al) construction
稳定剂 size stabilizer; stabilising [stabilizing] agent; stabilizator; stabilizer; antidecomposition additive < 润滑油中的 >
稳定剂基 stabilizer base
稳定剂加入数量 quantity of stabilizer
稳定剂用量 consumption of stabilizing
稳定剂种类 kind of stabilizer; type of stabilizer
稳定加速 firm acceleration
稳定加速过程 stable acceleration process
稳定夹 steadying bracket
稳定架 chock; steady; steady rest
稳定浆液 stabilizing grout; stable grout
稳定交通 stationary traffic
稳定交通流 stable flow; stable traffic flow; stationary traffic flow
稳定角 angle of repose
稳定阶段 range of stability; stage of stability; steady stage
稳定接箍 < 钻杆 > stabilizer sleeve
稳定接收机 regulated receiver
稳定接头 stabilized coupling
稳定结构 resistant structure; rock-steady structure; stable structure
稳定解 stable solution
稳定界限 margin of stability; phase margin
稳定襟翼 stabilizing spoiler
稳定进步 steady progression
稳定进场着陆 stabilized approach
稳定井水位 < 不抽水时的 > standing (water) level
稳定境 stable region
稳定就业 decasualization of labo(u)r
稳定矩 moment of stability
稳定矩阵 stability matrix; stable matrix
稳定绝缘子 stabilized insulator
稳定均匀流 steady uniform flow
稳定开放系统 stationary open system
稳定开关 stable switch
稳定可平行的 stably parallelizable
稳定可平行流 stable parallelizable flow
稳定空腔谐振器 stabilizer cavity
稳定控制 permanent control; stabilized control
稳定控制系统 stabilization control system
稳定控制信号 steering correction
稳定宽度 regime(n) width
稳定矿物 resistant mineral; stable mineral
稳定框 shakeless deckle
稳定框架 stable framework
稳定扩展 stable propagation
稳定拉杆 stabilizer bar
稳定离子 stabilizing ion
稳定理论 stabile theory; stability theory
稳定力 restoring force; stabilizing force
稳定力臂 arm of stability; lever of stability
稳定力矩 anti-tipping moment; moment of stability; restoring moment; righting moment; stabilization moment; stabilizing moment; strength

of moment
稳定力偶 righting couple
稳定沥青分散液 <即慢裂乳化沥青> stable bituminous dispersion
稳定砾石 stabilized gravel
稳定砾石路面 stabilized gravel surface
稳定砾石面层 stabilized gravel surface
稳定粒子 stable particle
稳定连接托架 stabilizing connecting bracket
稳定连接轴 stabilized coupling
稳定亮度 steady-state intensity
稳定裂缝 stable crack
稳定裂缝扩展 stable crack propagation
稳定零点 balanced null point
稳定流 constant current; permanent flow; stationary current; stationary flow; stationary stream; steady-state flow
稳定流变 steady flow
稳定流场 stationary flow field
稳定流抽水试验 steady flow pumping test
稳定流抽水试验法 method of steady pumping test
稳定流的不均匀性 steady-flow non-uniformity
稳定流动 steady flow
稳定流动成型 steady flow forming
稳定流动搅拌机 constant flow mixer
稳定流计算法 method of steady flow calculation
稳定流量 regime(n) flow; steady discharge; steady flow
稳定流量调节器 constant pressure flow controller
稳定流配线法 superimposed line method for steady flow
稳定流水井公式 well formula of steady state flow
稳定流态 steady flow state; steady-state of flow condition
稳定流图解法 graphic(al) method for steady flow
稳定流系数 steady flow coefficient
稳定流形 stable manifold
稳定硫同位素标准 standard of stable sulfur isotope
稳定馏出燃料 stable distillate fuel
稳定陆架相 stable-shelf facies
稳定滤波器 stable filter
稳定路边带 stabilized verge
稳定律 stable law
稳定率 coefficient of stability; coefficient of stabilization; index of stability
稳定论者 stabilist
稳定络合物 stable complex
稳定脉冲持继时间 stabilized pulse duration
稳定脉冲发生器 stable pulse generator; stable pulse generator clock
稳定脉动 steady pulsation
稳定镁橄榄石 stabilized forsterite
稳定密度 steady-state intensity
稳定面 stabilizing surface; standing face; surface of stability; unyielding surface
稳定能力 stabilizing power
稳定能量输入 steady energy input
稳定泥浆 <混有膨润土等> clay suspension
稳定泥结路面 stabilized soil-bound surface
稳定泥结面层 stabilized soil-bound surface; stabilized clay-bound surface

稳定凝固点 stable pour-point
稳定凝胶结构 stable gel structure
稳定排液 stable effluent
稳定片流 stable laminar current; steady laminar flow
稳定偏移 steady drift
稳定偏转 stabilized deflection
稳定漂移 erratic drift
稳定频率 established frequency; stabilized frequency
稳定频率建立时间 build-up time of frequency stabilization
稳定频率信号 stable frequency signal
稳定频闪放电管 stable strobe
稳定品种 distinct variety
稳定平衡 stable equilibrium
稳定平衡海岸线 stable equilibrium shoreline
稳定平衡相位 stable equilibrium phase
稳定平面爆震 steady planar detonation
稳定平台 stabilized platform; stable platform
稳定坡度 stabilized grade; stabilized slope; stabilizing slope; steady gradient
稳定坡角 stable slope angle
稳定期 stabilization period; stable period; stable phase; stationary phase
稳定期地震 silent earthquake
稳定鳍 fin stabilizer
稳定起动 regulated start
稳定气壳 stationary shell
稳定气流区 contour zone
稳定气团 stable air mass
稳定气柱 stable air column
稳定汽油 stabilized gasoline; stable gasoline
稳定器 anti-hunt means; ferrule; fluted coupling; holder; reactor; regulator; stabilization unit; stabilizator; stabilizer; stabilizer apparatus; stabilizing unit
稳定器摆动 stagger of stabilizer
稳定器弹簧 friction(al) spring
稳定器的倾斜角 incidence of stabilizer
稳定器杆 stabilizer bar
稳定器固定柱 stabilizer link assembly
稳定前馈 stabilizing feed forward
稳定前授 stabilizing feed forward
稳定强度 stability strength; steady-state intensity
稳定倾点 stable pour-point
稳定区范围 stability range
稳定区(域) stable region; stability domain; stability area; stable zone
稳定渠道 stabilized channel
稳定泉 constant spring; steady spring
稳定裙 stabilizing skirt
稳定燃烧 plateau burning; smooth burning; smooth combustion; stable burning; stable combustion; steady burning
稳定燃烧极限 stable combustion limit
稳定染色研究 steady-state dye study
稳定染色羽流 steady-state dye plume
稳定绕转 steady spin
稳定绕组 stabilized winding; stabilizing winding; tertiary winding
稳定热传导 steady heat conduction
稳定热流 steady flow of heat
稳定热水 regulated hot water
稳定人口 stable population
稳定蠕变 stationary creep; steady creep
稳定蠕变阶段 secondary creep stage
稳定蠕动 stationary creep; steady creep

稳定乳化液 stable emulsion
稳定乳状液 spontaneous emulsion; stable emulsion
稳定入渗率 limiting infiltration rate
稳定入渗能力 constant infiltration capacity; ultimate infiltration capacity
稳定入渗容量 constant infiltration capacity; ultimate infiltration capacity
稳定伞 drogue
稳定沙的植物 sand-stabilizing vegetation
稳定沙丘 arrested dune; stabilized dune; stable dune
稳定沙洲 stabilized bar
稳定砂砾 stabilized gravel
稳定砂砾土 stabilized gravel soil
稳定砂石路面 <即级配砂石路面或面层> stabilized granular surface
稳定砂石面层 stabilized granular surface
稳定上升 steady climb
稳定上同调运算 stable cohomology operation
稳定设施 stabilization works
稳定渗流 steady seepage; time-invariant seepage
稳定生成稳定 stable formation temperature
稳定生界限 stability locus
稳定剩余磁性 stable remanent magnetization
稳定湿度 constant humidity
稳定石屑 stabilized chip
稳定时间 settling time; stabilization time; stable time; steady time; settle (ment) time
稳定时期 stationary phase
稳定市场 stabilize the market
稳定示踪同位素 stable tracer isotope
稳定式半潜平台 stable semi-submersible platform
稳定式海洋平台 stable ocean platform
稳定势 stationary potential
稳定势能 stationary potential energy
稳定试验 stabilization test
稳定收敛 stable convergence
稳定输出 firm output
稳定束 stabilized beam
稳定数 stability number
稳定水舱 anti-rolling tank
稳定水道 stabilized channel
稳定水力断面 stable hydraulic section
稳定水流 stationary flow; steady flow
稳定水流阻力 steady drag force
稳定水舌 stable nappe
稳定水深 regime(n) depth
稳定水头高度 height of steady head
稳定水位 established water level; fixed level; permanent water level; stationary stage; steady water level
稳定水温度控制器 aquastat controller
稳定水源 permanent water
稳定水跃 stationary jump; steady jump
稳定伺服机构 stable servo
稳定伺服系统 stable servo
稳定速度 constant velocity; permanent speed; stabilized speed; stabilized velocity; steady speed; steady velocity
稳定速率 steady rate
稳定速率偏差 steady speed deviation
稳定碎石 stabilized macadam
稳定碎石土 stabilized crushed rock soil

稳定索 bull rope; stability cable; standing guy rope; steadying line; steadying rope; guy rope; standing line; stay rope
稳定塔 stabilizer column; stabilizer tower
稳定塔底产物 stabilizer bottoms
稳定塔顶排(出)气体 stabilizer overhead
稳定塔进料泵 stabilizer feed pump
稳定塔进油泵 stabilizer feed pump
稳定塔气体 stabilizer gas
稳定塔再沸器 stabilizing reboiler
稳定台地 stabilized platform; stable platform
稳定态 stable state; stationary state; tranquil state
稳定态培养 steady-state culture
稳定态特性 steady-state response
稳定态振动 steady-state vibration
稳定弹簧 stabilizing spring
稳定弹射 stabilized ejection
稳定塘 facultative pond; heter-aerobic pond; stability lagoon; stabilization lagoon; stabilization pond
稳定塘法 stabilization lagoon method; stabilization pond method
稳定塘-好氧塘 aerobic pond
稳定塘兼性厌氧塘 ampli-aerobic pond
稳定塘净化机理 purification mechanism of stabilization pond
稳定塘类型 type of stabilization pond
稳定塘-曝气塘 aerated pond
稳定塘设计 design for stabilization pond
稳定塘生态系统特征 ecosystem characters of stabilization pond
稳定塘厌氧塘 anaerobic pond
稳定特性 stability characteristic; steady-state characteristic
稳定特性曲线 stability curve
稳定体 <一种防波堤块体> stabit
稳定体系 stability system
稳定天然汽油 stabilized natural gasoline
稳定天线 stabilized antenna
稳定添加剂 stabilizing additive
稳定条件 stability condition; stabilization condition; stabilized condition; stable condition; steady condition
稳定调节 homothermism
稳定通风板 draught stabilizer
稳定通货 stabilize currency
稳定同量异位素 stable isobar
稳定同位素 stable isotope
稳定同位素比值 ratio value of stable isotope
稳定同位素标准 standard of stable isotope
稳定同位素地层学方法 stable isotopic stratigraphy
稳定同位素分析 stable isotope analysis
稳定同位素分析标准 standard of stable isotopic analysis
稳定同位素分析方法 method of stable isotopic analysis
稳定同位奏国际标准 stable isotope world-wide standard
稳定同位素示踪(法) stable isotope tracing
稳定同位素示踪剂 stable isotope tracer
稳定同位素稀释技术 stable isotope dilution technique
稳定同位素稀释质谱法 stable isotope dilution mass spectrometry
稳定同位素指示剂 stable isotope indicator

稳定同位素中国工作标准 stable isotope working standard in China
稳定统计模型 stable static model
稳定突水量 steady quantity of water bursting
稳定图 stability diagram
稳定图解 stability diagram
稳定图像 stabilized image;steady image
稳定土 hard compact soil;stabilised [stabilized] soil
稳定土拌和机 road-mix stabilizer;soil stabilizer;stabilized soil mixer
稳定土厂拌设备 stabilized soil mixing plant
稳定土成品贮仓 stabilization soil storage
稳定土基 stabilized soil base
稳定土基层 stabilized soil base course
稳定土搅拌机 pugmill type soil stabilization mixer;pulvimixer;soil stabilizer
稳定土路 stabilized soil road;stable soil road
稳定土路面 stabilized soil pavement
稳定土路修筑机 soil stabilizer
稳定土壤混凝土 stabilized-earth concrete
稳定土壤剂 earth stabilizer
稳定土设备 stabilizer
稳定土摊铺机 stabilized soil paver
稳定土用拌和机 stabilization mixer
稳定土筑路机 soil stabilizer
稳定湍流 steady turbulent flow
稳定推进器 stabilized feed
稳定退火 stabilizing annealing
稳定托架 stabilizer bracket
稳定陀螺仪 stabilizing gyroscope
稳定挖槽 stable dredge-cut
稳定网络 stabilization network;stabilizing network
稳定(维持)机能 maintenance function
稳定卫星 stabilized satellite
稳定位置 settling position
稳定位置偏差 steady position deviation
稳定温度 equilibrium temperature;immobile temperature;steady temperature
稳定涡流 stabilizing vortex
稳定污水流 stable effluent
稳定/无雾添加剂 stabilizer/anfog additive
稳定物价 price stabilization;stabilize commodity price
稳定稀释法 steady dilution method
稳定系(列)stability series
稳定系数 coefficient of fixation;coefficient of stability;coefficient of stabilization;environmental coefficient;margin of stability;stability coefficient;stability factor;stability margin;stability number;stabilization factor;stabilizing factor
稳定系统 anti-hunt circuit;stabilization system;stable system;steady-state system
稳定系统故障 stability malfunction
稳定下降 stabilized descent
稳定下来 settle out
稳定下渗能力 constant infiltration capacity;ultimate infiltration capacity
稳定下渗容量 constant infiltration capacity;ultimate infiltration capacity
稳定咸淡水界面近似方程 approximation equation for steady interface between salt and fresh water
稳定显示 stabilised [stabilized] presentation
稳定线路调整 stabilizing adjustment
稳定相 stable phase
稳定相角 phase-stable angle
稳定相区 phase-stable area
稳定相位 stable phase angle
稳定向量丛 stable vector bundle
稳定效率 stabilization efficiency
稳定效应 stabilizing effect;steadying effect
稳定卸荷 <指工程荷载> steady unloading
稳定型 stable form;stable type
稳定型水源地 stable water source
稳定型重力仪 stable-type gravimeter
稳定性 capacity to stand;constancy;fixedness;fixity;immobility;insensitiveness;insensitivity;rigidity;stability;stabilizing ability;stableness;stable stability;steadiness;sustainability
稳定性白云石耐火砖 stabilized dolomite brick
稳定性边坡 stabilized side slope
稳定性变化 stability change;stability variation
稳定性参数 stability parameter
稳定性测定 stability measurement
稳定性差 poor stable
稳定性差的 unsound
稳定性常数 stability constant
稳定性出价 stabilising bid
稳定性大气条件 stable atmospheric condition
稳定性导数 stability derivative
稳定性的实验室检验 laboratory tests of stability
稳定性的野外检验 field tests of stability
稳定性的增长 stability augmentation
稳定性调查 stability investigation
稳定性定形 <指干温热法> stability setting
稳定性范围 range of stability;stability boundary
稳定性方程式 stability equation
稳定性分析 stability analysis
稳定性改进 stability improvement
稳定性公式 stability formula
稳定性和适应性 <车辆在道路上行驶的> roadability
稳定性基层 stabilized base
稳定性极限 stability limit
稳定性计算 stability calculation;stability computation
稳定性加强系统 stability augmentation system
稳定性减小 stability reduction
稳定性交叉曲线 cross curves of stability
稳定性交换原理 stability exchange principle
稳定性理论 theory of stability
稳定性量测仪 stabilometer
稳定性氯素消毒剂 stable chlorine disinfectant
稳定性判据 stability criterion
稳定性曲线 curve of stability;stability curve
稳定性三角形 stability triangle
稳定性试验 breakdown test;stability test(ing)
稳定性试验机 stabilometer
稳定性试验模型 stability model;stability test model
稳定性试验台 stabilizing tester
稳定性试验仪 stability meter
稳定性数 stability number
稳定性条件 condition of stability;stability condition
稳定性调节 stability control
稳定性同位素 stability isotope
稳定性团聚体 stability aggregate
稳定性微污染地表水 stable micro-polluted surface water
稳定性系数 coefficient of stability;stability coefficient;stability factor
稳定性下降 stability drop
稳定性限度 limit of stability
稳定性消失点 vanishing point of stability
稳定性研究 stability investigation
稳定性因数 stability factor
稳定性因子 stability factor
稳定性阈值 stability threshold
稳定性元素 <阻止石墨化元素> stabilizing element
稳定性指标 stability index
稳定性指示器 stability indicator
稳定性指数 index of stability;stability index
稳定性准则 stability criterion
稳定性组构 stability fabrics
稳定性作用函数 stability function
稳定修正 stabilizing correction
稳定悬浮液 stable suspension
稳定旋转 steady rotation
稳定旋转特性 steady turning performance
稳定循环 constant circulation;cycle of stabilization
稳定压力 constant pressure;standing pressure;steady pressure
稳定压力阀 steady pressure valve
稳定压缩试验法 steady compression process
稳定压头 steady head
稳定延续时间 stably continuous period
稳定岩层 stable bed rock
稳定岩基 stable rock bed
稳定岩石 stable rock
稳定演变 steady progression
稳定焰燃烧器 flame-stabilized burner
稳定叶片 stabilizer vane
稳定液 clay suspension;slurry for stability;stabilizing bath
稳定液拌和机 mixer for stabilizing solution
稳定液逸散 circulation loss of stabilizing fluid
稳定一阶上同调运算 stable primary cohomology operation
稳定一种证券 stabilize a security
稳定仪 stabilometer
稳定因素 stability factor;stabilization factor;stabilizing factor
稳定因子 coagulation factor;stability factor;stabilization factor;stabilizing factor;stable factor
稳定音 stationary sound;stationary tone
稳定应变 steady strain
稳定应力 steady stress
稳定影响 stabilizing influence
稳定涌水 normal water flow
稳定用拉条 steady brace
稳定用调谐线圈 stabilizing turned coil
稳定域 stability domain;stability loci;stability locus;stability range;stable region
稳定裕度 margin of stability;stability margin
稳定裕量 margin of stability;stability margin
稳定元件 stable element
稳定元素 stable element
稳定原油 stabilized crude(oil)
稳定约化 stable reduction
稳定约束 scleronomic constraint;stable confinement
稳定运动 stable motion;steady motion
稳定运动状态 stable state of motion;steady-state of motion
稳定运输 stationary traffic
稳定运行 smooth operation;stabilization operation;stable operation;steady operation;steady run
稳定运转 steady running
稳定运转工况 steady running condition
稳定运转条件 steady running condition
稳定运转状态 steady running condition
稳定噪声 steady-state noise
稳定增长 standing growth
稳定照明 steady illumination
稳定振荡器 stabilized oscillator;stable oscillator
稳定振动 stable oscillation;steady oscillation;steady-state oscillation
稳定振动的纽马克分析法 Newmark's analysis method for steady state vibration
稳定振动状态 stable state of vibration
稳定振幅 stabilized amplitude;stable amplitude
稳定蒸发损失率 stationary evapo(u)ration loss rate
稳定整理 stabilised finish
稳定支撑 outrigger
稳定支脚 stabilizer foot
稳定支腿 stabilizing jack
稳定织物 stable fabric
稳定值 stable value;steady value
稳定值区 region of stable values
稳定植被 stable vegetation
稳定植物群落 stable phytocoenosium
稳定指数 index of stabilization;Ryzner stability index;stability index;index of stability
稳定中心 stable center [centre];metacenter [metacentre] <浮体的>
稳定中心高度 metacentric height
稳定种群 stable population
稳定重力仪 stable gravimeter
稳定周期 stable period
稳定注水量 steady quantity of water injection
稳定转动 stable running;steady rotation;steady running
稳定转速 stabilized speed
稳定桩 anchor spike
稳定装置 arresting gear;catcher;detent;draught stabilizer;stabilization plant;stabilized mount(ing);stabilizer apparatus;stabilizer plant;stabilizing device
稳定状况温度 steady-state temperature
稳定状态 case of stability;stability condition;stabilized condition;stabilized state;stable condition;stable state;stationary state;steady state
稳定状态的热交换 steady-state heat transfer
稳定状态阶段 steady state phase
稳定状态频率 steady-state frequency
稳定状态速度 terminal velocity
稳定状态条件 steady-state condition
稳定状态图 stable diagram
稳定状态系统 stead state system
稳定子群 stability subgroup;stable subgroup

稳定纵波 stabilized grade
稳定最大输出 firm peak capacity; firm peak output
稳定最高量 stable maximum
稳定最佳化 steady-state optimization
稳定作用 anti-hunt action; stability effect; stabilisation [stabilization]
稳度 degree of stability; stability
稳舵盒 deck stuffing box
稳风装置 draught stabilizer
稳幅 fixed amplitude
稳幅器 amplitude stabilizer
稳固 firmness; security; stability; stableness
稳固窗 fast window; fixed window
稳固的 stable
稳固底板 firm bottom
稳固地基 firm ground; solid foundation; steady foundation; terra firma
稳固顶板 firm top; hard roof
稳固件 firm ware
稳固设备 firm ware
稳固围岩 firm wall
稳固性 steadiness
稳固岩层 firm ground; tight formation
稳固岩石 firm rock; secure rock; self-sustaining rock
稳固支承 non-sinking support
稳管基座 equalizing bed
稳恒场退磁 steady field demagnetization
稳恒传号 steady mark
稳恒的 steady
稳恒电流 constant current
稳恒电阻 steadying resistance
稳恒负载 steady load
稳恒过程 steady process
稳恒荷载 steady load
稳恒拉力 steady pull
稳恒临界水流 constant critical flow
稳恒流 constant flow; steady stream
稳恒流动 steady flow
稳恒束 steady beam
稳恒态 steady state
稳恒态模型 steady-state model
稳恒态学说 steady-state theory
稳恒状态 steady state
稳恒状态边界层 steady-state boundary layer
稳弧 arc stabilization
稳弧剂 arc stabilizer [stabiliser]; ionizer
稳弧装置 arc stabilizer [stabiliser]
稳滑 stable sliding
稳滑方式 stable-slip behavio(u)r
稳化池 Emscher filter
稳(缓)变异构现象 mesomerism
稳剪切柔量 steady shear compliance
稳健估计 robust estimation
稳健原则 conservatism
稳静层 calm layer
稳静烟雾 calm smog
稳流 stationary flow; stationary stream; steady flow; tranquil flow; undisturbed flow; uniform flow
稳流板 stabilizer
稳流池 equalization pond
稳流冲击式喷气发动机 steady flow ram jet
稳流灯 barretter
稳流电感 stabilizing inductance
稳流电源 stabilized current supply
稳流电阻 ballast resistance; stabilizing resistance; steady resistance
稳流堆 steady flow reactor
稳流管 current regulator tube; regulator tube; anemostat <暖气或通风系统管路中的>
稳流过程 steady flow process
稳流搅拌 constant flow mixing
稳流螺旋式拌和器 constant flow screw-type mixer
稳流黏[粘]度 steady flow viscosity
稳流喷嘴流浆箱 steady flow nozzle headbox
稳流器 current regulator; current stabilizer
稳流燃烧 steady flow combustion
稳流设备 constant flow equipment
稳流式涡轮 steady flow turbine
稳流室 plenum chamber
稳流箱 tranquil(l)ing tank
稳流研磨 constant flow grinding
稳流液压缸 fluid-stabilized cylinder
稳流装置 current stabilizer
稳流纵剖面 steady-state profile
稳泡剂 air-retaining substance; bubble stabilizer; subs-stabilizing agent
稳频 frequency stabilization
稳频电源 constant frequency power supply
稳频二氧化碳激光器 frequency stabilized carbon dioxide laser
稳频氦氖激光器 frequency stabilized He-Ne laser
稳频计 frequency stabilometer
稳坡建筑物 grade stabilizing structure
稳谱器 spectrum stabilizer
稳水池 stabilization pond
稳水杆 rod for stilling water
稳水器 still well
稳水栅 steadying baffle; damping screen <水工试验的>
稳速试验 <发动机的> steady speed test
稳索 guy rope; standing rope
稳索绞辘 guy tackle; vang purchase
稳态 stable state; stationary state; steady state
稳态保持时间 steady-state residence time
稳态边界层 steady-state boundary layer
稳态边界条件 steady-state boundary condition
稳态变量 steady-state variable
稳态波动 steady-state wave motion
稳态测量 steady-state measurement
稳态测试方法 steady-state method of test
稳态常数 steady-state constant
稳态臭氧水平 steady-state ozone level
稳态触发电路 stable trigger circuit
稳态传导 steady-state transfer
稳态传热 steady-state heat transfer
稳态带张力 steady-state tape tension
稳态等离子体 steady-state plasma
稳态电抗 steady-state reactance
稳态电流 steady-state current
稳态短路 steady-state short-circuit
稳态堆 steady-state reactor
稳态法 steady-state method
稳态反应 <脉冲荷载下无阻尼振动之一部分> steady-state response
稳态反应增益 steady-state response gain
稳态方程 steady-state equation
稳态方位 steady bearing
稳态放电 steady-state discharge
稳态非均匀流 steady non-uniform flow
稳态分布 steady-state distribution
稳态分布方程 steady-state distribution equation
稳态分配 steady-state distribution
稳态分析 steady-state analysis
稳态复合 steady-state recombination
稳态概率 probability of stability
稳态高斯过程 stationary Gaussian process
稳态工作 quiescent operation
稳态功率系数 steady-state power coefficient
稳态共振曲线 steady-state resonance curve
稳态河口 steady-state estuary
稳态荷载 steady-state loading
稳态黑子 steady sunspot
稳态换向 steady-state commutation
稳态混响声 steady-state reverberation sound
稳态机制 homeostatic mechanism
稳态激励 steady stimulation; steady stimulus
稳态级 steady-state level
稳态极限功 steady-state power limit
稳态加速度 steady-state acceleration
稳态简谐振动 stationary simple harmonic motion
稳态解 <微分方程的> stable solution; steady-state solution
稳态近似 steady-state approximation
稳态晶体管 steady-state transistor
稳态聚变堆 steady-state fusion reactor
稳态均匀流 steady-state uniform flow
稳态控制 homeostatic control; stable control
稳态扩散 steady-state diffusion
稳态理论 steady-state theory
稳态链式反应 steady-state chain reaction
稳态量测 steady-state measurement
稳态流 stationary stream; steady-state current; steady-state flow
稳态流阻 flow resistance
稳态冒险 steady state hazard
稳态密度 steady-state concentration; steady-state density
稳态模拟 steady-state simulation
稳态模式分布 equilibrium mode distribution
稳态模型 steady-state model
稳态黏[粘]度 steady-state viscosity
稳态泡沫 steady-state froth
稳态偏差 eventual deviation; steady-state deviation
稳态频率控制 steady-state frequency control
稳态平衡 steady-state equilibrium
稳态前动力学 presteady state kinetics
稳态强迫振动 steady forced vibration; steady-state forced vibration
稳态燃料转换比 steady-state fuel conversion ratio
稳态燃烧 steady-state burning
稳态扰动 stationary disturbance
稳态热传导 steady-state conduction
稳态热交换器 steady-state heat exchanger
稳态热异常 steady-state thermal anomaly
稳态热应力 steady-state thermal stress
稳态柔量 steady-state compliance
稳态蠕变 steady-state creep
稳态蠕动 steady-state creep
稳态三相短路电流 permanent three-phase short circuit current; sustained three-phase short-circuit current
稳态渗透 steady-state permeation
稳态生长 steady-state growth
稳态生物膜 steady-state biofilm
稳态声 steady-state sound
稳态声压级 steady-state sound pressure level
稳态声源 steady-state sound source
稳态试验 steady-state test
稳态寿命 steady-state lifetime
稳态输出 steady-state output
稳态水面线 steady-state profile
稳态水跃 stationary jump
稳态说 steady-state theory
稳态速度调整率 steady-state speed regulation
稳态特性 steady-state characteristic; steady-state performance; steady-state response
稳态梯度 steady-state gradient
稳态条件 equilibrium condition; steady-state condition
稳态调节 steady-state regulation
稳态调整 steady-state regulation
稳态温度 steady-state temperature
稳态稳定储备量 steady-state stability margin
稳态稳定极限 steady-state stability limit
稳态稳定系数 steady-state stability margin
稳态稳定裕度 steady-state stability margin
稳态污染物浓度 steady-state concentration of pollutant
稳态误差 steady-state error
稳态系统 steady-state system
稳态现象 stationary phenomenon
稳态响应 steady-state response
稳态信号 steady-state signal; steady-state type of signal
稳态性能 steady-state behavio(u)r
稳态徐变 steady-state creep
稳态压力 steady-state pressure
稳态运动 stable motion; stable state of motion; steady motion
稳态运输 steady-state transport
稳态运行 steady-state operation
稳态运转 quiescent operation; quiet run
稳态噪声 stationary noise; steady-state noise
稳态噪声级 steady-state noise level
稳态噪声源 steady-state noise source
稳态增益 steady-state gain
稳态振荡 stable oscillation; stable state oscillation; steady-state oscillation
稳态振动 steady vibration
稳态振动的纽马克分析法 Newmark's method for steady state vibration
稳态振幅 steady-state amplitude
稳态蒸馏 steady-state distillation
稳态正弦波激振 steady-state sinusoidal excitation
稳态值 steady-state value
稳态滞后现象 stationary hysteresis
稳态中毒 steady-state poisoning
稳态种 equilibrium species
稳态周期 steady-state period
稳态转速 steady-state speed
稳态转速变化 steady-state speed variation
稳态转速变化增量 incremental permanent speed variation
稳态转速下降 incremental permanent speed droop
稳态转速增量 steady state incremental speed variation
稳态转弯 steady-state turn
稳态转向离心加速度 steady-state lateral acceleration
稳态最佳化 steady-state optimization
稳艇带 boat gripe
稳拖链 skag
稳妥的 reliable; safe; sound
稳妥的保险 sure-footed safety
稳妥的政策 prudential policy

W

稳妥价值 sound value
稳妥可靠的设计 adequate design
稳位桩 stay pile
稳显微组分 cryptomaceral
稳相参数 stable phase parameter
稳相加速 phase-stable acceleration
稳相加速器 phase-stable accelerator; synchro-cyclotron
稳相近似法 stationary phase approximation
稳相位法 stationary phase method
稳相原理 principle of stationary phase
稳相振荡 stable phase oscillation
稳斜钻具 angle maintenance tool assembly
稳心 metacenter [metacentre]; shifting centre
稳心半径 metacentric radius
稳心半径轨迹 locus of metacentric radii
稳心的 metacentric
稳心高度 metacentric height
稳心高度减少值 virtual loss of metacentric height
稳心轨迹曲线 metacentric involute
稳心距浮心高度 metacentric height above center [centre] of buoyancy
稳心曲线 curve of metacenter; locus of metacenter; metacenter curve
稳心曲线图 metacentric diagram
稳心图 metacentric diagram
稳心在浮心上的高度 the height of the transverse metacenter above the center of buoyancy
稳心在龙骨上的高度 height of metacentric above keel
稳心至龙骨距离 keel metacenter
稳芯垫砂 bedding a core
稳性 stability; stability of ship【船】
稳性测量仪 stabioga(u)ge
稳性范围 range of stability
稳性腐殖质 stable humus
稳性过大的船 stiff ship; stiff vessel
稳性和载重线及渔船安全分委员会 Subcommission on Stability and Load Lines and Fishing Vessel Safety
稳性力臂 lever-arm of stability; metacentric arm; righting arm; righting lever
稳性力臂及力矩 righting arms and moments
稳性力臂曲线 curve of righting arm
稳性力矩 moment of stability; righting couple; right moment; stability moment
稳性力偶 couple of stability
稳性曲线图 stability curve
稳性试验 stability test(ing)
稳性消失角 angle of vanishing stability
稳压 regulated voltage; stabilized voltage; voltage regulation; voltage-stabilizing; stabilivolt【电】; suppression of the load
稳压变压器 stabilizing transformer; voltage-stabilizing transformer
稳压变阻器系统 stabilizing resistor system
稳压层 deaerating layer; plenum space
稳压电路 mu-balanced circuit; stabilization circuit; voltage-stabilizing circuit
稳压电源 constant voltage power supply; regulated power supply; stabilized power source; stabilized voltage supply; voltage-stabilized source
稳压电源箱 regulated power pack
稳压二极管 breakdown diode; voltage regulator diode; voltage-stabilizing diode; Zener diode
稳压阀 pressure maintaining valve; unloading valve
稳压风缸 pneumatic cylinder under constant pressure
稳压管 regulation tube; regulator tube; stabilizer valve; voltage regulator tube; voltage-stabilizing tube; stabilivolt tube; stabilivolt valve【电】
稳压罐 compensator; surge tank
稳压极光红弧 stable auroral red arc
稳压流量控制器 constant pressure flow controller
稳压煤气工厂 pressure gas plant
稳压器 anti-fluctuator; constant voltage regulator [generator/ transformer]; manostat; potentiostat; pressostat; stabilizer; voltage-regulating equipment; volt(age) regulator; volt(age) stabilizer【电】; hydraulic pressure regulator; surge damper; pressure regulator; pressurizer; regulator; stabilizator
稳压器敏感性 sensibility of voltage regulator
稳压调节 constant voltage modulation
稳压系统 <土工试验用> self-compensating pressure system
稳压箱 pressurizer tank; surge tank
稳压用碳柱 carbon pile
稳压整流器 regulated rectifier; stabilized rectifier
稳压装置 pressure accumulator
稳压组件 stabipack
稳焰器 flame holder
稳翼 bilge keel
稳纸框 deckle
稳住航向 stand upon the course

问顶 roof sounding; roof tapping

问价 inquiry
问卷法 questionnaire
问卷油管下入深度 depth of coiled tubing
问事处 enquiry office; information bureau; inquiry office
问题长度 question length
问题程序 problem program(me)
问题单 queries; questionary; questionnaire
问题的关键 bottom line; crux of the matter
问题定式化 problem formulation
问题定义 problem definition
问题方式 problem mode
问题废物 problem waste
问题分析 problem analysis
问题构成 problem structuring
问题关键 key trouble spot
问题归约算子 problem reduction operator
问题归约图 problem reduction graph
问题环境 problem environment
问题回答系统 question answering system
问题解答策略 problem-solving strategy
问题解决 problem solving
问题解决行为 problem-solving behavio(u)r
问题决定 problem determination
问题求解 problem solving
问题求解程序 problem solver
问题数据 problem data
问题说明 problem definition; problem description
问题特征与火灾类型 problem characteristics and fire types
问题系统 problem system
问题叙述 problem description
问题应答长度 question length
问题诊断 problem diagnosis
问询处 enquiry office; inquiry desk; inquiry office
问询台 information board; information counter; information desk
问讯处 information bureau; information office; information room; inquiry office
问讯处柜台 enquiry desk
问讯台 information booth; information counter; information desk; information table
问讯厅 information hall

翁格那木 buck-eye

翁钠金云母 wonesite
翁氏孔菌属 <拉> Onnia

嗡鸣(进气)buzz

嗡嗡声 boom; singing
嗡嗡响 buzz
嗡嗡作响 churr
嗡音 hum note

瓮 basin; crock; jar; urn

瓮暗蓝 vat dark blue
瓮黄 vat yellow
瓮蓝 vat blue
瓮亮紫 vat brilliant violet
瓮染 vat colo(u)ring; vatting
瓮染料 vat colo(u)r
瓮染染料 vat dye(stuff)
瓮深黑 vat deep printing black
瓮声 tubby
瓮形的 urceolate
瓮印染辅剂 vat printing assistant

涡凹 <螺旋桨急转时产生的> cavitation(pocket)

涡凹反应器 cavitation reactor
涡凹气浮 cavitation air flotation
涡凹限度 cavitation limit
涡胞环流 cellular circulation
涡层 vortex sheet
涡场 vorticity field
涡成沙坝 eddy built bar
涡虫 flatworm; planaria
涡串 vortex lattice
涡道 street of vortex; vortex path; vortex street
涡道稳定性 vortex street stability
涡电流 eddy current; Foucault's current
涡电流测功机 eddy current brake
涡电流超声 eddy currents ultrasonics
涡电流法测厚仪 eddy current thickness meter
涡电流加热 eddy current heating
涡电流加热器 eddy current heater
涡电流扭矩仪 eddy current dynamometer
涡电流试验 eddy current test(ing)
涡电流探伤 eddy current inspection
涡电流阻尼器 eddy current damper
涡电式测功器 eddy current type dynamometer
涡动 eddy motion; eddy turbulence; swirl; swirling motion; turbulent motion; vortex; vortex motion; whirl; whirling motion
涡动泵 vortex pump
涡动传导率【气】 eddy conductivity
涡动传导性 eddy conductivity
涡动动力黏[粘]滞度 kinematic(al) eddy viscosity
涡动动能 eddy kinetic energy
涡动反应器 vortex reactor
涡动副 vortex pair
涡动剪应力 eddy shearing stress
涡动搅拌器 vortex agitator
涡动搅动器 vortex agitator
涡动扩散 eddy diffusion
涡动扩散率 eddy diffusivity
涡动扩散系数 coefficient of eddy diffusion
涡动能(量) eddy energy; energy of turbulence
涡动黏[粘](滞)度 eddy viscosity; kinematic(al) eddy viscosity
涡动黏[粘]滞系数 coefficient of eddy viscosity; eddy viscosity coefficient
涡动黏[粘]滞性 eddy viscosity
涡动气流 wind eddy
涡动热传导 eddy heat conduction
涡动输送 eddy transport
涡动速度 eddy velocity; fluctuation velocity; swirl speed
涡动速率 swirl speed
涡动通量 eddy flux
涡动通量仪 <直接测水库蒸发的仪器> evapotron
涡动通量蒸发公式 eddy flux evaporation formula
涡动相关 eddy correlation
涡动性 vorticity
涡动压力 eddy pressure
涡动应力 eddy stress
涡动有效势能 eddy available potential energy
涡动值 vortex value
涡动阻力 eddy resistance
涡动作用 eddy forcing
涡斗 vortex pot
涡度 eddy; vorticity
涡度场 field of vorticity
涡度方程 vorticity equation
涡度分量 component of vorticity
涡度扩散 eddy diffusion
涡度输送 vorticity transfer; vorticity transport
涡度输送理论 vorticity transport theory
涡度相关法 eddy correlation method
涡度中心 vorticity center [centre]
涡对 vortex pair
涡杆式开幅器 scroll expander; scroll opener
涡管式排沙孔 vortex ejector
涡管温度表 vortex thermometer
涡管型排沙道 vortex ejector
涡核 vortex core
涡核线 vortex trunk
涡核心 vortex core
涡痕 eddy marking
涡环 collar vortex
涡迹 <涡旋的尾迹> vortex trail; vortex path
涡激振动 vortex shedding excitation
涡桨式搅拌机 paddle-type mixer
涡桨转子 paddle rotor
涡街 vortex street; vortex trail
涡节流 turbulent throttle
涡卷 scroll
涡卷的 volute
涡卷花纹 swirl crotch figure
涡卷木纹 swirl grain
涡卷饰 cartouch(e); swirl

W

涡卷纹 swirl grain
涡卷线脚 floral scroll
涡卷心 eye of a volute
涡卷形线脚 swirls
涡卷形装饰 acanthus scroll
涡卷形装饰品 scroll
涡卷圆柱 twisted column
涡卷装饰 scroll work
涡卷准线 scroll directrix
涡壳 volute casing
涡壳混流泵 mixed flow volute pump
涡壳形通道 scroll-case access
涡空 cavitation
涡空度 degree of cavitation
涡量 vorticity
涡量方程 vorticity equation
涡量守恒 conservation of vorticity
涡量输运假设 vorticity transport hypothesis
涡列 street of vortex;vortex row;vortex street;vortex trail;vortex train
涡列引起的振动 vortex-induced vibration
涡流 vortex(eddy);vortex flow;vortexing;vortex motion;vortex type flow; vortical flow; vorticity (flow);whirl(ing current) bumpy flow; burble; churning; convolution;cross current;eddy flow;eddy water;erratic flow;Foucault current;gyrating current;helical flow;rotational flow;sinuous flow;spiral flow; stream swirl; swirl (ing flow); turbulence; turbulent current; turbulent flow; backset; bow wave;
涡流伴流 eddy wake
涡流泵 eddy pump; peripheral (turbine) pump; regenerative pump; turbulence pump;vortex pump
涡流捕尘板 eddy catching plate
涡流测功计 eddy current dynamometer
涡流测功器 eddy current dynamometer
涡流测厚仪 eddy current thickness meter
涡流测量仪器 eddy current instrument
涡流测速计 eddy current speed indicator
涡流层 backset bed;vortex band;vortex sheet;vorticity layer
涡流层理 backset bedding
涡流插芯 whirl insert
涡流常数 eddy current constant
涡流场 vortex field
涡流沉积淤泥 eddy-deposited silt
涡流沉沙池 vortex grit separator
涡流尺度 eddy scale;eddy size
涡流冲砂器 vortex grit washer
涡流除渣器 vortex cleaner
涡流传导 eddy conduction
涡流传导性 eddy conductivity
涡流传递 eddy transfer
涡流带 vortex band;vortex strip
涡流导流片 whirl vane
涡流地震检波器 eddy current seismometer
涡流电动机 eddy current motor
涡流电流 eddy current
涡流电凝聚-气浮-接触过滤组合工艺 vortex electrocoagulation-flo (a) tation-contact filtration process
涡流动力黏[粘]滞度 kinetic eddy viscosity
涡流动力黏[粘]滞系数 coefficient of kinetic eddy viscosity;kinetic eddy viscosity coefficient
涡流动能 eddy kinetic energy;turbulence energy
涡流度 turbulivity
涡流发电机 turbo-generator
涡流发散 vortex shedding
涡流发生器 vortex generator
涡流反复循环 vortex circulation and recirculation
涡流反应 eddy reaction;whirling reaction
涡流反应室 whirling reaction chamber
涡流方程 vortex equation;vorticity equation
涡流方向 set of eddy flow
涡流分布 vortex pattern
涡流分解炉 swirl furnace
涡流分解室 swirl calciner
涡流分离 eddying shedding;eddy separation;vortex separation
涡流分离流量计 eddy separation flowmeter;eddy shedding flowmeter;row shedding flowmeter
涡流分离器 eddy current separator
涡流分粒器 vortex classifier
涡流粉碎机 eddy mill
涡流附加损失 strand losses
涡流干扰 vortex interference
涡流干扰效应 vortex-interference effect
涡流隔离板 eddy current separator
涡流隔焰炉烧嘴 turbulence muffle burner
涡流鼓风机 turbofan
涡流固体分离器 swirl-flow separator
涡流管 vortex tube
涡流管喷射器 vortex tube ejector
涡流管制冷器 vortex tube refrigerator
涡流轨迹 vortex path
涡流痕 vortex sheet
涡流痕迹 eddy marking
涡流虹吸 volute siphon [syphon]
涡流环 vortex ring
涡流环叶栅 vortex-ring cascade
涡流缓行器 eddy current rail brake
涡流混合 eddy mixing
涡流混合反应器 turbulent mixer/reactor
涡流激动振荡 vortex-excited oscillation
涡流激励 vortex excitation
涡流激振区域 vortex excitation region
涡流计 eddy current ga(u)ge
涡流加热 eddy current heating
涡流假说 vortex hypothesis
涡流检测 eddy current examination
涡流检验 eddy current inspection
涡流界限 turbulence level
涡流卡片 eddycard
涡流卡片存储器 eddycard memory;eddycard store
涡流可调离心风机 vortex adjustable centrifugal fan
涡流空气滤净器 vortex air cleaner
涡流空蚀 vortex cavitation
涡流扩散 eddy diffusion; turbulent diffusion
涡流扩散度 eddy diffusivity
涡流扩散率 eddy diffusion coefficient;eddy diffusivity
涡流扩散系数 coefficient of eddy diffusion;eddy diffusion coefficient
涡流扩散项 eddy diffusion term
涡流扩散作用 eddy diffusion
涡流离合器 eddy current clutch
涡流理论 vortex theory
涡流联轴节 eddy current coupling
涡流流动 eddy flux;eddying flow
涡流流量计 eddy current flowmeter;turbine flowmeter
涡流流线 vortex filament
涡流笼式流量计 vortex cage meter
涡流密度 vortex density
涡流模式 turbulence mode
涡流黏[粘](滞)度 eddy viscosity
涡流黏[粘]滞度系数 coefficient of eddy viscosity;eddy viscosity coefficient
涡流黏[粘]滞性 eddy viscosity
涡流偶极子 vortex doublet
涡流耦合器 eddy current coupler
涡流盘 eddy current disc; spinning disk [disc]
涡流喷射 eddy effusion
涡流喷嘴 swirl nozzle
涡流片 swirl plate;vortex sheet
涡流片式喷嘴 shear-plate nozzle
涡流屏蔽 eddy current screen
涡流谱 eddy spectrum;vortex pattern
涡流强度 strength of vortex; vortex intensity;vortex strength;vorticity
涡流侵蚀作用 evorsion
涡流区 <沙漠中沙在空中悬浮区> eddy zone;eddy region;vortex cavity
涡流燃烧室 swirl burner;swirl combustion chamber
涡流热传导 eddy heat conduction; eddy heat flux
涡流热扩散系数 coefficient of eddy thermal diffusion;eddy thermal diffusivity
涡流熔化炉 vortex melting furnace
涡流砂粒分离器 vortex grit separator
涡流栅 vortex lattice
涡流声 vortex sound
涡流式拌和机 spiral-flow mixer
涡流式泵 turbine pump
涡流式沉砂池 churn flow grit chamber
涡流式传感器 eddy current transducer
涡流式二次冷凝器 cyclone-type after-condenser
涡流式反应室 whirling reaction chamber
涡流式感应图示仪 cyclograph
涡流式缓行器 eddy current type retarder
涡流式减速器 eddy current type retarder
涡流式检波器 eddy current type geophone
涡流式弥雾机 swirl-type atomizer
涡流式碾磨机 eddy mill
涡流式燃烧器 whirl burner
涡流式燃烧室 swirl combustion chamber; toroidal swirl chamber; vortex cavity
涡流式燃烧系统 vortex firing system
涡流式研磨机 eddy mill
涡流式转速表 eddy current revolution counter
涡流式转速计 eddy current tachometer
涡流事故 turbulence accident
涡流室 minor air cell;swirl chamber;turbulence chamber
涡流室式柴油机 swirl-chamber diesel
涡流束 vortex core
涡流衰减 vortex decay
涡流速度 eddy velocity; velocity of whirl;whirl velocity
涡流速度仪 vortex velocity meter
涡流速度指示表 eddy current speed indicator
涡流损耗 eddy current loss
涡流损耗异常 eddy current anomaly
涡流损失 eddy(current) loss;loss by swirls;vortex loss
涡流探伤 eddy current inspection;eddy current test(ing)
涡流探伤法 eddy current flaw detection
涡流探伤机 eddy current test(ing) equipment
涡流通量 eddy flux; moisture flux; turbulent flux
涡流图 vortex pattern
涡流脱落 vortex shedding
涡流位移电流 eddy displacement currents
涡流温度计 vortex thermometer
涡流稳定弧 vortex-stabilized arc
涡流误差 eddy error
涡流洗砂器 vortex grit washer
涡流系 vortex pattern
涡流系数 eddy current coefficient; eddy current factor
涡流效应 eddy current effect; eddy effect;Foucault's current effect;rotation effect; vortex shedding; vortex shedding effect
涡流形成 churning; eddy generation; eddying;eddy making
涡流形的 eddylike
涡流型放大元件 vortex fluid amplifier
涡流型分离器 eddy current separator
涡流型卸料方式 vortex-shaped discharge pattern
涡流旋风式选粉机 Van Tongeren type classifier
涡流旋流器 centriclone
涡流旋转制动机 eddy current rotating brake
涡流压力 eddy pressure
涡流压气机 drag compressor
涡流液体分离器 swirl-flow separator
涡流引起的 vortex-induced
涡流应力 eddy stress
涡流诱导的 vortex-induced
涡流预热器 turbulent flow preheater
涡流原理 vortex principle
涡流约束 eddy current confinement
涡流运动 eddying motion
涡流运动黏[粘]度 eddy(ing) kinematic viscosity
涡流运动学 vorticity kinematics
涡流闸 eddy current brake
涡流制动磁体 eddy current braking magnet
涡流制动(法) eddy current braking
涡流制动机 eddy current brake
涡流制动器 eddy current brake; Foucault's current brake
涡流制动器调速 speed regulation through eddy current braking device
涡流制冷效应 vortex refrigerating effect
涡流重差计 eddy current gravimeter
涡流重力分离器 cross-current gravity separator
涡流轴线 eddy axis
涡流转速计 drag-type tachometer; eddy current revolution counter
涡流阻力 eddy(ing) resistance;turbulence resistance;vortex drag
涡流阻尼 eddy current damping;eddy damping
涡流阻尼器 eddy current retarder
涡流作用 eddy current effect; vortex action
涡轮 impeller wheel;rear torus;scroll-wheel; turbine runner; turbine wheel;turbo-wheel;vortex ring
涡轮泵 roturbo; turbine pump; turbopump;volute pump
涡轮泵供油的 turbo-fed
涡轮泵压式供油 turbopump injection

涡轮泵组 gas turbine pump combination; pump unit; turbopump assembly
涡轮变流机 turbo-converter
涡轮变流器 turbo-converter
涡轮变压器 transformer turbine
涡轮操作的 turbocharged
涡轮柴油机 turbodiesel
涡轮敞炉 turbohearth
涡轮充气器 turbine aerator
涡轮冲压火箭 turboram rocket
涡轮冲压器接进气歧管的密封件 turbocharger to inlet manifold seal
涡轮冲压式喷气发动机 turboram jet engine
涡轮冲压(组合)喷气发动机 turboramjet
涡轮出口气体压力 turbine outlet pressure
涡轮出口温度 turbine-exit temperature
涡轮出口温度控制 turbine discharge temperature control
涡轮出气壳 gas outlet housing
涡轮传动 turbine-drive; turbo-transmission
涡轮传动泵 turbine-driven pump
涡轮船 turbine vessel
涡轮导流片 turbine diaphragm
涡轮导轮 turbine guide wheel
涡轮导向器 turbine stator
涡轮导向器气体流出速度 turbine nozzle exit velocity
涡轮导向器叶片 turbine nozzle blade; turbine nozzle vane
涡轮导向叶片 nozzle blade
涡轮导向装置 turbine nozzle
涡轮电动的 turbo-electric(al)
涡轮电动机车 turbine electric(al) locomotive
涡轮电力的 turbo-electric(al)
涡轮电力驱动 turbo-electric (al) drive
涡轮定轮 stator ring
涡轮定子 turbine stator
涡轮动力 turbo-power
涡轮动力滑翔机 turbine-powered glider
涡轮动力轴 turboshaft
涡轮动力装置 turbine power plant
涡轮发电动力装置 turbo-electric(al) propulsion
涡轮发电机 power turbine; turbine electric(al) generator; turbine generator; turbodynamo; turbo-generator
涡轮发电机舱 power turbine room; turbine generator room; turbo-generator room
涡轮发电机照明 turbo-generator lighting
涡轮发电机轴 turbo-generator shaft
涡轮发电机组 turbine generator set; turbo-generator set; turbo-generator unit; turbo-unit
涡轮发动机 turbine engine; turbo-engine
涡轮发动机的 turbo-electric(al)
涡轮发动机运输机 turbine transport
涡轮反用换流器 turbo-inverter
涡轮反作用度 turbine reaction
涡轮分配器 turbo-distributor
涡轮分子泵 turbomolecular pump
涡轮粉碎机 turbo-mill
涡轮风扇 turbofan
涡轮风扇动力装置 turbofan power plant
涡轮风扇式运输机 turbofan transport
涡轮浮选 turbine flo(a)tation
涡轮复式发动机 turbo-compound engine
涡轮盖螺旋帽 turbine cover screw cap
涡轮盖帽 turbine cover cap
涡轮干燥机 turbodrier
涡轮高压级 high-pressure turbine stage
涡轮功率计 turbine dynamometer
涡轮鼓风机 turbo-blower
涡轮固定叶轮 nozzle diaphragm
涡轮滚珠轴承盖 turbine ball bearing cover
涡轮换向齿轮 worm reversing gear
涡轮混合机 turbine mixer
涡轮混合器 turbo-mixer
涡轮活塞式混合发动机 turbo-compound aero-engine
涡轮火车 turbotrain
涡轮火箭发动机 turborocket
涡轮机 turbine; turbomachine
涡轮机保安设备 turbine protective device
涡轮机标志能力 turbine name-plate capacity
涡轮机舱 turbine room
涡轮机车 turbine locomotive
涡轮机出力 output of turbine
涡轮机传动 turbo-drive
涡轮机船 turbine ship
涡轮机导流隔板 turbine diaphragm
涡轮机底座 turbine foundation
涡轮机电力传动 turbine electric(al) drive
涡轮机电力机车 turbo-electric (al) locomotive
涡轮机电力推进船 turbo-electric(al) ship
涡轮机电力推进装置 turbo-electric(al) drive
涡轮机电力装置 turbo-electric (al) installation
涡轮机动力装置 turbo-power unit
涡轮机房 turbine hall
涡轮机放水口 turbine drain
涡轮机废气防护墙 blast fence
涡轮机功率 output of turbine; turbine output
涡轮机功率输出 power output of turbine
涡轮机固定部分 turbine stator
涡轮机级 turbine stage
涡轮机继动器 turbine servomotor
涡轮机监控设备 turbo-visory equipment
涡轮机减速推进装置 turbine-geared propulsion unit
涡轮机壳(体) turbine case; turbine cylinder; turbine housing
涡轮机客运列车 turboliner
涡轮机空放阀 turbine relief valve
涡轮机气蚀 turbine cavitation
涡轮机驱动 turbine-driven
涡轮机设备容量 installed wheel capacity
涡轮机设计性能 turbomachine design performance
涡轮机室 turbine house; turbine room
涡轮机损耗 turbine losses
涡轮机调整 turbine regulation
涡轮机外壳 turbine casing
涡轮机效率 turbine efficiency
涡轮机械 turbine machinery
涡轮机性能曲线 turbine performance curve
涡轮机性能图表 turbine performance chart
涡轮机压气机 turbomachinery compressor
涡轮机叶轮 turbine wheel; turbo-impeller
涡轮机叶片 turbine blade
涡轮机叶片振动 turbine bucket vibration
涡轮机油 turbine oil
涡轮机油压调速器 oil-pressure turbine governor
涡轮机运行曲线 turbine performance curve
涡轮机增压 turbocharging
涡轮机站 turbine station
涡轮机支承梁 turbine beam
涡轮机直接传动 turbine direct drive
涡轮机轴承 turbine bearing
涡轮机转鼓 turbine drum
涡轮机转轮 turbine disc [disk]; turbine rotor
涡轮机装置 turbine installation
涡轮机组 turbine set; turboset; turbomachinery
涡轮给水泵 turbo-feed pump
涡轮加速器 turbo-accelerator
涡轮减速器 worm reduction unit
涡轮交流发电机 turbo-alternator
涡轮搅拌机 turbine mixer; turbine stirrer; turbine-type agitator; turbomixer
涡轮进口温度 turbine inlet temperature
涡轮进口温度控制 turbine inlet temperature control
涡轮进气壳 gas inlet housing
涡轮进气速度 approach velocity
涡轮卷纬装置 turbo-winder
涡轮壳 turbine casing; turbine cylinder
涡轮壳环 turbine shroud ring
涡轮壳体 turbine shroud
涡轮冷气发动机 turboexpander
涡轮离心压缩机 turbo-compressor
涡轮力矩 turbine torque
涡轮励磁机 turboexciter
涡轮联轴节 turbo-coupling
涡轮流量计 revolving flowmeter; turbine flowmeter
涡轮螺桨发动机 turbo-propeller; turbo-propeller engine
涡轮螺桨发动机性能 turboprop performance
涡轮螺桨式客机 turboliner
涡轮螺桨组合 turbine-propeller combination
涡轮螺旋桨 turbine propeller
涡轮螺旋桨船身式水上飞机 turboprop boat
涡轮螺旋桨发动机 turbine-propeller engine; turboprop; turboprop engine; propeller turbine
涡轮螺旋桨系统 airscrew turbine system
涡轮螺旋桨运输机 turboprop transport
涡轮马达 turbine motor
涡轮秒流量 turbine flow per second
涡轮排气管 turbine exhaust pipe
涡轮排气机 turbo-exhauster
涡轮排水泵 turbo-extractor pump
涡轮盘 turbine runner disc
涡轮盘式干燥机 turbo-tray dryer
涡轮喷气发动机 fan jet; turbine jet; turbojet; turbojet engine
涡轮喷气机 turbine jet engine; turbojet engine
涡轮喷气型 turbojet version
涡轮喷嘴 turbine nozzle
涡轮膨胀机 turboexpander
涡轮起动机 turbo-starter
涡轮气体出口 turbine outlet
涡轮气体发生器 turbo-fed gas generator
涡轮气体净化器 turbine gas purifier
涡轮气体吸收器 turbine gas absorber
涡轮汽车 turbo-car
涡轮清垢 turbining
涡轮清管 tube turbining
涡轮驱动的 turbo-driven
涡轮驱动增压器 turbo-driven supercharger
涡轮燃烧炉 turbo burner
涡轮燃烧器 turbo burner
涡轮入口导管 turbine entry duct
涡轮扫气泵 turbo-scavenging blower
涡轮射流 turbojet
涡轮式 turbine-type; turbo-type
涡轮式拌和机 turbid mixer; turbine mixer
涡轮式包装机 turbo-packer
涡轮式泵 turbid pump; turbine pump
涡轮式风力选粉机 turbo-air separator
涡轮式风扇 turbofan
涡轮式风扇机 turbine fan
涡轮式鼓风机 turbine blower; turbo-blower
涡轮式环形交叉(道口)<车辆进入和离开交叉口时均按切线方向行驶> turbine-type rotary intersection
涡轮式混合器 turbine mixer; turbo-mixer
涡轮式混凝土搅拌机 turbo concrete mixer
涡轮式搅拌机 turbid stirrer; turbine mixer
涡轮式搅拌器 turbine agitator; turbine shape mixer; turbine-type mixer
涡轮式空压机 turbo-blower
涡轮式冷却器 reaction-type cooler
涡轮式离心泵 centrifugal pump of turbine type; turbine-centrifugal pump
涡轮式立体交叉 turbid(type) intersection; turbine-type interchange; turbine-type intersection
涡轮式流量计 turbine flowmeter
涡轮式喷灌机 turbine sprinkler
涡轮式喷头 turbine sprinkler
涡轮式气马达 turbine air motor
涡轮式深水泵 turbine-type deep well pump
涡轮式水表 inferential meter; turbine shape meter; turbine(-type) meter; turbine-type watermeter
涡轮式通风机 turbo-blower; turbofan
涡轮式消沫离心机 turbine-centrifugal foam breaker
涡轮式压气机 turbine compressor; turbo-compressor
涡轮式压缩机 turbine compressor; turbo-compressor
涡轮式增压器 turbo-blower; turbo-type supercharger
涡轮式制冷机 turbo-refrigerator
涡轮式钻(孔)机 vane borer
涡轮式钻土机 vane borer
涡轮体 turbine casing
涡轮调节器 turbo-regulator
涡轮通风机 radial fan; turbine-driven fan
涡轮推进 turbine propulsion
涡轮推进机 turboprop engine
涡轮推土机 turbo-dozer
涡轮往复蒸汽机联合装置 turbo-reciprocating engines
涡轮涡杆千斤顶 ratchet lever jack
涡轮涡杆最终传动 worm gear final drive
涡轮蜗杆传动机械 worm-geared machine
涡轮洗矿机 turbo-washer
涡轮型溶解机 turbo-dissolver
涡轮型叶轮搅拌器 turbine-type im-

W

peller agitator
涡轮性能 turbine performance
涡轮旋流器 turbocyclone
涡轮循环泵 turbo-circulator
涡轮压力比 turbine pressure ratio
涡轮压气活塞发动机 turbocharged piston engine
涡轮压气机 turbine-driven compressor
涡轮压缩机 turbo-driven compressor
涡轮研磨机 turbo-mill
涡轮叶 turbine blade
涡轮叶轮 turbine wheel; turbine wheel impeller
涡轮叶片 impeller blade; turbine bucket
涡轮叶片冷却 turbine blade cooling
涡轮叶片绕流损失 loss in turbine blade
涡轮叶片外箍 blade rim
涡轮油船 turbine tanker
涡轮圆盘 turbine disc [disk]
涡轮噪声 turbine noise
涡轮增压 turbocharge; turbocharging
涡轮增压柴油机 turboblown diesel engine; turbocharged diesel; turbo-compounded diesel
涡轮增压的 turbocharged
涡轮增压的脉冲变换系统 pulse-converter system of turbocharging
涡轮增压发动机 turbo-supercharged engine
涡轮增压后冷式发动机 turbocharged aftercooled engine
涡轮增压机 turbo; turbo-supercharged engine
涡轮增压器 turbo(super)charger
涡轮增压器接头 turbocharger coupling
涡轮增压器联轴节 turbocharger coupling
涡轮增压器配合 turbocharger matching
涡轮增压器匹配 turbocharger matching
涡轮增压器特性曲线 turbo-blower characteristics
涡轮增压器压缩比 turbocharger compression ratio
涡轮增压器压缩机 turbocharger compressor
涡轮增压器运行故障 turbocharger running defects
涡轮增压器转速 turbocharger speed
涡轮增压式柴油机 turbocharged diesel engine
涡轮增压式发动机 turbocharged engine
涡轮增压叶片 turbosupercharger bucket
涡轮增压装置 turbocharging installation
涡轮罩填密 turbine cover packing
涡轮振动继电器 turbine vibration relay
涡轮振动停机开关 turbine vibration shut-down switch
涡轮支柱 turboprop
涡轮直升机 turbocopter
涡轮制动器 turbine brake
涡轮轴 turbine shaft; turboshaft
涡轮轴发动机 turboshaft engine
涡轮轴流式扇风机 turbo-axial fan
涡轮轴压缩器组 turboshaft-compressor
涡轮转速 secondary speed
涡轮转速调节 turbine speed control
涡轮转子 turbo-rotor
涡轮转子轴 turbine rotor shaft
涡轮子午线剖面轮廓 meridianal contour
涡轮钻机 turbodrill; turbodrilling rig
涡轮钻机喷水枪 turbodrill monitor
涡轮钻机钻杆 turbine drill rod
涡轮钻进 turbine drilling; turbodrilling
涡轮钻具 turbine drill; turbodrill
涡轮钻具外壳 turbodrill housing
涡轮钻掘 turbodrilling
涡螺 volute
涡面 vortex sheet; vortex surface
涡面卷起 vortex plane rolling-up
涡黏[粘]度 vortex viscosity
涡偶 vortex pair
涡片 vortex sheet
涡强 vortex strength
涡区 vortex band
涡扇发动机 fan jet
涡舌 volute tongue
涡蚀穴 eversion hollow
涡式流量计 vortex shedding flowmeter
涡室 volute chamber
涡丝 vortex line
涡通量 vortex flux
涡头螺钉 socket head screw
涡纹 curl; running spiral design
涡纹饰镶面板 curly veneer
涡系 vortex system
涡线 line vortex; vortex line
涡线波导管 spiral coiled waveguide
涡线图谱 pattern of vortex line
涡线形状 pattern of vortex; pattern of vortex line
涡形齿轮 scroll-wheel
涡形定向器 vortex finder
涡形管 scroll; vortex pipe [piping]; vortex tube [tubing]
涡形花纹 scroll work design
涡形花样 volute
涡形剪床 scroll shear
涡形剪切机 scroll shear
涡形轮 snail
涡形器 eddy unit
涡形曲流 scroll meander
涡形沙坝 scroll bar
涡形饰纹镶面板 burl veneer; curly veneer
涡形凸轮 snail
涡形轴 scroll shaft
涡形装饰 cartouch(e); scroll work
涡形装置 cartouch(e)
涡旋 gyral; gyre; volute; volution; vortex; vorticity; whirlpool
涡旋场 eddy field; field of vorticity; rotational field; vortex field
涡旋澄清池 swirling clarifier
涡旋尺度 eddy size
涡旋传递 vorticity transport
涡旋次数 frequency of eddies
涡旋的 turbulent
涡旋度守恒 conservation of vorticity
涡旋分布法 vortex distribution method
涡旋风 eddy wind
涡旋风速计 vortex velocity anemometer
涡旋迹 vortex street
涡旋流动 vortex flow
涡旋密度 vortex density
涡旋喷注 swirl injection
涡旋频率 eddy frequency; frequency of eddies
涡旋频数 frequency of eddies
涡旋破碎 vortex breakdown
涡旋强度 vorticity
涡旋射流 swirling jet
涡旋射流空化 swirling jet-induced cavitation
涡旋时间 vortex time
涡旋式波导管 coiled waveguide
涡旋式密封垫 vortex type seal
涡旋式喷嘴 swirler
涡旋式压缩机 scroll compressor
涡旋饰 helix; volute
涡旋输移 vorticity transport
涡旋丝 vortex filament
涡旋体 swirl; vortex
涡旋脱落 vortex shedding
涡旋纹 swirl grain
涡旋紊流 eddying turbulence; turbulence
涡旋误差 swirl error
涡旋线 vortex filament
涡旋形 volute
涡旋形饰 volute
涡旋形弹簧 volute spring
涡旋叶片 swirl vane
涡旋雨 vortex rain
涡旋云系 vortex cloud system
涡旋运动 eddying motion; turbulent motion; turbulent movement; vortical motion; vorticity motion
涡旋中心 eye
涡旋状态 vorticity
涡旋阻力 eddy resistance
涡漩变形 eddy deformation
涡漩谱 eddy spectrum
涡穴 cahot
涡源 vorticity source
涡致振动 vortex-induced vibration
涡转 whirl(ing)
涡状流痕 vortex-like flow mark
涡状喷注 swirl injection
涡浊 cavitation damage
涡阻 vortex drag

窝

窝抱 brood

窝洞壁 cavity wall
窝洞充填 cavity filling
窝洞垫基 cavity base
窝洞消毒 cavity disinfection
窝洞修复 cavity restoration
窝洞暂封剂 cavity temporary sealant
窝洞制备 cavity preparation
窝工 idle; poor organization resulting in holding up the work; run(ning) idle; run-out of work
窝工的船舶时数 ship hours of idle time
窝工费用 expenses of idleness; idling charges
窝工工时 man-hour in idleness
窝工日 lay day
窝工日数 lay days
窝工时间 downtime; idle hours; idle period; idle time; time-out
窝工时期 bad time
窝工收费 idling charge
窝沟封闭剂 fissure sealant
窝管 honeycomb pipe; honeycomb tube
窝接 bell and spigot joint; socket and spigot joint(ing); socket joint; spigot and faucet joint; spigot joint
窝接杆件 socketed member
窝接口承端 socket end
窝接口大端 socket end; socket end of pipe
窝接口小端 spigot end of pipe
窝接式混凝土管 spigot and socket concrete pipe
窝接式接头 spigot and socket connection
窝接式接头管 spigot and socket pipe
窝接式陶管 spigot and socket stoneware pipe
窝模 snap
窝棚 booth; shack; shanty; shed; wickiup; dugout <采掘时搭的>; hunting box <打猎时住的>
窝头 snap
窝托横梁 backing
窝形的 nested
窝形柱 socketed column
窝穴 pot-hole
窝眼盘 cell plate
窝眼筒 pocket cylinder
窝眼圆盘 pocket disk
窝重 litter weight
窝钻 socket drill

蜗

蜗杆 endless screw; helicoid screw; hob; worm

蜗杆泵 screw pump; worm pump
蜗杆差动滑车 worm chain block
蜗杆齿弧式转向机 worm and sector steering device
蜗杆齿轮 gear on worm
蜗杆齿扇转向器 steering gear with worm sector
蜗杆齿条式 worm rack type
蜗杆齿条式驱动装置 worm rack type driving device
蜗杆传动 worm drive; worm-gearing
蜗杆传动齿条 worm rack
蜗杆传动的 worm-driven
蜗杆传动后轴 worm-driven rear axle
蜗杆传动绞车 worm-geared winch
蜗杆传动卷扬机 worm-geared winch
蜗杆传动马达 worm motor
蜗杆传动箱 worm box
蜗杆传动主动齿轮 worm driving pinion
蜗杆传动装置 worm gear drive
蜗杆单滚轮式转向机构 worm-and-single-roller steering gear
蜗杆给矿机 worm feeder
蜗杆滚齿机 worm hobbing machine
蜗杆滚刀 worm hob
蜗杆滚轮式转向机构 worm-and-roller steering gear
蜗杆滚轮式转向器 worm and roller-lever steering gear
蜗杆滑车 worm pulley block
蜗杆换向齿轮 worm reverse gear
蜗杆及扇齿付 worm and sector gear
蜗杆及扇齿轮结构 worm and sector gear structure
蜗杆挤压机 screw extrusion machine
蜗杆架 worm bracket
蜗杆减速齿轮箱 worm reduction gear unit
蜗杆减速传动装置 worm reduction gear
蜗杆减速装置 worm reduction gearing
蜗杆检查仪 worm tester
蜗杆节距 pitch of worm
蜗杆链滑车 differential chain block; worm chain block
蜗杆螺钉 worm screw
蜗杆螺母式转向 worm-and-nut type steering
蜗杆螺母转向器 steering gear with worm and nut
蜗杆螺丝 worm
蜗杆螺纹 thread worm; worm thread
蜗杆磨床 worm grinder
蜗杆啮合 worm mesh
蜗杆偏心调整套 worm eccentric adjusting sleeve
蜗杆起重机 worm hoist
蜗杆球 worm ball
蜗杆球导管夹 worm ball return guide clamp
蜗杆球回行导架 worm ball return

guide
蜗杆曲柄双销式转向机构 worm-and-twin-lever steering gear
蜗杆梢 screw tip
蜗杆式机械手动给进 worm-type mechanical hand feed
蜗杆式夹紧机构 worm clip
蜗杆式推进器 worm-type propeller
蜗杆式压缩机 worm compressor
蜗杆式转向机构 worm-type steering gear
蜗杆输送器 worm conveyer [conveyor]
蜗杆双滚轮式转向机构 worm-and-double-roller steering gear
蜗杆调节 worm adjustment
蜗杆凸轮辊式送料装置 worm-cam roller feeder
蜗杆推进器 worm feed
蜗杆推力螺钉 worm thrust screw
蜗杆往复球螺母 worm recirculating ball nut
蜗杆涡轮传动减速器 worm and gear reducer
蜗杆蜗轮传动装置 worm gear
蜗杆蜗轮 worm and wormwheel
蜗杆蜗轮组 worm and gear set
蜗杆蜗轮滑车组 worm and worm-wheel pulley block
蜗杆蜗轮卷扬机 worm gear machine
蜗杆蜗轮式驱动桥 worm-and-wheel axle
蜗杆蜗轮转向机构 worm-and-wheel steering gear
蜗杆铣床 worm milling machine
蜗杆铣刀 worm milling cutter
蜗杆箱 worm gear case
蜗杆销钉式转向器 peg-and-worm steering gear
蜗杆压榨机 forcer
蜗杆腰部 worm waist
蜗杆与滚轮 worm-and-roller
蜗杆与螺旋齿扇齿 worm-and-spiral-teeth sector
蜗杆与三齿式齿扇 worm-and-three-teeth sector
蜗杆与扇形轮 worm-and-sector
蜗杆与蜗轮 worm and worm gear
蜗杆罩 worm case;worm casing
蜗杆止动 worm lock
蜗杆止推轴承 worm thrust bearing
蜗杆制动器 worm brake
蜗杆制动器弹簧 worm brake spring
蜗杆制动器内制动锥体 worm brake cone
蜗杆制动装置 worm brake
蜗杆轴 worm shaft
蜗杆轴承 worm bearing
蜗杆轴承杯 worm bearing cup
蜗杆轴承的调整 worm bearing adjustment
蜗杆轴承的止推螺钉 worm bearing thrust screw
蜗杆轴承滚珠座圈 worm bearing ball race
蜗杆轴承调整螺母 worm bearing adjusting nut
蜗杆轴齿轮式 worm shaft and gear type
蜗杆轴隔套 worm shaft spacer tube
蜗杆轴螺母式 worm shaft and nut type
蜗杆轴扇轮式 worm shaft and sector type
蜗杆轴向间隙 worm shaft end play
蜗杆轴向节距 linear pitch
蜗杆转向 worm steer(ing)
蜗杆转向装置 worm steering gear
蜗杆锥棍轴承 worm shaft roller conical bearing
蜗管入口 spiral inlet
蜗卷式入口 helixal inlet
蜗卷叶梗 culiculus
蜗壳 scroll;scroll case;scroll casing;spiral case;spiral casing;spiral housing;volute;volute casing;volute housing;vortex chamber;wheel case
蜗壳安全泄水道 spiral-case relief sluice
蜗壳包角 nose angle of spiral casing;wrap of spiral case
蜗壳泵 volute pump
蜗壳鼻端 nose angle of spiral casing
蜗壳层 spiral-case floor
蜗壳的包角 nose angle
蜗壳隔舌 volute tongue
蜗壳喉部 volute throat
蜗壳角 volute angle
蜗壳接头 scroll-case connection
蜗壳进口 scroll-case access;spiral-case access
蜗壳进入孔 scroll-case access;spiral-case access;spiral-case manhole
蜗壳扩压器 volute diffuser
蜗壳排水管 scroll-cased drain pipe
蜗壳腔 volute chamber
蜗壳舌板 baffle vane of spiral case
蜗壳舌部 volute tongue
蜗壳式离心泵 screw-type centrifugal nozzle; volute-type centrifugal pump
蜗壳式收集器 volute collector
蜗壳(式)水轮机 spiral wheel;spiral-cased(water)turbine;spiral turbine
蜗壳通道 spiral-case access
蜗壳通气管 scroll-case vent
蜗壳尾端 nose angle of spiral casing
蜗壳形模型 cochlear model
蜗壳形排水管 scroll-case drain
蜗壳形通道 scroll-case access
蜗壳形转子 spiral-shaped rotor
蜗孔 helicotrema
蜗流螺旋起重器 worm gear screw jack
蜗轮 screw wheel;worm gear;worm wheel
蜗轮(标)度盘 worm-wheel dial
蜗轮传动 worm gear drive
蜗轮传动比 worm gear drive ratio
蜗轮传动绞车 worm-geared winch
蜗轮传动式螺母扳手 worm gear nut runner
蜗轮传动装置 worm gear drive mechanism; worm-gear(ing); worm-wheel gearing
蜗轮的轮缘 runner band
蜗轮副检查仪 worm gear conjugation tester
蜗轮鼓风机 turboblower
蜗轮管 worm pipe
蜗轮滚齿机 worm-wheel hobbing machine
蜗轮滚刀两用磨床 hob and worm grinding machine
蜗轮滚刀螺纹 worm-wheel hob thread
蜗轮滚(铣)刀 worm gear hob
蜗轮滚削 worm gear hobbing
蜗轮减速机 worm reduction gear;worm speed reducer
蜗轮减速器 worm gear reducer;worm speed reducer
蜗轮进给 worm gear feed
蜗轮卷扬机 worm-geared hoist
蜗轮抗弯强度 worm bending strength
蜗轮轮毂 worm-wheel hub
蜗轮螺杆机构 worm-and-wheel gear
蜗轮摩擦闸 worm-wheel friction brake
蜗轮摩擦制动器 worm-wheel friction brake
蜗轮磨床 worm grinding machine
蜗轮润滑油 worm-type lubricant
蜗轮润滑脂 worm gear grease
蜗轮式积极卷取运动 positive worm take-up motion
蜗轮式卷扬机 worm-geared machine
蜗轮式挖泥船 worm-wheel dredge(r)
蜗轮式闸门的启闭机 screw capstan head
蜗轮凸轮传动 worm gear cam drive
蜗轮蜗杆 worm and gear
蜗轮蜗杆传动 over-type worm gear;worm-gearing
蜗轮蜗杆传动装置 screw gearing
蜗轮蜗杆副 worm couple
蜗轮蜗杆减速箱 worm reduction box
蜗轮蜗杆千斤顶 ratchet lever jack
蜗轮蜗杆式差速器 worm gear differential
蜗轮蜗杆装置 worm-gear(ing)
蜗轮铣床 worm gear milling machine
蜗轮箱 worm gear case
蜗轮箱壳体 worm housing
蜗轮油 worm gear oil
蜗轮增压发动机 turbocharged engine
蜗轮增压机 turbocharger
蜗轮支架 worm-wheel bracket
蜗轮轴 worm-wheel shaft
蜗轮转向装置 worm steering gear
蜗轮状构造 turbine like structure
蜗牛壳状 cochlea(ry);cochleated
蜗牛线【数】cochleoid
蜗牛形曲线 limacon
蜗牛形柱墩 <古典建筑物的> apophyge
蜗牛状星云 snail-shaped nebula
蜗钎 worm screw
蜗室 spiral housing
蜗水管 aqueduct of cochlea
蜗水管内口 internal orifice of cochlear aqueduct
蜗水管外口 external aperture of cochlear aqueduct
蜗线尺 diminished scale
蜗线形外壳 volute casing
蜗形泵 worm pump
蜗形齿轮 <直径逐渐变化的> scroll gear
蜗形的 cochlear
蜗形管 worm pipe
蜗形机壳 volute casing
蜗形离心机 scroll centrifuge
蜗形轮 snail wheel
蜗形旁通风阀 scroll bypass damper
蜗形器 snail unit
蜗形沙坝 scroll bar
蜗形绳轮 fusee
蜗形弹簧 volute spring
蜗形体喉部 volute throat
蜗形腿狭台 console table
蜗形吸入室 volute suction
蜗形镶面板 curly veneer
蜗形重力介质选矿机 snail type heavy-medium separator
蜗压机 plodder
蜗缘逢 hem
蜗缘饰 hem
蜗支 cochlear branch

我触点继电器 non-contact relay

我触点开关 non-contacting switch
我方存款 nostro deposit
我方概不负责 without our responsibility
我方结存 balance in our favour
我方受益的余额 balance in our favour
我方无责任 without our responsibility
我方余额 balance in our favour
我方账户 nostro account
我国目前采用的干酪根分类 recognizable kerogen classification at present in China
我群意识 <个人与团体休戚相关的心理> we-group consciousness

沃波尔(目视)比色器 Walpole colo(u)rimeter

沃丹尔风 vaudaire;vauderon
沃德尔球形度系数 Wadell's sphericity factor
沃登重力仪 Worden gravimeter;Worden gravity meter
沃地 fertile land;fertile soil
沃丁神接待战死者英灵的殿堂 <北欧神话中的> Valhalla
沃尔对比 Wahl correlation
沃尔多硬度试验机 Waldo hardness tester
沃尔夫-伦德马克系统 Wolf-Lundmark system
沃尔夫式浮坝 Wolf's curtain
沃尔夫数 Wolf number
沃尔夫斯霍茨钻孔(灌注混凝土)桩 Wolfsholz bored pile
沃尔夫图 Wolf diagram
沃尔夫治河法 <用原木顺坝治河,常用于欧洲> Woolf system
沃尔哈德滴定 Volhard's titration
沃尔哈德硫氰酸钾溶液 Volhard's solution
沃尔康合金 <一种耐蚀铜合金> Vulcon metal
沃尔克型圆盘铸锭机 Walker casting wheel
沃尔曼盐 Wolman salt
沃尔曼盐剂 <木材防腐用> Wolman salt agent
沃尔喷漆枪 <一种油漆喷涂设备> Volspray
沃尔什变换 Walsh transformation
沃尔什-哈达马德变换 Walsh-Hadamard transformation
沃尔索牌工具 <一些机动、风动工具> Warsop
沃尔特冰阶【地】Warthe stage
沃尔特哈姆牌运斗 <装运砖块、灰浆等用> Wouldham hod
沃尔特曼流速仪 Woltman current meter
沃尔维斯海岭 Walvis ridge
沃发涕 <一种离子交换树脂> Wofatit
沃钒锰矿 vuorelainenite
沃格尔-奥萨格(毛细管)黏[粘]度计 Vogel-Ossag viscosimeter
沃格尔-福歇尔关系式 Vogel-Fulcher relation
沃格尔红 Vogel's red
沃科巴统 Waucobian
沃科普式起动器 Wauchope(type) starter
沃克(混凝土板)啮口接头 Walker interlocking joint
沃克-斯蒂尔摆杆硬度计 Walker-Steel's (swinging beam) hardness tester
沃拉莱板 <一种塑料贴面胶合板> Warerite
沃拉斯顿棱镜 Wollaston prism
沃拉斯顿辗压导线 rolling Wollaston wire
沃拉斯通偏光棱镜 Wollaston polarizing prism
沃兰孔隙水(气压式)压力计 Warlam piezometer

沃勒曲线 Wohler curve
沃硫砷镍矿 vozhminite
沃梅尔安全夹持器 Wommer safety clamp
沃硼钙石 volkovskite
沃赛特镍铬耐蚀合金 worthite
沃森编码 Watson code
沃斯列夫参数 Hvorslev parameter
沃斯列夫面 Hvorslev surface
沃斯列夫模型 Hvorslev soil model
沃斯塔试验法 Worstall heat test method
沃斯特氏红 Wurster's red
沃特勒奥那多控制 Ward-Leonard control
沃特勒奥那多控制系统 Ward-Leonard system
沃特勒奥那尔多拖动(电机) Ward-Leonard drive
沃特勒奥那尔多制 Ward-Leonard system
沃田 irrigate farmland
沃土 fat soil;fecund soil;fertile soil;generous soil;loam;loamy soil;rich clay;rich soil;top soil
沃屯牌窗 <一种专利天窗> Wotton
沃希托海槽 Ouachita marine trough
沃洲 oasis [复 oases];pioases
沃兹沃思装置 Wadsworth mounting

肟 oxime

肟基丙酮 isonitrosoacetone

卧

卧材 ledger;ledger plate

卧车 coach;night coach;sleeper;sleep(ing) car;sleeping carriage;snoozer;varnished car;wagon-lit <火车的>
卧车包房 compartment of sleeping car
卧车服务员 sleeping-car attendant
卧车内坐席间 sleeping car section
卧车票 sleeping car ticket
卧车上墙板 bunk panel
卧车随运乘客自用汽车的列车 auto-sleeper train;car-steeper train
卧城 bedroom city;bedroom community;bedroom town
卧城郊区 bedroom suburb;bed town suburb
卧城区 dormitory suburb
卧囱炉 horizontal-flued oven
卧倒门 falling(-leaf) gate;tilting door;tilting gate;tipping door;tumble gate;drop gate;flap gate
卧倒门拉缆 gate haulage rope
卧倒门前底板 gate apron
卧倒门支座 gate rest
卧倒(式)坞门 flap dock gate
卧倒闸门 flag gate;tumble gate
卧倒直立试验 recumbent-upright test
卧底 dinting;rock taking-up
卧底爆破 bottom-ripping shot
卧底链 ground cable
卧冬 lie up
卧冬港 winter harbo(u)r
卧佛 reclined Budda
卧佛寺 <北京香山> Sleeping Buddha Temple;Temple of the Sleeping Buddha
卧管 lying pipe
卧管槽 pipe chase
卧管式臭氧发生器 horizontal tube zonation generator
卧管蒸发器 horizontal tube evapo(u)rator;horizontal type evapo(u)rator
卧具 bedding
卧具等物 purser's stocking
卧具间 bed closet
卧孔菌属 <拉> Poria
卧梁 girth strip;seat beam;wall plate
卧炉焦 horizontal gas coke
卧模 <地坑造型> bed in
卧木 ground beam;ground brace;ground plate;ground sill;ledger;mud sill;sleeper plate
卧女枕 reclined woman pillow
卧盘式真空过滤器 horizontal filter
卧铺 <汽车、火车、轮船等的> berth
卧铺安全带 berth safety strap
卧铺安全带吊挂 berth safety strap hanger
卧铺安全带钩 berth safety strap hook
卧铺安全带联结扣 berth safety strap clamp
卧铺安全链挂 safety berth bracket
卧铺安全锁栓 safety berth latch
卧铺边框 berth side frame
卧铺车双人包房 double
卧铺车厢 sleeper
卧铺车厢小包房 roomette
卧铺灯 berth lamp
卧铺底板 berth bottom plate
卧铺垫 mattress
卧铺垫托 mattress seat
卧铺吊链 berth chain
卧铺端框 berth end frame
卧铺费 berth charges;sleeper charges
卧铺附加票 couchette supplement;sleeping-berth supplement
卧铺隔板 berth partition board
卧铺隔板卡铁 berth partition board latch;partition board latch
卧铺隔板套框 head board pocket
卧铺隔板折页 berth partition board hinge;partition board hinge
卧铺钩 berth hanger
卧铺固定绳 berth safety rope;safety rope
卧铺固定绳摩擦垫 berth safety rope liner
卧铺号码牌 berth number plate
卧铺间壁 berth partition board
卧铺开锁装置 unlocking apparatus
卧铺客车 couchette coach
卧铺帘 berth curtain
卧铺帘杆 berth curtain rod
卧铺帘杆卡子 berth curtain rod catch
卧铺帘杆支架 berth curtain rod arm
卧铺帘杆支架折页 berth curtain rod arm hinge
卧铺帘挂 berth curtain hanger
卧铺链轮 berth chain pulley
卧铺列车 car-sleeper train
卧铺票 berth ticket
卧铺容纳量 sleeping accommodation
卧铺上部隔板 head board
卧铺上铺挡板 bunk apron
卧铺收入 berth revenue
卧铺锁 berth lock
卧铺锁键托 berth lock bracket
卧铺锁托 berth lock bracket
卧铺腿 berth leg
卧铺位 passenger berth;sleeping berth
卧铺预定 advance berth reservation
卧铺折页 berth hinge
卧铺证 couchette voucher;sleeping-berth voucher
卧铺支铁 berth support
卧铺支铁套 berth support bracket
卧铺支铁托 berth support bracket
卧铺柱 berth post
卧式 horizon
卧式氨冷凝器 horizontal ammonia condenser
卧式氨蒸发器 horizontal ammonia evaporator
卧式拔丝机 horizontal wire-drawing machine
卧式拌和机 horizontal shaft mixer
卧式宝塔筒子 horizontal cone
卧式泵 horizontal type pump
卧式测微仪 horizontal micrometer
卧式插床 horizontal slotting machine
卧式缠绕机 horizontal winding machine;lathe type winding machine
卧式长管蒸发器 horizontal long-tube evapo(u)rator
卧式车床 horizontal lathe;lathe drill
卧式沉降螺旋卸料离心分离机 solid bowl type screw decanter
卧式充压型静电加速器 horizontal pressure generator
卧式冲床 horizontal punching machine
卧式冲击破碎机 horizontal impact crusher
卧式冲模插床 horizontal die slotting machine
卧式储罐 horizontal storage tank
卧式储浆池 horizontal stock chest
卧式串列静电加速器 horizontal tandem generator
卧式吹炉 barrel convertor
卧式锤碎机 horizontal hammer mill
卧式带锯机 horizontal band sawing machine
卧式单次拉丝机 single horizontal block
卧式单动冲床 horizontal single action presses
卧式单级泵 horizontal one-stage pump
卧式的 horizontal type
卧式底革滚压机 horizontal sole leather roller
卧式电动机 horizontal motor
卧式电机 horizontal machine
卧式电集尘器 horizontal electrofilter;horizontal electrostatic precipitator;horizontal flow electrostatic precipitator
卧式煅烧炉 horizontal calciner
卧式锻造机 Greenbat machine;horizontal forging machine;impactor
卧式对置发动机 horizontal opposed engine
卧式对置气缸发动机 pancake engine
卧式多室流化床干燥器 horizontal multicompartment fluidized bed drier [dryer]
卧式多轴夹盘车床 horizontal chucking multi-spindle
卧式发电机 horizontal shaft generator;horizontal type generator;horizontal type motor
卧式发动机 flat engine;horizontal engine
卧式分批搅拌装置 horizontal batch plant
卧式分条整经机 horizontal section warper
卧式釜 horizontal retort
卧式干馏釜 lying retort
卧式感应电动机 horizontal induction motor
卧式感应加热炉 horizontal induction furnace
卧式钢罐 horizontal steel cylinder
卧式钢筒 horizontal steel cylinder
卧式高速锯木机 horizontal high-speed log sawing machine
卧式高压消毒器 horizontal type high pressure steak sterilizer
卧式高压蒸气消毒柜 horizontal type high pressure sterilizer
卧式管壳式冷凝器 closed shell and tube condenser
卧式管式蒸发器 horizontal tube evapo(u)rator
卧式管形蒸发器 horizontal tube evapo(u)rator
卧式滚齿机 horizontal hobbing machine
卧式锅炉 horizontal boiler
卧式过热器 horizontal superheater
卧式豪猪开棉机 horizontal porcupine opener
卧式烘干滚筒 horizontal rotary dryer
卧式烘干机 horizontal drying machine;horizontal whirler
卧式烘砂滚筒 revolving drier [dryer];rotary sand drier
卧式滑净装置 horizontal cleaner
卧式回管锅炉 horizontal return tubular boiler
卧式回火管锅炉 horizontal return tubular boiler
卧式回转泵 horizontal rotary pump
卧式回转干燥器 horizontal rotary dryer
卧式回转高压釜 horizontal rotating autoclave
卧式回转工作台磨床 horizontal spindle rotary table grinding machine
卧式混凝土搅拌机 horizontal drum mixer;horizontal axis mixer
卧式火管锅炉 horizontal tubular boiler
卧式机床 horizontal type machine
卧式集气罐 horizontal air collector
卧式集中(采暖)炉 horizontal type central furnace
卧式给水加热器 horizontal type feed water heater
卧式挤压纺丝机 horizontal spinning extruder
卧式挤压机 horizontal extruder;horizontal extrusion presses
卧式加速器 horizontal type accelerator
卧式剪床 horizontal shear;horizontal shear machine
卧式碱洗 horizontal type alkaline cleaning
卧式胶带运输机 horizontal belt conveyer [conveyor]
卧式绞盘 gypsy
卧式搅拌机 horizontal mixer
卧式精镗床 horizontal fine boring machine
卧式静电加速器 horizontal generator
卧式锯床 horizontal saw mill
卧式锯胶机 horizontal rubber sawing machine
卧式卷取机 horizontal reel
卧式卷线筒 horizontal block
卧式壳管式蒸发器 closed shell and tube evapo(u)rator
卧式壳形的 horizontal shell-type
卧式框锯 horizontal frame saw
卧式框锯制材厂 horizontal frame saw mill
卧式扩散炉 horizontal diffusion furnace
卧式扩散器 horizontal diffuser
卧式拉床 horizontal broaching machine
卧式冷凝器 horizontal condenser
卧式冷气器 horizontal condenser
卧式离心泵 centrifugal pump with horizontal axis
卧式离心分选机 solid bowl centrifuge
卧式离心机 horizontal centrifuge
卧式离心浇铸机 horizontal centrifugal casting machine
卧式离心碾磨机 horizontal centrifu-

gal grinder
卧式离心筛 horizontal centrifugal screen
卧式离心水泵 horizontal centrifugal pump
卧式离心油泵 horizontal centrifugal oil pump
卧式连续动作螺旋压榨机 horizontal continuous-acting auger press
卧式硫铁矿回转炉 horizontal rotary pyrites burner
卧式炉 horizontal chamber furnace; horizontal chamber oven; horizontal furnace
卧式螺旋泵 horizontal screw pump; screw feed pump
卧式落地组合铣床 modular horizontal floor mill
卧式脉冲萃取器 horizontal pulse extractor
卧式密闭砂磨机 horizontal closed head bead mill
卧式膜蒸发器 horizontal film evapo(u)rator
卧式磨床 horizontal hone
卧式磨机 horizontal grinder
卧式内燃机 horizontal motor
卧式逆流式水冷却器 horizontal counterflow water cooler
卧式牛头刨床 horizontal shaper
卧式刨床 horizontal plane machine; horizontal planer
卧式喷雾凉水塔 horizontal spray chamber type cooling tower
卧式平板筛 horizontal flat screen
卧式平面磨床 horizontal surface grinder; horizontal surface grinding machine
卧式破碎机 horizontal crusher
卧式气泵 horizontal air pump
卧式气力提升机 horizontal air elevator
卧式气力压榨机 horizontal pneumatic press
卧式汽缸 horizontal cylinder
卧式氢气电阻炉 horizontal hydrogen resistance furnace
卧式清洗机 horizontal washing machine
卧式燃油锅炉 horizontal oil-firing boiler
卧式容器 horizontal vessel
卧式容器的鞍形支座 saddle supports for horizontal vessels
卧式杀菌锅 horizontal sterilizer
卧式砂磨机 horizontal sand grinder
卧式深孔钻床 horizontal deephole drilling machine
卧式升降台铣床 horizontal knee and-column type milling machine
卧式试样磨机 horizontal type sampling mill
卧式手摇泵 pump low down type
卧式双动油压机 horizontal double action oil hydraulic press
卧式双滚延压机 horizontal twin-roller machine
卧式双水轮机 horizontal shaft double turbine
卧式双轴对钻钻床 wheel-quartering machine
卧式双轴钻床 horizontal duplex drill
卧式水泵 horizontal shaft pump; horizontal(spindle)pump
卧式水管锅炉 horizontal water-tube boiler
卧式水轮机 horizontal turbine
卧式四缸发动机 flat-four engine
卧式塑料注射成型机 horizontal plastic injection mo(u)lding machine
卧式台钻 horizontal bench drill
卧式炭窑 lying charcoal kiln; lying meiler
卧式镗床 horizontal boring machine; horizontal boring unit
卧式镗铣床 horizontal boring and milling machine
卧式淘析器 horizontal elutriator
卧式套皮壳机 horizontal pulling on machine
卧式梯 horizontal ladder
卧式桶钩 horizontal barrel hook
卧式筒叶拌和机拌和配料 batch pugmill mixing
卧式土钻机 <用于横穿公路路面下层,安装管道等> horizontal auger
卧式外延炉 horizontal epitaxial furnace
卧式弯板机 horizontal plate bending machine
卧式往返工作台平面磨床 horizontal reciprocating-table surface grinder
卧式往复泵 parallel pump
卧式往复活塞压气机 horizontal type reciprocating compressor
卧式往复压缩机 horizontal opposed reciprocating compressor
卧式涡轮机 horizontal turbine
卧式蜗壳旋桨泵 horizontal volute propeller pump
卧式无钻座锤 horizontal counterblow hammer
卧式坞门 bottom-hinged box type flap gate; box falling gate; box gate; falling gate
卧式西门子衍射仪 horizontal Shemens diffractometer
卧式铣床 horizontal boring and milling machine; horizontal miller
卧式线材卷取机 horizontal wire rod reel
卧式箱形炉 horizontal box furnace
卧式斜撑闸门 bear-trap gate
卧式星形发动机 horizontal rotary engine
卧式蓄热室 horizontal regenerator
卧式悬挂 horizontal suspension
卧式旋风水膜除尘 horizontal water-film cyclone dust cleaning
卧式旋风水膜除尘器 horizontal water-film cyclone collector
卧式旋风筒 axial cyclone
卧式压床 horizontal press
卧式压缩机 horizontal(type) compressor
卧式研磨分散机 horizontal grinding dispersion machine
卧式窑 horizontal kiln
卧式液力压榨机 horizontal hydraulic press
卧式油罐 horizontal tank
卧式油压机 horizontal oil hydraulic
卧式油压千斤顶 horizontal hydraulic jack
卧式圆盘破碎机 horizontal disc [disk] crusher
卧式圆盘种子清选机 horizontal disc [disk] separator
卧式圆筒烘燥机 horizontal cylinder dryer
卧式圆筒形储罐 horizontal cylindric(al) tank
卧式圆形油罐 horizontal cylindric(al) tank
卧式圆形贮罐 bullet type tank
卧式运输带 horizontal belt conveyer [conveyor]
卧式运输筛 horizontal conveyer screen
卧式甑 horizontal retort
卧式轧边机 horizontal edger
卧式闸门 falling gate; flap gate; tilting gate; tipping door; tipping gate; tumble gate
卧式真空电阻炉 horizontal vacuum resistance furnace
卧式真空过滤机 horizontal vacuum filter
卧式振动离心机 horizontal vibrating centrifuge
卧式振动离心脱水机 horizontal vibrating screening centrifuge
卧式蒸发器 horizontal evapo(u)rator
卧式蒸罐 horizontal retort
卧式蒸罐制焦油 horizontal retort tar
卧式蒸汽机 horizontal steam engine
卧式蒸煮器 horizontal digester
卧式整经机 horizontal warper
卧式支脚 horizontal prop leg
卧式支座 horizontal support
卧式制版照相机 horizontal gallery camera
卧式煮炼锅 horizontal kier
卧式转炉 horizontal rotary furnace
卧式自动凿榫机 horizontal automatic mortising machine
卧式钻床 drill lathe; horizontal drill press
卧式钻机 horizontal drilling rig
卧式钻孔机 horizontal borer
卧式钻石镗床 horizontal diamond boring machine
卧式钻探机 horizontal boring machine
卧式钻镗两用机床 horizontal drilling and boring machine
卧室 bed chamber; bedroom; house room; retiring room; rooming occupancy; sleeping room; cubiculum <古罗马>
卧室壁橱 bedroom closet
卧室壁橱台架 bedroom closet bank
卧室窗(子) bedroom window
卧室灯具 bedroom lamps
卧室及起居两用的单间宿舍 bed-sitting flat
卧室兼起居间 bed sit(ter); bed-sitting room
卧室兼起居室 bed sit(ter); bed-sitting room
卧室空间 sleeping room space
卧室楼层 bedroom floor; bedroom stor(e)y
卧室门 bedroom door
卧室内壁橱 bedroom cupboard
卧室水盆 bedroom basin
卧室洗涤盆 bedroom basin
卧室用椅(子) bedroom chair
卧室装饰 bedroom decoration
卧榻 couch
卧筒式拌和机 horizontal drum mixer
卧筒式搅拌机 horizontal drum mixer
卧筒式淋灰机 drum-type lime watering treater
卧筒式叶片拌和机 pugmill type mixer
卧拖车 sleeping caravan
卧拖车停车位置 sleeping caravan standing
卧像 recumbent statue
卧椅 bed chair
卧镇 residential town
卧置 in the horizontal position; laid up
卧轴平面磨床 horizontal spindle surface grinder
卧轴式布置 <水轮机> horizontal shaft arrangement
卧轴式搅拌机 pugmill mixer
卧轴式离心泵 horizontal shaft centrifugal pump
卧砖 brick on bed
卧钻 horizontal auger

握

握把 grip; hand hold
握柄 grab handle; hand hold; lever
握柄把 lever handle
握柄扳动顺序表 lever pulling(manipulation) chart
握柄保险机 grip safety
握柄闭止把 lever latch
握柄闭止块 lever latch block
握柄表示灯 lever lamp; lever light
握柄操纵顺序表 lever pulling(manipulation) chart
握柄操纵员 leverman
握柄承架 lever shoe
握柄传动装置 lever actuator
握柄电路控制器 lever circuit controller
握柄电锁器 lever lock
握柄动程 lever movement; stroke of lever
握柄返回表示器 return indicator
握柄复位表示 lever return indication
握柄杆 lever arm
握柄回路管制器 lever circuit controller
握柄架 crankcase; lever base
握柄键 latch; lever latch
握柄接点 lever contact
握柄解锁号 lever release number
握柄空位 lever space
握柄控制 lever control
握柄联锁 direct lever interlocking
握柄名牌 lever name plate
握柄排列 lever arrangement
握柄式 lever type
握柄室 lever box
握柄锁闭 lever lock(ing)
握柄锁闭表示 lever lock indication
握柄锁闭控制 lever lock control
握柄锁闭器 lever lock
握柄台 crankcase; crank frame; lever apparatus; lever box; lever frame
握柄调整器 lever compensator
握柄停用卡圈 lever collar
握柄尾杆 lever tail
握柄止台 quadrant
握柄座 lever apparatus; lever box; lever stand
握持反射 grasp(ing) reflex
握持缆索 hold cable
握持力 durable grip; holding force; grip
握钉力 nail-holding ability; nail-holding power; nail withdrawal
握钉性 nail holding
握杆 holding rod
握固板 holding plate
握固长度 grip length
握固架 grip holder
握固千斤顶 <盾构工程用> gripper jack
握固头 gripping head
握管器 cylinder wrench
握裹 bind
握裹长度 <钢筋的> grip length; bond length
握裹耗损 loss of bond
握裹力 bondability; bond between concrete and steel; bond force; bond(ing) stress; bond resistance; clamping force; gripping force; holding power
握裹力劈裂破坏 bond split failure
握裹锚碇 anchorage by bond
握裹锚固 grip anchorage
握裹面积 bond area
握裹能力 gripping capability
握裹强度 bond strength

握裹失效 bond failure
握裹应力 anchorage bond stress; bond stress
握紧 clasp;grasp;grip;gripe
握力 grip strength;power of gripping
握力计 hand-dynamometer
握力减弱 grasp weakness
握式离合器 gripping clutch
握手 handle tornob;handshaking;knob
握手铁杆 grab iron
握手协议 handshake protocol
握索结 diamond knot;footrope knot; single diamond knot
握梯手 heel man
握团试验 test of grasping soil into a ball
握住 catch hold of;clasp;prehension

渥登重力仪 Worden gravimeter

渥拉斯顿棱镜 Wollaston prism
渥拉斯顿石英棱镜 quartz Wollaston prism
渥奇式滚水坝 ogee spillway
渥奇式屋顶 ogee roof
渥奇溢流坝 Ogee dam
渥太华砂 <美国美国水泥试验用标准砂> Ottawa sand;graded standard sand

硪 circular rammer;pummel;tamper

斡旋者 bridge builder

龌龊房间 dog hole

乌柏 Chinese sapium; Chinese tallow tree

乌柏木 tallow wood
乌柏皮油 Chinese vegetable tallow
乌柏油 stillingia oil;tallowseed oil
乌班吉河 Ubangi River
乌板树属 bilberry
乌伯娄德滴点 Ubbelodhe drop point
乌伯娄德黏[粘]度计 Ubbelodhe visco(si)meter
乌伯娄德熔点 Ubbelodhe melting point
乌伯娄德液化点 Ubbelodhe liquefying point
乌伯娄德液化点试验 Ubbelodhe liquefying-point test
乌布利希球 Ulbricht sphere
乌布利希球形光度计 Ulbricht globe; Ulbricht sphere
乌德硫磺回收过程 Uhde sulfur recovery process
乌尔比诺陶器 <意大利> Urbino ware
乌尔夫宋达陶器 <瑞典> Ulfsunda faience
乌尔曼防腐法 Wolman process
乌耳斯德统 <中泥盆世>【地】Ulsterian series
乌耳斯特阶【地】Ulsterian
乌干达桃花心木 Uganda mahogany
乌光釉的 matt-glazed
乌黑的 fuliginous
乌黑发亮 jet
乌黑发亮的 jet black
乌呼鲁 X 射线源表 Uhuru catalogue
乌金 babbit(t)
乌金釉 black bronze glaze; mirror black glaze;wu-jin glaze
乌金轴瓦 bearing pad
乌卡过程 Urca process
乌肯纳比色计 Ukena colo(u)rimeter
乌拉尔阶【地】Uralian(stage)
乌拉尔硼钙石 uralborite
乌拉尔山脉 Ural Mountains
乌拉尔统【地】Uralian series
乌拉尔祖母绿 Uralian emerald
乌拉圭河 Uruguay River
乌拉石 <多种石棉水泥制品> Urastone
乌拉坦 ethyl carbamate
乌拉梯斯期【地】Ulatisian
乌兰巴托 <蒙古首都> Ulan Bator
乌兰迪木 <热带美洲产的红棕色硬木> urunday
乌勒因 Uleine
乌里奈特便器 <一种公厕中女用便器> Urinettes
乌利水色计 Ule's scale of colo(u)r
乌利型染色标准 Ule's colo(u)r standard
乌洛托品 hexamethylene tetramine; hexamine;methenamine;urotroloine
乌洛托品废水 urotropine wastewater
乌煤 fusain; mineral charcoal; motherham;mother of coal
乌姆斯-伊特纳窑 <烧瓷釉黏[粘]土管的环窑> Ooms-Ittner kiln
乌木 ebony;ebony wood
乌木蜡 ebonite wax
乌纳尔 Uinal
乌尼维斯油 univis oil
乌硼钙石 uralborite
乌嘌呤石 guanine
乌萨烯酸盐 ursaenate
乌氏液化点 <即乌氏滴点> Ubbelodhe liquefying point
乌氏液化点试验 <即乌氏滴点试验> Ubbelodhe liquefying-point test
乌头酸 aconitic acid
乌鸦 crow
乌鸦黑色 corbeau
乌釉 black glaze
乌釉搪瓷 majolica enamel
乌釉陶器 majolica
乌贼墨(色)sepia
乌贼颜料 sepia
乌芝属 <拉> Amauroderma
乌兹别克斯坦 <亚洲> Uzbekistan
乌紫色 aubergine
乌嘴笔 curve pen;swivel pen

圬泵 dock pump

圬工 masonry;masonwork
圬工矮墙庙宇 masonry podium temple
圬工坝 masonry dam;massive dam
圬工白灰浆 white masonry mortar
圬工板 masonry plate
圬工板条 mason's lath
圬工边墙 masonry wall
圬工表面 masonry surface
圬工表面修整 regrate
圬工薄壳 masonry shell
圬工材料 masonry material
圬工层 masonry layer
圬工插锁 masonry lock
圬工沉井基础 masonry well foundation
圬工衬垫料 masonry lining material
圬工衬砌 masonry lining;masonry work lining
圬工衬砌导坑 masonry heading
圬工衬砌的 masonry-lined
圬工衬砌隧道 masonry-lined tunnel
圬工承包商 masonry contractor
圬工尺 mason's rule
圬工冲孔 dabbing
圬工除垢器 masonry cleaner
圬工储水坑 mason's trap
圬工处理 masonry cleaning
圬工船闸 masonry lock
圬工垂直缝 build
圬工锤 cutting hammer; mason's hammer
圬工存水井 mason's trap
圬工存水坑 mason's trap
圬工单层砖墙 masonry tier
圬工挡土墙 masonry retaining wall
圬工的砂浆找平层 bed joint
圬工的水平缝 bed joint
圬工的踏步形接合 horsed joint
圬工底层接缝 bed joint
圬工地牢 masonry dungeon
圬工垫块 mason's bolster
圬工定位销 masonry dowel
圬工洞室 masonry vault
圬工对接缝和层面修琢过的 scribbled
圬工墩 masonry pier
圬工墩座墙 masonry podium
圬工防潮密封 masonry seal
圬工防潮密封剂 masonry sealing agent
圬工防水水泥 brick cement
圬工风格 masonry work style
圬工缝 abre(a)uvior
圬工高层公寓 masonry high flat;masonry tall flat
圬工格板墙 masonry panel wall
圬工隔断 masonry partition
圬工隔膜 masonry diaphragm
圬工工程 masonry work
圬工工程公寓塔楼 masonry work apartment tower
圬工工程拱券 masonry work arch
圬工工程技术 masonry work technique
圬工工程学 mason(ry) construction
圬工工具 masonry tool
圬工拱 masonry arch
圬工拱的曲线形缝 collar joint
圬工拱桥 masonry arch(ed) bridge
圬工拱式重力坝 masonry arch gravity dam
圬工拱座 masonry shoulder
圬工勾缝 dabbing
圬工构件 masonry component; masonry unit
圬工构造学 masonry construction
圬工谷坊 masonry check dam
圬工管道 masonry duct; masonry work duct
圬工灌浆 masonry grout(ing)
圬工涵洞 masonry-stone culvert
圬工护岸 masonry revetment
圬工画线 scoring of masonry
圬工灰缝 masonry point
圬工灰浆 mason's mortar
圬工回填料 masonry backfill
圬工基础 foundation of masonry;masonry foundation
圬工基础墙 masonry footing wall;masonry foundation wall
圬工基座 paillasse
圬工技术 masonry
圬工加固 masonry reinforcing
圬工加筋 masonry reinforcing
圬工建基脚墙 masonry footing wall
圬工建筑 masonry architecture
圬工建筑构件 masonry building component; masonry component; masonry member
圬工建筑块材 masonry building block
圬工建筑师 mason-architect
圬工建筑物 masonry building
圬工建筑学 masonry construction
圬工建筑砖 masonry building tile
圬工角尺 mason's square
圬工脚手架 mason(ry)'s scaffold
圬工脚手台 siege
圬工接缝 mason's joint
圬工接合件 masonry fastener
圬工节制闸 masonry check
圬工节制闸阀 masonry check valve
圬工结构 masonry construction; masonry structure; masonry work structure
圬工结构系统 masonry structural system
圬工结合空心砖墙 masonry bonded hollow wall
圬工紧固件 masonry fastener
圬工井筒基础 masonry well foundation
圬工锯 masonry saw
圬工开口 masonry opening
圬工壳板 masonry skin
圬工块 masonry block;masonry mass; mass of masonry
圬工拦沙坝 masonry check dam
圬工廊道 masonry gallery
圬工联结原理 masonry bond principle
圬工梁 masonry beam
圬工领班 king master mason;master mason
圬工镘板 mason's float
圬工镘刀 mason's float;mason's trowel
圬工锚碇 masonry anchor
圬工锚具 masonry anchorage
圬工锚栓 raw bolt
圬工面料 masonry facing material
圬工磨床 rubbing bed
圬工抹子 mason's trowel
圬工墓 sepulcher
圬工墓穴 masonry tomb
圬工泥刀 masonry cutting blade; mason's trowel
圬工黏[粘]结剂 masonry bond
圬工黏[粘]结类型 masonry bond type
圬工女儿墙 parapet masonry(wall)
圬工平顶 <古希腊建筑> lacunaria
圬工平台 masonry platform
圬工平台庙宇 masonry platform temple
圬工平台圣堂 masonry platform temple
圬工平头凿 mason's flat-ended chisel
圬工屏幕 masonry screen
圬工破坏试验 masonry failure test
圬工砌层 masonry course
圬工砌缝 masonry joint;mason's joint
圬工砌合 <按砌块错缝方式而异的> masonry bond(ing)
圬工砌合类型 bond type
圬工砌体错缝 masonry stop
圬工砌体结合的空心墙 masonry bonded hollow wall
圬工砌体斜接缝 mason's stop
圬工砌体与地面的接缝 ground joint
圬工砌造 laying brick
圬工砌筑单元 masonry unit
圬工嵌缝工作 filler masonry work
圬工墙 flank masonry wall;masonry wall
圬工墙表皮 masonry wall skin
圬工墙材料 masonry wall material
圬工墙的安装 masonry wall installation
圬工墙的联结砖 masonry wall junction brick
圬工墙的突出部分 projection of a masonry wall
圬工墙的稳定性 masonry wall stability

圬工墙的斜交点 skewed junction of masonry walls
圬工墙的斜砌石贯入度 skew penetration of masonry walls
圬工墙顶 masonry wall top
圬工墙顶盖砖 masonry wall capping brick
圬工墙洞 masonry wall hollow
圬工墙钝角 birdsmouth quoin of masonry wall
圬工墙勾缝机 masonry wall-pointing machine
圬工墙构件 masonry wall element
圬工墙核心 masonry wall core
圬工墙后伸出的工作台 bench table
圬工墙厚度 masonry wall thickness
圬工墙基础 masonry wall footing
圬工墙技术革新 masonry wall break-through
圬工墙交叉 masonry wall crossing
圬工墙结构 masonry wall construction
圬工墙龛 masonry wall niche
圬工墙口 masonry wall opening
圬工墙块 masonry wall block
圬工墙拉杆 masonry wall tie
圬工墙梁 masonry wall beam
圬工墙裂口 masonry wall slot
圬工墙裂纹 masonry wall crack
圬工墙锚碇装置 masonry wall anchor
圬工墙面板 masonry wallboard
圬工墙内衬 masonry wall lining
圬工墙内的填料 moellon
圬工墙披水板构件 masonry wall flashing piece
圬工墙坡度 slope of a masonry wall
圬工墙砌层 masonry wall course
圬工墙强度 masonry wall strength
圬工墙塔 masonry wall tower
圬工墙体的金属防雨泛水板 metal masonry wall flashing piece
圬工墙体的金属防雨披水板 metal masonry wall flashing piece
圬工墙体的金属拉杆 metal masonry wall tie
圬工墙体的金属连接件 metal masonry wall tie
圬工墙体系 masonry wall system
圬工墙头 masonry wall head
圬工墙线 masonry wall line
圬工墙压顶 masonry wall coping
圬工墙翼 masonry wall wing
圬工墙直角交叉 rectangular masonry wall crossing
圬工墙中心线 masonry wall center line
圬工墙砖 masonry wall tile
圬工墙座 masonry wall base
圬工桥 masonry bridge
圬工桥墩 masonry bridge pier; masonry pier
圬工桥台 masonry abutment
圬工撬棒 setting bar
圬工清洁 masonry cleaning
圬工清洁剂 masonry cleaner
圬工穹隆 masonry vault
圬工渠道 masonry conduit
圬工砂 mason sand
圬工砂浆 masonry mortar; mason's mortar; pointing mortar
圬工渗流坑 masonry seepage pit
圬工施工 masonry construction
圬工石 masonry-stone; plums
圬工石灰膏 fixer's bedding
圬工石灰浆 fixer's bedding
圬工饰面 masonry work facing
圬工刷 mason's brush
圬工水槽 masonry flume
圬工水平仪 mason's level
圬工水闸 masonry lock
圬工随意砌法 random masonry bond
圬工随意砌筑 random masonry work
圬工碎片 masonry ruins
圬工体 masonry unit
圬工填充单元 masonry filler unit
圬工填(充)料 masonry fill
圬工填(充)料的 masonry-filled
圬工跳板 mason(ry)'s runway
圬工通道 masonry work duct
圬工筒仓 masonry silo
圬工涂料 masonry paint(ing); stone paint
圬工托架 masonry bracket
圬工瓦 masonry tile
圬工外墙衬里 external masonry wall lining
圬工外墙饰面 external masonry wall facing
圬工外墙柱 external masonry wall column
圬工下水道 masonry conduit-type sewer
圬工小锤 scutcher
圬工斜接面 mason's miter [mitre]; mason's stop
圬工烟囱隔板 masonry withe
圬工溢流堰 masonry weir
圬工用具 masonry tool
圬工用漆 masonry paint
圬工用切刀 masonry cutting blade
圬工用砂 masonry sand
圬工用熟石灰 mason's hydrated lime
圬工用水泥 masonry cement
圬工用水平尺 mason's level
圬工用小铁锤 mason's hammer
圬工油灰 mason's putty
圬工原理 masonry principle
圬工圆顶 masonry cupola
圬工凿 mason's chisel
圬工支墩 masonry buttress
圬工重力坝 masonry gravity dam
圬工主体 body of masonry
圬工住宅塔楼 masonry residence tower; masonry residential tower
圬工贮水池 masonry reservoir
圬工砖 masonry brick
圬工钻 masonry drill
圬土灰浆 masonry mortar

污斑 blot(ch); blur; clouding; contact stain; dirt spot; smear; smirch; smudging; soilure; stain; stare

污版 greasing; spewing
污布 soiled cotton
污布样 soiled swatch
污层 dirt bed
污底 foul bottom; fouling; green bottom
污底船 foul ship
污底裕度 fouling allowance
污点 black spot; blemish; blotch; blur; discoloration; grime; macula [复 maculae]; maculation; rust staining; slur; smirch; smudge; smutch; splodge; splotch; spot; stain; tache; taint; tarnish; mott(e)
污点法 stain test
污点清除 spot removal
污废物焚化处置法 crematory system of sewage disposal
污垢 dirt; filth; foul(ing); soiling
污垢沉淀池 mud pan
污垢沉积物 fouling deposit; scale; scale deposit
污垢荷载 soil loading
污垢监控 fouling monitoring; monitoring of fouling
污垢控制 fouling control
污垢控制剂 fouling control agent
污垢量 fouling amount
污垢热阻 fouling resistance
污垢物 contaminant; fouling product
污垢系数 dirty factor; foul(ing) coefficient; foul(ing) factor; scale coefficient; scale factor
污垢悬浮剂 soil-suspending agent
污垢悬浮能力 soil-suspending power
污垢抑制 fouling inhibition
污垢抑制剂 fouling inhibitor
污灌 irrigation with wastewater; sewage irrigation; wastewater irrigation
污灌农作 wastewater farming
污灌区 sewage irrigation region
污痕 taint
污化剂 activator
污黄毛 canary stained wool
污秽 dirtness; feculence [feculency]; pollution
污秽泵 sink evacuator
污秽的 dirty
污秽环境 dirty environment
污秽空气 foul air
污秽土地 foul land
污秽物 filth; foul(ing agent); grime
污迹 smear; smudge; stain; work-up <印刷物表面的>
污积带 dirt band
污经 soiled ends
污井 dirt well
污径比 ratio of wastewater discharge capacity to runoff
污媒疾病 pollution-related disease
污泥 mire; muck(soil); mud; sludge
污泥岸(滩) sludge bank
污泥保存 preservation of sludge
污泥泵 sludge pump; sludger; slurry pump; slush pump
污泥泵房 sludge pump house
污泥泵送 pumping of sludge; sludge pumping
污泥比生长率 specific growth rate of sludge
污泥比阻 sludge ratio friction
污泥变陈 ag(e)ing of sludge; age of sludge
污泥饼 sludge cake
污泥剥离机 sludge stripping machine
污泥驳 barging of sludge; sludge barge
污泥驳(船装)运 barging of sludge; sludge barging
污泥部分 sludge part
污泥采样 sludge sampling
污泥采样器 sludge damper
污泥槽 sludge channel; sludge sump; sludge trough
污泥槽车 sludge tank car; sludge tank wagon
污泥测定 sludge determination
污泥层 sludge blanket; sludge formation; sludge layer
污泥层反应池 sludge blanket reactor
污泥层滤池 sludge blanket filter
污泥掺和 sludge blending
污泥产量 sludge production; sludge yield
污泥产率 sludge yield
污泥产气量 gas production rate of sludge
污泥长期储存 sludge long-term storage
污泥场 sludge bed; sludge site
污泥沉淀 sludge deposit; sludge settling
污泥沉淀池 sludge lagoon
污泥沉积 sludge bank; sludge deposit(ion); sludge settling
污泥沉积室 sludge collector
污泥沉积效应 effect of sludge deposit; effect of sludge settling
污泥沉降 sludge settling
污泥沉降比 sludge settling ratio
污泥沉降池 sludge thickener
污泥沉降特性 settling characteristics of sludge; sludge settling characteristics
污泥成层沉淀速度 sludge stratified sedimentation velocity
污泥成分 sludge component
污泥成区沉降速度 zonal settling velocity of sludge
污泥池 lagoon for holding sludge; sludge chamber; sludge impoundment; sludge lagoon; sludge pond; sludge pool; sludge tank
污泥池灌气法 sludge tank activation
污泥池塘处理 sludge lagooning
污泥池塘堆储脱水 sludge dewatering by lagooning
污泥池塘法 lagooning
污泥池蓄处理 lagooning of sludge
污泥冲洗 sewage sludge washing
污泥抽取量 sludge pumpage
污泥、抽吸、装卸和输送系统 sludge pump, load and transfer system
污泥稠度 sludge thickness
污泥稠化 sludge thickening
污泥稠化器 sludge thickener
污泥出口 sludge outlet
污泥除杂粒 sludge degritting
污泥储存 <作为最后的处置方法> lagooning; sludge storage
污泥储斗 sludge hopper
污泥储量 sludge storage
污泥储留 sludge impoundment
污泥储留池 sludge impoundment basin
污泥处理 sludge conditioning; sludge disposal; sludge handling; sludge treatment
污泥处理厂 sludge treatment plant
污泥处理法 sludge handling process
污泥处理费 cost of sludge treatment
污泥处理工艺 sludge handling process
污泥处理设备 treatment equipment of sludge
污泥处理装置 sludge treatment plant
污泥处置 disposal of sludge; sludge disposal
污泥处置厂 sludge disposal plant
污泥处置费 cost of sludge disposal
污泥船 sludge tanker; sludge vessel
污泥船运至海(排放) sludge barged to sea
污泥床 sludge bed; sludge blanket
污泥床反应器 sludge bed reactor
污泥床厚度 sludge bed height
污泥床滤池 sludge bed filter
污泥纯氧好氧消化 sludge pure oxygen aerobic digestion
污泥的 sludgy
污泥的化学调理 chemical conditioning of sludge
污泥的农业价值 agricultural value of sludge
污泥的农业利用 agricultural utilization of sludge
污泥的燃烧值 fuel value of sludge
污泥的湿空气氧化 wet air oxidation of sludge
污泥的熟化 maturing of sludge
污泥的淘洗 elutriation of sludge
污泥的厌氧分解 anaerobic decomposition of sludge
污泥低温消化 psychrophilic sludge digestion

W

污泥斗 sludge conditioning pocket; sludge hopper; sludge pit
污泥堵塞 sludge clogging; sludge plug cock
污泥堆肥 sludge compost(ing)
污泥堆肥发酵处理 sludge composting and fermentation treatment
污泥堆积 sludge band
污泥堆积场 sludge dumping ground
污泥堆积地区 sludge dumping area
污泥堆积高度 sludge height
污泥多层床焚化炉 sludge multiple-hearth incineration
污泥多层床干燥器 sludge multi-hearth drier [dryer]
污泥恶氧消化 sludge anaerobic digestion
污泥发酵产酸 fermentation and acidogenesis of sludge
污泥发酵(法) sludge fermentation
污泥发泡势 foaming potential of sludge
污泥阀 sludge valve
污泥防护(设施) sludge proof
污泥肥料 fertiliser; sludge manure
污泥肥料价值 fertilizing value of sludge
污泥废弃 sludge wasting
污泥分布 sludge distribution
污泥分布器 sludge distributor
污泥分隔室 sludge compartment
污泥分级消化 stage digestion of sludge
污泥分解作用 sludge disintegration
污泥分离 sludge extract; sludge separation
污泥分离池 sludge extractor
污泥分离器 sludge separator
污泥分期消化 phase digestion
污泥焚化 sewage sludge incineration; sludge incineration
污泥焚化厂 sludge incineration plant
污泥焚化处理 crematory system of sewage disposal
污泥焚化炉 sludge incinerator
污泥焚烧 sewage sludge incineration; sludge incineration
污泥焚烧炉 sludge incinerator
污泥焚烧设备 sludge incineration facility
污泥粉碎机 sludge grinder
污泥浮选浓缩 sludge flo(a)tation thickening
污泥浮游物 sludge float
污泥腐化 sludge disintegration
污泥腐化法 sludge putrefaction
污泥腐殖质 sludge humus
污泥负荷比(率) sludge loading rate; sludge loading ratio; activated sludge loading; sludge lading rate
污泥复氧 sludge reaeration
污泥复氧法 sludge reaeration method
污泥覆盖物 scmutzdecke
污泥改善 sludge conditioning
污泥干负荷 sludge loading
污泥干化 drying of sludge; sludge drying
污泥干化场 open sludge drying bed; sand-drying bed; sludge bed; sludge dryer bed
污泥干化床 sludge drying bed
污泥干化床脱水 sludge dewatering by drying bed; sludge drying bed dewatering
污泥干化房 greenhouse for sludge drying
污泥干化机 band dryer
污泥干化器 sludge drier [dryer]
污泥干化装置 sludge drying facility
污泥干燥 sludge drying
污泥干燥场 sludge bed; sludge drying area; sludge drying bed
污泥干燥床 dry bed; sludge bed; sludge drying bed
污泥干燥器 sludge drier [dryer]
污泥高温分解 sludge pyrosis
污泥高温消化 thermophilic sludge digestion
污泥沟 muck ditch; sludge channel
污泥沟埋处置 sludge trenching
污泥固定 fixation of sludge
污泥固化 sludge solidify(ing)
污泥固体含量 solids concentration of sludge
污泥固体浓度 sludge solid concentration
污泥固体平衡 sludge solid balance
污泥固体(物) sludge solid
污泥固体物含量 sludge solid concentration
污泥刮板 squeegee
污泥刮泥机 sludge scraper
污泥管 sludge pipe; sludge tube; sludge withdrawal pipe
污泥管理 sludge management
污泥管理计划 sludge management program(me)
污泥管路输送 transportation sludge by pipe
污泥管线 sludge(pipe) line
污泥过滤 sludge filtration
污泥过滤器 sludge filter
污泥过剩 sludge excess
污泥海洋弃置 sludge dumping into ocean
污泥海洋倾卸 sludge dumping at sea
污泥含量 sludge content
污泥含水 sludge mould
污泥含水量 sludge moisture content
污泥含水量百分数 percentage of sludge moisture content
污泥含水率 sludge moisture; sludge moisture content
污泥含油 oil form sludge
污泥好氧消化 sludge aerobic digestion; sludge digestion aerobic
污泥合并高温分解 sludge co-pyrosis
污泥合并燃烧 sludge co-incineration
污泥和垃圾的卫生填埋 landfill
污泥荷载 sludge load
污泥荷载比 sludge loading ratio
污泥虹吸管 sludge siphon
污泥化学处理 chemical treatment of sludge
污泥化学处理方法 chemfix process
污泥化学加固(固定)处理法 chemfix process
污泥化学调理 sludge chemical conditioning
污泥灰法 sludge ash process
污泥回流 return sludge flow; sludge return; sludge water return
污泥回流比(率) return(ed) sludge ratio; sludge recirculation ratio; sludge recycle ratio; sludge return ratio
污泥回流量 amount of return sludge
污泥回流率 recycle ratio of sludge; sludge recycle flow rate
污泥回流速率 sludge return rate
污泥回流系统 sludge return system
污泥回流指数 inverse sludge index
污泥回流装置 sludge return apparatus
污泥回收 recycle(d) sludge; sludge reclamation
污泥回收产物 sludge product recovery
污泥回用 sludge reuse
污泥混合 sludge blending
污泥混合池 sludge blending tank
污泥混合堆肥 sludge co-composing
污泥活化 sludge activation
污泥活性 sludge activity
污泥机械脱水 mechanical dewatering of sludge; mechanical sludge dewatering
污泥积累 sludge accumulation
污泥加工 sludge processing
污泥加氯氧化 sludge chlorine oxidation
污泥加热处理 heat-treatment of sludge
污泥加热干化 heat drying of sludge; sludge heat drying
污泥加热干燥 heat drying of sludge; sludge heat drying
污泥减量 sludge reduction
污泥减容 sludge volume reduction
污泥检验 sludge examination; sludge test
污泥搅拌器 sludge stirrer
污泥搅动 sludge stirring
污泥接触处理法 sludge contact process
污泥接受器 sludge holding tank
污泥接种 sludge seeding
污泥结构 sludge structure
污泥结构曲线图 sludge structure curve
污泥井 sewer catch basin; sludge chamber; sludge well
污泥净化 sludge elutriation
污泥均化 homogenising of sludge
污泥均质池 sludge homogeneous basin
污泥菌分解 sludge digestion
污泥颗粒 sludge granule; sludge particle
污泥颗粒的 sludge granular
污泥颗粒化 sludge granulation
污泥颗粒粒径 particle-size of sludge
污泥坑 sludge pit; sludge sump
污泥块 sludge cake
污泥快速干燥器 sludge flash drier [dryer]
污泥老化 ag(e)ing of sludge
污泥累积 sludge accumulation
污泥冷冻法 sludge freezing process
污泥离心机 centrifuge of sludge
污泥离心机脱水 sludge dewatering by centrifuge
污泥离心脱水 centrifugal dewatering of sludge
污泥离心脱水机 sludge centrifuge
污泥利用 sludge utilization; utilization of sludge
污泥量 quantity sludge; sludge quantity
污泥量筒 sludge cylinder
污泥龄 age of sludge
污泥流 sludge flow
污泥流化床焚烧炉 sludge fluidized-bed incineration
污泥滤液 sludge filtrate
污泥埋藏(处理) sludge burial
污泥密度 density of sludge; sludge density
污泥密度指数 inverse sludge index; sludge density index
污泥免费管理 sludge free management
污泥面检测器 sludge level detector
污泥面探测器 sludge level sounder
污泥灭菌 sludge pasteurization; sludge sterilization
污泥磨碎 sludge grinding
污泥能源回收 energy recovery from sludge
污泥泥饼处置 sludge cake disposal
污泥年龄 sludge age
污泥浓度 concentration of sludge; sludge concentration; sludge thickening
污泥浓度指数 sludge density index
污泥浓缩 sludge concentration; sludge condensation; sludge thickening; sludge thickness; thickening of sludge
污泥浓缩池 sludge concentration tank; sludge thickener; sludge thickening basin; sludge thickening pond
污泥浓缩斗 sludge concentrated hopper
污泥浓缩机 sludge thickener
污泥浓缩器 sludge concentrator
污泥浓缩设备 sludge thickening equipment
污泥浓缩系数 sludge concentration factor
污泥耙 sludge rake
污泥排出口 sludge outlet
污泥排除 de-sludge; sludge removal; sludge wasting
污泥排放 sewage sludge disposal; sludge drain; sludge withdrawal
污泥排放泵 sludge draw-off pump
污泥排放管道 sludge discharge conduit
污泥培养 sludge cultivation
污泥喷雾干燥器 sludge spray drier [dryer]
污泥膨胀 sludge bulking; sludge expansion
污泥平衡池 sludge equilibrium pond
污泥破碎机 sludge shredder
污泥起沫 sludge foaming
污泥起泡 sludge foaming
污泥气顶盖 gas dome; sludge gas dome
污泥气管 sludge gas pipe
污泥气回流 gas recirculation; sludge gas recirculation
污泥气量计 sludge gas meter
污泥气热值 heat value of sludge gas
污泥气(体) sludge gas; sewage gas
污泥气体储存罐 sludge gas holder
污泥气体储柜 sludge gas tank
污泥气贮存罐 sludge gas holder
污泥倾卸 sludge dumping
污泥清除 sludge removal
污泥清扫机 clarifier
污泥区 sludge zone
污泥渠 sludge channel
污泥去除 sludge removal
污泥燃烧 combustion of sludge; sludge incineration
污泥燃烧炉 sludge incinerator
污泥热分解 thermal reduction of sludge
污泥热减缩 sludge thermal reduction
污泥热调节 thermal conditioning of sludge
污泥热值 sludge calorific value
污泥人工干燥 artificial sludge drying
污泥人工脱水 artificial sludge drying
污泥容积指数 sludge volume index
污泥蠕虫 sludge worm
污泥闪燃 sludge flash combustion
污泥上层液 supernatant
污泥上翻 rising sludge; sludge rising
污泥上浮 rising sludge; sludge rising
污泥上浮浓度 sludge rising concentration
污泥上涌 sludge boil
污泥设备 sludge equipment
污泥生产 sludge processing
污泥生长动力学 kinetics of sludge production
污泥生长指数 sludge growth index
污泥生成 sludge formation

污泥生成量 amount of sludge produced
污泥生成抑制剂 sludge inhibitor
污泥生物降解 sludge biodegradation
污泥生物量指数 sludge biomass index
污泥生物淋滤 sludge bioleaching
污泥湿气氧化 sludge wet-air oxidation
污泥湿式氧化法 sludge wet oxidation
污泥湿室焚烧 sludge wet combustion
污泥石灰处理 lime stabilization of sludge;sludge lime stabilization
污泥石灰稳定 lime stabilization of sludge;sludge lime stabilization
污泥试验 sludge examination;sludge test
污泥试样 sludge sample
污泥室 sludge chamber
污泥收集池 sludge collector
污泥收集器 sludge collector
污泥输送槽 sludge trough
污泥输送器 sludge conveyer [conveyor]
污泥熟化 maturing of sludge;sludge ripening
污泥衰减 sludge decay
污泥栓塞 sludge plug
污泥水 silty water;sludge water;slug water
污泥水分 sludge moisture
污泥水分含量 sludge moisture content
污泥水平带滤机脱水 sludge dewatering by horizontal belt filter
污泥水指示有机体 indicator organisms of polluted water
污泥丝状菌膨胀 sludge filamentation bulking
污泥缩变 sludge reduction
污泥摊铺 sludge spreading
污泥塘 lagoon for holding sludge;sludge lagoon;sludge pool
污泥淘析 elutriation of sludge
污泥淘洗 sludge elutriation
污泥特性 sludge characteristics;sludge property
污泥特性曲线 sludge characteristic curve
污泥提升机 sludge lifting machine
污泥体积 sludge volume
污泥体积比 sludge volume ratio
污泥体积缩小 sludge volume reduction
污泥体积与含量关系 volume-mass relationship for sludge
污泥体积指数 sludge volume index
污泥填地 sludge landfilling
污泥填埋 burial of sludge;sludge landfill
污泥调节 conditioning of sludge;sludge conditioning
污泥调理 sludge conditioning
污泥调理池 sludge conditioner
污泥调理剂 sludge conditioner
污泥调量 sludge conditioning
污泥调质 sludge conditioning
污泥停留时间 sludge residence time;sludge retention time
污泥投海 sludge sea disposal
污泥投配量 sludge dose
污泥土地应用 sludge land application
污泥土壤混合物 sludge-soil mixture
污泥脱水 dehydration of sludge;dewatering of sewage sludge;dewatering of sludge;sludge dewatering
污泥脱水床 sludge dewatering bed
污泥脱水机 sludge dewatering equipment;sludge dewatering machine
污泥脱水率 rate of sludge dewatering
污泥脱水热处理 heat conditioning and dewatering of sludge
污泥脱水设备 sludge dehydration facility
污泥脱水势 dewatering potential of sludge;sludge dewatering potential
污泥脱水系统 sludge dewatering system
污泥脱水系统工艺 sludge dewatering system process
污泥脱水性能 dewaterability characteristics of sludge
污泥完全停留 complete sludge retention
污泥稳定化 sludge stabilization
污泥稳定化处理 sludge stabilization process
污泥稳定化学处理法 sludge stabilization-chemical process
污泥稳定热处理法 sludge stabilization-thermal process
污泥稳定装置 sludge stabilization device
污泥消毒 sludge disinfection;sludge sterilization
污泥消化 digestion of sludge;slaking of sludge;sludge digestion
污泥消化槽 sludge digester
污泥消化池 digesting compartment;sludge digester;sludge digestion tank
污泥消化处理 sludge digestion
污泥消化处理罐 sludge digestion tank
污泥消化动力学 kinetics of sludge digestion
污泥消化发泡 foaming in sludge digestion
污泥消化法 sludge digestion method
污泥消化间 sludge digestion compartment
污泥消化气 sludge digestion gas
污泥消化气罐 digestive gas storage tank of sludge
污泥消化器 sludge digester
污泥消化热平衡 heat balance in sludge digestion
污泥消化室 sludge digestion chamber
污泥消解 sludge digestion
污泥消解槽 sludge digestion tank
污泥形态 sludge morphology
污泥需氧量 sludge oxidation demand
污泥絮凝物 sludge floc
污泥驯化 sludge acclimatization
污泥循环 recirculation of sludge;sludge(re)circulation;sludge recycle
污泥循环泵 sludge circulation pump
污泥循环澄清池 sludge circulation clarifier
污泥循环流 sludge recycle flow
污泥压干 sludge pressing
污泥压力 sludge pressing
污泥压滤 sludge press filtration
污泥压滤机 sludge press;sludge press filter
污泥压缩 sludge decreasing;sludge pressing
污泥压榨 sludge pressing
污泥厌氧微生物消化处理 anaerobic sludge digestion
污泥厌氧消化 sludge anaerobic digestion
污泥氧化 sludge oxidation
污泥氧化比 sludge oxidation ratio
污泥氧化定值 sludge oxidation demand
污泥氧化塘 sludge oxidation lagoon
污泥样本 sludge sample
污泥液 sludge liquor
污泥液体 sludge liquid
污泥移动 muck shifting
污泥淤积 accumulation of mud;accumulation of sludge
污泥淤积量 sludge storage
污泥预处理 sludge pretreatment
污泥运输 sludge conveyance
污泥运输成本 sludge hauling cost
污泥再煅烧 sludge recalcination
污泥再曝气 sludge reaeration
污泥再生 reactivate
污泥增长指数 sludge growing index
污泥增稠 sludge thickening
污泥增稠器 sludge thickener
污泥增活剂 sludge synergist
污泥增殖指数 sludge growth index
污泥渣 sludge silt
污泥闸门 sludge gate
污泥真空过滤 sludge vacuum filtration
污泥真空过滤脱水 sludge dewatering by vacuum filter
污泥值 sludge number;sludge value
污泥指数 sludge index
污泥质量 sludge quality
污泥质量保险细则 sludge quality assurance regulation
污泥质量干重 dry weight of sludge mass
污泥质量浓度 sludge mass concentration
污泥中的氨基酸 amino-acid in sludge
污泥中温消化 mesophilic sludge digestion
污泥重力浓缩 sludge gravity thickening
污泥重量降低 sludge weight reduction
污泥砖 bricks from sludge ash
污泥装运 muck shifting
污泥资源回收 resource recovery of sludge;sludge resource recovery
污泥自动氧化 autooxidation of sludge
污泥综合利用 comprehensive utilization of sludge
污泥组分 sludge composition
污泥最小化 sludge minimization
污泥最终处置 sludge ultimate disposal;ultimate disposal of sludge;ultimate sludge disposal
污气网纹 gas checking;gas crazing
污染 contaminate;contamination;defilement;dirt(i)ness;discoloration;pollute;pollution;taint
污染报警系统 pollution warning system
污染标度 pollution scale;scale of pollution
污染标准 pollution criterion;pollution norm
污染标准指标 pollution standard index
污染标准指数 pollution standard index
污染表面 contamination surface
污染病 pollution disease
污染波 waves of pollution
污染波衰减度 degree of pollution wave transformation
污染补偿政策 pollutant offset policy
污染补偿制度 pollution offset system
污染参数 parameters of pollution
污染测定 pollution determination;pollution measurement
污染测量 pollution measurement
污染测试器 contamination tester;pollution tester
污染层 dirt bench
污染产品收费 pollution product charges
污染产生量 pollutant generation quantity
污染产生源 pollution-creating source
污染沉积(物) marine sediment;pollution deposit
污染程度 amount of fouling;contamination degree;content of pollution;emission level;polluting strength;pollution degree;pollution level;scale of pollution;soilability;soiling procedure;degree of pollution;contamination level
污染程度测定器 contamination tester
污染澄清 sewage clarification
污染持续时间 duration of pollution
污染出水分析 analysis of sewage effluent
污染除去率 decontamination efficiency
污染处理 waste treatment
污染带 pollution belt;pollution zone;saprobic zone
污染带范围 pollutant band
污染带宽度 pollution zone width
污染当量 pollution(al) equivalent
污染的 contaminative;impure;pollutional
污染的阿发因子 alpha factor in wastewater
污染的沉积物 contaminant sediment
污染的大气 contaminated atmosphere;polluted atmosphere
污染的地下水 polluted groundwater
污染的法定最高限度 statutory ceiling on pollution
污染的法律问题 legal aspects of pollution
污染的房间 contaminated room
污染的放射性 contaminating radio activity
污染的海中沉积物 contaminant marine sediment
污染的河口 polluted estuary
污染的河流 polluted river;polluted stream
污染的河水 contaminated river water;polluted river water
污染的后果 pollution effect
污染的环境 contaminated environment;polluted environment
污染的径流 contaminant runoff
污染的浚挖物质 contaminant dredged material
污染的空气 foul air;contaminant air
污染的模式 pattern of pollution
污染的排水沟 polluted waterway
污染的全球影响 global effect of pollution
污染的散发物来源 emission source
污染的生物监测 biologic(al) monitoring of pollution
污染的食物 contaminated food
污染的水 polluted water
污染的水道 polluted waterway
污染的水域 polluted waters
污染的甜菜种子 contaminated beet
污染的土地 contaminant land;contaminated land
污染的土壤 contaminated soil
污染的压舱水 contaminated ballast
污染的烟羽 pollution plume
污染的淤泥 polluted ooze
污染的指标生物 index organism for pollution
污染等级 classes of pollution
污染底泥 contaminated sediment;polluted sediment
污染底泥管理对策 contaminated sediment management strategy
污染底泥浓度 sediment concentration of pollutant
污染地表水 polluted surface water
污染地带 contaminated zone;wastewater field;zone of pollution;

pollution zone
污染地区 contaminated area; contaminated region
污染地图集 pollution atlas
污染地下水 contaminated groundwater
污染地下水带 polluted groundwater zone
污染地下水修复 remediation of polluted underground water
污染地下水资源 contaminated groundwater resources
污染地质循环 geologic(al) cycle of pollutants
污染调查 pollution investigation; pollution survey
污染调查报告 pollution survey report
污染调查监测区 pollution research and monitoring area
污染毒度 pollution toxicity
污染毒物评估 toxicological assessment of pollution
污染毒性 pollution toxicity
污染度 degree of contamination; degree of staining; dirt(i)ness; dustiness
污染度检测 examination of pollution index
污染堆积 pollution accretion
污染对气候影响 pollutant effect on climate
污染对人类和社会的影响 pollution effect on man and society
污染对鱼类影响 pollutant effect on fishes; pollution effect on fishes
污染发散物 pollutant emission
污染法 sewage law
污染法规 pollution code
污染防护费用 cost of pollution protection
污染防止法 control of pollution act
污染防治 abatement pollution; contamination control; pollution control; pollution prevention; prevention and control of pollution
污染防治法 act of pollution control
污染防治工业 pollution prevention industry
污染防治技术 pollution prevention and control technology; pollution prevention technique
污染防治条例 pollution prevention and control ordinance; pollution prevention ordinance
污染放射性 contamination activity
污染分布 pollution distribution
污染分级标准 scale of pollution
污染分流系统 separate sewage system; separate system of sewage
污染风险 pollution risk
污染风险分析 contamination risk analysis; pollution risk analysis
污染风险管理 management of pollution risk
污染锋面位置 pollutional front location
污染负荷 <污水对河流等的> polluted load; pollution(al) load
污染负荷分配模型 pollution load allocation model
污染负荷量 air cleaning load; contamination load; pollution loading amount
污染负荷率 pollutant loading rate; pollution loading rate
污染负荷排放优化分配 optic(al) allocation for pollutant load discharged
污染负荷系数 pollution load coefficient
污染负荷影响 pollution load influence
污染负荷影响估算 estimation of pollution load influence
污染负荷指数 pollution loading index
污染工程 pollution engineering
污染工程技术 pollution engineering technique
污染工地 polluted ground
污染工业废水 polluted industrial wastewater
污染公害 pollution nuisance
污染公用事业 sewer utility
污染沟 muck ditch
污染关系 pollution relationship
污染管管顶 crown of sewer
污染管理 pollution management
污染管理费 costs of pollutant management
污染管制 contamination control
污染灌溉处置 irrigation sewage disposal
污染过程 pollution history
污染含水层 contaminated aquifer
污染河段 polluted reach
污染河口 polluted estuary
污染河流 contaminated river; contaminated stream; polluted river; polluted stream; pollutional river; pollutional stream
污染河水 polluted river water
污染河水处理 polluted river water treatment
污染痕迹 stain
污染后果 pollutional consequence; pollutional contribution
污染湖泊 polluted lake
污染化藓沼 pollutification bog
污染化学 pollution chemistry
污染环境 befouling environment; polluted environment; pollution of the environment
污染环境废物税 effluent tax
污染环境者 polluter
污染环境罪 crime against polluting environment
污染汇集地 pollution sink
污染混合带 pollution mixing zone
污染货物 contaminative cargo; polluting goods
污染机理 pollution mechanism
污染机制 pollution mechanism
污染积成物 pollution accretion
污染基准 pollution criterion
污染级别 pollution level
污染极限 pollution limit
污染疾病 pollution disease
污染计数管 contamination counter; pollution counter
污染计数计 contamination counter; pollution counter
污染计数器 contamination counter; pollution counter
污染剂量 contamination dose; pollution dose
污染剂量计 contamination meter
污染监测 contamination monitoring; pollution monitoring
污染监测器 contamination monitor
污染监测网 pollution monitoring network
污染监测仪 contamination monitor
污染监测与评价 pollution monitoring and assessment
污染监控船 pollution control ship
污染监视 pollution surveillance
污染茧 soiled cocoon
污染减轻 contamination abatement; pollution abatement
污染检测 pollution detection
污染经济学 pollution economics
污染阱 dirt trap
污染警报系统 pollution warning system
污染净化装置 pollution control equipment
污染径流 polluting runoff
污染径流模拟模型 pollutant runoff simulation model
污染颗粒 soiling particulate
污染空气 contaminated air; contamination air; polluted air
污染恐怖 pollution horror
污染控制 contamination control; pollution control
污染控制标准 standard of performance(of pollution)
污染控制船 pollution control ship
污染控制措施 pollution control measures
污染控制法规 pollution control regulation
污染控制法令 Pollution Control Act
污染控制管理 pollution control management
污染控制规划 pollution control planning; program(me) of pollution control
污染控制技术 engineering for pollution control; pollution control technology
污染控制建筑物 pollution control structure
污染控制局 Pollution Control Agency; Pollution Control Board
污染控制立法 pollution control legislation
污染控制目标 objective of pollution control
污染控制器 pollution controller
污染控制设施 pollution control facility
污染控制水平 level of pollution control
污染控制条例 pollution control regulation
污染控制系统 pollution control system
污染控制效能 pollution control function
污染控制协会 Association for Anticontamination
污染控制仪器 pollution control instrumentation
污染控制政策 pollution control policy
污染控制指标 pollution control index
污染控制指南 pollution control guide
污染控制指数 pollution control index
污染控制装置 pollution control facility; pollution control plant
污染控制作用 pollution control function
污染扩散 diffusion of pollution
污染类型 pollution type
污染历时 duration of pollution
污染量 pollutional load; quantity of pollution
污染了的覆盖层 contaminated overburden
污染了的淤泥 polluted ooze
污染零排放 zero discharge of pollutant
污染流 contaminant flow
污染流出液 contaminated effluent
污染流体 contaminant fluid
污染率 contamination rate
污染弥散 pollutant dispersion
污染密集货物 pollution-intensive goods
污染面积 contaminated area
污染敏感类型 pollution-sensitive typology
污染敏感时刻 pollution episode
污染敏感性分类 pollution-sensitive typology
污染模式 pollution model
污染膜 fouled membrane
污染能力 pollution capacity
污染泥浆池 dirty mud sump
污染凝聚 contamination condensation
污染农田 polluted agricultural land
污染浓度 pollution concentration
污染偶发事件 contamination accident
污染排放标准 pollution discharge standard
污染排放要求 emission requirement
污染评价 pollutant assessment; pollution assessment; pollution evaluation
污染评价指标 assessment index of pollution; pollutant assessment index
污染评价指数 assessment index of pollution; pollutant assessment index
污染起始值 pollutant start value
污染气象学 air pollution meteorology; pollution meteorology
污染潜势 pollution potential
污染强度 intensity of pollution; polluting strength; pollution intensity
污染区 blighted area; contaminated district; contaminated zone; polluted area; pollution area; pollution region; pollution zone
污染去污试验 contamination-decontamination experiment
污染全权交易市场 market-creation of pollution rights
污染权 market-creation of pollution rights; pollution right; rights to pollute
污染权利转让 pollution rent
污染容量 pollution capacity
污染容纳量 pollution-carrying capacity
污染溶液 foul solution
污染生理效应 pollution physiologic(al) effect
污染生态系统管理 management of polluted ecosystem
污染生态系统识别 identification of polluted ecosystem
污染生态学 pollution ecology
污染生物 fouling organism
污染生物监测 biologic(al) monitoring of pollution
污染生物学 pollution biology
污染生物指示器 biologic(al) indicator of pollution
污染生物指数 biologic(al) index of pollution; biotic index of pollution
污染食品 contaminated food products
污染食物 contaminated food
污染史 pollution history
污染事故 contamination accident; pollution incident
污染事件 contamination event; pollution episode
污染事件模型 pollution episodic model
污染事件统计 pollution incident statistics
污染势 pollution potential
污染试验 stain test; test for contamination; test for pollution
污染受害者 pollution-related victim
污染受体 pollution receptor
污染树脂物 resin poison
污染水 contaminated water; infected water; polluted water; pollution water; unclean water
污染水道 polluted waterway

污染水回收 polluted water recovery
污染水流 polluted flow
污染水平 level of pollutant; pollutant level; pollution level
污染水平与分类 contamination level and type
污染水体 polluted waters; pollution waters
污染水体水质 polluted waters quality; water quality of polluted waters
污染水文地质调查 contaminative-hydrogeologic(al) survey
污染水文地质剖面图 profile of pollution hydrogeology
污染水文地质图 map of pollution hydrogeology
污染水文学 contaminant hydrology
污染水修复 contaminated water remediation
污染水域 contaminated waters; polluted waters; pollution waters
污染水源 polluted water sources
污染水藻 polluted water algae
污染税 pollution tax
污染速度 rate of pollution
污染损害索赔(报告)书 claims documents of pollution damage
污染损失费用 pollution loss cost
污染特性 polluting property
污染特征 pollution characteristic; pollution feature
污染条件 pollutional condition
污染同化 pollution assimilation
污染图 pollutograph
污染图册 pollution atlas
污染途径 pollution pathway
污染土 polluted soil; contaminated soil
污染土壤的改良 improvement of contaminated soil
污染危害 contamination hazard; pollution hazard
污染危害的城市 pollution-plagued city
污染危险 pollution risk
污染危险频率 pollution risk frequency
污染威胁 pollution danger; pollution threat
污染微生物学 pollution microbiology
污染文摘 <期刊> Pollution Abstracts
污染物 contaminant; contaminator; impurity; polluted matter; polluted substance; fomes; fomite <如病人衣物床褥等>
污染物标准指标 pollutant standard index
污染物标准指数 pollutant standard index
污染物表征 pollutant characterization
污染物补偿政策 pollutant offset policy
污染物采样 contamination sampling
污染物残留质量 mass of pollutant remaining
污染物测定 measurement of contaminant
污染物产生量 pollutant generation quantity
污染物超标率 overstandard rate of pollutant
污染物-沉淀物相互作用 pollutant-sediment interaction
污染物成分 constituent of contamination; pollutant constituent
污染物持久性 persistency of pollutant
污染物处理 pollutant disposal
污染物处置 pollutant disposal
污染物传播 pollutant dispersion
污染物传播理论 contaminant migration theory
污染物单位负荷 pollutant load per unit
污染物的安全浓度 safe concentration of pollutant
污染物的长期效应 long-term effects of pollutant
污染物的持久性 persistence of pollutant; pollutant persistency
污染物的传播途径 pollutant pathway
污染物的地质大循环 geologic(al) cycle of pollutants
污染物的毒性 toxicity of pollutant
污染物的化学降解 pollutant chemical degradation
污染物的近地面浓度 ground-level concentration of contaminant
污染物的累积 build-up of pollutants
污染物的路径 pathway of a pollutant
污染物的浓度 concentration of pollutant
污染物的迁移 transport of pollutants
污染物的铅直梯度 vertical gradient of pollutant
污染物的侵蚀转移 transport of pollutant by erosion
污染物的生物地球化学循环 biogeochemical cycle of pollutants; biogeochemistry cycle of pollutant
污染物的生物降解作用 biodegradation of pollutant
污染物的生物评价 biologic(al) assessment of pollutant
污染物的释放 release of pollutant
污染物的危害 hazards of pollutant
污染物的物理吸附与物理沉淀 physical absorption and deposition of pollutant
污染物的形态 form of pollutant
污染物的影响 pollutant effect
污染物的转化 transformation of pollutant
污染物的转移 transfer of pollutant
污染物的自净作用 pollutant self-purification
污染物地址大循环 geologic(al) circle of pollutants
污染物动力学 pollutant dynamics
污染物短期浓度 short-term concentration of contaminant
污染物对气候的影响 pollutant effect on climate
污染物对人和社会的影响 pollutant effect on man and society
污染物对鱼类的影响 pollutant effect on fishes
污染物放出率 emission rate
污染物分布 pollutant distribution
污染物分解 decay of pollutant
污染物分类 classification of pollutants; contaminant classification
污染物分离 isolation of pollutant
污染物分析 pollutant analysis
污染物分析器 pollutant analyser [analyzer]
污染物分析仪 pollutant analyser [analyzer]
污染物负荷量 contaminant loading; pollutant burden; pollutant loading
污染物负荷模式 pollutant burden pattern
污染物感受性 pollutant susceptibility
污染物含量 contaminant content; pollutant content
污染物化学 pollutant chemistry
污染物环境标准 environmental standard of pollutant
污染物混合 combination of pollutant
污染物活度 activity of pollutant
污染物监测 pollutant monitoring; pollutant surveillance
污染物监视 pollutant surveillance
污染物检测 contaminant detection
污染物检出率 detective rate of pollutant
污染物鉴别 identification of pollutant; pollutant identification
污染物鉴定 identification of pollutant; pollutant identification
污染物降解作用 degradation of contaminant; pollutant degradation
污染物接受者 pollutant recipient
污染物控制技术标准 pollutant control technology standard
污染物扩散 contamination dispersal; diffusion of contaminant; diffusion of pollutant; dispersal of pollutant; dispersion of contaminant; pollutant dispersion
污染物扩散标准 pollutant emission standard
污染物扩散输移 dispersive pollutant transport
污染物扩散运动 dispersive movement of contaminant
污染物来源鉴别 pollutant source identification
污染物类 group of contaminants
污染物累积 contaminant accumulation; pollutant accumulation
污染物累积量 amount of pollutant accumulation
污染物量 contaminant loading; level of pollutant
污染物量控制 quantity control of pollutant
污染物流股 plume of contaminant
污染物流量 quantity of pollutant flow
污染物流束 plume of contaminant
污染物路径 pollutant pathway
污染物密度 density of pollutant
污染物目标 pollutant target
污染物目标暴露 pollutant target exposure
污染物浓度 pollutant concentration; pollutant level; pollutant load
污染物浓度场 pollutant concentration field
污染物浓度的每日容许摄取量 acceptable daily intake
污染物浓度动态特征 dynamics of pollutant concentration
污染物浓度控制 concentration control of pollutant; pollutant concentration control
污染物浓度水平 pollutant concentration level
污染物浓度限度 pollutant concentration limit
污染物浓度预报 pollutant concentration prediction
污染物排出 pollutant discharge
污染物排除 contaminant removal
污染物排放 blowdown; discharge of sewage; effluent discharge; pollutant discharge; pollutant emission
污染物排放标准 pollutant discharge standard; standard for discharge of pollutants
污染物排放标准总量 total amount of pollutant discharge standard
污染物排放负荷 pollutant discharge loading
污染物排放控制指标 guidelines of pollutant control for industries
污染物排放量 effluent level; quantity of discharge; quantity of pollutant discharged; discharge of pollutants
污染物排放率 emission rate
污染物排放浓度 pollutant discharge concentration
污染物排放申报登记 declaration and registration of pollutant discharge; discharge of reporting and registering
污染物排放申报登记制度 declaration and registration system of pollutant discharge; system of reporting and registering pollutant emission
污染物排放通道 pathways of pollutant discharge
污染物排放消除体系 pollutant discharge elimination system
污染物排放许可证 permit for pollutant discharge; pollution license
污染物排放许可制度 permit system for discharging pollutants
污染物排放质量 pollutant discharge quality
污染物评估 contaminant assessment
污染物迁移 convection of pollutants; transport of pollutants
污染物侵蚀转移 transport of pollutant by erosion
污染物清单 list of pollutant; list of polluting substance
污染物去除率 pollutant removal rate
污染物去除系数 pollutant removal efficiency
污染物散布 pollutant dispersal; pollutant dispersion
污染物生产量 pollutant generation quantity
污染物生物检测 contaminant biomonitoring
污染物生物降解 biodegradation of pollutant
污染物识别 pollutant identification
污染物释放 free of pollutant mass; pollution release
污染物输移 contaminant transport; pollutant transport; transformation of pollutant
污染物输移模拟模型 pollutant transport simulation model
污染物输移模型 pollutant transport model
污染物输移预报 pollutant transport prediction
污染物输移转化 pollutant transport transformation
污染物水平 contaminant level
污染物瞬时流场 contaminant transient flow field
污染物通量 pollutant flux
污染物途径 pathway of pollutants
污染物团 pollutant mass
污染物团释放 free of pollutant mass
污染物完全产生系数 complete production coefficient of impurities
污染物污染指标 the pollution quota of pollutant
污染物物料衡算 material balance calculation of pollutant
污染物削减量 quantity of pollutant reduced
污染物行径 pollutant pathway
污染物烟缕 contaminant plume; plume of contaminant
污染物烟缕预报 contaminant plume prediction
污染物易感性 pollutant susceptibility
污染物与环境的相互作用 pollutant interaction with environment
污染物远程输送 long-range transport of pollutant
污染物允许排放量 allowable quantity of pollutant discharged
污染物在大气中的总份额 total air-

borne fraction
污染物在土壤中的迁移 transport of pollutant in soil
污染物折纯碱量 pollutant quantity in pure base
污染物直接产生系数 direct production coefficient of impurities
污染物指标 pollutant index; pollutant target
污染物指数 pollutant index
污染物质 contaminating material; pollutant (substance); polluted matter; polluted substance; polluter; pollutional matter; fomes; fomite <如病人衣物床褥等>
污染物质处理 pollutant treatment
污染物质的地面浓度 ground concentration of pollutant; ground concentration of pollution material
污染物质极限容许浓度 permissible concentration limit of pollutant
污染物质降解作用 degradation of pollutant(substance)
污染物质量 mass of pollutant; pollutant mass
污染物质量连续性 continuity of pollutant mass
污染物质平衡 pollutant equilibrium
污染物质远期效应 long-term effects of pollutant
污染物转化 pollutant transformation
污染物综合指数 combined pollutant index
污染物总量控制 total amount control of pollutant
污染物最终处理 final treatment of pollutant
污染系数 contamination coefficient; contamination factor; pollution coefficient; pollution factor
污染险 risk of contamination
污染限度 contamination limit; pollution limit
污染限制 pollution restriction
污染消除 abatement of pollution; pollution(al) abatement
污染效应 contaminant effect; effect of pollution; polluting effect; pollutional effect
污染型能源 pollution energy
污染性货物 dirty cargo
污染性空中爆炸 contaminating air burst
污染性离子 pollution indication ion
污染性气溶胶 aerosol contaminant
污染性悬浮微粒 aerosol contaminant
污染性质 polluting property
污染许可证 pollution permit
污染循环 pollution cycle
污染烟缕 pollution plume
污染岩芯 <被泥浆污染的> foul the core
污染研究 pollution study
污染演化 evolution of pollution
污染氧化物 contaminating oxidant
污染要素数量 the number of pollution factors
污染液体 contaminated fluid
污染抑制 pollution abatement
污染因子 contamination factor
污染引起的疾病 disease caused by pollution; pollution-induced disease
污染应负的责任 pollution liability
污染影响 effect of pollution; pollution effect
污染油 polluting oil
污染油品处理设施 contamination oil processing facility
污染油品再注入设备 contamination oil reinjection device
污染鱼类 polluted fishes
污染与环境相互作用 pollutant interaction with environment
污染雨水 polluted rain
污染预报 pollution prediction
污染预测 pollution prediction
污染源 contaminated sources; contaminating sources; contamination sources; pollutant source; polluter; pollution source; sources of contamination; sources of pollution
污染源表征 contaminant source characterization
污染源采样 source sampling
污染源采样器 source sampler
污染源地 pollutant source region
污染源调查 investigation of contaminative sources; survey of pollution sources
污染源分布图 distribution map of pollution sources
污染源分布形状 arrange form of pollution sources
污染源概况 generalization of pollution sources
污染源贡献率 pollution source contribution ratio; source contribution ratio
污染源管理 management of pollution sources
污染源基准 pollutant source criterion; pollution source criterion
污染源鉴定 pollutant source identification
污染源控制 control of pollution sources; pollution source control
污染源控制对策 pollution source control strategy
污染源类型 pollutant source type; source type
污染源排放史 contaminant source release history; pollution source release history; release histories of contaminants; source release history
污染源排污物组分比率 source contribution ratio
污染源评价 pollution source assessment
污染源评价抽样系统 source assessment sampling system
污染源普查 general survey of pollution sources
污染源识别 identification of pollution sources; pollutant source identification; pollution source identification; pollution source recognition; source recognition
污染源位置 contaminant source location; pollution source location
污染源位置识别 identification of contaminant source locations; identification of pollution source locations
污染源详查 detailed survey of pollution sources
污染源与源强分析 pollution source and source strength analysis
污染运流 convection of pollutants
污染灾难 contamination accident
污染责任 pollution liability
污染责任保险 pollution liability insurance
污染毡状层 pollution carpet
污染者 discharger; polluter
污染者偿还原则 polluter pays principle
污染者付款原则 polluter pays principle
污染者负担原则 polluter pays principle
污染者负担政策 polluter pays policy
污染者赔偿原则 polluter pays-off principle
污染者支付原则 polluter pays principle
污染蒸汽 impure steam
污染指标 index of pollution; parameter of pollution; scale of pollution
污染指标物 indicator of pollution
污染指示剂 indicator of pollution; pollutional indicator
污染指示器 indicator of pollution; pollutional indicator
污染指示生物 pollutional indicating organism; pollution indication organism
污染指示物 indicator of pollution; pollution indicator
污染指数 contamination index; index of pollution; pollutant index; pollutional index
污染指数法 pollution index method
污染治理 pollution abatement; pollution control
污染治理费用 cost of pollution abatement
污染治理(经费)预算 budget for pollution policy
污染治理设备 pollution abatement equipment
污染周期 pollution cycle
污染转移 contaminant transport
污染状况 pollutional condition
污染状态 pollutional condition
污染准则 pollution criterion
污染自然净化 natural amelioration of pollution
污染综合控制 integrated pollution control
污染综合控制规划 integrated pollution control program(me)
污染总负荷 total loading of pollutant
污染作用 pollution interaction; soiling effect
污热阻 fouling thermal resistance
污塞层 fouling film
污色 contact stain
污色的 sordid
污砂 dirty sand
污水 befouled water; dirty water; drainage water; effluent; foul water; liquid filth; polluted water; refuse water; residuary water; sewage water; slop; slop water; soil sew(er)age; sullage; wasted water; wastewater <美>
污水岸边排放 sewage shore discharge
污水暗管 blind sewer
污水泵 dirt(y) water pump; dredge pump; non-clogging pump; non-clog type of pump; sanitary pump; sewage disposal pump; sewage handling pump; sewage pump; sewage water pump; sink evacuator; slurry pump; slush pump
污水泵房 sewage pump house; sewage pumping room; sewage pump plant
污水泵站 sewage pumping plant; sewage pumping station; wastewater pumping station
污水驳 sewerage barge
污水补给 sewage recharge
污水采样 sampling of effluent; sewage specimen; wastewater sampling
污水采样方法 sewage sampling method
污水参数 wastewater parameter
污水槽 septic tank; sewage tank; sink; slop sink; slop tank; sump; sump tank
污水槽研磨机 sink grinder
污水测量管 bilge sounding pipe
污水层 sewage layer
污水产生的气体 sewage gas
污水厂出水 sewage effluent; sewage outlet water; sewage works effluent
污水厂平面布置 plan of sewage treatment plants
污水厂入流格栅 grizzly feeder
污水厂设计年限 designing period of sewage(treatment) plants
污水厂位置 location of sewage treatment plants
污水厂总体布置 general layout of sewage treatment plants
污水场 wastewater field
污水超滲 sewage ultra-filtration
污水车 cesspoolage truck
污水沉淀 precipitation of sewage; settlement of sewage
污水沉淀池 detention of sewage; detention tank; effluent settling chamber; sewage settling basin; sewage tank
污水沉淀柜 sewage tank
污水沉淀室 sewage grit chamber
污水沉淀物 sewage solid
污水沉积的污泥 sewage sludge
污水沉降池 settlement tank
污水沉沙池 grit chamber; sewage grit chamber
污水沉沙室 detritus chamber; sewage grit chamber
污水成分 composition of sewage; sewage composition
污水澄清 clarification of sewage; sewage clarification
污水池 cesspool; foreyn; lagoon; septic tank; sewage fish pool; sewage lagoon; sewage pond; sewage reservoir; sewage tank; sewerage lagoon; small sewage tank; sump; tank sewer; wastewater basin
污水池格栅 grate of sink
污水池压顶石 coping stone of cesspool
污水冲洗 sewage flushing
污水抽射器 bilge ejector
污水出口 outlet of sewer; sewer outlet
污水出流 sewage effluent
污水出水分析 analysis of sewage effluent
污水初次沉淀槽 sewage preliminary settling tank
污水初次沉淀池 sewage preliminary basin; sewage preliminary settling tank
污水初级沉淀池 primary precipitation tank
污水初级沉降池 primary settlement tank
污水储池 effluent holding reservoir
污水储存 <指自然净化> lagooning
污水储存池 sewage stock pond
污水处理 disposal of sewage; effluent disposal; sewage disposal; sewage purification; sewerage and sewage treatment; treatment of sewage; treatment of wastewater; sewage treatment; wastewater treatment
污水处理标准 effluent disposal standard
污水处理残渣 sewage treatment residue
污水处理测定中心 wastewater treatment evaluation facility
污水处理厂 disposal plant; effluent plant; effluent treatment plant; sewage disposal plant; sewage disposal works; sewage plant; sewage purification plant; sewage treatment plant; sew(er)age treatment

works;treatment plant;wastewater disposal plant; wastewater treatment plant;sew(er)age works
污水处理厂厂主协会 Sewage Plant Manufacturing Association
污水处理厂出水 effluent from sewage treatment plant; sewage (treatment) plant effluent; treatment plant effluent; wastewater treatment plant effluent
污水处理厂的污泥 sewage sludge
污水处理厂废液排出管 outlet pipe
污水处理厂费用函数 cost function of wastewater treatment plant
污水处理厂模拟模型 sewage treatment plant simulation model
污水处理厂排除已处理污水的管道 outfall sewer
污水处理厂排放 sewage treatment plant discharge
污水处理厂设计 sewage treatment plant design
污水处理厂污泥 sewage treatment plant sludge
污水处理场 sewage disposal plant; sewage farm; sewage treatment plant;wastewater treatment unit
污水处理池 sewage treatment tank; treatment tank
污水处理船 sewage disposal vessel
污水处理单元过程最优化设计 optimization in design of unit processes in wastewater treatment
污水处理法 sewage treatment method;wastewater treatment method
污水处理方案 wastewater treatment scheme
污水处理工程 effluent treatment works; sewage disposal works; sewage treatment works
污水处理工艺 sewage disposal process;sewage treatment process; wastewater treating process; wastewater treatment process
污水处理工艺流程 sewage treatment technological process
污水处理构筑物 sewage treatment structure
污水处理过程 sewage disposal process; sewage treatment process
污水处理过的废水 sewage treated wastewater
污水处理后出水 sewage effluent; wastewater effluent
污水处理后出水分析 analysis of sewage effluent;analysis of wastewater effluent
污水处理后出水量 quantity of sewage effluent;quantity of wastewater effluent
污水处理后出水排放 discharge of sewage effluent; discharge of wastewater effluent
污水处理后出水水质 sewage effluent quality;wastewater effluent quality
污水处理后出水同化容量值 value of assimilative capacity of sewage effluent;value of assimilative capacity of wastewater effluent
污水处理及排放 sewage disposal
污水处理技术 sewage treatment technique
污水处理建筑物 sewage treatment structure
污水处理接触法 contact method of sewage treatment
污水处理结构 sewage treatment structure
污水处理利用 utilization of wastewater treatment
污水处理量 quantity of wastewater disposal; quantity of wastewater treatment
污水处理区 sewage disposal area
污水处理设备 foul water disposal facility;sewage treatment equipment
污水处理设备环境保护监督管理办法 guidelines of supervision and management of sewage treatment facility on environment(al) protection
污水处理设施 refuse water disposal facility;sewage disposal facility
污水处理水平 sewage treatment level
污水处理塔 digestion tower
污水处理系统 sewage disposal system; sewage treatment system; wastewater disposal system; wastewater treatment system
污水处理系统优化设计 optimization design of wastewater treatment system
污水处理效率 efficiency of wastewater treatment;wastewater treatment efficiency
污水处理优化模型 sewage treatment optimization model
污水处理余渣 sewage treatment residue
污水处理运作 wastewater treatment operation
污水处理站 sewage treatment station
污水处理装置 sewage treatment plant; waste-disposal plant; wastewater treatment equipment; wastewater treatment facility; wastewater treatment unit
污水处理最优化模型 wastewater treatment optimization model
污水处置 disposal of sewage;sewage disposal
污水处置措施 measures of waste liquid disposal
污水处置费 cost of sewage treatment
污水处置过程 sewage disposal;sewage disposal process
污水处置系统 <可将人粪便返回土地的> conservancy system; sewage disposal system
污水纯氧曝气工艺 unox process
污水次干道 submain sewer
污水次干管 submain sewer
污水萃取法 sewage extraction method
污水倒灌 backflow of sewage; backing-up of sewage
污水道 bilge way;foul sewer;limber passage; limber space; sewer; slops chute;wasteway
污水道附属物 sewer appurtenances
污水道盖板 limber board
污水道拱顶 sewer arch
污水道(秽)气 sewer gas
污水道检查井 sewer manhole
污水道排出口 sewer outfall
污水道排泄设备 sewer evacuator
污水道气体 sewer gas
污水道系统 sewer system
污水的逆渗透净化 reverse osmosis
污水的完全处理 <包括初级和二级处理> complete treatment of sewage
污水的卫生化学分析 sanitary chemical analysis of sewage
污水的一次混合量 sewage batch; sewage dose
污水地段 effluent farm;sewage farm
污水地下处置 wastewater underground disposal
污水地下渗漏场 leaching field
污水电解处理法 sewage electrolytic treatment
污水调查 sewage investigation
污水动物 saprobic animal
污水毒性 effluent toxicity;wastewater toxicity
污水毒性试验 effluent toxicity testing
污水二级处理 secondary treatment for sewage; secondary wastewater treatment
污水法规 sewage legislation
污水反渗透处理法 reverse osmotic treatment for sewage
污水反消化脱氮处理 nitrogen removal from wastewater by denitrification
污水方案 sewage scheme
污水肥料价值 fertilizing value of sewage
污水费率 sewage rate
污水费(用) wastewater charges;sewage charges
污水分解 decomposition of wastewater; sewage decomposition;wastewater decomposition
污水分解气体 sewage gas
污水分类 classification of wastewater
污水分离系统 effluent segregation system
污水分流 segregation of wastewater
污水分配管 distribution tile
污水分散处理 decentralized wastewater treatment
污水分析 analysis of polluted water; sew(er)age analysis
污水分析样 analysis sample of contaminated water
污水粪便系统 plumbing system
污水浮游生物 saproplankton
污水腐生生物 polysaprobe
污水负荷 sewage pollution value
污水负荷量 sew(er)age load(ing)
污水干道 main(sanitary)sewer
污水干管 main sewer;outfall sewer; sewer main;trunk sanitary sewer; trunk sewer
污水干线 sewer trunk
污水钢丝布 belt sewage screen
污水高负荷处理 high rate treatment
污水隔筛 sewage screen
污水工程 sewerage;sew(er)age engineering;sew(er)age works
污水工程法 sewerage law
污水工程师 sewage engineer
污水工程系统 sewerage system
污水公用事业 sewer utility
污水沟 dirty water ditch;foul water trench; muck ditch; sewage drain; sewer trench;side bilge
污水沟槽 sewer trench
污水沟抽水管系 bilge line;bilge pipe
污水沟道 wastewater sewer
污水沟盖板 bilge board;bilge limber board
污水沟内底板 bilge ceiling
污水垢 sump
污水观测点 observation point of waste water
污水管 cesspipe;foul sewer;foul water pipe; pipe sewer; refuse water pipe; sanitary pipe; sewage pipe; sewage piping; sewer; sewer pipe (drain); soil pipe; soil sewer; waste(water) pipe
污水管鞍座 sewer saddle
污水管保养 sewer maintenance
污水管冲刷 sewer scouring
污水管冲洗 sewer flushing
污水管冲洗阀 sewer sluice valve
污水管出口 outlet of sewer;sewer outlet;sewer orifice;sewer outfall; sewer outlet
污水管存水弯 sewer trap
污水管道 pipe sewer line; sanitary plumbing; sewage conduit; sewage line;sewer;sewerage conduit;sewer line;wastewater sewer;sanitary sewer
污水管道沉积物 sewer sediment
污水管道费用函数 cost function of sewage pipe
污水管道沟 foul water(pipe)trench
污水管道汇流井 junction chamber
污水管道排泄设备 sewer evacuator
污水管道设计 sanitary sewer design
污水管道网 network of sewers
污水管道系统 sewage conduit system
污水管道系统布置 plan of sewerage system
污水管道系统优化设计 optimization design of sewer pipe system
污水管道窨井 sanitary sewer manhole
污水管道中壅水 backwater in sewage duct
污水管干线 trunk sanitary sewer
污水管沟 refuse water pipe trench; sewer tunnel
污水管沟槽 sewage pipe trench
污水管管头成型机 jam socket machine
污水管基础 sewer bottom
污水管检查 sewer inspection
污水管检查井 sewer manhole
污水管检查孔 sewer manhole
污水管件 sewer fitting
污水管建造 conduit-type sewer construction
污水管尽端 dead-end sewer
污水管进户线 house sewer
污水管进口 sewer inlet
污水管理 sewage management; wastewater management
污水管理局 Sewer Authority
污水管理区 sewage management district
污水管理系统 wastewater management system
污水管流率 sewer rate
污水管路 plumbing drain
污水管黏[粘]垢 sewer slime
污水管排泄设施 sewer evacuator
污水管清管球 sewer pill
污水管清洗 sewer cleaning
污水管人孔 sewer manhole
污水管容量 sewer capacity
污水管设施 sewer facility;soil stack installation
污水管施工 sewer construction
污水管隧道 sewer tunnel
污水管条例 sewer ordinancc
污水管弯头 soil pipe elbow; waste trap
污水管网 sewer net(work)
污水管维修 sewer maintenance
污水管吸水装置 bilge injection
污水管系统 wastewater system
污水管线 sewage pipeline;sewer line
污水管修建方案 sewer-rehabilitation alternatives
污水管压制机 sewer pipe press
污水管闸阀 sewer sluice valve
污水管沼气 sewer gas
污水灌溉 irrigation with sewage;sewage farming;sewage irrigation;sewerage farm;wastewater irrigation
污水灌溉处理 broad irrigation;irrigation sewage disposal;sewage farming of land treatment
污水灌溉处理场地 irrigation sewage disposal field
污水灌溉处置 irrigation sewage disposal

污水灌溉地 irrigation sewage disposal field
污水灌溉定额 sewage irrigation norm
污水灌溉法 sewage farming of land treatment
污水灌溉(农)田 sewage farm;irrigated sewage field;irrigation sewage disposal farm
污水灌溉水质标准 quality standard of sewage irrigation
污水灌溉系统 irrigation system of sewage;sewage irrigation system
污水灌田 wastewater farming
污水规划 sewage scheme
污水柜 bilge tank;sewage tank
污水过渡段 sewer crossing
污水过滤 sew(er)age filtration
污水过滤池 bilge strainer;mud collector;sew(er)age filter
污水过滤器 bilge strainer;mud collector;sewerage filter
污水海湾排放 sewage discharge in harbo(u)r
污水海洋处置 ocean disposal of wastewater
污水涵管 sewer culvert
污水好氧生物处理 aerobic biological treatment of wastewater;biologic(al) aerobic treatment of wastewater
污水好氧生物处理反应器 aerobic biological treatment reactor of wastewater
污水耗氯量 chlorine demand of sewage
污水和污泥在农业中的应用 application of sewage and sludge in agriculture
污水湖 humic sewage lake;sewage lagoon
污水化学处理法 chemical methods of wastewater treatment;chemical treatment of sewage
污水化学分析 chemical analysis of sewage
污水化学需氧量 effluent chemical oxygen demand
污水环境科学 wastewater environmental science
污水回收 reclamation of sewage
污水回收利用 reclamation and utilization of sewage
污水回用 sewage reclamation
污水回注 wastewater re-injection
污水汇集 sewage collection
污水汇集系统 sewage collection system
污水混合水样 composite sample
污水活化池 sludge activation tank
污水机械处理法 mechanical sewage treatment;sewage mechanical treatment
污水机械脱水 sewage sludge dewatering
污水及压载管系 bilge and ballast system
污水及雨水合流制 combined system of sewerage
污水及雨水混合排水管 sewer for combined foul and surface water
污水及雨水两用排水管 sewer for combined foul and surface water
污水集水井 house inlet
污水集水管 collecting sewer;drainage collector;lateral sewer
污水集中处理 central waste water treatment
污水计量池 dosing chamber;dosing tank
污水剂量监测计 effluent monitor
污水加氯 chlorination of sewage;sewage chlorination
污水加氯消毒费 cost of sewage chlorination
污水间接排水管道 indirect drain;indirect waste piping
污水检测 examination of wastewater;sewage examination;test of sewage
污水检测探针 waste detection sounder
污水检查井 sewage control well;sewer manway
污水检验 sewage examination;test of sewage
污水建筑物 sewage construction
污水江心排放 sewage discharge in river central-line
污水截流干管 sewage main collecting pipe
污水截流管 interceptor;interceptor for sewage
污水截留井 house inlet;sewage interceptor
污水进行氧化的双平行渠 double oxidation ditch
污水进行氧化渠 oxidation ditch
污水经过完全处理后排放的液体 final effluent
污水井 <又称污水阱> absorbing well;bilge hat;bilge sump;bilge well;drainage hat;drainage pot;drainage well;drain port;dump well;dung hole;hat box;sink hole
污水净化 foul water purification;purification of sew(er)age;purifying of sewage;refuse water purification;sewage clarification;sewage purification
污水净化厂 sewage purification plant
污水净化池 sewage purifier;sew(er)age clarifier
污水净化法 clarification of sewage
污水净化器 sewage purifier
污水净化三级过程 ABC process of sewage purification
污水净化设备 sewage purifier
污水净化系统 waste(water) purification system
污水净化协会 Association for the Purification of Sewage
污水净化需氯量 chlorine demand of sewage
污水净化学会学报 Journal of the Institute of Sewage Purification
污水净化学会杂志 Journal of the institute of Sewage Purification
污水净化站自动化 automation of station for purification of sewage
污水净化装置 wastewater renovation plant;wastewater renovation unit
污水径流 wastewater runoff
污水径流调节池 wastewater flow equalization basin
污水聚集 sewage collection
污水卷流 sewage plume
污水菌 sewage fungus
污水颗粒 sewage particle;sewage particulate
污水坑 cesspit;cesspool;collecting sump;collection sump;dumb well;dung hole;foul water trench;leaching pit;sewage pit;sewage sump;sump;slop sink
污水坑道 sewage gallery;sewage tunnel;sewer gallery
污水控制 wastewater control
污水库 sewage reservoir
污水类型 type of wastewater
污水离析水流 wastewater isolation flow
污水立法 sewage legislation
污水利用 sewage utilization;utilization of sewage;wastewater use;wastewater utility;wastewater utilization
污水利用法 sewage utilization act
污水利用条例 sewage utilization act
污水连接井 sewage joining well
污水量 amount of sewage;sewage amount;sewage flow;sewage quantity;wastewater charges;wastewater flow
污水量分布 distribution of sewage
污水量配池 dosing chamber;dosing tank
污水流 effluent flow;liquid effluent;sewage stream;sewer flow
污水流出口 sewage outfall;wastewater outfall
污水流量 discharge of sewage;quantity of waste flow;sewage discharge;sewage flow(volume);sewage rate
污水流量计 dirty water meter;sewage flow meter
污水流率 rate-of-flow of sewage;sewage flow rate;sewage rate
污水流入量 sewage inflow;turbidity inflow
污水流入水道 transport by water of pollutant to water courses
污水垄沟灌溉 sewage ridge and furrow irrigation
污水滤池 filtration tank;sewage filter
污水滤器 sewage filter
污水滤网 sewage screen
污水滤网残渣 sewage screen residue
污水氯化处理 sewage chlorination
污水氯化(作用) sewage chlorination;chlorination of sewage
污水漫灌 broad irrigation of sewage;sewage flooding irrigation
污水漫淹灌溉 flood irrigation of sewage
污水明沟 open sewer
污水膜 sewage film
污水农灌定额 dosing rate of sewage irrigation
污水农灌价值 value of sewage irrigation
污水农灌土壤要求 soil requirements of sewage irrigation
污水农灌卫生要求 hygienic requirements of sewage irrigation
污水浓度 sewage concentration;concentration of sewage;sewage load(ing);sewage strength;strength of sewage
污水排出管 sewage discharge pipe;sewage discharge tube
污水排出口浓度 outfall concentration
污水排出量 discharge of sewage
污水排出器 sewage ejector
污水排除 sewage disposal
污水排除泵 effluent pump
污水排放 effluent;sewage discharge;sewage disposal;waste removal;wastewater discharge
污水排放标准 effluent-quality standard;sewage discharge standard;sewage drainage standard;sewage effluent standard
污水排放费 cost of sewage discharge
污水排放管 sewage discharge pipe;sewage discharge tube
污水排放口 sewage outlet
污水排放扩散管 diffuser for sewage discharge;sewage discharging diffusion pipe
污水排放量 discharge of waste water
污水排放量与径流量比 ratio of wastewater discharge capacity to runoff
污水排放浓度 concentration of waste water disposal
污水排放水力学 wastewater discharge hydraulics
污水排放位置 wastewater discharge location
污水排放系统 system of sewerage
污水排放装置 wastewater discharger
污水排海标准 ocean discharge criterion
污水排海口 ocean outfall;sea outfall
污水排海口水力学 sea outfall hydraulics
污水排气口 soil vent
污水排水管 soil drain
污水排水支管 soil branch
污水排泄 <处理后的> effluent sewage
污水排泄管 sewer pipe(drain)
污水排泄口 sewer outlet
污水排泄站 sewage discharging station
污水排泄总管 outfall sewer
污水旁管 lateral;lateral sewer
污水配水槽 distribution box
污水喷洒 sewage sprinkling
污水喷洒机 sewage distributor
污水喷射井 waste-injection well
污水喷射器 sewage ejector
污水盆 sewage sink;slop sink
污水盆格篦 grating of sink
污水曝气 aeration of sewage;sewage aeration;wastewater aerating
污水曝气池 aerated sewage lake
污水曝气处理 aeration of sewage
污水曝气处理法 activation
污水曝气湖 aerated sewage lake
污水曝气时间 aeration period of sewage;aeration time sewage
污水曝气周期 aeration period of sewage
污水畦灌 bed irrigation
污水气体 sewer gas
污水汽提 wastewater stripping
污水强度 strength of sewage
污水强度指数 index of sewage strength
污水情况 state of sewage
污水区 sewage district;sewage field
污水全部处理 complete treatment of sewage
污水人工降雨灌溉 sewage irrigation by sprinkler
污水人孔 sewage manhole
污水日流量 day flow of sewage
污水容量 sewage capacity
污水入渗通道 passage of waste water infiltration
污水软管 sewage hose
污水三级处理 tertiary sewage treatment;tertiary treatment for sewage;tertiary waste(water) treatment
污水三级处理法 advanced water treatment;tertiary sewage treatment method;tertiary wastewater treating process
污水杀菌 sterilization of sewage
污水筛网 sewage screen
污水筛渣掩埋 burial of sewage screenings
污水设备 wastewater appliance
污水设施 sewage facility
污水射流泵 sewage ejector;wastewater ejector
污水深度处理 advanced waste(water) treatment;complete treatment

W

of sewage
污水渗床 leaching bed
污水渗井 cesspit; cesspool; disposal well; infiltration well of sewage; leaching cesspool; leaching well; pervious cesspool; waste well; leaching pit
污水渗坑 infiltration pit of sewage; leaching cesspool; leach pit; leach pit for sewage; sewage sink; sink
污水渗库 infiltration reservoir of sewage
污水渗漏 effluent seepage
污水渗滤速率 infiltration rate to sewer
污水渗渠 infiltration ditch of sewage
污水生化处理 biochemical sewage treatment
污水生态学 sewage ecology
污水生物 saprobe; seprobia; wastewater organism
污水生物除磷 wastewater biological phosphorus removal
污水生物处理 biologic(al) sewage treatment; biologic(al) wastewater treatment; wastewater bio-treatment
污水生物处理厂 biologic(al) sewage disposal plant; biologic(al) sewage disposal works; biologic(al) sewage treatment plant; biologic(al) treatment plant
污水生物处理动力学 biologic(al) treatment kinetics of wastewater
污水生物处理系统 biologic(al) sewage treatment system; biologic(al) wastewater treatment system
污水生物带 saprobic zone
污水生物分解 biolysis of sewage; sewage biolysis
污水生物分类(法) saprobic classification; saprobien classification
污水生物分析 biologic(al) analysis of sewage
污水生物净化(法) biologic(al) sewage treatment; biologic(al) purification of sewage
污水生物三级处理 biologic(al) tertiary sewage treatment
污水生物系统 saprobes system; saprobic system; saprobien system
污水生物需氧量 effluent biological oxygen demand
污水生物学 biology of polluted water; sewage biology
污水生物学处理 biologic(al) sewage treatment
污水生物指示法 bioindication of polluted water
污水生物指数 saprobic index
污水实践 sewage practice
污水实施 sewage practice
污水试验 test of sewage
污水试样 sewage specimen
污水收集 collection of sewage; collection of wastes; sewage collection; wastewater collection
污水收集处理系统 wastewater collection and treatment system
污水收集管理系统 wastewater collection and management system
污水收集系统 wastewater collection system
污水收集装置 wastewater collection facility
污水输送 wastewater conveyance
污水输送管 sewer duct
污水竖管 sewer chimney; standing waste
污水水样 sewage sample
污水水样分析 analysis of sewage sample
污水水质 sewage quality; wastewater quality; water quality of sewage
污水水质标准 wastewater quality standard
污水水质分析 analysis of wastewater quality
污水水质监测 water quality monitoring of contaminated water
污水水质物理化学试验 physical and chemical examination of sewage
污水水质指标 wastewater quality index
污水水质指数 wastewater quality index
污水税 wastewater fees
污水隧道 sewage tunnel
污水隧洞 sewer tunnel
污水塘 lagoon; sewage lagoon; sewage pond
污水特性 characteristics of sewage; sewage characteristic; wastewater characteristic
污水特征 characteristics of sewage; sewage characteristic; wastewater characteristic
污水提升泵站 sewage lifting pump station
污水提升器 sewage ejector
污水提升站 sewage(re) lift station
污水提升装置 sewage raising plant
污水提送设备 Autaram
污水田 sewage farm
污水条件 sewage condition
污水条例 sewer ordinance
污水调节池 wastewater equalization pond
污水铁格栅 bar sewage screen
污水通道 foul water gallery; foul water tunnel
污水通风 sewage ventilation
污水通量 effluent flux
污水通气管 soil ventilation pipe
污水通气管路 venting loop
污水同化 wastewater assimilation
污水同化能力 wastewater assimilative capacity
污水同化容量 wastewater assimilative capacity
污水桶 slop pail
污水透气管回路 ventilator loop; venting loop
污水图 sewerage drawing
污水土地处理 sewage land treatment
污水土地处理系统 sewage disposal system by soil
污水土壤净化法 wastewater soil purification
污水脱色 sewage decolo(u)ration
污水瓦管 clay sewer pipe; sewer tile
污水完全二级处理 complete secondary sewage treatment
污水完全停留 complete sewage retention
污水网筛 mesh sewage screen; sewage screen
污水尾水处置费 cost of effluent disposal
污水稳定化 stabilization of sewage
污水稳定塘 wastewater stabilization pond
污水污泥 sewage sludge; wastewater sludge
污水污泥处理 sewage sludge disposal; sewage sludge treatment
污水污泥处置 sewage sludge disposal
污水污泥焚化 sewage sludge incineration
污水污泥焚烧 sewage sludge incineration
污水污泥干化 sewage sludge drying
污水污泥干化床 sewage sludge drying bed
污水污泥干燥床 sewage sludge drying bed
污水污泥培养 sewage sludge culture
污水污泥气体 sewage sludge gas
污水污泥清除 sewage sludge dumping
污水污泥施加负荷 sewage sludge application
污水污泥淘洗 sewage sludge washing
污水污泥脱水 dewatering of sewage sludge
污水污泥洗涤 sewage sludge washing
污水污泥消化 sewage sludge digestion
污水污染 polluted by sewage; sewage contamination; sew(er)age pollution
污水污染物 wastewater pollutant
污水污染物排放标准 wastewater pollutant discharge standard
污水污染值 sew(er)age pollution value
污水污染指示物 indicator of sewage pollution
污水物化处理 physico-chemical sewage treatment; physico-chemical treatment of wastewater
污水物化处理法 physico-chemical method of sewage treatment
污水物理分析 physical analysis of sewage
污水物理化学处理 physico-chemical sewage treatment
污水吸附法 adsorption process of wastewater
污水稀释 dilution of sewage; effluent dilution; wastewater dilution
污水系统 plumbing system; sewage stream; sewerage; sewage system; sewer(age) system
污水系统分布 sewage distribution
污水系统附属设备 sewer appurtenances
污水细菌 sewage bacteria
污水下水道 foul sewer
污水线网 sewer distribution line
污水消毒 disinfection of sewage; sewage disinfection; sewage sterilization; wastewater disinfection
污水消毒剂 wastewater disinfectant
污水消化塔 sewage digestion tower
污水消化作用 sewage digestion
污水硝化-反硝化脱氮处理 nitrogen removal from wastewater by nitri-denitrification
污水硝化脱氮处理 nitrogen removal from wastewater by nitrification
污水泄放管 relief sewer
污水泄水区 sewage outlet area
污水需氯量 chlorine demand of sewage
污水需氧量 effluent oxygen demand
污水窨井 sewer manhole; sewer manway
污水循环 recirculation of sewage
污水循环泵 sewage recirculating pump; wastewater recirculating pump
污水压气喷射器 sewage ejector
污水厌氧处理 anaerobic sewage treatment
污水厌氧生物处理 biologic(al) anaerobic treatment of sewage
污水养鱼 fish-farming with sewage; sewage fish culture
污水养鱼塘 sewage fish pool
污水养殖 sewage farming; wastewater aquaculture
污水氧化 sewage oxidation
污水氧化池 sewage oxidation pond
污水氧化处理 oxidized sewage
污水氧化沟 oxidation ditch
污水氧化塘 sewage lagoon; sewage oxidation basin; sewage oxidation pond; wastewater lagoon
污水一级处理 lower class treatment of wastewater; primary sewage treatment; primary treatment of sewage; primary wastewater treatment
污水溢出物 sewage effluent
污水引水池 sewage intake basin
污水蝇 sewage fly
污水油脂 sew(er)age grease
污水有机毒化学物浓度 effluent toxic chemical concentration
污水有机质 effluent organic matter
污水淤渣 sewage grit
污水预沉池 sewage preliminary basin; sewage preliminary tank
污水预处理 sewage pretreatment; wastewater pretreatment
污水预曝气 preaeration of sewage; preliminary aeration of wastewater
污水源 sources of wastewater; wastewater sources
污水远海排污口 long sea outfall
污水运输车 sewage lorry
污水再利用 effluent reuse
污水再生法 water renovation process
污水再生技术 wastewater regeneration engineering
污水再循环系统 wastewater recycling system
污水藻类净化 wastewater purification by algae
污水藻类生长处理 wastewater treatment by algal growth
污水渣抽出器 sewage solid extractor
污水渣滓抽出器 sewage solid extractor
污水闸阀 sewage sluice valve
污水栅栏 log-trash boom
污水真菌 sewage fungus
污水征收法令 wastewater tax law
污水征收费 wastewater charges
污水支沟 branch sewer
污水支管 branch sewer; collecting sewer; lateral sewer; sewer branch; soil branch
污水直接灌溉 direct irrigation
污水治理工程 sewerage implementation project
污水滞留沉淀池 sewage detention tank
污水滞留池 wastewater retention reservoir; detention tank
污水中除去砂砾 grit removal from sewage
污水中的粒子 sewage particulate
污水中的污泥 sewage sludge
污水中易爆物质 explosive substance in sewage
污水重复利用 sewage water recycling
污水重力分离 gravity separation of wastewater
污水主管 main(sanitary) sewer; trunk sewer
污水贮池 cesspit; cesspool
污水注入井 injection well
污水装置 sewage facility
污水状况 sewage condition
污水浊度 effluent turbidity
污水资源 wastewater resources
污水资源化 resourcelization of wastewater
污水自净作用 natural purification
污水综合处理 integrated sewage treat-

W

ment; integrated wastewater treatment
污水综合处理厂 integrated wastewater treatment plant
污水综合管理 comprehensive wastewater management
污水综合管理规划办法 comprehensive wastewater management planning process
污水综合排放标准 integrated wastewater discharge standard
污水综合排放一级标准 first grade of integrated wastewater discharge standard
污水总管 main sewer; sewage conduit; sewer main; trunk sanitary sewer; trunk sewer
污水组成 composition of sewage; wastewater composition
污水组分 composition of sewage; wastewater composition
污水最终沉淀池 sewage final settling basin; sewage final settling tank
污水最终沉降槽 contaminated oil settling tank; sewage final settling tank
污水最终沉降池 sewage final settling basin; sewage final settling tank
污损 blur(ring); contamination damage; defile; fouling; stain
污损本 soiled copy
污损复合体 fouling complex
污损膜 fouling film
污损群落 fouling community; fouling complex
污损生物 fouling organism
污土 muck soil
污物 dirt; fecula; feculence [feculency]; garbage; muck; off-scourings; smut(ch)
污物层 layer of dirt
污物处置 trash disposal
污物电梯 soiled lift
污物堆积 collecting dirt
污物焚化处理系统 crematory system of sewage disposal
污物滤池 dirt filter
污物滤器 dirt filter
污物排气管 soil and vent pipe
污物设备 soil fitments; soil fittings
污物桶 litter basket
污物下沉 down wash
污洗池篦子 sink grating
污洗池底板 sink mat
污洗池排水挡板 sink drain board
污下水道隧洞 sewer tunnel
污液 soiling solution
污液处理程序 slops disposal procedures
污液配方 soiling formula
污以煤烟 soot
污油 adulterated oil
污油泵 sump waste oil pump
污油槽 dirty oil tank
污油池 contaminated-oil basin
污油处理 effluent oil treatment
污油处理设备 effluent oil treatment equipment
污油管 contaminated oil pipe
污油柜 dirty ballast; sludge collecting tank; sludge tank
污油回收 effluent oil recovery
污油回收槽 oil skimming tank
污油回收船 oil skimmer; oil-spill collect vessel
污油净化装置 oil reconditioner
污油水舱 slop tank
污脏货物 dirty cargo
污浊 filthy; foul
污浊冰(夹)层 dirt band
污浊的 stagnant
污浊负荷量 pollution loading
污浊环境 sweated environment
污浊货作业津贴 dirt money
污浊空气 dirty air; foul air; leaving air; polluted air; rank(ed) air; stale air; vitiated air
污浊空气导管 foul air duct
污浊空气管道 foul air flue
污浊空气排出管 foul air duct
污浊排泄气 vitiated expired air
污浊物 foulant; fouling product
污浊物增溶剂 foulant solubility
污浊下水 foul sewer
污浊下水道 foul drain
污着 fouling
污着群落 fouling community
污着生物 fouling organism
污着生物群落 fouling organism community
污着速度 fouling velocity
污着物 fouling product
污渍 blemish; blot; smudge
污渍处理 blotter treatment

诬告 false accusation

钨靶 tungsten target

钨板 tungsten sheet
钨棒 tungsten rod
钨铋矿 russellite
钨的 tungstic; ungstenic
钨灯 tungsten lamp
钨灯丝 tungsten filament
钨电极 tungsten electrode; tungstic electrode; wolfram electrode
钨电极弧 tungsten arc
钨电极气体保护焊接法 gas tungsten-arc welding method
钨电阻加热元件 stratit element
钨发射体 tungsten emitter
钨粉 tungsten powder
钨覆盖层 tungsten coating
钨钢 Mushet('s) steel; tungsten steel; wolfram steel
钨钢牙钻 tungsten bur
钨钢钻头 tungsten drilling bit
钨膏灯丝 pasted filament
钨铬钢 tungsten chrome steel; tungsten-chromium steel
钨铬工具钢 tungsten-chromium tool steel
钨铬钴(硬质)合金 Stellite
钨工具钢 riffel steel
钨钴合金 tungsten-cobalt
钨钴钛系硬质合金 prolite
钨钴硬质合金 Igatalloy
钨硅酸 tungstosilicic acid
钨焊条电弧焊 tungsten arc welding
钨合金 tungsten alloy
钨弧焊 tungsten arc welding
钨弧熔化 tungsten arc melting
钨华 tungstic ocher; tungstite; wolframine; wolfram ocher
钨黄铜 tungsten brass; wolfram brass
钨基合金 heavy alloy
钨极 tungsten electrode
钨极电弧焊 tungsten arc welding
钨极电弧切割 tungsten arc cut
钨极惰性气体保护电弧焊 tungsten inert gas arc weld; tungsten inert gas welding
钨极惰性气体保护弧焊机 tungsten inert gas welding machine
钨极脉冲氩弧焊 pulsed tungsten argon arc weld(ing)
钨极气体保护焊 gas tungsten-arc welding
钨极氩弧焊 argon tungsten-arc welding
钨夹 tungsten clip
钨接点 tungsten contact; tungsten point
钨矿(石) tungsten ore
钨矿物 tungsten minerals
钨矿异常 anomaly of tungsten ore
钨粒 tungsten grain; tungsten particle
钨卤灯泡 tungsten halogen lamp
钨螺旋线 tungsten helix
钨镁酸铅陶瓷 lead magnesio-tungstate ceramics
钨锰矿 hu(e)bnerite
钨钼钢 tungsten-molybdenum steel
钨钼热电偶 tungsten-molybdenum thermocouple
钨钼酸系颜料 tungstomolybdic pigment
钨钼族元素 tungsten and molybdenum group elements
钨镍合金 tungsten nickel
钨盘 tungsten disc [disk]
钨铅矿 scheelitine; stolzite
钨青铜 tungsten blue; tungsten bronze; wolfram bronze
钨青铜结构 tungsten bronze structure
钨青铜型结构 tungsten bronze type structure
钨熔丝 tungsten fuse
钨砂 wolfram ore
钨生产商协会 Association of Tungsten Producers
钨丝 tungsten filament; tungsten wire; wolfram filament
钨丝白炽灯 tungsten lamp
钨丝灯(泡) osram lamp; tungsten lamp; tungsten bulb; tungsten filament lamp; tungsten light
钨丝电子管 bright emitter valve
钨丝弧光灯 tungsten arc lamp
钨丝加热器 tungsten heater
钨丝灵敏度 tungsten sensitivity
钨丝炉 tungsten coil furnace; tungsten wire furnace
钨丝卤化灯 tungsten halogen lamp
钨丝阴极 tungsten cathode
钨丝照明 tungsten lighting
钨酸 tungsten salt; tungstic acid; wolframic acid
钨酸铵 ammonium tungstate
钨酸钡 barium tungstate; barium white; barium wolframate
钨酸钙 artificial scheelite; calcium tungstate; calcium wolframate
钨酸镉荧光颜料 cadmium tungstate
钨酸根 tungstate radicle
钨酸钾 potassium tungstate; potassium wolframate
钨酸镁 magnesium tungstate
钨酸锰矿 manganese tungstate
钨酸钠 natrium wolframate; sodium tungstate; sodium wolframate
钨酸铅 lead tungstate; lead wolframate; plumbous tungstate
钨酸盐 tungstate; wolframate
钨酸盐玻璃 tungstate glass
钨酸盐基水处理 tungstate based water treatment agent
钨酸盐类 tungstates
钨酸盐陶瓷 tungstate ceramics
钨钛铬钴高合金高速钢 Tizit
钨钛碳化物 tungsten-titanium carbide
钨钛硬质合金 tungsten-carbide-titanium carbide-cobalt alloy
钨锑贝塔石 scheteligite
钨锑金石英脉矿床 tungsten-antimony-gold-quartz vein deposit
钨条 tungsten rod
钨铁 ferrotungsten; tungsten iron
钨铁矿 ferberite
钨铜 tungsten copper
钨铜电接触器材 tungsten-copper contacts
钨铜复合物 tungsten-copper composite
钨铜合金 <焊条用合金> Elconite
钨铜镍电接触器材 tungsten-copper-nickel contacts
钨铜镍铝合金 Romanium
钨铜镍制品 tungsten-copper-nickel composition
钨铜制品 tungsten-copper composition
钨钍铷氧化物 rubidium thorium tungsten oxide
钨锡矿化探 geochemical exploration for tin and tungsten
钨锡铍铌钽及稀土矿床模式 model of tungsten-tin-beryllium-niobium-tantalum and rate earth deposit
钨系硬质合金 tungalloy
钨锌矿 sanmartinite
钨氩充电器 tungar charger
钨氩管 tungar
钨氩管(吞加)充电机 tungar charger
钨氩管整流器 tungar rectifier; Rectigon rectifier <其中的一种>
钨氩整流管 tungar bulb
钨铀华 uranotungstite
钨赭石 tungstic ocher
钨质电接触器材 tungsten contact
钨质须簧 tungsten whisker
钨铸铁 tungsten cast iron

屋 house

屋板嵌灰泥 torching
屋椽式支护法 rafter timbering
屋顶 building roof; covering; house top; roof(ing); roof-top; dreadmought <片状沥青和预制浮石块的>
屋顶柏油 roofing flux
屋顶板 roof board(ing); roof deck(ing); roofing slate; roof slab; shingle
屋顶板背面沥青涂层 back coat(ing)
屋顶板层 shingle layer
屋顶板条 roof lathing
屋顶板岩石 roof stone
屋顶半桁架 <四坡屋顶的> jack truss
屋顶保温 roof insulation; thermal insulation of roof
屋顶保温空气层 air-space for roof insulation
屋顶保温填充料 insulating roof fill
屋顶崩落 roof fall
屋顶避雷线 roof conductor
屋顶避雷针 roof conductor
屋顶边泛水 head flashing
屋顶边缘 roof edge
屋顶便道 captain's walk; roof walk
屋顶标牌 roof sign
屋顶标志 roof sign
屋顶玻璃窗 roof glazing
屋顶玻璃装配 <不用油灰的> dry roofing glazing
屋顶薄钢板 roofing sheet
屋顶擦窗系统 roof-mounted window washing system
屋顶材料 <茅草、棕榈等> thatch
屋顶采光 daylighting design principle; roof lighting
屋顶采光窗 dormer
屋顶层 attic floor; attic stor(e)y; garret; garret gap-graded; half stor(e)y
屋顶层格栅 attic joist

屋顶层横梁 garret beam;loft beam
屋顶层楼面 roof deck(ing)
屋顶层面 roof deck(ing)
屋顶衬套 roof bushing
屋顶池 roof pond
屋顶抽风机 roof extract ventilator
屋顶出入孔 roof hatch;roof scuttle
屋顶出水口 roof outlet
屋顶出烟口 <古希腊或罗马的> opaion
屋顶除潮 humidity removal from roofs
屋顶窗 dormer window;exit opening;eye brow;lucarne;roof dormer;roof light;shed dormer;skylight window
屋顶窗侧壁 dormer cheek
屋顶存车场 roof-top car park
屋顶大梁 roof girder
屋顶挡雪板 roof guard
屋顶挡烟隔板 draft curtain;smoke curtain
屋顶稻草 roof straw
屋顶的交接 roof intersection
屋顶的金属覆盖层 roof covering sheet-metal
屋顶灯光照明 pyramidal light
屋顶底面 roof soffit
屋顶电扇 attic fan
屋顶垫片 roof substrate sheeting
屋顶吊杆 roof jib
屋顶吊钩 roof hook
屋顶端边 verge
屋顶断路器 roof disconnector
屋顶泛水 roof flashing
屋顶防护板 roof guard
屋顶防水层 roof covering
屋顶房间 appentice;attic room;pendice;penthouse;solarium [复 solaria/solariums]
屋顶风机 attic fan;roof ventilator
屋顶风帽 roof ventilator
屋顶封顶 roof seal(ing)
屋顶富丽装饰 roof enrichment
屋顶覆面油毛毡 asphalt felt roof covering
屋顶盖板 roof sheathing
屋顶盖瓦 tegula [复 tegulae]
屋顶高度 roof level
屋顶高跨比 pitch of a roof;roof pitch
屋顶高跨比角度 angle of roof pitch
屋顶格栅 roof joist
屋顶隔火屏 roof curtain
屋顶隔火山墙 roof screen
屋顶隔热 roof heat insulation;roof insulation;thermal insulation of roof
屋顶隔热板 roof insulation board
屋顶隔热层 roofing isolating layer;roof insulation
屋顶工程 roof work
屋顶工工作 roofer work
屋顶工用的钉子 roofer nail
屋顶工作 roof work
屋顶公寓 penthouse;penthouse apartment
屋顶拱 roof arch
屋顶拱腹 roof soffit
屋顶钩住铅板的钩子 thumb screw
屋顶钩状物 roof hook
屋顶构件 roof component;roof member
屋顶构造 roof construction
屋顶骨架 carcase roof(ing)
屋顶管道(系统)roof plumbing
屋顶锅炉房 roof boiler room
屋顶荷载 roof load(ing)
屋顶桁架 roof truss
屋顶桁架的主拉杆 main tie of roof truss
屋顶桁架的柱环与椽子的接棒 carpenter's boast
屋顶桁条 purlin(e) brace
屋顶花架 roof pergola;roof trellis
屋顶花样培植 roof garden planting
屋顶花园 roof gardening
屋顶花园餐厅 roof garden restaurant
屋顶花园酒吧 roof garden bar
屋顶花园绿化 roof garden planting
屋顶花园平台 roof-top terrace garden
屋顶缓斜的 low-pitched
屋顶活荷载 roof live load
屋顶活门 trap door on roof
屋顶机场 roof aerodrome;roof airdrome
屋顶机房 mechanical penthouse;penthouse
屋顶机械抽气机 mechanical roof extractor
屋顶基层片材 roof base sheeting
屋顶脊谷用瓦 flap tile
屋顶尖塔 roof spire
屋顶监测器 roof monitor
屋顶检查孔 inspection hole on roof
屋顶建造人员 roofer
屋顶胶贴物质 dressing composition
屋顶脚手架 roofer's scaffold(ing);roof staging
屋顶结构 roof structure
屋顶结构层找坡 roof structure to falls
屋顶金属网 roof lathing
屋顶景观 roofscape
屋顶卷材 storm sheet
屋顶绝热填料 roof fill insulation
屋顶开孔房屋 hypaethral
屋顶壳体 roof shell
屋顶空间 roof space
屋顶孔隙 roof void
屋顶跨度 roof span
屋顶框架 roof framing
屋顶框架结构 three-dimensional area-covering structure
屋顶肋条 roof rib
屋顶类型 roof type;type of roof
屋顶沥青保护层 coating asphalt of roof
屋顶梁 roof beam;roof girder
屋顶瞭望所 watching loft
屋顶瞭望台 covered spectator's stand;window's walk
屋顶檩条 roof purlin(e)
屋顶龙骨 roof joist
屋顶龙骨找坡 roof joists laid to falls
屋顶楼梯间 stair turret
屋顶绿化 roof greening
屋顶卵石 roof gravel
屋顶轮廓 roof contour
屋顶轮廓线 roof line
屋顶落水沟 roof gutter
屋顶茅草 roof straw
屋顶锚杆 roof bolt
屋顶锚固 roof anchorage
屋顶锚瓦条 close board
屋顶密封条 roof sealing strip
屋顶面 roof floor;roofing;roof sheathing
屋顶面层 roof coating
屋顶面墙 flying facade
屋顶木板 roof plank;shingle applicator;shingle cover(ing)
屋顶木板吊挂 shingle hanging
屋顶木板切断机 shingle cutter
屋顶木工活 roof woodwork
屋顶木条 roofing batten
屋顶木瓦 shingle applicator;shingle cover(ing)
屋顶木瓦板 roofing shingle
屋顶木瓦层 shingle layer
屋顶木瓦吊挂 shingle hanging
屋顶木瓦切断机 shingle cutter
屋顶木檐槽 wooden gutter
屋顶木雨水槽 wooden gutter
屋顶内衬 roof lining
屋顶内排水 interior drainage
屋顶黏[粘]合料 jew solder
屋顶女儿墙板 roof parapet slab
屋顶女儿墙通道 vamure
屋顶爬梯 roof ladder
屋顶排尘器 roof extract unit
屋顶排风机 roof extract unit
屋顶排风扇 roof fan
屋顶排气口 femerell;roof vent
屋顶排气笼 femerell
屋顶排气装置 roof exhaust unit
屋顶排水 roof drainage
屋顶排水暗沟 secret valley
屋顶排水沟 roof drain;roof gutter;roof valley;valley
屋顶排水沟的截面 section of a valley
屋顶排水口 roof drain
屋顶排水天沟 cullis
屋顶排烟窗 smoke and fire vent
屋顶排烟口 smoke and fire vent
屋顶跑道 roof trackway
屋顶喷射降温器 roof spray
屋顶平面 roof plan
屋顶平台 roof deck(ing);roof terrace
屋顶平台板 roof decking panel
屋顶平台薄板 roof decking sheeting
屋顶平瓦 plain roof tile
屋顶坡槽间隙盖板 gusset piece
屋顶坡度 angle of roof;fall of roof;inclination of roof;pitch of roof;roof pitch;roof slope;slope of roof
屋顶坡度陡的 high-pitched
屋顶坡度角 angle of roof pitch;angle of roof
屋顶剖面 roof cross-section
屋顶铺盖工 roof roofer
屋顶铺面集料 roof aggregate
屋顶铺石板瓦 roof slating
屋顶铺瓦 roof tiling;tiling
屋顶曝晒架 caisson
屋顶曝晒箱 caisson
屋顶漆 roof coating;roof(ing) paint
屋顶起重机 roof crane;roof hoist
屋顶气窗 bird house
屋顶铅垂器 roof plummet
屋顶戗脊 close-cut hip
屋顶墙面板涂料 shingle stain
屋顶倾斜角 angle of roof
屋顶穹隆 roof dome
屋顶热 roof heat
屋顶软木 roof cork
屋顶软木板 corkboard for heating lagging;corkboard for heat insulation
屋顶软木用板 cork slab for roof(ing)
屋顶散步地坪 roof-top promenade
屋顶散热器 cabane radiator
屋顶商店 roof shop
屋顶上梁庆宴 roof wetting party
屋顶上人孔 roof hatch
屋顶上小钟楼 bell cot
屋顶设备 roof-top equipment
屋顶设计 roof(ing) design
屋顶伸缩缝 roof expansion joint
屋顶伸缩缝盖板 roof expansion joint cover
屋顶升降机房 elevator penthouse
屋顶石板 slate board(ing)
屋顶石板瓦 <10 英寸 ×20 英寸> countess slate
屋顶式坝 roof weir
屋顶式筏道 bear-trap drift-chute
屋顶式放筏槽 bear-trap drift-chute
屋顶式空调器 roof-top air conditioner;roof-top air conditioning unit
屋顶式冷凝器 roof condenser
屋顶式路拱 roof-shaped crown
屋顶式堰 roof weir
屋顶式堰闸门 roof weir gate
屋顶式闸门 bear-trap gate;roof gate
屋顶式闸门槛 bear-trap sill
屋顶室 attic;sky parlor;tallet [tallot]
屋顶竖窗 luthern
屋顶水池 attic tank;roof tank
屋顶水泥 roof cement;jew solder <俚语>
屋顶水箱 attic tank;roof tank
屋顶塌落 roof collapse;roof fall
屋顶塔楼 fly tower;gazebo [复 gazebo(e)s]
屋顶太平走道 fire escape roof(walk) way
屋顶套管 roof bushing
屋顶梯吊钩 roof ladder hook
屋顶天窗 deck dormer;femerell;roof scuttle;trap door on roof;dormant window;dormer
屋顶天窗玻璃 pane of a roof
屋顶天窗玻璃格条 invincible
屋顶天窗盖板 roof dormer covering
屋顶天窗开口 roof-light opening
屋顶天窗配件 roof-light fittings
屋顶天沟 neck gutter
屋顶天沟挡雪板 gutter board
屋顶天线 loft antenna;overhouse aerial;roof aerial;roof antenna
屋顶天线杆 roof pole
屋顶铁覆盖层 iron roof cover(ing)
屋顶铁皮 roofing iron
屋顶亭子 mirador
屋顶庭园 roofscaping
屋顶停车场 car parking roof;roof parking area;roof(-top) car park;roof parking
屋顶停车坪 roof parking deck
屋顶通风 roof ventilation
屋顶通风窗 attic window;dormer-ventilator
屋顶通风窗口 dormer-ventilator opening;femerell
屋顶通风管 ventilation pipe on roof
屋顶通风机 roof ventilator
屋顶通风空心砖 attic vent block
屋顶通风孔 attic vent opening
屋顶通风口 roof terminal line
屋顶通风砌块 roof vent block
屋顶通风器 attic ventilator;roof extract ventilator;roof ventilator
屋顶通风器砌块 roof ventilator block
屋顶通风砖 roof ventilating tile
屋顶通风装置 attic ventilator tile
屋顶通气窗 ventilating eyebrow
屋顶通气孔 bird house;fumerell
屋顶通气装置 roof vent
屋顶透明板 roof-light sheet
屋顶透气孔 edge venting
屋顶凸饰 roof boss
屋顶突起物周边的泛水 head flashing
屋顶涂料 dressing paint;roof paste
屋顶托架 roof bearer;roof cradle
屋顶托梁 roof joist
屋顶瓦 roof(ing) tile
屋顶瓦下油毡 underslating felt
屋顶外围层 roof surround
屋顶网格 roof pane
屋顶网球场 roof-top tennis court
屋顶望板木材 roof sheathing lumber
屋顶桅杆 roof mast
屋顶维修保证书 roofing bond
屋顶位移 roof displacement
屋顶系统 roof system
屋顶下部结构系统 decking system
屋顶下储藏空间 loft
屋顶下封闭空间 blind attic
屋顶下面 underneath the roof
屋顶下热烟气隔板 draft curtain
屋顶线 roof line

屋顶线套管 roof bushing
屋顶消防栓 roof hydrant
屋顶小车 roof car
屋顶小车驱动的工作平台 roof-powered platform
屋顶小房间的入口 entrance roof overhang
屋顶小桁架 jack truss
屋顶小花园 alameda;roof garden
屋顶小梁 roof joist
屋顶小楼 penthouse
屋顶斜度 inclination of roof;pitch of roof;roof inclination
屋顶斜沟 close-cut valley
屋顶斜沟(排水槽)底板 valley board
屋顶斜角 angle of roof
屋顶信道 trapdoor
屋顶形的(建筑物) tectiform
屋顶形式的决定 determination of the roof shape
屋顶形状 roof shape
屋顶修理 roof repair
屋顶悬吊的荷载 roof-hung load
屋顶悬挂螺栓 roof suspension bolt
屋顶悬挑 roof cantilever;roof overhang
屋顶压力 roof pressure
屋顶烟囱 chimney stack;roof chimney
屋顶烟囱套筒 thimble
屋顶烟雾排出口 smoke scuttles
屋顶岩板 healing stone
屋顶檐沟采暖 roof gutter heating
屋顶阳台 roof terrace
屋顶以上的隔火墙 parapet
屋顶易碎,穿越小心! caution, fragile roof
屋顶用单层织物 single-layered fabric for roofing wall
屋顶用非金属柔性薄板 roofing felt
屋顶用空心黏[粘]土砖 clay pot for tiled roofs
屋顶用木料 roof timber
屋顶用砂浆 roofer's mortar
屋顶用油毛毡 sheathing paper
屋顶用直角尺 roofing square
屋顶(油)毛毡 roofing felt
屋顶油纸 roofing paper
屋顶游泳池 roof-top swimming pool
屋顶有盖出口 roof hatch
屋顶预制板 roof deck(ing)
屋顶载重 roof load
屋顶毡 roofing felt
屋顶找平面层 roof screed topping
屋顶整装空调器 roof-top type packaged air conditioner
屋顶支柱 roof column;roof stanchion
屋顶织布 roofing fabric
屋顶直升飞机停机场 roof helicopter airport;roof heliport
屋顶直升机降落场 roof-top heliport
屋顶纸板 roofing shingle
屋顶钟 tower clock
屋顶装饰 cresting;roof decoration;roof ornament
屋顶装饰部件 roof decorative feature
屋顶装饰处理 roof ornamental finish
屋顶装饰加工 roof decorative finish
屋顶装饰特点 roof decorative feature
屋顶装饰特征 roof ornamental feature
屋顶状的 tectorial
屋顶状堰 bear-trap weir
屋顶自然通风 fresh-air supply to roof
屋顶纵梁系 roof beam grillage
屋顶纵向支撑 vertical roof bracing
屋顶走道 roof walkway;widow's walk
屋顶走廊 roof gallery
屋顶作业防滑板 crawling board
屋盖 roof
屋盖板夹片 deck clip
屋盖构造 roof construction;roof framing
屋盖构造体系 roof framing system
屋盖网格板 lattice plate
屋盖支撑 roof bracing
屋盖周边气孔 edge vent
屋谷槽 valley gutter
屋谷黏[粘]土瓦 roof valley clay tile
屋谷瓦 roof valley tile
屋谷柱 roof valley post
屋桁 ridger girder
屋桁架小柱 princess
屋后(侧)服务性道路 rear service road
屋后(侧)辅助道路 rear service road
屋基 toft;tort
屋基石 foundation stone
屋脊 comb;fastigium;hip;ridge;roof ridge;fust <古称>;pien(d) <苏格兰>
屋脊鞍形脚手架 saddle scaffold
屋脊板 ridge board;ridge plate;ridging;saddle board;ligger <茅屋的>
屋脊抽气机 ridge extractor
屋脊抽气通风 ridge extract ventilation
屋脊窗 hip dormer
屋脊大梁 ridge girder
屋脊泛水 ridge stop
屋脊泛水片 ridge flashing sheet
屋脊防水铅皮 ridge stop
屋脊盖 ridgecap;ridge capping piece
屋脊盖板 hip sheet
屋脊盖条 hip capping
屋脊盖瓦 angle tile;arris tile;bonnet hip tile;bonnet tile;hip capping;hip tile;ridge capping
屋脊高度 ridge height
屋脊挂瓦钢条 hip hook
屋脊挂瓦铁 hip iron
屋脊节点 peak joint
屋脊卷筒形装饰 hip roll
屋脊卷形装饰 ridge roll
屋脊联结瓦 bullet roofing tile;ridge junction tile
屋脊梁 ridge beam;ridge piece;ridge pole;roof tree
屋脊梁的垂直锯口 ridge cut
屋脊梁上加钉的木条 ridge batten
屋脊梁系统 system of rafters
屋脊梁下支柱 post under ridge pole
屋脊檩条 ridge purlin(e)
屋脊木瓦 hip and ridge finishing pieces
屋脊披水铅皮 ridge stop
屋脊铅皮卷筒 hip roll;ridge roll
屋脊石 saddle stone
屋脊式路拱 gable crown
屋脊式天窗 ridge-type rooflight;ridge-type skylight
屋脊饰 ridge crest(ing);roof comb;roof crest(ing)
屋脊饰板 crest board
屋脊饰瓦 crest tile
屋脊水沟 ridge-runner
屋脊通风机 ridge ventilator
屋脊通风孔 ridge vent
屋脊筒瓦 tarus
屋脊瓦 bonnet hip tile;bonnet tile;book tile;crest tile;ridge tile
屋脊瓦层 ridge course
屋脊弯盖瓦 cone tile
屋脊弯瓦 bonnet hip tile
屋脊系杆 ridge binder
屋脊线 ridge;ridge line
屋脊线脚 hip mo(u)ld(ing)
屋脊小塔 ridge turret
屋脊形棱镜 roof prism
屋脊形式 ridge form
屋脊型 roofing
屋脊型烟羽 roofing plume
屋脊压顶片 ridge capping
屋脊压盖 hip capping
屋脊折叠缝 <金属板屋面的> ridge folding
屋脊止水铅皮 ridge stop
屋脊至屋脊间折叠缝 ridge-to-ridge folding
屋脊至屋面沟折叠缝 ridge-to-valley folding
屋脊装饰 roof crest(ing)
屋脊状的 roof-top
屋脊状支架 saddle-back
屋架 frame of roof;principal;rafter set;ridge framing;roof bent;roof frame;roof framing;roof principal;roof truss
屋架板 roof board(ing);roof slab
屋架的金属环箍 roof frame strap
屋架辅助杆件 secondary truss member
屋架覆盖 roof covering
屋架高跨相等时屋面坡度 full pitch
屋架桁架 principal
屋架横木 <古罗马建筑> transtrum [复 stranstra]
屋架间距 roof span
屋架间抗风剪刀撑 wind race
屋架间抗风力剪刀撑 wind brace
屋架结构 roof framing
屋架跨度 span of roof truss;span of truss;roof span
屋架拉杆 roof tie
屋架拉梁 footing beam;roofing beam
屋架拉条 collar tie
屋架木料 truss timber
屋架上弦杆 principal rafter
屋架水平撑条 batten of roof truss
屋架屋顶结构 lamellar roof structure
屋架系杆 stay piece
屋架系统 roof system
屋架斜撑 brace of roof truss;principal brace;strut of roof truss
屋架验收 acceptance of the carcass
屋架支托 roofing bracket
屋架中柱 crown post
屋架主撑 principal brace
屋角的尘埃 fug
屋角丁头砖 quoin header
屋角顶石 quoin bondstone
屋角顶砖 quoin bonder
屋角仿石(砌)体 block quoin
屋角构件 quoin
屋角石 quoin stone
屋角石块 angle quoin;angle tile;coin;quoin;stone angle quoin
屋角石块墙砌合 quoin bonding
屋角石镶边 return head
屋角束石 quoin bondstone
屋角斜瓦面 swept valley
屋角隅石 quoin
屋角砖 quoin;quoin brick;quoin header
屋角桩 quoin header
屋里的人 occupant
屋面 building roof;coin;roof;roof covering;top coat
屋面柏油 roofing asphalt
屋面柏油沥青 roofing asphaltic bitumen
屋面柏油纸 roof sheathing paper
屋面板 roof board(ing);roofing slab;roof panel;roof sheathing;roof slab;sheathing board;sheet roofing;shide;shingle;shingle panel
屋面板材 roof(ing) sheet
屋面板的凸露边 butt
屋面板钉 shingle nail
屋面板交接处天沟 valley
屋面板扭曲 tin-canning
屋面板起拱 tin-canning
屋面板上黑色砂砾条 top shadow line
屋面板太阳能收集器 solar collector roof panels
屋面板特性 characteristics of shingles
屋面板条 roof sheathing
屋面板瓦 shingle tile
屋面板下半嵌灰泥 shouldering half-torching
屋面板用钉 roof nail;slater's nails
屋面保护层 protected membrane;roof preservation coat;roof covering
屋面保护物 roof(ing) preserver
屋面保护装置 roof preserver
屋面保温层 roof insulation
屋面避雷网 air termination network
屋面避雷装置 roof conductor
屋面边饰瓦 roof curb tile
屋面边缘 roof curb
屋面边缘栏 deck curb
屋面边缘瓦 roof curb tile
屋面表层 roofing skin
屋面冰块挡板 snow board
屋面玻璃 drop dry
屋面薄钢板 light-ga(u)ge sheet;roofing iron
屋面薄膜 roof(ing) membrane
屋面薄膜防水体系 roof membrane system
屋面薄膜防水系统 roof membrane system
屋面薄铁皮 roofing iron
屋面材料 roofage;roof cladding;roofing application;roof(ing) material;roofing(product);bonded roof <厂商具保的>
屋面材料外露面积 exposed area of roof materials
屋面层骨料 roof aggregate
屋面衬板 sarking;sarking board
屋面衬垫油毡 underlining felt of roof
屋面衬毡 sarking felt;sarking felt underlining felt
屋面承包商 roofing contractor
屋面椽条 roof rafter
屋面挡水条 roofing waterstop
屋面吊 roof crane
屋面钉 slating nail
屋面泛水 apron flashing
屋面方块钢板 roofing square
屋面防潮纸 roofing paper
屋面防腐剂 roofing preservative
屋面防护栏 roof guard
屋面防漏条带 strip soaker
屋面防水接缝 roofing waterproofing joint
屋面防水系统 roofing waterproofing system
屋面防雨板 roof flashing
屋面防雨方法 roof flashing method
屋面防雨片 roof flashing strip
屋面防雨装置 roof flashing block
屋面附件 roofing accessories
屋面附配件 roof accessories
屋面覆盖层 roof cladding;roof cladding element;roof covering
屋面盖板 roof plate;roof terminal;shingle;roof deck(ing)
屋面盖以预制沥青材料 roof covering with asphalt ready roofing
屋面高跨比 pitch of roof
屋面隔热板 roofing insulation panel
屋面隔热材料 roof insulation
屋面隔热层 roof insulation
屋面隔热条 roof insulating strip
屋面隔热找平层 roof insulating screed
屋面工 building roofer;roofer;roof slater

屋面工程 roofing work
屋面工程造价 roofing cost
屋面工人使用的木斧 shingle hatchet
屋面构造 construction of roof
屋面固定 roofing fastening
屋面固定附件 roofing fixing accessories
屋面固定构件 roofing fastener
屋面挂瓦条 roof(ing) batten
屋面管套 roof flange
屋面管凸缘套 boot
屋面荷载 roof load
屋面桁条 roof purlin(e)
屋面滑移 slippage
屋面活荷载 roof live load
屋面积水 roof ponding
屋面基础片材 roofing substructure sheeting
屋面浆料 roofing paste
屋面浇注 roof pour(ing)
屋面接缝 roof joint
屋面截水条 ridge terrace
屋面金属箔 roofing foil
屋面金属箔层 roof foil
屋面金属薄板 roofing sheet-metal
屋面金属材料 roofing metal
屋面金属防水条 counter flashing
屋面金属片材 roof sheathing metal sheet
屋面金属片互搭接头 drip joint
屋面卷板 coiled roofing sheet
屋面卷材 coiled roofing sheet; rolled roofing; rolled roofing material; rolled strip roofing; roofing felt; roofing foil; roof rolls; shingle roll
屋面卷材面上凸起的花纹 cedartex
屋面绝热 roofing insulation
屋面绝缘 roofing insulation
屋面绝缘板 roof insulation board
屋面绝缘层 roof insulation
屋面颗粒材料 roofing granules
屋面腊克 roof lacquer
屋面栏杆 roof railing
屋面立缝 standing seam; stand-up welt
屋面沥青 roofing asphalt; roofing bitumen; dead level <含量极少或不含硬焦油脂的>; bonding compound <冷铺或热铺>
屋面沥青卷材 asphalt roll(ed-strip) roofing
屋面沥青天沟 check-fillet
屋面沥青油毡 asphalt roofing felt
屋面沥青纸 roofing paper
屋面粒渣 roofing granules
屋面凉亭 roof pergola
屋面梁 roof beam; roof girder
屋面流淌 slippage
屋面琉璃瓦 roofing terracotta
屋面露明卷材 <多层油毡屋面面层> cap sheet
屋面炉渣 roofing slag
屋面铝材 roof alumin(i)um
屋面麻丝板 millboard roofing material; roofing millboard
屋面马口铁覆盖板 tinplate roof cover(ing)
屋面面积 roof area
屋面木基层 wooden upper roof framing
屋面木条 roof(ing) batten; roof lath
屋面黏[粘]结剂 roofing cement
屋面女儿墙 roof parapet wall
屋面排气孔 stack vent
屋面排水 roof drainage
屋面排水沟 roof drain
屋面排水管上承口 cistern head
屋面排水口 roof outlet
屋面排水坡找平层 drainage slope fill
屋面披水板 roof flashing
屋面披水方法 roof flashing method
屋面披水片 roof flashing strip
屋面披水装置 roof flashing block
屋面片材 roofing membrane
屋面平板 flat roofing sheet
屋面平面图 roof plan
屋面平石板 table slate
屋面平台撑木 deck cant
屋面坡度 roofing pitch; slope of roof
屋面坡度角 slope of roof pitch; slope of roof
屋面铺敷 roofing application
屋面铺砂 roofing sand
屋面铺砂器 primary surfacing hopper
屋面铺瓦工程 rooftiling work
屋面铺细粒机 Norwood blender
屋面铺细粒机上料斗 Norwood hopper
屋面铅皮 roofing lead sheet
屋面倾斜角 angle of roof
屋面热拌石屑黏[粘]合料 hot chip(ping)s-bonding roofing cement
屋面热黏[粘]合料 hot bonding roofing cement
屋面洒水 roof water spray
屋面散步廊 roof promenade
屋面砂砾 roofing granules
屋面上釉材料 aluminex
屋面设计 roofing design
屋面施工 roofing application
屋面施工用支架 roofing bracket
屋面石板 arris ways; muffet; muffity
屋面石板瓦 princess; roofing rags; roof(ing) slate
屋面水 roof water
屋面水泥 roofing cement
屋面顺槽中的拉条 astel
屋面藤架 roof pergola
屋面体系 roofing system
屋面天沟 center gutter; open valley; valley
屋面天沟瓦 roof valley tile; valley roof tile
屋面天然漆 roof lacquer
屋面通风井 roofing shaft
屋面铜片 roofing copper
屋面突出物 roof protrusion
屋面涂层 roof coating
屋面涂料 liquid roofing; roof coating
屋面托座 roofing bracket
屋面瓦 roof(ing) tile; roof shingle
屋面瓦工 roof tiler
屋面瓦工厂 roofing tile factory
屋面瓦检验机 roofing tile tester
屋面望板 sarking board
屋面维修 maintenance of roofs
屋面系统 roof system
屋面下部结构片材 roof substructure sheeting
屋面纤维质材料 roofing fabric
屋面箱形下水沟的铅衬里 lead-lining of box roof gutter
屋面小石板 lady
屋面小碎石 roofing chip(ping)s
屋面斜沟支承板 valley board
屋面斜排水沟 V-gutter
屋面斜坡 roofing pitch
屋面行业 roofing trade
屋面檐板 roof cornice
屋面阳台 roof promenade
屋面用沥青胶泥 asphaltic-bitumen roofing cement
屋面用螺钉 roofing nail
屋面用软金属皮 flexible metal for roofing
屋面用砂 roofing sand
屋面用水泥 cement for roof; bull <俚语>
屋面用瓦楞钢皮 Robertson
屋面用油毛毡 ruberoid
屋面油灰 roofing putty
屋面油毛毡 roof sheathing paper; underlining felt; tarred roofing felt; roofing felt
屋面油毛毡的裂片 shive
屋面油毛毡卷材 sheet-roofing felt
屋面油毛毡在浇沥青前铺撒的一层砂或石屑 underlay mineral
屋面油毡 black roofing felt; Durex; fabric fiber [fibre]; roofer felt; roofing malthoid; tarred roofing felt; roofing felt
屋面油毡层 interlayment
屋面油毡防水层 roofing membrane
屋面油毡工厂 roof felt manufacture
屋面油毡胶黏[粘]剂 black roofing adhesive; roofing felt adhesive
屋面油毡老化 ag(e)ing of roofing felt; deterioration of roofing felts
屋面油毡铺砂 sand rubbing
屋面油毡铺设工 roofing felt layer
屋面油毡压条 counter battens
屋面油毡制造 roof felt manufacture
屋面油纸 roofing paper
屋面游戏场 roof playground
屋面预抛光 drop dry
屋面毡 fabric for roofing; felted fabric for roofing; roofing mat
屋面找平材料 roof screed material
屋面支撑 roof bracing
屋面支撑系统 roof bracing system
屋面支承 roof support
屋面总高度 total rise of a roof
屋内布线 domestic wiring
屋内车库 built-in garage
屋内厨房 inside kitchen
屋内大梁 inner girder
屋内管道 interior piping
屋内管线布置 service layout
屋内楼梯斜梁 inner string
屋内排水管 service layout
屋内人行交通 internal pedestrian traffic
屋内声响焦点 loud spot
屋内装修和设施 interior works
屋内自来水干管 building main
屋旁杂草 ruderal
屋前基地 front stead
屋前空地 frontage; frontage space; street front
屋前临街车道 frontage carriageway
屋前临街庭院 front yard
屋前小院 forecourt
屋水管闸门室 draft tube gate chamber
屋屯住宅 <设计相同的> tract house
屋瓦 building tile; roof tile
屋瓦接合法 tile hanging
屋瓦排列成鳞状的 scaled
屋外厕所 outhouse
屋外管子工 external plumber
屋外花园酒店 beer garden
屋外给水龙头 hydrant outside a building
屋外角柱 quoin post
屋外太平梯 exterior fire escape
屋外线脚 exterior trim
屋外消防栓 hydrant outside a building
屋外用油漆 exterior house paint
屋外照明 exterior lighting
屋外整修 outside finish
屋向 frontage
屋向天沟 open valley
屋檐 eaves; penthouse; principal cornice; weather check
屋檐暗水槽 secret gutter
屋檐板扇形饰 scallop
屋檐板条 eaves lath
屋檐单元 eaves unit
屋檐滴水 eaves drip [drop]
屋檐滴水槽 non-walking-way gutter
屋檐底板 soffit board
屋檐泛水 eaves flashing
屋檐高度 eaves height
屋檐模板 eaves mo(u)lding
屋檐排水沟 cullis
屋檐瓶口状滴水 bottle-nose drip
屋檐铺底板 starter board
屋檐铺底层 starter strip
屋檐铅垂线 eaves plummet
屋檐水槽 roof gutter
屋檐天线 eaves antenna
屋檐瓦 front tile
屋檐下的通风孔 eaves vent
屋檐详图 eaves detail
屋檐悬挑 eaves overhang
屋檐与屋瓦之间安置的交叉板条 ambrices
屋檐柱 eaves pole
屋隅 cant
屋隅角柱 quoin post
屋缘铺底层 starting strip
屋正面脚手架 facade scaffold(ing)
屋正面清洁用吊车 facade elevator
屋主住户 owner-occupant
屋柱的中楣 zoophorus

无 nought; zilch

无碍的冲刷浪 clear breach
无碍航物的港 clean harbo(u)r
无碍航物的海岸 clear coast; open coast
无碍航物锚地 clear anchorage
无碍航物水域 clear water
无安全岛分流的平交路口 unchannelized intersection
无安全栅栏的平面交叉 open crossing
无暗销伸缩缝 non-dowelled expansion joint
无暗销有榫的伸缩缝 non-dowelled joggle expansion joint
无凹槽的 unfluted; unnotched
无凹陷的土壤 soil without cave-ins
无把握的 unassured
无坝河段 undiked reach
无坝取水 intaking without dam
无坝式 reservoir with damless
无坝引水 damless intaking
无摆振起落架 shimmy-free landing gear
无斑的 stainless
无斑点的 immaculate
无斑非显晶质的 aphyric
无斑纹观察 speckle-free viewing
无斑隐晶质的 aphyric
无斑隐晶质结构 phenocryst-free cryptocrystalline texture
无板纲 Aplacophora
无版权的作品 work in public domain
无版权、无专利权状态 public domain
无伴奏的 unaccompanied
无瓣空气凿岩器 pulsator
无包层光纤 non-clad optic(al) fibre
无包层纤维 non-clad fiber [fibre]
无包皮电缆 uncovered cable
无包皮炸药 non-sheathed explosive
无包套热等静压 cladless HIP [hot isostatic pressure]
无包扎密封 packless seal
无包装的 loose; packless; unpacked
无包装的货物 unprotected cargo; unprotected goods
无包装货物 nude cargo; unpacked cargo; unpacked goods
无保持架的滚珠轴承 cageless ball bearing
无保持架轴承 cageless rolling bearing

W

无保持力的 unretentive
无保额保险单 no amount policy
无保护 unprotect
无保护层的路肩 unprotected shoulder
无保护层地面 open ground
无保护层电弧焊 unshielded arc welding
无保护的 naked; unprotected; unsheltered; unshielded
无保护颠倒温度表 unprotected reversing thermometer
无保护电弧焊 non shielded arc welding
无保护电路 unprotected circuit
无保护罩散热器 naked radiator
无保留的 outright; unconditional; unconditioned; unreserved
无保留条件承兑 absolute acceptance; clean acceptance; general acceptance; unconditional acceptance
无保留验收 unreserved acceptance; unreservedly acceptance
无保留意见 clean opinion
无保险驾驶人 uninsured motorist
无保险丝开关 no-fuse switch
无保养使用期 service-free life
无保障的投资 insecure investment
无保真性 infidelity
无保证贷款 blank credit
无保证的 unwarranted
无保证的应收应付 open credit
无保证合同 nude contract
无保证基金 non-guaranteed fund
无报 no message
无报酬 unfeed
无报酬的 rewardless; unremunerative; unrequited
无报酬的业务 unremunerative service
无报单 no advice
无报道 loss of information
无爆过程 non-knocking process
无爆燃烧 non-detonating combustion
无爆炸危险 nothingness explosive risk
无爆炸性 non-explosive
无爆震操作 knock-free operation
无爆震功率 knock-free power
无爆震敲击声的早燃 silent preignition
无爆震区 knock-free region
无爆震燃料 knock-free fuel
无备份油量的航程 no-reserves range
无背衬焊接法 non-backing process
无苯汽油 unbenzoled petrol
无绷绳固定的桅杆 free-standing mast
无泵反循环取芯钻具 reverse circulation core barrel without pump
无泵反循环钻进 reverse circulation drilling without pump
无泵体的充气式浮选槽 pneumatic cell without pump body
无泵钻进 pumpless drilling
无比较性的资料 non-comparable data
无比率 ratioless
无壁犁 mo(u)ld boardless plough
无边的 unbordered; unbounded; unlimited
无边际的 unmeasurable
无边际利润 unbounded profit
无边界框 borderless easel
无边框玻璃门 frameless glass door
无边女帽 bonnet
无边盘卷装 headless package
无边无际的 boundless
无边小圆软帽 bonnet
无边缘的 unskirted
无编号指令 unnumbered command
无编码输入系统 codeless input system
无变表示法 undeformed representation
无变化 monotony; uniformity
无变化的 monotonous; unconverted; unrelieved
无变量的 nonvariant
无变量器的 transless
无变量系统 invariant system
无变形的 undeformed
无变形钢 non-shrinkage steel; non-shrinking
无变形线 isoperimetric line; loxodrome; rhumb line
无变形状态 undeformed state
无变压器的 transformerless; transless
无变压器式 transformerless
无变压输出器 output transformerless
无变异增长 invariant increase
无变应性 anergia; anergy
无变应性的 anergic
无辩护的 undefended
无标尺视距仪 non-staff tachometer [tachymeter]
无标度性破坏项 scale-breaking term
无标号的 unlabel(l)ed
无标号公用块 blank common block
无标记 no mark
无标记的 unmarked
无标记元 no tag cell
无标牌的 unmarked
无标签车辆检测器 unlabel(l)ed car detector
无标签的 unlabel(l)ed
无标志 no mark
无标志的 unbeaconed; unsigned
无标志货物 unidentifiable cargo
无表决权的 nonvoting
无表决权股票 non-voting stock
无冰的 ice-free
无冰季节 ice-free season
无冰期 clearing period; ice-free period; ice-free season
无冰区域 ice-free area
无冰水道 ice-free waterway
无冰水面 open water
无柄的 sessile
无柄模具 shankless die
无柄水底植物 sessile benthos
无饼滑车 dumb sheave
无病毒的 avirulent
无波动 ripple disable
无波浪的 waveless
无波浪水面 undisturbed water level
无波纹的 unrippled
无玻璃集热器 unglazed collector
无玻璃体熟料 free-glass clinker
无玻璃相陶瓷 glass-free porcelain
无玻璃质瓷砖 non-vitreous tile
无铂玻璃 platinum-free glass
无补偿的 non-compensated; uncompensated
无补偿的延长工期 noncompensable delay
无补偿电离室 uncompensated ionization chamber
无补偿放大器 uncompensated amplifier
无补偿接触网 messenger wire and contact wire without balance
无补偿流量计 uncompensated flowmeter
无布料器的沥青路面平整机 spreader finisher without hopper
无擦痕 without striations
无彩色 achromatic colo(u)r; neutral colo(u)r
无彩色刺激 achromatic stimulus
无彩色视觉 achromatic vision
无参考价值的书 dead book
无参考目标视野 empty visual field
无参考线虚分量仪 imaging component instrument without referring wire
无参数检验 parameter-free test
无残留 noresidue at harvest time
无残留农药 non-persistent pesticide
无残值 zero salvage value
无舱盖集装箱船 hatchcoverless container ship
无操作 no-operation; no-op(s)
无操作码 source destination code
无操作位 no-operation bit
无操作者 operator less
无操作指令 no(n)-operation instruction; waste instruction; blank instruction
无糙面 absent shagreen surface
无槽 slotless
无槽垂直引上法 Pittsburgh process
无槽的 unslotted
无槽的门框 door frame without rebate
无槽电枢 smooth core armature
无槽杆件 unnotched bar
无槽机座 plain bed
无槽接骨螺钉 plain bone screw
无槽立柱 unfluted shaft of column
无槽汽轮发电机 no-slotted turbogenerator
无槽丝锥 fluteless tap
无槽凸缘垫圈 grooveless flange gasket
无槽引上玻璃 non-debiteuse drawing glass
无槽引上法 PPG-Pennvernon process
无槽柱身 unfluted shaft of column
无槽转子 unslotted rotor
无槽转子汽轮发电机 unslotted rotor turbogenerator
无草休闲地 bare fallow; clean fallow; weedfree fallow
无侧墙式车库 open-deck garage
无侧收缩 suppressed contraction
无侧(收)缩堰 suppressed weir
无侧推力岸壁 non-thrust quaywall
无侧限板式支座 unconfined pad bearing
无侧限变形模量 unconfined deformation modulus
无侧限不排水剪切试验 unconfined undrained shear test
无侧限单轴荷载 unconfined uniaxial loading
无侧限单轴压缩试验 unconfined uniaxial compression test
无侧限的 non-confined; unconfined
无侧限混凝土 non-confined concrete; unconfined concrete
无侧限剪切强度 unconfined shear strength
无侧限抗剪强度 unconfined shear strength
无侧限抗拉强度 unconfined tensile strength
无侧限抗压强度 unconfined compression strength; unconfined compressive strength
无侧限抗压强度试验 non-confined compression test; unconfined compression test; unconfined compressive strength test
无侧限膨胀 unconfined swelling
无侧限试件 unconfined specimen
无侧限压缩 unconfined compression
无侧限压缩强度 unconfined compression strength; unconfined compressive strength
无侧限压缩强度试验 non-confined compression test; unconfined compression test
无侧限压缩(强度)试验仪 unconfined compression test apparatus
无侧限压缩试验 non-confined compression test; unconfined compression test
无侧限压缩(试验)装置 unconfined compression apparatus
无侧限压缩仪 unconfined compression apparatus
无侧限应变试验 zero lateral strain test
无侧限圆柱体试样 unconfined cylindric(al) sample
无侧眼管子 blank pipe
无侧柱的 <古希腊、古罗马> apteral
无侧柱的神殿 apteral temple
无侧柱的寺庙 <古希腊> apteral temple
无侧柱的寺院 apteral temple
无测限压力试验 unconfined compression test
无层滑坡 homogeneous landslide
无层结湖 unstratified lake
无层理的 unstratified
无插拔力 zero insertion force
无插拔力插座 zero insertion force socket
无插口的管子 hubless pipe
无岔区段 switchless section
无差别 indifference; indifferentism
无差别成本 common cost; non-differential cost; sunk cost
无差别待遇 non-discriminatory
无差别的 indiscriminate; non-differentiated
无差别区间 indifference interval
无差别区域 indifference region
无差别曲面 indifference surface
无差别曲线 indifference curve
无差错的 error free
无差错运转周期 error free running period
无差错帧 error free frame
无差拍 dead beat
无差调节 isochronous control
无差调速器 zero droop governor
无差异分析 indifference analysis
无差异关系 indifference relationship
无差异拉力 indifferent tension
无差异曲线 indifference curve
无差异曲线的凸性 convexity of indifference curve
无差异水平 indifference level
无差异行销 undifferentiated marketing
无掺杂的 undoped
无掺杂气流系统 undoped flow system
无产权名 non-proprietary name
无颤噪效应 nonmicrophonic
无长石假白榴石微晶正长岩 feldspar-free pseudoleucite microsyenite
无长石假白榴正长岩 feldspar-free pseudoleucite syenite
无偿拨款 unpaid appropriation
无偿出口 unrequited export
无偿的 gratuitous; unpaid; unrequited; nude <契约等>
无偿调拨 unpaid appropriation
无偿调出 free allocated-out
无偿调入 free allocated-in
无偿付能力 bankruptcy; insolvency
无偿付债务能力 insolvency
无偿合同 contract without compensation
无偿还能力的 insolvent
无偿还义务 without recourse
无偿技术转与 voluntary conveyance
无偿列车 dead-heading train; deadhead train
无偿契约 gratuitous contract; naked

contract
无偿让与 voluntary conveyance
无偿使用 use without compensation
无偿试用 trial free
无偿受让人 voluntary grantee
无偿投资 gratuitous investment
无偿现金转移 unrequired current transfer
无偿行为 act without consideration
无偿援助 aid given gratis; gratuitous help; non-reimbursable assistance
无偿运距 free haul
无偿运输 deadhead traffic; free haul
无偿债能力的债务人 insolvent debtor
无偿转拨 unpaid transmission
无偿转让 voluntary conveyance
无常的 fugitive
无常平架惯性导航设备 gimballess inertial navigation equipment; strapped-down inertial navigation equipment
无场放射电流 field-free emission current
无超高的线路 track without cant
无超高路段 normal crown section
无超载式 non-overloading type
无潮 tideless
无潮波 amphidromic wave
无潮的 non-tidal; tide free; tideless
无潮点 amphidromic point; nodal point; no(n)-tide point
无潮海 tide-free sea; tideless sea
无潮海岸 tideless coast
无潮海滩 tideless beach
无潮河道 tideless river
无潮河段 non-tidal section
无潮河流 tide-free river; tideless river; non-tidal river
无潮区 amphidromic region; non-tidal compartment; tide-free region; tideless region
无潮水域 non-tidal waters
无潮体系 amphidromic system
无潮汐冲积河流 non-tidal alluvial river; non-tidal alluvial stream
无潮汐的 non-tidal; tide-free
无潮汐河流 non-tidal river
无潮汐区 non-tidal compartment
无潮汐影响的河流 fluvial stream
无潮系统 amphidromic system
无潮线 nodal line
无吵闹声的 rattle free
无车家庭 non car owning household
无车架车辆 frameless vehicle
无车架车身 frameless body
无车架结构 frameless construction
无车架小客车 frameless automobile
无车辆 wheelless
无车辆来往地区 traffic-free precinct
无车区 traffic-free area; traffic-free encave; traffic-free precinct; u-free area
无车站隧道 non-station tunnel
无车者 non-vehicle available
无尘 dustlessness
无尘操作 dust-free operation
无尘车间 dust-free plant; dust-free workshop
无尘的 dust-free; dustproof; free from dust
无尘粉末 free flowing powder
无尘工厂 Staublos plant
无尘环境 dust-free environment
无尘颗粒 free flowing granules
无尘空气 dust-free air; dust-free atmosphere
无尘矿山 non-dusty mine
无尘冷却 dust-free cooling
无尘路 dust-free road
无尘路面 dust-freed surface
无尘落料管 dustless loading chute [spout]; dust-free spout
无尘煤气 clean gas; dust-free gas
无尘耐磨路面 hygienic pavement
无尘区 dustproof enclosure
无尘石棉布 dustless asbestos cloth
无尘石棉手套 dustless asbestos gloves
无尘石屑 dustless screenings
无尘时间 dust-free time
无尘炭黑 dustless carbon black; free flowing black
无尘污连续卸船机 dust-free continuous ship unloader
无尘卸灰装置 dustless ash unloader
无尘卸料系统 dustless loading system
无尘凿岩 dust-free drilling
无尘装料 non-dusting charge
无尘装载 dustless loading
无尘钻进 dust-free drilling
无尘钻眼 dust-free drilling; dustless drilling
无沉淀物 deposit(e)-free
无沉淀物的油 non-break oil
无沉积水流 non-deposit flow
无沉积物水 sediment-free water
无沉降水流 non-deposit flow
无沉陷的 unyielding
无沉陷地基 unyielding foundation
无沉陷桥台 unyielding abutment
无衬垫的 packless
无衬垫纸型纸 no-pack
无衬里水池 unlined tank
无衬砌的 unlined
无衬砌渠道 canal with earth section; unlined canal
无衬砌隧道 rough tunnel; unlined tunnel
无衬砌隧洞 rough tunnel; unlined tunnel
无衬区 blank zone
无衬套阀 packless valve
无衬线的(铅字) sanserif
无撑杆的 stanchion-free
无承口管 no(n)-hub pipe
无乘务员列车 crewless train
无迟延通信[讯] no delay service
无迟延信道 undelayed channel
无尺寸参数 dimensionless parameter
无尺寸单位 dimensionless unit; non-dimensional unit
无尺寸的 dimensionless; non-dimensional
无尺度 zero dimension
无尺度量 dimensionless quantity; non-dimensional quantity
无尺度数 dimensionless number; non-dimensional number
无齿卷盘的 non-sprocketted
无齿轮机车 gearless locomotive
无齿轮马达 gearless motor
无齿轮磨机传动 gearless mill drive
无齿轮磨机电动机 gearless mill motor
无齿轮啮合的 ungeared
无齿轮球磨机传动装置 gearless ball mill drive
无齿轮式圆锥破碎机 gearless reduction gyratory
无齿耙 toothless drag harrow
无齿平土耙 smoothing harrow
无齿拖耙 toothless drag harrow
无齿隙齿轮 anti-backlash gear
无齿圆锯 friction(al) saw
无充填的 unfilled
无充填物的不连续面 clean discontinuities
无充填物的软弱结构面 clean discontinuities
无冲动继电器 surgeless relay
无冲击 absence of shocks
无冲击变速 faultless gear change
无冲击的 hammerless
无冲击工况 shockless entrance condition
无冲击振动器 non-impact vibrator
无冲刷流速 non-scouring velocity
无冲突多路存取 collision-free multi-access
无冲突相位 < 交通信号中,左、右转车辆同对向车流或行人没有冲突的相位 > protected phase
无冲突转弯 < 信号交叉口上左、右车辆同对向车流或行人不发生冲突的转弯 > protected turns
无冲洗液钻进 dry drilling
无重复正规列 normal series without repetition
无重影电视 ghost-free television
无抽气 zero-extraction
无酬时间 unoccupied time
无臭 odo(u)r free
无臭溶剂 odo(u)rless solvent
无臭石油溶剂 odo(u)rless mineral spirit
无臭味的 inodo(u)rous
无臭(味)涂料 odo(u)rless paint
无臭油 sweet oil
无臭油漆 odo(u)rless paint
无出口的 blind
无出口房间 dumb chamber
无出口湖 lake without outflow; lake without outlet; basinal lake
无出口水道 false channel
无出口通道 blind gallery
无出料 zero discharge
无出料系统 zero discharge system
无出流地区 endor(h)eic region
无出流湖 closed lake; drainless lake; lake without outflow
无处不在的 ubiquitous
无处稠密集 nowhere dense set
无触点臂的断电器 pivotless breaker
无触点传感器 contactless pickup; non-contacting pick-up
无触点厚度计 non-contacting thickness ga(u)ge
无触点记录 non-contact recording
无触点继电器 contactless relay; no touch relay
无触点检波器 no-contact pickup; non-contacting pick-up
无触点控制 static control
无触点控制元件 contactless control element
无触点起动 contactless starting
无触点式自动同步机 magslep; synchro magslep
无触点式自整角机 magslep
无触点同步机的解算器 magslip resolver
无触点同步机解算装置 magslip resolver
无触点纵向记录 non-contact longitudinal recording
无穿堂风 draft fan; draught-free; draughtless
无传导性 inconductivity
无传力杆接缝 undowelled joint
无传力设备的(混凝土)板角 unprotected corner
无船舶执照的船 interloper
无船承运人 non-vessel operation carrier
无船的运输业 passive commerce
无船机集装箱船 gearless container ship
无船位 no fix
无船闸水坝 non-navigable dam
无窗厂房 blackout plant
无窗的 windowless
无窗的层楼 blind stor(e)y
无窗房屋 black(out) building; windowless building; windowless house
无窗格窗 sashless window
无窗拱廊 blind arcade
无窗建筑 blackout building; windowless building
无窗结构物 windowless building
无窗框安装法 direct glazing; glazing without frame
无窗帘的 uncurtained
无窗楼层 blind stor(e)y; windowless story
无窗墙 blank(-out) wall; blind wall; dead wall; wall blank
无窗墙板 windowless panel
无唇护刃器 lipless guard
无疵病 freedom from defects; soundness
无疵(病)木材 clean(-cut) timber; clear(-cut) timber; clear stuff; clear wood
无疵材 clear lumber; clears; clear stuff; clear timber; clear wood; free stuff
无疵残 zero defect
无疵带 flawless tape
无疵的 unblemished
无疵锯材 clear lumber
无疵小试样 small clear specimen
无磁层的 uncoated
无磁场径迹 < 指云室或气泡室中 > no-field track
无磁偏线 agonic line
无磁性船 non-magnetic vessel
无磁性的 non-magnetic
无磁性钢 non-magnetic steel
无磁性异构体 non-magnetic isomer
无磁性铸铁 non-magnetic cast iron
无磁性钻头卡圈 non-magnetic drill collar
无磁修正的 unslaved
无磁滞 anhysteresis
无磁滞的 unslugged; unsluggish
无次品小组 zero defect team
无次序的 out-of-order; unmethodical
无刺激 disincentive
无刺激反应 vacuum response
无从探索的 unsearchable
无粗差的 blunder free
无促进剂的硫化胶 unaccelerated sulfur vulcanizate
无存储摄像装置 non-storage pick-up device
无存货 not in stock; out of stock
无存款 no funds; no effects < 银行在空头支票上的批语 >
无错操作 error free operation
无错程序 star program(me)
无错建立模式 correct-by-construction
无错秒 error free second
无错误 inerrancy
无错误编码 error free coding
无错误传输 error free transmission
无错误的 error free; errorless
无错误信道 error free channel; error free information
无错误信息 error free information
无答辩 nihil dicit; nil dicit
无答复 no-reply
无大害故障 fail passive; fail soft
无大梁结构 monocoque
无大门的 ungated
无大缺陷水泥 macrodefect free cement
无大修工作期间 overhaul period
无大修理运行期间 overhaul period

W

无代表性的 non-representative; unrepresentative
无代价的 gratuitous
无单交货 conversion of cargo
无单位的 dimensionless; non-dimensional
无单元 no-cell
无担保贷款 loan without(collateral) security; signature loan; straight loan; unsecured advance; unsecured loan
无担保的 naked; unsecured; unvouched
无担保的账户 unsecured account
无担保负债 unsecured liability
无担保合同 bare contract; naked contract
无担保契约 bare contract; naked contract
无担保融资 accommodation loan
无担保项目 general crossing; unsecured account
无担保信贷 open credit
无担保信托 naked trust
无担保(信用)贷款 open credit
无担保债券 debenture(bond); naked debenture; plain bond; unsecured bond
无担保纸币 uncovered money
无挡板的 unbaffled
无档的 unstudded
无档短环链 unstudded short link chain
无档链 open link chain
无档链环 open link; studless link
无档锚链 unstudded cable
无导管的 ductless
无导框式轴箱 guardless type axle box
无导流片扩压器 vanless diffuser
无导流片汽缸 unbaffled cylinder
无导向工业 foot-loose industry
无导向架的悬吊式打桩锤 free hanging pile hammer
无导向模 guideless die
无导向凸模 free punch
无到发站各次列车都准乘坐的客票 blank-to-blank ticket
无道口栏木的公路与铁路交叉 ungated level crossing
无道面的机场 unprepared field
无灯(光)航标 day beacon; day mark; unlit beacon
无等高线地图 planimetric(al) map
无/低费清洁生产方案 no/lose cost clean production option
无堤防河道 unleveed river
无滴痕涂料 drip-free paint
无滴口喷嘴 no-drip nozzle
无滴油漆 non-dripping oil paint
无敌的 unrivalled
无底 bottomless
无底板牵引车 skeletal trailer
无底的吊桶 bottomless bucket
无底洞 abime; abysm; abyss
无底坩埚 boot; hood; potette
无底筐格式结构 bottomless crib
无底量斗 <混凝土骨料计量用的> measuring frame
无底门铺料机 spreading machine without bottom doors
无底门摊铺机 spreading machine without bottom doors
无底门受泥船 mud barge without bootom door
无底木笼 open crib
无底片黑白照片 rotograph
无底瓶形建筑装饰砖 bootle brick
无底漆 primerless
无底套箱 bottomless case
无底网 fly net
无底匣钵 ringer
无底缘的 rimless
无抵触 no conflict
无抵押担保之债权人 unsecured creditor
无抵押借款 borrow without security; clean loan
无抵押资产 unmortgaged assets
无地表渗入 non-infiltration from surface
无地沥青的 non-asphaltic
无地沥青石油 non-asphaltic petroleum
无地下室 cellarless
无地下室的 basementless
无地线 freedom from ground connection
无地震地 aseptic region
无地震区 aseismatic region; aseismic region; non-seismic area; non-seismic region
无地址 zero address
无地质准备工作的钻孔 blind drilling
无缔合性的 unassociated
无点缀的 nude
无电沉积 electroless deposit(ion)
无电池电话 sound-powered telephone set; sound power telephone
无电导线 dead wire
无电的 without current
无电电路 dead circuit
无电顶锻留量 current-off upset allowance
无电顶锻时间 upset current-off time
无电顶锻余量 current-off upset allowance
无电镀镍钢 electroless nickel plated steel
无电段 dead section
无电敷镀 electroless plating
无电杆交叉 poleless transposition
无电感的 non-inductive
无电感电路 non-inductive circuit
无电感电容器 non-inductive capacitor
无电荷的 uncharged
无电荷分子 uncharged molecule
无电弧抽头换接器 arcless tap
无电极电导系统 electrodeless conductivity system
无电极电镀 electrodeless plating
无电极放电 electrodeless discharge
无电极放电灯 electrodeless discharge lamp
无电极紫外线 electrodeless ultraviolet
无电抗电阻 non-reactive resistance
无电缆交通信号 cableless linking of traffic signals
无电缆联动系统 cableless linking system
无电缆(信号)联 cableless linking system
无电流保护 no-current protection
无电流场 current-free field; non-current field
无电流的 currentless
无电码示像 no-code aspect
无电区 dead zone
无电区段 dead track section
无电区段警告信号 dead section warning signal
无电区段注意信号 dead section warning signal
无电容输入 input capacitorless
无电刷槽电动机 brush and slotless motor
无电刷电动机 brushless motor
无电刷发电机 brushless generator
无电梯大楼 walk-up
无电梯大楼的楼上房间 walk-up
无电梯的 walk-up
无电梯的大楼 walk-up building
无电梯的居住楼房 walk-up domestic building; walk-up residence building
无电梯的住房 walk-up domestic block
无电梯的住家楼房 walk-up residence block
无电梯的住宅楼 walk-up domestic block
无电梯多层楼房 walk-up
无电梯公寓 walk-up apartment
无电梯公寓楼 walk-up apartment house
无电梯楼层 walk-up stor(e)y
无电梯楼房 walk-up building
无电压 no-voltage
无电压保护试验 no-voltage protection test
无电压报警器 no-voltage alarm
无电压电磁式继电器 no-voltage electromagnetic relay
无电压断路器 no-voltage cut-out
无电压继电器 no-voltage relay
无电压状态 no-voltage condition
无电压自动断路器 no-voltage circuit-breaker
无电源电话 sound-powered telephone
无电子管放大器 tubeless amplifier; valveless amplifier
无电自动断路器 zero cutout
无电阻 non-resistance
无垫密封 seal(ing) without gaskets
无垫片式膜板 diaphragm packless
无垫压力偶密封 self-seal by pressure coupling
无顶的 roofless
无顶点 nonoriented vertex
无顶帆三桅纵帆船 ram schooner
无顶货柜 open top container
无顶棚楼板 open stor(e)y
无顶棚屋顶 open roof
无顶棚屋面 open-timbered roof
无顶平台 stoop
无顶树 pollard
无顶台井架 bald-headed derrick
无顶台钻塔 bald-headed derrick
无顶箱型集装箱 open-top box container
无顶压逆流再生【给】 atmospheric press bed counter current regeneration
无定额保险单 open policy
无定额的 open-end
无定额经费分配 indeterminate appropriation
无定湖 astatic lake
无定价的 unpriced
无定见的 ramshackle
无定期储存 indefinite storage
无定期的 undated
无定期喷泉 flowing geyser
无定位调节 astatic regulation; floating control
无定限的 indefinite
无定向摆 astatic pendulum
无定向摆放 random setting
无定向常数 astatic constant
无定向成网机 random web-laying equipment
无定向磁力计 astatic magnetometer
无定向磁力仪 astatic magnetometer
无定向磁力仪法 astatic magnetometer method
无定向磁偶. astatic couple
无定向磁针 astatic needle
无定向的 amphibolic; astatic; omnidirectional; unoriented
无定向地震触发器 omnidirectional seismic trigger
无定向电流计 astatic galvanometer
无定向电炉钢 non-oriented electric-(al) steel
无定向对 astatic pair
无定向风 baffle; baffling wind
无定向控制 astatic control
无定向流 unoriented current; vagrant current
无定向配边 unoriented cobordism
无定向平衡 astatic balance
无定向熔结凝灰岩 unoriented welded tuff
无定向式仪表 astatic instrument
无定向水系 astatic drainage; insequent drainage(system)
无定向弹性摆 astatic elastic pendulum
无定向调节 astatic governing; floating control
无定向调节器 astatic governor
无定向陀螺仪 astatic gyroscope
无定向微风 whiffling breeze
无定向无线电导航信号台 circular radio beacon
无定向纤维网 random(-laid) web
无定向性 astatism
无定向压力 directionless pressure
无定向洋流 variable current
无定向中心 astatic center
无定向重力计 astatic gravimeter; astatized gravimeter
无定向重力仪 astatic gravimeter
无定向作用 floating action
无定形半导体存储器 amorphous semiconductor memory
无定形不透明物质 amorphous opaque matter
无定形材料 amorphous material; among(st) material
无定形沉淀 amorphous sediment
无定形的 amorphous; structureless; unbodied
无定形二氧化硅 amorphous silica; soft silica
无定形固体 amorphous solid
无定形光学材料 amorphous optic(al) material
无定形硅石 amorphous silica
无定形硅石粉 amorphous dust
无定形结构 impalpable structure; undefined structure
无定形晶 crystal imperfection
无定形聚合物 amorphous polymer
无定形矿物 gel mineral
无定形蜡 among(st) wax; amorphous wax; ceresin(e) wax
无定形磷 amorphous phosphorus; red phosphorus
无定形磷酸盐 amorphous phosphate
无定形硫 amorphous sulfur
无定形泥炭 amorphous peat
无定形黏[粘]合剂 amorphous binder
无定形硼 amorphous boron
无定形区 amorphous region
无定形扫描信号 amorphous scanning signal; espews
无定形石墨 amorphous graphite
无定形碳 agraphitic carbon; amorphous carbon
无定形体 amorphous solid
无定形物质 among(st) substance; amorphous substance
无定形硒 amorphous selenium
无定形纤维素 amorphous cellulose
无定形现象 amorphism
无定形相 amorphous phase
无定形型 amorphous type
无定形性 amorphism
无定形雪 amorphous snow

无定形岩 amorphous rock
无定形氧化硅 amorphous silica
无定义符号 undefined symbol
无定住所的建筑师 architect errant
无定状高积云 alto cumulus informis
无定状云 informis
无锭轧机 direct rolling mill
无锭轧制 rolling from the molten condition
无动力船 non-powered ship
无动力的 motorless; non-motile; unpowered
无动力回转刮刀 undynamic(al) force rotary scraper
无动力进场着陆 unpowered approach
无动力上升 unpowered ascent
无动力输送机 unpowered conveyer [conveyor]
无动力速率 power-off speed
无动力推进的下滑 unpowered glide
无动力推进飞行 unassisted flight
无动力转向架 unpowered bogie
无动力自导引武器 unpowered homing weapon
无动区 agravic
无动作指令 ignore instruction
无逗点码 comma free code
无毒 asepsis; innocuity
无毒的 atoxic; avirulent; innocuous; innoxious; non-toxic
无毒害的 avirulent
无毒或低毒电镀 non-poisonous or less-poisonous electroplating
无毒焦油(沥青) non-toxic tar
无毒界量 limes zero
无毒流出物 innocuous effluent
无毒漆 non-toxic paint
无毒玩具喷漆 non-toxic toy lacquer
无毒污染物 non-toxic pollutant
无毒物质 non-poisonous material
无毒性 non-toxicity
无毒性的 non-toxic
无毒性反应的最高浓度 no-observed-adverse-effect level
无毒性涂料 non-toxic paint
无毒性作用剂量水平 non-toxic effect level
无毒盐类 non-toxic salt
无毒颜料 non-toxic pigment
无毒增塑剂 non-toxic plasticizer
无读数 non-metering
无独创性的 unoriginal
无堵塞爆破 open hole shooting
无镀层钢 black steel
无镀层钢板制品 uncoated flat product
无镀层管 black pipe
无镀层熟铁 black wrought iron
无端 sideless
无端穿绕 endless reeving
无端带 endless band; endless tape
无端回线 endless loop
无端胶片 endless film
无端接的 unterminated
无端接线 unterminated line
无端锯 endless saw
无端链 endless chain
无端链条拉木机 endless chain log haul-up
无端皮带 endless strap
无端铜网 endless wire
无端(循)环 endless loop
无端有根树 endlessly rooted tree
无短路接触开关 non-shorting contact switch
无断层的 unfaulted
无对的 azygous; impar; unpaired
无对流湖 non-convective lake
无惰性气体 inert free gas
无恶臭 odo(u)r free
无鲕石灰石 beer stone
无耳砖 lugless brick
无二次污染 without second-pollution
无二氧化碳蒸馏水 carbon dioxide free distilled water
无发的 acomous
无发动机的 motorless; unpowered
无阀泵 valveless pump
无阀冲击器 valveless air hammer
无阀的 valveless
无阀动作 valveless action
无阀工具 valveless tool
无阀计量泵 valveless metering pump
无阀滤池 non-valve filter; valveless filter(ing pool)
无阀门的 valveless
无阀式发动机 valveless engine
无阀型除锈器 valveless scaler
无阀重力滤池 valveless gravity filter
无阀作用 valveless action
无法辩护的 unwarrantable
无法表明 indemonstrability
无法表明的 indemonstrable
无法补救的 irremediable; irrepairable
无法操纵的 out of control
无法测定的损失 undeterminable loss
无法兑付的支票 uncollectible check
无法估计的 imponderable
无法核实的 unverifiable
无法后送的 unevacuable
无法回答的 unanswerable
无法回收的(木)支撑 non-recoverable timbering
无法计量的性质或状态 immensurability
无法计算的 beyond number; undeterminable
无法纪的社会状态 anomie
无法架桥的 unbridgeable
无法检验的 unverifiable
无法鉴别的 unidentifiable
无法交付 failure to deliver; non-delivery
无法交付的 undeliverable
无法交付的进口货 undeliverable import goods
无法交付的支票 undeliverable check
无法交付货物 non-deliverable goods; unable-deliverable cargo; unable-deliverable goods; undeliverable cargo; undeliverable goods
无法交付物品 non-deliverable articles; undeliverable articles
无法校正的 uncorrectable
无法校正的错误 uncorrectable error
无法解答的 unsolvable
无法解释 unaccountability
无法解释的 unaccountable
无法就地修复 beyond local repair
无法决策 non-decision-making
无法开垦的土地 unclaimable land
无法考证的 unverifiable
无法控制 out of control
无法控制的 beyond control; uncontrollable; ungovernable; uncontrolled
无法控制的通货膨胀 runaway inflation
无法兰的 flangeless
无法兰连接 jointless
无法兰盘接合件 junction piece without saddle
无法利用的 unserviceable
无法履行义务的情况 incapability of meeting obligation
无法律约束的 ruleless
无法论证的 undemonstrable
无法弥补的 irretrievable
无法平息的 unappeasable
无法确定的 indeterminable; undeterminable
无法实行 impracticability
无法收集的 uncollectible
无法收款的支票 uncollectible check
无法收现的应收票据 uncollectible note receivable
无法收现的应收账款 uncollectible receivable account
无法替换的 irreplaceable
无法投递 non-delivery
无法投递的邮件 nix
无法挽回的 irrepairable
无法挽救的 irrepairable
无法消化 indigestibility
无法销售的 unrealizable
无法修复 damaged beyond repair
无法修理的建筑物 structure beyond repair
无法预见的 unpredictable
无法征收的税款 uncollectible tax
无法证明的 indemonstrable
无法治理 unworkable harness
无帆船 dumb barge
无翻边轮胎 beadless tire [tyre]
无反冲 non-recoil
无反冲发射 recoilless emission
无反冲共振吸收 recoilless resonance absorption
无反冲力的 recoilless
无反冲转向 shock-proof steering
无反对的 unopposed
无反光玻璃 invisible glass
无反光层 non-reflection coating
无反光商店橱窗玻璃 invisible glazing
无反馈的 non-reactive
无反馈控制 open loop control
无反射 abolition of reflex; absent reflection; areflexia; non-reflection
无反射玻璃 non-reflecting glass
无反射层的 untamped
无反射层堆 unreflected reactor
无反射层装置 bare assembly
无反射橱窗 non-reflecting shop window
无反射的 non-reflective; unreflected
无反射空间 non-reflecting cavity
无反射墨水 non-reflective ink
无反射器测量技术 reflectorless measurement
无反射器式定向天线系统 beam primary aerial system; beam primary antenna system
无反射墙壁 reflection free wall
无反射区 non-reflecting cavity; reflection-free areas
无反射涂层 bloomed coating; non-reflecting coating
无反射终端 reflexless terminal
无反跳击锤 dead-stroke hammer
无反响的 anechoic
无反应 non-response
无反应的 non-reactive; reactionless; unresponsive
无反应骨料 unreactive aggregate
无反应力的 recoilless
无反应区 dead band
无反应性 anergy; non-reactivity
无反作用的伺服机构 reactionless servo-mechanism
无返回点 no return point
无方位声源 non-directional sound source
无方向射线 indirect ray
无方向天线 circular diagram aerial
无方向性的 direction-free; non-directional
无方向性硅钢片 non-oriented silicon steel
无方向性声源 non-dimensional sound source
无方向性天线 isotropic(al) antenna; unipole antenna
无方向性无线电信标 non-directional radio beacon
无防寒设备的发动机 unprotected engine
无防护部分 unshielded part
无防护道口 non-guarded crossing; non-protected crossing; unprotected (level) crossing
无防护的 non-protected; non-protective; undefended; unguarded; unshielded
无防护的海滩 nonprotective beach
无防护的河床 non-protected riverbed; unprotected river bed
无防护的混凝土衬垫 unprotected concrete pad
无防护发射场 unprotected site
无防护海滩 non-protected beach; unprotected beach
无防护机械 unfenced machinery
无防护金属结构 unprotected metal construction
无防护设备的 unprotected
无防护设备的平面交叉 unprotected level crossing
无防护站台 non-protective platform
无妨碍物的环境 nuisance-free environment
无房户 unallocated household; unallotted household
无纺玻璃布 non-woven glass fabric
无纺玻璃纤维网格布 non-woven scrim
无纺布 adhesive-bonded fabric; geotextile; non-woven cloth; unwoven cloth
无纺布滤材 non-woven media
无纺布伸缩装置 non-woven fabric expansion installation
无纺材料 non-woven material
无纺的 non-woven
无纺聚酯纤维 non-woven polyester fabric
无纺石棉布 non-woven asbestos cloth
无纺土工布 non-woven geofabric; non-woven geotextile
无纺型土工织物 non-woven geotextile
无纺织黏[粘]合织物 non-woven bonded fabric
无纺织物 bonded fabric; non-woven fabric
无放射性玻璃 radiation-free glass
无放射性尘埃 dead ash
无放射性原子 dark atom
无飞边模锻 flashless die forging; no-flash die forging
无废工艺 non-waste technology
无废技术 non-waste technique
无废物生产 wasteless production
无分布 distribution-free
无分步作用 stepless action
无分叉河流 unbranched river; unbranched stream
无分隔带道路 undivided road
无分隔带的车行道 undivided carriageway
无分隔带的双向车道 undivided two-way road
无分隔带公路 undivided highway
无分红权的(保险) non-participating
无分划调节器 non-indicating controller
无分回业务分保 non-reciprocal reinsurance
无分路的 unshunted
无分选 assortment
无分选的 non-sorted

无分异高原气候 non-differential plateau climate
无分异岩体 non-differentiated intrusive body
无分支电路 linear chain
无分支接头 straight joint
无分支连接 straight joint
无分支连接套管 straight joint
无粉尘的 non-dusting
无粉尘熟料 undusted clinker
无粉尘性颗粒 non-dusting granules
无粉腐蚀 powderless etching
无粉腐蚀机 powderless etching machine
无粪生活污水 gray water
无风 ash breeze;dead calm;draught-free;lull;quiet air
无风层 calm layer
无风带 calm belt;calm zone;dead belt;zone of silence
无风的 unwind;windless
无风点 no-wind position
无风洞 cave without wind
无风而下微雨 Irish hurricane;paddy's hurricane
无风帆 < 无动力驳船 > dumb barge
无风格 absence of style
无风海岸 windless coast
无风航程 still air range
无风化的曝晒 no weathering exposure
无风化露头 no weathering exposure
无风流靠泊 going alongside in calm weather
无风逆温 calm inversion
无风逆温污染 calm inversion pollution
无风期 calm;doldrum season
无风区 calm zone;decay area;windless region
无风扇电机 non-ventilated motor
无风艏向【航海】no-wind heading
无风天气 calm weather
无风位置 air position
无风位置指示器 no-wind position indicator
无风无浪 lull;calm
无风无流的推算船位 dead-reckoning position
无风无流靠码头 going alongside with no wind and current
无风险投资利率 risk free rate
无风眼 calm control eye
无封板的浅抽屉 tray
无封冰期 clearing period
无封缝料的接缝 unsealed joint
无封隔器测试 test without packer
无峰线路 < 驼峰区 > no hump track
无缝 no-seam
无缝不锈钢管 seamless stainless steel pipe
无缝道岔 continuously welded switches;continuously welded turnout
无缝的 drawn cylinder;joint free;unseamed;weldless;seamless
无缝底 seamless bottom
无缝地板 composition floor cover(ing);jointless floor(ing);seamless floor(ing)
无缝地板面层 seamless floor cover(ing)
无缝地板涂层 seamless flooring coating
无缝地板用的沥青膏 mastic asphalt for jointless floor(ing)
无缝地板装修 chipping
无缝地面 seamless floor(ing);seamless ground
无缝地面处理 seamless floor finishing
无缝地面做法 jointless floor(ing)
无缝吊杆 telescopic(al) derrick;weldless derrick
无缝顶盖 baldacohino
无缝封头 seamless head
无缝钢管 solid-drawn steel pile;solid-drawn steel tube;solid pipe;steel seamless tube;weldless steel conduit;no-hub pipe;seamless steel pipe
无缝钢管厂 seamless pipe mill;seamless steel tubing plant
无缝钢管的分段轧制设备 step-by-step type seamless tube rolling mill
无缝钢管的套管 seamless casing
无缝钢管校正器 straightener for seamless piping;straightener for seamless tubing
无缝钢管坯 seamless bloom
无缝钢管轧机 step-by-step type seamless tube rolling mill
无缝钢管桩 seamless steel tube pile
无缝钢轨 continuously welded rail
无缝钢套管 seamless casing
无缝高碳钢管 seamless carbon-steel pipe
无缝构造 jointless construction
无缝冠 seamless crown
无缝冠冲压机 seamless crown machine
无缝管 drawn pipe;seamless conduit;seamless pipe [piping];seamless tube [tubing];solid-drawn tube;weldless pipe [piping];seldless tube [tubing]
无缝管道 continuous pipe without joint
无缝管轧机 seamless pipe rolling mill
无缝管状织物 seamless tubing fabric
无缝光谱 slitless spectrum
无缝轨道 gapless rail;jointless track;seamless rail;seamless track
无缝锅炉筒 seamless shell course;seamless shell ring
无缝焊焊接 solderless joint
无缝环轧制 seamless ring rolling
无缝接头 blind joint
无缝结构 jointless structure
无缝金壳冠 seamless gold crown
无缝壳体 seamless shell
无缝空心球轴承 seamless hollow ball bearing
无缝控制钢管 seamless drawn steel pipe;seamless drawn steel tube
无缝冷凝器管 seamless condenser tube
无缝连接 join monolithically
无缝楼板 jointless floor(ing);seamless flooring
无缝楼面处理 seamless floor finishing
无缝楼面修饰 seamless floor finish
无缝路面 shut pavement
无缝路面构造 jointless construction
无缝铝管 seamless alumin(i)um
无缝铝护套 seamless alumin(i)um sheath
无缝门 seamless door
无缝铺地作业 fleximer
无缝气瓶 drawn cylinder
无缝柔性地面 fleximer
无缝软钢管 seamless mild-steel pipe
无缝软管 flexible seamless tubing
无缝舢板 seamless boat
无缝摄谱仪 slitless spectrograph
无缝石膏地面 gypsum floor
无缝水泥地坪 cement jointless floor(ing)
无缝套筒 seamless steel pipe [piping];seamless steel tube [tubing]
无缝铜水管 seamless copper tube
无缝筒体 seamless shell
无缝纹型软波导 seamless corrugated waveguide
无缝屋面表层 seamless roof skin
无缝隙 no fissure
无缝线路 continuous welded rail;continuous welded track;gapless track;jointless track;seamless track;welded rail
无缝型钢 weldless rolled steel;weldless tooled steel
无缝阴极 seamless cathode
无缝轧制钢管 seamless rolled tube
无缝轧制管 seamless rolled tube
无缝制品 dipped goods;seam-free product;seamless article
无缝装饰板 decorative in-situ floor(ing)
无缝紫铜管 seamless copper pipe;seamless copper tube
无缝紫铜水管 seamless copper water tube
无缝钻管 seamless drill pipe
无扶手沙发 armless chair
无扶手的梯子 individual-rung step ladder
无氟玻璃 fluoride-free glass
无浮子式液面控制器 floatless-type liquid level controller
无浮子液面控制器 floatless(liquid-) level controller
无符号的 unsigned
无符号(整)数 signless integer;unsigned integer
无符号整数格式 unsigned integer format
无辐散高度 level of nondivergence
无辐射产生 non-radiative generation;radiationless generation
无辐射共振 non-radiative resonance;radiationless resonance
无辐射过程 non-radiative process;radiationless process
无辐射衰变过程 non-radiative decay process;radiationless decay process
无辐射跃迁 non-radiative transition;radiationless transition
无辅助的 unaid
无辅助设备发动机 bare-engine
无腐蚀 freedom from corrosion
无腐蚀润滑 non-corrosive grease
无腐蚀物的 free from corrosive substances
无腐蚀性 non-corrosiveness;non-corrosivity
无腐蚀性的 non-corrosive
无腐蚀性焊剂 non-corrosive flux
无负荷 no-load(position);off-load
无负荷除氧 < 给水在非工作装置的除氧 > off-load deaeration
无负荷的 uncharge
无负荷电池 idle battery
无负荷构件 idle member
无负荷控制 no-load control
无负荷流量 no-load discharge
无负荷面砖 non-load-bearing tile
无负荷期 off-load period
无负荷起动 no-load starting
无负荷起动电磁阀 no-load starting electro-magnetic valve
无负荷切断器 no-load cut-out
无负荷试验 no-load test
无负荷速度 load-free speed;runaway speed
无负面环境影响 no-negative environmental impact
无负面效应 no-negative effect
无负弯矩值受拉钢筋 concordant tendon
无负载 absence of load(ing)
无负载的 no-load(ed)
无负载的传播延迟 unloaded propagation delay
无负载电缆 non-loaded cable
无负载电路 unloaded circuit
无负载电压 non-load voltage
无负载品质因数 unloaded quality factor
无负载释放动作 no-load release action
无负载速度 no-load speed
无负载特征 unloaded characteristic
无负载天线 unloaded antenna
无负载位 idle position
无负载位置 idle position
无负载运转 idle running;idling;no-load run(ning);non-loading run
无负载状态 no-load position
无负债 never indebted
无附带条件的提货单 clean bill of lading
无附加条件的合同 absolute contract
无附加载荷的运输 free haul
无覆层的 unpainted
无覆层钢 unpainted steel
无覆盖层 intectate-reticulate
无覆盖层地基 bare-base support
无覆盖的 unclothed;uncoated;uncovered
无覆盖地区 uncovered area
无覆盖甲板 unsheathed deck
无覆面层地基 < 如机场 > bare-base support
无改编通过旅客列车 unadaptable through passenger train
无钙胶结料 non-calcareous cement
无钙水泥 non-calcareous cement
无盖 top off
无盖板梳棉机 flatless card
无盖杯 open cup
无盖冰箱 uncovered refrigerator
无盖驳船 open cargo lighter
无盖敞口驳 open hopper barge
无盖车 open wagon
无盖的 non-operculate;uncowled
无盖二轮马车 < 英 > buggy
无盖货车 gondola car;truck;trundle
无盖井 open well
无盖漏斗车 open top hopper;open-top hopper car
无盖排水沟 water furrow
无盖容器 open-topped vessel
无盖箱式托盘 open box-pallet
无盖箱铸造 casting-in open
无盖轴箱 half box
无盖子也无保护的管子终端 plain end
无干扰 nil interference
无干扰波道 noiseless channel
无干扰的 undisturbed
无干扰的端间距 unperturbed end-to-end distance
无干扰地震图 noise-free seismogram
无干扰点 non-interference point
无干扰雷达设备 green gadget
无干扰瞄准系统 undisturbed line-of-sight system
无干扰频道 clear channel
无干扰式交通探测技术 non-intrusive traffic detection technologies
无干扰输出信号 undisturbed output signal
无干扰温度 non-interference temperature
无干扰无规线圈 unperturbed random coil
无干扰响应电压 undisturbed response voltage
无干扰信道 noiseless channel
无干扰运行 trouble-free operation
无杆电钻 rodless electrodrill
无杆角式压塑机 rodless angle [mo-

(u)lding] press
无杆锚 housing anchor; patent anchor; stockless anchor; swinging fluke anchor
无杆喷嘴 stemless nozzle
无杆首锚 stockless bower
无坩埚技术 crucibleless technique; pedestal technique
无感 non-sensation
无感导线 inductionless conductor
无感地震 feltless earthquake; not noticeable earthquake; unfelt earthquake
无感点 null; silent point
无感电路 non-inductive circuit
无感电阻 non-inductive resistance
无感分流器 non-inductive shunt
无感荷载 non-inductive load
无感觉的 impassible; insensitive
无感区 dead band
无感绕法 non-inductive winding
无感绕线灯丝 non-inductively wound filament
无感绕组 non-inductive winding
无感线圈 Ayrton-perrs winding; non-inductive coil
无感应的 inductionless; non-inductive
无感阻抗 non-reactive impedance
无橄方沸碱煌岩 fourchite
无刚性承台 non-rigid relieving platform supported on bearing piles
无钢筋的 non-reinforced; unreinforced
无钢筋混凝土 unreinforced concrete
无钢索内向挖土机 drag shovel
无港的 harbo(u)rless
无格滚筒式真空过滤器 single-chamber rotary drum vacuum filter
无格式存储容量 unformatted capacity
无格式的 unformatted
无格式读 unformatted read
无格式记录 unformatted record
无格式书写 unformatted write
无格式输入 formatless input
无格式输入输出 unformatted input-output
无格式文件 unformatted file
无格式写语句 unformatted write statement
无隔壁的 unseparated
无隔垫堆放的 dead-piled
无隔墙的 partition-free
无隔墙建筑物 loft building
无铬砖 chrome-free brick
无给料器的沥青路面平整机 spreader finisher without hopper
无根据的 groundless; unsound; untenable; unwarrantable; unwarranted
无根据的索赔 unwarranted claim
无根山 mountain without roots
无根树 unrooted tree
无跟部尖轨 <与固定式尖轨跟部相对应> heelless switch
无跟鞋 flattie; flatty
无更正 no correction
无耕种植 no-fill planting; zero till
无梗花栎 durmast(oak); sessile oak; stackless flowered oak
无梗肋板 unribbed slab
无工班时间 non-shift hours
无工伤的 injury-free
无公差 zero allowance
无公度 incommensurability
无公度的 incommensurable
无公海技术 non-nuisance technique
无公害 lesspollution; nuisance free
无公害闭环式工艺 nuisanceless closed loop technique
无公害闭路式工艺 nuisanceless closed loop technique
无公害的 nuisanceless
无公害工艺 nuisanceless technology
无公害能量 clean energy
无公害热处理 anti-pollution heat treatment
无公害食品 pollution-free food
无公害蔬菜 pollution-free vegetable
无公约数的 incommensurable
无功 idle work; idling; virtual work
无功百万伏安 million volt-ampere reactive
无功补偿 reactive-load compensation
无功补偿设备 reactive-load compensation equipment
无功部分 idle component; imaginary component; reaction component; reactive component; wattless component
无功的 reactive; wattless
无功电动势 reactive electromotive force
无功电力 reactive power
无功电流 idle current; idling current; reactive current; wattless current
无功电流瓦特计 idle-current wattmeter
无功电路 reactive circuit
无功电容性负荷 leading wattless load
无功电压 reactance voltage; reactive voltage
无功电压分量 reactive voltage component
无功电压降 reactive drop
无功而得的 unmerited
无功分量 idle component; quadrature component; reaction component; reactive component; wattless component
无功峰值限制器 reactive-peak limiter
无功伏安 <电抗功率单位,电压、电流和相角正弦的乘积> var [reactive volt(age)-ampere]; wattless volt-ampere
无功伏安表 idle-current wattmeter
无功伏安计 varmeter
无功伏安小时 var hour
无功伏安小时计 reactive volt-ampere-hour meter
无功负荷 reactive load; reactive termination; wattless load
无功负载 reactive load; wattless load
无功功率 blind power; reactance capacity; reactive kilovolt-ampere; reactive power; wattless power
无功功率补偿器 reactive power compensator
无功功率补偿装置 reactive power compensation equipment
无功功率成组调节装置 group reactive power regulating device
无功功率继电器 reactive power relay
无功功率控制 reactive power control
无功功率因数 reactive power factor
无功功率因数表 reactive factor meter
无功功率自动调整运行 automatic reactive power regulating operation
无功量 reactive value
无功率 inactivity
无功面积 lost area
无功能蛋白质 non-functioning protein
无功能的 non-functional; non-functioning
无功能量 quadergy
无功千伏安 kilovar; quadrature kilovoltampere; reactive kilovolt-ampere; wattless kilowatt
无功千伏安控制 reactive-KVA control
无功损耗 no-load loss
无功网络 quadrature network
无功线路 reactive line
无功效的 high idle
无功旋转 idling
无功因数 quadrature factor
无功元件 reactive element
无功运行 no-load operation
无功滞后负荷 lagging wattless load
无功周期 idle period
无供给的 unfurnished
无供应 non-available; unavailability
无汞船底漆 mercury free ship-bottom paint
无汞防霉剂 non-mercurial fungicide
无拱的 archless
无拱炉 archless kiln
无拱手的单椅 side chair
无拱窑 archless kiln
无沟地下铺设 tunnelless underground laying
无沟梁 unsewerage beam
无沟垄种植 fiat planting
无沟排水暗管铺设机 trenchless drainage machine
无沟渠的 unsewered
无构架的 frameless
无垢井孔 clean well
无箍的 collarless
无箍接头 unprotected tool joint
无股钢丝绳 non-spinning rope; non-stranded rope
无股票公司 non-stock corporation
无股息 dividend off; ex dividend
无骨架密封 unreinforced seal
无骨架式车身 monocoque body
无骨架式车体 monocoque coach body
无固定顶盖连续(作业)窑 archless continuous kiln
无固定时间 no fixed time
无固定体积 unsoundness
无固定体积的 unsound
无故的 gratuitous
无故缺席 absence without leave
无故障 test OK; zero defect
无故障操作 trouble-free operation
无故障的 fail safe; failure-free; fault free; faultless; trouble-free; trouble-proof
无故障的工作 unfailing service
无故障的工作期限 no-failure life
无故障的浸油离合器 trouble-free oil clutch
无故障概率 probability of non-failure
无故障工作 failure-free work; no-failure operation; trouble-free operation
无故障工作周期 instantaneous availability
无故障控制 fail-safe control
无故障使用 trouble-free service
无故障使用期 trouble-free life
无故障使用寿命 no-failure life
无故障试验 fall-safe test
无故障停车精度 fail-safe stop accuracy
无故障性能 trouble-free performance
无故障运行 failure-free operation; trouble-free operation
无故障运行时间 non-failure operation time
无故障运转 trouble-free operation; trouble-free running
无故障状态 unfaulty condition
无故障着陆 trouble-free landing
无挂车的载重车 sutruck
无关 independency
无关(测定的)运动 extraneous motion
无关成本 irrelevant cost
无关的 extraneous; independent; unconcerned; unrelated
无关的分割序列 independent sequence of partitions
无关的物质 foreign material
无关积分 independent integral
无关集 independent set
无关节的 inarticulate; jointless
无关解 independent solution
无关紧要 of no consequence
无关紧要的事物 inconsequential
无关颗粒凝集试验 inert particle agglutination test
无关空间 independent space
无关联 absence of correlation
无关联的 unattached; uncoupling
无关频率 unrelated frequencies
无关微分算子 independent differential operators
无关系的 irrelative; irrelevant; unallied
无关线性方程 independent linear equation
无关信号 extraneous wave
无关性 independence
无关要紧的 inessential
无关因子的消除 elimination of irrelevant factors
无关"与"门【计】 don't-care gate
无关元 independent element
无关知觉色 unrelated perceived colo(u)r
无关重要 count for nothing
无管的 pipeless
无管电钻 pipeless electric(al) drill
无管供暖与通风设备 pipeless heating and ventilation
无管锅炉 pipeless furnace
无管火炉 pipeless furnace
无管井 cavity well
无管漏斗 stemless funnel
无管轮胎 tubeless tire [tyre]
无管热风采暖 warm-air pipeless heating
无管热风供暖 warm-air pipeless heating
无管三通阀 pipeless triple valve
无管式采暖通风 pipeless heating and ventilation
无管式采暖系统 pipeless heating system; pipeless system of heating
无管式采暖与通风 pipeless heating and ventilation
无管式供暖系统 pipeless heating system; pipeless system of heating
无管式供暖与通风 pipeless heating and ventilation
无管式火炉 pipeless furnace
无管网气体灭火装置 extinguishing device without piping system
无管辖权抗辩 plea to the jurisdiction
无管制交叉口 uncontrolled intersection; unrestricted intersection
无管座电子管 baseless tube
无管座管 wired-in tube
无贯入 zero penetration
无惯例扣减条款 no customary deductions clause
无惯性控制系统 infinitely fast control system
无光 blackout; dead dull; dead flat; dulling; flatness; dullness <指釉面>
无光白漆 dead white paint
无光白釉 moonstone
无光百叶窗 black blind
无光斑点 flat spot
无光标志 non-luminous sign; non sign
无光表面 mat(te) surface; non-glossy surface
无光玻璃 dull glass; etched glass
无光薄板 dull-finish(ed) sheet
无光彩的 cold; dumb
无光层 aphotic layer
无光窗 blackout window
无光醇酸瓷漆 flat alkyd enamel
无光瓷漆 mat(te) enamel paint

W

无光道林纸 unglazed printing paper
无光的 aphotic; blind; flat; non-luminous; unglazed; cold <塑料表面>
无光的毛面 matted finish
无光放电 black discharge; dark discharge
无光浮标 unlighted buoy
无光光洁度 matt finish; stain finish
无光海洋环境 aphotic marine environment
无光黑漆 Berlin black paint; flat-back paint; unlighted black paint
无光化带 aphotic zone
无光化纤纱 pigment fibre yarn
无光挥发性漆 flat lacquer
无光(火)焰 non-luminous flame
无光浸洗液 matt dip; mat(te) dip
无光精整 <板材表面的> butler finish
无光蓝(色料) mat(te) blue
无光(毛面)镀锡薄钢板 dull plate
无光面 dull surface; matt surface
无光面层 mat(te) coat
无光面漆 flat finish
无光磨面层 rubbed finish
无光牛皮纸 pure kraft paper
无光漆 anti-glare paint; dull paint; frosted lacquer; frosting varnish; lusterless paint; mantle lacquer; matte lacquer; mat(t) varnish; non-glossy paint
无光漆层 flat coat
无光漆发花现象 ghosting; shiner
无光墙(壁)漆 flat wall paint; mat wall paint
无光墙面涂料 flat wall paint
无光清漆 mat(te) varnish; rubbing varnish; flat varnish
无光清漆粉饰 satin finish varnish
无光区 <海或湖中较深的不透光部分> aphotic zone
无光人造丝 pigment rayon
无光栅 no-raster
无光上釉 mat glazing
无光上釉的涂层 mat glazed coat(ing)
无光深水区 aphotic zone
无光饰面 flat finish; mat(te) finish(ing); satin finish
无光搪瓷 chalky enamel; mat(te) enamel
无光搪瓷釉 mat vitreous enamel
无光涂层 flat coat; mat(te) coating
无光涂料 flat paint; matt finish; mat(t) paint
无光线 absence of glare
无光线的 rayless
无光像片 mat(te) picture
无光像纸 mat(te) paper
无光硝基漆 flat lacquer
无光修饰 matt finish
无光研磨 dull polishing
无光焰 dark flame
无光印花布 mat(t) prints
无光油墨 dull black ink
无光油漆 blooming; mat(te) finish(ing); flat paint
无光油漆作业 flatting
无光油性漆 flat oil paint
无光釉 dull glaze; lusterless glaze; mat(te) glaze
无光釉的 matt-glazed
无光釉面砖 non-lustrous glazed tile
无光泽 flat gloss; mat(te); reluster; tarnish decolo(u) ration
无光泽白漆 dead white
无光泽板条 matting strip
无光泽表面 cloudy surface; frosted face; lusterless [lustreless] surface; mat(te) surface
无光泽表面打光 mat-surface glazing
无光泽表面抛光 mat-surface glazing
无光泽表面涂油 flat surface glazed coat(ing); flat surface glazing
无光泽玻璃 matted glass; matt-surface glass
无光泽彩料 mat vitrifiable colo(u) r
无光泽的 dim; flat; frosted; lacklustre; lusterless; mat(te); non-gloss; opaque; unpolished
无光泽的瓷釉 mat enamel
无光泽的亮煤 dull clarain
无光泽的裂化颜料 flat vitriflable colo(u) r
无光泽的墙漆 flat wall paint
无光泽的搪瓷 flat enamel
无光泽断口 mat fracture
无光泽断面 lusterless fracture
无光泽珐琅 flat vitreous enamel
无光泽蜡克 flat lacquer
无光泽类型 mat-pattern
无光泽面 mat(te) finish(ing); matt surface
无光泽面层 flat finish; mat coat
无光泽面层油漆 flat finish paint
无光泽抛光 dead finish
无光泽喷漆 flat lacquer
无光泽漆 mat paint
无光泽清漆 frosting varnish
无光泽饰面(材料) non-lustrous finish
无光泽饰面构件 matt-finish structural facing unit
无光泽搪瓷 flat porcelain enamel
无光泽搪瓷釉 mat porcelain enamel
无光泽涂层 opaque coat
无光泽涂料 dead white; frosted paint; mat(te) paint; non-lustrous finish
无光泽修饰 dull finish
无光泽银幕 mat screen
无光泽油漆 flat gloss oil paint; flat paint; frosted paint; mat(te) paint
无光泽油漆作业 flat spot
无光照片 mat print; matt(e) print
无光罩面 mat(te) coat
无光整理 matt finishing
无光纸 dull-finish(ed) paper; dull-surface paper
无龟裂 freedom from cracking (and crazing)
无规白噪声 random white noise
无规波 random wave
无规博奕 free game
无规不均匀媒质 randomly inhomogeneous medium
无规程性材料 cohesionless material
无规穿孔纤维板 random perforated cellulose fiber tile
无规存取分立地址 random access discrete address
无规的【化】 atactic
无规电子 random electron
无规定 no standard
无规定向纤维 randomly oriented fiber [fibre]
无规断链 random scission
无规分布背景天体 random background object
无规负载 random loading
无规格 non-specification
无规格材料 non-specification material
无规隔行扫描 random interlaced scanning
无规共聚物 random copolymer
无规共聚作用 random copolymerization
无规固溶体 random solid solution
无规过程 random process
无规划建造的房屋 haphazard building
无规划延伸的地区 <市区> sprawled area
无规划运动 random motion
无规交联 random crosslinking
无规聚丙烯 atactic polypropylene
无规聚丙烯改性沥青 atactic polypropylene modified asphalt
无规聚合物 random polymer; unregulated polymer
无规卷曲 random coil
无规卷曲构象 random coil conformation
无规卷曲聚合物 random coiling polymer
无规孔隙模型 random pore model
无规链 random chain
无规灵敏度 random sensitivity
无规律 disorder
无规律的 erratic; irregular
无规律的土壤 erratic soil
无规律地 at random
无规律荷载 random load
无规律曲线 erratic curve
无规律性 irregularity irregular
无规律运输 irregular traffic; irregular transport
无规律运行 irregular movement
无规排列 random array
无规碰撞 random collision
无规取向 random orientation
无规取样示波器 random sampling oscilloscope
无规绕(制)线圈 random coil
无规溶胶 atactose
无规入射 random incidence
无规入射隔声量 random-incidence transmission loss
无规入射灵敏度 random-incidence sensitivity
无规入射声场 random-incidence sound field
无规入射吸声系数 random-incidence sound absorption coefficient
无规入射响应 random-incidence response
无规散射 random-position scattering; random scatter(ing)
无规声 random sound
无规斯塔克效应 random Stark effect
无规速度 random velocity
无规碎裂模型 random fragmentation model
无规填料塔 randomly packed column
无规误差 random error
无规线团模型 random coil model
无规线型共聚物 random linear copolymer
无规相位 random phase
无规谐波 random harmonic
无规行走过程 random walk process
无规旋卷大分子 randomly coiled macromolecule
无规旋卷链 randomly coiled chain
无规源 random source
无规则安放 random-place
无规则编码 hash coding
无规则的 ruleless
无规则的斑点马赛克 dot random mosaic
无规则电流 random electron current
无规则分布 random distribution; randomization
无规则分布噪声 stochastic noise
无规则符合 stray coincidence
无规则干涉图样 random interference pattern
无规则建筑 sporadic building
无规则降解 random degradation
无规则结构网 random network
无规则紧密排列 random dense arrangement
无规则扩展现象 <市区> sprawl phenomenon
无规则裂纹 random crack(ing)
无规则排列 random arrangement
无规则取向 disordered orientation
无规则衰变 random degradation
无规则松散排列 random loose arrangement
无规则填充 random packing
无规则跳动 randomized jitter
无规则湍动 random turbulence
无规则瓦管排水 random tile drainage
无规则网络学说 random network theory
无规则询问 random interrogation
无规则移动扩散 random walk diffusion
无规则游动扩散 random walk diffusion
无规则运动 random motion
无规则运动能 random energy
无规则噪声 random noise; fluctuation noise
无规则振动 random vibration
无规则钻进 random drilling
无规则钻孔 random borehole; random drill hole
无轨采矿 trackless mining
无轨采区 trackless mining area
无轨车辆 trackless vehicle; freewheeled vehicle <美>
无轨道的 non-orbital; non-track
无轨道的隧道 trackless tunnel
无轨道电路区 non-track circuit territory
无轨道路 railless line
无轨的 freewheeled; tailless; trackless
无轨电车 railless tram; railless (trolley) car; trackless troll(e)y car; trackless troll(e)y car; trambus; troll(e)y bus; troll(e)y car; troll(e)y coach; electric(al) troll(e)y; troll(e)y <英>
无轨电车导电杆 troll(e)y jib
无轨电车电动机 troll(e)y-bus motor
无轨电车吊杆 troll(e)y boom
无轨电车吊运车 troll(e)y car
无轨电车架空天线接触杆 troll(e)y pole
无轨电车交通 flexible transport
无轨电车路线 troll(e)y line
无轨电车线 troll(e)y coach route
无轨滑模 non-railed slipform
无轨交通 trackless traffic; trackless transportation
无轨矿井 trackless mine
无轨列车 <由牵引车牵引一系列挂车搬运材料或货物> trackless train; tractor-trailer train
无轨气垫 gyrobus
无轨牵引 railless traction
无轨隧洞掘进(法) trackless tunnel(l)ing
无轨土方运输 railless earthmoving
无轨土方(作业) railless earthmoving
无轨挖运土 trackless earthmoving
无轨运输 flexible transport; rubber-tired haulage; trackless haulage; trackless transportation
无轨装料机 mobile-charging machine
无轨自行矿车 koalmobile
无滚球槽的轴承 unnotched type bearing
无滚子的 norol
无国籍 absent nationality
无国籍船舶 ships without nationality
无国籍人 stateless person
无过错责任 liability without fault
无过滤的管井 open-end well

无过失 no fault
无过失的 unimpeachable
无过失责任制 faultless responsibility
无过盈与间隙配合 metal-to-metal fit
无海岸国家 non-coastal state
无海绵体 no spongy body
无害 innocuity
无害出流水 innocuous effluent
无害的 harmless; innocuous; innoxious; non-deleterious; non-harmful; unharmful; unharming
无害的炭沉积 harmless carbon deposition
无害地段 harmless district
无害发展 development without destruction
无害废水 innocuous effluent; safe waste
无害废物 non-hazardous waste
无害航行 inoffensive navigation
无害化处理 innocent treatment
无害环境影响 no adverse environmental effect
无害排出物 innocuous effluent
无害坡度 harmless grade
无害区 innocuous zone
无害深度假说 harmless-depth theory
无害通过 innocent passage; inoffensive passage
无害通过权 <海洋法> right of innocent passage
无害物质 innocuous substance; innoxious substance
无害性 innocuousness
无害油田废物 non-hazardous oilfield
无害作用阈(值) threshold of no adverse effect
无焊布线 solderless wiring
无焊缝 non-fusion
无焊缝的 weldless
无焊缝钢管 seamless steel pipe; seamless steel tube; weldless steel pipe; weldless steel tube
无焊剂装配 solderless assembly
无焊接端头 pressure terminal; solderless terminal
无焊接合 solderless splice
无焊接头 solderless fitting
无焊连接 solderless connection; solderless joint
无焊料的 solderless
无航海经验的水手 fair weather sailor; flying-fish sailor; green sailor
无号码 no number
无号数的 numberless
无耗的 non-dissipative
无耗电缆 lossless cable
无耗滤波器 lossless filter
无耗散短线 non-dissipative stub
无耗散线 dissipationless line
无耗网络 lossless network; non-dissipative network
无耗元件 lossless element
无合理的替代方案 <废物处置> no reasonable alternatives
无河区 arheic area; arheic region; arheism; riverless
无河区的 ar(h)eic
无核国家 non-nuclear country
无核集 scattered set
无核区 nuclear-free zone
无荷弛度 unloaded sag
无荷弧垂 unloaded sag
无荷载 absence of load(ing); off-load; no-load
无荷载边界 free boundary
无荷载车轮径向偏心 unloaded radial wheel run-out
无荷载除气 off-load deaeration
无荷载的 unburdened; uncharged; unloaded
无荷载的横向阻力 unload lateral resistance
无荷载的稳定性 inequilibrium empty; stability unloaded
无荷载轮胎径向偏心 unloaded radial tire run-out
无荷载摩擦 no-load friction
无荷载喷嘴 no-load jet
无荷载膨胀 non-leaded swelling
无荷载膨胀率 non-leaded swelling rate
无荷载期 off-load period
无荷载下工作 idle motion
无荷载弦 unloaded chord
无荷载指示 no-load indication
无桁架的檩条屋顶 non-trussed purlin(e) roof
无桁架的小跨度屋盖 cottage roof
无横条痕 ring free
无横纹的 unstriated; unstriped
无横纹纤维 involuntary muscular fiber [fibre]; non-striated fiber [fibre]
无红利 ex dividend
无宏观缺陷结构材料 macrodefect free construction material
无宏观缺陷水泥 macrodefect free cement
无后效的 memoryless
无后效过程 Markov process
无后坐力的 recoilless
无后座力气锤 recoilless chipping hammer
无弧的 nonarcing
无弧折的球形储罐 plain spheroid
无护岸河床 unprotected river bed
无护堤的 unembanked
无护航的海上运输队 unescorted convoy
无护航飞行 unescorted flight
无护面的 unpaved
无护栅铁路交叉道 open railway crossing
无护卫的 unescorted
无护罩的 non-sheathed; uncanned
无花果蜡 gondang wax
无花果树 fig-tree; sycamore
无花果树蜡 fig-tree wax
无花果属树木 fig-tree
无花纹铺地瓷砖 plain floor(cover(ing)) tile
无花纹墙纸 plain paper
无花纹园艺用的压延玻璃 plain rolled horticultural glass
无华丽装饰的 inornate
无滑差牵引差速器 positive traction differential
无滑动 freedom from slip(page)
无滑动齿轮 slide [sliding] eliminated gear
无滑动点 zero slip point
无滑动切割 non-sliding cutting
无滑跑起飞 no-run takeoff
无滑跑着陆 no-roll landing
无滑驱动 non-slip drive
无滑润运转 run dry
无滑移传动 slip-free drive
无滑移的差速器 non-slip differential
无滑移理论 no(n)-slip theory
无滑移区 non-slip region; zone of zero slip
无滑移条件 no-slip condition
无滑制动系统 non-skid braking system
无环的 acyclic(al)
无环化合物 acyclic(al) compound
无环量势流 acyclic(al) potential flow; acyclic(al) potential motion
无环流的 circulation-free
无环烃 acyclic(al) hydrocarbon
无缓冲的 non-cushioned; unbuffered
无换热效应 zero heat transfer
无换向器电机 commutatorless machine
无患子【植】soap nut tree
无患子属【植】chinaberry; soapberry
无灰尘的 dustless; non-dusting
无灰的 ash-free; ashless; dust-free; dustless
无灰缝砌缝 non-bonded joint
无灰基 ash-free basis
无灰浆的 mortarless
无灰浆铺砌 dry paving
无灰浆砌石墙 dry stone wall
无灰浆砌筑 brick and brick
无灰焦 free-ash coke
无灰滤纸 ashless filter paper
无灰煤 ash-free coal; free-ash coal
无灰燃料 ash-free fuel
无灰砂浆 non-staining mortar
无灰筛屑 dustless screenings
无灰石屑 dustless screenings
无灰装玻璃 dry glazing; puttyless glazing
无挥发性的 non-volatile
无回答的 unanswered
无回力操纵系统 irreversible control
无回路设备的灌浆系统 open circuit grouting system
无回热系统 non-recuperative system
无回声的 anechoic
无回声房间 anechoic room
无回声区 echoless area; no echo area
无回声室 anechoic room
无回授控制 open cycle control
无回响的 unresounding
无回响房间 anechoic room; non-reverberant room
无回音 no-reply
无回音的 anechoic
无回音室 anechoic chamber
无回音厅 anechoic hall
无汇票 no orders
无混响的 non-reverberant
无活瓣的 valveless
无活动的 non-active
无活动关节的 no-hinged
无活动力的 inert
无活力 debility
无活力的 unvital
无活门的 valveless
无活塞杆的锤作用活塞 free running piston
无活性颜料 inert pigment
无火炊具 fireless cooker
无火的 fireless
无火花 no-spark
无火花的 nonarcing; non-sparking; sparkless; sparkproof
无火花断路 clean break
无火花工具 non-sparking tool
无火花换向 sparkless commutation
无火花区 no-spark zone
无火花性 non-sparkability
无火花整流 sparkless commutation; sparkless rectification
无火花整流器 sparkless commutator
无火花转动 sparkless-run
无火回送位 fireless running position
无火或无动力机车 cold locomotive
无火机车【铁】cold locomotive; dead locomotive; fireless locomotive; dead engine
无火机车装置 dead engine fitting
无火始动 fireless run
无火险的 fire safe
无火焰催化气体加热器 flameless catalytic gas heater
无火焰电离检测器 flameless ionization detector
无火焰发爆 non-flame blasting
无火焰发爆器 buster
无火焰原子化法 flameless atomization
无火焰原子吸收(法) non-flame atomic absorption
无火焰原子吸收分析 non-flame atomic absorption analysis
无火焰原子吸收光谱法 flameless atomic absorption spectrometry
无火蒸汽机 fireless steam loco(motive)
无火蒸汽机车 steam accumulator locomotive
无货车交通 non-commercial traffic; non-traffic
无货供应 nothing available
无货票车 <有车无货票> no-bill car
无机氨氮废水 inorganic ammonia nitrogen wastewater
无机薄膜 inorganic thin film
无机不溶试验 inorganic insoluble test
无机不溶物 inorganic insoluble substance
无机材料 inorganic material
无机沉淀(物) inorganic deposit; inorganic sediment
无机沉积(物) inorganic deposit; inorganic sediment
无机成分 inorganic constituent
无机成因气量 amount of abiogenetic gas
无机成因学说 inorganic origin theory
无机橙红 orange mineral
无机稠化润滑脂 inorganic gel-thickened grease
无机促进剂 inorganic accelerator
无机单磷酸盐 inorganic monophosphate
无机氮 inorganic nitrogen; mineral nitrogen
无机氮肥 inorganic nitrogenous fertilizer
无机的 inorganic; mineral; unorganic
无机毒物 inorganic toxic material
无机发光材料 phosphor
无机防锈剂 inorganic corrosion inhibitor
无机放射性碘 inorganic radio iodine
无机非金属材料 inorganic non-metal(lic) material
无机非金属材料的 ceramic
无机非金属材料学 ceramics
无机肥料 inorganic fertilizer; mineral fertilizer; mineral manure
无机废弃物 inorganic waste
无机废水 inorganic wastewater; mineral wastewater
无机分离膜 inorganic separation membrane
无机分析 inorganic analysis
无机粉沙 inorganic silt
无机敷层 inorganic coating
无机氟化合物 inorganic fluorine compound
无机复合体 inorganic complex
无机复合物 inorganic complex
无机副产物 inorganic byproduct
无机富锌打底料 inorganic zinc rich primer
无机富锌漆 inorganic zinc rich paint
无机富锌涂料 inorganic zinc rich coating; inorganic zinc rich primer
无机覆盖层 inorganic coating
无机高分子复合絮凝剂 inorganic polymer composite flocculant
无机高分子混凝剂 inorganic polymer coagulant
无机高分子聚合物 inorganic polymer
无机高分子絮凝剂 inorganic polymer flocculant

无机隔热材料 inorganic heat insulating material
无机汞 inorganic mercury
无机汞化合物 inorganic mercury compound
无机汞化学品 inorganic mercury chemicals
无机汞化学物 inorganic mercury chemicals
无机汞中毒 inorganic mercury poisoning
无机骨料 inorganic aggregate
无机固体 inorganic solid
无机过氧化物 inorganic peroxide
无机合成染料 inorganic synthetic(al) dye
无机合成实心电阻器 inorganic carbon solid composition resistor
无机合成颜料 inorganic synthetic(al) dye
无机化工 inorganic chemical industry
无机化合物 inorganic compound; mineral compound
无机化学 abiochemistry; inorganic chemistry
无机化学剂量计 inorganic chemical dosimeter
无机化学品污染 abiochemical pollution; inorganic chemical pollution
无机化学试剂 inorganic chemical reagent
无机化学污染物 abiochemical pollutant; inorganic chemical pollutant
无机化学物 inorganic chemicals
无机化学物污染 inorganic chemical pollution
无机化学药品 inorganic chemicals
无机环境 inorganic environment
无机环境因素 abiotic environment(al) factor
无机混凝剂 inorganic coagulant
无机混凝土骨料 inorganic concrete aggregate
无机集料 inorganic aggregate
无机碱 inorganic base
无机建筑材料 inorganic building material; inorganic constructional material; inorganic structural material
无机建筑涂料 inorganic architecture coating
无机胶结剂 inorganic cement
无机胶黏[粘]剂 inorganic adhesive
无机胶凝材料 inorganic cement; inorganic cementing material
无机胶凝复合材料 inorganic-bonded composite
无机胶凝木纤维材料 wood fiber inorganic composite
无机胶凝黏[粘]结层 coating by inorganic cementitious bonding
无机胶态 mineral colloid
无机胶体 inorganic colloid; mineral colloid
无机胶体凝胶润滑脂 inorganic colloid-gelled grease
无机胶质 mineral colloid
无机结合剂 inorganic bond
无机结合料 inorganic binder; inorganic binding agent
无机界 inorganic sphere; inorganic world
无机晶体 mineral crystal
无机晶体闪烁器 inorganic crystal scintillator
无机聚合物 inorganic polymer
无机绝缘 inorganic insulation
无机绝缘材料 inorganic insulation material
无机绝缘电缆 copper-sheathed cable; mineral-insulated cable
无机绝缘体 inorganic insulator
无机颗粒 inorganic particle
无机颗粒物 inorganic particulate matter
无机垃圾 inorganic refuse
无机离子 inorganic ions
无机离子含量 inorganic ion content
无机离子交换 inorganic ion exchange
无机离子交换剂 inorganic ion exchanger
无机离子交换膜 inorganic ion exchange membrane
无机离子交换纸 inorganic ion exchange paper
无机磷 inorganic phosphorus
无机磷光漆 inorganic phosphorescent paint
无机磷酸盐 inorganic phosphate
无机硫 inorganic sulfur
无机硫酸盐 mineral sulfate
无机铝盐防水剂 inorganic alumin(i)um salt waterproofing agent
无机络合物 inorganic complex
无机膜 inorganic membrane; mineral membrane
无机膜-生物反应器 inorganic membrane-bioreactor
无机能的 non-functioning
无机泥酸 mud acid
无机黏[粘]合剂 inorganic binder
无机黏[粘]合耐热氟金云母纸层压板 inorganic-bonded heat resistant fluorophlogopite plate
无机黏[粘]结剂 inorganic binder; mineral binder bond
无机黏[粘]土 inorganic clay
无机凝结剂 inorganic coagulant
无机农药 inorganic pesticide
无机膨润土 inorganic bentonite
无机漆 inorganic paint
无机漆基 inorganic binder
无机气体 inorganic gas
无机氢氧化物磷灰石 mineral hydroxyapatite
无机溶剂 inorganic solvent
无机溶解物 inorganic dissolved substance
无机溶解组分 inorganic dissolved component
无机溶质 inorganic solute
无机弱酸 inorganic weak acid
无机杀虫剂 inorganic insecticide
无机杀菌剂 inorganic fungicide
无机闪烁晶体 inorganic scintillation crystal
无机砷化合物 inorganic arsenic chemicals
无机生境 abiocoen
无机石油化学品 inorganic petrochemicals
无机树脂 inorganic resin; mineral resin
无机水化学 inorganic hydrochemistry
无机酸 inorganic acid; mineral acid
无机酸度 mineral acidity
无机酸酸度 inorganic acidity
无机酸营养湖 inorganic acidotrophic lake
无机酸酯 inorganic acid ester
无机燧石 inorganic chert
无机碳 inorganic carbon
无机碳酸 inorganic carbonic acid
无机碳酸钙 inorganic limestone; mineral limestone
无机碳体系 inorganic carbon system
无机陶瓷膜 inorganic ceramic membrane
无机填料澄清池 inorganic filler clarifier
无机涂层 inorganic coating
无机涂料 inorganic coating
无机土(壤) inorganic soil
无机微胶粒润滑脂 inorganic microgel grease
无机微颗粒 inorganic microparticle
无机污垢 inorganic foulant
无机污泥 inorganic sludge; mineral sludge
无机污染物 inorganic pollutant
无机污水 inorganic wastewater
无机物 inorganic matter; inorganic substance; mineral inorganic substance; minerals
无机物层面印痕 abiogloph
无机物沉积速率 rate of mineral precipitation
无机物稠化的润滑脂 inorganic base grease
无机物分析 mineral analysis
无机物复合型混凝剂 inorganic complex type coagulant
无机物工艺学 technology of inorganic chemicals
无机物护面 inorganic coating
无机物浓度 inorganic concentration
无机物溶解速率 rate of mineral dissolution
无机物水污染 water pollution by inorganic substance
无机物填充的电位器 inorganic solid composition potentiometer
无机物填料 inorganic filler
无机物外加剂 mineral admixture
无机物污染 inorganic pollution; mineral pollution
无机物指数 inorganic index
无机物质 dead-matter; inorganic material; inorganic substance; mineral matter; inorganic matter; mineral load <流水挟带的>
无机物组分 inorganic component; inorganic constituent
无机吸附剂 inorganic adsorbent; mineral adsorbent
无机吸着剂 inorganic sorbent
无机稀释剂 mineral thinner
无机锡 inorganic tin
无机纤维 inorfil; inorganic fiber [fibre]; mineral cord; mineral fiber [fibre]
无机纤维板 inorganic fiberboard; mineral fiber board
无机显微组分 inorganic microconstituent
无机橡胶 inorganic rubber
无机械致冷的冷藏车皮 refrigerated truck
无机新 advanced inorganic material
无机性粉尘 inorganic dust
无机絮凝剂 inorganic coagulant; inorganic flocculant
无机悬沙 inorganic suspended sediment
无机盐 inorganic salt; mine salt
无机盐生产 inorganic salt production
无机盐污染 inorganic salt pollution
无机颜料 inorganic pigment; mineral pigment
无机夜光漆 inorganic phosphorescent paint
无机营养 inorganic nutrition; mineral nutrition
无机营养物 inorganic nutrient
无机有毒物质 inorganic toxic material
无机有机高分子复合絮凝剂 inorganic-organic polymer composite flocculant
无机淤泥 inorganic silt
无机制剂 inorganic formulation
无机质 inorganic substance
无机质泥沙 inorganic silt
无机质谱法 inorganic mass spectrometry
无机致癌物 inorganic carcinogen
无机综合物 inorganic complex
无唧泥现象 <混凝土板> non-pumpers
无积沉底 hard ground
无基槽建筑物 trenchless construction
无基础的 groundless
无基金公债 unfunded debt
无基台模型 models without a sill
无基托的玻璃纤维卷材和垫片 sta-fit
无畸变 distortion-free; freedom from distortion; orthoscopy
无畸变波 undistorted wave
无畸变成像 distortion-free image formation; undistorted image formation
无畸变传输 distortionless transmission
无畸变的 orthoscopic; undistorted; distortionless
无畸变电路 distortionless circuit
无畸变观察 undistorted viewing
无畸变光具组 orthoscopic optic(al) system
无畸变记录 distortionless recording
无畸变校正仪 distortion-free correcting instrument; non-distortion correcting instrument
无畸变立体模型 orthoscopic stereomodel; undistorted stereomodel
无畸变立体像对 orthoscopic stereopair; undistorted stereopair
无畸变面 plane of no distortion
无畸变模式 undistorted mode
无畸变模型 undistorted model
无畸变目镜 orthoscopic eyepiece; orthoscopic ocular
无畸变全息图 distortion-free hologram; non-distortion hologram; undistorted hologram
无畸变视图 undistorted view
无畸变图像 orthoscope image; orthoscopic image
无畸变系统 orthoscopic system; rectilinear system
无畸变线 line of no distortion
无畸变线路 distortionless line
无畸变虚像 distortion-free virtual image
无激波 unshock
无激波的 unshocked
无激励检测继电器 no-excitation detection relay
无激励接触器 no-excitation contactor
无激励谐振器 unperturbed resonator
无级变量 infinitely variable
无级变量调节 infinite variability
无级变速 infinitely variable speed; infinite speed variation; infinite variation of ratio; infinitive stage transmission; progressive gear; progressive transmission; steplessly variable speed; stepless speed change
无级变速齿轮箱 infinitive variable gear box
无级变速传动机构 infinitely variable speed gearing
无级变速传动(装置) indefinitely variable transmission; infinitely variable speed drive; stepless variable drive; vari-speed drive; adjustable speed drive; infinitely variable speed transmission; stepless drive
无级变速范围 range of infinitely variable speeds
无级变速控制 stepless speed control
无级变速皮带 adjustable speed belt
无级变速器 buncher; speed variator; stepless transmission; variable drive; variable-speed unit; variator

无级变速驱动 stepless variable drive
无级变速调节 infinite variability
无级变速箱 stepless transmission
无级变速型 stepless speed changing type
无级变速液压传动 speed fluid drive
无级变速转动装置 vari-speed drive
无级变速装置 infinite variable speed mechanism; stepless speed change device; variable-speed unit
无级传动 stageless transmission
无级传动机构 stepless gear
无级船 unclassed ship
无级的 stepless
无级电压调整 stepless voltage regulation
无级加速 stepless acceleration
无级控制 stepless control
无级摩擦式传动 stepless friction transmission
无级配材料 ungraded material
无级配骨料 ungraded aggregate
无级配沥青混凝土 open-graded asphalt(ic) concrete
无级液力变矩器 variable-speed fluid drive
无级液力变速器 variable-speed hydraulic transmission
无级液体静压传动器 infinitely variable hydrodynamic(al) transmission
无极变速 variable speed
无极变速器 infinitely variable transmission; non-stage transmission
无极带 endless belt
无极带式抛掷充填机 endless thrower
无极带式洗矿槽 automatic strake
无极的 apolar; endless; neutral; poleless
无极端变形 free from extreme deformation
无极放电灯 electrodeless discharge lamp
无极分子 non-polar molecule
无极钢缆 endless cable
无极钢绳矿车运输 endless rope carhaul
无极钢索 endless rope
无极轨道电路 neutral track circuit
无极化合物 homopolar compound
无极继电器 neutral relay; non-polarized relay
无极加强接点继电器 neutral relay with heavy-duty contacts
无极价 homopolar valency
无极键联 homopolar binding
无极接点 neutral contact
无极缆道 endless cableway
无极链 endless chain
无极链斗式提升机 continuous bucket elevator
无极链给料机 endless chain feeder
无极链炉箅 endless chain grate
无极链式多斗挖掘机 endless chain trench excavator
无极链式给矿机 endless chain feeder
无极链式截煤机 endless chain coal cutter
无极偏极继电器 bias(s)ed-neutral relay
无极期的 acritical
无极绳 continuous rope; endless belt; endless rope
无极绳传动 continuous rope drive
无极绳挂链 lashing chain
无极绳耙斗搬运 endless rope scraper haulage
无极绳牵引 endless rope traction
无极绳运输机 endless rope haulage
无极绳运输系统 endless rope system
无极绳运输主要平巷 endless main
无级调节 infinitely regulation; infinitely variable control; stepless regulation
无级调节传动 stepless control gearing
无级调速 infinite speed variation; stepless speed control; stepless speed regulating; stepless speed regulation; stepless speed variation
无级调速传动装置 fully adjustable speed drive
无级调速器 stepless speed adjusting gear
无级调速伺服电动机 stepless servomotor
无级调压 stepless voltage regulation
无级调压器 stepless voltage regulator
无级调整 infinitely variable adjustment
无极衔铁 neutral armature; non-polarized armature
无极性 non-polarity
无极性归零记录方式 non-polarized return-to-zero
无极性化合物 non-polar compound
无极性键 homopolar bond; homopolar link
无极性胶体 homopolar colloid
无极性结合 non-polar binding
无极液体 non-polar liquid
无极音响机 non-polarized sound
无极运输机 endless conveyer [conveyor]
无脊椎动物的 invertebrate
无脊椎动物化石 non-vertebrate fossil
无脊椎动物区系 invertebrate fauna
无计划的 unplanned
无计划地延伸 <建筑物等> sprawl
无计划建筑 sporadic building
无计划钻探 wildcat
无记录 no record
无记名背书 general endorsement [indorsement]
无纪律 indiscipline
无技术的 unaccomplished
无季风期 non-monsoon period
无寄名受托保管收据 bearer depository receipts
无寄生图像的电视 ghost-free television
无加感线路 unloaded circuit
无加劲板 unstiffened plate
无加劲平板 unstiffened plate
无加劲式吊桥 unstiffened suspension bridge
无加力燃烧室的 unreheated
无加气剂混凝土 non-air-entrained concrete
无加强高的焊缝 flush weld
无加速条件 unaccelerated condition
无加温盘管罐车 no-piped tank car
无加温污泥消化槽 none heated sludge digestion tank
无夹圈滚动轴承 cageless anti-friction bearing
无家可归者 homelessness
无家游民聚集地区 skid-row
无甲板船 open boat; open vessel; undecked boat; undecked ship; undecked vessel
无甲板的 undecked
无价的 non-valent
无价样品 sample post
无价值 futility
无价值的 hungry; inutile; naught; no-(ac) count; unvalued; unworthy; useless; vain; valueless; weedy; worthless
无价值的东西 fluff; nought; rap
无价值矿地 bull pup
无价值土地 submarginal land
无价值物品 mud
无价值修剪 futility cut-off
无价值状况 anomie
无架乘式犁 frameless sulky plough; frameless sulky plow
无架挂车 frameless trailer
无架式油灌车 frameless tanker
无架凿岩机 unmounted drill
无尖头的 pointless
无尖削的 untapered
无间断的 unremitting
无间隔带 gapless tape
无间隔的 non-septate
无间隙 zero clearance; not free <机械零件之间>
无间隙的 gapless; unspaced
无间隙对接焊缝 unspaced butted weld
无间隙对接接头 tight butt joint
无间隙焊缝 closed weld
无间隙接头 blind joint; closed joint
无间隙结构 gapless structure
无间隙金属氧化物避雷器 metal oxide surge arrester without gaps
无间隙气门挺杆 zero lash valve lifter
无间隙扫描 scanning without gap
无间歇拌和(法) <指水泥混合料加水后立即拌和> zero-hour mixing
无间歇搅拌(法) zero-hour mixing
无监督操作 unattended operation
无监督分类法 unsupervised classification
无监控的 unmonitored
无监控系统 unmonitored control system
无监视的控制系统 unmonitored control system
无减速器发动机 ungeared engine
无减速伞着陆 unchuted landing
无检核点 <无多余观测交会的补点位置> non-check point; non-check position
无碱玻璃 non-alkali glass
无碱玻璃纤维 non-alkali glass fiber [fibre]
无见证人的手书遗嘱 holograph will
无建筑物的 unbuilt
无建筑物地区 unbuilt area
无建筑物区域 open space
无建筑翼部 transeptless
无鉴别力的 undiscriminating
无键插座 keyless socket
无键的夹具 keyless chuck
无键灯头 keyless socket
无键共振 no bond resonance
无键机构 keyless work
无键联轴节 keyless coupling
无键振铃 keyless ringing
无槛 sill loss
无浆的 unsized
无浆缝 dry joint; open joint
无浆干砌 dry jointed
无浆连接 ground joint
无浆砌墙 dry wall(ing)
无浆砌墙板 drywall
无浆砌体 dry masonry
无浆砌砖 steening
无奖金 ex bonus
无奖金计时工资 plain time rate
无交叉的 achiasmate
无交错突出的 nesting
无交换网络 non-switched network
无交流声 hum free
无交通往来的 trafficless
无胶的 unsized
无胶结 non-cementation
无胶结孔隙度 minus-cement porosity
无胶绝缘子 cementless insulator
无胶区 cementless area
无焦透镜 afocal lens
无焦系统 afocal system
无焦油煤气 clear gas
无角的 acerous
无角拱 fixed-end arch
无绞刀吸扬挖泥船 cutterless suction dredge(r)
无脚手架安装 erection without scaffolding
无铰的 no-hinged; without articulations
无铰刚构架 hingeless frame
无铰拱 arch without articulation; arc with articulation; fixed arch (at both ends); fixed-end arch; hingeless arch; no-hinged arch; non-articulated arch
无铰拱桥 fixed arch bridge; hingeless arch bridge
无铰拱桥梁桥 hingeless arch bridge
无铰构架 no-hinged frame
无铰(接)大梁 no-hinged girder
无铰接的 no-hinged
无铰接(发)券 no-hinged arch
无铰接拱 no-hinged arch
无铰接横梁 no-hinged girder
无铰接立柱 no-hinged support
无铰接梁 no-hinged beam
无铰接支撑 no-hinged support
无铰接支座 no-hinged support
无铰接柱 no-hinged column
无铰框架 no-hinged frame
无铰链的 hingeless; no-hinged
无铰支点 hingeless support
无搅拌装置的混凝土运料车 non-agitating truck; non-agitating unit; non-agitator truck
无搅动设备的混凝土运输车 non-agitating unit
无阶的 stepless
无阶段的全制动 non-progressive full (brake) application
无接触 non-contact; no touch
无接触测量法 non-contact measurement
无接触聪明卡 contactless smart card
无接触的 contactless
无接触干燥机 non-touch drier [dryer]
无接触激光传感器 contactless laser sensor
无接触控制元件 contactless control element
无接触量测法 non-contact measurement
无接触式记录 non-contact recording
无接触转速计 contactless tachometer
无接地搜索选择器 absence-of-ground searching selector
无接点部件 contactless unit
无接点插件 contactless plug unit
无接点单元 contactless unit
无接点的 contactless
无接点电子部件 contactless electronic component
无接点电子元件 contactless electronic component
无接点交流自整角机 telegon
无接点开关 contactless switching
无接点开关系统 contactless switching system
无接点逻辑电路 contactless logic circuit
无接点元件 contactless element
无接点转接 contactless switching
无接点转接系统 contactless switching system
无接缝的 jointless
无接缝接头 blind joint
无接缝铺地板法 Induroleum
无接缝水磨石地板 terrazzo jointless

floor(ing)
无接缝装饰地板 decorative jointless floor(ing)
无接箍套管 integral casing
无接头的 jointless
无接头管道 continuous pipe without joint
无接头环带 endless
无接头循环 endless loop
无接头油管 endless tubing
无节疤 free of knots
无节疤的 knot-free
无节疤木材 clear stuff; clear timber; clear wood
无节钢 plain bar(of reinforcement)
无节钢筋 plain bar (of reinforcement); plain reinforcement (bar); plain reinforcing bar
无节加强筋 plain reinforcement rod
无节木水管 continuous stave pipe
无节拍 clockless
无节新材 bodywood
无节制 extravagance
无节制的 inordinate
无节制地采伐森林 uncontrolled clearance
无节制地排放 uncontrolled emission
无节制地倾弃 fly tipping; uncontrolled tipping
无节制地清除 uncontrolled clearance
无结疤的木料 clean timber
无结的 knotless
无结构的 structureless; unorganized
无结构腐殖体 collinite; homocollinite
无结构胶凝体 collinite
无结构镜煤 collain; eu-vitrain
无结构镜质体 collinite; eu-vitrinite
无结构泥炭 amorphous peat
无结构式计算机 architecture free machine
无结构式信息处理机 architecture free processor
无结构土(壤) structureless soil
无结构纤维 unstructured fibre [fiber]
无结果的 unproductive
无结果的表决 inconclusive vote
无结果的投票 inconclusive ballot
无结果的选举 inconclusive ballot
无结合料层 unbound course
无结合料骨料 unbound aggregate
无结合料基层 unbound base
无结合料集料 unbound aggregate; untreated aggregate
无结合料路基 unbound(road) base
无解 non-solution
无介质粉磨 antogenous grinding
无介质磨 antogenous mill
无介质磨矿 run-of-mine-milling
无界变量 unbounded variable
无界的 unbounded
无界覆盖面 unbounded covering surface
无界函数 unbounded function; unlimited function
无界河流二维稳态水质模型 two-dimensional steady water quality model of non-boundary river
无界集 unbounded set
无界解【数】 unbounded solution
无界空间 unbounded space
无界实数集 unbounded set of real numbers
无界水流 unbounded stream
无金属丝的 unwired
无筋 non-reinforcement
无筋钢筋 plain reinforcement(bar)
无筋钢丝 plain wire
无筋混凝土 non-reinforced concrete; plain concrete; plane concrete; unreinforced concrete
无筋混凝土板 non-reinforced slab
无筋混凝土单个基础 plain concrete single base
无筋混凝土地基 plain concrete footing
无筋混凝土墩台基础 plain concrete foundation pier
无筋混凝土管 non-reinforced concrete pipe
无筋混凝土块体 plain concrete block
无筋混凝土路面 non-reinforced concrete pavement; unreinforced concrete pavement; unreinforced surface
无筋混凝土路面板 unreinforced concrete slab; non-reinforced concrete slab
无筋混凝土墙 plain concrete wall
无筋扩展基础 non-reinforced spread foundation
无筋路面 unreinforced surface
无筋路面板 non-reinforced concrete slab
无筋喷浆混凝土 unreinforced shotcrete
无筋平接缝 plain butt joint
无筋砌体 non-reinforced masonry; plain masonry; unreinforced masonry
无筋实心砖砌体 plain solid brick masonry
无筋砖建筑物 unreinforced brick building
无筋砖砌体 non-reinforced brick masonry; unreinforced brick masonry
无尽光栅 endless raster
无尽小数 infinite decimal; non-terminating decimal; unlimited decimal
无进位 no-carry
无进位加(法) false add
无经济价值气田 non-economic gas field
无经济价值油井 non-commercial well
无经济价值油气层 non-economic oil-gas bed
无经济价值油田 non-economic oil field
无经验的 callow; fresh water; half-baked; inexperienced
无经验的操作 green operation; McGee
无经验的驾驶员 untrained driver
无经验的人 greener; green hand
无经验的水手 lubber
无晶体振荡器 non-crystal oscillator
无晶锥 acone
无井管孔 open hole; uncased(bore) hole
无井楼梯 platform stair(case)
无径流区域 noncontributing area
无净化装置发动机 uncontrolled engine
无竞争的 uncompetitive
无竞争的出价 non-competitive bid(ding)
无竞争力的价格 non-competitive price; price not-competitive
无竞争能力的报价 non-competitive bid(ding)
无竞争能力的交通业务 non-competitive traffic
无静差摆 astatic pendulum
无静差环节 floating component
无静差控制 floating control; non-corresponding control
无静差控制器 floating controller
无静差调节 astatic governing; floating control
无静差调节器 astatic regulator
无静差元件 floating component
无静差作用 floating action; floating response; reset response
无静差作用速度 floating speed
无静电干扰的 non-static
无静电荷的 non-static
无救助利益 without benefit of salvage
无居民的岛 uninhabited island
无拘束区 <有害空间> unguided running section
无绝缘导线 uninsulated conductor
无绝缘电刷 non insolution brush
无绝缘电线 uninsulated wire
无绝缘节轨道电路 jointless track circuit
无绝缘节音频轨道电路 jointless audio frequency track circuit
无觉察到的毒性浓度 no detected toxic concentration
无觉察到的有害效应水平 no-observed-adverse-effect level
无均性的 heteropic(al)
无菌 asepsis; sterile
无菌操作 abacterial operation; asepsis; sterile working
无菌操作法 aseptic manipulation
无菌操作柜 aseptic manipulation cabinet; sterile cabinet
无菌操作室 aseptic manipulation room; sterile room
无菌产品 sterile product
无菌的 abacterial; aseptic; bacteria free; bioclean; germ-free; sterile
无菌动物 germ-free animal
无菌洞巾 aseptic hole-towel
无菌发生器 sterile generator
无菌房间 antiseptic room
无菌废水 sterilized wastewater
无菌隔离室 germfree isolator
无菌灌装 sterile filling
无菌过程 gnotobiosis
无菌过滤 aseptic filtration
无菌环境 gnotobasis
无菌技术 aseptic technique
无菌检验 steriling test
无菌胶乳 sterile latex
无菌净化室 aseptic clean room
无菌淋洗发生器 sterile elution generator
无菌培养 aseptic culture; axenic culture; sterile culture
无菌去离子水 sterile deionized water
无菌容器 sterile chamber
无菌溶液 sterile solution
无菌生产 aseptic production
无菌生物采样器 sterile biologic(al) sampler
无菌生物学 gnotobiology
无菌试验 sterility test
无菌室 bioclean bed; disinfection chamber; germless chamber; sterile room
无菌水 sterilized water
无菌填充 aseptic filling
无菌无离子水 sterile deionized water
无菌箱 sterile board
无菌橡胶膏 sterile adhesive plaster
无菌性 aseptic
无菌蒸馏水 sterile distilled water
无菌汁 processed juice
无开采价值的脉石 protore
无开关灯口 keyless socket
无开口舱壁 unbroken bulkhead
无开口的 astomous
无开口圬工墙 blank masonry wall
无看护的坝 dam without attendance
无抗电路 non-reactive circuit
无抗电路元件 non-reactive circuit element
无抗负载 non-reactive load
无抗倾斜交叉撑的框架 unbraced frame
无抗阻抗 non-reactive impedance
无靠背的 unbacked
无靠手木条的舷缘 open gunwale
无科学根据 rule-of-thumb
无颗粒型 agranular type
无壳套桩 uncased pile
无壳现浇桩 poured-in-place uncased pile
无壳桩 shell-less pile
无可比的 unexampled
无可比拟的 inapproachable; without equal
无可非议的 irrefragable; irrefutable; irreproachable; unquestionable
无可见有害作用水平 non-observable-adverse-effect level; non-observed-adverse-effect level
无可见作用水平 non-observable-effect level; non-observed-effect level
无可扣押之物 nulla bona
无可匹敌的 unapproachable
无可燃气体证书 gas free certificate
无可商讨的 non-negotiable
无可异议的推定 conclusive presumptions
无可争辩的 indisputable; unarguable; undeniable
无可争论 beyond dispute
无可争论的 indisputable
无可争议 no justification; without dispute
无可指摘的 unimpeachable
无刻度调节器 non-indicating controller
无刻度移液管 plain pipet(te)
无刻痕板材 clear stock
无坑木支撑的隧道 untimbered tunnel
无坑木支的 untimbered
无空白模件板 plain blank board
无空化性能 cavitation-free performance
无空化运行 cavitation-free operation
无空间电荷影响的情况 space-charge-free behavio(u)r
无空气的 air-free; airless; anaerobic
无空气火焰 non-aerated flame
无空气抛丸清理机 airless shot blasting installation
无空气抛丸清理装置 airless shot blasting installation
无空气喷漆 airless paint spraying
无空气喷漆涂装 airless spray painting
无空气喷枪 airless spray gun
无空气喷砂清理 airless blast cleaning
无空气喷涂 airless spraying
无空气喷涂装置 airless spraying equipment
无空气水 air-free water; de-aired water
无空蚀 free of cavitation
无空蚀性能 cavitation-free performance
无空蚀运行 cavitation-free operation
无空窝燃烧 non-raceway burning
无空隙 zero air void
无空隙的 void-free
无空隙浇注法 void-free cast method
无空隙密实度 <测定沥青混合料时用> zero air voids density
无空线(通信[讯]) no-lines
无孔表面 non-porous surface
无孔材 non-pored wood; non-porous timber; non-porous wood
无孔材料 pore-free material
无孔插件板 plain module board
无孔的 imporous; non-porous; porous-free; unpunched
无孔的状态 imperforation

无孔洞海墙 non-porous seawall
无孔法兰(盘) blank flange; blind flange; cast flange
无孔管 blank pipe; blind pipe
无孔花格窗 blank tracery; blind tracery
无孔混凝土 voidless concrete
无孔接头 joint without hole
无孔截头锥形筒离心机 bowl centrifuge
无孔颗粒 non-porous particle
无孔离心机 solid wall centrifuge
无孔密封垫 unpunched gasket
无孔膜 non-porous film
无孔桥 non-opening bridge
无孔碳 Karbate
无孔套管 blank casing
无孔凸缘 blank flange; blind flange; cast flange
无孔凸缘管堵 black flange
无孔稳定涂层 non-porous stable coating
无孔物质 voidless mass; voidless matter; voidless substance
无孔隙 imporosity
无孔隙的 imperforated; imporous; void-free; voidless
无孔隙复合材料 zero-porosity composite
无孔隙固体 non-porous solid
无孔隙体 pore-free mass
无孔屑纸带 chadded tape; chadless tape
无孔心轴 solid spindle
无孔性 imporosity; non-porosity
无孔纸带 virgin paper tape
无孔纸带卷 virgin paper-tape coil
无孔制品 non-porous article
无孔轴承 non-porous bearing
无孔砖 uncored brick
无孔转鼓 solid bowl
无孔转鼓离心机 solid bowl centrifuge
无控燃烧期 uncontrolled combustion phase
无控振荡 uncontrolled oscillation
无控制的发展 uncontrolled development
无控制的拦阻射击 uncontrolled barrage
无控制的裂纹 uncontrolled crack
无控制地区 uncontrolled area
无控制点镶嵌图【测】 uncontrolled mosaic; unchecked mosaic
无控制泛滥 uncontrolled flooding
无控制方式 non-control system
无控制交叉口 uncontrolled intersection
无控制交通系统 uncontrolled traffic system
无控制井 well out of control
无控制流(动) uncontrolled flow
无控制倾倒 uncontrolled dumping
无控制倾弃 fly tipping
无控制燃烧 uncontrolled combustion
无控制人行横道 <行人可优先于车辆通过的人行横道> uncontrolled pedestrian crossing; uncontrolled pedestrian cross walk
无控制设备的水库 uncontrolled storage
无控制通风 uncontrolled ventilation
无控制蓄水 uncontrolled storage
无控制堰 uncontrolled weir
无控制溢洪道 uncontrolled spillway
无口湖 lake without outlet
无扣留权保证金 no lien bond
无库存 not in stock
无库容的水电站 plant without storage
无库容电站 plant without storage

无快门照相机 shutterless camera
无矿地层 bare ground; dead ground
无矿地带 barren spots; barren zone
无矿基岩 <无锡砾石下的> kong
无矿物的 barren
无矿岩层 dead ground
无矿岩体 barren pluton
无矿钻孔 barren hole
无框玻璃 unframed glass
无框玻璃窗 frameless glass window
无框玻璃门 frameless glass door
无框扯窗 frameless sliding window
无框格窗 sashless window
无框(骨)架的 frameless
无框固定玻璃 fast sheet
无框架的窗玻璃 frameless glass
无框架底盘 frameless chassis
无框架汽车底盘 frameless chassis
无框架陀螺仪 gimballess gyroscope
无框镜 Venetian mirror
无框门 frameless door; unframed door
无框天窗 hipped skylight
无框推拉窗 frameless sliding window
无亏曲线 curve of deficiency zero
无亏损的 break-even
无亏损品位下限 break-even cut-off ore grade
无亏损最低品位 break-even cut-off ore grade
无困难 trouble-free
无扩散型相变 diffusionless transformation
无扩展疲劳裂缝 non-propagating fatigue crack
无括号表示法 Lukasawiez notation; parenthesis-free notation; Polish notation; prefix notation
无括号的 parenthesis-free
无拉杆板桩 cantilever sheet pile [piling]
无拉力材料 no-tension material
无拉力的 tension-free
无拉力分析 non-tension analysis
无拉平的着陆 unflared landing
无喇叭口的 unflared
无蜡的 wax-free
无蜡石油 wax-free (crude) oil
无蜡原油 wax-free crude
无来往 nothing between
无栏杆拖车 trunk platform
无栏杆旋梯的井孔 hollow newel
无栏木道口 unbarred road crossing; ungated level crossing
无缆常压潜水器 untethered atmospheric submersible
无缆式潜水 untethered diving
无缆式潜水器 cableless submersible; tetherless submersible; untethered vehicle
无缆遥控潜水器 untethered remotely-operated vehicle; untethered unmanned vehicle
无缆自航式潜水器 untethered self-propelled vehicle
无浪 calm(-glassy) sea; smooth sea
无劳动力能力 disability status
无雷声闪电 heat lightning
无肋多层薄壳 multiple ribless shell
无肋骨的 ecostate; ribless
无肋梁筏片基础 floating foundation non-costae girder
无肋片燃料元件 unfinned fuel element
无肋条的 unribbed
无棱角骨料 rounded aggregate
无棱角集料 rounded aggregate
无棱角砾石 rounded aggregate
无棱角砂 buckshot sand
无楞铁皮 plain sheet iron

无楞瓦 plain tile
无冷凝的 non-condensing
无冷凝房屋 condensation free dwelling
无冷却的 non-refrigerated
无冷却剂堆 uncooled reactor
无离析装料 non-segregated filling
无离子水 de-ionized water; ion-free water
无理不变式 irrational invariant
无理部分 irrational part
无理逮捕 unjustified arrest
无理单位制 irrational system of units
无理单项式 irrational monomial expression
无理的拖延 unreasonable delay
无理对合 irrational involution
无理方程 irrational equation; radical equation
无理根【数】 irrational root
无理函数 irrational function
无理合同 unconscionable contract
无理量 irrational quantity
无理实数 irrational real number
无理式 irrational expression
无理数【数】 irrational number; irrationality
无理数方程 irrational equation
无理性的 unreasonable
无理由的 unfounded
无理指数 irrational exponent
无力 impotence; inertia; weakness
无力偿付 insolvency; suspension
无力偿付债务 insolvency
无力偿还 insolvency
无力偿还者 insolvent
无力磁场 force-free field; force-free magnetic field
无力的 flabby; nerveless; powerless; unable
无力感 sense of powerlessness
无力还债 insolvency
无力还债的 bankrupt
无力还债者 insolvent
无力矩 moment-free
无力矩理论 shell theory without moments
无力矩油罐 moment free tank
无力支持 suspend
无力支付 financial in solvency; non-payment; suspend payment
无立木地 unstocked blank
无立柱支撑 <基坑支护用> flying shore
无利的 profitless
无利可图 detachment of interest
无利可图的 unremunerative
无利可图的条件 unremunerative terms
无利息 ex interest
无利息的 non-interest; non-interest bearing
无利益的 unprofitable
无沥青的 non-bituminous
无例外 cent per cent; without exception
无连接 no connection
无连接的过梁 loose lintel
无连接器 connectorless
无连接协议 connectionless protocol
无联结 non-bond
无联系的 unassociated; unattached
无联系梁的转向架 truck with no transversal connecting beam
无联运运输 no through traffic
无梁 beamless
无梁板 beamless slab; flat slab; capping slab
无梁板结构 flat slab construction; flat slab structure

无梁板结构中间板带 middle strip
无梁板柱结构 flat slab capital construction; flat slab column construction
无梁的 girderless
无梁地板 slab floor
无梁殿 beamless hall; brick-vault hall
无梁格栅楼面 bridging floor
无梁(桁架)平顶【建】 ceiling free of trussing
无梁楼板 flat slab floor; girderless floor; slab floor (cover(ing))
无梁楼板构造 flat slab construction
无梁楼板建筑 flat slab construction
无梁楼板结构 flat slab floor construction; mushroom slab construction
无梁楼板终饰 slab floor(ing) finish
无梁楼盖 <带柱帽的> flat slab floor; girderless floor; flab slab
无梁楼盖的柱帽 drop panel
无梁楼盖构造 girderless floor construction
无梁楼盖结构 girderless floor construction
无梁楼盖托板 <在无梁楼盖结构中围绕柱顶加厚的托板> drop panel
无梁楼盖系统 girderless floor system
无梁面板 girderless deck; slab decking; flat slab
无梁面板高桩码头 wharf of precast reinforced concrete slab without beam
无梁平板构造 flat slab beamless construction; flat slab construction
无梁平顶 ceiling without trussing
无梁天顶 ceiling free of trussing; ceiling without trussing
无梁天花板 flat slab ceiling
无粮上浆 non-grain sizing
无量 incalculability
无量纲 non-dimension; zero dimension
无量纲比 non-dimensional ratio
无量纲变量 dimensionless variable
无量纲变数 non-dimensional variable
无量纲参数 dimensionless group; non-dimensional parameter; dimensionless parameter
无量纲单位过程线 dimensionless unit hydrograph; non-dimensional unit hydrograph
无量纲的 dimensionless; non-dimensional
无量纲反应系数 dimensionless response factor
无量纲反应因数 dimensionless response factor
无量纲方差 dimensionless variance
无量纲分析 non-dimensional analysis
无量纲过程线 dimensionless hydrograph; non-dimensional hydrograph
无量纲亨利常数 dimensionless Henry constant
无量纲活度量 dimensionless activity quantity
无量纲量 dimensionless quantity; non-dimensional quantity
无量纲频率 dimensionless frequency; non-dimensional frequency
无量纲时间函数 dimensionless time function; non-dimensional time function
无量纲数 dimensionless number; non-dimensional number
无量纲特性 dimensionless characteristic
无量纲位移 dimensionless displacement
无量纲系数 dimensionless factor;

W

non-dimensional coefficient
无量纲形状矢量 dimensionless shape vector
无量纲因素 dimensionless factor; non-dimensional factor
无量纲因子 dimensionless factor; non-dimensional factor
无量纲应力 non-dimensional stress
无量纲值 dimensionless value
无量水设备的分水节制闸 non-metering offtake regulator
无料斗的沥青混凝土铺路机 asphaltic concrete paver without hopper
无料斗的沥青混凝土摊铺机 asphaltic concrete paver without hopper
无料箱沥青路面铺料机 blacktop spreader without hopper
无列 column-free
无裂变中子吸收 non-fission neutron absorption
无裂缝 free from flaw; free of cracks; no fissure; without cracks
无裂缝材料 unfissured material
无裂缝的 crackfree; flawless; uncracked
无裂缝岩石 unfissured rock
无裂缝状态 uncracked condition
无裂纹的 crackfree
无裂纹缝 flawless
无裂隙的黏[粘]土 intact clay
无裂隙固体 flawless solid
无裂隙黏[粘]土 non-fissured clay
无林草地 puoztas
无林地 non-stocked; unstocked forest land
无临床症状的 subclinical
无磷阻垢缓蚀剂 non-phosphorus scale and corrosion inhibitor
无零标度 set-up scale
无零点刻度仪表 set-up scale meter; suppressed-zero meter
无零点漂移放大器 drift-free amplifier
无零点仪表 < 刻度不是从零开始的仪表 > set-up scale instrument; set-up instrument; suppressed-zero instrument
无零点仪器 inferred-zero instrument; suppressed-zero instrument
无零数 roundness
无零位刻度(盘) suppressed-zero scale
无零位压力表 suppressed-zero pressure ga(u)ge
无零位仪表 suppressed-zero instrument
无领车轴 mul(l)ey axle
无流层 layer of no motion
无流道模具 runner-less mo(u)ld
无流动面 surface of no motion
无流动状态 no-flow condition
无流动资金 illiquidity
无流动资金的 illiquid
无流港口 slack-water harbo(u)r
无流关闭压头 no-flow shutoff head
无流区 arheic area; arheic region; arheism
无流区的 ar(h)eic
无流深度 depth no-motion
无流时水头 shut-off head
无流式压力盒 no-flow cell
无流水系 blind drainage
无流速运河 slack-water canal
无流位置 no-flow position
无硫的 sulphur-free
无硫腐蚀 sweet corrosion
无硫化氢的还原环境 reduction environment of non-hydrogen sulfide
无硫基 sulfur-free basis
无硫硫化 sulfurless cure
无硫硫化剂 sulfurless vulcanizing agent
无硫硫化作用 sulfurless vulcanization
无硫气 sweet gas
无硫燃料 sulfur [sulphur]-free fuel
无硫石油产品 sweet petroleum product
无硫熟料 sulfur-free clinker
无硫天然气 sulfur [sulphur]-free natural gas; sweet gas
无硫无氮基 sulfur and nitrogen free basis
无硫油 sweet oil
无楼梯踏板的 riser less
无楼座观众厅 single-floor type auditorium
无漏损的 leak tight
无漏泄 no-leak(age)
无漏泄储存 positive confinement
无漏嘴漏板 tipless bushing
无露头矿脉 blind lode; blind ore vein; blind vein
无炉箅煤气发生器 grateless producer
无炉衬冲天炉 liningless cupola
无卤 halogen free
无陆架国 shelf-less state
无路(可通)的 roadless
无路可通的土地 landlocked parcel
无路面道路 unmetallized road
无路面的 unmetallized; unsurfaced
无路面的土路 unsurfaced road
无路面路基 unsurfaced subgrade
无铝沸石 mountainite
无铝锌皂石 zincsilite
无氯防冻剂 non-chloride antifreezing admixture
无氯化物 free from chlorides
无氯杀虫剂 non-chlorinated insecticide
无氯型消毒剂 non-chlorodisinfectant
无氯盐 chloride free
无氯早强剂 non-chloride accelerator
无轮的 wheelless
无轮犁 wheelless plow
无轮缘车轮 blind wheel; flangeless wheel
无轮缘动轮 blind driving wheel
无论发生何事 haploid what may
无论据证实的 undefended
无论如何 at all cost; at any cost; in some way; rain or shine
无螺钉门把手 screwless knob
无螺杆挤出机 screwless extruder
无螺栓 boltless
无螺栓衬板 boltless liner; boltless lining
无螺栓固定系统 boltless fixing system
无螺栓接头 joint without fishbolt
无螺纹 threadless
无螺纹部分 < 螺钉、钻头等的 > shank; plain section; unthreaded portion
无螺纹的管接头 threadless coupling; Dresser coupling
无螺纹杆 blank bolt
无螺纹管 bare pipe; threadless pipe
无螺纹接器套筒 screwless adapter
无螺纹节 plain stem
无螺纹孔 unthreaded hole
无螺纹莲蓬头 screwless rose
无螺纹淋浴喷头 screwless rose
无螺纹栓 blank bolt; bolt blank
无马弗炉 unmuffled type furnace
无码 code absence
无卖方名称的包装清单 packing list in lain paper
无脉冲泵 pulse-free pump
无脉冲流 pulse-free flow
无脉冲色谱泵 pulse-free chromatographic pump
无脉冲信号 pulse-free signal
无脉动泵 flexible hose pump; pulseless pump
无脉动流 pulse-free flow
无脉羽叶 aphlebia
无慢化剂堆 unmoderated reactor
无芒草 bromegrass
无芒雀麦 bromegrass
无毛的 glabrous
无毛细的 non-capillary
无锚板桩 unanchored sheet piling
无锚单排板桩墙结构 cantilever single-wall sheet pile structure
无锚碇板桩 cantilever sheet pile
无锚碇板桩岸壁 cantilever sheet pile(d) quaywall
无锚碇挡土墙 cantilever(-type) retaining wall
无帽壁柱 lesene; pilaster strip
无帽螺钉 screw without nut
无煤区火力发电站 far from mine power plant
无门车身 flush-sided body
无门窗拱廊 blank arcade
无门窗墙 blank wall; blind wall; dead masonry wall; dead wall
无门的 doorless
无密封的 glandless; packingless
无密封的泵 packingless pump
无密封垫阀门 glandless valve
无密封环的泵 sealless pump
无密封填料伸缩接头 packless expansion joint
无密封压盖泵 glandless pump
无密级的 unclassified
无密押 no test
无面层的毛地板 uncovered floor
无面层的毛楼板 uncovered floor
无面值的 unvalued
无名的 anonymous; innominatal; innominate; unknown
无名高地 nameless height
无名公用块 blank common block
无名公用区 blank common block
无名控制段 unnamed control section
无名氏 anonym
无名数 abstract number; dimensionless number; obscured number
无名图段 unnamed segment
无名图纸 unnamed drawing
无名无因次群 anonymous dimensionless group
无名杂树 unknown trees; unspecified tree
无明暗差别的 flat
无明确定义的 ill-defined
无模板的 formworkless
无模板的混凝土块材筒仓 silo of concrete blocks acting as lost formwork
无模板混凝土 off-shuttering concrete
无模板浇注混凝土 off-formwork concreting; off-shutter concreting
无模板浇筑混凝土 off-formwork concreting
无模成型(法) off-hand working; mo(u)ldless forming
无模拉拔 dieless drawing
无模硫化 open cure
无模热压技术 mo(u)ldless technique for hot pressing
无模人工吹制 free blowing
无模人工吹制玻璃 free-blown glass
无模式标本 sine tipo
无模铸造 containerless casting
无摩擦 zero friction
无摩擦层面 frictionless plane
无摩擦的 friction-free; frictionless
无摩擦滚轴 frictionless roller
无摩擦河槽 frictionless channel
无摩擦河槽水流 frictionless channel flow
无摩擦铰(链) frictionless hinge; non-friction hinge
无摩擦流(动) frictionless flow
无摩擦流体 frictionless fluid
无摩擦模拟器 frictionless simulator
无摩擦液体 frictionless liquid
无摩阻的 frictionless
无摩阻流 frictionless flow
无摩阻流体 frictionless fluid; frictionless liquid
无磨耗 non-wear
无磨面的 unfacetted
无磨损性材料 non-abrasive material
无磨损旋转 wear-free rotation
无母数测验 non-parametric test
无母数统计 non-parametric statistics
无木撑隧道 untimbered tunnel
无木支撑隧洞 untimbered tunnel
无目标地 broadside
无目标排水 < 如低沼泽地带的 > aimless drainage
无目的的 pointless
无目录的 uncatalogued
无内聚性 incohesion
无内圈滚柱轴承 inner-ringless roller bearing
无内圈滚子轴承 inner ringless roller bearing; roller and outer ring assembly
无内胎轮胎 tubeless tire [tyre]
无内胎轮胎轮辋气门嘴孔 rim tubeless valve hole
无内凸肩车轴 < 火车 > raised wheel seat axle
无内柱建筑物 clear span building
无内座圈滚柱轴承 roller outer race assembly
无能偿付的抵押人 distressed mortgage
无能隙的超导(电)性 gapless superconductivity
无泥沙河流 sediment-free river; sediment-free stream; silt-free river; silt-free stream
无泥沙运动冲刷 scour without sediment motion
无逆流离心机 concurrent centrifuge
无腻子玻璃 puttyless glazing
无年轮树 endogen tree
无年轮限制的木材 open grain
无黏[粘]结的预应力筋 non-bonded tendon
无黏[粘]结钢筋丝(束) unbonded tendon
无黏[粘]结隔离 bond prevention
无黏[粘]结后张法 unbonded post-tensioning; unbonded pretensioning
无黏[粘]结剂团矿法 bindless briquetting
无黏[粘]结剂型砂 binderless sand
无黏[粘]结接缝 debonded joint
无黏[粘]结力筋 non-bonded tendon
无黏[粘]结力预应力 prestress without bond
无黏[粘]结力预应力钢索 non-bonded tendon
无黏[粘]结梁 unbonded beam
无黏[粘]结(梁)设计 unbonded design
无黏[粘]结体外预应力桥 unbonded externally prestressed bridge
无黏[粘]结性 incoherentness; non-tackiness
无黏[粘]结性煤 no caking coal
无黏[粘]结性土 non-cohesive soil
无黏[粘]结预加应力 prestressing without bond; unbonded prestressing
无黏[粘]结预应力 no-bond prestressing; prestress-free without bond
无黏[粘]结预应力钢筋 non-bonded

prestressed reinforcement; unbonded prestressed bar
无黏[粘]结预应力钢丝束 unbonded tendon
无黏[粘]结预应力构件 unbonded member
无黏[粘]结预应力混凝土 unbonded prestressed concrete
无黏[粘]结预应力筋 unbonded prestressing tendon; unbonded tendon
无黏[粘]结张拉 no-bond tensioning
无黏[粘]结张力 no-bond tensioning
无黏[粘]聚力材料 cohesionless material
无黏[粘]聚力的 cohesionless; incohesive; non-coherent; non-cohesive
无黏[粘]聚力土壤 cohesionless soil; non-coherent soil
无黏[粘]聚性的 cohesionless; non-cohesive
无黏[粘]流动 frictionless flow; ideal flow; inviscid flow
无黏[粘]土钻进冲洗液 clay-free drilling fluid; non-clay drilling fluid
无黏[粘]性 cohesionless; incoherence; non-stickness
无黏[粘]性材料 cohesionless material; cohesiveless material; incoherent material; non-cohesive material
无黏[粘]性沉积物 cohesionless sediment
无黏[粘]性冲积层 incoherent alluvium; non-cohesive alluvium; non-viscous alluvium
无黏[粘]性的 bondless; incoherent; inviscid; non-adhesive; non-coherent; non-cohesive; non-viscous
无黏[粘]性的生漆 lac without sticks
无黏[粘]性的紫胶 lac without sticks
无黏[粘]性流 inviscid flow; non-viscous flow
无黏[粘]性流体 frictionless fluid; inviscid fluid; non-cohesive fluid; non-viscous fluid
无黏[粘]性泥沙 cohesionless sediment; cohesiveless sediment; friction(al) sediment; non-cohesive sediment
无黏[粘]性填土 cohesiveless backfill
无黏[粘]性填土的主动土压力 a.e.p. [active earth pressure] of cohesionless backfill
无黏[粘]性土 cohesionless soil; cohesiveless soil; friction(al) soil; non-cohesive soil
无黏[粘]性土河岸 non-cohesive bank
无黏[粘]性土冲积河槽 noncohesive alluvial channel
无黏[粘]性土抗剪强度 shear strength of cohesionless soil
无黏[粘]性土壤 granular soil
无黏[粘]性土土压力 earth pressure of cohesiveless soil
无黏[粘]性土相对密度 relative density of cohesiveless soil
无黏[粘]性土压密度 compactibility of cohesiveless soil
无黏[粘]性土土样 cohesionless soil sample
无黏[粘]性土液化 liquefaction of cohesiveless soil
无黏[粘]性液体 non-viscous liquid
无黏[粘]着力的 non-adhesive
无捻粗纱 roving
无捻多股钢丝绳 no-throw wire rope
无捻钢丝绳 untwisted wire rope
无捻丝线 no-throw
无凝结水管热网 heating network without condensating pipe
无凝聚力的 cohesionless
无凝聚性 incoherence
无扭精轧机座 no-twist finished stand
无扭力螺距 zero torque pitch
无扭面 surface of no distortion
无扭曲 freedom from warpage
无扭曲的 non-twisted
无扭曲立体观察 strainless stereoscopic(al) viewing
无扭曲立体模型 parallax-free model; undistorted model
无扭曲模型 parallax-free model
无扭曲模型法原理 principle of undistortion-free model
无扭转的 torsion-free
无扭转运动 torque free motion
无偶极子 infinitesimal dipole
无排水的大便器 non-water-carriage toilet facility
无排水地区 endor(h)eic drainage; endor(h)eic region
无排水沟的路槽 undrained trench
无排水管路的 undrained
无排水抗剪强度 undrained shear strength
无排水区 drainless area
无排水设施的 undrained
无排水设施的路槽 undrained trench
无牌营业者 interloper
无旁路的 unbypassed
无炮泥爆破 unstemmed shot
无泡曝气 bubbleless aeration
无陪伴的 unescorted
无赔偿合同 naked contract
无赔款退货 return for no claim
无赔款折扣 no claim bonus; no claim discount
无配合子【植】 azygospore
无配件电动机 bare motor
无配筋的 unreinforced
无配筋的砖石幕墙 unreinforced masonry panel walls
无配线分界点 immediate train distance point without siding; train spacing point without distribution tracks
无配重平衡的臂架 jib balanced without counterweight
无喷管式喷雾器 boomless sprayer
无喷嘴的 noseless
无棚盖的 uncovered
无棚货车 open wagon
无棚站台 open platform
无硼玻璃 boron-free glass
无硼釉 boron-free glaze
无篷汽车 <美> breezer
无膨胀 zero dilatancy
无膨胀钙-碱凝胶 non-swelling lime-alkali gel
无膨胀性凝胶 rigid gel
无碰撞玻耳兹曼方程 collisionless Boltzmann equations
无碰撞冲击波 collisionless shock wave
无碰撞弓形激波 collision-free bow shock
无碰撞激波 collisionless shock wave
无碰撞撕裂不稳定性 collisionless tearing instability
无碰撞阻尼 collisionless damping
无皮边材 bright sap
无偏比估计量 unbias(s)ed ratio estimator
无偏差 zero deflection
无偏差的 agonic; unbias(s)ed
无偏差多项式 affectless polynomial
无偏差估计 unbias(s)ed estimation
无偏差估计量 unbias(s)ed estimate; unbias(s)ed estimation
无偏差检验 unbias(s)ed check; unbias(s)ed test
无偏差灵敏点 flat spot
无偏差统计 unbias(s)ed statistics
无偏差线 agonic line
无偏方差 unbias(s)ed variance
无偏估计量【数】 unbias(s)ed estimator
无偏估算 unbias(s)ed estimate
无偏航的 unyawed
无偏回归估计量 unbias(s)ed regression estimator
无偏见的评估 disinterested appraisal
无偏角线 <地图上把无磁偏角的点子联结起来的线> agonic line
无偏离散 unbias(s)ed variance
无偏临界区域 unbias(s)ed critical region
无偏平均数 unbias(s)ed mean
无偏试验 unbias(s)ed test
无偏统计量 unbias(s)ed statistics
无偏统计(值) unbias(s)ed statistics
无偏误差 unbias(s)ed error
无偏析 segregation-free
无偏析合金 non-segregation alloy
无偏析组织 non-segregated structure
无偏性【数】 unbias(s)edness
无偏样本 unbias(s)ed sample
无偏振粒子束 unpolarized particle beam
无偏置极化继电器 unbias(s)ed polarized relay
无偏置信区间【数】 unbias(s)ed confidence interval
无偏重性抽样 unbias(s)ed importance sampling; unbias(s)ed important sampling
无票乘车船者 stowaway
无票乘车人 fare-evader
无票乘客 <俚语> hoboes
无票运输 traffic without shipping documents
无(拼)缝地图 seamless map
无平方因子数 quadrat-free number
无平衡螺钉摆轮 no screw balance
无凭单项目处理系统 paperless item processing system
无凭证闭塞 tokenless block
无凭证闭塞制 <单线凭出发信号机显示信号行车的半自动闭塞> tokenless block system
无凭证的 tokenless
无屏蔽 unmask
无屏蔽箔 bare foil
无屏蔽的 unscreened; unshielded
无屏蔽计数管 bare counter
无屏蔽膜 no-screen film
无屏蔽双绞线对 unshielded twisted pair
无屏蔽水域 open water
无屏蔽望远镜 unshielded telescope
无屏胶片 no-screen film
无屏障泊位 exposed berth
无屏障地带 exposed area; exposed location
无屏障锚地 open roadstead
无屏障水域 exposed waters
无坡度的 non-sloping
无坡度姿态 unbanked attitude
无坡口 not beveled; plain end; square groove; unbeveled end
无坡口槽焊 square groove weld
无坡口对焊接 square butt welding
无坡口对接焊 plain butt weld
无坡口对接接头 straight butt weld
无坡口焊缝 square face weld
无坡口接头 unchamfered joint
无破裂性地壳运动 nofractured earth movement
无破损试验 non-destructive test(ing)
无铺板甲板 exposed deck
无铺面的 unpaved
无铺砌的 unpaved
无铺筑道面的简易机场 unpaved strip; unprepared strip; unsurfaced strip
无铺筑面机场 unpaved airfield
无期贷款 perpetual loan
无期徒刑 life imprisonment; life sentence
无欺骗意图 without fraudulent intent
无起化学反应化合物 non-reacting component
无气的 air-free
无气混凝土 air-free concrete
无气孔的 astomatal; imperforate; pore-free; void-free
无气孔混凝土 air-entrained concrete
无气流 draft fan; draught-free; draughtless
无气流的 draftproof; draught-proof
无气门的 valveless
无气门发动机 valveless engine; valveless motor
无气门轮胎 valveless tire
无气泡 free of blowholes; free of bubbles; freedom from concrete <混凝土>
无气喷漆机 airless paint spraying machine
无气喷漆枪 airless spray gun
无气喷射 airless injection; solid injection
无气喷射发动机 solid injection engine
无气喷射法 airless spray
无气喷射式柴油机 airless-injection diesel
无气喷射引擎 solid injection engine
无气蚀水流 cavitation-free flow
无气水 air-free water; degassed water
无气体保护的电弧焊 non-gas-shielded arc welding
无气体的 gasless
无气体无焊剂焊接法 non-gas nonflux process
无气味 freedom from odo(u)r; odo(u)r free; odo(u)rlessness
无气味的 inodo(u)rous; odo(u)rless
无气味的油漆稀释剂 odo(u)rless mineral spirit
无气味隔热毡 inodorous felt
无气味涂料 odo(u)rless paint
无气味油漆 odo(u)rless paint
无气味油毡 inodorous felt
无汽蚀水流 non-cavitation flow
无碛带 <冰期中的> driftless area
无迁移电池 cell without transference
无牵引压榨 no-draw press
无铅包电缆 non-lead-covered cable
无铅玻璃 crown glass
无铅掺和料 unleaded blend
无铅车用燃料 lead-free automobile fuel
无铅的 lead-free; leadless; unleaded
无铅抗爆添加剂 unleaded antiknock additive
无铅硼釉 boracic glaze
无铅漆 lead-free paint; leadless paint
无铅汽车燃料 no-lead automobile fuel
无铅汽油 lead-free gasoline; no-lead gasoline; unleaded gasoline
无铅燃料 lead-free fuel; unleaded fuel; unleaded gasoline; white gasoline
无铅色料 leadless colo(u)r
无铅搪瓷 leadless enamel
无铅涂料 lead-free paint
无铅玩具涂料 lead-free toy coating
无铅辛烷值 unleaded octane rating
无铅压电陶瓷 leadless piezoelectric (al) ceramics
无铅颜料 lead-free paint; lead-free pigment

无铅油漆 <适用于存放食物等且易受铅沾污之处> lead-free paint
无铅釉 lead-free glaze;leadless enamel;leadless glaze
无签证过境 transit without visa
无前后盘的叶轮 open impeller
无前例的 unprecedented
无潜水员技术 diverless technique
无潜水员井口 diverless well-head
无潜水员输油管连接 diverless flow-line connection
无潜水员水下修理工作 diverless underwater repair
无潜水员水下装调方法 method of diverless underwater intervention
无潜水员作业 diverless operation
无强力温度试验 no-strength temperature test
无墙壁的 wall-less
无墙筋金属板条粉刷隔墙 studless metal lath and plaster partition
无羟基表面 dehydroxylated surface
无翘曲 freedom from warpage
无切弧 arc without contact
无切口试样 unnotched specimen
无切线段 segment without contact
无切削成型 non-cutting shaping
无侵蚀性 not erodible
无倾伏褶皱 horizontal fold
无倾角的 aclinal;aclinic
无倾线 acline;acline line;aclinic line;magnetic equator
无倾斜的底片 untilted negative
无倾斜角的 aclinal;aclinic
无清偿能力 isolvency
无情地 remorselessly
无氰电镀 cyanideless electro-plating;non-cyanide plating
无穷 infinitude
无穷半可计算集 infinite semicomputable set
无穷变量 infinitely variable
无穷不连续点 infinite discontinuity
无穷插入法 infinite-pad method
无穷差 zero allowance
无穷长线 infinitely long line
无穷(乘)积 infinite product
无穷乘积表示 infinite product representation
无穷次可微函数 infinitely differentiable function
无穷大 infinite;infinitely great;infinity
无穷大插塞 infinity plug
无穷大数 infinite number
无穷大值 infinitely large quantity
无穷反衬法 infinite-pad method
无穷范围逼近 infinite range approximation
无穷分枝 infinite branch
无穷复合形 infinite complex
无穷号 sign of infinity
无穷和 infinite sum
无穷回归 infinite regress
无穷积分【数】 infinite integral
无穷(积分)限 <正的或负的> infinite limit
无穷级数 infinite series;infinitive progression
无穷级数的和 sum of infinite series
无穷集 infinite aggregate;infinite set
无(穷)尽的 endless
无穷尽集 infinite aggregate
无穷近 infinitely near
无穷矩阵 infinite matrix
无穷开区间 infinite open interval
无穷可分分布 infinitely divisible distribution
无穷可分律 infinitely divisible law
无穷离散群 infinite discrete group
无穷量 infinite quantity
无穷量数 zillion
无穷区间 infinite interval
无穷区域 infinite region
无穷群 infinite group
无穷三角级数 infinite trigonometrical series
无穷树 infinite tree
无穷数 infinitude
无穷数列 infinite series
无穷小 infinitely small;infinitesimality
无穷小变换 infinitesimal;infinitesimal transformation
无穷小常数 infinitesimal constant
无穷小单元 infinitesimal element
无穷小的 infinitesimal
无穷小的阶 order of an infinitesimal
无穷小典型变换 infinitesimal canonical transformations
无穷小法 infinitesimal method
无穷小核 infinitesimal nucleus
无穷小计算 infinitesimal calculus
无穷小结果 infinitely small resultant
无穷小量 dimensionless
无穷小邻域 infinitesimal area;infinitesimal neighbo(u)rhood
无穷小群 infinitesimal group
无穷小上限 infinitesimally small upper limit
无穷小生成元 infinitesimal generator
无穷小实数 infinitesimal real number
无穷小数 infinite decimal;infinitesimal value
无穷小算子 infinitesimal operator
无穷小特征标 infinitesimal character
无穷小位移 infinitesimal displacement
无穷小线性变换 infinitesimal linear transformation
无穷小相对位移 infinitesimal relative displacement
无穷小形变 infinitesimal deformation
无穷小旋转 infinitesimal rotation
无穷小幺正变换 infinitesimal unitary transformation
无穷小应变 infinitesimal strain
无穷小元素 infinitesimal element
无穷小圆 infinitesimal circle
无穷小运动 infinitesimal motion
无穷小增量 infinitesimal increment
无穷小值 infinitely small quantity
无穷序列 infinite sequence
无穷(循)环 infinite loop
无穷域 inferior field
无穷远 infinite distance
无穷远点 infinite point;point at infinite
无穷远点射投影 geometric(al) projection;perspective projection
无穷远极 pole at infinity
无穷增量 infinite increment
无穷滞后 infinite lag
无球粒顽辉陨石 aubrite
无球粒陨石 achondrite;shergottite
无曲柄泵 non-crank pump
无曲柄发动机 crankless engine
无曲柄压力机 crankless press
无取暖建筑空间 unheated building space
无取暖设备公寓 cold flat
无取向电工钢 non-oriented electric(al) steel
无圈曲线 dendrite
无权的 powerless
无权而定居公地者 squatter
无权码 non-weight(ed) code
无权追偿 without recourse
无权追索 without recourse
无缺点 freedom from defects;immaculacy
无缺点带 flawless tape
无缺点的 flawless
无缺点管理 zero defect management
无缺点计划 zero defect program(me)
无缺点运动 zero defect motion
无缺口的 unnotched
无缺陷 non-defect;soundness;zero defect
无缺陷表面 blemish-free surface
无缺陷材 clear lumber
无缺陷的 flawless;no defect
无缺陷管理 zero defect management
无缺陷计划 zero defect program(me)
无缺陷金属 sound metal
无缺陷料 clear stuff
无缺陷木材 <又称无节疤陷的木材> clear timber;clear stuff;clear wood
无缺陷漆膜 full coat
无缺陷区域 area free from defect
无缺陷铸件 zero defect casting
无确认式信息传递服务 unacknowledged information transfer service
无扰动的 non-turbulent
无扰动流 undisturbed flow
无扰动射流 undisturbed jet
无扰动试样 undisturbed sample
无扰动水流 free stream flow
无扰动转换 bumpless changeover
无扰控制 non-disturbance control
无扰流棒束 unbaffled rod bundle
无扰运动 undisturbed motion
无热玻璃退火炉 heatless lehr
无热的 afebrile;apyretic;apyrexial;athermal
无热活动渗眼 inactive thermal seepage
无热交换流动 zero heat current
无热流 zero heat current
无热期 apyrexia
无热溶液 athermal solutions
无热水供应的公寓 cold water flat
无热源的 pyrogen-free
无人变电所 non-attended substation;unmanned substation
无人驳 unmanned barge
无人舱 unmanned capsule;unmanned compartment
无人操纵的 driverless
无人操纵的挖掘机 unmanned excavator
无人操纵机舱 unattended engine-room
无人操纵监视仪器 unmanned surveillance equipment
无人操纵空中监视系统 unmanned aerial surveillance system
无人操作 unmanned operation
无人操作的 unmanned
无人操作台 unmanned station
无人测站 unmanned station
无人差测微器 impersonal micrometer
无人的 impersonal
无人地带 no man's land
无人地震监测站 unmanned seismic observatory
无人多用途轨道卫星 unmanned orbital multifunction satellite
无人多用途卫星 unmanned multifunction satellite
无人发射井 unmanned silo
无人放牧的 loose-grazed;unherded
无人飞机 unmanned plane
无人服务(商店) nonattendant service
无人浮标 robot buoy
无人工厂 unmanned factory
无人工作间 manless face
无人工作面 manless face
无人观测气球 unmanned observation balloon
无人观测站 non-attended station
无人管理 absentee control
无人管理泵站 unmanned pumping station
无人管理变电分站 nonattendant substation
无人管理的 non-agency;nonattendant;off-hand;unattended;unmanned;unwatched
无人管理的车站 non-agency station;unatteneded station
无人管理的会让站 unattended loop
无人管理的联轨点 unattended junction
无人管理的区间岔线路签机 unattended siding staff instrument
无人管理电站 unmanned power station
无人管理设备区域 unattended equipment area
无人管理站 unattended station;unmanned station
无人管理装置 unattended installation
无人管理自动泵站 unattended pumping station
无人回收运载工具 unmanned recovery vehicle
无人机舱 unattended machinery space;unmanned engine room;unmanned machinery space;zero-man engine room
无人机舱自动控制系统许可证 Automatic Control System for Unattended Engine-room Certificate
无人驾驶 manless drive;steer-away
无人驾驶补机 unmanned helper engine;unmanned pusher engine
无人驾驶舱 unmanned module
无人驾驶车辆 driverless vehicle
无人驾驶出租汽车 driverless taxi
无人驾驶船舶 drone
无人驾驶的 driverless;unpiloted;pilotless;unmanned
无人驾驶的调车机车 unattended switcher
无人驾驶的飞机 unmanned aircraft
无人驾驶的工程机械 unmanned equipment
无人驾驶的工业牵引车 driverless industrial tractor
无人驾驶的挖掘机 unmanned excavator
无人驾驶电子计算机控制的小车 driverless electronic-computer controlled truck
无人驾驶帆船 sail-assisted unmanned robot ship
无人驾驶飞机 drone;robot aircraft;robot plane
无人驾驶辅机 unmanned helper engine
无人驾驶机车 unmanned engine;unmanned locomotive
无人驾驶列车 crewless train;driverless train
无人驾驶设备 robot equipment
无人驾驶卫星 pilotless satellite
无人驾驶宇宙飞船 pilotless spaceship
无人驾驶自动避撞系统 automatic steer-away collision avoidance system
无人驾驶自动车 unmanned roving vehicle
无人驾驶自动折返 driverless reversal operation
无人监督的 unsupervised
无人监督配电站 unattended substation
无人监视台 non-attended station
无人居住的 uninhabited;unmanned;unpeople(d)
无人居住区 depopulated zone
无人看管的 unserviced

无人看守道口 non-guarded crossing; open crossing; unattended level crossing; unstaffed level crossing
无人看守道口的凭证机 unattended crossing place token instrument
无人看守的 unattended; unwatched; unmanned
无人看守的灯标 unwatched light
无人看守的灯船 unattended lightship; unmanned lightship; unwatched lightship
无人看守的灯塔 unattended lighthouse; unmanned lighthouse; unwatched lighthouse
无人看守的线路无线电台 unattended wayside radio
无人看守的作业 unattended operation
无人看守公路道口 unattended crossing; unmanned crossing
无人看守设备 unattended equipment
无人控制的 non-man control
无人控制观测气球 kitoon
无人控制系统 unmanned control system
无人口区 unpopulated area
无人流基线 zero inflow curve
无人名账号 impersonal account
无人配电站 unattended substation
无人潜水器 unmanned submersible; unmanned(underwater) vehicle
无人区 no man's land
无人认领的 unclaimed
无人认领的物品 unclaimed article
无人认领的行李 abandoned baggage; abandoned luggage; unclaimed luggage
无人认领货物 unclaimed cargo; unclaimed freight
无人认领行李 unclaimed baggage
无人深潜器 deep unmanned submersible
无人使用的 unoccupied
无人售货法 self-service
无人售货商店 self-service shop
无人售票公共汽车 conductorless bus
无人售书处 self-service bookstall
无人台站 unmanned station
无人为干扰 freedom from jamming
无人维护工作 unattended operation
无人烟的 wild
无人烟绿洲 uninhabited oasis
无人烟沃洲 uninhabited oasis
无人遥控潜水器 unmanned remotely-operated vehicle
无人运行 unattended operation
无人增音机 unattended repeater
无人增音站 unattended repeater station
无人照管的 unattended
无人值班操作 unattended operation
无人值班的 unattended; unstaffed
无人值班的变电所 non-attended station
无人值班的遥控电站 non-attended remotely controlled power station; unattended remotely controlled power station
无人值班的运行 unattended operation
无人值班设备 unattended equipment
无人值班台 non-attended station
无人值班信号楼 unstaffed tower
无人值守变电分所 nonattendant substation; non-attended substation; unattended substation
无人值守的运转 unattended operation
无人值守台 non-attended station; unattended station
无人值守站 non-attended station; unattended station
无人值守中继站 unattended relay station
无人主张的 unclaimed
无人住的 unoccupied
无人住的房子 vacant
无人自航式潜水器 unmanned free-swimming submersible
无人租住的 tenantless; untenanted
无任所大使 ambassador at large
无韧性的 immalleable; inviscid
无日期背书 undated endorsement
无日期的 timeless
无日期支票 undated check
无绒毛布片 fluffless rag
无容差 zero allowance
无溶剂的 solvent-free; solventless
无溶剂工业用涂料 solventless industrial finish
无溶剂环氧清漆 solventless epoxy varnish
无溶剂聚酯清漆 solventless polyester varnish
无溶剂漆 solventless paint
无溶剂清漆 solventless varnish
无溶剂涂料 solventless coating
无溶蚀现象 non-dissolution trace
无熔丝的 no-fuse
无熔丝断路器 no-fuse breaker; no-fuse circuit breaker
无熔丝开关 no-fuse switch
无熔丝配电盘 no-fuse panel
无熔线断路器 non-fuse breaker
无蠕变 freedom of creep
无蠕变的 creepless
无乳化剂的乳胶 emulsifier-free latex
无乳剂面 non-emulsion-coated face; non-emulsion-coated side
无润滑表面 unlubricated surface
无润滑摩擦 unlubricated friction
无润滑式压缩机 non-lubricated compressor
无润滑油的 greaseless
无润滑运转 running dry
无塞灌浆(法) non-packer grouting
无塞绳长途接续台 cordless toll switchboard
无塞绳电话交换机 cordless telephone switch board
无塞绳电话交换台 cordless telephone switch board
无塞绳式话传电报交换台 cordless panel phonogram equipment
无塞绳式交换台 cordless board
无散矢场 solenoidal field
无扫描 no-raster
无色 achromatic colo(u)r; achromaticity; achromatism; air colo(u)r
无色凹凸印 blind blocking
无色碧硒 achroite
无色玻璃 colo(u)rless glass
无色彩 achromatization
无色彩的 non-chromatic
无色差接物镜 achromatic objective
无色成色剂 colo(u)rless coupler
无色的 achromatic; air colo(u)red; colo(u)rless; leuco; lipochromous; uncolo(u)red
无色的粉剂 colo(u)rless powder
无色底釉 colo(u)rless base
无色点饰浮雕 clear stipple embossed
无色电气石 achroite
无色钙铝酸盐玻璃 colo(u)rless calcium aluminate glass
无色光学玻璃 colo(u)rless optic(al) glass
无色光学玻璃分类法 colo(u)rless optic(al) glass classification
无色光学玻璃牌号 designation of colo(u)rless optic(al) glass
无色滑油 bleach oil
无色化合物 leuco-compound
无色间隙 achromatic interval
无色晶体 clear crystal
无色空心玻璃制品 colo(u)rless hollow ware
无色孔雀石绿 leucomalachite green
无色零电平 no-colo(u)r zero-voltage level
无色煤油 water-white kerosene
无色瓶罐玻璃 flint container glass
无色染料 leuco dye
无色熔剂 colo(u)rless flux
无色砂浆 non-staining mortar
无色素的 non-pigmented
无色透镜 air colo(u)r lens; colo(u)rless lens
无色透明的 water white
无色无臭 no colo(u)r or smell
无色性 colo(u)rlessness
无色亚甲基蓝 leucomethylene blue
无色颜料 leucopigment
无色眼镜玻璃 colo(u)rless spectacle glass
无色油 bleached oil
无色釉 colo(u)rless glaze
无色阈值 achromatic threshold
无沙水流 sediment free flow
无砂大孔混合料 open stone mix
无砂大孔隙混凝土 no fines concrete
无砂河流 sediment-free stream
无砂混凝土 no fines concrete; non-fines concrete; popcorn concrete
无砂浆砌筑 dry masonry
无砂浆砌筑墙体 dry masonry wall
无砂浆砌砖 dry brick(building)
无砂孔轻质混凝土墙 no fines lightweight concrete wall
无砂目平板 grainless plate
无砂石膏灰粉刷 unsanded gypsum plaster
无砂石膏灰浆 unsanded gypsum
无砂石膏灰涂抹 unsanded gypsum plaster
无砂石膏抹灰料 unsanded gypsum
无山墙的 unpedimented
无闪光 absence of glare
无闪烁电路 flickerless circuit
无闪烁图像 flicker-free image
无栅的 ungridded
无栅天花板 contact ceiling
无扇出电路 fan-out-free circuit
无伤亡事故 non-injury accident
无商标 no mark
无商标汽油 unbranded gasoline
无商业价值 no-commercial value
无上限的 uncapped
无设备的 unaccommodated; unfurnished
无设备的跑道 uninstrumented strip
无设施的污泥干化床 unplanted sludge drying bed
无摄动轨道 unperturbed orbit
无摄轨道 undisturbed orbit
无摄运动 non-disturbed motion
无伸缩的旋转型桅式起重机 non-telescopic rotating mast crane
无伸缩剖面 plane of no-deformation
无伸缩性 non-elasticity
无伸缩性的 non-elastic
无渗碳体钢 oce steel
无升力冲角 angle of incidence of zero lift
无升力角 angle of no lift
无升力尾翼 floating tail
无生产力的 non-productive
无生代【地】 Azoic; Azoic era
无生的【地】 Azoic
无生化学 abiochemistry
无生纪【地】 Azoic period
无生界【地】 Azoic erathem; Azoic group
无生命 inanimation
无生命的 abiotic; Azoic【地】
无生命环境的相互作用 interactions with abiotic environment
无生命物质 non-living biomass
无生命组织 non-living tissue
无生气的 insipid
无生气时期 doldrums period
无生物降解能力的 non-biodegradable
无生物降解性 non-biodegradability
无生物界 inorganic sphere
无生物深度 depth zero of life
无生物时代的【地】 Azoic
无生物水域 azoic waters
无生物学 abiology
无生系【地】 Azoic system
无生意 absence of business
无生源论 abiogenesis
无生源说 abiogenesis; autogenesis; spontaneous generation
无生源新说 neo-abiogenesis
无声 aural null; aural zero; noiselessness; silence
无声白炽灯 anti-sing lamp
无声变速 quiet shifting; silent gear change
无声波放电 silent(electric) discharge
无声玻璃 flint glass
无声彩色循环幻灯片 silent technicolo(u)r loop
无声操作 noiseless action
无声槽 unmodulated groove
无声掣子 silent pawl
无声齿轮 quiet gear; silent gear
无声齿轮箱 all-silent gearbox
无声传动 noiseless drive
无声打桩法 silent pile driving
无声打桩机 silent pile-driver
无声打桩系统 silent pile driving system
无声的 dumb; dummy; inaudible; mute; noiseless; soundless; tuneless
无声(的第)三挡 quiet third
无声(的第)三挡速度 silent third speed
无声灯 noiseless lamp
无声发动机 silent-running engine
无声放电 effluvium [复 effluvia]; silent discharge
无声放电处理(法) voltolization
无声放气阀 silent blow off valve
无声放泄阀 silent blow off valve
无声工作速度 silent speed
无声航行 silent running
无声弧光灯 noiseless arc lamp
无声换挡 noiseless shift(ing); quiet gear; quiet speed
无声换挡变速器 silent-mesh gearbox
无声机车 <有凝汽器的> dummy
无声棘轮 silent ratchet wheel
无声铰链架置 silent-block mounting
无声进给 silent feed
无声警察 <十字路口的交通指挥装置> silent cop
无声链 laminated chain; leaf chain; silent chain
无声链传动 silent chain drive
无声链轮滚刀 silent chain sprocket hob
无声链式运输器 silent chain conveyer [conveyor]
无声料托 silent stock support
无声轮胎 silent tire [tyre]
无声铆接机 silent riveter
无声煤气灯 silent-flame gas burner
无声煤气炉 silent-flame gas burner

W

无声啮合 silent engagement;silent mesh
无声啮合变速器 silent-mesh transmission
无声啮合轮组 silent-mesh gear set
无声片 mute
无声起飞与着陆 quiet take-off and landing
无声擒纵机构 silent escapement
无声区 anacoustic(al) zone; dead space;zone of silence
无声三速传动齿轮箱 silent third gearbox
无声三速和四速传动齿轮组 silent third and fourth gear set
无声声道 unmodulated track
无声售货员 silent salesman
无声循环式幻灯片 silent film loop
无声影片 mute
无声运动 noiseless run(ning)
无声运行 silent running
无声运转 noiseless drive; quiet run; silent operation
无声运转曲线 quieting curve
无声振动 inaudible vibration
无胜负 break-even
无绳的 cordless
无绳电话 cordless telephone
无绳电话系统 cordless telephone system
无绳螺丝刀 cordless screwdriver
无绳塞头 cordless plug
无绳塞子 cordless plug
无绳通信[讯] cordless telecommunication
无剩余贷款 no surplus stock
无失真波 undistorted wave
无失真长途拨号 distortionless distance dial(l)ing
无失真传输 undistorted transmission
无失真的 distortionless;undistorted
无失真电路 distortionless circuit
无失真条件 distortionless condition
无失真图像 undistorted image;undistorted picture
无失真线路 distortionless line
无失真信号 undistorted wave
无施受的 aprotic
无湿胀状态 non-expansive condition; no-swell condition
无十字头发动机 trunk engine
无石膏水泥 gypsum free cement
无石棉板 asbestos-free slate
无石棉产品 asbestos-free products
无石棉硅酸钙板 asbestos-free calcium silicate board; xonotlite model insulating board
无石棉品 asbestos-free products
无石棉微孔硅酸钙 asbestos-free calcium silicate
无石棉制动制品 asbestos-free friction product
无石英的闪绿岩 quartz-free diorite
无石英质斑岩 quartz-free porphyry
无时代性形式 anti-period style
无时间限制的 hourless;timeless
无时间性 timelessness
无时间延迟 non-time delay
无时效处理 non-aging treatment
无时效的 non-delay
无时效钢 non-ag(e)ing steel
无实测资料的 unsurveyed
无实测资料地区 unsurveyed area
无实际价值的 barren of practical value
无实际意义的 intangible in value
无实体的 in substantial;unbodied
无实用意义的 ivy
无矢量 scalar
无矢量矩阵 scalar matrix
无使用价值商品 discommodity
无示踪物的 tracer free
无事故 zero defect
无事故的 accident-free
无事故运行 uninterrupted service
无事件概率 probability of none event
无事实根据的 ungrounded
无势运动 non-potential motion
无视差 no-parallel;parallax-free
无视差运动 parallax-free motion
无视界 without visibility
无视警告 ignore warming
无视信号 negligence of signal
无视油砂 miss an oil sand; overlook an oil sand
无试样 dry-out sample
无饰柱 rustic column
无适航性能 innavigability
无适应性 inadaptability; inelasticity; inflexibility
无适应性的 inelastic behavio(u)r
无收入的 non-revenue
无收入的吨公里 non-revenue ton-kilometers
无收入的运输 <如路务运输> non-revenue earning traffic
无收入货物 non-revenue freight
无收缩的 shrinkage-free
无收缩钢 non-shrinking
无收缩灌浆 nonshrinking grout
无收缩混凝土 non-shrinkage concrete; non-shrinking concrete; shrinkage-compensating concrete; shrinkless concrete
无收缩快凝波特兰水泥 non-shrinking rapid hardening Portland cement
无收缩快凝硅酸盐水泥 non-shrinking rapid hardening Portland cement
无收缩裂纹 free of shrinkage cracks
无收缩砂浆 non-shrink(ing) mortar
无收缩水泥 non-shrinkage cement; non-shrinking cement; shrinkage compensates cement
无收缩状态 no-shrink condition
无守车的不同运输方式联运列车 cabooseless intermodal train
无首系 singular set
无受力孔隙 unstressed pore
无受污渍的 non-staining
无受益条款 not to insure clause
无受追索权 without recourse
无输出变压器 output transformerless
无输出变压器电路 output transformerless circuit
无输出变压器功率放大器 output transformerless amplifier
无输入变压器 input transformerless
无熟料水泥 cement without clinker; clinker-free cement; clinkerless cement
无鼠的 free from rats
无束缚的 unbounded
无树大草原 savanna(h)
无树木的 treeless;un-wooded
无树区 hoogeveld
无树山地 wold
无树(叶)的 bald
无数 incalculability; innumerability; legion; myriad; wilderness; without number
无数的 incalculable; innumerable; innumerous; no end of; numberless; sumless; thousand and one; uncounted; unknown; unnumbered; unnumerable
无数据的 not available
无刷励磁 brushless excitation
无刷励磁系统 brushless excitation system
无衰耗波 undamped wave
无衰减 non-attenuating; zero decrement
无衰减波 non-attenuating wave; unattenuating wave; undamped wave
无衰减的 undamped
无衰减介质 nonattenuating medium
无衰减前向波 non-attenuating forward wave; unattenuated forward wave
无衰减振荡 undamped oscillation
无衰减振动 undamped vibration
无栓分隔栏 bow stall;loose-box
无双的 peerless;unparalleled
无双像场 horopter
无双像场曲线 horopter curve
无霜冰箱 frost-free refrigerator
无霜带 frostless zone;green belt;verdant zone
无霜的 frostless
无霜(冻) free of frost;frost-free
无霜季节 frost-free season; frostless season
无霜期 duration of frost-free period; frost-free growing season; frost-free period;frost-free season;frostless season;non-frost period
无霜期间 length of frost free period; length of frost free season
无霜日 day free of frost; day without frost
无霜生长期 frost-free growing period
无霜系统 no-frost system
无水 anhyetism
无水氨(液) anhydrous ammonia
无水白云母 anhydromuscovite
无水表的公寓 unmetered apartment
无水材料 water-free material
无水层 aquifuge
无水储气器 waterless gasholder
无水醇 absolute alcohol
无水带 anhydrite band
无水胆矾 chalcite
无水的 anhydric; anhydrous; moisture-free; non-aqueous; unhydrous; unwatered; water-free
无水的管阀门 dry valve
无水地层开挖 excavation in dry ground
无水煅烧石膏 hard-finished plaster
无水矾石 alumian
无水分的油 emulsifiable oil
无水粉饰 anhydrous plaster
无水高岭土 anhydrokaolin
无水高氯酸镁 anhydrone
无水高温处理厂 non-aqueous pyroprocessing plant
无水工艺 waterless technology
无水硅酸钙 anhydrous sulphate of silicate
无水硅酸铝 anhydrous alumin(i)um of silicate
无水硅酸盐 anhydrous silicate
无水黑云母 anhydrobiotite
无水回管 dry return pipe
无水基 moisture-free basis
无水甲醇 absolute methanol; absolute methyl alcohol
无水钾镁矾 langbeinite
无水钾锰矾 manganolangbeinite
无水钾盐镁矾 anhtdrokainite
无水碱芒硝 makite
无水建筑材料 anhydrous material
无水胶印 driography
无水井 dumb well;water-free well
无水酒精 absolute(ethyl)alcohol; anhydrous alcohol;raw spirit
无水立管 <消防用> dry riser
无水粒状高氯酸钡(干燥剂) desicchlora
无水磷酸 anhydrous phosphoric acid
无水硫铝酸钙 anhydrous calcium sulphoaluminate
无水硫酸钙 anhydrous calcium sulfate [sulphate]; anhydrous sulphate of calcium;dead plaster;drierite
无水硫酸钙粉饰 anhydrous calcium sulphate plaster
无水硫酸钾 anhydrous sulphate of potassium
无水硫酸钠 anhydrous sodium sulfate;anhydrous sulphate of sodium
无水硫酸铜 anhydrous copper sulfate
无水氯化氢 anhydrous hydrogen chloride
无水马力门限钻压 non-hydraulic threshold weight
无水芒硝 anhydrous salt cake; pyrotechnite; Sebastian salt; thenardite; verd(e) salt
无水醚 absolute ether
无水膜残迹的 water break-free
无水钠镁矾 vanthoffite
无水泥浇注料 cement free castable; no-cement castable
无水硼砂 borax anhydrous
无水平推力的拱形桁架 arched girder without horizontal thrust
无水溶剂 anhydrous solvent; dry solvent
无水沙漠区 xerochore
无水砂灰浆 <由石灰和砂组成的> mild mortar
无水生石灰 anhydrous lime
无水石膏 anhydrite; anhydrous gypsum; calcium sulphate anhydrate; dead-burnt gypsum; gypsum anhydrite; hard (-burnt) plaster; karstenite
无水石膏板 anhydrite sheet
无水石膏绷带 anhydrite bander
无水石膏粉饰 anhydrous gypsum plaster(ing)
无水石膏灰浆 anhydrous gypsum plastering
无水石膏灰泥 anhydrous calcium sulphate plaster; anhydrous gypsum plaster
无水石膏胶泥 anhydrous gypsum plaster
无水石膏胶凝材料 anhydrite cement
无水石膏(砌)块 anhydrite tile
无水石膏砌块隔墙 anhydrite tile partition
无水石膏石 anhydrite rock
无水石膏石灰砂浆 anhydrite lime mortar
无水石膏饰面 anhydrite surface
无水石膏水泥 anhydrite cement; anhydrous cement
无水石膏族 Middle Muschelkalk
无水石灰 anhydrous lime
无水石油 dry oil
无水时期 anhydrous period
无水试样 dry-out sample
无水苏打 calcinated soda;soda ash
无水酸 anhydrous acid
无水隧洞 non-water tunnel
无水碳酸钠 anhydrous sodium carbonate;natrium carbonicum;soda ash
无水天然气 dry natural gas
无水无灰基准 moisture and ash-free basis
无水物 anhydride
无水消防栓 dry hydrant
无水型砂 waterless mo(u)lding sand
无水羊毛蜡 anhydrous wool wax
无水羊毛脂 anhydrous wool fat;wool fat;wool grease;wool oil;wool wax
无水氧化铬绿色颜料 green anhydrous oxide chromium pigment
无水乙醇 absolute ethyl alcohol;anhydrous alcohol;dehydrated alcohol

无水运行＜关闭导叶后＞ nodischarge operation
无水藻的 weed-free
无水质污染 zero water pollution
无水重量 moisture free weight
无税 tax-free
无税的 tax-exempt
无税港口 free port
无税或退税 tax-free or tax-refund
无税货物 non-dutiable goods
无税离岸价 free on board stowed and trimmed
无顺序 out-of-order
无顺序随机规划 non-sequential stochastic programming
无瞬变过程的 transient-free
无瞬变现象的 transient-free
无说明分录 blind entry
无死伤事故 PDO accidental
无死锁的 dead-free
无死锁方案 deadlock-free scheme
无塑性 non-plasticity
无塑性的 non-plastic
无塑性断裂 non-ductile fracture
无塑性粉末 non-plastic powder
无塑性粉土 non-plastic silt
无塑性理想粗粒土 ideal coarse-grained soil of no plasticity
无塑性黏[粘]土 non-plastic clay
无塑性破坏 non-ductile fracture
无塑性土(壤) non-plastic soil
无塑性转变温度 nil-ductility (transition) temperature
无酸的 acid-free; acidless; free from acid; non-acidic
无酸油 acid-free oil
无随从的 unaccompanied
无髓心板 side cut
无髓心的锯材 side cut
无髓心木材 timber without pith
无损 no-wear
无损表征 non-destructive characterisation
无损材料试验 non-destructive material test
无损的 intact; non-destructive
无损读出 non-destructive reading; non-destructive readout
无损读出光存储器 non-destructive readout optic(al) memory
无损浮力 intact buoyancy
无损害 zero damage
无损耗 free of losses; non-loss
无损耗壁 loss-free wall
无损耗波导 lossless waveguide
无损耗材料 lossless material
无损耗的 loss-free; lossless
无损耗的筛分 screening without breakage
无损耗电介质 loss-free dielectric
无损耗电缆 lossless cable
无损耗电路 zero-loss circuit
无损耗反射机理 loss-free reflectivity mechanism
无损耗共振器 lossless cavity
无损耗结 lossless junction
无损耗介质 loss-free medium; lossless medium
无损耗滤波器 lossless filter
无损耗网络 ideal network
无损耗线 loss-free line; lossless line; no loss line
无损耗线路 ideal line
无损坏 damage-free; non-failure
无损坏的 undamaged
无损回收 undamaged recovery
无损技术 non-destructive technique
无损加法【计】 non-destructive addition
无损检测 non-destructive examination; non-destructive test(ing)
无损检测法 non-invasive method
无损检测器 non-destructive detector
无损检测设备厂 non-destructive testing equipment works
无损检查 non-destructive inspection
无损检验 harmless test; non-destructive examination; non-destructive inspection; non-destructive test(ing)
无损检验法 method of non-destructive examination; method of non-destructive inspection; non-destructive testing method
无损检验人员证明书 certification of non-destructive personnel
无损检验设备 non-destructive testing equipment
无损喷砂 non-erosive blasting
无损评估 non-destructive evaluation
无损评价 non-destructive evaluation
无损伤 zero damage
无损伤的黏[粘]土 intact clay
无损伤试验 non-destructive test
无损伤性圆体针 non-traumatic round-bodied needle
无损伤止血钳 non-traumatic hemostatic forceps
无损失 free of losses
无损失的 break-even
无损失电码 lossless code
无损失码 lossless code
无损失条件 loss-free condition
无损试验 non-destructive assay
无损试验技术 non-destructive testing technique
无损探伤 non-destructive evaluation; non-destructive (flaw) detection; non-destructive inspection; non-destructive test(ing)
无损探伤法 method of non-destructive inspection; non-destructive flaw-detecting method; non-destructive inspection method
无损探伤机 non-destructive inspection machine
无损探伤记录 record of non-destructive detection
无损探伤技术 non-destructive detection technique; non-destructive (testing) technique
无损探伤试验 non-destructive test(ing)
无损探伤术 harmless flaw detector; non-destructive testing technique
无损通道 lossless channel
无损稳定性 intact stability
无损压缩 lossless compression
无榫骨架 bare foot
无缩格排版 full out
无缩孔钢 non-piping steel
无索赔折扣 no claim discount
无索引的 indexless
无锁闭表示 lock-free indication
无锁定方式 lock-free system
无锁握柄 free lever
无胎沥青聚合物片材 unreinforced bitumen polymer sheet
无胎油毡 non-reinforced asphalt sheet
无胎缘轮胎 debeaded tire[tyre]
无台阶的金属化 no-step metallization
无太平门的建筑物 fire trap
无太平门等设施的建筑物 fire trap
无坍落度混凝土 dry-packed concrete; no-slump concrete; zero-slump concrete
无坍落混凝土 stiff concrete
无坍塌的土壤 soil without cave-ins
无摊铺斗的路面整铺机 paver-finisher without hopper
无弹簧的悬吊 spring suspension
无弹簧的悬挂 spring suspension
无弹簧门锁 dead latch
无弹簧锁闩 dead bolt(lock)
无弹簧托板转向架 spring plankless truck
无弹力橡胶 lifeless rubber
无弹力橡皮 lifeless rubber
无弹性 inelasticity; stereotype
无弹性的 unrecovered
无弹性供给 inelastic supply
无弹性轨道 dead track
无弹性扣件 untensioned rail fastening
无弹性碰撞 non-elastic collision
无弹性性能 inelastic behavio(u)r
无弹性性状 inelastic behavio(u)r
无弹性需求 inelastic demand
无弹性针钩 rigid hook
无碳不锈钢 carbon-free stainless steel
无碳的 carbon-free
无碳复写纸 no carbon (required) paper
无碳钢 carbon-free steel
无碳铝镍钴磁铁 Alcomax
无套管成井 casing less completion
无套管的 uncased
无套管混凝土灌注桩 shell-less pile
无套管井 uncased well
无套管绳索冲抓钻机 non-casing rope percussion-grab drill
无套管式混凝土桩 uncased concrete pile
无套管现场灌注桩 uncased concrete pile
无套管现浇(混凝土)桩 cast-in-situ uncased pile; poured-in-place uncased pile
无套管旋转钻探 rotary open hole drilling
无套管桩 uncased pile
无套管钻进 open hole drilling
无套管钻孔 uncased(bore) hole; uncased drill hole; uncased drill holing; barefooted; open hole
无套管钻孔灌注桩 shell-less pile; uncased-bored pile
无套管钻探 open hole drilling
无套节铸铁管 hubless cast-iron pipe
无套壳的 uncased
无套模 decapsulation
无套桩 shell-less type of pile
无特定功能 unspecialized [unspecialised]
无特定功能的 unspecialized [unspecialised]
无特色的 insipid
无特殊 unspecificness
无特殊装备的 unsophisticated
无特征的 undistinguishable
无梯级的 stepless
无梯级运河 stepless canal
无梯井两跑式楼梯 staircase of dog-legged type
无梯井式楼梯 dog-legged type stair(case)
无踢板 open riser
无踢板楼梯 open riser stair(case)
无提取物木材 extractive-free wood
无提升机跳汰机 bucket-less jig
无体积膨胀 free from volume increase
无体式钻头 bodiless bit
无体物权 property in action
无天电干扰的 static-free
无天花板屋顶 open roof
无添加剂的 non-additive
无添加剂的汽油 natural gasoline
无添加剂润滑油 non-additive oil
无填充的 packless
无填料泵 glandless pump; packless pump
无填料的 packless; unsized
无填料的平板梁 slab-and-beam without fillers
无填料阀 packless valve
无填料函的 glandless
无填料函式阀体 no box type valve body
无填料接合 packless joint
无填料散热器阀 packless radiator valve
无填料塔 no-padding tower
无填塞膜片阀 packless diaphragm valve
无条痕 absence of streaks
无条件背书 absolute endorsement; unconditional endorsement
无条件不等式 unconditional inequality
无条件承兑 absolute acceptance; unconditional acceptance
无条件承付 absolute acceptance; unconditional acceptance
无条件传送 unconditional transfer
无条件贷款 non-tied loan
无条件的 absolute; categorical; unconditional; unconditioned; unqualified; unreserved; without qualification
无条件的定货 discretional order
无条件的合同 bare contract
无条件的交易 unconditional operation
无条件的契约 bare contract
无条件的信用证 unconditional letter of credit
无条件的义务 unconditional obligation
无条件的银行保函 unconditional bank guarantee
无条件订购 open order
无条件反射 unconditioned reflex; unconditioning
无条件概率 unconditional probability
无条件供应 supply without cost [obligation]
无条件估计量 unconditional estimator
无条件继承的不动产 estate in fee
无条件交货 free delivery; unconditional delivery
无条件接受 absolute acceptance; clean acceptance
无条件劲度 unconditioned stiffness
无条件控制转移 unconditional transfer of control
无条件流动能力 unconditional liquidity
无条件契约 a bare contract
无条件收敛 unconditional convergence
无条件跳转 unconditional jump; unconditional transfer
无条件稳定 unconditional stability
无条件稳定的 unconditionally stable
无条件稳定的放大器 unconditionally stable amplifier
无条件稳定准则 unconditional stability criterion
无条件限制随机抽样 unrestricted random sampling
无条件销售 absolute sale
无条件信用书 clean credit
无条件性 unconditionality
无条件抑制 unconditioned inhibition
无条件语句【计】 unconditional statement
无条件指令 imperative statement
无条件转移 unconditional branch; unconditional jump; unconditional transfer
无条件转移指令 unconditional con-

trol jump instruction; unconditional control transfer instruction
无条件最惠国条款 unrestricted most favoured nation clause
无条件作用 unconditioning
无条款信用书 clean credit; open letter of credit
无条理 incoherence
无条纹板玻璃 cord-free plate glass
无条纹玻璃 plain glass
无调节池的径流式电站 run-of-river plant without pondage
无调节的 ungoverned
无调节喷头的喷灯 non-variable head torch
无调节水流 non-regulated flow
无调速器气马达 non-governed air motor
无调速状态 governor free state
无调头的往返动输 no-turn shuttle hauling
无调蓄演算 non-storage routing
无调中转车 transit car without resorting
无调中转车停留时间 detention time of car in transit without resorting
无调中转列车技术作业过程 operating procedure of transit train without resorting
无跳动运转 run true
无铁的磁系统 ironless magnetic system
无铁电枢 ironless armature
无铁交流伺服电动机 ironless alternative current servo motor
无铁鳞的 free from scale
无铁溶液 iron-free solution
无铁芯的 air-cored
无铁芯电感线圈 inductance without iron core
无铁心感应电炉 coreless-type induction furnace
无铁芯电枢 coreless armature
无铁芯扼流圈 no-core reactor
无铁芯感应线圈 inductance without iron core
无铁陨石 asiderite
无梃门 ledged door; unframed door
无通道的建筑物中央大厅 nave aisleless
无通道的教堂中部 nave aisleless
无通风 draft fan
无通过台车 blind-end car
无通航的 non-navigable
无通路开关 pathless switch
无同型减数分裂 apohomotypic meiosis
无铜光铁蓝 non-bronze blue
无铜硬铝 Aldray
无头槽塞 headless slotted plug
无头的 headless
无头钉 brad (nail); glazing sprig; headless nail; sprig; glazier's point <镶玻璃用>
无头钉(固定)圆线脚 wire brad bead
无头定位螺钉 headless locating screw
无头定位螺栓 headless set screw
无头钢钻 straight shank twist drill
无头螺钉 grub screw; headless screw
无头螺丝 grub screw
无头铆钉 headless rivet; straight-neck rivet; straight shank rivet
无头塞 headless plug
无头小鞋钉 sparable
无头轧制 endless rolling
无头止动螺钉 headless set screw
无头轴肩螺钉 headless shoulder screw
无投资款项 zero investment supply
无透镜照相机 pin-hole camera
无凸缘衬套 flush bushing
无凸缘衬筒 flangeless liner
无凸缘的 flangeless; unflanged
无凸缘管子 no-hub pipe
无凸缘轮胎 blind tyre; flangeless tire
无凸缘锚卡环 lugless anchor shackle
无突起 non-relief
无突缘的 flangeless
无图案水印辊 plain dandy
无图的 non-mapping
无图地区 uncharted area; uncovered area; unmapped area
无图零件切割面光洁度 cutting finish for parts without drawing
无涂层的 unplated
无涂层冷扎扁钢板 cold-rolled uncoated flat product
无涂层冷扎扁钢制品 cold-rolled uncoated flat product
无涂层纸 uncoated paper
无土栽培 growing plant without soil; soilless culture
无土覆盖的喀斯特 naked karst
无土种植 nutrient culture
无钍光学玻璃 thorium-free optic(al) glass
无湍流 turbulence-free flow
无推杆活塞 free running piston
无推力螺距 zero thrust pitch
无退还抽样 sampling without replacement
无托架的 bracketless
无托梁楼板 untrimmed floor
无托盘包装的砖块运输 palletless package brick transporting
无托运行李设备的车站 station not affording facilities for registered baggage
无拖车的载重汽车 single-unit truck; straight job
无拖斗卡车 single-unit truck
无拖期赔偿条款 no damages for delay clause
无瓦履带 shoeless track
无瓦斯矿山用机车 naked-flame locomotive
无外部电磁场线圈 fieldless coil
无外荷时的应力 inherent stress
无外汇出口 no-draft export
无外汇进口 no-draft import
无外加剂混凝土 plain concrete
无外壳的 non-sheathed; uncanned; uncased
无外力边界 force-free edge
无外力作用边 force-free edge
无外圈滚柱轴承 outer-ringless roller bearing
无外座圈滚柱轴承 roller inner race assembly
无弯矩 bending moment-less
无弯曲毛 broad wool
无网印刷 screenless printing
无危险的 non-hazardous; not dangerous
无微生物的 amicrobic
无围带的叶轮 unshrouded wheel
无围护井架 unboarded derrick
无围护钻塔 unboarded derrick
无桅帽的短上桅 stump mast
无维参数 dimensionless parameter; non-dimensional parameter
无维单位过程线 non-dimensional unit hydrograph
无维的【物】dimensionless
无维量 dimensionless quantity; non-dimensional quantity
无维稳定性 non-dimensional stability
无维因数 dimensionless factor
无位错 dislocation free
无位错单晶体 dislocation-free single crystal
无位错晶体 dislocation-free crystal
无位移花纹 rise-free design
无位移谱线 stationary line
无味 insipid taste
无味的 insipid
无纹的 unstriated
无纹钢筋 plain bar (of reinforcement)
无纹铬钢 plain chromium steel
无纹理 absence of streaks
无紊流 turbulence-free flow
无涡流 irrotational flow; non-eddy flow
无涡流的 non-vortex
无涡旋运动 vortex-free motion
无涡运动 irrotation(al) motion
无握裹的预应力 no-bond prestressing
无握裹力预应力 prestress without bond
无污点的 stainless; unstainable; unstained; virgin
无污染 non-pollution
无污染产品 green production
无污染的 contamination free; pollution free
无污染的生产废水 non-polluted industrial wastewater
无污染发动机 pollution-free engine
无污染工业废水 non-polluted industrial wastewater
无污染工艺 non-polluted technology; pollution-free technology
无污染供热 pollution-free heating
无污染河流 healthy stream
无污染或少污染工艺 non-polluting or pollution-reducing process
无污染抗氧剂 non-staining anti-oxidant
无污染控制 no pollution control
无污染能源 energy without pollution; non-polluting energy sources; pollution-free energy resources
无污染农药 non-contamination pesticide; pollution-free pesticide
无污染汽车 non-contamination automobile; non-polluting automobile; pollution-free automobile
无污染燃料 non-polluting fuel; pollution-free fuel
无污染设备 pollution-free equipment
无污染生产 pollution-free production
无污染塑料 pollution-free plastics
无污染涂层 non-polluting coat(ing)
无污染涂料 green paint
无污染制浆 pollution-free pulping
无污染装置 pollution-free equipment; pollution-free installation
无屋顶的 hyp(a)ethral; unroofed
无屋脊坡屋顶 cut roof
无钨硬质合金 tungsten-free cemented carbide
无坞墙的干(船)坞 dry dock without walls
无误差 errors excepted
无误差编码 error free encoding
无误差操作 error free operation
无误差的 error free; errorless; free from error
无误差的地图数据 clean map
无误差模拟像片 error free simulated photograph
无误差运行 error free running
无误差运转期 error free running period
无误差制 non-error system
无误的 unerring; unmistakable
无误运转周期 error free running period
无吸收能力的 non-absorbent
无吸收性的 non-absorbent
无吸收性的表面 non-absorbent surface
无吸引力的 insipid; unappealing
无息存款 interest free deposit; non-interest-bearing deposit
无息贷款 free loan; gift loan; interest free credit; interest free loan; loan without interest; non-interest loan
无息的 interest free
无锡钢板 tin-free steel
无锡青铜 tin-free bronze
无席垫的 non-matting
无系缆式潜水器 underwater untethered submersible
无系统的 unsystematic(al)
无细骨料 no fines aggregate
无细骨料混凝土 no fines concrete; popcorn concrete
无细集料 no fines aggregate
无细集料混凝土 no fines concrete
无细菌的 abacterial
无细孔 impunctate
无细孔结构 pore-free structure
无细扣方钻杆 non-thread kelly
无细料混凝土 non-fines concrete
无隙 zero clearance
无隙内接头 close nipple
无匣钵装窑法 open setting
无瑕疵的 irreproachable; unstained
无瑕疵的表面 blemish-free surface
无下标变量 non-subscripted variable; unsubscripted variable
无下标变量名 unsubscripted variable name
无下标的 unsubscripted
无下定义的 undefined
无下水道的 unsewered
无下水道的卫生设施 non-sewered sanitation
无先例的 unexampled
无纤道的运河岸 offside
无纤维状的 non-fibered [fibred]
无弦的 stringless
无现货 non-available goods
无现款 money out of hand; out of cash
无现款社会 cashless society
无线本地环路 radio local loop
无线传声器 wireless mic
无线传声器接收机 wireless microphone receiver
无线传输 radio communication
无线的 wireless
无线地震台网 radio linked seismometer network
无线电 radio; wireless
无线电安全信号 radio safety signal
无线电安全证书 wireless installation inspection certificate
无线电岸台 ground station
无线电保护通路 radio protection channel
无线电保护装置 radio link protection
无线电保密 radio security
无线电报 aerogram; marconigram; radio (tele) gram; radio telegraph; space telegraphy; telegraph radio; wireless message; wireless talker; wireless telegram; wireless telegraph
无线电报报警信号 radio telegraph alarm signal
无线电报测向 wireless telegraph direction finding
无线电报测向器 wireless telegraphy direction finder
无线电报传输 radio message transmission

无线电报船舶电台 radio telegraph ship station
无线电报的 wireless
无线电报发射机 radiotelegraph transmitter
无线电报发送 radio telegraphic (al) transmission
无线电报呼叫 radio telegraphic (al) call
无线电报机 aerograph; radio telegraph
无线电报交换台 wireless telegraphy board
无线电报警 radio warning
无线电报警和控制系统 radio alarm and control system
无线电报流水号数 wireless serial number
无线电报设备 radio telegraph installation
无线电报时信号 wireless time signal
无线电报室 radio telegraph operation room; wireless room
无线电报术 radio telegraphy; wireless telegraphy
无线电报台 wireless telegraph station
无线电报通信[讯] radio telegraph communication
无线电报通信[讯]程序 radio telegraph procedure
无线电报学 radio telegraphy; space telegraphy; wireless telegraphy
无线电报汛站 radio reporting river ga(u)ge
无线电报自动报警器 radio telegraph auto-alarm
无线电暴 radio storm
无线电变压器 wireless transformer
无线电标志 radio marker; radio signal
无线电标志信标 radio marker beacon
无线电波 air wave; radio wave
无线电波波前失真 radio wavefront distortion
无线电波测深 radio wave sounding
无线电波传播 radio (wave) propagation; radio (wave) transmission
无线电波传播公式 radio transmission formula
无线电波传播路径 radio path; radio ray path
无线电波传播情况预报 radio forecast
无线电波传播速度 velocity of radio waves
无线电波传播预报 radio propagation forecast
无线电波传播预测 radio wave propagation prediction
无线电波导 radio duct; radio waveguide
无线电波导航定位 radio range fix
无线电波道 radio canal; radio duct; radio (wave) channel
无线电波的散射 radio diffusion
无线电波法 radio wave method
无线电波反射法 radio wave reflection method
无线电波辐射 radio wave radiation
无线电波干涉法 radio wave interference method
无线电波检验 radioexamination
无线电波路径 radio (wave) path
无线电波倾斜传播 oblique radio transmission
无线电波散射 radio scattering
无线电波束偏离指示器 radio beam deviation indicator
无线电波束曲折 bending
无线电波束制导 radio beam guidance
无线电波衰落 fade out
无线电波探测 radio wave sounding
无线电波透视 radio wave penetration
无线电波透视法 radio wave penetration method
无线电波透视法视吸收系数等值线图 contour map of apparent absorption of radio wave penetration method
无线电波吸收 radio wave absorption
无线电波吸收材料 radio wave absorbing material
无线电波消失 radio fade-out
无线电播发 radiobroadcast(ing)
无线电播送 radio line
无线电舱 radio compartment; radio hold
无线电操纵 telemotion
无线电操纵泵站 radio-controlled pump station
无线电操纵船 wireless control vessel
无线电操纵的 radio-controlled
无线电操纵的调车机 rail-controlled shunter
无线电操纵的浮标 radio-controlled buoy
无线电操纵的起重机 radio-controlled crane
无线电操纵台 radio control box
无线电操纵推土机 radio control bulldozer
无线电操作 radio adaptation
无线电操作人员 radio operator; wireless operator
无线电测得的距离 radio range
无线电测地学 radio geology
无线电测定 radio determination
无线电测风观测 radio wind observation
无线电测风气球 radio pilot; rawin balloon
无线电测风(仪) radio wind; rawin
无线电测风站 radio wind station
无线电测高法 radar method; radio altimetry
无线电测高计 electronic altimeter; radio altimeter; reflection altimeter; terry
无线电测高仪 radio altimeter; radiometric altitude measuring apparatus
无线电测高仪和高差仪记录 records made by radio altimeter and statoscope
无线电测角器 radiogoniometer
无线电测角仪 radiogoniometer
无线电测距 radio distance-measuring; radio range finding; radio ranging measurement; radist [radio distance]
无线电测距波束 radio range beam
无线电测距定位 ranging radio positioning
无线电测距发射机 radio range transmitter
无线电测距计 tellurometer
无线电测距接收机 radio range receiver
无线电测距器 radio distance finder; radiosonde
无线电测距射束 radio range leg
无线电测距仪 radio distance-finding set; radio distance-measuring set; radiogoniometer; telegoniometer; tellurometer
无线电测距站 radio distance-finding station
无线电测量 radio geodesy
无线电测量器材 radio measuring equipment
无线电测量系统 telemetric system
无线电测量仪器 radio instrument
无线电测流仪 radio current meter
无线电测试设备 radio test set
无线电测位 radio position finding; wireless fixing
无线电测位电台 radio position finding station
无线电测位陆地电台 radio position finding land station
无线电测位行动电台 radio location mobile station
无线电测向 directional wireless; radio direction finding
无线电测向定向仪 radio direction finding receiver
无线电测向发射台 directional transmitter
无线电测向法 radio direction finder method
无线电测向计 radio direction (al) finder; radiogoniometer; wireless compass
无线电测向接收机 radio direction finding receiver
无线电测向控制台 radio direction finding control station
无线电测向器 bird dog; radio direction (al) finder; radiogoniometer; wireless compass
无线电测向术 radio goniometry
无线电测向数据库 radio direction finding data base
无线电测向台 direction finding station; radio direction finder (station); radio direction finding station; radio range station
无线电测向天线 direction finder antenna
无线电测向图 radio direction finding chart
无线电测向网 radio direction finding network
无线电测向系统 radio direction finding system
无线电测向仪 direction meter; radio direction (al) finder; radiogoniometer; radio setting apparatus; wireless compass; wireless direction (al) finder
无线电测向仪定位法 positioning by radio direction finder; positioning by radio goniometer; positioning by wireless compass
无线电测向仪上消除天线效应的装置 balancer
无线电测向仪误差 error of radio direction finder; radio direction finder error; radio goniometer error; wireless compass error
无线电测向站 radio direction finding station
无线电测位仪 radio position finder
无线电长途电话系统 radio long call system; radio trunk call system; radio trunk telephone system
无线电厂 radio factory
无线电场强 radio field intensity
无线电场强单位 radio field strength unit
无线电场强图 radio field strength map
无线电唱机 radiogram; wireless record player
无线电传播预测 radio propagation prediction
无线电传感潮位仪 radio linked tide ga(u)ge
无线电传感器 radiosensor
无线电传声器 radio microphono
无线电传送 radio; radio transmission
无线电传送功率 radio power
无线电传送机 radio transmitter
无线电传真设备 radio facsimile equipment
无线电传真室 facsimile room
无线电传真术 <用于车辆调度> photogrammetry
无线电传真图片 photoradiogram
无线电传真系统 radio facsimile system
无线电传真照片 radiophotogram; radio picture; radiophoto(graph)
无线电船位 radio fix; wire(less) fix
无线电船位线 radio position line
无线电磁航向指示器 radio magnetic indicator
无线电粗略定位 coarse radio location
无线电大地测量 radio triangulation
无线电大地测量学 radio geodesy
无线电大气 radio atmosphere
无线电单向定向器 unidirectional radio direction finder
无线电导标的航线 beacon course
无线电导标接收机 beacon receiver
无线电导航 aeronautical radio navigation; electronic navigation (al); radio guidance; radio navigation; radio range
无线电导航岸台 radio navigation land
无线电导航测标 radio beacon buoy; radio navigation buoy
无线电导航定向(法) radio range orientation
无线电导航飞行 radio flying
无线电导航浮标 sonobuoy; sono-radio buoy
无线电导航警报 wireless navigational warning
无线电导航静默 radio navigation silence
无线电导航陆地电台 radio navigation land station
无线电导航气球 radio pilot balloon
无线电导航设备 radio aid; radio aids to navigation; radio navigational aids
无线电导航式制导 radio navigation guidance
无线电导航台 radio direction finding station; radio range station
无线电导航图 enroute chart; radio facility chart; radio goniometric chart; radio navigation chart
无线电导航图网 radio navigational lattice
无线电导航卫星 radio navigation satellite
无线电导航系统 electronic navigation (al) system; radio direction finding system; radio navigation system; radio ranging system; radist [radio distance]
无线电导航系统接收机 radio navigation receiver
无线电导航信标 radio range beacon
无线电导航信号 radio beacon
无线电导航信号发射台 radio marker beacon
无线电导航信号浮标 radio beacon buoy
无线电导航信号区 quadrant
无线电导航信号台 radio beacon station
无线电导航仪 radio navigation aid instrument; radio navigation set; radio range
无线电导航移动电台 radio navigation mobile station
无线电导航站 direction finding station; radio direction finding station; radio range station
无线电导航装置 radio navigational device

无线电导向标 radio range beacon
无线电的 wireless
无线电灯船 radio light vessel
无线电低频 low radio frequency
无线电地波传播 radio ground-wave propagation
无线电地平(线) radio horizon
无线电地平线距离 radio horizon distance
无线电电标 radio telephony
无线电电传打字电报接收机 radioteletype receiver
无线电电传打字机 radio teletype writer; radioteletype
无线电电话机 radio telephone
无线电电离层传播 radio ionospheric propagation
无线电电路 radio channel; radio circuit
无线电电声测距法 ranging method by radio and sound wave
无线电电台定位 radio fix
无线电电子学 radioelectronics
无线电调度系统 radio dispatching system
无线电定位 radio detection and ranging; radio directional bearing; radio fix; radiolocate; radio location fixing; radio location <一种监测车辆运行位置的方法>
无线电定位地面电台 radio location land station
无线电定位点 radio range fix
无线电定位电台 radio determination station
无线电定位法 radio position fixing (method); radio positioning method
无线电定位量测仪 radio position plotter
无线电定位陆地电台 radio positioning land station
无线电定位器 radar; radio locator
无线电定位设备 radio fixing aids; radio location equipment
无线电定位天文学 radio locational astronomy
无线电定位图 radio location chart
无线电定位系统 radio-positioning system
无线电定位线 radio line of position
无线电定位仪 radiophare; radio phase; radio position finder; radio position fixing instrument
无线电定位移动电台 radio positioning mobile station
无线电定位站 radio location station
无线电定位装置 radar [radio detecting and ranging]
无线电定向 directional radio; radio bearing; radio direction finding; wireless direction finding
无线电定向标选择器 omnibearing selector; radial selector
无线电定向发射 directional radio radiation
无线电定向发射台 radio beam transmitting station
无线电定向法 radio direction finder method
无线电定向计 radio goniograph
无线电定向器 radiogoniometer
无线电定向设备 radio directive device
无线电定向台 radio compass station; radio direction finding station
无线电定向图 direction finding chart; radio chart; radio direction finding chart
无线电定向信标 directional beacon; directive beacon
无线电定向仪 radio bearing installation; radio direction (al) finder; wireless direction (al) finder; direction finder
无线电定向与测距 radio detection and ranging
无线电定向站 radio compass station
无线电对话机 two-way radio
无线电对讲接收机 radio control receiver
无线电发报机 radiotelegraph transmitter
无线电发报水文站 radio reporting river ga(u)ge
无线电发火装置 radio firing device
无线电发射 radio(-frequency) emission; radio radiation; radio transit; radio transmission; transmitting
无线电发射机 radio transmitter; radio transmitter equipment; radio transmitting set; transmitting set; X-emitter <俚语>
无线电发射机工作信号灯 radio output light
无线电发射控制系统 radio launch control system
无线电发射频率测量系统 radio transmission frequency measuring system
无线电发射室 radio transmitting room
无线电发射塔 broadcasting tower; radio tower; wireless tower
无线电发射台 radio transmitting station
无线电发射中心 radio transmission center [centre]; radio transmitting center [centre]
无线电发送 radiation transmission
无线电发送机 radio transmitting set
无线电发送器 radio transmitter
无线电法导航定向 radio range orientation
无线电反射点 apex [复 apices/apexes]
无线电反射镜 radio mirror
无线电反射信号 radio echo
无线电方位 radio bearing
无线电方位测定器 goniometer; radiogoniometer
无线电方位换算表 radio bearing conversion table
无线电方位角 radio angle
无线电方位偏差 radio bearing deviation
无线电方位图 radio bearing chart
无线电方向探测站 radio gonio station
无线电放大器 radioamplifier
无线电浮标 radio buoy; radio light vessel
无线电辐射 radio emission
无线电辐射激扰 radio noise burst
无线电辐射宁静部分 quiet component of radio radiation
无线电辐射温度 radio temperature
无线电干扰 barrage; radio countermeasure; radio disturbance; radio influence; radio interference; radio jamming; radio noise; radio suppression; spurious response; strays
无线电干扰波场强仪 radio interference field intensity meter
无线电干扰测量仪 radio interference measuring set
无线电干扰测量仪表 radio interference measuring instrument
无线电干扰场 radio interference field
无线电干扰的排除 radio interference suppression
无线电干扰电平 radio interference level
无线电干扰电压 radio interference voltage [RIV]
无线电干扰电压测量 RIV measurement
无线电干扰发生器 radio jammer
无线电干扰滤波器 radio interference filter
无线电干扰试验 radio interference test
无线电干扰水平 radio interference level
无线电干扰探测器 radio interference locator
无线电干扰消除器 clarifier; radio interference eliminator
无线电干扰信号发生器 radio interference generator set
无线电干扰抑制器 radio interference suppressor
无线电干扰源 radio noise source
无线电干涉测量 radio interferometry
无线电干涉仪 radio interferometer
无线电干抑消除器 radio interference suppressor
无线电杆 radio mast
无线电高导磁性合金 radio metal
无线电高度 radio altitude
无线电高度计天线 radio altimeter antenna
无线电高度计指示器 radio altimeter indicator
无线电高度指示器 radio altitude indicator
无线电高空测风仪 radio wind sounding; rawin
无线电高空测候 rawin
无线电高空测候器 radiosonde
无线电高空测候仪 radio wind sounding; rawinsonde
无线电高空仪 radio meteorograph
无线电高频测向仪 huff-buff
无线电跟踪 radio tracking
无线电跟踪站 radio tracking station
无线电跟踪装置 minitrack
无线电工程 radio engineering; wireless engineering
无线电工程师 radio engineer; wireless engineer
无线电工程师学会 Institution of Radio Engineers
无线电工艺 radio technology
无线电公电密语 radio service code
无线电功率 wireless power
无线电观测 radio observation
无线电惯性导航系统 radio equipped inertial navigation system
无线电惯性监控设备 radio inertial monitoring equipment
无线电惯性制导 radio inertial guidance
无线电惯性制导系统 radio inertial guidance system
无线电广播 broadcasting transmission; radiobroadcast (ing); radio cast; telediffusion
无线电广播车 radio car
无线电广播电台 radio broadcasting station
无线电广播发射机 radio broadcast transmitter
无线电广播节目 broadcasting program(me)
无线电广播设备 broadcasting equipment; radio broadcasting equipment
无线电广播通道 radio broadcasting channel
无线电广播网 radio broadcasting system
无线电广播站 radio broadcasting station
无线电广播者 <美> blaster
无线电广播转播 broadcast relaying
无线电广告广播员 spieler
无线电归航辅助设备 radio homing aid
无线电归航信标 radio homing beacon
无线电海流计 radio current meter
无线电海图室 radio and chart room
无线电航标 beacon light; navigational radio beacon; radio aid; radio beacon; radiophare; wireless beacon
无线电航标台组 grouping of radio beacon
无线电航海警告 radio navigational warning
无线电航行警告 radio navigational warning
无线电和电话通信[讯]系统 radio and telephone communication system
无线电呼号 radio call; radio call letters; radio call signal
无线电呼叫系统 radio paging system
无线电话 wireless; wireless telephone
无线电话报传输 radio voice transmission
无线电话报警信号 radio telephone alarm signal
无线电话发射机 radio phone transmitter; radio telephone transmitter
无线电话呼叫 radio telephone call
无线电话机 aerophone; bellboy; radio phone
无线电话交换机 wireless telephone exchanger
无线电话求救信号 mayday
无线电话术 radiophony; radio telephony
无线电话通信[讯] radio telephone communication; wireless telephonic communication
无线电话网 radio telephony network
无线电话务员 radioman
无线电话信号程序 radio telephone signal procedure
无线电话学 radiophony; radio telephony; wireless telephony
无线电话自动报警器 radio (tele) phone auto-alarm
无线电回波 radio echo
无线电回波法 radio echo method
无线电回波探测(法) radio echo observation; radio echo sounding
无线电回波探测器 radio echo detector
无线电回声测距法 radio acoustic (al) (sound) ranging; radio sound ranging
无线电火花制止器 radio spark suppressor
无线电火箭探空仪 radio sonde-rocket set
无线电机 radio apparatus
无线电机械室 radio cabin
无线电及电话管制 radio and telephone control
无线电极光 radio aurora
无线电技师 radioman
无线电技术 radio engineering; radio technics; radio technology
无线电技术通信[讯]工具 radio technical means of communication
无线电技术员 radio mechanician; wireless mechanician
无线电监测站 monitoring station
无线电监控 radio monitoring
无线电监控器 radio monitoring set
无线电监控系统 radio control system; radio monitoring system
无线电监听 wireless monitoring
无线电监听设备 radio monitoring e-

quipment
无线电检验器 radio tester
无线电交换台 radio exchange
无线电接力通信[讯] beam wireless
无线电接力系统 chain relaying;radio relay system
无线电接力线路 radio relay link;relay link
无线电接收 radio reception
无线电接收发送装置 radio reception and transmitting installation
无线电接收管 radio receiving tube
无线电接收机 radio receiver;radiosensor;radio set;wireless receiver
无线电接收机盒 receiver cabinet
无线电接收机壳 receiver cabin
无线电接收情形 radio reception condition
无线电接收室 radio receiving room
无线电接收站 radio receiving station
无线电接收装置 radio receiving set
无线电接线盒 radio terminal box
无线电截获站 radio intercept station
无线电截听单位 radio intercept unit
无线电截听或情报分析站 radio interception or intelligence analysis post
无线电截听控制设备 radio intercept control set
无线电截听站 radio intercepting post
无线电金属探测器 radio metal locator
无线电紧急信号 radio urgency signal
无线电近发引信 radio proximity fuse
无线电进场导航设备 radio approach aids
无线电经纬仪 radio theodolite
无线电警报 radio warning
无线电警报信号 radio alarm signal
无线电警标 radio beacon
无线电警告航标 warning radio beacon
无线电警戒 radio communication guard;radio detection;radio guard
无线电静默 radio listening silence;radio quiescence;radio silence;wireless silence
无线电静区 radio pocket;radio shadow
无线电救援信标 rescue radio beacon
无线电勘测 radio surveying
无线电勘探 radio prospection
无线电勘探法 radio method
无线电空中导航 radio aids to air navigation;radio avigation
无线电控空仪信号发送装置 radio sonde flight equipment
无线电控制 radio control;wireless control
无线电控制的 radio-controlled;wireless-controlled
无线电控制的交通 radio-controlled traffic
无线电控制的空中目标 radio-controlled aerial target
无线电控制的列车运行 radio-controlled train operation
无线电控制的调车机车 radio-controlled switch engine
无线电控制机车 radio-controlled locomotive
无线电控制继电器 radio control relay
无线电控制空中航行区 radio-controlled aerial-navigation
无线电控制雷达 radio-controlled radar
无线电控制盘 wireless panel
无线电控制气球 radio pilot balloon
无线电控制信号接收机 radio control receiver
无线电控制信号系统 radio-controlled signaling system
无线电控制运行 radio control operation
无线电控制站 radio-controlled station
无线电控制中心 radio-controlled center [centre]
无线电扩音器 radio speaker
无线电雷达电磁波吸收剂 radio radar absorber
无线电棱镜 radio prism
无线电立体声 radio stereophony
无线电连接 radio connection
无线电连接干涉仪 radio link interferometer
无线电联络 radio contact;wireless contact
无线电联络指挥系统 radio paging system
无线电联络中断时的处置办法 radio out procedure
无线电联系 wireless link
无线电链路 radio link
无线电领航 radio pilot
无线电六分仪 radiometric sextant
无线电录音机 wireless record player
无线电录制 radio transcription
无线电罗经 direction finder;radio compass;wireless compass
无线电罗盘 direction finder;radio compass;sense finder;wireless compass
无线电罗盘传感器 radio compass sensor
无线电罗盘定位 radio compass fix
无线电罗盘方位指示器 radio compass bearing indicator
无线电罗盘校准数据 radio compass calibration data
无线电罗盘接收机 radio compass receiver
无线电罗盘误差 radio compass error
无线电罗盘指示器 radio compass indicator
无线电盲目着陆设备 radio blind landing equipment
无线电盲区 radio skip zone
无线电瞄准器 radio sight
无线电瞄准线 radio line-of-sight
无线电瞄准装置 radio aiming device
无线电排 radio platoon
无线电判位 radio determination
无线电频道 radio-frequency channel
无线电频率 radio-frequency
无线电频率定位(法)<用于车辆定位系统> radio-frequency location
无线电频率分析仪 radio frequency analyser [analyzer]
无线电频率干扰 radio-frequency interference
无线电频率资源 radio-frequency resource
无线电频谱 radio-frequency spectrum;radio wave spectrum
无线电频谱分配 radio spectrum allocation
无线电频谱学 radio spectrography
无线电凭证行车制(系统) radio electric token system
无线电屏蔽 radio shielding
无线电气候学 radio climatology
无线电气象报告业务 radio meteorological service
无线电气象观测 radio meteorograph observation
无线电气象计 radio meteorograph;radiometeorograph
无线电气象探测 raob
无线电气象通信[讯] radio weather message
无线电气象图解 radio meteorogram
无线电气象学 radio meteorology
无线电气象仪 radio meteorograph
无线电汽车 radio car
无线电器材厂 radio appliance factory;radio equipment factory
无线电切换中心 radio switching centre
无线电窃听 radio intercept
无线电求救信号 radio alarm signal;radio distress signal
无线电求救信号频率 radio distress frequency
无线电扰动 radio storm
无线电人员 radioman
无线电散射 radio diffusion
无线电扇面标识浮标 radio fan marked beacon
无线电设备 radio;radio apparatus;radio equipment;radio facility;radio installation;radio set;wireless apparatus
无线电设备安全证书 safety radio telegraphy certificate
无线电设备舱 radio bay;radio equipment bay
无线电设备作用范围 radio coverage
无线电射束 radio(-frequency) beam
无线电摄像机 radio camera
无线电摄影术 radiophotography
无线电声波测距仪 radio acoustic(al) range finding equipment
无线电声的 radioacoustic
无线电声的浮标 radio sonobuoy;sonobuoy;sono-radio buoy
无线电声定位 radio acoustic(al) position finding
无线电声呐测距 radio sonic ranging
无线电声呐测距系统 radio sonic ranging system
无线电声呐浮标 radio sonic buoy
无线电声呐浮标与水下听音器 radio sonobuoy and hydrophone
无线电声(响)测距 radio acoustic(al)(sound) ranging
无线电声响定位 radio acoustic(al) position finding
无线电声学 radio acoustics
无线电声学法 radio acoustic(al) method
无线电声学遥测术 phonotelemetry
无线电时号 radio time signal
无线电时间码 radio time code
无线电时间信号 radio tick;radio time signal
无线电识别 radio recognition;radio recognition and identification
无线电识别发送信号 squawk
无线电识别设备 radio recognition equipment
无线电视 radio television;radiovision;wireless television
无线电室 radio cabin;radio house;radio room;wireless house
无线电收发报机 radio set;radio transceiver;radio transmitter and receiver;transceiver;wireless set
无线电收发报机组 transmitting-receiving radio set
无线电收发机 radio receiver set;radio receiver-transmitter;radio set;transceiver;wireless set
无线电收发两用机 transceiver;transmitter-receiver set
无线电收发信机 radio set;radio transceiver
无线电收发站 transmitting and receiving radio set
无线电收信机 radio receiver
无线电收音机 radio set;wireless
无线电收音困难的地区 dead spot
无线电手册 radio manual
无线电手机 radio handset
无线电数据 radio data
无线电数据传输线路 radio data link
无线电数据设备 radio data set
无线电数据通信[讯]设备 radio data link set
无线电数据系统 radio data system
无线电数据系统/交通信息通道 radio-data system/traffic message channel
无线电数字系统 radio digital system
无线电数字终端 radio digital terminal
无线电衰减测量 radio attenuation measurement
无线电双曲线定位 hyperbolic radio positioning
无线电双向通信[讯] radio two way communication
无线电搜索 radio search
无线电台 radio plant;radio post;radio set;radio station;wireless set;wireless station
无线电台干扰 radio station interference;station interference
无线电台广播网 broadcast transmission
无线电台呼号 radio station call sign
无线电探测 radio detecting and ranging;radio detection
无线电探测器 radar [radio detecting and ranging];radio detector;radiosonde
无线电探测器观测 radiosonde observation
无线电探测设备 radio detector equipment
无线电探测术 radio sondage technique
无线电探测仪 radio sounding apparatus
无线电探测与定位 radio detection and location
无线电探测资料 radar data;radio sonde data
无线电探空 radio sondage;radio sounding
无线电探空测风观测 rawinsonde observation
无线电探空测风仪 rawinsonde
无线电探空测风站 radio sonde and radio wind station
无线电探空观测 radio meteorograph observation;radio sonde observation;raob
无线电探空记录器 radio sonde recorder
无线电探空接收机 radio sonde receptor
无线电探空气球 radio sonde balloon
无线电探空器 radiosonde
无线电探空仪 radiosonde
无线电探空仪测风 radio sonde wind sounding
无线电探空仪和雷达测风<联合装置> radio sonde and radar wind sounding
无线电探空仪信号地面接收装置 radio sonde ground equipment
无线电探空仪转换开关 radio sonde commutator
无线电探空站 radio sonde station
无线电探向 radio homing;wireless direction finding
无线电探向器 radio locator
无线电陶瓷 radio ceramics
无线电天文导航系统 radio celestial navigation system
无线电天文学 radio astronomy
无线电天线 radio aerial;radio antenna;wireless aerial

W

无线电天线车 radio antenna truck
无线电天线杆 radio mast; radio tower; wireless mast; wireless tower
无线电天线塔 radio tower; wireless tower
无线电天线桅 radio mast
无线电铁塔 radio tower
无线电停止工作信号 silence signal
无线电通话方式 system of radio telephone
无线电通路 radio circuit
无线电通信[讯] radio; radio communication; telecommunication; wireless communication
无线电通信[讯]处 radio division
无线电通信[讯]传送 communication traffic
无线电通信[讯]船 radio ship
无线电通信[讯]电路 radio communication circuit
无线电通信[讯]电码 radio code
无线电通信[讯]发射机 radio communication transmitter
无线电通信[讯]干扰机 wireless jammer
无线电通信[讯]干扰台 radio channel jammer; radio link jammer
无线电通信[讯]工程 radio communication engineering
无线电通信[讯]工作程序 radio procedure
无线电通信[讯]规程 radio communication regulation
无线电通信[讯]规则 radio regulation
无线电通信[讯]记录本 radio log
无线电通信[讯]记录簿 radio log
无线电通信[讯]纪律 radio circuit discipline
无线电通信[讯]检查 radio check
无线电通信[讯]简语适用范围 radio code aptitude area
无线电通信[讯]舰 radio ship
无线电通信[讯]距离 radio communication range
无线电通信[讯]联络 communications traffic
无线电通信[讯]频率 radio communication frequency
无线电通信[讯]设备 radio communication equipment; radio communication set; telecommunication equipment
无线电通信[讯]试验船 radio research ship
无线电通信[讯]台 telecommunication station
无线电通信[讯]统计 radio communication statistics
无线电通信[讯]网 radio communication net; radio link network; radio network
无线电通信[讯]线路 link; radio communication diagram; radio link
无线电通信[讯]用国际标准时间 <格林尼治平均时> Z time; zulu time
无线电通信[讯]站 telecommunication station
无线电通信[讯]主任 wireless officer
无线电网 radio net
无线电网中心 network radio center [centre]
无线电微波搜索仪 radio microwave search set
无线电微波线路 radio microware link
无线电位置线 radio line of position
无线电雾号 radio fog signal; wireless fog signal; wireless telegraph fog signal
无线电吸收测量 radio absorption measurement
无线电系统 radio system
无线电线路 radio link; wireless link
无线电线圈 radio coil
无线电小组 wireless team
无线电信标 radio beacon; radio marker
无线电信标导航图 radio chart; wireless chart
无线电信标航向 radio beacon course
无线电信标机 radio beacon set
无线电信标监视站 radio beacon monitor station
无线电信标接收机 radio beacon receiver
无线电信标设备 radio beacon facility
无线电信标识别 radio beacon identification
无线电信标网 radio beacon network; radio range network
无线电标系信统 radio beacon system
无线电信标译码器 beacon decoder
无线电信标用滤波器 radio range filter
无线电信道 radio channel
无线电信管 radio detonator
无线电信号 radio-based signal; radio beeper; radio call; radio signal; tick; wireless signal
无线电信号报告码 radio signal reporting code
无线电信号标 radio marker beacon
无线电信号表 list of radio signals
无线电信号发射台 marker beacon; marker radio beacon
无线电信号截听器 radio signal interceptor
无线电信号输入 radio input
无线电信际 radiophare
无线电信汽车 radio trunk
无线电旋转指向标 rotating radio beacon
无线电寻呼 radio paging
无线电讯号消失 no radio
无线电验潮仪 radio tide meter
无线电遥测 radio telemetering
无线电遥测发射机 radio telemeter transmitter
无线电遥测法 radio telemetry
无线电遥测浮标 radio telemetering buoy
无线电遥测接收机 radio telemeter receiver
无线电遥测设备 radio telemetering equipment
无线电遥测摄影机 histogram recorder
无线电遥测水位计 radio remote-sensing fluviograph
无线电遥测系统 radio(-operate) remote-recording system; radio remote-sensing system
无线电遥测学 radio telemetry
无线电遥测与遥控 radio telemetry and remote control
无线电遥测站 radio telemetering station
无线电遥测装置 telemetering gear
无线电遥测资料的接收 telemetering reception
无线电遥感器 radiosensor
无线电遥控 radio telecontrol; radio telemetering; wireless control; wireless remote control
无线电遥控靶艇 radio-controlled target boat
无线电遥控机器人 radio robot
无线电遥控开关 on-off radio control
无线电遥控遥测站 radio control and metering service
无线电遥控自动装置 radio robot
无线电遥控自记地震仪 radio telerecording seismograph
无线电业务 radio service
无线电仪表板 radio escutcheon panel; radio panel
无线电仪器底板 radio chassis
无线电移动式定位电台 radio positioning mobile station
无线电抑制器 radio suppressor
无线电引信 radio bomb fuze; radio detonator
无线电用乙电池组 radio B battery
无线电有线电综合通信[讯] radio wire integration
无线电有线电综合通信[讯]台 radio wire integration station
无线电诱惑 radio deception
无线电与布板通信[讯]分排 radio and panel section
无线电与电视导航工具 radio and television aid to navigation
无线电与电视工程 radio and television engineering
无线电员 radiop
无线电噪声 radio noise
无线电噪声计 radio noise meter
无线电噪声图 radio noise map
无线电噪声指数 radio noise figure
无线电增益 radio gain
无线电侦察监听系统 radio reconnaissance monitoring system
无线电真方位 radio true bearing
无线电真空管 radio tube; radio valve
无线电振荡器 radio oscillator
无线电值班员 resident radioman
无线电指点标 radio marker station
无线电指点信标台 radio beacon marker
无线电指挥 radio flagging
无线电指挥波道 radio command channel
无线电指挥的调度集中 radio-activated centralized traffic control
无线电指挥控制系统 radio command control system
无线电指令 radio command; wireless order
无线电指令控制装置 radio command unit
无线电指令链路 radio command link
无线电指令系统 radio command system
无线电指令指示器 radio command indicator
无线电指令制导 radio command guidance
无线电指令制导系统 radio command guidance system
无线电指示台 radiophare
无线电指数 radio index
无线电指向标 directive radio beacon; radio beacon; range station
无线电指向标表 list of directive radio beacon; list of radio beacon
无线电指向标波束 radio course
无线电指向标分区图 radio beacon diagram
无线电指向标监视站 radio beacon monitor station
无线电指向标射程 radio beacon range
无线电指向标站 radio beacon station
无线电指引仪 radio director indicator
无线电制导波束 radio vector
无线电制导装置 electronic guidance equipment
无线电中断 radio blackout
无线电中断时间 radio blackout period
无线电中继段 radio relay section
无线电中继干涉仪 radio link interferometer
无线电中继台 relay base
无线电中继卫星 radio relay satellite
无线电中继系统 chain relaying
无线电中继线路 radio relay line; radio relay link
无线电中继增音器 radio relay repeater set
无线电中继站 radio relay; radio relay station
无线电中继制 radio relay system
无线电中继终端分排 radio relay terminal section
无线电中心 radio center [centre]
无线电中心台 central office
无线电中央台 radio central office
无线电终端设备 radio terminal; radio terminal equipment; radio terminal set
无线电终端组件 radio terminal assembly
无线电主天线 master radio antenna
无线电助航设备 radio aids to navigation
无线电助航设施 radio aid; radio navigation facility
无线电柱形天线 radio mast antenna
无线电转播 replete
无线电转播系统 radio re-broadcast system
无线电装备 radio equipment
无线电装配工 radio mechanician; wireless mechanician
无线电装置 radio plant; wire installation
无线电追踪 radio tracking
无线电追踪技术 radio tracking technique
无线电自动报警器 automatic radio alarm
无线电自动闭塞系统 wireless automatic block
无线电自动操纵 radio autocontrol
无线电自动测向仪 automatic radio direction finder
无线电自动发报水文站 radio reporting river ga(u)ge
无线电自动控制 radio autocontrol
无线电自动调整 radio autocontrol
无线电自动遥控水位计 radio robot level-meter
无线电自动遥控雨量计 radio robot ombrometer
无线电总局 radio central office
无线电阻尼器 radio suppressor
无线调度电话 dispatcher-controlled radio
无线调度防灾系统【铁】emergency radio dispatching system; radio dispatching system for train
无线防护报警装置 emergency radio alarming, protection equipment
无线杆座 mast base
无线和有线组合电路 combined radio and metallic circuit
无线话筒 portable radio sender
无线环路系统 wireless in the loop
无线回复 wireless reply
无线基台 radio base station
无线基站 wireless base station
无线接合 wireless bonding
无线接入 wireless access
无线卡 unruled card
无线收发转换装置 deplexing assembly
无线数据线路 wireless data link
无线索 no trace
无线通信[讯]车 radio car
无线通信[讯]电路 wireless communi-

W

cation line
无线通信[讯]天线安装 radio antenna installation
无线通信[讯]系统 radio communication system
无线通信[讯]业务 radio traffic
无线网关 wireless gateway
无线性畸变镜头 rectilinear lens
无线寻呼电话 radio paging service
无线寻呼台 radio pager station;radio paging station
无线寻呼系统 radio paging system
无线遥测地震仪 radio telerecording seismograph
无线遥测台网 radio telemetry network
无线移动电话 radio mobile telephone
无线移动电话系统 cellular mobile tele-communications
无线应用协议 wireless application protocol
无线与有线综合系统设施 radio and wire integration
无线远距离联合通信[讯] wireless telecommunication
无线载波服务区 radio signal carrier service area
无线重叠网 radio overlay network
无线装订 perfect binding;thermoplastic binding;threadless binding;unsewn binding
无线自动闭塞 wireless automatic block
无线自动闭塞系统 wireless automatic block system
无线自动闭塞制 wireless automatic block system
无线自动变速器 steplessly variable automatic transmission
无线组 wireless section
无限 endlessness;infinitude;infinity
无限安全几何条件 infinitely safe geometry
无限板状体的综合参数 synthetic(al) parameter of infinite sheet
无限棒束 infinite rod bundle
无限倍增系数 infinite multiplication factor
无限边界 infinite boundary
无限边坡 infinite slope
无限变化制动控制 infinitely variable braking control
无限变速调整 infinitely variable speed adjustment
无限波列 infinite wave train
无限层次模型 infinite hierarchy model
无限长 infinite length
无限长板条 infinite strip
无限长的 infinite
无限长梁 infinite beam
无限长线 infinite line
无限长锥形号筒 infinite conical horn
无限持续时间 infinite duration
无限冲淡 infinite dilution
无限冲激响应 infinite impulse response
无限词项 infinite term
无限磁场 infinite magnetic field
无限次映射 unlimited image number
无限存储过滤器 infinite memory filter
无限大【数】 infinitely great;infinity
无限大板 infinite plate
无限大比 infinite ratio
无限大的 infinite
无限大的数目 zillion
无限大点 infinity point
无限大电网 infinite power network
无限大膜 infinite membrane
无限大平板 infinite slab
无限大权 infinite weight
无限大容器 infinite reservoir
无限大衰减器 infinite attenuation network
无限担保 unlimited guarantee
无限的 boundless;endless;infinite;limitless;no end of;transfinite;unlimited;unmeasured
无限的范围 infinitude
无限地 ad infinitum
无限地带 infinite strip
无限典型群 infinite classical group
无限陡前沿 infinitely sharp leading edge
无限短时间 infinitesimal time
无限堆 infinite reactor
无限对策 infinite game
无限多值逻辑 infinite value logic
无限多重谱线 infinite multiplets
无限额的 open ended
无限反射层堆 infinitely reflected reactor
无限刚性 infinite rigidity;infinite stiffness
无限刚性的 infinitely rigid
无限刚性桩 infinitely rigid pile
无限高 infinite height
无限格拉斯曼流形 infinite Grassmann manifold
无限各向同性源 infinite isotropic(al) source
无限各向异性源 infinite anisotropic(al) source
无限公司 company of unlimited liability;unlimited company;unlimited partner
无限管束 infinite tube bundle
无限归纳 infinite induction
无限规划 infinite programming
无限含水层 infinite aquifer
无限含水带 infinite aquifer zone
无限号 sign of infinity
无限合伙 unlimited partnership
无限厚靶 infinitely thick target
无限厚的弹性层 elastic layer of infinite thickness
无限厚度含水层 unlimited thickness aquifer
无限厚样品 infinitely thick sample
无限互溶性 ultimate mutual solubility
无限滑动轴承 infinite journal bearing
无限回流 infinite reflux;infinite return
无限回流操作 infinite reflux operation
无限回流率 infinite reflux rate
无限活源 unlimited traffic source
无限基数 infinite cardinal number
无限级数 infinite series
无限集 infinite set
无限记忆滤波器 infinite memory filter
无限间断 infinite gap
无限阶 infinite order
无限接近点 infinitely near points
无限解 infinite solution
无限介质 infinite medium
无限介质倍增常数 infinite multiplication constant
无限介质倍增因子 infinite multiplication factor
无限介质谱 infinite medium spectrum
无限界含水层 infinite aquifer
无限竞争性投标 unlimited competitive bidding
无限竞争性招标 unlimited competitive bidding
无限静稳定性 infinite static stability
无限静止质量 infinite rest mass
无限跨度 infinite span
无限扩展 infinite expansion
无限扩张 infinite extension
无限力量 omnipotence
无限连分数 infinite continued fraction
无限链 infinite chain
无限量 infinitude
无限流动 unlimited flow
无限流动应变 unlimited flow strain
无限流体黏[粘]度计 infinite fluid viscometer
无限螺丝 endless screw
无限螺旋 endless screw
无限慢过程 infinitely slow process
无限媒质 infinite medium
无限平板堆 infinite slab reactor
无限平面 infinite plate
无限平面源 infinite plane source;infinite slab source
无限期 in perpetuity;unlimited duration
无限期罢工 strike of indefinite duration
无限期拨款 no-year appropriation
无限期的 dateless;indefinite
无限全域 infinite universe
无限权威 omnipotence
无限溶解 complete miscibility
无限栅格 infinite lattice
无限射气介质 infinite emanation medium
无限升限 unlimited ceiling
无限收敛 infinite convergence
无限寿命 infinite life
无限寿期 infinite lifetime
无限输入阻抗放大器 infinite input impedance amplifier
无限衰减 infinite attenuation
无限衰减滤波器 infinite rejection filter
无限衰减频率 frequency of infinite attenuation
无限素因子 infinite prime divisor
无限弹性 infinite elasticity
无限弹性带 infinite elastic strip
无限弹性体 infinite elastic body;infinite elastic solid
无限弹性楔型 infinite elastic wedge
无限弹性楔子 infinite elastic wedge
无限图 infinite graph
无限网状结构 infinite networks
无限维的 infinite dimensional
无限维概率分布 infinite dimensional probability distribution
无限维控制 infinite dimensional control
无限维线性系统理论 infinite dimensional linear systems theory
无限稳定度 infinite degrees of stability
无限(物)体 infinite body
无限稀度 infinite dilution
无限稀释 infinite dilution
无限稀释参比状态法 infinite dilution reference state means
无限稀释共振积分 infinite dilution resonance integral
无限稀释活度系数 infinite dilution activity coefficient
无限稀释价值 <材料> infinite dilution worth
无限稀释溶液 infinitely dilute solution
无限相容性 unlimited compatibility
无限小【数】 infinitely small
无限小变化 infinitesimal change
无限小的 infinitesimal
无限小合力 infinitely small resultant
无限小力 infinitesimal force
无限小偶极子 infinitesimal dipole
无限小数 unlimited decimal
无限小位错 infinitesimal dislocation
无限小位移 infinitesimal displacement
无限小涡量 infinitesimal vorticity
无限小误差概率 infinitesimal error probability
无限小形变 infinitesimal deformation
无限小应变 infinitesimal strain
无限小映射 infinitesimal mapping
无限小振幅波 wave of infinitely small amplitude
无限辛群 infinite symplectic group
无限型 infinite type
无限型茎 indeterminate stem
无限序列 infinite sequence
无限循环 endless loop;infinite loop
无限循环群 infinite cyclic group
无限酉群 infinite unitary group
无限浴比 infinite bath
无限域 infinite domain;infinite field
无限源 infinite source
无限远焦点 focus for infinity;infinity focus
无限远聚焦 focus for infinity
无限远能见度 unrestricted visibility
无限远调焦 focus at infinity
无限远直线 straight-line of infinity
无限责任 unlimited liability
无限责任公司 unlimited liability company
无限责任合伙企业 unlimited partnership
无限窄间隙 infinitely narrow gap
无限障板 infinite baffle
无限障板式扬声器 infinite-baffle loudspeaker
无限障隔式 infinite baffle
无限蒸发 infinite amount of vaporization
无限正交群 infinite orthogonal group
无限正则连分数 infinite regular continued fraction
无限直积 infinite direct product
无限制 unrestraint
无限制爆炸 unconfined explosion
无限制波 unbounded wave
无限制操作许可证 unrestricted operating licence
无限制的 unconditional;unrestrained;unrestricted;unconfined
无限制的否决权 absolute veto
无限制的免税仓库 unlimited duty-free storage
无限制抵押 unlimited mortgage
无限制额土地利用 unrestrained land-use
无限制灌溉 unrestricted irrigation
无限制灌浆 unrestricted grouting
无限制花型范围 unlimited pattern area
无限制混合物 unconfined mixture
无限制劲度 unconditioned stiffness
无限制空气喷口 unrestricted air-bleb
无限制流动 unrestricted flow
无限制能见度 unrestricted visibility
无限制塑性流动 unrestricted plastic flow
无限制随机采样 unrestricted random sampling
无限制随机样本 unrestricted random sample
无限制随机样品 unrestricted random sample
无限制投票 unrestricted ballot
无限制现场 non-restricted site
无限制优惠 non-restricted preference
无限质量 infinite mass
无限中止 indefinite suspension
无限桩 infinite pile
无限状态自动机 infinite state automata
无限资源 inexhaustible resource
无限自动机 infinite automata;infinite automaton

无限自由度 infinite degrees of freedom
无限总体 infinite population
无限阻抗检波 infinite impedance detection
无限阻抗检波器 infinite impedance detector
无相比性资料 non-comparable data
无相差觇标 phasedless target
无相关 absence of correlation; zero correlation
无相互影响控制 non-interacting control
无相互作用 non-interacting
无相位孔径综合 phaseless aperture synthesis
无相依变化 fully independent variation
无香味的 inodo(u)rous
无箱造型 boxless mo(u)lding; mo(u)ld in a snap flask
无箱铸型 flaskless mo(u)ld; snap flask mo(u)ld(ing)
无镶边装修的 untrimmed
无镶边装修楼板 untrimmed floor
无响应的 immune
无响应控制 non-corresponding control
无向传声器 astatic microphone; non-directional microphone
无向的 non-directional
无向地性 ageotropism; apogeotropism
无向点 nonoriented vertex
无向积 scalar product
无向检测器 non-direction(al) detector
无向孔隙 isotropic(al) porosity
无向量 scalar property; scalar quantity
无向量场 scalar field
无向量代谢 scalar metabolism
无向量的 scalar
无向量方程 scalar equation
无向量函数 scalar function
无向量积循环 scalar product cycle
无向量流 scalar flux
无向量通量 scalar flux
无向量域 scalar field
无向量轴 scalar axis
无向量阻抗 scalar impedance
无向通信[讯]网 non-oriented communication network
无向图 non-directed graph; undirected graph
无向性 isotropism; isotropy; scalar property
无向性点 isotropic(al) point
无项目 without project
无项目贷款 non-project loan
无像差 aberration-free
无像差的 aberrationless
无像差图像 unaberrated image
无像散摄谱仪 stigmatic spectrograph
无消耗 absence of wear
无销路 no market; sales resistance
无销路的 unsalable
无效 annul null; futility; idling; ineffectiveness; invalidation; nullity; of no effect
无效坝 dead dam
无效拌和料 inert aggregate
无效操作码 invalid op code
无效成本 idle cost
无效程序块 inactive block
无效穿孔 invalid punch
无效存水量 <贮水池的> dead storage
无效的 helpless; ineffective; invalid; powerless; reactive; sterile; unavailable; unprofitable; useless; vain; null and void

无效的(法律)行为 void act
无效地址 invalid address
无效点 ineffective point
无效电流 idle current; idling current
无效电路 idle circuit
无效分蘖 ineffective tillering; non-bearing tillering
无效分配 null allocation
无效符号 unblind
无效杆 inactive member
无效工时成本 idle time cost
无效公式 invalid formula
无效功 lost work
无效功率 reactance capacity; reactive power
无效构件 <适应外观要求而不承受荷重的构件> idle member
无效合同 contract void; invalid contract; void contract
无效呼叫 ineffective call
无效呼叫计数器 ineffective call meter
无效呼叫信号 signal for ineffective call
无效剂量 ineffective dose
无效假设 null hypothesis
无效键 invalid key
无效键条件 invalid key condition
无效降深 ineffective drawdown
无效交通 waste traffic
无效接收帧 invalid received frame
无效进尺 no effective footage
无效菌株 ineffective strain
无效孔隙度 inactive porosity
无效库容 inactive storage; non-effective storage
无效劳动 dis-utility of labo(u)r
无效类型 inefficient type
无效力 inefficiency
无效力的 non-effective; powerless
无效力期 adynamic(al) stage
无效量 ineffective dose
无效零位压缩 zero compression
无效率 inefficiency
无效码 invalid code
无效脉冲 idler pulse
无效锚固长度 ineffective length of anchorage
无效名 invalid name
无效命令 illegal command; invalid command
无效能 unavailable energy
无效炮眼 blown-out hole
无效票 invalid ballot
无效频率 idling frequency
无效凭单 invalid voucher
无效凭证 invalid voucher
无效期 adynamic(al) stage
无效区 dead space
无效圈 end coil
无效热耗 useless heat loss
无效杀虫剂 inefficient insecticide
无效上升 ineffective rise
无效升高 ineffective rise
无效时间 idle time; inactive time; ineffective time
无效食物 inert food
无效试呼 ineffective call attempt
无效输入 invalid input
无效数据 corrupted data
无效数位 invalid digit
无效水 unavailable water; unavailable soil water <土内的>
无效水分 echard; unavailable moisture
无效水量 <土粒固着水分> echard
无效顺序 invalid sequence
无效特许码 invalid authorization code
无效天线 antenna eliminator
无效条件检测 invalid condition detection

无效条款 clause of no effect; invalid clause
无效调整 idling adjustment
无效投资 make the investment inefficient
无效途径 invalid path
无效土壤水分 unavailable soil moisture
无效位 invalid bit
无效温度 ineffective temperature
无效文件 inactive file
无效线圈 dead coil; inactive coil
无效线匝 dead turn
无效行 inactive line
无效行程 back lash; lost motion
无效性 unavailability
无效序列 invalid sequence
无效页面分时 invalid page time-sharing
无效异常 ineffective anomaly
无效益耗用 non-beneficial consumptive use
无效益消耗 non-beneficial consumptive use
无效应变片 dummy ga(u)ge
无效应浓度 no effect concentration
无效油润滑 lost-oil lubrication
无效约束 inactive constraint
无效运动 lost motion
无效运输 ineffective traffic; invalid transport
无效运转 lost motion
无效整流片 dead segment
无效支票 dead cheque; invalid cheque
无效指令 ignore instruction; illegal command; invalid instruction
无效指针标志 null pointer indication
无效贮水量 dead storage
无效字符 idle character; invalid character
无效自变量 invalid arguments
无效钻探 unproductive boring
无效作废 null and void
无楔的混凝土方块 no-keyed concrete block
无斜度的 untapered
无斜度桩 untapered pile
无斜杆的桁架 truss without sloping members
无斜杆桁架梁 open frame girder
无斜坡的 non-sloping
无斜翼岸墩 breast abutment
无斜翼桥台 breast abutment
无谐振线 non-resonant line
无泄漏 no-leakage
无泄水区 endor(h)eic drainage
无屑车削 chipless turning
无屑成型 shaping without stock removal
无屑加工 chipless machining; chipless working
无懈可击的 unimpeachable; water-tight
无心棒拔制管材 sink drawing
无心材 free of heart center [centre]
无心的 coreless; non-central
无心二次曲面 noncentral quadric
无心二次曲线 conic(al) without center; non-central conic
无心珩床 centreless honing machine
无心精研机 centreless lapping machine
无心磨床 centerless [centreless] grinder; centerless [centreless] grinding machine
无心磨削 centerless grinding
无心磨削原理 centreless grinding machine principle
无心内圆磨 centreless internal grinder

无心内圆磨床 centreless grinding machine internal
无心外圆磨床 centreless grinding machine external
无心轴转向架 bogie with false pivot
无芯棒拔制 sinking
无芯编带 no-core braid
无芯工频感应电炉 line-frequency coreless induction furnace
无芯冷弯管 coreless cold elbowing
无芯头芯 touch core
无芯钻头 coreless bit
无锌环氧底漆 non-zinc epoxy primer
无薪假 leave without pay
无信号 absence of signal; no signal; zero signal
无信号电流 quiescent current; spacing current
无信号交叉口 non-signalized crossing; unsignalized intersection
无信号控制架空线系统 no-signal control line wire system
无信号控制架空线制 no-signal control line wire system
无信号区 blind area; dead space; zero signal zone
无信号损耗 spacing loss
无信号卫星 silent satellite
无信用 bad repute; discredited; not trustworthy
无行星齿轮分速器 gearless differential
无形飑 white squall
无形财产 incorporeal property; intangible property; non-visible property
无形财产权 intangible property rights
无形差额 invisible balance
无形成本 intangible cost
无形抽象画 metaphysical painting
无形出口 invisible export
无形出口和进口 invisible exports and imports
无形的 aeriform; immaterial; intangible; non-physical; non-visible
无形的标准 intangible standard
无形的飑风 white squall
无形动产 incorporeal chattel
无形费用 intangible cost
无形供给 invisible supply
无形股份 invisible stock
无形固定资产 intangible fixed assets
无形耗损 non-physical wear
无形洪水损害 intangible flood damage
无形价值 intangible value
无形交易 invisible transaction
无形进口 invisible import; invisible input
无形开发成本 intangible development cost
无形框架 invisible frame
无形浪费 invisible waste
无形利益 intangible returns; invisible gain
无形贸易 invisible trade
无形人工 immaterial labo(u)r
无形商品 intangible goods; intangible merchandise
无形生产 immaterial production
无形收入 invisible gain
无形收益 invisible earnings
无形收支项目 invisible item
无形手 intangible hand
无形损耗 immateriality wasting; invisible loss; moral depreciation; non-physical wear
无形损失 invisible loss; invisible waste; moral depreciation; non-material loss; non-physical loss
无形体 incorporeity
无形投资 intangible investment

无形项目 <用于国际收支表> invisible item
无形效果 intangible effect(iveness)
无形效益 intangible benefit;non-tangible benefit
无形因素评价 intangible analysis
无形应折旧财产 intangible depreciable property
无形资本 immaterial capital;incorporeal capital;intangible capital;invisible capital
无形资产 immaterial assets;incorporeal assets;intangible;intangible assets;invisible assets
无形资产的成本 cost of intangible;cost of intangible assets
无形资产取得审计 intangible assets acquisition audit
无形资产审计 intangible assets audit
无形资产摊销审计 intangible assets amortization audit
无形资产投资审计 intangible assets investment audit
无形资产项目 intangible item
无形资产转让审计 intangible assets transfer audit
无形资源 intangible resources
无形钻探费 intangible drilling cost
无性繁殖 asexual reproduction;clone
无性繁殖系 <植物> clone
无性系【生】clone
无休止的 non-stop
无修改指令 unmodified instruction
无锈的 rust-free;rustless
无锈皮的 scale free
无须补强 no reinforcement needed
无须动用现金的项目 items not requiring or providing cash
无须更正 no correction
无须加工的石板 self-faced stone
无须加工的石面 self-faced
无须加工石板 self-faced slab
无须调整的 a adjustment-free
无需保养的 service-free
无需偿还基金 non-reimbursable fund
无需答辩 no case to answer
无需付偿可享受之事物 <社会中的> public goods
无需批准手续 permit-free
无徐变的 creepless
无徐变混凝土 creepless concrete
无序 absent order;disorder
无序表 unordered table
无序材料 disordered material
无序场 disordered field
无序抽样 sampling without ordering
无序的 non-sequential;unordered
无序点阵 disordered lattice
无序度 degree of disorder
无序对 unordered pair
无序分布 random distribution
无序符号 unordered symbol
无序固溶体 random solid solution
无序合金 disordered alloy
无序化 disordering
无序基 unordered basis
无序结构 disordered structure
无序晶格 disordered lattice
无序连接 random splicing
无序区 region of disorder
无序取向 disordered orientation
无序扫描 non-sequential scanning
无序态 disordered state
无序网络结构假说 random network hypothesis
无序行为 chaotic behavio(u)r
无序性 randomness
无序褶皱 wild folds
无旋波 irrotational wave
无旋场 irrotational field
无旋键 irrotational binding
无旋流 turbulence flow
无旋流动 acyclic(al) motion;irrotational motion;irrotational flow
无旋流模型 irrotational flow model
无旋流速场 irrotational velocity field
无旋矢量场 irrotational vector field
无旋涡的 non-vortex;vortex free
无旋涡流 irrotational vortex;vortex-free flow
无旋涡流动 vortex-free motion
无旋涡室 quiescent chamber
无旋涡水流 eddy free flow;vortex-free discharge
无旋向量 irrotational vector
无旋向量场 irrational vector field;lamellar vector field
无旋形变 irrotational deformation
无旋性 irrotationality
无旋应变 irrotational strain;non-rotational strain
无旋运动 irrotational motion
无旋振动 irrotational vibration
无旋转的 non-rotational
无旋转滑动 translational slide
无旋转推进波 irrotational translation wave
无旋转向量场 irrotational vector field
无旋转形变 irrotational deformation
无旋转型运动 irrotational type of motion
无旋转原点 non-rotating origin
无选择的 indiscriminate
无选择吸收 unselective absorption
无选择性除草剂 non-selective herbicide
无选择性除莠剂 non-selective herbicide
无眩光 absence of glare;free from glare
无眩光玻璃 glare-free glass
无眩光的 glare-free;glare-reducing;non-glare
无雪冬季 open winter
无循环 loop free
无循环白水系统 no loop white water system
无循环编码 straight-line coding
无循环程序 straight-line code;straight-line coding
无循环算法 loop-free algorithm
无压暗渠 covered open channel
无压保护 no-voltage protection
无压差 no pressure difference
无压差钻速 non-differential-pressure penetration speed
无压成型 pressureless compacting;shaping without pressure
无压储箱 atmospheric storage tank
无压处理 non-pressure treatment
无压处理木材 nonpressure-treated timber
无压的 pressure-free
无压地下水 free groundwater;phreatic groundwater;unconfined groundwater
无压地下水含水层 phreatic aquifer
无压地下水流 flow with water table
无压地下水位 free ground water table;free water table;phreatic water table;unconfined water table
无压防腐处理木材 open-tank treatment
无压盖 non-bonnet
无压盖外螺纹 outside screw non-bonnet
无压供水泵 spout-delivery pump
无压固化树脂 zero pressure solidifying resin
无压管 non-pressure pipe
无压管道 high line conduit;non-pressure pipe
无压含水层 free aquifer;non-artesian aquifer;phreatic aquifer;unconfined aquifer
无压涵洞 free surface culvert;gravity culvert
无压继动阀 no-pressure relay valve
无压接触 potential free contact
无压接触输出 potential free contact output
无压浸渍法 infiltration process with vibration
无压开关 no-voltage cut-out
无压空气 <盾构施工> free air
无压力 zero pressure
无压力处理(法) <木材防腐> non-pressure treatment;non-pressure process
无压力的 non-pressure
无压力管道 free flow conduit
无压力管线 non-pressure pipeline
无压力式涵洞 inlet unsubmerged culvert;non-pressure culvert
无压力梯度流动 zero pressure gradient flow
无压力蒸汽 pressureless steam
无压流 flow in open air;free flow;free surface flow;gravity flow;non-pressure flow;unconfined flow
无压硫化 non-pressure cure
无压排水 non-pressure(d) drain(age)
无压排水工程 non-pressure drainage works
无压气机式喷气发动机 propulsive duct
无压容器 atmospheric vessel
无压烧结 pressureless sinter(ing)
无压烧结金钢石扩孔器 non-pressed sintering diamond reaming shell
无压渗流 free surface seepage;unconfined seepage flow
无压式水轮机 pressureless turbine
无压释放 no-voltage release
无压水槽 free flowing channel
无压水管 covered open channel;non-pressure pipe
无压水平面 free level
无压水头 free head
无压隧道 covered open channel;free flow tunnel;free level tunnel;non-pressure tunnel;tunnel under no pressure
无压隧洞 covered open channel;free flow tunnel;free level tunnel;non-pressure tunnel;tunnel under no pressure
无压线圈 no-voltage coil
无压印刷机 non-impact printer
无压载 unballast
无压贮箱 atmospheric storage tank
无押品货物 unsecured loan
无烟囱的煤气灶具 unvented gas appliance
无烟道采暖炉 flueless heater;unvented space heater
无烟道的 flueless
无烟道的燃具 unvented appliance
无烟道炉 flueless appliance
无烟道燃具 flueless appliance;type A appliance
无烟的 smoke-free;smokeless
无烟发射药 smokeless propellant
无烟工业 smokeless industry
无烟火箭发动机 smokeless rocket
无烟火炬 smokeless flare
无烟火药 ballistite;flameless explosive;smokeless powder
无烟火药的反应制止剂 deterrent
无烟列车 <电力或内燃机车牵引的列车> smoke-free train
无烟煤 anthracite;anthracite coal;blind coal;coal stone;glance coal;hard coal;kilkenny coal;malting coal;smokeless coal;stone coal;bird's eye <粒级 6 ~ 8 毫米>
无烟煤粉 anthracite duff;pulverised [pulverized] anthracite
无烟煤化作用 anthracitization
无烟煤级腐植煤 humanthracite
无烟煤矿 anthracite mine
无烟煤砾 culm
无烟煤炉 anthracite stove
无烟煤滤池介质 anthracite filter media
无烟煤煤砖 pelletized anthracite
无烟煤末 anthrafine
无烟煤筛分机 buckwheat screen
无烟煤细粒 anthracite fines
无烟煤渣 culm
无烟排气 smoke-free exhaust;smokeless exhaust
无烟区 smokeless zone
无烟燃料 smokeless fuel
无烟燃烧 smokeless combustion
无烟燃烧器 smokeless burner
无烟闪光粉 smokeless flash powder
无烟推进剂 smokeless propellant
无烟雾 smog-free
无烟雾化器 flameless atomizer
无烟(线状)火药 cordite
无烟延迟雷管 gasless delay detonator
无烟炸药 axite;smokeless powder
无延迟接续 no delay operation
无延迟双稳多谐振荡器 dynamic(al) flip-flop
无延迟通话业务 no-delay traffic
无延性的 inductile
无延性转变温度 nil-ductility(transition) temperature
无岩芯回次进尺 blank run
无岩芯进尺 non-cored footage
无岩芯式金刚钻头 non-coring type diamond bit
无岩芯钻进 non-core drilling;full hole drilling
无岩芯钻头 solid bit
无盐的 salt-free
无盐过程 salt-free process;saltless process
无颜料白漆 pittmentized paint
无颜料的 unpigmented
无颜料的底涂层 unpigmented base coat
无掩蔽的 exposed
无掩蔽地带 exposed area
无掩蔽海 open sea
无掩蔽海岸 open(-sea) coast;exposed shore
无掩蔽航道 exposed channel
无掩蔽外海施工方法 open sea construction method
无掩护的 uncovered
无掩护的滨线 exposed shoreline
无掩护港口 open harbo(u)r
无掩护港外锚地 open roadstead
无掩护港湾 open harbo(u)r
无掩护海岸 open coast
无掩护码头 unprotected terminal
无掩护锚地 open anchorage
无掩护水域 open water
无掩模 maskless
无掩模法 maskless process
无掩遮物的地带 exposed situation
无眼衬管 blank liner
无焰爆破 non-flame blasting
无焰催化燃烧 flameless catalytic combustion
无焰的 flameless;non-flammable

无焰法 flameless procedure
无焰火 smolder(ing);smo(u)lder
无焰火药 flameless powder;non-explosive agent
无焰闷热 smolder(ing)
无焰气体 non-luminous gas
无焰燃料 non-flammable fuel
无焰燃烧 flameless combustion;preaerated combustion
无焰燃烧器 flameless burner;radiant burner;surface combustion burner
无焰烧嘴 flameless burner
无焰式燃烧 flameless burning
无焰原子吸收法 flameless atomic absorption spectrophotometry
无焰原子吸收分光光度计 flameless atomic absorption spectrophotometer
无焰原子吸收光谱法 flameless atomic absorption spectroscopy
无阳离子固化 cationic-free curing
无养分的 dystrophic
无养护 zero maintenance
无养护道路 zero maintenance road
无养护方针 no-maintenance (policy);no-(policy)
无养护路面 zero maintenance pavement
无养护年限 maintenance-free life
无养护政策 no maintenance strategy
无氧的 oxygen free
无氧分解 anaerobic decomposition
无氧高导电性铜 oxygen-free high conductivity copper
无氧光细菌 anoxyphotobacteria
无氧过程 oxygen-free process
无氧焊剂 non-oxygen flux;oxygen-free flux
无氧化 non-oxidation
无氧化加热 scale free heating
无氧化加热炉 scale-free heating furnace
无氧化皮薄板 sheet-metal free from oxides
无氧化皮的 free from scale;non-scale;scale free
无氧化皮切削 clean-cut
无氧化气氛 non-oxidizing atmosphere
无氧环境 oxygen-free environment
无氧聚合止水材料 anaerobic sealant
无氧培养 anaerobic incubation
无氧培养器 anaerobic culture apparatus
无氧区 oxygen-free zone
无氧燃烧 anoxycausis
无氧态 anaerobic
无氧铜 oxygen-free copper
无摇动台转向架 rigid bolster truck
无摇枕转向架 bolsterless bogie
无药爆孔 unloaded hole
无药焊条 bare electrode;bare rod;uncoated electrode
无药雷管 blank cap
无要求 no requirement
无叶的 aphyllous;bald;vaneless
无叶扩压器 vaneless diffuser
无叶片的 vaneless
无叶片空间 vaneless space
无叶片离心泵 bladeless centrifugal pump
无叶片区 interspace
无叶片式扩散器 vaneless diffuser
无叶性的【植】aphyllous
无叶-有叶混合式扩压器 vaneless-vaned diffuser
无叶植物 aphyllous plant
无夜勤的执勤制 day manning
无液测高计 aneroid altimeter
无液的 aneroid
无液高度计 aneroid altimeter
无液气压表 aneroid;aneroid altimeter;aneroid barograph;aneroid barometer
无液气压测高计 aneroid altimeter
无液气压机 capsule aneroid
无液气压计 aneroid barograph;aneroid barometer
无液气压计膜片 aneroid diaphragm
无液气压记录器 aneroidograph
无液气压控制 aneroid control
无液气压器 aneroidograph
无液式气压自记仪 aneroidograph
无液压力传感器 aneroid capsule
无液压力计 aneroid manometer
无液自动气压计 aneroidograph
无液自记气压计 aneroid barograph
无一定尺寸限制的结构 tensegrity system
无依托翼侧 unsupported flank
无疑钢管式厚壁圆筒 thick-walled cylinder by seamless steel pipe
无以匹敌的事物 nonesuch
无异仪的 undisputed
无异议 by common consent;unanimity
无异议的 unanimous
无异议通过程序 no-objection procedure
无抑止剂汽油 uninhibited fuel
无疫健康证明书 clean bill of health
无疫通行证 certificate of health;certificate of pratique;pratique
无疫证书 <船只> bill of pratique;clean bill of health
无益健康的 unsalutary
无意的雷达干扰 unintentional radar interference
无意识的 unconscious;unintentional
无意识污染 unintentional pollution
无意识行为的表现 acting out
无意识性 automaticity
无意识选择 unconscious selection
无意污染 unintentional pollution
无意义相关 nonsenge correlation
无意中发生的 inadvertent
无意中发现 blunder(up)on;come actress
无意中泄漏 blunder out
无翼翅的 ribless
无翼飞行器 wingless aircraft
无翼昆虫 bristletail
无翼墙桥台 wingless abutment
无翼桥台 straight abutment;stub abutment
无翼缘主动轮 blind driver
无因次 non-dimension;zero dimension
无因次参数 dimensionless parameter;non-dimensional parameter
无因次常数 dimensionless constant
无因次单位 non-dimensional unit
无因次单位(水文)过程线 dimensionless unit hydrograph;non-dimensional unit(hydro)graph
无因次的 dimensionless;non-dimensional;zero dimensional
无因次过程线 dimensionless hydrograph
无因次量 dimensionless quantity;non-dimensional number;non-dimensional quantity
无因次量纲度 dimensionless measure
无因次流速分布 dimensionless current velocity distribution;dimensionless current velocity profile
无因次率定曲线 dimensionless rating curve
无因次曲线 dimensionless curve
无因次数 dimensionless number
无因次水位-流量关系曲线 dimensionless rating curve
无因次特性 dimensionless characteristic
无因次特性线 dimensionless performance
无因次弯沉盆 non-dimensional deflection basin
无因次系数 dimensionless coefficient;dimensionless factor;non-dimensional coefficient
无因次因数 dimensionless factor
无阴影的 unblanketed
无阴影区 unshaded area
无音槽 plain groove
无音调的声音 unpitched sound
无音信 silence
无银点 no-shadow point
无银感光材料 non-photographic(al) material
无银装置 non-silver unit
无引力开关 zero-gravity switch
无引线变换器 leadless inverted device
无引线钉头式接合 tailless nail head bonding
无引线接插头 patch-plug
无饮水宿营地 dry camp
无隐蔽的 naked
无应变 absence of strain
无应变的 undeformed;unstrained
无应变范围 level of no strain
无应变环 strainless ring
无应变介质 unstrained medium
无应变速度 zero strain velocity
无应答呼叫 no-reply call
无应力 free of stress;stress-free;unstressing
无应力变形 stressless deformation
无应力表 no-stress meter
无应力单晶 strain-free single crystal
无应力的 stressless;unstressed
无应力的孔隙 stress-free pore
无应力的细孔 stress-free pore
无应力端 <钢筋束> dead end
无应力杆件 <构架的> unstressed member
无应力混凝土 non-stressed concrete
无应力计 non-stress meter
无应力退火 stress-free annealing
无应力应变计 non-stress strain meter;stress-free strain meter
无应力状态 non-stress condition;zero stress state
无盈亏(的) break-even
无盈亏价格 break-even cost;break-even zone
无盈亏模型 breadthwise-even model;break-even model
无荧光的 non-blooming
无营养的 dystrophic
无营运业务的 unemployed shipping
无影灯 astral lamp;shadowless lamp
无影响剂量 ineffective dose;no-effect level
无影照明 concealed illumination;indirect illumination;shadowless illumination;shadowless lighting
无硬质路面的土路 unmetallized road
无涌 no swell
无涌浪的 surgeless
无用 all for naught;futility;good for nothing;non-utility;obsolescence;of no effect;out of condition
无用板坯 dummy sketch
无用玻璃 unusable glass
无用玻璃收集器 glass pocket
无用材料报废的权力 authority to scrap obsolete material
无用层 unavailable layer
无用存储单元 garbage;gibber(ish)
无用单元的收集 garbage collection
无用单元收集程序 garbage collector
无用单元压缩 garbage compaction
无用档案保管所 records holding area
无用的 exhausted;helpless;idle;inutile;naught;out of use;trumpery;unprofitable;unserviceable;useless;valueless
无用的人员 <俚语> dead wood
无用东西 vanity
无用功 idle work;waste energy
无用机械 fifth wheel
无用键 invalid key
无用空间 dead space
无用能 unavailable energy
无用区域 keep out area
无用人员 <俚语> fifth wheel
无用设备 waste appliance
无用束 idling beam
无用数据 garbage;gibber(ish);hash
无用数据输入 garbage in
无用土地 useless land
无用物 dross
无用信号 garbage signal
无用信息 garbage;gibber;gibberish;hash
无用振荡模的抑制 strapping
无用之物 dead wood
无用资产 dead assets
无用字 stop word
无优惠 ex privileges
无优先权的一方 innocent party
无油泵 oilless pump
无油超高真空镀膜设备 oil-free ultra-high vacuum coating equipment
无油超高真空排气台 oil-free ultra-high vacuum exhaust station
无油醇酸树脂 oil-free alkyd resin
无油的 oil-free
无油地段 barren gap
无油断路器 oilless circuit breaker
无油工业用压气 oil free industrial air
无油管完井 tubingless completion
无油过滤麻布 oilless hessian
无油灰玻璃木条木 wooden puttyless glazing bar
无油灰的 puttyless
无油灰的玻璃屋顶 puttyless glazing roof
无油灰缝镶装玻璃 patent glazing
无油灰木制玻璃窗心条 timber puttyless glazing bar
无油灰屋顶玻璃 patent roof glazing
无油灰镶玻璃法 patent glazing;puttyless glazing
无油灰装玻璃用的铝条 puttyless alumin(i)um glazing bar
无油灰装玻璃用钢条 puttyless steel bar
无油胶料 unextended compound
无油脚亚麻籽油 linseed oil without foots
无油井 barren well
无油聚酯 oil-free polyester
无油空气压气机 <不需要润滑气缸的> oilfree air compressor
无油蜡 oil-free wax
无油蜡膏 oil-free petrolatum
无油密封 dry seal
无油膜片泵 oil-free diaphragm pump
无油墨色带 uninked ribbon
无油润滑 oil-free lubrication
无油润滑空压机 non-lubricated air compressor
无油润滑迷宫式压缩机 oil-free labyrinth compressor
无油润滑气体压缩机 non-lubricated gas compressor
无油润滑压缩机 non-lubricated compressor
无油式的 oilless
无油压缩机 oil-free compressor;oil-

less compressor
无油页岩 barren shale
无油脂 grease free
无油轴承 oilless bearing
无游梁抽油设备 elephant
无有机物流出物 organo-free effluent
无有机物加速管 organic free tube
无有限伸缩线方位 orientation of finite longitudinal strain
无釉瓷 unglazed porcelain
无釉瓷砖 unglazed tile
无釉地砖 biscuit tile; bisque tile; quarry tile
无釉坩埚 unglazed crucible
无釉红炻器 <宜兴紫砂器皿> boccaro ware
无釉挤压砖 unglazed extruded tile
无釉面瓷砖 non-vitreous tile
无釉面轻集料 uncoated lightweight aggregate; unglazed lightweight aggregate
无釉面砖 unglazed tile
无釉铺地砖 paver; pavio(u)r
无釉陶 bisque
无釉陶瓷器皿 unglazed ware
无釉陶器 dry body
无釉瓦 unglazed roofing tile
无釉屋面瓦 unglazed roof tile
无余辉荧光屏 non-persistent screen
无雨的 rainless
无雨干旱带 rain shadow
无雨期 dry period; dry spell; period without rainfall; rain-free period; rainless period
无雨区 rainless region
无雨沙漠 rainless desert
无雨刷玻璃 wipeless windscreen
无雨天数 rain-free days
无语言连接词 metalinguistic connective
无浴缸的盥洗室 half-bath toilet room
无预定运行方向的专开旅游列车 mystery-trip train
无预应力的粗钢筋 non-prestressed bar reinforcement
无原球的 atrochal
无原位记录 no-home record
无原位仪表 apparatus without homing position
无原子测度空间 nonatomic measure space
无原子防护的设施 unhardened facility
无缘轮胎 plain tire [tyre]
无缘无故的 uncalled for
无源闭合环路 passive closed loop
无源部件 passive component
无源测距 passive localization
无源传感器 passive sensor; passive transducer
无源磁泡产生器 passive bubble generator
无源大地电阻 passive earth resistance
无源的 passive
无源电缆均衡器 passive cable equalizer
无源电路 passive circuit; passive electric(al) circuit
无源电阻 passive resistance
无源定位与导航装置 location and navigation device-passive
无源定向偶极子 director
无源反射器 passive reflector
无源辐射计 passive radio meter
无源伽马射线分析 passive gamma-ray analysis
无源跟踪系统 passive tracking system
无源固定信息应答器 passive fixed information transponder
无源光网络 passive optic(al) network
无源广播信道 passive broadcast channel
无源化学电离质谱法 passive chemical ionization mass spectrometry
无源换算电路 resistive matrix network
无源混频电路 resistive mixing pad
无源活化分析 passive activation analysis
无源基片 passive substitution
无源激光器 passive laser
无源检波 passive detection
无源减速器 inert retarder
无源矩阵网络 resistive matrix network
无源可变信息应答器 passive variable information transponder
无源流动 passive flow; source-free flow
无源滤波器 passive filter
无源器件 passive device
无源人造地球卫星 passive man-made satellite
无源设备 inactive component; passive component
无源声呐 passive sonar
无源式人造卫星通信[讯] passive satellite communication
无源收集 passive collection
无源探测 passive detection; passive sounding
无源探测器 passive detector; passive probe
无源天区 cold sky
无源天线 indirectly fed antenna; parasitic(al) antenna; passive antenna; reflecting antenna; reflector antenna
无源天线阵 parasitic(al) array
无源调谐线圈 inert-tuned coil
无源网络 passive(electric) network
无源网络站 passive network station
无源微波 passive microwave
无源微波传感的应用 application of passive microwave sensing
无源微波辐射计 passive microwave radio meter
无源微波遥感图像 passive microwave remote sensing image
无源系统 passive system
无源显示系统 passive display system
无源谐振腔 passive resonant cavity
无源信号标杆法 passive sign post
无源性 passivity
无源遥感 passive remote sensing
无源有源数据模拟 passive-active data simulation
无源有源探测定位 passive-active detection and location
无源元件 passive component; passive element
无源振子 parasitic(al) vibrator
无源支路 passive branch
无源指示装置 passive indication device
无源中继器 passive repeater; radio mirror
无源中继站 passive relay station
无源中子分析 passive neutron analysis
无源转发卫星 reflector satellite
无源总线 passive bus
无源阻抗 passive impedance
无源阻尼器 passive restraint
无远见 improvidence
无约束 absence of restrictions
无约束爆破 unconfined shot
无约束变量 unrestricted variable
无约束冰崩 uncontrolled disintegration
无约束不变式 unrestricted invariant
无约束车流 unrestricted traffic flow
无约束的 non-restraint; unconfined; unconstrained; unrestricted; unzoned
无约束的阻尼层 unconstrained damping layer
无约束构件 unrestrained member
无约束混凝土 unconfined concrete
无约束极小化 unconstrained minimization
无约束交通流 unrestricted traffic flow
无约束梁 unbonded beam; unconstrained beam; unrestrained beam
无约束流 unrestricted flow
无约束收缩 unconstrained shrinkage
无约束水流 unbounded stream; unrestricted flow; unconfined stream
无约束条件 unconfined condition
无约束系统 unconstrained system
无约束雪崩 unconfined avalanche
无约束药包 unconfined charge
无约束游丝 free spring
无约束最优化【数】unconstrained optimization
无月光期 dark of the moon
无月期间 interlunation
无钥匙表 keyless watch
无钥匙锁 keyless lock
无钥信号 keyless ringing
无越流补给 non-leaky recharge
无云 cloudlessness
无云区 cloud-free area
无云天空 clear sky
无运动面 level of no motion
无晕光的 non-halating
无杂草的 weed-free
无杂物的 unalloyed
无杂音信号 sure signal
无杂质半导体 intrinsic(al) semiconductor
无杂质的 uncontaminated
无杂质水 uncontaminated water
无杂质水泥 non-staining cement; white cement
无载饱和曲线 no-load saturation curve
无载波导 unloaded waveguide
无载测量法 no-load method
无载抽头 off-load tap
无载储备 cold reserve
无载垂度 unloaded sag
无载导线 dummy conductor
无载的 no-load
无载点 no-load point
无载电池 idle battery
无载电路 unloaded circuit
无载电压 no-load voltage
无载电阻 no-load resistance
无载端电压 no-load terminal voltage
无载短路法 no-load short-circuit method
无载阀门 no-load valve
无载法 no-load method
无载分支 return run
无载功 no-load work
无载功率 no-load power
无载合闸 no-load switching-in
无载荷 unload; zero load
无载荷的 non-loaded; unloaded
无载荷弦杆 unloaded chord
无载还原式 no-load restore type
无载机构 unloaded mechanism
无载继电器 no-load relay
无载剪力强度 no-load shear strength
无载励磁电压 no-load field voltage
无载流量 no-load flow; zero load flow
无载轮 unloading pulley
无载母线 dead main
无载能量 unloaded energy
无载喷嘴 no-load nozzle
无载品质因数 Q-unloaded
无载起动 no-load starting
无载起动指示灯 start light
无载切断 no-load switching off
无载情况 no-load condition
无载区 no-load zone
无载热耗 no-load heat consumption
无载时间 no-load time
无载时降落速度 unladen lower speed
无载时门架离地间隙 unladen-chassis ground clearance
无载时提升速度 unladen lift speed
无载试验 zero load test
无载释放 no-load release
无载释放电磁铁 no-load release magnet
无载释放动作 no-load release action
无载损耗 idling loss
无载弹簧 unloaded spring
无载特性(曲线) no-load characteristic; unload characteristic
无载体催化剂 unsupported catalyst
无载体的 unsupported
无载体胶粘剂 unsupported adhesive
无载体指示剂 carrier-free tracer
无载天线 unload antenna
无载调节 off-load regulation
无载调压抽头 no-load voltage regulator and tap
无载调压器 no-load voltage regulator
无载跳闸 no-load release
无载铜条 idle bar
无载托辊 idler roll
无载稳定性 stability unload; stable empty
无载弦 unloaded chord
无载运输 running no-load
无载运行 loose running; no-load operation; no-load run(ning); running no-load; vacant run(ning)
无载运转 idle run; no-load operation; no-load run(ning); running no-load; vacant run(ning)
无载运转的损失 unload loss
无载运转试验 no-load running test
无载正常转弯半径 normal unladen turning-radius
无载重量 empty weight; unladen weight; unloaded weight
无载转速 free running speed
无载转弯半径 unladen turning-radius
无载状态 no-load state
无再过热的 non-resuperheat
无再热的 non-reheat
无皂的 non-soap
无皂乳液 soap free emulsion
无皂润滑油 non-soap grease
无噪电路 silent circuit
无噪激光解调 noise-free demodulation of laser
无噪声 quietness
无噪声传动 silent running
无噪声打桩技术 noiseless piling technique
无噪声的 muting; noise-free; noiseless
无噪声电池 quiet battery
无噪声电动机 noiseless motor; quiet motor; silent motor
无噪声电路 noiseless circuit; quiet circuit; silent chain; silent circuit
无噪声电源 quiet power supply
无噪声反应堆 noiseless reactor
无噪声放大器 noiseless amplifier
无噪声风扇 silent fan
无噪声工作 noise-free operation
无噪声管 muting tube; muting valve
无噪声光 noise-free light
无噪声航速 noiseless speed
无噪声绞肉机 silent cutter

无噪声接收机 noise-free receiver
无噪声控制 mute control
无噪声路面 noiseless pavement
无噪声马达 quieter motor
无噪声脉冲 noiseless pulse
无噪声调谐 quiet tuning
无噪声调整 mute control
无噪声通风机 silent blower
无噪声图像 noise-free picture
无噪声狭缝 noiseless slit
无噪声线路 quiet line
无噪声相敏放大器 noiseless phase sensitive amplifier
无噪声信道 noise-free channel;noiseless channel
无噪声信号 noise-free signal
无噪声运转 noiseless run(ning);quiet running
无噪声载波 noise-free carrier
无噪调谐 quiet tuning;silent tuning
无噪音的 noise-free
无噪音快艇 quiet fast boat
无噪音路面 noiseless pavement
无泽面 mat(te)
无泽釉 mat(te) glaze
无责任的 exculpatory
无责任(交通事故) no fault
无增长经济 nil growth economy
无增强基材 unreinforced matrix
无增相位滞后 non-increasing phase lag
无增压装置的 unsupercharged
无增振幅 non-increasing amplitude
无增支成本 zero incremental cost
无渣板式轨道 ballastless slab track;ballastless track;unballasted track
无渣的 ballastless
无渣轨道 ballast-free track;ballastless track;unballasted permanent way;unballasted track
无渣轨道结构 ballast-free track structure;ballastless track structure
无渣梁 girder without ballast
无渣桥面 ballastless deck;bridge floor without ballast;unballasted floor;open bridge floor
无渣无轨桥面 bridge floor without ballast sleeper
无渣无枕梁 girder without ballast and sleeper
无渣油 sludgeless oil
无闸坝航道 free channel
无闸坝河段 free flowing stretch;free river reach
无闸坝河流 free flowing river;free river;free stream
无闸港池 open basin;open dock;coastal basin;tidal dock
无闸河段 lock-free stretch
无闸控制水流 ungated flow
无闸控溢洪道 uncontrolled spillway
无闸门的 ungated
无闸门溢洪道 ungated spillway;unobstructed spillway
无闸门溢流坝 self-spillway dam
无闸通海运河 open sea canal;stepless sea canal
无闸通航渠道 stepless canal
无闸堰顶 free crest of spillway;unobstructed crest of spillway
无闸运河 canal without locks
无栅栏 fenced-off
无债务纠纷 free from encumbrance;free from incumbrance
无展宽共振 unbroadened resonance
无展性的 immalleable
无绽边 no crack edge
无张力 no pull
无张力表面 traction-free surface
无张力的 tension-free
无张力刚度 no-tension rigidity
无张力环 strainless ring
无张力劲度 no-tension stiffness
无张力劲性 no-tension stiffness
无张力孔隙水 tension-free pore water
无张力硬度 no-tension rigidity
无张力轧制 tension-free rolling
无胀腐蚀 non-expansive corrosive
无胀性的 non-bloated
无障碍的 <对残疾人> barrier-free
无障碍的景观 unobstructed view
无障碍的入口 unobstructed access
无障碍的视野 unobstructed view
无障碍的通路 unobstructed access
无障碍环境 barrier-free environment
无障碍环境设计 barrier-free environment design
无障碍空间 headroom
无障碍设计 barrier-free design
无障碍设施 non-obstacle facility
无障碍视界 unobstructed vision
无障碍视线 unobstructed sight
无障碍水域 open waters
无障碍物地带 obstacle free zone
无障碍物锚地 clean anchorage
无障壁海岸层序 non-barrier-coastal sequence
无照导电性 dark conductivity
无照电流 dark current
无照电阻 dark resistance
无照明的教堂拱廊 unlit triforium (gallery)
无照营业 interlope
无罩的 uncowled
无罩灯 naked light
无罩弧光灯 open arc lamp
无罩燃烧器 uncovered burner
无遮蔽的 <窗等> unshaded
无遮蔽的土地 bare ground
无遮蔽港口 <不能挡风浪> roadstead
无遮蔽锚地 open(road) stead
无遮挡物的 unshielted
无遮盖的 uncovered
无遮盖的明渠 uncovered open channel
无遮盖光线 naked light
无遮盖火焰 naked flame
无遮光线 naked light
无遮喷砂法 free sand blasting
无遮喷丸清洗装置 free jet blast plant
无折边球形封头 spheric(al) head without folded edge
无折边锥体变径段 conic(al) reducer without straight flange
无折边锥形封头 conic(al) head without straight flange;conic(al) head without transition knuckle
无折叠链段 unfolded chain section
无折光性的 aclastic
无折射吸收 non-deviated absorption
无针弧面 plain segment
无砧座锤 counterblow hammer;double-faced hammer
无砧座锻锤 impacter [impactor];impacter forging hammer
无振 vibration-free
无振荡 oscillation-less
无振荡罗经 aperiodic(al) compass;deadbeat compass
无振动的 vibrationless
无振动冷却 non-shock chilling
无振动零部件 non-vibration parts
无振膜传声器 diaphragmless microphone
无振线圈 non-vibrating coil
无振压实遍数 static passes
无振运动 oscillation-free motion
无震板块 aseismic plate
无震冲击机 shockless jarring machine
无震带 non-seismic zone
无震地面变形 aseismic ground deformation
无震断层位移 aseismic fault displacement
无震概率 probability of none earthquake;probability of none event
无震海脊 aseismic ridge
无震海岭 aseismic ridge
无震湖 aseismic lake;non-seismic lake
无震滑动 aseismic slip
无震隆起 aseismic rise
无震前沿 aseismic front
无震区 aseismic region;non-seismic area;non-seismic region
无震形变 aseismic deformation
无争论的 undisputed
无争议性条款 non-contest clause
无蒸发养护 evaporation free curing
无正当理由的 unwarrantable
无证据的 naked
无支撑 no-support
无支撑采掘面 prop-free front
无支撑长度 unbraced length;unsupported length
无支撑的 unshored;unstayed;unsupported;untimbered
无支撑的柱长 unbraced length of column
无支撑叠合梁 unshored composite beam
无支撑高度 unsupported height
无支撑工作面 prop-free front
无支撑珩磨 free honing
无支撑间距 unsupported distance
无支撑开挖 excavate without timbering;excavation without timbering;no-support excavation;unbraced excavation
无支撑开挖工程能坚持的时间 stand-up time
无支撑宽度 unsupported width
无支撑框架 unbraced frame
无支撑膜 non-supported membrane
无支撑平封头和盖板 unstayed flat heads and covers
无支撑墙 unbraced wall
无支撑深度 unsupported depth
无支撑挖掘 excavate without timbering;excavation without timbering
无支撑挖土工程 excavation works without timbering
无支撑氧化铁陶瓷膜 unsupported ferric oxide ceramic membrane
无支撑柱 unbraced column
无支撑组合梁 unshored composite beam
无支承 support-free
无支承边 free edge
无支承长度 unsupported length
无支承的 non-supporting;unsupported
无支承切割 free cutting
无支付能力 bankruptcy
无支付能力的 unable to pay
无支付能力者 bankrupt
无支付期 no pay date
无支护顶板跨度 unsupported back span
无支护对角工作面 untimbered rill
无支护工作面 open face
无支护跨度 unsupported span;unsupported width
无支护倾斜分层充填法回采 untimbered rill stoping
无支护倾斜分层充填开采法 untimbered rill method
无支架安装(法) erection without scaffolding
无支架吊装法 erection with cableway
无支架施工 bridge erection without scaffolding;construction without trestle;erection without scaffolding
无支架天井 raw raise
无支链烃 unbranched chain hydrocarbon
无支腿作业 <起重机> on-rubber operation
无支柱 column-free
无支柱采掘面 prop-free front
无支柱的 stanchion-free;unsupported
无支柱工作面 prop-free front
无支柱体育馆 column-free sport facility
无枝的 clear
无枝干材 bodywood;clean-bole
无知的 uninformed
无知的高价购买 innocent purchase for value
无知的买主 good faith purchaser
无知因数 factor of ignorance
无织边宽带 wide tape without selvage
无脂的 fat-free
无脂密封阀 greaseless valve
无执照 ex warrants
无执照的 unlicensed
无执照的引航员 unlicensed pilot
无执照发射机 unlicensed transmitter
无执照海员 hoveller
无执照驾车罪 motoring offence
无执照酒店 shebeen
无直达交通 no through traffic
无职业的 jobless
无植被的 unvegetated
无植物区 aphytic zone
无止境的 never-ending
无纸的 paperless
无指挥器制 <步进式自动电话> non-director system
无制动辅助转弯半径 radius without steering brake turning
无制动机车 <无闸车> brake-free car
无制约塑性流动 unrestricted plastic flow
无秩序 chaos;out-of-course
无秩序的 chaotic;rough-and-tumble
无秩序行走 haphazard walking
无滞后过滤器 zero-lag filter
无滞性流 frictionless flow
无中断备用 no-break standby
无中继线 no trunks
无中间冷却压气机 non-intercooled compressor
无中间煤粉仓直接燃烧的煤磨 direct firing coal mill
无中间煤粉仓直接燃烧的碗煤磨 direct firing coal bowl mill
无中间煤粉仓直接燃烧系统 direct firing coal system
无中间支点的楼板 suspended floor
无中梁底架 center sillless underframe
无中心的 acentric;centerless [centreless]
无中心磨床 centerless grinder
无中心磨削 centerless grinding
无中心研磨 centerless grinding
无中心柱螺旋形楼梯 hollow newel stair(case);open newel stair-(case);open well stair(case)
无中柱的 astelic
无中柱螺旋梯 hollow newel stair-(case);open newel stair(case)
无中柱式 astely
无终止程序 endless program(me)
无重大故障的 fail passive
无重大事故的 uneventful
无重力 zero gravity
无重力的 agravic;non-gravitating
无重力分异的油藏 non-segregated reservoir
无重力区 agravic;zone of zero gravity
无重力状态 agravic state

无重量 imponderability；imponderableness；zero gravity
无重量流体 imponderable fluid；weightless fluid
无重量弹簧 weightless spring
无重状态 null-gravity state
无周期摆 aperiodic（al）pendulum；deadbeat pendulum
无周期波 aperiodic（al）wave
无轴线卷边 false wiring
无轴向应变固结 consolidation without axial strain
无肘板结构式 bracketless framing
无皱缩 pucker-free
无皱整理 non-creasing finish
无骤发环境风险 non-sudden environmental risk
无珠光体钢 pearlite free steel
无主材 waif
无主地 vacant succession
无主货物 abandoned merchandise；unclaimed goods
无主失物 waif
无主物 derelict
无主系 singular set
无主行李 astray baggage
无助力操纵 unpowered control
无助力操纵方向舵 unpowered rudder
无助推的 unboosted
无助洗剂的洗涤剂 unbuilt detergent
无柱 support-free；support-less
无柱的 column-free
无柱的楼梯 flying stair（case）
无柱地基 stereobate
无柱飞机库 column-free hanger
无柱式背面 astylar back
无柱式的（正面）astylar
无柱头飞机库 column-free hanger
无转换装置 non-changeover system
无转向架车 <即两轴车> bogie-less car
无转向架货车 <四轮货车> non-bogie wagon
无转向架客车 non bogie coach
无转向架四轮客车 non bogie coach
无转子硫化仪 rotorless curemeter
无桩底基 stereobate
无装备车辆 non-equipped vehicle
无装甲车体 unarmored body
无装甲的 unarmored
无装甲的轮式车辆 unarmored wheeled vehicle
无装饰的 bald；unadorned；undecked
无装饰的墙面板 plain panel
无装饰的镶板 plain panel
无装饰的柱 rustic column
无装饰格栅的楼板 untrimmed floor
无装饰或变化的方棱方角的性质 boxiness
无装卸设备的船 gearless ship；gearless vessel
无装卸设备的集装箱船 gearless container ship
无装载出发 start without a load
无撞击进口 shockless entrance
无追索权 drawn without recourse；non-recourse
无追索权背书 endorsement without recourse
无追索权的抵押 dry mortgage
无追索权的汇票 draft without recourse；non-protectable bill
无追索权的借据 non-recourse note
无追索权的借款 non-recourse debt
无追索权的欠款 non-recourse debt
无追索权汇票 non-protestable bill
无追索权信用证 without recourse letter of credit
无追索权资金筹措 non-recourse financing
无追索资金融通 non-recourse finance
无锥度桩 untapered pile
无准备的 extemporary；off-hand
无灼烧损失 ignition loss free
无资格 disability；disablement；disqualification；incapacity
无资格的 unqualified
无资格的证人 incompetent witness
无资料的 not available
无资料地区水文计算 hydrologic（al）computation of missing data region
无资料流域 unga（u）ged basin
无资助城市更新项目 non-assisted urban renewal project
无滋养的 dystrophic
无滋养湖 dystrophic lake
无滋养环境 dystrophic environment
无子女户 single-generation household
无子女双职工 dink
无子叶植物 acotyledon
无自备汽车家庭 non car owning household
无自动力的 inert
无自动调节的受控系统 controlled system without self-regulation
无自感线圈 Curtis winding
无自我程序设计 egoless programming
无自由度的 invariant
无自由面爆破 <除工作面外> grunnching
无自由氧的 anaerobic
无自转主序 zerorotation main sequence
无足够压舱物的 walt
无足轻重 count for little
无阻（碍）in the clear
无阻碍车流 free flowing traffic；unhampered flow of traffic
无阻碍车速 unimpeded speed
无阻碍车通 free flowing traffic
无阻碍道路通行能力 open road capacity
无阻碍的 free flowing；unobstructed
无阻碍交通 free（ly）moving traffic；free traffic
无阻碍空间 unobstructed space
无阻碍排料 unhindered（gravity）discharge
无阻碍容量 unobstructed capacity
无阻碍隧道 empty tunnel
无阻挡的 unobstructed
无阻铰链 frictionless hinge
无阻井喷 blowing in wild
无阻力横摇 unresisted rolling
无阻力卫星 drag-free satellite
无阻力卫星法 method of drag-free satellite
无阻尼摆 undamped pendulum
无阻尼波 undamped wave
无阻尼的 undamped
无阻尼的分析天平 undamped analytical balance
无阻尼的强迫振动 undamped forced vibration
无阻尼的受迫运动 undamped forced vibration
无阻尼电波 undamped electric wave
无阻尼固有频率 undamped natural frequency
无阻尼控制 undamped control
无阻尼模式 undamped mode
无阻尼频率 undamped frequency
无阻尼振荡 undamped oscillation
无阻尼振荡器 undamped oscillator
无阻尼振荡周期 undamped period
无阻尼振动 non-damping vibration；undamped vibration
无阻尼周期 undamped period
无阻尼自然频率 undamped natural frequency
无阻尼自由振动 undamped free vibration
无阻尼自振动频率 undamped free vibration frequency
无阻尼自振圆频率 undamped natural circular frequency
无阻尼自振周期 undamped natural period
无阻塞 congestion-free；non-blocking
无阻塞交换网络 non-blocking switching network
无阻塞网络 non-blocking network
无阻视线 unobstructed sight
无组织 inorganization
无组织的 uncoordinated；unmethodical
无组织进风 unorganized air supply
无组织排放 fugitive emission
无组织排风 unorganized exhaust
无组织排水 unorganized drainage
无组织自然通风 uncontrolled natural ventilation；unorganized natural ventilation
无钻杆反循环钻机 rodless reverse circulation rig
无钻座锻锤 impact forging hammer
无罪推定 presumption of innocence
无左右行车线的道路 undivided road
无作用代号 ignore code
无作用符号 ignore code
无作用密码 ignore code
无作用期间 inaction period
无作用区 neutral zone
无作用水平 no-effect level
无（作）用（字）符 ignore character
无坐标尺寸基准 three-dimensional reference
无座圈轴承 cageless bearer

吴

吴尔拉斯顿反射测角仪 Wollaston's reflecting goniometer

吴哥窟 <柬埔寨> Angkor vat
吴淞零点 Wusong horizontal zero；Wusong zero of elevation
吴吞式宽火箱 Wootten firebox

梧

梧桐 Chinese parasol tree；phoenix tree

梧桐科 Sterculiaceae
梧桐籽油 sterculic oil
梧子 gall

蜈

蜈蚣形植草渠 centipede grassed channel

五

五百年 quincentenary

五百年一遇洪水 returned flood of five hundred years
五瓣花 five-petaled flowers
五瓣花饰 cinquefoil
五瓣形拱 cinquefoil arch
五倍的 fivefold；quintuple；quintuplicate
五倍量 quintuple
五倍频器 quintuplet
五倍器 quintupler
五倍子 gall（nut）
五倍子丹宁酸 Chinese gallotannnic acid
五倍子的 gallic
五倍子酸 gallic acid
五倍子酸盐 gallate
五苯基乙烷 pentaphenylethane
五边形 pentagon
五边形的 pentagonal；quinquangular
五边形截面 pentagonal section
五边形石桥 five-sided stone bridge
五边形压痕器 pentagonal indenter
五边形支撑 pentagonal timbering
五标高点横断面 five-level section
五丙基 pentapropyl
五彩 full colo（u）r
五彩的 multicolo（u）r；technicolo（u）r
五彩拉毛粉刷 sgraffi（a）to
五彩拉毛陶瓷 sgraffi（a）to
五彩器皿 five-colo（u）red ware
五彩人物角瓜式瓶 polychrome figure curve vase
五彩砖 multicolo（u）r brick
五层板门 five-ply door
五层干燥机 five-tier drier
五层作（法）five ply
五钗松 Japanese white pine
五成摊销法 fifty percent amortization method
五成折旧法 depreciation-fifty percent method
五齿岩石抓斗 five-tine rock grapple
五触点塞孔 five-point jack
五床位的 five-bed
五次不尽根 quintic surd
五次代数曲线 quintic
五次的【数】quintic
五次方程 quintic equation
五次轮回选择方案 five cycle of a recurrent selection program（me）
五次曲线 quintic curve
五次谐波 quintuple harmonics
五次样条函数 quintic splines
五单体李晶 fiveling
五单位 five level
五单位码 five bit code；five-level code；five-unit alphabet
五单位起止式设备 five-unit start-stop apparatus
五单位数字保护电码 protected 5-unit numerical code
五单位字母 five-unit alphabet
五单元关节车 five-unit articulated car
五单元码 five-unit code
五档 five-range；five speed
五档变速器 five-speed transmission
五岛式泊位 five-island berth
五地址指令 five-address instruction
五点布井法 five-spot pattern
五点段面 five-level section
五点二次曲线平滑 quadratic smoothing with five point
五点法【航测】five-point method；five-spot method
五点拱 quint-point arch
五点井网 five-spot pattern
五点井网流动公式 five-spot flow formula
五点井网注水 five-spot water flooding
五点线性平滑 linear smoothing with five point
五点形的 quincuncial
五点移动平均 five points moving average
五点注水系 five-spot flood system
五碘化物 pentaiodide
五碘乙烷 pentaiodoethane
五电平码 five-level code
五电平起止操作 five-level start-stop operation
五斗橱 bachelor chest；chest of drawers；chiffonier
五斗柜 chest of drawers；chiffonier；commode
五二记数法 quibinary notation
五二码 quibinary；quibinary code
五方的 pentagonal
五分镍币 jitney
五分仪 quintant

W

五分之一对座 quintile
五分钟振荡 five-minute oscillation
五佛山群【地】Wufoshan group
五氟苯甲酸 pentafluorobenzoic acid
五氟化硫 sulfur pentafluoride
五氟化溴法 BrF_5 method
五氟氯乙烷 chloropentafluoroethane
五幅一联画 pentaptych
五幅一联图 pentaptych
五缸往复式泵 quintuple reciprocating pump
五个接触点 pentad
五个一套的 quinary
五个一组 pentad
五股编绳索 five-strand crabber's eye sennit
五鼓绞车 five-drum winch
五鼓卷扬机 five-drum winch
五辊滚压机 five-roller;five-roll mill
五辊矫直机 five-roll machine
五辊拉伸机 five-roller stretcher
五辊轧机 five-roller mill
五滚筒式研光机 five bowl universal calender
五毫米分划 half-centimeter division
五合 pentahapto
五合板 five plywood
五合板门 five-panel(led) door
五合透镜 five-element lens
五弧拱 cinquefoil arch
五花山墙 five-corbiestep gable with bargeboard;stepped gable wall
五滑轮冕形滑车 scatter sheave crown
五环三萜烷 pentacyclictriterpane
五簧片塞孔 five-point jack
五机座串列式轧机 five-stand tandem mill
五级博多码 five-level Baude code
五级成矿远景区 the fifth grade of minerogenetic prospect
五级成煤远景区 the fifth grade of coal-forming prospect
五级的船闸系统 a five-flight shiplock system
五级风 fresh breeze;wind of Beaufort force five
五级风浪 very rough sea
五级结构面 grade five discontinuity
五级结构体 grade five texture body
五级浪 force-five wave;rough sea
五级能见度 visibility poor
五级油气远景区 the fifth grade of oil-gas prospect
五极的 pentode
五极电子管 pentode tube
五极管 five-electrode tube; five-element tube;pentode;suppressor grid tube
五极管放大器 pentode amplifier
五极纵轴测深 quintic electrode longitudinal axis sounding
五极纵轴测深等断面图 contour section of quintic-dip longitudinal sounding
五极纵轴测深曲线图 diagram of quintic-dipole longitudinal sounding
五加仑装的汽油罐 jerrican
五夹板 five plywood
五甲基副品红 pentamethyl pararosaniline
五甲炔花青 pentamethine cyanme
五甲氧基红 pentamethoxyl red
五甲紫 pentamethyl violet
五价 pentad; pentavalence; quinquevalence [quinquevalency]
五价的 quinquevalent
五价铌的 niobic
五价钨的 tungstic
五价物 pentad
五价溴的 bromic
五架梁 five-purlin(e) beam
五件套浴室 five-fixture bathroom
五件一套 quint;quintet(te)
五角的 pentangular
五角枫 mono maple
五角棱镜 five-sided prism;pentagonal prism;pentaprism
五角棱镜直角器 five-sided square;pentagonal prism square
五角棱柱体 pentagonal prism
五角三八面体 gyroid
五角三四面体 tetartoid
五角十二面体 pentagonal dodecahedron;pyritohedron
五角石 pentagonite
五角数 pentagonal number
五角双锥 pentagonal bipyramid
五角星 five-pointed star;pentagram
五角星饰 pentacle
五角星形 pentacle;pentagram;pentalpha
五角形 pentagon
五角形堡垒 pentagonal bastion
五角形的 pentagonal;quinquangular
五角形地平面图 pentagonal ground plan
五角形防护幕 pentagonal screen
五角形截面 pentagonal section
五角形塔楼 pentagonal tower
五角形砖 pentagonal tile
五脚插头 five-point plug
五节环 five-membered ring
五节托辊 five roll idler
五金 hardware;metals
五金安装工程 ironmongery work
五金厂 hardware factory
五金电料 hardware and electric(al) appliance
五金店 hardware store;iron monger;ironmongery
五金工厂 hardware manufactory
五金工具 brightwork
五金光铁蓝 bronzeless blue
五金件 metallic fittings
五金零件工厂 hardware factory
五金零件装配机 hardware mounting machine
五金器具 brightwork;iron monger;iron ware;ironmongery
五金商人 hardware man
五金业 iron monger
五金制品工业 hardware industry
五金属光泽瓷漆 non-metallic enamel
五金属酞菁颜料 metal-free phthalocyanine
五金装饰品 ornamental hardware
五进码 quinary code
五进制的【计】quinary
五进制记数法 quinary notation
五进制数 quinary digit
五镜头空中照相机 five-lens aerial camera
五镜头摄影机 five lens camera
五居室单元 five-room(ed) dwelling unit
五聚物 pentamer
五开道岔 five-throw turnout
五开间的 five-bay
五块镶门板 five-panel(led) door
五类典型柱 five orders
五类典型柱石 fire orders
五棱镜 pentagonal prism;pentaprism
五棱锥 pentagonal pyramid
五力矩定理 theorem of five moments
五连形 pentagon
五联拱 cinquefoil
五零二速凝胶 five-zero-two ethyl quick setting adhesive
五零一一脲醛树脂胶 five-zero-one-one urea formaldehyde adhesive
五硫化二砷 arsenic pentasulfied
五硫化二锑 antimony pentasulfide; antimony red
五氯苯酚 pentachlorophenol
五氯苯酚钠盐 sodium pentachlorophenolate
五氯苯酚 74 <一种防虫剂> pentachlorophenol
五氯苯酚铜 copper pentachlorophenate
五氯酚钠 sodium pentachlorophenate; santobrite <抗真菌化学溶液>; Dowicide G <木材防腐剂>
五氯酚钠钠硼砂合剂 <水溶性木材防腐剂> noxtane
五氯酚钠乙基磷酸汞合剂 <木材防腐> melsan
五氯酚钠乙基磷酸汞硼砂合剂 <木材防腐> zylobrite
五氯酚石灰融合分析法 lime-fusion method;lime ignition method
五氯酚石油溶液 pentachlorophenol petroleum solution
五氯酚盐 pentachlorophenate
五氯酚中毒 pentachlorophenate poisoning
五氯化锑 antimony pentachloride
五氯化物 pentoxide
五氯联苯 pentachlorodiphenyl
五氯硫酚 pentachlorothiophenol
五氯硝基苯 pentachloronitrobenzene; quintozene
五氯乙烷 pentachloroethane
五轮滑车 five-sheave block
五轮列的 pentacyclic
五枚缎纹 five heddle satin
五枚斜纹 five leaf twill
五面镜 pentamirror
五面体 pentahedron
五年计划 five-year plan
五年间 pentad
五年时间 luster
五硼烷 pentaborane
五桥式底卸卡车 five axles bottom-dump
五曲柄曲轴 five-throw crankshaft
五人小组 quintet(te)
五日生化需氧量 five-day BOD [biological oxygen demand]
五日预报 five-day forecast
五蕊柳 bay-leaved willow;laurel-leaf willow
五色晕 pleochroic halo
五栅变频器 pentagrid converter
五栅电子管 pentagrid tube
五栅混频管 pentagrid mixer
五十公斤 centner
五十年一遇洪水 fifth-year flood
五十年一遇水位 fifty(50)-year recurrence interval stage
五十千克 centner
五十烷 pentacontane
五十周波整流器电动机 fifty cycle commutator motor
五十周波整流式电动机 fifty cycle commutator motor
五十周年的 semi-centennial
五十周年纪念 quinquagenary
五实和欠压实 compaction and under-compaction
五示像机车信号机 five-aspect cab signal
五水合硫代亚硫酸钠 sodium thiosulfate
五水合硫酸铜 blue copperas; blue jack;blue vitriol
五水化合物 pentahydrate
五水硫酸铜 blue stone;chalcanthite;cyanosite
五水锰矾 jokokuite
五水硼钙石 gowerite; pentahydroborite
五水碳钙石 pentahydrocalcite
五水碳镁石 lansfordite
五水泻盐 pentahydrite
五顺一丁或四顺一丁砌法 garden bond
五顺一丁砌法 five stretcher courses to one header course alternately
五速齿轮箱 five-range transmission;five-speed gear box
五羧酸 pentacarboxylic acid
五塔寺 Five Pagoda Temple
五台系【地】Wutai system
五台运动【地】Wutai movement
五屉柜 tallboy
五天期 pentad;penthemeron
五亭桥 five pavilion bridge
五通道扫描辐射计 five-channel scanning radio meter
五通塞门(取暖) five way cock
五涂层结构 five coat system
五弯矩定理 theorem of five moments
五桅船 five-master
五维空间 quintuple space
五位制 five-digit numbering;five digit system;five figure system
五位字节 quintet(te)
五弦桁架 five-chord truss
五显示闭塞信号机【铁】five aspect block signal
五显示机车信号机【铁】five-aspect cab signal
五线制 five-wire system
五相点 quintuple point
五小工业 five small industry
五心拱 arch of five center [centre]; five-centered [centred] arch
五心拱券 false ellipse
五心连拱 cinquefoil arch
五心柱变压器 five-legged transformer
五心装饰拱 false ellipse arch
五芯电缆 five-core cable
五行铆钉 quintuple riveting
五行铆接 quintuple riveting
五溴苯胺 pentabromoniline
五溴苯酚 pentabromophenol
五溴二苯醚 pentabromodiphenyl oxide
五溴化反应 pentabromization
五溴化磷 phosphorus pentabromide
五溴甲苯 pentabromotoluene
五溴一氯环己烷 pentabromochlorocyclohexane
五溴乙基苯 pentabromoethyl benzene
五亚乙基六胺 pentaethylene hexamine
五氧化二铋 bismuth pentoxide
五氧化二氮 nitrogen acid anhydride; nitrogen pentoxide
五氧化二钒 red cake;vanadic oxide;vanadium pentoxide
五氧化二钒尘粉 vanadium pentoxide dust
五氧化二钒烟 vanadium pentoxide fume
五氧化二磷 phosphoric anhydride; phosphor(o)us pentoxide
五氧化二磷含量 content of phosphorus pentoxide
五氧化二钼 molybdenum hemipentoxide
五氧化二镎 neptunium protoxide-oxide
五氧化二铌 niobium pentoxide
五氧化二砷 arsenic pentoxide
五氧化二钽 tantalum pentoxide
五氧化二钨 tungsten pentexide
五氧化物 pentoxide
五氧杂环十八烷 pentaoxacyclooctadecane
五叶窗花格 five-lobe tracery
五叶梅花饰 cincfoil
五叶爬山虎【植】Virginian creeper

五叶饰 quinquefoil
五叶推进器 five-blade propeller
五叶形截面 pentalobal cross-section
五叶形饰 cinq(ue)foil
五叶形饰的拱 five-foiled arch
五乙烯六胺 pentaethylene hexamine
五元 pental
五元醇 pentabasic alcohol
五元合金 quinary alloy
五元碱性的 pentabasic
五元码 quinary code
五元素铋矿床 bismuthic deposit of penta elements
五元素建造镍矿床 ore deposits of penta deposit formation
五元素型铀矿床 uranium deposit of penta element type
五元酸 pentabasic acid
五元系 quinary system
五元组 pental;quintuple
五元组图灵机 quintuple Turing machine
五元组图灵可计算函数 quintuple Turing computable function
五元组形式 quintuple form
五月金龟子 cockchafer
五趾吊 hang five
五中取二码 two-out-of-five code
五重的 fivefold
五重对称 fivefold symmetry
五重符合 five-fold coincidence
五重线 quintet(te)
五轴法 five-axis method
五轴式底卸卡车 five axles bottom-dump
五主轴颈曲轴 five bearing crankshaft
五柱式 five orders;pentastyle
五柱式建筑 pentastyle building;pentastylos
五柱式庙宇 pentastyle temple
五柱式铁芯 <变压器的> five limbs type core
五柱柱形 five orders
五组并排椭圆弹簧 quintet(te);quintuplet;quintuplet elliptic spring
五坐标程控机床 quintuple-control machine-tool

午标 noon-mark

午餐 dinner;luncheon;nooning
午餐盒 <管线工的> bait box
午餐会 luncheon
午餐招待会 luncheon
午潮 noon tide
午城黄土 <中国> Wucheng loess
午毒蛾 <拉> Lymantria dispar
午后 post meridian
午后班 afternoon watch
午后效应 <无线电> afternoon effect
午间休息 lunch-break;midday break
午门 Meridian Gate
午前 ante meridian
午前班 forenoon watch
午前的 antemeridian
午睡 nap
午台后部幕布 backdrop
午休 midday break;nooning
午宴 luncheon
午夜蓝 midnight blue
午夜倾倒者 midnight dumper
午夜噪声 midnight noise
午夜值班 middle watch;midwatch

伍德-安德逊地震仪 Wood-Anderson seismic instrument

伍德玻璃 Wood's glass
伍德(低熔)合金 Wood's metal;Wood's alloy
伍德法 <垂直拉玻璃管法> Wood's process
伍德光线 Wood's light
伍德华德黏[粘]度计 Woodward visco(si)meter
伍德华德调速器 Woodward speed governor
伍德华德调整器 Woodward regulator
伍德华德液压调速器 Woodward governor
伍德效应 Wood's effect
伍登粒级 Udden grade scale
伍登-温德华斯粒级标度 Udden-Wentworth grade scale
伍登-温德华斯(土)粒径分类表 Udden-Wentworth scale
伍尔顿砂岩 <一种暗红色砂岩建筑材料,英国兰开郡产> Woolton
伍尔夫静电计 Wulff electrometer
伍尔夫裂解炉 Wulff pyrolysis furnace
伍尔夫弦线静电计 Wulf string electrometer
伍尔黏[粘]土空心块 Wohl block
伍尔维奇层 Woolwich beds
伍拉维牌混凝土 <一种轻质混凝土> Woolaway
伍氏(低熔)合金 Wood's alloy
伍氏黏[粘]度计 Woodward visco(si)meter
伍氏网 Wulff net
伍兹·霍尔沉降分析仪(器) Woods Hole sediment analyser [analyzer]

庑殿【建】 hip roof

庑殿(式)屋顶 hipped roof;Chinese hipped roof;Wudian roof

武断行为 arbitrary action

武官 military officer
武木期【地】 Wurm
武器 arm(ament);ordnance;weapon
武器监控中心 weapon monitoring center [centre]
武器库 armory;arsenal
武器商船 defensively equipped merchant ship
武器碎片 weapon debris
武器系统库存控制站 weapon system stock control point
武士俑 warrior figures
武装部队技术情报局 <美> Armed Services Technical Information Agency
武装部队客票 forces ticket
武装检查船 armed boarding vessel
武装民船 privateer
武装商船 armed merchantman;armed vessel; defensively armed merchant; defensively equipped merchant ship

舞弊罚款 fraud penalty

舞弊行为 corrupt transaction
舞场 ballroom
舞池 dance floor;dancing floor
舞池地面 dance floor
舞池照明 dance floor lighting
舞鹤带 maizuru tectonic zone
舞瓶 <瓷器名> vase with dance design
舞台 scene; stage; theatre [theater]; pulpitum <罗马剧院里与乐队席相邻的>
舞台安灯天桥 light bridge
舞台安装 erection stage
舞台安装天桥 light bridge
舞台背后弧形背景幕布 cyclorama
舞台标高 stage level
舞台表演艺术 theatrical;theatrics
舞台布景 decor;scenery;stage scene(ry)
舞台布景存放处 scene dock
舞台布景设备 stage equipment
舞台布景升降机 sloat;slote
舞台布景升降结构 stage gridiron
舞台侧布景 coulisse
舞台侧面布景 cullis
舞台侧幕 tormentor
舞台侧墙 proscenium wall
舞台侧翼 flies
舞台储藏室 property room
舞台大桥上的栏杆 pin rail
舞台带形灯 stage strip light
舞台道 flyway
舞台的侧光 side light
舞台的台口以内部分的建筑 stagehouse
舞台的演戏面积 acting area
舞台的一部分但不起舞台作用 non-working stage
舞台灯 stage lamp
舞台灯光 acting area light;stage light(ing)
舞台灯光效果 stage lighting effect
舞台灯光耀眼 glare of the foot lights
舞台灯光装置 stage lighting units
舞台灯光总开关 blackout switch
舞台地板门 <演员从下面上升至舞台> star trap
舞台吊灯具和布景的钢管 lighting booth
舞台吊杆 batten;fly-bar
舞台吊幕 back drop;drop scene
舞台吊幕横管 pipe batten
舞台吊索 fly line
舞台防火帘 procenium
舞台防火幕(框) procenium
舞台附属用房 ancillary accommodations of stage
舞台高程 elevation of stage;stage level
舞台高度 stage level
舞台工场 stage shop
舞台工作间 working stage
舞台工作区 wings
舞台广告幕 advertisement curtain
舞台后部 upstage
舞台后方 back part of stage;upstage
舞台后台电源插销盒 stage pocket
舞台后台门 stage door
舞台后运布景门 loading door
舞台花道 <自舞台向观众席伸出> flyway
舞台幻灯机 cloud machine
舞台(活动)地板门电梯 trap elevator
舞台机械设备 stage machinery
舞台建筑 scene-building
舞台建筑的左右侧翼 versurae
舞台脚灯 footlight
舞台聚光灯 stage projector; stage spotlight
舞台卷幕 rolling curtain
舞台空间 stage space
舞台口 proscenium [复 proscenia]; proscenium arch
舞台口包厢 proscenium box;stage box
舞台口边框 architrave of proscenium
舞台口侧墙 proscenium wall
舞台口的幕(场间降下) act drop
舞台口门 proscenium door
舞台口上部灯光通廊 proscenium bridge
舞台两侧的窄幕 leg drop
舞台两侧耳房 <古希腊> paraskenion
舞台门孔 proscenium opening
舞台面 acting level; boards; scene(ry)
舞台幕布 stage curtain
舞台幕下垂线 curtain line
舞台旁特别包厢 stage box
舞台平衡锤系统 <用系绳法操作的> loose grid
舞台前部 downstage; inner proscenium; proscenium [复 proscenia]; stage-working
舞台前部的包厢 stage box
舞台前部空间 loft
舞台前部装置 proscenium [复 proscenia]
舞台前侧包厢 proscenium box
舞台前附台 apron stage
舞台前过道 cross-over
舞台前活动板门 corner trap
舞台前沿 front stage
舞台前缘灯 footlight
舞台墙上电源插板 stage wall pocket
舞台桥面系统 system of bridges of stage
舞台区 stage area
舞台入口 stage entrance
舞台上部布景控制处 flies
舞台上部横挂的照明灯 border light
舞台上空 flies;fly loft;upper part of stage
舞台设备 stage equipment; stage fittings
舞台设计 setting designing
舞台设计师 stage designer
舞台设计用的水色漆 designer colo(u)rs
舞台设在观众席中央的剧院 arena theatre [theater]
舞台升降机 console lift;stage lift
舞台升降控制器 console lift
舞台升降平台 grave trap
舞台疏散口 escape from stage; stage escape
舞台水下活动的幕 drop curtain
舞台索具 stage rigging
舞台塔 fly tower
舞台台唇 apron; forestage; proscenium [复 proscenia]
舞台台口 proscenium arch; proskenion;proscenium opening
舞台台口侧墙 proscenium wall
舞台台口灯光 border light
舞台台口排灯 first border
舞台天幕 <能横向卷动的> rolling cyclorama;backcloth
舞台天桥 fly floor;fly gallery
舞台天桥栏杆 working rail
舞台调光机 stage dimmer
舞台帷幕 surround curtain
舞台戏院 stage theater [theatre]
舞台下底层 mezzanine
舞台下活动门小室 trap cellar
舞台小五金 stage hardware
舞台效果 scenic effect;theatrics
舞台艺术 decor;histrionics
舞台硬件 stage hardware
舞台用房 stagehouse
舞台右侧 <以演员面对观众时为准> stage right
舞台造型 tableau [复 tableaus/tableaux]
舞台照明 dramatic lighting; general scene lighting; scenic lighting; stage illumination;stage light(ing)
舞台照明法 lightpot
舞台照明效应 stage lighting effect
舞台转盘 <施转舞台的圆形地板部分> disc turntable
舞台装置 decor;mise en scene

舞台左侧 <以演员面对观众时为准> stage left
舞厅 ballroom; dance hall; dancery; dancing hall; dancing saloon
舞厅休息室 ballroom foyer

勿倒置 keep upright

勿放顶上 do not stake on top
勿放湿处 do not store in damp place; keep dry
勿倾倒 not to be-tipped
勿受潮湿 keep dry
勿忘草【植】forget-me-not
勿用手钩 no hook
勿掷 don't cast

务实的方法 pragmatic approach

务实的态度 pragmatic attitude

戊胺 pentyl amine

戊醇 allyl alcohol alcohol; amylalcohol; pentanol
戊二胺 pentamethylenediamine
戊二醇 pentamethylene glycol; pentanediol
戊二酸 pentanedioic acid
戊二羧酸 pentane dicarboxylic acid
戊二烯 pentadiene
戊二烯橡胶 piperylene rubber
戊酚树脂 amylphenol resin
戊基醚 amyl ether
戊基纤维素 amyl cellulose
戊基乙炔 amylacetylene
戊腈 valeronitrile
戊硫醇 pentan-thiol
戊醚 amyl ether
戊内酰胺 valerolactam
戊内酯 valerolactone
戊硼烷 boron hydride; pentaborane
戊炔 pentyne
戊酸 valerianic acid; valerie
戊酸丁酯 butylvaleriate
戊酸戊酯 oilofapple
戊酸盐 valerate; valerianate
戊酮-3 diethyl ketone
戊烷 pentane
戊烷的溶解度 solubility of pentane
戊烷灯 pentane lamp
戊烷馏除塔 pentanizer; pentanizing column; pentanizing tower
戊烷三酮 triketopentane
戊烷碳稳定同位素组成 stable carbon isotopic composition of pentane
戊烷温度计 pentane thermometer
戊烯 amylene; valerene
戊烯醇 pentenol
戊烯二醛 glutaconaldehyde
戊烯酸 pentenoic acid
戊酰氯 valeric chloride; valeryl chloride

坞壁 dock wall; dockside; dock sidewall; drydock wall

坞壁顶部 dock coping
坞壁顶面 dock coping
坞壁顶面高程 dock coping level
坞壁活动脚手架的电流滑触器 electric pick-up for dock arm
坞壁平台 altar platform
坞壁梯阶 altar steps
坞壁悬臂脚手架 dock arm
坞边起重机 dockside crane
坞边起重机导轨 dockside crane rail
坞边墙 dock sidewall
坞长度 length of dock; drydock length
坞底 dock bottom; dock floor
坞底板 dock floor
坞底板剖面 dock floor profile
坞底板纵坡 longitudinal slope of dock floor
坞底高程 drydock floor level
坞底横向型灌水涵洞 transverse floor type flooding culvert
坞底静水压力 hydrostatic pressure on dock floor; hydrostatic underpressure
坞顶栏杆 coping railing
坞墩 dock (ing) (concrete) block; ship blocking
坞工坝 stone masonry dam
坞工的 masonic
坞灌水涵管 dock filling culvert
坞槛 dock apron; sill; drydock sill
坞槛高程 sill elevation
坞槛上水深 depth over the sill; drydock depth over sill
坞槛水深 depth of dock entrance
坞口 dock entrance
坞口宽度 width of dock entrance
坞口水深 depth on sill; depth of dock entrance
坞门 dock caisson; dock gate; entrance caisson; sluice of the dock; gate chamber
坞门舱室 dock gate chamber
坞门侧面龙骨 gate side keel
坞门的一端 gate end of dock
坞门顶部 dock gate crown
坞门干船坞 graving dock
坞门刚度 gate stiffness
坞门和坞墙的贴接面 meeting face of drydock
坞门槛 gate sill; dock sill
坞门槛标高 dock sill level
坞门开启 dock gate opening
坞门跨度 span of gate
坞门龙骨 gate keel
坞门枢轴 gate pivot
坞门支承 gate support
坞门止水 gate seal
坞门座 <浮坞门或滑动门的> dock seat; dock gate bed; gate seat; gate bed
坞内梁格结构 gridiron in dock
坞内临界吃水 critical docking draft
坞内龙骨垫 docking keel block
坞内水位标尺 dock water level indicator
坞内舾装 dock outfitting
坞墙 dock wall; sidewall of dock; drydock wall
坞墙顶高程 coping elevation
坞墙型灌水涵洞 sidewall type flooding culvert
坞群 dock group
坞深 depth of dock
坞式船闸闸室 dock-type lock chamber
坞式登陆舰 dock landing ship
坞式货船 dock cargo ship
坞室 dock chamber; interior of dry dock; dock basin
坞室长度 length of drydock
坞室宽度 width of drydock barrel
坞首 dock head; entrance of dry dock; headward end of dock; head of dock
坞首的一端 headend of dock
坞首端部 head end of drydock
坞首端墙 dock headwall
坞头槛 apron
坞修 docking repair; dry dock(ing)
坞修和上排 docking and slipping
坞修设施 dock facility
坞闸 sea gate
坞闸门 dock gate
坞站 docking station
坞长 dockmaster
坞中支撑 dock shore
坞组 grounded dock; grouped docks
坞座垫块 docking block
坞座墩木 docking block
坞座龙骨 grounding keel; docking keel

艻 fluorene

艻石 kratochvilite

物标 optic(al) marking

物标地理视距 geographic(al) range of an object
物标反舷角指示器 aspect indicator
物标录取 target acquisition
物标强度 target strength
物标视地平距离 horizon range from an object
物标搜索 target acquisition
物标舷角 aspect; target angle
物标正横 object abeam; target abeam
物标正横距离 distance of object when abeam
物产水火保险 fire and marine insurance
物产租金 <不包括设备和服务> economic rent
物的准动产 chattels real
物点 object point
物动视力 vision with object moving
物端 nose
物端光栅 objective grating
物端棱镜 objective prism
物端棱镜光谱 objective prism spectrum
物方 object space
物方焦点 object focal point; object focus
物方空间坐标系 object space coordinate system
物方物界 object space
物光束 object beach
物归原主 revesting
物候 phenology
物候的隔离 phenological isolation
物候仿真模型 phonological simulation model
物候观测 phenological observation
物候观察 phonological observation
物候季节 phonological season
物候历 phenological calendar
物候谱(系) phenological spectrum; phenospectrum
物候期 phenological date; phenological phase; phenophase; vegetation sampling period
物候日期 phenodate
物候生态谱系 phenoecological spectrum
物候图 phenogram; phenological chart
物候现象 phenology
物候学 phenology
物候演替 phenological succession
物化 materialize
物化参数 physical-chemical parameter; physico-chemical parameter
物化处理 physical-chemical treatment; physico-chemical treatment
物化的 physico-chemical
物化法 physical-chemical process; physio-chemical process
物化分析 instrumental analysis; physico-chemical analysis
物化检验 physical-chemical examination; physico-chemical examination
物化降解 physical-chemical degradation; physico-chemical degradation
物化净化(法) physical-chemical purification; physico-chemical purification
物化劳动 embodied labo(u)r; indirect labo(u)r; materialized labo(u)r
物化探综合异常图 synthetic(al) anomaly map of geophysics and geochemistry
物化脱氮处理 physical-chemical denitrification; physico-chemical denitrification
物化污水处理 physical-chemical sewage treatment; physico-chemical sewage treatment
物化系统 physical-chemical system; physico-chemical system
物化性质 physical-chemical property; physico-chemical property
物化预处理 physical-chemical pretreatment; physico-chemical pretreatment
物价 general price; prices of commodities
物价变动 fluctuation reserve
物价变动条款 escalation clause
物价变动准备 price fluctuation reserves
物价表 price list
物价波动 fluctuation in price; price fluctuation
物价补贴工资 escalating wage
物价飞涨 rocketing prices; skyrocketing prices; soaring price; steep increase in prices
物价浮动 price fluctuation
物价浮动系数 price fluctuation factor
物价管理 price control
物价管制 price control
物价结构 price structure
物价津贴 dearness allowance
物价局 pricing department
物价控制 price control
物价螺旋上升 price spiral
物价猛涨 breakthrough
物价膨胀 price inflation
物价上涨 inflation of prices; prices inflated; run high
物价上涨指数 escalation index
物价水平 level of price; price level
物价水准 level of price; price level
物价稳定 steadiness of commodity
物价下跌 decline in price
物价与所得政策 prices and incomes policy
物价折算指数 price converting index
物价政策 price policy; pricing policy
物价指数 cost of living indexes; index number of price; index of price; price index
物价指数表 tabular standard
物价指数化 indexation of prices
物价指数债券 indexed bond
物价总水平 general level of market prices; the overall price level
物价总指数 general price index
物价组 price group
物架 object carrier
物件 article; object
物件归还 restitution of property
物件所有权 title to property
物件之顶 <拉> caput
物焦点 object focal point

物角 object angle
物景深 depth of object space
物镜 field lens;objective;object(ive) glass;object(ive) lens
物镜测微计 object micrometer
物镜的光斑 flare
物镜反射光斑 flare
物镜放大 objective magnification
物镜放大率 objective lens magnification
物镜分辩率 resolving power of objective lens
物镜分解力 resolving power of lens
物镜盖 lens cap;objective cap
物镜光圆刻度 lens diaphragm scale
物镜光栅 objective grating
物镜滑筒 object slide
物镜环 object(ive)(glass) collar
物镜畸变差 distortion of lens
物镜架 objective carrier
物镜焦距 objective focal length
物镜孔 objective window
物镜孔径 objective aperture;object(ive) glass aperture
物镜快门 objective shutter
物镜框 cone;lens cone;lens mounting;lens panel;objective holder;objective mount
物镜类型 Aviar
物镜棱镜 objective prism
物镜前(后)节点 front-rear nodal point of lens
物镜散斑 objective speckle
物镜色差 chromatic aberration of lens
物镜视场 objective angular field;objective field of view
物镜视角范围 objective angular field
物镜视野 objective field of view
物镜调整 objective lens adjustment
物镜通光孔径 clear objective aperture
物镜筒 nosepiece;objective slide;objective tube
物镜透镜 objective lens
物镜象散校正器 objective astigmatic corrector
物镜遮光器 gobo flag
物镜遮光罩 counter-light lens hood;lens hood
物镜主光轴 optic(al) axis of lens
物镜主平面 central plane of objective
物镜转换盘 objective changer
物镜座 objective holder
物距 object(ive) distance
物空间 object space
物理摆 compound pendulum;physical pendulum
物理半衰期 physical half-life
物理爆炸 physical explosion
物理边界 physical boundary
物理变化 physical change
物理变数 physical variable
物理变性 physical modification
物理标准 physical criterion
物理剥蚀作用 physical exfoliation
物理不均匀性 physical heterogeneity
物理不相容性 physical incompatibility
物理布局 physical layout
物理参数 physical parameter
物理操作 physical operations
物理测定 physical determination;physical measurement
物理测井 geophysical (well-) log(ging)(for oil prospecting)
物理测量 physical measurement
物理层 physical layer;physical level
物理层理形式 physical stratification pattern
物理缠结 physical entanglement
物理常数 physical constant
物理场变化 physical field change
物理成熟 physical ripening
物理尺度 physical size
物理除尘 physical aspirating;physical dedusting
物理处理单元操作 physical treatment unit operation
物理处理(法) physical treatment
物理催化剂 physical catalyst
物理大地测量学 physical geodesy
物理大气压 physical atmosphere
物理单元 physical location
物理单元操作 physical unit operation
物理单元处理法 physical unit process
物理当量伦琴 Rontgen equivalent physical;tissue roentgen
物理的衰期 physical half-life
物理地层学 physical stratigraphy
物理地貌学 physical geomorphology
物理地平 physical horizon
物理地球化学 physical geochemistry
物理地址 physical address
物理地质现象 physical geologic(al) phenomenon
物理地质现象调查 investigation of physico-geologic(al) phenomena
物理地质学 physical geology
物理地质作用 physio-geologic(al) function
物理点数 sum total of physical point
物理电工计算机与自动控制情报服务中心 Information Services of Physics, Electrotechnology, Computers and Control
物理电路 physical circuit
物理电子学 physical electronics
物理定律 physical law
物理定向 physical orientation
物理定则产生器 physical law generator
物理毒剂 physical poison
物理对象 physical object
物理发泡 physical blowing;physical foaming
物理发泡剂 physical blowing agent;physical foamer;physical foaming agent
物理法 physical method
物理方法测地质年代 physical time
物理防老化剂 physical antioxidant
物理防水 physical waterproofing
物理防治(法) physical control;physical prevention and treatment
物理放大因数 physical amplification factor;physical magnification factor
物理非线性 physical non-linearity
物理分类 physical classification
物理分离 physical separation
物理分析 physical analysis;physical assay
物理分选 physical separation
物理分选装置 physical sorting equipment
物理风化 physical deterioration;physical weathering
物理风化残留物 physical residue
物理风化作用 physical weathering
物理辐射效应 physical radiation effect
物理腐蚀 physical corrosion;physical deterioration
物理概念 physical concept(ion)
物理干旱 physical drought
物理干扰 physical interference
物理干燥 physical dryness
物理干燥法 drying by evapo(u)ration
物理刚玉 physical corundum
物理格式 physical format
物理隔离 physical separation
物理根数 physical element
物理公式 physical equation
物理共振性 physical resonance
物理故障 physical fault
物理关系 physical relationship
物理光度计 physical photometer
物理光学 physical optics
物理过程 physical process
物理过程研究 studies of the physical processes
物理海洋学 physical oceanography
物理化学 physical chemistry;physico-chemistry
物理化学参数 physico-chemical parameter
物理化学常数 physico-chemical constant
物理化学处理(法) physico-chemical treatment
物理化学单元过程 physico-chemical unit process
物理化学的 physico-chemical
物理化学地质学 physico-chemical geology
物理化学法 physical-chemical process
物理化学方法 physico-chemical method
物理化学分离法 discrete process of physical chemistry
物理化学分析 physico-chemical analysis
物理化学过程 physical-chemical process;physico-chemical process
物理化学结合水 physico-chemical bound water
物理化学净化 physico-chemical purification
物理化学力学 physico-chemical mechanics
物理化学屏障 physico-chemical barrier
物理化学特性 physico-chemical characteristic
物理化学污水处理 physico-chemical sewage treatment
物理化学吸附作用 physical chemistry adsorption
物理化学性质 physico-chemical property
物理化学指标 physical and chemical indices;physico-chemical indices
物理环境 physical environment
物理火灾 physical fire
物理火灾模型 physical fire model
物理机理 physical mechanism
物理机械性能 physical and mechanical property
物理及物理化学分析 physical and physico-chemical analysis
物理记录 physical record
物理记录定位 physical record positioner
物理加工过程 physical refining process
物理加固土 mechanical stabilization for soil
物理检验 physical check;physical examination
物理检验法 physical test(ing)
物理检验试样 physical testing sample
物理降解 physical degradation
物理接触连接器 physical contact connector
物理接口 physical interface
物理节点 physical node
物理结构 physical structure
物理结构煤化作用 physics-structure coalification
物理净化作用 physical purification
物理勘探(法) physical exploration;physical logging;physical prospecting
物理抗性 physical resistance
物理科学 physical science
物理控制 physical control
物理矿物学 physical mineralogy
物理老化 physical ageing
物理理论 physical theory
物理力学 physical mechanics
物理力学特性 physico-mechanical characteristic
物理力学性质 physico-mechanical property
物理连接 physical connection
物理连续表 physically-contiguous list
物理连续性 physical continuity
物理联编 physical binding
物理量 physical quantity
物理量值 magnitude of physical quantity
物理疗法 modality;physiatrics;physical therapy;physiotherapy
物理伦琴当量 Roentgen equivalent physical
物理模块 physical module
物理模拟 physical analog(ue);physical modelling;physical simulation
物理模拟方法 physical analog(ue) method
物理模拟系统 physical simulation system
物理模式 physical mode
物理模型 physical model(ling)
物理目视双星 physical visual binary
物理内存 physical memory
物理耐性 physical tolerance
物理能 physical energy
物理浓集 physical upgrading
物理配对 physical pairing
物理平衡 physical equilibrium
物理气候 physical climate
物理气候系统 physical climate system
物理气候学 physical climatology
物理气相沉积 physical vapour deposition
物理气象学 physical meteorology
物理侵蚀 physical attack
物理清洗法 physical cleaning method
物理驱动力 physical driving force
物理取样 physical sampling
物理人工模拟 physical manual simulation
物理溶剂 physical solvent
物理软化剂 physical softener
物理砂粒 physical sand
物理伤害 physical injury
物理上可实现性 physical realizability
物理设备 physical device;physical equipment
物理设备表 physical device table
物理设备号 physical device number
物理设计 physical design
物理设计决定 physical design decision
物理生热作用 physical thermogenesis
物理生态学 physical ecology
物理声学 physical acoustics
物理湿度计 physical hygrometer
物理实体 physical entity
物理实验设备 physical experimental equipment
物理实验室 physics laboratory
物理示踪 physical tracing
物理示踪剂 physical tracer
物理式气体分析器 physical gas analyser [analyzer]
物理势态 physical situation
物理试验 physical test(ing)
物理试验室 physics laboratory

W

物理寿命 physical life
物理树结构 physical tree structure
物理数据 physical data
物理数据独立性 physical data independence
物理数据库设计 physical database design
物理数据描述语言 physical data description language
物理数学模型 physico-mathematical model
物理衰减 physical depreciation
物理双星 binary;physical double star
物理水文因素 physical hydrographic factor
物理顺序存取 physical sequential access
物理塑限 physical plastic limit
物理损伤 physical impairment
物理探查 physical prospecting
物理探查记录 <音波、导电、密度、水分、放射能等> physical logging
物理特性 physical characteristic;physical property
物理特性保持年限 physical life
物理特征 physical property
物理天平动 physical libration
物理天气分析 physical weather analysis
物理天文学 physical astronomy
物理条件 physical condition
物理通道 physical channel
物理通路 physical path
物理统计预报 physico-statistic(al) prediction
物理蜕变 physical disintegration
物理脱硫 physical desulfurization
物理脱色 physical decolo(u)rization
物理外观 physical appearance
物理网络结构 physical network arrangement
物理网络资源 physical network resource
物理位置 physical position
物理文件格式 physical file format
物理文件结构 physical file structure
物理文件名字 physical file name
物理文件系统 physical file system
物理稳定性 physical stability
物理污染 physical pollution
物理污染指数 physical pollution index
物理误差 physical error
物理吸附 physical adsorption
物理吸附物种 physically adsorbed species
物理吸附作用 physical absorption; physisorption
物理吸着 physisorption
物理系统 physical system
物理系统设计 physical system design
物理显影 physical development
物理现象 physical phenomenon
物理限制 physical limit
物理相互作用 physical interaction
物理相容性 physical compatibility
物理效应 physical effect
物理行为 physical behavio(u)r
物理性崩解 physical disintegration
物理性剥蚀 physical degradation
物理性防老剂 physical antioxidant
物理性分隔物 physical barrier
物理性干旱 physical drought
物理性假同晶(现象) physical pseudomorphy
物理性检验 physical examination
物理性能 physical aspect; physical characteristic;physical property
物理性能试验 physical property test
物理性黏[粘]粒 physical clay
物理性破坏 physical damage
物理性试验 physical test(ing)
物理性损坏 physical damage
物理性损伤 physical injury
物理性污染物 physical pollutant
物理性异构 physical isomerism
物理性应力 physical stress
物理性质 physical appearance;physical aspect;physical behavio(u)r; physical characteristic; physical feature; physical nature; physical property
物理性状 physical behavio(u)r
物理修复 physical remediation
物理选矿 physical upgrading
物理学 natural philosophy
物理学家 physicist
物理学界 community of physicists
物理研究 physical research
物理要求 physical requirement
物理冶金的 physical metallurgical
物理冶金学 physical metallurgy
物理异构 physical isomerism
物理异构物 physical isomer
物理异性 physical isomerism
物理意义 physical significance
物理因素 physical agent;physical factor
物理因子 physical factor
物理因子数 number of physical factor
物理硬化 physical hardening
物理有机化学 physical organic chemistry
物理指标 physical index
物理指示剂 physical tracer
物理治疗 naturopathy
物理属性 physical behavio(u)r
物理转化 physical transformation
物理装置 physical unit
物理状态 physical condition;physical state
物理状态混合物 physical mix(ture)
物理子女指示符 physical child pointer
物理自净化作用 physical self-purification
物理组织 physical organization
物力 material resources
物力论塔 dynamicist
物力资源 material resources;physical resources
物量指数 volume index
物料 article;material;material balancing calculation;matter;stores;supplies
物料安全数据表 materials safety data schedule
物料搬运 materials handling
物料搬运单 move order
物料搬运工程师 materials handling engineer
物料搬运机 materials handling machinery
物料搬运设备 material handling equipment
物料崩落 material avalanche
物料比重 material density
物料编号 stock number
物料表 material statement
物料参数 parameter of material
物料层 bed of material
物料储备 materials reserve
物料储存 material store
物料船 store ship;store vessel
物料单 store list
物料的膨胀 material swell
物料的重复操作 double-handling of material
物料等级 material gradation
物料费用 supplies expenses
物料分级法 material gradation
物料分配器 stock distribution
物料分配小仓 material distributing vessel
物料负荷率 material loading factor
物料供应 materials handling
物料供应工程师 materials handling engineer
物料管理 handling of goods and materials
物料衡算 balance of materials; balance of matters
物料衡算报告 material balance report
物料衡算法 material balance method
物料衡算量 material accountability
物料衡算区 material balance area
物料回收 material retrieval
物料积聚 build-up of material
物料架拱 arching of material
物料间 storeroom
物料检查 survey of stores
物料结块 build-up of material
物料结团 material agglomeration
物料进出量核算 material balance
物料(进出)平衡 material balance
物料净化 decontamination of materials
物料库费用 storeroom expenses
物料库容量 silo content
物料扩散流动 diffusive flow
物料量 inventory
物料流(程) material flow
物料流量 material rate
物料流速 mass flow rate
物料密度 material density
物料名称 name of materials
物料黏[粘]稠 material consistency
物料盘存 supplies inventory
物料盘存制 balance of stock system; balance-of-stores system
物料膨胀 swell of materials
物料平衡 balance of materials; balance of matters; burden balance; mass balance; reconciliation of inventory
物料平衡关系 material balance relationship
物料平衡计算 material balancing calculation
物料平衡限制条件 material balance constraint
物料迁移机理 mass transport mechanism
物料人工搬运 man-handling of materials
物料生产 material produce
物料输送泵 material handling pump
物料输送装置 material handling device
物料顺窑移动 travel down the kiln
物料特性 material characteristic
物料提升机 material lift
物料通过额 throughput capacity
物料通过量 throughput; throughput capacity
物料通过速率 rate of throughput; throughput rate
物料通过体积 throughput volume
物料通过窑所需时间 time of passage through kiln of material
物料通过重量 throughput weight
物料现状报告 material status report
物料消耗计算 material consumption calculation
物料性质 material property
物料循环 circulation of materials
物料与负荷因数 materials and load factors
物料与空气的接触面 material air interface
物料员 store keeper
物料运动速度 material rate
物料运输 material handling; material transport
物料在窑内停留时间 time of passage through kiln of material
物料装卸 materials handling
物料自动堆垛台 unscrambler
物料总存量 material inventory
物流 flow of material;materials flow; physical distribution
物流链 logistics chain
物流平台 logistics platform
物流数据库 logistics database
物流外部经营的 logistics outsourcing
物流网络 logistics network
物流系统 logistics system
物流园 logistic park
物流中心 logistics center [centre]
物流作业 logistics operation
物面 hill plane
物面效应 surface effect
物品 article;matter;stores;supplies
物品被损坏或损坏其他货物的风险 risk of article being damaged or causing damage to other commodities
物品本身自然属性 natural property inherent to articles
物品长度 length of article
物品袋(卧铺) pocket
物品单 inventory
物品的相对重量 relative weight of article
物品密度 density of article
物品清单 <零担车> list of articles
物品清单处理程序 bill of material processor
物品网 hammock;suspended net
物品网挂 net hook
物品叙述 description of property
物品种类 item
物平面 object plane
物平面扫描仪 object plane scanner
物权 ownership of property; real right;title
物权保留条款 retention of title clause
物权担保 security for interest;security interest
物权的准据法 applicable law for real rights
物权法 law of property
物权契据 title deed for land
物权诉讼 action in rem;real action
物权债券 document of title
物权证书 certificate of title; document of title
物深 object depth
物态 physical state; state (ment) of matter
物态边界面 state boundary surface
物态变化 change of states
物态方程式 equation of state
物态连续性 continuity of state
物态逆变 reversibility of states
物探 geophysical exploration; geophysical prospecting
物探报告 geophysical exploration report
物探成果图 resultant maps of geophysical exploration;result maps of geophysical prospecting
物探方法类型 type of geophysical prospecting method
物探工作布置图 layout of geophysical prospecting work
物探屏蔽区 blind zone
物探重力二级基点 second-grade gravity base station for exploration geophysics
物探重力三级基点 third-grade gravity base station for exploration geo-

physics
物探重力一级基点 first grade gravity base station for exploration geophysics
物探(钻孔)用喷射式钻头 geophysical jetting bit
物体 body;object;substance
物体表面凝结的薄冰 glaze ice
物体波 object wave
物体的尖端 the tip of anything
物体堵塞 solid block(age)
物体方位【测】object space
物体高度 object height
物体光波 object wave
物体光束 object beam
物体后部 afterbody
物体盲 object blindness
物体绕流 flow around a body
物体色 object colo(u)r
物体条件配色 object metamerism
物体下部 underbody
物体限制孔径 object defining aperture
物体选择 object selection
物体轴线 body axis
物体自由度 body freedom
物体阻力 body resistance
物物交换 barter; dicker; made-off; trade-off
物物交易条件 barter terms of trade
物物型算子 parabolic(al) operator
物系 system
物相 phase
物相变化 phase transformation
物相定律 phase rule
物相分析 phase analysis; physical phase analysis
物像变换 object-image transform
物像不等 aniseikonia
物像测量 image measurement
物像重合法 object-image coincidence method
物像柱 xat;totem-pole <北美印第安人屋前刻的图腾像柱>
物性 properties of matter
物性测试样品 sample testing physical property
物业不动产 real estate
物业法法案 Law of Property Act
物业管理 estate management;property management;property manager
物业管理协议 property management agreement
物业交付文据制作 conveyancing
物业开发 real estate development
物业水火保险 fire and marine insurance
物业税 property tax
物影照片 photogram
物有所值 good value for money
物证 act and deed;material evidence
物质 mass; materiality; matter; substance
物质波 material wave;matter wave
物质波移 mass wasting
物质不灭定律 law of conservation of matter; law of indestructibility of matter;principle of conservation of matter
物质财产 material property
物质产品 material products
物质储备 fund
物质传递 mass transfer; material transfer
物质传递机理 mass transfer mechanism
物质刺激 incentive
物质刺激制度 incentive scheme
物质导数 individual derivative;material derivative; substantial derivative
物质的 physical
物质的量 quantity of mass
物质的溶液态 solution state of matter
物质的运输 transport of hazardous materials
物质地平 physical horizon
物质分离机理 separation mechanism of substance
物质分配 allocation of materials
物质辐射器 mass radiator
物质改造 astration
物质管理 goods handling; material handling
物质和生物环境 physical and biotic environment
物质化 materialization
物质环境 physical environment
物质技术基础 material and technical basis
物质奖励 material incentives;material reward
物质交换 interchange of matter;mass exchange
物质结构 physical structure
物质刺激 material incentives
物质力量 physical force
物质利益 material benefit; material gains;material interest
物质流 material flow
物质能量 matter energy
物质浓度 material concentration;substance concentration
物质赔偿 physical make-up
物质平衡 mass balance;material balance
物质平衡法 material balance method
物质平衡方程式 equation of material balance;material balance equation
物质坡移 mass wasting
物质期 matter era
物质迁移方向 migrational direction of materials
物质迁移机理 material transport mechanism
物质权益 material equity
物质三态 three states of matter
物质设备 physical equipment
物质设施 physical facility
物质生产 physical production
物质生产部门 branched of material production
物质生产领域 sphere of material production
物质生产资料 material instrument of production
物质时代 matter era
物质世界 material(istic) world;natural world;physical world
物质收获 material yielding
物质收获表 material yield table
物质守恒 conservation of matter
物质守恒定律 law of conservation of matter; law of indestructibility of matter;principle of conservation of matter
物质授予能量 energy imparted to matter
物质损耗 body waste
物质损伤 material damage
物质损失 material damage; material loss;mass wasting <滑坡整体移动中的>【地】
物质脱水的重量 dry weight
物质文明 material civilization
物质吸收强度 mass absorption intensity
物质形态 physical form
物质性 materiality
物质性蓄积 material accumulation
物质性质 physical property
物质需求 material demand
物质循环 mass cycle;material cycle; substance cycle
物质循环作用 role in cycle of matter
物质因素 physical factor
物质援助 material assistance
物质再生系统 regenerative system of matter
物质在大气中的存在时间 atmospheric life of a substance
物质占优期 matter dominated era
物质张量 matter tensor
物质折旧 physical depreciation
物质证据 physical evidence
物质质量 material mass
物质主义 materialism
物质状态 physique
物质资本 material capital; physical capital
物质资产 physical assets
物质资源 material resources;physical resources
物质自我 material self
物质组成 material composition
物种 species
物种产卵地 spawning bed for species
物种的破坏性开发 destructive exploitation of species
物种的自然环境 national environment of a species
物种地层分带 stratigraphic(al) range of species
物种多样性 species diversity
物种多样性指数 species diversity index
物种反应性 species' reactivity
物种分布 species distribution
物种分布地区 species distribution area
物种丰度 abundance of species;species abundance
物种丰富 abundance of species;species abundance
物种丰富的生物群落 species-rich biomes
物种活度 species' activity
物种间相关估算 interspecies correlation estimation
物种均匀度 species evenness
物种流 species flow
物种描术鉴定要点 diagnosis in species description
物种灭绝 species extinction;eradication
物种浓度 species' concentration
物种起源 origin of species
物种生态学 genecology
物种消失 loss of species
物种形成 speciation
物种资源 species resource
物种组成 species composition
物主 owner
物主记号 owner's mark
物主身份 ownership
物主提供 supplied by the owner
物主提供的 provided by the owner
物资 goods and materials; material; material assets
物资保管 storekeeping
物资采购权 purchasing right of materials
物资储备 material reserves;stock of materials
物资储备定额 material stocking quota
物资储备量 stocking quantity of materials
物资倒流 retrograde materiel action
物资调拨 allocation of materials
物资调拨方案 material allocation scheme
物资调配 material adjustment
物资调用计划 physical planning
物资调运 distribution of material; physical distribution
物资调运的全过程 through distribution
物资调运规划 physical planning
物资调运计划 physical planning
物资调运库 material distribution building
物资调运中心 material distribution center [centre]
物资分配 allocation of materials;distribution of material;material allotment
物资分配制度 material allocation scheme
物资供应 material supply
物资供应成本 material supply cost
物资供应处 material supply department
物资供应船 material supply ship;store ship;supply ship
物资供应管理体制 administrative system of material supply
物资供应计划 material supply plan
物资供应网的联系 connection of the supply networks
物资供应中心 support center [centre]
物资管理 handling of goods and materials;inventory control;inventory management; material management;resource management
物资合理调拨 rational distribution and supply of products
物资回收 material salvage
物资计划 material plan(ning)
物资技术供应代理人 purchasing agent
物资交流 interflow of commodities; material exchange
物资领退制度 material requisition return system
物资流通 flow of material; material flow
物资码头 supply service quay;supply service terminal; supply service wharf
物资目录 material catalog(ue)
物资盘存 material inventory
物资配送 material distribution
物资平衡 material balance
物资平衡表 material balance sheet
物资平衡分配计划 distribution plan of material balance
物资申请计划 material application plan
物资损坏 material damage
物资损失 material damage
物资提前运输 advanced transport of products
物资统计 material statistics
物资统配 unified allocation of materials
物资投放 material release
物资消耗 consumption of materials; material consumption
物资消耗定额 material consumption quota; norm of material consumption
物资需求量 demand quantity of material
物资需要量规划 material requirement programming
物资与供应 material and supply
物资政策 material policy
物资质量发展目标 qualitative materiel development objective
物资资源 physical resources
物资总存量 material inventory

误报 misdeclaration;misrepresentation
误报的火警 false alarm of fire
误报警概率 false alarm probability
误被注销 cancelled in error
误比特概率 bit error probability
误比特率 bit error probability;bit error rate
误拨 misassignment
误拨货车 < 对运货人 > missassignment of cars
误操动作 misoperation
误操作 improper operation;incorrect manipulation;malfunction;Maloperation;mishandling;incorrect switching < 开关 >
误操作故障 mishandling failure
误操作失效 misuse failure
误插 misplug
误查 false drop
误差 aberration;blunder;discrepancy [discrepance];error;fault;inaccuracy;jitter;mistake;oversight;relative accuracy;straggling
误差百分率 error percentage;percentage error
误差百分数 error percentage
误差棒 error bar
误差比 error rate;error ratio
误差变量 error variance
误差标记 error flag
误差表 table of errors
误差补偿 deviation compensation;error compensation
误差补偿器 error compensator
误差测定 error determine
误差测量函数 error measuring function
误差测量环节 error measuring element
误差测量计 error meter
误差测量系统 error measuring system
误差测量装置 error measuring means
误差产生 error production
误差常数 error constant
误差场 error field
误差乘法器 error multiplier
误差抽样控制系统 error sampled control system
误差出现率 rate of occurrence of errors
误差处理 error processing;error treatment
误差传播 error propagation;propagation of errors
误差传播定律 law of error propagation;law of propagation of errors
误差传播率 law of propagation of errors
误差传递 propagation transfer
误差传递函数 error transfer function
误差传感器 error pick-up;error-sensing element
误差传感装置 error sensing device
误差大小 error magnitude
误差带 error band
误差单位 unit of error
误差导出文法 error induced grammar
误差的估计量 estimate of error
误差的消除 elimination of error
误差的主要成分 primary component of error
误差电压 error voltage
误差电压极性 error voltage polarity
误差定律【数】 law of errors;error law
误差定位码 error locating code
误差发送器 error pick-off
误差法 error method
误差范围 error band;error limit;error norm;error range;extent of error;limit of error;range of error
误差方差 error variance
误差方程式 error equation;observation equation
误差方程系数 error equation coefficient
误差方向 direction of error;misalignment
误差方向图 error pattern
误差防护 error protection
误差放大器 error amplifier
误差分布 distribution of errors;error distribution
误差分布定律【数】 law of distribution of errors
误差分布原理 error distribution principle
误差分量 component error
误差分配 error distribution
误差分配码 error distributing code
误差分析 error analysis;erroneous analysis
误差缝 joint for allowance
误差符号 error character;error symbol
误差幅度 error band
误差改正的 error correcting
误差改正电路 error correction circuit
误差概率 error probability;probability of error
误差概率函数 error probability function
误差概率图 error probability diagram
误差更正的伺服机构 error correction servo
误差共振 error resonance
误差估计 error evaluation;estimation of error;error estimate
误差函数 error function
误差函数的补函数 complementary error function
误差函数积分 error function integral
误差函数试验 error function test
误差合成定律 law of combination of errors
误差呼叫 error call
误差恢复 error recovery
误差积分 error integral
误差积分器 error integrator
误差畸变串 error-deformed string
误差畸变链 error deformation string
误差极限 error limit;limit of error;maximum allowable error;margin of error
误差极小 < 99.999999999% > eleven nines
误差计算 error calculation;error calculus
误差计值多项式 error evaluator polynomial
误差记号 error character
误差记录程序 error logger
误差记录器 error register
误差监督程序 error monitor
误差检测 error detection;error inspection
误差检测程序 error detection routine
误差检测代码 error detecting code
误差检测电路 error detector circuit
误差检测电码 error detecting code
误差检测及校正码 error detecting and correcting code;Hamming code
误差检测码 error detection code
误差检测器 error detector
误差检测设备 error detecting facility
误差检测系统 error detection system
误差检测仪 error detector
误差检测与校正 error detection and correction
误差检查 error check;error detecting
误差检出方式 error detecting system
误差检验 error checking;error inspection;error test(ed)
误差检验码 error checking code
误差检验与校正 error checking and correction
误差角 error angle;error bound
误差校验 error check(ing)
误差校验符 error check character
误差校验码 error checking code
误差校验装置 error checking arrangement
误差校正 adjustment of errors;compensating of error;correction of error;error correcting;error correction;error of correction
误差校正操纵台 error correction console
误差校正程序 error correcting program(me)
误差校正(代)码 error correction code
误差校正符号 error correction code
误差校正过程 error recovery procedure
误差校正逻辑 error correction logic
误差校正剖析 error correction profile
误差校正系统 error correcting system
误差校正信息 error checking information
误差校正训练步骤 error correction training procedure
误差校正译码器 error correcting decoder
误差校准 calibrate for error
误差校准器 error corrector device
误差界限 error threshold;margin of error;bounds on error;error bound
误差纠错码组 error correcting code
误差矩阵 error matrix
误差距离 error distance;error span;miss-distance
误差控制 error control
误差控制记号 error control character
误差控制技术 error control technique
误差控制码 error control code
误差控制系统 error actuated system
误差跨度 error span
误差框 error box
误差来源 source of errors;error source
误差类型 error type
误差累积 accumulation of error;error accumulation
误差离散 error variance
误差理论 error theory;theory of error
误差量 extent of the error;margin of error
误差灵敏度 error sensitivity
误差菱形 diamond of error
误差率【数】 error rate;error ratio;rate of deviation
误差率测量设备 error rate measuring equipment
误差敏感度数 error sensitivity number
误差敏感元件 error-sensing element;error sensor
误差模型 error model
误差能量 error energy
误差判据 error criterion
误差配赋 adjustment of errors
误差偏差 error variance
误差平方 error square;square error
误差平方和 error sum of squares
误差平方积分法 integral square-error method
误差平方准则 error squared criterion
误差平行四边形 error parallelogram
误差谱密度 error spectral density;error spectrum density
误差清除 error dump(ing)
误差区间 error burst;error interval
误差曲线 curve of error;deviation curve;error curve
误差曲线图 graph of errors
误差容限 error margin;error tolerance
误差容许限 error allowance
误差三角形 error triangle;triangle of error
误差数据 error information
误差特性 error characteristic;error performance
误差条件 error condition
误差条线 error bar
误差调相 error phasing
误差调整 adjustment of errors;compensation of errors;error adjustment
误差统计估计 statistic(al) estimate of error
误差图解 diagram of errors
误差图形 error figure
误差图样 error pattern
误差椭圆 ellipse of errors;error ellipse;error ellipsoid
误差椭圆单元 error ellipse element
误差系数 error coefficient;error constant
误差限度 error limit;limit of error
误差限制值 values of error limit
误差相对极限 relative limit of error
误差向量 error vector
误差向量值 values of error vector
误差项 error term
误差消除 error cancelling
误差消息 error message
误差效应 error effect
误差协方差 error covariance
误差信号 error signal
误差信号编码器 error signal encoder
误差信号变换器 error transformer
误差信号产生 error signal generation
误差信号发生器 error signal generator
误差信号放大器 error amplifier
误差信号检测 error signal detection
误差信号检测器 error signal detector
误差信号解码器 error signal decoder
误差信号脉冲 error pulse
误差信息 control information;error information;error message
误差修正 error correction
误差修正剖析 error correcting parsing
误差修正剖析器 error correcting parser
误差序列 error sequence
误差选择 error option
误差已消除 errors excepted
误差影响 effect of errors
误差与精度 error and accuracy
误差预算 error budget
误差原因 source of error
误差圆 circle of error;circle of uncertainty;error circle
误差圆半径 error circular radius
误差诊断 error diagnostics
误差振动 jitter
误差指示 error indication
误差指示电路 error indicating circuit
误差指示器 error indicator
误差指示设备 error indication facility
误差指示系统 error indicating system
误差指数律 exponential law of error
误差中断 error interrupt
误差重算 error retry

误差转换 error transformation
误差状态 error state
误差准则 error criterion;error norm
误差自动检测(程序) automatic error detection
误差自动校正 automatic error correction
误差自动校正装置 automatic error request equipment
误差综合(法) error synthesis(method)
误差总改正 total correction of error
误差总校正 total correction of error
误长记录 wrong length record
误称 misnomer
误乘 taking wrong train
误乘列车 take a wrong train
误触发 false firing
误传送 misfeed
误传信号 false call
误错比 error ratio
误导 misorient;misrepresentation
误点的 <汽车火车> overdue
误点列车 overdue train
误调入 false call
误动 false operation
误动作 false action;inadvertent operation;incorrect operation;misoperation
误动作概率 malfunction probability
误读 misread
误发警报 false alarm
误放 misplacement
误符定位子 errata locator
误符计值子 errata evaluator
误付 payment by mistake of fact
误工 loss of working time
误供给 misfeed
误购 buying wrong ticket
误绘 out of drawing
误记 misdescription
误记日期 misdate
误检(索) false drop
误交付 delivery mistake
误接 misconnect(ing)
误解 misapprehension;misconception;misconstruction;misunderstanding
误解事物 mistake of facts
误拒绝 false rejection
误块秒比 errored second ratio
误馈送 misfeed
误录【计】 drop in
误码 error code
误码比 error ratio
误码测试仪 code error tester
误码率 bit error rate;code error rate;error rate
误码秒 errored second
误码增殖 error multiplication
误码增殖因子 error multiplication factor
误码字 error code word
误派 misassignment
误判断 misinterpret;misjudge
误配 mismatch
误配材料 mismatched material
误期 astern of one's reckoning;behind schedule;detention period;overdue
误期表 overdue list
误期船只 overdue vessel
误期的 belated
误期罚款 liquidated damages for delay;penalty for delay
误期费用 cost of delay;demurrage
误期赔偿费 damages for delay;liquidated damages;liquidated damages for delay
误期责任 liability for delay
误燃 false firing
误认 misread
误认信号显示 misread signal aspect;misunderstand signal aspect
误时 out of time
误识别风险 risk of misrecognition
误售 selling wrong ticket
误输送 misfeed
误送 miscarriage
误算 miscalculation;miscount
误调 misadjustment
误调节 misregulation
误调整 misregulation
误跳闸 incorrect trip
误像 false image
误写 clerical error
误卸 mislanding;unloaded error
误选 false drop
误译 misinterpretation
误引 misquotation
误印 misprint
误用 abuse;misapplication;misuse
误运包裹 astray package
误运货票 astray waybill
误运免费送至到达站 <对误运或误卸货物免费运送至到达站> free-astray
误置 misplace
误置缺陷 freckle defect;mislocation defect
误装 misloading
误装卸 wrong handling
误字率 character error rate;word error probability
误组合自防止码 self-demarking code

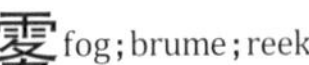
雾 fog;brume;reek

雾岸 fog bank
雾斑 haze;tarnish
雾标 fog buoy;fog signal
雾标志 fog mark
雾笛 fog reed
雾层顶 fog horizon
雾尘 spray dust
雾持续时间 fog duration
雾穿透 haze penetration
雾带 fog belt;mist belt
雾岛效应 fog-island effect
雾道 fog channel
雾的消散 dissipation of fog
雾灯 fog-lamp
雾灯光信号 mist light-signal
雾堤 fog bank
雾滴 cloud drip;fog drip;fog drop;nebulizer drop
雾滴分离器 spray trap
雾滴分离装置 mist separation equipment
雾笛 diaphone;fog horn;fog signal;fog siren;fog whistle;reel horn
雾点 cloud point
雾风 fog wind
雾峰中明亮部分 fog-dog
雾灌 fine-spray irrigation
雾光 <汽车前灯防雾光束> fog-beam;gloss haze
雾害 fog pollution
雾航音响信号 fog alarm;sound signal
雾号 fog horn;fog signal
雾号发声器 fog signal emitter
雾号浮标 fog buoy;horn buoy;trumpet buoy
雾号间歇 silence of fog signal
雾号角 foghorn
雾号声音异常传播 <因通过不同密度大气而产生的差错> mohn effect
雾号台 fog signal station
雾号站 fog signaling plant
雾号自动控制器 automatic fog signal control
雾虹 false white rainbow;fog bow;fogeaster;white rainbow
雾化 aerosolize;atomise [atomize];atomizing;comminution;condensation;fogging;atomization;pulverization
雾化板 pulverizer plate;sprayer plate
雾化程度 degree of atomization
雾化淬火 spray quenching
雾化的 atomizing
雾化度 degree of atomization
雾化法 atomization method
雾化粉化 atomize
雾化粉(末) atomized powder
雾化风机 atomizing air blower
雾化干扰 nebulization interference
雾化干燥 spray drying
雾化干燥剂 atomized drier [dryer]
雾化干燥器 atomized drier [dryer]
雾化剂 aerosol;atomising [atomizing] concentrate
雾化角 spray angle
雾化介质 atomization medium;atomizing medium
雾化空气 atomizing air
雾化沥青封面 fog coat
雾化粒度 atomized particle
雾化粒径 atomized particle size
雾化良好的 well-atomized
雾化铝粉 atomized alumin(i)um powder
雾化膜喷嘴 fan-spray nozzle
雾化黏[粘]度 atomizing viscosity
雾化盘 atomizing disc [disk]
雾化喷射器 spray nozzle
雾化喷头 atomizer;spraying nozzle
雾化喷涂 atomizing spraying
雾化喷油器 oil atomizer
雾化喷嘴 atomization burner;atomizer;atomizing nozzle;spraying apparatus;spraying nozzle
雾化喷嘴套管 atomization nut;atomizing nut
雾化期 atomization period
雾化气 atomizing air
雾化器 atomization device;atomizer;diffuser;micromizer;nebulizer;pulverizer;spraying atomizer;spraying gun;vapo(u)rizer
雾化区 range of atomization
雾化燃料 atomized fuel
雾化燃料喷洒 atomization fuel spray;atomized fuel spray
雾化燃烧器 atomizer burner;atomizing burner;vapo(u)rizing burner
雾化润滑 atomized lubrication
雾化上釉 atomization glazing
雾化设备 atomization plant;atomizing unit
雾化式泵 atomizing pump
雾化式喷嘴 atomized spray injector
雾化式燃烧室 atomizer combustion chamber
雾化室 atomizer chamber
雾化水 atomized water
雾化水喷头 water atomizing nozzle
雾化水喷嘴 atomized water jet;atomized water nozzle
雾化铁粉 Mannesmann iron powder
雾化头 spray(ing) head
雾化尾迹 vapo(u)r trail
雾化细度 atomized particle size;atomizing fineness
雾化效率 nebulization efficiency
雾化性 sprayability
雾化悬浮技术 atomized suspension(al) technique
雾化悬浮体 atomized suspension
雾化悬物燃烧技术 atomized suspension technique of combustion
雾化压力 atomizing pressure
雾化液体 atomized liquid
雾化用蒸汽 atomizing steam
雾化油 atomized oil;fogging oil
雾化运用 atomization application;atomizing application
雾化蒸汽 atomization steam
雾化装置 atomization device;atomization plant;spraying apparatus
雾化状态 spray pattern
雾化锥 atomizer cone;atomizing cone
雾化作用 nebulization
雾灰色 mist grey
雾级标度 fog scale
雾检测器 fog detector
雾角 fog horn
雾角浮标 horn buoy
雾晶 fog crystal
雾警 fog warning
雾警报装置 fog alert system
雾警设备 fog warning apparatus
雾径迹 fog track
雾镜 fog filter
雾喇叭 fog horn;fog trumpet
雾冷堆 fog-cooled reactor
雾冷反应堆 fog reactor
雾量计 fogmeter
雾林 fog forest
雾铃 fog bell;fog gong
雾流 fogging;mist flow
雾锣 fog gong
雾霾 fog haze
雾密度 fog density
雾密度指示器 fog-density indicator
雾沫 entrainment
雾沫分离器 demister(of vehicle window);entrainment separator;entrainment trap
雾沫分离装置 entrainment separation
雾浓的 soupy
雾浓度 fog density
雾浓度指示器 fog-density indicator
雾炮 fog gun
雾喷心 atomizing core
雾喷嘴 atomizing nozzle
雾气 mist;nebula;vapo(u)r
雾气溶液 fog solution
雾气洗涤器 fume scrubber
雾气养护 fog curing
雾汽笛 fog siren
雾情探测灯 fog detecting light;fog detector light
雾情信号 fog signal
雾区 fog belt;fog region
雾日 day with fog;fog(gy) day
雾哨 fog whistle
雾生 fog formation
雾式的 fog-type
雾式喷嘴 fog-type nozzle
雾室 concrete moist curing;damp room;fog cabinet;fog room <混凝土养护的>;humidity chamber;humid room;moist chamber;moist closet;moist(ure) room
雾室养护 fog cure;fog curing;moist curing;fog-room curing
雾室养护的 fog-cured
雾凇【气】 glazed frost;freezing fog;rime;soft rime
雾探测 fog detection
雾天 day with fog;thick weather
雾天灯 fog light
雾天警告信号 fog warning sign
雾天行车灯 fog-lamp
雾天音响信号喇叭 fog horn
雾系数 coefficient of haze
雾消 burn(a);burn-off
雾消散 burn-off;fog dispersal;fog dissipation

雾信号 fog bell;fog gong;fog signal
雾信号复示器 fog repeater
雾信号器 megafog;nautophone
雾信号台 fog signal station
雾信号员 fogman;fog signalman
雾烟 fog dust
雾烟机 fog machine
雾液 spray film
雾用探照灯 mist projector
雾雨 fog drip; fog precipitation; fog rain;fog shower
雾障演替顶极 fog climax
雾遮的 turbid
雾阵 fog bank
雾中航行设备 anti-fog device
雾中升起的满月 fogeaster
雾中拖标 fog buoy;fog spar;position buoy;towing buoy;towing spar
雾中信号笛 diaphone
雾中行车 fogday traffic
雾钟 fog bell
雾状薄膜 cloudy film
雾状带 nepheloid zone
雾状的 nepheloid;vapo(u)rous
雾状封层 flush seat;fog seal coat
雾状封面 <路面> fog seal;fog seal coat
雾状花纹 <钢铁表面燃烧后出现的> damascene
雾状空化 cloud cavitation
雾状空气 atomized air
雾状冷却器 vapo(u)r coolant
雾状两相流动 fog two-phase flow
雾状流 mist flow;spray flow
雾状燃料 pulverised [pulverized] fuel
雾状云 nebulosus cloud
雾浊 blushing
雾阻 fog bound

W

X

夕阳产业 declining industry; sunset-industry
夕阳反照 sun wake
夕照 after glow; after-light

汐 evening tide

西阿拉黄檀 kingwood
西埃列兹柯单位 Ciereszko unit
西岸海洋性气候 west-coast marine climate
西岸气候 west coast climate
西澳大利亚海流 west Australian current
西澳大利亚海盆 west Australian basin
西巴基斯坦水利电力开发局 Water and Power Development Authority of west Pakistan
西班牙白 Spanish white
西班牙彩砖 azulejo
西班牙草 esparto
西班牙草纤维 Spanish grass fiber[fibre]
西班牙港 <特立尼达和多巴哥首都> port-of-Spain
西班牙高哥特式建筑 Spanish high Gothic(style)
西班牙哥特式建筑 Spanish type of Gothic
西班牙黄 Spanish yellow
西班牙货物集装箱运输网 Spanish freight container network
西班牙景天 Spanish stonecrop
西班牙卡达兰的新艺术形式 modernism
西班牙栎 Spanish oak
西班牙栗木 Spanish chestnut
西班牙绿 <浅暗绿色> Spanish green
西班牙摩尔建筑 Hispano-Moresque architecture
西班牙皮革 Spanish leather
西班牙式百叶窗 Spanish blind
西班牙式窗帘 Spanish blind
西班牙式建筑 Spanish architecture; Spanish style
西班牙式庭园 Spanish garden
西班牙式屋面瓦 Spanish roofing tile
西班牙式圆瓦 mission tile
西班牙式遮帘 Spanish blind
西班牙松节油 Spanish oil of turpentine
西班牙桃花心木 Spanish mahogany
西班牙屋顶瓦 mission roofing tile
西班牙香椿 Spanish cedar
西班牙氧化铁红 Spanish red oxide
西班牙赭石 Spanish ocher
西班牙主教堂 capilla mayor
西班牙棕 Spanish brown
西半球 occident; the Occident; western hemisphere
西半球波束 west hemibeam; west hemispheric beams
西北 northwest
西北大风 northwester
西北的 northwestern
西北东 east-northwest[ENW]
西北风 northwester; northwesterly wind; northwest wind
西北光线法 northwest lighting
西北光斜照法 northwest lighting
西北-华北半干旱亚热带 northwest and north China semiarid subtropical zone
西北季风 northwest monsoon
西北角法 <一种线性规划方法> northwest corner method
西北偏北 northwest by north
西北偏西 northwest by west
西北区 northwest China
西北太平洋 northwest pacific ocean
西北西 west-northwest[WNW]
西北蛀船虫 northwest shipworm
西比驳 sea-bee barge; sea-bee lighter
西比船 <海蜂式载驳货船> sea-bee carrier
西比型护岸混凝土块体 seabee unit
西波雷克斯板 Siporex slab
西波雷克斯材料 <一种轻质绝缘材料> Siporex
西波雷克斯墙板 Siporex wall slab
西波雷克斯轻质绝缘集料混凝土 Siporex concrete
西波列蒂堰 Cippolletti weir
西伯电气石 siberite
西伯利亚 Siberia
西伯利亚板块 Siberian plate
西伯利亚大陆冰盖 glacial sheet of Siberia
西伯利亚地台 Siberian platform
西伯利亚地台南缘深断裂系 Southern Marginal deep fracture zone of the Siberian platform
西伯利亚地台碳酸盐岩滩三叶虫地理大区【地】Siberian platform carbonate bank trilobite region
西伯利亚反气旋 Siberian anticyclone
西伯利亚高压 Siberian high
西伯利亚古陆 Siberia paleocontinent
西伯利亚古气候期 Siberia paleoclimate epoch
西伯利亚极性巨带 Siberia polarity hyperzone
西伯利亚极性巨时 Siberia polarity hyperchron
西伯利亚极性巨时间带 Siberia polarity hyperchronzone
西伯利亚冷杉 Siberian fir
西伯利亚落叶松 Siberian larch
西伯利亚五针松 Siberian cedar; Siberian pine
西伯利亚杏 Siberian apricot
西伯利亚鹦鹉螺地理大区 Siberian nautiloid region
西伯利亚有孔虫地理大区 Siberian foraminiferal region
西伯利亚云杉 Siberian spruce
西伯利亚植物地理大区 Siberian floral region
西伯特三相电弧炉 Siebert furnace
西伯值 Sieber number
西博氏胡桃 Japan walnut
西部 westward
西部刺桐 western coralbean
西部房屋正面 western facade
西部分等(货运) <美> Western Classification
西部幅度差异性活动区 differential intensive active area in West China
西部红杉 Shinglewood
西部红雪杉 western red cedar
西部黄松 <美国和加拿大的> western pine
西部建筑翼部 western transept
西部落叶松 <美> western larch; western tamarack
西部门廊 western porch
西部砌砖法 western method
西部润滑油 <美国加利福尼亚润滑油> western lubricating oil
西部山地白松 <美> western white pine
西部山地森林土区 the west mountain forest soil area
西部山墙 western pediment
西部式框架 western framing
西部塔楼 western tower
西部铁杉 western hemlock
西部外廊 western gallery
西部香脂冷杉 western grand fir; western white fir
西部云杉 Western spruce
西部柱廊上三角形或弧形檐饰 western pediment
西部走廊 western gallery
西餐馆 European restaurant; western food restaurant
西藏柏木 Himalayan cypress
西藏红杉 Himalaya larch
西藏式(建筑)Tibetan style
西侧 western side
西侧门座 western portal
西侧墙 west-facing wall
西侧入口 western portal
西赤道逆流 Western Equatorial countercurrent
西窗 west-facing window
西大距 western elongation
西德船级社 Germanischer Lloyds
西点【天】west point(of the horizon)
西端 <中世纪教堂的> west end
西端唱诗班席位 western quire
西端(多层)廊屋 <德国与荷兰的塔楼教堂> westblock
西端塔楼 west end tower
西尔艾洛伊铋合金 Sealalloy
西尔弗莱克斯镀锌法 Silflex process
西尔福斯铜银合金 Silfos
西尔玛雷克铝硅镁合金 Silmalec
西尔式带填充丝钢丝绳 Seale filler wire rope
西尔斯比定律 Silsbee rule
西尔斯比效应 Silsbee effect
西尔瓦灵顿混合式钢丝绳 Seale Warrington wire rope
西尔威斯特工艺 Sylvester process
西尔(旋转圆筒式)黏[粘]度计 Searle visco(si)meter
西范托风 Siffanto
西方 the west; westward
西方拜占庭(建筑)风格 western Byzantine style
西方拜占庭(建筑)格式 western Byzantine style
西方侧柏 northern white cedar
西方鹅耳枥 Common hornbeam
西方风格 western style
西方古典园林 western classical garden
西方国家 the Occident
西方红杉木瓦 royal
西方桦 black birch; western birch
西方桧柏 western cedar; western juniper
西方建筑 western architecture
西方教堂唱诗班席位 western choir
西方落叶松 hackmatack
西方人 occidental; the Occident
西方式房屋正面 western facade
西方位标 west cardinal mark
西方文化的 occidental
西方文明 occident; the Occident
西非橄仁树 frake
西非国家经济共同体 Economic Community of West African States
西非合欢木 African walnut
西非花梨木 African rosewood
西非黄麻 West African jute
西非黄檀木 bubinga
西非吉纳树胶 W. African kino
西非青龙木 African rosewood
西非桃花心木 cababa
西非天然沥青 libolite
西非乌木 West African ebony
西非樱桃木 African cherry
西非荚檀木 African rosewood
西非硬木 <其中的一种> denya
西非(硬)树脂 west African Copal
西非洲酒椰纤维 West African bass
西非紫檀 West African padauk
西菲克斯定位系统 Sea-fix system
西风 wester(ly); west wind
西风波 wave in westerlies; westerly wave
西风槽 trough in westerlies; westerly trough
西风吹流 westerly drift current; west wind drift
西风带 band of westerlies; prevailing westerlies; subpolar westerlies; subtropic (al) westerlies; temperate westerlies; westerly belt; west wind belt; zonal westerlies; zonal wind; westerlies
西风多雨带 westerlies rain belt
西风漂流 west wind drift
西风气流 westerly current
西风型 westerly type
西夫青铜 Sifbronze
西弗吉尼亚 <美国州名> West Virginia
西弗吉尼亚红云杉 West Virginia Spruce
西戈壁盆地 west Gobi basin
西格巴恩标志 siegbahn notation
西格测温锥 pyrometric cone
西格玛焊接 sigma welding
西格玛桨式混合机 Sigma blade mixer
西格玛铝基合金 sigmalium alloy
西格玛相 sigma phase
西格玛叶片式捏合机 Sigma blade kneader
西格玛因子 sigma factor
西格三角锥耐火板 <高温测量用> pyrometric Seger cone equivalent
西格示温锥 pyrometric cone
西格瓦特式 Siegwart
西格瓦特式楼板 <一种预制空心钢筋混凝土板楼盖> Siegwart floor
西根阶 <晚泥盆世>【地】Siegenian (stage)
西谷椰子树属【植】sago palm
西瓜皮绿 water-melon green
西红粉 ferric oxide
西红柿 love apple
西加云杉 silver spruce; sitka spruce
西郊 western suburbs
西界标 west cardinal mark
西进 westing
西经 west longitude(of Greenwich)
西距 westing
西距角 west elongation; western elongation
西科板 <一种芯材为黏[粘]结碎木的胶合板> Syncore
西科斯温度计 Six's thermometer
西兰阶【地】Seelandian(stage)
西乐尼丝 <商品名> Celanese
西勒维斯特惯性律 Sylvester's law of inertia
西勒维斯特零性律 Sylvester's law of nullity
西雷科 <一种锻造镁合金> Selektron
西立面 west elevation
西隆酮 <一种水泥液体硬化剂> Sealontone
西露蒂课题 Cerruti problem

西罗防水剂 Sealocrete
西罗克灌浆材料 <地基处理用> Siroc grout
西罗克净 <一种混凝土表面清洁剂> Sealoclean
西马特砂滤池 Simater sand filter
西美拉尼西亚海沟 west Melanesian trench
西门子 <电导单位> Siemens
西门子电阻单位 Siemens unit
西门子法 Siemens Method
西门子格子体 Siemens straight packing
西门子功率计 Siemens dynamometer
西门子静电伏特计 Siemens static voltmeter
西门子炉 Siemens furnace
西门子-马丁钢 Siemens-Martin steel; open-hearth steel
西门子-马丁平炉 Siemens-Martin furnace
西门子煤气发生炉 Siemens producer
西门子平炉钢 Siemens steel
西门子式半自动闭塞 Siemens block
西门子式电功率计 Siemens electrodynamometer
西门子式格子砖 Siemens setting checker
西门子锌基轴承合金 Siemens alloy
西蒙・古德温图表 Simon-Goodwin charts
西蒙得木蜡 Jojoba wax
西蒙扩散 Simon diffusion
西蒙尼公式 Simonys' formula
西蒙森现象 Simonsen phenomenon
西蒙石 Simonellite
西蒙斯型破碎机 Symons crusher
西蒙斯型圆锥破碎机 Symons cone crusher
西蒙钻进理论 Simon's drilling theory
西面 western side
西面墙 west-facing wall
西姆卡曝气器 Simcar aerator
西姆林清操纵法 Heimlich maneuver
西内部盆地 west inland basin
西内双壳类繁盛中心 western interior bivalve endemic center[centre]
西南 southwest[SW]
西南成矿区 <中国> southwestern China metallogenetic province
西南大风 sou's wester; southwester; sou-wester
西南的 southwestern
西南方向的 southwestward
西南风 southwester; southwesterly wind; southwest wind
西南高地 southwest China upland
西南季风 southwest monsoon
西南偏南 southwest by south
西南偏西 southwest by west
西南区 southwest China
西南日本内带 inner zone of southwest Japan
西南日本外带 outer zone of southwest Japan
西南塔里木盆地坳陷区 southwest Tarim downwarping region
西南太平洋 Southwest Pacific Ocean
西南西 west southwest
西南亚雪杉 cedar of Lebanon; Lebanon cedar
西欧共同市场 European Common Market
西欧经济合作与发展组织 Organization for Economic Cooperation and Development
西欧联盟 <北大西洋公约组织> Western European Union
西欧内河航道分级标准 West-European inland waterway classification scale
西欧内河流分级标准 West-European inland waterway classification scale
西欧铁路定期客票 Eurailpass
西欧铁路机车车辆投资公司 European Company for the Financing of Railway Rolling Stock
西硼钙石 sibirskite
西偏北 west by north
西偏南 west by south
西墙 west wall
西撒哈拉 West Sahara
西萨摩亚 Western Samoa
西沙尔地毯 sisal carpeting
西沙尔麻 sisal hemp
西沙尔麻加强的 sisal reinforced
西沙尔麻绳 sisal hemp rope
西沙尔麻索 sisal rope
西沙尔麻织物 sisal fiber[fibre]
西沙尔牛皮纸 sisal kraft paper
西沙尔纤维 sisal fiber[fibre]
西斯丁学派 <11 世纪法国早期哥特式建筑的> Ecole Cistercienne
西太平洋 West Pacific Ocean
西太平洋岛弧毕乌夫带深断裂系 deep fractural zone representing the Benioff zone of the West Pacific Island arcs
西太平洋岛弧地槽褶皱区 geosynclinal fold region of Island Arcs of the western pacific
西太平洋地槽系 Western Pacific geosyncline system
西太平洋活动区 active area in West Pacific
西太平洋软体动物地理区系 west Pacific molluscan realm
西锑砷铜铅石 theisite
西烃石 Simonellite
西图廓 left-hand border; left-hand edge; westerly limit
西妥僧侣修道会哥特式建筑 Cistercian Gothic(style)
西妥僧侣修道会教堂 Cistercian church
西妥僧侣修道院 Cistercian abbey; Cistercian monastery
西妥僧侣修道院会堂 Cistercian abbey-house
西瓦度 <一种木材防腐剂> Sylvadure
西屋 <仿罗马式教堂的> westwork
西屋电弧焊 <惰性气体保护金属极弧焊> Westing-arc welding
西武式护舷 fender Seibu
西西北 west northwest
西西伯利亚盆地 west Siberian basin
西西里大理石 <意大利> Sicilian marble
西西里漆树 Rhus coriaria; Sicilian sumac
西西里蔷薇 Sicilian rose
西西南 west southwest
西喜马拉雅构造段 Western Himalaya tectonic segment
西下的 westering
西向 west orientation
西向漂移速率 westward drift rate
西新地岛海槽 West Wovaya Zemlya trough
西行的 westbound
西行航程 westing
西行航线 westward course
西行列车 westbound train
西雅图港 <美> Seattle Port
西雅图-温哥华-日本航线 Seattle-Vancouver-Japan route
西亚建筑 West Asian architecture
西洋丁香 Syringa vulgaris
西洋红 crimson lake
西洋建筑史 history of Western architecture
西洋接骨木 European elder
西洋杉板条顶棚 cedar slatted ceiling
西印度毒漆树 Rhus metopium
西印度红木 carapa; carapa oil; West Indian mahogany
西印度黄花椒木 West Indian satinwood
西印度群岛建筑 West Indies architecture
西印度杉木 Havana wood
西印度乌木 West Indian ebony
西印度洋 West Indian Ocean
西中大平洋渔业委员会 Western Central Atlantic Fishing Commission

吸

吸铵量 ammonium absorption
吸板 suction disc; suction plate
吸瓣 suction clack
吸杯 suction cup
吸泵 suction pump
吸槽 suction cell
吸槽式溢洪道 suction-slot spillway
吸潮 absorbed moisture; absorption of moisture; humidity absorption; moisture absorption
吸潮器 <变压器用> air-breather; breather
吸潮试验 porosity test
吸潮性 absorptivity; hygroscopicity
吸尘 control of dust; dust absorption; dust arrest(ment); dust collecting; dust collection; dust exhaust; dust extraction; dust suction; Hoover; lade; sweep-up
吸尘袋 dust bag; dust-collecting sleeve
吸尘电极 collecting electrode; collector electrode
吸尘管 sweep-up pipe
吸尘机 suction machine
吸尘凝集器 aspiration condenser
吸尘器 aspirator; cleaner; dirt collector; dust arrester[arrestor]; dust catcher; dust-collecting fan; dust collector; dust concentrator; dust exhauster; dust extractor; dusting brush; dust precipitator; dust suctor; exhaustor; suction cleaner; suction plant; suction sweeper
吸尘器电源插座 power point for cleaners
吸尘清洁 suction cleaning
吸尘设备 dust exhaust plant
吸尘式抛光机 dust absorption polishing machine
吸尘式清扫车 suction sweeper
吸尘系统 dust-collecting system; dust suction system
吸尘箱 dust bin
吸尘罩 dust-collecting hood; dust exhaust hood; dust hood; suction cap; suction hood
吸尘装置 aspiration; dust arrester plant; dust exhaust apparatus; dust-separation equipment; dust suction installation; suction device; suction plant
吸尘作用 aspiration
吸持 holding; locking; caging; confinement; sticking
吸持棒 holding bar
吸持沉降 adhesion settling
吸持磁铁 holding magnet; locking magnet
吸持电磁铁 portative electromagnet; pulley magnet; holding magnet
吸持电流 retaining current
吸持电路 holding circuit; locking circuit
吸持电压 holding voltage
吸持功率 holding power
吸持继电器 holding relay; sticking relay
吸持力 holding force
吸持螺线圈 sucking solenoid
吸持塞孔 latch retainer
吸持时间 <继电器的> holding period
吸持位置 <继电器的> holding position; pull-up position
吸持衔铁 holding armature
吸持线圈 holding(-on) coil; holding out coil; hold on coil; retaining coil
吸持线圈导线 locked coil conductor
吸持装置 holding device
吸抽 aspiration
吸抽加压泵 sucking and forcing pump; sucking pump
吸出 aspirate; aspiration; drain off
吸出阀 aspirating valve; bleeder valve
吸出高度 draft height; draught height; suction head; static draft head <水轮机的>
吸出管 draft pipe; draft tube
吸出管损失 draft tube loss
吸出集流管 suction manifold
吸出孔 suction eye
吸出喇叭管外缘 suction umbrella
吸出率 rate of draft
吸出落差 draft head; draught head
吸出器 aspirator
吸出式通风 exhausting-type ventilation; exhaust system of ventilation
吸出式通风方式 portal-to-exhaust shaft; suction ventilation mode
吸出式凿岩 suction drilling
吸出室 suction chamber
吸出术 exsuccation; exsuction; suction extraction
吸出水头 draft head; draught head
吸出物 aspirate
吸出性能 suction performance
吸除式附面层控制 suction boundary layer control
吸除系统压气机 suction compressor
吸除系统压缩机 suction compressor
吸除作用 resporption
吸粗 rude respiration
吸导轮 suction guide wheel
吸电子 electron-withdrawing
吸电子的 electron-attracting; electrophilic
吸垫 mat
吸掉 removal by suction
吸顶暗灯 flush light
吸顶灯 ceiling light; ceiling mounted lamp; coffered light panel
吸顶灯具 recessed fixture; surface-mounted luminaire
吸顶小灯 downlight
吸动电流 operate current; operating current; working current
吸动电平 keying level
吸动时间 operate time; pull-out time; reacting time
吸动仪表 suction operated instrument
吸阀控制杆 air lever
吸风 indraft; induced draft
吸风道 suction air duct
吸风风扇 inlet fan; suction fan
吸风管 aspirating air pipe; aspiration channel; uptake
吸风管道 aspiration channel
吸风管线 <隧道通风系统中的> foul air duct
吸风机 draft fan; exhaust blower; exhauster; suction blower; suction

fan; suction ventilator; induced draft fan
吸风口 air suction intake;exhaust inlet;exhaust opening;inlet scoop
吸风冷却塔 induced-draft cooling tower
吸风器 air catcher
吸风烧结 down-draft sintering
吸风烧结机 down-draft sintering machine
吸风式牵拉 suction take-down
吸风式清洁装置 pneumatic suction cleaning system
吸风式散热风扇 suction type radiator fan
吸风水冷却器 induced draught water cooler
吸风塔 aspiration leg
吸风罩 suspended hood
吸风锥 suction cone
吸附 adsorption;adsorb;sticking
吸附板 adsorption plate
吸附泵 adsorbent pump; sorption pump
吸附泵送 sorption pumping
吸附比 adsorption ratio;suction ratio
吸附变化传感器 sorption-deformation sensor
吸附表面 adsorption surface;adsorptive surface
吸附表面积 adsorption surface area
吸附表面膜 adsorbed surface film
吸附波 adsorption wave
吸附薄膜 adsorptive hull
吸附材料 adsorbed material;adsorbed matter; sorbed material; sorbing material
吸附槽 adsorption tank
吸附层 ad-layer;adsorbed layer;adsorbing layer;adsorption layer;sorbate layer; sorbed layer; sorbing layer
吸附层厚度 thickness of the adsorbed layer
吸附层析 adsorption chromatography
吸附差示脉冲伏安法 adsorption differential pulse voltammetry
吸附常数 adsorption constant
吸附沉淀 adsorption precipitation;adsorptive precipitation
吸附沉淀剂 adsorption precipitant
吸附澄清剂 adsorption clarifier
吸附池 adsorption tank
吸附处理法 adsorption process
吸附床 adsorption bed
吸附催化 adsorption catalysis
吸附催化剂 adsorptive catalyst
吸附萃取法 sorption-extraction
吸附带 adsorption band;suction belt
吸附导体阴极 adconductor cathode
吸附的 adsorbed; adsorbent; adsorptive;sorptive;suctorial
吸附的可逆性 reversibility of adsorption
吸附的阳离子 adsorbed cation
吸附的正离子 adsorbed cation
吸附等容线 adsorption isostere
吸附等温等势线 adsorption isotherm isopotential
吸附等温方程 adsorption isotherm equation
吸附等温式 isothermic equation for adsorption
吸附等温线 adsorbed isotherm; adsorption isotherm;suction isotherm
吸附等压线 adsorption isobar
吸附点 adsorption point; sorption point
吸附电荷 adsorbed charge
吸附电流 adsorption current
吸附电位 adsorption potential
吸附定位 adsorbed orientation
吸附氡法 adsorption radon technique
吸附氡时间 adsorption radon time
吸附动力学 adsorption dynamics;adsorption kinetics;sorption kinetics
吸附动力学方程 adsorption kinetics equation
吸附动力学曲线 adsorption kinetics curve
吸附度 adsorptivity; degree of adsorption
吸附段 adsorption section
吸附法 adsorption method;adsorptive process
吸附法脱氮 adsorption denitrification
吸附反应 adsorption reaction; sorption reaction
吸附反应器 adsorption reactor
吸附方法 adsorption method
吸附方式 adsorption pattern
吸附非方程式 adsorption pattern
吸附分光计 adsorption spectrometer
吸附分离 adsorption separation;fractionation by adsorption
吸附分析 adsorption analysis; sorption analysis
吸附分子 adsorbed molecule
吸附分子接合 adsorbed molecule bind
吸附粉末 adsorbing powder;adsorptive powder
吸附负荷曲线 adsorption load curve;adsorptive load curve
吸附改性剂 adsorbent modifier
吸附干燥过程 adsorption drying process
吸附公式 adsorption equation
吸附固体-溶液界面 adsorptive solid-solution interface
吸附罐 adsorption tank
吸附过程 adsorption process; adsorbent process
吸附过滤 adsorption filtration
吸附过滤介质 adsorbent filtering medium
吸附含水量 adsorption water content
吸附焓 enthalpy of adsorption
吸附化合物 adsorption compound
吸附活度 adsorption activity
吸附机理 adsorption mechanism
吸附基 adsorbing group
吸附及脱附平衡 adsorption-desorption equilibrium
吸附及脱附周期 adsorption-desorption cycle
吸附计 adsorption ga(u)ge
吸附剂 adsorbentia; adsorbing substance; adsorbent; adsorber; adsorbing agent; adsorbing material; adsorbing substance; adsorption agent; adsorptive agent; sorbed agent; sorbent; sorbing agent; sorption agent
吸附剂比表面积 specific surface area of adsorbent
吸附剂表面积 adsorbent surface area
吸附剂床 adsorbent bed
吸附剂过滤介质 adsorbent filtering medium
吸附剂过滤器 adsorbent filter
吸附剂活性 adsorbent activity
吸附剂剂量 adsorbent dose
吸附剂粒子 adsorbent particle
吸附剂粒子表面 surface of adsorbent particles
吸附剂膜系统 adsorbent-membrane system
吸附剂浓度 adsorbent concentration
吸附剂气固色谱 active solid
吸附剂吸附质对 adsorbent-adsorbate pair
吸附剂相 adsorbent phase
吸附剂效率 efficiency of the adsorbent
吸附剂性能 performance of the adsorbent
吸附剂再生 adsorbent regeneration; regeneration of adsorbent
吸附剂质量 mass of adsorbent
吸附减活剂 adsorbent deactivator
吸附检测器 adsorbed detector
吸附件 suction attachment
吸附胶体浮选 adsorbing colloid flo(a)tation
吸附胶团 admicelle
吸附/解吸作用 sorption/desorption
吸附介质 adsorptive media
吸附精制 adsorption refining;refining with adsorbents
吸附精制吸附速率 adsorption rate
吸附阱 adsorption trap
吸附壳 adsorptive hull
吸附空间 adsorption space
吸附离子 adion;adsorbed ion
吸附理论 adsorption theory
吸附力 adsorption affinity;adsorption power; adsorptive attraction; adsorptive force; power of adsorption;sorptive attraction
吸附力场 adsorptive force field
吸附量 adsorbance;adsorbed amount; adsorbing capacity; adsorption amount
吸附磷 adsorbed phosphorus
吸附滤池 adsorbent filter
吸附滤器 adsorbent filter
吸附率 adsorptivity
吸附密度 adsorption density;sorption density
吸附面 adsorption plane
吸附模拟 adsorption modeling
吸附模式 adsorption model
吸附膜 adsorbed film; adsorbed film of water; adsorption film; adsorption film of water;adsorption membrane;sorbed film; sorbed film of water;sorption film of water
吸附能 adsorbed energy
吸附能力 adsorbability; adsorption capacity; adsorptive capacity; adsorptive power; power of adsorption;sorption capacity
吸附能量 adsorption energy
吸附黏[粘]土 adsorbing clay
吸附浓度 adsorbent concentration; adsorbed concentration
吸附平衡 adsorption equilibrium
吸附平衡常数 adsorption equilibrium constant; adsorptive equilibrium constant
吸附平衡常数分布 distribution of adsorptive equilibrium constant
吸附平衡理论 adsorption equilibrium theory
吸附平衡时间 adsorption equilibrium time
吸附谱带 adsorption band
吸附气泡分离法 adsorption bubble separation
吸附气泡技术 adsorptive bubble technique
吸附气提 adsorption stripping
吸附气体 adsorbed gas; adsorption gas;sorbed gas
吸附气体法 gas adsorption method
吸附气相色谱法 adsorption gas chromatography
吸附汽油 adsorption gasoline
吸附器 adsorber
吸附器面积 area of adsorption device
吸附强度 adsorption strength; sorption strength
吸附区 adsorption zone; adsorptive region
吸附区高度 height of adsorption zone
吸附曲线 adsorption curve
吸附热 adsorption heat; heat of adsorption
吸附热检测器 adsorption heat detector;heat of adsorption detector
吸附热力学 adsorption thermodynamics
吸附容量 adsorption capacitive;adsorptive capacity;sorption capacity
吸附容量相关 adsorption correlation
吸附容量指数 adsorption capacity index
吸附溶质 adsorbed solute
吸附色层分离法 adsorption chromatography
吸附色谱法 adsorption chromatography
吸附色谱分离(法) adsorption chromatography
吸附设备 adsorbent aggregate
吸附生物降解法 adsorption biodegradation process
吸附生物降解活性污泥法 adsorption biodegradation activated sludge method
吸附生物氧化法 adsorption bio-oxidation process
吸附湿度表 adsorption hygrometer
吸附时间 adsorption time;adsorptive time
吸附式减湿器 sorbent dehumidifier
吸附式减湿装置 adsorption dehumidifier
吸附式净化 adsorption cleaning
吸附式制冷 adsorption refrigeration
吸附式制冷系统 adsorption refrigeration system
吸附式制冷循环 adsorption refrigeration cycle
吸附势 adsorption potential
吸附势理论 adsorption potential theory;adsorptive potential theory
吸附试验 adsorption test
吸附适用性 adsorption applicability
吸附术语 adsorption terminology
吸附树脂 adsorbent resin;adsorption resin; adsorptive resin; polymeric adsorbent
吸附衰减稳态 adsorption-decay steady state
吸附水 adsorbed moisture;adsorption water; adherent water; adhesive water;adsorbed water;adsorption-moisture; adsorptive water; attached water;bound water;hydration water; hydroscopic water; pellicular water;planar water
吸附水分 adsorbed moisture; entrapped moisture
吸附水分校正系数 hygroscopic moisture correction factor
吸附水膜 adsorbed film of water; capillary film
吸附速度 adsorption velocity
吸附速度理论 adsorption velocity theory
吸附速度系数 coefficient of adsorption velocity
吸附速率 rate of adsorption;adsorption rate
吸附速率常数 adsorption rate constant
吸附速率理论 adsorption rate theory
吸附塔 adsorption column;adsorption

X

tower
吸附特性 characterization of (ad)-sorption
吸附特征 adsorption characteristic; sorption characteristic
吸附梯度 adsorbent gradient
吸附提取 sorption-extraction
吸附体 adsorbent
吸附体积 adsorption volume
吸附天平 adsorption balance
吸附天然有机质 adsorbed natural organic matter
吸附推动力 adsorptive thrust force
吸附脱苯 debenzolization of adsorber
吸附脱臭 adsorption deodo(u)rizing
吸附脱附再生再吸附 adsorption-desorption-regeneration-readsorption
吸附瓦斯 adsorption gas
吸附位置 adsorption site
吸附温度 adsorbent temperature
吸附物 adsorbate; adsorptive matter; sorbate
吸附物密度 adsorbate density
吸附物体积 adsorbate volume
吸附物质 adsorbed substance; adsorbent; adsorption substance; sorbed mass
吸附物种 adsorbed species
吸附物种活度系数 activity coefficient of adsorbed species
吸附洗提 adsorption elution
吸附系数 adsorption coefficient; coefficient of adsorption
吸附系统 adsorption system; adsorptive system
吸附现象 adsorption phenomenon; sorption phenomenon
吸附限 adsorption edge
吸附相 adsorbed phase
吸附相浓度 adsorbed phase concentration
吸附效率 adsorption efficiency
吸附效应 adsorption effect
吸附型钝化剂 adsorption inhibitor
吸附型滤池 adsorbent type filter; adsorptive type filter
吸附型滤器 adsorbent type filter
吸附型抑制剂 adsorption inhibitor
吸附性 adsorbability; adsorbency; adsorptivity
吸附性材料 adsorbability material; adsorptive material; sorptive material
吸附性复合体 adsorption complex
吸附性过滤 adsorptive filtration
吸附性过滤器 adsorptive filter
吸附性钠 adsorbed sodium
吸附性能 adsorbing performance; adsorption performance; adsorption property; adsorptive performance; adsorptive property
吸附性阳离子 adsorbed cation
吸附性载体 adsorptive support
吸附选择 adsorbent selection; adsorption selection
吸附选择性 adsorption selectivity
吸附循环 adsorption cycle
吸附压力 adsorbent pressure; adsorption pressure
吸附氩 adsorbed argon
吸附厌氧序批间歇式反应器 adsorption-anaerobic sequencing batch reactor
吸附引力 adsorptive attraction
吸附油 adsorbed oil
吸附原理 adsorption principle
吸附原子 adatom
吸附杂质 adsorbing impurity
吸附再生法 adsorption regeneration process; biosorption process; contact stabilization process
吸附再生活性污泥法 adsorption regeneration activated sludge process
吸附再生曝气 adsorption regeneration aeration
吸附障 adsorption barrier
吸附罩 adsorption trap; adsorptive hull
吸附值 adsorptive value
吸附指示剂 adsorption indicator
吸附指数 adsorption index; adsorption exponent
吸附制冷 adsorption refrigeration
吸附质 adsorbate; sorbate
吸附质分子 adsorbate molecule
吸附质分子分布 distribution of adsorbate molecule
吸附质活度 adsorbate activity
吸附质类型 type of adsorbates
吸附质累积 adsorbate accumulation
吸附质量 amount of adsorbate
吸附质浓度 adsorbate concentration
吸附置换 adsorptive displacement
吸附周期 adsorption cycle
吸附柱 adsorbing column; adsorption column
吸附柱色谱法 adsorption column chromatography
吸附装置 adsorbent equipment; adsorber; adsorption equipment
吸附状态 adsorbed state
吸附着水分校正系数 hygrocsopic moisture correction factor
吸附组分迁移距离 transfer distance of adsorption component
吸附作用 adsorption; adsorptive action
吸干 blotting
吸杆 sucker rod
吸杆扳手 sucker rod wrench
吸杆钩 sucker rod hook
吸杆蜡 sucker rod wax
吸杆系统 system of span pieces
吸汞分析 mercury intrusion analysis
吸垢 dirt pickup
吸管 pipet(te); sucker; sucking pipe; sucking tube; suction attachment; suction conduit; suction pipe; suction tube; withdrawal tube
吸管法 pipet(te) method
吸管分析法 pipet(te) analysis
吸管盒 pipet(te) box
吸管架 pipet(te) stand; pipet(te) support; suction tube ladder
吸管接头 pipet(te) connection; pipet(te) joint; suction attachment
吸管滤网 suction basket
吸管脉冲 suction pulsation
吸管式 hydrocone type
吸管式比重计 syringe hydrometer
吸管式风速表 suction anemometer
吸管式挖泥船 suction dredger
吸管损失 suction pipe loss
吸管托架 suction pipe bracket
吸管弯头 suction bend(ing)
吸管位置监控 suction tube position monitoring
吸管洗涤吹干橡皮球 pipet(te) blower
吸管橡胶套 reinforced rubber suction hose; suction sleeve
吸管消毒器 pipet(te) sterilizer
吸管自动控制器 automatic suction tube controller
吸罐 cucurbitula; cupping glass
吸光 <无光漆表面明确不均> ghosting
吸光百分率 percent absorption
吸光材料 light absorbent
吸光测定法 absorptionmetry; absorption photometry
吸光度 absorbence [absorbency]; extinction
吸光度比值导数分光光度法 ratio spectrum-derivative spectrophotometry
吸光分析 absorptionmetric analysis; absorptionmetry
吸光光度法 optic(al) method
吸光计 absorptiometer
吸光率 absorbance; absorptance
吸光升降机 sucker rod elevator
吸光系数 absorptivity; extinction coefficient; specific extinction coefficient
吸光性 light absorbency
吸光眼镜玻璃 absorbing spectacle glass
吸光指数 absorbance[absorbancy] index
吸合 pull-in
吸合电流 actuating current
吸合时间 pick-up time; response time
吸合位置 operating position
吸合值 pull-on value
吸红试验 dye adsorption test; fuchsin(e) test; ink test
吸红外和紫外高硅氧玻璃 infrared and ultraviolet adsorbing Vycor glass
吸红外(线)玻璃 infrared ray absorbent glass
吸灰泵 ash pump
吸回效应 resorptive effect
吸混作用 persorption
吸积 accretion
吸积假说 accretion hypothesis
吸积理论 accretion theory
吸积盘 accretion disc[disk]
吸钾量 volume of potassium absorption
吸浆量 absorption mud; acceptance of grout; grout acceptance
吸浆率 absorption rate; rate of adsorption grout
吸浆能力 acceptance of grout
吸浆损失 loss of returns
吸浆堰板 nozzle type slice
吸胶布 bleeder cloth
吸进凸轮 admission cam; inlet(-valve) cam
吸进氧气 take in oxygen
吸空气 suck
吸空气装置 respiratory device
吸口 mouthpiece; suction port
吸口衬环 suction mouth liner; suction mouth wearing ring
吸口关闭阀 suction inlet shutoff valve
吸口深度计 suction mouth level indicator
吸口深度指示仪 suction depth indicator
吸蜡炉 wax-absorbing furnace
吸蓝量 methylene blue absorption
吸浪器 wave absorber
吸力 force of attraction; portative force; suction; suction force; suction pressure
吸力泵 <地下连续墙> suction pump
吸力泵井 well pumped by suction
吸力侧 suction side
吸力冲洗 suction washing
吸力冲洗器 suction washer
吸力反循环钻井法 portadrill reverse circulation method
吸力计 suction ga(u)ge
吸力加油 suction oiling
吸力进给杯 suction feed cup
吸力密封 suction seal
吸力面 suction surface
吸力排水 drainage by suction
吸力强度 suction strength
吸力输送器 suction conveyer[conveyor]
吸力水头 suction water head
吸力水柱(高度) suction water column
吸力特性曲线 <继电器> pull curve
吸力头 suction hose
吸力挖泥机 pump dredge(r)
吸力洗净器 suction washer
吸力效应 suction effect
吸力漩涡 suction eddy
吸力运输机 suction conveyer[conveyor]
吸力钻井法 portadrill method
吸利管架 pipet(te) support
吸粮机 marine elevator; pneumatic grain elevator; pneumatic grain handling machine
吸粮机橡皮管头上的嘴 camel
吸量管 pipet(te); plunging siphon; valinche
吸料模 gathering mo(u)ld
吸料模的钢套 nose bushing
吸料头 suction head
吸流变压器 <交流电气化铁道系统中> booster transformer
吸流变压器供电方式 booster transformer feeding system
吸留 occlusion
吸留化合物 occlusion compound
吸留气体 occluded gas
吸留水 occluded water; occlusion water
吸留物 occlusion
吸硫 sulfuring; sulphuring
吸滤漏斗 filter funnel
吸滤瓶 filter(ing) flask; suction bottle; suction flash
吸滤器 nutsch[filter]; suction filter; suction strainer; vacuum filter
吸滤纸 blotting paper
吸滤作用 imbibition
吸氯特性 chlorine absorptive property
吸氯性 chlorine absorbability
吸棉嘴 pick-up capturer
吸磨机 suction mill
吸墨粉 anti-setoff powder; pounce
吸墨试验 blotting test
吸墨水试验 ink test
吸墨水纸 blotter
吸墨水纸滚台 blotting pad
吸墨性 absorbence[absorbency]; ink receptivity
吸墨用具 blotter
吸墨纸 blotter; blotting paper
吸纳成本法 absorption costing
吸能边界 energy absorbing boundary
吸能材料 energy absorbing material
吸能代谢反应 endogenic reaction
吸能的 endoergic
吸能反应 <需要输入能量的生化反应> endergonic reaction
吸能构件 energy absorbing element
吸能过程 endoergic process
吸能能力 energy absorption capacity
吸能器 power absorber
吸能潜力 energy adsorption potential
吸能容量 energy absorbing capacity
吸能元件 power-absorbing; power-absorbing element
吸能装置 energy absorber
吸泥泵 dredge pump; dredging pump; excavating pump; hydraulic excavator; mud (suction) pump; scum pump; sludger; diaphragm pump; mud sucker
吸泥泵额定功率 dredge pump capacity
吸泥泵流量 dredge pump flow rate

X

吸泥采样 dredge sampling
吸泥船 drag boat; hydraulic suction dredge(r); pump dredge(r); sand pump dredge(r); suction dredge-(r)
吸泥船吹填 <筑堤、填岸、填地> barged-in fill
吸泥船锚桩 dredge(r) spud
吸泥管 dredging pipe(line); dredging tube; mud suction pipe; suction tube; suction pipe
吸泥管吊架 suction pipe gantry
吸泥管水头损失 head loss of suction pipe
吸泥管速度 suction tube velocity
吸泥管套管 suction sleeve
吸泥管位置监控仪 suction tube position monitor
吸泥管线 suction pipeline
吸泥管直径 suction tube diameter
吸泥机 dredge pump; excavating pump; hydraulic suction dredge(r); pump dredge(r); suction dredge(r)
吸泥机作业区 barrow area
吸泥浆罐 mud suction tank
吸泥口 suction head; suction inlet; suction mouth
吸泥能力 dredge capacity
吸泥器 hydraulic suction dredge(r); ooze sucker; suction dredge(r)
吸泥切割头 dredging cutting head
吸泥软管 suction hose
吸泥砂机 <吸水质土用的> sand-sucker
吸泥式挖泥船 drag-suction dredge-(r); hydraulic pipe line dredge(r)
吸泥套管 suction sleeve
吸泥头 suction head
吸排气过程 suction and exhaust process
吸排气水软管 discharge and suction hose
吸排烟道 suction flue
吸盘 chuck; holdfast; sucking disc [disk]; suction cup; suction disc
吸盘花纹的轮胎 suction cup tread tyre
吸盘架 sucker frame
吸盘静电计 attracted-disc[disk] electrometer
吸盘式摞包机 suction nozzle palletizer
吸盘天线 sucker antenna
吸盘头 dustpan head
吸盘挖泥船 dustpan(suction) dredge-(r)
吸盘用工作垫块 chuck block
吸盘状陷窝 concave sucking disc
吸皮匣 skin suction box
吸起 pick-up
吸起的继电器 relay picked up
吸起电流 pick-up current
吸起电路 operating circuit; pick-up circuit
吸起电压 attracting voltage; holding voltage; operating voltage; operational voltage; pick-up voltage; working voltage
吸起接点 pick-up contact
吸起绕组 pick-up winding
吸起时间 operating time; operation time; pick-up time; time of operation
吸起位置 on position
吸起线 pick-up wire
吸起线圈 pick-up coil
吸起值 pick-up value
吸气 admission of air; air admission; air admittance; air drawing; aspirate; aspiration; aspire; inhaling; inspiration; inspire; suction gas
吸气泵 aspirating pump; aspirator pump; getter pump; inhaler; suction pump
吸气测量计 inspirometer
吸气冲程 air suction stroke; induction stroke; inspiration stroke; suction stroke
吸气储(备)量 inspiratory reserve volume
吸气导管 intake guidance; intake guide
吸气导管装置 intake guide unit
吸气的 air-breathing; inspiratory
吸气发动机 aspirating engine
吸气阀 air induction valve; air suction valve; inhalation valve; inspiratory valve; sniffing valve; sniffle valve; snifter valve; snifting valve; air admittance valve
吸气阀杆 choke lever
吸气方式 aspiration
吸气分析 analysis by absorption of gases
吸气风道 air intake duct
吸气风扇 suction fan
吸气风箱 suction ventilating fan
吸气盖 suction cowl
吸气给料装置 suction feeder
吸气管 air entry tube; air intake duct; air intake tube; air suction pipe; induction pipe; inhalation tube; sucking tube; suction duct; suction nozzle; suction tube; wind catcher
吸气管道 air inlet duct; aspiration piping; cold gas duct
吸气管线 suction line
吸气管阻焰器 induction flame damper
吸气管嘴 intake nozzle
吸气过滤泵 aspirator filter pump
吸气回扬 back-pumping with air absorption
吸气机 air sniffer; aspirated engine
吸气积尘器 aspirator combined with dust collector
吸气集尘器 aspirator combined with dust collector; aspirator with dust collector
吸气计 inspirometer
吸气剂 air getter; degasser; getter (material); air intake
吸气剂溅射设备 getter sputtering equipment
吸气剂快速蒸发 getter flash
吸气剂托 getter tab
吸气剂蒸发 getter evapo(u)ration
吸气剂作用 getter action
吸气节流阀 suction throttling valve
吸气壳 air suction casing
吸气孔 air intake opening; aspirating hole; aspirating mouth; suction hole
吸气口 absorption air inlet; air entry; air suction inlet; air suction port; aspirating hole; aspirating mouth; aspiration inlet; exhaust port; fresh-air inlet; inlet scoop; suction inlet; air suction
吸气框 suck holder
吸气离子泵 getter-ion pump
吸气力 inspiratory force
吸气量 inspiratory capacity
吸气流速 inspiratory flow rate
吸气喷镀 getter sputtering
吸气瓶 aspirator bottle
吸气期 intake period
吸气歧管 induction manifold
吸气器 aspirator; getter; inhaler; inspirator; sniffler
吸气取样 snift
吸气取样探针 sniffer probe
吸气燃烧器 aspirating burner
吸气容器 gettering container
吸气软管 inhaling hose
吸气扇 vent fan
吸气湿度计 aspirated hygrometer; aspirating psychrometer; aspiration psychrometer
吸气式抽风机 extraction fan
吸气式发生器 suction producer
吸气式分选机 air suction separator
吸气式高温计 aspirating pyrometer
吸气式激光器 air-breathing laser
吸气式集装箱 air suction container
吸气式冷凝器 aspiration condenser
吸气式清洗 suction cleaning
吸气式输送系统 pneumatic suction conveying system
吸气室 aspirating chamber; induction chamber; suction(air) chamber
吸气塔 aspiration column
吸气探针 sniffer
吸气套筒 inlet sleeve
吸气停顿 inspiratory standstill
吸气通风 suction draft; suction ventilation
吸气通风扇 suction ventilating fan
吸气通风系统 input system
吸气筒 aspiration column
吸气凸轮 air inlet cam
吸气温度 suction temperature
吸气涡流 suction eddy
吸气系统 suction system
吸气相 inspiratory phase
吸气箱 gettering container
吸气行程 suction stroke
吸气型化油器 suction type carburet-(t)or[caruret(t)er]
吸气旋塞 suction cock
吸气压力 aspirating pressure; pressure of inspiration; suction pressure
吸气压力计 suction pressure ga(u)ge
吸气压力控制 aspiration pressure control
吸气压力调节阀 crankcase pressure regulator; hold-back valve
吸气压力调节器 suction pressure regulator
吸气叶片 suction vane
吸气罩 draft hood; exhaust hood; suction hood
吸气装置 aspirator; breath apparatus; gas suction plant; getter device; suction system
吸气总量 getter capacity
吸气嘴 vacuum nozzle
吸气作用 suction effect
吸汽缸 aspirating cylinder; suction cylinder
吸器盘 haptor; sucker
吸枪 suction gun; suction pistol
吸取 aspiration; take suction
吸取管 breather hole
吸取速度 capture velocity
吸取先进技术 import advanced technology; utilizing advanced technology
吸取液体 imbibition liquor
吸取转轮 suction runner
吸取装置 pick-up gear
吸去 blotting
吸去沟渠中泥沙的卡车 gull(e)y emptier
吸热 absorption of heat; decalescence; heat absorption; heat pick-up; heat sink; reception of heat
吸热泵 absorption heat pump
吸热变化 endothermal change; endothermic change
吸热玻璃 anti-actinic glass; anti-solar glass; heat-absorbing glass; heat-absorptive glass
吸热部件 dissipater[dissipator]; heat sink
吸热材料 heat-sink material
吸热层 heat-sink shell
吸热窗纱 solar screening
吸热挡风玻璃 heat-absorbing glass; heat-absorbing windshield
吸热的 endothermal; endothermic; heat consuming; thermonegative; heat-absorbing
吸热的水 hottest water
吸热反应 endothermal reaction; endothermic reaction; heat-absorbing reaction; thermonegative reaction
吸热反应的 endothermal; endothermic
吸热峰 endothermic peak
吸热钢化玻璃 tempered heat-absorbing glass
吸热谷 endothermic valley
吸热过程 absorption; endoergic process; endothermic process
吸热化合物 endothermic compound
吸热剂 heat absorbent
吸热降解反应 endothermic degradation reaction
吸热冷却系统 heat-sink system
吸热量 heat absorption; heat absorption capacity
吸热滤光片 heat-absorbing filter
吸热率 absorptivity
吸热面 heat-absorbent surface; heat-absorbing surface; heat adsorption surface
吸热面积 heat absorption area
吸热能 endothermic energy
吸热能力 heat absorption capacity; heat carrying capacity
吸热能量 heat absorption capacity
吸热气氛 <光亮退火用的中性无二氧化碳炉气> endothermic atmosphere
吸热器 heat absorber[abstractor]; heat dump; heat sink; heat trap; thermal absorber
吸热强度 heat absorption rate
吸热燃料 endothermic fuel
吸热容量 heat adsorption capacity; heat carrying capacity; heat-sinking capacity
吸热设备 heat sink; thermal sink
吸热式冰箱 absorption refrigerator
吸热速率 heat absorption rate
吸热特性 endothermic character
吸热体 heat carrier; heat receiver
吸热铜 heat-sink copper
吸热线 endotherm
吸热效率 heat adsorption efficiency
吸热效应 endothermal effect; endothermic effect
吸热型气体 endogas; endothermic gas
吸热型气体发生设备 endogas unit
吸热性 heat absorptivity
吸热罩 heat-absorbing shield; heat-sink shield
吸热转化 endothermic conversion; endothermic disintegration
吸热转换 endothermic transition
吸热装置 heat sink
吸热装置设计 heat-sink design
吸热作用 endothermal effect; endothermic effect; heat-absorbing action
吸入 aspirate; aspiration; entrain; imbibe; imbibition; indraft; indraught; ingestion; inleakage; inspiration; inspire; soaking in; suction

吸入瓣 inlet clack
吸入泵 aspiring pump; drawing pump; suction pump; drawlift <扬程不超过吸高>
吸入比转速 suction specific speed
吸入边 suction side
吸入波 suction wave
吸入补充阀 suction replenishing valve
吸入侧 upstream side
吸入侧筒形接头 suction side cartridge
吸入侧压头 submergence head; suction head
吸入侧真空 vacuum head
吸入层 blot coat
吸入冲程 indoor stroke; induction stroke; inlet stroke; intake stroke; suction stroke
吸入穿流式 draw-thru
吸入穿流式风机系统 draw-through fan system
吸入穿流式供暖机组 draw-through heater
吸入粗滤器 suction strainer
吸入单向阀 suction check valve
吸入导管 suction lead
吸入的 drawn; entraining; inductive; inspiratory
吸入的空气 draw-in air
吸入毒性 toxicity on inhalation
吸入端 inlet side; suction end; suction side
吸入端泵盖 suction cover
吸入端法兰 suction flange
吸入阀 clack valve; induction valve; inflow valve; inlet valve; intake valve; suction valve
吸入阀簧 suction valve spring
吸入阀螺栓及开尾销 suction valve bolt and split pin
吸入阀片 suction valve disc
吸入阀片提举器 suction valve lifter
吸入法 inhalation; inhaling method
吸入分叉管 suction branch
吸入负压 negative suction
吸入干管 suction main
吸入缸 suction cylinder
吸入高度 intake head; suction head; suction height
吸入管 draft tube; induction pipe; sucker; sucking tube; suction chute; suction duct; suction intake; suction pipe; suction tube; drawing lift <泵的>
吸入管道 intake line; suction conduit
吸入管道系统 suction piping
吸入管过滤器 suction strainer
吸入管接头 suction pipe joint
吸入管滤网 suction line screen
吸入管滤油器 suction line filter
吸入管路 suction line
吸入管逆止阀 suction check valve
吸入管湿空气的绝对压力 absolute wet center manifold pressure
吸入管系统 suction piping
吸入管线 inlet line; suction line; suction pipeline; suction piping; upstream line
吸入管压力调节阀 suction pressure regulating valve
吸入管嘴 suction nozzle
吸入罐 suction tank
吸入集管 suction header
吸入剂 inhalant
吸入剂量 inhalation dose; suction dose
吸入加热器 suction heater
吸入接触 suction contact
吸入开关 suction cock
吸入空气 air suction; sniffing; suction air
吸入空气室 suction air chamber
吸入孔 inlet hole; suction eye; suction port
吸入孔道 suction passage
吸入口 inlet port; intake opening; sucker hole; sucking port; suction chute; suction inlet; suction mouth; suction opening; suction port
吸入口底板 suction manifold
吸入口法兰 intake flange
吸入口接头 input connection; suction connection
吸入口静压 static pressure of inlet
吸入口连接 suction connection
吸入口调节器 suction damper
吸入口外壳 suction cover
吸入口罩 mask for inhalation
吸入口真空度 suction water head
吸入口直径 diameter of inlet; inlet diameter
吸入口轴承压盖 suction side bearing cover
吸入喇叭 suction bell
吸入喇叭口 suction cone
吸入冷凝器 aspiration condenser
吸入连接短管 suction connection stub(pipe)
吸入连接管 suction connecting pipe; suction connection; suction piece
吸入量 intake; soakage; suction volume
吸入流 inlet flow; input flow; suction current
吸入流动 intake flow
吸入漏斗 suction funnel
吸入滤网 intake screen; intake strainer; suction strainer; strum <防止固体物进入泵或吸水管>
吸入煤气式烧嘴 aspirating burner
吸入密封 suction seal
吸入能力 inlet capacity; suction capacity
吸入逆止阀 suction check valve
吸入喷射泵 sucking jet pump
吸入喷嘴 inlet nozzle
吸入歧管 intake manifold
吸入气 inspired air
吸入汽化器 suction type carburet(t)or[caruret(t)er]
吸入器 inhalator; inhaler; inspirator; sucker
吸入器械 inhalation apparatus
吸入腔 suction chamber
吸入区 suction zone
吸入容量 inlet capacity
吸入软管 suction hose
吸入射流泵 sucking jet pump
吸入式采砂船 suction dredge(r)
吸入式电磁铁 plunger electromagnet
吸入式分级机 suction sorter
吸入式浮油回收装置 suction skimmer
吸入式鼓风机 suction blower
吸入式继电器 plunger relay
吸入式空气冷却 draft type air-cooling
吸入式空气滤清器 aspirated cleaner
吸入式滤清器 suction filter
吸入式煤气发生炉 suction gas producer
吸入式煤气机 suction gas engine
吸入式泥沙采样器 suction type sediment sampler
吸入式喷枪 suction feed type gun
吸入式撇油器 suction skimmer
吸入式起动注水器 suction primer
吸入式气动涂料喷射机 suction paint sprayer
吸入式气体过滤器 suction gas strainer
吸入式取样器 suction type sampler
吸入式砂泵 suction bailer
吸入式收集器 suction sweeper
吸入式输运器 vacuum blower
吸入式通风 suction draught; suction (type) ventilation
吸入式通风机 suction blower
吸入式通风调节器 suction type draught regulator
吸入式通风系统 suction plenum; suction ventilation system
吸入式液压油滤器 suction hydraulic filter
吸入式钻泥提取器 suction boiler
吸入试验 inhalation test
吸入室 inlet chamber; suction bell
吸入水 imbibition moisture; sucking water; suction water
吸入水分 entrapped humidity; imbibitional moisture
吸入水口 suction mouth
吸入水流 suction current
吸入水头 suction water head
吸入速度 suction speed; suction velocity
吸入损失系数 intake loss coefficient
吸入塔 suction stack
吸入套管 suction bush
吸入通风 suction draft; suction venting
吸入通风器 suction ventilator
吸入通风装置 suction ventilator
吸入筒 suction drum
吸入途径 inhalation route
吸入危害试验 inhalation hazard test
吸入危害性 inhalation hazard
吸入温度 inlet temperature; suction temperature
吸入物 inhalation
吸入系统 intake system; suction system
吸入箱 inlet box; inlet chest; suction box
吸入效率 suction efficiency
吸入效应 suction effect
吸入行程 suction stroke
吸入性 injectivity
吸入旋涡 suction vortex
吸入压力 back pressure; suction pressure
吸入压力表 back-pressure ga(u)ge; low-pressure ga(u)ge; suction ga(u)ge
吸入压力计 induction pressure ga(u)ge
吸入压力调节 back-pressure regulation
吸入压力调节阀 back-pressure regulator; evaporator pressure regulating valve
吸入压头 inlet head; suction head; suction pressure head
吸入扬程 <泵的> static draft head
吸入氧气 oxygen intake
吸入溢流 suction overfall
吸入硬化的 air set
吸入闸门 inlet sluice
吸入支管 suction manifold
吸入值 pull-in value
吸入中毒度 inhalation toxicity
吸入中毒几率系数 index of potential inhalation toxicity
吸入钟口 suction bell
吸入转换阀 suction reversing valve
吸入装置 suction apparatus
吸入状况 suction condition
吸入锥 suction cone
吸入资金 outside fund
吸入总管 suction main
吸入阻力 resistance of suction; suction drag; suction resistance
吸入嘴 suction nozzle
吸入作用 imbibition
吸扫【疏】suction
吸色高温计 colo(u)r-extinction pyrometer
吸砂泵 <又称吸沙泵> sand pump
吸砂船 cutterless sand-sucker; sand pump dredge(r)
吸砂机 sand-sucker; suction type dredge(r)
吸砂机构 suction cutter apparatus; suction ladder
吸砂头 suction cutter
吸砂用的吸泥机 sand-sucker
吸上变压器 boosting transformer booster
吸上高度 suction head; suction lift
吸上高度调节 submergence control
吸上式注水器 lifting injector
吸上线【电】boosting wire
吸升高度 suction head; suction height; suction lift
吸升极限 suction limit
吸升力 suction head; suction lift
吸升水头 suction(water)head; vacuum head
吸升损失 suction loss
吸升扬程 suction lift; suction water head
吸升中途增压设备 suction booster
吸声 deadening; noise deadening; sound absorption; sound trapping
吸声板 abat-voix; absorbent board; absorbing board; absorptive lining; acoustic(al)(filter)board; acoustic(al) panel; acoustic(al) tex; acoustic(al) tile; coffer; noise barrier; sound-absorbent board; sound-deadening board
吸声板条 absorbing sheet
吸声背衬 acoustic(al) backing; sound-absorbent backing
吸声背衬材料 absorbent backing; absorbing backing
吸声表面 sound-absorbent surface [surfacing]
吸声玻璃 absorbent glass; absorbing glass; absorptive control glass; absorptive glass; sound-absorbent glass
吸声玻璃棉板 glass wool acoustic(al) board
吸声箔 absorptive foil; acoustic(al) foil; sound-absorbent foil
吸声薄板 absorbent sheet; absorptive waffle; sound-absorbent sheet
吸声材料 absorbent; absorbent material; absorbing material; absorptive material; acoustic(al) absorbent; acoustic(al) absorptive material; acoustic (al) material; celotex (board); draping; sound-absorbent; sound-absorbent material; sound-absorbing material; sound absorption material; sound-deadening material; Cabot's quilt <墙壁或地板的>; Corkoustic <一种专卖的>
吸声材料施工法 absorbent material construction(method)
吸声测量 acoustic(al) absorptivity measurement
吸声层 absorbent panel
吸声尘劈 absorbing wedge; acoustic(al) absorption wedge; sound absorption wedge
吸声衬板 absorbent lining; absorptive backing
吸声衬层 absorbent lining
吸声衬垫 acoustic(al) lining; sound-absorbent lining; sound-absorbent pad

吸声衬砌 sound-absorbing lining
吸声处理 sound-absorbent treatment; sound-absorbing treatment; sound absorptive treatment
吸声带宽 bandwidth of sound absorption
吸声单位 sound absorption unit
吸声单元 absorbent unit; sound-absorbent unit
吸声的 sound absorbing; sound absorptive; sound damping; sound-deaden
吸声的蛭石灰膏 vermiculite sound absorbent plaster
吸声垫 absorbent blanket; absorbing pad; absorptive facing; absorptive pad; acoustic(al) blanket
吸声垫板 absorptive backing
吸声吊顶 absorbent hung ceiling; absorbent suspended ceiling; absorbing hung ceiling; absorbing suspended ceiling; acoustic(al) ceiling; suspended acoustic(al) ceiling
吸声吊顶棚 hung acoustic(al) ceiling
吸声吊顶砖 acoustic(al) ceiling tile
吸声顶板 absorbent ceiling; noise-absorbing ceiling
吸声顶棚 absorbing ceiling(board); absorptive ceiling; acoustele; acoustic(al) ceiling board; sound-absorbent ceiling(board); suspended acoustic(al) ceiling
吸声顶棚薄板 absorptive ceiling sheet; sound-absorbent ceiling sheet
吸声顶棚体系 sound-absorbent ceiling system
吸声顶棚涂料 absorbing ceiling paint; acoustic(al) ceiling paint; sound-absorbent ceiling paint
吸声顶棚系统 acoustic(al) ceiling system
吸声顶棚砖 acoustic(al) ceiling tile
吸声方形块 absorbing waffle
吸声粉末 absorbent powder
吸声粉刷 sound-absorbent plaster; sound absorptive plaster
吸声粉刷层 absorbent plaster; absorbing plaster; acoustic(al) plaster
吸声粉刷骨料 absorbent plaster aggregate; sound-absorbent plaster aggregate
吸声粉刷灰浆骨料 absorbing plaster aggregate
吸声粉刷集料 absorbent plaster aggregate
吸声粉刷天花板 absorbent plaster ceiling
吸声蜂窝表面 sound-absorbent waffle
吸声附件 unit absorber
吸声覆盖层 sound-absorbent cover(ing)
吸声隔板 absorptive blanket; acoustic(al) coffer; acoustic(al) waffle; celotex board; absorptive board
吸声隔层 absorbent pad; acoustic(al) blanket
吸声隔墙 acoustic(al) partition
吸声构件 sound-absorbent unit; unit absorber
吸声构造 absorbing construction; acoustic(al) construction; sound-absorbing construction
吸声构造材料 absorbing construction material; absorptive construction material
吸声构造方法 sound-absorbent construction method
吸声管道 absorbing duct
吸声管道衬砌 acoustic(al) duct lining
吸声护壁板 acoustic(al) cassette
吸声护面 absorptive cover(ing); absorptive(sur) facing
吸声护墙板 absorbent cassette
吸声灰膏 absorbing plaster; absorptive plaster; acoustic(al) plaster
吸声灰浆 acoustic(al) plaster
吸声灰泥 absorbing plaster; sound-absorbent plaster
吸声灰泥顶棚 sound-absorbent plaster ceiling
吸声灰泥骨料 sound-absorbent plaster aggregate
吸声剂 acoustic(al) absorbent; sound absorber
吸声夹层板 acoustic(al) coffer
吸声尖劈 wedge absorber
吸声建筑 acoustic(al) construction
吸声建筑材料 absorbent construction material; absorbing construction material; absorptive construction material; sound-absorbent construction(al) material
吸声建筑单元 absorbent building unit
吸声建筑构件 sound-absorbent building unit
吸声降噪 noise reduction by absorption
吸声结构 sound-absorbing structure
吸声介质 acoustic(al) absorbing medium
吸声金属箔 absorptive foil; absorbent foil; absorbing foil
吸声金属顶棚 absorbent metal ceiling; absorbing metal ceiling; absorptive metal ceiling
吸声金属天花板 absorbent metal ceiling; absorbing metal ceiling; absorptive metal ceiling
吸声空心砖 absorbing tile
吸声空心砖顶篷 absorbing tile ceiling
吸声空心砖天花板 absorbing tile ceiling
吸声块 absorbent wall block
吸声力 acoustic(al) absorptivity; sound-absorbing power; sound absorption
吸声量 equivalent absorption area; sound absorption capacity
吸声料 noise absorption
吸声率 absorptivity; acoustic(al) absorptivity; noise absorption factor; sound absorption coefficient
吸声毛毯 balsam wool
吸声门 acoustic(al) door
吸声面 sound-absorbing surface
吸声面层 absorbent cover(ing); absorbent facing; absorbent surfacing; absorbing facing; absorbing surfacing; absorptive cover(ing); absorptive(sur) facing; acoustic(al) lining
吸声面砖 absorbent tile; absorbing tile; acoustic(al) tile
吸声模板 absorbent shutter; sound-absorbent formboard
吸声抹灰顶棚 acoustic(al) plaster ceiling
吸声木丝板 absorbent wood fiber board; sound-absorbent wood-wool board
吸声木条 absorptive sheet
吸声木质纤维板 absorbing wood fiber[fibre] board; absorptive wood fiber[fibre] board; acoustic(al) wood fiber[fibre] board
吸声内衬 absorbing lining
吸声能力 absorbent capacity; absorption power; absorption quality; absorptive capacity; sound absorption capacity; sound absorption power
吸声盘 sound-absorbent pan
吸声泡沫玻璃 sound adsorbing foam glass
吸声喷涂材料 acoustic(al) sprayed-on material
吸声喷涂粉刷 absorbent sprayed-on plaster; absorbing sprayed on plaster; absorptive sprayed on plaster; sound-absorbent sprayed-on plaster
吸声喷涂灰泥 absorbing sprayed on plaster; absorptive sprayed on plaster; sound-absorbent sprayed-on plaster
吸声平顶 absorbent hung ceiling; absorbent suspended ceiling; absorbing hung ceiling; absorbing suspended ceiling; acoustic(al) ceiling; sound-absorbent ceiling(board)
吸声平顶镶板 sound-absorbent coffer
吸声屏幕 acoustic(al) screen
吸声屏障 absorbent lined barrier
吸声漆 absorbent paint; anechoic paint; sound-absorbing paint
吸声砌块 absorbing wall block; acoustic(al) block; sound adsorption block
吸声器 acoustic(al) damper; filter plexer; sound absorber; sound damper
吸声墙 absorbent wall; absorptive wall; sound-absorbent wall
吸声墙板 absorbent wall block; absorptive panel
吸声墙空心砖 absorbent hollow tile; absorbent wall tile
吸声墙面砖 acoustic(al) wall tile
吸声墙砌块 sound-absorbent wall block
吸声墙纸 absorbent wallpaper; absorbing wallpaper; acoustic(al) wallpaper; sound-absorbent wallpaper
吸声墙砖 absorbent wall brick; sound-absorbent wall brick
吸声穹隆 acoustic(al) vault
吸声设备 absorbent unit
吸声设施 acoustic(al) absorbing device
吸声石膏装饰板 decorative acoustic(al) gypsum
吸声试验 sound absorption test
吸声室 absorbent chamber; absorbing chamber; absorptive chamber; anechoic chamber; dead room; sound-absorbent chamber
吸声塑料 acoustic(al) plastics
吸声损失 acoustic(al) adsorption loss
吸声毯 absorbing felted fabric; sound-absorbent quilt; sound-absorbing blanket
吸声毯毡 absorbent blanket
吸声特性 acoustic(al) performance; sound adsorption characteristic
吸声体 noise reducing splitter; sound-absorbent; sound absorber
吸声体系 sound-absorbent system
吸声天花板 absorbent ceiling; absorbing ceiling(board); absorptive ceiling; acoustele; acoustic(al) ceiling board; acoustic(al) ceiling panel
吸声天花板板条 absorbent ceiling sheet
吸声天花板涂料 absorbing ceiling paint; acoustic(al) ceiling paint
吸声天花贴面板 absorbing ceiling tile
吸声贴面 absorbing lining
吸声贴面板 acoustic(al) cassette
吸声贴面层 absorbing cover(ing); sound-absorbent facing
吸声贴砖 acoustic(al) tile; acoustolith tile
吸声通风顶棚 acoustic(al) ventilating ceiling
吸声通风天花板 acoustic(al) ventilating ceiling
吸声涂层 absorptive cover(ing); sound-deadening coat(ing)
吸声涂料 absorbent paint; acoustic(al) coating; acoustic(al) paint; anti-noise paint; sound-absorbent paint; sound-absorbing paint; sound-deadening paint
吸声涂面 absorbing lining
吸声涂面层 absorbing cover(ing)
吸声瓦板 absorbent tile; acoustic(al) tile
吸声圬工墙 sound-absorbent masonry wall
吸声系数 acoustic(al) absorption coefficient; acoustic(al) absorptivity; acoustic(al) adsorption coefficient; acoustic(al) adsorption factor; acoustic(al) factor; coefficient of acoustics; coefficient of sound adsorption; noise absorption factor; normal incident absorption coefficient; sound-absorbing coefficient; sound absorption coefficient
吸声系统 absorbent system; absorbing system; sound-absorbing system; sound absorption coefficient
吸声纤维板 absorbent wood fiber board; absorbing fiber board; absorptive fiber[fibre] board; acoustic(al) fiber board; sound-absorbent fiber[fibre] board
吸声箱 sound box
吸声镶板 absorbent coffer; absorbent panel; absorbing panel; absorbing waffle; absorptive panel; absorptive waffle; acoustic(al) cassette; acoustic(al) coffer; acoustic(al) lay-in panel; acoustic(al) panel; sound-absorbent cassette; absorbing coffer
吸声镶面板 absorbing cassette; absorptive cassette; absorptive coffer
吸声镶面层 absorbing cover(ing)
吸声小块 absorptive cassette
吸声效率 sound absorption efficiency
吸声楔 acoustic(al) wedge
吸声性能 absorption capacity; absorption quality; absorptive property; acoustic(al) behavio(u)r; acoustic(al) property
吸声性质 sound absorption property; sound absorption quality
吸声悬挂顶棚 sound-absorbent hung ceiling; sound-absorbent suspended ceiling
吸声因素 sound adsorption factor
吸声油漆 absorbing paint; absorptive paint; anti-noise paint; sound-absorbing paint; acoustic(al) paint
吸声毡 absorbent felt(ed fabric); absorptive blanket; absorptive felt(ed fabric); acoustic(al) felt; baffle blanket; felt deadener; sound-absorbent blanket; sound adsorbing blanket; sound attenuation blanket
吸声毡料织物 absorptive felt(ed fabric)
吸声毡织物 sound-absorbent felt(ed fabric)
吸声毡织物顶棚 sound-absorbent felted fabric ceiling
吸声罩 sound-absorbing cover
吸声织物 acoustextile
吸声蛭石粉刷 vermiculite acoustic-

(al) plaster
吸声蛭石灰泥 vermiculite acoustic(al) plaster
吸声砖 absorbent brick; absorbing brick; absorptive brick; absorptive tile; acoustic (al) brick; acoustic(al) tile; sound-absorbent brick; sound-absorbent tile
吸声砖顶棚 acoustic(al) tile ceiling; sound-absorbent tile ceiling
吸声砖平顶 acoustic(al) tile ceiling
吸声砖砌墙 absorbing masonry wall; absorptive masonry wall
吸声砖石墙 absorbent masonry wall; absorbing masonry wall; absorptive masonry wall; sound-absorbent masonry wall
吸声砖天花板 acoustic(al) tile ceiling
吸声装置 absorbent unit; absorbing system; absorptive system
吸声琢面 deadening dressing
吸声阻力 acoustic(al) resistance
吸声组合设备 absorbing building unit
吸声组合物 sound-deadening composition; sound-deadening compound
吸声组装部件 absorptive building unit
吸声组装构件 absorptive building unit
吸声作用 sound absorption
吸湿 absorbed moisture; absorption of moisture; moisture absorption; moisture regain
吸湿不稳定性 hygro-instability
吸湿材料 hygroscopic material; moisture absorption
吸湿材料除潮 dehydration
吸湿尘 hygroscopic dust
吸湿衬垫 absorptive liner
吸湿持水度 specific retention of moisture absorption
吸湿的 hygroscopic; moisture-retentive
吸湿度 hygroscopic degree; hygroscopic humidity; hygroscopicity; wettability
吸湿干燥 hygroscopic desiccation
吸湿含水量 hygroscopic moisture content; hygroscopic water content; hygroscopicity; hygroscopic moisture
吸湿含水率 hygroscopic moisture
吸湿机制 hygroscopic mechanism
吸湿计 hygrometric moisture meter
吸湿剂 desiccant; desiccating agent; hygroscopic agent; moisture absorbent; moisture-retentive chemicals; moisture absorption
吸湿开裂 moisture crazing
吸湿力 moisture absorption
吸湿量 damping capacity; hygroscopic moisture content; hygroscopic water content
吸湿率 hygroscopicity; rate of moisture absorption
吸湿模板衬垫 absorptive form lining

吸湿能力 damping capacity; hygroscopic capacity; moisture absorption power
吸湿膨胀 hygroscopic expansion; moisture expansion; hygrometric expansion
吸湿膨胀性 hygroexpansivity
吸湿平衡 hygroscopic equilibrium; moisture equilibrium at dry side
吸湿器 dehumidifier; moisture absorber
吸湿潜力 hygroscopic potential
吸湿容量 hygroscopic capacity; moisture absorption capacity
吸湿柔软剂 hygroscopic softener
吸湿式湿度计 absorption hygrometer
吸湿势 hygroscopic potential
吸湿试验 fuchsin(e) test; hygroscopic test; moisture absorption test; desiccator test
吸湿数 hygroscopic number
吸湿水 hygrcosopic water
吸湿水分 hygroscopic moisture
吸湿水系数 hygroscopic water coefficient
吸湿土壤水 hygroscopic soil water
吸湿物 hygroscopic substance
吸湿物质 hygroscopic substance
吸湿系数 coefficient of moisture adsorption; hygroscopic coefficient
吸湿纤维板 absorbent fiber [fibre] board
吸湿现象 hygrometric phenomenon; hygroscopic effect; hygroscopic phenomenon
吸湿效应 hygroscopic effect
吸湿型水分测定计 hygroscopic moisture meter
吸湿性 absorptivity; hygroscopicity; hygroscopic property; hygroscopic effect; hygroscopic property; moisture of absorption; moisture regain; moisture retention; water-absorbing quality
吸湿性材料 hygroscopic material
吸湿性的 hygrometric
吸湿性核 hygroscopic nucleus
吸湿性核素 hygroscopic nuclei
吸湿性货物 humidity absorbing goods; hygroscopic cargo
吸湿性粒子 hygroscopic particle
吸湿性能 hygroscopic behavio(u)r; hygroscopicity; hygroscopic property
吸湿性物质 hygroscopic matter
吸湿性盐 deliquescent salt; hygroscopic salt
吸湿压力 hygroscopic pressure
吸湿盐 hygroscopic salt
吸湿盐法 hygroscopic salt method
吸湿盐类 <如氯化钙、氯化镁> hygroscopic salts
吸湿硬膏 <即湿熟石膏> dreston
吸湿运动 hygrometric movement
吸湿作用 hygroscopic absorption; hygroscopic effect
吸石机 rock washing machine
吸式清扫机 suction cleaner
吸式挖泥船 suction cutter dredge(r); suction dredge(r)
吸式挖泥头 suction cutter head
吸式挖土机 suction cutter dredge(r); suction dredge(r)
吸势 suction potential
吸室 suction chamber
吸收 absorb; co-opt; digest; drink(ing); imbibition; uptake
吸收板 absorber plate
吸收杯 absorption cell
吸收杯池 sample cell
吸收本领 absorbence [absorbency]; absorbing power; absorption power; absorptive power; absorption capacity
吸收泵 absorption pump; clean-up pump
吸收比 absorbed ratio; absorptance; specific absorption
吸收比率 absorption rate
吸收比色计 absorptiometer
吸收壁放大管 resistive-wall amplifier tube
吸收边 absorption edge; absorption rim
吸收边能量 absorption edge energy
吸收边缘 absorption edge
吸收表层 absorbing cover(ing)
吸收表面 absorbing surface
吸收表面膜 absorbed surface film
吸收波长计 absorptive wavemeter
吸收波段 absorption band
吸收波谱 absorption spectrum
吸收薄膜 absorbing membrane; absorptive hull
吸收材料 absorbent material; absorbing material
吸收操作线 operation line of absorption process
吸收槽 absorption cell
吸收测定器 absorptiometer
吸收测工器 absorption dynamometer
吸收测功仪 absorbing dynamometer
吸收测量分析 absorptiometric analysis
吸收测量学 absorptionmetry
吸收层 absorbed bed; absorbed layer; absorbent bed; absorbing zone; absorption bed; absorption region; trapping layer
吸收常数 absorption constant
吸收场地 absorption field
吸收车间 absorption plant
吸收衬度像 absorption-contrast image
吸收成本(计算)法 absorption costing
吸收池 absorption cell; absorption tank
吸收抽提 absorptive extraction
吸收初速 initial rate of absorption
吸收处理 absorption treatment
吸收床 absorbed bed; absorbent bed; absorption bed
吸收代谢作用 assimilatory metabolism
吸收带 absorption band; absorption zone
吸收带边缘 absorption edge
吸收带长度 absorption zone length
吸收带强度 absorption band intensity; intensity of absorption bands
吸收的 absorbent; sorptive; absorbing
吸收的复合材料 absorbable composite material
吸收等温线 absorption isotherm
吸收电磁波的特性 radar absorption characteristic
吸收电抗器 soaking reactor
吸收电流 <非完全介质中> absorption current
吸收电路 absorbing circuit; absorption circuit; trap circuit
吸收电容器 absorption capacitor
吸收电子图像 absorption electron image
吸收电阻 dead resistance
吸收垫 suction pad
吸收定律 absorption law
吸收度 absorbance; absorptance
吸收度比值 absorbance ratio
吸收度试验 mounting test
吸收端 absorption edge; foot valve
吸收发射比 absorptivity-emissivity ratio
吸收法 absorptive method; absorption (method); absorption process; soak-up method; equity method <即产权净值法>
吸收法测定 absorptionmetric determination
吸收法分离器 absorbent separator
吸收法回收的汽油 absorption gasoline
吸收法脱氮 absorption denitrification
吸收范围 absorption limit; absorption region
吸收废物容量 waste assimilation capacity; waste assimilative capacity
吸收分布 absorption distribution
吸收分光光度测定(法) absorption spectrophotometry
吸收分光光度计 difference spectrophotometer
吸收分光镜 absorption spectroscope
吸收分光摄像仪 absorption spectrograph
吸收分光学 absorption spectroscopy
吸收分离 absorption extraction
吸收分析法 absorption approach
吸收份额 absorbed fraction
吸收峰(值) absorption peak
吸收敷层 darkflex
吸收俘获 absorption capture
吸收(辐射)剂量 absorbed dose
吸收干燥 drying by absorption
吸收干燥法 absorptive drying
吸收隔板 absorbing septum
吸收功率 absorbed power; absorbent power; absorption power
吸收功率计 absorption dynamometer
吸收功能 absorption function
吸收沟槽 absorption trench
吸收管 absorbent; absorber; absorption tube; absorption pipe
吸收管法 absorption tube method
吸收罐 tourie; tourill
吸收光 absorbed light
吸收光带 absorption hand
吸收光度法 absorption photometry
吸收光度计 absorptiometer; absorption meter
吸收光屏 absorption screen
吸收光谱 absorbance spectrum; absorption spectrum
吸收光谱(测定)法 absorption spectrometry
吸收光谱带 absorbing band; absorption band
吸收光谱分析 absorption spectroanalysis; absorption spectroscopy
吸收光谱分析仪 absorption spectrometer
吸收光谱化学分析 absorption spectrochemical analysis
吸收光谱检查 absorption spectroscopy
吸收光谱术 absorption spectrometry
吸收光谱图 absorption spectrogram
吸收光谱学 absorption spectroscopy
吸收光谱仪 absorption spectrometer
吸收光线的 light-absorbing
吸收光楔 absorbing wedge; absorption wedge
吸收过程 absorbent process; absorption process
吸收过渡 absorptive transition
吸收过滤介质 absorbent filtering medium
吸收函数 absorption function
吸收合并 consolidation by merger
吸收和分配 uptake and partitioning
吸收横断面 absorption cross-section
吸收红外线玻璃 heat-absorbing glass; infrared-absorbing glass; infrared ray absorbent glass
吸收红外(线)气体 infrared-absorbing gas
吸收后状态 post-absorptive state
吸收化合物 absorption compound
吸收还原法脱氮 control of NOx by absorption-reduction process
吸收灰尘 absorbing dust
吸收机 absorption machine
吸收极限 absorption edge; absorption limit
吸收极限频率 absorption limiting frequency
吸收集 absorbing set
吸收技术 suck

吸收剂 absorbefacient; absorbent; absorbentia; absorbent material; absorber; absorbing material; absorbing medium; absorption agent; carrier; getter; sorbing agent
吸收剂量 dosage; dose; uptake dose
吸收剂量比值 specific absorbed dose
吸收剂量分布 absorbed dose distribution
吸收剂量量热器 absorbed dose calorimeter
吸收剂量率 absorbed dose rate
吸收剂量指标 absorbed dose index
吸收剂溶液 absorbent solution
吸收肩 absorption shoulder
吸收阶段 absorption stage
吸收截面 absorption cross-section
吸收截止频率 absorption limiting frequency
吸收晶体光谱 absorption crystal spectrum
吸收精馏塔 reboiled absorber
吸收井 absorbing well; inverted well; negative well; suction well
吸收阱 absorbing trap
吸收坑 absorption pit
吸收孔 absorbing hole
吸收控制 absorber control; absorption control
吸收口 suction nozzle
吸收库 reservoir
吸收库中的各种物质 reservoir species
吸收亏损 absorb loss
吸收劳动力 absorption of labo(u)r power
吸收理论 absorption approach
吸收量 absorbability; absorbing capacity; absorption capacity; absorptive capacity; absorptivity; loading
吸收料 dope
吸收滤波器 trap circuit
吸收滤光镜 absorbing filter
吸收滤光片 absorption filter; barrier filter
吸收律 absorption law
吸收率 absorbence[absorbency]; absorption coefficient; absorption index; absorption rate; absorptive index; absorptive power; absorptivity; absorption coefficient; rate of absorption; ratio of absorption; specific absorption; suction rate
吸收率计 absorptiometer
吸收马尔可夫链 absorbing Markov chain
吸收媒介 absorbing medium
吸收媒质 absorbing medium
吸收面 absorption surface; capture area; absorption cross-section < 天线的 >
吸收面积 absorption area
吸收模板衬 absorptive form lining
吸收能级 absorption level
吸收能力 absorbability; absorbence [absorbency]; absorbent capacity; absorbing ability; absorbing capacity; absorbing power; absorptance; absorption capacity; absorption power; absorption property; absorption quality; absorptive power; absorptivity; absorption capacity; absorption power; assimilative capacity; assimilatory power; inverted capacity; receptivity; suction capacity; intake capacity < 钻孔的 >
吸收能力稳定性 stability of absorption
吸收能量 absorbed energy; energy absorbing capacity
吸收扭矩 absorbed torque

吸收排水井 inverted drainage well
吸收盘 absorption tray; absorption pan
吸收盘管 absorption coil
吸收劈 absorbing wedge
吸收片的吸收限 absorption limit of absorbed flat
吸收频带 absorption band
吸收频率 absorption frequency
吸收频谱 absorption spectrum
吸收平衡 absorption equilibrium
吸收平衡线 equilibrium curve of absorption process
吸收瓶 absorption bottle; absorption bulb
吸收谱 absorption spectrum
吸收谱带 absorption band
吸收谱段 absorption band
吸收谱线 absorption line
吸收期 absorption stage
吸收齐 dope
吸收汽油 absorption gasoline
吸收器 absorber; absorption vessel; annihilator; suction header
吸收器泵 absorber pump
吸收器电路 absorber circuit
吸收器冷却器 absorber cooler
吸收器液面计 level ga(u)ge for absorber
吸收强度 absorption strength
吸收球管 < 气体的 > absorbing pipet(te)
吸收区(域) absorbing zone; absorption region; absorption zone; uptake zone
吸收曲线 absorption curve; sorption curve
吸收曲线斜率 slope of absorption curve
吸收全息图 absorption hologram
吸收权益性投资 absorbing beneficial interest investment
吸收热 absorption heat; heat of absorption
吸收热量 ingress of heat
吸收容量 inverted capacity
吸收容器 absorption vessel
吸收溶出伏安法 absorption stripping voltammetry
吸收溶剂作用 lyosorption
吸收溶液 absorption solution
吸收色 absorption colo(u)r
吸收色层法 absorption chromatography
吸收色谱法 absorption chromatography
吸收色谱学 absorption chromatography
吸收上的差异 differential absorption
吸收社会劳动力 absorption of labo(u)r power
吸收射线 absorption ray
吸收摄谱仪 absorption spectrograph
吸收湿度 moisture of absorption
吸收湿度表 absorption hygrometer
吸收湿度计 absorption hygrometer; chemical hygrometer
吸收式冰箱 absorption refrigerator
吸收式波长计 absorption type wavemeter
吸收式测功器 absorption dynamometer
吸收式的 absorption type
吸收式分光光度计 absorption spectrophotometer
吸收式供冷 absorption cooling
吸收式光学高温计 absorption pyrometer
吸收式过滤器 absorptive type filter
吸收式净化 absorption cleaning
吸收式空调设备 absorption air conditioning
吸收式冷冻 absorption refrigeration
吸收式冷冻机 absorbent refrigerator; absorption refrigerator; absorption refrigerating machine; absorption refrigeration machine
吸收式频率计 absorption type frequency meter
吸收式燃气制冷机 absorption gas refrigerator
吸收式湿度表 absorption hygrometer
吸收式衰减器 absorption attenuator; absorptive attenuater[attenuator]
吸收式探伤器 absorption flaw detector
吸收式调温机 absorption unit
吸收式消音器 absorption silencer
吸收式盐水冷气设备 absorption brine chilling unit
吸收式液体冷却装置 absorption type liquid chiller
吸收式制冷 absorption cooling; absorption refrigeration
吸收式制冷机 absorbent refrigerator; absorption refrigerating machine; absorption refrigerator; absorption type refrigerating machine
吸收式制冷机组 absorption type refrigerating unit
吸收式制冷剂 absorption type refrigerating machine
吸收式制冷器 absorption chiller
吸收式制冷系统 absorption(refrigeration) system
吸收式制冷循环 absorption refrigeration cycle
吸收式制冷装置 absorption refrigerating plant
吸收试剂 absorbing chemical
吸收试验 absorption test; sorptivity test
吸收试验池 absorption cell
吸收试验仪 absorption apparatus
吸收室 absorption cell; absorption chamber
吸收衰减 attenuation by absorption
吸收衰减补偿 absorption attenuation compensation
吸收衰减补偿值 absorption attenuation compensation value
吸收衰减器 resistive padding
吸收衰竭 assimilatory depletion
吸收衰落 absorption fading
吸收水 absorbed water; absorption water; occluded water; water of imbibition; hydroscopic water
吸收水层 absorption field
吸收水分 absorbed moisture; absorption moisture; take-up water; imbibition < 岩石土壤孔隙的 >
吸收水分当量 suction moisture equivalent
吸收水膜 absorbed film of water; absorption film of water; sorbed film of water; sorption film of water
吸收水头 suction water head
吸收速度 infiltration rate
吸收速率 absorption rate; rate of absorption
吸收速率比值 specific absorption rate
吸收损耗 absorption loss
吸收损失量 absorption loss
吸收塔 absorber; absorbing column; absorbing tower; absorption column; absorption tower
吸收塔板 absorption tray
吸收塔冷却器 absorber cooler
吸收碳 carbon pick-up
吸收特性 absorption property; absorptive character
吸收体 absorbent; absorbent material; absorber
吸收天平 sorption balance
吸收填充塔 packed column for absorption
吸收调制 absorbing modulation; absorption modulation; absorptive modulation
吸收通量 absorbed flux
吸收头 power termination
吸收透镜 absorption lens
吸收涂层 absorptive coat
吸收外国游客的旅游业 inward tourism
吸收外国资金 attract foreign pound
吸收外商直接投资的方式 incorporating foreign direct investment
吸收外援的能力 absorptive capacity for external assistance
吸收外资 absorb foreign capital; absorption of foreign investment; foreign capital inducement; foreign capital intake
吸收完毕 post-absorption
吸收污染度 assimilative capacity
吸收污染量 assimilative capacity
吸收物 absorbate; absorbent
吸收物煅烧 absorbate incineration
吸收洗涤器 absorber washer
吸收系数 absorbance [absorbancy]; absorbing coefficient; absorptance; absorption coefficient; absorption factor; absorptivity; absorptivity coefficient; coefficient of absorption; specific absorption
吸收系数的提取 extraction of absorption coefficients
吸收系数剖面 absorption coefficient section
吸收系数剖面法 method of absorption coefficient section
吸收系统 absorption system
吸收匣 absorption cell
吸收限 absorption edge; absorption limit
吸收限能量 absorption edge energy
吸收陷波电路 absorption(wave) trap
吸收陷阱 absorption(wave) trap
吸收消声器 absorbing silencer
吸收消音器 absorbing silencer
吸收效率 absorption efficiency
吸收效应 absorption effect; sink effect
吸收效应修正 absorption effect correction
吸收新人员 recruitment
吸收型的 absorptive type
吸收型过滤器 absorptive type filter
吸收型消声器 silencer of the muffler type
吸收型轴承合金 absorbent metal
吸收性 absorbability; absorbance; absorptance; absorptive character; absorptive quality; absorptivity
吸收性材料 absorbing material; absorptive material; attenuating material
吸收性的 absorbefacient; absorbent; absorptive; spongy
吸收性粉末 absorbing powder; absorptive powder
吸收性缝线 absorbable suture
吸收性复合体 absorbing complex; absorptive complex
吸收性过滤 absorbent filtration
吸收性过滤器 absorbent filter
吸收性介质 absorbing medium
吸收性滤清器 absorbent filter; absorptive type filter
吸收性媒质 resistive medium

吸收性明胶海绵 absorbable gelatin sponge
吸收性模板 absorptive form
吸收性能 absorption capacity; absorption property; absorptive property; absorption behavio(u)r; receptivity
吸收性能试验 absorbency test
吸收性热泵 absorption heat pump
吸收性炭 absorbent charcoal
吸收修正量 absorption correction
吸收选择器 absorbing selector
吸收循环 absorption cycle
吸收烟灰设备 plant for smoke absorption
吸收氧化法脱氮 control of NOx by absorption-oxidation process
吸收样品池 sample cell
吸收液 absorption liquid
吸收仪 absorbing apparatus
吸收异向性 absorb anisotropy
吸收因数 <被物体吸收的光通量与照射于该物体的光通量之比> absorption factor
吸收因子 absorption factor
吸收引力 absorptive attraction
吸收营养 absorption of nourishment
吸收油 absorbent oil; absorber oil
吸收游资 absorb idle funds
吸收跃迁 absorptive transition
吸收增强效应 absorption enhancement effect
吸收沾染 absorption impurity
吸收障碍 malabsorption
吸收振动 vibration-absorption
吸收震式凿岩机 shock absorber type drill
吸收值 absorption value; absorptive value
吸收纸 absorption paper
吸收指数 absorptive index; absorption index; index of absorption
吸收制动器 absorption brake
吸收中心 absorption centre
吸收中子 intercept neutron
吸收轴 absorption axis
吸收柱 absorbing column; absorption column
吸收装置 absorber; absorbing apparatus; absorption equipment; absorption device; absorption plant; absorption unit
吸收状态 absorbing state; absorption state
吸收着色 absorption colo(u)ring
吸收着色力 absorption tinting strength
吸收资金 bring in funds
吸收紫外线玻璃 ultraviolet absorbent glass; ultraviolet (ray) absorbing glass
吸收总系数 overall absorption coefficient
吸收组织 absorption tissue
吸收最大值 absorption peak
吸收作用 absorption (effect); imbibing; sorption
吸收作用试验 absorption test
吸水 absorption; moisture absorption; take-up water; water regain; water sucking; water uptake; water absorption
吸水泵 pulsating pump; suction pump; water suction pump
吸水比率 absorption ratio
吸水布 absorbent cloth
吸水槽 suction sump
吸水槽式溢洪道 suction-slot spillway
吸水层 absorbed layer
吸水场地 absorption field
吸水衬里 absorptive liner
吸水程度 level of water being sucked
吸水池 priming reservoir; suction bay; suction tank
吸水的 hygroscopic; water-absorbed; water-absorbing; water receptive; water-retaining
吸水底阀 suction floor valve; suction foot valve
吸水地层 absorbed ground; absorbent formation; absorbent ground
吸水度 water-absorbing capacity; water absorption quality
吸水发胀黏[粘]土 bloated clay
吸水阀 suction valve; foot valve; induction valve
吸水阀座 suction valve seat
吸水范围 suction range
吸水覆盖层 suction mat unit
吸水干管 sucking main; suction main
吸水高度 suction height; suction length; suction lift; suction water head
吸水高度调节 suction height controlling
吸水骨料 absorbent aggregate; hydrophilic aggregate
吸水管 induction manifold; intake manifold; rattlehead; sucking pipe; sucking tube; suction conduit; suction duct; suction manifold; suction pipe; suction tube; suction water pipe; water adsorption tube
吸水管道 suction pipeline; suction piping
吸水管道沟槽 absorption trench
吸水管端 snore piece
吸水管滤网 suction pipe strainer
吸水管路 suction pipeline
吸水管水头损失 suction pipe loss
吸水管损失 suction pipe loss
吸水管弯头 suction bend(ing)
吸水管系统 suction piping system
吸水管线 suction pipeline; suction piping
吸水过程 hygroscopic absorption
吸水过滤器 suction strainer
吸水河湾 suction bay
吸水汇管 suction manifold
吸水机 suction machine; water scooping machine
吸水及放水阀座 suction and discharge valve seat
吸水极限 suction range
吸水集料 absorbent aggregate; hydrophilic aggregate
吸水剂 water absorbent
吸水胶管 pump suction hose
吸水进水头 suction header
吸水进水主干 suction header
吸水井 absorbent well; absorbing well; absorption well; collecting well; dead well; entrance well; inlet well; inverted drainage well; inverted well; negative well; recharge well; sucking water well; suction well; wet well; suction pool <泵站的>
吸水静压 suction static pressure
吸水坑 adsorption pit; pumping pit; suction pit
吸水口 sucking hole; suction mouth
吸水口连接 suction connection
吸水口条件 suction hole condition
吸水里衬 <混凝土模板的> absorptive lining
吸水力 water-absorbing force
吸水量 hydroscopic capacity; infiltration; moisture of absorption; soakage; water-holding capacity; water regain; water retaining capacity; moisture pickup
吸水量测定 measurement of water adsorption
吸水滤网 suction screen
吸水率 absorption rate; coefficient of water adsorption; degree of water adsorption; specific adsorption; specific water absorption; suction rate; water absorption; water absorptivity; water intake rate
吸水率试验 water absorptivity test
吸水模板 absorbent shutter; moisture-absorbing formwork; vacuum form; vacuum shutter(ing); water-absorbing form
吸水内底 water-absorbent insole
吸水能力 absorbent power; absorbility; absorptive capacity for water; infiltration capacity; water-absorbed capacity; water-absorbing capacity; water-absorptive capacity; water retaining capacity
吸水膨胀 adsorption swelling; imbibition; water expansion
吸水膨胀倍数 expansion multiple after water absorption
吸水平管 suction pipe
吸水器 aspirator; water aspirator
吸水曲线 water suction curve
吸水容量 water-absorbent capacity
吸水软管 suction hose; water suction (rubber) hose
吸水纱布 absorbent gauze
吸水砂岩 water sensitive sandstone
吸水石 water-absorbing stone
吸水时间 absorbing time
吸水式梯田 absorption-type terrace
吸水试验 absorption test; water absorption test; sorptivity test
吸水受水器 suction header
吸水水池 break cistern
吸水水头 draft head; draught head; suction head; suction lift
吸水损失水量 absorption loss
吸水速率 absorption rate
吸水特性 suction characteristic; water absorption character
吸水梯地 absorptive terrace
吸水头 sucking head; sucking water head; suction head; suction water head
吸水弯管 suction elbow
吸水网 suction grid
吸水系数 adsorption coefficient; coefficient of moisture absorption; hygroscopic water coefficient
吸水狭长开口槽式溢洪道 suction-slot spillway
吸水现象 water soaking
吸水箱 suction box; suction tank; vacuum box
吸水橡胶管 water suction (rubber) hose
吸水性 absorption of water; hygroscopicity; hygroscopic property; hygroscopy; water-absorbing quality; water absorptivity; water absorption
吸水性的 absorptive
吸水性聚合物 water-absorbing polymer
吸水性模板衬里 absorptive form lining; absorptive lining
吸水性能 absorptive capacity for water
吸水性能测定仪 bibliometer
吸水旋塞 suction cock
吸水漩涡 suction vortex
吸水压力 suction pressure; suction tension
吸水压榨 suction press
吸水压榨辊 suction press roll
吸水岩层 water sensitive formation
吸水扬程 suction head
吸水页岩 water sensitive shale
吸水液化 deliquescent
吸水与放泄阀簧 suction and discharge valve
吸水与放泄阀座 suction and discharge valve seat
吸水纸 absorbent paper; bibulous paper; drinking paper
吸水纸模型 blotter model
吸水指数 injectivity index
吸水肘管 suction elbow
吸水总管 main suction; suction main
吸水作用 absorption of water; water absorption
吸送式风力输送器 pneumatic conveyer[conveyor] with suction
吸送式气力输送机 suction type pneumatic conveyer[conveyor]
吸送式气力输送器 suction pneumatic conveyer[conveyor]
吸提暗渠 suction culvert
吸铁吊具 lifting magnet
吸铁石 lodestone; magnet
吸铁式电动机 attracted-iron motor
吸筒 sucking tube; suction tube
吸头 sucker
吸污植物 pollution-absorbing plant
吸物 absorbate
吸行波 suction wave
吸压泵 combined suction and force pump; suction and force pump
吸压混合式气力输送 suction and compressed air conveying
吸压混合式通风 exhausting and blowing combined ventilation system
吸压力 combined suction and force pump; suction pressure
吸压两用泵 sucking and forcing pump; combined suction and force pump
吸压两用传送系统 combined suction and pressure conveying system
吸压式通风机 suction and-pressure fan
吸压双作用泵 double-acting pump
吸烟 smoking
吸烟车 smoking car
吸烟车厢 <火车上的> smoker; smoke car(riage)
吸烟道 suction flue
吸烟间 <铁路客车> smoking compartment; smoking room
吸烟器 smoke absorbing device
吸烟舌 smoker's tongue
吸烟设备 smoking accommodation
吸烟室 smoke room; smoking room; swing room
吸烟室排出物 smoke house emission
吸烟眩晕 dizziness caused by smoking
吸烟站 fag station
吸烟者 smoker
吸焰器 flame trap
吸扬暗渠 suction culvert
吸扬泵 aspiration pump; lift pump; suction pump
吸扬采砂船 sand suction dredge(r)
吸扬涵洞 suction culvert
吸扬能力 suction capacity
吸扬设备 suction equipment; suction plant; suction unit
吸扬式 suction type
吸扬式泵 aspiration pump
吸扬式采砂船 aggregate suction dredge(r)
吸扬式开底挖泥船 suction hopper dredge(r); trailing suction hopper dredge(r)

吸扬式排水 suction method drainage
吸扬式疏浚机 hydraulic dredge(r)
吸扬式挖泥船 cutter-head dredge(r); cutter-head suction dredge(r); pump dredge; discharging dredge(r); hopper dredge(r); hydraulic(pipe line) dredge(r); hydraulic suction dredge(r); pipeline dredge(r); pump dredge(r); reclamation dredge (r); sand pump dredge (r); sand-sucker; suction dredge(r)
吸扬式挖泥机 cutter-head dredge(r); discharging dredge(r); hydraulic pipe line dredge(r); suction cutter dredge(r); suction dredge(r); pump dredge(r)
吸扬式挖泥沙装置 suction cutter apparatus
吸扬装舱挖泥船 hopper suction dredge(r); suction hopper dredge(r)
吸氧 oxygen uptake
吸氧剂 oxygen absorbent
吸氧减压(法) oxygen decompression
吸氧量 < 水样在 27℃ 时 4 小时自过锰酸盐中吸取的氧量 > oxygen absorbed
吸氧率 oxygen uptake rate
吸氧器 oxygen inhaler
吸氧装置 respiratory device
吸液 imbibition
吸液泵 aspirator
吸液点 entry point
吸液管 pipet(te)
吸液管架 pipet(te) rack; pipet(te) stand
吸液量 liquid absorption
吸液喷射器 lifting injector
吸液膨润 imbibe
吸液膨胀 imbibe
吸液器 syringe
吸液速度 rate of liquid aspiration
吸液印相法 imbibition process
吸液罩 fluid aspirator
吸液针 aspirating needle
吸液作用 imbibing; wicking < 光缆 >
吸移管 pipet(te); volumetric(al) pipet(te)
吸移管法 < 试验土级配的 > pipet(te) method
吸移管管理器 pipettor
吸音 deadening; noise-absorbing; noise absorption; noise deadening; sound absorption; sound deadening
吸音板 abat-voix; absorptive lining; acoustic(al)(filter) board; acoustic(al) tile; coffer
吸音材料 acoustic(al) absorbent; acoustic(al)(absorber) material; damping material; sound-absorbing material
吸音单位 unit of acoustic(al) absorption
吸音的 acoustic(al)
吸音吊顶 acoustic(al) ceiling
吸音顶棚 acoustic(al) ceiling; acoustic(al) ceiling board
吸音顶棚体系 acoustic(al) ceiling system
吸音粉刷 acoustic(al) plaster
吸音格子板 sound-absorbent form-board
吸音构造 acoustic(al) construction
吸音灰膏 acoustic(al) plaster
吸音灰泥 acoustic(al) plaster
吸音技术 acoustic(al) technology
吸音路面 noise-absorbing pavement
吸音率 acoustic(al) absorptivity; sound absorption coefficient
吸音门 acoustic(al) door
吸音面 sound-absorbing surface
吸音模型板 sound-absorbent form-board
吸音抹灰工作 acoustic(al) plaster-work
吸音能力 acoustic(al) absorptivity
吸音喷涂材料 acoustic(al) sprayed-on material
吸音平顶 acoustic(al) ceiling
吸音屏幕 acoustic(al) screen
吸音漆 acoustic(al) paint; anechoic paint
吸音嵌板 acoustic(al) panel
吸音墙粉 acoustic(al) plaster
吸音墙面砖 acoustic(al) wall tile
吸音穹隆 acoustic(al) vault
吸音设施 acoustic(al) absorbing device
吸音式消声器 arrestor muffler
吸音试验 sound absorption test
吸音室 anechoic chamber; dead room
吸音体 acoustic(al) absorbent; acoustic(al) absorber
吸音天花板 acoustic(al) ceiling board; acoustic(al) pay-in board
吸音贴面砖 acoustolith tile
吸音涂料 acoustic (al) paint; anti-noise paint
吸音系数 coefficient of sound absorbing; coefficient of sound adsorption
吸音纤维板 deadening board
吸音镶板 acoustic(al) panel
吸音油漆 acoustic (al) paint; anti-noise paint
吸音毡 acoustic(al) felt; baffle blanket; deadening felt
吸音砖 acoustic(al) tile
吸音阻力 acoustic(al) resistance
吸引 attractiveness; magnetize; pull; sucking; siphon off < 指吸引过境交通到另一条路 >
吸引半径 attraction radius; attractive radius
吸引比引导比 entrainment ratio
吸引导液法 aspiration drainage; suction drainage
吸引地区 catchment area; territory served
吸引点 attractor
吸引端 suction side
吸引范围 domain of attraction; traffic catchment area【铁】; attractive sphere < 设施服务 >
吸引高度 suction head; suction height; suction lift
吸引刮匙 suction curette
吸引管 draught tube; suction tube; sucker
吸引管线 suction line; upstream line
吸引交通 attracted traffic
吸引交通量 absorbed traffic volume; attracted traffic volume
吸引开关 suction shaft
吸引力 absorptivity; attraction(force); attractive force; attractive power; pick-up force; traction; allure
吸引率 absorptivity
吸引镊 suction forceps
吸引钮 suction button
吸引喷射器 lifting injector
吸引喷嘴泵 sucking jet pump
吸引瓶 suction bottle
吸引器 aspirator; suction apparatus
吸引器瓶架 suction bottle frame
吸引器头 suction aspirator tip
吸引圈 attractive circle; sphere of attraction
吸引设备 suction apparatus
吸引升力 suction lift
吸引时间 pull-up time
吸引式输送机 suction scavenging machine
吸引式输送设备 suction conveyer [conveyor]
吸引式输送装置 suction conveyer [conveyor]
吸引术 aspiration; suction
吸引瓦斯 suction gas
吸引外国资金 attracting foreign funds
吸引现象 suction phenomenon
吸引线圈 sucking coil
吸引相 suction phase
吸引型电动机中的涡流 eddy currents in attraction type motor
吸引型直线悬浮电动机 attraction type linear suspension motor
吸引压力 suction pressure
吸引引流法 aspiration drainage; suction drainage
吸引域 domain of attraction
吸引质量 attracting mass
吸引中心 attraction pole
吸引注射器 aspirating syringe
吸引装置 suction device
吸引资本标准 capital-attracting standard
吸引阻力 attraction resistance; suction resistance
吸引阻力场 attraction force field
吸引作用 sucking action; suction action; absorbing effect < 交通客、货流的 >
吸油 blot
吸油泵 drainage pump; oil scavenge pump; oil suction pump
吸油材料 blotting material; oil absorption material; oil adsorption material
吸油层 blotter coat
吸油处理 blotter treatment
吸油的 oil-absorbing; oil absorption; oil-attracting
吸油骨料 blotter aggregate
吸油管 intake pipe; oil suction pipe; oil suction tube; sucking pipe
吸油管连接器 oil suction pipe adapter
吸油管路 inlet line; intake line
吸油管系 suction line
吸油合金 oilite
吸油后白度 brightness in oil
吸油后毛体积比重 < 集料的 > bulk impregnated specific gravity
吸油集料 blotter; blotter aggregate
吸油井管 well head
吸油口 inlet port; oil suction
吸油量 oil absorption(value); oil absorption volume; oil adsorption content; oil number
吸油能力 blotting capacity; oil absorption
吸油器 blotter; oil header
吸油栅 absorbing boom
吸油绳 wick
吸油绳润滑器 wick lubricator
吸油试验 oil absorption test
吸油系统 suction line; suction system
吸油橡胶管 oil suction rubber hose
吸油性聚合物 oil-absorbing polymer
吸油性能 oil absorbency; oil absorptiveness
吸油性能试验 oil absorption test
吸油性黏[粘]合剂 oil absorbent adhesive
吸油值 oil factor
吸釉 < 窑具吸取釉中挥发性成分 > sucking
吸鱼泵 fish pump
吸涨水 imbibition water
吸涨体 imbibant
吸涨压 imbibitional pressure
吸胀 imbibition
吸胀水 imbibitional water
吸胀体 imbibant
吸胀压 imbibition pressure
吸针 suction spindle
吸振基础 vibration-absorbing base
吸振能力 damping capacity
吸振器 bump level(l)er; vibration absorber
吸震能力 damping capacity
吸震器 bump level(l)er; shock suppresser[suppressor]
吸枝(植物)的 suckering
吸纸台 suction table
吸锥 suction cone
吸着 absorption; sorbing
吸着层 absorbed layer
吸着等温线 sorption isotherm
吸着点 sorption site
吸着分离法 separation by sorption
吸着和解吸作用 sorption and desorption
吸着剂 sorbent
吸着剂面浮油回收装置 sorbent surface skimmer
吸着剂面撇油装置 sorbent surface skimmer
吸着剂原料 sorbent feed
吸着检测器 sorption detector
吸着介质 sorptive medium
吸着界限 absorption limit; adsorption limit
吸着能力 absorptive capacity; sorptive power
吸着平衡 sorption equilibrium
吸着强度 sorption strength
吸着热检测器 heat of sorption detector
吸着时间 hold(ing) time
吸着水 absorbed water; absorption water; attached(ground) water; hygroscopic moisture; hygroscopic water; unfree water; held water < 土壤内地下水位以上的 >; retained water < 土壤岩石内的 >
吸着水分 entrapped moisture; hygroscopic moisture
吸着水含量 hygroscopic water content
吸着水校正系数 hygroscopic moisture correction factor
吸着水结构 structure of absorbed water
吸着水量 hygroscopic moisture content
吸着速率 sorption rate
吸着损失 adsorption loss
吸着梯度 sorption gradient
吸着物 sorbate
吸着稀薄化作用 sorption exhaust
吸着性能 sorption property
吸着油 occluded oil
吸着质 sorptive
吸着滞后现象 sorption hysteresis
吸着作用 sorption
吸紫外玻璃 ultraviolet absorbing glass
吸阻力 suction drag
吸嘴 suction mouth; suction nozzle
吸嘴滤网 suction strainer

希柏登结 Heberden's node; Heberden-Rosenbach node

希宾石 khibinskite
希宾式轻巧家具 < 英国 18 世纪的 > Chippendale

希宾岩 chibinite
希伯尼-沙莱里有铰腕足动物地理大区 Hiberno-Salairian articulate brachiopod region
希伯特标准 Hibbert standard
希伯特电池 Hibbert cell
希得波兰特萃取机 Hildebrandt extractor
希德鲁阶【地】Chideruan
希尔阿丹斯止门滑轮 Hill adams
希尔伯恩探测器 Hilborn detector
希尔伯特变换式 Hilbert's transform(ation)
希尔伯特变压器 Hilbert's transformer
希尔伯特不变积分 Hilbert's invariant integral
希尔伯特超平行体 Hilbert's parallelotope
希尔伯特多项式 Hilbert's polynomial
希尔伯特概型 Hilbert's scheme
希尔伯特基 Hilbertian basis
希尔伯特基定理 Hilbert's basis theorem
希尔伯特矩阵 Hilbert's matrix
希尔伯特空间 Hilbert's space
希尔伯特立体 Hilbert's cube
希尔伯特零点定理 Hilbert's null theorem
希尔伯特模函数 Hilbert's modular function
希尔伯特模群 Hilbert's modular group
希尔伯特模形式 Hilbert's modular form
希尔伯特特征函数 Hilbert's characteristic function
希尔伯特型问题 Hilbert-type problem
希尔德布兰德电解池 Hildebrand cell
希尔德布兰德函数 Hildebrand function
希尔方程式 Hill's equation
希尔令型浮选槽 Hearing cell
希尔绿 cupric arsenite
希尔斯大厦 Sears Building
希尔特定律 Hilt's law
希尔行列 Hill determinant
希尔兹曲线 < 启动条件的 > Shields diagram
希耳伯特空间 Hilbert space
希夫碱 Schiff base
希格比模型 Higbie model
希格斯玻色子 Higgs bosons
希格斯机制 Higgs mechanism
希硅锆钠钙石 hiortdahlite
希金斯电炉 < 一种熔制耐火材料的电炉 > Higgins furnace
希克森黏[粘]度计 Hickson visco(si)meter
希拉克利绝缘材料 < 一种厚度不大于3英寸的检验材料,1英寸 = 0.0254米 > Heraklith
希拉克利菱苦土木屑建筑板 Heraklith magnesite-bound excelsior building slab

希腊爱奥尼柱式 Grecian Ionic order;Greek Ionic order
希腊彩条粗布 Greek stripes
希腊船级社 Greek Register;Hellenic Register(of Shipping)
希腊复古式 Greek revival;neo-Greek
希腊复兴式 Greek revival;neo-Greek
希腊古建筑正面上方的三角墙 eagle
希腊化 Hellenization
希腊化巴洛克 Hellenistic Baroque
希腊化巴西利卡 Hellenistic Basilica
希腊化长方形会堂 Hellenistic Basilica
希腊回纹饰 Green key
希腊回纹装饰 Ala Grecque;Grecian fret;Grecian key pattern;Greek key pattern;Greek fret
希腊桧 Greek juniper
希腊或罗马体育训练馆 palestra[复 palestrae]
希腊建筑 Hellenic architecture
希腊建筑花饰 < 以棕叶饰为基本装饰 > honeysuckle ornament
希腊建筑形式 Grecian architectural style;Greek architectural style
希腊教堂洗礼池 colymbethra
希腊教堂洗礼室 colymbethra
希腊教堂至圣所旁的房间 parabema
希腊剧院舞台 logeion
希腊剧院舞台中央通乐队的扶梯 Charonian steps
希腊剧院舞台中央通乐队的踏步 Charonian steps
希腊拉丁方 < 即拉丁方 >【数】 Graeco-Latin square
希腊拉丁方方法 Greco-Latin square method
希腊冷杉 Greek fir
希腊卵形装饰线脚 Greek ovolo mo(u)lding
希腊罗马式 Grecian-Roman(style);Greco-Roman style; Green-Roman style
希腊罗马式瓦 Greek(-and)-Roman tile
希腊罗马式戏院 odeom
希腊盆地 Hellas
希腊燃烧剂 Greek fire
希腊神殿的边墙 pteromata
希腊神话中的大力士 sam(p)son
希腊神庙 Grecian temple;Greek temple
希腊十字形 Greek cross
希腊石碑 Greek stela
希腊石灰石 sarcophagus
希腊式的格子花 aligreek
希腊式建筑 Grecian architecture
希腊式十字架 Grecian cross
希腊式挖花花边 Greek lace
希腊式屋面坡度 < 15° > Grecian pitch
希腊式屋面瓦 Greek roof tile
希腊式叶形装饰 Grecian acanthus leaf
希腊式柱 Grecian column;Greek column
希腊式装饰艺术 Grecian ornamental art;Greek decoration art;Greek ornamental art
希腊陶立克式 Grecian Doric order;Greek Doric order
希腊陶立克式神庙 Grecian Doric temple;Greek Doric temple
希腊陶立克式柱头 Grecian Doric capital;Greek Doric capital
希腊西南焚风 Sirocco di levante
希腊艺术 Greek art
希腊议院建筑 Greek Senate building
希腊原始木雕神像 xoanon
希腊正教会教堂 katabasis
希腊正教堂祭台下放圣骨和圣物处 catabasis
希腊柱式 Grecian order;Green order;Greek order
希腊柱型 Green order;Greek order
希腊砖石网状砌体 dictyotheton
希腊装饰 < 多在柱顶盘上用对称的树叶形和玫瑰花形装饰 > Greek ornament
希腊装饰线脚 Greek mo(u)lding
希腊装饰线条 Greek mo(u)lding
希腊字母 Greek alphabet;Greek letter
希腊字母第二十个字母 upsilon
希腊字母第十九个字母 tau
希莱尔定位法 Hilaire method
希兰导航系统 Shiran system
希立-肖模型 Hele-Shaw model
希列(打桩)公式 Hiley's formula
希列公式估算单桩承载力 estimating pile bearing capacity by Hiley formula
希鲁枢轴 Schiele's pivot
希南动态测功计 Heenan dynamic dynamometer
希南液力扭矩计 Heenan hydraulic torque meter
希佩克(海底)沉积土取样器 Shipek sediment sampler
希柔公式 Hero's formula
希萨尔反向极性带 Hissar reversed polarity zone
希萨尔反向极性时 Hissar reversed polarity chron
希萨尔反向极性时间带 Hissar reversed polarity chronzone
希氏公式 < 一种设计混凝土路面厚度的古典公式 > Sheet's formula
希望路线 hope line
希望路线图 hope line draft
希望寿命 expectation of life
希望线 desire line
希沃特 < 符号 Sv,剂量当量单位 > sivert

昔

昔日 once upon a time

析

析冰作用 ice segregation

析出 separate out;separation
析出点 freezing point
析出段 elutriation leg
析出挥发物 distillation
析出粒子 precipitation particle
析出粒子的取向 orientation of precipitate
析出量 flush out amount;settling out amount
析出气体 bubbling;liberation of gases;product gas
析出强化 precipitation strength
析出石墨 indigenous graphite
析出物 educt
析出相 precipitated phase
析出硬化 precipitation hardening
析得干性油 segregated oil;separated oil
析晶 crystallization;devitrification;devitrify
析晶结石 devitrification stone
析晶晶核 devitrification nuclei
析晶路程 crystallization path
析晶面 crystallization face
析晶倾向 tendency towards devitrification
析晶热 heat of crystallization
析晶温度 recrystallization temperature
析晶温度范围 crystalline range;temperature range of crystallization
析晶温度下限 lower limit of crystallization temperation
析离 isolation
析离体 dialyse;schlieren
析离体构造 segregation schlieren structure
析流器 diverter
析取 disjunction;extraction;scarify(ing);scorification
析取范式 disjunctive normal form
析取概念 disjunctive concept
析取合取目标树 disjunctive-conjunctive goal tree
析取检索 disjunctive search
析取克立格法 disjunctive Kriging method
析取克立格方差 disjunctive Kriging variance
析取克立格方程组 system of disjunctive Kriging equations
析取克立格估计量 disjunctive Kriging estimator
析取克立格估值直方图 histogram of disjunctive Kriging estimator value
析取项 disjunct
析取指令 extract instruction
析取字 extracter[extractor]
析色器 colo(u)r analysing filter
析水 bleeding
析水量 separate water volume
析水率 sweating rate
析水速率 drainage rate
析稀 vacuolate
析稀胶粒 vacuole
析稀作用 vacuolation
析相温度 phase separation temperature
析相作用 phase separation
析像 image analysis;image dissection
析像倍增管 image dissector multiplier
析像管 dissector;image dissector tube
析像管倍增器 dissector multiplier
析像管摄影机 image dissector camera
析像力不足 lack of resolution
析像能力 resolution power;resolving power
析像盘 disk scanner
析像器 image dissector
析像摄像机 image dissection camera
析像系数 resolution ratio;resolution response
析像线 resolution line
析像圆盘 disk scanner
析盐蒸发器 salting-out evapo(u)rator
析液时间 drainage time
析因 factorial;factotization
析因设计 factorial design
析因实验 factorial experiment;
析因实验分析 analysis of a factorial experiment
析因实验设计 factorial experiment design
析因试验 factorial test;factorial trial
析因数据的变换分析 transformational analysis of factorial data
析因相关矩阵 factoring correlation matrix
析银法 quartation
析皂 graining
析枝纹 < 陶瓷纹饰之一 > separate sprays

矽

矽尘 silicious dust

矽尘吸入 inhalation of silica particles
矽尘作用 action of silicium dust
矽肺 pulmonary silicosis;silicosis
矽肺病 anthracosilicosis;pneumonoultramicroscopic siliccovolcanoconiosis;silicosis
矽钢 silicon steel
矽钢片 magnet steel sheet
矽胶 silica gel
矽卡岩 skarn
矽卡岩白钨矿矿床 scheelite skarn deposit
矽卡岩化 skarnzation
矽卡岩化作用 skarnization
矽卡岩矿床 skarn ore deposit
矽卡岩矿田构造 skarn orefield structure
矽卡岩矿物 skarn mineral
矽卡岩硼矿床 boron-bearing skarn deposit

矽卡岩型方钍石矿床 skarn-thorianitic deposit
矽卡岩型硫铁矿床 skarn-type pyrite deposit
矽卡岩型钼矿床 skarn molybdenum deposit
矽卡岩型铅锌矿床 lead-zinc deposit in skarn
矽卡岩型铜矿床 skarn-type copper deposit
矽卡岩型锡矿床 skarn-type tin deposit
矽铝锌铅石 sauconite
矽镁带 sima sphere
矽镁石 humite
矽酸盐 silicate
矽酸盐沉着病 silicatosis
矽酸盐肺 silicate pneumcconiosis;silicatosis
矽炭银 agysical
矽铁肺 silicosiderosis
矽土 silica soil
矽线黑云片麻岩 sillimanite biotite gneiss
矽线石 fibrolite;sillimanite
矽线石白云母片麻岩 sillimanite muscovite gneiss
矽线石白云母片岩 sillimanite-muscovite-schist
矽线石白云母石英片岩 sillimanite muscovite quartz schist
矽线石二云母片岩 sillimanite dimicaceous schist
矽线石二云母石英片岩 sillimanite two mica quartz schist
矽线石二云片麻岩 sillimanite two mica gneiss
矽线石黑云母片岩 sillimanite-biotite schist
矽线石黑云母石英片岩 sillimanite biotite quartz schist
矽线石矿床 sillimanite deposit
矽线石矿石 ore grade sillimanite
矽线石榴二长片麻岩 sillimanite garnet potash-feldspar and plagioclase gneiss
矽线石榴钾长片麻岩 sillimanite garnet potash-feldspar gneiss
矽线石石英片岩 sillimanite-quartz schist
矽线石斜长正长麻粒岩 sillimanite plagioclase orthoclase granulite
矽藻土 bergmeal;moler

息

息差年率 interest differential

息粗 raucous breathing
息单 interest statement
息角 angle of repose
息漏失 spillover
息灭脉冲电平 pedestal
息票 certificate of interest;coupon;coupon sheet;interest coupon;interest warrant;talons
息票到期日 due date of coupon
息票登记簿 coupon register
息票调换券 talons
息票(公司)债券 coupon bond
息票利率 coupon rate
息票收益率 coupon yield
息票支付 coupon payments
息前税前收益 earnings before interest and tax
息钱 interest on money
息氏黏[粘]度 Sengler viscosity
息债 loan on interest
息折 interest passbook
息止位 rest position

牺

牺牲财产 sacrifice property

牺牲船舶 sacrifice of ship
牺牲的 sacrificial
牺牲电极 corrosion target
牺牲腐蚀保护法 sacrificial corrosion
牺牲金属 sacrificial metal
牺牲金属保护层 sacrificial metal coating
牺牲墓穴 sacrificial pit
牺牲品 sacrifice
牺牲戗台 sacrificial berm
牺牲涂层式防护 sacrificial protection
牺牲阳极 sacrificial anode;galvanic anode
牺牲阳极(保护)法 sacrificial anode protection
牺牲阳极防蚀法 cathodic protection by galvanic anodes
牺牲阳极腐蚀 sacrificial corrosion
牺牲者 victim

悉

悉尼港桥 <澳> Sydney Habo(u)r Bridge

悉尼自适应交通(控制)系统 <澳大利亚开发的的一种自适应交通信号网络控制系统> Sydney Coordinated Adaptive Traffic System

桸

桸榁 California redwood;coast redwood;redwood

淅

淅沥声 spatter

烯

烯丙胺 allylamine

烯丙醇 allyl alcohol;propenol;propenyl alcohol
烯丙醇聚合物 allyl alcohol polymer;allyl polymer
烯丙醇酯 allyl alcohol ester
烯丙基 allyl(group)
烯丙基丙酮 allylacetone
烯丙基的 allylic
烯丙基淀粉 allyl starch
烯丙基化 allylation
烯丙基化硫 allyl sulfide;diallyl sulfide;oil garlic;thioallyl ether
烯丙基聚合物 allyl polymer
烯丙基硫醇 allyl sulfhydrate
烯丙基硫醚 allyl sulphide
烯丙基卤 allyl halide
烯丙基氯 allyl chloride
烯丙基醚 allyl ether
烯丙基氰 allyl cyanide
烯丙基树脂 allyl resin
烯丙基树脂类 allyl resins
烯丙基塑料 allyl plastics
烯丙基缩水甘油醚 allyl glycidyl ether;glycidyl allyl ether
烯丙基溴 allyl bromide;bromoallylene
烯丙(类)塑料 allyl plastics
烯丙硫醇 allyl mercaptan
烯丙卤型 allylic halides type
烯丙取代 allylic substitution
烯丙树脂 allyl plastics;allyl resin
烯丙酯聚合物 allyl ester polymer
烯丙重排作用 allylic rearrangement
烯醇 enol;olefin(e) alcohol;olefinic alcohol
烯醇化程度 enolizability
烯醇化物 enolate
烯醇化作用 enolization
烯醇磷酸酯 enol phosphate
烯醇式 enol form
烯醇式丙酮酸 enol pyruvic acid
烯醇型 enol form
烯醇型脂 enolic ester type
烯化 alkylene
烯化氧 alkylene oxide
烯化氧聚合物 alkylene oxide polymer
烯化作用 olefination
烯键 olefinic bond;olefinic link
烯醛 olefin(e) aldehyde
烯炔 eneyne
烯炔化合物 enyne
烯炔烃 enyne
烯酸 olefin(e) acid
烯碳 olefinic carbon
烯烃 olefinic hydrocarbon;olefin(e) hydrocarbon
烯烃的水化作用 hydration of olefines
烯烃的转化 conversion of olefines
烯烃叠合合成润滑油 olefin(e) polymer oil
烯烃共聚物 olefin(e) copolymer
烯烃含量 olefin(e) content;olefinic content
烯烃化合物 olefin(e) compound
烯烃基 alkylene
烯烃聚合 olefinic polymerization
烯烃聚合油 olefin polymer oil
烯烃配位化合物 olefin(e) complex
烯烃燃料 olefinic fuel
烯烃树脂 olefin resin
烯烃塑料 olefin(e) plastics
烯烃纤维 olefin(e) fiber[fibre]
烯烃转化法 olefin(e) conversion process
烯烃族的 olefinic
烯酮 keten(e)
烯亚胺 alkyleneimine
烯属磺酸酯 olefin(e) sulfonate
烯属聚合作用 olefinic polymerization
烯属酸 olefinic acid
烯属碳水化合物 olefinic carbohydrate
烯属烃 olefin(e)

硒

硒钯矿 palladseite

硒宝石红玻璃 selenium ruby glass
硒钡铀矿 guilleminite
硒铋矿 guanajuatite
硒铋铅汞铜矿 petrovicite
硒铋银 bohdanowiczite
硒不锈钢 selenium stainless steel
硒层 selenium layer
硒的 selenic
硒的危害 selenium hazard
硒碲 selen-tellurjum
硒碲铋矿 Kawazulite
硒碲铋铅矿 poubaite
硒碲镍矿 kitkaite
硒堆 selenium pile;selenium stack
硒镉红颜料 selenium red
硒镉矿 cadmoselite
硒汞矿 tiemanite
硒钴矿 cobaltomenite
硒鼓复印 xerox
硒光电池 selenic photoelectric(al) cell;selenium cell;selenium photocell
硒光电池继电器 selenium-cell relay
硒光电管 selenic photoelectric(al) cell;selenium cell
硒光度计 selenium photometer
硒光敏电阻 selenium conductive cell
硒害 selenium hazard
硒合剂 selenium mixture
硒化 selenizing
硒化镉光电管 cadmium selenide cell
硒化镉颜料 cadmium selenide pigment
硒化合物 selenium compound
硒化钾 potassium selenide
硒化钠 sodium selenide
硒化镍 nickelous selenide
硒化铅光敏元件 lead selenide cell
硒化氢 hydrogen selenide
硒化物 selenide
硒化银 silver selenide
硒黄铜矿 eskebornite
硒金银矿 flschesserite
硒静电复印 xerox(copy);xerographic(al) printer;xerox machine
硒静电复印机 xerographic(al) printer
硒氪法 selenium-krypton method
硒矿 selenium ore
硒矿床 selenium deposit
硒矿指示植物 indicator plant of selenium
硒硫铋铅矿 weibullite;wittite
硒硫铋铅铜矿 proudite
硒硫铋铜铅矿 junoite
硒硫碲铋矿 csiklovaite
硒硫镍钴矿 selenio-siegenite
硒硫砷矿 jeromite
硒钼矿 drysdallite
硒镍矿 blockite
硒铅矾 olsacherite
硒铅矿 clausthalite
硒铅铜镍矿 petroseite;peuroseite
硒砷硫黄 jeromite
硒酸 selenic acid
硒酸钾 potassium selenate
硒酸钠 sodium selenate
硒酸镍 nickelous selenate
硒酸铷 rubidium selenate
硒酸盐 selenate
硒酸盐碲酸盐 selenates tellurates
硒铊硐矿 sabatierite
硒铊铁铜矿 bukocite
硒铊银铜矿 crookesite
硒锑铜矿 permingeatite
硒调色剂 selenium toner
硒铁 ferro-selenium
硒铁矿 achavalite
硒铜 selenium copper
硒铜钯矿 oosterboschite
硒铜辉铅铋矿 selenocosalite
硒铜矿 athabascaite;berzelianite;krutaite
硒铜蓝 klockmannite
硒铜镍矿 penroseite
硒铜铅铀矿 demesmaekerite
硒铜银矿 eucairite
硒铜铀矿 marthozite
硒铜钴矿 tyrrellite
硒土 selenium soil
硒污染 selenium contamination;selenium pollution
硒银矿 naumannite
硒黝铜矿 hakite
硒与铀矿指示植物 indicator plant of selenium and uranium
硒整流堆盒 selenium pile case
硒整流焊机 selenium rectifier welder
硒整流片 selenium cell
硒整流器 selenium rectifier;seletron
硒整流器盘 selenium rectifier board
硒质光电管 selenium cell
硒中毒 selenium intoxication;selenium poisoning;selenosis

犀

犀角 rhinoceros horn

犀角线 capricornoid

稀

稀矮植物区 fell-field

稀氨溶液 liquor ammoniae dilutus
稀拌 mushy consistence[consistency];wet consistence[consistency]

X

稀表性浮游生物 spanipelagic plankton
稀薄 rarefaction; rarity; subtilization; under-pressure; wateriness
稀薄波 rarefaction wave
稀薄大气 rarefied atmosphere
稀薄得很的 well-thinned
稀薄的 rare; subtle; watery; weak
稀薄度 tenuity
稀薄灌注 fine pouring
稀薄极限 <混合气燃烧的> lean-limit; limit of inflammability
稀薄空气 light air; rare(fied) air
稀薄气体 low-density gas; rarefied gas; rare gas
稀薄气体等离子体 rare gas plasma
稀薄气体动力学 rarefied gas dynamics
稀薄气体放电 discharge through rarefied gas
稀薄气体空气动力学 super-aerodynamics
稀薄手感 thin feel
稀薄细麻布 sheer lawns
稀薄液体 wash
稀薄因数 tenuity factor
稀薄因子 tenuity factor
稀薄纸浆 lapped pulp
稀薄状态 <土、混凝土的> sloppy condition
稀布阵列天线 thinned array antenna
稀的 thin
稀的浓度 dilute concentration
稀堆 <每层用垫木隔开> open piling
稀垛 open stacking
稀垛法 open stacking method
稀化 desaturation
稀化泥 rarefied mudflow
稀灰泥喷射机 fine plaster throwing machine
稀混合气 lean mix(ture); poor mix(ture)
稀混合气的位置 lean position
稀混合气发动机 lean-burn engine
稀混合气极限 lean mixture limit
稀混合气强度 lean mixture strength
稀混合气式净化排气发动机 lean-burn emission controlled engine
稀混合物 lean(fuel) mix(ture); sloppy mix(ture)
稀混凝土 sloppy concrete; slurry (-type) concrete
稀碱 dilute alkali
稀碱金属元素 rare alkaline metals
稀碱式醋酸铅溶液 liquor plumbi subacetatis dilutus
稀碱液 sig water
稀见品种 rare species
稀浆 slurry
稀浆传运法 water slurry transportation
稀浆封层 slurry seal; slurry seal coat
稀浆管道 slurry pipeline
稀浆结度 thin flowing consistency
稀浆进料 slurry feed
稀浆喂料 slurry feed
稀浆炸药 slurry explosive
稀浆制备 slurry preparation
稀空法 spacing loosened method
稀料 diluent; diluting agent; paint thinner; thinner
稀料浆 fluid slurry; thin slurry
稀流草层 loose grasses
稀流输送 lean phase conveying
稀硫酸 dilute sulphuric acid
稀硫酸贮槽 dilute sulfuric acid storage tank
稀路由电话通信[讯] thin-route telephone message
稀路由电信 thin-route telecommunication
稀氯化钙回流泵 diluted CaCl2 solution return pump
稀泥 mud; ooze
稀泥浆 clay emulsion; liquid slurry; mud slush; slurry; slush; thin mud
稀泥浆自动排出船外装置【疏】automatic light mixture overboard-discharging device; automatic overboard-dump system for light mixture
稀泥砂 slurry
稀泥炭 muskeg
稀黏[粘]程度 sliminess
稀黏[粘]稠度 thin consistency
稀黏[粘]度 microviscosity
稀黏[粘]性 sliminess
稀黏[粘]滞度 thin consistency; thin viscosity
稀铺屋面板 open boarding
稀漆 diluted paint
稀漆剂 lacquer diluent; lacquer thinner; thinning additive; thinning agent
稀奇的 strange
稀气层 deaerating layer
稀墙泥 thin plaster
稀缺程度高 high degree of scarcity
稀缺规律 law of scarcity
稀缺价值 scarcity value
稀缺商品 bottleneck commodity; exclusive
稀缺性 scarcity
稀燃 lean-burn
稀燃料混气比 lean fuel-air ratio
稀燃料空气比 lean fuel-air ratio
稀溶液 dilute solution; weak liquor; weak solution
稀散波 depression-type wave; expansion fan; expansion shock
稀散金属矿产 disperse metals commodity
稀散元素 disperse element
稀纱布 cheese cloth
稀砂浆 grout; slurry; thin mortar
稀少交通 little traffic
稀石灰浆 lime slurry
稀释 attenuate; attenuation; cutting back; desaturation; dilution; liquefaction; liquefy; sleak (ing); thinning; thin out
稀释安定性 dilution stability
稀释柏油 cutback road tar
稀释倍数 dilution factor; dilution ratio
稀释比 <河水水量与入河污水之比> dilution factor
稀释比例 dilution rate
稀释比率 dilution ratio; thinning ratio
稀释比色法 dilution colo(u)rimetry
稀释槽 thinning tank
稀释测定 dilution metering
稀释测量 dilution ga(u)ging
稀释测流法 dilution ga(u)ging
稀释产品 cutback product
稀释产物 cutback; cutback tar
稀释池 diluting tank
稀释稠度 thin consistency
稀释处理 treatment by dilution
稀释处理法 method of dilution
稀释处置 disposal by dilution
稀释的 diluted; liquefied
稀释滴度 dilution titer
稀释地沥青 cutback asphalt
稀释定律 law of dilution
稀释度 degree of dilution; dilutability; dilution
稀释度估算 estimation of dilution
稀释法 dilution approach; dilution method; method of dilution
稀释法测流 dilution ga(u)ging
稀释法排放污染物 disposal by dilution
稀释废水 dilute waste; dilute wastewater
稀释分离 separating by dilution
稀释分析 dilution analysis
稀释管 <测爆仪用> dilution tube
稀释罐 thinning tank
稀释好的 well-thinned
稀释(后)水量 diluted discharge
稀释糊 reduction thickening
稀释回配 cutting back
稀释机制 dilution mechanism
稀释极限 dilution limit
稀释剂 attenuant; desaturator; diluent (material); diluter; diluting agent; dilution agent; flux(ing agent); flux paste; liquefier; liquifier; thinner; thinning agent; flux oil <沥青的>
稀释剂容器 diluting receiver
稀释浆 reduction paste
稀释胶剂 diluting the latex
稀释焦油 cutback tar
稀释焦油沥青 cutback road tar
稀释空气 diluent air; dilution air
稀释矿物酸 diluted mineral acid
稀释扩散 dilution and diffusion
稀释扩散规律 rule of diffusion with dilution
稀释力 dilute power; diluting power
稀释沥青 cutback (asphaltic) bitumen; doped cutback <掺添加剂的>; air blow asphalt
稀释沥青黏[粘]合料 cutback asphalt binder
稀释沥青乳化剂 cutback asphaltic (bitumen) emulsion
稀释沥青乳胶 cutback asphalt(bitumen) emulsion
稀释沥青乳浊液 cutback asphalt emulsion
稀释沥青蒸馏装置 cutback asphaltic bitumen distillation apparatus
稀释量 dilute volume
稀释量试验 dilution test
稀释料浆 diluted slurry
稀释流动点 dilute pour point
稀释率 dilution ratio; rate of dilution
稀释灭火 extinguishment by dilution
稀释能力 dilution capacity
稀释黏[粘]度计 dilution visco(si)meter
稀释浓度 diluted concentration
稀释浓缩试验 dilution and concentration test
稀释排放 dilution discharge
稀释气体 diluent gas; dilution gas
稀释倾点 dilute pour point
稀释区 diluent zone; dilution zone
稀释燃料油 fuel dilution
稀释燃烧产物 diluted combustion product
稀释热 differential heat of dilution; dilution heat; heat of dilution
稀释溶剂 diluent solvent; fluxing medium; reducing solvent; retarder thinner
稀释溶液 diluted solution
稀释乳化油 diluted soluble oil
稀释润滑剂 diluted lubricant
稀释熵 entropy of dilution
稀释石油 thinned oil
稀释水 dilute water; dilution water
稀释水溶液 dilute aqueous solution
稀释水系统 dilute aquatic system
稀释速度 dilution velocity
稀释速率 dilution rate
稀释通风 dilution ventilation
稀释微分热 partial heat of dilution
稀释稳定性 dilution stability
稀释污泥体积指数 diluted sludge volume index
稀释污水 dilute sewage; dilute wastewater
稀释无机酸 diluted mineral acid
稀释系数 coefficient of dilution; dilution factor
稀释现象 dilution phenomenon
稀释限度 dilution limit; thinning limit
稀释效率 dilution efficiency
稀释效应 dilution effect; effect of dilution
稀释性能试验 reducibility test
稀释性试验 dilution property test
稀释需水量 dilution requirement
稀释氧化塘 dilution oxidation pond
稀释要求 dilution requirement
稀释液 dilution solution
稀释因子 diluting factor; dilution factor
稀释因子法 diluted factor approach
稀释用石灰 fluxing lime
稀释用水 thinned water
稀释用印浆 reduction clear
稀释油 flux(ing) oil
稀释有毒气体需风量 air requirement by diluting harmful gas
稀释有机物水混合物 dilute organic-water mixture
稀释与接种法 dilution and seeding method
稀释预测 dilution prediction
稀释制冷机 dilution refrigerator
稀释终点 dilution end point
稀释猪猡污水 dilute swine wastewater
稀释作用 diluting effect; dilution process
稀释作用的扩散规律 rule of diffusion with dilution
稀疏傍管薄壁组织 scanty paratracheal parenchyma
稀疏冰 broken ice; loose pack ice; open pack ice; slack ice
稀疏波 decompression wave; expansion wave; rarefaction wave; suction wave
稀疏的 rare; sparse
稀疏的植物 sparse vegetation
稀疏浮冰 sailing ice; scattered ice
稀疏浮冰群 open drift ice; open pack ice
稀疏化 rarefaction
稀疏混合物 rarefied mixture
稀疏集居 dispersed settlement
稀疏交通 sparse traffic
稀疏矩阵 spare matrix; sparse matrix
稀疏控制 distance ground control
稀疏林地 sparse wood
稀疏流冰群 loose pack ice; open drift ice; open pack ice; sailing ice; slack ice
稀疏逆矩阵 sparse inverse
稀疏气体 rarefied gas
稀疏区 rarefaction
稀疏群丛 open association
稀疏群落 open community
稀疏群系 open formation
稀疏索引 sparse index
稀疏天线阵 thinned array
稀疏微波型 rare minute echoes
稀疏纹理 semi-open texture
稀疏细雨 thin rain
稀疏向量 sparse vector
稀疏像限 rarefaction quadrant
稀疏隐蔽 sparse concealment
稀疏阵列天线 thinned array antenna
稀疏植被 sparse vegetation
稀疏状态 rarefaction state
稀疏作用 rarefaction
稀树草原 park land; savanna(h)
稀树草原气候 savanna(h) climate

稀树草原疏林 savanna(h) woodland
稀树干草原 serrados;tree savanna
稀水泥浆 cement(-water) grout;thin cement grout; thin cement slurry; wet cement paste
稀松窗帘用布 scrim
稀松织物背衬 double back
稀松织物衬 scrim back
稀酸 acidic dilute;dilute(d) acid
稀酸洗涤 dilute acid washing
稀酸液 pickle
稀土 rare earth;tombarthite
稀土贝塔石 rare earth betafite
稀土玻璃 rare earth glass
稀土掺杂光纤 rare earth doped optic(al) fiber
稀土掺杂剂 rare earth dopant
稀土除去系统 rare earth removal system
稀土磁体 rare earth magnet
稀土催化双氧水氧化工艺 rare earth catalysis-H_2O_2 oxidation process
稀土发光材料 rare earth luminescent material
稀土锆石 hagatalite;oyamalite;riberirite
稀土钴磁钢 rare earth cobalt magnet
稀土钴磁铁 cobalt rare earth magnet
稀土光学玻璃 rare earth optic(al) glass
稀土合金 rare earth alloy
稀土化合物类 rare earth compounds
稀土金属 rare earth metal;rare metal
稀土金属催干剂 rare earth drier[dryer]
稀土金属合金 remalloy
稀土金属化合物 rare earth compound
稀土金属混合物 mischmetal
稀土金属矿产 rare earth commodity
稀土金属矿物 rare earth mineral
稀土金属氧化物 rare earth oxidation
稀土金属氧化物阴极 rare earth oxide-coated cathode
稀土金属与镁的合金 kunheim metal
稀土精矿分解废水 rare earth concentrate decomposition wastewater
稀土矿(石) rare earth ore
稀土类荧光粉 rare earth phosphor
稀土离子 rare earth ion
稀土沥青 carbocer
稀土磷铀矿 lermontovite
稀土络合物 rare earth complex
稀土镁合金 magnesium-rare earth
稀土钠硬硅钙石 rare earth miserite
稀土石榴石 rare earth garnet
稀土添加剂 rare earth addition
稀土铁石榴子石 rare earth iron garnet
稀土盐 rare earth salts
稀土氧化物 rare earth oxide
稀土氧化物荧光粉 rare earth oxide phosphor
稀土元素 rare earth element; rare earth metal
稀土元素分析 rare earth element analysis
稀土正铁氧体 rare earth orthoferrite
稀土总量 total rare earth content
稀纹 fast groove
稀污泥 slurry
稀污水 dilute sewage; dilute sewerage;weak sewage
稀雾 mist or thin fog
稀析胶 <油漆用> glue-size
稀相 dilute phase;lean phase
稀相床 dilute phase bed; lean phase bed
稀硝酸 weak nitric acid
稀性泥石流 diluted debris flow;liquid mud-stone flow
稀性泥石流沉积 thinned-debris-flow deposit
稀盐 dilute salt
稀盐酸 dilute hydrochloric acid
稀液 thin liquid
稀油 thin oil
稀油墨 soft ink
稀油漆 lean paint
稀油润滑系统 oil lubricating system
稀有 rarity
稀有暴风雨 <在美国是 10 年或 25 年一遇的大暴风雨> occasional storm
稀有暴雨 occasional storm
稀有本 rare issue
稀有潮 abnormal tide
稀有的 infrequent;phenomenal;rare; scarce; uncommon; unusual; unwonted
稀有动物 rare animal
稀有度 rarity
稀有分类群 rare taxa
稀有贵金属 rare noble metal
稀有花饰贴面 rare veneer
稀有混合物 rare mixture
稀有碱基 minor base;rare base
稀有碱土金属 rare alkaline earth metal
稀有金属 exotic metal; rare metal; scarce metal
稀有金属工业 rare metals industry
稀有金属矿产 rare earth commodities
稀有金属矿石 rare-metal ore
稀有金属矿物 mineral of rare metals
稀有金属矿异常 anomaly of rare metals ore
稀有金属热电偶 rare-metal couple
稀有金属温差电偶 rare-metal thermocouple
稀有矿物 rare mineral
稀有类型 infrequent types
稀有品种 rare variety
稀有气体 noble gas;rare gas
稀有气体法 rare gas method
稀有气体管 rare gas tube
稀有气体聚集法 rare gas polymerization method
稀有气体笼形包合物 rare gas clathrate
稀有气体笼形物 rare gas clathrate compound
稀有轻金属 rare light metal
稀有燃料 exotic fuel
稀有通货 scare currency
稀有图书馆 rare book library
稀有文献资料 rare materials
稀有稳定同位素 rare stable isotope
稀有物 rarity
稀有(物)种 rare species
稀有物资 critical material
稀有性 rarity
稀有性程度图 rare degree figure
稀有氧化物 rare oxide
稀有元素 rare element
稀有元素地球化学 geochemistry of rare elements
稀有元素分析 rare element analysis
稀有元素化学 rare element chemistry
稀有原料 scarce raw materials
稀淤泥 muskeg
稀遇潮 abnormal tide
稀遇洪水 rare flood
稀遇洪水频率 frequency of rare floods
稀遇水位 exceptional water level
稀植 spaced planting;thin planting
稀族 rare earths

舾装 apparel;apparel and tackle;equipment;fitting out;outfit

舾装泊位 <在船坞附近的水区> fitting-out basin; fitting-out berth; outfitting berth
舾装不全的 out of rig
舾装车间 fitting-out shop;outfitting shop
舾装船舶 fitting-out vessel
舾装船坞 fitting-out dock
舾装港口 fitting-out port
舾装工 outfitter
舾装工程 outfitting
舾装码头 fitting-out dock;fitting-out pier; fitting-out quay; fitting-out wharf; outfitting pier; outfitting quay
舾装品 outfit
舾装齐备 all found
舾装起重机 fitting out crane
舾装设备 outfitting
舾装突堤码头 outfitting pier

溪 beck; brooklet; creek; runnel; springlet

溪沟拦河坝 rubble-intercepting dam in brook
溪谷 dale;dean;dene;gole;linn;vale
溪谷口 embouchure
溪谷小坝 check dam
溪河群落 namatium
溪涧 mountain brook;mountain stream
溪口 brook outlet
溪口导坝 training dike at mountain brook
溪口导堤 rubble-guiding jetty at brook-outlet
溪口急滩 brook outlet rapids
溪流 arroyo;beck;bourn(e);brook; ravine stream; rill (e); riverlet; stream;rivulet
溪流涵洞 brook culvert
溪流截夺 stream piracy
溪流口 mouth of a stream
溪流侵蚀 rill erosion;stream erosion
溪流袭夺 stream piracy
溪水采样 sampling of stream water
溪线 river valley line

锡拔染法 tin discharge

锡钯矿 stannopalladinite
锡白 tin oxide;tin white
锡斑 tine plaque
锡板 sheet tin;tin sheet
锡包线 solder-covered wire
锡钡钛石 pabstite
锡病 tin plague
锡铂钯矿 atokite
锡箔 tin foil;tin leaf
锡箔电极 tin-foil electrode
锡箔电容器 tin-foil capacitor;tin-foil condenser
锡箔片 tin-foil plate
锡箔条 strip of tin foil
锡箔条回波 tin-foil echo
锡槽 tin bath
锡槽边封 side sealing of tin bath
锡槽底砖 tin bath bottom block; tin bath siege
锡槽空间分隔墙 tin bath partition wall
锡槽砖 tin bath block
锡尘病 stannosis
锡尘肺 stannosis
锡衬垫 tin liner
锡醇 tin spirit
锡锉 tin file
锡的 stannic;tinny
锡的电镀 tin electroplating
锡的多因复成矿床 polygenetic compound tin deposit
锡滴坑 <浮法玻璃缺陷> drip crater
锡钉 tin tack
锡锭 ingot tin; pig tin; tin ingot; tin slab
锡尔河 <为亚洲中部的内陆河,源于天山山脉,流经图兰低地后注入咸海> Syr Darya River
锡尔特盆地 Sirte basin
锡粉 glass putty;tin ash;tin powder
锡弗诺岛人宝库 <位于希腊德尔菲> Treasury of the Siphaians
锡覆盖层 tin coating
锡刚玉 nigerite
锡钢板 bright tin plate
锡钢皮容器 tinplate container
锡铬红 mineral lake
锡工 tinsmith
锡工厂 tinwork
锡工场 tinwork
锡汞合金 tin amalgam
锡管 tin-tube
锡管充药器 tin tube filler
锡管封口机 tin tube shutter
锡罐 tin can
锡光泽彩 tin lustre[luster]
锡汗 tin exudation;tin sweat(ing)
锡焊 soft soldering; soldering; tin burning;tin soldering;tin-lead bonding;tin welding
锡焊衬套 soldering bushing
锡焊搭接 plumb joint; soldered lap joint
锡焊的 soldered
锡焊灯 soldering lamp
锡焊缝 soldering seam
锡焊膏 tinol
锡焊管 soldered pipe
锡焊剂 tin solder
锡焊接 plumb joint;solder(ed) joint
锡焊接头 solder(ed) jointing
锡焊连接 solder joint
锡焊料 pewter solder;tinsmith's solder;tin solder
锡焊配件 soldered fitting
锡焊平接缝 flat soldered seam
锡焊铁 solder(ing) iron
锡焊咬合缝 flat-lock soldered seam
锡焊药 soft soldering flux
锡焊液 soldering fluid
锡合金 Ashbury metal;tin alloy
锡化物 stannide
锡黄铜 Chamet bronze;tin brass
锡灰 <锡与铅的混合氧化物> grey-tin;tin ash
锡基巴比合金 tin-base babbit; tin-base babbit metal
锡基白合金 ocpan
锡基白合金衬层 tin-base white-metal linings
锡基高强度轴承合金 Mota metal
锡基合金 kaiserzinn; Kamash alloy; queen's metal; tin-base (d) alloy; trifle
锡基密封合金 nickeline
锡基锑铅铜合金 Parson's brass
锡基锑铜轴承合金 Parson's metal
锡基轴承 tin-base bearing
锡基轴承合金 Hoyt's metal; Karmarsch alloy;tin-base babbit
锡匠 tinsmith;whitesmith
锡晶料 tin frit
锡精炼厂 tin refiner
锡精炼设备 tin-refining plant
锡卡西亚胡桃木 <欧洲南部> Circassian walnut
锡科人宝库 Treasury of Sikyon
锡块 block tin;pig tin;tin bar
锡矿 stannary;tin ore;wheal
锡矿工 tinner
锡矿矿工 tinner

锡矿脉 tin lode
锡矿区 stannary
锡矿山 tin mine
锡矿石 tin ore
锡矿异常 anomaly of tine ore
锡蜡 pewter
锡蜡制的 pewter
锡蜡制器皿 pewter
锡兰锆石 Matura diamonds
锡兰黑纹棕色硬木 coromandel
锡兰桃花心木或杉木 lunumidella
锡锂大隅石 brannockite
锡林格勒矿 xilingolite
锡林滚筒 cylinder;swift
锡磷青铜带 tin phosphorus bronze band
锡瘤 list edge
锡铝矿 nigerite
锡伦盘 Thelen pan
锡伦蒸发器 Thelen evapo(u)rator
锡罗科风 Sirocco
锡罗科式扇风机 sirocco fan
锡锰钽矿 wodginite
锡内穆阶 <晚侏罗世>【地】Sinemurian
锡镍 tin-nickel
锡镍镀层 tin-nickel coating
锡片 tin metal sheet;tin sheet
锡器 tinware
锡钎料 tin solder
锡铅铋易熔合金 Malott metal
锡铅各半焊料 half-and-half solder
锡铅焊料 tin-lead solder
锡铅合金 pewter;tinsel
锡铅合金焊料 fine solder
锡铅矿石 Sn-bearing lead ore
锡铅钎料 tin-lead solder
锡铅青铜 tin leaded bronze
锡铅软焊料 tinsmith solder
锡铅锑合金 mischzinn
锡铅轴承合金 ley
锡青铜 tin bronze
锡青铜合金 tin bronze alloys
锡热析 tin sweat(ing)
锡熔炼炉 tin smelter
锡砂 stream tin
锡砂矿基岩 shelf
锡石 black tin cassiterite;cassiterite;tin spar;tin stone
锡石-白钨-石英脉建造【地】cassiterite-scheelite-quartz vein formation
锡石含量 cassiterite
锡石矿石 cassiterite ore
锡石硫化物矿床 cassiterite-sulfide deposit
锡石石英脉建造 cassiterite-quartz vein formation
锡石石英脉矿床 cassiterite-quartz vein deposit
锡石云英岩 zwitter
锡试验 tin test
锡斯坦地块 Sistan massif
锡斯坦风 Seistan
锡酸 stannic acid
锡酸钡 barium stannate
锡酸铋 bismuth stannate
锡酸钙 calcium stannate
锡酸钙陶瓷 calcium stannate ceramics
锡酸钴蓝 cerulean blue
锡酸钾 potassium stannate
锡酸镁 magnesium stannate
锡酸钠 preparing salt;sodium stannate
锡酸铌尼奥斯坦(超导材料)Niostan
锡酸镍 nickel stannate
锡酸铅 lead stannate
锡酸锶 strontium stannate
锡酸盐 stannate
锡酸盐陶瓷 stannate ceramics
锡淘洗盘 tin buddle
锡特 <气流流量单位每小时英尺> scid
锡特加云杉 tideland spruce
锡锑焊料 queen metal
锡锑合金 tin pewter
锡锑合金板 plate pewter
锡锑铅青铜 Reith's alloy
锡锑铜合金 Babbitt(alloy);Britannia metal;Tutania alloy
锡锑系轴承合金 Alg(i)er alloy
锡条 tin bar;bar tin
锡铁合金的机械性混合物 scruff
锡铁山石 xitieshanite
锡铁钽矿 ixiolite
锡铜 tinned copper
锡铜钯矿 cabritte
锡铜合金 gun metal
锡铜锑合金 Britannia
锡铜锑合金焊接 Britannia joint
锡铜轴承合金 Fahry metal;Fahry alloy
锡涂层 tin coating
锡污板 scruff(y)plate
锡污染 pollution by tin
锡细晶石 stannomicrolite
锡锌 tin-zinc
锡锌合金 red brass
锡锌铅铜焊料 Soluminium
锡锌青铜 gun metal
锡盐 pink salt
锡液 <氯化亚锡溶液> tin liquor
锡疫 tin disease;tin pest;tin plague
锡铟薄膜 tin-indium film
锡油 tin oil
锡黝铜矿 colusite
锡釉 tin enamel;tin glaze
锡釉彩陶 <意大利> maioliaca;majolica
锡釉彩陶用色料 majolica colo(u)r
锡釉彩陶釉 majolica glaze
锡釉陶瓷 stanniferous glaze
锡釉陶器 faience;stanniferous faience;Tervueren faience
锡釉瓦 tin glaze tile
锡浴 tin bath
锡增量 tin weighting
锡增量处理 tin finish
锡渣 dross(spot);scruff;tin dross
锡渣带 dross band
锡纸 silver paper;tin foil;tin-foil paper
锡制品 tinwork
锡中毒 tin poisoning
锡珠 tin sweat(ing)
锡紫 tin violet
锡组 tin group

熄灯 light out

熄灯器 extinguisher
熄弧 arc blow-out;quenching of arc
熄弧磁铁 blowout magnet
熄弧隔板 arc deflector
熄弧沟 arc chute
熄弧脉冲电平 pedestal
熄弧能力 arc-rupturing capacity
熄弧器 arc arrester;arc suppressor
熄弧器触点 blowout contact
熄弧室 expulsion element
熄火 annihilate; annihilation;brennschluss;extinction of a flame;fire off; flame failure; flame-out;quenching of a flame
熄火安全装置 flame failure device;flame safeguard
熄火保护装置 flame failure device;flame-out trip device; flame safeguard
熄火点 burnout;cut-off point
熄火电位 extinction potential
熄火后的稳定 post cut-off stabilization
熄火滑行 coasting with engine off
熄火脉冲 extinguishing pulse;turn-off pulse
熄火山 dormant volcano;extinguished volcano
熄火时间 burn-out time
熄火时刻 brennschluss time
熄火阴极电流 off cathode current
熄火蒸汽吹入管 steam inlet for snuffing
熄火直径 quenching diameter
熄火装置 blanket
熄焦废水 coke quenching effluent
熄焦机 coke quenching machine
熄焦水 coke quenching water
熄灭 amortize; blackout; blanketing;blanking; blankoff; blowing-out;blowout;burn-out;cancellation;extinct;extinguishing;extinguishment;flame-out;go-out;put out;quench;stifle;suffocating
熄灭波 blanking wave
熄灭灯光 blackout
熄灭电路 blanking circuit;killer circuit;quench circuit
熄灭电平 blanking level
熄灭电势 extinction potential
熄灭电压 blackout voltage;blanking voltage;extinction voltage <闸流管>
熄灭过程 blanket process
熄灭含水量 extinction moisture content
熄灭火花 quench
熄灭火花装置 spark quench device
熄灭极限 extinction limit
熄灭率 extinctivity
熄灭脉冲 blackout pulse; blanking pulse;pedestal pulse
熄灭脉冲电平 blanking pedestal;pedestal level
熄灭脉冲放大器 blanking(pulse)amplifier
熄灭脉冲黑带 blanking bar
熄灭器 annihilator; extinguisher;quencher
熄灭设备 blowdown facility
熄灭时间 fall time;fire out time
熄灭式火花发报机 quenched spark transmitter
熄灭握手 stop-engine button
熄灭信号持续时间 blanking time
熄灭信号电平 blanking level
熄灭者 extinguisher
熄灭装置 blanker

熙和园 Summer Palace

熙和园乐寿堂 Hall of Happiness and Longevity
熙和园仁寿殿 Hall of Benevolene and Longevity
熙和园玉澜堂 Hall of Jade Ripples
熙提 <表面亮度单位> stilb

膝部 knee

膝部向外弯的 bandy
膝动式独立悬架 knee action suspension
膝杆 knee lever
膝曲褶皱【地】knee fold
膝上型计算机 laptop computer
膝形杆 knee
膝形杆动作 knee action
膝形构造 knee-type structure
膝形夹头 knee clamp
膝形铰接作用 knee action
膝形拉条 knee brace
膝形犁刀 knee colter
膝形托架 knee bracket
膝形悬臂托架 knee-braced bracket
膝折带 kink band
膝折带产状 attitude of kink band
膝折带宽度 width of kink band
膝折褶皱 kink fold(ing)
膝褶皱 kink fold
膝状的 geniculate;genual
膝状剪 angled scissors
膝状铰链 knee butt
膝状曲线弯曲电平 knee level
膝状曲线弯曲压缩 knee compression
膝状双晶 knee twin
膝状体 corpus geniculum;geniculate body
膝状物 knee

榍头 gable wall head

熹微光 first light

习得反应 acquired response

习得特性 acquired characteristic
习得行为 learned behavio(u)r
习惯 convention;custom;habit;knack;wont
习惯包装 conventional packing;customary packing
习惯保有 customary tenure
习惯保有财产 customary tenure estates
习惯程序 customary procedure
习惯持续性假设 habit persistence hypothesis
习惯持续性模型 habit persistence model
习惯订价法 customary pricing
习惯法抵押 common law mortgage
习惯法(规)common law;customary law;tacit law;unwritten law
习惯法信托机构 common law trust
习惯法责任 common law liability
习惯规则 customary rule
习惯国际法 customary international law
习惯航线 customary route;customary voyage
习惯化 habituation
习惯货币 customary money
习惯居所 habitual residence; usual residence
习惯快速装卸 customary quick despatch[despatch]
习惯面 habit plane
习惯命名法 common nomenclature;customary nomenclature;trivial nomenclature
习惯磨耗 customary deductions
习惯皮重 customary tare
习惯期限 <支付外国汇票的> usance
习惯期限的汇票 bill at usance;time draft
习惯商法 custom of merchant
习惯特征的测验 test of customary characteristics
习惯响度级 habitual loudness level
习惯形成 habituation
习惯形成原理 habit formation principle
习惯性 habituation

习惯(性的)价格 customary price
习惯因素 habit factor
习惯用的材料 orthodox material
习惯用漆 conventional paint
习惯与惯例 custom and practice
习惯运费单位 customary freight unit
习惯装卸速度 customary dispatch; customary quick dispatch[despatch]
习惯作用 habituation
习惯做法 accepted practice; conventional method; conventional practice; practices prevailing; regular practice
习生地 habitat; natural abode
习俗 convention; custom; habitude
习俗的 orthodox
习性 habit(us); propensity
习性变化 habit modification
习性面 habit face; habit plane
习性谱 ethogram
习性学 ethology
习艺所 work-house
习用表示法 conventional representation
习用的 conventional
习用滴滤池 conventional trickling filter
习用阀 conventional valve
习用方法 conventional approach; convention(al) method
习用符号 conventional sign; conventional symbol
习用荷载 conventional load(ing)
习用活性污泥法 conventional activated sludge process
习用集料 conventional aggregate
习用计量单位 conventional measurement unit
习用夹套 conventional jacket
习用掘进法 conventional drivage method
习用扩散曝气池 conventional diffused aeration tank
习用配合法 arbitrary proportioning; arbitrary proportion method
习用配料法 arbitrary proportioning; arbitrary proportion method
习用润滑油 conventional type lubricant
习用投影 conventional projection
习用图表 conventional diagram
习用型 conventional type
习用性 convention
习用凿井方法 conventional shaft sinking method

席

席包 mats

席播 mat seeding
席不暇暖式的管理 seat of the pants management style
席冲断层 sheet thrust
席夫碱 Schiff base
席工 mat work
席间电路 interposition circuit
席间联络电路 order wire circuit
席卷 engulf
席勒绿 Scheele's green
席勒塑性仪 <测泥浆流动性用> Shearer plastometer
席理 sheeting
席梦思床垫 box spring
席式防波堤 hovering breakwater
席式供暖系统 cable mat heating system
席式供热系统 cable mat heating system
席位 gallery
席纹地板 basket weave parquetry flooring; inlaid parquet; parquet floor; parquetry flooring in basket weave pattern
席纹地面 inlaid parquet; parquet
席纹马赛克 basket weave mosaic
席纹木地板 herringbone parquetry
席纹铺砖 basket weave pattern brick
席纹砌法 basket weave bond
席纹砌合 basket weave bond; diaper bond
席纹式 herringbone pattern
席纹图案 basket weave pattern; basket work pattern
席纹图案的磨砖对缝墙 basket weave pattern brick wall
席纹织物 basket weave
席纹柱 natte
席纹砖墙 basket weave pattern brick wall
席形基础 mat foundation
席状 sheeted; sheet-like
席状沉积 blanket deposit
席状断层地区 sheeted ground
席状构造 sheet structure
席状护岸 blanket revetment
席状交代矿床 manto
席状节理【地】sheet joint
席状晶体 mat crystal
席状矿床 sheeted ground; sheeted (vein) deposit
席状矿带 sheeted zone
席状矿体 sheeted ore body
席状脉 sheeted vein
席状脉群 sheeted veins
席状泥炭 sheet peat
席状砂 sheet sand
席状砂体 sheeted sand body
席状砂岩 sheet sandstone
席状岩墙群 sheet diked swarm
席状杂岩 sheeted complex
席子 mat(ting)

袭

袭夺 beheading; derangement; robber; abstraction <河流>

袭夺河 captor river; captured river; capturing river; capturing stream; diverter; mutilate driver; pirated stream
袭夺(河)湾 elbow of capture
袭夺作用 abstraction
袭击 beam down (up) on; onset; rampage
袭击报告 raid report
袭击时间 raid period
袭击性巡逻 raid patrol
袭来波 incoming wave

洗

洗氨器 ammonia scrubber; ammonia washer

洗板机 slab washer
洗杯机 glass washing machine
洗苯器 benzol(e) scrubber
洗不掉的 indelible
洗擦用法兰绒布块 <英> flannel
洗菜池 vegetable preparation tank; vegetable sink
洗菜盆 vegetable sink; vegetable sink unit
洗餐具室 basin room
洗舱 hold washing; ship cleaning; washing of tanks; tank cleaning
洗舱费 cleaning tank charges
洗舱设备 tank-cleaning equipment
洗舱系统 tank-cleaning system; tank-washing system
洗舱系统泵 tank-cleaning pump
洗舱站 cleaning station
洗舱装置 tank-cleaning plant
洗槽 launderer; tank washer; wash bath; wash trough; wash tub; tin disk <洗岩粉样品用>
洗槽振动筛 sluice-mounted screen
洗车场 car wash yard; vehicle washdown(yard); vehicle washing area; wash-down yard for cars
洗车处 car wash; wash stand
洗车床 car cleaning plant
洗车房 car wash; car washing room
洗车废水 car washing wastewater; wastewater from car washer
洗车沟 car washing canal
洗车机 carriage cleaner; car washer; coach washing machine; motor flusher; motor water car; railcar washer【铁】
洗车架 service rack
洗车库 car cleaning plant
洗车棚 car wash booth
洗车台 car washing stand; service rack
洗车线 car washing line; car washing track; washing line; washing track
洗车厢 car wash booth
洗尘 air washing
洗尘器 dust scrubber
洗出 leaching-out
洗出法 lavage; lavation
洗出骨料 exposed aggregate
洗出回路 eluant circuit
洗出母液 eluent; elutriant
洗出溶液 wash solution
洗出图形法【测】water-cote method
洗出液 eluate
洗出液浓度 eluate grade
洗出移液管 washout pipet(te)
洗除 wasout
洗窗机 window cradle machine
洗窗台架 window-washing cradle
洗床单房 <旅馆> linen room
洗涤 ablution; clean-up; eluate; flush away; lavation; rinsing; scouring; scrubbing; swill; syringe; toilet; washery; washing
洗涤凹槽 washing recess
洗涤本领 washing power
洗涤泵 washer pump
洗涤比 wash ratio
洗涤步骤 launchering procedure
洗涤残余物 debris
洗涤槽 launder; rinse bath; sink; Swill bath; wash channel; washing tank; washing-trough
洗涤槽下料筒 washer boot
洗涤沉淀 washing precipitation
洗涤沉降罐 washing-settling tank
洗涤池 ablution fountain; kitchen sink; sink
洗涤池存水弯 sink trap
洗涤池排水 sink drain
洗涤除尘(法) dedusting by washing; washing dedusting
洗涤除尘器 scrubber
洗涤处理 aqueous desizing; washing treatment
洗涤打浆机 washing beater
洗涤的 abluent; abstergent; detergent
洗涤底漆 active primer; etching primer; self-etch primer; wash primer
洗涤段 washing section
洗涤法 washing method
洗涤坊废水 laundry waste
洗涤废水 kitchen waste; washes; wash waste
洗涤废物 washery waste
洗涤废液 scrubbing raffinate
洗涤废渣 washery refuse
洗涤分级机 launder classifier; washing classifier
洗涤分级器 launder classifier; washing classifier
洗涤分散剂 detergent dispersant
洗涤分析试验 <确定混凝土骨料配合比> washing analysis test
洗涤粉 cleaning powder
洗涤干燥机 scrubbing-and-drying unit
洗涤管 sluicing pipe
洗涤罐 scrubber tank; washing tank
洗涤过的萃取液 scrubbed extract
洗涤和预加工转鼓 washing and conditioning drum
洗涤混合液 cleaning solution; cleaning mixture; washing solution
洗涤机 cleanser; flow washer; scouring machine; washer; washing machine; washing roller; wet mill; launderer
洗涤集尘器 scrubber collector
洗涤集管 wash header
洗涤剂 abluent; abstergent; cleaner's solvent; cleaning agent; cleaning compound; cleaning mixture; cleaning solution; cleanser; cleansing agent; detergent; detergent agent; detersive; laundry detergent; lotion; scouring agent; washing agent
洗涤剂废水 detergent waste
洗涤剂化合物 detergent compound
洗涤剂喷射 shampoo spray
洗涤剂喷射拱门 shampoo spray arch
洗涤剂污染 detergent pollution
洗涤剂吸附 adsorption of detergent
洗涤剂助剂 builder
洗涤间 washery; washing room; wash-room
洗涤间设备 wash room equipment
洗涤碱 sal soda
洗涤阶段 washing stage
洗涤蓝 laundry blue
洗涤冷凝器 scrubber condenser
洗涤沥干物重 washed drained weight
洗涤滤液 wash filtrate
洗涤滤液真空接受器 wash filtrate vacuum receiver
洗涤木盆 trough wash basin
洗涤能力 washability
洗涤碾磨机 washing mill
洗涤浓密机 washing thickener
洗涤浓缩器 wash thickener
洗涤盘 dry sink
洗涤盆 sink(bowl); washing bowl; wash(ing) sink; wash (ing) tub; wash-up sink(unit)
洗涤盆存水弯 sink(water) trap
洗涤盆龙头 sink faucet
洗涤盆排水道 sink drain
洗涤盆水龙头 sink bib
洗涤盆污水 <厨房> slop water
洗涤瓶 wash(ing) bottle
洗涤器 abstergent; clean(s)er; scrubber; syringe; washer; washing apparatus; washing device
洗涤器板 scrubber plate
洗涤器废水 washer wastewater
洗涤器刷辊 scrubber brush roll
洗涤器淤渣 scrubber sludge
洗涤泉 ablution fountain
洗涤溶液 washing solution
洗涤砂砾石 sand and gravel wash
洗涤设备 scrubber; scrubbing plant; wash fixture; washing equipment
洗涤设施 washing device
洗涤石墨 graphite water
洗涤时间 wash time
洗涤式滤尘器 scrubber filter
洗涤式压滤机 washing press
洗涤试验 scrubbing test; washing test; decantation test

洗涤室 washing chamber
洗涤室挡水板 eliminator plate
洗涤室阻水板 eliminator plate
洗涤收集器 scrubber collector
洗涤水 flush water; scouring water; scrubbing water; wash(ing)-water
洗涤水泵 wash(ing) water pump
洗涤水池 kitchen sink
洗涤水管 washing pipe
洗涤水回收系统 wash-water recovery system
洗涤水截止阀 stop valve for washing
洗涤水入口 entrance of washing water; washing water inlet
洗涤水系统 wash-water system
洗涤水止逆阀 check valve for washing water
洗涤速率 rate of washing
洗涤损失 washing out loss
洗涤塔 scrubber; scrubber wash tower; scrub column; tower washer; washer; washing tower; washover string
洗涤塔盘 wash tray
洗涤桶 washing beck; wash tub
洗涤筒 cylindric(al) washer
洗涤温度 wash temperature
洗涤污水 laundry wastewater
洗涤物 washings
洗涤箱 wash(er) box
洗涤效果 scrubbing effect
洗涤性能 scourability
洗涤液 cleaning solution; cleaner; cleaning fluid; cleaning liquor; cleansing fluid; cleansing solution; scrubbing solution; washer solvent; washing liquid
洗涤液出口管 washing liquor exhaust pipe
洗涤液喷嘴 washing nozzle
洗涤液贮槽 washing liquor tank
洗涤用锅炉 wash boiler
洗涤用碱 washing soda
洗涤用喷枪 wash gun
洗涤用水 bathing water; lavation
洗涤用稀释剂 wash thinner
洗涤用轴承 bearing for washing
洗涤油 detergent oil; flushing oil; washing oil
洗涤渣 washed-residue
洗涤折痕 <整理疵点> washer breaks
洗涤者 washer
洗涤蒸馏釜 wash oil still
洗涤周期 washing cycle
洗涤助剂 washing assistant
洗涤柱 column washer; washing column
洗涤转筒筛 wash(ing) trommel
洗涤装置 sink unit; wash mill
洗涤装置电动机 washing device motor
洗点式土壤触探计 wash-point soil penetrometer
洗掉 wash down
洗碟池 dish pan; dish washing sink(unit); washing-up sink
洗碟机 dish washer
洗碟剂 dish washing agent
洗碟盆 dish pan; dish washing sink(unit); washing-up sink
洗碟器 plate washer
洗碟室 plate scullery
洗碟子水 dish water
洗发盆 shampoo bowl
洗干净的混凝土表面 scrubbed concrete surface
洗杆 ramrod
洗工 fuller
洗鼓 washing drum
洗管 tube cleaning
洗管机 tube cleaner
洗管器 pipe cleaner; tube cleaner
洗盥槽 launder
洗罐【铁】tank car washing
洗罐机 can rinser
洗罐库 tank car washing shed
洗罐棚 tank car washing shed
洗罐设备 tank-cleaning equipment
洗罐线 tank vehicle washing line
洗过的沉淀 washed precipitate
洗过的混凝土平板纸 washed concrete slab paper
洗过的砾石 wash gravel
洗过的煤 washed coal
洗灰 ash washing
洗积 wash
洗剂 lotion
洗甲板器 wash deck gear
洗甲板软管 wash deck hose; wash deck pipe
洗甲板水泵 wash deck pump
洗浆池 drainer; pulp washer
洗浆机 pulp washer; stock washer; wash hollander; washing engine
洗焦废水 coke-washing wastewater
洗脚池 foot bath
洗街机 street washer
洗洁剂 detergent
洗金属淡酸液 pickle
洗井 borehole flushing; clean-out of well; hole flushing; washing the well; well cleaning; well flushing; well washing; flushing <石油钻探时>
洗井方法 method of well cleaning
洗井工作数据采集 data collection of well cleaning
洗井管柱 wash pipe
洗井后的井深 depth of well after well cleaning
洗井后的水井出水量 water yield after well cleaning
洗井前的井深 depth of well before well cleaning
洗井前的水井出水量 water yield before well cleaning
洗井日期 date of well cleaning
洗井设备 development equipment
洗井水 wash-down water
洗井液 drill fluid; flushing fluid
洗井液压力值 pressure value of well cleaning liquid
洗井用的时间 time of well cleaning
洗井装置 development equipment
洗净 scour(ing); washing off; wash-(ing)-out; abstersion; cleaning; depurate; deterage; eluate; elution; elutriation
洗净玻璃 elutriating glass
洗净材料 washed material
洗净的 ablutionary; detergent; washed-up
洗净的河卵石 wash river run rock
洗净的碎煤 washed slack
洗净放水开关 washing out plug and drain cock
洗净和脱脂 cleaning and degreasing
洗净机 scrubber
洗净剂 abluent; cleaner; detergent; remover; scourer
洗净剂喷布车 detergent spray truck
洗净空气通道 washed air duct
洗净砾石 gravel wash; washed gravel
洗净毛 clean wool
洗净器 washer
洗净砂 washed sand
洗净石屑 washed chip(ping)s
洗净树脂 eluted resin
洗净性 detergency[detergence]
洗净性能 washing-off property
洗净液 detergent; remover; ablution
洗净装置 cleaning equipment
洗客车场 coach cleaning yard
洗客车员 carriage cleaner
洗孔 borehole flushing; clean-out of hole; hole flushing
洗孔水 wash-down water
洗口瓶 vase with washer shaped month
洗矿 dress; elutriate; washed ore; wash(ing)
洗矿槽 buddle; hut(ch); launder; log washer; sluice(box); strake; trough washer; wash box
洗矿槽脱水传送器 dewatering flight conveyer[conveyor]
洗矿厂产品 washery product
洗矿厂泄出水 washery effluent
洗矿厂用水 washery water
洗矿工 mineral washer
洗矿滚筒筛 rotary washing screen
洗矿机 log washer; mineral washer; washing apparatus; washing plant
洗矿机械 washing machinery
洗矿精矿 wash up
洗矿泥沙沉积堆 mine wash
洗矿浓密机 wash thickener
洗矿喷水 washing spray
洗矿筛 washing screen
洗矿水 hutch water; ore washer; washwater
洗矿损耗 washing-loss
洗矿台 rag-frame; washing table
洗矿提升机 washing elevator
洗矿桶 keeve
洗矿转筒筛 revolving washing screen
洗礼池 baptistery; hagiasterium
洗礼盘 font
洗礼盆 piscina
洗礼室 baptismal room
洗礼堂 baptismal room; baptistery
洗砾厂 gravel washing plant
洗砾机 gravel washer
洗砾筛 gravel washing screen
洗砾石道砟 washed-gravel ballast
洗砾石机 gravel washing screen
洗脸架 commode
洗脸间 washing hand
洗脸盆 basin; wash basin; wash bowl
洗脸盆混合龙头 mixing tap in lavatory
洗脸盆托架详图 details of basin support
洗脸台 commode
洗料池 diffuser
洗料机 washer
洗流 wash
洗流角 angle of crab; rake angle
洗炉 boiler washout; prepurging; slugging
洗炉堵 washout plug
洗炉堵孔 washout plug hole
洗炉堵孔法兰 washout plug hole flange
洗炉后点火 lighting-up after washout
洗炉机 boiler washer
洗炉金属液 wash metal
洗炉盘 rack
洗炉塞 washout plug
洗炉橡胶管 boiler washout rubber hose
洗滤 filter(ing) wash(ing)
洗滤器 filtering washer; washing filter
洗滤水 filter washing water; filter wastewater
洗滤液 washing filtrate
洗路车 street sweeper with washer
洗路机 street flusher
洗轮机 wheel washer
洗麻石刷面【建】depreter[depeter]
洗毛槽 scouring bowl
洗毛废水 wool scouring wastewater
洗毛废液 wool scouring waste liquor
洗毛工业 wool scouring industry; wool washing industry
洗毛工业废水 wool scouring manufacturing wastewater
洗毛机 wool rinsing machine; wool scouring machine; wool washing machine
洗毛生产废水 wool scouring manufacturing wastewater
洗毛污水 wool scouring effluent
洗煤 coal washing; washed coal; washing of coal
洗煤厂 coal-cleaning plant; coal washery; coal-washing plant
洗煤厂废水 washery wastewater
洗煤厂建设 coal-cleaning plant building
洗煤场 coal-washing plant
洗煤废弃物渗沥液 coal-cleaning waste leachate
洗煤废水 coal-dressing wastewater; coal-washing wastewater
洗煤机 coal washer; washer
洗煤设备 coal-washing plant
洗煤渗沥液 coal-cleaning leachate
洗煤效率 efficiency of separation
洗面盆 lavatory
洗面器 lavatory basin; wash basin; wash bowl; wash-hand basin; wash stand
洗面器放出管联结器 drip coupling
洗面器放水塞门 washstand bibcock
洗面器水龙头 basin faucet
洗萘氨水 naphthalene and ammonia washing water
洗泥斑 washer mark
洗泥机 wash mill; washing dolly
洗黏[粘]土 washed clay
洗耙 <淘泥机的> washing gates
洗盘池 dish washing sink(unit)
洗盘间 dish washing room; plate scullery
洗盆 washing basin
洗皮转鼓 wash wheel
洗片 develop
洗片挂 wet film hanger
洗片罐 film washer; print washer
洗片机 developing machine; processing machine
洗片桶 processing tank
洗平板机 panel washer
洗瓶废水 bottle washing waste
洗瓶机 bottle washer; bottle washing machine; bottle washing plant
洗瓶刷 bottle brush
洗瓶水 bottle rinse water; bottle washing water
洗漆工 paint scrubber
洗漆剂 paint scrubber; remover
洗漆器 paint scrubber
洗漆刷 paint scrubber
洗气 gas washing; scrub【化】
洗气池下冷水龙头 undersink cold tap unit
洗气池下面 undersink
洗气废水 gas washing wastewater
洗气管 sluicing pipe
洗气罐 scrubber tank; scrubber wash tower
洗气机 gas cleaner
洗气盆排水管 sink drain
洗气瓶 gas(washing) bottle; wash-(er) bottle
洗气器 gas washer
洗气器废水 washer wastewater
洗气器液体 scrubber liquid; scrubber

liquor
洗气溶液 scrubbing solution
洗气设备 gas washing plant
洗气式压滤机 washing press
洗气室 flashback chamber; washing chamber
洗气收集器 scrubber collector
洗气水 gas water
洗气塔 aeration tower;gas tower;gas wash tower;scrubber wash tower
洗气用水 gas water
洗气装置 air washer
洗汽 steam washing
洗清 wash(ing) down
洗球机 marble washer
洗去 flush away; wash(ing)-off; wash(ing-) out
洗染店 cleaners and dyers;laundering and dyeing shop
洗扫舱 cleaning and sweeping holds
洗砂 cleaning of sand;grit washing
洗砂厂 sand washer plant;sand-washing plant
洗砂场 sand washer plant;sand-washing plant
洗砂沟 sluiceway
洗砂机 grit washer; sand scrubber; sand washer; sand-washing machine
洗砂设备 sand washer;sand-washing equipment
洗砂石均匀程度 uniformity of washing aggregate
洗砂跳汰机 sand-cleaner jig
洗砂装 grit washer
洗筛机 washing and screening machine
洗绳机 rope washing machine
洗圣器池 piscina
洗石机 gravel washing plant; scrubber;stone scrubber;stone washer; winding apparatus;log washer
洗石面 <混凝土表面> exposed aggregate
洗石筛 gravel washing screen
洗石筛分机 washing screen
洗石设备 aggregate washing plant
洗石子粉刷 granitic plaster
洗石子面 washed finish
洗手不干的 washed-up
洗手间 lavatory; rest room; washing hand;washroom;loo <英俚语>
洗手间成套器皿 toilet set
洗手间器皿 toilet ware
洗手盆 hand(wash) basin; lavatory (basin); wash basin; wash bowl; scrub sink <外科手术用>
洗手盆排水栓 basin plug
洗手器 hand washbasin;wash bowl
洗手刷 hand brush
洗刷 scour;scrub
洗刷风化物 efflorescence cleaning
洗刷锅炉 washing out boiler
洗刷过的混凝土表面 washed concrete surface
洗刷甲板的人 deck scrubber
洗刷器 scourer
洗刷系统 wash down system
洗刷下来的水 washing down water
洗刷线 cleaning siding;cleaning track; washing track
洗刷作用 scrubbing action;sheet wash
洗水仓 wash-water sump
洗水澄清 water clarification
洗水澄清槽 wash-water settling tank
洗水沟 wash-water channel
洗水回收 wash-water reclamation
洗水式凿岩机 wash boring drill
洗水式凿岩设备 wash boring rig
洗碎石场 gravel washing plant
洗碎石饰面 depreter[depeter]
洗烫 laundering
洗陶土 washed kaoline
洗提 elute
洗提常数 elution constant
洗提段 elution section
洗提法 elution method
洗提分级 elution fractionation
洗提峰 eluting peak
洗提剂 eluant; eluting reagent; scrub solution
洗提能力 eluting power
洗提气体色层分离法 elution gas chromatography
洗提器 elutriator;stripper
洗提曲线 elution curve
洗提容积 elution volume
洗提设备 stripping apparatus
洗提时间 elution time
洗提顺序 eluting order; eluting sequence;elution order
洗提塔 elution column;stripping column
洗提条件 elution requirement
洗提温度 eluting temperature
洗提效应 eluting effect
洗提循环 elution cycle
洗提液 eluant;elutriant
洗提液成分 eluant composition
洗提液强度 eluant strength
洗提液强度梯度 eluant strength gradient
洗提液清洗柱 eluant stripper column
洗提液组分 eluant component
洗提展开 elution development
洗提周期 elution period
洗铁 washed metal
洗桶机 can rinser;drum rinser
洗桶水 wash-water
洗筒 washing cylinder
洗筒之润滑 washing cylinder lubrication
洗拖把池 housemaid's sink
洗脱 elute;eluting;elution;elutriation
洗脱程序 elution program(me)
洗脱带 elution band
洗脱法 elution method
洗脱方式 elution mode; type of elution
洗脱分析 elution analysis
洗脱峰 eluting peak
洗脱过程 elution process
洗脱混合液 eluent mixture
洗脱剂 eluent;eluting agent;elutriant
洗脱能力 eluting power;elutive power
洗脱浓度 eluent concentration
洗脱气体 eluent gas
洗脱器 eluotropic
洗脱强度 eluotropic strength
洗脱色谱法 elution chromatography
洗脱时间 elution time
洗脱顺序 eluting order; eluting sequence
洗脱体积 elution volume
洗脱条件 elution requirement
洗脱温度 eluting temperature
洗脱物 eluate
洗脱系统 elution system
洗脱效率 elution efficiency
洗脱效应 eluting effect
洗脱性质 eluent property
洗脱序 eluotropic series
洗脱液 eluent; elutriant; spent regenerant
洗脱液除气 degassing of eluent
洗脱液浓度 eluent strength
洗脱液强度 eluant strength
洗脱液强度梯度 eluant strength gradient
洗脱液清洗柱 eluant stripper column
洗脱液体积 effluent volume
洗脱液组成 eluent composition
洗脱液组分 eluant component
洗脱用气体 eluant gas
洗碗槽 scullery;scullery basin
洗碗池 dish washing sink(unit)
洗碗碟机 dish washing machine
洗碗机 bowl washer; bowl-washing machine;dish washing machine
洗碗间 bowl-washing room
洗碗盆 kitchen sink;pot sink
洗碗器 plate washer
洗碗室 plate scullery
洗碗刷 mop
洗污 contaminate
洗箱 tank washer
洗鞋解卡法 <用装有管鞋钻头的冲洗管柱钻透被卡钻具> washover fishing operation
洗选 dressing by washing;elutriation; washing
洗选厂 washery
洗选法 washing process
洗选费 washing cost
洗选格 washing cell
洗选过程 washing process
洗选机 wet washer
洗选间 wash station
洗选矿泥坡面板 jagging board
洗选流矿槽 sluice box
洗选煤 cleaned coal;separation coal; washed coal
洗选设备 washing appliance;washing machinery
洗选尾矿 washing refuse
洗选系统 washing circuit
洗选效果 washing performance
洗选效率 washing efficiency
洗选旋流器 washing cyclone
洗岩 rock awash
洗盐 washing salt
洗盐水量 leaching requirement
洗眼杯 eyecup
洗眼设备 <劳动保护用> eye washing bath
洗液 cleansing solution; flushing oil; lotion;washing liquid;washing liquor;washing solution;washing water
洗液废水处置 sullage disposal
洗衣板 scrub board;wash board
洗衣板玻璃 washboard glass
洗衣板式波浪的形成 <路面> formation of washboard waves
洗衣板式道路 washboard course
洗衣槽 clothes chute;laundry tray
洗衣池 laundry sink; laundry tray; laundry tub
洗衣的碱水 buck
洗衣店 laundry
洗衣房 laundry; laundry room; laundry wash house;wash house;washroom
洗衣房废水 laundry wastewater;wash house effluence;wash house waste; wash room waste
洗衣房热风炉 laundry stove
洗衣房设备 laundry equipment
洗衣废水 laundry waste
洗衣粉 laundry soap powder
洗衣干燥柜 <住宅中的> cabinet for dry washing
洗衣工 launderer;laundryman
洗衣锅炉 wash boiler
洗衣烘干室 drying room
洗衣机 flow washer; laundry machine;washer;washing machine
洗衣机用搅拌棒独轮小车 dolly
洗衣间 laundry room;washing place
洗衣筐 buck basket
洗衣蓝 laundry blue
洗衣篓 wash basket
洗衣盆 laundry tray;laundry tub;set tub <较深的一种>
洗衣日 wash day
洗衣室 laundry
洗衣水槽 laundry sink
洗衣桶 dolly tub
洗衣脱水机 washer-drier[dryer]
洗衣皂 laundry solid soap
洗印车间 film laboratory
洗用碱水 washing lye
洗油 oil washing;scrubbing oil;wash(ing) oil <一种煤馏油,相当于重油>
洗油脱苯 debenzolization of oil; oil stripping
洗油蒸馏釜 wash oil still
洗余的水 scourage
洗余水 wash-water
洗余液 scrubbing raffinate
洗浴 bath
洗浴废水 bathing waste(water)
洗浴更衣室 bath closet
洗浴水 bathing water
洗浴水洗衣水 bath(ing) laundry water
洗浴洗衣粉 bath-laundry water
洗熨(衣) launder
洗澡 bath
洗澡废水 bath(ing) waste(water)
洗澡间 bathroom cabinet; shower bath room
洗澡盆 wash tub
洗澡水 bath(ing) water
洗澡水处理 bath(ing) water treatment
洗澡污水 bath(ing) waste(water)
洗渣 washery slag
洗濯机 washing machine
洗濯盆 wash tub
洗濯室 washing chamber

徙动 migration

徙动常数 migration constant

铣背槽 milling back flutes

铣边机 edge milling machine
铣边翼形螺钉 milled edge thumb screw
铣槽 channeling;milling flute;milling of grooves
铣槽尺 key-seat rule
铣槽刀 channel(l)ing cutter;grooving cutter
铣槽刀具 grooving tool
铣槽刀轴 saw arbor
铣槽附件 slotting attachment
铣槽机 slotter
铣槽锯 grooving saw
铣槽装置 slotting attachment
铣车床 mill-turn
铣成边 milled edge
铣成的 milled
铣成的产品 milled product
铣成的木材 milled wood
铣齿 gear milling;milled tooth
铣齿侧面 flank of tooth
铣齿刀架 milling slide
铣齿机 gear milling machine
铣齿轮 cut gear
铣齿轮齿 milling gear teeth
铣齿密封喷射钻头 sealed jet milled bit
铣齿钻头 milled bit
铣床 mill(er);milling machine

铣床抽风罩 milling machine hood
铣床顶尖 miller center[centre]
铣床分度头 index
铣床附件 milling machine accessory
铣床工作台 milling machine table
铣床夹具 milling jig
铣床立柱 milling machine upright
铣床零件 milling machine parts
铣床轧头 milling machine dog
铣床主轴 milling machine spindle
铣床转速刻度盘 milling machine speed dial
铣刀 cutter;fraise;hobbing;mill;milling cutter;milling-tool
铣刀柄 milling cutter shank
铣刀齿节 pitch of cutter teeth
铣刀齿数 cutter tooth number
铣刀齿形 milling cutter tooth form
铣刀端面后角 face clearance
铣刀附件 fraise adapter
铣刀杆 cutter arbor; cutter spindle; fraise arbor; milling(cutter) arbor; milling cutter spindle
铣刀夹具 milling fixture
铣刀铰刀磨床 cutter-and-reamer grinder
铣刀卡盘 cutter chuck
铣刀量隙规 cutter clearance ga(u)ge
铣刀磨床 milling cutter grinding machine
铣刀盘 facer;facing cutter
铣刀盘刃磨机 cutter-head grinder
铣刀片 cutter blade
铣刀式穿孔器 star perforator
铣刀头 milling head
铣刀头锯 <开槽锯> dado head saw
铣刀修磨机 milling cutter sharpening machine
铣刀直径 milling cutter diameter
铣刀轴 cutter spindle;milling(cutter) arbor;milling machine arbor
铣捣 wheel cutter
铣掉 mill off
铣法 milling
铣工 miller(hand); milling machine operator
铣工车间 milling machine shop
铣工工作 milling work
铣光边 milled edge
铣轨搬运夹钳 rail tongs
铣轨机 rail milling machine
铣环 milling ring
铣加工导座 milling shoe
铣键槽 key seat
铣键槽机 keyseater
铣角 milling angle
铣轮式挖泥机 cutter-head dredge(r)
铣轮式挖土机 cutter-head dredge(r)
铣螺纹 screw-thread milling; thread milling
铣螺旋线 scroll milling
铣模机 die sinker
铣磨路面机 road milling machine
铣刨 cold planing
铣刨法 milling
铣刨机 road planer
铣刨精度 milling accuracy
铣平面 face milling;milling flat
铣切深度 milling depth
铣切式挖掘机 milling excavation
铣去 milling off
铣刃式钻头 <消减孔内钻具用> drill mill
铣铁 pig;pig iron
铣铁压块 kentledge
铣头 cutter head; cutting head; mill head
铣头的偏转角 heeling
铣纹机 threader
铣锨 shovel
铣削 gang milling;mill(ing)
铣削操作 milling machine operation; milling operation
铣削刀痕 milling mark
铣削刀架 milling slide
铣削动力头 fraise unit; milling unit head
铣削法 milling
铣削附件 milling attachment
铣削革屑 trimming dust
铣削过的外底边 trimmed edge
铣削加工 milling
铣削夹具 milling fixture
铣削深度 milling depth
铣削装置 milling attachment
铣鞋 rotary shoe
铣直槽 milling straight flute
铣制螺母 milled nut
铣制螺栓 milled bolt
铣制钻头 cut drill
铣锥 milling tap
铣锥形槽 milling cone groove
铣钻联合机床 combined drill and mill machine

喜冰雪的 chionophilous

喜采集奇石的人 rock hound
喜草原的 psilophilus
喜池沼的 tiphophilus
喜氮植物 nitrogen loving plant;nitrophile
喜低温 psychrophilic
喜低温的 cryophilic
喜低温微生物 cryophilic microorganism
喜钙植物 calcicole;calcicolous plant; calcipete;calciphile
喜光的 photophilic;photophilous
喜光树 sun-loving plant
喜光植物 heliophile;heliophilous plant; sun plant
喜光种 sun-loving species
喜旱的 xerophile;xerophilous
喜旱性 xerophile
喜旱植物 xerophile;xerophilous crop; xerophilous plant
喜河流的 potamophilus
喜急流的 rhyacophilus
喜碱的 basophilous
喜拉马雅山山脉【地】Himalayan mountains;Himalayas
喜拉马雅山运动【地】Himalayan movement
喜冷植物 psychrophilic plant
喜马拉雅柏木 Himalayan cypress
喜马拉雅大区 Himalaya region
喜马拉雅地槽褶皱区【地】Himalayan geosynclinal fold region
喜马拉雅构造带 Himalaya tectonic zone
喜马拉雅海 Himalaya sea
喜马拉雅海退 Himalaya regression
喜马拉雅海相动物地理区 Himalaya marine faunal province
喜马拉雅继承性断褶带 Himalaya inheriting fault-fold zone
喜马拉雅继褶隆起 Himalaya fault-fold swell
喜马拉雅菊石地理大区 Himalaya ammonite region
喜马拉雅冷杉 Himalayan fir
喜马拉雅南边缘坳陷系 South Himalayan trough system
喜马拉雅-南岭构造带【地】Himalaya-Nanling tectonic zone
喜马拉雅南缘近期拗陷 neoid depression in the south marginal of Himalaya
喜马拉雅期地槽【地】Himalayan geosyncline
喜马拉雅山 Himalaya mountains
喜马拉雅山南麓断裂构造带 Himalayan southern foot fault belt
喜马拉雅山区 Himalayas
喜马拉雅山杉木 Himalaya cedar
喜马拉雅山雪杉 deodar
喜马拉雅台褶带 Himalayan platform folded belt
喜马拉雅腕足动物地理大区 Himalayan brachiopod region
喜马拉雅型造山带 orogenic zone of Himalayas type
喜马拉雅旋回【地】Himalayan cycle
喜马拉雅亚旋回 Himalayan subcycle
喜马拉雅银(冷)杉 Himalayan silver fir
喜马拉雅云杉 Himalayan spruce
喜马拉雅造山运动 Himalayan orogeny
喜马拉雅辗掩断裂系 Himalayan Nappe fracture zone
喜马拉雅褶皱系 Himalayan fold system
喜马拉亚松 Himalayan pine
喜马樱 Himalayan cherry
喜庆日 festival
喜热生物 thermophilous organism
喜沙的 ammophilous
喜沙丘的 thinophilus
喜沙植物 psammophile
喜山期 Himalayan period
喜生于木材上的 xylogenous;xylophilous
喜湿菌 psychrophile
喜湿细菌 hydrophilic bacteria
喜湿植物 hygrophilous plant
喜水(生)的 hygrophile;hygrophilous
喜水物 hydrophile
喜水小核菌 Sclerotium hydrophilum
喜水植物 hydrophilous plant; hydrophyte;water-loving plant
喜酸的 acidophilic;acidophilous;oxyphilous
喜酸植物 oxylophyte; oxyphiles; oxyphilic plant
喜土植物 geophilous plant
喜温的 thermophilic
喜温生物 thermophilic organism
喜温树种 thermophilic species
喜温细菌 mesophilous bacteria; thermophilic bacteria
喜温性 thermophily
喜温植物 thermophile; thermophilic; thermophilous;therophyte
喜盐的 drimophilous;halophilic
喜盐生物 halobiont;halophile
喜盐微生物 halophile
喜盐植物 halophile
喜阳的 heliophilous
喜阳植物 heliad; heliophile; heliophilous plant
喜氧呼吸带 aerobic respiration zone
喜氧细菌 aerobic bacteria
喜氧细菌降解 aerobic bacteria degradation
喜氧细菌降解作用带 aerobic bacteria degradation zone
喜阴的 shade demanding
喜阴植物 ombrophyte;sciophile;shade-loving plant;shade plant;skiophyte
喜雨的 ombrogenous; ombrophilic; ombrophilous
喜雨植物 ombrophile; ombrophilous plant
喜嶂矿 xifengite
喜沼泽的 limnophilous
喜中温的 mesophilic

屣痕 heel mark

戏剧 theatre[theater]

戏剧表演 histrionics
戏剧性冲击 dramatic impact
戏剧演出 theatrical;theatrics
戏台 stage
戏院 theatre[theater]
戏院楼厅 balcony
戏院内的前排包厢 dress circle
戏院正厅后排 parquet circle
戏院最下层包厢 baignoire

系 department; faculty; series; system【地】

系岸沙洲 tombolo
系岸试车 dockside trial;dock test
系板 stay plate;strap(ping);tie plate
系比试验 clone test;clone trial
系标签的 label(l)ed
系泊 tie-down <专指拖车住房在驻地的系固>;mooring【船】
系泊臂架 mooring boom
系泊部位 docking station
系泊船舶 docked vessel;moored ship; moored vessel
系泊船处 point of laying up
系泊大型船 large moored ship
系泊岛 mooring island
系泊岛式码头 open wharf mooring island
系泊的 moored
系泊地 mooring berth;tying-up place; mooring place
系泊点 mooring point
系泊动作 moorage
系泊方式 mooring system
系泊费 berthage; moorage; mooring fee
系泊浮筒 anchor(age) buoy;mooring buoy
系泊钢缆 mooring wire
系泊港池 mooring basin
系泊荷点 mooring point
系泊荷点结构的设计 design of mooring point structure
系泊荷点结构物 mooring point structure
系泊荷载 mooring load
系泊荷载计算 evaluation of mooring load
系泊绞车 mooring winch
系泊绞盘 mooring capstan
系泊块体 concrete block for mooring post
系泊缆(索) hawser; mooring cable; mooring fast; mooring hawser; mooring line;mooring rope
系泊链 mooring chain
系泊码头 mooring place;mooring terminal
系泊锚 mooring anchor
系泊锚链 mooring chain
系泊锚位 mooring anchorage
系泊模式 mooring pattern
系泊平台 mooring platform
系泊牵引两用桩 combined mooring and warping bollard
系泊区 mooring zone
系泊权 mooring right
系泊设备 berthing accommodation; mooring arrangement; mooring facility;mooring fittings
系泊设施 berthing facility; docking facility
系泊试车 dock trial; mooring trial;

trial at anchor; dock test; dockside trial; quay trial
系泊试车码头 quay for material trial
系泊试验 basin trial; dockside trial; dock test; dock trial; mooring trial
系泊试运转 mooring trial
系泊栓 mooring cleat
系泊水域 dock basin
系泊索具 mooring tackle
系泊无线电声呐浮标 anchored radio-sono-buoy
系泊系统 mooring system
系泊小船 moored boat
系泊小船的浮筒绳索 trot
系泊用具 moorings
系泊用栓桩 mooring peg
系泊转环 mooring swivel
系泊桩 anchor strut; mooring stump
系泊装置 fastening device; mooring gear; mooring device
系泊作业 moorage
系材 accouplement
系成员 set member
系船 laid up; moor
系船臂架 mooring boom
系船标锚碇 buoy mooring
系船驳(船) dummy barge; mooring barge
系船沉箱 mooring caisson
系船池 marina
系船处 moorage; mooring
系船触岸索 breast(ing) line
系船传铃钟 docking telegraph
系船船首索 bowline
系船船尾索 stern line
系船船坞 wet dock
系船簇桩 dolphin pile moorings; mooring dolphin
系船墩 mooring dolphin; dolphin; mooring pier
系船浮标 anchorage buoy; anchor buoy; dolphin; mooring buoy
系船浮标使用税 buoyage; buoy dues
系船浮筒 anchorage buoy; dolphin; mooring buoy
系船浮筒脱缆钩 release hook for mooring buoy; release hook for mooring buoy
系船港池 mooring basin
系船钩 line hook
系船钩环 mooring shackle
系船荷载 mooring load
系船环 mooring ring
系船混凝土块 mooring block
系船建筑(物) mooring structure
系船绞车 <靠码头用> docking winch
系船结构(物) mooring structure
系船卡环 mooring shackle
系船靠岸索 breasting line
系船拉力 mooring pull
系船缆 mooring fast; mooring line
系船缆导轮 fairlead
系船缆绳 mooring line
系船缆索 mooring line; mooring rope; mooring hawser
系船缆张力 mooring line tension
系船力 bollard pull; mooring force; tie-up force; bollard force
系船力分布 distribution of mooring force
系船链索 mooring chain
系船码头 mooring pier; mooring terminal; mooring wharf
系船锚 mooring anchor
系船木桩 mooring post
系船能力 berthing capacity
系船区 docking area
系船设备 berthing facility; mooring accessories; mooring arrangement; mooring equipment; mooring facility; mooring gear; ship mooring; tie-up facility
系船设施 berthing facility; mooring accessories; mooring arrangement; mooring equipment; mooring facility; mooring gear; tie-up facility
系船栓 mooring bolt
系船索 bridle; hawser; mooring guy; mooring hawser; mooring rope; painter
系船塔架 mooring tower
系船系统 mooring system
系船纤维缆 mooring fiber[fibre] rope
系船卸扣 docking shackle; mooring shackle; mooring swivel
系船柱 anchorage spud; bitt; bollard; deadhead; dolphin; guard post; make fast; mooring bitt; mooring bollard; mooring dolphin; mooring post; nigger head; ship-mooring bollard; warping bollard; mooring bollard; fender post
系船柱荷载 bollard loading
系船柱混凝土块体 concrete block for mooring post
系船柱架 spud frame
系船柱拉力 bollard pull
系船柱拉力试验 bollard pull trial
系船柱力度 bollard strength
系船柱索 spud line
系船柱线 spud line
系船柱柱头形式 type of bollard heads
系船桩 bollard; dolphin; mooring pile [piling]; pile dolphin; mooring post
系船桩拉力 bollard pull
系船桩缆 spud cable
系船桩式挖泥船 spud-type suction dredge(r)
系船桩式吸泥船 spud-type suction dredge(r)
系船桩索 spud cable
系船桩柱 mooring dolphin
系船桩组 pile moorings
系椽 collar beam; collar rafter; top beam
系存放方式 set location mode
系带 brace band; bridle; chalaza; frenulum; ligament; tie band; vinculum [复 vinculums/vincula]
系单浮 secure to single buoy
系定板 anchor plate
系定杆 pull-off pole
系定理 corollary
系定位方式 set location mode
系定序准则 set ordering criterion
系定柱 pull-off pole
系定桩 fastening pile
系耳 attaching lug
系帆索 lacing line
系浮 buoy mooring
系浮标的投弃货物 lagan; lagend
系浮筒 securing to a buoy
系杆 binder(lever); brace rod; bracing rod; bridle bar; collar beam; collar bracing; collar rafter; lace bar; lacing bar; sag rod; span piece; straining piece; structuring piece; strutting piece; tie bar; tie beam; tie bolt; tie member; tie rod; tie-strut; tirant; top beam
系杆臂架 bowstring trussed boom
系杆轭 tie-rod yoke
系杆拱 bowstring arch; tied arch
系杆拱大梁 tie arched girder
系杆拱拱弦 choppy of bowstring arch; chord of bowstring arch
系杆拱梁 bowstring girder
系杆拱桥 bowstring arch bridge; bowstring(girder) bridge; tied-arch bridge
系杆桁拱 braced tie-arch
系杆桁架 bowstring truss
系杆加劲柱 tie column
系杆间距 tie distance
系杆节点 <人字木屋架> collar joint
系杆结构 tied structure
系杆拉条 tie-rod bracing
系杆联结桩 braced pile
系杆梁 bowstring beam
系杆球端 tie-rod ball
系钢 bridle iron
系拱 strainer arch
系固物 fastener
系固装置 tie-down fitting
系机地锚 ground anchor
系集平均值 assembly average
系枷牛舍 stanchion barn
系间跃迁 intersystem transition
系件 tie; tie member
系节 set section
系结 hitch
系紧 anchor; fastening; tie-down
系紧板 lacing; lacing board
系紧的绳 hatched knot
系紧钢丝 <模板> racking wire
系紧构件 tie member
系紧滑轮 tie-down sheaves
系紧夹 <指在热状态安上的> shrink link
系紧链条 tie chain
系紧梁 tie piece
系紧螺钉 captive screw
系紧螺母 anchor nut
系紧螺栓 anchorage bolt; anchored bolt; box closure; coupling bolt; drift bolt; holding-down bolt; pinch bolt; tie bolt; tie-down bolt
系紧线 stay wire
系紧用具 lacer
系紧装置 lashing device; tie-down fitting
系靠船台 dolphin
系孔 lashing eye; tie hole
系拉杆 tie rod
系缆 brace rope; warping hawser
系缆壁洞 bollard niche; bollard recess
系缆泊位 mooring dolphin
系缆点 mooring point
系缆点上的荷载 load on mooring point
系缆墩 cleat; mooring dolphin
系缆耳 mooring cleat
系缆浮标 dolphin; mooring dolphin
系缆附件 mooring accessories
系缆工作船 line boat
系缆钩 mooring hook; chock
系缆环 lashing ring; ring fast
系缆活钩 sliphook for connecting rope
系缆活结 slip racking
系缆绞车 mooring winch
系缆卷筒 warping barrel
系缆力 mooring force
系缆木桩 deadhead
系缆设备 mooring arrangement
系缆受力量测仪 mooring force gage
系缆栓 belaying cleat
系缆索 mooring rope
系缆铁件 mooring accessories; mooring fittings; mooring hardware
系缆艇 mooring craft
系缆眼板 eyeplate for mooring; pad eye for mooring
系缆柱 bitt; dolphin; knight; landfast; mooring bitt; mooring post; riding bitt; snubbing post; timber head
系缆柱杆 bitt pin
系缆柱滚筒 ninepin block
系缆柱基础 bollard foundation
系缆柱拉力 bollard pull
系缆柱支承板 bitt bracket
系缆柱支柱 bitt standard
系缆柱柱头 bitthead
系缆桩 anchor pile; bitt; bollard head; dolphin; mooring bitt; mooring bollard
系缆桩横杆 cross bitt
系缆装置 mooring accessories
系缆组 mooring gang
系缆座 bollard; cleat
系牢艇首缆 secure a painter
系类型 set type
系连海底的浮式护舷 buoyant fender moored to sea bed
系链 link chain; riding chain; stay-chain; tether
系链钩 line hook
系链扣座 cable clench; cable clinch
系梁 collar(beam); collar brace; collarino; collar rafter; connected yoke; dormant tree; joining balk; joining beam; landing beam; spanner; span piece; spar piece; spring beam; straining beam; stringer; tie-back; tie beam; tie girder; tie piece; tirant; top beam
系梁横(撑)木 collar stretcher
系梁人字木屋顶 collar rafter roof
系梁人字木屋架 collar beam roof truss
系梁人字屋顶 collar(beam) roof
系梁人字屋面 collar(beam) roof
系梁三角屋架 collar(-and)-tie roof
系梁屋顶 tie-beam roof
系梁屋盖 tie-beam roof
系梁屋架 tie-beam roof
系梁下的格栅 joist under collar beam
系列 bank; family; series; corollary
系列泵 pump in series
系列编号 series designation
系列布置 series arrangement
系列产品 series production
系列成图 systematic mapping
系列澄清池 banks clarifier
系列澄清器 banks clarifier
系列冲模 follow die
系列船模 serial ship model; series of ship model
系列法 rosette method
系列杠杆 lazy jack
系列工程模型 serial engineering model
系列规格 series specification
系列焊(接) series welding
系列航次 series of voyage
系列号 serial number
系列号的前缀 serial number prefix
系列号码 serial number[S/N]
系列合同 serial contract
系列河工模型 series of river engineering model
系列化 normalization; serialization; seriation
系列化产品 serialization products
系列离散 serial variance
系列片 family chip
系列平均值 serial mean
系列铺路机械 paving train
系列期权 series of options
系列渠化 continuous canalisation[canalization]
系列圈闭的油气聚集 oil and gas accumulation of series trap
系列热解 series pyrolysis
系列射线照相术 serial radiography
系列生产 serial production; series manufacture
系列生产的 serially produced
系列数据分析(法) panel analysis
系列投标 serial tendering
系列外的 off-line
系列相关 sequential correlation; serial

correlation
系列相关系数 sequential correlation coefficient;serial correlation coefficient
系列小地震 earthquake swarm
系列型号 serial model No.
系列性表格 chained list
系列学习 serial learning
系列压模 follow die
系列应力损失 sequence-stressing loss
系列预测法 series forecasting method
系列圆盘线脚 <诺曼底式建筑> byzant
系列运行 series operation
系列增压 series supercharging
系列债券 serial bond;series bond
系列招标文件 <美> bidding documents
系列照片 serial-gram
系列照相术 serialography;seriography
系列照相装置 serialograph;seriograph
系列制图 systematic mapping
系留 mooring
系留椿 mooring dolphin
系留灯 riding lamps
系留地 tying-up place
系留趸船 captive barge;captive float;captive pontoon
系留杆 mooring mast
系留港池 anchorage basin;tie-up basin
系留观测网 mooring network
系留观察气球 captive balloon survey
系留监测系统 moored monitoring system
系留紧固件 captive fastener
系留缆 retaining rope
系留力 tie-up force
系留联结 hitching
系留码头 lay berth;tying-up place
系留气球 captive balloon;kite balloon;kitoon[kytoon];tethered balloon
系留气球勘测 captive balloon survey
系留气艇 kite-air ship
系留式监视系统 moored surveillance system
系留水雷 moored mine
系留索 mooring guy;mooring hawser;mooring wire;stay wire
系留塔 mooring mast;mooring tower
系留探空仪 wire sonde
系留柱 hitching post;mooring post
系留装具 mooring harness
系论 corollary
系锚板 anchor strap
系锚短柱 cathead
系锚杆 cathead
系描述项 set description entry
系模钢条 wall spacer
系木 accouplement;bond timber;chain timber;lacing;lacing bar;timber tie
系木绳套 timber hitch
系木索环 timber hitch
系谱 genealogy;lineage;line of descent
系谱表 genealogical table
系谱分类 phylogenetic classification
系谱关系 genealogical relation;genealogical relationship
系谱树 family tree;genealogical tree
系墙 toggle bolt
系墙螺栓 toggle bolt;toggle mechanism
系墙索 wall tie
系墙铁 metal wall tie;tie iron;wall tie
系圈 collar beam
系绕钢索 bend a cable
系上(绳索) turn in
系绳 guy rope;lashing;tether
系绳板条 cleat
系绳处 belay
系绳环垫板 lashing ring pad
系绳角铁 cleat;transom cleat
系绳链 tether
系绳栓 belaying pin;cleat
系绳索 hitch
系绳铁角 <加固货物装载用> cleat
系绳柱 lashing post
系绳桩 bollard
系绳座 holdfast
系石 bonder brick;bond header;bond stone;parpend(stone);perpeyn;through bonder;through stone
系石层 bonder course;bondstone course;course of bondstones
系兽板 head board
系数 coefficient;course of exchange ratio;factor;modulus[复 moduli];multiplier;quotient;quotiety;rating of exchange;ratio
系数 A <孔隙压力> A-parameter
系数 K 值 values of factor K
系数部件 coefficient unit
系数乘法器 coefficient multiplier
系数带 coefficient tape
系数的变换 transformation of coefficient
系数的观察趋势 observed trend in coefficients
系数的可估性 estimability of coefficient
系数电势计 coefficient potentiometer
系数电位器 coefficient potentiometer
系数分布 coefficient distribution
系数给定部件 coefficient unit
系数给定单元 coefficient unit
系数矩阵 coefficient matrix;matrix of coefficient;system matrix【数】
系数式电位计 coefficient potentiometer
系数统计推论 parametric(al) statistical inference
系数向量 coefficient vector
系数行列式 determinant of coefficient;system determinant
系数修正法 level(l)ing factor method
系数域 coefficient domain;field of coefficient
系数值 coefficient value
系数组 coefficient unit
系栓 drift bolt
系双浮 secure to double buoys
系丝 tie wire
系索 dragline;grab rope;lacing;lasher;lashing wire;tie line;guy wire <挖掘机的>
系索耳 belaying cleat;cleat;range cleat
系索棍 batonet
系索环 lacing eye;lacing grommet;lashing ring
系索栓 belaying pin
系索眼环 lashing eye
系索柱 guy stub;nigger head
系索桩 belaying pin;guy anchor;guy stake;toggle
系塔索 yaw guy;yaw line
系天幕杆 jack rod
系条目 set entry
系铁螺钉 yoke bolt
系铁销 gudgeon
系艇杆 boat boom;guest warp boom;lower boom;riding boom;swinging boom
系艇缆 boat line;chest rope;guess rope;guess warp;guest rope;guest warp
系艇索 painter line
系统 formation;lineage;network;system;tract
系统安全 system safety
系统安全程序 system assurance program(me)
系统安全工程 system safety engineering
系统安全压力 system relief pressure
系统安装 system assemble;system install;system installation
系统百科全书 system encyclopedia
系统保存文件 system save file
系统保护 system protection
系统保险程序 system assurance program(me)
系统保证出力 assured system capacity
系统背压 system backpressure
系统本征值 system eigenvalue
系统变化 systematic variation
系统变换 system change
系统变量符号 system variable symbol
系统辨识 system identification
系统辨识理论 system identification theory
系统标准 system standard
系统标准标号 system standard label
系统补偿 system balance
系统不变量 system invariant
系统不精确性 systematic inaccuracy
系统不确定度 systematic uncertainty
系统不确定性 systematic uncertainty
系统不同时率 system diversity factor
系统布局 topology of the system
系统布置 system layout
系统采样 systematic sampling
系统采样法 systematic sampling method
系统采样格式 systematic sampling scheme
系统参数 system parameter
系统操纵压力 line pressure
系统操作 system operation
系统操作员 sysop;system operator
系统测定 systematic measurement
系统测量程序 system measurement routine
系统测试 system test(ing);testing the system
系统测试程序 system measurement routine;system tester;system testing program(me)
系统测试方式 system test mode
系统测试机 system tester
系统测试全套设备 system test complex
系统测试仪 system tester
系统插板 system board
系统查错程序 system error routine
系统常数 system constant
系统常驻 system residence
系统超松弛 systematic overrelaxation
系统成分 system composition
系统程序 system program(me)
系统程序操作员 system program(me) operator
系统程序常驻区 system residence
系统程序错误 system program(me) error
系统程序号 system programmer
系统程序计划员 system programmer
系统程序库 component library;system library
系统程序设计 system programming
系统程序设计语言 system programming language
系统程序设计员 system programmer
系统程序图 system chart
系统程序营运人员 system program(me) operator
系统池 system pool
系统充液量 system refill fluid capacity
系统重新启动 system restart
系统重新运行 system roll-back
系统重置 system reset
系统重组 system reconfiguration
系统抽样(法) system(atic) sampling
系统抽样方案 systematic sampling schematization;systematic sampling scheme
系统抽样概型 systematic sampling scheme
系统抽样模型 systematic sampling schematization;systematic sampling scheme
系统抽样样品 systematic sample
系统储备 system reserves
系统储能 system stored energy
系统处理程序 system service program(me)
系统处理单位 system unit
系统处理命令 system action command
系统处理误差 system processing error
系统传递函数 system transfer function
系统存储量 system storage capacity
系统存储器 system storage;system memory
系统错后复原 system error recovery
系统错误 system mistake
系统带 system tape
系统带宽 system bandwidth
系统导纳 system admittance
系统导向的变革 system-centered [centred] approach to change
系统的 scientific;systematic;systemic
系统的变量 system variable
系统的布置管理 systematic layout planning
系统的测度 systematic measure
系统的产生代换品系 systematic production of substitution lines
系统的传递函数 system transfer
系统的社会经济因素 systematic socioeconomic factors
系统的瞬态响应 system's transient response
系统的调制传递锐度 system-modulation-transfer acutance
系统的协调 system coordination
系统的选择方法 system alternatives
系统的自我调节 systematic self-adjustment
系统的自我组织 systematic self-organization
系统等待 system wait(ing)
系统等价 system equivalence
系统地阐述 formulation
系统地计划 formulize
系统颠簸 churning
系统电功率摆动 system electrical power swing
系统电压 system voltage
系统电压降 system voltage drop
系统电压稳定器 system voltage stabilizer
系统电源 system power source;system power supply
系统电阻 system resistance
系统调查研究 systematic investigation and study
系统调度 system call
系统调度室 system control room
系统调用 system call
系统调用中断 system call interrupt

X

系统定向 systematic orientation
系统定义 system definition
系统定义信息 system definition information
系统动力学 dynamic(al) system; system dynamics
系统动态学 system dynamics
系统动态学模拟模型 system dynamics simulation model
系统动态学预测法 system dynamics forecasting method
系统动态预示模型 system dynamics predictive model
系统堆栈 system stack
系统多路转换 system multiplex
系统发生 phylogeny
系统发生带 phylozone
系统发生的 phylogenetic
系统发生图 phylogenetic chart
系统发生学 phylogenetics
系统发育 phylogeny
系统发展 systems development
系统发展公司 systems development corporation
系统发展过程 systems development process
系统发展要求 systems development requirement
系统方法 system approach; systems method
系统方位 systematic orientation
系统仿真 system simulation
系统分辨能力 system resolution
系统分布 systematic distribution
系统分类 genealogical classification; phyletic classification; phylogenesis classification; system classification
系统分离 systematic separation; system splitter
系统分列 system separation; system splitting
系统分析 circuit analysis; harmonic analysis; system(atic) analysis
系统分析(方)法 system approach
系统分析工程学 system analysis engineering
系统分析和综合模型 system analysis and integration model
系统分析技术 <利用计算机等数学方法来研究问题并提出多种解决办法以资选择> systems analysis(technique)
系统分析经理 system analysis manager
系统分析模型 system analysis model
系统分析人员 systems analysis staff; systems analyst
系统分析仪 system analyser
系统分析员 system analyst
系统分析专家 system analyst
系统分组 hierarchic(al) classification; systematic grouping
系统风机 system fan
系统风险 system risk
系统封闭 system lock
系统封锁 system lock
系统服务程序 system service program(me)
系统服务控制点 system service control point
系统服务申请 system service request
系统辅助控制单元 system auxiliary control element
系统负荷 system loading
系统负荷系数 system load factor
系统负异常 systematic negative anomaly
系统负载 system load
系统负载特性 system load characteristic
系统负载状况 system loading condition
系统复位 system reset
系统复原 system reset
系统改进时间 system improvement time
系统改进组 system improvement group
系统概念 system concept
系统干管 system main
系统跟踪 system trace
系统跟踪标记 system trace tab
系统更换 system conversion
系统工程 system(atic) engineering; systems enrichment
系统工程方案 system engineering program(me)
系统工程方法论 systems engineering methodology
系统工程工具 system engineering tool
系统工程师 system engineer
系统工程学 <运筹学的相邻学科,使用模拟、数理统计、概率论、排队论、信息论等数学方法来处理工程中的系统问题> system engineering
系统工程学评价 system engineering evaluation
系统工程与技术指导 system-engineering and technical direction
系统功能 system function
系统功能设计 system function design
系统功能失灵 system malfunction
系统功能图 systems function diagram
系统功效 system behavio(u)r
系统功效模型 system behavio(u)r model
系统供电 mains power supply
系统供电变压器 system service transformer
系统构成 configuration of system
系统估计 systems estimation
系统固件 system firmware
系统固有延迟 system delay
系统故障 system down; system failure; system malfunction
系统故障分析 system trouble analysis
系统故障监测 fault monitoring
系统故障排除 recovery from system failure
系统故障探测器 fault finder
系统故障允许程度 fault tolerance
系统关闭 system shutdown
系统关联 <指信号在系统中有时间上的关联> coordination
系统观 systematic perspective
系统观测 systematic observation; system measurement
系统观测法 method by series
系统观察 systematic observation
系统观点 systematic point of view
系统管理 system management
系统管理程序 system supervisor
系统管理功能 system management function
系统管理学派 systems management school
系统管理员 system operator
系统光谱分析 systematic spectral analysis
系统规定参数 constraint
系统规范 system specification
系统规划 system planning; system programming
系统规划程序 system program(me) planning
系统规则 system convention
系统过程 systematic procedure
系统过调量 system overshoot
系统函数 system function
系统合成 system synthesis
系统和程序 systems and procedures
系统核心容量 system residence volume
系统荷载因数 system load factor
系统宏指令 system macro instruction
系统后备 system backup
系统后援通道 system support channel
系统互换性 system compatibility
系统互连 system interconnection
系统化 regiment; schematization; systematization; systemize; sytematize
系统化采样 systematic sampling
系统化信息 systematic fashion information
系统化支护 systematic support
系统划分 system partition; zoning
系统环境 system environment
系统环境仿真 system environmental simulation
系统环境记录 system environment recording
系统缓冲单元 system buffer element
系统缓冲区 system buffer
系统恢复 system recovery
系统恢复管理 systems recovery management
系统会聚误差 systematic convergence error
系统火力发电厂 system thermal power plant
系统机械化 system mechanization
系统基本原理 system philosophy
系统畸变 systematic distortion
系统及其环境相互作用的数学模型 mathematic(al) models of interaction between system and its environment
系统级模拟 system level simulation
系统级芯片 system on chip
系统集成 system(atic) integration
系统集成技术 system integration technique
系统集中控制 system centralized control
系统几何尺寸 system configuration
系统计划 system planning
系统计划人员 system planner
系统计时 system time stamp
系统记录功能 system log function
系统记录器 system log
系统记录数据集 system log data set
系统记录文件 system record file
系统记时器 system timer
系统记时员 system timer
系统技术 systems technology
系统技术顾问 systems technical consultant
系统技术规格书 system specification
系统技术说明 system specification
系统尖峰负荷 system peak load
系统间的 intersystem
系统间通信[讯] intersystem communication
系统兼容性 system compatibility
系统监督程序 system monitor
系统监控程序 system monitor
系统监视装置 system monitor unit
系统检查 systematic inspection; system inspection
系统检错码 systematic error checking code
系统检验 system check; system testing(ing)
系统检验模块 system check module
系统检验器 system checker
系统建筑方法 system building approach
系统鉴别 system identification
系统鉴定 systematic identification
系统降级 system degrade
系统交叉 system transposition
系统校验 system check
系统校验模件 system check module
系统校正像片 system-corrected image
系统校准 system calibration
系统接地 system earth; system ground
系统接地导体 system grounding conductor
系统接地法 system grounding method
系统接口模件 system interface module
系统接口设计 system interface design
系统节理 systematic joint
系统结构 structure of a system; system architecture; system configuration; system organization; system structure
系统结构改变 system reconfiguration
系统解列 system detachment
系统近似平衡 system approaches equilibrium
系统经理 system manager
系统精度 system accuracy
系统矩阵 system matrix
系统聚类 system cluster
系统卷宗 system volume
系统决策模式 systematic decision model
系统开发 system development
系统开发和分析程序 system development and analysis program(me)
系统开始 system initiation
系统科学 system science
系统可变符号 system variable symbol
系统可靠度 system reliability
系统可靠性 fail-safety; system reliability
系统可靠性分析 systems reliability analysis
系统可用性 system availability
系统空间平均车速 space mean speed in system
系统控制 control of system; system control
系统控制板 system control panel
系统控制记录点 control-metering point of systems
系统控制接口 system control interface
系统控制块 system control block
系统控制面板 system control panel
系统控制器 system controller
系统控制台 system console
系统控制中心 system control center [centre]
系统框架 system sketches
系统框图 system chart; system flowchart
系统扩充 system expansion
系统类目 systematic category
系统离差 system deviation
系统理论 system(atic) theory
系统理论法 systematic theory method
系统力矩系数 moment coefficient of combination
系统利用度 availability system; system availability
系统利用率 availability system; system availability; system utilization; system utilization factor
系统利用率记录 system utilization logging
系统利用率记录程序 system utilization logger
系统利用率记录器 system utilization logger
系统连贯 systems linking

系统连接 system connection
系统连接线 tie conductor
系统联调 system integration
系统龄 systems age
系统流程 system flow
系统流程图 system chart; system flowchart
系统论 systematology; system theory
系统(论)方法 systems approach
系统逻辑 system logic
系统码 system(atic) code
系统锚杆 systematic rock bolt
系统锚固 pattern bolting; system bolting
系统描述问题 system description problem
系统名(称) systematic name
系统命令 system command
系统命令解释 system command interpretation
系统命令执行 system command executive
系统命名法 system nomenclature
系统模件 system module
系统模拟 system analog; system model(l)ing; system simulation
系统模拟模型 system simulation model
系统模拟网络测试方法 systematic analog network testing approach
系统模拟语言 system analog(ue) language
系统模式 system mode
系统模型 system model
系统模型化 systems modelling
系统模造 system model(l)ing
系统母线 system busbar
系统目标 aims of system(s)
系统内的 intrasystem
系统内期望人数 expected number in the system(s)
系统内制冷剂容量 refrigerant charge
系统能 system energy
系统能力的恢复 system recovery
系统耦合 system linking
系统排水 system de-watering
系统排水渠 systematical drainage canal
系统配套 system support
系统配置 system configuration
系统匹配 system matching
系统偏差 systematic deviation; system deviation
系统漂移速率 systematic drift rate
系统频率 system frequency
系统频率偏移 system frequency excursion
系统频率特性 system response; system transient output response
系统平衡 system balance; system equilibration
系统评价 system evaluation; systems assessment
系统评价程序 systematic evaluation program(me)
系统评价与决策 assessment and decision of system
系统屏蔽 system mask
系统破坏 system failure
系统企业 system house
系统启动 system start-up
系统启动命令 system action command
系统起动 system start-up
系统切换 system switching
系统侵染 systemic infection
系统情报 system information
系统情报管理 system information management
系统区 system area; system pool
系统区段平衡 system reaches equilibrium
系统区域 system realm
系统取样 systematic sampling
系统扰动 system disturbance
系统任务 system task
系统任务保存区 system task save area
系统任务区 system task partition
系统任务优先数 priority of system task
系统容积 system volume; volumetric-(al) content of system
系统容量 system capacity; volume content of system
系统软件 system software
系统软件包 system software package
系统软件成分 system software component
系统软设备 system software
系统扫描程序 system scanner
系统熵 system entropy
系统设备 system device
系统设计 system design; system layout
系统设计标准 system design criterion
系统设计的最佳化 optimization of system design
系统设计方法 systematic design method
系统设计工具 system design aids
系统设计规范 systematic design discipline
系统设计模拟器 system design simulator
系统设计手段 system design aids
系统设计说明书 systems design specification
系统设计员 system designer; system planner
系统设计最优化 optimized system(s) design
系统审计 operational auditing; systems audit
系统生成 system generation
系统生成时间 system generation time
系统生命周期 systems life cycle
系统生态学 systems ecology
系统生物学 system biology
系统失灵 system down
系统失效 thrashing
系统失真 systematic distortion
系统施工法 systematic building; system building; system construction method
系统施力 systematic application
系统时钟 system clock
系统时钟控制 system clock control
系统时钟控制器 system clock controller
系统时钟控制信号 system clock control signal
系统识别 system identification
系统识别与参数估计 system identification and parameter estimation
系统实施 systems implementation
系统实用程序 system utility program(me)
系统实用设备 system utility device
系统式管理组织结构 systems structure
系统试验 system test(ing)
系统试验器 system exerciser[exercisor]; system tester
系统试验站 system test station
系统试验装置 system test set
系统手册 system handbook; system manual
系统寿命周期 system life cycle
系统输出值 output value of system
系统输出装置 system output unit
系统输入值 input value of system
系统输入装置 system input device; system input unit
系统树 systematic tree; system tree
系统树图 dendrogram
系统数据收集 system data acquisition
系统数据总线 system data bus
系统数学分析法 systematic mathematical analysis
系统数学模型 systematic mathematical model; system mathematical model
系统水平 system level
系统水平衡 system water balance
系统水头 system head
系统水文学 system hydrology
系统顺序 systematic order
系统瞬时输出频率特性 system transient output response
系统说明 formulate
系统思路 systems approach
系统思维 system thinking
系统死锁 system deadlock
系统松弛步骤 systematic relaxation procedure
系统速度 system speed
系统算法 system algorithm
系统随机样本 systematic random sample
系统损耗 system loss
系统特性 system performance
系统特性参数 system specific parameter
系统特性化学因子 system specific chemical factor
系统条件 system condition
系统调试 system debug
系统调优 system optimization
系统停工 system shutdown
系统停机 system down; system halt
系统停止页 system stop page
系统通信[讯] system communication
系统通信[讯处理 system communication processing
系统统计分析 system statistical analysis
系统统计量 systematic statistic
系统统计信息 system statistical information
系统投配量 system dose
系统透盐率 system salt permeability
系统突变 systematic mutation
系统图 system(atic) diagram; system drawing; system map
系统途径 system approach
系统退化 system degradation
系统完成码 system completion code
系统完整性 system integrity
系统网络 grid; system network
系统网络结构 system network architecture; system network structure
系统网络体系 system network architecture
系统维护 system maintenance
系统维修手册 system maintenance manual
系统稳定性 stability of a system; system stability
系统稳定性分析 system stability analysis
系统污垢物 system foulant
系统误差 bias error; constant error; regular error; systematic error; systematic residuals; system error
系统误差检验码 systematic error checking code
系统细部设计 system detail design
系统细目概念 systematic isolate idea
系统现场记录 system environment recording
系统现状的保存 system description maintenance
系统线路 system line
系统线路网络 network of system lines
系统相容性 system compatibility
系统响应 system response
系统消光 system extinction
系统小时最大负载 system maximum hourly load
系统小组 systems group
系统效率 system effectiveness; system efficiency
系统效应 systematic effect
系统协调 communicating the system
系统协调性 system compatibility
系统心理学 system psychology
系统信号控制机 system signal controller
系统信号组合法 combination method system(of traffic signal)
系统信息 system information
系统信息数据集 system message data set
系统行为 systems behavio(u)r
系统形成 system generation
系统性 systematicness
系统性变化 system variation
系统性差的 ill-structured
系统性程序 systematic procedure
系统性的 systematic
系统性风险 general market risk; marketwide risk; systematic risk
系统性感觉丧失 systematic desensitization
系统性季节偏差 systematic seasonal deviation
系统性能 system performance
系统性能监视器 system performance monitor
系统性能降低 system degradation
系统性能校正化合物 system performance calibration compound
系统性能评价 system performance evaluation
系统性能数据 system performance information
系统性能有效度 system performance effectiveness
系统性能指标 system performance index
系统性硬化 systemic sclerosis
系统性直线离差 systematic deviation from linearity
系统性组合 schematism
系统修正 systematic correction
系统虚拟地址 system virtual address
系统选择 systematic selection; systemic selection; system option
系统选择方案 systems alternatives
系统学 genealogy; systematics; systematology
系统学派 systems school
系统询问 system interrogation
系统循环 systemic circulation
系统压力 system pressure
系统压头 system head
系统压头特性 system head characteristic
系统压头特性曲线 system head curve
系统研究 system approach; system research; system study; research-on-research <对研究方法和发展过程的>
系统研究法 systematic approach; systems approach; systems approach to management; systems research
系统研制 system development
系统研制步序 development system step

系统验收试验 system acceptance tests
系统样本 systematic sample
系统要件说明书 systems requirements specification
系统要求标准 systematic requirement criterion
系统液体容量 system fluid capacity
系统移交 system conversion
系统因素 system factor
系统因子 system factor
系统引导偏差 miss-distance bias
系统应急保护 system emergency protection
系统应用程序的提供 system utility program(me) support
系统硬件 system hardware
系统用户 system user
系统优化 system optimize
系统优选 system optimization
系统油液灌注容量 system refill fluid capacity
系统有效性 availability;system effectiveness
系统语言 system language
系统预置 system initialization
系统元件 system component;system element
系统约束 system constraint
系统运行 systems operation
系统运行程序 system log
系统运行费用 system running cost
系统运行机制 systems operational mechanism
系统运行记录 system log
系统运行试验 systems implementation test
系统运行文件 system log
系统再起动 system restart
系统赞助人 system sponsor
系统噪声 systematic noise; system noise
系统噪声指数 system noise figure
系统增益漂移 systemic gain drift
系统展示监控器 system display monitor
系统诊断 system diagnosis
系统诊断程序 system diagnostics
系统振荡 system oscillation
系统整合【交】system integration
系统支持 system support
系统支援程序 system support program(me)
系统支援功能 system support function
系统支援机械 system support machine
系统支援设备 system support
系统执行语言 system implementation language
系统指令 system command
系统制束 system constraint
系统中的问题 intersystem problem
系统中断 system interrupt; system outage
系统中断调度程序 system interrupt dispatch routine
系统中断动作 system interrupt action
系统中断请求 system interrupt request
系统中心 system center[centre];system point
系统中心线 system center[centre] line
系统中性点接地方式 earthing method of system neutral
系统主管 system supervisor
系统驻留容积 system residence volume
系统转储 system dump
系统转换 system switching
系统装机容量 system installed capacity
系统装入程序 system loader
系统状况 system status
系统状况询问 system status interrogation
系统状态 system state;system status
系统状态表 system state table
系统状态指令 system mode instruction
系统准备程序 system preparation routine
系统咨询部 system advisory board
系统资料管理 system resource management
系统资源 system resources
系统资源管理 system resource management
系统子成分 system subcomponent
系统子程序 system subroutine
系统综合 system integration; system synthesis
系统总开销 system overhead
系统总体积 total volume of system
系统总线 system bus
系统总线负载 system bus loading
系统总线接口 system bus interface
系统总线宽度 system bus width
系统总线仲裁器 system bus arbitrator
系统阻抗 system impedance
系统阻抗比 system impedance ratio
系统阻力 system resistance
系统阻力曲线 system resistance curve
系统阻力特性 system resistance characteristic
系统阻尼 system damping
系统组成 system composition
系统组件 system constituent
系统组织 system organization
系统最佳化 system optimization
系统最优 system optimum
系统最优分配原则 system-optimized assignment principle
系统最优化 system optimization
系统最优交通分配模型 system optimum traffic assignment;system optimum traffic assignment model
系瓦绳 tile tie
系网环 net ring;ring for safety net; ring to safety ring
系物横杆 lash rail
系线 anchor line string; conode; tie line
系线测量法 tie line method
系线法 <越过障碍物的测量方法> tie line method
系项 set entry
系序 set order
系选择 set selection
系选择格式 set selection format
系圆片 tie disc[disk]
系在浮筒上的链或短缆 <供系泊时使用> mooring bridle
系在缆桩的绳 dolphin fast
系者 tier
系值 set occurrence;train value
系值选择 set selection
系主 set owner
系主任 director of department;dean
系住 anchor;moor
系柱 bollard cleat
系柱试验 <在岸边试验推进性能> bollard test
系子登记项 set subentry
系子系目 set subentry
系综 <统计力学的> ensemble; assemblage
系综密度 coefficient of probability
系综平均(值) ensemble average
系族 strain

细

细白垩 whiting

细白垩胶 clairecolle;clearcole
细白砂 crust
细白石膏 alabaster
细白土 fine clay
细柏油混凝土 fine tar concrete
细斑状 minophyric
细棒材 pencil rod
细胞表面疏水性 cell-surface hydrophobicity
细胞状结构 cellular construction;cellular structure
细碧角斑岩 spillite-keratophyre
细碧角斑岩建造 spilitic-keratophyre formation
细碧结构 spilitic texture
细碧岩 spilite
细碧岩化作用 spilitization
细碧岩角斑岩系 spilite-keratophyre sequence
细碧岩套 spilitic suite
细碧岩质玻璃 spilitic glass
细碧岩组 spilitic suite
细碧岩组合 spilitic association
细臂 microarm
细扁木锉 smooth flat wood rasp
细冰针 frazil ice
细波花纹 fiddleback
细波纹 ghost wires;insertion waves
细玻棒 fine glass rod
细玻璃管 mat(t)ras(s)
细薄布 cambric
细薄防水布 jaconet
细薄干煤 blind coal
细薄毛织物 bastiste
细薄棉布 muslin
细薄平纹毛织物 muslin-delaine
细布 fine cloth;percale
细部 detail;fine details
细部布置 layout of details
细部测量 detail(ed) survey(ing)
细部测图 detail mapping;detail plotting
细部地形 fine texture topography
细部点 detail point
细部构造 detail construction
细部规划 detailed planning; detailed program(me)
细部结构 detailed construction;detailing
细部截面 detailed section
细部清晰度 sharpness of detail
细部三角测量 detailed triangulation
细部设计 detail(ed) design; detailing (design)
细部设计阶段 detailed design stage
细部设计联合体 stage of detailed planning
细部设计员 detailer
细部图 detail chart; detail drawing; detailed plan
细部详图 detail drawing
细部详细 richness of details
细部再现 rendering of detail
细部再现力 detail reproduction
细部装配 detail assembly
细部装配样板 detail assembly template
细部作业计划 detailed operating[operation] schedule
细部坐标 coordinate of details
细部坐标点 detail coordinate point
细擦痕 stria[复 striae]
细采伐剩余物 light slash
细槽 cannelure
细槽沟 groove
细槽光栅 fine-grooved grating
细槽铸型 delicate flute cast
细层 laminated bedding;microlayer
细层纹的 finely laminated
细层纹岩石 finely laminated rock
细察 scrutinization;scrutiny
细长 slenderness
细长比(例) ratio of slenderness;slenderness proportion;slenderness ratio;slender proportion
细长刀 bistoury
细长的 slab-sided;slender
细长度 degree of slenderness; slenderness(degree); slenderness proportion
细长杆式变速 cane type shift
细长工件 slender piece
细长构件的横截面 slender section
细长骨料 elongated aggregate
细长骨料指数 elongation index
细长管式荧光灯 slimline type fluorescence lamp
细长集料 elongated aggregate
细长颗粒 elongated particle; slender particle
细长粒料 elongated aggregate
细长粒子 elongated particle; slender particle
细长梁 slender beam
细长内廊 <古希腊> pastas
细长片 strip
细长片状骨料 flaky and elongated aggregate
细长平底的船 canal boat
细长气泡 blibe;cat's eye <玻璃中的>
细长墙 slender wall
细长三角旗 pennant
细长体 slender configuration
细长体理论 slender body theory
细长条纹 air line;hairline
细长效应 slenderness effect
细长楔 slender wedge
细长形 elongate; slender form; slender type;slimline type
细长型 slender form
细长延长药包 long narrow cartridge
细长药卷 <引发雷管> string load instant cap
细长液柱 slender stream
细长荧光灯 slimline fluorescent lamp
细长荧光灯装置 slimline lamp fluorescent luminaire fixture
细长凿(刀) ripping chisel
细长指数 elongation index
细长柱 slender column;long column
细长锥体 slender cone
细尘岩 pulveryte
细沉积物 fine sediment
细齿 serration
细齿锉 saw file
细齿滚刀 serration hobber
细齿滚铣刀 serration hob
细齿环规 serration ring ga(u)ge
细齿环形扳手 serrated ring spanner
细齿锯 fine-toothed saw;half-sip saw
细齿锯条 fine-toothed saw blade
细齿可逆棘轮 fine-toothed reversible ratchet
细齿拉刀 serration broach
细齿连接 serration joint
细齿螺母 serrated nut
细齿木锯 half rip-saw
细齿塞规 serration plug ga(u)ge
细齿树花 Ramalina denticulata
细齿纹 tooth
细齿铣刀 fine-toothed cutter
细齿形 serrate profile
细齿轴 serrated shaft;serration shaft
细充填料 fine fill
细瓷 fine porcelain;refined porcelain

X

细瓷化炻器 fine porcelainized stoneware
细瓷状断口 fine pottery fracture
细刺壳属 <拉> Aphauostigme
细粗骨料比 sand-coarse aggregate ratio
细粗集料比 fine coarse aggregate ratio
细锉 fine(cut) file; polishing file; slim file; smooth-cut file; smooth file
细锉刀 fine file
细大理石渣 fine marble chip
细带 band(e)let; strapping
细带饰 band(e)let
细的 light-ga(u)ge; slim
细的板材 light-ga(u)ge
细的钢丝 light-ga(u)ge
细的光面圆钢筋 pencil rod
细的铁丝 light-ga(u)ge
细等高线 light-line contour; thin contour
细滴灌溉 trickle irrigate
细滴乳液 fine emulsion
细碲粉 <商品名> Telloy
细点荫罩 fine dot shadow mask
细雕 elaborate carved work; entail
细动作研究 micromotion study
细读数 fine reading
细度 degree of fineness; fineness degree
细度比 fineness ratio
细度参数 fineness parameter
细度测定 determination of fineness; fineness determination
细度测定仪 fineness meter
细度成熟度试验仪 fineness-maturity tester
细度规格 fineness specification
细度极限 fineness limit
细度计 fineness ga(u)ge; grindometer
细度模量 fineness modulus; modulus of fineness
细度模数 Abram's fineness modulus; fineness modulus; modulus of fineness
细度试验 fineness test
细度系数 fineness factor; fineness modulus; coefficient of fineness <悬浮固体与细度的比率>
细度相同的颗粒 grains of equal size
细度因子 fineness factor; surface factor
细度指数 <重塑土某一强度时的含水量> fineness number
细端 short end
细端直径 diameter at smaller end
细段 thin segment
细盾霉属 <拉> Leptothyrium
细剁斧面 dabbed finish
细剁石面 fine axed stone; smoothed-axed stone face
细方石 ashlar[ashler]; square stone

细分 detailed final sorting; subdivide; subdivision
细分度的分布 fine-scaled distribution
细分分级器 classifier for fine separation
细分解 fine resolution
细分类 subclassification
细分类法 close classification
细分类住宅小区 subdivision
细分离器 fine separator
细分散程度 degree of subdivision
细分散淡水泥浆 dispersed fresh water mud
细分散树脂体 diffuse resinite
细分散系统 fine dispersed system
细分筛 sifting screen
细分市场 segment market
细分水流 subdivided flow
细分选 fine separation
细粉 burgy; fines; fine breeze; fine dust; fine powder
细粉产量 fines output
细粉尘 fine dust
细粉的百分率 percentage of fines
细粉分离器 pulverised coal collector; pulverized coal collector
细粉粒分析 fine silt analysis
细粉粒黏[粘]结 fine silt bond
细粉料 fines
细粉率 acceptance efficiency; percentage of fines; percent fines
细粉磨 fine comminution; regrind mill; fine grinding
细粉磨的 finely ground
细粉磨机 finishing mill
细粉末 flower
细粉砂 <又称细粉沙> fine silt; fine silty sand
细粉砂结构 fine slit texture
细粉砂屑 fine silty clast
细粉砂岩 fine siltstone
细粉石墨 fine graphite
细粉碎 fine reduction
细粉碎机 slimer
细粉土 fine silt
细粉细度 fineness of powder
细粉卸出 fine discharge
细粉状的 finely powdered
细粉状黏[粘]土 fine clay
细缝 feint; fine draw; hairline; stylolite
细缝开裂 crazing
细缝筛 needle-slot screen
细浮石 pumicite
细腐熟腐殖质 fine mull
细腐殖质 mull
细干卷 cirrus[复 cirri]
细杆 pin
细杆螺栓 through bolt
细钢筋 concrete reinforcement wire; pencil rod; single-wire <常指直径 3～7 毫米>; distribution bar reinforcement <与主筋成直角的>
细钢缆扫海 fine-wire sweep
细钢丝 fine steel wire; thin steel wire
细钢丝绳 seizing wire
细高的 drawn-up
细格钢丝网 wire-mesh with close grid; wire-mesh with fine grid
细格花窗 <伊斯兰教建筑> mushrabeyed work
细格筛 finer mesh
细格栅 fine ga(u)ge screen; fine rack; fine screen
细工 fine workmanship
细工锯 fret saw
细工镶嵌的 tessellated
细工用材 small timber
细沟 rill(e); stria[复 striae]
细沟冲蚀 rill wash; shoestring washing
细沟冲刷 rill wash(ing); shoestring washing; rill erosion
细沟阶段 rill stage
细沟侵蚀 microchannel erosion; rill erosion; rill washing; rillwork
细沟装饰柱 striated column
细沟状冲刷 rill erosion
细谷 ravine
细骨料 fine aggregate; fines; mineral dust; small aggregate
细骨料百分比 percent fines
细骨料粉砂含量试验 silt content test for fine aggregate
细骨料粉土含量试验 silt content test for fine aggregate
细骨料灌浆料 sand(ed) grout
细骨料含量 sand percentage
细骨料含量试验 organic matter test for fine aggregate
细骨料混凝土 concrete made with fine aggregate; fine aggregate concrete; fine concrete; mierobeton
细骨料混凝土混合料 fine concrete mix
细骨料区 fine zone
细管 microtubule; tubule
细管供暖系统 small pipe system
细管黏[粘]度计 caplastometer
细管热水集中供暖设备 small pipe system
细管热水集中供暖系统 small bore system; small pipe system
细管束薄膜反渗器 <成水的> permasep membrane permeator
细管荧光灯 slimline lamp
细光栅扫描 microscanning
细光束聚光灯 pin spot; pin spot-light
细辊 smooth roll
细过滤器 fine filter
细焊料 fine solder
细号钢丝网 chicken wire
细号线 light-ga(u)ge wire
细河砂 fine stream sand
细花岗岩 microgranite
细花键连接 serration
细滑石(粉) fine talc
细滑移带 fine slip band
细化 refine; refinement; refining; thinning
细化温度 refining temperature
细灰缝 fine stuff
细灰浆 fine stuff
细灰浆机 fine plaster machine
细混合骨料 fine aggregate mixture
细混合集料 fine aggregate mixture
细混和物 fine admixture
细活 neat work
细级配 fine grading; fine size grading
细级配材料 <水泥与细集料的总称> fine-graded material
细级配的 fine-graded
细级配地沥青混合料 fine asphalt mix(ture)
细级配地沥青混凝土 fine asphalt concrete; fine-graded asphaltic concrete
细级配地沥青混凝土路面 fine asphalt surfacing; fine-graded asphaltic concrete pavement
细级配地沥青混凝土磨耗层 fine-graded asphaltic concrete carpet
细级配地沥青混凝土瓦 fine-graded asphaltic concrete tile
细级配地沥青沥青砖 fine asphalt tile
细级配地沥青磨耗层 fine asphalt carpet
细级配骨料 finely graded aggregate; soft sand
细级配集料 finely graded aggregate
细级配沥青混凝土 finely graded bituminous concrete
细级配沥青碎石 fine-graded macadam
细级配马克当路面 fine-graded macadam
细级配(石油)沥青混凝土瓦 fine-graded asphaltic concrete tile
细级压碎 small crushing
细集料 fine aggregate; mineral dust; mineral filler; small aggregate; fine concrete aggregate <混凝土的>
细集料混合物 fine aggregate mixture
细集料混凝土 fine aggregate concrete; fine concrete; microbeton
细集料混凝土楼面找平层 fine concrete floor screed
细集料混凝土楼面找平层材料 fine concrete screed material
细集料棱角 fine aggregate angularity
细集料式道路 fine aggregate type of road
细加工 fine finish; fining; fining-off; upper cut
细加工条石 finished small stone
细尖 apiculus
细尖熔灰岩 pulverulite
细尖三角锉 extra slim taper file
细尖水翼 slender pointed hydrofoil
细尖塔 needle spire; slender spire
细讲 give particulars
细焦点 fine focus
细焦粉 fine coke breeze; finely ground coke
细节 detailing; details; fine details; fine end; minor details; particular; pin knot; ramification; specialties; the ins and outs
细节对比度 detail contrast
细节放大器 detail enhancer
细节分解力 resolution detail
细节距影孔板 fine-pitch mask
细节清晰度 fine detail resolution; resolution detail
细节设计 detailed design; detailed engineering
细节失落 detailloss
细节再现 detail rendering
细结构土 fine-textured soil
细介壳灰岩 microcoquina
细金刚砂 fine emery powder
细金刚砂粉 emery flour; flour of emery; powdered emery
细金属丝 fine wire
细金属丝网 fine wire netting; wire gauze
细金属网 <覆盖墙板接缝用的> fly wire
细金属网眼 fine metal mesh
细进刀 fine feed
细进给 micrometer feed
细晶 fine-grained
细晶白云岩 fine crystalline dolomite
细晶超塑成型 fine crystal superplastic forming
细晶错综状的【地】autallotriomorphic
细晶的 fine crystalline
细晶低碳钢 plastalloy
细晶锭 fine-grained ingot
细晶断口 fine-grained fracture
细晶刚玉 microlite
细晶花岗石 fine crystalline granite
细晶花岗岩 aplite-granite
细晶花岗质的 aplite-granitic
细晶化 grain refinement; grain refining
细晶灰岩 fine crystalline limestone
细晶结构 aplitic texture; fine crystalline texture
细晶粒 close grain; fine grain
细晶粒的 compact grained; fine granular
细晶粒度 fineness of grain
细晶粒钢 fine-grained steel; grain refined steel; grain refining steel
细晶磷灰 carbonate-fluorapafite
细晶磷灰石 francolite
细晶石 microlite
细晶石含量 microlite content
细晶石矿石 micromite ore
细晶岩 alaskite; aplite; diabase; haplite
细晶岩类 aplite group; rhenopalites
细晶岩脉 aplitic dike[dyke]
细晶质粗面岩 aplitic trachyte
细晶质的 cryptomerous; finely crystalline

细颈大坛 demijohn
细颈钢瓶 flask
细颈瓶 narrow(ed)-neck(ed) bottle; ampulla <古希腊罗马的,有双耳的>
细颈器皿 narrow neck ware
细颈盛水瓶 <有玻璃塞的> decanter
细颈形衔铁 isthmus armature
细颈柱 <风蚀、浪蚀> pedestal; rock pedestal
细净化 fine cleaning
细净化区 fine clarification
细距 fine pitch
细锯 fine saw; jig saw; smoothing saw
细锯锉 slim saw file
细锯纹区 fine hackle
细聚集体 fine agglutinant
细聚焦 fine focusing
细聚焦 X 射线管 fine-focus X-ray tube
细菌 bacteria; germ; microbe
细菌变异 bacterial variation
细菌病毒含量超标 overstandard of bacterial virus content
细菌测量 bacterial survey
细菌测量方法 bacteria survey
细菌沉淀 bacteria precipitation
细菌成因 bacteriogene
细菌处理 bacteria(l) treatment
细菌床 <生物滤池的> bacteria bed
细菌的 bacterial
细菌电池 bacteria cell
细菌毒素 bacteriotoxin
细菌繁殖数 bacterial count
细菌肥料 bacteria(l) fertilizer; bacteria(l) manure
细菌分解作用 bacterial decomposition; microbial decomposition
细菌分析 bacteria(l) analysis; bacteriological analysis
细菌腐蚀 bacteria(l) corrosion
细菌过滤器 biofilter
细菌含量 bacterial content
细菌化学 bacterial chemistry
细菌化验 bacterial analysis; bacteriological analysis
细菌化验室 bacteriology laboratory
细菌还原作用 bacteria reduction
细菌活化剂 bacteria activator
细菌活性 bacteria activity
细菌计数 bacterial count; bacteriological count; counting of bacteria; enumeration of bacteria
细菌计数器 bacteriological counting apparatus; counting apparatus for bacteria
细菌减少 bacterial reduction
细菌检验 bacteria(l) examination; bacteria(l) test
细菌检验室 bacteriological laboratory
细菌鉴定 identification of bacteria
细菌降解法 bacterial degradation
细菌浸出 bacteria leaching
细菌净化 bacterial treatment
细菌镜检法 bacterioscopy
细菌菌苗 bacterial vaccine
细菌勘察 bacterial prospecting
细菌抗力 bacterial resistance
细菌抗药性 bacterial resistance
细菌沥滤 bacterial leaching
细菌滤池 bacterial bed
细菌滤除法 sterilization by filtration
细菌滤器 bacterial filter
细菌密度 bacterial density
细菌黏[粘]合 adhesion of bacteria
细菌黏[粘]液 bacteria(l) slime
细菌浓度 bacterial concentration
细菌培养 bacterial culture; pure culture
细菌培养法 bacterial cultivation
细菌培养基 bacterial culture medium
细菌培养物 bacterial culture
细菌培养液 inoculums[复 inocula]
细菌平衡 bacterial equilibrium
细菌平衡指数 bacterial balance index
细菌迁移距离 transfer distance of bacteria
细菌侵蚀 bacterial attack
细菌生长曲线 bacterial growth curve; growth curve of bacteria
细菌生存时间 survival period of bacteria
细菌示踪物 bacterial tracer
细菌试验 bacteria examination; bacteriological examination
细菌衰减 bacterial decay
细菌死亡 bacterial mortality
细菌同化 bacterial assimilation
细菌脱蜡 bacterial dewaxing
细菌污染 bacterial contamination; bacterial pollution; bacteriological contamination; germ contamination
细菌污染指数 bacteriological pollution index
细菌武器 germ weapon
细菌吸附作用 adsorption of bacteria
细菌吸收 bacterial assimilation
细菌消化 bacterial digestion
细菌性痢疾 bacillar dysentery
细菌性杀虫剂 bacterial insecticide
细菌性水质 microbial water quality
细菌性污染物 bacterial contaminant
细菌性有机碳 bacterial organic carbon
细菌学 bacteriology; microbiology
细菌学的 bacteriological
细菌学分析 bacteriological analysis
细菌学检验 bacteriological analysis
细菌学指标 bacteriological index
细菌氧化作用 bacterial oxidation
细菌冶金 bacterial metallurgy
细菌再生 bacteriological aftergrowth
细菌指示物 bacterial indicator
细菌指数 bacteriological index
细菌种群 bacterial population
细菌转化 bacterial conversion
细菌状的 bacteroid
细菌总数 bacteria amount; total amount of bacteria; total-count of bacteria colonies
细菌作用 bacteria(l) action
细砍砖 fair cutting
细看 scrutiny
细颗粒 fine grain; fine particle
细颗粒部分 fine fraction
细颗粒材料 fine material
细颗粒沉积物 fine-grained sediment
细颗粒的 fine angular; fine-grained; fine(ly) granular; close grained
细颗粒分析 fine analysis; fine grain analysis
细颗粒管型 fine granular cast
细颗粒混合物 fine grain mixture
细颗粒间相互排斥作用 <土内> interparticle repulsion
细颗粒泥沙 fine sediment
细颗粒土 fine-grained soil
细颗粒物 fine particulate mass
细颗粒有机物 fine particulate organic matter
细颗粒织构 close texture
细颗粒状的 fine granular
细刻磨 fine cutting; smoothing <空心玻璃>
细孔 osculum; pore
细孔壁 porous wall
细孔的 fine-pored
细孔电解 electrostenolysis
细孔垫布 finely porous mat
细孔分布 pore size distribution
细孔格栅 fine rack
细孔隔膜电解 electrostenolysis
细孔硅胶 Kiselgel A
细孔金属丝筛网 fine wire screen
细孔径柱 narrow-bore column
细孔扩散 pore diffusion
细孔拦污栅 filter screen
细孔漏斗 sintered glass filter
细孔滤网滤清器 fine-mesh filter
细孔黏[粘]度计 orifice-type visco(si)meter
细孔凝胶 small porosity gel
细孔喷油嘴 fine nozzle
细孔容积 pore volume
细孔筛 bolter; close-meshed screen; fine ga(u)ge screen; fine grain screen; fine-mesh sieve; fine screen; hair sieve; small mesh sieve
细孔筛布 fine-mesh screen cloth
细孔网 fine-meshed screen
细孔污水滤网 fine sewage screen
细孔隙 fine pore
细孔隙的 finely porous
细孔小喉型 small throat with connecting small pore
细孔中喉型 middle throat with connecting small pore
细孔钻 rat hole drilling
细口的 narrow-mouth(ed)
细口瓶【化】narrow-neck(ed) bottle; narrow-mouth(ed) bottle
细口烧瓶 narrow-mouth(ed) flask
细筘 fine reed
细矿仓 fine ore bin; fine ore storage
细矿粉 fine ore
细矿脉 scun; stringer vein; thread
细矿脉脉道 stringer
细矿脉脉道源头 stringer-head
细矿石 fine ore
细矿团化 pelletization[pelletisation]
细矿物面材 fine mineral surfacing
细矿物撒布料 fine mineral surfacing
细拉毛饰面 fine stipple finish
细浪 choppy sea; lipper; ripple; rippling sea; smooth sea; cat's paw
细类 subclass
细沥青 refined asphalt
细砾 fine gravel; grit gravel; pea gravel; pebble
细砾波痕 granule ripple
细砾薄表层 fine surface mulch of gravel
细砾角砾岩 granule breccia
细砾结构 granule texture
细砾矿床 fine gravel ore deposit
细砾泼撒抹面 pebble-dash plaster
细砾石 bird's eye gravel; fine gravel; grail; pea gravel
细砾石滤池 fine gravel filter
细砾土 fine-grained soil; fine gravel
细砾镶嵌沙漠面 pebble mosaic
细砾岩 granulestone
细粒 caliche; riddlings
细粒百分率 percent fines
细粒板 fine particle board
细粒变晶结构 fine granular crystalloblastic texture
细粒薄层砂岩 sand flag
细粒部分 fine component
细粒材料 fine-sorted material
细粒残留物 fine residue
细粒产物 fine product
细粒沉积岩 pulverite; pulveryte
细粒充填料 fine fill
细粒冲积物 fine-grained alluvial deposit
细粒大理石罩面水磨石 berliner
细粒的 close(d)-grained; fine; fine-graded; fine-grained
细粒的过渡层 fine transition
细粒的小砖 pinhead tile
细粒地沥青混凝土联结层 fine-grained asphaltic binder course
细粒度 fine graininess
细粒度破碎腔室 fine chamber
细粒度油石 microstone
细粒断裂 smooth fracture
细粒反滤料 fine transition
细粒分级 fine(size) grading
细粒分散 fine break-up
细粒粉末 fine-grained powder
细粒钙质软泥 drewite
细粒构造 fine-grained structure
细粒骨料 fine-grained aggregate
细粒含量 fines content; percent fines
细粒含量百分率 percentage fines
细粒黑云花岗岩 Oglesby blue granite
细粒花岗石 fine crystalline granite; Cornish granite <产于英国康沃尔郡>
细粒花岗岩 fine-grained granite
细粒花岗岩脉 fine-grained granite dike
细粒化 grain refinement; sliming
细粒灰浆 fine-grained mortar
细粒灰岩 fine-grained limestone
细粒混合料 fine mix
细粒混凝土 fine-grained concrete
细粒混凝土保护层 protective layer of fine-grained concrete
细粒级 fine fraction
细粒级的 fine-graded; fine-grained
细粒级范围 fine fraction range
细粒级沥青混凝土 fine-graded asphaltic concrete
细粒级曲线 fine fraction curve
细粒集合体 fine aggregate
细粒集料 fine-grained aggregate
细粒结构 close-grained structure; fine(-grained) texture; fine granular texture
细粒金刚石 small whole stone
细粒金刚石粉 bort(z) powder
细粒金刚石钻头 bort(z)-set bit; diamond particle bit; multistone bit; small stone bit
细粒金刚石钻头抛光 <钻进极硬细粒岩石时> bit polishing
细粒精矿 slime concentrate
细粒径 fine grain size
细粒聚结 coalescing of fine particles
细粒类土 fine-grained soil
细粒沥青混凝土 fine-graded asphalt concrete
细粒砾石 grit gravel
细粒砾石骨料 fine-grained gravel aggregate
细粒砾石集料 fine-grained gravel aggregate
细粒料 fine-grained material; mineral dust <混凝土用的>
细粒料填缝 seal(ing) with fines
细粒料止水 seal(ing) with fines
细粒裂纹 fine-grained fracture
细粒炉黑 fine furnace black
细粒磨石 turkey stone
细粒泥灰石 fine-grained
细粒泥质岩 clunch
细粒黏[粘]土 fine clay; ultra-clay
细粒凝灰岩 fine tuff
细粒凝聚 coalescing of fine particles
细粒喷砂面 fine-stippled sandblasted finish
细粒片地沥青混合料 fine sheet asphalt mixture
细粒轻质混凝土骨料 fine-grained lightweight concrete aggregate
细粒轻质混凝土集料 fine-grained lightweight concrete aggregate
细粒曲流带沉积 fine-grained mean-

der belt deposit
细粒人造石 fine-grained artificial stone; fine-grained patent stone; fine-grained reconstituted stone
细粒砂 fine(-grained) sand; finished sand; packsand
细粒砂浆 fine-grained mortar
细粒砂金矿 flour gold
细粒砂石 fine-grained sandstone
细粒砂岩 fine-grained sandstone; flint; post stone
细粒砂状结构 fine granular psamitic texture
细粒筛分 fine sizing
细粒石 bird's eye gravel
细粒石灰石石屑 fine limestone chip(ping)s
细粒石屑 fine-grained chip(ping)s
细粒石油沥青混凝土 fine(-graded) asphalt(ic) concrete
细粒式沥青混凝土路面 fine-graded bituminous concrete pavement
细粒树脂 finely divided resin
细粒松散材料 fine bulk material
细粒碎屑 fine-grained clastics
细粒碳 fine-grained carbon
细粒陶瓷 fine grain ceramics
细粒跳汰 fine jigging
细粒图像 fine-grained picture
细粒土(壤) fine earth; fine-grained soil; fines; fine soil; fine-textured soil
细粒无烟煤 culm
细粒物料 fine
细粒物料淘洗盘 fine pan
细粒雾化的燃料 finely pulverized fuel
细粒玄武岩 fine-grained basalt
细粒岩块 fine-grained monolith
细粒盐 fine-grained salt
细粒摇床 slimer
细粒杂质的过滤 fine particle filtration
细粒再加工石料 fine-grained reconstituted stone
细粒炸药 fine-grained powder
细粒珍珠岩 fine-grained perlite
细粒铸石 fine-grained cast stone; fine-grained patent stone
细粒状的 finely granular
细粒状断口 fine-grained fracture
细粒状岩石 gruss
细粒子 fines
细粒子热裂炉黑 fine thermal black
细粒组分 fine component
细粒组织 compact-grain structure
细粮 small grain
细梁 slender beam
细料 fine stuff; smalls
细料表面层 rough wall backing
细料成分 fine ingredient
细料冲洗分级脱水机 fine material washer-classifier-dehydrator
细料的 fine-grained
细料含量 fine content
细料含量百分数 percentage fines
细料混合料 fine mix
细料浆 fine slurry
细料浆溢流 fines overflow
细料进给 fine feed
细料进给范围 fine feed range
细料进给曲线 fine feed curve
细料轧碎机 fine breaker; fine crusher; fine crushing machine
细料轧制 fine breaking
细料轧制机 fine breaker; fine breaking machine
细料纸板 Bristol board; Bristol paper
细料最佳百分率 optimum fine aggregate percentage
细劣煤 duff
细裂 checking
细裂缝 check crack; fine crack; fine flaw; hair check(ing); hair crack; hair-like crack; hairline crack; hairline cracking; hair seam; minute crack
细裂纹 checking; feather check(ing); feather cracking; hair check(ing); fine fissure; hair(line) crack; map crack(ing); shelling; craze <陶瓷、混凝土的>
细裂纹面 fine-grained fracture
细裂纹釉 crazing glaze; glaze crazing
细邻近 fine proximity
细鳞白云母 damourite
细鳞片状 finely squamose
细鳞云母 cookeite
细流 arroyo; beck; bourn(e); brooklet; dribble; fillet; rill(e); runlet; runnel; streamlet; thread of stream; trickle; tube of flow
细流冲蚀 rill erosion
细流冲刷 rill washing; shoestring washing
细流动作 dribble action
细流法制造钻粒 shotting
细流灌溉 trickle irrigation; trickle irrigation system
细流痕 fringe rill mark; rill mark
细流模 rill mo(u)ld
细流排水 rill drainage
细硫砷铅矿 gratonite; hatchite
细炉黑 fine furnace black
细滤 fine filtering
细滤池 fine filter
细滤器 fine cleaner; fine filter; fine strainer; secondary filter
细滤清器 final filter; fine cleaner
细滤网 fine-mesh filter; fine screen
细铝粉 atomized alumin(i)um
细卵石 fine gravel
细卵石过滤器 gravel filter
细罗纹机 fine rib machine
细罗纹织物 fine-ribbed fabric
细螺纹 fine thread
细麻布 bastiste; cambric; fine linen; lawn
细麻布绝缘 cambric insulation
细麻布绝缘套管 spaghetti tubing
细麻带 cambric tape
细麻缆绳 cambric cable
细麻绳 fine cordage
细麻索缆 cambric cable
细马氏体 hardenit
细马氏硬化体 hardenite
细脉 lead vein; string; stringer vein
细脉浸染状金矿石 netted-disseminated Au ore
细脉浸染状矿石 netted-disseminated ore
细脉浸染状铅锌矿石 netted-disseminated Pb-Zn ore
细脉侵染矿床 ore deposits of fine intrusive veins
细毛玻璃 satin finish glass; velvet-finish glass
细毛毛雨 thin drizzle
细毛面玻璃 velet-finish glass
细毛面饰 satin finish
细毛面酸蚀 satin etch
细毛面酸蚀法 satin etching of sand-blasted surface
细毛线 fingering
细貌 finely dissected topography; fine texture topography
细煤 duff
细密缝 fine draw
细密级配土(壤) closely graded soil
细密结构 close-grained structure; fine-grained texture; fine granular texture; fine texture
细密拦污栅 strainer rack; trash rack
细密扫描 close scanning
细密组织的 fine textured
细棉布 cotton shirting
细磨 final grinding; fine comminution; finish grinding; smooth grinding; smoothing <平板玻璃>; polish grind
细磨白垩 ground chalk
细磨白漆 mill white
细磨边 smoothed edge
细磨材料 impalpable flour
细磨仓 fine grinding chamber; fine grinding compartment; finish department; secondary grinding compartment
细磨刀石 hone; oil stone
细磨的 finely ground; super-fine ground
细磨分隔仓 fine grinding compartment
细磨高炉矿渣 finely pulverized blast furnace slag
细磨过的 fine ground
细磨回路 finish grinding circuit
细磨机 atomizer mill; fine grinding mill; finish mill; grinding mill; mechanical fine-grader; pulverizing mill; regrinding mill; secondary grinding mill
细磨加工面 fine-rubbed finish; honed finish
细磨颗粒 finely ground grain
细磨矿石 finely ground ore
细磨矿物掺和料 finely divided mineral admixture
细磨料 finer abrasive
细磨磨料 levigated abrasive
细磨木浆 finish mill
细磨石 fine grinding stone; hone; razor stone
细磨石灰 finely ground lime
细磨石料 meal
细磨石条 hone stick
细磨石英 finely ground quartz
细磨熟料 finely powdered clinker
细磨水泥 finely ground cement
细磨碎石 finely ground rock
细磨圆角 pencil round
细磨圆角的 pencil-rounded
细磨重晶石 finely ground barite[baryte]
细末 dead small; smalls
细末里运动 Cimmerian orogeny
细末土(壤) heavy soil
细木 fine wood
细木板 block board
细木壁饰 cabinet finish
细木车间 joiner's shop; joiner's workshop
细木锉 cabinet rasp
细木工 cabinet maker; casework; finishing carpentry; joinery; woodwork
细木工板 block board; stave core; stripboard; strip core; veneered strip-glued board
细木工场 joiner's shop
细木工车间 joinery
细木工程 joiner(y) work; woodwork construction; cabinet work
细木工程施工 joinery work
细木工程准备工作 first fixings
细木工锉 cabinet file
细木工的最后加工 cleaning up
细木工规尺 grasshopper ga(u)ge; spider gage
细木工花格挑窗 mushrabeyed work
细木工画线规 joiner's gauge
细木工件 joinery component; joinery member; joinery unit
细木工胶 wood glue
细木工料 small timber
细木工领班 foreman carpenter
细木工木槌 joiner's mallet
细木工涂装 cabinet finish
细木工艺 cabinet making; cabinet wood; joinery(making)
细木工用材 cabinet wood; joinery timber; joinery wood
细木工用锤 Warrington hammer
细木工用的圆规 joiner's compasses
细木工用工具箱 joiner's kit
细木工用木料 cabinet wood
细木工制品 millwork
细木工装饰 cabinet finish
细木工装修 carpenter's finish
细木工作 cabinet wood; cabinet work; joinery
细木工作业 finish carpentry
细木刮砌 scraper
细木管楦 boxwood tampin; turning pin
细木横割锯齿 joiner's cross-cut saw bench
细木横切锯齿 joiner's cross-cut saw bench
细木护壁板 <房间墙壁的> boiserie
细木活材质规定 joinery
细木家具工 cabinet work
细木匠 joiner
细木锯 compass saw; half ripper; keyhole saw
细木刨 cabinet scraper
细木片 matchwood
细木器 finishing carpentry
细木饰面工作 finishing off
细木镶嵌工 intarsist
细木镶嵌装饰 intarsia; taria
细木屑 fine; wood flour; wood meat
细木芯板结构 lumber-core construction
细木用胶水 joiner's glue
细木作 fine wood-work; joiner's work; woodwork
细目 detail; detailed catalogue; enumeration; particularity; subheading; subsection
细目表 detail item; schedule; subsection
细目打印 detail printing
细目带 detail tape
细目分类(法) breakdown; close classification; exact classification
细目卡片 detail(ed) card
细目筛 micromesh sieve
细目式计划 form of detailed schemes
细目丝网 gauze wire
细目网 fine-mesh wire
细目文档 detail file
细目文件 detail file; transaction file
细目预算 detail budget
细耐火泥 finely ground fire clay
细泥 <原砂的> fine silt; fine clay
细泥浆 fine clay slurry
细泥沙 fine deposit
细霓霞岩 tinguaite
细拈绳 nettle stuff
细黏[粘]土 fine clay; fine deposit; fine sediment; montmorillonoid
细黏[粘]土悬浮物 fine clay slurry
细碾水泥 finely ground cement
细凝灰岩 politic tuff
细偶线期 lepto-zygenema
细盘条 light rod
细抛光 fine polishing
细刨 smoothing plane; smoothing planing; smooth plane; truing plane; trying plane
细刨床 smoothing plane machine; smoothing planing machine; smoot-

hing planner
细刨花 excelsior; fine flake; wood wool
细刨花板 excelsior board
细喷砂面 fine-stippled sandblasted finish
细喷雾枪 fine water spray nozzle
细片 shred
细片状珠光体 fine pearlite
细漂白土粉 fine earth
细平 fine grading
细平布 broadcloth;fine plain
细平衡 fine balance
细平耙 smoothing harrow
细平齐头锉 equaling file; equalizing file
细破碎 fine crushing
细漆布 cambric
细铅丝 spun lead
细切地形 finely dissected topography
细切割 fine cut
细切削 smooth cut;upper cut
细青铜丝网 fine bronze wire netting
细轻质混凝土骨料 fine lightweight concrete aggregate
细轻质混凝土集料 fine lightweight concrete aggregate
细泉 springlet;weeping spring
细热裂黑 fine thermal black
细人造砂 fine manufactured sand
细熔岩流 drib(b)let
细如毛发的 hair-thin
细乳状液 fine emulsion
细散粉土 finely divided silt
细散固体 finely divided solid
细扫描 close scanning
细纱 spun yarn
细砂 <又称细沙> mo;fine-grained sand; fines; graining sand; silver sand;soft sand
细砂百分含量 percentage of fine sand
细砂布 crocus cloth
细砂层 fine sand loam;layer of fine sand
细砂地基 fine sand foundation
细砂分类 classification of fine sand
细砂粉刷 sand plaster
细砂过滤 fine sand filter
细砂海滩 fine sand beach
细砂含量达一半的土(壤) soils with fines content up to 50%
细砂混凝土 fine sand concrete
细砂混凝土骨料 fine sand concrete aggregate
细砂混凝土集料 fine sand concrete aggregate
细砂浆 fine mortar
细砂沥青 mineral-filled asphalt
细砂粒 fine sand
细砂垆坶 fine sand loam
细砂轮 close-grained wheel;fine grinding wheel;fine grit wheel;fine plain emery wheel;fine wheel
细砂磨光的 smoothly sanded
细砂磨砖 Bristol stone
细砂壤土 fine sand loam;fine sandy loam
细砂陶器 fine sandy clay wares
细砂土 mo;decomposed granite
细砂土壤 floury soil
细砂屑 fine sandy clast
细砂岩 fine sandstone;gritstone;pack-sand
细砂岩储集层 fine sand reservoir
细砂质垆坶 fine sandy loam
细砂质磨石 buhrstone;grinding stone;millstone
细砂注入机 fine sand feeder
细筛 closed-mesh screen; finishing screen;lawn;sifting screen;under-size sieve
细筛绢网 fine silk net
细筛孔 fine mesh
细筛孔的 close-meshed;fine-meshed
细筛料 hutching
细筛目载体 fine-mesh solid support
细筛网 fine screen
细筛眼 narrow mesh
细山线条 baguette
细射束 narrow beam
细绳 binder twine; bobbin; cordage; hamber;hambrolin(e);hambrough;marlin(e);small stuff
细湿罗音 fine moist rale
细石 fines;roe stone
细石工 dressed masonry
细石灰 finishing lime; flour lime (stone)
细石灰粉 kemidol
细石灰膏 <抹灰工用的> plasterer's putty
细石混凝土 fine aggregate concrete; pea gravel concrete
细石混凝土路石饰面 granitic finish
细石混凝土面层 granolithic screed
细石料 fine metal;rock fines
细石抹灰饰面 rock dash
细石器时代的 <中石器阶> Microlithic
细石器文化 microlithic culture
细石屑 blinding;fine chip(ping)s;quarry dust;quarry fines;guttings <路面用>;blinding(material) <填充表面孔隙的>
细石渣 fine ballast;hoggin
细实线 fine line
细炻器 fine stoneware
细数 net amount
细数据通道 fine data channel
细水泥浆 fine-cement grout
细丝 filament;thread
细丝钢丝绳 fine wire rope
细丝焊(接) fine wire welding
细丝间基质 interfilamentous matrix
细丝静电计 filament electrometer
细丝盘条 wire coil
细丝石棉 amiant;amianthin(it)e;amianthus;amiantos;amiantus
细丝网筛 fine wire screen
细丝质的 filamentous
细丝状的 filamentous
细碎 comminution;milling grinding
细碎玻璃 finely crushed glass
细碎的 fine-crushed
细碎的金属丝 shredded wire
细碎对辊机 grinding roll
细碎工段 fine crushing department
细碎辊碎机 fine crushing rolls
细碎机 fine crushing machine; small aggregate crusher
细碎级配砾石 fine-crushed graded gravel
细碎金属丝底线轮胎 shredded wire underthread tire[tyre]
细碎破碎机 tertiary crusher; fine crusher
细碎燃料 hogged fuel
细碎石 chippings; chip stone; fine-crushed rock;finely broken stone;fine stone <AASHO 规定粒径 3/8 英寸细 10 号筛,1 英寸 =0.254 米>
细碎石块 fine broken rock; fine-crushed rock
细碎石料 fine broken stone; fine-crushed stone
细碎石路 fine-crushed road; fine-crushed rock road
细碎室 <碎石机的> fine crushing chamber
细碎屑基质 finely detritus groundmass
细碎屑结构 fine clastic texture
细碎屑土 fine clastic soil;fine-grained soil
细碎旋回破碎机 fine reduction gyratory crusher
细碎研磨机 mill disintegrator grinder
细碎用颚式破碎机 fine jaw crusher; secondary jaw crusher
细碎用旋回圆锥式破碎机 fine gyratory crusher;fine reduction gyratory crusher
细碎圆锥破碎机 gyro-granulator
细碎作用 fine reduction
细索 marlin(e)
细索类 small cordage
细索引 fine index
细索油麻绳 spun yarn
细弹簧 hair spring
细弹簧分规 hairspring dividers
细弹簧两脚规 hairspring dividers
细炭粉末 fine coke
细探针 mandrin
细探子 stylet
细陶瓷 fine ceramics
细陶粒 fine ceramisite
细填充料 fine filler;finely divided materials
细填料 fine stopper; fine stuffing; flour filler;knifing filler
细条 strand
细条尺 <制造钢梁用的> pole strip
细条带状结构 finely banded structure
细条带状煤 finely banded coal
细条灯芯绒 fine needle corduroy
细条片 sliver
细条纹 brush line; fine cord; fine line;fine ream;fine streak;string;thread;silking <涂层上的>
细条效应 pin stripe effect
细条子 pin stripe
细条子帆布 pin-stripe duck
细调 fine regulation;fine set(ting);fine tuning control
细调焦距螺旋 fine focusing adjustment knob
细调节 fine regulation;fine adjustment
细调节器 fine adjustment knob
细调控制 fine control
细调料槽 fine feed adjustment tank
细调擒纵叉 adjusting element
细调谐 fine tuning
细铁粉磁芯 sirufer(core)
细铁矿 fine iron ore
细铁丝钉 wire nail
细铁丝绳 wire line
细铁屑 swarf
细头电热丝加热器 finend strip(electric)heater
细头二联管 finend duplex tube
细头管 finend pipe;finend tube
细图 detail
细土 fine earth;soil fines
细土覆盖层 dust mulch
细土粒 fine soil grain
细团块 fine crumb
细团粒的 fine granular
细团粒状结构 finely granular structure
细拓扑 fine topology
细网 fine structure mesh
细网点印版 fine-screen halftone
细网格 finer mesh;refined net
细网格式车行道铺面 fine-meshed carriageway grid
细网目片 fine dot raster
细网筛 hair sieve
细网栅 fine-mesh grid;fine screen
细网眼的 close-meshed
细网(眼)滤器 fine-mesh filter
细网抑制栅 fine-mesh barrier grid
细网振动筛 fine-screen shaker
细微 fine;fineness;impalpability
细微差别 faint difference; nuance; shading
细微成分 fine constituent
细微的 impalpable;microfine;microscopic;minute;pinpoint
细微的差别 fine distinction
细微地震 earthquake tremor
细微动作研究 micromotion study
细微构造 fine structure;minute structure
细微结构 fine texture
细微孔隙 micropore
细微粒体 fine-grained micrinite
细微裂缝 crazing crack;microcrack(ing);minor crack;hairline cracking <钢铁内部的>
细微裂纹 crazing crack;microcrack(ing);minor crack;hairline cracking <钢铁内部的>
细微马赛克 micromosaic
细微区别 subtlety
细微纹理 fine grain
细纬 fine filling;fine pick
细纬档 fine filling bar
细纹 fine grain; fine thread; smooth cut
细纹板材 close(d)grain
细纹标准 close-grained rule
细纹锉 smooth-cut file;super-cut file
细纹大理石 alabaster;onyx marble
细纹的 fine;fine-grained <木材>
细纹合成饰面板 fine-line compound veneer
细纹继裂面 fine-grained fracture
细纹建筑石料 fine-grained building stone
细纹精刻的马赛克 vermiculated mosaic
细纹开裂 crazing
细纹刻石刀 fine cut burr
细纹理 close(d)grain
细纹理石料 fine-grained stone
细纹裂面 fine-grained fracture
细纹螺钉 fine thread screw
细纹螺丝板牙 fine thread die
细纹螺丝钢板 fine thread die
细纹螺旋 fine screw
细纹面琢石 fibrous-stroked dressing stone
细纹木(材) close-grained wood;fine-grained wood; fine wood; jack wood;narrow-ringed timber;fine-textured wood
细纹木丝板 fine-grained wood wool board
细纹修琢(石面) fibrous-stroked dressing
细纹凿面的 seam-faced
细纹钻 fine bur
细雾 mist
细雾滴喷嘴 fine-spray nozzle
细雾化喷嘴 high dispersion nozzle

细矽线石岩 fibrolit rock
细隙 areola
细隙的 areolar
细纤丝 microfiber[microfibre];microfibril
细纤维 fibril;fine fiber[fibre]
细线 fine thread; fraction line; geographic(al)limit;neat line;pencil(l)ing;thin-wire
细线道 brush line;fine line
细线刻针 fine-line graver
细线理状煤 finely striated coal
细线体 fine type;lean type;light type
细线体字 light face letter
细线条饰 bandlet

细线条图 hatched drawing
细线通 brush line
细销 hairpin
细小裂纹 microfissure
细小铅条 lead wool
细小石器 microlith
细小线条 bagnette
细斜纹布 jean(ette); reversed jeanette
细屑 fines
细屑镜煤 anthraxylon
细屑粒 fine crumb; fine sieve
细屑丝炭 attrital anthrxylon
细屑体 attrinite
细屑岩 lutite; lutyte
细屑质的 microclastic
细心琢磨成圆 cut true and square
细芯电缆 small capacity cable
细芯铜轴电缆 pencil coaxial cable
细虚线 fine dotted line
细玄岩 anamesite
细悬浮固体 fine suspended solid
细悬浮泥沙 fine suspended material; fine suspended sediment
细悬浮物 fine divide suspended matter
细悬移质 fine-sediment load; fine suspended load; wash load(ing)
细选粉 fine separation
细选者 sifter
细牙 closely pitched
细牙槽刨铁 fine indented cut plating iron
细牙螺钉 fine-pitch screw
细牙螺丝攻 fine screw tap
细牙螺纹 fine(-pitched) thread; Swiss screw-thread; Thury screw-thread
细牙普通螺纹 fine plain thread
细牙丝锥 fine thread tap
细亚麻布 lawn
细亚砂土 fine sandy loam
细岩脉 stringer vein; veinlet; veinule
细岩砂的相对体积 fractional volume of silt
细研的石英与玻璃的混合物 Gaspar
细研磨 fine lapping
细研磨膏 fine paste
细研磨机 fine grinder
细眼的 fine-meshed
细眼钢丝网 light-mesh steel fabric
细眼筛 close-meshed sieve; small mesh sieve; fine sieve
细眼丝网 gauze wire
细眼网 minnow net
细眼网目细孔 fine mesh
细腰管 venture tube; Venturi
细腰形试块 waisted specimen
细叶桉 gray gum; horncap eucalyptus
细油麻绳 spun yarn; tarred marline
细淤泥 fine silt
细雨 drizzle; drizzling (rain); soft rain
细圆齿状的 crenulate
细圆锉 rat tail(ed) file
细阅 scrutinize; scrutiny
细云母布 fine mica cloth
细云母片岩 chocolate
细云母石 fine talc
细凿边框 drafted margin; margin draft
细凿石 fine picked stone; fine-pointed stone
细凿石锤整修 fine bush hammered
细凿石面 fine picked stone finish; fine-pointed stone finish
细凿修整 fine-pointed finish
细凿琢面 dabbed finish
细则 by(e)-law; detailed rules and regulations; technical regulation
细轧 fine crushing
细轧室 < 轧石机的 > fine chamber; fine crushing chamber
细账 itemized account
细褶裥 fine gathering
细褶皱【地】plication; crenulate; crenulation; minute folding
细针状结晶 fine needles
细珍珠岩 fine perlite
细震颤 fine tremor
细枝 twig
细枝的 twiggy
细直棂分隔式双窗 ajimez
细至粗粒的 fine-to-coarse-grained
细至中粒的 fine-to-medium-grained
细质地土壤 fine-textured soil
细质木材 fine-textured wood
细质丝 cytoplasmic filament
细致的黏[粘]土悬浮物 fine clay slurry
细致地摊平路面 fine grading
细致度 fineness
细致路面拉坡 < 纵坡 > fine grading
细致平衡 detailed balance
细致体质 fine constitution
细致调整 close regulation
细致纹理 fine grain; fine texture; smooth grain
细致型 fine type
细重晶石 croylstone
细珠光体 nodular troostite
细柱 buttress shaft; slender column
细柱柳 Salix gracilistyla
细柱型 small tube type
细筑路碎石 fine road-metal
细砖工 ga(u)ged brickwork
细桩 buttress pile; buttress shaft; fine pile
细琢 fine-pointed finish
细琢方块石 < 上边缘企口接合 > channel(l)ed quoin
细琢方石 pick-dressed ashlar
细琢花岗岩贴接面 dressed granite meeting face
细琢料石 dressed stone
细琢石 dressed ashlar; fine-pointed dressing
细琢石料 stroked work
细琢石路面 dressed stone pavement
细琢石面 fine-pointed dressing
细琢石砌体 dressed masonry
细琢石圬工 dressed masonry
细琢修整 fine-pointed dressing
细紫柳 slender purpleosier willow
细钻 jig boring; jig drill

隙长 gap length

隙动差 back lash; lost motion
隙缝 chink; crack; crevice; fissure; slit; slot
隙缝辐射器 slot radiator
隙缝宽度 aperture width
隙缝滤水管 slotted pipe
隙缝侵蚀 < 土壤的 > internal erosion
隙缝式沉砂室 slot grit chamber
隙缝式空气扩撒器 slot sir diffuser
隙缝式锚固 slot anchor
隙缝式模型 slot pattern
隙缝体积 interstitial volume
隙缝天线 aperture antenna; slot antenna
隙缝天线阵 slot array; slotted antenna array
隙缝调整 play adjustment
隙缝系数 gap coefficient
隙缝泄漏 slot leakage
隙规 gap ga(u)ge
隙灰比 void-cement ratio
隙灰比法配合混凝土成分 proportioning by void-cement ratio
隙灰比法配料 proportioning by void-cement ratio
隙灰比配料法 void-cement ratio method(of proportioning material)
隙间边 interlacunar marginal
隙间部分 interstitial fraction
隙间腐蚀 contact corrosion; crevice corrosion
隙间结露 interstitial condensation
隙间液体 interstitial liquid
隙角 clearance angle; cutting clearance
隙径 clearance diameter
隙孔 lyriform pore
隙跨比 fissure spacing-span ratio
隙宽 gap length
隙裂 crazing
隙囊 clearance pocket
隙片 feeler; feeler blade
隙偏效应 gap tilt effect
隙泉 fracture spring
隙扫描 gap scanning
隙钽矿 alumotantite
隙填充 gap filling
隙透 effusion
隙透法 effusion method
隙透计 effusiometer
隙透冷却 effusion cooling
隙压比 gap factor
隙压系数 gap factor
隙状器 lyriform fissure

虾池 pound

虾罐头加工废物 shrimp canning waste
虾蟆夯 pummel
虾艇 prawn boat

瞎 blindness

瞎的暗井 blind
瞎缝 bastard joint; blind joint; closed joint【铁】
瞎孔 < 爆破的 > failed hole
瞎炮 blowout shot; cut-off shot; failed hole; fast shot; misfire(d charge); misfired detonation; missed hole; miss-fire shot; unexploded charge
瞎炮处理 handling misfire; handling misfiring; handling of misfire; misfire handling
瞎炮孔 dead hole; misfired hole; missed hole; miss shothole; unexplosive hole; unfired hole
瞎炮炮眼 bootleg
瞎跑 misfiring
瞎跑眼 miss-fire shot

匣钵 saggar[sagger]

匣钵垫板 crank
匣钵垛 cell
匣钵回转成型机 rotary sagger machine
匣钵坯体 saggar body
匣钵圈 < 无底匣钵 > ring
匣钵上集附的渣 encrustation
匣钵土 seggar clay
匣钵修补用耐火泥料 saggar cement
匣钵压机 saggar press
匣钵渣 saggar grog
匣钵柱 bung
匣盖 < 盒式百叶窗 > cover flap
匣户 the sagger makers
匣内软肖叶 box shutter
匣升降机 coffin hoist
匣式采样器 box type sampler
匣式窗框 box frame; cased window frame
匣式磁盘【计】cassette cartridge
匣式电桥 box bridge
匣式回照器 box heliotrope
匣式胶卷【计】cassette film
匣式取样器 box type sampler
匣式取样器水样 < 美国泥沙测验用语 > box sample
匣式软百叶里衬 back-flap
匣式升降机 coffin hoist
匣式榫头 boxed tenon
匣式凸榫 boxed tenon
匣式消音器 cell-type muffler
匣式檐口 closed eaves
匣式中竖框 double boxed mullion
匣式钻模 box jig
匣形窗架 box frame of window; cased window
匣形窗框 boxed frame(of window); cased(sash) frame; cased window frame
匣形的 boxed
匣形飞檐 closed cornice
匣形沟渠 box drain
匣形构架 box frame
匣形结构 box-frame(d) construction
匣形截面的水落管 box section leader
匣形梁 box beam; box girder; cased beam
匣形(门窗)竖框 boxed mullion
匣形起重机 coffin crane; coffin hoist
匣形水槽 box gutter; parapet gutter
匣形水落管 box section leader
匣形榫 box tenon
匣形天沟 parapet gutter
匣形柱 case(d) column
匣型 box type
匣装百叶 boxing shutter

峡 dalles

峡部 isthmus
峡道 defile
峡道口 channel firth[frith]
峡的 isthmian; isthmic
峡沟 flume
峡谷 cajon; canyon; clough; clove; combe; deep valley; dell; dough; ghyll; glen; gulch; gullet; kloof; kluf; linn; narrow defile; nulla(h); poort < 内部非洲的 >; ravine; rock gorge; water gap; gorge; quebrada < 美国西部的 >; barranca < 美 >; donga < 南非的 >
峡谷坝 valley dam
峡谷壁 canyon wall; gorge wall
峡谷边 valley side
峡谷边碛 valley train
峡谷飑 gull(e)y squall
峡谷城市 gap town
峡谷出口 boca
峡谷陡壁 canyon side; canyon wall
峡谷风 fall wind; ravine wind; canyon wind
峡谷风景区 valley scenic spot
峡谷滚石堆 fan
峡谷河段 canyoned river reach
峡谷河流 canyoned stream
峡谷急流(险滩) dalles
峡谷阶地 canyon bench
峡谷街道 canyon street
峡谷宽阔河段 embouchure
峡谷邻近的宽阶地 esplanade
峡谷流 canyon current
峡谷轮廓 canyon profile
峡谷名称 name of canyon
峡谷剖面 canyon profile
峡谷桥 nullah bridge
峡谷峭壁 dalls
峡谷侵蚀 ravine erosion

峡谷水坝式电站 valley dam(power) plant
峡谷填充圈闭 canyon-fill trap
峡谷蜿蜒河道 intrenched meander
峡谷效应 <高层建筑的> canyon effect(of high buildings)
峡谷形状 canyon shape
峡谷型水库 gorge type reservoir
峡谷因数 canyon factor
峡谷淤积 canyon fill
峡谷状湖 glen loch
峡口 narrows
峡口城市 gap town
峡路 col
峡门 embouchure
峡切面 sectiones isthmi
峡区 gorge area;gorge district
峡湾 fiord[fjord]
峡湾谷 fiord[fjord] valley
峡湾海岸 fiord[fjord] coast
峡湾海面 fiord[fjord] coast
峡湾里的多年冰 <一部分由雪组成> sikussak
峡湾型的 fjord type;fjord like
峡湾型海岸 fjord type shore;fjord-type coast
峡湾型海岸线 fjord type shoreline;fjord type coastline
峡中急流 dalles

狭隘的住所 close quarter

狭隘刻板的 one track
狭凹槽 quirk
狭板 batten;stave;planchettes <宽度在4.5英寸以下的,1英寸=0.254米>
狭板道 <泥地上铺板> duck board
狭边(钩齿)粗木锯 whip saw
狭槽 slot
狭槽接合 slot(ted) joint
狭槽连接 slot joint
狭槽刨 quirk router
狭槽榫 open mortise
狭槽榫接 slot mortise
狭槽装置螺钉 slot screw
狭长背斜 elongated anticline
狭长背斜圈闭 elongated anticline trap
狭长槽 <混凝土中的> reglet
狭长草地 strake
狭长城市 linear city
狭长窗 slit window
狭长道路 strip road
狭长的暗礁 spit of land
狭长的海湾 ria;coastal inlet
狭长的旗 streamer
狭长的浅滩 spit of land
狭长的山脊 hogback
狭长的土地 strip
狭长的纸带 streamer
狭长地带 corridor;pan handle;strip of ground
狭长沟槽 slit trench
狭长海港 arm of the sea
狭长基地 strip of ground
狭长脊岭状沙丘 saif dune;seif(dune)
狭长料子 slit fabrics
狭长流冰区 stream ice
狭长隆起物 rampart
狭长木框罩 <窗帘等挂棒上的> pelmet
狭长深海槽 foredeep
狭长式填土 sliver fill
狭长台布 <装饰用> table runner
狭长台地 bench
狭长投影 pan handle
狭长小孔 <开在墙上的> loophole
狭长形市区地带 <美国,由两相邻城市逐步形成> strip city
狭长钻 ensiform file;slitting file
狭车行道 narrow street
狭带 narrow band;narrow tape;tape
狭道 berm(e);bottleneck;narrow pass;throat
狭底水沟 canch
狭地带 tang
狭地峡 dumb bell
狭丁字刀架 <木工车床> narrow T rest
狭多盐生物 polystenohaline
狭帆布 narrow duck
狭范围控制器 narrow band controller
狭分布种 stenotopic species
狭缝 slit
狭缝灯 slit lamp
狭缝法 flat-gap process
狭缝法试验 flat jack slot test
狭缝方向滤波 crack directional filtering
狭缝分光镜 slit spectroscope
狭缝宽度 slit width
狭缝滤水管 slit filter
狭缝黏[粘]性液模型 Hele-Shaw model
狭缝式换热器 slit recuperator
狭缝式荧光灯 aperture type fluorescent lamp
狭缝形孔隙 slit-shaped pore
狭缝堰 notched weir
狭幅床单布 narrow sheeting
狭幅地毯 narrow carpet
狭幅饰带 narrow braid
狭幅织物 narrow cloth
狭钢条 bar
狭沟 gull(e)y
狭谷 coom;coombe;gull;narrow gorge;narrow valley
狭谷式热室 canyon
狭谷式水库 gorge type reservoir
狭谷小平原 valley flat
狭管现象 funnel(l)ing
狭管效应 funnel(l)ing;Venturi effect
狭光性动物 stenophotic animal
狭轨 narrow rule
狭轨距 narrow-ga(u)ge
狭轨铁路 narrow-ga(u)ge
狭轨(小)机车 dolly car
狭海湾 calanque
狭海峡 gut;narrow waters
狭航道 gat
狭壕 slit trench
狭河槽 neck channel
狭间隙焊 narrow gap welding
狭角 narrow angle
狭角照相机 narrow angle camera
狭颈裁弯 loop cutoff;neck cut-off
狭径 defile
狭孔 narrow opening;slit;slot
狭口 narrow orifice
狭口流速 jaw speed
狭口送风速度 slot air velocity
狭矿脉 slicking
狭棱条 narrow wale
狭梁 narrow beam;slender beam
狭路 bottleneck road;gullet;pass;pass road
狭路出土挖方(法) gulletting
狭木条 fillet
狭木条楼面 batten flooring
狭木条拼成的水管 wood-stave pipe
狭年轮的 narrow-zoned
狭频波 sharp wave
狭谱放大器 narrow band amplifier
狭栖性 stenoky
狭浅河槽 restricted channel
狭浅水域 confined waters;restricted waters
狭桥标志 narrow bridge sign
狭筛眼 narrow mesh
狭扇面记录仪 narrow-sector recorder
狭生长带 narrow zone
狭湿性动物 stenohygric animal
狭束辐射 narrow beam radiation
狭束回声测深仪 narrow beam echo sounder
狭水道 gut;narrow channel;narrow water;sluit <南非>;swash way
狭水道的开口处 issue
狭水道掉头 cast in a narrow channel
狭水道航行 channel navigation
狭缩 narrowing;pinch;pinch-out
狭条 ribbon;slat;strap
狭条板屋面 batten seam roofing
狭条板屋面覆盖 batten roof cladding
狭条扁柱 strip pilaster
狭条船壳板 strip planking
狭条地板 strip flooring
狭条工地 ground strip
狭条荷载 strip loading
狭条花边 tape lace
狭条锯 pattern-maker's saw
狭条抗拉 strip tensile
狭条拉伸试验 strip tension test
狭条连线 bar
狭条排水槽 fillet gutter
狭条水槽 fillet gutter
狭通带滤波器 narrow band pass filter
狭通带轴 narrow band axis
狭通过台 narrow vestibule
狭凸缘的 narrow-flanged
狭温生物 stenothermal organism
狭温微生物 microphilic
狭温细菌 microphilic bacteria
狭温性 stenotherm
狭温性的 stenothermic;stenothermy
狭温性动物 stenothermal animal
狭温种 stenotherm
狭小地方用的单向风钻 non-reversible close-quarter pneumatic drill
狭小地方用的风钻 reversible close-quarter pneumatic drill
狭小汇率幅度 narrow band
狭盐度性 stenohaline
狭盐分生物 stenohaline organism
狭盐性 stenohalinity
狭盐性的 stenohaline
狭盐性动物 stenohaline animal
狭盐性鱼类 stenohaline fishes
狭盐性藻类 stenohaline algae
狭叶的 narrow-leaved
狭叶榕 slimleaf fig
狭义货币 narrow sense currency
狭义建筑学 <不包括销管道、雕刻装饰等> architecture proper
狭义量纲分析 restricted dimensional analysis
狭义市场 narrow sense marketing
狭义拓扑学 narrow topology
狭义网论 special net theory
狭义相对论 special relativity;special relativity theory;special theory of relativity
狭闸门 narrow gate
狭窄 coarctation;slenderness
狭窄场地 tight quarter
狭窄场工作 close work
狭窄场所作业 close quarter work
狭窄处 close quarter
狭窄的 slender;stenotic
狭窄的出入口 mouse;mouse hole
狭窄的高架人行道 catwalk
狭窄的航道入口 gullet
狭窄的居住空间 close quarter
狭窄的入口 jaw
狭窄的山脊 knife edge
狭窄的深层快速海洋流 stream current
狭窄的通道 strait
狭窄的小海湾 creek
狭窄的装饰物 gimp
狭窄地采掘 strait work
狭窄地带 neck of land
狭窄段 bottleneck
狭窄段道路 bottleneck road
狭窄海岬 neck
狭窄海湾 firth[frith]
狭窄海峡 kyle
狭窄航道 gat;narrow channel;navigation pass
狭窄河槽 restricted channel
狭窄河段 restricted reach
狭窄急流 neck current
狭窄裂纹 narrow fissure
狭窄区 bottleneck
狭窄人行道 catwalk
狭窄水道 narrow channel;narrow pass;narrow waterway;restricted waterway;gut
狭窄水域 confined waters;narrow waters;restricted waters
狭窄隧道 narrow tunnel
狭窄通道 ginnel;slype
狭窄通道用的叉车 narrow-aisle fork truck
狭窄通航水道 navigable pass
狭窄通路 narrow gap
狭窄小道 catwalk
狭窄型 stenotic type
狭窄烟道 cramped flue
狭窄拥挤的 bottleneck
狭窄正面 narrow front
狭直条饰 tringle
狭字 elongated

瑕疵 blemish;blot;bug;disfigurement;flaw;gall;mote;speck;stain;tache;vice

瑕疵修补 repair of blemishes

辖区 precinct;prefecture

霞白斑岩 arkite

霞辉二长斑岩 allochetite
霞辉碳酸盐伟晶岩 kasenite pegmatite
霞辉碳酸盐岩 kasenite
霞辉斜长岩 modumite
霞磷岩 neapite
霞鳞石英 christensenite;tridymite
霞榴正长岩【地】borolanite
霞闪正煌岩 sannaite
霞石 eleolite;nepernepheline;nepheline;nephelite;phonite
霞石安山岩 nepheline andesite
霞石白榴石方沸石碧玄岩 nepheline leucite analcime basanite
霞石白榴石方沸石碱玄岩 nepheline leucite analcime tephrite
霞石白榴石黝方石方钠石兰方斑岩 nepheline leucite nosean sodalite hauynophyre
霞石斑岩 nepheline-porphyry
霞石碧玄岩 nepheline leucite analcime basanite
霞石粗安岩 tautirite
霞石粗面岩 nepheline trachyte
霞石二长碉 nepheline monzonite
霞石二长岩 nepheline monzonite
霞石方沸碱煌岩 nepheline monchiquite
霞石方钠石微晶正长岩 nepheline sodalite microsyenite
霞石方钠石正长岩 nepheline sodalite syenite
霞石光斑岩 nepheline(eleolitic) porphyry
霞石化 nephelinization

X

霞石黄长岩 nepheline melilitite
霞石辉长岩 nepheline gabbro
霞石辉石岩 nepheline-pyroxene rocks
霞石碱玄岩 nepheline tephrite
霞石闪长岩 nepheline diorite
霞石条纹 nepheline worm
霞石响岩 nepheline phonolite
霞石玄武岩 nepheline basalt
霞石岩 nephelinite
霞石岩-白榴岩类 nephelinite-leucitite group
霞石岩结构 nephelinitic texture
霞石黝方石方钠石白榴斑岩 nepheline nosean sodalite leucitophyre
霞石正长伟晶岩 nepheline syenite pegmatite
霞石正长细晶岩 nepheline syenite-aplite
霞石正长岩 nepheline syenite
霞石正长岩类 nepheline syenite group
霞石正长岩族 midalkalite
霞响岩 nephelinitoid phonolite
霞斜石 theralite
霞斜岩 theralite
霞斜岩类 theralite group
霞斜岩质厄塞岩 theralitic essexite
霞岩 nephelinite

下

下按按钮 downward button

下凹 concave-down(ward); downwarp; notching
下凹垂线 notching curve
下凹的 concave-down(ward); notching
下凹的中间分隔带 depressed median
下凹锻模 bottom swage
下凹交错层理 concave cross-bedding
下凹面 sunk face
下凹曲线 notching curve
下凹式水舌 depressed nappe
下凹铁路线 sunken track
下凹效用曲线 concave downward utility curve
下凹形坡 concave slope
下奥陶纪【地】Lower Ordovician
下奥陶统【地】Lower Ordovician series
下白边 bottom margin
下白垩统【地】Lower Cretaceous series
下摆动颚板面 lower swing jaw face
下班 be off duty; come off work; knock off; turn in; off-duty
下班高峰时间 work-to-home peak hour
下班时间 close time; closing time; quitting-time
下半部 lower half; lower part
下半部瓶身缩小 insweep
下半部曲轴箱 lower half crankcase
下半部组装 lower half assembly
下半端罩 lower end shield; lower shield
下半断面 bench
下半格 lower semi-lattice
下半功率频率 lower half-power frequency
下半空间 lower half-space
下半连续函数 lower semi-continuous function
下半连续性 lower semi-continuity
下半模格 lower semi-modular lattice
下半平面 lower half-plane
下半旗 half mast
下半汽缸 lower cylinder half; lower half casing
下半球 lower semisphere
下半圈 lower branch
下半箱 mo(u)ld drag
下半箱体 bottom half
下半型 mo(u)ld drag
下半轴瓦 low brass
下包络线 lower envelope curve; minimum envelope curve
下包络原理 lower envelope principle
下雹 hail storm
下雹子 sleet
下北美生物带 Lower Sonoran life zone
下贝氏体 lower bainite
下背 lower back
下辈 inferior
下边 bottom margin; lower band; lower margin
下边板 lower edge board
下边部分被削掉的 undercut
下边带 down-sideband; low(er) sideband
下边带频谱 lower sideband spectrum
下边带上变频器 lower sideband upconverter
下边带特性 lower sideband characteristic
下边带通信[讯] lower sideband communication
下边观测高度 < 天体 > observed altitude of lower limb
下边界 lower boundary
下边滩 downstream sand bank; lower side flat
下边线 lower border
下边缘 lower limb
下变差 lower variation
下变频 down-conversion
下变频器 down converter
下标 footnote; subscript(index); suffix
下标变量 index variable; subscript(ed) variable
下标表 subscript list
下标表达式 subscript expression
下标成分 indexed component
下标范围 subscript range
下标界 subscript bound
下标名 index name
下标数据名 subscripted data name
下标位置 subscript position
下标限制名 subcripted qualified name
下标形式 form of subscript
下表层套管井眼 conductor hole
下表面 bottom surface
下表面蒙皮 bottom skin
下表面套管的钻孔 surface borehole
下冰风 off-ice wind
下剥 under spall
下部 bottom; lower; lower curtate
下部凹陷 undercut
下部板块 lower plate
下部半断面 bench; bottom drift; bottom section
下部边缘区 lower marginal area
下部表面假接缝 bottom surface dummy joint
下部侧撑 lower lateral
下部超前工作面 bottom heading
下部承口 bottom socket
下部承载结构 load-carrying substructure
下部冲刷河岸 undercut bank
下部穿孔 underpunch
下部穿孔区 lower curtate
下部传动 underneath drive
下部传动式搅拌机 underdriven mixer
下部窗页扇 bottom leaf
下部挡板 under shield
下部导洞 bottom drift
下部导坑 bottom drift
下部导向轮 < 铲土机 > lower idler
下部的上排钢筋 < 梁的 > upper bottom layer
下部地层 substratum[复 substrata]; understratum[复 understrata]
下部定位垫铁 soffit spacer
下部堆芯板 lower core(support) plate
下部堆芯围筒 lower core support barrel
下部副气口 lower auxiliary steam port
下部覆盖层 undercloak
下部隔膜板及阀杆 lower diaphragm plate and valve stem
下部给料燃烧方式 underfeed combustion
下部构架 bottom boom
下部构造层次 lower tectonic level
下部滚轮 < 铲土机 > lower idler
下部含水量 lower aquifer
下部护板 underguard
下部滑轮 < 铲土机 > lower idler
下部机械滑动部分 lower carriage
下部集水装置 underdrain system
下部加(燃)料炉 underfeed furnace
下部间隙 underclearance
下部建筑 base unit; infrastructure; substructure; understructure; underwork
下部建筑成本 infrastructure cost
下部建筑费用 infrastructure charges
下部脚手架 lower falsework; lower scaffold
下部铰接的框格窗 bottom-hinged sash window
下部结构 base unit; infrastructure; lower structure; substruction; substructure; support(ing) structure; undercarriage; underpart; understructure; underwork
下部结构灯光 beam lower
下部结构工程 substructure work
下部结构体积 volume of substructure
下部壳体 lower housing
下部空隙 underclearance
下部矿脉 underset
下部框架撑条 lower frame bracing
下部拉板 < 路牌机 > bottom slide
下部拉紧装置 lower tension carriage
下部联结系 sub-connecting system
下部链轮 bottom chain sprocket
下部陆源建造 lower terrigenous formation
下部履带 lower run of track
下部门页扇 bottom leaf
下部排气式集中采暖炉 downflow-type central furnace
下部排烟锅炉 down-draft boiler
下部喷管 bottom bullnose
下部平底相纹理构造 lower plane-bed facies lamellar structure
下部气泡式鼓包 < 机身 > lower bubble
下部气室 down-plenum
下部驱动式压床 underdrive press
下部燃烧炉 underfired furnace
下部手制动轴导架 lower brake shaft bearing
下部水流动态 lower flow regime(n)
下部送纸 lower feed
下部台阶 bottom bench
下部掏槽 bottom cut; lower cut; underhole
下部淘空 bottom cut; undermining
下部淘刷 undercutting
下部套管 < 套管柱中的 > casing starter
下部提升筒 lower lift drum
下部弯曲 < 挖泥船身的 > lower knuckle
下部围护侧板 toe board
下部围岩 underlying wall rock
下部维护侧板 toe bead; toe board
下部温带板 lower ring
下部吸气阀 lower suction valve
下部吸入阀 lower suction valve
下部线圈 lower coil
下部泄水隧道 lower discharge tunnel
下部行走部分 base carrier
下部压紧胶辊 nipple roll lower
下部支撑 under-bracing
下部支承轴承 lower bearing
下部枝条 under branch
下部滞水层 hypolimnion
下部滞水层掺混 hypolimnetic mixing
下部中段的 low-level
下部主汽口 lower main steam port
下部装罐水平 lower decking level
下部装载运输机 bottom-loading conveyer[conveyor]
下舱孔 hatch; hatchway
下侧 underside; downside
下侧板 lower side panel
下侧的 downside; subjacent
下侧拉紧的皮带 belt driving under
下侧梁 bottom chord; bottom side rail < 集装箱 >
下侧梁矫直 < 集装箱 > straight bottom rail
下侧梁嵌补 < 集装箱 > inserting bottom rail
下侧门 bottom side door
下侧门板压铁 bottom side door washer plate
下侧门搭扣座 bottom side door hinge plate stop bracket
下侧门挡 bottom side door stop
下侧门钩链 bottom side door hanger chain
下侧门扣铁 bottom side door hinge plate stop
下侧门折页 bottom side door hinge plate
下侧面 downthrow side; lower side
下侧牵引 underdrag
下侧条 hypotrematic
下测晶体分辨率 resolution of down measured crystal
下测晶体体积 down survey crystal volume
下测站 downstation
下层 bottom bed; bottom layer; low coat; lower bed; lower course; lower layer; sublayer; subterrane; undercourse; underlayer; understratum[复 understrata]
下层(矮生)植物 understor(e)y
下层材料 subsurface material
下层城堡 lower citadel
下层道碴 subballast
下层道路 subsurface highway; subsurface road
下层的 lower; subjacent
下层地板 blind floor; subfloor(ing); false floor < 冷藏车双层地板的 >
下层地层 substratum[复 substrata]; understratum[复 understrata]
下层地层构造 substrata formation; understratum formation
下层地基 subjacent bed
下层地窖 subcellar
下层地面板 subfloor
下层地下室 subbasement; subcellar
下层断面测绘 subbottom profiling
下层浮游生物 hypoplankton
下层构造 infrastructure; subsurface structure; understructure
下层海流 submarine current
下层河岸阶地 lower bench
下层基础 subfoundation

下层甲板 lower deck;orlop deck
下层甲板设备 below deck equipment
下层驾驶台 lower bridge
下层建筑 substruction;substructure
下层节点 lower level node
下层结构 understructure
下层居民 underclass
下层矿脉 underset
下层林 lower storey of forest
下层林丛 underb(r)ush;under growth;underwood(growth)
下层林木 underwood
下层流 subsurface current;undercurrent
下层楼 lower floor
下层路面 underpavement
下层密灌丛 underscrub
下层面层 lower coat
下层木 underwood
下层逆流 underset;undertow
下层逆流漩涡 sea-pouce;sea push
下层排水道 subdrain
下层平面图 ground plan
下层破裂面 lower failure plane
下层铺面 underpavement
下层桥楼 lower bridge
下层桥面 lower bridge deck; lower deck
下层清液 subnatant
下层绕组 bottom slot layer
下层乳剂 lower emulsion
下层筛 lower screen
下层试样 lower sample
下层受光区 dysphotic zone
下层树林 undergrove
下层水 subnatant liquid;water below oil reservoir
下层水域 subjacent waters
下层土 bottom soil;buried soil;lower-horizon soil;subsoil;substratum [复 substrata];undersoil
下层土测量 subsoil survey
下层土沉降 subsidence of the subsoil
下层土冲刷破坏 failure by subsurface erosion
下层土恶化 subsoil deterioration
下层土勘探 subsoil exploration
下层土流失 migration of subsoil
下层土面 subsurface
下层土排水 subsoil drainage
下层土排水沟 half-socket pipe;subsoil drain
下层土排水管 subsoil pipe
下层土壤排水 subsoil drainage
下层土压实 subsoil compacting;subsoil compaction
下层托盘 bottom deck
下层洗流 down washing
下层线棒 <电机> lower bar
下层新红砂岩 lower new red sandstone
下层烟囱 lower funnel;lower stack
下层用料 subsurface material
下层预应力标高 lower prestressing level
下层站台 under platform
下层滞水带 <湖、海的> hypolymnion [hypolimnion]
下层滞水区 <湖、海的> hypolymnion [hypolimnion]
下层滞水区湖 bypolimnium
下层自游生物 subnekton
下叉连杆 lower wishbone link
下差 allowance below nominal size
下车 alighting; debark; disentrain; drop-off; landing; set-down; step down; undercarriage; lower structure <起重机>
下车灯光 beam lower
下车客流量 alighting passenger volume
下车人数 alighting passengers
下车月台 landing platform
下车站 debarkation point
下车支架 base frame
下扯窗盖条 deep bead;sill bead;ventilating bead
下沉 dip; downward plunging; gravitating; sagging; setting; sink(age); sinking; slump settlement; subside; subsidence;subsiding;sunk;swag
下沉岸 coast of subsidence;submergence coast
下沉岸线 shoreline of depression; submerged coastline; submerged shoreline
下沉凹槽 trench of subsidence
下沉板 settlement plate; subsiding sheet
下沉板块 downgoing plate
下沉板片 subsiding sheet
下沉比 settling ratio
下沉变化 settlement change
下沉滨线 shoreline of submergence; submergence shoreline;shoreline of depression
下沉不匀 uneven settlement
下沉槽池 subsidence trough
下沉测量仪 convergence ga(u)ge
下沉沉井 setting caisson
下沉的 submerged
下沉范围 depression range
下沉风 under the wind;down-wind
下沉改正 <标尺或脚架的> correction for settling
下沉过程线图 yield-time diagram
下沉海岸 depression coast; plunging coast; submerged coast; coast of submergence;sinking coast
下沉海岸线 shoreline of submergence
下沉礁 submerged reef
下沉校正 correction for settling
下沉坑 dig-down pit;sunken pit
下沉力 negative buoyancy; sinking force
下沉梁法 sagging beam method
下沉量-时间曲线 convergence time curve
下沉流 downward current
下沉率 rate of sinking
下沉面 surface of subsidence
下沉面积 area of subsidence;subsidence area
下沉逆温 subsidence inversion;subsidence temperature inversion
下沉盆地 subsidence basin; subsidence trough
下沉气流 descending air current; down current; down draft; down draught;down wash
下沉区 zone of subsidence
下沉曲线 sinking curve; subsidence curve
下沉深度 <浮坞的> sinkage depth
下沉时间图 yield-time diagram
下沉式地板 depressed floor
下沉式动臂起重机 sinking derrick
下沉式花园 sunken garden
下沉式流液洞 drop-throat
下沉式天窗 sinking skylight;sunk skylight
下沉式圆辊闸门 submersible roller sluice gate
下沉试验 <浮坞的> sinkage trial
下沉水 cascading water
下沉水位 sunken water level
下沉水域 sunken waters
下沉说 subsidence theory
下沉速度 downwash velocity;settling rate; sinking speed; sinking velocity;subsiding velocity;terminal velocity;velocity of subsidence
下沉速率 fall(ing) rate
下沉物 hypostasis
下沉系数 coefficient of settlement
下沉箱 sinking of caisson
下沉性 setting quality
下沉延迟 fall delay
下沉与控制 sinking and control
下沉缘琢 sunk draft
下沉增温 subsidence inversion
下沉支座 yielding seat;yielding support
下沉阻力 <沉箱的> sinking resistance
下沉作业 sinking operation
下衬 lower gasket;underlay;underlie
下撑 under-bracing
下撑式 strut-framed
下承矮桁架 half-through truss
下承矮墙桁架 half-through bridge
下承板梁 half-through girder;through girder;through plate girder
下承层 subjacent layer
下承的 through
下承拱 through arch
下承桁 through girder bridge
下承桁架 through truss
下承跨 through span
下承梁 through beam; through girder;upstand beam
下承梁桥 through girder bridge
下承锚底座 lower anchor bracket
下承桥 through bridge
下承桥大梁 through girder
下承桥跨 through space; through spalling
下承桥面 through bridge roadway above boom
下承式 through-type
下承式板梁 through plate girder
下承式低桁架桥 low truss bridge
下承式叠板弹簧 underhung laminated spring
下承式公路桥 bottom-layer(road) bridge;bottom road bridge
下承式桁架 bottom supporting truss; through truss
下承式桁架桥 through truss bridge
下承式桥 through bridge
下承式栓焊梁桥 bolted and welded through truss bridge
下承式悬臂桥 through cantilever bridge
下承台 lower pile cap
下承行车道 roadway below
下承行车路 roadway below
下承悬臂梁 through cantilever
下承支座 yielding-type support
下承座 lower seat;lower socket
下池 lower pool
下齿 lower tooth
下冲 down rush; downthrust; undercut;underscouring;undershoot
下冲板块 underthrust plate
下冲程 stroke down
下冲的 undershot
下冲的物质 <从山上或高处> down wash
下冲点 plunge point
下冲断层【地】underthrust
下冲杆 lower plunger;lower punch
下冲气流 downflow;down wash
下冲失真 undershoot distortion
下冲式 undershot
下冲式废水渠 undershot-type wasteway
下冲式水道 undershot-type waterway
下冲式水轮 undershot wheel
下冲式水轮机 undershot turbine;undershot water wheel
下冲式水压机 down stroke hydraulic press
下冲式闸门 underpour type gate;undershot gate
下冲水轮 undershot water wheel
下冲头压力 lower punch pressure
下冲信号 underswing
下抽炉 underdraft furnace
下出料 bottom discharge
下除渣口 bottom tap
下穿 fly under
下穿道 underway
下穿交叉 fly under crossing
下穿交叉道 underpass
下穿交叉路 undercrossing;underpass
下穿路 <在跨线桥下通过> depressed road(way);depressed highway
下穿式道路枢纽 fly under interchange
下穿式(立体)立交 undercrossing;fly under; fly under interchange; underpass; vehicular undercrossing; underpass grade separation
下穿式桥 <立体交叉的> underbridge
下穿式桥桥下净空 underbridge clearance
下穿式人行过道 pedestrian underpass
下穿式自行车道 cycle underpass
下穿水底通道 underway
下穿污水管 depressed sewer
下穿线 underpass
下传动 underdrive
下传动的立辊轧机 underdriven vertical edger
下传链路 down line
下船 debark; disembark; disembarkation;disembarkment;unship
下船港 port of debarkation
下船货物 off-load cargo
下船台 float out;undocking
下吹的 katabatic
下吹风 canyon wind;katabatic wind;fall wind
下吹式暖风机 downward discharge unit heater
下吹式通风装置 downflow unit
下垂 buckling; cocking-down; dipping; hang-down; nutation; preshoot;sag(ging)
下垂长度 sag length
下垂的 flagging; hanging; pendant; sagged;lop
下垂的拱顶石 pendant keystone
下垂的缆索 slack rope
下垂的遮阳篷 drop awning
下垂的指针 sagitta
下垂点 sagging point
下垂度 slump;droop
下垂改正 catenary correction
下垂钩 drop hook
下垂管 dip pipe
下垂焊缝 sagged weld
下垂桁架 dropped girder

下垂力矩 sagging moment
下垂流淌试验 sag-flow test
下垂龙骨 drop keel;sliding keel
下垂披叠板 novelty;novelty siding
下垂曲线 drop-down curve;sag(ging) curve
下垂趋向 sagging tendency
下垂式抽屉拉手 drop drawer pull
下垂式信号装置 pendant signal
下垂式钥匙孔盖 drop escutcheon
下垂式钻机 pendent drill
下垂试验 sag test
下垂索 sagging cable
下垂特性(曲线)droop characteristic
下垂体 pendant

下垂天线 drag antenna;trailing antenna;trailing wire
下垂托板 dropped panel
下垂托板造型 drop mo(u)lding
下垂拖板 dropped panel
下垂物 lappet
下垂物装饰 pendant
下垂形洒水喷头 pendant sprinkler
下垂枝 descending branch
下垂状态 hang
下锤头 lower ram
下次决算 next closing of account
下错窗 drop tracery
下达 convey(ing)
下达简令 brief briefing
下大雨 drench(ing);rain cats and dogs
下蛋式块料摊铺机 egg layer
下挡栏杆 bottom transom
下挡墙 lower retaining wall
下刀架 bottom tool
下导洞法 < 隧道施工的 > bottom heading method;bottom-drift method
下导洞推进法 bottom-drift method
下导管 downcomer;downtake tube
下导轨 lower guide;lower guide rail;lower guide track;lower rail
下导函数 lower derived function
下导坑 bottom drift;bottom heading;conductor hole
下导坑超前先拱后强法 bottom-drift method;bottom heading method
下导坑核心支承开挖法 bottom-drift and ring-cut method
下导坑开挖法 bottom-drift excavation method;bottom heading method of excavation
下导坑漏斗棚架法 bottom-drift excavation method
下导坑先墙后拱法 bottom-drift excavation method
下导轮 bottom tumbler;lower tumbler
下导轮轴承 bearing of lower tumbler shaft
下导模法 inverted Stepanov technique
下导气管 downtake
下导数 lower derivative;lower differential coefficient
下导叶环 lower guide-vane ring
下导轴承 bottom guide bearing;lower guide bearing
下导轴承支架 lower guide bearing bracket
下导轴瓦 lower guide metal
下道 lower track
下道车 cars on-line
下的 suballern
下的中甲板 lower tween deck
下等高线面积 area of lower contour
下等酒店 shebeen
下等客舱 steerage
下等客栈 doss house
下等品 low quality
下底板 lower plate;lower shoe;lower wall
下底梁结构 low-sill structure
下地壳层 lower crust
下地壳硅镁 lower crust sima
下地幔 inner mantle;lower mantle;mesosphere
下第三纪【地】Eocene period
下第三系【地】Eocene system;Eogene;Pal(a)eogene system
下点 subpoint
下电极臂 bottom arm
下垫板 lower bolster
下垫层 base course;lower course;underlayer;underlying bed
下垫面 underlying surface
下垫面地形 underlying topography
下垫岩石 underlying rock
下吊式 underslung type
下跌 come down;dip;drop
下跌的桩锤 uncoupling hammer
下跌平衡法 declining balance method
下跌趋势 downward tendency
下顶料 bottom knockout
下定付款 cash with order
下定决心 make-up one's mind
下定向管操作 drive piping operation
下定义的 undefined
下动颚式破碎机 Blake(-type jaw) breaker;Blake(-type jaw) crusher
下犊牛栏 maturity pen
下堵 bottom plug
下端 lower end;lower extreme;soffit;underbed;underside end < 给进液压缸的 >
下端插入 bottom entry
下端点 lower extreme point
下端阀座 lower value seat
下端盖 bottom end cover
下端固定支承 fixed earth support
下端局设备 downset
下端梁 < 集装箱 > bottom end rail;bottom end transverse member
下端试水位旋塞 lower ga(u)ge cock
下端调整阀 lower adjusting valve
下端万向节 lower cardan
下端向上升 bottom lift
下端桩 reach post
下端自由支承 free earth support
下段 < 地层露头 > lower bench
下锻模 anvil tool
下堆栈 push-down stack
下蹲值 squat value
下多雨的 weeping
下舵杆 lower stock;main rudder piece;main stock
下舵枢 heel brace;heel gudgeon
下颚 lower jaw
下颚板 < 破碎机的 > lower cheek plate
下二层舱 lower tween deck
下二叠统【地】lower Permian series
下阀杆密封 lower stem seal
下阀片 lower disc
下法兰 lower flange
下翻梁 downstand beam
下反角 cathedral angle;inverted dihedral;negative dihedral
下方【疏】bed measure;in-situ volume of dredged material;volume in-situ
下方产量 in-situ production;output in bed measure
下方穿过 fly under crossing
下方的 downward
下方的测定 in-situ measurement
下方给煤机 underfeed stoker
下方加煤机 underfeed stoker
下方渗流 underseepage
下方值 lower valve
下风防波堤 lee(ward) breakwater
下枋 lower fillet and fascia
下放的管子 run-down pipe;run-down tube
下放点 setting point
下放矸石井筒 rock shaft
下放矿石 ore drawing
下放物体的垂直运输机 lowering conveyer[conveyor]
下飞机 deplane
下飞机道 < 机场 > deplaning road
下分层 bottom bench;lower leaf;lower slice
下分岔 down-splitting
下分法 subdivision method
下分解 lower decomposition
下分泌说 descending secretion theory
下分配板 lower distribution plate
下分式双管系统 double pipe up-feed system
下分式系统 upfeed system
下风 alee;lee
下风岸 lee shore;leeward bank
下风板 lee board
下风侧 lee side;leeward side
下风潮 lee tide;leeward tide
下风的 down-wind;leeward
下风舵 helm alee
下风方向 leeward side
下风防波堤 lee breakwater;leeward breakwater
下风井 downcast
下风距离 down-wind distance
下风满舵 hard alee;hard down
下风锚 lee anchor
下风面 airside face;lee face;lee side;leeward side
下风面护面 < 防波堤 > rear face armo(u)r
下风面块体 < 防波堤 > rear face armo(u)r
下风坡 lee slope
下风沙丘 lee dune
下风弦(杆)leeward chord
下风舷 disengaged side;lee side
下风漩涡 lee eddy
下封闭层 negative confining bed
下峰机车 pulldown engine
下峰信号【铁】hump trimming signal
下伏 underlie
下伏饱和黏[粘]土层 underlying saturated clay strata
下伏材料 underlying material
下伏层 underlying layer
下伏沉积 underlying deposit;underlying sediment
下伏冲积层 underlying alluvium
下伏冲积土 underlying alluvial soil
下伏的 subjacent;underlying
下伏地层 substratum[复 substrata];underlying stratum
下伏地基土 underlying foundation soil
下伏地质层组 underlying geologic(al) formation
下伏基岩 subterrane;underlying bedrock;underlying rock
下伏基岩面坡度 slope of underlying bed rock
下伏喀斯特 subjacent karst
下伏面 underlying surface
下伏泥砂 underlying soil
下伏潜流 underlying phreatic flow
下伏体 underlying mass
下伏填土层 underlying fill
下伏土(壤)underlying soil
下伏围岩 underlying country rock
下伏系统 underlying system
下伏系统地层 strata of underlying system
下伏岩层 underlayer;underlying stratum;underlying bed
下伏岩溶 subjacent karst
下伏岩石 underlying rock
下浮 floating downward
下俯 nutation
下附数字 inferior figures
下覆辊 lower couch(roll)
下覆岩石 feu
下覆岩土层 D-horizon
下盖 base cup;lower cover
下盖板 bottom cover plate
下盖梁 pile bent sill
下干 inferior trunk;truncus inferior
下杆 lower beam
下工作辊 bottom working roll
下工作台 bottom mounting plate
下供立管 drop riser
下供燃料炉膛 underfeed furnace
下供式 down feed
下供式分配 down-feed distribution
下供式立管 down-feed riser
下供式立管系统 down-feed riser system
下供式系统 down-feed system
下供竖管 drop riser
下拱板 < 拱板转向架 > bottom arch bar
下拱壁 arch buttress
下拱杆 lower arch bar
下沟 lower(ing) in
下构造层物质上涌 infrastructural upwelling
下古生代【地】lower Paleozoic Era
下古生界【地】Lower Paleozoic Erathem
下固定颚板面 lower fixed jaw face
下挂梁 drop in beam
下冠层 lower canopy
下冠长 lower crown length
下冠高 lower crown height
下管 pipe installation;pipe setting;pipe sinking
下管后的钻孔水位 water-level in well after casing
下管滑道 exit ramp
下管理限 lower control limit
下管前的钻孔水位 water-level in well before casing
下管钳 backup tongs;backup wrench
下管钳操作工 < 卸钻杆时的 > backup man
下管取样钻探 tube sample boring
下规定限 lower specification limit
下跪人像 kneeling figure
下辊 bottom roll;lower roll
下辊轮 bottom roller
下滚子旁承 side bearing roller
下锅管 water drum
下海产卵的 catadromous
下海绿石砂 lower greensand
下函数 hypofunction;minor function
下寒带 lower frigid zone
下寒武统【地】Lower Cambrian series
下合 inferior conjunction
下合潮 inferior tide
下河 watering
下河斜坡道 dike ramp
下盒 lower casing
下桁材 bottom side rail;lower side rail
下桁条 inferior purlin(e)
下横撑 lower transverse strut
下横档 bottom rail;bottom door rail
下横档护板 < 门的 > kicking piece
下横桁吊索 yard slip
下横梁 lower beam;lower cross beam;lower separator
下横木 lower rail
下横支撑 lower lateral bracing
下弧面 lower camber side
下弧试验 flashover test
下滑 downslide;glide;gliding;letting down;slide down
下滑车 lower block
下滑道 glide slope;undocking【港】
下滑道天线 < 机场 > glide slope antenna
下滑道无线电波束 < 机场 > glide slope radio beam
下滑道无线电波束发射机 < 机场 > glide slope beam radio transmitter
下滑道信标 glide-path beacon
下滑锋【气】catafront;katafront
下滑轨道 glide path

下滑轨迹 glide path;glide trajectory
下滑角 angle of descent;glide angle;glide slope;gliding angle
下滑轮组 lower block
下滑面 downslide surface;glide slope
下滑漂移 downdrift
下滑坡度 glide slope
下滑速度 launching speed;launching velocity
下滑位移 downslope displacement
下滑信标 glide-path localizer
下滑指向标发射机 glide-path transmitter
下滑着陆系统 glide-path landing system
下环 bottom cover;bottom ring;shroud <水轮机转轮的>
下灰岩群 lower limestone group
下灰再燃送风机 return cinder fan
下回转塔式起重机 low-level slewing tower crane
下火车 detrain;disentrain
下击式水轮 undershot water wheel
下机架 lower bearing bracket;lower spider
下机壳 lower cover
下积分 lower integral
下级 inferiority;lower levels;subordinate
下级的 inferior
下级机构 subordinate body
下级欠缴运输进款 transport revenue receivable from subordinate unit
下级职员 underclerk
下极 lower pole
下极部 lower pole piece
下极面 lower pole face
下极限 floor margin;inferior limit
下极限事件 inferior limit event
下棘轮 <手制动> bottom ratchet wheel
下棘轮掣子 foot pawl
下集油箱 lower oil header
下给单立管 down-feed one-pipe riser
下给立管 down-feed riser;drop riser
下给料式燃烧 underfeed combustion
下给式单管系统 down-feed one-pipe system
下给式供暖系统 drop system
下给式热水系统 underfeed system
下给式双管供暖系统 two-pipe hot water downfeed system
下给式双管热水供暖系统 two-pipe direct water downfeed system
下给式双管热水系统 twin-pipe hot water downfeed system
下给式双管系统 down-feed two-pipe system
下给式系统 down-feed system
下给重力式供暖系统 down-feed gravity heating system
下加热式炉 underfeed furnace
下加线字符 underlined character
下夹板 lower plate
下夹送辊 bottom pinch roll
下甲板 under deck
下甲板吨位 underdeck tonnage
下价货 low priced goods
下间隔块 slab spacer
下槛 bottom sill;cill;mud sill;sill (plate)
下槛锚固螺栓 plate anchor;sill anchor
下浆板块 descending plate
下降 coastdown;deadening;decaying;declining;decrease;degreasing;degression;derating;descend;downbeat;downdrift;downfall;downslide;drawdown;droop(ing);drop-off;drop(out);fall(off);going down;let down;lowering;step down
下降板块 descending plate;subduction plate
下降比 lapse rate;suppression ratio
下降边 lagging edge;back edge;following edge;trailing edge
下降标高 drawdown level
下降波 descending wave
下降部分 sloping portion
下降操作 drawdown operation
下降齿轮系 drop train gear
下降传动装置 drop train gear
下降次序 descending order
下降到冰点以下的温度 a lowering of temperature below the freezing point
下降的 descendant;descending;supergene
下降的特性曲线 negative characteristic
下降点 <温度等> drop(ping) point
下降电流 dropout current
下降段 descending branch;descent trajectory;lowering limb
下降断块【地】downthrow block
下降法 descent method
下降方向 descent direction
下降风 canyon wind;fall wind;gravity wind;katabatic wind;mountain breeze;mountain wind
下降伏安特性 falling volt-ampere characteristic
下降高度 falling head;falling height;load-lowering height
下降高速度【机】high load lowering speed
下降割面 descending face
下降管 downcomer;downstream tube;downtake
下降管环路 downflow circuit
下降管回路 downcomer circuit
下降海岸 depression coast;sinking coast
下降海滨线 shoreline of emergence
下降函数【数】decreasing function
下降极限限位器【机】lowering height limiter
下降几率 decreasing hazard rate
下降角 angle of descent
下降阶段 subsiding stage
下降径流 descending flow
下降卡子 falling pawl
下降空气 descending air
下降流 descending current;downflow;downwelling(current)
下降漏斗 <地下水面的、井内抽水形成的> cone of pressure relief;cone of exhaustion;cone of influence;cone of depression;crater of depression
下降路程 <电梯> down trip
下降路径 descent path
下降率 rate of decline;rate of descend
下降率保持 rate-of-descent hold
下降率传感器 rate-of-descent sensor
下降率控制 rate-of-descent control
下降率指示器 rate-of-descent indicator
下降盘 <断层的> downthrow side;downthrow wall
下降盘沉降幅度 subsiding amplitude of downthrow side
下降盘沉降速度 subsiding velocity of downthrow side
下降盘地层厚度 strata thickness of downthrow side
下降平顶 drop ceiling
下降平原 subsided plain
下降坡度 descending grade;down grade;falling gradient
下降气流 air pocket;descending air current;downflow;downflow draft;down-wind
下降气流的 katabatic
下降气流区 hyperbar
下降区 depressed region
下降曲线 decline curve;depression curve;descending curve;drop-down curve;droop line
下降趋势 downswing;downturn;downward trend;downtrend <经济方面的>
下降泉 descending spring;gravity spring
下降溶液成矿说 descension theory
下降深度限位器 lowering limiter
下降深度指示器 load-lowering height indicator
下降时间 fall time
下降式活门 hinged flash gate
下降式火道 downflow flue
下降式进水口 drop inlet structure
下降式竖式烟道 descending vertical flue
下降式双管系统 double pipe dropping system
下降式通风 downward ventilation
下降式烟气 downflow flue
下降水 descending water;supergene water
下降水面曲线 falling surface curve
下降水平距离关系 <地下水位> distance-drawdown relationship
下降水平面 drawdown level;drawdown water level
下降水头 depression head;fall head of water;dropping head
下降水位 falling level;falling stage
下降顺序 descending order
下降速度 descending velocity;fall(ing) speed;fall(ing) velocity;lowering speed;load-lowering speed
下降速度调节手柄 lowering rate control lever
下降速率 rate of descent <飞机>;lowering rate
下降索道 fall way
下降特性 fall characteristic
下降特征曲线 drooping characteristic curve;falling characteristic(curve)
下降梯度 downward gradient
下降调速器 droop governor
下降温度 descending temperature;squatting temperature
下降温度计 Kata thermometer
下降系数比拟法 analogy method of drawdown coefficient
下降相 descending phase
下降斜率 descending slope
下降行程 down stroke;down trip <电梯>
下降序列 decreasing sequence
下降压力 falling pressure
下降烟道 downward flow flue
下降延迟 fall delay
下降沿 negative-going edge;trailing edge
下降预制(混凝土)隧道筒管部分 lowering precast tunnel sections
下降云 fall cloud
下降运动 bathygenesis;lowering motion;subsiding movement
下降制动开关 lowering brake switching
下降制动器控制(装置) lowering brake control
下降周期 falling period
下降主管 falling main
下降转矩 falling torque
下降转为上升区 converting area from depression into uplifting
下降锥(体) cone of depression;cone of drawdown;cone of influence
下交叉 crossing below
下浇口 peg gate;submarine gate
下胶塞 bottom plug
下角 Cornu inferius;inferior horn;underhorn
下脚管 off-cut pipe
下脚混凝土 rejected concrete
下脚焦炭 scrap coke
下脚料 leftover;leftover bits and pieces;mill cull;off-cut;reject;revert;scantling;wastes;foots
下脚煤 refuse coal
下脚(木)材 mill culls
下脚切口 foot cut
下脚油 sump oil;waste grease
下接某页 continued on…sheet
下接头 lower contact
下节 lower segment
下节点 descendant;lower node
下截盘 undercutting jib
下界 lower bound
下界表达式 lower bound expression
下金枋 lower purlin(e) tiebeam
下金桁 lower principal purlin(e)
下金檩 lower principal purlin(e)
下近似值 lower approximate value
下近中天 ex-meridian below pole
下进汽口 lower steam inlet port
下井管 sinking
下井检查 down-the-hole inspection
下井速度 downhole speed
下井筒 well sinking
下井筒速度 sinking velocity
下井仪模块开展器 extender downhole ga(u)ge module
下锯 cutting
下锯法 sawing procedure
下锯口 undercut
下卷式 underwind
下卷式卷绕 underwinding
下开口 under shed
下开式车钩提杆 bottom operation uncoupling lever
下开式车钩提杆装置 rotary operating mechanism
下壳 lower casing
下刻度盘 <经纬仪> lower circle
下刻机 undercutter
下客 off-load;set-down
下客港 port of debarkation;port of disembarkation;port of landing
下客时间 alighting time
下孔法 downhole method;downhole shooting
下孔法锤击钻进 downhole hammer drilling
下孔速度 downhole speed
下孔型 lower pass
下控制杆 lower control arm
下控制限 lower control limit
下口 birdmouth;undercut
下库 lower pool
下跨零点 downward zero crossing;zero down-crossing
下矿井时间 breaking-in period
下框架 lower frame
下馈式加煤机 underfeed stoker
下馈式烧煤机 underfeed stoker
下扩大室 <调压井> lower expansion chamber
下拉 drag down;pulldown
下拉把手 pulldown handle
下拉法 down-draw process
下拉杆 bottom connecting rod;bottom truck connection;brake strut;

X

lower link; brake lever coupling bar <内侧悬挂制动>
下拉杆安全托 bottom rod guide; safety support
下拉荷载 downdrag; negative friction of pile; downdrag load
下拉力 down pull; drag-down force
下拉式菜单 pulldown menu
下拉式搁板 pulldown utility shelf
下拉条(轴箱) bottom brace
下来 <从车、飞机等处下来> alight
下栏门窗 bottom rail
下栏索 footline
下类 lower class
下连接子卡 lower anchors
下联(结系) under-bracing
下联锁继电器 lower interlock relay
下联箱 subheader
下链条 lower chain
下梁 beam lower; lower bolster; underbeam
下料 baiting; blanking; lay(ing) off
下料表 <钢筋或木材等> cutting list
下料不顺 sticky
下料长度 fabrication length
下料工 blanker; layout man
下料公差 blanking tolerance
下料管 feed downpipe; slick line; vertical trunking <混凝土浇注用>
下料环 cutting-off bushing
下料机 blanking machine; rod shearing machine
下料间隙 blanking clearance
下料口 throat opening
下料力 blanking pressure
下料利用率 utilization rate for the semiprocessed materials
下料馏槽 discharge duct
下料模 blanking die
下料清单 cutting list
下料生产线 blanking line
下料压力机 cutting machine
下料重量超差 miss-cropping
下列 lower row
下列的 following
下列签字人 undersigned
下列土地分级法 following system of land classification
下临界点 lower change point
下临界冷却速度 lower critical cooling speed
下临界冷却速率 lower critical cooling rate
下檩 pole plate
下流 downflow
下流槽 downtank
下流固定床反应器 downflow fixed bed reactor
下流固定膜生物反应器 downflow fixed film bioreactor
下流管 down pipe; downspout conductor; downtake
下流河岸(因冲积)的自然长成 colmatage
下流护床 fore apron
下流量 down-off
下流气流 <立管的> down draught
下流式单位散热器 downward discharge unit heater
下流式固定床 downflow fixed bed
下流式接触床 downflow contact bed
下流式快砂滤池 rapid downward flow sand filter
下流式滤池 downflow filter
下流式烧结机 down-draft sintering machine
下流稳态膜 downflow stationary film
下流斜板式沉淀池 inclined plank settling tank of downflow
下流悬挂海绵 downflow hanging sponge
下流厌氧紊动床 downflow anaerobic turbulent bed
下楼机 lowerer
下漏式模 push-through die
下炉腹线 lower bosh line
下绿砂层 <白垩纪>【地】lower greensand
下掠式排气 downswept exhaust
下轮 lower whorl
下螺母 lower nut
下落 descending; downcast; downdrop; downswing; lowering; sink; whereabouts; downthrow
下落地块【地】downthrow block
下落点 setting point; west point (of the horizon)【天】
下落断层 dipper; downcast fault; downthrow fault; jump-down; thrown fault; trap-down
下落断层作用 down faulting; throw faulting
下落管 down pipe; drop pipe
下落滑动 thrown slip
下落(滑)断层【地】down-slip fault
下落机构 lowering mechanism
下落快门 drop shutter
下落盘【地】downthrow block
下落盆地断层 down-to-basin fault
下落期 decline phase
下落时间 downtime; fall time; lowering time
下落式窗玻璃腔 window pan
下落式顶棚散流器 step-down ceiling diffuser
下落式顶填泥芯 wing core
下落式(棘轮)掣子 falling latch; falling pawl
下落式立筒烘干机 dropping shaft dryer
下落式取土器 drop sampler
下落式拖车 drop bed trailer
下落式悬挂 drop suspension
下落输送 drop delivery
下落搜索 whereabouts search
下落搜索策略 whereabouts strategy
下落速度 falling speed; falling velocity; lowering rate; rate of drop
下落位置 lowering position
下落芯 drop core; wing core
下落芯头 drop print
下落翼 downcast side; downthrow
下落终速(度) terminal fall velocity
下落重量 falling weight
下马石 horse block
下马氏点 martensite finish(ing) point
下毛毛雨 drizzle
下锚 at anchor; bottom anchor; moor; top drop anchor; drop anchor
下锚的驳船 anchored barge
下锚处 haven
下锚船 anchor barge; anchoring ship
下锚段衬砌 anchor-section liner; anchor-section lining
下锚浮标 anchor buoy
下锚固定角钢 anchor fixing angle steel
下锚跨 anchor span
下锚埋入杆 rod for anchor
下锚平衡板 balance sheet for anchor
下锚楔套 wedge sleeve for anchor
下冒头 <门的> bottom door rail; bottom rail (of door); lower door rail
下楣(柱) architrave; epistyle
下煤舱 bunker hold
下煤减压装置 throat inside burner
下门 bottom door
下密度 lower density
下密封 lower seal
下面 felt side; lower side; underside; undersurface
下面部分 lower half; lower part; lower section
下面传动的 underdriven
下面的物件 subjacent body
下面盾状 hypopeltate
下面光 lower light
下面接触式导电钢轨 under-contact conductor rail
下面结构 infrastructure
下面支承 subjacent support
下模 bed die; body mo(u)ld; bottom force; counter-die; lower die; lower mo(u)ld half; negative die; pattern draw(ing)
下模板 lower bolster
下模冲 bottom punch; lower punch
下模剂 mo(u)ld release agent
下模具 bottom die
下模型 drag
下模座 die shoe
下磨 head of mill
下磨盘转动的磨粉机 under runner
下木 underb(r)ush; under growth; underplant; underroof; understor(e)y; underwood
下木塞 <偏斜楔支座用的> base plug
下木塞(注水泥)法 bottom-packer method
下木栽植 undergrowth planting
下挠 downwarping
下挠轴 sagging axle
下泥盆统【地】Lower Devonian series
下枭【建】reversa; lower cyma
下拧 screw-down
下排污 lower drain
下排障器踏级 lower pilot step
下盘 bottom wall; foot side; foot wall; heading side; lower plate【测】; lying wall; lower wall【地】; lower circle <经纬仪>; heading wall
下盘层 hypothecium
下盘的【地】underlying
下盘的地层 strata of foot wall
下盘微动螺丝 lower plate slow motion screw
下盘斜井 footwall shaft; underlay shaft; underlier; underlie shaft; underlying shaft
下盘晕 lower wall halo
下盘制动螺丝 <经纬仪的> lower clamp
下盘制动螺旋 <经纬仪的> lower clamp
下盘主动下投 footwall active downthrow
下盘转动 <复测经纬仪的> lower motion
下旁承 truck safety bearer; truck side bearing
下旁承钢 side bearing iron
下旁承盒 side bearing pocket
下旁承螺栓 truck side bearer bolt
下喷水器 bottom water sprays
下皮 bottom; soffit
下皮标高 plancier level; soffit level
下偏差 lower deviation; lower variation of tolerance
下偏耳 low parry arc
下偏光镜 lower Nicol; polarizer
下偏置 below-center offset
下漂浮生物 hypoenuston
下平榑 eaves purlin(e)
下平联 lower lateral
下平巷底板结构 low-sill structure
下平纵联 bottom lateral bracing
下坡 aft slope; decent; declivity; descending; descending slope; falling gradient; negative grade; sag grading
下坡安全速度 safe downhill speed
下坡铲土法 declining earth shoveling process
下坡传送带 decline conveyer
下坡道 descending grade; down grade; downhill grade; falling grade; minus grade; descending gradient; descending ramp; down gradient; downhill gradient; down ramp; failing gradient; falling gradient; falling slope; gug <采矿、隧道、坑道的>
下坡道防护电路 protection circuit for heavy down grade approaching
下坡的 declivitous; down grade; downslope
下坡度 descending gradient; down gradient; downhill gradient; falling gradient; down grade
下坡段 decline section; falling portion
下坡段落 downhill section
下坡方向 direction of fall
下坡防滑机构 <汽车的> hill holder
下坡风 downslope wind; fall wind; settle draft
下坡焊 downward welding in the inclined position
下坡开挖 downgrade excavation
下坡路 downhill path; downhill road
下坡路的 downhill
下坡锚 anchor cast on downward slope
下坡坡道 down ramp
下坡气流 downslope slow
下坡速率路标 grade severity rating sign; grade severity rating system
下坡推土法 declining earth pushing process
下坡行程 downhill journey; downwards journey; downwards run
下坡运输带 decline conveyer
下铺 lower bed; lower berth; lower bunk
下棋机 chess playing machine
下棋模式 chess pattern
下起重滑轮 load lower block
下气道 downtake; gas downtake
下气管 downcomer; gas downtake
下气筒垫密片 lower air cylinder gasket
下气烟道 downcomer
下汽包 lower drum; water drum
下汽车 detruck
下汽缸 lower cylinder
下汽缸垫密片 lower steam cylinder gasket
下潜涵洞 dive culvert
下潜架 submerged cradle
下潜平台 submerged platform
下潜深度指示器 depth indicator
下潜时间 descent time
下潜式工作舱 submerged work
下潜式减压舱 submerged decompression chamber; submersible decompression chamber
下潜式绞车 submersible winch
下潜信号 diving alarm
下潜型载驳货船 submergible type barge carrier
下潜压力 submergence pressure
下潜压载 diving ballast
下潜制动装置 dive brake
下浅海地带 lower sublittoral zone; outer sublittoral zone
下墙板 wainscot panel
下墙板上镶条 upper wainscot rail
下墙板下镶条 lower wainscot rail
下墙板镶条 wainscot(ting) rail

下桥塞 setting a bridge plug
下翘 downwarp(ing)
下切 basal sapping;cutting;downcut;regressive erosion;sapping;vertical erosion
下切法 downcutting method;undercutting method <隧道工程>
下切河(流) corrading river;corrading stream;degrading river;degrading stream;downcutting river;downcutting stream;corroding stream
下切冷锋 undercutting cold front
下切侵蚀 downcut;downward erosion
下切式剪切机 downcut shears;up-and-down cut shears
下切水流 cutting jet
下切削位置 lower cutting position
下切作用【地】downcutting;incision;vertical erosion
下揿按钮 downward button
下倾 declination;decline;declining;declivity;descend;dipping;down dip;fall over
下倾板块 descending plate
下倾的 cataclinal;catacline;downhill
下倾的输送管线 downhill pipeline
下倾断块 downdip block
下倾角 angle of declination
下倾前缘 drooped leading edge
下倾输送 downhill conveying
下倾输送机 downhill conveyer[conveyor]
下倾型 catacline
下倾岩巷 dipping stone drift
下区 lower curtate
下区段 lower curtate
下曲 downwarping
下曲拐 lower crank
下曲拐铜套 lower crank bushing
下曲面 lower surface camber
下曲线 lower curve
下曲轴箱 lower crankcase
下屈服点 lower yield point
下屈服应力 lower yield stress
下确界 greatest lower bound;infimum【数】
下绕 underwind
下热水箱 lower hot water tank
下容差限 lower tolerance limit
下入槽内的扁键 sunk key
下入沟内 <管道的> lowering in
下入孔内 sent down the hole
下塞深度 plugging depth
下塞位置 plugging site
下三叠统【地】Lower Triassic series
下三角形矩阵 lower triangular matrix
下三角洲平原沉积 lower delta-plain deposit
下三角座 trigonid
下三角座底板 bottom cam box plate
下三角座箱 bottom cam box assembly
下扫描线 lower tracer
下砂箱 bottom mo(u)lding box;drag box;drag flask
下筛 lower sieve;shoe sieve
下筛斑 inferior perforated spot
下山 board-down;diphead
下山车场 incline landing
下山坡 downhill
下山巷道 bord-down
下栅板 lower grid plate
下扇 lower fan
下舍入 round down
下设排水系统的坞底板 under-drained drydock floor
下射波 down-coming wave
下射的 undershot
下射喷燃 downshot firing
下射喷燃器 downshot burner
下射曲叶水轮 Poncelet wheel
下射燃烧炉膛 downward fired furnace
下射式 undershot
下射式水轮机 undershot water wheel
下射式引水闸门 undershot head gate
下射式闸门 underpour type gate;undershot gate
下伸小柱 <中世纪英国屋架的> pendent post
下身盆 bidet
下身洗盆 bidet
下深槽 downstream deep;downstream pool;lower deep;lower pool
下渗 infiltration;sumping
下渗过程 infiltration process
下渗理论 infiltration theory;theory of infiltration
下渗率 infiltration rate;intake rate
下渗能力 infiltration capacity
下渗强度 infiltration intensity
下渗区 infiltration area;infiltration zone
下渗容量 infiltration capacity
下渗容量方程 infiltration capacity equation
下渗容量曲线 infiltration capacity curve
下渗实验区 infitrometer plot
下渗水 infiltrated water;infiltration water
下渗速度 infiltration velocity;intake speed
下渗损失 infiltration loss
下渗系数 infiltration coefficient
下渗演算 infiltration routing
下渗值 infiltration value
下渗指数 infiltration index
下蜃景 inferior mirage
下升船机 float out;undocking
下声发射器 underwater sound projector
下湿地 <北非> dambos
下石炭统【地】Lower Carboniferous series
下蚀作用 vertical erosion
下视图 bottom view
下室 <教堂的> a(u)mbry
下首闸门 lower gate
下枢轴 lower pintle
下枢轴浇铸 lower pivot casting
下枢轴铸件 lower pivot casting
下枢轴铸模 lower pivot casting
下枢轴铸造 lower pivot casting
下输管 down pipe
下鼠笼 lower cage
下鼠笼条 bottom bar;lower bar
下蜀黏[粘]土 Xiashu clay
下述的 undermentioned
下竖撑 bottom vertical brace
下刷 down rush
下水 downbound;launching;launch (out)
下水驳船队 downbound tow
下水车 launching troll(e)y
下水程序 launching procedure
下水池 downstream bay
下水尺 lower ga(u)ge
下水出闸时间 downbound exit time of lockage
下水船道 launching way
下水船台 launching platform
下水道 cloaca;culvert;down(flow) pipe;effluent sewer;gull(e)y drain;kennel;outfall sewer;sanitary sewer;sewage drain;sewer;sew(er)age;sewer culvert;sewer tunnel;soil pipe
下水道布置 plan of sewerage
下水道沉泥井 sewer catch basin
下水道沉沙池 sewer trap
下水道沉沙洗涤机 sewage grit washer
下水道沉渣池 detritus tank
下水道尺寸 sewer size
下水道冲洗器 sewer flusher
下水道出水口 outlet of sewer
下水道出水总管 main outlet of sewer
下水道次干管 submain sewer
下水道粗格栅 coarse sewage screen
下水道倒拱 sewer invert
下水道地区 sewer zone
下水道跌落出口 outfall
下水道跌落井 drop manhole
下水道短程海底出口 short sea outfall
下水道范围 sewer zone
下水道防臭弯管 sewer trap
下水道废气 sewer gas
下水道分流检修孔 diversion manhole
下水道附属物 sewage appurtenant;sewer appurtenances
下水道改流装置 underpass deflector
下水道盖板 channel mo(u)ld
下水道干管 main sewer;main sewer;trunk sewer
下水道工程 sewage works;sewerage
下水道工程系统 sewerage system
下水道工程学 sewerage engineering
下水道工人 sewer man
下水道拱顶 sewer arch
下水道管 sewer line;sewer pipe (drain)
下水道管材料 sewer pipe material
下水道管道截流堰 separating weir
下水道管道系统 sewer(age) system
下水道管道汇合处检修井 junction manhole
下水道管理区 sewer district
下水道海底出口 sea outfall;submarine outfall
下水道海底出口的扩散器 outfall diffuser
下水道海中出口扩散器 outfall diffuser
下水道互通式交叉 sewer interchange
下水道汇流井 function chamber
下水道基础 sewer bottom
下水道检查井 sanitary sewer manhole;sewer manhole;inspection chamber
下水道检查孔 sewer manhole
下水道检查口 sewer manhole
下水道建筑物 sewer construction
下水道接头 sewer connection
下水道接头油灰 sewer jointing compound
下水道截面 sewer section
下水道尽端 dead-end sewer
下水道进口截污井 gull(e)y trap;yard trap
下水道进口截污设备 yard trap
下水道进入孔 sewer manhole
下水道口 inlet of sewer
下水道拦截油的设施 oil interceptor
下水道内底 sewer invert
下水道排放场地 absorption field
下水道排水口 sewer outfall
下水道排水构筑物 <入海、湖或河流的> sewer outfall
下水道排水建筑物 <入海、湖或河流的> outfall
下水道平面图 plan of sewerage system
下水道坡度 sewer grade;sewer slope
下水道气体 sewer(age) gas
下水道清除 sewer cleaning
下水道清孔绞刀 sewer reamer
下水道清理工 sewer rat
下水道清理设备 channel scraper and elevator
下水道清掏机 sewer evacuator
下水道清洗 sewer cleaning
下水道清洗设备 sewer cleaning device
下水道区(域) sewer(age) district;sewer(age) territory
下水道容量 sewer capacity
下水道入口 inlet of sewer
下水道散发气味 sewage odo(u)r
下水道上游端的检修孔 end manhole
下水道设备 sanitation
下水道设备系统 sanitation system
下水道设计 sewer design
下水道竖向支流检查井 drop manhole
下水道水系 drainage
下水道抬高 raising of sewage
下水道陶管 sewer tile
下水道通风 sewerage ventilation
下水道图 map of sewers;sewer map
下水道污泥 sewage sludge;sewer sludge
下水道污水 effluent sewerage
下水道污水容许最大浓度 instantaneous effluent limitation
下水道系统 drainage system;sewage system;system of sewerage;system of sewers
下水道系统布置 plan of sewerage system
下水道系统的支流与主流汇合段 junction chamber
下水道系统连接污水处理厂或容纳水域的末端管道 outfall sewer
下水道系统评估调查 sewer system evaluation survey
下水道系统通风管 utility vent
下水道系统形式 pattern of sewage system
下水道系统溢流 sewage system overflow;sewerage system overfall
下水道系统装置 sewer appurtenances
下水道楔形砖 wedge-type sewer brick
下水道修建工 drainer
下水道窨井 sewer catch basin
下水道溢出口建筑物 outlet structure for closed drain
下水道溢流 sewerage(system) overflow
下水道溢流堰 <溢泄超量的雨水> leaping weir
下水道用管 sewer pipe
下水道用上釉防水陶管 sewer tile
下水道用陶器 sewer stoneware
下水道用砖 sewer brick
下水道淤渣 sewage grit;sewerage grit
下水道远程海底出口 long sea outfall
下水道支沟 branch sewer
下水道支管 sewer connection;branch discharge pipe;branch sewer
下水道支渠 branch sewer
下水道制品 sewer article
下水道筑堰挡水 stanking off
下水道综合征 sewer syndrome
下水道纵剖面 profile of sever
下水吨数 downstream tonnage
下水防臭阀 gull(e)y trap
下水分析 sewage analysis
下水工程 sewerage
下水工地 <船舶、沉箱、管道等> launch site
下水沟 sewage drain;sewer culvert
下水管 downcomer;downflow pipe;down-take pipe;sewer(pipe)
下水管道 pipe sewer line;sewage line;sewer line
下水管道承插式内平弯路 recessed fitting
下水管道承插式内平弯头 recessed fitting

下水管道连接井 junction chamber
下水管道排水能力 sewer capacity
下水管道清扫机 sewer pipe cleaner
下水管道系统 network of conduit-type sewers;sew(er)age system
下水管户线 service drain
下水管漏泄试验器 asphyxiator
下水管线 sewer line
下水柜或下舱装水的压载 hold water ballast
下水航驳 descending barge
下水航程 downbound journey
下水航驶 proceed down the river
下水滑道 ground ways;launching way;slipway;standing way
下水滑道倾斜度数 declivity of ways
下水滑脂 launching grease
下水(货)运量 downbound traffic;downstream traffic
下水计算 launching calculation
下水架 puppet
下水架用楔 slice
下水口 gull(e)y hole
下水口及存水弯 waste and shower trap
下水库 <抽水蓄能电站的> tail reservoir
下水了的(船舶)off the stock
下水排水 sewage discharge
下水排水口 waste outlet
下水坡道 launching slide
下水坡度 launching grade
下水牵索 launching cable
下水曲线 launching curve
下水日期 date of launching; launch date
下水设备 launching facility
下水时船舶重量 launching weight
下水式 launching ceremony
下水水位 downstream level;level of tail water
下水隧道 sewer tunnel
下水隧洞 sewer tunnel
下水台 shipway
下水条款 launching clause
下水通气竖管 stack vent
下水图纸 launching drawing
下水瓦管 crock
下水稳定性 launching stability
下水污泥的地面处理 land disposal of sewage sludge
下水污泥干化 sewage sludge drying
下水箱 lower header;lower tank
下水箱给水装置 lower tank system water supply apparatus
下水斜坡道 launching ramp
下水行驶 downstream sailing
下水仪式 launching ceremony
下水仪式台 launching platform
下水淤渣 sewage grit
下水运量 downbound traffic volume
下水运输 downstream traffic;downstream transportation
下水支船架 launching cradle
下水支管 branch sewer
下水支架 puppet
下水重量 launching weight
下水装置 boat launch;launching gear;launching installation
下水作业 launching operation
下死点 bottom dead-center[centre];lower dead center[centre]
下死点标记 bottom dead-center indicator
下四分位长度 lower quartile length
下饲式炉排 <柱塞进煤的> ram and retort grate
下饲 underfeed
下饲式加煤机 underfeed stoker
下饲式加煤装置 retort stoker
下饲式炉排 retort grate;retort-type stoker
下饲式炉膛 underfeed furnace
下送变换器 down converter
下塑限 lower plastic limit
下梭口 lower shed
下索 lower cable
下锁臂 lower lock arm
下锁销杆 <车钩> bottom lock lifter bar;bottom lock lifter lever
下锁销杆吊 bottom lock lifter bar bracket
下台 step down
下台阶 bottom bench;lower bench
下滩 descend a rapid
下套管 case in; casing installation; hang inside; insert the casing; land casing; pipe driving; running casing;sinking pipe casing
下套管部位 casing point
下套管程序设计 casing string design
下套管的井 cased bore hole
下套管的钻孔 cased hole
下套管滑车 casing block
下套管记录 casing record
下套管深度 casing point;landing depth
下套管时间 setting casing time
下套管用工具 casing appliances
下套管钻孔 cased bore hole
下套筒 lower sleeve
下梯段 lower bench
下梯段超前掘进 heading overhand bench
下提杆座 uncoupling rod foot
下提式(车钩) bottom uncoupling type
下填法 underfill method
下挑丁坝 downstream-angled spur dike[dyke];downward-point groin
下调压池 lower surge basin
下调压井 lower surge basin
下调压室 lower surge chamber
下通式塔式起重机 underrunning stacker crane
下同调类 homology class
下统【地】Lower Series
下筒 doff(ing);removal of cake
下筒体 lower shell
下筒体法兰 flange of lower shell
下投侧 downthrow side
下投断层 drop fault
下投式探空仪 dropsonde
下投梭 underpick
下凸轮轴驱动齿 lower camshaft drive gear
下凸轮轴蜗轮 lower camshaft worm gear
下凸模 lower punch
下突重荷模 torose load cast
下图廓 bottom border; bottom margin;lower edge;lower margin;lower border
下涂 bodying coat;key coat
下推 push-down
下推表 push-down list
下推队列 push-down queue
下推排队 push-down queue
下推式存储器 push-down storage
下托板 ram bolster
下托辊 lower supporting roller;snub pulley <胶带运输机的>
下托架 bottom bracket
下托牙 lower
下拖式潜水工作舱 downhaul utility capsule
下脱模 bottom knockout
下挖 sinking;undercut
下挖的 subexcavated
下挖断面 subexcavated section
下挖坑 dig-down pit;sunken pit
下挖式多斗挖掘机 bucket excavator for downward scraping
下挖式多斗挖土机 bucket excavator for downward scraping
下挖式链斗挖掘机 bucket excavator for downward scraping
下挖式链斗挖土机 bucket excavator for downward scraping
下外包线 minimum envelope curve
下外侧面 facies inferolateralis;inferolateral surface
下弯 decurvature;kick down;sagging
下弯桥轴 cambering axle
下桅 lower mast
下桅的左右支索 lower rigging;lower shrouds
下桅帆 heavy sail
下桅帽 mast yoke;yoke of mast
下桅盘 round top
下桅盘护绳 futtock rigging
下纬度 lower latitude
下位 inferior
下位潮 inferior tide
下位机 slave
下位锁 down lock
下温带 lower temperate zone
下卧饱和黏[粘]土层 underlying saturated clay strata
下卧薄层的 thin-bedded
下卧层【地】subjacent bed; subjacent layer; sublayer; underlayer; underlying bed; underlying layer; underlying stratum; substratum[复 substrata]
下卧层土 underlying layer of soil
下卧层应力增量 subsurface stress increment
下卧地基 underlying ground
下卧地基土 underlying foundation soil
下卧黏[粘]土层 underlying clay layer;underlying clay stratum
下卧软土层 soft underlying soil
下卧砂层 underlying sand layer; underlying sand stratum
下卧砂岩 underlying sandstone
下卧体 subjacent body
下卧土 subsoil
下卧土层 underlying soil stratum
下吸 down draft;down draught
下吸式抽风罩 down-draft hood
下吸式化油器 down-draft carbureter [carburetor]
下吸式煤气发生炉 down-draft gas producer
下吸式汽化器 down-draft carbureter [carburetor]; down-draught carbureter[carburetor]
下吸式通风 down draught
下洗角 angle of downwash
下洗(流)down wash
下系带 frenulum
下细雨 drizzle
下弦 bottom boom;bottom chord;dichotomy; lower chord; low(er) boom <桁架的>; lower camber; lower string; third quarter; last quarter【天】
下弦板 lower boom plate
下弦承重悬臂桥 cantilever-through bridge
下弦风撑 bottom chord wind bracing
下弦杆 bottom boom bar; bottom boom member; bottom boom rod; bottom chord bar; bottom chord member; lower boom rod; lower chord bar; lower chord rod; lower truss
下弦杆部件 lower boom member
下弦杆钢条 bottom boom bar
下弦杆连接板 bottom boom junction plate;bottom chord junction bar
下弦杆纵向钢条 bottom chord longitudinal bar
下弦钢条 bottom chord bar
下弦构件 lower chord member
下弦横撑 lower lateral bracing
下弦横向水平支撑 bottom lateral bracing;lower lateral bracing
下弦剪刀撑 cross bracing of the bottom chord
下弦接板 lower boom junction plate
下弦节点 lower chord panel point
下弦节点连接板 bottom boom junction plate
下弦联板 lower chord junction plate
下弦小潮 third quarter neap tide
下弦应力 bottom chord stress
下弦月 last quarter of the moon;third quarter of the moon;waning moon
下弦支撑 lower chord bracing
下弦纵杆 bottom boom longitudinal bar;lower boom longitudinal bar
下弦纵向钢条 bottom boom longitudinal bar
下现晨景 inferior mirage
下线工具 inserting tool
下线角 inferior angle
下线装置 coil assembling apparatus;coil inserting apparatus
下限 low(er) bound; lower boundary;lower extreme;lower limit;low extreme;low limit;prescribed minimum
下限尺寸 low limit dimension
下限的 limited to the left
下限定理 lower bound theorem
下限公差 low limit of tolerance
下限估计 lower bound estimate
下限函数 lower limit function
下限寄存器 lower limit register
下限接受值 lower acceptance value
下限截止频率 lower cut-off frequency;lower limiting frequency
下限解 lower-bound solution
下限控制 lower operating range control
下限累加器 lower accumulator
下限临界流速 lower critical velocity
下限浓度 concentration limit;threshold concentration
下限稳定性 lower limit of stability
下限限额 lower limit
下限信号器 low alarm
下限原理 lower bound theorem
下限越界 off-normal lower
下限直径 diameter limit
下限值 lower limit value;lower range value;value of inferior boundary
下陷 bogging down; cave-in; caving; joggle; sag(ging); sinkage; slump settlement;subsidence
下陷包层光纤 depressed cladding fiber[fibre]
下陷边 <石工> sunk draft
下陷的 downcast
下陷地面 subsiding ground
下陷改正 correction for sag
下陷口 <放置蹭鞋垫的> mat sinkage
下陷量 sinkage
下陷率 subsidence rate
下陷面 subsided surface;sunk face
下陷区 caving zone
下陷型模 bottom swage
下陷性 subsidability
下箱 lower box;lower tank
下向 down the dip;lower quadrant
下向臂板 lower quadrant arm;lower quadrant(semaphore)blade

下向臂板信号机【铁】lower quadrant (semaphore) signal
下向采掘 underhand work
下向单翼回采工作面 underhand single stope
下向方框支架安装法 method of underhand square setting
下向分层崩落采矿法 radial top-slicing;top slicing mining
下向过滤 downward filtration
下向回采 underhand stope
下向回采梯段 underhand stope bench
下向加荷 downhill loading
下向扩孔钻头 under-reaming bit
下向配水管 down distribution
下向气流 down draft;down draught
下向倾斜分层崩落采矿法 top slicing with inclined slices
下向扫描 down sweep
下向式气腿 downward air leg
下向输送 downhill conveying
下向输送机 downhill conveyer[conveyor]
下向双翼回采工作面 underhand double stope
下向水平分层崩落采矿法 top slicing and caving; top slicing combined with ore caving
下向水平分层崩落法 top-slice method
下向水平分层崩落矿房法 top slicing by rooms
下向水平分层顶板崩落采矿法 top slicing cover caving
下向梯段 underhand bench
下向梯段爆破 underhand blasting
下向梯段回采 underhand work
下向梯段掘进 underhand work
下向梯段开挖 underhand work
下向梯段式长壁回采 underhand longwall
下向梯段式的 underhand
下向梯段凿岩 underhand work
下向通道 downpass
下向通风 down (cast) ventilation; down draft;down draught
下向通风机 downcast ventilator
下向通风系统 downward system of ventilation
下向引出管 downcomer
下向凿岩机 downhole drill;sinker
下向钻眼 downhole
下项线 inferior nuchal line;linea nuchae inferior
下象限 lower quadrant
下小钢轮 undercrown wheel
下楔 hyposphene
下斜撑 bottom diagonal brace
下斜(坡)declivity
下斜直线 straight downward-sloping lines
下泄 down draught
下泄热交换器 let-down heat exchanger
下泄式渠首闸门 undershot head gate
下泄式水闸 under sluice
下泄式闸门 underpour type gate;undershot gate
下泄水流 discharge flow
下泄水位 drawdown level
下泄物 spigot
下心盘 bogie center[centre] plate; center[centre] bearing;female center[centre] plate;truck center[centre] plate
下心盘螺栓 truck center[centre] plate bolt
下心盘支承面 center[centre] bearing
下芯 core setting
下芯机 core setter
下芯夹 core positioner
下芯区 coring-up station
下芯样板 core setting ga(u)ge
下信标灯 lower beacon light
下行 descend
下行波反褶积剖面 deconvolution section of downwave
下行波剖面 downgoing wave section
下行驳船 descending barge
下行驳船队 downbound tows
下行车排队 downstream queue
下行程 stroke down
下行冲程 down(ward) stroke
下行出闸时间 downbound exit time
下行船(舶) descending boat; descending vessel; downbound boat; downbound ship; downbound vessel
下行船队 downbound fleet
下行的 descending; downbound; downstream; underhand; underrunning
下行顶推船队 downbound push-train
下行多普勒 down-Doppler
下行法 descending method
下行方面 downside
下行方向 down direction
下行分层崩落采矿法 top slicing
下行感染 descending infection
下行沟通 downward communication
下行海岸 downcoast
下行货流 downbound cargo flow
下行给水 down-feed distribution
下行交通 <从市镇中心区向边缘的交通> down(bound) traffic
下行进(船)闸时间 downbound entry time of lockage
下行开采 descending mining; downward mining
下行控制 descending control
下行列车 down(bound) train
下行流 downward flow
下行流速 descending velocity
下行桥式起重机 underrunning bridge crane
下行色层分离法 descending chromatography
下行色谱法 descending chromatography
下行上给式 upfeed distribution
下行上给式蒸汽供暖系统 steam-heating up-feed system
下行式葫芦 bottom running hoist
下行式热水系统 underfeed system
下行束 descending tracts
下行竖管 falling main
下行水流 let-down stream
下行梯段回采法 underhand stoping
下行系统 downward system
下行线 down line;down track
下行线路 down line;down link;turndown wiring
下行效应 top-down effect
下行斜坡 aft slope
下行信息交流 downward communication
下行性的 descending
下行咽喉 down throat
下行烟道 downpass;downtake(chamber)
下行液流 descending flow of sap
下行液体 descending liquid
下行抑制径路 descending inhibitory path
下行抑制性网状投射 descending inhibitory reticular projection
下行抑制作用 descending inhibitory action
下行易化性网状投射 desending facilitory reticular projection
下行运量 downbound commerce; downbound traffic(volume);downbound transport
下行匝道 down ramp
下行展开法 descending developing method; descending development method
下行正线 down main track
下行制气 down-run
下行主管 <下水道> falling main
下型 mo(u)ld bottom half
下型锤 hardle
下型模 anvil swage
下型箱 drag;drag box;nowel
下型箱板 drag plate
下型芯头 drag point
下悬 bottom hung
下悬窗 bottom-hinged window; bottom hung window;hopper window
下悬的 underhung
下悬挂臂 lower suspension arm
下悬管 down pipe
下悬内开窗 <医院> hospital window
下悬喷管 boom downpipe;drop pipe
下悬起重机 underhung crane
下悬式搬运车 over-the-load carrier
下悬式车架 underslung car frame
下悬式窗 bottom hung window
下悬式的 underslung
下悬式电葫芦 undersling hoist; underslung hoist
下悬式起重机 underslung hoist
下悬式散热器 underslung radiator
下悬式输送机 underslung conveyer [conveyor]
下悬索的 sub-cabled
下悬索连续吊桥 sub-cabled continuous suspension bridge
下旋内开气窗 hopper vent
下旋钮 turned-down button
下雪 snow;snowfall
下雪的 snowy
下雪天 snowy weather
下压 underdraught
下压板 lower platen
下压模 dip mo(u)ld
下压式压机 inverted ram press
下压式造型机 mo(u)lding machine with down sand frame
下亚高山 lower subalpine
下延的 decurrent
下延火成岩体 subjacent igneous body
下延羽状的 decursively pinnate
下摇摆轮 lower wig-wag
下摇架 bottom cradle
下药 dosage
下曳 drag down
下曳根 pull-root
下曳力 downpull force
下曳气流 down current;down draft; down draught
下页 overleaf
下液管长度 length of down-comer
下液管高度 height of downcomer
下一层 lower layer
下一出口标志 next exit sign
下一代 coming generation; younger generation
下一代的 follow-on
下一动作 next move
下一记录 next record
下一进路开通 clear for next route
下一纳税年度 subsequent taxable year
下一卫星 <GPS 用语> next constellation
下一位 low order
下一位数 low-order digit
下一页 following page;next page
下一语句【计】next statement
下一指令寄存器 next-instruction register
下移 down shift;shift down
下移波痕 downslope ripple
下移河曲 sweeping meander
下移河湾 sweeping meander
下移距离 downstream moved distance;downward shift distance
下移曲流 sweeping meander
下移曲线段 sweeping meander
下移烧结炉 lowering furnace
下议院 lower house
下意识 subconsciousness
下意识的 subconscious
下溢 underflow
下溢中断 interrupt on underflow
下翼 bottom wing;lower wing;lower limb【地】
下翼片 lower panel
下翼缘 bottom flange;lower flange
下翼缘板 lower flange plate
下引航向 dropped outward pilot
下涌浪 down surge
下油管 running the tubing string into well
下油管记录 tubing record
下油盘 lower oil pan
下游 afterbay; lower course; lower river;under water
下游坝壳 <土石坝的> downstream shell
下游坝面 air face;downstream face
下游坝坡 downstream batter
下游坝体 downstream shell
下游坝趾 downstream toe(of dam)
下游坝趾棱体 downstream toe prism
下游边 downstream side;lower reach
下游侧 downstream side; down-draft side <漂沙的>
下游产品 <制造业> downstream products
下游承推墙 lower thrust wall
下游池 downstream bay
下游冲刷 downstream erosion
下游船闸 downstream lock
下游船闸的上游河段 lower head-water section
下游船闸进水口 lower lock entrance
下游船闸闸门 lower lock gate
下游导堤 lower(guide)wall
下游导墙 downstream guide wall
下游导向叶片 downstream guide vane
下游的 downriver;downstream;lower
下游底板 downstream floor
下游电站 downstream station;lower station
下游段 downstream course; low course;lower reach;lower river
下游断面 downstream cross section; downstream range
下游墩尖 downstream pier nosing
下游发电厂 downstream power plant
下游发电站 downstream power plant
下游法 downstream method
下游反向曲线石板 <保护桥墩用> cyma downstream slab
下游防冲铺砌 downstream apron
下游防护措施 downstream protection
下游防护槛 lower guard sill
下游管 downstream pipe;downstream tube
下游管道 sewer downstream
下游管线 down line;downstream line
下游过度段 downstream transition
下游海岸 downcoast
下游河床 downstream bottom
下游河道 lower-river course; lower stream course
下游河段 alluvial tract; downstream reach; downstream waters; lower course;lower river;lower stretch;

plain tract; tail bay; lower reach; lower basin; lower pond; lower pool <船闸的>; downstream water <堤坝的>; downstream bay <临近建筑物的>
下游河流 lower reach
下游河湾 aft-bay; afterbay
下游护岸 downstream protection; downstream revetment
下游护坡 downstream protection
下游护墙 lower guard wall
下游护坦 <有时包括海漫在内> downstream floor; downstream apron
下游混凝土防冲铺砌 downstream concrete apron
下游混凝土护坦 downstream concrete apron
下游渐变段 downstream transition
下游控制 downstream control
下游口门 <港池的> downstream entrance
下游立视图 downstream view
下游临时坝 downstream temporary dam
下游流域 low river basin
下游门扉 lower leaf
下游门叶 lower leaf
下游面 downstream side
下游面坡度 downstream batter; downstream slope
下游面坡脚 downstream toe
下游面应力 downstream stress
下游排水 downstream drainage
下游排水井 relief well
下游坡 downstream slope
下游坡脚 downstream toe of dam
下游区 downstream region
下游区段 downstream section
下游设备 downstream equipment
下游受益 downstream benefit
下游水 downstream water; tail water; under-stream water
下游水(边)线 downstream water line
下游水池 lower pool
下游水库 downstream reservoir; lower reservoir
下游水面 downstream surface
下游水深 <坝闸> tailwater depth
下游水深标尺 downstream ga(u)ge
下游水塘 lower pond
下游水位 downstream stage; downstream(water) level; level of tail water; lower pool elevation; low-water level; tailwater elevation; tailwater level
下游水文测验断面 downstream measuring section
下游水域 downstream waters; tail waters
下游水源 downflow water source
下游顺流限制 downstream limit
下游塘 afterbay; lower pool
下游梯级 downstream stage; lower development
下游调压池 lower surge basin
下游调压井 downstream surge chamber
下游调压室 downstream surge chamber
下游停泊区 downstream anchorage; downstream berthing area; downstream garage
下游围堰 downstream cofferdam
下游围堰工程地质剖面图 engineering geological profile downstream cofferdam
下游子围堰轴线工程地质剖面图 engineering geological section along axis of downstream cofferdam
下游尾水渠墙 tailrace wall
下游喂料 downstream feeding
下游物流 downstream
下游衔接段 downstream transitional section
下游向S形板 ogee downstream slab
下游楔形体 <土石坝的> downstream fill
下游泄水闸门 lower sluice gate
下游蓄水池 lower storage basin
下游压力 <液压元件> downstream pressure
下游堰 aft-bay
下游翼墙 downstream wing wall
下游引航道 lower approach; lower approach channel; lower lock approach
下游闸门 <船闸的> aftergate; aft gate; tail gate; lower(lock) gate
下游闸门凹壁 lower gate recess
下游闸门门槽 lower gate recess
下游闸门门扇 downstream leaf
下游闸室 downstream lock chamber
下游闸首门槛 downstream gate sill
下游最低水位 downstream minimum water level
下游最低通行水位 downstream lowest navigable water level
下游最高通行水位 downstream highest navigable water level
下有吊顶的楼板 double floor
下有吊顶的楼板饰面 double floor covering
下有吊顶天花板 double floor
下隅撑 lower angle brace
下雨的 rainy
下雨可能性 rainfall probability
下雨时间 rainy spell
下雨天 day of rain; day with rain; rainy day; wet day
下雨形成的小河 wet weather rill
下雨预兆 rain shadow
下元古代【地】Lower Proterozoic era
下缘 bottom edge; lower margin
下缘纤维 lower fiber[fibre]
下源的 anogene
下月 last quarter
下越流层 down leakage layer
下运河段 lower canal reach
下载 down load(ing)
下楂 undercut
下轧槽 bottom pass
下轧道 lower pass
下轧辊 bottom roll
下闸门 lower lock gate; tail gate
下闸室 lower chamber; bottom chamber
下闸首 downstream gate head; lower gate bay; lower head; tail bay
下闸首墩 lower gate block
下闸首人字闸门 lower miter[mitre] gate
下闸首翼墙 lower wing wall
下闸装置 gate lowering mechanism
下涨式火箱 extended fire box; wide fire box
下折式 downfolding
下阵雨 shower
下支承座 bottom pivot
下支架 subframe; undercarriage
下止点 bottom dead-center[centre]; bottom dead point; lower dead center[centre]; lower dead point
下止点附近汽缸壁的磨损 lower cylinder bore wear
下止点指示器 bottom dead-center[centre] indicator; lower dead-center [centre] indicator
下志留【地】Lower Silurian
下志留统【地】Lower Silurian series
下制动螺丝 lower clamp
下置车架底盘 underslung chassis
下置杠杆 under-lever
下置式 underneath type
下置式击锤 underhammer
下置式蜗杆传动 underslung worm drive
下置蜗杆 underslung worm
下置信区间 lower confidence interval
下置信限 lower confidence limit
下中级品 low middling
下中天 inferior transit; lower culmination; lower transit
下中天高度 altitude below pole; altitude of lower transit; meridian altitude below pole
下种 seeding; seed shedding
下轴 bottom spindle; lower shaft
下轴承盖 lower ball cover
下轴承架 lower bearing spider
下轴承箱 lower bearing housing
下轴承箱盖 lower bearing housing cap
下轴承支架 lower bracket
下轴承座 bottom chock
下轴瓦 lower bearing bush; lower brass; lower half
下侏罗纪【地】Lower Jurassic
下侏罗纪蓝黏[粘]土 blue lias clay
下侏罗统【地】Lower Jurassic series
下属机构 agency
下注 bottom pouring
下注的 bottom poured
下注空气(泡)接触曝气池 downflow bubble contact aerator
下柱列 lower colonnade
下铸(法) bottom casting; bottom pouring; uphill casting; uphill teeming; indirect casting
下转 downturn
下转后退闭锁机 rolling block
下桩 stabbing
下装 down(-line) load(ing); lower level loading
下坠 straining; tenesmus
下坠的 tenesmic
下浊点 lower cloud point
下子午圈 ante meridian; lower branch
下子午线 inferior meridian; lower branch of the meridian
下子座 hypostroma
下纵标集 lower ordinate set
下走 <磁性单元由于连续部分驱动脉冲的作用而磁化强度降低> walk down
下钻 going down; going in; go into the hole; runback into the hole; running-in
下钻杆速度 sinking velocity
下钻杆(下套管)时间 make-up time
下钻杆遇阻 slacking off
下钻管 stabbing
下钻速度 running speed
下钻所需时间 time of running in
下作用水阀 under lever faucet
下作用洗手塞门 under lever bibcock
下坐【船】squat
下座圈 lower race ring
下座套 lower bushing
下座椅连杆 lower seat connecting arm

夏半年 summer half year
夏半球 summer hemisphere
夏贝纤维 Sharpey's fibre
夏布 grass cloth; grass linen
夏材 late wood; summer tree; summer wood
夏潮 summer tide
夏初 early summer
夏初或春末 early summer or late spring
夏堤 summer dike[dyke]; summer levee
夏飞尔安全玻璃 Chauvel safety glass
夏干区 summer dry region
夏干温暖气候 warm climate with dry summer
夏港 summer port
夏耕休闲 clean summer fallow
夏宫 summer palace
夏灌 summer irrigation
夏灌冬用 recharge in summer and utilizing in winter
夏湖 <月球的> lacus aestatis
夏季 <印度6~10月> Kharif
夏季(白)油 summer white oil
夏季插 summer cutting
夏季车轴油 summer axle oil
夏季吃水线 summer draft
夏季厨房 summer kitchen
夏季淡水满载水线 freshwater load in summer
夏季的 aestival
夏季的冷负荷 summer cooling load
夏季等温线 isotheral
夏季度时 summer degree hour
夏季堆积 summer accumulation
夏季放牧 summer pasturing
夏季服务 summer service
夏季浮标装置 summer buoyancy
夏季浮游生物 summer plankton
夏季干舷 summer freeboard
夏季干舷标志 summer freeboard mark
夏季海滨阶地 summer berm
夏季洪水 summer flood
夏季洪水期 summer flood season
夏季黄油 summer yellow oil
夏季混凝土施工 concreting in hot weather
夏季级 summer grade
夏季级汽油 summer grade gasoline
夏季季风 summer monsoon
夏季季风海流 summer monsoon current
夏季剪修 summer-pruning
夏季交通 summer traffic
夏季浇筑的混凝土 hot-weather concreting; summer-placed concrete
夏季径流预报 summer runoff forecast
夏季绝对休闲 clean summer fallow
夏季空气调节 summer air-conditioning
夏季空气调节室外计算日平均温度 outdoor design mean daily temperature for summer air conditioning
夏季空调 summer air-conditioning
夏季空调室外计算干球温度 outdoor design dry-bulb temperature for summer air-conditioning
夏季空调室外计算湿球温度 outdoor design wet-bulb temperature for summer air-conditioning
夏季空调室外计算逐时温度 outdoor design hourly temperature for summer air-conditioning
夏季空调系统 summer air-conditioning system
夏季枯水型(河流) Sequanian type
夏季垃圾 summer refuse
夏季冷却容量 summer capacity
夏季路 summer road
夏季落叶 summer leaf-drop
夏季满载吃水线 summer load(water) line
夏季漫水冬季干燥的土地 turlough

夏季木材 summer wood
夏季木材载重线 summer timber load line
夏季牧场 summer range
夏季喷药 summer spray
夏季平均日交通量 summer average daily traffic
夏季平均水位 mean summer level
夏季汽油 summer gasoline
夏季区带 summer area;summer zone
夏季容许最大吃水 maximum allowance summer draught
夏季润滑 summer lubrication
夏季施工 hot-weather construction
夏季时间 <夏季把钟点拨快一小时的制度> daylight saving time;summer time
夏季市场回稳 summer rally
夏季水边低地 summer polder
夏季水面 summer water level
夏季水位 summer water level
夏季停滞 summer stagnation
夏季停滞期 summer stagnation period
夏季通风室外计算温度 outdoor design temperature for summer ventilation
夏季通风室外计算相对温度 outdoor design relative humidity for summer ventilation
夏季驼峰【铁】low hump;summer hump
夏季限制峰高 limited hump height at summer
夏季相 aestival aspect
夏季型吃水 summer mo(u)lded draft
夏季休耕 summer fallow
夏季休闲地 summer fallow;summer fallow field
夏季修整 summer patching
夏季蓄水库 summer reservoir
夏季一年生植物 aestival annual plant
夏季用柴油 summer diesel oil
夏季用齿轮油 summer compound oil
夏季用润滑油 summer lubricating oil
夏季用润滑油品级 summer grade
夏季用石油产品 summer grade
夏季油 summer oil
夏季油舱 summer tank
夏季运行 summer service
夏季运行图和时刻表 summer train working diagram and timetable
夏季载重水线吃水 summer draft
夏季载重线 summer load(water)line
夏季载重线标志 summer draft mark
夏季植物 summer annual
夏季住宅 summer residence
夏季装载线 summer load(water)line
夏季载重线 summer load(ing)line
夏枯草属 self-heal
夏令电能 summer electric(al)energy
夏令电影院 summer motion picture theater[theatre]
夏令供水龙头 summer hydrant
夏令给水龙头 summer hydrant
夏令剧场 summer theater[theatre]
夏令剧院 straw hat theater;summer theater[theatre]
夏令垃圾 summer refuse
夏令时 daylight-saving
夏令时间 <夏季把钟点拨快一小时的制度> summer time;daylight saving time
夏令时中午 summer noon
夏令停滞 summer time stagnation
夏令载货吃水线 summer loading;summer on loadline
夏绿灌木林 aestifruticeta
夏绿灌木群落 aestifruticeta
夏绿木本群落 aestilignosa
夏绿乔木林 aestatisilvae;aestisilvae
夏绿乔木群落 aestisilvae
夏绿硬叶林 aestidurilignosa;aestilignosa
夏绿硬叶木本群落 aestidurilignosa
夏马风 Shamal
夏眠 aestivate
夏木 summer wood
夏木材 late wood
夏生植物 therophyte
夏湿区 summer wet region
夏时制 daylight saving time
夏氏冲击试验机 resilience testing machine
夏氏塔冷杉 Shasta fir
夏天的 aestival
夏威夷海岭 Hawaiian ridge
夏威夷群岛 Hawaiian Islands
夏威夷式火山活动 Hawaiian activity
夏威夷式喷发 Hawaiian type eruption
夏雾 summer fog
夏西统【地】Chazyan
夏汛 summer flood;summer season
夏汛期 summer flood season
夏用黑色润滑油 summer black oil
夏用汽油 summer petrol
夏用润滑液面高度 summer oil level
夏用油 summer oil
夏用(油)品级 summer grade
夏蛰【动】aestivate
夏至 June solstice;midsummer
夏至潮 solstitial tide
夏至点 first point of Cancer;summer solstice
夏至日 summer solstice
夏至线 the Tropic(al)of Cancer

罅

罅隙 breach;chink;cranny

罅釉 <陶瓷> crazing glaze;glaze crazing

仙

仙后(星)座【天】Cassiopeia

仙境 wonderland
仙人球 cactus
仙人掌 cactus;Indian-fig;nopal
仙人掌式抓斗 cactus grab
仙人砖 figurine
仙人走兽【建】ridge ornamentation of celestial-being and beasts
仙食菌 ambrocia fungus

先

先报 previous notice

先报次序 <遍历二叉树的方法之一> preorder
先辈过滤器 ancestry filter
先闭后断换向开关接点 make-before-break changeover switch contact
先闭后开 make-before-break
先闭后开触点 make-before-break contact
先潮 <日潮先于月潮> acceleration of tide;priming of tide
先成的 antecedent
先成地台 antecedent platform
先成谷 antecedent valley
先成河 antecedent drainage;antecedent river;antecedent stream
先成晶体 phantom crystal
先成水系 antecedent drainage
先成顺向河 antecedent consequent river
先成峡 antecedent gorge
先成现象 prothetely
先成学说 preformation theory
先成作用 antecedention
先冲 subpunch(ing)
先冲后淤(河床)cut-and-fill
先储存后发送制 store-and-forward basis
先存断裂网络 previous fracture network
先存组构 preexisting fabric
先打一根短桩 spud pile
先贷后存的利差 <英> round-tripping
先导波 preceding wave
先导操纵的 pilot-operated
先导操作阀 pilot-operated valve
先导产业 forerunner industry;leading industry
先导车标志 pilot car sign;pilot ear sign
先导冲突理论 theory of antecedent conflicts
先导的 pilot
先导阀 pilot valve
先导阀控制的阀 pilot-actuated valve
先导工作面 leading surface
先导管道 pilot line
先导管路 pilot line
先导化合物 lead compound
先导回摆液压缸 pilot swing cylinder
先导级 pilot stage
先导价格 price leader
先导价格制 price leadership
先导孔 pilot hole
先导控制 pilot-operated control valve
先导控制阀 piloted control;pilot-operated control
先导控制压气系统 pilot-operated air system
先导阀的先导管道 pilot valve pilot line
先导流量 pilot flow
先导锚固 pilot anchor
先导面 leading surface
先导模型 pilot model
先导气流 pilot air
先导闪击 leader stroke
先导闪流 pilot streamer
先导升降液压缸 pilot lift cylinder
先导式压力控制阀 pilot-operated pressure control
先导伺服阀 pilot valve
先导系统 pilot system
先导压 pilot pressure
先导压头 pilot ram
先导液压缸 pilot cylinder
先导油路 pilot circuit;piloting
先导指标数字 leading indicator
先导桩 pilot pile;spud pile
先岛群岛隆起地带 Sakishima Shoto uplift region
先到货物 forward arrivals
先到期先做规则 first in system-first served rule
先到先办制 first come first-served system
先到先服务 first-in first-out
先到先做原则 principle of first-in first-served
先到者先接受服务原则 first come first-served
先电离 preionization
先定变量 predetermined variable
先动触点 early contact
先断后合接点 break-before-make contact
先断后接式 break make system
先发言权 <一方的律师在法庭上的> preaudience
先发样品 advance sample
先发制人 forestall
先放粗料后灌砂浆 intrusion grouting
先放后核制 valuation after release system
先锋 pioneer;vanguard
先锋阶段 pioneer stage
先锋群落 initial community;pioneer community;prodophytium
先锋生物 pioneer
先锋植物 pioneer plant
先锋作物 pioneering crop
先付 cash in advance;make advance
先付定金 layaway plan
先付费用 front-end fee
先付规则 pay first rule
先付后接制公用电话站 pay station;prepay station
先付货款 cash with order
先付票据通知 advice of bill accepted
先付运费 prior carriage charges
先拱后墙法 archaic[arched] roof in advance of wall tunnel(l)ing method;Belgian method;flying arch method;inverted lining method
先拱后墙施工法 upside-down construction
先共晶奥氏体 pro-eutectic austenite
先共晶的 proeutectic
先共晶渗碳体 pro-eutectic cementite
先共析 pro-eutectoid
先共析渗碳体 pro-eutectoid cementite
先共析体 pro-eutectoid
先共析铁素体 pro-eutectoid ferrite
先光期 prephoto-phase
先寒武【地】Pre-Cambrian
先寒武系的 Infracambrian
先合后断接点 close-before-open contact
先合后断开关 make-before-break switch
先合后离接点 make-before-break contact
先后次序 precedence[precedency]
先后关系 precedence relationship
先后顺序表 precedence table
先后张拉混凝土 <预应力的> pre-post-tensioned concrete
先后张拉结构 <预应力的> prepost-tensioned construction
先后张拉结合法 prepost tensioning
先加工面 first side
先加速 pre-acceleration
先接后断 make-before-break
先接后断触点 make-before-break contact
先接后付制公用电话站 post-pay station
先接后离 make-before-break
先接后离触簧 make-before-break contact spring
先结束先送(出) <文件等的> first-ended first-out
先进 advancement
先进安全车辆 advanced safe vehicle
先进玻璃 advanced glass
先进材料 advanced material
先进船舶操纵控制系统 advanced ship's operation control system
先进导坑 pilot drift;pilot hole;pilot-tunnel heading
先进道路交通信息技术 advanced road transport telematics
先进的车辆控制系统 advanced vehicle control system
先进的回收处理 advanced reclamation treatment
先进的驾驶员信息系统 advanced driver information system
先进的交通管理系统 advanced traffic management system
先进的交通管制系统 progressive sys-

tem of traffic control
先进的旅行者信息服务 advanced traveler information services
先进的旅行者信息系统 advanced traveler information system
先进的汽车控制系统 advanced vehicle control system
先进的设计公式 upgraded design formula
先进的停车管理系统 advanced parking management system
先进的污水处理 advanced sewage treatment
先进的无人调查系统 advanced unmanned search system
先进的乡村交通系统 advanced rural transport system
先进的逸出气体控制设备 advanced air emission control device
先进地面交通 advanced ground transport
先进地面运输 advanced ground transport
先进定额 advanced norm
先进房屋 advanced house
先进废水处理技术 advanced water technology
先进复合材料 advanced composite material
先进个人或集体 pacesetter
先进更新法 advanced reproduction
先进工程塑料 advanced engineering plastics
先进工业本底污染浓度 pre-industrial background concentration
先进工艺 advanced process
先进工作法 advanced ways of working
先进工作方法 advanced method of working
先进工作者 advanced worker
先进公共交通系统 advanced public transportation system
先进构思列车 <美> advanced concept train
先进固体逻辑技术 advanced solid logic technology
先进管理系统 advanced administrative system
先进光导摄像管摄像系统 advanced vidicon camera system
先进国家 advanced country
先进后出 first-in last-out
先进后出交易 back-to-back escrow
先进环境控制系统 advanced environment(al) control system
先进计算处理机 advanced computational processor
先进计算环境 advanced computing environment
先进记录系统 advanced record system
先进技术 advanced technique; advanced technology

先进技术电路模块 advanced technique circuit module
先进技术企业 enterprise with advance technology
先进技术卫星 advanced technology satellite
先进技术转让 transfer of advanced technology
先进建筑 advanced architecture; progressive architecture
先进交互调试系统 advanced interactive debugging system
先进交通管理系统 progressive system of traffic control
先进交通运输信息技术 advanced transport telematics
先进经验 advanced experience
先进科学计算机 advanced scientific computer
先进雷达航站系统 advanced radar terminal system
先进列车控制系统 <美> advanced train control system
先进列车控制与安全系统 advanced train control safety system
先进留库盘存法 first in still here
先进配置和电源管理界面 advanced configuration and power interface
先进平均定额 norm of advanced average
先进平均数 advanced average
先进企业的劳动生产率 best-practice labo(u)r productivity
先进气冷反应堆 advanced gas-cooled reactor
先进汽车交通信息与通信[讯]系统 advanced mobile traffic information and communication system
先进汽车信息系统 advanced mobile information system
先进潜水系统 advanced diving system
先进设备 advanced equipment; advanced plant; sophisticated equipment
先进设计 advanced design
先进甚高频通信[讯]系统 advanced VHF[very high frequency] communication system
先进数据库系统 advanced data base system
先进水平 advanced stage
先进司机信息系统 advanced driver information system
先进陶瓷 advanced ceramics; new ceramics
先进铁路电子系统 <美> advanced radar traffic control system
先进网络系统体系 advanced network system architecture
先进污水处理(法) advanced waste (water) treatment
先进先出 <等待处理方法> first-in first-out; push-up
先进先出法 first-in first-out method
先进先出栈 first-in first-out stack
先进先服务 first come first-service
先进型 advanced type
先进型电气化高速旅客列车 <英> advanced passenger train
先进仪器 advanced instrument
先进用户信息系统 advanced user information system
先进转换堆 advanced converter
先据者 predecessor
先决论 predeterminism
先决条件 condition (of) precedent; postulate; precondition; prerequisite; prior condition
先决问题 preliminary question
先开顶部 cut straps first and remove top
先开阀 jockey valve; pilot
先开后合 break-before-make
先开后合触点 break-before-make contact
先开后合接点 break-before-make contact
先开汇票 advance bill
先开挖隧道上半部 top heading
先客后货 high-speed train prior than low speed train; passenger train prior than goods train
先来先服务 first come first-served; first come first-service
先例的 antecedent; precedent(ial)
先例约束力的原则 rule of precedent
先令 <英国银币> shilling
先令舵 schilling rudder
先罗马式建筑形式 Carolingian architecture
先买的 preemptive
先买后付 buy-now-later
先买权 preemption; preemptive right
先慢后快 positively acceleration
先铺法 prebedding
先期报告 advance report
先期材料申请 advance material request
先期点火 premature ignition
先期固结 preconsolidation; previous consolidation
先期固结土 over-consolidated soil; preconsolidated soil
先期固结压力 preconsolidation pressure
先期加荷 prior loading
先期胶凝 premature gelation
先期教会 previous church
先期教堂 previous church
先期结构 <预应力混凝土旋工过程中转换体系前的结构> structure at initial stage; initial structure
先期违约 anticipatory breach
先期蓄水 preimpoundment
先期压密土 preconsolidated soil
先期应力-应变历史 prior stress-strain history
先启门扇 active door
先砌拱后砌墙法 inverted lining
先前报价和日期 previous quotation and date
先前的 antecedent
先前河 <澳大利亚用词,指现在河以前,古河以后的水系> prior River
先前输出态 previous output state
先前输入态 previous input state
先前吸取的收入 precollected revenue
先前值 preceding value
先遣人员 advance agent
先墙后拱衬砌法 ordinary lining process (from bottom to top)
先墙后拱法 wall in advance of arched roof tunneling method
先驱 pioneer; precursor; vanguard
先驱产品 pioneer product
先驱工业 pioneer industry
先驱企业 pioneer enterprise
先驱群体 pioneer population
先驱物 precursor
先驱者 forerunner
先驱植物 pioneer plant
先驱种 pioneer species
先趋网络图 <计划协调技术的一种方法> precedence network
先趋涌 ahead swell
先取 preempt
先取权 law of priority
先入先出表 push-up list
先入先出单 push-up list
先入之见 preconceived idea; preconception
先生产后基建 production comes before capital construction
先蚀再淤阶地 fillstrath terrace
先试验 pilot test
先收敛再发散 come-and-go; convergence-and-divergence
先受偿债权人 senior creditor
先天禀赋 native endowment
先天的 connate; inherent
先天畸形 congenital malformation
先天倾向 congenital disposition
先天释放机制 innate releasing mechanism
先天行为 innate behavio(u)r
先贴好瓷砖的隔墙 pre-tiled partition
先头波 preceding wave
先头舰 flotilla leader
先挖后填式地下铁道 dug and covered underground railway
先挖后填式地下铁路 dug and covered underground railway
先挖后支法 postponed installation of tunnel support
先污染后治理说 theory of treatment after pollution
先写 preempt
先写后记录 log-write-ahead
先卸批准单 request note
先行 precession
先行部分 undercarriage
先行策略 look-ahead strategy
先行车 preceding vehicle
先行处理机 preprocessor
先行传送 anticipatory staging
先行的 preceding
先行定址 one-ahead addressing
先行方式 anticipation mode
先行费 front money
先行工程 pioneering work
先行工作 prerequisite activity
先行规则 antecedent rule
先行缓冲 anticipatory buffering
先行活动 prior activity
先行获取 advance acquisition
先行加法器 look-ahead adder
先行进位发生器 look-ahead carry generator
先行进位(法) carry lookahead
先行进位方式 look-ahead system
先行进位加法器 anticipated carry adder; carry look ahead adder
先行进位输出 look-ahead carry output
先行经济指标 leading indicator
先行经济指标综合指数 composite index of lagging indicators
先行经济指数 anticipated economic index
先行矩阵 predecessor matrix
先行控制 advanced control; lock-ahead control
先行控制部件【计】lock-ahead unit
先行块 predecessor block
先行列车 leading train; preceding train
先行零 leading zero
先行脉冲 precursor pulse
先行目标 antecedent goal
先行请求 anticipated request
先行申报制度 prelodgement system
先行试销 first put on trial sale
先行试钻 pilot boring
先行调页 anticipatory paging
先行跳跃进位 look-ahead skip
先行脱扣器 preferential trip
先行下沉 preconsolidation settlement
先行形式 anticipation mode
先行性行业 pioneer industry
先行性行业减税 relief for pioneer industry
先行研究 prior study
先行者 forerunner
先行指数 leading indicator
先行装置 look-ahead facility
先验标准差 a priori standard deviation
先验的 a priori <拉>; transcendental
先验分布 a priori distribution
先验概率 a priori probability; theoretic(al) probability
先验估计 a priori estimate
先验估值 a priori estimate
先验后验分析 a prior posterior analysis

先验结构模型 a prior structural model
先验精度分析 a prior accuracy analysis
先验均方根差 a priori standard deviation
先验论 apriorism;metempiricism;metempirics
先验模型 a prior model
先验权 a prior weight
先验权中误差 mean error of prior weight
先验误差界 a priori error bound
先验信息 apriori information;a prior information
先验值 a prior value
先验主义 metempiricism;metempirics
先在断层 preexisting fault
先在破裂岩 prefractured rock
先在山带 preexisting mountain belts
先张的 pretensioned;pretensioning
先张法 pretensioned system; pretensioning method; pretensioning process;pretensioning type
先张法长线台座 long bed for pretensioning
先张法池座 pretensioning bed
先张法和后张法综合并用的预应力技术 pre/post-tensioning
先张法混凝土 pretensioned concrete
先张法混凝土的长线法 long line method of pretensioning concrete
先张法台座 pretensioning bed
先张法预(加)应力 pretensioned prestress(ing)
先张法预应力构件 pretensioned member
先张法预应力混凝土 pretensioned concrete; pretensioned prestress(ing) concrete
先张法预应力混凝土管桩 pretensioned spun concrete pile
先张法预应力混凝土梁 pretensioned prestressed concrete beam
先张法预应力台座 pretensioned bed; pretensioning bench
先张法预应力桩 pretensioning prestressed pile
先张法张拉台座 pretensioning bed
先张钢丝 pretensioned wire
先张钢丝束 pretensioned tendon
先张管道 < 即先张法预应力混凝土管道 > pretensioned pipe
先张后张法 prepost tensioning
先张混凝土构件 pretensioning concrete member
先张技术 pretensioning
先张拉 prestretching;pretension
先张拉的钢绞索 pretensioning strand
先张拉钢丝 pretensioned wire
先张拉钢丝束 pretensioned tendon
先张拉结构 pretensioned construction
先张拉设备 pretensioned system
先张拉线束 pretensioned tendon
先张梁 pretensioned beam
先张预应力构件 pretensioned member
先张预应力混凝土 pretensioned concrete
先张预应力混凝土板 Hoyer slab
先张预应力混凝土空心楼板的填料 Hoyer hollow floor filler
先张预应力梁 pretensioned beam
先张预应力砼 pretensioned concrete
先张预应力桩 pretensioned prestressed pile
先兆 forerunner; monition; premonition;premonitory
先兆的 aural
先兆预感 aura[复 aurae/auras]
先支付部分贷款 < 分期付款购货 > pay down
先质 precursor
先组式 precoordination
先组式标引法 precoordinate indexing system
先组式索引 precoordinate index
先钻 < 钻孔前 > subdrilling
先钻后扩法 reaming-after-boring method

纤

钡锂石 balipholite

纤铋铀矿 uranosphaerite
纤船的机车 shore-bound locomotive
纤道 strip(e) road; tow(ing) path; track road;trackway
纤度 titer[titre]
纤钒钙石 fernophrastite
纤方解石 tartufite
纤沸石 gonnardite
纤工 boat tracker;tracker
纤硅钙石 hydrowollastonite; riversideite
纤硅锆钠石 elpidite
纤硅碱钙石 rhodesite
纤硅铜矿 plancheite
纤黑蛭石 bardolite
纤滑石 asbestine;beaconite
纤钾明矾 potash alum
纤居 nanocurie
纤克 nanogram
纤孔菌属 < 拉 > Inonotus
纤拉的船 track boat
纤蜡石 neurolite
纤磷钙铝石 crandallite
纤磷铝石 vashegyite
纤磷铝铀矿 ranunculite
纤磷石 phoshofibrite
纤磷铁矿 kerchenite
纤硫铋铅矿 goongarrite
纤硫锑铅矿 robinsonite
纤绿铜锌矿 zincrosasite
纤毛 cilia
纤毛虫 ciliate
纤毛原虫 ciliate
纤镁柱石 magnesiocarpholite
纤锰闪石 carpholite
纤锰柱石 carpholite
纤米 nanometer
纤秒 nanosecond
纤钠海泡石 loughlinite
纤钠明矾 mendozite
纤钠铁矾 sideronatrite;urusite
纤硼钙石 bakerite
纤硼镁石 ascharite
纤硼石 stassfurtite
纤弱的 slender
纤闪黑玢岩 uralitophyre
纤闪辉绿岩 ophite;uralite-diabase
纤闪石 byssolite;uralite
纤闪石化【地】uralitization
纤蛇纹石 chrysolite;chrysotilite
纤砷钙铝石 arsenocrandallite
纤砷钴镍矿 forbesite
纤砷铁锌石 ojuelaite
纤绳 cordelle;tow line;tracking rope
纤石膏 fibrous gypsum
纤手 tracker
纤水滑石 nemalite
纤水磷铍石 moraesite
纤水绿矾 glockerite;pitticite
纤水镁石 nemalite
纤水碳镁石 artinite
纤丝 fibril;fibrilla[复 fibrillae]
纤丝的 fibrillar;fibrillary
纤丝角(度) fibril angle
纤碳铀矿 rutherfordine
纤铁矾 fibroferrite;siderotil
纤铁矿 lepidocrocite
纤铁矿岩 lepidocrocite rock
纤铁闪石 montasite;rhodusite
纤网补偿装置 fiber[fibre] web compensation device
纤维 fiber[fibre];staple
纤维按长度分级 stapling
纤维百分比 fiber percentage
纤维板 beaver board; cellulose fiber[fibre] tile; fiber[fibre](building) board; fiber[fibre] sheet; fiberwood;fiberboard[fibreboard];fibreboard panel;fibrolite;fibrous slab; pressed-fiber[fibre] board; presswood board; softboard; celotex board < 木质纤维毡压制的 >
纤维板厂废水 fabric board mill wastewater;fiber[fibre] board mill wastewater
纤维板吊顶 fiber[fibre] board ceiling
纤维板集装箱 fibreboard container
纤维板模具 masonite die
纤维板平顶 fiber[fibre] board ceiling
纤维板强度试验机 tester for fiberboard strength
纤维板贴面【建】 fiberboard [fibreboard] finishing;textile finishing
纤维板桶 fibercan
纤维板箱 fiber(board)[fibre(board)] box;fiberboard[fibreboard] case;fiber[fibre] cam
纤维板箱包装 fiber[fibre] box package
纤维板箱罐 fiberboard[fibreboard] can
纤维板有网纹的背面 screen side
纤维包裹 fibrous encapsulation
纤维包膜 fibrous capsule
纤维饱和点 < 木材的 > fiber[fibre] saturation point; fiber[fibre] saturated point
纤维饱和度 fiber[fibre] saturated level
纤维保温材料 fiber[fibre] insulating material
纤维保温层 fiber[fibre] insulation layer
纤维保温体 fibrous insulator
纤维爆裂 burst
纤维背衬 fiber[fibre] backing
纤维壁板 fiber[fibre] building board
纤维壁厚 fiber[fibre] wall thickness
纤维编包 fibrous braiding
纤维编织 fibrage
纤维编织保护层 braiding
纤维变晶结构 fibrous blastic texture; nematoblastic texture
纤维变晶状的 fibroblastic
纤维变性 fibrosis; fibrous degeneration
纤维辫 fiber[fibre] strand
纤维表示 fibring
纤维波导管 fiber[fibre] guide
纤维玻璃 fiberglass;fibrous glass
纤维玻璃池 fibrous tank
纤维玻璃钢瓦 fiberglass tile
纤维玻璃增强塑料 fibreglass reinforced plastics
纤维薄板绝缘垫片 fiber[fibre] insulating spacer
纤维薄壁 fiber[fibre] membrane
纤维补强材料 fiber-reinforced materials
纤维补强的 fiber-reinforced
纤维材料 fiber[fibre] material;fibrous material
纤维材料齿轮 fiber[fibre] gear
纤维材料过滤器 fibrous filter
纤维彩色图像器 fiber[fibre] scope
纤维层 fibrage; fiber[fibre] mat; fibrous coat; fibrous layer; fibrous stratum
纤维层的 fibrolaminar
纤维层过滤除尘器 fibrous mat dust filter
纤维层内纤维 fibers in the fibrous layers
纤维层曲 fiber[fibre] buckling
纤维缠结 fiber[fibre] entanglement; fiber[fibre] matting
纤维缠结作用 fiber[fibre] entanglement
纤维缠绕法玻纤增强热固性树脂管 filament-wound glass fiber[fibre] reinforced thermosetting resin pipe
纤维长度 fiber[fibre] length
纤维长度变化 fiber[fibre] length variation
纤维长度测试仪 fiber[fibre] sorter
纤维长度的控制 fiber[fibre] length control
纤维长度分布图 fiber[fibre] diagram
纤维长度分配 fiber[fibre] length distribution
纤维长度分析仪 fiber[fibre] diagram machine
纤维长度排列 fiber[fibre] length array
纤维长度试验仪 fiber[fibre] length tester
纤维长间距 fibrous long spacing
纤维唱针 fiber[fibre] needle
纤维超滤机 Polcon Varge superfilter
纤维沉降区 fiber[fibre] collecting zone
纤维沉降室 fiber[fibre] collection hood;fiber[fibre] laydown hood
纤维衬底 fiber[fibre] base
纤维衬垫 < 石棉或软木衬垫 > fiber[fibre] packing; fiber-reinforced washer;fiber[fibre] gasket
纤维衬套 fiber[fibre] bush
纤维成分 fibrated composition; fibrous composition
纤维成型 fiber[fibre] forming
纤维赤铁矿 fibrous red iron ore
纤维冲击强力试验仪 fiber[fibre] impact tester
纤维抽出力 fiber[fibre] withdrawal force
纤维处理剂 < 润滑抗静电 > fiber[fibre] finish
纤维吹制法 fiber[fibre] blowing
纤维丛 fiber[fibre] bundle
纤维丛的约化 reduction of fiber[fibre] bundle
纤维丛理论【数】fiber[fibre] bundle
纤维粗度 fiber[fibre] coarseness
纤维带 cellulose tape;fabric belt;fibrous zone;Scotch tape
纤维导管 fiber[fibre] duct
纤维导管模具 fiber[fibre] duct
纤维导光灯 fiber[fibre] lamp
纤维的多孔性 fiber[fibre] porosity
纤维等同周期 fiber[fibre] identity period
纤维底层 fiber-reinforced underlay; fiber[fibre] underlay
纤维地毯 fiber[fibre] rug
纤维电池 fiber[fibre] cell
纤维垫板 fiber[fibre] building mat;fiber[fibre] washer
纤维垫过滤器 fibrous mat filter
纤维垫块 fiber[fibre] block
纤维垫片 fiber[fibre] washer
纤维垫圈 fiber-reinforced washer;fiber[fibre] washer;fibrous composition washer
纤维定量喂给控制装置 fibremeter
纤维定向 fiber[fibre] orientation
纤维定形机 fiber[fibre] setter

纤维端 fiber[fibre] end
纤维端密度 fiber[fibre] end density
纤维断口铁 fibrous iron
纤维断裂 fiber[fibre] breakage
纤维二糖 cellobiose
纤维方向 machine direction
纤维防潮的 fibrated dampproofing
纤维废水 fibrous waste
纤维沸石 fibrous zeolite
纤维分布 fiber[fibre] distribution
纤维分解 cellulosis
纤维分解化学制浆机 defibrator-chemipulp
纤维分离 defibering;defibration
纤维分离的木材 defibrated wood
纤维分离度 fiber[fibre] separation index
纤维分离机 defibering machine;defibrator;fluffer
纤维分析 fiber[fibre] analysis
纤维分选器 fiber[fibre] separator
纤维酚醛树脂 fibrous bakelite
纤维复合材料 fiber[fibre] composite;fibrous composite(material)
纤维复合层压材料 fiber[fibre] composite laminate
纤维改性 fiber[fibre] modification
纤维钙化 fibrocalcification
纤维钙化的 fibrocalcific
纤维钙石凝胶 tobermorite gel
纤维干涉图 fiber[fibre] interference figure
纤维干湿饱和点 <木材的> fiber[fibre] saturation point
纤维隔离层 fibrous insulation
纤维隔热层 fibrous insulation
纤维隔声板 acoustic(al) celotex(tile)
纤维隔音板 acoustic(al) celotex board;acoustic(al) celotex(tile);acoustic(al) fiber[fibre] board
纤维根 fiber[fibre] root
纤维构造 fiber[fibre] structure;fibrous structure
纤维管(道) fiber[fibre] conduit;fiber[fibre] pipe;fiber[fibre] tube;fiber[fibre] conduit; fiber-reinforced wiring conduit; fiber-reinforced pipe[tube]
纤维罐 fiber[fibre] can
纤维光笔 fiber[fibre] pen
纤维光导 fiber[fibre] light guide
纤维光点 fiber[fibre] optic(al) dot
纤维光电子学 fiberoptronics
纤维光块 fiber[fibre] optic(al) block
纤维光缆 fiber[fibre] optic(al) cable
纤维光腔 optic(al) fiber[fibre] cavity
纤维光束记录器 fibrooptic(al) recorder
纤维光学 fiber[fibre] optics;photonics
纤维光学编码-解码装置 fiber[fibre] optics image-encoding-de-coding device
纤维光学编码器 fiber[fibre] optics code apparatus
纤维光学传感器 fiber[fibre] optic(al) sensor
纤维光学放大器 fiber[fibre] optics amplifier
纤维光学光导管 fiber[fibre] optic(al) catheter
纤维光学集束 fiber[fibre] optic(al) array
纤维光学记录示波管 fiber[fibre] optic(al) oscilloscope recording tube
纤维光学技术 fiber[fibre] optic(al) technique
纤维光学监控 fiber[fibre] optic(al) monitoring
纤维光学镜 fiber[fibre] scope
纤维光学面板 fiber[fibre] optics faceplate
纤维光学耦合 fiber[fibre] optics coupling
纤维光学屏面管 fiber[fibre] optic(al) faced tube
纤维光学屏幕 fiber[fibre] optic(al) screen
纤维光学扫描变换器 fiber[fibre] optic(al) scan converter
纤维光学扫描器 fiber[fibre] optic(al) scanner
纤维光学摄影机 fiber[fibre] optics camera
纤维光学图像传递装置 fiber[fibre] optic(al) image transfer device
纤维光学元件 fiber[fibre] optic(al) member
纤维光学增强器 fiber[fibre] optics intensifier
纤维光学照明器 fiber[fibre] optic(al) illuminator
纤维光学照样机 fiber[fibre] optics camera
纤维光学装置 fiber[fibre] optic(al) device
纤维过滤器 fabric filter;fiber(-pad) filter;fibrous filter
纤维合成层板 fiber[fibre] covered plywood
纤维合成物 fibrated compound;fibrous compound
纤维和金属衬里板式 fabric and metal lined plate type
纤维褐煤 board coal
纤维花岗变晶结构 fibrous granoblarblastic texture
纤维滑动轴承 fiber[fibre] glide bearing
纤维滑溜性 fiber[fibre] slippage
纤维滑石 agalite
纤维化 fibrosis
纤维化的 fibrated
纤维化沥青乳液 fibrated asphalt emulsion
纤维化石棉 fiberized asbestos
纤维环 annuli fibrosi;fiber[fibre] ring;fibrous ring
纤维灰浆 fibrous plaster;staff
纤维灰浆花饰 staff floriation
纤维灰浆磨面 fibrous plastering
纤维灰浆抹灰工 fibrous plasterer
纤维灰浆抹面 fibrous plastering
纤维灰浆抹面工作 stick and rag work
纤维灰泥 fibred plaster;fibrous plaster
纤维灰泥浇铸 fibrous plaster cast
纤维灰泥浇铸成品 fibrous plaster molding
纤维灰泥模型 fibrous plastering mo(u)ld
纤维灰泥模子 fibrous plastering mo(u)ld
纤维回潮测试仪 fiber[fibre] moisture regain tester
纤维回收 fiber[fibre] recovery;fibrous recovery
纤维回收机 fiber[fibre] recovery machine
纤维混凝土 fiber-reinforced concrete
纤维混凝土板 fibrated concrete slab
纤维混凝土管 fibrated concrete tube
纤维混凝土制品 fibrated concrete product
纤维活性染料 fiber[fibre] reactive dye
纤维火山渣 thread-lace scoria
纤维级 fiber[fibre] grade
纤维计算直径 calculated diameter of fiber
纤维加筋结构 fibre-reinforced structure
纤维加筋塑料 fibre-reinforced plastics
纤维加气混凝土 faircrete[fair-air concrete]; fiber[fibre] air-entrained concrete
纤维加强玻璃管 fiber-reinforced glass pipe
纤维加强材料 fiber-reinforced materials
纤维加强的塑料筋 fibre reinforced plastic reinforcement
纤维加强混凝土 fiber[fibre]-reinforced concrete;fibrous concrete
纤维加强金属 filament-reinforced metal
纤维加强(筋) fibrous reinforcement
纤维加强喷射混凝土 fiber reinforced shotcrete
纤维加强塑料 fiber-reinforced plastics
纤维加强土 fiber reinforced soil
纤维加速方式 fiber[fibre] accelerating mode
纤维夹层玻璃 ply glass
纤维夹持器 lint retainer
纤维钾明矾 kalinite
纤维间的 interfibrous
纤维间分离阻力 fiber[fibre] drag
纤维间隔 fibrous septum
纤维间接触 fiber[fibre] to-fiber[fibre] contact
纤维间结合键 fiber[fibre] to-fiber[fibre] bond
纤维间距机理 fiber[fibre] spacing mechanism
纤维间摩擦 fiber[fibre] friction
纤维间黏[粘]合 inter-fiber[fibre] bond
纤维检验 examination of fibers
纤维建筑板发展组织 <英> Fiber[fibre] Building Board Development Organization
纤维鉴别 fiber[fibre] identification
纤维浆料 fiber[fibre] stuff
纤维胶 <人造丝、赛璐珞等的原料> viscose fiber;vissilk
纤维胶合板 compo board;composition board
纤维胶木 fabroil
纤维胶木齿轮 fabroil gear
纤维胶泥 fibrous plaster
纤维胶质 fibroglia;inoglia
纤维胶质原纤维 fibroglia fibrils
纤维接触点 fiber[fibre] contact point
纤维结构 fiber[fibre] texture;fibrous texture
纤维结晶状 fibrocrystalline
纤维结块 fiber[fibre] balling
纤维介质 fiber[fibre] media
纤维金属 fiber[fibre] metal
纤维金属材料 metallic fiber[fibre]
纤维进入点 fiber[fibre] entry point
纤维浸解坑 retting pound
纤维镜 fiber[fibre] scope
纤维聚合物 fiber[fibre] forming polymer
纤维聚糖 cellulosan
纤维绝热材料 fiber[fibre] insulating material;fibrous insulant
纤维绝热层 fiber[fibre] insulation layer
纤维绝缘 fiber[fibre] insulation fibrous insulation
纤维绝缘材料 fiber[fibre] insulating material;fibrous insulant
纤维绝缘层 fibrous covering;fiber[fibre] insulation layer;
纤维绝缘垫板 fiber[fibre] plate
纤维绝缘体 fiber[fibre] insulator
纤维绝缘线 fiber[fibre] insulated wire
纤维绝缘座 fiber[fibre] base
纤维抗拉强度 fiber[fibre] tensile strength;tensile strength of fiber
纤维空间 fiber[fibre] space;pseudo-fiber[fibre] space
纤维孔 fibrous pore
纤维矿石 fibrous mineral
纤维拉伸 fiber[fibre] in tension
纤维拉伸强力试验仪 fiber[fibre] tensile strength tester
纤维蜡 fiber[fibre] wax
纤维篮 fiber[fibre] basket
纤维缆(索) fiber[fibre] rope
纤维离解机 fibrator
纤维沥青合成物 fibrous asphalt[bituminous] compound
纤维沥青路面 fibered[fibred] asphalt pavement
纤维粒状变晶结构 fibrous granularblastic texture
纤维连接材料 fibrous jointing material
纤维帘布 fibrecord
纤维量 fiber[fibre] weight
纤维裂缝 fiber[fibre] fracture
纤维裂痕 fibrous fracture
纤维裂面 fibrous fracture
纤维流均匀度 fiber[fibre] flow uniformity
纤维流量计 flockmeter
纤维流形 fibred manifold
纤维滤纸 fiber[fibre] filter paper
纤维绿铜锌矿 rosasite
纤维毛细作用 capillarity of fiber[fibre]
纤维密度 fiber[fibre] density
纤维密封垫 fiber[fibre] packing
纤维密封环 fibrous composition seal(ing) ring
纤维面层板 fiber[fibre] covered plywood
纤维模板 fiber[fibre] pan
纤维膜 fibrous membrane
纤维膜束 fiber[fibre] film bundle
纤维泥灰 fibrous plaster
纤维泥浆 fibrous peat;fibrous slurry;fibrous turf
纤维泥煤 fibrous turf
纤维泥炭 fiber[fibre] peat;fiber[fibre] turf;fibrous turf
纤维黏[粘]合 fiber[fibre] bonding;fibrous union
纤维黏[粘]合织物 bonded fiber[fibre] fabric
纤维黏[粘]胶 viscose
纤维凝聚 fiber[fibre] deposition
纤维排列 fiber[fibre] alignment;fiber[fibre] array;fiber[fibre] pattern
纤维排列图 fiber[fibre] array diagram
纤维配比 fiber[fibre] composition;fiber[fibre] furnish;fibrous composition
纤维配筋 fiber[fibre] reinforcement
纤维配置 fiber[fibre] placement
纤维喷浇混凝土 fibrous shotcrete
纤维皮带 fabric belt
纤维片 fiber[fibre] plate
纤维品 fabric
纤维平均强度 mean fiber[fibre] strength
纤维铺装板 fibrous-felted board
纤维漆 cellulose lacquer
纤维强度 fiber[fibre] intensity;fiber(-reinforced) stress; fiber[fibre] strength;fibrous strength
纤维强化的 fiber[fibre] strengthened
纤维强化复合材料 fiber[fibre]-reinforced composite material
纤维强化金属 fibre[fibre]-strength-

ened metal
纤维墙板 fiber[fibre](-reinforced) wallboard
纤维墙布 fiber[fibre] wall paper
纤维切断 fiber[fibre] cut(ting)
纤维切断机 fiber[fibre] cutting machine;stapler;stapling machine
纤维切割器 fiber[fibre] cutter
纤维球过滤 fiber[fibre] ball filtration
纤维球滤料 fiber[fibre] ball filter material
纤维取向 fiber[fibre] orientation
纤维取样器 fibrosampler
纤维热绝缘层 fibrous thermal insulation
纤维绒 fiberflock
纤维容器 fiber[fibre] container
纤维熔合 fiber[fibre] fusion
纤维软化剂 softening agent for fiber
纤维润滑 fiber[fibre] lubrication
纤维润滑脂 fiber[fibre] grease
纤维散失 fiber[fibre] wash
纤维纱团 fiber[fibre] yarn
纤维砂轮 fiber[fibre] grinding wheel;fiber[fibre] sanding disk
纤维筛分仪 fiber[fibre] classifier
纤维闪石 uralite
纤维闪烁体 fiber[fibre] scintillator
纤维扇面 fan of filaments
纤维伸长法黏[粘]度计 fiber[fibre] elongation visco(si) meter
纤维绳 cordage;fibercord;fiber[fibre] rope
纤维绳长插接 fiber[fibre] rope long splicing
纤维绳短插接 fiber[fibre] rope short splicing
纤维绳索 fibrecord
纤维绳眼环插接 fiber[fibre] rope eye splicing
纤维石 cebollite;inolith;satin stone;satin spar
纤维石膏 fiber[fibre] plaster;fibrous gypsum;sericolite;striated gypsum
纤维石膏板 fiber[fibre] gypsum board;fibrous gypsum board;fibrous plaster
纤维石膏工 fibrous plasterer
纤维石膏工作 stick and rag work
纤维石膏制品 stick and rag work
纤维石膏制造的模型 fibrous plaster model
纤维式观测器 fiber[fibre] scope
纤维式预塑缝 fiber[fibre] type premoulded joint
纤维收集器 fabric collector
纤维束 fiber[fibre] assembly;fiber[fibre] bundle;tow
纤维束材料 towed material
纤维束粗纺 roving
纤维树脂施敷器 fiber[fibre] resin depositor
纤维栓塞 fiber[fibre] plug
纤维水 funicular water
纤维水泥 fiber[fibre] cement
纤维水泥板 cement fibrolite plate;fiber[fibre] cement board[slab]
纤维丝 fiber[fibre] yarn;thread
纤维撕裂 fiber[fibre] tear
纤维素 cellulose
纤维素保温材料 cellulosic insulation material
纤维素壁 cellulose wall
纤维素薄板 cellulose sheet
纤维素薄膜条 slit cellulose film
纤维素薄片 cellulose foil
纤维素材料的阻燃 fire-retardant for cellulosic material
纤维素瓷漆 cellulose enamel;enamel lacquer
纤维素发酵废水 cellulose zymolytic wastewater
纤维素发酵细菌 cellulose fermenting bacteria
纤维素凡立水 cellulose varnish
纤维素废(旧)料 cellulosic waste
纤维素废品 cellulosic waste
纤维素分解能力 cellulose decomposing capacity
纤维素分解细菌 cellulose decomposing
纤维素分解作用 cellulose decomposition
纤维素粉 cellulose powder
纤维素封口 cellulose seal
纤维素改性丙烯酸漆 cellulose modified acrylic lacquer
纤维素过滤器 cellulose filter
纤维素胶 cellulose glue
纤维素胶带 cellulose tape
纤维素胶片 cellulose film
纤维素胶黏[粘]剂 cellulose adhesive
纤维素介质 cellulose medium
纤维素绝热材料 cellulosic insulation material
纤维素类交换剂 cellulosic exchanger
纤维素(类)塑料的 cellulosic
纤维素离子交换剂 cellulosic ion exchanger
纤维素媒介 cellulose vehicle
纤维素酶 cellulase
纤维素醚 cellulose ether
纤维素膜 cellulose film;cellulose membrane
纤维素膜过滤 cellulose membrane filtration
纤维素木屑 cellulose debris
纤维素黏[粘]合剂 cellulose adhesive;cellulose cement
纤维素喷漆 cellulose lacquer
纤维素片基 cellulose support
纤维素漆 cellulose nitrate lacquer;cellulose paint;cellulosic varnish
纤维素清漆 cellulose varnish
纤维素溶剂 cellulose solvent
纤维素三钠 trisodium cellulose
纤维素树胶 cellulose gum
纤维素树脂 cellulosic resin
纤维素塑料 cellulosic plastics;cellulose plastics
纤维素填充料 cellulose stopper
纤维素填料 cellulose filler;cellulosic sizing agent
纤维素涂料 cellulose coating;cellulosic varnish
纤维素稀释剂 cellulose thinner
纤维素纤维 cellulose fiber[fibre]
纤维素纤维贴墙板 cellulose fiber tile
纤维素纤维瓦 cellulose fiber[fibre] tile
纤维素型(电)焊条 high cellulose type electrode
纤维素型焊条 cellulose-coated electrode
纤维素性的 fibrinous
纤维素性渗出物 fibrinous exudate
纤维素衍生物 cellulose derivative
纤维素样变性 fibrinoid degeneration
纤维素增稠剂 cellulose thickener
纤维素毡 cellulose mat
纤维素纸 cellulose paper
纤维素酯 cellulose ester
纤维素质 cellulosic
纤维塑料 compo
纤维随机分布的预浸板材 random fiber[fibre] prepreg sheet
纤维随机排列 random fiber[fibre] array
纤维碎片 fiber[fibre] debris
纤维损耗 fiber[fibre] loss
纤维损伤 fiber[fibre] damage;fiber[fibre] tendering
纤维锁结织物 fibrelock fabric
纤维弹性的 fiber[fibre] elastic
纤维弹性试验 fiber[fibre] elasticity test
纤维炭 peaty fibrous coal
纤维陶瓷 fiber[fibre] ceramics
纤维体积比 fiber[fibre] volume fraction
纤维体积率 volume percentage of fiber
纤维填充密度 fiber[fibre] packing density
纤维填充物 fiber[fibre] fill
纤维填充系数 fiber[fibre] packing fraction
纤维填料 fiber[fibre] filler;fiber[fibre] packing;fibrefill;fibrous filler;fibrous packing;wadding
纤维填塞物 fiber[fibre] fill
纤维头石灰岩 fibrous limestone
纤维图 fibrogram
纤维涂层技术 fiber[fibre] coating technique
纤维推移 fiber[fibre] shuffling
纤维外皮 fiber[fibre] incrustation
纤维完整性 fiber[fibre] integrity
纤维网 fiber[fibre] netting;fiber[fibre] sheet;fiber[fibre] web;fibrous fleece;fibrous reticulum
纤维网的 fibroreticulate
纤维网过滤器 fabric filter
纤维网络纱 fiber[fibre] network yarn
纤维物质 fibrous matter
纤维雾消除剂 fiber[fibre] mist eliminator
纤维细度 fiber[fibre] fineness
纤维细度气流测定仪 arealometer
纤维细度指示器 fiber[fibre] fineness indicator
纤维细菌 cellulose bacteria
纤维箱 fiberboard box
纤维项 fiber[fibre] term
纤维屑 <塞缝隙用> flocks and hards;fiber[fibre] tow;lint
纤维芯 fiber[fibre] core
纤维锌矿 wurtzite
纤维形成 fibration
纤维形成的 fibroplastic
纤维形态测视仪 fiber[fibre] morphometer
纤维形态学 fiber[fibre] morphology
纤维形状比 aspect ratio of fibers[fibres]
纤维型 fibrous type
纤维型滑石粉 fibrous talc
纤维性颤动 fibrillation
纤维性尘埃 fabric dust;fiber[fibre] dust;fibrous dust
纤维性粉尘 fabric dust;fiber[fibre] dust;fibrous dust
纤维性灰尘 fiber[fibre] dust
纤维性混凝土 <加入纤维质填充料如石棉、锯屑的混凝土> fiber[fibre] concrete;fibrous concrete;fibrated concrete
纤维性泥岩 fibrous peat
纤维性强直 fibrous ankylosis
纤维性收缩 fibrillary contractions
纤维性塑料砖 fibroplastic tile
纤维性土(壤) <如泥炭土、腐植土、沼泽地土、高度有机土等> fibrous soil
纤维悬浮液 fiber[fibre] suspension
纤维压力线法 funicular pressure line method
纤维压力线拱 funicular pressure line arch
纤维压力线拱顶 funicular pressure line vault
纤维压榨机 fiber[fibre] press
纤维延伸度 fiber[fibre] extent
纤维冶金 fiber[fibre] metallurgy
纤维应变 fiber[fibre] strain
纤维应力 fiber[fibre](-reinforced) stress
纤维应力公式 fiber[fibre] stress formula
纤维硬化 fibrosclerosis
纤维硬结 fibroid induration
纤维硬纸匣 fibreboard box
纤维预浸料 fiber[fibre] prepreg
纤维原料 fibrous raw material
纤维灶窑 hair kiln
纤维增强 fiber[fibre](-reinforced) reinforcement;fibrous reinforcement
纤维增强丙烯酸酯 fiber[fibre]-reinforced acrylate
纤维增强材料 fiber[fibre]-reinforced materials; fibrous reinforcement material
纤维增强的 fiber[fibre]-reinforced
纤维增强低碱度水泥建筑平板 fiber[fibre]-reinforced low-pH cement construction board
纤维增强复合材料 fiber[fibre] -reinforced composite material
纤维增强混凝土 fiber[fibre]-reinforced concrete;fibrous concrete
纤维增强金属 fibre(glass) reinforced metal
纤维增强金属材料 fiber-reinforced metal
纤维增强金属复合材料 fiber[fibre]-reinforced metal composite
纤维增强金属基复合材料 fiber[fibre]-reinforced metal matrix composite
纤维增强聚合物混凝土 fiber[fibre]-reinforced polymer concrete
纤维增强聚酯 fiber[fibre]-reinforced polyester
纤维增强聚酯片材 fiber[fibre]-reinforced polyester sheet
纤维增强耐火材料 fiber[fibre] reinforcement refractory
纤维增强热塑性塑料 fiber[fibre]-reinforced thermoplastics
纤维增强石膏板 fiber[fibre]-reinforced gypsum board
纤维增强石膏模 fibrous plaster
纤维增强水泥 fiber[fibre] cement composite; fiber[fibre]-reinforced cement
纤维增强塑料 fiber[fibre]-reinforced plastics
纤维增强塑料产品 fiber[fibre]-reinforced plastic product
纤维增强塑料风扇 fiber[fibre]-reinforced plastic fan
纤维增强塑料鼓风机 fiber[fibre]-reinforced plastic blower
纤维增强塑料筋 fiber[fibre]-reinforced plastic rod
纤维增强陶瓷 fiber[fibre]-reinforced ceramics
纤维增强陶瓷基复合材料 fiber[fibre]-reinforced ceramic material composite
纤维增生 fibroplasia
纤维毡 fibrofelt
纤维毡板 fibrous-felted board;fibrous-felted panel
纤维毡型管套 fiber[fibre] blanket pipe
纤维张力线 funicular tension line
纤维照明系统 fiber[fibre] optics illuminator
纤维织品 scrim
纤维织物除尘器 fabric filter

纤维织物(片材) fabric sheet(ing)
纤维直径 fiber[fibre] diameter
纤维直径指数 fiber[fibre] diameter index
纤维纸 fiber[fibre] paper; fibrous paper
纤维纸板 fiber[fibre] sheet
纤维纸板垫 fiber[fibre] gasket
纤维质 nemaline
纤维质材料 fibrous material
纤维质的 cellulosic; fibrous
纤维质废物 cellulosic waste
纤维质骨料 fibrous aggregate
纤维质骨料混凝土 fiber aggregate concrete
纤维质过滤嘴 fibrous filter
纤维质灰浆混凝土 gryptcrete
纤维质胶体 fibrous tactoid
纤维质煤 board coal
纤维质木粉 hardwood flour
纤维质清漆 cellulosic varnish
纤维质填料 cellulosic sizing agents
纤维质涂料 cellulosic coating; cellulosic paint; cellulosic varnish
纤维重量率 fiber[fibre] concentration by weight
纤维周边 fiber[fibre] periphery
纤维轴线 fiber[fibre] axis
纤维轴向不均匀性 fiber[fibre] taper
纤维转移 fiber[fibre] migration; fiber[fibre] transfer
纤维状 nemaline; sinewy
纤维状保温材料 fibrous heat insulator
纤维状表面 fibrous surface
纤维状冰 acicular ice; satin ice
纤维状的 fibrous; filiform; nemaline
纤维状的耐火材料 fibrous refractory
纤维状的森林粗腐殖质 fibrous mor
纤维状断口 fibrous failure; fibrous fracture; fibrous rupture
纤维状断裂 ductile fracture; fibrous fracture
纤维状方解石 fibrous calcite
纤维状分布 funicular distribution
纤维状粉末 fibrous powder
纤维状腐蚀 filiform corrosion
纤维状共晶合金 fibrous eutectic alloy
纤维状构造 bacillar structure; fibrous structure; filamentary structure
纤维状硅酸盐 fibrous silicate
纤维状焊缝 fibrous weld
纤维状集合体 fibrous aggregate; fibrous assembly
纤维状结构 fibrous texture; nerve structure
纤维状结晶 fibrous crystal; whisker
纤维状金属 fibrous metal
纤维状绝缘材料 fibrous insulating material
纤维状颗粒 fibrous particle
纤维状裂缝 craze; fibrous fracture
纤维状裂纹 fibrous fracture
纤维状煤 needle coal
纤维状泥炭 fibrous peat
纤维状黏[粘]合剂 fiber[fibre] adhesive
纤维状泉华 fibrous sinter
纤维状绒毛 fibrous wool
纤维状润滑脂 fiber[fibre] grease; fibrous grease
纤维状蛇纹岩 picrolite
纤维状石灰岩 fibrous limestone
纤维状碳 fibrous carbon
纤维状铁 fibrous iron
纤维状网络结构 fibrous network structure
纤维状纤维素 fibrous cellulose
纤维状形成 stringer formation
纤维状岩石 fibrous rock
纤维状氧化铝 fibrous alumina
纤维状组织 bacillar fabric
纤维着色 fiber[fibre] staining
纤维纵切面 fiber[fibre] profile
纤维纵视图 fiber[fibre] longitudinal view
纤维组织 fiber[fibre] pattern; nerve structure
纤维作物 fiber[fibre] crop
纤锡矿 dneprovskite; wood tin
纤细的 slight
纤心 fiber core
纤心直径 fiber core diameter
纤锌矿 fleischerite; wurtite
纤锌矿结构 wurtzite structure
纤锌矿型氮化硼 wurtzite BN
纤锌锰矿 woodruffite
纤锌锰石 woodruffite
纤铀铋矿 uranosphaerite
纤重钾矾 minesite
纤柱晶石 kryptotil
纤状变晶结构 nematoblastic texture
纤状变晶质的 nematoblastic
纤状变晶状 fibroblastic
纤状蛇纹石 chrysotile

氙

氙灯 xenon vapo(u)r lamp

氙灯老化(试验)机 xenon(arc) weatherometer; xenotest apparatus
氙碘法 xenon-iodine method
氙法 xenology method
氙航标灯 xenon aid lamp
氙弧光灯 xenon arc lamp
氙弧老化机 xenon arc weatherometer
氙气标准白(色)光源 xenon standard white light source
氙气灯 xenon(arc) lamp; xenon vapo(u)r lamp
氙气灯(耐候)试验机 xenotest apparatus
氙气管 xenon lamp

掀

掀板 opening; parting

掀动 tilting
掀动按钮 actuated button
掀断盆地【地】tilt-block basin
掀起 washing off
掀斜 tilt(ing)
掀斜地块【地】tilted block
掀斜断块 tilted block; tilted fault block
掀斜构造 tilting structure
掀斜山 tilted mountain
掀斜式断块山 tilted fault-block mountain

酰

酰氨基黄 acylamino yellow

酰胺 acidamide; acylamide
酰胺粉 amidpulver
酰胺化作用 amidation
酰胺萘酚红 amido naphthol red
酰胺染料 amido colo(u)r
酰胺纤维 nylon
酰胺-酰亚胺树脂 amide-imide resin
酰胺-酯聚合物 amide-ester polymer
酰胺族 amide group
酰二胺 acid diamide
酰化 acidate
酰化的 acidylable
酰化聚合物 acylated polymer
酰化作用 acidation; acidylation; acylation
酰基 acyl
酰亚胺树脂 imide resin

锨

锨式挖沟机 trench hoe

鲜

鲜贝壳松脂 bush kauri

鲜橙色 bright-orange
鲜纯色 saturate(d) colo(u)r
鲜果船 fruit carrier; fruit ship
鲜果汁 squash
鲜红 <明代早期铜红釉> fresh red
鲜红的 puniceous
鲜红色 bright red; coceine; madder; scarlet; cerise
鲜红色的 cherry; florid; madder
鲜红银朱颜料 scarlet vermilion pigment
鲜黄色 canaria; canary
鲜黄色玻璃 canary glass
鲜活货物 fresh and live freight; fresh and live goods
鲜活商品 fresh goods
鲜货 green ball freight; live cargo; perishable cargo
鲜货船 perishable cargo ship
鲜桔红(色) nacarat
鲜蓝色 amparo blue
鲜绿的 bright green
鲜绿色 emerald; emerald green; smaragd; vivid green
鲜绿氧化铬 emerald oxide of chromium
鲜明 vividness
鲜明表象 eidetic imagery
鲜明的 bright; vivid; trenchant <指轮廓等>
鲜明的颜色 strikingly colo(u)r
鲜明度 boldness
鲜明度光泽计 visibility glossmeter
鲜明色 assertive colo(u)r; bright colo(u)r
鲜明棕色的 high-brown
鲜污泥 green sludge
鲜血胶 blood adhesive
鲜艳 brilliance
鲜艳的橘红色 nacarat
鲜艳的颜色 technicolo(u)r
鲜艳色 bright colo(u)r; bright vivid colo(u)r
鲜映性 depth of image; distinctness of image; gloss distinctness of image
鲜鱼舱 fresh fish hold
鲜渔船 fresh fish carrier
鲜重 fresh weight; green weight
鲜贮法 fresh keeping method
鲜紫红色 solferino

闲

闲摆 idle pendulum

闲触点 idle contact
闲荡 potter
闲导条 idle bar
闲故障时间 downtime
闲季 off-season
闲接点 dead contact; idle contact
闲轮 idler
闲频 idle frequency
闲频信号 idler
闲圈 idle coil
闲散 stand-off
闲散劳动力 idle labo(u)r
闲散生产力吸收 absorption of idle capacity
闲散土地 scattered plots of unutilized land
闲散现金 idle money
闲散资金 idle capital; idle fund; idle money; inactive money; scattered funds; spare capital
闲栅 free potential grid
闲时交通 leisure traffic
闲市后市场 after market
闲田 vacant field; vacant land
闲暇工业 leisure industry
闲暇间 dodge time
闲暇消遣设施 leisure facility
闲线 disengaged(free) line; idle line
闲线路【数】idle channel
闲匝损失 dead-end loss
闲职 sinecure
闲置 depose; idling; laid up; leave unused; lie idle; set aside
闲置泊位 lay berth
闲置不用的船 laid-up ship; laid-up tonnage
闲置车道 free track
闲置船(舶) idle shipping; laid-up vessel; lay boat; lay ship; lay vessel; mothball ship; unemployed ship
闲置船舶停靠的码头 lay berth
闲置船锚地 laid-up vessel anchorage
闲置船只 laid-up ship
闲置的 on the shelf; standing idle
闲置的车辆 idle vehicle
闲置的货币资本 idle money-capital
闲置的机器 idle machine
闲置的建设用地 vacant building land
闲置地 idle land
闲置电池 idle battery
闲置吨位 laid-up tonnage
闲置费用 <指设备> expense of idleness; idle time cost; idling charges
闲置腐蚀 shelf corrosion
闲置货币 idle money
闲置机器 idle machine
闲置机器时间报告 idle machine time report
闲置机组 idle unit
闲置空间 <仓库> idle space
闲置率 vacancy rate
闲置能力 idle capacity
闲置能力成本 idle capacity cost
闲置能力损失 idle capacity loss
闲置能量 <温度较高时的韧度> shelf energy
闲置能量差异 idle capacity variance
闲置能量成本 idle capacity cost
闲置能量损失 idle capacity loss
闲置跑道 idle runway
闲置人工成本 idle labo(u)r cost
闲置人年 idle man-year
闲置人时数 idle man-hours
闲置人员 idle hand
闲置设备 dormant equipment; idle equipment; idle unit; stand-by equipment
闲置设备清单 idle equipment report (list)
闲置设施 idle facility
闲置生产能力 spare capacity; idle capacity
闲置生产能力差异 idle capacity variance
闲置时间 idle time; inactive time; shelf life; stand-by time; stand-by unattended time
闲置时间补充率 idle time supplementary rate
闲置时间差异 idle time variance
闲置时间和利用时间百分比 percentage of idle time to utilization time
闲置时间开支 idle time expenses
闲置试验 shelf test
闲置寿命 shelf life
闲置损耗 idling loss
闲置损失 idling loss
闲置土地 vacant land
闲置现金 idle cash
闲置线路 deadline
闲置余额 idle balance

闲置账户结存 dormant account balance
闲置折旧 shelf depreciation
闲置资本 disposable capital;idle capital;unemployed capital
闲置资产 dormant assets;idle assets
闲置资金 dead money;disposable capital;idle capital;idle fund;idle money;laid-up capital;volatile money
闲置资源效果 idle resources effect

弦

弦 chord; hypotenuse; latus[复 latera];subtense

弦板 boom plate;chord plate
弦波 waves on a string
弦材 bastard lumber
弦测法 chord measurement
弦长 chordal height;chord length
弦长比例尺 line of chord
弦长归算 reduction to chord length
弦潮 neap tide;quadrature tide
弦尺 <量测桶板材的单位> chord-foot
弦齿高 chordal addendum; chordal height
弦齿厚(度) chordal thickness
弦带 chordal band
弦的对弧【数】arc subtended by a chord
弦对弧 subtend
弦杆 boom chord; boom element; boom rod; chord bar; chord member;chord rod;suspension chord
弦杆端孔眼 <弦杆端头加大便于插销钉> chord head
弦杆构件 chord element
弦杆桁架 chord truss
弦杆加固 boom stiffening
弦杆加劲 boom bracing;flange bracing
弦杆角钢 angle iron of the chords
弦杆角铁 angle iron of the chords
弦杆接合板 chord splice
弦杆螺栓 chord bolt
弦杆面积 chord area
弦杆模量 chord modulus
弦杆拼接板 chord splice
弦杆平面 chord plane
弦杆式钻孔机 chord boring machine
弦杆支柱 boom support
弦杆组合构件 <型钢的> packed chord
弦构件 chord member
弦环形网络 chordal ring network
弦角 angle of chord;subtense angle
弦角法 chord-angle method
弦接运行 chordal running
弦节距 chordal pitch
弦截 flat-sawn
弦距 middle ordinate
弦距校正 correction for chord distance
弦锯 backsaw(ing);bastard sawing;slab cut
弦锯板 bastard sawed board
弦锯材 bastard sawed;bastard sawed lumber; bastard sawn lumber; plain-sawn lumber
弦锯的 plain-sawed;tangential sawn
弦锯法 plain-sawing
弦锯精选橡木材 plain-sawed select oak;plain-sawn select oak
弦锯面 plain grain face
弦锯木材 plain-sawed lumber
弦锯木料 plain-sawn timber
弦锯无节疤橡木材 plain-sawed clear oak;plain-sawn clear oak
弦控指针 string pointer
弦脉 angry pulse
弦面 hypotenuse face
弦面材 bastard lumber; flat-grain lumber; flat-sawed lumber; plain-sawed lumber
弦面花纹 slash figure
弦面环裂 slash figure check
弦面木纹 slash grain
弦面纹理 bastard grain; flat grain; slash grain
弦偏角 choppy deflection angle;chord deflection angle
弦偏距 chord deflection
弦切 flat cut;side cut
弦切板 flat-grained lumber;flat-sawed lumber;plain-sawed lumber
弦切法 chord contact method
弦切角 chord-tangent angle
弦切距 bowstring tangent distance
弦切面 bastard grain;flat grain;plain grain face;tangential section
弦切纹理 flat-sawn grain
弦绕组 chord(ed) winding
弦声 waves on a string
弦式测力计 string dynamometer
弦式测压机 dynamometer
弦式电流计 string galvanometer
弦式遥控温度计 vibrating tele-hygrometer
弦式引伸仪 strain wire
弦式应变计 string extensometer;wire strain ga(u)ge
弦式钻孔伸长计 wire-type borehole extensometer
弦束曲线 string curve
弦丝传动 string drive
弦纹 bowstring pattern; raised line design
弦系拱 <有拉杆的> tied arch
弦线 chord line; chord wire; music wire;string
弦线测角仪 subtense instrument
弦线齿顶高 chord addendum
弦线齿厚度 chordal tooth thickness
弦线电流计 Einthoven's galvanometer;string galvanometer
弦线电位计 string potentiometer
弦线法标定曲线 pegging out a curve from the chord
弦线绘角法 plotting angle by chords;plotting of angle by chords
弦线检流计 string galvanometer
弦线控制方式 <自动沥青混凝土摊铺机控制高程的一种方式> string-line system
弦线量角器 milrule
弦线模量 chord modulus
弦线偏距 chord deflection offset
弦线偏距法 method of chord deflection distance;method of chord deflection offsets
弦线偏向角 chord deflection angle
弦线式检流计 string galvanometer
弦线式静电计 string electrometer
弦线式示波器 string oscillograph
弦线式仪表 string-shadow instrument
弦线支距【测】chord(deflection) offset
弦线支距法 <曲线测量用的> choppy offset method;offset from chord method;chord offset method
弦线中点 mid-chord
弦线重力仪 linear gravimeter
弦线桩定法 chord stationary method
弦响器 organum chordotonale
弦向 chord direction
弦向薄壁组织 tangential parenchyma
弦向的 chordwise
弦向分力 chord force;chordwise force
弦向分量 chord component; chordwise component
弦向距离 chordwise distance
弦向锯材 bastard sawn
弦向锯切的 flat-sawed;flat-sawn
弦向模量 chord modulus
弦向切削 flat cut
弦向弯曲 chordwise bending
弦音计 sonometer
弦月潮水 quadrature tide
弦月窗 lunette;lunette window
弦月形 mososeries
弦中距 mid-ordinate;mid-ordinate of chord
弦中距测量 midordinate-to-chord measurement
弦中距间坐标 middle-ordinate for chord
弦轴 chord axis

贤

贤良名人祠 Hall of Fame

咸

咸潮倒灌 intrusion of tide saltwater

咸淡水 brackish water
咸淡水动物区系 brackish water fauna
咸淡水混合区 brackish water zone
咸淡水交接面 salt-water front
咸淡水界面 fresh-salt water interface
咸淡水界面埋深 buried depth of salt-fresh water interface
咸淡水界面在海平面以下 depth of salt-fresh water interface under sea level
咸淡水养殖 brackish water culture
咸的 saliferous;saline;salty
咸度 salinity
咸风 salt breeze
咸海 Aral Sea;saline sea
咸湖内海水 saline lake and interior sea water
咸化的 brackish
咸化泻湖相 saline lagoon facies
咸货 salt provisions
咸卤水溢出带 saline and brine water overflow zone
咸蓬 suaeda glauce
咸泉 saline spring
咸水 brine; saline water; salt(y) water
咸水冰 salt-water ice
咸水淡化 conversion of saline water;saline water conversion;saline water reclamation
咸水动物区系 brackish water fauna
咸水反渗淡化 brackish water reverse osmosis
咸水浮游生物 haliplankton[haloplankton]
咸水港口 saline harbo(u)r; saline port
咸水隔层 salt-water barrier
咸水光 <追踪测流> salt-water light
咸水湖 lagoon water;saline lake;salt lake;salt-water lake
咸水湖水 <内陆的> athalassohaline water
咸水环境 salt-water environment
咸水冷流变 salt water creep
咸水流 saline current
咸水屏障 salt-water barrier
咸水前锋 salt-water front
咸水潜流 salt-water underrun
咸水入侵 saline encroachment;saline intrusion; salinity intrusion; salt-water encroachment;salt-water intrusion
咸水入侵地下水 saline water intrusion groundwater
咸水生态系统 brackish water ecosystem;saline water ecosystem
咸水生物 halobiont;halobios
咸水下移 saline water descending
咸水楔 saline wedge;salt brine wedge;salt-water wedge
咸水泻湖 salt-water lagoon
咸水异重流 salt-water underrun
咸水沼泽 saline bog
咸水沼泽沉积 salt-water swamp deposit
咸味 saline taste;salty taste

涎

涎布 mokador

涎流冰 salivary flow ice

舷

舷 board;chine

舷边 gunnel
舷边安全网 cargo saving net;side net
舷边构架 side framing
舷边和水面相平 gunwale down;gunwale to
舷边计程仪 sidestreaming log;viking log
舷边角钢 gunwale(angle) bar
舷边栏杆 gunwale rail
舷边列板 gunwale strake
舷边没入水面以下 gunwale under
舷边缘 gunwale
舷边作业吊架 flying scaffold
舷材 balk
舷舱 wing compartment;wing space
舷侧 broadside;shipboard;topside
舷侧波 sea abeam
舷侧出灰口 chute
舷侧带缆柱 side bitt;side bollard
舷侧导水板 spurn water;wash board;wash strake;water board;weatherboard(ing)
舷侧地 broadside
舷侧顶部外板 strake below sheer strake;topside strake
舷侧独立舱 independent tank
舷侧阀 hull valve;hull ventilator
舷侧防摇水舱 stabilizing bilge
舷侧工作木排 floating stage
舷侧工作平底小船 painting punt
舷侧合板 plywood side planking
舷侧厚板 sheer plank; sheer plate; sheer strake
舷侧浪 athwart sea
舷侧披水板 bilge board
舷侧漆 broadside paint
舷侧燃料舱 side bunker
舷侧水柜 cantilever tank;corner tank;gunnel tank; gunwale tank; outboard tank;topside tank;wing tank
舷侧踏板 flake
舷侧通海阀 flood valve
舷侧突出部分 sponson
舷侧突出台 sponson
舷侧外板 side strake;topside planking;topside plating
舷侧污水道 side water course
舷侧系艇杆 boat boom; guest warp boom;riding boom
舷侧相向 beam on;broadside on
舷侧卸货 side discharge
舷侧卸货跳板 side loading ramp
舷侧圆窗 bull's eye
舷侧装货门 cargo port;side port
舷侧装卸方式 sideboard system;side port system

X

舷侧纵板 riser
舷侧纵材 side longitudinal
舷侧纵通角钢 <在甲板边板上> stringer angle;stringer bar
舷侧坐板 side bench
舷窗 port light;air port;illuminator;porthole; scuttle; side light; side scuttle;dead light <船上的>
舷窗扳手 scuttle key
舷窗玻璃 airport glass;port (hole) glass;deadlight
舷窗法兰 port flange
舷窗风暴盖 dead light
舷窗风斗 subscuttle;wind catcher
舷窗盖 port buckler; scuttle blind; scuttle cover
舷窗盖板吊钩 port hook
舷窗滚筒 side roller
舷窗厚玻璃 port lens
舷窗护板 port apron
舷窗槛 sill of side-scuttle
舷窗框 scuttle frame
舷窗帘 airport screen
舷窗楣板 port brow
舷窗内盖 dead light;side scuttle blind
舷窗水密盖 scuttle key;scuttle lid
舷窗铁盖 battle light; blind cover; blinder; iron cover; scuttle hatch (cover)
舷窗凸缘 port flange
舷窗外盖 dead light; side light plug; side scuttle plug
舷窗橡胶 port rubber
舷窗橡胶垫圈 scuttle rubber washer
舷窗檐板 port brow
舷窗遮水楣 eyebrow of side scuttle
舷灯 running light;side light
舷灯玻璃 port light glass
舷灯插头 side light plug
舷灯光源宽度 filament width of side-light
舷灯架 side light tower
舷灯遮板 screen of sidelight
舷灯遮光板 screen for side light;side light screen
舷顶列板 sheer strake
舷弧 sheer;sheer of gunwale
舷弧度 bilge circle
舷弧高 sheer height
舷弧模 sheer mo(u)ld
舷弧线 sheer curve;sheer line
舷弧样板条 sheering batten
舷角 angle on the bow;bow bearing; relative bearing
舷靠舷 chine to chine
舷孔 air port;porthole;side scuttle
舷孔半截盖 half port
舷孔纱窗 airport screen
舷栏杆 accommodation rail; guard rail
舷肋骨 shell frame;side frame
舷门 port buckler; port door; port gangway; accommodation ladder; entrance port; gangway; gangway door; gangway ladder; gangway port;side port
舷门槛 port sill
舷门栏 port sill
舷门索 man rope
舷门斜跳板 side ramp
舷门装卸方式 side porter system;si-porter system
舷内的 inboard
舷墙 bulwark;weather-board(ing)
舷墙板 bulwark plating
舷墙导缆钩 bulwark chock
舷墙等急转折部分 hance
舷墙扶手 main rail
舷墙栏杆 bulwark rail;waist rail
舷墙排水口 bulwark freeing port; bulwark port; freeing port; wash port
舷墙网 bulwark netting
舷墙线 bulwark line
舷墙斜撑 bulwark brace
舷墙支柱 bulwark stay
舷墙柱 bulwark stanchion
舷墙纵材 bulwark strake
舷桥 wharf ladder
舷伸梁 sponson beam
舷台 sponson
舷梯 accommodation ladder;bulwark ladder; gangway (ladder); railing ladder; wharf ladder; sea steps 【船】;side ladder; accommodation net【船】
舷梯侧板 stairway stringer
舷梯吊柱 accommodation ladder davit
舷梯扶手 ladder rail;side rail
舷梯绞车 accommodation ladder winch
舷梯绞辘 gangway tackle
舷梯栏杆 ladder rail
舷梯平台 gangway platform
舷梯梯级 side step
舷梯围布 ladder screen
舷梯型橡胶护舷 ladder-type fender
舷梯值班 <船靠码头时> gangway watch
舷梯值班人员 gangway man
舷拖(网)渔船 side trawler
舷外 outside board
舷外的 outboard;ovb[overboard]
舷外吊杆 yard boom;yard derrick
舷外发动机 outboard engine;outboard motor
舷外浮材 outrigger
舷外挂机 kicker; outboard engine; outboard motor
舷外链节 outboard shot
舷外排出管 overboard discharge line
舷外排水阀 overboard discharge valve
舷外推进机 outdrive set
舷外推进器 outboard motor
舷外消磁绕组 outboard coil
舷外斜桁 burton boom; outboard boom
舷外斜木 outrigger
舷外支架 outrigger
舷外作业 outboard work;stage
舷外作业板架 outside staging; stage board
舷围 rubber
舷线护缘 sheer mo(u)lding; sheer rail
舷缘 gunwale
舷缘板 covering board;plank sheer
舷缘材 plank sheer
舷缘衬板 gunwale plate
舷缘第二列板 landing strake
舷缘内倾 tumble home
舷缘上列板 berthing
舷缘着水 gunwale down;gunwale to
舷遮 side awning

衔轨 <高架快速道路用> armature rail

衔接 connector;join;link-up
衔接臂 engaging arm
衔接的 end-to-end
衔接的运输单位 connecting carrier
衔接复生 syntaxial overgrowth
衔接合并(铁路)end-to-end merger
衔接轮 side roller
衔接砌法 <两相互垂直墙的> racking
衔接器 adapter [adaptor] (connector);engager
衔接深度 depth of engagement
衔接丝杆 engagement screw
衔接(涂)层 tie coat
衔接误差 bridging error
衔套 lug
衔铁 anchor cone;armature;armature iron;keeper;reed;yoke
衔铁比 armature ratio
衔铁臂 <继电器的> armature arm
衔铁槽 armature slot
衔铁颤动 <继电器> armature bounce
衔铁超程 armature overtravel
衔铁触点 armature contact
衔铁传动比 armature lever ratio
衔铁挡 armature stop
衔铁动程 armature stroke; armature travel
衔铁端 armature end
衔铁端缓动铜环 armature end slug
衔铁反跳 armature bounce
衔铁(防黏[粘])余隙 armature residual gap
衔铁杆 armature lever
衔铁后挡 <继电器的> armature back stop
衔铁回跳 armature rebound
衔铁架 <继电器的> armature bracket
衔铁间隙 armature gap
衔铁绝缘架 <继电器的> armature card
衔铁控制装置 reed control system
衔铁拉簧 armature tensioning springs
衔铁上升式继电器 tractive armature type relay
衔铁式继电器 armature relay
衔铁释放 armature releasing
衔铁释放超行程 armature dropout overtravel
衔铁瞬时延迟 relay armature hesitation
衔铁铁芯 armature core
衔铁头 armature stub
衔铁推杆 armature lifter
衔铁弯曲线 armature bending tool
衔铁吸合过度 armature pickup overthrow
衔铁吸合式继电器 attract armature relay
衔铁吸入式断电器 clapper relay
衔铁线圈 armature coil
衔铁行程 armature travel; relay armature travel
衔铁游隙 armature play
衔铁振动 armature chatter
衔铁滞缓 armature hesitation
衔铁轴 armature shaft
衔铁组 armature assembly

嫌冰雪的 chionophobous

嫌风植物 anemophobe
嫌钙植物 calcifuge;calcifugous plant; calciphobe
嫌光的 photopholic
嫌寒植物 frigofuge
嫌旱的 xerophobous
嫌忌点 dislike point
嫌忌值 dislike value
嫌碱植物 basifuge
嫌菌的 mycophobic
嫌气分解 anaerobic decomposition
嫌气腐烂 anaerobic decay
嫌气过程 anaerobic process
嫌气环境 anaerobic environment
嫌气接触(池)anaerobic contact
嫌气菌 anaerobic bacteria
嫌气菌分解 anaerobic decomposition
嫌气菌还原处理 anaerobic treatment
嫌气生物 anaerobic organism
嫌气(生物)处理 anaerobic treatment
嫌气塘 anaerobic pond
嫌气条件 anaerobic condition
嫌气土壤微生植物 anaerophytobiont
嫌气微生物 anaerobe;anaerobia
嫌气细菌 anaerobe; anaerobic bacteria
嫌气消化 anaerobic digestion
嫌气消化污泥 anaerobically digested sludge
嫌气性的 anaerobic
嫌气性接触法 anaerobic contact process
嫌气性消化槽 anaerobic digester
嫌气性消化池 anaerobic digester
嫌气性消化污泥负荷 sludge loading of anaerobic digester
嫌色的 chromophobe;chromophobic
嫌色性 chromophobia
嫌水植物 hygrophobe
嫌酸的 acidofuge;acidophobous
嫌酸植物 oxyphobe
嫌污水真菌 lymaphobe
嫌雪植物 chionophobous;chionophobous plant
嫌盐植物 halophobe
嫌阳植物 heliophobe; heliophobous plant;skiophyte
嫌氧环境 anaerobic environment
嫌氧菌 anoxybiotic bacteria
嫌氧生物处理 anaerobic biological treatment
嫌氧塘 anaerobic pond
嫌氧消化 anaerobic digestion
嫌疑犯 culprit
嫌雨植物 ombrophobe

冼砂机 clay washer

冼砂网 washing screen

显斑时间 presentation time

显参数 explicit parameter
显程序 explicit program(me)
显出电子枪 reading gun
显出火石的墙面 polled face
显出母本的品质 qualities of the female parent
显出现 explicit occurrence
显出一个侧面的 half-faced
显磁极 projecting pole;salient pole
显磁极发电机 salient pole generator
显的 explicit
显定合一浆 monopaste
显定义 explicit definition
显而易见 exposed to view
显而易见的 visible
显二次函数 explicit quadratic function
显光管 arcotron
显光性 gloss developing
显含成本 explicit cost
显函数 explicit function
显焓 sensible enthalpy
显号接续制 coded call display working
显花植物 carpophyte; flower plant; phanerogam
显花植物式木材 phanerogamic wood
显化偏好 revealed preference
显画器 camera lucida
显迹 tracing
显迹实验 tracer experiment
显迹试验 tracer experiment

X

显极 salient pole
显极电机 salient pole machine
显极式转子 salient pole type rotor
显极性 salience[saliency]
显见面 visible surface
显解 explicit solution
显晶结构 phaneric texture; phanerocrystalline texture
显晶碎屑岩 macroclastic rock
显晶岩 phanerite; phanerocrystalline rock
显晶岩的 phaneritic
显晶质的 phaneric; phanerocrystalline; phaneromer
显晶质结构 phanerocrystalline texture
显冷负荷 sensible cooling load
显粒岩 phaneromere
显露 emergence; exposition; exposure; relieving; revealment; revelation; stick-out; unfold
显露出来 come to light
显露气孔 phaneropore
显露式垂直归档制度 visible-vertical filing system
显露式记录设备 visible record equipment
显露叶轮 open impeller
显名委托人 named principal
显明的 phanerous; phanic
显气孔率 apparent porosity
显然相反 contradistinction
显热 sensible heat; sensible heat content
显热比 sensible heat ratio
显热得热量 sensible heat gain
显热负荷 sensible heat load; dry ton <美>
显热空气冷却器 sensible heat air cooler
显热冷却 sensible heat cooling
显热冷却器 sensible heat cooler
显热冷却效果 sensible cooling effect; sensible heat cooling effect
显热量 sensible heat capacity; sensible heat quantity
显热流 sensible heat flow
显热损失 sensible heat loss
显热系数 sensible heat factor
显热因数 sensible heat factor
显热增量 sensible heat gain
显色 colo(u)ration; colo(u)r development; development; development of colo(u)r
显色板 colo(u)r plate
显色对比度 development contrast
显色法 development process
显色反应 colo(u)r-producing reaction; colo(u)r reaction
显色反应板 colo(u)r reaction plate
显色管 chromoscope
显色剂 developer; developing agent; photograph developer
显色鉴定 colo(u)r identification
显色染料 developed dye; diazo dye; ingrain dye
显色试剂 colo(u)r-producing reagent; spray reagent; colo(u)r reagent
显色试验 colo(u)r test
显色性 colo(u)r rendering
显色盐染料 diazol colo(u)r
显色指数 colo(u)r rendering index
显熵 sensible entropy
显生代【地】Phanerozoic time
显生代地质年代表 Phanerozoic Time-Scale
显生宇【地】Phanerozoic Eonothem
显生宇的 Phanerozoic
显生宙【地】Phanerozoic Eon
显生宙的【地】Phanerozoic
显生宙地质年代表 Phanerozoic geochronologic(al) scale
显示 discovery; exhibition; presentation; show; unroll; video picture
显示板 display panel; visual tablet【计】; indicator panel <警报器的>
显示报警器 display alarm
显示部件 display unit
显示菜单 display menu
显示场 display field
显示程序 display routine
显示尺寸 display size
显示处理机 display processor
显示处理器 display processing unit
显示窗口 display window
显示磁鼓 display drum
显示存储板 display store card
显示存储管 display storage tube
显示存取 display access
显示存取变址板 display access and modify card
显示单元 display unit
显示到地表的孔内爆炸作用 shothole disturbance
显示灯 display lamp
显示灯盘 display lamp panel
显示电键 display key
显示电路 display circuit; indicating circuit; indication circuit
显示电子束 reading beam
显示法 explicit representation; method of presentation
显示范围 display range; indication range
显示方法 display packing
显示方式 display mode; mode of signal indication
显示访问 display access
显示分析控制台 display analysis console
显示复印机 display copier
显示改变 aspect change; aspect modification; indication change
显示格式 display format
显示更新速度 display refresher rate
显示管 display tube; indicator tube; telltale pipe; telltale tube <溢水指示>
显示和编辑软件 software for display and edit
显示缓冲器 display buffer
显示缓冲区 display buffer
显示缓冲区存储器 display buffer storage
显示幻象 indication phantom
显示绘图仪 display plotter
显示机构 indication mechanism
显示激励器 display actuator
显示极明显 excellent clear displayed
显示记号 display mark
显示技术 display technique; display technology; techniques of display
显示剂 colo(u)r developing agent; developer; spray reagent
显示寄存器 display register
显示监视器 display monitor
显示检查器 visual detector
显示检索曲线 overt-retrieval plot
显示检验 visual examination
显示角 visual angle
显示较隐瞒 comparative obscure displayed
显示接口装置 display adapter unit
显示距离 indication range; range of a signal; visibility distance; visibility range; visible distance; vision distance; visual distance; visual range; sighting distance <信号灯光>
显示距离不足 lacking visibility; poor visibility
显示距离测试 visibility test
显示开关 display switch
显示拷贝器 display copier
显示刻度 display scale
显示空间 display space
显示控制 display control
显示控制盘 visual control panel
显示控制器 display controller
显示控制台 display console
显示类型 display category
显示冷却 sensible cooling
显示量 display item
显示列 display column
显示灵敏度 display sensibility
显示绿灯时间 display(ed) green time
显示逻辑装置 display logic unit
显示门闩 <显示厕所有没有人> indicator bolt
显示门栓 indicating bolt
显示面 visible face; visible side
显示名称 aspect name
显示明显 clear displayed
显示模件 display module
显示模拟 displace simulation
显示模式 display mode
显示模型 display model
显示凝聚力 apparent cohesion
显示牌价荧光屏 teleregister
显示盘 display board; display panel
显示偏好 revealed preference
显示屏 display glass; display panel; display screen; scope face; visual display board
显示其重要性 expression of interest
显示器 cathode-ray tube; display (station); display unit; indicator; scope unit; visual display terminal
显示器标度 display scale
显示器储存容量 display memory
显示器的上回波信号 scope return
显示器端 silencing end
显示器基座 display pedestal
显示器荧光屏 scope face
显示器组 display bank
显示轻微开裂 show slight breaks
显示清单 display menu
显示区 display place; display space
显示区域 display region
显示软件 display software
显示扫描 reading scan
显示设备 display device; display equipment
显示时间 viewing time
显示时钟 read clock
显示式示波器 indicating oscillograph
显示适配部件 display adapter; display adapting unit
显示适配器 display adapter
显示数据分析 display data analysis
显示数据模块 display data module
显示数据询问器 video data interrogator
显示数据终端 video data terminal
显示顺序 <色灯信号> colo(u)r sequence
显示特性 display feature
显示条件不良 poor visibility condition
显示图像 display image; screen image
显示位置 display position
显示文本 display text
显示文件 display file
显示物 indicator
显示误差 indication error
显示系统 demonstration system; display system
显示线 display line
显示小室 display booth
显示心理学 display psychology
显示信号 display signal
显示信息处理机 display information processor
显示行 display line
显示行为 display behavio(u)r
显示形式 display format
显示型指示器 display indicator
显示选通门 indicator gate
显示仪表 display device; display instrument; indicating instrument
显示隐瞒 obscure displayed
显示与控制系统的设计 display and control design
显示与控制装置 display and control unit
显示域 display field; display frame
显示元件 indicator element
显示站 display station
显示帧 display frame
显示指令 display command; readout command
显示中心 display center[centre]
显示重复存储器 refresh memory
显示周期 display cycle
显示柱 display column
显示转接器 display adapter
显示装置 display device; display panel; display unit; readout device; visual display unit
显示字段 display field
显示字符 display character; graphic(al) character
显示组 display group
显示组件 display module
显式饱和度计算 explicit saturation calculation
显式辨识 explicit identification
显式标识 explicit identification
显式表达式 explicit expression; explicit formulation
显式表示法 explicit expression; explicit representation
显式参数 explicit parameter
显式差分格式 explicit difference scheme
显式差分公式 explicit difference formula
显式程序库 explicit library
显式地址 explicit address
显式法 explicit scheme
显式方法 explicit method
显式格式 explicit scheme
显式公式 explicit formula
显式公式化 explicit formulation
显式估计法 explicit estimation technique
显式关系 explicit relation
显式加权 explicit weighting
显式交替方向 alternative direction explicit
显式解 explicit solution
显式类型 explicit type
显式类型结合 explicit type association
显式说明 explicit declaration
显式松弛 explicit relaxation
显式网格 network for the explicit method
显式一元算符 explicit unary operator
显式引用 explicit reference
显式有限差分(法) explicit finite difference
显式有限差分格式 explicit finite difference scheme
显式有限差分计算 explicit finite difference calculation
显式属性 explicit attribute
显式自校正器 explicit self-tuner
显数学模型 explicit mathematical model

显似凝聚力 apparent cohesion
显似弹性极限 apparent elastic limit
显图处理机 graphic(al) display processor
显图器 <铺画列车运行图,用阴极射线管表示> graphic(al) display apparatus
显微 X 射线法 microradiology
显微 X 射线照相术 microradiography
显微暗煤 microdurain
显微斑晶的 mediiphyric
显微半圆角度规 vernier protractor
显微包裹体 microinclusion
显微比较镜 comparascope
显微比重法 microspecific gravity method
显微变晶结构 microcrystalloblastic texture
显微玻基斑状结构 vitriphyric texture
显微薄片 microsection
显微不均匀性 microinhomogeneity
显微不平度 microinregularity
显微操纵 micromanipulation
显微操纵器 micromanipulator
显微操作 micromanipulation
显微操作叉 microfork
显微操作器 micromanipulator
显微操作术 micrurgical technique; micrurgy
显微(操作)针 microneedle
显微测焦法 microfocus(s)ing
显微测角仪 microgoniometer
显微测径器 micrometer cal(l)ipers
显微测量 microdetermination; micrometry
显微测密术 densi(to)metry
显微测谱术 microspectrometry
显微测微计 microscope micrometer
显微层状构造 microstratified structure
显微差热分析仪 microdifferential thermal analyser[analyzer]
显微尘粒 microscopic dust particles
显微处理 microprocessing
显微传真电报 micrograph; microphotogram
显微传真电报术 micrography
显微纯洁(度) microcleanliness
显微单位 micrometer unit; microscopic unit
显微的 micrographic; microscopic
显微电泳(法) microelectrophoresis; microscopic electrophoresis
显微电子探针分析 microprobe analysis
显微电阻焊 microresistance welding
显微度谱术 microspectrometry
显微断口分析 microfracturegraphy
显微断口检验法 microfracturegraphy
显微断面 microsection
显微断谱学 microfractography; microfracturegraphy
显微法折射率测量 microrefractometry
显微反应 microreaction
显微放大(技)术 micrurgy
显微放大器 micrograph
显微放射显影术 microradiography
显微放射照片 microradiogram
显微放射照相的 microradiographic
显微放射照相术 microradiography
显微放射自显术 microradiography
显微放射自显影 microradiography
显微放射自显影术 microautoradiography
显微放射自显影相片 microautoradiogram
显微放射自显影照相 microautograph
显微放映机 microprojector
显微分光光度 microspectrophotometry
显微分光光度法 microscopic spectrophotometry
显微分光光度计 microspectrophotometer
显微分光光度计法 microphotometer method
显微分光镜 microspectroscope
显微分光镜检查 microspectroscopy
显微分析 microanalysis; microanalyze
显微分析器 microanalyzer
显微浮游生物 microplankton
显微辐射计 microradiometer; radiomicrometer
显微腐蚀 interdendritic corrosion; microcorrosion
显微干涉仪 microinterferometer
显微高温计 micropyrometer
显微共生 implication
显微构造 microscopic structure; microstructure
显微构造尺度 microscopic scale
显微构造和组构 microstructure and fabric
显微构造类型和性质 type and feature of microstructure and fabric
显微构造学 microtectonics
显微观察 microexamination
显微光波干涉仪 profilograph; profilometer; surface analyser[analyzer]
显微光度计 microdensitometer; microphotometer
显微光度计比较器 microphotometer comparator
显微光度术 microphotometry
显微光刻技术 microphotolithographic technique
显微光密度 microphoto density
显微光密度计 microphoto densitometer
显微光谱图 microphotogram
显微光谱学 microspectroscopy
显微焊接 microwelding
显微化学 microscopic chemistry
显微化学法 microscopic chemical method
显微化学分析 microchemical analysis
显微划痕硬度 microcharacter hardness
显微划痕硬度计 microcharacter sclerometer
显微幻灯 megascope
显微灰化 spodography
显微绘图术 micrography
显微机械加工 micromachining
显微极谱仪 micropolarograph
显微技术 microtechnic; microtechnique
显微加工 microprocessing
显微检查 microexamination; micrographic(al) examination; micrography
显微检验 microexamination; micrographic(al) examination; microtest
显微检验设备 micromanipulator
显微胶卷阅读器 film reader
显微角砾岩 microbreccia
显微结构 fine texture; microscopic texture; microstructure; microtexture
显微结晶 microscopic crystal
显微结晶学 microcrystallography
显微解剖 microdissection
显微晶体学 microcrystallography
显微晶质结构 microcrystalline texture
显微镜 binoculars; microscope
显微镜操作 microscope work
显微镜操作设备 micromanipulator
显微镜测微器 microscope micrometer
显微镜抽筒 microscope draw tube
显微镜传真 microphotograph
显微镜的 microscopic
显微镜的工作距离 working distance of microscope
显微镜的换镜旋座 nosepiece; revolving nosepiece of microscope
显微镜电视 microscopic television
显微镜电泳池 microscope electrophoresis cell
显微镜读数 microscopic reading
显微镜法 microscopic method
显微镜分辨率 reduction of the microscope
显微镜分析 microscopic analysis
显微镜分析法 microscopic analysis method
显微镜盖玻片 cover glass
显微镜观察 microscopic examination
显微镜管嘴 nosepiece
显微镜光度计 microphotometer; microscope photometer
显微镜光谱描记法 microspectrography
显微镜换镜旋座 revolving nosepiece
显微镜计数 microscopic count(ing)
显微镜计数法 microscope count method
显微镜计算 microscopic count
显微镜加温台 microscopic warming table
显微镜检查 microscopic examination
显微镜检查法 microscopy
显微镜检查用薄片 microsection
显微镜检验 microscopic examination
显微镜鉴定 microscopic identification
显微镜镜鼻 nosepiece
显微镜粒度法 microscope sizing
显微镜煤 microvitrain
显微镜描记法 micrography
显微镜目测法 eye survey by microscope
显微镜目镜 microscopic ocular
显微镜摄影用闪光灯 electroflash for microphotography
显微镜视野 field of microscope; microscopic field
显微镜试验 micrographic(al) test; microscopic test
显微镜试样制片技术 mounting technique
显微镜数菌法 microscopic bacterial count
显微镜台 substage
显微镜台下灯 substage microlamp
显微镜筒 microscope tube
显微镜外镜 epimirror
显微镜物镜 microobjective; microscope objective; microscopic objective
显微镜下 microscopically
显微镜旋转台 microscope turn table
显微镜学 microscopy
显微镜学家 microscopist
显微镜荧光灯源 fluorescent light source
显微镜用灯 microlamp
显微镜用盖玻片 microscope slide glass
显微镜用载玻片 microscope slide
显微镜用载片 microglass
显微镜载玻片 microslide; microscope slide
显微镜载片台 microscope stage
显微镜载体玻璃 microscope cover glass
显微镜载物玻璃 object carrier
显微镜载(物)片 microscope slide
显微镜载物台 microscope carrier; microscope stage; microscopic stage; microstat
显微镜照明装置 microscope illuminator
显微镜照相机 microscope camera
显微镜直接观测 direct observation through microscope
显微镜转接器 microscope adapter
显微镜座 microscope base
显微镜座内装置式显微镜灯 in-base illuminator
显微镜座上装置式显微镜灯 on-base illuminator
显微抗拉强度 microtensile strength
显微(科)学 micrology
显微可熔性 microsolubleness
显微空隙 microscopic void
显微空穴聚结 microvoid coalescence
显微孔隙 microscopic void
显微矿物学 micromineralogy
显微扩视镜 euscope
显微拉力强度 microtensile strength
显微拉制仪 microforge
显微粒径 microscopic size
显微粒状结构 micrograined texture
显微粒子计数器 microparticle counter
显微亮煤 microclarain
显微量尺 micrometer[micrometre]
显微量热测定 micrithermometric(al) measurements
显微裂缝 microcrack
显微裂纹 microcracking; microcrazing; microfissure; microflaw
显微裂隙 microcrack
显微磷片结构 microscaly texture
显微流动点 microflow point
显微孪晶 microtwin
显微轮廓仪 microprofilometer
显微煤岩类型 microlithotypes of coal
显微煤岩类型分类 microlithotype classification
显微煤岩组分 maceral
显微煤岩组分分类 maceral classification
显微密度测定 microdensitometry
显微密度计 microdensitometer
显微密度计测定线 microdensitometer tracing
显微描绘器 camera lucida
显微描绘仪 camera lucida
显微模型 microscope model
显微磨片 microsection
显微培养 microculture
显微疲劳裂纹 microscopic fatigue crack
显微偏析 microsegregation
显微品质 microcrystalline
显微气孔结构 microvesicular texture
显微器具 microinstrument
显微切面 microsection
显微切片 microsection
显微切片刀 microtome
显微切片浮雕抛光 relief polishing
显微切片机 microtome
显微侵蚀 microetching
显微清洁度 microcleanliness
显微全息电影照相术 cineholomicroscopy
显微全息摄影 microholography
显微全息图 microhologram
显微全息照片 microhologram
显微全息照相术 microholography
显微熔点 microfusible point
显微熔点测定器 micromelting point apparatus
显微蠕变 microcreep
显微扫描 microscan
显微筛状变晶结构 microdiablastic texture

显微射线照片 microradiogram
显微射线照相机 microradiography unit
显微射线照相术 microradiography
显微射线自动摄影 microautoradiograph
显微摄谱仪 microspectrograph
显微摄影 micrograph; photographic(al) micrograph; photographic(al) micrography
显微摄影测量(法) microphotogrammetry
显微摄影测量术 photomicrometrology
显微摄影机 microphotographic(al) camera; photomicrographic camera; photomicroscope
显微摄影术 microphotography; photomicrography
显微摄影业 micrographic
显微示波器 microoscillograph
显微收缩 microshrinkage
显微疏松 microporosity
显微术 microscopy
显微树脂学 resinography
显微数据 <录在胶片上的数据> microimage data
显微丝炭 microfusain
显微损伤 microdamage
显微缩孔 micropipe; microshrinkage; microshrink-hole
显微探针 microprobe
显微体积法 microvolume method
显微投影(法) microprojection
显微投影器 microprojection apparatus
显微投影仪 microprojection; microprojector; microscopic projector
显微透镜 microlens
显微图 microgram; micrograph
显微图书 filmbook
显微图形放大器 megalograph
显微图形放大装置 megalograph
显微威达凝集试验 microscopic Widal agglutination test
显微维氏硬度计 microvicker
显微伟晶结构【地】micropegmatitic structure
显微文象结构【地】micrographic(al) texture
显微纹影仪 microschlieren
显微物理性质 microphysical property
显微系统 microscopic system
显微纤维结构 microfibrous texture
显微相 microfacies
显微效应 microeffect
显微岩石学 microlithology
显微岩相 microlithofacies
显微岩性学 microlithology
显微样品 microobject
显微隐晶斑岩 aphanophyre
显微隐晶质 microaphanitic; microcryptocrystalline
显微隐晶质的 aphaniphyric
显微隐晶质结构 microaphanitic texture; microcryptocrystalline texture
显微应力 microscopic stress; microstress
显微荧光测定 microfluorometric(al) determination
显微荧光分光度(测定)法 microspectrofluorometry
显微荧光分光计 microspectrofluorimeter
显微荧光分析 microfluorescence analysis
显微荧光光度法 microfluorophotometry
显微荧光光度技术 microfluorophotometry
显微荧光计 microfluorometer
显微荧光镜 microfluoroscope
显微映象 microprojection
显微映象器 microprojector
显微硬度测定法 microhardness determination method
显微硬度计 microhardness instrument; microhardness scale; microhardness tester; microhardometer
显微硬度试验 microhardness test
显微硬度(值) microhardness
显微硬度锥头 microhardness head
显微阅读器 microfilm reader
显微载片 micromount
显微载片染色缸 Naples jar
显微照片 fotomicrograph; microcopy; micrograph; microphotograph; photomicrograph
显微照片的 megascopic
显微(照片)阅读器 microreader
显微照相 microphotograph; photomicrogram; photomicrograph
显微照相的 megascopic
显微照相机 photomicrographic camera; photomicroscope
显微照相卡片 microfiche
显微照相设备 photomicrographic apparatus
显微照相术 micrography; microphotography; photomicrography; photomicroscopy
显微照相图 microphotogram
显微折射计 microrefractometer
显微折射率测量法 microrefractometry
显微褶皱 microscopic fold
显微制图 micrography
显微组分 maceral
显微组分变种 maceral variety
显微组分含量 maceral content
显微组分亚组 submaceral group
显微组分组 maceral group
显微组织 microscopic construction; microscopic structure; microstructure
显微组织学 microhistology
显现 phanerosis
显现光泽 glossy up
显现时间 presentation time
显相纸 development paper
显像 developing process; development; discovery; display; photodevelopment; presentation; video picture; visualization
显像不足 underdevelopment; underexposure
显像存储示波器 storage oscilloscope
显像电话 video telephone
显像对比度 reproduced image contrast
显像法 visualization method
显像范围 development range
显像分辨力 reproduced image resolution
显像粉 toner
显像固定 display lock
显像管 Braun tube; converter tube; charactron; display tube; image reproducer; image tube; imaging tube; kinescope; oscillight; picture tube; teletron; teletube; television picture tube
显像管玻璃 funnel glass; teletube glass
显像管厂 kinescope factory
显像管框架 magic veil
显像管录像 teletranscription; kinephoto; kinescope
显像管上图像的录制 kinescope recording
显像管外壳 kinescope bulb
显像管阳极 picture tube anode
显像管荫罩 aperture mask; planar mask; shadow mask
显像管有九次幂调制特性的电视接收机 n-th power-law receiver
显像管罩 picture tube hood
显像记录 video recording
显像记录装置 video recording device
显像剂 developer; photograph developer; photographic(al) developer; replenisher
显像密度测定 densi(to)metry
显像密度计 densi(to)meter
显像面 picture plane
显像屏 picture screen
显像器 visualizer
显像清晰度 reproduced image resolution
显像三基色 receiver primary; display primary
显像时间 time of developing
显像误差 development error
显像雾翳 development fog
显像形式 kinoform
显像液槽汲取者 dipper
显像荧光屏 display screen
显像噪声 picture noise
显像帧面 picture frame
显像纸 developing-out paper
显形式 explicit form
显型 phenotype
显性度 degree of dominance
显性分析 proximate analysis
显性性状 dominant character
显压强度 wet-compressive strength
显要席 <古罗马教堂> tribunal
显因律 law of vividness
显影 developing; developing process; development
显影不足 underdevelopment
显影不足的 underdeveloped
显影参数 photographic(al) parameter
显影槽 development tank; processing tank
显影反差 development contrast
显影废水 development wastewater
显影缸 developing tank
显影管 development tube; image tube
显影罐 development tank
显影过程 development process; development stage; stage of development
显影过度 overdevelop(ment)
显影机 developer; developing machine
显影及定影 developing and fixing
显影剂 developer; developing agent; photographic(al) developer
显影剂条纹 developer streaks
显影胶片 developed film
显影粒子 developed grain
显影蒙翳 chemical fog
显影敏感性 photographic(al) sensitivity
显影墨 developing ink
显影盘 developing tray
显影屏打样法 cathode-ray tube proof; display roof
显影器 developer; development mechanical kit
显影设备 developing outfit; development equipment
显影时间 development time
显影速度 developing speed; developing velocity; emulsion speed; emulsion velocity; rate of development
显影条件 development condition
显影桶 developing tank
显影图 developed pattern
显影完毕 develop to finality
显影温度 development temperature
显影像过度 overdeveloping; overdevelop(ment)
显影效果 developing effect
显影效应 development effect; processing effect
显影液 developer; developing solution; photographic(al) developer; soup
显影液槽 dipper
显影仪 visualizer
显影纸 development paper
显影指南 developing instruction
显影中心 development center
显域土 mature soil; zonal soil
显约束 explicit constraint
显真饰面 natural finish
显震气压表 air barometer
显著 evidence; predominance; prominence[prominency]
显著差别 marked difference; significant difference
显著差异 marked difference
显著的 dominant; dramatic; effective; notable; noticeable; observable; protuberant; remarkable; salient; splendent; visible
显著的白色网状图案 striking network pattern of white colo(u)r
显著的疏忽 gross negligence
显著的水硬性石灰浆 eminently hydraulic lime mortar
显著地形 topographic(al) feature
显著点 significant point
显著方位树 prominent surveyed tree
显著浮雕装修 anaglyph
显著共振 prominent resonance
显著雷达目标 radar conspicuous object
显著目标 conspicuous object
显著目标一览表 list of conspicuous object
显著起伏 strong relief
显著实惠 tangible benefit
显著水平 conspicuous level; level of significance【数】
显著特征 outstanding feature
显著天气 significant weather
显著物标 conspicuous object
显著物标一览表 list of conspicuous object
显著误差 appreciable error; conspicuous error; significant error; gross error
显著下降 greater drop in yield
显著相关 significant correlation
显著性 significance
显著性比率 significance ratio
显著性测定 significance test
显著性测验 significant test; test of significance
显著性差异 significant difference
显著性分析 significance analysis
显著性概率 significance probability
显著性检验 significance test; test of significance
显著性检验组合 combination of significance tests
显著性界限 significance limit
显著性试验 significance test; test of significance
显著性水平 level of significance; significance level; significant level
显著性水准 level of significance; significance level
显著性条件 significant condition
显著性序贯检验 sequential significance test
显著因素图 marked factor figure

显著增长 substantial increase
显字灯 figure lamp
显字管 character display tube;charactron;typotron
显字示波管 charachtron
显踪同位素 label(l)ed isotope
显坐标 palpable coordinate

险波 scarp

险段 dangerous section;traffic hazard;vulnerable spot <堤防>
险恶的天气现象 threatening appearance of weather
险恶地 foul area;foul patch
险恶地方 nasty place
险恶海面 ugly sea
险恶气候 ugly and threatening weather
险恶天气 fiendish weather;foul weather;hostile weather;thundery sky
险工 <堤防> vulnerable spot
险礁 dangerous reef;dangerous rock;danger rock;vigia
险峻的 arduous;precipitous
险峻土地 steep land
险库 hazardous reservoir
险坡 dangerous hill;scarp
险区 danger area;dangerous zone;danger zone;foul area
险滩 cascade;critical bar;dangerous passage;dells;hazardous rapids;rapids
险滩段 cascade portion
险滩警灯 warning light
险象 danger sign
险性事故 bad accident;dangerous accident
险肇事故 near injuries
险阻的 bold

蚬蛤壳铲斗液压缸 clam cylinder

蚬壳铲斗 clamshell
蚬壳铲斗取料半径 scoop radius

藓纲【植】Masci

藓类 Musci
藓类地衣植物 mosslichen vegetation
藓类沼泽 high moor;moss moor
藓类植物 music
藓沼 black bog;bog

县 county;prefecture

县城 county seat;county town
县城规划 county-region planning
县道 county highway;county road
县电话中心局 county telephone central office

县法院书记官 clerk of county court
县干道 trunk road
县公路 county highway;county road <美>
县管理委员会 <美> board of supervisors
县规划道路 county planning
县级道路 county road
县级公路 county road
县给排水管理局 county water and sewer authority
县给水质量管理局 county regional water quality control authority
县界 county boundary;county line
县局 county telephone central office
县内自动电话局 community automatic exchange
县署大厦 prefectural edifice
县图 county map
县行政官 county manager
县域规划 country regional planning
县长 chief executive;county government
县镇地方道路 county road;county route
县镇功能和组织 county functions and organization
县镇规划 county planning
县镇建筑规范 country budding code
县镇总体规划 county master plan
县政府办公楼 court house
县政府所在地 county seat

现拌的裸露混凝土 exposed field concrete

现拌法 mixed-in-place method;mixing-in-place method
现拌混凝土 field concrete;job-mix(ed)concrete;site-mixed concrete
现拌机 mix-in-place machine
现拌石膏底灰灰浆 gypsum job-mix(ed)basecoat plaster
现拌现浇混凝土 concrete mixed and placed in site
现场 field engineering;job site;locality;locus in quo;scene;site;working site
现场安全 site security
现场安全监督 safe supervisor
现场安全检验单 job site safety checklist
现场安全员 site safety officer
现场安装 erection on site;fabricated on site;field connection;field erection;field installation;on-site installation
现场安装的 field-erected;field-installed;field-mounted;site-erected
现场安装的格笼 erected-in-place crib
现场安装费用 field-labor cost
现场安装工作 on-site installation work
现场安装和维修条件 installation and maintenance condition at field
现场安装螺栓 field bolt(ing)
现场安装容器 field assembly of vessels
现场按钮 field push button
现场搬运车 site handler
现场办公室 on-site office
现场拌和 field mix;in-place mixing;job mix;mixed-in-place;mixing at site;mixing-in-place;on-site mixing
现场拌和材料 materials-mixed in situ
现场拌和厂 site mixing plant
现场拌和的 site mixed
现场拌和法 mixed-in-place method;mixing-in-place method
现场拌和防渗墙 mix-in-place wall
现场拌和混凝土 job-mix(ed)concrete
现场拌和沥青 bituminous mix-in-place
现场拌和配方 job mix formula
现场拌和配合比 job mix
现场拌和试验 field mixing test
现场拌和土 mix-in-place
现场拌和装置 site mixing plant
现场拌灰板 spot board
现场拌制灰泥 job plaster
现场拌制混凝土 field concrete;job-mix(ed)concrete;site-mixed concrete
现场保险费 construction site insurance premium
现场保养 in the field
现场爆炸试验 field blasting test
现场边界 site boundary
现场边坡图 field slope chart
现场变异条件 differing site conditions
现场辨别 field discrimination
现场标志 site-marking
现场表面再生 in-place surface recycling
现场布局 site layout
现场布线 field wiring
现场布线图 field wiring diagram
现场布置任务通知单 field work order
现场布置任务通知书 field work order
现场布置图 site layout plan
现场部件 field unit
现场材料管理员 site checker
现场操纵 on-site handling
现场操作 site operation
现场操作测试 field operational test
现场操作规范 field work standards
现场操作条件 on-the-job condition
现场操作站 field station
现场草图 field sketch
现场测定 field measurement;in-place measurement;on-the-spot determination
现场测定加州承载比 in-site CBR[California bearing ratio]
现场测定应变 in-situ strain
现场测定应力 in-situ stress
现场测绘图 field sheet
现场测量 field measure;field measurement;field survey(ing);in-situ measurement;on-the-job measurement;original form survey;site measuring;site survey;locate survey
现场测量队长 party chief
现场测量法 field measuring technique
现场测量记录本 chain book
现场测量模量 field modulus;in-situ modulus
现场测试 field experiment;field measurement;field test(ing);in-situ test(ing);site test
现场测图 field mapping
现场查勘 site exploration;site reconnaissance;field investigation
现场场地勘察 field site reconnaissance
现场车间 site shop
现场车辆 on-site vehicle
现场衬里 in-situ lining
现场衬砌 in-situ lining
现场承载试验 field bearing test
现场承重试验 field bearing test
现场程序编制 field programming
现场尺寸 site dimensions
现场抽查 snap check;spot check(ing)
现场抽水试验 field pumping test
现场初始压缩曲线 field initial compression curve
现场处理 site treatment;field treatment
现场处理单元 local processing unit
现场处置 site disposal
现场触探试验 field sounding test
现场淬火 field quenching
现场存货 field warehousing
现场存料 material at site
现场打桩 impact cast-in-situ pile
现场大气污染 field atmospheric pollution
现场代表 field representative;projective representative
现场代理人 site agent;site representative
现场单件生产车间 job working shop
现场单元 field unit
现场道路 site road
现场的 field;in place;local;on-job;on-site;on-the-spot
现场灯具 site lantern
现场地势图 site topography
现场地形 site contour
现场地形图 site topography
现场地质特征 field geology feature
现场点 field location;field station
现场点选择按钮 location(selection)button
现场电码设备 field code station;field coding station
现场电码设备室 field code station housing
现场电码匣 field code station;field coding station
现场电码装置 code field installation
现场电视广播 live television broadcast(ing)
现场电视监测生产法 television monitored production
现场电视录像 field television video recording
现场电阻率测定 field resistivity measurement
现场雕凿的石材线脚 revale
现场调查 field investigation;field method;field survey;on-site examination;on-the-spot investigation;site inspection;site investigation;spot investigation;spot survey
现场调查设计 field survey design
现场调查研究 field study
现场调查员 field worker
现场调度 on-site dispatching;spot dispatch
现场钉的铆钉 field rivet
现场定测 layout in field
现场定量 in-situ quantification
现场定线 field location;field staking-out;on-site staking-out
现场定线工作 field location work;staking out work
现场动力试验 in-situ dynamic(al)test
现场对象 field function
现场发电站 site power plant
现场发电装置 site power plant
现场发泡 foaming in place;foaming in situ;foam-in-place;froth formation in situ;in-situ foaming
现场发泡隔热材料 site-foamed insulation
现场发泡绝热 foamed-in-place insulation;foamed-in-situ insulation
现场发泡塑料 foamed-in-situ plastics
现场发射 field emission
现场法 site method
现场反应 field feedback
现场范围 site coverage
现场仿真 field simulation
现场访问 facility visit
现场费用工程师 field cost engineer
现场分析 in-site analysis;in-situ analysis
现场焚毁 on-site incineration
现场服务 field service;on-the-spot service
现场服务车 on-the-job service unit
现场服务设施 field service unit
现场复查 field review
现场复核 field check
现场改装阀 field conversion;field conversion valve
现场更动通知 field order
现场工厂 on-site plant;portable factory

现场工程控制工程师 site control engineer
现场工程设计更改 field engineering design change
现场工程设计更改计划表 field engineering design change schedule
现场工程师 field engineer; site engineer
现场工程修改通知 field order
现场工棚灯 site hut lamp
现场工棚供热 site hut heating
现场工人 on-site labo(u)r
现场工作 filed work; on-site work; site work
现场工作记录本 field book
现场工作人员 field personnel; site staff
现场工作性能 field performance
现场工作指南 field guide
现场功能 field performance
现场供热 site heating
现场供应 site accommodation
现场估工 figuring-production on-the-job
现场估价 field evaluation
现场估算生产量 figuring-production on-the-job
现场估算师 field estimator
现场顾问 field adviser
现场观测 field observation; site observation
现场观察 field observation
现场观察员 field observer
现场管理 field control; field management; on-site management; on-the-spot administration; site management; worksite management
现场管理费 site administration expenses; site expenses
现场灌浆 site grouting
现场灌注混凝土 concrete in at site
现场灌注球基桩 Franki pile
现场灌注桩 caisson pile; site pile; bored pile; cast-in-place pile; in situ pile
现场灌注桩墙 bored pile wall
现场灌筑 cast-in-place; poured-in-place
现场灌筑的 job-placed; job-placed forming
现场灌筑混凝土 concrete in situ
现场广播 live broadcast
现场规章 site regulation
现场含水当量 field moisture equivalent
现场含水量密度试验 field moisture-density test
现场焊接 field weld(ing); on-site welding; site weld(ing)
现场焊接的 field welded; site welded
现场航测图 field sheet
现场核查 on-site verification
现场核数员 field auditor
现场荷载试验 field bearing test; field loading test; field plate loading test; in-situ loading test
现场划线 site-marking
现场回放装置 field playback
现场回收重新使用 in-place recycling
现场会 clinic
现场会计 field accountant; site accounting
现场会议 on-the-spot conference; on-the-spot meeting; site meeting
现场绘制草图 field sketching
现场混合 mixed-in-place
现场混合装置 spot-mix plant
现场混凝土构件厂 insley plant
现场混凝土立方体强度试验 field cube test
现场混凝土强度 strength in situ
现场活动 site activity
现场活化分析 in-situ activation analysis
现场机修工 site mechanic
现场及其通道占用权 possession of site and access thereto
现场极谱仪 field polarograph
现场计量 site measuring; field measurement; site measurement
现场计数 field enumeration
现场计算 field calculation; field computation
现场记录 field note; field record(ing); record on spot; site records
现场记录保留继电器 field stick register relay
现场记录簿 field book
现场技工 site mechanic
现场技术传授 on-site transfer of technology
现场继电器 field relay
现场继电器匣 field relay group
现场继电器组 field relay group
现场加工 field processing; on-site handling
现场加工车间 job workshop
现场加工钢筋 field bending
现场加工弯钢管 field bend
现场加工弯钢筋 field bend
现场加工制造的 field fabricated
现场架模 on-the-job
现场架设 field erection
现场监测 field monitoring; in-situ monitoring; on-site monitoring
现场监督 field supervision; site inspection
现场监督员 job superintendent
现场监工员 project supervisor <美>; clerk of work <英>
现场监控量测技术要求 specifications for in-situ monitoring measurement
现场剪切触探仪 in-situ shear sounding apparatus; iskymeter
现场剪切试验 in-situ shear test
现场检测 in-situ inspection; on-site inspection
现场检测方法 in-situ check and test method
现场检查 field examination; field inspection; on-site inspection; on-spot check; site inspection; spot checking; spot inspection
现场检查制度 system of on-site inspection
现场检验 field check; field inspection; floor inspection; in-situ inspection; inspection on spot; job site check; observation of the work; on-site inspection; spot inspection
现场建造 in-place construction
现场建筑师 site architect
现场鉴定 field evaluation; field identification; on-the-spot appraisal
现场交付 delivery on field; spot delivery
现场交货 delivery on(the) spot; ex-works; spot delivery
现场交货价格 ex point of origin
现场交通 <施工时> site traffic
现场浇捣的 cast-in-site; cast-in-situ
现场浇捣钢筋混凝土 cast-in-situ reinforced concrete
现场浇捣混凝土 job-placed; job-placed concrete; job-poured concrete
现场浇捣混凝土管 concrete pipes cast in place
现场浇捣加气混凝土 cast-in-situ aerated concrete
现场浇灌 cast-in-place; job-placed; poured-in-place
现场浇灌工程 in-situ casting work; site casting work
现场浇灌混凝土 cast-in-place concrete; job-placed concrete; job-poured concrete; site cast concrete
现场浇灌墙壁 wall cast in situ
现场浇灌桩 cast-in-place pile; site cast concrete pile
现场浇混凝土 cast-in-situ concrete; concrete in situ; in-situ reinforced concrete; poured-in-place concrete
现场浇混凝土板 cast-in-situ concrete slab; concrete slab in situ; in-situ reinforced concrete slab; poured-in-place concrete slab
现场浇混凝土胸墙 cast-in-situ concrete parapet wall
现场浇制场 site casting yard
现场浇制带壳桩 cast-in-place shell pile
现场浇制钢筋混凝土 reinforced concrete cast on site
现场浇制钢壳混凝土桩 cast-in-place cased pile
现场浇制构件 cast-in member
现场浇制混凝土 site cast concrete
现场浇制混凝土桩 site cast concrete pile
现场浇制桩 cast-in-place(concrete) pile; cast-in-situ(concrete) pile; mo(u)lded-in-place pile
现场浇制桩打桩机 cast-in-situ piling machine
现场浇注 cast-in-place; cast-in-situ; job-place; poured-in-place
现场浇注的混凝土桩 cast-in-situ concrete pile
现场浇注的箱桩 cast-in-place cased pile
现场浇注工程 in-situ casting work; site casting work
现场浇注混凝土 cast-in-place concrete; cast-in-site; job-placed concrete; job-poured concrete; cast-in-situ concrete
现场浇注加气混凝土 in-situ aerated concrete
现场浇注楼梯间 in-situ-cast stair(case)
现场浇注桩 cast-in-place pile; site cast concrete pile
现场浇筑 cast in situ; in-situ-cast; poured-in-place
现场浇筑大体积混凝土墙 in situ mass concrete wall
现场浇筑的 cast-in-place; cast-in-site; job-placed
现场浇筑的混凝土 in-situ concrete
现场浇筑的碾压混凝土 field placed roller-compacted concrete
现场浇筑钢筋混凝土墙 in situ reinforced concrete wall
现场浇筑工程 in-situ casting work; site casting work
现场浇筑灰浆 cast-in-situ mortar
现场浇筑混凝土 cast-in-place concrete; cast-in-situ concrete; job-placed concrete; job-poured concrete; site cast concrete; concrete cast in position; concrete placing in site; poured-in-place concrete; field concrete; in situ concrete
现场浇筑混凝土墙 cast-on-place concrete wall
现场浇筑楼板 in situ floor
现场浇筑墙 walling cast in situ
现场浇筑水磨石 cast-in-situ terrazzo
现场浇筑桩 site cast concrete pile; cast-in-place pile
现场浇筑钻孔桩 in-situ pile
现场浇铸 cast-in-place
现场浇桩场 on-site pile casting yard
现场胶黏[粘] field gluing
现场搅拌 field mix; in-place mixing; mixed-in-place; mixing at site; on-site mixing
现场搅拌厂 mix-in-place mixing plant; site mixing plant
现场搅拌法 mixed-in-place method
现场搅拌混凝土 job-mix(ed) concrete; site-mixed concrete
现场搅拌机 mix-in-place machine; site mixer; worksite mixer
现场搅拌装置 site mixing plant
现场校核高程 checked spot elevation at site
现场校正 spot calibration
现场校准 spot calibration
现场接点 site joint
现场接合 field joint; site joint
现场接线 field connection
现场结合 field joint
现场解决 be to deal with on the spot
现场紧急工程命令 emergency field order
现场进度计划 job site schedule
现场进度计划师 job site scheduler
现场进路 site access road
现场进入权 site entrance
现场浸渍 field impregnation
现场经费 site expenses
现场经纪 floor broker; pit broker
现场经理 field service manager; project manager; site manager
现场就地混合 site mixing in place
现场聚合 in-situ polymerization
现场决标 award at tender opening
现场决定 be decided on job-site; be determined on job-site; determination on site
现场开发 site development
现场勘测 site investigation; site survey; site reconnaissance
现场勘察 field investigation; inspection of site; on-the-spot investigation; site exploration; site inspection; site investigation
现场勘察测试 site surveying and test
现场勘察前准备工作 desk study
现场勘探 site exploration
现场考查 on-site examination
现场考察 site inspection
现场可更换部件 field-replaceable unit
现场刻度 wellsite calibration
现场刻度器 wellsite calibrator
现场客 floor trader
现场空间 site space
现场控制 field control; local control
现场控制电路 field control circuit
现场控制工程师 field control engineer
现场控制盘 field control panel
现场控制试验 field control testing
现场控制台 field control station
现场矿石质量管理 site ore supervision
现场拉伸 job site stretching
现场冷弯管子 field bend
现场冷再生 cold in-place recycling
现场立模 on-job forming; on-the-job forming
现场连接 field connection; site connection; site joint
现场联结 field connection; on-site connection
现场梁抗弯试验 field beam test
现场梁试验 works beam test
现场量测 in-place measurement; in-situ measurement

现场料斗 site bin
现场领导 field superintendent
现场录音 live recording
现场路面再生 in-place surface recycling
现场螺栓连接 site bolting
现场铆钉 site rivet
现场铆钉对接接头 site riveted butt joint
现场铆合 site riveting
现场铆接 field rivet(ing); site riveting
现场铆接的 site riveted
现场密度试验用取芯钻法 corduroy cutter method for field density determinations
现场密(实)度 <现场压实的密实度> field density; in-situ density
现场密(实)度试验 field density test
现场灭火活动 field operation
现场模架 on-the-job forming
现场模拟试验 site simulation test
现场模塑填缝料 field-molded sealant
现场模制 mo(u)lded-in-place
现场模制泡沫 in-situ-cast foaming
现场模制桩 mo(u)lded-in-place pile
现场抹灰打底 in-situ plasterwork
现场抹灰粉饰 in-situ plastering
现场碾压试验 field compaction test
现场培训 on-site training
现场配合 field mix
现场配合比公式 <沥青混合料等的> job mix formula
现场配合比设计 job mix design
现场配合公式 job mix formula
现场配料 field mix
现场配料设计 site mix formulation
现场配料组成 site mix formulation
现场配线 field wiring
现场配制泡沫隔绝层 foam(ed)-in-place insulation
现场喷洒绝缘材料 sprayed-in-place insulation
现场膨胀 expanding in situ
现场拼接 field connection; field splice
现场拼装 field assembly
现场频率可编程序音调传感器 field frequency programmable tone transmitter
现场平衡 field balancing
现场平衡法 field balancing technique
现场平面图 locate plan; site plan
现场平整 site preparation
现场评价 field evaluation
现场铺设的 job-placed
现场铺瓦 site tiling
现场砌筑砖工 in-situ brickwork
现场潜水系统 field diving system
现场强度 in-situ strength; situ strength
现场轻便铁路 field railway
现场清理 clean-up and move out; clearing of site; preparation of site; site cleaning; site clearance
现场清洗 in-place cleaning
现场情报 field intelligence
现场情况 field situation
现场取样 spot sample; spot sampling
现场燃烧 <采掘石油时在地下的> in-situ combustion
现场人员 field staff; site personnel; site staff
现场人员训练 field personnel training
现场入口 site approach
现场设备 field apparatus; field equipment; field plant; site accommodation; site plant
现场设备平面图 site plant plan
现场设计 field design; site design; site development
现场设计负责人 job captain
现场设施 site accommodation; site facility; site installation
现场设施计划 site facility program(me)
现场设施平面图 site facility plan; site installations plan
现场摄像机 field camera; live camera
现场渗透抽水试验 pumping test of field permeability
现场渗透试验 field permeability test; field seepage test
现场渗透试验法 field permeability test method
现场生产性能测定 on-farm performance
现场失效 field failure
现场施工 at-site construction; field construction; in-place construction; on-site construction; site operation; field coating <指涂料方面的施工>; field works
现场施工工程师 field control engineer
现场施工会议 field construction meeting; field production meeting
现场施工监督 field supervision of construction
现场施工检查 field check
现场施工维修涂料 site-applied maintenance coating
现场施工准备 preliminary operation
现场施工组织 site construction organization; site organization
现场湿度当量 field moisture equivalent
现场湿法事后处理 wet job-site aftertreatment
现场十字板剪切试验 field vane shear test
现场十字板试验 field vane test
现场实测浚挖数量 <俗称"下方">【疏】in situ volume
现场实践 site practice
现场实施 site operation
现场实习 workshop practice
现场实验 field experiment; on-the-site experiment; on-the-spot experiment
现场实验分析 analysis of field testing
现场实验室 on-job laboratory
现场实样 site sample
现场食品 ready foods
现场食堂 site canteen
现场使用 field operation; field service; field usage; field use
现场使用可靠性 on-the-job dependability
现场使用性能 field performance
现场事故 on-the-job accident; work accident
现场事故断开 emergency off local
现场视察 field inspection; observation of site; observation of the work; site inspection
现场试坑浸水试验 in-situ collapsibility test
现场试块试验 works cube test
现场试验 field experiment; field experimentation; field investigation; field test(ing); field test in place; field trial; full-scale experiment; inplace test; in-situ test(ing); in situ test in place; in-place test; on-job test; on-site test; site test; test in place; site trial; spot test; test in site; test on site
现场试验程序 field test procedures
现场试验的 field tested
现场试验法 field test method
现场试验室 field laboratory; site laboratory
现场收购 on-the-spot collection
现场手册 field manual
现场手绘 field sketching
现场数据 field data; field evidence; in-site data; site data
现场数据收集 field data collection
现场数字处理系统 on-site digital processing system
现场刷(油)漆 field painting
现场栓接 field bolted; site bolting
现场栓接的 field bolted
现场水文观测 field hydrological observation; field hydrological survey
现场说明 field description
现场松散体积配合比 mix proportion by loose volume
现场踏勘 field exploration; field inspection; field reconnaissance; reconnaissance trip; site inspection; site investigation
现场坍落试验 in-place slump test
现场提供 job offer
现场弹性系数测定 coefficient of elasticity test in situ
现场填充 in-situ filling
现场条件 field condition; on-site condition; site condition
现场调试 field survey
现场调整装置 field adjusting device
现场调制涂料 job-mix(ed) paint
现场贴面砖隔墙 ready-made tiled partition
现场贴砖 site tiling
现场铁路 site railway
现场通信[讯] on-scene communication
现场统计 field statistics
现场筒仓 site silo
现场投标 instant tender; on-site tender
现场图 site map
现场涂胶黏[粘]合 field gluing
现场涂漆的 field-coated
现场涂装 field application; field coating; field painting
现场土 in-situ soil; soil in place; soil in-situ
现场土木工程 civil works on site
现场土强度 in-situ soil strength
现场土壤 site soil
现场土壤剪切触探仪 iskymeter
现场土壤剪切计 iskymeter
现场土壤剪切仪 iskymeter
现场土壤试验 in-situ soil test
现场推销经理 field sales manager
现场挖出的散石 fieldstone
现场外的生产检查 off-site production
现场外培训 off-site training
现场外预制 off-site fabrication
现场弯筋作业 field bending
现场完成工作(项目) job made
现场围墙 site enclosure
现场维修 field maintenance; on-site maintenance; on-site repair
现场位置平面图 location plan
现场温度 temperature in situ
现场无需的设备或人员 <俚语> fifth wheel
现场显示 status display
现场协调 coordination on site; site coordination
现场信号施工队 field signal construction gang
现场信息 field intelligence
现场行驶速度 on-the-job travel speed
现场修建的 site built
现场修理 current repair; field repair; repair in-situ; running repair; spot-light repair; spot repair
现场修理车间 one spot; spot shop
现场修理线路 repair, inspect and paint track
现场压力计 in-situ pressiometer; in-situ pressure meter
现场压实度 field compacted density
现场压实曲线 field compaction curve
现场压实试验 field compaction test
现场压水试验 pressure test of field permeability
现场压缩曲线 field compression curve
现场研究 workplace study
现场验收 site acceptance
现场验收试验 field acceptance test
现场验证 field validation
现场验证过的 field proven
现场验证过硬件 field proven hardware
现场养护 field curing; job curing
现场养护的 job-cured
现场养护的(混凝土)圆柱试体 field-cured cylinder; job-cured cylinder
现场养护的(混凝土)圆柱体试块 field-cured cylinder
现场养护混凝土 job-cured concrete
现场养护圆柱体试件 field-cured cylinders
现场养生 job curing
现场业务 field service
现场仪表板 local panel
现场仪器操作 field instrumentation
现场以外的训练 vestibule training
现场因素 site factor
现场应力测量 in-situ stress measurement
现场应力测试 in-situ stress measurement
现场用短截套管作成的岩芯管 poorboy core barrel
现场用分级机 mill-type classifier
现场用间接法测定混凝土强度 estimated in situ cube strength
现场用湿度计 in-situ moisture meter
现场用压缩机 accessible compressor; field service compressor
现场油漆 field painting
现场油漆的 field-painted
现场预加应力 job site prestressing; site prestressing
现场预浇 site precasting; site prefabrication
现场预浇混凝土厂 field site precast concrete factory
现场预应力张拉 job stretching of tension; job tensioning
现场预制 site precasting; site prefabrication
现场预制吊装法 site prefabricated method
现场预制桩 in situ pile
现场预制装配工厂 on-site prefabricating plant
现场预装配 site prefabrication
现场运输 site transport; transportation on the site
现场运行 field operation
现场运行测试 field operational test
现场运行人员 field operator; operational site staff; site operation staff
现场载重试验 field loading test
现场再熔化 re-melting on site
现场再压缩曲线 field recompression curve
现场张拉 job site tensioning; site stretching; site tensioning
现场照明 site lighting
现场震中 field epicentre
现场支模 on-the-job forming
现场直剪实验 in-situ shear test

现场指导 field guide
现场指挥 on-scene commander
现场指挥员 field director
现场制 L 形短管 field elbow
现场制备混凝土 site concrete
现场制砂 in-situ sand
现场制弯头 field bend;field elbow;field pipe bend
现场制造 custom-built;field manufacturing
现场制作 fabrication on site;field fabrication;field work;site fabrication
现场制作的 site built
现场制作装配 built-on-the-job
现场质量管理 site supervision
现场质量控制 field quality control
现场质量控制试验 on-the-job control test
现场注水渗透试验 pouring-in permeability test
现场装接 on-site connection
现场装配 field assembly;field connection;field erection;field fabrication;on-the-job assembly;site assembly
现场装配的 field fabricated;site assembled
现场装配胶合板 field-applied plywood panel
现场装卸 on-site handling
现场装卸指挥人员 spotter
现场装载 circus loading
现场装置 site installation
现场装置平面图 site plant plan
现场准备工作 site preparation
现场准备计划 site preparation programme
现场资料 field data;site data;site information
现场自然状况调查 physical survey
现场总平面设计 site planning
现场总线 field bus
现场组织 site organization
现场组织设计 site planning
现场组装 dress up;field assembled;field assembly;field connection;onsite assembly;on-the-job assembly;site assembly
现场组装的 site assembled
现场组装的混凝土模板 built- in place form
现场组装式转轮 site fabricated runner
现场钻孔灌注桩 in-situ concrete pile
现场作业 field operation;field work;site operation
现场作业自动化 field automation
现钞 <尤指美钞> cash;currency note;long green
现钞汇率 cash rate
现成备用容量 ready reserves
现成参考资料 ready reference
现成产品 off-the-shelf
现成程序 canned program(me)
现成船舶 spot ship
现成的 hand-me-down;in stock;made-up;off the peg;off-the-rack;off-the-shelf;pick-up;ready-made;ready-to-wear;bought <指非定制的>
现成的成套待装配板料 shook
现成的物质生产条件 existing material condition of production
现成服装店 slop shop
现成副食品店 delitessen store
现成杆柱 grown spar
现成构件 fabricated parts
现成集成电路 off-the-shelf integrated circuit
现成解决办法 off-the-shelf solution
现成零件 off-the-shelf part
现成配料 ready-made mixture
现成品 ready-made articles
现成设备 off-the-shelf gear;off-the-shelf hardware
现成石膏抹灰拌料 gypsum ready-mixed stuff
现成食品 convenience foods;fast foods
现成食品店 delicatessen
现成水分 readily available moisture
现成制品 ready-made goods
现存 on hand;stock on hand
现存财产 existing assets
现存的工程 existing works
现存腐蚀 existing corrosion
现存公司 existing company
现存价值 carrying value
现存建筑物 existing building
现存量 standing crop;stock on hand
现存临时性住宅 existing residential building-temporary
现存企业 existing company;existing enterprise
现存燃料 net fuel
现存生物量 standing crop biomass;standing population
现存数量 quantity on hand
现存特惠关税 existing preferential tariff
现存物资 goods and materials in stock;stock in storage;stock on hand
现存永久性住宅 existing residential building-permanent
现存值 existing value
现存植被 actual vegetation
现存制成品 finished goods on hand
现存资本商品 existing capital goods
现存总数 amount on hand
现代 present generation;present period;recent epoch;recent period
现代板块 modern plate
现代冰川 contemporary glacier;live glacier;living glacier;present ice
现代冰川作用 contemporary glaciation;recent glaciation
现代剥蚀区分布图 distribution map of recent erosion area
现代部门 modern sectors
现代车用汽油 modern motor spirit
现代沉积 made ground
现代沉积期 period of modern deposition
现代沉积区分布图 distribution map of recent sedimentation area
现代沉积物 modern deposit;recent sediment
现代成熟的技术和设备 state-of-the-art technology and equipment
现代成盐作用 present salt precipitation
现代承重墙 contemporary bearing wall
现代城市 current city
现代城市轻轨快速交通系统 advanced light rapid transit
现代城市轻轨铁路交通系统 advanced light rail transit
现代冲积物 recent fluvial sediments
现代垂直运动图 map of recent vertical movement
现代大洋名称 name of recent ocean
现代的专业化大生产 modern specialized large-scale production
现代底部沉积(物) recent bottom sediments
现代地壳等升线图 recent isoanabase map of earth crust
现代地壳运动 modern crust movement
现代地貌 recent landform
现代地球铅 modern terrestrial lead
现代地形 live form;live landfill;live landform
现代电气设备 modern electrical equipment
现代动力与工程 <期刊> Modern Power Engineering
现代方法 modernism
现代非均衡增长理论 modern theory of uneven growth
现代风格 modernity;modern style
现代风格瓷器 modern style faience
现代风格的 modernistic
现代福利经济 modern welfare economy
现代复合材料 advanced composite material
现代高烟囱施工法 modern system
现代工业 modern industry
现代工艺 contemporary technique
现代工艺水平 the state of the art
现代工艺艺术 state-of-the-art
现代公路 <美国期刊名> Modern Highways
现代构造应力场 current tectonic stress field
现代构造应力场研究方法 study method of recent tectonic stress field
现代构造运动 recent tectonic movement
现代管理 modern management
现代国防 modern national defense
现代海洋底部铁锰铜矿床 manganese-iron-copper deposit on the modern ocean floor
现代河床 recent channel
现代河谷剖面图 profile of recent valley
现代河流沉积物 recent fluvial sediments
现代河流泥沙 recent fluvial sediments
现代核估计 modern kernel estimate
现代化 aggiornamento;modernization;update
现代化程度 up-to-dateness
现代化大工业 modernized large industry
现代化的 state-of-the-art;updated;up-to-date;up to the-minute;down-to-date
现代化的型式 modern version
现代化港口 modern port
现代化工具 <生活方面的> modern conveniences
现代化工艺过程 modernized technological processes
现代化管理 modernization management
现代化计划 modernization plan;modernization program(me);updating plan;updating program(me)
现代化技术 current technique
现代化建筑 modern architecture
现代化炼油厂 modern refinery
现代化趋势 modernization drive
现代化设备 modern equipment;sophisticated equipment;state-of-the-art facility;modern conveniences <住房的>
现代化设施 state-of-art facility
现代化水质模拟 state-of-the-art water quality modeling
现代化驼峰编组场 modern hump yard
现代化训练 updating training
现代化野营场地 modern campground
现代化运输 modern transport
现代化轧机 present-day mill
现代化之前 pre-modern
现代绘画 modern painting
现代机械化土方工程 modern mechanized earthwork
现代计算方法 modern computing method
现代技术 current technology;modern technology
现代技术(发展)水平 state of art;state-of-the-art
现代间歇泉 active geyser
现代建筑 contemporary architecture
现代建筑形式 style modern
现代经济管理 modern economic management
现代经济计量学 modern econometrics
现代经济学 modern economics
现代科学技术 modern science and technology
现代控制理论 modern control theory
现代类型 modern type
现代淋余层 neo-eluvium
现代农业 modern agriculture
现代派 modernism;modernist school
现代派式样 modern style
现代企业管理 modern enterprise management
现代气候学 neoclimatology
现代汽车 modern car
现代铅 modern lead
现代乔治式建筑 modern Georgian architecture
现代桥梁 modern bridge
现代桥梁建筑技术 modern bridge building technique
现代轻轨快速运输系统 advanced light rapid transit
现代珊瑚礁 biostrome
现代商业概览 <美> Survey of Current Business
现代式 modernism
现代数据输入技术 modern data entry techniques
现代数字计算机 modern digital computer
现代数字系统 modern digital system
现代水平运动图 map of recent horizontal movement
现代水系图 recent drainage map
现代税收制度 modern taxation system
现代陶瓷 advanced ceramics;contemporary ceramics;modern ceramics
现代体 modern face
现代通信[讯]技术 modern communications technology
现代统计学 modern statistics
现代文明病 modern civilized illness
现代系统理论 modern system theory
现代信号处理技术 modern signal processing technique
现代型模型及型芯制造法 modern mo(u)ld and core making process
现代型式 late-model;recent model
现代性 modernity;modernness
现代岩溶 modern karst
现代颜料 modern pigment
现代样式 modern style;up-to-date style
现代一般平衡理论 modern general equilibrium theory
现代艺术 modern art
现代英格兰建筑 modern English architecture
现代优化交通感应(控制) modern optical vehicle actuated
现代与当代建筑 modern and contemporary architecture

现代运输 modern transport
现代运输部门 modern transport sector
现代植被 recent vegetation
现地调查 on-the-site inquiry; on-the-spot inquiry
现地对照地形 checking the terrain at the actual ground
现地控制单元 local control unit
现地址 current address
现点 position of spectator
现付 cash payment; down payment; money down; pay on the spot; spot payment
现付成本 current outlay cost
现付费用 out-of-pocket cost; out-of-pocket expenses
现付金额 down payment
现付票 cash order
现购 cash purchase
现购发票 cash invoice
现购自运 cash-and-carry
现灌砂浆 in-situ-cast mortar
现焊三通 wrap-round tee
现患调查 prevalence survey
现汇 cash remittance; convertible foreign exchange; spot exchange
现汇汇率 spot exchange rate
现汇交易 spot exchange transaction
现汇结算 cash settlement
现货 actual goods; actual stuff; goods in stock; merchandise on hand; off-the-shelf; prompt goods; spot cargo; spot commodity; spot goods; spots; stock in hand; stock on hand
现货工作项目及内容 tally items
现货供应 off-the-shelf
现货购进 spot purchase
现货和期货购进 buying for both spot and forward delivery
现货价(格) price on spot; spot price
现货交付 spot delivery
现货交换 against actual
现货交易 over-the-counter trading; spot deal; spot trading; spot transaction
现货交易掮客 spot broker
现货买家 spot buyer
现货卖家 spot seller
现货贸易确认函 confirmation memorandum
现货贸易主合同 master sale & purchase agreement for spot trading
现货抛补 spot cover
现货期权 American option; spot option
现货契约 spot contract
现货日期的次日 spot-next
现货升水 backwardation
现货时价 spot quotation
现货市场 actual market; cash market; physical market; spot market
现货市场价格 actual market price
现货销售 cash sale; sell on the spot; spot sale

现货样品 actual sample; sample of existing goods
现货溢价 backwardation
现价 current price; current rate; current selling price; going rate; present price; ruling price
现价估算值 current-price estimate
现价美元估算值 current dollar estimate
现浇 in-situ-cast
现浇板缝 poured-in-place joint; site cast joint
现浇包壳(混凝土)桩 in-situ shell pile
现浇薄壳混凝土烟囱 shell-cast concrete chimney
现浇大体积混凝土岸壁 mass-in-situ concrete wall; mass wall
现浇带 pour strip
现浇的 casting; cast-in-place; cast-in-situ
现浇的混凝土面层 concrete bay
现浇地板 in-situ flooring
现浇地面 in-situ floor
现浇地坪 cast-in-situ floor
现浇电缆沟 in-situ cable duct
现浇墩 cast-in-situ pier
现浇墩子 in-situ pier
现浇防波堤 in-situ pier
现浇钢筋混凝土 cast-in-situ reinforced concrete; field reinforced concrete; in-situ reinforced concrete; poured-in-place reinforced concrete; reinforced in-situ concrete; site-placed reinforced concrete
现浇钢筋混凝土结构 cast-in-place reinforced concrete structure; cast-in-situ reinforced concrete structure
现浇钢筋混凝土肋 reinforced cast in-situ rib
现浇钢筋混凝土楼板 in-situ reinforced concrete floor
现浇钢筋混凝土楼面 field reinforced concrete floor
现浇钢筋混凝土螺旋桩 reinforced in-situ screw pile
现浇钢筋混凝土墙 cast-in-situ reinforced concrete wall
现浇钢筋混凝土桥梁 in-situ reinforced concrete bridge
现浇钢筋混凝土桩 shell pile Aba-Lorenz
现浇钢壳混凝土桩 cased pile
现浇构件 cast-in member
现浇管道 in-situ pipeline
现浇灰浆 cast-in-situ mortar; in-situ mortar
现浇混凝土 cast-in-place concrete; concrete cast in place; cast in situ concrete; concrete placed in situ; concrete placing in site; field architectural concrete; field concrete; in-situ-cast concrete; in-situ concrete; job-placed concrete; on-site concrete; poured-in-place concrete; site-placed concrete
现浇混凝土板 in-situ concrete slab
现浇混凝土薄壳 field concrete shell; in-situ concrete shell
现浇混凝土侧石梁 in-situ concrete curb beam
现浇混凝土衬砌外轮廓周边 extrados of cast-in-situ concrete lining
现浇混凝土带肋楼面 field concrete rib(bed) floor
现浇混凝土电缆管道 field concrete cable duct
现浇混凝土垫层 cast-in-situ concrete base course
现浇混凝土顶板 cast-in-place concrete ceiling
现浇混凝土顶层 cast in-place topping
现浇混凝土钢壳桩 metal-cased pile
现浇混凝土工程(施工) poured construction
现浇混凝土构架 field concrete frame
现浇混凝土管 cast-in-situ concrete pipe
现浇混凝土结构 field concrete structure; in-situ concrete structure; poured concrete structure; structural cast-in-place concrete
现浇混凝土框架 poured-in-place concrete frame
现浇混凝土肋 in-situ concrete rib
现浇混凝土楼板 cast-in-place floor
现浇混凝土楼面 field concrete floor
现浇混凝土楼梯 field concrete stair(case); in-situ-cast stair(case)
现浇混凝土楼梯斜梁 in-situ-cast string
现浇混凝土路面 cast-in-situ concrete pavement
现浇混凝土面层 in situ concrete pavement
现浇混凝土铺面 cast-in-situ concrete pavement
现浇混凝土施工 cast-in-situ construction
现浇混凝土填充 field concrete filling
现浇混凝土填料 in-situ concrete filling
现浇混凝土屋檐 field concrete eaves unit
现浇混凝土檐口构件 poured-in-place concrete eaves unit
现浇混凝土阳台 field concrete balcony; poured-in-place concrete balcony
现浇混凝土桩 cast-in-place concrete pile; cast-in-situ concrete pile; in-situ concrete pile; in-situ pile
现浇混凝土桩墙 cast-in-situ concrete piled wall; in situ concrete piled wall
现浇混凝土钻孔桩围堰 bored cast-in-place piling cofferdam
现浇加气钢筋混凝土 in-situ-cast aerated reinforced concrete
现浇加气混凝土 cast-in-situ aerated concrete; field aerated concrete; in-situ aeration concrete; site-placed aerated concrete
现浇加气轻质混凝土 in-situ-cast aerated light concrete
现浇加气砂浆 in-situ-cast aerated mortar
现浇建筑(工程) in-situ construction
现浇建筑混凝土 cast-in-situ architectural concrete; site-placed architectural concrete
现浇结构混凝土 structural field concrete; structural in-situ cast concrete
现浇锚固 cast-in-place anchor
现浇帽梁 cast-in-situ capping
现浇面层 in-situ-cast topping
现浇平台 in-situ landing
现浇桥台 in-situ pier
现浇轻骨料混凝土 in-situ lightweight-aggregate concrete
现浇轻质混凝土 field lightweight concrete
现浇砂浆 poured-in-place mortar; site-placed mortar
现浇施工方法 in-situ-cast construction method
现浇实心混凝土楼板 in-situ solid concrete floor
现浇式钢筋混凝土结构 monolithic reinforced concrete structure
现浇饰面混凝土 fair-faced cast-in-place concrete; fair-faced cast-in-situ concrete; fair-faced field concrete; fair-faced site-placed concrete
现浇水磨石 cast-in-situ terrazzo; in-situ terrazzo; monolithic terrazzo
现浇隧道 in-situ tunnel
现浇无接缝耐火材料 monolithic refractory
现浇胸墙 cast-in-situ breast wall
现浇预应力混凝土桩 in-situ-cast prestressed concrete pile
现浇柱 cast-in-situ pier
现浇筑地板 in-situ floor finishes
现浇桩 in-situ pile; situ-cast pile
现浇桩帽 in-situ cap; in-situ pile cap
现阶段 the present stage
现今地应力场 present ground stress field
现今构造体系 present tectonic system
现今构造应力场 present tectonic stress field
现金保证 cash guarantee
现金保证金 dollar margin
现金(报)表 cash statement
现金及现金等价物增加额 net increase in cash and cash equivalents
现金及银行结存 cash and bank balance
现金流动表 cash flow statement
现金流量图 cash flow diagram
现金流入 cash inflow
现金账 cash accounting
现金折扣 cash discount; discount cash
现净值法 net present value method
现觉显微光度计 visual microphotometer
现款 cash; cash on the bank; currency note; hard cash; money down; ready cash; ready money; spot cash
现款保险库 cash vault
现款采购 buy for ready money
现款订货 cash in order; cash with order
现款购买 cash purchase; purchase on cash
现款价格 spot price
现款奖金 cash reward
现款交割 cash delivery
现款交货 cash on delivery
现款交易 bargain on the spot; cash account; for love or for money; no credit; ready money business; for money < 伦敦证券所用语 >
现款交易市场 cash market
现款九折 make an allowance of 10 percent for cash payment
现款售价 cash price
现款提货 cash-and-carry
现款销售 sale by real cash; sales in hard cash
现款余额 cash balance
现款支付 net cash; payment by cash; payment in cash; ready payment
现款资金 cash fund
现况 existing circumstance
现况报告 status report
现况调查 inventory survey; prevalence survey
现况图 existing state plane map
现买现卖 hand-to-mouth buying
现卖 cash sale
现期 current period
现期船 spot ship
现期存货 existing stock
现期汇率 demand rate; spot exchange rate
现期汇票 demand bill
现期交易部 < 美国粮食交易所 > pit
现期纳税法 current tax payment act
现期期刊 current serial
现期书刊 < 指最近出版的一期 > expiry issue
现期外汇 spot foreign exchange
现期小麦交易部 < 美交易所 > wheat pit
现期债务 demand obligation
现扫比 blip-scan ratio
现色性 colo(u)r rendering
现色指数 colo(u)r index
现时安排 current arrangement

现时产量 current yield
现时成本 current cost
现时底泥污染 present-day pollution of bottom sediments
现时方位 present azimuth
现时高低角 present angular height
现时购买力 current purchasing power
现时环境 present environment
现时价格 up-to-date price
现时交通量 current traffic
现时角坐标 present angular coordinate
现时距离 present distance
现时可达标准 currently attainable standard
现时耐用性分类 present serviceability rating
现时耐用性指数 present serviceability index
现时生产的产品 current production
现时市价 current market price
现时数据 current data
现时水平 current level
现时水体状况 present-day state of water bodies;present-day state of waters
现时水污染 present-day pollution of water
现时提单 current bill of lading
现时天然水状况 present-day regime of natural water
现时烃污染 present-day hydrocarbon pollution
现时统计数字 current statistics
现时投入 current input
现时信息 current information
现时性能评级 present performance rating
现时盈余 current earnings
现时用户 active user
现时运量 current traffic
现时值 current value
现时指令 current order
现时重置成本 current replacement cost
现时资本价值 present capital value
现时资料 current data;current information
现时资料处理 current data processing
现时资源 current resources
现实的产生活动 real productive activity
现实的价值 actual value
现实的生产限度 practical capacity point
现实地质学 actuogeology
现实分析 factual analysis
现实感 feeling of reality;presence;reality sense
现实感丧失 derealization
现实购买力 current purchasing power
现实古生物学 actuopaleontology
现实解体 derealization
现实界系统 real-world system
现实经济学 applied economics
现实林 real forest
现实挑战 actual challenges
现实威胁 existing threat
现实性 actualism;feasibility;realizability
现实性条件 reality condition
现实蓄积量 actual growing stock
现实验证能力 capacity to test reality
现实仰角 present elevation
现实意义 immediate significance;practical significance
现实原则 reality principle
现实主义 realism
现实主义标准 criterion of realism
现实主义的 realistic
现实主义态度 realism attitude
现实资本 actual capital
现实最佳航向 best available true heading
现世的 secular;terrestrial
现世纪 recent period
现势图 advance chart;correction map;correction sheet
现势资料 current information;current intelligence
现收现付制 cash basis
现图法 cartographic(al)representation
现亡率 mortality
现象的 phenomenal
现象分析 phenomenological analysis
现象环境 phenomenal environment
现象特征 pattern of events
现象学 phenomenology
现象(学的)心理学 phenomenological psychology
现象学模型 <用于评价路面性能> phenominological model
现象自我 phenomenal self
现销 cash sale;sale by real cash
现销发票 cash sale invoice
现销价格 cash sale price
现销票 cash tickets
现行安全规程 applicable safety procedures
现行版 current edition
现行办法 current practice
现行比率 current ratio
现行标准 active standard;actual standard;current criterion;currently effective standard;current required standard;current standard;standard in force;working standard
现行标准成本 current standard cost
现行材料标准 applicable material standard
现行材料规格 applicable material specification
现行成本 current cost
现行成本会计 current cost accounting
现行船舶维修计划 current ship's maintenance project
现行的 actual;effective
现行抵押(契据)existing mortgage
现行地图 current map
现行地下水水质风险评价法 current groundwater quality risk-assessment method
现行定额 current rating
现行动向 current trend
现行兑换率 current rate
现行法律 current law;existing law;law in force;law in operation
现行法律条款 Existing Legislation Clause
现行方法 current method
现行费率 current rate;prevailing rate
现行费用 current cost
现行工资 prevailing wage
现行工作制度 current practice
现行估计值 current estimate
现行观测 current observation
现行惯例 current practice
现行规定 current criterion
现行规范 current code;currently effective code;current specification;existing code
现行规范和标准 applicable codes and standards
现行规则 current regulation;regulations in force
现行规章 current regulation;regulations in force
现行规章制度 rules and regulations in force
现行国内价格 current domestic value
现行汇率 current foreign exchange rate;current rate;prevailing rate;going rate
现行汇率法 current rate method
现行汇率折算法 current rate translation method
现行寄存器 actual register
现行价格 current cost;current price;existing price;financial price;going price;present price;prevailing price
现行价格法 going rate pricing
现行价率 current rate
现行价值 going value
现行建筑规范 present building code
现行交通量 current traffic
现行结构 existing design
现行决议 standing resolution
现行理想的标准成本 current ideal-standard cost
现行利率 current interest;current rate of interest;prevailing rate of interest
现行目录【计】current directory
现行平均日交通量 current average daily traffic
现行驱动器 current driver
现行趋向 current trend
现行日本工业标准 applicable JIS standard
现行市(场)价(格)current market price
现行市(场)价值 current market value
现行售价 current selling price
现行数据 active data;current data
现行数据处理 current data processing
现行水质标准 current water quality standard
现行税率 tariff in force
现行税制 current tax system
现行特惠关税 existing preferential duties
现行体系 current system
现行条例 current regulation
现行条令 current regulation
现行通令 standing order
现行限度 currency margin;current margin
现行行情股票 active stock
现行要求 current requirements
现行用途 existing use
现行优先次序指示器 current priority indicator
现行运价表 rate quotations
现行运量 current traffic
现行折合率 ar ruling rate
现行政策 present policy
现行支出机构 the current pattern of spending
现行值 current value
现行指令 current order
现行指令码 current instruction code
现行指令字 current instruction word
现行指示符 current indicator
现行重置成本 current replacement cost
现行转移指令 current transfer order
现行资料 current data
现已开采深度 nowadays mining depth
现役 active duty;active service
现役表 active list
现役船 commissioned ship;ship in commission
现役的 commissioned
现役舰队 active fleet
现役区 active area
现役数据区 active data area
现役栈 active stack
现役主项 active master item
现用表 active list
现用采购 buying for current needs
现用操作符 active operator
现用单元参考 current location reference
现用的 off-the-shelf
现用的地面支援设备 operational ground support equipment
现用放射性仪器设备 operational radiation instrumentation equipment
现用废物处理场 active waste disposal site
现用废物处置场 active waste disposal site
现用符号 current operation symbol
现用呼号 operational call sign
现用机场 active aerodrome;air port
现用基地 operational base
现用计算机 active computer
现用卡片 active card
现用量水装备 <生活用水的> active meter service
现用跑道 active runway;live runway
现用气象卫星系统 operational meteorological satellite system
现用区 active area
现用燃料 present-day fuel
现用摄影机 off-the-shelf camera
现用数据区 active data area
现用文档【计】active file
现用文件 active file
现用系统 off-the-shelf system
现用现买 buy as required
现用项目 off-the-shelf item
现用循环 active DO-loop
现用主文件 active master file
现用主项 active master item
现用资源调查系统 operational resource inventory system
现用钻机 active rig
现有 on hand
现有泊位 available berth;existing berth
现有材料 available material;existing material;materials now available;materials on hand
现有产品 existing products
现有车(辆)wagon on hand
现有车计划 plan of number of wagons on hand
现有车数 cars on hand
现有成本 present cost
现有成果 on-going results
现有船 existing ship
现有存货 stock inventory;stock on hand
现有道路 existing road
现有道砟 existing ballast
现有的 available;off-the-shelf
现有的和建造中的船舶 ships afloat and building
现有的抗震能力 existing seismic-resistant capacity
现有的权利 vested in possession
现有地面高程 existing grade
现有电话局 existing exchange
现有电量 available capacity
现有定货 order on hand
现有定货量 backlog of business;backlog of orders
现有房屋 existing house
现有风险 availability risk
现有服务能力 present serviceability
现有高程 existing grade
现有工厂能力 existing plant force
现有公路 existing highway

现有公用(事业)设施 existing utility
现有功能评定 present serviceability rating
现有功能指标 present serviceability index
现有供多住户居住的出租房屋 existing multifamily rental housing
现有管网 existing pipework
现有航道通过能力 available channel capacity;available waterway traffic ability
现有合流管网系统 existing combined sewerage system
现有和潜在的优越性 available and latent superiorities
现有河道 existing waterway
现有货物运量通知 advice of traffic on hand
现有机车 existing locomotive
现有机器的原价值 original cost existing machine
现有机械 existing machinery
现有基金 funds on hand; funds in hand
现有建筑 existing construction
现有建筑结构 existing structure
现有建筑物 existing building
现有交通 existing traffic
现有交通量 existing traffic volume
现有交通系统 existing traffic system
现有结构图 existing construction map
现有经验 present experience
现有井 existing well
现有抗震设计规定 existing seismic-resistant design provision
现有流量 existing flow
现有(路况)服务能力 present serviceability
现有路面 in service pavement
现有路线 existing route
现有路由 existing route
现有内存储量 available memory
现有能力 available capacity
现有平均日交通量 current average daily traffic
现有企业的扩充 expansion of established enterprises
现有契约限额 balance of contract
现有潜水器 existing submersible
现有桥 existing bridge
现有情报 existence information
现有人口 actual population; current population; present-in-area population
现有设备 available equipment;available facility;existing facility;existing equipment
现有设备安装率 ratio of installed equipment available
现有设备实际使用率 rate of actual use of available equipment
现有设计 existing design
现有设施 available equipment;available facility; existing equipment; existing facility;existing utility

现有生产能力 existing production capacity
现有生产设备 equipment facility;existing equipment
现有数据 available data
现有水位 existing water table
现有条件 existing condition
现有铁路 existing railroad
现有外汇额 exchange position
现有文件 active file
现有圬工 existing masonry
现有污水处理厂 existing wastewater treatment plant
现有污水系统扩建 expansion of existing sewerage system
现有物种 live species
现有物资 material on hand
现有线路改善 improvement of existing lines
现有项目 off-the-shelf item
现有型式 existing type
现有余款 balance on hand
现有雨量计网 existing rain gauge networks
现有运量 existing traffic
现有住房 existing housing
现有住房量 existing housing stock; housing inventory;housing stock
现有状态 available mode; standing state
现有资产扩建 addition to existing assets
现有资产增值 addition to existing assets
现有资料 available information;existing data
现有资料来源 existing data sources
现有资源 available resources
现有资源总量 total available resources
现有总收入 gross available income
现在代替当时 nunc pro tunc
现在高低角 present elevation
现在航向 present heading
现在价值 present worth
现在进行中的吹填工程 reclamation works presently underway
现在距离 present range
现在模式法 present pattern method
现在气象条件 present meteorological condition
现在收入 present income
现在水平距离 present ground range
现在天气 current weather; present weather
现在贴现值 present discounted value
现在位置 present position
现在位置距离 range at present position
现在斜距离 present slant range
现在蓄积 present yield
现在坐标 present coordinate;present-position coordinate
现值 current terms; current value; original cost to date;present value; present worth
现值比较法 present worth comparison method
现值标准 present value standard
现值法 <经济分析中的一种方法> present worth method;present value method
现值分析 present value analysis;present worth analysis
现值分析法 discounted cash flow analysis;present value approach
现值率 present value rate
现值美元 current dollar
现值投资费用 present worth capital cost
现值系数 single-payment present worth factor
现值因子 present worth factor
现值折旧 present value depreciation
现值指数 present value index
现职 current job; current post; present employment
现制楼地面 in-situ flooring
现制美术水磨石 in-situ artistic terrazzo
现制泡沫保温 site-foamed insulation
现制泡沫保温材料 foamed-in-place insulation material
现制水磨石 in-situ terrazzo
现制水磨石楼地面 in-situ terrazzo flooring
现制水磨石面 in-situ terrazzo facing
现制水磨石踢脚 in-situ terrazzo skirting
现制水磨石镶边 in-situ terrazzo border
现制塑料楼地板 in-situ plastics flooring
现制贴面 applied trim
现状报表 reports-status
现状道路 existing road
现状法 <用于交通预测> present pattern method
现状分区 actual zoning
现状分析 analysis of existing conditions
现状荷载 updated load
现状剪切强度 updated shear strength
现状条款 as they are;telquel
现状与动向 state-of-the-art

线 thread;wire;yarn;line <长度单位,1 线 = 1/12 英寸,1 英寸 = 0.0254米>

线凹盆 <岛弧区> nuclear basin
线把【数】bundle of lines
线摆 string pendulum
线斑状 linophyric
线板 spool
线棒保护层 bar armo(u)r
线棒绝缘 bar insulation
线棒扭弯器 bar cropper
线孢盾壳属 <拉> Linospora
线比例尺 bar scale
线闭合差 linear closure; linear discrepancy
线变位 linear deflection
线变位分量 component of linear deflection
线变形 <混凝土路面的> linear change
线变形模量 modulus of linear deformation
线标方程 line coordinates equation
线标准 line standard
线不相交道路 line-disjoint paths
线材 wire bar;wire rod;wire stock
线材成型机 wire forming machinery
线材尺寸 wire size
线材粗细(直径) thickness of wire
线材导板 wire guide
线材反复缠绕试验 snarl(ing) test
线材复绕机 respooling machine
线材工业 wire industry
线材规格 wire ga(u)ge
线材(滚)轧机 looping mill;rod mill
线材号数 ga(u)ge number
线材火焰喷枪 wire flame spray gun
线材火焰喷涂 wire flame spraying
线材剪切机 wire shears
线材矫直机 wire-straightening machine
线材卷 rod coil
线材卷架 wire-rod reel
线材卷盘条 wire-rod rolls
线材卷取和绕卷机 wire-coiling and winding machine
线材孔型 wire-rod pass
线材拉拔机 coil winder
线材拉后直接退火 direct patenting
线材拉直切割机 wire-straightening and cutting machine
线材拉制试验 drawing test
线材扭转试验机 wire torque tester
线材坯 rod stock;wire bar
线材漆 wire enamel;wire lacquer
线材清净作业线 wire-cleaning line
线材弯曲和成型机 wire bending and forming machines
线材无扭精轧机组 no-twist finishing block
线材压扁机 wire flattening mill
线材用电动机 wire motor
线材用硅青铜 cusiloy
线材轧辊 wire roll
线材轧机 rod-rolling mill;wire mill;wire-rod mill
线材轧制 rod rolling; wire milling; wire-rod milling;wire rope rolling
线材整直器 wire straightener
线材直径 ga(u)ge of diameter of wire
线槽 raceway; trunking; wire chase; wire duct;wire troughing;wireway
线槽盖 capping
线槽节距 slot pitch
线槽楼面 raceway floor
线槽漏泄 slot leakage
线槽脉动 slot ripple
线槽系数 slot factor
线槽占空系数 slot space-factor
线测标 linear measuring
线测验 line test
线叉 turnout
线叉线夹 overhead line crossing
线岔【电】cross-over;overhead crossing
线产生器 line generator
线长闭合差 <导线坐标闭合差平方和的根> linear closure
线长度 line length
线长改正【测】linear reduction
线长误差 linear error
线长延(伸)器 line stretcher
线超松弛 line overrelaxation
线超载 line surcharge load
线成像穿透计 wire image penetrameter
线尺比较 wire comparison
线尺寸 linear dimension
线尺度 linear content; linear dimension;linear dimensioning
线齿菌属 <拉> Grammothele
线虫的防治 control of nematode
线储量法 line reserve method
线穿透测厚仪 penetron
线丛 linear bundle;line bundle
线丛的次 degree of complex
线丛蓝面 complex surface
线丛曲线 complex curve
线粗 line weight;line width
线簇 line family
线存储器 linear memory;linear store
线代码扫描器 bar code scanner
线带 tape
线带卷支持器 coil holder
线带装置 belting wire;wire belting
线担 ally arm;cross arm;cross member;pole-arm <电杆>
线担撑脚 vertical brace
线道 drawing line; hole; line; string; thread;wire line
线的 lineal;linear
线的产状 attitude of line
线的交叉 crossing of wires
线的空间分布特征 spatial distribution characteristics of lines
线的宽度 line weight
线的平行踪迹 parallel trace of lines
线的指向 sense of line
线地电压 line-to-ground voltage
线地间电容 wire-to-earth capacity
线点对偶 higher pair
线电定向器 radio direction(al) finder
线电荷 line charge
线电流 line current
线电流计 single-string galvanometer
线电压 circuit voltage; line voltage; voltage between lines

线电压分布 line voltage distribution
线电压降电压表补偿器 line drop voltmeter compensator
线电压控制 line voltage controller
线电压起动 line voltage start
线电压限制器 line voltage limiter
线垫 latchet;tingle;lead tack < 砌砖用 >
线雕(刻)line cut;line engraving
线迭代 line iteration
线迭代法 line iteration method
线碟 saucer
线钉 wire nail
线定存储器 wired-in memory
线锭 wire bar
线独立数 line independence number
线度 dimension
线端 line end;terminal;terminal end
线端接线柱 terminal binding post
线端扎接 terminating
线段 line section;line segment;section of line;segmenting
线段编号 segment number
线段标识符 segment identifier
线段布线算法 line routing algorithm
线段裁剪 line clipping
线段裁剪算法 line clipping algorithm
线段近似法 broken-line analysis
线段轮廓面积 segments bound area
线段拟合 line-fit(ting)
线段误差 line error
线段相交 lines intersect
线段中心 centroid of length
线段字符发生器 stroke character generator
线段组 segment set
线断面样本 line transect sample
线堆模型 line-reactor model
线对 line pair;pair;wire pair
线对称【数】line symmetry
线对称图 line symmetric graph
线对错接 pair split
线对分裂 pair split
线对间电容 pair-to-pair capacity
线对接错 pair split
线对线故障 line-to-line fault
线对增容系统 pair gain system
线发射 line emission;line reflection
线反演 line inversion
线方位 course bearing
线分辨率 linear resolving power
线分布力 force per unit length
线分划 line division
线分裂 line splitting
线分析(法)line(ar)analysis
线风区 line fetch
线缝 linear slit
线缝探测器 seam detector
线伏组织 line structure
线幅 line width
线辐 wire spoke
线辐轮 wire spoke wheel
线辐射阻止本领 linear radiative stopping power
线负载 linear load
线覆盖 line covering
线覆盖数 line covering number
线刚度 flexural rigidity
线钢丝锯 jig saw
线跟踪 line-following
线功率 linear heat generation rate;linear heating power;linear heat rating;linear(power)rating
线沟漏斗 ribbed funnel
线估计方差 line term of estimation variance
线箍 binding clip;clamp;wire ferrule
线固定式 wire mounting
线挂水准器 string level
线管 spool
线管壳 conduit seal
线光谱 line spectrum[复 spectra]
线光顺 line smoothing
线光源 linear lighting source;line light source;line source
线规 divider;line width ga(u)ge;standard wire;wire ga(u)ge;wire size
线规板 wire-ga(u)ge plate
线规(编)号 wire-ga(u)ge number
线含矿率 linear ore ratio
线焊机 seam welder
线焊(接)line weld(ing);seam weld(ing)
线号 wire ga(u)ge;wire marking;wire size
线合理性 line justification
线荷模型 line charge model
线荷载 line load(ing)
线荷重 line load(ing)
线盒 joint box
线盒熔丝 fixture cut-out
线弧 bank;cross connecting field;terminal assembly;terminal block
线弧触排 link frame
线弧电缆 < 选择器架 > bank cable
线弧复式电缆 bank multiple cable
线弧盖 bank cover
线弧接头 line bank contact
线弧清拭器 bank cleaner
线弧容量 bank capacity
线弧组件 bank assembly
线化 linearize
线化法 linearized method
线化空气动力学 linearized aerodynamics
线化理论 linearized theory
线划板【测】detail plate;line copy plate;line plate
线划串 line string
线划粗度 line weight
线划地图修测 revision of line maps
线划端点 line ending
线划端正 line ending
线划符号 line symbol
线划复照片 line photograph
线划复照图 line photograph
线划跟踪 line-following;line tracing
线划跟踪器 line-follower;line tracer
线划跟踪装置 line-following device
线划光洁度 line sharpness
线划刻图 line engraving
线划宽度 line thickness;line width
线划平直度 line straightness
线划清绘 line drawn in ink;line work
线划清晰度 line acutance
线划数字化器 linear digitizer
线划填料 line filler
线划图 line drawing
线划图像 graphic(al)image;line image
线划图形 line figure;line image
线划拓扑结构 line topology
线划位移 plumb line deflection
线划无关【数】linear independence
线划形状 shape of line
线划阳片 linework positive
线划要素 line(ar)feature;line element
线划原图 line copy;line original;line print;line work;outline draft
线划质量 quality of line drafting
线划自动跟踪 automatic line-following;automatic line tracing
线划自动跟踪器 automatic line-follower automatic line tracer
线划综合 line generation
线环 eyelet;wire eye;wire eyelet;wire loop;wire thimble
线换向变流器 line commutated inverter
线簧继电器 wire spring relay
线簧式插头 formed wire contact
线汇 congruence of lines;line congress;line congruence
线汇的秩 rank of a congruence
线绘图机 line plotter
线"或"【计】wired-OR
线积分 curvilinear integral;line integral
线级数 line series
线计算 line computation
线夹 amphenol connector;cable clamp;cable cleat;clamp clip(terminal);cleat;contact wire clamp;crimped lock;fastener;terminal block;wire clamp;wire clip;wire connector;wire holder;wiring clip
线夹布线 cleat wiring
线夹端子 clip terminal
线架 coil holder;swift
线间电容 mutual capacitance;wire-to-wire capacitance;wire-to-wire capacity
线间电压 circuit voltage;voltage between lines;line-to-line voltage
线间短路 line-to-line short circuit;short-circuit between conductors
线间距 distance between tracks;midway between tracks;track center [centre] distance
线间距加宽 widening of line
线间耦合传播方式 line coupling transmission system
线间噪扰电压 line-to-line noise influence voltage
线剪 wire cutter;wire snips
线渐屈线 filar evolute
线渐伸线 filar involute
线交换集中器 line switching concentrator
线角 line angle;mo(u)lding
线角刨【建】beading plane;beading tool
线绞 skein
线脚 architrave;garnish mo(u)lding;mo(u)ld a form;mo(u)lding;skintle;skintled brickwork
线脚边缘 mo(u)lded edge
线脚材料 mo(u)lding material
线脚槽 < 室内陈设瓷器的 > plate rail
线脚槽刨 quirk router
线脚测绘器 cymograph
线脚成型 shaping of mo(u)lding;sticking
线脚成型机 sticker machine;sticker mo(u)lder
线脚的放大 enlarging mo(u)lding
线脚钉夹钳 brad setter
线脚端部 return ell
线脚端头 return head
线脚工作 mo(u)lded work
线脚固定块 skirting block
线脚灰泥 mo(u)lding plaster
线脚模板 mo(u)ld plate
线脚抹灰 running off
线脚刨【建】banding plane;beading plane;beading tool;bull-nose;mo(u)lding plane
线脚刨床 mo(u)lding planer
线脚切割器 mo(u)lding cutter
线脚砂光机 variety sander
线脚饰 nulling
线脚饰件样本 mo(u)lding pattern
线脚条 < 绕柱身的 > cymbia;cincture
线脚托座 foot block
线脚压割机 mo(u)lding machine
线脚压制机 mo(u)lding machine
线脚样板 horse;horsed mo(u)ld
线脚样模 horsed mo(u)ld
线脚支架 mitre shoot
线脚止端 stop mo(u)lding
线脚中的扇形饰 nulling
线脚终饰 running off
线脚转弯处 return head
线脚装饰 < 护墙板的 > base mo(u)lding
线脚装饰处理 entail
线接触 line contact
线接触钢丝绳 linear contact lay wire rope
线接触压力 line contact pressure
线接触轴承 line contact bearing
线接收器 line receiver
线接头 wire terminal
线接与门电路 implied AND circuit
线结 < 书脊顶带的 > beadroll
线截 line-intercept
线解法 nomography
线解图 alinement chart
线界限长度标准 line standard of length
线金属量 linear metal productivity
线劲度 linear stiffness
线经度 linear longitude
线径 linear diameter;wire diameter
线径测量计 line width ga(u)ge
线径测量装置 wire ga(u)ging equipment
线径规 stubs ga(u)ge
线矩阵 wire matrix
线距 interval between lines;line distance;line width;wire distance
线飓 line squall
线锯 bow saw;coping saw;fret saw;jig saw;scroll saw;trepan;trephine;wire saw
线锯床 scroll sawing machine
线锯导引器 wire saw guide
线锯导子 wire saw guide
线卷 coil of wire;rope reel
线卷弹簧 coil spring
线卡 wire ga(u)ge
线卡子 guy clip
线开光源 linear light source
线抗拉强度试验机 single yarn strength tester
线刻技术 line cutting technique
线控振荡器 line-controlled oscillator
线控制 line control
线宽 line spacing
线宽比例因子 line width scale factor
线宽度 line width
线宽化 line broadening method
线宽控制 line width control
线宽控制技术 line width control technique
线宽特性 line width characteristic
线框 line boundary
线捆扎 line tying
线扩充器 line stretcher
线括弧 vinculum[复 vinculums/ vincula]
线拉伸 line drawing
线拉条 wire bracing
线缆皮 wire covering
线缆涂料 wire coating
线缆最大弛度 maximum sag of cable
线缆最大垂度 maximum sag of cable
线勒子 cutting ga(u)ge;marking ga(u)ge
线离子密度 linear ion density
线理 lineation
线理成因类型 genetic(al)type of lineation
线理的方位【地】orientation of lineation
线理的世代 generation of lineation
线理构造 lineation structure
线理观测点 observation point of line-

ation
线理类型 type of lineation
线理特征 feature of lineation
线理与褶皱轴关系 relationship between lineation and fold axis
线理运动类型 kinematic(al) type of lineation
线理状结构 striated structure
线理状煤 striated coal
线连接 wire joint
线连通度 line-connectivity
线连通图 line-connected graph
线连续性系数 coefficient of linear continuity
线链 linear chain
线链架 line-link frame
线链控制器 line-link controller
线量误差 linear discrepancy
线列 alignment
线列传感器 line-array sensor
线列传声器 line microphone; machine-gun microphone
线列声源 linear array of simple source
线列水听器 line hydrophone
线列陷波电路 line trap
线列(阵)接收机 linear array receiver
线裂隙率 fissure ratio along a line
线流 filament of water; filamentous flow; linear flow; streaming flow; streamline flow; filamental flow
线炉 line oven
线路 line(link); route; track; network; wiring; circuitry; circuit【电】; railway line【铁】; permanent way
线路安全标准 track safety standard
线路百米标 picket post
线路班 < 保养路轨工作班 > track maintenance gang
线路板 breadboard
线路板厂废水 circuit board plant wastewater
线路保险开关 line protection breaker
线路保养 line service
线路闭塞 block occupied; route locking
线路闭锁 track blocking
线路避雷器 leakage conductor; line gap
线路避雷针 line lightning arrestor
线路边侧 side relative to track
线路边沟 track ditch
线路变形公差 distortion tolerance of track
线路变压器 line transformer
线路标 marker
线路标志 route sign; track post; track sign; wayside marker; wayside sign
线路标桩 track marking
线路别疏解 track oriented decrossing; untwining for leading lines
线路病害 track damage; track defect; track deterioration; track fault
线路病害情况 description of track damages
线路不通 service interrupt(ion)
线路布置 circuit layout; track arrangement
线路布置图 track plan; wiring diagram; schematic wiring diagram
线路布置图照明盘 track diagram panel
线路参数 circuit value; line parameter
线路残余电流 line residual current
线路侧 line side; wayside
线路测定 staking of track
线路测绘 wire plotting
线路测绘面积 geologic(al) mapping area of road
线路测量 line survey(ing); route survey(ing)
线路测试器 circuit tester
线路测试入口点 line test-access point
线路测试终接器 line test connector
线路测试子系统 line testing subsystem
线路测验车 line inspection car
线路插头 wiring plug
线路差错 line error
线路拆除 removal of track
线路拆移 removal of track
线路长度 length of line; track line
线路常规检查 route inspection
线路常数 line constant
线路抄平 track level(l)ing
线路潮流方程式 line flow equation
线路车辆识别点 wayside identification point
线路乘客总量 line ridership
线路充电 line charging
线路充电电压 line charge voltage
线路充电容量 line charging capacity; line charging power
线路出口 line outlet
线路出清 line clear
线路出清程序 line clear procedure
线路传输 line transmission
线路传输频带 line transmission frequency band
线路传输频谱 line transmission frequency spectrum
线路传输群 line transmission group
线路传输误差 line transmission error
线路传送速度 line speed
线路垂直交叉 route right-angle intersection
线路瓷夹 split knob
线路错接 misconnecting
线路打扰 line hit
线路大修【铁】track overhauling; major repair of track; overhaul and upgrading of track; track renewal
线路代数 circuit algebra
线路带电 line charging
线路单段平衡 single section balancing
线路单元 line element
线路档距 line span
线路导管 circuit conduit
线路导纳 line admittance
线路导频报警 line pilot alarm
线路导体 line conductor
线路导线 line conductor; line wire; route traverse【测】
线路导线起始张力 initial conductor tension
线路导行方法论 route guidance methodology
线路导行微机 route guidance microcomputer
线路倒换环 line-switch ring
线路到列车间的无线通信[讯]频率 wayside-to-train frequency
线路道钉 track spike
线路的 permanent way
线路的接通 continuity of the line
线路灯 line lamp
线路等级 class of track; line classification; track classification
线路等值电容 equivalent line capacitance
线路低通滤波器 line low-pass filter
线路地面设备 wayside equipment
线路地图 route map
线路点 field location; field station
线路电池 line battery
线路电导 line conductance
线路电杆 line pole
线路电感 line inductance
线路电抗器 feed(er) reactor
线路电缆 line cord
线路电流 line current
线路电流变换器 line current transducer
线路电流表 line milliammeter
线路电流计 line current tester
线路电路 line circuit
线路电码匣 line code station
线路电码装置 code field installation
线路电纳 line admittance; line susceptance
线路电平 line level
线路电气常数 linear electric(al) constant
线路电容 line capacitance
线路电容器 line capacitor
线路电容器电压变换器 line capacitor voltage transducer
线路电压 line voltage
线路电压变化 line voltage variation
线路电压变换器 line voltage transducer
线路电压降补偿 line drop compensation
线路电压降(落) line drop
线路电压调整 line regulation
线路电压调整器 line voltage regulator
线路电晕 line corona
线路电阻 line resistance
线路吊牌信号 line drop signal
线路调度员 route controller
线路定期防护设备 track protected equipment
线路定位 line location
线路定向 determination of line direction
线路端 line terminal
线路端接容量 line termination capacity
线路断开 line break; line release; circuit loop break < 在靠近一对绝缘子处的 >
线路断开装置 cut-out
线路断流器 line circuit-breaker
线路断路器 circuit breaker; line breaker; line circuit-breaker
线路对地 line-to-ground
线路对地电压 line-to-earth voltage
线路对地短路 line-to-ground short circuit
线路钝角交叉 obtuse crossing
线路多段平衡 multisection balancing
线路扼流圈 line choking coil; screening protector
线路发码 line coding
线路发送部件 line transmitting unit
线路发送单元 line transmitting unit
线路发送环节 line transmitting link
线路繁忙时 busy time of line
线路繁忙时间 route busy time
线路反接电键 line reverse key
线路方程式 equation of line
线路方向 track alignment
线路仿真器 line simulator
线路放大器 line amplifier
线路放大器架 line amplifier bay
线路放样 positioner of a line; positioning of line
线路非直线系数 line non-linear coefficient; route non-linear coefficient
线路分段 line sectionalizing
线路分割器 line splitter
线路分类 classification of line; line classification
线路分配器 link allotter
线路分析 circuit analysis
线路分向滤波器 line separating filter
线路分支 line tap
线路分组 grouping of tracks
线路封闭 closure of track
线路封闭标志 track-closed post
线路封锁 blocking up of track; line closure; track closure
线路封锁时间 blockage time of track
线路敷设 wiring layout
线路敷设方式 laying mode of route
线路敷设图 wiring plan
线路符号示意图 schematic diagram
线路辐射 line radiation
线路负荷 line load
线路负荷强度 line load intensity
线路附加网络 line building-out network
线路复测 repetition survey of existing; resurvey of line
线路复用 channel compression
线路改建 line reconstruction; track reconstruction
线路改进 line development
线路改坡 line regrading; regrading of line
线路改善 betterment to the lines; track improvement
线路改移 alignment deviation
线路干扰 line hit
线路感温器 line heat detector
线路感应器 track inductor; wayside coil; wayside indicator
线路刚度 track rigidity
线路高程控制测量 route plane control survey
线路高通滤波器 line high-pass filter
线路隔波器 line trap
线路隔离继电器 line isolation relay
线路隔离开关 line disconnecting switch
线路隔离器 line pad
线路更换 track replacement
线路更新 relaying of track; renewal of track; track renewal
线路工 line(s)man; plate layer; trackman; wire man
线路工程 line work
线路工程地质测绘 engineering geology mapping of road
线路工程队 line crew
线路工程师 line engineer; permanent way engineer
线路工具 track tool
线路工区 permanent way gang
线路工区工长 ganger
线路工区工具房 section tool house
线路工区工人 line(s)man; plate layer; section man; trackman; trackwalker
线路工区工人摇车 platelayer's troll(e)y
线路工区领班 ganger
线路工区领工员 gang foreman
线路工区用车 gang car
线路工人检电器 lineman's detector
线路工作 line work; permanent way work; track work
线路工作程序 permanent way program(me)
线路工作计划 permanent way program(me)
线路公里里程 track kilometrage
线路功率 line power
线路功率谱 line power spectrum
线路共享 wire sharing
线路共享系统 wire-sharing system
线路固定标桩 track fixed stake
线路固定的 hard-wired
线路固定使用 specialization of tracks
线路故障 line failure; line fault; track failure; track obstruction
线路故障探测装置 line fail detection

set;line fault anticipator
线路关闭 closure of line
线路规程 line discipline
线路和建筑物 permanent way and structures
线路和中继线群 line and trunk group
线路荷载 load of roadway
线路横向水平 cross level of track
线路弧刷 line wiper
线路环节 line link
线路环路 line loop
线路缓冲器 line buffer
线路换态过程 switching process
线路回路 line loop
线路回路操作 line loop operation
线路机械化更新 mechanical track renew
线路畸变 line distortion
线路激励放大器 line driving amplifier
线路激励器 line driver
线路集中单元 line concentrating unit
线路集中器 line concentrator
线路集总平衡 balancing by balancing elements
线路几何尺寸 route geometry
线路几何学 route geometry
线路记录 line record
线路继电器 line relay
线路继电器室 wayside housing
线路加长器 line lengthener
线路加固 track strengthening
线路加速段 acceleration section
线路架设 construction of line; line construction
线路间距 midway between tracks
线路监控设备 line monitoring equipment
线路监视器 actual monitor;line monitor
线路监听器 line monitor
线路检查 inspection of line; inspection of track; line inspection; route inspection
线路检查员 permanent way inspector
线路检索 route search
线路检修工 line man
线路建筑长度 length of constructed line; length of construction line; length of line constructed
线路交叉 crossing; crossing of lines; crossing of tracks; track intersection; transpose
线路交叉方案 transposition plan
线路交叉平衡 balancing by transposition
线路交叉形式 route intersection type
线路交换 circuit switching; line switching
线路交换集中器 line switching concentrator
线路交换网络 circuit switching network
线路交换系统 line switching system
线路交换型 line switching type
线路交换制 line switching system
线路交织 <在标准轨与窄轨或宽轨之间> interlacing of tracks
线路矫正 rectification of alignment
线路接触器 line contactor
线路接地 line-to-ground
线路接地冲击危险 line-to-ground shock hazard
线路接地故障 line-to-ground fault
线路接地事故 line-to-earth fault
线路接法 wiring process
线路接口 line interface
线路接口部件 line interface unit
线路接口控制器 line interface controller
线路接口硬件 line interface hardware
线路接收部件 line receiving unit
线路接收单元 line receiving unit
线路接收环节 line receiving link
线路接收机 line receiver
线路接帚 line wiper
线路节流阀 line choke
线路解锁 track release
线路解锁电路 line freeing circuit
线路尽头站 <缆索铁路和架空索道> line-end station
线路浸水 inundation on tracks
线路经常维修 current track maintenance; track current maintenance
线路净衰减 overall line attenuation
线路净损耗 circuit net loss
线路距离 route distance
线路决定 route determination
线路绝缘 line insulation
线路绝缘子 line insulator
线路绝缘自动测试器 automatic line insulation tester
线路均衡器 line equalizer
线路喀啦声 line scratch
线路开关 line breaker; line switch(ing); wiring switch
线路开关网 circuit switched network
线路开通 block cancel(l)ed; block cleared; block released; line clear; releasing track; track clear
线路开通表示 track release indication
线路开通电报 line clear telegram
线路开通电路 line freeing circuit
线路开通解锁 line-clear release
线路开通示像 line clear aspect
线路开通通知 track release advice
线路开通显示位置 proceed position
线路开通信号示像 line clear signal aspect
线路开通证 line clear message
线路开销 line overhead
线路勘查 line location; location of line
线路勘察 investigation of linear road
线路可靠性 circuit reliability
线路可用率 link availability
线路客流周转量 line passenger circulation
线路空闲 line clear; line free; line idle
线路控制 line control
线路控制程序 line control program(me); line control routine
线路控制分程序 line control block
线路控制符号识别器 stunt box
线路控制计算机 line control computer
线路控制继电器 line(impulse) control relay
线路控制块 line control block
线路馈电 line feed
线路馈电表 line feeder
线路扩建 track expansion
线路拉线 line guy
线路雷电特性 line lightning performance
线路里程 route kilometrage; route mileage; trackage
线路理论通过能力 theoretic(al) line capacity; theoretic(al) track capacity
线路连接 track junction
线路连接命令 connect data set to line
线路连接器 line connector
线路连接设备 line connection equipment
线路连结 track connection
线路链路控制分站 line-link sub-control station
线路链路控制站 line-link control station
线路零件 circuit element
线路领工员 track-foreman
线路流通量 line flow
线路留车表示器 detached vehicle indicator
线路滤波器 line filter
线路滤波器平衡 line filter balance
线路路基 infrastructure
线路路由 routing of route
线路螺栓 track bolt
线路码 line code
线路码型 line code pattern
线路码组 line block
线路埋深 buried depth of route
线路忙音 line-busy tone
线路模拟器 link simulator
线路模型 circuit model
线路末端交换台 line terminal switchboard
线路难行地段 hard point of track
线路能力 line capacity
线路排水 track drainage
线路旁扫描器 track-side scanner
线路配件 line fittings
线路配线架 circuit patch bay
线路偏置 line bias
线路频带分隔滤波器 line separating filter
线路频率 line frequency
线路平衡 line balance; line balancing
线路平衡变换器 line balance converter
线路平衡度 line balance
线路平均负荷度 average road carrying intensity
线路平面 line plan; route plan
线路平面及纵断面 line plan and profile
线路平面控制测量 route plane control survey
线路平面图 line plan; track plan; wiring diagram【电】
线路平行错移 tracks by parallel shifting apart
线路平行交织 route parallel weaving
线路平纵面 route plan and profile
线路坡度 track gradient
线路坡度图 grade diagram of line; gradient diagram of line
线路坡度整修 line regrading; regrading of line
线路铺设测量 track-laying measurement
线路普查 reconnaissance survey
线路起动电动机 line-start motor
线路器材 line material
线路器具 line hardware
线路桥隧养护 maintenance of way and structures
线路清理班组 brush gang; brush party; brush team
线路区段 track section
线路区间 railroad section
线路曲率半径 curvature radius of track; track curvature radius
线路曲折因数 line non-linear factor
线路全长 total track length
线路容车辆数 car capacity of track
线路容车量 car capacity of track
线路容量 circuit capacity【电】; line of capacity; line capacity; line carrying capacity; track capacity of line【铁】
线路容许速度 permissible speed of a line
线路入口 line inlet
线路塞孔 line jack
线路塞绳 line link
线路塞绳架 line-link frame
线路扫描 line scanning
线路上部建筑 superstructure of track
线路上油库 line depot; line tank farm
线路设备 ground equipment; line equipment; line facility; line plant; roadside apparatus; roadside equipment; roadway equipment; track facility; track set; track-side equipment; wayside equipment
线路设备继电器 field relay
线路设计 wiring layout
线路设施 track facility
线路伸延器 line stretcher
线路生产法 line production method
线路失真 line distortion
线路施工机械 <指铺轨、铺碴等机械> permanent way construction machinery
线路施工现场速度信号 speed signal for track construction site
线路时间曲线 track-time curve
线路识别连接器 line marker connector
线路识别装置 line identification equipment
线路使用程度 line utilization
线路使用权 <铁路与铁路间> trackage rights
线路使用系数 line-use ratio
线路示意图 conspectus; diagrammatic map
线路试验车 track testing troll(e)y
线路试验器 line tester
线路试验中继线 line test trunk
线路适配器 line adapter[adaptor]
线路释放线夹 free-center-type clamp
线路输出 line output
线路输出变压器 line output transformer
线路输入 line input
线路数 number of lines
线路数据 <计算列车运行时分用> track data
线路衰减 line attenuation
线路衰减器 line pad
线路双交叉 scissors track
线路水准 track level
线路水准测量 route level(l)ing
线路松动 decomposition of track; strength failure of track
线路速度 circuit speed
线路速度曲线 track-velocity curve
线路速率 line speed
线路损耗 line loss
线路损失 line loss
线路损失率 rate of electricity loss from transmission line
线路所 block house; block office; block post; block signal box; block station; intermediate block post; token station
线路所管理员 block pose keeper
线路所值班员 block operator; post keeper; signalman; station operator
线路锁定 anchor(ing) of track
线路踏看 route reconnaissance
线路弹性 resilience of track
线路特性 line characteristic
线路特性失真 line characteristic distortion
线路调节 line conditioning
线路调谐 line tuning
线路调整 circuit conditioning
线路调整率 line regulation
线路停车站 line station
线路停电 line outage
线路停电率计算器 line-outage calculator
线路停用 line out of service
线路通过能力 capacity of track; carrying capacity of track; line throughput capacity; rail capacity; track ca-

pacity; line carrying capacity【铁】; road capacity【道】
线路通量 line flux
线路通路点 line access point
线路通信[讯]量 line traffic
线路同步 line synchronization
线路图 plan of wiring; scheme of wiring; circuit (diagram)【电】; layout chart; line chart; line diagram; line drawing; line(-route) map; network chart; route map; branching program(me); schema[复 schemata]; track map【铁】
线路图寻迹 tracing
线路图追索 tracing
线路网 line network
线路网密度 network density
线路维护 line upkeep
线路维护工 plate layer
线路维护规程 line rule
线路维修 line maintenance; maintenance of tracks; maintenance of way
线路维修标准 standard of track maintenance
线路维修规则 line rule
线路维修考核图 control chart for track maintenance
线路维修设备 track maintenance equipment
线路维修统计表 track maintenance statistic chart
线路维修(总)登记簿 ledger for track maintenance
线路无线电台 wayside radio
线路无载电流 line charging current
线路系统控制功能块 line system control block
线路系统重力仪 gravimeter with linear system
线路匣 track side cubicle
线路下沉 subsidence of track
线路衔接器 line adapter[adaptor]
线路限界信号 <警冲标> limit signal of track
线路陷波器 line trap
线路箱 panel board box; panel board cabinet; panel box
线路协议 line protocol
线路斜交 route skew intersection
线路信号 lineside signal; track-side signal; wayside signal(ling); line signal
线路信号编码 line signal code
线路信号标 wayside sign
线路信号单元 line signal element
线路信号点 wayside signal location
线路信号电路 lineside signalling circuit
线路信号发送连接器 line signal sender connector
线路信号机【铁】lineside signal; trackside signal
线路信号示像 wayside signal aspect
线路性能 link performance
线路修复 line repair
线路蓄电池 line accumulator
线路选定 route planning
线路选通监察器 line strobe monitor
线路选位 line location
线路选线 route selection
线路选择 selection of route
线路选择器 line selection unit; line selector
线路巡查员 ridge-runner
线路循环试验 loopback test
线路压差(率) line regulation
线路压降补偿器 line drop compensator
线路延长 extension of line; extension of track; track extension
线路延展 route development
线路延展长度 extended length of line
线路研究 corridor study
线路养护 maintenance of way; track maintenance
线路养护班 track maintenance gang
线路养护账目 maintenance of way account
线路业务 line activity; line work
线路业务调度 line traffic coordination
线路业务量 line traffic
线路影响 line influence
线路有车表示器 vehicle-on-line indicator
线路有效长度 effective length of track; effective track length
线路与标高 line and level
线路与高程 line and level
线路元件 circuit element
线路元件测试 roadway element test
线路元件点 roadway element location
线路元件故障 roadway element failure
线路园林化 landscape gardening
线路远端 far end; rear end of line
线路允许速度 permissive speed on track; railway permitted velocity
线路运营长度【铁】operating distance of track
线路杂音 line noise
线路载客量 line capacity
线路再生中继机 line regenerative repeater
线路噪声 circuit noise; line noise
线路噪声电平 circuit noise level
线路增压器 in-line booster
线路增音机 line repeater
线路占用 track occupancy; track occupation
线路占用标 D sign
线路占用率 line fill
线路占用时刻 busy time of line
线路占用图 track occupation diagram; track occupation graph
线路障碍 obstacle on the track; track obstruction
线路遮断 obstacle on the track; track blockade
线路遮断器 scotch block
线路折断 obstruction of track
线路争夺状态 contention mode
线路正常 line normal
线路支架 line tower
线路支线 lateral
线路执行控制功能块 line execution control block
线路值班员 route controller; track controller
线路质量 quality of track
线路中的 in-circuit
线路中断 interruption on track; line interruption; track outage
线路中断原因 reason for outage
线路中继操作 hook operation
线路中继站 route relay
线路中间配线架 line intermediate distributing frame
线路中线测量 center[centre] line survey(ing)
线路中心间距 distance between centers of lines
线路中心线 center[centre] line of track; track alignment; track center[centre] line
线路中修 intermediate repair of track; track intermediate repair
线路终点 end of track
线路终点站 line-end station
线路终端 line terminal; line termination
线路终端负载网络 line terminating network
线路终端设备 line terminal equipment; line terminating equipment; line termination equipment
线路终端升压器 tail-end booster
线路终端网络 line terminating network
线路终接控制器 line termination controller
线路重复因数 line overlap factor
线路重执参数 line retry parameter
线路柱形绝缘子 line-post insulator
线路专门化 specialization of tracks
线路转换 line switching; system switching
线路转换技术 line switching technique
线路转接 circuit switching; line switching
线路转接器 line adapter[adaptor]
线路桩 line stake
线路装置 wayside device
线路状况 condition of line; status of line
线路状态 condition of line
线路子系统 line subsystem
线路自动解锁 automatic track-release
线路自动信号设备 automatic wayside signal(l)ing
线路自动转换 automatic line switching
线路总长 total track length
线路总工程师【铁】chief permanent way engineer
线路总延长 road miles; route miles
线路纵断面 route profile of the line
线路纵断面的简化 condensation of route profile
线路纵断面试验 route profile test
线路纵断面图 line longitudinal profile; line profile; track profile
线路纵横断面测量 route column-cross profile survey
线路纵剖面图 batter board
线路纵向差动保护装置 line longitudinal differential protection device
线路纵向水平 longitudinal track level
线路走向 line of alignment; strike of alignment; trend of railway line
线路阻波器 line trap
线路阻抗 line impedance
线路阻抗匹配 line impedance matching
线路阻塞 obstruction of track
线路最大通过能力 maximum line capacity
线路最小通过能力 minimum line capacity
线路最终通过能力 resultant line capacity
线麻籽油 hempseed oil
线密度 < 法定单位为特(克斯) > line-(ar) density
线密封 linear sealing
线描 line drawing
线木集材机 feller-skidder
线内 on-line
线内距离 intraline distance
线能量转移 linear energy transfer
线扭折 kinking of a wire
线耦 line pair
线耦合 line coupling
线耦合装置 line stretcher
线盘 wire coil
线盘千斤顶 drum jack
线盘卸料器 stripping spider
线盘卸下机 wire-coil stripping machine
线旁 trackside; wayside
线膨胀 linear dilatation; linear expansion
线膨胀测量法 linear dilatometry
线膨胀测试仪 linear dilatometer
线膨胀率 linear expansion rate
线膨胀系数 linear expansibility; linear expansion factor linear; line(ar) expansion coefficient; coefficient of linear expansion
线膨胀限界 linear-expansion limit
线膨胀型继电器 linear expansion relay
线偏光 linearly polarized light
线偏振 linear polarization; plane polarization
线偏振波 linearly polarized wave
线偏振光束 linearly polarized light beam
线平均径 linear mean diameter
线、平面或立体在另一平面上的投影 auxiliary projection
线谱鉴定 line spectrum test
线器 scriber
线钳 wire cutter
线强度比 line intensity ratio
线桥建筑折旧费 line and bridge building depreciation expenses
线切割废水 line-cutting wastewater
线切割机床 linear cutting machine
线切术 seriscission
线球 skein; spireme
线球变换 line-sphere transformation
线渠 culvert
线圈 bobbin; coil(curl); coiler; electric(al) coil; fake; inductance; inductance coil; inductor; loop; parallel circle; pipe coil; plot; spool; winding
线圈包扎层 coil covering
线圈比较器 coil comparator
线圈边 coil-side
线圈变压器 coil transformer
线圈布置 coil arrangement
线圈测试器 coil comparator
线圈长度 loop length; winding length
线圈长度变化 slurgalls
线圈常数 coil constant
线圈抽头 coil tap
线圈出线端 finish lead
线圈(磁电)效应 solenoid effect
线圈磁力计 coil magnetometer
线圈磁漆 coil enamel
线圈的 coiled
线圈的包扎层 coil covering
线圈第二节距 second pitch of coil
线圈电感 coil inductance
线圈电流 coil current
线圈电流计 coil galvanometer
线圈电容 coil capacity
线圈端部 end turn
线圈端部绑扎 coil lashing
线圈端部换位 inverted-turn transposition
线圈端漏磁 coil-end leakage
线圈端漏电抗 coil-end leakage reactance
线圈短路试验仪 growler
线圈段 coil segment
线圈分段 coil grading
线圈分支 path
线圈管 coil form; ferrule; inductor form
线圈盒 coil box; magnet case < 电磁阀上的 >
线圈盒盖 magnet coil case cap
线圈加感 coil loading

线圈架 bobbin;coil form;dowel;former;inductor form;wire frame
线圈架开槽式线圈 slotted-form winding
线圈架心 bobbin core
线圈间击穿 coil-to-coil breakdown
线圈间距 coil space;coil span
线圈间绝缘 intercoil insulation
线圈间隙系数 space factor
线圈检测器 coil detector
线圈接头 coil joint
线圈节距 coil pitch;coil span;winding pitch
线圈节距系数 coil span factor
线圈结构 loop construction;loop structure
线圈浸渍清漆 coil impregnating varnish
线圈绝缘 coil insulation
线圈绝缘测试器 coil insulator tester
线圈绝缘层 coil covering
线圈宽度 coil width
线圈框架 magazine
线圈拉型机 coil spreading machine
线圈连接(线)coil tie
线圈脉冲发生器 coil pulser
线圈密封式继电器 sealed coil relay
线圈内引线 start lead
线圈盘 coil panel
线圈品质因数 coil-Q
线圈清漆 coil varnish
线圈圈数计数器 coil turn counter
线圈绕组 coil winding
线圈试验器 coil comparator
线圈体积 winding volume
线圈调节器 coil adjuster
线圈铁芯 magnetic bobbin core
线圈涂料 coil coating
线圈温度 zonal temperature
线圈温度指示器 winding temperature indicator
线圈形天线 coil antenna
线圈有效边 active side of coil
线圈占空系数【电】coefficient of charge;coil space factor
线圈质量因数 coil constant
线圈中和 coil neutralization
线圈轴线 coil axis
线圈组 coil assembly;coil group;coil unit
线圈组件 coil assembly;coil kit;coil pack
线权 line weight
线群 group of lines;group of tracks;set of tracks
线群出站灯光信号 common starting light signal
线群出站灯光信号机【铁】common starting light signal
线群出站信号 group starting signal
线群出站信号机【铁】group starting signal
线群占线继电器 group marking relay
线绕变压器 wire-wound transformer
线绕变阻器 wire rheostat;wire-wound rheostat
线绕部件 wire-wound component
线绕磁极 wire-wound pole
线绕电极计数管 wire counter
线绕电极计数器 wire counter
线绕电炉 wire-wound furnace
线绕电容器 wire capacitor
线绕电枢 wire-wound armature
线绕电位计 wire-wound potentiometer
线绕电位器 wire-wound potential meter;wire-wound potentiometer
线绕电阻 resistance coil;resistance winding; restraining coil; wire-wound resistance
线绕电阻器 wire-wound resistor
线绕非线性电位器 non-linear wound potentiometer
线绕管接头 wire-wound pipe[piping] joint; wire-wound tube [tubing] joint
线绕和板极计数管 wire and plate counter
线绕计数器 wire counter
线绕接头 wire-wound joint
线绕可变电阻 frame resistance
线绕冷子管 wire-wound cryotron
线绕式 wire-wound
线绕式感应电动机 wire-wound induction motor;wound-rotor induction motor; wound-type induction motor
线绕式脉冲传感器 peaking strip
线绕式转子 phase wound rotor;wound rotor
线绕转子 coil wound rotor
线绕转子式电动机 wound-rotor type motor
线热膨胀率 linear thermal expansibility
线热膨胀系数 linear coefficient of thermal expansion
线热胀系数 coefficient of heat transference
线热阻 linear thermal resistance
线三角 wire triangle
线散射系数 line scattering coefficient
线扫描示波器 line strobe monitor
线扫描速度 linear scanning speed
线扫描周期 line period;line scanning period
线色散率 linear dispersion
线色数 line-chromatic number
线色素 mitochrome
线栅 grate;grating;wire grating;wire grid
线伤痕防护板 set on wound preventing plate
线上标(准)地 linear sample plot
线上分析 on-line analysis
线上监听器 on-line monitor
线上(项目)above the line
线上(项目预算)支出 above-the-line expenditures
线上预算 <英国预算中的经常性项目> above the line
线上综合物料管理系统 on-line integrated materials management system
线上作业 on-line operation; on-line processing
线射波矢 diffracted wave vector
线声源 line source of sound
线声源模型 line source model
线示加工留量 line showing finishing allowance
线示图 nomogram
线式变形测定仪 wire strain ga(u)ge
线式参变管 wire parametron
线式觇标 linear target
线式打印机 stylus
线式混凝土铰支座 linear type concrete hinge-bearing
线式结晶器 linear crystallizer
线式脉冲器 line type pulser
线式偏振光 linearly polarized light
线式示波器 string oscillograph
线式天线阵 linear antenna array
线饰进口带层 mo(u)lded-intake belt course
线收缩 linear contraction; shrinkage of line;linear shrinkage
线收缩率 ratio of linear shrinkage
线收缩系数 coefficient of linear shrinkage
线收缩限度 linear-shrinkage limit
线束 bale of wire;bight;bunch;combined main; hank; pencil of lines; set of tracks; group of lines; track group【铁】; cluster of sidings; group of tracks <铁路编组场>
线束包线 beam envelope
线束对 bunched pair
线束对合 line involution
线束缓行器 group retarder
线束缓行器出口端 leaving end of group retarder
线束缓行器控制 group retarder control
线束缓行器以远的道岔阻力 switch resistance for switches beyond group retarder
线束减速器 group retarder
线数 line number;number of lines
线刷 wire brush
线刷清洁器 wire brush cleaner
线水准测量 line level(l)ing
线水准仪 line level
线松驰 line relaxation
线搜索 line search
线素几何学 line geometry
线素向量 line element vector
线速度 line speed
线速率 linear rate; linear speed; line rate;line speed
线塑性理论 simple plastic theory
线损 line loss
线缩 <沿长度方向的收缩> lineal shrinkage
线缩率 linear shrinkage rate
线缩试验 lineal shrinkage test
线索 clue; scent; thread; finger post <解决问题的>
线索表 threaded list
线索调查 trace survey
线索习得等同性 acquired equivalence of cues
线索习得明显性 acquired distinctiveness of cues
线索向量 line element vector
线锁环 wire locking ring
线锁设备 wire lock equipment
线铊 plumb
线滩堡礁 bank barrier
线体加强金属 whisker reinforced metal
线条 cornice; dessin; line; lineation; linellae;stria[复 striae]
线条板座块 architrave block
线条彩色印件 line-colo(u)r
线条测试图 bar pattern
线条的 lineal;linear
线条的接续 wire connection
线条断续器 line interrupter
线条分辨检验图 line resolution
线条光谱 line spectrum
线条和成型机 mo(u)lding and shaping machine
线条画 delineation;line drawing
线条机 mo(u)lding machine; pantograph
线条检测 line detection
线条件 line condition
线条交叉 cross connection
线条进度表 bar chart
线条进度图 Gantt progress chart
线条刻蚀 line etching
线条快门开关 cable release; flexible wire release;wire release
线条扩散 bleed line
线条类型 line style
线条轮廓平滑的 flowing
线条刨 mo(u)lding plane
线条平滑的 flowing
线条区 fringe area
线条示踪器 bar tracer
线条式样 line pattern
线条调和 harmony of lines
线条凸版 line block;line cut
线条图表 bar chart; bar graph; Gantt chart;string diagram
线条图形 bar graph;line pattern
线条网目混合版 line-halftone combination
线条网屏 line-screen
线条网线混合版 composite block
线条细化 line thinning
线条形图形 bar pattern
线条一致 harmony of lines
线条印刷机 line printer
线条原理 principles of line
线条状 linear
线调制器 line modulator
线头 stub;thrum
线头焊片 chape
线头接栓 terminal post
线头脱落 wire lead drop out
线头匣 outlet box
线图 line(ar) chart; line(ar) graph; line(ar) map;string diagram
线图处理 line drawing processing
线团 ball;skein
线外编码 out-of-line coding
线外车站 off-line station
线外绘图仪 off-line plotter
线外计算【计】off line computation
线外装置 off-line installation
线网 line net;reticle;wire mesh
线网层率 ply-rating
线网长度 line net length;network line length
线网乘行人次总数 total network travel passenger/time
线网除雾器 wire-mesh demister
线网方案 line network plan;line network scheme
线网分隔层 wire-mesh section
线网服务水平 service level of line network
线网负荷强度 carrying capacity of line network; carrying intensity of line network
线网感光胶混合机 silk screen paste mixer
线网隔层 wire gauze diaphragm
线网管理系统 line network management system
线网规划 line net planning
线网规模 line network scale
线网过滤器 wire gauze filter
线网护罩 wire netting protector
线网绘图仪 wire-mesh plotting device
线网搅拌器 wire stirrer
线网结构 line network structure
线网客流 line network passenger volume
线网炉 wire-mesh oven
线网滤波器 line network filter
线网密度 density of line network
线网平均负荷度 average carrying intensity of line network
线网平均换乘系数 average transfer coefficient of line network
线网屏(蔽)wire-mesh screen
线网实施规划 line network execution planning; line network execution program(me)
线网适应性 line net adaptability
线网填充料 wire gauze packing
线网用地控制规划 land-use control planning for line network
线网总长(度)line net total length;total length of line network
线尾 wire tail

线纬度 linear latitude
线位 line location
线位错 line dislocation
线位移 linear displacement
线位移测量计 linear displacement ga-(u)ge
线位移传感器 linear movement pick-up;rectilinear transducer
线位移分量 translational component
线位移感应式传感器 linear inducto-syn;Nultrax
线位移和角位移传感器 linear-and-angular-movement pickup
线纹 cord design
线纹笛鲷 red sea lined snapper
线纹计数器 thread counter
线纹饰玻璃 thread glass
线纹装饰法 threading
线稳定 line stabilization
线稳振荡器 line-stabilized oscillator
线涡 line vortex
线涡流 line volume;line vortex
线污染源 line pollution source;line source;line source of pollution
线污染源模型 line source model
线务员 line man;line mechanician;line(s)man;wire man
线务员工具包 lineman tool case
线吸收 line absorption
线吸收系数 linear attenuation coefficient
线系极限 series limit
线系绝灭 phyletic extinction
线系物种形成 phyletic speciation
线隙 line gap
线下式客运站 <站房与站台标高高低不同> combined high and low level station
线下收支 below-the-line item;payments and receipts
线下项目 below-the-line;below-the-line item
线下预算支出 below-the-line expenditures
线向 alignment;alinement
线向标 alignment marker
线向不正 misalignment;disorder of line <轨道的>
线向控制 alignment control
线向量 line vector
线向偏差 alignment deviation
线向收缩试验 lineal shrinkage test
线向调整 trued for alignment
线向振动 straight vibration
线芯 wire core
线芯钩 core hitch
线芯试验 test on core
线芯线圈 wire-core coil
线形 alignment;line font;line shape
线形变化 geometric(al)variable
线形变形 linear deformation
线形标准 geometric(al)standard
线形波 linear wave
线形城市 strip city
线形道路 by-road
线形的 linear;nemeous
线形低聚物 linear oligomer
线形动量 linear momentum
线形分布 linear distribution
线形分解 linear resolution
线形峰点 line peak
线形浮雕 alaglyph
线形符号 line symbol
线形辐射器 linear radiator
线形刮斗式 catenary type unloader
线形光 linear light
线形光源 linear light source
线形规划分析 linear programming analysis
线形函数 line shape function
线形荷载 linear load
线形桁架 linear truss;polygon truss
线形结构 lineation
线形聚酯 linear polyester
线形刻划板 scribe board
线形控制 alignment control
线形良好的河流 well-aligned river
线形流 streamline flow
线形龙骨 rockered keel
线形黏[粘]弹性 linear viscoelasticity
线形缺陷 line defect
线形三角链(锁)【测】linear triangulation chain;linear chain of triangulation
线形三角网 linear triangulation network
线形沙丘 linear dune
线形设计 alignment design;design of alignment;geometric(al)design
线形声波延迟线 wire-type acoustic-(al)delay-line
线形蚀刻 linear etching;line etching
线形示意图 linear diagram;line diagram
线形塑性理论 linear plastic theory
线形缩聚物 linear condensation polymer
线形锁 linear triangulation chain
线形填缝 string packing
线形填料 string packing
线形图 line plan;string diagram
线形图显示 line drawing display
线形细裂 line checking
线形协调 alignment coordination;harmony of alignment
线形样方 linear sample plot
线形样条 line transect
线形摇座 linear rocker bearing
线形要素 alignment element
线形(引起的)延误 geometric(al)delay
线形载重系统 linear weight-carrying system
线形炸药 linear explosives
线形折旧法 straight-line depreciation
线形褶皱 linear fold
线形支承体系 linear supporting system
线形至椭圆形 varies in shape from linear to oval
线形装饰 linear decoration;line of ornament
线形阻力 alignment resistance
线形组合 linear array
线型 alphabet of lines;line style;type of line
线型闭合 closure
线型变感元件 wiretron
线型变化 variation of linear type
线型大分子 linear macromolecule
线型地震 linear earthquake
线型叠加 linear superposition
线型定义 line type definition
线型酚醛环氧黏[粘]合剂 epoxy novolac adhesive
线型酚醛清漆 novolak
线型酚醛树脂 novolac[novolak] resin
线型浮雕 anaglyph
线型高聚物 linear high polymer;linear superpolymer
线型构件 linear element
线型构造 lineament
线型函数 linear function
线型火灾探测器 line type fire detector
线型浇注 cord-pour
线型聚合物 linear polymer
线型聚酰胺 linear polyamide
线型聚乙烯 linear polyethylene
线型开裂 line type cracking
线型控制 proportional control
线型刨床 mo(u)lding plane
线型谱带 line spectrum[复 spectra]
线型气动感温探测器 pneumatic line type heat detector
线型蚀刻 line etching
线型损耗 linear depletion
线型缩合 linear condensation
线型缩合环 linear condensed rings
线型调制器 line type modulator
线型图 body lines;lines
线型图型. line style
线型外壳 volute casing
线型源加热器 line source heater
线型阵列 linear array
线型酯 linear ester
线型注入系统 linear injection system
线性摆动 rectilinear oscillation
线性半对数趋势线 linear semilogarithmic trend line
线性半序空间 linearly semi ordered space
线性伴随群 linear adjoint group
线性倍增时间 linear doubling time
线性逼近【数】linear approximation
线性比 linear ratio
线性比尺 linear scale ratio
线性比功率 linear specific power
线性比较 linearity check
线性比较器 continuous comparator;linear comparator
线性比例尺 distance scale;linear scale
线性比例系数 linear scale factor
线性比色法 linear colo(u)rimetric method
线性闭包 linear closure
线性边缘算子 linear boundary operators
线性边值问题 linear boundary value problem
线性编码 uniform encoding
线性编码器 linear encoder
线性变参数系统 linear variation parameter system
线性变分 linear variation
线性变化 lineal change;linear change;linear variation
线性变化电位器 rectilinear potentiometer
线性变化分布荷载 uniformly varying distribution load
线性变化荷载 linearly varying load;linear variation load
线性变化型季节变动 linear seasonal variation
线性变化应力场 linearly varying stress field
线性变换 linear transformation
线性变换差动变压器 linear variable differential transformer
线性变换的核 kernel of a linear transformation
线性变换矩阵 matrix of a linear transformation
线性变换器 linear quantizer
线性变换式 linear transform
线性变量 linear variable
线性变频 multiplicative mixing
线性变形 linear deformation
线性变形模量 modulus of linear deformation
线性变压器 linear transformer
线性标尺 linear scale
线性标度 linear scale
线性表 linear list
线性表示 linear expression
线性表示法 linear representation
线性波 linear wave
线性波导加速器 linear waveguide accelerator
线性波理论 linear wave theory
线性波形 linear waveform
线性波形失真 linear waveform distortion
线性波形响应 linear waveform response
线性薄壳理论 linear shell theory
线性补偿器 linearity compensator
线性不等式 linear inequality;linear unequality
线性不等式约束 linear inequality constraint
线性不等式组 linear inequalities
线性不可分函数 linear unseparable function
线性不稳定运动 linearly unstable motion
线性不相关 linear independence
线性不相交性 linear disjointedness
线性布尔递归 linear Boolean recursion
线性部分 linear segment
线性部件 linear unit
线性材料性能 linear material behavio(u)r
线性采样 line sampling
线性彩色解调 linear colo(u)r demodulation
线性参数 linear parameter
线性参数放大器 linear parametric amplifier
线性残余膨胀 linear after expansion
线性残余收缩 linear after contraction
线性操作机构 linear actuator
线性测度 linear measure
线性测量 linear measure;linear measurement
线性测量装置 linear measuring assembly
线性测试卡 linearity test card
线性测试器 linearity checker
线性测试信号 linearity test signal
线性测试振荡器 linearity test generator
线性测斜仪 line clinometer
线性策略规则 linear strategy rule
线性插补器 linear interpolater
线性插值 linear interpolation
线性插值法 linear interpolation method
线性查找 linear search
线性差动传感器 Linear variable displacement transducers
线性差分方程 linear difference equation
线性差分微分方程 linear differential-difference equation
线性觇标系统 linear target system
线性长度 linear length
线性偿付 linear repayment
线性常数 linear constant
线性车辆跟随模型 linear car-following model
线性车辆容量 <每一车辆单位长度(米)内的客位数> linear vehicle capacity
线性车速-密度模型 linear speed-density model
线性沉降速率 linear settling rate
线性成像扫描系统 linear imaging scanner system
线性程序 linear program(me)
线性程序控制 linear programming control
线性程序设计 linear programming
线性尺寸 linear dimension;linear size
线性尺寸稳定性 linear dimensional stability
线性尺度 linear dimension

线性传感器 linear transducer
线性串音 linear crosstalk
线性磁异常 linear magnetic anomaly
线性磁致伸缩系数 linear magnetostriction factor
线性大规模集成电路 linear large scale integrated circuit
线性代换 linear substitution
线性代数 linear algebra
线性代数方程 linear algebraic equation
线性代数方程组 system of linear algebraic equations
线性代数计算算法 computational method of linear algebra
线性代数群 linear algebraic group
线性代数系 linear system of algebra
线性单回路控制系统 linear single-loop control system
线性单群 linear simple group
线性单位过程线理论 linear unit hydrograph theory
线性单一方程式模式 linear single-equation model
线性单元 linear element
线性导体 linear conductor
线性的 lineal; linear; one-dimensional; unidimensional
线性等价类 linear equivalence classes
线性等价因子 linearly equivalent divisor
线性等距 linear isometry
线性等温线 linear isotherm
线性低密度聚乙烯 linear low density polyethylene
线性滴定 linear titration
线性迪奥番廷方程 linear Diophantine equation
线性地震 linear earthquake
线性递归 linear recurrence
线性递归关系 linear recurrence relation
线性递归序列 linear recurring sequence
线性递推关系 linear recurrence relation
线性点集 linear point set
线性点集系 linear system of point groups
线性电动机 electric(al) linear motor; linear motor
线性电动机驱动的 PM 电车 linear-motored "network" peoplemover
线性电动机驱动的轨道交通网系统 linear-motored network system
线性电动机驱动的悬浮式短途往返列车 linear-motored hover-shuttle
线性电动机驱动运行 linear-motored operation
线性电动机型减速器 linear-motor type retarder
线性电感耦合存储器 linear inductive coupling storage[store]
线性电光效应 linear electro-optical effect
线性电荷密度 linear charge density
线性电介质 linear dielectric
线性电抗器 linear reactor
线性电离 linear ionization
线性电流 linear current; linear reactance; rectilinear current
线性电流范围 linear current range
线性电流环 linear electric(al) current loop
线性电流密度 linear current density
线性电路 linear circuit; linearity circuit
线性电路试验 linear circuit test
线性电平 linear level
线性电气常数 linear electric(al) constant
线性电气传动 linear motion electric(al) drive
线性电容 linear capacitance
线性电势计 simple potentiometer
线性电位 linear potential
线性电位计 linear potentiometer
线性电位器 linearity potentiometer; linear potentiometer
线性电压 linear voltage
线性电压波形 linear voltage wave form
线性电压差分传感器 linear voltage differential transformer
线性电压位移传感器 linear voltage displacement transducer
线性电子加速器 linac; linear electron accelerator
线性电子矩阵化 linear electronic matrixing
线性电阻 linear resistance; linear resistor
线性电阻补偿 line resistance compensation
线性电阻分布特性 linear taper
线性电阻流量计 linear-resistance flowmeter
线性电阻器 linear resistor; ohmic resistor
线性电阻元件 linear resistive element
线性叠加 linear superposition
线性叠加原理 principle of linear superposition
线性定律 linear law
线性动力特性 linear dynamic(al) property
线性动力系统 linear dynamic(al) system
线性动力学 linear dynamics
线性动量 linear momentum
线性动态范围 linear dynamic(al) range
线性动态模型 linear dynamic(al) model
线性动态系统 linear dynamic(al) system
线性独立 linear independence
线性独立函数 linearly independent function
线性独立解法 linear independent solution
线性度 linearity
线性度量空间 linear metric space
线性段 linearity range
线性断裂力学 linear fracture mechanics
线性对合 linear involution
线性对数变换器 linear-to-log converter
线性对数放大器 linear log(arithmic) amplifier
线性对数接收机 line log receiver
线性对数量化器 linear log quantizer
线性对应的 colinear
线性多变量控制 linear multivariable control
线性多变量系统 linear multivariable system
线性多步法 linear multistep method
线性多参数系统 linear multiparameter system
线性多项式 linear polynomial
线性多元回归模型 linear multiple regression model
线性额定功率 linear power
线性二极管 linear diode
线性二维结构 linear two-dimensional structure
线性发电机 linear generator
线性反馈控制 linear feedback control
线性反馈系统 linear feedback system
线性反演法 linear inversion method
线性反应 linear response
线性泛函 linear function(al)
线性泛函微分方程 linear functional differential equation
线性泛函序列 sequence of linear functionals
线性范围 linear range; proportional band; proportional range
线性方差 linear equation
线性方程解算装置 linear equation solver
线性方程式 linear equation
线性方程组 linear system of equations; system of linear equations
线性方程组程序 linear equation system program(me)
线性方程组的相容性 consistency of linear equations
线性方程组解法 solution of linear equation
线性放大(倍数) linear amplification; linear magnification
线性放大器 linear amplifier
线性放大系数 lateral magnifying power; linear magnification coefficient
线性非理想色谱法 linear non-ideal chromatography
线性非平稳地质统计学 linear non-stationary geostatistics
线性非时变滤波器 linear time invariant filter
线性费用函数 linear cost function
线性分布 linear distribution
线性分布理论 straight-line theory
线性分层器 linear quantizer
线性分级结 linearly graded junction
线性分级杂质分布 linearly graded impurity distribution
线性分类器 linear classifier
线性分离器 linear separator
线性分离系统 linear split system
线性分流系统 linear split system
线性分配等温线 linear partition isotherm
线性分散 linear dispersion
线性分式变换 linear fractional transformation
线性分式规划 linear fractional programming
线性分式函数 linear fractional function
线性分式群 linear fractional group
线性分析 linear analysis
线性分支系统 linear path system
线性分组码 linear block code
线性风浪相互作用 linear wind-wave interaction
线性负荷 linear load
线性负荷体系 linear load-bearing system; linear load-carrying system; linear loaded system
线性复合 linear recombination
线性概率模型【数】linear probability model
线性感应泵 linear induction pump
线性感应电动机 linear induction motor
线性感应马达 linear induction motor
线性刚度 linear rigidity; linear stiffness
线性高速试验台 linear high-speed test bench
线性格 linear lattice
线性跟踪 linear tracking
线性功率放大器 linear power amplifier
线性供应承数 linear supply function
线性拱 linear arch
线性共振 linear resonance
线性共振系统 linear resonant system
线性构造 aligned structure; lineament
线性构造图 lineament map
线性估计 linear estimation
线性估计量 linear estimate
线性关税减让法 linear reduction of tariffs
线性关系 linear dependence; linear relation(ship); straight-line relation
线性关系的假设检验 testing hypothesis of linear relationship
线性管理法规 linear regulation control law
线性光滑 linear smoothing; line smoothing
线性光密度计 linear densitometer
线性光学 linear optics
线性光学系统 linear light system; linear optical system
线性光栅 striated pattern
线性归一化 linear normalization
线性规划 linear plan; line(ar) program(me); linear programming; mathematic(al) programming
线性规划程序 linear programming program(me)
线性规划的对偶性 duality in linear programming
线性规划的对偶性定理 duality theorem in linear programming
线性规划的退化 degeneracy in linear programming
线性规划的应用 application of linear programming
线性规划法 linear programming method; linear programming technique; linear programming theory
线性规划反褶积 linear programming deconvolution
线性规划分配法 linear programming distribution method
线性规划模型 linear programming model
线性规划问题 linear programming problem
线性规划问题的代数解法 algebraic(al) linear programming
线性规划问题的矩阵结构 matrix formulation of linear programming problem
线性规划问题的可行解区域 feasibility of linear programming problem
线性规划问题解 solution of linear program(me)
线性规划中的单纯形法 simplex method of linear programming
线性规划中的分解 decomposition in linear programming
线性规划最优化理论 mathematic(al) programming
线性过程 linear process
线性过滤模型 linear filter model
线性过滤器 linear filter
线性函数 linear function
线性(核)放射性 linear activity
线性荷载 collinear load(ing); linear load(ing); knife-edge load(ing)
线性荷载体系 linear load-bearing system; linear load-carrying system
线性恒定 constant of linearity
线性厚度 linear thickness
线性滑动电位器 linear sliding potentiometer
线性化 linearisation; linearization
线性化单元 linearizer
线性化电路 linearity circuit; linearizer

线性化多项式 linearized polynomial
线性化法 method of linearization
线性化方程 linearized equation
线性化概念 linearization concept
线性化观测方程 linearized observation equation
线性化函数 linearized function
线性化极大似然估计量 linearized maximum likelihood estimator
线性化理论 first-order theory
线性化流动 linearized flow
线性化网络 linearized network
线性化系统 linearized system
线性环节 linear element
线性换能器 linear transducer
线性回归 linear regression
线性回归法 linear regression method
线性回归方程式【数】equation of linear regression
线性回归分析 linear regression analysis
线性回归估计值 linear regression estimate
线性回归技术 linear regression technique
线性回归模型 linear regression model
线性回火 linear tempering
线性回游 linear migration
线性混合估计法 linear mixed estimation method
线性机 linear machine
线性积分方程 linear integral equation
线性积累破坏准则 linear accumulative damage criterion
线性基本图形 linear fundamental figure
线性基础结构 linear infrastructure
线性畸变 linear distortion
线性极大似然法 linear maximum likelihood method
线性极化 linear polarization
线性极化波 linearly polarized wave
线性极化法 linear polarization method
线性极化关联测量 linear polarization correlation measurement
线性极化输出 linearly polarized output
线性极小化 linear minimization
线性极小化问题 linear minimization problem
线性极坐标图 linear-polar coordinate plot
线性集成电路 linear integrated circuit
线性计算 linear computation
线性加权细胞模型 linear-weighted cell model
线性加速表 linear accelerometer
线性加速度 linear acceleration
线性加速度法 linear acceleration method
线性加速器 linear accelerator
线性加速自记器 linear accelerograph
线性加速自记仪 linear accelerograph
线性假设 linear hypothesis
线性假设的检验 linear hypothesis testing
线性假设模型 linear hypothesis model
线性检波 linear detection; linear rectification; straight-line detection
线性检波器 linear detector; linear rectifier; line rectifier
线性检验 linear test; test for linearity
线性简单点阵 linear simple lattice
线性鉴别器 linear discriminator
线性鉴频器 linear discriminator
线性交变梯度系统 linear alternative gradient system
线性角编码器 linear angle encoder
线性校平 linear smoothing
线性校正 linearity correction
线性校正电路 linearity correction circuit
线性校正放大器 linearity correcting amplifier
线性校准曲线 linear calibration curve
线性阶式函数发生器 linear staircase function generator
线性结 linear junction
线性结构 linear organization; linear structure
线性结构方程 linear structural equation
线性结构方程组 linear structural equation system
线性截止 linear cut-off
线性截止低通滤波器 linear cut-off low-pass filter
线性解释经济模型 linear explanatory economic model
线性解调器电路 linear demodulator circuit
线性紧子范畴 linearly compact subcategory
线性近似 linear approximation
线性进程 linear flow
线性径向扫描 linear radial sweep
线性矩 linear moment
线性矩阵 linear matrix
线性锯齿波 linear saw-tooth wave
线性聚焦透镜 linear focus lens
线性决策法则 linear decision rule
线性决策律 linear decision rule
线性决策模型 linear decision model
线性均衡器 linear equalizer
线性均聚物 linear homopolymer
线性均匀功能 linear homogeneous function
线性均匀生产函数 linear homogeneous production function
线性开关 linear switch(ing)
线性抗噪声传声器 linear noise-cancelling microphone
线性可变电阻 linearly variable resistance; linear variable resistor
线性可分函数 linearly separable function
线性可分离 linearly separable
线性可调差动变压器 linear variable differential transformer
线性刻度 linear scale
线性刻度欧姆表 linear scale ohmmeter
线性空间 linear space
线性空间不变光学系统 linear space-invariant optical system
线性空间桁架 linear space truss
线性空间滤波器 linear spatial filter
线性空气动力特性 linear aerodynamic characteristics
线性控制 linear control; linearity control
线性控制电动机构 linear control electromechanism
线性控制电路 linearity control circuit
线性控制范围 range of linearity control
线性控制过程 linear control process
线性控制理论 linear control theory
线性控制器 linearity controller
线性控制系统 linear control system
线性块码 linear block code
线性亏格 linear genus
线性扩散 linear diffusion
线性扩展 linear expanding; linear stretching
线性扩张 linear extension
线性累积损坏准则 linear accumulative damage criterion
线性离散规划 linear discrete programming
线性离散值系 linear discrete valued system(s)
线性理论 linearized theory; linear theory; straight-line theory
线性理论法 linear theory method
线性理论模型 linearized theoretical models
线性理想色谱 linear ideal chromatography
线性力 linear force
线性力矩 linear moment
线性连续集 linear continuous set
线性连续统 linear continuum
线性联立方程 linear simultaneous equation
线性联立方程组 linear simultaneous equations
线性联络 linear connection
线性链 linear chain
线性量化 equal interval quantizing
线性量化器 linear quantizer
线性流 linear flow; linear stream
线性流动 linear fluidity
线性流量特性 linear flow characteristic
线性流速 linear flow rate
线性流形【数】linear manifold
线性滤波 linear filtering
线性滤波器 linear filter
线性逻辑阶段建设程序 linear construction
线性马达 linear motor
线性码 linear code
线性脉冲放大器 linear pulse amplifier
线性脉冲滤波器 linear pulse filter
线性脉冲调制器 linear quantizer
线性门 linear gate
线性密度计 linear densimeter
线性密封 linear sealing
线性敏感度测量法 linear sensitivity measures
线性模拟控制 linear analogue control
线性模式 linear mode; striated pattern
线性模型 linear model; liner model
线性摩擦 linear friction
线性目标 linear objective
线性目标函数 linear objective function
线性内插 linear interpolation; proportional interpolation
线性内插法 linear interpolation method
线性能量传递 linear energy transfer
线性拟合 linear fitting
线性黏[粘]弹特性 linear viscoelastic behaviour
线性黏[粘]弹微极性材料 linear viscoelastic micropolar material
线性黏[粘]弹性 linear viscoelasticity
线性黏[粘]弹性材料 linear viscoelastic material
线性黏[粘]弹性介质 linear viscoelastic medium
线性黏[粘]弹性流变模型 four parameter model
线性黏[粘]性阻尼 linear viscous damping
线性黏[粘]性阻尼系数 linear viscous damping coefficient
线性黏[粘]滞阻尼器 linear viscous dashpot
线性耦合动态论 linear coupled dynamic theory
线性耦合器 linear coupler
线性耦合装置 linear coupler
线性排列 linear arrangement; linear array; linear permutation
线性判别函数 linear discriminant function
线性抛线型方程 linear parabolic equations
线性膨胀 line(al) expansion; polar expansion; linear expansion
线性膨胀系数 coefficient of linear expansion; linear expansion coefficient
线性偏微分方程 linear partial differential equation
线性偏微分算子 linear partial differential operator
线性偏振 linear polarization
线性偏振光 linear polarized light
线性偏振光输出 linearly polarized light output
线性偏振极化模 linearly polarized mode
线性漂移 linear drift
线性频率电容器 straight-line frequency capacitor
线性频率响应 linear frequency response
线性平衡线 linear equilibrium line
线性平滑 linear smoothing
线性平滑算法 linear smoothing algorithm
线性平均直径 linear mean diameter
线性平稳地质统计学 linear stationary geostatistics
线性谱 linear spectrum
线性谱空间 linear spectral space
线性谱模型 linear spectral model
线性齐次方程 linear homogeneous equation
线性齐次方程组 linear homogeneous system of equations
线性齐次群 linear homogeneous group
线性气流分布 linear air distribution
线性器件 linear unit
线性嵌入 linear imbedding
线性侵蚀 linear erosion
线性倾向 linear trend
线性求积仪 linear planimeter
线性区域 range of linearity
线性曲面系 linear system of surfaces
线性曲线 linearity curve
线性曲线系 linear system of curves
线性驱动 linear drive
线性趋势 linear trend
线性全身扫描机 linear whole body scanner
线性群 linear group
线性燃耗 linear burn up
线性燃烧速率 linear burning rate
线性扰动 linear perturbation
线性扰动理论 linear perturbation theory
线性热膨胀 linear thermal expansion
线性热(膨)胀系数 coefficient of linear thermal expansion; linear thermal expansion coefficient
线性热源 line heat source
线性容许函数 linear admissible function
线性溶剂化能相关法 linear salvation energy relationship
线性柔度 linear compliance
线性蠕变 linear creep
线性三维技术 linear three-dimensional technique
线性散列 linear hash
线性散射 linear scattering
线性扫描 linear scan(ning); linear (time base) sweep; linear trace; rectilinear scan(ning)
线性扫描点 line scanning spot
线性扫描伏安法 linear scan voltam-

metry;linear sweep voltammetry
线性扫描率 linear sweep rate;uniform sweep rate
线性扫描器 rectilinear scanner
线性扫描延迟电路 linear sweep delay circuit
线性扫描振荡器 linear sweep generator
线性色度失真 linear chroma distortion
线性色谱(法)linear chromatography
线性色散 linear dispersion
线性商法 linear quotient method
线性上升电压 linear increase voltage
线性摄动 linear perturbation
线性摄像管 linear tube
线性摄像器件 linear image pick-up device
线性伸长 linear stretching
线性伸长率 linear elongation rate
线性渗透定律 <即达西公式,水头损失与渗流速度成线性关系> linear law of seepage flow;law of linear permeability
线性生产函数 linear production function
线性生产活动 linear production activity
线性生长速率 linear growth rate
线性生成 linear spanning
线性声源 linear sound
线性剩余码 linear residue code
线性失误率分布 linear failure rate distribution
线性失真 linear distortion; linearity distortion
线性失真的 linear distorted
线性时变参数网络 linear time-varying network
线性时变信道 linear time-variant channel
线性时不变系统 fixed-time linear system
线性时基 linear base;linear time base
线性时基振荡器 linear time-base oscillator
线性时间不变系统 linear time-invariant system
线性时间算法 linear time algorithm
线性时间整量化系统 linear time-quantized control system
线性时序电路 linear sequence circuit
线性矢量函数 linear vector function
线性矢量空间 linear vector space
线性视差 linear parallax
线性视频放大器 linear video amplifier
线性试样偏差 linear sample bias
线性收敛 linear convergence
线性收缩 lineal shrinkage; line(ar) shrinkage
线性收缩率 linear shrinkage ratio
线性守恒系统 linear conservation system
线性受力变形 linearly varying strain
线性输出 linear output
线性输出地平仪 linear output horizon sensor
线性输出电流 linear anode current
线性输出反馈 linear output feedback
线性输入带 linear input tape
线性输入形式策略 linear-input from strategy
线性束 linear pencil
线性束流密度 linear beam density
线性树码 linear tree code
线性数 linear number
线性衰减 linear attenuation
线性衰减量 linear decrement
线性衰减器 line pad
线性衰减系数 linear attenuation coefficient;linear extinction coefficient
线性双折射 linear birefringence
线性水库 linear reservoir
线性水库模型 linear reservoir model
线性水库群 linear reservoir system
线性水库系统 linear reservoir system
线性瞬时滤波 linear temporal filtering
线性瞬时值 linear instantaneous value
线性瞬态分析 linear transient analysis
线性斯塔克效应 linear Stark effect
线性伺服机构 linear servomechanism
线性伺服系统动力学 linear servosystem dynamics
线性伺服执行机构 linear servo actuator
线性速度 linear speed;linear velocity
线性速度测量法 linear speed method
线性速度-密度模型 linear speed-concentration model
线性速率常数 linear rate constant
线性速率定律 linear rate law
线性塑料光纤 linear plastic optical fiber
线性算符 linearity operator; linear operator
线性算子 linearity operator; linear operator
线性随车模型 linear car-following model
线性随机回归模型 linear stochastic regression model
线性随机假设 linear stochastic hypothesis
线性随机系统 linear stochastic system;line stochastic system
线性损伤法则 linear damage law
线性缩放仪 linear pantograph
线性台阵 linear array
线性弹簧 Hookean spring;linear spring
线性弹性 linear elasticity
线性弹性变形 linear elasticity
线性弹性变形的 linearly elastic
线性弹性动力学 linear elastodynamics
线性弹性断裂力学 linear elastic fracture mechanics
线性弹性方式 linear elastic manner
线性探查 linear probe
线性特性 linear characteristic
线性特征 line(ar) feature
线性特征标 linear character
线性特征提取 linear feature extraction
线性特征物 lineament
线性梯度 linear gradient;linear grading
线性梯度结 linear gradient junction
线性梯级信号 linearity staircase signal
线性体长度 length of lineament
线性体长度方位频数直方图分析 length-azimuth-frequency histogram analysis of lineament
线性体长度频数直方图分析 length-frequency histogram analysis of lineament
线性体单交叉控矿 ore-forming control of monolinearment of lineament
线性体多交叉控矿 ore-forming control of multilinearment of lineament
线性体方位 azimuth of lineament
线性体方位偏差分析 azimuthal deviation analysis of lineament
线性体方位频数直方图分析 azimuth-frequency histogram analysis of lineament
线性体方位异常度分析 azimuthal anomalous degree analysis of lineament
线性体分析方法 method of lineament analysis
线性体光学傅立叶分析 optic(al) Fourier analysis of lineament
线性体环形体交切控矿 ore-forming control of linear-circular intersection
线性体密度分析 density analysis of lineament
线性体模式分析 modal analysis of lineament
线性体频数 frequency of lineament
线性体统计分析 statistic(al) analysis of lineament
线性体系 linear system
线性体形态 shape of lineament
线性体中心对称度分析 central symmetric degree analysis of lineament
线性体种类 category of lineaments
线性条件 linear condition; linearized condition
线性调节器 linear regulator
线性调节器问题 linear regulator problem
线性调频 linear frequency modulation
线性调频雷达 chirp radar
线性调频脉冲 chirp;linear pulse
线性调频转发器 serrodyne
线性调整 frame linearity control;linearity control
线性调整器 linearity controller
线性调制 linear modulation
线性同步电(动)机 linear synchronous motor; linear synchronous machine
线性同余 linear congruence
线性统计抽样模型 linear statistical sampling model
线性透视 linear perspective
线性图 linear graph
线性图案 linearity pattern
线性图形 linear figure
线性推进 linear propulsion
线性退火 linear annealing
线性外推(法)linear extrapolation
线性外推距离 linear extrapolation distance
线性外延(法)linear extrapolation
线性弯曲理论 linear bending theory
线性烷基磺酸酯(或盐)<一种可代替 ABS 的可生化降解的表面活性剂> linear alkyl-sulfonate
线性网 linear net
线性网络 linear circuit; linear network
线性网络法 linear network method
线性网络理论 linear-network theory
线性往复扫描 linear reciprocating sweep
线性微电路 linear microcircuit
线性微分方程 linear differential equation
线性微分算符 linear differential operator
线性微分算子 linear differential operator
线性微密度计 linear microdensitometer
线性微扰分析 linear perturbation analysis
线性微型系统 linear microsystem
线性维数 linear dimension
线性伪度量空间 linear pseudometric space
线性位移 linear displacement; linear transformation; linear translation; tendon transformation
线性温度梯度 linear temperature gradient
线性稳定性分析 linear stability analysis
线性涡轮机 linear turbine
线性无定形聚合物 linear amorphous polymer
线性无关 linearity independence
线性无关的 linearly independent
线性无关的向量 linearly independent vector
线性无关积分 linearly independent integral
线性无关集 linearly independent set
线性无关解 linearly independent solution
线性无关量组 linearly independent quantities
线性无关性 linear independence
线性无关元 linearly independent elements
线性无关族 linearly independent family
线性无偏估计 linear unbiased estimation
线性无旋流 linear irrotational flow
线性无缘性 linear disjointure
线性无源耦合网络 linear passive coupling network
线性误差 linear error;linearity error; miss-distance
线性吸附 linear adsorption
线性吸附系数 linear adsorption coefficient
线性吸收系数 linear absorption coefficient
线性洗提力梯度 linear eluant strength gradient
线性洗脱液强度梯度 linear eluant strength gradient
线性系数 linearity factor;time scale factor
线性系(统)linear system
线性系统的技术指标 specifications of linear system
线性系统的马尔柯夫矢量法 Markov vector approach for linear systems
线性系统分析 linear system analysis
线性系统理论 linear system theory
线性系统统计函数 linear systematic statistics function
线性系统统计量 linear systematic statistic
线性下降电压 linear decrease voltage
线性显示 conventional display;linear indication
线性现象 linear phenomenon
线性线丛 linear complex; linear line complex
线性线丛系 linear system of complex
线性线汇 linear line congruence
线性相关 linear correlation; linear(ity) dependence
线性相关的向量 linearly dependent vector
线性相关度量 measure of linear correlation
线性相关系数 linearly dependent coefficient
线性相关性关系 linear dependence relation
线性相箝位 linear phase sweep
线性相位比较器 linear phase comparator
线性相位滤波器 linear phase filter
线性相位失真 linear phase distortion
线性相位特性放大器 phase linear

amplifier
线性相位特性工作 linear phase operation
线性相位特性接收机 phase linear receiver
线性相位特性系统 linear phase system
线性相位系统函数 linear phase system function
线性相位运用 linear phase operation
线性相移 linear phase shift
线性响应 linear response
线性向量函数 linear vector function
线性向量空间 linear vector space
线性项 linear term
线性削波 linear clipping
线性削波器 linear clipper
线性消解 linear resolution
线性斜率延迟滤波器 linear-slope delay filter
线性斜坡电压 linear ramp
线性谐振加速器 linear resonator accelerator
线性信号 linear signal
线性信息处理语言 linear information processing language
线性形变 linear deformation
线性形式 linear form
线性形状函数 linear shape function
线性型 linear model;linear type
线性性质 linear behavio(u)r
线性修匀 linear smoothing
线性需求函数 linear demand function
线性序 linear order
线性序列机 linear sequential machine
线性旋转变压器 linear resolver
线性旋转流 linear rotational flow
线性旋转弹簧 linear rotational spring
线性选择 linear selection
线性循环序列 linear recurrent sequence
线性压曲理论 linear buckling theory
线性延迟失真 linear delay distortion
线性延伸的城市 linear town
线性延伸电路 linear delay circuit
线性延时电路 linear time delay circuit
线性延展系数 coefficient of linear extensibility
线性氧化速率 linear oxidation rate
线性摇臂轴承 linear rocker bearing
线性仪表 linear metre
线性移位寄存器 linear shift register
线性异步结构 linear asynchronous structure
线性因数 linearity factor
线性因子 linear factor
线性应变 linear strain
线性应变地震仪 linear strain seismograph
线性应变图 linear strain diagram
线性应力 linear stress(ing)
线性应力场 linear stress field
线性英寸 linear inch
线性影像相移 linear phase
线性映射 linear mapping
线性映射的秩 rank of linear mapping
线性优先函数 linear precedence function
线性游标 linear vernier
线性有界自动机 linear bounded automat
线性有限自动机 linear bounded automation
线性有序集 chain; linearly ordered set
线性有序链 linearly ordered chain
线性有序模 linearly ordered module
线性语言 linear language
线性预报值 linear predictor
线性预测 linear prediction
线性预测编码 linear predictive coding
线性预测理论 linear prediction theory
线性预加应力 < 钢筋混凝土 > linear prestressing
线性元件 linear element
线性元素 linear element
线性约束 linear constraint;linear restriction
线性匀滑转换 crossfade linearity
线性运动 linear motion; motion of translation
线性运动滑块 linear motion slides
线性运算 linear operation
线性运算部件 linear operational element
线性运算器 linear operator
线性载荷 line load
线性再散列法 linear rehash method
线性再收缩 linear after contraction
线性责任图 linear responsibility chart
线性增长 arithmetic(al) growth; linear growth; simple interest
线性增加荷载 linear increasing load
线性增益 linear gain
线性增益接收机 linear gain receiver
线性展开 linear expansion
线性张弛法 line relaxation method
线性张量 linear tensor
线性阵列 linear array
线性阵列扫描仪 linear array scanner
线性阵列网络 linear array network
线性振荡 linear oscillation; rectilinear oscillation
线性振荡器 linear oscillator
线性振动 linear vibration
线性振子 linear oscillator
线性正应变 linear normal strain
线性支出系统 linear expenditure system
线性执行机构 linear actuator
线性直读式温度计 linear direct reading thermometer
线性直流饱和控制 linear direct current saturation control
线性秩 linear rank
线性致动器 linear actuator
线性滞后 linear hysteretic; linear lag
线性转发器 linear repeater
线性装置 linear unit
线性子空间 linear subspace
线性子流形 linear submanifold
线性子图 linear subgraph
线性自由能关系 linear free-energy relationship
线性自由能相关 linear-free-energy-related
线性自由能相关法 linear free-energy relationship
线性阻抗 linear impedance
线性阻抗继电器 linear-impedance relay
线性阻尼 linear damping
线性阻止能力 linear stopping power
线性组合 collinearity; linear combination
线性钻进 line drilling
线性最佳化 linear optimization
线性最佳随机系统 linear optimal stochastic system
线性最佳系统 linear optimal system
线性最小平方法 linear least square method
线性最优化 linear optimization
线选存储器 linear selection storage; two-dimensional memory
线选法 linear selection method; linear selection system
线选开关 linear selection switch
线选开关存储器 Olsen memory
线选择 line selection
线寻找器 line finder
线压 line(al) pressure
线压力 linear pressure; line intensity
线压增加器 in-line booster
线延伸 linear extension
线岩溶率 karst factor counted along a line
线样板 line templet
线样征 string sign
线应变 linear strain; longitudinal strain; unit elongation; unit extension
线应变率 linear strain rate; linear strain ratio
线应力 linear stress
线英里 wire mile
线硬件 < 架空明线的金属附件 > line hardware
线有根图 line-rooted graph
线有向图 line digraph
线"与"【计】wired-AND
线与弧相交 line and arc intersect
线元 linear element
线元法 line element method
线元素 line element
线圆锥 line-cone
线源 linear source
线源结构 line source structure
线源模式 line source model
线晕 tint
线匝 coil of wire; conductor turn; wire turn
线匝比 turn ratio
线匝绝缘 turn insulation
线匝试验器 interturn tester
线展宽 line broadening
线张力 line tension
线张应变计 wire strain ga(u)ge
线胀率 linear expansion
线胀系数 coefficient of linear expansion
线胀性 linear expansibility; linear expansivity
线障脉冲测试器 pulse echo fault locator
线阵扫描 linear array scanning
线振荡器 line oscillator; phase shift oscillator
线振式换能器 vibrating wire transducer
线振应变仪 vibrating wire strain ga(u)ge
线振质谱计 farvitron
线震源 linear source
线正则图 line-regular graph
线之间虚线 interline
线支承 line bearing; line support
线支点处力矩 moment about point of support
线支座 linear support
线织物 cord woven fabric
线制品 wire work
线质系数 quality factor
线中电压 line-to-neutral voltage
线轴 bobbin; quill; reel; spool(er); wire reel
线轴式瓷绝缘器 spool-type porcelain insulator
线轴式滑阀 spool valve
线轴式绝缘子 spool insulator
线轴形沙坝 spool bar
线轴形圆柱床 spool bed
线主管道 trunk line
线转换器 line transformer
线转迹线 line roulette
线装药密度 linear charge concentration
线状 line form; striation
线状斑岩 linophyre
线状保险丝 wire fuse
线状背斜带 linear anticline zone
线状层理 linear stratification
线状的 filiform; linear
线状地物 align feature
线状地震 linear earthquake
线状电晕 line corona
线状调查 linear survey
线状风暴 line squall
线状风化壳 linear residuum
线状符号 line symbol
线状辐射源 linear radiation source
线状腐蚀 filiform corrosion
线状腐蚀坑 lineage
线状钢材 long steel product
线状更新抽样法 linear regeneration sampling
线状构造 lineation; lineation structure
线状光谱 line spectrum[复 spectra]
线状焊 string beading
线状焊道 stringer bead
线状焊料 yarn solder
线状花纹 line effect
线状火药 cordite
线状基线尺 base measuring wire; measuring wire
线状夹杂物 line inclusion
线状夹渣 line inclusion; slag stringer
线状胶体 linear colloid
线状焦点 line-focus
线状礁 ribbon reef
线状礁油气藏趋向带 linear reef pool trend
线状浸入 linear intrusion
线状流动构造 linear flow structure
线状目标 linear target; line target
线状排列 linear parallelism
线状喷出 linear eruption
线状劈理 linear cleavage
线状片理 linear schistosity
线状(频)谱 line spectrum[复 spectra]
线状频谱光源 line spectrum source
线状侵蚀 linear erosion
线状缺陷 linear discontinuities; line defect
线状熔断器 wire fuse
线状沙脊 sand streak
线状沙丘 linear dune
线状闪电 linear lightning
线状烧烙 line-firing
线状蚀变带 linear alteration zone
线状水体 linear water body; linear waters
线状天线 wire antenna
线状通道 filamentary region
线状图 linear graph
线状纹孔 linear pit
线状物 thread
线状烟缕 line plume
线状延伸的城市 lineal town
线状要素 align feature; linear element
线状叶理 linear foliation
线状云 line cloud
线状震源 line source of earthquake
线状装饰 linear decoration
线状装药 string loading
线状组构 linear fabric
线状组织 filum; ice-flower-like structure < 焊接的 >
线状钻孔 line drilling
线坠 bob
线着色 line colo(u)ring
线族 family of lines
线阻应变仪 wire resistance strain ga(u)ge
线组 line set
线组电阻 winding resistance
线坐标 line(ar) coordinates

限

限变 boundedness

限变器 deglitcher
限波控制器 chopper controller
限波系统 chopper system
限差(范围) limit of error;tolerance
限产超雇 featherbedding
限长钻臂 fixed-length boom
限程 range line
限程杆 distance bar
限程器 arrester[arrestor];guard
限程约束 range constraint
限带宽度 width of root zone
限带频谱 band-limited frequency spectrum
限带随机过程 band-limited random process
限带信号 band-limited signal
限当年有效的(预算或拨款)授权 annual authority;one-year authority
限当日 good the day
限当月 good this month
限地承兑 local acceptance
限定 conditioning; delimit; finitude; limitation; peg; prescribe; qualification; restrict; terminate
限定版 small impression
限定保险人责任的条款 clause limiting insurer liability
限定报价 limited offer
限定背书 qualified endorsement; restricted endorsement
限定变长编码 restricted variable length code
限定标准 peg
限定表达式 qualified expression
限定捕获量 restricting catch
限定财产继承权 tail
限定查账证明书 qualified audit certificate
限定产权财产 qualified property
限定尺寸 limit size
限定臭气浓度 threshold odo(u)r
限定出行(交通方式) captive trip
限定词 determiner;qualifier
限定的 terminative
限定的频带宽度 limited bandwidth
限定地点承兑 acceptance qualified as to place
限定订货 closed order
限定段名 qualified paragraph name
限定方法与装备的规范 <与最终产品规范对比> method and equipment specification
限定符 qualifier
限定符号 qualifier
限定雇主劳工 contract labo(u)r
限定环 limit collar;stop collar
限定汇率 pegging
限定货币 definitive money
限定集装箱 captive container
限定继承权 entail
限定继承人 heir in tail
限定价格 pegged price;price ceiling
限定金额及支付期的支票 limited check
限定(进口)订单 closed indent
限定静载荷 deadman load
限定开关时间 finite switching time
限定块 stop block
限定力 constraining force
限定流动基金 restricted current funds
限定名 qualified name
限定器 delimiter;stunt box
限定取款额度的可转让账户 negotiable order of withdrawal
限定时间 limiting time;time limit
限定时间承兑 acceptance qualified as to time
限定收购银行信用证 restricted credit
限定输出功率 limited output
限定随机抽样法 restricted random sampling
限定外包轮廓 bulk envelope
限定位置 limiting position
限定物 terminator
限定效率 restriction efficiency
限定性季节模型 definite seasonal pattern
限定性特征 determinant attributes
限定性指令 limit order
限定意见报告书 qualified opinion report
限定意义 specialize meanings
限定用途的公积金 restricted surplus
限定用途的现金 restricted cash
限定用途基金 restricted fund
限定在道路以外行驶的车辆 off-road vehicle
限定在道路以外行驶的自卸车 off-road dump truck
限定在公路以外行驶的车辆 off-highway vehicle
限定在公路以外行驶的卡车 off-highway truck
限定重量 design weight
限定准备金 qualifying reserves
限定资产 defined assets
限定最高价格 fixed ceiling price
限动齿 control tooth
限动挡 limit stop
限动环 stop collar
限动机构 stopper mechanism
限动件 limiter;limiting member;stopper
限动块 stop block
限动框 framing mask
限动螺帽 back nut
限动螺母 back nut;binding nut
限动器 arrester[arrestor]; bridle; debooster;limiter
限动销 banking pin;stopper pin
限动油门 gated throttle
限度 limit(ation); limited range; restriction;thresholding
限度波长 threshold wavelength
限度尺寸 size at limit
限度等级 grade of limits
限度厚度 design thickness
限度计算 limit calculation
限度控制 limit control
限度排污水质 water quality based effluent limitation
限度试验 limiting test
限度调节器 limit of regulator
限度稀释传代 limit dilution passage
限额 amount of limitation; limit; norm;quota;stint
限额表 limit table;line sheet;table of limits
限额拨款 closed-end appropriation
限额采伐制度 system of cutting quota
限额的 normed
限额抵押 closed mortgage
限额抵押债券 closed mortgage bond
限额对策 quota game
限额发放物资制度 quota system for distributing materials
限额发料 issuance of materials on a norm basis
限额分配 ration
限额基金 closed-end fund
限额价格 rationing price
限额交易 rationed exchange
限额结算 normal settlement; settlement within quota
限额进出口制 quota system
限额控制 limit control
限额领料 material requisition on quota
限额领料单 limited material requisition;materials requisition on quota
限额领料制 materials requisition on quota
限额内联邦债务 Federal debt subject to limitation
限额配给 rationing
限额赊销账户 revolving charge account
限额设计 quota design
限额输出 export quota; rationed export
限额输入 import quota
限额项目 line item
限额信贷 rationed credit
限额型抵押 closed-end mortgage
限额循环周转信贷 revolving credit
限额循环周转信贷额度 revolving line of credit
限额循环周转信贷合同 revolving credit agreement
限额循环周转信贷计划 revolving credit plan
限额循环周转信贷账户 revolving credit account
限额样本 quota sample
限额以上 above norm;above quota
限额以上的 above norm
限额以上的合同 above-norm project
限额以下 below-norm
限额与不限额移民 quota and non-quota immigrants
限额政策 quota policy
限额支票 limited check;limited cheque
限额支票簿 limited check book
限额制 closed system;quota system
限幅 amplitude limit; chop; clipping; cut ridge;limiting;slicing
限幅比 limit ratio
限幅边际 limiting margin
限幅波 clipped wave
限幅道 clipped trace
限幅电路 amplitude limiter circuit; clipped circuit; clipper circuit; clipping circuit; limiter circuit; limiting circuit; peak-clipping circuit; slice circuit
限幅电平 clip level;limiting level
限幅电瓶 clip level
限幅电压 limiting voltage
限幅二极管 bootstrap diode; clipper diode;limiter diode
限幅反馈 limiting feedback
限幅放大器 clipper amplifier;clipping amplifier;limiting amplifier
限幅管 limiter tube
限幅级 limiter stage
限幅控制 chopper control
限幅起始电平 threshold level of amplitude limiter
限幅器 amplitude limiter; clipper; clipper circuit; clipper-limiter; clipping circuit;debooster;slicer
限幅器转移 tertiary interference
限幅前馈 limiting feedforward
限幅系数 clipping factor;limiting figure
限幅信号 limited signal
限负荷运行 load limit operation
限光遮板 screening for light
限海上范围 sea-borne only
限航区 restricted area
限滑差速器 limited slip differential
限火蔓延措施 fire limit
限极 thresold
限价 ceiling price; ceiling rate; check price; forced quotation; frozen price; lid on price; limit prices; price fixing; price limit; vaiorisation;valorize
限价补进或卖出 stop-loss order;stop order
限价订单 limit(ed) order; stop-limit order
限价平仓单 stop-limit order
限价商店 limited price store
限价停止订单 stop-limit order
限角转向 lock steering
限界 boundary line; circumscription; clearance; gabarite; ga(u)ge; limited range; limiting dimension; margin;clearance limit
限界包络 clearance envelop
限界测验帘 <俚语> tell-tales
限界沉降速度 terminal dropping velocity;terminal fall velocity;terminal setting velocity
限界的 marginal
限界电平调整 cut-off adjustment
限界对 bound pair
限界法 bound method
限界分析 marginal analysis
限界高度 clearance height;headroom
限界加宽 clearance widening;demarcation widened
限界架 loading ga(u)ge;clearance ga(u)ge
限界检测 profile checking
限界检测车 clearance car
限界检测器 height-width contour detector;high and wide load detector
限界检查框 clearance profile
限界检查器 clearance detector;clearance treadle;limit treadle
限界建筑物 limiting structure
限界聚焦 confined focus(s)ing
限界开关 limit switch
限界框 clearance limit frame
限界利润 marginal profit; marginal return
限界轮廓 limiting outline
限界轮廓 clearance diagram
限界门 vehicle ga(u)ge device
限界上下文 boundary context
限界上下文文法 bounded context grammar
限界失调警告 clearance disorder alarm
限界收益 marginal gain
限界调整 threshold adjustment
限界图 clearance diagram
限界推移力 limiting tractive force; limiting tractive power
限界线 clearance line;margin line
限界消费 marginal expenditures
限界效用 marginal utility
限界信贷 marginal credit
限界以下的 submarginal
限界以下的耕作土地 submarginal land
限界营业 marginal business
限界值 threshold value
限紧装置 stop work
限矩型液力偶合器 load limiting type coupling
限距 range line
限菌区系 gnotobiota
限菌生物学 gnotobiology; gnotobiotics
限控富营养化区 eutrophication control area of specialization
限亏止单 stop-loss order
限累模式 limited space-charge accumulation mode
限力扳手 torque wrench
限力装置 force-limiting device
限粒格筛 limiting grizzly
限粒筛 limiting screen
限量 limited range

限量泵 measuring pump; metering pump
限量放牧 rational grazing; rationed grazing
限量供水计划 use-restriction plan
限量灌瓶器 filling limit device
限量积分器 limited integrator
限量控制 limit control
限量运输 capacitated transportation
限流保护 current-limiting protection
限流电感 current-limiting inductance
限流电抗器 current-limiting reactor; limiting reactor
限流电抗线圈 current-limiting reactance
限流电路 current-limited circuit; current-limiting circuit
限流电阻 current-limiting resistance
限流电阻器 current-limiting resistor; limiting resistor
限流阀 limiting-flow valve
限流分路器 limiter shunt
限流过载保护装置 current-limiting overcurrent protective device
限流环 restrictor ring
限流继电器 current-limit(ing) relay
限流开关 current limit switch
限流孔 metering hole
限流孔口 limited orifice
限流控制 metering control
限流淋浴器 limiting-flow shower head
限流率 metering rate
限流喷管 metering nozzle
限流起动器 current-limiting starter
限流器 amperite; current limiter; flow snubber; restrictor
限流熔断器 limiting fuse
限流熔丝 current-limiting fuse
限流设备 water-restricting equipment
限流设施 limiting-flow device
限流水 bound water
限流通路 flow-limiting passage
限流系统 metering system
限流线圈 current-limiting inductor
限流信号 metering signal
限流信号交叉口 metering signal intersection; throttling signal intersection
限流因子 orificing factor
限流阴极 robber
限流匝道 metering pump
限流闸板 tweel
限流指示器 limit indicator
限流嘴 flow-limiting nozzle; metering jet
限流作用 metering function
限某区域的合作 closed regional cooperation
限内发行 intra-issue
限扭矩式离合器 torque limiting clutch
限坡坡率 rate of ruling grade
限坡选定 choice of ruling grade
限期 date of expiry; deadline; time limit
限期保险单 time policy
限期偿还贷款 payback loans within a set time
限期付款信用证 time letter of credit
限期环境污染治理制度 system of treating environmental pollution within a prescribed time
限期汇票 tenor draft
限期缴费人寿保险 limited-payment life insurance
限期缴费人寿保险单 limited-payment life policy
限期票据 demand bill
限期影响 dead influence
限期有效合同 term contract
限期整顿 consolidation within a specified time
限期治理 undertake treatment within a prescribed limit of time
限期治理环境污染制度 system of treating environmental pollution within a prescribed time
限期装船 timed shipment
限期租用的地产 term
限燃材料 <药柱铠装用> inhibiting material; inhibitor material
限燃层 restricting
限热 thermal relief
限筛尺寸 limiting screen size
限上限下项目 upper and lower investment limit project
限深规 depth stop
限深器 depth stop
限时报警器 time alarm
限时断路继电器 time cut-off relay
限时工作方式 operation time-limit system
限时继电器 time limit relay; time relay
限时解锁 definite time limit release; time release
限时解锁继电器 time release relay
限时解锁器 definite time limit release; time release
限时解锁图 time release scheme
限时解锁装置 timing releasing device
限时器 timer
限时人工解锁 manual time release
限时熔线 time fuse
限时式最高需量指示器 restricted hour maximum demand indicator
限时释放器 time release
限时元件 timing element
限时专递 special delivery
限使用一次注射器 disposable syringe
限束光阑 beam-defining jaw; beam-defining slit
限束器 beam-defining clipper
限束小孔 beam-defining aperture
限速 permitted speed; restricted speed; restrictive speed; slow order; speed limit; speed restriction
限速板 speed limitation board; speed restriction board
限速闭塞区段 speed restrictive block
限速标 restriction board; speed limitation board; speed limit notice
限速标记 speed limit marking
限速标志 speed-limit(ed) sign; state-(d)-speed sign
限速步(骤) rate-limiting step
限速颠簸槛 <美> speed control bump
限速断路器 speed limiting circuit breaker
限速反应 rate-limiting reaction
限速防坠装置【机】drop preventing device for carriage
限速进行 proceed at restricted speed
限速开关 speed limiting switch; governor switch; overspeed switch
限速路 limited speed road
限速路标 speed limit road sign
限速路槛 road hump; speed-control hump; road hump
限速模式 restricted mode
限速牌 restriction board; speed limitation board; speed restriction board
限速器 emergency governor; maximum speed governor; overrunning governor; over-speed governor; over-speed limit device; safety governor; speed limitator [limiter]; speed limiting device; speed limiting governor
限速丘 speed control hump
限速区 restricted speed area; restricted speed zone; restrictive speed area; restrictive zone; speed limit zone
限速区段 speed restricted section
限速区间 speed restrictive block
限速区域 speed restrictive area
限速设备 speed limit(ing) device
限速示像 restricted speed aspect; restricting aspect; restrictive aspect
限速输送机 retarding conveyer [conveyor]
限速调节器 speed limiting governor
限速驼峰 speed control hump; speed hump; road hump
限速信号 restricting signal; restrictive signal; speed-limited signal; speed restriction signal; speed signal
限速因素 rate-limiting factor
限速因子 rate-limiting factor
限速运行 proceed at restricted speed; speed limit; speed limitation; speed restriction
限速指示标 restricting speed indicator sign
限速指示器 restricting speed indication sign
限速制动器 speed limiting brake
限速装置 over-speed limit device; speed limit(ing) device
限外发行 extra-limit issue
限外条款 exclusion clause
限外温度计 ultra-thermometer
限位 spacing
限位板 fix plate; stopper
限位挡 limit stop
限位挡块 limit stop; positive stop
限位钉 banking pin
限位阀 limit valve
限位杆 gag lever post
限位工作台 spacing table
限位环 check ring; spacing ring; stop collar
限位机构 stop mechanism
限位检验 limit check
限位角转向泵 lock steering pump
限位开关 limit switch; over-travel-limit switch
限位开关保护的连接模 limit switches guarding progressive die
限位控制 limit control
限位块 stopper
限位链 anchoring chain; check chain; restraining device
限位螺钉 banking screw
限位器 limiter; limiting device; retainer; stop
限位圈 spacing collar
限位套筒 spacing collar
限位填片 spacing shim
限位销 spacer pin
限位指示器 cut-off level ga(u)ge
限位柱塞 stop plunger
限位装置 caging device; kickout; stop device
限温继电器 temperature limiting relay
限下剂量 sublimited dose
限心 limiting zone
限压半导体二级管 stabistor
限压阀 compression relief valve; pressure limitation valve; pressure limiting valve; pressure-relief valve; relief valve
限压阀杆 pressure limiting valve stem
限压阀钢球 pressure-relief valve ball
限压阀弹簧 pressure valve spring
限压器 voltage limiter
限压燃烧 limiting pressure combustion
限压设施 pressure limiting device
限压缩力密封 confinement-controlled seal
限压装置 pressure limiting device
限用胶黏[粘]剂 restricted adhesive
限油喷嘴 metering jet
限油器 oil-limiter
限于吃水的船 vessel constrained by her draught
限于计算的任务 compute-bound task
限于水上范围 water-borne only
限于微扰动运动 motion limited to small disturbances
限于温带气候 be limited to temperate climates
限阈 limiting value
限员投标 closed bid
限约流 bound flow
限运货物 restricted freight; restricted goods
限载保险丝 current limiting fuse
限载调速器 load limiting governor
限载转换器 load-limit changer
限在某些列车上有效 limited availability in certain trains
限照灯光 screened light
限值 limitation; threshold value
限值排布 limited value assignment
限止 limitation
限止阀 limit stop valve
限止流 restricted flow
限止器 eliminator; limitator
限止器盖 limiting device cover
限止线 limiting line
限止叶片 limit blade
限制 astriction; circumscribe; circumscribing; circumscription; clipping; confine(ment); constraint; contract; crimp; limit; limitation; narrowing; qualification; regulate; restrain; restrict; restriction; stint; tether
限制半径 limiting radius
限制比 restriction ratio
限制变形 restraint deformation
限制标志 restriction sign
限制表面 controlled surface; limiting surface
限制冰冻深度的路基(标高)设计 limited subgrade frost penetration design
限制步骤 rate-limiting step; rating-limiting step
限制部位 restriction site
限制财产处理权证书 backbond
限制层 confining layer; limiting layer
限制产量者 restricter
限制场 limiting field
限制超车视距 restricted passing sight distance
限制车速 regulation speed
限制车速的道路 <由人口密集区连接到边区> speed road
限制承兑 qualified acceptance
限制程度 degree of restraint
限制吃水 limited draught
限制吃水船 restricted-draft ship
限制尺寸 limit(ing) dimension; limit(ing) size
限制充气 restricted charger
限制大气干扰设备 X-stopper
限制贷款 credit squeeze
限制的 limiting; restrictive
限制的随机抽样法 restricted random sampling
限制等级 class of restriction
限制地层 confining stratum
限制地下水运动地层 confining zone
限制点 limiting point
限制电路 limiter circuit; limiting cir-

cuit
限制电平 cut-off level; slice level
限制电平范围 limiting level range
限制堵 choke plug
限制墩 restraining dolphin; mooring dolphin
限制额 limited amount
限制发行 restricted circulation
限制阀 inhibitor valve; limiter valve; limiting valve
限制阀盖 limiting valve cap
限制阀簧 limiting valve spring
限制阀体 limiting valve body
限制反馈 limiting feedback
限制反应 limiting response
限制范围 limit range
限制房屋高度的区划规定 height density
限制放大器 limiting amplifier
限制非生产性投入 limitation of nonproductive input
限制费用 restriction of expenditures
限制分路器 limit shunt
限制分配的材料 critical material
限制峰高【铁】limiting height of hump
限制缝 limiting slit
限制符 delimiter
限制幅器 limiter; limiting device
限制负荷电阻器 load limiting resistor
限制杆 limit rod
限制高度 restricted height
限制格式 restricted format
限制根 restricted root
限制工业化 deindustrialization
限制工资增长 limiting wage increases
限制供水 restrictive water supply
限制管 threshold tube
限制灌浆 containment grouting
限制航槽 restricted channel; restricted waterway
限制航道 restricted channel
限制航道阻力 restricted water resistance
限制航线 limited service route
限制航行 restricted voyage
限制和反限制 restriction versus opposition to restriction
限制荷载 limit load
限制弧 restriction arc
限制环 limit collar
限制级 limiter stage
限制剂量当量 dose-equivalent limit
限制价格 check price; lid on price
限制价格订货 limited price order
限制交通 restricted traffic
限制交通街道 traffic restricted street
限制接地保护 restricted earth protection
限制接地故障保护 restricted earth fault protection
限制进出的道路 limited way
限制进入 entrance restriction
限制进入的道路 limited access road
限制进入的桥梁 limited access bridge
限制进入的区域 restricted area
限制净距 restricted clearance
限制净空 restricted clearance
限制竞争的协议 combined trade restrictions
限制距离 limiting distance
限制开关 limit switch
限制可变长代码 restricted variable length code
限制空间 restricted quarter
限制空气喷口 restricted air-bled
限制空气压力 limited air pressure; partial air pressure
限制孔 choke
限制孔径 limiting aperture
限制孔口 restriction orifice
限制控制器 restriction controller
限制框 framing mask
限制扩散 restricted diffusion
限制扩散层析法 restricted diffusion chromatography
限制冷缩 restrained cooling shrinkage
限制力 restraining force
限制力矩 limiting moment
限制粒径 constrained diameter; controlled diameter
限制量计 limit ga(u)ge
限制流量 limiting discharge
限制流通 restricted circulation
限制路基冰冻深度设计法 limited subgrade frost penetration method
限制贸易 restrain trade; restrictive trade
限制贸易实施法庭 Restrictive Practices Court
限制膜片 limiting diaphragm
限制(某些车辆)进入的公路 limited access highway
限制浓度 limiting concentration
限制配船部位的条款 restrictions on shipping space clause
限制膨胀 confined expansion
限制频率 limiting frequency
限制坡道 ruling grade
限制坡度 constrained grade; limiting grade; limiting gradient; maximum resistance grade; operating grade; ruling grade; ruling gradient; rolling gradient <铁路、道路的>
限制齐次完整群 restricted homogeneous holonomy group
限制汽车行驶区域 automobile restricted zone
限制器 arrester[arrestor]; boundary member; chopper; clipper; debooster; killer; killer restrainer; limitator; limiter; limit stop; restrainer; restrictor; safety dog; slicer; stopper
限制器电极 limiter electrode
限制器特性 limiter characteristic
限制前馈 limiting feedforward
限制(倾)斜度 limiting gradient
限制区间 carrying capacity limiting section; critical section; speed restrictive section; limiting section
限制曲线 limiting curve
限制日期 limiting date
限制塞 limit plug
限制筛孔 effective screen cut-point
限制上坡坡度 ruling up-gradient
限制奢侈品进口 restriction of luxury imports
限制生产定额 gold-bricking
限制生产函数 limitational production function
限制声明 declaration of restrictions
限制石墨化 restricting graphitization
限制时间 binding hours
限制时间的停车 time limit parking
限制使用 restricted use
限制使用财产协定 restrictive covenant
限制使用的胶黏[粘]剂 restricted adhesive
限制使用失去效率 loss of containment
限制驶入 entrance restriction
限制市场 restricted market
限制示像 restricting aspect; restrictive aspect
限制示像区 zone of restrictive aspect
限制式适应控制 adaptive control constraint
限制视距 restrictive sight distance
限制水道 restricted waterway; confined waters
限制水路效应 effect of confined waters
限制水面 limit water surface
限制水深 limiting depth
限制水域 restricted waters; confined waters
限制速度 limitation velocity; limited speed; limiting speed; restricted speed; restrictive speed
限制速度区 restricted speed area; restrictive speed area
限制速率 speed limit
限制速率基质 rate-limiting substance
限制塑性流动 contained plastic flow
限制弹簧 restraining spring
限制条件 boundary condition; limiting condition; qualification; restrictive condition
限制条件下运输货物 <如易燃品、易爆品等> restricted freight
限制条款 proviso[复 proviso(e)s]; qualifying clause
限制停车视距 restricted stopping sight distance
限制通过 restricted passage
限制通过能力的道路 limited road
限制通货膨胀 arrest inflation
限制通频带的放大器 bandwidth-limiting amplifier
限制通行街道 close up street
限制投标 restricted tender
限制陀螺 constrained gyro
限制弯矩 constraining moment
限制弯曲 forced meandering
限制完整群 restricted holonomy group
限制位置 restrictive position
限制污水排放量 limit drainage amount of polluted water
限制污水排放浓度 limit drainage concentration of polluted water
限制物 bridle; limiter
限制物品 restricted articles
限制下坡坡度 ruling down-gradient
限制显示 restrictive indication; restrictive position
限制显示区 zone of restrictive aspect
限制线 barrier line; line of demarcation
限制消费 limited consumption
限制销售 restrict sale
限制效率 restriction efficiency
限制信号 restricting signal; restrictive signal
限制信号机【铁】stick signal
限制行车速度的道路 restricted speed way
限制性背书 qualified endorsement; restrictive endorsement
限制性措施 restrictive practice
限制性贷款 tied loan
限制性的 limitative
限制性定价 limit pricing
限制性分配 restricted distribution
限制性航道 restricted channel
限制性环节 critical link
限制性寄生 restrictive host
限制性交易措施 restrictive trade practices
限制性经营办法 restrictive business practice
限制性劳工措施 restrictive labo(u)r practice
限制性内切位点 restriction site
限制性契约 restrictive covenant
限制性权利边界 margin of restrictive rights
限制性燃烧 restricted burning
限制性商业措施 restrictive business practice; restrictive commercial practice
限制性商业惯例 restrictive business practice
限制性商业做法 restrictive business practice; restrictive commercial practice
限制性随机抽样 restricted random sampling
限制性条款 proviso[复 proviso(e)s]; restrictive covenant; restrictive clause
限制性投标 limited tender; restricted tender
限制性投票 limited ballot; restricted ballot
限制性温度 restrictive temperature
限制性效应 restrictive effect
限制性信贷政策 restrictive credit policy
限制性许可证 restrictive license
限制性援助 tied aid
限制性运转 restricted operation
限制性债权 claims subject to limitation
限制性政策 restrictive policy
限制性资源 constrained resources
限制性自行 restricted proper motion
限制修正系统 restriction modification system
限制序贯程序 restricted sequential procedure
限制压力 control pressure
限制议付 restricted negotiation
限制议付信用证 restricted (negotiable letter of) credit
限制因素 limiting factor; rating-limiting factor
限制因子 limiting factor; restriction factor
限制因子律 law of limiting factor
限制营业合同 contract in restraint of trade
限制营业税率 restrictive business tariff
限制用电 brown-out
限制用户 limited subscriber
限制用途的现金 restricted cash
限制用途的资产 restricted assets
限制优先级 limit priority
限制于小范围 focalize
限制渔船数 limiting the number of fishing vessels
限制与非限制随机走动 restricted and unrestricted random walk
限制运动 constrained motion
限制在公路以外使用的自卸车 off-highway dump truck
限制责任 limited liability
限制占有 restricted occupancy
限制障碍物的地面 <在飞机场周围,用以保障航空安全> obstruction restriction surface
限制褶皱 non-sequent fold(ing)
限制振动 constrained vibration; restricted vibration
限制正态性假定 restrictive normality assumption
限制直积群 restricted direct product group
限制主义 restrictionism
限制转弯 turn regulation
限制装车 restriction in wagon loading
限制资格的背书 qualified endorsement
限制资源 resource scheduling method
限制总捕捞量 limiting total amount of fishing
限制纵坡 limiting gradient
限制钻进工作量 curtailment of drilling

限制作用下的拉应力 tensile stresses under restraint
限制座席的列车 <如特快> train with limited accommodation
限制座席和停车站的特别快车 limited express train
限重标志 weight limit sign

宪

宪法 constitutional law

宪法(上的)规定 constitutional provision;constitutional prescription
宪章 charter

陷

陷凹 cul-de-sac(street);pouch

陷凹镜 culdoscope
陷凹镜检查 culdoscopy
陷波 trapped wave
陷波电路 trap;trap circuit;wave trap
陷波放大器 trap amplifier
陷波开关 trap switcher
陷波理论 trapping theory
陷波滤波器 notching filter;trap filter
陷波频率 notch frequency;trap frequency
陷波频率抑制 trap rejection
陷波器 trap filter;trapper;wave filter;wave trap
陷波器陷除频率范围 notch filter frequency range
陷波线圈 trap coil
陷捕物 birdlime
陷出 trap out
陷点 trapping spot
陷光器 light trap
陷获层 trapping layer
陷获常数 trapping constant
陷获级 trapping level
陷获中心 trapping center[centre]
陷阱 dead hole;entrap;fall trap;gin;semi-conductor trap;swallet hole;trap(-hole);pitfall
陷阱操作 trap operation
陷阱地址 trap address
陷阱方式 trapping mode
陷阱理论 trap theory
陷阱门 trap door
陷阱模型 trap model
陷阱能 trap energy
陷阱能级 trapping level
陷阱屏蔽 trap mask
陷阱位 trap bit
陷阱效应 trap effect
陷阱型 trap pattern;trapping
陷阱指令 trap instruction
陷坑 chasm;crater;pitfall;sink;sinkage;sinkaging
陷孔 crater
陷口几何关系 <指冲击钻进时,陷口容积、钻头几何形状同岩石阻力三参数关系> crater geometry relations

陷窟 chasm
陷落 cave;cave-in;collapse;come down;depress;downcast;downfall;falling-in;founder(ing)
陷落侧 downcast side;downthrow side
陷落地层 caving formation;caving ground;falling ground
陷落地块 depressed block;downthrow;downthrow block
陷落地震 collapse earthquake;depression earthquake;depressive earthquake
陷落范围 hold range;retention range
陷落湖 <地壳变动形成的> diastrophic lake
陷落火山口 ca(u)ldron;sunken caldera
陷落角 angle of draw;collapse angle;drawdown angle
陷落脉 collapsing pulse
陷落喷火口 depression crater
陷落盆地 subsidence basin
陷落区 caving area
陷落效应 slumping
陷落性地震 depressive earthquake
陷频滤波器 notch filter
陷球壳属 <拉> Trematosphaeria
陷扰指数 trapping index
陷入 emboly;fall into;plunge;trap in
陷入僵局 deadlock
陷入泥潭 mire down in mud;swamp
陷入泥中 bog;bogging down
陷入沼泽 swamp
陷声 sound trapping
陷损成辙 worn into ruts
陷窝 dimple
陷窝状水系模式 lacunate mode
陷型锤 set hammer
陷型模 hollow swage;swage
陷型模锻 swage process
陷型模锻机 swager
陷型模压铸件 die pressed casting
陷型桩 swage pile
陷穴 concave;sink hole
陷穴侵蚀 sinkhole erosion
陷油陷 trap
陷于困境 stalemate
陷噪器 noise trap

馅

馅饼 pie

献

献祭的 sacrificial

献身墓穴 sacrificial pit
献身于桥梁事业 bridge dedication

腺

腺病毒 AD virus

腺甾烷 conane

霰

霰 graupel;ice rain;sago snow

霰弹补给速度 pellet rate
霰弹冲击钻进 pellet impact drilling
霰弹冲击钻头 pellet impact bit
霰弹袋试验 shot-bag test
霰弹法 pellet system
霰弹功率 pellet power
霰弹计数回路 pellet counting circuit
霰弹聚集 pellet cloud
霰弹粒 shot grain
霰弹喷口比 pellet-to-nozzle ratio
霰弹喷射速度记录器 pellet rate recorder
霰弹碰撞 pellet interference
霰弹直径比 pellet-to-diameter ratio
霰弹制备设备 pelletizer
霰弹钻头结构 pellet bit texture
霰弹钻头喷射泵 pellet bit jet pump
霰弹钻头设计 pellet bit design
霰弹钻头研制 pellet bit development
霰弹作用 pellet bit
霰石 aragonite;chimborazite;conchite
霰石华 flos ferri;flower of iron
霰石灰岩 aragonitic limestone
霰雪丸 soft hail

乡

乡 <美国、加拿大> township

乡村 country;countryside;village
乡村暴雨污染 rural stormwater pollution
乡村别墅 villa rustica
乡村步行小道 country foot path
乡村车站 rural station
乡村城堡住宅 country house 'castle'
乡村城市化 village urbanization
乡村大厦 country mansion
乡村道路 back road;drift;dustroad;hay road;rural road;cart road;cart track;cart way;country road
乡村道路交叉口 rural intersection
乡村道路交通的季节性特点 seasonal character(of rural traffic)
乡村道路桥梁 village road bridge
乡村的 rural;rustic;sylvan;villatic
乡村的舒适 amentities of the countryside
乡村地 rural section
乡村地带 rural area
乡村地区 rural area;rural district;rural section
乡村电话 rural telephony
乡村电话线 rural distribution wire
乡村房舍 rustic home
乡村房屋 rustic home
乡村风味 rusticity
乡村风味的 rustic
乡村高速公路 rural motorway
乡村工业 rural industry
乡村公路 rural highway;rural road
乡村公园 country park
乡村供水 rural water supply
乡村广场 village square
乡村规划 rural planning
乡村国家 rural country
乡村和市镇建筑规范 country and township building code
乡村给水 rural water supply
乡村家具 cottage furniture
乡村建设者 country builder
乡村建筑 rural architecture
乡村建筑地段 village building lot
乡村交通调查 rural traffic survey
乡村教堂 country church;rural church;village church
乡村景色 village scene
乡村俱乐部 country club
乡村开发 rural development
乡村疗养院 village hospital
乡村林 village forest
乡村林地 village forest field
乡村路 carriage road;farm road;village road
乡村庙宇地段 village house lot
乡村农用工业 rural agricultural industry
乡村配电线 rural distribution wire
乡村朴实瓦片 rustic tile
乡村企业 rural enterprise
乡村区域规划 country planning;rural planning <已开发建设区域以外的>
乡村人口 rural population
乡村人口移入城市 rural-urban migration
乡村山庄 manison
乡村上门接送业务 country collection and delivery service
乡村社区 rural community
乡村式 rusticity
乡村水平 country level
乡村特点 rusticity
乡村土路 country road
乡村委员会 countryside commission
乡村污水处理 rural sewage disposal
乡村污水灌溉田 rural sewage disposal farm
乡村污水系统 rural sewage
乡村小屋 box
乡村医院 cottage hospital;rural hospital
乡村银行 country bank
乡村邮政路线 rural mail delivery route
乡村运量观测 rural traffic survey
乡村支票 country check
乡村中公共绿地 village green
乡村住房贷款 rural housing loan
乡村住房地基贷款 rural housing site loan
乡村住房联合会 Rural Housing Alliance
乡村自动电话局 community automatic exchange
乡道 country road
乡公路 <即乡道> township road
乡间别墅 country house;country villa
乡间道路 country road;rural road;secondary road;rural route
乡间的 rustic
乡间的店铺 village shop
乡间公路 rural motorway
乡间汽车道 rural motorway
乡间损失 country damage
乡间土路 country road;rural road;secondary road
乡间小路 country road
乡间住宅 country seat
乡路 country road
乡区 rural area;rural district
乡趣园 rustic finish;rustic garden
乡土保护 local environmental protection
乡土材料 indigenous materials
乡土的 indigenous;native;vernacular
乡土覆盖植物 indigenous cover
乡土化 provincialization
乡土环境教材 leaching materials of local environmental science
乡土建筑 vernacular architecture;vernacular construction
乡土教材 regional teaching materials
乡土品种 indigenous variety;indigenous species
乡土味造型 vernacular form
乡土艺术 vernacular arts
乡土植被 native vegetation
乡土植物 indigenous plant;native plant
乡土植物群 indigenous flora
乡土种 autochthon(e)[复 autochthon(e)s];native species
乡下的 rural
乡下人 countryman
乡野家具 rustic woodwork
乡镇 rural community;rural township
乡镇财政 village and town finance
乡镇道路 communal road;township road;village road
乡镇工业 rural industry;village and town run industry
乡镇公用设施规划图 community facility plan
乡镇规划 community planning
乡镇规划道路 county planning
乡镇接壤地带 rurban
乡镇企业 enterprise in townships(and towns);rural and small town enterprises;rural enterprise;township and village enterprise;township enterprise;township(-run) enterprise
乡镇企业环境管理 environmental management for enterprise of village and town
乡镇企业集团化 grouping of township enterprises
乡镇企业开发区 township and village

enterprise development zones
乡镇桥 town bridge
乡镇委员会 county council
乡镇运输点 country shipping point
乡镇长 chief executive;township government

相伴 attach to

相伴变量 associated variate;concomitant variable
相伴对称有向图 associated symmetric directed graph
相伴矩阵 adjoin matrix; associated matrix;conjugate matrix
相伴数 associate number
相伴图 associated graph
相伴无向图 associated undirected graph
相伴系数 coefficient of association
相伴有向图 associated directed graph
相伴张量 associated tensor
相伴阵列 associated array
相补事件 complementary event
相不平衡 phase unbalance
相参 coherent
相参波 coherent wave
相乘 multiplication
相乘混频 multiplicative mixing
相乘信道 multiplicative channel
相乘噪声 multiplicative noise
相持【统】tie
相持秩 tied rank
相斥的 repellent
相斥相 repulsion phase
相斥性 repellence[repellency]
相斥原理 exclusion principle
相除 dividing
相传动产 heirloom
相从变动 concomitant variation
相从变量 concomitant variable
相错式接头 alternate joint; broken joint;staggered joint
相错轴 crossed axis;skew axis
相当 correspondence
相当长 at some length
相当吹风距离 equivalent fetch length
相当大的 appreciable;tidy
相当大的差异 substantial difference
相当大的幅度 considerable latitude
相当大的少数 sizable minority
相当大的深度 a considerable depth
相当的 equivalent; reasonableness; tantamount
相当的火灾烈度 equivalent fire severity
相当桁 equivalent girder;hull girder
相当截面 equivalent section
相当均匀的 well-proportioned
相当流量 corresponding discharge
相当流量曲线 curve of corresponding discharge
相当湿度 equivalent moisture
相当水位 equivalent water level
相当卫生的水 more sanitary water
相当位温 equivalent potential temperature
相当温度 equivalent temperature
相当详细 at some length
相当应力 equivalent stress
相当于 amount to;contain
相当圆截面 equivalent circular section
相当直径 equivalent diameter
相当准确的近似值 close approximation
相导体 phase conductor
相导图 derivation graph
相的变化 facies change;phase change
相等 balance with; equality; equipollence;equivalency;parity
相等波纹性质 equal-ripple property
相等成本 equivalent cost
相等的 equal; equivalent; identical; tantamount; uniform; equiphase【物】
相等电路 equality circuit
相等高度 double altitudes;equal altitude
相等或超过概率 exceedance probability
相等集 equal set
相等截距约束 equal intercept restrictions
相等课税收益率 equivalent taxable yield
相等力 identical force
相等连接操作 equi-join operation
相等频数曲线 isofrequency curve
相等平价 parity of representation
相等评价法 peer rating
相等深度 even depth
相等数 equal number
相等物 equivalent
相等向量 equivalent vector
相等性 equality; equalization; equation
相等样品 matched samples
相等转移 branch on equality
相滴定 phase titration
相抵 balance;counter-balance;offset
相抵错误 errors of compensation
相抵的 abutting
相抵的错误 offsetting error
相抵减项 offsetting deduction
相度损伤 examination of wound
相对 end for end
相对安定度 relative stability
相对安置角 < 倾斜摄影中倾角分析的 > relative setting angle
相对百分数 relative percentage
相对板块运动 relative plate motion
相对饱和 relative saturation
相对饱和差 relative saturation deficit
相对饱和度 relative saturation degree
相对保持 relative retention
相对保持率 relative retention ratio
相对保留 relative retention
相对保留时间 relative retention time
相对保留体积 relative retention volume
相对保留值 relative retention value
相对曝光量对数 relative logH(E)
相对本构造区的方向 direction opposite this tectonic region
相对比 relative ratio
相对比电离 relative specific ionization
相对比较 relative comparison
相对比例(尺)relative scale
相对比例误差 proportional error; fractional error
相对比值法 relative ratio method
相对闭合差 ratio of closure;relative closing error
相对闭集 relatively closed set
相对边际价值 relative marginal value
相对边界 relative boundary
相对编号 relative number(ing)
相对编号法 relative number
相对编码 relative coding
相对编址分时 relative addressing time-sharing
相对变差函数 relative variogram
相对变动 relative variability
相对变化 relative change; relative variable;relative variation
相对变位 relative displacement
相对变形 relative deformation
相对变异度测度 measure of relative variability
相对变异度量 measure of relative variation
相对标高 relative elevation; relative level
相对标绘 relative plot
相对标志权重 relative index weight
相对标准 relative standard
相对标准偏差 relative standard deviation
相对标准偏差椭圆 relative standard deviation ellipse
相对标准误 relative standard error
相对标准误差 relative standard deviation
相对表面 apparent surface
相对表面活度 relative surface activity
相对表面积 relative surface area
相对表面积浓度 relative concentration of surface area
相对表面磨损 relative surface wear
相对表示法 relative presentation
相对波动 relative fluctuation
相对波段 relative band
相对补格 relatively complemented lattice
相对补救损失 relative saving loss
相对不变测度 relatively invariant measure
相对不变泛函 relatively invariant functionals
相对不变式 relative invariant
相对不对称性 relative asymmetry
相对不可溶性固体 relatively insoluble solid
相对不易疲劳性 relative indefatigability
相对不应期 relative refractory period
相对参数 relative parameter
相对参数灵敏度函数 relative parameter sensitivity function
相对糙度 relative roughness
相对糙率 relative roughness
相对侧 opposite side
相对测定 relative determination
相对测光 relative photometry
相对测量 relative measurement
相对插入损耗 relative insertion loss
相对查全率 relative recall
相对差异量 measure of relative dispersion
相对差阈 relative difference limen
相对产额 relative yield
相对产量 fractional yield
相对长度 relative length
相对超前 relative advance
相对超声波流速仪 contrapropagating ultrasonic flowmeter
相对潮 opposite tide
相对沉降量 relative settlement; relative compressibility
相对沉降系数 coefficient of relative settlement
相对沉陷 relative settlement
相对成本控制 relative cost control
相对成本系数 relative cost coefficient
相对成本向量 relative cost vector
相对程度 relative magnitude
相对程序 relative code; relative program(me);relocatable code
相对程序设计 relative programming
相对持久性 relative persistence
相对尺寸 relative dimension; relative size
相对尺度 relative scalar;relative scale
相对充气量 relative charge
相对冲击 relative impact
相对冲击能量 relative impact energy
相对稠度 relative consistence[consistency]
相对稠度曲线 relative consistence [consistency] curve
相对传输电平 relative transmission level
相对传输响应 relative transmitting response
相对传送电平 relative transmission level
相对垂线偏差 astrogeodetic deflection;relative deflection;relative deviation of vertical
相对磁导率 relative permeability
相对磁道 relative track
相对磁化率 relative magnetic susceptibility
相对磁力吸引性 relative magnetic attractability
相对磁强计 relative magnetometer
相对次数 relative degree
相对粗糙(度)relative roughness
相对粗糙(度)系数 coefficient of relative roughness; relative roughness factor
相对粗度 relative coarseness
相对催化活性 relative catalytic activity
相对脆性 relative brittleness
相对大小 relative magnitude;relative size
相对代码 relative code
相对代数数域 relative algebraic number field
相对代谢率 relative metabolic rate
相对带宽 relative bandwidth
相对当量 relative equivalent
相对导磁率 relative permeability
相对导函子 relative derived functor
相对的生物效应 relative biologic(al) effect
相对等高线 relative contour
相对等级 relative magnitude
相对等效值 relative equivalent
相对地 phase-to-earth;relatively
相对地势 relative relief
相对地形 relative topography
相对地形图 relative relief map
相对地应力测量 relative ground stress measurement
相对地址 relative address
相对地址标号 relative address label
相对地址代码 position independent code
相对地质年代表 relative time scale
相对地质时期 relative geologic time
相对电极化率 relative electric(al) susceptibility
相对电离比度 relative specific ionization
相对电离层吸收仪 riometer
相对电流电平 relative current level
相对电平 relative level
相对电平点 relative level point
相对电容率 relative permittivity
相对电压电平 relative voltage level
相对电压反应 relative voltage response
相对电压降 relative voltage drop
相对电子束加速器 relativistic electronic beam accelerator
相对电阻 relative resistance
相对订单 < 外汇订单 > counterpart
相对定位 relative positioning
相对定位法 < 卫星大地测量 > translocation mode
相对定向 relative orientation
相对定向不定性 uncertainty of relative orientation
相对定向函数 function of relative orientation

相对定向误差 error of relative orientation
相对定向像片 relative oriented photo; relative oriented picture
相对定向元素 elements of relative orientation
相对定向照片 relative photograph
相对动量 relative momentum
相对毒度 relative toxicity
相对毒力比 relative toxic ratio
相对毒性 relative toxicity
相对独立 relative independence
相对独立性 relative independentability
相对短距系数 relative pitch-shortening value
相对对比度 relative contrast
相对对比率 relative contrast
相对对数曝光时间 relative log exposure
相对多度 relative abundance
相对多样性 relative diversity
相对多余度 relative redundance [redundancy]
相对发光度 relative luminosity
相对发光效率 relative light efficiency; relative luminous efficiency
相对发挥度 relative volatility
相对反应速率 relative reaction rate
相对反应物量 relative amounts of reactants
相对反应性 comparative reactivity
相对范数 relative norm
相对方差 relative variance
相对方位 reciprocal orientation; relative azimuth; relative orientation
相对方位角 relative azimuth; relative bearing
相对方位角指示器 relative bearing indicator
相对方位平面位置指示器 heading upward plan position indicator
相对方位显示 heading upward presentation
相对方位元素 elements of relative orientation
相对方向 opposite direction; relative direction
相对方向盘转角进行动力转向的系统 proportional demand steering system
相对飞行高度 relative flying height
相对非紧子区域 relatively noncompact subregion
相对费用 relative cost
相对分布 percentage distribution; relative distribution
相对分布函数 relative distribution function
相对分量 relative component
相对分馏系数 relative fractionation coefficient
相对分歧指数 relative ramification index
相对分子质量 relative molecular mass
相对分子质量分布 relative molecular mass distribution
相对份额 relative share
相对丰度 abundance ratio; relative abundance
相对风 apparent wind; relative wind
相对风量 relative blowing rate
相对风速 relative wind velocity
相对风险 relative risk
相对风险系数 relative risk factor
相对峰面积 relative peak area
相对幅值 relative amplitude
相对辐射强度特征 relative radiance strength signature
相对负荷容量 relative load capacity
相对富集系数 relative enrichment factor
相对覆盖度 relative cover
相对覆盖率 relative coverage ratio
相对伽马场图 relative gamma field map
相对概率 relative probability
相对干涉作用 relative interference effect
相对干遮盖力 relative dry hiding power
相对感受性 relative sensitivity
相对感应 relative induction
相对刚度 relative rigidity; relative stiffener; relative stiffness
相对刚度半径 radius of relative stiffness
相对刚度比 ratio of relative stiffness; relative stiffness ratio
相对刚性 relative rigidity; relative stiffness
相对刚性半径 radius of relative stiffness
相对刚性比 ratio of relative stiffness; relative stiffness ratio
相对高差 relative height difference
相对高程 relative elevation
相对高度 relative altitude; relative height; free board <水面与陆地的>
相对高度位置 relative altitude position
相对隔水层 relative aquifuge
相对隔水层厚度 thickness of relative aquifuge
相对隔水层界线 boundary of relative confining bed
相对隔水层岩性 rock type of relative aquifuge
相对工资 relative wages
相对功率 relative power
相对功率增益 relative power gain
相对功效 relative potency
相对共振积分 relative resonance integral
相对构型 relative configuration
相对购买力 relative purchasing power
相对鼓风强度 relative blowing rate
相对固定费用 relative fixed cost; relative fixed expenses
相对惯性 relative inertness
相对光度测量 relative photometry
相对光谱定向反射比 relative spectral directional reflectance
相对光谱功率分布 relative spectral power distribution
相对光谱灵敏度 relative spectral sensitivity
相对光谱能量 relative spectral energy
相对光谱能量分布 relative spectral energy distribution
相对光泽计 relative glossmeter
相对轨道 relative orbit; relative trajectory
相对过剩人口 relative overpopulation; relative surplus-population
相对过调 relative overshoot
相对海底速度 <航海> ground speed
相对海平面的绝对航高 flying height above mean sea level
相对含冰量 relative ice content
相对含量 comparative content; relative content
相对含量分析 ratio analysis
相对含量分析器 content ratio analyser[analyzer]; ratio analyser[analyzer]; relative content analyser[analyzer]
相对含水量 relative water content
相对含水率 relative water content
相对航高 flying altitude above ground; flying height above terrain; flying height above the ground; relative flight height; relative flying altitude; terrain clearance
相对航向 relative course
相对航向指示器 relative heading indicator
相对黑度 relative blackness
相对横摇 relative rolling
相对厚度 relative depth; relative thickness
相对滑动 relative sliding; relative sliding movement; relative slip
相对滑移理论 relative slip theory
相对滑移运动 relative sliding movement
相对化 relativization
相对环境价值 relative environment value
相对环境指数 relative environment index
相对环向(压)力 relative hoop force
相对缓解 relative remission
相对挥发度 relative volatility
相对混响级 relative reverberation level
相对活度 relative activity
相对活度测量 <放射性> relative activity determination
相对活性 relative activity
相对积分不变式 relative integral invariant
相对基点重力差 gravity difference to base station
相对极 anti-pole; relative pole
相对极大或极小 relative maximum or minimum
相对极大值 relative maximum
相对极线 relative extremal
相对极限 relative limit
相对极小点 relative minimum point
相对极小模型 relatively minimal model
相对极小值 relative minimum
相对极性 relative polarity
相对极值 relative extreme (value); relative extremity
相对极值曲线 relative extremal curve
相对加溶能力 relative solubilizing power
相对加速度 relative acceleration
相对价格 relative price
相对价格差别 relative price difference
相对价格效应 relative price effect
相对价值 relative value; relative worth
相对价值的比例表 scale of relative values
相对价值的同样量的变化 same change of magnitude in relative value
相对价值递减 diminishing relative value
相对价值量 magnitude of relative value
相对价值形式 relative form of value; relative value form
相对间距 relative spacing
相对剪切 relative shear
相对焦距 relative focal length
相对校正因子 relative correction factor
相对阶次 relative order
相对接近航速 relative speed approach
相对接收响应 relative receiving response
相对结果 relative result
相对介电常数 relative dielectric(al) constant; relative inductivity; relative permittivity
相对紧半群 relatively compact semigroup
相对紧度 relative firmness
相对紧域 relatively compact domain
相对紧致 relatively compact
相对劲度 relative stiffener; relative stiffness
相对劲度半径 radius of relative stiffness
相对劲度比 relative stiffness ratio
相对近似值 relative value approximation
相对精度 relative accuracy; relative precision
相对精确性 relative precision
相对净精确度 relative net precision
相对径向磨损 relative radial wear
相对径向速度 relative radial velocity
相对镜面光泽度 relative specular glossiness
相对距离 relative distance
相对均差 relative mean deviation
相对均衡 relative equilibrium
相对均匀收敛性 relative uniform convergence
相对均匀性 relative homogeneity
相对开集 relatively open set
相对抗滑性 relative antiskid characteristic
相对抗滑值 <以肯脱基岩地沥青试样作为1> relative resistance value
相对可见度 luminosity coefficient
相对可见度曲线 luminosity curve; relative visibility curve
相对可见度因数 relative visibility factor
相对可靠性 relative reliability
相对可透性 relative perviousness
相对克分子响应 relative molar response
相对空气密度 relative air density
相对孔径 aperture ratio; relative aperture; relative opening
相对孔隙 relative pore space
相对孔隙度 fractional porosity; relative porosity
相对孔隙率 fractional porosity; relative porosity
相对控制 relative control
相对控制范围 relative control range; relative range of control
相对控制条件方程 condition equation of relative control
相对跨度 relative advance
相对跨距 relative pitch
相对亏损 relative deficit
相对扩散 relative diffusion
相对累积频率曲线 relative cumulative frequency curve
相对离差 relative scatter
相对离散空间 relative discrete space
相对力矩 relative moment
相对利益 comparative advantage; relative advantage
相对连接的二极管 opposed diode
相对链复形 relative chain complex
相对亮度 relative brightness; relative luminosity
相对亮度曲线 relative luminance curve
相对亮度系数 relative luminosity factor
相对量 relative quantity
相对烈度 relative intensity
相对邻近 relative proximity
相对邻域 relative neighbo(u)rhood
相对林地 relative forest land
相对灵敏度 relative quickness; relative response; relative sensitivity
相对灵敏度系数 relative sensitivity coefficient; relative sensitivity factor
相对灵敏黏[粘]合剂 relative sensitive

X

相对流 relative current
相对流出角 relative exit angle
相对流动 relative flow
相对流量 relative discharge
相对流平性 relative level(l)ing
相对流速 relative current speed;relative velocity
相对流线 relative line of flow
相对流形 relative manifold
相对硫酸含量 relative acid content
相对漏洞 relative photographic(al) gap
相对旅程时间 relative travel time
相对旅运成本 relative travel cost
相对旅运服务 relative travel service
相对率 relative rate
相对绿时差【交】relative offset
相对论 relativism;relativity theory;theory of relativity
相对论等离子体 relativistic plasma
相对论电动力学 relativistic electrodynamics
相对论极限 extreme relativistic limit
相对论简并 relativistic degeneracy
相对论力学 relativistic mechanics
相对论偏折 relativistic deflection
相对论位移 relativity shift
相对论性波动方程 relativistic wave equation
相对论性场 relativistic field
相对论性改正 relativistic correction
相对论性粒子 relativistic particle
相对论性能量 relativistic energy
相对论性能量范围 relativistic energy range
相对论性能量损失率 relativistic rate of energy loss
相对论性情况 relativistic case
相对论性热力学 relativistic thermodynamics
相对论性速度 relativistic velocity
相对论性效应 relativistic effect
相对论性协变量 relativistic covariant
相对论性修正 relativistic correction
相对论性因子 relativistic factor
相对论性直线加速器 relativistic linac
相对论性质量 relativistic mass
相对论性质量方程 relativistic mass equation
相对论修正 relativity correction
相对论运动学 relativistic kinematics
相对论质量 relativistic mass
相对螺距 relative pitch
相对埋入深度 relative embedment
相对埋深 relative embedment
相对脉冲高度 relative pulse height
相对密度 density index;density ratio;relative compaction;specific density
相对密度计 gravi(to)meter;pycnometer
相对密度试验 relative density test
相对密切圆 relative osculating circle
相对密实度 relative compaction;relative compactness;relative density
相对免税率 franchise
相对面积响应 relative area response
相对敏感性 relative sensitivity
相对命令 relative command
相对摩擦系数 relative friction coefficient
相对摩尔量 relative molecular weight
相对摩尔热含 relative molar heat content
相对磨耗率 wear rating
相对磨损率 relative wear rate
相对内效率 internal efficiency ratio
相对耐久性 durability ratio
相对耐磨性 relative wear resistance
相对能见度 relative visibility
相对能力 relative capacity
相对年代 relative age
相对年代测定 relative dating
相对年代学 relative chronology
相对年龄 relative age
相对黏[粘]度温度数(值)relative viscosity temperature number
相对黏[粘]性 relative tack
相对黏[粘](滞)度 relative viscosity
相对捻度 relative twist
相对浓度 relative concentration
相对浓缩系数 relative enhancement coefficient
相对排放量 relative discharge
相对排放量指数 relative emission index
相对排架法 relative location;relative location system
相对排列货物起卸口系统 bank of elevators
相对排气速度 relative outlet velocity
相对排污量 relative discharge
相对判别式 relative discriminant
相对膨胀 relative expansion
相对偏比熵 relative partial specific entropy
相对偏差 relative deviation
相对偏度 relative skewness
相对偏离 relative departure
相对偏心率 relative eccentricity
相对贫困化 relative pauperization
相对频率 relative frequency
相对频率变差 relative frequency draft
相对频率变差率 relative frequency variation rate
相对频率分布 relative frequency distribution
相对频率函数 relative frequency function
相对频率偏差 relative frequency deviation
相对频率图 relative frequency diagram
相对频率误差 relative error of the frequency
相对频率直方图 relative frequency histogram
相对频偏 phase deviation
相对频谱曲线 relative spectrum curve
相对频数 relative frequency
相对平衡 relative balance;relative equilibrium
相对平静 relative quiet
相对平均二级差 relative average difference of second order
相对平均离差 relative mean deviation
相对平均数 average of relatives
相对平均信息量 relative entropy
相对平移 relative translation
相对坪斜率 relative plateau slope
相对起伏 relative relief
相对气流 relative wind
相对气流速度 relative air speed
相对牵引力 relative traction
相对强度 relative intensity;relative strength
相对强度输入/输出值 relative intensity input/output value
相对强度系数 coefficient of relative strength
相对强吸附 relative strong adsorption
相对切削速度 relative cutting speed
相对切削性 relative machinability
相对倾角 decalage;relative inclination;relative tilt angle
相对倾斜 relative tilt
相对倾斜计 relative inclinometer
相对倾斜仪 relative inclinometer
相对清晰度 relative articulation
相对曲度 <弯道曲度与间宽之比> relative curvature
相对曲率 relative curvature
相对曲率半径 relative radius of curvature
相对曲面拟合 relative surface fitting
相对全距 relative range
相对全损 relative total loss
相对权重 relative weight
相对燃耗 fuel utilization;relative burnup
相对热应力模数 relative thermal stress modulus
相对认址【计】relative addressing
相对日斑数 relative sunspot number
相对日照 relative sunshine
相对容量 relative capacity
相对容许河流污染 comparative acceptable river pollution
相对溶剂化作用 relative solvation
相对冗余(度)relative redundance[redundancy]
相对三倍精度 relative triple precision
相对散射力 relative scattering power
相对散射强度 relative scatter intensity
相对色彩 relative hue
相对色度电平 relative chroma level
相对色度时间 relative chroma time
相对色散 relative dispersion
相对色散的倒数 reciprocal relative dispersion
相对色散偏离值 deviation of relative dispersion from normal
相对扇形搜索 relative sector search
相对熵 relative entropy
相对上链复形 relative cochain complex
相对上升区 positive area;positive element
相对上同调 relative cohomology
相对上同调论 relative cohomology theory
相对上同调群 relative cohomology group
相对伸长 elongation per unit length;relative elongation;specific elongation;ultimate set
相对深度 relative depth
相对渗入量 relative infiltration
相对渗透度 relative permeability
相对渗透率 relative permeability
相对渗透率与饱和率的关系(曲线) relative permeability ratio
相对渗透率之比 relative permeability ratio
相对渗透系数 coefficient of relative permeability
相对渗透性 relative permeability;relative perviousness
相对生产量 relative productivity
相对生产率 relative productivity
相对生产密度 relative production density
相对生长 relative growth
相对生长率 relative growth rate
相对生物效率 relative biologic(al) efficiency
相对生物效应 relative biologic(al) effectiveness
相对生物学效能 relative biologic(al) effectiveness
相对生物有效性 relative biologic(al) effectiveness
相对剩余 relative surplus
相对失调 relative detuning
相对湿度 comparative absolute humidity;relative degree of humidity;relative humidity;relative moisture
相对湿度百分率 percentage relative humidity
相对湿度度计 absorption hygrometer
相对湿度换算器 relative humidity converter
相对湿度计 absorption hygrometer
相对湿度显示器 relative humidity indicator
相对湿度指示器 relative humidity indicator
相对湿润性 relative wettability
相对时差 relative time difference
相对时代 relative age
相对矢量 relative vector
相对式对接【铁】opposite joint;square joint
相对式接头【铁】opposite joint;square joint
相对视差 relative parallax
相对视向速度 relative radial velocity
相对适应性 relative adaptability
相对收入 relative income
相对收入差别 relative income difference
相对收入分配 relative distribution of income
相对收入假设 relative income hypothesis
相对收益和绝对收益 relative and absolute returns
相对寿命 comparative lifetime
相对疏伐强度 relative thinning intensity
相对疏水性 relative hydrophilicity
相对输出量 relative output value
相对竖直线偏差 relative deviation of vertical
相对数 proportion;relative number
相对数据 relative data
相对衰减 relative attenuation;relative damping ratio
相对双反射率 relative bireflectivity
相对双支管 double branch
相对水深 relative depth;relative water depth
相对瞬心 relative instantaneous center[centre]
相对死亡率指数 relative mortality index
相对搜索 relative search
相对速度 relative speed;relative velocity
相对速率 relative speed
相对塑性指数 relative plasticity index
相对酸度 relative acidity
相对损伤因数 relative damage factor
相对损失 fractional loss
相对缩短 relative reduction;specific compression
相对所得 relative income
相对弹性恢复 relative elastic recovery
相对弹性回弹 relative elastic recovery
相对弹性挠度 <管道的> flexibility factor
相对探伤灵敏度 overall flaw detection sensitivity
相对特异性 relative specificity
相对梯度 relative gradient
相对梯级 relative step
相对体积 relative volume
相对体积变形 relative volumetric deformation
相对体积法 relative volume method
相对天体物理学 relativistic astrophysics
相对调和测度 relative harmonic measure
相对调节 relative regulation
相对停留时间 relative resident time;

relative retention time
相对停滞阶段 stage of a relative standstill
相对同调代数【数】relative homological algebra
相对同调理论 relative homology theory
相对同调群 relative homology group
相对同伦群 relative homotopy group
相对同位素丰度测量 relative isotope abundance determination
相对投资 counter investment
相对透气率 relative permeability
相对透射比 transmittancy
相对透射率 relative transmittance
相对凸模直径 specific punch diameter
相对推力 relative force
相对弯曲 relative bending
相对弯曲半径 relative bending radius
相对往来账户 reciprocal account
相对危害性 relative hazard
相对危害因数 relative hazard factor
相对危险度 comparative risk; relative risk
相对危险性 comparative risk; relative risk
相对微分几何学 relative differential geometry
相对微分自由能 relative partial free energy
相对位偏 malalignment
相对位移 relative displacement
相对位移方式 mode of relative displacement
相对位移量 relative position movement
相对位置 relative location; relative position
相对温度 relative temperature
相对文件 relative file
相对稳定地带 relatively steady range
相对稳定地块 relative stable area
相对稳定度 relative stability
相对稳定阶段 relative stability stage
相对稳定期 relatively steady period
相对稳定性 relative stability
相对涡度 relative vorticity
相对涡流 relative eddy
相对污染负荷 relative pollutant loading
相对污染水平 relative contamination level
相对污染指数 relative pollutant index; relative pollution index
相对误差 fractional error; percentage error; proportional error; relative error
相对误差椭圆 relative error ellipse
相对吸附量 relative adsorbed amount
相对吸附系数 relative adsorption coefficient
相对吸收 relative absorption
相对吸收物 relative absorbance
相对吸收项 relative-absorption term
相对系数 relative coefficient
相对下沉量 quantity of relative subsidence; relative subsidence
相对下沉系数 coefficient of relative settlement
相对相容性 relative consistence[consistency]
相对相(位) relative phase
相对响应 relative response
相对响应因子 relative response factor
相对响应值 relative response value
相对向上移动 positive movement
相对项(目) relative term
相对消耗量 relative consumption
相对销价法 relative sales value method
相对销售价值法 relative sales value method
相对效率 generic efficiency; relative efficiency
相对效率检验 relative efficiency inspection
相对效能 relative efficiency
相对效应比 relative effect ratio
相对协方差 relative covariance
相对谐波含量 relative harmonic content
相对信用证 counter letter of credit
相对形变 relative deflection
相对形变梯度 relative deformation gradient
相对性 relative property; relativism; relativity
相对性能 relative performance
相对性燃烧指标 relative flammability index
相对性原理 principle of relativity; relativity principle
相对性原则 relativization principle
相对性状 relative character
相对休眠 relative dormancy
相对休眠期 relative dormant period
相对虚分量剖面平面图 profile-plan figure of relative imaginary component
相对虚分量剖面图 profile figure of relative imaginary component
相对旋角 relative swing
相对旋涡 relative eddy
相对旋转 counter-rotating; counter-rotation
相对旋转元件 relatively rotating elements
相对选择性 relative selectivity
相对眩光 relative glare
相对寻址 relative addressing
相对循环 relative cycle
相对压力 relative pressure
相对压力传感器 differential pressure transducer
相对压力系数 relative pressure coefficient
相对压强 relative intensity of pressure
相对压实度 relative compaction; relative compactness
相对压实作用 relative compaction
相对压缩度 relative compressibility
相对压缩量 relative compressibility
相对压缩性 relative compressibility
相对压下量 relative reduction
相对延迟 relative delay
相对延伸率 relative elongation
相对延时 relative time delay
相对要素成本 relative factor cost
相对要素价格 relative factor price
相对一致收敛 relative uniform converge
相对一致星收敛【数】relative uniform star convergence
相对一致性 relative uniformity
相对移动 relative shift
相对移位速率 rate of travel
相对异常 relative anomaly
相对异常剖面曲线 profile curve of relative anomaly
相对异常衰减曲线 attenuation curve of relative anomaly
相对应的人(或物) flip side
相对硬度 relative hardness
相对优点 relative merit
相对优势 comparative advantage; relative odds
相对有利 comparative advantage
相对有利条件原则 principle of comparative advantage
相对阈限 relative limen; relative threshold
相对圆 relative circle
相对约束 relative restraint
相对运动 relative motion; relative movement
相对运动方向 direction of relative movement
相对运动速度 speed of relative movement
相对运动显示 relative motion display
相对运动线 relative movement line
相对运动行程 distance of relative movement
相对运动作图 relative plot
相对运价 relative rate
相对运转特性 relative operating characteristic
相对噪度 relative noiseness
相对噪声功率 relative noise power
相对噪声温度 relative noise temperature
相对增量 relative increment
相对增益 relative gain
相对张量 relative tensor
相对涨落 relative fluctuation
相对账目 opposite account
相对照 cross reference
相对折射率 relative index of refraction; relative refraction index
相对振幅处理 relative amplitude processing
相对振幅剖面 relative amplitude section
相对震速 relative seismic velocity
相对蒸发 relative evapo(u)ration
相对蒸发量 evaporation opportunity
相对蒸气密度 relative vapo(u)r density
相对蒸气压 relative vapo(u)r pressure
相对蒸腾 relative transpiration
相对正方形搜索 relative square search
相对正截面 reciprocal normal sections
相对支付能力 relative capacity to pay
相对值 relative magnitude; relative quantity; relative size; relative value【数】
相对值系统 per-unit system
相对指标 relative indicator; relative indicatrix; relative indices
相对指数 relative index
相对制约 relative restraint
相对质量 mass ratio
相对质量过剩 relative mass excess
相对质量亏损 relative mass defect
相对置信区间 relative confidence interval
相对中毒 relative poisoning
相对中误差 relative mean error
相对中误差椭圆 relative standard deviation ellipse
相对重力 relative gravity
相对重力测定 relative gravity determination
相对重力测量 relative gravity measurement
相对重力数据 relative gravity data
相对重量 relative weight
相对重要性 relative importance
相对周期误差 relative error of the period
相对周期运动 relative periodic (al) movement
相对主义 relativism
相对专一性 relative specificity
相对转动 relative rotation
相对转换比 relative conversion ratio
相对转角 mutual rotation
相对转角半径 relative corner radius
相对转弯 counter-rotation
相对转移指令 jump relative instruction
相对装入程序 relative loader
相对准确度 relative accuracy
相对着色力 relative colo(u)r strength; relative tinting strength
相对自行【天】relative proper motion
相对总基点重力差 gravity difference to main base station
相对纵坡 <即路面边缘纵坡和实际超高纵坡之间的差值> relative longitudinal slope
相对纵轴的角位移 roll displacement
相对走时 relative travel time
相对阻力 relative resistance
相对阻尼系数 relative damping factor
相对阻止本领 relative stopping power
相对阻滞 relative retardation
相对最大值 relative maximum
相对最小面积 relative minimum of area
相对最小值 relative minimum
相对作业水准 relative activity level
相对坐标 incremental coordinates; relative coordinates
相对坐标系 relative coordinate system
相反 contradiction; contrariety; reverse; reversed polarity
相反表达式 reciprocal expression
相反步骤 reverse procedure
相反的 abhorrent; antagonistic; backward; counter; inverse; inverted; opposite; reciprocal; reverse
相反的极 reciprocal pole
相反的计划 cross purpose
相反的事物 contra
相反地 per contra; the other way
相反方向 opposite direction; right-about
相反符号 contrary sign; reversed sign
相反关系 inverse relationship
相反航向 opposite course
相反极 anti-pole
相反价值 opposing value
相反命题 contrary proposition
相反频率 reverse-frequency
相反溶质 opposite-type solute
相反色 opposite colo(u)r
相反万向的 criss-cross
相反系统 reciprocal system
相反效果 boomerang effect
相反需求类型 inverse demand pattern
相反旋转 counter-rotating
相反应 phase reaction
相反应力 counter stress
相反折弯 opposite folding
相反值 inverse value
相反转 reversal of phase
相反组断层 antithetic fault
相反作用 opposing reaction
相符褶皱 congruous fold
相辅的 explementary
相辅相成的发展 develop in a complementary way
相辅相成的经济结构 complementary economic structure
相辅需求 complementary demand
相负荷 phase load
相富集系数 phase enrichment coefficient
相干 cohere; coherent
相干背景 coherent background
相干波 coherent wave
相干波列 coherent wave train; interacting wave train

相干参考 coherent reference
相干侧视雷达 coherent side looking radar
相干长度 coherent length
相干成像 coherent image
相干成像系统 coherent imaging system
相干传递函数 coherent transformation function
相干次波 coherent secondary wave
相干单色波 coherent monochromatic wave
相干度 coherent degree;degree of coherence
相干断续波 coherent interrupted waves
相干发射 coherent emission
相干发射机 coherent transmitter
相干发生器 coherent oscillator
相干反射 coherent reflecting
相干范围 coherence range; coherent range
相干分光计 coherent spectrometer
相干辐射 coherent radiation
相干光 coherent light
相干光成象 imaging with coherent light
相干光处理 coherent optical processing
相干光处理方法 processing method with coherent/partial coherent light
相干光定位器 coherent optical locator
相干光光束 coherent light beam
相干光雷达 coherent light radar; coherent optical radar; colidar; optic(al) coherent radar
相干光雷达系统 coherent optical radar system
相干光平行处理 coherent optical parallax processing
相干光谱 coherent spectrum
相干光全息术 coherent holography
相干光束 coherent beam
相干光调制 modulation of coherent light modulation
相干光通信[讯] coherent light communication
相干光信息处理 coherent optical information processing
相干光学 coherent optics
相干光学测图 coherent optical mapping
相干光学处理系统 coherent optical processing system
相干光学计算机 coherent optical computer
相干光源 coherent(light) source
相干光自适应技术 coherent optical adaptive technique
相干函数 coherence function
相干红外激光 coherent infrared laser
相干红外雷达 coherent infrared radar
相干回波 coherent echo
相干积分 coherent integration
相干技术 coherence technique
相干加强 coherence emphasis;coherent enhancement
相干加强剖面 coherent enhancement section
相干检波 coherent detection
相干检波器 coherent detector
相干接收 coherent reception
相干接收器 coherent receiver
相干结构 coherent structure
相干解调 coherent detection
相干矩阵 coherence matrix
相干喇曼光谱法 coherent Raman spectrometry
相干雷达 coherent radar
相干雷达像片 coherent radar photograph
相干滤波 coherence filtering
相干孪晶间界 coherent twin boundary
相干脉冲法 coherent-impulse method
相干脉冲雷达 coherent-pulse radar
相干频率 coherent frequency;coincidence frequency
相干散射 coherent scattering
相干散射区 coherently scattering region
相干时间 coherence time
相干式微密度计 coherent microdensitometer
相干视频信号 coherent video
相干调制 coherence modulation; modulation coherence
相干通信[讯]信道 coherent communication channel
相干投影 coherent projection
相干系统 coherent system
相干相移键控制 coherent phase shift keying
相干效应 coherence effect
相干信号 coherent signal
相干性 coherence[coherency]
相干应变 coherency strain
相干影像 coherent video
相干运转 coherent operation
相干载波 coherent carrier
相干载波系统 coherent carrier system
相干噪声 coherence noise; coherent noise
相干噪声滤波 coherence noise filtering
相干照明 coherent illumination
相干振荡 coherent generation
相干振荡器 coherent generator
相干震荡器 coherent oscillator;coho
相干转发器 coherent transponder
相干组合 coherence array
相隔信道干扰 alternate channel interference
相关 interlocks; interrelate; mutuality;rapport
相关半径 correlation radius
相关比方差分析 analysis of variance for correlation ratio
相关比例法 correlation ratio method
相关比(率) correlation ratio; related ratio;relevance ratio
相关变换 correlation transformation
相关变量 correlated variables
相关变率 related rate
相关变异 correlated variation
相关变异性 correlated variability
相关标识符 relevant identifier
相关标引 multiple-aspect indexing; relational indexing
相关表 correlation table; dependence list
相关部分 relevant portion
相关参数 correlation parameter
相关参照 reciprocal reference
相关测量仪 correlation-measuring instrument
相关测向台 correlation direction finder
相关产量范围 relevant volume range
相关产品 related product
相关产业助推 complementary related industries
相关长度 correlation length
相关沉积法 method of correlative sediments
相关成本 related cost;relevant cost
相关成本计算 relevant costing
相关乘积 relative product
相关程度 correlativity;degree of correlation
相关程序 dependent program(me); relative program(me)
相关尺寸标注 associative dimensioning
相关抽样 correlated sampling
相关存储器 relational memory; relational storage
相关带宽 correlation bandwidth
相关单位 coherent units
相关当局 appropriate authority
相关的 coherent;correlative
相关的合并估计 combined estimates of correlation
相关的设备 relating device
相关地 relatively
相关地形 relative topography
相关(定)律 law of correlation
相关定向仪 correlation direction finder
相关动作 relevant action
相关对比法 correlation method
相关法 method of correlates
相关反射 coherent reflecting
相关范围 relevant range
相关方程(式) correlation equation; correlate(d) equation; dependent equation; equation of correlation
相关方法 correlation method; correlation technique
相关方面 correlation aspect
相关方位 relative bearing
相关费用 correlative charges
相关分类法 relative classification
相关分析 analysis of correlation; relative analysis; relevance analysis; correlation analysis
相关分析法 correlation analysis method
相关风险 relevant risk
相关复合类 phased class
相关概念系列 series of concept
相关干涉仪 correlation interferometer
相关感色 related perceived colo(u)r
相关跟踪测距系统 correlation tracking and range system
相关跟踪三角测量系统 correlation tracing and triangulation system
相关跟踪系统 correlation tracking system
相关工程 associated works
相关工艺 related trades
相关公式 correlation formula
相关构件 correlated element
相关估计量【数】estimate of correlation
相关故障 dependent failure
相关关系 dependency [dependence] relation;related relation
相关观测 correlated observation;correlation observation
相关观测平差 adjustment of correlated observations
相关光谱 correlation spectrum
相关光纤束 coherent bundle
相关过程 correlated process
相关函数 coherence function; correlation function; dependence function;function of correlation;related function
相关函数分析 correlation function analysis
相关函数分析仪 correlation function analyser[analyzer];correlator
相关函数计算记录器 correlatograph; correlatogram
相关函数组 dependent functions
相关横道图 linked bar chart
相关活度 related activity
相关货物 complementary goods
相关机构网 institutional framework
相关计 correlometer
相关计算 correlation calculation
相关计算机 correlation computer
相关技术 correlation technique
相关检波 coherent detection; cross-correlation detection
相关检波器 correlation detector
相关检测 correlation detection
相关检测器 correlation detector
相关检索 correlative indexing
相关角 related angle
相关接收 correlation reception
相关精度 correlation accuracy;correlation precision
相关矩阵 correlation matrix
相关距离 correlation distances
相关控制 relevant control
相关控制服务单元 association control service element
相关雷达 correlation radar
相关类型 correlation type
相关理论 correlation theory
相关联设备 associated equipment;associated outfit
相关流动准则 associated flow rule
相关率 correlative ratio
相关论 theory of correlation
相关螺旋 relational coiling;relational spiral
相关码体系 code-dependent system
相关脉冲 coherent pulse
相关脉冲法 coherent-impulse method
相关脉冲雷达 coherent-pulse radar
相关门宽度 correlation gate width
相关面 correlation surface
相关模型 correlation model
相关内插器 related interpolator
相关频数比值法 correlation sequence method
相关平均信息量 joint entropy
相关剖面 correlation section
相关器 correlating device;correlation device; correlation-type receiver; correlator;correlator device
相关曲线 correlation curve; correlation diagram; correlative curve; curve of relation; relation curve; relative curve;scattergram
相关曲线图 correl(at)ogram
相关曲线图分析 correlogram analysis
相关取向跟踪测距系统 correlation orientation tracking and range system
相关散点图 correlation scatter diagram
相关散射 coherent scattering
相关色温 correlated colo(u)r temperature;correlated temperature
相关商品 related product
相关熵 joint entropy
相关声码器 relevant vocoder
相关时窗时间 length of correlation time window
相关时间 correlation time
相关式 related expression; relational expression
相关事件 dependent event
相关视频信号 cohered video
相关视要素 correlation apparent feature
相关试验 test for correlation
相关受益与成本 relevant benefits and cost
相关树 relevance tree
相关数 correlation numbers; depend-

ency number
相关数据 related data
相关数据处理 associative data processing
相关数据处理机 correlated data processor
相关数据库 relational database
相关数列 correlated series
相关水权 correlative water rights
相关水位 correlation water level; correlative ga(u)ge; relation stage
相关水位曲线 stage relation curve
相关伺服回路 correlator-servo loop
相关损失 correlation loss
相关索引 coordinate indexing; correlative indexing; manipulative indexing
相关索引法 coordinate indexing
相关特性 relation property
相关条件 correlated condition
相关通道程序 related channel program(me)
相关投资 correlative investment
相关图 correlation chart; correlation curve; correlation diagram; correlatogram; correlatograph; scattergram
相关图分析 correlogram analysis
相关图形 relation graph
相关微分方程 related differential equation
相关位置 correlated site
相关文件 associated file
相关问题 interrelated issues
相关物 correlate; correlative
相关误差 dependent error
相关系 mutual relationship
相关系数 coefficient of correlation; correlation coefficient; correlation factor; index of correlation; related coefficient
相关系数测试仪 measuring set for correlation coefficient
相关系数的可靠性 reliability of correlation coefficient
相关系数的一致性 consistency of coefficients of correlation
相关系数法 correlation coefficient method
相关系数检验 correlation coefficient test
相关系数矩阵 matrix for the correlation coefficient
相关系数显著性 significance of correlation coefficient
相关显著水平 significant level of correlation
相关显著性实验 test of significance for correlation
相关线端 related terminal
相关线段 relevant segment
相关线性 dependent linearity
相关线性方程 dependent linear equation
相关相 relation phase
相关项(目) relevant item; related term
相关项目效果 effect of relative project
相关形体 related feature
相关型检漏仪 correlator-type leak detection device
相关性 correlation; dependence[dependency]; correlativity; reciprocity
相关性测度 similarity measure
相关性程度 correlativity
相关性分析 analysis of relationship
相关性模型 correlative model
相关性试验 correlation test
相关性状 correlated character; correlated traits
相关学科 allied discipline; allied subject; interfacing discipline
相关研究 correlation studies
相关掩模图像 correlation mask image
相关掩膜技术 correlation mask technique
相关仪 correlator
相关因素 correlation factor; correlative factor
相关因子 correlation factor
相关与回归 correlation and regression
相关与回归分析 correlation and regression analysis
相关与回归技术 correlation and regression technique
相关噪声 correlated noise
相关张量 correlation
相关账户 reciprocal account
相关振动 coupled vibration
相关知觉色 related perceived colo(u)r
相关指标 index of correlation
相关指数 correlation index; index of correlation
相关质量 correlation quality
相关轴(线) axis of reference; reference axis
相关主题 related subject
相关专利优先权服务 patent family service
相关状态 correlation behaviour
相关资料 correlation data; related data
相关资料的表达 disclosure of relevant information
相关坐标 dependent coordinates
相贯的组合体 intersecting assembly
相贯体 intersecting body; intersection of solids
相合 fall-in; in harmony with
相合变换 congruent transformation
相合矩阵 congruent matrix
相互安排 reciprocal arrangement
相互拌和 intermixing
相互保险 mutual insurance
相互保险社 mutual insurance society
相互保险协会 Mutual Insurance Association
相互备用 backup each other
相互比较 intercomparison
相互补偿约定 mutual indemnification agreement
相互补充 complementation
相互采购 reciprocal purchase
相互参考 cross reference
相互参照 cross reference
相互掺混 interblend
相互超车 mutual overtaking; mutual passing
相互沉淀作用 reciprocal precipitation
相互承兑 mutual acceptance
相互承付 cross acceptance
相互出借协定 interloan agreement
相互穿插 interfingering
相互穿透聚合物网络 interpenetrating polymer network
相互传染 cross infection
相互传染性 intercommunicability
相互串音 interaction crosstalk
相互垂直 both perpendicular; orthogonality; orthogonalization
相互垂直的 mutual perpendicular; orthogonal
相互垂直的钢筋网 <混凝土墙板结构内的> curtain reinforcement
相互垂直偏转 orthogonal deflection
相互促进作用 interaction facilitation
相互存款保险公司 mutual deposit premium
相互代理 mutual agency
相互的 commutative; mutual; reciprocal
相互抵消 repeal by implication
相互抵消的债权 claims set-off against each other
相互抵消的债务 cross debt
相互电抗器 mutual reactor
相互调用 cross call
相互叠砌 interbed
相互订货 cross order
相互定线 mutual alignment
相互定向 coorientation
相互独家垄断 bilateral monopoly
相互独立的随机变量 mutually independent random variables
相互对角比 co-diagonalization
相互对立的利益 antagonistic interest
相互对照 cross reference
相互对照表 cross reference list
相互反复选择 recurrent reciprocal selection
相互反射 interreflexion
相互反射能力 interreflectance
相互反射系数 interreflectance
相互分散 reciprocal dispersion
相互分摊额 reciprocal apportionment
相互干扰 mutual interference
相互购买 reciprocal buying
相互购买信用交换所 counter-purchase credit exchange
相互关联 cross-relate; interdependence[interdependency]
相互关联目标 interacting goal
相互关系 correlation; correlativity; cross connection; cross-correlation; interdependence[interdependency]; interface; interrelation; interrelationship; reciprocity; relation(ship)
相互关系的一致性 interface compatibility
相互关系和谐的需要 need for relatedness
相互关系约束条件 interrelationship constraint
相互贯穿 interfinger(ing)
相互贯穿的构件 interfingering member
相互广播公司 <美> Mutual Broadcasting System
相互合作 mutual cooperation
相互合作地 in tandem
相互核对 cross check; mutual check
相互回交 recurrent backcross
相互汇率 reciprocal rate
相互混合 intermix
相互混频 reciprocal mixing
相互混(溶)性 intermiscibility
相互寄生物 reciprocal parasite
相互间多重依存关系 multiple interdependency
相互间通信 reciprocal communication
相互监督 mutual check
相互检查 cross check; cross proving
相互检验 cross check
相互交叉 intercross
相互交叉的 criss-cross
相互交叉筋束 tendon crossing each other
相互交换 reciprocal interchange
相互校核 cross check
相互校验 cross check
相互校准 intercalibration
相互抉择的稳定 discretionary stabilization
相互抉择利润 discretionary profits
相互开发抵用票据 cross drawing
相互可操作(运行)的 interoperable
相互可见的 intervisible
相互控股 reciprocal holdings
相互扩散 interdiffusion
相互来往 intercommunication
相互理解 mutual understanding
相互利用 interavailability
相互连接 backflow connection; interconnection
相互连通的孔隙 interconnected void
相互联锁 reciprocal interlocking
相互联系 intercommunication; interrelate; interrelation
相互联系的 interrelated
相互联系控制机构 interconnected control
相互联系图 interconnection diagram
相互两家垄断 bilateral duopoly
相互满足的状态 <指人际关系方面> mutual comfort
相互贸易协定法 Reciprocal Trade Agreement Act
相互贸易信用 mutual trading credit
相互免除 reciprocal exemption
相互啮合的齿轮 pitch wheel
相互凝结作用 intercoagulation
相互凝聚作用 intercoagulation
相互耦合 interconnection; intercoordination; intercoupling; interdependence[interdependency]
相互排斥 mutual exclusion
相互配合 interact; interaction
相互匹配的一部分 counterpart
相互迁移 intermigration
相互确定 interfix
相互赊欠 swap credit
相互渗透 intercoagulation; interpenetrate; interpenetration; mutual coagulation
相互失谐级联电路 staggered circuit
相互式接头 alternate joint; staggered joint
相互适应 interadaptation
相互锁住的瓦片 <英国制的> Bridgewater tile
相互特许 cross license[licence]; cross licensing[licencing]
相互条件 conditions of reciprocity
相互条约 reciprocal treaty
相互调制 cross modulation; intermodulation
相互调制法 intermodulation method
相互调制分量 intermodulation product
相互调制频蛊 intermodulation frequency
相互通货持有 mutual currency holding
相互通货账户 mutual currency account
相互通信[讯] intercommunication
相互通信[讯]制 intercommunication system
相互同步系统 mutually synchronized system
相互同化 reciprocal assimilation
相互同意 mutual assent
相互投资联营公司 reciprocal affiliations
相互透支 mutual swing credit
相互透支额度 mutual serving credit
相互推斥 mutual repulsion
相互往来账户 mutual current account
相互委托销售 reciprocal consignment
相互位移 mutual displacement
相互位置 mutual alignment
相互吸附 adsorptive interaction
相互吸引 interattraction; mutual affinity; mutual attraction
相互销售 reciprocal sale
相互效应 interaction effect

相互楔交 interfinger
相互协商 mutual consultation
相互协调 intercoordination
相互协调的 mutually compatible
相互行为 interbehavio(u)r
相互需求 reciprocal demand
相互需求法则 law of reciprocal demand
相互研磨 intergrind(ing); interground
相互一致条件 concurrent terms
相互依存 correlation dependence; interdependence [interdependency]; mutual dependence; reciprocal interdependence
相互依存权 right of solidarity
相互依存条件 concurrent condition
相互依赖 interdependence [interdependency]
相互依赖的 interdependent
相互抑制 reciprocal inhibition
相互抑制竞争 mutual inhibition competition
相互易位 reciprocal translocation
相互引力 mutual attraction
相互影射 homology
相互影响 interact; interaction; interaction effect; interrelationship; mutual effect
相互影响车流 interacting traffic stream
相互影响的 interactive
相互影响分析 cross impact analysis; interaction analysis
相互影响分析法 mutually affecting analysis method
相互影响技能 interactive skills
相互影响矩阵 interaction matrix
相互影响矩阵分析预测法 interaction matrix analysis forecasting
相互影响模型 interactive model
相互影响心理学 transactional psychology
相互影响行为 interactive behavio(u)r
相互诱导 reciprocal induction
相互约束 reciprocal bond
相互运动 mutual movement
相互责任 cross liability; mutual responsibility
相互赠与 mutual gift
相互债务 cross liability; mutual debt
相互沾染 cross contamination
相互账户 reciprocal account
相互账户的协调 reconciliation of reciprocal accounts
相互支援的建设 mutual-aid construction
相互制约 interact; interaction; mutual check
相互制约的 interactive
相互制约活动 interacting activity
相互重叠 interlap
相互重叠的枝条 touching each other
相互重组 reciprocal recombinant
相互注销 reciprocal cancellation
相互转股 mutual share transfer
相互转化 interconversion; reciprocal transformation
相互转换 interconversion
相互转接干线 interswitch trunk
相互状态 reciprocity
相互自化作用 mutual assimilation
相互自然选择 reciprocal natural selection
相互组合 inter-combination
相互作用 coaction; correlation; cross coupling; event; interacting; interaction effect; interactivity; intercoupling; interplay; mutual action; mutual effect; reciprocity; interworking <机械零件的>
相互作用参数 interaction parameter
相互作用测绘制图软件系统 interactive graphics package
相互作用场 interacting field
相互作用的分支系统 interacting subsystem
相互作用的影响 interaction effect
相互作用对 interaction partners
相互作用范围 interaction volume
相互作用过程 interaction process
相互作用哈密顿函数 interaction Hamiltonian
相互作用核 interacting nuclei
相互作用截面 interaction cross section
相互作用空间 interacting space
相互作用力 interaction force; interactive force; mutual force
相互作用粒子 interacting particles
相互作用率 interaction rate
相互作用面 interface
相互作用模型 interactance model; interaction stress; interactive model
相互作用能 interacting energy; interaction energy
相互作用平均自由程 interaction mean free path
相互作用区域 interaction zone
相互作用圈 sphere of antagonistic effect
相互作用软件系统 interactive software system
相互作用时间 interaction time
相互作用式显示器 interactive display
相互作用势 interaction potential
相互作用说 interactionism
相互作用图表 interaction diagram
相互作用图景 interaction picture
相互作用图像 interaction picture; interaction representation
相互作用物 interactant
相互作用系数 interaction coefficient
相互作用型 coactive pattern
相互作用性质 interactive property
相互作用序列 encounter sequence
相互作用研究 repercussion study
相互作用应力 interaction stress
相互作用状态 interacting state
相会点 <美> passing point
相混 interblend
相混合 mingle
相机反光镜 view finder
相机记录仪 camera
相机快门 drop shutter
相积分 phase integral
相际沉淀 interphase precipitate
相继 in sequence
相继操作 operation in tandem
相继层 successive layers
相继差值 successive difference
相继场 successive field
相继的 sequent
相继电器 phase relay
相继动作 cascade operation
相继对比 successive contrast
相继各层 successive layers
相继两低潮的高度差 low-water inequality
相继平滑值 running mean
相继平均法 method of successive average
相继平均值 consecutive mean; overlapping mean; running mean
相继数据组 generation data group
相继文件 consequential file
相继行 consecutive line; sequential lines
相继运送人 on-carrier
相继值 consecutive values
相继总和 successive summation
相加 additive; totalling
相加的联合作用 additive joint action
相加的效用函数 additive utility function
相加电路 added circuit
相加法 additive method
相加规划 addition rule
相加合成 additional combining
相加合成补色 additive complementary colo(u)r
相加合成基色 primary additive colo(u)rs
相加混频 additive mixing
相加级 adder stage
相加脉冲 add pulse
相加门 add gate
相加器 summator
相加数据 summarized information
相加问题 adding up problem
相加误差 additive error
相加效应 additive effect; summation effect
相加信号 sum signal
相加性 additive property; additivity
相加性常数 additive constant
相加作用 additive action; additive effect; summation action
相夹带剔除 phase disentrainment
相减电路 subtraction circuit
相减合成基色 primary subtractive colo(u)r; subtractive primaries
相减脉冲 subtract pulse
相交 bisect; intersect(ion)
相交成锐角的力 forces intersecting at an acute angle
相交道 <交叉口的> intersection legs
相交道路 intersecting roads
相交的 crossed; intersecting
相交点 crossing point; cross(-over) point; intersecting point; junction point; point of intersection
相交点噪声 crosspoint noise
相交定理 intersection theorem
相交法 method of intersection
相交钢箍 interlocking stirrup
相交滑移带 intersecting slip bands
相交环 intersecting ring
相交角 angle of intersection; intersecting angle; intersection angle; skew angle
相交街道 crossed street
相交理论 intersection theory
相交理论法 intersection theory method
相交链 intersecting chain
相交路段 intersection legs
相交路线 intersecting routes
相交平面 intersecting plan; intersecting plane; intersection plane
相交坡度 <交叉道路的> intersection grade
相交剖面的综合显示 display of combination of intersections
相交数 intersection number
相交数据 intersection data
相交通 intercommunicate
相交位错 intersecting dislocations
相交线 intersecting line; intersection line; line of intersection
相交因素 intersectional elements
相交于一点的线 convergent lines
相交轴 concurrent axis; intersecting shaft
相接触的 contagious
相接环形山 contiguous crater
相接环形山链 contiguous chain
相接面 abutted surface
相接漂移管 consecutive drift tube
相截性 transversality
相近的 related
相近性 proximity
相近选址 hashing addressing
相近寻址 hashing addressing
相距很近或紧靠着树立两根立柱 accouplement
相距很远 wide apart
相类似的 analogous
相连 attach to
相连单元 contiguous location
相连的 conjoint; conterminal; conterminous
相连的城市 strip city
相连工程 associated works
相连接的漏斗节 <浇注混凝土用> articulated drop chute
相连曲线点 curve to curve
相连式独户住宅 single family attached dwelling
相连项 contiguous item
相连住宅 attached housing
相联 association
相联变动 associated variation
相联并行计算 associative and parallel computation
相联查表 associative table looking
相联成对比较设计 linked paired comparison designs
相联处理机 associative processor
相联存储符合 associative memory match
相联存储寄存器 associative storage register
相联存储器 annex(e) memory; annex(e) storage; associative memory; associative storage; catalog memory; content-addressable memory; content-addressed memory; content-addressed storage; multiaccess associative memory; parallel search storage; searching memory
相联度 degree of association
相联度量 measure of association
相联对 association pair
相联关键码 associative key
相联关键字 associative key
相联寄存器 associative register
相联检索 associative retrieval; associative search
相联结的电力系统 interconnected system
相联数据处理 associative data processing
相联条款 tying clause
相联系的指示物 tied indicator
相联系数 coefficient of association
相联性的象限检验法 corner test of association
相联语言 associative language
相联阵列处理机 associative array processor
相量 phasor
相量电流 phasor current
相量方程式 phasor equation
相量幅角 phasor argument
相量功率 complex power; phasor power
相量功率因数 phasor power factor
相量函数 phasor function
相量积 phasor product
相量图 Blondel diagram; phasor diagram
相量圆图 clock diagram
相邻板 adjacent plank
相邻臂 alternate arm
相邻边 adjacent edge
相邻边界上的 limitrophe
相邻波道干扰 adjacent channel interference

相邻测线线距 neighbo(u)ring line distance
相邻层面间防止移动的锁键 mechanical bond
相邻车道 abutting lane;adjacent traffic lane
相邻道岔 adjoining points
相邻道叠加 adjoining traces stack
相邻的 adjacent;closely spaced
相邻的科学 contiguous branches of science
相邻的科学分支 contiguous branches of science
相邻的墙 adjacent wall
相邻的桥跨 adjacent opening;adjacent span
相邻的正交线 adjacent orthogonal
相邻等高线的高度差 contour interval
相邻等级 adjacent rank
相邻地产 abutting property;adjacent premises
相邻地段 abutting lot
相邻地界线 neighbo(u)ring property
相邻地区 adjacent country;immediate vicinity
相邻点 contiguous stations;continuous point
相邻点间相对中误差 relative mean square error of adjustment points
相邻点平均化 neighbo(u)rhood averaging
相邻读数 consecutive reading
相邻断面 adjacent section
相邻二地亮块体 a pair of neighboring massives
相邻方格 adjacent square
相邻房地产 adjoining premises
相邻房地产分界线 abuttal
相邻(房地产)业主 adjacent owner
相邻房屋 adjacent blocks;adjacent building;adjoining blocks
相邻分析 neighbo(u)rhood analysis
相邻构件接缝面的间距 joint clearance
相邻构造单元名称 name of adjacent tectonic region
相邻股道 adjacent track;neighbo(u)ring track
相邻关系 neighbo(u)ring relations
相邻轨道 adjacent track;adjoining rail;neighbo(u)ring track
相邻海区 adjacent sea
相邻含水层 neighbo(u)ring aquifer
相邻含水层越流补给 leakage form adjacent aquifer
相邻航(空)线 adjacent flight lines;adjacent flight strips;adjoining flight strips;neighbo(u)ring flight lines
相邻航线测量 adjoining survey
相邻航线间隔 <航测> distance between adjacent flight lines;spacing of adjacent strips
相邻化合物 adjacent compound
相邻混凝土浇筑层 adjacent lift;adjoining lift
相邻极值点 adjacent extreme point
相邻建筑物 adjacent building;adjacent structure
相邻建筑物业主 adjoining owner
相邻交叉口 adjacent intersection;adjacent junction
相邻浇筑块 adjacent blocks
相邻街坊 adjacent blocks
相邻节点 adjacent node
相邻矩阵 adjacency matrix
相邻刻度的间隔 scale span
相邻跨度 neighbo(u)ring opening;neighbo(u)ring span
相邻立体模型 adjacent stereomodel
相邻两高潮之间的时间 tidal period
相邻两柱间距离等于柱直径之二倍的 systyle
相邻落程的水头差 head difference between neighbouring drawdown
相邻面 adjoining plane
相邻模式间拍频 adjacent mode beat frequency
相邻模型 adjacent model;adjoining model;contiguous model
相邻排列不重叠现场灌注桩挡土墙 contiguous bored pile wall
相邻排列现场灌注桩 contiguous bored pile;contiguous bored pile
相邻排列钻孔灌注桩挡土墙施工法 contiguous bored piling
相邻排列钻孔灌注桩墙 contiguous pile wall
相邻炮点点距 neighbo(u)ring shot point gap
相邻期 adjacent periods
相邻桥孔 adjoining opening
相邻切线 adjacent tangent
相邻区 adjacent region
相邻区段 adjoining district;adjoining section
相邻区间 adjoining section
相邻曲线 adjacent curves
相邻溶液 adjacent solution
相邻设备 adjacent accommodation
相邻数 consecutive numbers
相邻水域 adjacent waters
相邻特许(租借)地 adjoining concession
相邻通道 adjacent channel
相邻通道干扰 adjacent channel interference
相邻通路干扰 adjacent channel interference
相邻图幅 adjacent map;adjacent (map) sheet;adjoining (map) sheet;contiguous map sheet;contiguous sheet;continuation map sheet;continuation sheet;neighbo(u)ring (map) sheet
相邻图幅名称 adjoining sheet names
相邻图形 contiguous graphics
相邻位置 adjacent location;adjacent position
相邻线 adjacent track;neighbo(u)ring track
相邻线匝 adjacent turns
相邻项 adjacency;contiguous item
相邻像对 adjacent pair;consecutive pair
相邻像片 adjacent photograph;adjoining photograph
相邻小站 adjacent wayside stations
相邻效应 neighbo(u)rhood effect
相邻信道 adjacent channel
相邻信道干扰 adjacent channel interference
相邻信道共振 adjacent resonance
相邻信道频率 adjacent channel frequency
相邻信道衰减 adjacent channel attenuation
相邻信道选择性 adjacent-channel selectivity
相邻信道载波 adjacent carrier
相邻学科 interfacing discipline
相邻影响 adjacent effect
相邻影响曲线图【岩】adjacent effects chart;settlement effects chart
相邻域 adjacent domain
相邻站 continuous stations
相邻照片 adjacent photograph
相邻帧 consecutive frame
相邻值 consecutive values
相邻指数 consecutive indexing
相邻周节误差 adjacent pitch error
相邻字符 adjacent character
相拟定律 model law
相啮合齿轮 engaged wheel
相偶极子 phase dipole
相拍干扰 beat interference
相旁地(段)adjacent contract section
相配 assort;match;match up;suit;suitability
相配安装 match-fitting
相配的 assorted;suitable
相配订单 matched orders
相配阶段 conjugate stages
相配取穴 principle of point association
相配预制 match-casting
相匹配 phase match
相配线 <航测拼图的> match lines
相谱 phase spectrum
相前 phase front
相嵌接合 halved joint;halving;scarfing;straight scarf joint
相嵌连接 interdigitated junction
相切 tangency;tangential
相切变换 contact transformation
相切圆 tangent circle
相切柱面 tangent cylinder
相容 consistent
相容逼近(法)consistent approximation
相容程度 degree of consistency[consistence]
相容程序 inclusive routine
相容次序 consistent order
相容次序矩阵 consistently ordered matrix
相容存储 compatible storage
相容的 compatible
相容度 degree of compatibility
相容段 inclusive segment
相容方程式 compatibility equation;compatible equation;consistency equation;consistent equation
相容刚度矩阵 consistent stiffness matrix
相容估计 consistent estimate;consistent estimation
相容估计量 consistent estimator
相容关系 compatibility relation
相容观察 compatible observation
相容极限 limit of compatibility
相容几何刚度 consistent geometric(al) stiffness
相容检定 consistent calibration
相容检验 consistent test
相容节点荷载 consistent nodal load
相容矩阵 consistent matrix
相容模型 compatible mode
相容情况 compatibility condition
相容生物再生 consistent biologic(al) regeneration
相容事件 compatible event
相容条件 condition of compatibility;consistency condition
相容图 compatible chart
相容性 compatibility;consistence [consistency]
相容性定律 law of compatibility
相容性法 compatibility method
相容性法则 compatibility law
相容性方程 equation of compatibility
相容性检验 consistency check
相容性试验 compatibility test
相容性条件 compatibility condition
相容性证明 consistency proof
相容元 conforming element;consistent element
相容元素 compatible element
相容质量矩阵 consistent mass matrix
相容置换 substitute consistently
相溶度分析 phase solubility analysis
相溶解度 phase solubility
相溶解度分析 phase solubility analysis
相溶物 solutrope
相渗透率 phase permeability
相生(现象)synergism;synergy
相适应流动法则 associative flow rule
相疏 alienation
相疏系数 alienation coefficient;coefficient of alienament
相思树 acacia;rich acacia
相思树装饰性硬木 <产于澳大利亚> Gidgee
相似 equiform;resemblance;semble;similarity;similitude
相似比 ratio of similitude;similarity ratio;similitude ratio
相似比值法 likelihood ratio method
相似比值判别 Gaussian
相似变换 equiform transformation;similarity transformation;transformation of similitude
相似变换群 equiform group
相似变量 similar argument;similarity variable
相似标准 similarity criterion
相似表示 similar representation
相似材料模拟试验 model test of similar material
相似参数 similarity parameter;similar parameter
相似程度 degree of similitude
相似导线 similars
相似的 equiformal;homoplastic;similar
相似的船舶 similarity ship
相似的联合作用 similar joint action
相似电路 analogous circuit
相似定理 correspondence theorem
相似定律 affinity law;analogous law;law of similarity;scaling law;similarity law;similarity rule;law of similitude
相似多边形 similar polygons
相似多面体 similar polyhedrons
相似二次曲面 similar quadrics
相似二次曲线 similar conics
相似法 analog(ue) method;likelihood method;similarity method
相似法则 similar law
相似方阵 similar square matrices
相似风巷 similar airway
相似工况 similar condition;similarity operation condition;similar operating condition
相似估计 likelihood estimation
相似关系 similarity relation(ship)
相似规则 rule of similarity
相似轨道 similar orbits
相似化合物 analog(ue) compounds
相似几何学 equiform geometry;similar geometry
相似价值函数 similar cost function
相似检定 similar calibration;similar test
相似检定法 likelihood ratio method
相似检验 similar test
相似矩阵 similarity matrices;similar matrix[matrices]
相似颗粒混凝土 like-grained concrete
相似扩大 homothety
相似类目 similar class
相似理论 similarity theory;similitude theory
相似流(动)similar flow
相似流域 analog(ue) basin;similar basin
相似流域比拟法 transposition method
相似律 law of similarity;similarity;

similarity law;similarity rule;similar law;similitude law
相似率 percentage similarity
相似论 theory of similarity
相似面 similar face
相似模拟 analog(ue) simulation;analogy simulation
相似模拟模型 analog(ue) simulation model
相似模型 analog(ue) model(ing);analogy model;scale model;similarity model;similar model
相似模型涡轮机 scale model turbine
相似年(份) analog(ue) year;similar year
相似排列 similar permutation
相似判据 criterion of similarity
相似判数 criterion of similarity
相似偏离 similarity deviation
相似三角形 similar triangles
相似色调的色泽 related shades
相似试验序列 series of similar experiments
相似树 similar tree
相似数 similarity number
相似数系 similar system(s) of numbers
相似双生 concordant twin
相似水轮机 similar turbine
相似特性法 similitude method
相似条件 condition of similarity;similarity consideration
相似同构的 similarity isomorphic
相似图形 similar figures
相似屋顶 homogeneous roof
相似物 resemblance
相似系数 coefficient of conformity;coefficient of similarity;similarity coefficient;similarity factor
相似效应 similitude effect
相似形式 <符合异物同形概念的> homologous forms
相似性 analogy;likelihood;likeness;resemblance;similarity;similitude
相似性测度 similarity measure
相似性定理 similarity theorem
相似性法则 similarity rule
相似性化简 similarity reduction
相似性假说 similarity hypothesis
相似性理论 similarity theory;theory of similarity
相似性判别 similarity discrimination
相似性判据 criteria of similarity;criteria of similitude
相似性试验 similarity test
相似性水平 similarity level
相似性条件 condition of similitude
相似性原理 principle of similitude
相似性指数 similarity index
相似性准则 similarity criterion;similitude criterion
相似有序集 similar ordered aggregates
相似域 similar region
相似原理 analog(ue) principle;principle of similitude;similarity principle;similitude principle
相似褶皱 similar fold(ing)
相似指数分析 similarity-index analysis
相似中心 center[centre] of similarity;center[centre] of similitude;homothetic center[centre]
相似种 sibling species
相似属性 like attribute
相似准则 similarity criterion
相通 interpenetrate
相同 come to the same thing;identical congruent
相同比例尺地图系列 common-scale strip
相同变元表 identical argument list
相同标准 identical standard
相同表 identical table
相同产品 like products
相同成分 identical component
相同处理机 identical processor
相同的 alike;congruent;discontinuous;identical
相同的广土类 same broad soil groups
相同对 tied pair
相同分布 same distribution
相同覆盖 uniform coverage
相同干旱气候 arid homoclimate
相同固定误差 same constant error
相同函数引用 identical function reference
相同计算 identical computation
相同类型的养分 similar kinds of nutrients
相同片段 identical sections
相同频率 same frequency
相同平衡条件 same equilibrium condition
相同期望 equating expectation
相同气候 homoclimate;homoclime
相同气候区 homoclime
相同使用期 equal life
相同竖向应变 equal vertical strain
相同水分 considerable moisture
相同顺序 same sequence
相同网络 identical network
相同位穿孔 batten;cordonnier
相同位穿孔检查 peek-a-boo check
相同位点 same loci
相同位置 same position
相同项 identical entry
相同英尺烛光图 isofoot candle diagram
相同钥匙的圆筒弹子锁 keyed-alike cylinders
相同运算 identical operation
相同质量 equal in quality
相同烛光图 isocandela diagram
相同子表 identical sublist
相同子树 identical subtree
相向 opposite direction
相向车辆 opposing vehicle
相向交通 opposing traffic
相向运行 opposing movement
相消干扰 destructive interference
相消干涉 destructive interference
相消律 cancellation law
相咬 <齿轮等> tooth
相依 contingence
相依变数 interdependent variable
相依表 <质量管理> contingency table
相依程度 correlativity
相依关系 dependence relation;dependency relationship
相依函数 dependent function;interdependent function
相依契约 dependent covenants
相依强度 strength of dependence
相依时滞 dependent time lag
相依事件【数】dependent event
相依随机事件 dependent random event
相依系数 association coefficient;coefficient of contingency
相依效用函数 interdependent utility function
相依行动 dependent action
相依性 dependence
相依样本 dependent sample
相宜 suitability
相宜的 congenial
相倚因素 interdependence factor
相异 dissimilarity;dissimilation;diversity
相异的 alien;divergent
相异段 differential segment
相异根 distinct root
相异关系 relation of diversity
相异号【数】contrary sign
相异零点 distinct zero
相异品系 divergent strains
相异性指数 diversity index
相易的 reciprocal
相因而生的 consequential
相因数 phase factor
相引 coupling
相引线 phase terminal
相应 correspondence
相应变异性 corresponding variability
相应齿廓 counterpart profile
相应的 appropriate;corresponding;homologous
相应的百分比 corresponding percentage
相应的地基土设计参数 corresponding subgrade soil design parameter
相应的设备 adequate provision
相应的数据 pertinent data
相应的资料 corresponding data;pertinent data
相应地增长 corresponding increase
相应点 corresponding point;homologous point;relevant point;homogeneous point
相应电路 correspondence circuit;corresponding circuit
相应电子衍射 relative electron diffraction
相应发生 <如权利与责任> accrue
相应发生的事件 accrual
相应反应 relevant reaction
相应峰顶水位 corresponding crest;corresponding crest stage
相应洪峰水位 corresponding crest;corresponding crest stage
相应互惠依存 reciprocal interdependence
相应阶段 corresponding stage
相应节点位移 relevant nodal displacement
相应流动法则 associative flow rule
相应流量 corresponding discharge
相应流量关系 flow relation
相应流量曲线 curve of corresponding discharge
相应平衡 relevant equilibrium
相应实地尺寸 corresponding ground dimension;equivalent ground dimension
相应水位 concordant water-level;corresponding water stage;corresponding water level;equivalent water level
相应水位读数 equivalent ga(u)ge reading
相应水位关系 ga(u)ge relation;stage relation
相应水位曲线 stage relation
相应速度 corresponding speed
相应投影 corresponding projection;homolographic(al) projection
相应图面位置 proper map-place
相应位置 relevant position
相应稳定性常数 relevant stability constant
相应物 homologue
相应像片 homologous photograph
相应压力 relevant pressure
相应于有效波高的波浪周期 wave period associated with effective wave heights
相应赠款 matching grant
相应责任 correspond responsibility duty
相应照片 homologous photograph
相应值 relevant value
相应状态定律 law of corresponding states
相应最高水位 corresponding crest;corresponding crest stage
相遇 meetion;rendezvous
相遇局面 meeting situation
相遇时距曲线 reverse hodograph
相遇时距曲线系统 reverse control hodograph
相约 cancel;cancellation
相约取消期 cancelling date
相助串联 series-aiding connection
相撞 barge against
相撞航线 collision course

香

香柏 arborvitae
香柏木 <尼日利亚产> Moboron
香柏木箱 cedar chest
香柏油 cedar oil
香草 vanilla
香草醛 vanillin
香肠 sausage
香肠构造 boudinage structure
香肠构造特征【地】feature of boudinage
香椿 Chinese cedar;Chinese mahogany
香椿属 <拉> Cedrela
香刺柏 incense juniper
香的 sweet
香豆树脂 cumar gum;cumar resin
香豆酸 c(o)umaric acid
香豆酮 c(o)umarone
香豆酮树脂 coumarone-indene resin;coumarone resin;paracoumarone resin
香豆酮茚树脂 coumarone-indene resin;polycoumarone resin;paracoumarone-indene resin
香榧 Japanese torrey
香粉 powder
香港工程师学会 Hongkong Institution of Engineers
香港公路和铁路结构设计指南 Hongkong Structures Design Manual for Highways and Railways
香港海员工会 Hong Kong Seamen's Union
香港桥梁美学监视委员会 Watchdog Committee on the Appearance of Bridges, H.K.
香港石色油漆 Hong Kong stone
香港铁路公共交通公司 Mass Transit Railway Corporation of Hong Kong
香港土木工程指南 <根据 BS5400 结合香港具体条件制订的设计标准> Hongkong Civil Engineering Manual
香港银行同业拆放利率 Hongkong Interbank Offered Rate
香膏 balm;balsam
香菇属 <拉> Lentinus
香果树 Henry emmenopterys
香旱芹油 ajawa oil;ajowan oil
香花石 hsianghualite
香桦 fragrant birch
香槐属 yellow wood;Cladrastis <拉>
香胶 balsam;Benjamin;Cologne glue
香蕉插孔 banana jack
香蕉插口 banana jack
香蕉插头 banana jack;banana pin;banana plug;split plug
香蕉串 banana cluster
香蕉国 <指只有单一经济作物的拉丁美洲国家> banana republic
香蕉苹果 banana apple

香蕉水 amyl acetate; banana oil; lacquer thinner; oil of bananas
香蕉卸船机 banana elevator; banana unloader
香蕉形座位 banana seat
香蕉油 amyl acetate; banana oil
香蕉运输船 banana carrier
香精油 essential oil
香料 aromatics; fragrance; odo(u)rant; perfume; spice
香料厂 perfumery
香料缸 lavender jar
香料固定剂 perfume fixative
香料化合物 flavorful compound
香料贸易 spice trade
香料植物 perfume plant; spice berry
香料渍 spice curing
香料作物 spice crop
香炉 censer
香茅酸 rhodinic acid
香茅油 mahapengiri oil
香木 incense wood
香气 odo(u)r; scent
香气污染 fragrance pollution
香芹酮 carvonone
香杉 incense cedar
香杉木 aromatic cedar
香树精 amyrin
香栓菌 sweet odor conk
香水 perfume
香水盒 perfume box
香水月季 tea rose
香水制造厂 perfumery
香松树胶 gum Sandarac; sandarac(h); sandarac(h) gum
香桃木 myrtle; myrtus
香桃木蜡 myrtle wax
香味 fragrance; perfume; scent
香味桃花心木 sapele mahogany; scented mahogany
香味印刷 ag(e)ing printing
香味油墨 perfumed ink
香肖楠 bastard cedar; incense cedar; post cedar
香肖楠属 incense cedar
香烟点火器 lighter
香烟店 cigar store
香烟烧痕 cigarette burn
香烟烟雾 smoke from cigarette
香烟烟熏 cigarette smoking
香叶醇 geraniol
香叶醛 geranial
香叶树子油 spice bush seed oil
香叶烯 myrcene
香叶油 bay oil; geranium oil; myrcia oil
香油 balm; sesame oil
香橼 <颜色为黄绿色> citron
香橼树 citron
香皂 fancy soap; toilet soap
香獐子 musk deer
香樟木 camphor wood
香脂 balm; balsam
香脂的 balmy
香脂冷杉 balsam fir
香脂松节油 balsam turpentine
香脂杨 balsam poplar

厢车 box car

厢底净空 <电梯> bottom car clearance
厢房 adjacent accommodation; appentice; flank; levecel; wing room
厢房拱顶 aisle-vault
厢房过道 aisle passage
厢埽工 sunken fascine
厢式搬场汽车 furniture van
厢式车身 station wagon
厢式车厢 van body
厢式挂车 trailer wagon; van(-type body) trailer
厢式货车 van truck; closed-top van
厢式货物运输车 panel van
厢式家具运送汽车 furniture van
厢式食品运送汽车 food container van
厢式送货车 delivery van
厢式送货汽车 delivery truck; delivery van
厢式拖车 van trailer
厢式邮政车 mail wagon
厢式载货汽车 parcel van
厢式载重汽车 panel-body truck; wagon truck
厢堂桁幅 nave aisle bay
厢堂廊台 nave aisle gallery

箱 box; store holder; bell housing <如飞轮壳等>

箱板 box board
箱板凹 case plank board dented
箱板材 case wood; packing case wood
箱板焦 case plank board scorched
箱板裂 case plank board split cracked
箱板料 box shook
箱板木材 box timber; boxwood
箱板破 case plank board broken
箱板塌 carton plank dented
箱板条破裂 cases with boards and battens broken
箱板凸出 case plank board bulged
箱板凸起 carton board bulged
箱板弯翘 case plank board warped
箱板性能测定 primitive test of container board
箱板修补 carton board patched; case plank board patched
箱板有钉眼 cases with nail holes
箱板撞破 cases with boards staved in
箱半空 case half empty
箱被封妥 case sealed
箱边压坏 cases jammed at the side
箱变形 carton deformed
箱擦花 carton chafed
箱材 box lumber
箱长度 body length
箱车 box car; house-car
箱衬 box casing
箱衬板 flask liner
箱衬纸板 gusset felt
箱池 tank
箱尺 box staff
箱脆弱 case frail
箱带 crossbar; flask bar
箱的几何形状 tank geometry
箱底 hopper base; floor <集装箱>
箱底板 box board
箱底承载力 floor loading capability
箱底加热器 tank bottom heater
箱底焦 bottom skid broken
箱底结构 base structure
箱底破裂 cases broken at bottom
箱底托盘 box pallet
箱底脱落 cases bottoms off
箱顶 <集装箱> roof
箱顶强度 <集装箱> strength of roof
箱顶强度试验 <集装箱> roof strength test
箱顶压坏 top of case crushed
箱斗倾翻装置 box-tipping equipment
箱斗式分布机 box-hopper spreader
箱斗式饲槽 hopper type feeder
箱斗式摊铺机 box-hopper spreader
箱端裂开 cases with ends split
箱法扩散 box diffusion
箱封条脱落 case seal off
箱盖 tank cover
箱格导柱 cell guide
箱格导柱口 entry of cell guide
箱格防波堤 crib breakwater
箱格钢板桩防波堤 cellular sheet pile breakwater
箱格基础 coffered foundation
箱格式坝 cellular dam
箱格式挡土墙 cellular retaining wall
箱格式货舱 cellular hold
箱格式集装箱船 cellular container vessel
箱格式集装箱轮 cellular container ship
箱格式重力坝 cellular gravity dam
箱格形结构 cellular structure
箱格形梁 cellular girder
箱格形心墙 cellular core wall
箱拱 regular arm
箱共鸣 boominess resonance
箱箍 case band
箱箍断脱 cases with bands broken
箱管 container management
箱柜钩眼 hook ring
箱涵 box frame in jack-in method; culvert box; slab culvert
箱涵排水 box drain
箱涵刃角 cutting edge of box frame
箱汗 <集装箱> container sweat; sweat on container
箱号 case number; crate number
箱盒 case; chamber
箱盒用材 short end
箱盒运送板架 case pallet
箱货绳扣 box sling
箱基 box foundation
箱夹 clamp
箱架 box stand
箱架堆装方式 rack system
箱角导向块 corner guide
箱角配件 corner fittings
箱角破裂 cases broken at corner
箱筋 crossbar of the mo(u)lding box
箱开断面 box section
箱开截面 box section
箱壳 box casing
箱空 case empty
箱框 box frame; upset frame <砂的>
箱梁 box beam
箱梁刚架 box rigid frame
箱梁桥 box girder bridge; box type beam bridge
箱梁型护栏 box girder fence
箱量 <集装箱> container through-put/ volume
箱龄 age of container
箱流 <指集装箱> container flow
箱门搭扣件 <集装箱> door holder
箱门封条 door seal gasket
箱门横梁 <集装箱的> door cross member
箱门卡住 seized door
箱门开口尺寸 dimensions of door opening
箱门密封垫 door seal gasket
箱门密封垫黏[粘]合 bonding door gasket
箱门强度 door strength
箱门强度试验 door strength test
箱门支撑杆 box strut
箱末端板条裂开 cases with end battens split
箱内气体压力 loading pressure
箱内容物有破碎声 carton contents rattling
箱内退火 box annealing; pot annealing
箱内旋管 coil in box
箱内旋管式冷却器 coil in box cooler
箱旁板破裂 cases with side boards broken
箱旁板条破裂 cases with side batten broken
箱旁破裂 carton broken at side
箱皮撕裂 carton wrapper torn
箱破损 case broken
箱圈 false cheek; raising middle flask; sand frame
箱容利用率 utilization factor of container volume
箱容量 tankage; volume of tank
箱容系数 container specific volume capacity coefficient; volume coefficient of container
箱渗碳 box hardening
箱式 cabinet-type
箱式板桩 boxed sheet piling
箱式变阻器 box type rheostat
箱式标尺 box staff
箱式表盘 trunk dial
箱式布线法 case way wiring system
箱式测潮表 box ga(u)ge
箱式铲斗 scraper box
箱式潮位计 box ga(u)ge
箱式车 house-car; van
箱式沉箱 box caisson
箱式船坞 box dock; chamber dock
箱式船闸 chamber lock; chamber navigation lock
箱式窗架 cased frame
箱式窗框 cased window frame
箱式大梁悬臂桥 box girder cantilever bridge
箱式电炉 cabinet-type electric(al) furnace; electric(al) box furnace
箱式电桥 post office bridge
箱式电梯 boxed stair(case)
箱式电阻炉 box type resistance furnace
箱式定量器 sliding box feeder
箱式翻斗 box dumper
箱式翻砂模 box mo(u)ld
箱式分布机 box-hopper spreader
箱式风机 cabinet fan
箱式风扇 cased type fan
箱式浮(船)坞 box dock
箱式干燥器 box drier; hopper drier [dryer]; loft drier[dryer]; tray and compartment drier
箱式干燥窑 box kiln
箱式钢板桩 box steel sheet piling; steel box pile
箱式钢桩 steel box pile
箱式锅炉 box type boiler; tank boiler
箱式过滤器 box filter
箱式涵洞 box culvert
箱式桁架桥 trussed-box-girder bridge
箱式烘干器 chamber drier[dryer]
箱式呼吸器 cabinet respirator
箱式滑阀 three-port slide
箱式活门 three-port slide
箱式货车 box car; box wagon
箱式货运车辆 box freight car
箱式基础 coffered foundation
箱式给料机 box feeder
箱式计量给料器 chop feeder
箱式加料器 box feeder
箱式加热炉 box type heater
箱式夹具 box jig
箱式家具 blanket chest
箱式剪力试验 box shear test
箱式建筑 cellular building
箱式结构 tank construction
箱式截面板桩 box section sheet pile
箱式截面纵梁 box section string
箱式截面桩 box section pile
箱式金属片芯撑 sheet-metal chaplet
箱式救生筏 pontoon raft
箱式卷线机 tank winding machine
箱式空调器 cabinet-type air conditioner

箱式框架构架 box-frame(d) construction
箱式框架建筑 box-frame(d) construction
箱式冷却器 box cooler
箱式连接器 box connector
箱式流明计 box lumen meter
箱式楼梯 boxed stair(case);box stair(case);closed stair(case)
箱式炉 batch furnace; box type furnace; box type oven; chamber furnace;oven-type furnace
箱式炉退火 batch annealing
箱式模板 box form(work)
箱式模型 box model
箱式泥芯撑 perforated chaplet;sheet-metal chaplet
箱式排水沟 rectangular gutter
箱式配料法 box batching method
箱式棚车箱 box shed car
箱式硼扩散 boron capsule diffusion
箱式起重爪 box lewis
箱式气净化器 box gas purifier
箱式桥 box bridge
箱式桥台 box abutment
箱式取样器取样 sampling by box
箱式热水系统 tank hot-water system
箱式上料机 box type feeder
箱式伸臂 box apron
箱式渗碳炉 case-hardening furnace
箱式石料摊铺机 box(stone) spreader
箱式手(推)车 box barrow
箱式摊铺机 box-hopper spreader;box type spreader;paving box;spreader box
箱式弹簧床 box spring bed
箱式搪烧炉 box type enamelling furnace
箱式天沟 box roof gutter;rectangular gutter
箱式天然对流器 cabinet-type natural convector
箱式调速器 cabinet-type governor
箱式退火窑 box annealing furnace
箱式坞门 caisson gate; closing pontoon
箱式斜接 box scarf
箱式蓄热室 box type regenerator
箱式压滤机 chamber filter press
箱式檐沟 rectangular gutter
箱式扬声器 cabinet loudspeaker
箱式窑 box furnace;box kiln
箱式液冷却器 tank-type liquid cooler
箱式造型 box mo(u)ld
箱式褶裥 box pleating
箱式真空吸尘器 can(n)ister
箱式振动器 casing oscillator
箱式整流器 tank rectifier
箱式铸铁工作台 box-like casting table
箱式自然对流器 cabinet-type natural convector
箱式组合钢柱 box-form built steel column
箱式钻模 box type jig
箱摔碎 case crushed
箱撕破压扁 carton torn, dented
箱套 encasement
箱体 case
箱体底部 bottom half
箱体底座 box type bed
箱体端盖 housing plate
箱体加油口弯管 housing filler elbow
箱体容量 tank capacity
箱体塞子 housing plug
箱体套筒 housing sleeve
箱体支架 anti-nose-dive leg
箱体轴承座 housing bearing seat
箱体轴套 housing bushing
箱体座 housing seat
箱头窗 box-head window
箱外紧固件 outer-tank attachment
箱位 slot
箱位号<集装箱> slot number
箱位检测器 position detector
箱位数 number of slots
箱位租赁权<集装箱> slot charter
箱物存放处 cloakroom
箱匣 box
箱镶板裂开 cases with panels split
箱销 box pin
箱谐振 boominess
箱形 box shape;egg shape
箱形暗沟 box drain
箱形坝 box dam
箱形百叶窗 boxing shutter; folding shutter
箱形板<桥梁> box slab
箱形板梁 box plate girder; tabular girder
箱形板梁桥 tabular bridge
箱形泵 box pump
箱形臂 box type jib
箱形臂架 box boom
箱形臂流动式起重机 mobile crane with box section
箱形边沟 box gutter
箱形剥离法 box cut
箱形薄板 box slab
箱形薄壁梁桥 box girder bridge
箱形舱口盖 pontoon hatch cover; pontoon type hatch cover
箱形层压板门式刚架 box-plywood portal frame
箱形铲(运)斗 box-scraper
箱形潮位计 box ga(u)ge
箱形潮位仪 box ga(u)ge
箱形沉箱 America caisson; box caisson
箱形出口 box out
箱形船闸 box gate; chamber gate; coffer lock
箱形窗框 box casing; cased window frame
箱形床脚 box typed leg
箱形大梁 box(ed) girder; cased girder; hollow box girder; hollow web girder; pontoon girder
箱形单梁龙门起重机 single-box girder gantry crane
箱形单梁起重机 monobox crane
箱形刀架 cutter box
箱形导坑 box heading
箱形的 boxed;box-shaped;cased
箱形的刮土戽斗 box-shaped scraper bucket
箱形底架 box section under frame
箱形底座 bed box bed;box bed
箱形吊杆 box type boom
箱形吊索 box sling
箱形钉固的输送器 box stapling conveyer[conveyor]
箱形渡槽 box flume
箱形断路器 tank truck circuit-breaker
箱形断面 box profile
箱形断面的横向机构 box cross member
箱形断面底脚 box section foundation
箱形断面钢梁 steel box beam
箱形断面基础 box section foundation
箱形断面基脚 box section foundation
箱形断面结构 box section construction
箱形断面肋 box sectioned rib
箱形断面索 box section string
箱形断面摇枕 box section bolster
箱形断面柱 box section column
箱形锻造风箱 chest bellows
箱形堆垛法 box-end stacking
箱形翻斗车 box body dump car
箱形防波堤 box breakwater
箱形飞檐 box cornice
箱形风筝<测气象用> box kite
箱形浮(船)坞 box floating dock;box type floating caisson
箱形钢 box steel
箱形钢板桩 box steel sheet piling
箱形钢结构系统 box steel structure system
箱形钢梁 steel box girder
箱形钢柱 box(ed) steel column
箱形钢桩 box pile;steel box pile
箱形格构大梁 box lattice(d) girder
箱形给料器 sliding box feeder
箱形拱桥 box arch bridge;box ribbed arch bridge
箱形拱座 box abutment
箱形构架 box frame
箱形构造 box type construction;box-work
箱形构造类型 box construction type
箱形刮泥器 cabinet scraper
箱形管 box pipe;box-shaped pipe
箱形管道 box conduit
箱形管片 box segment
箱形光度计 box photometer
箱形涵(洞) box culvert;slab culvert
箱形和趸船型防波堤 box and pontoon breakwater
箱形桁 box girder
箱形桁架 box truss
箱形桁架梁 box truss girder
箱形桁架桥 trussed-box-girder bridge
箱形横截面 box cross-section
箱形混凝土摊铺机 box type concrete spreader
箱形活塞 box type piston
箱形货舱口盖 cargo hatch pontoon cover
箱形货物 boxed cargo
箱形机座电动机 box frame motor
箱形基础 box foundation;coffer foundation
箱形基脚 box footing
箱形脊 box-shaped ridge
箱形集装箱 box type container
箱形夹具 box type jig
箱形甲板船 trunk deck vessel
箱形架 box type frame
箱形减速器 box reducer
箱形减压器 box reducer
箱形胶合板龙门架 box-plywood portal frame
箱形胶合板门式框架 box-plywood portal frame
箱形教堂凳 box-pew
箱形结构 box construction;box structure; box type construction; cased structure;tubular construction
箱形结构的环绕肋 wrap-around rib
箱形结构系统 cubicle system
箱形结构形式 box construction type
箱形截面 box section
箱形截面槽钢柱 column of channel box section
箱形截面大梁 box section girder
箱形截面的框架 box section frame
箱形截面梁 box girder; box section girder
箱形截面梁楼板 box beam floor
箱形截面梁式起重机 box girder crane
箱形截面楼梯斜梁 box section stringer
箱形截面设计 box section design
箱形截面柱 box section column
箱形进水口 box inlet;box intake
箱形开关 box switch
箱形开挖法 box cutting method
箱形靠背长椅 box-pew
箱形空气过滤器 box type air filter
箱形孔型 box pass
箱形框 cased frame
箱形框架 box(ed) frame;box section frame;cellular framing
箱形框架钢筋混凝土结构 box frame type reinforced concrete construction
箱形框架结构 box-frame(d) construction
箱形框架施工法 box frame construction
箱形框架式抗震墙 box frame type shear wall
箱形框纱窗 box screen
箱形肋断面 box rib profile
箱形肋拱桥 box ribbed arch bridge
箱形肋骨 box frame
箱形肋下承式拱 box-rib-through-arch
箱形冷凝器 box type cooler
箱形联结器 box coupling
箱形联轴节 box coupling
箱形梁 box(-shaped) beam; cased beam;hollow box beam
箱形梁的内模板 void former
箱形梁桁架式龙门起重机 box girder truss gantry crane
箱形梁楼板 box beam floor
箱形梁模板 box former
箱形梁桥 box girder bridge
箱形梁曲线桥 curved box girder bridge
箱形梁式悬臂桥 box girder cantilever bridge
箱形料架 boxing-in;pocket setting
箱形流槽 box launder
箱形龙骨 duct keel
箱形楼梯 box(ed) stair(case);closed stair(case)
箱形路沟 box gutter
箱形路面排水沟 box rainwater gutter
箱形螺母 box nut
箱形门阶 box stoop
箱形门框 box casing
箱形面具 box respirator
箱形模板 box form(work);box shuttering
箱形内龙骨 box keelson
箱形排水沟 box drain
箱形排水渠 box drain
箱形墙垛 boxed buttress wall
箱形桥 box bridge
箱形桥台 box(type) abutment
箱形曲流 box-shaped meander
箱形商店 box store
箱形设计 box design
箱形升降机 casing elevator
箱形石料撒布机 spreader box
箱形石料摊铺机 spreading box;spreader box
箱形石笼 box gabion
箱形石棉水泥天沟 asbestos-cement box roof gutter
箱形输水道 box conduit
箱形双梁龙门吊车 double box girder gantry crane
箱形双梁龙门起重机 double box girder gantry crane
箱形水槽 box(ed)(rainwater) gutter; box flume; parapet gutter; trough gutter
箱形水道 box conduit
箱形水沟 box gutter
箱形水力分级机 water scalping tank
箱形隧道 box section; box tunnel; rectangular shaped tunnel
箱形台 tank table
箱形体系 box system
箱形天线 box antenna;box loop
箱形挺杆 box jib

箱形通道 < 施工用 > box heading
箱形通风管 < 散装货船用的 > box ventilator
箱形凸起甲板 trunk deck
箱形托盘 box pallet
箱形拖车 box trailer
箱形挖槽 box cut
箱形挖泥法 dredging box cut
箱形围堰 box cofferdam; box dam; box weir
箱形屋顶水沟 boxed roof gutter
箱形屋顶雨水沟 special box roof gutter
箱形坞门 dock caisson
箱形舞台 proscenium frame stage
箱形物 box
箱形峡谷 box canyon
箱形舷缘 box gunwale
箱形小车 box car
箱形型孔 < 用箱形模板在混凝土中形成的孔口或洞 > box out
箱形岩芯取样器 box corer
箱形檐槽 parapet gutter
箱形檐沟 box(ed) gutter
箱形验潮计 box ga(u)ge
箱形验潮仪 box ga(u)ge
箱形叶轮 cased impeller
箱形油断路器 tank-type oil circuit breaker
箱形雨水槽 box gutter; parallel gutter
箱形雨水沟 box rainwater gutter
箱形预应力混凝土结构 box(-shaped) prestressed concrete structure
箱形轧槽 box groove
箱形枕 box-shaped pillow
箱形整体结构件 box type integral structure
箱形支撑 box brace
箱形支撑杆 box strut
箱形支座 < 手推车马道的 > box horse
箱形柱 box(-shaped) column; cased column
箱形桩 box pile
箱形字盘 trunk dial
箱形座 box stand
箱穴 Box Hole
箱移法 casing method
箱用钉 box nail
箱载(重)利用率 coefficient of utilization of container's capacity; loading factor of container
箱植法 box planting
箱中退火 close annealing
箱主代号 < 集装箱 > owner code
箱注入口 tank filler
箱柱 box column
箱装 in boxes; incase
箱装部件 pack unit
箱装货吊钩 case hook
箱装货(物) boxed cargo; cargo in boxes; cargo in cartons; cargo in case; cargo in chests; cargo in crates; case goods; case cargo
箱装式调节器 case-mounted controller
箱装油 case oil
箱状 box-like
箱状(插齿刀)刀架 cutter box
箱状构造 boxwork
箱状建筑物 box building; box structure
箱状峡谷 box canyon
箱状褶皱【地】box fold
箱撞碎 case smashed
箱子 box; case; cassette; chest; trunk
箱子闭锁 box closure
箱子变形 case deformed misshaped
箱子不牢固 frail case
箱子储藏室 trunk room
箱子脆弱 case fragile
箱子倒置 case upside down arrow mark down
箱子翻钉 case re-nailed
箱子间 boxroom
箱子经修补 case repaired
箱子设计 box design
箱子摔破重钉 cases cracked and re-nailed
箱子压坏 case crushed
箱座 box base; box seat; box stand; tank block
箱座接合 tank base joint

襄

襄理 assistant manager
襄审官 < 国际法院 > assessor

镶

镶安全玻璃 security glazing
镶板 boiserie; fitch; lining board; panel(board); panel(l)ing; plate panel; shingle panel; veneer; veneer board; veneer(ed) panel; wall panel; wood veneer; spandrel < 两层窗之间的 >; crotch veneer < 取自树杈的 >
镶板安排 panel arrangement
镶板壁板 panel siding
镶板玻璃槽口 rebate of glazing
镶板薄壳 panel shell
镶板布置 panel arrangement
镶板材 panel(l)ing lumber
镶板材料 planking stuff
镶板的 veneered
镶板的可互换性 panel interchangeability
镶板的切割面 loose side
镶板的直观面 primary panel
镶板底面 panel soffit
镶板地板 surround slab
镶板钉 panel pin
镶板顶棚 lacunar; panel ceiling
镶板隔断 paneled partition
镶板隔墙 panel curtain wall; plate panel partition(wall)
镶板桁架结构系统 panel-truss-structure system; Lu-Re-Co
镶板花式模型 panel pattern
镶板或护墙板分格线脚 applied mo(u)lding
镶板机 match boarding machine
镶板夹杆 sett
镶板建筑系统 panel construction system
镶板(胶)合板 plypanel
镶板结构 veneer construction
镶板结构系统 panel structural system
镶板锯 veneer saw
镶板框 paneled framing
镶板离缝 crossband gap
镶板立面 panel facade
镶板门 mo(u)lded matching door; panel(ling) door; panelled door
镶板门框架 panelled framing
镶板门下沉线脚 drop mo(u)lding
镶板模 panel mo(u)ld
镶板模板工程 panel formwork
镶板幕墙 panel curtain wall
镶板内顶板 panel ceiling
镶板配套装修 panel lining
镶板拼装法 pattern match
镶板平顶 pan ceiling; panel ceiling
镶板墙 veneered wall
镶板墙拉杆 veneer tie; veneer wall tie
镶板切割 veneer cutting
镶板切制 veneer slicing
镶板设计 pan design
镶板施工 panel construction
镶板施工方法 panel construction method
镶板式模板 post-and-panel form
镶板式柱墙 column-and-panel wall
镶板饰墙面 wall veneering
镶板天花板 pan ceiling; panel ceiling
镶板天棚 pan ceiling
镶板外貌 panel facade
镶板外墙 external panel wall
镶板围篱 panel fence
镶板细工 panel covering; panel(l)ing
镶板线条饰 panel mo(u)lding
镶板小分格 subdivided panel
镶板油漆 panel-painting
镶孢属 < 拉 > Coniothecium
镶壁家具 building in furniture; built-in furniture
镶边 edge lining; ending; flange up; infix; inset; lipping; list; overshoot; purfle; purfling; selvage [selvedge]; trimming
镶边板 edge plate; edging board
镶边板条 cant strip; chamfer strip; eaves strip
镶边材料 fringe material
镶边的 enchased; purfled
镶边短锯 back saw
镶边干板 edged plate
镶边构造 veneered construction
镶边花边 insertion lace
镶边机 welter
镶边角板 corner board
镶边结构 rimmed texture
镶边框 trim; trimming
镶边框的孔口 trimmed opening
镶边楼板 surround slab
镶边陆架 rimmed shelf
镶边面砖 marginal tile
镶边平板 edged plate
镶边石 border stone; trim stone
镶边手据 back saw; bead saw
镶边手套 insert glove
镶边陶块 < 屋顶 > curb clay tile
镶边条 banding; edging strip
镶边细工 purfled work
镶边线条 edge mo(u)lding; edge strip; edging strip
镶边压缝条 cant strip; chamfer strip
镶边压条 tilting fillet; tilting piece
镶边砖石 angle closer
镶玻璃 glaring
镶玻璃扁头钉 glazier's sprig
镶玻璃部件 glazing unit
镶玻璃材料 glazing material
镶玻璃裁口 glazing rebate; rebate of glazing
镶玻璃槽口 glass rabate; glass rabbet; glazing mo(u)lding; glazing rebate
镶玻璃的 glassed
镶玻璃的舱口 window hatch
镶玻璃的车顶 observation room
镶玻璃的钢芯铅条 reinforced glazing bar
镶玻璃的檩条 glazing purlin(e)
镶玻璃的木门 glazed timber door
镶玻璃的外廊 glassed-in veranda(h)
镶玻璃的型材 glazing profile
镶玻璃的阳台 glassed-in veranda(h)
镶玻璃垫条 glazing gasket
镶玻璃定位块 glazing block
镶玻璃法 glazing mo(u)lding
镶玻璃方法 glazing technique
镶玻璃隔墙 glazed partition
镶玻璃工人 glazier
镶玻璃技术 glazing technique
镶玻璃夹 glazing clip
镶玻璃门 glazed door; sash door
镶玻璃密封料 glazing compound
镶玻璃密封条 glazing gasket
镶玻璃面料团 glazing compound
镶玻璃木窗棂条 wooden glazing bar
镶玻璃铅条 canes; window lead
镶玻璃条 glazing bead
镶玻璃铁条 iron glazing bar
镶玻璃行业 glazing trade
镶玻璃翼板 wing
镶玻璃用扁头钉 glazing sprig
镶玻璃用扁头针 glazier's sprig
镶玻璃用角状压条 glazing angle
镶玻璃(用)油灰 glazier's putty; fixing compound; glazing putty; sash putty
镶补板 panel(l)ing
镶补混凝土 dental concrete
镶补料 filling material
镶补木料 graving piece
镶侧石的步行道 kerbed footway
镶衬 lining
镶衬梁 joggled beam
镶成棋盘花纹 tessellate
镶齿 inserted tooth; tungsten-carbide insert
镶齿扳牙 inserted die
镶齿刀头 inserted blade
镶齿的 tipped
镶齿端铣刀 inserted blade end mill
镶齿滚刀 inserted tooth hob
镶齿铰刀 inserted blade reamer; inserted reamer
镶齿锯 inserted tooth saw
镶齿扩孔钻 inserted blade core drill
镶齿拉刀 inserted broach; inserted tooth broach
镶齿螺纹梳刀丝锥 inserted chaser tap
镶齿密封喷射滑动钻头 sealed jet tungsten carbide insert sliding bit
镶齿密封喷射钻头 sealed jet tungsten carbide insert bit
镶齿平面铣刀 inserted tooth facing milling cutter
镶齿三面刃铣刀 inserted side milling cutter
镶齿(式)铣刀 inserted-tooth(milling) cutter; inserted fraise
镶齿丝锥 inserted tap
镶齿硬质合金刀片 inserted carbide
镶齿圆锯 segmental saw
镶齿钻头 tungsten-carbide insert bit
镶窗玻璃金属条 metal astragal
镶唇 eyeleting
镶刀钻头 insert bit
镶的 lined
镶地板用胶合板 plywood square
镶舵入座 gulleting of rudder
镶盖 veneer
镶钢片 < 镶嵌针槽的 > inserted trick
镶高速钢齿的阔面铣刀 slap milling cutter with high speed steel inserted teeth
镶格工作 panel work
镶焊 built-up welding; welding on
镶航空像片 mosaic air photo; mosaic photo strip
镶合腹接 side veneer-grafting
镶合浇制 match casting
镶合浇制法 match casting method
镶合浇铸 match casting
镶合梁 joggle(d) beam
镶合砂型 sectional mo(u)ld
镶合条件 fitting condition
镶护壁板 wainscot(t)ing
镶护墙板 wainscot(t)ing; wall paneling
镶花 enchase
镶花边 lace; lacework
镶花地板 mosaic fingers; mosaic parquetry; parquet

镶花地板材料 parquet flooring
镶花木工 parquet work
镶花细工 parquetry
镶辑复照图 photo index; index photography
镶夹条 jib
镶尖榫接头 tusk and tenon joint
镶尖装置 tipper
镶胶合板门 veneered door
镶接 hatched grafting; splice; whip grafting; board and brace <木材的>
镶接扳手 insertion spanner
镶接板 fish plate; splice plate
镶接长度 length of restraint
镶接的指针 splicing needle
镶接工作 joggled work; joggle work
镶接木板工作 board-and-brace work
镶接木板作业 board-and-brace work
镶接头 fish joint
镶金刚石锯盘 diamond slitting wheel
镶金刚石钻头 bort(z)-set bit
镶进 infix
镶进线饰带层 mo(u)lded-intake belt course
镶镜用螺钉 mirror screw
镶刻 applique
镶空心玻璃砖墙 glass concrete construction
镶孔环 eyelet
镶块 glut(ting)
镶块锻模 inserted forging die
镶块模 assembled die; insert die
镶块拼花地板 inlaid parquet
镶块式模 sectional die
镶框 bezel
镶框架 framing
镶框镜 framed mirror
镶铝装饰用胶合板 plymax
镶轮滑车 mortised block
镶马赛克的建筑物的立面 mosaic facade
镶马赛克的建筑物的正面 mosaic facade
镶面 cladding; dressing; facet(te); veneer(ing)
镶面板 cover slab; face slab; facing slab; lining board; panel(l)ing; veneer
镶面板包装 package of veneer
镶面板锯 veneer saw
镶面板面层 face veneer
镶面板黏[粘]结剂 veneer adhesive
镶面板刨刀 veneer knife; veneer shaver
镶面板切割机 veneer cutting machine
镶面板系铁 veneer tie
镶面板装饰 package of veneer
镶面薄板 facing sheet
镶面材料 cladding material; dressing material; facework material
镶面层板 curly veneer
镶面瓷砖 paving tile
镶面的 veneered
镶面方法 facing system
镶面工作 face work
镶面构造 veneered construction
镶面混凝土板 facing concrete slab
镶面机 facing machine
镶面胶合板 veneered plywood
镶面结构 veneered construction; veneered structure
镶面门 veneered door
镶面模板 <钢筋混凝土> shell formwork
镶面刨花板 veneered chipboard
镶面砌体 dressed masonry
镶面墙 face(d) wall; veneer(ed) wall
镶面墙系铁 veneered wall tie; veneer tie
镶面石 dressing stone
镶面石板 veneer stone
镶面石层 opus lithostratum
镶面圬工 faced masonry
镶面系墙件 veneer tie; veneer wall tie
镶面砖 veneer(ed) brick
镶模板 panel mo(u)ld
镶木 parquet; veneer; veneer wood
镶木板 wood panel(l)ing
镶木厂 parquet plant; parquet work
镶木地板 inlaid parquet; mosaic parquetry; overlay flooring; parquet floor; parquetry; wood block floor
镶木地板打光 parquet polishing
镶木地板打光机 parquet polishing machine
镶木地板的板块 parquetry block
镶木地板的密封 parquetry sealing
镶木地板的拼镶 parquetry composite
镶木地板覆盖 block parquetry floor covering
镶木地板工 parquet floor layer
镶木地板清漆 parquet varnish
镶木地板条块覆盖 block parquetry floor covering
镶木地板细工的木块 parquetry block
镶木地板细工木料 parquetry wood
镶木地板用木材 parquetry timber
镶木法 parquetry
镶木工 mosaic
镶木工艺 wood mosaic
镶木工作 parquetry; parquet work
镶木机 machine for parquetry work
镶木块 wood mosaic
镶木楼板 wood block floor
镶木马赛克 wood mosaic
镶木条的甲板捻缝法 splining
镶木条甲板 splined deck
镶木条形地板 parquet strip flooring; strip flooring
镶木涂料 strip coating
镶木细工 parquet; parquetry; parquet work
镶木细工的拼镶 parquetry composite
镶木细工桶板 parquetry stave
镶木制品 tarsia
镶配地板 matched floor
镶配天花板 matched ceiling
镶配屋面采光玻璃 king's glazing
镶片刀具 tipped tool
镶片锯切 rift saw(ing)
镶片磨轮 segment grinding wheel
镶片木材 pieced timber
镶片式翅片管 imbedded fin tube
镶片圆锯 rift saw; wing saw
镶片钻头 D-bit
镶拼板 matching
镶拼底板 <钉在墙筋或格栅上的> tight sheathing
镶拼花板 match boarding
镶铺瓷砖 terra-cotta
镶铅玻璃 leaded glass
镶铅装饰胶合板 plymax
镶嵌 damascene; inlay(ing); laydown; marriage; matching; montage; mounting; set in
镶嵌巴比合金 Babbitting
镶嵌板 elbow board; mosaic panel; panel board
镶嵌玻璃 glass inlaying; incrusted glass; mosaic glass; Roman mosaic
镶嵌玻璃窗 mosaic window
镶嵌玻璃瓷砖 glass mosaic tile
镶嵌玻璃的槽口 rebate of glazing
镶嵌玻璃用的油灰 glazier's putty
镶嵌玻璃用铅 glazier's lead
镶嵌玻璃用支架 glazier's rack
镶嵌材料 inlay
镶嵌层 mosaic coating; mosaic surface; parquetry layer; parquet sealing
镶嵌衬里的 mosaic-lined
镶嵌成花纹的路面 tessellated pavement
镶嵌成像 mosaic imaging
镶嵌齿 inserted tooth
镶嵌窗玻璃有檐铅条 lead came
镶嵌锤 driving mallet
镶嵌大理石片的水磨石 berliner
镶嵌的 enchased; inlaid; interbedded; mosaic; planted
镶嵌的玻璃 incrusted glass
镶嵌的航摄照片带 mosaic photo strip
镶嵌的家具 boulle
镶嵌地板 inlaid flooring; inlaid parquet; mosaic floor(ing); mosaic parquetry; parquet floor
镶嵌地块 felder; mosaic structure
镶嵌地面 inlaid flooring
镶嵌地砖 mosaic tile
镶嵌电极 mosaic electrode
镶嵌垫块 setting block
镶嵌雕刻品 superimposed carving
镶嵌缝隙 mosaic fissure
镶嵌覆盖层 mosaic clad
镶嵌工 inlayer; mosaicker; mounter
镶嵌工程 inlaid work; mosaic work
镶嵌工具 setting tool
镶嵌工艺 inlay; marquetry; mosaic; tessellated work
镶嵌工艺品 buhl
镶嵌拱顶 mosaic vault
镶嵌构造 inlaid structure; mosaic structure
镶嵌光电元件 mosaic
镶嵌规 <金刚石钻头> setting ga(u)ge
镶嵌航片 airphoto mosaic
镶嵌花边 mosaic lace
镶嵌花坛 mosaic bed
镶嵌花纹 inlay pattern
镶嵌花纹油毡 inlaid linoleum
镶嵌花样 mosaic
镶嵌机 pointing machine
镶嵌机具 inserted implement
镶嵌技术 marquetry
镶嵌加工 inlaid work
镶嵌夹具 setting block
镶嵌家具 built-in furniture
镶嵌件 inserts
镶嵌胶结 mosaic cementation
镶嵌接头 housed joint
镶嵌结构 mosaic texture; interlocked structure
镶嵌金刚石的制品 diamond-impregnated composition
镶嵌金刚石扩孔器 set reaming shell
镶嵌金属包覆 inlay cladding
镶嵌晶体 inserted crystal; mosaic crystal
镶嵌锯 inlaying saw
镶嵌块 mosaic block
镶嵌块件 mosaic building block
镶嵌框架 mosaic frame; mosaic panel
镶嵌粒间孔隙 mosaic intergranular pore
镶嵌粒状变晶结构 mosaic granoblastic texture
镶嵌楼梯 mosaic stair(case)
镶嵌螺栓 reamer bolt
镶嵌马赛克装饰特色 mosaic ornamental feature
镶嵌马赛克装饰特征 mosaic ornamental feature
镶嵌门挡 planted stop
镶嵌门面 mosaic facade
镶嵌面 mosaic area
镶嵌面板 mosaic panel
镶嵌面层 surface mounting
镶嵌模型法 inlaid model method
镶嵌模样 mosaic pattern
镶嵌木 inlaid wood
镶嵌木板条块 parquet block
镶嵌木工 parquet work
镶嵌木块 parquet block
镶嵌木料 boule; parquet timber
镶嵌内径 <金刚石钻头的> set inside diameter
镶嵌能手 mosaic artist
镶嵌排列 arrangement of mosaic
镶嵌胚乳 mosaic endosperm
镶嵌拼花板 mosaic parquet panel
镶嵌拼接 scarf splice
镶嵌品 boutique
镶嵌屏薄膜 mosaic screen film
镶嵌屏幕 mosaic screen
镶嵌铺面砖 inlaid flooring brick
镶嵌漆器 inlaid lacquer ware
镶嵌砌块墙 coffered wall
镶嵌器 photoisland grid
镶嵌钎头 insert rock bit
镶嵌铅条 calm; came
镶嵌墙板 lay-in panel
镶嵌穹隆 mosaic dome
镶嵌群落 mosaic community
镶嵌任意块圬工 mosaic random rubble masonry
镶嵌石块圬工工程 mosaic masonry work
镶嵌式 mosaic type
镶嵌式电路图 mosaic circuit diagram
镶嵌式构造 mosaic
镶嵌式红外系统 mosaic infrared system
镶嵌式楼梯间 mosaic stair(case)
镶嵌式绒线刺绣 mosaic wool work
镶嵌式陶瓷 champleve enamel
镶嵌式外包装 mosaic cover(ing)
镶嵌式外壳 mosaic clad
镶嵌式圆管 cylindric(al) interlocking pipe
镶嵌式圆屋顶 mosaic cupola; mosaic vault
镶嵌式轴衬 insert liner
镶嵌式轴瓦 insert bearing
镶嵌饰 inlay; intarsia; parquetry
镶嵌饰物的家具 buhl work
镶嵌室外表面的修整 mosaic outdoor finish
镶嵌竖窄条玻璃的门 vertical-vision-light door
镶嵌说 mosaic theory
镶嵌碎裂结构 interlocked texture
镶嵌索引图 <一种用于索引的像片略图> index mosaic
镶嵌碳化钨合金片的活钻头 detachable tungsten carbide insert bit
镶嵌陶瓷细片 ceramic tessera
镶嵌套管鞋 set casing shoe
镶嵌条带 inserted strips
镶嵌贴面 mosaic surface
镶嵌图 mosaic diagram; mosaic figure; mosaic map; sheet assembly
镶嵌图案 mosaic; mosaic pattern
镶嵌图案玻璃 picture forming glass
镶嵌图案的玻璃 picture glass
镶嵌图板 mosaicking board; panel base
镶嵌图片 panel base
镶嵌外表最后加工 mosaic exterior finish
镶嵌外径 <金刚石钻头的> set outside diameter
镶嵌卫片 <遥感地质> space image mosaic
镶嵌物 incrustation; inlaid material; inlet; tessera[复 tesserae]
镶嵌物似的 tesseral
镶嵌细工 inlaid work; marqueterie; mosaic work; parquetry; tessellated work; marquetry

镶嵌细木工 mosaic woodwork;reignier work;wood(en) mosaic
镶嵌细木工地板 mosaic fingers
镶嵌细作木板 board parquetry
镶嵌现象 mosaicism
镶嵌线条 inlaid mo(u)lding;laid-in mo(u)lding
镶嵌像片排系列 mosaic photo series
镶嵌像片重叠边剔薄 featheredging
镶嵌小方石 mosaic paving sett
镶嵌小圆顶 mosaic cupola
镶嵌形式 mosaic pattern
镶嵌型板 clincher pattern plate
镶嵌性 mosaicism
镶嵌乙烯材料 inlaid vinyl goods
镶嵌乙烯制品 inlaid vinyl goods
镶嵌艺术 art of inlaying
镶嵌艺术家 mosaic artist
镶嵌用材 mosaic wood
镶嵌用彩色玻璃 tessera[复 tesserae]
镶嵌用乳白玻璃 flashed opal glass
镶嵌用小块 mosaic tessera
镶嵌油毡 inlaid linoleum
镶嵌造型 insert mo(u)lding
镶嵌照片系列 mosaic photo series
镶嵌者 inlayer;mosaicker
镶嵌直径 <金刚石钻头的> set diameter
镶嵌制品 inlaid goods;inlaid work
镶嵌制作工 mosaicsit
镶嵌重量 <钻头上金刚石的> set weight
镶嵌珠宝的日用品 boutique
镶嵌砖 mosaic window
镶嵌装订 mosaic binding
镶嵌装饰 boule;boulle;incrustation;inlaid decoration;mosaic decoration;mosaic enrichment
镶嵌装饰槽 lacuna[复 lacunae]
镶嵌装饰法 mishima
镶嵌装饰品 marqueterie;marquetry
镶嵌装饰性整修 mosaic decorative finish
镶嵌状 mosaic shape
镶嵌状沙漠面 desert mosaic
镶嵌钻头 insert bit;slug bit
镶嵌作业 inlaid work
镶墙板单元 wall unit
镶墙边的半顶砖 half header
镶墙(边)的砖石 closer;wall tie closer
镶墙面 veneer of wall
镶切刃岩芯钻头 insert set bit
镶刃刀胚 insert blank
镶刃刀片 insert bit
镶刃钎头 tipped steel
镶刃钻头 inserted drill;tipped bit
镶入 inlay;insert;trim in
镶入板 filler
镶入齿 inserted blade
镶入刀头 bit insert
镶入的 built-in
镶入的衬套 inserted sleeve
镶入块 insert
镶入配件 built-in parts
镶入砌块 coffered block
镶入砌体 coffered block
镶入器件 built-in fitting
镶入刃口 bit insert
镶入设施 built-in items
镶入式 inset type
镶入式白炽灯 recess incandescent lamp
镶入式半圆件 mortise(d) astragal
镶入式缸套 inserted liner;insertion liner
镶入式模具 insert die
镶入式钎头 insert bit
镶入式荧光灯 recessed fluorescent fixture
镶入式荧光灯带 recess continuous row fluorescent fixture
镶入式浴盒 built-in bathtub
镶入线条 laid-in mo(u)lding
镶色玻璃 cased glass;casing glass
镶石工作 face work
镶石护面码头 quay wall with stone facing
镶石铺面 Roman mosaic;tessellated paving
镶式板柱墙 column-and-panel wall
镶饰 braze;encrust;incrust;inlaid
镶饰玻璃面 glass facing
镶饰玻璃面工作 glass face work
镶饰建筑物立面的玻璃块 glass-facade block
镶饰建筑物正面的玻璃块 glass-facade block
镶饰青铜 sheathing bronze
镶双锯齿形花边 engrail
镶塑料板 veneered plastics board
镶塑料的 plastic tined
镶塑料头的小锤 plastic tip hammer
镶榫 mortise and tenon
镶榫接合 mortice[mortise]-and-tenon joint;trimmer joint
镶榫接头 mortice[mortise]-and-tenon joint;trimmer joint
镶锁铺砌(路面) interlocking paving
镶碳化钨钻头 tungsten insert bit
镶套 bush fitting
镶套衬垫 slip-in liner
镶套套管 slip-in liner
镶条 fit strip;liner;mo(u)lding;stripe
镶条地板 inlaid-strip flooring
镶条地面 inlaid-strip flooring
镶贴 encrustation;incrustation
镶贴金属板(防水) me-tyl-wood
镶铁路缘石 armo(u)red curb
镶图(底)板 mosaic mount
镶新月形斜肋 crescent oblique rib
镶新月形斜棱 crescent oblique rib
镶压性拱形支架 yielding arch
镶压性金属地架 yieldable steel arch
镶压性金属支架 yielding steel arch
镶以木板 boarding
镶硬质合金刀片的麻花钻头 twist drill with hard alloy tip
镶硬质合金切削刃 tipped edge
镶硬质合金钻杆 tungsten-carbide-tipped drill rod
镶硬质合金钻头 tungsten-carbide-tipped bit
镶有玻璃的建筑正面 glass facade
镶有环圈的 ringed
镶有金刚石的 diamond set
镶有金刚石的工具 dia-tool
镶有金刚石的套管靴 diamond-set casing shoe
镶有金刚石条的扩孔器 insert reaming shell
镶有硬质合金的钻头 hard-faced bit
镶有装饰框的窗 architecturally enframed window
镶釉漆 glazing putty
镶在框子内 incase
镶在墙上 immure
镶注油管 cast-in oil lead
镶柱的 brick-encased
镶铸的 cast-in
镶砖 brick setting
镶砖的 brick-encased
镶砖房屋 brick clad building
镶砖混凝土板 brick-lined concrete slab
镶砖建筑物 brick clad building
镶砖圬工 brick-lined masonry
镶装刀片 insert chip
镶装喷管 insert nozzle
镶钻头工具 bit setting tools
镶钻头环 bit setting ring
镶钻头夹具 bit setting block

详

详表 full edition
详测 close mapping;detailed survey
详测曲线图 detail log
详查 detailed investigation;detailed prospecting;detailed survey
详查报告 detailed prospecting report
详查阶段 stage of detailed investigation
详查者 sifter
详查最终 detailed prospecting for construct
详函 covering letter
详记 specify
详检 through check
详见附表 details as per attached list;for details see attached table
详尽分录 covering entry
详尽规划过的 well-planned
详尽说明书 cookbook
详尽细节 elaboration
详尽研究 exhaustive research
详尽指示 in-depth briefing
详尽作业程序表 detailed operating [operation] schedule
详究 scrutinization;scrutiny
详勘 main investigation
详勘报告 detailed exploratory report
详勘钻孔 development borehole
详论 dwell on
详情随后邮寄 details to follow
详式提单 long form bill of lading
详述 dilatation;expound;give particulars;particularize;enlarge on
详述技术标准 descriptive specification
详探井 detailed exploratory well;development well
详图 comprehensive chart;comprehensive map;congested map;detail(ed) drawing;detailed map;detailed sketch;part by part
详图工程 detail engineering
详图设计 detail(ed) design
详图设计员 detailer
详细 part by part
详细办法 detailed ways and means
详细报告 circumstantial report;detailed report;details
详细背书 endorsement in full
详细标记 detail designator
详细布置计划 detailed layout planning
详细布置图 detail layout plan
详细步骤 detailed procedure
详细测图 close(d) mapping
详细查账报告 long-term report
详细的 at length;detailed;particular
详细的技术资料 technical data
详细的目的 purposes stated in detail
详细的实地观察 close up
详细的试验目的 detailed test objective
详细的图纸 detail drawing
详细调查 close investigation;detailed investigation;detailed prospecting
详细调查阶段 detail survey stage;particular investigation stage
详细断面图 removed section
详细分类账 detailed ledger
详细分析 detailed analysis;multianalysis
详细概算 detailed cost estimate;detailed estimate
详细概算书 detailed estimation
详细格式 long form
详细工程 detail engineering
详细工程计划 detailed program(me)
详细工程设计 definite project plan
详细工作清单 exhaustive work
详细供应范围 detailed scope of supply
详细故障分析 detailed failure analysis
详细规定项目 specified-in-detail item
详细规范 closed specification;detail(ed) specification
详细规格(书) detail(ed) specification;detail(ed) requirement
详细规划 blueprint;detail(ed) planning
详细规划条件 condition for site planning
详细规划图 detailed plan
详细海岸线 detailed coastline;detailed shoreline
详细核对 complete checking
详细绘图 detailed sketch
详细计划 detailed plan
详细记录 detail(ed) record
详细记载保险单 specification policies
详细技术经济评价 detailed evaluation of technical economics
详细技术说明书 detailed technical specification
详细技术要求 detailed technical requirement
详细价目表 detailed price list
详细检查 canvassing;detailed examination;detailed inspection
详细进度计划表 detail schedule
详细经费 detailed appropriation
详细勘测 detailed survey;detail surveying
详细勘察 detailed exploration;detailed (geotechnical) investigation;detailed prospecting
详细勘探 detailed exploration;detailed (geotechnical) investigation;detailed prospecting
详细勘探阶段 detail prospecting stage
详细勘探钻井 development drilling
详细勘探钻孔 development borehole
详细可选性试验 detailed test of mineral dressing ability
详细零件图 detail drawing of parts
详细流程图 detail chart;detail flowchart
详细论述 detailed coverage
详细面谈法 detailed interview
详细目录 inventory
详细内容后告 details to follow
详细平面 detailed plan
详细剖面 detailed section
详细清单 detailed list;exhaustive list
详细情节 ins and outs
详细情况 detail;further information
详细情况报告 detail status report
详细日程计划 detail schedule
详细设计 design in detail;detail(ed) design;detailing
详细设计报告 definite plan report
详细设计阶段 detailed design stage
详细设施 facility detail
详细审查 sifting
详细审计 detailed audit
详细审计报告 detailed audit report;long form report
详细时间 detail time
详细数据 detailed data;particulars
详细说明 detailed description;detailed explanation;detailed instruction;detailed particulars;extended explanation;particular;specification;specify

详细说明的 specified
详细说明理论 expound theory
详细说明书 closed specification;detail(ed) specification
详细索引标志 detail index mark
详细讨论 disquisition
详细条款 detailed provisions
详细推理 fine draw
详细网络图 detail network
详细网示 <工程进度表> detail network
详细文件资料 detail file
详细文摘 extended abstract
详细系统说明 detailed system description
详细现场资料 detailed field data
详细项目单 itemized list
详细叙述 recount
详细研究 detailed study;dissect;scrutiny;workover
详细要求 detail requirement
详细用途 detailed use of commodity
详细预计的经营损益表 stabilized operating statement
详细预算 detailed budget;detailed estimate(of construction cost)
详细造价分析 detail cost analysis
详细账目 itemized account
详细装箱单 detailed packing list
详细资料 detailed data;detail file;particulars
详细纵断面图 detailed profile
详细作业(进度)计划 detailed operating[operation] schedule

享

享受保险赔偿者 beneficiary
享受递减律 law of diminishing returns
享受年金权利者 annuitant
享受年金人 annuitant
享受优惠的资格 preferential entitlement
享受资料 means of enjoyment
享用 beneficial use
享有 holding
享有期间 tenure
享有全部可保利益 full interest admitted
享有声誉 bear reputation
享有所有权的 titular
享有用益权的人 usufructuary

响

响板 sound board;sounding tester
响导船 leader
响导者 conductor
响导桩 pilot pile
响度 loudness;sonority
响度标度 loudness scale;sone scale
响度单位 loudness unit;phon
响度单位计 volume unit indicator
响度的感觉 loudness sensation
响度等级 loudness scale
响度范围 volume range
响度分析仪 loudness analyser[analyzer]
响度复原 loudness recruitment
响度函数 loudness function
响度级 level of loudness;loudness level;volume level
响度级调整装置 vogad;voice-operated gain adjusting device
响度计 phon(o)meter;program(me) meter
响度鉴定 loudness evaluation
响度降低 loudness reduction
响度均衡控制 loudness equalization control
响度控制 loudness control
响度控制电路 loudness control circuit
响度控制开关 loudness control switch
响度扩大 volume expansion
响度密度 loudness density
响度平衡 loudness balance
响度评定 loudness rating
响度试验 loudness test
响度图 loudness pattern
响度限制器 volume limiter
响度效率因数 loudness efficiency factor
响度指数 loudness index
响度指向性指数 loudness directivity index
响度重振 loudness recruitment
响墩 detonating cartridge;torpedo;detonator <雾信号>
响墩安置机 detonator machine;detonator placer;torpedo placer
响墩安置机握柄 detonator placer lever
响墩信号 detonating signal;torpedo
响弧 crackling arc
响亮 vibrance
响亮的 vibrant
响裂 clinks
响铃 ring-down
响铃浮标 bell buoy
响区 loud spot
响哨浮标 whistle buoy;whistling buoy
响声 blare;tantara
响岩 clinkstone[klinkstone];phonolite[phonolyte]
响岩巴黎白 cliffstone Paris white
响岩结构 phonolitic texture
响岩类 phonolite group
响岩质粗面斑岩 phonolitic trachyte porphyry
响岩质粗面岩 phonolitic trachyte
响叶杨 <山白杨> Chinese aspen
响应 answer back;receptance;respond;response
响应比 response ratio
响应变换 response transform
响应表面 responding surface
响应场 response field
响应持续时间 response duration
响应带宽 responsive bandwidth
响应的 responsive
响应法 response method
响应范围 responding range;response range
响应方差 response variance
响应幅相图 Bode diagram;log magnitude and phase diagram
响应函数 response function
响应级 sensitivity level
响应极限 response limit
响应计划的撤销 deactivation of a response plan
响应率 responsivity
响应率的效应 effect in response ratio
响应慢 low-response
响应面法 response surface method
响应模式 response mode;responsive mode
响应频率 response frequency
响应频率特性 response-frequency characteristic
响应频率图 response-frequency diagram
响应谱 response spectrum
响应器 responder;responser;responsive device
响应曲面 response surface
响应曲线 responding curve;response curve
响应设计 responsive design;responsive planning
响应时间 respond time;response time;time response
响应时间常数 responsive time constant
响应数据 response data
响应速度 response speed;screen speed;speed of response
响应速率 response rate
响应探测器 response detector
响应特性 response characteristic
响应特性曲线 response curve
响应图 response diagram
响应位 response position
响应系数 response coefficient
响应信号 answering signal;response signal
响应信息 response message
响应形状 response shape
响应性 responsibility;responsiveness;responsivity
响应性投标 responsive bid
响应因数 response factor
响应因子 responding factor;response factor
响应阈 threshold of response
响应者 respondent;responder
响应值 response value
响应指令 response instruction
响应滞后 response lag
响指示器 audible indicator
响质岩 phonolitoid

想

想象方式 imaginative type
想象力 imagination
想象实验 thought experiment
想象型 imaginative type
想象者 visualizer
想要投资的开发商 would-be developer

向

向……方向的交通 traffic bound for…
向岸 ashore
向岸冰压力 shoreward thrust of ice
向岸冰涌 ice shore
向岸大风 onshore gale;onshore wind
向岸的 onshore;shoreward
向岸的大量运输 shoreward mass transport
向岸风 inshore breeze;inshore wind;off-sea wind;onshore wind
向岸机车 pilot engine
向岸-离岸流 onshore-offshore current
向岸-离岸输沙 onshore-offshore transport
向岸流 indraft;inshore current;onshore current
向岸倾斜 tilt to coast
向岸上 ashore
向岸上抛泥 onshore disposal
向岸水流运动 indraft;inshore current;onshore current
向岸水团运动 shoreward mass transport
向岸推力 shoreward thrust
向岸拖的浅水围圈 beach seine
向岸涌浪 landing swell
向暗性 skototaxis
向保险公司索赔 claim against insurance company
向保险人索赔 claim against underwriter
向保证人提供的担保 counter guarantee
向北的 northward
向北海岸 upcoast
向北航行 northbound
向北开的 northbound
向北面海岸的 <美> upcoast
向北驶 northing
向北驶的 northbound
向岔线送车 placing of cars at sidings
向厂商直接进货的零售商 desk jobber;drop shipper
向场透镜 field lens
向车内 aboard
向承包商下达指令 instructions to contractors
向承运人索赔 claim against carrier
向赤道方向航行 depress the pole
向触性 thigmotropism
向船外 overboard
向船外排水 overboard discharge
向船尾 abaft;astern
向磁极性 verticity
向磁性 magnetropism
向大地排放污水或处理过的污水 land disposal
向大气排放 atmospheric exhaust
向大西洋方向迁移 migration toward Atlantic
向导 cicerone;pilot;usher
向导标 marker post
向导车 pilot car
向导机车 pilot engine
向导信号 go ahead
向导压力 pilot pressure
向低频调谐 tuning downward
向地根 geotropic root
向地球方向 earthward
向地球移动的 earthbound
向地下岩层注水 water infusion
向地下运送垃圾装置 crib
向地性 geotropism
向第三方索赔 claim compensation from a third party
向第三债务人下达扣押令 garnish;garnishee
向第三者租用 third party leasing
向电性 electrotropism
向定义链 forward definition chain
向东 eastward
向东北 northeastward
向东方 due east;orientation
向东航程 easting
向东航行 easting
向东南突出边缘弧 southeast facing marginal arc
向东(行)驶的 eastbound
向盾尾形成的空隙压注砾石 <盾构施工> gravel(l)ing
向法院起诉 access to court
向法院申诉权利 access to court
向反行车信号法 against current-of-traffic signal(l)ing
向风 aweather;fetch to
向风岸 windward bank;lee shore;weather shore
向风的 windward
向风海岸 windward coast
向风桁架 windward truss
向风面 exposed side;stoss side;windward face;windward side;windy side
向风漂流 windward drift
向辐射性 radiotropism
向个人直接销售 direct marketing by personal solicitation
向各处 hither and thither
向公众公开拍卖出售的财产 public sale
向顾客提供信息 customer communication
向光的 lucipetal

X

向光性 phototropism
向光性的 phototropic
向光移动技术 photomigration technique
向光运动 phototropic movement
向国外托收出口票据 foreign collections
向海岸的 coastward
向海岸上的 upcoast
向海边 down the country
向海边的 downcountry
向海的 oceanward;offshore;seaward
向海的一面 sea side
向海方向 offward;seaward
向海防波堤 seaward breakwater
向海风 land wind;offshore wind
向海港池 seaward basin
向海关申报出口 enter out
向海关申报进口 enter in
向海关申请出口或进口 apply to the customs
向海航行 sea going;stand to sea
向海洄游 seaward migration
向海疆界 seaward boundary
向海界限 seaward boundary
向海陆坡 sea slope
向海坡面 seaward slope
向海倾斜 tilt to sea
向海区 offshore area
向海三角洲前积模式 seaward deltaic pregradational mode
向海上 outside;seaward
向海滩输送的离岸带沉积物 backpassing
向海凸出 seaward-convex
向海推进岸线 prograding shoreline
向海推进法 end-on system;over-end system
向海一边 outshore
向海一面 seaward side
向河口 downstream;down the country
向河口处 downriver
向河口的 downcountry;downriver
向河坡 riverward slope
向红(基)团 bathochrome
向红效应 bathochromatic effect;bathochromic effect
向红移 bathochromatic shift;bathochromic shift
向后 backward;rearward
向后差分算子 backward difference operator
向后串视 back range
向后的 astern;backward
向后调车信号 shunt-back signal
向后方的 rearward
向后焊 backhand welding
向后合并 backward integration
向后滑行 reverse coast
向后划桨 backward stroke
向后进位 carry back
向后连接 backward link
向后连锁率 backward linkage
向后连锁效应 backward linkage effect
向后内插公式 backward interpolation formula
向后拍摄航摄仪 rear-shooting camera
向后气流 slip stream
向后牵引滑车轮 backhaul sheave
向后倾倒 backfall
向后倾翻的位置 tilt-back angle
向后倾覆稳定性 <起重机> backward stability
向后倾斜 rake aft;recede
向后倾斜供给曲线 backward-sloping supply curve
向后倾卸 rear-dumping
向后散射 back-scattering;backward scatter
向后散射体 back-scatterer
向后收起 retract rearward
向后衰竭 backward failure
向后推进 backrunning
向后弯 recurve
向后弯的 recurvate
向后弯曲 curved backward
向后弯曲供给曲线 backward bending supply curve
向后弯叶片通风机 backward tip fan
向后卸料刮土机 backward-dumping (wheel) scraper
向后延伸的 rearward extend
向后引用 backward reference
向后游动 walk back
向后运动 astern motion;setback
向后折叠 fold back
向后整合 backward integration
向后转 about-face;right-about;right-about-face
向后追踪技术 back trace technique
向后座 rear-facing seat
向湖心倾斜 tilt to lake center[centre]
向化性 chem(i)otaxis
向回拉缆索的导向滑轮 tailblock
向基底部 basad
向基底的 based
向近郊 uptown
向径 vector;radius vector
向空(安全)泄放阀 atmospheric relief valve
向蓝移的 hypsochromic
向/离岸泥沙运动 onshore-offshore sediment movement
向/离岸漂沙 on-off shore drift
向里 inward;pullback
向里偏距 track in
向里弯曲 turn(ing)-in
向里斜展 <窗框侧壁,平面成八字形> flu(e)ing
向两侧移动 sideshift
向量 vector;vector quantity
向量比 vector ratio
向量变换 vector change
向量变量 vector variable
向量表示 vector representation
向量不变式 vector inequality;vector variant
向量测度 vector measure
向量产生 line generation;vector generation
向量产生器 vector generator
向量长(度) length of vector;vector length
向量场 vector field;vectorial field
向量场位 potential of vector field
向量处理 vector processing
向量磁位 magnetic vector potential
向量丛【数】vector bundle
向量代数 vector algebra
向量代谢 vectorial metabolism
向量的 vectorial
向量的分解 resolution of a vector
向量的合成 composition of vectors
向量的绝对值 absolute value of a vector;magnitude of a vector
向量电位 electric(al) vector potential
向量多边形 vector polygon
向量范数 norm of vector
向量方程 vector equation
向量方式 vector mode
向量分布 vector distribution
向量分量 component of vector
向量分析 vector analysis
向量格的根 radical of a vector lattice
向量公式 vector equation
向量过程 vector process
向量函数 vector function
向量和 sum of vectors;vector sum
向量化 vectorization
向量环 vector loop
向量绘图法 vector drawing method
向量积 cross product;outer product;vector product
向量计算机 vector computer
向量计算器 vector calculator
向量寄存器 vector registor
向量加法 addition of vectors;vector addition;vectorial addition
向量减法 vector subtraction
向量交换律 commutative law of vector
向量角 vector angle;vectorial angle
向量径 radius of vector;vector radius
向量矩 moment of a vector;vector moment
向量矩阵 vector matrix
向量空间 vector space
向量空间的范畴 category of vector space
向量空间的维数 dimension of a vector space
向量控制 vector control
向量马尔可夫过程 vector Markov processes
向量描述符 vector descriptor
向量瞄准器 vector gunsight
向量模 norm of vector
向量平差 vectorial adjustment
向量屏蔽寄存器 vector mask register
向量区写入封锁 write lockout
向量群 vector group
向量势 vector potential
向量算符 vector operator
向量算子 vector operator
向量投影图 vector projection
向量图 arrow diagram;hodograph;vectogram;vectograph;vector diagram;vector drawing
向量图形学 vector graphics
向量微积分 vector calculus
向量位 vector potential
向量误差 vectorial error
向量显示器 vectorscope
向量相关系数 vector correlation coefficient
向量相加 addition of vectors;vectorial addition
向量心磁计 vector magnetocardiograph
向量性 vector-oriented
向量旋度 curl of a vector
向量循环 vector loop
向量演算 vector calculus
向量元 element vector
向量运输 vectorial transport
向量运算 vector arithmetic;vector operation
向量运算方式 vector mode
向量正交分解(法) orthogonal decomposition of vector
向量值函数 vector-valued function
向量值函数的散度 divergence of a vector-valued function
向量中断 vectored interrupt
向量子空间 vector subspace
向列化合物 nematic compound
向列结构 nematic structure
向列态 nematic state
向列调整 nematic alignment
向列相 nematic;nematic phase
向列型 nematic
向列型结构 nematic structure
向列型液晶 nematic liquid crystal
向列型中间相 nematic mesophase
向流面 stoss face;stoss side
向流性 rheotropism
向陆风 landward wind
向陆突出 landward-convex
向某点集中 funnel
向南 southward
向南的 austral;meridional;southern
向南海岸 <美> downcoast
向南航行 south bound;southing
向内 entad
向内拨号 inward dial(l)ing
向内侧移动 shift inboard
向内筹借资金 funds secured from internal sources
向内的 inward
向内开(窗) inward opening
向内开的窗扇 in-swinging casement
向内开启的侧悬窗 side-hung casement window opening inwards
向内扩散 indiffusion
向内流 indraft
向内平开窗 <门窗的> in ward-swinging
向内倾斜 tumbling-in
向内倾斜的 inward tipping
向内渗漏 influent seepage;inward seepage
向内渗透 endosmosis
向内生长 ingrowth
向内式城市 intropolis
向内凸胀 belling in;bellying in
向内突入 intrusion
向内挖 cut into
向内弯曲 incurvation;inflect;inflexion;introflection;introflexion;necked-in <边缘>
向内下 intero-inferiorly
向内行驶 inward bound
向内褶曲 folding inward
向内转动 inboard turning
向你方买进 book with you
向农田地下渗透污水 land filtration
向旁边 sidewise
向盆地一侧下落的断层 down-to-basin
向平面外的 out-of-plane
向气性 aerotropism
向签票人本人支付 pay to self
向前边 forward edge
向前边锁率 forward linkage
向前插值 forward interpolation
向前差分法 forward difference method
向前冲程 forward stroke;head reach
向前出口的混凝土搅拌机 mud slinger
向前的 forward
向前电压 forward voltage
向前调车臂板 draw-ahead arm
向前调车信号 draw-ahead signal;shunt-ahead signal
向前调车信号机【铁】draw-ahead signal
向前方位 forward bearing
向前焊(接) forehand welding;forward welding
向前合并 forward integration
向前滑行 forward coast;forward skidding
向前绞 haul forward
向前进 make headway
向前看 look ahead
向前看的 forward looking
向前连锁反应 forward linkage effect
向前连系效果 forward linkage effect
向前流动 forward flow
向前孪生指示符 forward twin pointer
向前迈进 forge ahead
向前猛冲的 plunging
向前内插 forward interpolation
向前跑 forward run
向前倾倒 forward tipping
向前倾斜的 forward tipping
向前燃烧 forward burning
向前散射 forward scatter
向前散射角 forward-scattering angle

向前散射粒子计数管 forward-angle counter
向前推进 push ahead
向前弯叶片通风机 forward-tip fan
向前微分法 forward difference method
向前行驶 forward travel
向前移动 make way
向前引用 forward reference
向前运动 forward motion;travel forward
向前运行 ahead running;forward running
向前整合 forward integration
向前直滑 straight-ahead skid
向曲轴箱加满油 refilling crankcase
向全换算 interconversion of directions
向热气流发射试验 hot shot
向热性 thermotaxis;thermotropism
向日葵 Helianthus annulus;oonopsis sop;sunflower
向日葵油 sunflower(seed)oil
向日性 heliotropism
向日仪 sun seeker
向色性 chromotropism;chromotropy
向上 turn up;uprising
向上凹的 concave-up(ward)
向上奔流 uprush
向上变薄层序 thinning-upward sequence
向上变粗层序 coarsening-upward sequence
向上变动性 upward mobile
向上变厚层序 thickening-upward sequence
向上变换器 up-converter
向上变细层序 fining-upward sequence
向上冲程 upstroke
向上抽风 updraft
向上抽风窑 updraft kiln
向上出料落料模 return blanking die
向上传布(扩展)的裂缝 crackings propagating upwards
向上垂准 plumb up
向上打井 shaft raising
向上打扇型眼的钻车 fan jumbo
向上打眼 uphole drilling
向上打眼凿岩机 overhead driller
向上大气辐射 upward atmospheric radiation
向上挡阻的机械装置 rising warded mechanism
向上的 anadromous;antrorse;ascending;skyward;upward
向上的坡面 uphill side
向上的斜坡 uphill slope
向上地球辐射 upward terrestrial radiation
向上地形 positive landform
向上顶出坯料 push-back blank
向上顶蚀作用 overhead stopping
向上动作 up movement
向上断层 upcast fault
向上对流式 overhead convection type
向上对流式管式炉 overhead convection type of pipe still
向上翻的 upturned
向上翻的梁 upturned beam
向上翻转 tip up wards
向上翻转竖立(施工)方法 tilt-up method
向上反作用力 upward reaction
向上方 upward;upwards
向上分类法 ascending system
向上分量 upward component
向上分支 ascending branch
向上风 aweather;windward
向上风航行 in the wind
向上浮起的 assurgent
向上根 standing root
向上弓【地】upper bend
向上攻击 upward attack
向上供送系统 upfeed distribution
向上拱 upper bend
向上拱曲的 hog-backed
向上焊接 uphill welding;upward welding
向上横流式通风 upward transverse ventilating
向上横向式系统 upward transverse system
向上回采矿石法 overhead stopping
向上及向下比敦器 upward and downward comparator
向上集中器 hyperconcentrator
向上兼容的 upward compatible
向上兼容性 upward compatibility
向上静水压力 upward hydrostatic(al)pressure
向上静液压 upward hydrostatic(al)pressure
向上卷 acock
向上掘进 stopping
向上开采 upward mining
向上开的门 upward-acting door
向上开挖 raising;stooping-down roof <隧道的>
向上开挖施工法 raising
向上扩散云室 upward diffusion cloud chamber
向上拉索 <斜架车或缆车的> uphaul cable
向上累积 cumulated upward;upward cumulation
向上力 uplifting force;upward force
向上立焊 upward welding in the vertical position;vertical up welding
向上连续积分 upward continuation integral
向上流 upwell
向上流动 upflow;upward flow
向上挠曲 upward deflection;upwarp(ing)
向上排代 upward displacement
向上排风罩 upward plenum
向上排气 air updraft;overhead exhaust;updraft;up-draught
向上偏斜 upward bias
向上偏转 upward deflexion
向上坡度 uphill grade
向上铺衬 under-drawn
向上气流 updraft stream
向上牵引链条 inhaul chain
向上翘曲 upwarp deflection;upwarping
向上切割 high cut
向上倾斜 acclivity;up-dip
向上倾斜的 acclivitous
向上倾斜的炮眼 uphole
向上倾斜的弯道 <从内侧至外侧,以保障行车安全> banked bend
向上趋势 up-trend
向上趋势分析 upward trend analysis
向上全辐射 upward total radiation
向上渗透 upward percolation;upward seepage
向上升起的 assurgent
向上式侧向凿岩机 offset stoper
向上式凿岩机 stoper;stoper drill;stoping drill;stopper hammer
向上式凿岩机支架 stoper leg
向上水流反滤层 upward flow filter
向上水压力 uplift;upward hydraulic pressure
向上送风口 upward air outlet
向上抬起 <船受浪推动时的> scend
向上替代影响 upward substitution effect
向上调整 upward adjustment
向上通风 updraft;upward draft;upward ventilation
向上通风炉 updraft furnace
向上通风竖井 up-shaft
向上通风筒 upcast ventilator
向上通风系统 upward system of ventilation;upward ventilation system
向上投 upthrow
向上突出 upthrust
向上推进 raising feed
向上推力 high thrust
向上弯 bending up
向上弯梁 <打桩架底下的> moonbeam
向上弯曲 camber(ing);hogging bending;kick-up;turn up
向上弯曲的构件 camber piece
向上弯曲度 convexity
向上弯曲钢筋 bent-up bar
向上弯曲缆索 bending up cables
向上相容性 upward compatibility
向上削土 upward soil cutting
向上斜的 acclivitous;acclivous
向上斜坡 acclivity
向上行驰的坡面 uphill
向上行动 up movement
向上旋开的 swing up
向上压力 buoyancy pressure;uplift(ing)pressure;upward pressure
向上延拓 upward continuation
向上延续法 upward continuation integral
向上移动 move-up
向上引用 upward reference
向上游的 downriver;upcurrent;upstream
向上游洄游 upstream migrating;upstream migrate
向上游倾斜的 slanted upstream
向上运的运输机 upward conveyer [conveyor]
向上运动 up movement;upward movement
向上运输 upward transport
向上凿井 shaft raising
向上增长 <沉积物的> vertical accretion
向上折算 reduction ascending
向上褶皱 upfold
向上帚形 upward broom
向上转换型磷光体 up-conversion phosphor
向上转移符号 shift out character
向上转移频率 roll-off frequency
向上钻进 uphole drilling;upward drilling
向上钻孔 uphole;upward borehole;upward drill-hole;upward-pointing hole
向上钻孔侵蚀 upward drilling
向上作用的负载 upload
向上作用的荷载 upload
向神供品 <古希腊> agalma
向渗性 osmotropism
向声性 phonotropism
向湿性 hygrotropism
向实体性 stereotaxis;stereotropism
向市中心区的交通 downtown terminal traffic
向收货人交付货物 delivery of freight to consignee
向首加速度 cephalad acceleration
向水性 hydrotaxis;hydrotropism
向四方倾斜的褶皱【地】quaquaversal fold
向太平洋方向迁移 migration toward pacific
向天空的 skyward
向外 turnout
向外长波辐射 outgoing long-wave radiation
向外的 outward
向外发展政策 outward-looking development policy
向外翻的 upturned
向外浮运 floating-out
向外鼓 outward bulging
向外拐弯 outcurve
向外回旋门 out-opening door
向外交通 outbound traffic
向外开出 outward-bound
向外开出的路 outward-bound road
向外开的门 upward-acting door
向外开的门窗 out-opening
向外开启的侧悬窗 side-hung casement window opening outwards
向外扩散 outdiffusion
向外扩伸 flare
向外流 outset
向外流涡轮机 outward flow turbine
向外排放 out-emission
向外排水 pump out
向外翘曲 buckle outward
向外渗出 outward seepage
向外疏散 decentralization
向外挑出 corbel out
向外凸出 to buckle outward
向外凸的半个型板 oddside pattern
向外凸胀 belling out;bellying out
向外突出 upthrust
向外推力 outward thrust
向外弯出的 pillowed
向外弯曲 necked out;to buckle outward;turnout
向外弯曲边 necked out
向外弯曲的 bandy
向外张开 flare
向外张开的 splay
向外张开的双导流堤 divergent jetties
向外周 peripherad
向尾部的 astern
向尾部倾斜的长坡度汽车顶 fast-back
向尾加速度 caudad acceleration
向位 bearing
向位换算 conversion of course or bearing
向位圈 chart compass;compass rose
向温性 thermotropism
向西的 toward west;westward;westering
向西航行 westbound;westing
向西南突出边缘弧 southwest facing marginal arc
向西坐标 westing
向下 downturn;regressive
向下凹陷 undercut
向下变换器 down converter
向下并合 downstair(case)merger
向下齿轮系 drop train gear
向下冲程 down stroke
向下冲击式凿岩机 hammer sinker
向下冲刷 downward erosion;downward scour;vertical erosion
向下抽风 down draught
向下抽风间歇式砖窑 down-draft intermittent brick kiln;down-draught intermittent brick kiln
向下穿孔 down drilling
向下传播的裂缝 crackings propagating downwards
向下传动装置 drop train gear
向下垂准 plumb down
向下打竖井 shaft sinking
向下打眼钻爆法 flat cut
向下的 decurrent;descending;downcast;underhand;downward
向下的重量 downward load
向下的走向 downside
向下调用 downward call
向下顶替 backtrack

X

向下断层 downcast fault; down-slip fault
向下发送 down-line load; down-load
向下翻倒 tip downwards
向下方向 downdip
向下风 alee; leeward (ly); make lee way
向下风横漂 crab drift
向下风化 downward weathering
向下风面漂移【船】sag
向下风偏转 fall to leeward
向下风漂流 sag to leeward
向下辐射 downward radiation
向下弓形弯曲 downward bowing
向下给料系统 down-feed system
向下兼容的 downward compatible
向下兼容性 downward compatibility
向下开挖 downward drifting
向下拉 pulldown; down-haul
向下拉缆索 < 斜架车或缆车的 > downhaul cable
向下累积 cumulated downward
向下力 downward force
向下立焊 downward welding in the vertical position; vertical down welding
向下立焊条 electrode for vertical down welding
向下流的 defluent
向下流动 devolution
向下流动的液体 descending liquid
向下流推移的沙波 progressive sand wave
向下挠曲 downward deflection
向下派生器 infrageneralizer
向下炮孔 downward hole
向下炮眼 plugger hole
向下喷射燃烧 downshot firing
向下偏斜 downward bias
向下偏倚 biased downward; downward bias
向下气流 down draft; down draught
向下翘曲 downsagging
向下侵蚀 deepening; downcut (ting); downcutting erosion; vertical erosion
向下倾斜 downward sloping
向下倾斜的 declivitous
向下倾斜交叉 < 道路的 > dip crossing
向下倾卸 tip downwards
向下全辐射 downward total radiation
向下式凿岩机 sinker
向下输送 downward conveying; down wash
向下输送机 downward conveyer [conveyor]; decline conveyer
向下水流 downward flow
向下调制 downward modulation
向下通风 down draft
向下通风发生器 down draught producer
向下通风系统 downward system of ventilation; downward ventilation system
向下凸出 downward bulge
向下挖 cut down
向下弯的 decurved; downcurved
向下弯曲 down buckling; downward deflection; downwarping
向下压力 down pressure; downward pressure
向下压强 downward pressure
向下延拓 downward continuation
向下移位 infraplacement
向下涌出 downwelling
向下游 downriver; downstream; downdrift
向下游漂流 downdrift
向下游漂移 downdrift
向下游推进沙波 progressive sand wave
向下运动 downstream; downward movement
向下凿岩 downward drilling
向下凿岩机 downhole driller
向下照光 downlight
向下折算 reduction descending
向下蒸发 inverted evapo (u) ration
向下支管 drop pipe
向下帚形 downward broom
向下转形 apostrophe
向下转移符号 shift in character
向下锥型 tapered down
向下钻进 put down a borehole; subdrilling
向下钻孔 downhole drilling; downward drilling
向下钻眼 sunk
向下钻眼机 downhole driller
向相反方向地 criss-cross
向斜【地】down fold; syncline
向斜坳陷湖 synclinal down-warped lake
向斜闭合 canoe fold
向斜闭合度 synclinal closure
向斜槽【地】down fold; trough of syncline; mulde; trough of a syncline
向斜层【地】swally; syncline; mulde
向斜储水构造 storage structure of syncline
向斜大陆海 synclinal continental sea
向斜到背斜的过渡区 transitional area from syncline to anticline
向斜的向心倾伏末端 plunging end area of syncline
向斜底线 trough line
向斜地层 synclinal formation; synclinal stratum
向斜断层 synclinal fault
向斜谷 synclinal valley; valley of subsidence
向斜含水带 water-bearing zone of syncline
向斜河 (流) synclinal river; synclinal stream
向斜泉 synclinal spring
向斜山 synclinal mountain; syncline mountain
向斜山脊 synclinal ridge
向斜枢 (纽) hinge of syncline; synclinal hinge
向斜水系 insequent drainage (system)
向斜隧道 syncline tunnel
向斜翼 trough limb; synclinal limb
向斜油藏 synclinal oil pool
向斜油气田 synclinal oil-gas field
向斜褶皱 synclinal fold
向斜中心 cener [centre] of syncline; core of syncline; trough core
向斜轴 axis of syncline; synclinal axis; trough axis
向斜轴向河 axial stream
向斜状抬升高原 moela
向心 centrocline
向心爆炸 implosion
向心泵 centripetal pump
向心的 centripetal
向心断层 centripetal fault
向心辐流式水轮机 radial inflow wheel
向心辐内流式水轮机 radial inflow turbine
向心滚子轴承 radial roller bearing
向心化 centralization
向心加速度 acceleration centripetal; centripetal acceleration
向心交代【地】centripetal replacement
向心交通 centripetal movement
向心结构 centric texture
向心进汽 radial-inward admission
向心径流叶轮 < 液力耦合器 > inflow wheel
向心卡车 radial truck
向心力 centripetal force
向心面 inside surface
向心排水 centripetal drain (age)
向心迁徙 centripetal migration
向心倾角 centroclinal dip
向心倾斜 centripetal tendency; centroclinal dip
向心球轴承 annular ball bearing; radial ball bearing
向心神经 centripetal nerve
向心市场 centripetal market
向心式 radial inflow
向心式的 radial inward
向心式燃气轮机 radial-inward flow gas turbine
向心式水泵 centripetal pump
向心式水轮机 centripetal turbine; inward-flow turbine
向心式透平 inward flow radial turbine
向心式涡轮 inward flow turbine
向心式涡轮机 radial inflow turbine; radial-inward (flow) turbine
向心式压气机 radial inflow compressor
向心式压缩机 radial inflow compressor
向心式叶轮 < 液力传动的 > inflow wheel
向心收缩 concentric (al) contraction
向心水系 centripetal drainage
向心推力滚动轴承 radial-thrust bearing
向心推力球轴承 angular (contact) ball bearing; radial-thrust ball bearing
向心型 centric pattern; centripetal pattern
向心性 centrality
向心性缩小 concentric (al) contraction
向心选择 centroclinal selection
向心叶轮 centripetal wheel
向心移动 centripetal movement
向心圆锥滚子轴承 annular tapered roller bearing
向心展开 centripetal development
向心褶皱 centroclinal fold
向心止推滚动轴承 angular contact bearing
向心止推滚珠轴承 angular contact ball bearing
向心止推轴承 angular bearing; radial-thrust bearing
向心轴承 radial bearing; transverse bearing
向心轴颈轴承 radial journal bearing
向心作用 centripetal action
向新灌灰缝中填碎石 gal (l) et
向形 synform
向形背斜 synformal anticline
向形的 synformal
向形盆地 synformal basin
向性 tropism
向性弯曲 tropic (al) curvature
向压性 barotropism
向阳病房 sun parlo (u) r
向阳侧 day side
向阳间 sun parlo (u) r
向阳客厅 sun parlo (u) r
向阳性 heliotropism
向阳植物 heliotrope
向洋迁移 migration in direction of ocean
向氧性 aerotropism; oxygenotropism
向翼 guide vane
向银行提款 make a draft on a bank
向银行透支 bank overdraft; overdraft from the national bank
向右 < 手势 > guide right
向右侧 dextrad
向右的 dexiotropic; dexiotropous; rightward
向右对齐 right (-hand) justify
向右焊 backward welding
向右横移 swing to starboard
向右缓转 easy starboard; starboard easy
向右倾斜的 negatively skewed
向右舷 astarboard
向右旋运动 right-lateral motion
向右旋转 dextrorotation
向右旋转的 dextrorotary; dextrorotatory
向右移位开启 shift right open
向右转 deasil; rotation to the right
向右转舵 turn to starboard
向源冲刷 headwater erosion; regressive erosion
向源的 retrogressive
向源底层流 upstream bottom current
向源堆积 retrogressive accumulation aggradation
向源滑坡 retrogressive slide
向源切割 head cut; headcutting
向源侵蚀 backward erosion; headward erosion; headwater erosion; regressive erosion; retrogressive erosion
向远侧 distad
向月潮 superior tide
向月球发射 moon shot
向运送人交货条件 free carrier
向责任第三方追偿 recourse against the responsible third party
向震性 seismotropism
向震中 kataseism
向震中的 kataseismic
向正东 due east
向枝松 creeping pine
向指数 zonal index
向中的 centripetal; mesial
向中端 mediad end
向中路遮断 afferent blockage; afferent hindrance
向中线 medially
向中心倾斜 centrocline
向中央延伸的 medial
向终点速度 velocity towards destination
向周围 peripherad
向轴的 adaxial
向轴猪毛菜 solsola nitraria
向着湖泊的 lakeward
向着两端地 endway; endwise
向紫增色基 hypsochrome
向自性 autotropism
向左 < 手势 > guide lift
向左侧运动 left-lateral motion
向左的 leftward
向左对齐 left justify
向左焊 forehand welding
向左横移 swing to port
向左抢风调向 port tack
向左舷 aport
向左舷的 larboard
向左旋转 left-handed
向左移 (位) left shift
向左/右转舵 put the rudder to port/starboard
向左转舵 turn to port
向左转向 alter course to port

巷道 alley way; heading; intake; room neck

巷道壁 wall

巷道边墙 gallery wall
巷道底【矿】floor
巷道顶板 tunnel roof
巷道堆垛机 aisle stacker
巷道堆垛起重机 aisle stacking crane
巷道盖顶 roof
巷道高度 headroom
巷道工作面 roadhead
巷道贯通 holing
巷道横梁 stemple
巷道横木 stemple
巷道进尺成本 footage cost of tunneling
巷道掘进 drivage;heading advance
巷道掘进测量 adit digging survey
巷道掘进工程 head driving
巷道掘进机 boring machine; roadheader
巷道掘进通风 heading ventilation
巷道掘通 work through;thirling <隧道的>
巷道开拓 development drifting
巷道口 access adit;heading collar
巷道两侧背板 side lagging
巷道面通风 heading ventilation
巷道排水 drainage by drift
巷道刷帮 flitching
巷道梯段掘进法 heading and bench method
巷道梯段开挖法 heading and bench method
巷道突水 bursting water in mining tunnel
巷道托座 way bracket
巷道位置 opening location
巷道形式 type of heading
巷道腰线 grade line
巷道隅撑 way bracket
巷道支护费用 support cost of tunnel
巷道支架 excavation support
巷道纵剖面图 drift profile
巷道钻孔结合疏干 dewatering with gallery and borehole
巷顶钻机 ceiling drilling machine
巷计数 lane count
巷空微分能谱测量 aerodifferential spectrum survey
巷门 lane gate
巷识别 lane identification

项部 napex

项次号 item number
项的大小 item size
项迭代因子 iteration factor
项链式麦克风 Lavalier microphone
项码 item code
项目 article;entry;project;subject of entry;item
项目办公室 project office
项目报告 project paper; project report
项目备选方案 project alternative
项目边界 project boundaries
项目编制阶段 project preparation phase
项目标记 project mark
项目表 chart of accounts
项目财务估价 financial evaluation of projects; project financial evaluation
项目财务评价 financial evaluation of projects; project-financing evaluation
项目采购 project purchasing
项目策划 job design
项目参数 project parameter
项目成本 item cost;project cost
项目程序 project procedure
项目筹备融通资金 project preparation facility
项目储备 backlog
项目处理 item processing
项目代表 project representative
项目贷款 loan of project; project finance; project-financing loan; project lending;project loan
项目单 menu
项目单区 menu area
项目的参与方 project participant
项目的核准 authorization for project;project approval
项目的技术标准 project specification
项目的列入 inclusion of items
项目的拟订 project formulation
项目的批准 authorization for project;project approval
项目的任务及限界 project definition
项目的审定 authorization for project;project approval
项目的使用年限 useful life of the project
项目的所有者 owner of the project
项目的属性 attribute of items
项目的总评价 overall appraisal of a project
项目的总投标价值 total tender value of project
项目定期贷款 term loan
项目发起人 project promoter;project sponsor
项目发展周期 project development cycle
项目法 item method;project method
项目法人责任制 project corporation responsibility system
项目范围 scope of project
项目方案 project alternative
项目费用 project cost
项目分隔符 item separation symbol
项目分类 classification of items;project classification
项目分配 allocation of items
项目分析 item analysis;project analysis
项目分析研究 job analysis;job study
项目分项分析 job breakdown
项目服务面积 catchment area
项目负责人 principal-in-charge;principal investigator;project director; project manager
项目概算 project estimate
项目工程 <建设项目统盘实施的技艺,包括美学、功能、技术、经济和各项有关问题的解决> project engineering
项目工程师 project engineer
项目工程师负责制 project engineer system
项目工程师制度 project engineer system
项目工作计划完成率 planned performance rate of project work
项目工作组 project group; project team
项目估价 item quotation
项目估算师 project estimator
项目管理 project control; project management
项目管理承包合同 management contract
项目管理承包商 management contractor
项目管理工具 project management tool
项目管理人 project manager
项目管理问题 project management problem
项目管理组织 project organization
项目规划 project planning
项目规划编制阶段 project preparation phase
项目规模 project size;item size
项目号 item number;project number
项目核对法 check list
项目核准权 approval authority for projects;projects approval authority
项目后评估 post project evaluation
项目后评价 post project evaluation
项目后评价和后继行动 post project evaluation and follow-up
项目化组织 projectized organization
项目划分 segregation of projects;segregation of items
项目环境改善 project environmental improvement
项目环境评价 environment assessment of project
项目货款额度 project line
项目级短期路面性能监测 short-term project monitoring
项目级模型 site-specific adaptive model
项目计划 project plan(ning);project programming
项目计划编制 project planning
项目计划核准 project approval
项目计划设计图 project planning chart
项目记录 item record
项目技术评价 project technology evaluation
项目加权 item weighting
项目间接费用 indirect expenses of project
项目间接效益 indirect benefit of project
项目监督人 project superintendent; project supervisor
项目监控 project monitoring
项目监理 project supervision
项目检验回归曲线 item-test regression curve
项目简介 project brief;project introduction
项目建设费用 cost of project implementation
项目建设进度表 project implementation schedule; project implementation scheduling
项目建议书 project proposal(report); proposals for the projects;proposed task of project
项目建筑师 project architect
项目鉴定 project appraisal;project identification
项目鉴定标准 standard of project appraisal
项目进度计划师 project scheduler
项目进行情况 operation of a project
项目经济估价 economic evaluation of a project;project economic evaluation
项目经理 project manager
项目经理负责制 manager responsibility system(of a key project under construction)
项目经理违约 default by project manager
项目经营 project management
项目矩阵组织 project matrix organization
项目竣工 completion of project;project completion
项目开发招标 inviting bids of project development
项目开工 kick-off of a project;project commencement; start-up of a project
项目可行性分析 project feasibility analysis
项目控制 project control
项目块 entry block
项目历史 history of project
项目利润评价 project cost-benefit assessments
项目链 item chain
项目论证 project justification
项目名 item name
项目名称 name of item;project title
项目目标管理 project management by objectives
项目拟订 project formulation
项目年限 project age
项目批准 project approval
项目匹配法 matching item
项目评估 project appraisal;project evaluation
项目评估法 project evaluation and review technique
项目评估及复核技术 project evaluation and review technique
项目评估及复审计划进度表 project evaluation and review technique schedule
项目评估技术 technique for project evaluation
项目评估框架 assessment framework
项目评审技术 project evaluation and review technique
项目企业经济评价 enterprise's economic appraisal of a project
项目区 project area
项目区灌溉需水量 project irrigation requirement
项目全部记录 project record documents
项目全部完成 project closeout
项目全面管理 overall project management
项目确定 project identification
项目人员 project staff
项目认定 project identification
项目日志 project diary
项目融资 project finance
项目删除 deletion of items
项目上端 item top
项目设备 project equipment
项目设计 item design;project design; project engineering
项目设计工程师负责制 project engineer responsibility system
项目设计师 project designer
项目设计文件 project design documents
项目设计者 project designer
项目设想 project conception
项目社会评价 project's social value appraisal
项目审查 project appraisal
项目审批制 the system of examination and approval of projects
项目生产 project production
项目生态环境评价 environmental appraisal of a project
项目识别附注 project identification notes
项目识别号 item identification number
项目实施 project execution; project implementation
项目实施表 project implementation schedule
项目实施贷款 seed money loans
项目实施单位 project implementation agency
项目实施费用 seed money
项目实施计划 project scheduling; plan of works

项目寿命 project life
项目数据库 project database
项目说明 description of items; item description; job description
项目说明书 specification of an item; term of reference
项目特定技术标准 job specification
项目投资评估 investment appraisal
项目投资评价 project investment evaluation
项目投资(时)期 investment phase of a project
项目图与布置图 project charts and layouts
项目完成报告 completion report; project completion report
项目完成并移交给业主后或者工程弃置后的意外保险 completed operations insurance
项目完成后审核 post evaluation
项目网络法 project network technique
项目网络分析 project network analysis
项目网络计划分析 project network analysis
项目网址 project web site
项目文件 item file
项目细账 job account
项目小组 project team organization
项目协调 project coordination
项目信息 project information
项目行 item line
项目型援助 project type assistance
项目修理通知单 item repairing order
项目选定 project identification
项目选择 project selection
项目训练(法) project training
项目研究编目 item study listing
项目研究小组 project study group; project study team
项目业绩 project performance
项目业务 activities of project
项目一次采购法 project purchasing
项目一览表 itemized schedule
项目预算 program(me) budget; project budget
项目预算编制 program(me) budgeting
项目预算方案 project budget requirement
项目暂停 project suspension
项目政治评价 political appraisal of a project
项目支助业务 project supporting services
项目执行 project implementation
项目执行评估报告 project performance assessment report
项目执行情况 project performance
项目执行情况审计报告 project performance audit report
项目指标 project indicator
项目质量考核 project quality test
项目中的经济效益 financing package
项目周期 project cycle
项目助理员 project assistant
项目转让 project assignment
项目转移 item advance
项目准备 preparation of project
项目准备融通资金 project preparation facility
项目准则 project criterion
项目资本金 project capital
项目资金筹措 project financing
项目资金的规划 planning of project finance
项目资金的制订 planning of project finance
项目资金回收率 project rate of return
项目资助业务 project supporting services
项目自然资源评价 appraisal of project from the view of natural resources
项目综合评价 overall appraisal of a project; project comprehensive evaluation
项目组成 item design; set-up of project
项目组成部分 project component
项数 number of items
项图 term diagram
项系数法 partial factor method
项相关矩阵 term-term matrix
项值 term value

相【地】facies

相背 end for end
相比 phase ratio
相比率 phase ratio
相变 change of phase; phase transformation; phase transition; facies change【地】
相变材料 phase change material
相变超塑成型 phase changing superplastic forming
相变成因 phase transformation genesis of earthquake
相变的级 order of phase transition
相变点 transformation point; transformation temperature
相变点测定器 critical point tester
相变范围 phase transformation range
相变化 phase change
相变换 phase conversion
相变换行制彩色电视 phase alternating line colo(u)r television
相变级数 order of phase transition
相变理论 phase change theory
相变量 phase variable
相变能量 phase transition energy
相变平衡条件 phase equilibrium condition
相变前沿 transition front
相变区 phase change zone
相变热 critical heat; heat of transformation
相变热阻 thermal resistance to phase transition
相变塑性 transformation plasticity
相变位错 transformation dislocation
相变温度 phase transition temperature; temperature transformation; transformation temperature
相变系数 phase change coefficient
相变型温度 phase inversion temperature
相变性 interconvertibility
相变异 phase variation
相变应力 transformation stress
相变诱发的超塑性 transformation induced super plasticity
相变诱起塑性 transformation induced plasticity
相变增韧 transformation toughening
相变增韧陶瓷 phase transformation toughened ceramic
相变增韧氧化锆 phase transformation toughened zirconia; transformation-toughened zirconia
相变制冷 refrigeration of phase change
相变滞后 hysteresis of transformation
相并 side-by-side
相波 phase wave
相补角 phase margin
相不平衡 phase unbalance
相参 coherent
相参波 coherent wave
相参光 coherent light
相参散射 coherent scattering
相参振荡器 coherent generator; coho
相册 album
相差 phase contrast; phase difference
相差保护 phase comparison protection
相差保护装置 phase comparison protection unit
相差衬托 phase contrast
相差电影摄影术 phase contrast cinematography
相差电子显微镜 phase contrast electronic microscope
相差高频电流保护 carrier current phase-differential protection
相差计 phase meter
相差继电器 phase comparison relay
相差接物镜 phase objective
相差聚光镜 phase contrast condenser
相差显微镜 phase contrast microscope
相差显微镜检查 phase contrast microscopy
相差显微术 phase contrast microscopy
相差载波保护装置 carrier current phase differential protection device
相差指示器 phase difference indicator
相长的 constructive
相长干涉 constructive interference
相衬 phase contrast
相衬定则 phase contrast rule
相衬显微镜 phase contrast microscope; phase microscope
相衬显微术 phase contrast microscopy
相称 due proportion; equipoise; proportionality; proportionment; symmetry
相称比例 scale of ratios
相称的 assorted; proportional; proportionate(d); proportioning; symmetric(al)
相称性 commensurability
相成分 phase constituent
相冲突 contradict
相重出版物 parallel publication
相重合 phase coincidence
相重数 multiplicity
相重项 multiplet
相重性 multiplicity
相带 facies zone; phase belt
相带宽度 belt span
相点 phase point; representation point; specific phase
相电流 phase current
相电压 phase-to-phase voltage; phase voltage; star voltage; Y-voltage <在星形连接中的>
相电压调整器 phase voltage regulator
相电阻 phase resistance
相叠区间 interval overlapping one another
相端子 phase terminal
相发电机 diphase
相方程式 phase equation
相分布 phase distribution
相分解 phase decomposition
相分界面 phase contacting area
相分离 phase disengagement; phase separation
相分离纺丝 phase separation spinning
相分离工艺 phase separation process
相分离互穿聚合物网络 phase separated interpenetrating polymer network
相分离级 phase splitting stage
相分离器 phase separator
相分离区 phase separation region
相分配法 phase partition
相分析 phase analysis
相幅鉴别器 phase amplitude discriminator
相负荷 phase load
相富集系数 phase enrichment coefficient
相共振 phase resonance
相故障保护 phase failure protection
相故障继电器 phase-fault relay
相轨道 phase trajectory
相过滤方法 phase filtering method
相函数 phase function
相间 phase-to-phase
相间边界 interphase boundary
相间边界结 interphase boundary junction
相间边界能 interphase boundary energy
相间变压器 interphase transformer
相间的 interphase
相间电感器 interphase reactor
相间电位 interphase potential
相间电压 interlinked tension; phase-to-phase voltage; voltage between phases
相间短路 phase fault
相间短路继电器 phase-fault relay
相间分界面 interphase interface
相间隔板 phase partition
相间绝缘 phase insulation
相间离开 alternate disjunction
相间连杆 interphase connecting rod
相间连接线 interphase connection
相间黏[粘]结力 interphase bond
相间耦合 interphase coupling
相间速断过流保护 interphase quick-break overcurrent protection; phase-to-phase quick-break overcurrent protection
相间缩聚 interphase polycondensation
相间相互作用 interphase interaction
相间信道干扰 alternate channel interference
相间作用 interaction phase; phase interaction
相角 phase angle
相角差 angular phase difference; phase angle difference
相角电压表 phase angle voltmeter
相角鉴别 phase angle discrimination
相角误差 phase angle error
相角指示器 phase angle indicator
相结构 phase structure
相结构文法 phase structure grammar
相界 phase boundary
相界电势 phase boundary potential
相界交联 phase boundary crosslinking
相界面 interphase; phase contacting area; phase interface
相界面传质 interphase mass transfer
相界面积 phase contact area
相界曲线 boundary curve
相界线 limiting line
相空间 gamma space; phase space
相空间光学 phase space optics
相空间积分 phase space integral
相空间因子 phase space factor
相空间元件 phase space cell
相控天线 phased antenna
相控天线阵 phased array
相控阵干扰机 phased array jammer
相控阵空间跟踪雷达 phased array space tracking radar
相控阵列雷达 phased array radar
相控阵三坐标雷达 phased-array 3-D radar
相控阵扫描 phase array scanning
相控阵式扇形扫描显像仪 phased ar-

X

ray sector scanner
相控阵天线 phased-array antenna
相控整流器 phase-controlled rectifier
相类 type of facies
相流失 phase bleed
相滤波器 phase filter
相律【物】phase rule
相律平衡 phase rule equilibrium
相逻辑 phase logic
相面积 phase area
相敏 phase sensitive
相敏参量地震计 phase sensitive parametric(al)seismometer
相敏触发器 phase sensitive trigger
相敏电路 phase sensitive circuit
相敏放大器 phase sensitive amplifier
相敏检波器 phase sensitive rectifier
相敏鉴流 phase depending on rectification
相敏解调器 phase sensitive demodulator
相敏调制器 phase sensitive modulator
相敏元件 phase sensitive element
相敏整流 phase depending on rectification
相敏整流器 phase sensitive detector; phase sensitive rectifier
相模式 facies model
相片 finished print
相片比例尺 photograph scale
相片参考点 reference point of picture
相片冲洗 photographic(al)processing
相片地图 photomap
相片调绘 annonation classification of pictures
相片方位元素 element of orientation for picture
相片纠正仪 photo transformer; transforming printer
相片控制点 control point of picture
相片内方位元素 data of inner orientation
相片判读者 photointerpreter
相片缩小仪 photoreducer; reduction printer
相片索引图 photo index
相片外方位元素 data of outer orientation; element of exterior orientation
相片镶嵌图 universal photo
相片中点 photograph center[centre]
相偏析 phase segregation
相偏移 phase deviation
相频畸变 phase frequency distortion
相频失真 phase frequency distortion
相频特性 phase frequency characteristic
相频特性曲线 phase shift frequency curve
相频响应 phase frequency response
相平衡 phase equilibrium; transfer equilibrium
相平衡计算机 phase equilibrium computer
相平衡曲线 phase equilibrium line
相平衡同位素效应 phase equilibrium isotope effect
相平衡图 phase equilibrium diagram
相平均 phase average
相平面 phase plane
相平面法 phase plane method
相平面分析 phase plane analysis
相平面曲线 phase-plan curve
相平面图 phase plane diagram
相谱 phase spectrum
相前 phase front
相绕式电动机 phase wound motor
相绕组 phase winding
相栅 phase grating
相时 phase duration
相式车辆检测器 phase mode vehicle detector
相数 number of phases; phase number
相速(度)phase speed; phase velocity
相速度法 phase speed method; phase velocity method
相锁 phase in-lock
相态 phase state
相态关系 phase relationship
相态稳定性 phase stability
相态转化热 heat of transformation
相套 facies suite
相特性 phase behavio(u)r
相体积 phase volume
相体积比 phase volume ratio
相体积理论 phase volume theory
相调制 phase module
相图 facies map; phase diagram; equilibrium diagram <合金的>
相位 phase position
相位摆动 phase hunting; phase swing
相位比较 phase comparison
相位比较电路 phase-comparing network
相位比较器 phase comparator; phase detector
相位比较式保护 phase comparison protection
相位编码【计】phase encode; phase coding; phase encoding
相位编码技术 phase encoding technique
相位编码脉冲压缩 phase-coded pulse compression
相位编码器 phase encoder
相位变化 phase variation
相位变化过程 phase history
相位变化率 rate of change of phase
相位变换 phase transformation
相位变换机 phase converter
相位变换开关 phase change switch
相位变换器 phase converter; phase transformer
相位变换适配器 phase adapter
相位表 phase measuring equipment; phase meter; phasometer
相位波动 phase fluctuation; phase moiré【计】
相位波前 phase front
相位补偿 phase compensation
相位补偿器 phase compensator
相位补偿装置 phase compensating device
相位不等 phase inequality
相位不稳定的 phase-unstable
相位不重合 out phase; put-phase
相位布置 phasing arrangement
相位采样 phase sample; phase sampling
相位测距法 method of distance measurement by phase
相位差 dephasing; difference of phase; out-of-phase; phase defect; phase difference; phase displacement; phase shift
相位差测量 phase difference measurement
相位差法 phase difference method
相位差剖面曲线 profile curve of phase difference
相位差相移 phase shift
相位差仪 phase meter
相位差影响 out-of-phase difference effect
相位常量 phase constant
相位常数 phase constant
相位超前 leading in phase; phase advance; phase lead(ing)
相位超前补偿机 phase advancer
相位超前补偿器 advancer; phase advancer
相位超前角 leading phase angle
相位超前网络 phase advance network
相位超前运行 leading power factor operation
相位迟滞 retardant of phase
相位储备 phase margin
相位传递函数 phase transfer function
相位导前 phase lead
相位倒转 phase inversion; phase reversal
相位地址系统 phase address system
相位抖动 phase jitter
相位抖动测量仪 phase jitter tester
相位对比 phase correlation
相位对相位的 phase-to-phase
相位多值性 phase ambiguity
相位法 <信号配时的一种方法> phase related method; phase method
相位法测距 method of distance measurement by phase
相位反差 phase contrast
相位反应 phase response
相位分辨率 phase resolution
相位分辨能力 phase resolution
相位复矢量 complexor; phasor
相位改变 180 度 phase reversal
相位改正 phase equalization
相位干扰 phase disturbance
相位跟踪器 phase tracker
相位共振 phase resonance
相位关系 phase relation; phase relationship
相位规则【化】phase rule
相位轨迹 phase locus
相位后移的 dephased
相位畸变 phase distortion
相位畸变系数 phase-distortion coefficient
相位激电仪 phase instrument for IP
相位计 phase measuring equipment; phase meter; phaser; phasometer
相位加速器 fazotron
相位间的 phase-to-phase
相位监视器 phase monitor
相位检测 phase-detecting; phase detection
相位检测器 phase detector
相位检查 phase check
相位鉴别 phase discrimination
相位鉴别器 phase discriminator; phase shift discriminator
相位交叉 phase crossover
相位交叉频率 phase crossover frequency
相位交叠 phase crossover
相位角 hour angle; phase angle
相位校正 correction for phase; phase correction
相位校正级 phase-correcting stage
相位解调器 phase demodulator
相位均衡 phase equalization
相位均衡网络 phase equalizing network
相位控制 phase control
相位控制电路 phase control circuit
相位控制机 phase controller
相位控制键 phase control key
相位控制器 phase controller
相位控制装置 phase control device
相位亏损 phase defect
相位连续键控 phase-continuous keying
相位量测 phase measuring
相位量测系统 phase measuring system
相位灵敏的 phase sensitive
相位灵敏度 phase sensitivity
相位滤波 phase filtering
相位落后 phase lag(ging)
相位落后角 <应变与应力的相位差角> phase lag angle
相位脉冲 phase pulse
相位弥散 phase debunching
相位密度谱 phase density spectrum
相位模块 phase module
相位模拟器 phase simulator
相位偏移 phase deviation; phase displacement; phase drift
相位偏移减小 phase compression
相位偏移值 phase pushing figure
相位漂移 phase drift
相位频率畸变 phase frequency distortion
相位(频率)响应特性 phase response characterization
相位平衡 phase balance; phase equilibrium
相位平衡继电器 phase-balance relay
相位平面 phase plane
相位谱 phase spectrum
相位谱曲线 phase spectrum curve
相位曲线 phase curve
相位取样器 position sampler
相位全息术 phase holography
相位全息图 phase hologram
相位确定 phase determination
相位日变化 diurnal phase change
相位容量 phase margin
相位散乱 phase debunching
相位失真 phase distortion; phase shift
相位失真补偿 phase compensation
相位时间比 phase split
相位时间调制 phase time modulation
相位识别脉冲 phase identify pulse; p-pulse
相位矢量指示 vector phase indication
相位式定位器 phase localizer
相位数 <交流电> number of phases
相位水平剖面 horizontal phase section
相位伺服系统 phase servo system
相位速度 phase speed; phase velocity; wave velocity
相位特性 phase characteristic
相位特性控制 phasing characteristic control
相位特征 phase characteristic
相位梯形 phase keystone; phase trapezoid
相位提前 phase lead
相位调节 phase modulation
相位调节器 phase regulator
相位调节装置 phase modulation unit
相位调整 phase adjustment; phasing
相位调整叠加剖面 phase adjusted stack section
相位调整器 phase shifter
相位调制 phase modulation
相位调制包线 phase modulation envelope
相位调制常 phase modulation constant
相位调制信号 phase-modulated wave
相位调置 phase setting
相位调准 phase adjustment
相位停车 phasing stop
相位同步 phase lock(ing); phase synchronism
相位同步控制 phase-locked control
相位突变 phase jump
相位突然异常 sudden phase anomaly
相位图 phase diagram; phase portrait; constitutional diagram
相位推延 phase retardation
相位位移 phase displacement
相位稳定性 phase stability
相位误差 phase error
相位系数 phase coefficient

相位显示器 phase monitor
相位显微镜技术 phase microscopy
相位相干的 phase-coherent
相位响应特性 phase response
相位谐振 phase resonance
相位信息 phase information
相位选择 phase selection
相位选择钮 phase selector
相位延迟 phase-delay;phase shift delay
相位延迟测定 phase-delay measurement
相位延迟测量 phase-delay measurement
相位延迟电路 phase-delay network
相位延迟网络 phase-delay network
相位延时失真 phase-delay distortion
相位延误 phasing delay
相位一致的 in-step
相位一致条件 in-step condition
相位仪 phasometer
相位移(动) phase displacement;phase drift
相位移动的 out-of-phase
相位移角 angle of phase displacement
相位移偏移 phase shift migration
相位异常 phase anomaly
相位因数 phase factor
相位余量 phase margin;phase reserve
相位预矫 phase predistortion
相位预均衡 phase pre-equalization
相位裕度 phase margin
相位裕量 phase margin
相位月潮不等 phase inequality
相位跃变 phase change;phase jump;phase step
相位运行(行列)表 phase-movement matrix
相位噪声 phase noise
相位障碍 phase fault
相位指示器 electrogoniometer;phase indicator
相位指示器平衡控制 phase indicator balancing control
相位滞后 lagging of phase;lag in phase;phase-delay;phase lag;phase-tag;retardant of phase;retardation of phase
相位滞后电路 phase lag network
相位滞后角 phase lag angle
相位周期 phase cycle
相位周值 phase revolution
相位转换 phase conversion;phase transformation
相位转换附加器 phase adapter
相吻合齿轮 engaged wheel
相稳定性 phase stability
相物镜 phase objective
相物体 phase object
相线 phase line
相响应 phase response
相序 phase order;phase rotation;phase sequence
相序表 phase-sequence meter
相序倒转 reversal of phase sequence
相序继电器 phase-sequence relay
相序控制 control of phase sequence
相序形成图 phase-sequence generation diagram
相序指示计 phase-sequence meter
相序指示器 phase-sequence indicator
相延迟 phase-delay;phase retardation
相延迟温度系数 thermal retardation coefficient
相延多普勒测图 delay Doppler mapping
相延失真 delay distortion
相延续时间 phase duration
相氧化 phase oxidation
相液解度 phase solubility
相移 dephasing;misphasing;phase displacement;phase shift
相移编码 phase shift coding
相移编码器 phase shift coder
相移变换器 shift converter
相移变压器 phasing transformer
相移常数 phase shift constant
相移传声器 phase shift microphone
相移磁强计 phase shift magnetometer
相移电路 phase shift circuit
相移电容器 phase shifting capacitor
相移发生器 phase generator
相移话筒 phase shift microphone
相移鉴频器 Foster-Seeley discriminator;phase shift discriminator
相移键控 phase shift keying
相移角 phase angle
相移滤波器 all-pass filter
相移片 phase plate
相移频率曲线 phase shift frequency curve
相移器 phaser
相移全向无线电指向标 phase shift omnidirectional radio range
相移失真 phase shift distortion
相移失真试验 differential phase test
相移特性 phase shift characteristic
相移特征曲线 phase shift characteristic
相移调制器 phase shift modulator
相移突变 phase shift mutation
相移网络 phase shift network
相移响应特性 phase shift response
相移压缩 phase compression
相移延迟 phase shift delay
相移译码器 phase shift decoder
相移振荡器 phase shift oscillator
相移装置 phase changer
相移阻抗 phase shifting impedance
相倚接点 dependent contact
相域 phase region
相阵 phased array
相振荡频率 phase oscillation frequency
相纸漏过光的 light-struck
相制电路 interaction circuit
相滞 lagging of phase
相中点接地 mid-phase grounding
相转变 phase conversion;phase shift;phase transformation;phase transition;phase variation
相转变图 phase diagram
相转化法 phase inversion
相转换 phase transition
相转换线 line of phase transformation
相转移法 phase transfer method
相转移模型 phase transfer model
相状态 phase behavio(u)r
相阻 phase resistance
相组 facies group;phase bank;phase group
相组成 phase composition
相组合 facies association
相坐标 phase coordinates
相座 phase site

象

象白蚁 Nasutitermes parvonasutus
象白蚁属 <拉> Nasutitermes
象鼻 snout
象鼻虫 weevil
象鼻管 articulated conduit;articulated(drop) chute;trunk;elephant trunk <浇混凝土用的>
象鼻架伸臂 double lever jib
象鼻式混凝土泵 elephant concrete pump
象鼻式卸废料管 elephant's trunking
象鼻状装置 elephant trunk
象波道 picture channel
象差 quadrantal deviation
象大理石的 marmoraceous
象甲 weevil
象甲科 <拉> Curculionidae
象角 quadrant
象角船首 club foot
象蜡树 elephant ash
象沥青的 pitchy
象毛的 hair-like
象模标本 Icotype
象气体的 gassy
象软木的 corky
象树脂的 resinaceous
象铜的 cupreous
象锡的 tinny
象限 quad;triangular space
象限差 quadrantal error
象限潮 neap tide
象限乘法器 quadrant multiplier
象限尺 range of quadrant
象限的 quadrantal;quadratic
象限点 half cardinals;intercardinal point;quadrantal point
象限法 quadrant method
象限方位(角) quadrantal bearing
象限方向 quadrantal heading
象限分布 quadrantal distribution
象限分(线)规 quadrant divider
象限光电倍增器 quadrant photomultiplier
象限含水层 quadrant aquifer
象限航向 intercardinal heading
象限横摇误差 quadrant intercardinal rolling error
象限弧 bearing arc;quadrantal arc
象限检核 <校核各象限函数与角度的关系> quadrant test
象限角 bearing angle;quadrant(al) angle;quadrant(al) bearing
象限静电计 quadrant electrometer
象限类型 quadrant type
象限罗盘仪 bearing compass;quadrant compass
象限瞄准具 quadrant sight
象限偏差 quadrantal deviation
象限偏差校正器 quadrantal deviation corrector
象限球 quadrantal sphere
象限球面三角形 quadrantal spherical triangle;quadrantal triangle
象限四分仪 quadrant
象限图 quadrantal diagram
象限误差 quadrantal error;quadrant error
象限误差成分 quadrantal component of error
象限误差修正器 quadrantal error corrector
象限线 quarter section;quarter section line
象限线圈 quadrant coil
象限相关 quadrant dependence
象限修正 quadrantal correction
象限仪 quad;range quadrant
象限仪水准器臂 quadrant arm
象限仪座 quadrant mount
象限圆规 wing compasses
象限圆饰 quarter round
象限值 quadranture
象限轴 quadrature axis
象限自差 quadrantal deviation
象星的 stallate;stallated
象形单位图 pictorial unit chart
象形符号 representational symbol
象形锅炉 elephant boiler
象形图 figurative diagram;figurative graph;ideograph chart;pictogram;pictorial chart;picture graph;shape resembling chart
象形文结构 runic texture
象形文字 hierograph;pictograph;picture-writing
象形字 pictograph
象形字母 <古代北欧文字的字母> runic character
象牙 ivory
象牙白 <一种釉色品> ivory white
象牙白光泽纸板 ivory bristol
象牙白纸板 ivory board
象牙白装饰 ivory finish
象牙匙 ivory spoon
象牙瓷 ivory china;ivory porcelain
象牙雕刻 sculpture in ivory
象牙雕刻品 ivory sculpture
象牙工艺 ivory work
象牙海岸 Ivory Coast
象牙黑 abaiser
象牙黑颜料 ivory black pigment
象牙黄 <一种釉色品> ivory yellow
象牙螺 ivory shell
象牙墨 ivory black
象牙木 <产于南美,淡白带褐黄色有时微绿色> ivory wood
象牙色 ivory tint
象牙色表层 ivory glazed finish
象牙色的 eburnean;ivory-colo(u)red
象牙色浸蜡印像法 ivory type
象牙色釉 ivory glaze
象牙色釉表层 ivory glazed coat(ing)
象牙(炭)黑 ivory black
象牙屑 ivory chip(ping)s
象牙样的 eburneous
象牙椰子 carozo nut
象牙椰子树 ivory palm
象牙艺术雕刻品 artistic object in ivory
象牙之塔 ivory tower
象牙制品 ivory work
象牙棕榈 vegetable ivory
象烟灰的 fuliginous
象银的 argentine
象征 badge;emblem;symbolize;token;typify
象征常数 figurative constant
象征的 figurative
象征的设计 figurative design
象征的实体 antitype
象征符号 stylized symbol
象征化作用 symbolization
象征码 interpreter code;symbolic code
象征手法 symbolism
象征速度 nominal speed
象征速率 nominal speed
象征特性 figurative character;figurative element
象征图 symbolic diagram
象征误用 misuse of symbol
象征性保费 nominal premium
象征性偿付 token payment
象征性的 symbolic;token
象征性雇佣 tokenism
象征性交付 symbolic delivery
象征性交货 symbolic delivery
象征性进口 token import
象征性赔偿 nominal damage
象征性设计 figurative design
象征性输出 token export
象征性输入 token import
象征性思维 symbolic thought
象征性损失 nominal damage
象征性租金 pepper-corn rent
象征学 symbology
象征意义 symbolic meaning
象征元 figurative element
象征主义 symbolism
象足(迹)形裂缝 elephant crack

像

像斑 image patch
像倍增器 image multiplier
像变形 deformation of image

像标 image scale
像差 aberration(defect);astigmation;image error;image fault;optic(al)aberration
像差符号 aberration mark
像差改正 aberration correction
像差计 aberrometer
像差角 aberration angle
像差容限 tolerance for aberration
像差系数 aberration coefficient
像差源 aberrant source
像长石的 feldspathic
像场 image field;picture-field
像场变化 change of image area;image area change;image size change
像场分割取景器 split field
像场角 angle of coverage;objective angle of image field;coverage angle
像场曲率 curvature of image field
像场曲面 field curvature
像场缩小 picture-field warning
像场弯曲 field aberration;field curvature
像场修正透镜 image flattening lens
像衬比 image contrast
像衬比技术 image contrast technique
像瓷 Parian Paros
像大理石的 marbly;marmoreal
像倒置 image inversion
像底点 nadir point;photograph plumb point;photo nadir point
像底点三角测量 plumb point triangulation
像点 image point;image spot;photo point;picture dot;picture element;picture point;point image
像点辨认 point identification
像点尺寸 image spot size
像点定位 image point setting
像点交错 picture-dot interlacing
像点控制 picture point control
像点扩散 diffusion of the point image
像点面积 elementary area
像点模糊 diffusion of image point;diffusion of the point image
像点投影线 image ray
像点位移 displacement of images;image displacement
像点坐标 coordinates of image point;coordinates of photo point;image (position) coordinates;photographic(al) coordinates;picture coordinates;plate coordinates
像点坐标改正 correction of image coordinates
像电荷 image charge
像电流 image current
像电流密度 image current density
像电影的 filmic
像动稳定补偿 image motion compensation
像对【测】airphoto pair;image pairs;pair of pictures;photopair;double image
像对定向 orientation of pairs
像对共面(条件)<像片平行于摄影基线> basal coplane
像对(水)平共面 horizontal coplane
像法测图 photographic(al) mapping
像方 image space
像方焦点 image focal point;rear focus
像方景深 depth of image space
像方位 image aspect
像肥皂的 soapy
像幅 dimension of image;dimension of picture;frame area;frame size;image ratio;image shape picture;image size;picture area;picture format;picture ratio;picture size;plate size;size of image
像幅大小 frame size
像高修正 height-of-image adjustment
像古希腊盛大节日般的状况<祭神庆典> Panathenaic way
像管 charactron
像海员的 seamanlike
像含数据 image-contained data
像函数 image function
像合成 image synthesis
像横线 parallel to the axis of tilt;photograph parallel
像灰一样的 cinereour
像畸变 anamorphosis;image distortion
像畸变的 anamorphoser
像集 image set
像间的 interimage
像间效应 interimage effect
像角 angle of image;image angle;picture angle
像节点 node of emission
像晶石的 sparry;spathic;spathose
像距 image distance
像空间 image space
像空间辅助坐标系 photo space auxiliary coordinate system
像空间坐标系(统)image space coordinate system;photo space coordinate system;picture space coordinate system
像宽 image width
像框 picture frame;plate frame
像框光轴点 optic(al)centre of registering frame
像框距 frame pitch
像框面 plane of registering frame
像扩散 image spread
像蜡的 waxen
像蓝宝石的 sapphirine
像力 image force
像力模型 image force model
像亮化器 image intensifier
像录制 image transcription
像弥散 image confusion
像面 image side;image surface;picture plane;plate plane
像面朝向 image direction
像面辐照度 image plane irradiance
像面平整镜 field flattener
像面曲率 image surface curvature
像面扫描 image plane scanning
像面照度的均匀性 image illumination uniformity
像面坐标 image coordinates
像片 image picture;paper print;photo;photograph;photographic(al)image
像片曝光放大表 photoexposure enlarging meter
像片比例尺 image scale;photographic(al)scale;photo scale;picture scale
像片比例尺数字 image scale figure;photo scale figure
像片边缘 margin of image;margin of picture;white margin
像片边缘清晰度 marginal definition
像片编号 exposure number
像片编辑员 photo editor;picture editor
像片编图 mapping from photograph;photocharting
像片变形 picture distortion
像片标准点 photographic(al)normal point;standard point of photo
像片簿 album
像片测量 picture measurement
像片测倾仪 topoangulator
像片测图 mapping from photograph;photogrammetric(al)mapping;photographic(al)mapping;photomapping
像片测图仪 image mapping apparatus;photoplotter
像片尺寸 image ratio;image size;picture ratio
像片垂线 photograph perpendicular
像片导线 photo polygon;photo-polygonometric(al)traverse
像片导线测量 photopolygonometry
像片等角点 photograph is center
像片地平线 horizon trace;image horizon;picture horizon
像片地质解译 geologic(al)interpretation of photograph
像片地质判读 geologic(al)interpretation of photograph
像片定向 image orientation;photo-point control
像片对 image pairs;photopair
像片对假定平面的相对倾斜<不一定是水平面的>【航测】relative tilt
像片方位角 azimuth of photograph
像片放大 photo enlargement;photographic(al)enlargement
像片分辨率 image resolution;photographic(al)resolution;picture resolution
像片分析 photo analysis;photographic(al)analysis
像片改正 correction of photo deformation;rectification of aerial photograph
像片感光面 photosensitive plane
像片高程控制 vertical photo-control
像片高程控制点 elevation photo control point;picture height control point;vertical photo-control point
像片光轴点 foot of optic(al)axis
像片归心 centering of photograph;photo centering
像片烘干机 print drier[dryer]
像片基线 photo(graph)base(line);photographic(al)image base
像片基线长度 base length on the photo
像片基线方向 photo-base orientation
像片基准面 photographic(al)datum
像片几何学 photo geometry
像片计数器 frame counter
像片迹线 picture trace
像片减薄 reduction of image
像片检索 photo indexing
像片角 photo corner
像片角锥体 photograph pyramid
像片校正 photoproof
像片接收 photo-reception
像片解释 photographic(al)interpretation
像片纠正 photo(graphic)rectification;rectification of photograph
像片纠正仪 photo transformer
像片拷贝 photographic(al)copy
像片空中三角测量 photogrammetric(al)aerial triangulation
像片控制 image control;photo-control
像片控制测量 control surveying of photograph;photo-control survey
像片控制点 control point of photograph;photo-control point;picture control point
像片控制点辨认 photoidentification
像片控制点略图 photo-control diagram
像片库 photographic(al)library
像片框 picture carrier;pix carrier;vision carrier
像片框标 fiducial marks on photograph
像片离心 photograph decentration
像片连测 photo-control
像片连测略图 photo-control diagram
像片连接点 picture contact point
像片联测 photographic(al)extension;photo-point control;plate conjunction
像片联测略图 photo-control diagram
像片量测 image measurement;measurement of photographic;picture measurement;picture measuring
像片量测仪 image measuring apparatus;picture measuring apparatus
像片量角法 photogoniometric method
像片量角仪 photogoniometer
像片量角仪编图 photoalidade compilation
像片量角仪编图法 photoalidade method
像片略图 mosaic assembly;photographic(al)sketch;serial mosaic;strip mosaic;unchecked mosaic;uncontrolled mosaic;unrectified mosaic
像片麻岩的 gneissoid
像片内方位元素 elements of interior orientation;interior orientation elements
像片盘 photo carriage;photo carrier plate;picture carrier plate
像片判读 annotate a photograph;photographic(al)interpretation;photographic(al)reading;photoidentification;photointerpretation
像片判读标志 photointerpretation key
像片判读技术 photo interpretative technique
像片判读手册 manual of photographic(al)interpretation
像片判读仪 photointerpreter;photopret
像片判读员 photographic(al)interpreter
像片判解 photointerpretation
像片偏心 photoexcenter
像片平高控制点 full control point;horizontal and elevation picture control point;horizontal and vertical control point of photograph;horizontal and vertical photo control point
像片平面 photoplan
像片平面控制 horizontal photo-control
像片平面控制点 horizontal photo-control point;picture horizontal control point
像片平面图 annotated(photographic)mosaic;photomap;photoplan;picture plan;photographic(al)plan
像片倾(斜)角 tilt angle of photograph
像片清晰度光泽 distinctness-of-image gloss
像片三角测量 photogrammetric(al)triangulation;phototriangulation
像片三角量测仪 phototriangulation instrument
像片扫描系统 photoscan system
像片上的距离 photographic(al)distance
像片视准量角仪 photoalidade
像片数据压缩 photographic(al)data reduction
像片数字化 photodigitizer
像片水平控制点 picture horizontal control point
像片水平线 image horizon;photograph parallel;picture horizon
像片顺序 sequence of photograph
像片缩小 photoreduction

像片缩小仪 photoreducer; reduction printer
像片索引图 index photography; index to photograph; photo index; plot map
像片调绘 field photo interpretation; photograph annotation; topographic-(al) identification
像片天底点 photographic(al) nadir; photograph nadir; plate nadir
像片投影仪 photo projector
像片图 aerophotographic(al) sketch; photoplan; picture plane
像片外方位元素 elements of exterior orientation; exterior orientation elements
像片镶嵌 adjustment of images; mosaic assembly; mosaicking; mount of the picture; photographic(al) compilation; photograph montage; serial mosaic
像片镶嵌胶 photomountant paste
像片镶嵌略图 trimming and mounting diagram
像片镶嵌索引图 mosaic index; photo mosaic index
像片镶嵌图 imagery mosaic; mosaic map; photomontage; photomosaic
像片镶嵌仪 mosaicker
像片修测图 photo-revised map
像片旋角 swing angle of photograph
像片旋偏角 swing of photo
像片选点 picture point selection; point selection
像片野外控制点【航测】outdoor picture control point
像片有效面积 photographic(al) effective area
像片照明电路 picture illumination circuit
像片整饰框 retouching frame
像片质量 image quality; picture quality
像片质量改善 improvement of photograph
像片中心 optic(al) picture center [centre]; photocenter[photocentre]; photograph center [centre]; plate center[centre]
像片重叠 photograph overlap; picture overlap
像片重叠调节器 photographic(al) coverage regulator
像片重叠指示器 overlap indicator
像片轴 photograph axis
像片主点 optic(al) picture center [centre]
像片主横线 principal parallel
像片主距 principal distance of photo
像片主总线 principal vertical(line)
像片主纵线 photograph meridian; principal meridian
像片转绘仪 camera lucida; camera obscure; sketch master

像片锥形法 photographic(al) pyramid
像片子午线 photograph meridian
像片自动镶嵌仪 automatic mosaicker
像片组 block of photographs
像片坐标 image (position) coordinates; photo (graph) coordinates; photographic(al) coordinates; picture coordinates; plate coordinates
像片坐标系 photo coordinate system
像片坐标轴 fiducial axis; photograph axis
像频 image frequency; picture frequency
像频干扰 image frequency interference
像频抗拒比 image ratio
像频率选择性 image frequency selectivity
像频调制 image modulation
像频响应 image frequency response; image response
像频抑制比 image frequency rejection ratio
像平面 image plan; image plane; plane of delineation
像平面全息摄影 image plane holography
像平面图 photoplan
像平面坐标系 image plane coordinate system; photo plane coordinate system; picture plane coordinate system
像平面坐标系统 photo coordinate system
像青玉色的 sapphirine
像清晰度 image sharpness
像曲线 image curve
像圈 image circle
像全息图 image hologram
像容数据 image-contained data
像三角形的 deltaic
像散 astigmatic aberration; astigmation
像散差 astigmatic aberration; astigmatic difference
像散成像 astigmatic image
像散的 astigmatic
像散光束 astigmatic bundle
像散计 astigm(at)ometer
像散焦点 astigmatic focus
像散校正 astigmatic correction; astigmation correction
像散校正器 astigmatic corrector; astigmatism corrector
像散校正装置 astigmator
像散镜 astigmatizer; astigmatoscope
像散器 astigmatizer
像散调节 astigmatic accommodation; binocular accommodation
像散透镜 astigmat(ic) lens
像散现象 astigmatism
像散性 astigmation; astigmatism
像散装置 astigmatic mounting; astigmatizer
像砂粒大小的尺寸 sandy size
像蛇的 snake-like
像石 image-stone
像束 image beam; video beam
像束电视 video beam television
像水平线 photograph parallel
像水手的 seamanlike
像素 image element; mirror element; pel; picture dot; picture element; picture point; pixel
像素比例尺 pixel scale
像素充实法 picture element replenishment method
像素传送时间 picture point time
像素缓冲区 pixel array cache
像素控制 picture point control
像素密度法 dot density method
像素面积 elemental area; elementary area; pixel area
像素扫描 pixel scan
像素输入 pixel loading
像素数组大小 pixel array dimensions
像素显示时间 picture point time
像素信号 picture element signal
像素阵列 pixel array
像台 <承雕像的平台> entablement
像陶瓷的 ceramic-like
像天然石材 stone-like
像图 image pattern
像弯曲 image curvature
像位规定 image position specification
像位置 image position
像相常数 image phase constant
像信号板 image plate
像信号中频变压器 image intermediate frequency transformer
像形图 image chart
像形印痕 hierogloph
像形状变换器 image shape converter
像移 image motion; image movement; image drift
像移补偿 image motion compensation; image-movement compensation
像移补偿器 image motion compensator
像移动函数 image motion function
像移因素 image motion factor
像印花棉布一样的墙纸 chintz
像元 image element; mirror element; pel; picture element; pixel
像元大小 pixel size
像源 image source
像源网络 network of image sources
像增强管 image intensifier tube
像增强光敏薄膜 image intensifying photosensitive film
像增强器微通道板 image intensifier-microchannel plate
像增强式分流直像管 image intensifier isocon
像增强析像管 image intensifier orthicon tube
像增强系统 image intensifying system
像帧 picture frame
像纸 development paper; photocopying paper; photographic(al) paper; photosensitive paper; printing-out paper; sensitive paper; sensitized paper
像纸反差等级 paper grade; printing grade
像质 picture element
像质计 image quality indicator
像中心 iconocenter
像重现 image reconstruction
像重现装置 image reproducer
像主点 image center[centre]; optic-(al) picture center[centre]; photograph center[centre]; plate center [centre]; principal point of photograph
像主点横坐标 abscissa of principal point
像主点落水 principal point of photograph on water
像主点三角测量 principal point triangulation
像主点位置 principal point location
像主点纵坐标 ordinate of principal point
像主距 photo-principal distance; principal distance
像转换管 image converter tube
像转换器 image converter
像转换器扫描照相机 image converter streak camera
像坐标校正 correction of image coordinates
像座 entablement

橡

橡浆 latex[复 latices/ latexes]; rubber latex
橡浆地板 rubber latex floor
橡浆废水 latex wastewater
橡浆改性沥青 latex-modified asphalt
橡浆胶结软木嵌缝料 rubber-bound cork filler
橡浆沥青混合物 latex bitumen mixture
橡浆清 serum
橡浆水泥 rubber cement
橡浆涂料 latex paint
橡胶 rubber(glue); elastica; elastic caoutchouc; gum (elastic); India rubber
橡胶安全踏板 rubber safety tread
橡胶安装垫 rubber mount
橡胶按钮 rubber knob
橡胶凹模成型 rubber pad forming; Guerin process
橡胶把手 rubber hand grip
橡胶坝 flexible dam; rubber dam
橡胶板 rubber panel; rubber plate; rubber sheet(ing); sheet rubber
橡胶板式泥浆输送器 rubber slat mud conveyer[conveyor]
橡胶板压形法 rubber pad forming process
橡胶版 rubber plate
橡胶版油墨 rubber plate ink
橡胶包带 rubber insulating tape
橡胶包的 rubber-sheathed
橡胶包面的 rubber-surfaced
橡胶包皮电缆 rubber-sheathed cable
橡胶包套 rubber bag
橡胶泵阀 rubber pump valve
橡胶边侧密封 rubber side seal
橡胶边缘密封 rubber side seal
橡胶扁肩钎 rubber collar drill steel
橡胶表层 rubber skin
橡胶玻璃密封条 rubber glazing channel
橡胶箔 rubber foil
橡胶薄垫片 rubber shim
橡胶薄膜 rubber film; rubber membrane
橡胶薄片 rubber sheet(ing)
橡胶布 rubber cloth; rubberized cloth; rubberized fabric; rubber sheet(ing); rubber tissue
橡胶布辊筒 rubber-blanket cylinder
橡胶布外罩 rubberized-fabric envelope
橡胶部件 rubber unit
橡胶擦 rubber eraser
橡胶槽 rubber trough
橡胶草 grass rubber plant
橡胶测厚计 rubber ga(u)ge
橡胶层 rubber layer
橡胶插塞 rubber plug
橡胶长靴 rubber boots; sea boots
橡胶厂 rubber factory; rubber plant
橡胶厂废水 rubber mill wastewater; rubber processing plant wastewater
橡胶车轮护舷 rubber-tired truck-wheel fender
橡胶衬 rubber gasket
橡胶衬板 rubber lining plate
橡胶衬垫 neoprene washer; rubber liner; rubber lining; rubber packing; rubber grommet
橡胶衬垫接缝 <混凝土路面> rubber-gasketed joint
橡胶衬里 lining rubber; rubber lining; rubber sheet lining
橡胶衬里的 rubber-lined
橡胶衬里的齿耙 rubber-lined rake
橡胶衬里管 rubber-lined pipe
橡胶衬里水龙带 rubber-lined fire hose
橡胶衬里消防软管 rubber-lined fire hose
橡胶衬料 rubber paving
橡胶衬砌 armo(u)rite rubber lining
橡胶衬球磨机 rubber-lined ball mill
橡胶衬套 rubber bushing
橡胶成分 rubber constituent
橡胶成型 rubber pad forming

橡胶成型法 rubber pad forming process
橡胶承垫块 rubber bearing pad
橡胶承受周期性弯曲的寿命 flex life
橡胶承座 rubber bearing
橡胶冲裁 rubber pad blanking
橡胶除冰带 rubber-boot deicer
橡胶传动带 rubber transmission belting
橡胶传输带 rubberized conveyor belts
橡胶传送带 rubber conveyor belt
橡胶锤 rubber hammer;rubber mallet
橡胶磁铁 ferrogum
橡胶促进剂 rubber acceleratant;rubber accelerator
橡胶促进剂废水 rubber acceleratant wastewater
橡胶打捞浮筒 rubber salvage pontoon
橡胶代用品 rubber substitute
橡胶带 gum band;rubber belt(ing);rubber tape
橡胶袋 rubber bag;rubber bulb
橡胶挡块 rubber stop
橡胶挡泥板 rubber fender; rubber mudguard;rubber wing
橡胶挡水埂 rubber waterbar
橡胶挡水条 rubber waterbar
橡胶刀 rubber tapping knife
橡胶导管 rubber catheter;rubber tremie tube;rubber conduit
橡胶的底脚 thickness mo(u)lding
橡胶登陆艇 rubber landing craft;rubber landing ship
橡胶底层 rubber-based coating
橡胶底胶黏[粘]剂 rubber-based adhesive
橡胶地板擦 squeegee[squilgee]
橡胶地板覆盖层 rubber floor covering
橡胶地沥青 bituthene;rubber asphalt;rubberized asphalt
橡胶地沥青防水膜 bituthene waterproofing membrane
橡胶地沥青嵌缝板 rubberized asphalt sealing strip
橡胶地面 rubber flooring; rubber sheet flooring
橡胶地毯 kamptulicon
橡胶电线 India-rubber wire
橡胶垫 rubber blanket;rubber insert(ion); rubber mat(ting); rubber mattress; rubber mounting; rubber pad(ding); rubber seat; rubber sheet(ing)
橡胶垫板 rubber bearing pad
橡胶垫布 rubber sheet(ing)
橡胶垫层 rubber cushioning; rubber spacer
橡胶垫底层 rubber underlay
橡胶垫环 rubber gasket
橡胶垫架 rubber mount
橡胶垫块 rubber bearing;rubber pad
橡胶垫块护舷 rubber cushion fender
橡胶垫料 rubber packing
橡胶垫履带板 rubber pad shoe
橡胶垫密封的 rubber sealed
橡胶垫密片 rubber gasket
橡胶垫片 rubber distance piece;rubber ring; rubber washer; rubber gasket
橡胶垫片支座 <桥梁的> rubber pad bearing
橡胶垫钎肩式钎尾 shrunk rubber collar shank
橡胶垫圈 gum washer; packing rubber; rubber gasket; rubber grommet; rubber packing ring; rubber washer;rubber ring
橡胶垫圈接口 rubber gasket joint
橡胶垫圈密封 rubber-ring seal
橡胶垫式成型法 rubber pad process
橡胶垫条 rubber weather-strip
橡胶垫压力机 rubber pad press
橡胶垫支承 rubber cushion supporter
橡胶垫座 rubber seat retainer
橡胶顶盘 rubber support plate
橡胶顶撞块 <码头的> rubber bumper
橡胶定位销 rubber dowel
橡胶定子 elastomer stator
橡胶毒物 rubber poisons
橡胶端套 rubber pot-head
橡胶鳄皮开裂现象 alligatoring of rubber
橡胶耳垫 rubber ear cushion
橡胶阀座 rubber valve seat
橡胶帆布软管 rubber canvas hose
橡胶帆布艇 faltboat;fold(ing) boat
橡胶帆布制品 rubber and canvas article
橡胶防冲装置 rubber fender
橡胶防风条 rubber wind seal
橡胶防护板 rubber apron
橡胶防护层 rubber apron
橡胶防护圈 rubber apron
橡胶防护罩 rubber gaiter
橡胶防老剂 rubber antioxidant
橡胶防水带 rubber strip
橡胶防水封口 rubberized waterproof seal
橡胶防水片材 rubber waterproofing sheet
橡胶防水条 rubber strip
橡胶防振制品 rubber pad; rubber spring;vibration-proof rubber
橡胶防震装置 rubber vibroinsulator
橡胶防撞垫 rubber fender
橡胶防撞块 <码头的> rubber bumper
橡胶防撞设施 rubber fender
橡胶废料 rubber waste
橡胶废水 rubber wastewater
橡胶废物 rubber waste
橡胶粉(末) rubber powder; powdered rubber
橡胶风挡清洁器 rubber windshield cleaner
橡胶封套 rubber cuff
橡胶敷面 rubber-based coating
橡胶覆盖层 rubber facing
橡胶覆盖地面 rubber floor
橡胶覆面挤干辊 rubber drying rolls
橡胶改性沥青 rubber-modified bitumen
橡胶改性沥青油毡 rubber-modified asphalt felt
橡胶改性砂浆 latex-modified mortar
橡胶改性塑料 rubber-modified plastics
橡胶盖 rubber serum cap
橡胶膏 rubber adhesive plaster
橡胶膏涂料机 spreading machine for adhesive plaster
橡胶隔板 rubber separator
橡胶隔膜 rubber diaphragm separator
橡胶隔膜泵 rubber diaphragm pump
橡胶隔振垫层 rubber mounting
橡胶隔振器 rubber isolator; rubber vibration insulator
橡胶隔振体 rubber-type vibration isolator
橡胶隔震器 rubber shock absorber
橡胶工业 rubber industry
橡胶工业机械 rubber machinery
橡胶工业用黏[粘]土 rubber clay
橡胶箍稳定器 rubber-sleeve stabilizer
橡胶骨料 rubber aggregate
橡胶刮板 rubber flap; rubber grab; squeegee;squilgee
橡胶刮板式升运器 rubber-flight elevator
橡胶刮刀 rubber squeegee
橡胶刮管器 <从提升钻杆上清除泥浆的> stripper rubber
橡胶刮水板 rubber squeegee
橡胶管 hose; rubber canvas hose; rubber pipe;rubber tube
橡胶管卡圈 rubber hose clamp
橡胶管式水准器 rubber tube level
橡胶管式水准仪 rubber tube level
橡胶管套 rubber sleeve
橡胶辊筒 rubber-coated pressure bowl
橡胶辊(子) rubber roller
橡胶滚轮 rubber-covered roller
橡胶滚筒 roller squeegee
橡胶滚轴 rolling rubber
橡胶滚子 rubber squeegee;squeegee
橡胶过滤板 rubber screen plate
橡胶过早硫化 scorching
橡胶海绵 cellular rubber; expanded rubber;foam rubber;rubber foam; rubber sponge
橡胶号码印 rubber number stamp
橡胶合成树脂混合物 rubber-synthetic resin mix(ture)
橡胶护面的 rubber-lagged
橡胶护墙板 sheeting rubber
橡胶护舷 <码头用> rubber fender; elastomeric fender
橡胶护舷失去弹性变脆的温度 glass-transition temperature of rubber fender
橡胶化焦油铺路合剂 rubberized-tar paving mixture
橡胶环 rubber collar; rubber grommet;rubber ring
橡胶缓冲挡 rubber bumper; rubber buffer
橡胶缓冲垫 rubber buffer
橡胶缓冲器 India-rubber cushion; rubber buffer;rubber bumper;rubber cushion;rubber draft gear
橡胶缓冲装置 rubber buffer;rubber bumper
橡胶簧 India-rubber spring
橡胶灰浆桶 rubber hod
橡胶混合物 rubber composition;rubber compound;rubber stock
橡胶活动坝 collapsible rubber dam
橡胶活塞 rubber piston
橡胶机器带 rubber machine belting
橡胶基密封剂 rubber-based sealant
橡胶基涂料 rubber-based paint
橡胶基质 rubber mass;rubber matrix
橡胶级滑石 rubber-grade talc
橡胶集料 rubber aggregate
橡胶集装袋 rubber container
橡胶计 rubbermeter
橡胶加工 rubber processing
橡胶加工厂 rubber mill
橡胶加工设备 rubber tooling
橡胶加工用油 rubber processing oil
橡胶夹层 channel rubber
橡胶夹层地板 floating floor
橡胶夹衬软管 rubber-lined hose
橡胶夹手 rubber-lined claw
橡胶夹心弹簧 rubber sandwich spring
橡胶间隔片 rubber spacer
橡胶减震 rubber shock absorption
橡胶减震垫 rubber cushion
橡胶减震器 rubber bumper; rubber damper;rubber shock absorber
橡胶减震筛 rubber-supported screen
橡胶减震支座 rubber-block support
橡胶减震柱 rubber shock absorber strut
橡胶减震装置 rubber shock absorber
橡胶件 rubber element
橡胶建筑胶黏[粘]剂 rubber building mastic
橡胶建筑胶黏[粘]水泥 rubber building mastic
橡胶浆 rubber latex
橡胶浆沥青乳液 rubber latex-asphalt emulsion
橡胶浆黏[粘]接剂 rubber latex cement
橡胶浆水泥砂浆 rubber latex cement mortar
橡胶胶 rubber mastic
橡胶胶带 rubber tape
橡胶胶合 rubber bond
橡胶胶合剂 rubber cement
橡胶胶结软木垫 rubber-bonded cork pad
橡胶胶乳 rubber latex
橡胶胶水 rubber cement;rubber solution
橡胶胶水滤器 rubber solution strainer
橡胶接缝 rubber joint
橡胶接管 rubber cone
橡胶接头 elastomer connector; rubber joint;rubber union
橡胶结构垫圈 rubber structural gasket
橡胶结合剂 gum cement;rubber bond
橡胶结合砂轮 rubber wheel
橡胶截水条 rubber waterstop
橡胶金属叠合垫块 rubber-bonded-to-metal pad
橡胶金属叠合垫片 rubber-bonded-to-metal pad
橡胶金属铰节 silent block
橡胶金属结合法 plioweld
橡胶金属结合件 rubber-metal assembly
橡胶金属结合组装配件 rubber-metal assembly
橡胶紧密卷(车门) rubber backing roll
橡胶绝缘 rubber insulation
橡胶绝缘板 rubber insulating board
橡胶绝缘编包风雨线 rubber-covered braided
橡胶绝缘导线 rubber conductor
橡胶绝缘的 rubber-covered; rubber-insulated
橡胶绝缘的编包风雨线 rubber-covered braided weather proof wire
橡胶绝缘电缆 Indian rubber cable; India-rubber cable;india-rubber insulated cable; rubber-insulated cable;rubber cable
橡胶绝缘电线 rubber insulation wire;rubber-insulated wire
橡胶绝缘软线 cab-tyre cord
橡胶绝缘双层编包线 rubber-covered double braided
橡胶绝缘体减震架 rubber-insulated body mount
橡胶绝缘外壳 rubber-covered outer casing
橡胶绝缘外罩 rubber-covered outer casing
橡胶绝缘线 India-rubber wire
橡胶绝缘子 rubber-insulator
橡胶铠装电缆 rubber-sheathed cable
橡胶颗粒改性沥青 crumb rubber modified asphalt
橡胶颗粒改性沥青混合料 crumb rubber modified mixture
橡胶壳 <电瓶的> rubber jar
橡胶控制条 neoprene control strips
橡胶块 rubber block
橡胶块减震器 rubber-block shock absorber
橡胶块联轴器 coupling with rubber pad
橡胶块履带 rubber-block track

橡胶块（铺砌）路面 rubber-block pavement
橡胶快速接头 rubber quick-joint
橡胶拉伸模量 rubber extension modulus
橡胶拉深 rubber pad drawing
橡胶缆 rubber cable
橡胶老化 ag(e)ing of rubber
橡胶类黏[粘]结剂 rubber cement
橡胶沥青 rubber bitumen
橡胶沥青层面材料 ruberoid
橡胶沥青底基 rubberised bitumen underlay
橡胶沥青封缝料 Thormaseal
橡胶沥青合成料 rubber-bitumen composition
橡胶沥青混合物 latex bitumen mixture;rubber-bitumen composition
橡胶沥青接缝料 Thorma-joint
橡胶沥青密封 rubber bitumen seal
橡胶沥青密封剂 rubber asphaltic bitumen sealing compound
橡胶沥青密封条 rubberized asphalt sealing strip
橡胶沥青乳胶 rubber(asphaltic)bitumen emulsion
橡胶沥青填缝 rubber-bitumen joint
橡胶沥青填缝料 rubber-bitumen sealing compound
橡胶粒 rubber crumb
橡胶连接层 rubber bonding layer
橡胶连接器 rubber coupling
橡胶帘布缓冲层 rubberized breaker cord
橡胶联结器 rubber coupling
橡胶联轴节 rubber coupling
橡胶劣化 rubber deterioration
橡胶裂化 rubber cracking
橡胶溜管 rubber elephant trunk
橡胶硫化 vulcanization of rubber; vulcanize
橡胶硫化器 vulcanizer
橡胶硫化设备 vulcanizing equipment
橡胶硫化作用 vulcanization
橡胶楼板覆盖层 rubber floor covering
橡胶履带 rubber(belt)track;rubber crawler
橡胶履带板 rubber track shoe
橡胶履带车辆 rubber-tracked vehicle
橡胶履带片 rubber track shoe
橡胶绿 rubber green
橡胶轮 rubber wheel
橡胶轮卡车 rubber-tyred truck
橡胶轮圈 rubber rim
橡胶轮手推车 Georgia buggy
橡胶轮胎 rubber tire[tyre]; rubber wheel tyre[tire]
橡胶轮胎（车辆组成的）列车 rubber-tyred train
橡胶轮胎的搬运汽车 rubber-tired carrier
橡胶轮胎高架起重机 rubber-tyred overhead crane
橡胶轮胎护舷 rubber-tired fender
橡胶轮胎缓冲装置 rubber-tired fender
橡胶轮胎卡车起重机 rubber tyred truck crane
橡胶轮胎内带 rubber wheel tyre inner tube
橡胶轮胎式车轮 rubber-tyred wheel
橡胶轮胎式的 rubber-tired
橡胶轮胎压路机 rubber-tired[tyred] roller
橡胶轮胎辗压机 rubber-tired roller
橡胶轮拖拉机 rubber-mounted tractor
橡胶轮压路机 rubber wheeled roller
橡胶帽 rubber cap
橡胶门 rubber door
橡胶门扇 rubber door leaf
橡胶密封的 rubber sealed
橡胶密封的柔性接头 rubber sealed flexible joint
橡胶密封垫 gum gasket;rubber gasket;rubber grommet;rubber seal
橡胶密封垫圈 rubber sealing gasket
橡胶密封阀 rubber sealed valve
橡胶密封环 rubber sealing ring
橡胶密封环钎尾 seal ring shank
橡胶密封件厂 sealing parts factory
橡胶密封圈 rubber grommet; rubber packing ring;rubber sealing ring
橡胶密封套 rubber sealing boot
橡胶密封条 rubber sealing strip
橡胶面层 rubber facing
橡胶（面层）砖 rubber tile
橡胶面衬垫 rubber-faced pad
橡胶面钢板 rubber-faced steel plate
橡胶面炉垫 rubber-faced pad
橡胶面铺地砖 rubber-faced tile
橡胶面铺砌块 rubber-surfaced paving block
橡胶面石棉水泥面砖 cement asbestos rubber tile
橡胶面砖地面 rubber tile floor(ing)
橡胶模 rubber die
橡胶模成型 rubber pressing
橡胶模压力机 rubber die press
橡胶膜 rubber skin
橡胶膜板 rubber diaphragm
橡胶膜校正 rubber membrane correction
橡胶膜片曝气器 rubber film aerator
橡胶摩擦传动 rubber friction drive
橡胶摩擦带 rubber friction tape
橡胶摩擦缓冲器 rubber friction draft gear
橡胶磨 rubber mill
橡胶内衬泥沙泵 rubberized sand pump
橡胶内胎 rubber inner tube
橡胶泥 rubber dough
橡胶黏[粘]合剂 gum cement; rubber adhesive
橡胶黏[粘]结剂 gum cement; rubber bond;rubber cement
橡胶黏[粘]结料 rubber bond
橡胶黏[粘]结砂轮 rubber-bond wheel
橡胶捻缝 marine glue ca(u)lking
橡胶凝块 rubber lump
橡胶扭力轴承 rubber torsional bearing
橡胶排水管 rubber drain
橡胶盘根 rubber packing
橡胶盘减震器 rubber disk shock absorber
橡胶盘减震支柱 rubber disk shock strut
橡胶抛光杯 rubber polish cup
橡胶跑道 rubber track
橡胶泡胀试验液 rubber-swelling test fluids
橡胶配方 rubber compounding
橡胶配合剂 rubber ingredient
橡胶喷嘴 rubber pinch tip
橡胶膨润试验 rubber swelling test
橡胶膨胀节 rubber expansion joint
橡胶碰垫 rubber fender
橡胶碰门垫 bumper;rubber silencer
橡胶皮带 flat rubber belting
橡胶皮带传动 elastic drive
橡胶皮带输送机 rubber belt conveyer[conveyor]
橡胶皮模 rubber mo(u)ld
橡胶皮碗 packing rubber;rubber cup
橡胶片 rubber tissue;rubber washer; sheet rubber; rubber tile <铺地面用>
橡胶片涂漆 padding
橡胶铺地砖 rubber flooring tile;rubber tile
橡胶铺垫 rubber carpet
橡胶铺块 rubber paving block
橡胶铺面 rubber paving
橡胶铺面地板 rubber flooring
橡胶蹼 flipper
橡胶漆 rubber-based coating; rubber paint;rubber solution
橡胶气垫 air mattress;rubber cushion
橡胶气囊 rubber balloon
橡胶气囊防舷材 pneumatic rubber fender
橡胶气球法 <测密度> rubber balloon method
橡胶钎肩 rubber collar
橡胶嵌缝材料 rubber joint filler
橡胶嵌入片 rubber insert(ion)
橡胶嵌入效应 rubber penetration effect
橡胶切片机 rubber sheet cutter
橡胶轻制沥青 rubberized cutback
橡胶清漆 rubber varnish
橡胶球 rubber ball
橡胶球台面 rubber-ball deck
橡胶圈 India-rubber ring;rubber band
橡胶圈减震器 rubber-ring shock absorber
橡胶圈连接 rubber-gasketed joint
橡胶圈密封 seal with elastomeric washer
橡胶圈式缓冲托辊 rubber-ring-type buffer idler
橡胶热补机 rubber vulcanizing machine
橡胶热炼机 rubber warmer
橡胶容器 flexible cell; rubber container
橡胶溶剂 rubber solvent
橡胶溶液 rubber solution
橡胶乳 latex emulsion
橡胶乳剂涂料 rubber-emulsion paint
橡胶乳胶漆 rubber-emulsion paint
橡胶乳液 latex liquefaction;latex liquid;rubber emulsion;rubber latex
橡胶乳液地面 rubber latex floor
橡胶乳液黏[粘]合剂 rubber latex cement
橡胶乳液黏[粘]结剂 rubber latex cement
橡胶乳液水泥 rubber latex cement
橡胶乳液涂料 rubber-emulsion paint
橡胶乳汁 rubber latex
橡胶软管 flexible rubber pipe;rubber rotary hose;rubber hose
橡胶软管连接 rubber hose connection
橡胶软线 flexible rubber-insulated wire
橡胶塞 rubber cork;rubber stop(per); soft rubber ball;go-devil <美国橡胶塞俗称>
橡胶三角垫密封 seal by elastomeric delta gasket
橡胶三角皮带 rubber V belt
橡胶砂轮 elastic grinding wheel
橡胶筛板 rubber screen plate
橡胶伸缩管 rubber bellows
橡胶伸缩接头 neoprene section expansion joint
橡胶绳 rubber cord
橡胶石板瓦 Roberoid slab
橡胶石块路面 rubber sett paving
橡胶石棉板 rubber asbestos plate
橡胶石棉垫料 rubber asbestos pad; Klingerite
橡胶式聚束 rubber-banding
橡胶试塞 <试验管的密实性> rubber test plug
橡胶拭子 rubber swab
橡胶手柄 rubber grip;rubber hand grip
橡胶手揿泵 rubber hand pump
橡胶手套 rubber gloves
橡胶熟化剂 rubber-curing agent
橡胶树 gumwood;rubber tree;pana-ma rubber tree <中南美产的>
橡胶树材 gumwood
橡胶树脂 gum lac;gum resin;rubber resin
橡胶树脂并用胶 rubber-resin alloy; rubber-resin blend
橡胶树脂沥青密封条 para-plastic rubberized asphalt sealing strip
橡胶树脂黏[粘]合剂 rubber-resin adhesive
橡胶树子油 rubber-seed oil
橡胶双O形环密封 double O-ring rubber seal
橡胶水封 rubber stop;rubber water-stop
橡胶水管 rubber water pipe[piping]; rubber water tube[tubing]
橡胶水龙带 Indian rubber hose;rubber rotary hose
橡胶水泥 latex cement; rubber cement
橡胶水银探条 rubber mercury bougie
橡胶素炼 masticate
橡胶索 rubber cord
橡胶态 rubbery state
橡胶态高弹区 rubbery plateau zone
橡胶态剪切模量 rubbery shear modulus
橡胶态拉伸模量 rubbery tensile modulus
橡胶态流动区 rubbery flow zone
橡胶态转变 rubber transition
橡胶弹簧 rubber spring
橡胶弹簧钩环 rubber spring shackle
橡胶弹簧缓冲盘 side buffer with rubber spring
橡胶弹性 rubber elasticity
橡胶弹性链 rubber spring chain
橡胶套 rubber case;rubber jacket
橡胶套电缆 rubber jacket cable
橡胶套管 rubber bush
橡胶套辊 rubber-covered roll
橡胶套手柄 rubber-covered handle
橡胶套筒 rubber sheath;rubber sleeve
橡胶套弹性联轴器 rubber packed coupling
橡胶套鞋 <美> gumshoe
橡胶添加剂 rubber additive
橡胶填充料 rubber filler
橡胶填料 gum filler; India-rubber packing;rubber packing
橡胶填密圈 rubber-ring packing
橡胶填圈 rubber seal
橡胶填实接缝 rubber gasket joint
橡胶填隙片 gum gasket
橡胶条 rubber strip
橡胶贴面 rubber tiling
橡胶贴面的 rubber-faced
橡胶艇 rubber boat;rubber dingey
橡胶头锤 rubber headed hammer
橡胶透声窗 gum window
橡胶凸版油墨 flexographic(al)ink
橡胶凸缘 rubber flange
橡胶图章 rubber stamp
橡胶图章印花装饰法 rubber stamp decoration
橡胶涂层 rubber covering
橡胶涂层地板 rubber floor covering
橡胶涂层织物 rubber-coated fabric
橡胶涂料 rubber coating
橡胶涂面导管 rubber-lined pipe
橡胶涂面管道 rubber-lined pipe
橡胶涂面输送管 rubber-lined pipe
橡胶涂面瓦 rubber-faced tile
橡胶土 rubbery soil
橡胶托轮 rubber wheel
橡胶拖把 rubber swab

X

橡胶(外包)电缆 rubber-sheathed cable
橡胶外包线 rubber-sheathed wire
橡胶碗 rubber bowl;rubber cup
橡胶碗打捞器 packer rubber grab
橡胶万向联轴节驱动 rubber swivel-joint cardan drive
橡胶围裙 rubber apron
橡胶污染 rubber contamination
橡胶屋面 <一种屋面材料> Roberoid roofing
橡胶屋面油毡 Roberoid roofing felt
橡胶线 rubber-covered wire; rubber thread
橡胶相容性 rubber-compatible
橡胶镶框 rubber trim
橡胶镶面金属履带铰节 rubber-faced track
橡胶消声器 rubber silencer
橡胶效应 rubber effect
橡胶芯 rubber core
橡胶型材 rubber profile
橡胶型缓行器 rubber-type retarder
橡胶型胶黏[粘]剂 rubber adhesive
橡胶性物质 rubbery substance
橡胶悬挂弹簧 <座位的> rubber suspension spring
橡胶悬架 rubber suspension
橡胶靴套 rubber boots
橡胶压垫圈 rubber seat holder
橡胶压缝 rubber compression joint
橡胶压缩式止水条 rubber compression type seal
橡胶压条 rubber hold-down strip
橡胶压制 rubber pressing
橡胶衍生物涂料 rubber derivate paint
橡胶样 rubber-like
橡胶样的 rubbery
橡胶叶片 rubber blade;rubber paddle
橡胶异构体 rubber isomer
橡胶硬度 rubber hardness(degree)
橡胶硬度测试器 rubber hardness tester
橡胶硬度计 rubbermeter
橡胶用品制造机 rubber goods making machinery
橡胶用溶汽油 gum turpentine oil
橡胶用润滑剂 rubber lubricant
橡胶用炭黑 rubber grade carbon black
橡胶用陶土 rubber clay
橡胶油封 rubber oil seal
橡胶油灰 rubber mastic
橡胶与金属相结合 attaching rubber to metal
橡胶与塑料研究协会 <英> Rubber and Plastics Research Association
橡胶元件 rubber element
橡胶园 rubber estate;rubber plant
橡胶原批 masterbatch
橡胶圆盘 puck
橡胶增强材料 rubber reinforcement
橡胶增塑剂 rubber plastizing agent
橡胶闸瓦 rubber shoe
橡胶毡 rubber felt
橡胶胀缝装置 rubber expansion joint device
橡胶胀形模 rubber bladder
橡胶障穿孔器 rubber dam punch
橡胶障吊锤 rubber dam plumb
橡胶障夹 rubber dam clamp
橡胶障夹夹持钳 rubber dam clamp forceps;rubber dam clamp holding forceps
橡胶真空吸盘 rubber vacuum cup
橡胶枕簧 rubber-encased bolster spring
橡胶整理 racking of rubber
橡胶支承 rubber bearing
橡胶支座 elastomeric pad bearing; rubber seat; neoprene pad; rubber bearing;rubber plinth
橡胶直浇口 rubber gate stick
橡胶植物 rubber bearing plant; rubber producing plant; rubber yielding plant
橡胶止水 rubber(water) stop;rubber seal;water sealing with rubber
橡胶止水带 rubber waterbar;rubber waterstop strip
橡胶止水环 neoprene seal(ing) ring; rubber sealing ring
橡胶止水条 rubber sealing strip; staunching rod <堰顶闸门用>
橡胶指套 rubber finger-cot; rubber finger-stall
橡胶制(混凝土)输送管 rubber elephant trunk
橡胶制品 rubber goods;rubber item; rubber product
橡胶制品承受动力弯曲寿命的试验 flex life test
橡胶制止器 rubber accelerator
橡胶质黏[粘]合剂 rubber-based adhesive
橡胶种植园 plantation of rubber;rubber plantation
橡胶轴承 rubber bearing;rubber shaft bearing;rubber-supporting bearing
橡胶皱纹管 rubber bellows
橡胶助剂 rubber compounding ingredient
橡胶爪式输送器 rubber-fingered conveyer[conveyor]
橡胶砖地板 rubber tile floor(ing)
橡胶装修 rubber trim
橡胶状材料 rubber-like material
橡胶状的 elastomeric
橡胶状共聚物 rubbery copolymer
橡胶状聚合物 rubbery polymer
橡胶状流动 rubbery flow
橡胶锥形交通路标 rubber traffic cone
橡胶阻尼器 rubber cushion assembly
橡胶组分 rubber composition
橡胶组合 building rubber compound
橡胶座平板振动器 rubber based plate vibrator
橡胶座扫灰刷 rubber-set jamb brush
橡筋绳牵引起飞 shock cord takeoff
橡筋弹射起飞 rubber-cord start
橡木 oak;Tasmanian oak <澳大利亚塔斯马尼亚产>
橡木板条地板 strip oak flooring
橡木窗台板 oak sill
橡木地板 oak floor
橡木垫块 oak block
橡木护壁板 wainscot(ting) oak
橡木槛 oak threshold
橡木篱桩 oak picket
橡木拼花地板 oak parquet(ry)
橡木清漆 oak varnish
橡木条编成的篱笆 woven-oak fencing
橡木条栅栏 oak slat fence
橡木条篱笆 oak slat fence
橡木屋顶板 oak shingle
橡木镶板 oak panel
橡木栅栏 oak fence
橡木栅柱 oak stake fence
橡木枕(木) oak sleeper
橡木桩 oak post
橡皮 argtum
橡皮板 sheet rubber
橡皮包角搅拌叶片 rubber edged blade
橡皮包面铺砌块 rubber-surfaced paving block
橡皮包线 rubber-covered wire
橡皮布 blanket
橡皮布夹头 blanker damp
橡皮布紧轴 blanker wind
橡皮擦 kneaded eraser;kneaded rubber
橡皮层石棉盖板 asbestos rubber tile
橡皮衬垫接缝 <混凝土路面> rubber-gasketed joint
橡皮衬垫水力压接 rubber gasket connection
橡皮衬里 channel rubber
橡皮衬砌 armorite rubber lining
橡皮冲气艇 pneumatic boat
橡皮船 rubber dinghy
橡皮锤 rubber hammer;rubber mallet
橡皮带 elastic;rubber belt(ing)
橡皮带绝缘 belt insulation
橡皮带式生成线 rubber-banding
橡皮底帆布鞋 plimsoll
橡皮垫 rubber packing;valve rubber; rubber cushion
橡皮垫附件 rubber pad attachment
橡皮垫密片 rubber gasket
橡皮垫片 rubber gasket
橡皮垫圈 gum washer;rubber gasket; rubber-ring packing;rubber washer
橡皮垫支架振动筛 tyroc screen
橡皮筏 inflatable raft
橡皮帆布艇 collapsible lifeboat
橡皮帆布鞋 <美> gumshoe
橡皮防冰套 rubber deicer boot
橡皮防冲击装置 rubber fender system
橡皮防冲物 rubber fender
橡皮防水带 waterproof rubber strip
橡皮防水条 waterproof rubber strip
橡皮封套 rubber cuff
橡皮膏 adhesive plaster;adhesive tape; sticking plaster;strapping
橡皮隔振器 rubber vibration insulator
橡皮刮板 squeegee;pipe wiper <清除管外泥浆的>
橡皮刮泥板 <清除管内泥浆用> pipe wiper rubber
橡皮管 rubber tubing
橡皮滚筒 blanket cylinder;offset blanket
橡皮滚(子) squeegee
橡皮缓冲器 die cushion
橡皮夹层 channel rubber
橡皮减震器 rubber shock absorber
橡皮剪切块车轴悬挂 rubber shear block axle suspension
橡皮胶 rubber cement
橡皮胶轮胎 rubber tire[tyre]
橡皮胶水 cementing rubber
橡皮筋 bungee
橡皮救生艇 inflatable life-boat
橡皮绝缘编包风雨线 weatherproof wire
橡皮绝缘插座 cab-tyre connector
橡皮绝缘导线 rubber conductor
橡皮绝缘的 cable tyre;cab-tyre
橡皮绝缘电缆 cab-tyre cable
橡皮绝缘软电缆 cab-tyre cable
橡皮绝缘软线 cab-tyre cord
橡皮绝缘软性电缆 cab-tyre rubber
橡皮块 rubber block
橡皮块铺砌 rubber-block pavement
橡皮沥青 dopplerite
橡皮裂纹深度法 <确定空气污染程度的> rubber cracking method
橡皮轮胎缓冲物 rubber-tired truck-wheel fender
橡皮密封垫 gum gasket
橡皮模成型法 Wheelon forming process
橡皮模压制成型法 Marform process
橡皮膜 rubber membrane
橡皮膜校正 <土的三轴试验> membrane correction
橡皮膜嵌入效应 <三轴试验> membrane penetration effect
橡皮膜顺变性 <三轴试验> membrane compliance
橡皮奶头 nipple
橡皮泥 model(l)er's clay
橡皮铺面 rubber pavement; rubber paving
橡皮铺面块 rubber paving block
橡皮圈 quoit;rubber band
橡皮软管 rubber hose
橡皮塞 rubber bung
橡皮塞子 rubber plug;rubber stopper
橡皮扫帚 squeegee
橡皮树 <三叶树胶> para-rubber tree
橡皮栓 rubber plug
橡皮套 boot; rubber sleeve; rubber sheath <用于大型三轴试验>
橡皮填密片 gum gasket
橡皮艇 dinge(y);inflatable boat
橡皮头铅笔 abrasive pencil
橡皮土 cheesy; rubbery clay; sponge soil;spongy soil;mud wave
橡皮吸气球 rubber suction ball
橡皮吸水泵 force cup
橡皮吸水器 force cup
橡皮芯 rubber core
橡皮芯线 rolled latex
橡皮压印模 cavity-embossing rubber die
橡皮油箱 flexible bag
橡皮制品 rubber goods
橡皮制止器 rubber stopper
橡皮座 ebonite base
橡乳胶 latex emulsion
橡乳胶黏[粘]合剂 latex emulsion adhesive
橡乳胶黏[粘]合料 latex emulsion adhesive
橡实管 acorn tube;shoe-button tube
橡实夹 acorn
橡实形电池 Acorn cell
橡实形电子管 acorn tube
橡树 buck-eye;oak
橡树果 acorn
橡树子 acorn
橡塑防水片材 rubber-plastic waterproofing sheet
橡苔油 oakmoss oil
橡碗 acorn cup;valonea;valonia
橡子 acorn

肖

肖伯抗拉试验仪 Schopper's tensile tester
肖伯抗张强度试验机 Schopper's tensile strength machine
肖伯耐折度测定仪 Schopper's folding tester
肖伯扭力仪 Schopper's torsion meter
肖伯织物顶破强力机 Schopper's bursting strength tester
肖伯织物耐磨试验仪 Schopper's cloth abrasion tester
肖德贝里铜镍矿床 Sudbury copper-nickel deposit
肖恩喷射器 Shone ejector
肖尔曼热风冲天炉 Schuermann cupola;Schuermann furnace
肖兰波路径 shoran wave path
肖兰穿线飞行测距法 shoran-line crossing
肖兰导航系统 shoran system
肖兰定位 shoran position fixing
肖兰法 shoran method
肖兰反算问题 inverse shoran problem
肖兰精密导航装置 <电子测距体系,测自空中站到两个地面站的距离> shoran[short-range navigation system]
肖兰距离校正 shoran reduction

肖兰控制 shoran control
肖兰控制航空摄影 shoran-controlled photography
肖兰控制摄影 shoran-controlled photography
肖兰控制网 shoran net; shoran network
肖兰三边测量 shoran trilateration
肖兰三角测量 shoran triangulation
肖兰问题 shoran problem
肖兰像片定位 shoran control
肖兰直线指示器 <帮助飞行员作直线飞行用> shoran straight line indicator
肖兰作用距离 shoran range
肖钠长石 pericline
肖钠长石律 pericline law
肖钠长石双晶律 pericline twin law
肖瑞金属喷涂法 Schori process
肖瑞抗腐雾 Schori
肖氏防护罩 Shaw guard
肖氏回弹硬度测定 scleroscopic hardness test
肖氏回跳硬度器 Shore scleroscope
肖氏回跳硬度试验 scleroscopic hardness test; Shore scleroscope hardness test
肖氏精密铸造法 Shaw process
肖氏弹跳硬度试验 rebound Shore hardness test
肖氏窑 Shaw kiln
肖氏硬度 durometer hardness; rebound hardness; scleroscope hardness; Shaw hardness; Shore hardness; Shore scleroscope hardness
肖氏硬度测定 shore test
肖氏硬度级 Shore hardness scale
肖氏硬度计 hardness drop tester; sclerometer; scleroscope; Shore durometer; shore hardness tester; Shore scleroscope
肖氏硬度实验 Shore hardness test
肖氏硬度实验仪 shore hardness tester
肖氏硬度试验 drop hardness test; rebound test; Shore hardness test
肖氏硬度试验方法 Shore hardness testing method
肖氏硬度试验计 shore hardness tester
肖氏硬度试验仪 shore hardness tester; Shore scleroscope
肖氏硬度数 Shore hardness number
肖氏铸造法 Shaw process
肖特基放射 Schottky emission
肖特基缺陷 Schottky defect
肖特基势垒 Schottky barrier
肖特基位错 Schottky disorder
肖特钟 free-pendulum clock; Shortt clock
肖沃特稳定指数 Showalter stability index

枭混线【建】cima recta; cyma recta; ogee mo(u)lding

X

枭混线脚 cyma reversa; sima reversa

削笔刀 pen knife

削壁 escarpment
削壁采准 resuing development
削边板 chamfered board; cloven board
削边(尺) feather-edge
削边粗石工(作) chamfered rustic work
削边刀 edge tool
削边踏步 chamfered step
削边镶板 chamfered panel
削边砖 feather-edged brick; side cut brick
削波 clipping; slicing
削波电路 chopping circuit; clipper circuit; clipping circuit
削波电平 clip level
削波二极管 clipper diode
削波放大器 clipping amplifier
削波管 clipper tube
削波畸变 clipping distortion
削波控制 chopper control
削波频率 chopping frequency
削波砌块的岸壁 quay wall of sliced blockwork
削波砌块的岸墙 quay wall of sliced blockwork
削波器 chopper; clipper; slicer; wave clipper
削波器时间常数 clipping constant
削波时间常数 clipping time constant
削薄轮缘 thinned wheel flange
削薄木板 shingle board
削成钝角 blunt
削成碎片 splintering off
削成斜面 <木端> flinching
削除事项 deletion
削磁开关 field-breaking switch
削刀 broach; sharpener
削刀式剥皮机 cutter-head barker
削低山 subdued mountain
削掉 chipping off
削顶脉冲 chopped pulse
削陡作用 oversteepening
削短叶片 shortened blade
削断山嘴 truncated spur
削发融资 <美国术语> hair cut finance
削方 square yard; squaring
削峰 crest truncation; despiking; peak clipping
削峰背斜 scalped anticline
削峰电路 despiker circuit; peak-clipping circuit
削峰电阻 despiking resistance
削峰高(度) crest truncation
削峰器 despiker
削光面 faced surface
削痕 sheeter lines
削价 cut price; lower the price; markdown; price cutting; undercut price
削价出售 undercut
削价处理商品 selling goods at reduced prices
削价竞争 price-cutting competition; price war
削价市况 discount market
削价运费 distress freight
削尖 fine away; pointing; sharpen(ing)
削尖的 sharp-pointed
削尖脉冲 peaking pulse
削尖山嘴 sharpened spur
削减 chop; curtail(ment); cutback; minify; pare; put down; undercut
削减带 subduction zone
削减贷款 curtail credits to enterprises
削减的生产量 curtailed production
削减的消费量 curtailed consumption
削减非生产性开支 apply the ax to nonproductive expenditures
削减费用 abridgment of expenses; deduct expenses
削减工资 rate cutting; wage cut
削减关税 reduction in tariff
削减管理费 decreased overhead
削减或消除账面价值 write off
削减进口 import curtailment
削减经费 ax(e) [复 axes] outlay
削减军需储备 military reserve decrease
削减预算 chop the budget; trim a budget; trim budget
削减支出 put down expenditures
削剪 clipping
削角【建】angle of chamfer; angle of bevel; angle of skew; chamfer(ing); cope; top rake
削角边 chamfered edge
削角边板 canted bearing plate
削角沉箱防波堤 sloping top caisson breakwater
削角的 chamfered
削角电刷 bevel(l)ed brush
削角垫块 angle block
削角防波堤 chamfered gravity breakwater
削角角度 chamfer angle
削角刨 chamfering plane
削角砌缝 chamfered rustication
削角砌块 chamfered block
削角式防波堤 gravity breakwater with cutaway cross-section
削角条 cant strip; plane strip
削角直立堤 chamfered vertical breakwater
削角柱 chamfered column
削角钻 chamfer bit
削节 <木材的> knobbling; knobble
削截圈闭 truncation trap
削具 sharpener
削砍石块 knobble
削孔模 broaching die
削宽榫 tease tenon
削梁 cope in steel beam
削裂 stripping
削面 bevel
削面凹凿 paring gouge
削皮 paring
削皮刀 paring knife
削皮工 spudder
削皮器 scratcher; spudder
削片 shave; splintering; spall
削片打磨机 chip grinder
削片机 chipper; shaving machine
削片压痕 chip marking; chip marks
削平 bevelment; bulldozing; dubbing; level(l)ing; scapple; truncation
削平波 chipped wave
削平补强的焊缝 flush weld; weld machined flush
削平槽规 adzing ga(u)ge
削平的 faced
削平焊道 flush welding
削平焊缝 weld machined flush
削平洪峰 smoothing off peak discharge
削平石面 plucked finish
削平作用 bevel(l)ing
削坡 bank grading; grading; slop cutting
削坡减载 slope-cutting for lightening load
削铅笔刀 pen knife
削峭作用 oversteepening
削切(山)坡 facetted spur
削去 chip(ping) off; detruncate; pare; scrape off; thinning
削去屋脊的坡屋顶 cut roof
削弱 abate; abating; de-emphasis; rebate; weakening
削弱磁场的步级 step of field weakening
削弱磁场阀 field-weakening valve
削弱磁场控制 field weakening control
削弱磁场系数 field weakening factor
削弱磁场运转控制 control of weak field operation
削弱的 weaken
削弱区 weak zone
削弱系数 coefficient of extinction
削蚀背斜 truncated anticline
削蚀的岬角 truncated headland
削蚀作用 erosional truncation
削榫 beaking joint
削榫机 dovetailing machine
削台 shooting block
削土能力 earth cutting ability
削下来 chip-off
削下来的皮 paring
削屑 shaving
削圆角 cavetto
削圆转角 truncated corner
削匀机 shaving machine
削錾 paring chisel
削凿(刀) cape chisel; paring chisel; cross-cut chisel
削枕机 adzing machine
削整 adz(e); adzing
削正 shaving
削砖砌层 split course

消摆龙骨 bilge chock

消雹 hail mitigation
消冰 deicing
消冰剂 deicing agent
消波 beach reflection; wave reduction
消波岸坡 wave absorbing beach
消波沉箱堤 wave dissipating caisson breakwater
消波池 surge basin; wave basin; wave trap
消波丁坝 wave reflecting spur wall
消波工程 wave breaking works; wave dissipating works
消波构筑物 works for dissipating wave
消波海滩 expending beach; spending beach
消波护岸 wave absorbing revetment; wave dissipation revetment
消波混凝土块体 hollow square; wave dissipating concrete block
消波建筑物 wave absorbing structure; wave breaking works
消波结构 wave absorbing structure
消波井 stilling well
消波阱 wave trap
消波空气帘 air-bubble breakwater
消波块覆盖堤 upright breakwater covered with wave dissipating concrete blocks
消波块(体) wave energy dissipating concrete block; wave breaker block; wave breaking block
消波器 water breaker; wave absorber; wave damper; wave subducer; wave suppressor
消波设备 wave absorber; wave eliminater
消波设施 wave absorbing construction; wave breaker; wave breaking facility; wave damping device; wave trap
消波室 wave chamber
消波滩 spending beach; wave absorbing beach; wave trap
消波噪声 clipped noise
消波装置 wave absorber; wave damper
消波作用 water damping effect; wave damping effect
消侧音 anti-side tone
消侧音电话机 anti-sidetone set
消侧音电路 anti-sidetone circuit
消侧音感应线圈 anti-sidetone induction coil
消层 de-stratification
消长 growth and decline; regression
消超载台 overload relief bed
消尘剂 dust laying agent

消冲电路 smearer
消冲击器 shock suppresser[suppressor]
消除 cancel(lation);clearing;efface;eliminate;elimination;erase;extinguish(ing);slake;backing-off < 应力的 >
消除……间的相互影响 decouple
消除白口退火 chill removing annealing
消除背景噪声 anti-ground noise
消除本机干扰 clutter(ing)rejection
消除壁垒和堵塞 remove barriers and blockades
消除表面漏泄法 guarding
消除病源 getting rid of the source of infection
消除不确定性与变化 offset uncertainty and change
消除步骤 removal process
消除残余应力退火 relief annealing
消除残余应力退火炉 stress-relieving furnace
消除测站偏心 allow for an eccentric set
消除差错 elimination error
消除长期趋势 detrending
消除常数 elimination constant
消除程序修改 unwind(ing)
消除错误原因 error cause elimination;error cause removal
消除氮的氧化物 denoxing
消除地面影响 anti-ground
消除电离作用 de-ionization
消除电路 cancel circuit;wipe circuit
消除毒气 degas
消除毒气的毒性 degassing
消除毒性 toxicity elimination
消除堵塞 unblocking
消除短路(现象)unshorting
消除多余的线 < 计算机绘图 > drop line
消除多余运算 remove redundant operation
消除恶臭 neutralization of odor
消除法 method of elimination;null method
消除放大器 erase amplifier
消除放射性 radioactive decontamination
消除放射性措施 contamination control
消除放射性污染 deactivation;detoxification
消除放射性循环 decontamination cycle
消除放射性沾染 desactivation;radioactive decontamination
消除放射性质点 removing of radiological agents
消除分配的缺点 remove the defects of distribution
消除干扰 denoise
消除感染 eliminating infection
消除公害 abatement of nuisance
消除功能 obliterating power
消除故障 trouble shoot(ing)
消除故障的人 trouble-shooter
消除过程 removal process
消除灰尘 abatement of dust(ing)
消除回波 echo elimination;echo suppression
消除回声 echo elimination
消除火花 arc control;extinction of spark
消除火花用整流器 rectifier for contact protection
消除机 canceling machine
消除积炭 removing carbon
消除激励 depriming
消除激励电路 deenergizing circuit
消除疾病 eradication of disease
消除剂 remover
消除季节变动后的数据 deseasonalized data
消除季节变动后数列 deseasonalized series
消除季节变动影响 deseasonalizing
消除季节性因素 deseasonalization
消除加工硬化 release of work hardening
消除键 cancel key
消除(交通)危险事故计划 < 美 > hazard elimination program(me)
消除结块 de-blocking
消除静电 electrostatic elimination;elimination of static electricity
消除静电作用 destaticization
消除局部应力热处理 local stress relief heat treatment
消除开关 defeat switch
消除空气污染 air pollution abatement
消除浪费 waste elimination
消除亮度 deaden
消除临时工 decasualize
消除隆起 abatement of swell
消除率 elimination factor;elimination rate
消除码 blanking code
消除脉冲 extinguishing pulse
消除铆接漏缝 stop riveting
消除冒口残根 pad
消除内残余应力退火 stress-relief annealing
消除内应力 relieving of internal stress;removal of internal stress
消除内应力退火 stress-relief annealing
消除年月日 purge date
消除黏[粘]结 debond(ing)
消除黏[粘]结的预应力筋 debonded tendon
消除偏差 deviation compensation
消除平交道口 elimination of grade crossings
消除气泡器 bubble eliminator
消除气体 removal of gases
消除器 allayer;annihilator;canceller;eliminator;eradicator;eraser;extinguisher;suppressor
消除氰化物质 cyanicide
消除区组效应法 method of eliminating block effects
消除缺点 debug
消除热量 elimination of heat
消除沙槛 bar removal
消除失业 eliminate unemployment
消除视差 parallax clearance
消除收缩的热处理 heat-treatment for shrinkage relief
消除数据波动 smoothing
消除水 elimination of water
消除水气 < 玻璃等 > defog
消除水污染 abandonment of water pollution;abatement of water pollution
消除速率 elimination rate
消除速率常数 elimination rate constant
消除坍方 removing landslide
消除通货膨胀指数 deflator
消除外汇风险 eliminate foreign exchange risk
消除危石 chopping;scaling;trimming
消除污点 spotting out
消除污染 abatement of pollution;abating pollution;decontamination;depollution;pollution abatement
消除污染贷款 anti-pollution loan
消除污染流体 decontamination fluid
消除污染区域 decontamination area
消除污染设施 decontamination device
消除污染室 decontamination chamber
消除污染投资 anti-pollution investment
消除污染物 removal of pollutants
消除污染循环 decontamination cycle
消除物 erasure
消除误差 debug
消除信号 erasure signal
消除压力 neutral pressure
消除烟尘 abatement of smoke
消除烟雾设施 anti-fog and anti-smog device;anti-fog antismog
消除岩尘 rockdusting
消除叶片气流分离 recovery from the blade stall
消除隐患 hazard clean-up
消除应力 relaxation of stress;relieving stress;stress-relieving
消除应力处理裂纹 stress-relief annealing crack
消除应力的 stress-relieved
消除应力的热处理 stress-relief heat treatment;stress-relieving
消除应力钢丝 stress-relieved wire
消除应力回火 strain relief tempering
消除应力加热温度 stress-relief temperature
消除应力热处理 stress-relief heat treatment
消除应力热处理炉 heat-treatment furnace for stress relieving
消除应力时效 stress ag(e)ing
消除应力退火 recovery;stress annealing;thermal relief
消除应力岩石 relaxed rock
消除涌浪 abatement of swell
消除用具 eraser
消除油轮残余气体工具 gas devourer
消除有害事物的成本 abatement cost
消除余应力回火 stress-relief tempering
消除杂音的 anti-jamming
消除噪声 abatement of noise;noise elimination
消除噪声电容器 anti-hum capacitor
消除者 eliminator
消除指令【计】purge
消除指印防锈油 fingerprint remover
消除指印纹(防锈)试验 fingerprint removal test
消除肿胀 abatement of swell
消除种族隔离 desegregation
消除装置开关 eraser gate
消磁 degaussing;de-magnetization;depolarization;erasement;field discharge;washout;wiping
消磁按钮 erase key
消磁泊位 wiping berth
消磁程度 erasability
消磁船 degaussing ship;degaussing vessel
消磁带 degaussing belt
消磁电缆 degaussing cable
消磁电流 erasing current
消磁电阻 field-breaking resistance;field-discharge resistance
消磁观测场 degaussing range
消磁剂 depolarizer[depolorizer]
消磁开关 field-breaking switch;field-discharge switch
消磁力 de-magnetizing force
消磁率 de-magnetizing factor
消磁脉冲 erasing pulse
消磁能力 erasing ability
消磁频率 erase frequency
消磁器 degaussing gear;eraser;magnetic eraser
消磁曲线 de-magnetization curve
消磁绕组 degaussing coil;killer winding
消磁设施 degaussing facility
消磁时间 erasing time
消磁头 erase head;eraser;erasing head
消磁系数 de-magnetizing factor
消磁系统 eraser system
消磁线圈 anti-magnetized coil;degaussing coil
消磁效果 erasure effect
消磁效应 demagnetising effect
消磁因素 erasing factor
消磁元件 degaussing component
消磁站 degaussing range
消磁振荡器 erase oscillator
消磁作用 de-magnetizing action;demagnetizing effect
消电离 de-ionization
消电离电位 extinction potential
消电离电压 de-ionization potential
消电离断路器 de-ion circuit breaker
消电离剂 de-ionizing agent;ionization buffer
消电离灭磁器 de-ionization de-exciter
消电纳网络 susceptance-annulling network
消毒 anti-poisoning;decontamination;despumate;detoxify;disinfect;sterilize;pasteurize < 法国巴氏法 >
消毒杯 disinfectant cup
消毒槽 disinfecting chamber;sterilizing chamber
消毒厕所 chemical closet
消毒厕座 chemical closet
消毒测定器 sterilometer
消毒厂 disinfection plant
消毒池 disinfecting pool;disinfecting tank;sterilizing chamber
消毒纯水 sterilized pure water
消毒措施 measure for disinfection
消毒的 antiseptic(al);aseptic;disinfectant;sterile
消毒灯 sterilamp
消毒动力学 disinfection kinetics
消毒法 disinfection;sterilization
消毒方法 means of disinfection
消毒房间 antiseptic room
消毒废水 disinfected wastewater
消毒废物 disinfected waste
消毒费 disinfection charges
消毒副产品 disinfection byproduct
消毒副产物 disinfection byproduct
消毒坩埚 sterilized crucible
消毒工艺 disinfection process
消毒剂 disinfector;sterilizing agent
消毒剂残留 disinfectant residual
消毒剂副产物 disinfectant byproduct
消毒剂浓度 disinfectant concentration
消毒剂评价 disinfectant evaluation
消毒剂与消毒副产品法令 Disinfectant/ Disinfection Byproducts Rule
消毒净水(法)javellization
消毒漏斗 sterilized fennel
消毒滤器 filter-sterilizer
消毒滤纸 sterilized filter paper
消毒灭菌 sterilization
消毒喷雾器 dynalysor
消毒器 sterilizer;sterilizing apparatus;sterilizing outfit
消毒器具 disinfector
消毒钳 sterilizing forceps
消毒杀菌剂 disinfectant
消毒纱布 antiseptic gauze
消毒烧杯 sterilized beaker
消毒烧瓶 sterilized flask
消毒设备 decontaminating apparatus;disinfection equipment;disinfection

plant; equipment of disinfection; sterilizer instrument; sterilizing apparatus; sterilizing equipment
消毒设施 disinfection equipment; disinfection facility
消毒试管 sterilized test tube
消毒室 disinfection chamber; disinfection room; sterilizing chamber; sterilizing room; decontamination chamber
消毒水 sterilized water; water sterilization
消毒塔 sterilizing tower
消毒土壤 disinfection soil; pasteurized soil
消毒箱 disinfecting chamber
消毒效应 sterilizing effect
消毒烟剂 fumigant
消毒药 shell sanitizer
消毒用电热器 electric(al) heater for sterilization
消毒员 disinfectioner; sterilizer
消毒皂 disinfected soap
消毒者 disinfector
消毒蒸锅 autoclave
消毒证明文件 phytosanitary certificate
消毒证书 certificate of disinfection
消毒纸尖 sterilized absorbent points
消毒种子 disease-treated seed; disinfection seed
消毒组合工艺 disinfection combination process
消毒作用 disinfection; disinfecting action; sterilizing effect
消多色差的 apochromat
消多色差物镜 apochromatic objective
消扼流效应 unchoking effect
消扼流作用 unchoking effect
消反射敷层 anti-reflecting coating
消反射膜 anti-reflecting film
消防 fire control; fire fighting; fire-protection; fire suppression
消防爱好者 fire fan
消防安全标志 fire-protection safety sign
消防安全部门 fire-fighting safety service
消防安全出口 fire escape
消防安全措施 means of escape; method of mean
消防安全服 fire safety clothing
消防安全合格证书 fire safety certificate
消防安全检查员 fire runner
消防安全教育 fire safety education
消防安全例行检查 fire security routine inspection
消防安全门 fire exit
消防安全通道 means of escape
消防安全系统法 fire safety system approach
消防安全周 fire safety week
消防包装 fire-protection encasement
消防泵 fire-extinguishing pump; fire-fighting pump; fire (-proofing) pump
消防泵房 fire-fighting pumping room; fire pump room
消防泵检验合格证 pump certification
消防泵接水口 fire service connection
消防泵站 fire pump station
消防泵组 fire-fighting pump set
消防标志 fire symbol
消防标准 fire-protection criterion
消防博物馆网络系统 fire museum network system
消防部门 fire agent
消防部门的设备 facilities of fire department
消防部门工作人员 fire department personnel
消防部门数据 fire department system
消防部门系统 fire service system
消防部门业务 fire department operation
消防部门主管人 fire chief
消防部署 fire bill; fire-fighting station; fire quarters
消防产业 fire-protection industry
消防车 fire brigade vehicle; fire control car; fire engine; fire-extinguishing tanker; fire fighter truck; fire-fighting vehicle; fire pumper; fire-tank wagon; fire tender; fire truck; motor fire brigade vehicle; motor fire brigade vehicle
消防车道 fire driveway
消防车胶管卷 fire brigade hose reel; fire truck reel
消防车接头 pumper outlet nozzle
消防车库 fire engine house; fire engine room; fire house
消防车领班 engine boss
消防车上的液压起重机 snorkel
消防车通道 fire-fighting vehicle access; fire lane
消防车行程图 fire-protection map
消防出口楼梯 fire-fighting exit staircase
消防出入口 fire department access
消防储水 fire storage
消防储水池 fire cistern
消防处 fire brigade; fire department
消防处接头 fire department connection
消防船 fire(-fighting) boat; fire-fighting ship; fire float; fire tender; fire vessel; pump ship
消防窗(户) fire-fighting window; fire window; escape window
消防措施 fire-fighting procedure; fire precaution; fire-proofing measure; fire-proofing protection
消防大接头 pumper outlet nozzle
消防带喷口 fire hose nozzle
消防等级 fire-fighting grading
消防电话线 fire telephone line
消防电梯 emergency elevator; fire elevator; fire fighter's elevator; fire-fighting elevator; fire lift
消防电梯开关 fire lift switch
消防电梯优先开关 fire lift priority switch
消防调度 fire despatch[dispatch]
消防队 fire brigade; fire-fighting crew; fire-fighting gang; fire-fighting party; fire-fighting team; fire service; firing gang; hose company; salvage corps; station house; fire company <美>; firing crew
消防队长 fire chief; fire marshal; marshal
消防队驾驶员 fire department vehicle driver
消防队进入点 fire department access point
消防队控制室 fire brigade control room
消防队水管车 hose cart
消防队锁匙箱 fire department key box
消防队员 fire fighter; fire patrolman; pompier
消防队之友 fire buff
消防队专业资格 fire fighter professional qualification
消防队组织 fire brigade organization
消防阀 fire valve
消防法规 fire legislation; fire statute
消防飞机 fire airplane
消防费 fire rate
消防分队 engine company
消防分区 fire limit; fire district
消防分区控制机 fire detection system control unit
消防封锁线 fire line
消防服 fire-protection clothing; fire suit
消防服务表彰日 fire service recognition day
消防斧 fire(man's) axe
消防负荷 fire load(ing)
消防负责人 fire boss
消防干管 fire main(pipe)
消防感温式报警探测器 fire alarm detector of thermal type
消防岗位 fire combat station; fire service
消防隔墙 fire partition wall
消防工程 fire-protection engineering
消防工程师 fire-protection engineer
消防工作 fire service; fire work
消防工作专家 fire service expert
消防供水 fire-fighting supply; fire water supply
消防供水持续时间 fire water supply duration
消防供水设备 waterworks for fire-fighting
消防供水系统 fire system
消防供水线 fire water line
消防钩 fire hook
消防管(道) extinguishing pipe; fire main(pipe); fire annihilator pipe; fire(service) pipe
消防管理 fire management
消防管理备忘录 fire management note
消防管理计划 fire management plan
消防管理路系统 fire main system
消防管理系统 fire management system
消防管理员 fire warden
消防管线 fire line
消防规范 fire code; fire-protection code
消防规则 fire regulation
消防柜 fire hose cabinet
消防护拱 fire-protection haunching
消防护目镜 fire goggles
消防机 fire engine
消防机构 fire department
消防机构位置软件包 fire service location package
消防及火警设备证书 certificate for fire extinguishing and detecting apparatus
消防及救护站 fire and ambulance station
消防及融雪系统 fire and snow melting system
消防给水 fire supply; fire water supply
消防监督 fire superintendent
消防监视站 fire monitor
消防检查体系 fire service inspection system
消防鉴定 fire rating
消防胶管 fire engine hose; fire-fighting hose
消防教官 fire service instructor
消防教育 fire education
消防界 fire community
消防界线 fire limit
消防紧急电梯 fire emergency elevator
消防紧急管理 fire emergency management
消防井 hydrant well
消防警报装置 fire alarm
消防警戒线 fire line
消防救生部署表 fire and rescue bill
消防局 fire administration; fire service
消防开关 fireman's switch
消防科学 fire science
消防孔 fire-fighting window
消防控制 fire control
消防控制盘 fire service control panel
消防控制设备 fire control equipment
消防控制室 fire service control room
消防控制系统 fire control system
消防力量 fire cover
消防立管 standpipe for fire-fighting
消防立管储水管 standpipe
消防连接管 fire service connection
消防连接口 fire department connection
消防练习塔 fire drill tower
消防瞭望台 fire tower
消防列车 fire-fighting train
消防流量 fire discharge; fire flow
消防流量测定 hydrant-flow test
消防流量计算 hydrant-flow calculation
消防龙头 fire cock; fire hose; fire plug; hydrant bend; wall hydrant; water hydrant
消防龙头阀 hydrant valve
消防龙头护罩 hydrant bonnet
消防龙头套筒 hydrant barrel
消防龙头箱 hydrant box
消防楼梯 fire-fighting staircase; fire-fighting stairway
消防楼梯间 fire staircase
消防路线 fire-extinguishing line
消防帽 fire hat
消防门 fire door
消防面具 fire mask; smoke helmet
消防灭火设施 fire-fighting equipment
消防灭火设备 fire-fighting equipment
消防炮 fire-fighting monitor; fire monitor
消防喷淋系统 fire-protection sprinkler system
消防喷射器 fire-fighting monitor
消防喷射水管 fired pipe
消防喷射水流 fire stream
消防喷水系统 fire sprinkler system
消防喷头 applicator
消防喷嘴 fire nozzle
消防汽车接头 connection of steam; steamer connection
消防器 fire engine
消防器材 extinguisher; fire apparatus; fire-fighting appliance; fire service implement
消防器材储藏处 fire cache
消防器材设备 fire-fighting equipment
消防器械 fire-fighting appliance
消防枪 fire branch; fire hose nozzle
消防勤务 fire-fighting service
消防情报实地调查 fire information field investigation
消防区 fire-protection district
消防区划 fire zoning
消防区图 protected compartment
消防人员 fire control operator; fire fighter; fire guard; fire man; hoseman
消防入口 access for fire-fighting
消防软管 fire hose
消防软管架 fire hose rack
消防沙 fire sand
消防设备 fire apparatus; fire appliance; fire control unit; fire device; fire equipment; fire extinction equipment; fire-extinguishing appliance; fire-extinguishing equipment; fire-extinguishing installation; fire-extinguishing plant; fire-fighting apparatus; fire-fighting appliance; fire

installation; fire prevention equipment; fire-protecting arrangement; fire-protection equipment; fire protection facility
消防设备标志 fire equipment sign
消防设备布置图 fire-fighting plan
消防设备洞 niches for fire-fighting equipment
消防设备管理员 equipment officer
消防设备制造商协会 Fire Equipment Manufacturers Association
消防设施 extinguishments equipment; fire-fighting device; fire-protection service; fire service
消防设施门 fire control access door
消防设施组件 fire assembly
消防石膏板 fire fighter gypsum board
消防实践 firemanship
消防手套 fire fighter's glove
消防疏散训练 fire exit drill
消防竖管 fire standpipe
消防栓 fire cock; fire hydrant; fire plug; hydrant (stem); water hydrant; water plug
消防栓毕托计 fire Pitometer; hydrant pitometer
消防栓尺寸 fire size; hydrant size
消防栓阀座口径 hydrant size
消防栓管嘴帽 fire cap; hydrant cap
消防栓井 fire hydrant chamber
消防栓连接器 fire connector; hydrant connector
消防栓流速计 hydrant pitometer
消防栓帽 hydrant bonnet
消防栓竖管 hydrant standpipe; standpipe of a hydrant
消防栓筒 hydrant barrel
消防栓箱 fire(hydrant) box; fire hydrant cabinet; fire hydrant chamber
消防栓泄水阀 fire drain valve; hydrant drain valve
消防水泵 fire(service) pump
消防水表 fire service meter
消防水槽 fire cistern; fire water pond
消防水车 water tender
消防水池 emergency pit; fire cistern; fire-fighting pool; fire water pond
消防水带架 fire hose rack
消防水带接口 fire hose coupling
消防水带卷盘 fire hose reel
消防水带配件 fire hose fitting
消防水道 water work for fire-fighting
消防水管 fire hose; fire service main; hydrant pipe; steel pipe for fire-fighting water
消防水管网 fire-fighting water pipe network
消防水管系统 fire main system; standpipe system
消防水管线 fire water line
消防水罐 fire-protection water tank
消防水喉 hose reel
消防水井 fire well
消防水流 fire stream
消防水龙 turret
消防水龙带 fire(-fighting) hose
消防水龙带接头 fire hose coupling
消防水龙带晾干塔 hose-drying tower
消防水龙带喷嘴 fire hose nozzle
消防水龙带屋 hose house
消防水龙带箱 fire hose box; fire hose cabinet
消防水龙头 fire hydrant; hose tap
消防水枪 fire branch; fire-fighting lance; fire gun; fire nozzle
消防水栓 hydraulic hydrant
消防水栓三通 hydrant tee
消防水桶 fire bucket
消防水箱 fire water box; fire water tank
消防水需要量 fire water requirement
消防水压力 fire pressure; fire water pressure
消防水用户收费 <按水管直径(英寸)乘长度(英尺)计算> service charge for water inch foot
消防水源 fire-protection water supply; fire service water supply
消防水直接压力 direct fire pressure
消防水自动增压泵 fire booster
消防水总管 fire main(pipe); fire-water main
消防损失条款 civil authority clause
消防塔 fire tower
消防毯 fire blanket
消防梯 aerial ladder; fire-fighting ladder; fire ladder; fire service ladder; scaling ladder
消防梯脚板 heel plate
消防条例 fire control regulation; fire ordinance; fire service act
消防艇 fire boat; fire-fighting craft; fire launch
消防艇消防用泵 fire float pump
消防通道 fire department access; fire-extinguishing run; fire-fighting access; fire passage; fire road
消防通气孔 fire venting
消防通信[讯]调动中心 fire alarm headquarter
消防通信[讯]调度室 fire call receiving
消防桶 fire bucket
消防筒 fire drencher
消防头盔 fire fighter helmet
消防图 fire-protection drawing
消防拖车 fire-fighting trailer
消防拖轮 fire-fighting tug
消防网 fire-fighting monitor
消防文摘 <期刊> Fire Control Digest
消防污水两用泵 fire and bilge pump
消防无线电通信[讯]网 fire radio communication network
消防吸水管 fire suction hose
消防系统 fire control system; fire-extinguisher system; fire-extinguishing system; fire-fighting system; fire-protection system
消防线 fire limit
消防箱 hydrant cabinet
消防效能试验 effectiveness test of fire-fighting installation
消防信号系统 fire-protection signal(l)ing system
消防需水量 fire demand; fire demand of water
消防需水流量 fire demand rate; required fire flow
消防需水率 fire demand rate
消防宣传 fire-protection propaganda
消防靴 fire fighter's boots; fire-fighting footwear
消防蓄水池 fire cistern
消防学校 fire-fighting school
消防学院 fire academy
消防巡查 fire watch
消防巡查制度 fire patrol system
消防巡逻 fire patrol
消防巡逻队员 fire patrolman
消防巡逻员 fire walker
消防训练 fire service training
消防训练设施 fire training facility
消防研究 fire research
消防研究机构 fire research station
消防演习 fire drill; fire exercise; fire practice
消防业 fire service
消防业务信息 fire service information
消防应急服务 fire emergency services
消防应急通道 fire escape route
消防营地 fire camp
消防用泵 fire-extinguishing pump
消防用窗 fire brigade access window
消防用的 pompier
消防用斧 fire axe
消防用检测水表 fire service detector check meter
消防用具 fire-protection appliance; sundries for fire-fighting and protection
消防用具支架 fire-extinguisher bracket
消防用泡沫喷射器 foam applicator
消防用品 fire service inventory
消防用软管 fire hose; water hose
消防用砂 fire-extinguishing sand
消防用水 fire fighting water; fire flow; water for fire-fighting purposes; fire water
消防用水储罐 fire water storage tank
消防用水费 service charge for water inch foot
消防用水量 fire consumption(of water); fire demand
消防用水系统 fire water system
消防员 fire man
消防员的台架 fireman's platform
消防员电梯 firefighting lift
消防员防护服 protective fire fighter's clothing
消防员制服 fire fighter uniform; fireman's uniform
消防员装备 fireman's outfit
消防云梯 extending ladder; extension ladder; telescopic(al) ladder
消防战斗服 fire fighter's protective clothing; fire-fighting clothing; fire-fighting tunic; fire-fighting turnout
消防战斗员 fire fighter
消防站 fire engine house; fire hose station; fire house; fire station
消防织物 fire-fighting fabric
消防值班 fire watch
消防值班员 fire watcher
消防指挥车 fire-fighting command car
消防指挥台 fire command station
消防指挥系统 fire command system
消防指挥站 fire command station
消防指挥中心 fire command center [centre]
消防制度 fire-fighting system
消防制品 fired product
消防秩序 fire control order
消防贮水池 fire cistern
消防柱 hydrant
消防专业人员 fire-protection professional
消防专用管道 fire line
消防装备箱 fire equipment cabinet
消防装置 fire-fighting unit; fire installation; fire plant; fire-proof installation; fire service equipment
消防总管 fire main system
消防走廊 fire lobby; fire corridor
消防组织 fire-protection organization
消防作业 fire-fighting operation; fire-fighting process; fire-fighting work
消费 consume; expend; expense; spending
消费包装 consumer package
消费保证 consumption guarantees
消费保证金 deposit(e) for consumption
消费标准 consumption standard
消费不足 inadequate consumption; limited consumption; underconsumption
消费部门 consumer sector
消费采购 buying for ultimate consumption
消费参数 consumption parameter
消费层次 hierarchy of consumption
消费产品 consumption product
消费产业部门 consuming industries
消费场合 demand point
消费场所 consumption point
消费城市 consumer city; consuming city; tyrannopolis <专制统治的>
消费乘数 consumption multiplier
消费贷款 consumer loan; consumption loan; loan for consumption
消费单位 spending unit
消费的 consumptive
消费的对偶性 duality in consumption
消费的发展中国家 consumer developing country
消费的经济顺序法则 law of economic order of consumption
消费的量度 measure of consumption
消费地 place of consumption
消费点 consuming point; demand point
消费额 amount of consumption
消费方程 consumption equation
消费方面的耗费增加 consumption diseconomies
消费方面的节约 consumption economies
消费方式 consumption pattern; pattern of consumption; spending pattern
消费费用 consumption charges
消费服务 consumer service
消费高涨 consumption boom
消费公众 consuming public
消费构成 composition of consumer demand; structure of consumer demand; structure of consumption
消费管线 lateral service piping
消费国 country of consumption
消费函数 consumption function
消费函数模型 consumption function model
消费合作社 consumer's cooperative society; user's cooperative
消费后的垃圾 postconsumer waste
消费后的物质回收 postconsumer recycling
消费户 consumer
消费化 consumerization
消费基金 consumption fund; funds for consumption
消费基金来源 consumption fund source
消费集 consumption set
消费技术 consumption technology
消费价格水平 the general price level
消费价格指数 consumer price index number
消费价值趋势 consumption development trends
消费结构 consumption structure
消费借贷合同 contract of loan for consumption
消费决策 consumer's decision
消费均衡 expending equilibrium
消费可能边缘 consumption possibility frontier
消费可能线 consumption possibility line
消费控制 control of consumption
消费宽减额 consumption allowance
消费力 consumptive power
消费量 amount used; consumption
消费量大用户 large-scale consumer
消费率 consumption rate
消费模拟法则 law of consumption imitation
消费模式 consumption model; consumption pattern
消费内容 content of consumption
消费能力 consuming capacity; consumption capacity

消费膨胀 inflation of consumption
消费品 articles of consumption; consumer's goods; consumer commodity; consumer items; consumption goods
消费品安全委员会 Consumer Product Safety Commission
消费品产量 output of consumer goods
消费品工业 consumer goods industries
消费品购买 consumer purchase
消费品和劳务 consumer goods and services
消费品价格指数 consumer price index
消费品零售税 purchase tax
消费品评审组 consumer jury
消费品市场 consumer goods market
消费品市场研究 consumer research
消费品输入 consumption-related import
消费品物价膨胀 consumer price inflation
消费品物价指数 consumer price index
消费品销售 consumer sale
消费品需要结构 composition of consumer goods
消费品总账 consumable ledger
消费平衡 expending equilibrium
消费倾向 propensity to consume
消费倾向图表 propensity-to-consume schedule
消费曲线 consumption curve
消费趋向性工业 industries close to consumption areas
消费群众 consuming public
消费商品 consumables; consumer lines
消费社会 consumer society; mass consumption society
消费伸缩性 elasticity of demand
消费剩余废料 consumption residue
消费失调 imbalance of consumption
消费市场 consumer market
消费收入 income for consumption
消费收入比 consumption-income ratio
消费收入表 consumption-income schedule
消费收入关系 consumption-income relation
消费收入序列 consumption-income sequence
消费(数)量指数 consumption quantity index
消费水平 consumption level; level of consumption; standards of consumption
消费税 consumption duty; consumption tax; excise duty; excise tax; tax on consumption; tax on expenditures
消费通货膨胀 consumption inflation
消费投资成本 outlay cost
消费图表 consumption schedule
消费途径 consumption path
消费外部性 consumption externality
消费无差异曲线 consumption indifference curve
消费系数法 consumption coefficient method
消费箱 expendable box
消费项目 article of consumption
消费效用函数 consumption utility function
消费协会 consumption guild
消费信贷 consumer credit; consumption credit; loan for consumption; unproductive credit
消费信贷保护法 consumer credit protection act
消费信贷公司 consumer finance company
消费信贷管理 consumer credit control; control on consumer credit
消费信贷专业机构 consumer credit agency
消费行为理论 consumer theory
消费形态 consumption pattern
消费性的物质生活 consumption part of material life
消费性积累 consumptive accumulation
消费性开支 consumer spending
消费性劳务 consumer service; consumption service
消费性用途 consumptive use
消费性用途系数 consumptive use coefficient
消费性支出 non-productive expenditures
消费需求 demand of consumption
消费需求方程 consumer demand equation
消费需求膨胀 inflation of consumption demand; swollen consumption demand
消费需求研究 consumer study
消费压制 underconsumption
消费余量 consumer surplus
消费预算分析 consumer budget analysis
消费增长率 consumption increase rate
消费者 consumer
消费者保护 consumer protection
消费者部门 consumer service
消费者差额受益 consumer's rent; consumer surplus
消费者产品调查 consumer producer research
消费者超需函数 consumer's exceed demand function
消费者的购买力 consumer's purchasing power
消费者的特征 consumer characteristics
消费者的需求 consumer's requirements; consumer's wants
消费者的意见和要求 needs and opinions of consumers
消费者抵押信用 consumer mortgage credit
消费者地区差价指数 index of regional difference of consumer prices
消费者调查 consumer's survey; market research
消费者动机 consumer motivation
消费者对产品质量的反应 quality information feedback
消费者额外负担 consumer excess burden
消费者方便物品 consumer convenience goods
消费者分期付款信贷 consumer instalment credit
消费者风险 consumer's risk
消费者风险点 consumer risk point
消费者购买力 consumer purchasing power
消费者合作社 consumer cooperative
消费者货物 consumer goods
消费者价格 consumer price
消费者价格指数 consumer price index
消费者阶层 consumer strata
消费者借贷 consumer lending
消费者均衡 consumer equilibrium
消费者开支 consumer's expenditures; consumer spending
消费者可支配收入 consumer's disposable income
消费者垃圾 consumer's waste
消费者联合会 Consumer Association
消费者联盟 <美> Consumer Union
消费者满足极大化 maximization of consumer satisfaction
消费者目标市场 consumer orientated market
消费者偏好 consumer's preference
消费者剩余(额) consumer's surplus
消费者使用试验 consumer use test
消费者损失 consumer loss
消费者所付小额税款 nuisance tax
消费者态度 consumer attitude
消费者投资 consumer investment
消费者无差别曲线 consumer indifference curve
消费者物价指数 consumer price index
消费者希望购买额 consumer buying expectations
消费者协会 consumer's cooperative society; consumers' association
消费者心理因素 customer mentality
消费者信任 consumer confidence
消费者信用保险 consumer credit insurance
消费者信用卡保条 consumer credit card slips
消费者行为 consumer's behavio(u)r; consumer action; consumer behavio(u)r
消费者行为论 theory of consumer's behavio(u)r
消费者行为研究 consumer behavio(u)r research
消费者需求 consumer demand
消费者需求表列的性质 nature of schedule of consumer demand
消费者需求的决定因素 determinants of consumer demand
消费者需求理论 consumer demand theory
消费者需要 consumer wants
消费者选择 consumer's choice
消费者选择论 theory of consumer's choice
消费者研究 consumer research
消费者要求条件 consumer requirements
消费者有奖竞赛 consumer contest
消费者预算 consumer budgets
消费者预算论 theory of consumer's budget
消费者运动 consumer movement
消费者诈骗处理科 consumer fraud division
消费者债务利息 interest on consumer debt
消费者征信机构 consumer reporting agencies
消费者支出 consumer expenditures
消费者至上 consumer's sovereignty
消费者至上论 consumer orientation
消费者主权 consumer's sovereignty
消费者资本 consumption capital
消费者租金 consumer's rent
消费支出 consumer spending; consumption expenditures
消费支出分类 classification of consumption expenditures
消费支出收入模式 consumer expenditure-income pattern
消费支出税 expenditure taxes
消费支出统计 statistics of consumption expenditures
消费支出账户 consumption expenditure account
消费指数 index of consumption
消费质量 consumption quality
消费滞差 consumption lag
消费周期 consumption cycle
消费主义 consumerism
消费状况 condition of consumption
消费资产 expense assets
消费资金 consumption capital
消费资料 consumer goods; consumption goods; means of consumption; means of subsistence
消费资料的生产 production of consumer goods
消费总量 total quantity consumed
消盖水洞 estavel
消干扰的 anti-jamming
消感 anti-induction
消感导线 anti-induction conductor
消感网络 anti-induction network
消根【数】rationalize
消光 deluster(ing); extinction; short finish
消光表面 frosting
消光衬比 extinction contrast
消光处理 matting operation
消光度 degree of flatting
消光法 felting down; light extinction method; photosedimentation <粒径分析方法之一>
消光法曝光表 extinction meter
消光方向 direction of extinction; extinction direction
消光规则 <结晶面的> extinction rule
消光计 <又称悬移质测量计> light extinction meter
消光剂 deadening agent; delusterant; dull agent; flatt(en)ing agent; frosting agent; gloss reducer; matting agent
消光角 angle of extinction; extinction angle
消光角补角 angle of isocline
消光距离 extinction distance
消光率 extinctivity
消光轮廓 extinction contour
消光面 mat(te) surface
消光尼龙 delustered nylon
消光漆 flat varnish
消光试验 extinction test
消光涂层 delustred coating
消光涂料 delustring coating material; material
消光位置 extinction position
消光系数 coefficient of extinction; extinction coefficient; extinguishing coefficient; light extinction coefficient
消光效率 flatting efficiency
消光效应 flatting effect
消光楔 neutral wedge
消光修饰 dull finish
消光修整 mat finish
消光颜料 delustering pigment; flatting pigment; matting pigment
消光液 deglossing liquid
消光影 isogyre
消光油 flatting oil
消光罩面漆 mat finish paint
消光整理 mat finish
消光整理喷漆 dull-finish(ed) lacquer
消光值 extinction value
消光浊度法 photoextinction method
消光浊度计 extinction turbidimeter
消过毒的 sterile
消号 disannul a call
消耗 consume; consumption; attrition; decrement; dissipation; diverging; draining; drawdown; exhaustion; expenditure; wastage; wear away; wear off; wear out
消耗百分比 percentage depletion
消耗标准 quota of expenditures
消耗不足 underconsumption
消耗材料 consumable material; con-

sumables
消耗臭氧物质 ozone depleter
消耗的材料 lost material
消耗的等待时间 spent waiting time
消耗的总蒸发水量 consumptive use
消耗递减率 ratio of consumer decreasing progressively
消耗递减主要因素 principal factors of consumer decreasing progressively
消耗递增率 ratio of consumer increasing progressively
消耗递增主要因素 principal factors of consumer increasing progressively
消耗电阻 dead resistance
消耗掉 peter out;use up
消耗定额 consumption norm;norm of consumption; quota of consumption;rate of expenditures
消耗动力的 power-consuming
消耗费用 consumption cost
消耗高峰期 consumption peak
消耗工具和设备 consumable tool and equipment
消耗工时 lost time
消耗功 consumed work; dissipative work;work input
消耗功率 consumed power;dissipated power
消耗故障 wear-out failure
消耗过高 exorbitant expenditures
消耗冷藏集装箱 consumptive refrigerated container
消耗量 amount of consumption; amount used; consumptive use; quantity of consumption;wastage
消耗量记录 consumption record
消耗量试验 consumption test
消耗量所居位数 number of consumption to arrange in order
消耗量占百分比 annual consumption at percentage
消耗零件 expendable parts
消耗率 consumption rate;rate of consumption;specific consumption;use rate
消耗马力【机】friction(al)horsepower
消耗满足率 ratio of full consumption
消耗能 dissipated energy
消耗能量 consumed energy;lost energy
消耗品 consumable article;consumable item; consumables; expendables; expendable supply; expenditure item
消耗品(物料)库 consumable store
消耗气驱油藏 depletion-drive pool
消耗器 customer
消耗器材备份 bench stock
消耗曲线 use condition
消耗热 chargeable heat;hectic fever
消耗失效 wear-out failure
消耗使用 consumptive use
消耗式载体 expendable carrier
消耗数量 quantity consumed;consumption
消耗水量 water loss
消耗缩减 consumption decrease
消耗土壤养分 use up soil nutrient
消耗污泥 ripe sludge
消耗物 expendables
消耗物料 spent material
消耗物品 non-durable goods;non-durables
消耗型水源地 depleting water source
消耗性包装 expendable packaging
消耗性材料 expendable material
消耗性的 consumptive;expendable
消耗性地图 consumption map
消耗性工具 perishable tool
消耗性经费 exhaustive expenditures
消耗性配件 consumable supplies
消耗性器材 consumable supplies
消耗性用品 expendable supplies
消耗性用水 consumptive use of water;consumptive water supply;expendable water supply
消耗性用途 consumptive use
消耗性支出 exhaustion expenditures; exhaustive expenditures
消耗性装置 expendable equipment
消耗需量 consumptive requirement
消耗因数 consumption factor
消耗用水 water for consumption
消耗与补充 waste and repair
消耗载荷 consumable load
消耗增长 consumption increase
消耗值 consumption figure
消耗指数 consumption index
消耗装置 consumer
消耗状况 use condition
消耗资产 wasting assets
消耗资产折耗 depletion for wasting assets
消和槽 slake tank;slaking tank
消和池 slaking apparatus
消和器 slaking apparatus
消弧 arc blow-out; arc extinction; arc-extinguishing; arc suppressing; extinction of arc
消弧电路 crowbar circuit
消弧电压 extinction voltage of arc
消弧环 arcing ring
消弧角 arcing horn;extinction angle
消弧器 arc arrester [arrestor]; arc catcher;extinguisher;spark absorber;spark killer
消弧圈 extinguisher
消弧栅 arc chute
消弧室 arc extinguish chamber;explosion chamber
消弧线路 crowbar circuit
消弧线圈 arc-extinguishing coil; arc suppressing coil; arc suppression coil; extinction coil; extinguishing coil;Peterson coil
消化 digest;slaking
消化槽 digester;digestion cell
消化池 digester [digestor]; digestion cell; digestion chamber; digestion tank;slaking pan
消化池保温 insulation of digester
消化池保温层 insulating blanket of digester
消化池爆炸 explosions of digestion tank
消化池出入人孔 digester access manhole
消化池的圆顶盖 gas dome
消化池发动机 gas engines for digester gas
消化池浮盖 floating digester cover
消化池加热 digester heating
消化池加热盘管 heating coils in digester tank
消化池搅拌 digester mixing
消化池气(体)爆炸危险 explosion hazards of digester gas
消化池气(体)比重 specific gravity of digester gas
消化池气(体)产量 digester gas production
消化池气(体)成分 composition of digester gas
消化池气(体)存储 storage of digester gas
消化池气(体)利用 digester gas utilization;utilization of digester gas
消化池气(体)热值 fuel value of digester gas
消化池气(体)生产 production of digester gas
消化池气(体)收集 collection of digester gas
消化池气(体)特性 characteristics of digester gas
消化池气(体)循环搅拌器 digester gas circulation mixer
消化池热交换器 digester heat exchanger;heat-exchange in digestion tank
消化池酸性发酵 digester acid fermentation
消化池循环唧泥管 digester recirculation suction
消化池液体排除管 digester liquor draw off
消化池液体溢流 digest liquor overflow
消化池引流搅拌器 digester draft tube mixer
消化处理 digestion process;digestive treatment
消化道 alimentary canal;digestive canal;enteron
消化罐 digestion tank
消化过程 digestion process
消化过的污泥 digested sludge
消化剂 digestant; digester [digestor]; digestive
消化率 digestibility;rate of digestion
消化盘管 digester coil
消化期 digestion period
消化气体 digestive gas
消化器 digester[digestor];slaker
消化器官 digestive apparatus
消化热 slaking heat
消化石灰 limoid
消化石灰螺旋输送器 slaking screw
消化时间 digestion period; digestion time;time of slaking
消化试验 digestion trial
消化室 digesting compartment; digestion chamber; digestion compartment
消化室容积 volume of digestion chamber
消化水池 digestion water tank
消化污泥 digested sludge;ripe sludge
消化污泥排除 digested sludge draw off
消化污水污泥 digest sewage sludge
消化系数 digestion coefficient
消化系统 alimentary system; digestive system;systematic digestorium
消化系(统)疾病 digestive disease
消化箱 slaking box
消化效率 digestive efficiency
消化性 digestibility
消化性能 slaking behavio(u)r
消化药 digestant;digester;digestive
消化液 digest;digestive solution
消化作用 digestion
消晃板 antisloshing baffle
消火(给水)栓筒 hydrant barrel
消火花磁铁 blowout magnet
消火花的 anti-spark
消火花电路 antispark circuit;quenching circuit
消火花角隙 blowout horn
消火花线圈 blowout coil
消火花箱 blowout box
消火器 fire-extinguisher
消火栓 fire cock;fire hydrant;hydrant;penstock;fire plug <老式的>
消火栓毕托计 hydrant pitometer
消火栓尺寸 hydrant size
消火栓出水口 hose outlet
消火栓管嘴帽 hydrant cap
消火栓柜 fire hose and fire plug cabinet
消火栓接消防泵的出口 pumper outlet
消火栓井 fire hydrant chamber; hydrant pit
消火栓帽 hydrant bonnet
消火栓箱 fire cabinet; fire hydrant cabinet;hose cabinet
消火栓泄水阀 hydrant drain
消火水枪 fire nozzle
消火旋塞 fire cock
消迹 erasion
消迹电流 erasing current
消迹放大器 erase amplifier
消迹墨水 canaigre ink
消迹能力 erasing ability
消迹振荡器 erase oscillator
消极安全性 passive safety
消极保证 negative assurance;negative guarantee
消极保证条款 negative pledge clause
消极的 negative;passive
消极防火 passive fire defence
消极复原法 pessimistic recovery technique
消极干预 passive intervention
消极隔振 negative vibration isolation; passive insulation;passive isolation
消极贸易 passive trade
消极施工法 <施工前保留地基内冻层的> passive method
消极适应 passive adaption
消极态度 negative attitude
消极效果 negative effect
消极信托 passive trust
消极性 negativity
消极性城市化 negative urbanization
消极性管制 negative control
消极因素 minus factor;negative factor
消极约束 passive constraint
消极债券 passive bond
消极状态 passivity
消极自感 negative self-feeling
消极作用 passive role
消加常数 <视距测量的> anallatism
消加常数(透)镜 anallatic lens
消加常数望远镜 anallatic telescope
消减 cutback;cut down
消减板块 consumering plate;subducting plate
消减板块边缘 consumed plate boundary
消减边界 consumed boundary
消减带 consumering zone; extinction zone;subduction zone
消减断面 attenuation cross section
消减声响 dead sounding
消减污物处置 decrease the disposal of pollutants
消减型陆外下陷盆地 subduction downwarped extracontinental basin
消减烟雾(法)abatement of smoke
消减涌浪设备 surge suppressor
消减噪声 noise reduction
消减作用 consumption;subduction
消解 digestion; resolution; slacking <石灰等的>
消解白云石质石灰浆 dolomite plaster
消解法 resolution method
消解过程 slaking process
消解含水量 slaking water content
消解率 slaking modulus
消解黏[粘]土 slaking clay
消解热 slaking heat
消解石膏块 core of gypsum
消解石灰筛块 corduroy of lime;core of lime
消解时间 digestion time
消解式 resolvent
消解水管真空设施 vacuum breaker
消解速率 rate of slaking

消解温度 digestion temperature
消解系数 slaking modulus
消尽指示器 null detector
消控火势 corralling a fire
消浪措施 wave breaking measure
消浪海滩 expending beach; spending beach
消浪块体 wave breaker block; wave breaking block
消浪乱石堆 rubble wave absorber
消浪栅 <水工模型用的> steadying baffle
消浪滩 spending beach
消力 baffle; baffling
消力坝 stilling dam
消力板 baffle plate
消力槽溢流 stilling-pond overflow
消力池 stiller; still(ing) basin; stilling pond; stilling pool; toe basin; water apron; water cushion; absorption basin; absorption pond; absorption pool; cushion basin; cushion pond; cushion pool; plunge pool; spillway basin
消力池溢流 stilling-pond overflow
消力齿 baffler teeth; dental; dentil; dragon's tooth [复 teeth]; energy-dissipating dents
消力齿槛 dentated sill
消力洞穴 stilling cavern
消力墩 baffle; baffle block <静水池或河道中的>; baffle pier; control block; dragon's tooth[复 teeth]; energy dispersion baffle; energy-dispersion baffle; energy dispersion block; floor block; impact block
消力格栅 stilling grid
消力戽 bucket; energy-dissipating bucket
消力戽护坦 bucket apron
消力戽内的水辊 bucket roller
消力戽消能 energy dissipation of roller bucket
消力槛 anti-scour lip; anti-scour sill; baffle sill; baffle threshold; clap(ping) sill; end baffle; end sill; energy absorber; energy-dissipating sill; floor sill; stilling baffle; dentated sill
消力结构 baffle structure
消力坎 bucket; stilling baffle
消力孔洞 stilling cavern
消力块 baffle block
消力器 baffler
消力墙 baffle wall
消力设施 baffle facility; baffle structure
消力塘 absorption pond; stilling basin; stilling pool; stilling pond
消力箱 stilling box
消力锥 baffle prong
消链 destroy chain
消零 <在印刷前将无效零位消去>【计】zero suppression; null suppression; suppress zero; zero blanking
消零电路 zero blanking circuit
消零法 zero elimination
消零图像字符 zero suppression picture character
消零字符 zero suppression character
消落 degrading; drawdown; regression
消落段 falling limb; falling segment; falling stage
消落率 <洪水> rate of subsidence
消落期 falling period; falling stage
消落区 drawdown area; drawdown zone; drawdown land
消落时间 time of recession
消落水位 <水库> drawdown level
消没 vanishing
消没角 vanishing angle
消没通量 vanishing flux
消灭 annihilate; annihilation; blow over; destroy; eradicate; extinction; extinguish; extirpate; knock; obliteration
消灭赤字 shortage control
消灭虫害 eliminating pest
消灭地表火 extinguishing grass fire
消灭点 <透视> point vanishing
消灭概率 extinction probability
消灭痕迹 obliterate
消灭花系统 blowout system
消灭火花 blowout; spark quenching
消灭火花电容器 spark quenching capacitor; spark quenching condenser
消灭林火 extinguishing forest fire
消灭难以根除的杂草 eliminating troublesome weeds
消灭器 annihilator; extinguisher
消灭时间 extinction time
消灭时效 bar of statute of limitation; extinctive prescription
消灭危害性大的杂草 eliminating ill weeds
消灭效果 extinction effect
消灭噪声装置 noise blanker
消灭者 extinguisher
消磨时间 kill time
消抹射束 scan-off beam
消沫剂 defoamer agent; defoaming agent
消能 baffle; baffling; dissipation energy; dissipation of energy; dissipation of kinetic energy; energy dissipation; energy of dissipation; spending of energy
消能槽 diffusion trench
消能池 baffler; surge basin; tumble bay; tumbling bay; stilling basin
消能齿 baffler teeth; dragon's tooth[复 teeth]; energy-dissipating dents; pool cushion
消能的 anti-scour
消能丁坝 baffle groin; baffler prong
消能墩 baffle block; baffle pier; dragon's tooth [复 teeth]; energy-dispersion baffle; energy dispersion block
消能范围 area of dissipation
消能防冲设施 energy dissipating and anti-scour facility
消能工 dissipator; energy absorber; energy dissipator
消能工程 energy dissipating works; energy dissipator
消能沟 diffusion trench
消能构件 baffler module
消能构筑物 energy dissipator
消能柜 baffle box
消能海滩 expending beach
消能护岸 energy dissipation revetment
消能戽 roller bucket
消能混凝土块体 energy-dissipating concrete block
消能建筑 arrestment
消能建筑物 dissipator; energy dissipation structure
消能槛 annihilator of energy; baffle threshold; energy-dissipating sill
消能结构 energy dissipator
消能井 stilling well
消能阱 wave trap
消能坎 end sill; energy-dissipating sill
消能块 baffle block
消能棱体 anti-scour prism
消能率 rate of energy dissipation
消能面积 area of dissipation
消能能力 energy absorbing capacity
消能坡 breakwater glacis
消能器 annihilator of energy; energy absorber; energy dissipator
消能墙 baffle wall
消能区 area of dissipation; energy dispersion block
消能设备 energy absorber; energy-dissipating device; energy dissipator; water-stilling device
消能设施 dissipator; energy dispersion device
消能试验 dissipation test
消能室 stilling chamber
消能台阶 velocity reducing steps
消能滩 breakwater glacis; dummy beach; spending beach
消能梯级 velocity reducing steps
消能系数 coefficient of performance
消能箱 baffle box
消能桩 buckling column fender; energy-dissipating pile
消能装置 energy disperser; energy dissipator; energy killer
消能作用 damping effect; stilling action
消浓器 deconcentrator
消泡 defoam; foam fractionation
消泡剂 air-detraining admixture; anti-blowing agent; anti-bubbling agent; anti-foam(er); anti-foaming agent; defoamant; defoamer agent; defoaming agent; defrother; degassing agent; foam control agent powder; foam depressant; foam inhibitor; foam killing agent; foam suppressant; foam suppressor; air-detraining compound
消泡器 deaerator; defrother; froth breaker
消泡添加剂 anti-foam additive
消泡作用 defoaming
消偏光振镜 depolariser[depolorizer]
消偏振镜 depolariser[depolorizer]
消偏振器 depolariser[depolorizer]
消偏振因素 depolarization factor
消偏振作用 depolarization
消气 gettering
消气感受器 deflation receptor
消气剂 getter
消气剂的胶合剂 getter binder
消气剂涂层 coating getter
消气剂用黏[粘]合剂 getter binder
消气器 getter
消碛 recessional moraine
消遣 distraction; refection
消遣的 recreational; kill time
消遣活动 recreational activity
消遣性出行 recreational trip
消遣用水 recreational use of water
消球差板 <航测制图用> aspheric(al) plate; compensating plate
消球差放大镜 aspheric(al) magnifier
消球差校正 aspheric(al) correction
消球差镜 aplanat
消球差透镜 aplanat; aplanatic lens; aspheric(al) lens
消球差望远镜 aplanatic telescope; aspheric(al) telescope
消球差性 asphericm
消去 cancellation
消去电子枪 erasing gun
消去法 elimination method; method of elimination; subtractive process
消去反应 elimination reaction
消去光泽 flatting down
消去记录 erasing record
消去加成机制 elimination-addition mechanism
消去离子 de-ionization
消去零 null suppression
消去律 cancellation law
消去率 elimination factor
消去前补零 leading zero suppress
消去取代基 elimination of group
消去试验 elimination test
消去水 elimination of water
消去系数 elimination factor
消去信号输入 erase input
消去者 annihilator
消去作用 elimination action
消热罩 heat-sink shield
消溶礁 resorbed reef
消溶胀作用 deswelling
消熔 resorption
消融 dissipation; wastage
消融冰 disappearing ice; melting of ice
消融冰碛 ablation drift; ablation till; wastage moraine; ablation moraine
消融层 ablation zone
消融带 <冰川> ablation zone; zone of ablation; zone of wastage; ablation belt
消融地形 ablation form
消融范围 area of dissipation
消融角砾岩 ablation breccia
消融理论 <冰川> theory of ablation
消融量 ablation amount
消融率 ablation factor; ablation rate; rate of ablation
消融面 ablating surface; ablation surface
消融面积 area of dissipation
消融期 ablation period; ablation season
消融强度 ablation intensity
消融区 <冰川> ablation zone; area of dissipation; wastage zone; zone of ablation; zone of waste; ablation area; zone of wastage; zone of dissipation
消融热 heat of ablation
消融系数 ablation factor
消融仪 ablatograph
消融沼泽 ablation swamp
消融作用 ablation
消弱系数 extinction coefficient
消散 breakaway; clear away; dissipating; dissipation; slake; slaking; vanish
消散波 evanescent wave
消散的 discursive; discutient; resolvent
消散剂 resolvent
消散能量 dissipated energy
消散器 dissipator
消散时间 resolution time
消散试验 <孔隙压力的> dissipation test
消散速度 dissipation rate
消散尾迹 dissipation trail
消散系数 dissipation factor
消散效应 dissipative effect
消散性 evanescence
消散作用 dissipative effect
消色 colo(u)r buffing; colo(u)r reduction; decolo(u)rization
消色测定法 achromic method
消色差 achromation; achromatization
消色差磁色谱计 achromatic magnetic mass spectrometer
消色差的 achromatic
消色差点 achromatic point
消色差干涉条纹 achromatic interference fringe
消色差感觉 achromatic sensation
消色差光 achromatic light
消色差集光镜 achromatic convex lens
消色差镜系 achromatic system
消色差聚光镜(头) achromatic condenser
消色差聚光器 achromatic condenser

消色差聚光透镜 achromatic condenser
消色差棱镜 achromatic prism
消色差目镜 achromatic eyepiece; achromatic ocular
消色差色线 achromatic locus
消色差双合透镜 achromatic doublet lens
消色差天线 achromatic aerial; achromatic antenna
消色差条纹 achromatic fringe
消色差透镜 achromat(ic lens); colo(u)r corrected lens
消色差透镜组 applique
消色差望远镜 achromatic telescope; anaberrational telescope
消色差位相版 achromatic phase plate
消色差物镜 achromatic objective
消色差像 achromatic image
消色差性 achromaticity; achromatism
消色差颜色 achromatic colo(u)r
消色点 achromic point
消色电路 colo(u)r killer circuit; killer stage
消色光 achromatic light
消色糊精 achrodextrin
消色力 achromatic power; lightening power; reducing power; tint-reducing power; whitening power
消色灵 eradicator
消色期 achromic period
消色器 colo(u)r killer
消色器电路 killer circuit
消色时间 achromic period
消色指示剂 achromatic indicator
消升华作用 desublimation
消声 acoustics insulation; amortization; damping; deafen; muffle; muffling; noise abatement; noise-damping; noise deadening; noise elimination; noise reduction; silencing; silencing of noise; sound attenuation; sound insulation; sound-proofing; sound reduction
消声板 muffler plate; sound deadening board
消声被 sound-absorbing quilt; sound-deadening quilt
消声本领 obliterating power
消声比 extinction ratio
消声玻璃 acoustic(al) control glass
消声材料 sound-absorbent material; sound-absorbing material; sound-damping material
消声槽 anechoic trap
消声程序 noise abatement procedure
消声储水器 silent cistern
消声处理 deadening
消声措施 noise countermeasure; noise elimination measure
消声打桩(法) muffler piling
消声打桩设备 silent pile driving system
消声的 anechoic; anti-hum; sound absorptive; sound damping; sound-muffling
消声灯 noiseless lamp
消声垫 noise absorption mat
消声段 muffler section; silencer section
消声法 sound deadening
消声方法 silencing method
消声防逆阀 silent check valve; sound-damping check valve
消声废气锅炉 silencer-boiler
消声粉刷材料 acoustic(al) plaster aggregate
消声粉刷骨料 acoustic(al) plaster aggregate
消声粉刷集料 acoustic(al) plaster aggregate
消声隔板 sound gobo
消声隔片 tormenter[tormentor]
消声功能 obliterating power
消声管 sound-absorbing duct
消声胶合板 sound-deadening plywood
消声金属箔 acoustic(al) foil
消声金属顶棚 acoustic(al) metal ceiling
消声金属天花板 acoustic(al) metal ceiling
消声静压箱 sound-absorbing and static pressure box
消声坑 silence pit
消声量 sound-deadening capacity
消声滤气器 silencer-filter
消声滤气器型空气进口 silencer-filter type air intake
消声门 sound attenuating door; sound door; sound-rated door
消声排气 muffled air exhaust
消声盘 noise-deadened wheel
消声喷管 suppressor nozzle
消声片 sound splitter
消声谱 non-echo spectrum
消声漆 acoustic(al) paint; anti-noise paint; sound-absorbing paint
消声器 acoustic(al) clarifier[damper/filter]; anti-rattler; anti-rumble; anti-speaker; anti-squeak; attenuator; auto-muffler; buffer; damp(en)er; deadener; deadfender; dissipative muffler; intake silencer; muffler; noise abatement device; noise canceller [cover/ damper/ deadener/ reducer/ snubber/ suppresser/ suppressor]; silencer; silencing pot; snubber; sound absorber [attenuator/ damper/ eliminator/ splitter/ suppressor]
消声器爆声 muffler explosion
消声器测量 measurement of muffler performance
消声器出口 silencing end
消声器出气管 muffler outer pipe; muffler outlet piper
消声器出气管托架 muffler outlet pipe bracket
消声器带条 muffler strap
消声器反火 muffler explosion
消声器隔断器 muffler cut-out
消声器进口 muffler inlet
消声器内爆音 backshot
消声器排气管 muffler tail pipe
消声器腔 silencer chamber
消声器切断阀 muffler cutout valve
消声器外壳 muffler shell
消声器尾管夹 muffler tail pipe clamp; muffler tail pipe support clip
消声器尾管支架 muffler tail pipe support
消声器修理套 muffler repair jacket
消声器支架 muffler bracket; silencer support
消声器组 muffler group
消声球形龙头 silent falling ball tap
消声锐度 null sharpness
消声设备 silencing equipment
消声试验 silence test
消声室 anechoic chamber; anechoic room; baffle chamber; dead room; free field chamber; free field room; non-echo chamber; sound attenuation room
消声碎石墙 acoustic(al) masonry wall
消声筒 silencing pot
消声涂料 anti-noise paint; sound-deadening paint
消声弯头 bend muffler; lined bend; lined elbow
消声圬工墙 acoustic(al) masonry wall
消声系数 acoustic(al) reduction factor; erasing factor; extinction coefficient; extinction factor; extinction ratio; sound-reduction factor
消声箱 attenuator box; silence chamber; sound-absorbing box; sound attenuating box
消声效果 silence effect
消声因数 acoustic(al) reduction factor; erasing factor; extinction factor
消声罩 acoustic(al) enclosure; silencing hood; acoustic(al) box <打桩用>
消声砖石墙 acoustic(al) masonry wall
消声装置 muffler device; noise abatement equipment; silencing device; silencing unit; sound absorber; sound arrester; sound-damping device; sound-damping equipment; sound-damping unit; sound suppression attachment
消声锥面 quieting ramp
消声作用 acoustic(al) blanking; sound deadening
消失 abolition; deadening; die away; die out; disappear; disappearance; evaporation; merge; occultation; off air; submerge; vanish; wear off; thin out【地】
消失板块 consumering plate
消失波 evanescent wave
消失潮 vanishing tide
消失带 consumering zone; consuming zone; extinction zone
消失殆尽 die out
消失的天然放射性核 extinct natural radionucleus
消失的天然放射性核素 extinct natural radionuclides
消失点 vanishing point
消失反应 reaction of disappearance
消失方向 direction of extinction
消失河(流) disappearing river; disappearing stream
消失空化 disappearing cavitation
消失块 consuming plate
消失率 rate of dissipation
消失面 vanishing plane
消失区 fade area; fade zone
消失时间 elapsed time
消失系统 loss system
消失线 vanishing line
消失性 evanescence
消失性乳膏基质 vanishing cream base
消失制服务系统 mass service system with rejection
消失作用 consumption; disappearance; extinction; obliteration
消石 niter; nitrum
消石灰 air-slaked quick lime; calcium hydroxide; drowned lime; hydrated lime; hydrate of lime; killed lime; lime hydrate; slaked lime; white lime
消石灰槽 slake tank
消石灰的碳化 carbonation of hydrated lime
消石灰加料仓 slaked lime feeding tank
消石灰器 slaker
消蚀 ablation
消蚀河流 degrading stream
消蚀聚合物 ablative polymer
消蚀率 ablating rate
消蚀涂层 ablation coating
消蚀涂料 ablative coating
消蚀性能指数 ablative performance index
消视差平面位置显示器 virtual plan-position reflectoscope
消视点 vanishing point
消视线 vanishing line
消逝 dissolve
消逝时间 elapsed time
消衰延程 decay distance
消水池 still pool; water apron
消速 stilling
消速池 stilling basin; stilling box
消损 wear and tear
消滩水力指标 hydraulic parameter of rapids abating
消退 extinction; subsidence
消退的 extinctive
消退法 repercussion
消退率 depletion rate
消退排放量 recession discharge
消退曲线 desorption curve
消退水量 depletion rate
消退型 degenerated type; regressive type
消退性实践法 negative practice
消退性抑制 extinctive inhibition
消拖尾电路 smearer
消亡 dissolution; extinction; withering-away
消亡湖 extinct lake
消位 digit absorption
消位器 digit absorber
消涡池 vortex suppressing basin
消涡室 stilling chamber
消污润滑油 <内燃机用的> detergent oil
消雾 dissipation of fog; fog dispersal
消雾剂 fog dispeller; fog disperser
消雾滤光器 haze filter
消雾器 demister(of vehicle window)
消息 intelligence; poop; tidings
消息包 packet
消息编号 message numbering
消息编辑作业 message edit tasking
消息标题 message header
消息处理作业 message process tasking
消息传递 message passing
消息传递时间 message transfer time
消息答复 message acknowledgment
消息的准确度 message accuracy
消息防护 message protection
消息格式 message format
消息缓冲区 message buffer
消息监控作业 message monitoring task
消息交换 full-time message switching
消息交流渠道 communication channel
消息类型 type of message
消息灵通人士 informed source
消息漏失 spillover of message
消息路由表 message routing table
消息率 message rate
消息内插 message interpolation
消息首部 message header
消息宿 message sink
消息通信[讯]功能 message communication function
消息尾 message ending
消息位 message digit
消息优先等级 message precedence
消息源 message source
消息指示符 message indicator
消息组 block of the message
消隙弹簧 anti-backlash spring; backlash spring
消隙装置 anti-backlash device
消闲中心 leisure center[centre]
消向 disorientation
消像差反射望远镜 anaberrational reflector
消像差望远镜 anaberrational telescope
消像反射望远镜 anastigmatic reflector

消像散射器 anastigmator
消像散射束 anastigmatic beam
消像散透镜 anastigmat(ic lens)
消像散物镜 anastigmatic objective
消像散性 anastigmatism
消旋混合物 racemic mixture
消旋缆 de-spin cable
消旋索 de-spin cable
消旋天线 despun antenna
消旋装置 despin mechanism
消旋作用 racemation;racemization
消压管 pressure-relief pipe
消压荷载 decompression load
消压阶段 decompression stage
消压状态 state of decompression
消烟 smoke abatement;smoke elimination
消烟材料 fumivorous material;smoke consuming material
消烟除尘 abatement of smoke and dust;eliminate smoke and dust;smoke prevention and dust control
消烟除雾设备 smoke abatement device
消烟防护 smoke protection
消烟过滤器 smoke filter
消烟剂 anti-smoke agent;smoke suppressant;smoke suppressor
消烟前室 smoke-stop anteroom
消烟设备 smoke elimination equipment
消焰器 flame catcher;flame damper;flame trap;flash hider
消摇液体舱 anti-rolling tank
消抑 killing
消音 dampen;deadening;deafen;noise abatement;noise-damping;noise elimination;noise isolation;silencing;sound damping;sound deadening;sound reduction
消音板 abat-voix
消音材料 acoustic(al) absorbent;acoustic(al) absorber;deadening;quieter material
消音方法 silencing method
消音放大器 erase amplifier
消音管 hush pipe
消音机械装置 damping mechanism
消音键 ventil
消音胶合板 sound-deadening plywood
消音孔 bloop
消音量 noise reduction
消音路面 <行车噪声较低> quieter pavement
消音盘 noise-deadened wheel
消音器 absorption damper;acoustic(al) damper;amortisseur;antisqueak;attenuator;baffler;bumper;deadener;intake silencer;muffler;noise damper;noise deadener;noise killer;noise suppresser[suppressor];quencher;quieter;silencer;snubber;sound suppressor;sourdine
消音器安全阀 muffler safety valve
消音锐度 null sharpness
消音式喷管 noise reducer nozzle
消音试验 anechoic test
消音室 anechoic chamber;anechoic room;dead room;sound attenuating chamber;sound damping chamber
消音贴片 bloop
消音筒 hush pipe
消音涂层 muffle coating
消音效果 erasure effect;silence effect
消音元件 noise-absorbing elements
消音振荡器 obliterating oscillator
消音装置 noise killer;silencer
消音锥度 quieting ramp
消隐 blackout;blanking;blankoff
消隐波 blanking wave
消隐电路 blanking circuit
消隐电平 blanking level;pedestal;pedestal level
消隐电平变化 variation of blanking level
消隐电势谱 disappearance potential spectroscopy
消隐放大器 blackout amplifier
消隐光导管回描 blank vidicon retrace
消隐混合器 blanking mixer
消隐技术 blanking technology
消隐控制 blanking control
消隐脉冲 blackout pulse;blanking impulse;blanking pulse;blanking signal;quench pulse
消隐脉冲电平 blanking pedestal;pedestal level
消隐脉冲电平调整管 pedestal control tube
消隐脉冲发生器 blanking impulse generator;blanking-pulse generator
消隐脉冲放大器 blanking amplifier
消隐脉冲高度 pedestal height
消隐脉冲管 blanking pulse tube
消隐脉冲混频管 blanking mixer tube
消隐脉冲消除器 blank deleter
消隐脉冲选通电路 blanking gate
消隐门 blanking gate
消隐时间 blanking time
消隐信号 blanking signal;blanking wave
消隐装置 blanker
消应处理 stress-relief treatment
消应回火 stress-relief annealing;stress-relief tempering
消元法 elimination(method);elimination of unknowns;method of elimination
消元规则 rule of elimination
消元式 eliminant;resultant
消元问题 problem of elimination
消晕作用 anti-halation
消杂音电池 noise-killing battery
消杂音设备 noise suppression equipment
消载荷 removal of load
消噪 noise-absorbing
消噪耳套 ear muffs
消噪抗扰 noise immunization
消噪声电池 noise-killing battery
消噪声电路 noise balancing circuit
消噪装置 noise blanker
消振 oscillation damping;shock absorption;weakening
消振材料 oscillation damping material
消振层 vibration-absorbing layer
消振措施 vibration-absorbing measure
消振管 flexible tube
消振平衡 vibration balancing
消振器 dashpot;shock absorber;shock eliminator;vibration absorber
消振箱 damp box
消振装置 vibrate absorber
消震 absorb shock
消震镜头 dynalens
消震炮孔 snubber hole
消震炮眼 snubber hole
消震器 dynamic(al) damper;shock absorber
消震橡胶 shock-absorbing rubber
消震橡皮 shock-absorbing rubber
消震液 shock-absorbing fluid
消震油 shock absorber oil
消字头 eraser

萧

萧条 bad time;depression;stagnancy;turn down
萧条的劳动市场 slack labo(u)r market
萧条地区 depressed area;special area
萧条膨胀 slumpflation
萧条期 slack time;winter
萧条市场 depressed market;flat market
萧条市面 sluggish market

硝

硝氨基甲酸乙酯 nitrourethane
硝氨基甲酸酯 nitrocarbamate
硝铵 aerolite;amidpulver
硝铵安全炸药 ammonium nitrate permissible explosive
硝铵二硝基萘炸药 ammonite
硝铵基半凝胶炸药类 semi-gelatins
硝铵胶质炸药 ammonia gelatin(e) dynamite;ammonium nitrate gelatin(e)
硝铵锯屑 monobel
硝铵铝粉炸药 alumatol
硝铵燃料油 ammonium nitrate and fuel oil
硝铵炭炸药 dynamitron
硝铵油炸药 ammonium nitrate-fuel oil mixture;ANFO[ammonium nitrate and fuel oil] mixture
硝铵铀矿 uramphite
硝铵炸药 ammonal;ammon-dynamite;ammonia dynamite;ammonium nitrate explosive;dynamite;nitramon;schneebergite;raschite
硝铵炸药废水 ammonia dynamite wastewater
硝铵炸药装药软管 ammonium nitrate hose
硝胺 nitramine;tetralite;tetrye
硝胺化合物 nitroamino-compound
硝饼 niter cake
硝代 niter
硝代程度 degree of nitration
硝甘炸药卷 dynamite stick
硝酐 dinitrogen pentoxide;nitric anhydride
硝化 nitrate;nitrating;nitration;nitrify
硝化程度 degree of nitration
硝化处理 process of nitrification
硝化床 nitriary
硝化的 nitrifying
硝化滴滤池 nitrifying trickling filter
硝化地沥青 nitrated asphalt
硝化地沥青石 nitrated asphaltite
硝化淀粉 nitrostarch;starch nitrate
硝化淀粉炸药 nitrostarch explosive
硝化法 nitrification process
硝化反硝化生物去除剩余磷 nitrification denitrification biological excess phosphorus removal
硝化反硝化作用强度 nitrification-denitrification intensity
硝化反应 nitration reaction;nitrification reaction
硝化方法 nitriding process
硝化分离器 nitrator-separator
硝化釜 nitrating pot
硝化甘醇 nitroglycol
硝化甘油 angioneurosin;devil's brew;dynamite glycer;explosive gelatin(e);explosive oil;glycerol trinitrate;monobel;nitroglycerol;soup;blasting oil
硝化甘油代用品 niteoglycerin substitute
硝化甘油火药 nitroglycerine powder
硝化甘油酰胺炸药 nitroglycerine-amide powder
硝化甘油硝铵炸药 nitroglycerine-ammonium nitrate powder
硝化甘油炸药 dynamite;gelatin(e) dynamite;gelinite;nitrogelatine;nitroglycerine dynamite;nitroglycerine explosive;nitroglycerine powder;nitroglycerin(e)
硝化甘油炸药包 dynamite cartridge
硝化杆菌 nitrobacter
硝化工艺 nitrification process
硝化过程 nitrification process
硝化糊精 nitro-dextrin
硝化还原作用 nitrate reduction
硝化火药 nitroexplosive;nitropowder
硝化级苯 nitration benzol;nitration grade benzene;nitro-grade benzene
硝化级二甲苯 nitration grade xylene
硝化级甲苯 nitration grade toluene
硝化剂 nitrating agent
硝化胶质炸药 nitrogelatine
硝化酒精 nitrated alcohol
硝化聚甘油 nitrated polyglycerin
硝化菌 nitrobacteria
硝化菌活性 nitrifying bacterium activity
硝化离析器 nitrating separator
硝化离心机 nitrating centrifuge
硝化力 nitrifying power
硝化沥青 nitrated asphalt
硝化了的 nitrated
硝化磷酸盐 nitrophosphate
硝化滤池 nitrifying filter
硝化煤焦油(沥青) nitrated coal-tar
硝化棉 cellulose nitrate;collodion cotton;guncotton;nitrocellulose;nitrocotton;pyroxylin(e)
硝化棉涂料 pyroxylin(e) lacquer
硝化棉致发白 cotton blushing
硝化明胶炸药 nitrogelatine
硝化木材 nitro-timber
硝化木素 nitrolignin
硝化能力 nitrification ability;nitrifying capacity
硝化器 nitrator
硝化强度 nitrifying capacity
硝化生物滤池 nitrifying biofilter
硝化生物膜 nitrifying biofilm
硝化试验 nitrating test
硝化丝 nitrosilk
硝化速率 nitrification rate;nitrifying rate;rate of nitrification
硝化速率常数 nitrification rate constant
硝化酸 nitrating acid
硝化酸磷酸钾 nitrophoska
硝化涂料 nitrodope
硝化温度 nitriding temperature
硝化污泥 nitrifying sludge
硝化细菌 nitrification bacteria;nitrifier;nitrifying bacteria;nitrifying organism;nitrobacteria <使亚硝酸盐氧化为硝酸盐的细菌>
硝化纤维 nitrocellulose;cellulose nitrate
硝化纤维表面涂层 nitrocellulose surface coat
硝化纤维法 nitrocellulose process
硝化纤维胶 <黏[粘]贴应变片的胶合剂> nitrocellulose cement;nitrocellulose glue
硝化纤维面漆 nitrocellulose surface coat
硝化纤维漆 nitrocellulose paint
硝化纤维清漆 cellon lacquer;zapon (lacquer);zapon varnish;nitro-lacquer
硝化纤维人造丝 nitrosilk
硝化纤维丝 nitrocellulose silk
硝化纤维素 guncotton;nitrated cellulose;nitrocellulose;nitrocotton
硝化纤维素瓷漆 zapon enamel
硝化纤维素地板 cellulose flooring;nitrocellulose flooring
硝化纤维素火药 nitrocellulose powder

硝化纤维素胶泥 < 即杜卡胶,黏[粘]贴应变片的胶合剂 > nitrated cement;Duco cement
硝化纤维素黏[粘]合剂 nitrocellulose adhesive
硝化纤维素喷漆 nitrocellulose lacquer
硝化纤维素漆 cellulose enamel;nitrocellulose lacquer
硝化纤维素漆片 nitrocellulose chip
硝化纤维素清漆 nitrocellulose varnish
硝化纤维素塞子 nitrocellulose stopper
硝化纤维素塑料 cellulose nitrate plastics;pyroxylin(e) plastics
硝化纤维素涂料 nitrocellulose dope;nitrocellulose paint
硝化纤维素油漆 nitrocellulose paint
硝化橡胶 nitrite elastomer
硝化厌氧氨氧化 nitrification-anaerobic ammonia oxidation
硝化液回流 nitrification liquor recycling
硝化抑制剂 inhibitor for nitrification;nitrification inhibitor
硝化用产品 nitration grade products
硝化油 nitrated oil;nitro-oil
硝化炸药 nitroexplosive;nitropowder
硝化蔗糖 nitrosaccharose
硝化纸 nitrated paper;nitro-paper
硝化作用 niter;nitrification;nitration
硝化作用的抑制 inhibition of nitrification
硝基 nitro;nitro-group
硝基胺 nitra-amine;nitramine
硝基苯 mirbane oil;nitrobenzol
硝基苯胺 nitroaniline;nitrophenylamine
硝基苯二胺 nitrophenylene diamine
硝基苯废水 nitrobenzene wastewater
硝基苯废水处理 nitrobenzene wastewater treatment
硝基苯酚 dinitrophenolate;nitrophenol
硝基苯酚废水 nitrophenol wastewater
硝基苯酚盐 nitrophenolate
硝基苯基 nitrobenzophenone
硝基苯甲腈 nitrobenzonitrile
硝基苯甲醛 nitrobenzaldehyde
硝基苯甲酸 nitrobenzoic acid
硝基苯甲酸盐 nitrobenzoate
硝基苯甲酰胺 nitrobenzamide
硝基苯甲酰苯胺 nitrobenzanilide
硝基苯降解 nitrobenzene degradation
硝基苯醌 nitrobenzoquinone
硝基苯乙醚 nitrophenetol
硝基苯乙酸 nitrophenyl-acetic acid
硝基苯乙烯 nitrostyrolene
硝基苯中毒 nitrobenzene poisoning
硝基丙烷 nitropropane
硝基铂酸根 nitroplatinate
硝基产品 nitrocellulose product
硝基醇 nitroalcohol
硝基瓷漆 lacquer enamel;nitroenamel
硝基(代)苯 nitrobenzene
硝基丁烷 nitrobutane
硝基二苯胺 nitrodiphenylamine
硝基二苯基甲烷 nitrodiphenylmethane
硝基二甲苯 nitroxylene
硝基二甲基苯胺 nitrodimethylaniline
硝基芳酸 nitroaromatic acid
硝基芳香化合物 nitroaromatic
硝基废水 nitro-group wastewater
硝基酚 nitrophenol
硝基酚染料 nitrophenolic dye
硝基氟过程 Nitrofluor process
硝基腐殖酸 nitro-humic acid
硝基甘油 glyceryl trinitrate;nitroglycerin(e)
硝基胍(炸药) nitroguanidine
硝基航空涂料 nitrate dope
硝基化合物 nitro-compound
硝基环己烷 nitrocyclohexane
硝基环己烷法 nitrocyclohexane process
硝基磺酸 nitrosulfonic acid
硝基甲苯 < 一种烈性炸药 > nitrotoluene
硝基甲苯胺 nitro-methylaniline;nitrotoluidine
硝基甲酚 nitrocresol
硝基甲烷 nitrocarbol;nitromethane
硝基甲烷氧气混合气 nitromethane-oxygen mixture
硝基假象牙 celluloid
硝基间苯二甲酸 nitroisophthalic acid
硝基金属 nitrometal
硝基酒石酸 nitrotartaric acid
硝基聚合物 nitropolymer
硝基枯烯 nitrocumene
硝基联苯胺 nitrobenzidine
硝基绿 nitrate green
硝基氯苯 nitro-chlorobenzene
硝基氯苯胺红 nitro-chloraniline red
硝基氯仿 chloropicrin;nitrochloroform
硝基蒙皮漆 nitrate dope
硝基面漆 finishing lacquer
硝基萘 nitronaphthalene
硝基脲 nitrourea
硝基喷漆 spraying lacquer
硝基漆 celluloid paint;lacquer;nitrate paint;nitrocellulose paint;nitro-lacquer;pyroxylin(e) lacquer
硝基漆漆膜擦平 pulling over of nitrate paint
硝基漆稀料 lacquer thinner
硝基漆稀释剂 thinner for nitrocellulose finishes
硝基清漆 clear lacquer;nitrate varnish;nitrocellulose lacquer;nitrodope;zapon(lacquer);zapon varnish
硝基染料 nitro dyestuff
硝基塞 nitrocellulose stopper
硝基石蜡 nitroparaffin
硝基涂料 nitrocellulose paint
硝基烷 nitroparaffin
硝基系颜料 nitropigment
硝基纤维二道浆 lacquer surfacer
硝基纤维封闭底漆 lacquer sealer
硝基纤维漆稀释剂 lacquer thinner
硝基纤维清漆 xylonite-solution
硝基纤维素 nitrocellulose
硝基纤维素漆 lacquer
硝基纤维(素)漆用溶剂 boiler
硝基盐酸 nitro-muriatic acid
硝基乙烷 nitroethane
硝基脂肪酸 nitrofatty acid
硝镪离子 nitronium
硝棉 cordite;guncotton
硝棉胶 collodion
硝棉漆 pyroxylin(e) lacquer
硝棉塑料 pyroxylin(e) plastics
硝棉制动器 pyroxylin(e) stopper
硝皮 tannage
硝皮工人 tanner
硝镪水 nitric acid
硝石 niter[nitre];nitrokalite;saltier;saltpeter[saltpetre]
硝石层 caliche
硝石的 nitric
硝石洞 saltpeter cave
硝石矿床 nitrate bed
硝石炉 niter[nitre] oven
硝石球 niter[nitre] ball
硝石土 saltpeter earth
硝石盐 saltpeter salt
硝水 salt water
硝酸 aqua fortis;azotic acid;hydrogen nitrate;nitric acid
硝酸铵 ammonium nitrate;ammonium nitrite
硝酸铵高含量炸药 high-ammonium-nitrate-content dynamite
硝酸铵胶质炸药 ammonia gelatin(e)
硝酸铵燃油 ammonium nitrate fuel oil
硝酸铵与柴油燃料混合物 igdanit
硝酸铵炸药 < 爆破用 > ammonium nitrate powder;ammonium explosive
硝酸钡 nitrate of baryta
硝酸苯胺 aniline nitrate
硝酸苯汞 phenylmercuric nitrate
硝酸铋 bismuth nitrate;Spanish white
硝酸丙酯 n-propyl nitrate
硝酸厂 nitric plant
硝酸醋酸纤维素 nitroacetyl cellulose
硝酸淬火 nitric acid quenching
硝酸发动机 nitric acid unit
硝酸法浆粕 nitric acid pulp
硝酸分解 decomposition with HNO3
硝酸钙 nitrate of lime
硝酸钙溶洞 saltpeter cave
硝酸钙土 saltpeter earth
硝酸甘油 britonite;carbonite;grisounite;monarkite;monobel;nitrobid;nitroglycerin(e);nitroglycerol;nitroglyn;nitrolingual;nitromel;nitrong;nitrorectal;nitroretard;nitrostat;nitrozell retard;Nobel's explosive;nysconitrine;trinalgon
硝酸甘油等混合炸药 carbonification
硝酸甘油火药 straight dynamite
硝酸甘油胶质炸药 gelignite
硝酸甘油溶液 nitroglycerine solution
硝酸甘油炸药 axite
硝酸根 nitrate;nitrate radical
硝酸根的 nitric
硝酸根离子 nitrate ion
硝酸汞 mercuric nitrate
硝酸汞滴定法 mercuric nitrate titration
硝酸汞容量法 mercuric nitrate volume method
硝酸合氢离子 nitriacidium ion
硝酸和硫酸的混酸 nitration mixture
硝酸化过磷酸 nitrate superphosphate
硝酸还原酶 nitrate reductase
硝酸基 nitrato
硝酸基胶片 nitrate-base film
硝酸基络合物 nitrato complex
硝酸加热试验 nitration acid heat test
硝酸甲醇溶液 nital
硝酸甲脂和酒精的混合物 myrol
硝酸钾 aerolite;amidpulver;britonite;niter;nitrate of potash;nitrate of potassium;nitre;potassic nitrate;potassium nitrate;saltpeter[saltpetre]
硝酸钾钠 nitropo
硝酸钾炸药 grisounite
硝酸胶片基 nitrate film base
硝酸浸蚀试验法 spot test
硝酸菌 Nitromonas
硝酸类 < 硝酸及亚硝酸的统称 > nitrose
硝酸灵 nitron
硝酸硫酸混合酸浸渍液 < 电镀用 > ackey
硝酸铝 alumin(i)um nitrate
硝酸绿 nitrate green
硝酸镁 magnesium nitrate
硝酸锰 manganese nitrate
硝酸棉炸药 axite
硝酸钠 Chili saltpeter;cubic(al) niter;nitrate of soda;nitre;saltier;sodium biphosphate;sodium niter;sodium nitrate
硝酸钠溶液浸渍处理法 nitralising
硝酸钠溶液浸渍净化法 nitralising
硝酸尿素 urea nitrate
硝酸镍 nickel nitrate;nickelous nitrate
硝酸铅 lead nitrate;plumbic nitrate
硝酸氢氟酸分解 decomposition with mixture of HNO_3 and HF
硝酸烧伤 nitric acid burn
硝酸试剂 nitron
硝酸试验 nitric-acid test
硝酸双氧铀 uranium nitrate;uranyl nitrate
硝酸铁 ferric nitrate;iron nitrate
硝酸铜 copper nitrate;cupric nitrate
硝酸吸收塔 nitric acid absorber
硝酸析银法 quartation
硝酸细菌 nitrate bacteria
硝酸纤维 nitric acid fibre
硝酸纤维素 cellulose enamel;nitrocellulose;cellulose nitrate
硝酸纤维素胶片 nitrate film;nitrocellulose film
硝酸纤维素(木粉)浆 plastic wood
硝酸纤维素黏[粘]结剂 cellulose adhesive
硝酸纤维素漆 cellulose nitrate paint;nitrocellulose lacquer;zapon lacquer
硝酸纤维素清漆 zapon varnish
硝酸纤维素塑料 cellulose nitrate plastics
硝酸纤维素填料 cellulose nitrate stopper
硝酸锌 zinc nitrate
硝酸亚汞 mercurous nitrate
硝酸亚铁 ferrous nitrate
硝酸岩矿床 nitrate deposit
硝酸盐 azotate;nitrate
硝酸盐氮 nitrate nitrogen
硝酸盐的淋溶 leaching of nitrate
硝酸盐废水 nitrate wastewater
硝酸盐和氯化物浓度 nitrate and chloride concentration
硝酸盐还原法 nitrate reduction
硝酸盐还原细菌 nitrate-reducing bacteria
硝酸盐还原作用 nitrate reduction
硝酸盐矿物 nitrate mineral
硝酸盐冷却 nitrate cooling
硝酸盐浓度 nitrate concentration
硝酸盐污染 nitrate pollution
硝酸盐细菌 nitrate bacteria
硝酸盐氧化溶液 nitric acid oxidation solution
硝酸盐阴性杆菌 nitrate-negative bacillus
硝酸盐与硫酸盐还原 nitrate and sulphate reduction
硝酸盐再生作用 nitrate regeneration
硝酸盐植物 nitrate plant
硝酸盐中毒 nitrate poisoning
硝酸乙酸纤维素 cellulose nitroacetate;nitroacetyl cellulose
硝酸异丙酯 isopropylnitrate
硝酸银 common caustic;lunar caustic;silver nitrate
硝酸银滴定法 silver nitrate titrimetry
硝酸银分光光度法 silver nitrate spectrophotometry

硝酸酯 nitrate
硝态氮 nitrate nitrogen
硝态氮负荷 nitrate nitrogen load
硝态氮浓度 concentration of nitrate nitrogen
硝纤象牙 celluloid
硝烟 nitrous fume

销

销板 feather tongue;pin plate;spline joint
销棒 spike
销棒装置 mechanical dowel installer

销闭式离合器 locking clutch; lock-up clutch
销采比 ratio of marketing divided by output
销槽 cotterway; keyhole slot; key seat(ing); key slot; keyway; pick-up slot
销槽刨床 keyway planer
销槽铣刀 cotter mill
销衬套 <活塞销的> pin bush(ing)
销承压强度 pin-bearing strength
销齿条 pin rack
销除 elimination
销除法 working-off process; writing-off process
销除债务准备 extinguishment reserve
销地市场 markets in sales areas
销钉 brad; core pin; dowel; dowel bar; dowel pin; foot lug; key bolt; nail dowel; pin bolt; set pin; spike knot; spud; wire nail
销钉扳手 pin wrench
销钉板 dowel plate
销钉打入工具 dowel driver
销钉定位 pin registration
销钉断面剪力 dowel shear
销钉封接机 stud pin inserting machine
销钉杆 pin shank
销钉管 stud tube
销钉辊 spike roller
销钉铰链 pin hinge
销钉接合 dowel(l)ed joint; dowel(ling)(pin) joint; pin connection
销钉接头 dowel(l)ed joint; dowel(ling)(pin) joint; pin connection
销钉孔(眼) cotter hole; dowel pin hole; pin-hole
销钉连接 bayonet; pin connection; pin joint; pinning; pin seal
销钉帽 dowel cap
销钉拼接板 pin splice
销钉汽笛 tag gun
销钉枪 stud gun
销钉式擒纵摆轮 pin pallet escapement balance
销钉式擒纵叉 pin pallet; pin pallet fork
销钉式擒纵机构 pin pallet(lever) escapement
销钉式擒纵轮 pin pallet escape wheel
销钉锁式开关 pin tumbler
销钉套筒 dowel sleeve
销钉头 stem knob
销钉歪斜 dowel deflection
销钉用黄铜 pin metal
销钉桩 pin pile
销钉桩在基岩内的承插孔 pile socket
销钉钻 pin drill; dowel bit <木工用>
销钉作用 dowel action
销定位 finger setting
销耳式接口 pin lug coupling
销阀式喷油嘴 pintle-type atomizer; pintaux nozzle <带辅助喷孔的>

销负荷 pin load
销杆 pin rod
销钩 choker hook; pintle hook
销固定 fixed pin
销规 pin ga(u)ge
销辊 pin roll
销号 cancellation of a call; logoff
销后买回 sale-and-repurchase
销后租回 sale and leaseback
销簧 dog chart
销毁 demolishing; demolishment
销毁库存 destruction of stock
销毁式过油管射孔器 expendable through-tubing perforator
销毁式射孔器 expendable gun
销毁许可 demolition permission
销毁炸药 destruction of explosives
销毁证明书 cremation certificate
销货 merchandise sales
销货搬运费 cartage-out
销货补偿 allowance for sale
销货车费 cartage-out
销货成本 cost of goods sold; cost of merchandise sold; cost of sales
销货成本汇总表 summary of cost of goods sold
销货纯利 net profit on sales; net trading profit
销货单 bill of sale
销货登记簿 sales register
销货点 point of sale
销货点终端机 point-of-sale terminal
销货订单 sales order
销货发票 invoice for sales; sales invoice; sales ticket
销货发票簿 invoice-book outwards; invoice outward book
销货法 sales system
销货费(用) selling expenses; cost of sales; sales expenses
销货合同 sales contract
销货回扣 sales rebate
销货回扣及折让 sales rebate and allowance
销货会谈 sales convention
销货记录 record of goods sold
销货净额 net sales
销货净额对资产总额的比率 ratio of net sales to total assets
销货净利 net profit on sales
销货净收入 net proceeds from sales
销货净值法 net sales-value method
销货客户 trade creditor
销货毛利 gross margin; gross profit on sales; gross trading profit; trading profit
销货毛利法 gross profit method
销货毛利分析 gross profit analysis
销货毛利与销货比率 ratio of margin on sales to net sales
销货凭证 sale voucher
销货清单 account sale
销货渠道 marketing channel
销货日记账 sales daybook; sales journal
销货收入 avails of the sale; income from sales of products; proceeds of sale
销货收益率 rate of return on sales; return on sales
销货损益 sales profit and loss
销货退回 return inward; return sales; sales returns
销货退回报告单 returned sales report
销货退回通知单 credit memo for sales returns; credit memorandum for sales return; return sales memo
销货退回账 returns book inward
销货退回账户 returned sales account
销货退回折扣 discount on returned sales
销货现金 sales dollar
销货要约 offer to sell; selling offer
销货佣金 selling commission
销货与管理费用 selling and administrative expenses
销货预测技术 sales forecasting technique
销货预算 sales budget
销货员推销 personal selling
销货运费 carriage outward; delivery expenses; freight out(war)d; outward freight; transportation out
销货账 account of goods sold; merchandise sales account; sales account
销货账簿 sales book
销货账目 account of(goods) sales
销货折价 discount allowed; discount granted
销货折扣 sales discount
销货知识 sales knowledge
销货总成本 gross cost of merchandise sold
销货总额 gross sales
销货总额法 total sales method
销键 pin key
销铰 pin hinge
销铰轴芯铰链 pin hinge
销接 pin-connected joint; pin joint; pin-keyed
销接的 pin-connected
销接端 pin(ned) end
销接拱梁 pin arched girder
销接合 bayonet joint; cotter joint; pin connection
销接桁架 pin-jointed truss
销接框架 pin-connected frame; pin frame
销接链 pin chain
销接式夹桩器 pin connecting chuck
销接头 hinged joint; pin connection; pin joint
销接支柱 pin-ended strut
销紧卡环 check ring
销紧开关 <陀螺测斜仪> switch for caging
销壳 pin boss
销孔 matching hole[holing]; pin-hole [holing]; drawbore
销孔钉 drawbore
销孔磨床 pin-hole grinder
销孔式插锁 pin-in-hole lock
销孔钻 pin drill
销口(接头) lock joint
销连接 cottering; dowel(l)ed joint; pin connection; pin coupling
销联桁梁 pin-jointed truss
销联结 pin coupling
销路 consumption; debouche; market(ability); outlet; salability; salableness
销路不佳 decline of sale; dull of sale; poor sales
销路呆滞 narrow outlay; sale resistance
销路调查 market research
销路好 easy of sale
销路好的 marketable
销轮 pin wheel
销轮的销座 shroud
销轮销座(钟表) shroud
销螺母 pin nut
销起子 key driver
销钳 pin pliers
销入 pinning-in
销式安全接头连接 pin type safety joint
销式连接 pin type attachment
销式联结钻头 <凿岩机> pin type attachment
销售 distribution; market(ing); merchandising; selling
销售百分比法 percentage of sales approach; percentage of sales method
销售百分比分析 percentage of sales analysis
销售百分率分析 percentage of sales analysis
销售包装 distribution packaging; sales package
销售保证 sales warrant
销售比较法 sales comparison approach
销售边标 margin on sales
销售表 trading statement
销售不畅 dull sale
销售部 sales department
销售部经理 sales manager
销售部门 marketing agencies; marketing department; sales department
销售采购 buying for resale
销售策略 marketing strategy
销售场所 merchandising location
销售成本 cost of distribution; cost of goods sold; cost of marketing; cost of sales; distribution cost; distributive cost; marketing cost; selling cost
销售成本分析 marketing cost analysis
销售成本顾问 marketing cost consultant
销售成本预算 selling cost budget
销售处 sales department; sales office
销售纯利率 net profit rate to sales; sales margin
销售刺激因素 sales stimulation factor
销售促进 sales promotion
销售代表 sales representative
销售代理商 sales agent; selling agent
销售代理业务 sales commission
销售贷款公司 sales finance company
销售单位 marketing unit
销售淡季 period of slack sales
销售导向 sales orientation
销售的边际利润 margin-profit on sale
销售的产品 product sold
销售的毛收入 gross income from sales
销售底货 clearance sale
销售地区权利条款 territory rights clause
销售点系统 point of sale
销售点终端机 point-of-sale terminal
销售调查 distributive trade survey; marketing research
销售订单 sales order
销售订单处理 sales order processing
销售定额 sales quota
销售短视 marketing myopia
销售堆 sale lot
销售对策 marketing game
销售额 amount of sales; sales; sales amount; sales value; sale volume
销售额百分比预测法 percentage of sales forecasting method
销售额分类 classification of sales
销售额极大化原理 sales-maximization principle
销售额统计预测 statistic(al) forecast of sales
销售额与利润的比例 sales and profit comparisons
销售额预测 sales forecast
销售额增加 upswing of sales
销售额最大化假设 sales maximization hypothesis
销售发票 outgoing invoice; sales invoice
销售方式 modes of sale
销售方式多样化 diversification of sales
销售方针 marketing policy; sales policy
销售费用 cost of marketing; cost of sales; distribution charges; sales expenses; selling charges; selling cost; selling expenses
销售费用差异 selling expenses variance
销售分处 branch sales office
销售分类账 sales ledger
销售分析 sale analysis
销售份额 market share; sales quota
销售服务组织 sales service organization
销售概念 marketing concept; selling concept

销售纲领 selling platform
销售工厂 distribution plant
销售工程 sales engineering
销售工程师 marketing engineer;sales engineer
销售公司 sales company
销售构成 sales mix
销售估计 sales estimate
销售管理 marketing management;sales management
销售惯例 marketing practice
销售规模 sales volume
销售国 country of sale
销售国外的商品 commodity sold overseas
销售过程 distribution chain;marketing process
销售合同 contract of sales;sales agreement;sales contract
销售合同应收账款 sales contract receivable
销售合作社 marketing cooperative
销售核查 sales audit
销售化企业观念 marketing-oriented business concept
销售回扣及折让 sales rebate and allowance
销售会计 sales accounting
销售或退货 on sale or return
销售机构 marketing mechanism
销售机会 sales opportunities
销售激励研究 sales and marketing motivation study
销售及管理成本 selling and administrative cost;selling and administrative expenses
销售及管理费用 selling and administrative expenses
销售及管理费用分析表 selling and administrative expense analysis sheet
销售集团 selling group
销售集团承销部分 selling group pot
销售计划 marketing plan(ning);sales plan
销售计价基准 basis for recording sales
销售记录 act of sales;sales records
销售技术 marketing technique;sales technique
销售价(格) selling price;retail price
销售价格差异 selling rice variance
销售价值 realization value;sale value
销售价值法 sales value method
销售减衰率 ratio of marketing decreasing
销售减衰主要因素 principal factors of marketing decreasing
销售检核人 marketing controller
销售结构 sales structure
销售截止日期 <印在食品等包装上> pull date
销售金额计划 dollar merchandise plan
销售金融公司 sale finance company
销售经纪人助理 sales associate
销售经理 marketer;sales manager
销售竞争 sales contest
销售局 Marketing Boards
销售决策变数 marketing decision variable
销售科 sales department
销售控制研究 sales control research
销售-库存比率 sales-inventory ratio
销售力量 sales potential
销售利润 profit on sales;salable profit;sales profit
销售利润边标 margin-profit on sale
销售利润中心 marketing profit center [centre]
销售链系 sales chain
销售量 quantity sold;sales volume;volume of sales
销售量估计 estimation of sales volume
销售量所居位数 code of marketing in order
销售量责任额 sales volume quota
销售量占百分比 market share at percentage
销售率 sale rate;selling rate
销售毛利 gross profit on sales;gross trading profit
销售模型 marketing model
销售目标矩阵表 sales objective matrix
销售目录 sales catalogue
销售能力 ability to sell;selling capacity
销售年度 marketing year
销售配额条款 quota clause
销售情报 marketing information
销售情报系统 marketing information system
销售情况 sales status
销售区 commercial territory
销售区试验 sales area test
销售区域 sales territory
销售渠道 channel of distribution;marketing channel
销售渠道长度 the length of distribution channel
销售权 power of sale
销售确认书 confirmation of sales;confirmatory sales note;sales confirmation
销售人员薪金 sales salaries
销售人员综合市场预测法 sales-force-composite method
销售日记账 sales journal
销售商 distributor
销售商风险 seller's risk
销售商网广告 dealer tie-in
销售设施 distribution facility
销售深度检查 sales depth test
销售审计 sale auditing
销售收入 proceeds of sale;revenue from goods sold
销售税 sale tax
销售提供价值 sales yield value
销售网 dealer network;distribution net(work);sales network
销售物价指数 consumer price index
销售限额 sales quota
销售协议 agreement of sales;sales agreement
销售循环 market cycle
销售义务 obligation sale
销售意愿 willingness to sell
销售佣金 sale commission
销售预测 market forecast(ing)
销售运费 transportation out
销售增长率 ratio of marketing increasing
销售增长主要因素 principal factors of marketing increasing
销售占有率 capture rate
销售账 account sale
销售折扣 discount on sales
销售折扣提成 allowance for discounts available
销售折让 allowance on sales
销售证书 certificate of sale
销售状况 condition of sale
销售总额 gross proceeds;gross sales
销售组合 sales mix
销售组合计划 marketing mix program(me)
销售组织 marketing organizing
销枢 drift pin
销闩 catch lever
销栓 foxtail
销栓挂钩 pintle hook
销栓连接 pin connection
销栓作用 dowel action
销榫接缝 dowel tongue joint
销榫接头 pin joint
销套 guide;pin bush(ing)
销头扳手 pin spanner
销心 pin center[centre]
销眼 eye
销约条款 cancellation clause
销账 cancel a debt;cancel from an account;charge off;crossing of account;cross-off an account;remove from an account;write off;release of record <解除抵押品的留置权>
销中心线 pin center[centre]
销轴 clevis pin;pin roll;ripper pin
销轴拔卸器液压缸 pin puller cylinder
销轴承 pin bearing
销轴锁紧 lock clasp
销轴装卸器 pin puller
销住 pinning
销住的 pin-locking
销子 bolt pin;drift pin;stud(bolt)
销子槽 pin groove;pin slit;pin slot
销子连接 pegging
销子起子 key driver
销子套筒 dowel sleeve
销子轴承 pin bearing
销子座 bracket
销座 key seat;keyway;pin boss

小

小V形槽 veiner
小矮窗 <教堂圣坛右侧的> lychnoscope;offertory window
小安装角的叶片 low stagger blade
小鞍 back pad
小暗谷 rincon
小暗井 <由上而下开凿的> winze
小凹 cove
小凹槽 dado
小凹坑 pit hole
小凹圆饰 quarter hollow
小凹圆线脚 gorge cut
小八开图纸 <13 英寸×16 英寸,1 英寸=0.0254 米> small cap
小把 wisp
小坝 minor dam;mill dam <为磨坊蓄水用>;anicut <印度南方为灌溉而筑成的>
小白点 small particles
小白果 Chinese hazel
小白铁钉 small white nail
小百货 small articles of daily use
小摆锤 small pendulum
小摆设 bauble;bric-a-brac;knick-knack
小班 subcompartment
小斑 speckle;stigma;stigmata
小斑点 fleck
小搬小运 parcel collection;transportation of small goods
小板 <金刚石形的,用于装饰线脚> lozenge
小板方形玻璃板 quarrel
小板簧片 minor spring leaf;small leaf spring clip
小板簧套 small spring bushing
小板簧支销 small leaf spring pin
小板块 microplate;small plate
小板坯 coke bar
小板条 small strip
小板条地板 parquet strip flooring
小半径 sharp radius
小半径的环形工程 quick sweep
小半径的平曲线和竖曲线 sharp horizontal and vertical curvatures
小半径的转弯 short turning
小半径盘旋下降 narrow-spiral glide
小半径曲线 sharp curve;sharp-radius curve;short-radius curve;small-radius curve;tight curve
小半径曲线黏[粘]降【铁】reduction of adhesion at small radius curve;reduction of adhesion on minimal radius curve
小半径曲线坡度折减 compensation of sharp curve
小半径弯桥 sharp-radius curved bridge
小半径弯头 sharp bend
小半径转角 sharp corner
小半径作业 quick sweep
小半砖接头 queen closer
小拌(砂浆)板 <搅拌混凝土或砂浆用的> spot board
小包 parcel;pouch;subcontractor;sub-subcontractor <从大承包商中转包部分工程>
小包房 <卧车的> roomette
小包工 subcontractor
小包工头 lumper
小包裹加 bag rack
小包货 ballot
小包货样 small sample
小包机 bundling press
小包体 bleb
小包厢 <剧院的> cubiculo;family circle
小包运价 parcel rate
小孢绿盘菌 <拉> Chlorosplenium aeruginascens
小孢子 microspore
小孢子体 microsporinite
小苞片 bracteole
小雹 ice hail;small hail
小保险箱 coffret;pyx
小堡 fortlet
小堡垒 fortalice
小堡眼 crenelet;oillet(te)
小报 tabloid
小抱箍 small guide strap
小爆破 light blast(ing);light shot
小爆炸 light shot
小杯 pocill;pocillum;small cup
小杯菌属 <拉> Phialea
小杯状体 calicle
小杯状穴 calicle
小杯子 noggin
小北斗 Little Dipper
小本经营 business with a small capital;do business in a small way;hand-to-mouth operation;shoe-string trading
小比例(尺) small scale
小比例尺测量 small-scale survey
小比例尺测图 small-scale survey
小比例尺测图仪 small-scale plotter
小比例尺的 small scale
小比例尺地图 small-scale map
小比例尺地形图 small-scale topographic map
小比例尺地质测量 small-scale geologic(al) survey

小比例尺海图 small-scale chart
小比例尺模型 small-scale pilot model
小比例尺模型试验 small-scale model test
小比例尺水工模型 small-scale hydraulic model
小比例尺图 small-scale drawing
小比例柱体试验法 small-scale cylinder test method
小笔刷 fitch liner
小闭环 loop stop
小闭曲线 small closed curve
小壁橱 cubby
小壁石 bullset

小编队 small formation
小扁豆层(体) lentil
小扁漆刷 lining tool
小匾额 plaquette
小便槽 trough urinal; urinal; urinal channel; urinal gutter; urinal trough
小便槽冲水阀 urinal flush valve
小便槽冲洗管 sparge pipe; sparge pipe for urinal
小便槽的挡板 urine-repellent
小便池 trough urinal; urinal; urinal gutter; urinary
小便池冲洗 urinal flushing
小便斗 bowl urinal; urinal
小便斗分隔间的排列 range of urinal stalls
小便隔间 urinal stall
小便间 stall urinal
小便器 stall urinal; urinal(bowl); urinal spreader; urinal stall
小便器存水弯 urinal trap
小便器滤网 urinal strainer
小便所 urinal; urinary
小便污垢 urine stain
小变更 minor change
小变化 microvariation
小变位 small deflection
小变形锻造 saddening
小变压器亭 kiosk
小辫 pigtail
小标题 cross head(ing); subhead(ing); subtitle
小别墅 cottage; small villa; villanette
小滨海湖 lochan
小冰雹 ice pellets
小冰川 bergy bit; glacieret
小冰川脊 < 高 10 ~ 30 米 > paha
小冰堆 screw ice
小冰河 glacieret
小冰块 cake ice; ice cube; small ice piece; calf(ice) < 从冰山上崩落的冰块 >
小冰块崩落 calving
小冰盘 small ice floe
小冰期 little ice age
小冰球 ice pellets
小冰山 floeberg; floe ice; growler
小冰隙 small fracture
小冰原 small ice-field
小病 ailment
小波 small echo; small wave; smooth sea; wavelet
小波分析 wavelet analysis
小波纹板 short-pitch corrugated sheet
小玻璃瓶 phial; vial; ampo(u)le < 装一次用量药剂 >
小玻璃球 < 在路面上反光用的 > ballotini
小薄片 flakelet
小檗 barberry
小檗属 barberry; Berberis < 拉 >
小补偿空间爆破 small compensation space blasting
小不平度 small irregularities
小部分 a fraction of; snippet
小材 small wood
小材形数 small wood form factor
小彩色胶片 colo(u)r patch
小参数法 method of small parameter
小参数稳定准则 small parameter stability criterion
小餐馆 bistro; coffee shop; estaminet; luncheonette; lunch room
小餐间家具 dinette
小餐室 buffet; dinette; cenaculum < 古罗马 >
小餐厅 grill(e) room; private dining room; small dining room
小餐厅桌椅 < 厨房旁边的 > dinette(set)
小残丘 pimple mound
小舱单 cargo list
小舱口 scuttle
小舱口盖 scuttle cover
小舱室 flap; small cabin
小操纵台 cabinet
小操作室 minor-operating theatre [theater]
小糙 small slugs
小槽 fuller; minor trough; small groove; vale
小槽车 bowser
小槽试验资料 flume data
小槽线脚 quirk mo(u)lding
小草地 grass plot
小草坪 grass plot
小册子 booklet; brochure; chapbook; pamphlet; tabloid
小侧窗 bulleye window
小层 chamberlet
小层序 minor sequence
小叉车 little forklift
小插图 vignette
小插销 draw bolt
小碴 < 四分之一砖 > quarter closer
小差距汇率幅度 narrow band; narrow exchange rate margin
小差距汇率幅度安排 narrow margin arrangement
小柴捆 cord wood
小产品 miscellaneous goods
小铲 paddle
小长旗 awheft
小长气泡 elongated blister; elongated bubble
小潮 dead tide; neap; neap tide
小潮层序 neap sequence
小潮差 microtidal range; tidal range at neaps
小潮差海岸带 < 潮差小于 2 米 > microtidal belt
小潮潮差 mean(neap) range; tidal range at neaps
小潮潮流 neap tidal current
小潮潮升 neap rise
小潮迟后 lagging of neap
小潮到达河流上游界限 upstream limit of neap tide
小潮的最高低潮位 highest low water of neap tides
小潮低潮 low-water neap-tide; mean low water neaps
小潮低潮高 low-water neaps; neap low water
小潮低潮面 low-water neaps
小潮低潮位 low water neaps
小潮低水位 neap low water
小潮高潮面 high water neaps; high water neap-tide; neap high water
小潮高潮(位) high water neaps; high water neap-tide; neap high water
小潮高水位 neap high water
小潮高度 neap rise
小潮流速 neap rate; neap velocity
小潮平均潮差 mean neap range
小潮平均潮升 mean neap rise
小潮平均低潮面 low-water ordinary neap tide; mean low water neaps; mean low water of ordinary neap tide
小潮平均低潮位 low-water ordinary neap tide; mean low water neaps; mean low water neap tides; mean low water of ordinary neap tide
小潮平均高潮面 height of mean high-water of neap tide; high water of ordinary neap tide; mean high water neaps
小潮平均高潮位 height of mean high-water of neap tide; high water of ordinary neap tide; mean high water neaps
小潮期 neap season
小潮升 neap rise
小潮时平均低水位 mean low water neaps
小潮提前 priming of neap
小潮汛 lesser flood
小潮涨落差 neap range
小潮最高潮位 maximum high water level of neaps
小车 barrow; bogie [bogey/ bogy]; buggy; cab(in); cart; dan; dilly; hand barrow; hand cart; hut(ch); larry; lorry; pony truck; trolly; wagon; wheel barrow; pipe trolley < 从钻塔中拖出管子的 >
小车变幅的臂架 < 起重机 > luffing jib
小车变幅水平式臂架 troll(e)y boom
小车错车装置 < 施工隧道内 > car passer
小车大梁 troll(e)y girder
小车道板 < 脚手架上的 > cart way panel
小车轨道 troll(e)y track
小车轨距 troll(e)y span
小车荷载 troll(e)y load
小车滑行导轨 troll(e)y sliding guide rail
小车浇铸 buggy casting
小车卷扬机 troll(e)y hoist
小车卡车专用运输船 pure car-truck carrier
小车库 small garage
小车式道岔 sliding switch
小车式起落架 bogie landing gear; bogie undercarriage
小车式输送机 car conveyer[conveyor]
小车式退火窑 pan lehr
小车式铸型输送机 car-type mo(u)ld conveyer[conveyor]
小车式自动落纱机 mobile autodoffer
小车输送机 troll(e)y conveyer[conveyor]
小车塔式起重机 troll(e)y-jib tower crane
小车弹簧 bogie bolster
小车膛式炉 bogie hearth furnace
小车箱系统 small cabin system
小车行程 < 桥式起重机的 > troll(e)y travel
小车行驶大梁 < 桥吊的 > troll(e)y girder
小车运行速度 cross travel speed; traverse speed; troll(e)y(travel(l)ing) speed
小车走行机构 troll(e)y running mechanism
小尘暴 devil; dust devil
小沉淀池 small pit
小沉箱 leech; limpet dam
小陈列室 cabinet
小撑 jack shore
小撑杆 toe dog
小承包商 lumper; subcontractor
小承大小头 small end bell reducer
小城堡 castelet; chatelet; small castle
小城区 zonule
小城市 small city; small urban
小城镇 small town; tank town; townlet
小城镇建设 construction of small towns
小程序 applet; small routine
小程序块 blockette
小吃 collation; refection
小吃部 buffet; cafeteria; kiosk; refreshment room; snack counter
小吃店 beaufet; confectaurant; lunch room; snack bar
小吃饭间 < 厨房边的 > dinette
小吃摊 concessionaire; refreshment stand
小池 cuvette
小池塘 dill
小尺寸 small dimension; subsize; undersized bearing
小尺寸的 light-sized; midget; minor sized; minute-sized; pony-size; small scale; small-sized
小尺寸的零件 undersize part
小尺寸的异形钢材 < 槽钢 shaped bar
小尺寸风洞试验数据 small-scale tunnel data
小尺寸管子 < 指直径 24 英寸及以下的管子, 1 英寸 = 0.0254 米 > little inch
小尺寸桁架 jack truss
小尺寸模型 small-scale model
小尺寸木材 second-growth timber
小尺寸平板 small-sized slab
小尺寸陶瓦锦砖 stoneware small-size mosaic
小尺度 microscale; small scale
小尺度气候 small-scale climatology
小尺度气候学 microclimatology
小尺度天气 microweather
小尺度紊动 small-scale turbulence
小齿 denticle; dentil
小齿距齿 close-toothed
小齿锯 jig saw
小齿轮 miniature gear; pinion(gear); pinion wheel; step-down pinion
小齿轮传动 pinion drive; pinion stand gear
小齿轮传动轴 pinion drive shaft
小齿轮传动装置 pinion gearing
小齿轮导承 pinion guide
小齿轮的一个齿叶 pinion leaf
小齿轮对 pinion mate
小齿轮和齿条转向装置 pinion-and-rack steering
小齿轮架 pinion stand
小齿轮刨齿机 pinion-gear shaper
小齿轮润滑脂 pinion grease
小齿轮室 pinion barrel
小齿轮套筒轴 pinion quill shaft
小齿轮外侧轴承 pinion shaft outside bearing
小齿轮外罩 pinion housing
小齿轮铣刀 pinion cutter
小齿轮线坯 pinion wire
小齿轮箱 pinion stand
小齿轮轴 pinion shaft
小齿轮轴承 pinion(shaft) bearing
小齿轮组 crowd(ed) pinion
小齿轮座 pinion stand
小齿铣刀 fine-pitch cutter
小齿状突起 denticle; denticulation
小充电电流 trickle charging current
小冲沉桩 pile jetting(method)
小冲床 bear
小冲击参数碰撞 close collision
小冲角 low incidence
小冲天炉 small cupola
小冲头 pin punch
小冲子 small punch
小虫孔 < 直径 1/16 ~ 1/8 英寸, 1 英寸 = 0.0254 米 > shothole; pin(worm) hole
小虫孔健全材 sound wormy
小出租车 minicab
小厨房 k'ette; kitchenet(te); package kitchen; small kitchen
小橱 cupboard
小橱橱门闩 cupboard turn
小橱房 wanigan
小储藏室 den
小储水池 cistern

小川 brooklet;streamlet
小船 cog; dinge(y); dinghy; small craft;cockboat <大船的供应船>
小船避风港 boat harbo(u)r
小船舶地 boat basin
小船舶位 small craft berth
小船厂 boatyard
小船池 camber
小船船厂 boatyard
小船船闸 recreational lock
小船船长 skipper
小船登岸处 small boat landing
小船队 flotilla
小船浮码头 floating pontoon for small craft
小船港 small craft harbo(u)r; boat harbo(u)r;small boat harbo(u)r
小船港池 boat basin;boat dock
小船过冬停泊水域 harbo(u)r for lagging up river craft
小船航道 boat channel
小船横舵柄绳 twiddling line
小船桨手 wherryman
小船警报 small craft warning
小船码头 boat dock; boat harbo(u)r; boat landing; boat wharf; small boat landing
小船锚地 anchorage for small vessels; small craft anchorage; small vessel anchorage
小船升船滑道 boat lift railway
小船停泊处 harbo(u)r for lagging up river craft
小船坞 camber;marina
小船系缆桩 boat cleat
小船下水设施 launching installation for small vessels
小船业 boatbuilding
小船运输 boatage;boat hire
小船运输费 boatman charges
小船闸 boat lock
小椽 dwarf rafter; jack rafter; jack timber;sprocket piece
小椽的斜切面 bevel(l)ing cut of jack rafter
小椽子的垂直切割 perpendicular cut of jack rafter
小窗 fenestrule; little window; loop; small window; lucarne <尖塔上的>;low side window <老教堂内高坛墙上的>;leper window <中世纪教堂外墙上的>
小窗洞 fenestella;fenestral
小窗口 wicket
小窗扇 <纱窗上的,供人手伸出> screen flap
小床 bell cot;couch
小锤 crandall;small hammer
小蠹虫 bark beetle
小丛壳 <拉>属 Glomerella
小丛林 bosk(et);bosquet;grove
小脆柄菇属 <拉> Psathyrella
小村 small village
小村庄 hamlet
小锉 cabinet file
小错 lapse
小/大环境 micro/macro environment
小大陆桥 mini-land bridge
小带 zonule
小带间隙 Petit's canal
小带锯 pony hand-saw;table band saw
小带纤维 fibrae zonulares
小袋 pouch
小单位 microscale
小弹丸 pellet
小挡泥板 small mud guard
小刀 dirk;knife;pen knife
小刀头 tool bit
小导管 condulet;ductile;ductule
小导管预注浆 pregrouting with mi-cropipe
小岛 ait;isle(t);holm(e) <河或湖中的>
小岛屿 small island
小捣锤 mash hammer
小道 ai(s)le;foot path;path(way); promnard;track
小的步行商业街 mini-mall
小的锻造部件 small forged parts
小的静电干扰声 clicking
小的矩形排水井 box drain
小灯船 beacon boat
小灯光 dimmed illumination
小灯泡 miniature bulb;small bulb
小凳 stool
小低(气)压 microdepression
小滴 dripping;water droplet
小底部阻力喷管 low base-drag nozzle
小地 bavli
小地槽 minor geosyncline
小地块 small block
小地理品种 microgeographic race
小地区填补测量 patchwork survey
小地松 lodgepole pine
小地毯 <房间用> area rug
小地物 underfeature
小地形 microrelief;minor feature
小地形区域 region of little relief
小地形学 microtopography
小地旋回【地】deuterogaikum
小地榆 small burnet
小地震 minor earthquake;tremor;minor shock
小地震动 small magnitude earthquake
小地震区 minor seismic area
小地址存储区域 low-order memory area
小点 dot;speckle;tick
小电池 baby battery;cell
小电动机 small-size motor
小电机 small machine
小电流 low current;trickle
小电流充电器 trickle charger
小电流等离子弧焊 low-current plasma arc welding
小电偶 small electronic couple
小电器设备 small electric(al) appliance
小电影馆 cinematheque
小电珠 bulb;flash lamp
小电阻测量表 evershed ducter
小垫布 doily;tidy
小垫片 minipad
小雕像 figurine;statuette
小吊车 grass hopper
小吊杆 small suspension rod
小吊绳 timenoguy
小吊艇 davit craft
小吊桶 windlass bucket
小钓船 troller
小跌水潭 plunge pool
小钉 fine nail;lead tack
小钉书机 stapler
小定额灌溉 deficit irrigation
小东西 pinhead
小动物缸 mouse jar
小动物群 faunule
小动物饲养场 small animal reservoir
小动物饲养园 terrarium
小洞 fenestra;orifice;small cave
小洞穴 bug hole
小洞织补法 small hole darning
小斗车 buggy;decauville truck;wagon
小豆石 pea gravel;pea grit
小豆石灰浆 pea gravel grout
小蠹虫卵孔道 brood tunnel
小蠹科 <拉> Scolytiolae
小蠹属 <拉> Seolytus
小端 small end;tip end
小端齿接触 toe contact
小端面 small end face
小端铜瓦 little-end brass
小端销 little-end pin
小端轴承 little-end bearing
小段 subparagraph;subsection
小断层【地】minor fault;small fault; thurm
小断面 light section;small cross-section
小断面道坑 small-section gallery
小断面的 light-ga(u)ge
小断面隧道 micro-tunnel; small section tunnel;tiny tunnel
小断面隧道掘进机 micro-tunnel(l)er
小断面线路 small profile line
小断面型钢 bar size section
小断面钻进 drilling in cramped quarter
小堆 little heap
小队 crew;squad
小队长 platoon commander
小队工作 team work
小对船 small pair boats
小吨 short ton
小吨位船舶 light-tonnage vessel
小吨位货汽车 pick-up
小吨位运货汽车 pick-up truck
小墩 <充当支座> pendicule
小墩支撑的 pendiculated
小盾片 scutcheon
小额 groat;petty sum
小额保险 petty insurance
小额补贴 fringe benefit
小额财产的管理 administration of estates of small value
小额储存 small lot storage
小额储蓄者存款证 small saver's certificate
小额存款 petty deposit
小额贷款 petty loan
小额贷款公司 licensed lender; small loan company
小额股票 stock scrip
小额耗用财产 petty sum wasting property
小额活期存款 petty cash current deposit
小额货币 money of small denominations
小额交易汇价 retail rate of exchange
小额款项 hay
小额零星存款 retail deposits
小额免费发行证券 fractional free issue
小额免费分配 fractional free distribution
小额批发 jobbing
小额欠款单据 chit;chitty
小额损害赔偿 nominal damage
小额损失 small damage
小额损失赔偿 nominal damage
小额索赔法庭 small claims court
小额通货 scrip
小额现金 petty cash
小额现金支付 petty cash payments
小额消费品税 <美> nuisance tax
小额银行券 fractional bank note
小额银行业务市场 retail banking market
小额优惠 fringe benefit
小额优惠工资 fringe wage
小额预付款 petty imprest
小额预算 baby budget
小额债券 baby bond;small bond
小额支票 small check
小额资本 shoestring
小儿科 department of paediatrics; paediatrics
小而不值钱的东西 flivver
小而长的海湾 armlet
小而舒适的客厅 snug
小二乘估计量【数】least squares estimator
小发动机 puffer
小阀(门) bib valve;pet valve
小帆布袋 satchel
小帆船 jolly boat; sailing craft; smack;yawl
小翻斗车 spoil car
小反射面目标 small reflective area target
小饭店 chop-house; coffee-tavern; cook shop;inn
小范围点检测器 small area point detector
小范围屈服 small-scale yield
小范围屈服理论 yield theory of small scope
小贩 pack man;padlar;pedder <沿街兜售的>; chapman[复 chapmen] <英>
小贩船 bum boat
小贩的货物 peddlery
小贩法例 Hawkers Act
小贩经营 peddlery
小方案 offshoot program(me)
小方材 <2~4 英寸见方,1 英寸见方 = 0.0006 平方米> scantling; baby square;batten;small square
小方锉 small square file
小方地毯 carpet tile
小方格 small check
小方格网 close-meshed rectangular grid
小方尖塔 pyramidion
小方角材 <1 英寸×1 英寸以下,1 英寸 = 0.0254 米> baby scantling
小方块 dice
小方块石砌墙 snecked wall
小方木 baby square;scantling lumber
小方山 butte;meseta
小方石 ashlar;cube sett;sett;shoddy work;blockage <美>;small stone block <嵌花式铺砌用>
小方石的石料 <美> blockage stone
小方石辐射形地面 radial small stone sett paving
小方石辐射形路面 radial small stone sett paving
小方石辐射形铺面 radial small stone sett paving
小方石块 Kleinpflaster; small sett; stone sett
小方石块机 sett machine
小方石块路面 small-sized sett paving;stone-sett pavement
小方石块铺路面 sett paving
小方石块制作 sett-making
小方石路 sett-paved road
小方石路面 small sett paving; stone-sett pavement; stone sett paving; sett paving
小方石路面下的砂垫层 sand layer under a sett paving
小方石面层 sett paving
小方石拼花路面 mosaic paving sett
小方石铺路 paving with setts
小方石铺路砂 sand for sett paving
小方石铺面 pegtop paving
小方石铺砌 pavement in sets; paving in setts; paving in stone blocks; random pavement; sett paving; small sett paving; stone sett paving
小方石铺砌路面 durax-cube pavement
小方石制造机 sett-making machine
小方厅 antechamber
小方线条 riglet
小方形开口缝 bar fagoting
小枋 baby square;scantling

小房 chamberlet
小房间 booth; cell; cellula; cubicle; cuddy; pigeon hole; cabinet; roomette < 卧车的 >
小房间壁橱 mouse hole
小房内的 intralocular
小房屋 maisonette; pill-box
小房子 house let
小飞机 douser
小飞机场 air park; air strip
小飞机库 hangarette; small hangar
小费 beer-money; bribe; gratuity; palm grease; perquisite; tip
小分包人 sub-subcontractor
小分包商 sub-subcontractor
小分包者 sub-subcontractor
小分队 contingent
小分隔间 stall
小分类 minor; subclass; subclassification
小分数功率 subfractional rating
小分支 subbranch
小分子 micromolecule; small molecule
小风暴 storm in a teacup
小风速计 air(o)meter
小浮标 dan buoy
小浮标敷设艇 dan layer
小浮冰 small ice floe
小浮冰块 calf
小浮沉样 small float-and-sink sample
小浮码头 floating boat dock
小浮游生物 net plankton
小幅度贬值 mini-devaluation
小幅度波动 fluctuation within a narrow range
小幅度递增分配方式 marginal increase allocation method
小幅度调整汇率 crawling peg
小幅度调制 small amplitude modulation
小蜉蝣 caenis
小斧 < 救生艇配置品 > hatchet
小斧刃 bit of hatchet
小辅圆 minor auxiliary circle
小负荷管式 small duty pipe-still
小负荷运转 light running
小改革 minor tune up
小改进 minor betterment
小改正 small correction
小盖板 little covering plate
小盖皿【化】capsule
小概念 minor concept
小杆材树 small pole
小杆钩 pike staff
小坩锅 monkey
小钢轨 baby rail
小钢轨便道 narrow steel track
小钢梁 bar joist
小钢坯加热炉 billet furnace
小钢凿 gad
小港 creek; outport; portlet
小港道 channel creek
小港口 standing harbo(u)r
小港湾 small creek
小港运费附加费 local port surcharge; outport surcharge
小港站 pick-up and drop off point
小高地 microknoll
小高尔夫球场 putting green
小高台 < 瞭望用的 > crow's nest
小疙瘩 spicule; tit; wart
小搁板 berm
小格 minor division; subcell
小格栅 sleeper joist
小格栅固定夹 sleeper clip
小格栅墙 sleeper wall
小格通风扇 night vent
小格型围堰 small cellular cofferdam
小格型消声器 cell-type sound absorber
小格子 subbox
小隔离圈 small isolation circle
小更动 minor change
小埂 hawse
小梗 sterigma
小工 hod carrier; hodman; tupper; unskilled labo(u)r; builder's labo(u)rer < 建筑施工的 >
小工程项目 minor structure
小工具 go-devil
小工商业者 small craftsmen and traders
小工业者 small industrialist
小工艺品 fancy goods
小工作室 minor-operating theatre [theater]
小工作台 snap bench
小公牛 bullock
小公司 < 实力不足的 > gyppo
小公寓房屋 flat building
小功率 low-power; miniwatt
小功率波导终端负载 low-power waveguide termination
小功率测量 low-power measurement
小功率齿轮 low-power gear
小功率船 low-powered vessel
小功率的 light duty; low-duty; low-powered
小功率灯 low-power lamp; small light
小功率电动机 fractional electric(al) motor; low-power motor; small power motor
小功率电机 low-power machine
小功率电台 low-power station; miniwatt station
小功率发电机 < 小于 1 马力的,1 马力 = 735.50 瓦 > fractional horsepower motor
小功率发动机 low-powered engine
小功率反应堆 low-power output reactor
小功率放大器 low-power amplifier; miniwatt amplifier
小功率沸腾式反应堆 low-power water boiler reactor
小功率负载 low-power load
小功率工作状态 light duty; low-duty
小功率机车 small unit
小功率机床 light-duty machine
小功率计算机 small power computer
小功率晶体管 low-power transistor
小功率空气压缩机 fractional horsepower air compressor
小功率全向无线电信标 low-powered non-directional radio beacon
小功率绕组 low-power winding
小功率扇形指点标 low-powered fan marker
小功率调制系统 low-power modulation system
小功率图像发射机 low-power video transmitter
小功率拖拉机 light tractor; small tractor
小功率信号 low-power level signal
小功率信号参量放大器 low-level parametric amplifier
小功率询问器 low-power interrogator
小功率压缩机 baby compressor
小功率研究性反应堆 low-power research reactor
小功率因数电路 low-power factor circuit
小功率源 low-level source
小功率振铃器 low-powered ringer
小功率转播台 low-power repeater station
小宫殿 chatelet
小拱廊 arcature; small arcade
小沟 grindle; grip; gullet; gutter; minor groove; rill(e); shallow ditch
小沟铸型 gutter cast
小钩 crotchet; hamula
小钩子 hooklet
小孤山 butte
小菇属 < 拉 > Mycena
小菇状的 mycenoid
小谷 dale; dol
小谷坊 burrock
小股东 minority stockholders
小故障 teething trouble
小挂尺 small hanging tape
小关 < 双杆装卸舱口吊杆 > hatch boom
小关吊货索 up-and-down fall
小观察孔 oillet(te)
小冠巴西棕蜡 ouricury wax
小冠椰子油 ouricury oil
小管 incher; tubule
小管采暖系统 small bore heating system; small bore system
小管径集中式采暖系统 microbore heating system
小管径隧道 pipe small-diameter tunnel
小惯量电动机 minertial motor
小惯性传感器 fast-response transducer
小惯性热电偶 fast thermocouple; high-velocity thermo couple; quick-response thermocouple
小惯性压力表 lightly damped ga(u)ge
小灌木 arbuscle; arbuscula; small shrub
小灌木材 copse
小灌木丛 undershrub
小灌木林 coppice(wood); copse (wood); spinn(e)y
小罐 canister; cannikin
小罐包装漆 shelf goods
小光灯 dim light; dimmed illumination; dimmer; fender lamp; low beam
小光灯泡 dim light bulb
小光栅 subraster
小光栅偏转系统 subraster deflection yoke
小广场 small square
小广告 < 美俗语 > adlet
小规格 small format
小规格材料 small dimension stock
小规格平板 slab of small format
小规格铜管 light-ga(u)ge copper tube
小规模 microscale; pilot scale; smallness
小规模的 miniature; pygmy; small scale
小规模地 on a small scale
小规模地震 miniature earthquake
小规模工厂 semi-work
小规模工程计划 small project
小规模灌溉 microscale irrigation
小规模集成电路 small-scale integrated circuit; small-scale integration
小规模计划 small project
小规模建筑场地 small-scale building site
小规模批量生产 small scale serial production
小规模企业管理 small business management
小规模区域地质调查 small-scale regional geologic(al) surveying
小规模燃烧 small-scale combustion
小规模生产 production on small scale; small lot manufacture; small lot production; small-scale production
小规模施工场地 small-scale building site
小规模试验 bench-scale test(ing); pilot(-scale) test; pilot-scale trial; small-scale test
小规模试验工厂 pilot plant; small-scale pilot plant
小规模栽植 small-scale commercial planting
小规模制造 small lot manufacture
小规模租屋经营者 small-scale tenant operator; Mrs. Murphy
小轨道 baby track
小锅炉 arcola; small boiler
小果蔷薇 small fruit rose
小孩卧室壁橱 bedroom closet for children
小海角 cusp
小海口 zawn
小海损 accustomed average
小海图 chartlet
小海湾 armlet; cove; coving; creek; fleet; sea inlet; small bay; voe < 设得兰群岛和奥尼克群岛的 >; pocket beach
小海峡 breach; kill
小旱 partial drought
小航线进场 tight circling approach
小号的 light-sized
小号钉 nail of small ga(u)ge; small ga(u)ge nail
小号发动机 jack engine
小河 arroyo; beck; bourn(e); brook(let); burn(a); burnene; fleet; font; lesser river; rill(e); riverlet; rivulet; runlet; runnel; streamlet
小河冲蚀 rill wash
小河冲刷 brook-clearing
小河床 minor river bed
小河道网 anabranch
小河段 subreach
小河港引航员 huffler
小河沟 rill(e); rillet
小河谷 coulee; nant
小河河堤 creek wall
小河口 cove; hope; runnel; streamlet; creek
小河拦水坝 pen
小河流 small creek; coulee
小河桥 brook bridge
小河调节池 brook retention basin
小河弯 minor meander; submeander
小核 micronucleus
小核体 nucleolus
小荷载轴承 spigot bearing
小盒 capsule
小盒子 small box; pyx; etui < 放针、牙签的 >
小黑座壳属 < 拉 > Phaeocryptopus
小桁架 pony set; pony truss; small truss
小桁条 subpurlin(e)
小横桁 monkey spar
小横梁 trave
小弧 small arc
小弧齿轮 small helical bevel gear
小胡同 close alley
小湖 laguna; lakelet; lochan; mar; mere
小槲树林 chaparral
小虎钳 lock smith's clamp; vice clamp
小花 floret; floweret
小花盘 small face plate
小花蔷薇 small-flower rose
小花饰【建】floweret
小花束 nosegay
小花香槐 Chinese yellowwood
小花形装饰 fleuret
小滑车 jigger; small sheave
小滑车护挡板 small sheave nest guard
小滑车支承架 small sheave bearing

小滑车组 small sheave block
小滑车组限程器 small sheave nest guard
小滑车组支架 small sheave carrier
小滑车组支座 small sheave carrier
小滑轮 monkey wheel;small pulley
小滑轮吊车 whip
小滑轮组 small pulley block; small sheave block;small sheave nest
小滑轮组护挡板 small pulley nest guard
小滑轮组限程器 small pulley nest guard
小滑轮组支承架 small pulley carrier
小滑轴 small sheave
小划艇 scull
小划子 boat
小环 annulet;circlet;ringlet;zonule
小环化合物【化】small ring compound
小环礁 faro reef
小环境 microenvironment; sub-environment
小环裂 slight shake
小环路 minor loop
小环形珊瑚礁 atoll(on)
小环藻属 cyclotella
小换工 minor change
小回归潮差 small tropic range
小回路 inner loop;minor loop
小汇流条 small bus-bar
小汇水面积洪水公式 flood formula of small collective area
小活塞 petcock;pocket piston
小火车 douser
小火车头 pony engine;pug
小火道 branch canal
小火花 sparklet
小火口 craterkin; craterlet; hydrothermal crater let
小火轮 steam cutter;steam launch
小火山口 craterkin;craterlet;maar
小火山(丘)monticule
小火山锥 conelet
小货车 buggying
小货船 small cargo boat
小货物发单 bill of parcel
小机车 dolly;gadgetry;mule;dink(e)y < 运货、林场及调车用 >
小机动车 doodlebug
小机工 tinker
小机件 dingus;gadgetry
小机具 widget
小机器匠 tinker
小机械 widget
小基层单位 small establishment
小集材道 rack
小集合 small set
小集落 microcolony
小集水沟 micro-catchment
小集水区 small watershed;subwatershed
小集团 clique;coterie
小集中 bunching
小脊 < 俚语 > thank-you-madam
小脊齿脉菌属 < 拉 > Lopharia
小计 abstract;minor total;small count;subtotal;sum
小记号 tick
小纪 < 15 年 > indiction
小剂量照射 low dose irradiation;low dose of exposure
小季节风 little monsoon
小祭坛 altalet
小夹 clip
小夹板 small splint;splintlet
小夹板固定法 small splint immobilization
小家具 knickknack
小家庭 nuclear family
小岬 cusp;point
小岬角 point
小尖锤 cavil axe
小尖刀 stylet
小尖顶 pinnacle;spirelet
小尖山 butte
小尖塔 pinnacle;spirelet
小间 booth;closet;cubicle
小间断【地】diastem
小间隔取样 close sampling
小间隔样品 close sample
小间隙 close clearance; interstice; small gap
小检查 minor inspection
小检修 line break;line check
小件 package
小件包装商品 packed parcel
小件储存系统 miniload system
小件古玩 bibelot
小件货(物) shipping parcel; smalls; parcel cargo
小件货载舷门 cargo port
小件寄存处 little piece checkroom
小件升降机 dumb waiter
小件物品寄存处 check room
小件行李 small luggage
小件行李架 package shelf; parcel shelf
小件行旅 packet
小件邮递 small packets post
小件轧制部件 small rolled parts
小件作物干燥设备 small crop drying equipments
小建筑物 < 用作庙宇的 > aedicule
小舰队 flotilla;squadron
小僵块 cakes of cement
小交换机 private branch exchange; private branch exchange switchboard
小交换机复接线弧 private branch exchange arc
小交易 dicker
小礁岛 cay[kay]
小角边界 small angle boundary
小角度 X 射线散射 small angle X-ray scattering
小角度 X 射线衍射 small angle X-ray diffraction
小角度边界 small angle boundary
小角度测定仪 < 光学的 > micrometer ga(u)ge
小角度俯冲 slight dive
小角度晶间界 small angle boundary
小角度倾斜 low oblique
小角度驶靠 going alongside with small angle
小角度衍射 small angle diffraction
小角度照明配件 narrow angle lighting fittings
小角度转舵 small helm
小角度转弯 sharp angle turning
小角法 method of small angle measurement;minor angle method
小角光散射 low-angle light scattering
小角晶粒间界 low-angle grain boundary;small angle grain boundary
小角散射(法) low-angle scattering; small angle scattering
小角弹性散射 small angle elastic scattering
小角砧 beakiron
小角锥体 pyramidion
小绞车 jigger;puffer;windlass
小绞辘 handy billy;jig
小绞棉纱 ley;lea
小绞丝 light skein
小脚轮 castorite;swivel-type castor; trundle;caster[castor] < 家具下面装的 >
小搅拌板 < 苏格兰 > ligger
小轿车 light car;motorcar;sedan car
小教堂 angle chapel;chapel;domestic chapel;parekklesion;transept chapel < 从教堂十字交叉处进入的 >; prothesis < 古希腊教堂 >;chantry < 捐献的 >
小阶地 terracettes
小阶梯光栅 echelette grating
小接砖 queen closer
小街 by-pass street
小节 small knot < 直径为 0.5 ~ 0.75 英寸的,1 英寸 = 0.0254 米 >;knobble;subsection
小节点 nodule
小节端 point of hock
小节距齿铣刀 fine-pitch cutter
小节距管壁 tangent tube wall
小节律 microrhythm
小节筛 sliver screen
小结 brief sum-up;interim summary; nodule
小结节 nodule
小截面 small bore
小截面材料 light material
小金属锅 < 有横柄的 > piggin;pipkin
小金属圆片 planchet
小金字塔 pyramidion;small pyramid
小经济模型 microeconomic model
小晶粒 fine grain
小晶面 facet(te)
小晶片 platelet
小晶体 minicrystal;small crystal
小井距 close well spacing
小井掘进 pit sinking
小井眼完成 tubingless completion
小颈大瓶 demijohn
小景观 microlandschaft
小景气 boomlet
小径 alley(way);byway;foot path; footway; pathway; trail; unmade path
小径材 small dimension wood
小径基准尺寸 basic minor diameter
小径汽缸发动机 small bore engine
小径围绕 path surround
小镜煤 microvitrain
小酒吧 cantina;dram shop
小酒吧间 crib
小酒杯 shot glass
小酒店 ale-house;dram shop;inn;pot house;pub;tavern
小矩型计算机 mini-supercomputer
小剧场 little theatre
小卷蛾科 < 拉 > Olethreutidae
小绝缘子 pony insulator
小咖啡馆 estaminet;small café
小卡 gram-calorie;lesser-calorie;mean calorie;small calorie;therm
小卡车 barney;larry;motor wagon
小卡口灯座 small bayonet cap
小开本书 dwarf book
小开空隙率 little open void ratio
小开型孔隙 small open void
小康社会 affluent society
小康社会发展规划 comparatively well-off social development planning
小康社会指标体系 indicator system for moderately well-off society
小康(生活) relatively comfortable life;relatively comfortable standard of living
小康型消费结构 affluent consumption structure
小糠草 bentgrass;red top
小颗粒 granule
小颗粒的 short-grained;small-grained
小颗粒聚集体 microaggregate
小颗粒硬质合金 < 粒径为 3 ~ 4 毫米 > clustered carbide
小刻线刀 scratch
小客车 minibus; passenger car; passenger vehicle
小客车出行 car trip
小客车换算当量值 passenger-cars equivalent unit;passenger car equivalence
小客车(交通量)单位 < 用于计算道路通行能力 > passenger car unit [pcu]
小客轿车 limousine;sedan
小客栈 herberge;posada;rest house
小坑 craterlet;dig < 光学玻璃缺陷 >
小空隙的 areolat(ed)
小孔 aperture; bug hole; eye hole; eyelet; foramen; micropore; minipore; orifice; ostiolum; oylet; pin-hole;pin-holing;pink;posthole; small hole; snoot; stoma; ventage; hungry spot < 新浇混凝土表面的 >
小孔单向阀 orifice check valve
小孔的 narrow meshed
小孔阀 orifice valve
小孔法 orifice method
小孔规 small hole ga(u)ge
小孔畸变 aperture distortion
小孔集中加热 mini-bore central heating
小孔径 small bore
小孔径活塞 small bore piston
小孔径宽视场镜头 Topogon lens
小孔径透镜 slow lens
小孔径系统 small bore system
小孔径凿岩 slim-hole drilling
小孔径钻进 slim-hole drilling; small diameter drilling
小孔径钻凿 slim-hole drilling
小孔菌属 < 拉 > Microporus
小孔螺旋状排列 spiral pattern of aperture
小孔耦合 hole-coupling
小孔膨胀消声器 expansion chamber micropore muffler
小孔熔焊 eyelet welding
小孔熔接 eyelet welding
小孔塞子 orifice tap
小孔砂眼 dit
小孔释放法 < 测应力 > Mathar method
小孔调节阀 orifice control valve
小孔网 fine structure mesh
小孔隙 fine pore
小孔隙带大孔隙 small pore and big pore
小孔消声器 micropole diffusor; micropore muffler
小孔效应 keyhole effect
小孔型等离子弧焊 keyhole-mode welding
小孔折流板 orifice baffle
小孔止回阀 orifice check valve
小孔钻 star drill
小口 microstome
小口大玻璃瓶 carboy
小口尖底瓶 < 瓷器名 > vase with small mouth and pointed base
小口径 minor caliber;small bore
小口径测斜仪 small diameter inclinometer
小口径缸 small bore cylinder
小口径管道 small bore pipe
小口径管线 small pipeline
小口径机动泵 small bore mobile type pump
小口径胶管水带 first-aid hose
小口径配管方式 small bore piping system
小口径三通 narrow tee
小口径水带线 hand hose line
小口径钻管 small bore pipe
小口径钻机 slim-hole drilling machine;slim-hole rig;small hole drill;

small hole rig
小口径钻进 slim-hole drilling; small diameter hole drilling; small hole boring; small hole drilling
小口径钻井 slim well; small well
小口径钻孔 slim hole; small diameter borehole; small hole
小口径钻孔法 small hole method
小口瓶【化】narrow-mouth(ed) container
小口瓶压吹成型法 narrow neck press-blow process
小口试剂瓶 narrow mouth reagent bottle
小口双耳壶 < 瓷器名 > ewer with small mouth and double handles
小库锁 < 耐火保险箱内的 > sub-treasury lock
小跨度 short span
小跨度结构 short-span structure
小跨度屋盖 short-span roof; cottage roof < 无桁架的 >
小块 blob; blockette; button; gobbet; lump; nub(ble); pat; tablet
小块扁平石 pennystone
小块冰 calf
小块草地 a pocket-handkerchief lawn; grass plot
小块冲积地 patches of alluvium
小块大理石 tessella
小块地 patch
小块地区 < 死胡同 > pocket
小块地毯 scatter rug; rug
小块方形玻璃 quarry glass
小块废钢 small-sized scrap
小块浮冰 brash ice; patch; glacon
小块灌丛 mogote
小块焦炭 egg coke
小块金属 biscuit metal
小块锦砖 small mosaic; small-sized mosaic
小块茎 tubercle
小块可见的天空 patch of visible sky
小块料 reduct; smalls; dead smalls < 煤、矿石的 >
小块林 woodland
小块林地 woodlot
小块绿地 vest-pocket park
小块马赛克 small mosaic; small-sized mosaic
小块煤 chestnut coal; dead stall; nut coal; small-sized coal
小块木材 stud
小块木料 scantling
小块拼合模板 paneled form; panel form
小块平面法 facet method
小块铺路方块石 small-sized paving sett
小块切割机 dicing cutter
小块石 dornick; finger stone; pinner
小块石灰 small lime
小块石料 scantling
小块石砌体 ragwork
小块碎砖 brick hardcore
小块体 biscuit
小块图像 image subsection
小块土地 a pocket-handkerchief of land; parcel (land); plot; outlot; outparcel < 抵押没包括进去的 >; subdivision of land < 供出售的 >
小块土地合并后的增值 plottage increment
小块土地所有制 proprietorship of land parcels
小块无烟煤 culm
小块镶嵌物 tessera[复 tesserae]
小块雨区 rain patches
小块琢石 small ashlar
小快艇 cog
小矿车 barney; jutty
小矿体 small orebody; squat
小窥视孔 eyelet
小捆 packet; wisp
小阔斧 broad hatchet
小拉手 small pull handle
小蜡孔 < 介壳虫中的 > micropore
小蜡烛 taper
小缆 cablet
小浪 < 海浪二级 > slight sea; smooth sea
小浪海况 low wave condition
小浪花 leaper; lipper
小老虎窗 small dormer
小雷暴 slight thunderstorm
小雷艇 yawl
小肋材 jack rib
小类 minor sort; subclass; subdivision; subregime
小冷库 walk-in box; walk-in refrigerator
小礼拜堂 Lady Chapel; oratory
小礼堂 < 殡仪馆的 > antechapel
小力增压器 low-duty supercharger
小立柱 < 篱笆用的 > stoa
小沥青撒布机 dandy
小砾(石) pea gravel; small gravel
小笠原(副热带)高压 ogas awara high
小粒 pellet
小粒黄色琥珀 muckite
小粒黄色树脂 muckite
小粒圆石 granule roundstone
小联管节螺母 small union nut
小联管节柱螺栓 small union stud
小链轮起重机 coffin hoist
小梁 bridging joist; joist; junior beam; spar; stringer; trabecula[复 trabeculae]
小梁侧封板 apron facing; apron lining
小梁和大梁的连接 beam-to-girder connection
小梁间加固的剪力撑 solid bridging
小梁脚手架 needle beam scaffold
小梁结构 small structure
小梁空心管楼盖 armo(u)red tubular floor
小梁空心砖楼盖 precast beam and hollow-tile floor
小梁楼板 beam floor; joist slab
小梁托撑基础 needle beam underpinning
小量 small quantity
小量不一致 minor inconsistence
小量成批生产 small lot production
小量程仪表 minimum meter
小量当地零发料 small quantity for local delivery
小量的折扣 shading
小量镦粗 shallow draught
小量混凝土试验性拌和 trial batch
小量贸易 small trade
小量生产 low production; production on small scale; short-run production; small-scale production; small volume production
小量试拌 trail batch
小量试样 meso sample
小量危险品限额 allowable limited amount of dangerous cargo
小量位移 inching; small displacement
小疗养所 small rest home
小疗养院 small rest home
小料 minor ingredient; scantling
小料浆池 sump
小裂缝 small check
小林 grove
小鳞茎 clove
小灵猫 lesser civet cat
小菱面 minor rhombohedral face
小菱形头钢线钉 small rose head steel wire nail
小零件 midget
小流量 low discharge
小流量采样器 small flow sampler
小流量温泉 small volume spring
小流域 minor watershed; small drainage basin; small watershed
小流域洪水公式 flood formula of small watershed
小流域治理 small valley treatment; taming small rivers
小瘤 nodule; nub
小六角较扁螺母 small thinner hexagonal nut
小龙骨 cross furring; furring channel; quartering
小龙头 draw cock; petcock; sill cock; small cock; small faucet; stop cock
小漏 dribbling; pin-hole leak; trivial leak
小炉脖底砖 block beneath port neck
小炉脖拱 port cap rake; port crown rake
小炉脖碹 port cap rake; port crown rake
小炉侧墙 cheek; port jamb; portside wall
小炉侧墙砖 edger block
小炉口 drum
小炉口框架 framing of port mouth
小炉上升道 upcast
小炉下口燃烧器 underport burner
小炉砖 burner block
小炉子 chauffer
小陆块 microcontinent
小鹿 fawn
小鹿毛色 fawn brown
小路 foot path; footway; alley(way); by-path; by-road; byway; driveway; lesser road; minor road; narrow pass; notch; path(way); piste; sideway; trail (road); unmade path; berm(e); bridle path; angiportus < 古罗马 >
小旅店 herberge; inn; minch house; posada
小旅馆 club-fraternity-lodge; hostel; inn; lodge; pub; public house; tap house
小卵石 gravel pebble; kidney; ordinary pebble; ovals; pea gravel; pebble
小卵石花纹 armure
小卵石涂抹 pebble dash
小轮 annulet; truckle; trundle
小轮车架 roll back
小罗 small gross
小罗经 baby compass; box and needle
小罗盘 baby compass
小螺距 fine pitch
小螺口茄形指示灯泡 small egg-plant screw indicating lamp
小螺栓 stove bolt
小螺栓扳手 spud wrench
小螺丝刀 small screw driver
小螺丝钻 gimlet
小螺旋 minor spiral
小螺旋伞齿轮 helical bevel pinion
小螺旋扇回旋加速器 low-spiral cyclotron
小落潮流 lesser ebb
小落水洞 daya
小麻把 strick
小麻绳 knittle; nettle
小麻籽油 hempseed oil
小马 pony
小马达船 runabout
小马力船 low-powered vessel
小马力电动机 fractional horsepower motor
小马力水轮机 tiny horsepower hydraulic turbine
小马力拖拉机 low-powered tractor
小麦淀粉 wheat starch
小麦糊浆 wheat paste
小麦筛渣 wheat screenings
小麦穗 ears of wheat
小麦田 wheat paddock
小卖部 buffet; canteen; petty store; retail department; retailing counter; service pantry
小卖者 vendor
小脉 small pulse; small vein
小镘刀 spoon and square
小盲谷【地】hope
小毛病 teething trouble
小毛刷 fitch
小茅屋 < 墨西哥及美国西南印第安人的杆柱、泥草茅屋 > jacal
小锚 grapnel; grappling iron; kedge (anchor); kelleg; kill(i)ck; killock [kellock]; stream anchor
小锚绳 kedge rope
小帽桉 tallow wood eucalyptus
小煤车 < 煤井内用 > coal tub
小煤块 nuts
小煤炱属 < 拉 > Meliola
小煤窑名称 small coal mine name
小煤柱 stump
小门 ostiole; ostiolum; wicket; wicket gate; dwarf door < 活动门的下半截门 >; wicket < 正门旁的 >
小门道 small gateway
小门闩 cottage latch
小米 millet
小密室 closet
小面包车 van pool
小面积洪水 small area flood
小面积流冰群 ice patch
小面积闪烁 small area flicker
小面积试涂 brushout
小苗 plantlet
小庙 basilica julia
小皿 capsule
小明沟 small ditch
小模数齿轮 fine-pitch gear; micron gear
小模数滚刀 micron hob
小抹刀 small spatula
小木板 keeler; strip
小木槌 gavel; wood mallet
小木方 scantling
小木节 cat eye; pin knot
小木锯 trim saw
小木块 little joiner
小木块拼接 parquet
小木料 scantling
小木刨 block plane
小木条 batten
小木桶 kid
小木屋 bangalow; log cabin
小木箱 box
小木楔 angle block; little joiner; nose key; page
小木桩 peg; piquet; setting-out stake; spile; stake; wooden peg
小木作 joinery work
小目标探测器 small object detector
小牧场 paddock; small livestock farming
小牧场主 ranchette
小内测千分卡尺 small inside micrometer cal(l)ipers
小内齿圈 small internal gear
小内管径 < 半英寸管，用于水泵助力的集中供热系统 > small bore
小内径管 small bore tubing
小难题 glitch
小囊【生】saccule
小挠度 small deflection

小挠度理论 small deflection theory
小挠度弯曲 bending with small deflection
小挠艇 cockle
小脑山坡 clivis
小能力驼峰 small capacity of hump
小泥铲 trowel
小泥门 upper hopper door
小泥塘 boglet
小年 off year
小牛 calf
小牛肝菌属 <拉> Boletinus
小牛栏及产房 calf and maternity quarters
小牛皮 calf;calfskin
小牛腿 cantilever block; cantilever vault
小农场 croft;small farm;steading
小农场主 ranchero
小农具 small farm tools
小弄 alley way
小耙 rastellus
小排 float
小排量汽车 light car
小排列 minispread
小排气孔 small vent
小排水沟 dingle;small trough
小排水管 weeper
小盘 pannikin
小盘状的 kneepan-shaped
小炮孔 popshot
小炮眼 block hole;plug shot
小炮眼爆破 pop-blasting;pop-shoot(ing)
小泡 alveolus[复 alveoli];bullule;vesicle
小泡的 vesicular
小泡状海绵体 small bubble spongy body
小配电亭 kiosk
小配电站 distribution substation
小配件 gadget
小盆子齿轮 bevel pinion
小棚 booth
小棚车 cariole
小棚架 pony set
小棚屋 humpy;small hut
小篷车 carriole
小批 job lot;sublot
小批成组工艺 small batch technology
小批合同 job lot contract
小批化学液体专用船 chemical parcel tanker
小批货物 small lot
小批货物托运人 smaller stripper
小批量 short run;small lot
小批量发货人(托运人) low-volume shipper
小批量货 small lot cargo
小批量生产 packaged production; short production run; short-run production; small lot production; small batch production
小批量生产的车辆 low-volume vehicle
小批散货 bulk parcel
小批生产 ascertainable production; low production; preproduction; small lot manufacture; small-scale manufacturing; small scale serial production; small serial production;small-size production
小批生产机械 preproduction machine
小批生产用模具 short-run die
小批试制 limited-run trial production; trial manufacture on a small batch basis; trial-produce on a small batch basis
小批转接 packet-switch
小批装油 bulk reduction
小皮板下区 infracuticular region
小皮带 belt lace
小片 crumb; dice; flake; flakelet; fleck; pellet; scrip; slot; snatch; snip;tablet
小片安置机 chip positioner
小片彩色玻璃 smalto
小片彩色珐琅 smalto
小片处理机 chip handler
小片地 odd area
小片冻融作用地形【地】solifluction lobe
小片堆垛机 small-size sheet stacker
小片焊合 chip bonding
小片接合 chip bonding
小片接头 chip bond
小片连接 die attachment
小片林 woodlot
小片水面 patch
小片物 patch
小片雨区 rain patches
小片状体 platelet
小偏心 small eccentricity
小偏心受压 small eccentric compression
小偏转角显像管 narrow angle picture tube
小漂砾 boulderet
小频数 small frequencies
小平板【测】traverse table;tablet
小平板法 microplating method
小平板仪 traverse plane table
小平板仪与经纬仪联合测图 joint mapping with plane-table and transit
小平底拖网船 dory trawler
小平底(鱼)船 dory
小平顶丘<高原面残余> tent-hill
小平房 bunkhouse
小平均日潮差 small mean diurnal range
小平面<多面体的> facet(te)
小平面分类 facet-classification
小平面立体测图仪 stereo facet plotter
小平台 berm(e)
小平凿 snap
小瓶 flasket;vial
小坡度 gentle bank;light grade;light gradient;low gradient
小坡度坡道 ramp with a slight pitch
小坡度倾斜玻璃窗 slightly pitched glazing
小坡度倾斜屋顶 slightly pitched roof
小坡度人字屋顶 slightly pitched gable roof
小剖面 small section
小瀑布 cascade;small falls;water cascade
小瀑布潭 plunge pool
小漆桶 pail
小祈祷处<教堂内的> aedicule
小旗 awheft; fanion【测】; bannerol <葬礼用>
小企业 peanut
小企业产品 small enterprise products
小企业贷款 small business loan
小企业金融机构 financial institutions for small businesses
小企业融资 small business finance
小起伏 microrelief;microtopography
小起伏地形 underfeature
小起伏微地形 gilgai
小起重机 donkey crane;grass hopper <俚语>
小气窗 night vent
小气道阻力测定 small airway resistance determination
小气候 local climate;microclimate
小气候测定 microclimate measurement
小气候测量 microclimatic survey
小气候的模拟 simulation of microclimates
小气候的人工调控 artificial modification of microclimates
小气候评定 micrometeorologic(al) evaluation
小气候热岛 microclimatic heat island
小气候效应 microclimatic effect
小气候学 microclimatology
小气候异常的遥感 remote-sensing of microclimatic stress
小气候因素 microclimatic factor
小气候资料 microclimate data
小气孔 pin-holing;spike hole;spile-hole
小气泡 air marking;bubblet;fine air bubble; fine gas bubble; minute bubble;bleb;vesicle <矿物或岩石中的>;fine seed <小于 0.2 毫米>
小气泡多的玻璃 very seedy glass
小气泡麻孔 air marking;air marks
小气泡曝气 fine air bubble aeration
小气泡曝气系统 fine air bubble aeration system
小气球 ballonet
小气田 small gas field
小气涡 microcyclone
小气象学 micrometeorology
小汽车 cyclecar; douser; motorcar; passenger car; road louse[复 lice] <美国汽车俗称>
小汽车吊具 automobile sling; autosling;car sling
小汽车间 small garage
小汽车旅行住房 pick-up camper
小汽车停车区 car park(ing) area
小汽车停放场地<送客换乘火车的> kiss-and-ride parking provision
小汽车限制区【交】auto-restricted zone
小汽船 boat;pinnace
小汽机 donkey engine
小汽艇 cutter; launch; motor boat; motor launch;pinnace
小汽艇租费 launch hire
小器具 gadget
小器具设计 gadgeteering
小器皿箱 canteen
小器窑 kiln for small pieces; small ware kiln
小千斤 topping lift bull rope
小铅柱试验 small lead block test
小前提 subsumption
小前题 minor proposition
小钳 small forceps
小钳子 nippers;tweezers
小戗角屋顶 hipped gable roof
小枪 lancet
小枪眼 oillet(te)
小腔的 locular
小腔室 capacitor
小腔形成 loculation
小墙 wallette
小桥 foot bridge; minor bridge; pinock;small bridge
小巧精致的 unobstructive
小窃 petty pilferage
小寝室 wanigan
小青瓦 blue roofing tile
小丘 hillock;hummock;hump;hurst; microknoll; monticule; mound of earth;rising ground
小球 buke; bulb; pellet; pill; small ball;small sphere;globule <汞、油等>
小球测定法 pill test
小球过滤 glomerular filtration
小球磨 small ball-mill
小球腔菌属<拉> Laptosphaeria
小球清洗器 ball cleaner
小球饰 pellet mo(u)lding
小球体 spherule
小球形成 pellet formation
小球形恒温器 pellet-type thermostat
小球藻素 chlorellin
小球藻(属) chlorella
小球状的 globular
小球状镶嵌物 button-like insert
小区 blockette; dwelling district; housing area;mini-area;minor subdivision;plot;subzone
小区测验<径流的> plot measurement
小区观测<尤指径流的> plot observation
小区灌溉 basin irrigation
小区规划 housing estate planning; plot planning;sector planning
小区划分 microzonation
小区划化 microregionalization
小区机误变量 plot error variance
小区间 minizone
小区交通调查 cordon count
小区径流 plot runoff
小区面积 plot area
小区内部街道 interior street
小区排列 plot arrangement
小区热力站 area thermal substation
小区设计 plot plan;plot sign
小区生活污水处理 community domestic wastewater treatment
小区试验 plot experiment;plot test
小区形心 district centroid
小区域 subblock;zonule
小区域划分 microzoning;microzonation
小区域气候 small-scale climatology
小区域区划 microregional plotting
小区域市场 small region market
小区中心 branch center[centre];center of housing estate;neighbo(u)rhood center[centre];zone centroid
小曲流 submeander
小圈 circlet
小圈圆规 drop compasses
小圈子 coterie
小权分散 decentralize power on minor issues
小泉 springlet
小缺点 mote
小缺陷 blemish;minor defect
小群 grouplet
小群丛 microassociation
小群落 microcoenosium; microcommunity;minor community;society
小群落植物 assembly plant
小群体 microcommunity
小扰动法 linear perturbation theory
小扰动法计算 perturbation calculus
小扰动水面 perturbed water surface
小人造卫星 moonlet
小日潮差 small diurnal range
小日程计划 detail operating schedule;detail schedule
小日用品 small retail items
小容积发动机 small displacement engine
小容积罐 volume tank
小容量 low-capacity
小容量存储器 small capacity memory
小容量的 low-duty
小容量低压灭草喷雾器 low-volume low-pressure weed sprayer
小容量电缆 low-capacity cable;small capacity cable
小容量电站 low-capacity plant
小容量混凝土搅拌机 small capacity concrete mixer

小容量搅拌机 small capacity mixer
小容量喷雾器 low volume sprayer
小容量水泵 low-capacity pump
小容量水轮机 tiny horsepower hydraulic(al)turbine
小容量系列气枪 small capacitance series air-gun
小容量液压系统 low-capacity hydraulic system
小容器 small container
小熔深焊条 low penetration electrode
小入口 ostiole
小塞门 <放泄用> petcock
小塞子 spile;spill
小三度 minor third
小三和弦 minor chord
小三角测量 microtriangulation;minor triangulation; subsidiary survey; subsidiary triangulation
小三角锉 balance arm file
小三角点 minor triangulation point
小三角旗 pencel;pennant
小三角旗的旗杆 pennant rod
小三角形瓦块 shouldering half-torching
小三角洲 <冲刷形成的> washover; wave delta
小伞齿轮 cone pinion; transmission pinion
小伞齿轮传动 bevel pinion gearing
小伞形齿轮 bevel pinion
小伞形花序 umbellule
小扫描法 mini-scanning method
小沙波 sediment ripple
小沙岛 sand cay;sandkey
小沙洲 sand cay;sandkey
小纱窗 <纱窗上的> wicket screen
小砂坝 sandkey;sand cay
小砂层 sand sheet
小砂丘 sand drift
小砂原 sand sheet
小砂砖 <擦甲板转角用的> prayer book;sand brick
小筛分样 small-size sample
小山 bhur(land); carn; knap; knoll; kopfe; kyr; loma; mound; pinnacle; single hill
小山的顶 knap
小山岗 monticule
小山岭 wold
小山坡 hillside
小山墙 gablet
小山丘 hill;koppie
小山山顶 hill top
小珊瑚礁 faro
小舢板 cockboat;cocket;cockle boat; cog;dingey;dinghy
小商船船长 skipper
小商店 babyshark; corner shop <住宅区附近的>
小商品 small commodities;small wares
小商品价格 prices of miscellaneous goods
小商品经济 petty commodity economy;small commodity economy
小商品批发 wholesale of small articles
小商品生产者 small commodity producer
小商品自动售货机 vending machine
小商人 tradesman
小商小贩 small tradespeople and peddlers
小商业 small business under individual ownership; small enterprises in commerce;small shop
小勺 bail
小舌片 strap
小舍 booth
小设备 midget
小升降绳 lazy halyard
小生产 small-scale production; test run(ning)
小生境 ecologic(al)niche;microhabitat;niche
小生境多样化 niche diversification
小生境多样性 alpha diversity
小生态系统 small ecosystem
小生物 minimus
小生物群落 minor biome;minor community
小生意 dicker
小绳 laniard [lanyard]; marlin(e); small stuff;tarred fittings;twine
小绳缠扎钩口的钩 mousing hook
小圣堂 <古罗马小神庙中的> cellula
小湿室 moisture cell
小十字符 tick mark
小十字形 crosslet
小石 agger arenal;gravel
小石板 lady;smalls
小石板瓦 little slate
小石壁 bullset
小石堆 barrans
小石块 domick;finger stone
小石块路面上沥青表面层 bituminous topping on set paving
小石块铺面的浇注混合剂 pouring compound for stone sett paving
小石块铺砌 pegtop paving
小石块铺砌沟渠 ditch sett paving
小石块砌筑陀螺形路面 pegtop paving
小石屋 <麦加大清真寺中的> Cabba
小石柱 <石灰石等中的> stylolite
小石子 handstone;pebble
小石子底座 handstone subbase
小石子混凝土 fine aggregate concrete
小时变化量 hourly variation
小时变化系数 hour variation coefficient
小时变量 variable per hour
小时表 hour meter
小时产量 hourly capacity;hourly tonnage;hourly output
小时成本 hourly cost
小时出力 hourly efficiency
小时定额 hourly rating;hour target; one-hour rating
小时费率 hourly rate
小时风速 hourly wind speed
小时风速系数 hourly speed factor
小时工资 hourly pay; hourly wage; per hour wage;wage per hour
小时工资率 hourly wages rate; pay rate per hour;time on daywork
小时工资收入 hourly earnings
小时功率 hourly rating;one-hour rating
小时固定工资率计划 hourly rate wage plan
小时呼 call hour
小时基数 hourly basis
小时计 hour meter
小时间隔 hourly interval
小时奖金制 hourly premium system
小时降雨量 hourly rainfall(depth)
小时交通量 hourly traffic capacity; hourly traffic volume
小时均值 hourly average
小时平均 hourly average
小时平均进尺 average footage per hour
小时平均收入 average hourly earnings
小时生产率 hourly efficiency;hourly production;output per hour
小时生产能力 hourly capacity;hourly production capacity
小时数 hourage
小时损失率 hourly loss rate
小时用水量 hourly consumption
小时雨量 one-hour rainfall
小时增殖率 hourly growth rate
小时值 hourly value;one-hour value
小时制功率 hourly output
小时制牵引力 hourly tractive effort
小时最大用水量 hourly maximum water consumption
小时作业产量 output per hour
小食堂 luncheon room;lunch room
小食亭 refreshment kiosk
小矢高拱 depressed arch
小市场 small market
小市镇 small town
小式 wooden frame without dougong
小事故 minor accident; minor disaster;minor failure
小视场辐射计 small field radiometer
小视界 low coverage
小试 bench-scale test(ing); small-scale test
小试件 small specimen
小试投机 dabble speculation
小试验 bench run
小饰板 plaquette
小室 booth; cab(in); cabinet; cell; chamberlet; cope; cubicle; cuddy; a(u)mbry <教堂>
小室隔断 cabinet filler
小室供暖 cabinet heating
小室冷库 locker plant
小手柄 miniature lever
小手工业者 small handicraftsman
小手锯 chest saw
小手磨 quern
小手术室 minor-operating theatre [theater]
小手摇泵 one-armed Johnny
小书斋 <修道院的> carol(le)
小鼠洞径 size of mouse hole
小鼠洞深 depth of mouse hole
小鼠洞斜 inclination of mouse hole
小束 fascicle
小树 arboret;sapling;small tree
小树丛 coppice(wood);copse wood; underbrush
小树节 <直径小于 0.75 英寸,1 英寸 = 0.0254 米> small knot;pin knot
小树林 bosk(et);grove
小树林的 nemoral
小树柳 little tree willow
小树梢料 brush wood
小树枝状相思树 slim wirilda acacia
小树脂囊 <宽 1/8 英寸以下,长 4 英寸以下,1 英寸 = 0.0254 米> small pitch pocket
小数 broken number; decimal fraction; decimal number; decimals; small number
小数表示 fractional representation
小数部分 decimal part; fixed-point part; fractional part; mantissa <浮点数中的>
小数的 decimal;fractional;fractionary
小数的循环节 period of a circulating decimal;recurring period
小数的整数部分 higher decimal
小数点 arithmetic(al)point;base point; decimal point;dot;point;radix point
小数点定位器 decimal point positioner
小数点对位 decimal point alignment
小数点区分符 point specifier
小数点调准 decimal point alignment
小数点位置 scaling position
小数点自动定位计算机 automatic decimal point computer
小数点自动定位运算 automatic decimal point arithmetic(al)
小数定点制 fractional fixed point
小数定律 law of small numbers
小数定位器 decimal point positioner
小数砝码 fractional weights
小数计算机 <小数点在最前面的定点计算机> fractional computer
小数目 peanut
小数位【数】decimal place
小数位等量 decimal equivalent
小数循环节 period of circulating decimal; period of repeating decimal; repetend
小数运算 fractional arithmetic
小衰退 minirecession
小栓 spile;spill
小双尖拱 two-cusped arch
小水坝 stank
小水池 pondlet; overnight pond <灌溉用>
小水滴 droplet
小水电 mini hydro
小水电站 midget water power station; small hydro-power station; small-scale hydroelectric(al) station
小水龙头 bibcock;bib tap;sill cock; sink bib
小水龙头弯嘴 bib nozzle
小水轮机 mini hydro turbine
小水泥厂 local cement plant
小水泥工业 mini cement industry
小水塘 livestock reservoir
小水桶 breaker
小水头 small head
小水团 bolus of water
小水洼 calicheras
小水位差 small head
小水螅 microhydra
小水眼钻头 restricted bit
小水堰 mill-weir
小睡 nap
小丝带圈 <装饰女帽的> coque
小四边形构件 quarrel
小四开图纸 sheet-and-third foolscap; sheet-and-half cap <11.5 英寸×24.5 英寸,1 英寸 = 0.0254 米>;flat cap <14 英寸×17 英寸,1 英寸 = 0.0254 米>
小松 poor pine
小苏打 baking soda; dicarbonate; sodium bicarbonate;sodium hydrogen carbonate
小苏打灭火干粉 dry chemical extinguisher
小塑像 figurine;statuette
小碎粒 comminutions; small broken particles
小碎石 finely crushed rock
小碎石块 small clod
小碎石料 hand-broken metal
小碎石子 hand-broken stone
小碎土块 small clod
小穗状花【植】spicule
小损害 microlesion
小损失互保协会 small damage club
小损益 petty loss and profit
小榫 small tenon
小塔 small tower;baby tower;diminishing tower <建筑装饰用>
小塔楼 <墙上凸出的> bartizan
小台服务员 steward
小台架 <放讲稿的> lectern
小台钻 little beakiron
小台座 piedouche
小太阳轮 small sun gear
小太阴椭圆日分潮 small lunar elliptic diurnal component
小弹簧 little spring

小探巷 monkey drift
小碳粒电阻炉 carbon resistance furnace
小镗杆 quill
小塘 pulk
小套公寓房间 <有厨房及卫生设备的> efficiency apartment(or unit)
小特厚板 <宽 8 英寸,厚 5 英寸以下,1 英寸 = 0.0254 米> scantling
小藤壶 balanid
小藤石 kotoite
小梯田 terracettes;terrwicket
小提单 delivery order
小提花织物 dobby weave
小提升闩 cottage latch
小提箱 suitcase
小体 corpusc(u)le
小体重 volume weight by small-size sample
小天窗 dwarf skylight
小天地 microcosm(os)
小天鹅 whistling swan
小天井 light well;open shaft;jump-up <由平巷向上开掘的>
小天气 microweather
小天文卫星 small astronomical satellite
小田地 croft
小填角焊 light fillet weld
小填角焊缝 light filler
小调变度磁铁 low-flutter magnet
小调式 minor mode
小调整 minor adjustment;minor tune up
小笤帚 whisk
小铁锤 adz(e)-eye hammer;mash hammer;lump hammer
小铁皮钻 stake
小铁梃 heaver
小铁研式抗裂试验 small Tekken type cracking test
小铁砧 small anvil;stake;bench stake
小亭 alcove;kiosk
小停顿 dwell
小艇 boat;cockboat;dingey;small boat;small craft;pinnace <大船附带的>
小艇车间 boat shop
小艇港 marina
小艇(的港湾)基地 marina
小艇护舷材 belting
小艇甲板 awning deck;hurricane deck;hurricane roof;promenade deck
小艇驾驶术 boating
小艇空气箱 boat air tank
小艇锚 boat anchor;killagh
小艇索具 boat's rig
小艇停靠区 marina
小艇推进器 boat propeller gear
小艇尾室 dick(e)y
小艇尾台 stern sheet
小艇系锚绳 rode
小艇舷侧栏杆顶围绳 ridge rope
小艇属具箱 boat box
小艇作业 boat work
小通道 passage aisle
小通气孔 small vent
小同轴电缆 small diameter coaxial cable
小铜钉 <固定屋顶铅皮的> lead nail
小桶 kilderkin;rundlet;small cask;keg;firkin <英国容量单位,1 小桶 = 9 加仑或 40.9 升>
小偷小摸 pilferage
小头 little end;short end;small end;tip end <原木的>;bottom end <桩或圆木的>
小头衬套 small end bushing
小头钉 casing nail;lost-head nail;sprig;brad nail
小头直径 small-end diameter;small-top diameter;tip diameter
小凸壁缘 astragal frieze
小凸嵌线玻璃 reeded glass
小凸嵌线脚 reed mo(u)ld
小凸嵌线饰 reed mo(u)ld;reeds
小凸圆体花饰 bagnette
小凸缘线脚 reeding
小凸装饰线脚 astragal cornice
小突起 monticule
小图像 image subsection
小土坝 farm dam
小土地出租者 lessor of small plots;petty lessor;small land lessor
小土堆 rideau;earth hummock
小土墩 Mima mound
小土滑阶坎 terracettes
小土脊 rideau
小土阶 sheep-tracks;terracettes
小土块 small clod
小土粒 ped
小土丘 rideau
小土山 <威尔士> twyn
小团粒结构土壤 fine soil
小推车 dilly;plain trolley
小推锄 small thrust hoe
小推力面 minor thrust face
小腿 calf
小托梁 binding joist;needle beam
小拖车 small tractor;towing dolly
小拖轮 marine tractor;small tugboat;water tractor
小洼地 microdepression;daia <沙漠中的>
小洼泉 dimple spring
小瓦 galleting tile
小瓦锅 <有横柄的> piggin;pipkin
小外环弹簧 <缓冲器的> preliminary outer ring
小弯椽 jack rib
小弯度曲流 small curved meander
小弯头 eyesight elbow
小弯缘 curved jack rafter
小湾 armlet;bight;coving;creek
小湾或河口 liman
小丸 pellet;pillet
小玩意 gadgetry
小挽缆桩 sheet bitt;small bitt
小碗柜 locker
小网眼 fine mesh
小望远镜 spyglass
小围墙 enclosing wall
小尾孢属 <拉> Cercosporella
小卫星 moonlet
小紊流度 small-scale turbulence
小稳性船 tender ship;tender vessel
小问题 subproblem;teething problem
小窝棚 dog house
小窝器 small pit organ
小卧室 control cubicle;cubicle;cubiculo
小卧室客车 chambrette car
小屋 cabana;cab(in);cottage;house let;hovel;hut(ch);lodge;maisonette;small house;benab <几内亚>;shieling <牧场附近的>
小屋顶 rooflet;small roof
小屋野营 cabin camp
小屋子 <美国印第安人的> wickiup
小无花果树 sycamore
小无极绳牵引机 small endless rope tractor
小五金 <建筑用> hardware;architectural metal;furniture;metal fittings
小五金安装工程 hardware work
小五金材料 ironmongery materials
小五金门锁 cabinet latch
小五金配件 hardware fittings
小五金商 ironmonger
小五金饰面 ironmongery finishes
小五金项目 hardware item
小五金制品厂 hardware product factory
小五金装饰件 trim hardware
小物件打捞篮 <钻孔内的> red-and-yellow basket;reed basket
小物件打捞器 <钻孔内的> junk retriever;pick-up grab
小物品打捞篮 <钻孔内的> red-and-yellow basket;reed basket;open-end basket
小物品打捞器 <钻孔内的> junk retriever;pick-up grab
小溪 arroyo;beck;bourn(e);breach;brooklet;burn(a);burnene;creek;dingle;rill(e);rillet;rivulet;rivus;runlet;runnel;sike;small brook;streamlet;brook <以泉水为源的小溪、小河>
小溪谷 dell
小溪缆桩 dollie
小溪流 arroga;brooklet stream
小溪流域 beck
小溪群落 namatium;rhoium
小溪调节池 brook retention basin
小系统风机控制箱 control box for small system fan
小细锉 drill file
小细矿脉 boke
小隙缝分离系统 small aperture separator system
小匣 casket
小峡谷 hawse
小纤维 <组成木纤维的细小纤维> fibril
小舷门 half port;port sash
小香菇属 <拉> Lentinellus
小箱 camera;casket
小镶板 small panel
小镶板构造 small panel construction
小巷(道) alley(way);by-lane;cundy;pedestrian lane;streetlet;access lane;side street;throughер
小巷住房 alley dwelling
小像幅航空摄影 small format aerial photography
小像幅摄影机 small format camera
小像幅投影仪 miniature projector
小橡皮轮车 dolly
小橡皮条 <浇混凝土前放在模板上的> ke-it
小削角 <方桩四角> small chamfer
小写金额 amount in figures
小写体 lower case
小写字符 lower case alphabetic character
小写字母 lower case letter;minuscule
小写字体 lower case type
小心 care;precaution
小心搬运 handle with care
小心爆炸 caution, risk of explosion
小心标志 care mark;cautionary mark
小心超车 <交通警告标志用语,美国> pass with care
小心超高 caution, limited overhead height;limited overhead height
小心触电 caution, risk of electric(al) shock;electric(al) shock
小心瓷器 porcelain with care
小心电击 caution, risk of electric(al) shock
小心电离辐射 caution, risk of ionizing radiation
小心掉落 do not drop
小心防火 precaution against fire
小心非电离辐射 caution, non-ionizing radiation
小心腐蚀物质 caution, corrosive substance
小心工厂车辆 caution, industrial trucks
小心滑倒 caution, slippery surface
"小心火车"标志 "railroad warning" sign
小心火烛 caution, risk of fire
小心激光 caution, laser beam
小心警犬 guard dog
小心落物标 beware of falling sign
小心前进(准备停车)的显示 proceed under CAUTION
小心强磁场 caution, strong magnetic field
小心轻放 handle with care;no rough handling
小心生物感染 caution, biological hazard
小心使用 care in applications
小心悬空重物 caution, overhead load
小心运行 cautious running;running under caution
小心正面朝上 right side up with care
小心中毒 caution, toxic hazard
小信号 small signal
小信号参数 small signal parameters
小信号分析 small signal analysis
小信号功率增益 small signal power gain
小信号晶体管 small-signal transistor
小信号理论 small signal theory
小信号性能 small signal behavio(u)r;small signal characteristic;small signal performance
小信号增益 small signal gain
小信号柱 signal doll
小星群 asterism
小星形的 stellular;stellulate
小行星 asteroid;minor planet;planetoid
小行星带 asteroid belt
小行星环内的内行星 inner planet
小行星环外的外行星 outer planet
小型 bench scale;compact type;miniature type;minitype
小型爱迪生螺旋灯头 small Edison screw-cap
小型爱迪生式螺旋灯头 miniature Edison screw-up
小型安全开关 baby switch
小型百货商店 junior department store
小型扳手 pocket wrench
小型板块 small-size plate
小型版框 card chase
小型办公室与家庭办公 small office home office
小型半甲板驳船 monkey boat
小型半圆室 <教堂的> absis
小型包层 miniature clad
小型堡寨 <中世纪英国> peel
小型爆破 small explosion
小型本生灯 micro Bunsen burner
小型变压器 miniature transformer;ouncer transformer

小型别墅 cottage;cottage suburb
小型别墅区域(单元) cottage community
小型玻壳管制造机 miniature glass tube machine
小型玻璃器皿 small-size glass ware
小型采暖炉 room heater;space heater
小型采暖燃具 space heating appliance
小型槽钢 channel steel bar;junior channel
小型草耙 handle hay rake
小型(侧翻式)货车 jubilee wagon
小型侧卸式货车 jubilee truck;jubilee wagon

小型测厚仪 pocket-size thickness ga(u)ge
小型测距仪 mini-ranger
小型测试器 pocket tester
小型测微计 small test
小型测振仪 microsound scope
小型插头 miniplug
小型柴油电机车 small-size diesel electric(al) locomotive
小型铲斗 small profile bucket
小型常压灭菌器 small atmospheric sterilizer
小型超级市场 small supermarket;superette
小型车 midget car
小型车床 bench lathe
小型车辆 dilly
小型承载板试验 mini-type loading plate test
小型城堡 chatelet
小型城市 compact city
小型城市自动化有轨电车交通系统 personal rapid transit system
小型冲击式采样吸收管 midget impinger
小型冲孔机支承 bear
小型冲天炉 cupolette
小型抽水泵 drip pump
小型出海船舶 sea craft
小型出纳机 mimiteller
小型出租马车 fiacre
小型出租汽车 minicab
小型厨房 small-sized kitchen
小型处理机 mini-processor
小型处理系统 small treatment system
小型处理装置 package treatment plant
小型船舶 floating craft;mini-ship;mini-vessel
小型窗 fenestella;fenestral
小型磁鼓 minidrum
小型存储器 minimum storage
小型措施费 small measure expenses
小型打桩机 ringing engine
小型带泵灭火器 pump tank extinguisher;pump tank fire extinguisher
小型带锯机 pony band saw;pony rig
小型单相变压器 small single-phase transformer
小型单相感应电动机 small single-phase induction motor
小型导航雷达 small navigational radar
小型的 bantam;bench scale;compact;downsized;midget;miniature;pocket;pony;pony-size;small scale;small-sized;vest-pocket
小型灯 pygmy lamp
小型灯标 minor light
小型灯标船 light-float
小型灯泡 midget bulb;miniature lamp
小型灯座 candelabrum[复 candelabra] base;miniature base;miniature lampholder;miniature lamp-socket
小型登陆车 weasel
小型低温制冷器 cryomite
小型低压槽 minor trough
小型狄法尔试验 <试验小尺寸集料抗磨值方法之一>【道】micro-Deval test
小型底片摄影 miniature negative photography;miniature photography
小型底栖生物 meiobenthos
小型底卸式铲运机 pan
小型地雷场 small mine block
小型地貌 small-scale landform
小型地面终端 small earth terminal
小型地震仪 geophone;seismic timer
小型电车 minitram
小型电池 compact battery
小型电动工具 microelectric(al) tool;portable electric(al) tool
小型电动机 light power motor;low-power motor;midget motor;miniature motor;pony motor;small type motor
小型电动机车 mule
小型电动运货升降机井道 power dumbwaiter hoistway
小型电感比较仪 minicom
小型电感器 midget inductor
小型(电解)溶解器 bench-scale dissolver
小型电路 miniature circuit
小型电路试验器 midget circuit tester
小型电气路签 miniature electrical train staff
小型电器支回路 small appliance branch circuit
小型电容器 button capacitor;midget capacitor;midget condenser
小型电台 low-power station
小型电影摄影机 cine camera;cinekodak
小型电源插头 attaching plug
小型电站 midget plant
小型电子管 micropup;miniature tube;peanut tube;midget tube
小型电子计算机 minicomputer
小型雕塑座 <山尖或山花端部的> acroter
小型调查 minisurvey
小型叠合装置 midget polyunit
小型动力装置 miniature power unit
小型冻结间 walk-in freezer
小型斗车 decauville wagon
小型斗式提升加料机 small-size bucket elevator loader
小型断电器 miniature circuit breaker
小型对拖网 pareja;pareja trawl
小型墩台沉井 Chicago caisson;Chicago well
小型多用设备 small-scale multiple-use equipment
小型耳机 insert earphone
小型发报机-接收机 miniature transmitter-receiver
小型发电厂 small power station
小型发电机 small motor;electric(al) set
小型发电机组 small generating set
小型发电站 mini power plant
小型发动机 compact engine;puffer;small engine
小型阀(门) pet valve
小型翻斗车 jubilee skip;small dumper
小型防毒器 small anti-gas apparatus
小型仿真程序 miniemulator
小型放大镜 loup(e)
小型放大器 mini amplifier
小型放大器调制器 mini-amplifier modulator
小型放线阀 petcock
小型飞船 blimp
小型飞机 light aeroplane;aviette <体育运动用>
小型废水处理厂 small wastewater treatment plant
小型焚化炉 small-scale incinerator
小型伏特计 pocket voltmeter
小型服务结构物 minor service structure
小型浮游动物 microzooplankton
小型浮游生物 microplankton
小型辅助机械 donkey engine
小型钢 light(weight) section
小型钢材 light shape;small steel shape
小型钢材剪断机 bar shear
小型钢铁厂 mini steel plant
小型钢轧机 jobbing mill;light section mill
小型钢制翻斗车 steel jubilee skip
小型钢制料车 steel jubilee skip
小型港口拖船 small harbo(u)r tug
小型高尔夫球场 miniature golf course
小型高速客运 personal rapid transit
小型高速客运系统 personal rapid transit system
小型高温计 micropyrometer
小型高效城市交通控制系统 compact urban traffic control system
小型工厂 midget plant;portable factory
小型工程 small project
小型工程计划 small project
小型工地 microsite
小型工具 small tool
小型工具箱 glove compartment
小型工业 back yard industry
小型公共汽车 baby bus;minibus;minimum bus;shuttle bus;jitney(bus)<美>
小型公寓 small-sized flat
小型公寓住宅 studio apartment
小型公园 vest-pocket park
小型供暖器 space heater
小型供热器 space heater
小型供水系统 small water-supply system
小型拱 small arch
小型拱廊 arcature
小型构造 minor structure
小型构造尺度 minor structure scale
小型鼓状剥皮机 bag barker;pocket barker
小型管 miniature tube;peanut tube
小型管底 miniature base
小型管管脚 bantam stem
小型管帽 miniature cap
小型贯入仪 pocket penetrometer
小型惯性陀螺 miniature inertial gyro
小型灌溉 microscale irrigation
小型灌溉坝 farm dam
小型灌溉工程 minor irrigation
小型灌溉水塘 farm pond
小型光电倍增管 miniature photomultiplier
小型光电测距仪 microranger
小型辊式破碎机 monkey roll crusher;monkey rolls
小型焊枪 midget
小型航标 minor aids-to-navigation
小型航道整治工程 small-scale channel regulation project
小型航空母舰 escort carrier
小型核电池 miniature nuclear battery
小型核电站 small-size nuclear power plant
小型盒式磁带机 minicartridge
小型横洞 fox hole
小型红外线干燥机 small infrared drier
小型护航航空母舰 jeep
小型滑车 dan
小型滑车支承架 small pulley bearing
小型滑动切片机 small sliding microtome
小型滑轮式起重机 pole derrick
小型滑轮支承架 small pulley bearing
小型滑坡 small landslide
小型滑翔机 aviette
小型滑雪提升机 small type ski hoist
小型滑脂加注器 mini lube
小型化 miniaturization
小型化的 midget
小型化电路 miniaturized circuit
小型化记录仪 miniaturized recorder
小型化系统 miniaturized system
小型话筒 lapel microphone;microtelephone
小型环流 minor circulation
小型缓冲器 miniature retarder
小型缓行器 miniature retarder
小型回转窑 mini rotary kiln
小型会计机 monorobot
小型绘图台 graphic(al) tablet
小型混凝土拌和机 baby concrete mixer;half-bag mixer
小型混凝土搅拌机 baby concrete mixer;half-bag mixer
小型活塞<辅助的> slave piston
小型火箭 pencil rocket
小型火箭发射活动控制中心 small rockets launch operation controlled center[centre]
小型火箭升降装置 small rocket lift device
小型货车 baby truck;dilly;jubilee wagon;light van;roadster;jubilee truck<一种车身窄而深的侧卸式>
小型货物电梯 small goods elevator
小型货物升降机 small goods elevator
小型货运电梯 small type goods lift
小型机场 air park;small airport
小型机场跑道指示器 small airport runway indicator
小型机车 jack engine;junior machine;motorscooter
小型机动车 cyclecar;cycling-car
小型机动摩托车 motorscooter
小型机库 hangarette
小型机器机械化 mechanism with small machines
小型机械 small-size machine;light plant
小型机械化 power choring
小型机械手 micromanipulator
小型积分陀螺 miniature integrating gyro
小型集会场所 minor place of(outdoor) assembly
小型集装船 baby container
小型集装箱 small container;portable container
小型给水器 water tender
小型给水系统 small water-supply system
小型计数管 miniature counter
小型计算机 miniature computer;minicom;small computer;small-size computer
小型计算机操作系统 minicomputer operating system
小型计算机程序设计 minicomputer programming
小型计算机的通信[讯]处理机 minicomputer communication processor
小型计算机的外围设备 minicomputer peripherals
小型计算机控制的终端 minicomputer controlled terminal
小型计算机软件 minicomputer software
小型计算机输入输出设备 minicomputer input and output
小型计算机调谐器 personal computer front end
小型计算机网络 microcomputer network
小型计算机系统 minicomputer system
小型计算机系统接口 small computer system interface
小型计算机指令 minicomputer instruction
小型计算机终端控制器 minicomputer terminal controller
小型技术措施贷款 loan for technical facility
小型继电器 midget relay;miniature

relay
小型家庭神位 <罗马> lararium
小型家用设备 small domestic appliance
小型架式风钻 pom-pom
小型架式气钻 pom-pom
小型监测器 pocket monitor
小型减速机 packaged reducer
小型检测器 portable detector
小型建设项目 small construction item; small construction project
小型交通车 minibus
小型角锥玻璃 small pyramid glass
小型轿车 pony car
小型接收机 midget receiver; miniature radio receiver; vest-pocket receiver
小型接线器 miniswitch reed matrix
小型结构(物) small structure; minor structure
小型近海航船 <印度尼西亚一带> paraos
小型晶体 minicrystal
小型晶体管 minitransistor
小型晶体管式测微指示表 pulcom
小型晶体座 button mounting
小型静电陀螺 miniature electrostatic gyro
小型纠正仪 small type rectifier
小型救生艇或供应艇 cockboat; dinghy
小型救援拖船 small rescue tug
小型居住单元 small dwelling unit
小型剧院 <古希腊或罗马的> ode(i)on
小型聚光灯 baby can; baby light; baby spot(light)
小型卷扬机 winch
小型绝缘试验器 meg
小型掘进机 minimole
小型军用潜水器 miniature military submersible
小型卡口灯座 miniature bayonet base
小型卡片 minicard
小型卡型盒式磁带机 minicassette
小型开沟犁 scooter; small gang plough
小型开关 minor switch; switchette
小型开口沉井 leech
小型科研卫星 small scientific research satellite
小型客车 microbus; minibus; station wagon; wagon
小型客车帆船 sail board
小型客机 air-taxi
小型空气取样器 minihead air sampler
小型空气压缩机 baby air compressor
小型空中目标 small-size air target
小型控制台 consolette
小型快递交通系统 people move transit system; personal rapid transit system
小型快速扫描电子显微镜 mini-rapid scan electron microscope
小型矿车 cocopan; corf; jubilee truck
小型矿床 small-size ore deposit
小型矿山 miniature-tonnage mine
小型垃圾焚化炉 small incinerator
小型垃圾箱 individual waste container
小型蜡台形灯座 candelabrum base
小型冷藏间 walk-in refrigerator
小型立体坐标量测仪 microstereocomparator
小型沥青喷布机 dandy
小型沥青喷洒机 dandy
小型沥青再生搅拌机 mini-recycler
小型连拱 arcature
小型链式起重机 coffin hoist
小型链式升降机 coffin hoist
小型疗养院 nursing home
小型料车 decauville wagon
小型列车 minitrain
小型临时工棚 casern
小型临时设施 small-scale temporary facility
小型流动机械库 small-size travel(l)ing crane and car shed
小型流动式发电机 small-size travel(l)ing power generator
小型流速计 miniature current meter
小型流速仪 diminutive current meter; miniature current meter
小型六分仪 stadi(o)meter
小型陆桥运输 <集装箱> mini-land bridge service
小型陆桥运销 mini-land bridge service system
小型旅宿车 mini-motor-home
小型旅游公共汽车 baby bus
小型履带吊车 small crane on tractor track; small crawler crane
小型履带式拖拉机 baby track-layer
小型罗马祭坛 <露天的> sacellum
小型螺丝千斤顶 <顶脚手架用> crick
小型螺旋起重器 small screw jack
小型落地式唱机 consolette
小型落地式电视机 consolette
小型落地式接收机 consolette
小型马桶 private stable
小型煤气发生器 gasogene
小型模件 minimodule
小型模型 bench model
小型摩托车 scooter
小型碾磨钢段 cylpeb
小型爬梯 abbreviated ladder
小型排锯 sash gang saw
小型排气阀 draw cock
小型排水阀 draw cock
小型配电板 small distributor
小型棚屋 small-sized hut
小型拼装污水厂 package sewage-treatment plant
小型平板车 lift truck
小型平底渔船 fishing dory
小型破冰船 pocket icebreaker
小型破碎机 small breaker; small breaking machine; small crusher; small crushing machine
小型企业 small business; small lot producer; small-sized enterprise
小型起重机 minor crane; dink(e)y <俚语,起重能力小于196千牛>
小型汽车 baby car; compacted car; kart; minicar
小型汽车库 small motor garage
小型砌块 small block
小型牵引车 small-size tractor
小型潜水器 miniature submersible
小型桥台 mini-abutment
小型侵入体 minor intrusion
小型轻便货车 mini pickup
小型轻便望远镜 prospective glass
小型轻轨翻斗车 jubilee skip
小型倾卸车 jubilee wagon
小型球面反射镜 mirror ball
小型全断面隧洞掘进机 mini fullfacer
小型燃油锅炉 oil burning package boiler
小型热带风暴 midget tropical storm
小型热水器 sink heater; small water heater
小型软磁盘 flippi; minidiskette
小型软盘机 minifloppy disk drive
小型塞孔 pup jack
小型三角架 pocket tripod
小型散货船 mini-bulker
小型散货轮 mini-bulker
小型散货平舱机 calf-dozer
小型散装 minibulk
小型散装(货)船 minibulk carrier; mini-bulker
小型沙发椅 settee
小型设备 bantam; compact unit; midget plant; small appliance; small plant; mini-plant <试生产用的>
小型设备机械化 mechanism with small machines
小型摄影机 miniature camera
小型升降机 dumb waiter
小型生产 small-scale production
小型(十二脚)电子管 compactron
小型十字板剪切仪 pocket shear meter; torvane
小型十字板仪 pocket shear meter; torvane
小型实验装置 mini-plant
小型食物链 microscopic food chain
小型示波器 pocketscope
小型试件 subsized specimen
小型试验 bench(-scale) test(ing); mini-test; pilot test; small test
小型试验槽 small laboratory flume
小型试验机 small-scale tester
小型试验设备 pilot plant unit
小型试验水槽 miniature open flume
小型试验性工厂 pilot plant
小型室内电力供热器 electric(al) space heater
小型室外集合场所 minor place of (outdoor) assembly
小型收集器 minicollector
小型手柄 miniature lever
小型手动穿孔机 portable hand punch
小型手绞车 gypsy winch
小型手摇车床 throw lathe
小型双人潜水器 diving saucer
小型水池 service reservoir; service tank
小型水处理系统 small water treatment system
小型水电 small hydropower
小型水电站 small hydro-power station; small water power station
小型水罐消防车 first-aid outfit
小型水库 service reservoir
小型水力发电厂 microhydraulic station
小型水力发电站 small hydroelectric(al) plant
小型水力旋风(分粒)器 miniature hydrocyclone
小型水利工程 small hydraulic project; small-scale water conservancy project
小型水陆两用车 jeep
小型水轮机 diminutive turbine
小型水泥厂 mini cement plant; small-size cement plant
小型水平仪 pocket level
小型水枪 little giant
小型水源地 small-sized water source
小型私人医院 <英> nursing home
小型送话器 microtelephone
小型酸性转炉 baby Bessemer converter
小型酸性转炉钢 little Bessemer steel
小型随车起重装置 <一种带绞盘的> cherry pick
小型隧道 minitunnel; microtunnel
小型隧道施工技术 microtunnelling
小型隧道挖掘机 microtunneller; microtunnelling machine
小型隧洞掘进机 miniborer
小型隧洞掘进机挖掘施工 moling
小型台卡导航仪 Decca Hi-Fix equipment
小型台秤 small platform scales
小型套房 flatlet
小型套节 midget socket
小型条信号 minibar
小型调车机车 dink(e)y; pony engine
小型调谐器 minituner
小型调制解调器 minimodem
小型铁道车 jubilee wagon
小型通信[讯]终端设备 small communications terminal
小型投影测图仪 aerovelox
小型透平 pony turbine
小型突水 small-scale of bursting water
小型图书馆 small library
小型推土机 baby bulldozer; calf-dozer; minibul; skipdozer
小型拖拉机 baby tractor; compact tractor; midget tractor; miniature tractor; small-size tractor; small tractor
小型拖轮 light tug; water tractor; bumboat <装备有卷扬机、起锚机、顶推及拖带设备等的>
小型驼峰编组场 mini-hump yard
小型挖掘机 mini-excavator
小型挖泥船 mini-dredge(r)
小型万用表 midget tester
小型望远镜 spyglass
小型桅杆式起重机 pole derrick
小型温切斯特磁盘机 mini-Winchester disk drive
小型问题 small-scale problem
小型涡轮机 pony turbine
小型握柄式动力联锁架 miniature-lever power frame
小型污泥处理厂 small sludge treatment works
小型污水处理厂 small sewage treatment works; small wastewater treatment works
小型污水处理装置 small sewage treatment plant
小型无线电收发机 pack unit
小型物 midget; miniature
小型物镜 microobjective
小型物像 microform
小型系列生产 small scale serial production
小型项目 small project
小型消防艇 pup boat
小型小客车 small car
小型信号机【铁】dwarf signal
小型星光镜 small starlight scope
小型型材轧机 light section mill
小型型钢 bar stock; light(er) section; merchant bar; small section
小型型钢卷取机 thin-stock reel
小型型钢轧机 small section mill
小型型芯座 pocket plum
小型旋塞(阀) petcock; pet valve
小型选择器 minor switch
小型循环 small round
小型压力传感器 miniature pressure cell
小型压力喷雾器 small pressure sprayer
小型压力撒雾器 small pressure sprayer
小型压路机 baby roller; skipdozer; portable roller <装橡胶胎后轮的>
小型压缩机 baby compressor; fractional compressor; light-duty compressor
小型压缩机冷凝机组 fractional condensing unit
小型压土器 baby roller
小型烟雾图 miniature smoke chart
小型沿岸运输船 small coastal transport
小型檐托 abaciscus
小型眼底镜 microphthalmoscope
小型氧消耗比色计 bench-scale oxygen consumption colo(u)rimeter
小型样芯 small coring

小型摇床 miniature table
小型移动式电站 packaged power plant
小型移动式起重机 portable crane
小型移动式装置 packaged equipment
小型音乐厅 odeum
小型应用技术卫星 small application technology satellite
小型涌泉 pocket spring
小型用户 small user
小型油船 mail boat;packed boat
小型油轮 small-size tanker
小型元件 matrix component;miniature component;miniaturized component;minicomponent
小型圆角光子 egg sleeker
小型圆盘耙 small disc harrow
小型越野汽车 jeep
小型运货车 barrow truck
小型运货卡车 baby truck
小型运货汽车 mini lorry;motor wagon;panel truck;pick-up truck
小型运输货车 haul wagon
小型载货篷车 panel truck
小型载客轮渡 water-bus
小型载人潜水器 small manned submersible
小型载重汽车 small truck
小型再生道路重铺机 miniremixer
小型凿井用泵 running lift
小型藻类废水处理 microbial wastewater treatment
小型轧材 black bar;merchant bar
小型轧钢厂 bar mill
小型轧机 bar mill;merchant bar mill
小型轧机中间机座 pony
小型闸刀开关 baby knife switch
小型招牌 small professional or announcement signs
小型照片 microphotograph
小型照相机 Kodak;miniature camera;minicam;minicam(era)
小型照相术 microphotography
小型折叠式集装箱 small-size collapsible container; small-size folding container
小型褶皱 mesoscopic fold
小型侦察车 beep
小型真空泵 minipump
小型真空泵系统 packaged vacuum pump system
小型(真空)电容器 peanut capacitor
小型真空设备 packaged vacuum unit
小型振动台 small-scale vibration table;small-size vibration table
小型蒸馏塔 baby tower
小型蒸汽超重机 donkey crane
小型蒸汽机 jack engine
小型支架 pony support
小型制冷机 miniature refrigerator
小型制冷技术 miniature refrigerating technology
小型制冷器 small cooler
小型中继站 satellite studio
小型中子加速器 small neutron accelerator
小型钟壳 miniature clock case
小型重力翻斗车 gravity type small dumper
小型轴承 small-size bearing
小型烛台灯座 candelabrum base
小型主机 minihost
小型注油器 cadger
小型柱 colonnette;columella;small column
小型铸件 small casting
小型砖 economy brick
小型转向架 pony truck
小型桩 <英国通常指直径小于 300mm 的桩> micropile;mini-pile
小型装配单元 minor package unit
小型装卸铺设机 small loading and laying machine
小型装药 small charge
小型装置 midget plant;miniature set;packaged plant;packaged unit
小型装置单元 package unit
小型资源 make full use of limited resources
小型自动电话交换台 unit automatic exchange
小型自动化电车快速轨道交通(系统)personal rapid transit
小型自动化电车式(城市)轨道交通(系统)automatic group transit
小型自动售货商店 superette
小型自动数据系统 personnel automated data system
小型自卸车 power barrow
小型阻抗联结器 wee-Z bond
小型组匣 minimodule
小型组装污水处理厂 package sewage-treatment plant
小型钻机 pocket borer;sack borer
小型钻机钻杆 sack borer stem
小熊星座 Ursa Major
小熊星座的七颗主星 Little Dipper
小熊座 cynosure
小休 vacationette
小修 light(running)repair;operating maintenance; routine repair; running repair;temporary repair;transient service;turnaround
小修保养 routine maintenance
小修补 minor maintenance; routine maintenance
小修改 minor change
小修和安装 light maintenance and installation truck
小修理 current repair;easy servicing;light maintenance; light overhaul;light repair;minor overhaul;minor repair; periodic(al) inspection;slight repair
小修小补 dribbing
小修业务 running repair service
小旋风 eddy wind;whirly
小旋回 epicycle
小旋塞 petcock;draw cock <放气用>
小旋转接头 small swivel
小旋转轮 small pivoted wheel
小漩涡 dimple
小穴 <矿物或岩石中的> vesicle
小学 elementary school;grade school;primary school
小学环境教育 environmental education in primary school
小学生 pupil
小学校舍 elementary school building
小雪 flurry;light snow;spit
小雪暴 slight thunderstorm
小循环 lesser circulation;minor cycle
小循环法 partial circulation process
小循环运行 partial circulation operation
小压辊 lump roll
小压(力)机 subpress
小压下量 light reduction
小压下量轧制 saddening
小牙轮 pinion
小牙饰 denticulation
小崖【地】nip;scarplet
小亚细亚 <黑海与阿拉伯地区的总名> Asia Minor
小亚细亚建筑 architecture of Asia minor
小亚细亚型爱奥尼克亚柱式 Ionic order of the Asia Minor type
小烟道 bottom flue
小烟管 tube
小岩脉 dikelet
小岩墙 dikelet
小岩石 rocklet
小眼接头 slim-hole joint
小眼井钻进 slim-hole drilling
小眼孔 eyelet
小眼温度 temperature at orifice
小眼凿深孔法 small bore deep-hole method
小演替系列 microsere
小宴会厅 function room;small banquet hall
小焰管钟形管头加工工具 small tube belling tool
小燕尾旗 pennant
小阳春【气】gossamer
小样本 hand sample;small sample
小样本法 small sample method;small sample technique
小样本时序研究 small sample time-series study
小样品 hand sample;small sample
小样品测定系统 small sample assay system
小样品分布 small sample distribution
小样品散射 small sample scattering
小样品选择器 small sample chopper
小样取样 increment sampling
小窑调查 investigation of small coal mine
小药片 tabloid
小药瓶 phial
小叶 leaflet
小叶病 little leaf disease
小叶的 foliolate
小叶榉树 Chinese zelkova
小叶六道木 small leaf abelia
小叶杨 Simon poplar
小页岩 little slate
小液滴 droplet
小衣橱 commode
小医院 cottage hospital
小翼 winglet
小应变 small strain
小应变幅 small strain amplitude
小应力 small stress
小应用程序 applet
小樱桃 little cherry
小硬颗粒 corn
小用具 <如蘸药的小棉花棍,涂漆的小刷等> applicator
小油壶 cadger
小油门慢车 low idle
小油漆桶 paint kettle
小油绳 small stuff; tarred fittings;yacht marline
小油田 small oil field
小油桶 oil-can(ning)
小油嘴 oil nipple
小游艇船坞 marina
小游泳池 minor swimming pool
小游园 petty street garden;vest-pocket park
小幼树 <高 3~10 英尺,1 英尺 = 0.3048米> small sapling
小于标准直径的岩芯 undersize core
小于常规的尺寸 undersize
小于符合 less than match
小于或等于 less than or equal to
小于 90 度的方位角 quadrantal bearing;reduced bearing
小于两万分之一的 ultra-micro
小于某粒径的颗粒百分率 percent finer
小于某粒径的颗粒百分数 percent finer
小于某筛号的材料 minus No. material
小于筛孔半径的粒度比 half-size ratio
小于条件 less than condition
小于原子的 subatomic
小于正常折射 less than normal refraction
小鱼 fingerling
小鱼船 beach boat;cog;cutter;fishing smack;tosher;smack
小鱼艇 coracle
小鱼围网 minnow seine
小宇宙 microcosm(os)
小雨 fine rain;flurry;gull(e)y;light rain; low intensity storm; small rain;soft rain;sprinkle
小浴盆 short bath tub
小园地 spade farm
小原木 batten
小圆材 <胸径大于 4 英寸,小于 8 英寸,1 英寸 = 0.0254 米> small pole;batten;round billet
小圆齿 crenulation
小圆窗 bull's eye;bull's eye window;bullion;bulls eye;dead eye;oculus;roundel;round window;small circular window
小圆锉 tongue file
小圆构造带 small circle belt
小圆刮刀 bead slicker
小圆规 bow compasses;bow pens;pen bow;spring bow compasses
小圆航路 small circle
小圆环带 small circle girdle
小圆井 small circular shaft
小圆井展开图 extending map of small cylindrical pit
小圆锯 burr;pad saw
小圆块 cob;cob walling
小圆棱 eased edge
小圆螺母 small round nut
小圆木 small log
小圆刨 spokeshave
小圆丘 hump;mamelon
小圆石 gravel pebble;gravel stone;kidney;small round stone;water-worn pebble
小圆石子 pebble
小圆饰线 mo(u)lding fillet
小圆塔楼 round small tower
小圆头 bull head
小圆头栏杆 bull-head(ed)rail
小圆凸 bosselation;torulus
小圆凸线 baguette
小圆弯 knuckle bend
小圆屋顶 cupola
小圆屋顶间 intercupola
小圆线脚 baguette;bead
小圆形物 roundlet
小圆岩块 rognon
小圆凿 quick gouge
小月亮 moonlet
小云滴 cloud droplet
小云朵 cloudlet
小运输线 small transfer line
小运转机车 transfer engine
小运转交路【铁】transfer routing
小运转列车 exchange train;removal or transfer train; transfer train;transship train
小运转列车取送车行程 local tripping for collection and distribution
小运转列车行程 transfer run
小韵律 small rhythm
小杂物间 cubby
小杂质 small impurity
小灾害 minor disaster
小凿 small chisel
小赠品 <商人给顾客的> lagniappe
小轧型机 small section rolling mill
小闸门 wicket <大闸门的一部分>;handstop
小闸作用管 application pipe of independent brake valve
小斋堂 misericord(e)
小债券 baby bond;small piece bonds

小债务法庭 small debts court
小站 minor station; minor stop; way station <列车快车不停的>; halt <无建筑物的>
小站电气集中联锁 all-relay interlocking for way-station
小张 <纸条、纸片等的> scrip
小涨潮流 lesser flood
小帐顶床 bonnet bed
小账 beer-money
小照片 midget
小照相机物镜 miniature camera lens
小折刀 pocket-knife
小折凳 X-stool
小折叠桌 servette
小褶皱【地】minor fold; puckering
小枕木 crown-tree
小阵雨 light shower; passing shower; slight shower of rain
小振荡法 small oscillations method
小振动 small vibration
小振幅 small amplitude
小振幅波 small amplitude wave; wave of low amplitude; infinitely small-amplitude wave
小振幅波理论 small amplitude wave theory
小振幅剪切振荡 small amplitude shear oscillation
小振幅脉冲极谱法 small amplitude pulse polarography
小镇 hicktown; small town; townlet
小震 minor earthquake; tremor
小震带 minor seismic belt
小震动 small shock
小正齿轮 spur pinion
小正厅 <古罗马> atriolum
小支流 pup; sprout; subdivided flow
小支重架 minor bogey
小枝 branchlet; ramulus; sprig; twig
小枝的 ramular
小枝条 small stems
小直径 minor diameter
小直径齿轮 small diameter gear
小直径封隔器 pony packer
小直径工件 small diameter workpiece
小直径管 narrow-bore tube; narrow tube
小直径管线 small pipeline; thin(pipe) line
小直径暖气管 small bore heating pipe
小直径钎钢 small ga(u)ge steel
小直径深孔崩矿法 small bore deep-hole method
小直径隧道 micro-tunnel; small size tunnel
小直径隧道掘进机 micro-tunnel(l)er
小直径同轴电缆 small diameter coaxial cable
小直径凸模 pin punch
小直径钻进法 slim-hole system
小直径钻孔 slim hole; small diameter drill hole
小植物群落 microstrand
小指标计分奖 point system on small quotas
小指标竞赛 hundred-point emulation
小指示灯 bezel
小制动器 inertia brake
小质点 fine particle
小质量质点 low-mass particle
小中继站 minor relay station
小中取大法则 minimax criterion
小中庭 <古罗马> atriolum
小钟杆平球架 bell bearing
小重力加速度离心机 low-g centrifuge
小舟桨手 whiff
小周期 minor cycle; word time
小周期脉冲发生器 minor cycle pulse generator
小周天 a small circle of the evolutive
小皱纹 reticulation
小珠 bead
小珠状端壁 nodular end wall
小烛树蜡 candelilla wax
小主平面 minor principal plane
小主应变 minor principal strain
小主应力 minor principal stress
小主应力面 minor principal stress plane
小住屋 doll-house
小住宅 bungalow <带回廊的>; bastide <法国南部的农村>
小住宅区的观测 observation of settlements
小柱 cancelli; colonnette; columella; pillaret; princess post; small column; stake; subpost
小柱冠 abaciscus
小柱栏杆 baluster supporting a railing
小柱屋架 princess post truss
小砖 Calculon
小转发站 minor relay station
小转弯 steep turn
小转运商 small shipper
小转运站 minor relay station
小庄园 minifundium
小桩 small pile; spile; sapling <加固货物装载>
小桩子【测】ground peg
小装饰品 doodad; knickknack
小装饰物 affix
小装卸车 <英> jubilee wagon
小装置 dingus
小锥齿轮 bevel pinion
小锥体 bocca
小锥(子) bradawl; sprig bit; conelet
小宗货物 parcel
小宗批发商 jobber
小纵梁 jack stringer; stringer
小阻力 slight drag
小阻力物体 low-drag body
小阻尼 underdamping
小阻尼振子 low-damped oscillator
小组 panel; squad; subcommittee; subgroup
小组办公楼 prefecture
小组测试法 panel method
小组调查会 panel
小组访谈 group interview
小组负责人 section leader
小组管理 management of working groups; team management
小组会议 panel meeting; panel session
小组活动 group action
小组集体工作 group work
小组监工 walking ganger
小组间冲突动态性 dynamics of intergroup conflict
小组奖金工资制 group-bonus wage plan
小组奖励计划 group incentive plan
小组进位 group carry
小组讨论会 panel discussion; panel session
小组委员会 subcommittee
小组信息 blockette; subblock
小组岩脉 dike set
小组自主生产法 autonomous group approach
小钻 baby drill; first bit; pocket borer
小钻机 <矿层钻进专用> drilling-in unit
小作坊 individual workshop
小作用区域 low coverage

效

效度准则 validity criterion
效果 effect; effectiveness; impact; payoff; result; returns on investment
效果的衡量 efficiency measure; measure of effectiveness
效果监察器 effect monitor
效果检验 validity check
效果鉴定 impact assessment
效果滤光片 effects filter
效果律 law of effect
效果照明装置 brancart
效价递增 increasing titer gradient
效价强度 potency
效力 effect; effectiveness; efficacy; potency
效力递减律 law of relativity
效力可加性 additive of effect
效力准则 validity criterion
效率 ability; coefficient of efficiency; coefficient of performance; coefficient of useful effect; effectiveness; workpiece ratio
效率半衰期 efficiency half life
效率比 efficiency ratio
效率测定 efficiency test(ing)
效率测定机 efficiency testing machine
效率测量 measurement of efficiency
效率差异 efficiency variance
效率成本 efficiency cost
效率-出力曲线 efficiency-power curve
效率单位 efficiency unit
效率的契约 efficiency contract
效率等价 efficiency equivalence
效率低 inefficiency
效率低于标准 below standard performance
效率递减 diminishing efficiency
效率定员法 efficiency-based personnel allocation method
效率费用比 benefit-cost ratio
效率费用分析 benefit-cost analysis
效率分析 analysis of effect; efficiency analysis
效率负荷范围系数 efficiency load-range factor
效率高的 efficient; expeditious
效率高低 level of efficiency
效率工程师 efficiency engineer
效率工资制 efficiency wage system
效率公式 <群桩的> efficiency formula
效率估计 efficiency estimation
效率谷点 efficiency dip
效率观点 view of efficiency
效率轨迹 efficiency locus
效率合约 efficiency contract
效率很低的 inefficient
效率换算 <模型效率换算成真机效率> efficiency step up
效率价格 efficiency price
效率减少 reduction of efficiency
效率检查 efficiency audit
效率降低系数 loss of efficiency factor
效率量度 measures of effectiveness
效率频带 efficiency band
效率频率特性 efficiency frequency characteristic
效率评价 efficiency rating
效率期 effector phase
效率曲线 efficiency curve
效率曲线的平均高 planimetric(al) average efficiency
效率人 efficiency man
效率审计 efficiency audit
效率升降机 continuous elevator
效率十二原则 twelve principles of efficiency
效率市场 efficiency market
效率示踪法 efficiency tracer method
效率试验 efficiency test
效率试验机 efficiency testing machine
效率收入 efficiency earnings
效率水平 level of efficiency
效率特性 efficiency characteristic
效率提高 increase of efficiency
效率调制 efficiency modulation
效率图 efficiency chart; efficiency diagram
效率外推法 effectiveness factor; efficiency extrapolation method
效率系数 efficiency factor; output efficiency
效率系数方程式 efficiency equation
效率限界 efficiency of frontier
效率因数 efficiency factor
效率因子 efficiency factor
效率与公平 performance and fairness
效率预算法 efficiency budgeting
效率增进运动 efficiency movement
效率值 efficiency value
效率指标 efficiency index
效率指示 efficiency indication
效率指数 effectiveness index; efficiency index; index of performance
效率制 efficiency system
效率专家 efficiency engineer; efficiency expert
效能 effectiveness; efficacy; efficiency; performance; potency
效能测定 potency assay
效能单位 potency unit
效能分析【交】effectiveness analysis
效能试验 approval test; effectiveness test; potency test
效能指数 efficiency index; index of performance
效坛下官窑 Jiaotan xia official kiln
效验 efficacy
效益 achievement; benefit; effectiveness; efficiency; efficiency earnings; performance; usefulness
效益成本比(率) benefit-cost ratio
效益成本比率法 benefit-cost ratio method
效益成本分析 benefit-cost analysis
效益成本关系 benefit(-to)-cost relationship
效益费用比(率) benefit-cost ratio
效益费用分析 benefit-cost analysis
效益费用关系 benefit-to-cost relationship
效益分析 benefit-risk analysis
效益浮动工资 floating wages linked with performance
效益工程 benefit project
效益函数 benefit function
效益流程 flow of benefits
效益流量 benefit flow
效益评估 benefit assessment
效益评价程序 benefit-evaluation procedure
效益输出 performance output
效益现值 present value of benefit
效益预测 benefit forecast; benefit prediction; benefit projection; ex ante measurement of benefit

效益指标 efficiency target; measure of performance; performance indicator
效益最大的货币区 optimum currency areas
效应 effect
效应玻璃 effect glass
效应部 effector
效应递减 diminished utility
效应光阀管 light valve tube
效应机理 effector mechanism
效应阶段 effective stage
效应局限 locality of effect
效应可加性 additive of effect
效应力 efficacy

效应量 effect quantity
效应滤色器 effect filter
效应浓度 effect concentration
效应评估 effect assessment
效应器 effector
效应外推模型 effect extrapolation model
效应问题工作组 working group on effects
效应因子 effector
效用 effectiveness; influence; usefulness; utility
效用报偿 utility payment
效用测度实验 utility-measurement experiments
效用成本分析 utility cost analysis
效用尺度 utility scale
效用抽象 abstraction of utility
效用单位 utility unit
效用等高线 equal-utility contour
效用等级 degree of utility
效用递减 diminishing utility
效用递减律 law of diminishing utility
效用分析 utility analysis
效用估价 effectiveness evaluation
效用固定分析法 fixed effectiveness approach
效用函数 effectiveness function; utility function
效用极大化 utility maximization
效用价值论 utility theory of value
效用可能性曲线 utility possibility curve
效用亏损与货币亏损比较 utility versus money loss
效用理论 utility theory
效用理论的对偶性定理 duality theorem in utility theory
效用率 utility ratio
效用论 theory of utility
效用模型 utility model
效用曲线 utility curve
效用树 utility tree
效用水平 utility level
效用损失率 percentage of utility loss
效用投资对偶性 utility expenditure duality
效用系数 effectiveness factor
效用消失 lost usefulness
效用支出对偶性 utility expenditure duality
效用值 utility measure
效用指标 effectiveness indicator; utility index; measures of effectiveness <描述交通设施服务质量的参数，如车速、延误、油耗等>
效用指数 utility index
效用指数的唯一性 uniqueness of utility index
效用最大化 utility maximization
效用最大化法则 utility maximizing rule

校办工厂 school plant

校车 school bus
校舍 school building; schoolhouse
校舍建筑规划 school building program(me)
校外的 extramural
校董事会董事 trustee; board of trustees
校园 campus; school garden; school yard
校园规划 campus planning
校园建筑 campus building
校园平面布置图 <大学> campus plan
校园网【计】campus (computer) network
校长 chancellor; director; master; president; rector

笑容条件 conditions of compatibility

笑靥花 bridal wreath

啸鸣 singing

啸鸣器 howler
啸声 hiss; howl; squealing; whistler; whistling sound; screeching <飞机发动机的>
啸声电弧 hissing arc
啸声干扰 whistle interference; whistler
啸声衰减 howling damping
啸声信号 whistling signal
啸声抑制 squelch
啸音 whistler-type noise

些微 smallness

楔 forelock; gate wedge; glut; gore

楔板 wedge plate
楔闭 wedge clothing
楔边 <用以劈石> feathers
楔边板 <板材缺陷> feather-edged board
楔边板安装 feather-edged boarding
楔表面钢丝绳 keystone strand wire rope
楔槽式绳头卡子 wedge socket fitting
楔冲断层 wedging thrust
楔冲式冲断层 wedge thrusts
楔出 edge away
楔锤 wedge hammer
楔唇角 lip angle
楔刀 froe; frow
楔垫抗拉强度试验 <螺钉等的> wedge tensile strength test
楔垫密封 wedge gasket closure
楔阀 block valve
楔缝(堵漏法) wedging method
楔缝螺栓 slit and wedge bolt
楔缝锚杆 sliding wedge roof bolt
楔缝锚头 spring loaded wedge
楔缝式 slot-and-wedge type
楔缝式杆柱 slot-and-wedge bolt
楔缝式锚 split wedge anchor
楔缝式锚杆 slit bolt; slit-rod-and-wedge-type bolt; slit-wedge type rock bolt; slot-and-wedge bolt; split bolt; wedge-type rock bolt
楔缝式锚杆支护 slot-and-wedge fastening
楔缝式岩石锚杆 slotted rock bolt
楔杆 tapered bar
楔杆阀冲水厕所 wedge valve closet
楔固 keying; wedging
楔光片测温计 wedge pyrometer
楔焊 wedge bond(ing)
楔合 conjunction; wedging
楔合板 coupling plank
楔合的 coped
楔合块 wedging block
楔合梁 keyed beam
楔合用弓形支撑 <盾构的> key segment
楔合作用 wedging action
楔横轧 cross wedge rolling
楔横轧模 wedge rolling tool
楔厚 wedge thickness
楔击缺口冲击试验 notch wedge impact
楔积 wedge product
楔尖劈 nib
楔尖沙坝 wedge-tip bar
楔尖式钻孔定向器 whipstock
楔键 taper key; wedge key
楔角 angle of throat; angle of wedge; insert angle; lip angle; wedge angle
楔角量 volume of tapering
楔接板 wedge boarding
楔接锻接 crotch weld
楔接焊接 crotch weld
楔接(合) wedge (d) joint (ing); wedge grafting
楔紧 ca(u)lk; wedge-caulking; wedging
楔紧钢板 wedging plate
楔紧作用 wedge action; wedging action
楔聚焦 wedge focusing
楔开 wedged off
楔壳式锚杆 wedge-and-sleeve bolt
楔壳式锚栓 wedge-nut bolt
楔口冲击韧性 notch impact toughness
楔口焊接 scarf weld(ing)
楔块 wedge (block); angle quoin; chock; feathers; voussoir; wedging block; coign(e)
楔块分置式厚壁圆筒 thick-walled cylinder by disperse tapered blocks
楔块拱 voussoir arch
楔块扩孔 expanding with wedge block
楔块理论 wedge theory
楔块锚 wedge anchor(age)
楔块劈裂作用 wedging
楔块式推力轴承 Kingsbury-type thrust bearing; Michell thrust bearing
楔块调整 wedge adjustment
楔牢 wedge
楔连接 cottering; foxtail wedge joint
楔裂 wedging
楔裂法 wedging
楔流 wedge flow
楔滤因子 wedge factor
楔螺母胀壳锚杆 wedge-nut expansion shell bolt
楔落 wedging down
楔面 ramp; wedge surface
楔面断面 wedge section
楔面接头 scarf joint(ing)
楔木 stow wood
楔劈石机 gadding machine
楔(劈)作用 wedge action
楔片 key
楔起 ramming up; wedge up; wedging up
楔嵌接 wedge clamp joint
楔嵌压 wedge pressure
楔取法 wedge method
楔圈 keying
楔入 coping; incuneation; wedge; wedging
楔入板 wedging plate
楔入化合物 wedging compound
楔入剂 wedging agent
楔入锚固 wedge anchor(age)
楔入效应 wedging effect
楔入褶皱 wedging folds
楔入作用 wedging action
楔石 clavel; clavis; clawel
楔实碾压 <碎石路的> key-rolling
楔式传动 wedge drive
楔式队形 wedge formation
楔式接合器 wedge bonder
楔式开挖 wedge cut
楔式离合器 sprag clutch
楔式锚具 wedge anchor(age)
楔式切割 wedge cut
楔式热模锻压力机 wedge press
楔式提芯器 wedge core lifter
楔式握裹 wedging bond
楔式斜嵌(接) oblique scarf with wedge
楔式桩夹 wedge pile chuck
楔栓 forelock; linchpin
楔榫 wedge(d) tenon
楔锁 wedge yoke
楔锁式支柱 wedge prop
楔弹簧 wedge spring
楔套 adapter sleeve; taper clamping sleeve
楔体法 wedge method
楔体分析法 wedge analysis
楔体拱 cuneatic arch; wedge arch
楔体结构 wedge design
楔体理论 <土压力理论的> wedge theory
楔体破坏 wedge failure
楔条 firring; spline
楔铁 fid
楔线 wedge line
楔镶板 wedge boarding
楔削 taper; gib and cotter <立柱的>
楔削厚度 thickness taper
楔削榫 tapered tenon
楔效应 keystone effect; wedge effect
楔形 dovetail form; sugar-loaf fashion; wedge shape
楔形安全臼 <轧机> wedge-type breaking piece
楔形安全装置 wedge grip gear
楔形暗销 apple ring dowel
楔形板 clapboard; feather-edged board; tapered plate; wedge(-shaped) plate
楔形板材 taper plate
楔形板桩 wedge(-shaped) sheet pile
楔形棒条筛 tapered grizzly bar; wedge-bar screen
楔形碑 cuneiform
楔形贝 wedge shell
楔形比例尺 wedge scale
楔形比色计 wedge-colo(u)rimeter
楔形闭锁机 wedge type
楔形壁座 wedge curb; wedging curb
楔形边 <一边厚、一边薄> tapered bevel; tapered edge; tapered rim; feather-edge
楔形扁材轧机 taper mill
楔形扁销 wedge cotter
楔形表层面砖 quoin veneer split tile
楔形波 wedge wave
楔形薄板 tapered sheet
楔形不锈钢丝滤网 stainless steel wedge-wire screen
楔形槽 chase wedge; key groove; lewis hole; wedge-shaped groove; wedge-shaped slot
楔形槽沟 wedge gap
楔形槽活塞环 wedge channel grooved piston ring
楔形(槽)轮摩擦传动 friction(al)-grooved gearing
楔形侧部 dovetailed side piece
楔形测针 wedge probe
楔形插条 wedge cutting
楔形铲刀 V-shaped
楔形长槽 keyhole slot
楔形长钉 brob
楔形超复层【地】interfinger
楔形超声延迟线 wedge-type ultrasonic delay line
楔形潮沟口 tidal wedge
楔形衬垫 razor-edge liner
楔形衬环 tapering plate
楔形承插口 key type socket
楔形承口装置 wedge socket fitting
楔形尺 measuring wedge

楔形齿皮带 cog belt
楔形储池 wedge reservoir;wedge storage
楔形传动带 wedge-shaped belt
楔形锤片 <粉碎机的> hacked-type hammer
楔形搭接 tapered lap
楔形带 triangular belt;wedge belt
楔形挡板 <板宽逐渐减少> bevel siding
楔形导板 wedge slide
楔形导向器 wedge director
楔形的 arrowheaded; cuneate; cuneiform; sphenoid; V-shaped; V-type; wedged;wedge-form;wedge-shaped
楔形等高线 wedge isobars
楔形底板指示灯 wedge-base lamp
楔形底唇金刚石钻头 wedge-set bit
楔形地脚螺栓 lewis bolt;wedge lewis bolt
楔形地块 wedge-shaped block
楔形点 wedge point
楔形电极 wedge electrode
楔形电刷 wedge-shaped brush
楔形垫 chock
楔形垫板 razor-edge liner
楔形垫块 chock
楔形垫密封 seal by wedge-like gasket
楔形垫密封的计算荷载 calculation load for wedge gasket
楔形垫木 bevel(l)ed sleeper
楔形垫片 skim liner;tapered shim
楔形垫圈 bevel(l)ed washer
楔形钉 cut nail;clasp nail
楔形钉尖 chisel nail point
楔形定位装置 wedge setting gear
楔形定型模板 wedge-shaped form
楔形断裂 wedge-shaped fracture
楔形断面 wedge section
楔形断面筛条 double-tap reeled grizzly
楔形堆垛 <仓库中> wedge storage
楔形阀 wedge valve
楔形法兰 wedge-riser flange
楔形防爬器 wedge-type anti-creeper
楔形防水墙 wedge-shaped dam
楔形分级衬板 tapered classifying liner plate
楔形缝 key(ed)joint(ing)
楔形盖顶石 wedge coping
楔形杆件 tapered member
楔形钢板 <加固基础用> wedging plate
楔形钢筛 wedge wire screen
楔形钢丝 wedge-section wire
楔形高(气)压 wedge-shaped high
楔形高压线 wedge line
楔形工字钢梁 tapered-flange beam
楔形工作缝 keyed construction joint
楔形弓形支撑 <盾构的> taper segment
楔形拱 voussoir arch
楔形拱石 arch block; arch stone; voussoir
楔形拱砖 ga(u)ged brick;voussoir brick
楔形沟 tapering gutter
楔形管 taper pipe
楔形罐道 wedge guide
楔形规 wedge ga(u)ge
楔形轨式脱鞋器 wedge-rail type brake shoe escapement
楔形滚动闸门 wedge roller gate
楔形过滤器 wedge filter
楔形焊接 wedge bond(ing)
楔形荷载 wedge-shaped load
楔形后缘 trailing-edge wedge
楔形滑刀式开沟器 wedge-shaped runner
楔形滑阀 wedge-type slide valve
楔形滑塌 wedge failure
楔形环 cambridge ring; taper ring <管片衬砌>
楔形活门 wedge(-type) gate valve
楔形技术 wedged-shaped technique
楔形夹 wedge clip;wedge grip
楔形夹板 wedge clamp
楔形夹层 interfinger(ing)
楔形夹层部件 interfingering member
楔形夹层构件 interfingering member
楔形夹持器 holding dog
楔形夹块 <预应力锚具> anchor grip
楔形夹盘 wedge chuck
楔形夹头 <拉力试验用> Templin chuck
楔形尖 wedge grip
楔形渐窄梁 tapered-flange beam
楔形键 voussoir key;wedge key
楔形交错层理构造 wedge-shaped cross bedding structure
楔形浇道 wedge gate
楔形浇口 knife gate;wedge gate
楔形角 key groove;V-angle
楔形角度块规 wedge(block) ga(u)ge
楔形接缝 splayed joint(ing);wedge joint(ing)
楔形接合 foxtail joint; wedge bond(ing);wedge joint(ing)
楔形接头 wedge(-shaped)joint(ing)
楔形节插条 nodal wedge type cutting
楔形结构 wedge structure
楔形结合 dovetailing
楔形截面环 wedge ring
楔形金属丝 wedge-wire
楔形金属丝筛网 wedge-wire screening
楔形进气边 wedged leading edge
楔形晶 wedge crystal
楔形晶间断裂 wedge-type fracture
楔形晶体 cumeat
楔形井框 wedging crib
楔形举扬盘 taper lifter
楔形开口环接件 wedge-type split ring connector
楔形开裂加荷 wedge-opening-loading
楔形开挖 V-cut;wedge cut
楔形刻针尖 wedge-shaped point
楔形空气过滤器 expanded type air filter
楔形孔 wedge slot
楔形孔口 wedge-shaped opening
楔形扣板 wedge clip; wedge-shaped clip
楔形块 cathead;cupola block
楔形块规 wedge block ga(u)ge
楔形块检验 wedge test
楔形块数据 wedge data
楔形扩孔器 wedging reamer
楔形扩孔钻头 bull-nose bit; wedge reaming bit;wedging reamer
楔形棱镜 prism wedge
楔形笠石 cuneiform
楔形连接器 wedge-type connector
楔形联结套筒 wedge socket
楔形联结装置 wedge coupler
楔形联轴器 wedge coupling
楔形梁 key beam; tapered beam; wedge beam
楔形裂缝 wedge-shaped fracture
楔形零件 wedge piece
楔形流 wedge flow
楔形楼梯踏步 wedge-shaped step
楔形楼梯踏步板 wedge-shaped tread
楔形路缘石 wedge-shaped curb
楔形绿地 green wedge
楔形轮摩擦传动 wedge friction gearing
楔形螺栓 wedge bolt
楔形螺栓垫圈 bevel(l)ed washer
楔形螺纹 tapered thread
楔形锚固 wedge anchor(age)
楔形锚固系统 wedge anchoring system
楔形锚具 wedge anchor(age); anchor wedge
楔形密封 wedge-type seal
楔形磨擦轮 wedge friction wheel; wedge gear
楔形木钉 sprocket
楔形木块 bolt
楔形内接 wedging inscribing
楔形喷管 wedge nozzle
楔形坯架 wedge-stilt
楔形偏导器 wedge-shaped deflector
楔形平瓦 tapered plain tile
楔形破坏 wedge(-type) failure
楔形破裂面分析法 <用于土壤稳定计算> soil trial wedge method
楔形破裂面试算法 <土体稳定计算的> soil trial wedge method
楔形砌层 key course
楔形砌拱 cuneatic arch; cuneiform arch;wedge cuneatic arch
楔形砌块 wedge-shaped block
楔形千分尺 wedge micrometer
楔形嵌接 keyed scarf
楔形嵌入件 wedge insert
楔形墙压顶 wedge coping
楔形切割 wedge cut
楔形取样器技术 wedge sampling technique
楔形燃烧室 combustion chamber of wedge form
楔形塞 wedge plug
楔形塞垫 taper liner
楔形三角洲 wedge-like delta
楔形伞伐作业 wedge system
楔形伞伐作业方式 wedge-shaped shelter-wood system
楔形砂浆接头 wedge-shaped mortar joint
楔形筛条 tapered grizzly bar; wedge bar
楔形设计 wedge design;wedge plan
楔形摄谱仪 wedge spectrograph
楔形施工缝 keyed construction joint
楔形石 angle quoin; coillon; quoin (stone); voussoir (stone); wedge (-shaped) stone;arch stone
楔形石砌隧洞 wedge block tunnel
楔形试块 Y-block
楔形试片 wedge test piece
楔形试验 chill test
楔形手杆 wedge buckle
楔形束 wedge beam
楔形水体 wedge flow
楔形丝 wedge-shaped hair
楔形丝筛面 wedge-wire deck
楔形丝筛网 wedge-wire screen
楔形丝织筛布 wedge-wire cloth
楔形丝织网 wedge-wire mesh
楔形松弛函数 wedge relaxation function
楔形松弛谱 wedge-shaped relaxation spectrum
楔形榫 culvertail; wheelwright's tenon;tapered tenon
楔形榫槽接合 tongue-and-lip joint
楔形榫键 dovetail feather joint
楔形榫接合 joint dovetail
楔形榫结合 cocking
楔形榫头 dovetail;tapered tenon
楔形踏步 wedge-shaped tread
楔形掏槽 V-cut; wedge cut; wedge-shaped cut;wedging cut
楔形掏槽爆破 wedge shot; wedging shot
楔形掏槽孔 wedge cutter
楔形梯级 step wedge
楔形体 wedge
楔形体储存 wedge storage
楔形体堆垛 wedge stacking
楔形体贮存 wedge storage
楔形天线 wedge antenna
楔形填隙料 wedge filler
楔形填隙片 dutchman; shim; tapered shim
楔形条干化床 wedge-wire drying bed
楔形条色谱 wedge-strip chromatography
楔形条网 wedge-wire screen
楔形条状筛板 wedge-bar screen deck
楔形通海阀 wedge gate flooding valve
楔形透镜聚光镜 wedge lens condenser
楔形凸凹榫接合 tongue-and-lip joint
楔形图 wedge pattern
楔形土地 gusset;wedged-shaped lot
楔形脱轨器 scotch block derail
楔形网过滤器 wedge-shaped dust filter
楔形物 wedge
楔形物激波 wedge shock
楔形误差 prismatic(al) error; wedge error
楔形吸收器 wedge absorber
楔形系杆 wedge buckle
楔形系统 <锚具或夹具> collet system
楔形隙 wedge-shaped gap
楔形隙避雷器 wedge-shaped lightning arrester
楔形狭槽 wedged slot
楔形线夹 wedge clamp
楔形线脚 <诺曼底式建筑特征> nail-head(ed) mo(u)lding
楔形线圈 wedge-shaped coil
楔形销钉 wedge stud
楔形小木条 centering slips
楔形斜层理 wedge-shaped cross-bedding
楔形斜接缝 oblique scarf with wedge
楔形斜嵌 oblique scarf with wedge
楔形斜嵌接 oblique-angled scarf(joint) with wedge
楔形泄水阀 wedge sluice valve
楔形芯头泥芯 wing core
楔形蓄水体 wedge storage
楔形压 wedge pressure
楔形压板 wedge clamp
楔形压焊 wedge bond(ing)
楔形压力联结装置 wedge compression coupler
楔形牙尖 wedge-shaped cusp
楔形岩芯提断器 slip-type core lifter; wedge core lifter
楔形验孔规 differential hole ga(u)ge
楔形堰 wedge-type weir
楔形堰板 wedge weir panel
楔形翼型 wedge aerofoil
楔形翼缘梁 tapered-flange beam
楔形阴沟砖 wedge sewer brick
楔形油膜 wedge-shaped oil film
楔形云母 wedge-shaped mica
楔形栽植 wedge planting
楔形轧制 wedge rolling
楔形闸门 wedge gate
楔形闸(门)阀 wedge action valve; wedge(sluice) valve;wedge(-type) gate valve
楔形遮光栅照明窗 parawedge louver
楔形整合【地】sphenoconformity
楔形支持物 quoin
楔形纸 wedged-shaped paper
楔形制动器 key-operated brake; V-block brake;wedge brake
楔形制销 wedge cotter
楔形柱 tapered column; wedge col-

umn;wedge post
楔形砖 arch brick;brick for wedge use;compass brick;crown brick;dome brick;feather-edged brick;king closer;radial brick;radiating brick;radius brick;voussoir;wedge (-shaped) brick;wedge (-shaped) tile;voussoir brick
楔形砖建烟囱法 custodis
楔形桩 wedge pile;cuneiform pile
楔形装甲 wedge-shaped armor
楔形装置法 wedge method of attachment
楔形钻杆夹持器 knife-dog
楔形钻头 wedge bit;wedging bit
楔形座 wedge mount
楔叶类植物 sphenophsida
楔住 chocking-up;choke;keying action;lashing
楔砖 key brick
楔桩 Coignet pile
楔装曲柄 keyed-on crank
楔状的 cuneiform;sphenoid;wedge-shaped
楔状地块【地】wedge-shaped block
楔状渐伐作业法 wedge system
楔状礁 wedge-form reef
楔状密封片 wedge seal
楔状伞伐作业 shelter-wood wedge system
楔状扇形块 wedge segment
楔状铁栓 fid
楔状隙 embrasure
楔状岩层 wedge bed
楔状岩体 wedge-shaped rock body
楔状岩芯自卡 wedging
楔子 wedge;block;chock;cleat;packing;quoin;spike;trig
楔子劈入面 cleat plane
楔子强度 cleat strength
楔座 master sheet;wedge block

歇

歇工 knock off;stop work
歇光 occult light
歇后降低 <桩的承载力> relaxation
歇后增长 set-up;freeze <桩承载力的>
歇火 burn-out;stop firing
歇脚板 foot rest
歇脚地 stepped stone;stepping stone
歇山 gable and hip roof
歇山(式)屋顶 hip and gable roof;shread-head roof
歇山屋顶山墙 hipped gable
歇山屋面 clipped gable
歇市 close market
歇斯底里性格 hysterical personality
歇业 close a business;close door;go out of business
歇业者 retiree
歇振频率 quench frequency
歇振调制射频 quench-modulated radiofrequency

蝎

蝎毒 scorpion venom
蝎毒液中毒 scorpion venom poisoning
蝎螫中毒 scorpionism
蝎尾形结构 scorpion
蝎蜇伤 scorpion sting
蝎蜇中毒 scorpion sting poisoning
蝎子 scorpion

协

协编单位 assistant compiled unit
协编人 assistant editor
协变(差) covariance
协变导数 covariant derivative
协变分量 covariant component
协变计算 covariant calculation
协变矩阵 covariance matrix
协变量 covariant
协变式的 covariant
协变式方程 covariant equation
协变微分法 covariant differentiation
协变性 covariability;covariance;covariation
协变性函数 covariance function
协变性原理 principle of covariance
协变张量 covariant tensor
协变指标 covariant index
协不变量 co-invariant
协步变换 cogredient transformation
协订单一税则 convention single tariff
协定 agreement;concordat;deed;pact;treaty;convention
协定保险价额 agreed value
协定草案 draft convention
协定的损害赔偿金 liquidated damages
协定费率 tariff rate
协定关税 agreed tariff;conventional customs duties;conventional tariff;fixed tariff duties
协定国家 Countries under the Agreement
协定价格 agreement price;conventional price;price agreed upon;stipulated price;striking price
协定价值 <承运人与托运人互相协定的货物价值> agreed valuation
协定结束 termination of agreement
协定利率 conventional interest rate;conventional money rate
协定内容 treaty contents
协定年度 agreement year
协定配额 bilateral quota
协定期满 termination of agreement
协定全损 compromised total loss
协定日期 date of agreement
协定书 letter of agreement;memorandum of agreement
协定税额 conventional tax
协定税率 conventional tariff
协定税则 agreement tariff;conventional tariff
协定弹限 conventional elastic limit
协定条件 treaty condition
协定条款 treaty articles
协定文本 treaty wording
协定向量 covariant vector
协定运费率 agreed rate
协定运价率 <按货物运量的增加而减低价率> agreed rate
协定摘要 treaty particulars
协动潮 cooperating tide
协方差 covariance
协方差比例 covariance proportion
协方差参数 covariance parameter
协方差的效率 efficiency of covariance
协方差反应 covariance response
协方差分析 analysis of covariance;covariance analysis
协方差分析表 analysis of covariance table
协方差分析模型 analysis of covariance model
协方差函数 covariance function
协方差矩阵 covariance matrix
协方差律 covariance law
协方差模型 covariance model
协方差平稳过程 covariance stationary process
协方差原理 principle of covariance
协方差组分 components of covariance
协合剂 synergist
协合曲线 synergistic curve
协合效应 synergistic effect
协合效应曲线 synergistic effect curve
协合作用 synergism;synergistic action
协和 consonance
协和汇流 accordant junction
协和界面胶结物结构 compromise boundary cement texture
协和面 compromise faces
协和山脉 harmonic mountain range
协和式飞机 Concorde plane
协和式客机 Concorde airliner
协和万能绘图头 concord universal ruling head
协和褶皱【地】accordant fold
协和振动模型 consistent vibration model
协和指数 synergetic index
协和作用 synergism;synergy
协会 academy;association;institute;institution;organization;society;union
协会罢工险条款 <货物> Institute Strikes Clause
协会保险单 Institute policy
协会保险条款 institute clause
协会本部 association headquarters
协会的 associative
协会航空货物保险条款 Institute Air Cargo Clause
协会货物平安险条款 Institute Cargo Clauses
协会货运保险条款 Institute Cargo Clauses
协会全险条款 all risks;Institute Cargo Clauses
协会危险药品条款 Institute Dangerous Drugs Clause
协会章程 memorandum of an association;articles of association
协会重置条款 Institute Replacement Clauses
协会总部 association headquarters
协力 coaction;collective efforts;combined effort;concerted effort
协力创新法 synectics
协力弹簧 <漏斗车底门> assistance spring
协力作用 synergism
协联 on cam
协联飞逸转速 on-cam runaway speed
协联工况 combination operating condition;on cam operating condition
协联关系 <导叶开度与桨叶角度的> cam relationship;on cam relationship
协联机构 runner blade control mechanism;combinator <转桨式水轮机的>
协联控制 joint control
协联特性曲线 combination curve
协联凸轮 copying cam
协联系统 copying system
协商 agreement;confer;consultation;negotiation
协商标 negotiation tendering
协商采购 negotiated procurement
协商的 give-and-take
协商费用 charge negotiable;negotiation charges
协商管理 consultative management
协商合同 negotiated contract;negotiation contract
协商和解 settlement by agreement
协商会(议) conference;consultative conference;round table
协商机构 machinery for consulsations;machinery of consultation
协商价格 negotiated price
协商阶段 <承包工程合同谈判> negotiation phase
协商解决 compromise settlement;resolved by negotiation;settlement by agreement
协商市场 negotiated market
协商手续费 negotiation commission
协商条款 negotiated terms
协商投标 negotiated tender
协商委员会 consultative committee
协商一致 consensus
协商一致通过 adoption by consensus
协商招标 negotiated bidding
协商者 negotiator
协调 accordance;chime;concordance [concordancy];concordantly;coordination;in synchronism;reconcile;reconciliation;tune
协调比例 scale of ratios
协调标准 harmonized standard
协调不能 incoordination
协调程序 coordinator
协调传动 coordinated drive
协调的 concordant;congruous
协调的调节 coupled control
协调的工资政策 coordinated wage policy
协调发展 coordinated development
协调方程(式) equation of compatibility;compatibility equation
协调防治 integrated control
协调分析 trade-off analysis
协调服务 communication
协调钢束 <在预应力混凝土结构中不产生次反力的钢束布置> concordant cable
协调工作 <建筑施工充分配合各项工种,抢时间开工> free float
协调功能 coordinating function;mediation function
协调功能测试器 ataxiameter
协调关系 rapport
协调规划 coordinate(d) planning
协调过程 process of coordination
协调荷载 matched load
协调滑移 cooperative slip
协调机构 coordinating body;coordinator
协调机理 concerted mechanism
协调机能 coordination function
协调基准面 coordinating plane
协调计划 <资源计划之一> liaison plan
协调计算【计】coordinated computing
协调距离 coordinated distance;coordination distance
协调决策 coordinated decision
协调控制系统 coordinated control system
协调联运 intermodal transport
协调联运公路 intermodal road
协调联运技术 intermodal technology
协调流量 concordant flow;coordinated flow
协调扭转 compatibility torsion
协调器 synchronizer;coordinator <双扇门关闭的>
协调人 coordinator
协调设备 mediation device
协调生境 harmonic habitat
协调世界时 coordinated universal time
协调世界时系统 universal time coordinated system
协调式 coordination form
协调水系 accordant drainage
协调条件 compatibility condition
协调委员会 coordinate commission;coordination committee

协调系数 cooperation index
协调小组 conciliation panel; coordinating group
协调信贷 blending credit
协调性 compatibility; compatibleness; harmony
协调性条件 compatibility condition
协调样本 concordant sample
协调一致 reconcilement
协调应用功能 mediation application function
协调元 conforming element; consistent element
协调员 coordinator
协调运动 coordinated movement; coordinate exercise
协调运行的交叉口(体)系 chain of coordinated crossings
协调者 coordinator
协调褶皱 harmonic fold
协调指标 coordinate index
协调中心 coordinating center [centre]; coordination center[centre]
协同操作 cooperating
协同沉淀抗体 coprecipitating antibody
协同程序 coroutine
协同处理 coprocessing
协同处理机 coprocessor
协同处理器 coprocessor
协同萃取 synergistic extraction
协同存取数组 coordinate access array
协同的 synergistic
协同(调)运行交叉口 cooperative intersection
协同动作 concerted action; team work
协同毒性效应 synergistic toxic effect
协同工作 team work
协同共栖 synergism
协同驾驶 cooperative driving
协同检索 coordinate retrieval
协同降解 synergistic degradation
协同克立格法 intermodal Cokriging method
协同例程 coroutine
协同联运 intermodal transportation
协同凝集 coagglutination
协同区域化变量 coregionalized variables
协同式管理制度 cooperative-type management
协同收缩 cocontraction
协同索引 cooperation index
协同通信[讯]联络 cooperative communication
协同系数 coefficient of concordance
协同效应 synergetic effect; synergistic effect
协同型开发环境 convert development environment
协同学 synergetics
协同演化说 coordinated evolutionalism
协同一贯联运 intermodal transportation
协同指数 synergistic index
协同作业 team play
协同作用 synergism; synergistic action; synergy
协相谱 co-spectrum
协议 agreement; negotiation
协议板 agreed-dimension slab
协议表格 agreement form
协议草案 draft form of agreement
协议地区 agreement area
协议范围 extent of agreement; scope of agreement
协议格式 agreement form
协议工程 protocol engineering
协议公司 agreement corporation
协议关税 bargaining tariff
协议管理 consultative management
协议规格 model agreement
协议货物 conference cargo
协议价格 agreed-upon price; agreement price; negotiated price
协议解决 settlement by agreement
协议力量 bargaining power
协议利益 benefit of agreement
协议面积 agreement area
协议年度 agreement year
协议配额 negotiated quota
协议期满日 terminal of an agreement
协议签订生效日期 date of agreement
协议区域 agreed territory
协议全损 arranged total loss; compromised total loss
协议日期 date of agreement
协议生效 agreement in force
协议石料 agreed-dimension stone
协议书 agreement form; letter of agreement; written agreement; protocol
协议书表格 form of agreement
协议书的语言 language of the agreement
协议书格式 form of agreement
协议书规定的货币 currency of agreement
协议书号码 agreement number
协议书生效 agreement effective
协议水费 agreement water rate; lease water rate
协议条款 agreement form; provisions of the agreement; terms of an agreement
协议条文 articles of agreement
协议细节 detail of treaty
协议效力 force of agreement
协议有效期(限) duration of agreement
协议与补偿 accord and satisfaction
协议运费率 conference freight rate
协议运价 arranged rate
协议栈 protocol stack
协议证书 deed of settlement
协议中有关开脱责任的条款 exculpatory clause
协议终止(日)期 terminal of an agreement
协议重量 agreed weight
协议装货日期 stemming date
协约法 conventional law
协张量 co-tensor
协振潮 cooperating tide; co-oscillating tide
协振分潮 co-oscillating partial tide
协助 assistance; help; lend a hand; take part with
协助保管设备档案 aiding in keeping equipment records
协助查明当地规章 assistance with local regulation
协助电解物 indifferent electrolyte
协助法 <影响他人改变行为> collaborative method
协助计划材料流程 aiding planning materials flow
协助计划公共设施 aiding planning for utilities
协助清理债务人 debt-consolidators; debt-pooler
协助推销人 dealer aids
协助作用 synergism
协作 collaboration; cooperation; coordination; team work
协作部门 collaboration organization
协作出版 coordinated publishing
协作单位 collaborator
协作的 synerg(et)ic; synergistic
协作的成就 team-work achievement
协作方案 team-work project
协作关系 cooperative relationship
协作规划 team-work project
协作合同 contract of cooperation
协作集体 co-acting group
协作计划 cooperation plan
协作剂 synergist
协作价格 coordinated price
协作件检验 inspection of contracted-out parts
协作建筑师 associate architect; coordinator architect
协作贸易 cooperative international trade
协作排水 cooperative drainage
协作器 synergist
协作设计 team-work project
协作试验 coordinated trial
协作试验研究 interlaboratory study
协作网络 synergetic network
协作委员会 cooperative committee; coordinate commission
协作系统 coordinated system
协作型作风 collaborating style
协作研究 team study
协作研究方案 cooperative research program(me)
协作运输 coordinated transportation
协作者 coadjutor; collaborator; team-worker
协作指数 cooperation coefficient
协作作用 synergism

胁

胁变 strain
胁变观察器 colmascope; strain viewer
胁从犯 coerced or induced offender
胁迫法 <影响他人改变行为> coercive method
胁强 stress
胁强计 stressometer

挟

挟 pinch
挟冰水流 ice-bearing current
挟持能力 carrying power
挟带 entrainment
挟带大量泥沙的 sediment-laden
挟带大量泥沙的河流 heavy burdened river; heavy burdened stream
挟带的泥沙 entrained sediment
挟带力 dragging force; entrainment force
挟带能力 carrying capacity; carrying power
挟带泥沙 sediment entrainment
挟带泥沙的河流 burdened river; burdened stream
挟带泥沙的水流 burdened stream
挟带泥沙河 competent river
挟带泥沙能力 sediment carrying power; sediment-carrying capacity
挟角 angle of bite
挟力 pinching force
挟漂浮物河流 debris-laden stream
挟气 entrapped air
挟气量 entrained gas content
挟沙 sanding; silt-carrying
挟沙潮流 sediment-laden tidal flow
挟沙风 sand-bearing wind; sand-driving wind
挟沙港道 sediment-laden stream
挟沙河口 sediment-carrying estuary
挟沙河流 competent river; competent stream; loaded river; loaded stream; sediment-bearing stream; sediment-carrying river; sediment-laden river; sediment-laden stream; silt-carrying river
挟沙量 sand-carrying capacity; sediment-carrying capacity
挟沙能力 sand-carrying capacity; sediment-carrying capacity; sediment transport capacity; silt carrying capacity
挟沙水 drift water; sediment-laden water; water-sediment complex; water-sediment mixture
挟沙水流 debris-laden flow; sediment fluid flow; sediment-laden flow; silt-carrying flow; silt-laden flow
挟沙水流动黏[粘]滞度 kinematic(al) viscosity for sediment-laden flow
挟水沙层 water-bearing sand; water-holding sand
挟屑冰河流 frazil river; frazil stream
挟岩屑河流 debris-laden stream
挟运能力 carrying capacity; carrying power; sediment transport capacity; sediment transport competence [competency]
挟者 nipper
挟子 tweezers

偕

偕胺肟 amidoxime
偕二甲基 gem-dimethyl
偕行人数 accompanying number

斜

斜 T 形管 sweep tee
斜埃尔米特型 skew Hermite form
斜岸 glacis; sloping bank
斜岸墩 diagonal abutment pier
斜拔 oblique pull
斜拔力 oblique pull
斜白霞斑岩 plagifoyaitarkite
斜斑粗安岩【地】vulsinite
斜板 gradient plate; inclined plate; pitch plate; skew(ed) slab; skew plate; sloping panel
斜板沉淀 sloping plank-settling; sloping plate settling
斜板沉淀池 inclined board type precipitation tank; inclined plate settler; plate clarifier; plate settler; settler clarifier; skew plate settler; sloping plank-settling tank
斜板法 tilting plate method
斜板分离器 tilted-plate separator
斜板浮选法 sloping plate floatation process
斜板隔油沉淀池 oil trap with sloping plank
斜板管沉淀池 sloping plank tube settling tank
斜板脚踏板 splayed baseboard
斜板结构 tilted-slab structure
斜板曝气器 inclined plane aerator
斜板桥 skew(ed) slab bridge
斜板穹隆 tilted-slab dome
斜板扇形体 tilted-slab segment
斜板式沉淀池 inclined plank settler; inclined plank settling tank
斜板式分离器 tilted-plate separator
斜板式隔油池 inclined plank isolating-oil pool
斜板式瞄准具 ramp sight; ramp-type sight
斜板填土机 angle blade; angle filler
斜板条 diagonal strip
斜板凸轮 swash plate

斜板推土机 angledozer; angling dozer; dozer with angling blade; grade builder; trail builder
斜板屋顶 tilted-slab roof
斜板箱 tilting box
斜板小圆屋顶 tilted-slab cupola
斜板桩 batter sheet piling
斜半砖 miter[mitre] half
斜背式车身 fast-back
斜背堰 inclined weir
斜铋钯矿 froodite
斜壁 diagonal siding; skew wall
斜壁板 bevel(led) side[siding]
斜壁沟 ca(u)nch
斜壁谷 <美> coulee
斜壁通道 slant-wall duct
斜臂起重机 angle crane
斜臂式挖沟机 inclined boom type trench excavator
斜边 arrissed edge; bevel siding; chamfered edge; mitre[miter]; sloped edge; snipped edge; steep bevel; stop scallop's chamfered edge; hypotenuse <直角三角形的>
斜边板条 chamfer strip
斜边玻璃 bevel(led) glass
斜边玻璃门 bevel(led) glass door
斜边彩画玻璃 <铅条镶的> leaded beveled plate
斜边差 side set
斜边锉 taper file
斜边底层 splayed ground
斜边肋骨 bevel frame
斜边门 bevel(led) door
斜边抛光机 bevel polishing machine
斜边刨床 plate edge bevel(l)ing machine
斜边墙 bevel wall
斜边切割 bevel cut(ting)
斜边凿 bevel chisel
斜边砖 bevel brick; bevel tile; edge skew
斜边装饰 raking cornice
斜扁栓 tapered gib
斜扁销 tightening key
斜扁销调整楔 tightening key
斜变换 slant transformation
斜变截面梁 inclined haunched beam
斜标尺 inclined staff ga(u)ge; slope ga(u)ge
斜表面 skewed surface
斜柄刷 oblique brush
斜玻璃窗 inclined glazing
斜玻璃幕墙 inclined glass curtain wall
斜薄板 <木板墙> angle board
斜裁布抛光轮 bias buff
斜裁机 bias cutter
斜舱壁 sloping bulkhead
斜操纵台 bench board
斜槽 bevel groove; chute; flume(chute); inclined shoot; inclined trough chute; lander; oblique slot; skewed slot; sloping platform; sloping trough; spout; tilted trough; slanting chute <泄水的>
斜槽板 chute board
斜槽插接 bevel(l)ed housing
斜槽出料 chute(d) discharge
斜槽磁极 obliquely slotted pole
斜槽跌水 inclined drop
斜槽废水 flume wastewater
斜槽沟 skew slot
斜槽进料焚化炉 chute-fed incinerator
斜槽进料器 chute feeder
斜槽进料装置 chute feed
斜槽距 sloping slot pitch
斜槽口 oblique notch; skew notch
斜槽溜混凝土 chuting concrete
斜槽漏磁通 skew leakage flux
斜槽炉排 chute grate
斜槽刨 skew rebate plane; spout plane
斜槽上料器 chute feeder
斜槽式电动机 obliquely slotted motor
斜槽式筏道 chute raft
斜槽式溢洪道 chute spillway
斜槽式溢流道 chute spillway
斜槽式鱼道 Denil fish pass; diagonal baulk fish pass; graded-channel fish pass; skew fish pass
斜槽式重力进料 gravity chute feed
斜槽输送 chuting
斜槽鼠笼式转子 skewed slot squirrel cage rotor
斜槽眼 bevel(l)ed housing
斜槽运料 chuting
斜槽运料设备 chuting plant
斜槽运料系统 chuting system
斜槽运输 chuting
斜槽运输设备 chuting plant
斜槽运输系统 chuting system
斜槽装料系统 chuting system
斜侧接 oblique side grafting
斜侧面 prism
斜侧面镜头 angle shot
斜侧墙 taper side
斜侧照明 oblique lighting
斜侧支撑 diagonal side bracing
斜层 oblique bedding
斜层顶天窗 sloping roof skylight
斜层介质 obliquely layered medium
斜层理【地】oblique bedding; inclined bedding
斜层理构造 diagonal bedding structure
斜层砌合 sloping bond
斜叉 T 形管节 skew tee
斜叉钉法 toed nailing
斜叉钉法的 toe-nailed
斜叉管接头 breeching fitting
斜叉三通 skew tee; Y-pipe
斜叉三通管 pipe branch
斜叉四通 double Y
斜叉直线 skew line
斜插 oblique cutting
斜插板阀 inclined damper; inclined slide valve
斜插俯冲 oblique subduction
斜茬 racking; racking back
斜茬施工缝 ranking work joint
斜铲 <推土机的> tilt blade
斜铲犁雪机 angling plough
斜铲式平地机 angle blade scraper
斜铲(式)推土机 angled blade bulldozer; angling blade (bull) dozer; angledozer; angling dozer; side dozer; tilt dozer; three-side(d)(bull) dozer; tilting dozer
斜铲作业 angle-dozing
斜长斑岩 plagiophyre
斜长粗面岩 plagiotrachyte
斜长二辉麻粒岩 piriklazite
斜长方格地板 diagonal rectangular grid floor
斜长花岗斑岩 plagiogranite porphyry
斜长花岗岩 plagioclase granite; plagiogranite
斜长花岗岩类 pragiogranite group
斜长煌斑岩 spessartite
斜长辉长岩 plagioclase gabbro
斜长角闪片岩 plagioclase hornblende schist
斜长角闪岩 amphibolite; plagioclase amphibolite
斜长角闪岩相组 plagioclase amphibolite facies group
斜长片麻岩 plagioclase gneiss
斜长石 anorthose; plagioclase feldspar
斜长石砂碎屑岩 plagioclase-arenite
斜长石砂岩 plagioclase arkose
斜长碎屑岩 plagioclastic-arenite
斜长细晶岩 plagiaplite
斜长岩 anorthosite; plagioclase rock; plagioclasite
斜长英安岩 plagiodacite
斜场管 inclined field tube
斜衬线 oblique serif
斜撑 arm tie; back stay; batter brace; batter post; brace strut; bracing piece; bracing strut; brail; corner brace; corner bracing; crippling; diagonal bar; diagonal bracing; diagonal bridging; diagonal member; diagonal strut; inclined piece; inclined shore; inclined strut; knee brace; knee bracing; needle shoring; outrigger; raker; raking element; raking prop; rider; riding shore; slanted strut; sprag; strut bracing; strut leg; strutting piece; backshore
斜撑臂 diagonal brace
斜撑长度计算表 brace table
斜撑长度刻度表 brace scale
斜撑的贴墙横木 bottom shore
斜撑的支座 sole plate
斜撑底部垫块 solepiece
斜撑垫底板 sole piece
斜撑顶杆 tilt brace
斜撑帆杆 sprit(sail)
斜撑腹板 braced web
斜撑杆 <俗称八字架> brail; diagonal stay; strut tenon; sway rod; diagonal bar; diagonal member; diagonal brace
斜撑构成门 framed and braced door
斜撑桁架 scissors truss
斜撑加强支架 reinforced stull
斜撑角 brace angle
斜撑脚 rafter foot
斜撑脚垫 sole piece
斜撑节段 braced panel
斜撑靠船墩 raker dolphin
斜撑块 sway brace; thrust block
斜撑框架 brace(d) frame[framing]; full framing
斜撑梁 braced beam; brail
斜撑梁腹 braced web
斜撑梁桥 strutted beam bridge
斜撑帽 rider cap
斜撑木 raking timber; rance; rider shore; raking shore
斜撑起来 sheer up
斜撑墙 diagonal buttress
斜撑式单向离合器 sprag type one way clutch
斜撑式离合器 sprag clutch
斜撑式满布木拱架 full-span wooden inclined strut centering[centring]
斜撑式起重机 derrick crane
斜撑式桅杆起重机 rigid braced derrick crane
斜撑式卧倒门 strutted flap gate
斜撑条 diagonal stay
斜撑头 diagonal member head
斜撑脱断 kickout
斜撑桅杆起重机 stiffleg
斜撑屋架 knee-brace roof
斜撑系船柱 raker dolphin
斜撑系统 shoring
斜撑型钻塔 sway braced derrick
斜撑意外松动 kickout
斜撑闸板坝 bear-trap dam
斜撑闸板门 bear-trap gate
斜撑闸板堰 bear-trap weir
斜撑闸筏道 bear-trap drift-chute
斜撑支柱 breast timber
斜撑柱 rider shore; shore; spur shore
斜撑自卸车 raker dump car
斜撑座 brace bracket
斜承石板 skew table
斜承重椽 valley jack rafter
斜橙黄石 plagiocitrite
斜齿 helical tooth; oblique tooth; skewed gear tooth
斜齿轨 inclined rack
斜齿筘 angle reed
斜齿离合器 bevel(l)ed claw-clutch
斜齿轮 angular gear; bevel(led) gear; bevel wheel; helical gear; helical spur; miter gear; oblique tooth gear; rocket wheel; skew gear; skew wheel; spiral gear; spiral wheel
斜齿轮传动 bevel gear drive
斜齿轮传动灯泡式水轮机 bevel geared bulb turbine
斜齿轮起重器 bevel(l)ed gear jack
斜齿轮式泵 helical gear pump; spiral gear pump
斜齿轮闸门启闭机 bevel gear gate lifting device
斜齿轮主传动(机构) bevel gear main drive
斜齿伞齿轮 spiral bevel gear
斜齿伞形齿轮传动 spiral bevel gear drive
斜齿式分级机 rake classifier
斜齿条 screw rack
斜齿纹 diagonal pitch
斜齿型 plagiodont
斜齿锥齿轮 skew bevel gear
斜齿锥齿轮付 skew bevel gear pair
斜冲击波 oblique shock wave
斜触 <船与码头> glancing blow
斜穿过 athwart
斜传动齿轮 bevel wheel
斜船首 raked stem; raking stem
斜船首柱底部曲率半径 rake radius
斜船尾 raked stern
斜窗 batement light; oblique window
斜吹转炉 Kaldo furnace
斜唇形入口 bevel(l)ed lip entrance
斜磁黄铁矿 clinopyrrhotite
斜磁流体冲击波 oblique hydromagnetic shock wave
斜错缝满条砌法 raking stretcher bond
斜搭接 scarfing joint
斜挡板 bevel siding
斜挡风玻璃 sloping windshield
斜挡圈 separate thrust collar
斜挡水板 raking flashing(piece)
斜挡土墙 skew retaining wall
斜挡栅 oblique boom
斜导程 bevel lead
斜导电壁发电机 diagonal conducting wall generator
斜导杆 inclined guide
斜导轨 slant rail
斜导架(式)打桩机 batter leader pile driver
斜导线导轮 angle pulley
斜道 chute; rampway
斜的 askew; bias; oblique; off-angle; sideway; sidewise; skewed
斜的楼梯背面 sloping soffit of stair(case)
斜的木垫 <教堂跪拜用> kneeling board
斜的剖面线 hatching
斜的掏槽炮眼 angling snubbing hole
斜的突缘 oblique nosing
斜等轴测图 cavalier drawing
斜低压槽【气】titled trough
斜堤 glacis; esplanade <外岸的>
斜底 heavy tap; heel tap; sloping bed; sloping bottom
斜底板 sloping panel
斜底仓 slanting-bottom bin

斜底层 splayed ground
斜底车 inclined bottom car
斜底储存器 sloping bottom bin
斜底船坞 slip dock
斜底的 slant-bottom scoop
斜底罐 sloping bottom tank
斜底加热炉 roll-over type furnace
斜底料仓 slanting-bottom bin
斜底料斗 sloping bottom bin
斜底漏卸车 inclined bottom car
斜底面板 <漏斗车的> slope sheet
斜底式炉 longitudinal sloping hearth furnace;roll down furnace
斜底座 sloping bottom
斜地 sidelong ground
斜地盘 sloping site
斜地坪 sloping floor
斜碲钯矿 telluropalladinite
斜点阵 oblique lattice
斜垫 taper liner
斜垫板 canted bearing plate; wedge plate
斜垫块 angle block; angle washer; flare block;inclinator
斜垫圈 angle washer; bevel(led) washer;tapered washer
斜垫轴承 tilting bearing; tilting pad bearing
斜垫自位轴承 tilting bearing
斜吊斗 engine plane
斜吊杆 inclined hanger
斜吊杆式挖沟机 slanting boom type trenching machine
斜吊杆悬索桥 suspension bridge with inclined suspenders
斜吊索 angulated roping;cable stay; inclined hanger cable
斜叠键嵌接 oblique tabled scarf with key
斜叠嵌接 oblique tabled scarf
斜钉 concealed nail(ing);skew nail; toe(d) nail;tusk nail
斜钉材料 skew nailed material
斜钉钉子 toe nail
斜钉法 skew nailing; tusk nailing; toed nailing
斜钉拼板(法) edge toe nailing
斜顶 wagon top
斜顶板 sloping top panel
斜顶板铺板 sloped bulkhead plating
斜顶板纵骨 sloped bulkhead stiffener
斜顶撑 <隧洞> batter brace
斜顶格栅 bevel(l)ed joist
斜顶拱顶 wagon-headed vault
斜顶锅炉 wagon-top boiler
斜顶炉 sloping roof furnace
斜顶式管束单元 inclined tube bundle unit
斜顶踢脚板 splayed baseboard;splayed skirting
斜顶形老虎窗 wagon-headed dormer
斜顶形天花板 wagon-headed ceiling
斜顶形屋顶 wagon roof
斜顶堰 weir with inclined crest
斜动小车 skewed trolley;skewing trolley
斜洞 inclined adit
斜洞门 skew portal
斜斗输送机 bucket inclined conveyer [conveyor]
斜度 angularity; crab; falling gradient; inclination; incline batter; obliquity; pendence; skewness; slant angle;slope;taper
斜度变化 slope deviation
斜度测定 slopes test
斜度测量 measurement of gradient; measurement of tilt;slope level(l)ing
斜度尺 adjustable triangle
斜度分压器 bearing potentiometer
斜度改正 slope correction
斜度管 slope indicator tube
斜度规 slope ga(u)ge;tilt ga(u)ge
斜度过载噪声 slope overload noise
斜度基点压缩 slope-keypoint compaction
斜度计 gradienter[gradiometer];tilt-(o)meter
斜度校正 slope correction
斜度卷尺 slope tape
斜度量规 sloping ga(u)ge
斜度灵敏度 slope sensitivity
斜度磨光装置 bevel grinding attachment
斜度偏差 slope deviation
斜度曲线 slope curve
斜度调整 slope control;slope rectification
斜度调整杆 tilting lever
斜度仪 clinometer; gradienter; grading instrument;slope meter;tilt(o)meter
斜度增益 slope gain
斜度转折点(角)slope discontinuity
斜端 splay end
斜端墙 taper end
斜端砖 end skew brick
斜断 oblique fracture
斜断层【地】oblique fault;semi-longitudinal fault;semi-transverse fault; diagonal fault
斜断面 oblique section
斜堆 diagonal stacking
斜堆法 inclined stacking
斜对称 skew symmetry
斜对称变换 skew-symmetric(al) transformation
斜对称并矢式 skew symmetric(al) dyadic
斜对称的 anti-symmetric(al);skew-symmetric(al)
斜对称荷载 obliquely symmetric(al) load
斜对称矩阵 skew matrix;skew-symmetric(al) matrix
斜对称行列式 skew determinant
斜对称性 anti-symmetry
斜对称张量 skew-symmetric(al) tensor
斜对焊 oblique butt weld
斜对接 bevel(l)ed halving; oblique butt joint;oblique halving
斜对接缝焊 slanting butt seam welding
斜对接接头 scarf butt joint
斜对切 bevelment
斜二轴测图 cabinet drawing
斜发沸石 clinoptilolite
斜发送 emphasis transmission; slope transmission
斜阀 pitch valve
斜法 slant method
斜帆 angulated sail;fore-and-aft sail
斜钒铋矿 clinobisvanite
斜钒铅矿 chervetite
斜反射 oblique reflection
斜反射照明法 oblique reflected lighting
斜方白云石 rhombic(al) spar
斜方板 bevel board
斜方板晶石 epididymite
斜方钡锰闪叶石 orthoericssonite
斜方(单)锥 rhombic(al) pyramid
斜方的 trimetric
斜方底面 rhombic(al) base
斜方碲钴矿 mattagamite
斜方碲金矿 krennerite
斜方碲铅石 plumbotellurite
斜方碲铁矿 erohbergite;frohbergite
斜方短柱 rhombic(al) branchy-prism
斜方短锥 rhombic(al) branchy-pyramid
斜方对称 orthorhombic(al) symmetry;rhombic(al) symmetry
斜方鲕绿泥石 orthochamosite
斜方矾石 felsobanyaite
斜方钙沸石 gismondite
斜方格钢板 raised diamond checker plate
斜方汞银矿 paraschacherite
斜方硅钡钛石 barrio-orthojoaquinite
斜方硅钙石 felite;kilckoanite
斜方硅灰石 belith
斜方硅钠钡钛石 orthojoaquinite
斜方硅钠锶钡钛石 strontio-orthojoaquinite
斜方辉橄岩 harzburgite
斜方辉铅铋矿 cosalite
斜方辉砷钴矿 orthocobaltite
斜方辉石 enstenite;orthorhombic(al) pyroxene
斜方辉石橄榄岩 harzburgite
斜方辉石花岗岩 orthopyroxene granite
斜方辉石类 enstenite;orthopyroxene
斜方辉石岩 orthopyroxenite
斜方辉锑铅矿 meneghinite
斜方回纹饰 diamond fret
斜方角闪石 authophyllite
斜方金铜矿 auricupride
斜方晶 prismatic(al) crystal
斜方晶的 orthorhombic;rhombic(al)
斜方晶结构 orthorhombic(al) structure
斜方晶体 rhomboidal
斜方晶系【地】rhombic(al) system; trimetric system; orthorhombic crystal system
斜方晶系排列 <金属的> orthorhombic(al) packing
斜方孔格栅 diagonal square grid
斜方块砌体防波堤 sliced blockwork breakwater
斜方块织物 diamond cloth; diamond fabric
斜方蓝辉铜矿 anilite
斜方硫 rhombic(al) sulfur; rhombic(al) sulphur
斜方硫铋铅矿 bonchevite
斜方硫镍矿 godlevskite
斜方硫锑铅矿 geocronite
斜方硫铁铜矿 haycockite
斜方硫锡矿 ottemannite
斜方六面体 rhomb
斜方铝矾 khademite rostite
斜方氯砷铅矿 heliophyllite
斜方镁黑镁铁锰矿 rhombomagnojacobsite
斜方锰矿 ramsdellite;rasorite
斜方锰顽辉石 donpeacorite
斜方钠钙锆石 ortholovenite
斜方钠锆石 gaidonnayite
斜方钠铌矿 lueshite
斜方硼镁锰矿 orthopinakiolite
斜方硼砂 kernite;rasorite
斜方砌法 reticulated work
斜方铅铋钯矿 polarite
斜方全面体晶系 rhombic(al) holohedral crystal system
斜方砷 arsenolamprite
斜方砷钴矿 safflorite
斜方砷镍矿 rammelsbergite
斜方砷铁矿 loellingite
斜方砷铜矿 paxite
斜方十二面体 rhombododecahedron
斜方双楔 rhombic(al) disphenoid
斜方双锥 rhombic(al) dipyramid
斜方水锰矿 groutite
斜方水硼镁石 perobrazhenskite
斜方四面体 rhombic(al) disphenoid
斜方钛铀矿 orthobrannerite
斜方锑镍矿 nisbite
斜方锑铁矿 seinajokite
斜方锑银矿 dyserasite
斜方体 corpus trapezoideum; trapezoid body
斜方投影图 plan oblique drawing
斜方顽辉石 orthoenstatite
斜方网状构造 rhombic(al) net structure
斜方硒镍矿 kullerudite
斜方硒铜矿 athabascaite
斜方锡钯矿 paolovite
斜方线 linea trapezoidea; trapezoid line
斜方向运动 diagonal crossing
斜方楔 rhombic(al) sphenoid
斜方楔体类 rhombic(al) sphenoidal class
斜方形 diamond shape;rhomb;rhombus[复 rhombuses/rhombi]
斜方形的 rhombic(al);trapezial;trapeziform;trapezoid
斜方形浮雕花饰 diamond fret
斜方形轨道(交)叉点 diamond point
斜方形砖 rhomb brick
斜方柱 rhombic(al) prism
斜方组构 orthorhombic(al) fabric
斜分比例尺 diagonal scale
斜分辨率 <侧视雷达> slant resolution
斜分尺 transverse graduation
斜风 oblique wind
斜风道 dip switch;slope air course
斜峰 skewed peak
斜缝 inclined joint;skew joint;splayed joint(ing); struck joint; weathered joint
斜缝假平顶 false ceiling with slanting joints
斜缝浇筑 <混凝土的> inclined joint placing;inclined joint pouring
斜扶壁 diagonal buttress
斜扶垛 diagonal buttress
斜扶柱 diagonal buttressing pier
斜浮射流 inclined buoyant jet
斜副杆 counter
斜腹板 angle web; inclined web; tapered web
斜腹板 Z 形钢 acute zee
斜腹板箱梁 box girder with sloping exterior webs
斜腹杆 diagonal web member
斜腹杆桁架 <上下弦平行,三角形腹杆系统或并附加竖杆的> Warren truss
斜腹杆桁架桥 Warren truss bridge
斜腹杆三角形桁架 Belgian roof truss
斜腹杆系 diagonal web system
斜钙沸石 wairakite
斜盖瓦 diagonal slating; drop-point slating
斜盖盐箱形房屋 saltbox
斜杆 batter post; braced strut; bracing;bracing strut;diagonal bracing; diagonal rod;strut;sway rod
斜杆阀 valve with inclined stem
斜杆加劲 stiffening by diagonals
斜杆球形阀 inclined stem globe valve
斜杆钻架 drill jib
斜钢箍 inclined stirrup
斜钢筋 diagonal bar;inclined bar;inclined reinforcement;inclined steel; oblique reinforcement;sloping steel
斜杠杆 angle lever
斜高 elevation slope line; inclined height;slant height
斜高测量仪 oblique height finder
斜锆石 baddeleyite;Brazilite

斜锆石砾 zirkite
斜割 miter[mitre] cut
斜割边 feather-edge
斜割法 slanting cut
斜割机 mitring machine
斜割理 oblique cleat
斜割刨 miter[mitre](cutting) plane
斜格 diagrid
斜格锉 double cut file
斜格排 skew grillage girder bridge
斜格排梁桥 angle web; skew grillage girder bridge
斜格式绕组 lattice winding
斜格式线圈 lattice-wound coil
斜格形绕组 lattice winding
斜格形四端网络(架) X-quadripole
斜格栅 inclined screen
斜工作面 slanting face
斜工作面上向采矿法 inclined rill system
斜拱 askew arch; inclined arch; oblique arch; raking arch; skew(ed) arch; sloping arch
斜拱顶 rampant vault; rising vault; screw vault
斜拱角 skew angle
斜拱筒 inclined barrel arch
斜勾缝 weathered joint; weathering pointing; weather(-struck) joint(ing)
斜沟 chute; skewed slot; slope ditch; tapered slot; tapering gutter
斜沟槽 center gutter; valley gutter
斜沟(处的)椽子 valley rafter
斜沟处抬高的椽子 valley jack rafter
斜沟椽 angle rafter
斜沟泛水 valley flashing piece
斜沟木瓦 valley shingle
斜沟黏[粘]土瓦 <屋顶排水的> tapered valley clay roof(ing) tile
斜沟式阶田 graded-channel type terrace
斜沟陶土瓦 roof valley clay tile; valley clay roof(ing) tile
斜沟瓦 valley tile
斜沟小椽条 valley jack; valley jack rafter
斜沟小屋顶 cricket; saddle piece; chimney cricket <烟囱后面排水用的>
斜沟掩埋 sloping-trench landfill
斜构架 diagonal frame; skewed frame
斜谷 diagonal valley
斜鼓形混凝土搅拌机 tilting drum concrete mixer
斜鼓形搅拌机 tilting(drum) mixer
斜刮板 diagonal screed
斜刮板式修整器 diagonal screed finisher
斜刮缝 struck joint
斜刮线脚 struck mo(u)lding
斜挂角度 blade angle
斜挂推土板 angledozer; angling bulldozer
斜挂推土板的推土机 angling blade bulldozer
斜管 diagonal tube; inclined tube; pipe chute; pitch pipe[piping]; pitch tube [tubing]; sloped pipe; sloped tube
斜管沉淀 inclined tube precipitation
斜管沉淀池 inclined tube settler; inclined tube settling tank; sloping pipe settling tank; tube clarifier; tube settler
斜管沉降器 inclined tube settler
斜管澄清池 tube clarifier
斜管风压计 inclined tube ga(u)ge
斜管黏[粘]度计 inclined tube visco(si)meter
斜管气压计 inclined tube ga(u)ge
斜管式测压计 inclined tube manometer; slanting leg manometer
斜管式测压器 inclined tube manometer
斜管式沉淀池 inclined tube precipitation tank
斜管式压力计 inclined tube manometer; slanting leg manometer
斜管微压计 inclined tube ga(u)ge; inclined tube micromanometer; slanting leg manometer
斜管蒸发器 inclined tube evapo(u)rator
斜光(线) skew ray
斜光锥 oblique pencil
斜硅钙石 larnite
斜硅钙铀矿 lambertite
斜硅灰石 belite
斜硅铝铜矿 ajoite
斜硅镁石 clinohumite
斜硅锰石 sonolite
斜硅钠钙石 istisuite
斜硅酸石 larnite
斜轨道 oblique orbit
斜轨上滑行的吊杆 sliding boom on tilting track
斜辊 cross roll; skew table roll; web roll
斜辊道工作台 skew roller table
斜辊横矫直机 cross-roll straightening machine
斜辊制管法 Mannesmann process
斜滚式整平梁 <修整混凝土路面用> diagonal finishing beam; diagonal roller beam
斜涵(洞) skew(ed) culvert
斜焊 bevel weld; inclined weld; skew weld
斜焊接缝 incline weld; skew weld
斜航 loxodrome; rhumb line
斜航法 loxodromics; loxodromy
斜航(海)线 loxodrome
斜号 virgule
斜合帽 mitred cap
斜合水门 mitered[mitred] gate
斜荷载 inclined load; oblique loading)
斜痕 bias
斜桁 gaff; gavelock
斜桁撑杆后支索 backrope
斜桁横帆 lug sail
斜桁架式外筒 diagonal truss tube
斜桁仰角 steeve
斜桁支索 vang
斜桁纵帆 trysail
斜横 crosswise
斜横挡 angle web
斜红磷铁矿 phosphosiderite
斜红铁矾 kornelite
斜弧 oblique arc
斜弧度测量 oblique arc measurement
斜护板 diagonal sheathing
斜护墙 sloping apron
斜护坦 inclined apron; sloping apron
斜滑板 taper slide
斜滑断层 diagonal slip fault; oblique-slip fault
斜划桨 slope oar
斜划(线) shilling-stroke
斜环带 oblique girdle
斜环索线 oblique strophoid
斜煌岩 spessartine; spessartite; kersantite
斜辉铅铋银矿 diaphorite
斜辉石 augite; clinopyroxene
斜击式水轮机 angular impulse turbine; Turgo impulse turbine
斜积 skew product; slant product
斜激波 oblique shock; oblique shock wave
斜极化 inclined polarization; oblique polarization
斜极靴 skewed pole-shoe
斜脊椽角的边棱 backing hip rafter
斜脊梁 mitred hip
斜脊瓦盖 bonnet tile; hip tile
斜脊屋顶 hipped roof
斜脊锥形筒瓦 cone hip tile
斜钾铁矾 yavapaiite
斜架 cant frame
斜架车 cradle equipped with wheels
斜架车滑道 marine railway
斜架车拉索 carriage cable
斜架滑道 slipway and wedged chassis
斜架下水车【船】beaching wedged chassis; launching cradle; sliding cradle
斜架轴承 angle pedestal bearing
斜间距 diagonal pitch
斜肩 bevel(l)ed shoulder
斜剪机 bevel shearing machine
斜剪力 skew shearing force
斜检测 slope detection
斜碱沸石 amicite
斜碱铁矾 clinoungemachite
斜渐近线 slanting asymptote
斜键 oblique key; taper key
斜桨式搅拌器 inclined paddle type agitator
斜交 cornerwise; obliquity; skew crossing
斜交板 skewed plate; skew(ed) slab
斜交不整合 angular unconformity; clino-unconformity
斜交层理【地】diagonal bedding
斜交叉 oblique crossing; skewed crossing
斜交叉道路 oblique crossing; skew crossing
斜交叉管 skew branch
斜交磁化 oblique magnetization
斜交道 skew crossing
斜交道路 diagonal street
斜交的 heterotropic; oblique; skewed
斜交地震反射结构 oblique seismic reflection configuration
斜交渡口 stream crossing in skew angle
斜交反铁磁性 canted antiferromagnetism
斜交方格 diagonal square grid
斜交方形花纹 ballerina check
斜交格构体系 diagonal-grid system; diagrid system
斜交拱 oblique arch; skew arch
斜交拱桥 skew arch bridge
斜交轨迹 oblique trajectories
斜交涵洞 skew(ed) culvert
斜交激震波面 oblique shock front
斜交激震波前 oblique shock front
斜交交叉口 oblique intersection; skew intersection
斜交角(度) angle of skew; skew angle; skew bevel
斜交角焊缝 oblique fillet weld
斜交接管 oblique nozzle
斜交接合 skew joint
斜交截面 skewed section
斜交镜 inclined mirror
斜交矿脉 pee; slant vein
斜交肋 diagonal rib
斜交帘布层轮胎 bias ply tire
斜交裂隙 diagonal fissure
斜交轮胎 cross-ply
斜交木纹 interlocked grain; interlocking grain
斜交平交道 skewed crossing
斜交前积模式 oblique progradational mode
斜交浅滩 reticulated bar
斜交墙 cant(ed) wall
斜交桥 oblique bridge; skew(ed) bridge
斜交桥板 skew bridge slab
斜交桥跨度 skew span
斜交球 oblique sphere
斜交三通管接 skew T
斜交伞齿轮 angular bevel gear
斜交升交点赤经 oblique ascension
斜交石 <支持拱脚的> kneeler; kneestone; skew table
斜交水道 stream crossing in skew angle
斜交条纹 diagonal bars tread
斜交网格型楼板 diagrid floor
斜交纹理 oblique lamination
斜交轴 oblique axis
斜交轴伞齿轮 angle gear; angular gear
斜交轴锥齿轮 angular bevel gear
斜交坐标 oblique coordinates
斜角 angle of inclination; bevel(corner); bevel(led) corner; inclination angle; oblique angle; skewed angle; slope angle; mitre bevel <玻璃的>
斜角鞍形连接件 oblique-angled saddle junction piece
斜角板 bevel board; bevel sheet
斜角板桩 bevel sheet pile
斜角半砖 mitered bat
斜角棒 miter[mitre] rod
斜角边 mitred border
斜角(补强)铁 knee iron
斜角部件 angling
斜角采光光源 angle lighting luminaire
斜角槽 mitre shoot
斜角撑杆 diagonal strut; knee bracing
斜角尺 joint rule; miter[mitre] square; square miter[mitre]
斜角椽 angle rafter; valley rafter
斜角存车 diagonal parking; parking at an angle
斜角导轨 angle guide
斜角的 oblique-angled
斜角丁字接头 inclined tee joint
斜角钉 miter[mitre] brad
斜角定线板 bevel board
斜角堵头砖 mitered closer
斜角端部翘起 oblique-angled end cocking
斜角端部榫接 oblique-angled end cogging
斜角伐树刀 angling shearing blade
斜角分支接头 angle tee[T]
斜角缝 mitre[miter]
斜角缝线 miter[mitre] line
斜角覆盖层 diagonal sheathing
斜角杆 miter rod
斜角格构 lattice
斜角格栅 bevel(l)ed joist
斜角拱 miter[mitre] arch
斜角拱座 skewback
斜角规 angle protractor; bevel ga(u)ge; bevel protractor; bevel scale; bevel square; miter[mitre] square; sauterelle; set square; tee bevel
斜角轨枕 bevel(led) rectangular sleeper; bevel(led) rectangular tie
斜角焊 bevel weld(ing)
斜角焊缝 oblique fillet
斜角焊接 oblique fillet weld
斜角焊接缝 oblique fillet welding
斜角横切面铺地板木块 bevel-corner end-grain wood block
斜角汇合 oblique-angled junction
斜角机头向内(停放) angled nose-in
斜角机头向外(停放) angled nose-out
斜角夹板 corner clamp
斜角架 angular mount
斜角铰刀 angular reamer
斜角接缝 mitered joint; mitred joint

斜角接缝尺 joint rule
斜角接箍 turned-down coupling
斜角接头 franking; miter(ed) joint; oblique-angled joint; skew joint; splayed joint(ing)
斜角接砖 mitered closer
斜角孔 angular aperture
斜角棱镜 oblique prism
斜角离心机 angle centrifuge
斜角立体投影 <工程画> cabinet projection
斜角连接 angular contact; miter[mitre] joint; oblique-angled junction; splayed joint(ing)
斜角连接管子 oblique-angled pipe junction
斜角联结 splayed joint(ing)
斜角连接 miter joint
斜角裂缝 angled crack; oblique-angled cracking
斜角裂开 oblique-angled cracking
斜角檩条 angle purlin(e)
斜角面 bevel face
斜角木板 bevel board
斜角木支撑 racked timbering
斜角黏[粘]合 oblique bond
斜角黏[粘]接夹具 miter clamp
斜角平行六面体 inclined parallelopiped
斜角铺砌 diagonal bond
斜角砌合 diagonal bond
斜角启动锤 bevel start hammer
斜角砌合 diagonal bond; oblique-angled bond
斜角嵌接 oblique scarf
斜角强度 oblique-angled strength
斜角墙用砖 squint brick
斜角翘起 oblique-angled cocking
斜角切割机 mitering machine
斜角球面三角形 oblique spheric(al) triangle
斜角三通 angle tee[T]
斜角设计 bevel design
斜角摄影 oblique photography
斜角石 skew
斜角时差校正 dip-moveout correction
斜角竖起 oblique-angled cocking
斜角榫 mitered joint
斜角缩放仪 skew pantograph
斜角探头 angle probe
斜角跳板【船】angled rampway
斜角铁板 mitered sheet
斜角停车 angle parking; diagonal parking; parking at an angle
斜角投影 oblique-angled projection
斜角推土铲 angle bulldozer
斜角推土法 oblique angle earth pushing process
斜角推土机 angle(blade) dozer; angle bulldozer; angling dozer; side dozer
斜角圬工模样 oblique-angled masonry pattern
斜角圬工形式 oblique-angled masonry pattern
斜角铣床 chamfering bit
斜角铣刀 angle milling cutter; angular cutter
斜角系 oblique system
斜角镶玻璃条 bevel(l)ed bead
斜角镶条 quirk
斜角小带 diagonal band
斜角咬口连接件 oblique-angled saddle junction piece
斜角隅砖 squint brick; squint window
斜角支撑 racked timbering
斜角支管 Y-pipe; Y-tube
斜角轴 oblique-angled axis
斜角砖 angle brick; bevel brick; squint
斜角转盘 oblique turntable
斜角转弯 oblique-angle turn
斜角装帧 mitered decoration
斜角坐标 oblique coordinates
斜角坐标系 oblique coordinate system; oblique axis
斜绞辘 cant fall; cant purchase
斜脚刚架桥 rigid frame bridge with inclined(slant) legs
斜铰刀 angle drift; angle reamer; angle sweep
斜铰接接头 sloping hinge
斜阶跃波 ramp step wave
斜接 angular contact; bevel(led) joint; chamfered joint; diagonal joint; mitering[mitring]; mitre[miter] joint; oblique-angled joint; oblique joint; scarf(ing); scarf together; skew joint; juxtaposition【地】
斜接 T 形接头 oblique T joint
斜接波纹连接片 mitre brad
斜接插销 donkey's ear
斜接尺 joint rod
斜接触 angular contact
斜接的 scarfed
斜接丁字接头 inclined T joint
斜接端 miter[mitre] end
斜接阀 miter[mitre] valve
斜接封闭 lock miter[metre]
斜接缝 diagonal joint; miter[mitre] joint; inclined joint; bevel joint
斜接缝帆 diagonal cut sail
斜接缝加固件 mitre brad
斜接缝角钉 miter[mitre] brad
斜接缝三角帆 diagonal jib; mitered jib
斜接扶手弯折处 mitred knee
斜接拱 miter[mitre] arch
斜接规 miter[mitre] ga(u)ge
斜接海棠角【建】quirk-mitered corner
斜接焊接 miter[mitre] welding
斜接合 inclined joint; oblique joint; raking bond; skew joint; mitered [mitred] joint
斜接夹 miter[mitre] clamp
斜接角 miter[mitre] angle
斜接角焊缝 miter weld
斜接接缝 <地板的> splayed heading
斜接接头 inclined joint; scarf joint(ing)
斜接近段的长度 length of the oblique exposure
斜接口 inclined joint; miter[mitre] joint
斜接框架 miter[mitre] framing
斜接帽盖 mitered[mitred] cap
斜接面 miter[mitre] plane
斜接排水沟 <45 度的连接> miter [mitre] drain
斜接刨 miter[mitre] plane
斜接戗角 mitered[mitred] hip
斜接式活塞环 miter[mitre]-cut piston ring
斜接式铺板 diagonal planking
斜接式铁芯 mitered[mitred] core
斜接榫 oblique-angled scarf(joint)
斜接天沟 mitered[mitred] valley
斜接头 bent sub; mason's miter[mitre]; oblique compression joint; oblique joint; toe joint; splayed heading joint <地板的>
斜接头焊接 miter[mitre] welding
斜接瓦屋面沟 mitered[mitred] valley
斜接弯管 mitered[mitered] bend; mitered[mitred] elbow
斜接弯头 mitered[mitred] knee; miter [mitre] elbow
斜接线 miter[mitre] line
斜接燕尾榫 miter[mitre] dovetail
斜接褶皱 miter[mitre] folding
斜接正方形 miter[mitre] square
斜接支承 miter[mitre] bearing
斜接柱 miter[mitre] bar; miter[mitre] column; mitering[mitring] post; mitered[mitred] post
斜街 diagonal street
斜节 <木材的> splay knot; spike knot
斜节理 angular jointing; clinoclase; diagonal joint
斜截 bevel; snape
斜截棱柱 truncated prism
斜截棱锥 truncated pyramid
斜截面 oblique section; sloping section
斜截式 slope-intercept form
斜截筒柱 truncated cylinder
斜截头 miter[mitre] cut(ting)
斜截头角锥 truncated pyramid
斜截头屋顶 hip(ped) roof
斜解理 plagioclastic
斜筋 bent-up bar; diagonal reinforcement
斜筋骨 cant frame
斜晶石 clinohedrite
斜晶系 oblique system
斜井 deflecting well; inclined opening; inclined shaft; slant(ing) well; slope mine; slope shaft; sloping chamber; sloping shaft
斜井垂直深度 vertical depth of inclined well
斜井挡车器 incline stop
斜井的长度 length of inclined well
斜井吊桥 lifting bridge for inclined shaft
斜井吊桥调车场 drawbridge inset of slope
斜井工作面 backwall
斜井罐笼 incline bogie; slope cage
斜井箕斗 inclined shaft skip
斜井箕斗提升机 incline(d) skip hoist
斜井井壁 slope wall
斜井井底车场 incline inset; slope bottom
斜井井口 slope collar
斜井卷扬机 incline engine
斜井掘进 incline driving; slope driving; slope sinking
斜井掘进工作面 slope face
斜井开采矿 slope mine
斜井开拓 slope development
斜井开挖 slope driving; slope sinking
斜井口 headman; slope mouth
斜井口把钩工 headman
斜井矿柱 slope pillar
斜井木支护 inclined shaft timbering
斜井皮带提升机 hoist belt
斜井平车道结构 composition of slope bottom
斜井平车道调车场 plane inset of slope
斜井坡度 gradient of inclined well
斜井上下把钩 <采矿> hook tender; rope cutter
斜井式鱼闸 inclined shaft fish lock
斜井用车道参数 swing parting parameter of slope
斜井用车调车场 swing parting switchyard of slope
斜井提升车架 slope carriage
斜井提升钢丝绳 incline rope
斜井提升机 incline hoist
斜井通风 inclined shaft ventilation
斜井拖运系统 dilly
斜井维修工 incline trackman
斜井巷 dook
斜井展开图 extending map of slop mine
斜井抓岩机 incline grab
斜井装卸台 incline landing
斜井装岩机 inclined shaft mucking apparatus
斜井字 projecting diamond
斜径 clino-diagonal
斜径木纹 comb grain
斜径刨切 rift cutting
斜鸠尾【建】oblique dovetail
斜矩形 rhomboid
斜矩阵 skew matrix
斜距 inclined distance; oblique distance; slant distance; slant range; slope distance
斜距变换器 slant-range converter
斜距读数 slope distance readout
斜距分辨率 slant range resolution
斜距高度转换开关 slant-range-altitude switch
斜距校正 slant range correction
斜距离 inclined range
斜距图像 slant range image
斜距显示 display of slope distance
斜距丈量 slope taping
斜距指示器 slant range marker
斜锯 taper sawed; tapersawn
斜锯板 miter[mitre] board; resawed tapered shake; resawn tapered shake
斜锯槽 mitre[mitre] board
斜锯导引的工具 miter[mitre] box; miter[mitre]-saw cut; miter[mitre]-sawing hoard
斜锯辅助块 miter[mitre] block
斜锯架 bevel jack; miter[mitre] block
斜锯面 cheek cut
斜锯切 miter[mitre] cutting
斜掘机 sloper
斜卡 miter[mitre] clamp
斜开采面 beat away
斜开接头 bevel(l)ed halving
斜开口 diagonal cut joint
斜开口活塞 bevel joint piston ring
斜开口活塞环 bevel joint piston ring
斜砍 miter[mitre] cut
斜靠的 recumbent
斜坑 inclined adit; inclined shaft
斜坑道 sloping adit
斜孔 angle hole; angling hole; angular hole; crooked hole; inclination borehole; incline hole; skew hole; slant hole; slanting well; downward borehole <坑道钻进>
斜孔掏槽 inclined hole cut
斜孔弯斜地区 crooked hole country
斜孔网状克特尔塔板 Kittel tray
斜孔小窗 squint(window); hagioscope <十字形教堂的>
斜孔预制块 raking aperture block
斜孔钻进 slant hole boring; slant hole drilling
斜孔钻进法 off-angle drilling
斜孔钻进技术 slant hole technique
斜口 sinking; splay; dipped finish <玻璃制品缺陷>
斜口扳手 alligator wrench; angle wrench; offset wrench
斜口侧壁 splayed jambs
斜口侧石 sloping curb
斜口对接焊 angle butt weld(ing)
斜口环 miter[mitre] cut ring
斜口接缝 scarf(ing) joint(ing)
斜口接合 scarf(ing) joint(ing)
斜口接头 scarfed joint
斜口阔边刨 badger
斜口门窗洞 splayed reveal
斜口木 miter[mitre] timber
斜口刨床 scarf planer
斜口平接板 angle butt strap
斜口钳 diagonal cutting pliers
斜口四通 oblique cross; slanting crossing
斜口榫 angular tenon
斜口转角缝 birdsmouth corner joint
斜库容 slope storage

斜跨叠加褶皱 oblique superposed folds
斜跨度 skew span;slope span
斜跨结构 skew span
斜跨桥 stream crossing in skew angle
斜块 <直立防波堤用> sloping block
斜块拱座 skewback
斜框格 skew grid
斜框架 diagonal frame
斜拉 oblique pull
斜拉板梁桥 cable-stayed girder bridge
斜拉大幕 tableau curtain
斜拉导线 lateral conductor
斜拉杆 diagonal(draw) bar;diagonal tie;link arm;tension diagonal;angle brace;angle tie;batter brace
斜拉杆头 diagonal rod head
斜拉钢缆 <预应力混凝土中的> diagonal cable
斜拉桁架 stayed truss
斜拉结构 cable-stayed construction;cable-stayed structure
斜拉力 <混凝土受弯构件剪应力引起的> diagonal tension
斜拉力破坏 <混凝土受弯构件的> diagonal tension failure
斜拉裂缝 diagonal tension crack
斜拉破坏 diagonal tension failure
斜拉桥 backstayed bridge;bridle chord bridge;cable-stayed bridge;stayed-cable bridge
斜拉桥垂直双面索纵向布置 vertical double plane transverse cable arrangement of cable-stayed bridge
斜拉桥单索面横向布置 single-plane transverse cable arrangement of cable-stayed bridge
斜拉桥辐射形纵向布置 radiation longitudinal cable arrangement of cable-stayed bridge
斜拉桥桥塔 pylon of cable-stayed bridge;tower of cable-stayed bridge
斜拉桥扇形纵向布置 fan longitudinal cable arrangement of cable-stayed bridge
斜拉桥竖琴形纵向布置 harp longitudinal cable arrangement of cable-stayed bridge
斜拉桥斜双索面横向布置 sloping double plane transverse cable arrangement of cable-stayed bridge
斜拉桥星形纵向索布置 star longitudinal cable arrangement of cable-stayed bridge
斜拉索网型 cable net type
斜拉条 diagonal brace;diagonal tie
斜拉悬臂 cable-stayed cantilever
斜拉悬索组合体系桥 hybrid cable-supported bridge system
斜拉应力 diagonal tensile stress;diagonal tension stress;inclined tension stress
斜拉桩 batter tension pile
斜拉桩板桩(结构) sheet-pile supported by batter piles
斜兰硒铜矿 clinochalcomenite
斜蓝铜矾 wroewolfeite
斜缆道 inclined cableway;inclined ropeway
斜缆桥 diagonal cable bridge
斜缆(索) inclined cable;backstay cable;sloped cable;springing line;stay cable
斜廊 inclined gallery
斜浪 oblique sea
斜肋 diagrid;oblique-angled rib;oblique rib
斜肋板 cant floor;cant timber
斜肋构架(网格) diagrid
斜肋骨部 cant body
斜肋骨(框架) cant frame
斜肋楼板 diagrid floor
斜肋条 diagonal rib
斜肋瓦 diagonal ribbed tile
斜肋网格屋顶 diagrid floor
斜棱 slant edge;sloped edge
斜棱滚轮 bevel(l)ed wheel
斜棱角 slanting edge
斜棱柱 oblique prism;slant prism
斜棱锥(体) oblique pyramid;slant pyramid
斜锂闪石 clinoholmquistite
斜力 incline(d) force
斜连接 skewed connection;toe joint
斜链线结构 inclined catenary construction;inclined catenary structure
斜链形悬挂 inclined catenary;skew catenary
斜链形悬挂的架空接触线 inclined overhead contact line
斜梁 bracing piece;bridge board;cant beam;diagonal beam;inclined beam;miter [mitre] sill;raking beam;ramp beam;ramp carriage;sloped beam;sloping beam
斜梁地板 diagonal beam floor
斜梁楼盖结构 diagonal beam floor
斜梁有侧板的楼梯 closed string stair(case)
斜两跨板 skew two-span slab
斜量法 gradient method
斜量角规 bevel protractor
斜料斗 bin
斜列丁砖 cant header
斜列断层 echelon fault
斜列节理 en echelon joints
斜列沙坝 diagonal bar
斜列式布置 oblique arrangement
斜列式构造 en echelon structures
斜列式矿脉 echelon-like veins
斜列式停车 diagonal parking
斜列式停车道 diagonal parking lane
斜列式植针法 twill set
斜列停车 angle parking;inclined parking;saw-tooth parking arrangement
斜列停车道 angle parking lane;inclined parking lane
斜列停放 angle parking
斜列停放的 angle parked
斜列系统 diagonal system
斜列驻车 inclined parking
斜列砖 cant stretcher
斜裂 oblique segmentation;slant crack
斜裂缝 angular crack;inclined crack;oblique cracking;diagonal crack
斜裂纹 diagonal crack(ing)
斜磷钙铁矿 mitridatite
斜磷硅钙钠石 clinophosinaite
斜磷铝钙石 montgomeryite
斜磷铝铁石 clinobarrandite
斜磷锰矿 stewartite
斜磷锰铁矿 saropide
斜磷铅铀矿 parsonsite
斜磷锌矿 spencerite
斜菱碱铁矾 clinoungemachite
斜溜槽 inclined chute;slant chute
斜溜井 slant chute
斜流 oblique flow
斜流可逆式水泵水轮机 Deriaz turbine
斜流式沉淀池 inclined flow precipitation tank
斜流式水轮机 Deriaz turbine;diagonal flow turbine;inclined flow turbine;mixed flow variable pitch turbine
斜流式水轮机转轮 feathering Francis runner
斜流式压缩器 diagonal flow compressor
斜流通风式发电机 diagonal flow generator
斜流涡轮机 diagonal turbine
斜流箱 spiral-flow tank
斜硫砷汞铊矿 christite
斜硫砷钴矿 alloclasite
斜硫砷银矿 smithite
斜硫锑铋镍矿 parkerite
斜硫锑铅矿 plagionite
斜硫锑铊矿 parapierrotite
斜六面体 oblique parallelepiped
斜楼座 inclined gallery
斜炉箅 angular grate;inclined grate
斜炉箅垃圾焚化炉 inclined grate incinerator
斜炉底 sloping hearth
斜路 cross-cut;ramp;rampway;slade
斜路牙子 sloped curb
斜路缘 sloped curb;sloping curb
斜铝矾 jurbanite
斜铝石 tanatarite
斜率 gradient;offsetting;rate of grade;rate of slope;slope;amount of inclination <井身的>
斜率比 slope ratio
斜率比测定 slope ratio assay
斜率比法 slope ratio method
斜率常数 gradient constant
斜率窗口 window of the slope
斜率放大 slope amplification
斜率过载 slope overload
斜率过载误差 slope overload error
斜率函数 slope function
斜率计 slope meter
斜率计算 slope calculation;slope computation
斜率检测 slope detection
斜率检测器 slope detector
斜率鉴定法 slope detection method
斜率鉴频法 slope detection method
斜率鉴频器 slope detector
斜率灵敏度 slope sensitivity
斜率滤波器 slope filter
斜率偏移 slope deviation
斜率容限 slope tolerance
斜率调整 slope regulation
斜率系数 slope coefficient
斜率信号 slope signal
斜绿泥石 clinochlore;clinochlorite
斜氯硼钙石 sologoite
斜氯铜矿 botallackite
斜螺旋齿轮 spiral wheel
斜螺旋面 oblique helicoid
斜螺旋线 skewed helix
斜码头 inclined wharf
斜脉 rake
斜锚 inclined anchor
斜锚桩钢板桩岸壁 steel sheet pile wall with inclined anchor pile
斜铆距 diagonal pitch
斜镁川石 clinojimthompsonite
斜镁锂闪石 magnesioclinoholmquistite
斜锰针钠钙石 schizolite
斜面 angular surface;bevel surface;cant;chamfer;declining;droop;gradient;inclined plane;inclined surface;oblique plane;pendence;ramp;scarf;skew surface;sloping face;splay
斜面板 cant board
斜面板壁 bevel siding
斜面半搭接 bevel halving
斜面半迭接 bevel halving
斜面报架 sloping newspaper rack
斜面壁柱 battered pilaster
斜面补给流 slope-supply flow
斜面槽形榫 bevel cocking;bevel cogging;bevel corking
斜面齿轮 bevel(led) gear
斜面齿轮传动 bevel gear drive
斜面齿轮传动装置 bevel gearing
斜面窗洞 embrasure
斜面锉 cant file;taper(ed) file
斜面导板 inclined guide plate
斜面的 sloping
斜面的外墙覆面木板 bevel(l)ed siding
斜面地基 sloping foundation;sloping ground;sloping site
斜面垫板 canted tie plate
斜面垫圈 angle washer;bevel(l)ed washer
斜面跌水 inclined drop
斜面丁坝 dip groin;dipping dike[dyke]
斜面丁砖 cant header;splay header
斜面顶砖 bevel(l)ed closer
斜面定线板 bevel board
斜面断口 angular fracture;shear fracture
斜面对接 bevel(l)ed halving
斜面墩 batter pier;miter[mitre] pier
斜面阀 miter[mitre] valve
斜面法 <测试用> inclined plane method
斜面方钢 chamfered square bar
斜面防浪墙 shaped sea wall
斜面防雨板 skew flashing
斜面封口砖 bevel(l)ed closer
斜面封檐板 bevel(l)ed siding
斜面缝 miter joint
斜面拱 fluing arch;sluing arch;splaying arch
斜面箍筋 inclined stirrup
斜面刮板 sloping screed
斜面刮土铲运机 slope scraper
斜面海堤 shaped sea wall
斜面海塘 shaped sea wall
斜面焊接 scarf weld(ing)
斜面弧形板 inclined segment
斜面护坦 sloping apron
斜面鸡笼 rollaway nest
斜面积 inclined area
斜面加工 scarf
斜面键 taper key
斜面键盘 sloping keyboard
斜面胶合 scarfed joint
斜面角 angle of chamfer;bevel angle
斜面校正(法) slope correction
斜面接缝 splayed joint
斜面接合 bevel(l)ed halving;birdsmouthing jointing;miter[mitre] joint;skew joint
斜面接头 bevel(led) halving;bevel(l)ed joint;miter[mitre] joint
斜面镜 flat inclined mirror
斜面控制盘 sloping control panel
斜面栏杆 raking baluster;raking balustrade
斜面肋骨 <船首尾> frame bevel
斜面累积洗矿槽 building buddle
斜面连接 mitre jointing
斜面梁 hip rafter
斜面露头石 bevel header
斜面路牙 sloped curb
斜面路缘石 rolled curb
斜面密封阀 bevel seat(ed) valve
斜面密集补强 concentrated reinforcement with inclined surface
斜面磨削 angular surface grinding
斜面黏[粘]度计 inclined plane visco(si)meter
斜面黏[粘]接 scarf;scarfed joint
斜面刨削 angle planing;angular planing
斜面培养(法) slant culture
斜面披水板 raking flashing
斜面坡度分段变化桩 step taper pile

斜面铺面 slope pavement
斜面铺砌 slope paving;sloping bond
斜面砌缝石 cant bondstone
斜面砌缝砖石 cant bonder
斜面砌墙石 bevel bonder;bevel bondstone;splay bonder
斜面砌砖 cant stretcher
斜面墙 battered wall;battering wall;talus wall;embrasure
斜面墙帽 feather-edged coping
斜面桥墩 batter pier;leaning pier
斜面切边刀具 bevel trimming cutter
斜面切边机 bevel trimmer
斜面轻型屋顶 sloping light weight roof
斜面散水 raking flashing
斜面设计 bevel design
斜面升船道 boat lift railway;craft railway
斜面升船机 inclined ship lift;ship incline;ship lifter with sloping desk;ship railway;inclined plane <船舶由一条水道转到另一条的>
斜面式墩 batter pier
斜面式防波堤 sloping faced breakwater
斜面式桥墩 batter pier
斜面式溢流堰 bevel surface overflow
斜面式鱼道 inclined plane fish way
斜面书架 sloping shelf
斜面书桌 reading desk
斜面束石 splay bondstone
斜面双晶 Baveno twin
斜面顺砖 cant stretcher;splay stretcher
斜面体 beveloid;clinohedral
斜面铁条 ramp
斜面头钎子 splayed drill
斜面凸凹榫接合 tongue miter
斜面推力轴承 tapered-land thrust bearing
斜面挖土填埋垃圾技术 ramp technique
斜面弯管 miter[mitre] bend bondstone
斜面圬工墙 sloping masonry wall
斜面屋顶 sloping roof
斜面屋顶面积 sloped roof area
斜面铣刀 bevel milling cutter
斜面铣削 angular milling
斜面系石 splay bondstone
斜面下部挡土墙 skirt retaining wall
斜面线脚 bevel mo(u)lding;cant mo(u)lding;rake(d) mo(u)lding;raking mo(u)lding
斜面线条 bevel mo(u)lding;cant mo(u)lding;rake(d) mo(u)lding;raking mo(u)lding
斜面销 wedge pin
斜面小齿轮 bevel gear pinion;bevel pinion
斜面小齿轮罩 bevel pinion housing
斜面型(结晶) plagiohedral
斜面修整机 bevel trimming cutter;mitring machine
斜面压顶(石) raking capping;raking coping
斜面研磨机 mitering[mitring] grinding machine;miter[mitre] grinding machine
斜面雨水板 bevel(l)ed siding
斜面圆筒 oblique cylinder
斜面运输 chuting
斜面闸室墙 batter chamber wall
斜面折叠 canted folding
斜面钟 inclined plane clock
斜面轴承 tapered(-land) bearing
斜面柱 <柱径收缩的> contracture;battered post
斜面砖 bevel(l)ed brick;cant brick;neck brick;slope brick;splay(ed) brick
斜面砖层收尾砖 bevel(l)ed closer
斜面转车台 inclined traverser
斜面装车台 livestock ramp
斜面状(装)饰 cant mo(u)lding
斜面桌式控制台 sloping desk control panel
斜面作业基地 sloping site
斜摩擦齿轮 bevel friction gear
斜摩擦轮 bevel friction gear;bevel wheel
斜磨 bevel grinding
斜墨卡托投影 skew Mercator projection
斜木板屋顶 wooden tilted-slab roof
斜木撑 inclined timber shore;racked timbering
斜木托条 splayed ground
斜木纹 cross grain;oblique grain;sloping grain;diagonal grain
斜木纹压力 compression inclined to grain
斜木纹压缩 compression inclined to grain
斜目镜 diagonal eyepiece
斜内推土板 angledozer blade
斜钠锆石 elpidite
斜钠明矾 tamarugite
斜钠鱼眼石 natroapophyllite
斜能见度 oblique visibility;slant visibility
斜捻沟 <木甲板接缝的斜坡口> outgage
斜镍矾 nickel-hexahydrite
斜扭因数 skew factor
斜爬模板 skew lifting formwork
斜排条板 luffer-boarding
斜排泄眼注水泥管鞋 whirler shoe
斜盘 sloping cam plate
斜盘电机 swash plate motor
斜盘角度 swash plate angle
斜盘式搅拌机 inclined drum type mixer
斜盘式压缩机 swash plate compressor;wobble plate compressor
斜盘式轴向活塞(油)泵 cam plate type axial piston pump;swash plate axial piston pump
斜盘式轴向柱塞泵 swash plate axial cylinder pump;swash plate type axial plunger pump
斜盘式轴向柱塞电动机 cam-type axial piston motor
斜盘轴 wobbler shaft
斜刨 miter plane
斜刨法 angle planimeter;angle planing;angular planing
斜刨计 angular planimeter
斜炮洞 angling hole;grip shot;oblique hole
斜炮孔 angling hole;grip shot;oblique hole
斜炮眼 angling hole;grip shot;oblique hole
斜跑尺寸 pitch dimension;rake dimension
斜配孔型轧制法 diagonal flange method
斜硼钙石 calcioborite
斜硼镁钙石 clinokurchatovite
斜硼钠钙石 probertite
斜劈 sloping wedge
斜劈接 oblique cleft grafting
斜片百叶窗 abatjour
斜片簧 tapered leaf spring
斜偏心荷载 oblique eccentric load(ing)
斜拼合 tongue miter[mitre]
斜平面三角形 oblique plane triangle
斜平头接合 oblique butt joint
斜平镶舢板 diagonal carvel built
斜平巷 cross gate
斜平行垫铁 degree parallel
斜平行四边形板 skewed parallelogram plate
斜坡 slope(ground);backfall;berm(e) shoulder;clino;declining;declivity;glacis;inclined plane;inclined ramp;inclined slope;incline grade;incline gradient;mountain slope;ramp;side slope
斜坡板 tapersplit
斜坡板块 grade slab
斜坡比率 slope ratio
斜坡边缘 slope edge
斜坡标桩 gradient peg;gradient stake
斜坡表面 sloping surface
斜坡表面处理 slope dressing
斜坡表面排水系统 face drain
斜坡波形 ramp waveform
斜坡部分 slope portion;slope section
斜坡长度 length of slope
斜坡车站 station on an incline
斜坡沉积 clinoform;clinothem
斜坡沉陷 slope caving
斜坡冲蚀 slope erosion
斜坡冲刷 slope wash
斜坡粗糙度系数 roughness coefficient of slope
斜坡带 slope zone
斜坡道 inclined approach;inclined plane;inclined roadway;on-board ramp way;raised approach;ramp(road);slope ramp;sloping way
斜坡道管理工 motion driver
斜坡道开采 ramp mining
斜坡道路 slope course
斜坡道无轨掘进 slope railless driving
斜坡的 clival
斜坡的覆盖物 covering of a slope
斜坡的护坡 slope lining
斜坡的换算水平距离 horizontal equivalent
斜坡的空中索道 <运重物的> inclined ropeway
斜坡的中点 midpoint of slope
斜坡堤堤顶高程 elevation of rubble mound breakwater crest
斜坡堤堤顶宽度 width at the rubble mound breakwater crest
斜坡堤基床垫层 bedding layer of mound type breakwater
斜坡底部 toe of slope
斜坡地 sloping ground;sloping land
斜坡地板结构 slab-on-grade construction
斜坡地道 inclined subway
斜坡地洞窑 ground-hog kiln
斜坡地基 sloping site
斜坡地形 <如大陆坡、三角洲前积层> clinoform
斜坡地用拖拉机 slope field tractor
斜坡电压 ramp voltage
斜坡垫板 canted plate
斜坡顶沉箱防波堤 sloping top caisson breakwater
斜坡定位装置 sloping ga(u)ge
斜坡陡度 slope gradient
斜坡陡削度 steepness of setting;steepness of slope
斜坡度 ramp grade
斜坡端 <屋顶的> hipped end
斜坡对流 sloping convection
斜坡发生器 ramp generator
斜坡法 ramp method
斜坡法灌浆 grout slope
斜坡改正 slope correction
斜坡割草附件 slope mowing attachment
斜坡割草机 slope mower
斜坡割草装置 slope mowing attachment
斜坡谷仓 bank barn
斜坡轨道 tilting rail
斜坡海漫坝 sloping apron
斜坡海漫闸 sloping apron
斜坡函数 ramp function
斜坡函数发生器 ramp function generator
斜坡函数响应 ramp function response
斜坡河底 shelving bottom
斜坡护底 sloping apron
斜坡护面 slope pavement
斜坡护面防波堤 sloping faced breakwater
斜坡护坡 slope revetment
斜坡护墙 slope revetment;sloping apron
斜坡滑道 inclined slip
斜坡毁损 slope failure
斜坡混凝土 sloping concrete
斜坡混凝土板 slab-on-grade
斜坡混凝土铺料机 slope concrete paver
斜坡基础 sloped foundation
斜坡计算 slope calculation;slope computation
斜坡继电器 ramp relay
斜坡角 slope angle
斜坡角度 degree of slope
斜坡校正 slope correction
斜坡进口 ramp entrance
斜坡进口道 ramped approach
斜坡距离 grade distance;slope distance
斜坡卷扬开拓 development of slope cable lift
斜坡卷扬提升 slope hoisting
斜坡开挖或回填的权利 slope easement
斜坡块石铺面 slope sett paving
斜坡块石铺砌 slope sett paving
斜坡梁 grade beam
斜坡料堆 <集料的> ramped pile
斜坡流 slope current
斜坡隆起 slope crown
斜坡楼板结构 slab-on-grade construction
斜坡路 <牵拉船或货物的> canal incline;incline
斜坡路面 <供老弱,残疾人使用的> sloped pavement;slip road
斜坡路面填缝混合物 slope paving joint sealing compound
斜坡绿化 slope greening;slope planting
斜坡码头 sloping faced wharf;sloped wharf
斜坡码头护岸 pavement of the wharf slope
斜坡面积 sloping area
斜坡面排水 slope drain
斜坡面铺盖排水 blanket drain
斜坡模型 slope model
斜坡磨损 slope erosion

斜坡抹面机 slope finisher
斜坡墨卡托投影 oblique Mercator projection
斜坡排除 slope removal
斜坡排水 batter drainage;slope drainage
斜坡喷射混凝土施工缝 board butt joint
斜坡平整机 slope trimmer
斜坡坡度 gradient of slope;ramp grade;slope gradient
斜坡坡率 gradient of slope
斜坡破坏 slope failure
斜坡破坏形式 failure mode of slope
斜坡铺路石块 slope sett

斜坡铺面 pitching of slope
斜坡铺面机 slope facing machine; slope-lining machine
斜坡铺砌 pitching of slope; slope paving
斜坡铺砌机 slope paving machine
斜坡侵蚀 slope wash
斜坡倾斜角 slope inclination
斜坡区段 slope portion; slope section
斜坡蠕动 slope creep
斜坡山墙尖屋顶 jerkin head
斜坡上位置 location on a slope
斜坡升船机 inclined plane
斜坡(升或降)百分比 percent of slope
斜坡失稳主导因素 principal factors of slope failure
斜坡施工 slope construction; slope work
斜坡施工设备 slope construction equipment
斜坡式底脚 sloped footing
斜坡式防波堤 breakwater with sloping faces; inclined breakwater; mound(-type) breakwater; rubble-mound breakwater; sloping breakwater; sloping faced breakwater; sloping mound breakwater; slope-surface breakwater
斜坡式防波堤堤面坡度 armo(u)r-course slope
斜坡式防波堤的圆锥形堤头 round-head
斜坡式海堤 sloping seawall
斜坡式护墙 sloping apron
斜坡式护坦 inclined floor; sloped floor; sloping apron
斜坡式基础 sloping footing; sloping foundation; sloped footing
斜坡式冷床 rake cooling bed
斜坡式临水结构物 sloped waterfront structure
斜坡式路缘石 sloped curb
斜坡式码头 inclined dock; inclined terminal; inclined wharf; sloping quay; sloping wharf
斜坡式输送机 inclined conveyer[conveyor]
斜坡式踏步 sloped step
斜坡式堰 slope weir
斜坡输入 ramp input
斜坡输入信号 ramp input signal
斜坡双晶 Baveno twin
斜坡水标尺 inclined ga(u)ge
斜坡水尺 inclined (staff) ga(u)ge; sloping ga(u)ge
斜坡隧道 slope tunnel
斜坡踏步板 ramped steps
斜坡台 ramp
斜坡梯田 graded terrace
斜坡停车制动装置 hill holder
斜坡通道 access ramp
斜坡凸面 slope crown
斜坡推进法 <一种水平移动式浇注隧道衬里方法> advancing slope method
斜坡推进灌浆(法) <一种作预填骨料混凝土的施工方法> advancing slope grouting
斜坡挖成台阶形 bench the slopes
斜坡挖土 slope cut(ting)
斜坡挖土及整平机 gradall
斜坡稳定系数 stability factor of slope
斜坡稳定性 slope stability; stability of slope; over-the-side stability【机】
斜坡屋顶 sloping roof; pitched roof
斜坡屋顶桁架 pitched truss
斜坡屋顶门窗 internal dormer
斜坡误差电压 ramp error voltage
斜坡系数 grade factor
斜坡线 profile of slope; slope line
斜坡效率 slope efficiency
斜坡斜度 slope inclination
斜坡信号 ramp signal
斜坡形成器 ramp former
斜坡修整 slope trimming; trimming of slope
斜坡修整机 slope trimmer
斜坡压辊 slope roller
斜坡压实机 slope compactor
斜坡岩层 clinothem
斜坡岩体变形观测 monitoring of deformation of rock mass of slope
斜坡养护 slope maintenance
斜坡样板 batter(ing) ga(u)ge; batter templet
斜坡移动 slope removal
斜坡移动速度 velocity of slope
斜坡引道 ramped approach
斜坡引桥 ramp bridge
斜坡匝道 slope ramp
斜坡栽培 slope culture
斜坡造林 slope planting
斜坡轧制 taper rolling
斜坡张应力带宽度 width of tension zone
斜坡丈量 slope chaining
斜坡整面机 slope finisher
斜坡整平 slope flattening
斜坡整平压实机 slope level(l)ing and compacting machine
斜坡整修 slope dressing
斜坡整修机 slope trimming machine
斜坡植草坪 slope sodding
斜坡重力作用 gravitational processes on slope
斜坡主缆车道 inclined slip
斜坡装卸台 loading ramp
斜坡自动控制器 automatic slope control
斜坡纵断面 slope longitudinal section
斜坡纵断面图 profile of slope
斜坡纵剖面图 grade profile; profile of slope
斜坡走道 inclined walk
斜坡阻力系数 grade resistance factor
斜坡作业推土板 stoper dozer blade
斜破板 taper-split shake
斜剖面 angular section; oblique profile
斜剖面法 taper sectioning
斜剖视图 oblique section view
斜铺板 diagonal slating; drop-point slating
斜铺板材屋面 diagonal roofing; French method roofing
斜铺衬板 diagonal sheathing
斜铺地板 diagonal flooring
斜铺法 diagonal built
斜铺方砖地面 pointel(le)
斜铺楼板 slab-on-grade
斜铺毛地板 diagonal subfloor; oblique scarfing
斜铺面砖 diagonal tile
斜铺瓦 diagonal tile
斜铺屋面板 diagonal slating
斜企口接缝 chamfer-tongued jointing
斜起重臂 inclined gib; inclined jib
斜砌 racking; sloping bond
斜砌壁洞 cant bay
斜砌层 inclined course; sliced layer; sloping layer
斜砌法 sliced work
斜砌方块 edgewise cube; sliced block; blocks in sloping bond
斜砌方块工程 sliced blockwork
斜砌方块砌体 sliced blockwork; sloping blockwork
斜砌合墙 slope bonder
斜砌合石 slope bondstone
斜砌块 mitre block
斜砌块体 sliced blockwork
斜砌石 skew
斜砌石角度 angle of skew
斜砌式方块岸壁 quay wall of sliced blockwork; quay wall of sloping blockwork
斜砌式方块码头 quay of sloping blockwork; quay of sliced blockwork
斜砌图案 diagonal pattern
斜砌筑 sloping bond
斜砌砖 skew brick
斜砌砖层 rack(ing) course; tumbling course <与水平层相交的>
斜砌砖工 sloping bond
斜牵板 diagonal tie plate
斜牵引杆 diagonal draw bar
斜前沿 skew front
斜前沿边缘 sloping front boundary
斜嵌 oblique halving
斜嵌槽 scarf(ing)
斜嵌槽接 French scarf(joint); oblique scarf joint
斜嵌接 oblique(-angled) scarf(joint); oblique halving
斜嵌接头 oblique scarf joint; skew scarf joint
斜嵌连接 scarf connection
斜嵌式窗台板 bevel(l)ed-rabbeted window stool
斜嵌条 skew fillet
斜嵌砖层 tumbled-in course; tumbling course
斜墙 battered wall; battering wall; diagonal siding; die wall; miter[mitre] wall
斜墙坝 dip wall dam
斜墙式防波堤 sloping wall type breakwater
斜羟氯铅矿 paralaurionite; rafaelite
斜羟砷锰石 allactite
斜桥 askew bridge; bridge of skew (plan) form; hoist incline; oblique bridge; skew bridge
斜切 angle cut; bevel cut; bevel(l)ing; bevelment; chamfer(ing); oblique cut(ting); raking cutting; slanting cut
斜切边 chamfered edge
斜切材料 bevel material
斜切窗槛 mitred sill
斜切的 chamfered
斜切顶 diagonal topping
斜切割 miter[mitre] cutting
斜切机 angle cutter; bias cutter
斜切尖轨 chamfer cut
斜切渐长海图 oblique cylindric(al) orthomorphic projection; oblique Mercator projection
斜切角 angle of chamfer; chamfered angle; rake angle; bevel angle
斜切锯 miter[mitre] saw
斜切口 <活塞环> miter[mitre] joint
斜切门槛 mitered[mitred] sill
斜切面 scarf; sloping section
斜切砌缝 cant bonder
斜切戗脊 cut-and-mitered[mitred] hip
斜切式砌合法 clip bond
斜切手钳 diagonal cutting pliers
斜切投影赤道 oblique equator
斜切削刃 oblique cutting edge; side cutting edge
斜切楔形砖 feather brick
斜切心射投影 oblique gnomonic projection
斜切形 chamfer shape
斜切辙叉轨头 chamfer cut
斜切柱 mitered[mitred] post
斜切砖 cut splay
斜轻砷铜矿 clinoclasite
斜倾 squint
斜倾断层 semi-transverse fault
斜倾型 apsacline
斜穹隆屋顶 wagon vault
斜球面投影 oblique stereographic(al) projection
斜球三角形 oblique spheric(al) triangle
斜球体 oblique sphere
斜曲柄 oblique crank; oblique crank web
斜曲尺 bevel ga(u)ge; bevel square
斜曲面 skew surface
斜曲线 grading curve; skew curve
斜屈光性 plagiophototropism
斜趋光性 plagiophototaxy
斜渠 diagonal drain; diagonal drainage
斜圈绕组 skew coil winding
斜刃割法 oblique cutting
斜刃剪切机 inclined throat shears
斜刃面 basil
斜刃刨 skew plane
斜刃切削 oblique cutting
斜刃凸模 bevel(l)ed punch
斜刃凿 bevel(l)ed-edge chisel
斜入射 oblique incidence
斜入射传播 oblique incidence transmission
斜入射传输 oblique incidence transmission
斜入射反射率 oblique incidence reflectivity
斜入射涂敷 oblique incidence coating
斜入射吸收系数 oblique incident absorption coefficient
斜三角形 oblique-angled triangle; oblique triangle
斜三跨板 skew three-span slab
斜三通 double Y; skew T; sweep tee; wye
斜三通管 bevel arm piece
斜三通管接 oblique T
斜伞齿轮 skew bevel gear; skew bevel wheel
斜伞形 skew bevel
斜散热器 sloping radiator
斜散热器护栅 sloping radiator grille
斜沙洲 <与海岸线成交角的> reticulated bar
斜砂轮 tapered emery-wheel
斜筛 inclined screen; slanted screen; uphill screening
斜山墙尖两坡屋顶 <山墙尖呈斜坡的> jerkin head roof
斜闪石 aenigmatite
斜梢铆钉 taper rivet
斜射 oblique fire
斜射法 oblique incidence method
斜射光 skew ray
斜射角 squint angle
斜射投影法 clinographic(al) projection
斜射线 oblique ray
斜射线效应 plant path effect
斜射影 oblique projection
斜射照明 oblique illumination
斜伸缩仪 skew pantograph
斜砷钯矿 palladoarsenide
斜砷镁钙石 irhtemite
斜砷钴矿 clinosafflorite
斜升 deflected ascent
斜升高门铰 rising butt
斜升合页 rising hinge
斜升门铰链 rising(butt) hinge
斜升式储气罐 spirally guided holder
斜生单干形 oblique cordon
斜生多干形 oblique palmette
斜生石笋 heligmite
斜十字撑 herringbone bracing

斜石 inclined stone
斜石墙压顶(石)splaying coping
斜驶变换 loxodromic transformation
斜驶代换 loxodromic substitution
斜驶靠泊 oblique berthing
斜驶靠拢 sheer alongside
斜驶螺线 loxodromic spiral
斜驶曲线 loxodrome curve
斜驶线 loxodrome;loxodromic line
斜驶循环群 loxodromic cyclic(al) group
斜式沉降器 inclined clarifier
斜式导井 inclined shaft
斜式的 splayed
斜式盖顶 splayed coping
斜式钢琴 oblique pianoforte
斜式构件 sloping member
斜式路缘(石)sloping curb;splayed curb;mountable curb
斜式抛物线拱 sloping parabolic arch
斜式筛 sloping screed
斜式水泵 inclined shaft pump
斜式缘石 sloping curb[kurb]
斜式造形 splay mo(u)lding
斜式甑 inclined retort
斜式折叠 splayed folding
斜式(蒸馏)罐 inclined retort
斜视 inclined sight;side glance;strabismus
斜视的 squint
斜视钩 squint hook;strabismus hook
斜视光楔 strabismus wedge
斜视计 deviometer;strabismeter;strabometer
斜视校正 correction of strabismus;squint correction
斜视校正棱镜 squint-correcting prism
斜视剪 strabismus scissors
斜视角 angle of squint;squint
斜视图 oblique drawing;oblique view
斜视误差 squint error
斜收敛角 oblique convergence
斜艏 raked stem
斜受拉钢筋 diagonal tension bar
斜束 angle beam
斜栓 pintle
斜栓节链 pintle chain
斜水尺 inclined staff ga(u)ge;slope ga(u)ge;sloping ga(u)ge
斜水钙钾矾 gorgeyite
斜水沟 batter drainage;diagonal drain(age)
斜水沟缝 weathered pointing
斜水硅钙石 killalaite
斜水硼镁石 preobrazhenskite
斜水跃 oblique hydraulic jump
斜顺轨 <第三轨起点的> ramp rail
斜丝框 inclined reel
斜四通 oblique crossing
斜榫 mitre[miter];oblique tenon;tapered tenon
斜榫接 bridle joint;skew table;tongued miter[mitre]
斜榫锯 miter saw
斜榫拼接 tongued miter[mitre]
斜缩图仪 plagiograph
斜索 chorda oblique;oblique cord;stay cable
斜索道 inclined ropeway
斜索面 inclined cable plane
斜索面斜拉桥 cable-stayed bridge with inclined cable plane;inclined plane cable-stayed bridge
斜索桥 bridle chord bridge
斜塔 leaning tower;inclined derrick <指钻塔>
斜塔式闸门 tilting fillet;tilting gate
斜踏步 <楼梯的> wheel(ing)steps;turn tread;winder
斜踏步楼梯 winder stair(case)
斜台 sloping bench;sloping platform
斜台阶 bevel(l)ed shoulder
斜台拉伸 tapered extrusion
斜毯式投料机 skew blanket feeder
斜探头 angle beam probe;angle beam searching unit;angle probe;angular sensor
斜躺椅 leg-rest reclining seat
斜掏槽(炮)眼 angled snubbing hole
斜套 shank alidade
斜梯式装车台 step ramp loading chute
斜梯形纹螺钉 buttress thread screw
斜踢面 raking riser
斜体波 inclined body wave
斜体的 italic
斜体罗马字 sloped roman
斜体文字 Italic letters
斜体字 inclined letter;italic lettering;italics;oblique letter;sloping type
斜体字注记 oblique name
斜天沟 valley gutter <上大下小的>;tapered valley <上小下大的>
斜天沟椽 valley jack rafter;valley rafter
斜天沟底板 valley board
斜天沟泛水 valley flashing
斜天球 oblique sphere
斜填角 oblique fillet
斜挑檐 rake cornice;raking cornice
斜条 bevel(l)ed edge;slanted bar
斜条格构 lattice
斜条格(构)窗 lattice window
斜条砌合 raking stretcher bond
斜条形调制盘 slanted bar reticle
斜条状叶片 slanted-bar blade
斜调 slope regulation
斜调网络 slop regulation network
斜调谐器 tilt tuner
斜跳板 angled type rampway;dock board
斜贴(角)焊缝 oblique fillet weld
斜贴(角)焊接 oblique fillet weld
斜铁 gate wedge
斜铁辉石 clinoferrosilite
斜铁锂闪石 ferroclinoholmquisite
斜通道 winze <连接矿内各层的>;rampway <如汽车运输船的车辆甲板斜道等>
斜同步 non-horizontal simultaneous
斜铜泡石 clinotyolite
斜筒 inclined shaft
斜筒拱 skewed barrel arch;inclined barrel arch
斜筒搅拌机 tipping trough mixer
斜筒穹顶 inclined barrel vault
斜筒式拌和机 tilting drum mixer
斜筒式混凝土拌和机 tilting drum concrete mixer
斜筒式混凝土搅拌机 tilting drum concrete mixer
斜筒形拱 skew-barred arch
斜筒形结构 tapered-cylindric(al) structure
斜筒形穹顶 inclined barrel vault;skew barrel vault
斜头 plagiocephalia
斜头对接焊 butt-welded[welding] with chamfered ends
斜头管 bevel(l)ed pipe
斜头轨枕 tapped sleeper
斜头平焊接 butt weld with chamfered ends
斜头平接焊 butt-welded[welding] with chamfered ends
斜头原木 joggled timber
斜投射 oblique incidence
斜投射探测 oblique incidence sounding
斜投影 cavalier projection;inclined projection
斜投影极 oblique projection pole
斜投影图 oblique projection drawing
斜透视 diagonal perspective;oblique perspective
斜透视图 angular perspective
斜钍石 huttonite
斜推法 oblique bed process;inclined shearing method <现场剪力试验的>
斜推杆式升料机 inclined pusher bar elevator
斜推式推土机 tilt dozer
斜腿 <井架的> sloping leg
斜腿的车行路拱 splay-legged roadway arch
斜腿刚构 frame with slant legs;inclined legged frame structure;slant-legged rigid frame structure
斜腿刚构桥 slant-legged rigid frame bridge
斜腿刚架 frame with raking struts;inclined legged frame
斜腿刚架桥 frame bridge with raking struts;inclined legged frame bridge;rigid frame bridge with inclined legs;slant-legged rigid frame bridge
斜托石 summer stone
斜托座 angle bracket
斜拖 oblique pull
斜拖试验【船】oblique towing test
斜瓦式推力轴承 tilting pad thrust bearing
斜瓦式轴承 tilting pad bearing
斜歪面墙 <播音室用,避免连续声反射> skew wall
斜歪倾伏褶皱 inclined fold
斜歪水平褶皱 inclined horizontal fold
斜歪系数 skew factor
斜歪褶皱 inclined fold
斜弯翘曲 diagonal warping
斜弯曲 skew bending
斜弯头 miter[mitre] bend;miter[mitre] knee <楼梯扶手>
斜顽辉石 clinoenstatite
斜顽辉石瓷 clinoenstatite porcelain
斜顽火石 clinoenstatite
斜网 inclined screen
斜网格楼板 diagrid floor
斜微商 oblique derivative
斜桅三角帆 cap jib
斜桅仰角 steeve
斜纬 oblique weft
斜纬线 oblique parallel
斜位 loxosis;oblique position
斜位移 oblique displacement
斜温层 metalimnion;thermocline(layer)
斜纹 interlocking grain;oblique(to) grain
斜纹背织物 twill backing
斜纹编织 twill(ed) weave
斜纹布 diamond cloth;diamond woven fabric;drill weave
斜纹材 diagonal wood
斜纹衬里 twill backed
斜纹齿 loxodont
斜纹绸 faille
斜纹锉 float-cut file;single cut file
斜纹的 comb-grained;cross-grained;diagonal-grained
斜纹断续的织物 broken twill weave
斜纹方平织物 twilled mat
斜纹花型 diagonal pattern
斜纹胶合板 diagonal plywood
斜纹角度 twill angle
斜纹接合 rake bond;raking bond
斜纹理 cross grain;diagonal grain;oblique grain
斜纹帘布层 bias ply
斜纹滤筛网 twill shape screen;twill type screen
斜纹木材 cross-fibered wood;cross grain wood;diagonal-grained wood
斜纹刨切单板 endy veneer
斜纹砌合 rake bond;raking bond
斜纹砌块 oblique(masonry)bond
斜纹筛网 twill type screen
斜纹式滤网 twill screen
斜纹顺砖砌法 raking stretcher bond
斜纹图案 diagonal pattern
斜纹织物 diamond fabric;twill fabric;twills
斜纹缀面 broached work;punched work
斜纹组织 twill weave
斜窝接合 <木结构> bevel(l)ed housing
斜卧管式进水口 inclined penstock intake
斜卧位 recumbent position
斜卧褶皱 reclined fold
斜污水网筛 inclined sewage screen
斜屋顶 double pitch roof;pitched roof;tilt roof
斜屋顶板窗 flat window on pitched roof
斜屋顶内凹窗 internal dormer
斜屋顶天窗 sloping roof skylight
斜屋顶屋脊板 saddle board
斜屋脊 hip;mitered[mitred] hip
斜屋脊椽 hip rafter
斜屋面 pitched roof;sloping roof;tilt roof
斜钨铅矿 raspite
斜吸式化油器 slanting carburetor
斜硒镍矿 wilkmanite
斜铣轴 tilting milling spindle
斜系船缆 spring line
斜系杆 diagonal tie
斜系缆线 spring line
斜隙 skew joint;splayed joint(ing)
斜霞正长岩 nosykombite
斜纤维蛇纹石 clino-chrysotile
斜弦杆 inclined chord
斜弦杆桁架 inclined chord of truss
斜弦支撑 diagonal chord bracing
斜舷板 waist stroke
斜线 bias;hatching line;skew line;slanting;slash
斜线比例尺 diagonal scale;oblique scale
斜线尺 clinograph
斜线道 oblique line;oblique stroke;diagonal dross scar <浮法玻璃缺陷>
斜线电压 ramp voltage
斜线读微尺 transverse scale
斜线分离符号 <表示分数、日期等的> solidus[复 solidi]
斜线号 slant;solidus[复 solidi];virgule
斜线脚 raked mo(u)lding;ranking mo(u)lding
斜线距离能见度 slant range visibility
斜线刻绘仪 <阴像刻图用> spacing device
斜线圈 skew coil
斜线圈仪表 inclined coil instrument
斜线深度 slant depth
斜镶板地板 diagonal cassette slab floor
斜镶板楼盖 diagonal cassette slab floor;diagonal coffer(slab) floor;diagonal waffle floor
斜镶舢板 diagonal planking
斜向 oblique course;slant;toeing
斜向波 oblique wave
斜向场 angled field
斜向传动碾磨机 angle drive grinder
斜向磁化 oblique magnetization

斜向磁迹 skewed track
斜向错动 oblique dislocation
斜向的 insequent;semi-transverse
斜向地性 plagiogeotropism
斜向电刷 diagonal brush
斜向钉钉子 skew nailing;tusk nailing
斜向定位装置 sloping ga(u)ge
斜向断层【地】diagonal fault;semi-longitudinal fault
斜向对称的 obliquely zygomorphic
斜向对流 slantwise convection
斜向对位 angular spotting
斜向发展 diagonal expansion
斜向反力 inclined reaction
斜向泛水 raking flashing(piece)
斜向辐射天线 skew antenna
斜向钢筋 diagonal bar;diagonal reinforcement;sloping reinforcement
斜向构件 dagger;diagonal member
斜向箍筋 inclined stirrup
斜向谷 insequent valley
斜向光性 plagiophototropism
斜向光泽 angular sheen
斜向滚齿 oblique hobbing
斜向海岸 oblique coast
斜向海岸线 oblique coastline
斜向航测 oblique aerial photography
斜向航空摄影 oblique aerial photography
斜向河(流)insequent river;insequent stream
斜向荷载 inclined load
斜向横挖法【疏】obliquely transverse dredging method
斜向滑动断层 oblique-slip fault
斜向滑坡 insequent landslide
斜向环榫 skewed ring dowel
斜向剪切作用 oblique shear(ing)
斜向交叉(口)skew crossing;skew intersection
斜向接(合)joining on skew
斜向接头 mitre joint
斜向街道 diagonal street
斜向节理【地】oblique joint
斜向开裂 diagonal crack(ing);sloping crack
斜向抗剪钢筋 inclined shear reinforcement
斜向抗拉钢筋 diagonal tension bar
斜向靠泊 angular berthing;oblique berthing
斜向扩散波 oblique expansion wave
斜向拉力 diagonal tension
斜向拉伸 oblique extension
斜向力 oblique force;skew force
斜向裂缝 diagonal crack(ing);inclined crack
斜向裂缝发展 diagonal cracking development
斜向路线 diagonal route
斜向锚桩 raking anchor pile
斜向耙地 diagonal harrowing
斜向排除 slope removal
斜向排水沟 batter drainage;oblique gutter
斜向配筋 sloping reinforcement
斜向碰撞 oblique collision
斜向坡道 diagonal ramp
斜向铺砌 diagonal pavement;diagonal paving
斜向砌合 diagonal bond raking bond
斜向强度 oblique strength
斜向翘起的 oblique cocking
斜向上弦杆 sloping top chord
斜向驶靠 oblique berthing
斜向式导水机构 inclined guide apparatus
斜向受拉 diagonal in tension
斜向受力桩 oblique bearing pile
斜向受弯 oblique bending
斜向受压 diagonal in compression
斜向水流 oblique flow
斜向水系 insequent drainage(system)
斜向探测 oblique incidence sounding
斜向探测电离图 oblique ionogram
斜向停车 angle parking
斜向停车布置 <停车场> inclined parking arrangement
斜向圬工砌合 diagonal masonry bond
斜向圬工图案 diagonal masonry pattern
斜向系列 clinosequence
斜向下游 slanted down stream
斜向线 line of dip
斜向相撞 angle collision
斜向性 plagiotropism
斜向压力 diagonal compression
斜向压力灌浆 advance slope grouting
斜向压应力 diagonal compression stress
斜向移动 slope removal
斜向应力 oblique stress
斜向凿割带刷的模板 miter templet
斜向凿岩 slant drilling
斜向轧辊 slope roller
斜向蒸镀 oblique evapo(u)ration plating
斜向蒸发 oblique evapo(u)ration
斜向支撑 sway bracing
斜向直移 oblique translation
斜向肘材 dagger knee
斜向驻波 diagonal standing wave
斜向转动碾磨机 angle drive grinder
斜向钻进 down pointing borehole
斜向钻孔 inclined boring
斜向作用力 inclined force
斜像差 oblique aberration
斜像散 oblique astigmation;oblique astigmatism
斜削 bead cut;bevel(ling);splay-(ing)【建】
斜削板 feather-edged board
斜削边 bevel(led)edge;chamfered edge;feather edging;feathering;splayed edge;tapered edge
斜削边墙板 tapered(-edge)wallboard
斜削波形瓦 tapered roll pantile
斜削插条 whip cutting
斜削的 bevel(l)ed;taper
斜削端 bevel(l)ed end
斜削浮雕装饰【建】bossage without bevel
斜削角 angle of taper;bevel angle;chamfered angle;hollow chamfer
斜削角的端部 bevel(l)ed end
斜削接 bevel(led)halving
斜削接缝 bevel(l)ed joint;chamfered joint;miter[mitre] joint
斜削接合 bevel(l)ed joint;scarfed joint
斜削接口螺纹 Higbee cut
斜削接头 bevel(led)joint;chamfered joint;miter[mitre] joint;skew joint;diagonal joint
斜削铆钉 tapered rivet
斜削面 sloping section
斜削木块 bevel(l)ed block
斜削销 feather tongue
斜削型端接缝 tapering end joint
斜削砖 bevel(l)ed brick;bevel(l)ed closer;feather-edged brick;skew brick
斜消光 inclined extinction
斜销 taper pin
斜楔 sloping edge
斜楔块 tapered wedge
斜楔石 trigonite
斜楔榫 miter[mitre] dovetail
斜楔体法 inclined wedge method
斜楔作用确定回程成型模 positive-return cam action forming die
斜心距 diagonal pitch
斜芯墙 sloping core
斜芯墙坝 dam with inclined core
斜芯墙土石坝 inclined core earth-rockfill dam
斜星形线 oblique asteroid
斜行 squint
斜行进 oblique march
斜行开降机 inclined lift
斜行纹理 <木材的> cross grain
斜形波 ramp
斜形舱壁板 sloped bulkhead plating
斜形槽 drift slot
斜形齿轮 bevel gear
斜形磁极 skewed pole
斜形基础 sloped footing
斜形架空接触线 inclined overhead contact line
斜形交叉 oblique crossing;skew crossing
斜形坡度 diminution
斜形砌块 skew block
斜形狭圆锯 compass saw;keyhole saw;pad saw
斜形小窗洞 <教堂内墙上的> loricula
斜形悬链线 inclined overhead contact line
斜形砖 squint brick
斜形座阀 bevel seat(ed)valve;conic-(al)seat valve
斜型圆钢 bevel bar
斜悬臂多斗挖掘机 slanting ladder ditcher
斜旋转因子解 oblique rotation factor solution
斜碹 skew arch;sloping arch
斜压波 baroclinic wave
斜压不稳定 baroclinic instability
斜压场 baroclinic field
斜压的 boroclinic
斜压杆 inclined strut
斜压过程 baroclinic process
斜压海洋 baroclinic ocean
斜压力 diagonal compression
斜压流 baroclinic flow
斜压流体 baroclinic fluid
斜压模式 baroclinic model
斜压破坏 diagonal compression failure
斜压情况 baroclinic condition
斜压缩 oblique compression
斜压涡动输送 baroclinic eddy transform
斜压涡流 baroclinic eddy
斜压涡旋 baroclinic vortex
斜压性 baroclinicity;barocliny
斜盐层 halocline
斜檐 rake cornice
斜眼 incline hole
斜眼格筛 diagrid
斜眼格栅 diagrid
斜眼凿 <打斜榫用> bevel chisel
斜堰 oblique weir;skew weir
斜燕尾榫 miter[mitre] dovetail
斜摇座 inclined rocker
斜叶涡轮式搅拌器 pitched turbine type agitator
斜腋 tapered haunch
斜移存储 skewed storage
斜移断层 diagonal-slip fault
斜移方案 skewed scheme;skewing scheme
斜移距离 skewed distance
斜倚 lean against
斜倚的 accumbent
斜倚卧铺 reclining berth
斜翼墙 flare wall;flare wing wall;oblique wing wall
斜翼墙桥台 abutment of flare wing walls;flare wing wall abutment
斜翼式法兰西斯转轮 feathering Francis runner
斜翼式转轮 feathering vane runner
斜翼叶片 <水轮机的> feathering vane
斜翼状轮叶的水轮机 feathering vane wheel
斜黝帘石 clinozoisite
斜隅石 squint quoin
斜鸳鸯榫头 bevel(led)cogging
斜圆拱 imperfect arch
斜圆筒形拱 skew-barred arch
斜圆柱体 oblique cylinder
斜圆柱正形投影 oblique cylindric-(al)orthomorphic projection;oblique Mercator projection
斜圆柱正形投影图 oblique cylindric-(al)orthomorphic chart;oblique Mercator chart
斜圆锥(体)oblique circular cone;oblique cone
斜缘 bevel(led)edge
斜凿 inclined flat chisel;pitching chisel;skew chisel
斜凿面 pitch face
斜凿石 charred stone
斜凿纹 diagonal pitch
斜轧 cross rolling;oblique rolling;skew rolling;slant rolling
斜轧厂 skew rolling mill
斜轧穿孔机 cone mill;Mannesmann piercing mill
斜轧机 cross-rolling mill;skew rolling mill;slant-rolling mill
斜轧扩径机 plug expander
斜轧式扩管机 rotary tube expanding machine
斜轧式轧机 skew mill;slant mill
斜斩砖 bevel(l)ed closer
斜展角 <护轮轨的> flare bevel
斜张钢缆 stay cable
斜张角 flare angle;open-angle
斜张桥 cable-stayed bridge;stayed-cable bridge
斜张天幕 sloping awning;tenting;tenting an awning
斜照法 oblique illumination(method)
斜照光 oblique light
斜照型照明器 angle lighting fitting;angle lighting luminaire
斜照晕 oblique hill shading
斜照晕宣法 shading with oblique lighting
斜罩 stagger
斜罩角 <叶片的> stagger angle
斜遮雨板 raking flashing(piece)
斜折合接 oblique joint
斜折线形屋顶 gambrel roof
斜褶皱 oblique fold
斜震 oblique shock
斜支撑 batter brace;bearing diagonal;diagonal brace[bracing];inclined strut;needle shoring;raking strut;ranking shore;sway brace;sway bracing;sway strut
斜支撑法 raking shoring;shoring
斜支撑木 inclined shore;raker;raking shore
斜支承 inclined bearing;raking support
斜支墩 diagonal buttressing pier
斜支杆 diagonal bridging
斜支距 oblique offset
斜支索 diagonal stay
斜支腿 sloping leg
斜支线 splay branch
斜支柱 back leg;batter post;diagonal bracing;diagonal pillar;diagonal

stanchion;diagonal strut;raker;raking prop;raking shore;sprag
斜支座(止推)轴承 oblique pillow-block bearing
斜直闪石 slino-anthophyllite
斜置 diagonal laying;diagonal placed;diagonal-wire placed;tilt
斜置安全阀 tilt relief valve
斜置丁砖皮 diagonal bond;herring-bone bond
斜置发动机 inclined engine
斜置阀 inclined valve;slanting set valve;sloping valve
斜置钢梁孔型 diagonal-beam pass
斜置箍筋 inclined stirrup
斜置桁架 diagonal trussing
斜置锯子桌 canting saw table;tilting saw table
斜置孔型 diagonal pass
斜置孔型系统 diagonal system
斜置冷却管 canted radiator tube
斜置模型 tilted model
斜置气门 inclined valve
斜置汽缸 oblique cylinder
斜置砂轮座 angular wheel slide
斜置栅(筛) inclined grizzly;sloping grizzly
斜置式舱盖 tilt stowing hatchcover
斜置式六汽缸(发动机) slant six(cylindered engine)
斜置式(蒸汽)绞车 diagonal winch
斜置水锤泵 tilt ram
斜置轧辊 cross roll
斜置装甲板 inclined armor
斜置撞锤 tilt ram
斜轴 clino-axis;oblique case;oblique position;skew axis;sloping shaft
斜轴拌和机 high discharge mixer;inclined axis mixer;inclined axis truck mixer
斜轴测射影法 oblique axonometry
斜轴测投影 oblique axonometric projection
斜轴承 angle pedestal bearer;oblique pillow-block bearing
斜轴传动 sloping-shaft drive
斜轴的 clino-diagonal
斜轴阀 valve with inclined stem
斜轴管式水轮机 inclined shaft tubular turbine
斜轴贯流式水轮机 inclined shaft tubular turbine
斜轴环流 clino-axis circular current
斜轴搅拌机 inclined axis truck mixer
斜轴锯 drunken saw blade
斜轴面 clinopinacoid
斜轴墨卡托 oblique Mercator
斜轴墨卡托投影 oblique Mercator projection
斜轴墨卡托投影地图 oblique Mercator projection chart
斜轴坡面 clinodome
斜轴式拌和机 high-discharge mixer;inclined-axis mixer
斜轴式机组 slant unit
斜轴式搅拌机 inclined axis mixer
斜轴式轴向活塞马达 angle type axial piston motor
斜轴式轴向柱塞泵 angle type axial piston pump;bent-axis type axial plunger pump
斜轴式柱塞泵缸体 inclined cylinder block
斜轴双晶 inclined twin
斜轴投影 oblique projection;skewed projection
斜轴投影赤道 oblique equator
斜轴投影地图 oblique projection chart
斜轴投影格网 oblique graticule;oblique projection graticule
斜轴投影极点 oblique pole
斜轴投影经度 oblique projection longitude
斜轴投影经纬网 oblique graticule
斜轴投影经线 oblique meridian
斜轴投影纬度 oblique projection latitude
斜轴投影纬线 oblique meridian;oblique parallel
斜轴投影无变形线 oblique rhumb line
斜轴透视圆柱投影 perspective oblique cylindric(al) projection
斜轴图 oblique chart
斜轴线 oblique axis
斜轴圆柱正形投影地图 oblique cylindric(al) orthomorphic chart
斜轴柱 clinoprism;oblique column
斜轴锥 clinopyramid
斜轴坐标系 skew coordinate system
斜轴座架 oblique axis mount
斜注切割电熔耐火材料 diamond cut lug fusion-cast refractory
斜柱 batter post;inclined post;juggler;oblique cylinder;spur post
斜柱船首 rake of bow
斜柱杆式起重机 stiffleg derrick
斜柱排架 diagonal bent
斜柱塔架 batter leg tower
斜爪 tapered jaw
斜砖 bevel brick;skew brick
斜砖砌合 raking course
斜砖砖层 raking course
斜转窗 tilt and turn window
斜桩 angle(d) pile;batter(ed) pile;brace pile;dip pile;inclined pile;oblique pile;pile pitching;raked pile;raker pile;raking pile;raking prop;sloped pile;taper pile;spur pile(for dredging)【疏】
斜桩打桩机 batter leader pile driver;oblique pile driver
斜桩敷设 out-batter pile driving
斜桩机 batter leader pile driver
斜桩框架 raking frame
斜桩缆索 pile pitching
斜桩锚碇设施 raking-piles anchorage
斜桩锚碇式板桩码头 sheet pile quay-wall with batter anchor piles
斜桩群 batter pile cluster;group of ranking piles
斜桩突堤码头 batter pile pier
斜桩斜度 pitching of pile
斜桩栈桥式码头 trestle type wharf with battered piles
斜桩支撑板桩堤岸 sheet-pile bulkhead with batter-pile support
斜装 angle mount;toeing
斜装推土板 angledozer blade
斜撞锤 battering ram
斜锥 oblique cone;oblique pencil
斜锥度 tapering
斜锥式 beaver type
斜准器 slope level
斜着 bevelways;sideway
斜琢边石 pitched stone
斜自动楼梯通道 inclined escalator shaft
斜自然硫 rosickyite
斜钻 slant drilling
斜钻孔 angled hole;angling well;deviating hole;inclined(bore)hole;oblique hole
斜嘴钳 diagonal cutting nippers
斜坐标 skew coordinates
斜座阀 bevel seat(ed) valve;slanting set valve
斜座起重机 angle crane
斜座石 club skew;skew corbel;skew putt;skew table
斜座台 skew corbel
斜座止推轴承 oblique plummer block bearing
斜座轴承 angular contact bearer

谐

谐波 harmonics;harmonic wave;harmonization;overtone
谐波百分比 percent harmonic
谐波闭锁继电器 harmonic blocking relay
谐波变换 harmonic conversion
谐波变换换能器 harmonic conversion transducer
谐波变换器 harmonic converter
谐波变频 harmonic conversion
谐波变频器 harmonic converter
谐波标度盘 harmonic dial
谐波波导管 harmonic waveguide
谐波补偿 harmonic compensation
谐波部分 harmonic component
谐波产生 harmonic generation
谐波长 resonance wave length
谐波常数 harmonic constant
谐波场 harmonic field
谐波齿轮 harmonic gear
谐波传动 harmonic motion drive
谐波次数 order of harmonic
谐波的 harmonic
谐波电动机 harmonic motor
谐波电话振铃器 harmonic telephone ringer
谐波电流 harmonic current
谐波电路失谐 detuning of resonant circuit
谐波电压 harmonic voltage
谐波定向天线 harmonic wire projector
谐波发射 harmonic emission
谐波发生器 harmonic generator;harmonic oscillator;harmonic producer;harmonic source
谐波放大器 harmonic amplifier
谐波分量 harmonic component;harmonic constituent
谐波分量发生器 harmonic component generator
谐波分析 Fourier analysis;harmonic analysis;harmonic reduction
谐波分析表 harmonic-analysis schedule
谐波分析器 harmonic analyser[analyzer];harmonic wave analyser[analyzer]
谐波辐射 harmonic radiation;radiation of harmonics
谐波干扰 harmonic disturbance;harmonic interference
谐波干扰力 harmonic disturbing force
谐波干涉 harmonic interference
谐波功率 harmonic power
谐波共轭体系 harmonically conjugate system
谐波共振 harmonic resonance
谐波过滤 harmonic filtration
谐波含量 harmonic content
谐波含有率 rate of harmonic
谐波函数 harmonic function
谐波合成 harmonic synthesis
谐波环流 harmonic circulating current
谐波回声 harmonic echo
谐波混频器 harmonic mixer
谐波畸变 harmonic distortion
谐波畸变法 harmonic distortion method
谐波激励 harmonic drive
谐波级次 harmonic number;harmonic order
谐波级数 harmonic progression;harmonic series
谐波级数展开 harmonic series expansion
谐波继电保护 harmonic relaying
谐波加速 harmonic acceleration
谐波减速装置 harmonic speed changer
谐波检波器 harmonic detector
谐波校正 harmonic correction
谐波阶数 harmonic order number
谐波接收器 harmonic receiver
谐波近似 harmonic approximation
谐波晶体 overtone crystal
谐波聚束 harmonic bunching
谐波雷达 harmonic radar
谐波励磁 harmonic excitation
谐波励磁发电机 harmonic excited generator
谐波励磁系统 harmonic excitation system
谐波漏电抗 harmonic leakage reactance
谐波滤波器 harmonic filter
谐波滤除 harmonic filtration
谐波滤除电路 harmonic filtering circuit;harmonic suppressor
谐波脉动 harmonic pulsation
谐波脉动率 harmonic regulation
谐波偏流 harmonic bias
谐波偏置 harmonic bias
谐波频率 harmonic frequency;multiple frequency;overtone frequency
谐波频率特性 harmonic frequency characteristic
谐波频率响应 harmonic frequency response
谐波频率转换器 harmonic frequency converter
谐波平衡 harmonic balance
谐波平衡曲轴 harmonically balanced crankshaft
谐波平衡仪 harmonic balancer
谐波曲线 harmonic curve
谐波群 harmonic group
谐波扰动 harmonic disturbance
谐波扰动扭矩 harmonic exciting torque
谐波绕组 harmonic winding
谐波容限 harmonic tolerance
谐波失真 harmonic distortion
谐波失真测量设备 harmonic distortion measuring set
谐波失真电平 harmonic distortion level
谐波失真衰减 harmonic distortion attenuation
谐波失真因数 harmonic distortion factor
谐波试验 harmonic test
谐波输出功率 harmonic output power
谐波衰耗 attenuation of harmonic product
谐波衰减 decay of harmonics;harmonic attenuation
谐波损耗 harmonic loss
谐波特性 harmonic response
谐波天线 harmonic aerial;harmonic antenna
谐波调谐 harmonic tuning
谐波调谐偏转电路 harmonically tuned deflection circuit
谐波位移 harmonic displacement
谐波吸收器 harmonic absorber
谐波系(列) harmonic series
谐波系数 harmonic coefficient;harmonic wave factor
谐波线性化 harmonic linearization
谐波相角 harmonic phase angle
谐波相位 harmonic phase
谐波响应 frequency response;harmonic response

谐波响应法 harmonic response method
谐波响应函数 frequency response function
谐波响应特性(曲线) frequency response characteristic; harmonic response characteristic
谐波响应图 harmonic response diagram
谐波消除 harmonic cancellation
谐波消除器 harmonic eliminator; harmonic excluder
谐波效应 harmonic effect
谐波形 harmonic forms
谐波型磁放大器 magnettor
谐波型晶体 harmonic crystal; harmonic mode crystal
谐波选择信号 harmonic selective signal(l)ing
谐波选择振铃 harmonic selective ringing
谐波压电晶体 overtone crystal
谐波压缩器 harmonic compressor
谐波异步转矩 harmonic asynchronous torque
谐波抑制 harmonic rejection; harmonic suppression
谐波抑制器 harmonic shutter; harmonic suppressor; harmonic trap
谐波抑制网络 harmonic suppression network
谐波因数 harmonic wave factor
谐波源 harmonic source
谐波运动 harmonic motion
谐波运动凸轮 harmonic motion cam
谐波运行 harmonic operation
谐波展开 harmonic development
谐波振荡 harmonic cycling; harmonic oscillation
谐波振荡器 harmonic generator
谐波振动 harmonic vibration
谐波振幅 harmonic amplitude
谐波振型 overtones mode
谐波指示器 harmonic detector
谐波指数 harmonic index
谐波周期 harmonic period
谐波转矩 harmonic torque
谐波综合 harmonic synthesis
谐波综合器 harmonic synthesizer
谐单力点源 harmonic single-force point source
谐函数 harmonic function
谐和 concord; consonance; harmonization; harmony
谐和摆动 harmonic oscillation
谐和边界 compromise boundary
谐和颤动 <火山喷发时发生的地壳轻微持续的震动> harmonic tremor
谐和常数 harmonic constant
谐和传动 harmonic drive
谐和的音调 euphony
谐和地压采煤法 harmonic coal mining method
谐和点 harmonic point
谐和分潮 tidal harmonic constituent
谐和复量 complex harmonic quantity
谐和感 harmony
谐和滚动 harmonic roll
谐和函数 harmonic function
谐和荷载 harmonic load
谐和回声 harmonic echo
谐和力 harmonic force
谐和频率 frequency harmonic; harmonic frequency
谐和平衡器 harmonic balancer
谐和趋势分析 harmonic trend analysis
谐和山脉 harmonic mountain range
谐和衰减 harmonic decline
谐和元素 harmonic element
谐和运动 harmonic motion
谐和运动凸轮 harmonic motion cam
谐和褶皱 harmonic fold
谐和褶皱作用 harmonic folding
谐和振荡 harmonic oscillation
谐和振动 harmonic vibration
谐量 harmonic quantity
谐量分析 Fourier analysis
谐频 harmonic frequency; harmonics
谐频放大器 harmonic amplifier
谐频分量 harmonic component
谐腔四极管 resnatron; resonator-tron
谐调 harmony
谐调潮流常数 harmonic current constant
谐调生境 harmonic habitat
谐音 chimb(e); harmonic; harmonic tone; symphony
谐音阶 order of harmonic
谐音联想 clang association
谐音汽笛 chime whistle
谐音容器 harmonical vase
谐音推进器 singing propeller
谐音系列 harmonic series of sound
谐音选择呼叫 harmonic selective signal(l)ing
谐音音栓 harmonic stop
谐运动 simple harmonic motion
谐振 harmonic motion; harmonic oscillation; harmonic vibration; resonance(oscillation); resonant vibration; resonate; sympathetic(al) vibration; synchronous vibration; syntony; syntonization
谐振比(率) resonance ratio
谐振臂 resonance arm
谐振变化法 resonance variation method
谐振变换 resonant transformation
谐振变压器 resonance transformer; resonant transformer
谐振波(浪) resonance wave; synchronous wave
谐振波线圈 resonance wave coil
谐振槽路 resonant tank; resonant tank circuit
谐振测量法 resonance measurement
谐振测试法 resonance test method
谐振场 resonance field
谐振触发器 resonant flip-flop
谐振传输线 resonance transmission line
谐振窗 resonant window
谐振荡 resonance oscillation
谐振的 resonant; syntonic
谐振的电压升高 resonant voltage step-up
谐振点 resonance point; resonant point
谐振电流 resonance current
谐振电路 acceptor; resonance circuit; resonant circuit; resonating circuit; tank circuit
谐振电路常数 resonance constant
谐振电路阻抗 resonant circuit impedance
谐振电桥 resonance bridge
谐振电容器 resonant capacitor
谐振电位 resonance potential
谐振电压 resonance potential; resonance voltage; sinusoidal voltage
谐振电子管 resonator-tron
谐振电阻 resonance resistance; resonant resistance
谐振动 resonance vibration
谐振扼流圈 resonant choke
谐振法 <测混凝土强度的> resonance method
谐振反谐振法 resonance-antiresonance method
谐振范围 resonance range
谐振放大率 resonance ratio
谐振放大器 resonant amplifier
谐振放大系数 resonance amplified coefficient
谐振放电器 resonant discharger; soft rhumbatron
谐振分路 resonant shunt
谐振峰(值) harmonic peak; resonance peak
谐振蜂鸣器 resonance hummer
谐振缝 resonant slit
谐振辐射 harmonic radiation
谐振辐射计 resonance radiometer
谐振干扰 resonance interference
谐振感应器 resonance inductor
谐振管 resonatron
谐振荷载 resonant load
谐振环 resonant ring
谐振簧片继电器 resonant-reed relay
谐振簧片式转速测量计 resonant-reed tachometer
谐振簧片式转速计 resonant-reed tachometer
谐振回路 acceptor
谐振回路阻抗 resonant circuit impedance
谐振混频器 resonant mixer
谐振激励 resonant excitation
谐振记录器 harmonograph
谐振继电器 discriminating relay; frequency responsive relay; resonant relay; selecting relay; selective relay
谐振角频率 resonance angular frequency
谐振接地系统 resonant earthed system; resonant grounded system
谐振截面 resonance cross-section
谐振空腔 resonance chamber; resonant cavity
谐振空隙 resonant gap
谐振控制式晶体管 resonant-gate transistor
谐振馈线 resonant feeder
谐振离子 harmonic ion
谐振蜜 resonance ratio
谐振模 mode of resonance; resonant mode
谐振模式 mode of resonance; resonance frequency
谐振模数 resonant mode number
谐振膜片 resonant diaphragm; resonant iris
谐振膜转换开关 resonant-iris switch
谐振能级间距 resonance-level spacing
谐振耦合 resonant coupling
谐振匹配 resonance matching
谐振频率 frequency of resonance; harmonic frequency; resonance frequency; resonant frequency; resonant vibration frequency
谐振频率计 reso-meter; resonance efficiency meter; resonance frequency meter
谐振频率调制 resonant frequency modulation
谐振频率温度系数 temperature coefficient of resonance frequency; temperature coefficient of resonant frequency
谐振频率响应 resonant frequency response
谐振频谱 resonance spectrum
谐振平衡试验机 below resonance balancing machine
谐振瓶 syntonic jar; syntony jar
谐振器 cavity of resonator; harmonic oscillator; resonant cavity; resonant element; resonator; syntonizer
谐振器磁铁 resonator magnet
谐振器固定夹 resonator fixing clamp
谐振器护罩 resonator guard
谐振器间隙 resonator gap
谐振器绝缘板 resonator insulator
谐振器栅极 resonator grid
谐振器止块 resonator banking stop
谐振腔 resonant chamber; resonator
谐振腔波长计 cavity resonator wavemeter; cavity wavemeter
谐振腔磁控管 resonator magnetron
谐振腔反射镜准直 cavity reflector alignment
谐振腔频率计 cavity frequency meter
谐振腔天线 cavity antenna
谐振区 resonance range; resonance region
谐振曲线 resonance curve; response curve
谐振曲线峰值 resonance peak
谐振曲线面积法 resonance curve area method
谐振圈天线 resonance coil antenna
谐振锐度 resonance sharpness; sharpness of resonance
谐振舌片式测速发电机 resonant-reed tachometer
谐振升压 resonance step-up
谐振示波器 resonoscope
谐振式波长计 resonance frequency wavemeter; resonance wavemeter
谐振式对接焊 resonance butt weld
谐振式绝缘装置 resonant insulating apparatus
谐振式频率表 resonant frequency meter
谐振式频率计 resonance frequency meter
谐振式频率指示器 resonant circuit type frequency indicator
谐振式调压器 resonance type voltage regulator
谐振式消声器 resonator type silencer
谐振式仪表 resonance instrument; resonant type instrument
谐振式阻抗连接变压器 resonated impedance bond
谐振式阻抗联结器 resonated impedance bond
谐振式阻容振荡器 resonance type capacitor resistance oscillator
谐振试验 resonance test
谐振试验机 resonant tester
谐振四端网络 resonant four-terminal network
谐振探测 resonance probe
谐振特征 resonance characteristic
谐振天线 resonant aerial; resonant antenna
谐振吸减器 resonant absorber
谐振吸声体 resonant sound absorber
谐振吸收 resonance absorption
谐振隙 resonant gap
谐振现象 resonance effect; resonance phenomenon
谐振线 resonant line; transmission line
谐振线圈 resonance coil; resonance inductor
谐振线调谐器 resonant line tuner
谐振线振荡器 resonant line oscillator
谐振箱 resonant chamber
谐振响应 resonance response
谐振消声器 resonator type absorber
谐振效率 resonance efficiency
谐振效应 resonance effect
谐振性激增 resonance step-up
谐振压电石英片 resonating piezoid
谐振抑制器 resonant choke
谐振用石英晶体片 resonating piezoid

谐振用石英片 resonating piezoid
谐振元件 resonant element
谐振增强器 resonance intensifier
谐振振荡电路 resonance oscillatory circuit
谐振指示器 resonance indicator
谐振装置 resonance device
谐振状态 resonance condition
谐振子 harmonic oscillator; harmonic vibrator; simple oscillator
谐振阻抗 resonance impedance
谐振阻尼器 resonator damper
谐振作用 resonance action
谐震 harmonic tremor

携播 phoresy

携布植物 deamochore
携带 carry; carry-over; tote
携带包装 carrier pack
携带大量泥沙 silt-laden
携带大量岩屑的泥浆液 cuttings laden mud fluid
携带电磁铁 portative electromagnet
携带机理 entrainment mechanism
携带介质 carrying agent
携带泥沙 picking-up of sediment
携带泥沙的 sand-bearing; sediment-entraining
携带泥沙的河流 sand-bearing river
携带设备 portable testing set
携带式 portable type
携带式 X 射线分析仪 portable X-ray analyser[analyzer]
携带式 X 射线设备 portable X-ray equipment
携带式 X 荧光仪 portable fluorescence apparatus
携带式安培计 pocket ammeter
携带式暗室 <户外用> tent
携带式标准仪表 portable standard meter
携带式冰箱 portable refrigerator
携带式测试器 routine tester
携带式测试设备箱 portable test kit
携带式超声波测厚仪 audiga(u)ge
携带式灯 extension lamp
携带式地下水(位)测定仪 portable dipmeter
携带式电话 portable telephone
携带式电视机 walkie-lookie
携带式放射线剂量计 pocket dosimeter
携带式放射性测井仪 portable radioactive logging apparatus
携带式辐射监测器 portable radiation monitor
携带式辐射能测定仪 cutie
携带式高温计 portable pyrometer
携带式光导摄像管摄像机 creepie-peepie
携带式红外线探测器 metascope
携带式剂量计 pocket chamber
携带式检测器 portable detector
携带式检电器 lineman's detector
携带式接收机 mobile receiver; portable receiver
携带式离心机 portable centrifuge
携带式麻醉机 portable anesthesia apparatus
携带式脉率指示表 portable pulse rate indicator
携带式黏[粘]度计 portable visco(si)meter
携带式起搏器 portable pacemaker
携带式气动工具 portable air tool
携带式前轮定位仪 portable alignment ga(u)ge
携带式栅陷振荡器 portable dipmeter
携带式示波器 portable oscillograph
携带式试验装置 portable test set
携带式水平仪 pocket level
携带式无线电话机 walkie-talkie; walky-talky
携带式无线电台 portable radio station
携带式吸尘器 portable suction fan
携带式显微镜 portable microscope
携带式线性计数率计 portable linear ratemeter
携带式消防栓流量计 portable hydrant flowmeter
携带式氧气瓶 carry-around oxygen cylinder
携带式野外能谱仪 portable field spectrometer
携带式液化石油气采暖炉 mobile space heater
携带式仪表 portable instrument
携带式仪器箱 carrying case
携带式译意风 <俚语> walkie-hearie
携带式硬度计 portable hardness tester
携带式油压千斤顶 portable oil jack
携带式(游标)卡尺 pocket slide cal(l)pier
携带式直流 pH 计 portable battery pH-meter
携带式钟表 portable timekeeper
携带式浊度计 portable turbidimeter
携带水分 moisture carry-over
携带水分能力 moisture carrying capacity
携带速度 entrainment velocity
携带土壤水分 carry-over soil moisture
携带系数 entrainment coefficient
携带岩屑能力 <泥浆的> carrying power; solids-carrying capacity
携带用小型收发报机 handie-talkie
携带运动 dragging motion; dragging movement
携带者 carrier
携泥量 silt carrying capacity
携气剂 air-entraining agent
携沙量 silt carrying capacity
携沙能力 moving capacity
携沙液 sand carrier
携手 hand in hand
携污容量 soil carrying capacity
携污性 soil carrying property
携氧能力 oxygen carrying capacity
携运箱 carrying case

缬氨酸 valine

鞋带菌 honey fungus; shoestring fungus

鞋带状 shoestring-form
鞋带状砂层 shoestring sand
鞋带状油气藏趋向带 shoestring sand pool trend
鞋底铁掌 calkin
鞋店 shoe shop; shoe store
鞋钉 ca(u)lking nail
鞋盖 gaiter
鞋工 cordwainer
鞋匠板凳 cobbler's bench
鞋扣形电池 shoe-button cell
鞋面压延机 upper calendar
鞋钮管 shoe-button tube
鞋襻 bootstrap
鞋套 shoe cover
鞋形连接件 oblique saddle junction piece
鞋形耙头 shoe-shaped drag head
鞋形物 shoe
鞋印 heel mark
鞋状砂矿层 shoestring sand

写保护 write lockout

写保护环 write protection ring
写报告 paper work
写出指令 order-writing
写穿透电压 write-breakthrough voltage
写错日期 misdate
写读头 write head; writing head
写放大器 write amplifier
写封锁 write lockout
写过程 write procedure
写后读 read-after-write
写后读检查 read-after-write check
写后干扰 post-write disturb
写后干扰脉冲 post-write disturb pulse
写恢复 write recovery
写回【计】write-back
写景地图 illustration map
写景地质学 scenographic geology
写景法 scenography
写景图 pictorial map
写控制 write gating
写控制逻辑 write control logic
写论文 dissert(ate)
写明 stipulate
写命令线 write enable line
写区 writing field
写驱动绕组 write drive winding
写任务 writing task
写入总线 write bus
写生薄 sketch book
写生画 paint from nature; sketch
写时读 read-while-writing
写时间 write time
写实报告 factory report; factual report
写实记录 factory record; factual record
写述电路 writing circuit
写数策动器 writer driver
写数据 write data
写数驱动器 write driver
写数头 write-read head
写锁定 write lockout
写头 write head; writing head
写信室 writing room
写寻址 write addressing
写意 liberal style
写意画 painting in free-sketch style
写字间楼层 office stor(e)y
写字楼 office building
写字台 knee desk; lectern; secretary; writing desk; writing table; bureau [复 bureau/ bureaus] <英>
写字仪 writing apparatus
写字椅 writing-chair
写作室 writing room

泄爆 explosive venting

泄爆窗 blowout window
泄爆门 explosion relief port
泄爆屋顶 blowout roof
泄爆装置 explosion relief provision
泄冰 ice discharge
泄冰槽 ice chute
泄冰道 ice chute
泄冰建筑物 ice pass structure
泄冰渠 ice escape channel
泄冰闸 ice sluice; slush ice sluice
泄冰闸门 ice gate
泄槽 chute(channel)
泄差输送 air pulse conveying
泄出 blow-off; drain away; effluvium [复 effluvia]; outflowing; shedding; spill
泄出存油 defueling
泄出的 emulgent
泄出管 bleeder
泄出量 leakage
泄出水 discharged water
泄出水流 discharge flow
泄出物 discharge; evacuation
泄出装置 draining provision
泄地电流 earth current
泄地电流表 earth current meter
泄电指示器 leak detector
泄阀球 drain valve ball
泄放 bleed-off; flashing
泄放波 surge wave
泄放点 discharge point
泄放电流 bleeder current; leakage current
泄放电路 bleeder chain; bleeder circuit; bleed-off circuit; leak circuit
泄放电阻 leak resistance
泄放扼流器 bleed choke
泄放阀 bleed(er) valve; bleeding cock; discharge(service) valve; release valve
泄放管 discharge tube
泄放管道 discharge conduit
泄放活门 blowdown valve
泄放孔 discharge hole; escape orifice
泄放流量 draw-off discharge
泄放浓度值 leakage concentration magnitude
泄放器 bleeder
泄放曲线 release curve; release diagram
泄放塞 drain plug
泄放水流 discharging current
泄放线圈 leakage coil; leak coil
泄放旋塞 delivery cock; drain cock; draw(-off) cock; waste cock
泄放装置 relief device
泄愤效应 spite effect
泄管 exit tube
泄洪 flood discharge; flood relief
泄洪坝段 spillway dam section
泄洪道 flood gate; flood way; freshet canal; rapid flood passage
泄洪底孔 bottom outlet; bottom sluice
泄洪阀门 flood relief valve
泄洪工程 flood relief works
泄洪涵管 spillway culvert
泄洪建筑物 flood discharge structure; flood-releasing work
泄洪孔 flood opening
泄洪口 diversion cut
泄洪量 overflow discharge
泄洪能力 flood carrying capacity
泄洪前缘 flood-releasing front
泄洪桥 relief bridge
泄洪区 flood way
泄洪渠 freshet canal; overflow channel; flood relief channel
泄洪渠道 flood relief channel
泄洪水道 flood channel
泄洪隧道 flood discharge tunnel; sluice tunnel
泄洪隧洞 flood discharge tunnel; sluice tunnel; spillway tunnel
泄洪堰 waste weir
泄洪闸(门) flood gate; flood diversion sluice; flood relief sluice; sluice; spillway gate
泄降 drawdown
泄降区 drawdown zone
泄降曲线 drawdown curve; drop-down curve
泄降曲线纵断面 drop-down section
泄降时间 time of drawdown
泄降水位流量曲线 drawdown-dis-

charge curve
泄降涌浪 drawdown surge
泄空 emptying
泄空阀 emptying valve
泄空时间 emptying time; time of emptying
泄力 stress-relieving
泄量高程关系 discharge-elevation relation
泄量高程关系曲线 discharge-elevation relation curve
泄料 blowdown
泄料池 blowdown pit
泄料阀 blowdown valve
泄料管路 blowdown piping
泄料罐 blowdown container
泄料桶 blow tank
泄料系统 blowdown system
泄料箱 blowdown tank
泄料烟道 blowdown stack
泄凌道 ice sluice
泄流 aerial drainage; cascading water; discharge flow
泄流板 draw-off pan
泄流变压器 draining transformer
泄流冲刷速度 downwash velocity
泄流底涵 lower discharge tunnel
泄流段 discharge bay; discharge range
泄流阀 eduction valve
泄流范围 discharge range
泄流方法 method of discharging
泄流管 expansion pipe; leak-off pipe; water discharge pipe; water discharge piping
泄流河 exit river; exit stream
泄流孔 dewatering opening; dewatering orifice; discharge hole; discharge opening; discharge orifice; drain hole; outlet entrance
泄流量 water discharge
泄流能力 capacity of spillway; discharge capacity; discharging capacity; flow capacity
泄流排水管 relief sewer
泄流器 bleeder
泄流软管 discharge hose
泄流室 effusion cell
泄流水孔 weep hole
泄流速度 discharge velocity
泄漏 blow by; breakthrough; dispersion; draining; efflux; leakage; leaking; let out; oil leak(age); runoff; scatter; severity sew
泄漏槽 leakage tank
泄漏程度 leakiness
泄漏处 weep
泄漏的阀 leaky valve
泄漏点 leakage point
泄漏电导 scatter admittance
泄漏电导率 leakage conductivity
泄漏电流 leakage current
泄漏电流检测器 leakage current detector
泄漏电路 leakage path
泄漏电平 leak level
泄漏电阻 bleeder; bleeder resistance; leakage resistance
泄漏电阻器 bleeder resistor[resister]
泄漏阀 leak valve
泄漏放电 leakage discharge; spillover
泄漏峰 escape peak
泄漏辐射 leakage radiation
泄漏辐射剂量 leakage radiation dose
泄漏功率 leakage power
泄漏管路 leakage line
泄漏罐 leakage tank
泄漏光谱 leakage spectrum
泄漏和滴失 leakage and drip
泄漏剂量 leaking dose
泄漏检测器 leakage indicator
泄漏检查 leak detection
泄漏检验 leak check
泄漏进风 leakage intake
泄漏距离 leakage distance
泄漏控制 leakage control
泄漏量 amount of leakage; leakage amount; leakage rate; spillage
泄漏流 leakage flow
泄漏流量 leakage flow rate
泄漏路径噪声 leakage-path noise
泄漏率 leakage rate
泄漏容量 breakthrough capacity
泄漏试验 air-tight test; leak(age) test
泄漏数字积分 digital leakage integration
泄漏损失 leakage loss; leakage ullage
泄漏探测器 leak detector; leak finder
泄漏通道 leakage path
泄漏通量 leakage flux
泄漏系数 coefficient of leaking; leakance
泄漏现场 leakage scene
泄漏信号孔 telltale hole
泄漏液点 weep point
泄漏抑制 leakage reduction
泄漏指示孔 leakage indicator hole
泄漏至地 leak to ground
泄漏中子 leakage neutron
泄漏阻抗 leakage impedance
泄露 disclose; disclosure; leak out
泄露点 breakthrough point
泄露辐射 compromising emanation
泄露机密 breach of confidence
泄密 squeal
泄气 aerofluxus; air escape; air leak(age)
泄气阀 air release valve; escape valve; release valve; snuffle valve
泄气管 gas outlet
泄气轮胎 deflated tire[tyre]
泄气旋塞 air escape cock
泄气状态 deflated condition
泄砂阀 sand flush valve
泄砂孔 <又称泄沙孔> sand outlet
泄砂量 discharge of solids
泄砂堰 flush weir
泄砂闸(门) sand escape; sand sluice; sand sluicing gate
泄束磁铁 beam-dump magnet
泄水 dewatering; discharging; draining (off the water); flash; sluice; sluicing; unwater(ing); water release
泄水坝 flush dam; flush weir; spillway dam
泄水板 ablution board; drain(ing) board
泄水泵的排水涵洞 pump discharge tunnel
泄水波 release wave
泄水槽 discharge trough; weeping channel; wriggle
泄水池 discharge basin; discharge bay
泄水冲沙 sluicing and scouring
泄水冲刷 slugging; sluicing; jet scour
泄水处 off-take
泄水带 belt of discharge; zone of discharge
泄水道 by-passing; discharge conduit; downcomer; flashway; outlet conduit; outlet sluice; scour outlet; sluice; sluiceway; spillway; tailrace; water escape
泄水道操作廊道 sluice operating gallery
泄水道操作室 sluice operating chamber
泄水道截水环 <坝内用的> sluice collar
泄水道进口 outlet entrance; sluice entrance
泄水道坡度 flow gradient
泄水道闸 sluiceway gate
泄水道闸门室 sluice-gate chamber
泄水底管 bottom-discharge pipe
泄水底孔 bottom outlet; bottom sluice; bottom sluice gate; deep outlet; dewatering conduit; low-level outlet; unwatering conduit
泄水底孔闸门 bottom sluice gate; gate ground
泄水点 draw-off point
泄水斗 drainage bunker
泄水断面面积 discharge area
泄水阀(门) bleed valve; discharge valve; drain cock; drip valve; emptying valve; outlet valve; release valve; washout valve; draw-off valve; sluice valve
泄水分离器 drain separator
泄水复氧 reoxygenation by discharging
泄水工程 outfall works; outlet works
泄水沟 catch-drain; drain ditch; feed ditch; leak-off; outlet drain; relief ditch; weeper drain
泄水沟盖 gull(e)y cover
泄水构造 water-escape structure
泄水构筑物 discharge structure; outlet works; release structure; release works; escape works
泄水关系曲线 water discharge relation curve
泄水管(道) discharge conduit; discharge pipe; discharge tube; discharging tube; draw(n)-off pipe; escape pipe; lower leg; outlet pipe; runoff conduit; runoff pipe; sluice pipe[piping]; sluice tube[tubing]; tap pipe; water discharge pipe[piping]; water drainage pipe[piping]; bleeder pipe; downcomer; drain(age) pipe; outlet conduit; water drain pipe; weep(ing) pipe; spout <阳台等处的>
泄水管道(水头)损失 discharge-line loss
泄水管进口 outlet entrance
泄水管闩 drain cock
泄水管线 drainage line; drainage pipeline; sluice line
泄水柜 drainage tank; drain cistern
泄水涵洞 discharge culvert; discharge opening; draw-off culvert; emptying culvert; sluice culvert; dewatering culvert
泄水滑槽 trash chute
泄水机构 <水库的> outlet element
泄水建筑(物) discharge structure; escape works; outlet structure; outlet works; release works; sluice works; water release structure
泄水节点 drained joint
泄水结构 water outlet
泄水截门 washout valve
泄水井 catch pit; drainage shaft; drainage well; dry well; negative well
泄水阱 drainage sump; pocket well
泄水孔 discharge opening; discharge orifice; drainage hole; drainage opening; outlet entrance; weep drain; weeper; relief hole
泄水(孔)坝段 sluice section
泄水孔底 unwatering conduit
泄水孔过渡段 outlet transition
泄水孔渐变段 outlet transition
泄水孔进口 outlet entrance
泄水孔流量 outlet discharge
泄水孔容量 outlet capacity
泄水孔塞 bleed(er) cock; bleeder plug; bleeding plug; bottom plug; drain bolt; drain cock; drain pin; drain plug; docking plug
泄水控制建筑(物) outlet control structure
泄水控制闸板 outlet regulating gate
泄水控制闸门 outlet regulating gate
泄水口 discharge opening; discharge outlet; discharge port; drainage opening; drainage outlet; drain hole; drain(ing) opening; off-take; opening for drainage; outlet port; output port; scupper; waste outlet
泄水口结构 outfall
泄水廊道 discharge culvert; emptying culvert; unwatering gallery
泄水冷却器 drain cooler
泄水量 escapage; discharge quantity
泄水龙头 draining tap; draw-off cock
泄水路 drain way
泄水门 drain gate
泄水面积 drainage area; runoff area
泄水能力 discharge capacity; discharging capacity; flow capacity; flow discharge; outlet capacity
泄水能量 flow capacity
泄水排除 draw-off
泄水盘 drain pan
泄水坡度 grade of discharge
泄水器 drainer
泄水球阀 drain ball-valve
泄水渠(道) discharge canal; discharge channel; escape canal; outflow channel; outlet channel; sluice channel; sluiceway canal; sluiceway channel; sluicing canal; sluicing channel; dewatering channel
泄水渠进口 outlet entrance
泄水软管 discharge hose
泄水塞 draining plug; drain stopper; docking plug
泄水塞体 drain cock body
泄水设备 draining equipment; draining installation; emptying device; sluice equipment; sluice installation
泄水时间 emptying time; dewatering time; time to dewater
泄水栓 clearing plug
泄水速度 discharge velocity; velocity of discharge
泄水隧道 drainage tunnel; outlet tunnel; water discharge tunnel
泄水隧洞 discharge tunnel; draw-off tunnel
泄水塔 draw-off tower
泄水台 water-table
泄水条件 discharge condition
泄水瓦管 bleeder tile; flashing tile
泄水弯管顶部通风口 crown-vent
泄水弯头 downspout elbow
泄水涡流 outflow vortex
泄水系统 drainage system; emptying system; hydraulic emptying system; water release system; dewatering system
泄水箱 sluice box
泄水旋塞 drainage cock; draining cock; draw-off cock
泄水眼 bleed hole
泄水堰 escape weir; flush weir; overfall weir; overflow weir; sluice weir
泄水闸 discharge sluice; discharging sluice; emptying gate; escape sluice; gole; outlet sluice; release sluice; sluice; sluice valve; water release gate
泄水闸板 flashboard; floodboard; sluice board
泄水闸口 gate outlet

泄水闸流量 discharge of sluice;sluicing capacity
泄水闸门 flood gate;outlet gate;release(d) gate;sluice gate;waste gate;penning gate
泄水闸门下缘 sluice-gate lip
泄水闸室 sluice chamber
泄水闸室滚动闸门 rolling gate of sluice chamber
泄水闸通道量 sluice capacity
泄水闸引水 sluicing
泄水主管 main drain
泄水锥 discharge cone;hub extension;runner cone;runner boss <桨式的>;hub cone <轴流式水轮机的>
泄水总管 drain manifold
泄水钻孔 discharge borehole;release(bore) hole
泄水嘴 <阳台、平台等向外排水的> downspout
泄完水的船坞 dewatered drydock
泄污道 trashway
泄蓄(关系) outflow storage
泄蓄因数曲线 <水库> outflow storage factor curve
泄压 pressure relief
泄压槽 pressure-relief slot
泄压阀 decompression valve;pressure escaping valve;pressure-release valve;pressure-relief damper;pressure-relief valve;relief valve
泄压风门 pressure-relief damper;relief damper
泄压杆 compression release lever
泄压拱 auxiliary arch
泄压管线 relief(pipe) line
泄压开关 compression relief cock
泄压孔 pressure port
泄压口 pressure-relief opening
泄压门 pressure port
泄压旁通阀 relief bypass valve
泄压通气管 relief vent
泄压旋塞 compression relief;relief cock
泄压装置 pressure-relief device;pressure-relief provision
泄烟道 vent stack
泄油 oil discharge
泄油槽 drainage pan;drain pan
泄油阀 gate gurgle valve;oil discharge valve;oil drain valve
泄油管 drainage pipe;oil discharge pipe
泄油流道 leakage passage
泄油器 drip
泄油塞 oil drainer
泄油嘴 leak-off nozzle
泄走 venting

泻

泻槽式溢洪道 chute spillway

泻出 outpour
泻湖 barachois;barrier lagoon;bayou;border lake;etang;lagoon;laguna
泻湖岸边土地 lagoon-side
泻湖边缘 lagoon margin
泻湖滨面沉积 lagoon shoreface deposit
泻湖沉积 lagoonal deposit
泻湖沉积土 lagoon sedimentary soil
泻湖岛 lagoon island
泻湖复合体 lagoon complex
泻湖港 bay harbo(u)r;lagoon harbo(u)r;lagoon port
泻湖港湾 lagoon harbo(u)r
泻湖海湾 lagoon harbo(u)r
泻湖河口 lagoon mouth
泻湖建造 lagoonal formation
泻湖口 lagoon mouth
泻湖类型 type of lagoon
泻湖坡 lagoon cliff;lagoon scarp;lagoon slope
泻湖区 lagoon area
泻湖沙坝沉积 lagoon sand-bar deposit
泻湖砂 lagoon sand
泻湖水 lagooned water
泻湖水道宽 width of lagoon channel
泻湖水盐度 salinity of lagoon
泻湖滩 lagoon beach
泻湖湾 lagoon harbo(u)r
泻湖相 lagoon facies
泻湖淤泥 lagoon mud
泻湖沼泽演替系列群丛 lagoon marsh association
泻剂 purgative prescription
泻利盐 epsomite
泻流 effusion
泻流冷却 effusion cooling
泻鼠李 Common buckthorn
泻水槽 watering groove;weather groove;weathering trough
泻水场地 absorption field
泻水的 weathered
泻水缝 weathered joint;weather(ing) joint(ing)
泻水勾缝 stuck point pointing;weathered joint;weathering pointing;weather-struck joint
泻水沟 flash
泻水假屋顶 <烟囱后的> cricket
泻水井 absorbing well;waste well
泻水坡 <烟囱或桥墩的> weathering back
泻水坡度 <窗台等> weathered slope
泻水倾斜面板 canting strip
泻水台 offshoot;water-table
泻水线脚 weather mo(u)lding
泻水线条 weathering mo(u)lding
泻水斜板 weathering
泻水斜度 weathering slope
泻水斜角 weathering
泻水斜面 weathering
泻水闸 flushing sluice
泻性水 aperient water
泻盐 <硫酸镁结晶> Epsom salt;magnesium sulfate
泻盐矿 epsomite

卸

卸岸 landing;unloading ashore

卸岸货账册 landing book
卸岸价格 landed price
卸岸日期 date of landing
卸岸与交付 landing and delivery
卸板垛机 slab unpiler
卸板机 unloader
卸材机 off-loader
卸材台 log deck
卸槽 shoot
卸车 unload(ing)
卸车场 unloading yard
卸车吨数 tons unloaded
卸车费 unloading charges
卸车工 tripper man;unloader
卸车工艺系统 car unloading technology system
卸车机 car unloader;unloader;wagon unloader
卸车机械 wagon shaker
卸车及安装成本 unloading and erecting cost
卸车计划 unloading plan
卸车坑 dump hopper;train unloading pit;dump pit
卸车能力 capacity of unloading
卸车器具 unloading device
卸车清单 list unloaded wagons
卸车数 car unloadings;number of wagons unloaded;wagon unloadings
卸车台 car unloading platform;incoming(loading) space
卸车线 unloading siding
卸车站台 unloading platform
卸车装备 unloading installation
卸车作业 car unloading operation;unloading operation
卸成部件 take to piece
卸成长堆 dump in a windrow
卸成行 dump in windrow
卸出分析 dump analysis
卸出列表 dump list
卸出数字 outturn quantity
卸出损失 discharge loss
卸出线 discharge line
卸出斜铁 drift key
卸出支管 discharge manifold
卸出装置 discharge equipment
卸除 disengage;unload
卸除道砟机 ballast unloader
卸除舾装 unship
卸除压载 unballast
卸船 stevedore;unship;unloading of ship
卸船费 discharging charges;unloading charges
卸船港 unloading port
卸船工艺系统 ship-unloading technology system
卸船后不负风险 no risk after discharge
卸船机 ship unloader;unloader
卸船率 discharging rate
卸船码头 discharge terminal
卸船桥式起重机 unloading bridge
卸船设备 ship-unloading plant
卸船时间 unloading time
卸船速率 unloading rate
卸船(突堤)码头 unloading pier
卸船用斗式提升机 barge-unloading elevator
卸船重量 landed weight
卸船装火车 ex-ship to rail
卸船作业 ship-discharging operation;ship-unloading operation;unloading operation
卸船作业区 discharge terminal
卸道砟 ballast unloading
卸掉 take-off
卸掉螺栓的 unbolted
卸斗时间 dump time
卸斗液压缸 dump cylinder
卸垛 destacking
卸垛机 unpiler
卸舵钩 rudder pit;rudder well
卸阀器 valve removal;valve remover
卸放装置 discharger
卸负荷 load removal;load-shedding
卸负荷孔 equalizing hole
卸管手把 breakout deuce
卸轨起重机 rail unloading crane
卸荷 decompression;discharge;down load(ing);lay-off;load off;relief
卸荷板 relieving floor;relieving platform;relieving slab
卸荷爆破 relieving shot
卸荷带宽度 width of relaxed zone
卸荷阀 relieving valve;unloading valve;valve removal
卸荷拱 discharging arch;relief arch;relieving arch
卸荷回路 relief circuit
卸荷活塞 labyrinth piston
卸荷结构面滑坡 landslide of unload structural plane
卸荷孔 balancing hole
卸荷裂隙 relaxed fracture;unloading fissure
卸荷龄期 load removal age
卸荷模量 decompression modulus;unloading modulus
卸荷平台 relief platform;relieving platform
卸荷期 off-load period
卸荷气环 relieved compression ring
卸荷器 capacity unloader;unloader
卸荷强度 loosening strength
卸荷曲线 decompression curve;unloading curve
卸荷时间 time of unloading
卸荷试验 unloading test
卸荷手柄 relief lever
卸荷台式码头 relieving platform wharf
卸荷调节 off-load regulation
卸荷旋塞 relief cock
卸荷运转 running unloaded
卸荷装置 relief arrangement
卸荷钻 unloading auger
卸灰道 <蒸汽机车的> ashpit road
卸灰阀 cinder valve
卸灰管 cinder pipe
卸灰口 ash dump
卸灰炉排 dump grate
卸灰器 dust discharger
卸混凝土的溜槽 flow trough
卸活塞器 piston removal
卸货 discharge;discharging;clear a ship;disburden;landing of cargo;outloading;uncharge;unlade;unload cargo;unload(ing);unship;off-load(cargo)
卸货班 discharging crew;discharging gang;discharging party;discharging team;unloading crew;unloading gang;unloading party;unloading team
卸货报告 landing account;outrun report
卸货泊位 discharging berth;ship unloading berth
卸货产品 discharge product
卸货场 lighter's wharf;unloading yard
卸货场废水 wastewater from debarking operations
卸货车 car dump;dump car;dump cart;dumper
卸货车位 <汽车> unloading bay
卸货尘埃控制 tipple dust control
卸货成本 discharging expenses;landed cost
卸货处 discharging place;landing place;unloading place;unloading yard
卸货次序 discharge rotation
卸货代理商 landing agent
卸货单 landing order;unloading certificate
卸货的 off-load
卸货的一侧 delivery part
卸货地点 discharging place;unloading place;unloading point;landing place
卸货点 break-bulk point;clearance point;debarkation point;discharging point;tipping point;unloading point;unloading terminal
卸货斗 discharge hopper
卸货费(用) unloading cost;discharging expenses;discharging charges;landing charges;landing expenses;landing hire;unloading charges
卸货费用及栈租 receiving storage delivery
卸货费在内的到岸价格条款 cost, in-

surance, freight land terms; freight land terms
卸货风险 unloading hazard
卸货港(口) unloading port; port of debarkation; port of discharge; discharge port; discharging port; port of debarkation; port of delivery; port of discharge; port of unloading; port of discharging; port of disembarkation; port of landing
卸货港标志 discharging port mark; port mark
卸货港订运 home booking
卸货港口 port of discharging
卸货港序 calling for order
卸货工具 discharger
卸货轨 unloading track
卸货河弯 discharging bay; unloading bay
卸货后不负责 non-risk after discharge
卸货机 decrater; discharger; elevator; ship unloader【港】; unloader
卸货机泵 unloader pump
卸货机构 unloading gear
卸货机塔 unloader tower
卸货记录 cargo boat note
卸货记录簿 landing account
卸货间 delivery room
卸货减轻船的负载 lightening a ship
卸货卡 unloading card
卸货口【船】discharging hatch
卸货率 unloading rate
卸货螺旋 load displacing screw
卸货码头 port of discharging; unloading quay; unloading pier; discharge berth; discharge jetty; discharge quay; discharge wharf; discharging berth; discharging quay; discharging wharf; landing wharf; lighter's wharf; port of discharge; unloading dock; unloading terminal; unloading wharf; tipping jetty <散货的>
卸货面 unloading front
卸货能力 discharging capacity; takeaway capacity; unloading capacity; discharge capacity
卸货牌 unloading card
卸货棚 <一般适用于卸煤漏斗车> unloading shed
卸货品质条件 landed quality term; landing quality term
卸货平台 discharging platform; landing platform; unloading platform
卸货期 discharging period; unloading period
卸货器 unloader
卸货区 unloading area
卸货曲线 discharging curve; unloading curve
卸货人 discharger
卸货日 lying days
卸货日期 date of discharge; date of discharging; date of landing; discharging day
卸货上岸 landing
卸货设备 unloading aid; delivering device; discharging equipment; discharging plant; unloading equipment; unloading plant
卸货设施 discharging facility
卸货申请书 application to unload
卸货时间 discharge time; discharging period; discharging time; unloading period; unloading time
卸货时间表 discharging time list
卸货时外表明显损坏 apparent damage on discharge
卸货事故报告书 report on exception to cargo discharged
卸货收据 discharge receipt
卸货授受单 boat note
卸货数量条件 landed quantity term
卸货顺序单 discharging list
卸货速度 rate of discharging
卸货速率 rate of discharging; unloading speed; unloading rate
卸货台 receiving area
卸货跳板 unloading gang plank; unloading ramp
卸货停机坪 unloading apron
卸货通知 notice of readiness
卸货通知单 landing order
卸货推板 unloading bulkhead
卸货推料器 discharge pusher
卸货完毕 discharged finish
卸货网 cargo net; cargo network
卸货效率 discharge rate; discharging efficiency; discharging rate; rate of discharging; unloading efficiency
卸货斜台 unloading ramp
卸货许可单 bill of sufferance
卸货许可证 discharging permit
卸货样品 outturn sample
卸货用螺旋输送机 unloading auger
卸货在驳船上 lighterage
卸货栈桥 unloading bridge
卸货站 discharge station; tipping point; unloading station
卸货站台 discharge platform; discharging platform; unloading platform
卸货折减 discharge reduction
卸货者 unloader
卸货证 landing card
卸货证明 landing certificate; landing permit
卸货证书 unloading certificate
卸货重量 delivered weight; discharging weight; landed weight; landing weight; outturn weight
卸货重量条件 arrival weight terms; landed weight term
卸货重量为准 landed weight final
卸货装置 discharging equipment; discharging gear; discharging plant; easing gear; lading discharge device; shedder
卸货准备完成通知书 notice of readiness
卸货准单 permit to discharge; permit to unlade; permit to unload
卸货作业 break-bulk operation; discharging; discharging cargo work
卸货作业台 discharging platform; unloading platform
卸件装置 shedder
卸焦炭推杆 coke discharging ram
卸卷机 ejecting gear
卸卷器 stripper plate
卸开 dismantle; unlink
卸开接头 hook-off joint
卸空 clear; empty; unstow
卸空车计划 plan of empty wagons after unloading and transshipment
卸空车数 empty rolling stock amount of discharging
卸空的 unladen
卸空集装箱 emptying container
卸空日期 date of complete of discharging; final landing date
卸空时间 dump time
卸扣 shackle
卸扣滑车 shackle block
卸扣开口 mouth of shackle
卸扣扭矩 breakout torque
卸扣栓 shackle bolt
卸扣弯部 bow of shackle
卸矿点 draw point
卸矿机 ore unloader
卸矿塔 unloading rig
卸矿信号灯 dump light
卸矿栈桥 unloading gantry
卸捆器 sheaf discharger
卸粮机 grain unloader; marine elevator
卸粮螺旋离合器接合杆 unloading auger clutch lever
卸粮伸出管 unloader arm
卸粮升运器 delivery elevator
卸粮拖车 header box
卸料 discharging; dumping; off-load; unlade; unload(ing)
卸料(阀)闸门 discharge gate
卸料按钮 discharge button
卸料板 floor opening; stripper
卸料板式输送机 drawing plate conveyer[conveyor]
卸料泵 discharge pump
卸料篦板 discharge grate
卸料臂 unloading arm
卸料仓 discharge chamber; live silo
卸料槽 blow pit; blow tank; discharge launder; discharge trough; dumping trough; reject chute; unload chute
卸料侧 discharge side
卸料场 discharge place; dumping ground; dumping place; dumping position; dumping site
卸料车 bottom-discharge tractor; dummy car; throw-off carriage
卸料车轨道 dump track
卸料带 discharge conveyer[conveyor]; outlet zone
卸料导管 blowdown line
卸料地道 escape tunnel
卸料点 emptying point; point of discharge
卸料吊车 unloading crane
卸料吊斗 <船用的> marine leg
卸料吊杆 discharge boom
卸料吊柱 discharge boom
卸料斗 discharge bucket; discharge hopper; dump skip; tipping hopper; dump bucket
卸料端 delivery end; discharge end; exhaust end; outlet end; throw-off end
卸料端空心轴(颈) discharge-end trunnion; outlet trunnion
卸料端磨头 outlet head
卸料端中空轴 discharge-end trunnion
卸料端轴颈 outlet trunnion
卸料堆 dump pile
卸料阀 discharge flow grate; discharge valve; emptying valve
卸料翻斗 discharge cradle
卸料方法 method of dumping
卸料钢板 steel dump sheet
卸料高度 dumping height; truck height <对卡车装载时的>
卸料格栅 discharge grid
卸料工 dumpman
卸料刮板 drainage separator; drain separator
卸料刮刀 unloader knife
卸料管 discharge conduit; discharge duct; discharge pipe; discharge tube; draw(n)-off pipe
卸料管法兰 discharge flange
卸料管子弯头 discharge pipe elbow
卸料罐阀 relief tank
卸料滚动阀门 discharge rotary valve
卸料滚筒 discharge roll
卸料滑槽 delivery chute; emptying chute
卸料活门开闭机构 discharge gate operating mechanism
卸料机 kick-off mechanism; off-loader; stripper machine; stripping machine; unloader; unloading device; unloading machine
卸料机构 shedding mechanism
卸料架 discharge frame; throw-off shelf
卸料间 unloading bay
卸料胶带输送机 discharge belt conveyer[conveyor]
卸料角 angle of discharge; dump(ing) angle
卸料绞刀 discharge screw; extracting screw
卸料距离 spoil discharge reach
卸料开口宽度 breadth of discharge opening
卸料坑 discharging pit
卸料空气斜槽 discharge air slide
卸料孔 discharge hole; discharge opening; discharge outlet; spigot discharge
卸料口 discharge opening; discharge outlet; discharge port; product outlet
卸料口颈圈 outlet collar
卸料口临界直径 critical outlet diameter
卸料口面积 discharge area
卸料溜槽 discharge chute; dump(ing) apron; drop chute
卸料溜子 discharge chute; dump(ing) apron
卸料漏斗 discharge funnel; dump hopper
卸料炉栅 discharge grating
卸料螺旋 discharge auger; discharge screw; emptying auger
卸料螺旋的操纵 discharge auger control
卸料螺旋输送机 discharge screw conveyer[conveyor]; extracting screw; extracting screw conveyer[conveyor]
卸料皮带输送机 drawing belt conveyer[conveyor]
卸料平底船 discharge scow; dump scow
卸料平台 drawing floor
卸料起重机 dumping crane
卸料器 discharger; shedder; unloader; tripper <皮带运输机中途卸料用的>
卸料强度 discharge intensity
卸料裙板 discharge apron
卸料容器 discharge cash; discharge vessel
卸料设备 unloading assembly; unloading device
卸料时间 discharge time; discharging time; dumping time; time of unloading
卸料时间表 unloading schedule
卸料输送机 discharge conveyer[conveyor]
卸料竖管 discharge shaft
卸料速度 discharge rate; discharging speed; mass rate of emission
卸料塔 dump tower
卸料台 dump platform; throw-off shelf; unloading platform
卸料台架 unloading skid
卸料调节阀 adjustable discharge flow gate
卸料筒 drop chute
卸料桶 blow tank
卸料推杆 discharge ram
卸料位置 discharge position; dump position
卸料系数 coefficient of discharge
卸料箱 discharge box
卸料小车 discharge cart; dump buggy; tripper; tip-up troll(e)y <输送的>

卸料斜槽 discharge chute
卸料悬臂 discharge boom; jib-end <输送机卸料端安装的>
卸料用板式输送机 discharging plate conveyer[conveyor]
卸料用胶带输送机 discharging belt conveyer[conveyor]
卸料闸板 discharge shutter
卸料站 discharge terminal
卸料站台 discharging platform; dump platform
卸料装备 unloading device
卸料装置 discharge apparatus; discharge device; drawing mechanism; dumping device; empty device; tripper; unloading device
卸料锥(体) discharge cone; relief cone
卸轮器 wheel puller
卸轮辋器 rim expander
卸螺丝 unscrew
卸煤场 coal unloading yard
卸煤工人 coal whipper
卸煤机 coal drop; coal unloader; coal whipper
卸煤码头 coal unloading dock; coal unloading terminal; coal unloading wharf
卸煤门 coal dump
卸煤器 coal tip
卸煤筒 coal drop
卸煤系列化 <船舶> systematized coal unloading
卸模 mo(u)ld unloading
卸泥 dredged material dumping
卸泥标志 dumping mark
卸泥船 dumping vessel
卸泥区 dumping area; dumping ground; dumping place
卸泥区测量 dumping area survey
卸丕槽 pressure-relief groove
卸气臂 unloading arm
卸钎器 jumper detacher
卸清船上货物 clear a ship
卸清货车 clear the lorries
卸去负荷 unloading
卸去负载保护继电器 unloading protecting relay
卸去负载压力继电器 unloading pressure relay
卸去索具 strip
卸燃料 discharge of the fuel
卸任董事长 outgoing chairman
卸任理事长 outgoing chairman
卸任主席 outgoing chairman
卸砂输送机 spill sand conveyer[conveyor]
卸石梯 rock ladder
卸式运输车 bottom-dump haul(i)er
卸水泥机 cement dump
卸水泥设备 cement unloading equipment
卸土 discharge
卸土板 ejector
卸土板导向滚轮 <铲运机的> guide roller of ejector
卸土堆 tipped fill
卸土厚度 <铲土机> depth of spread
卸土距离 spoil discharge reach
卸土器杠杆 <铲土机> ejector lever
卸完时支付 payable on completion of discharging
卸物装置 easing gear
卸下 back-off; detach(ing); disboard; dismount; haul down; knock down; off-load(ing); take-down; unload
卸下船货 hoist down a cargo
卸下底片 unloading the film
卸下钢绳 unreeve
卸下螺丝 unscrew
卸下皮带 throw-off a belt
卸下转向架 detruck
卸线钩 coil stripper
卸箱分舱箱位图 discharging bay plan
卸箱清单 container unloading list
卸箱指示 <集装箱> general discharging instructions
卸修 overhauling
卸压 de-stress; pressure relief
卸压安全阀 pressure-relief valve
卸压舱水 deballasting
卸压阀 pressure-relief valve; relief valve; safety valve
卸压杆 compression release lever; compression release shaft
卸压活门 pressure-relief valve
卸压孔 pressure-relief vent
卸压弹性模量 decompression elasticity modulus
卸压载 deballasting
卸压装置 pressure relieving device
卸窑机 kiln unloading unit
卸液软管 unloading hose
卸液作业 liquid discharging operation
卸油 oil discharge; oil unloading
卸油泵 discharge pump
卸油臂 unloading arm
卸油导管 unloading line
卸油阀 oil drain cock; oil drain valve
卸油管 oil discharge pipe
卸油管路 unloading line
卸油平台 landing platform
卸油软管 discharge hose
卸油损耗 discharge loss
卸油台 platform for unloading tank car
卸油效率 oil discharging efficiency; oil discharging rate
卸鱼机械 fish unloading machinery
卸鱼及鱼加工区 fish landing and products processing area
卸鱼码头 fish landing
卸鱼棚 unloading fish shelter
卸运港 unloading port
卸运装置 extractor transport unit
卸载 break the bulk; debark; disburden; discharge; discharging; load off; load removal; off bear; overboard-dump; relief load; relieve(d load); removal of load; uncharge; unlade; unload(ing); unship【港】; off-load
卸载(减轻)作用(效应) relieving effect
卸载半径 discharging radius; dumping radius
卸载保护继电器 unloading protecting relay
卸载爆破 relieving shot; stress-relief blast
卸载爆破法 relief blasting; relief method
卸载泊位 discharge berth
卸载舱口 unloading hatch
卸载槽 discharge launder; stress-relief grooves; unloading trough
卸载场 unloading area
卸载车 self-discharger
卸载传动装置 unloading gear
卸载传送带 discharge belt
卸载传送机 unloading conveyer[conveyor]
卸载导线下垂量 final unloaded sag; final unloading sag
卸载的 balanced; off-load
卸载点 discharge point; disposal point; off-loading point; unloading point
卸载端 out end
卸载端驱动装置 <输送机> delivery end drive
卸载阀 equalizing valve; feather valve; unloading valve
卸载法 relaxation method; stress-relief method
卸载方式 way of dumping
卸载杆 unloader lever
卸载高度 discharge height; dump(ing) clearance; dump(ing) height; height of load removal
卸载拱 discharging arch; relieving arch; unloading arch
卸载拱顶 discharging vault
卸载管 discharge pipe
卸载过程 unloading path
卸载荷载 lock off load
卸载回路 relief circuit
卸载货的 off-load; off-load
卸载机 off-loader; unloader; unloading machine
卸载机构 unloader power element; unloading mechanism; dumping mechanism
卸载机切碎螺旋推运器 unloader's cutting auger
卸载及转弯时间 <铲土机> dump and turn time
卸载角 discharge angle; dump(ing) angle
卸载距离 dump clearance; dumping reach
卸载拉绳 inhaul cable
卸载梁 chock release
卸载溜槽 dumping apron
卸载螺旋运输机 unloading auger
卸载锚地 discharge anchorage
卸载门 balanced door
卸载面积 <负或正面积> relieving area
卸载能力 unloading capacity
卸载皮带输送机 unloading belt conveyer[conveyor]
卸载期 off-load period; timing of surcharge removal
卸载起动 unloaded start
卸载起重机 unloader lifter
卸载器 deloader; emptier
卸载切碎吹送机 unloader blower chopper
卸载曲线 unloading curve
卸载容量 unloading capacity
卸载设备 unloading equipment; unloading gear
卸载升运器 delivery elevator; discharge elevator; unloading elevator
卸载时间 dump time; emptying time
卸载时自重 unladen weight
卸载式码头 relieving quay; relieving wharf
卸载式平台 relieving platform
卸载式载重汽车 discharge vehicle
卸载输送机 conveyer unloader; discharge conveyer[conveyor]
卸载输送链 discharging chain
卸载输送器 conveyer unloader; discharge conveyer[conveyor]
卸载速度 discharge rate; unloading speed
卸载速率 rate of debarkation
卸载套 unloader sleeve
卸载通道 relief passage
卸载途径 unloading path
卸载脱浅 discharging for refloat
卸载完毕 completion of discharge
卸载位置 unloading position
卸载温度 unloading temperature
卸载系统 emptying system
卸载箱 unloader box
卸载巷道 relief roadway
卸载循环 unloading cycle
卸载岩体 relaxed rock
卸载油缸 unloading cylinder
卸载月台 relieving platform
卸载月台式码头 relieving platform type wharf
卸载运输带 throw-off belt
卸载再压环 decompression and recompression loop
卸载装置 discharge mechanism; emptying device; shedding mechanism; unloading device; unloading mechanism
卸载装置离合器 unloader clutch
卸渣 trapping muck; ballast unloading; dumping
卸渣场 slag dump; dumping area
卸渣阀 blow valve
卸渣炉箅 drop plate; tipping grate
卸渣炉排 dump plate
卸渣炉栅 dumping grate
卸渣门 discharge cover
卸渣门垫圈 packing for discharge cover
卸渣器 ballast unloader
卸砖机 unloading machine
卸装 undressing
卸装方法 discharging method
卸钻杆 rod uncoupling
卸钻套楔 drill drift
卸钻头器 bit puller

屑

屑冰 acicular ice; frazil; frazil ice; slack ice

屑冰的形成 frazilization
屑冰块 frozen frazil slush
屑痕 chip scratch
屑粒 crumb
屑粒土 crumby soil
屑粒状结构 crumble texture; crumb structure
屑片 flaky particle
屑伤 <磨光玻璃缺陷> block rack
屑蚀地形 truncated landform
屑物 leftover
屑重 chip weight

谢

谢贝利河 Shebelle River

谢才方程 Chezy equation
谢才公式 Chezy formula
谢才速度系数 Chezy velocity factor
谢才系数 Chezy coefficient
谢澈尔滚动式竖旋桥 Scherzer roller bascule bridge
谢代特炸药 <蓖麻油、过氯酸铵、二硝基甲苯的混合物> Cheddite
谢尔贝薄壁取土器 Shelby tube sampler; thin-walled Shelby tube sampler
谢尔贝薄壁取样器 Shelby tube sampler
谢尔贝取土器 Shelby soil sampler
谢菲尔特系统 Sheffield system
谢绝订货 decline an order
谢绝发盘 decline offer
谢拉格锁 <一种筒状锁> Schlage lock
谢勒绿色 Seheele's green
谢勒绿颜料 Seheele's green
谢利得高频感电炉 Schneider furnace
谢利得推进器 Schneider propeller
谢林哈式铰链调节瓣通风装置 Sheringham valve
谢尼克金属疲劳试验机 Schenick machine
谢泼德断口淬透试验 Shepherd pene-

tration-fracture test
谢泼德校正估计量 Shepherd-corrected estimator
谢氏系数 coefficient of Chezy formula
谢瓦尔特控制模型 Shewhart control model
谢瓦尔特控制图 Shewhart control chart
谢瓦尔特图 Shewhart chart
谢语 acknowledgement

榍石 sphene;titanite

榭pavilion on terrace;terraced pavilion

懈怠 omission

蟹壳青 < 色釉名 > crust green

蟹耙式清岩机 rake-up ballast remover;rake-up rock remover
蟹青釉 crab green glaze
蟹式转向 crab steer
蟹行 crabwise motion
蟹行式转向 four-wheel crab steering
蟹形窑 crab-shape tank furnace
蟹眼结 crabber's eye knot;cross running knot
蟹眼式望远镜 hyposcope
蟹爪 crab claw
蟹爪敞口式取样器 crab claw open-type sampler
蟹爪裂纹釉 crab's foot crackle
蟹爪式起重机 crab;crab crane;crab derrick
蟹爪式装岩机装岩 loading by crab's paws loader
蟹爪式装载机 crab gathering arm loader;gathering arm loader
蟹子草 crab grass
蟹足状 crab-like

心板 cored plate

心板材 centre board
心棒 axle;mandrel bar;triblet
心棒杆 mandrel bar
心边交界材 intermediate wood
心不在焉的 absent-minded;unthinking
心部 hearting
心部裂纹 < 木材的 > heart check
心部木材 heart wood
心部热处理的 < 高中频淬火件 > core-treated
心材 comb-grained wood;core stock;heart wood;sound heartwood;true-wood;duramen;heart
心材板 heart plank
心材变色 heart stain
心材腐蚀 heartwood rot
心材腐朽 heart rot;heartwood rot;stem-rot
心材含有物 heart substance
心材面 inner surface
心材面板 heart-face board
心材树木 heartwood tree
心材树种 heartwood tree;tree with colo(u)red heartwood
心材枕木 heart tie;heartwood sleeper
心电图 electrocardiogram
心电图室 electrocardiography room
心电图仪 electrocardiograph
心对心法 center-to-center method
心腐 centre wood rot;heartwood rot;stem-rot
心腐材 punky wood
心骨 rodding of cores
心管 inner cylinder
心轨 nose rail;point rail
心轨缺口 point notch
心果山核桃 bitternut
心合板 < 以刨花等夹心 > cordboard
心环 thimbler
心环索眼插接 thimbler eye splice
心迹线 centroid
心绞痛 angina pectoris
心径 core diameter
心境平静 equilibrium[复 equilibria/equilibriums]
心孔铰刀 center[centre] reamer
心力 central force
心裂 < 木材的 > box heart check;heart shake;star check;growth shake;heart check;rift crack
心裂木材 quaggy timber
心流涡轮机 inward flow turbine
心木 Comanchic-grained wood;comb-grained wood;heart wood
心捻油杯 wick-feed oil cup
心盘 center[centre] plate;pivot bearing
心盘复原杆 center plate centering stud
心盘荷载 center[centre] plate loading force
心盘间距离 bogie centers
心盘锁销 locking center pin
心盘支重 center pivot loading
心盘座块 center plate block
心坯 core stock
心皮 carpel
心砌合 heart bond
心区 central zone
心砂 core sand
心射极平投影 gnomonic projection
心射切面投影 gnomonic projection
心射投影 central projection
心射投影尺 gnomonic ruler;gnomonic scale
心射投影基准子午线 principal meridian of gnomonic chart
心射图中央子午线 central meridian of gnomonic chart
心身疾病 psychosomatic disorder
心数 number of cores
心算 mental arithmetic;mental calculation
心滩 batture;centre bar;mid-channel bar;middle bar;middle ground shoal
心滩沉积 channel bar deposit
心滩地 batture land
心滩相 channel bar facies
心土 bottom soil;subsoil
心线 corduroy wire;core wire
心线和金属壳间的电容 shunt capacitance
心销 center[centre] pivot
心形单眼木拼 heart block
心形的 caroid
心形方向图接收 heart shape reception
心形卡环 heart-shaped shackle
心形镘刀 heart and square
心形砌合法 heart bond
心形曲线 cardioid
心形饰 < 线脚中的 > open heart mo(u)lding
心形凸轮 heart cam
心形图案 cloven leaf pattern;cardioid pattern
心形卸扣 heart shackle
心形样品的嵌入 < 用于试验路面防滑性能的 > corduroy insertion
心血管(系统)的 cardiovascular
心压力 eccentric compression
心叶椴 small-leaved lime
心叶水杨梅 Haldu;heart leaf adina
心因性障碍 psychogenic disorder
心影 podoid
心影曲线 podoid of a curve
心载荷 centrifugal load(ing)
心脏地带 heartland
心脏式 cardioid
心脏形的 cordate
心脏形曲线 cardioid
心脏形线【数】cardioid
心证 evidence through inner conviction
心轴 arbor;central spindle;drift pin;monkey;pivot;spindle;trib(o)let
心轴打桩法 mandrel driving
心轴减震器 spindle shock absorber
心轴孔 arbor hole
心轴套 spindle sleeve
心轴托 spindle holder
心轴支承 pivoting bearing
心轴轴承 spindle bearing
心轴主销 vertical pivot pin
心轴组件 spindle assembly
心轴座 spindle seat
心柱断面 stem section

芯板 central piece;core plate

芯板材 core board;core stock;particle board
芯板料 core stock
芯板裂缝 core gap
芯棒 core rod;mandril;piercer;piercing point bar;plug;mandrel < 桩工 >
芯棒抽出机构 stripper mechanism
芯棒划痕 plug lines
芯棒切断机 core rod cutter
芯棒式无缝管轧机 mandrel mill
芯棒旋转穿孔机 rotary piercing mill
芯棒轧管法 plug rolling;plug roll process
芯玻璃 core glass
芯部伸出的火花塞 projected-core spark plug
芯部硬度 core hardness
芯部周长 core circumference
芯材 core material;sandwich material
芯材胶合板 core board
芯层 central layer;core sheet;interlayer;sandwich layer
芯层发泡层压制品 foam-cored laminate
芯层物料 core material
芯层毡 coremat
芯撑 stud
芯撑支撑面积 chaplet area
芯锻 core forging
芯飞边缺肉 < 铸件缺陷 > core fin
芯杆 core bar;core pin;plug
芯杆拔管法 bar drawing
芯钢丝 core wire
芯钢线 corduroy wire
芯给润滑 wick lubrication
芯骨 arbor;core bar
芯骨架 core arbor;core crab
芯骨校直机 core wire straightening machine
芯管 core barrel;core pipe
芯管支承的 corduroy-supported
芯合板 core board
芯盒 core box
芯盒翻转机 core drawer;core drawing machine
芯盒密封垫 core box sealer
芯盒排气槽 core box vent
芯块胶合板 block board
芯框 buck frame;core frame
芯缆 core cable
芯模 core pattern;mandrel
芯模转角 rotating angle of mandrel
芯皮黏[粘]合 fillet
芯片 chip;slug
芯片测试 chip testing
芯片工艺 chip technology
芯片焊接机 die bonder
芯片微处理器 chip microprocessor
芯片组 chipset
芯拼合胶黏[粘]剂 core splice adhesive
芯墙 core wall < 用于堤坝或井点滤管等 >;hearting;heart wall;corduroy-wall < 用于堤坝或井点滤管等的 >;central zone < 土坝的 >
芯墙坝 core wall dam;dam with core wall;heart wall dam
芯墙材料 < 土石坝的 > core material
芯墙顶(部) < 土坝的 > core crest;core crown
芯墙沟 core trench
芯墙基槽 core trench
芯墙截水槽 core trench
芯墙截水沟 core trench
芯墙倾斜(度) core slope
芯墙式堆积坝 vertical core rockfill dam
芯墙式堆石坝 core wall type rockfill dam
芯墙式土坝 core type dam;core earth dam
芯绕组 core winding
芯砂 core sand
芯砂搅拌机 corduroy sand mixer;core mixer;core sand mixer
芯石 core stone
芯式泵 cartridge-type pump;plug-in pump
芯式变压器 winding core type transformer
芯水 core water
芯碳棒 cored carbon
芯填充系数 core packing fraction
芯铁 core bar;core grid
芯铁损失 core loss
芯头标记 core register
芯头间隙 clearance;print clearance
芯吸润滑器 oil siphon
芯吸作用 wicking
芯线 component wire;core wire
芯线绞距 lay of wire
芯线绝缘 core insulation
芯形件 spool element
芯型 core pattern
芯型样品的嵌入 core insertion
芯样 core sample
芯样抗压强度 core compressive strength
芯样试验 core test
芯油 core oil
芯轴 mandrel[mandril]
芯轴拔长 fullering with the core bar
芯轴密封 stem seal
芯轴压机 mandrel press
芯轴造形机 spindle mo(u)lder
芯柱 button stem;central screw;core of column;stem
芯柱机 button stem machine;stem machine
芯柱烧结炉 stem sintering furnace
芯子 body core;cartridge;mandrel [mandril];core
芯子凹陷 core depression
芯子半径 radius of the core
芯子分离 core separation
芯子拼接 core splicing

芯子压环 core crush
芯子窑 core drier
芯座 core print seat;mandrel holder

辛

辛胺 octylam(in)e
辛变换 symplectic transformation
辛醇 capryl(ic)alcohol;octanol;octyl alcohol
辛的 symplectic
辛迪加 <企业的联合组织> syndicate
辛迪加贷款 syndicate loan
辛迪加解除限制 syndicate release
辛迪加投标 syndicated tender
辛迪加限制 syndicate restriction
辛迪加限制终止 syndicate termination
辛迪加协议 syndicate agreement
辛迪加银行业务 syndicate banking
辛迪加组织 syndication
辛二醇 octamethylene glycol
辛二酸 octane diacid;suberic acid
辛二酸盐 suberate
辛二酸酯 suberate
辛二酰 octanedioyl
辛基 octyl
辛基苯 octylbenzene
辛基酚 octyl phenol
辛基化合物 capryl compounds
辛基硫酸钠 sodium octyl sulfate
辛克莱式压力黏[粘]度计 Sinclair pressure visco(si)meter
辛苦而令人讨厌的工作 drudgery
辛苦工作 toil work
辛辣的 tart
辛硫磷 phoxim
辛硫砷铜矿 sinnerite
辛普发拉格条件 <在直接投影体系中,物体、透镜、像平面必须是共线的,才能得到清晰的焦点> Scheimpflug condition
辛普拉斯表面曝气系统 Simplex surface-aeration system
辛普拉斯曝气器 Simplex aerator
辛普拉斯套管连接 Simplex joint
辛普拉斯型舵 Simplex rudder
辛普拉斯桩 <可抽出外壳套管的混凝土桩> Simplex pile
辛普拉斯钻孔灌注桩 Simplex concrete pile
辛普森多样性指数 Simpson's diversity index
辛普森法则 Simpson's rule
辛普森公式 Simpson's formula
辛普森抛物线法则 Simpson parabolic rule
辛普森抛物线总计法则 Simpson's rule
辛普森三分之一公式 Simpson's 1/3 formula
辛普森指数 Simpson's index
辛羟砷锰石 synadelphite
辛群 symplectic group
辛酸 caprylic acid
辛酸乙酯 ethyl caprilate
辛酮 octanone
辛烷 octane
辛烷单位 octane unit
辛烷额定值 octane rating
辛烷二甲酸 octane dicarboxylic acid
辛烷基 octyl
辛烷曲线 octane curve
辛烷溶解度 solubility of octane
辛烷值 octane level;octane number;octane rating;octane ratio;octane value
辛烷值标度 octane scale
辛烷值测定 octane number determination
辛烷值的要求 octane number requirement
辛烷值感应 octane number response
辛烷值校正器 octane corrector
辛烷值列线图 octane number nomogram
辛烷值实验室测定法 octane laboratory method
辛烷值试验机 octane number test engine
辛烷值特性 octane performance
辛烷值提高剂 octane enhancer
辛烷值提级 octane upgrading
辛烷值下降 octane fade
辛烷值选择器 octane selector
辛烷值研究法 octane research method
辛烷值增进剂 octane number improver
辛烷值指数 octane rating index
辛烯 octylene
辛辛那提-篮岭山字型构造体系【地】 Cincinnati-Blue epsilon structural system
辛辛纳提阶 <美国晚奥陶世>【地】 Cineinnatian stage

欣

欣德利(球面)蜗杆 Hindley worm;hour-glass worm
欣德利蜗轮 Hindley worm gear
欣古河 Xingo River

锌

锌 zinc
锌凹凸槽 zinc rebate
锌白 China white;Chinese white;zincolith;zinc oxide;white zinc;zinc white
锌白磁漆 Chinese white stand oil enamel
锌白粉 zinc oxide powder
锌白漆 white zinc paint;zinc white
锌白铅矿 iglesiasite
锌白熟油瓷漆 zinc white stand oil enamel
锌白调合漆 ready mixed zinc white paint
锌白铜 German silver;packfong
锌摆轮 zinc balance
锌板 rolled tin;sheet zinc;zinc plate;zinc sheet;zinc slab
锌板防蚀 zinc cure
锌板厚度 zinc thickness
锌板厚度规 zinc ga(u)ge
锌板平屋顶系统 <表面无钎焊和钉头> roll-cap system
锌板条 zinc strip
锌板条屋顶 zinc batten roof
锌版 zincograph;zinc plate
锌版防蚀处理 cronak process
锌版复制 papyrotype
锌版画 zincograph
锌版剩余药层去除法 post-Cronak
锌版(印刷)术 zincography
锌版转印法 anastatic process
锌版装饰 zinc plate finish
锌半电池 zinc couple
锌棒 spelter;zinc bar;zinc rod
锌饱和 zincification
锌钡白 <一种用于油漆的白色颜料> enamel white;lithosphere;lithopone;pearl white
锌钡白级 lithopone grade
锌钡白锌白调合漆 prepared lithopone-zinc white paint
锌钡铅玻璃 zinc-barium-lead glass
锌玻璃砖 zinc glaze tile
锌箔 zinc foil
锌薄板 zinc sheet
锌尘 zinc fume
锌衬里 zinc liner
锌橙色 zincorange
锌赤铁矾 zincobotryogen
锌催干剂 zinc drier
锌带 strip zinc
锌的 zincky;zincous
锌的浓度 zinc concentration
锌点 zinc point
锌电极 zinc electrode
锌电解设备 zinc electrolyzing plant
锌电流 zinc current
锌电偶 zinc couple;zinc half-cell
锌钉 zinc nail
锌锭 zinc ingot(metal);zinc pig
锌毒性 zinc toxicity
锌堵 zinc plug
锌镀层 zinc coating
锌镀层火焰加固处理 flame scaling
锌二乙基二硫代氨醛甲酸盐 <一种沥青改性掺加剂> zinc-diethyl-dithiocarbamate
锌矾 zinc vitriol;zinkosite
锌矾石 zinc aluminite
锌方解石 zincocalcite
锌肥 zinc fertilizer
锌粉 powdered zinc;zinc ash;zinc dust;zinc powder
锌粉沉降 precipitation with zinc dust
锌粉底漆 zinc dust primer
锌粉防锈漆 zinc dust anticorrosive paint
锌粉防锈涂料 zinc dust anticorrosive paint
锌粉膏 zinc dust paste
锌粉精制 zinc dust purification
锌粉漆 zinc dust paint
锌粉热镀 sherardise[sherardize]
锌粉涂层 metallic zinc coating
锌粉涂料 zinc paint
锌粉颜料 zinc dust pigment
锌粉蒸馏 zinc dust distillation
锌浮渣 zinc ash
锌腐蚀 zinc corrosion
锌腐蚀试验 zinc corrosion test
锌钙白 sulphopone
锌钙铜矾 serpierite
锌盖板 zinc shingle
锌干料 zinc drier
锌膏 calamine cream
锌铬黄 zinc chromate;zinc yellow;zinc chrome
锌铬黄底漆 zinc chromate primer
锌铬黄防锈漆 zinc chromate anti-rusting paint
锌汞电池 zinc-mercuric oxide cell;zinc mercury battery;zinc mercury cell
锌汞合金 zinc amalgam
锌汞齐 zinc amalgam
锌管 zinc pipe[piping];zinc tube[tubing]
锌管渠 zinc conduit
锌罐 zinc can
锌硅酸盐涂层 zinc silicate coating
锌辊轧花整理 zinc finish
锌锅炉板 zinc boiler plate
锌焊料 spelter;zinc solder
锌焊药 spelter solder
锌合金 alloyed zinc;kirksite <模具用>;zinc alloy
锌合金衬板 zinc alloy lining plate
锌合金涂层钢板 zinc alloy coated sheet steel
锌黑 zinc black
锌黑锰矿 zinc hausmannite
锌护板 zinc protector
锌花 <镀锌薄钢板表面的> spangles
锌华 flowers of zinc;zinc bloom
锌华凝结器 devanture
锌化 zinc impregnation
锌化单宁酸 zinc-tannin
锌化合物 zinc compound
锌化杂酚油 zinc creosote
锌还原法 zinc reducing method
锌黄 buttercup yellow;zinc chromate hydroxide
锌黄长石 hardystonite
锌黄粉 zinc yellow powder
锌黄锡矿 kesterite
锌黄颜料 citron yellow;zinc chromate
锌灰 zinc ash;zinc grey[gray]
锌灰油(漆)zinc gray
锌基合金 binding metal;zinc-base alloy
锌基合金模 zinc alloy die
锌基糊状铸造合金 zinc-base slush casting alloy
锌基铅铜焊料 Mouray's solder
锌基模铸件表面机械抛光 fadgenising
锌基泡沫活化稳定剂 zinc-based foam activator/ stabilizer
锌基润滑脂 zinc base grease
锌基压铸合金 reversed brass;zinc-base die casting alloy
锌基轴承合金 Fenton bearing metal;Fenton's metal;Ledebur bearing alloy;Pierott metal;zinc-base bearing metal
锌极 <电池的> zincode
锌尖晶石 zinc aluminate;zinc spinel
锌精炼厂 zinc refiner
锌壳 cadmia;zinc crust
锌空气电池 zinc-air battery;zinc-air cell
锌孔雀石 rosasite
锌块 spelter;zinc slab;zinc spelter
锌矿 zinc mine
锌矿泥 zinc sludge
锌矿石 zinc ore
锌矿指示植物 indicator plant of zinc
锌冷凝器 zinc condenser
锌离子 zinc ion
锌粒 zinc granule
锌硫磷 phoxim
锌铝矾 dietrichite
锌铝合金 allumen;E alloy;E-metal;zinc alumin(i)um;alumin(i)um zinc
锌铝尖晶石 gahnite
锌铝蛇纹石 fraipontite
锌绿 zinc green
锌绿松石 faustite
锌绿铁矿 zinc-rockbridgeite
锌氯电池 zinc-chlorine battery
锌氯化铵 zinc ammonium chloride
锌氯化银原电池 zinc-silver-chloride primary cell
锌镁胆矾 zinc-magnesium chalcanthite
锌镁电镀合金 zinc-magnesium galvanizing alloy
锌镁矾 mooreite
锌镁铜铬锰铁硅合金 zircal
锌蒙脱石 sauconite
锌锰电池 zinc-manganese battery
锌锰橄榄石 roepperite
锌锰红辉石 zinc-schefferite
锌锰辉石 fowlerite;keatingine
锌锰碱性干电池 crown cell
锌锰矿 hetaerolite
锌锰泻盐 zinc-fauserite
锌棉 zinc sponge
锌冕玻璃 zinc crown glass;zinc silicate glass
锌面层 zinc coating
锌明矾 zinc aluminite
锌模 zinc templet

锌钼矿石 Zn-bearing Mo ore
锌钠硅玻璃 zinc silicate glass
锌镍铜合金 copper nickel
锌镍涂层薄钢板 zinc-nickel coated sheet
锌镍蓄电池 zinc-nickel storage battery
锌硼酸盐玻璃 zinc-borate glass
锌皮 rolled tin;zinc sheet
锌皮标号 zinc ga(u)ge
锌皮泛水 zinc flashing
锌皮披水 zinc covering
锌皮平屋顶 zinc covered flat roof
锌皮屋顶 zinc roof;zinc sheet roofing
锌皮屋面 zinc covering;zinc(sheet) roof(ing)
锌皮油灰 zinc putty
锌皮雨水槽沟 <匣形的> zinc box rainwater gutter
锌片 sheet zinc;zinc(metal) sheet; zinc slab
锌漆 zinc paint
锌起爆雷管 zinc detonator
锌铅矿石 Zn-Pb ore
锌铅锡合金 Calamine
锌铅蓄电池 zinc-lead accumulator
锌青铜 ormolu;zinc bronze
锌日光榴石 genthelvite
锌熔池 molten spelter tank
锌熔炼炉 zinc smelter
锌熔喷层 zinc spray coating
锌熔液槽 spelter pot
锌乳石 voltzite
锌闪烁 zinc flash
锌石 zincite
锌试剂 zincon
锌水绿矾 zinc melanterite
锌水砷铜矿 zinclavendulan
锌酸 zincic acid
锌酸钴 cobalt green
锌酸盐 zincate
锌酸盐处理 zincate treatment
锌酸钴 cobalt zincate
锌条 strip zinc;zink rod
锌铁 zinc iron
锌铁电池 zinc iron cell
锌铁矾 bianchite
锌铁橄榄石 stirlingite
锌铁黄长石 justite
锌铁尖晶石 franklinite
锌铁磷酸盐涂层 anchorite
锌铁披水 zinc flashing
锌铁皮 tutanaga
锌铜矾 glaucokerinite
锌铜焊接合金 spelter soldering alloy
锌铜焊料 spelter;spelter solder
锌铜合金 brazing metal;copper-zinc alloy;platine;spelter
锌铜矿石 Zn-bearing copper ore
锌铜铝矾 glaucokerinite
锌铜铝合金 Alneon
锌铜水绿矾 zinc-copper melanterite
锌铜钛合金 zinc-copper-titanium alloy
锌凸版 zinc engraving;zinco;zinc etching
锌凸版印件 zincograph
锌涂层 spelter coating
锌涂层钢板 zinc-coated sheet steel
锌涂料 spelter coating
锌稳定性时间试验 zinc stability time test
锌污染 pollution by zinc;zinc pollution
锌污染物 zinc pollutant
锌锡焊条 half-and-half solder
锌锡铜合金 ormolu
锌锡轴承合金 Fenton bearing metal
锌(系)颜料 zinc pigment
锌箱 zinc box
锌蓄电池 zinc battery;zinc storage battery
锌压铸 zinc die casting
锌烟罐 zinc smoke candle
锌盐 zinc salt
锌盐变性剂工艺 zinc-modifier system
锌阳极 zinc anode
锌氧电池 zinc-oxygen battery
锌氧粉 zinc oxide powder
锌叶绿矾 zincocopiapite
锌银合金 zinc-silver alloy
锌银蓄电池 zinc-silver accumulator
锌引爆管 zinc blasting cap;zinc detonator
锌釉 Bristol glaze;zinc glaze
锌釉砖 zinc glaze tile
锌云母 hendricksite
锌皂 zinc soap
锌皂石 sauconite;zinc-saponite
锌渣 cadmia;zinc dross;zinc slag
锌蒸馏罐 zinc retort
锌制摆轮 zink balance
锌质炉瘤 cadmia
锌质嵌线 zinc rule
锌中毒 zinc poisoning
锌铸件 zinc casting
锌着色 zinc flash

新

新安石 xinanite

新凹陷 neo-concave area
新坳陷 neo-depression
新奥(地利隧道)法 <一种隧洞掘进和支护方法> new Austrian tunnel(l)ing method
新奥地利(隧道)开挖法 new Austrian tunnel(l)ing method
新奥地利隧洞掘进法 new Austrian tunnel(l)ing method
新奥地利学派 Neo-Austrian school
新奥尔林港 <美> New Orleans Port
新巴洛克式 Neo-Baroque style
新巴洛克式建筑 Neo-Baroque architecture
新坝地 polder
新白颜料 new white pigment
新拜占廷建筑 neo-Byzantine architecture
新板 newly-made slab
新版 new edition;redaction
新版本 reissue
新版海图 new edition chart
新版日期 date of new edition
新办工业 infant industry
新拌的 green
新拌的贫混凝土 lean freshly mixed concrete
新拌和的 freshly mixed
新拌灰浆 green mortar
新拌灰泥 freshly mixed mortar
新拌混凝土 as-mixed concrete;as-placed concrete;fresh concrete; freshly mixed concrete;green concrete;immature concrete;unset concrete;wet concrete
新拌混凝土稠度试验 ball test
新拌混凝土的密度 freshly mixed concrete density
新拌混凝土的碳化 carbonation of fresh concrete
新拌混凝土的运送 haul of fresh concrete
新拌混凝土快速分析 rapid analysis of fresh concrete
新拌混凝土密度 green concrete density
新拌砂浆 freshly mixed mortar;green mortar
新拌水泥砂浆 fresh cement mortar
新保护主义 new protectionism
新暴露顶板 fresh roof
新北区 Neoarctic region
新变量 new variable
新变量的构造 construction of new variables
新变量求积分法 integration by new variables
新变态 neomorphosis
新冰 newly-frozen ice;young ice
新冰川作用 neoglaciation
新冰皮 skin
新兵训练中心 recruit training center [centre]
新不列颠海沟 new Britain trench
新部件 new parts
新材料 fresh material;innovative material;new material;new substance
新材料的批准 approval of new materials
新材料评价 evaluation of new materials
新采矿物 fresh mineral
新采砂 green sand
新采石料(所含)水分 quarry sap
新彩 new colo(u)rs;polychrome painting
新残积土 immature residual soil
新草覆盖 fresh mulch
新测 new survey
新层型 neostratotype
新碴起道 raising on new ballast
新产品 later products;new face;newly-made products;new products; novelty;offspring;outlay;pioneer products
新产品车间 new product division
新产品发展 new product development
新产品发展过程 new product development process
新产品发展计划 new product development plan
新产品估价 new product evaluation
新产品经济预测 economic forecasting of new products
新产品经营策略 strategy for new product development
新产品开发 initial series of production;new product development
新产品开发程序 procedures of new product development
新产品开发方案评价 evaluation of development program(me) of new products
新产品年度 model year
新产品年度差异 model-year variation
新产品渗透理论 diffusion of innovations
新产品试销 trial sale of new product
新产品试验 development test
新产品试制 pilot;trial production of new products
新产品试制程序 procedure of trial production of new products
新产品试制费 expenses for trial manufacture of new products;expenses in the trial manufacture of new products;fees for trial production of new products;funds for trial manufacture of new products;money for new product development; subsidize trial manufacture of new products
新产品试制基金 new product trial production fund
新产品推介方案 new product introduction plan
新产品专利 new product monopoly
新厂房和新设备投资额 expenditure on new plant and equipment
新场所境 new habitat
新潮体系 current system
新车交接 new car delivery service
新车轮 new wheel
新车展览 motorama
新沉积物 recent sediment
新陈代谢 metabolic rate
新陈代谢记录器 metabograph
新陈代谢说 metabolism
新陈代谢污染物 metabolic pollutant
新成本 neosome
新成冰 fresh ice;freshly frozen ice; newly-formed ice
新成河岸 newly-formed bank;newly-forming bank
新成矿作用 neomineralization
新成体 metasome;neosome
新成土 entisol
新城内城镇 new-town-in-town
新城(市) new town
新城镇 new town
新尺寸 new dimension
新冲积层 newly-accumulated alluvium
新冲积土 young alluvial soil
新冲积物 recent deposits of alluvium
新出矿湿度 pit moisture
新出窑砖 kiln-fresh brick
新传输方案 new transmission plan
新船的试验航行 shakedown cruise
新船名录 new register book of shipping
新磁带 raw tape
新粗野主义 neo-brutalism;New Brutalism
新催化剂 raw catalyst
新村建设 housing estate development
新村开发 housing estate development
新村(庄) new village
新达尔 <一种铝基合金> Cindal
新导电棒 clean stub
新到地图目录 map accessions list
新到货(物) fresh arrival;new arrival
新到图书 accession;new acquisitions
新到资料索引 accession index
新道碴 fresh ballast
新德里 <印度首都> New Delhi
新的操作人员 novice operator
新的充填带 green pack
新的管理决策科学 new science of management decision
新的借款办法 new technique of borrowing
新的栖息地 new habitat
新的污泥 green sludge
新的最低价位 new low
新的最高价位 new high
新地巨旋回【地】Neogaic megacycle
新地时期 epoch of neogenicum
新地旋回【地】neogaikum
新第三纪【地】Neogene(period)
新第三纪气候分带 Neogene climatic zonation
新第三纪夷平面 Neogene planation surface
新第三系【地】Neogene system
新雕像揭幕仪式 unveil of a statue
新订货单 new order
新定线道路 newly-located road
新定义 redetermination
新动向 recent trend
新度系数 newness coefficient of vehicle
新段落 new paragraph
新断层 neo-fault
新堆反应性 clean reactivity
新堆砌体 green pack
新发明 gadgetry;invention
新发明的 neoteric

新发特许状 recharter
新发现 discovery;revelation
新发现的 newfound;newly discovered
新发行的股票 coming out
新发行的资本股票 new capital issue
新发展 recent advance;recent development
新发展的技能 newly-developed technique
新发展计划 new development plan
新伐 fresh cut
新伐材 green wood
新伐的 <木材> new fallen
新伐木材 fresh wood;live wood;unseasoned lumber [timber/ wood];green lumber;green timber;green wood
新伐树杆 green pole
新伐原木 green log
新翻地 new ploughed field
新范畴 neocategory
新方案 new departure
新分贝 new decibel
新分类号 new classification number
新分类学 new systematics
新风 fresh air;make-up air;outside air;replacement air
新风百叶窗 fresh-air louvers
新风比 fresh-air ratio
新风比例 fresh-air proportion
新风补偿量 fresh-air make-up
新风采暖器 fresh-air warming appliance
新风采暖设备 fresh-air warming appliance
新风处理 fresh-air operation
新风道 fresh-air duct;fresh-air flue;fresh flue
新风阀 outside air damper
新风风道 outside air intake duct
新风负荷 outdoor air load
新风格 latest style;new style
新风供应 fresh-air supply
新风管道系统 fresh-air duct pump
新风过滤器 make-up air filter
新风换气 ventilation with fresh air
新风机 fresh-air blower;fresh-air fan
新风机组 fresh-air handling unit;outside air unit
新风加热 fresh-air warming
新风加热器 fresh-air heater;fresh-air heating appliance;fresh-air warmer
新风加热设备 fresh-air heating appliance
新风进风塔 fresh-air inlet tower;fresh-air intake tower
新风进口竖管 fresh-air intake stack
新风空气进口 outside air opening
新风冷负荷 cooling load from outdoor air;cooling load from ventilation
新风量 fresh-air requirement;fresh-air volume
新风平衡进口 fresh-air balanced intake
新风系统 fresh-air system;primary air system
新风循环 fresh-air circulation
新风预热器 fresh-air preheater
新敷设管道 new-laid pipe;new-laid tube
新服役船 raw ship
新腐烂 green rot
新干线 new trunk;Shinkansen <日本高速列车路线>
新感觉涂层 coating with soft
新感觉涂料 suede coating
新港 new harbo(u)r
新港区浚挖 newly-built port dredging
新港区开挖 new work dredging
新高岭石 neokaolin(e)
新哥特式 neo-Gothic
新哥特式建筑 neo-Gothic architecture
新割干草 new mown hay
新耕土 freshly plowed soil
新工程 new construction
新工人培训学校 vestibule school
新工业国 new industrial state
新工业区 new industrial district
新工艺 innovative technology;new process
新工艺试行 shakedown
新工艺试验 shakedown
新工作项目建议 new work item proposal
新公司 new incorporation
新构成主义 neo-construction
新构造单元 neotectonic elements
新构造分析图 neotectonic analytical map
新构造骨架【地】neotectonic frame
新构造合成图 synthetic(al) neotectonic map
新构造活动区 neotectonic active region
新构造类型 neotectonic type
新构造类型图 map of neotectonic type
新构造区划图 neotectonic zoning
新构造体系 new tectonic system
新构造图 neotectonic map
新构造图件 neotectonic map
新构造稳定区 neotectonic stabile region
新构造学 neotectonics
新构造研究法 research method of neotectonic movement
新构造运动 neotectonic movement;neotectonics
新构造运动的继承性 neotectonic inheritance
新构造运动调查 investigation of new tectonic movement
新构造运动分类 type of neotectonic movement
新构造运动活动地区 active area of new tectogenesis
新构造运动活跃程度 mobile level of new tectogenesis
新构造运动极活跃地区 very active area of new tectogenesis
新构造运动阶段 stage of neotectonic movement
新构造运动类型 movement type
新构造运动强度图 intensity map of neotectonic movement
新构造运动区 neotectonic movement division
新构造运动时期 period of neotectonic movement
新构造运动速度 velocity of neotectonic movement
新构造运动停顿地区 stable area of new tectogenesis
新构造运动一般地区 ordinary area of new tectogenesis
新构造综合图 comprehensive neotectonic map
新构造作用的时间分类 classification on the age of neotectonism
新古典建筑 neoantique architecture
新古典建筑式 beoclassic
新古典经济学 neoclassical economics
新古典式建筑 neoclassical architecture
新古典学派 neoclassical school
新古典学派生产函数 neoclassical production function
新古典增长理论 neoclassical growth theory
新古典主义 neoclassicism
新古典主义建筑 neoclassical architecture
新古典综合派 neoclassical synthesis
新股份 new stock
新雇用者 <指在任何一个时期中进入就业的总人数> engagements
新刮清干管 newly scraped main
新国际回合 new international round
新国际经济秩序 new international economic[economy] order
新国际正常重力公式 new international normal gravity formula
新国际(烛)光(单位) candela [ed];new candle
新海特拉明 neohetramine
新海西(造山)运动 Subhercynian orogeny
新罕布什尔 <美国州名> new Hampahire
新合伙人 junior partner
新河谷桥 <位于美国> new River Gorge Bridge
新荷兰式系船浮标 New Dutch mooring buoy
新赫布里底板块 New Hebrides plate
新红砂岩 new red sandstone
新红砂岩统 <早三叠世>【地】New Red Sandstone series
新花式 new pattern
新花样 dodge
新花样设计 dodge
新华夏构造体系【地】Neocathaysian system
新华夏式【地】Neocathaysian
新华夏系 Neocathaysian
新华夏系构造体系 Neocathaysian tectonic system
新环境适应 postadaptation
新环境主义 <1945 年后美国出现的> new environmentalism
新换旧 new for old
新换旧条款 new for old clause
新黄土 neo-loess;young loess
新黄颜料 new yellow pigment
新灰 fresh lime
新灰坑 live pit
新灰水 new lime liquor
新汇票 <代替退票另加手续费在内的> redraft
新会员 entrant
新活力论 neovitalism
新火山岩 neovolcanic rock
新火山作用 <第四纪>【地】neovolcanism
新火岩体 neosome
新货币 new money
新货币借款 new money loan
新货物装卸泊位 new cargo handling berth
新机器磨合 break-in
新机器跑合 break-in
新机器首次起动 initial start-up
新几内亚海槽 new Guinea trough
新己烷 neohexane
新计算元素 neo-computing element
新记录 new record
新纪元 epoch;new epoch;new era
新技术 innovative technology;new technique
新技术采用期 implementation period of new technique[technology]
新技术风险投资 venture capital of new technology
新技术革命 new technologic(al) revolution
新技术鉴定 appraisal of new technology
新技术开发 development of new technology;new technologic(al) development
新技术开发的推动力 an impulse for new technologic(al) developments
新技术开发研究费 new technique exploration expenses
新技术评价 evaluation of new technology
新技术推广 dissemination of new technology
新技术有偿转让 remunerative transfer of new technology
新技艺 Art Nouveau
新加坡港 Singapore Port
新加坡国际货币交易所 Singapore International Monetary Exchange
新加坡银行同业拆放利率 Singapore Interbank Offered Rate
新加坡元 Singapore dollar
新加入者 entrant
新价值再生产成本 cost of reproduction new value
新驾驶员 fresh driver
新见解 neodoxy
新建 newbuild;new mine
新建的 newbuilt;newly-built;new made
新建房屋 newbuilt house
新建废水管理工程 new wastewater management project
新建港口 newbuilt port;newly-built port
新建工程 new erection;newly(-built) construction
新建工程的危险 hazards of new construction
新建工业 infant industry
新建设项目 new construction
新建铁路 new constructed railway;new-laid line;newly-built railway
新建项目 new construction item;newly-built project
新建住宅开工数 new dwelling construction started
新建筑 new building
新建筑材料 new construction material
新建筑配玻璃的挂门 closing in
新建筑完工量价值 value of new construction put in place
新剑桥学派 neo-Cambridge school
新疆圆柏 savin(e)
新降雪 newborn snow;young snow
新交通系统 new transport(ation) system
新浇的 freshly placed;new-laid
新浇灌的 newly laid
新浇混凝土 <未完全凝固的混凝土> green concrete;as-placed concrete;fresh concrete;freshly mixed concrete;freshly placed concrete;newly laid concrete;newly-placed concrete;unset concrete
新浇混凝土边缘圆角抹刀 arrissing tool
新浇混凝土对模板的压力 pressure developed by concrete on framework
新浇混凝土棱边圆角修整器 arrissing tool
新浇混凝土上部浮浆层 laitance layer
新浇混凝土桩 pile of freshly placed concrete
新浇水泥板 newly-made slab
新浇制楼板 green slab
新浇制砌块 green block
新浇制墙板 green panel
新浇筑的 newly laid
新浇筑的混凝土 as-cast concrete
新胶 fresh rubber
新街坊 new block
新结冰 fresh ice;freshly frozen ice

X

新结构 new construction
新结构设计 development engineering
新结晶作用 neocrystallization
新结硬混凝土 green concrete
新借款办法 new technique of borrowing
新金刚石 new diamond;virgin diamond
新近沉积黏[粘]土 young clay
新近沉积黏[粘]性土 recent deposited cohesive soil
新近沉积土 recent deposited soil;recent soil;recent deposit
新近堆积黄土 recent deposited loess
新近堆积土 recently deposited soil
新近降雨 recent rain(fall)
新近资料 current data
新进入淡水河的海鱼 fresh run
新经济体制 new economic system
新经济政策 new economic policy
新经验主义 <指北欧斯堪的纳维亚人的1940年代的建筑风格> New Empiricism
新井投产 bring in a well
新旧过渡工程 switchover works
新居民区 area for new settlements
新局面 new departure
新开地 plantation;thwaite
新开发区 new developed area
新开工(程)项目 newly-commenced project
新开垦地 broken ground
新开垦土地 cull-land
新开挖边坡 bare cut slope
新开账户 new account
新凯恩斯理论 neo-Keynesian theory
新科技 new science and technology
新颗粒 new grain
新颗粒大小 size of new grain
新客观性 neo-objectivity;new objectivity
新垦地 broken ground;reclaimed land
新垦荒地 newly broken;young clearing virgin soil
新款设备 stylistic apparatus
新矿产地发现率 new mine field discovery rate
新矿物化 neomineralization
新矿物生成作用 neoformation;neogenesis
新来的人 new arrival;newcomer
新蓝 new blue
新浪漫主义 neo-Romanticism
新老混凝土接缝 stoppage joint
新老混凝土之间的结合层 bonding layer
新老砌体接缝 sliding joint
新老砖墙咬接 keying-in
新理性主义 neo-rationalism
新历 new style
新历史主义 neo-historicism
新立账户 new account
新料 virgin material
新料压缩机 make-up compressor
新淋洗过的树脂 freshly eluted resin
新磷钙铁矿 neomesselite
新零件 new parts
新领域 frontier;new area
新露顶板 green roof
新陆地 <海水退减后的> derelict
新路由 new route
新罗马(建筑)风格 neo-Romanesque
新落的(雪) new fallen
新马尔萨斯主义 neo-Malthusianism
新煤贫属 <拉> Neocapnodium
新蒙片 neomask
新面貌 new look
新模式 neotype
新墨西哥 <美国州名> New Mexico
新木材 green lumber
新木料 green lumber;green timber;green wood
新那水 <俗称> thinner
新能源和可再生能源 new and renewable sources of energy
新能源开发技术 developmental technology of new energy resources
新泥沙沉积期 period of modern deposition
新年赠品 handsel
新娘的 bridal
新镍试剂 furildioxime
新凝(结)砂浆 freshly set mortar
新凝泥浆 freshly set mortar
新诺帕尔石 <一种合成岩石> Sinopal
新派建筑 modern architecture
新喷出岩 kainolithe;neoeffusive
新批地段 new grant lot
新品性能 initial performance
新品种光学玻璃 new type optical glass
新铺的 newly laid
新铺轨道 newly laid track
新铺混凝土 newly laid concrete;newly-placed concrete
新铺面层 green surface
新铺线路 newly laid track
新奇的 unprecedented
新奇商品 fancy goods
新企业经营集团 new business venture group
新起点 new departure
新汽 live open steam;live steam;working steam;direct steam
新汽阀 live steam valve
新汽供暖 live steam heating
新汽管 live steam pipe
新汽温度 initial steam temperature
新汽压力 initial steam pressure;live steam pressure
新汽压力调节器 initial steam pressure regulator
新汽再加热 live steam reheating
新砌的砖砌体 green brick work
新砌砌体 green masonry
新砌圬工 green masonry
新钱 fresh money
新嵌板 newly-made panel
新区 newly-built district
新区段 new block
新区普查孔 blue sky exploratory well
新取砂 green sand
新全球构造学 new global tectonics
新全型 neoholotype
新全责拨款授权 <美> new obligation(al) authority
新燃料 fresh fuel
新热带动物地理区 neotropical zoogeographic region
新热带间断分布 neotropical disjunction
新热带区 neotropical region
新热带区的 neotropical
新热带植物地理区系 neotropical floral realm
新热带植物区 neotropical region
新热区 neotropics
新人的 neoanthropic
新三水氧化铝 nordstrandite
新砂 fresh sand;new sand
新设备 development
新设备操作培训 new equipment training
新设备跑合阶段 break-in period
新设备跑合期 break-in period
新设备试运转 shakedown
新设干管 newly laid main
新设施 innovation
新设置的 newly-installed
新升 <体积单位,1 新升 = 10^{-3}立方米> new liter[litre]
新升地面湖 newland lake
新生 entrant;palingenesis
新生变形作用 neomorphism
新生表面 fresh surface
新生部分的侵染 reinfection of newer portions
新生材料 virgin material
新生层 cambium[复 cambi]
新生产部门 new lines of production
新生产堆 new production reactor
新生产方式 new productive system
新生代【地】Caenozoicus;Cainozoic era;Cenozoic era;Kainozoic era;Neozoic era
新生代海洋 Cenozoic ocean
新生代盆地 kainozoic basin
新生代褶皱 Cenozoic fold
新生代植物 cenophyte
新生的 juvenile;nascent;neoformative;neogenic;neonatal;newborn
新生地槽 rejuvenate geosyncline
新生地幔 juvenile mantle
新生工业 infant industry
新生沟谷 gull(e)y
新生界 Cainozoic group;Cenozoic erathem;Cenozoic group;Cenozoic system;Kainozoic group
新生矿物 neogenic mineral
新生氢 nascent hydrogen
新生事物 neocomer;newborn thing;newcomer
新生水 juvenile water;nascent water
新生态 nascent state
新生态玻璃纤维强度 strength of virgin glass fibre
新生态铝 nascent alumin(i)um
新生态氧 nascent oxygen
新生土 recent soil
新生岩浆 neomagma
新生氧 nascent oxygen
新生沼泽 new marsh
新生枝 innovation
新施工管道清扫 new construction cleanup
新湿式浸煮法 novel wet digestion procedure
新石 neolite
新石器 neolith
新石器时代【地】Neolithic Age;New Stone Age
新石器时代的 Neolithic
新石器时代文化 Neolithic culture
新石器时期 Neolithic period
新时代风格 epoch of style
新世界 new world
新世界联盟 the new world alliance
新市镇 new town
新式的 modern;neoteric;new-fashioned
新式动物园 wildlife park
新式发动机 new work engine
新式机床 junior machine
新式机动车 new motor vehicle
新式汽车 new motor vehicle
新式设备 <住房的> modern conveniences
新式铜灵 neocuproin
新式吸收板仪器 new suction plate apparatus
新式悬浮预热器 new suspension preheater
新式样 new design
新式样设计师 stylist
新式住宅 <郊外的> cottage
新事物 coming thing;development;offspring
新事业开发群 new venture group
新饰品 novelty
新手 freshener;green hand;inexperienced operator;neophyte;new face;new hand;novice;noviciate;raw hand;tyro
新手操作 neophyte operation;neophytic operation
新书目录 accession catalogue
新书样本 advance copy
新输入队列 new input queue
新输入排队 new input queue
新树立的标牌 newly-made panel
新数据标帜 new data flag
新刷清干管 newly brushed main
新水 new water
新水手 green sailor;green seaman;landlubber
新塑型主义 neoplasticism
新塑性图 new plasticity chart
新塑造主义 neoplasticism
新索利迪契特水泥 <一种混合硅酸盐水泥> neosolidizit cement
新台升地面湖 newland lake
新陶瓷 new ceramics
新添工程 additional work;extra work
新添加燃料 green fuel
新填路堤 freshly made fill;green embankment
新填土 recent fill
新条款 new terms
新调整 new adjustment
新同步 new sync
新投资 new investment
新投资发生后一年内之利益 next year advantage from project
新投资净额 net investment required
新投资收益报酬率 rate of return on new investment
新投资所得率 rate of return on new investment
新投资添置成本 installed cost of project
新涂未干的油漆 fresh paint
新土 new soil
新土力学参数 new soil parameter
新土力学系数 new soil coefficient
新钍 mesothorium
新挖河槽 newly-cut channel
新挖路堑 freshly made cut
新完工建筑物 new-building completed
新玩意 dofunny
新维修轨道 freshly maintained rail
新位置 new location
新文艺复兴式 neo-Renaissance
新文艺复兴式建筑 neo-Renaissance architecture
新闻报道 reportage
新闻报纸板 news board
新闻编辑室 newsroom
新闻部门 information service
新闻采访工作 legwork
新闻处 press department
新闻传真 pressfax
新闻电报 press telegram
新闻电话 press call
新闻电讯 press dispatch
新闻电讯社 wire agency;wire service
新闻电影院 news cinema;news theatre
新闻发布会 news release conference
新闻发布室 briefing room
新闻封锁 blackout;box out
新闻服务器 news server
新闻稿 news release
新闻工作者 journalist;pressman
新闻公报 press releases
新闻广播 newscast
新闻广播员 commentator
新闻记者 journalist;newsman
新闻记者席 press gallery
新闻简报 bulletin;news briefing;news summary

新闻节目时间 information hour
新闻媒介 medium of communication
新闻社 news agency
新闻文化参赞 press and cultural counselor
新闻影片 newsreel
新闻纸 newspaper;newsprint
新污染区 zone of recent pollution
新污染源执行标准 new source performance standard
新戊二醇 neopentyl glycol
新戊烷 neopentane
新物质 new substance
新西兰阿尔卑斯转换断层 New Zealand Alpine transform fault
新西兰白松 Kahikatea
新西兰标准学会 New Zealand Standards Institute
新西兰地槽 New Zealand geosyncline
新西兰红木 Podocarpus totara
新西兰绿软玉 New Zealand greenstone
新西兰罗汉松 Kahikatea;totara
新西兰气象局 New Zealand Meteorological Service
新西兰双壳类地理亚区 New Zealand bivalve subprovince
新西兰柚木 New Zealand teak
新希腊式 <1840 年在法国开创的建筑形式> neo-Grec; neo-Greek; New Grecian type
新希腊式建筑 New-Greek architecture
新下的雪 fresh snow
新鲜 freshness;verdure;viridity
新鲜白土 sweet clay
新鲜充量 fresh charge
新鲜出炉热水泥 hot cement
新鲜催化剂 fresh catalyst
新鲜催化剂料斗 fresh catalyst hopper
新鲜的 eye-opening; unweathered <指岩石的风化程度>
新鲜度 freshness
新鲜废水 fresh wastewater
新鲜废物 fresh waste
新鲜干草垛 grass cock
新鲜花岗岩 fresh granite
新鲜灰浆 green plaster
新鲜混凝土 as-mixed concrete;immature concrete
新鲜基岩 fresh bedrock
新鲜胶乳 fresh latex
新鲜空气 fresh air;outdoor air
新鲜空气补充量 fresh-air make-up
新鲜空气导入装置 fresh-air unit
新鲜空气风道 fresh-air duct
新鲜空气鼓风机 fresh-air blower
新鲜空气管道 fresh-air duct;fresh-air flue
新鲜空气过滤层 fresh-air filter
新鲜空气环流 fresh-air circulation
新鲜空气加热 fresh-air warming
新鲜空气加热器 fresh-air warmer
新鲜空气进口 fresh-air inlet
新鲜空气量 ventilation requirement; volume of fresh air
新鲜空气流通 ventilation with fresh air
新鲜空气面具 fresh-air mask
新鲜空气入口 fresh-air inlet; fresh-air intake
新鲜空气输送装置 fresh-air unit
新鲜空气送风机 fresh-air fan
新鲜空气吸入口 outdoor air intake
新鲜空气循环 fresh-air circulation
新鲜裂化气 fresh cracked gas
新鲜漏油 fresh oil spills
新鲜露头 fresh outcrop
新鲜露头面【地】freshly exposed surface
新鲜气流 intake air
新鲜气体 live gas
新鲜气团 fresh
新鲜入风 fresh intake
新鲜润滑油系统 fresh oil system
新鲜润滑油箱 fresh oil tank
新鲜砂浆 green plaster
新鲜蔬菜 fresh vegetables
新鲜水 fresh water
新鲜水果 fresh fruit
新鲜污泥 fresh sludge;green sludge
新鲜污水 fresh sewage
新鲜血液 new blood
新鲜岩石 fresh rock;unaltered rock; unweathered rock
新鲜岩芯技术 fresh-core technique
新鲜样品 fresh sample
新鲜原料 fresh feed
新鲜原料收集器 fresh feed surge drum
新鲜原料转化率 fresh feed conversion
新鲜原油 fresh crude oil
新鲜蒸汽 fresh steam;live steam;working steam
新鲜状态 fresh state;green state
新现实主义 neo-realism
新线建设 new line construction
新相 kainotype
新相火山岩 cenotypal rock
新相岩 kainotype rock
新巷道开挖 new work dredging
新项目工程 green field project
新项目施工 new construction activity
新写实主义 neo-realism
新心硬木 <造船用> green heart
新信息 fresh information
新兴产业和高技术产业 emerging and high-tech industries
新兴城市 boom(ing) city;boom(ing) town;born city;mushroom town; newly developing;newly rising city
新兴工业 growth industry;infant industry
新兴工业城市 new industrial city
新兴工业国家 young industrial country
新兴工业化国家 newly industrialized country
新兴工业化国家和地区 newly industrialization country and region
新兴工业区 new industry area
新兴国家 emerging countries;emerging nation
新兴科技领域 new scientific and technological undertakings
新兴力量 newly emerging forces
新兴市镇 <美> boom town
新兴物质生产部门 newly developing material production department
新兴行业 growth industry
新星形成【天】astration
新行为主义 neo-behaviorism
新形成 new formation
新形成的 new made
新形成的冰 fresh ice
新形成体 neomorph
新形式主义(建筑)New Formalism
新形体形成 neomorphism
新型 modern version; neotype; new pattern;new style;utility model
新型材料 new material
新型超积极喂纱装置 new ultra positive feed
新型成圈方法 new-loopforming system
新型的 late-model;neotype;new model
新型斗轮挖泥船 new type bucket wheel dredge(r)
新型发动机 new work engine
新型发动机原型 pilot engine
新型方法 innovative approach;innovative method
新型干法水泥生产 new dry process cement production
新型工业陶瓷 new industry ceramics
新型号 new model;new style
新型号投产 model launching
新型环氧树脂 neo-epoxy resin
新型技术装备 new process equipment
新型控制设备 modern control equipment
新型连接件 modern connecter[connector]
新型木墙板 novelty siding
新型轻轨快速运输系统 advanced light rapid transit
新型设计 new design
新型设计控制 new design control
新型生物混合系统 novel biological hybrid system
新型陶瓷 new-ceram;new ceramics
新型陶瓷纳滤膜 new ceramic nanofiltration membrane
新型污染物 emerging pollutant
新型悬浮预热器快速分解炉 new suspension preheater with flash calciner
新型旋转桥式装船机 novel slewing-bridge type shiploader
新型一体化 3R(除碳、氮脱磷)生物反应器 new combined 3R(carbon, nitrogen and phosphorus removal) bioreactor
新型运输工具 new mode
新修 rebuild;renewal
新学说 neodoxy
新学院派形式主义 Neo-Academic Formalism
新雪 freshly fallen snow;fresh snow; new snow
新雪冰 <落在水面的雪所形成的> new snow-ice
新雪层 fresh snow layer
新芽 sprout
新亚甲蓝 new methylene blue
新亚铜试剂 enocuproine
新岩芯钻头 new core bit
新样品 fresh sample
新样式 new mode;new style pattern
新叶黄素 heofucoxanthin
新一代 new generation
新一代自动绘图系统 new generation AutoCAD system
新艺术风格的(上釉)陶瓷 Art Nouveau faience
新艺术派 Art Nouveau
新艺术运动 <19 世纪末法国和比利时的> Art Nouveau
新引种植物 neophyte
新印象派 neo-impressionism
新英格兰砌砖法 New English method;pick and dip
新英格兰式 <美国初期建筑> New English colonial
新英格兰殖民式建筑 New England colonial architecture; New English colonial architecture
新颖 novelty;originality
新颖的 new face;originative
新颖花纹地板 novelty flooring
新颖花样 novelty pattern
新颖拼花地板 novelty flooring
新颖设计 novel design
新颖性 <指专利产品发明> novelty
新用途开发概率 probability of new usage develop
新用途影响程度 affected degree of new usage
新油 clean oil;fresh oil;new oil
新油漆 fresh paint
新有效温度 new effective temperature
新淤积泥沙 modern deposit;modern sediment; newly-accumulated sediment
新淤泥 bungum
新余额 new balance
新预制的混凝土块 prefabricated concrete green block
新域 neofield;new world
新元古巨旋回【地】Neo-Algonquian megacycle
新月 crescent moon;demilune;meniscus[复 menisci/meniscuses]; new moon
新月大潮 new moon spring tide
新月灰光 earth light
新月饰 crescent;meniscus[复 menisci/meniscuses]
新月体 crescent
新月体形成 crescent formation
新月湾 bight
新月效应 meniscus effect
新月形 concavo-convex; crescent-shaped;lunar;lune
新月形冰碛层 crescentic moraine
新月形冰丘 ice barchan
新月形波痕 crescent ripple;washover crescent
新月形擦痕 crescentic scour mark
新月形冲刷痕 crescentic scour mark
新月形的 bicorn;crescent(ic);half-moon-shaped;lunate;meniscoid
新月形地 lunate bone
新月形叠加褶皱 crescent superposed folds
新月形废河弯 meander crescent;old meander crescent
新月形拱 crescent arch;sickle arch; sickle-shaped arch
新月形海滩 crescent beach
新月形河湾 lunate bend
新月形湖 crescentic lake;oxbow lake
新月形礁 crescentic reef
新月形砾滩 shingle barchan
新月形三角洲 lunate delta
新月形沙坝 lunate bar
新月形沙埂 lunate bar
新月形沙垄 crescentic sand-ridge
新月形沙丘 barchan(e); crescentic dune;meniscus dune;parabolic(al) dune;barkhan
新月形沙丘地 barkhan sand
新月形沙丘链 barchan[barkhan] chain;continuous barchan;continuous crescent dune
新月形沙滩 sickle beach
新月形沙土 barkhan dune; crescent dune
新月形沙洲 crescentic-shaped bar;lunate bar
新月形水流痕 crescent current marks; current crescent
新月形水下沙洲 crescentic submarine bar
新月形条斑 crescentic formed penciling
新月形凸岸边滩 crescent point bar; crescent-shaped point bar
新月形弯顶 meander crescent
新月形物 meniscus[复 menisci/ meniscuses]
新月形雪丘 snow barchan(e)
新月形照相机 meniscus camera
新月形状拱顶 lunette vault
新月型交错原理构造 crescent cross-bedding structure
新月型胶结物结构 meniscus cement texture

新月状 crescent
新月状物 crescent;lunute
新运动区的 neokinetic
新灾变论 neocatastrophism
新造 newbuild
新造的 newbuilt; newly-built; newly made
新造型主义 neoplasticism
新泽西 < 美国州名 > new Jersey
新泽西式护墙 new Jersey barrier
新增 creation
新增成本 additional cost;extra cost
新增储量 new additional reserves
新增的工作 additional work
新增风险 additional risks
新增购买力 newly-increased purchasing power
新增固定资产 added fixed assets;new fixed assets
新增固定资产费用 cost of new fixed assets
新增雇员税收减免 new jobs credit
新增关税 new additional duty
新增机械设备支出 new plant and equipment expenditures
新增机组 additional unit
新增积累资金 newly-increased accumulation funds
新增价值 newly-increased value
新增交通量 < 由于新地区开发而引起的 > generated traffic
新增就业人员 new entrants to the labo(u)r force
新增矿山生产能力 annual output of new mining
新增生产能力 added capacity; new productive capacity
新增消费资金 newly-increased consumption funds
新增效益 new efficacy
新增资产 new assets
新债券 new bond
新债务人 expromissor
新褶皱 new fold
新阵地地域 new position area
新蒸汽 fresh steam;live steam
新蒸汽注水器 live steam injector
新政策 new deal
新值 new value
新指标 give new set
新制板材 freshly made slab
新制产品 green product
新制成的产品 newly-made products
新制成品 freshly made product;fresh product
新制的 fire-new
新制地图样张 pilot sheet
新制度经济学家 neo-institutional economics
新制构件 green member;green unit
新制阶段 green state
新制块材 green unit
新制黏[粘]土砖 green clay brick
新制品 green unit
新制平板 newly-made slab
新制砖瓦 green tile
新中古时代的 neo-mediaeval
新中世纪的 neo-mediaeval
新中世纪式建筑 neo-mediaeval architecture
新烛光 candela;new candle
新住处 new habitat
新住宅区 settlement
新住宅区职能 function of settlement
新筑混凝土 freshly placed concrete
新筑砌体 green masonry
新筑圬工 green masonry
新专款授权 new obligation(al) authority
新装玻璃 first glazing
新着丝点 neocentromere
新资本所得率 rate of return on new investment
新资金 new funds
新资料 current information
新资源 new resources
新子午卫星 nova transit satellite
新自由派风格的 neo-liberty
新自由型 new freedom type
新租船契约 new charter
新租船契约方 new charter party
新租户 ingoing tenant
新组织的 neoblastic
新钻工 boll weevil
新钻进方法 novel drilling method
新钻头试验过程 bit break-in procedure
新坐标(系)格网 fresh grid
新做的 new made

薪

薪材 energy wood; fire wood; fuel wood; kindling wood; short end; yule log

薪材造林 fuelwood planation
薪额 amount of salary
薪俸 pay
薪俸税 salaries tax
薪给报酬所得税 income tax on salaries and remunerations
薪工 payroll
薪工标准 pay scale
薪工成本 payroll cost
薪工登记簿 payroll register
薪工分配 payroll distribution
薪工分配簿 payroll distribution book
薪工分析 payroll analysis
薪工管理 payroll control
薪工管理部门 payroll department
薪工管理部门费用 payroll department expenses
薪工会计 payroll accounting
薪工结算员 payroll clerk
薪工结算账户 payroll clearing account
薪工清单 payroll sheet
薪工审计 payroll audit
薪工税 payroll tax
薪工通知单 payroll slip
薪工账户 payroll account
薪工支票 payroll check
薪工制度 payroll system
薪工周转金 payroll fund
薪工总额 gross pay
薪级表 salary scales;salary table
薪金 emoluments; pay; pay cheque; wages
薪金保险 salary insurance
薪金比率 comparative salary ratio
薪金标准 wage level
薪金标准化 salary standardization
薪金表 salary scales;salary schedule
薪金等级 pay grade;salary class
薪金调查 salary survey
薪金费用 salaries expenses
薪金分配表 salary distribution sheet
薪金管理 salary administration
薪金和津贴 salaries and allowances
薪金计算 salary accounting
薪金加奖金制 salary plus bonus
薪金结构 salary structure
薪金、津贴和酬金 salaries, allowance and emoluments
薪金俱乐部 salary club
薪金率 salary rate
薪金清册 salary roll
薪金税 salaries bat;salaries tax
薪金线制度 track system
薪金增长曲线图 salary progression curve
薪金中预扣所得税法 < 英 > pay-as-you-earn
薪金自动增长 salary increment
薪津冻结 pay pause
薪水 compensation;pay;wage
薪水阶层 salariat
薪水账 pay list;payroll
薪水支票 pay check
薪炭材 fag(g)ot wood;fire wood
薪炭林 firewood forest;fuel forest
薪饷制 system of regular pay
薪资 salary
薪资保险 salary insurance
薪资标准 pay scale
薪资单 payroll
薪资冻结 pay pause
薪资规程 regulations governing salaries and wages
薪资曲线 wage curve
薪资体系 salary system
薪资与福利 compensation & welfare
薪资照付的病假 paid sick leave
薪资照付的假期 leave with pay;paid leave
薪资照付的例行假日 paid public holidays
薪资照付的年度休假 paid annual vacation
薪资总额 gross salaries

囟

囟钠矾 schairerite

囟钠石 suphohalite

信

信标 beacon;marker

信标测向仪 beacon direction finder
信标处理系统 beacon processing system
信标导航 beacon
信标灯 beacon lamp;beacon light
信标灯光 marker light
信标电台 beacon station
信标分布密度 beacon density
信标跟踪 beacon tracing; beacon tracking
信标接收器 beacon receiver
信标雷达 beacon radar
信标码 beacon-code
信标频率 beacon frequency
信标识别数据 beacon identification data
信标塔 beacon tower
信标天线 marker antenna
信标延迟 beacon delay
信标遗失 beacon stealing
信标有效距离 beacon range
信标指示器 marker beacon indicator
信槽 mail slot
信槽孔 letter slot
信秤 letter balance;letterweight
信尺长度 multiple lengths
信串比 signal-crosstalk ratio
信贷 bill finance; credit; credit and loan
信贷安全系数 safety factor
信贷保函 credit guarantee
信贷保密保付代理 confidential factoring
信贷保证 credit guarantee
信贷部(门) credit department
信贷承诺 credit commitment
信贷乘数 credit multiplier
信贷储备 credit reserve
信贷担保 credit guarantee; credit guaranty;guarantee for loans
信贷的层层加大 pyramid of credit
信贷的计划偿还 scheduled repayments on credits
信贷额 credit amounts;credit volume
信贷额度 credit limit;credit line;line of credit
信贷发行 credit issue;issuing credit
信贷方针 credit extending policy
信贷费用 finance charges
信贷分类 credit rating
信贷份额 credit tranche
信贷风险 credit risk
信贷封锁 credit blockade
信贷杠杆 credit leverage
信贷公司 finance company;financial corporation
信贷供应者 supplier's credit
信贷关系 credit relation
信贷管理 credit control;credit management
信贷管制法 credit control act
信贷合同 finance contract
信贷核对 credit verification
信贷机构 credit agency;credit institution;credit society
信贷基金 credit fund
信贷价值 credit standing;credit worthiness
信贷监督 credit supervision
信贷交易 credit transaction
信贷接受人 credit receiver
信贷节制 credit restraint;credit restriction;restriction of credit
信贷金额 credit amount
信贷紧缩 credit squeeze
信贷经理 credit manager
信贷竞争 credit competition; credit race;loan competition;loan race
信贷决策 credit decision
信贷可获量 credit availability
信贷控制 credit control
信贷扩张 credit expansion
信贷联合 credit association
信贷配给 credit rating
信贷膨胀 credit inflation
信贷凭证 credit memo
信贷期以前 precredit
信贷人寿保险 credit life insurance
信贷申请表 credit application
信贷使用 availment of credit
信贷市场 credit market
信贷收支 credit receipts and payments
信贷收支平衡 balance between credits and payments
信贷所 credit society
信贷条件 credit terms
信贷条款 terms of credit
信贷同业公会 credit guild
信贷文件 credit documents
信贷系统 credit system
信贷限额 credit limit;high credit;line of credit
信贷限额保证 cover for lines of credit
信贷限制 credit restriction
信贷协定 credit agreement
信贷协定的期限和条件 terms of loan agreements
信贷协会 credit union
信贷协议 credit agreement
信贷需要 credit requirement
信贷续期 renewal of facility
信贷延期 dating
信贷业务 credit operation
信贷银行 credit bank; Kredietbank NV < 比利时 >
信贷员 loan officer;loan teller
信贷政策 credit policy
信贷政策调节论 theory of regulation by credit policy
信贷支持的信贷 credit supported by

other credit
信贷值 value of credit
信贷质量 quality of credit
信贷中心 credit center[centre]
信贷资本 credit capital
信贷资金 credit fund
信贷总额指标 indicator for total credit
信贷最高限额 credit ceiling
信道 channel;signal channel
信道比 channel ratio
信道编号 channel number
信道编码 channel coding
信道编码器 channel encoder
信道变换器 channel translator
信道波导管 channel waveguide
信道处理单元 channel bank
信道触发电路 channel trigger circuit
信道串音 channel crosstalk
信道带 channel strip
信道带宽 channel bandwidth
信道单束光缆 channel single-bundle cable
信道单线光缆 channel single-fiber cable
信道(的频率)间隔 channel spacing
信道等级 channel grade
信道地址 channel allocation
信道发射机 channel(l)ized transmitter
信道放大器 channel amplifier
信道分路器 channel-subdivider
信道分配 channel allocation; channel assignment
信道分用 channel demultiplexing
信道服务单元 data service units
信道合并 channel packing
信道化 channelizing
信道划分 channelize
信道获得延迟 channel acquisition delay
信道架 channel bank
信道间插入频率 slot frequency
信道间串音 interchannel crosstalk
信道间的 interchannel
信道间干扰 cross fire; interchannel interference
信道间距 channel separate; channel separation
信道间噪声 interchannel noise
信道矩阵 channel matrix
信道可靠性 channel reliability
信道控制 channel control
信道宽度 channel width
信道扩展 channel spread
信道利用指数 channel utilization index
信道连接器 channel connector
信道脉冲 channel pulse
信道门 channel gate
信道命令 channel command
信道频带利用率 data rate-to-bandwidth ratio
信道频宽变化 channel-width variation
信道容量 channel capacity
信道入口光学控制 optic(al) access control
信道失真 channel distortion
信道时隙 channel time slot
信道识别 channel identification
信道使用率 channel utilization
信道输入 channel input
信道数 number of channels
信道衰减 path attenuation
信道双工通信[讯] double channel duplex
信道速率 channel efficiency; channel speed
信道探头 signal channel probe
信道同步器 channel synchronizer
信道稳定性 channel stability
信道陷波电路 channel trap
信道相位调整 channel phasing
信道序号 channel designator; channel sequence number
信道选择电路 channel selectivity circuit
信道选择器 channel selector
信道压缩 channel compression
信道移动器 channel shifter
信道译码器 channel decoder
信道英里 channel mile
信道载频滤波器 channel carrier filter
信道占用 channel busy
信道指配 channel assignment
信道质量指数 quality index of a channel
信道终端 channel terminal
信道终端编码器 channel terminal coder
信道终端解码器 channel terminal decoder
信道种类 class of channel
信道转换 channel switching
信道组 channel bank; channel group
信得过标 bona fide bid
信得过产品 trustworthy product
信得过信贷 deemed paid credit
信的附言 postscript
信度 degree of confidence; inter-scorers reliability
信度解二次方程法则 Hindu method for quadratic equations
信风 trade(wind)
信风赤道槽 trade-wind equatorial trough
信风带【气】trade-wind zone; trade-wind belt; zone of trade-wind
信风锋 trade-wind front
信风海流 trade-wind current
信风环流 trade-wind circulation
信风积云 trade cumulus; trade-wind cumulus
信风空气 trade air
信风逆温 trade inversion; trade-wind inversion
信风漂流 trade drift
信风区 trade-wind region
信风沙漠 trade-wind desert
信风系统 trade-wind system
信封式成型帘 envelope type forming hood
信管 fusee; fuse tube
信管电路 detonator circuit
信管口 eye; fuse cavity
信管装置 fuse mechanism
信函 letter
信函传真 facsimile mail; post fax
信函传真机 mail-facsimile apparatus; postfax apparatus
信函订单 letter orders
信函求偿 mail reimbursement
信函自动分选 automatic partition of letters
信号 semaphore; signal(ling); wig-wag(signal)
信号矮柱 doll
信号按钮 signal button; signal(l)ing key
信号摆幅 signal swing
信号板 backplate; signal panel; signal plate
信号保养员 signal maintainer
信号报警 signal alarm
信号报警设备 signal(l)ing alarm equipment
信号倍增器 multiplex(er); signal multiplier
信号比 actuating signal ratio; signal ratio
信号比较法 signal-comparison method
信号笔 sign pen
信号闭塞 non-passage of signal
信号闭塞装置 signal block system
信号臂 semaphore arm
信号臂板 arm semaphore; blade; semaphore arm; signal blade; signal arm
信号臂板复示器 signal arm repeater; signal blade repeater
信号臂板夹 blade grip
信号臂板平衡锤 semaphore signal counterbalance
信号臂板位置 signal position
信号臂板位置检查 signal arm proving
信号臂板选别器 electric(al) coupling of signal arm
信号臂板重锤 semaphore signal counterbalance
信号臂板轴 axle of signal blade
信号边频带 signal frequency side band
信号编码 signal coding
信号编码容量 signal code capacity
信号编码装置 signal coding equipment
信号变换 signal conversion; signal transformation; signal translation
信号变换存储器 signal converter storage tube
信号变换器 signal converter; signal inverter; signal transformer
信号变换设备 signal conversion equipment
信号变换装置 chromacoder
信号变压器 signal transformer
信号标 beacon; semaphore; signal beacon
信号标杆加路码表法 signpost and odometer
信号标志 signal; signal mark(er)
信号标志灯 signal marker light
信号标准化 signal normalization; signal standardization
信号表示 indication of signal; signal indication
信号表示灯 signal indication lamp; signal indication light; signal indicator
信号表示法 signal representation
信号表示镜 roundel; signal spectacle
信号(表示)面 <信号室的一面,控制一个方向的交通> signal face
信号表示器 signal indicator
信号波 signal wave
信号波包络 signal-wave envelope
信号波形 keying waveform; signal waveform
信号波形成分 signal waveform elements
信号波形校正网络 signal-shaping network
信号波形监视 signal monitoring
信号波形修整 signal conditioning
信号玻璃 signal glass
信号玻璃制品 signal glassware
信号补偿器 signal compensator
信号捕获 signal capture
信号捕捉装置 signal acquisition unit
信号不定性 signal ambiguity
信号不稳 swinging of signals
信号不稳定性 jitter
信号布板 signal(l)ing panel
信号布板码 panel code
信号布条 signal strip
信号布置 arrangement of signals; signal(l)ing arrangement; signal(l)ing plan
信号布置图 diagram of signal(l)ing layout
信号部队 buzzer
信号参量 signal parameter
信号操纵 signal operation
信号操纵导线 signal operating wire
信号操纵盘 signal(l)ing panel
信号操作 signal operation
信号测量器 signal measuring apparatus
信号测试 signal testing
信号测试设备 signal-testing apparatus
信号层 signal plane
信号察觉理论 theory of signal detectability
信号产生站 signal generating station
信号长度 signal length
信号场 signal field; signal ground
信号场强度 signal field strength
信号场周期 signal field period
信号车间 signal(work)shop
信号成分 signal component
信号迟延时间 signal delay time
信号持续时间 signal duration
信号冲撞 signal bumping
信号处理 signal process(ing)
信号处理单元 signal processing unit
信号处理放大器 processing amplifier
信号处理机 signal processor
信号处理机控制器 signal processor controller
信号处理天线 signal processing antenna
信号处理系统 signal-handling equipment; signal processing system
信号处理装置 signal-handling apparatus
信号触点 signal(l)ing contact
信号传播速度 signal velocity
信号传递 signal transmission
信号传递函数 signal transfer function
信号传动机构 signal operating mechanism
信号传动装置 signal(l)ing gear
信号传感器 signal transducer
信号传输 signal propagation; signal transmission; signal(l)ing
信号传输方式 signal transmission form
信号传输时间引起的失真 transit time distortion
信号传输通道 signal channel
信号传送 signal transmission
信号传送测试 signal(l)ing test
信号传送延迟 signal transferring lag
信号传送滞后 signal transferring lag
信号锤 knocker
信号刺激 signal stimulus
信号存储器 signal storage
信号错误开放 false clearing of signal
信号代码 signal code
信号代码表 signal code table
信号带宽 signal bandwidth
信号单元 signal element
信号单置点 single signal location
信号弹 signal projectile
信号弹发射器 pyrotechnic generator
信号导线 signal wire
信号导线导轮座 signal wheel base
信号的 semiotic
信号灯 alarm lamp; alert lamp; approach light; call lamp; flash lamp; headlight; indicating lamp; lantern; marker light; pilot lamp; pilot light; signal flare; signal lamp; signal lantern; signal light (ing); telltale lamp; winker
信号灯按钮 signal(l)ing lamp push button
信号灯插口 signal lamp socket
信号灯电话交换台 lamp switchboard
信号灯工 lampman
信号灯光 code light; identification

light; pilot light; signal light
信号灯光检查 signal light proving
信号灯光显示距离测试 signal light visibility test
信号灯红玻璃 red signal lamp glass
信号灯换相间隔 change interval
信号灯架 signal lamp bracket
信号灯降压控制 dimmer control; dimming control
信号灯交叉口 signal control intersection; signalized intersection
信号灯控制人行横道 light-controlled pedestrian crossing
信号灯控制行人过街道 Pelicon crossing
信号灯框 signal lantern case
信号灯蓝玻璃 blue signal lamp glass
信号灯器 optic(al) unit
信号灯闪烁器 signal lamp flicker
信号灯室 signal lantern case
信号灯调压 voltage regulation of signal lamps
信号灯显示时间序列 interval sequence
信号灯相调制 <按电子计算机控制中心命令的> phase module
信号灯周期分配 cyclic distribution of signal lights; split cycle
信号灯柱 switch tower
信号灯紫玻璃 purple signal lamp glass
信号灯座 signal lamp housing; signal lamp socket
信号等强线 equal signal strength line; equisignal strength line
信号地 signal ground
信号地线 signal ground
信号点 signal location
信号电池 signal(l)ing battery
信号电动机 signal motor
信号电极 pick-up electrode
信号电缆 bell cable; signal cable; cab-tyre cable <地下连续墙>
信号电流 signal(ling) current
信号电路 signal(l)ing circuit
信号电路被阻塞或不能工作的表示 indication of interrupted or non-functioning signal circuits
信号电路控制器 signal circuit controller
信号电码 signal code
信号电平 signal level
信号电平比调整电位计 fader potentiometer
信号电平表 signal level meter
信号电平波动 signal level fluctuation
信号电平分布 signal level distribution
信号电平固定放大器 clamper amplifier
信号电平检测器 signal level detector
信号电平控制 level control; signal level control
信号电平试验片 signal level test film
信号电线路 signal circuit line
信号电压 signal voltage
信号电压发生器 signal voltage generator

信号电源 signal power source
信号电源闸 power source room for signal
信号吊牌 signal drop
信号吊绳滑车 <船的> jewel block
信号叠加 superposition of signals
信号定时 signal timing
信号定时键盘 signal timing dial
信号丢失 loss of signal
信号动作 signal operation
信号读出 signal reading
信号度盘 dial; dial unit
信号度盘时间键定位 dial key setting
信号度盘转速 dial speed
信号渡越时间 signal transit time
信号对比度 signal contrast
信号对噪声和失真比 signal-to-noise and distortion ratio
信号对准指示符 sector alignment indicator
信号多路传输装置 signal multiplexing unit
信号多重性 signal ambiguity
信号遏制 signal curbing
信号二极管 signal diode
信号发射规则 signal regulation
信号发生 signal generation; signal production
信号发生法 signal generating means
信号发生器 signal generating device; signal generator
信号发生器环形天线 signal-generator loop
信号发生源 signal generating source
信号发送程序 signal(l)ing procedure
信号发送存储器 sending signal memory
信号发送器 sender unit; signal transmitter
信号发送速度 signal(l)ing rate; signal(l)ing speed
信号发送台 sender transmitting station
信号阀 signal valve
信号法 signal(l)ing
信号反射 signal reflex
信号反射镜 signal reflector
信号反应器 signal repeater
信号反褶积 signal deconvolution
信号方案 signal(l)ing scheme
信号方牌 square signal
信号方式 aspect; signal aspect
信号方位显示 signal azimuth display
信号方向 sense
信号防护范围 signal limit
信号防护区段 signal limit
信号防止重复锁闭器 one-pull signal lock
信号房 signal box; signalman's cabin
信号放大 signal amplification; signal multiplication
信号放大器 signal amplifier
信号放映机 signal projector
信号分类机 signal classifier
信号分离 demultiplex; separation of signals; signal separation
信号分离带通滤波器 signal-separation filter
信号分离电路 demultiplexing circuit
信号分离器 demultiplexer; signal separator
信号分量 signal component
信号分配存储器 signal distribution memory
信号分配放大器 distribution amplifier
信号分配盘 signal distribution panel
信号分配器 signal distributor; signal splitter
信号分配装置 signal distribution equipment; signal distribution unit
信号分析 signal analysis
信号分析器 signal analyser [analyzer]; signalyzer
信号分析与测量仪表 signal analysing and measuring instruments
信号风笛 signal whistle
信号风缸 signal reservoir
信号风管 signal pipe
信号风管滤尘器 signal pipe strainer
信号风管遮断塞门 signal pipe cut-out cock
信号浮标 dan buoy; flag float; signal buoy
信号浮标释放装置 marker buoy release unit
信号符号 signal code; signal(l)ing symbol
信号符合 signal correspondence
信号幅度 signal level
信号幅度分布 signal amplitude distribution
信号附属品 signal appendant
信号复示器 signal repeater
信号复示器箱 signal repeater box
信号复位器 signal replacer; signal reverser
信号复原 signal replacement; signal restoring
信号复原机械接触器 mechanical replacement treadle
信号复原机械踏板 mechanical replacement treadle
信号复原开关 recall switch
信号覆盖范围 signal cover
信号概率 signal probability
信号干扰比 signal-to-interference ratio; signal-to-noise ratio
信号杆 signal bar; signal lever; signal mast; signal post; signal rod; signal staff; telltale rod <桩工>
信号感受 sensing
信号跟踪 tracing
信号跟踪仪器 signal-tracing instrument
信号更换 signal replacement
信号工 flagman; signalman
信号工厂 signal shop
信号工程师 signal engineer
信号工程学 signal(l)ing engineering
信号工具 signal means; signal tool
信号工区 signal section
信号工作 signal working
信号功率 signal power
信号故障 signal failure; signal(l)ing fault
信号故障积累 accumulation of signal faults
信号故障检寻器 signal tracer
信号故障累积 accumulation signal fault
信号关闭 signal at stop; signal on
信号关闭表示 stop signal indication
信号管 signal pipe; signal piping
信号管理控制器 signal supervisory control
信号管线 signal pipe line
信号光线 signal ray
信号光学 signal optics
信号规则 signal book; signal code; signal instruction; signal regulation
信号轨道电路 signal track circuit
信号过阀指示器 sector alignment indicator
信号号角 signal horn
信号和进路控制中心 signals and routes control center
信号(滑)杆 signal slide
信号化 signalization; signal(l)ing
信号化环形交叉 signalized roundabout
信号还原 de-emphasis
信号恢复 signal reconstruction
信号回路管制器 signal circuit controller
信号混合 signal mixing
信号混合器 signal mixer
信号混合装置 signal mixer unit
信号混响 signal reverberation
信号混淆 blurring
信号活动 activity; signal activity
信号机【铁】numerator; ringing machine; semaphore; signal (generator); teleseme
信号机背板 signal unit back
信号机道岔联锁检查器 signal point detector
信号机的间隔 space interval; spacing interval; spacing of signals
信号机的设备箱 location case
信号机(底)座 signal base
信号机点灯 signal lighting
信号机点灯电路 signal lighting circuit
信号机点灯电源 signal lighting power source
信号机电操作 electric(al) working of signals
信号机电动气动阀 electropneumatic signal valve
信号机电空操作 electropneumatic signal working
信号机构 signal head; signal machine
信号机号码 signal number
信号机后方 advance of a signal; ahead of a signal; back of a signal; in advance of a signal; in rear of a signal; inside of a signal
信号机间隔 signal headway
信号机间隔距离 distance between signals
信号机间隔研究 signal spacing study
信号机接近亮灯(法) approach signal lighting
信号机经常关闭的闭塞制 block system with signals normally at danger
信号机经常开放的闭塞制 block system with signals normally at clear
信号机具 signal(l)ing apparatus
信号机绿灯时间 controller green time
信号机内方 advance of a signal; ahead of a signal; back of a signal; in advance of a signal; inside of a signal
信号机前方 in advance of a signal; outside of a signal; rear of a signal
信号机塞孔 annunciator jack; enunciator jack
信号机顺序 sequence of signals; succession of signals
信号机头 signal head
信号机头部装置 signal head assembly
信号机头支持装置 signal support
信号机涂漆 signal painting
信号机外方 outside of a signal; rear of a signal
信号机无效标志 signal-not-in-use sign
信号机系统 annunciator system; succession of signals
信号机箱 signal case
信号机械 signal(l)ing gear
信号机-信号机的进路排列【铁】signal-to-signal route lineup
信号机移位 signal relocation
信号机照明 signal lighting
信号机柱 signal mast; signal post
信号机柱顶 signal pinnacle
信号机组 controller assembly
信号积累 integration of signals; signal integration
信号畸变 signal distortion
信号极 pick-up electrode
信号极性 signal polarity
信号极性变换 signal inversion
信号极性变换器 signal inverter
信号极性转换开关 signal polarity switch
信号集合 signal integration
信号计划 signal(l)ing scheme
信号计数器 event counter; signal counter
信号记录电报术 signal-recording telegraphy
信号记录器 signal recorder
信号技工 signal mechanician
信号技术 signal(l)ing art; signal(l)

ing technique
信号技术员 signal mechanician
信号继电器 enunciator relay; indicating relay; indication relay; signal relay
信号寄存器 sign register
信号寄存器电源 signal register power source
信号加工 signal conditioning
信号加密系统 signal encrypt system
信号架 < 多层平面车道上的 > sign gantry
信号假说 signal hypothesis
信号间隔 momentum range; signal space; wave spacing
信号间距 signal distance
信号间距离 span of signals
信号监督电路 signal supervisory circuit
信号监视 signal monitoring
信号监视器 picture monitor; signal monitor
信号检测 signal detection
信号检测阀 signal detection threshold
信号检测概率 signal detection probability
信号检测理论 signal detection theory
信号检测器 signal detector
信号检查 signal check
信号检查器 signal monitor
信号检修所 signal maintenance depot; signal repair station
信号建设 signal construction
信号鉴别 signal discrimination
信号降低 signal degradation
信号交互线 alternate line
信号交换 handshaking
信号交换输入输出控制 handshake control
信号绞车 signal winch
信号校正 signal calibration; signal correction
信号校正电路 peaking circuit
信号校准 signal calibration
信号教学设备 signal teaching equipment
信号阶段转换时间 signal stage change time
信号接合 signal bond
信号接近控制电路 signal approach control circuit
信号接口装置 signal interface unit
信号接力 signal relay
信号接收存储器 receiving signal memory
信号接收电极 signal receiving electrode
信号接收电路 signal receiving circuit
信号接收分配器 signal receiver and distributer
信号接收器 signal receiver
信号结 signal node; Troisier's node; Virchow's node
信号结构 sign structure
信号解释 signal interpretation
信号解调器 signal demodulator; sound demodulator
信号经济学 signal(l)ing economics
信号距离 Hamming distance; signal distance
信号卡 signal card
信号开放 clearing of signal; signal at clear; signal off
信号开放表示 clear signal indication
信号开环系统 open loop system
信号可辨度界限 signal threshold
信号可辨阀 signal threshold
信号可见度 signal view
信号空间 signal space
信号孔 signal hole; telltale hole
信号控制 control of signal; level control; signal control; signal operation
信号控制的存储 signal control storage
信号控制电路 signal control circuit
信号控制继电器 signal control relay
信号控制交叉 controlled intersection junction
信号控制交叉口 signalized intersection
信号控制距 signal spacing
信号控制模式 signal control mode
信号控制器 signal controller; signal monitor
信号控制区 signal control area
信号控制人行道 controlled pedestrian crossing
信号控制手钮 signal control knob
信号控制线 signal control wire
信号控制旋钮 signal control knob
信号控制中心 signal(l)ing center
信号宽度编码 width coding
信号拉绳 signal cord
信号拉绳挂圈 signal cord guide
信号拉绳接头 signal splice
信号拉绳套管 signal cord bushing
信号喇叭 signal horn
信号雷管 < 铁路警报用 > torpedo
信号联动方案 signal coordination schematization; signal coordination scheme
信号联锁 interlocking of signals; signal interlock
信号联锁图 signal interlocking chart
信号量化 signal quantization
信号瞭望距离 signal view; signal visibility
信号灵敏度 signal sensitivity
信号铃 alarm; alarm bell; annunciator bell; call bell; jingle bell; signal bell; signal hammer
信号铃锤 knocker
信号铃符号 bell character
信号铃流发生器 tone and ringing generator
信号流 traffic flow
信号流程图 signal flow graph
信号流跟踪器 trace flow
信号流图 flow graph; signal flow chart; signal flow graph
信号楼 cab(in); control box; control cabin; control tower; interlocked signal tower; interlocking(control) tower; interlocking station; signal cabin; signals and routes control center; signal box; signal tower
信号楼扳道员 tower switchman
信号楼表示器 tower indicator
信号楼操纵员 tower switchman
信号楼轨道表示盘 tower track diagram
信号楼合并方案 signal box amalgamation scheme
信号楼合并计划 signal box amalgamation scheme
信号楼控制地区 area under control of signal cabin
信号楼控制范围表示盘 signal box diagram
信号楼控制区 signal cabin zone
信号楼设备 tower equipment
信号楼锁闭器械 tower lock instrument
信号楼值班员 interlocking operator
信号滤波 signal filtering
信号滤波器 traffic filter
信号率 signal(l)ing rate
信号论 signal theory
信号码 code word
信号码元 signal element
信号脉冲 marker pulse; signal impulse
信号脉冲间隙 signal pulse interval
信号脉冲重复频率 signal pulse repetition frequency
信号门 signal gate
信号瞄准器 periscope; signal alignment device; signal sighting device
信号命令 signal command
信号模件 signal module
信号模块 signal zone adapter module
信号模拟 signal imitation
信号能级 signal level
信号霓虹灯 signal neon lamp
信号逆变器 signal inverter
信号钮 order button
信号耦合环 signal coupling loop
信号牌 signal board
信号盘 signal panel
信号炮 signal gun
信号配时 signal setting; signal timing; signal timing dial
信号配时方 signal plan generation system
信号配时方案 signal plan
信号配时方案产生系统 signal plan generation system
信号配时计算机程序 computer signal timing program(me)
信号配时图 signal timing diagram
信号配置 interlocking layout
信号偏差比 signal-to-deviation ratio
信号偏压 signal bias
信号偏压计 bias meter
信号偏移 signal bias; signal excursion
信号偏置法 offset signal method
信号漂移 signal drift
信号频率 signal(ling) frequency
信号频率范围 signal frequency range
信号频率放大器 signal frequency amplifier
信号频率偏移 signal frequency shift
信号频谱 spectrum of signal
信号平均法 signal averaging
信号平均器 signal averager
信号屏蔽接地技术 signal-shield ground
信号旗 code flag; flag; pennant; signal-(ling) flag; signal pennant; staff flag; track flag; waft[weft]; wig-wag
信号旗方向 direction of weft
信号旗手 flagger; flagman; signal flagman
信号旗停车 flag halt; flag stop
信号旗站 signal station
信号旗组 signal kit
信号起伏 signal fluctuation
信号起落变化 swinging of signals
信号气球 signal balloon
信号器 signal(l)ing apparatus; advertiser; annunciator; enunciator; ringer; signal head < 一种信号装置,可向一个或几个方向显示信号 >
信号器交叉摆动警告(火车即到)装置 wig-wag
信号前沿 signal leading edge
信号枪 ground signal projector; pistol; pyrotechnic generator; pyrotechnic pistol; signal gun; signal pistol
信号强度 signal intensity; signal strength
信号强度变动 swinging
信号强度测量器 s-meter
信号强度等级 signal strength scale
信号强度计 signal strength meter; s-meter; S-unit meter
信号强度调整 signal strength adjustment
信号强度限制器 volume limiter
信号强度与天电强度之比 signal static ratio
信号强度指示器 s-meter
信号桥 signal bridge
信号桥柱 bridge mast
信号切断 signal cut
信号清尾时间【交】clearance time of signal
信号球 signal ball
信号区间 signal spacing
信号区域 signal area; signal range
信号取样 sample of signal
信号取样迭代法 signal sampling iteration method
信号取样网络 sampling network
信号群 ensemble; signal group
信号绕组 signal winding
信号冗余度 signal redundancy
信号软管 signal hose
信号扫描 signal scan
信号扫描器 signal scanner
信号色玻璃框 signal spectacle
信号色灯 signal colo(u)r light
信号色光 signal colo(u)r light
信号栅极 signal grid
信号闪光灯 blinker; signal flasher
信号设备 arrangement of signals; signal equipment; signal facility; signal(l)ing apparatus; signal(l)ing arrangement
信号设备包 package of signal(l)ing equipment
信号设备的改造 remodel(l)ing of signal(l)ing installation
信号设备现代化 modernization of signal(l)ing
信号设备用线 signal(l)ing wire
信号设计 interlocking layout; signal design
信号设计图 signal(l)ing plan; signal(l)ing scheme; interlocking plan
信号设施 signal equipment
信号设置 signal setting
信号声带 click track
信号绳 draw cord; pull rope; signal halyard; signal line
信号失落补偿器 dropout compensator; signal dropout compensator
信号失真 signal(ling) distortion
信号失真发生器 signal distortion generator
信号失真率 signal distortion rate
信号施工 signal construction
信号施工人员 signal construction man
信号时差 signal offset
信号识别 signal identification; signal recognition
信号示像 signal aspect
信号示踪器 signal tracer
信号式线路故障寻找器 signal tracer
信号试验器 signal tester
信号室 signal cabin; signal box
信号室电源切换箱 signal cabin power source changeover box
信号手 buzzer
信号手柄 signal handle; signal lever
信号手柄表示灯 signal lever light
信号手柄继电器 signal lever relay
信号手枪 flare gun
信号书 signal book
信号输出 signal output
信号输出电流 signal output current
信号输出均匀性 signal output uniformity
信号输入 signal input
信号输入电极 signal receiving electrode
信号输入通道 signal input channel
信号输入中断 signal input interrupt
信号数据处理变换器 signal data con-

verter
信号数据处理器 signal data processor
信号数据记录器 signal data recorder;signal data recorder unit
信号数据转换器 signal data converter
信号衰减 attenuation; loss of signal strength;signal attenuation
信号衰落 signal fadeout;signal fading
信号双置点 double signal location
信号水平 signal level
信号顺序 sequence of the information
信号顺序跳越 slip sequence
信号顺序图 diagram showing signal sequence
信号搜索接收机 signal-seeking receiver
信号速度 signal speed
信号所 signal box
信号索 signal stay
信号锁闭 signal locking
信号锁闭器 signal lock
信号锁杆 signal locking bar
信号锁条 signal locking bar
信号塔 beacon; signal tower; switch tower
信号踏板 signal pedal
信号台 beacon;gazebo[复 gazebo(e)-s];lantern;signal bridge;signal station
信号探照灯 signal search light
信号特征分析 signature analysis
信号提取 signal extraction
信号调节 signal conditioning
信号调节器 signal conditioner
信号调零 signal balancing
信号调谐放大器 signal-tuned amplifier
信号调整 signal conditioning
信号调整器 signal-conditioning unit
信号调制 signal modulation
信号调制成分 signal modulation component
信号调制器 signal modulator
信号调制(深度)百分率 picture modulation
信号跳越现示顺序 <每隔一定时间,在原定信号现示顺序中跳去一个或两个,以适应不同方向车流量的不平衡状态> skip sequence
信号停止开关 signal shutdown switch
信号通带 signal pass-band
信号通道 signal(l)ing line; signal path
信号通地 signal earth
信号通过能力 communication capacity
信号通路 signal(ling) channel;signal passage;signal routing
信号通路的各种干扰 signal channel miscellaneous interference
信号通信[讯] signal communication; wig-wag
信号通信段 signal and communications district
信号通信[讯]距离 signal(l)ing distance
信号通信[讯]系统 signal communication system
信号通知的列车调动 signaled train movement
信号同步 signal synchronization; signal timing
信号同步系统 signal synchronizing system
信号透镜 signal lens
信号图解法 graphic(al) method of signal
信号图像 signal image;signal pattern
信号托架 signal bracket; signal cantilever
信号托架柱 signal bracket post
信号网 signal net
信号网络管理信号 signal(l)ing-network-management signal
信号往返时间滞后 signal roundtrip time delay
信号桅 signal mast
信号桅杆 signal mast
信号维修 signal maintenance
信号维修队 signal maintenance force
信号维修工 signal maintainer
信号维修员 signal maintainer
信号尾部 signal tail;wave tail
信号稳定放大器 signal stabilization amplifier
信号握柄 signal lever
信号握柄表示灯 signal lever light
信号握柄操纵台 signal lever platform
信号握柄电锁器 electric(al) signal lever lock
信号握柄拐肘 signal crank
信号无效标 signal out of use sign;St. Andrew's crossing
信号系统 colo(u)r system; signal(ling) system; system of signs and signals
信号系统协调 system coordination
信号显示 indication of signal; phase; signal aspect and indication; signal display;signal indication;signal position
信号显示板 indicator panel
信号显示次序 interval sequence
信号显示检验 signal indication checking
信号显示距离 signal range;signal view; signal visibility
信号显示名称 wording of signal indication
信号显示时间序列 interval sequence
信号显示用语 wording of signal indication
信号显示装置 phase equipment
信号线 order wire; signal line; signal wire
信号线抗扰性 signal-line noise immunity
信号线路 marker line; signal(l)ing line
信号线路控制电路 signal line control circuit
信号线务员 signal line(s)man
信号限值器 signal limiter
信号限制 signal constraint
信号相 <交通信号周期中分配给一定交通或交通组合的通行时间> singal phase
信号相变换 phase conversion; signal phase conversion
信号相时间分配 phase split
信号相位 signal phase;signal phasing
信号相位比 phase split
信号相位分配 signal phase split
信号相位起伏 signal phase fluctuation
信号相位顺序 phase sequence
信号相移 signal phase shift
信号相运行图 <色灯信号每相的交通运行图> phase diagram
信号箱 indicator panel; signal box; signal chest;signal locker
信号消失 blackout
信号消息 signal message
信号效果 signal effect
信号形成部件 shaping unit
信号形成器 shaping unit
信号形态 signal aspect
信号序列 burst;signal train
信号选别器 signal replacer;signal reverser;signal slot
信号选择继电器 signal selector relay
信号选择器 signal selector;signal slot
信号选择器脉冲 signal selector patch
信号学 signal(l)ing
信号压缩 compression of signal
信号延长 signal carry-over
信号延迟 code delay; coding delay; signal delay
信号延迟时间 signal delay time
信号延滞 signal delay
信号颜色 signal(ling) colo(u)r
信号遥控 long-distance signal operation
信号仪器 signal(ling) apparatus
信号抑制 signal curbing
信号译码 signal interpretation
信号译释 signal interpretation
信号溢出 signal overflow
信号音接收器 tone receiver
信号音中继线 tone trunk
信号引发器 signal initiator
信号引起的噪声 signal generated noise
信号引线 signal lead
信号印刷电路板 signal printed board
信号拥挤 traffic congestion
信号用接地 signal earth; signal ground
信号用整流器 signal rectifier
信号油 signal oil
信号与量比噪声比 signal-to-quantizing noise ratio
信号与所排进路符合 signal correspondence
信号语言 sign language
信号预告距离 presignal(l)ing distance
信号阈 signal threshold
信号员 signal(l)er; signalman; signal operator; bellboy <俚语>; tower operator <美>;tower man <信号楼的>
信号圆牌 target
信号圆盘 signal disc[disk]
信号源 information source; signal source;supply oscillator
信号源电阻 signal resistance
信号源方框符号 source block
信号源阻抗 generator impedance
信号钥匙 signal(l)ing key
信号跃迁 signal transition
信号载波调频记录法 signal-carrier FM recording
信号载波频率 signal-carrier frequency
信号载线 signal carrying line
信号再生 signal regeneration; signal reshaping
信号噪声 signal noise
信号噪声比 signal(-to)-noise ratio
信号噪声比性能 signal(-to)-noise performance
信号噪声特性 spurious response
信号增强 signal enhancement
信号增强型轻便地震仪 portable signal enhancement seismograph
信号增益 signal gain
信号站 signal station
信号遮光板 signal back light blinder
信号振荡器 signal oscillator
信号振幅 signal amplitude
信号整流器 signal rectifier
信号整形 signal regeneration; signal-(re)shaping;signal standardization
信号整形网络 signal-shaping network
信号支架 signal support
信号支柱 signal support
信号值 signal value
信号指挥的运行 signal(l)ed movement
信号指示阀 indicator valve
信号指示器 signal indicator; signal meter
信号制 signal(ling) system
信号质量代码 signal reporting codes
信号质量检测器 signal quality detector
信号滞后 signal lag
信号滞留时间 signal delay time
信号中的噪声 noise in signal
信号中继船 repeating ship
信号中继器 signal repeater
信号中心 signal center[centre]
信号中央控制 signal central control
信号中央控制室 signal central (control) room
信号中站 signal relay point
信号终端 signal terminal
信号终端单元 signal(l)ing terminal unit
信号终端区域雷达 signal terminal area radar
信号终端装置 signal(l)ing terminal unit
信号钟 signal bell;signal clock
信号周期 signal cycle;signal period
信号周期交通量 signal cycle traffic volume
信号属性 signal attribute
信号注入 signal injection
信号注入器 signal injector
信号柱 signal head assembly; signal post
信号柱顶 pinnacle
信号柱帽 pinnacle
信号柱旁安装 side-of-mast mounting
信号转换 signal conversion; signal transformation
信号转换存储管 signal converter storage tube
信号转换开关 signal transfer switch
信号转换器 signal adapter; signal converter; signal shifter; signal transducer
信号转换装置 signal conversion equipment
信号转子 signal rotor
信号装置 advertiser;annunciator;signal device; signal fixture; signal installation;signal(l)er;signal(l)ing apparatus;signal(l)ing device;signal(l)ing installation; telltell device; telltale device
信号状况 signal(l)ing condition
信号状态 signal condition
信号追赶 signal race
信号追踪 signal tracing
信号资料处理机 signal data processor
信号字 signal letters
信号字数 block of information
信号自动化的 automatically signalled
信号自动同步交换 automatic handshake
信号自相关 signal autocorrelation
信号组 signal set
信号组成 signal make-up
信号组规范 <美国铁道协会的> Signal Section Specifications
信号组合设备 signal combining equipment
信号组匣 signal module
信号最优化程序 signal optimization program(me)
信号作用 signalization
信号坐标 signal coordinates
信汇 letter of remittance;letter transfer;mail remittance;mail transfer
信汇汇率 rate of mail transfer
信汇委托书 mail transfer advice
信笺 letter form
信笺上端所印文字 letterhead
信件 missive
信件滑槽 letter chute;mail chute
信件误投 miscarriage
信据特征 information content

信孔盖 door tidy
信赖 accreditation; on the strength of; reliance; rest on; trust
信赖程度 confidence level
信赖度 reliability
信赖度工程 reliability engineering
信赖级 confidence level
信赖区间 confidence interval
信赖性购买 trust purchasing
信令 signal(l)ing
信令点 signal(l)ing point
信令链路 signal(l)ing link
信令路由 signal(l)ing route
信令时隙 signal(l)ing time slot
信令数据链路 signal(l)ing data link
信令网功能 signal(l)ing network function
信令系统 signal(l)ing system
信令虚通路 signal(l)ing virtual channel
信令音 signal(l)ing tone
信令终端 signal(l)ing terminal
信令终端设备 signal(l)ing terminal equipment
信路 signal channel
信念 morale
信频放大器 octamonic amplifier
信任 accredit; confide; confidence; credence; credit; reliance; trust
信任案 vote of confidence
信任代理人 accredited agent; credit agent
信任理事会 board of trustee
信任区 trust region
信任危机 crisis of confidence
信任要素 element of confidence
信石 arsenic
信使 bearer of despatches; courier; messenger
信手剖面图 freehand profile
信宿 message sink; sink
信天翁海渊 albatross deep
信筒 mailbox; pillar post
信头 cell header
信头差错控制 header error control
信托 dependability; entrust; trust
信托保险 trust insurance
信托保证金 trust deposit
信托部 corporate trust
信托财产 fiduciary estate; trust estate; trust property
信托财产的经管及支出情况 charge and discharge statement
信托财产的指定继承人 trust remainder
信托财产授与人 settlor
信托仓库 safe deposit
信托储蓄银行 trust and savings bank; trustee savings bank
信托存款 deposit(e) in trust; trust deposit
信托代理商 trust agent
信托贷款 outside loan
信托单位 trust unit
信托的 fiduciary
信托动产 movable property in trust
信托放款 outside loan
信托费 trust charges; trust fee
信托服务 fiduciary service
信托服务公司 trust service corporation
信托服务系统 trust service system
信托公司 loan and trust company; trust company
信托股份公司 trust stock company
信托合约 trust indenture; trust obligation
信托和储蓄银行 trust and savings bank
信托会计 fiduciary accounting; trust account
信托机构 fiduciary institution; trust institution
信托机关 fiduciary institution
信托基金 fund-in-trust; pool of fiduciary; trust fund
信托基金拨款 trust fund appropriation
信托基金方案 trust fund program(me)
信托基金分类账 trust fund ledger
信托基金户间往来 trust inter-fund transaction
信托基金账户 trust fund account
信托及代理基金 trust-and-agency fund
信托金库 trust fund bureau
信托契据 deed of trust; trust deed
信托契约 deed of trust; fiduciary contract
信托人 truster[trustor]
信托商店 commission agent; commission house; commission shop; trust shop
信托商店管理 management of commission shops
信托收据 trust receipt
信托手续 trust process
信托受益人 cestui que trust
信托书 covenant of seizin; letter of trust; trust deed; trust indenture; trust letter
信托投资 fiduciary contribution; trust investment
信托投资业务 trust and investment business
信托协议 declaration of trust; trust agreement
信托业务 fiducial business; fiduciary business; fiduciary work; trust business
信托业务会计 trust accounting
信托银行 trustee bank
信托银行公司 trust and banking company
信托银行业务 trust banking
信托与介入 commitment and involvement
信托债券 trust bonds
信托账户 account in trust; closing the trustee's books; trust account
信托证券 securities in trust
信托证书 trust certificate; trust instrument
信托职能 trust function
信托周转基金 trust revolving fund
信托咨询公司 trust consultancy corporation
信托资产 fiduciary estate; trust assets
信托资产本值 trust corpus
信托资金 money in trust; trust fund
信息 communication; information
信息安全 information security
信息按等级层次扩散 hierarchic(al) diffusion of information
信息板 informatory board
信息包 information package; information packet; packet
信息包交换 packet-switch
信息包随机存取存储器 packet random access memory
信息保护 protect file; protective file
信息保密 information security
信息爆炸 information explosion
信息备忘录 placement memorandum
信息编辑 message editing
信息编码 information encoding; message coding
信息编排器 message composer
信息编组 message blocking
信息变换 information conversion
信息标号应答询问 cue-response query
信息标号属性 keyed attribute
信息标记 information flag
信息标头 information heading
信息标志 information mark
信息标桩 information stake
信息表 information sheet
信息表示 information representation; present of information
信息并行传输 parallel transmission of information
信息波道 information channel
信息不灵 access to information in short
信息材料 information material
信息采集 information acquisition
信息测度 information measure
信息测量 information measurement
信息策略 informational strategy
信息插入 message interpolation
信息查询设施 information inquiry facility
信息查询系统 referral information system
信息查询装置 kiosk
信息产业 information industry
信息长度 message length
信息场 information field
信息成本 information cost
信息成本函数 information cost function
信息城市 information city
信息程序 information program(me)
信息储备 information fund
信息储存 information accumulation; information storage
信息处理 data handling; information-(al) processing; information handling; information treatment; message processing; processing of information
信息处理编码 information processing code
信息处理程序 information-handling program(me); information processing program(me); information treatment program(me); message processing program(me)
信息处理分配系统 message processing distribution system
信息处理分析 information process analysis
信息处理机 handler; information-handling machine; information processing machine; information processor; information treatment machine; processor
信息处理能力 information processing capability; information processing capacity
信息处理器 information processor; message handler
信息处理容量 data-handling capacity
信息处理速率 information-handling rate
信息处理系统 information-handling system; information processing system; message handling system; message processing system
信息处理循环 information-processing cycle
信息处理语言 information processing language
信息处理中心 information processing center[centre]
信息处理装置 information-handling apparatus; information processing apparatus; information processing system
信息传播 information dissemination
信息传递 information transfer
信息传递反馈 communication feedback
信息传递速率 message data rate
信息传递线 information-carrying wire
信息传视系统 videotext
信息传输 information transmission
信息传输设备 information transmission equipment
信息传输速度 information transmission speed
信息传输速率 rate of information transmission
信息传输系统 data transmission system; information transmission system
信息传输线 transmission line of information
信息传输效率 information transfer efficiency
信息传输性质 quality of information transfer
信息传送率 rate of information transmission
信息传送时间 message transfer time
信息传送速度 rate of information throughput
信息传送速率 rate of information throughput
信息传送系统 information delivery system
信息传送线路 information circuit
信息传送信道 information transfer channel
信息传送循环 information-processing cycle
信息串属性 string attribute
信息存储 information storage; memorise; storage of information; data storage; digital recording
信息存储方法 information storage means
信息存储方式 information storage means
信息存储管 information storage tube
信息存储和检索 information storage and retrieval
信息存储和检索系统 information storage and retrieval system
信息存储器 information storage; information storage unit; information storing device; message memory
信息存储体 storage medium[复media]
信息存储系统 information storage system
信息存储与检索自动化 automation of information storage and retrieval
信息存储装置 information storing device
信息存取时间 access time
信息代数 information algebra
信息带宽 information bandwidth; intelligence bandwidth
信息单位 bit of information; code bit
信息单元 information unit; piece of information
信息道 information track
信息道格式 track format
信息道索引 track index
信息的 informative; informatory
信息的边标成本 marginal cost of information
信息的存取 access to information
信息的多余部分 redundant
信息的访问 access to information
信息的恢复 repair of information
信息的实用性 practicality of information
信息的顺序 sequence of information
信息的消费化 consumerization of information
信息的选择性传播 selective dissemination of information

信息的自然对数单位 <等于 1.443 二进制单位> neutral unit
信息点阵结构 information lattice
信息电路 information circuit
信息电平 message level
信息叠加 superimpose
信息丢失 bit drop-out; information dropout
信息读出时间 information readout time
信息读出线 information-read-wire
信息段 information field; message section; message segment
信息队列 message queue
信息队列数据集 message queue data set
信息多项式 message polynomial
信息多余部分 redundant
信息发生器 information generator
信息发送 message routing
信息反馈 information feedback; massage feedback
信息反馈系统 information feedback system
信息反馈制环 information feedback control loop
信息反馈中心 center of information feedback; information feedback center[centre]
信息反射 information reflection
信息方面的角色 informational role
信息方式 information pattern; informationwise
信息访问 information access; message reference
信息分隔符 information separator; information separator character
信息分开 unpack
信息分块 block sort
信息分类 information classification
信息分类编码标准 standard for classified code of information
信息分离符 information separator
信息分离器 information separator
信息分配器 information distributor
信息服务 information service
信息符号 information symbol; message digit
信息覆盖 overlaying of information
信息(改变区)转储 change dump
信息概率 informational probability
信息感受 information sensing
信息感受能力 information sensing capability
信息感应 information sensing
信息港 infoport
信息高速公路 information (super) highway
信息革命 information revolution
信息格式 information format
信息更换 change dump
信息工程 information technology
信息工程师 information engineer
信息工程学 information engineering
信息工具包 information kit
信息功能单元 information functional unit
信息功效 informationally efficient
信息共同体 information community
信息沟通 information channel(l)ing
信息管理 information control
信息管理检索和传播系统 information management; retrieval and dissemination system
信息管理系统 information management system
信息互换文件 file for information interchange
信息化社会 informationizationed society; information society
信息化时代 information age

信息环境保护者 information environmentalist
信息环境学 information environment science
信息缓冲 message buffer(ing)
信息恢复 information retrieval
信息恢复系统 information retrieval system
信息恢复与存储系统 system for information retrieval and storage
信息混入 bit drop-in; information drop-in
信息或资料来源 quarry
信息获取 information acquisition
信息获取活动 informational activity
信息获取要求 intelligence requirement
信息机 information machine
信息积累 information accumulation
信息基础 information infrastructure
信息极限 information limitation
信息集 information set
信息集合 ensemble of communication
信息记录组 interblock
信息记录组间隙 interblock gap
信息记时 message time stamping
信息技术 information technique; information technology; telematics
信息技术辅助(公共汽)车站信息显示 telematics-aided bus stop information display
信息技术革命 information technique revolution
信息寄存器 information register; message register
信息加工 information processing; information tailoring
信息加工(处理)与决策 information processing and decision
信息加工元件 information processing element
信息加下标 coordinate indexing
信息价格函数 information cost function
信息减缩变换 data reduction
信息检索 information retrieval; information search
信息检索技术 information retrieval technique
信息检索系统 information retrieval system
信息检索语言 information retrieval language
信息简化 reduction of data
信息简缩变换 reduction of data
信息交换 full-time message switching; information interchange; message exchange; message switching
信息交换标准代码 standard code for information interchange
信息交换存储媒体 volume for information interchange
信息交换格式 message exchange format
信息交换技术 message switching technique
信息交换网(络) message switching network; switched message network
信息交换系统 message switching system
信息交换用代码 code for information interchange
信息交换站 clearing house
信息交换中心 message switching center[centre]
信息交换装置 message exchanger
信息交流 information exchange
信息交流程序 communication process
信息交流功能 communication function

信息交流渠道 channel of communication
信息交流网络 communication network
信息交流运动概念 communication campaign concept
信息交流中心 information clearing house
信息交流组合 communication mix
信息接收机 intelligence receiver
信息接收器 information sink; message sink; recipient
信息接收速度 information rate
信息节点 information node
信息结构 information structure; message structure
信息结构模型 information structure model
信息结构设计 information structure design
信息结束 end of message
信息结束代码 end-of-message code
信息结束(字)符 end-of-message character
信息介体 information medium
信息介质 information medium[复 media]
信息经济学 information economics
信息矩阵 information matrix
信息卡片 informative card
信息开始 start of message
信息开始标志 beginning-of-information marker
信息(科)学 informatics; information science
信息科学技术 information science technology
信息科学家协会 Institute of Information Scientists
信息科学讨论会 information science institute
信息可加性 information activity; information additivity
信息可用线 information-available line
信息空间 information space
信息控制 information control; message control
信息控制程序 information control program(me); message control program(me)
信息控制块 message control block
信息控制论 informatics
信息控制任务 message control task
信息控制系统 information control system; management information system
信息库 bank of information; information bank; information base; data bank
信息库系统 information bank system
信息块 block of information; information block; message block
信息块长度 information block length
信息块传送结束 end-of-transmission block
信息块结束 end of block
信息来源 sources of information
信息来源表示法 informant representation
信息理论 communication theory; information theory
信息联系 informational linkage
信息链路 information link
信息量 amount of information; content of information; information quantity; quantity of information; information content
信息量单位 information unit
信息量度 measure of information
信息量法 quantity of information method

信息量值 information magnitude
信息量自然单位 natural unit of information content
信息列队 message queue
信息流 flow of information; information stream; message flow
信息流程图 information flow chart; information flow diagram
信息流的影响 impact of information flow
信息流分析 information flow analysis
信息流跟踪程序 flow tracer
信息流控制 flow control; information flow control
信息流速 information flow-rate
信息流通 information flow
信息漏失 spillover signal; spillover (of message)
信息漏失符号 escape character
信息录象带 information video tape
信息路径选择码 message routing code
信息率 data rate; informational rate
信息轮询 message polling
信息论 information theory
信息逻辑机 information logical machine
信息码 character code; information code
信息码组 block of information; information block
信息脉冲 information pulse; intelligence carrying pulse
信息媒介 information medium
信息媒体 information medium[复 media]
信息门电路 information gate
信息密度 information density
信息密集工业 information-intensive types of industries
信息模式 information pattern
信息模型 information model
信息内插法 message interpolation
信息内容 information content
信息排队 information queue; message queue(ing)
信息评价 information appraisal
信息奇偶性 message parity
信息企业 information firm
信息起始点 load point
信息起始点标记 load point indicator; load point marker
信息前馈 feed forward of information
信息清除 erasing of information
信息区 block of information; information region
信息区分 data separation
信息渠道 line of information
信息取样 message sample
信息日志功能 message journalizing function
信息容量 data capacity; information (al-handling) capacity
信息融合(技术) information fusion
信息冗余 information excess
信息丧失 loss of information; walk down
信息熵 comentropy; entropy of information; information entropy
信息社会 information society
信息剩余度 redundancy of information
信息时代 information age
信息矢量 information vector
信息使用 information utilization
信息始端标志 beginning-of-information marker
信息市场 information market
信息视窗 info-window
信息收发自动化 automatization of sending and receiving operation of

messages
信息收集 information collection; information gathering
信息收集和储存 data capture
信息收集器 information collector
信息收集装置 information collection unit
信息首部 information heading
信息输出 information output; output information
信息输出方式 message stream mode
信息输出公式 output format
信息输出区 message area
信息输入 information input
信息输入输出的时间【计】access time
信息数 information number
信息数据 information data; intelligence data
信息数据集 information data set; message data set
信息数位 message digit
信息数字处理 digital processing
信息搜集 information gathering
信息素 pheromone
信息速率 information rate
信息损失 information loss; loss of information
信息提供商 information offering
信息提供者 informant
信息提取 information extraction
信息提取能力 information extraction ability
信息提要 information abstract
信息体 informosome
信息体系结构 information architecture
信息调节 accommodation of information
信息调制 message modulation
信息通道 information bus; information channel; information path
信息通道容量 channel capacity
信息通过量 throughput
信息通路 information channel; information path; message routing
信息通信[讯] message communication
信息通信[讯]系统 information communication system
信息统计机 file computer
信息图像 frame; information image
信息吞吐量 rate of information throughput
信息吞吐率 rate of information throughput
信息网(络) information network; data network
信息网络系统 information network system
信息位 information bit
信息位传送速率 transfer rate of information
信息稳定性 information stability
信息污染 information pollution
信息无损计算机 information lossless machine
信息误差 information error
信息系统 information system; intelligence system
信息显示 character display; informational representation; message display; presentation of information; present of information
信息显示板 message panel
信息显示控制台 message display console
信息显示器 content indicator; information display
信息显示速度 information display rate
信息显示系统 information display system(s)
信息线 information wire
信息向量 dope vector; information vector
信息项 item of information
信息效益 efficiency brought about by information
信息写入线 information-write-wire
信息心理学 information psychology
信息信道 information channel
信息信号 information signal
信息形式 information form
信息性摘要 informative abstract
信息序列 information sequence
信息选择传布 selective dissemination of information
信息选择系统 information selection system
信息学 informatics
信息循环 information circulation
信息压缩 information compression
信息延迟 message delay
信息要求 information requirement
信息译码 information decoding
信息引导 offer guidance by supplying information
信息隐蔽 information hiding
信息用户 information user
信息优先级 message priority
信息语言 information(-oriented) language
信息预报 information prediction
信息预选器 preselector of information
信息域 information field
信息元 information word
信息源 information generator; information source; message source; source data
信息源编码 information source coding
信息源的平均信息量 information source entropy
信息载体 information carrier; semantide
信息载体产物 episemantic
信息再现 information display
信息再现速率 information display rate
信息噪声比 message to noise ratio
信息增益 information gain
信息闸门 information gate
信息摘要 information abstract
信息整理 data recovery
信息正文 message text
信息证实代码 message authentication code
信息指定通道 message routing
信息指示器 message indicator
信息中心 information center[centre]
信息终端 intelligence terminal
信息终了符 end of message
信息重复循环 recirculation
信息重获 recovery of information
信息周期 information cycle
信息周转率 rate of information throughput
信息主通路 highway circuit
信息转储 dump; memory dump; storage dump
信息转储程序 dump routine
信息转换 information conversion; message switching
信息转换功能 information conversion function
信息转换器 transcriber
信息转换体 informofer
信息转接 message switching; store-and-forward
信息转接计算机 message switching computer
信息转接网络 message switching network; switched message network
信息转接系统 message switching system
信息转接中心 message switching center[centre]
信息转移通路 message transfer channel
信息资源 information resources
信息资源共享 information resources sharing
信息子 informofer
信息子系统 information subsystem
信息字 information word
信息字段 information field
信息字节 information byte
信息自动处理 automated processing of information
信息自动检索系统 automatic information retrieval system
信息综合 convergence of evidence
信息总工作量 message throughput
信息总量 informational capacity
信息总线 <电子交换机的> bus
信息组 block; byte; field; message block; message group
信息组标记 field mark; group mark
信息组标识 field identification
信息组差错率 block error rate
信息组长度 block length; field length
信息组传 block transfer
信息组传输结束符 end-of-transmission block character
信息组传输终止符 end-of-transmission block character
信息组传送 block transfer
信息组存储器 byte storage
信息组错误率 block error rate
信息组代码 block code
信息组定义 field definition
信息组多路控制位 block multiplexing control bit
信息组间隙 block gap
信息组检查字符 block check character
信息组结束 end of block
信息组结束信号 end of block signal
信息组控制 field control
信息组名 block name; field name
信息组前缀 block prefix
信息组前缀字段 block prefix field
信息组顺序指示符 block sequence indicator
信息组作废符号 block cancel character; block ignore character
信箱 letter box; letter drop-off; mailbox; pillar-box; postbox
信箱板 <安装在大门上的信槽孔板> letter-box plate
信箱公司 letter-box company
信箱蓝 letter-box blue
信箱区 mailbox
信箱投口金属板 letter drop plate
信箱投信口开缝板 letter plate
信用 credence; credibility; credit
信用摆动额 swing credit
信用保单 policy proof of interest; wager policy
信用保险 credit insurance; insurance of credit guarantee
信用保险单 credit policy; hono(u)r policy
信用保险费 credit insurance premium
信用保证 credit security
信用保证金额 credit amount
信用保证书 letter of guarantee
信用保证债券 guaranteed bond
信用保证制度 credit guarantee system
信用报告 credit report
信用报告机构 credit reporting agencies
信用备谘 credit reference
信用比率 credit ratio
信用标准 credit standard
信用部分贷款 <国际货币基金组织> credit tranche
信用差距 credibility gap; credit gap
信用产生 credit creation
信用产生的界限 limitation on credit creation
信用承兑 acceptance for honour
信用程度 creditability
信用创造 credit creation
信用措施 credit facility
信用贷款 blank credit; character loan; credit loan; debt of hono(u)r; fiduciary debt; fiduciary loan; loan on credit; open credit; open fiduciary loan; straight loan; uncovered advance; unsecured loan
信用贷款股份 debenture stock
信用贷款申请单 credit application form
信用单证 credit documents
信用担保 del credere
信用担保代理人 del credere agent
信用的可供应情况学说 availability doctrine
信用地位 credit standing
信用第一 credit first
信用电话 credit telephone
信用调查 credit inquiry; credit information
信用调查部 credit department
信用调查档案 accountant file; credit file
信用调查机构 credit agency
信用调查录 opinion book
信用调查员 credit man
信用定额 credit rating
信用冻结 credit freeze
信用多倍扩大 multiple expansion of credit
信用额度 credit line; limit of credit; line of credit
信用发行 fiduciary issue
信用发行的纸币 fiduciary money
信用范围 fiduciary capacity
信用方式 form of credit
信用分类 credit classification
信用分析 analysis for credit; credit analysis
信用分析公司 credit analysis firm
信用份额提款 drawing in the credit branches
信用风险 credit risk
信用风险保险 credit risk insurance
信用根据 basis of credit
信用工具 credit instrument
信用公司债券 debenture bond
信用购买契约 credit contract
信用股票 debenture stock
信用管理 credit control; credit management
信用管制 credit control
信用管制手段 instrument of credit control
信用合作社 credit cooperative; loan society; credit union
信用合作社投资总汇 credit union investment pools
信用核实 credit verification
信用换汇 credit swap
信用汇款 credit remittance
信用汇票 credit bill
信用货币 credit currency; credit money; faith money; fiduciary money
信用货币化 monetization of credit
信用机构 fiduciary institution
信用基础 basis of credit
信用交易 credit transaction; deal on

credit; on credit; transaction with credit
信用结论 confidence inference
信用借贷资本 debenture capital
信用借款 character loan; charter loan; confidence debt; credit loan; debt of hono(u)r; fiduciary loan; open fiduciary loan
信用紧缩 credit contraction; credit screw; credit squeeze; credit squeezing; tight credit; tight money
信用经济 credit economy
信用卡 access card; charge card; credit card
信用卡财务费 credit card finance charges
信用卡电话 credit card call
信用卡记账法 card system
信用卡系统 credit card system
信用恐慌 credit crunch
信用扩张 credit expansion
信用流通 credit circulation
信用论 credit theory
信用买卖 credit sale; sale by credit
信用媒介 fiduciary media
信用名誉 credit standing
信用能力 credit rating
信用膨胀 bloated credit; credit expansion; credit inflation; swollen credit; truth worthiness inflation
信用票据 bill of credit; blank credit; clean bill; credit instruments; credit note; credit paper
信用评级 credit rating
信用期限 credit period
信用欺诈 credit fraud
信用清算组 credit clearing division
信用情报交换 credit interchange
信用让与人 grantor of credit
信用赊账 on credit account
信用设施 credit facility
信用社 credit association; credit cooperative
信用通货 credit currency
信用投资 fiduciary contribution
信用透支 credit facility; fiduciary overdraft
信用往来 open credit
信用危机 credit crisis
信用线 line of credit
信用限额 credit line; limited credit
信用循环 credit cycle
信用要素 element of credit
信用银行 credit bank
信用余额 credit balance
信用债券 debenture; debenture certificate; debenture stock; debenture trust deed; fidelity bond
信用债券持有人 debenture holder
信用证 letter of credit[L/C]
信用证保兑 letter of credit confirmation
信用证部分转让申请书 application for partial transfer of letter of credit
信用证的过户与转让 transfer and assignability of credit
信用证的开立 establishment of letter of credit
信用证的条件 condition in the letter of credit
信用证费 letter of credit charges
信用证格式 specimen of letter of credit
信用证汇票 bill drawn on a letter of credit
信用证结算 settlement of account by letter of credit
信用证(截止)有效日期 expiration date of letter of credit
信用证开发方 accrediting party
信用证开证申请人 accountee
信用证开证银行 issuing bank
信用证券 credit documents; credit paper; trust bonds
信用证软性条款 credit with soft clauses
信用证申请人 applicant for letter of credit
信用证申请书 application for letter of credit
信用证收益人 credit beneficiary; credit user
信用证受益人 beneficiary of transferable credit
信用证书 credit documents; instrument of credit
信用证条件 letter of credit terms; terms of credit
信用证通知书 advice of letter of credit
信用证统一惯例 uniform customs for credits
信用证未用部分 unused portion of letter of credit
信用证未用余额 unused balance of letter of credit
信用证修改申请书 application for alteration of credit
信用证延期 credit extension
信用证有效期 validity of credit
信用证有效性 expiry date of L/C; validity of credit
信用证逾期 credit expired
信用证约定书 commercial credit agreement
信用证正本 original letter of credit
信用证转让申请书 application for (advice of) transfer of L/C
信用证总限额 aggregate amount of letter of credit
信用证总账 letter of credit ledger
信用政策 credit policy
信用纸币 fiduciary currency
信用制度 credit system
信用质量控制 qualitative credit control; selective credit control
信用转让 credit transfer
信用转账 credit transfer
信用状况 credit standing; credit status
信用咨询公司 credit agency; credit bureau
信用咨询组 credit inquiry division
信用资本 use capital
信用资料 credit information
信用资料社 credit bureau
信用组合 credit union
信誉 credibility; credit standing; credit worthiness; goodwill; reputation
信誉保险单 hono(u)r policy
信誉标 bona fide bid
信誉不好者 dead beat
信誉承包商合格预选 <指工程招标> prequalification of (prospective) bidders
信誉承兑 act of hono(u)r
信誉借款 debt of hono(u)r
信誉可靠最低标价 lowest responsible bid; lowest responsive bid
信誉可靠最低标价的投标人 lowest responsible bidder
信誉可靠最低标价的投标商 lowest qualified bidder; lowest responsible bidder
信誉良好的企业 business of good standing
信誉良好的银行 reputable bank
信元 cell
信元差错比 cell error ratio
信元传递延时 cell transfer delay
信元定界 cell delineation
信元丢失比 cell loss ratio
信元速率解耦 cell rate decoupling
信元延时变化 cell delay variation
信源编码 source encoding
信源编码器 source encoder
信源符号集 source alphabet
信源换能器 source transducer
信源率 source rate
信源字母集 source alphabet
信杂比 signal-to-noise ratio
信噪比 signal-to-mask ratio; signal-to-noise performance; signal(-to)-noise ratio; speech-noise ratio
信噪比改善因数 improvement factor of signal to noise ratio
信噪比改善阈值 improvement threshold
信噪比极限 limit of signal to noise ratio
信噪特征 signal-to-noise characteristic
信纸 notepaper; writing paper
信中附件 enclosure

兴波伴流 wave wake

兴波马力 wave horsepower
兴波阻力 wave-forming resistance; wave(-making) drag; wave-making resistance
兴奋纯度 excitation purity
兴奋剂 excitant; stimulant; stimulative
兴奋剂用作物 stimulant crop
兴奋扩散 irradiation of excitation
兴奋输入 excitatory input
兴奋性 excitability
兴奋作用 stimulation
兴建的沿岸挡水堤 raised coast(al) barrier
兴利 beneficial use
兴利计算 beneficial purpose calculation
兴利库容 utilizable capacity
兴利目标 <需蓄水的> conservation purpose
兴利调度 utilizable regulation
兴利蓄水库容 conservation storage
兴利蓄水目标 conservation purpose
兴起时间 rise time
兴趣测验 interest test
兴趣调查 interest inventory
兴趣缺乏 interest blank
兴斯堡反应 Hinsberg reaction
兴旺时期 boom period
兴业区 enterprise zone
兴叶扁柏 white cedar

星爆式 star burst

星币 star note
星标 asterisk; star target
星表 durchmusterung; star catalog(ue); star list
星表号数 catalog(ue) number
星表示 star representation
星表误差 ephemeris error
星彩 asterism
星彩宝石 asteria
星彩红宝石 star ruby
星彩蓝宝石 star sapphire
星彩石 star-stone
星彩石英 aventurin(e) quartz; star-quartz
星彩性 asterism
星场 stellar field
星场跟踪仪 star field tracker
星传感器 star sensor
星窗 <飞机的> astral
星代数 star algebra
星带 star braid
星的 astral; sidereal
星的判别 star identification
星灯 star lamp
星等 stellar magnitude
星等比 magnitude ratio
星等系统 magnitude system
星点 asterion; stardust; star point
星点间的 interasteric
星点试验 star test
星点釉 starred glaze
星鲽 halibut
星对法 <天文观测> star pair method
星盾贝属 <拉> Asterina
星多边形 starred polygon
星峰岩群 star peak group
星跟踪式定位器 star tracker
星观测 star observation
星冠 astral crown
星光 starlight
星光宝石 star gem
星光镜 starlight scope
星光望远镜 starlight telescope
星光照度 starlight illumination
星光中程观察装置 starlight medium range observation device
星号 asterisk; star; star number; star symbol
星号保护 asterisk protection
星号程序 blue ribbon program(me); star program(me)
星号记法 asterisk notation
星号图像字符 asterisk picture character
星火规划 catch-up program(me)
星火计划 catch-up program(me)
星际 interspace
星际尘埃 interstellar dust
星际导航 interplanetary navigation
星际的 astral; intergalactic; interplanetary; interstellar
星际法 interplanetary law
星际航行 astronavigation; interplanetary flight; interplanetary navigation; interplanetary travel
星际间吸收 interstellar absorption
星际监视台 interplanetary monitoring platform
星际监视卫星 interplanetary monitor satellite
星际空间 outer space
星际旅行 grand tour; interplanetary travel
星际探测人造卫星 satellite for interplanetary probes
星际通信[讯] interstellar communication
星际站 satellite station
星绞 star quad stranding
星绞电缆 star quad cable
星绞结构 star quad construction
星绞四线组 spiral quad; star quad
星绞四线组电缆 star quad cable
星绞四芯软电缆 spiral four cable; spiral quad cable
星界的 astral
星径曲率改正 correction for curvature of star image path; curvature correction
星壳孢属 <拉> Asteromella
星空背景 star background; stellar background
星空测绘 star-field mapping
星空图 sky chart
星肋穹顶 stellar vault
星历表 astronomic(al) ephemeris; ephemeris[复 ephemerides]
星历表计算 ephemeris table computation

星历表计算同步卫星 ephemeris table computation synchronous satellite
星历表时间 ephemeris time
星历表数据 ephemeris data
星历秒 ephemeris second
星历日 ephemeris day
星历时 ephemeris time
星量 nebulosity
星裂 star check;star shake
星裂节 star-checked knot
星流 star drift;star-streaming
星轮 spider;sprocket;star wheel
星轮机构 star-wheel mechanism
星轮进给 star feed
星轮排渣器 star-wheel extractor
星轮升运器 star elevator
星轮式 star-wheel type
星轮式控轴器 <控制矿车运行用> star-wheel axle controller
星轮式输送器 star conveyer[conveyor]
星轮式提升机 star elevator
星轮式镇压器 star roller;star-wheeled roller
星芒 asterism
星芒图 star figure
星芒图案 star burst pattern
星毛栎 box white oak;brash oak;iron oak;post oak
星盘 astrolabe
星匹配 stellar map matching
星期服务 weekly service
星期六条款 Saturdays clause
星期日班 Sunday duty
星期日和假日除外 Sundays and holidays excepted
星期日和假日在内 Sundays and holidays included
星期日旅行 Sunday excursion
星期日停运的列车 daily-except-Sunday train
星期日以外的任何一天 weekday
星球表面的阴暗区 mare
星球仪 celestial globe;star globe
星球自动定向器 star seeker
星区 constellation
星三角换接开关 star-delta reversing switch
星三角启动器 star-delta starter
星三角形接法 delta connection;star-delta connection
星散薄壁组织 diffuse parenchyma;scattered parenchyma
星散聚合薄壁组织 diffuse-in-aggregates parenchyma
星散线 stelloid
星色 star colo(u)r
星蛇纹石 radiotine
星式搜索法 <优化算法中的> star-search method
星式斜拉桥 star cable-stay bridge
星式斜缆桥 star cable-stay bridge
星式信号弹 star flare
星式信息交流网络 star communication network
星式装药 star-type charge
星收敛 star convergence
星丝 astral fiber
星宿仪 astroscope
星体 aster;cytaster
星体大气物理学 aeronomy
星体的 stellar
星体的次级粒子 secondary particle of star
星体的跟踪 star tracking
星体地貌 planetary landform
星体地物 extraterrestrial feature
星体高度总改正量 total correction of star's altitude
星体跟踪天文导航 star-tracking guidance
星体跟踪望远镜 star-tracking telescope
星体跟踪系统 star-tracking system
星体装拱顶 stellar vault
星天牛 <拉> Anoplophora chinensis
星图 celestial atlas;celestial chart;sky diagram;star atlas;star chart;start map
星图集 star atlas
星图学 uranography
星团 clustre;star cluster;stardust;stellar cluster
星位导航 bearing by stars
星位角 angle of parallax;angle of position;angular parallax;parallactic angle;position angle
星系 galaxy;star galaxy
星系际的 intergalactic
星下点 subcelestial point;subsatral point;substellar point
星下位置 substellar position
星象 star(like) image
星象跟踪仪 astrotracker;star tracker
星象迹线 trail
星象罗盘 astrocompass
星象罗盘定向 star compass orientation
星象宁静度 seeing
星协 stellar association
星星形接线 star-star connection
星星形连接 star-star connection
星形 radial;star(pattern);stellar pattern
星形摆动装置 stirring star
星形棒 star-shaped slug
星形臂 spider arm
星形边撑 star temple
星形标记 asterisk
星形布局 star topology
星形布置图 star-shaped plan;stellar plan
星形仓 intecell
星形测膛规 star ga(u)ge
星形长凿 plugging chisel
星形齿轮 spider gear;star gear
星形齿轮架 range carrier
星形齿轮铣刀 sprocket cutter
星形磁铁 <用于磁倾仪> spider magnet
星形导向装置 star guide
星形的 ast(e)roid;starlike;star-shaped;stellar;stellated;stelliform
星形电流 star current
星形电路 Y-shaped network
星形垫圈 star washer
星形定心器 star centralizer
星形发动机 radial cylinder engine;radial engine;radial motor
星形阀 star valve
星形分布系统 star distributed system
星形蜂巢形 star honeycomb
星形扶正器 star centralizer
星形复形 star complex
星形纲 Stellaroidea
星形高聚物 star polymer
星形格 star lattice
星形隔仓板 star-shaped dispenser
星形工具 spider kit
星形拱 stellar vault
星形航空发动机 radial aeroengine
星形红宝石 star ruby
星形弧坑裂纹 star crater crack
星形环裂 star shake
星形活塞式气马达 radial piston air motor
星形机架 spider mounting
星形集 star-shaped set
星形给料闸门 star gate
星形计量给料器 star computation batcher
星形加料器 star feeder
星形加细 star refinement
星形架 starframe
星形架染色法 star dyeing
星形架蒸化机 star ager
星形架蒸汽箱 star steamer
星形建筑 Y-shaped building
星形建筑物 star-shaped building
星形交叉 junction of the star type;star junction;Y-junction
星形绞合 star twisting
星形铰刀 rose reamer
星形搅拌机 star-shaped agitator
星形搅动装置 stirring star
星形接法 star-connected system;star connection(system);star-star connection;Y-connection
星形接头 star connector
星形接线 star connection
星形接线法 wye wire connection
星形接线相电压 star voltage
星形结构 star quad formation;star structure
星形截面 star cross section
星形进料器 rotary feeder;star batcher;star feeder
星形卷取 starred roll
星形开裂 star shake
星形空冷式发动机 radial air-cooled engine
星形控制器 star controller
星形扩孔器 star reamer
星形肋拱顶 star-ribbed vault;star vault;stellar vault
星形连接 star connect;star connection;Y-connection
星形连接的 star-connected
星形连接电路 star circuit;star-connected circuit
星形连接电阻箱 star box;Y-box
星形连接法 star connection;star connection method;wye connection
星形连接器 star coupling
星形连接制 star-connected system
星形联结 Y-junction
星形联结的 Y-connected
星形裂纹 star shake;star fracture <平板玻璃>;star mark <搪瓷表面缺陷>
星形炉栅 star grid
星形滤油片 star-shaped filter element
星形轮 four-arm spider;spider(assembly);spider wheel;spoked assembly;star wheel;geneva cam <十字轮机构的>
星形轮的滚针轴承 spider needle bearing
星形轮毂 spider center[centre]
星形轮式给料机 star-wheel type feeder
星形轮运动 star-wheel motion
星形轮转动 input of planetary
星形马达 radial motor
星形脉冲串增量 star burst increment
星形脉冲群试验样图 star burst test pattern
星形脉冲试验带 star burst test tape
星形脉冲组 star burst
星形锚具 star-notched anchorage
星形模型 star-shaped pattern;stellar pattern
星形母线 star bus bar
星形捏手 star knob
星形耦合器 star coupler
星形排列 star-shaped disposition;starwise disposition
星形配料器 star batcher
星形片 star plate
星形平面 star-shaped plan
星形平面的塔式住宅 star house
星形汽缸 radial cylinder
星形钎头 star-shaped bit
星形曲线 asteroid;asteroid-shaped curve
星形曲折接法 interconnected star connection;star-zigzag connection
星形曲折连接 star-zigzag connection
星形全波整流器 wye rectifier
星形燃烧器 star burner
星形扰动装置 stirring star
星形三角变换 delta-Y transformation;pi-T transformation;Y-delta transformation
星形三角接线 star-delta connection
星形三角开关 star-delta switch
星形三角连接法 star-delta connection method
星形三角接线 star-delta connection wire
星形三角起动 <感应电动机> star-delta control;star-delta starting
星形三角起动继电器 <感应电动机> star-delta starting relay
星形三角形起动法 Y-delta starting method
星形三角形起动器 star-delta starter;Y-delta starter
星形三角形切换起动器 star-delta switching starter
星形三角形转换开关 star-delta switch
星形三角转换 star-delta switching
星形饰【建】starlike ornamentation
星形手轮 star handle;star hand wheel
星形手钮 palm grip hand knob
星形枢纽 <道路> junction of the star type
星形说 nebular theory
星形四线扭绞 star quad
星形四线组扭绞 <电缆> star quad twist
星形四心线组 star quad
星形索 star cable
星形套筒扳手 spider spanner;spider wrench
星形套筒铰刀 rose shell reamer
星形体 star body
星形天线 star aerial
星形调制盘 star reticle
星形铁 mill star
星形图 star graph
星形脱模器 stripping star
星形拓扑 star topology
星形网 stellated reticulum
星形网络 star connection;star network;stelliform network;Y-connection
星形网眼 star net
星形网眼变换 star-mesh transformation
星形往复式压缩机 reciprocating compressor with radial cylinders;semi-radial reciprocating compressor
星形物 star
星形线 astroid
星形线脚 star mo(u)lding
星形相交路线 star route
星形斜拉桥 star type cable-stayed bridge
星形斜张桥 star type cable-stayed bridge
星形信标链 star beacon chain
星形星形接线 star-star connection
星形修整工具 star dresser
星形旋钮螺旋 star knob screw
星形叶桨架 <强制式搅拌机的> paddle star
星形有限的 star finite
星形有限复合形 star finite complex
星形有限覆盖 star finite covering

星形有限空间 star finite space
星形有限性 star finite property
星形域 star domain
星形凿 stared drill
星形支撑 star strut
星形中点接地的 star-grounded
星形中心器 star centralizer
星形装饰(品) star ornament;stellated ornament
星形撞伤 star-shaped bruise
星形着陆装置 star-shaped landing gear
星形琢磨花样 star cut
星形钻(头) rose bit;six-point bit;star drill;star(-shaped) bit
星形钻头导向器 rose bit pilot
星掩蔽 eclipse
星叶石 astrophyllite
星夜 starry sky
星云 nebula;star cloud
星云红移 nebular red shift
星云假说 nebular hypothesis
星云母 halite;star mica
星云谱线 nebular lines
星云摄谱仪 nebular spectrograph
星云线脚 nebula mo(u)lding
星云岩 nebulite
星云跃迁 nebular transitions
星云状的 nebular;nebulitic;nebulous
星云状物质【天】nebulosity
星载遥感系统 satellite-borne remote sensing system
星载转播站 spaceborne repeater station
星盏 sparkling glaze
星盏釉 star flashing glaze
星支 star prong
星质学 planetary geology
星蛭石 hallite
星轴承合金 star alloy
星状 stellar;stellated
星状冰川 star glacier
星状的 ast(e)roid;astral;astrotorus
星状断口 rosette fracture
星状耳石 asteriscus
星状隔膜 stellated diaphragm
星状拱顶 stellar vaulting
星状构造 stellated structure
星状光圈 star-shaped diaphragm
星状花 star flower
星状花的 star flowered
星状结构 stellated texture
星状结晶釉 starred glaze
星状截面 star-shaped profile
星状晶体 stellar crystal
星状肋拱顶 star vault
星状肋穹顶 stellar vault
星状肋穹隆 star(-ribbed) vault
星状沥青煤 stellarite
星状连接 Y-connection
星状裂纹 star(like) crack(ing);star fracture
星状毛 stellated hair
星状帽 star-cap

星状耦合器 star coupler
星状曲线 starlike curve
星状闪电 stellar lightning
星状水系 stellated drainage
星状碳极电池 star-shaped carbon cell
星状图 star diagrams
星状网络 star network
星状物 asterisk;star
星状域 starlike domain
星状装配 spider assembly
星状装饰 astreated
星状组合 spider assembly
星字炸药 <硝铵、硝酸甘油、三硝基甲苯炸药> Astralite
星族 stellar population
星座(花饰) constellation

猩红栎 scarlet oak

猩红磷 scarlet phosphorus
猩红热 scarlet fever
猩红色 scarlet;scarlet red
猩红色淀 geranium lake
猩红酸 scarlet acid
猩猩草子油 garcia nutans seed oil

刑事制裁 penal sanction

行包仓库 luggage and parcel depository

行包承运 acceptance of luggage consignment
行包到达作业收入 baggage arrival operation income
行包的承运 acceptance of luggages and parcels
行包发送、到达、中转量 volume of parcel dispatch and arrival and transshipment
行包房 baggage office;luggage and parcel house
行包快运专列 luggage express train
行包流线 baggages and parcels flow paths
行包事故 luggage and parcel traffic accident
行包事故案卷 files of luggage and parcel accident
行包事故查询 inquiry of luggage and parcel accident
行包事故等级 class of luggage and parcel accident
行包事故立案 register of luggage and parcel accident
行包事故种类 classification of luggage and parcel accident
行包托运 consigning of luggages and parcels
行包邮政地道 tunnel for(transporting) luggage and postbag
行包运价 price of luggage and parcel traffic
行包站台 luggage and parcel platform
行包中转作业收入 baggage transshipment income
行包周转量 turnover of parcel transportation
行包专列 special train for luggage and parcel
行包专列统计 parcels train statistics
行边喷射 marginal spray
行边效应 bounding effect
行编号 line number
行编号数据 line-numbered data
行编号文件 line numbered file
行编辑程序 line editor
行变程 marching
行变换 line translation
行变换器 line transformer
行标 rower
行标号 line label
行标记 line mark
行波 constant wave;moving wave;progress(ive) wave;propagating wave;running wave;travel(l)ing echo;travel(l)ing wave
行波保护 travel(l)ing-wave protection
行波比 travel(l)ing-wave ratio
行波波导 travel(l)ing waveguide
行波参量放大器 travel(l)ing-wave parametric amplifier
行波场 travel(l)ing-wave field
行波传送 rippling trough
行波传送进位 ripple through;ripple through carry
行波传送进位装置 ripple through carry unit
行波多束速调管 travel(l)ing-wave multiple beam klystron
行波放大器 travel(l)ing-wave amplifier
行波分离器 travel(l)ing-wave separator
行波功率放大器 travel(l)ing-wave power amplifier
行波共振腔 travel(l)ing-wave cavity
行波管 travel(l)ing-wave tube
行波管放大器 travel(l)ing-wave amplifier;travel(l)ing-wave tube amplifier
行波管振荡器 travel(l)ing-wave tube oscillator
行波光电倍增管 travel(l)ing-wave photomultiplier
行波光电管 travel(l)ing-wave phototube
行波加速器 travel(l)ing-wave accelerator
行波结构 travel(l)ing-wave structure
行波进位 ripple carry
行波进位方式 ripple carry system
行波进位加法器 ripple carry adder
行波螺旋波导 travel(l)ing-wave helix
行波脉泽 travel(l)ing-wave maser
行波示波管 travel(l)ing-wave oscillograph;wamoscope;wave-modulated oscilloscope
行波示波器 travel(l)ing-wave oscillograph;travel(l)ing-wave oscilloscope
行波式绕组 progressive wave winding
行波速调管 travel(l)ing-wave klystron;twystron
行波探测器 travel(l)ing detector
行波天线 progressive wave antenna;travel(l)ing-wave aerial;travel(l)ing-wave antenna;wave antenna
行波微波激射器 travel(l)ing-wave maser
行波系数 travel(l)ing-wave coefficient
行波系统 travel(l)ing-wave system
行波线 travel(l)ing-wave line
行波线性加速器 travel(l)ing-wave linear accelerator
行波谐振器 travel(l)ing-wave resonator
行波型参量放大器 travel(l)ing-wave type parametric amplifier
行波型铁磁放大器 travel(l)ing-wave ferromagnetic amplifier
行波型振荡 travel(l)ing-wave type oscillation
行波学说 travel(l)ing-wave theory
行波约束 travel(l)ing-wave confinement
行波振荡 travel(l)ing-wave oscillation
行波直线加速器 travel(l)ing-wave linac;travel(l)ing-wave linear accelerator
行播 sowing in line;sowing in row
行播机 row planter
行播器 row-seeder
行播小区 row plot
行步拖车 baggage towing tractor
行差 <测微器> run error
行车 bridge crane;crane;driving;hall crane;transporter crane;travel(l)-er;overhead(traveling) crane <行车俗称>;running of trains;train running;train working【铁】
行车安全 safe of traffic;safety of operation;safety of running;safety of traffic;safety running;traffic safety
行车报告 traffic report
行车报警 travel alarm
行车闭塞法 block(block) working;train block system
行车闭塞系统 block working system;train block system
行车闭塞制 block working system;train block system
行车贬值 road depreciation
行车标志 operation sign
行车表 service diagram
行车部分 travel(l)ed portion
行车操作 driving operation
行车畅通 circulation of traffic
行车车道 driving lane;running lane;travel(l)ed lane
行车冲击(力) traffic blow;vehicle impact;impact of traffic
行车大梁 crane girder;runaway girder;travel(l)er
行车带 traffic strip;traffic zone
行车导向箭头 direction marking
行车道 carriageway;driveway;running lane;stripe;traffic lane;travel(l)ed way;wagon road
行车道变动 lane change
行车道变更 lane change
行车道变形缝 carriageway expansion joint
行车道标志线 carriageway marking
行车道分隔带 divisional strip
行车道分隔线 stripe
行车道净宽 clear width of carriageway
行车道宽度 driving lane width
行车道连接缝 carriageway expansion joint
行车道伸缩缝 carriageway expansion joint
行车道纵梁 deck stringer
行车道租金 lane rental
行车的 trafficked
行车灯光 driving light
行车调度 traffic control
行车调度集中 centralized traffic control
行车调度命令空白格式 train order blank
行车调度通信[讯]电路 train dispatching circuit
行车调度系统 traffic control system
行车调度中心 traffic control center[center]
行车饭费 travel(l)ing allowance
行车范围 traffic area
行车方法 method of operation
行车方向 direction of movement;direction of traffic;driving direction;running direction;traffic direction
行车方向分析器 directional analyser
行车方向控制手柄 mode direction handle
行车费(用) cost of vehicle operation;running charges;running cost;running expenses;travel cost;operating cost;operating expenses;motor vehicle running cost <包括燃料、轮胎、修理、润滑、折旧等与车辆营运有关的费用,但不包括与车辆里程无关的费用,如执照费、利息等>
行车分隔 <包括对间分隔,快慢分隔及转变交通分隔等> segregation of traffics
行车钢轨 <与护轮轨相对应> running rail
行车公寓 train crew dormitory

行车规程 operating regulation
行车规则 operating rule;rule of road
行车轨迹 wheel path
行车荷载 traffic load; vehicular live load
行车横向分布系数 transverse vehicle distribution factor
行车横向净空 vehicular horizontal clearance
行车滑道 crane runway
行车滑道柱 crane runway column
行车机构 undercarriage
行车机件 running gear
行车激振 vibration excited by moving truck
行车记录簿【铁】train record book
行车记录设备 train movement recording equipment
行车驾驶室 <起重机> transport station cab
行车间隔 headway;running interval
行车监察员 operating supervisor
行车监督系统 traffic monitoring system
行车箭头标 direction arrow
行车奖金 running premium
行车交叉【铁】traffic intersection
行车交通 moving traffic
行车交通线 thoroughfare track
行车交织段 weaving section
行车脚闸 travel service brake
行车接触导线 crane trolley wire
行车进路 train route
行车净距 traffic clearance
行车开灯时间 lighting-up
行车控制 control of train operation; driving control
行车栏杆 traffic railing
行车里程 car mileage; mileage; vehicle-miles of travel
行车梁 crane beam; crane runway girder;runway beam
行车梁滑道 crane runway
行车量 traffic load
行车路面 road carriageway; running surface of rail
行车路线 driving route; travel path; travel way
行车旅程 car mileage
行车门座 crane portal
行车密度 rate of traffic flow
行车命令 train order
行车命令继电器 train order relay
行车命令信号机 <对司机>【铁】 train order signal
行车命令运行法 train order operation
行车模拟器 traffic simulator
行车能见度 driving visibility
行车能量 trafficability
行车碾压 traffic bound
行车碾压路面 traffic-bound road
行车碾压筑路法 traffic-bound method(of construction)
行车偶然发生事件 operating contingence
行车判断能力 road sense
行车频率 frequency of service; service frequency
行车平稳性 <路面的> smooth-riding quality
行车平稳质量 riding quality
行车凭证 running token;token
行车坡道 <停车场的> driving ramp
行车(前后)颠簸 vehicle pitching
行车强度 intensity of traffic
行车强度函数 function of traffic intensities
行车桥 bridge overhead travel(l)ing crane
行车桥面 traffic-carrying deck
行车升值 road appreciation
行车时间 running time;travel time
行车时间表 time schedule(chart)
行车时刻 running time; scheduled time
行车时刻表 timetable
行车时刻表的全部修订 blank sheet revision
行车式料斗 bopper-on-rail
行车式漏斗 hopper-on-rail
行车式抛砂机 gantry slinger
行车事故 auto accident; movement accident; operating accident; operating trouble; traffic accident; train operation accident【铁】
行车事故件数 number of train operation accident
行车事故率 fatality rate; traffic accident rate
行车视距 running sight distance
行车试验 road performance test;road test;traffic test
行车试验期 traffic testing period
行车室 traffic operation office
行车适应性 roadability
行车手势 arm signal
行车舒适性 riding comfort;roadability
行车竖向净空 vehicular vertical clearance
行车速度 <以实际行驶时间除全程距离> running speed
行车速率 driving speed;speed of travel
行车隧道 vehicular tunnel
行车隧洞 vehicular tunnel
行车条件 driving condition; operating condition;traffic condition
行车调整 traffic arrangement; train operation adjustment
行车通告 train running notice
行车通知 traffic advice; traffic notification
行车通知单 traffic notification
行车统计 operating statistics; train operating statistics
行车图形显示器 graphic(al) display
行车紊乱 kneading of traffic; pounding of traffic
行车稳定性 roadability
行车稳定性和适应性 roadability
行车显示 traffic(movement) phase
行车线 running line;running track
行车限制 traffic limitation
行车辛烷值 road octane-value
行车信号登记簿 train signaling record book
行车信号(机)【铁】running signal; train signal
行车行列 moving traffic lane; autocade <美>
行车性能 operating performance
行车压实的 traffic bound
行车压实碎石路 traffic-bound macadam
行车拥挤 congestion of traffic
行车优先权 traffic priority
行车责任事故间隔里程 kilometres interval of running responsible accident
行车障碍 driving obstacle
行车振动 traffic variation
行车震颤 road shocks
行车支出 running expenses
行车指令 driving instruction
行车指向箭头 direction marking
行车制动器 service brake;travel brake
行车质量 driving quality
行车质量等级 riding quality grade
行车中断 traffic interruption
行车主任 <主管全段行车和调车工作,调度和配车工作除外> train master
行车装置 travel(l)ing gear
行车字符显示器 character display
行车自动打印机 automatic train movement printer
行车阻力 resistance to traction
行车组织 train operation organization
行车作业 traffic operation
行车作用 traffic action
行程 array pitch; distance of run; flight; haul cycle; length of run; length of travel;travel;trip(ping)
行程编码 run-length encoding
行程不匀 wabble
行程参数 travel parameter
行程操纵阀 range selector valve
行程差 progressive error;progressive inequality;path difference <直达声与反射声的>
行程长度 haul distance; length of stroke;stroke length;travel length; trip length
行程超限 excess of stroke
行程车速 journal speed;journey speed; travel speed
行程次数 number of strokes
行程次数利用系数 <压力机的> utilization coefficient of strokes
行程挡块 stroke dog
行程的无级调节 variable stroke control
行程的下死点 bottom end of stroke
行程的展长 lengthening of run
行程定位器 range finder
行程动机 journal motive
行程范围 travel(ling) range
行程放大器 stroke multiplier
行程分布 trip distribution
行程改变 stroke alteration
行程杆 tripping arm
行程换向 return of stroke
行程极限 extreme limit of travel
行程计 stroke counter
行程计划 itinerary
行程检验法 runs test
行程交换模型 trip interchange model
行程开关 limit(ing) switch; overtravel-limit switch;position switch; stop limit switch;terminal stopping device;travel switch
行程控制阀 range selector valve
行程控制捏手 stroke control knob
行程利用率 <又称里程利用率,为一定时期内汽车重载行驶公里数与总行驶公里数之比> travel utilization ratio
行程模型 travel pattern
行程排量 stroke volume
行程频率 travel frequency
行程起点 start of a run
行程时间 journal time;journey time; travel time
行程时间比 <用于高速干道和一般道路> travel-time ratio
行程时间矩阵 travel time matrices
行程时间曲线 travel-time curve
行程时间图 travel-time map
行程时间系数 coefficient of travel time
行程时间延误 <实际行驶的总行程时间与同车辆在自由行驶时平均车速计算的行程时间之差> travel-time delay
行程速度 no-load speed
行程天线 path-length antenna
行程调节 stroke control
行程调整 stroke adjustment
行程调整轴 stroke adjusting shaft
行程位置调整 adjustment for position of stroke
行程限度 limitation of length of path; limitation of length of stroke
行程限位开关 lead limit switch
行程限位器 end stop
行程限制器 arresting device; arresting gear;limit stop;overstroke safety device;stop piece
行程限制销 stroke limiting pin
行程延长 extension of journey
行程英里数 miles of travel
行程终点 stroke end;trip end
行船风 junk wind
行船通知 <码头、港口的> dock returns
行导位置 guide position
行道灯上的棱形玻璃(镜) paved prism;pave prism
行道树 alee-tree; alley tree; avenue tree; border tree; roadside tree planting;shade tree;sidewalk tree; street tree
行道栽植 avenue planting
行得通的竞争 workable competition
行灯 cable lamp;portable lamp;portable light(er);service lamp
行灯灯泡 portable lamp bulb
行灯灯座 portable lamp plug socket
行点旗礼 dip the flag
行电平 line level
行调 traffic controller
行调交叉【铁】traffic-switching intersecting
行动 behavio(u)r; deed; deport; ongoing
行动不能【医】akinesia
行动参数 action parameter
行动操纵 finger control
行动的范畴 category of behavio(u)r
行动地区 movement area
行动点 action limit
行动电台 mobile station
行动吊车 bridge crane;mobile station
行动方案 action plan
行动方案计划 action program(me) plan
行动方向的设想 action-designed assumption
行动负荷 moving load
行动纲领 program(me) of action
行动荷载 travel(l)ing load
行动计划 blueprint;plan of action
行动技能 action skill
行动检测器 motion detector
行动开始时刻 zero hour
行动控制 finger control
行动派 Tachism
行动绳 distance line
行动失调 dystaxia
行动时滞 action lag
行动式机构 walking mechanism
行动式拉铲挖掘机 walking dragline
行动式碎石设备 walking crushing plant
行动式索斗挖掘机 walking dragline excavator
行动式挖泥机 walking dredge(r)
行动式挖土机 walking excavator
行动式装置 walking mechanism
行动调整 finger control
行动系统 behavio(u)ral system
行动训练 action training
行动研究 action research
行动椅 bath chair
行动载重 rolling load
行动债券 performance bond
行动装置 running gear
行动准则 operative norm
行动自由的资本 liberated capital
行宫 <古建筑> temporary palace

行轨式制动闸 carracing retarder
行洪河槽 flood channel
行洪河道 flood channel
行洪区 floodway district
行贿 bribe(ry);commit bribery;corrupt practice
行贿购买 buying off
行贿人 briber
行贿收买 buying off;buying over
行贿受贿 offering or accepting bribes
行贿受贿分子 corruptionist
行贿受贿罪 offence of bribery
行贿物 bribe
行贿者 briber
行贿资金 slush fund
行近流 approach flow
行近流速 approach(ing)(of)approach
行近流速水头 approach velocity head
行近平面 arrival level
行近水流 approaching flow
行近水头 head of approach
行近速度 approaching velocity;velocity of approach
行近速率 approach speed
行近信号 approach signal
行进 locomote;marching
行进表面波 progressive surface wave
行进波 advanced wave;progressive wave;running wave;travel(l)ing wave
行进功率谱 running power spectrum
行进空化 travel(ling)cavitation
行进扰动 travel(l)ing disturbance
行进时间 advance time;time of travel <洪水波等的>
行进式潮汐波动 progressive type of tidal oscillation
行进水流 progressive flow
行进水头 head of approach
行进速度 approach velocity;travel(ling)speed;travel(l)ing velocity
行进速率 travel(l)ing rate;travel(l)ing speed
行进涌浪 travel(l)ing surge
行进振幅波 progressive oscillatory
行进中卸载 non-stop dumping
行进重力波 progressive gravity wave
行经曲线 <列车> taking of curves
行军虫 armyworm
行军床 camp bed;cot;tent bed
行李 baggage;belongings;dunnage;impedimenta;luggage
行李搬运 baggage handling;luggage handling
行李搬运车 baggage troll(e)y;baggage truck;luggage troll(e)y;luggage truck;package delivery truck
行李搬运工 luggage porter
行李搬运设施 baggage handling facility
行李包裹承运 acceptance of luggages and parcels
行李包裹交付 delivery of luggages and parcels
行李包裹托运 consigning of luggages and parcels
行李、包裹、邮运收入 baggage, parcel and post service revenue
行李包裹运价 tariff of luggages and parcels
行李包裹运输成本 transport cost of luggages and parcels
行李包裹运送量 luggage and parcel traffic volume
行李包裹中转加给收入 luggage and parcel transfer increased revenue
行李保价费 luggage and parcel security value charges
行李保险 baggage insurance;luggage insurance

行李报关 luggage declaration
行李报关单 baggage declaration form
行李便餐车 baggage buffet car
行李标签 luggage label
行李舱 baggage cabin;baggage compartment;baggage hold;boot compartment;luggage compartment;luggage hold
行李舱底板 boot floor;luggage floor panel
行李舱地板 luggage compartment floor
行李舱地板衬垫 luggage compartment trimming
行李舱地板垫毯 luggage floor mat;trunk mat
行李舱口 luggage hatch
行李车 baggage car(t);baggage van;baggage wagon;blind car;luggage car(t);luggage van;luggage wagon
行李车拉门护门 door guard
行李车拉门护铁 door guard
行李车门 baggage car door
行李车厢 freight car;luggage van
行李储藏室 luggage store
行李处理柜台 baggage handling counter
行李处所 baggage space
行李传送带 baggage conveyer[conveyor]
行李存放处 check room;cloakroom;luggage depositary
行李存放室 baggage room;luggage room
行李袋 cargo container
行李登记费 luggage registration charges
行李地下通道 baggage way
行李电梯 baggage elevator;luggage elevator
行李定义 definition of luggage
行李动车 motor luggage van
行李发出 despatching of luggage
行李发运 dispatching of baggage;dispatching of luggage
行李房 baggage(check)room;baggage office;baggage registration office;baggage room;luggage office;luggage room;trunk room
行李费 baggage charges;baggage fee;luggage charges
行李格架 baggage grid;luggage grid
行李工【港】porter
行李管理 baggage handling
行李和包裹的保管 storage of luggages and parcels
行李和包裹的承运 accept the conveyance of luggages and parcels
行李和包裹的支付 delivery of luggages and parcels
行李和包裹的中转 transfer of luggages and parcels
行李环形道 baggage roundabout
行李计量器 luggage counter;luggage sticker
行李寄存处 check room;left-luggage office
行李寄存室 luggage store
行李架 baggage carrier;baggage holder;baggage rack;baggage shelf;bag rack;luggage carrier;luggage rack;luggage shelf;package rack;parcel rack;rack
行李架可放倒的延长部分 luggage carrier extension
行李架栏杆 luggage rail(ing)
行李间 baggage compartment;baggage hall;luggage compartment;luggage locker;luggage room;trunk room
行李间灯 luggage compartment lamp;luggage compartment light

行李间地席 luggage compartment lining;luggage compartment mat
行李间钥匙 luggage-boot key;trunk lid key
行李检查 luggage inspection
行李检查大厅 baggage examination hall
行李卡车 baggage truck
行李框栏杆 <车顶的> luggage rail on top
行李领取处 luggage claim;baggage claim
行李领取室 baggage delivery office
行李流程路线 baggage flow route
行李路程单 luggage waybill
行李牌 baggage check;placard
行李棚 baggage shed
行李篷车 baggage van
行李票 baggage check;baggage ticket;baggage voucher;luggage ticket;luggage voucher
行李认领处 baggage claim area
行李容量 baggage capacity
行李入口处 baggage entrance
行李申报单 baggage declaration
行李升降机 baggage lift;luggage elevator
行李室 baggage room;luggage room
行李收据 luggage receipt
行李收入 baggage revenue
行李收运 acceptance of luggage
行李手推车 baggage barrow;block truck;luggage barrow
行李输送系统 baggage conveyance system
行李通关单 baggage sufferance
行李铜牌 <免费行李发牌不起票> metal check
行李推车 baggage truck
行李托运 registration of luggage
行李托运室 baggage registration office
行李托运台 baggage(check)counter;luggage(check)counter
行李拖车 baggage tractor
行李网固定卡扣 luggage-net bracket
行李网架 <车上> basket rack;luggage net
行李网支架 luggage-net support
行李箱 cargo container;luggage carrier
行李箱盖 luggage-boot lid
行李小车 baggage truck
行李小推车 luggage carrier
行李邮政车 luggage and mail van
行李邮政守车 mail-luggage-guard's van
行李预先登记(托运) advance registration of luggage
行李员 baggage clerk;baggage man;luggage clerk;luggage van guard
行李运价表 baggage tariff
行李运输 baggage traffic;luggage traffic
行李运输经理 baggage traffic manager
行李暂存箱 luggage locker
行李责任事故发生率 occurrence rate of liability accident for luggage
行李责任事故件数 number of liability accident for luggage
行李站台 baggage platform
行李支架 baggage holder;luggage holder
行李重量限度 baggage allowance
行李装卸线 baggage load and unload track
行李走廊 baggage corridor
行前【交】pre-trip
行前出行计划 pre-trip planning
行人 foot passenger;passer-by;pedestrian;walker
行人安全措施 protection of pedestrians
行人安全岛 pedestrian(refuse)island
行人安全防护 protection of pedestrians
行人安全设施 pedestrian safety devices
行人按钮控制 pedestrian push-button control
行人按钮控制交通信号 pedestrian push-button control signal
行人按钮信号标志 pedestrian-actuated signal sign
行人扳动的交通信号 pedestrian operated traffic signal
行人避车处 refuge manhole
行人车祸 pedestrian motor vehicle traffic accident;pedestrian vehicle accident
行人车辆冲突点 pedestrian-vehicular conflict
行人传动控制器 <一种自动控制器,其中的部分信号,特别是行人"行走"信号,是由行人感应器引动的> pedestrian-actuated controller
行人传送带 pedestrian conveyer[conveyor]
行人道化 pedestrianize
行人道交通管制 pedestrian control
行人道体系 gallery system
行人的 pedestrian
行人的保护 pedestrian protection
行人地带 pedestrian zone
行人地带标(志)线 pedestrian zone marker
行人地道 passenger subway;pedestrian tunnel;subway for pedestrians
行人地区标志线 pedestrian zone marker
行人调查 pedestrian survey
行人动态 pedestrian behavio(u)r
行人渡口旅客 pedestrian ferry passenger
行人分隔带 pedestrian barrier panel
行人感知器 pedestrian detector
行人隔栏 pedestrian separation fence
行人隔离栏 pedestrian separation fence
行人观测计数 pedestrian count
行人管理 pedestrian control
行人管制栅栏 pedestrian control fence
行人过街标志 crossing sign;pedestrian crossing sign
行人过街道 pedestrian crossing
行人过街清尾时间 <从允许行人过街信号终止到不允许过街信号出现之间一段空隙时间> pedestrian clearance interval
行人过路线 pedestrian crossing
行人荷载 pedestrian load
行人护栏 pedestrian barrier;pedestrian guardrail
行人活动的安全措施 safe accommodation of pedestrian movements
行人计数 <一定时间内通过一定地点的行人数> count pedestrian
行人检测器 pedestrian detector
行人交通 foot traffic;pedestrian flow;pedestrian traffic
行人交通安全措施 protection of pedestrians
行人交通管制 pedestrian control
行人交通量 pedestrian traffic volume;pedestrian volume
行人交通信号 pedestrian signal
行人街道 pedestrian street
行人金属防护栏 pedestrian metal guardrail
行人净空 pedestrian space
行人靠街左走标志 walk on left sign

行人控制设施 pedestrian control device
行人栏杆 pedestrian rail
行人廊道体系 gallery system
行人流 flow of pedestrians; pedestrian flow
行人流率 pedestrian flow rate
行人流通量 < 车站 > pedestrian circulation
行人楼梯 pedestrian stair(case)
行人轮渡 pedestrian ferry
行人密集 pedestrian congestion
行人面积模数 pedestrian area module
行人桥 pedestrian bridge
行人揿钮信号 pedestrian-actuated signal
行人揿钮信号标志 pedestrian-actuated signal sign
行人请求绿灯 pedestrian recall
行人区 pedestrian zone
行人群 pedestrian group
行人色灯(信号)控制 pedestrian light control
行人上下车地带 pedestrian traffic zone
行人上下车门道 pedestrian traffic door
行人随意行走 haphazard walking
行人探测器 pedestrian detector
行人天桥 overhead walkway; pedestrian bridge; pedestrian skyway
行人调控信号 pedestrian-actuated signal
行人通道 passenger tunnel
行人通过量 people throughout
行人通行 pedestrian traffic
行人通行控制器 pedestrian control device
行人往来交通 < 车站 > pedestrian traffic
行人无秩序行走 haphazard walking
行人先走信号显示 leading pedestrian phase
行人小道 by-walk
行人心理 pedestrian psychology
行人信号显示 pedestrian phase; pedestrian signal display
行人信号相 < 分配给行人交通的信号相 > pedestrian phase
行人信号相位 pedestrian signal phase
行人行车半专用信号显示 semi-exclusive pedestrian-vehicle phase
行人行程 person trip
行人性状 pedestrian behavio(u)r
行人拥挤 pedestrian congestion
行人与车辆分流 pedestrian segregation
行人与车辆碰撞事故 pedestrian motor vehicle traffic accident
行人与自行车道 pedestrian and cycle track
行人与自行车事故 pedestrian and cycle accident
行人运送系统 pedestrian distribution system
行人栅栏 pedestrian railing
行人占地面积 pedestrian space
行人专用道 pedestrian road; pedestrian way < 英国道路用地外的 >
行人专用区 pedestrian precinct
行人专用商业区 pedestrian precinct
行人专用显示系统 scramble system
行人专用信号相 exclusive pedestrian phase
行人走的小路 footpath
行人阻挡栅栏 pedestrian barrier panel
行使功能 functionating
行使假文件 uttering forged documents
行使扣押权的债权人 attaching creditor; creditor's equity
行使期权 exercise option
行使权力 exercise
行使权利 enjoyment of right; exercise power
行使权利文书 enabling declaration
行使伪币 uttering false coin
行使职权 administer; discharge one's duties; discharge one's functions and powers; exercise of authority; exertion
行使职务 officiate
行使职责 functionate
行使专利权 use of patent
行驶 exercise; riding
行驶操纵装置 roading function control
行驶车辆观测法 floating car method
行驶车流 moving traffic
行驶车速 < 设计的 > running speed
行驶车座 travel(l)ing pedestal
行驶点 running point
行驶方式 mode of travel; travel mode
行驶方向 traffic movement
行驶方向稳定性 stability of travel direction
行驶规则 rules of the road
行驶过度的 overriding
行驶换向 directional change
行驶机构 travel(ling) mechanism
行驶净空 man(o)euvering space
行驶控制 travel control
行驶里程 miles operated; road haul
行驶灵活性 road mobility
行驶路线 driving route
行驶履带 tread caterpillar
行驶轮 travel(l)ing wheel
行驶平顺性 ride performance; riding comfort
行驶时间 running time
行驶时外形尺寸 < 起重机 > overall travel dimension
行驶时噪声低的机器 quiet-riding
行驶舒适性指数 riding comfortable index
行驶速度 running speed; transport speed; travel(l)ing speed
行驶速度控制手柄 travel speed lever
行驶速率 speed of travel
行驶特性 operational characteristic
行驶稳定器 ride stabilizer
行驶性能 rideability; running performance
行驶性能试验 travel performance test
行驶振动传递特性 ride transfer characteristic
行驶(职权的)人 exerciser
行驶制动器 service brake; travel brake
行驶质量 driving quality; ride quality; riding characteristic
行驶质量指数 riding quality index
行驶中称重 weigh in-motion
行驶中换挡 on-the-go shifting
行驶中调整 on-the-go adjustment
行驶装置 travel(l)ing gear
行驶状态 travel(l)ing condition
行驶状态下的轴荷分配 transport axle distance
行驶总速度 overall travel speed
行驶阻力 driving resistance; running resistance
行驶阻力测定 moving resistance measurement
行停信号 stop-and-go signal; stop-and-proceed signal
行为背景 behavio(u)r setting
行为变革的过程 process of behavio(u)r change
行为错乱 fragmentation
行为地理学 behavio(u)ral geography
行为动力学 behavio(u)r dynamics
行为范型 behavio(u)r pattern
行为方案 course of an action
行为方程 behavio(u)r equation
行为工程师 behavio(u)ral engineer
行为功能 action function
行为环境 behavio(u)ral environment
行为矫正精神病学 orthopsychiatry
行为决策 behavio(u)r decision making
行为决策函数 behavio(u)ral decision function
行为科学 behavio(u)r science; behavio(u)ral science
行为空间 behavio(u)r space
行为量表 behavio(u)r rating scale
行为论 behavio(u)rism
行为模式 behavio(u)r pattern
行为模式的表义作用 semantization
行为模型 behavio(u)r model
行为税 act tax
行为势能 behavio(u)r potential
行为修正 behavio(u)r modification
行星 planet
行星摆轮混砂机 rotation mixer
行星摆线针轮减速机 cycloidal planetary gear speed reducer
行星摆转混砂机 rotation mixer
行星边界层 planetary boundary layer
行星变速轮 planetary transmission (gear)
行星变速器 planetary transmission
行星表面分析 planetary surface analysis
行星表面探测 planet surface exploration
行星表面学 planetography
行星表面坐标 planetographic coordinates
行星波 planetary wave
行星测图 planetary mapping
行星差动齿轮 planet differential gear
行星齿轮 epicyclic(al) gear; planet; planetary gear; planetary wheel; planet differential; planet gear(ing); satellite gear; star gear; sun-and-planet wheel
行星齿轮变速箱 planetary gear type transmission
行星齿轮差速器 satellite differential
行星齿轮传动 epicyclic(al) gear transmission; planetary drive; planetary gearing; planetary gear transmission; sun-and-planet gearing
行星齿轮传动比 planetary gear ratio
行星齿轮传动单向离合器 planetary overrunning clutch
行星齿轮传动滚筒 planetary geared drum
行星齿轮传动机构 planetary gear mechanism
行星齿轮传动提升机 planetary geared hoist
行星齿轮传动系 epicyclic(al) gear train; planetary train
行星齿轮传动线 epicyclic(al) driveline
行星齿轮传动装置 spur planetary gearing
行星齿轮的 planetary
行星齿轮定时装置 planetary timing gear
行星齿轮动力驱动 planet power drive
行星齿轮分配机构 planetary tuning gear
行星齿轮后轴 planetary rear axle
行星齿轮混合机 planetary gear mixer
行星齿轮机构壳体 planet cage
行星齿轮架 pinion frame; planet(ary)(gear) carrier
行星齿轮架的小齿轮 carrier pinion
行星齿轮减速 epicyclic(al) gear reduction; planetary gear reduction
行星齿轮减速机 planet gear speed reducer
行星齿轮减速装置 epicyclic(al) reduction gear unit; planetary reducer; planet gear speed reducer
行星齿轮轮毂 planetary hub gear
行星齿轮轮毂减速 epicyclic(al) hub reduction; planetary hub reduction
行星齿轮驱动系统 planetary drive system
行星齿轮驱动轴 planetary drive axle
行星齿轮驱动装置 planetary drive set
行星齿轮式变速器 epicyclic(al) gear transmission
行星齿轮式吊车 planetary geared hoist
行星齿轮系 epicyclic(al) train; planetary gear train
行星齿轮箱 epicyclic(al) gearbox; planetary transmission box; planetary cage
行星齿轮型变速器 planetary gear type inverter
行星齿轮支座 planet carrier
行星齿轮终传动 planetary final drive
行星齿轮终减速 planetary final reduction
行星齿轮轴 planetary axle; planetary pin
行星齿轮轴架 planetary wheel carrier
行星齿轮转向 planetary steer(ing)
行星齿轮转向机构 planetary steering device; planetary traversing mechanism
行星齿轮转向轴 planetary steering axle
行星齿轮转向装置 planetary power steering
行星齿轮装置 epicyclic(al) gearing; planetary gear apparatus
行星齿轮组 epicyclic(al) gearing; spool pinion; sun-and-planet gears
行星齿轮最终传动 planetary reduction final drive
行星传动 epicyclic(al) gearing; epicyclic (al) transmission; planetary drive; planetary transmission
行星传动齿轮 planet(ary) pinion
行星传动第一级固定齿轮 fixed primary annulus
行星传动箱 planet cage
行星传动装置箱 planetary cage
行星传动装置液压控制阀 range transmission hydraulic control valve
行星大气 planetary atmosphere
行星大气环流 planetary circulation
行星的 planetary
行星的涡度效应 planetary vorticity effect
行星跟踪仪 planet tracker
行星工程 planetary engineering
行星光行差 planetary aberration; planetary light deviation
行星轨道 planetary orbit
行星轨道内空间 cisplanetary space
行星轨道倾角 inclination of planetary orbits
行星环境 planetary environment
行星环流 planetary circulation
行星回动齿轮组 epicyclic(al) reversing gear set
行星急流 planetary jet
行星几何学 planetary geometry
行星际测量卫星 interplanetary measurement satellite
行星际场 interplanetary field

行星际尘埃 interplanetary dust
行星际冲击波 interplanetary shock wave
行星际磁暴 interplanetary magnetic storm
行星际磁场 interplanetary magnetic field
行星际辐射 extraterrestrial radiation
行星际光 extraterrestrial light
行星际轨迹 extraterrestrial trajectory
行星际航行 interplanetary navigation
行星际环境 extraterrestrial environment
行星际监视台 interplanetary monitoring platform
行星减速齿轮 planetary reducer; planetary reduction gear
行星减速齿轮装置 epicylic reduction gear unit; planetary reducer
行星减速电动机 epicycle motor
行星减速器 epicyclic(al) reduction gear; planetary reduction gear
行星减速装置 planetary reducer
行星空间 planetary space
行星空间导航系统 planetary space navigational system
行星框架 planetary frame
行星流星雨 planetary stream
行星轮 planet; planetary wheel
行星轮传动桥轴 planetary axle
行星轮架 planet carrier
行星轮架的小齿轮 spider gear
行星轮减速桥轴 planetary axle
行星轮结构 planetary configuration
行星轮式动力换挡 planetary power shift
行星轮式终传动 planetary final drive
行星轮式转向机构 planetary steering device
行星轮输入 input of planetary
行星轮他动箱 epicyclic(al) transmission
行星轮转向装置 planetary steer(ing)
行星面的 planetographic
行星平行轴二级减速齿轮 planetary-parallel gear
行星平行轴减速齿轮箱 planetary-parallel reduction gear
行星上登陆 planetary landing
行星摄动 planetary perturbation
行星生态系统 planetary ecosystem
行星式 planetary type
行星式拌和机 planetary compulsory mixer; planetary stirrer
行星式变速箱 planetary transmission
行星式缠绕机 planetary type winding machine
行星式车轮传动 planetary wheel drive
行星式齿轮 sun-and-planet gear
行星式齿轮传动 planet gearing; planetary transmission
行星式齿轮传动系 planetary gearing
行星式齿轮减速 epicyclic(al) reduction gear of planetary type
行星式齿轮系 planetary gear system
行星式齿轮组 planetary gear set; planetary wheel set
行星式齿系 planetary set gear
行星式传动装置 planetary transmission
行星式电子 planetary electron
行星式动力变速箱 planetary(type) power shift transmission
行星式粉碎机 planet type grinding mill
行星式钢球冲击缩管机 planetary ball swaging
行星式钢丝绳捻股机 planetary stranding machine
行星式高频振动器 planetary high frequency vibrator
行星式混合机 orbital-motion mixer; planetary type mixer
行星式混合器 planetary mixer
行星式混凝土搅拌机 planetary (type) concrete mixer
行星式减速齿轮系 planetary reduction gear
行星式桨状混合器 planetary paddle mixer
行星式搅拌机 orbital-motion mixer; planetary stirring machine; planetary(type) mixer
行星式搅拌捏和机 planetary mixing and kneading machine
行星式搅拌器 planetary agitator; planetary beater; planetary compulsory mixer; planetary mixer; planetary stirrer; planetary stirring machine
行星式冷却机 planetary cooler
行星式冷却筒 planetary tube
行星式冷轧管机 planetary tube cold-rolling mill
行星式螺杆挤压机 planetary screw extruder
行星式螺纹铣床 planetary thread miller; planetary thread milling machine
行星式磨机 planetary mill
行星式磨筒 planetary tube
行星式内圆磨床 planetary internal grinder
行星式碾磨机 planetary mill
行星式牵引传动系统 planetary traction transmission
行星式球磨机 planetary ball mill
行星式圈条器 planetary coiler
行星式热轧机 planetary hot mill
行星式摄影机 planetary camera
行星式铣床 planetary milling machine
行星式旋转发动机 planetary rotation engine
行星式旋转叶片 <搅拌机的> planetary rotating paddles
行星式研磨机 planetary type grinding machine
行星式液压换挡 planetary power shifting
行星式轧机 planetary(rolling) mill
行星式振动器 planetarium type vibrator
行星式制动器 planetary brake
行星式转向轮系 planetary steering gear
行星式转向装置 differential steering
行星式组轮 planetary set gear
行星岁差 planetary precession; planetary precession of equinoxes
行星天文 planetary astronomy
行星凸轮 planetary cam
行星托架 planetary carrier
行星位置测量仪 planet sensor
行星位置图 planet location diagram
行星温度 planetary temperature
行星铣法 planetary milling
行星系 planetary system
行星相对位置 planetary configuration
行星相互位置图 planetary configuration
行星小齿轮 planet(ary) pinion gear
行星小齿轮架 planet pinion carrier
行星小齿轮轴 spider pin
行星心轴 planetary spindle
行星信息 planetary information
行星旋转机 planetary rotation machine
行星旋转运动 planetary rotary movement
行星研究仪器舱 planetary capsule
行星研磨机 star lapping machine
行星运动 epicyclic(al) motion; planetary motion; sun-and-planet motion
行星增速器 planet gear speeder
行星轧机 planetary rolling
行星针轮减速机 cyclo reducer
行星中心的 planetocentric
行星中心坐标 planetocentric coordinates
行星中心坐标系 planetocentric coordinate system
行星重力场 planetary gravitational field
行星转向轮 planetary steering gear
行星锥齿轮装置 bevel planet gearing
行凶者 perpetrator
行邮守车 mail-luggage-guard's van
行载 moving load
行遮光 line vignetting
行针 needle conveying; needle transmission
行政办公建筑 administration complex
行政办公楼 administration block; administration building; administration unit; administrative building
行政办公室 administration office; administrative building
行政报告 executive report
行政边界 administration boundary
行政编制 executive preparator
行政部门 administration; administrative division; administrative section; executive; executive branch
行政处罚 administrative penalty; administrative sanction
行政措施 administrative measure; execution process
行政单位 administrative unit; political unit
行政当局 executive authority
行政的 administrative
行政的薪金和津贴 executive salaries and bonuses
行政等级 administrative hierarchy
行政地区 administrative zone
行政调查 administrative census
行政定价范围 range of administered price
行政罚款 administrative fine
行政法规 administrative law; administrative regulation; administrative rule; business law
行政法(令) administration law; administrative law
行政法庭 administrative tribunal
行政法院 administrative tribunal
行政方法 administrative means
行政费表 schedule of administrative fee
行政费用 administrative cost; administrative expenses; organizational cost
行政否决权 executive veto
行政负责人 administrator
行政干预 administrative interference; administrative intervention
行政工作人员 administrative officer
行政功能 administrative function
行政官 executive
行政管理 administrative arrangement; administrative management; executive management; management engineering
行政管理办公用品 office of executive management use
行政管理部门 administration authority; administration section; intendance
行政管理处 administrative department; administrative management service
行政管理当局 administrative authority
行政管理的 administrative
行政管理多道程序设计 executive control multiprogramming
行政管理费 administration and supervision cost; administration cost; administrative cost; administrative expenditures; administrative expenses; administrative spending; spending on administration and management
行政管理费用 overhead cost
行政管理负责人 administrative authority
行政管理工程信息管理系统 administrative engineering information management system
行政管理和预算局 <美> Office of Management and Budget
行政管理机构 administrative body; administrative organ
行政管理机关 administrative body
行政管理理论 administrative management theory
行政管理楼 administrative building; directors building
行政管理人才的就业市场 market for executive talent
行政管理人员 administration staff; administrative personnel; administrative staff; legislators & administrators
行政管理数据处理 administration data processing; administrative data processing
行政管理数据处理机 administrative data processor
行政管理学 administrative science
行政管理学会 Administration Management Society
行政管理员 administrator
行政管理支出 administrative expenditures
行政管理终端系统 administrative terminal system
行政管制 administrative control
行政规范 administrative rule
行政规章 rule of administration
行政和财务处 administrative and financial services
行政会议 assize
行政机构 administrative agency; administrative machinery; civil service
行政机关 administrative authority; administrative office; administrative organ
行政监督 administrative supervision
行政建筑 administrative building
行政建筑群 administration complex
行政建筑物 administration block
行政解剖 administrative autopsy
行政界 administrative boundary
行政经理 administrative manager
行政决策 administrative decision; executive decision
行政决定 administrative action
行政开支 administrative expenses; cost of supervision
行政控制 administrative control
行政理事会行政管理费 administrative council
行政立法 administrative legislation
行政领导 executive directors
行政楼 bouleuterion
行政命令 administrative orders
行政批准 administrative approval
行政区 administrative area; administrative district; administrative divi-

sion;civil town;district;political unit
行政区划 administrative delimitation;administrative division
行政区划分 regionalism
行政区划界 administrative boundary;civil(division) boundary;political boundary
行政区划略图 administrative diagram
行政区划示意图 administrative index;civil parish index;index to boundaries
行政区划图 administrative map;boundary map;political map
行政区划资料 administrative data;administrative detail;administrative information
行政区位置 position of administrative division
行政区域 administrative delimitation;administrative division;administrative region
行政区镇 civil township
行政人员 administrative personnel;administrative staff;administrative worker;administrator;executive;executive worker;official
行政人员决策模式 model of decision making by administrative staff
行政上的标准 standard of an executive
行政审计 administration audit;administrative audit
行政事务 civil service;government service
行政事务工作 administration duty
行政事务主任 administration manager
行政手段 administrative means;administrative measure
行政诉讼 administrative action;administrative litigation
行政诉讼程序 administrative proceeding
行政特区 special administration area
行政体制 administrative set-up
行政统计资料 administrative statistics
行政协定 executive agreement
行政行为 act of administration;administrative behavio(u)r
行政性保护 administrative protection
行政性会议 meeting in camera
行政性业务 administrative services
行政预算 administration budget;administrative budget
行政责任 administrative responsibility
行政支持 administrative backstopping
行政支出 administrative expenditures
行政支援 administrative support
行政执行 administrative execution
行政职员 executive staff
行政指导 administrative guidance
行政指挥 administrative command
行政制度式组织 bureaucratic form of organization
行政中心 administration center[centre];administrative center[centre]
行政主管 office manager
行政走私 administrative smuggling
行政组 administration division
行政组织理论 theories of bureaucracy
行走部分变速箱 traction gear box
行走部分传动系 propulsion gear
行走部分传动装置 propulsion drive
行走部分的传动皮带 traction drive belt
行走部分无级变速器操纵杆 traction variable-speed control
行走部件 traction member
行走传动 travel drive
行走传动离合器 traction clutch
行走的 travel(l)ing
行走底盘的司机室 <起重机> transport station cab;chassis cab
行走电动机 travel(l)ing motor
行走吊车 moving crane
行走方式 travel mode
行走方向的独立控制 independent direction control
行走横轴 horizontal traction shaft;horizontal travel(l)ing shaft
行走滑轮 travel(l)ing wheel
行走机构 bogie[bogey/ bogy];running mechanism;travel mechanism;undercarriage
行走机构制动器 travel(l)ing gear brake
行走机械脚的沉陷量 walking foot sinkage
行走驾驶台 transport station
行走架 running support
行走结构驾驶室 chassis cab
行走距离 travel(l)ed distance
行走履带 tread caterpillar
行走轮 road wheel;travel(l)ing wheel
行走轮轮面 tread of travel(l)ing wheel
行走轮驱动 traction drive
行走轮驱动的割捆机 bull wheel driven binder
行走轮驱动的喷粉机 traction duster
行走轮驱动的喷雾机 traction sprayer
行走轮驱动机构 ground drive mechanism
行走轮驱动式割草机 traction driven mower
行走轮摇动臂 wheel swing arm
行走轮轴 live axle
行走马达 tramming motor
行走设备 walking arrangement
行走式拉索脱模机 walking dragline stripper
行走式模板 traveling formwork
行走式破碎机 walking crusher
行走式起重机 walking crane
行走式砂处理机 sand kicker
行走式索斗挖掘机 walking dragline excavator
行走式索斗挖泥船 walking dredge(r)
行走式挖土机 walking excavator
行走式悬臂吊车 cantilever walking crane
行走式支架 walking prop
行走式自升式施工驳船 walking jack up
行走司机室 transport station
行走速度 speed of travel;travel(l)ing speed;walking speed
行走速率 speed of travel
行走系统 running gear;running system
行走线 line of travel;walking line
行走信号 walk signal
行走支架 undercarriage frame
行走装置 running gear;travel(l)ing gear;walking arrangement;walking gear
行走装置液压回路 travel circuit
行走阻力 running resistance;travel(l)ing resistance

形

形板 shaped plate
形板测验 form board test
形变 shape change
形变地震仪 deformation seismograph
形变二次式 quadric of deformation
形变法 deformation method
形变方程 deformation equation
形变分析 deformation analysis
形变功 work of deformation
形变回火 tempforming
形变结构 deformation structure
形变矩阵 deformation matrix
形变力矩 moment of deformation
形变模量 modulus of deformation;stress-deformation modulus;stress-strain modulus
形变能 deformation energy
形变能量守恒原理 theory of constant energy of deformation
形变能守恒理论 theory of constant energy of deformation
形变热处理(法) ausform(ing);marworking;thermomechanical treatment
形变热处理钢丝 ausformed steel wire
形变速率 rate of deformation
形变系数 coefficient of deformation
形变硬化 strain-hardening
形变支架 <膨胀岩中隧道衬砌用> yielding support
形参【计】formal parameter
形常数 <计算杆件系统的杆件变形时用> shape coefficient;shape constant
形成 build-up;forming
形成凹面 dishing
形成白浪 cockle
形成半径 formative radius
形成杯状凹陷 cupping
形成标准之前的 pre-standard
形成薄膜的 film-forming
形成操作 forming operation
形成层 cambium[复 cambi];formation【地】
形成层带 cambial zone
形成层环 cambium ring
形成层理【地】encrustation;encrustment
形成常数 formation constant
形成车辙的 rutted
形成齿轮 formative gear
形成串的 bobbed
形成大块 formation of lump
形成大理石纹路 marbling
形成的 formative
形成地址 calculated address;generated address
形成电弧 arcing
形成电压 formation voltage;formative voltage
形成电子对 pair creation
形成方法 method of formation
形成方式 generation type;mode of formation
形成蜂窝表面 honeycombing surface
形成港湾 embayment
形成拱状 arching
形成骨架 skeletonize
形成规则 formation rule;production rule
形成过程 establishment;forming process
形成和发展 formation and development
形成花边 <板材深冲时> earing
形成花纹 patterning
形成环路 looping in
形成环状沟槽的钻头 <由于金刚石损坏在端部> ringed out bit
形成回路 looping in
形成机理 mechanism of formation
形成脊状 ridge
形成价值的实体 value-creating substance
形成胶态 gum forming
形成胶体 gel forming
形成角度 angulation
形成角状地 cornerways;cornerwise
形成阶地 terracing
形成结节 tubercle formation
形成结晶 crystal formation
形成晶核 nucleate
形成晶核能力 nucleating power
形成矩形脉冲 squaring
形成颗粒 granulation
形成坑洞 pitting;potting
形成坑塘 dishing
形成空泡 vacuolate;vacuolation;vacuole formation;vacuolization
形成孔洞的型芯 <混凝土构件内> hole-forming core
形成喇叭状 splay
形成粒状 granulation
形成裂缝 crack growth;fissuration
形成裂缝弯矩 cracking moment
形成垄断 create a monopoly
形成率 formation rate;speed of formation
形成螺旋状细沟 <岩芯上> spiral grooving
形成码 generated code
形成面 forming fare
形成膜状的黏[粘]合剂制动器 membrane-forming type bond braker
形成蘑菇形断口 cupping
形成木纹表面 board-marked
形成泥包 <空气吹洗钻进时湿岩粉所形成的> bridge and ball
形成凝块 clustering
形成气孔 blistering
形成气泡 bubble formation;vesiculate;vesiculation
形成气穴 cavitation erosion
形成器 former
形成球状 ball up
形成区 formative region
形成渠道系统 canalization
形成群落的植物 gregarious plant
形成热 heat of formation
形成热裂缝 hot tear(ing)
形成十字形交叉 criss-cross
形成时间 formation time
形成时滞 formation time lag
形成熟料的放热过程 exothermic clinkering process
形成水堵 water blocking
形成水舌 cusping
形成水锥 cusping;water coning
形成速度 speed of formation
形成条纹 streaking
形成凸凹的条纹 raised stripe
形成土楔 wedging
形成外门的门栏 bar post
形成弯状 embayment
形成网眼状空隙 areolation
形成温度 formation temperature
形成纹理 veining
形成物质 formative substance
形成细颈 neck down
形成狭窄通道 laning
形成斜纹的双向翘曲 twilled double warp
形成压力 build the pressure
形成氧化皮 scale forming;scaling effect
形成窑皮能力 coating-forming capacity
形成因素 formation factor;formative factor
形成于山麓的 piedmont
形成原纤维 fibrillation
形成圆锥形 coning
形成蒸气的 vapo(u)rous;vapo(u)rific
形成支晶 Christmas tree formation
形成周边流幕 external flow vehicle
形成皱纹 gof(f)ering
形成珠状 beading

形成组织 formative tissue
形齿轮 generating gear
形点 form point
形点高 form-point height
形定义 shape definition
形而上学 metaphysics
形高 form height
形函数 shape function
形河曲 oxbowu
形核位置 nucleation site
形换热器 needle recuperator
形迹 evidence;feature;vestige
形迹滑距 trace slip
形迹滑距断层 trace-slip fault
形级 form class
形率 form quotient
形貌学 morphology
形面阻力 profile drag
形模 shape model
形模成型 contour forming
形实替换程序【计】thunk
形示信号 visual signal
形式 form(at);modality;mode shape;pattern
形式伴随算子 formally adjoint operator
形式逼近 formal approximation
形式变化 change of style
形式变换 formal transformation
形式变量 formal variable
形式变元 formal argument
形式标志 formal denotation; formal notation
形式表达法 formal representation
形式不可判定命题 formal undecidable proposition
形式参数 formal parameter
形式参数表 formal parameter list
形式参数部分 formal parameter part
形式参数调用 formal parameter call
形式操作 formal operations
形式产品 formal product
形式抄袭 imitation of style
形式程序验证【计】formal program verification
形式处理 formal treatment
形式次数 formal degree
形式代码 formal code
形式代数 formal algebra
形式的 formal
形式的概念 concept of form
形式的首项系数 formal leading coefficient
形式地址 formal address
形式定义 formal definition
形式对称的立面图 formal symmetrical elevation
形式发票 pro forma invoice
形式繁多的 multiformed
形式方法 formal approach; formal method
形式分析 formal parsing
形式分析算法 formal parsing algorithm
形式概念 formal notion;form concept
形式概型 formal scheme
形式更新 refashion
形式公理学 formal axiomatics
形式公司 corporate shell
形式构造 formal construction
形式规则 formal rule
形式化 formalization;formalize
形式化计算机 formalized computer
形式化计算机程序 formalized computer program(me)
形式化模型 formalized model
形式化算法 formalized arithmetic
形式环 formal ring
形式积分法 formal integration
形式记录 formal record
形式间可靠性系数 interform reliability coefficient
形式键 formal bond
形式解 formal solution
形式近似 formal approximation
形式决算表 pro forma statement
形式控制 formal control;format control
形式类 formal class
形式离散误差 formal discretization error
形式理论 formal theory
形式论 formalism
形式逻辑学 formal logic
形式码 form code
形式美 beauty in form
形式幂 symbolic power
形式幂级数 formal power series
形式幂级数环 formal power series ring;ring of formal power series
形式幂级数域 formal power series field
形式描述 formal description
形式模仿 imitation of style
形式模型 formal model
形式谱 formal spectrum
形式群 formal group
形式人口学 formal demography
形式上 pro forma
形式上的 modal
形式上的差别 phenomenon difference
形式上限 formal upper bound
形式设计说明 formal design specification
形式实参对应 formal-actual parameter correspondence
形式实域 formally real field
形式矢量 symbolic vector
形式说明 formal specification
形式说明词 formal declarer
形式算式 algorithmic form
形式体系 formalism
形式体系化 formalization
形式统一 unity of form
形式推导 formal deduction
形式推理 formal inference
形式系统 formal system
形式下界 formal lower-bound
形式下推自动机 formal pushdown automaton
形式效用 form utility
形式信号 form signal
形式行 formal row
形式性 formality
形式训练 formal discipline
形式有效性 formal validation
形式与空间概念 formal and spatial concept
形式语言 formal language
形式语义学 formal semantics
形式原因 formal cause
形式蕴涵 formal implication
形式证法 formal proof
形式证明 formal verification
形式重建 grimthorpe
形式主义 formalism
形势 position; state of affairs; status of affairs
形势规律 law of the situation
形势图 situation map
形势险恶的 squally
形数 figurate number
形似 similarity in appearance; take the form of
形似层理 apparent bedding
形似的 apparent;shaped
形似动物之花 animal flower
形似竖琴的物件 harp
形似性 paralogy
形态 conformation;feature;modality;shape
形态比 aspect ratio
形态变化 conformational change;morphodifferentiation;morphologic(al) change;morphologic(al) differentiation
形态变异 morphologic(al) change
形态变种 botanical variety
形态标志 morphologic(al) index
形态不美的 unshaded
形态糙度 form roughness
形态测定法 morphometry
形态测量参数 morphometric(al) parameter
形态测量学 morphometry
形态测量研究 morphometric(al) research
形态层组 form-set
形态差异 morphologic(al) differences
形态成因的 morphogen(et)ic
形态单元 <地形分析中的> morphologic(al) unit
形态调查(河流的) geomorphologic(al) survey(of river)
形态发生 morphogenesis
形态分化 morphodifferentiation;morphologic(al) differentiation
形态分类 morphologic(al) classification
形态分离法 form separation method
形态分离仪 form separator
形态分区 form and structure zoning
形态分析 morphologic(al) analysis
形态分析法 shape analysis
形态复杂程度综合指标 comprehensive index of morphologic(al) complexity
形态复杂的 complex morphologic(al)
形态感 form sense
形态构成 morphosis
形态观察 morphologic(al) observation
形态函数 <包括一个或多个不定位移参数的函数,在有限单元法中为求节点位移的函数> shape function;morphic function
形态很复杂的 extremely complex morphologic(al)
形态价值 form value
形态检查 inspection of shape
形态简单的 simple morphologic(al)
形态鉴定 morphologic(al) identification
形态较简单的 less simple morphologic(al)
形态结构 morphologic(al) structure;morphosis;morphostructure
形态矿物学 morphologic(al) mineralogy
形态量侧 morphometry
形态硫量 form sulphur content
形态论 gestaltism
形态面 form surface
形态视觉 form vision
形态特性 morphologic(al) property
形态特征 morphologic(al) character(istic)
形态天文学 morphologic(al) astronomy
形态稳定性 morphologic(al) stability
形态物理学 morphophysics
形态误差 morphologic(al) error
形态系数 shape factor
形态相关 morphologic(al) correlation
形态效用 form utility
形态心理学 gestalt psychology
形态形成 morphogenesis
形态形成的 morphogenetic
形态性质 morphologic(al) property
形态性状 morphologic(al) characters
形态学 morphology
形态学报 Journal of Morphology
形态学分类 morphologic(al) classification
形态学检查 morphologic(al) examination
形态学(上)的 morphologic(al)
形态研究法 morphologic(al) research
形态演变过程 morphologic(al) process
形态因素 morphology factor
形态运筹法 game theory
形态正常 normomorph
形态指标 form index
形态制图语言 morphologic(al) mapping language
形态属 form genus
形体的刚度 rigidity of modulus
形体风格 body style
形体感觉 stereognosis
形体规划 physical planning
形体排列 dimensional orientation
形同虚设的规定 dead letter
形位变体 allomorph
形位测量 surface relationship measurement
形位误差 form and position error
形稳性 dimensional stability
形稳性稳定电极 dimensionally stable anode; dimensionally stable electrode
形相 configuration
形象 profile
形象尺寸 size of image
形象符号 drawn symbol;pictorial symbol
形象工程 figurative engineering
形象构思的 plastically conceived
形象化 visualisation[visualization]
形象化报表 pictorial statement
形象化分类 classify in a figurative sense
形象化算法 visualization algorithm
形象进度 graphic progress
形象模型 iconic model;physical model(ling)
形象扭曲的镜子 shock mirror
形象失真的镜 shock mirror
形象图库 image library
形象显示 display present
形象艺术 figurative art
形象专柜 image counter
形心 center[centre](of) figure;center[centre](of) form;figure of form
形心距离 centroidal distance
形心频率 centroid(al) frequency
形心曲线 centroid;poid
形心凸轮 frog cam
形心轴 centroidal axis
形心主轴 principal axis of centroid
形质指数 quality index(number)
形装载 I-type loading
形状 appearance; configuration; figuration;form;shape
形状保持不变的 form-retentive
形状保持不性的 form-retentiveness
形状保持性 shape retention; shape-retentiveness
形状比 aspect ratio
形状编号 <钢筋等> code better
形状编码 shape coding
形状变量 configuration variable
形状不对称 configurational asymmetry
形状不规则的 out of shape
形状不规则金刚石 random-shaped stone
形状不同的 difform;diveriform
形状参数 form parameter;shape parameter

形状糙率 form roughness
形状测定器 shapometer
形状测量 shape measure
形状常数 shape constant
形状尺寸信息 shaped size information
形状代码 shape code
形状的概念 concept of shape
形状的恢复 recovery of form
形状的精确度 accuracy of shape
形状的真实性 trueness of shape
形状的准确性 accuracy of shape
形状分比 morphologic(al) differentiation
形状分解 shape decomposition
形状分配 shape distribution
形状分析 shape analysis
形状复杂矿体 complex-shape orebody
形状概念 shape concept
形状公差 form tolerances; tolerance of form
形状共振 dimensional resonance
形状规则的舱室 squared-off cargo space
形状函数 form function; shape function
形状函数矩阵 shape function matrix
形状号码 shape number
形状恢复 recovery of shape; shape recovery
形状畸变 shape distortion
形状计 shapometer
形状记号 symbolism of form
形状记忆合金 shape memory alloy
形状记忆合金机器人 shape memory robot
形状记忆聚合物材料 shape memory polymer material
形状记忆陶瓷 shape memory ceramics
形状记忆效应 shape memory effect
形状检定 shape detection
形状检验装置 shape check device
形状矩阵 shape matrix
形状描述 shape description
形状模型 shape model
形状扭曲 distorted shape
形状偏差 deviation from exact shape
形状偏离的 off the form
形状识别 shape recognition
形状双折射 structural birefringence
形状水头损失 form loss of head
形状损失 form loss
形状损失系数 form-loss coefficient
形状弹性 elasticity of form
形状弹性散射 shape-elastic scattering
形状特性 geometric(al) characteristic
形状调整 shape control
形状稳定性 shape stability; stability of shape
形状系数 displacement coefficient; shape coefficient; shape factor; form factor <共振试件的>
形状相同的盆地 equant-shaped basin
形状效果 shape effect
形状效应 form effect
形状信号 form signal
形状选矿(法) separation by shape
形状寻迹器 form tracer
形状要素 form element
形状因数 configuration factor; shape factor; form factor
形状因素 <某种断面梁的断裂模量和标准断面梁的断裂模量的比值> form factor; shape factor
形状因子 form factor; shape factor
形状应力因数 form stress factor
形状与空间的概念 concept of form and space
形状振荡 shape oscillation
形状正确的 true-to-shape
形状知觉 form perception
形状知觉常性 form constance
形状指数 shape index
形状阻力 form drag; form resistance; shape resistance
形状组构 shape fabric
形状作用 shape effect
形阻 form drag

型

型板 lagging; match board; mold-board; mo(u)ld(ing) board; shaping plate; stencil (plate); template

型板布置 template layout
型板回送机 return conveyor for caulphater
型板犁 mo(u)ld board plough
型板弯曲试验 guided bend test
型板镶块 pattern plate insert
型板印刷 stenciling
型板印刷机 wire printer
型板造型 plate mo(u)lding
型变 morphotropy
型变异 form variation
型别 type
型冰 sized ice
型材 bar section; bar stock; patterned lumber; profile; profiling; sectional bar; sectional material; section bar; section material; section wood; shaped material; shape wood
型材钢丝 section wire
型材辊式矫正机 gagger
型材剪切机 angle shears
型材矫直机 section-straightening machine
型材水道 waterway
型材水路 waterway
型材铜 copper section
型材弯曲机 section-bending machine
型材轧辊 section roll
型材轧机 bar and shape mill; bar mill; section rolling mill
型吃水 draft mo(u)lded; form draft; geometric(al) draft; mo(u)lded draft; mo(u)lded draught
型船 type ship
型锤 dollie; setter; setting hammer; swaging hammer
型钉 sprig
型锻 stamp forging; swage; swaged forging; swaging
型锻机 swage engine; swaging machine
型锻模 swage die
型锻芯轴 swaging mandrel
型缝 groove part
型干舷 form freeboard; geometric(al) freeboard
型钢 bar iron; bar shape; bar steel; billet bar; commercial steel; fashioned iron; figured bar iron; formed steel; iron section; lightweight section; merchant steel; profile(d) iron; profile(d) steel; rolled(steel) section; rolled(steel) shape; roller steel section; section(al) iron; section(al) steel; section(al) bar; shape(d) steel; steel bar; steel profile; steel section; steel shape; structural steel; swage
型钢板 pressed sheet metal; shaped steel plate
型钢标准断面 standard steel section
型钢标准截面 standard steel section
型钢材 structural section
型钢大梁 compound girder
型钢底座 formed steel base
型钢断面 rolled profile; steel section
型钢分选机组 bar rolled stock sorting unit
型钢钢筋 rigid armo(u)ring
型钢格栅 rolled steel joist(beam)
型钢构架 structure-steel frame[framing]
型钢桁架 bar girder
型钢混成柱 combination column
型钢混凝土 steel reinforced concrete
型钢混凝土房屋 concrete-steel building
型钢混凝土混合结构 composite construction in steel and concrete
型钢混凝土结构 section steel concrete structure
型钢剪切机 joist shears
型钢建筑物 formed-steel construction
型钢矫正压力机 bull press
型钢矫直机 section straightener; shape straightener
型钢校直压力机 straightening press for steel sections
型钢结构 formed steel construction
型钢截面 steel section
型钢截面积 area of structural steel
型钢框架 rolled section frame; section frame
型钢连接 bar link
型钢梁 rolled steel joist(beam)
型钢梁桥 rolled steel beam bridge; shape steel beam bridge
型钢配筋 steel shape reinforcement; section reinforcement
型钢喷砂车间 profile-blasting shop
型钢数量表 quantity of section steel; quantity of shaped steel
型钢水池 sectional tank
型钢轧槽 sectional groove
型钢轧辊 shaped roll
型钢轧机 mill for rolling sections; section mill
型钢轧制 rolling of sectional iron; rolling of sections
型钢转向架 pressed steel bogie
型构件 shaped block
型号 model; model number; type (number); version
型号标志 mark
型号表示法 model designation
型号的多样化 diversity of types and size
型号改变 model change
型号名称 model designation
型号铭牌 nomenclature plate
型号说明 demonstration of the type
型号系列 range of models
型盒 flask
型盒压机 dental flask press
型基线 mo(u)ld base line
型架经纬仪 transit square
型检测管 <检测管系是否漏气设施> water-gauge
型件 shaped block
型交叉 type crossing
型焦 form-coke
型卷线圈 former-wound coil
型壳 shell
型宽 breadth-mo(u)lded; mo(u)lded beam; mo(u)lded breadth(of vessel); mo(u)lded width
型梁 shaped beam
型料 mo(u)lding mixture
型铆 snap riveting
型煤 section mo(u)ld
型面 mo(u)lding surface
型面阻力 form drag
型模 joint mo(u)ld; mo(u)lded form; pattern die; section mo(u)ld; swage die
型模钢 mo(u)ld steel
型模靠模机床 mo(u)ld and die copying machine
型模块 swage block
型排水量 mo(u)lded displacement
型刨 bead plane; profiler
型坯 parison
型坯工 parison maker
型坯膨胀 parison swell
型片 matrix[复 matrixes/matrices] (band)
型腔 dies cavity; die space; impression; mo(u)ld cavity
型腔的充填 cavity fill
型腔模 cavity die
型砂 casting sand; foundry sand; mo(u)lding sand; plasticine; sand mix
型砂斑点 embedded grit
型砂比重 sand density
型砂标准抗弯试块 standard transverse test core
型砂标准拉力试块 standard tensile-test core
型砂成分 mo(u)lding sand composition
型砂处理 sand conditioning
型砂挡肋 sand edge
型砂捣锤 pegging rammer
型砂捣实器 sand rammer
型砂翻新 rebonding of mo(u)lding sand
型砂分选装置 sand separator
型砂覆膜 sand coating; sand precoating
型砂回性 sand maturing
型砂混合料 mo(u)lding sand mixture
型砂挤压机 squeezer
型砂加入剂 sand addition
型砂减泥 weakening of mo(u)lding sand
型砂紧实度 sand compaction
型砂浸湿 tempering
型砂空隙度 sand porosity
型砂控制 sand control
型砂流动性 sand flowability
型砂密度 sand density
型砂模型 plasticine model
型砂黏[粘]合剂 sand binder
型砂黏[粘]结剂 mo(u)lding sand binder
型砂配方 sand formulation
型砂配制 mo(u)lding-sand preparation
型砂配置机 sand conditioner
型砂膨胀 sand expansion
型砂强度 sand bond; sand strength
型砂强度试验机 sand strength testing machine
型砂湿度 mo(u)lding sand moisture
型砂试验 sand test
型砂试验室 sand laboratory
型砂试样 sand test specimen
型砂熟化 sand maturing
型砂水分 sand humidity; sand moisture content
型砂水分控制器 mo(u)ldability controller
型砂水泥 sandy casting cement
型砂调匀 sand maturing
型砂透气性 sand permeability
型砂稳定剂 sand stabilizer
型砂性能 mo(u)lding-sand property
型砂运送 sand conveyance
型砂再生 sand reclamation
型砂制备 sand conditioning; sand preparation
型砂质量调整 sand control
型深 depth mo(u)lded; mo(u)ld(ed) depth(of vessel)

型式 mode; model; mode shape; portrait; type; type of structure
型式标准化 standardization of types
型式的概念 concept of style
型式典型化 standardization of types
型式规格化 standardization of types
型式技术 type technology
型式批准 type approval
型式评价 type evaluation
型式试验 type approval test; type test
型式选择及设计 select type and design
型数比 aspect ratio
型态溃散性 core breakdown property
型套 mo(u)ld jacket; slip jacket
型特殊性 type specificity
型条边 trim-edge
型铁 dolly; fashioned iron; figured iron; mo(u)ld pig iron; profile(d) iron; rolled steel member; section(al) iron; shape(d) iron; swage; swedge
型铁架 section iron chassis
型铁索端配件 swaged cable-end-fittings
型铁楔形锚 swage-wedge anchor
型铁楔型锚碇 swage-wedge anchorage
型头铆钉 set-head rivet
型线 frame line; frame station; mo(u)lded line
型线孔口 shaped orifice
型线图 drawing line diagram; mo(u)lded line diagram
型线坐标 profile coordinate
型箱 casting box; mo(u)lding flask; mo(u)ld(ing) box
型箱用钩头钉 hook foundry nail
型芯 core; cove box bit; mandrel
型芯板 core board; core plate
型芯表面硬度测定 core testing
型芯材料 core material
型芯车 core oven car
型芯车床 core turning lathe
型芯撑 chaplet
型芯撑钉 chaplet nail
型芯成型工具 core former
型芯吹干机 core blowing machine
型芯吹砂机 core blower
型芯错位 mismatch in core
型芯顶部 crown of core
型芯定位座 core indicator; core locator; core marker; core register
型芯飞边 joint flash
型芯干燥机 core dryer
型芯骨架 core skeleton
型芯盒 core box
型芯盒刨 core box plane
型芯烘干 core baking
型芯烘干炉 core-baking oven
型芯烘炉 core drier; core drying furnace
型芯混合料 core mixture
型芯机 core former; core machine
型芯记号 core mark; top print

型芯夹 core setting jig
型芯夹具 core jig
型芯架 core rack
型芯胶 core gum
型芯开边 core parting line
型芯烤炉 core oven
型芯孔 cored hole
型芯溃散性 core collapsibility
型芯拉出器 core puller
型芯内浇口 core gate
型芯耐火性 core refractiveness
型芯黏[粘]合法 core slurring; slurring
型芯黏[粘]合膏 core paste
型芯黏[粘]合剂 corduroy binder; core adhesive; core binder; core gum
型芯黏[粘]合液 core slurry
型芯排气塞 core vent(ing)
型芯披缝 core joint flash
型芯片 cake core
型芯破碎 core breakdown
型芯染料 core dye
型芯砂 core sand; foundry core sand
型芯绳 core rope
型芯铁 core bar; core iron
型芯铁的 iron core
型芯铁机 corduroy iron
型芯铁条整直机 core rod straightening machine
型芯头 core bedding frame
型芯透气性 cored density
型芯涂料 core coating; core cover; core dressing; core wash
型芯涂料浆 core paint
型芯修整 core dressing
型芯样板 core template
型芯移出 core pull
型芯移出程序 core pull sequence
型芯硬化 core hardening
型芯造型 core sand mo(u)lding
型芯掌 sand-bearing
型芯制造机 core-making machine
型芯轴 corduroy barrel; core rod; spindle; core barrel
型芯钻杆整直机 core rod straightening machine
型芯钻机 core cutting machine
型芯座 core print
型穴 push-up
型造品 fictile
型砧 swage block; swage die
型砧座 swage block stand
型值表 offset table; table of offset
型砖 encallow; feather-edge brick; shaped brick
型状稳定性 form stability
型状阻力 form drag
型组件 shaped block
型钻 swage

醒目标志 visible marker

醒目程度 boldness
醒目广告 display advertisement; eye-catcher
醒目色 assertive colo(u)r; attractive colo(u)r
醒目条款 conspicuous clause
醒目效应 catch eye effect
醒目性 highlighting

杏红色 almond pink

杏黄 imperial yellow
杏黄色 apricot
杏绿 almond green
杏仁 almond; stone of apricots
杏仁桉 gaint gum
杏仁孔 amygdule
杏仁体 amygdale; amygdaloidal; amygdaloidal body
杏仁细碧岩 dunstone
杏仁岩 amygdaloid; amygdaloidal lava
杏仁油 almond oil; apricot kernel oil; peach kernel oil; persic oil
杏仁状的 amygdaloidal
杏仁状构造 amygdaloidal structure
杏仁状辉绿岩 dunstone
杏仁状结构 amygdaloidal texture
杏仁状结构辉绿岩 amygdaloidal diabase
杏仁状玄武岩 amygdaloidal basalt
杏仁子 amygdule
杏树 apricot; apricot tree
杏圆形装饰板 mandorla

姓 surname

姓名地址录 directory
姓名索引 name index

幸存法 survival method; survivorship method

幸存概率 probability of survival
幸存者 near miss; survival; survivor
幸免于难 near miss

性别 sexuality

性别年龄结构 sex-age structure
性能 behavio(u)r; characteristic; functional performance; performance; worth
性能保证 performance guarantee; quality assurance
性能保证书 performance bond
性能标志板 duty plate
性能标准 performance criterion; quality specification; quality standard; performance standard
性能标准化法 standardizing performance method
性能表 capability chart; performance chart; performance table; table of standard performance; work sheet
性能不匀 spread in performance
性能参数 performance parameter; performance statistics
性能测定 performance measurement
性能测量 performance measure
性能测试 performance measurement
性能测验 aptitude test
性能差 poor performance
性能尺寸 characteristic dimension
性能单位 unit of performance
性能的一致性 consistence of performance
性能等级 <SHRP 沥青分级标准> performance grade
性能范围 functional range; performance range
性能分类 property sort
性能分析 performance analysis
性能负载 performance load
性能更好的 performance-plus
性能估定 performance evaluation
性能估计 assessment of performance; performance estimation; performance prediction
性能估价 performance evaluation
性能故障 functional failure; malfunction; performance bug
性能故障概率 probability of failure performance; probability of performance failure
性能规定 performance provision
性能规格 operating specification; operational specification; performance specification; specification
性能函数 performance function
性能合格试验 performance qualification test
性能荷载关系 performance loading relationship
性能换算 performance reduction
性能恢复试验 recovery test
性能集中分析 qualitative investigation
性能计算 performance calculation; performance computation
性能记载 record of performance
性能技术标准 performance specification
性能价格比 cost performance; performance cost ratio; price performance
性能监察 performance monitoring
性能监视器 performance monitor
性能检查 performance checking
性能检验 performance examination
性能检验周期 performance proof cycle
性能鉴定试验 performance evaluation test
性能考核 performance test
性能可靠性与价格 performance-cost-dependability
性能控制传感器 quality control pick-up
性能良好 high performance; high performance characteristic
性能良好的 well-behaved
性能描写 characterization
性能评估 performance evaluation
性能评价 performance assessment; performance evaluation; rating of merits
性能评价模拟 performance evaluation simulation
性能评价要素 performance evaluation element
性能曲线 characteristic curve; performance chart; performance curve
性能上的缺陷 defect in property
性能设计 functional design; performance design
性能失灵 unperformance
性能试验 aptitude test; characteristic test; functional test; functioning test; performance test; performance trial; service; service test
性能试验模型 service test model
性能试验台 performance test stand
性能手册 performance manual
性能术语 performance term
性能数 performance number
性能数据 mathematicals; performance data; performance figure
性能数字 performance figure
性能说明书 performance specification; statement of performance
性能损失 performance loss
性能特点 functional characteristics
性能特性 performance characteristic
性能特征 characteristic features; performance characteristic
性能提高 performance build-up
性能调试 performance adjuster; performance adjustment
性能统计 performance statistics
性能图 performance chart
性能系谱 performance pedigree
性能系数 characterization factor; coefficient of performance; performance coefficient
性能下降 degradation of performance; performance reduction
性能下降法 performance reduction method
性能相抵触货 cargo of contradictory nature
性能相互关系 performance correlation
性能研究 performance study
性能验证测试 compliance test
性能要求 performance requirement
性能因数 figure of merit; performance factor
性能因素 performance factor
性能优化 performance optimization
性能优良 high performance; satisfactory performance
性能与成本估价 performance and cost evaluation

性能预测 performance prediction
性能阈值点 performance point
性能指标 figure index;figure of merit;index of performance;measure of performance;performance criterion;performance figure;performance guarantee;performance index;performance level;performance measure;quality characteristic
性能指标计 performance index meter
性能指数 figure index;index of performance;measure of performance;performance figure;performance index;performance level;performance measure
性能准则 performance criterion
性能最佳性 availability;reliability;serviceability
性态数 condition number
性质 belongings;character;nature;property
性质变化 qualitative change
性质不明露头 unknown
性质上的 qualitative
性质未确定故障 indeterminate fault
性状 behavio(u)r
性状模型 behavio(u)ral model

凶年 bad year

凶器 lethal tools;lethal weapon;weapon
凶险 bad risk
凶险风 travesier

兄弟节点 brother node

兄弟任务表 brother task table

匈牙利柔性路面设计法 Hungary flexible pavement design method

匈牙利式建筑 Hungarian architecture;Hungary architecture

汹涛 confused sea;heavy sea;phenomenon sea

汹涛海面 very heady seaway
汹涌<风或浪的> bluster;popple
汹涌的波浪<河道中的> boiling water
汹涌的波涛 boiling waves
汹涌的长浪 heavy swell
汹涌的涌浪 heavy swell
汹涌海面 heavy sea;peaky sea;surging sea
汹涌海洋 chopping sea;choppy sea
汹涌急流 flashy flow
汹涌碎波 surging breaker

胸板 breast board;breast plate

胸部 breast
胸部压重物 breast lead weight
胸高<树木的> breast height;chestheight
胸高护墙面饰 breast(ing) lining
胸高护墙线 window back
胸(高直)径<木材> breast-height diameter;chest-height diameter
胸(高周)围<木材> breast-height girth
胸挂电话机 breast telephone
胸挂送话器 breast plate microphone;breast transmitter;chest transmitter
胸辊 breast roll
胸径<树木离地 1.37 米处的直径> diameter breast high
胸口顶板 breast plate
胸腔千斤顶 breasting jack
胸墙 altar;annular dam;breast (wall);breast work;crown wall;dwarf wall;face wall;fascia wall;head wall;parapet (wall);parapet railing;parapet wall;protected wall;protecting wall;skimmer wall;trash baffle;front wall<桥台的>;cap wall;cope wall
胸墙标志 parapet signs
胸墙高度 breast height
胸墙设计 design of crest structure
胸墙式桥台 breast abutment
胸墙索 breasting line
胸墙下的人字闸门 truncated miter [mitre] gate
胸墙下闸门 truncated gate
胸墙作业 breast work
胸射式水轮机 breast-shot water wheel;half breast water wheel
胸式桥台 breast abutment
胸水轮 breast water wheel
胸索 breast line
胸台 terminal pedestal
胸像 bust;sculptured bust
胸像台 terminal pedestal;terminus[复 termini]
胸像台座的上部 vagina
胸像柱 terminal figure;terminal statue;term
胸压手摇钻 breast borer;breast brace;breast drill;breast plate borer
胸压钻孔器 breast brace
胸钻 breast drill
胸座 terminal pedestal

雄厚资金 abundant fund

雄黄 arsenic sulphate;realgar;red arsenic glass;red orpiment;sandarac(h)
雄榫 cog(-tongue);cut joint;tenon;tongue;lewis<铁制的>
雄榫肩部受压表面 abutment cheek
雄榫接合 cogged joint;cogging;cross cocking
雄榫嵌接 coak scarf joint
雄榫嵌装 cogging
雄榫上的斜肩 gain
雄榫斜口接合 coak scarf joint;cog scarf joint
雄榫装入 cogging
雄伟的窗 monumental window
雄甾烷 andtostane

熊本水俣病<日本> Kumamoto-Minamata disease

熊葱<一种植物> ramson
熊果酸 ursolic acid
熊阱式坝 bear-trap dam
熊阱式筏道 bear-trap drift-chute
熊阱式放筏槽 bear-trap drift-chute
熊阱式堰 roof weir
熊阱式闸门 bear-trap gate
熊阱式闸槛 bear-trap sill
熊阱式自动堰 automatic bear-trap weir
熊阱堰 bear trap;bear-trap weir;roof weir
熊阱堰闸门 roof weir gate
熊阱闸门 bear-trap gate
熊狸 binturong
熊猫 panda
熊猫斑纹(人行)横道 panda crossing
熊猫车<涂成熊猫斑纹的警察巡逻车,英国> panda car
熊猫瓶<瓷器名> vase with panda design
熊市 bear market
熊形灯 bear-shaped lamp
熊熊燃烧 blaze

休班船员 standing off crew

休病假 on sick leave
休伯特<一种家用机器人> Hubot
休产日 pause days
休风 damping down
休复的 refractary
休耕地 fallow;fallow field;fallow ground;fallow land;idle land
休耕时间 fallow time
休耕水田 fallow paddy field
休耕与不休耕地 fallowed and nonfallowed land
休工培训日 day release
休会 adjourn
休假 vacate;vacation
休假地 vacation land
休假费 leave cost
休假工资 holiday pay;pay for holidays
休假津贴 vacation allowance
休假列车 leave train
休假旅行 pleasure trip
休假日 day off;off-day
休假中心 leisure center[centre]
休静 resting phase
休克<俗称> shock
休利特电动卸料机 Hulett electric(al) unloader
休林格方程 Heulinger equation
休仑系【地】Huronian system
休曼曲线<粒径分配> Schumann curve
休眠 dormancy
休眠的 dormant
休眠地热系统 dormant geothermal system
休眠后期 later rest
休眠火山 dormant volcano
休眠季节灌溉 dormant-season irrigation
休眠间歇泉 dormant geyser
休眠期 dormant period;dormant season;dormant stage;period of dormancy;quiescent stage;repose period
休眠期喷射 dormant spraying
休眠期施用 dormant application
休眠期修剪 dormant pruning
休眠泉 dormant spring
休眠型 resting form
休眠状态 resting state
休谟管<离心混凝土管> Hume pipe
休谟管道 Hume duct
休谟黄金流动平衡作用 Hume gold flow equilibrating mechanism
休憩室(俗称) den
休斯敦港<美> Port Houston
休士电报机 Hughes' telegraph
休斯铅基轴承合金 Hughes' metal
休特顿公式<计算斜面上风压的公式> Hutton's formula
休乌特型装载机 Huwood loader
休乌特振荡剥煤机 Huwood slicer
休息长凳 lay bench;easy bench<井场的>
休息场地 halting place
休息场所 rest area;nymphaeum<古罗马的>
休息车 lounge
休息车棚<火车的> lounge car
休息池<鱼道上的> resting pool
休息处 breathing place;breathing space;couch;nest;resting place
休息地方 camp
休息减让时间 relaxation allowance
休息平台 half-space;landing elevator;stair platform;footpace<楼梯的>
休息区 rest(ing) area
休息日 day off;rest-day
休息设施 leisure facility
休息时间 respite;rest period;time of rest
休息室 angle pavilion;canteen;day room;den;drawing room;green room;lobby (area);lounge;parlo(u)r;recreation room;recroom;rest room;retirement room;retiring room;waiting room;withdrawing room;schola<古希腊体育场的>
休息室入口 entrance lobby
休息所 minch house;pull-up
休息厅 lobby;lounge;lounge hall;arrival lounge<机场的>
休息娱乐区 recreation area
休息与问讯处标志 rest and information area sign
休闲<土地> fallowness
休闲草场 spelled pasture
休闲草地 amenity grass land
休闲船 leisure craft;recreational craft
休闲船闸 recreational lock
休闲稻田 fallow paddy field
休闲的 fallow
休闲地 fallow;fallow field;fallow land;idle land
休闲地作物 fallow crop
休闲海滩 recreational beach
休闲季节 off-season
休闲及居住用的海滩 beach for recreation and residential purpose
休闲轮牧 rest-rotation grazing
休闲设备 dormant equipment
休闲设施 recovery facility
休闲水田 fallow paddy field
休闲田 fallow field
休闲艇 recreational craft;pleasure boat;pleasure boat
休闲土地 fallow land
休闲土壤 fallow soil
休闲制 fallowing
休闲制度 fallow system
休闲转数 rest-rotation grazing
休闲状态 fallowness
休闲作物 fallow crop
休养 convalescence;recreate;recreation;recuperation
休养城市 resort city;resort town
休养城镇 resort town
休养的 recreational
休养地 health resort;sanitarium
休养环境 recreational environment
休养假 convalescent leave
休养林业 recreation forestry

休养区 leisure area;recreational (focal) area;recreational precinct
休养区公园 health resort park
休养区旅馆 health resort hotel
休养生息 bring recovery;recuperate and multiply;rehabilitation;rest and build up strength;revitalize
休养胜地 resort place
休养室 recovery room
休养所 convalescent home;health spa;rest house;rest room;sanatorium
休养用水 recreational use of water
休养娱乐区 recreation(al) area
休养娱乐用地 recreation(al) area
休养院 convalescent home;rest home

休业 business of vacation; lockout; shut-down; suspension of business
休业工厂 idle plant
休战 truce
休战协定 truce
休整 rest and recuperation
休整地区 rehabilitation area
休整计划 rehabilitation program(me)
休止 diapause; dormancy
休止的 dormant
休止辅助触头 normally closed interlock
休止角 angle of repose; angle of rest; critical slope; repose angle; slope of repose
休止平衡 equilibrium at rest
休止坡度 slope of repose
休止期 resting phase
休止时期 idle period
休止状态 inactivity; state of rest

修版 retouching

修版笔 colo(u)red pencil; crayon pencil
修版机 retouching machine
修版汽笔 aerograph
修版台 light table
修版颜料 retouch colo(u)rs
修版液 film opaque
修杯机 cup trimming machine
修边 deburr(ing); edging; trimming
修边铲斗 profile bucket
修边车床 slicing lathe
修边冲模 ripper die
修边刀 edge knife; rib; trimming knife
修边刀具 knife tool
修边的木板 edge-shot board
修边工 mender; ribbon-mender
修边工具 deburring tool; edger; edge tool
修边和角 flat edge and bevel
修边机 deburring machine; edger; edge sponging machine; flat edge trimmer; hedge trimmer; straight-line edger; trimming knife machine
修边夹具 shooting board
修边剪刀 edging shears
修边胶 finish cement; finish compound
修边锯 jointer saw; skinner saw; straight-line edger
修边镘刀 <用于新鲜混凝土的> edging trowel
修边模 comb die; flush trim die; pinch-off die; trimming die
修边模板 margin templet
修边抹子 margin trowel
修边木板 edge shot
修边刨 joining plane; jointing plane; shooting plane
修边刨床 edge planer
修边炮孔 popshot
修边炮眼 rib hole
修边器 edger; edging trowel
修边饰器 bead sleeker
修边压力机 trimming press
修边眼 trim hole
修边余量 shaving allowance; trimming allowance
修边准备 edge preparation
修表螺旋 jeweler's screw
修补 burling; fettle; make good; mend(ing); patch(ing repair); repairing good; repatching (repairing); revamp; tinker
修补板料 patch plate
修补材料 patching material
修补袋 repaired mended bag
修补单元 patching unit
修补刀 fettling knife
修补的 remediable; remedial
修补的坑洞 patching holing
修补段区 repair patch
修补费用 renewal expenses
修补膏 gelling agent; healing agent
修补工 patcher; tinker
修补工厂 patching plant
修补工程 making good; patching operation; patch work; remanent work; remedial work
修补工具 repair outfit
修补工修补 tinker
修补工作 making good; patching operation; patch(ing) work; remanent work; remedial work
修补工作专家 patchwork specialist
修补过的木材 pieced timber
修补焊 repairing welding
修补灰浆 patching plaster
修补混合料 patching mix(ture)
修补混合物 patching compound
修补混凝土 patching concrete
修补机 patching machine
修补技术 patching technique
修补剂 healant; healing agent; patch compound
修补嫁接 repair grafting; saving grafting
修补胶 repair sheets; repair gum
修补坑洞 patching hole
修补沥青路面小队 bituminous-patching crew
修补料 patching material
修补裂缝 mud daub
修补裂纹 mended crack
修补漏缝 remedy a leak
修补炉壁 fettling
修补炉衬 daub; fettle; fettling
修补路面坑槽 patch repair
修补轮胎 retread
修补配合料 patching compound; patching mix(ture)
修补漆 refinishing paint; touch-up paint
修补汽车轮胎 retread
修补缺陷 remedying defect
修补乳化液 patching emulsion
修补乳胶 patch(ing) emulsion
修补砂浆 patch mortar
修补设备 patching plant; repair outfit
修补石膏 patch plaster
修补受损的索具 mend the service
修补术 neoplasty
修补损坏边坡面 refilling of damaged slope surface
修补物 prosthesis; restorer
修补性 toolability
修补用板片 panel patch
修补用瓷漆 finish enamel
修补用粉刷 repair plaster
修补用灰泥 repair mortar
修补用混凝土 repair concrete
修补用泥浆 mending slip
修补用砂浆 patching mortar; repair mortar
修补用小锅炉 patch kettle
修补用组合物 patching composition
修补元件 patching unit
修补着色 inked
修补作业 patch work
修布 burling
修布钳 burling-irons
修测 maintain; revision survey
修测图 revised map
修测图板 deletion model
修车厂 car shop
修车场 repair yard
修车车间 car shop
修车灯 inspection lamp
修车工 fixer
修车工具 engine tool
修车股道 repair track
修车坑 drop pit; inspection pit
修车库 car repair(ing) shed
修车棚 car repair(ing) shed; maintenance shed; repair shed
修车起重机 garage jack
修车器具 garage equipment
修车器械 garage equipment
修车千斤顶 garage jack
修车台位 position for repairing a rolling stock
修车台位长度 position length for repairing car
修车线【铁】car repair track; repair siding; repair track; vehicle repair track
修车用灯 inspection lamp
修车用起重器 garage jack
修成直角 square up
修程 class of repairs; course of repair; repair schedule
修程范围 scope of repair course; scope of repair programmes
修齿 fit
修船班 boat repair gang
修船舶位 repair berth
修船拆船厂 repair and dismantling yard
修船厂 broadside railway dry dock; repairing dock; repairing yard; repair shipyard; ship-repair yard
修船厂辅助小船码头 yard craft pier
修船场 broadside railway dry dock
修船处 careenage
修船船台 repairing slipway; dry-docking and repairing berth; repair berth
修船船位 <船台的> repair station
修船船坞 dry dock; graving (dry) dock; repairing dock
修船的格子船台 gridiron
修船方责任 ship repairer's liability
修船分队 boat repair gang
修船干船坞 repairing drydock; graving drydock
修船港 repair port
修船港池 careening basin; ship repair basin
修船工业 ship repair industry
修船滑道 hauling-up slip; marine railroad; marine railway; slipway
修船基地 repairing base
修船架 gridiron
修船架滑道 ship slipway
修船码头 repair(ing) quay; ship-repair pier; ship-repair quay
修船木工 shipwright
修船区 ship-repair yard
修船湿坞 slip dock
修船台 railway dry dock
修船位 repairing berth
修船坞 slip dock; basin drydock; graving dock
修船坞池 repairing basin
修船小组 boat repair group
修船突堤码头 repair pier
修窗吊篮 window cradle machine
修锉叶轮 impeller shaping
修挡土墙 making retaining wall
修道士 monk
修道院 abbey; hermitage; monkery; religious house; monastery <男的>; priory <小的>
修道院花园或公墓 paradise
修道院会客室 locutorium; locutory
修道院教堂 abbey church; monastic church
修道院墓地 cloister cemetery
修道院世俗斋堂 frater for lay brethren
修道院所在地 abbeystead
修订 recension; re-edit; revisal; revise
修订版 correcting plate; expurgated edition; new print; revised edition; revision
修订报价 change in offer
修订本 adapted edition; redaction; revised edition; revision
修订标准版 revised standard version
修订单体法 revised simplex method
修订的麦卡利地震烈度表 modified Mercalli scale
修订的最优分配 revised optimum allocation
修订规范 code revision
修订后的拨款 revised appropriation
修订计划 remake a plan; reschedule
修订列车时刻表 modify a train schedule; timetable revision
修订列车运行图 train diagram revision
修订说明 revision note
修订小组 amendments group
修订者 expurgator
修订最低限度标准模型 revised minimum standard model
修房木料 house bote
修房下料 house bote
修缝凿 dumb iron; jerry iron; opening iron
修复 damage control; instauration; make [making] good; recondition(ing); rehabilitate; rehabilitation; reinstatement; renovate; renovation; reparation; restoration; restore; retirement
修复保养时间 corrective maintenance time
修复标签 recovery slip
修复车辆 reconditioned vehicle
修复臭氧层空洞 repair the ozone hole
修复措施 reclamation activities
修复的 reparative
修复的零件 reconditioned parts
修复方案 regulation scheme; rehabilitation schematization
修复费用 rehabilitation charges; rehabilitation expenses
修复工程 regulation works; rehabilitation works; remedial works
修复工作 reclamation works; recovering; rehabilitation work; reinstatement work; restoration work
修复合成 repair synthesis
修复后性能 post-repair behavio(u)r
修复计划 rehabilitation program(me)
修复技术 regulation technique; remediation technology; renewal technique
修复路面 redeck(ing); resurfacing
修复履带板筋 regrouser
修复轮胎 rebuild a tyre
修复平均时间 mean time to repair
修复漆面 inpaint
修复前等候时间 waiting time to repair
修复桥面 redecking
修复湿地 restore wetland
修复时间 maintenance downtime; repair time
修复受损的图形 recover damaged drawing
修复术 prosthesis; prothesis
修复水毁设施 repair the facility
修复说明 rebuilding instruction
修复体 dummy
修复条款 replacement clause
修复图 restoration drawing

修复性 rebuildability;recoverability
修复学 prosthetics
修复叶片 reblading
修复用零件 repair parts
修复用陶瓷 prosthetic ceramics
修复有效度测试 rehabilitation effectiveness testing
修复原状 repair to original shape
修复-运营-移交 rehabilitation-operate-transfer
修复者 renovator
修复指数 maintainability index
修复组 prosthetic group
修复作用 repair
修改 amend(ment);correction;modify;revision;rework;treatment;update;updating
修改布置 revisions in layouts
修改操作 retouching operation
修改程序 update routine; updating program(me);updating routine
修改带 change tape;transaction tape
修改单纯形法 revised simplex method
修改到期日承兑 qualified acceptance as to time
修改的 updated
修改等时法测试 modified isochronal test
修改地址 modification address;modified address
修改反馈 modifying feedback
修改方案 modified scheme; updated version
修改方式 alter mode
修改分配法 modified distribution method
修改干砌(法) modified loose lay
修改工程 alternation;alternation work
修改合同 amendment of contract; modification of contract; revise a contract;revision of contract
修改后的法规 <美> revised statutes
修改后的设计 revised design
修改计划 revision of plan
修改记录图 record drawings
修改寄存器 modifier register
修改量 index word;modifier
修改(了的)设计 revised design
修改流 revised flow
修改命令 modifier command
修改平均数 modified mean
修改前馈 modifying feedforward
修改区域 modifier area
修改权 power of amendment
修改设计 alter a design;change in design
修改设计的 derated
修改设计建议 engineering change proposal
修改事项 particulars of amendment
修改手续费 amendment commission
修改水泵曲线 modified curve
修改松铺(法) modified loose lay
修改通知单 amendment advice;change order;variation order
修改通知(书) amendment advice; change order;variation order
修改图纸 revision on drawing
修改位 modify bit
修改文件 amendment file;updating of files
修改箱位单 restowage list;restowage plan
修改信号 resignal(l)ing
修改信号计划 resignal(l)ing scheme
修改信用证 alter the letter of credit; amend the letter of credit
修改信用证金额 amend the amount of the letter of credit
修改信用证申请书 application for amendment of letter of credit
修改信用证条件 amend the terms of the letter of credit
修改信用证条款 amend the terms of the letter of credit
修改型 modified form
修改性承兑 qualified acceptance
修改液 correcting fluid; liquid opaque;opaquing fluid
修改意见 comment on revision; suggested amendment
修改者 modifier
修改值 modified value
修改资料档案 updating of files
修改作业计划 rescheduling
修干枝 dry pruning
修高枝 high pruning
修刮 combining
修管工具 pipe tool
修管箍 pipe repair clamp
修管夹箍 patch clamp
修管器 pipe twist
修光 dress smooth;sleek(ing)
修光阀面 valve refacing
修光工具 sleeker; slick; smoother; striking tool
修光器 slicker
修光石面的工具 comb
修光钻头 slick bit
修过边的金属丝网布 gauze wire cloth with cropped edges
修壕沟防护 entrenchment
修好 fettle
修合 backfit
修机车坑 engine pit
修机器坑 engine pit
修剪 cropping; cutting back; dress; lop;prune;scissoring;trim(ming)
修剪的丛篱 trimmed hedge
修剪高度 cutting range;pruning range
修剪工具 clippers;pruning implement
修剪工人 cropper
修剪灌木的 topiary
修剪机 cropper;hedge trimmer;trimmer
修剪锯 pruning saw;trimming saw
修剪枯枝 dry pruning
修剪苗 stripling
修剪器 clipping machine
修剪人 trimmer
修剪融资 hair cut finance
修剪树冠 pollarding
修剪树木 lop;pruning trees
修剪树枝 amputation; disbranch; pruning trees
修剪桃树 pruning peaches
修剪下来的小树枝 lop
修剪向里生长的枝条 cutting out inward growing branches
修剪艺术 topiary;topiary art
修剪整形 <树木的> topiary work
修建 rehabilitation
修建第二线 double tracking;laying of second track;track doubling
修建第三线 laying of third track
修建队 maintenance and construction brigade
修建路基单价 grading price
修建路基造价 grading price
修建人员 builder
修建设计 construction design
修建烟囱、尖塔等高空作业工人 steeplejack
修建者 restorer
修建中 under construction
修角工具 corner tool
修井 well conditioning; well repair; well workover
修井队 clean-out crew
修井工作 clean-out job
修井机 service rig;workover rig
修井机械 clean-out machine(ry)
修井井架 workover derrick
修井平台 workover platform
修井设备 repair drill hole assembly; servicing unit;well pulling machine
修井钻杆 clean-out string
修井钻柱 clean-out string
修旧利废 repair and utilize old or discarded things
修旧利废材料 reclaimed material
修锯 filing
修锯锉 saw file
修锯工 saw doctor
修锯机 gummer
修均法 graduation
修竣车数 number of repaired wagon
修篱笆用的钉子 fence nail
修理 chipping; docking【船】; fixing; fix up; make[making] good; make-up; overhaul; patching; put in repair; recondition; redress; refit; renovate; reparation; revamp; servicing
修理包 maintenance kit; maintenance package; maintenance packaging
修理保险 repair risk insurance
修理备件的申请单 deadline requisition
修理标签 bad order card;repair label
修理标准值 repair standard value
修理驳 repair barge;workshop barge
修理部分 repair part
修理部件 repair parts
修理厂 backshop; depot repair; maintenance depot; maintenance shop; rehabilitation shop; repair depot; repair factory; repair shop; salvage shop;service shop
修理场 repair depot
修理车 breakdown van; machine shop wagon; maintenance car; mobile repair shop; recovery vehicle; repair truck; repair wagon; shop truck; shop wagon; tender; tool wagon; window van;workshop wagon
修理车间 backshop; machine assembly department; maintenance shop; repair shop; repair workshop; salvage shop; service shop; jobbing shop <工地现场的>
修理车辆 repair wagon
修理成本 cost of repairs;repairing cost
修理程序 procedure of repairing
修理尺寸 repair(ing) size
修理船 factory vessel; floating workshop;repair ship
修理窗扇 window trim
修理单 repair list;repair tag
修理的 reparative
修理店 fix-it shop; mendery; repair shop
修理队 maintenance party;repair crew; repair team
修理阀座 reseating
修理范围 scope of repair(ing) course; scope of repair(ing) programmes
修理方案 repair forecast
修理费均衡准备 repairs equalization reserves
修理费(用) cost of repairs; overhaul charges; repair charges; repair(ing) cost; repair(ing) expenses
修理分间 repair room
修理服务台 repair clerk's desk
修理服务行业 repair and service trades
修理复杂系数 repair complex coefficient
修理港(口) repair port;port of repair
修理工厂 repair plant;salvage shop
修理工场 repair shop
修理工程 repairing and maintaining works;repair works
修理工程车 breakdown tender; machine shop wagon; mobile shop; mobile workshop; repair truck; repair wagon; shop wagon; tool wagon;workshop wagon
修理工段 repair section
修理工具 maintenance tool; repair outfit; repair tool; service equipment
修理工具包 maintenance kit; repair kit
修理工(人) repairer; repairman; fettler; fitter; mender; serviceman; donkey doctor <俚语>
修理工艺 repairing technology
修理工艺过程 technical process of repair
修理工组 repair team
修理工作 remedial work; repairing; repair(ing) work; repair services; workover
修理工作船 repair ship; mending vessel
修理工作队 black gang
修理工作量定额 repair work norm
修理工作通知单 repair work order; repair production order
修理管线前先堵塞管口 cramming
修理管线区域 firing-line
修理规程 field manual
修理焊接容器中的缺陷 repair defects in welded vessel
修理和保养 repairing and keeping
修理和改良 repairs and betterment
修理和更换 repairs and replacement
修理和技术保养时间 repair and servicing time
修理和维护 repair and maintenance
修理后试验 repair test
修理基地 depot repair; repair base; repair depot
修理基金 repair funds
修理基金银行存款 bank deposit of major repair fund
修理计划 repair forecast; repair program(me); repair schedule
修理夹 leak clamp
修理价格 repair cost
修理间 rehabilitation shop; repair bay; repairing room; repair work; service shop; workshop
修理间堆场 workshop storage area
修理间隔平均时间 mean time before failure
修理间隔期 inter-repair time; repair interval
修理检验 survey of repair
修理可能性 repairability
修理孔 access for repair
修理口 access for repair
修理拉杆 repair rod
修理劳动量定额 quota for repair labor
修理类别 class of repairs
修理列车 maintenance train
修理零件存货 inventory of repair parts
修理螺纹 rethread(ing)
修理码头 repair jetty; repair quay
修理木支架 retimber
修理棚 repair shed; car repair(ing) shed <车的>; dock hangar <船的>; maintenance hangar <飞机的>
修理棚厂 dock hangar
修理平均时间 mean time to repair
修理平台 repair platform

X

修理期间用的替换转向架 replacement bogie used during repairs
修理期限 time limit of repair
修理勤务 maintenance service
修理情况 maintenance status
修理人员 repairman
修理任务 repair assignment
修理任务单 repair order
修理设备 repair facility; repairing equipment; repair installation; workover equipment
修理设施 repair facility
修理时间 downtime; out-of-service time; repair time
修理室 repair room
修理手册 repair manual
修理守则 repair guide
修理说明书 repair guide
修理所 repair shop
修理台 repair bench; repair platform; repair stand
修理镗气缸 rebore
修理梯 ladder for repairing
修理填单 < 路外车 > billing repair card
修理通知单 repair order
修理图 repair drawing
修理屋面用防水塑性石棉合成物 hermetex
修理线 repair track
修理项目 items of repair job; repairing item
修理小组 maintenance section
修理行业的技术人员 maintenance technician
修理须知 servicing hint
修理延迟时间 repair delay time
修理业 repair services
修理业务 maintenance service
修理用备件 repair parts
修理用部件 service parts
修理用工具 repair outfit
修理用航天器 repair space vehicle
修理用机床 repair machinery
修理用具 repair outfit; service set
修理用具箱 repair box
修理用零件 repair parts; service parts
修理用配件 repair parts
修理用乳剂 repair emulsion
修理用乳胶 repair emulsion
修理用设备 repair equipment; repair facility; repair installation; repair outfit
修理用丝锥 patching tap
修理用整套器具 repair kit
修理预测 repair forecast
修理预算 repairing budget
修理凿 fettling chisel
修理闸门 guard gate
修理站 breakdown service; service center [centre]; service station; servicing center [centre]; servicing depot
修理账目 repairing account
修理指南 repair guide
修理指南说明书 repair instruction
修理中 under repair
修理中的机器 machine under repairs
修理中机车 locomotive under repair
修理周期 cycle of repairs; repair cycle
修理周期结构 repair cycle structure
修理专业化 specialization of repairs
修理装置 repair facility
修理准备 provision for repairs; reserve for repairs
修理准备时间 administrative time
修理组 maintenance team
修理钻头 dress(a) bit
修理钻头用吊架 bit dressing crane
修理作业 reconditioning work; remedial operation
修炉 fettle; furnace repair; patching
修炉衬喷枪 patching gun
修炉浆料 patching mix(ture)
修炉口 cleaning door
修炉搪料 patch
修路 road mending; road repair
修路边沟 ditching of road
修路灯号 road beacon (for highway construction)
修路工 road surfaceman
修路工程 road building; road making
修路工人 road-man
修路绕道 traffic diversion
修轮胎工具包 tire repair kit
修毛刺 shaving
修面机 finisher
修面錾 paring chisel
修描 retouch
修描过的照片 retouch
修模车间 mo(u)ld-repair shop
修模锉 riffler file
修磨 coping; retipping
修磨横刃 chisel edge thinning; web thinning
修磨机 sharpening machine
修磨钎杆 drill sharpening
修呢 burling
修女唱诗班 nun's choir
修刨 shaping
修配 overhaul(ing)
修配玻璃工人 repairing glazier
修配厂 maintenance depot; repair plant; repair shop
修配车 mobile workshop
修配车间 auxiliary shop; repair and spare parts workshop; repair shop; machine shop
修配车间设备 equipment for maintenance shop
修配船坞厂 repair shipyard shop
修配的设备 reconditioning equipment
修配服务 repair services
修配工 fitter; repairman
修配工厂 repair workshop
修配工组 repair group
修配间 repair workshop
修配业市场 after market
修配站 maintenance depot
修坯 fettle; fettling; repaired biscuit; trim(ming)
修坯承座 chuck; chum
修坯刀具 fettling knife
修坯机 fettling machine; finishing machine; peeler; trimming machine
修坯毛刷 edging brush
修坯台 scouring table
修平 flattening; level(l)ing; slicing; smoothing
修平板 level(l)ing beam; screed board
修平尺 level(l)ing beam
修平刀 trowel
修平地面 screed floor
修平法检验 test of smoothness
修平机 trimmer
修平滤波器 roof filter
修平器 flattening hammer
修平曲线图 smoothed curve chart
修平用长镘刀 skipfloat
修坡 gradation; grading; sloping; slope finishing
修坡机 siding machine
修坡作业 bank work
修钎 retapering
修钎车间 drill repair shop; drill sharpening shop
修钎工 bit dresser; bit setter; sharpener
修钎机 bit dresser; bit grinding machine; bit sharpener; dressing machine; dressor; drill sharpener; sharpening machine; jackmill < 凿岩机的 >
修钎间 redressing shop
修钎炉 jack furnace
修桥梁石块 bridge stone
修切边缘 scarf
修切山嘴 trimmed spur
修缮 fettle; make good; refit; betterment
修缮部 maintenance department
修缮成本 repair cost
修缮房产的贷款保险 property improvement loan insurance
修缮费 maintenance charges
修缮费账 repair and maintenance account
修缮工程 betterment works; repair work; restoration work
修缮工作 renovation work
修缮和扩建 betterment and extension
修缮经费 betterment; betterment cost; renovation cost; repairing cost; repairing expenses
修缮期间保险 repair risk insurance
修缮通知单 betterment order
修缮用锤 fettling hammer
修缮用品 operating and maintenance supplies
修缮用物料 operating and maintenance supplies
修缮准备 reserve for repairs
修士医院 infirmary for lay brethren
修饰 adorn; ceiling trim; dressing; embellish(ment); finishing; garnish; modify; picked dressing; retouch(ing)
修饰斑点 spotting-in
修饰边缘工具 edge tool
修饰表面法 method of facing
修饰材料 setting stuff
修饰层 finishing coat
修饰打光 polished dressing
修饰刀 casing knife
修饰丁(头)砖 clipped header
修饰工 patcher
修饰工具 dresser
修饰工作 finishing work
修饰过的材料 dressed stuff
修饰过的隧道 trimmed tunnel
修饰过的橡木家具 limed oak
修饰过的小室 dressing cab(in)
修饰过梁 clipped lintel
修饰剂 finishing agent
修饰抹子 finishing trowel
修饰平整的 smooth-finished
修饰隧道 trimmed tunnel
修饰台 retouching stand
修饰涂层 finish coat
修饰涂料 grout
修饰性处理 modifying processing
修饰因子 modifier; modifying factor
修饰阴角的镘板 inside-angle tool
修饰用布料 trimming cloth
修饰用油漆 dressing paint
修饰用组成物 dressing composition
修饰照相 retouched photograph
修饰字 index word
修饰作用 modification
修树工作 topiary work
修树剪 secateurs
修树枝用的长柄刀 long hafted billhook
修水龙带用套夹 hose mender
修锁工人 locksmith
修梯田机 terracer
修通段 section of restored traffic
修桶工人 cooper
修涂参考图 blueprinting plate; cyan printing plate
修土样器 soil lathe
修网线刻刀 lining tool
修圩 impound
修屋面用步级板 duck board
修屋面用临时步级板 duck board
修下残枝 trimming
修下树枝堆 brash
修线扑火 parallel fire suppression
修像框 retouching frame
修削 dubbing; tease
修形砌块 former block
修型 finish the mo(u)ld
修型工具 mo(u)ld mending tool
修型镘刀 sleek; slick(er)
修型小勺 spoon slicker
修样器 soil lathe
修圆 cavetto
修圆的 rounded-off
修圆角 filled corner
修圆角工具 rounding tool
修圆镘刀 round sleeker
修圆转角 truncated corner
修缘锯 edge-trimming saw
修缘刨 edge trimmer; edge-trimming plane
修匀 smoothing
修匀常数 smoothing constant
修匀的经验公式 empiric(al) formula in smoothing
修匀的理论公式 theoretic(al) formula in smoothing
修匀法 smoothing method
修匀公式 smoothing formula
修匀过程 smoothing procedure; smoothing process
修匀回归分析 smooth regression analysis
修匀理论动态数列 smoothed theoretical dynamic series
修匀频数曲线 smoothed frequency curve
修匀曲线 smoothed curve
修匀时序数据 smoothing time-series data
修匀效果 effect of smoothing
修匀值 smoothed value
修凿 boasted work; chisel(l)ing
修造船厂 combination ship yard; ship repairing and building yard
修造船滑道 drawdock; slipway < 英 >
修造船脚手架 scaffolding
修造船设施 drydocking facility
修造船水工建筑物 hydraulic structure for shipyard
修整 conditioning; cropping; decorate; decoration; dressing; fairing; fettling【冶】; finishing; finish work; making good; reclamation; reconditioning; rectification; regrade; rounding-off; scalping; searing; shaping; shaving; touch (ing) up; trimming; truing; working up; workover
修整岸线 rectification of bankline
修整板 finishing board
修整爆破 trimming blast
修整边缘 bannering
修整标桩 finishing stake
修整表面 refacing; regrading skin; shaving
修整材料 trimming material
修整层 surface layer
修整车间 finishing room
修整成斜面 taper trimming
修整程序 finishing procedure
修整冲孔 drifting
修整冲模压力机 die spotting press
修整锤 bush hammer; dressing hammer; facing hammer; roughening tool

修整刀具 finishing knife
修整道路凹凸 dribbing
修整道路坑凹 dribbing
修整的挤压锭 scalped extrusion ingot
修整的屋顶层房间 finished attic
修整锭面机 deseaming machine
修整阀座 reseat
修整缝 clip joint
修整工 fabricator;finisher
修整工段 conditioning department;dressing room
修整工具 cleaning means;dresser;dressing tool;resurface;trimming tool;finishing tool
修整工具用金刚石(颗粒) diamond grains for metal bond dressing tool
修整工作 finishing operation;finishing process;finishing work;reinstatement work
修整沟槽平地机 trench grader
修整刮板 finishing screed
修整规板 finishing board
修整过的边坡 rounded slope
修整过的地面高度 finished ground level
修整过的楼面 finished floor
修整过的坡度 finished grade
修整好的面层 finished grade;finish grade
修整好的木材 dressed lumber
修整混凝土路面用的台架 float bridge
修整机 burring machine;dresser;dressing machine;finishing machine;scalper;trimmer;trimming machine;finisher
修整机框架 <路面> finisher frame
修整机料斗 finisher hopper
修整剪切机 trimming shears
修整角 angle of trim(ming)
修整接合 clip joint
修整接合砌砖层 clip course
修整锯 trim(mer) saw
修整锯齿 gulleting
修整均值 <舍去一些边值后得到的> trimmed mean
修整孔洞 trim hole
修整跨 scalping bay
修整宽度 graded width
修整类型 styles of dressing
修整料 dressing material
修整轮 freeing wheel
修整镘板 finishing float
修整铆钉梢 gardening up;tomahawking
修整铆钉梢的铁锤 tomahawk
修整模 restriking die
修整坯体边缘 towing
修整漆面污斑 spot finishing
修整器 dresser [dressor];dressing tool;finisher;reseater;trimmer
修整钎子 bit dressing
修整切削 fair cutting;fair raking cutting
修整砂轮 trimming wheel;trueing of grinding wheel
修整砂轮机 emery wheel dresser
修整石板用凿刀 sax
修整时间 tooling time
修整手续 finishing procedure
修整树态 topiary work
修整树形 topiary work
修整树形的 topiary
修整数据 trimming data
修整速度 trimming speed
修整统计量 trimmed estimator
修整圬工表面 regrating
修整误差 rounding error
修整系数 adjustment factor
修整斜坡 ramping
修整压力机 trimming press
修整样板 <混凝土表面> lute
修整用车床 trimming lathe
修整用金钢刀架 truing diamond holder
修整用金钢石刀 truing diamond
修整用抹子 finishing trowel
修整预算 revised estimate
修整圆盘 fettling disk
修整錾 fettling chisel
修整指令 round-off order
修整铸件 dressing
修整铸件锤 chasing hammer
修整装置 trimming device;truing device
修整钻头 bit dressing;dressing bit;retip a bit
修正 adjust;amend;meliorate;readjusting;readjustment;reduction;reformation;revisal;update
修正案 amendment
修正贝塞耳函数 modified Bessel function
修正标尺 deflection scale
修正标准 amended standard
修正表 calibration card;correction card;correction chart;correction data
修正补码 modified complement
修正布置 revision in layouts
修正草案 amended draft
修正(测定骨料磨耗量的)道氏试验 modified Dorry test
修正测量 revision survey
修正承载比 modified bearing ratio
修正齿 corrected tooth
修正齿轮 corrected gear
修正处理 corrective treatment
修正的 updated
修正的爱因斯坦法 modified Einstein method;modified Einstein procedure
修正的布格异常 modified Bouguer anomaly
修正的传热系数 corrected heat transfer coefficient
修正的单纯方法 modified simplex method;revised simplex method
修正的二项分布 modified binomial distribution
修正的固定模型变形 modified version of fixed model
修正的贯击数 modified blow count
修正的虎克定律 modified Hooker's law
修正的回归自变量 modified regressor
修正的季节指数 modified seasonal index
修正的净填方量 net corrected fill
修正的马蹄型 modified horseshoe
修正的麦卡利地震烈度表 modified Mercalli intensity scale;modified Mercalli scale
修正的美国各州公路工作者协会击实试验 modified AASHTO compaction test
修正的目标函数 modified objective function
修正的频数表 modified frequency table
修正的平截头棱锥体法 modified prismoidal method
修正的儒略日 modified Julian day
修正的三角形成果表 revision point list
修正的设计 modified design;revised design
修正的水油比曲线法 corrected water-oil ratio curve method
修正的太平洋型岸线 modified Pacific type coastline
修正的提前高度 predicted and corrected altitude
修正的现金余额方程式 revised cash balance equation
修正的仰角 adjusted elevation
修正的正规方程组 modified system of normal equations
修正的正态分布 modified normal distribution
修正的指数曲线 modified exponential curve
修正等距抽样 modified systematic sampling
修正地震烈度表卒 revised earthquake intensity scale
修正地震影响系数 modified design seismic coefficient
修正地震影响系数法 modified design seismic coefficient method
修正电流 correcting current
修正短路棒 modified short-bond
修正对数成败优势比 modified log odds
修正多尔利氏试验 <测定集料磨耗量的> modified Dorry test
修正方案 draft amendment;readjustment plan
修正方位角 corrected azimuth
修正分配法 modified distribution method
修正分析算法 revised parsing algorithm
修正弗兰德利希模型 modified Freundlich model
修正浮动汇率制 modified system of flexible exchange rate
修正盖勒金矢量 modified Galerkin vector
修正概率 correction probability
修正刚度指标 reduced rigidity index
修正高度 corrected altitude
修正格式 amended form
修正公式 modified formula
修正功率 corrected output
修正光楔 correction wedge
修正汉克尔函数 modified Hankel functions
修正合同 amendment of contract;revised contract
修正后尺寸 neat size
修正后的地基承载力特征值 revised characteristic value of subgrade bearing capacity
修正后的价格指数 corrected price indices
修正后的控制界限 modified control limit
修正(后的)预算 revised budget
修正后价格 revised price
修正(后)深度 <井的> corrected depth
修正换算后的结果 modified result
修正回路 corrective loop
修正击实试验 modified Proctor test
修正机构 correction mechanism
修正计划 adjustment plan(ning);amended plan
修正计算 correction computation
修正记录 amendment record;change record
修正加州承载比 modified California bearing ratio
修正角 correction angle
修正劲度 modified stiffness
修正均方连续差 modified mean square successive difference
修正均数 adjusted mean;corrected mean
修正空化数 modified cavitation number
修正空速 calibrated air speed
修正孔 trimming hole
修正孔用丝锥 screw chaser
修正雷诺数 modified Reynolds number
修正力矩 corrective moment
修正量 amendment;correction
修正流速 modified velocity
修正率 adjusted rate;corrected rate;revised rate
修正麦氏地震烈度表 modified Mercalli intensity scale
修正孟塞尔表色制 Munsell renovation system
修正面 balancing plane;truing face
修正偏心受压法 modified eccentric compression method
修正平均值 modified mean
修正坡度 <实际坡度加曲线阻力折减的相应坡度> compensated grade
修正普罗克托击实试验 <又称修正普氏击实试验,一般指重锤击实法> modified Proctor's compaction test
修正普罗克托密实度 modified Proctor density
修正普罗克托密度试验法 modified Proctor density method
修正器 corrector
修正曲线 correction curve;fair curve
修正权责发生制 modified accrual basis
修正缺陷 corrective pitting
修正柔度矩阵 modified flexibility matrix
修正深度 modified depth
修正时间 correction time
修正识别算法 revised recognizer
修正数 correction number
修正数据 correction data;processed information;specified data
修正水头历时曲线 modified head duration curve
修正死亡率 adjusted death rate;revised death-rate
修正速度 modified velocity
修正条款 amending clause;bisque clause
修正条文 amended reading
修正条约 revised treaty
修正条约的协定 amending agreement
修正统一责任制 modified uniform liability system
修正图 amended plan;correction map;revised drawing
修正椭圆公式 modified ellipse equation
修正网络 corrective net
修正文件 revised documents
修正系数 auxiliary value;coefficient of correction;compensation factor;correction coefficient;correction factor;corrective coefficient;modified coefficient
修正线 modified line
修正限 expurgated bound
修正项 correction term
修正协定 amending agreement
修正信息 update information
修正性能等级 modified performance level
修正液 liquid paper;modified liquid
修正因数 correction factor;correctness factor;factor of correction
修正因素 corrective moment
修正印鉴 amendment to authorized signatures
修正有效温度 corrected effective temperature
修正账单 adjustment of account
修正折射指数 modified index of re-

fraction; modified refractive index; refractive modulus
修正值 amendment; corrected value; correction; correction value; modified value
修正指标 correction index
修正指令 revision directive
修正指数 modified index; revised index
修正质量 modified mass
修正逐次超松弛法 modified successive over relaxation
修正转速 corrected speed
修正总吨 corrected gross tonnage
修正总重合度 modified contact ratio
修正阻抗继电器 modified impedance relay
修枝 prune; trimming
修枝刀 pruning knife
修枝工具 pruning implement
修枝工作 topiary work
修枝剪 pruning scissors; pruning shears; tree shears
修枝剪刀 pruner; tree shears
修枝锯 pruning saw; trim(ming) saw
修枝手锯 pruning saw
修枝者 pruner
修筑程序 construction sequence
修筑池塘 pond building
修筑底脚 underpinning
修筑隔墙 damming
修筑灌水田埂 ridging for irrigation
修筑路基 subgrading
修筑路肩 shouldering
修筑路面 road surface
修筑面层 surface coating
修筑梯田 terracing
修筑田埂 ridging
修筑围堰 cofferdamming
修砖錾 brick chisel
修准钻 finishing drill
修琢 dressing; knobbing; regrate
修琢方石 quarry dressed
修琢过的 dressed
修琢后的尺寸 dressed size
修琢加工过的石方工程 milled stonework
修琢加工过的石块 milled natural stone
修琢加工过的石料 milled stonework
修琢面 trimmed surface
修琢石 dressed stone; dressing stone
修琢石板的尖锤 sax
修琢石块 <采石场> kibble
修琢石块表面 dabbing
修琢铁 break iron
修钻头机 bit sharpener

羞

羞明 photophobia

髹

髹漆 lacquering

髹漆底涂 paint primer

朽

朽病 timber rot

朽材 powder post
朽坏的 decayed; rotten
朽坏区 decay area
朽节 dead knot; decayed knot; unsound knot
朽木 dead wood; decayed knot; dote; frowy; mo(u)ldered wood; punk; rotten wood; touchwood
朽木节 rotten knot
朽壤 rotten earth
朽松木烷 fichtelite
朽心木材 druxey; druxiness

绣

绣花地毯 embroidered carpet

绣花丝线 filoscelle
绣球花 Chinese snowball
绣像 tapestry portrait

袖

袖阀灌浆管 sleeved grout pipe

袖阀套管法灌浆 tube-a-manchette grouting
袖阀套管灌浆法 <法国 Soletanche 公司开发的一种砂砾石层灌浆方法> Soletanche method
袖口 cuff
袖套状浸润 cuffing infiltration
袖网浮子 wing shape float
袖珍 rest-pocket
袖珍版 miniature edition; pocket edition; vest-pocket edition
袖珍报警器 pocket alerter
袖珍报警器充电器 pocket alerter charger
袖珍报警器发送器 pocket alerter transmitter
袖珍本 diamond edition; miniature book; pocket book; pocket edition
袖珍变压器 ouncer transformer
袖珍步谈机 handle talkie
袖珍测厚仪 miniature thickness ga(u)ge
袖珍测试计 pocket meter
袖珍潮流图集 pocket tidal stream atlas
袖珍词典 pocket dictionary
袖珍磁力仪 pocket magnetometer
袖珍带盘 pocket-size reel
袖珍的 compact; midget; pocket(-size); vest-pocket <照片、书等的>
袖珍灯 pocket flashlight; pocket lamp
袖珍地图 pocket map
袖珍地图集 pocket atlas; small atlas
袖珍电动槽刨机 centec
袖珍电动刻纹器 centec
袖珍电离室 pocket chamber
袖珍电流计 battery ga(u)ge
袖珍电视 microtelevision
袖珍电视机 minitelevision; minitube
袖珍发报机 <救生艇属具> Gibson girl; pocket transmitter
袖珍放大镜 hand viewer; pocket magnifier
袖珍风速标 pocket anemometer
袖珍伏特计 pocket voltmeter
袖珍个人计算机 wallet personnel computer
袖珍含气量测定仪 chace air meter
袖珍活塞 pocket piston
袖珍计数机 pocket calculator
袖珍计数器 junior counter; vest-pocket calculator
袖珍计算机 pocket calculator; pocket computer
袖珍计算器 vest-pocket calculator
袖珍剂量计 pocket chamber
袖珍监测仪 pocket monitor
袖珍接收机 pocket receiver
袖珍经纬仪 pocket-sized transit; pocket theodolite; pocket transit
袖珍立体镜 pocket stereo view; pocket stereoscope
袖珍量测立体镜 pocket measuring stereoscope
袖珍六分仪 pocket sextant; box sextant
袖珍录音机 pocket recorder
袖珍罗盘(仪) pocket compass; Bouguer compass
袖珍气压计 pocket barograph
袖珍摄像头 compact camera
袖珍示波器 pocketscope
袖珍式表 pocket watch
袖珍式存储器 pocket memory
袖珍式电子闪光装置 pocket-size electronic flash unit
袖珍式断路器 miniature circuit breaker
袖珍式放射线测量仪 pocket chamber
袖珍式贯入器 pocket penetrometer
袖珍式贯入仪 pocket penetrometer
袖珍式混凝土贯入仪 pocket concrete penetrometer
袖珍式计算器 pocket calculator
袖珍式精确测距定位仪 Syledis
袖珍式喷蒸汽设备 <贴墙纸用> steam-stripping appliance
袖珍式收音机 pocket radio set
袖珍式无线电设备 pocket radio
袖珍式(肖氏)硬度计 hard Scope
袖珍式仪表 pocket meter
袖珍式阅读器 hand viewer
袖珍手册 pocket-sized manual
袖珍双金属温度计 pocket bimetal thermometer
袖珍水平仪 pocket level
袖珍体视镜 pocket stereoscope
袖珍天文表 pocket chronometer
袖珍土壤试验仪 pocket-sized soil testing instrument
袖珍万用电表 pocket multimeter
袖珍温度计 pocket thermometer
袖珍无线电传呼机 pocket bell
袖珍显微镜 pocket microscope
袖珍型 compact type
袖珍蓄电池 pocket accumulator
袖珍仪表 pocket instrument
袖珍凿子 pocket chisel
袖珍照相机 pocket camera; vest-pocket camera
袖砖 sleeve brick
袖子 sleeve

锈

锈疤 incrustation

锈斑 aerugo; flash rusting; pitting(due to cavitation); rust mark; rust spot; rust staining; rusty spot; splodge; splotch; splotches of rust; tarnish; rust stain <混凝土表面的>
锈斑病 rubigo
锈孢锈菌属 <拉> Aecidium; Roestelia
锈孢子 aeci(di)ospore
锈层 rust layer; rusty scale
锈点 rust spot
锈浮包 rust blister
锈痕 iron mo(u)ld
锈和瘪损 rusty and dented
锈和漏损 rusty and leaky
锈红色 rust red
锈化处理 sull coat(ing)
锈坏 eat away
锈迹 rust creep; rust print; rusty stain
锈接 rust joint
锈结 rust joint
锈结核 nodule of rust
锈金 rusty gold
锈疽 incrustation
锈坑 fretting
锈块 scale crust
锈鳞铲除 removal of scale
锈瘤 tubercle
锈膜 film of rust; rust film; tarnish film
锈疱 rust blister
锈皮 debris of oxide; mill scale; scale
锈皮层 scale layer
锈蠕变 rust creep
锈色 patina; rust
锈色的 rubiginous
锈蚀 corrosion; rust; rust-eaten; rustiness; rusting(corrosion); staining
锈蚀产生的易损性 corrosion embrittlement
锈蚀程度 degree of rusting
锈蚀的 corrodent; corrosive; iron stained
锈蚀等级 rust grade scale; rusting grade
锈蚀度 rusting degree
锈蚀过程 rusting process
锈蚀厚度 corrosion allowance
锈蚀机理 mechanism corrosion
锈蚀控制 corrosion control
锈蚀鳞片 rust scale
锈蚀麻点 pitting corrosion
锈蚀疲劳试验 corrosion-fatigue test
锈蚀容许量 corrosion allowance
锈蚀深度 corrosion penetration
锈蚀试验 corrosion test(ing); tarnishing test
锈蚀速率 corrosion rate; rusting rate
锈蚀透 rust through
锈蚀危险 risk of rusting
锈蚀小坑 pitting corrosion
锈蚀质量 rusty quality
锈蚀状态 rusty state
锈蚀作用 action of rust
锈霜 rust bloom
锈水效应 red water effect
锈损险 risk of rust
锈桶 drums rusty
锈纹 rust streak
锈污点 iron stain
锈屑 residual rust
锈徐变 rust creep
锈氧化皮 scale
锈印 rust mark
锈渍 rust stain
锈棕色的 rusty-brown

溴

溴氨酸废水 bromoamine acid wastewater

溴处理 bromation
溴处理 bromination; bromization
溴代丙烯酸 bromoacrylic acid
溴的 bromic
溴碘比值系数 bromine-iodine ratio
溴碘水 bromine-iodine water
溴碘锶水 bromine-iodine-strontium water
溴仿 bromoform; methenyl tribromide; tribromomethane
溴氟碳化合物 bromofluorocarbons
溴化铵 bromide of ammonium
溴化白金纸 platinum bromide paper
溴化催化剂 bromination catalyst
溴化镉 cadmium bromide
溴化铬 chromium bromide; chromous bromide
溴化硅 silicon bromide
溴化剂 bromizating agent
溴化钾 potassium bromide
溴化金 auric bromide; gold bromide
溴化锂 bromide of lithium; lithium bromide
溴化锂水 lithium bromide water
溴化锂水溶液 lithium bromide water solution
溴化锂吸收式直燃机 direct-fired absorption LiBr refrigerating machine
溴化锂吸收式制冷 lithium bromide absorption refrigeration
溴化锂吸收式制冷机 lithium bromide absorption-type refrigerating machine
溴化锂制冷机 lithium bromide chiller; lithium bromide refrigerating

machine
溴化钠 bromide of sodium
溴化氢 hydrogen bromide
溴化氰 bromine cyanide
溴化容量法 bromination volumetric(al) method
溴化试验 bromine test
溴化铊 thallium bromide
溴化钽 tantalum bromide
溴化铽 terbium bromide
溴化铜 copper bromide
溴化物 bromide
溴化橡胶 brominated rubber
溴化锌 zinc bromate
溴化亚铂 platinous bromide
溴化亚锡 stannous bromide
溴化乙烯 ethylene bromide;ethylene dibromide
溴化乙酯 ethyl bromide
溴化银 silver bromide
溴化银乳剂 bromide emulsion
溴化银像纸 bromide paper;bromide print
溴化银纸照片 bromide print
溴化荧光素钠盐 eosin
溴化油 bromine oil
溴化纸像片 photographic(al) bromide picture;photographic(al) bromide print
溴化作用 bromation
溴甲苯 toluene bromide
溴甲酚蓝 bromocresol blue
溴甲酚绿 bromocresol green
溴甲酚紫 bromocresol purple
溴矿床 bromine deposit
溴锂水 bromine-lithium water
溴量法 bromometry
溴硫磷 bromophos
溴氯酚蓝 bromchlorphenol blue
溴氯甲烷 bromochloromethane
溴氯乙腈 bromochloroacetonitrile
溴氯乙酸盐 bromochloroacetate
溴硼锂水 bromine-boron-lithium water
溴硼水 bromine-boron water
溴水氧化法 bromine water oxidation method
溴素纸 bromide paper
溴酸钾 potassium bromate
溴酸镧 lanthanum bromate
溴酸锌 zinc bromate
溴酸盐 bromate
溴钨丝灯 bromine tungsten filament lamp
溴乙酸 bromoacetic acid
溴乙烷 bromoethane
溴乙烷磺酸 bromoethanesulfonic acid
溴银矿 bromargyrite;bromyrite
溴原子 bromine atom
溴值 bromine number
溴指数 bromine index
溴中毒 bromine poisoning;bromism

嗅 smell;sniff

嗅出 smell out
嗅觉 olfaction;scent;smell(ing)
嗅试法 <测挥发性物质密封程度> scent test method
嗅味试验 scent test

须 爆破取材的工程 rock job

须茬儿 bristle
须穿工作服并系安全带 wear safety harness/belt
须带口罩 respiratory protection must be worn
须戴面罩 wear face shield
须付附加费 subject to additional charges
须付关税 subject to customs duty
须付加价票 subject to payment of supplement
须更改的已更改 <拉> mutatis mutandis
须购附加票的列车 train on which a supplement is payable
须经批准/同意 subject to approval
须经声明 subject to declaration
须经预先核准 subject to prior approval
须两次操纵的离台器 double-clutching
须签名盖章的期票 note to order
须签字照付的票据 bill to order
须藤石 sudoite
须小心调车的货车 wagon to be shunted with care
须用可调护板 use adjustable guard
须用现金支付 terms cash
须征税 subject to duty
须知 guideline
须状晶体 whisker crystal

虚 摆 fictitious pendulum;hypothetical pendulum

虚半轴 imaginary semi-axis
虚包金属门 hollow metal door
虚报 false declaration; false reporting; misrepresentation; misstatement
虚报的账单 padded bill
虚报价 illusory offer
虚报开支账目 padding
虚报亏损 falsification the account; falsifying the account
虚报冒领 make a fraudulent application and claim
虚报账目 cook an account;padding
虚本初子午线 prime fictitious meridian
虚比降 virtual slope
虚比特 dummy bit
虚边坡 virtual slope
虚变量 imaginary variable
虚变位 virtual deformation; virtual displacement
虚变元 virtual argument
虚标志 virtual mark
虚波阵面 imaginary wave front
虚波阻抗剖面 pseudo-impedance section
虚补功原理 principle of virtual complementary work
虚部 idle component;imaginary component;imaginary part <复数的>;reaction component;reactive component
虚部算子 imaginary part operator
虚部运算 imaginary part operation
虚参数 virtual parameter
虚测标 virtual mark
虚长(度) virtual length
虚常数 imaginary constant
虚场边界 virtual field boundary
虚超球面 imaginary hypersphere
虚齿数 virtual number of teeth
虚冲程 empty stroke
虚触发 false triggering
虚存策略 virtual memory strategy
虚存概念 virtual memory concept
虚存管理 virtual(memory) management
虚存机构 virtual memory mechanism
虚存计算机 virtual memory computer;virtual memory machine
虚存技术 virtual memory technique
虚存结构 virtual memory structure
虚存系统 virtual memory system
虚存硬件 virtual memory hardware
虚存执行系统 virtual memory executive system
虚存指示字 virtual memory pointer
虚的 dummy;vain
虚等待时间【数】virtual waiting time
虚等斜线 fictitious loxodrome; fictitious loxodromic curve; fictitious rhumb line
虚地磁北极 virtual geomagnetic north pole
虚地磁北极位置 virtual geomagnetic north pole position
虚地磁极 virtual geomagnetic pole
虚地磁极类型 type of virtual geomagnetic pole
虚地磁极位置 virtual geomagnetic pole position
虚地磁南极 virtual geomagnetic south pole
虚地磁南极位置 virtual geomagnetic south pole position
虚地图文件 virtual map file
虚地址 dummy address; virtual address
虚地址空间 virtual address space
虚点 mathematic(al) point; virtual point;imaginary point
虚点源模式 virtual point source model
虚点指令【计】dummy order
虚电抗 fictitious reactance
虚电势 virtual potential
虚电阻 virtual resistance
虚调用 virtual call
虚调用控制 virtual call control
虚调用手段 virtual call facility
虚陡度 virtual steepness
虚段 virtual segment
虚段结构 virtual segment structure
虚对 virtual pair
虚额定值 virtual rating
虚二次超曲面 imaginary quadric hypersurface
虚二次域 imaginary quadratic field
虚反射 ghost
虚反射速度剖面 pseudo-reflection coefficient section
虚分量 imaginary component; reaction component; reactive component
虚分量剖面平面图 profiling-plan figure of imaginary component
虚分量相对异常曲线 relative anomaly curve of imaginary component
虚负载 artificial load;dummy load;imaginary load(ing);phantom load
虚负载法 dummy load method;phantom loading method
虚刚度 fictitious stiffness
虚高(度) virtual height; equivalent height
虚根【数】imaginary root
虚工作 dummy activity
虚功 idle work; useless work; virtual work
虚功定理 theorem of virtual work
虚功定律 law of virtual work;virtual work law
虚功法 method of virtual work;virtual work approach
虚功方程 virtual work equation
虚功(功)率 fictitious power; virtual power
虚功互等定理 reciprocal virtual work theorem
虚功理论 virtual work theory
虚功原理 law of virtual work;principle of virtual work; virtual work principle
虚功原理法 virtual work method
虚拱 false arch
虚构 fiction
虚构成本 fictitious cost
虚构赤道 fictitious equator
虚构存在 fictitious existence
虚构的 feigned;fictitious;imaginary
虚构的价格 fictitious quotation
虚构的索赔 fabricated claim
虚构股息 fictitious dividend
虚构荷载法 method of fictitious loads
虚构恒向线 fictitious loxodrome;fictitious rhumb line
虚构极 fictitious pole
虚构经度 fictitious longitude
虚构梁 conjugate beam;hollow-trussed beam
虚构年 Besselian year;fictitious year
虚构图网 fictitious graticule
虚构纬度 fictitious latitude
虚构需求 fictitious use
虚构因素 fictitious factor
虚构预应力 fictitious prestressing
虚构账目 fictitious account
虚构资本 fictitious capital
虚构资产 fictitious assets
虚惯性 artificial inertia
虚光 vignette
虚光子场 virtual photon field
虚轨道 virtual orbital
虚过程 dummy activity;virtual process
虚焊接头 dry joint; cold joint; cold solder joint;rosin joint
虚焊接线 rosin connection
虚行 virtual row
虚耗人工 lost labo(u)r
虚耗时间 lost time
虚荷载 dummy load; imaginary load(ing);virtual load(ing)
虚荷载因素 imaginary load factor
虚厚度 nominal thickness
虚呼叫功能 virtual call capability
虚呼叫设备 virtual call facility
虚呼叫设施 virtual call facility
虚幻 phantom
虚幻灯光 phantom light
虚幻示像 ghost aspect; phantom aspect
虚幻显示 phantom indication
虚幻现象 airy appearance
虚幻信号 ghost signal;phantom signal
虚换位 void transposition
虚记录 virtual record
虚加应力 imaginary stressing
虚价 fictitious price; nominal price; shadow price
虚价社会价值理论 false social value theory
虚假 false hood
虚假变数 dummy variable
虚假代理 ostensible agency
虚假弹道 pseudo-trajectory
虚假的 deceptive;false;ostensible
虚假的陈述 false statement
虚假的生产过剩 temporary overproduction
虚假的所得税申报书 fraudulent income tax returns
虚假低标 artificially low tender
虚假定位 false bearing
虚假动作 false operation
虚假供需系统 bogus demand and supply system
虚假光 <灯塔的> false light
虚假海底 phantom bottom
虚假河底 phantom bottom
虚假活塞行程 difference between running and standing travel; false

X

piston travel
虚假交易 fictitious transaction
虚假接地 virtual ground
虚假利润 fictitious profit
虚假轮廓线 false contouring
虚假脉冲 ghost image; ghost pulse; ghost signal
虚假贸易 false trading
虚假耦合 virtual coupling
虚假谱带 spurious band
虚假设 null hypothesis
虚假声明 <重量或品类的> false declaration
虚假收入 spurious revenues
虚假数字 meaningless figures; questionable figures
虚假四位二进制 dummy tetrad
虚假所有权 apparent time
虚假索赔 false claim
虚假梯度 false gradient
虚假填报 <货物运单> false billing
虚假物 dummy
虚假线圈 dummy coil
虚假信道 wrong channel
虚假信号 false signal; ghost signal; spurious signal
虚假需求 fictitious use; imaginary demand
虚假因素 fictitious factor
虚假硬度 false brinelling
虚假远程耦合 virtual long-range coupling
虚假指示 phantom indication
虚假资本 watered capital
虚交点 apparent intersection; imaginary intersection point; imaging intersection point
虚焦点 imaginary focus; virtual focus
虚铰 imaginary hinge
虚接地 bad earth; virtual earth
虚接地隔离器 virtual-earth buffer
虚节 dummy section
虚结构 virtual structure
虚结果 virtual result
虚井 image well
虚警 false alarm
虚警概率 false alarm probability
虚警蛊 false alarm rate
虚警时间 false alarm time
虚距 false distance
虚距离 virtual distance
虚绝对速度剖面 pseudo-absolute velocity section
虚空间 imaginary space
虚夸 vanity
虚亏格 virtual deficiency
虚拉伸 imaginary stretching
虚力 virtual force
虚力矩 virtual moment
虚力原理 principle of virtual forces
虚立方根 imaginary cube root
虚利 nominal profit
虚连接 virtual connection
虚梁 imaginary beam
虚梁法 conjugate beam method; virtual beam method
虚量 imaginary quantity
虚量子 virtual quantum
虚零 false zero
虚零点 virtual zero point
虚零点法 false zero method
虚零检验 false zero test
虚流量 virtual discharge
虚漏 false leak; virtual leak
虚路由 virtual route
虚率 false alarm rate
虚螺距 virtual pitch
虚描线 inactive line
虚模型 virtual model
虚能 virtual energy

虚能级 virtual energy level; virtual level
虚拟变量 dummy variable; invented variate
虚拟变量回归 dummy variable regression
虚拟变量模型 dummy variable model
虚拟变量陷阱 dummy variable trap
虚拟变数 dummy variable
虚拟变形 virtual deformation
虚拟变元 dummy argument
虚拟操作 pseudo-operation
虚拟操作系统 virtual operating system
虚拟产业部门 dummy industry
虚拟场 virtual field
虚拟赤道 fictitious equator
虚拟处理 dummy treatment
虚拟处理机 virtual processor
虚拟磁盘 virtual disk
虚拟存储法 virtual access method
虚拟存储器 hypothetical memory; virtual storage; virtual memory
虚拟存储器地址 virtual memory address
虚拟存储器分配 virtual memory allocation
虚拟存储系统 virtual storage system
虚拟存储寻址 virtual memory addressing
虚拟存取法 virtual access method
虚拟打印机 virtual printer
虚拟单坡 <将复坡虚拟为当量的单坡> hypothetical single slope
虚拟的增长 suppositional growth
虚拟地址 virtual address
虚拟点 virtual point
虚拟点污染源 virtual point source of pollution
虚拟电路 virtual circuit
虚拟反应堆 virtual reactor
虚拟方程 fictitious equation
虚拟方式 virtual mode
虚拟分区存取法 virtual partitioned access method
虚拟分子量 fictitious molecular weight
虚拟杆件 dummy member; dummy element
虚拟公用存储器 virtual common memory
虚拟观测 fictitious observation
虚拟观测方程 fictitious observation equation
虚拟观测量 fictitious observable; quasi-observable
虚拟观测量平差法 adjustment with fictitious observed quantities
虚拟观测值 dummy observation
虚拟惯性 virtual inertia
虚拟荷载 fictitious load
虚拟荷载法 <或称弹性荷重法,计算静定梁与桁架的挠度的方法> method of fictitious loads
虚拟呼叫 virtual call
虚拟化 virtualization
虚拟环境 virtual environment
虚拟回归 dummy regression
虚拟活动 dummy activity
虚拟机 virtual machine
虚拟机安全性 virtual machine security
虚拟机核心 virtual machine kernal
虚拟机技术 virtual machine technique
虚拟机械 virtual machine
虚拟级 virtual level; virtual stage
虚拟计算机 host computer; virtual computer
虚拟技术 virtual technique
虚拟校正 fictitious correction
虚拟接口 virtual interface
虚拟介电常数 fictitious dielectric(al) constant
虚拟井 image well
虚拟空间 virtual space
虚拟控制台 virtual console
虚拟力 fictitious force
虚拟立体模型 imaginary stereoscope
虚拟连接 virtual connection
虚拟连线 dummy link
虚拟模型 virtual model
虚拟内存【计】virtual memory
虚拟内存管理单元 virtual memory management unit
虚拟排水量 virtual displacement
虚拟屏幕 virtual screen
虚拟谱密度 fictitious spectral density
虚拟前期固结压力 virtual preconsolidation pressure
虚拟设计荷载 fictitious design load
虚拟设计荷载系数 fictitious design load factor
虚拟湿球温度 pseudo-wet-bulb temperature
虚拟实体 pseudo-entity
虚拟市场 dummy market
虚拟释放 virtual unbundling
虚拟首子午线 prime fictitious meridian
虚拟数据项 virtual field
虚拟水平 dummy level
虚拟顺序存取法 virtual sequential access method
虚拟速度 pseudo-velocity
虚拟随机 pseudo-random
虚拟索引顺序存取法 virtual indexed sequential access method
虚拟通道逻辑 virtual channel logic
虚拟通道网络 virtual channel network
虚拟通路 virtual path
虚拟图 virtual map
虚拟图像 virtual mapping
虚拟外生变量 dummy exogenous variable
虚拟网络 virtual network
虚拟位移原理 <即虚功原理> principle of virtual displacement
虚拟文件 virtual file
虚拟文件库 virtual file store
虚拟下标指标 dummy subscript index
虚拟现实方法 virtual reality method
虚拟现实建模语言 virtual reality modeling language
虚拟线路 virtual circuit
虚拟线路控制程序 virtual circuit control program(me)
虚拟线栅 fictitious graticule
虚拟效率系统颠簸 churning
虚拟形式 virtual form
虚拟仪器系统设计 design of virtual instrument system
虚拟源 fictitious source; virtual source
虚拟远程通信[讯]存取方法 virtual telecommunications access method
虚拟终端 virtual terminal
虚拟终端网络 virtual terminal network
虚拟终端协议 virtual terminal protocol
虚拟桩 dummy pile
虚拟资本 fictitious capital
虚拟字段 virtual field
虚拟作业 dummy activity
虚捻 false twist
虚凝 hesitation set(ting)
虚耦 dummy coupling
虚牌价 nominal quotation
虚盘 offer without engagement; unconfirmed offer
虚判定值 virtual decision value
虚判决值 virtual decision value
虚抛 wash sale
虚炮点 image shotpoint
虚频率 imaginary frequency
虚平面 imaginary plane
虚平面图 virtual plane
虚平移运动 virtual translation
虚坡(度) <考虑动能作用的坡度,即换算坡度> virtual grade; virtual gradient; virtual slope
虚枪涂层 fogged coat
虚球 imaginary sphere
虚球圆 imaginary circle at infinity
虚曲面 imaginary surface
虚曲线 broken curve; dashed curve(line); dotted curve; imaginary curve
虚荣性牌照 <号码及字母由汽车主自选> vanity plate
虚容器 virtual container
虚弱 asthenia
虚弱公司 thin corporation
虚色 false colo(u)r
虚上界 virtual upper bound
虚设公司 bogus company; bubble company
虚设荷载 dummy imaginary load
虚设结合能 fictitious binding energy
虚设模型 dummy model
虚设物 dummy
虚设线圈 dummy coil
虚设信号 signal fictive
虚声源 virtual sound source
虚时间坐标 imaginary time coordinate
虚实对比 contrast between solid and void; solid-and-void contrast
虚实分量法 imaginary-real component method
虚实徘徊势 stepping lightly and heavily alternately
虚实线 broken and solid line
虚势 imaginary potential
虚饰门面 papier-mâche facade
虚数 imaginary(quantity); imaginary number
虚数变换 imaginary transformation
虚数部分 imaginary component; imaginary part; reaction component; reactive component; wattless component
虚数单位 imaginary unit
虚数的 imaginary
虚数累加器 imaginary accumulator
虚数项 imaginary terms
虚数折射率 imaginary refractive index
虚数轴 axis of imaginaries
虚税 false tax
虚说明词 virtual declarer
虚说明符 virtual declarator
虚四元数 imaginary quaternion
虚素因子 imaginary prime divisor
虚速度 pseudo-velocity; virtual velocity
虚速度定律 law of virtual velocity
虚抬利润 inflated profits
虚态 virtual state
虚条纹图 phantomatic fringe pattern
虚通路 virtual channel
虚同步 false synchronization
虚图 imaginary circle; virtual mapping
虚脱 prostration
虚椭圆 imaginary ellipse
虚椭圆面 imaginary ellipsoid
虚椭圆柱面 imaginary elliptic cylinder
虚外围设备 virtual peripheral
虚弯矩 dummy moment
虚伪的法律行为 hypocritical legal act
虚伪显示 phantom indication
虚伪账户 garage account
虚纬线 fictitious parallel

虚位 fictitious potential;vacancy
虚位法 regulus false
虚位移<指在静力平衡下,由其他力引起的位移> virtual displacement
虚位移场 virtual displacement field
虚位移定理 theorem of virtual displacement; virtual displacement theorem
虚位移定律 law of virtual displacement;virtual displacement law
虚位移法 method of virtual displacement
虚位移原理 law of virtual displacement;virtual displacement principle
虚温 virtual temperature
虚圬工墙 hollow masonry wall
虚无假设 null hypothesis
虚无限素因子 imaginary infinite prime divisors
虚无主义 nihilism
虚下界 virtual lower bound
虚线 break line;broken curve;broken line;dash(ed) line;dot(ted) line;dotted rule;hidden line;imaginary line;intermittent line;interrupted line;invisible line;phantom line;pointed line;short dashed line;vacant line
虚线笔 dotting pen
虚线等高线 dashed contour;dashed contour line;dashed land line
虚线等高线图 dashed-line contour plot
虚线箭头 dotted arrow
虚线矩阵设计 dot matrix description
虚线曲线 broken curve
虚线圆 broken circle
虚相对速度剖面 pseudo-relative velocity section
虚相交 imaginary intersection
虚想边界 fictitious boundary
虚像 false image;virtual image
虚像井 image well
虚像立体模型 virtual image stereomodel
虚像全息立体模型 virtual image holographic(al) stereomodel
虚销 holding latch
虚循环 virtual cycle
虚循环域 imaginary cyclic(al) field
虚压 virtual pressure
虚样本点 imaginary sample points
虚异常 pseudo-anomaly
虚因子 imaginary factor
虚阴极 virtual cathode
虚阴极电镀 dummying
虚应变 virtual strain
虚应力 virtual stress
虚盈实亏 false profit
虚影 double image
虚预加应力 imaginary prestress
虚域 imaginary field
虚元素 imaginary element
虚圆 virtual circle
虚圆点 circular point;focoid;imaginary circular point
虚圆面 spurious disk
虚源干涉 image interference
虚载(荷) dummy load;fictitious load;imaginary load(ing);virtual load
虚造干扰设备 meacon(ing)
虚增资本 watered capital
虚张拉 imaginary tensioning
虚账户 nominal account
虚直径 nominal diameter
虚指数 imaginary exponent
虚质量 apparent mass;virtual mass
虚质量效应 virtual mass effect
虚重力 virtual gravity
虚重心 virtual center of gravity
虚轴 axis of imaginaries;dummy axis;false shaft;free axis;imaginary axis;virtual axis
虚注入器 virtual injector
虚转动惯量 virtual moment of inertia
虚锥 imaginary cone
虚子午线 fictitious meridian;virtual meridian
虚自变数 imaginary argument
虚纵断面 virtual profile
虚阻抗 virtual impedance
虚作业 dummy activity

嘘

嘘声 hiss
嘘嘘响 zip

需

需报量 reportable quantity
需补气区 air shed
需氮量 nitrogen requirement
需电额控制 electric(al) demand control
需电量 power requirement
需电量过程线 demand hydrograph
需电量累积曲线 demand mass curve
需方 purchaser
需废弃的 obsolescent
需付费的 pay
需付利息的债务 active debt
需功量 power requirement
需光度 intolerance of shade;light requirement
需光量 light requirement
需加催硬剂的无水煅烧石膏 hard-finished plaster
需加固的构件 distressed structure
需价 call for bids
需经批准 subject to approval
需款即付 bill on demand
需量标准 demand rate
需量表 demand meter
需量记录器 demand recorder;demand register
需量累积曲线<水、电等的> demand mass curve
需量率 demand rate
需量系数 demand;demand factor
需量限制器 demand limiter
需氯量 chlorine demand;chlorine required;chlorine requirement
需能量 energy requirement
需气的 aerobic
需气量 air demand;air requirement;load;requirement of air;requirement of gas
需气生物处理 aerobic biological treatment
需气细菌 aerobic bacteria
需气性 aerophilic;aerophilons
需气性生物处理 aerobic biological treatment
需求饱和 demand saturation
需求报告 statement of requirements
需求变动 change in demand;demand shift
需求变量 demand variable
需求波动 demand-side fluctuation;swings of demand
需求不足 scant demand
需求不足型失业 demand-deficient unemployment
需求部门 sector of demand
需求参数 demand parameter
需求层次 hierarchy of needs
需求层次理论 hierarchy of needs theory;need hierarchy theory
需求差异订价法 demand differential pricing
需求出行 demand trips
需求促成的通货膨胀 demand-pull inflation
需求带动性的通货膨胀 demand-pull inflation
需求的单一弹性 unitary elasticity of demand
需求的点弹性 point elasticity of demand
需求的弧弹性 arc elasticity of demand
需求的互补 need complementarity
需求的价格弹性 price elasticity of demand
需求的交叉弹性 cross-elasticity of demand
需求的特性 demand characteristics
需求的吸入弹性 income density of demand
需求的周期性波动 cyclic(al) swings in demand
需求递减法则 law of diminishing demand
需求点 demand point
需求顶点 top-out
需求定价法 demand-oriented pricing;demand pricing
需求定律 law of demand
需求分析 demand analysis;needs analysis
需求供给的转变 shift in demand supply
需求供给量的增加 increase in demand supply
需求估计 demand estimation
需求估计量 demand estimation
需求管理 demand management
需求过多引起的通货膨胀 excessive demand inflation
需求过剩 demand surplus
需求过旺 excessive demand
需求函数 demand curve;demand function
需求货物 demand goods
需求季节 demand season
需求价格 demand price
需求价格弹性 pirce elasticity of demand
需求价格弹性系数 coefficient of price elasticity of demand
需求价格灵活性 price elasticity of demand
需求减缩率 ratio of demand contraction
需求减缩主要因素 principal factors of demand contraction
需求检测器 demand detector
需求交通量 demand volume
需求结构 demand structure
需求紧缩 demand retrenchment
需求竞争 demand competition
需求开始点 need trigger
需求扩张性 demand expansibility
需求拉动 demand pull
需求累积曲线 demand mass curve
需求量 quantity demanded
需求量变动 change in quantity demanded;change in quantity supplied
需求流量过程曲线 demand hydrograph
需求率 demand factor;demand ratio
需求论 theory of demand
需求满足率 ratio of demand full
需求密度 demand density
需求模式 demand model
需求膨胀 demand inflation;demand pull
需求批准符号 requirements clearance symbol
需求疲乏 weakening in demand
需求疲弱 weakening in demand
需求平衡系统 demand balance system
需求牵引力 demand pressure
需求曲线 demand curve
需求曲线的移动 shift in demand curve
需求日期 date of acquisition
需求-容量比 demand-capacity ratio
需求伸缩性 demand elasticity;elasticity of demand
需求收入的弹性 income demand elasticity
需求收入弹性系数 coefficient of income elasticity of demand
需求缩减 demand depletion
需求弹性 demand elasticity;elasticity of demand
需求弹性的持久性 durability for demand elasticity
需求弹性的倾向性估计 biased estimation of demand elasticity
需求弹性为一 unitary elasticity of demand
需求特点 characteristics of demand
需求通行能力型控制 demand-capacity control
需求稳定性 demand stability
需求系数 demand factor;service demand factor
需求下降 faltering demand;recession in demand
需求项目 demand element
需求效果 effect of demand
需求学派 demand school;demand side
需求压力 demand pressure;demand pull
需求要素变动 change in factor demand
需求一览表 demand schedule
需求因素 demand factor
需求引起的通货膨胀 demand-pull inflation
需求与供给 demand and supply
需求与供给模型 demand and supply model
需求与供应 demand and supply
需求预测 demand forecast(ing);forecasting of demand
需求预告 demand notice
需求阈限 need trigger
需求约束条件 demand constraint
需求增长 demand growth;demand increase;growth in demand
需求增长率 ratio of demand increase
需求增长主要因素 principal factors of demand increase
需求增加 increased demand
需取得进口许可证的交易 subject to import license
需热量 caloric requirement;calorific requirement;heat demand;heat requirement;thermometric constant
需时 take time;take uptime
需时最长的各项连续工作 total of longest consecutive jobs
需水 water need
需水高峰时间 peak demand time
需水供水比 demand-supply ratio
需水量 amount of water required;duty of water;water consumption;water demand;water requirement
需水量过程线 demand hydrograph
需水量累积曲线 demand mass curve
需水量曲线 demand curve;water-demand curve
需水性 water requirement
需选矿石 milling ore
需氧 aerobic oxidation

需氧处理 aerobic treatment; oxygen-demanding pollution
需氧处理效率 aerobic treatment efficiency
需氧代谢 aerobic metabolism
需氧的 aerobian; aerobic; oxybiontic
需氧二次处理 aerobic secondary treatment
需氧二级处理 aerobic secondary treatment
需氧法 aerobic process
需氧废水 oxygen-demanding waste (water)
需氧废物 oxygen-demanding waste
需氧分解 aerobic breakdown; aerobic decomposition
需氧固氮细菌 aerobic nitrogen-fixing
需氧过程 aerobic process
需氧含碳物质 oxygen demanding carbonaceous material
需氧呼吸 aerobic respiration
需氧环境 aerobic atmosphere
需氧净化 aerobic purification
需氧菌 aerobe; aeromicrobe
需氧菌的 aerobic
需氧菌接触碎石基床 contact bed
需氧菌塘 aerobic lagoon
需氧量 oxygen demand; oxygen requirement
需氧量指数 oxygen demand index
需氧酶 aerobic enzyme
需氧培养 aerobic culture
需氧生活 aerobiosis; oxybiosis
需氧生物 aerobic organism
需氧生物处理 aerobic biological treatment
需氧生物处理法 aerobic biological process
需氧生物滤池 aerobic biological filter
需氧生物学处理法 aerobic biological treatment process
需氧生物氧化 aerobic biooxidation
需氧塘 aerobic pond
需氧条件 aerobic condition
需氧土壤微生物 aerophytobiont
需氧微生物 aerobe; aerobian
需氧微生物污水处理基床 percolating filter
需氧污泥消化 aerobic sludge digestion
需氧污染 oxygen-demanding pollution
需氧物质 oxygen-demanding material
需氧细菌 aerobacteria; aerobe bacteria; aerobic bacteria
需氧细菌处理 aerobic bacteria treatment
需氧细菌学 aerobacteriology
需氧消化池 aerobic digester
需氧消化(法) aerobic digestion
需氧性 aerobism
需氧性生物 oxybiotic organism
需氧性生物处理 aerobic biological treatment
需氧氧化作用 aerobic oxidation
需氧异养菌 aerobic heterotrophic bacteria
需氧有机沉积物 aerobic organic sediment
需氧指数 oxygen index
需氧组分 oxygen-demanding constituent
需要标明尺寸和钢筋间距的正常钢筋网 scheduled fabric
需要层次 need hierarchy
需要抽力 required draft
需要存储量 storage requirement
需要大翻修的房产 handyman's special
需要大量劳动力的货物 labor-intensive goods
需要大量劳力的工业产品 labor-intensive industrial product
需要挡位 required gear
需要的创造 need creation
需要的双方吻合 double coincidence of need
需要的水库供水量 reservoir demand
需要的一致 coincidence of need
需要点 demand point; destination <运输问题中心需要点>
需要订立协议 requirement
需要分等级理论 need-ranking theory
需要风量 required air volume
需要峰高【铁】necessary height of hump
需要附加补强 requiring additional reinforcement
需要改良土壤的地区 area requiring soil improvements
需要更新换代的商业区 commercial improvement area
需要工作面 area requirement
需要功率 required power
需要构成 pattern of need
需要技巧的 tricky
需要加固的结构 distressed structure
需要加固土壤的地区 area requiring soil improvements
需要价格 demand price
需要价值 value of demand; value on demand
需要进行大量研究工作的 research-intensive
需要经费 investment funds needed
需要库容 storage requirement
需要两人共同工作的大块石 two-man riprap
需要量 demand quantity; requirement
需要量估计 demand estimation
需要量累积曲线 demand mass curve
需要量曲线 demand curve
需要流量 discharge required; flow demand
需要流量过程线图 demand hydrograph
需要面积 area requirement; required area; space required
需要伸缩性 elasticity of demand
需要声明书 statement of necessity
需要时进行公断者 referee in case of need
需要数量 quantity required
需要水量 required amount of water
需要弹性 elasticity of demand
需要性 desirability
需要修理的 in need of repair
需要医治的工伤 doctor's case
需要溢价 call premium
需要因数 requirement factor
需要因素 demand factor; requirement factor
需要预测 requirements forecasting
需要在工地外完善和完成的部分项目 off-site improvement
需要照度 needed illuminance
需要证书 certificate of necessity
需要住宿的 live in
需用窗户面积 required window
需用的水文图 hydrograph of demand
需用电量过程线 demand hydrograph
需用功 work requirement
需用功率 power requirement; required (horse) power
需用脚手架的工程高度 scaffold-high
需用量<水、电、气、燃料等的> utility requirement
需用量曲线 demand curve
需用率 demand factor
需用马力 horsepower requirement; required (horse) power
需用日期 required date
需用时限 demand interval
需用水量过程线 demand hydrograph
需用台数 number required
需用系数 demand factor
需用因素 demand factor
需用硬通货偿还的贷款 hard loan
需用资料的次数 access frequency
需助铲的铲运机 push loaded scraper

墟

墟集市 kermis

徐

徐变 creep (age); gradual change; slow change; time-yield

徐变比 creep ratio
徐变变形 creep deformation; time deformation; time-dependent deformation
徐变差 creep variation
徐变(产生的)屈曲 creep induced buckling
徐变沉降 creep settlement
徐变成熟系数 creep maturity coefficient
徐变的损失 loss due to creep
徐变断裂 creep runway; creep rupture
徐变断裂强度 creep breaking strength; creep rupture strength
徐变复原 creep recovery; elastic aftereffect
徐变观测 creep observation
徐变过程 creep process
徐变后效 creep recovery
徐变滑动 creep slide
徐变恢复<卸载后徐变随时间而减小> creep recovery
徐变机理 creep mechanism; mechanism of creep
徐变机制 creep mechanism
徐变极限 creep limit; limit of creep
徐变快速试验法 acceleration creep test
徐变理论 creep theory; theory of creep
徐变力学 theory of creep
徐变量 magnitude of creep
徐变裂缝 creep crack
徐变流动度<变形与时间关系曲线的斜度> creep fluidity
徐变模量 creep modulus; modulus of creep
徐变挠曲 creep deflection; creep deflexion
徐变能力 creep capacity
徐变破坏 creep failure; creep rupture
徐变破坏强度 creep rupture strength
徐变破坏试验 creep rupture test
徐变强度 creep (ing) strength
徐变曲线 creep curve
徐变屈服 creep compliance
徐变失稳 creep instability
徐变时间曲线 creep time curve
徐变试验 creep test
徐变试验机 creep test (ing) machine
徐变收缩 creep shortening
徐变速度 creep speed
徐变速率 creep rate; rate of creep
徐变速率系数 creeping speed factor
徐变损失 plastic flow loss; plastic loss; loss of creeping
徐变特性 creep behavio(u)r; creep characteristic
徐变梯度 creep gradient
徐变系数 coefficient of creep; creep coefficient; creep factor
徐变效应 creep effect
徐变修正系数 creep correction factor
徐变压力 creeping pressure
徐变压曲理论 theory of creep buckling
徐变仪 creep meter
徐变引起的预应力损失 prestressing loss due to creep
徐变应变 creep strain
徐变应变回复 creep strain recovery
徐变应力 creep stress
徐变增长 creep growth; creep increase
徐变终值 final value of creep
徐动应变 creep strain
徐沸 simmer
徐冷区 annealing zone
徐流河 sluggish river; sluggish stream
徐燃 smolder (ing)
徐热 soaking heat
徐升式起重机 creeper travel(l)er
徐舒损失<预应力损失> loss of relaxation
徐塑系数 yielding flow coefficient
徐行 worm
徐徐倾斜 shelve
徐徐移动 edge

许

许多分子组成的 supramolecular

许多平均数间的差数 differences among means
许多人 multitude
许多实际措施 many practical decisions
许购定额 open-to-buy
许购定额报告 open-to-buy report
许购定额的估计 open-to-buy estimate
许可 all-clear; approval; approve; clearing; concession; permission; permittance; sanction
许可标记 label(l)ed
许可标准 approval standard; permissible criterion
许可产量 permissible yield
许可沉降 allowable settlement
许可程序 licensing procedure; licensing program(me)
许可的 allowable; permissible; permissive
许可的环境标准 allowable environmental condition
许可的环境条件 allowable environmental condition
许可的漏损率 leakage
许可的水准 permissible point
许可方 licensor
许可负载 allowable load
许可工作强度 allowable working strength
许可合同 license[licence] contract
许可荷载 allowable load; charge of surety
许可寄主 permissive host
许可力矩 moment allowance
许可流速 permissible velocity; safe velocity
许可挠度 permissible flexibility
许可挠性 permissible flexibility
许可浓度 allowable concentration
许可配额混合制 license[licence] and quota system
许可强度 allowable strength
许可使用 permitted use
许可使用策略 acceptable use policy
许可收缩量 allowable for shrinkage
许可调节 permissible control; permissive control
许可停车的街道 authorised[authorized] street parking
许可通商 license to trade

许可温度 permissible temperature; permissive temperature
许可信号 enabling signal
许可性寄主 permissive host
许可应力 permissible stress
许可应力范围 safe range of stress
许可营业商店 licensed house
许可运输漏损率 transportation leakage
许可证 certificate(of authority);certificate of authorization; charter; clearance; permit; warrant; licence [license];clearance paper;clearance permit <指船出海用的>
许可证持有者 licencee[licensee]
许可证的颁发 licensing;licensure
许可证费(用) license[licence] fee;license expenses;permit-fee
许可证及执照 license[licence] and permit bond
许可证交易 licencing
许可证接受方 licencee[licensee]
许可证接受方不得推翻原议 licencee [licensee] estoppel
许可证接受方不得违约 licencee[licensee] estoppel
许可证接受人 licencee[licensee]
许可证贸易 license[licence] trade;license [licence] transaction; licensing;licensure
许可证贸易联盟 license[licence] trade union
许可证贸易协议 licensing trade agreement
许可证申请书 application for licence [license]
许可证生产 license[licence] production
许可证书 <驻在国发给领事或商务人员的> exequatur
许可证税 excise
许可证条件 licence[license] condition
许可证协定 licence[licensing] agreement
许可证协议 licence[license] agreement
许可证业务 licensing operations
许可证有效 license[licence] in place
许可证制度 license[licensing] system; permit system; system of licencing [licensing]
许可制贸易 trading under license
许可重量 safety weight
许可转让的技术 licensed technology
许可转数 permissible revolutions
许可总鱼获量 total allowable catch
许诺 commitment
许诺条款 grant clause
许偏差 permissible variation
许瓦尔兹不等式 Schwarz inequality
许瓦尔兹-克里斯托弗尔变换 Schwarz-Christoffel transformation
许瓦尔兹引理 Schwarz's lemma
许许多多 a world of
许用 admissible;allowable
许用安全应力 safe allowable stress
许用材料 approved material
许用长度 permissible length
许用承载 allowable bearing capacity
许用承载力 allowable bearing pressure
许用承载能力 safe bearing load;safe bearing power;safe carrying capacity(property); safe load carrying capacity
许用单位安全应力 safe allowable unit stress
许用的 permissible
许用工作压力 allowable working pressure
许用工作应力 allowable working stress
许用荷载 permissible load; safe allowable load; safe working load(ing);working load
许用加载 allowable force
许用间隙 safety clearance
许用接触应力 allowable contact stress
许用矿山设备 permissible mine equipment
许用拉应力 allowable tensile stress
许用脉动扭矩 allowable pulsary torque
许用磨损量 allowable wear
许用能力 allowable capacity
许用扭振应力 allowable torsional vibration stress
许用偏差 allowable deviation
许用频率 allowed frequency
许用强度 working strength
许用切应力 allowable shear stress
许用倾侧力矩 allowable heeling moment
许用设计应力 allowable design stress
许用塑料填料 approved plastic filler
许用弯曲应力 allowable flexural stress
许用温度 allowable temperature
许用误差 allowance error
许用系数 allowance factor
许用形式 permissible shapes
许用压力 allowable pressure;set pressure
许用应力 admissible stress;allowable stress; design strength; permissible stress; safe stress; safe working stress;working stress
许用应力法 working stress method
许用应力折减系数 reduction coefficient of allowable stress
许用永久扭曲 allowable permanent twist
许用余隙 allow clearance
许用炸药 permitted explosive
许用胀隙 allowed space for expansion
许用值 allowable value
许用最大扭矩 allowable maximum torque

旭

旭日式谐振系统 rising-sun resonator system
旭日型磁控管 rising-sun magnetron

序

序编码 out-of-line coding
序变能力 system's ability for orderly change
序磁性 ordered magnetism
序对 ordered pairs
序贯编码 sequential coding
序贯策略 sequential strategy
序贯程序 sequential procedure
序贯抽样 sequential sampling
序贯抽样检查方案 sequential sampling inspection plan
序贯单形法 sequential simplex method
序贯调查观测值 sequential investigation observation
序贯对比指数 sequential comparison index
序贯法 sequential method
序贯反应 sequential response
序贯分析 sequential analysis
序贯风险函数 sequential risk function
序贯概率比检验 sequential probability-ratio test
序贯构码 sequential encoding
序贯估计 sequential estimation
序贯观测 sequential observation
序贯观测者 sequential observer
序贯光放大器 sequential light amplifier
序贯假设检验 sequential hypothesis testing
序贯检测 sequential detection
序贯检定 sequential test
序贯检验的平均样本容量 average sample size in sequential tests
序贯解码 sequential decoding
序贯决策 sequential decision
序贯决策过程 sequential decision process
序贯判别 sequential discrimination
序贯设计 sequential design
序贯识别 sequential recognition
序贯试验 sequential trial
序贯试验设计 sequential trial design
序贯试样 sequential sample
序贯数据 serial data
序贯搜索(法)sequential search
序贯随机规划 sequential stochastic programming
序贯选优法 sequential search
序贯选择 sequential selection
序贯寻优(法)sequential search
序贯样本 sequential sample
序贯译码 sequential decoding
序号 code number;order number;ordinal number;serial number
序级制 rank system
序景 order of sceneries
序理检测 sequence check
序粒层 graded bed
序列 line up;order;ranking;sequence;suite;train
序列编码 sequential coding
序列变差 serial variation
序列变换 sequence transformation
序列表 sequence table
序列不相关扰动 serially uncorrelated disturbance
序列采样 sequential sampling
序列参数 sequential parameter
序列槽 sequential tank; sequential trough
序列长度 sequence length
序列长度分布 sequence length distribution
序列程序 sequencer program(me)
序列程序机 sequence program(me) machine
序列池 sequential basin
序列抽样 sequential sampling
序列抽样法 sequential sampling plan
序列抽样检验 sequential sampling inspection;serial sampling inspection
序列的 intersequential;sequential
序列的极限 limit of a sequence
序列的相加 addition of series
序列调用 sequence calling
序列独立 serial independence
序列断点 sequence break
序列对比 matching of successions
序列发生器 sequencer
序列方差 serial variance
序列分类法 method of sequential sorting
序列分析 sequence analysis;sequential analysis
序列分析器 sequential analyser[analyzer]
序列分析仪 sequential analyser[analyzer]
序列改变器 sequence alternator
序列概率比 sequential probability ratio
序列概率比试验 sequential probability-ratio test
序列港池 sequential harbo(u)r basin
序列估计 sequential estimation
序列过滤器 sequence filter
序列焊接时间调节器 sequence weld timer
序列号 serial number;series number
序列合同 serial contract
序列互相依存 sequential interdependence
序列极限 limit of sequence
序列记录 subsequent record
序列监视器 sequence monitor
序列检测 sequential detection
序列检查 sequence checking
序列检验 sequence check;sequential test
序列检验程序 sequence check(ing) routine
序列交流发电机 sequence alternator
序列校验 sequence check
序列校验程序 sequence checking routine
序列接收 sequential reception
序列结构 sequential structure
序列结束 end of sequence
序列解码 sequential decoding
序列空间 sequence space
序列控制 sequence control;sequential control
序列控制带 sequence control tape
序列控制计算器 sequence control calculator
序列控制寄存器 sequence control register
序列滤波 sequential filtering
序列码 sequence code
序列脉冲 train pulse
序列判据 sequence criterion
序列平均值 serial mean
序列熵 sequence entropy
序列设计 serial design
序列生产 serial production
序列生产过程 serialization
序列生成程序 sequence generator
序列施工法 sequential construction
序列时间调节器 sequence timer
序列式分布 serial distribution
序列式增压器 in-line booster
序列数 sequence number
序列搜索 sequential search
序列随机性检验 test for serial randomness
序列天波 train of skywave
序列图表 sequence chart
序列完备空间 sequentially complete space
序列无约束极小化方法 sequential unconstrained minimization technique
序列误差 sequence error;sequential error
序列系统 sequential system
序列相关 correlation between series;sequential correlation;serial correlation
序列相关检验 serial correlation test;test for serial correlation
序列相关模型 serial correlation model
序列相关误差过程 serially correlated error process
序列相关系数 sequential correlation coefficient;serial correlation coefficient
序列信号 sequence signal
序列信息数据 sequence information data
序列修匀 smoothing of series
序列应力损失 sequence-stressing loss
序列债券 serial bond
序列战略 sequential strategy

序列支付 ranked payoff
序列值 sequential value
序列滞后相关 serial lag correlation
序列自动化 sequential automation
序轮机 sequence switch
序码发生器 sequence generator
序脉 serial vein
序幕 prolog(ue)
序偶 ordered pairs
序批式反冲洗滤池 sequential backwashing filter
序批式反应器 sequencing batch reactor
序批式反应器活性污泥法 sequencing batch reactor activated sludge process
序批式好氧污泥层反应器 sequential aerobic sludge blanket reactor
序批式活性污泥法 sequencing batch activated sludge process
序批式活性污泥反应池 sequencing batch activated sludge reactor
序批式氯化消毒工艺 sequential chlorination disinfection technology
序批式膜反应器 sequencing batch membrane reactor
序批式膜生物反应器 sequencing batch membrane bioreactor
序批式内循环反应器 inner loop sequencing batch reactor
序批式气提反应器 sequencing batch airlift reactor
序批式人工湿地 sequential constructed wetland
序批式生物膜反应器 biofilm sequencing batch reactor; sequencing batch biofilm reactor
序批式厌氧生物反应器 anaerobic sequencing batch bioreactor; sequencing batch anaerobic bioreactor
序批式移动床生物膜反应器 moving bed sequencing batch biofilm reactor; sequencing batch moving bed biofilm reactor
序时记录 chronological record
序时平均数 chronological average
序时延迟 sequential time delay
序时账簿 book of chronological entry; chronological books
序数 ordinal; ordinal number; serial number
序数尺度 ordinal scale
序数和 ordinal sum
序数积 ordinal product
序数论 ordinalism
序数幂 ordinal power
序数效用 ordinal utility
序数效用论 theory of ordinal utility
序文条款 introductory provisions
序系 introduction; series
序有界的 order-bounded
序转变 order-disorder transition

叙词 descriptor

叙词网络 descriptor network
叙词语言 descriptor language
叙级标准 grading standard
叙拉古城的雅典城神庙 Temple of Athena at Syracuse
叙利亚地沥青 Syrian asphalt
叙利亚拱顶 Syrian vault
叙利亚建筑 Syrian architecture
叙利亚式 Syria type
叙述目标 statement of objectives
叙述器 describer
叙述式报表 explanatory statement; narrative form
叙述式损益表 narrative form of profit and loss statement
叙述特性 characterize
叙述统计 descriptive statistics
叙述性决策理论 descriptive decision theory
叙述性情报 descriptive information
叙述者 delineator; describer
叙永石 endellite; halloysite

恤金 pension

恤金准备 reserve for pensions
恤养基金 pension fund
恤养基金准备 pension fund reserves
恤养金 pension

绪丝 floss

续保 renewal(of insurance)

续保保(险)费 renewal premium
续保费 renew premium
续保收费率 reissue rate
续保收据 renewal receipt
续保险 renew of insurance
续次拉延模 redrawing die
续存公司 surviving company
续订合同 renewal of contract
续订货 renew order
续订权 renewal option
续订通知书 renewal notice
续发玻璃状体 secondary vitreous
续发地震 ensuing earthquake
续发晶状体纤维 secondary lens fibers
续发率 secondary attack rate
续灌 continuity flow irrigation
续航 continuation of the journey
续航半径 cruising radius; steaming radius; steaming range
续航发动机 sustainer motor
续航距离 cruising range; radius of steaming range
续航力 cruising radius; endurance; range; sea endurance; steaming radius; steaming range
续航力测定航次 endurance run
续航力大的舰船 long legged
续航力试验 endurance test; endurance trial
续航时间 cruise duration; cruising duration; endurance; flight endurance
续航性能 duration performance
续会 resumed session
续加燃料 refueling
续建大中型项目 large and medium-sized project to be continued
续建项目费用 expense for continued project
续接 splicing wire
续料 payoff
续流 after flow
续流泵 continuous flow pump
续流二极管 flywheel diode
续馏液 subsequent distillate
续滤液 subsequent filtrate
续生成本 recurring cost
续线 splicing wire
续行列车 following train
续页入口 page entry
续源流行 continuing sources epidemic
续运 on-carriage; reforwarding
续运合同 <集装箱> contract of on-carriage
续展注册 renewal of registration
续纸器 automatic sheet feeder
续纸装置 feeder unit
续租 relet
续阻条款 prolongation clause
续钻小一号尺寸钻孔 follow-up hole

酗酒违章 alcohol involvement

絮化 flocks

絮化点 flock point
絮化剂 flocculating agent
絮化作用 flocculation
絮聚剂 flocculant
絮棉梳理机 wadding card
絮凝 coagulation
絮凝层 flocbed
絮凝沉淀 flocculating sedimentation; flocculating settling; flocculation deposition; flocculation sedimentation; flocculation settling; flocculent settling
絮凝沉淀层 flocbed; floc blanket; flocculent layer
絮凝沉淀处理 coagulation sedimentation filtration
絮凝沉积物 flocculated sediment
絮凝沉降 flocculating settling; flocculent settling
絮凝沉降池 flocculation settler; flock settler
絮凝沉降单位 flocculating unit
絮凝成分 flocculating constituent
絮凝澄清池 clariflocculator; flocculation basin
絮凝澄清器 flocculater clarifier; flocculator settler
絮凝池 floc basin; flocculating chamber; flocculating tank; flocculation basin; flocculation tank
絮凝处理 coagulation treatment; floccing
絮凝促进剂 flocculation accelerator
絮凝促凝剂 flocculation accelerator
絮凝单位 flocculation unit
絮凝的 flocculent; flocky
絮凝点 floc(culation) point
絮凝动力学 flocculation kinetics
絮凝度 degree of flocculation
絮凝法 flocculation process
絮凝反应 flocculation reaction; flocculoreaction
絮凝反应池 flocculation basin; flocculation reaction tank
絮凝分散现象 deflocculation
絮凝浮选 flocflo(a)tation
絮凝浮选法 flocculation-flo(a)tation
絮凝化 flocculating
絮凝化比 <测定沥青胶体稳定性的一种指标> flocculation ratio
絮凝化限度 flocculation limit; flocculence limit
絮凝活性 flocculating activity
絮凝活性生物固体 flocculated activated biosolid
絮凝机理 flocculating mechanism; flocculation mechanism
絮凝极限 flocculation limit; limit of flocculation
絮凝剂 coagulant; flocculant; flocculating reagent; flocculation agent; flocculator; flocculent; floc-forming chemical reagent; floe-forming chemical reagent; leiocome
絮凝剂产生菌 flocculant-producing microbe
絮凝剂投加量 flocculant dosage
絮凝剂用量 dose of flocculant
絮凝加速剂 flocculation accelerator
絮凝胶 flocculent gel; leiocome glue
絮凝胶乳浆 flocculated latex crumb
絮凝胶体 flocculated colloid; flocculating colloid
絮凝搅拌 flocculated mixing
絮凝搅拌机 flocculator
絮凝搅拌器 flocculator
絮凝结构 flocculated structure; flocculent structure
絮凝粒 floccule
絮凝滤清池 clariflocculator
絮凝率 flocculation ratio
絮凝气浮法 flocculation-flo(a)tation method
絮凝气浮吸附 flocculation-flo(a)tation adsorption
絮凝区 flocculating chamber; flocculation zone
絮凝生物接触氧化法 flocculation-biological contact oxidation process
絮凝石膏 flocculent gypsum
絮凝时间 flocculation time
絮凝试验 floc(culation) test
絮凝室 flocculating chamber
絮凝水泥 flocculated cement
絮凝体 flocculating constituent; flocculation body; floccule body; flock
絮凝体形成 floc formation
絮凝体组分 flocculating constituent
絮凝添加剂 flocculating admixture
絮凝土 flocculent soil
絮凝团 cluster of flocculates
絮凝脱色剂 flocculating-decolo(u)rizing agent
絮凝脱水技术 flocculation dewatering technology
絮凝脱油法 flocculation oil removing method
絮凝物 clump; floccing; floccules; flocs; flocculate
絮凝物结构 floc structure
絮凝物生成菌 floc-forming bacteria
絮凝物特性 floc characteristic
絮凝物形成 floc formation
絮凝系数 flocculation factor; flocculation ratio
絮凝限度 limit of flocculation
絮凝相 phase of flocculation
絮凝效果 flocculated effect
絮凝效率 flocculating efficiency; flocculation efficiency
絮凝效应 flocculating effect
絮凝形成 flocculant formation; flocculant forming
絮凝形态特征 flocculation morphologic property
絮凝性 flocculability
絮凝性能 flocculating property; flocculation property
絮凝性污泥 flocculated sludge; flocculator sludge; flocculent sludge
絮凝序批间歇式反应器活性污泥法 flocculation-sequencing batch reactor activated sludge process
絮凝因素 flocculation factor
絮凝渣 flocculated sludge
絮凝值 coagulation value; flocculation value
絮凝助沉 flocculation-sedimentation aids
絮凝助沉剂 sedimentation aid
絮凝助剂 flocculating agent
絮凝装置 flocculating unit; flocculation plant; flocculation unit
絮凝状沉淀 flocculant deposit; flocculent deposit
絮凝状沉积物 flocculated sediment; flocculent deposit
絮凝状况 flocculating condition
絮凝状泥沙 flocculated sediment
絮凝状黏[粘]土 flocculated clay
絮凝作用 flocculation

絮片 floc(culu)s
絮球 blowball
絮散 deflocculation
絮体 floc
絮体带出 floc carryover
絮体核心 floc nucleus
絮体加重剂 floc weighing agent
絮体破碎 floc break-up
絮体强度 floc strength
絮体生成 floc formation
絮体拦栅 floc barrier
絮体水 floc water
絮团 flock
絮羽化 feathering
絮云【气】cirrocumulus
絮状沉淀 clump;flocculent(precipitate);floccules;flock;flocs
絮状沉淀槽 pre-flock chamber
絮状沉淀单位 flocculating unit;limit of flocculation
絮状沉淀法 flocculence
絮状沉淀反应 flocculation precipitation
絮状沉淀试验 flocculation test
絮状沉淀物 flocculent precipitate;flocky precipitate
絮状沉降 flocculent settling
絮状的 floccose;flocculent;flocky
絮状反应 flocculoreaction
絮状滚胶法 cottoning
絮状混浊 flocculent turbidity
絮状活性污泥 activated sludge floc
絮状结构 flocculent structure
絮状凝集现象 flocculating phenomenon
絮状试验 flocculation test
絮状体 floccule;flocculus;flocs
絮状污泥 flocculated sludge;flocculent sludge
絮状物 floc;floccules;floccus[复 flocci];floss
絮状抑制试验 flocculation inhibition test
絮状阴影 patchy shadow
絮状云 cotton ball clouds;floccus[复 flocci]
絮状支托黏[粘]结 flocculated clay buttress bond
絮状作用 flocculation
絮浊反应 flocculation and turbidity reaction
絮浊试验 flocculation-turbidity test

畜草平衡 balance between pastures and livestock

畜产品 animal products;livestock products
畜产品产量 output of livestock product
畜队 team
畜队交通 team track
畜队(拖拉)荷载 team load
畜副产品 animal by-products
畜牧 cattle breeding;farming;raise animal;raise livestock;rear poultry
畜牧场 animal farm;cattle farm;livestock farm;stock-farm
畜牧场废物 waste from stockfarming
畜牧场桥式起重机 stockyard transporter
畜牧场输送器 stockyard transporter
畜牧场运输机 stockyard transporter
畜牧副产品 animal by-products
畜牧经营 animal industry
畜牧生产 animal production
畜牧生产体系 livestock production system
畜牧兽医科学研究所 institute of zootechnics and veterinary science
畜牧水产业 livestock and aquaculture industries
畜牧水源开发 stock water development
畜牧饲养业废水 waste form stock-raising production
畜牧围场 paddock
畜牧学 animal husbandry;animal science;zootechnical science
畜牧学报 Journal of Animal Science
畜牧业 animal agriculture;animal husbanding;animal husbandry;graziery;husbandry;live(stock) breeding;livestock farming;livestock husbandry;pastoral farming;pasturage;stock farming;stocking raising
畜牧业生产工厂化 factory-like production in livestock husbandry
畜牧业者 grazier;stock farmer
畜牧业总产量 gross output of cattle breeding
畜牧用水 stock water
畜牧用水池 stock pond
畜牧用水开发 stock water development
畜牧者 farmer
畜疫 murrain

蓄产量 animal yield

蓄潮 tidal storage
蓄潮池 tidal pool
蓄潮冲游塘 scouring basin
蓄潮湾 storage bay
蓄存槽 stored tank
蓄存系数 storage factor
蓄电 condense;electric(al) power storage;storage
蓄电池 accumulator;accumulator cell;electric(al) accumulator;electric(al) battery;electric(al) power storage;secondary cell;secondary generator;storage battery;storage cell;Exide accumulator <一种牵引设备用的>
蓄电池安装托架 battery carrier
蓄电池搬运车 electric(al) battery truck
蓄电池薄片 battery separator
蓄电池舱 accumulator room
蓄电池操作的工业卡车 battery operated industrial truck
蓄电池槽 accumulator box;accumulator container
蓄电池车 accumulator vehicle;battery car(t);battery truck;battery vehicle;storage battery car
蓄电池车间 accumulator plant
蓄电池沉淀物 battery mud
蓄电池充电 accumulator charging;battery charge;battery charging;battery input
蓄电池充电插塞 battery charging plug
蓄电池充电插座 battery charging receptacle
蓄电池充电电流表 battery charging ammeter
蓄电池充电电流调整 battery current regulation
蓄电池充电电阻器 battery charging resistor
蓄电池充电机 accumulator charger;battery charger
蓄电池充电及检修车间 battery charging and overhaul shop
蓄电池充电间 accumulator charging room;accumulator plant;battery house
蓄电池充电接触器 battery charging contactor
蓄电池充电器 accumulator charger;battery charger;charger
蓄电池充电设备 battery charging equipment
蓄电池充电小车 accumulator charging pushcart
蓄电池充电用变阻器 battery charging rheostat
蓄电池充电站 battery charging station
蓄电池充电整流器 accumulator rectifier
蓄电池充电指示灯泡 battery charger bulb;battery control lamp bulb
蓄电池充放电安时计 battery meter
蓄电池充放电记录仪 storage battery meter
蓄电池充放电装置 charging and discharging equipment for accumulator
蓄电池出电量 battery output
蓄电池串联法 battery series connection
蓄电池单位 battery cell
蓄电池地线 battery ground cable
蓄电池点火系统 battery ignition system
蓄电池电动车 battery-electric truck
蓄电池电极 battery terminal
蓄电池电解液槽 battery case
蓄电池电解质 battery electrolyte
蓄电池电缆 battery cable
蓄电池电流测量分流器 battery current measuring shunt
蓄电池电路 battery circuit
蓄电池电压表 excel tester
蓄电池电液泡 storage battery gassing
蓄电池发火 battery ignition
蓄电池放电指示器 battery discharge indicator
蓄电池浮充率 floating rate
蓄电池负载开关 battery load switch
蓄电池盖 battery cover
蓄电池缸 accumulator vessel
蓄电池隔板 accumulator separator;battery separator;electrode separator
蓄电池隔离器 battery isolator
蓄电池供电 storage battery supply
蓄电池供电磁场接触器 battery field contactor
蓄电池故障指示器 battery fault indicator
蓄电池轨道车 accumulator rail-car
蓄电池柜 battery cubicle
蓄电池过量充电 battery over charge
蓄电池机车 battery(-driven) locomotive;storage battery locomotive
蓄电池机车牵引 battery-powered haulage
蓄电池机车运输 battery-powered haulage
蓄电池极板 accumulator plate;battery element
蓄电池极板隔离板 accumulator separator
蓄电池极板合金 accumulator metal
蓄电池极板间的胶木片 accumulator side piece
蓄电池加热器 battery heater
蓄电池加注酸液 priming of battery
蓄电池架 accumulator stand;battery rack;battery stand;bearer for accumulator
蓄电池间 accumulator workshop
蓄电池检查票插 battery check card holder
蓄电池接头 battery connector
蓄电池接线端子 battery terminal
蓄电池接线头 battery connector
蓄电池接线柱 battery terminal
蓄电池金属极板 accumulator metal
蓄电池绝缘子 accumulator insulator;battery insulator
蓄电池卡车 battery truck;electric(al) battery truck;storage battery truck
蓄电池开关 battery switch
蓄电池壳 battery container;battery jar;jar
蓄电池馈电 accumulator feeding
蓄电池流出量 battery flux
蓄电池冒气 storage battery gassing
蓄电池木隔板 battery wood separator
蓄电池配电板 battery switchboard
蓄电池配电盘 accumulator switchboard;battery distribution board
蓄电池配电盘调节器 accumulator switch regulator
蓄电池起重机 battery crane
蓄电池牵引 accumulator traction;battery traction
蓄电池牵引车 battery tractor
蓄电池牵引机 battery tractor
蓄电池牵引机车 accumulator loco(motive)
蓄电池铅板 battery grid;battery lead plate
蓄电池铅合金 lead battery metal
蓄电池清洁棒 accumulator cleaning stick
蓄电池驱动 battery drive
蓄电池容量 accumulator capacity;battery capacity;capacity of storage battery;storage capacity;store capacity;battery cell current capacity
蓄电池容器 accumulator box;accumulator container;accumulator jar;battery jar
蓄电池溶液 battery electrolyte;battery solution
蓄电池栅板 accumulator grid
蓄电池失效 battery cell failure
蓄电池实验器 battery tester
蓄电池式电机车 accumulator loco(motive)
蓄电池式焊机 battery welder
蓄电池式机车 battery locomotive;battery-type locomotive;electric(al) battery locomotive
蓄电池室 accumulator house;accumulator plant;accumulator room;battery house;battery room
蓄电池酸液 accumulator acid;battery acid
蓄电池同极连接片 strap
蓄电池同性极板汇流条 terminal yoke
蓄电池外壳 battery case
蓄电池线 battery cable
蓄电池线夹 battery clip
蓄电池线接头 battery terminal
蓄电池箱 accumulator box;accumulator case;accumulator tank;battery box;battery cupboard
蓄电池箱吊 battery box hanger;battery box support
蓄电池橡皮隔板 battery rubber separator
蓄电池液槽 battery sump
蓄电池阴极板 negative battery plate
蓄电池载运车 accumulator carriage
蓄电池载重车 battery truck;electric(al) battery truck;storage battery truck
蓄电池再充电 battery recharging
蓄电池站 accumulator plant
蓄电池照明 battery lighting
蓄电池支架 battery support

蓄电池重新充电 refreshen
蓄电池转换开关 accumulator switch; battery cut-over switch
蓄电池转换调节装置 accumulator switch regulator
蓄电池组 accumulator; accumulator battery; accumulator storage battery; battery (pack); galvanic battery; rechargeable battery; secondary battery; storage battery; battery bank
蓄电池组放电 battery discharge
蓄电量 <电池的> charge capacity
蓄电瓶 accumulator jar
蓄电瓶废水 battery wastewater
蓄电瓶汇流条 battery bus
蓄电瓶极板网栅 accumulator grid
蓄电瓶室 battery room
蓄电器 condenser
蓄电装置 electric(al) storage device
蓄洪 flood retention; flood storage; storage of flood
蓄洪坝 flood-control dam; flood dam
蓄洪低地 washland
蓄洪防旱 store flood-water for use in a drought
蓄洪工程 flood retention work; flood storage project; flood storage work
蓄洪构筑物 detention structure; floodwater retarding structure
蓄洪河槽 flood-control channel
蓄洪建筑物 detention structure; floodwater retarding structure
蓄洪垦殖 flood storage with land reclamation
蓄洪垦殖工程 flood storage and reclamation works
蓄洪库 flood basin
蓄洪能力 flood absorption capacity; flood(storage) capacity
蓄洪区 flood basin; flood-control reservoir; flood storage area; flood retention area; pondage land
蓄洪设施 detention structure
蓄洪水库 detention reservoir; flood basin; flood-control reservoir; flood detention reservoir; flood reservoir; flood retention basin; flood storage basin; flood storage reservoir; retention basin
蓄换热式空气预热器 recuperative air preheater
蓄汇波 filling release wave
蓄积 accumulate; stock
蓄积倍数 accumulation multiple
蓄积毒性 cumulative toxicity
蓄积毒性试验 cumulative toxicity test
蓄积粮食 store up grain
蓄积量 stockpile
蓄积率 rate of accumulation
蓄积疲劳 accumulated fatigue
蓄积试验 accumulation test
蓄积水 intercepted water
蓄积系数 cumulative coefficient
蓄积性毒物 accumulative toxicant; cumulative poison(ing)
蓄积性中毒 cumulative intoxication; cumulative poisoning; retention toxicosis
蓄积作用 accumulative action; cumulative action; cumulative effect
蓄加充电阀门 accumulator charging valve
蓄加充电开关 accumulator charging valve
蓄栏 bawn
蓄冷 accumulation of cold; cold accumulation; hold-over
蓄冷板 hold-over plate
蓄冷介质 cool storage medium
蓄冷库 storage rate
蓄冷盘管 hold-over coil
蓄冷系统 hold-over system
蓄力器 accumulator
蓄量 storage volume
蓄量常数 <马斯京根法的 *K* 值> storage constant
蓄量出流曲线 storage-discharge curve
蓄量方程式 storage equation
蓄量系数 coefficient of storage
蓄能 energy storage accumulation; storage energy
蓄能泵 storage pump
蓄能电厂 storage plant
蓄能电机车 electrogyro locomotive
蓄能电容器 energy storage capacitor
蓄能电站 storage energy power station; storage plant
蓄能焊 stored energy welding
蓄能合闸 store energy closing
蓄能器 accumulator; energy storage
蓄能器壳体 accumulator housing
蓄能器预充 precharge of accumulator
蓄能器增压 accumulator charging
蓄能曲线 stored energy curve
蓄能式发电站 storage plant; storage station
蓄能式(水力)发电厂 storage plant
蓄能水泵 storage pump
蓄能系统 storage system
蓄泥沙水库 debris storage basin
蓄泥石库 debris storage basin
蓄泥石洼地 debris storage basin
蓄气瓶 gasometer
蓄气试验 accumulation test
蓄汽器 steam accumulator
蓄汽桶 receiver tank
蓄清排浑 impounding clear water and discharging sediment-laden water
蓄热 heat recuperating; heat retention; heat storage; hot storage; stored heat; thermal storage; heat-stored curing <水泥混凝土的>
蓄热材料 heat storage material
蓄热的 regenerant; regenerative
蓄热法 thermos method
蓄热负荷 storage load
蓄热负荷系数 storage load factor
蓄热坩埚炉 regeneration crucible furnace
蓄热锅炉 heat storage boiler
蓄热块组放热器 block storage radiator
蓄热量 stored heat; thermal storage capacity
蓄热炉 regenerative furnace
蓄热面积 heating surface
蓄热能力 heat retention capacity; heat storage capacity; heat storage power; heat storing capacity; thermal storage capacity
蓄热器 accumulator; heat accumulator; heat reservoir; heat storage; regenerator; storage-type calorifier; storage water heater; themophore; Thermofor; thermophore
蓄热器连续渗滤 Thermofor continuous percolation
蓄热器排污阀 accumulator blowdown valve
蓄热器汽轮机 accumulator turbine
蓄热器砖格子砌体 checkerwork
蓄热墙 thermal storage wall
蓄热式 heat accumulating type
蓄热式采暖 thermal storage heating
蓄热式池窑 regenerative tank furnace
蓄热式供暖器 storage heater
蓄热式供热系统 thermal storage heating system
蓄热式换热 regenerative heat exchange
蓄热式换热器 regenerative heat exchanger
蓄热式加热法 regenerative firing
蓄热式绝热操作 regenerative adiabatic operation
蓄热式均热炉 Amco soaking pit; regenerative soaking pit
蓄热式空气循环系统 regenerative air cycle system
蓄热式空气预热器 regenerative air preheater
蓄热式炉 regenerative furnace; regenerative oven
蓄热式热风炉 regenerative hot blast stove
蓄热式窑 regenerator kiln
蓄热式窑炉 regenerative furnace
蓄热室 checker chamber; heat regenerator; regenerative chamber; regenerator chamber
蓄热室布置 regenerator setting
蓄热室出口 checker port
蓄热室顶部空间 regenerator top space
蓄热室格子砖堆砌 regenerator packing
蓄热室烟道 regenerator flue
蓄热室支撑 bracing of regenerator
蓄热水池 heat storage tank; thermal storage tank
蓄热特性 heat storage capacity; heat storage characteristic; thermal storage characteristic
蓄热体 heat accumulator; heat retainer; heat retaining mass
蓄热系数 coefficient of accumulation of heat; coefficient of heat accumulation; coefficient of heat storage; coefficient of thermal storage; thermal admittance; thermal storage coefficient
蓄热系统 heat accumulator system
蓄热效应 heat storage effect
蓄热性能 heat storage property; heat storage quality
蓄热烟道 regenerative flue
蓄热养护 <混凝土> heat-stored curing; heat storage maintenance
蓄热养生 heat-stored curing
蓄热指数 heat storage index
蓄热装置 heat storage device
蓄热作用 accumulation of heat; thermal storage effect
蓄容曲线 storage capacity curve
蓄沙池 sediment storage basin
蓄砂坑 sand storage pit
蓄势器 pressure reservoir
蓄水 damming; impound(age); impounded water; impoundment water; pondage; ponding; retaining; stock of water; storage of water; stored water; water accumulation; water impoundment; water storage; water conservancy; water conservation
蓄水坝 impounded dam; impounding dam; retaining dam; storage dam; water storage dam
蓄水保水 water reserves
蓄水陂塘 water pan
蓄水泵 storage pump
蓄水比 storage ratio
蓄水材料 water-holding material
蓄水槽 catch basin; catchwork irrigation; ice bank tank; water cistern; water storage tank
蓄水层 aqueous stratum; aquifer; aquiferous layer; nappe aquifer; reservoir bed; water-bearing layer; water carrier
蓄水层抽水及注水试验 aquifer test
蓄水层渗透能力 permeability of aquifer
蓄水层输水能力 transmissibility
蓄水常数 storage constant
蓄水程度 percentage of storage
蓄水池 body of water; cistern; conservation pool; deluge collection pond; feeding reservoir; hydraulic accumulator; impounded basin; impounded body; impounded reservoir; impounding pond; impounding reservoir; impoundment; lasher; pondage reservoir; receiving basin; reserve pool; reservoir; retaining basin; retention basin; retention pond; retention reservoir; storage basin; storage bay; storage pond; storage pool; storage reservoir; storage tank; stored-water tank; water cistern; water pond; water reservoir; water storage pond; water storage reservoir; water (storage) tank; water supply; earth tank <用于牲畜用水的>
蓄水池出流曲线 storage-outflow curve
蓄水池顶盖 reservoir roof
蓄水池管理 pond management
蓄水池容量 reservoir capacity
蓄水池式发电厂 pondage type power plant
蓄水池式水力发电厂 hydropower plant with reservoir
蓄水池式水力发电站 reservoir type power plant
蓄水池中可能的最高水位 capacity level
蓄水的 aquiferous; impounded; water-bearing; water carrying; water holding
蓄水度 percentage of storage
蓄水发电厂 dam power plant
蓄水发电站 storage power station
蓄水法 <确定土渠渗透损失的> pondage method
蓄水方案 impounding scheme; storage scheme
蓄水放出 release of stored water
蓄水港池 impounded(harbo(u)r) basin; impoundment basin
蓄水港池闸门 impounded dock gate
蓄水高程 impounding elevation; impounding water level; storage level
蓄水高度 level of storage
蓄水工程 retaining works; storage works; water conservancy works; water storage project
蓄水供水箱 feed cistern
蓄水灌溉系统 storage irrigation system
蓄水罐 catchwork irrigation; water storage tank
蓄水河湾 storage bay
蓄水后活动性 post-impounding activity
蓄水湖 accumulation lake
蓄水集水坑 storage sump
蓄水建筑物 water-retaining structure
蓄水结构 water-retaining structure
蓄水井筒 storage shaft
蓄水开放 release of stored water
蓄水坑 storage sump
蓄水库 conservation (storage) reservoir; impounded reservoir; impounding reservoir; storage basin; storage reservoir
蓄水库容 conservation storage
蓄水库泄水曲线 storage-outflow curve
蓄水廊道 storage gallery
蓄水累积曲线 cumulative curve of

storage;mass curve of storage
蓄水量 impoundment; impoundage; pondage; storage content; water stock;water storage(capacity)
蓄水量变率 rate of storage change
蓄水量方程 storage equation
蓄水量改正 pondage correction
蓄水量校正 pondage correction
蓄水量限度 filling limit
蓄水量泄水量关系曲线 storage-outflow curve
蓄水量与开挖量之比 storage/excavation ratio
蓄水量与土方量之比 water-to-earth ratio
蓄水量与挖方量之比 water-to-earth ratio
蓄水率 percentage of storage;storage coefficient; storage factor; water-holding rate
蓄水面标高 impounded water level
蓄水面积 impounded area;storage area
蓄水能力 pondage capacity; water-holding capacity
蓄水平衡 storage balance
蓄水期 impoundment stage; storage period
蓄水器 accumulator;storage vessel
蓄水前池 headwater pond
蓄水前地震活动性 preimpounding seismicity
蓄水区 accumulator;water reserves
蓄水区域图 pondage topographic(al) map
蓄水曲线 storage curve
蓄水渠 Boezem;storage canal
蓄水容积 storage volume
蓄水容量 storage capacity
蓄水设备 storage facility;water storage facility
蓄水时间 storage period; storage time;water storage time
蓄水式港池 impounded basin; impounded dock
蓄水式水电厂 reservoir type power plant
蓄水式水电站 dam type hydroelectric(al) power plant;dam type power station;reservoir power plant
蓄水式雨量计 storage rain ga(u)ge
蓄水竖井 storage shaft
蓄水水库 water conservation reservoir
蓄水水位 retained water level
蓄水塔 draw-off shaft
蓄水梯田 absorption(-type) terrace
蓄水调峰 storage for peak production
蓄水调节 storage balance
蓄水围坑双壁开沟犁 basin lister
蓄水位 impounded level
蓄水稳定性 storage stability
蓄水系数 coefficient of storage;storage coefficient
蓄水箱 cistern tank;storage tank;water storage tank
蓄水箱顶部的通气孔 sniff hole
蓄水泄放 release of stored water
蓄水泄水曲线 storage-outflow curve
蓄水压力 retained water pressure
蓄水压力计 storage ga(u)ge
蓄水溢流 storage over-flow
蓄水与供给系统 storage and water supply system
蓄水运行 storage operation
蓄水闸 damming lock; flood-control dam;pound-lock;retaining lock
蓄水闸门 pound-lock
蓄水周期 cycle of storage;storage cycle;storage period
蓄水总量 storage volume
蓄丝式拉丝机 cumulative type wire-drawing
蓄污洼地 storage sewage depression
蓄泄 <水库的> outflow storage
蓄泄曲线 storage-draft curve
蓄泄因数曲线 <水库的> outflow storage factor curve
蓄雪保墒 snow retention
蓄压柜 accumulator
蓄压器 accumulator;high-pressure gas container; oil-pressure reservoir; pressure accumulator
蓄压器分隔活塞 piston separator
蓄盐器 salt accumulator
蓄养池 holding pond;stock pond
蓄液器 hydraulic accumulator
蓄意的违法行为 wilful misconduct
蓄意造成的损失 wilfully caused loss
蓄油表层 oil-absorbing coating
蓄油器 fuel accumulator
蓄油器系统 oil accumulator system
蓄鱼箱 cauf
蓄雨水池 rainwater retention basin

轩

轩 windowed veranda

宣

宣布 announce(ment); divulge; proclaim; proclamation; promulgate; pronounce

宣布撤回(声明等) recant
宣布分配股息 declaration of dividend
宣布共同海损 declare general average
宣布日 announcement day
宣布日期 declaration date;date of declaration
宣布无效 declare defeasance;overrule
宣布选择港口 declare port option
宣布装卸港 declare discharging/loading port
宣传处 publicity department
宣传地图 advertising map; poster map
宣传费 propaganda expenses; publicity expenses
宣传工具 <指书刊、电影、广播等> information medium [复 media]; mass media
宣传公司政策 publicize company policy
宣传画 poster
宣传节目时间 information hour
宣传品 promo
宣德窑 Xuande ware
宣告 adjudication; enunciation; pronounce
宣告合同无效 avoidance of contract; to declare the contract avoided
宣告判决 pronouncement of judgment
宣告破产 adjudication of bankruptcy; be declared bankrupt suspend; declare bankruptcy; suspend; suspend payment
宣告无力偿债 declaration of inability to pay
宣告无效 annul(ment) <法令、合同等的>; declaration of avoidance; voidance
宣告无效的债券 invalidated bonds
宣告无效条款 denunciation clause
宣告无罪 acquittal
宣礼塔 minaret
宣判无罪 acquit
宣誓度量人 sworn measurer and weigher
宣誓发票 sworn commercial invoice; sworn invoice
宣泄流量 discharge rate
宣纸 rice paper

喧

喧闹地区 hive

喧嚣潮汐急流 roust
喧噪 rumpus

玄

玄玻凝灰岩 palagonite tuff

玄精石 selenitum
玄块凝灰岩 basalt-agglomerate tuff
玄砂石 wacke
玄闪石 basaltic hornblende; oxyhornblende
玄武斑岩 basal-porphyry; basalt-porphyry; toadstone
玄武波浪质地层 basaltic glassy substratum
玄武玻璃【地】sordawalite; tachylyte; basalt obsidian; hyalobasalt; jaspoid; tachylite; wichtisite
玄武玻璃质底层 basaltic glassy substratum
玄武黑曜岩 tachylite
玄武湖公园 Xuanwu lake garden
玄武火山凝灰岩 basaltic tuff
玄武角闪石 lamprobolite
玄武黏[粘]土 basalt clay; basaltic clay; bole
玄武凝灰岩 basaltic tuff; basalt tuff; trap tuff
玄武球颗结构 variolitic texture
玄武熔岩 basaltic lava
玄武石 basaltic stone; graystone
玄武石土 basal wacke
玄武土 basalt clay; basaltic wacke; basalt (gray) wacke; whine-rock; wacke
玄武无球粒陨石 basaltic achondrite
玄武岩 basalt; lava basalt
玄武岩板 basaltic slab
玄武岩玻璃 basalt glass
玄武岩采石场 basalt quarry
玄武岩层 basaltic layer; basaltic stratum
玄武岩的 basaltic; basaltine
玄武岩的羊毛状物 basaltic wool
玄武岩地板 basalt floor tile
玄武岩地壳层 basaltic crustal layer
玄武岩豆砾石 basalt pea gravel
玄武岩盾 shield basalt
玄武岩粉末 basaltic meal; basaltic powder
玄武岩风化物 wacke
玄武岩高原 basalt plateau
玄武岩构造【地】basalt structure
玄武岩辉长岩石 basalt-gabbro rock
玄武岩混凝土 basaltic concrete
玄武岩混凝土板 basalt concrete slab
玄武岩火山石 basaltic lava
玄武岩基质 basal matrix
玄武岩浆 basaltic magma; basalt magma
玄武岩角闪石 basaltic hornblende; basaltine; lamprobolite; oxyhornblende
玄武岩节理 basaltic jointing; basaltic parting
玄武岩壳 basaltic shell
玄武岩矿棉 basalt wool
玄武岩亏损机制 basalt depletion mechanism
玄武岩类 basalt group; basaltic rocks; basanitoid
玄武岩类岩石 whinstone
玄武岩流 basalt flow; basaltic flow
玄武岩楼面瓷砖 basaltic floor tile
玄武岩脉 basalt dike
玄武岩黏[粘]结材料 basal cementing material)
玄武岩黏[粘]结料 basal matrix
玄武岩铺路小方石 basalt(ic) sett
玄武岩铺路小方石的热处理 heat-treatment of basalt paving setts
玄武岩(铺路用)小方石 basalt paving sett
玄武岩墙 basalt face
玄武岩区 malpais
玄武岩熔岩 basaltic lava
玄武岩砂 basaltic sand
玄武岩砂石 basaltic sandstone
玄武岩石板 basalt slab
玄武岩石粉 basalt meal; basalt powder
玄武岩炻器 basalt stoneware
玄武岩蚀变红黏[粘]土 erinite
玄武岩水层 basalt water
玄武岩水泥板 basalt-cement tile
玄武岩水泥砖 basalt-cement tile
玄武岩碎石 crushed basalt stone
玄武岩碎屑 basalt chip(ping)s
玄武岩碎屑混凝 basaltic chip(ping)s concrete
玄武岩碎屑混凝土 basalt chip(ping) concrete
玄武岩铁矿石 basaltic iron ore
玄武岩土 basaltic wacke
玄武岩细砾 basalt pea gravel
玄武岩纤维 basalt fiber[fibre]
玄武岩小方石路面 basalt sett paving
玄武岩小砾石 basaltic pea gravel
玄武岩屑堆 basaltic debris; basaltic scree
玄武岩压碎砂 basaltic stamp sand
玄武岩渣棉 basalt wool
玄武岩制品 basalt
玄武岩质底层 basaltic substratum
玄武岩质泛滥喷发 basaltic flood eruption
玄武岩状 <柱状> basaltiform
玄武质地壳层 basaltic crustal layer
玄武质火山角砾岩 basaltic volcanic breccia
玄武质集块岩 basaltic agglomerate
玄武质科马提岩 basaltic komatite
玄武质凝灰岩 basaltic tuff
玄武质熔岩 basaltic lava
玄武质岩 basaltic rock
玄学 metaphysics

悬

悬岸 overhanging bank; sheer cliff; steep

悬摆 dangle
悬摆稳定器 pendulum stabilizer
悬摆物 dangler
悬摆仪 <测定坝、桥梁和基础等的水平位移用> hanging pendulum
悬摆振动 pendulous vibration
悬板桥 stress ribbon bridge
悬板桌 drop-leaf table
悬帮 hanging wall; superincumbent bed; superjacent bed
悬帮平峒 hanging adit
悬壁板墙结构 cantilevered sheet wall structure
悬壁托梁 ancon(e)
悬壁支架 outboard support
悬臂 ally arm; bracket; cantilever(ing); column arm; jib(arm); jut; outreach; over arm; overhang; pole bracket(cantilever); pylon; top ladder
悬臂摆动机构 whipping gear
悬臂板 back slab; cantilever plate; cantilever slab; corbel back slab; projecting slab
悬臂板桩 unanchored sheet piling; cantilever sheet pile

悬臂板桩岸壁 cantilever sheet pile-(ed) quaywall; unanchored sheet pile quaywall
悬臂板桩式突堤码头 cantilever sheet pile jetty
悬臂绷绳 boom line
悬臂部分 cantilever arm; cantilever segment; cantilever portion <航线加密网设有野外控制点检核的伸延部分>; bracketed part <混凝土梁的>
悬臂侧伸长 side overhang
悬臂长柄挖斗掘土机 <每斗 0.4 ~ 60 立方米> steam navvy
悬臂长度 boom reach; cantilever-(ing) length; jib length; length of cantilever; overhang length
悬臂超出履带外缘的距离 overhang over track
悬臂超出轮胎或履带外缘的距离 overhang beyond tire/track
悬臂超出轮胎外缘的距离 overhang beyond tire
悬臂冲床 overhanging press
悬臂冲击试验机 Izod impact machine
悬臂冲击试验值 Izod impact value
悬臂传动机构 luffing mechanism
悬臂传动装置 luffing gear
悬臂打桩设备 cantilever pile driving plant
悬臂大梁 cantilever(ing) girder
悬臂单排板墙 cantilever(ed) single-wall
悬臂单排板墙结构 cantilever(ed) single-wall structure
悬臂单元 cantilevering unit
悬臂挡墙 flexible wall
悬臂挡土墙 cantilever retaining wall
悬臂的 bracketed; cantilevered; free-standing
悬臂的车站雨棚 cantilever platform roof
悬臂电动机 luffing motor
悬臂吊车 arm crane; boom hoist; bracket crane; cantilever crane; jib crane
悬臂吊杆 cantilever jib; luffing boom
悬臂吊杆铲土机 luffing-boom shovel
悬臂吊机 arm crane; jib crane; rotating jib
悬臂吊索 arm sling; luffing line <起重机的>
悬臂吊拖车 working trailer with jib crane
悬臂顶梁 cantilever bar
悬臂顶柱 top column
悬臂定位器 cantilever positioning device
悬臂端 cantilevered end
悬臂端外伸 overhang of cantilever ends
悬臂段 cantilever segment
悬臂法 cantilevering method
悬臂法施工 free cantilever method
悬臂方法 cantilever method
悬臂方式 cantilever mode
悬臂风应力分析法 cantilever method of wind stress analysis
悬臂扶垛 hanging buttress
悬臂俯仰机构 boom hoisting mechanism
悬臂负荷 cantilever loading
悬臂杆式喷灌机 boom sprinkler
悬臂钢板桩丁坝 cantilever sheet pile groyne
悬臂钢桁架桥 cantilever steel truss bridge
悬臂格栅 cantilever grizzly
悬臂弓形架 cantilever arched bridge
悬臂弓形梁 cantilevering curved girder

悬臂拱 cantilever arch
悬臂拱桁桥 cantilever arch truss bridge
悬臂拱梁 cantilevering arched girder
悬臂拱桥 cantilever arched bridge
悬臂拱式大梁 cantilever arched girder
悬臂拱式桁架 cantilever arch truss
悬臂拱形大梁 cantilever arched bridge
悬臂拱形桁架 cantilever arch truss
悬臂拱形桁桥 cantilever arch truss bridge
悬臂拱形桥 cantilever arched bridge
悬臂构件 cantilevering component
悬臂构件固定铁 tail(ing) iron
悬臂构件嵌入墙内 tailing in
悬臂构造 lookout
悬臂固定钢构件 tailing iron
悬臂固定式卸料塔 cantilever fixed unloading tower
悬臂辊 open-end roll
悬臂辊式轧机 overhang roll type mill
悬臂航线三角测量 cantilever(strip) triangulation
悬臂荷载 cantilever(ing) load(ing); overhung load
悬臂桁架 cantilever truss
悬臂桁架梁 cantilever girder; overhanging girder
悬臂桁架桥 cantilever truss bridge
悬臂后端伸出长度 rear overhang
悬臂后伸长 rear overhang
悬臂回转式排风罩 cantilever hood
悬臂活动半径 jib head radius
悬臂基础 cantilever foundation; cantilever footing
悬臂基脚 cantilever footing
悬臂加撑杆的屋顶桁架 cantilever strutted roof truss
悬臂加载 cantilever loading
悬臂架 console; freehand tool holder
悬臂架设 cantilever erection; erection by protrusion
悬臂架设法 cantilever erection method; cantilever method of erection
悬臂架设施工法 cantilever erection construction
悬臂剪力墙 cantilever shear wall
悬臂浇注 free cantilever casting
悬臂浇筑 cantilever casting
悬臂浇筑法 cast-in-place cantilever method; free-span method
悬臂脚手架 hanging scaffold(ing); hanging stage; needle scaffold; outrigger scaffold
悬臂铰接梁 Gerber's beam
悬臂结构 arm structure; cantilever structure
悬臂距 cantilever arm
悬臂距离 cantilever spacing; horn spacing
悬臂锯机 line-bar resaw
悬臂卡车 cantilever truck
悬臂跨度 cantilever span
悬臂跨径 cantilever space; cantilever span
悬臂跨(距) cantilever spalling
悬臂块 cantilever block
悬臂框架 cantilever frame
悬臂拉杆 boom tie
悬臂肋骨 cantilever frame
悬臂力矩 cantilever moment
悬臂连续梁 Gerber beam
悬臂梁 beam with one overhanging end; beam with overhanging ends; butt cap; cantilever(ing) beam; cant lever; clamp-free beam; hanging beam; outrigger; outrigger beam; overhanging beam; projecting beam; semibeam; semi-girder; socle beam

悬臂梁长度 length of cantilever
悬臂梁底板 outrigger base
悬臂梁截面 cantilever section
悬臂梁桥 cantilever beam bridge; cantilever girder bridge; Gerber bridge; simple supported girder bridge
悬臂梁(施工的)混凝土浇注 pouring in cantilever work
悬臂梁式冲击机 Izod impact machine
悬臂梁式冲击试验 cantilever beam impact tester; Izod(impact) test
悬臂梁式冲击试验机 cantilevered beam impact tester
悬臂梁式挡土墙 cantilever retaining wall
悬臂梁式碰撞试验 Izod impact test
悬臂梁屋顶 hammer-beam roof
悬臂梁转动角 angle rotation of cantilever beam
悬臂楼板 cantilevering floor
悬臂楼梯 bracketed step
悬臂楼梯平台 cantilevered landing
悬臂锚碇 cantilever anchorage
悬臂模板 cantilever formwork; cantilever shuttering
悬臂末端 cantilevering end
悬臂牛腿 cantilever bracket
悬臂拼装 erecting by overhang; free cantilever erection
悬臂拼装法 cantilever erection method; cantilevering assemble; erection by protrusion; free-span method
悬臂平板坝 overhanging dam
悬臂平板梁 overhanging beam
悬臂平台 cantilever platform
悬臂铺管机 boom cat
悬臂起重杆 luffing jib
悬臂起重机 boom crane; boom derrick; boom hoist; jib crab; jibcrane; overhang crane; rotating derrick; swing boom crane; cantilever crane
悬臂起重机船 floating derrick
悬臂起重机架 boom cat
悬臂起重机台架 cantilever(ed) gantry
悬臂砌筑 cantilever masonry(work)
悬臂牵索滑车 <挖掘机> luff tackle
悬臂墙 cantilever(ed) wall
悬臂桥 cantilever bridge
悬臂桥墩 jutting-off-pier
悬臂桥面板 cantilever flange
悬臂桥台 cantilever abutment
悬臂倾斜角 <起重机> boom pitch
悬臂人行桥 cantilever pedestrian bridge
悬臂扇形体 <桥梁结构上的> cantilever segment
悬臂设计法 cantilever method of design
悬臂伸出 cantilever out
悬臂伸距 handling radius
悬臂伸缩门式起重机 gantry crane with retractable boom
悬臂升降机构 boom hoisting mechanism
悬臂施工(法) cantilever construction
悬臂施工架 cantilever construction carriage
悬臂式(C形)压床 gap-frame press
悬臂式安装 overhung mounting
悬臂式岸壁 cantilevered quay wall
悬臂式板桩 cantilever sheet pile[piling]
悬臂式板桩岸壁 cantilever sheet pile bulkhead
悬臂式板桩码头 cantilever sheet pile-(ed) quaywall
悬臂式板桩墙 cantilever sheet piling wall
悬臂式棒条筛 cantilever grizzly

悬臂式壁灯 bracket light
悬臂式标志 cantilever sign
悬臂式铲斗 boom bucket
悬臂式打桩机 cantilever driving plant; overhanging pile driver
悬臂式单侧信号桥 cantilever half bridge
悬臂式单排板桩墙 cantilevered single-wall
悬臂式挡土墙 cantilever(-type) retaining wall; cantilever wall
悬臂式的 overhung
悬臂式电子皮带秤 arm-suspended belt scale; hang-arm electronic belt scale
悬臂式钢坝 steel cantilever dam
悬臂式钢板弹簧 cantilever spring
悬臂式钢桁架 cantilever steel truss
悬臂式钢桁架桥 steel cantilever truss bridge
悬臂式拱桥 archaic cantilever bridge; arched cantilever bridge; cantilever arched bridge
悬臂式构架 cantilever(ed) frame [framing]
悬臂式构件 cantilevered element
悬臂式桁架 cantilever truss; overhanging truss
悬臂式桁架桥 cantilever lattice girder bridge
悬臂式花格大梁桥 cantilever lattice girder bridge
悬臂式基础 cantilever foundation; cantilever footing
悬臂式加料吊车 underslung charging crane
悬臂式加密 cantilever bridging; cantilever extension <指像片的角测量由有控制地区扩展到无控制地区>
悬臂式交通标志 cantilever(-type) traffic sign
悬臂式脚手架 bracket scaffold(ing); scaffold on bracket
悬臂式脚手架走道 cantilevered gantry
悬臂式结构 cantilever construction
悬臂式井架 cantilever derrick
悬臂式刻图仪 floating arm engraver
悬臂式龙门起重机 cantilever gantry crane
悬臂式楼梯 bracketed stair(case); bracketed step; cantilevered steps
悬臂式螺旋 open-end auger
悬臂式码头岸壁 cantilever bulkhead quay; cantilever bulkhead wharf
悬臂式锚碇设施 cantilever anchorage
悬臂式门吊 cantilever gantry crane
悬臂式模板 cantilever form(work)
悬臂式膜窗 cantilever diaphragm
悬臂式喷灌机 boom sprinkler
悬臂式平板坝 cantilever-deck dam
悬臂式起重机 arm crane; boom hoist; bracket crane; cantilever(ed boom) crane; gib hoist; jib crane
悬臂式千斤顶 outrigger jack
悬臂式墙 cantilever-type wall
悬臂式峭壁 overhanging cliff
悬臂式轻便钻架 cantilever portable drill rig
悬臂式穹顶 cantilever dome
悬臂式人行道 cantilever(ed) footway; cantilever for footway; overhanging footway; cantilevered foot path
悬臂式人行桥 cantilever foot bridge
悬臂式输送机 rocker conveyer[conveyor]
悬臂式挑檐 bracketed cornice
悬臂式通风机 overhung-type fan
悬臂式透平 overhung turbine
悬臂式挖掘装载机 swing loader

悬臂式卧倒门 cantilever flap gate
悬臂式屋顶 cantilever roof
悬臂式坞壁小车 dock arm
悬臂式压床 overhanging press
悬臂式压力机 horn press
悬臂式圆屋顶 cantilever cupola
悬臂式闸墙 cantilever lock wall
悬臂式支墩坝 cantilever buttress dam
悬臂式走道 overhanging footway
悬臂输送带 boom belt conveyer[conveyor]
悬臂输送机 arm conveyer[conveyor];boom conveyer[conveyor]
悬臂双曲线抛物面屋顶 cantilever-type hyperbolic paraboloidal roof
悬臂水翼 cantilever hydrofoil
悬臂踏板 hanging steps
悬臂踏步 cantilever steps
悬臂梯级 cantilever steps
悬臂提升机 cat's head
悬臂天平 cantilever construction carriage
悬臂挑梁 corbel arm
悬臂条屏 cantilever grizzly
悬臂条筛 cantilever grizzly
悬臂条网屏 cantilever bar screen
悬臂条子铁栅筛 cantilever screen of bars
悬臂铁栅筛 cantilever bar screen
悬臂托架 cantilever bracket
悬臂托梁 ancon(e);hammer beam
悬臂托梁桁架 hammer-beam truss
悬臂托柱 hammer post
悬臂托座 cantilever bracket;jib bracket
悬臂挖泥船 boom dredge(r)
悬臂圬工 cantilevering masonry
悬臂下降超速限制器 cantilever descending anti-overspeed device
悬臂限位装置 cantilever caging device
悬臂向前施工法 cantilevering forward
悬臂销 drawbar pin
悬臂斜脊屋顶 cantilever hipped-plate roof
悬臂芯 balanced core; short square core
悬臂行车梁 cantilevering crane(runway)girder
悬臂型平台 outrigger platform
悬臂延伸(段)cantilever extension;overhang extension
悬臂引伸 cantilever extension
悬臂应变仪 strain-ga(u)ged cantilever
悬臂应力分析 cantilever analysis
悬臂圆锯 radial saw
悬臂运动 luffing motion
悬臂再锯机 line-bar resaw
悬臂折板薄壳屋顶 cantilever prismatic shell roof
悬臂折板屋面 cantilever folded plate roof
悬臂支承 arm rest
悬臂支承桁架 cantilever braced truss
悬臂支承架 cantilever supporting strut
悬臂支承销 arm base pin
悬臂支架 hanging bracket;pole bracket(cantilever);cantilever outrigger
悬臂支托 cantilever bracket
悬臂支托脚手架 needle scaffold
悬臂支柱 outrigger shore
悬臂轴 overhung shaft
悬臂轴承架 stay bracket
悬臂肘板 bracket of cantilever
悬臂柱 cantilever column; cantilever mast;cantilever post
悬臂砖 corbel brick
悬臂桩 cantilever pile
悬臂装料杆 peel
悬臂装料机 jib loader
悬臂装卸吊杆 loading boom
悬臂总伸距 overall overhang
悬臂钻床 drilling machine with jointed arm
悬臂作用 cantilever action;cantilever effect
悬臂座 cantilever support
悬臂座板 cantilever saddle
悬冰 crystal ice;hanging ice;ice fall
悬冰川 cascading glacier; cornice glacier;hanging glacier
悬冰斗 hanging cirque
悬仓 suspension bunker
悬撑槽钢 sling stay channel
悬撑角座 sling stay T
悬尺测距 catenary taping
悬冲积扇 hanging alluvial fan
悬出的构件卸荷作用 relieving effect of overhanging element
悬窗滑轮窗梃 sash run
悬垂 lop;overhang(ing);pendency
悬垂把手 drop handle
悬垂插座 hanging socket
悬垂导管导轮 hanging pipe carrier; suspended pipe carrier; suspending pipe carrier
悬垂导线导轮 suspended wire carrier
悬垂的 dependent; overhung; pendant;pendulous
悬垂的叶片(装饰)swag leaf
悬垂的照明电线 drop cord
悬垂吊着 pend
悬垂腹 pendulous abdomen
悬垂拱 catenary arch
悬垂构造 pendant formation
悬垂回波 overhang echo
悬垂汇流 hanging junction
悬垂绝缘子 suspension insulator
悬垂绝缘子串 link suspension insulator;suspension string
悬垂门 overhung door
悬垂石膏 suspension plaster
悬垂式按扭 pendent switch; suspension push
悬垂式按钮站 pressel switch
悬垂式加速计 pendulous accelerator; pendulous accelerometer
悬垂式水尺 chain ga(u)ge; wire-weight ga(u)ge
悬垂式信号灯 pendent signal
悬垂式信号装置 pendant signal; pendent signal
悬垂式照明 catenary lighting
悬垂体 <洞顶的> horst
悬垂位 vertical position
悬垂物 overhang
悬垂线 catenarian;catenary
悬垂线测距 measuring in catenary
悬垂线夹 suspension clamp
悬垂心 drop heart;pendulous heart
悬垂信号机【铁】hanging signal; suspended signal
悬垂形绝缘子 tie-down insulator
悬垂型 suspended pattern
悬垂性(能)draping property
悬垂直曲线 sag vertical curve
悬锤 bob; hanging load; plumb bob; suspension weight
悬锤测量法 plumb bob method
悬锤机构 plumb bob section
悬锤(式)水(标)尺 wire weight ga(u)ge
悬锤式水位计 wire weight ga(u)ge
悬锤陀螺仪 pendulous gyroscope
悬锤支座 overhanging support
悬带 suspensor(y)
悬带桥 stress ribbon bridge; suspended ribbon bridge
悬的 suspensory
悬灯 ceiling lamp
悬灯梁 candle beam
悬滴 hanging drop
悬滴标本 hanging drop preparation
悬滴槽 hanging drop cell
悬滴法 direct transmission; pendent drop method;sessile drop method
悬滴汞电极 hanging mercury drop electrode
悬滴试验 hanging drop test
悬滴术 hanging drop technique
悬滴涂布观察法 spread suspension method
悬滴雾化器 hanging drop atomizer
悬滴装置 hanging drop preparation
悬点 point of suspension; suspension center[centre]; suspension point; center of suspension
悬吊 a' cockbill;overhead suspension
悬吊鞍形屋顶 cable roof with saddle shape
悬吊按钮 pendant push
悬吊按钮站 pendant box
悬吊壁 hanging wall
悬吊部件 suspension member
悬吊槽 hanging gutter
悬吊槽式运输机 suspended tray conveyer[conveyor]
悬吊车辆 suspended car
悬吊窗扇 <用于双悬窗> hung sash
悬吊磁铁 suspended magnet
悬吊大梁 independent girder
悬吊带 sling
悬吊的 hanging;pendent;suspended
悬吊灯具 pendant light fitting
悬吊底模的拉杆 formwork hanger tie
悬吊地板 suspended tread floor
悬吊点 point of suspension
悬吊电磁铁 suspended electromagnet
悬吊顶 suspended ceiling
悬吊法 <测质的或测重心的> suspension method
悬吊翻转电动机 pivoted motor
悬吊杆 suspension rod
悬吊钢丝 suspension wire
悬吊高度 overhung height
悬吊隔声板 hanging baffle
悬吊工具 hanging harness;suspension attachment
悬吊构造 suspension construction
悬吊骨架 suspended framing
悬吊管 hanged pipe; hanged tube; hanging pipe;pendant tube;pendent pipe; pendent tube; sling conduit; sling pipe; sling tube; suspended pipe; underslung conduit; underslung pipe[piping];underslung tube[tubing]
悬吊管道 suspended pipeline
悬吊管子的钢带 pipe strap
悬吊辊式粉碎机 suspended roller mill
悬吊桁架 drop in truss
悬吊护板 <上轧辊出口侧的> hanging guide
悬吊滑车式支架 pulley rig
悬吊滑轮 hanging sheave;pulley with suspension;sling block
悬吊环 suspension bail
悬吊货物 suspended cargo
悬吊机构 suspension gear
悬吊件 pendant fitting
悬吊绞车 suspension winch
悬吊脚手架 hanging scaffold(ing); suspended scaffold(ing); suspended staging
悬吊接头 swing joint
悬吊结构 suspended structure; suspension structure
悬吊截盘 underslung jib
悬吊进漏斗 suspension bunker
悬吊开关 pear-switch;pendant push; pressel switch;pendant switch
悬吊开关盒 pendant control box
悬吊壳体 hung shell
悬吊跨度 suspended spalling
悬吊梁 hanger beam;suspension beam
悬吊楼板 hanging floor
悬吊螺栓方法 hanging bolt method
悬吊锚碇索 suspension anchor cable
悬吊面 suspended floor
悬吊模板工程 suspended formwork; suspended shuttering
悬吊耐火钢模夹紧装置 beam clamp
悬吊耐火墙 suspended refractory wall
悬吊平顶 <起隔声绝热作用的> counter ceiling; drop ceiling; suspended ceiling
悬吊平台 hanging platform
悬吊牵条 sling stay
悬吊潜水钟 suspended diving bell
悬吊墙面板 hanging shingling
悬吊取样管 swing sample pipe
悬吊绳 pendant cord
悬吊式按钮 pear-push
悬吊式板屋顶 plate suspension roof(ing)
悬吊式壁面铺盖层 suspended wall lining
悬吊式簸动运输机 suspended shaker
悬吊式操作台 suspended platform
悬吊式打桩导向架 free-hanging lead
悬吊式地面板 suspended ground floor
悬吊式堵塞器 hole-down packer; hook wall packer
悬吊式堵头 <水下浇筑混凝土导管的> traveling plug
悬吊式飞机库 aircraft hanger
悬吊式钢板弹簧 <位于车轴下的> underslung laminated spring
悬吊式钢筋混凝土屋顶 suspended reinforced concrete roof
悬吊式工作笼 work cage
悬吊式挂车 hanging trailer;tug
悬吊式锅炉 suspended boiler
悬吊式过热器 pendant superheater
悬吊式桁架 suspension truss; underslung truss;hanging truss
悬吊式喉镜 suspension laryngoscope
悬吊式火力钻机 suspension-piercing machine
悬吊式火力钻进 suspension piercing
悬吊式脚手架 ship scaffold; suspended scaffold(ing); swinging staging;swinging scaffold
悬吊式框架堰 suspended-frame weir
悬吊式扩散板 suspended diffuse plate
悬吊式拉边机 hanging top roller
悬吊式料斗 swing(ing) hopper
悬吊式楼板 suspended floor
悬吊式楼梯 suspended stair(case)
悬吊式漏斗 swinging hopper
悬吊式炉 suspended type of furnace
悬吊式模板 pendent forms; suspended formwork;suspended forms
悬吊式平顶漫射照明灯盒 suspended light diffusing ceiling panel
悬吊式平顶漫射照明灯框 suspended light diffusing ceiling panel
悬吊式平台 suspended platform
悬吊式穹顶 pendant vault
悬吊式升降机 carriage elevator
悬吊式水轮发电机 suspension hydraulic generator
悬吊式水平仪 string level
悬吊式踏板 suspended pedal
悬吊式天顶 suspension ceiling
悬吊式天花板 suspended ceiling

悬吊式跳板 flying gangway
悬吊式屋顶 suspended roof
悬吊式吸声器 suspended absorber
悬吊式吸声天花板 suspended acoustical ceiling
悬吊式吸音器 suspended absorber
悬吊式照明器具 suspended luminaire
悬吊式照明设备 suspended luminaire
悬吊式照明装置 pendant luminaire
悬吊式中间再热器 pendant reheater
悬吊术 suspension method
悬吊台 suspension yoke
悬吊体系 suspended system
悬吊天花板的光源格栅 luminaire grid suspension ceiling
悬吊天花板系统的光源格栅 luminaire grid suspension ceiling system
悬吊托架 pendentive bracketing
悬吊屋顶 cable roof;suspension roof
悬吊物 handing object;suspended object;suspender;suspensory
悬吊物件用的梁 suspended beam
悬吊系杆 hanging tie
悬吊系统 suspension system
悬吊系统参数 coefficient of suspense system
悬吊纤维板 hanging fiber
悬吊弦 pendant cord
悬吊线圈 suspension coil
悬吊效果 suspension effect
悬吊岩 pendent rock
悬吊元件 drop-in unit
悬吊载荷 suspended load
悬吊在顶板上的钢绳支架输送机 roof-supported wire rope conveyer [conveyor]
悬吊振动筛 aerovibro screen
悬吊支承 suspended support
悬吊质量 suspended mass
悬吊重量 suspended weight
悬吊重物搬运的装置 suspended load handling device
悬吊轴承 drop bearing;hanger bearing;hanging bearing
悬吊柱 <门的> swinging post
悬吊砖 suspension brick
悬吊装饰 drop ornament
悬吊装置 hanger attachment;pendant fixture
悬吊作用 hanging effect
悬顶 hanging arch
悬顶支护 suspension roof support
悬动钻床 overhead travel(l)ing drilling machine
悬斗秤 suspension bucket scale;suspension hopper scale
悬而未决的 pending;suspensive;unresolved
悬而未决的问题 open question
悬阀 hanging valve
悬扶壁 hanging buttress
悬扶垛 hanging buttress
悬浮 dispersion;flo(a)tation;hover;suspend
悬浮百分率 suspension percentage
悬浮摆轮 floating balance
悬浮搬运 suspension transport
悬浮焙烧 flash roasting;suspension roasting
悬浮焙烧炉 flash roaster;suspension roaster
悬浮冰 hanging ice
悬浮材料 suspended material
悬浮残渣 non-filtrable residue
悬浮层 floating blanket;superposed layer;suspension layer
悬浮车 levitated vehicle
悬浮车辆 levitation vehicle
悬浮尘埃 suspended dust;suspending dust
悬浮尘埃浓度指数 concentration index of suspended dust
悬浮尘埃微粒 suspended particles of dust
悬浮沉积(物)suspended sediment
悬浮澄清池 sludge blanket clarifier;sludge blanket filter
悬浮床反应器 suspended bed reactor
悬浮床沙 suspended bed sediment
悬浮床沙质 suspend bed sediment
悬浮磁分离器 suspended magnet separator
悬浮单体 suspending monomer
悬浮的 suspended
悬浮的粒状物质 suspended particulate
悬浮底质 suspended bed sediment
悬浮电极 suspension electrode
悬浮煅烧窑 suspension kiln
悬浮法 suspension method;suspension process
悬浮反应堆 slurry reactor
悬浮肥料 suspension fertilizer
悬浮废液雾化技术 atomized suspended oxidation technique
悬浮分级 levigation
悬浮分离法 suspension separation method
悬浮分离理论 theory of suspended separation
悬浮浮子 composite float
悬浮负载 suspended load;suspension load
悬浮共聚 suspension copolymerization
悬浮固体 suspended solids
悬浮固体的去除 suspended solid removal
悬浮固体含量 suspended solid content
悬浮固体接触澄清池 suspended solid contact clarifier
悬浮固体接触反应池 suspended solid contact reaction basin;suspended solid contact reactor
悬浮固体接触反应器 suspended solid contact reactor
悬浮固体颗粒 solid particle in suspension;suspended solid particle
悬浮固体颗粒去除 suspended solid removal
悬浮固体量 quantity of suspended solid
悬浮固体浓度 suspended solid concentration
悬浮固体水位图 suspended solid hydrograph
悬浮固体微粒层 fluidized bed
悬浮固体污染物 suspended solid pollutant
悬浮固体物质 solid material in suspension;suspended solid matter;suspended solid material
悬浮固体有机碳 suspended solid organic carbon
悬浮固体总量 total suspended solids
悬浮烘燥机 air lay drying machine
悬浮灰分 suspended ash
悬浮挥发性固体 suspended volatile solid
悬浮混合物 suspended mixture
悬浮火车 aerotrain
悬浮剂 deflocculant;deflocculater;deflocculation agent;deflocculator;dispersing agent;suspended agent;suspending agent;suspending medium;suspension agent
悬浮加热法 levitation heating
悬浮夹带 entrain
悬浮胶(体)suspended colloid;suspension colloid
悬浮结构 suspension structure;suspended structure
悬浮介质 disperse medium;dispersion medium;dispersive medium;suspending medium
悬浮介质床 suspended media blanket
悬浮金属粉末 leafed powder
悬浮聚合(法)bead polymerization;pearl polymerization;suspension polymerization
悬浮颗粒 particle in suspension;suspended particle;suspended particulate;suspended solids
悬浮颗粒沉降速度 setting velocity of suspended particles
悬浮颗粒量测器 hi-volume sampler
悬浮颗粒生物膜法 suspended biological film enveloped particle method
悬浮颗粒污泥床反应器 suspended granular sludge bed reactor
悬浮颗粒物采样装置 suspended particulate sampling equipment
悬浮颗粒物质 suspended particulate matter
悬浮颗粒物自动采样器 automatic suspended particulate sampler
悬浮颗粒总量指数 total suspended particulate index
悬浮快速内燃预热器 suspension flash preheater
悬浮力 floating power
悬浮粒状物质 suspended particle
悬浮粒子 suspended particle
悬浮流 suspension current
悬浮流速 suspended flow velocity;sustaining velocity
悬浮硫酸盐 suspended sulfate
悬浮炉 shower furnace
悬浮率 rate of pickup;suspension percentage
悬浮煤粉尘 coal-in-air suspension
悬浮能力 suspending power
悬浮泥沙 suspended load;suspended material;suspended sand;suspended sediment;suspended silt
悬浮泥沙浓度 suspended sediment concentration
悬浮泥沙总量 total suspended sediment
悬浮泥渣层 sludge blanket
悬浮泥渣层反应池 sludge blanket reactor
悬浮气化煤气发生炉 suspension gasifier;suspension producer
悬浮气力输送机 floating pneumatic conveyer[conveyor]
悬浮气体中的微粒 gas-borne particles
悬浮区熔生长法 floating zone melting growth
悬浮燃料 suspension fuel
悬浮燃烧 suspension burning;suspension combustion;suspension firing
悬浮染色 suspension dyeing
悬浮溶剂的黏[粘]滞度 viscosity of suspending medium
悬浮熔炼 smelting in suspension;suspension smelting
悬浮熔融 levitation melting
悬浮沙 sand in teeter;suspended sediment;teetering sand
悬浮闪烁体 suspended scintillator
悬浮生长处理法 suspended-growth process
悬浮生长反应器 suspended-growth reactor
悬浮生长生物法 suspended-growth biological process
悬浮生长脱氮法 suspended-growth denitrification process
悬浮生长微生物 suspended-growth microorganism
悬浮生长污泥 suspended-growth sludge
悬浮生长硝化法 suspended-growth nitrification process
悬浮石墨 deflocculated graphite
悬浮式顶棚 floating ceiling
悬浮式短途往返列车 hover-shuttle
悬浮式分解炉 suspension calciner
悬浮式后桥轴 oscillating axle
悬浮式混凝土地板 suspended concrete floor
悬浮式混凝土基础 suspended concrete base
悬浮式混凝土基底 suspended concrete base
悬浮式混凝土毛地板 suspended concrete(slab)subfloor
悬浮式列车 hovertrain
悬浮式桥轴 free floating axle
悬浮式陀螺仪 floated gyro(scope)
悬浮式预热器 suspension(type)preheater
悬浮式直方图 suspended histogram
悬浮式直接脱硫 suspension type direct desulfurization
悬浮输送 fluidized conveying
悬浮输送机 fluidized conveyer[conveyor]
悬浮水 suspended water
悬浮水带 suspended-water zone
悬浮水分 suspended water
悬浮速度 speed of suspension;suspended velocity;suspending velocity;sustaining velocity
悬浮塑料填料 suspended plastic filler
悬浮隧道 floating tunnel;submerged tunnel
悬浮态燃料 suspension type fuel
悬浮探空仪 suspended sonde
悬浮碳 suspended carbon
悬浮体 disperse solid;slurry;soliquoid;suspended substance
悬浮体分选法 suspensoid process
悬浮体粒子薄层 platelet
悬浮体流变学 rheology of suspensions;suspension rheology
悬浮体流量计 suspended body flowmeter
悬浮体浓度(测定)仪 Tyndallometer
悬浮体散射术 nephelometry
悬浮体散射仪 nephelometer
悬浮体系 suspension system
悬浮体系斜拉桥 cable-stayed bridge in floating system
悬浮体渣脚 suspended sedimentation
悬浮体渣脚取样器 suspended sediment sampler
悬浮体轧染(法)pigment padding;pigment pad dyeing
悬浮添加剂 suspension additive;dispersant additive
悬浮填充 suspension packing
悬浮填料 suspended filler
悬浮填料床 suspended packed bed
悬浮土沙 suspended sand;suspended load
悬浮土沙采取器 suspended load sampler
悬浮团粒去除 suspended aggregate removal
悬浮陀螺仪 floating gyro;floating type gyroscope
悬浮微粒 aerosol particle;suspended microparticle
悬浮微粒分离 <大气的> aerosol separation
悬浮微粒控制 particulate control;suspended microparticle control

悬浮微粒污染物 suspended microparticle contaminant; aerosol contaminant <空气中的>
悬浮微粒总量 total suspended particulates
悬浮微生物 air-borne microbial
悬浮尾矿排出沟 tailrace
悬浮稳定性 suspension stability
悬浮污泥 float(ing) sludge
悬浮污泥层 suspended sludge blanket
悬浮污泥层法 suspended sludge blanket process
悬浮污泥澄清池 suspended sludge clarifier
悬浮污泥技术 suspended sludge technology
悬浮污泥浓度 suspended sludge concentration
悬浮污泥质取样器 hi-volume sampler
悬浮无机物 suspended inorganic matter; suspended mineral
悬浮物 matter in suspension; seston; suspended load; suspended matter; suspended solids; suspended substance
悬浮物沉积 sedimentation of suspension
悬浮物分析仪 suspended solid analyser[analyzer]
悬浮物流变 rheology of suspension
悬浮物浓度 suspended matter concentration
悬浮物浓度分布 distribution of suspended matter concentration
悬浮物体分离试验 <废水的> jar test
悬浮物质 suspended load; suspended matter; suspended substance; suspended material
悬浮物质的颗粒 particle of suspended matter
悬浮物中的沉淀颗粒 sedimentary particles in suspension
悬浮物中的非生物部分 tripton
悬浮雾化技术 atomized suspension(al) technique
悬浮系 disperse system; dispersion system
悬浮细胞 suspension cell
悬浮现象 suspension phenomenon
悬浮相 suspended phase
悬浮相似性 suspension similarity
悬浮效应 levitation effect
悬浮性 anti-settling property; suspension property
悬浮性固体 suspended solids
悬浮性固体量 floatable solid content
悬浮性粒子 suspended particulate
悬浮旋 suspended liquid
悬浮烟灰和烟的浓度指数 concentration index of suspended smoke and soot
悬浮研磨法 suspension liquor grinding
悬浮液 dispersion; soliquoid; suspending liquid; suspension dispersion; suspension fluid; suspension liquid
悬浮液处理 preconditioning
悬浮液(反应)堆 suspension reactor
悬浮液密度 density of suspended liquid
悬浮液体 suspending liquid
悬浮液体颗粒组成电导测定仪 co(u)lter counter
悬浮液雾化技术 atomized suspension(al) technique
悬浮异重流 suspended density flow; suspended turbidity current
悬浮油 suspended oil; suspension oil
悬浮有机物 suspended organic matter
悬浮淤泥 suspended mud
悬浮淤渣 suspended sludge
悬浮预热 suspension preheating
悬浮预热器 suspension preheater
悬浮预热器干法窑 dry process kiln with suspension preheater
悬浮预热器和分解炉 suspension preheater and flash furnace
悬浮预热器窑 suspension preheater kiln
悬浮预热系统 suspension preheating system
悬浮运载工具 hovercraft
悬浮载体 suspended carrier
悬浮载体生物膜反应器 suspended carrier biofilm reactor
悬浮在水中的 neutrally buoyant
悬浮质 suspended load; suspended matter; suspensate
悬浮质搬运 suspended load transport
悬浮质的固体颗粒 solid particle in suspension
悬浮质点 particles in suspension
悬浮质流 turbidity current
悬浮质泥沙 sediment in suspension
悬浮质移动率 rate of transfer of suspended particles
悬浮质运动 movement of floating debris
悬浮质中沉淀颗粒 sedimentary particles in suspension
悬浮中的沉淀物 sedimentary material in suspension
悬浮轴承 suspension bearing
悬浮转变 suspended transformation
悬浮转变作用 suspended transformation
悬浮桩 floating pile
悬浮桩基 floating pile foundation
悬浮状态 disperse state; dispersion state; suspended state; suspension; quick condition <土的>
悬浮状态的 in suspension
悬浮状态的泥沙 sediment in suspension
悬浮状油 suspended oil
悬浮浊流 suspended turbidity current
悬浮着的 in suspension
悬浮总体 suspension population
悬浮作用 suspension effect
悬浮座位 suspension seat
悬干 hang dry
悬杆 hanger; hanging bolt; hanging stick; loft; suspender; suspension chord; suspension link; suspension rod
悬杆堆料机 boom stacker
悬杆法 hanger method
悬杆式测流仪 rod-suspended current meter
悬杆式流速仪 rod-suspended current meter
悬杆式水文缆道 rod-suspended hydrometric(al) cableway
悬杆水文缆道 hydrometric(al) cableway with rod suspension
悬拱 sprung arch; sprung roof; suspended arch; suspension arch; hanging arch
悬拱窑 catenary kiln
悬沟 hanging valley
悬钩法 hanger method
悬钩子 bramble
悬谷 hanging valley; suspended valley
悬挂 append; drape; draping; hanging-up); mount (ing); suspend; suspense
悬挂摆 inverted pendulum
悬挂摆动篦床式熟料篦冷机 pendulum clinker cooler
悬挂棒 hanger rod
悬挂臂 mounting arm; suspension arm
悬挂部件 sprung parts
悬挂采油封隔器 retainer production packer
悬挂抽油杆到摇杆端部的装置 mule-head hanger
悬挂垂球 freely hanging plummet
悬挂单轨系统 underslung monorail system
悬挂导杆 hanging guide
悬挂导架 hanging guide; hanging leader
悬挂的 hanging; suspended; suspensory; underslung
悬挂的灯座 pendant lampholder; pendant lamp-socket
悬挂的楼梯 <防火用> cantilever steps
悬挂滴水槽 hung gutter
悬挂点 hitch point; landing top; pendant point; point of suspension; suspension point; terminal vertex
悬挂段 drop-in section
悬挂垛盘的木框 hanging set
悬挂砝码 suspended counterweight
悬挂辅助件 suspension-auxiliary components
悬挂干燥 loft-dried; loft drying
悬挂干燥器 loft drier[dryer]
悬挂杆 suspension post
悬挂给料斗 overhead feed hopper
悬挂工具 mounted implement
悬挂钩环 suspension shackle
悬挂构造 suspension construction
悬挂罐笼的保险链 bridle chain
悬挂轨道 underslung track
悬挂滚翻中心 suspension roll center [centre]
悬挂荷载 suspended load; hanging load <钻塔等的>
悬挂桁架 hanging truss
悬挂护舷 hanging fender
悬挂滑车 hanging block; hanging sheave
悬挂滑车螺栓 suspension tackle bolt
悬挂机架 mounting frame; mounting group
悬挂夹具 suspension chord
悬挂家具 suspension furniture
悬挂架 headstock
悬挂架销钉 suspension spike
悬挂减振器 suspension damper
悬挂建筑(物) suspension building; suspension construction; slung-span construction; suspended structure
悬挂键 dangling bond
悬挂脚手架 suspended catwalk
悬挂脚手通道 suspended catwalk
悬挂铰接轴 linkage pitman
悬挂接头 hanger coupling
悬挂结构 slung-span construction; suspended structure; tension structure
悬挂经纬仪 suspension theodolite
悬挂经纬仪横托架 arm for theodolite; arm for transit
悬挂开关 cord switch; pendant switch
悬挂抗侧倾刚度 roll stiffness of suspension
悬挂跨度 suspended span
悬挂棱镜补偿器 compensator with suspended lens
悬挂犁 mounted plow
悬挂犁犁轮 mounted-plow wheels
悬挂联结梁 <可代替常规的枕梁> 【铁】suspension adapter
悬挂帘 suspended screen
悬挂链条 suspension link
悬挂链移动曝气技术 suspended chain moving aeration technology
悬挂梁 suspension beam
悬挂料斗 suspension bunker
悬挂裂纹 hanger crack
悬挂楼板 hung floor
悬挂炉顶 suspended arch
悬挂罗经 hanging compass
悬挂螺栓 hanging bolt; suspension bolt
悬挂毛细水 suspended capillary water
悬挂锚链 pendant chain
悬挂门滑机 barn-door hanger
悬挂模板 suspended shuttering; suspended formwork
悬挂模型 suspended model
悬挂膜结构 suspended membrane structure
悬挂木瓦 hanging shingling
悬挂盘式提升机 suspended tray elevator
悬挂平衡重 suspended counterweight
悬挂屏幕 suspended screen
悬挂起重机 overhang crane
悬挂器 hanger
悬挂桥式堆垛起重机 suspended stacking crane
悬挂泉 suspended spring
悬挂洒水器 pendant sprinkler; pendent sprinkler
悬挂三通 drop tee
悬挂砂轮机 suspended grinder
悬挂筛 suspended screen; suspension screen
悬挂设备 suspension equipment; suspension fixture
悬挂升程 lifting range
悬挂式 suspended type; suspension type; suspension ultimate
悬挂式安装法 underslung suspension
悬挂式安装管路 pendant mounting channel
悬挂式按钮 pendant control switch
悬挂式玻璃窗 suspended glazing
悬挂式薄壳 suspended shell
悬挂式传感器 pendulous pick-up
悬挂式传声器 suspended microphone
悬挂式打桩导(向)架 hanging leader
悬挂式单轨铁路 suspended monorail(way)
悬挂式单梁吊 underhung crane
悬挂式单位散热器 suspended type unit heater
悬挂式导向体 suspended guide
悬挂式的 direct connected; direct-mounted; linkage-mounted
悬挂式灯具 pendant light fitting; suspension light fitting
悬挂式底层地板 suspended ground floor
悬挂式地板 suspension floor
悬挂式地板梁 suspended floor beam
悬挂式点积采样器 cable-suspended point-integrating sampler
悬挂式电(动)机 fully suspended motor; gimbaled motor; suspended motor
悬挂式电铃按钮 pressel
悬挂式电流计 suspended coil galvanometer
悬挂式电气设备 pendant
悬挂式电梯缆绳架 underslung car frame
悬挂式吊车 underhung crane
悬挂式顶棚 grid ceiling
悬挂式锭翼 suspended flyer
悬挂式对接 suspended butt-joint
悬挂式防冲装置 suspend fender
悬挂式防撞物 suspended fender

X

悬挂式防撞装置 suspended fender
悬挂式封隔器 hold-down packer
悬挂式干粉灭火器 hanged powder extinguishing equipment
悬挂式干燥 festooning
悬挂式干燥器 hanging drier
悬挂式干燥室 festoon drier[dryer]
悬挂式钢筋混凝土屋顶 suspended reinforced concrete roof
悬挂式割草机 direct-connected mower
悬挂式格栅体系 suspension grid system
悬挂式隔墙 suspended partition(wall)
悬挂式合页 gudgeon
悬挂式桁架 suspended truss
悬挂式桁架桥 hanging truss bridge
悬挂式烘布机 loop drier
悬挂式烘燥机 hanging dryer
悬挂式护木 hanging timer fender
悬挂式护舷 suspended fender; hanging fender
悬挂式护舷木 hung wood fender
悬挂式滑翅运动 hang gliding
悬挂式灰墁顶棚 suspended plaster ceiling
悬挂式灰墁平顶 suspended plaster ceiling
悬挂式机具 <拖拉机上的> inserted implement
悬挂式机具架 mounted toolbar
悬挂式积点采样器 cable-suspended point-integrating sampler
悬挂式基阵监视系统 suspended array surveillance system
悬挂式加料起重机 underslung charging crane
悬挂式建筑 suspended building
悬挂式浇包 troll(e)y ladle
悬挂式绞车 linkage winch; link winch
悬挂式脚手架 boat scaffold(ing); cradle scaffold; flying falsework; flying scaffold; flying shelf; hanging scaffold(ing); scaffold cradle; suspended scaffold(ing); suspended staying; swinging scaffold; hung scaffold(ing); scaffold suspended; suspended shuttering; suspended staging
悬挂式脚手架的悬吊梁 needle beam
悬挂式结构 suspension structure
悬挂式结构体系 suspension structural system
悬挂式金属顶棚 suspended metal ceiling
悬挂式金属平顶 suspended metal ceiling
悬挂式经纬仪 hanging theodolite
悬挂式聚光灯 suspension spotlight
悬挂式绝缘器 suspension type insulator
悬挂式绝缘子 insulated hanger; suspension insulator
悬挂式开关 pendant(control) switch; suspension switch
悬挂式可拆卸封隔器 hook wall flooding packer
悬挂式可调座椅 adjustable suspension seat
悬挂式控制踏板 pendant control pedal
悬挂式框架堰 suspended-frame weir
悬挂式离心机 suspended centrifuge
悬挂式犁 mounted plough
悬挂式料斗 suspension banker
悬挂式溜槽 suspended chute
悬挂式流速仪 cable-suspended current meter
悬挂式楼板构造 hung floor construction
悬挂式楼层 suspension stor(e)y
悬挂式炉 suspended furnace
悬挂式落砂架 suspended shake-out device
悬挂式门 overhung door
悬挂式灭火装置 hanged extinguishing equipment
悬挂式木护舷 hanging timer fender; hung-wood fender
悬挂式暖风机 suspended type unit heater
悬挂式排水沟 suspended gutter
悬挂式抛砂机 swing slinger
悬挂式抛丸清理机 hanger abrator
悬挂式喷气烘燥机 hanging nozzle drier
悬挂式喷水器 pendant sprinkler
悬挂式喷雾机 hanging sprayer
悬挂式碰垫 hung fender; suspended fender
悬挂式起重机 suspended crane; troll(e)y crane; underslung crane
悬挂式起重小车 troll(e)y hoist
悬挂式起重装置 overhead lifting gear
悬挂式气动吊车 pendent-type air hoist
悬挂式牵引割草机 hanging power hay mower
悬挂式桥板 suspended catwalk
悬挂式桥式起重机 underslung bridge crane
悬挂式轻轨系统 suspended light rail system
悬挂式倾斜仪 hanging clinometer
悬挂式散热器(件) suspended type unit heater
悬挂式砂轮机 swing-frame grinder
悬挂式上部结构 suspended superstructure
悬挂式石膏板 suspension gypsum slab
悬挂式书架 suspension shelving
悬挂式输送机 overhead(trolley) conveyer[conveyor]; underslung conveyer[conveyor]
悬挂式双动油压机 suspension type oil hydraulic press
悬挂式双向犁 mounted alternate plow
悬挂式水准管 string level
悬挂式水准器 string level
悬挂式弹簧 suspended spring
悬挂式天沟 hanging roof gutter
悬挂式通用犁 mounted general purpose plow
悬挂式推拉门 underhung door
悬挂式托辊 suspended idler
悬挂式托辊结构 suspended idler structure
悬挂式挖泥机 suspension dredge(r)
悬挂式挖土铲斗 mounted earth scoop
悬挂式帷幕 hanging curtain; partial curtain
悬挂式喂料机 chain feeder
悬挂式屋顶 tension roof
悬挂式屋盖 hanging roof; suspended roof
悬挂式屋架 hanging roof truss
悬挂式舞台 suspension stage
悬挂式信号机【铁】pendant signal
悬挂式行走机构 bogy undercarriage system
悬挂式旋臂起重机 underhung slewing jib crane
悬挂式旋转割草机 suspended rotary mower
悬挂式仪表 hook-on instrument
悬挂式雨水槽 suspended gutter
悬挂式运输机 overhead trolley conveyer[conveyor]; underslung conveyer[conveyor]
悬挂式照明 suspension light
悬挂式照明器 pendant(-type) luminaire
悬挂式照明设备 pendant luminaire fixture
悬挂式照明装置 suspension lighting fixture; suspension luminaire fixture
悬挂式振动筛 hanger supported shaking screen
悬挂式重力防撞装置 suspended gravity fender
悬挂式重力护舷 suspended gravity fender
悬挂式装载车 mounted load carrier
悬挂式装载机 mounted load carrier
悬挂式自动整平器 floating screed
悬挂式走道 pendant walkway
悬挂式座椅 suspension seat
悬挂式座椅套 suspension seat cover
悬挂水 hanging water
悬挂水准仪 string level
悬挂索 suspension line
悬挂锁定器 linkage stabilizer
悬挂锁紧机构 suspension lock device
悬挂锁紧手柄 suspension locking handle
悬挂弹簧 pendulum spring; suspension spring
悬挂弹性件 hanging elastic member
悬挂体系 suspension system
悬挂跳板 suspended catwalk
悬挂铁脚式交叉 drop bracket type disposition
悬挂铁条 bridle iron; stirrup strap
悬挂托架 suspension bracket; trapeze hanger
悬挂拖运 suspended tow
悬挂网格 suspended screen
悬挂稳定性控制 suspension stability control
悬挂屋盖 suspended roof
悬挂屋架 hanging truss roof
悬挂物 suspender
悬挂吸声体 suspended sound-absorber
悬挂吸收器 suspended absorber
悬挂系固有频率 suspension natural frequency
悬挂系统 partially separated system; suspension system
悬挂系统侧倾效应 suspension roll geometry effect
悬挂系统垂直刚度 suspension vertical stiffness
悬挂细栅 suspended screen
悬挂线 hanger wire; suspension line
悬挂线夹 suspension clamp
悬挂小车 underslung trolley
悬挂型 mounted model; suspension type
悬挂型有轨巷道堆垛机 suspended S/R machine
悬挂岩溶 perched karst
悬挂于屋顶上的管道作业 roof-hung ductwork
悬挂载荷 suspended load
悬挂在窗子下半部的窗帘 brise-bise
悬挂在顶部轨道的滑动门或折叠门 overhung door
悬挂在墙上的脚手架 needle scaffold
悬挂在水中 suspended in water
悬挂在拖拉机中部的农机具 mid-mounted tool
悬挂在载重汽车上的 lorry-mounted
悬挂照明体 suspended luminaire
悬挂支撑 suspension stay
悬挂支承 hanger bearing; hanging support
悬挂支墩 hanging buttress
悬挂织物 antique satin
悬挂质量 sprung mass; suspended mass
悬挂中心线 center[centre] line of support
悬挂重力式防撞装置 gravity type fender
悬挂重力式护舷 suspended-gravity fender
悬挂重量 suspended weight
悬挂轴 suspension pin; suspension shaft
悬挂轴承盒 hanger box
悬挂柱 suspension post
悬挂转向架 suspension bogie
悬挂桩 suspension peg
悬挂装置 bogie[bog(e)y]; hanger; hookup; lifting linkage; linkage hitch; linkage levelling screw; linkage mounting; mounted equipment; suspension arrangement; suspension device; suspension gear; suspension linkage; suspension mounting; suspension system
悬挂装置的锁定器 linkage stabilizer
悬挂装置等级 linkage category
悬挂装置缸筒 suspension cylinder
悬挂装置滑油泵 suspension lubrication pump
悬挂装置滑油泵联杆 suspension pump linkage
悬挂装置滑油管 suspension lubrication pipe
悬挂装置滑油箱 suspension lubrication tank
悬挂装置减震器 suspension cylinder
悬挂装置检查孔 suspension access
悬挂装置铰接轴 linkage pitman
悬挂装置类别 linkage category
悬挂装置锁定杆 linkage check rod
悬挂装置限位链 linkage check chain
悬挂装置油缸 lifting cylinder
悬挂锥体 suspended cone
悬挂总成 mounting group
悬管轴承组合 hanger bearing assembly
悬轨 suspension rail
悬轨式移动称量配料器 suspension rail type travel(l)ing weighbatcher
悬轨运输系统 suspension rail conveying system
悬辊 overhang roll
悬辊法 <制管工艺> Rocla process; roller suspension process
悬辊磨 suspension roller mill
悬河 <河床高出地面的河流> ceiling river; hanging river; suspended river
悬荷 suspended load
悬荷搬运 suspended load transport
悬化室 hanging room
悬环式烘布机 loop drier
悬簧 suspended spring; suspension spring; underhung spring
悬记账 clearing account
悬架 hanger(bracket); suspension fork
悬架臂 suspension lever; suspension link
悬架布置 suspension arrangement
悬架车轮 sprung-hub wheel
悬架传递振动的性能 suspension transmissibility
悬架大梁 suspended girder
悬架吊车 overhang crane
悬架吊耳 suspension shackle
悬架刚度 suspension rate; suspension stiffness
悬架杠杆销轴 suspension lever pivot pin
悬架固定点 suspension attachment points
悬架回弹限位器 rebound stop

悬架机 suspension
悬架前伸长 front overhang
悬架桥面 suspended floor
悬架式按钮台 pendant
悬架式电闸 pendant switch
悬架式拉木机 suspension type log haul-up
悬架式运输机 suspension type conveyer[conveyor]
悬架弹簧 bearing spring; suspension spring
悬架铁道 suspended railroad; suspended railway
悬架托架 suspension bracket
悬架系统 suspension system
悬架元件 suspension member
悬架支臂 hanger arm
悬架支承 suspension support
悬架支承加强板 suspension mounting reinforcement
悬架轴 hanger shaft
悬架轴距 suspension wheelbase
悬浇 free cantilever casting
悬浇法 cantilever placing method
悬胶 hanging paste; suspension colloid
悬胶荷载 suspension load
悬胶溶解荷载 suspension-dissolve load
悬胶态 suspensoid state
悬胶体 suspensoid
悬胶液 suspension; suspensoid
悬脚手架 suspended scaffold(ing)
悬接 bridge joint
悬接式接头 suspended joint
悬节 overhead suspension
悬空 hanging; overhang
悬空板 floating slab
悬空部分 < 悬臂掌子面的 > overhang
悬空操作台 suspended platform
悬空车 suspended car
悬空的 overhung; pending
悬空的工作架 flying shelf
悬空地板 sprung floor cover(ing); suspended floor
悬空地板之下的空间 solum
悬空顶撑 dog shore; flier; flying shore; horizontal shore
悬空端 free end
悬空高度 unsupported height
悬空工作架 flying; flying falsework
悬空管 overhanging pipe
悬空号志 suspended signal
悬空横撑 < 两建筑物外墙之间的 > flying shore
悬空基极 floating base
悬空加热法 levitation heating
悬空架 flying shelf
悬空架管 overhanging pipe
悬空架设 overhead suspension
悬空键 dangling bond
悬空脚手架 flying falsework; flying scaffold; hanging scaffold(ing); hanging stage; overhanging scaffold; swing scaffold
悬空接头 bridge joint
悬空距离 unsupported distance; unsupported space
悬空宽度 unsupported width
悬空缆 aerial spud
悬空缆车 cable car; cable carriage
悬空楼梯 hanging stair(case); overhanging stairs
悬空煤 hanging coal
悬空模板 flying form
悬空式面板 suspended deck
悬空式面板结构 suspended deck structure
悬空式梯台 hanging steps
悬空索道 aerial conveyer[conveyor]
悬空踏步 hanging steps
悬空铁路 suspended railroad
悬空烟道 bridging over flue
悬空园 hanging garden
悬空支撑 shoring
悬空支架 overhead suspension
悬孔 suspended span
悬控式软管 hanging hose
悬跨 slung-span; suspended space; suspended spalling; suspended span
悬跨钢轨接头 overhanging rail joint
悬跨管线 span line
悬跨建筑 slung-span construction
悬跨连续格构梁 slung-span continuous lattice(d) beam
悬跨连续梁 slung-span continuous beam
悬跨梁 hung-span beam
悬筐式输送设备 suspended tray conveyer[conveyor]
悬筐式蒸发器 basket evapo(u)rator
悬篮 hanging basket
悬篮栽培 basked planting
悬篮植物 basked plant
悬缆 messenger
悬缆挂灯线 cable suspension light support
悬缆平衡器 suspension rope equalizer
悬缆索 cable messenger
悬缆线 cable messenger; messenger cable; messenger wire; stranded wire; suspension cable; suspension rope
悬粒灰岩 float stone
悬链 catenarian suspension; chain sling; hanging chain; suspension chain; catenary suspension
悬链吊架 catenary suspension; chain hanger
悬链斗式卸机 catenary unloader
悬链钢丝绳 catenary wire
悬链拱 catenarian arch
悬链构造 catenary construction
悬链(回转)面 catenoid
悬链挠度 catenoid
悬链桥 braced-chain of suspension bridge
悬链曲面【数】catenoid
悬链曲线 catenary curve
悬链曲线拱 catenary arch; Ctesiphon arch; ktesiphon arch
悬链式脚手架控制垂度的提升器 vertical pickup
悬链式连续抛丸清理机 shot hanger blast
悬链式连续卸船机 catenary continuous unloader
悬链式锚腿系泊 catenary anchor leg mooring
悬链式锚腿系泊设施 catenary anchor leg mooring
悬链式(清仓)链斗机 bucket of catenary-mode
悬链式筛 chain hung type screen
悬链式输送机 rocker conveyer[conveyor]
悬链式卸船机 catenary chain unloader
悬链式装船机 catenary chain loader
悬链输送机 aerial conveyer[conveyor]
悬链腿式锚泊装置 catenary anchor leg mooring
悬链线 catenarian; catenary line; contact line; funicular curve
悬链线不对称改正 correction for unsymmetry of catenary
悬链线测距 measuring in catenary
悬链线垂度 catenoid
悬链线的 catenary
悬链线电压 catenary voltage
悬链线改正 catenary correction; correction for sag
悬链线拱 catenarian arch; catenary arch
悬链线拱桥 catenary arch bridge
悬链线校正 catenary correction
悬链线检查车 catenary inspection vehicle
悬链线结构 catenary construction
悬链线桥 catenary bridge
悬链线作用 catenarian action; catenary action
悬链形渡槽 catenary flume
悬链状波痕 catenary ripple
悬链状拱 catenary arch; tatenerian arch
悬梁 drop in girder; over arm; over beam; suspension girder
悬梁秤 suspending beam scale
悬梁拱 bowstring arch
悬梁式振动 cantilever vibration
悬料 bridging; hang-up
悬铃木 < 法国梧桐 > chenar[chinar](tree); plane tree; platan(e); oriental plane-tree
悬铃木属 plane tree; sycamore; Platanus < 拉 >
悬轮混合机 Muller mixer
悬门门柱 swinging post
悬模法 Rocla process
悬膜 suspensory membrane
悬木防护器 hung wood fender; suspended fender
悬木缓冲器 hung wood fender; suspended fender
悬起 hang-up
悬欠账户 swing account
悬墙 curtain wall(ing); overhanging wall; suspended wall
悬桥 hanging bridge; suspension bridge
悬桥变位论 deflection theory
悬桥固位论 elastic theory
悬桥桁链 braced chain; braced-chain of suspension bridge
悬桥简化理论 elastic theory
悬桥面 suspended deck; suspended floor
悬桥桥跨 suspended span
悬桥索 suspension bridge cable
悬球浮子 ball-and-line float
悬球状云 pocky cloud
悬区熔化 suspended zone melting
悬圈式电流计 suspended coil galvanometer
悬燃炉 suspension fire furnace
悬溶胶体 suspensoid
悬熔 drip melting
悬熔法 levitation melting
悬熔设备 levitating melting apparatus
悬融系 suspension system
悬沙 hydraulic suspension; suspended sediment
悬沙沉积 deposit(e) of suspended sediment
悬沙模型律 law of suspended load model; law of suspended sediment model
悬沙输送率 transportation rate of suspended load
悬山 overhanging gable roof
悬山屋顶 Chinese overhung gable-end roof
悬山屋檐板 verge course
悬栅 open grid
悬栅门 hinged barrier
悬舌式弹簧销 hinged latch bolt
悬伸部分 overhang
悬伸船尾 counter stern; overhanging stern
悬伸的 overhanging
悬伸甲板 overhanging deck
悬伸架设 erecting by overhang
悬伸脚手平台 projecting scaffold
悬伸式人行道 cantilever over-hanging footway
悬伸艉 counter stern
悬绳线 catenary
悬式板 suspended slab
悬式棒形绝缘子 overhead lines insulator
悬式标尺 hanging rod
悬式玻璃窗 hung glazing
悬式测量器 hanging measuring instrument
悬式测斜器 hanging(in) clinometer; suspension clinometer
悬式测斜仪 hanging(in) clinometer; suspension clinometer
悬式撑臂 hanging brace
悬式导架 hanging leader
悬式舵 hinged rudder
悬式发电机 suspended type generator; suspension type generator
悬式防护木 hung wood fender
悬式扶垛 hanging buttress
悬式格架 hanging shelf
悬式隔墙 hung partition(wall)
悬式骨架 hung shell
悬式刮板输送机 suspended flight conveyer[conveyor]
悬式桁架 hanging truss; suspended truss
悬式桁架桥 suspension truss bridge
悬式护舷 hung fender
悬式护木 hung fender
悬式混凝土桥面 suspended concrete floor
悬式集电弓架 suspension pantograph
悬式加劲桁架 suspended stiffening truss
悬式脚手架 suspended scaffold(ing)
悬式接头 suspended joint; suspension joint
悬式结构 suspension structure
悬式金属天花板 suspended metal ceiling
悬式金属条板 suspended metal lath
悬式经纬仪 hanging theodolite; hanging transit; suspended theodolite; suspended transit; suspension theodolite; suspension transit
悬式锯 overhead saw
悬式绝缘子 suspension insulator; tie-down insulator
悬式绝缘子串 suspension insulator string
悬式梁 suspended beam
悬式楼板 hanging floor; hanging stor(e)y; hung floor; hung story
悬式楼面 hanging floor; hanging stor(e)y; hung floor; hung story
悬式罗盘(仪) hanging compass; suspension compass
悬式模板 suspended shuttering
悬式排水瓦管 hung tilework
悬式平衡重 suspended counterweight
悬式人行桥 suspended pedestrian bridge
悬式散热器(件) suspended type unit heater
悬式水力发电机 suspension type hydro-generator
悬式水轮发电机 suspended hydrogenerator
悬式水平仪 hanging level; suspended level
悬式水准仪 hanging level; suspended level
悬式缩放仪 suspension pantograph
悬式天花板 suspended ceiling
悬式温度计 sling thermometer

悬式屋顶 suspension roof
悬式吸音天花板 hung acoustic(al) ceiling
悬式线路绝缘子 overhead line insulator
悬式雨水的檐槽 hanging rainwater gutter
悬式预热器 suspension type preheater
悬式支承 hanging support
悬式支承装置 suspension support installation
悬式装置 suspension mounting
悬式灯座 hanging socket
悬式座架 suspension mounting
悬饰 pendant
悬饰浮雕 pendant boss
悬饰柱 pendant post
悬殊 disparity
悬殊的 intervallic
悬水面湖 perched lake
悬水准管 hanging level
悬丝 suspension fiber[fibre]; suspensory wire
悬丝静电计 fiber electrometer
悬丝式磁力仪 torsion-type magnetometer
悬索 fly(ing) wire; pendant(wire); span wire; suspended cable; suspender; suspension cable; suspension cord; suspension rope; wire suspension
悬索操纵 pendant control
悬索道 <料场运石用> runway; blondin
悬索的下垂 sag of cable
悬索吊装法 cable erection
悬索公路桥 suspended highway bridge
悬索拱 <由索和拱组合的结构> suspend-arch
悬索构件 suspension member
悬索和斜拉混合体系 <指桥梁结构> mixed suspension and cable-stayed system
悬索桁架 cable truss
悬索桁架桥 suspension truss bridge
悬索加劲桁架 suspended stiffening truss
悬索接头 cable connection; cable joint
悬索结构 cable-supported construction; cable(-suspended) structure; catenarian construction; suspended cable structure; suspension structure
悬索卷轴 <流速仪的> suspension reel
悬索开关 pendant switch
悬索控制 pendant control
悬索梁 overhanging beam
悬索锚碇 <悬索桥缆索各式的锚固> suspended cable anchor; suspension cable anchor
悬索锚碇基础 suspension-cable anchor
悬索锚固 suspension cable anchor
悬索偏角改正 correction for deflection of a suspension cable
悬索平衡器 rope suspension equalizer
悬索平屋顶 cable flat roof
悬索桥 cable(cantilever) bridge; cable-supported bridge; cable suspension bridge; flying bridge; hanging bridge; rope suspension bridge; suspended bridge; suspension bridge; wire(suspension) bridge
悬索桥的桁架链 braced chain
悬索桥吊桥桥塔 pylon of suspension bridge; tower of suspension bridge
悬索桥钢索 bridge cable; suspension bridge cable
悬索桥加劲桁架 stiffening truss of suspension bridge
悬索桥抗风缆 wind cable
悬索桥空气动力稳定性 air dynamic(al) stability of suspension bridge
悬索桥跨度 suspended span
悬索桥缆索下面的走道 catwalk
悬索桥桥塔 suspension bridge tower
悬索曲度 cable bent
悬索筛 cable-suspended screen
悬索式水文缆道 cable-suspended hydrometric(al) cableway; hydrometric(al) cableway with cable suspension
悬索双曲抛物面屋顶 cabled hyperbolic paraboloidal roof
悬索铁道 cable railroad; cable railway; suspended railroad; suspended railway
悬索铁路 cable railroad; cable railway; hanging railway; suspended railroad; suspended railway; suspension cable railway
悬索网 cable network
悬索屋顶 suspended(cable) roof; suspension roof
悬索屋顶施工 suspension cable roof construction
悬索屋盖 cable roof; hanging roof; suspended cable roof
悬索屋面 suspension roof
悬索系统 suspended cable system; suspension cable system
悬索线 catenarian
悬索悬臂屋顶 rope-suspended cantilever(ed) roof
悬索悬挂屋顶 cable-suspended cantilever(ed) roof
悬索悬挑屋顶 cable cantilevering roof
悬索支承结构 cable-supported structure
悬索作用 overhanging action
悬梯 hanging ladder
悬体测定法 tyndallimetry
悬体测定计 tyndallimeter
悬体桥 swing-on bridge
悬挑 cantilevering; corbelling; corbel out; overhanging
悬挑板 cantilevering plate; cantilevering slab; corbel back slab
悬挑板结构 bracket board construction
悬挑标志 projecting sign
悬挑标志牌 armal sign
悬挑薄板 cantilevering sheet
悬挑长度 overhanging length
悬挑窗扇 overhang sash
悬挑的扇形踏步 overhanging angle(type) step
悬挑段 cantilevering segment
悬挑扶垛 hanging buttress
悬挑复折屋顶 cantilevering hipped-plate roof
悬挑钢筋混凝土板 cantilever reinforced concrete slab
悬挑拱顶 cantilever vault
悬挑构件 cantilevering member
悬挑件 encorbel(l)ment
悬挑脚手架 builder's jack
悬挑结构 suspended structure
悬挑壳体 cantilevering shell
悬挑楼层 cantilevered stor(e)y
悬挑楼面板 cantilevering floor slab
悬挑楼梯 cantilevered stairway; cantilevering stair(case); hanging steps; suspended stair(case)
悬挑楼梯平台 cantilevering landing
悬挑露台 cantilevering terrace
悬挑平台 cantilevering terrace
悬挑倾斜板屋面 cantilevering tilted-slab roof
悬挑式虹吸管溢洪道 overhanging siphon spillway
悬挑式建筑 jettied construction
悬挑式脚手架 projecting scaffold
悬挑式楼梯 cantilever stair(case); overhanging stair(case); hanging stair(case)
悬挑式门枢轴 gate hook
悬挑式屋顶 overhanging roof
悬挑踏步 cantilever(ing) steps; hanging steps
悬挑台阶 cantilever(ing) steps
悬挑体系 cantilevering system
悬挑铁件 corbelling iron; corbel pin
悬挑铁支撑 corbelling iron
悬挑屋顶 cantilevering roof; overhanging roof
悬挑屋顶板 cantilevering roof slab
悬挑斜列砖层 dogtooth course
悬挑压顶 projecting coping
悬挑折板屋顶 cantilevering folded plate roof; cantilevering folded slab roof
悬挑支架 lookout
悬停 hovering
悬停持续时间 hovering endurance
悬停阶段 hovering phase
悬停转弯 hovering turn
悬停状态功率 hovering power
悬凸部 after overhang
悬突 overhang
悬腿架 swing
悬托拱 suspended arch
悬托架 hanging bracket
悬网坝 suspended-net dam
悬网式谷坊 suspended-net check dam
悬网式节制坝 suspended-net check dam
悬污能力 soil-suspending ability
悬污作用 soil-suspending action
悬物杆 hanger rod
悬线链垂度 catenoid
悬线链挠度 catenoid
悬线链曲面 catenoid
悬线曲面 catenoid
悬心 centre of suspension
悬崖 beetling cliff; bluff; brow; cleve; cliff; crag; escarpment; hanger; hanging side; impending cliff; klip; linn; mountain scarp; overhanging cliff; palisade; pendant cliff; precipice; promontory; rock cliff; scarp
悬崖壁 scarp side
悬崖冰川 cliff glacier
悬崖岛 skerry
悬崖陡岸 high coast
悬崖防护 cliff protection
悬崖海岸 cliffed coast
悬崖绝壁 hanging wall
悬崖狂涛 precipitous
悬崖坡 front slope; inface; scarp face; scarp slope <与倾向坡相对的>
悬崖峭壁海岸 cliff coast
悬崖峭壁加固工程 cliff stabilization works
悬崖泉 cliff spring
悬崖下的屑堆 scree
悬崖线 scarp line
悬崖小路 corniche
悬岩 beetling cliff; bluff; brow; cleve; cliff; greben; overhanging cliff; rock bastion; rock cliff; rocky spur; spur
悬岩壁 cliffside
悬岩绝壁 sheer precipice and overhanging rock; steep precipice and cliff
悬岩泉 cliff spring
悬叶泵 swinging-vane pump
悬叶式水泵 swinging-vane pump
悬液 suspension
悬液灌浆 suspension grouting
悬液浆 suspension grout
悬液稳定性 suspension stability
悬液效应 suspension effect
悬移荷载 suspension load
悬移流 suspension current
悬移泥沙 silt suspension; suspended sediment
悬移沙 silt suspension
悬移质 hydraulic suspension; silt load; solid material in suspension; suspended load; suspended matter; suspended sediment; suspended solid material; suspended solids; suspension load; wash load(ing)
悬移质比率 specifically suspended load
悬移质采样 suspended load sampling
悬移质采样器 suspended load sampler; suspended sediment sampler; suspension load sampler
悬移质测定 suspended load measurement
悬移质测验 suspended load measurement
悬移质的超声测量 acoustic(al) measurement of suspended solid
悬移质动床测量 movable-bed with suspended load
悬移质干重 dry suspended solids
悬移质含沙量 concentration of suspended load; suspended content; suspended sediment concentration
悬移质含沙浓度 concentration of suspended load
悬移质河工模型 river model with suspended load
悬移质积点采样器 point-integrating suspended sediment sampler
悬移质积深采样器 depth-integrating suspended sediment sampler
悬移质流动量 suspended sediment flux
悬移质泥沙 sediment in suspension; suspension sediment; suspended load
悬移质泥沙连续测验记录 continuous suspended sediment record
悬移质泥沙量 suspended load
悬移质泥沙模型试验 suspended load model test
悬移质浓度 suspended sediment concentration
悬移质平衡 suspended load budget
悬移质取样器 suspended load sampler; suspension load sampler
悬移质输沙 sediment in suspension; suspended sediment
悬移质输沙率 rate of transportation of suspended load; suspended load; suspended sediment discharge; transport rate of suspended load
悬移质输沙率测定 suspended load discharge measurement
悬移质输沙能力 transport capacity of suspended load
悬移质输送 suspended load transport
悬移质输送率 rate of transportation of suspended load
悬移质挟沙能力 transport capacity of suspended load
悬移质泄放 release of suspended solid
悬移质淤积 deposition of suspended load; deposition of suspended material
悬移质运动 movement of floating debris
悬移质总量 total suspended material; total suspended sediment; total suspended solids
悬移作用 suspension

悬游固体接触凝集沉淀池 suspended solids contact reactor
悬鱼筛 fish
悬园 hanging garden
悬账 outstanding account; suspension account; unsettled account
悬置 suspension
悬置大梁 suspension girder
悬置的 suspended
悬置动力学 suspension dynamics
悬置阀 suspension valve
悬置桁架 suspended truss
悬置绞车 suspension winch
悬置跨 suspended span
悬置式电功率计 suspension electro-dynamometer
悬置式凸轮轴 overhead camshaft
悬置弹簧 suspension spring
悬置系统 suspension system
悬置中心 center[centre] of suspension
悬置轴承 suspension bearing
悬钟山墙 ringing-loft
悬钟式采样器 suspension clock sampler
悬轴式回转碎矿机 suspended-spindle (type) gyratory crusher
悬轴式破碎机 suspended-shaft crusher
悬轴式旋回破碎机 suspended-spindle (type) gyratory crusher
悬柱 hanging pillar; pendant post
悬爪 dew claw
悬桩 suspension pile
悬装式电动机 overhung-type motor
悬装钻机 ceiling drilling machine
悬浊固体 suspended solids
悬浊率 suspension percentage
悬浊物质定量分析 quantitative analysis of suspension
悬浊液 suspension
悬浊作用 suspension effect
悬着墙 hanging wall
悬着水 hanging water; hang retention water; pendular water; vadose water; wandering water; suspended water <地下水位以上的土壤含水>
悬着水层 hanging layer; layer of suspended water
悬着水带 hanging zone; zone of suspended water
悬着水分 suspended moisture

旋

旋摆颚板轴 swing jaw shaft
旋板换热器 spiral plate exchanger
旋板机 veneer lathe
旋棒 <螺丝等的> tommy bar
旋杯 bell; rotary cup
旋杯风速仪 cup anemometer
旋杯式风速计 cup anemometer; cup-cross anemometer
旋杯式海流计 cup (-type current) meter; Price current meter; rotating cup type current meter
旋杯式静电喷涂机 cup-type electrostatic paint sprayer
旋杯式流速仪 cup (-type current) meter; Price current meter; rotating cup type current meter; vertical-axis cup type current meter
旋壁 spirotheca
旋壁起重机 wall jib crane
旋臂 jib arm; radial arm; radius arm; radius rod; spiral arm; swing arm; swing lever; swivel (ling) arm; swivel(l)ing jib
旋臂壁钻 wall radial drill
旋臂吊车 wall jib crane; whirler(y) crane; whirley
旋臂吊杆 luffing davit
旋臂吊机 jib crane
旋臂锯 line-bar resaw; radial arm saw
旋臂裂土器 radial ripper
旋臂起重机 boom hoist; slewing crane; swing crane; turntable crane
旋臂牵引杆 swivel draw bar
旋臂曲柄 shutter jib winch
旋臂升降机 arm (conveying) elevator
旋臂升降机构 arm elevating mechanism
旋臂式定线机 radial arm routing
旋臂式加料机 underslung charger
旋臂式起重机 all-round crane; jib crane; luffing crane; swing jib crane; swing lever crane; turning crane; revolving boom crane; rotary boom crane; slewing crane
旋臂松土器 radial ripper
旋臂镗床 radial arm boring machine
旋臂支承 radial arm bearing
旋臂装货机 jib loader
旋臂装料机 jib loader
旋臂钻床 drilling machine with jointed arm; radial arm boring machine; radial arm drill; radial drilling machine
旋边工具 bead tool
旋变角度 swing angle
旋冰堆 screw ice
旋冰群 screwing pack
旋冰壅塞 screwing
旋柄快开入孔 swing-type quick opening manhole
旋柄快开手孔 swing-type quick opening handhole
旋场 curl field; nutation field
旋车盘 turntable
旋车台 tarn table
旋车主轴 live spindle
旋成螺栓 turned bolt
旋尺支距法 swing offset
旋冲 spin
旋冲钻 churn drill
旋冲钻的卷扬机 spudding drum
旋出 back-out; screw off; swing out
旋出螺丝 unscrew
旋锄机 rotary hoe
旋窗 pivoted window
旋窗窗扇 pivoted sash
旋床 turning lathe
旋床工人 spinner
旋床加工的木配件 <如栏杆柱等> turned wooden articles
旋锤式磨机 swing hammer mill
旋锤式破碎机 split hammer rotary granulator; swing hammer crusher
旋锤式碎石机 split hammer rotary granulator; swing hammer crusher
旋磁比 gyromagnetic ratio
旋磁材料 gyromagnetic material
旋磁共振 gyroresonance
旋磁介质 gyromagnetic medium
旋磁频率 gyromagnetic frequency
旋磁铁氧体 gyromagnetic ferrite
旋磁效应 gyromagnetic effect
旋磁学 gyromagnetics
旋大 turn up
旋带蒸馏柱 spinning band column
旋道 spiral(l)ing road
旋笛 siren
旋电介质 gyroelectric (al) medium
旋动石屑摊铺机 spinner chip spreader; spinner-type distributor for chip(ping)s
旋动石子摊铺机 spinner-type distributor for stones
旋动式干湿球湿度计 whirling psychrometer
旋度 rotation; rotor; strength of vortex; vorticity; curl【数】
旋度波 rotation wave
旋度方程 curl equation
旋度算符 curl operator
旋锻 swage
旋锻机 rotary swager; rotary swaging machine; swager
旋颚式破碎机 jaw gyratory crusher
旋阀 stop cock
旋分路签 composite staff; divisible staff
旋分器 cyclone separator
旋风 cyclostrophic wind; cyclonic wind; land spout; revolving storm; swirl; tourbill(i)on; turbulent wind; whirlblast; whirlwind; wind spout; tornado【气】
旋风尘柱 dust whirl
旋风除尘 cyclone precipitation
旋风除尘器 centrifugal dust separator; cyclone dust catcher; cyclone scrubber; cyclone separator; dust collector cyclone
旋风除尘器钢支架 steel support of cyclone dust collector
旋风除尘器组 battery cyclone
旋风除尘装置 dust-collecting cyclone
旋风除雾器 cyclone mist eliminator
旋风储存器 cyclone collector
旋风的 cyclonic
旋风涤气器 cyclone scrubber
旋风分尘器 cyclone dust separator
旋风分级机 whirlwind classifier
旋风分级器 cyclone classifier
旋风分离管 cyclone tube
旋风分离器 cyclone separating device; cyclone (settler); cyclopneumatic separator; turbo-separator
旋风分离器阻 multicyclone
旋风分选器 cyclone classifier
旋风管 tornadotron
旋风过滤器 cyclone filter
旋风集尘器 cyclone dust collector; cyclone scrubber; dedusting cyclone; dust-collecting cyclone
旋风集尘器效率 cyclone efficiency
旋风集尘器性能 cyclone performance
旋风聚尘器 cyclone dust collector
旋风立筒预热器窑 cyclone shaft preheater kiln
旋风流动机构 vortex shedding mechanism
旋风炉 cyclone burner; cyclone combustor; cyclone furnace
旋风炉膛 turbofurnace
旋风磨碎机 cyclone mill
旋风喷淋塔 cyclonic spray tower
旋风喷淋洗涤器 cyclonic spray scrubber
旋风喷雾塔 cyclonic spray tower
旋风气流干燥器 vortex conveyer drier[dryer]
旋风气选机 whirlwind air separator
旋风汽水分离器 cyclone steam separator
旋风器 cyclone
旋风曲度 cyclonic curvature
旋风燃烧 cyclonic combustion
旋风燃烧锅炉 cyclone-fired boiler
旋风燃烧器 vortex burner
旋风燃烧室 cyclone combustion chamber
旋风燃烧系统 cyclonic combustion system
旋风式除尘器 cyclone dust collector; cyclone dust extractor; cyclone receiver
旋风式分离器 cyclone separator
旋风式集尘机 cyclone-type collector
旋风式集尘器 tornado dust collector
旋风式集料器 cyclone receiver
旋风式计算机 whirlwind computer
旋风式静电喷涂装置 cyclone-type electrostatic spraying apparatus
旋风式颗粒层集尘器 cyclone granular bed filter
旋风式空气分离机 whirlwind air separator
旋风式空气滤清器 cyclone air cleaner
旋风式螺纹铣床 thread peeling machine; thread whirling machine
旋风式气选机 whirlwind air separator
旋风式燃烧室 cyclone combustion chamber; cyclone-type of combustion
旋风式撒布器 cyclone spreader
旋风式收尘器 cyclone scrubber
旋风式收集器 cyclone receiver
旋风式悬浮预热器 cyclone suspension preheater
旋风式选粉机 cyclone (-type) (air) separator
旋风式循环空气选粉机 cyclone (re) circulating air separator
旋风室 cyclone chamber
旋风收尘器 cyclonic collector; rotoclone collector
旋风收集器 cyclone collector
旋风输送管 cyclone tube
旋风水膜除尘器 cyclone water-film scrubber
旋风筒堵塞 cyclone blocking
旋风筒分装的颗粒层集尘器 granular bed filter with separate cyclone
旋风微滴收集器 cyclonic droplet collector
旋风位置测定仪 barocyclonometer
旋风涡 tourbill(i)on
旋风吸尘器 cyclone dust collector; dust-collecting cyclone; rotoclone collector
旋风洗涤器 cyclone scrubber; cyclone washer
旋风洗衣机 cyclone washer
旋风形涡流 cyclonic whirl
旋风型计算机 whirlwind computer
旋风型气流 cyclonic flow
旋风眼【气】eye of cyclone; bull's eye
旋风预热器 cyclone preheater
旋风预热器干法窑 dry process kiln with cyclone preheater
旋风预热器窑 cyclone preheater kiln
旋风增稠器 cyclone thickener
旋风炸药 cyclonite; cyclotrimethylene trinitramine; hexahydro-1,3,5-trinitro-symtriazine; sym-trimethylene trinitramine; trinitrotrimethylenetriamine
旋风制冷防热服 vortex tube heat-protective clothing
旋耕 rotary tillage
旋耕机 gyrotiller; rotary cultivator; rotary tiller; rotavator; rotocultivator; rototiller

旋耕碎土机 ground tilling mill
旋耕装置 rotary cultivator attachment
旋工 lathe hand; lathe turner; spinner; turner
旋工厂 turnery
旋工程序 site procedure
旋工工作 turnery
旋工和易性 constructability
旋工荷载 operating load
旋工后排水设施 post-construction drainage establishment
旋工勘查 investigation during con-

struction
旋工缺陷 construction deficiency
旋工制品 turnery
旋刮刀 skiving cutter
旋管 coiler; coil pipe; pancock coil; pipe coil
旋管过滤器 coil filter
旋管加热器 spiral heater
旋管冷凝器 coiled condenser; worm condenser
旋管冷却 coil cooling
旋管冷却器 coiled condenser
旋管式波导管 coiled waveguide
旋管式储液器 coil-type accumulator
旋管真空滤器 coil-type vacuum filter
旋管蒸发器 coil evapo(u)rator
旋光 optic; rotation
旋光本领 rotatory power
旋光玻璃 rotation glass
旋光测定法 polarimetry
旋光测定计 polarimeter
旋光的 optically active; rotatory
旋光度 optic(al) activity; optic(al) rotation
旋光对映体 optic(al) antipode; optic(al) enantiomorph
旋光分光法 spectropolarimetry
旋光分光计 spectropolarimeter
旋光分散 optic(al) rotatory dispersion
旋光分析 polarimetric analysis; polarimetry; rotational analysis
旋光改变作用 mutarotation
旋光光楔 rotating wedge
旋光计 polarimeter; polariscope
旋光角 angle of rotation
旋光镜 polariscope
旋光镜的 polariscopic
旋光镜检查 polariscopy
旋光离子交换剂 optically active exchanger
旋光率 specific rotary power; specific rotation
旋光器 light rotator; polarization apparatus
旋光色觉镜 ophthalmoleukoscope
旋光色散 optic(al) rotatory dispersion
旋光试验灯 strobolamp
旋光体 optically active form
旋光物 optically active substance
旋光显微镜 polarimicroscope
旋光性 optic(al) activity; optic(al)-rotation; rotary polarization; rotation property
旋光性降低 optic(al) depression
旋光性聚合物 optically active polymer
旋光性气体 optically active gas
旋光性色散 rotary dispersion; rotatory dispersion
旋光性增强 optic(al) exaltation
旋光仪 polarimeter
旋光异构 optic(al) isomerism; rotational isomerism
旋光异构体 mutamer; optic(al) isomer
旋光中心 rotophore
旋光轴 rotatory axis
旋光转化作用 Walden inversion
旋滚流 hydraulic roller
旋衡 cyclostrophic
旋衡风 cyclostrophic wind
旋衡流 cyclostrophic flow
旋花油 rhodium oil
旋滑 rotational slide
旋滑除渣器 whirling runner
旋环 knurled ring
旋环次序 circular order
旋簧键 coiled key
旋回半径 swinging radius; turning radius
旋回层【地】cyclothem
旋回层理 cyclic(al) bedding; cyclic(al) layering
旋回层理构造 cyclic(al) bedding structure
旋回层序 cyclic(al) sequence
旋回沉积 cyclic(al) deposit
旋回沉积作用 cyclic(al) sedimentation; cyclothermic sedimentation
旋回道路 lacet road
旋回浮标 swinging buoy
旋回固定侧 pivot flank
旋回弧形断裂 encircling arcuate faulting
旋回阶地 cyclic(al) terrace
旋回裂隙 cyclic(al) fracture
旋回面 rotation surface
旋回破碎机 reduction gyratory breaker
旋回球面破碎机 gyrasphere crusher
旋回圈 turning circle
旋回圈半径 radius of turning circle
旋回式破碎机 gyratory crusher
旋回式一次破碎机 gyratory primary crusher
旋回数目 number of cycles
旋回条纹 cyclic(al) stria
旋回涡流 toroidal swirl
旋回涌水量 cycle water yield
旋回圆锥破碎机 gyracone crusher
旋回运动 gyratory motion; gyratory movement
旋回周期 turning interval
旋脊 choma
旋架 swing frame
旋架磨床 swing-frame grinder; swing-frame grinding machine
旋桨泵 propeller pump
旋桨流速仪 rotor-type current meter; horizontal shaft current meter [metre]
旋桨曝气器 paddle aerator
旋桨式拌和机 propeller-type mixer; axial trough mixer
旋桨式海流计 propeller-type current meter
旋桨式加料器 rotary-paddle feeder
旋桨式搅拌器 propeller mixer
旋桨式流速计 propeller current meter
旋桨式流速仪 helical current meter; propeller(-type) current meter; screw current meter; horizontal-axis propeller type current meter
旋桨式排气装置 propeller release
旋桨式曝气池 propeller aerator
旋桨式曝气器 propeller aerator
旋桨式扇风机 propeller fan
旋桨式水轮机 propeller(-type) turbine
旋桨式挖泥船 cutter dredge(r); propeller-type dredge(r)
旋桨式转轮 propeller-type impeller
旋桨吸泥机 cutter-suction dredge(r); suction cutter dredge(r)
旋桨吸泥疏浚 suction cutter dredging
旋桨修整机 revolving paddle finisher
旋桨叶式料位指示器 rotating paddle type material-level indicator
旋角 angle of swing; swing angle
旋角校正 swing adjustment
旋角校准 swing calibration
旋角近似法 approximation for swing
旋角钮 swinging knob
旋角误差 swing error
旋角指示器 indicator of swing
旋角轴 swinging axis
旋节分解 spinodal decomposition
旋节溶线 spinodal curve
旋节现象 spinodal phenomenon
旋节线 spinodal
旋节线分解 spinodal decomposition
旋紧 handling tight; screw up
旋紧管子接头 handling tight
旋紧螺钉 screw(ing)
旋紧螺帽 spanner nut
旋紧器 coupling device
旋进 precess; precession
旋进磁铁 precessor
旋进流量计 vortex precession flow meter
旋进式热轧机 hot reeling machine
旋进照相法 precession photograph method
旋进照相机 precession camera
旋进照相图 precession photograph
旋进轴 precession axis
旋晶衍射图 rotating crystal pattern
旋镜 rotating mirror
旋矩电机 torque motor
旋卷 convolution
旋卷层理 convolute bedding
旋卷构造 convolute structure; vortex structure
旋卷纹理 convolute lamination
旋卷物 curling stuff
旋卷柱 wreathed column
旋开 swivel
旋开窗 pivoted casement
旋开阀 swinging valve
旋开孔 pivot span
旋开跨 pivot span; swing span
旋开螺栓 unbolt
旋开桥 pivot bridge; swing bridge; swivel bridge; turning bridge
旋开桥的转动机械 turning machinery of swing bridge
旋开桥平衡锤 end block of swing bridge
旋开式快泄板 shutter
旋开式闸阀 screw-down valve
旋开式闸门 hinge(d) gate
旋壳几何形态 coiling geometry
旋壳旋转方向 coiling direction
旋扣螺丝 turn(buckle) screw
旋扩 under-ream(ing)
旋离 whiz
旋量 curl; rotation; spinor
旋量空间 spin space
旋量群 spinor group
旋裂 spiral cleavage
旋流 eddy flow; rotational flow; spiral flow; swirling flow; vertiginous current; whirl(ing) pool; whirlwind
旋流泵 cyclone pump
旋流沉淀器 cyclone settler
旋流沉砂池 rotational flow grit chamber; vortex flow grit
旋流澄清器 cyclone clarifier
旋流池 gyractor
旋流除尘器 cyclone dust collector
旋流除尘器组 multicyclone collector
旋流除砂器 <锤击钻进用的> cyclone
旋流短节 whirled sub
旋流反应池 spiral-flow reaction basin
旋流沸腾式 SF 型预分解炉 swirling fluidization type of SF preheater
旋流分级机 cyclone classifier; cyclone separator
旋流分级器 cyclone clarifier; cyclone separator; vortrap
旋流分离 cyclone separation
旋流分离器 cyclone(hydraulic) separator; gun barrel
旋流风口 spiral-flow port; swirl diffuser; twist opening; twist outlet
旋流浮选机 cyclo-cell
旋流回收式撇油器 cyclone-recovery skimmer
旋流混合 whirling mixing
旋流混浆器 vortex flow jet mixer
旋流集尘器 dust-collecting cyclone
旋流加气池 spiral-flow tank
旋流搅拌器 whirling mixer
旋流净化器 cyclone clarifier
旋流扩散器 swirl diffuser
旋流流量计 swirling meter
旋流滤箱 spiral-flow tank
旋流喷射 swirl injection
旋流喷射洗涤器 cyclonic spray scrubber
旋流喷嘴 swirl(ing) nozzle
旋流喷嘴型喷射器 swirl-nozzle type injector
旋流片 spinning disk[disc]
旋流曝气 spiral-flow aeration
旋流曝气池 spiral-flow aeration tank
旋流曝气器 spiral-flow aerator
旋流曝气系统 spiral-flow aeration system
旋流曝气箱 spiral-flow tank
旋流器 cyclone; swirler; swirl vane; whirlcone
旋流器组 cyclone cluster
旋流区 eddy zone; vortex zone
旋流燃烧室 swirl combustion chamber
旋流塞 whirling insert
旋流式 spiral-flow system
旋流式 SF 型预分解炉 swirling type of SF preheater
旋流式反应池 spiral-flow reaction basin
旋流式空气散流器 vortex air diffuser
旋流式喷燃器 whirl burner
旋流式喷嘴 rotary burner; spin-cup burner; swirl jet
旋流式气体洗涤器 cyclone gas washer
旋流式清洗机 spiral-flow washer
旋流式燃料喷嘴 fuel swirler
旋流式燃烧器 cyclone burner; swirl-flow burner; vortex burner
旋流式燃烧室 air swirl combustion chamber
旋流式烧嘴 whirl burner
旋流式雾化器 swirl atomizer
旋流室 spin chamber; vortex chamber
旋流收尘器 cyclone settler
旋流套管鞋 whirler shoe
旋流通气器 cyclone breather
旋流窝 vortex cell
旋流卧式滚筒 whirling horizontal drum
旋流洗矿机 washing cyclone
旋流洗砂器 cyclone grit washer
旋流系统 spiral-flow system
旋流箱 spiral-flow tank
旋流信标 rotation beacon
旋流叶片 swirl(ing) vane
旋流叶片式喷燃器 intervane burner
旋律 melody
旋轮 <旋压用> spinning roller
旋轮安装角 adjusting angle; position angle
旋轮线 cycloid; roulette; trochoid
旋轮线短轴 minor trochoidal axis
旋轮线形缸体 trochoidal rotor housing
旋轮线转子 trochoidal rotor
旋轮线转子机 trochoidal type machine
旋轮圆角半径 roller working radius
旋门柱 hanging post; hinging post
旋密特混凝土试锤 <用于无破损试验> Schmidt concrete test hammer
旋木工件 turned works
旋木工作 turned work
旋木用车刀 turning gouge
旋扭构造体系【地】rotational shear

structural system; rotational shear tectonic system
旋扭构造体系组成 compose rotational shear structural system
旋扭调节器 knob
旋扭轴 axis of rotation
旋钮 button; knob; rotary knob; turn button; turn knob
旋钮灯口 key socket
旋钮开关 knob switch
旋钮控制 knob-operated control
旋盘 knurled disc
旋盘比色器 disc[disk] comparator
旋盘铺砂机 spinner gritter
旋盘铺砂器 spinner disk spreader; spinner gritter
旋盘式起重机 revolving disc crane
旋刨木片 rotary cut
旋喷 jet grouting
旋喷法 chemical churning process; rotary sprinkling method
旋喷法加固地基 jet grouting with cement
旋喷灌浆 jet grouting
旋喷桩 <即喷射桩> chemical churning pile; churning pile; jet-grouted pile
旋坯 biscuit throwing
旋坯成形 jiggering
旋坯工 jigger operator
旋坯用泥料 jigger body
旋片式直联真空泵 rotary-vane and directly connected vacuum pump
旋偏光度 rotation
旋偏振光 rotatory polarization
旋偏振光角 rotatory polarization angle
旋启浮桥 pontoon swing bridge
旋启式 swing type
旋启式止回阀 swing check valve
旋钳 wrench
旋桥 swing bridge; turn(ing) bridge
旋桥信号 swing bridge signal
旋切 peel
旋切层板 rotary cut veneer
旋切单板 rotary cut veneer
旋切的 rotary
旋切木段 peeler block
旋切用原木 peeler
旋切用圆木 peeler log
旋切原木 peeler(lathe)log
旋圈 volution
旋绕 convolute; convolution; curl up; wind around
旋绕层理 convolute bedding
旋绕管子的线 line spinning
旋绕机 spinning machine
旋绕式钢丝绳 spiral wound wire
旋绕饰【建】circumvolution
旋入 screw in
旋入长度 screw-in length
旋入螺钉 screwing
旋入深度 screw-in depth
旋入式保险丝 screw-plug fuse
旋入式滤清器 spin-on filter
旋入型温度计 screw-in type thermometer
旋塞 ball-cock tap; bib(-cock); cock tap; drain cock; faucet; plug cock; plug valve; swivel plug; tap; waterway
旋塞扳手 cock key; cock spanner
旋塞柄 cock-plug head; shank
旋塞触止 cock stop
旋塞阀 plug tap; stop cock
旋塞阀体 housing of stopcock
旋塞开关 cock; plug cock; plug valve
旋塞开展 opening of cock
旋塞壳体 barrel
旋塞孔洞盖 cock hole cover
旋塞龙头 cock stop; plug tap; tap cock
旋塞式阀(门) plug valve
旋塞式滑阀 rotary plug valve
旋塞式开关 cock key; cock tap; plug key
旋塞式龙头 plug tap
旋塞手柄 crutch key
旋塞体 cock body
旋塞头 cock-plug head; shank
旋塞污泥浓缩器 rotoplug sludge concentrator
旋塞芯 plug
旋塞钥匙 cock key
旋塞嘴 bib nozzle
旋筛塔 spiral-screen column
旋上 screw on
旋上钮 turned-up button
旋上凸缘 screwed flange
旋式灯 rotary lamp
旋式交叉 rotary intersection
旋式锥形贯入计 rotary cone penetrometer
旋枢起重机 swing crane
旋枢栅门 swing gate
旋束放大管 spiraling-beam amplifier
旋束振荡管 spiraling-beam oscillator
旋刷刮板式清 rotary broom-slat sweeper
旋速计 tach(e)ometer
旋速记录器 tach(e)ograph
旋梭 rotating shuttle
旋梭缝 rotary hook
旋锁 twist lock
旋锁封盖系统 roto-lock closure system
旋锁开锁时间 twistlock unlocking time
旋锁锁紧时间 twistlock locking time
旋梯 circular newel stair(case); helical stair(case); revolving ladder; screw stairs; spiral stair(case); winding stair(case); caracol(e)
旋梯的筒状中空柱 hollow newel
旋梯中柱 newel(post); solid newel
旋梯轴管 newel tube
旋梯柱井孔 hollow newel
旋条带 rifle tie
旋调管 cyclophone; Sychlophone
旋铁发电机 inductor generator
旋筒风力推进船 wind-driven rotor ship
旋筒筛 revolving screen
旋筒式风力推进船 rotor ship
旋筒式水膜除尘器 horizontal cyclone water-film scrubber
旋涂 spin coating
旋涂器 spinner
旋围喷水器 revolving distributor
旋涡 churning(of water); eddy; nuclear pool; rotational vortex; swirl; vortex; whirl(ing); whirling pool
旋涡泵 peripheral pump; vortex pump
旋涡侧浇口 whirl gate
旋涡侧滤渣浇口 whirl gate
旋涡产生的力矩 moment due to vortices
旋涡冲击式磨机 Hametag impact mill
旋涡出流 outflow vortex
旋涡初生 vorticity initiation
旋涡除尘器 centrifugal collector
旋涡道 street of vortex
旋涡笛 vortex whistle
旋涡底谷 inverted cusp
旋涡地区 eddy region
旋涡点 vortex point
旋涡发生器 vortex generator
旋涡放大器 vortex amplifier
旋涡分离 cyclone separation; vortex shedding
旋涡分离频率 vortex shedding frequency
旋涡分离器 roto-clone separator
旋涡沟槽洞 swirl and though cave
旋涡核(心) whirl core
旋涡虹吸式冲洗功能 siphon-vortex flushing action
旋涡虹吸式坐便器 siphon-vortex water closet
旋涡虹吸式冲洗功能 siphon-vortex flushing action
旋涡花形装饰 torsel
旋涡花样 scroll; spiral scroll; torsel
旋涡环形构造 spiral ring structure
旋涡混合曝气法 swirl mix aeration process
旋涡集渣浇口 spinner runner
旋涡计 swirlmeter
旋涡搅拌器 vortex agitator
旋涡净气器 vortex box
旋涡空化 vortex cavitation
旋涡列 street of vortex
旋涡流 bumpy flow; whirling current
旋涡流动 vortex motion
旋涡流量计 swirling meter; vortex meter
旋涡螺栓 vortex trunk
旋涡模 vortex cast
旋涡磨机 Hametag mill
旋涡偶 vortex pair
旋涡喷嘴 swirling nozzle; swirl sprayer; whirl jet nozzle
旋涡喷嘴型喷射器 swirl-nozzle type injector
旋涡破除器 vortex breaker
旋涡破坏器 vortex breaker
旋涡起始区 region of vorticity initiation
旋涡强度 vortex intensity
旋涡区 eddy region; vortex shedding; vortex zone
旋涡缺陷 swirl defect
旋涡熔炼法 cyclone smelting method
旋涡矢量 vortex vector
旋涡式 turbulent pattern
旋涡式喷嘴 swirl atomizer; swirler; swirl nozzle; swirl-type burner
旋涡式清洁机 eddy washing machine
旋涡式燃烧器 swirl-type burner
旋涡式燃烧室 vortex combustion chamber
旋涡式水泥活化器 vortical activator of cement
旋涡式摊铺机 <铺松散材料用> whirl spreader
旋涡饰 scroll; scroll ornament
旋涡饰盖顶 roll capping
旋涡室 <内燃机的> whirl chamber
旋涡收尘器 cyclonic collector; rotoclone
旋涡衰减 vortex decay
旋涡水 eddy current; eddy water; whirlpool
旋涡速度 whirling speed; whirl velocity
旋涡损失 churning loss
旋涡条纹 vortex streak
旋涡尾 vortex trail
旋涡尾迹 vortex shedding
旋涡纹 curly grain
旋涡纹理 whirled lamination
旋涡线 spire; vortex line
旋涡消衰时间 decay time of eddy
旋涡效应 whirlpool effect
旋涡形 helicoid
旋涡形彩纹 floral scroll
旋涡形仓斗 whirlpool-type hopper
旋涡形的 scrolled; scroll-shaped; voluted
旋涡形盖顶瓦 roll-capped
旋涡形回转 gyrate
旋涡形晶体 scroll crystal
旋涡形抹灰饰面 swirl finish
旋涡形山墙 scroll(ed-shaped) gable
旋涡形山墙饰 volute gable
旋涡形陶槽 swing cut
旋涡形线脚 roll mo(u)lding; scroll mo(u)lding
旋涡形线饰 volute helix
旋涡形柱顶 scroll-shaped capital; volute capital
旋涡形柱头 volute capital
旋涡形装饰 volute ornament
旋涡研磨 eddy milling
旋涡浴槽 whirlpool bath tub
旋涡运动 eddy motion; stirring motion; swirling motion; vortex motion; vortical motion
旋涡中心 vortex core
旋涡周期 vortex period
旋涡轴线 eddy axis; vortex axis
旋涡铸型 vortex cast
旋涡转移 vorticity transfer
旋涡状构造 whirl structure
旋涡状混合片麻岩 swirled amphogneiss
旋涡状排列 whirllike arrangement
旋涡锥 vortex cone
旋涡阻力 eddy resistance
旋涡作用 eddy action; swirling action; vortex action; vorticity
旋下钮 turned-down button
旋向性 handedness
旋像 image rotation picture
旋削 rotary cut
旋小 turn down
旋屑 turning
旋心 center[centre] of gyration
旋星体 spiraster
旋形凿井法 screw-type sinking method
旋形钻孔器 circular bit
旋旋传动 auger drive
旋压 rotary extrusion; spinning; spin-on
旋压车床 bulging lathe
旋压车床用夹头 spin chuck
旋压成型 mo(u)ld pressing; spin
旋压成型法 flow forming; spin forming
旋压成型机 spinning former
旋压成型机床 spin forming machine
旋压成型模 final spinning block
旋压电钮 turn-push button
旋压阀 screw-down valve
旋压封头 head formed by spinning
旋压盖 <螺钉上的> screw-down cap
旋压工具 spinner's chisel; spinning tool
旋压机 spinning machine
旋压机床 spinning lathe; spinning machine
旋压加工 spinning
旋压螺丝钢板 spınning die
旋压螺纹 threading
旋压铆钉机 spin riveter
旋压铆钉连接 spun rivet connection
旋压模 spinning block; spinning chuck
旋压喷砂机 spinning gritter
旋压式除尘器 centrifugal dust arrester; centrifugal dust remover
旋压式放水龙头 screw-down pattern draining tap
旋压芯模 spinning mandrel
旋窑轮带支座 chair of kiln tyre
旋叶铲 rotary blade
旋叶给料器 rotary-vane feeder
旋叶式风速仪 revolving vane anemometer
旋叶式挖泥船 swing-vane dredge(r)
旋叶式屋顶风扇 radial roof fan
旋叶饰【建】foliation
旋叶送料器 rotary-vane feeder
旋液澄清器 cyclone clarifier
旋液分离器 centriclone; Dutch cy-

clone; hydraulic cyclone; hydrocyclone; wet cyclone
旋翼 aerovane
旋翼(飞)机 gyro(co)pter; rotaplane; rotorcraft; gyroplane; rotor aircraft; rotor plane; autogiro[autogyro]
旋翼桨尖灯 rotor-tip lights
旋翼桨叶荷载 blade loading
旋翼铰链 rotor hinge
旋翼拉力 rotor thrust
旋翼洗流 rotorwash
旋翼叶盘 rotor disc[disk]
旋翼叶片 main-rotor blade
旋翼直升飞机 giro
旋翼周期变距 feather
旋印照片 rotograph
旋圆 rounding
旋圆加半径 round turn and half hitch
旋錾 turning chisel
旋凿 screw chisel; screwdriver; turnscrew
旋轧法 flow-turn
旋闸 wicket gate
旋支 circumflex branch
旋制薄板 rotary veneer
旋制层板 rotary veneer
旋制的 spun
旋制灯柱模子 spun lighting column mo(u)ld
旋制电杆模子 spun lighting column mo(u)ld
旋制管 spun pipe
旋制混凝土 spun concrete
旋制混凝土法 spun concrete process
旋制混凝土管 spun concrete conduit; spun concrete pipe; spun concrete tube
旋制混凝土管放置车间 spun concrete pipe laying plant
旋制混凝土管铺设车间 spun concrete pipe laying plant
旋制混凝土涵管 spun concrete pipe culvert
旋制沥青衬里管 spun bitumen-line pipe
旋制铆钉 spun rivet
旋制内衬 spun lining
旋制品 turnery; turnery-ware
旋制水泥衬里管 spun cement-line pipe
旋制铁泵泵送总管 spun iron pumping main
旋制涂层铁管 coated spun iron pipe
旋制制品 spunware
旋制桩 turned pile
旋制桩模 spun pile mo(u)ld
旋轴 swivel
旋轴剪草机 spindle mower
旋轴密封 rotary shaft seal
旋轴线轨迹 trochoidal track
旋帚式扫路车 rotary sweeper
旋转 rotation; circumgyrate; circumgyration; convolution; gyrate; revolution; revolve; revolving; slew; spin; swing; swivel; swivel(l)ing motion; turning; twirl; wheeling
旋转安全 spin safe
旋转按钮 rotary button
旋转靶 rotary target; rotating target
旋转百叶式干燥机 rotary louvre drier[dryer]; roto-louvre drier[dryer]
旋转板 flap panel; revolving panel; swivel plate
旋转板块 rotated plate
旋转板式给料机 rotary plate filler
旋转半径 radius of rotation; radius of turn; radius rod; swinging radius; turning radius
旋转半桥式刮泥器 rotating half-bridge scraper
旋转拌制混凝土 spun concrete
旋转棒 radial bar; radius rod < 抹灰用的 >
旋转棒的刮板 radius shoe
旋转棒空气采样器 rotated bar air sampler
旋转棒空气取样器 rotorod air sampler
旋转报警灯 rotating warning light
旋转曝光法 rotational exposing method
旋转曝光图像 rotational exposing image
旋转备用(容量) spinning reserves
旋转焙烧炉 rotating furnace
旋转泵 hurling pump; pump rotary type; rotary(type) pump; rotating pump; rotor pump
旋转箅式立窑 rotary grate shaft kiln
旋转臂 pitman arm; rotating arm; slewing arm; whirling arm; boom of slewing; swing jib
旋转边带测试法 rotation sideband testing method
旋转边峰 rotating side peak
旋转编码器 rotary encoder
旋转扁椭球 oblate ellipsoid of rotation
旋转变斑晶 helicitic phenocryst
旋转变幅起重机臂 pivoted luffing jib
旋转变幅式起重机 luffing and slewing type crane
旋转变换 rotation transformation
旋转变换角 rotation transform angle
旋转变换器 rotary converter[converter]; rotation transducer
旋转变量器 rotary transformer
旋转变流机 rotary converter[converter]; rototory converter
旋转变流器 rotary converter[converter]; rotating converter
旋转变流器机车 rotary converter locomotive
旋转变相机 rotating phase converter
旋转变形 rotational deformation
旋转变压器 magslep; magslip; rotary transformer; rotatable transformer; rotating transformer; slewing service
旋转变压器线 slewing service line
旋转变阻器 revolving rheostat
旋转标志灯 rotating beacon; rotating beacon lighting
旋转表面冲刷器 rotary surface washer
旋转表面淬火法 spin hardening; spinhard heat treatment
旋转波 rotary wave; rotating wave; rotational wave
旋转波导可变衰减器 rotary waveguide variable attenuator
旋转波束 rotary beam
旋转波束发射机 rotating beam transmitter
旋转波束无线电信标台 revolving beam station
旋转波纹材 curl wreath
旋转播料器 swivel feeding spout
旋转铂电极 rotating platinum electrode
旋转薄层色谱法 rotating thin layer chromatography
旋转薄层色谱仪 rotating thin layer chromatograph
旋转薄壳 revolutional shell; rotational shell
旋转不良 malrotation
旋转不平衡质量 rotating unbalanced mass
旋转不稳定度 rotational instability
旋转不稳定性 rotational instability
旋转布料胶带运输机 radial spreader (conveyor)
旋转布料器 revolving top
旋转布水器 revolving distributor; rotary distributor; rotating water distributor; swinging distributor; swivel water distributor
旋转步进继电器 rotary stepping relay
旋转部分 < 机器上的 > rotating part; rotor
旋转部件 rotating element; rotating member; turning unit
旋转部件支承装置 supporting device for slewing part
旋转参数 rotation parameter
旋转餐厅 revolving (floor) restaurant; rotating restaurant
旋转舱 rotatable cabin; rotating cabin
旋转舱口 revolving hatch
旋转操纵杆 swivel lever
旋转槽 rotation groove
旋转测度器 gyrograph
旋转层理 convolute bedding
旋转层状位移 rotational laminar displacement
旋转插销 swing plug
旋转铲斗 rotating(grab) scoop
旋转铲斗装置 rotating dipper unit
旋转铲掘机 rotary spading machine
旋转铲土机 revolving shovel
旋转长椭球 prolate ellipsoid of rotation
旋转场 revolving field; rotating field; rotational field
旋转场磁极 rotating-field magnet
旋转场电磁泵 rotating-field electromagnetic pump
旋转场理论 revolving field theory
旋转场天线 rotating-field antenna
旋转场型 rotating-field type
旋转场型交流发电机 rotating-field type alternator
旋转场型交流发动机 rotating-field type engine
旋转潮 amphidromic tide; rotary tide; whirl tide
旋转潮波系统 rotary tidal waves system
旋转潮流 amphidromic current; rotary stream; rotary (tidal) current; tidal rotary current
旋转车床 bore-and-turning mill; boring and turning machine
旋转车钩 rotary coupler
旋转沉箱 rotating caisson
旋转衬里机 spun line machine
旋转衬套 rotation sleeve bushing
旋转成型 rotational forming
旋转成型法 rotoforming
旋转成型机 rotational mo(u)lding machine
旋转齿 rotary teeth
旋转齿轮泵 rotary gear pump
旋转冲击工具 rotary percussive tool
旋转冲击架式钻机 rotary percussive drifter drill
旋转冲击式凿岩机 roto-percussive hammer
旋转冲击式(凿岩)钻车 rotary percussion drill jumbo
旋转冲击式钻机 rotary-percussion drill
旋转冲击式钻进 rotary-percussion drilling
旋转冲击式钻眼 rotary-percussion drilling
旋转冲击应力 cyclic(al) stress
旋转冲击钻 rotary-percussion boring rig
旋转冲钻 churn drill(er)
旋转抽气机 rotary exhaust machine
旋转除草器 spin weeder
旋转除雪机 snow caster
旋转锄 revolving spade; rotary hoe; soil miller
旋转储藏架 rotating storage rack
旋转穿孔 rotary piercing
旋转传动装置 rotating drive; slewing drive
旋转传动装置制动器 slewing gear brake
旋转传感器 ring laser rotation sensor; rotation sensor
旋转传送机料斗 swivel conveyer bucket loader
旋转船首 swinging the ship
旋转窗 in-swinging window; pivoted casement; pivot(ed) window; revolution window; revolving window; rotating sector
旋转窗扇 pivoted sash; swivel sash
旋转床 revolving bed
旋转床接触器 rotating bed contactor
旋转床式焚化炉 rotary hearth incinerator
旋转吹管 rotating blowpipe
旋转锤 roto-hammer
旋转锤锻机 rotary swager
旋转锤磨机 rotary hammer mill
旋转锤式破碎机 rotary hammer type breaker
旋转磁场 circular magnetic field; revolving magnetic field; rotary field; rotating(magnetic) field
旋转磁场变压器 rotating-field transformer
旋转磁场电动机 rotary field motor
旋转磁场交流发电机 revolving field alternator
旋转磁场式同步发电机 revolving field type synchronous alternator
旋转磁场式仪表 rotating-field instrument
旋转磁放大器 rotary magnetic amplifier
旋转磁力仪 spinner magnetometer
旋转磁力仪法 spinner magnetometer method
旋转磁铁式磁选机 rotating-magnet separator
旋转次数 number of revolutions
旋转粗糙法 rotary roughening
旋转淬火 rolling quenching
旋转萃取器 rotating extractor
旋转锉 rotary file
旋转打包机 revolving packer head
旋转打号槌 revolving die-hammer
旋转打击钻孔 rotary-percussion drill
旋转打桩机 swivel(l)ing pile driver
旋转代换 rotational substitution
旋转单耳 swivel with eye
旋转挡板 butterfly; rotary baffle; swinging damper; swivel damper; turning valve
旋转刀 spinning blade
旋转刀架 swivel(l)ing tool-holder; swivel(l)ing tool post
旋转刀具圆周速度测定 rotary tools circular velocity measurement
旋转刀盘 rotary head
旋转刀片 rotary bit
旋转导板 rotating guide
旋转导航标 rotating lens beacon
旋转导流罩 rotating fairing
旋转导向叶片 rotational guide vane
旋转导叶 rotating guide vane
旋转导柱 rotary guide
旋转倒反轴 inversion axis; rotation-

inversion axis
旋转到过热状态 running hot
旋转道路起重机 slewing road crane
旋转的 aswivel;gyratory;pivoted;rotary;rotational;rotative;rotatory;spinning;turning;vertiginous;whirly
旋转的臂 pivot arm
旋转的竖直轴 vertical axis of revolution
旋转的铁槽锯 swivel(l)ing grooving saw
旋转的心脏形方向性图 rotating cardioid pattern
旋转灯(光) revolving light; rotating light
旋转灯塔仪 rotating lighthouse system
旋转滴汞电极 rotated dropping mercury electrode
旋转底座 rotating base
旋转地板 rotary floor
旋转地板的餐馆 rotary floor restaurant
旋转地板的饭店 rotary floor restaurant
旋转地震仪 rotation seismograph
旋转点 pivoting point; point of rotation;turning point
旋转电场 rotating electric(al) field
旋转电磁铁 rotary electromagnet
旋转电动工具 rotary power tool
旋转电动势 revolving electro-motive force
旋转电镀桶 rotary plating barrel
旋转电弧 rotating arc
旋转电弧焊管机 rotating arc pipe welding machine
旋转电弧焊(接) rotating arc welding
旋转电弧炉 Stassano furnace
旋转电机 electric(al) rotating machinery; rotating electric(al) machine; rotating machine; slewing motor
旋转电极 rotary electrode
旋转电流 rotatory current
旋转电容器 rotating capacitor
旋转电势 rotational voltage
旋转电枢 rotating armature
旋转电枢交流发电机 revolving-armature alternator
旋转电枢区 revolving-armature type
旋转电枢式电机 revolving-armature type machine
旋转电枢式发电机 revolving-armature type alternator
旋转电枢式同步电动机 inside-out motor;rotating-armature machine
旋转电刷 rotary brush;rotating brush
旋转电伺服机构 rotary electric(al) servo
旋转电压 rotational voltage
旋转电子 rotating electron
旋转垫衬 V 形压气浮选槽 rotating-mat V-type cell
旋转吊车 rotating crane
旋转吊杆 slewing boom;slewing jib
旋转吊杆起重机 rotary boom crane
旋转吊货杆 slewing derrick
旋转吊货杆的牵绳 slew line;slue line
旋转吊货钩 swivel cargo hook
旋转吊锚架 spinning cathead
旋转吊钳 swivel wrench
旋转吊桥 bascule bridge;drawbridge; revolving draw bridge
旋转顶部 slewing crown
旋转定位读出 rotational position sensing
旋转定向天线 rotary beam antenna
旋转定向无线电导向台 revolving directional radio beacon
旋转动作 whirling action
旋转动作阀 rotary action valve
旋转斗轮式回收设备 rotary bucket wheel reclaimer
旋转斗式挖掘机 rotary bucket excavator
旋转度 swing
旋转断层【地】pivotal fault; rotary fault;rotational fault;scissors fault
旋转断面 <表示细长物体的断面,平面图中心轴旋转 90°> revolved section
旋转断续继电器 rotary interrupter relay
旋转锻打机 rotary swaging machine
旋转锻机 spin-forging machine
旋转锻烧窑 rotary calciner
旋转锻造 rotary swaging forging
旋转对称 rotation symmetry
旋转对称的 rotationally symmetrical
旋转对称双曲面 hyperboloid of rotational symmetry
旋转对称性 rotational symmetry
旋转对称轴 symmetric(al) axis of rotation;symmetry axis of rotation
旋转对话式信标 rotating talking beacon
旋转惰性 rotational inertia
旋转扼流圈 rotating choke; rotating choke-coil
旋转发电装置转子 rotating generating unit rotor
旋转发动机 revolving dynamo; spin engine;spin motor
旋转发射器 revolving expeller
旋转发射装置 rotary launcher
旋转阀分配机构 revolving valve mechanism
旋转阀(门) rotary valve; rotating valve; rotation valve; swing cock; swing faucet;swivel valve;change-over valve;swing valve;turning valve
旋转法 rotation method;rotary process <井筒下沉的一种施工方法>
旋转法安装 erection by swing
旋转法架设 erection by swing method
旋转法施工 erection by swing method
旋转法则 <顺时针转为正,反时针转为负> corkscrew rule
旋转法钻进钻孔 borehole drilling
旋转翻笼 rotary end tip
旋转反射炉 rotator
旋转反射器 rotoflector
旋转反伸轴 rotoinversion axis
旋转反向变流机 rotary inverter
旋转反向开关 rotary reverse switch
旋转反映轴 rotation-reflection axis; rotoreflection axis
旋转方向 direction of rotation; hand of rotation; rotation(al) direction; rotation sense; sense of rotation; sign of rotation
旋转房屋 rotary house;rotating house
旋转放大器 rotating amplifier
旋转放大器式励磁系统 rotating amplifier excitation system
旋转放电器 rotary discharger;rotary gap
旋转分级机 gyrotor classifier
旋转分量 rotational component
旋转分流器 rotary flow divider
旋转分流梭 rotating spreader
旋转分路式流量计 rotary shunt-type flowmeter
旋转分配器 rotary distributor
旋转分像盘 spiral disk
旋转风 cyclostrophic wind
旋转风暴 revolving storm; rotary storm
旋转风挡 swivel damper
旋转风速表 rotation anemometer
旋转封接 rotating bearing seal
旋转浮冰群 screwing pack
旋转辐射仪 rotary irradiation apparatus
旋转辐照 rotatory irradiation
旋转俯仰堆料机 slewing luffing stacker
旋转负荷 rotary load
旋转复制 swivel replication
旋转副翼 spinneron
旋转副轴 swivel(l)ing counter shaft
旋转盖 rotating cap;rotocap
旋转干扰 rotary jamming
旋转干燥器 rotary drier;rotatory drier[dryer]
旋转杆 pitman arm; swinger; swivel rod;whirling arm
旋转杆式沟管清 rotating rod sewer cleaner
旋转杆仪 rolling beam apparatus
旋转感 sense of rotation
旋转感应式磁电机 rotating inductor-type magneto
旋转感应式电动机 rotating induction motor
旋转刚度 rotational rigidity;rotational stiffness
旋转缸 rotating cylinder
旋转缸体 rotary cylinder-block
旋转缸体径向泵 rotary block radial pump
旋转缸体式发动机 revolving-block engine
旋转缸体式径向柱塞泵 rotary block radial pump
旋转钢丝绳 rotating wire rope
旋转搁板 revolving shelf
旋转格筛 rotary grizzly
旋转格网 revolving screen
旋转隔板 rotating barrier
旋转隔板电动机 rotary abutment motor
旋转隔板马达 rotary abutment motor
旋转给料斗 rotary feed reservoir
旋转给料器 rotary feeder
旋转耕耘机 rotavator;rototiller
旋转工作台 revolving table;turnable table;swivel table
旋转工作台式立轴平面磨床 vertical spindle surface grinder with rotary table
旋转功能 spinfunction
旋转供料槽 swivel chute
旋转供料台 revolving feed table
旋转汞泵 rotary mercury pump
旋转钩 rotary hook;swivel hook
旋转钩身 swivel shank
旋转构造 convolute structure;vortex structure
旋转鼓 going barrel
旋转鼓风机 positive blower
旋转鼓风机增压器 rotary blower supercharger
旋转鼓轮 rotating drum
旋转鼓式干燥器 rotary drum drier [dryer]
旋转鼓式洗涤器 revolving drum washer
旋转故障 rotary fault;rotational fault
旋转刮板 rotor segment; strickle board;strickle sweep;rotating paddle <供料机的>
旋转刮板真空泵 rotary moving blade vacuum pump
旋转挂片试验 rotary coupon test
旋转管 swivel(ling) conduit;swivel(ling) pipe;swivel(ling) tube
旋转管接头 swivel joint;swivel(l)ing hose connection
旋转管帽 rotocap
旋转管式炉 rotary tube furnace
旋转管式液位计 turning tube type liquid level ga(u)ge
旋转管头 spool-type casing head
旋转管柱 slewing tubular column
旋转灌溉 centre-pivot irrigation
旋转光标 rotary beacon
旋转光束 rotating beam
旋转光楔 rotating wedge
旋转光楔对【测】diasporometer
旋转光闸法 rotating sector method
旋转轨道起重机 revolving track crane
旋转辊凸缘 swivel(l)ing-roller flanging die
旋转滚筒清洗法 barrel cleaning
旋转滚柱 swivel(l)ing idler
旋转锅炉 revolving boiler; rotary boiler
旋转锅体用手轮 handle for pan position
旋转过滤鼓筒 rotating filter drum
旋转焊接 spin welding
旋转航标 rotary beacon;rotating beacon
旋转航标灯 rotating lens beacon
旋转航天站 rotating space station
旋转黑灰炉 revolving black ash furnace
旋转桁车 revolving crane
旋转桁架木板(活动)坝 wicket dam
旋转弧 arc of rotation
旋转虎钳 rotary vice[vise]; swivel vice[vise]
旋转戽斗式脱水机 rotoscoop
旋转滑槽 slewing chute
旋转滑动 rotational slide; rotational slip
旋转滑动叶片式流量计 rotary-sliding-vane flowmeter
旋转滑动叶片式压缩机 rotary-sliding-vane compressor
旋转滑阀 revolving valve;rotating pilot valve
旋转滑阀式发动机 rotary slide valve engine
旋转滑块曲柄机构 turning slider crank mechanism
旋转滑块式曲柄导杆机构 turning slider crank mechanism
旋转滑坡 rotational landslide; rotational slide;rotational slip
旋转环 slewing ring; swivel(l)ing ring
旋转环境 rotating environment
旋转环盘电极 rotating ring disk electrode
旋转环破碎机 rotary ring crusher
旋转环形天线 rotable loop; rotating loop antenna
旋转环形天线发射机 rotating loop transmitter
旋转环形天线信标台 rotating loop station
旋转环状天线 rotary loop aerial; rotary loop antenna;rotating loop
旋转环状天线指向标 rotating loop aerial beacon
旋转环状无线电信标 rotating loop radio beacon
旋转环状无线电指向标 rotating loop radio beacon
旋转换流机 dynamo-electric(al) motor;dynamomotor
旋转灰板 revolving ash table
旋转灰盘 revolving ash pan
旋转回臂高压整流器 snook rectifier

旋转回归线 gyrotropic
旋转回流式塔盘 rotational current tray
旋转回止阀 swing check valve
旋转混合器 impeller
旋转活动桥 swing bridge
旋转活动式扩散器 rotating active diffuser
旋转活塞 rotary piston; rotating piston
旋转活塞侧边 rotor side
旋转活塞柴油机 rotary piston diesel engine
旋转活塞传动装置 rotary piston drive
旋转活塞发动机 rotary internal combustion engine; rotary piston engine
旋转活塞宽度 rotor width
旋转活塞马达 rotary piston motor
旋转活塞内燃机 rotary piston internal combustion engine
旋转活塞式鼓风机 rotary piston blower
旋转活塞式正压排代流量计 rotary-piston-type-positive-displacement flowmeter
旋转活塞压气机 oscillating piston compressor
旋转活塞压缩机 rotary piston compressor
旋转活塞蒸汽机 rotary piston steam engine
旋转火花放电器 rotary spark gap
旋转火花隙 rotary(spark)gap
旋转火焰 swirl flame
旋转货架 swing clip
旋转机 gyroscope; revolving machine; rotating machine; rotator; whirler
旋转机动性 slewability
旋转机构 revolving gear; rotary mechanism; rotating mechanism; rotation mechanism; slewing device; slewing mechanism; swivel(l)ing mechanism; traversing mechanism; turning gear; turn(ing) mechanism
旋转机构表 tourbill(i)on watch
旋转机键 rotary switch
旋转机头 rotating head
旋转机械 rotary machinery
旋转机械接头 swivel mechanical joint
旋转机械输出 rotary mechanical output
旋转机制 swing mechanism
旋转唧水筒 helical rotor pump
旋转畸变 rotational distortion
旋转畸形 rotation deformity
旋转激光测量仪 rotating laser
旋转激光束 rotating laser beam
旋转极 pole of rotation
旋转极化 rotary polarization
旋转计 tropometer
旋转计数器 revolution counter; revolution meter
旋转计算器 rotary calculator
旋转记录器 rotation recorder
旋转记录仪 rotation recorder
旋转技术 rotation technique
旋转继电器 rotary-movement relay
旋转加热炉 rotary furnace
旋转加热器 rotary regenerative heater
旋转加速度 rotary acceleration
旋转加油器 rotary oiler
旋转架 gudgeon; revolving shelf; rotary riddle
旋转架式钻机 rotary drifter
旋转监视器 rotary warded mechanism
旋转减慢 spin down
旋转减速 rotational delay
旋转减速齿轮 revolving reduction gear
旋转剪板机 rotary clippers
旋转剪床 rotary shear
旋转剪力法 gyratory shear method
旋转剪切 rotational shear
旋转剪切滑动 rotational shear slide
旋转剪切机 rotary cutter
旋转剪切运动 rotary shearing motion
旋转检测器 rotation detector
旋转检出机构 rotation detection mechanism
旋转件 revolving part
旋转键离合器 rolling key clutch
旋转桨叶搅拌机 revolving blade mixer
旋转交轴剃齿法 rotary crossed axis shaving
旋转浇筑混凝土 spun casting concrete
旋转角 angle of rotation; rotary angle; rotation angle; slewing angle; swing angle; swivel(l)ing angle; twist angle
旋转角传感器 rotation angle sensor
旋转角度 angle of roll; angle of swing
旋转角反射器 rotary reflector
旋转角速度 spine velocity
旋转绞接管 articulated pipe; articulated tube
旋转铰接 swing hinge
旋转铰联结 swing joint
旋转铰链 Harmon hinge
旋转搅拌 rotary agitation
旋转搅拌机 gyratory shaker; whirling mixer
旋转搅拌器 impeller; rotating stirrer
旋转阶段 rotary stage
旋转接触器 rotary connector; rotary contactor
旋转接触装置 rotary connection
旋转接点 slewing point
旋转接合 swing joint; swivel connection; swivel coupling; swivel joint; turning joint
旋转接力器 rotary type servomotor
旋转接头 air inlet swivel; revolving joint; slack puller; swing joint; swivel connection; swivel coupling; swivel joint; turning joint; stay tightener <拉索的>
旋转接头管 hinged pipe; hinged tube
旋转节 slack puller
旋转节理 rotatory joint
旋转节中的球体 swivel ball
旋转结构 rotary texture; rotational scheduling
旋转截止阀 shut-off rotary valve
旋转紧线器 swivel tight; swivel tightener
旋转进给 swivel feed(ing)
旋转进料器 rotary table feeder
旋转进气口 air inlet swivel
旋转浸煮器 rotary digester
旋转晶粒 rotational grain
旋转晶体 X 射线照相 rotating crystal X-ray photograph
旋转晶体法 rotating crystal method
旋转晶体衍射图 rotating crystal pattern
旋转精度 running accuracy
旋转镜 pivoting mirror; revolving mirror; rotary mirror; rotating mirror; rotation mirror
旋转镜扫描器 rotating mirror analyser[analyzer]
旋转镜扫描系统 rotary mirror scan system
旋转镜闪频仪 revolving mirror stroboscope
旋转就位法 imposed rotation method
旋转矩性阵列 rotated rectangular array
旋转矩阵 rotation matrix; spin matrix
旋转矩阵元素 element of rotation matrix
旋转锯 rotary saw; turning saw
旋转卷布压辊 swivel(l)ing pressure cloth beam
旋转卡环 swivel shackle
旋转开关 rotary switch; rotating switch; rotative switch
旋转抗渣试验法 rotary slag-resistance test method
旋转壳 revolutionary shell; rotary shell
旋转刻点仪 spiral-shaft dotting machine
旋转刻针 swivel pen point
旋转空气分布器 swing diffuser
旋转空心轴 revolving hollow spindle
旋转控制 spin control
旋转控制阀 rotary control valve
旋转控制机 rotatrol
旋转控制装置 rotating control assembly
旋转快门 rotary shutter; rotating shutter
旋转框 rotating frame
旋转扩孔器 rotary underreamer
旋转扩口机 flaring machine
旋转扩散器 rotating diffuser; rotative diffuser; swing diffuser
旋转扩散体 rotating diffuser; rotative diffuser; swing diffuser
旋转扩散系数 rotational diffusion coefficient
旋转拉紧器 swivel tightener
旋转类型 rotation type
旋转棱镜 rotary prism; rotating prism
旋转棱镜全景照相机 rotating prism panoramic camera
旋转棱子式缝纫机 rotary sewing machine
旋转冷凝器 rotary condenser
旋转离位接(触)点 rotary off-normal contact
旋转离位弹簧 rotary off-normal spring
旋转离心设备 spinner equipment
旋转犁 rotary plough; rotary tiller
旋转犁式给料机 rotary plough feeder
旋转力 revolving force; rotary force
旋转力矩 axial torque; moment of rotation; rotary moment; rotating moment; rotating torque; turning moment; twisting couple; twisting moment
旋转力矩测量器 torquemeter
旋转力偶 rotary couple; rotating couple; rotation couple; turning couple; turning pair
旋转励磁机 rotating exciter
旋转粒子 rotating particle
旋转连接 rotary joint; rotatory joint; slewing point
旋转连接器 rotating joint
旋转连续过滤机 rotary continuous filter
旋转连续过滤器 rotary continuous filter
旋转联杆 rotating link
旋转联管节 swivel union; union swivel
旋转连接 rotary joint
旋转联锁 rotation interlocking
旋转联锁开关 rotary interlocked switch
旋转镰刀式割草机 rotary scythe grass cutter
旋转链 rotary chain
旋转链式卷扬机 rotary chain draw works
旋转链式提升机 rotary chain draw works
旋转链式张紧装置 rotary chain tightener
旋转铃流机 rotary ringing generator
旋转溜槽 swivel(l)ing chute
旋转流 gyrating current; rotational swirling flow; spindrift; gyratory current
旋转流量 rotary current
旋转流速计 screw current meter
旋转六角棱镜 rotating hexagonal prism
旋转龙头 swing cock; swing faucet
旋转楼板餐厅 revolving loading floor restaurant
旋转楼梯踏步 turret step
旋转楼梯筒柱 barrel newel
旋转漏斗脱水机 rotary hopper dewaterer
旋转炉 revolving burner; revolving furnace; rotary oven; rotating furnace; rotary furnace
旋转炉箅 revolving grate; rotary grate
旋转炉箅发生炉 revolving grate producer
旋转炉箅式立窑 rotating grate shaft kiln
旋转炉顶电炉 swing-roof furnace
旋转炉排 rotary grate
旋转滤波器 convolutional filter
旋转滤鼓 rotating filter drum
旋转滤机 rotary filter
旋转滤色盘 rotating filter
旋转滤网 rotary sieve
旋转路刷 revolving brush; rotary broom; rotating broom; rotating brush
旋转路刷附件 rotating brush attachment
旋转路帚 revolving brush; rotary broom
旋转率 turn rate
旋转孪晶 rotation twin
旋转轮 swivel(ling) wheel
旋转轮闪频仪 revolving drum stroboscope
旋转轮叶式水泵 rotary-vane pump
旋转螺管式继电器 rotary solenoid relay
旋转螺帽方法 turn-of-nut method
旋转螺栓 rotary bolt; spagnolet
旋转螺纹钢板盘 rotary screwing chuck
旋转螺线管 rotary solenoid
旋转螺旋刮片输送机 swivel flight conveyer[conveyor]
旋转螺旋压缩机 rotary screw compressor
旋转毛刷 rotating brush
旋转锚栓 spinning cathead
旋转帽 turncap
旋转门 circulating door; pivoted door; pivot gate; revolution door; revolving door; roller door; rotary door; rotating door; swing gate
旋转门入口 swing-type entrance
旋转门式送料机 rotary gate feeder
旋转迷宫 turning labyrinth
旋转密封 rotary seal; rotating seal
旋转密封环 rotating seal ring
旋转面 rotating face; rotating surface; rotation surface; surface of revolution; surface of rotation
旋转面壳(体) shell of revolution
旋转模板 rotating temple
旋转模锻 impact swaging; swage machining
旋转模塑 rotational mo(u)lding
旋转模型 rotating model; rotational model

旋转磨面机 rotary float
旋转抹子 rotating trowel
旋转木马 car(r)ousel; giddy-go-round; merry-go-round (equipment)
旋转能 energy of rotation; rotational energy
旋转能量子 roton
旋转逆变器 rotary inverter
旋转黏[粘]度计 rotating visco(si)meter; rotational visco(si)meter; stirring visco(si)mete; torque(-type)visco(si)meter
旋转黏[粘]度计式真空计 rotating visco(si)meter vacuum ga(u)ge
旋转耙 rotary harrow
旋转排气台 rotary exhaust machine
旋转盘 rotary disk[disc]; swash plate
旋转盘式拌和机 rotating pan mixer
旋转盘式加料器 rotary table feeder
旋转盘式装车机 rotary disk loader
旋转抛光 rotary finishing
旋转抛物面 paraboloid of revolution; rotational paraboloid
旋转抛物面反射器 paraboloidal antenna; paraboloidal reflector
旋转刨 rotary planning
旋转炮塔 cupola; rotating turret
旋转配流阀 rotary distributing valve
旋转配水池 rotary distributor; swinging distributor
旋转配水器 rotary distributor; swinging distributor
旋转喷撒器 rotary diffuser
旋转喷管 swivel(l)ing nozzle
旋转喷口冷却器 rotary jet cooler
旋转喷洒器 rotating sprinkler
旋转喷洒头 rotary head
旋转喷射过渡 rotating spray transfer
旋转喷水 rotary distribution
旋转喷水器 revolving distributor; rotary distributor; rotating distributor
旋转喷水钻头 power swivel
旋转喷水嘴 girandole
旋转喷头 revolving nozzle
旋转喷头式喷洒器 rotating-head sprinkler
旋转喷注燃油器 rotary cup oil burner
旋转喷嘴 swing diffuser; swivel(l)-ing nozzle
旋转坯料冷却器 rotary billet cooler
旋转偏光 rotation polarization
旋转偏向 rotary deflection
旋转偏心式起振机 rotating eccentric weight exciter
旋转频率 gyro frequency; rotational frequency
旋转平口钳 swivel vice
旋转平台 rotary platform; rotary stage; swiveling platform
旋转平整机 rotatory finisher
旋转屏蔽塞 rotating shield plug
旋转屏蔽体 rotating shield
旋转破坏试验 spin burst test
旋转破碎机 gyratory breaker; rotary breaker; rotary crusher
旋转剖视 revolved section
旋转剖视图 revolving section view
旋转铺毡机 rotary felt layer
旋转曝气器 swirl diffuser
旋转起重船 slewing crane barge; slewing floating crane
旋转起重机 swing(ing)crane; whirler
旋转气动工具 rotary air tool; rotary pneumatic tool
旋转汽缸 revolving cylinder
旋转汽缸式发动机 revolving cylinder engine
旋转器 gyrator; revolver; rotator; spinner; turner
旋转千斤顶 traversing jack
旋转钎头 attack drill
旋转枪管 rotating barrel
旋转强磨钻眼法 rotary abrasive drilling
旋转桥 pivot bridge; turn bridge
旋转桥式散货装船机 slewing bridge type ship loader for bulk cargos
旋转桥台 pivot pier
旋转切出层板 rotating-cut veneer
旋转切刀 revolving knife; rotary cutter; rotary on-demand cutter; rotating cutter
旋转切断刀 rotary cut-off knife
旋转切断机 rotary cutter
旋转切割机 rotary knife cutter; rotary on-demand cutter; rotary(type) cutter; rotating cutter
旋转切割器 rotary(on-demand) cutter; rotating cutter
旋转切碎机 rotocycle
旋转切碎器 rotospeed
旋转切削 peeling
旋转擒纵机构 revolving escapement
旋转倾卸 rotary end tip
旋转倾卸作业 rotary dumping operation
旋转清管器 rotating go-devil; rotating pig; rotating scraper
旋转清雪机 snow caster
旋转清淤机 revolving sludge scraper
旋转穹隆 revolving dome
旋转秋千 <公园内的> giant's stride
旋转球 rotating sphere
旋转球柄 <门锁的> turn knob
旋转球体模拟试验 model experiment with a rotating sphere
旋转球头 rotating ballhead
旋转驱动器 rotating drive
旋转驱动装置 slewing drive
旋转取土器 displacement auger
旋转取样器 rotating sampler
旋转圈 slew a full circle
旋转群 rotation group
旋转燃油泵 rotary fuel oil pump
旋转扰动 rotational disturbance
旋转容量 <热备用容量> spinning capacity
旋转熔接 spin welding
旋转蠕动 spin creepage
旋转洒水机 rotary distributor
旋转洒水器 rotating sprinkler
旋转三角槽块 slewing V-block
旋转三棱镜 axicon lens
旋转三通 swivel tee fitting
旋转伞 rotafoil parachute; rotary parachute; rotating parachute
旋转扫描法 rotating scanner method
旋转扫描镜 rotary scan mirror
旋转扫描轮 rotating scanning wheel
旋转扫描器 rotary scanner; rotating scanner
旋转扫描线 rotating trace
旋转扫雪机 snow blowing machine; snow caster
旋转扫雪犁 snow blowing machine
旋转杀菌器 rotary sterilizer
旋转砂轮 revolving wheel
旋转筛 gyratory screen; revolving grid; trammel; revolving screen; rotary screen
旋转筛磁头 rotary screen magnetic head
旋转筛式挖掘机 rotary screen digger
旋转筛选机 rotary screen
旋转扇形齿轮 rotating sector
旋转烧结窑 rotary sintering kiln
旋转勺斗钻机 rotary bucket drill
旋转设备 rotating equipment
旋转设备制动器 slewing gear brake
旋转设计 rotary design; rotatable design
旋转射束 rotating beam
旋转射束云高计 rotating beam ceilometer
旋转伸臂起重机 revolving boom crane
旋转伸缩悬臂挂物架 swing concertina arm
旋转升降桥 gyratory lift bridge
旋转失速 rotating stall
旋转失稳速度 whirling speed
旋转施工起重机 slewing construction crane
旋转湿磨机 wet rotary grinder
旋转石灰窑 rotary lime kiln
旋转石屑摊铺机 spinner chip spreader
旋转石油喷灯 rotary oil burner
旋转食品架 dumb waiter
旋转矢量 rotating vector; rotation vector; circuitation vector
旋转矢量图 circle of vector diagram
旋转式 rotary type
旋转式按钮 turn button
旋转式板阀 rotary flat valve
旋转式半自动闭塞机 rotary interlocking block
旋转式拌和管 swivel(l)ing mixing chamber
旋转式拌和机 <就地拌和稳定土或路面混合料用> rotary type mixer
旋转式包装机 rotary packer
旋转式泵 rotary pump
旋转式变相机 rotary phase converter
旋转式变相器 rotary phase changer
旋转式标志 circling mark
旋转式表面冲洗器 rotary washer
旋转式布告板 revolving tackboard
旋转式布水器 rotary water-spreader
旋转式步进电动机 rotary stepping motor
旋转式步进电机 rotary stepped machine
旋转式擦洗机 revolving scrubber
旋转式采样器 rotary type sampling tube
旋转式测向仪 rotating direction finder
旋转式铲 revolving shovel; rotary shovel
旋转式铲刀 rotary blade
旋转式铲斗 rotating bucket
旋转式铲斗车 full circle shovel
旋转式铲斗挖沟机 rotary scoop trencher
旋转式车窗 swivel window
旋转式匙型刮料机 rotating spoon
旋转式冲击钻机 rotary impact drilling machine
旋转式冲击钻孔 rotary impact drilling
旋转式冲击钻探 rotary impact drilling
旋转式冲洗器 rotary washing screen
旋转式出土斗 rotary bucket
旋转式除草机 rotary weeder
旋转式除芒器 rotary awner
旋转式除雪机 blower-type snow plough; rotary snow plough; rotary snow plow
旋转式除雪犁 rotary plow
旋转式传感器 rotating detector
旋转式传送带清洁器 rotary belt cleaner
旋转式船尾引桥 slewing stern ramp
旋转式窗扇 swing sash
旋转式锤 rotary hammer
旋转式磁电机 rotary magneto
旋转式磁放大器 rotary magnetic amplifier; rotating magnetic amplifier
旋转式磁力仪 spinner magnetometer; spinning magnetometer
旋转式粗粉分离器 rotating classifier
旋转式打桩机 swivel(l)ing pile driver
旋转式打桩设备 rotary type pile driving plant
旋转式单层寻线 rotary hunting
旋转式单色器 rotating monochromator
旋转式单向洗瓶机 come-back bottle washing machine
旋转式刀盘 rotating cutter head
旋转式刀头 revolving type of cutter head
旋转式导向架 swivel lead
旋转式道岔 rotary type points
旋转式灯标 rotating beacon; rotating light
旋转式灯塔 rotating beacon
旋转式电焊台 rotating jig for welding; rotation jig for welding
旋转式电话分局 rotary substation
旋转式电力铲 electric(al) revolving shovel
旋转式电位计 rotating potentiometer
旋转式电位器 rotating potentiometer
旋转式吊机 slewing crane; swing crane; swivel crane
旋转式吊艇柱 common davit; ordinary davit; radial davit; rotary davit; rotating davit; rotation davit; round bar davit
旋转式叠料器 rotary stacker
旋转式定位桩【疏】rotary spud
旋转式短切机轮 rotary chopper wheel
旋转式断束器 rotating beam chopper
旋转式断续器 rotary interrupter
旋转式堆垛机 rotopiler
旋转式堆积机 rotary stacker
旋转式堆料机 revolving stacker
旋转式对称膜片 rotary symmetric(al) diaphragm
旋转式盾构 rotary shield
旋转式多路接头 rotary manifold
旋转式发动机 rotary engine
旋转式发射装置 rotating launcher
旋转式阀门 <船闸输水廊道的> rotation gate
旋转式翻车机 revolving dumper; rotary(car) dumper
旋转式翻斗 rotating skip
旋转式翻斗车 rotary dumper; rotary skip; rotary tippler; rotating tipper
旋转式翻土机 rotary tiller
旋转式翻卸矿车 rotary dump car
旋转式防喷器 revolving blow-out preventer
旋转式防撞信号灯 rotating beacon light
旋转式放大机 rotating amplifier
旋转式放大器 rotary amplifier
旋转式飞机 gyropter
旋转式飞机库 revolving air plane hangar; rotary airplane hangar
旋转式分布机 rotary distributor; spin-spreader
旋转式分度夹具 rotary indexing fixture
旋转式分级 roto-sort
旋转式分级机 rotary grader
旋转式分粒器 rotary classifier
旋转式分路开关 rotary branch switch
旋转式焚化炉 rotary incinerator
旋转式风扇 rotary fan
旋转式伏特计 rotary voltmeter
旋转式干草摊晒机 gyrotedder hay conditioner
旋转式干馏炉 rotary retort
旋转式干湿表 sling psychrometer; whirling psychrometer; whirling

hygrometer
旋转式干湿球干湿表 sling psychrometer
旋转式干湿球湿度计 sling psychrometer
旋转式干式真空泵 rotary dry vacuum pump
旋转式干衣机 rotary clothes drier [dryer]
旋转式干燥机 rotary drier [dryer]; spin-drier[dryer]
旋转式干燥炉 rotary drier[dryer]
旋转式干燥器 rotary drier [dryer]; spin-drier[dryer]
旋转式钢管矫直机 rotary tube straightener
旋转式割草机 reel-type relay; rotamower; rotary grass cutter; rotary mower; rotary scythe
旋转式给料槽 swivel charging chute
旋转式给料机 rotary feeder
旋转式给料器 roll feeder
旋转式给湿机 rotoconditioner
旋转式公寓塔楼 rotary apartment tower; rotary residence tower; rotating apartment tower
旋转式供料器 roll feeder
旋转式供油泵 rotary type fuel feed pump
旋转式勾缝机 rototrowel
旋转式骨料撒布机 whirl aggregate spreader
旋转式鼓风机 rotary blower
旋转式鼓形栅网 rotary drum screen
旋转式固定架 rotary jig
旋转式固定排量发动机 rotary fixed displacement motor
旋转式刮刀 rotary blade
旋转式刮刀加料器 rotary knife feeder
旋转式刮泥机 rotating mud scraper; rotating sludge scraper
旋转式刮土机 rotary scraper
旋转式管接头 swivel pipe joint
旋转式灌木铲除机 rotary scrub slasher
旋转式灌木切除机 rotary brush cutter
旋转式光阑 disk diaphragm
旋转式辊筒 revolving roll
旋转式过滤器 rotary filter
旋转式焊接变压器 rotary welding transformer
旋转式焊(接)机 rotating soldering machine; rotary welding set; rotating welding machine
旋转式夯具 rotating ram
旋转式横臂 rotating boom
旋转式弧焊机 rotating arc welder; rotating arc welding machine; rotating arc welding set
旋转式护舷 rotating fender
旋转式护罩 rotating housing
旋转式滑片泵 rotary-sliding-vane
旋转式滑钳 rotary slide tongs
旋转式环碎机 rotary ring crusher
旋转式环形萃取器 rotary annular extractor
旋转式环形无线电导标 rotating loop radio beacon
旋转式换热器 rotary heat exchanger
旋转式混合机 revolving mixer; rotary mixer
旋转式混炼机 rotary mixer
旋转式活塞泵 rotary piston pump; rotopiston pump
旋转式活塞泵 rotating piston pump
旋转式机铲 revolving shovel
旋转式机动路刷 rotary power broom
旋转式机器 rotating machinery
旋转式激光水平仪 rotating laser
旋转式激励器 rotary actuator
旋转式集尘器 dust-collecting cyclone
旋转式挤芯棒机 rotary core machine
旋转式计数器 rotary counter
旋转式继电器 rotary relay
旋转式给料器 revolving batch charger; rotary-paddle feeder
旋转式供料器 revolving batch charger; rotary-paddle feeder
旋转式加料机 roll feeder
旋转式加料器 rotary feeder
旋转式加热炉 rotary type heating furnace
旋转式加压水洗机 rotary squeegee washer
旋转式捡拾器 rotary pickup
旋转式建筑施工起重机 rotating building crane
旋转式交换机 rotary switch board
旋转式交流换热器 rotary regenerator
旋转式绞车 swing winch
旋转式绞刀 rotary cutter
旋转式搅拌机 rotary agitator; rotary mixer; rotating mixer; rotating vessel
旋转式接力器 rotary servomotor
旋转式接生模型 rotary midwifery phantom
旋转式节流滑阀 rotary barrel throttle
旋转式金刚石钻机 rotary diamond drill
旋转式金刚石钻头 rotating diamond bit
旋转式浸渍涂漆 roto-dipping
旋转式浸渍涂漆法 roto-dip process
旋转式精馏柱 rotary thermal rectifying column
旋转式居住塔楼 rotary dwelling tower; rotating dwelling tower
旋转式聚光灯 swivel(l)ing spotlight
旋转式卷扬机滚筒 rotary draw works drum
旋转式掘进机 rotary header
旋转式掘土器 rotary cutter
旋转式卡片档 revolving card file; rotation card file
旋转式开采机 rotary miner
旋转式可变电容器 vane capacitor
旋转式可变电阻箱 rotating rheostat box
旋转式空气过滤器 rotary air filter; rotating type air filter
旋转式空气压缩机 rotary air compressor
旋转式空压机 rotary air compressor
旋转式快门 rotary type shutter
旋转式矿浆分配器 revolving pulp distributor
旋转式扩幅辊 revolving expander
旋转式垃圾筛 travel(l)ing screen
旋转式拉弯机 rotary draw bender
旋转式喇叭口 revolving bell mouth
旋转式拦污栅 revolving screen
旋转式栏路栅 swing gate
旋转式冷凝器 rotary condenser
旋转式冷凝作用 rotating condensation
旋转式犁雪机 rotary snow plough
旋转式连接 swing joint
旋转式联锁闭塞机 rotary interlocking block
旋转式流化床燃烧 rotating fluidized bed combustion
旋转式流量计 rotameter[rotometer]
旋转式流速仪 rotary meter; rotating current meter
旋转式龙门吊 revolving portal-type crane
旋转式搂草机 rotary rake
旋转式漏斗 revolving hopper; rotary hopper; swivel charging chute
旋转式炉筒 rotating furnace drum
旋转式滤尘器 rotary strainer
旋转式滤清器 rotary strainer
旋转式路基整平板 rotating grading screed
旋转式路刷 rotary brush; rotating broom
旋转式路刷附件 rotary brush attachment
旋转式路帚 rotating broom
旋转式螺旋犁 rotating worm plow
旋转式落料管 rotation spout
旋转式镘浆机 rototrowel
旋转式锚链舱 revolving chain locker
旋转式煤气表 turbine meter
旋转式磨料磨损试验机 rotary abrasive tester
旋转式抹子 rotary float
旋转式内燃机 rotary internal combustion engine
旋转式泥铲 rotating sludge rake
旋转式泥浆冲洗钻进 rotary mud flush drilling
旋转式泥耙 rotating sludge ranger
旋转式逆变机 inverted rotary converter
旋转式黏[粘]度计 rotary visco(si)meter
旋转式耙集机构 rotating rake mechanism
旋转式排水沟挖掘机 rotary drain cutter
旋转式炮孔钻机 rotary blasthole rig
旋转式炮眼钻机 rotary shot drill
旋转式炮眼钻进 rotary blasthole drilling
旋转式培土机 rotary ridger
旋转式配水 rotary distribution
旋转式喷动雾 rotating spray
旋转式喷粉器 rotary duster
旋转式喷灌机 revolving sprinkler; rotary sprinkler; rotating sprinkler
旋转式喷灌器 revolving sprinkler; rotary rainer; rotary sprinkler; rotating sprinkler
旋转式喷淋机 rotary sprinkler
旋转式喷洒机 rotary sprinkler; spin-spreader
旋转式喷洒器 rotary sprinkler
旋转式喷砂 blasting with round shot
旋转式喷射泵 rotary injection pump
旋转式喷水器 revolving sprinkler
旋转式喷丝嘴 rotating nozzle
旋转式喷头 rotary sprinkler
旋转式喷涂机 rotary spraying machine
旋转式喷雾器 rotary atomizer; rotating sprayer
旋转式喷嘴 rotary type sprinkling nozzle; rotating nozzle
旋转式偏振滤光片 rotating polarizing filter
旋转式频道选择器 turret tuner
旋转式平地机 rotating level(l)er
旋转式平面配流阀 rotary disc valve
旋转式破碎机 gyratory crusher
旋转式起重机 revolving crane; rotary crane; turning crane; revolving derrick; rotating crane; rotating derrick; slewing crane; whirl(er) crane
旋转式起重小车 rotating trolley
旋转式气动马达 rotary pneumatic actuator
旋转式气动执行机构 rotary pneumatic actuator
旋转式汽管接头 rotary steam joint
旋转式汽化器 rotary evapo(u)rator
旋转式前灯 swivel(l)ing head lamp
旋转式切割器 revolving sickle bar; rotating cutting mechanism
旋转式切茎器 rotary beater
旋转式切碎机 rotary chopper
旋转式切削 rotary cutting
旋转式切削盾构 rotary cutting shield
旋转式切削头 revolving type of cutter head
旋转式倾倒高边敞车 rotary dump gondola
旋转式倾卸机 rotary dumper
旋转式清理机 rotary cleaner
旋转式清粮筒 rotocleaner
旋转式清洗机 rotary washer
旋转式球磨机 rotary ball mill
旋转式取岩芯钻探 rotary core drilling
旋转式取芯钻头 rotary core bit
旋转式燃烧器 circular burner; rotary burner
旋转式热交换器 rotary heat exchanger
旋转式热空气发动机 rotary air engine
旋转式软管架 swing hose rack
旋转式软管接头 rotary hose coupling
旋转式软盘组 rotating flexible disk pack
旋转式撒布机 spinner broadcaster; spinner distributor; spin(ner-type) spreader; whirl spreader
旋转式洒水机 rotary distributor
旋转式洒水器 revolving sprinkler; whirling sprinkler
旋转式三棱镜 rotary triple prism
旋转式扫地机 rotary broom
旋转式扫路机 rotary(power) broom; rotating broom; rotary sweeper
旋转式扫描器 cyclonome
旋转式扫雪机 rotary snow plough
旋转式筛 gyratory sifter; revolving screen
旋转式筛分机 rotary screen
旋转式筛砂器 rotary sand screen
旋转式栅 revolving screen
旋转式栅门 bascule gate; passimeter; turnstile
旋转式上煤机 rotary feeder
旋转式烧嘴 rotary burner
旋转式摄影机 rotary camera
旋转式深耕铲 rotary subsoiler
旋转式升降机停车库 merry-go-round garage
旋转式生物表面处理 rotating biologic(al) surface process
旋转式生物接触反应器 rotating biologic(al) contact reactor
旋转式生物接触器 rotating biologic(al) contactor
旋转式生物滤器 rotating biologic(al) filter
旋转式生物培养基 rotating biologic(al) media
旋转式施工起重机 rotary construction crane
旋转式湿度计 whirling hygrometer; whirling psychrometer
旋转式湿潜热交换器 rotary sensible and latent heat exchanger
旋转式石板磨光机 jenny lind
旋转式手制动机 turn hand brake
旋转式输送机 fan conveyer[conveyor]; pivoted conveyer[conveyor]
旋转式输送器 pivoted conveyer[conveyor]
旋转式水泵 rotating pump
旋转式水表 rotary water meter
旋转式水冲钻探设备 rotary wash boring equipment
旋转式水轮布水器 rotating water wheel distributor
旋转式水轮曝气器 rotating turbine a-

erator
旋转式水泥活化器 rotational activator of cement
旋转式水泥窑 rotary cement kiln
旋转式水枪 distributor nozzle
旋转式水银转速计 rotating mercury tachometer
旋转式松紧螺旋扣 swivel(l)ing turnbuckle
旋转式松土机 rotary tiller;rotating soil loosener;rotovator
旋转式送风口 rotary supply air outlet
旋转式碎石机 gyratory breaker;gyratory crusher;rotary crusher
旋转式碎土机 harrow
旋转式塔吊 rotary-tower crane
旋转式烫轧机 rotary ironing calendar
旋转式提升机滚筒 rotary draw works drum
旋转式填料盖 rotating gland
旋转式调相机 rotary phase modifier
旋转式艇架 swing boat davit
旋转式通风机 rotary blower
旋转式通风帽 swivel cowl
旋转式同步器 rotary synchronizer
旋转式同步示波器 rotary synchroscope
旋转式凸轮驱动压片机 rotary type cam-driven tabletting press
旋转式涂布 whirl-on
旋转式土壤拌和机 rotary soil mixer
旋转式推进装置 rotofeed
旋转式脱棉器 rotary doffer
旋转式脱水机 spin-drier[dryer]
旋转式椭圆活塞泵 rotary elliptical piston pump
旋转式挖沟机 rotary trencher;rotor digger
旋转式挖掘铲 rotary share
旋转式挖掘法 rotary digging method
旋转式挖掘机 revolving excavator;revolving shovel;rocker-arm shovel;rotary digger;rotary excavator
旋转式挖掘抛掷机 rotating lifting spinner
旋转式挖土机 hook roller;revolving shovel
旋转式弯管机 rotary pipe bending machine
旋转式位置指示灯 rotating beacon warning lighting
旋转式温度表 sling thermometer;whirling thermometer
旋转式温度计 sling thermometer;whirling thermometer
旋转式污泥清除机 revolving sludge scraper
旋转式无护孔钻探 rotary probe drilling
旋转式无线电导标 rotating radio beacon
旋转式无线电信标台 rotating radio beacon station
旋转式舞台 revolving stage
旋转式坞门 swinging caisson
旋转式雾化喷头 rotary atomizer
旋转式吸尘器 rotary dust collector
旋转式洗涤器 revolving scrubber;rotary scrubber;rotating scrubber
旋转式铣头 revolving type of cutter head
旋转式系船臂 rotating mooring boom
旋转式细料轧碎机 fine gyrasphere crusher
旋转式细筛 rotary fine screen
旋转式显微探针台架 revolving microprobe stage
旋转式相位变换器 rotary phase converter
旋转式消石灰器 rotary slaker;rotating slaker
旋转式卸车吊门 swiveling unloading flap
旋转式卸料槽 swivel(l)ing discharge chute
旋转式信标 rotating beacon
旋转式修整机 rotary finisher
旋转式絮凝作用 rotating condensation
旋转式悬臂 rotating boom
旋转式悬臂吊车 pivoted jib crane
旋转式选粉机 rotary type air separator
旋转式选矿机 rotary concentrator
旋转式选择开关 uniselector
旋转式选择器 rotary selector;uniselector
旋转式选择器配线架 uniselector distribution frame
旋转式雪犁 rotary snow plough
旋转式寻线器 uniselector
旋转式循环水泵 rotary circulating pump
旋转式压路机 rotary press
旋转式压片机 rotary tablet machine
旋转式压实机 gyratory compactor
旋转式压缩成型机 rotary compression mo(u)lding press
旋转式压缩机 rotating compressor;rotary compressor
旋转式压缩空气振动器 rotary air vibrator
旋转式压制机 revolving press
旋转式哑铃形人造重力航天站 rotating dumbbell pseudo-gravity station
旋转式烟囱帽 chimney jack;chimney stacker
旋转式岩石钻 rotary rock drill
旋转式岩石钻孔 rotating rock drilling
旋转式岩芯管 swivel barrel
旋转式岩芯机钻取的试样 rotary core sample
旋转式岩芯钻探 rotary core boring;rotary drilling
旋转式岩芯钻头 rotary core bit
旋转式研磨机 rotary honing machine
旋转式叶轮给料机 revolving plow reclaimer
旋转式叶片泵 rotary-vane type pump
旋转式叶片电动机 rolling vane motor
旋转式液动机 rotary hydraulic motor
旋转式液压马达 rotary hydraulic motor
旋转式移动门式起重机 revolving gantry crane
旋转式油泵 displacement oil pump;rotary oil pump
旋转式油封 rotary oil seal
旋转式油封机械泵 rotary oil sealed mechanical pump
旋转式油烧嘴 rotary oil burner
旋转式余摆线发动机 rotary trochoidal-type engine
旋转式鱼栅 revolving fish screen
旋转式预选器 rotary line switch
旋转式原动机 rotary prime mover
旋转式圆筒筛 trammel(screen)
旋转式远射程喷灌装置 rotary rain gun
旋转式运动 rotational type of motion
旋转式运送机 fan conveyer[conveyor];rotary fan conveyer[conveyor]
旋转式运送装置 car(r)ousel
旋转式再生加热器 rotary regenerative heater
旋转式再生器 rotary regenerator
旋转式凿岩机 rotary jack hammer;rotary machine
旋转式凿岩钻车 rotary drill jumbo
旋转式闸门 pivot gate;revolving gate;rotary gate;wicket-type gate;rotating gate;swinging caisson
旋转式遮光器 rotary shutter
旋转式遮光罩 revolving light screen;revolving light shade
旋转式折页刀 flying tuck
旋转式真空泵 rotary vacuum pump
旋转式真空干燥器 rotary vacuum drier[dryer];rotating vacuum drier[dryer]
旋转式真空过滤 rotary vacuum filtration
旋转式真空过滤器 rotary vacuum filter
旋转式振动机 rotary type vibrator
旋转式振动器 rotary type vibrator
旋转式振动细筛 rotary vibratory fine screening
旋转式蒸发器 rotary evapo(u)rator;rotating evapo(u)rator
旋转式蒸气挖土机 revolving steam shovel
旋转式整平器 rotating level(l)er
旋转式直流焊机 welding motor generator
旋转式止回阀 rotary type check valve
旋转式制粒机 rotary granulator
旋转式重力浓缩器 rotating gravity concentrator
旋转式住宅塔楼 rotating housing tower
旋转式柱塞泵 rotary-plunger pump
旋转式抓斗 rotary grab;rotating grapple;rotator grapple
旋转式抓斗起重机 rotary grab crane
旋转式转筒筛 trommel screen
旋转式装船机 slewing shiploader;slewing loader
旋转式装袋器 rotary bagger
旋转式装卸起重机 revolving loading and unloading crane
旋转式装载机 rotary loader;swing-type loader
旋转式撞槌 rotating ram
旋转式锥形破碎机 gyratory cone breaker
旋转式自动电话制 rotary automatic system;rotary dial system
旋转式自动调节器 rototrol
旋转式自动同步机 rotating synchro
旋转式钻车 rotary drill jumbo
旋转式钻床 rotary drill
旋转式钻锤 rotary drill hammer
旋转式钻机 rotary drill;rotary(drilling)rig;rotary machine;swivel(l)ing drill;rotary drilling machine
旋转式钻机系统 rotary drilling system
旋转式钻机自动推进装置 rotafeed
旋转式钻进 boring by rotation
旋转式钻井法 rotary drilling system
旋转式钻孔 rotary drill hole;rotating drilling
旋转式钻孔机 rotary drill;rotary(drilling)rig;shot-coring machine
旋转式钻孔器 rotary drill
旋转式钻模 rotary jig
旋转式钻石机 rotary rock drill;rotatory rock drill
旋转式钻探 rotary boring;rotary drilling;rotating boring;rotating drilling
旋转式钻探最优化 rotary drilling optimisation[optimization]
旋转式钻探绳索 rotary drilling line
旋转式钻探装置 rotary drilling outfit;rotary drilling unit
旋转式钻头 rotary(drilling)bit
旋转式钻岩机 rotary rock drill;rotatory rock drill
旋转视图 rotating view
旋转试验 rotation test;whirling test;spin test <杯式流速仪的>
旋转试验机 gyratory testing machine;rotary test rig
旋转试验台 spin-test rig;whirling test stand
旋转室 rotating room
旋转释放 rotary release
旋转手柄 live handle;winding handle;tommy bar <套筒扳手的>
旋转手风琴式挂物架 swing concertina arm
旋转手锥 twist gimlet
旋转书架 revolving bookstand;rotating bookcase
旋转枢轴 swivel pin
旋转枢轴轴承 slewing bearing
旋转输出 rotary output
旋转输送机 Rotatruder
旋转输送机臂 revolving conveyer boom
旋转输送系统 rotary induction system
旋转竖杆的悬臂架 elevating boom
旋转刷 revolving brush;rotary brush;rotary broom <街道清扫机的>
旋转刷式播种机 rotary brush drill
旋转衰荡 spin fading
旋转衰减器 rotary attenuator
旋转甩击式撒布机 rotating flail spreader
旋转双晶 rotation twin
旋转双曲面 hyperboloid of revolution;rotating hyperboloid
旋转双曲型 hyperbola of revolution
旋转水流 rotational flow;rotative flow;rotary current
旋转水门 rotary water-valve
旋转水泥头 swivel cement head
旋转水平的总体 population of rotational levels
旋转伺服阀 rotary servo valve
旋转伺服马达 rotary servomotor
旋转送风口 rotary supply outlet;rotating air outlet with movable guide vanes
旋转送料 swivel feed(ing)
旋转搜索 rotary search
旋转速度 rotary velocity;rotational velocity;rotative velocity;slewing speed;speed of turn;spinning speed;swing speed;swing velocity;velocity of rotation;velocity of whirl;rotary speed;rotation speed;rotative speed;speed of rotation;swivel(l)ing speed
旋转速率 rate of revolution
旋转碎石机 gyrating crusher
旋转损耗 rotational loss
旋转缩分器 spinning riffler
旋转锁风喂料机 rotary air lock feeder
旋转锁簧 rotary locking spring
旋转锁闩 rotary latch
旋转塔 rotary column;spinner column

旋转塔式起重机 revolving tower crane;slewing column crane;slewing pillar crane
旋转台 revolving platform;rotary table;rotating plate;rotating stage;rotating table;turnplate;universal stage;whirler
旋转弹塑性系数 gyratory elasto-plastic idex
旋转套管 spinning cathead
旋转套管架 revolving tube carrier
旋转套筒式磁电机 rotating sleeve type magneto
旋转梯 swivel ladder
旋转梯式多斗挖掘机 slewing bucket-

ladder excavator
旋转体 axially symmetric(al) body; revolving solid; rotator; solid of revolution; solid of rotation; swivel
旋转天窗 hinged skylight
旋转天线 revolving antenna
旋转天线测向仪 Adcock directional finder
旋转天线监视雷达 rotating surveillance radar
旋转天线探向器 Adcock direction finder
旋转天线位置指示刻度盘 antenna repeat dial
旋转天线系统 rotary antenna system
旋转天线罩 rotodome
旋转调节 rotation regulation
旋转调节器 rotating control; rotational control
旋转调头支点 pivoting point
旋转调整 swivel adjustment
旋转调整器 swivel adjustor; swivel regulator
旋转调制盘 rotating chopper
旋转挑选台 picking disk
旋转跳板【船】slewing rampway; slewed ramp
旋转铁芯 rotary core
旋转铁芯式继电器 revolution core type relay; rotating core type relay
旋转铁芯型 revolving-core type
旋转挺杆起重机 swing jib crane
旋转同步 rotation synchronization
旋转同步机 rotating synchro
旋转同步励磁机 rotating synchronous exciter
旋转同步装置 rotosyn
旋转同心管分馏柱 rotating concentric tube distilling column
旋转同轴圆筒测黏[粘]法 rotating coaxial-cylinder viscometry
旋转筒 mandrel; rotating cylinder
旋转筒式磁化焙烧炉 rotary tube type magnetic roasting furnace
旋转筒式粉碎机 rotary drum pulverizer
旋转筒形节流阀门 rotary barrel throttle
旋转头 swivel(ling) head
旋转透镜 relay lens
旋转透镜筒 rotating lens drum
旋转凸轮式自动电压调整器 rotary cam type automatic voltage regulator
旋转图(解) rotating photograph; rotation diagram
旋转图指标化 rotating photograph indexing
旋转涂布 spin coating
旋转涂布器 spinning coater
旋转涂敷 spin coated
旋转涂覆法 whirl coating
旋转涂漆法 spinning
旋转推力 rotary thrust
旋转脱离 rotating stall
旋转脱水机 rotary dewaterer
旋转椭球 spheroid of revolution
旋转椭球面 ellipsoid of rotation; rotation ellipsoid
旋转椭球偏心率 eccentricity of spheroid of revolution
旋转椭球体 ellipsoid of revolution; ellipsoid of rotation
旋转椭圆球面 ellipsoid of revolution
旋转椭圆体 ellipsoid of revolution
旋转挖掘机 revolving shovel
旋转挖掘机的冲压件 dipper stick ram
旋转挖掘器 rotary lifter
旋转挖土机 revolving excavator; revolving shovel
旋转弯曲疲劳试验 rotary bending fatigue test
旋转弯曲试验机 rotary bending tester
旋转腕臂【电】hinged cantilever
旋转腕臂底座 hinged cantilever bracket
旋转网 rotation net
旋转往复式发动机 gyro-reciprocating engine
旋转往复油泵 radial plunger oil pump; radial plunger spinning pump
旋转微细结构 rotational fine structure
旋转微型铂电极 rotated platinum microelectrode
旋转桅杆式起重机 rotating mast crane
旋转位移 swing offset
旋转位置 revolving position
旋转温度计 rotating thermometer
旋转稳定的有效载重 spin-stabilized payload
旋转稳定指数 gyratory stability index
旋转涡流 whirl vortex
旋转卧式鼓轮 revolving horizontal drum
旋转无线电航标 revolving radio beacon; rotating radio beacon
旋转无线电指向标 revolving radio beacon; rotating radio beacon
旋转无线电指向标台 rotating radio beacon station
旋转无线电指向标站 rotating radio beacon station
旋转舞台 trick scene
旋转物 gig; rotating body; whirl
旋转误差 rotational error
旋转雾化器 rotary atomizer
旋转吸管式挖泥船 rotating suction pipe dredge(r)
旋转洗涤喷嘴 rotary washing nozzle
旋转洗片机 rotary print washer
旋转洗砂机 rotating sand washer
旋转铣刀头 revolving cutter head
旋转系泊臂架 rotating mooring boom
旋转系船臂架 rotation mooring arm
旋转系数 coefficient of rotation; revolution coefficient; rotation factor
旋转衔铁 rotating iron
旋转衔铁式过流继电器 rotating-iron type over-current relay
旋转衔铁式过压继电器 rotating-iron type over-voltage relay
旋转衔铁式欠压继电器 rotating-iron type under-voltage relay
旋转线 rotational line
旋转线圈 moving coil; revolving coil; rotating choke
旋转线圈法 rotating coil method
旋转线圈式磁通计 rotating coil fluxmeter
旋转线圈指示器 rotating coil indicator
旋转相量 rotating phasor
旋转相位超前补偿器 rotary phase advancer
旋转相位调整变压器 rotatable phase adjusting transformer
旋转向量 rotary vector; rotating vector
旋转向量场 rotated vector field
旋转向下流动悬浮液光催化反应器 rotating falling slurry photocatalytic reactor
旋转效应 rotation effect
旋转斜槽 slewing chute
旋转斜盘 swash cam; swash plate
旋转斜盘泵 swash plate pump
旋转斜盘滑履 swash plate slipper
旋转斜盘机构 swash plate mechanism
旋转斜桩打桩机 slewing and raking pile driving plant
旋转斜桩打桩装置 slewing and raking pile driving plant
旋转卸料车 revolving dump car
旋转卸料管 swing arm discharge pipe
旋转心轴 live spindle
旋转信标 rotary signal
旋转信标灯 revolving beacon light
旋转信标台 rotating beacon station
旋转信号 sign of rotation
旋转星形给料器 rotating segment feeder
旋转星座图 revolving planisphere
旋转型操作器 rotor-type operator
旋转型的 torsion type
旋转型浮坞门 swinging floating caisson
旋转型格网 rotary type screen
旋转型过滤器 rotary filter
旋转型滑坡 rotational landslide
旋转型热成型机 rotary type thermoforming machine
旋转形调制盘 rotating reticle
旋转型容积式压汽机 rotary displacement compressor
旋转型双板热成型机 rotary type twin sheet thermoforming machine
旋转型无线电信标 rotating pattern radio beacon
旋转型无线电指向标 rotating pattern radio beacon
旋转型小型翻斗车 swivel skip dumper
旋转型蓄热式 heat accumulating of revolution type
旋转型蓄热式换热器 accumulating type heat exchanger of revolution
旋转型运动 rotational type of motion
旋转性的 rotary; rotating
旋转旋工法 erection by swing method
旋转选线器 rotary switch
旋转压紧器 rotary packer
旋转压片机 rotary pelleting machine; rotary tablet presser
旋转压气机 rotary compressor
旋转压实法 gyratory compaction
旋转压实机 superpave gyratory compactor Superpave
旋转压碎机 rotary breaker; rotary crusher
旋转压缩器 rotary compressor
旋转压榨机 rotary squeezer
旋转压砖机 revolver press
旋转亚能级 rotational sub-level
旋转延迟 rotational delay
旋转阳极 rotating anode
旋转阳极 X 射线管 rotating anode X-ray tube
旋转阳极管 rotating anode tube
旋转窑 rotary calciner; rotary kiln
旋转叶刀 rotary blade
旋转叶轮 rotating paddle
旋转叶轮式料位指示器 level indicator with rotating paddles
旋转叶轮式正压排代流量计 rotary-vane-type-positive-displacement flowmeter
旋转叶片 revolving vane; rotary blade; rotating paddle
旋转叶片泵 rotating vane pump
旋转叶片机 rotary-vane machine
旋转叶片器 rotary-vane machine
旋转叶片式粉碎机 rotating blade-type disintegrator
旋转叶片式快门 revolving disc shutter; rotary disk shutter
旋转叶片式流量计 rotary bucket-type flowmeter
旋转叶片式配料机 rotary-vane batcher
旋转叶片式配料器 rotary-vane batcher
旋转叶片型稳定器 rotary blade stabilizer
旋转叶片装置 rotary-vane arrangement
旋转叶栅 rotating cascade
旋转液面计 rotary ga(u)ge
旋转液体 rotating liquid
旋转液压传动装置 rotary fluid power device
旋转液压机械 rotary fluid(pressure) machine
旋转液压千斤顶 rotary hydraulic jack
旋转移动 swivel(l)ing movement
旋转移动门式起重机 revolver gantry crane; revolving gantry crane
旋转移位 rotation displacement
旋转椅 stacking chair
旋转翼流量计 rotary-vane meter
旋转阴极 rotating cathode
旋转隐斜视 cyclophoria
旋转应变 rotational strain
旋转应变椭球体 rotational strain ellipsoid
旋转应力 rotary stress; rotational stress
旋转硬化 rotational hardening
旋转硬化热处理 spin-hard heat treatment
旋转油缸 slewing ram
旋转油膜 oil whirl
旋转油盆 rotating sump
旋转余弦 rotation cosine
旋转与冲击分动式钻机 independent rotation drill
旋转与冲击联动式钻机 fixed rotation drill
旋转元件 rotating element
旋转圆板 armature
旋转圆盘 revolving metal table
旋转圆盘刀 revolving disc[disk]
旋转圆盘电极 rotating disk electrode
旋转圆盘混合器 rotary disk contactor
旋转圆盘式黏[粘]度计 rotating disk visco(si)meter
旋转圆筒 rotating cylinder; rotor
旋转圆筒干燥器 rotary drum drier [dryer]
旋转圆筒混合机 rotating tumbling-barrel-type mixer
旋转圆筒筛式粗选机 rotary screen type precleaner
旋转圆筒式黏[粘]度计 rotating cylinder visco(si)meter
旋转圆筒窑 rotary cylindrical kiln
旋转圆屋顶 revolving dome
旋转圆锥给料机 rotating conical feeder
旋转远射式喷嘴 rotating far-throwing nozzle
旋转约束 rotational restraint
旋转跃迁 rotational transition
旋转运动 gyroscopic motion; movement of rotation; pivoting motion; rotary motion; rotational motion; slewing motion; turning movement; whirling motion; rotational movement
旋转再生空气加热器 rotary regenerative air heater
旋转凿岩 rotary drilling; rotary rock drilling
旋转噪声 rotational noise
旋转张量 rotation tensor
旋转照明设备 swivel luminaire fixture
旋转照相机 rotation camera
旋转罩 swivel housing
旋转罩型雷达天线 rotodome radar antenna
旋转罩型天线 rotodome antenna

旋转遮光器 revolving shutter;rotary shutter
旋转折流板 rotary baffle
旋转针齿破碎机 rotary-pick breaker
旋转针法 rotation spindle method
旋转针台 spindle stage
旋转真空干燥泵 rotary dry vacuum pump
旋转真空蒸发器 rotary vacuum evapo(u)rator
旋转振动铆钉机 rotary vibrating riveter
旋转振动筛 circle throw vibrating screen;rotor-vibrating screen
旋转振动试验 rotating vibration test
旋转蒸发 rota-evapo(u)ration;rotary evapo(u)ration
旋转蒸发器 rotatory evapo(u)rator
旋转整流机 rotary rectifier
旋转整流器 rotary rectifier;rotating rectifier
旋转整流器发电机 rotating rectifier alternator
旋转整流器励磁机 rotating rectifier exciter
旋转整平机 rotary float
旋转支承基面 swivel(l)ing base
旋转支距 swing offset
旋转支座 pivotal bearing;turnable support
旋转执行器 rotary actuator
旋转直径 turning diameter
旋转止回阀 swing check valve
旋转止漏环 rotating seal ring
旋转指示器 rotation indicator
旋转指向标 rotating beacon
旋转制动杆 swing lock lever
旋转制管 spun pipe
旋转制交换机 rotary switch board
旋转制品 spunware
旋转制终接器 rotary connector
旋转质量 gyrating mass;rotating mass
旋转质量惯性 steadying effect
旋转质量起振器 rotating mass shaker
旋转质量型激振 rotating mass type excitation
旋转中心 center[centre] of gyration;center[centre] of rotating movement;center[centre] of rotation;moment pole;pivot center[centre];rotary center[centre];rotating center[centre];rotation center[centre]
旋转中心法 method of center[centre] of rotation
旋转中心柱萃取器 rotating concentric cylinder extractor
旋转中心钻 rolling center drill
旋转重力偶 rotational gravity force couple
旋转重量 rotating weight
旋转周期 period of rotation
旋转轴 axis of revolution;axis of rotation;axis of swing;fulcrum pin;main pivot;pivot pin;revolution axis;revolving shaft;revolving spindle;rotary axis;rotational axis;swivel bearing;symmetry axis;transit axis;turning axle
旋转轴承 pivoted bearing;slack puller;swivel bearing
旋转轴接合 pivot knuckle joint
旋转轴颈 pivot journal;swivel neck
旋转轴颈接合 pivot knuckle joint
旋转轴颈轴承 sluing bearing
旋转轴线 rotation(al) axis;spin axis;axis of rotation
旋转轴支点 pivot point
旋转主动掣子 stop for rotary driving pawl
旋转注浆成型 whirlering
旋转柱 slew post
旋转柱塞液压马达 rotary piston hydraulic motor
旋转柱塞油泵 radial plunger oil pump
旋转柱体 rotating cylinder
旋转铸塑 rotational casting;rotational mo(u)lded
旋转轴线 centerline of rotation;swing axis
旋转爪 rotary pawl
旋转转换开关 rotary sampling switch
旋转转换开关凸轮 spinning cathead
旋转转轴 pivot
旋转装载机 swing loader
旋转装置 rotating device;slewing device;slewing unit;swivel(l)ing gear;swivel(l)ing mechanism;swivel(l)ing unit;turning gear
旋转装置制动器 slewing device brake;slewing unit brake;swivel(l)ing hear brake;swivel(l)ing mechanism brake;swivel(l)ing unit brake
旋转状态 rotational movement;slew mode
旋转子 gyrator
旋转自动计数器 self-registering revolution counter
旋转自锁式安全吊货钩 swivel safety cargo hook
旋转阻尼 damping due to rotation;rotary damping
旋转钻 rotary drill;rotary hammer
旋转钻成孔现场灌注桩 rotary bored pile
旋转钻管卡瓦 rotary slip
旋转钻机 rotation drill;rotary drill
旋转钻机打炮眼 rotary blasthole drilling
旋转钻机地下全套钻具组 drill string
旋转钻进 rotary boring;rotary drilling
旋转钻进绞车 draw works
旋转钻进钻管 rotary drill pipe
旋转钻井机 rotary;rotary rig
旋转钻井装备 rotary drilling rig
旋转钻孔 rotary boring;rotary drilling
旋转钻孔机 circular bit
旋转钻孔器 circular bit
旋转钻探 rotary boring;rotary drilling
旋转钻探方法 rotary drilling(method)
旋转钻探固定钻绳 dead-line
旋转钻探机 rotary rig
旋转钻头 rotary bit;rotary cutter
旋转钻岩 rotary rock drilling
旋转作用 rotative action;turning effort;whirling
旋转坐标 rotational coordinates
旋转座 rotary seat;rotating seat
旋转座老虎钳 swivel vice
旋装式挖泥船 cutter dredge(r)
旋状白鼓丁 polycarpea roberti
旋锥式破碎机 swinging hammer breaker;swinging hammer crusher
旋锥式碎石机 swinging hammer breaker;swinging hammer crusher
旋锥式轧碎机 swinging hammer breaker;swinging hammer crusher
旋锥体 vortex cone
旋子 spinor;roton <一种粒子>
旋子彩画 tangent circle pattern;whirling flower decorative painting
旋子团花图案 whirling flower pattern
旋钻甲虫 auger beetle
旋钻钻井 rotary drilling
旋座 nosepiece

漩

漩风气流通路 cyclonic path
漩流 helical flow
漩流式燃烧室 toroidal swirl chamber
漩流损失 eddy current loss
漩水 eddy
漩涡 burble;river boil;spiral vortex;swirl;vortex
漩涡沉积泥沙 eddy-deposited silt
漩涡尺度 eddy scale
漩涡的切线分速度 velocity of whirl
漩涡虹吸式溢洪道 vortex-siphon spillway
漩涡花饰 cartouch(e)
漩涡扩散作用 eddy diffusion
漩涡流量计 vortex flowmeter
漩涡磨机 eddy mill
漩涡饰 scroll
漩涡饰盖顶 roll-capped
漩涡饰盖顶脊瓦 roll-capped ridge tile
漩涡误差 swirl error
漩涡形 scroll
漩涡形饰 scroll work
漩涡形线脚 roll mo(u)lding;scroll mo(u)lding
漩涡抑制器 vortex suppression
漩涡转速 swirl speed

选

选拔 cull(ling)
选拔孔 aperture
选拔孔隙 mask aperture
选拔器 extractor
选拔隙 aperture
选坝 selecting damsite
选标 selection of bid;selection of tender
选别采矿法 resuing method;reuse method of mining
选别充填 selective filling
选别杆 slot bar
选别公差 assorting tolerance;sorting tolerance
选别回采 reuse stoping
选别器 disengager;replacer
选别器表示器 slot indicator
选别器电磁铁 slot magnet
选别器电路 reverser circuit
选别器复示器 slot repeater
选别器控制的机械信号机【铁】slotted mechanical signal
选别器握柄 selecting lever;selector lever;slot lever
选别筛 separating screen
选别作业 separating operation
选布间 rag sorting room
选材 select suitable material
选材绠 sorting jack
选材机 sorter
选材口 sorting gap
选材台 sorting table
选材员 grader
选层锚 selective zone anchor
选层型 lectostratotype
选厂测量 surveying for site selection
选厂用水 mill water
选厂址 location of industry
选场 selected scenes
选尘器 dust selection chamber
选弛渐近法 relaxation method
选池 scavenger
选出 filtering;pick out;single out;sorting out;voting in
选出木材 picking wood
选出文物 cull
选词分区 wording area
选档手柄 selecting lever;selector lever;slot lever
选帝后的宫殿 electoral castle
选点 point selection;reconnaissance point;reconnaissance survey;running survey;set point
选点法 method of selected points;selected-point method
选点略图 reconnaissance diagram
选点梯 reconnaissance ladder
选点条件 location condition
选点图 reconnaissance sketch
选点正交 orthogonal for finite sum
选点组 reconnaissance party
选定 decide
选定标准地层剖面 lectostratotype
选定波长法 selected ordinate method
选定材料 selected material
选定长度 designated length
选定尺寸 dimensioning
选定传动比 optional gear ratio
选定道路尺寸 road dimensioning
选定的方案 selected alternative;selected program(me);selected scheme
选定的函数群 selected function groups
选定的航迹 selected track
选定的航向 selected heading
选定的计划 selected plan;selected program(me)
选定的商品 selected merchantable
选定的设计方案 selected design alternative;selected project alternative
选定的试验 optional test
选定的特性 optional feature
选定的投标商 selected bidder
选定点值 set point value
选定多层焊次序 selective block sequence
选定港(口) picked port
选定(构件)断面 dimensioning
选定航向 selected heading
选定基准 selected reference
选定截流方案 selected closure scheme
选定径路操纵台 route-setting control console
选定粒度 designated size
选定脉冲 selected pulse
选定目标 selected reference
选定区电子衍射 selected area electron diffraction
选定深度 selected depth
选定时间 selected time
选定水雷 selected mine
选定速度 selected speed
选定太阳能表面 designated solar surface
选定温度 selected temperature
选定(项目的)试验 optional test
选伐 selective felling;selective logging
选分仓 separating chamber;separator chamber
选分机 classifier;rejector
选分率 precision of separation
选分筛 classifying screen;separating screen;sizing screen
选分筛片性的 classifying screen
选分铁栅筛 sorting grizzly
选分仪 kicksorter
选粉机 classifier air separator
选粉机粗粉 separator tailings;tailings of separator
选粉机负荷 separator load
选粉机和声响信号控制磨机喂料 mill feed control by separator and acoustics
选粉机外壳 outer separator casing
选粉机外锥 outer separator cone
选粉机细粉 separator fines
选粉机效率 separator efficiency
选粉机效率曲线 Tromp curve
选粉机叶片 separator blade
选粉气流 separating air
选粉效率 classification efficiency;separating efficiency;separation efficiency
选港费 option charges

选港附加费 optional additional charges
选港交货 option delivery
选港提单 optional bill of lading
选港装载 optional stowage
选购 choose;option
选购产品 shopping goods
选购的附加设备 optional equipment
选购品 free choice of goods
选购权交易所 option exchange
选购权买卖 option trading
选购权市场 option market
选购权证 option warrant;warrant
选购商品 commodities purchased by choice;free choice of goods;selective purchase
选购设备模式 equipment selection model
选好度变化 preference change
选号 numeric(al) selection
选号箱 selector box
选呼电话 selective call line telephone
选呼电话主机 central equipment for selective call line telephone
选呼调度系统 selective dispatching system
选呼调度制 selective dispatching system
选集 enriching;florilegium[复 florilegia]
选价预算 <建筑> account transaction
选件 option
选叫键 selective call key
选叫通话箱 selective calling device and talking set
选叫通信[讯] selective communication
选接单元 adjustable interconnection block
选接器放大器 pick-up amplifier
选接完成信号 end-of-selection signal
选局机架 office frame
选局器 zone selector;office selector
选局中继器 switching selector repeater
选举程序 electoral process
选矿 beneficiation; dressing; milling (of ore); mine dressing; mineral beneficiation; mineral concentration;mineral dressing;mineral separation; ore beneficiation; ore cleaning;ore dressing
选矿比 concentration ratio; ratio of concentration
选矿槽 concentration basin
选矿槽析流板 apron
选矿产量 dressing output
选矿产率 dressing yield
选矿产品 ore-dressing product
选矿厂 beneficiation plant;concentrating mill; concentration plant; dressing work; mineral dressing plant;ore-dressing plant;ore separation plant;washing plant
选矿厂厂址 millsite
选矿厂废石 mill chats
选矿厂接收处理的矿石 run-of-mill
选矿厂尾矿 mill tailings
选矿场 battery
选矿车间 enrichment plant
选矿成本 dressing cost
选矿程度 dressing procedure
选矿床 picking table
选矿单位 cleaning unit
选矿凳 mill bench
选矿法 treating process;treatment process
选矿方案 scheme of concentration
选矿方法 mineral separation process
选矿废水 ore-dressing wastewater
选矿废水处理 treatment of ore-dressing wastewater; wastewater treatment of ore-dressing
选矿废渣 reject
选矿费 treatment charges
选矿分离机 cobbing separator
选矿浮选废水 flo(a)tation wastewater
选矿工 dresser;millman
选矿工程师 mineral preparation engineer;mineral processing engineer; ore-dressing engineer
选矿后的废石 lean material
选矿后作业 adverse separation practice
选矿回收率 concentration recovery; ore dressing recovery percentage
选矿机 concentrating machine; concentration machine; concentrator; dresser;mineral dressing machine; preparatory;beneficiation plant
选矿技术经济指标 indices of mineral dressing
选矿精度 precision of separation
选矿流程 beneficiation flow;flow sheet of mineral dressing;mineral dressing flow; mineral processing flowsheet; ore-dressing flow; ore-dressing scheme
选矿流程图 flowchart of ore processing
选矿流程研究 study of ore-dressing scheme
选矿前作业 milling prophase operation
选矿取样 head sampling; mill sampling
选矿筛子 ore preparation screen
选矿设备 cleaning unit; dressing equipment;preparation equipment
选矿水 dress water
选矿台 cleaning table; concentrating table;sorting table
选矿筒 trommel;trommel screen
选矿尾矿 ore-dressing tailings
选矿系数 rate of enrichment
选矿摇床 concentrating table
选矿用水 mill water
选矿站 mineral separation station;ore dressing station
选矿指标 metallurgic(al) goal
选矿中间产物 chats
选矿装置 dressing plant;ore-dressing plant
选矿综合费用 processing cost
选矿作业 milling operation;ore-dressing practice
选粒机 classer
选粒器 Dorr classifier
选粒(铁栅)筛 sorting grizzly
选链 select chain
选料 select material
选料层 selected material course
选料斗 hopper feedback
选料器 mix selector
选录 excerpt
选路 route selection
选路电路 route calling circuit;route selecting circuit
选路开关 routing switchboard;selector switch
选路控制 routing control
选路器 route selector
选路式信号 route signal(l)ing
选路式信号系统 route signal(l)ing system
选路式信号制 route signal(l)ing system
选路完成 route completion
选路网络 route selection network; routing network
选路信号 splitting signal(ling)
选路信号机【铁】splitting signal
选码器 code selector
选码振铃 code call
选煤 coal dressing;coal preparation; coal separation
选煤厂 coal-dressing plant;coal-separating plant;coal washery;concentrator
选煤车间 coal preparation plant
选煤废水 coal-dressing wastewater
选煤机 coal picker;coal preparation machine;concentrator
选煤站 coal dressing station;coal separation station
选民 constituent
选模标本 lectotype
选模器 mode selector
选模式 lectotype
选磨 selective grinding
选木打枝 selective high pruning
选派 appointment
选配 assortative mating;fitting;matching;selected mating;selecting and fitting;selective assembly;selective pairing
选配的 apolegamic
选配电路板 option board
选配多项式 fitting polynomial
选配法 matching method; trial-and-error procedure
选配骨料 aggregate grading
选配合 selective fit
选配集料 aggregate grading
选配量规 selector ga(u)ge
选配零件副 matched pair
选配轮胎 type sizing
选配曲线 curve fit(ting);fitting curve
选片 option panel
选片控制器 reject control
选票 ballot
选频 frequency selection
选频带滤波器 band selective filter
选频电平表 selection level meter;selective level meter
选频电平指示器 frequency selective level indicator
选频放大器 accentuator; frequency selecting amplifier;frequency selective amplifier
选频伏特计 frequency selective voltmeter
选频连接器 frequency selecting connector
选频铃 harmonic ringer
选频式电平表 frequency selective level indicator
选频网络 frequency selective network
选频系统 frequency selective system
选频振铃 harmonic ringing;harmonic selective ringing
选频终接器 frequency selecting connector
选频阻尼材料 frequency selective damping material
选器 selecting unit
选区 ward
选区的居民 constituency
选区机 district selector
选区机架 district frame
选区衍射 selected diffraction
选取 access
选取标准 criterion of selection;selection standard
选取的样品 outgoing material
选取电路 selecting circuit
选取法 pull-up method
选取控制 access control
选取码 access code
选取线 alternative line
选取指标 index of selection
选权式迭代法 interaction method with variable weights
选任制 system of election of cadres
选色电极系统 colo(u)r-selecting-electrode system
选色电极系统透过率 colo(u)r-selecting-electrode transmission
选色镜 dichroic filter
选色用偏转系统 colo(u)r-selector deflection system
选砂 sand-sifting
选砂试验 test of sand selection
选筛台 picking table
选石机 stone picker
选手 player
选数管 selectron
选数器 selector
选速装置 pick-off gear
选台呼叫 general call to two or more specific(al) stations
选台旋钮 station selector
选题情报服务 selective dissemination of information
选通 gating;strobe;unit sampling
选通触发电路 gated flip-flop
选通单元 gating unit
选通道 gate tube
选通电路 gate circuit;gating circuit; selective gate;strobe circuit;strobotron circuit
选通定标器 gated scaler
选通多谐振荡器 gate multivibrator
选通阀 gate valve
选通方波发生器 gated square wave generator
选通放大器 gated amplifier
选通辐射计 gated radiometer
选通跟踪滤波器 gated tracking filter
选通跟踪滤波器系统 gated tracking filter system
选通关断可控硅整流器 gate turnoff thyristor
选通管 gate tube
选通级 gating stage
选通计数器 gated counter
选通继电器 relay gate
选通开关 gating switch
选通控制 gate control
选通控制器具 gate controlling device
选通码 gating code
选通脉冲 gate(im) pulse;gating(im) pulse;pulse gate;selective impulse; strobe(pulse)
选通脉冲边缘 strobe edge
选通脉冲标志 strobe marker
选通脉冲发生器 gate(pulse) generator;gating pulse generator;strobing pulse generator; switching pulse generator
选通脉冲放大器 gated amplifier
选通脉冲幅度 gated phase amplitude
选通脉冲宽度 gate length;gatewidth; key-pulse width;trigger-gate width
选通脉冲宽度调整 gatewidth control
选通脉冲门 strobe gate
选通脉冲输入 strobe input
选通脉冲(转换)开关 strobe switch
选通门 strobing gate
选通门检测器 gate detector
选通能力 gating capability
选通切断电路 gate-off circuit
选通全加器 gated full adder
选通绕组 gate winding
选通扫描 gated sweep
选通时间 gating time
选通时钟 gated clock
选通微型器件 gating module
选通系统 gating system

选通线路 gating line
选通线圈 gate winding
选通信[讯]号 gate signal;gating signal
选通行扫描 gating line-scan
选通装置 strobe unit
选通阻断电路 gate-off circuit
选通作用 gate action
选图设备 pick device
选线 reconnaissance(survey);route selecting;route selection;selection of route;railway location【铁】
选线测量 reconnaissance survey
选线方案 variant projects of location
选线工程师 locating engineer;location engineer
选线接线器 line choice connector
选线开关 selector switch
选线控制 selection control;selective control
选线器 selector
选线器系统 line-finder system
选线自动化 automation of route selection
选相交流电码轨道电路 phase selective AC coded track circuit
选相器 phase selector
选星 choosing star;selection of stars
选行监视器 line strobe monitor
选型 lectotype;proplasm;type selection
选型负荷 type selection load
选修科 minor
选修课 optional course;selective course
选修课程 alternate course
选穴法 point selection
选岩粉样(品)sludge sampling
选样 collection;sample selection;sampling action;selective sampling
选样器 dip can;sampler
选样认可 acceptance sampling
选样作用 sampling action
选冶试验报告名称 title of test report of mineral dressing ability and metallurgy
选冶问题未解决 mineral separative and metallurgic(al) problems unsettled
选用 adoption
选用标准化尺寸时的合理数字 preferred number
选用材料 select material
选用尺寸 preferred dimension;preferred size
选用的类表 preferred scheme
选择的投标者名单 invited bidders
选用的资料 selected data;selected information
选用概率的设计方法 probabilistic design method
选用钢筋 preferred reinforcement
选用恒星 selected stars
选用建筑<推荐采用的建筑形式> architecture elect
选用经理 manager selection
选择开挖 selective digging
选用零件表 preferred parts list
选用名 alternate name
选用顺序 preferred order;preferred sequence
选用通用图 typical drawings adopted
选用位置 preferred site
选用折旧方法声明 notification of depreciation method
选用值 preferred value
选优获益 opportunity cost
选余的 cull
选余的东西 culls
选余之物 cull
选源盘 source-selector disk
选择 choice;finding action;make choice of;option;preference;selecting
选择搬运 selective transportation
选择板 option board;option panel
选择报价 variation in bidding conditions
选择比 selection ratio;selectivity ratio
选择笔 selector pen
选择变量 choice variable
选择标度 selectivity scale
选择标记插入 selective mark insertion
选择标商阶段 negotiation phase
选择标准 option standard;selection criterion;selective standard
选择表 option table
选择波长法 selected ordinate method
选择补充库存管理制度 management system of choosing supplement stock
选择不稳定度 selective instability
选择布线法 discretionary wiring method
选择部件 alternative pack
选择菜单 selection menu
选择参入 selective administration
选择参数 selection parameter
选择操作 selecting operation
选择测量系统 selective measuring system
选择插件板 selection plugboard
选择差别 selection differential
选择产次率 select issue rates
选择常数 selectivity constant
选择厂址论 location theory
选择场址勘察阶段 detailed planning stage
选择车辆检测 selective vehicle detection
选择成本 alternative cost;discretionary cost
选择成熟 selective maturation
选择程序 option program(me)
选择程序计算机 selective sequence computer
选择程序计算器 selective sequence calculator
选择持有人 option holder
选择齿轮传动 selective gear drive
选择齿轮式变速器 selective gear transmission
选择冲洗厕所 alternatives to the flush toilet
选择抽样(法)choice-based sampling;selective sampling
选择穿透性 selective permeability
选择船位 assumed position;chosen position
选择磁铁 select(ing)magnet
选择磁芯矩阵 selection core matrix
选择次优方案 selection of the second best course of action;selection of the second best scheme
选择催化还原法 selective catalystic reduction process
选择催化裂化 selective catalystic cracking
选择催化转化 selective catalystic conversion
选择淬火 selective quenching
选择代码 selective code
选择带 select tape
选择单 menu
选择单位 selection unit
选择道 selective track
选择的 optional;selected;selective
选择的出行 choice trip
选择的初期阶段 initial stages of selection
选择的个体 selected individual
选择的进路(或径路)route chosen
选择的填置层<指路基最上部,或指下层基层> selected filling layer
选择的准则 criterion of choice
选择地址及对比操作 select address and contrast operate
选择点 choice point
选择电磁铁 selection magnet
选择电极 selective electrode
选择电键 selection key
选择电路 selecting circuit;selection circuit;voting circuit
选择定则 selection principle;selection rule
选择定址 selective addressing
选择毒物 selective poison
选择毒性 selective toxicity
选择度 selectance
选择镀 selective plating
选择断路 selective tripping
选择断路系统 selective tripping system
选择对象 alternative
选择发酵 selective fermentation
选择发射频率 select transmit frequency
选择发生 elective genesis
选择发送频率 select transmit frequency
选择阀 selector valve
选择法 back-and-forth method;selection method
选择法庭条款 choice of forum clause
选择反射 selective reflection
选择反射能力 selective reflectivity
选择反应 choice reaction;selection response
选择范畴 selective category
选择方案 option;selection scheme
选择方法 decision-making process;selective method;system of selection
选择方式 choice mode
选择放大 selective enlargement
选择放射 selective emission
选择放松 relaxation of selection
选择分类 selection sort
选择分馏 selective rectification
选择风缸 selector volume
选择风化作用 selective weathering
选择俘获 selective capture
选择辐射 selective radiation
选择辐射体 selective radiator
选择伽马-伽马测井 selective gamma-gamma log
选择伽马-伽马测井曲线 selective gamma-gamma log curve
选择杆 selecting lever;selection lever;selector lever;slot lever
选择港 optional port;port of option
选择港货物 optional cargo
选择港提单 optional bill of lading
选择高产品种 selection of a high yielding variety
选择跟踪 selective trace[tracing]
选择公理【数】axiom of choice;axiom of selection
选择功能 option function;selection function
选择供货单位 choice sources of supply;choose sources of supply
选择供应的附件 optional units
选择共鸣 selective resonance
选择共振 selective resonance
选择关税 selectionary tariff
选择关系 preference relation
选择管 gate tube
选择广播发射台 selective broadcast sending station
选择规格化变换 select normalization transformation
选择规则 choice rule;selective rule
选择过程 selective process
选择函数 choice function;selection function
选择合并 selective adversity combining
选择桁架木板坝 leaf dam
选择呼号 selective call
选择呼叫 multidrop;selected calling;selection call;selective calling;selective ringing
选择呼叫拒绝 selective call rejection
选择呼叫设备 selective call equipment
选择呼叫系统 selected calling system
选择呼叫振铃 selective call ringing
选择呼叫制 selective calling system
选择呼叫装置 selcall device
选择互换性 selective interchangeability
选择挥发 selective volatilization
选择混合相裂化 selective mixed-phase cracking
选择混凝土配合比 selective concrete batching
选择活门 selector valve
选择货币贷款 optional currencies loan
选择机构 selecting mechanism;selection mechanism;selector mechanism
选择机能 optional function
选择基 selectophore
选择激发法 selective excitation method
选择激发机理 selective excitation mechanism
选择级 stage of selection
选择极限 selection limit
选择继电器 discriminating relay
选择寄存器 mask register
选择加热 selective heating
选择检验 selection check;selective check;select verify
选择建筑基地 option of building field
选择鉴别特征 selective identification feature
选择键 selective key
选择交叉 selective transposition(of cable quad)
选择交代作用【地】selective metasomatism;selective replacement
选择校验 selection check
选择阶段 choice phase
选择接卡箱 selective stacker
选择接口 option interface
选择接收 selective reception
选择接收机 selective receiver
选择节段 selected segment
选择结晶 selective freezing
选择进料器 selective feeder
选择进路 selection route
选择浸出 selective leaching
选择经度 assumed longitude;chosen longitude
选择精馏 selective rectification
选择精制 selective finishing
选择矩阵 selection matrix
选择聚合 selective polymerization
选择聚合过程 selective polymerization process
选择聚合物 selective polymer
选择聚焦 selective focus
选择开关 option switch;selection switch;selective switch;selector
选择开关旋钮 selector knob
选择开挖法 selective excavation
选择砍伐 selective cutting;selective logging
选择砍伐稀疏 selective thinning
选择砍筏 selective cutting
选择砍筏稀疏 selective thinning

X

选择可用性 selective availability
选择刻蚀 selective etching
选择控制 alternative control; selection control
选择控制系统 selective control system
选择控制制 selective control system
选择框 choice box
选择扩散 selective diffusion
选择扩散式路径确定 selective flooding routing
选择理论 choice theory
选择连接法 discretionary wiring (method)
选择量 quantity of selection
选择列表 selective listing; select list
选择裂化 selective cracking
选择裂化过程 selective cracking process
选择滤波器 selective filter
选择滤色片 selective screen
选择路线 selection schemer
选择率 selectivity
选择氯化(法) selective chlorination
选择码 option code; selector code
选择脉冲 selection(im) pulse; selective pulse
选择脉冲发生器 selector-pulse generator
选择命令 select command
选择模型 preference pattern
选择内容印出 selective dump
选择能力 selective power; selectivity
选择排除 preferential elimination
选择排流法 selective drainage method
选择培养基 selective medium
选择配份 choice share
选择配合 selective fit
选择偏差 selection bias
选择频率控制 selective frequency control
选择平均值 selective mean
选择器 chooser; designator; finder; mix selector; selecting unit; selector
选择器按钮 selector(push) button
选择器标度盘 selector dial
选择器标记 selector mark
选择器步进磁铁 selector stepping magnet
选择器叉 selector fork
选择器的 a、b 线弧接线排【铁】link bank
选择器的 c 线弧【铁】private bank
选择器的触排 field of selection
选择器的寄存器 selector register
选择器电键 selector key
选择器电路 selector circuit
选择器电刷 pilot brush
选择器阀 selector valve
选择器负荷容量 selector carrying capacity
选择器杆 selector rod
选择器机架 selector frame
选择器级 rank of switches
选择器继电器 selector relay
选择器架 selector panel; selector shelf
选择器开关 multiple contact switch; selector switch
选择器刻度 selector scale
选择器量程 selector range
选择器列 rank of selectors
选择器塞头 selector plug
选择器线弧 translation field
选择器箱 selector set
选择器寻线时间 selector hunting time
选择器增音机 selector-repeater
选择器轴的导槽 selector shaft guide
选择器装置 selector installation
选择器自由回转起始接点 normal post contact
选择强度 intensity of selection; selection intensity
选择清除 <有选择地将一个信息从一个存储单元转移到另一个存储单元>【计】selective dump
选择曲线 trade-off curve
选择权 option(al) right
选择权变化率 option delta
选择权合同 option contract
选择权交易 option dealing
选择权买方 option buyer
选择权卖方 option seller; option writer
选择权清算公司 options clearing corporation
选择权外汇契约 option exchange contract
选择权协议 option agreement
选择权有效期限 option period
选择染色 selective staining
选择绕组 selection winding
选择人才 select good persons
选择溶剂提取 selective solvent extraction
选择熔化 selective fusion
选择锐度 sharpness of selection
选择色调校正 selective tint correction
选择射击阵地 selection of firing position
选择渗透膜 perspective membrane
选择渗透性 permeaselectivity; selective permeability
选择生命表 select life table
选择声频放大器 selective audio amplifier
选择识别 selective identification
选择使用 alternative use
选择使用的编组线 selection of classification tracks to be used
选择式变速器 selective transmission
选择式间苗机 selective gapper
选择式接近控制 selective approach control
选择收费制 optional rate
选择手段 selection approach
选择数据 selecting data
选择数字发射器 selective digit emitter
选择数字发送器 selective digit emitter
选择水源 selective water source
选择税 alternative duty
选择顺序 preference ranking; selecting sequence; selective sequential
选择搜索 selective searching
选择算法 selection algorithm
选择特性(曲线) selectivity characteristic
选择替换技术 selection replacement technique
选择填料 selected fill
选择条款 selectivity clause
选择停机指令 optional half instruction; optional stop instruction
选择通道 selector channel
选择通话 brokers call; multidrop
选择通路 selection path; selector channel
选择通透性 permselectivity; selective permeability
选择通信[讯] selective communication
选择统计量 selection statistic
选择投标方法 selective bid method
选择投弃 selective jettison
选择透过性 perspectivity; selective permeability
选择退火 selective annealing
选择外延 selective epitaxy
选择网络 selective network
选择纬度 assumed latitude; chosen latitude
选择位 select bit; subdevice bit
选择位置 assumed position; chosen position
选择问答法 multiple choice question
选择问题 routing problem
选择无效 invalid selection
选择物 candidate
选择物标 select target
选择吸附 selective absorption
选择吸声 selective absorption of sound
选择吸收 discrete absorption
选择吸收器 selective absorber
选择吸收性能 selective absorbability
选择吸收作用 selective absorption
选择系数 selectance; selection coefficient
选择系统 selective system
选择狭缝 selection slit
选择显示 selective display
选择线 selection wire; select line
选择项目 alternative item; option
选择消磁 selective erase
选择谐振 selective consonance
选择卸港货物 optional cargo
选择卸货港费 additional for optional port of discharge; optional port charges; optional surcharge
选择卸货港和附加费 optional destination and option fee
选择卸货港交货 optional delivery
选择卸货港所增运费 optional charges
选择信标雷达 selective beacon radar
选择信号 selection signal; selective signal
选择信号装置 selection signal(l)ing installation
选择信息 readout
选择信息路线 message routing
选择行为 choice behavio(u)r
选择性 selectivity
选择性螯合树脂 selective chelating resin
选择性搬运 selective removal
选择性保护 selective protection
选择性保护法 selective protection method
选择性保留 selective retention
选择性被动反射器吊舱 optional passive reflector pods
选择性变速 selective gear shift(ing)
选择性标记 selected marker
选择性标引法 selective indexing
选择性拨款 discretionary funding
选择性擦除 selective erasing
选择性采样 selective sampling
选择性采样器 selective sampler
选择性层位封堵 selective plugging
选择性差动继电器 selective differential relay
选择性沉淀 selective precipitation
选择性沉积 selective deposition
选择性除草剂 selective herbicide
选择性除莠 selective weeding
选择性除莠油 selective type herbicidal oil
选择性处理 selective treatment
选择性催化还原法脱氮 control of NOx by selective catalytic reduction
选择性存货管理 selective inventory control
选择性的 alternative; selective
选择性的电离压力计 selective ionization ga(u)ge
选择性地层试验 straddle testing
选择性电极法 selective electrode method
选择性电路 selective circuit
选择性电平测量设备 selective level measuring equipment
选择性顶替 organocalie; selective displacement
选择性定位 selective localization
选择性二叠合 selective dimerization
选择性法律条款 choice of law clause
选择性反萃取 selective stripping
选择性反调制 selective demodulation
选择性反应 selective reaction
选择性放大器 selective amplifier
选择性放牧 selective grazing
选择性分离 selective separation
选择性分配 selective distribution
选择性分配资金 selective distribution of fund
选择性风化 selective weathering
选择性浮选 selective flo(a)tation
选择性浮游分选法 Bradford preferential separation process
选择性辐射计 selective radiometer
选择性腐蚀 selective corrosion
选择性负反馈 selective inverse feedback
选择性富集 selective enrichment
选择性干扰 selective interference; selective jamming
选择性干扰机 selective jammer; spot jammer
选择性固定性成本 discretionary fixed cost
选择性光电效应 selective photoelectric(al) effect
选择性光电子发射 selective photoelectric(al) emission
选择性光学自动跟踪 selective optical lock-on
选择性广播接收台 selective broadcast receiving station
选择性函数 selectivity function
选择性还原 selective reduction
选择性换能器 selective transducer
选择性恢复压力 selective repressuring
选择性计算 optional calculation
选择性继电器 selecting relay; selective relay
选择性加热 selective heating
选择性检测 selective detection
选择性接触 selective contact
选择性结构 selective structure
选择性结晶 preferential crystallization
选择性聚集 selective aggregation
选择性决策 alternative decision
选择性开采(法) selective mining; high grading
选择性控制 selective control; selectivity control
选择性扩散器 selective diffuser
选择性离子电极 ion selective electrode
选择性离子交换 selective ion exchange
选择性灵敏度 selective sensitivity
选择性滤光镜 selective filter
选择性滤光片 selective absorbent; selective absorber
选择性滤光器 selective filter
选择性螺旋给进 selective screwfeed
选择性膜 selective membrane
选择性磨矿 selective grinding
选择性黏[粘]附 selective attachment
选择性扭曲 selective distortion
选择性农药 selective pesticide
选择性培养基 selective culture medium
选择性配额 selective quota
选择性破坏 selective damage

X

选择性破碎 selective breaking; selective crushing
选择性亲和力 selective affinity
选择性侵蚀 selective erosion
选择性清除累加器 selectivity clear accumulator
选择性球状团聚 selective spheric(al) agglomeration
选择性曲线 selectivity curve
选择性燃烧 selective combustion
选择性燃烧法 selective combustion method
选择性染色翻版法 selective dyeing process
选择性溶剂 selective solvent
选择性溶剂抽提 selective solvent extraction
选择性溶剂抽提油 selective solvent extracted oil
选择性溶剂萃取法 selective solvent extraction method
选择性溶剂过程 selective solvent process
选择性溶剂精制油 selective-solvent-refined oil
选择性溶剂提取法 selective solvent extraction method
选择性溶解 preferential solubility; selective solubility
选择性溶解扩散 selective solubility diffusion
选择性溶湿润 preferential wetting; selective wetting
选择性塞子 <可以插入任何一排塞孔> wander plug
选择性散射 selective scattering
选择性杀鼠剂 selective rodenticide
选择性渗透 selective permeation
选择性渗透的 permselective
选择性渗透膜 permselective membrane
选择性渗透作用 selective osmosis
选择性生长 selective growth
选择性生物强化 selective bioaugmentation
选择性试剂 selective reagent
选择性试验 selective test
选择性收集 selective collection
选择性衰落 selective fading
选择性水化 selective hydration
选择性酸处理 selective acidizing
选择性酸化 selective acidizing
选择性酸碱催化 specific acid base catalysis
选择性提取 selective extraction
选择性提取技术 selective extraction technique
选择性调节系统 selective control system
选择性调整 selectivity control
选择性同位素吸收 selective isotope absorption
选择性投标 selective bid(ding); selective tender(ing)
选择性突变 selective mutation
选择性突变性 selective mutant
选择性图 selectivity diagram
选择性土 selected soils
选择性挖掘 <按照不同土类进行挖掘> selective digging
选择性网络 selective network
选择性吸附 preferential adsorption
选择性吸附催化法 selective adsorption-catalytic process
选择性吸附剂 selective adsorbent
选择性吸光 selective absorption
选择性吸气剂 selective getter
选择性吸收 preferential absorption; selective absorption
选择性吸着剂 selective sorbent
选择性洗脱 selective elution
选择性系数 coefficient of selectivity; selective coefficient; selectivity factor; selectivity coefficient
选择性消除 selective erasure
选择性消费 optional consumption; selective consumption
选择性消灭 selective elimination
选择性携带剂 selective entraining agent
选择性信贷控制 selective credit control
选择性信道 selective channel
选择性信息 selective information
选择性信息更新 selective updating
选择性信息提供 selective dissemination of information
选择性信用管理 selective credit control
选择性絮凝 selective flocculation
选择性絮凝剂 selective flocculator
选择性压裂 selective fracturing
选择性延时动作 time selective action
选择性研磨 differential grinding; selective grinding
选择性阳极过程 selective anodic process
选择性抑菌作用 selective bacteriostasis
选择性抑制剂 selective depressant
选择性优惠关税制度 selective system of preference
选择性有机化合物 selected organic compound
选择性再结晶 selective recrystallization
选择性再吸收 selective reabsorption
选择性债务 alternative obligation
选择性招标 selected bidding; selective bid(ding); selective tender(ing); limited tendering
选择性蒸发 selective evapo(u)ration
选择性知觉 selective perception
选择性指示剂稀释曲线测定 selective indicator dilution curve determination
选择性指数 selectivity index
选择性重吸收 selective reabsorption
选择性转储 selection dump
选择性转化 selective conversion
选择性资本 optional capital
选择性子弹射孔器 selective bullet gun perforator
选择性自动控制 automatic selectivity control
选择性自动雷达识别装置 selective automatic radar identification equipment
选择性作用 selective action
选择序列 selective sequential
选择学说 selectionism; selection theory
选择寻迹程序 snapshot program(me)
选择寻址 selective addressing
选择寻址存储器 selective addressing memory
选择压力 selection pressure; selective pressure
选择氧化 selective oxidation
选择氧化工艺 selective oxidation process
选择移圈 selective loop transferring
选择因数 selectivity factor
选择硬化 selective hardening
选择泳移 selective migration
选择优势 selective advantage
选择优先权 selecting priority
选择优先吸附 selective preferential adsorption
选择有利性 selective advantage
选择语句 case statement; selection statement
选择育种 selective breeding
选择元(件) selection element; alternative element
选择原则 selection principle
选择圆盘 selector disc[disk]
选择远期外汇契约 optional forward exchange contract
选择运销通路 selection of channel
选择再活化性 selective reactivation property
选择噪声检波电路 selective noise-detection circuit
选择增效 selective advance
选择斩波辐射计 selective chopper radiometer
选择招标 selected bidding; selective tender(ing)
选择照明 selective illumination
选择折射 selective refraction
选择者 selector
选择阵列法 discretionary array method
选择振铃 selective ringing
选择振铃信号 individual signal
选择值 selective value
选择指 <一种用来探查穿孔卡或纸带系统孔洞的探头>【计】finger; selecting finger
选择指点标 selector marker
选择指令 selection instruction; select order
选择指令组 optional instruction set
选择指示器 selector marker
选择指数 selection index
选择指数法 selection index method
选择指针 select finger
选择制 selective system
选择制菌作用 selective bacteriostasis
选择种 selective species
选择住所 domicile of choice
选择转储 selective dump; snapshot dump
选择转换开关 selective switchgear
选择装配 selective assembly
选择装配式制造 selective assembly manufacturing
选择装载 selective loading
选择装置 strobe unit
选择追踪程序 selective tracing routine
选择纵坐标法 selected ordinate method
选择组 select set
选择最优方案 choose the best alternatives; select the best alternatives
选择作用 selection
选站 layout of station; station selection
选站按钮 station selection(push) button
选站电码 station selection code
选站继电器 station selector relay
选站脉冲 station selection impulse
选针 needle selection
选针集圈 selective tucking
选针片 selection piece; selector bit
选针纹板 selecting card
选址 choosing the location; location selection; selection of plant location; site selection
选址比较方案 alternative site
选址不受限制的工业 foot-loose industry
选址查勘 site selection investigation
选址单元 selected cell
选址定线问题 location routing problem
选址分析 siting analysis
选址规划 siting plan
选址阶段 siting stage
选址勘察 siting investigation; siting survey
选址(理)论 location theory
选址受限制的工业 foot-tight industry
选址术 selecting address technique
选址条件 location condition
选址研究 siting analysis; siting study
选址意见书 site designation memorandum
选主元 pivoting
选组 framing; grouping selection
选组标志器 selector stage marker
选组级 selector stage
选组计数器 batch(ing) counter
选组器 group selector
选组器机架 group selector rack
选组总数 batch total

炫

炫光 <令人一时模糊看不清的> veiling glare
炫目的砂 blinding sand
炫目危险 risk of glare
炫耀 flare
炫耀光栅 blazed grating

眩

眩感 dazzle
眩光 blinding glare; glare
眩光带 glare zone
眩光灯 glaring light
眩光恢复时间 return time of vision
眩光控制 glare control
眩光屏 glare screen
眩光视力 glare vision
眩光危险 risk of glare
眩光效应 glare effect
眩光指数 glare index
眩目 dazzle; glare
眩目白色 dazzling white
眩目的 blinding
眩目灯光 dazzle light(ing); glaring light
眩目效应 glare effect
眩耀光 glaring light
眩晕 dizziness; giddiness; staggers; vertigo
眩晕的 vertiginous
眩晕症 megrims; staggers

渲

渲染 rendering
渲染技术 rendering technique

碹

碹承台 skew angle bearer
碹碹角钢 skewback angle
碹碹砖 skew brick
碹滴 crown drop; smelter drippings
碹顶 apex of arch
碹顶高度 spring arch
碹高 arch rise; rise of arch
碹拱中心角 included angle of arch
碹环 arch ring
碹脚 skew; springer
碹口 quarl(e)
碹跨 arch span
碹跨比 rise to span ratio
碹肋 arch rib
碹圈 arch ring
碹胎 arch centering[centring]
碹砖 arch brick; springer
碹座 arch support

镟

镟床 lathe; turning machine
镟木家具 all-turned furniture

镟切 rotary cut
镟削孔 turning hole
镟圆 rounding
镟制螺栓 turned bolt
镟制用短原木段 billet

靴 boot

靴板 shoe
靴带方案 bootstrap scheme
靴耳属 < 拉 > Crepidotus
靴梁 boot-beam plate
靴袢动力学 bootstrap dynamics
靴式开沟器 shoe colter; shoe runner; shoe-type furrow opener
靴式舯板 half decked boat; half decker
靴筒 stove pipe < 沸腾钢缺陷 >; bootleg < 轧制缺陷 >
靴头 toe
靴形 boot-shaped
靴形锅炉 boot boiler
靴形滑车 shoe block
靴形心 boot-shaped heart
靴形制动器 shoe brake
靴状吸嘴 < 挖泥船的 > shoe-shaped drag head

薛泊式球笼等速万向节 Rzeppa (constant velocity) universal joint

薛佛地板 Schaffer floor
薛佛酸 Schaffer's acid
薛氏温度计 Six's thermometer
薛氏最高最低温度计 Six maximum and minimum thermometer

穴 cavity; delve; excavation; pocket; sinus

穴播 bunch planting; dropping in hill; hole seeding; hole sowing; sowing in holes
穴洞 grotto
穴洞网 network of burrows
穴腐 < 木材内 > pocket rot
穴灌 hill spacing; hole irrigating[irrigation]
穴居 cave dwelling
穴居的 troglobiotic
穴居动物 cave animal; troglodyte
穴居人 cave-man; troglodyte
穴居生物学 biospeleology
穴居时代 cave period
穴居型 fossorial
穴居者 burrower
穴距 distance between hills; distance between holes
穴孔网 spongework
穴泉 pocket spring
穴施 hole application
穴施法 application in planting hole; perforated method
穴蚀 cavitation (erosion); pit corrosion
穴蚀磨损 cavitation wear
穴式墓 trench-grave; trench-tomb
穴移植 pit transplanting
穴植 dibble; hill planting; hole planting; hole transplanting; pit planting
穴植法 hole-method; hole-planted method
穴中栽植 centre hole planting
穴珠 cave pearl
穴珠粒级 grain-size of cave pearl
穴状栽植 hole planting; pit planting

学

学报 academic(al) journal; journal; proceedings; transaction

学部 workshop
学部主任 division chairman
学费 premium; schooling
学会 academy; association; college; institute; institution; learned society; society
学会会员 academician
学会会员资格 fellowship
学会建筑 institutional building
学会条款 institute clause
学界 academic(al) community
学究式的 ivy
学科 course; department; discipline; subject
学科带头人 leaders in their chosen field of learning
学科的分支 subdiscipline
学科之间的 interdisciplinary
学历 educational qualification; education background; academic(al) history < 大专以上的 >
学历与专业能力 academic(al) and professional ability
学历与资历 academic(al) and professional ability
学龄儿童组 child of school age
学龄前儿童游戏场地 preschool children's playground
学名 nomen; scientific name
学年 academic(al) year; school year
学派 school
学期 school term; session; term; term time
学前教育 preschool education
学舍 lyceum
学生 t 分布 student's t distribution
学生 t 检验【数】student's t test
学生合作住房 student cooperative housing
学生及实习生 students and apprentices
学生季票 student's season-ticket
学生寄宿舍 students' hostel
学生居住单元 students' dwelling unit
学生联合会 student union
学生票价 student fare
学生评价 student assessment
学生容量 pupils' capacity
学生食堂 pupils' dining room; students' dining hall
学生宿舍 students' dormitory; study-bedroom; students' residential hostel < 校外的 >
学生用商品 back-to-school goods
学生寓所 students' quarters
学生阅览室 pupils' reading room; students' reading room
学生招待所 students' hostel
学士学位 bachelor's degree; bachelorship
学术报告 academic(al) report; colloquium; learned report; memoir
学术成就 academic(al) achievement; scholastic achievement; scholastic attainment
学术城 academic(al) city
学术代表作 portfolio
学术的 scientific
学术动态 academic(al) trends
学术环境 academia
学术会议 academic(al) conference; academic(al) meeting; institute; symposium
学术活动 academic(al) activity
学术机构 academic(al) institute; academic(al) institution
学术讲演 disquisition
学术讲演(或讨论)会堂 lyceum
学术交流 academic(al) exchange
学术界 academia; academic(al) world
学术论坛 academic(al) market
学术论文 academic(al) dissertation; academic(al) writing; disquisition; dissertation; memoir; research paper; scientific paper
学术权威 academic(al) authority
学术水平 level of scholarship
学术水平较高的报刊 highbrow press
学术水平较高的出版物 highbrow publication
学术讨论 academic(al) discussion
学术讨论会 colloquium; seminar; symposium
学术讨论会参加者 clinic delegate
学术团体 academese; academic(al) body; academic(al) society; learned organization; learned society
学术委员会 academy committee
学术文章风格 academese
学术行话 academese
学术性的 academic(al)
学术研究 academic(al) research
学术研究的绪论 isagoge
学术研究会 seminar
学术演讲 disquisition; dissertation
学术座谈会 conversazione
学说 theory
学童过街标志 school crossing sign; students crossing sign
学童交通安全值勤(员) school safety patrol
学徒 apprentice; improver
学徒工 prentice
学徒合同 contract of apprenticeship
学徒年限 apprenticeship
学徒培训 apprenticeship training
学徒期间 apprenticeship
学徒期限 tirocinium
学徒契约 indenture
学徒身份 apprenticeship; tirocinium
学徒实习车间 apprentice(work) shop
学徒实习工场 apprentice(work) shop
学徒式训练 apprentice training
学徒条例 apprenticeship act
学徒制 apprenticeship
学徒制度 apprenticeship system
学位 academic(al) degree; degree
学位论文 dissertation
学位授予典礼 commencement
学位证书 diploma
学位制度 academic(al) degree system
学习辨识 learning identification
学习程序 learning program(me)
学习传输功能 learning transfer function
学习的经济效果 economics of learning
学习动机 academic(al) motive
学习高原现象 plateau[复 plateaus/plateaux]
学习活动 learning activities
学习机 learning machine
学习机制 learning mechanism
学习计划 curriculum[复 curricula/curriculums]
学习驾驶员 driving pupil; learner driver
学习就业序列 learning-employment sequence
学习矩阵 learning matrix
学习控制 learning control
学习控制系统 learning control system
学习理论 learning theory
学习年数 years of schooling completed
学习期限 period of schooling
学习曲线 learning curve
学习容量 learning capacity
学习实践极限 limit of practice
学习适应能力 learning and adapting capability
学习司机 apprentice driver; learner driver; trainee driver; student engineer【铁】
学习算法 learning algorithm
学习特性 learned characteristics
学习网络 learning network
学习系统 learning system
学习效果律 law of learning efficacy
学习心理学 psychology of learning
学习因素说 factor theory of learning
学习者 learner
学习中心 study center[centre]
学习资料中心 learning resource center[centre]
学衔 academic(al) rank
学校 school; seminary
学校标志 school sign
学校操场 school playground
学校村 school village
学校大礼堂 school auditorium
学校大楼 educational block
学校儿童季票 school child's season ticket
学校管理体制改革 reform of school administrative system
学校环境教育 school environmental education
学校会堂 school assembly hall
学校建筑 school construction; school structure
学校建筑学 school architecture
学校街坊 school block
学校课程 school course
学校体育馆 school gymnasium
学校图书馆 school library
学校卫生 school health
学校卫生学 school hygiene
学校用公共汽车 school bus
学校园 school garden
学校噪声 noise in school
学校智力 academic(al) intelligence
学校注册人数 school enrollment
学校综合建筑 school complex
学演机器人 playback robot
学员 trainee
学园城市 school town
学院 academic(al) institution; academy; college; faculty; institute; seminary < 尤指私立女子学院 >
学院城市 academic(al) city
学院的 ivy
学院风格建筑 collegiate architecture
学院副院长 subdean
学院功能 university function
学院建筑 institutional building
学院派风格 academic(al) style
学院派绘画 academic(al) painting
学院式 academi(ci)sm
学院式建筑风格 academic(al) style of building
学院院长 dean of college
学者 bookman; scholar

雪 snow

雪岸 snow bank
雪凹 niche
雪凹冰川 niche-glacier
雪白 albedo of snow
雪白的 niveous; snowy; snow white
雪白云石 gurhofite
雪报春 snow primrose
雪暴 blizzard; Buran; drift storm; snow broom; snow storm; drifting snow
雪暴频率 blizzard frequency
雪被 snow cover

雪崩 avalanche; billow; snow avalanche; snowquake; snowslide; snow slip
雪崩报警器 snowslide alarm
雪崩冰块 avalanche ice
雪崩冰碛 avalanche moraine
雪崩槽沟 avalanche chute
雪崩摧毁 avalanche breakdown
雪崩电离 cumulative ionization
雪崩电路 avalanche circuit
雪崩电压 avalanche voltage
雪崩二极管 avalanche diode
雪崩二极管振荡器 avalanche diode oscillator
雪崩防护 avalanche defense[defence]; avalanche protection
雪崩防护林 avalanche preventing forest
雪崩防护棚 avalanche shed
雪崩防护设施 avalanche protecting facility
雪崩防廊 avalanche gallery
雪崩防御 snowslide defense[defence]
雪崩防御墙 avalanche defense[defence] wall
雪崩防止林 avalanche preventing forest
雪崩防止设备 check facility for snow slide
雪崩防治 snow slide protection
雪崩风 avalanche wind; flurry
雪崩复合低温开关 cryosar
雪崩感应徙动 avalanche-induced migration
雪崩光电二极管耦合器 avalanche photodiode coupler
雪崩光电检测器 avalanche photodetector
雪崩光二极管 avalanche photo diode
雪崩击穿 avalanche breakdown; breakdown avalanche
雪崩晶体管 <利用雪崩现象而放大电流的面结合型晶体管> avalanche transistor
雪崩警报 avalanche alarm
雪崩警报器 avalanche alarm device
雪崩扩展常数 avalanche-development constant
雪崩流 channeled avalanche
雪崩气浪 avalanche wind
雪崩区 avalanche range; Zener region
雪崩效应 <电子的> avalanche
雪崩形半导体 avalanche type semiconductor
雪崩源 avalanche source
雪崩噪声 avalanche noise
雪崩振荡器 avalanche oscillator
雪崩阻挡物 avalanche baffle
雪飑【气】snow squall
雪冰 slob; snow ice
雪冰崩落 firncrash
雪冰化作用 firnification
雪冰界线 firn edge
雪冰区 firn zone
雪冰线 firn line
雪波 sastruga; zastruga
雪层 snow deposit
雪层上径迹测量 track survey on snow layer
雪层上汽车伽马测量 gamma survey on snow with truck
雪铲 snow shovel
雪铲模型 snow-plough model
雪铲形反射器 snow-shovel reflector
雪车 sleight
雪成熟 snow ripening
雪成云 snow cloud
雪秤 snow balance
雪尺 snow roller; snow scale
雪冲刷 snow scavenging
雪吹程 fetch of snow
雪带 nival belt; nival zone; snow belt; snow zone
雪挡 snow baffle; snowbreak
雪的 nival
雪的保护系统 snow protection system
雪的变态 snow metamorphosis
雪的变形 snow metamorphosis
雪的变性 snow metamorphosis
雪的变质 snow metamorphosis
雪的吹积 snow drifting
雪的反射性 albedo of snow
雪的含水量 snow density
雪的累积 snow accumulation
雪的密度 snow density
雪的漂动 snow drifting
雪的软化 snow ripening
雪的升华 sublimation of snow
雪的熟化 <开始融化前> snow ripening
雪的水当量 snowpack water equivalent; water equivalent of snow
雪的研究 snow research
雪的蕴藏量 quantity of snow
雪的蕴水量 <以雪体积的百分数计> quantity of snow
雪的滞留 retention of snow
雪堤 snow bank; snow dike[dyke]
雪堤冰川 nivation glacier
雪堤掘进 snowbank digging
雪地车 ski mobile
雪地车辆 snow-going vehicle
雪地钓钟柳 snowland penstemon
雪地防滑链 snow chain
雪地履带式车辆 snow-cat
雪地履带与驱动链轮 snow sprocket
雪地起落架 snow gear
雪地汽车 over-snow vehicle
雪地牵引车 snow tractor
雪地驱动链轮 snow sprocket
雪地伪装涂料 snow reflecting camouflage paint
雪地用履带式汽车 snow mobile
雪地用汽车轮胎 snow tire[tyre]
雪地用拖拉机 snow tractor
雪地植物群落 chionophytia
雪地装夹 snow set
雪地着陆 snow landing
雪点 snow point
雪堆 drifting; snow bank; snow blockade; snow drift; snowpack; snow wreath
雪耳 earing
雪幡 snow virga
雪反照 albedo of snow
雪费尔焊缝组织图 Schaeffer's diagram
雪封 snow blockade
雪封的 snow bound; snowy
雪腐病 snow mo(u)ld
雪覆盖度 areal coverage of snow
雪盖 snow cover; snow mantle
雪盖的形成 accumulation of snow
雪糕 ice lolly
雪冠 snow cap
雪规 snow ga(u)ge
雪硅钙石 crestmoreite; hydrowollastonite; tobermorite
雪硅钙石胶 tobermorite gel
雪害 snow damage; snow drifting; snow hazard
雪荷载 snow load(ing)
雪荷载应力 snow load stress
雪荷载值 value of snow load
雪洪 snowmelt flood
雪厚 snow depth
雪花 snow flake
雪花玻璃 alabaster glass; frosted glass
雪花瓷器 alabaster ware
雪花覆盖冰 flower ice
雪花干扰 snow interference
雪花环 snow garland; snow wreath
雪花莲属 snowdrop
雪花面饰【建】frosted finish
雪花石膏 compact gypsum; gypseous alabaster; gypsum-lime mortar; onychite
雪花石膏采石场 alabaster quarry
雪花石膏色调 alabaster hue
雪花石膏柱 alabaster column
雪花式形成室 snow box forming hood
雪花饰面 caustic etch
雪花形干扰 snow storm
雪花属 Galanthus
雪滑道 snowslide
雪季 falling weather
雪夹冰粒 coin snow
雪浆 snow sludge
雪窖 firn basin
雪界 snow limit
雪晶 snow crystal
雪景 snow-scape
雪卷 snow roller
雪壳 snow crust
雪坑 nivation cirque
雪坑冰川 cliff glacier
雪栏 snow screen; snow board; snow breaker; snow guard
雪雷暴 snow thunder storm
雪犁 snow plough; snow plow
雪犁翼 snow wing
雪链 <防滑用的> snow chain
雪链轮 snow sprocket
雪量 snowfall
雪量计 nivometer; snow ga(u)ge
雪量雨量计 snow-rain ga(u)ge
雪量雨量器 snow-rain ga(u)ge
雪龄 age of snow
雪盲的 snowblind
雪盲症 snow blindness
雪密度 density of snow; snow density
雪面 nival surface
雪面波纹 as skavl; sastruga; zastruga
雪面逆温【气】spring inversion
雪面蒸发 evaporation from snow; snow evapo(u)ration
雪尼尔地毯 chenille axminster carpet; chenille carpet
雪尼尔线织物 chenille
雪泥 slush; snezhura; snow slush
雪喷泉 snow geyser
雪片 flake; snow block; snow flake
雪飘地带 snow belt
雪坡 ramp
雪橇 bobsled; carriole; over-snow machine; skier; sled; sledge; sleight; ski
雪橇路 bobsleigh run
雪橇棚屋 ski hut
雪橇汽车 sledge car
雪橇式绞车 ski-hoist
雪橇式提升机 ski-hoist
雪桥 snow bridge
雪撬列 <履带拖拉机牵引的> cat train
雪茄烟型河谷 cigar-shaped valley
雪丘 snow dune
雪球 snowball
雪球构造 snowball garnet structure
雪球结构 snowball texture
雪球式抽样 snowball sampling
雪区 snow limit; snow-patch; snow region
雪区突然陷落 snow tremor
雪人 snowman
雪日 day of snow; snow day; day with snow
雪融化 snow melt
雪融水 snow melt
雪栅 snow fence
雪深 snow depth
雪深尺 snow scale
雪生藻类 cryophilic algae; snow algae
雪蚀 nival erosion; snow erosion; snow-patch erosion
雪蚀凹地 nivation cirque; nivation hollow; snow niche
雪蚀冰川 nivation glacier; snowbank glacier
雪蚀坑 nivation cirque; nivation hollow
雪蚀型冰川 nivation type glacier
雪蚀作用 nivation
雪试样 snow sampler
雪霜蚀作用 snow-patch erosion
雪水 lolly; slosh; slush ice; snow-broth; snow melt; snow water
雪水补给河 snow-feed river
雪水冲刷 nival erosion
雪水储量 snow storage
雪水当量 snow water equivalent
雪水含量 liquid water content of snow
雪水径流 snowmelt runoff
雪水流域 snow shed
雪水文学 snow hydrology
雪水蓄灌 liman irrigation
雪松 cedar; cedar tree; cedrus; deodar; Indian cedar
雪松醇 cedrol
雪松坚果油 cedar-nut oil
雪松胶 cedar gum
雪松木材 deodar cedar
雪松木油 cedar wood oil
雪松树岩盖 cedar-tree laccolith
雪松叶油 cedar leaves oil
雪松油 cedar oil
雪松属 <拉> cedar; Cedrus
雪天作业 snow work
雪丸 graupel; soft hail; tapioca snow; snow pellets <包括软雪和霰>
雪污染 snow pollution
雪屋 snowhouse; snow igloo
雪席 snow mat
雪线 accumulation line; firn limit; snow limit; snow line
雪线变动 snowline fluctuation
雪线高程 snowline elevation
雪线高度 snowline height
雪线升高 snowline rise
雪线以上植物区 nival flora
雪限 snow limit
雪陷 snow tremor
雪鞋 <在新浇混凝土面上工作时使用> snow shoes
雪蓄水 snow storage
雪檐 cornice
雪样 snow core; snow sample
雪野 snow field
雪一般地落下 snow
雪壅回水 backwater from snow
雪雨霰 sleet
雪原 chionic; snow field
雪原气候 snow climate
雪源河 snowfed river; snowfed stream
雪蕴水量 snow retention
雪灾 snow emergency
雪载 snow load(ing)
雪载应力 snow load stress
雪载值 snow load value
雪载重 snow load
雪障 snowbreak; snow fence
雪照云光 snow blink; snow sky
雪折 snowbreak
雪辙 wheel track on snow
雪阵 snow flurry
雪质 quality of snow; thermal quality of snow
雪中绝对含水量 absolute water con-

tent of snow
雪中污染物 pollutant in snow
雪中装卸 snow discharging; snow handling; snow work
雪珠 sago snow; tapioca snow
雪柱收集器 snow sampler; snow tube
雪桩 snow stake
雪状物 snow

鳕

鳕鱼肝油 cod-liver oil
鳕鱼油 cod oil

血

血白蛋白胶水 blood albumin glue
血白朊胶结剂 blood albumin glue
血本 hard-earned capital
血蛋白胶结剂 blood albumin glue
血滴石 bloodstone; heliotrope
血管 vein
血汗制度 sweating system
血红蛋白 h(a)emoglobin
血红胶皿菌 Patellaria sanguine
血红木 blood wood
血红色 sanguine
血红色的 blood red
血浆 plasma
血胶 blood albumin; blood glue
血库 blood bank
血块 thrombus
血蓝蛋白 hemocyanin
血流速度计 rheometer
血皮槭 paperbark maple
血清 lymph
血清保存中心 serum reference bank
血清蛋白 serum albumin
血球 corpusc(u)le; globule; hemocyte
血球识别 blood cell identification
血球畜栏 corpuscle
血石 bloodstone
血栓 thrombus
血统公民资格 citizenship by descent
血统关系 genealogical relation
血吸虫 blood fluke; schistosome
血吸虫病 bilharziasis; schistosomiasis
血吸虫皮炎 schistosome dermatitis
血吸虫属 Schistosoma; Schistosomum
血细胞 hemocyte
血压 blood pressure
血压表 manometer
血压过低 hypotension
血压计 sphygmomanometer
血液检验室 blood examinating room
血液循环失调 blood circulation disorder
血雨 blood rain
血中酒精浓度 <用以测定驾驶员酒醉程度> blood alcohol concentration
血朱栓菌 Trametes cinnabarina var. sanguinea

勋

勋章 decoration; medal

熏

熏 smo(u)ldering
熏舱 deratization; fumigation; fumigation of ship's holds; hold fumigation
熏舱费(用) fumigation expenses
熏舱设备 fumigation plant
熏舱证书 certificate of fumigation; deratization certificate; fumigation certificate
熏船 fumigation
熏船燃料 fumigant; match
熏干 smoke dry(ing); smoke seasoning
熏干的 smoke dried
熏干木材 smoke-dried lumber
熏黑 dingy; soot(ing)
熏黑的 sooty
熏硫箱 sulphur box
熏气消毒场 site for fumigation
熏肉房 smoke house
熏杀剂 fumigant
熏晒图 ozalid print
熏晒图重氮方法 diazo process
熏烧 smo(u)lder(ing)
熏香剂 fumet(te)
熏烟 smoking; smudge
熏烟玻璃 smoked glass
熏烟草 fumeroot; fumewort
熏烟干燥窑 smoke kiln
熏烟剂 smoke generator
熏烟消毒 fumigation
熏烟纸 smoked paper; smoked sheet
熏烟纸滚筒记录器 smoked paper drum recorder
熏烟纸滚筒系统 smoked paper drum system
熏烟装置 aerator
熏窑 smother kiln
熏衣草醇 Lavandulol
熏衣草油 lavender oil; oil of lavender
熏衣类油 Lavandine oil
熏蒸 fumigating; fumigation
熏蒸场 fumatorium; fumatory
熏蒸法 fumigating system
熏蒸机 gassing machine
熏蒸剂 fumigation agent
熏蒸剂中毒 fumigant poisoning
熏蒸煤 steaming coal
熏蒸气消毒场 site for fumigation
熏蒸器 <用来消灭害虫> fumatorium
熏蒸杀虫剂 fumigating insecticide
熏蒸室 fumatory; fumigation chamber
熏蒸消毒(法) fumigate; fumigation
熏蒸消毒剂 fumigant
熏蒸消毒器 fumigator
熏蒸消毒室 fumatorium
熏蒸作用 fumigation action
熏制 bloat
熏制房 smoke chamber
熏制图 dyeline

薰

薰衣草 lavender

寻

寻北器 north-finding instrument; polar finder
寻波 ordinary wave
寻查 chase
寻常标 ordinary tower; ordinary beacon【航海】
寻常波 ordinary wave
寻常的 common; vulgaris
寻常符号 ordinary symbol
寻常光波分量 ordinary-wave component
寻常光线 ordinary light ray
寻常射线 ordinary ray
寻常叶 foliage leaf
寻常折射率 ordinary index of refraction
寻出并排除窃听器 debug
寻的系统误差 seeker error
寻地级【计】addressing level
寻读函数 search-read function
寻峰 peak-seeking
寻根法 location of root
寻呼电话【船】bank paging call; paging telephone set
寻呼机 pager
寻呼接收机 paging receiver
寻呼区 paging zone
寻呼信道 paging channel
寻获陨石 find meteorite
寻迹器 finder; beam finder
寻极仪 north-seeking instrument; polar attachment
寻检器 searcher
寻矿 ore-search
寻漏器 leakage finder; leak finder
寻觅器 hunter; searcher
寻觅速度 hunting speed
寻觅泄漏点 leak hunting
寻求 finding
寻热阻 hunt group
寻线 finding; hunting
寻线动作 finding action
寻线机 call finder; finder switch; hunting switch; line finder; line switch; rotary hunting connector
寻线机电路 line finder circuit
寻线机架 line switch board; line switch shelf
寻线机系统 line-finder system
寻线继电器 step forward relay
寻线接点 hunting contact
寻线开关 hunting switch
寻线期间 hunting period
寻线起动继电器 hunting start relay
寻线器 beam finder; finder; line finder
寻线器架 line-finder shelf
寻线器开关 line-finder switch
寻线群 hunt group
寻线时间 hunting time
寻线速度 hunting speed
寻相器 phase hunter
寻向测向 direction finding
寻像护罩 finder mask
寻像监视器 viewfinder monitor
寻像屏 finder screen
寻像器 finder; picture finder; view finder(unit)
寻像器图像 viewfinder picture
寻像器遮光罩 viewfinder hood
寻星度盘 finder-circle; setting circle
寻星镜 finder
寻星仪 planispheric astrolabe; star finder; star identifier
寻优法 search method
寻找 bird dogging; fish for; hunting; seek
寻找故障 trouble-hunting; trouble shoot
寻找平衡 hunting
寻找事故 trouble shoot
寻找算法 find algorithm
寻找天然金块 nuggeting
寻找途径 find one's way
寻找最大值 maximize
寻找最小值 minimize
寻址 addressing
寻址操作 addressing operation
寻址单元 selected cell
寻址点 addressable point
寻址电路 addressing circuit
寻址技术 addressing technique
寻址能力 addressing capability
寻址字符 addressing character

巡

巡测 scanning; tour ga(u)ging
巡测仪 survey meter
巡查班 visiting gang; visiting party; visiting team
巡查人员 patrol
巡查下水道 sewer patrolling
巡察电话 patrol telephone
巡察电话装置 patrol telephone device
巡道 patrol(ling)
巡道车 patrol car; platelayer's troll(e)y【铁】
巡道工 patrolman; track inspector; track patrolman; track walk; trackwalker
巡道工作 patrol work
巡道员 patrolman; trackwalker
巡堤员 dike keeper; dike reeve; levee patrol
巡航 cruise; cruising
巡航半径 cruising radius; cruising range
巡航范围 cruising range
巡航航程 cruising range
巡航快艇 cruising yacht
巡航深度 cruising depth
巡航速度 cruise speed; cruising speed; cruising velocity
巡航速度航程 range at cruising speed
巡航速率 cruising rate[rating]
巡航状态发动机 cruise propulsion unit
巡航状态发动机推力 cruising thrust
巡护员 patrol
巡回 itinerate; menagerie; rove
巡回班长 walking ganger
巡回保养 patrol maintenance
巡回测流断面 visiting gauging section
巡回查账员 travel(l)ing auditor
巡回大使 ambassador at large; roving ambassador
巡回的 roving; travel(l)ing
巡回法庭 <英> eyre
巡回法院 circuit court
巡回服务车 service patrol
巡回画廊 artmobile
巡回稽核 itinerating auditor
巡回记者 roving correspondent
巡回监测器 sequence monitor
巡回检测 circulation detection; data scanning; data-logging【计】
巡回检测系统 data-logging system
巡回检测装置 cyclic(al) measuring device; scanning device
巡回检查 inspection tour; patrol inspection; round inspection
巡回检查任务 itinerary mission
巡回检查员 drifter; itinerary personnel
巡回检修工 roving maintenance man
巡回潜水作业 excursion diving operation
巡回清扫 patrol cleaning
巡回区 patrolling section
巡回审理 circuit and on-spot trial
巡回使节 itinerant envoy
巡回式安排 round-robin scheduling
巡回图书馆 itinerating library
巡回推销员 travel(l)ing salesman
巡回戏剧 <美> road show
巡回系统 turnaround system
巡回小修 running maintenance; running repair
巡回养护【道】patrol maintenance
巡回养路 patrol maintenance
巡回(养路)制 patrol system(of road maintenance)
巡回医疗队 travel(l)ing dispensary
巡回医疗水上飞机 seaplane ambulance
巡回展览 travel(l)ing exhibition
巡检报告 inspection report
巡检速度 polling rate
巡检用标准仪器 travel(l)ing standard
巡检周期 polling period
巡警 patrolman
巡路车 <道路养护用> motor patrol
巡路平地机 patrol grader
巡路人员 trackwalker
巡逻 patrol
巡逻车 paddy wagon; patrol car; places; scout car
巡逻船 guard boat; guard ship; picket

X

ship;revenue boat;watch boat;patrol vessel
巡逻队 patrol gang
巡逻工 patrolman
巡逻救助艇 patrol-rescue boat;rescue cruiser
巡逻路线 patrol route
巡逻炮舰 patrol sloop
巡逻平路机 autopatrol grader
巡逻区 beat;patrol area
巡逻艇 guard boat;guard ship;patrol boat;patrol vessel;picket ship;watch boat
巡逻养护制 patrol system(of road maintenance)
巡逻养路 patrol maintenance
巡逻用照相机 patrol camera
巡逻员 patrolman
巡逻站 patrol station
巡逻者 patrol;patrolman
巡视 cruise;patrol;perambulation
巡视工长 travel(l)ing ganger;walking ganger
巡视工队 patrol gang
巡视箱制度 patrol box system
巡视员 inspector/monitor
巡视者 perambulator
巡守时间 patrol time
巡线 line inspection;line walking;patrol;pipeline walking
巡线电话 patrol telephone
巡线工 line(s)man
巡线员 line attendant;line walker
巡线站 maintenance station
巡行车辆 cruising vehicle
巡行的 ambulatory
巡行控制系统 cruise control system
巡行速度 cruising speed
巡行(稳定行驶)车速 cruise speed
巡行(稳定行驶)时间 cruise time
巡行修路机 autopatrol
巡洋班轮 cruise liner
巡洋舰 cruiser
巡洋舰型船尾 cruiser stern
巡夜人 roundsman
巡游 cruise;itinerate
巡游车道 cruising way
巡游船 cruise vessel

旬报 ten-day report

旬报表 ten days report
旬计划 ten days' plan
旬间装车计划 ten-day's wagon loading plan
旬降水量 precipitation in ten-day periods
旬平均湿度 mean decade humidity;mean dekad humidity;ten-day's average humidity
旬平均温度 mean decade temperature;mean dekad temperature;ten-day's average temperature
旬平均值 ten-day mean
旬星 ten-day star
旬要车计划表 ten days' wagon requisition plan

询标 bid inquiry

询答装置 transactor
询价 ask the price;call for offers;enquiry[inquiry]
询价单 inquiry list;inquiry sheet
询价函件 letter of inquiry
询价文件 enquiry document
询价信 letter of inquiry
询问 inquiry;interrogation;query
询问编码 interrogation coding
询问程序 interrogator
询问处理 inquiry processing
询问处理机 interrogating processor
询问电码 interrogation code[coding]
询问调查员 interviewer
询问定价 pricing information
询问反应标志 cue-response query
询问符次序 enquiry character sequence
询问呼叫 enquiry call
询问机 interrogation unit;transponder
询问机应答器 responser
询问记录和定位系统 interrogation recording and location system
询问间隔 interrogation spacing
询问交货 delivery information
询问距离 interrogating range
询问孔 interrogation hole
询问雷达信标辅助装置 inquisitor
询问脉冲 interrogation pulse
询问脉冲间隔 interrogation pulse spacing
询问判定算法 query evaluation algorithm
询问频率 interrogation frequency
询问器 challenger;interrogator(set);interrogator-transmitter;transponder
询问绕组 interrogation winding
询问设备 inquiry unit
询问生成语言 query generation language
询问台 enquiry station;inquiry desk;inquiry station
询问天线 interrogation antenna
询问无效率 countdown
询问系统 inquiry system
询问显示终端 inquiry display terminal
询问线路 interrogation link
询问信道 interrogation link
询问信号 enquiry signal;interrogating signal;interrogation signal;request signal
询问信号发生器 interrogate generator
询问延迟 interrogator delay
询问应答 inquire response
询问应答方式 interrogator-responder system
询问应答机 interrogator-responder
询问应答器 interrogator-responder
询问用户电报呼叫 request for information telex call
询问用终端设备 inquiry station
询问与通信[讯]系统 inquiry and communication system
询问语言 query language;question signal
询问语言研制 query language development
询问栈 inquiry station
询问站 inquiry station
询问者 interrogator
询问中断 interrogate interrupt
询问终端显示器 inquiry terminal display
询问字符 enquiry character;who-are-you

循轨波 guided wave

循环 circular track;circulate;circulation;cycle;loop around;loop cycle;period;recurrence;revolve;rotate;rotation;round;whirligig;looping【计】;recur(sion)【数】
循环拌和制 circulating mixing system
循环报表 cycle billing
循环倍率 circulating ratio;circulation factor;circulation rate
循环本体 loop body
循环泵 circulating water pump;circulation pump;ebullator;pump in recirculation;recirculating pump;recirculation pump;recycle pump
循环泵水头 circulating pump head
循环比 recirculating ratio;recirculation ratio;recycle ratio
循环编号 numbering cycle
循环编译 loop compilation
循环变动 cyclic(al) fluctuation;cyclic(al) movement
循环变动调整 cyclic(al) adjustment
循环变化 cyclic(al) change
循环变换 cyclic(al) transformation
循环变量 cyclic(al) variable;loop variable
循环变流器 cycloconverter
循环表 loop table
循环表示 cyclic(al) representation
循环表元素 loop table element
循环波动支配月曲线 month for cyclic(al) dominance curve
循环波动支配月数 month for cyclic(al) dominance
循环波峰 cyclic(al) peak
循环补给水加热器 circulation feed water heater
循环不变式 loop invariant
循环不均匀度 cyclic(al) irregularity factor
循环不稳性 cyclic(al) sterility
循环不止的 endless
循环采购 circular buying
循环采暖系统 hydronic heating system
循环参数 cycle condition;loop parameter
循环操作 cycle operation;cycling;loop function;loop operation;recirculation operation
循环操作状态 state of cycle operation
循环槽 circulating ditch
循环查点存货 cycle count
循环长度 length of the cycle
循环场 cyclic(al) field
循环超滤膜生物反应器 recirculated ultrafiltration membrane bioreactor
循环沉淀池 circulating sedimentation tank;circulation sedimentation tank
循环沉积 cyclothem
循环沉积作用 cyclic(al) sedimentation
循环承销融资 revolving underwriting facility
循环乘积码 cyclic(al) product code
循环程序 cycle program(me);loop code;looping routine;loop program(me)
循环程序计数器 cycle program(me) counter
循环程序控制 cycle program(me) control
循环池 circulatory pool
循环充气 return air
循环抽取 circular pumping;cyclic(al) pumping
循环初始化 loop initialization
循环初置 <初始状态的> loop initialization
循环除霜系统 cycle defrost system
循环储存 circular storage
循环储存器 circular storage;circulating memory;cyclic(al) storage
循环处理过程 treatment cycle
循环穿吊法 repeating tie
循环传送 cyclic(al) transfer
循环传送带 endless belt conveyer[conveyor]
循环磁带放音机 tefiphone
循环磁化 cyclic(al) magnetization
循环磁化状态 cyclically magnetized condition
循环次数 cycle index;number of cycles
循环次数计数器 cycle index counter
循环次序 circular order;cycle index;cyclic(al) order
循环存储 circulating memory
循环存储器 circulating memory;circulating storage;circulating store;cyclic(al) memory;cyclic(al) storage;recirculating loop memory
循环存储器存取 cyclic(al) storage access
循环存取 cyclic(al) access
循环错误 loop error
循环打捞筒 circulating over-shot
循环代码 loop code
循环带 endless belt;endless loop;tape loop
循环贷款 revoluting loan;revolving credit
循环贷款法 revolving loan
循环单剪试验 circular simple shear test;cyclic(al) simple shear test
循环单体 recycle monomer
循环当量比重 equivalent circulating mud weight
循环倒换 loop reversal
循环倒置 loop reversal
循环的 circuiting;circular;circulatory;cyclic(al);periodic(al) recurrent;recurrent;recursive;rotational;rotatory;round-robin;circulating;cycling
循环的回洗水 recycled backwash water
循环点群 circular point group;cyclic(al) point group
循环电芬顿反应器 circulating electro-Fenton reactor
循环电路 circulation circuit;circulator;cycling circuit
循环吊链 endless sling
循环吊运式(停)车库 rotopark car parking system
循环调度 round-robin scheduling
循环调度法 cyclic(al) dispatching method
循环迭代法 cyclic(al) iterative method
循环定理 circulation theorem;recurrence theorem
循环定时器 cycle timer
循环动 cycle event
循环读出 cyclic(al) readout
循环段 circulation section
循环对称函数 cyclosymmetric function
循环多斗挖沟机 endless chain trench excavator
循环二进制码 cyclic(al) binary code

循环发送 dispatch loop
循环阀 circulation valve;recycle valve;cycling valve
循环法 cyclic(al) method;recirculation process;round-robin
循环法测镭含量 radium content determined by cycle method
循环反演 loop inversion
循环反应 circular reaction;circular response;cyclic(al) reaction
循环反应器 recirculation reactor
循环范围 cycle range
循环方程 cyclic(al) equation
循环方式 round robin discipline
循环方向 circulating direction

循环防止 prevention of cycling
循环飞灰 circulating dust
循环废钢铁 circulating scrap
循环分配 loop distribution
循环分配算法 loop distribution algorithm
循环分批式干燥机 repeated batch drier[dryer]
循环分数 circular fraction;circulating fraction
循环风 circulating air; recirculated air;return air
循环风机 circulating fan
循环伏安法 cyclic(al)voltammetry
循环浮床反应器 circulating floating bed reactor
循环负荷 circulating load; cycling load;recirculating load
循环负荷率 circulating load ratio
循环负荷试验 cycling test
循环负载 repeated load(ing)
循环附加流量 additional circulating flow
循环附加流率 additional circulating flow rate
循环复位 cycle reservoir;cycle reset
循环干扰 circulating disturbance;circulation disturbance
循环干湿法 circular wetting and drying;cyclic(al)wetting and drying
循环干燥 circular drying; cyclic(al) drying
循环干燥机 recirculation drier[dryer]
循环干燥器 recirculation drier[dryer]
循环给料 circuit feed
循环工况 state of cycle operation
循环工质 cycle fluid
循环工作 periodic(al)duty
循环公式 recurrence formula; recurrent formula;recursion formula
循环公式法 recurrence formula method
循环功能 circulatory function
循环供给系统 circulation supply system
循环供暖系统 circulation supply system; recirculating heating system; recirculation heating system
循环供热 circulation heating
循环供热装置 circulation heating installation
循环供水 periodic(al)feeding;supply of circulating water
循环供油系统 circulation supply system
循环沟 circulating ditch
循环关系 recursion relation
循环管 circular pipe;circulation pipe; circulation tube[tubing]; cyclic(al) pipe;recirculation pipe[piping]
循环管道 circulating line;pipe circulation line
循环管路 circulating pipe; circulation line
循环管线 circulation line; pipe loop; recirculation piping line
循环灌溉 circular irrigation
循环光合磷酸化作用 cyclic(al)photophosphorylation
循环滚珠导管 recirculating ball guide
循环滚珠式转向盘 recirculating ball type steering wheel
循环过程 cycle process; cyclic(al) process;cycling process
循环过滤 circulating filtration
循环过滤活动床 fluidized bed recycling filtration
循环海水 recycle brine
循环函数 cyclic(al)function
循环行列式 circulant determinant;cyclic(al) determinant; recurrent determinant;circulant【数】
循环行列式设计 circulant design
循环号码 rotation number
循环荷载 circular load(ing); cyclic(al)load(ing);recurring load
循环恒等式 cyclic(al)identity
循环呼叫 recursive call
循环呼叫装置 signal call device
循环缓冲 circular buffering
循环缓冲器 cyclic(al)buffer
循环换流器 cycloconverter
循环回流 circulating reflux;circulation reflux;flow circuit
循环回水 circulation return
循环回水管 circulation return pipe
循环回线存储器【计】recirculating loop memory
循环回用水 water recycled
循环回用水系统 water recycle system
循环混合 recycle mixing
循环混合机 circulating mixer
循环活化 cyclic(al)activation
循环货物列车 shuttle freight train
循环机能检查 circulatory function test
循环积分器 cyclic(al)integrator
循环基金 revolving fund;rotary fund
循环级数 recurring series
循环极限 cycle limit
循环集 cycle set
循环给水 circulation supply;cyclic(al) supply
循环给水管 cyclic(al)supply pipe
循环给水加热器 cyclic(al)feed water heater
循环计时器 cycle timer
循环计数 cycle count
循环计数器 cycle counter;loop counter
循环计数器复原 cycle reset
循环计算操作 loop computing function
循环寄存器 circulating register
循环加荷 circular load(ing); cyclic(al)load(ing)
循环加热 heat cycling
循环加热冷却系统 hydronics
循环加热器 circulating heater; recirculation heater
循环加热系统 circulating heating system;hydronic heating system;loop heating system
循环加湿 circular wetting;cyclic(al) wetting
循环加速 loop speed-up
循环加速层次 loop speed-up hierarchy
循环加油 circulating oiling
循环加油润滑 circulation oiling
循环加载 loading cycle
循环加载试验 cyclic(al)test
循环间隔 intercycle
循环剪切试验 circular shear test;cyclic(al)shear test
循环剪切阻力 cyclic(al)shear resistance
循环剪应力比 circular shear stress ratio;cyclic(al)shear stress ratio
循环检测 loop detection
循环检查 cycle check
循环检查程序 loop checker; loop tester
循环检查系统 cyclic(al) detection system;cyclic(al)scanning system; loop checking system
循环检查制 cyclic(al)detection system; cyclic(al) scanning system; loop checking system
循环检索 chaining search
循环检验 circular test; cyclic(al) checker;round inspection
循环碱 circulating alkali
循环浆 recycle slurry
循环交换 revolving swap
循环交路 loop-type of locomotive routing system
循环胶片 endless film
循环胶体捕集 recirculation gel trap
循环校验 cycle check; cyclic(al) check
循环阶段 cycle stage;phase of the cycle
循环节 loop body; recurring period; repetend <循环小数的>
循环结束 loop ends;loop termination
循环解法 cycle solution
循环介质 circulating medium
循环借位 end-around borrow
循环进程测试 process loop test
循环进尺 footage circulation
循环进位【计】chain carry; end-around carry;cyclic(al)carry
循环进位移位 end-around carry shift
循环浸出 recycle leaching
循环经济 circular economy
循环纠错码 BCH[Bose-Chaudhuri] code
循环矩阵 circulant matrix; circular matrix;cycle matrix;cyclic(al)matrix
循环决定模型 cyclic(al)deterministic model
循环开采 cyclic(al)mining
循环开单制 cycle billing system
循环抗体 circulating antibody
循环空间 cyclic(al)space
循环空气 circulating air;recirculated air;return air
循环空气分级器 turbo-classifier
循环空气加热 heating by circulating air
循环空气进口 recirculated air intake
循环空气流通 ventilation with recirculated air
循环空气旁管 recirculated air by-pass
循环空气入口 recirculated air intake
循环空气式选粉机 circulating air separator; separator operated by circulating air;turbo-separator
循环控制 cycle control;loop control
循环控制变量 loop control variable
循环控制结构 loop control structure
循环控制开关 cycle control switch
循环控制算法 loop control algorithm
循环控制装置 cycler
循环库 cycling pool
循环类型 cyclic(al)patterns
循环类型测定与识别 measuring and identifying cyclic(al)patterns
循环冷却 circulating cooling; cooling by circulation; cool recycle; recirculating cooling
循环冷却处理系统 cooling system for circulating water
循环冷却排水 circulating cooling discharged water
循环冷却器 recirculation cooler
循环冷却水 circulating cooling water;recirculating cooling water
循环冷却水柜 holy-water tank
循环冷却水流 circulating cooling fluid
循环冷却水系统 recirculating cooling water system; recirculating water cooling system
循环冷系统 recirculating cooling system
循环冷却装置 cooling back installation
循环离差 cyclic(al)deviation
循环理论 recursion theory
循环力 circulating force; cyclic(al) force
循环连分数 recurring continued fraction
循环连通图 cyclically connected graph
循环连通性 cyclic(al)connection
循环连通有向图 cyclically connected directed-graph
循环连续 recycling continuum
循环链 endless chain;gearing chain
循环链炉算 endless chain grate
循环链式输送机 endless chain conveyer [conveyor]; gearing chain conveyer[conveyor]
循环链式挖沟机 endless chain trench excavator
循环量 circulating quantity; circulation quantity
循环裂点 circular nick; cyclic(al) nick
循环裂化操作 recycle cracking operation
循环淋浴 circulating shower
循环流程 circulation process
循环流(动) circulation flow; circular flow; circulating flow; circulatory flow
循环流动计数器 loop-line flow counter
循环流动式电解浴 cycle mobile-electrobath
循环流动性 circular mobility; cyclic(al)mobility
循环流化床 recycled fluidized bed
循环流化燃烧室 circulating fluidized bed combustor
循环流量 circulation discharge; circulation volume
循环流速 rate of circulating flow
循环流体 circulation fluid
循环馏分 recycle fraction
循环炉 circular oven;cyclic(al)oven
循环滤油器 circulating oil strainer
循环轮盘系统 cyclic(al) carousel system
循环论法 <以甲证乙复以乙证甲的骗人的论证法> vicious circle
循环码 chain code; cyclic(al) code; recurrent code;reflected code
循环脉冲发生器 rotary pulse generator
循环门 circulating door
循环描述 looping description
循环模 cyclic(al)module; module of cycles
循环模式 material cycle model
循环模数转换器 circulation analog(ue)/ digital converter
循环凝胶渗透色谱 recycle gel permeation chromatography
循环凝结剂 recycled coagulant
循环扭剪试验 circular torsional shear test;cyclic(al)torsional shear test
循环排队网络 cyclic(al)queuing network
循环排列【数】circular permutation; cyclic(al) permutation; cyclic(al) queues
循环排列码 circular permutation code
循环排污水 circulating discharged wastewater
循环排序 cyclic(al) ordering; rotational ordering
循环盘存 cycle count;cycle inventory
循环盘点 cycle count
循环判处 cycle criterion
循环判据 cycle criterion; cyclic(al) criterion
循环喷雾器的喷杆 circulating spray bar
循环喷油管 circulating spray bar
循环疲劳 cyclic(al)fatigue

X

循环片 loop-film
循环频率 cycle frequency
循环平地机圆圈 circlet
循环平均法 cyclic(al)average method
循环曝气池 circulated aeration tank
循环期 circulation period;recurrence interval
循环气 recycle gas
循环气鼓风机 recycle gas blower
循环气体炉 recycle gas furnace
循环气预热器 circulating gas preheater
循环汽油 recycle gasoline
循环器 recirculator
循环器脉塞 circulator maser
循环球形转向器 recirculating ball type steering gear; steering gear with circulating ball
循环区 circulation zone
循环区间 intercycle
循环曲线 cyclic(al)curve
循环趋势 cycle tendency;cyclic(al) trend
循环取样法 rotated sampling
循环群 cyclic(al)group
循环燃料反应堆 circulating fuel reactor
循环燃烧器 cyclic(al)burner
循环热气炉 circulation oven
循环热水 circulating hot water
循环热水器 side arm water heater
循环热水系统 recirculating heating system
循环容量 circulation volume
循环溶液反应堆 circulating-solution reactor
循环冗余 cyclic(al)redundance
循环冗余校验 cyclic(al)redundancy check
循环冗余校验码 cyclic(al)redundancy check code
循环冗余检验符号 cyclic(al)redundancy check character
循环冗余码符号 cyclic(al)redundancy character
循环入口 loop head
循环润滑(法)circulating lubrication; circulation lubrication; oil circulation lubrication
循环润滑系统 circulating oil system; circulation lubrication system
循环润滑油系统 oil circulation lubricating system
循环洒布系统 circulating spreading system
循环三轴试验 circular triaxial test; cyclic(al)triaxial test
循环扫描 scan-round
循环色谱法 recycle chromatography; recycling chromatography
循环砂滤池 recirculating sand filter; sand recirculation filter
循环上涨 circular rise
循环设备 circulating equipment;recycle unit
循环深度 depth of round
循环施工 cyclic(al)construction
循环时间 circulation time; cycle length;cycle time;cyclic(al)time; cycling time;loop time
循环时间测定 circulation time determination
循环使用的材料 recycled material
循环式安排 round-robin scheduling
循环式包销协议 revolving underwriting facility
循环式闭路灌浆(法)closed circuit grouting
循环式冲洗系统 through flushing system
循环式伏安污泥 cyclic(al)voltammetric sludge
循环式活性污泥法 cyclic(al)activated sludge process;cyclic(al)activated sludge technology
循环式活性污泥反应器 cyclic(al)activated sludge reactor
循环式活性污泥系统 cyclic(al)activated sludge system
循环式火灾探测器 fire cycle detector
循环式架空索道 circulating aerial ropeway
循环式冷却方式 circulating cooling system
循环式冷却系统 circulating cooling system
循环式冷却装置 circulating cooling system
循环式密封 recirculated seal
循环式皮带输送机 endless belt conveyer[conveyor]
循环式热力系统 cyclic(al)thermomechanical system
循环式热水器 circulating water heater
循环式室内加热器 circulating type room heater
循环式输送机 endless belt conveyer [conveyor]
循环式水管锅炉 circulation boiler
循环式水浴锅 circulating water bath
循环式网络 cycle-type network
循环式预应力张拉设备 merry-go-round equipment
循环式直接炼钢法 cyclosteel process
循环式中断 cycled interrupt
循环事件 recurrent event
循环试验 circle test;circular test;cycle test;cyclic(al)test;round-robin test
循环室 circulating chamber
循环寿命 cycle life
循环寿命试验 cycling life test
循环输入码(法)cyclic(al)feeding code
循环输送带 endless apron
循环输送机 endless belt conveyer [conveyor]
循环输送链 endless conveyor chains
循环数 circular number;cycle number;recurrent number
循环数码 cyclic-digit code
循环数模/数转换器 cyclic(al)analog/ digital converter
循环数字滤波器 recursive digital filter
循环衰竭 circulatory failure
循环水 backwater;circulating water; circulation water;recirculated water; recirculating water; recirculation water; recycle water; vadose water
循环水泵 circulating pump;water circulating pump; water circulation pump
循环水泵房 pump for circulating water
循环水泵站 pump station of recirculation water; water circulation pumping station
循环水舱 circulating tank
循环水槽 annular tank;recirculating flume;recirculation flume
循环水阀 circulating water valve
循环水分析 circulated water analysis
循环水管 circulating conduit; circulating pipe;circulating tube
循环水回用 circulating water reuse
循环水给水管 circulation supply pipe
循环水检漏装置 circulation water leakage detector
循环水冷却 cooling water circulation
循环水冷却法 cooling back
循环水流 circulated flow;circulating current;circulating flow;recirculation current;recirculation(water) flow
循环水滤网 circulating water screen
循环水路 water circuit
循环水清洗器 recirculating water washer
循环水入口阀 circulation water inlet valve
循环水头 circulating head;circulating pressure
循环水温继电器 circulating-water temperature relay
循环水系统 circulating water system; reclaimed water system;water circulation system; circulation system;recirculation system
循环水下水 vadose water
循环水箱 circulating tank
循环水养殖系统 recirculating aquaculture system
循环水柱 circulating water head
循环送风 circulating air supply
循环速度 circulating speed;circulating velocity; cycle speed; rate of circulation
循环速率 circulated rate;circulating rate;circulation rate;rate of circulation;speed of circulation
循环算法 recursion algorithm;round-robin algorithm
循环损失 loss of circulation;loss of returns
循环索 circulating cable;rope circuit
循环索系 moving cable system
循环塔盘 recycling tray
循环特性 cycle performance
循环特征 cycle specificity
循环提升机 circulating elevator
循环调节 cycle control;regulation of circulation
循环调用 recursive call
循环停机 loop stop;stop loop
循环停滞 circulatory stasis
循环通道 circulation passage(way)
循环通风 circulating ventilation;loop vent
循环通风系统 recirculating system of ventilation
循环通路 peripheral passage
循环同步 around synchronization
循环同步法 periodic(al)resynchronization
循环头 loop head;hydraulic circulating head <冲击钻的>
循环图 cycle diagram
循环图案 allover;repeat design
循环图像 chain image
循环退火 cycle annealing;cyclic(al) annealing
循环瓦斯油 recycle gas oil
循环网带提升机 continuous wire-belt elevator
循环网路 cycle-type network
循环尾气 recycled off-gas
循环位移 circular shift;end-road shift
循环谓词 loop predicate
循环文件 circular file
循环紊乱 disturbance of circulation
循环污泥 circulating sludge
循环物质 recycled material
循环误差 cyclic(al)error
循环吸附解吸试验 cyclic(al)load-elution test
循环系 recycle system
循环系平均充盈压 mean filling pressure of circulatory system
循环系数 circulation factor;circulation ratio;recirculating ratio;recirculation ratio; recycle ratio; systematic circulatorium
循环系统 circulating system;circulatory system;systematic circulation
循环系统补充水 make-up water of circulation system
循环系统多导记录仪 circulatory system polygraph
循环系统接口 circulation connection
循环系统压降系数 coefficient of circulating pressure drop
循环系统总体积 total volume of cycle system
循环线路 recycle circuit
循环限制 loop restriction
循环相关 circular correlation
循环相关函数 circular correlation function
循环消去回代法 recursive method
循环硝化滤池 cyclo-nitrifying filter
循环小数 circulating decimal;circulator; period fraction; periodic(al) decimal; periodic(al) fraction; recurrent determinant;recurring decimal;repeater;repeating decimal
循环小数循环节 recurring period
循环效率 cycle efficiency;cyclic(al) effect;efficiency of cycle
循环效应 cyclic(al)effect
循环信贷限额 revolving line of credit
循环信用证 revolving credit;revolving letter of credit
循环型 circular form;circulation pattern
循环性的 circulative
循环性精神病 circular psychosis;cyclothymia
循环性能 cycle performance
循环性侵染 cyclic(al)infection
循环性情感 cyclothymia
循环性振荡 cyclic(al)oscillation
循环性支付差额 cyclic(al)disequilibrium
循环序列 periodic(al)sequence
循环蓄水量 cyclic(al)storage
循环悬浮体 recycle slurry
循环选粉机 recirculating air separator
循环学说 cycle theory
循环寻址 cyclic(al)addressing
循环压降 circulation pressure loss
循环压力 circulating head;circulating pressure;circulation pressure
循环压缩机 booster compressor;recycle compressor
循环压头 circulating head
循环延迟选择器 cycle delay selector
循环盐水 circulating brine
循环盐雾紫外线照射 cyclic(al)salt/ fog/ ultraviolet expose
循环液 circulating fluid; circulation fluid
循环液泵 recycle liquid pump
循环液流 circulating current
循环液漏失 circulation loss
循环液压力 circuit pressure
循环移动的破碎板 travel(l)ing breaker plate
循环移位 circuit shift;cycle shift;cyclic(al)shift; non-arithmetic shift; ring shift;circular shift;end-around shift
循环移位寄存器 cyclic(al)shift register;end-around-shift register
循环因素 cyclic(al)factor
循环应变 cyclic(al)strain
循环应变软化 cyclic(al)strain softening
循环应力 cyclic(al)stress;pulsating

stress
循环应力比 cycle stress ratio
循环应力序列 sequence of cyclic(al) stress
循环用水 recycled water; water for use
循环用水系统 water system of circulation
循环优化 circular optimization; loop optimization
循环优化法 circular optimization method; loop optimization method
循环优先 rotation priority
循环油 circulating oil; recycle oil; recycle stock
循环油泵 circulating oil pump; oil circulating pump
循环油阀 oil circulation valve
循环油管 circulating conduit; circulating pipe; circulating tube; oil circulating pipe
循环油冷 oil circulation cooling
循环油料 cycle stock
循环油路 oil circulation
循环语句【计】do statement; loop statement; for statement
循环预置 loop initialization
循环元素表 for list
循环圆 circle of circulation; circulating circle
循环圆有效直径 effective diameter of circulating circle
循环约束条件 cyclic(al) constraint condition
循环运动 circulatory motion; cycle motion; cycle movement; cyclic (al) movement
循环运输机 endless belt conveyer [conveyor]
循环运行 circuit working; shuttle service【铁】
循环运转 merry-go-round
循环运转列车 shuttle train
循环运转制【铁】circular system of locomotive running
循环载荷装载法 cycle load(ing); cyclic(al) load(ing)
循环再补给 recurring issue
循环再生 cyclic(al) regeneration
循环增压 loop pressurization
循环整流器 cyclorectifier
循环执行 looping execution
循环直达列车 shuttled block train
循环指标 cycle index; cyclic(al) indicator
循环指令 loop command; recursion instruction
循环指示器 circulating indicator
循环指数 circular index; cycle index; cyclic(al) index
循环制导法 circuit rider approach
循环制导原理 circuit rider concept
循环制冷 refrigerated cycle
循环制冷剂 circulating refrigerant
循环制约 cyclic(al) constraint
循环置换 cyclic(al) permutation
循环置换标题索引 permuted-title index
循环置换码 cyclic(al) permuted code
循环中断 circulation loss; cyclic(al) interrupt
循环重整 reforming with recycle
循环周期 cycle duration; cycle period; cycling period
循环骤停 circulatory arrest
循环专用单元 loop box
循环转移 loop jump
循环装置 circulating device; circulator; cycling plant
循环状态 loop state; recurrent state
循环准则 cycle criterion
循环子句 for clause; loop clause
循环子空间 cyclic(al) subspace
循环总量 circulation flow
循环钻液损失 loss of circulation
循环作业 circular operation; cycle operation; cyclic(al) operation
循环作业定额 circular operation quota
循环作业时间 cycle time
循环作用 cyclic(al) action
循环坐标 cyclic(al) coordinates; ignorable coordinates
循迹表 tracking meter
循迹反光镜 tracking mirror
循迹误差 tracking error
循经选穴法 corresponding channel point selection
循序而来 come in its turn
循序计算 sequential computation
循序渐进 progressive procession
循序渐进法 step-by-step method
循序前进法 progressive method; step-by-step method
循序提取方法 sequential extraction method
循序性集聚需求模型【交】sequential aggregate demand model
循序性离散需求模型【交】sequential disaggregate demand model
循溢水管 flow pipe

训

训导长 warden

训练 apprenticeship; breed; education; exercitation; training
训练靶场 training range
训练班 training class
训练报告 training report
训练备忘录 training memorandum
训练标准 training standard
训练部 training command; training department
训练场 drill ground; training area; training field
训练程序 training program(me)
训练大纲 training bill
训练迭代法 training iterative method
训练发射控制中心 training launch control center[centre]
训练分类器 training classifier
训练分析员 training analyst
训练辅助器材 training aid
训练负荷 training load
训练公报 training bulletin
训练馆 practice hall
训练管理 training management
训练机长 training captain
训练基地 training base; training site
训练集 training set
训练计划 training plan(ning); training program(me)
训练计算机 training computer
训练进度表 training schedule
训练立体镜 stereoscopic (al) training; training stereoscope
训练命令 training order
训练目标 training objective
训练年度 training year
训练配属 training attachment
训练评定 training evaluation
训练期 apprenticeship; training period
训练气球 training balloon
训练器材 trainer; training device
训练器材保管员 training device man
训练区 training area
训练区文件 training area files
训练人 trainer
训练任务 training mission
训练设备 exercise equipment; training device
训练设施 training facility
训练时间 training time
训练实验组 training experimental group
训练手册 drill manual; training manual
训练塔 drill tower
训练条令 training regulation
训练图 training chart
训练现场 training site
训练学校 drill school; training school
训练有素的班组人员 well-trained crew; well-trained gang; well-trained party
训练员 drillmaster
训练中心 training center[centre]
训数点 arithmetic(al) point

讯

讯号灯 indicating lamp

汛

freshet

汛光灯 flood light
汛光照明 floodlighting
汛后期 post-freshet period
汛期 flood period; flood season; freshet period; high flow season; high water period; high water season
汛期次数 frequency of flooding period
汛期径流量 runoff in flood season
汛期流量 freshet discharge; freshet flow
汛期末 ending of flood
汛期时间 flooding period
汛前库容减少 preflood decrease of storage
汛前期 prefreshet period
汛前水库放水 preflood decrease of storage
汛前水库泄水 preflood decrease of storage
汛前预泄 preflood decrease of storage; prefreshet decrease of storage
汛情 flood regime(n)

迅

迅变干涉图 flash figure

迅达普鲁防潮液<一种由沥青橡胶配制的,用于胶黏[粘]木块> Synthaprufe
迅发反应性 prompt reactivity
迅发构造效应 prompt tectonic effect
迅发裂变能 prompt of fission energy
迅接制 dial system working
迅烈放电 disruptive discharge
迅猛前进 pelt
迅速摆动 galloping oscillation
迅速闭合 rapid closing; snap
迅速变成 flash to
迅速变化的 vertiginous
迅速不停地钻进 swivel bailing
迅速测定 rapid test
迅速测定土壤法 quick test for soil; rapid test for soil
迅速掣住 snub
迅速处理 expedite
迅速传播的 fast-spreading
迅速冻结 quick freezing
迅速二相反应 prompt biphasic reaction
迅速发送 customary quick dispatch [despatch]
迅速发运 instant dispatching
迅速发展 great strikes
迅速发展投资基金 growth fund
迅速分析法 rapid analysis method
迅速加衬施工法 concrete as you go method
迅速降压 rapid depressurization
迅速绞缆 heave in lively
迅速结算 make a prompt settlement
迅速解决 make a prompt settlement
迅速进行 hand-over hand
迅速拉出编组线上的车辆 prompt removal of cars from classification tracks
迅速磨损 rapid wearing
迅速燃烧 deflagrate; rapid combustion
迅速燃烧期 rapid combustion period
迅速熔断 quick break
迅速渗透 rapid permeability
迅速失压 rapid decompression
迅速调整 quick adjusting
迅速退回 prompt return
迅速形成 rapid formation
迅速修垫圈 quick repair washer
迅速应答 rapid answer
迅速增加 run-up
迅速增长 dramatic growth; explosion; explosive increase; run-up
迅速展开 rapid deployment
迅速涨水 flush of water
迅速作用恒温控制阀 snap-acting thermostatic valve

驯

驯化 acclimatization; acclimation <美>

驯化河流 civilized river; civilized stream
驯化活性污泥 acclimated activated sludge
驯化阶段 domesticated stage
驯化菌种 acclimated strain
驯化培养 acclimated culture
驯化时间 acclimatization time
驯化微生物 acclimated microorganism
驯化污泥 acclimation sludge
驯化研究所 Acclimatization Institute
驯化园 garden of acclimatization
驯化作用 acclimation; acclimatization
驯马术 manege
驯马围场 paddock

狗

狗伏革菌属<拉> Laeticorticium

狗孔菌属<拉> Laetiporus

逊

逊沸蒸馏器 subboiling still

逊径 undersize
逊径骨料 undersize aggregate
逊原子的 subatomic

殉

殉爆 flashover; sympathetic detonation

殉爆距离 explosive coupling distance
殉爆试验 gap test
殉爆药包 remaining cartridge
殉道堂 martyrium
殉葬陶瓷器 sepulchral pottery

蕈

mushroom

蕈环 annulus[复 annuli/ annuluses]
蕈空气旁管 recirculated air by-pass
蕈头式支墩坝 mushroom-head buttress dam
蕈线阀 poppet valve
蕈形阀 open chamber needle valve
蕈形锚 mushroom anchor
蕈状混合器 mushroom mixer

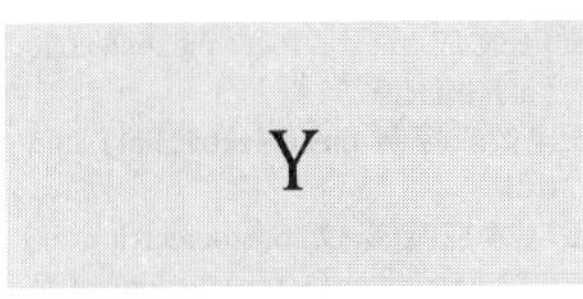

压凹 dint; indentation; shallow recessing
压凹试验 cupping test
压凹凸印刷 die stamping
压凹硬度试验 indentation test
压拔套管构造 casing tube pressing and extracting mechanism
压拔套管机构 casing tube pressing and extracting mechanism
压摆动轴 oscillating shaft
压扳机 trigger squeeze
压板 clamp(ing bar); clamp(ing) plate; clip plate; collar band; compression plate; hold-down; holding-down plate; jacking platen; platen; press head; presspahn; press(ure) back; pressure plate; set plate
压板垫 follower gasket
压板风速表 pressure plate anemometer
压板滚柱 pressure plate roller
压板货叉 keep plate fork
压板间距 <压机的> daylight opening
压板开档 daylight opening
压板螺栓 follower bolt
压板片 presser blade
压板式风速计 pressure plate anemometer
压板式滑片 presser slide
压板弹簧 rag iron spring
压板铁片 clip tingle
压板销 pressure plate stud
压板运动 presser motion
压板装置 presser device
压包机 cargo jack
压背法 back-pressing method
压泵 press pump
压比控制系统 pressure ratio control system
压边机 beading machine; flanging machine; flanging press
压边浇口 cornop gate; edge gate; knife gate; lap gate
压边浇铸 lip casting
压边接头 selvage joint
压边力 pressure-pad-force
压边内烧口 kiss gate
压边条 fillet
压扁 bruise; bruising; crimp; flatten; squash; squelch; flaser【地】
压扁层理【地】 flaser bedding
压扁构造【地】 flaser structure
压扁辊 conditioner roll
压扁辉长岩 flaser gabbro
压扁机 bruiser; flatting mill
压扁率 ratio of flattening
压扁试验 crushing test; flattening test
压扁丝 flattened wire
压扁褶皱 flattening fold
压扁状态下的弹簧 collapsed spring
压扁作用 flattening
压变形吸收能 energy absorbed in compression
压瘪 deflation
压瘪管 <断气方法> buckle; buckled pipe
压柄轴 lever pin
压饼 pressed pellet
压饼吹杯机 tumbler cup blowing machine
压饼吹制成型机 tumbler blowing machine
压波钢丝 crimped wire
压波纹机 creasing machine
压箔 blocking
压箔机 blanking press; stamping press
压薄缝焊 mash seam weld(ing)
压薄滚焊 mash seam weld(ing)
压薄机 thinning machine
压不碎的 uncrushable
压舱材料 ballasting; ballasting material
压舱袋 ballast bag
压舱费 ballastage
压舱柜 ballast tank
压舱海水 sea ballast
压舱货航程 ballast passage
压舱货(物) ballast cargo; bottom cargo
压舱货载 bottom cargo
压舱料 kentledge goods
压舱砂 ballast sand; sand ballast
压舱水 ballast water; ship ballast; water ballast
压舱水泵 ballast pump
压舱水舱 ballast tank; water ballast tank
压舱水舱室 ballast compartment
压舱水管线 ballast(pipe) line
压舱水罐 water ballast tank
压舱水总管 ballast(-water) main
压舱物 ballast; ballast cargo; base cargo; bottom cargo
压舱物料 ballasting material
压舱载(船舶) ballasting
压槽 indent; pressed recess; rolling groove
压槽窗玻璃 fluted sheet
压槽锤 fuller; fullering tool; Fuller's tool
压槽锤开槽 fullering
压槽模 fuller
压槽片玻璃 fluted sheet
压槽子 suppressing debiteuse down
压槽子的砖 plug
压草机 weed breaker
压侧(面) compression side
压差 differential pressure; delivery head; drop in pressure; pressure difference; pressure differential; pressure drop; pressure head
压差变送器 pressure difference transmitter; pressure differential transmitter
压差测量 measurement of difference pressure
压差乘子 differential pressure multiplier
压差传感器 differential pressure pick-up; differential pressure switch; differential pressure transducer; pressure difference sensor
压差传声器 gradient microphone; pressure gradient microphone; velocity microphone
压差传送器 differential pressure transmitter
压差阀 differential pressure regulator
压差法 pressure differential method
压差范围 pressure differential range
压差分区 pressure difference sub-area
压差隔膜 differential diaphragm
压差激励器 differential pressure producer
压差计 differential ga(u)ge; differential manometer; differential pressure ga(u)ge; differential pressure meter
压差计负压升高度 manometric(al) suction lift
压差计负压柱高度 manometric(al) suction head
压差记录器 differential recorder
压差卡钻 differential pressure sticking
压差开关 differential pressure switch; pressure difference switch
压差控制器 differential pressure controller
压差料位控制装置 pressure differential level controller
压差领航 pressure differential navigation
压差流量计 construction type flowmeter; differential pressure flowmeter; head meter
压差流速计 differential pressure flowmeter
压差密度计 gradiomanometer
压差能换器 pressure gradient transducer
压差式底河床质采样器 pressure difference bed material sampler
压差式底沙采样器 pressure difference (type) bed-load sampler
压差式流量计 pressure differential meter
压差式推移质取样器 pressure difference bed-load sampler
压差式真空计 differential vacuum ga(u)ge
压差水听器 gradient hydrophone; pressure difference hydrophone; pressure differential hydrophone
压差调节阀 pressure difference valve
压差调节器 differential pressure regulator; differential pressure controller
压差温度计 manometric(al) thermometer
压差油阀 differential pressure fuel valve
压差指示计 differential pressure indicator
压差指示器 differential pressure indicator
压差阻力 pressure drag
压差阻力系数 pressure drag coefficient
压成的轮廓 pressed profile
压成的黏[粘]土砖坯 pressed raw clay brick
压成的形状 pressed shape
压成倾斜 canting
压成声盘 pressing
压成丸形 pellet(ize)
压程曲线 pressure-travel curve
压尺 nose bar
压尺垂距 pressure bar lead
压尺间距 pressure bar gap
压出 extrudation; extrude; extruding; extrusion; force off; pinch(ing)-out; spewing; squeeze-out
压出板材 extruded sheet
压出半成品 extrudant
压出玻璃瓦 extruded glazed tile
压出的 extrusive
压出的型材 extruded shape
压出阀 discharge service van
压出干燥法橡胶 extrusion-dried rubber
压出辊 extraction roll
压出痕迹 dint
压出机 extruder; extruding machine
压出机机膛 extruder bore
压出机机筒内衬 extruder cylinder liner
压出机机头 extruder head; extruding-head
压出机机头芯型 extruder-head core
压出机机座 extruder base; extruder stand
压出机口型 extruder die
压出机螺杆 extruder-screw
压出机芯型 extruder core
压出机型号 extruder size
压出可能性 extrudability
压出冷却联动装置 extruder and cooling train
压出力 extrusion pressure
压出量 extruder output
压出流水线 extrusion line
压出泥浆 displace mud
压出坯料 extruder blanks
压出式注压机 extrusion type injection machine
压出水 water of compaction
压出速度 extruder rate
压出物 extrudate
压出型材 extrudate
压出压力 extruder pressure
压出压延机 extruder-calender
压出液 press juice
压出釉面瓦 extruded glazed tile
压出釉面砖 extruded glazed tile
压出指数 extrusion index
压出质量 extrusion quality
压出助剂 extruding aids
压杵 stamp
压穿方向 direction of pressing
压船货用的铁块 kentledge
压床 machine press; press; press machine
压床打料棒 knockout
压床垫板 bolster plate
压吹法 <玻璃> press-and-blow process
压吹机 press-and-blow machine
压磁材料 piezomagnetic material
压磁的 piezomagnetic
压磁法 piezomagnetic method
压磁化 piezomagnetization
压磁矩 piezomagnetic moment
压磁灵敏度常数 piezomagnetic sensitivity constant
压磁系数 piezomagnetic coefficient
压磁效应 piezomagnetic effect; piezomagnetism
压磁性 piezomagnetic; piezomagnetic effect
压磁应力计 piezomagnetic stress ga(u)ge
压淬机 quenching press
压带辊 belt hold roller
压带机 belt press
压带轮 guide pulley; lay-on roller; pinch roller
压带输送机 double belt conveyer [conveyor]
压刀 hook; push-type broach
压刀板 hold-down clip
压倒 override; stoop
压倒的多数 on the overwhelming majority
压倒的力量 force majeure
压倒多数 overwhelming majority
压得过密的 overcompacted
压的非常紧密 strong compacting
压低 abatement; depress; drive down
压低出厂价格 put down the price ex-works
压低到以前水平 roll back
压低等级 graded down
压低基数 lower base figure
压低需求措施 demand-curbing measure
压底舱费 charges for ballast
压地毯条 carpet rod
压点 pressure point
压电半导体 piezoelectric(al) semiconductor
压电泵 piezoelectric(al) pump
压电变换器 piezoelectric(al) sender

压电变形常数 piezoelectric(al) deformation constant
压电变压器 piezoelectric(al) transformer
压电材料 piezoelectric(al) material
压电常数 piezoelectric(al) constant
压电充电 piezoelectric(al) charging
压电传递 piezoelectric(al) transmission
压电传感器 piezoelectric(al) sensor
压电传声器 piezoelectric(al) microphone
压电的 piezoelectric
压电等效网络 piezoelectric(al) equivalent network
压电地震检波器 piezoelectric(al) seismometer
压电点火 piezoelectric(al) ignition
压电点火器 piezolighter
压电电池 piezocell
压电电压 piezoelectric(al) voltage
压电电压常数 piezoelectric(al) voltage constant
压电电子 piezoelectron
压电电阻 piezoresistance
压电电阻测量 piezoresistance measurement
压电电阻加速度计 piezoresistive accelerometer
压电电阻率 piezoresistivity
压电电阻器 piezoresistor
压电电阻拾声器 piezoresistive pick-up
压电电阻系数 piezoresistance coefficient
压电电阻效应 piezoresistance effect
压电堆 piezoelectric(al) stack
压电发电机 piezoelectric(al) generator
压电发光 piezoluminescence
压电方程 piezoelectric(al) equation
压电蜂鸣器 piezoelectric(al) buzzer
压电高温计 piezoelectric(al) pyrometer
压电共振器 piezoelectric(al) resonator
压电光调制器 piezoelectric(al) light modulator
压电话筒 piezoelectric(al) microphone
压电换能晶体 transducing piezoid
压电换能器 piezoelectric(al) transducer; piezometric transducer
压电换能石英片 transducing piezoid
压电活性 piezoelectric(al) activity
压电极化率 piezoelectric(al) susceptibility
压电计 piezoelectric(al) ga(u)ge; piezometer
压电记录加速度计 piezoelectric(al) recording accelerometer
压电加速表 piezoelectric(al) accelerometer
压电监视器 piezoelectric(al) monitor
压电检测器 piezoelectric(al) detector
压电角速度传感器 piezoelectric(al) angular rate sensor
压电校准器 piezoelectric(al) calibrator
压电结型换能器 piezojunction transducer
压电介质的 piezodielectric(al)
压电劲度常数 piezoelectric(al) stiffness constant
压电晶片 crystal plate; piezoelectric(al) crystal
压电晶体 piezocrystal; piezoelectric(al) crystal; piezoquartz
压电晶体部件 piezoelectric(al) crystal unit
压电晶体点火 piezoelectric(al) ignition system
压电晶体电路 piezoelectric(al) crystal circuit
压电晶体加速度计 piezoelectric(al) accelerometer
压电晶体滤波器 piezoelectric(al) crystal filter
压电晶体片 piezoelectric(al) crystal plate
压电晶体频率 piezoelectric(al) frequency
压电晶体送受话器 piezophony
压电晶体微音器 adp microphone
压电晶体谐振器 piezoresonator
压电晶体扬声器 crystal loudspeaker; piezoelectric(al) loudspeaker
压电晶体元件 piezoelectric(al) crystal element
压电矩阵 piezoelectric(al) matrix
压电聚合物换能器 piezoelectric(al) polymer transducer
压电刻纹头 crystal cutter; crystal cutting; piezoelectric(al) cutter; piezoelectric(al) cutterhead
压电控制 piezoelectric(al) control
压电量测系统 piezoelectric(al) measuring system
压电灵敏度 piezoelectric(al) susceptibility
压电流量计 piezoelectric(al) flowmeter
压电滤波器 piezoelectric(al) filter
压电模量 piezoelectric(al) modulus
压电模数 piezoelectric(al) modulus
压电耦合 piezoelectric(al) coupling
压电耦合常数 piezoelectric(al) coupling constant
压电耦合器 piezocoupler; piezoelectric(al) coupler
压电耦合系数 piezoelectric(al) coupling coefficient
压电频率 piezoelectric(al) frequency
压电器件 piezoelectric(al) device
压电三十度截法 thirty-degree piezoelectric(al) cut
压电声强计 piezoelectric(al) phonometer
压电石英 piezoelectric(al) quartz; piezoid; piezoquartz
压电石英法飘尘监测仪 piezoelectric(al) quartz airborne dust monitor
压电石英法飘尘取样器 piezoelectric(al) quartz airborne dust sampler
压电石英检测器 piezoelectric(al) quartz detector
压电石英校准器 piezocalibrator
压电石英晶体 piezoelectric(al) quartz crystal
压电石英晶体共振器 piezoelectric(al) quartz crystal resonator
压电石英晶体压力计 piezoquartz manometer
压电石英矿床 piezoelectric (al) quartz deposit
压电石英滤波器 piezoelectric(al) quartz filter
压电石英片 piezoelectric(al) quartz plate
压电石英稳频器 piezoelectric(al) stabilizer
压电石英质量敏感器 piezoelectric(al) quartz mass sensor
压电拾声器 piezoelectric(al) pick-up
压电拾音器 crystal pickup; piezoelectric(al) pick-up
压电示波器 piezoelectric(al) oscillograph
压电式测量器 piezoelectric(al) ga(u)ge
压电式传感器 piezoelectric(al) cell; piezoelectric(al) transducer
压电式传声器 crystal microphone; piezoelectric(al) microphone
压电式地震计 piezoelectric(al) seismometer; piezometric seismometer
压电式地震仪 piezoelectric(al) seismometer
压电式换能器 piezoelectric(al) transducer
压电式继电器 piezoelectric(al) relay
压电式加速度计 piezoelectric(al) accelerometer
压电式检波器 pressure detector
压电式接收器 piezoelectric(al) receiver
压电式拾音头 piezoelectric(al) cartridge
压电式示波器 moving-iron oscillograph; piezoelectric (al) oscillograph
压电式探测器 piezoelectric(al) detector
压电式探纬装置 piezoelectric(al) weft-detector system
压电式微音器 piezoelectric(al) microphone
压电式谐振器 piezoelectric(al) resonator
压电式压力表 piezoelectric(al) pressure indicator
压电式压力盒 piezoelectric(al) cell
压电式压力计 piezoelectric(al) pressure ga(u)ge
压电式扬声器 piezoelectric(al) loudspeaker
压电式振荡器 piezoelectric(al) oscillator
压电式振动计 piezoelectric(al) type vibration ga(u)ge
压电受话器 piezo receiver
压电树脂 piezoelectric(al) resin
压电水晶 piezocrystal; piezoelectric(al) quartz
压电水晶矿床 piezoelectric(al) quartz deposit
压电顺度 piezoelectric(al) compliance
压电送受话器 piezophony
压电探针 piezoelectric(al) probe
压电陶瓷 piezoelectric(al) ceramics
压电陶瓷材料 piezoelectric(al) ceramic material
压电陶瓷传声器 piezoelectric(al) ceramics microphone
压电陶瓷放大器 ceramic amplifier
压电陶瓷环 piezoelectric(al) ceramic ring
压电陶瓷鉴频器 piezoelectric(al) ceramic discriminator
压电陶瓷抗噪声传声器 piezoceramic noisecancelling microphone
压电陶瓷气体火花塞 piezoelectric(al) ceramic gas sparking plug
压电陶瓷拾音器 ceramic pickup
压电陶瓷受话器 piezoelectric(al) ceramic receiver
压电陶瓷微音器 ceramic piezoelectric(al) microphone
压电特性 piezoelectric(al) property
压电体 piezoelectrics
压电天平式粉尘计 piezo-balance type dust monitor
压电推进器 piezoelectric(al) driver
压电陀螺仪 piezoelectric(al) gyroscope
压电稳频器 piezoelectric(al) stabilizer
压电系数 piezoelectric(al) coefficient; piezoelectric(al) constant; piezoelectric(al) modulus
压电系统 piezoelectric(al) system
压电现象 piezo-electricity
压电效应 piezoeffect; piezoelectric(al) effect
压电效应控制 piezoelectric(al) control
压电效应应变仪 piezoelectric(al) effect strain ga(u)ge; piezoelectric(al) strain ga(u)ge
压电胁变常数 piezoelectric(al) strain constant
压电胁强 piezoelectric(al) stress
压电胁强常数 piezoelectric(al) stress coefficient; piezoelectric(al) stress constant
压电谐振器 piezoelectric(al) resonator; piezoresonator
压电性 piezo-electricity
压电学 piezo-electricity; piezoelectrics
压电压力计 piezoelectric(al) ga(u)ge; piezoelectric(al) pressure meter
压电扬声器 piezoelectric(al) speaker
压电仪 piezoelectric(al) apparatus; piezoelectric(al) ga(u)ge
压电应变仪 piezoelectric(al) strain ga(u)ge
压电圆片式水下检波器 disc hyophone
压电源 piezoelectric(al) source
压电跃变 piezoelectric(al) transition
压电张量 piezoelectric(al) tensor
压电振荡 piezoelectric(al) oscillation
压电振荡器 piezoelectric(al) oscillator; piezooscillator
压电振动 piezoelectric(al) vibration
压电振动机 piezoelectric(al) vibration machine
压电振动器 piezoelectric(al) vibration generator; piezoelectric (al) vibrator
压电振子 piezoelectric(al) vibrator
压电振子等效电路 equivalent circuit of piezoelectric(al) resonator
压电整频 piezocontrol
压电指示器 piezoelectric(al) indicator
压电滞后 piezoelectric(al) hysteresis
压电轴 piezoelectric(al) axis
压电轴向计 rodometer
压电转换器 piezoelectric(al) transducer
压电阻 compressive resistance
压钉杆 nail set
压钉机 stapling machine
压钉器 brad punch
压顶【建】 cope; coping; blocking course; capping; capping piece; projecting cover; tabling
压顶板 cap(ping) piece; capping slab; cap plate; coping slab; slab cap(ping)
压顶层 cap sheet
压顶防护层 capping cover layer
压顶灰浆 <圆柱试件抗压试验时,手抹面找平用> plaster for capping
压顶块 cope block; cap plate; capping piece
压顶梁 capping beam; capping piece; coping beam
压顶料 ballast
压顶锚具 capped anchorage
压顶内泛水 coping with inside wash
压顶砌块 coping block
压顶嵌条 <砖石山墙的> skew fillet
压顶石 capping stone; capstone; copestone; coping stone; topstone
压顶石座 skew butt
压顶双向泛水 coping with outside wash; coping with two-way wash
压顶物 capping mass
压顶线脚【建】 cap mo(u)ld(ing); lorymer

压顶砖 capping brick;coping brick
压定 compression set
压定木 staypack
压端连接 press bond
压短比 compressibility ratio
压锻 press forging;tamp forging
压锻模 steel die
压舵 carrying a rudder; meeting a rudder
压耳 clamp lug
压发拉发混合式发火装置 pressure-pull firing device
压发引信 push detonator
压发装置 pressure firing device
压阀垫 stop plate
压粉 press-powder
压粉磁铁 powder magnet
压粉过程 pressed powder process
压粉铁芯 dust core;magnetic powder core
压风 forced air;pulsated air
压风顶把 pneumatic hold-on
压风管 pneumatic tube;pressure ventilation pipe
压风机 blowing fan;forcing fan;pressure fan;pressure ventilator
压风机的高压级 high-pressure stage
压风机活塞 pneumatic piston
压风加热 forced air heating
压风胶管 pneumatic hose
压风冷却 forced air cooling
压风设备 pneumatic plant
压封 compression seal;press seal
压封式防喷器 blowout preventer of the pressure packer type
压峰 voltage crest
压峰剂 depressor
压缝 tooled joint;tooling
压缝板 batten plate;covering of joint; joint cover(ing); joint fillet cover strip
压缝法 joint pressing-in method
压缝滚子 <裱糊墙纸用> seam roller
压缝焊 cover weld
压缝器 jointer
压缝嵌条 <顶棚、天花板及墙板接缝处> cover fillet
压缝石条 lithistid
压缝条 batten;bead;cover fillet;covering of joint; cover plate; cover slab; cover strip; fillet; joint(ing) strip; pad; panel strip; plain flat strip mo(u)lding;wind stop;wood batten;welting strip <屋顶与墙的>
压缝条封缝 joint sealed by cover strip
压缝条贴边 cover mo(u)ld(ing)
压缝线脚 cover mo(u)ld(ing)
压盖 capping;gland;gland bush;gland cap;junk ring <填料盒的>
压盖保险 gland lock
压盖衬层 glandless lining
压盖衬套 gland liner
压盖的 overlying
压盖垫圈 gland washer
压盖(沸腾)钢 capped steel
压盖罐 friction(al) top can
压盖机 capper;closing machine
压盖螺母 gland nut
压盖螺栓 gland bolt
压盖密封 gland packing;gland seal
压盖密封环 gland seal ring
压盖密填 gland packer
压盖填缝料 sealer
压盖填料 gland packing
压盖旋塞 gland cock
压盖柱塞 gland plunger
压干 pressing
压干草机 hay press
压杆 compressed bar; compression bar;compression member;compression rod;compression strut;depression bar; member in compression; plunger; pressure bar; pressure lever;push brace;strut;strut bracing; strut frame
压杆长细比 slenderness ratio of compression member
压杆导架 presser bar spring bracket
压杆断裂 compression stalk rupture
压杆和拉杆法 strut-and-tie method
压杆破裂 compressive stalk rupture
压杆式室外消防栓 compression-type post fire hydrant
压杆式室外消火栓 compression-type post fire hydrant
压杆式消防栓 compression fire hydrant
压杆稳定性 stability of compression member
压杆轴 pressure shaft
压钢 compressed steel
压钢转向架 pressed steel bogie
压港 port congestion
压供 forcing
压汞法 mercury injection method
压汞分析 mercury intrusion analysis
压汞仪 mercury porosimeter
压钩坡【铁】coupler compressing grade
压管风速计 anemobiagraph;pressure tube anemometer
压管接头 compression sleeve
压管式风速表 pressure tube anemometer
压管式风速计 pressure tube anemometer
压管式风速仪 pressure tube anemometer
压灌混凝土 guncreting
压光 burnishing;press(ed) finish-(ing);rolling
压光的 hard finished;piezooptic(al)
压光辊 felt wrapped roll
压光机 calender;mangle
压光面 trowel finish
压光面饰 pressed finish
压光磨面 hard-finished plastering
压光抹子 finish trowel
压光饰面 hard-finished plastering
压光系数 piezooptic(al) coefficient
压光效应 photoelastic effect;piezooptic(al) effect
压光性质 piezooptic(al) property
压光装饰 calender finish
压轨机 rail press
压辊 compression roll(er); ductor; expression roll;press roll;pressure roll(er)
压辊砝码 press roll weight
压辊杠杆 press roll lever
压辊加热装置 drum heating device
压辊条纹 sea line
压辊线性静载荷 static linear drum load
压辊罩 crushing roll shell
压辊转向 drum steering
压滚 nip drum;press roll
压焓曲线 pressure-enthalpy curve
压焓图 pressure-enthalpy chart
压焊 bonding;pressure welding;push weld(ing)
压焊机 press(ure) welder
压合 forcing fit;press fit;pressure in; stitching
压合胶粘剂 pressure-sensitive adhesive
压合连接 compression joint
压合铁轮箍 pressed-on iron tyre
压痕 brinelling; dint; impression; indentation;press mark(ing);pressure mark(ing);rut
压痕变形钢丝 indented wire
压痕的长对角线 long diagonal of indentation
压痕的短对角线 short diagonal of indentation
压痕反应 constrictive mark reaction
压痕钢筋 indented bar
压痕钢丝 indented wire
压痕钢索 indented strand
压痕各向异性 indentation anisotropy
压痕机 marking press
压痕螺栓 indented bolt
压痕面积 <试验硬度时用> area of cup;area of indentation
压痕蠕变 indentation creep
压痕深(度) depth of cup;depth of impression; depth of indentation; penetration depth
压痕深度计 indentation depth ga(u)ge
压痕试验 indentation test
压痕硬度 durometer hardness;indentation hardness; penetration hardness
压痕硬度机 indentation machine
压痕硬度计 indentation hardness tester;indentation machine
压痕硬度计的压头 indenter [indentor]
压痕硬度试验 indentation hardness test
压痕硬度试验机 indentation hardness tester
压痕硬度指数 indentation hardness number
压痕直径 impression diameter
压痕指数 indentation index
压痕阻力 resistance to indentation
压后冷却式 aftercooled
压互 potential transformer; voltage transformer
压花 coining; depressed design; die stamping;emboss(ment);figurine; impression; knurl(ing); lettering; pattern embossing;ragging;stain on firing;stamping
压花板 embossed sheet; patterned sheet
压花板材料 embossing plate
压花玻璃 cathedral glass;configurated glass; figured glass; figured rolled glass;figured roller glass;figured roll glass;figured sheet glass; floral pattern glass; hammered glass; obscured glass; patterned glass;rough plate glass
压花玻璃墙 wall of patterned glass
压花薄膜 embossed film
压花布 embossed cloth
压花冲头 forcer
压花刀具 roulette
压花垫 embossing pad
压花雕饰 <颗粒屋面板上> cedartex
压花法 embossing
压花钢板 sun steel
压花工具 knurling tool
压花辊 embossing roll;knurl roll
压花辊子 embossing roller
压花滚刀 pinking roller
压花滚轮 knurling tool
压花滚筒 emboss roller;knurl roller; impression cylinder
压花机 coining mill;embosser;embossing machine;knurling machine;knurling tool;knurlizer;marking device
压花夹丝玻璃 wired patterned glass
压花金属薄板 embossed sheet
压花螺纹预应力钢丝 helically crimped wire
压花模 coining die;embossing die
压花片材 embossed sheeting
压花平板玻璃 figured plate glass
压花漆布 embossed linoleum
压花墙纸 <厚纸压花制成的> anaglyptic wallpaper;embossed paper
压花设计 embossed design
压花塑料 embossed plastic material; embossed plastics
压花铁 embossing iron
压花头 embossing die
压花图案 embossed pattern
压花涂饰剂 embossed finish
压花纹 knurling
压花效应 embossed effect
压花修复活塞 knurled piston
压花硬木板 embossed hardboard
压花硬质纤维板 embossed hardboard
压花织物 embossed texture
压花铸币 coin
压花装订布 embossed calico
压坏 collapsing; crush; compression failure
压坏应力 crushing stress
压环 compressed ring; female adapter;grip socket; junk ring; pressure ring;retaining ring;verge ring
压环机 ring squeezer
压环钳 collar crushing forceps
压簧 compression spring;pull-out piece spring;setting bolt spring
压簧螺钉 pull-out piece spring screw; screw for setting lever spring
压毁 collapsing; compression failure; crushing
压毁荷载 crushing load
压毁力 crushing force
压毁模量 modulus of crushing
压毁强度 crushing strength
压毁强度试验 crushing strength test
压毁试验 crushing test
压毁应变(度) crushing strain
压毁应力 crushing stress
压火 bank a fire;banked(coal) fire; banking-up;damp;damping down
压火锅炉 idle boiler
压火损失 banking loss
压机 press
压机闭合时间 press cooling time
压机工 presser
压机构造 press arrangement
压机冷却时间 press cooling time
压机旁研磨机 beside-the-press grinder
压机压板间距 daylight opening
压机压力 pressure of press
压机座 press bed
压迹 impression
压迹形成 track-forming
压积的雪 compacted snow
压积的雪堆 compacted snow formation
压挤 extruding
压挤成型 extrusion mo(u)lding
压挤的地面砖 extruded floor(ing) tile
压挤的铝合金 extruded alumin(i)um alloy
压挤的乙烯小五金 extruded vinyl hardware
压挤断面 extrusion section
压挤阀 squeeze valve
压挤方法 extrusion process
压挤钢模 extrusion die
压挤灌浆 squeeze
压挤机 extruder;extruding machine
压挤节理 compression joint
压挤截面 extruded section
压挤绝缘 extrudable insulation
压挤密封 crush seal
压挤模 compression die

压挤熟铁块 nobbing
压挤送料机 compression feeder
压挤型材 extrusion shape
压挤运行 run-in
压挤桩 compressed pile
压脊 pressure ridge
压加剩余磁化强度 pressure remanent magnetization
压夹虎钳 grip vice [vise]
压夹紧环 ring clamp
压价 beat-down price;demand a lower price;force down price;force price down;price squeeze;rollback price;send down price
压价政策 roll-back policy
压架 press table
压尖 pointing
压尖机 bar pointer;pointer;push pointer
压坚硬的雪 hard snow
压剪 compressed shear
压剪边界 compresso-shear boundary
压剪试验机 compresso-shear testing machine
压剪性构造面 compresso-shear structure plane
压剪性节理 compresso-shear joint
压剪性深断裂 compresso-shear deep fracture
压件 casting die
压浆 grouting;mortar intrusion;mud jack;pressure grout(ing)
压浆泵 boogie box;boogie pump;grout(ing) pump;mud jack pump <填充混凝土路面下空隙用的>
压浆稠度 grout consistency [consistence]
压浆处理 artificial cementation;artificial solidification
压浆法 mud jacking;prepakt method
压浆法浇灌水下混凝土 prepakt concrete
压浆方法<填充混凝土路面下的空隙用> mud jack method;mortar grouting method
压浆方式 shot method
压浆混凝土 grouted aggregate concrete;grouted concrete;intrusion concrete;prepackaged concrete;prepakt concrete
压浆技术 post-pressure grouting technology
压浆孔 grout hole;grout vent;injection hole
压浆排水 water-exclusion by grouting
压浆设备 grouting equipment;grout injection apparatus
压浆水 pulp press water
压浆套管 grout socket
压浆填充混凝土路面下空隙并顶升下沉路面 mud jacking
压浆填料 grout filler
压浆压力 grout pressure
压降 drop of pressure;falling off;fall of pressure;loss of pressure head;pressure differential;pressure drop;pressure loss;pressure reduction;voltage drop【电】
压降法 pressure drop method
压降检验 check and drop
压降曲线 drawdown curve;fall-off curve
压降水流报警器 drop in pressure water flow alarm
压降系数 pressure drop coefficient
压降效应 pressure drop effect
压降压力计 pressure drop manometer
压降振荡 pressure drop oscillation
压降值 pressure drop
压胶辊 dandy roll
压(胶片的)平板 pressure plate
压角机 angle mo(u)lding press
压角条 comer strip;sprig<镶玻璃用>
压脚 presser foot
压脚提升高度 presser lifting amount
压脚提升器 presser foot lifter
压脚调节翼形螺钉 pressure regulating thumb screw
压脚子 peav(e)y
压接 pressure welding
压接端子 pressure connector
压接对偶 pressure pair
压接环 compression thimble
压接接头 compression joint
压接连接 crimp connection
压接模 pressing die
压接栅板 pressure-locked grating
压接式柔性导体接头 compressed flexible rail bond
压接线夹 compression conductor clamp
压接型连接 compression-type of connection
压结结晶方式 piezocrystallization way
压结晶作用 piezocrystallization
压结前烧结 presenter
压金属箔机 foil laminating machine
压筋 beading;pressed rib;ribbing
压筋车体墙板 mounded wall
压筋模 beading die;embossing die
压紧 buckle;compact(ing);compaction;compress(ing);force against;holding-down;impact(ion);jam;pack;pinch
压紧板 hold-down;pressure strip
压紧包 dump bale
压紧部分 pressing section
压紧传感器 jam sensor
压紧锤 shingling hammer
压紧的 compact
压紧的石棉胶板 compressed asbestos sheet
压紧垫圈 packing washer
压紧法兰 supported flange;supporting flange
压紧盖 clamping cap;friction(al) top
压紧盖板 compacting cover plate;hold-down plate
压紧杆<打捆装置> compressor arm;holding-down rod
压紧杆扳手 presser bar lifter
压紧箍 packing clamp
压紧管箍 compression coupling
压紧辊 friction(al) roll;hold-down roller;pressure roll(er)
压紧环 clamp ring;hold-down ring;pressure ring;set collar
压紧机构 hold-down mechanism
压紧夹板 hold-down clip
压紧检测器 jam sensor
压紧力 packing force;pressing force
压紧联轴节 compression coupling
压紧轮 contact roller;pinch roller
压紧螺钉 check screw;clamp screw;forcing screw;housing pin;housing screw;screw tightener;thrust screw
压紧螺帽 compression nut;gland nut
压紧螺母 compression nut;gland nut;hold-down nut;packing nut
压紧螺栓 gland bolt;hold(ing)-down bolt;press fit screwing;thrust bolt
压紧螺丝 check screw;thrust screw
压紧螺旋 hold(ing)-down screw;housing pin;housing screw
压紧密封 pinch-off seal
压紧密封圈 positive seal
压紧面 seal face
压紧模 pressure die
压紧器 compactor;densener;squeezer
压紧钳 crimper
压紧区域 pinch area
压紧式配水龙头 compression-type hydrant
压紧式输送器 pressure conveyer [conveyor]
压紧式水龙头 compression-type cock
压紧式填料 jam type packing
压紧式消防龙头 compression-type hydrant
压紧式消防栓 compression-type hydrant
压紧弹簧 hold-down spring;retainer spring
压紧套箍<用以压入取样器> drive head yoke
压紧系数 compacting factor
压紧楔 drag wedge
压紧型消防栓 compression(type) fire hydrant
压紧研齿<无齿隙的> cramp lapping
压紧装置 gag;hold-down;hold-down gear;holding apparatus;hold-down gag<剪切机的>
压进 pressing in;ride down
压进心轴 press mandrel
压晶 piezocrystal
压井 killing the well
压井放喷管线 choke and kill line
压井管汇 kill-line manifold
压井管线 kill line
压井管柱 kill string
压井技术 well killing technique
压井泥浆比重 killing mud weight
压井液输送管线 kill and choke line
压具 hold(ing)-down
压锯机 stretcher
压(卷)边机 beading machine
压觉 pressure sensation
压觉痛觉模式 pressure-pain pattern
压开螺钉 press-off bolt
压开螺栓 press-off bolt
压开锁 Canadian latch;Norfolk latch
压坑 button;compressed pit;indentation;knockout
压控变容器 varicap
压控放大器 voltage-controlled amplifier
压控晶体振荡器 voltage-controlled crystal oscillator
压控可变延迟线 voltage-controlled variable delay line
压控石英晶体振荡器 voltage-controlled quartz crystal oscillator
压控衰减器 voltage-controlled attenuator
压控性盆地 compresso-shear basin
压控移相器 voltage-controlled phase shifter
压控振荡器 voltage-controlled oscillator
压块 briquetting;compact;hold-down;pressed compact;squeeze head;wafer
压块法 briquetting process
压块机 block-making machine;block-press machine;briquet(te) press;briquetter;briquetting machine;briquetting press;cuber;waferer;wafer machine
压块机构 forming mechanism
压块料 briquetting batch
压块密度 compact density;pressure density
压块用阴模 briquetting die
压矿 unminable protective safety territory
压溃 bruise;flowed head
压捆 baling
压捆机 bale press;baler;press-baler;pull plunger press
压捆密度调节器 bale density adjuster
压捆室 bale channel;baling chamber;press channel
压捆室的定切刀 bale chamber blade
压扩编码 compand coding
压扩律 companding law
压扩式短波电话终端机 linked compressor-expander system radio telephone terminal
压拉模材料 die material
压拉强度比 compressive-tensile strength ratio
压拉双作用预应力 prestressing with subsequent compression and tension
压蜡机 wax injection machine;wax injection press;wax injector
压阑石 rectangular stone slab
压牢装置 hold-down fixture
压力 compression;compression force;intermediate discharge pressure;piezo;pression;pressure;pressure force
压力安全阀 pressure-relief valve;pressure safety valve
压力安全壳 pressure containment
压力安全装置 pressure limiting device;pressure-relief device
压力按钮弹簧 pressure button spring
压力摆动 pressure pendulation
压力板 holding-down clip;pressure plate
压力板出流法 pressure plate outflow method
压力半径 pressure radius
压力瓣 pressure circulation;pressure clack;pressure flap
压力棒 press bar
压力保持 maintenance of pressure;pressure maintenance
压力保持部件 pressure retaining member
压力保持阀 pressure retaining valve
压力保持方法<油层> pressure maintenance method
压力保持回路 pressure holding circuit
压力保持器 retainer
压力保持装置 pressurizer
压力保护阀 pressure protection valve
压力保护开关 pressure head switch
压力保险开关 pressure safety switch
压力保险塞 pressure fuse plug
压力报警器 pressure alarm
压力报警装置 pressure warning unit
压力倍数 pressure multiplication factor
压力倍增器 pressure multiplier
压力泵 compression pump;force lift;force pump;press pump;suction and force pump
压力泵输送管 pressure pump delivery pipe
压力泵送 squeeze pumping
压力泵送混凝土 precrete
压力泵吸油管 pressure pump suction pipe
压力比 compression ratio;pressure ratio
压力比感传器 pressure ratio detector
压力比降 pressure gradient
压力比较开关 comparative pressure switch
压力比控制 pressure ratio control
压力比控制器 pressure ratio controller
压力比例混合器 pressure proportio-

ning tank
压力比容图 pressure and specific volume diagram
压力边界 pressure boundary
压力变动 pressure oscillation
压力变动情况 pressure history
压力变化 pressure change; pressure variation
压力变化率 pressure change rate
压力变化曲线 pressure history
压力变化速度 rate of pressure change
压力变换器 pressure converter; pressure transducer
压力变宽 pressure broadening
压力变送器 pressure transmitter
压力-变形曲线 pressure-displacement curve
压力变质作用 piezometamorphism; pressure metamorphism; pressure pattern metamorphism
压力表 compression ga(u)ge; manometer; manometer pressure ga(u)ge; piezometer; pressure ga(u)ge; pressure indicator; pressure meter
压力表板总成 pressure ga(u)ge dash unit
压力表当量 monometric(al) equivalent
压力表读数 ga(u)ge reading
压力表管 ga(u)ge line
压力表缓冲器 ga(u)ge saver; pressure ga(u)ge damper
压力表减振器 pressure ga(u)ge snubber
压力表检验器 pressure ga(u)ge tester
压力表校正仪 pressure ga(u)ge tester
压力表校准 manometer calibration
压力表接口 pressure ga(u)ge connection; pressure tap; pressure tap(ping) hole
压力表接口和螺塞 manometer tap
压力表接头 pressure tap(ping)
压力表开关 pressure ga(u)ge switch
压力表塞门 ga(u)ge cock
压力式温度计 manometric(al) thermometer
压力表式液位指示器 manometer liquid level indicator
压力表试验器 deadweight tester
压力冰 pressure ice; rough ice
压力病【救】 compressed-air illness; compressed-air sickness
压力波 compressive wave; force(d) wave; piestic wave; pressure shock; pressure wave; compression wave
压力波传播 pressure wave emission; pressure wave propagation
压力波传播时间 pressure wave travel time
压力波动 fluctuation of pressure; pressure fluctuation; pressure pulsation; pressure surge; surge
压力波动阀 pressure wave valve
压力波动极限 pressure surge limit
压力波动速度 pressure wave velocity
压力波反射 pressure wave reflection
压力波腹 pressure wave antinode
压力波节点 pressure wave node
压力波前 pressure wavefront
压力波调节 surge control
压力波调压 surge control
压力波增压 pressure wave supercharge; pressure wave supercharging
压力波阻尼器 pressure snubber
压力玻璃纤维袋集尘器 pressure-type glass baghouse
压力薄膜 pressure membrane
压力薄膜仪 pressure film instrument
压力补偿 pressure compensation
压力补偿的 pressure compensated
压力补偿阀 pressure-compensated valve
压力补偿控制 pressure-compensated control; pressure-compensating control
压力补偿流量调整阀 pressure-compensated flow control valve
压力补偿瓶 pressure-compensation bottle
压力补偿器 compensator; pressure compensator; pressure equalizer
压力补偿式泵 pressure-compensated pump
压力补偿式流量计 pressure-compensated flowmeter
压力补偿调节 pressure-compensation control
压力补偿系统 pressure compensating system
压力补偿转子流量计 balanced-pressure rotameter
压力补偿装置 pressure-compensation device
压力不足 under-pressure
压力擦亮 pressed polish
压力仓式输送机 pressure vessel conveyer [conveyor]
压力仓式输送机系统 pressure vessel conveying system
压力舱 barochamber; pressure cabin; pressure chamber
压力操纵的 pressure-operated
压力操纵阀 pressure-operated valve
压力操纵开关 pressure-operated switch
压力操作装置 pressure-operated device
压力槽 head tank; overhead tank; pressure tank
压力测定 compression test; piezometry; pressure determination; pressure ga(u)ging; pressure measurement
压力测井 pressure log
压力测力计 compression dynamometer
压力测量 measurement of pressure; pressure measurement; pressure survey
压力测量泵 pressure measurement pump
压力测量表 pressure meter
压力测量孔 pick-up hole
压力测量器 pressometer
压力测量系统 pressure-measuring system
压力测量仪器 pressure-measuring instrument
压力测深仪 pressure sounder
压力测深针 pressure sounder
压力差 difference in pressure; differential; differential head; pressure difference; pressure differential; pressure sink; differential pressure
压力差调节机构 differential pressure control mechanism
压力差诱导进入的空气 induced air
压力差指示控制器 pressure difference indicating controller
压力场 field of pressure; pressure field
压力超增 pressure override
压力潮位表 pressure tide ga(u)ge
压力潮位计 pressure tide ga(u)ge
压力潮汐计 pressure fide ga(u)ge
压力沉附物 pressure coat
压力成型 compression mo(u)lding; pressure forming
压力成型机 compression mo(u)lding press
压力成型装置 pressure former
压力程序设计 pressure programming
压力持续 pressure sustaining
压力齿轮泵 pressure gear pump
压力冲程 pressure stroke
压力冲击 compression shock; pressure percussion; pressure shock; water hammer
压力冲击波 pressure shock wave; pressure surge
压力冲洗 blow wash; pressure flush; pressure wash(ing)
压力冲洗阀 pressure flush valve
压力冲洗作业 pressure washing operation
压力抽风机 forced draught fan
压力抽风井 forced draft shaft
压力除节机 pressure knotter
压力除气 pressure deaeration
压力处理(法) <木材防腐> pressure treatment
压力触点 pressure contact
压力穿透 pressure penetration
压力传播 pressure propagation; pressure transmission
压力传导系数 coefficient of pressure conductivity
压力传递 pressure propagation; pressure transfer; transmission of pressure
压力传递介质 pressure transmission medium
压力传动的 pressure-actuated
压力传感 pressure sensing
压力传感阀 pressure-sensing valve
压力传感盒 pressure-sensitive cell
压力传感块 pressure-sensitive pad
压力传感器 load cell; pressure capsule; pressure cell; pressure detector; pressure ga(u)ge; pressure pick-up; pressure probe; pressure-sensing device; pressure sensor; pressure transducer; pressure transmitter; pressure unit; electric (al) pressure transducer; load cell
压力传感器系统 pressure capsule system
压力传感式定量自动秤 load cell weighing equipment
压力传感装置 pressure-sensing device
压力传声器 pressure microphone
压力传送 transmission of pressure
压力传送器 pressure transmitter
压力吹风器 pressure blower
压力吹气机 pressure blower
压力吹洗 blow wash
压力粗滤器 pressure strainer
压力袋 pressure bag
压力袋成型 pressure bag mo(u)lding
压力袋集尘器 pressure-type baghouse
压力袋模塑成型 pressure bag mo(u)lding
压力单位 pressure unit
压力挡风雨条 pressure weather stripping
压力导出孔 pressure tapping
压力道钉 pressure spike
压力的当量分布 equivalent distribution of pressure
压力的高度 discharge head
压力的交变分量 alternating component of pressure
压力的均衡 equalization of pressure
压力的上限 upper limit of pressure
压力的温度灵敏度 temperature sensitivity of pressure
压力等级 pressure rating
压力等式 equality of pressure
压力地面 ground pressure
压力递减 pressure decline; pressure depletion
压力点 pressure point; pressure spot
压力点火 pressure ignition
压力电缆 pressed-core cable
压力电位计 pressure potentiometer
压力电阻效应 piezoresistance
压力垫 pressure pad
压力垫顶出 pressure-pad lift-out
压力垫圈 pressure disc [disk]
压力垫枕 pressure pillow
压力定额 pressure rating
压力定形 compression set
压力定型 compression set; compressive set
压力动量曲线 pressure-momentum curve
压力动态 barokinesis
压力动作开关水流报警器 pressure-operated switch water flow alarm
压力动作开关水流启动装置 pressure-operated switch water flow initiating device
压力洞室试验 pressure chamber test
压力陡度 pressure gradient
压力读数 pressure reading
压力断开装置 pressure trip
压力断路器 pressure differential cut-out
压力锻造 press forging
压力对流加热器 forced convection heater
压力多边线法 funicular pressure line method
压力发送器 pressure capsule; pressure cell
压力发展(过程) development of pressure
压力阀(门) compression valve; outlet valve; pressure valve
压力阀开始通过流体的压力 cracking pressure
压力法 pressure method <测量混凝土中空气量的>; pressure application
压力范围 compression stress field; limit of pressure; pressure limit; pressure range; pressure schedule; pressure span
压力方程 pressure equation
压力防腐木材保存法 pressure process of timber preservation
压力防腐木料 pressure preserved
压力防护装置 pressure safeguard
压力放大 pressure amplification; pressure boost; pressure gain
压力放大器 pressure amplifier
压力放大系数 pressure amplification factor; pressure gain factor
压力放空阀 pressure vent valve
压力飞溅合用润滑 combined force-feed and splash lubrication
压力沸水器 pressure boiler
压力分布 distribution of pressure; pressure distribution; pressure field structure
压力分布测定 pressure distribution measurement; pressure exploration
压力分布规律 pressure law
压力分布环 pressure distribution ring
压力分布计算 pressure distribution calculation
压力分布模型 pressure-plotting model
压力分配器 pressure distributor
压力分布图 pressure distribution chart; pressure distribution pattern;

pressure-plotting; pressure profile
压力分布图绘制 pressure-plotting
压力分布系数 coefficient of pressure distribution
压力分布研究 pressure investigation
压力分量 pressure component
压力分配 pressure dividing
压力分配阀 pressure distributing valve; pressure dividing valve
压力分配控制阀 pressure dividing control valve
压力分配器 pressure distributor
压力分散 pressure dispersion
压力分散层 pressure-relieving joint
压力风缸 pressure reservoir
压力风扇 pressure fan
压力风速表 pressure anemometer
压力风速计 pressure anemometer
压力封盖 pressure seal cover
压力封口 pressure seal
压力峰值 pressure peak
压力缝 pressure slot
压力浮动 pressure fluctuation
压力浮选(法) pressure flo(a)tation
压力浮选器 pressure flo(a)tation equipment
压力浮子 pressure float
压力幅度 pressure amplitude
压力腐蚀 pressure corrosion
压力负荷 pressure load(ing)
压力复原 pressure reestablishment
压力干管 force(d) main; pressure main (pipe)
压力干扰 pressure disturbance; pressure interfere
压力干燥 drying by pressure
压力感传膜片 pressure-sensitive diaphragm
压力感生带 pressure-induced band
压力感生光谱 pressure-induced spectrum
压力感受盒 pressure-sensitive cell
压力感受器 barocepter [baroceptor]; pressoreceptor; pressure ga(u)ge pick-up; pressure head
压力感受器反射 pressoreceptor reflex
压力感受性 pressosensitivity
压力感应开关 pressure-sensitive switch
压力感知触点 mano-contact
压力感知检数器 pressure-sensitive vehicle detector
压力刚性飞艇 pressure rigid airship
压力缸 pressure cylinder
压力钢管 penstock; pressure steel pipe; steel penstock; steel pipe penstock
压力钢管阀(门) penstock valve
压力钢管进口阀 valve at inlet to penstock
压力钢管镇墩 anchor of steel penstock
压力钢筋 negative reinforcement
压力钢进水管 steel penstock
压力高度 pressure altitude; discharge head <压缩机的>
压力割理 pressure cleat
压力给油 forced fuel feed; force feed
压力给油汽化器 pressure fed carburetor
压力给油润滑 pressure-feed lubrication

压力供给 forced feed
压力供给泵 pressure maintaining pump
压力供气 forced air supply
压力供气设备 forced system
压力供水 pressure water supply
压力供应 pressure supply
压力供油法 force-feed oiling
压力拱 pressure arch
压力鼓风 forced air blast
压力鼓风机 pressure blower
压力固定 pressure fixation; pressure fixing
压力故障 pressure disturbance
压力管 flow pipe; manometer tube; pressure pipe; pressure tube; pressure tubing, rising pipe
压力管道 conduit under pressure; penstock; penstock pipe; power conduit; pressure piping(-line); pressure conduit
压力管道断面 penstock section
压力管道法 pressure line method
压力管道反射时间 penstock reflection time
压力管道分叉管 penstock manifold
压力管道管段 penstock section
压力管道管节 penstock course
压力管道截面 penstock section
压力管道进水口 penstock intake
压力管道流 closed conduit flow
压力管道排水 penstock drain
压力管道排水总管 penstock drain header
压力管道旁通管室 penstock bypass chamber
压力管道水力学 penstock hydraulics
压力管道调压井 penstock reducer
压力管道闸门 penstock gate
压力管道支撑 penstock bracing
压力管道支墩 penstock pier
压力管道支管 penstock manifold
压力管道直线段 penstock tangent
压力管反应器 pressure tube reactor
压力管风速表 pressure tube anemometer
压力管接头 pressure connection
压力管井 penstock shaft
压力管理 stress management
压力管路 pressure line; pressure piping(-line)
压力管渠 conduit under pressure; pressure conduit; pressure pipeline
压力管塞 manometer tube gland; monometer tube gland
压力管式反应堆 pressure tube reactor
压力管线 pressure flow pipeline; pressure (pipe) line; pressure piping(-line)
压力贯入 pressure penetration
压力贯入试验 pressure penetration test
压力灌缝锅 bougie
压力灌浆 injection grout(ing); pressure cementing; squeeze cementing; impregnation <土壤防水处理>; permeation grouting; pressure grouting
压力灌浆泵 boogie box; boogie pump
压力灌浆的 pressure-grouted
压力灌浆法 cementation process; pressure grout(ing)
压力灌浆管 grout pipe; pressure grout(ing) pipe; pressure grout(ing) tube
压力灌浆管桩 pneumatic caisson pile; pneumatic pressure pile; pressure pile
压力灌浆贯入 pressure grout penetration
压力灌浆盒 grout box
压力灌浆混凝土 concrete for pressure grouting; guncreting; intruded concrete; prepacked aggregate concrete; pressure grouted aggregate concrete
压力灌浆机 boogie pump; pressure grouting machine
压力灌浆深度 pressure grout penetration
压力灌浆筒 pressure grouting pan
压力灌浆外加剂 fluidifier
压力灌浇水泥浆 pressure cement grouting
压力灌射水泥浆 cement injection
压力灌水管 garden hose
压力灌注(法) pressure injection
压力灌注玛瑙脂 <接缝中防水> ca(u)lking tool
压力灌注(石)灰浆法 lime slurry pressure injection
压力灌注桩 pressure pile
压力罐 overhead tank; pressure pot; pressure tank
压力罐装水泥 bottled cement
压力光曲拱 pressure line arch
压力硅化法 compression silicified process; pressure silicified process
压力硅化加固法 pressure silicification method
压力柜 head tank; pressure tank
压力柜喷洒机 tank sprayer
压力辊 pressure roll(er)
压力滚筒 pressure bowl
压力滚子 puck
压力锅 pressure cooker
压力过大 overpressure
压力过低 hypotension
压力过低警告灯 pressure-low warning light
压力过高 hypertension; hypertonia
压力过滤 pressure filtration
压力过滤层 pressure filter
压力过滤池 pressure filter tank
压力过滤法 pressure filtration process
压力过滤缸 pressure filter tank
压力过滤机 <污泥脱水方法之一> filter press
压力过滤器 compression filter; pressure filter
压力函数 pressure function
压力涵洞 pressure culvert
压力焓图 pressure-enthalpy diagram
压力焊 pressure welding
压力焊接 pressure welding; solid-phase welding
压力和飞溅润滑 force and splash lubricating
压力和真空释放阀 pressure-and-vacuum release valve
压力荷载 compression load; compressive load(ing); pressure load(ing)
压力盒 head box; load cell; pressure capsule; pressure cell
压力盒盒压 cell pressure
压力盒(示)压力 cell pressure
压力盒试验 cell test; pressure cell test
压力盒系统 <测定应力的设备> pressure capsule system
压力盒装置 <测定土中应力的设备> pressure cell installation
压力横向分布 pressure traverse
压力化学 piezochemistry
压力环 compression ring; press(ure) ring
压力环境 forced circulation; pressure environment
压力环流 forced circulation
压力缓冲罐 pressure surge tank
压力缓冲回路 shock-absorbing circuit
压力缓冲器 pressure snubber
压力换能器 pressotransducer; pressure transducer
压力簧座 pressure spring seat
压力恢复 build-up of pressure; pressure recovery; pressure restoration
压力恢复方法 pressure restoration method
压力恢复曲线 build-up curve
压力恢复特性 pressure-recovery characteristic
压力恢复系数 pressure-recovery factor; repressuring coefficient
压力回路 pressure circuit
压力回授控制 pressure feedback control
压力回水 pressure return water
压力活塞 pressure piston
压力机 forcing machine; machine press; press; pressing machine; punch; squeezer; compression machine; mechanical press
压力机安装 press erection
压力机的液压缸 press cylinder
压力机垫木 press pan
压力机吨位 press tonnage
压力机滑块 press ram; press slide
压力机滑块上的模柄孔 recess of press ram
压力机活塞 ram piston
压力机上横梁 press crown
压力机调整不良 press misalignment
压力机行程 stroke of a press
压力机压头 press ram
压力机压制 press compacting; press compaction
压力机油滴 press drip
压力机柱塞 press plunger
压力机自动生产线 automated press line
压力机自动装置 press automation
压力积聚 build-up of pressure
压力积累 accumulation of pressure
压力级 pressure level; pressure stage
压力级冲动式汽轮机 pressure-stage impulse turbine
压力级制 pressure range
压力极限 limit of pressure; pressure limit
压力集度 pressure intensity
压力计 manograph; manometer pressure ga(u)ge; piezoga(u)ge; piezometer; pressiometer; pressure capsule; pressure ga(u)ge; pressure indicator; pressure manometer; pressure meter; compression ga(u)ge
压力计测量 manometry
压力计导管 ga(u)ge pipe
压力计感应塞 ga(u)ge probe
压力计管(压)盖 manometer tube gland
压力计接管 nozzle for pressure ga(u)ge
压力计接口 pressure tap
压力计量 pressure measurement; measurement of pressure
压力计量器 manometer
压力计平面 piezometric level
压力计示压力 ga(u)ge pressure
压力计式波高计 pressure meter type of wave recorder
压力计式温度计 pressure ga(u)ge type thermometer
压力计水面 piezometric water level
压力计托筒 recorder carrier
压力计压力 manometer pressure
压力计指针 ga(u)ge hand
压力计总压头 manometric(al) total head
压力记录计 pressure recorder
压力记录控制器 pressure recording controller
压力记录器 manograph; pressure recorder

压力记录系统 pressure-recording system
压力记录仪 pressure recorder; recording pressure ga(u)ge
压力继电器 pressure differential cut-out; pressure relay; pressure switch
压力加工 pressed work(ing); press-work
压力加工法 press-working method
压力加工费用 press-working cost
压力加料 forced feed
压力加氯机 pressure chlorinator
压力加油 pressure filling; pressure fuel feed; pressure oiling; pressure refueling
压力加油杯 drive-type oil cup
压力加油法 pressure-feed oiling
压力加油器 drive-type oil cup; pressure-feed filler
压力夹钳风阀 pressure grip throttle
压力夹钳风门 pressure grip throttle
压力尖峰 pressure spike
压力监测器 pressure monitor
压力监视器 pressure monitor
压力减低 hypotension
压力减弱 release of pressure
压力减振 pressure damping
压力检波器 pressure detector
压力检测器 pressure detector
压力检漏(测量) pressure survey
压力检验 pressure check
压力舰体 pressure hull
压力降 differential pressure; draw-down; pressure drop; pressure fall; pressure sink; pressure pumped down <因孔内油水而造成的>
压力降低 loss of pressure; lowering of pressure; pressure decay; pressure decrease; pressure drop; pressure unloading
压力降低漏斗 pressure-relief cone
压力降落 head fall; pressure decrease
压力降系数 pressure decreasing coefficient
压力降增加 pressure drop increase
压力交替作用 pressure replacement
压力浇注 pressure pouring
压力浇铸 pressure casting
压力浇铸机 pressure applicator
压力胶管 pressure hose
压力焦滴 pressure tar
压力角 angle of pressure; pressure angle
压力矫正机 gag press; garment press; ga(u)ge press
压力矫直机 ga(u)ge straighter
压力脚 pressure foot
压力校正 pressure correction
压力校正因子 correction factor for pressure; pressure correction factor
压力校准 pressure calibration
压力校准装置 pressure ga(u)ge
压力阶段 pressure stage
压力接触板 pressure contact board
压力接触法 pressure contact method
压力接触焊 pressure contact welding
压力接点 press(ure) contact
压力接点可控硅整流器 pressure contact thyristor
压力接缝 pressure joint
压力接合 compression joint; press bond
压力接口 pressure joint
压力接收器 pressure receiver
压力节理【地】 induced cleavage; pressure cleat; pressure joint
压力结构 pressure texture
压力介质 pressure medium
压力紧箍 press shrinking
压力进给 forced feed; positive-geared feed; pressure feed
压力进给加油器 force-feed oiler
压力进给系统 pressure-feed system
压力进料 force(d) feed
压力进水口 pressure intake port
压力浸渍树脂处理的木材 compregnated timber
压力井 penstock shaft
压力井道 pressure shaft
压力警报器 pressure alarm indicator
压力警报系统 pressure alarm system
压力剧变 pressure jump
压力绝缘 pressure insulation
压力均衡法 pressure balancing
压力均衡化 pressure equalizing
压力均衡(现象) isostasy; isostatic(al) balance
压力均匀化 pressure equalization
压力开关 pressure switch
压力开关组件 press switch assembly
压力壳层 pressure shell
压力壳顶盖 pressure vessel head
压力可调节弹簧 pressure-setting spring
压力坑道 pressure gallery
压力孔 pressure port
压力孔隙比 pressure-void ratio
压力孔隙比(关系)曲线 pressure-void ratio curve
压力孔隙比图(表) pressure-void ratio diagram
压力控制 control of pressure; pressure(-dependent) control
压力控制潮汐计 pressure-operated tide ga(u)ge
压力控制电路 pressure control circuit
压力控制阀 pressure-controlled valve; pressurizing valve
压力控制焊接 pressure-controlled weld
压力控制回路 pressure control circuit; pressure-control loop
压力控制继电器 pressure-control relay
压力控制开关 pressure switch
压力控制器 pressure controller; pressure governor
压力控制设备 pressure control equipment
压力控制式装置 pressure-controlled plant
压力控制顺序阀 pressure maintaining valve; priority valve
压力控制伺服阀 pressure-control servovalve
压力控制装置 pressure control device; pressuretrol
压力盔形图 helmet; pressure-helmet
压力老化容器 <一种沥青老化仪器> pressure [pressurized] ag(e)ing vessel
压力累积 accumulation of pressure
压力冷却 pressure cooling
压力离合器 pressure clutch
压力连接器 pressure connector; solderless connector
压力炼油 pressure rendering
压力量纲 pressure dimension
压力量雪袋 pressure pillow
压力裂缝 diaclase; induced cleavage
压力裂隙 induced cleavage; pressure cleat
压力灵敏度 pressure sensitivity
压力灵敏元件 pressure cell
压力流 confined flow; flow under pressure; pressure flow
压力流浆箱 pressure-type head box
压力流量计 differential pressure flowmeter
压力流量自动控制阀 automatic pressure and flow control valve
压力硫化 press cure
压力龙头 compression faucet
压力滤池 pressure filter; pressure filter tank
压力滤罐 pressure filter
压力滤器 filter press; pressure filter
压力滤网 pressure strainer
压力滤油 pressure filter
压力滤油机 oil filter press
压力滤油器 oil filter press
压力率 pressure rating
压力螺钉 pressure screw
压力螺钉防松螺母 pressure screw check nut
压力落差 differential pressure; pressure drop
压力脉冲 pressure pulse
压力脉冲阻尼器 pressure oscillation damper; pressure pulsation damper
压力脉动 pressure fluctuation; pressure oscillation; pressure pulsation
压力脉动波 pressure pulse wave
压力铆接机 riveting press
压力帽口 pressure riser
压力猛增 pressure surge
压力弥雾机 pressure atomizer
压力密封 pressure seal
压力密封车 pressure-sealed car
压力密封的 pressure-tight
压力密封环 pressure ring
压力密封件 compression seal
压力密封料 compression seal
压力密封帽 pressure seal cap
压力密封压盖 pressure seal bonnet
压力面 driving face; face of propeller; pressure(sur)face <压头所形成的面>; pressure side; thrust face
压力面图 pressure surface map
压力灭菌 pressure sterilization
压力敏感阀 pressure-sensitive valve
压力敏感块 <压力量测器> pressure-sensitive pat
压力敏感效应 pressure-sensitive effect
压力敏感性 pressure sensitivity
压力敏感性能 pressure-sensitivity characteristic
压力敏感元件 pressure-sensing device; pressure-sensitive device; pressure sensor
压力敏感纸 pressure sensitive paper
压力模版 pressure stencil
压力模式 pressure pattern
压力模数 pressure modulus
压力膜法 pressure membrane method
压力膜析装置 pressure membrane apparatus
压力磨光 pressed polish
压力磨浆 pressurized refining
压力磨浆机 pressurized refiner
压力木材防腐法 pressure process of timber preservation
压力幕涂法 pressure curtain coating
压力挠曲试验 pressure deflection test
压力能 <流体的> pressure energy
压力泥浆浇注 pressure slip casting
压力逆变 reversal of the pressure
压力黏[粘]度 pressure viscosity
压力-黏[粘]度比 pressure-viscosity ratio
压力黏[粘]度计 pressure viscosimeter
压力黏[粘]度特性 pressure viscosity characteristics
压力黏[粘]度指数 pressure viscosity exponent
压力黏[粘]结 pressure bonding
压力排水 pressure drainage
压力排水系统 pressure drainage system
压力盘簧 helical compression spring
压力抛光 press polish; pressure polishing
压力抛物线分布 parabolic(al) distribution of pressure
压力泡 <摩擦桩或桩群周围的> bulb of pressure; pressure bulb
压力喷布 pressure distribution
压力喷浆 gunite; guniting; gunning; pneumatic mortar applying; pressure mortar
压力喷浆法衬砌 gunite lining
压力喷浆机 gunite machine
压力喷浆面层 gunite covering
压力喷洒 pressure feed; pressure spray
压力喷洒机 pressure distributor; tank pressure distributor
压力喷洒器 pressure sprayer
压力喷洒蛭石(轻骨料) pressure-gun type vermiculite
压力喷射 pressure injection
压力喷射沥青的程序 bitumen gunite process
压力喷射沥青的设备 bitumen gunite equipment
压力喷射腻子 pressure-gun type putty
压力喷射枪 pressure gun
压力喷射燃烧器 pressure jet burner
压力喷射式燃烧器 gun-type burner
压力喷射式燃烧装置 pressure injection burner
压力喷射雾化器 pressure jet atomizer
压力喷射型玛琋脂 pressure-gun type mastic
压力喷射油灰 pressure-gun type putty
压力喷射造桩基法 pressure-injected footing
压力喷水 pressure spray
压力喷水系统 pressure water-spraying system; pressurizing water system
压力喷涂 air spraying
压力喷涂机 pressure sprayer
压力喷雾 press(ure) atomization; pressure spray
压力喷雾器 compression sprayer
压力喷雾燃烧器 pressure atomizing burner
压力喷雾燃烧室 pressure atomizing burner
压力喷雾式加湿器 pressure spray type humidifier
压力喷雾式燃烧器 pressure spray type burner
压力喷雾箱 pressure spray tank
压力喷洗 pressure spray
压力喷油燃烧器 pressure burner; pressure jet burner; pressure oil burner
压力喷注 pressure injection
压力喷注法 pressure injection(method)
压力喷注桩基法 pressure-injected footing
压力喷嘴 delivery cone; drive nozzle; pressure jet apparatus; pressure nozzle
压力喷嘴燃烧器 pressure jet burner
压力膨胀接头 pressure expanded joint
压力膨胀曲线 pressure expansion curve
压力片理 pressure schistosity
压力偏离额定值 pressure excursion
压力偏置负载 pressure bias load
压力偏置荷载 pressure bias load
压力偏转 pressure deflection
压力漂淋 pressure-rinse
压力平方差 quadratic pressure drop
压力平衡 equalization of pressure; pressure balance; pressure equalization; pressure equalizing; pressure equilibrium

压力平衡槽 relief groove
压力平衡常数 pressure equilibrium constant
压力平衡的燃料元件 pressure equalized fuel element
压力平衡阀 equilibrated valve; pressure balance valve
压力平衡管 equalizing pipe; pressure balance line; pressure equalizing pipe; pressure equalizing tube
压力平衡管道 pressure equalizing passageway
压力平衡器 pressure balancer; pressure equalizer
压力平衡装置 pressure balance adapter; pressure balancing device
压力评定 pressure evaluation
压力瓶装水泥 bottled cement
压力坡降 pressure gradient
压力破坏 compression failure; fail in compression
压力剖面 pressure traverse
压力曝气装置 pressure-type aeration device
压力启动器 pressure starter
压力起伏 pressure fluctuation
压力气氛 pressure atmosphere
压力气化 pressure gasification
压力气化煤气 pressurized gas
压力气化器 pressure gasifier
压力气流 artificial draught; blast draft; forced draft
压力气流式输送器 pressure pneumatic conveyer [conveyor]
压力气体瓶 pressure gas vessel
压力器 pressure ga(u)ge
压力前池 intake basin
压力枪 pressure gun
压力枪喷洒型墙外粉刷 pressure-gun type exterior plaster(ing)
压力枪喷射型沥青 pressure-gun type asphalt
压力枪喷射型软木塞 pressure-gun type cork
压力枪喷射型水泥浆 pressure-gun type cement
压力枪(注入的)润滑脂 pressure-gun grease
压力腔 pressure chamber
压力强度 compressive strength; intensity of compression; intensity of pressure; pressure intensity; pressure strength
压力强度测定仪 plunger tester
压力清洗装置 pressure washer
压力清洗作业 pressure washing operation
压力穹隆 pressure dome
压力球囊浮扬打捞法 salvage by pressurized-sphere-injector method
压力区 compression zone; pressure area; pressure zone
压力区钢筋 reinforcement in compression
压力区域 pressure span
压力曲线 pressure curve
压力曲线图 pressure-graph
压力驱动不稳定性 pressure driven instability
压力驱动的报警开关 pressure-actuated alarm switch
压力驱动密封垫圈 pressure-actuated seal
压力驱动膜分离技术 pressure driven membrane separation technology
压力屈服点 compression yield point
压力取芯工具 pressured coring equipment
压力去除 <从容器、管线中用压力去除液体或固体> blowdown
压力燃料供给 forced fuel feed
压力燃料进给 forced fuel feed
压力燃料箱 pressure-feed tank
压力燃烧器 pressure burner
压力扰动 pressure disturbance
压力热风供暖 forced air heating
压力热水供暖 hot-water pressure heating
压力热致发光 piezothermoluminescence
压力容积关系 pressure-volume relationship
压力容积图 pressure-volume diagram
压力容积温度关系(曲线) pressure-volume-temperature-relation
压力容量图 pressure-volume diagram
压力容器 pressure container; pressure reservoir; pressure tank; pressure vessel; pressurized reserve
压力容器安全监察规程 safety supervision regulation of pressure vessel
压力容器规范 pressure vessel code
压力容器级钢板 flange quality plate
压力容器结构 construction of pressure vessel
压力容器结构规范 pressure vessel construction code
压力容器可达性 pressure vessels accessibility
压力容器密封 seal(ing) for pressure vessels
压力容器上的安全阀 pop valve
压力容器式(反应)堆 pressure vessel reactor
压力容器塔式反应器 pressure vessel tower reactor
压力容器用钢 steels for pressure vessel use
压力容器制造厂 pressure vessel manufacturer
压力溶化 pressure fusing
压力熔焊 pressure thermit(e) welding
压力入口 pressure port
压力软管 forcing hose; pressure hose
压力润滑 forced lubrication; pressure-feed lubrication; pressure lubricating
压力润滑泵 forced lubrication pump
压力润滑的 pressure-lubricated
压力润滑法 force-feed lubrication; pressure lubrication
压力润滑器 force-feed lubricator; pressure lubricator
压力润滑系统 force-feed lubrication system; pressure lubrication system
压力润滑油泵 oil-pressure lubrication pump
压力润滑脂 pressure grease
压力润滑作用 pressure lubrication
压力洒布机 pressure distributor
压力三角形 triangle of maximum pressure
压力扫描激光器 pressure-scanned laser
压力砂浆 pneumatic mortar
压力砂滤池 pressure sand filter; pressure-type filter; pressure-type sand filter
压力砂滤法 pressure sand filtration method
压力砂滤器 pressure sand filter
压力筛 pressure strainer; pressurized screen
压力上胶 pressure-glued
压力上升 overpressure; pressure build-up
压力上升速度 rate of pressure rise
压力上网 pressure approach onto wire
压力上限 upper pressure limit
压力设定 pressure setting
压力射流 pressure jet
压力射油燃烧器 pressure jet (oil) burner
压力渗析法 piezodialysis; pressure dialysis
压力升(高) build-up of pressure; pressure lifting; pressure rise; pressurizing; pressure build-up; pressure increase
压力升高比 rate of explosion
压力升高(速)率 rate of pressure rise
压力施加法 pressure of application
压力时间(测流)法 pressure-time method
压力时间关系曲线 pressure-time history
压力时间(曲线)图 pressure-time diagram
压力时间示功图 pressure-time indicator diagram
压力式比例混合器 pressure proportioner; pressure side proportioner
压力式波高计 pressure-type wave meter
压力式布滤集尘器 pressure type
压力式潮位计 pressure-operated tide gauge
压力式车辆检数器 pressure detector
压力式储存筒 pressure-type storage cylinder
压力式地音计 pressure geophone
压力式防水注油器 pressure-type waterproof lubricating device
压力式风速计 anemobiagraph; pressure anemometer
压力式过滤器 pressure filter; upflow filter
压力式过鱼闸 pressure fishlock
压力式涵洞 outlet submerged culvert; pressure culvert
压力式缓行器 pressure-type retarder
压力式静电加速器 pressure-type electrostatic(al) accelerator
压力式空气冷却器 pressure-type air cooler
压力式流量计 pressure-type backwater system; pressure-type flowmeter; pressure-type return system
压力式滤池 pressure filter
压力式滤清器 upflow filter
压力式排气器 pressure bleeder
压力式起阀器 pressure-type valve lifter; compression-type valve lifter
压力式取样筒 pressure-type core barrel
压力式热水供暖 forced hot water heating
压力式射流喷嘴 pressure jet nozzle
压力式生物接触氧化法 pressurized bio-contact oxidation process
压力式水加热器 pressure-type water heater
压力式水力旋流分离器 pressured hydrocyclone separator
压力式水位计 pressure-type ga(u)ge
压力式水箱 pressure tank
压力式水银温度计 pressure-type mercury-filled thermometer
压力式通风 force-in ventilation
压力式温度计 capillary thermometer; piezometer type thermometer; pressure(-type) thermometer
压力式消防栓 compression-type hydrant; pressure-type hydrant
压力式消火栓 compression-type hydrant; pressure-type hydrant
压力式验潮仪 pressure-type tide ga(u)ge
压力式液肥洒布机 pressurized spreader
压力式饮用水冷却器 pressure-type drinking water cooler
压力式蒸煮器 pressure cooker
压力式支持器 pressure-type holder
压力式装药器 pressure-type apparatus
压力式钻芯采样机 pressure corer
压力式钻芯取样机 pressure corer
压力势 pressure potential
压力试验 compression test; pressure test(ing); tightness test
压力试验表 test pressure ga(u)ge
压力试验机 compression test(ing) machine; pressure test(ing) machine
压力试验机校准仪 calibrator for compressive testing machine
压力试验机座板 compression plate; pressure plate
压力试验计 pressure test ga(u)ge
压力试验接头 pressure test fittings
压力试验设备 pressure testing device; pressure testing instrument
压力试验证(明)书 certificate of pressure test
压力试验装置 pressure-testing unit
压力室 altitude chamber; balancing gate pit; compression chamber; depression box; pressure chamber; pressure reservoir
压力室压力 ambient pressure; cell pressure
压力释放 pressure discharge; release of pressure; pressure release
压力释放阀 pressure-release valve
压力释压阀 pressure-relief valve
压力输泥管 pressure sludge pipe
压力输水道 pressure aqueduct
压力输水干管 water pressure main
压力输水管 water pressure pipe [piping]; water pressure tube [tubing]; penstock
压力输水管线 water pressure line
压力输水隧洞 hydro tube; hydrotunnel
压力输送 hydraulic feed; positive delivery; pressure feed; pressurization
压力输送管 steel pipe penstock
压力输送管道 conduit under pressure; pressure tubing; head conduit
压力输送汽油 pressure-feed gasoline
压力输送燃料系统 pressure fuel system
压力输送系统 pressure flow system; pressurized feeding system
压力竖井 pressure shaft
压力衰减 pressure decline
压力栓 pressure lock
压力水 compressed water; power water; pressure water; pressurized water; press water
压力水泵 force(d) pump; forcing pump; pressure pump
压力水槽 head tank
压力水池 head reservoir; pressure pond; pressure pool
压力水出口阀 outlet valve
压力水道 pressure conduit; pressure tunnel
压力水缸 cylinder
压力水缸固定板 cylinder setting plate
压力水管 force pipe; forcing pipe; forcing tube; head conduit; pressure conduit; pressure water pipe; pressure water piping; pressure water tube; pressure water tubing; steel pipe penstock
压力水管管节 penstock course

压力水管管座 penstock footing
压力水管路 pressure water (pipe) line
压力水管时间常数 time constant of water passage
压力水管中压力波反射时间 reflection time of penstock
压力水管中压力波往复时间 reflection time of penstock
压力水进口阀 inlet valve
压力水冷反应堆 pressurized-water reactor
压力水面 pressure water surface
压力水幕 pressure shower screen
压力水喷射器 pressure water jet
压力水射流 jet of pressure water
压力水头 pressure head; working pressure head
压力水位计 pressure fluviograph
压力水箱 compression tank; elevated tank; head tank; pressure box; pressure head tank; pressure reservoir
压力水箱冲洗大便器 pressure tank water closet
压力水箱供水方式 pressure tank water supplying
压力水闸 pressure sluice
压力顺序阀 priority valve
压力顺序控制阀 pressure sequence (-controlled) valve
压力松弛试验 compression relaxation experiment
压力送风 forced draft
压力送风机 forced draft fan; forced draught fan; forced fan
压力送风井 forced draft shaft
压力送料 force(d) feed; pressure feed
压力送料机 gravity head feeder
压力送料润滑 pressure-feed lubrication
压力送水总管 water pressure main
压力速度级复合式汽轮机 pressure velocity compounded turbine
压力速度联合分级法 pressure and velocity stages
压力酸处理 pressure acidizing
压力酸化 pressure acidizing
压力隧道 hydrotunnel; power tunnel; pressure tunnel; pressurized tunnel
压力隧洞 hydrotunnel; power tunnel; pressure tunnel
压力损耗 loss of pressure; pressure loss
压力损耗系数 pressure loss coefficient; pressure loss factor
压力损失 draught loss; loss of head; loss of pressure; loss of pressure head; pressure drop; pressure loss
压力损失动作的雨淋阀 pressure-loss operated deluge valve
压力损失换算表 pressure loss conversion chart
压力损失系数 pressure drop coefficient; pressure loss coefficient; pressure loss factor
压力锁结格栅 pressure-locked grating
压力弹程曲线 force-pitch curve
压力弹簧 compression spring; pressure spring
压力弹簧隔热垫圈 pressure spring insulating washer
压力弹簧帽 pressure spring cap
压力探测器 pressure detector
压力探头 pressure probe
压力套筒 pressure house; pressure sleeve
压力特性曲线 pressure characteristic
压力梯度 baric gradient; barometric(al) gradient; gradient of pressure; pressure gradient
压力梯度校正因子 baric gradient correction factor; pressure gradient correction factor
压力梯度相似 pressure gradient analogy
压力梯度效应 pressure gradient effect
压力提拉法 pressure balance method
压力体积关系 pressure-volume relation
压力体积曲线 pressure-volume diagram
压力体积曲线图 pressure-volume chart
压力体积温度叠加 pressure-volume-temperature superposition
压力体积温度关系(曲线) pressure-volume-temperature-relation
压力填缝枪 pressure gun
压力填塞(接)缝 compression joint
压力调定螺塞 pressure-setting plug
压力调节 pressure adjustment; pressure control; pressure governing; pressure regulation; pressure setting
压力调节的薄膜 pressure-controlled diaphragm
压力调节阀 distribution valve; pressure-controlled valve; pressure regulating valve; pressure regulator valve; pressure-relief valve
压力调节罐 <管道系统的> surge vessel
压力调节螺丝 pressure-adjusting screw
压力调节器 boost controller; differential pressure controller; pressure controller; pressure governor; pressure regulator; pressure stat
压力调整 pressure adjustment; pressure setting; regulation of pressure
压力调整阀 pressure modulation valve
压力调整螺钉 pressure-adjusting screw
压力调整器 manostat; pressure regulator; pressure scheduler; relief damper
压力调整弹簧 pressure-adjusting spring
压力调整叶片 pressure-adjusting blade
压力调制辐射仪 pressure modulated radiometer
压力调制室 pressure-defined chamber
压力通风 forced air ventilation; forced draft; forced draught; forced ventilation; plenum ventilation; positive ventilating; positive ventilation; pressure ventilation
压力通风法 plenum method of ventilation; plenum ventilation method
压力通风风扇 forced draft fan; forced fan
压力通风机 compressor fan; forced draft blower; pressure fan; pusher-type fan
压力通风集尘器 forced draught filter
压力通风冷却塔 forced draught cooling tower
压力通风炉 forced draught furnace
压力通风扇 force-draft fan
压力通风系统 forced air ventilating system; plenum [复 plenums/plana] system
压力通气式曝气器 forced draught aerator
压力桶 pressure pot
压力筒 pressure cylinder
压力头 head pressure; pressure head
压力投配 pressure dosing
压力透镜体 pressure lenses
压力凸缘 pressure flange
压力突变 jump in pressure; pressure jump
压力突降 explosive decompression; sudden drop in pressure
压力突然波动 rapid pressure fluctuation
压力突跃 pressure discontinuity; pressure jump
压力突增 pressure jump
压力图 pressure chart; pressure diagram
压力图表 pressure schedule
压力团体 pressure group
压力推力 pressure thrust
压力脱水机 water mangle
压力外级冲动式汽轮机 Rateau turbine
压力弯曲方法 pressing bend method
压力维持 pressure maintenance
压力位移变换器 pressure-displacement transducer
压力-位移曲线 pressure-displacement curve
压力位置线 pressure line of position
压力喂料 pressure feed
压力喂料机 pressure feeder
压力温度补偿式阀 pressure-temperature compensated valve
压力温度范围 pressure-temperature region
压力温度系数 pressure-temperature coefficient
压力温度仪表 pressure thermometer
压力稳定 pressure level-off; pressure stability; pressure stabilization
压力稳定的 pressure-stabilized
压力稳定器 manostat
压力稳定上升 steady rise of pressure
压力稳定性 pressure stability
压力涡轮机 pressure turbine
压力污水管 sewage force main
压力误差 pressure error
压力雾化 pressure atomization
压力雾化喷枪 pressure atomizing lance
压力雾化喷嘴 pressure atomized fog jet; pressure jet apparatus
压力雾化器 pressure atomizer
压力雾化油烧嘴 pressure atomizing oil burner
压力吸收塔 pressure absorption tower
压力吸扬器 hydraulic ram
压力系数 coefficient of pressure; pressure coefficient; pressure factor
压力系统 pressure system
压力系统自动调节器 pressure system automatic regulator
压力下产出量 yielding under pressure
压力下出水量 yielding under pressure
压力下降 decline in pressure; drop in pressure; drop of pressure; loss of pressure; pressure breakdown; pressure decay; pressure drop
压力下降速度 rate of pressure drop
压力下进样 sample introduction under pressure
压力下陷 pressure sinkage
压力纤维 fibre in compression
压力线 line of pressure; pressure grade line; pressure line
压力线拱 pressure line arch
压力线拱顶 pressure line vault
压力线斜率 slope of pressure line
压力限度 pressure limitation
压力限位装置 pressure ga(u)ge
压力箱 pressure box; pressure tank
压力箱喷洒机 tank sprayer
压力箱容器 pressure tank container
压力箱式过滤机 pressure tank filter
压力箱式输送机 pressure tank conveyer [conveyor]
压力响应数据 pressure-response data
压力消减 pressure damping
压力消散 pressure dissipation
压力小管 forced pipe; pressure pipe
压力效率 pressure efficiency
压力效应 pressure effect
压力斜面 pressure slope
压力斜向灌浆 advance slope grouting
压力泄放装置 pressure relieving device
压力卸货 forced discharge
压力形成(过程) development of pressure
压力型测波仪 pressure-type wave ga(u)ge
压力型风速仪 pressure-type anemometer
压力型化油器 pressure-type carburetor
压力型自记测波仪 pressure type wave recorder
压力旋涡 pressure eddy
压力眩晕 alternobaric vertigo
压力循环 force-circulation; pressure circulation
压力循环锅炉 forced-flow boiler
压力循环润滑系统 pressure circulation lubricating system
压力循环式热水供暖 hot-water heating with forced circulation
压力循环水 pressure cycle water
压力循环系统 forced circulation system; pressure circulation system
压力循环蒸发器 forced circulation evapo(u)rator
压力迅速下降 flash down
压力压缩率 compression pressure ratio
压力扬吸机 hydraulic ram
压力氧化容器 pressure oxygen vessel
压力遥控系统 remote pressure control circuit
压力叶轮机 pressure turbine
压力液 hydraulic liquid
压力依存性 pressure dependence
压力仪 piezometer orifice; pressure ga(u)ge
压力移动中心 centre of pressure travel
压力异常 pressure anomaly
压力抑制系统 pressure suppression system
压力抑制型安全壳 pressure suppression type containment
压力因次 pressure dimension
压力引水系统 pressure diversion system
压力影 pressure shadow
压力油 oil under pressure; pressure fluid; pressure oil
压力油缸 oil-pressure cylinder
压力油管 pressure oil pipe
压力油管道 pressure oil (pipe) line
压力油管线 pressure oil (pipe) line
压力油罐洒油车 tank pressure distributor
压力油柜 pressure oil tank
压力油浸 pressure creosoting
压力油流量 pump supply oil flow
压力油滤网 hydraulic screen
压力油千斤顶 pressure oil jack
压力油线路 pressure oil line
压力油箱 pressure oil tank
压力油嘴 push-type lubricating fitting
压力与密度关系图 diagram showing the relation between pressure and density
压力元件 pressure element
压力跃变 pressure jump
压力再分布 pressure redistribution

压力增长 overpressure;pressure build-up;pressure rise
压力增长度(比) pressure ratio
压力增长曲线 pressure build-up curve
压力增大 pressure build-up
压力增加 pressure rise
压力增量 pressure excess;pressure increment;pressure override
压力增量比 pressure increment ratio
压力增益 pressure gain
压力闸门 pressure sluice
压力真空表 compound pressure and vacuum ga(u)ge;pressure vacuometer
压力真空呼吸阀 pressure-and-vacuum breather valve
压力枕 Freyssinet jack;pressure cushion
压力振荡 pressure oscillation
压力振动 compression shock
压力振幅 amplitude of pressure
压力蒸发器 pressure evaporator
压力蒸炼(法) pressure treatment
压力蒸馏物 pressure distillate
压力蒸汽锅 autoclave
压力正常 normotension
压力支撑 pressure mounting
压力值 pressure value
压力纸 pressure sensitive paper
压力指示灯 pressure indicator light
压力指示记录器 pressure indicator and recorder
压力指示控制器 pressure indicating controller
压力指示器 pressure indicator
压力致宽 pressure broadening
压力滞后 pressure delay
压力中继器 pressure repeater
压力中心 center [centre] of pressure;center [centre] of thrust;center [centre] of compression;pressure center [centre]
压力中心系数 center of pressure coefficient
压力中心移动 center of pressure shift;shift of pressure center [centre]
压力钟 <即钟形压力空间> pressure bell
压力周期 pressure cycle
压力轴 pressure axis
压力骤增 pressure jump
压力主管(道) pressure main (pipe);force main
压力注浆 pressure casting;pressure grout(ing);slip casting by pressure
压力注入 pressure impregnation;pressure injection
压力注射枪 pressure gun
压力注油器 compression grease cup
压力铸锭 pressure pouring
压力铸造 die casting;press (die) casting;pressure casting;pressure-die-casting
压力铸造法 injection mo(u)lding
压力铸制器 die-casting machine;pressure casting machine
压力转换吸附法 pressure-swing adsorption system
压力转换循环 pressure-swing cycle
压力转移 pressure transfer
压力桩 non-uplift pile
压力装料 pressure feed
压力装配 press fit;force fit
压力装置 pressure apparatus;pressure setting;pressure unit
压力锥 stress cone
压力锥印 <坯块缺陷> pressure cone
压力着色试验 pressure dye test
压力自动补偿系统 self-compensating pressure system
压力自记计 pressure recorder
压力自记器 manograph;pressure-graph;pressure register;recording manometer
压力自记仪 pressure-graph;pressure register
压力自紧力 energized force by pressure
压力自上而下的 superincumbent
压力总管 force main;pressure header;pressure main (pipe)
压力阻力 pressure resistance
压力阻尼 pressure damping
压力组装 forced fit
压力钻 friction(al) jewel;press drill
压力钻进 pressure drilling
压力钻孔 thrust boring
压力钻孔机 thrust borer
压力最大值 pressure peak
压力作动阀 pressure-actuated valve
压力作用 pressure action;pressure effect
压力作用开关 pressure-operated switch
压力作用下加宽 pressure broadening
压链锤 sentinel
压梁 pressure bar;reaction beam
压梁机 beam bender
压料垫 pressure pad
压料辊 nip rolls
压料机 jumper;swage
压料矫正器 swage sharper
压料(锯齿) swage set
压料器 eccentric swage
压料塞 pot plunger
压料式喷枪 pressure-feed type spray gun
压裂 compression fracture;pressure crack;pressure parting;slip crack
压裂泵车 fracturing truck
压裂泵压 pump pressure of fracturing
压裂层段 interval fractured
压裂处理 <地层水力> fracture treatment
压裂缝 pressure-break
压裂构造 cataclastic structure
压裂焦油(沥青) <石油加压热裂后的副产品> pressure tar
压裂设备 fracturing unit
压裂试验 crushing test
压裂纤维 punched fabric
压裂液 fracture fluid;fracturing fluid
压裂液泵排量 pumping rate of fracturing fluid
压裂液罐车 fracturing fluid tank truck
压裂液体 <水力压裂油层的> break-down fluid
压裂液添加剂 fracturing fluid
压裂液总用量 amount of fracturing fluid
压流 baric flow
压流计 pressure flowmeter
压流冷却(法) pressure cooling
压流润滑法 pressure fed lubrication
压流循环润滑法 self-contained lubrication
压垄辊 ridge roller
压滤 filter pressing;pressing;press(ure) filtration
压滤布 filter press cloth
压滤成型 slurry pressing
压滤干化 press drying
压滤机 blotter press;filter press (machine);press(ure) filter
压滤机板 filter press plate
压滤机滤饼 cake of filter-press
压滤机滤脱水 dewatering in filter press
压滤机卸料装置 press dumper
压滤机组 battery of presses
压滤馏分 press(ure) distillate
压滤(泥)饼 filter press cake;press cake
压滤泥浆 press mud
压滤器 press(ure) filter;ram filter
压滤去蜡油 pressed distillate
压滤式电池 filter press cell
压滤脱蜡 press dewaxing
压滤脱水 dewatering in filter press;filter pressing
压滤型电化学反应器 filter press type electrochemical reactor
压滤制 filter pressing
压滤作用 filter pressing;press filtration
压路辊 compaction roller
压路机 bulldozer;compaction roller;compactor roller;pavement roller;road drag;road grader;road level(l)ing machine;road roller;roller;street roller
压路机垫 shutter mat
压路机发动机 shutter engine;shutter motor
压路机滚轮 roller drum;roller wheel
压路机滚筒 compression roll
压路机加油器 shutter oiler
压路机驾驶员 roller man;roller operator
压路机静重 nonvibrating weight
压路机模型 roller pattern
压路机碾轮 road roller wheel
压路机碾压混凝土 roller-compacted concrete
压路机前轮 guide roll
压路机曲柄轴套 shutter handle sleeve
压路机驱动轮 compression roll
压路机拖运车 roller trailer
压路机旋臂 shutter jib
压路机压实 roller compaction
压路机引擎 shutter engine
压路机圆形碾压路线 circular roller path
压路机主动轮 drive roller wheel
压路机组 train of rollers
压路机作业区 roller operating zone;rolling zone
压路碾 road roller
压铝 jewelling
压轮 pinch roll
压轮机 wheel set press
压麻子油清漆 bunghole oil
压脉波 pressure pulse wave
压脉器 tourniquet
压铆机 riveting press;squeezer;squeeze riveter
压铆器 squeezer
压帽坯机 harder
压煤砖机 briquet(te) press
压密 compacting;consolidation
压密百分比 percent compaction
压密变形阶段 compaction deformation stage
压密不足 under-compacted
压密不排水三轴压缩试验 consolidated-undrained triaxial compression test
压密沉降 consolidation settlement
压密沉陷 consolidation settlement
压密度 consolidation ratio
压密封环 junk ring
压密机 baler
压密极限 compaction limit
压密理论 consolidation theory
压密率 compaction rate
压密黏[粘]土 stiff clay
压密器 consolidation press
压密试验 consolidation test
压密注浆 compaction grouting
压敏 pressure-sensitive;voltage-sensitive;voltage-dependent
压敏薄膜 pressure-sensitive film
压敏材料 pressure-sensitive
压敏潮位计 pressure-sensing tide ga(u)ge
压敏传感波浪记录仪 pressure-sensitive wave-recording device
压敏的 pressure-sensitive
压敏电池 bimorph cell
压敏电容器 compression capacitor
压敏电阻 piezoresistance;piezoresistor;pressure-sensing resistor;varistor [varister]
压敏电阻材料 piezoresistive material
压敏电阻传感器 piezoresistive pick-up
压敏电阻器 piezoresistor;voltage-dependent resister [resistor]
压敏换能器 vibrotron
压敏绘图垫板 pressure-sensitive sketch pad
压敏继电器 pressure-sensitive relay
压敏检波器 pressure-sensitive geophone
压敏键盘 pressure-sensitive keyboard
压敏胶 contact adhesive
压敏胶带 dry tape;pressure-sensitive adhesive tape
压敏胶带试验 pressure-sensitive tape test
压敏胶粘剂 contact adhesive;contact-bond adhesive;pressure-sensitive adhesive;dry-bond adhesive
压敏流向测定管 pressure-sensitive direction probe
压敏黏[粘]合 pressure-sensitive adhesion
压敏黏[粘]合剂 contact-bond adhesive;pressure-sensitive cement
压敏黏[粘]结剂 contact-bond adhesive;pressure-sensitive adhesive
压敏染料 pressure sensitive dye
压敏式测定管 pressure-sensitive direction probe
压敏式测针 pressure-sensitive direction probe
压敏式方向测针 pressure-sensitive direction probe
压敏式检数器 pressure-sensitive detector
压敏水听器 pressure-sensitive hydrophone
压敏透明注记 pressure-sensitive transparency lettering
压敏效应 pressure-sensitive effect;voltage-sensitive effect
压敏性黏[粘]合剂 pressure-sensitive adhesive
压敏性黏[粘]结带 pressure-sensitive adhesive tape
压敏元件 pressure cell;pressure-sensing device;pressure-sensitive element;pressure-sensitive pick-up
压敏纸 impact paper
压敏注记 pressure-sensitive adhesive lettering
压敏装置 pressure responsive device
压模 compression mo(u)ld;die cast;mo(u)lding-die;pressed film;pressing die;pressing mo(u)ld(ing);stamper
压模板条 mo(u)lding lath
压模车间 stamping shop
压模衬里 die liner
压模衬片 die lining
压模衬筒 die barrel

压模成型 compression mo(u)lding
压模承台 < 压床的 > press bolster
压模充填比 die fill ratio
压模冲头 plunger
压模淬火 die quenching
压模定位器 mo(u)ld positioner
压模法 die pressing
压模附件 die accessories
压模钢链板 < 板式给料机中的 > formed mild steel pan
压模滑架 die yoke
压模机 mo(u)ld(ing) press
压模机压型机 press mo(u)lding machine
压模铝合金 alumin(i)um die-casting alloy
压模摩擦 die friction
压模内衬 die liner
压模器 swager
压模嵌入件 die insert
压模润滑剂 die lubricant
压模台板 die plate
压模套管 die sleeve
压模印刷 die-press printing
压模元件 die segment
压模制模(法) hubbing
压模铸件 die casting
压模铸造 die casting
压模装料 die-filling
压模装配 die assembly
压模装置 die arrangement;die assembly
压膜 press film
压膜轴承 squeeze film bearing
压能 pressure energy
压泥机 clay press
压泥浆 < 填充混凝土路面下的空隙 > mud jacking
压泥片装置 batting out device
压黏[粘]土机 clay press
压黏[粘]云母块 micalex [mycalex]
压黏[粘]云母石 micalex [mycalex]
压黏[粘]云母石绝缘 micalex [mycalex] insulation
压凝汽油 casing-head gasoline
压扭性断层 compression contortion fault
压扭性断裂 compressive and torsion fracture
压扭性结构面【地】compresso-shear structural plane;structural plane of compresso-shearing origin
压扭性盆地 compresso-shear basin
压排沉排 ballasting and sinking of mattress
压盘 press disc;press plate;pressure disc [disk]
压盘毂 presser-plate hub
压盘套 pressure disk sleeve
压刨机 surfacer;thicknesser;thicknessing machine
压配合 force fit;interference fit;press fit
压配耐磨导套 press-fit wearing bushings
压膨胀 compression swelling
压膨胀水泥 shrinkage-compensated cement
压坯 compact(ing);green compact;pressed compact;pressed green compact;pressed shape
压坯厂 briquetting plant
压坯冲头 compacting punch
压坯机 briquet(te);briquetting machine
压坯密度 green density
压坯强度 green strength
压坯用阳模 hump mo(u)ld
压皮机 bark press
压皮片 skin graft spatula
压片 preform;presser bit;tabletting
压片玻璃 contact glass;cover glass
压片充气锤 inflatable pneumatic platen cushion
压片法 pressed disc method
压片分散法 chip dispersion method
压片机 briquet(te) press;pelletizer;rubber sheeter;sheeter;tabletting machine;tabletting press
压片技术 pressed-disc technique;squash technique
压片轮 presser wheel
压片凸轮 presser cam
压片用阳模 hump mo(u)ld
压平 flatten;gag;ironing;nipping;smashing;squeegee
压平板 platen;locating back;backing;pressure back < 航摄仪的 >
压平表面隆起 knobbling
压平玻璃检查格网 register-glass reseau
压平布 pressure cloth
压平布面方法 prein
压平的 depressed
压平机 flatter;flatting mill;nipping machine;smashing machine
压平精度 accuracy of flatness
压平力 flattening pressure
压平器 flattener;smoothing board
压平铁碾 < 平整沥青路面用 > smoothing iron
压平突出物 knobbing
压坡 counterpoising
压坡度 banking
压坡脚 loading foot slope
压坡棱体平台 supporting berm
压迫 pinch;stress
压破 implosion
压谱级 pressure spectrum level
压气 compressed air;compressive air
压气爆破 air shooting
压气爆破筒 airdox;shell receiver
压气爆破筒采煤系统 airdox system
压气泵 compression pump;pneumatic pump;pressure pump;pusher pump;suction and force pump
压气泵排水 compressed-air pump drainage
压气泵(输)送 pneumatic pumping
压气病 compressed-air sickness
压气波浪计 pneumatic wave recorder
压气舱沉井 pneumatic shaft sinking
压气操纵 air control
压气操作 pneumatic operation
压气槽 pneumatic transport
压气沉箱 compressed-air caisson;pneumatic caisson
压气沉箱基础 compressed-air foundation
压气沉箱基础工程 pneumatic foundation work
压气沉箱掘进工作 pneumatic work
压气沉箱凿井 pneumatic shaft sinking
压气沉箱凿井法 pneumatic method of sinking
压气冲洗式钻进 air flush drilling
压气抽水 pneumatic pumping
压气抽水泵 air-powered pump
压气抽运系统 pressurized gas pumping system
压气传动 air drive;compressed-air drive;pneumatic transmission
压气传送器 pneumatic transmitter
压气吹刷 air flushing
压气吹洗 air flushing;pneumatic cleaning
压气锤 pneumatic power hammer
压气打桩 pneumatic piling
压气电灯 pneumatic-electric(al) lamp
压气电缆 gas compression cable
压气垫座 pneumatic pillow
压气顶把 pneumatic hold-on
压气锻钎机 compressed-air dressing machine
压气发动机 air motor;air-pressure engine;compressed-air engine
压气法 compressed-air method
压气翻路机 pneumatic scarifier
压气反洗 backing blowing
压气缸 air cylinder
压气钢制沉箱 pneumatic steel caisson
压气供水(设备) pneumatic water supply
压气供水系统 pneumatic water supply system
压气供应 air-feed(ing);compressed-air supply
压气管 air manifold;air pipe
压气管道 air-pressure duct;compressed-air duct
压气灌浆 pneumatic injection
压气灌筑 < 混凝土 > pneumatic placing
压气夯 pneumatic ram
压气缓冲室 dead cushion space
压气活塞 air piston
压气机 air blower;air compressor;air pump;blower;compression pump;compressor;compressor machine;gas booster;gas compressor
压气机扳钮开关 compressor toggle switch
压气机厂 air compressor plant
压气机出口 blower outlet
压气机出口温度 compressor delivery temperature
压气机出口压力 compressor outlet pressure
压气机传动装置 compressor drive
压气机(多孔)放气平台 compressor platform
压气机房 compressor plant
压气机放气调节器 compressor bleed governor
压气机功 compressor work
压气机功率 compressor power
压气机鼓轮 compressor drum
压气机机壳 compressor casing
压气机进风装置 compressor air intake device
压气机进口防冰装置 compressor intake anti-icing system
压气机进口压力 compressor intake pressure
压气机进气预旋 compressor inlet pre-whirl
压气机起动电阻 compressor starting resistor
压气机起动联锁 compressor start interlock
压气机气缸 compressor cylinder
压气机汽缸 compressor housing
压气机容量 compressor capacity
压气机失速 compressor stall
压气机特性线图 compressor map
压气机调节器 compressor governor
压气机涡轮配合 compressor-turbine matching
压气机涡轮匹配 compressor-turbine matching
压气机械搅拌式浮选机 mechanical air flo(a)tation machine
压气机叶轮 blower impeller;compressor impeller
压气机叶片 compressor blade
压气机叶栅 compressor blade row;compressor cascade
压气机罩壳 blower casing
压气机真空泵 compressor vacuum pump
压气减震器 pneumatic shock absorber
压气浇筑 pneumatic placing
压气浇筑混凝土 air placed concrete
压气绞车 pneumatic winch
压气搅拌(法) pneumatic stirring
压气搅拌器 air agitator
压气锯 air saw
压气掘岩机 pneumatic excavator
压气掘凿机 pneumatic excavator
压气控制 air control
压气连接(法) pneumatic connection
压气氯化法 aerochlorination
压气落煤 compressed-air blasting
压气密封的 air-tight
压气排水打捞法 method of dewatering with compressed air
压气喷浆机 air injector;pneumatic placer
压气喷漆枪 painting compressor
压气喷漆装置 compressed-air painting apparatus
压气喷砂机 air blaster
压气喷射 air injection;pneumatic fed < 混凝土 >
压气喷射器 compressed-air ejector;compressed-air jet;pneumatic ejector;Shone ejector
压气喷涂法 air spray finishing
压气喷雾 atomization by pressure air;atomized by compressed air
压气喷雾器 atomizer by compressed air
压气曝气 pneumatic aeration
压气起动器 air-starter
压气千斤顶 pneumatic jack
压气曲线 compression curve
压气驱动泵 compressed-air drive pump
压气驱动的 air-driver
压气入口 pressure port
压气设备 compressor;gas booster
压气设备折旧摊销及大修费 compressor equipment depreciation apportion and overhaul charges
压气升降机 pneumatic elevator
压气升液器 montejus
压气施工法 pneumatic excavation;pneumatic method
压气式备用密封 inflatable stand-by seal
压气式潮汐发生器 pneumatic tide generator
压气式沉降仪 pneumatic settlement cell
压气式防波堤 pneumatic barrier;pneumatic breakwater
压气式夯具 compressed-air tamper
压气式检修密封 inflatable stand-by seal
压气式均衡风缸 < 机车的 > pneumatic surge chamber
压气式路面破碎机 pneumatic paving breaker
压气式密封 inflatable seal
压气式取样器 compressed-air sampler
压气式调压室 compressed-air surge chamber
压气试验 pneumatic test(ing)
压气试验器 compressometer
压气室 compression chamber;plenum chamber
压气输送 pneumatic transport
压气输送机 pneumatic conveyer [conveyor]
压气输送距离 transportation distance of compressed air
压气输送器 air-slide conveyer [con-

veyor]
压气输送水泥混合室 turbulence chamber
压气输送系统 pneumatic conveying system
压气竖井 pressurized shaft
压气水位计 pneumatic level ga(u)ge;pneumatic water ga(u)ge
压气提升法挖泥 airlift excavating
压气提升气罐顶 lifter roof
压气提升油罐顶 lifter roof
压气提水桶 pneumatic water barrel
压气调节器 compressor governor
压气通风 ventilation by forced draft
压气推进 air-feed(ing)
压气推进伸缩式凿岩机 air feed stoper
压气推进式凿岩机 air-feed drill
压气推进式钻机 air-feed drill
压气脱水调节阀 air regulating valve for dewatering
压气挖掘机 pneumatic excavation
压气污水泵 pneumatic sewage ejector
压气系统 pneumatic system
压气箱 pneumatic tank
压气行程 compression stroke
压气修整机 pneumatic finisher
压气修整器 pneumatic finisher
压气叶轮 compressor impeller
压气油漆喷涂器 compressed-air painting apparatus
压气运送设备 <混凝土在管道中的> pneumatic transport placer
压气凿 pneumatic chisel
压气凿孔机 pneumatic perforator
压气站 compressor station
压气轴承 compressed-air bearing
压气助力器 air-booster
压气总管 air manifold
压气钻井 air bar
压气钻孔 pneumatic drilling
压气钻芯取样器 pneumatic core sampler
压器 depressor
压铅 ballasting lead weight;diver's lead weight
压嵌组件 press fit package
压强 <单位面积上的压力> pressure intensity; intensity of compression; intensity of pressure; pressure; pressure per unit area; pressure strength; strength of pressure
压强测量仪 pressure-measuring set
压强差 pressure difference
压强传声器 pressure-operated microphone
压强单位 pressure unit; unit of pressure
压强耳机 pressure headphone
压强共振 pressure resonance
压强计 compression manometer; piezometer; pressure ga(u)ge; tens(i)ometer
压强器 pressure ga(u)ge
压强深度 pressure depth
压强式传声器 pressure microphone
压强水听器 pressure hydrophone
压强水头差降 drop in pressure head
压强梯度 baric gradient
压强中心 center [centre] of pressure; pressure center [centre]
压墙木 spur beam
压切点 point of attack
压切掘进机 <一种软地层隧道掘进机械> pressure face machine
压切(岩石) attack
压青 green manuring; ploughing in green
压球法 ball method
压球机 pelletizer
压区挡板 nip guard
压曲 buckling
压曲安全 buckling safety
压曲板 buckle(d) plate
压曲标准 buckling criterion
压曲不稳定性 buckling instability
压曲长度 buckling length
压曲幅度 buckling amplitude
压曲钢板 buckled steel plate
压曲高度 buckling height
压曲公式 buckling formula
压曲拐点 buckling point
压曲荷载 buckling load(ing)
压曲荷载板 buckling load plate
压曲荷载强度 buckling load strength
压曲计算长度 buckling effective length
压曲劲度 buckling stiffness
压曲铠装包装 buckling reinforcement
压曲抗(阻)力 buckling resistance
压曲理论 buckling theory
压曲力 buckling force
压曲临界荷载 buckling load(ing); crippling load; critical buckling load
压曲临界值 buckling value
压曲模量 buckling modulus
压曲模数 buckling modulus
压曲破坏 blow-up; breaking-in bulking; buckling breaking; buckling failure; buckling rupture; failure by buckling; failure in buckling
压曲破裂 buckling rupture
压曲强度 buckling strength
压曲强度分析 buckling analysis
压曲强度极限 buckling limit
压曲强度计算 calculation of the buckling strength
压曲情况 buckling case
压曲试验 buckling investigation; buckling test
压曲撕裂 buckling breaking
压曲特性 buckling behavio(u)r
压曲条件 buckling condition
压曲图形 buckling configuration
压曲危险 buckling risk
压曲稳定性 buckling safety; buckling stability
压曲问题 buckling problem
压曲系数 buckling coefficient; buckling factor
压曲限度 buckling limit
压曲效应 buckling effect
压曲形式 buckling mode
压曲形状 buckling configuration
压曲压力 buckling pressure
压曲应变 buckling strain
压曲应力 buckling action; buckling stress
压曲值 buckling value
压曲柱形护舷 buckling column fender
压曲状态 buckling condition
压曲阻力 buckling resistance
压曲作用 buckling action
压屈 buckle; buckling; lateral deflection
压屈变形 buckling deformation; buckling strain
压屈荷载 buckling load; crippling load
压屈抗力 buckling resistance
压屈模式 buckling mode
压屈破坏 failure by buckling
压屈强度 buckling strength
压屈构件 buckling member
压屈稳定性 buckling stability
压屈系数 buckling coefficient
压屈阻力 resistance to crippling
压屈作用 buckling action
压圈 clamping ring; junk ring; pressing ring; snap ring
压燃式(柴油)发动机 compression ignition engine
压燃式柴油机 full diesel
压燃式发动机燃料 pressure combustion motor fuel
压染机 padding machine
压染试验 pressure dye test
压热的 piezocalorie
压热硫化锅 autoclave press
压热器 autoclave
压热器法 autoclave process
压热试验 autoclave test
压热系数 piezocaloric coefficient
压热效应 piezocaloric effect
压刃板 knife bracket
压容图 pressure-volume chart; pressure-volume diagram
压容应变计 piezoelectric(al) condenser strain ga(u)ge
压溶 pressolution; pressure solution
压溶劈理 pressure-solution cleavage
压溶作用 pressure solution
压入 embedment; force on; forcing; pressing in
压入暗销 press-in connector; spike dowel
压入冲洗法 jacking jetting process
压入的 bulged in
压入端 push head
压入法 indentation method; pressing in method <测定硬度时采用的>; jacking method; plunging
压入法黏[粘]度计 penetration viscometer
压入法取样 drive sample; drive sampling
压入法取样器 drive sampler
压入法施工垫板 jacking plate
压入法通风 blowing method of ventilation
压入接头 <木材联结器> press-in connector
压入抗力 penetration resistance
压入料 injection mix
压入螺钉 drive screw
压入能力 embedability
压入泥浆的路面 slurry-mud jacked pavement
压入配合 force fit; forcing fit; hand fit; press fit
压入配合表壳 press fit case
压入配合试验 press fit test
压入配合余量 crush allowance
压入润滑(法) forced feed lubrication
压入石屑 rolled-in (stone) chip(ping)-s
压入式 pressure type; shrunk in
压入式打桩 pneumatic piling
压入式骨料处治法 <路面防滑、防软等> rolled-in treatment
压入式灌注桩 injection grout pile
压入式轨缝连接器 pressed type bond
压入式轨隙连接器 pressed type bond
压入式加油嘴 drive-type grease fitting
压入式拉杆 push-through tie
压入式连杆 push-through tie
压入式模 push-through die
压入式旁压仪 push-in pressuremeter
压入式取土器 pressed sampler
压入式取样器 pressed sampler
压入式 blowing system
压入式通风 blowing system of ventilation; blowing ventilation; forcing-in ventilation
压入式通风管 blowing ventilation pipe
压入式通风系统 pressure system of ventilation
压入式阳模 male tool
压入式桩 jacked(-in) pile
压入试验 indentation test; push-in test
压入水量 volume of the water injected
压入榫(钉) press-in connector; press-in dowel; spike dowel
压入头 bell jar; indenter
压入心轴 press mandrel
压入型 press fit type
压入性 embedability
压入盐水 intake brine
压入逸出式(通风) louver system; punka(h)
压入桩 jacked pile; displacement pile; pressed pile; pressure(-in) pile
压入桩托换 jacked pile underpinning
压入装料机 force-feed loader
压入阻力 penetration resistance
压塞 injection piston
压塞机 corker
压纱板 schlagblech
压砂 sand flow
压砂法 <沉管隧道用> sand flow method
压上 forcing on
压上的盖 press-on cap
压舌板 tongue depressor
压射成型 injection mo(u)lding
压射冲程 shot stroke
压射(汽)缸 injection cylinder
压射室 injection chamber; shot chamber
压射压力 injection pressure
压射周期 shot cycle
压伸器 compander [compandor]; compressor-expander
压深率 indentation ratio
压渗膜 piezodialysis membrane
压渗土料 soil for landside seepage berm
压升 voltage rise
压升高度 delivery lift
压绳盘 retainer disk [disc]
压绳器 rope guard
压剩余磁化 piezoremanent magnetization; pressure remanent magnetization
压石机 stone press
压实 compact(ing); compaction; compaction by compression; compression; ramming; rolling
压实百分数 percent compaction
压实板 squeeze board
压实板木门 high-density wood door
压实背斜 compaction anticline
压实比 compacting ratio; compaction ratio
压实边缘 compact edge
压实遍数 compaction pass; compactor pass
压实不足的 poorly compacted
压实步骤 compaction process
压实部分 consolidation section
压实材 staypack
压实层 compacted layer; compacted lift
压实层厚度 compacted thickness; thickness of compacted layer
压实层内密度 in-place density
压实差的回填土 poorly compacted backfill
压实厂房设施 compaction plant
压实程度 compaction rate; deep of compaction
压实稠化 densification
压实次数 compactor pass
压实粗砂 compacted coarse sand
压实导向装置 compaction guidance equipment

压实的 compacted; consolidated; rolled
压实的材料 pressed laminated wood
压实的积雪 hard-packed snow
压实的透水性填土 compacted pervious fill
压实的雪 hard(-packed) snow
压实的雪跑道 packed snow runway
压实底土 compacted subgrade
压实(底)土衬砌 compacted earth lining
压实底土基层 compacted subsoil base
压实度 amount of compaction; compact(ed)ness; degree of compaction; degree of compression; ratio of compression
压实度控制法 compaction control method
压实度试验 compactness test
压实度调节器 compression regulator
压实度自动检测装置 automatic compactometer
压实堆石 compacted rockfill; rolled rockfill
压实阀 squeeze valve
压实方 compacted measure
压实方法 compacted method; compaction process; method of compaction
压实防渗填料 compacted impervious fill
压实废物 milled refuse
压实分层厚度 compacted lift
压实风化砂 compacted weathered sand
压实赶光 float finish
压实缸 squeezing cylinder
压实高度 <弹簧> solid height
压实工程 compaction work
压实工具 compactor
压实工艺 compaction technology
压实工作 compacting work; compaction work
压实功 compacting effort; compactive effort
压实功能 compactive effort
压实光面机 compacting and finishing machine
压实辊 compaction roller; compression roll
压实过程 compaction process; process of compaction
压实很差 poorly compacted
压实后的方量 compacted cubic yard
压实后立方码 <土料的> fill yard
压实厚度 compacted depth; compaction depth; finished thickness
压实回填 rolled backfill
压实回填料 inbed
压实回填土 compacted backfill
压实混凝土 compacted concrete; compressed concrete; stamped concrete
压实混凝土桩 compacted concrete pile
压实活塞 squeeze piston
压实机 compacting machine; compacting machining; compaction wheel
压实机本体 basic compactor
压实机具 compacting equipment; compaction equipment; compactor
压实机附属机具 compactor attachment
压实机械 compacting machinery; compacting plant; compaction equipment; compaction machinery; compactor
压实机械试验 compaction machine test
压实机械种类 compaction machine type
压实机主机 basic compactor
压实基层 rolled base
压实计 compactometer
压实记录仪 compaction recorder
压实技术 compaction technique; compaction technology
压实技术标准 compaction specification
压实加强 consolidation
压实胶合板 densified plywood
压实焦渣填层 compacted cinder fill
压实脚扳宽度 pad width
压实宽度 compacting width; overall rolling width
压实宽度测定 compacting width measurement; compaction width measurement
压实类型和分带 compaction type and zone
压实力 compacted force; compactive effort; compactive force
压实立方码数 <土的> compacted yard
压实梁 tamping beam
压实路面 compacted surface
压实率 compaction rate; specific compaction
压实轮齿尖 compactor tips
压实密度 compacted density; compactness
压实模型 compaction mo(u)ld
压实木材 compressed wood; high-density wood
压实能 compaction energy
压实能力 compaction capacity
压实能量 compacting [compaction/compactive] energy
压实黏[粘]土 compacted clay
压实黏[粘]性土 compacted cohesive soil
压实凝结 pack set
压实平面 compaction plane
压实器 compactor
压实强度 compaction strength
压实曲线 compaction curve; densification curve
压实容重 compaction unit weight; compaction weight; compression bulk density
压实砂和砾 compacted sands and gravels
压实砂土 compacted sand
压实砂桩 compaction sand pile; sand compaction pile
压实砂桩法 <改善地基用> sand compaction pile method
压实设备 compacted equipment; compacting equipment; compacting plant; compaction equipment
压实深度 compacted depth; compacted thickness; compaction depth; depth of compaction
压实深度测定 compaction depth measurement
压实石块填方 compacted rockfill
压实式垃圾车 compactor collection vehicle
压实式造型机 squeeze molding machine
压实试验 compacting test; densification test; compaction test
压实水平 level of compaction
压实松密度 compression bulk density
压实速度 compacting speed
压实速度测定 rolling speed measurement
压实速率 compaction rate
压实体积 compacted volume
压实天然沥青铺面 compressed rock asphalt surfacing
压实填石 compacted rockfill
压实填土 compacted (earth) fill; rolled earth fill
压实通道 compacting pass
压实土 compacted earth; compacted soil; hard compact soil
压实土方量 compacted yard
压实土基础 compacted soil foundation
压实土壤(法) spotting
压实(土壤时挤出的)水 water of compaction
压实土桩 compacted earth pile
压实土状态 compacted soil condition
压实温度 compaction temperature
压实稳定 stabilization by compaction
压实无机质砂与粉砂混合物 compact inorganic sand and silt mixture
压实系数 coefficient of compaction; coefficient of consolidation; compaction coefficient; compaction factor; compacting factor
压实系数试验 compacting factor test
压实系数试验仪 compacting factor apparatus
压实细砂 compacted fine sand
压实纤维墙面板 tileboard
压实效果 compacting effect; compaction effect; compactive effort; effectiveness of compaction
压实效率 compaction efficiency
压实效应 compactive effort
压实(型)路堤 compact embankment
压实性 compact(ed)ness; compactibility
压实修整机 compacting and finishing machine; compacting and finishing machining
压实雪层 snowpack
压实雪地机场 compacted snow-field
压实压力 compaction pressure
压实轧轮 compaction roll
压实仪 compaction device
压实因数 compacting factor
压实引导装置 compaction guidance equipment
压实造型 squeeze mo(u)lding
压实造型机 squeeze mo(u)lding machine
压实褶皱 compaction fold
压实桩 compacting pile; compaction pile
压实桩施工法 compaction pile method
压实作业 compacting
压实作用 compacting action; compaction effect; compactive effort; densification
压书板 pressing board
压栓 holding-down bolt
压水板 air damper
压水泵 suction and force pump
压水槽 pressure tank
压水冲洗 blow wash
压水堆 pressurized-water reactor
压水法 setting out
压水反应堆 pressured-water reactor; pressurized-water reactor
压水辊 squeeze roll
压水核反应堆 pressured-water nuclear reactor; pressurized-water nuclear reactor
压水冷却 forced water cooling
压水喷水机 pressurized-water distributor
压水燃烧法 pressurized aqueous combustion
压水试验 Packer (permeability) test; injection test; hydraulic pressure test; Lugeon test; pressurizing water test; pump(ing)-in test; water pressure test; water pressurizing test
压水试验方法 the method of Lugeon test
压水试验封(闭)隔器 pressurizing water test packer; water pressurizing test packer
压水试验孔 water pressure test hole
压水试验孔数 number of water pressure test hole
压水试验总长度 total length of water pressure test segment
压死的炸药 dead-pressed explosive
压送 forcing; squeeze pump
压送泵 compressing pump; force lift
压送导管 supply conduit
压送管(道) conduit under pressure; discharge pipe; force pipe
压送管路 pressure line; pressure piping(-line)
压送罐容量 pot capacity
压送式风力输送器 pneumatic conveyer with pressure
压送式喷雾机 pressure atomiser
压送式喷雾器 pressure atomizer
压送式输送带 compression web
压送水头 delivery head
压送压力 pressurization pressure
压塑 compact(ion); compression mo(u)lding
压塑机 plasticator
压塑料 compression mo(u)lding compound
压塑性 compactibility
压塑性不良 poor compactibility
压碎 crush(ing); quassation; scrunch; spalling; stamp breaking; stamp crushing
压碎板 crushing pad
压碎变形 crushing strain
压碎变质 rupture metamorphism
压碎粗骨料 crushed coarse aggregate
压碎粗集料 crushed coarse aggregate
压碎大褶皱 big crush fold
压碎带 crushed zone
压碎的 crushed
压碎的混凝土 crushed concrete
压碎法 crushing method
压碎负荷 crushing load
压碎复褶曲 crushing fold
压碎工具 crushing tool
压碎滚筒 crushing roll
压碎含硅石 crushed tripolite
压碎荷载 crushing load
压碎机 bruiser; crusher; crushing machine; crushing mill; pan roll; roll crusher; runner; chip-breaker <用于压实木片、石片的>
压碎机的锤击 throw of the crusher
压碎机用油 crusher oil
压碎角砾结构 crust brecciatic texture
压碎角砾岩 crush breccia
压碎阶段 crushing stage
压碎结构 crush texture; pressure texture
压碎力 crushing force
压碎砾石骨料 crushed gravel fines
压碎砾岩 crush conglomerate; tectonic conglomerate
压碎面 crush plane
压碎破坏 crushing failure
压碎器 crushing apparatus
压碎强度 crushing strength
压碎强度试验 crushing strength test
压碎砂 stamp sand
压碎砂砾 crushed gravel sand
压碎石块骨料 crushed rock fines
压碎试验 crushing test
压碎台 crushing stage

压碎岩（石） crushed rock; rock crushing
压碎样品 crushing sample
压碎应力 crushing stress
压碎站 crushing station
压碎值 crushing strength; crushing value
压碎值试模 crushing strength tester
压碎值试验 crushing strength test
压碎值试验仪 crushing value tester
压碎指标值 crushing index value
压碎装置 crushing device
压损 crushing; damage of pressure; damaged owing to crushing
压损系数 pressure drop coefficient
压缩 boil down; capsule; compress; compression set; constricting; constriction; constringe; contraction
压缩安全阀 compression relief valve
压缩百分率 percentage of consolidation
压缩板 flake board
压缩爆破 compression blasting
压缩爆燃 compression knock
压缩泵 compression pump
压缩比（率） compressibility ratio; compression ratio; compressive ratio; ram ratio; ratio of compression; pressure ratio; degree of compaction; degree of compression; efficiency of compression; fill ratio; compression volume ratio <汽缸等>
压缩边缘 compressed edge
压缩变定 compression set
压缩变形 compression [compressive] deformation; compression set
压缩变形阻力 resistance to compression
压缩表 compaction table
压缩饼干 <救生艇用> hard bread; ship's bread
压缩波 compression(al) wave; compressive wave; longitudinal wave; pressure wave; stress wave; wave compression; waves of compression
压缩不足 insufficient compression
压缩材料 compression material
压缩残渣 pressed residue
压缩草纤维板 compressed straw slab
压缩层 compressed layer; compression layer; compression region; compression stratum; compression zone
压缩层厚度 <地基计算的> compression zone depth
压缩车间 compression plant
压缩沉淀 compressive settling
压缩沉降 compression settling; compressive settling; oedometer settlement; settlement due to compression
压缩沉降阶段 compression settling stage
压缩成本 cost squeeze
压缩程度 compression ratio; intensity of compression
压缩程序 condensing routine; packing routine; reduction program(me)
压缩程序穿孔卡片叠 squeeze pack
压缩冲程【机】 compression stroke
压缩冲击 compression shock
压缩储存【计】 compressed memory
压缩储气罐 air reservoir
压缩处理 compression treatment
压缩穿孔卡片叠 condensed pack; squeeze pack
压缩打包 shrink-wrapping
压缩稻草 pressed straw
压缩的 compact; compressed; compressive; encapsulated

压缩的材料 compressed material
压缩的泥炭（土） compressed peat
压缩点 compression point; point of compression
压缩点火 compressed [compression] ignition
压缩点燃 ignition due to compression
压缩定理 compress theorem
压缩度 amount of compression; compressed limit; degree of compression
压缩度盘 suppressed scale
压缩断裂 compression failure
压缩发火 ignition by compression
压缩阀 compression valve
压缩法 compression method; compression process; condensation method; method by condensation; method of condensation
压缩反差 contrast reducing
压缩范围 reduction range
压缩方式 compress mode
压缩房 compressor house
压缩放大器 compression amplifier; compressor amplifier
压缩非线性 compressive non-linearity
压缩废钢 pressed scrap
压缩废气 blow-by gas
压缩分类表 reduction of schedules
压缩负荷 compression load; compressive load(ing)
压缩复原率 compression recovery rate
压缩改正的调整大地水准面 cogeoid of the condensation reduction
压缩刚度 compressive rigidity
压缩格式 compressed format
压缩功 workdone during compression
压缩构件 compressional member in compression; compression component
压缩固结试验 compression consolidation test
压缩关系曲线 e versus log p curve
压缩管 compressed pipe
压缩管件接头 compression fitting joint
压缩过程 compression process
压缩海绵 compressed sponge
压缩荷载 compression load; compression pressure; compressive load(ing)
压缩厚度 compressed thickness
压缩化学战剂 pressure charge
压缩环 compression ring; piston ring; pressure ring
压缩回弹型护舷 recoiling fender
压缩回弹性 compressive resilience
压缩机 blower; compressing apparatus; compression engine; compression pump; compressor
压缩机本体 compressor body
压缩机部件 compressor component
压缩机车间 compressor plant
压缩机冲程 compressor stroke
压缩机抽水井 gas lift flowing well
压缩机单元 compressor section
压缩机的压缩端 downstream side of the compressor
压缩机端盖 compressor head
压缩机阀 compressor valve
压缩机房 compressor plant; compressor room
压缩机附件 compressor accessories; compressor attachment
压缩机工作容量 compressor displacement
压缩机活塞 compressor piston
压缩机活塞的行程容积 displacement of compressor

压缩机集气室 compressor manifold
压缩机加热器 compressor heater
压缩机架 compressor cradle
压缩机减压装置 compressor relief device
压缩机类型 compressor type
压缩机密封 compressor seal
压缩机排量 compressor displacement
压缩机歧管 compressor manifold
压缩机汽缸 compressor cylinder
压缩机曲轴箱 compressor crankcase
压缩机驱动的工具 air tool
压缩机容量 compressor capacity
压缩机入口 suction port of compressor
压缩机润滑油 compressor luboil
压缩机设计 compressor design
压缩机式液冷机 compressor-type liquid chiller
压缩机式液体冷却器 compressor-type liquid chiller
压缩机输出量 compressor output
压缩机特性曲线 operating envelope of compressor
压缩机调节器 compressor governor
压缩机调速器 compressor governor
压缩机调压器 compressor governor
压缩机透平 compressor turbine
压缩机吸入压力 compressor suction pressure
压缩机效率 compressor efficiency
压缩机效能 compressor performance
压缩机型垃圾车 compactor vehicle
压缩机性能 compressor performance
压缩机压比 compressor pressure ratio
压缩机压力 compressor pressure
压缩机叶轮 compressor impeller
压缩机油 compressor oil
压缩机运行 compressor operation
压缩机站 compressor station
压缩机重 weight of compressor
压缩机转子 compressor drum; compressor rotor
压缩机装置 compression plant; compressor installation; compressor plant
压缩机自动调压器 automatic compressor-adjuster
压缩机组 compressor assembly; compressor set; compressor unit
压缩基建规模 cutting down capital construction scale
压缩激波 compression shock wave
压缩级 compression stage; stage of compression
压缩极限 compressed limit; compression limit; limit of compression
压缩计 <测量压缩变形的仪器> compressometer; piezometer
压缩技术 compress(ion) technique
压缩加料滑脂杯 compression grease cup
压缩加强件 compression reinforcement
压缩加热 compressional heating
压缩加油杯 compression cup
压缩减湿器 compression dehumidifier
压缩减湿装置 compression dehumidifier
压缩检查 compression check
压缩绞线 compact-stranded wire
压缩接头 compression joint
压缩结构 pressure texture
压缩界限 compressed limit
压缩卡片组 condensed deck
压缩开支 cut-down expenses; reduce expenses; retrench
压缩刻度 suppressed scale

压缩空气 compressed air; compression of air; compressive air; heavy air; pressed air; pressure air; pressurized air
压缩空气安全规则 safety rules in compressed air
压缩空气拌和机喷射器 compressed-air mixer-injector
压缩空气爆破 bursting by compressed air
压缩空气泵 air-driven pump; compressed-air pump
压缩空气病 compressed-air illness; compressed-air sickness
压缩空气操纵的配料设备 pneumatically operated batching plant
压缩空气操纵器 compressed-air control
压缩空气插入式振捣器 compressed-air internal vibrator
压缩空气沉箱 air caisson
压缩空气冲井钻进 air drilling
压缩空气冲压机 compressed-air ram
压缩空气除尘 compressed-air cleaning; pneumatic dedusting
压缩空气储存罐 compressed-air reservoir
压缩空气储存器 compressed-air receiver; compressed-air reservoir
压缩空气储罐 air accumulator; air reservoir
压缩空气储气罐 compressed-air reservoir
压缩空气储气器 compressed-air reservoir
压缩空气储蓄器 compressed-air accumulator
压缩空气传动 compressed-air transmission; pneumatic transmission
压缩空气传送 conveyer by compressed air; air slide
压缩空气传送机 compressed-air conveyer [conveyor]
压缩空气吹风管 air lance
压缩空气吹管 compressed air blow pipe
压缩空气吹洗 compressed-air cleaning
压缩空气锤 pneumatic forging hammer; compressed-air hammer
压缩空气淬火 compressed-air quenching
压缩空气存储器 compressed-air accumulator
压缩空气打桩机 compressed-air pile driver; compressed-air pile hammer
压缩空气捣固机 compressed-air tamper
压缩空气低压指示灯 low air pressure light
压缩空气电动机 compressed-air motor
压缩空气电弧气刨 arc air gouging
压缩空气电机 air motor
压缩空气电容器 compressed air capacitor
压缩空气钉箱机 compressed-air nailing machine
压缩空气动力 compressed-air power
压缩空气断路 compressed-air engine
压缩空气断路器 compressed-air circuit breaker
压缩空气盾构开挖 air shield driving
压缩空气盾构推进 pneumatic shield driving
压缩空气多向接头 compressed-air junction manifold
压缩空气发动机 compressed-air engine; pneumatic engine
压缩空气法 pressed air method
压缩空气防止土方坍塌的挖土法 ple-

num method
压缩空气分配器 compressed-air distributor
压缩空气分配总管 air manifold
压缩空气风道 compressed-air tunnel
压缩空气风镐 compressed-air pick
压缩空气辅助的 pneumatically assisted
压缩空气干线 compressed-air main
压缩空气干燥 compressed-air drying
压缩空气高速锤 petro-forge machine
压缩空气给进支腿 air feed leg
压缩空气工程 compressed-air work
压缩空气工具 compressed-air tool
压缩空气工作条件下的气闸 airlock in compressed-air work
压缩空气供给 compressed-air supply; pressure supply
压缩空气供气系统 compressed air supply system
压缩空气供应 air supply; compressed-air supply; pressure supply
压缩空气供应线 compressed-air line
压缩空气鼓风机 compressed air blower
压缩空气管 air-pressure duct; compressed-air hose
压缩空气管道 air line; air pipe; air piping compressed; compressed-air pipe(line); compressed-air (pipe) line; compressed-air piping; pneumatic tube; pneumatic tubing; pressure piping(-line); compressed-air tube
压缩空气管道系统 air tube system
压缩空气管道装置 air tube installation
压缩空气管路 pneumatic line
压缩空气管系 compressed-air piping
压缩空气罐 compressed-air cylinder; compressed-air receiver; compressed-air tank
压缩空气夯 compressed-air ram; compressed-air tamper
压缩空气夯锤 compressed-air ram
压缩空气夯实 compressed-air packing
压缩空气夯实机 compressed-air rammer
压缩空气厚度 compressed-air thickness
压缩空气(滑板)输送机 air-slide conveyer [conveyor]
压缩空气滑车 compressed-air pulley block
压缩空气滑动门驱动(装置) compressed-air sliding door drive
压缩空气或蒸汽发音器 tyf(h)on
压缩空气机 air compressor; compressed-air unit
压缩空气机车 compressed-air locomotive
压缩空气加压机构 pneumatic pressure device
压缩空气浇筑混凝土 concrete placed by compressed air
压缩空气绞车 compressed-air winch
压缩空气搅拌 compressed-air agitation
压缩空气搅拌机 compressed-air mixer
压缩空气进料电动机 compressed-air feed motor
压缩空气进入管 compressed-air inlet pipe
压缩空气掘进 compressed-air drive
压缩空气卡盘 compressed-air chuck
压缩空气开关 compressed air switch
压缩空气开挖隧道法 compressed-air method of tunnel(l)ing
压缩空气开挖隧洞法 compressed-air method of tunnel(l)ing
压缩空气开凿法 compressed-air method
压缩空气开凿隧道法 compressed-air method of tunnel(l)ing
压缩空气控制 compressed-air control; pneumatic control
压缩空气控制的 air-actuated; pneumatically controlled
压缩空气控制的骨料料仓闸门 pneumatically controlled aggregate bin gates
压缩空气控制器 pneumatic controller
压缩空气控制系统 compressed-air control system; pneumatic control system
压缩空气扩张器 compressed-air expander
压缩空气冷却器 after-cooler; compressed-air cooler
压缩空气连接总管 air junction manifold
压缩空气量 air quantity; air supply
压缩空气流量 air delivery
压缩空气马达 compressed-air motor
压缩空气排水 drainage by compressed air
压缩空气喷漆器 air-painting sprayer
压缩空气喷漆器械 compressed-air painting apparatus
压缩空气喷漆设备 air-painting equipment
压缩空气喷枪 air spray pistol
压缩空气喷射器 air sprayer pistol; compressed-air ejector
压缩空气喷涂 compressed-air spraying
压缩空气喷涂器 paint blower
压缩空气喷雾 atomize by compressed air
压缩空气喷雾法 compressed-air atomization
压缩空气喷雾机 compressed-air type sprayer
压缩空气平衡装置 pneumatic counter balance
压缩空气瓶 air bottle; compressed-air bottle; high-pressure air tank; pneumatic cylinder
压缩空气曝气 compressed-air aeration
压缩空气曝气法 air diffusion method
压缩空气起动机 compressed-air starter; air injection starter; pneumatic starter
压缩空气起动器 compressed-air starter
压缩空气起动阀 pneumatic starting valve
压缩空气起动设备 air starting system
压缩空气起动系统 air starting system
压缩空气起动装置 pneumatic starting device
压缩空气起锚绞车 compressed-air anchor winch
压缩空气起重机 compressed-air hoist
压缩空气起重器 compressed-air jack
压缩空气气缸 compressed-air cylinder
压缩空气潜水 compressed-air diving
压缩空气潜水器 compressed-air diver
压缩空气潜水方案 compressed-air diving profile; compressed-air diving schedule
压缩空气潜水方法 compressed-air diver procedure
压缩空气潜水服 compressed-air suit
压缩空气枪 air-pressure gun; compressed-air gun
压缩空气敲锈锤 compressed-air scaling hammer
压缩空气切断 compressed gas cutout
压缩空气清孔【地】 air flushing
压缩空气清洗 compressed-air cleaning
压缩空气驱动 air propelling; air sliding; compressed-air drive; pneumatic drive
压缩空气驱动泵 air-operated pump
压缩空气驱动的 air-actuated; air-powered
压缩空气取土器 compressed-air sampler
压缩空气全起动装置 compressed-air starter
压缩空气容器 compressed-air container; compressed-air vessel
压缩空气入口 compressed-air inlet
压缩空气软管 compressed-air hose; flexible air line
压缩空气润滑器 air lubricator
压缩空气润滑装置 air lubricator
压缩空气塞 compressed-air lock
压缩空气设备 compressed-air equipment; compressed-air installation; pneumatic plant
压缩空气射流 compressed-air jet
压缩空气升降机 compressed-air elevator; pneumatic elevator
压缩空气升液器 pressure-air montejus
压缩空气式拌和机 compressed-air mixer
压缩空气式搅拌机 compressed-air mixer
压缩空气试验 air-pressure test
压缩空气室 delivery air chamber
压缩空气疏松器 compressed-air expander
压缩空气输送 pneumatic conveying
压缩空气输送管 blow pipe; compressed-air piping; pneumatic tube; compressed-air feed pipe
压缩空气输送机 air(-slide) conveyer [conveyor]
压缩空气输送器 air(-slide) conveyer [conveyor]
压缩空气输送系统 compressed-air conveying system
压缩空气水泥运送 pneumatic cement handling
压缩空气水泥装卸 pneumatic cement handling
压缩空气瞬时喷出 blowout of compressed air
压缩空气伺服设备 air-booster
压缩空气伺服装置 air-booster
压缩空气提升 compressed-air lifting
压缩空气提升机 pneumatic hoist
压缩空气提升挖泥船 air lift dredger
压缩空气条件下施工 compressed air working
压缩空气调节器 compressed-air governor; air governor
压缩空气筒 accumulator
压缩空气推动的 air-operated; compressed-air-assisted
压缩空气挖掘设备 air excavation equipment
压缩空气外振动器 compressed-air external vibrator
压缩空气温度 compressed-air temperature
压缩空气吸泥进气管 <地下连续墙> nozzle for compressed air
压缩空气吸入器 compressed-air inspirator
压缩空气系统 compressed-air system
压缩空气下沉 air sinking
压缩空气下的基础工程 foundation work under compressed air
压缩空气箱 pressure accumulator; pressurized container
压缩空气消耗量 compressed-air consumption
压缩空气卸车系统 <用于粒状和粉状货物> air compression unloading system
压缩空气需要量 compressed-air demand; demanding quantity of compressed air
压缩空气压力 compressed-air pressure
压缩空气压力表 air-pressure ga(u)ge; compressed-air ga(u)ge
压缩空气扬水泵 air lift; air lift pump
压缩空气扬水器 air lift
压缩空气遥控 pneumatic remote control
压缩空气油漆机具 compressed-air painting apparatus
压缩空气油漆械具 compressed-air painting apparatus
压缩空气油漆装置 compressed-air painting apparatus
压缩空气凿尖 compressed-air point
压缩空气闸 compressed-air brake; compressed-air lock
压缩空气闸门 pneumatic lock
压缩空气站 air starting station; compressed-air plant; compressed-air station
压缩空气振动器 compressed-air vibrator; pneumatic vibrator
压缩空气支承作用 supporting effect of compressed air
压缩空气制动 pneumatic braking; pressurized air braking
压缩空气制动器 compressed-air brake
压缩空气轴承 compressed-air bearing
压缩空气主管路 compressed-air main
压缩空气助推的 compressed-air-boosted; pneumatically boosted
压缩空气贮箱 ballasted accumulator; compressed-air accumulator
压缩空气桩锤 compressed-air pile hammer
压缩空气桩夹 compressed-air pile chuck
压缩空气装置 air equipment; air plant; blowing plant; compressed-air installation
压缩空气钻进 pressure-air drilling
压缩库存 reduce stocks
压缩扩展 companding
压缩扩展器 compander [compandor]; compressor(-and)-expander
压缩扩张器 compander
压缩垃圾储藏箱 compacted waste container
压缩垃圾箱 compacted waste container
压缩类型 compression type
压缩冷凝机组 condensing unit
压缩力 compressing force; compressure; draught pressure; force of compression; compression stress; compressive force; compressive stress
压缩利润 squeeze on profit
压缩联轴节 compression coupling
压缩量 amount of compression; decrement
压缩列车停站时间 decrease train dwelling time on station
压缩裂缝 pressure fissure
压缩裂隙 compression fissure; compressive fissure
压缩流 compressible flow; compressive flow

压缩流扩大 compressible flow divergence
压缩流量能量 compressible flow energy
压缩流量能量方程 compressible flow energy equation
压缩流收缩 compressible flow convergence
压缩漏气 compression release
压缩率 bulk factor; compressibility; compressibility coefficient; compressibility factor; compression rate; compression ratio; rate of compression; specific compression
压缩码 compressed code
压缩煤气 high-pressure gas; pressurized gas
压缩煤气罐 gas cylinder
压缩煤气瓶 gas cylinder
压缩煤气瓶储藏室 gas cylinder room
压缩煤气装置 compressed gas installation
压缩密度函数 compressed density function
压缩密封垫 compression gasket
压缩密封件 compression seal
压缩密封接头 compression gasket joint
压缩模量 compressibility modulus; compression modulus; compressive modulus; modulus of compressibility; modulus of compression; oedometric modulus
压缩模数 modulus of compression
压缩模塑 compression mo(u)lding
压缩模塑力 compression mo(u)lding pressure
压缩木 lignostone; staypack
压缩木材 compregnated wood; compressed wood; compression wood
压缩能量 compression energy
压缩黏[粘]度 compressional viscosity
压缩黏[粘]土层 compressible clay layer
压缩碾压 breakdown rolling
压缩凝结 compaction
压缩盘【计】compact disc
压缩喷嘴 constricting nozzle
压缩膨胀循环 compression-expansion cycle
压缩疲劳 compression failure; endurance in compression; fatigue in compression
压缩疲劳强度 compressive fatigue strength
压缩疲劳试验 repeated compression test
压缩平行板黏[粘]度计 compressive parallel plate viscometer
压缩坡 bunching grade
压缩破坏 compression failure; fail in compression
压缩起始压力 initial pressure of compression
压缩气管 compressed-air pipe(line); compressed-air (pipe) line
压缩气夯 compressed-air tamper
压缩气喷嘴 airgun
压缩气瓶 compressed gas cylinder
压缩气体 compressed gas; pressure gas; stored gas
压缩气体成型机 compressed gas forming machine
压缩气体电缆 gas compression cable
压缩气体断路器 compressed gas cutout
压缩气体罐 pressure gas tank
压缩气体和液化气体 compressed gas and liquid gas
压缩气体控制的 air-operated
压缩气体容器 pressurization gas cascade
压缩气体制造业协会 <美> Compressed Gas Association
压缩气体最初重量 initial storage gas weight
压缩气筒 compressed-air container; compressed gas cylinder
压缩气钻 compressed-air drill
压缩器 compressor; constrictor; impeller
压缩器的第一级 inlet compressor stage
压缩强度 compression strength; compressive strength; crushing strength; intensity of compression
压缩翘曲 compression deflection
压缩氢装瓶装置 compressed hydrogen unit
压缩区 compression(al) zone
压缩区间运行时间 decrease train transit time in section
压缩曲率 <喷管的> compressive curvature
压缩曲线【岩】compression curve; compression diagram; compressive curve; oedometer curve; pressure-void ratio curve; e-logp curve; pore ratio-pressure curve <孔隙比与压力关系曲线>
压缩曲线类型 type of compression curve
压缩曲线转折点 compression intercept
压缩屈服点 compressive yield point
压缩燃料 compact fuel
压缩热 compression heat; heat of compression
压缩容积 compression space
压缩蠕变 compressive creep
压缩软木 compressed cork
压缩软木板 compressed cork slab
压缩软木厂 compressed cork factory
压缩设备 compression system; gas booster <气体的>
压缩十进制 packed decimal
压缩十进制数格式 packed decimal number format
压缩时间 compression time; condensed time
压缩时屈服强度 yield compression strength
压缩实用程序 condense utility
压缩始点温度 initial temperature of compression
压缩式避雷器 compression-type lightning arrestor
压缩式成型密封 squeeze-type mo(u)lded seal
压缩式防震垫 compression pad
压缩式均匀度试验机 compression-type regularity tester
压缩式冷冻机 compression-type refrigerating machine
压缩式冷冻系统 compression-type refrigeration system
压缩式冷水机 compression chiller unit
压缩式冷水机组 compression-type water chiller
压缩式密封垫 compression seal gasket
压缩式燃气制冷机 compression gas refrigerator
压缩式热泵 compression heat pump
压缩式十进制数 packed decimal
压缩式系统 compression system
压缩式压力计 compression manometer
压缩式真空规 McLeod ga(u)ge
压缩式真空计 compression manometer
压缩式制冷 compression refrigeration
压缩式制冷机 compression refrigerating machine; compression refrigerator; compression-type refrigerating machine; compression-type refrigerator; compressor chiller; compressor type refrigerator
压缩式制冷机组 compression-type refrigerating unit; compression-type refrigeration unit
压缩式制冷系统 compression-type refrigerating system
压缩式制冷循环 compression refrigeration cycle; compression-type refrigerating cycle
压缩试件 compressed specimen
压缩试验 compression test; compressive test; oedometer test; oedometric test
压缩试验机 compression tester; compression testing machine
压缩试验器 compressometer
压缩试验仪 compression tester
压缩试样 pressurized sample
压缩室 compression chamber; delivery chamber; discharge chamber; plenum chamber
压缩属性 packed attribute
压缩树脂 press-resin
压缩数据 compressed data
压缩数据存储系统 compressed data storage system
压缩速率 compression rate
压缩算法 compression algorithm
压缩损失 compression loss
压缩弹簧 compression spring; pressure spring
压缩弹簧钩 compression shackle
压缩弹性 elasticity of compression
压缩弹性极限 compressive elastic limit
压缩弹性模量 modulus of elasticity in compression
压缩特性的转变 compression intercept
压缩天然气 compressed natural gas
压缩头线 headline; pressure grade line
压缩图 compression diagram
压缩退火热解石墨 compression-annealed pyrolytic graphite
压缩温度 compression temperature
压缩系数 coefficient of compressibility; coefficient of compression; compressibility (factor); compression coefficient; compression efficiency; compression factor; compression index; contraction coefficient; efficiency of compression; pressing factor; super-expansibility factor; contraction ratio <气流的>
压缩系统 compression system
压缩纤维 fibre under compression
压缩纤维板 compressed fiberboard
压缩线 compression line
压缩线圈 collapse coil
压缩限度 compression limit; limit of compression
压缩效率 compression efficiency; manometric(al) efficiency
压缩效应 compressibility influence
压缩信用 credit squeeze
压缩行程 compression travel; pressure stroke
压缩形式 compressed format
压缩性 coercibility; compressibility; contractibility; contractibleness
压缩性材料 compressible material
压缩性地基 compressible foundation
压缩性垫密材料 compressible packing material
压缩性滤渣 compressible cake
压缩性模数 compressibility modulus; modulus of compressibility
压缩性泡流 compressibility burble
压缩性土层 compressible formation; compressible ground; compressible strata; compressive formation
压缩性土(壤) compressible soil
压缩性误差 compressibility error
压缩性系数 compressibility coefficient
压缩性阻力 compressibility drag
压缩徐变 compressive creep
压缩旋塞 compression cock
压缩循环 compression cycle
压缩压力 compression pressure
压缩压力表 compression ga(u)ge
压缩压力计 compression ga(u)ge; crusher ga(u)ge
压缩延迟振铃 abbreviated and delayed ringing
压缩液态毒气 Mace
压缩仪 compression tester; compressive apparatus; compressometer; o(e)dometer
压缩仪器 compression apparatus
压缩因数 bulk factor; compressibility factor; deviation factor; gas deviation factor; shrinkage factor; supercompressibility factor
压缩因素 bulk factor; compressibility factor; shrinkage factor
压缩因子 compressibility; compressibility factor; shrinkage factor
压缩应变 compressing [compression/compressive] strain; compression deformation; compressive deformation
压缩应力 compression(al) stress; compressive stress; stress under compression
压缩应力波 compressive wave
压缩应力范围 compression stress field
压缩应力强度 intensity of compressive stress
压缩应力松弛 compression stress relaxation
压缩影响 compressibility effect
压缩永久变形 compression set; permanent compression set
压缩振荡 compressive oscillation
压缩振动 compressional vibration
压缩指标 compressed limit
压缩指令汇卡 condensed instruction deck
压缩指令卡片组 condensed instruction deck
压缩指示器 compression indicator
压缩指数 cake compressibility; compressibility index; compression exponent; compression index
压缩制冷 compression-type refrigeration
压缩制冷系统 compression refrigeration system
压缩制冷循环 compression-type refrigeration cycle
压缩制模木材 compressed pattern lumber
压缩终点温度 final temperature of compression
压缩主支曲线 main compression curve; virgin compression curve
压缩/转移模压机 compression/transfer mo(u)lding machine
压缩装袋机 compression-bagging machine
压缩装置 compression equipment; compression plant
压缩状态 incompression; state of

compression
压台 press table
压探头 piezoprobe
压烫拉幅定形机 press tenter
压套 junk ring
压填料栅板 hold-down grid
压条 batten (bar);batten strip;bead-(ing); cornice; depression bar; hatch securing bar; hold-down strip;mo(u)lding; mo(u)lding joint cover; mound layer; muck bar;patand;treadle bar
压条衬板 bar liner;liner with pressing bar
压条法 layerage;marcottage;marcotting layerage; planting of layers; layering
压条繁殖法 layerage
压条架 batten seat
压条扦插区 layer cuttage division
压条区 layering plot
压条镶玻璃法 bead glazing
压条枝 layer
压条装配玻璃法 bead glazing
压弹控制 push-pop control
压弹式堆栈 push-pop stack
压铁 clamping plate;foundry weight; iron ballast;iron weight;kentledge; mo(u)ld weight
压头 head of delivery;head of pressure;hydraulic pressure head;pressure differential; pressure head; ram(mer);squeeze head;indenter <硬度试验机的>
压头变化 head variation
压头差 fall head of water; head difference;head differential
压头沉落 head sink
压头反向 reversal of the head
压头范围 head range;range of head
压头高度 fall head of water
压头恢复 head recovery
压头机 heading machine
压头计 head meter
压头减小 reduction of pressure head
压头流量曲线 head-capacity curve
压头排量曲线 head-capacity curve
压头偏差损耗 head misalignment loss
压头式流量计 head flowmeter
压头式弯机 ram-type bending machine
压头损失 head loss;loss in head;loss of fall;loss (of) head;lost head
压头损失计 loss of head ga(u)ge
压头损失仪 loss of head ga(u)ge
压头系数 head coefficient
压头下降式压机 down stroke press
压头线 pressure grade line
压头相对落差 relative fall
压头箱 head box;head tank
压头镶玻璃 bead glazing
压头移动速率 rate of head movement
压头油缸 ram cylinder
压头油箱 overhead reservoir
压头再分布 redistribution of pressure head
压凸冲模 embossing die
压凸印刷 raised printing
压凸印刷墙纸 ink-embossed paper
压图块 chart weight
压涂法 extrusion method
压涂焊条 extruded electrode
压土机 compactor;packer;press wheel; soil compaction roller
压土碾 tamping roller
压瓦钉 tile pin
压瓦机 tile press
压瓦垅 corrugation
压弯 braking;press bending
压弯成型机 brake
压弯成型(法) gag process
压弯机 bending brake;bending press; bulldozer; camber jack; forming press;press brake
压弯机模 press brake dies
压弯抗力 buckling resistance
压弯模 bending die
压丸 pelleting
压网 pressure screen
压位差 head pressure;pressure head
压温湿记录器 barothermohydrograph
压温图(表)【气】 pressure-temperature chart
压纹 emboss(ing)
压纹板 embossed sheet; graining board;patterned sheet
压纹的 embossed;intagliated
压纹辊 embossing roller
压纹滚 graining roller
压纹机 embosser; embossing calender;gauffer machine;gof(f)er machine;grainer;graining machine
压纹瓦 intaglio tile
压纹效应 grain effect
压纹装饰 calender crush finish
压窝 indenting
压吸效应 pressure and suction effect
压熄(烟火等) dinch
压洗 pressure wash(ing)
压下 cogging; depress(ion); force down;press down
压下的 depressed
压下的接触轨 depressed conductor-rail
压下的踏板 depressed pedal
压下垫块 pressure block
压下机构 screw-down
压下控制 screw control
压下螺钉 housing screw
压下螺丝 housing screw; pressure screw
压下失效弹簧 dead spring
压下弹簧 dead spring
压下调整螺钉 housing adjusting screw
压下系统 press down system
压下引砖 depressing of the draw bar
压下装置 screw-down gear; screw-down structure
压陷 matting
压向 beam down (up)on
压向下风 fall of;sag
压向正偏 forward bias
压胁强 pressure stress
压心电缆 pressed-core cable
压型 embossing;pressing;profiling; tamping
压型板 contour plate
压型玻璃 extruded glass
压型粉末 compact powder
压型钢板 profiled steel sheet; steel sheet
压型钢材 profile steel
压型管 pressed pipe
压型机 cambering machine;mo(u)lding press;tamping machine
压型毛坯 pressing
压型模 compression mo(u)ld
压型坯料 mo(u)lded blank
压型熔模 pattern die
压型设备 tamping plant
压型体 formed body;pressed body
压型制品 pressed product
压性 piezotropy
压性断层 compression fault;compressive fault
压性方程 equation of piezotropy
压性结构面【地】 compressive structural plane
压性流体 piezotropic fluid
压性深断裂 compressional deep fracture
压性追踪构造 compression track structure
压袖板 sleeve board
压雪机 snow roller
压延 calendering;drawing;flattening; malleation;rolling
压延板材 flat rolled;rolled sheet
压延边缘 running edge
压延(扁)钢条 flat rolled steel bar
压延玻璃 plain rolled glass; rolled glass;cirrus glass <有特殊光照效果的>
压延玻璃板 rolled sheet glass
压延冲模 drawing die
压延锤痕玻璃 double-rolled cast
压延磁各向异性 roll magnetic anisotropy
压延淬火 rolled hardening
压延的 flat rolled
压延的刨花板 extruded chipboard
压延的碎胶合板 extruded chipboard
压延法 calendering process; rolling process
压延钢 rolled steel
压延钢板 pressed steel plate
压延钢梁 rolled steel beam
压延工 flattener
压延辊 forming roll(er);casting roller <平板玻璃的>
压延黄铜型材 extruded brass section
压延黄铜制品 extruded brass
压延机 calender; calender stack; flattener
压延机侧辊 offset roll of calender
压延机旁滚筒 offset roll of calender
压延机中辊 centre roll of calender
压延件 draw piece
压延金属 extruded metal
压延排水 rolling effluent
压延平板玻璃 rolled plate glass
压延润滑 rolling lubrication
压延铁 rolled iron
压延头 spreading chest
压延凸模 draw punch
压延系数 cupping ductility value
压延效应 calender grain
压延形变热处理 ausrolling
压延性能 drawability;drawing quality
压延用黄铜 brass for rolling and drawing
压延轧辊 reduction roll
压延装置 draw gear
压盐 lowing the salt location in soil
压檐梁 capping beam;capping piece
压檐木 capping piece;parapet piece
压檐墙 parapet wall
压檐石 barge stone
压檐瓦 capping tile
压檐砖 capping brick
压样法 compaction sample process
压液辊 squeezer roll
压抑 repression
压抑的 depressing
压抑作用 repression
压印 autogram;impress(ing);impression; indent; seal press; stamp(ing);tire scuffs
压印标记 seal
压印锤 set hammer
压印底圈的模型 footer mo(u)ld
压印垫板 baren
压印工作 stamp work
压印管 embossed tube
压印滚筒 impression cylinder
压印过程 mo(u)lding process
压印花纹 embossed decoration;stamping
压印机 blocking press;coining press; marking device;stamping machine; stamping press
压印加工 coining;embossed work
压印力 force of impression
压印密度控制 impression control
压印模 coining die;embossing die; marking stamp;stamping die
压印模板 impression block
压印墨 stamping ink
压印黏[粘]接 print bonding
压印偏心调节法 impression eccentric
压印器 imprinter
压印深度 depth of indentation
压印图案模糊 faint impression
压印系统 impression system
压印压力机 coining press
压印硬度 pressure hardness
压应变 crushing strain
压应力 compressive resistance;crush(ing) stress
压应力分量 compressive stress component
压应力封接 compression seal
压应力轨迹线 compressive trajectory
压应力计 compressive stress ga(u)ge; pressure cell; Ritter pressure cell
压应力釉 compression glaze
压应力张量 tensor of stress
压硬沥青成块 briquetting pitch
压泳(现象) barophoresis
压油泵 pressure oil pump
压油机 force-feed lubricator; grease press
压油器 lubricating press
压油润滑 forced lubrication;force feed; full force feed lubrication; oil-pressure lubrication
压油润滑的 force-lubricated
压油润滑法 force-feed lubrication (method)
压油润滑器 forced lubricator;force-feed lubricator
压油室 pumping chamber
压有车辙的泥地 rutted mud
压鱼机 gag press
压原 reduction
压源 potential source
压载【船】 ballast; bottom stowage; light load
压载泵 ballast pump
压载边舱 side ballast tank
压载布置 ballast configuration
压载舱 ballast chamber;ballast compartment
压载舱通海水管道 sea delivery
压载舱吸水管 ballast suction pipe
压载吃水 ballasted draft
压载吃水线 ballast (load-) line; ballast waterline
压载出港 ballast departure
压载到港状态 ballast arrival condition
压载趸船 ballasted pontoon
压载浮船 ballasted pontoon
压载杆 ballast bar;sinker bar
压载管系 ballast line; ballast piping system
压载航次 ballast run;ballast voyage
压载航行 ballast passage;ballast sailing;sailing in ballast;ballast voyage
压载航行船 ship in ballast
压载混凝土 ballast concrete
压载货 ballast cargo;kentledge;kentledge goods;limber kentledge
压载块 ballast weight
压载龙骨 ballast keel
压载滤水体 loaded filter
压载凝结器 ballast condenser

压载排水量 ballast displacement; light load displacement
压载试航 ballast trial
压载栓 ballast bolt
压载双层底舱 ballast tank
压载水 ballast water; water ballast
压载水泵 ballast pump
压载水泵舱 ballast pump room
压载水泵系统 ballast pumping system
压载水舱 ballast (water) tank; trimming tank; water ballast space; water ballast tank
压载水舱换水 freshen the ballast
压载水管 ballast pipe
压载水管接头 ballast connection
压载水柜 ballast (water) tank; shifting tank
压载水排尽 ballast blown out
压载水线 ballast waterline
压载水箱 ballast tank
压载水总管 ballast main
压载说明 ballasting instruction
压载调整 ballasting up; ballast trimming
压载调整装置 ballast movement arrangement
压载铁(块) kentledge; limber kentledge
压载投放 ballast release
压载图 ballast plan
压载物 ballast(er); ballasting material; ballast weight
压载物驳船 ballast lighter
压载物舷门 ballast port
压载物质量 ballast mass
压载物装置 ballasting
压载系统 ballasting system
压载舷孔 ballast hole
压载箱 ballast box; ballast pocket; ballast tank; weight basket; weight box; weight tray
压载重量释放试验 ballast weight release test
压载装置 ballaster
压载状态 ballasted condition; in ballast; light condition; no-cargo condition
压在……上 bear on
压在上面的 incumbent; superincumbent; superjacent
压轧 nip
压轧表皮效应 alligator skin effect
压轧机 crushing force
压轧机架 reduction stand
压轧伤 crush injury
压轧芯材 rolled core
压榨 crush; mechanical expression; pinch; pressing; squeezing
压榨板 rammer board; ramming board
压榨泵 press pump
压榨比 press ratio
压榨饼 press cake
压榨部 press part; press section
压榨的 compressive
压榨法 expression
压榨辊 press roll
压榨辊的合成网套 press fabrics
压榨辊外壳 press shell
压榨机 fulling board; presser; press-plate machine; ramming board; squeezer
压榨间 press house
压榨螺旋 propelling worm
压榨镊(钳) expression forceps
压榨器 expresser; presser
压榨施胶 press coating
压榨碎木机 pocket grinder
压榨筒 press cylinder
压榨脱蜡过程 press dewaxing process
压榨性能 pressability
压榨液 pressed liquor
压榨渣 press residue
压榨纸板 pressboard
压毡 press felt
压栈 pop down
压栈操作 push operation
压张钢丝 post-tensioning wire
压张疲劳 compression-tension fatigue
压折 upset; back edge; backfin <轧制中金属反向流动形成的>
压折缝的器具 creaser
压折器 crimper
压砧 anvil
压砧缸式超高压装置 anvil-cylinder ultra high pressure apparatus
压砧同步 anvil synchronizing
压枕木 bunkload
压蒸材料 autoclaved material
压蒸处理 autoclave treatment; autoclaving
压蒸法 autoclave expansion test
压蒸粉煤灰砖 autoclave fly-ash brick
压蒸釜 autoclave
压蒸釜成型 autoclaved mo(u)lding
压蒸构件 autoclaved unit
压蒸混凝土 autoclave-cured concrete
压蒸混凝土板 autoclaved concrete slab
压蒸加气混凝土 autoclaved aerated concrete
压蒸加气砌块 autoclaved aerated block
压蒸膨胀(率) expansion in autoclave test
压蒸膨胀试验 autoclave expansion test
压蒸膨胀值 autoclave expansion value
压蒸轻质混凝土 autoclaved lightweight concrete
压蒸石膏 autoclave plaster
压蒸试验 autoclaved test
压蒸养护 autoclave curing; autoclaving
压蒸养护混凝土 autoclave-cured concrete
压蒸养护周期 autoclave cycle
压直 straightening
压直机 gag press; straightening press
压止层 <地下室柔性防水层的混凝土覆盖层> loading coat
压纸板 pressboard
压纸尺 paper-holder
压纸滚轮凸轮 pressure roller cam
压纸卷轴 platen
压指 press(ure) finger
压指标 forcing down the targets; forcing up the target
压制 compact(ing); compress(ion); extrude; mo(u)lding; pressing; quash
压制板 forced board; pressboard; pressed tile; presswood; stamped plate
压制棒坯 pressed bar
压制波纹 corrugating
压制玻璃 pressed glass; cross reeded <两面肋条花纹交叉的>
压制玻璃基底 pressed glass base
压制玻璃器皿 pressware
压制薄钢板 pressed sheet steel
压制薄片木 pressed material
压制材料 pressed material
压制成波浪形 embossing
压制成的刹车块 pressed blocks lining
压制成块 briquetting
压制成块的配合料 consolidating batch
压制成块方法 briquetting method
压制成品 presswork
压制成型 compaction; press forming
压制成型过程 compacting process
压制雏形 pressing blank
压制磁铁 pressed magnet
压制粗黏[粘]土砖 pressed raw clay brick
压制稻草板隔墙 compressed straw slab partition (wall)
压制的 pressed; rolled; seamless
压制的板材 pressed sheet
压制的玻璃容器 pressed container glassware; pressed hollow glassware
压制的玻璃制品 pressed glassware
压制的晶质玻璃器皿 pressed crystal ware
压制的平瓦 repressed plain tile
压制的软木板 compressed corkboard
压制的炭电极 green electrode
压制的瓦 repressed tile
压制的屋面瓦 pressed tile
压制的制品 pressed ware
压制的砖 repressed brick
压制地沥青 compressed asphalt; stamp asphalt
压制地沥青烘烤机 compressed asphalt roaster
压制地沥青块 compressed asphalt block
压制地沥青砖 compressed asphalt tile
压制钉 stamped nail
压制多层板 compressed laminated wood
压制多次波 multiple suppression
压制防护板 pressed shield
压制钢 pressed steel
压制钢板 dished plate; pressed steel plate
压制钢结构建筑构件 pressed steel building component
压制锆(英石)砖 pressed zircon block
压制跟踪 neutralized track
压制工 presser
压制工件 spun work
压制工具 compaction
压制工艺 press process
压制钩环 pressed shackle
压制管 extruded pipe
压制管理 management by drive
压制轨枕 pressed sleeper
压制过程 pressing process
压制过的氧化钚 pressed plutonium oxide
压制禾杆 compressed straw
压制混合料 compressed mixture
压制混凝土 compressed concrete; pressed concrete
压制混凝土管 pressed concrete pipe
压制混凝土型材 extruded concrete profile; extruded concrete section; extruded concrete shape
压制混凝土桩 compressed pile
压制机 briquetting machine
压制机的模环 pressing ring
压制机械 pressing machinery
压制件 mo(u)lding; rolled-up stock
压制件缺陷 mo(u)lding fault
压制焦 form-coke
压制截面 pressed section
压制金属 pressed metal
压制金属薄板 pressed sheet metal
压制金属薄板控制盘 embossed sheet metal panel
压制金属薄板面板 embossed sheet metal panel
压制金属接头 pressed metal node
压制金属型材 pressed metal section
压制井喷 killing wells; kill the well
压制开裂 pressing crack
压制梁 pressed girder
压制裂纹 pressure check
压制零件 compression fittings
压制毛坯 slug press
压制毛毡 overfelt
压制煤块 pressed fuel
压制煤砖 briquetting of coal
压制密度 pressed density
压制模具 compacting tool set
压制模制数 modulus of pressing
压制木材 compressed wood; pressed wood
压制泥片 batting out
压制泥炭 mo(u)ld ed peat
压制黏[粘]土瓦 pressed clay-tile
压制黏[粘]土屋面瓦 pressed clay roof(ing) tile
压制黏[粘]土砖 pressed clay brick
压制片 compressed tablet
压制品 compact; presswork; stamping
压制坡莫合金粉 pressed permalloy powder
压制器皿 pressed ware
压制强度 press(ing) strength
压制燃料 pressed charge; pressed fuel
压制软木 compressed cork
压制润滑剂 pressing lubricant
压制砂 crusher sand
压制烧结熔化机 pressing-sintering-melting machine
压制烧结熔炼设备 pressing-sintering-melting unit
压制失控定向井 killer well
压制石棉水泥板 compressed asbestos-cement panel; compressed asbestos-cement sheet
压制石棉纸板 compressed asbestos sheet
压制时间 press time
压制试验 briquetting test
压制水泥矿渣砖 compressed cement-slag brick
压制速度 pressing speed
压制陶瓷多孔管 stamped ceramic porous tube
压制陶土咬接瓦 extruded interlocking clay roof(ing) tile
压制陶土咬接瓦屋面覆盖层 extrusion interlocking clay tile roof(ing) cover(ing)
压制体积系数 bulk factor
压制条件 pressing condition
压制透镜 pressed lens
压制透镜毛坯 pressing lens blank
压制图案模型 pattern mo(u)ld
压制外壳 drawn-shell case
压制外形 pressed profile
压制位置 pressing position
压制温度 press temperature
压制纤维板 compressed fiberboard [fibreboard]; compressed straw fibre building slab; fibre building board; pressed-fiber board
压制相干噪声 suppression of coherent noise
压制效应 depressing effect
压制型材 extruded section; extruded shape
压制型钢 pressed steel; pressed steel shape
压制型钢水槽 pressed steel sectional tank
压制型钢水箱 pressed steel sectional tank
压制型件 pressed section
压制性 briquettability
压制压力 pressing pressure
压制岩地沥青(块) compressed rock asphalt
压制岩沥青 compressed natural rock asphalt
压制阴模(法) hubbing; die hubbing
压制用粉末 pressing powder
压制云母 pressed mica

压制炸药 pressed explosive
压制毡 compressed felt;pressed felt
压制毡的制造 pressed felt manufacture
压制植物纤维 compressed vegetable fiber [fibre]
压制纸板 compressed straw slab; strawboard;fuller board
压制砖 pressed brick;struck brick
压制阻力 pressure drag
压制作用 suppression
压致变色 piezallochromy
压致电离 pressure ionization
压致移动 pressure shift
压致增宽 pressure broadening
压重 ballast weight;mo(u)ld weight; weight counterbalance
压重层(填土) counterweight fill
压重氮 pressure diazo
压重反滤层 loaded filter; weighted filter
压重飞机 ballasted aeroplane
压重混凝土 ballast concrete
压重料 kentledge
压重式手推滚筒 ballast type hand roller
压重填土 counterweight fill
压重箱 ballast tank
压皱钢纤维 crimped steel fiber [fibre]
压住 hold-back;hold-down;suppress
压住井喷 harness a well
压注 pressure penetration
压注泵 injection pump
压注化学溶液 chemical injection
压注井 pressure well
压注式润滑枪 force-feed gun;pressure gun
压注树脂 injection resin
压注水泥 snubbing cementing
压注油浸防腐法 pressure creosoting
压注油嘴 force filling oil nozzle
压注装置 injection unit
压柱 compression leg
压柱公式【建】column formula
压铸 compression casting; pressure casting; transfer mo(u)lding; die casting
压铸玻璃 cast(ing) glass
压铸玻璃板 cast glazed tile
压铸玻璃隔墙 cast glass partition (wall)
压铸玻璃门 cast glass door
压铸玻璃瓦 cast glass tile
压铸玻璃阳台护墙 cast glass balcony parapet
压铸玻璃砖 cast glass tile
压铸产品 die-casting material
压铸的玻璃天篷 cast glass canopy
压铸的玻璃挑出屋顶 cast glass roof overhang
压铸底盘 die cast chassis
压铸法 die casting;press mo(u)lding method; pressure casting; pressure casting process
压铸工艺 extrusion process
压铸过程 extrusion process
压铸合金 die-casting alloy
压铸黄铜 brastil
压铸机 casting machine;die-casting machine;pressure-die-casting machine
压铸件 die cast(ing); die pressed casting;pressure (die) casting
压铸金属 die-casting metal
压铸金属纡管 pressure-die-casting pirn
压铸滤热玻璃 cast glass calorex
压铸铝 alumin(i)um die casting
压铸铝合金 alumin(i)um die-casting alloy;pack alloy
压铸模 casting die; compression mo(u)ld;die cast dies;press mo(u)ld;steel die
压铸模具合金钢 pressure-casting die steel
压铸模铸造 pressure mo(u)lding
压铸温度 casting temperature
压铸锌 zinc die casting
压铸锌合金 zinc die casting alloys
压铸轴承合金 die-casting metal
压砖机 brick mo(u)lding press;brick press; brick-pressing machine; tile press
压砖机润滑油 brick oil
压砖计数器 stroke counter
压桩 pile jacking
压桩法 press pile driving method
压桩机 pile press machine; press-in pile driver
压桩力 press force
压桩力测量系统 pile press force measuring
压桩力传递系统 pile press force transmitting system
压桩速度 press-in speed
压装齿圈 pinion rings
压装吨位 mounting tonnage
压装压力 mounting pressure
压装压力机 arber press
压装应力 stress due to press mounting
压锥头 piezocone
压足 presser foot
压阻压力变送器 piezoresistive pressure transducer
压钻机钻床 drill press

押标保金 bid bond

押标保证 bid bond;tender bond
押标金 bid bond; guaranty bond; initial bond
押船借款 bottomry
押船契约 bottomry bond; hypothecation
押房 mortgage a house
押汇 bill of draft;bill of exchange;negotiation
押汇中心 bill center [centre]
押汇总质押书 general letter of hypothecation
押货 mortgage goods
押解出境 deportation;reconduction
押借 borrow money on
押借限额 loan value
押金 antecedent money;cash deposit (as collateral); deposit(e) for security deposit;deposit(e) in security; guarantee deposit; purchase money mortgage;quarantee deposit;retention;retention money;security deposit
押金存款 margin deposit
押金购买 buy on margin
押金回扣 financing fee
押金缺少时的价格 exhaust price
押金收据 certificate of deposit
押金退款制度 deposit(e)/refund system
押款 borrow money on security;loan on security;sum lent on a mortgage
押款的受押人 loan holder
押款协议书 servicing agreement
押码 authenticating code;cypher;test key;test word
押没品价 foreclosure value
押品的赎回 redemption
押契 mortgage deed
押切法 nail-pressure
押手 finger tip pressing technique; hand-pressing
押运 convoy;escort;escort goods in transportation;escorting of parcel
押运车 drover's car
押运费 escort fee
押运人 convoyer;convoy man
押运人票价 caretaker ticket rate
押运员 escort;freight clerk;supercargo
押账 leave something as security for a loan
押租 deposit(e) with landlord; deposit money;foregift;rent deposit

垭口 bealock; cross-over; nek; notch; col; saddle-back; poort <南非的>

鸦片 opium

鸭蛋青 duck's-egg green;pale blue

鸭蛋青瓷器 duck-egg porcelain
鸭蛋圆 oval
鸭脚铲 duckfoot sweep
鸭脚锄铲 duckfoot shovel
鸭脚树 maidenhair tree
鸭绿色 duck green
鸭屁股船尾 <小船> lute stern
鸭绒垫 eiderdown
鸭式布局 canard configuration
鸭式飞机 canard
鸭式构图 canard layout
鸭式设计方案 canard layout
鸭式水翼简图 canard
鸭式图形 canard configuration
鸭尾艄 fantail
鸭掌式锄铲 duck foot
鸭掌式中耕机 duckfoot cultivator
鸭足式除草器 duckfoot sweep
鸭嘴笔 border pen; crow-quill pen; drafting pen; drawing pen; duckbill pen;ruling pen
鸭嘴钉 duckbill nail
鸭嘴(斗)装载机 duckbill
鸭嘴钩 <铺管用> duck hook
鸭嘴剪 duckbill snips
鸭嘴钳 duckbill pliers;duckbill snips; flat-nose pliers;gripping tongs
鸭嘴式装载铲 duckbill blade
鸭嘴式装载机 duckbill loader
鸭嘴兽型堰顶 duckbill sill shape
鸭嘴形的 duckbill
鸭嘴形堰 duckbill weir
鸭嘴形溢洪道 duckbills type spillway
鸭嘴形钻(头) duckbill bit
鸭嘴堰 <用于渠道调节水量> duckbill weir
鸭嘴直线笔 chart drawing pen

牙白色 cream white

牙板 serrated slips; hold-down slips <夹持器的>
牙板夹头 gripper
牙板头 clamp head
牙病防治所 dental clinic
牙齿 tooth
牙齿的 dental;dentary
牙齿分离 demeshing
牙齿花饰 crenel(l)ated mo(u)lding
牙齿磨损 bit-tooth wear
牙齿磨损高度 tooth wearing height
牙齿磨损级别 grade of tooth wearing
牙齿磨损模式 tooth wearing mode
牙齿磨损速度 tooth wearing rate
牙齿磨损系数 factor of tooth wear
牙齿箱 change-speed gear;transmission case
牙齿状三角洲 cuspate delta
牙齿最优磨损量 optimum tooth wearing height
牙齿最终磨损量 final tooth wearing height
牙雕 ivory wares
牙雕素瓶 carved ivory vase
牙氟中毒 fluorosis of the teeth
牙垢 calculus [复 calculuses/calculi]
牙行 intermediary business
牙黑 ivory black
牙黄色环 yellow cadmium fringe on teeth
牙尖雌雄榫 keyed mortise and tenon
牙科 dental surgery;dentistry;department of dentistry
牙科材料 dental material
牙科瓷 dental porcelain
牙科室 dental room
牙科手术间 dental treatment cubicle
牙科诊室 dental clinic; dental treatment room
牙科中心 dental center [centre]
牙轮 cone
牙轮标准直径 ga(u)ge tip diameter of the cone
牙轮齿排互插深度 depth of interfit
牙轮的背面 rear flank
牙轮轮齿 cutter teeth
牙轮钎头 steel-toothed roller bit
牙轮切削具 rolling cutter
牙轮式扩孔器 pin-and-eye type reamer;rock bit type hole opener
牙轮销 cutter pin
牙轮中间一排齿 intermediate row teeth
牙轮轴 cone axis
牙轮锥顶 <钻头的> nose of cone
牙轮钻机 rotary drilling machine
牙轮钻进 roller bit drilling; toothed roller drilling
牙轮钻头 cone bit;cone rock bit;indented chisel;riffler;rock drill bit; rock roller bit;roller bit;roller cutter bit; rolling cutter (rock) bit; steel-toothed roller bit; toothed roller bit; true-rolling bit; simplex rock bit <钻硬岩石用>
牙轮钻头打捞套筒 rock bit cone fishing socket
牙轮钻头结构 rock bit structure
牙轮钻头偏斜极限 cone-type bit deviation limits
牙轮钻头试验钻机 rock bit testing rig
牙轮钻头镶嵌块 rock bit insert
牙轮钻头爪 sheet of the bit
牙轮钻头中的油壶润滑器 weevil
牙轮钻头锥形轴承 rock bit cone bearing
牙轮钻头钻进 rock bit drilling;roller boring
牙买加苦树 bitter wood
牙买加罗汉松 yacca
牙排水口 curb outlet
牙钳 die tongs
牙嵌离合器 claw coupling;coupling clutch;jack clutch;jaw clutch;positive clutch
牙嵌联轴节 clutch coupling
牙嵌式离合器 dog clutch
牙刷架 tooth-brush holder
牙形石绝灭 conodont glacial stage
牙形石类 conodontophoridia
牙医室 dentist office
牙状突石 tusses
牙状物 bud

芽孢 spore

芽胞 microspore

芽胞杆菌 bacillus
芽胞油页岩 bitumenite
芽层 sheaf of germs
芽接刀 budding knife

蚜虫 aphid

蚜虫属 <拉> Aphis

崖柏 giant arbor-vitae

崖壁 rock wall
崖岛型海岸 skerry-type coast
崖道 non-working platform
崖底堆积物 cliff debris
崖底侵蚀 scarping
崖堆 slide rock; talus deposit
崖复活断层 revived fault scarp
崖公窑 Cui-gong ware
崖脚缓坡 haldenhang; wash slope
崖柳 Salix xerophila
崖面 sheer
崖坡后退 recession of cliffs
崖坡悬崖 scarp
崖上砂洞 zawn
崖石 cliffstone
崖坍 cliff landslide
崖头山顶 brow
崖下阶地 undercliff

哑变量 dummy variable; invented variate

哑变量名 dummy variable name
哑变元 dummy argument; dummy variable
哑变元数组 dummy argument array
哑参数 dummy parameter
哑槽 blank groove; plain groove
哑串 dummy string
哑的 dumb
哑点 blind spot; dead center [centre]; dead point; dead spot; null point
哑点角 angle of silence
哑点宽度 range of silence
哑段 dummy section
哑光漆 mat(t) paint
哑过程 dummy procedure
哑角 reentrant angle
哑节 dummy section
哑节点 dummy node
哑金属 high damping metal
哑控制段 dummy control section
哑口【道】 pass; saddle-back
哑矿 <含矿很少的> deaf ore
哑铃 dumb bell
哑铃式轮胎防舷材 floating wheel rubber fender
哑铃式桥墩 dumbbell pier
哑铃形槽 dumbbell slot
哑铃形的 dumbbell-shaped
哑铃形卷装 dumbbell-shaped package
哑铃形立体交叉口 dumb-bell junction
哑铃形零速度线 dumbbell-shaped curves of zero velocity
哑铃形桥墩 dumbbell pier
哑铃形试验片 dumbbell specimen
哑铃形样板 dumbbell specimen
哑铃形窑 dumbbell kiln
哑铃柱端 die
哑铃状 dogbone shape
哑罗经 dumb card; dumb compass; dummy compass; ectropometer; pelorus
哑罗经座 pelorus stand
哑炮 blown-out shot; cut-off shot
哑色的 flat
哑纹 blank groove; virginal groove
哑下标 dummy suffix
哑音 dumb sound
哑音区 <混凝土> dummy area
哑指标 dummy index
哑指标记号 dummy-index notation
哑终端 dumb terminal
哑自变量 dummy argument

雅丹地形 jardang; yardang; yarding

雅碲锌石 yafsoanite
雅典的列雪格拉德文体纪念碑 monument of Lysikrates at Athens
雅典的奈基神庙 Temple of Nike Apteros at Athens
雅典的宙斯神庙 Temple of Zeus at Athens
雅典列柱式 attic order
雅典娜神殿 athen(a)eum
雅典式窗 <希腊> Athenian window
雅典式柱基座 attic base
雅典卫城 Acropolis of Athens
雅典卫城精致的门道 propylaeum [propylaea]
雅典宪章 <1933 年确定城市规划原则的，法国> Athens Charter; Charte d' Athenes
雅各布式 <十七世纪英国的> Jacobean style
雅各振动台 <一种混凝土振动台> Jagger table
雅加达 <印度尼西亚首都> Djakarta
雅可白特时代的玻璃 Jacobite glasses
雅可比变换 Jacobi's transformation; Jacobian transformation
雅可比变换行列式 Jacobi's transformation determinant
雅可比簇 Jacobian variety
雅可比第二解法 Jacobi's second method of solution
雅可比定理 Jacobi's theorem
雅可比多项式 Jacobi's polynomial; Jacobian polynomial
雅可比法 Jacobi's method for determining; Jacobi's method
雅可比方程 Jacobi's equation
雅可比公式 Jacobi's formula
雅可比函数 Jacobian function
雅可比行列式 Jacobian; Jacobian determinant; Jacobian determination
雅可比合金 Jacobi's alloy
雅可比恒等式 Jacobi's identity
雅可比积分 Jacobi's integral
雅可比级数 Jacobi's series
雅可比记号 Jacobi's symbol
雅可比井公式 Jacobian well formula
雅可比矩阵 Jacobi's matrix; Jacobian matrix
雅可比矩阵法 Jacobi's matrix method
雅可比括号 Jacobi's bracket
雅可比曲线【数】 Jacobian curve
雅可比软梯 Jacob's ladder; Jacobian ladder
雅可比条件 Jacobi's condition
雅可比椭球 Jacobi's ellipsoid
雅可比椭圆函数 Jacobi's elliptic function; Jacobian elliptic function
雅可比微分方程 Jacobi's differential equation
雅可比系 Jacobi's system
雅可比虚变换 Jacobi's imaginary transformation
雅可比扎头 Jacobi's drill chuck
雅可比锥 Jacobi's taper
雅可布港冰川 Jacobshaven Glacier
雅可布森流向仪 Jacobsen direction meter
雅可布森重排 Jacobsen rearrangement
雅硫铜矿 yarrowite
雅明补偿器 Jamin compensator
雅明干涉显微镜 Jamin interference microscope
雅明干涉仪 Jamin intreferometer
雅明棱镜 Jamin prism
雅明链 Jamin's chain
雅明条纹 Jamin fringes
雅明效应 Jamin effect
雅明折射计 Jamin refractometer
雅努斯四面凯旋门 Arch of Janus Quadrifrones
雅温得 <喀麦隆首都> Yaounde
雅西尔金属棒 <一种检查进水情况的金属棒> Jasil water bar
雅致色泽 art shade

亚氨基二羧酸 iminodicarboxylic acid

亚氨基碳酸 iminocarbonic acid
亚氨基碳酸酯 iminocarbonic ester
亚胺 imine
亚胺改性醇酸树脂 imide modified alkyd
亚澳太平洋大区系 Asian-Australian-Pacific realm
亚巴拿马型集装箱船 sub-Panamax container ship
亚北部的 subboreal
亚北方的 subboreal
亚北极的 subarctic
亚贝壳属 <拉> Bulliardella
亚贝壳状 subconchoidal
亚倍频程滤波器 suboctave filter
亚苯基苯胺黑 benzanil black
亚苯基苯胺蓝 benzanil blue
亚苯基苯胺染料 benzanil colo(u)r
亚苯基二异氰酸酯 phenylene diisocyanate
亚苄基氨基苯 benzalaminophenol
亚苄基苯胺 benzalaniline
亚苄基二胺 tolylene diamine
亚变化泉水 subvariable spring
亚变群落 lociation
亚变种 subvariety
亚表层 subsurface stratum
亚表层空气 subsurface air
亚表面波 subsurface wave
亚表面缺陷 subsurface defect
亚表土 subsurface field; subsurface soil
亚表土径流 subsurface runoff
亚表土镇压器 subsurface land roller
亚滨海带 sublittoral zone
亚滨湖带 sublittoral zone
亚冰期 stadial
亚冰期的 subglacial
亚冰碛谷 sub-drift valley
亚丙基二醇 trimethylene glycol
亚伯丁花岗岩 <产于英国的灰色或淡红色的花岗岩> Aberdeen granite
亚伯拉罕稠度计 Abraham's consistometer
亚伯拉罕（混凝土）水灰比定律 Abraham's law
亚铂的 platinous
亚侧耳属 <拉> Hohenbuehelia
亚层 subgroup; subshell
亚层土 subsoil
亚层土排水 subsoil drainage
亚层土钻进钻头 spudding bit
亚长石砂岩 subarkose
亚长石质岩屑砂屑岩 subfeldspathic lithic arenite
亚超重元素 subsuperheavy elements
亚琛大教堂 Aix-la-Chapelle cathedral
亚琛的查理曼礼拜堂 Chapel of Charlemgne at Aachen
亚成分 subcomponent
亚成熟形态 immature form
亚程序 metaprogram(me)
亚承压地下水 subartesian groundwater
亚赤道的 subequatorial
亚船壳属 <拉> Gloniella
亚纯的 meromorphic
亚纯函数 meromorphic function
亚磁体 metamagnet
亚大气压放电 subatmospheric(al) discharge
亚大气压力 subatmospheric (al) pressure
亚代 subage
亚带 subzone
亚单位 subunit
亚单元 subunit
亚氮的 nitrous
亚当地图投影 Adams projection
亚当风格 Adam's style
亚当建筑纸 <英乔治三世时亚当建筑师采用的罗马亚历山大纸> Alexandrian paper
亚当式建筑 Adam architecture
亚当斯·尼克尔逊 LAB 色差方程式 ANLAB colo(u)r difference equation
亚当斯玻璃砂除铁法 Adams process
亚当斯方法 Adams method
亚当斯-尼克尔逊彩色空间 Adams-Nickerson space
亚当斯-尼克尔逊-斯塔尔茨色差方程 Adams-Nickerson-Stultz colo(u)r difference equation
亚当斯色度值 Adams chromatic value
亚当斯色度值定量指示法 Adams chromatic value system
亚当斯色度值色差方程式 Adams chromatic value colo(u)r difference equation
亚当斯颜色定量指示法 Adams chromatic value system
亚当斯砖墩实用公式 Adam's formula
亚得里亚板块 Adriatic plate
亚得里亚海 Adriatic Sea
亚的斯亚贝巴 <埃塞俄比亚首都> Addis Ababa
亚低频 <0.3～3 千赫> infralow frequency
亚碲的 tellurous
亚碲酸盐 tellurite
亚碲酸盐亚硒酸盐 selenites tellurites
亚碲铜矿 rajite
亚点阵 sublattice
亚丁基【化】 butylene
亚丁湾 Gulf of Aden
亚顶极 subclimax
亚蒽基 anthrylene
亚尔登线圈 Ayrton-perrs winding
亚二甲苯基二胺 xylylene diamine
亚二甲苯基二异氰酸酯 xylylene diisocyanate
亚钒的 vanadous
亚非法律咨询委员会 Asian-African Legal Consultative Committee
亚砜【化】 sulfoxide
亚干旱的 subarid
亚纲【生】 subclass
亚高山草甸 subalpine meadow
亚高山带 subalpine belt; subalpine zone
亚高山的 subalpine
亚高山地区 subalpine region
亚高山黑色石灰土 subalpine rendzina
亚高山泥炭 subalpine peat
亚高山森林 orohylile
亚高山植物区系 subalpine flora
亚高山植物群落 orophytia
亚镉化合物 cadmous compound
亚铬酸 chromous acid

亚铬酸铜 copper chromite
亚铬酸盐 chromite
亚功率 subpower
亚汞盐类 mercurous salts
亚共晶的 eutectiferous;hypoeutectic
亚共晶合金 hypoeutectic alloy
亚共晶生铁 hypoeutectic cast iron
亚共析的 hypoeutectoid
亚共析钢 hypoeutectoid steel;hyposteel
亚共振 subresonance
亚构造层 tectonic sublayer
亚构造和位错观测方法 observation method of subfabric and dislocation
亚固线 subsolidus
亚固线数据 subsolidus data
亚灌木的 suffrutescent
亚光频 suboptic(al) frequency
亚光频衍射图样 suboptic(al) diffraction pattern
亚光栅偏转线圈 subraster deflection coil
亚光速 subvelocity of light
亚规格水 subspecification water
亚海西运动【地】Subhercynian orogeny
亚寒带 subfrigid zone;subpolar zone;subpolar region
亚寒带的 subarctic
亚寒带低压 subpolar low
亚寒带低压带 subpolar low pressure belt
亚寒带反气旋 subpolar anticyclone
亚寒带极限线 subpolar convergence
亚寒带气候 climate of Taiga
亚寒带区 subpolar region
亚寒带温湿气候 humid subarctic climate
亚寒带针叶林灰化土带 coniferous forest illimerized soil zone in subfrigid zone
亚寒带针叶林气候 climate of coniferous forest on the subfrigid zone
亚毫米 submillimeter
亚毫米波 submillimeter wave
亚毫米波段 submillimeter region
亚毫米波天文学 submillimeter astronomy
亚毫米波振荡器 Teratron
亚毫米光子计数器 submillimeter photon counter
亚毫微秒 subnanosecond
亚衡钢筋比 balanced percentage of reinforcement
亚化学计量萃取 substoichiometric extraction
亚化学计量的 substoichiometric
亚化学计量法 substoichiometry
亚化学计量分离 substoichiometric separation
亚化学计量分析 substoichiometric analysis
亚化学计量化合物 substoichiometric compound
亚化学计量同位素稀释 substoichiometric isotope dilution
亚化学计量同位素稀释分析 substoichiometric isotope dilution analysis
亚化学计量学 substoichiometry
亚磺酰 sulphinyl
亚灰瓦克岩 subgraywacke
亚基 subunit
亚极地冰川 subpolar glacier
亚棘盘菌属 < 拉 > Erinellina
亚加米农陵墓 Tomb of Agamemnon
亚镓的 gallous
亚甲基 methylene
亚甲基丁二酸 itaconic(al) acid
亚甲基丁二酸酐 itaconic(al) anhydride
亚甲基二苯基二异氰酸酯 methylene diphenyl-4
亚甲基二硅 disilmethylene
亚甲基消除力 methylene removal of stress
亚甲基应力缓解 methylene stress relieving
亚甲蓝 methylene blue
亚阶 < 地层的 > substage
亚结构 metastructure;substratum [复 substrata]; substruction; substructure
亚界【生】subkingdom
亚金的 aurous
亚金属的 submetallic
亚经济性地热田 latent economical geothermal field
亚晶胞 subcell
亚晶格 sublattice
亚晶界 subboundary
亚晶界结构 subboundary structure
亚晶粒 subgrain
亚晶粒边界 subgrain boundary
亚晶粒结构 subgrain structure
亚晶粒组织 subgrain fabric
亚晶态结构 subcrystalline structure
亚晶网 subnet(work)
亚静预燃室 quiescent pre-combustion chamber
亚科 subfamily
亚颗粒 subgrain
亚颗粒大小 size of subgrain
亚空泡水翼船 subcavitating craft
亚拉巴马州(柔性路面)设计法 < 美 > Alabama design method
亚蓝闪石 carinthine
亚类 subcategory
亚离子激光 argon ion laser
亚历山大半岛灯塔 < 古代七大奇观之一 > pharos of Alexandria
亚历山大港 < 埃及 > Port Alexandria
亚历山大凝胶强度试验器 Alexander tester
亚历山大式大理石镶嵌铺面 Alexandrian work
亚历山大统 < 下志留纪 >【地】Alexandrian
亚历山大早期报警系统 Alexander early warning system
亚利桑那 < 美国州名 > Arizona
亚利桑那大理石 Arizona marble
亚利桑那松 Arizona pine
亚临界 precritical
亚临界奥氏体 subcritical austenite
亚临界参数 subcritical condition
亚临界处理 subcritical treatment
亚临界的 close-to-critical;subcritical
亚临界的体积 subcritical volume
亚临界度 subcriticality
亚临界发电技术 subcritical power technology
亚临界过滤 subcritical filtration
亚临界开采面积 subcritical area of extraction
亚临界开采区 subcritical area of extraction
亚临界矿化 subcritical mineralization
亚临界扩展 subcritical crack growth;subcritical propagation
亚临界离心机 subcritical centrifuge
亚临界裂纹增长 subcritical crack growth
亚临界流(动) subcritical flow
亚临界流速 subcritical velocity
亚临界缺陷增长 subcritical flaw growth
亚临界水 subcritical water
亚临界水流 subcritical flow
亚临界水氧化 subcritical water oxidation
亚临界速度 subcritical speed;subcritical velocity
亚临界条件 subcritical condition
亚临界温度 subcritical temperature
亚临界温度退火 subcritical annealing
亚临界压力 subcritical pressure
亚临界压力锅炉 subcritical pressure boiler
亚临界压涡轮机 subcritical pressure turbine
亚临界质理 subcritical mass
亚临界转子 subcritical rotor
亚临界状态 below critical
亚临界状态的测量 subcritical measurement
亚磷的 phosphorous
亚磷酸 phosphorous acid
亚磷酸苯基二癸酯 phenyl didecyl phosphite
亚磷酸苯基新戊酯 phenyl neopentyl phosphite
亚磷酸二苯基十三烷酯 diphenyl tridecyl phosphate
亚磷酸二苯基一癸酯 diphenyl decyl phosphite
亚磷酸二乙酯 diethyl phosphite
亚磷酸二油醇酯 dioleyl phosphite
亚磷酸铅 lead phosphite
亚磷酸三丁酯 tributyl phosphite
亚磷酸三甲苯酯 tricresyl phosphate
亚磷酸盐 phosphite
亚磷酸酯 phosphite
亚磷酸酯类抗氧剂 phosphite antioxidant
亚琉态的 subvileous
亚硫的 sulphurous
亚硫酐 sulfurous anhydride
亚硫酸 sulfurous [sulphurous] acid
亚硫酸铵 ammonium sulphide
亚硫酸铵废水 ammonia base sulfite waste(water)
亚硫酸铋琼脂 bismuth sulfite agar
亚硫酸废液 sulphite spent liquor;sulphite waste liquor
亚硫酸钙 calcium sulfite
亚硫酸酐 sulfurous [sulphurous] acid anhydride
亚硫酸化塔 sulfitation tower
亚硫酸化作用 sulfitation
亚硫酸钾 potassium sulfite
亚硫酸钾法 potassium sulfite [sulphite] process
亚硫酸钠 sodium sulfite [sulphide]
亚硫酸钠法烟气脱硫 Wellman-Lord process
亚硫酸钠分光光度法 sodium sulfite spectrophotometry
亚硫酸钠工艺过程 sodium sulfite process
亚硫酸气 sulfurous [sulphurous] acid gas
亚硫酸氢铵 ammonium hydrogen sulphite
亚硫酸氢铵法 ammonium bisulfite process
亚硫酸氢钙 calcium bisulfite
亚硫酸氢钾 potassium acid sulfite; potassium bisulfite; potassium hydrogen sulfite
亚硫酸氢钠 bisulfite sodium; hydrosulfite of sodium;sodium bisulfite; sodium hydrogen sulfite
亚硫酸氢铷 rubidium bisulfite
亚硫酸氢盐 bisulfite [bisulphite]; bisulphate; dithionite; hydrosulfite; hyposulfite
亚硫酸氢盐蒸煮 bisulfite cooking
亚硫酸纤维素 sulfite cellulose
亚硫酸锌 zinc sulfite
亚硫酸盐 sulfite [sulphite]
亚硫酸盐厂废水 sulfite [sulphite]-mill effluent
亚硫酸盐处理 sulphuring
亚硫酸盐法 sulfite [sulphite] process
亚硫酸盐废液 sulfite [sulphite] spent liquor;sulfite [sulphite] waste liquor;waste sulfite [sulphite]
亚硫酸盐废液副产品 sulfite [sulphite] waste liquor by-product
亚硫酸盐固体 sulfite [sulphite] solid
亚硫酸盐还原法 sulfite [sulphite] reduction
亚硫酸盐还原活性 sulfite [sulphite] reduction activity
亚硫酸盐还原菌 sulfite [sulphite]-reducing bacteria
亚硫酸盐碱液 sulfite [sulphite] lye
亚硫酸盐碱液胶粘剂 sulfite [sulphite] lye adhesive
亚硫酸盐离子 sulfite [sulphite] ion
亚硫酸盐路用结合料 < 即亚硫酸废液,磺化木质素 > sulfite [sulphite] road binder
亚硫酸盐木浆 sulfite [sulphite] wood pulp
亚硫酸盐木质素 sulfite [sulphite] lye adhesive
亚硫酸盐赛璐珞 sulfite [sulphite] cellulose
亚硫酸盐松节油 sulfite [sulphite] turpentine
亚硫酸盐纤维素 sulfite [sulphite] cellulose
亚硫酸盐纤维素萃取物 sulfite [sulphite] cellulose extract
亚硫酸盐纤维素提出物 sulfite [sulphite] cellulose extract
亚硫酸盐蒸煮液 sulfite [sulphite] cooking liquor
亚硫酸盐纸浆 sulfite [sulphite] pulp
亚硫酸盐纸浆厂废水 sulfite [sulphite] pulp mill wastewater
亚硫酸盐纸浆法 sulfite [sulphite] pulping process
亚硫酸盐纸浆废碱液 sulfite [sulphite] waste lye
亚硫酸盐纸浆废液 spent sulfite [sulphite] liquor;sulfite [sulphite] cellulose liquor; sulfite [sulphite] spent liquor;waste sulfite [sulphite] liquor
亚硫酸盐纸浆浆化 sulfite [sulphite] pulping
亚硫酸盐纸浆浆化工艺 sulfite [sulphite] pulping process
亚硫酸盐纸浆制造工业污水 sulfite [sulphite] pulp manufacture industry sewage
亚硫酸盐渍土 sulphurous acrid saline soil
亚硫酸纸浆 sulfite [sulphite] pulp
亚硫酸纸浆废水 sulfite [sulphite] pulp mill wastewater
亚硫酸纸浆废液 lignosulphonate
亚硫酸酯 sulfite [sulphite]
亚硫碳铅石 macphersite
亚硫碳树脂 durinite
亚硫酰(二)氯 thionyl chloride
亚硫酰阱 thionyl hydrazine
亚氯酸 chlorous acid
亚氯酸化 chloritization
亚氯酸钾 potassium chloride
亚氯酸钠 sodium chloride
亚氯酸盐【化】chlorite
亚氯盐渍土 chlorine saline soil
亚伦杖 < 绕蛇杆形饰 > Aaron's rod
亚麻 common flax
亚麻拔取脱粒机 puller-deseeder
亚麻被单布 linen sheeting
亚麻布 flax burlap;linen;linen cloth;lint

亚麻布的防水处理 water-repellent treatment for lien
亚麻布卷尺 linen-tape
亚麻布料 linen goods
亚麻布碾磨轮 linen cloth grinding wheel
亚麻布水龙带 linen hose
亚麻布纹高级书写纸 linen bank
亚麻布纹饰面 linen finish
亚麻布帐帘 linen drapery
亚麻布纸 linen paper
亚麻布坐垫 flax burlap mat
亚麻厂废水 flax mill wastewater
亚麻尘 flax dust
亚麻除籽工艺 ruffing
亚麻除籽叶工艺 rippling
亚麻粗布 flax hessian; linen crash
亚麻粗布刮路机 flax hessian drag
亚麻粗布刮路机整修混凝土路面 flax hessian drag finish
亚麻粗纤维 flax hards
亚麻带 linen-tape
亚麻袋布 flax sacking
亚麻地毯 flax carpet
亚麻地毯底布 linen rug
亚麻电木 linen bakelite
亚麻短纤维 flax hards; flax tow
亚麻帆布 flax canvas; linen canvas; linen duck
亚麻干茎 flax stock
亚麻杆 falx straw
亚麻杆碎片 shive
亚麻稿秆 flax straw
亚麻割捆机 flax buncher
亚麻根腐病 foot rot of flax
亚麻工业 linen industry
亚麻厚油 blown linseed oil; stand linseed oil
亚麻黄质 linoxanthine
亚麻芥 butch flax; false flax
亚麻精 linolenin; trilinolenin
亚麻捆 bobbin
亚麻腊 flax wax
亚麻联匹 linen pot
亚麻刨花板 flaxboard
亚麻皮 flax shives
亚麻漂白 crofting
亚麻破布 linen rags
亚麻清油 boiled linseed oil; linseed oil varnish
亚麻仁 flaxseed; linseed
亚麻仁油 flaxseed oil; haarlem oil; stand oil; oil of flaxseed; linseed oil
亚麻仁油含量 stand oil content
亚麻仁油合成树脂釉瓷 stand oil synthetic resin enamel
亚麻仁油壶 stand oil kettle; stand oil pot
亚麻仁油灰 linseed oil putty
亚麻仁油混合试验 linseed oil combination test
亚麻仁油混合性试验 miscibility test in linseed oil
亚麻仁油混溶性试验 miscibility test in linseed oil
亚麻仁油漆 stand oil paint
亚麻仁油气味试验 linseed oil odor test
亚麻仁油清漆 linseed oil varnish
亚麻仁油酸 linoleic acid
亚麻仁油硝酸试验 linseed oil nitric acid test
亚麻仁油锌白釉瓷 stand oil white enamel
亚麻仁油型清漆 stand oil type varnish
亚麻仁油荧光试验 linseed oil fluorescence test
亚麻仁油皂 linseed oil soap
亚麻仁油毡 lino(leum)
亚麻仁油脂肪酸 fatty acid of linseed oil; linseed oil fatty acid
亚麻仁油中国白釉瓷 stand oil Chinese white enamel
亚麻仁脂肪酸 linseed fatty acid
亚麻(色)的 flaxen
亚麻纱 linen yarn
亚麻绳 flax rope; linen rope
亚麻梳籽装置 deseeder
亚麻属 flax; Linum < 拉 >
亚麻束 beet
亚麻双面梳麻机 duplex flax hackling machine
亚麻酸 linolenic acid
亚麻酸和亚油酸 linolenic and linoleic
亚麻酸盐 linolenate
亚麻碎茎辊 flax-breaking roller
亚麻碎料板 flaxboard
亚麻碎屑 flax-waste
亚麻填料 seed bag
亚麻条格布 linen check
亚麻脱粒辊 linseed roller
亚麻脱粒机 deseeding machine
亚麻脱酸 flax degumming
亚麻脱籽机 whipper deseeding machine
亚麻网眼布 linen mesh
亚麻纤维 flax; flax fiber [fibre]; linen fiber
亚麻纤维板 flax fiber board
亚麻线 flax
亚麻屑 flax shives; pouce
亚麻羊毛交织物 linsey
亚麻油 flax oil; lino(leum) oil
亚麻油定油 linseed stand oil
亚麻油凡立水 linseed oil varnish
亚麻油厚油 tekaol
亚麻油(清)漆 linseed oil varnish
亚麻油桐油定油 linseed-tung stand oil
亚麻油橡胶 factis
亚麻油盐催干剂 linolenate drier
亚麻油毡 lino(leum)
亚麻油毡浮雕版 linocut
亚麻油毡铺地 lino(leum) flooring
亚麻油毡铺块 Linotile
亚麻油脂 linol(e)in
亚麻毡 flax felt
亚麻织机筘号 linen reed count
亚麻织物 flax burlap; linen; linen cloth
亚麻纸 linen
亚麻制的 flaxen
亚麻籽 flaxseed; linseed
亚麻籽饼 linseed cake
亚麻籽粉 linseed meal
亚麻籽油腻子 linseed oil putty
亚麻籽熟油 linseed stand oil
亚麻籽油 flaxseed oil; linseed oil
亚麻籽油醇酸树脂 linseed oil alkyd
亚麻籽油打底剂 linseed oil priming agent
亚麻籽油底涂 prime coating with linseed oil
亚麻籽油干燥 linseed oil drying
亚麻籽油灰 linseed oil putty
亚麻籽油基 linseed oil base
亚麻籽油漆布 lino(leum)
亚麻籽油清漆 linseed oil varnish; kettle boiled linseed oil
亚麻籽油溶液 linseed oil solution
亚麻籽油乳剂 linseed oil emulsion
亚麻籽油酸 linseed oil acid
亚麻籽油酸盐 linoleate
亚麻籽油涂料 linseed oil paint
亚麻籽油油漆 linseed oil paint
亚麻渍漆 linseed oil varnish
亚马孙地槽 Amazon geosyncline
亚马孙-哥伦比亚腕足动物地理大区 Amazon-Colombian brachiopod region
亚马孙古陆【地】Amazonia
亚马孙合作条约组织 Tratado de Cooperation Amazonia
亚马孙河 Amazon River
亚马孙深海扇 Amazonas abyssal fan
亚马托铅锡锑轴承合金 Yamato metal
亚毛细管的 subcapillary
亚毛细管孔隙率 subcapillary porosity
亚毛细孔 subcapillary openings
亚毛细孔隙 subcapillary interstice
亚煤贫属 < 拉 > Hypocapnodium
亚美尼亚红 Armenian red
亚美尼亚红土 Armenian bole
亚美尼亚建筑 Armenian architecture
亚美尼亚式 Armenian style
亚门 subphylum
亚锰的 manganous
亚锰酸盐 manganite
亚眠大教堂 < 为 1200 ~ 1269 年法国哥特式建筑 > Amiens cathedral
亚模型 submodel
亚目 suborder
亚南极带 subantarctic zone
亚南极中层水 subantarctic Intermediate Water
亚铌的 niobous
亚黏[粘]土 clay(ey) loam; clayey soil; dauk [dawk]; donk; lam; loam; mild clay; sabulous clay; sandy clay; subclay
亚黏[粘]土的 loamy
亚黏[粘]土堵壁 loam seal(ing)
亚黏[粘]土化作用 loamification
亚黏[粘]土截水墙 loam seal(ing)
亚黏[粘]土围堰 loam seal(ing)
亚欧地震构造图 Asia-Europe seismotectonic map
亚偏晶 hypomonotectic
亚偏晶合金 hypomonotectic alloy
亚平流层 substratosphere
亚平宁山脉 Montes Apennines
亚期 substage
亚脐菇属 < 拉 > Omphalina
亚千电子伏 subkeV
亚铅酸盐 plumbite
亚潜像 latent sub-image
亚浅海底的 ellitoral
亚乔木 < 树高 9 ~ 30 米 > mesophanerophyte
亚秦岭山地 sub-Qinling mountain
亚球腔菌属 < 拉 > Metasphaeria
亚区 subregion
亚区带土壤 intrazonal soil
亚区域 subblock
亚全球性巨成矿带 subglobal metallogenic belt
亚全息图 subhologram
亚群 subgroup
亚群系 paraclimax
亚壤土 clayey loam
亚壤土砂 loamy light sand
亚热带 semitropics; subtropic(al) belt; subtropic(al) zone; subtropics
亚热带常绿林黄壤带 subtropic(al) evergreen forest yellow cinnamon earth zone
亚热带的 semitropic(al); subtropic(al)
亚热带地区 subtropic(al) region
亚热带动物 subtropic(al) animal
亚热带辐合带 subtropical convergence
亚热带干旱气候 dry subtropic(al) climate
亚热带高压带 subtropic(al) high belt
亚热带混交雨林 mixed subtropic(al) rain forest
亚热带季风气候 subtropic(al) monsoon climate
亚热带季雨林红壤带 subtropic(al) monsoon forest red soil zone
亚热带林 subtropic(al) forest
亚热带曝晒试验 subtropic(al) exposure test(ing)
亚热带气候 subtropic(al) climate
亚热带森林 subtropic(al) forest
亚热带生态系统 subtropic(al) ecosystem
亚热带生态学 subtropic(al) ecology
亚热带湿润气候 moist subtropic(al) climate
亚热带式冰箱 subtropic(al) class refrigerator
亚热带土壤 subtropic(al) soil
亚热带无风带 subtropic(al) calm belt; subtropic(al) calms
亚热带亚带 subtropic(al) subzone
亚热带岩溶 subtropic(al) karst
亚热带雨林 laurilignosa; subtropic(al) rain forest
亚热带针叶树 subtropic(al) conifer
亚热带植物 subtropic(al) crop; subtropic(al) plant
亚热的 subthermal
亚软质 quasi-soft
亚砂土 clayey loam; clayey sand; loam; loam(y) sand; mild sand; sandy loam (soil); subsand
亚砂土的 loamy
亚筛 subsieve
亚筛范围 subsieve range
亚筛分析 subsieve analysis
亚筛粉(末) subsieve fraction; subsieve powder
亚筛份额 subsieve fraction
亚筛颗粒 subsieve particle
亚筛粒度 subsieve; subsieve size
亚筛粒度范围 < 小于 325 目的粉末 > subsieve size range
亚筛粒度分析仪 subsieve size apparatus
亚筛粒子 subsieve particle
亚筛料末 subsieve fraction; subsieve powder
亚筛细度 < 不能用筛子分级的微粒 > subsieve fineness
亚杉 taiwania
亚砷酸 arsenious acid
亚砷酸钙 calcium arsenite
亚砷酸酐 arsenous acid anhydride
亚砷酸钾 potassium arsenite
亚砷酸钾溶液 potassium arsenite solution
亚砷酸钠 arsenite sodium; sodium arsenite
亚砷酸氢铜 Scheele's green
亚砷酸铜 copper arsenite; cupric arsenite
亚砷酸铜氨(溶)液 < 一种木材防腐剂 > ammonical copper arsenite; chemonite; ammoniated copper arsenite
亚砷酸铜复盐 Schweinfurt(h) green
亚砷酸锌 zinc arsenite
亚砷酸盐 arsenite; nickel arsenite
亚砷酸盐农药 arsenite pesticide
亚砷铜氨液 < 木材防腐用 > ammoniacal copper arsenate
亚声 infra-acoustic; infra-audible; infrasonic sound; infrasound
亚声波 subaudio wave
亚声波振荡 subaudio oscillation
亚声频率 infra-acoustic(al) frequency; infrasonic frequency
亚声频振荡器 subaudio oscillator
亚声频振动 subsonic vibrations
亚声破裂 subsonic rupture
亚声破裂速度 subsonic rupture speed
亚声哨 subsonic whistle
亚声速 infra-audible sound; subsonic speed
亚声速边界层 subsonic boundary layer

亚声速的 infrasonic;subaudio;subsonic
亚声速空气动力学 subsonic aerodynamics
亚声速空气动力学系数 subsonic force coefficient
亚声速流 subsonic flow
亚声速马赫数 subsonic Mach number
亚声速喷气飞行器 subsonic jet aircraft
亚声速气流 subsonic flow
亚声速区 subsonic area
亚声速升力 subsonic lift
亚声速稳定性 subsonic stability
亚声速压缩 subsonic compression
亚声速压缩机 subsonic compressor
亚声速运动 subsonic motion
亚声探测 infrasonic detection
亚湿土 subhumid soil
亚石墨 subgraphite
亚氏孢粉 afzelia
亚世 subepoch
亚属 subgenus
亚数据库 subdata base
亚松森 < 巴拉圭首都 > Asuncion
亚松森桥 < 巴拉圭 > Asuncion Bridge
亚速尔反气旋 Azores anticyclone
亚速尔高压 Azores high
亚速海 Azor sea
亚损伤 subdamage
亚铊的 thallous
亚太地区国际贸易博览会 Asia-Pacific International Trade Fair
亚太地区小水电研究培训中心 < 中国 > The Asia-Pacific Regional Network for Small Hydro Power
亚太经济合作组织 Asia Pacific Economic Cooperation
亚太经济和社会委员会 Economic and Social Commission for Asia and the Pacific
亚太经济圈 Asia-Pacific economic circle
亚太西洋期 subAtlantic epoch
亚态 metastate;substate
亚钛的 titanous
亚弹性 hypoelasticity
亚弹性的 hypoelastic
亚碳 carbene
亚锑酸盐 antimonite
亚体系 subsystem
亚条纹 subfringe
亚条纹技术 subfringe technique
亚铁 ferrous iron
亚铁磁材料 ferro-magnetic material
亚铁磁共振 ferrimagnetic resonance
亚铁磁体 ferrimagnetics; ferrimagnetism
亚铁磁性 ferrimagnetism
亚铁磁性矿物 ferrimagnetism mineral
亚铁磁性限幅器 ferrimagnetic limiter
亚铁的【化】ferro(us)
亚铁电晶体 ferro-electrics
亚铁离子 ferrous ion
亚铁钠闪石 arfvedsonite
亚铁氰化钾 potassium ferrocyamde; prussiate of potash; yellow prussiate
亚铁氰化钠 sodium ferrocyanide; yellow prussiate of soda
亚铁氰化铁 Berlin blue; ferric ferrocyanide; iron ferrocyanide; Prussian blue
亚铁氰化物 ferrocyanide
亚铁氰化物法 ferrocyanide process
亚铁氰酸 ferrocyanic acid
亚烃基 alkylene
亚同步 metasynchronism;subsynchronous
亚同步卫星 subsynchronous satellite
亚同温层 substratosphere
亚桐定油 linseed-tung stand oil
亚铜的 cuprous;molybdous
亚头顶钢线钉 flat smooth head steel wire nail
亚土层 subsoil;undersoil
亚土层勘探 subsoil exploration
亚土层排水 subsoil drainage
亚土层水 subsoil water
亚网络 subnet(work)
亚微粉末 submicro powder
亚微观的 submicroscopic
亚微观的大小 submicroscopic size
亚微观断裂 submicrofracture
亚微观构造 submicroscopic structure
亚微观结构 submicrostructure
亚微观粒子 submicroscopic particle
亚微观束状簇 submicroscopic clustering
亚微观损伤 submicroscopic damage
亚微观特征 submicroscopic feature
亚微观形态特征 submicro-morphological feature
亚微管【生】submicrotubule
亚微颗粒排放 submicro-particulate emission
亚微颗粒物质 submicro-particulate matter
亚微克 submicrogram
亚微克量 submicrogram quantity
亚微孔隙 submicroscopic porosity
亚微粒 amicron; hypomicron; submicron; submicro particle
亚微粒气溶胶 submicronic aerosol
亚微粒体 submicrosome
亚微粒子 subparticle
亚微量 submicro
亚微量法 submicro method
亚微量分析 submicro analysis
亚微量试样 submicrosample
亚微裂纹 submicroscopic cracking
亚微米 submicron
亚微米波 submillimeter wave
亚微米薄膜 submicron film
亚微米空传微粒 submicron airborne particle
亚微米粒子 submicron particle
亚微米气溶胶 submicron aerosol
亚微米微粒排放 submicron particulate emission
亚微米微粒物质 submicron particulate matter
亚微米细粒 submicron
亚微秒 submicrosecond
亚微微秒 subpicosecond
亚微微秒脉冲 subpicosecond pulse
亚微细粒 submicron particle
亚微细粒的分析 size analysis of submicron particles
亚微细粒的筛析 size analysis of submicron particles
亚微型 submicron
亚微型栅 submicron gate
亚微震 submicroearthquake
亚维耳水 Javelle water
亚位点 sublocus
亚温带的 subtemperate
亚稳 metastable
亚稳电平 metastable level
亚稳(定)的 metastable
亚稳(定)结构 metastable structure
亚稳(定)平衡 metastable equilibrium
亚稳定平衡图 metastable diagram
亚稳定土 metastable soil; substable soil
亚稳定型 metastable type
亚稳定性 metastability
亚稳度 metastability
亚稳分解 metastable decomposition
亚稳分界线 spinodal line
亚稳峰 metastable peak
亚稳固体 metastable solid
亚稳过饱和 metastable supersaturation
亚稳过渡 metastable transition
亚稳氦磁强计 metastable helium magnetometer
亚稳核 metastable nucleus
亚稳极限 metastable limit
亚稳境 metastable region
亚稳区 metastable range; metastable region
亚稳缺陷 metastable defect
亚稳溶液 metastable solution
亚稳态 semi-stable state
亚稳态白云方解石 metastable dolomitic calcite
亚稳态存储元件 metastable memory element
亚稳态方程 metastable equation
亚稳态峰 metastable peak
亚稳态粒子数 metastable population
亚稳态能级 metastable energy level
亚稳态跃迁 metastable-state jump; metastable-state transition
亚稳相 metastable phase
亚稳相生长 metastable growth
亚稳相图 metastable phase diagram
亚稳中间体 metastable intermediate
亚稳状态 metastable state
亚戊基二醇 pentamethylene glycol
亚戊基四胺 pentamethylene tetramine
亚西利亚建筑 < 公元前 1275 ~ 公元前 538 年 > Assyrian architecture
亚西利亚装饰 Assyrian ornament
亚硒的 selenious
亚硒酸钠 sodium selenite
亚硒酸盐 selenite
亚锡的 stannous
亚锡酸盐 stannite
亚系 subsystem
亚系分化 subline differentiation
亚系统 subsystem
亚细胞结构 subcellular structure
亚细亚酸 asiatic acid
亚显微的 submicroscopic
亚显微结构 submicroscopic structure; submicrostructure; ultra-structure
亚显微毛细现象 submicroscopic capillarity
亚显微缺失 submicroscopic deletion
亚显微细纤维 submicroscopic microfiber; submicroscopic microfibril
亚显微组分 submaceral
亚线性的 sublinear
亚线性振动 sublinear vibration
亚相 parafacies;subfacies
亚相变处理 subcritical treatment
亚相变的 subcritical
亚硝胺 nitrite amine
亚硝氮浓度 nitrous nitrogen concentration
亚硝的 nitrous
亚硝化菌 nitrosomonas
亚硝化作用 nitrosofication
亚硝基苯胺铵 cupferron
亚硝基胍(炸药) nitrosoguanidine
亚硝基系颜料 nitroso pigment
亚硝酸 nitrous acid
亚硝酸氮 nitrite nitrogen
亚硝酸二环己胺 dicyclohexyl amine nitrite
亚硝酸二环己基铵 ammonium nitrite
亚硝酸钙 calcium nitrite
亚硝酸根 nitrite
亚硝酸根合高钴酸钾 cobalt yellow
亚硝酸钴钾 cobalt potassium nitrite
亚硝酸红色烟雾 nitrous fume
亚硝酸钠 sodium nitrite
亚硝酸锶 strontium nitrite
亚硝酸细菌 nitrosomonas
亚硝酸烟雾 nitrous fume
亚硝酸盐 nitrite
亚硝酸盐氮 nitrite nitrogen; nitrous nitrogen
亚硝酸盐累积 nitrite accumulation
亚硝酸烟雾 nitrous fumes
亚硝酸银 silver nitride
亚硝酸酯 nitrite
亚硝态氮 nitrite nitrogen
亚形铜的【化】cuprous
亚型 hipotype [hypotype]
亚烟煤级腐殖煤 humodite
亚岩屑瓦克岩 sublithwacke
亚沿岸的(湖) sublittoral
亚乙硅烷基 disilanylene
亚乙基二氨基甲酸乙酯 ethylidene diurethane
亚乙基二醇 ethylene glycol
亚乙基二氯 ethylidene dichloride
亚乙基缩二脲 ethylidene biuret
亚乙烯氯化树脂 vinylidene chlorite resin
亚乙烯树脂 vinylidene resin
亚音频 infrasonic undersonic frequency
亚音频的 infra-acoustic; infrasonic; subaudible; subaudio; subsonic; undersonic
亚音频电报 infra-acoustic(al) telegraphy
亚音频电报机 sub-audio telegraph set
亚音频级的 subvoice-grade
亚音频级信道 subaudio-grade channel; subvoice-grade channel
亚音频音调 subaudible tone
亚音频振荡 subaudio oscillation
亚音频振荡器 subaudio generator
亚音速 subsonic speed; subsonic velocity
亚音速冲压喷气发动机 subsonic ramjet
亚音速的 infrasonic; subaudio; subsonic
亚音速风洞 subsonic wind tunnel
亚油基醇 linoleyl alcohol
亚油精 linol(e)in
亚油酸 linol(e)ic [linolek] acid
亚油酸钙 calcium noleate
亚油酸钴 cobalt linoleate
亚油酸钾 potassium linoleate
亚油酸锰 manganese linoleate
亚油酸铅 lead linoleate
亚铀的 uranous
亚阈 subthreshold
亚阈浓度 subthreshold concentration
亚原子的 subatomic
亚缘的 submarginal
亚杂砂岩 subgraywacke
亚折射 subrefraction
亚锗的 germanous
亚锗酸盐 germanite
亚正常水面 subnormal pressure
亚值 subthreshold
亚植物群落 subassociation
亚致死剂量 sublethal dose
亚致死接触 sublethal exposure
亚致死量 sublethal amount
亚致死浓度 sublethal concentration
亚致死水平 sublethal level
亚致死损伤 sublethal damage
亚致死效应 sublethal effect
亚致死照射 sublethal exposure
亚中孔型 submesothyrid
亚中温热液矿床 leptothermal deposit
亚中心 subcenter [subcentre]
亚肿锌石 leiteite
亚种 subspecies
亚重力 subgravity
亚重力波 infra-gravity wave

亚周期型 subperiodic type
亚洲大陆 continent of Asia
亚洲大米贸易基金 Asian Rice Trade Fund
亚洲的东南季风 junk wind
亚洲地面天气分析图 analysis of surface on Asia
亚洲工学院 <泰国> Asian Institute of Technology
亚洲工业发展理事会 Asian Industrial Development Council
亚洲管理组织协会 Asian Association of Management Organization
亚洲和太平洋国际气象电信网 International Meteorological Telecommunication Network in Asia and Pacific
亚洲回归热 Asiatic relapsing fever
亚洲基础设施投资银行 <简称亚投行> The Asian Infrastructure Investment Bank
亚洲及太平洋地区经济社会和文化委员会 Economic, Social and Cultural Commission for Asia and the Pacific
亚洲及太平洋地区经济与社会委员会 <联合国> Economic and Social Commission for Asia and the Pacific
亚洲及西太平洋地区承包商协会国际联合会 International Federation of Asian and Western Pacific Contractors Association
亚洲开发基金 Asian Development Fund
亚洲开发研究所 Asian Development Institute
亚洲开发银行 Asian Development Bank
亚洲开发中心 Asian Development Center [Centre]
亚洲内部清算中心 Clearing Center for Intra-Asia
亚洲清算联盟 Asian Clearing Union
亚洲清算同盟 Asian Payment Union
亚洲人 oriental
亚洲生产力组织 Asian Productivity Organization
亚洲式厕所 Asiatic closet
亚洲式柱座 Asiatic base
亚洲蓑蛾 <拉> Canephora asiatica
亚洲太平洋理事会 Asian-Pacific Council
亚洲象【动】elephant maximus
亚洲岩土工程情报中心 <泰国> Asian Information Center for Geotechnical Engineering
亚洲以西全部国家 occident; the Occident
亚洲与远东经济委员会 <ESCAP 的旧名> Economic Commission for Asia and the Far East
亚洲再保险公司 Asia Reinsurance Company
亚轴节 paracardo
亚壮年期 submature
亚族 subtribe
亚组 subunit
亚组构 subfabric
亚组织 substructure
亚祖黏[粘]土 <美国，为一种膨胀性黏[粘]土> Yazoo clay
亚祖式河川 <支干流平行的> Yazoo stream
亚祖式(河流) Yazoo
亚最佳的 suboptimal
亚最佳可控线路系统 suboptimally controlled linear system

研

研光 calender(ing); calender run; glaze; planish
研光薄钢板 planished iron; Russia iron
研光的 glossy
研光度 glossiness
研光辊 calender roll
研光滚 calender bowl
研光机 calender; calendering machine; calender stack; calender train; float plate; glazing machine; mangler
研光机脂 calender grease
研光上色 calender colo(u)ring
研光铁 planishing hammer
研光铁皮 planished iron; Russia iron
研光效应 calender grain
研光用云母 talc(um)

氩

氩39/氩40测年法 Ar39/Ar40 dating method
氩39氩40等时线 39Ar-40Ar isochron
氩39氩39年龄 40Ar/39Ar ages
氩40氩39法 40Ar/39Ar method
氩(保护)气氛 argon atmosphere
氩除气的自来水 tap water degassed with argon
氩等离子体 argon plasma
氩电离检测器 argon ionization detector
氩丢失量 lost argon
氩光灯 argon-filled lamp
氩弧 argon arc
氩弧焊机 argon-arc welder
氩弧焊(接) argon-arc weld(ing)
氩弧焊炬 argon-arc torch
氩弧切割 argon-arc cutting
氩护电弧焊 argon shielded arc welding
氩辉光灯 argon glow lamp
氩激光拔道 argon laser lining
氩激光校直(轨道) argon laser lining
氩激光凝固 argon laser photocoagulation
氩激光(器) argon laser
氩钾法 argon-potassium method
氩检测器 argon detector
氩离子化鉴定器 argon ionization detector
氩气 argon gas
氩气保护层 argon shield
氩气保护烧结 Ar-sintering
氩气保护铸造 argon cast
氩气灯 argon lamp
氩气闪光灯 argon flash
氩气体激光器 argon gas laser
氩稳定同位素组成 stable isotopic composition of argon
氩氧脱碳 argon oxygen decarburization
氩整流器 argon rectifier

咽

咽喉 bottleneck; swallow; throat
咽喉道岔【铁】throat points; throat switch; throat turnouts
咽喉管 uptake
咽喉轨道 throat track
咽喉宽度 throat width
咽喉利用率 the utilization ratio of throat
咽喉区 bottleneck; bottleneck section; gorge; throat area
咽喉区长度 length of switching area; throat length
咽喉区道岔通过能力 carrying capacity of throat points
咽喉区平行作业 simultaneous operation at throat
咽喉设计 design of throat
咽喉线 throat track
咽喉信号机【铁】signal in throat section
咽喉状部分 throat
咽料 <俗称> souring
咽门 fauces
咽气平衡炉具 balanced flued appliance

烟

烟 reek
烟斑 smoke spot; smoking patch
烟波 smoke wave
烟播散 smoke spread
烟草 tobacco
烟草仓库 tobacco warehouse
烟草尘肺 tobaccosis
烟草工业 tobacco industry
烟草货物 tobacco goods
烟草焦油 tobacco tar
烟草烟雾 tobacco smoke
烟草制品废水 tobacco waste(water)
烟草籽油 tobacco-seed oil
烟层顶 smoke horizon
烟层云 smoke stratus
烟尘 flue dust; smog; smoke; smoke (and) dust; smother; soot and dust; stack emission
烟尘测量计 smokemeter
烟尘层顶 smoke horizon
烟尘尘着病 tabacosis
烟尘沉积 deposit(e) of flue dust
烟尘肺 tabacosis pulmonum
烟尘负荷 smoking load
烟尘过滤器 smoke filter
烟尘计 smokemeter [smokometer]
烟尘记录器 smoke recorder
烟尘监测仪 smoke monitor
烟尘检测器 smoke detector
烟尘降落 ash fall
烟尘颗粒 smoke particle
烟尘扩散 emission of smoke
烟尘密度测定器 smokescope
烟尘密度计 smoke density indicator; smoke density meter
烟尘浓度测定 measurement of soot and dust concentration
烟尘浓度测定器 smokescope
烟尘浓度测量 measurement of soot and dust concentration
烟尘浓度计 dust monitor; smoke density meter
烟尘排放 soot emission
烟尘热 fume fever
烟尘室 <壁炉与烟道之间的> smoke chamber
烟尘探测器 smoke detector
烟尘雾 smoke fog
烟尘指示器 smoke indicator
烟尘浊度计 umbrascope
烟尘自动采样器 automatic smoke sampler
烟橱 fume cupboard; fume hood; fuming cupboard
烟橱通风罩 draught hood
烟囱 chimney shaft; chimney stack; lum; smoke chimney; smoke flue; smokestack; smoking pipe; stack; stack chimney; stove pipe
烟囱拔风 chimney draft; stack draft
烟囱拔风调节器 chimney draught regulator
烟囱本体 stack shell
烟囱壁 chimney wall
烟囱标志 funnel mark
烟囱标志灯 funnel light
烟囱侧壁 chimney cheek; chimney jambs
烟囱侧壁底脚 codding
烟囱侧壁基础 codding
烟囱超高 chimney superelevation
烟囱衬壁 chimney back; chimney lining
烟囱衬壁的预制混凝土空心块体 chimney block
烟囱衬砌 chimney lining
烟囱抽风 chimney draught; chimney ventilation; stack draught
烟囱抽风罩 chimney aspirator
烟囱抽力 chimney draft; stack draft; stack draught
烟囱抽力操作系统 chimney operated system
烟囱抽气器 chimney aspirator
烟囱抽气罩 chimney aspirator
烟囱抽吸作用 stack effect
烟囱出灰门 soot door of chimney
烟囱出口 chimney outlet
烟囱出气速度 stack exit velocity
烟囱出屋面 chimney above roof
烟囱出烟孔 stack outlet
烟囱除尘器 arrester [arrestor]
烟囱处置 stack disposal
烟囱单层隔墙 one-withe wall
烟囱挡板 stack damper
烟囱导板 chimney slide
烟囱倒灌风 stack downdraft
烟囱到屋顶的烟道 uptake
烟囱道 chimney flue; gas offtake
烟囱灯罩 chimney
烟囱(底部)进口 chimney intake
烟囱底部烟道进口 chimney intake at base
烟囱地点烟垢 chimney spot
烟囱地基勘察 exploration of chimney foundation
烟囱顶(部) chimney capital; chimney head; chimney top
烟囱顶泛水 fla(u)nching
烟囱顶盖 chimney pot
烟囱顶管 chimney can; chimney cowl; chimney pot
烟囱顶(上的瓦)筒管 tallboy
烟囱顶陶管 clay pot
烟囱顶罩 chimney cap; funnel apron; funnel bonnet; funnel cap; funnel umbrella; funnel hood
烟囱发散物 stack emission
烟囱筏基 chimney raft
烟囱泛水 chimney flashing
烟囱防雨板 chimney flashing
烟囱飞灰 grit
烟囱风挡 chimney damper
烟囱风帽 baffle board; chimney aspirator; hood; wind cowl; tallboy
烟囱风门 chimney damper; damper of chimney
烟囱风雨帽 colt
烟囱风罩 draft hood
烟囱封堵 fire stop
烟囱盖 funnel cover
烟囱盖顶 chimney mantle
烟囱盖头 chimney lid
烟囱高度 chimney height; height of chimney; stack height
烟囱高度效应 chimney altitude effect
烟囱隔板 withe [wythe]
烟囱隔网 <阻拦火星> chimney netting
烟囱工业 smoke stack industries
烟囱鼓风机 chimney fan
烟囱管 chimney fondu; funnel pipe
烟囱管道 chimney flue; chimney pipe
烟囱管帽 chimney can; chimney pot
烟囱管型井筒 stove pipe well casing
烟囱横断面面积 funnel area
烟囱喉部 stack throat
烟囱后倾角 angle of funnel
烟囱滑道 chimney slide
烟囱灰 flue dust
烟囱灰门 soot door of chimney
烟囱混凝土 chimney concrete
烟囱或烟囱管 chimney or vent
烟囱基础 chimney pad
烟囱基座 funnel base
烟囱集灰槽 soot pocket

烟囱检查 stack survey
烟囱铰链板 chimney flap
烟囱节气门 chimney register
烟囱截面 stack cross-section
烟囱空气阻滞器 barometric(al) damper
烟囱口 stack nozzle
烟囱口取样 sampling fo a stack
烟囱口网络 spark arrester [arrestor]
烟囱拉索 stack guy
烟囱喇叭口 stack flare
烟囱缆 stack guy
烟囱冷却器 chimney cooler
烟囱连接点 chimney junction
烟囱连接段 connected stack
烟囱连接管(道) chimney connector
烟囱连通管 vent connector
烟囱帽 bonnet; bonnet hood; chimney cap; chimney head; chimney hood; chimney terminal; chimney top; cowl; funnel hood; smoke jack; smoke stack cap; smoke stack hood; stack cap
烟囱帽顶 chimney capital
烟囱帽盖 funnel cap
烟囱门洞 checkdraft
烟囱面积 funnel area
烟囱模板 shaft formwork
烟囱内衬 chimney liner; chimney lining; flue lining
烟囱内的尖塔形通风设备 stack funnel
烟囱内多层隔墙厚度 multiple withe [wythe]
烟囱内抹面 perget
烟囱内套 chimney liner
烟囱内罩帽 cap; cravat
烟囱能力 chimney capacity
烟囱爬梯 chimney ladder
烟囱排放 chimney emission; smoke stack emission
烟囱排放参数 stack condition
烟囱排放量 stack emission
烟囱排放物 stack effluent
烟囱排水边沟 back gutter
烟囱排水槽 chimney gutter
烟囱排烟 chimney emission; stack draft
烟囱排烟道 stack vent
烟囱喷出 vomit
烟囱喷烟速度 stack exit velocity
烟囱漆 funnel paint; stack paint
烟囱气流 chimney draft; chimney draught; stack draft
烟囱气流调节器 chimney draught regulator
烟囱气体 funnel gases; stack gas
烟囱砌合 chimney bond
烟囱砌块 chimney block
烟囱砌筑 hovelling; smoke pipe bonding
烟囱砌筑成茅舍状 <四周开窗> hovelling
烟囱砌砖法 chimney bond
烟囱牵索圈 funnel ring
烟囱牵条 funnel guy
烟囱倾斜 stack leaning
烟囱群板泛水 chimney apron
烟囱群板突裙 chimney apron
烟囱热耗 chimney loss
烟囱扫除 chimney sweep; chimney sweeping
烟囱上部 chimney head
烟囱上口直径 diameter of stack opening
烟囱上升道 funnel uptake
烟囱身 chimney shaft; shaft; tun
烟囱施工脚手架 chimney construction scaffold(ing)
烟囱施工卷扬机 chimney construction hoist
烟囱式码法 closed chimney packing
烟囱式排水 <土石坝的> chimney drain
烟囱式气流 chimney current
烟囱式云 chimney cloud
烟囱收分 chimney taper
烟囱收缩侧壁 chimney wing
烟囱刷 funnel wire brush
烟囱水平连接部 breeching
烟囱体 chimney stack
烟囱调风门 chimney damper; chimney register
烟囱调节板 chimney back plate; chimney lid; chimney register
烟囱调节(风)门 funnel damper; stack damper
烟囱调节器 chimney damper
烟囱铁连接件 iron chimney connector
烟囱铁爬梯 access hook for chimneys
烟囱通风 chimney draught; chimney ventilation; stack draft
烟囱通风力 chimney draft
烟囱通风帽 <烟囱用的金属旋转通风器> chimney coul
烟囱通风罩 chimney cowl
烟囱筒 smoke pipe
烟囱筒帽 chimney can
烟囱筒身 chimney shaft
烟囱筒身用砖(圬工) masonry brick for chimney shaft
烟囱筒腰 hench
烟囱头 chimney head; smoke stack head
烟囱外壳 chimney casing; chimney shell; funnel shaft
烟囱外壳墙 chimney shell wall
烟囱外套 funnel casing
烟囱桅 mack; stack mast
烟囱温度 stack temperature
烟囱稳索 funnel guy; funnel shroud; funnel stay
烟囱吸力 chimney effect
烟囱效率 chimney efficiency
烟囱效应 chimney effect; flue effect; high stack effect; stack effect
烟囱修建工人 steeplejack
烟囱旋帽 chimney jack
烟囱旋转帽 turncap
烟囱碹 chimney arch
烟囱压顶 creasing course
烟囱咽喉 funnel uptake
烟囱烟道 chimney flue
烟囱烟灰 chimney soot
烟囱烟羽状况 chimney plume behavio(u)r
烟囱因素 stack factor
烟囱有效高度 effective stack height
烟囱有效吸力 available draught of flue
烟囱雨 stack rain
烟囱雨水槽 back gutter; chimney gutter
烟囱闸板 check damp; chimney flap; flue damper; flue shutter; stack damper
烟囱障碍灯 signal lamp on chimney
烟囱罩 bonnet; smoke hood; smoke jack; smoke pipe cap; watch cap
烟囱罩布 funnel cover
烟囱遮盖 chimney hood
烟囱支索 funnel guy; funnel shroud; funnel stay; smoke stack guy
烟囱中热交换器 stack heat exchanger
烟囱砖(块) chimney brick
烟囱状衰减器 chimney attenuator
烟囱状物 chimney
烟囱锥度 draft angle
烟囱纵向分隔板 wythe
烟囱阻力 stack friction
烟囱最高高度 maximum stack height
烟囱作用 stack action
烟囱座 chimney foot; sconce
烟囱座基 chimney base
烟窗 femerell
烟挡 smoke shelf
烟道 air flue; breeching; chimney flue; chimney neck; chimney pipe; combustion flue; ducting; flue; flue gas dust; flue pipe; flueway; foul air duct; funnel; furnace flue; gas circuit; gas conduit; gas duct; gas flue; gas outlet chamber; smoke canal; smoke channel; smoke flue; trunk way; uptake; vent line; vent stake; chimney intake <由锅炉到烟囱间的>
烟道表面 area of flue
烟道布置 flue arrangement
烟道采暖炉 duct furnace
烟道采样 stack sampling
烟道尘 flue dust; quick ash
烟道沉尘室 flue dust pocket
烟道衬块 chimney pot; flue liner
烟道衬里 flue lining
烟道衬砖 flue liner; flue lining
烟道出口 flue outlet
烟道出口抽力 flue outlet draft
烟道除尘器 flue dust collector
烟道挡板 damper; flue baffler; flue shutter; stack damper
烟道挡风板 checkdraft
烟道地板 flue baffler
烟道洞口 checkdraft
烟道阀 chimney valve
烟道废气 flue gas
烟道风帽 baffle board
烟道干燥器 flue drier [dryer]
烟道隔板 furnace bridge
烟道隔墙 baffler; mid-feather; withe
烟道工 flue man
烟道工作段 area of flue
烟道拱 chimney arch; flue arch; stack arch
烟道箍 flue collar
烟道管 flue pipe
烟道管壁 flue side
烟道横截面面积 gas passage area
烟道滑挡 chimney sliding damper
烟道滑动风阀 sliding flue damper
烟道滑动风门 sliding flue damper
烟道灰 flue dust
烟道汇集口 throating
烟道集合 flue grouping
烟道集流 flue gathering
烟道接火 start-up flue
烟道接口 flue collar
烟道接头 <锅炉上的> flue collar
烟道节气门 chimney register; chimney valve
烟道结露 condensation in flues
烟道截面 stack cross-section
烟道截面面积 stack area
烟道进口 chimney intake
烟道进料焚化炉 flue-fed incinerator
烟道进料型公寓焚化炉 flue-fed apartment-type incinerator
烟道净空 flueway
烟道孔 chimney hole; floss hole; flue opening
烟道口 flue opening
烟道冷冻作用 <穿过大涵管上方路面因管道通风促进土基冻胀隆起而破损> chimney action
烟道连接 flue connection
烟道流通截面 free gas passage area
烟道煤灰的清除 flue gas dust removal
烟道门 chimney flap; flue door
烟道面积 flue area
烟道内衬 flue lining
烟道耐火拱顶 flue bridge
烟道排风机 flue exhauster
烟道排气 flue gas
烟道排气净化废水 wastewater from wet scrubber
烟道气 combustion gas; effluent gas; stack gas
烟道气采样方法 stack sampling procedure
烟道气采样器 flue gas sampler
烟道气处理 flue gas treatment
烟道气的放空口 flue gas vent
烟道气的冷却 flue gas cooling
烟道气的凝结物 flue gas condensate
烟道气阀门 flue gas valve
烟道气废热锅炉 flue gas waste boiler
烟道气分析 flue gas analysis
烟道气分析器 flue gas analysis meter
烟道气分析仪 flue gas tester
烟道气分析装置 flue gas analysing apparatus
烟道气粉尘 mean dust concentration
烟道气粉尘平均浓度 mean dust concentration in stack gas
烟道气鼓风机 flue gas blower
烟道气管 flue pipe
烟道气锅炉 flue gas boiler
烟道气回收系统 fume recovery system
烟道气检验仪 flue gas tester
烟道气净化器 flue gas purifier
烟道气净化系统 flue gas cleaning system
烟道气开关 flue gas valve
烟道气流速 velocity of flue gas
烟道气气柜 flue gas holder
烟道气清洗器 flue gas washer
烟道气热损失 flue gas loss
烟道气入口 waste gas inlet
烟道气脱硫 desulfurization of flue gas; flue gas desulfurization
烟道气脱硫过程 flue gas desulfurization process
烟道气脱硫技术 stack-gas desulfurization technique
烟道气脱硫系统 flue gas desulfurization system
烟道气脱硫装置 flue gas desulfurization facility; stack-gas desulfurization facility
烟道气温度计 flue gas thermometer
烟道气洗涤器 flue gas scrubber; flue gas washer
烟道气循环 flue gas recirculation; flue gas return
烟道气循环炉 flue gas recirculation furnace
烟道气引风机 flue gas fan
烟道气再热 flue gas re-heating; reheating of flue gas
烟道气中的固体 stack solid
烟道气中粉尘含量测量法 method of measuring dust content in stack gas
烟道气重量 flue gas weight
烟道砌块 flue block
烟道墙 flue wall
烟道墙壁 hot wall
烟道桥 flue bridge
烟道切面积 stack area
烟道入口 flue inlet
烟道式格子体 smooth plain packing; solid chimney flue checker work
烟道受热面 flue heating surface; flue surface
烟道刷 flue brush
烟道损失 flue loss
烟道炭黑 impingement black
烟道套筒 flue collar
烟道调节板 chimney back; smoke damper
烟道调节挡板 flue damper
烟道调节闸门 flop damper
烟道通风器 uptake ventilator

烟道通条 flue auger
烟道凸缘 wind shelf
烟道瓦筒 flue tile
烟道弯管 flue bend
烟道吸尘装置 flue dust retainer
烟道系统 flue arrangement; flue assembly
烟道碹 flue arch; stack arch
烟道碹拱 flue arch
烟道咽喉 chimney throat(ing); chimney waist
烟道有效面积 free gas passage area
烟道预制块 flue block
烟道闸 flue damper; stack damper
烟道闸板 flue shutter
烟道闸门 flue damper
烟道砖 flue block; flue brick
烟道自动风门 smoke damper
烟道纵隔 withy
烟道纵向分隔板 wythe
烟道纵向分隔墙 mid-feather
烟道阻力 flue resistance
烟道组合 flue grouping
烟道座 flue block
烟的毒性 smoke toxicity
烟的光学密度 optic(al) density of smoke
烟的可探测量 detectable quantities of smoke
烟的控制 smoke control
烟的生产 tobacco-curing
烟斗点火钳 tobacco tongs
烟斗菌属 < 拉 > Solenia
烟斗泥 Indian pipestone
烟斗石 pipestone
烟度测量 smoke measurement
烟度计 smokemeter; smokescope; smoke telegram
烟风洞 smoke wind tunnel
烟感报警器 smoke alarm
烟感检测器 smoke detector
烟感器 smoke receptor
烟感探测器 smoke detector
烟感系统 smoke detection system
烟拱 smoke arch
烟垢 chimney soot; chimney spot; crock
烟垢吹净装置 soot blowing equipment
烟管 fire tube; flue header; smoke pipe; smoke tube
烟管板 flue tube sheet
烟管刮灰器 go-devil
烟管滚子 flue roller
烟管锅炉 flue tube boiler; Scotch boiler; smoke tube boiler
烟管焊机 flue welding machine
烟管焊接机 flue welder
烟管菌属 < 拉 > Bjerkandera
烟管群 nest of smoke pipes
烟管扫除 blowout
烟管式火灾报警系统 smoke pipe fire alarm system
烟管式火灾报警装置 smoke detecting arrangement; smoke indicator; smoke pipe fire detector; smoke vent system
烟管刷 go-devil
烟管型圆筒锅炉 smoke tube type cylindrical boiler
烟管族 nest of smoke pipes
烟罐 smoke can
烟光遮蔽高 smoke defilade
烟柜 fume chamber; fume hood; fuming cupboard
烟锅 smoke pan
烟害 damage by fume; fume damage; fume nuisance; fume pollution; smelter injury; smoke damage; smoke nuisance
烟褐色 infuscate
烟黑 gas black; jet black; smoke black
烟黑色 soot; sooty black
烟黑颜料 vegetable black
烟黑沾污了的 soot-laden
烟花爆竹污染 fire crackers pollution
烟化 fuming; smo(u)lder
烟化点 fuming-off point
烟灰 flue cinder; fly(ing)-ash; grime; soot; soot and dust
烟灰沉积 smoke deposition; soot deposit(ion)
烟灰沉降 soot fall
烟灰吹除 soot blowing
烟灰的 sooty
烟灰碟 ash tray
烟灰反应 pyrotechnic reaction
烟灰分离器 flue-dust separator
烟灰缸 ash tray
烟灰盒 ash receptacle; ash tray
烟灰降落 ash fall
烟灰浓度 soot concentration
烟灰排放标准 emission standard of smoke and soot
烟灰盘 ash receptacle
烟灰清除机 soot blower
烟灰色 smoke
烟灰色玻璃 smoked glass
烟灰收集器 flue dust collector
烟灰水泥 fly-ash cement
烟灰水泥混凝土 fly-ash cement concrete
烟灰台 smoking stand
烟灰状构造 smoky structure
烟挥发指数 smoke volatility index
烟火 fire works; pyrotechnics; smudge
烟火报警器 pyralarm
烟火冲击 pyrotechnic shock
烟火的 pyrotechnic
烟火点火器 pyrotechnic(-type) igniter
烟火发火剂 pyrotechnic igniter
烟火浮子 flag float
烟火剂 pyrotechnic composition; pyrotechnic compound
烟火检验器 rocket tester
烟火喷射装药 pyrotechnic blow-off charge
烟火气体发生器 pyrotechnic gas generator
烟火器材储藏室 pyrotechnic locker
烟火施放法 pyrotechnics; pyrotechny
烟火使用技术 pyrotechnics
烟火式充气器 pyrotechnic inflator
烟火探测 fire-smoke detection
烟火探测器 pyrotector
烟火效果物质 pyrotechnic effect material
烟火信号 pyrotechnic light; pyrotechnic signal; self-igniting light
烟火信号弹 pyrotechnics; signal(l)ing rocket
烟火信号码 pyrotechnic code
烟火信号通信[讯] pyrotechnic communication
烟火信号延时线 pyrotechnic delay train
烟火信号装置 pyrotechnic installation
烟火制造术 pyrotechnics; pyrotechny
烟迹 smoke puff; smoke trail; smoking trace
烟剂农药 smoky pesticide
烟碱 nicotine
烟胶 < 熏干的生橡胶片 > smoked sheet
烟晶 morion; smoke quartz; smokestone; smoky quartz
烟精 nicotine
烟扩散 smoke diffusion
烟粒 soot
烟流 plume; smoke plume
烟流技术 smoke technique
烟流向 drift
烟楼 plume
烟缕 smoke plume
烟缕边界 plume boundary
烟缕不透明度 plume opacity
烟缕高度 plume height
烟缕漂移时间 plume travel time
烟缕上升 plume rise
烟缕上升高度 rising height of smoke plume
烟缕上升模式 plume rise model
烟缕伸展范围 plume spread
烟缕湿度 plume moisture
烟缕污染 plume contamination
烟缕型式 type of stack gas plume
烟缕中线 plume center line
烟霾 smaze; smoke haze
烟煤 bitumenite; bituminous coal; bitumite; boghead(ite) coal; bright coal; fat coal; pitch coal; soft coal
烟煤级腐殖煤 humanthracon
烟煤焦 bituminous coke
烟煤结块 bituminous coal caking
烟煤块 bituminous lump coal
烟霉病 dark mildew; sooty mold
烟霉属 < 拉 > Fumago
烟密度 smoke density
烟密度计 smokometer
烟密度指示器 smoke density indicator
烟墨 lamp black
烟幕 artificial smoke; barrage; camouflage; curtain of smoke; screening smoke; smoke curtain; smoke obscuring screen; smoke screen
烟幕剂【化】 nitrocelluose mixture
烟幕施放喷管 smoke discharge manifold
烟幕施放器 smoke generator
烟幕示踪法 method of smoke screen tracing
烟拟水流照片 < 气流模型 > smoke photograph of flow
烟凝结物 smoke condensate
烟浓度 smoke concentration; smoke density
烟浓度计 smoke photometer
烟浓度检验计 smoke tester
烟浓度指示器 smoke density indicator
烟棚 barn
烟片 smoke sheet
烟气 combustion gas; flue gas; fume; gas of combustion; smoke; stack gas
烟气报警器 smoke alarm
烟气爆炸 flue gas explosion
烟气比例调节挡板 gas proportioning damper
烟气捕收管 flue collector
烟气测定 smoke determination
烟气测试探头 gas probe
烟气冲刷 sweep of gases
烟气抽提 smoke extract
烟气抽提管 smoke extract duct; smoke extract shaft
烟气出口 exhaust gas outlet; gas outlet; smoke outlet
烟气出口管 smoke outlet duct; smoke outlet shaft
烟气出口温度 outlet gas temperature
烟气除尘器 flue dust collector
烟气处理 flue gas treatment
烟气处理设备 smoke treatment plant
烟气单位数 smoke unit
烟气挡板 flue gas baffle; gas baffle; smoke damper
烟气导向器 < 蒸汽机车 > smoke deflector
烟气道抹面 parg(et)ing
烟气毒害 fume damage
烟气发生器 smoke generator
烟气防火装置 flue gas fire extinguisher; flue gas system
烟气分离器 fume separator
烟气分析 flue gas analysis
烟气分析器 flue gas analysing apparatus; flue gas analyser [analyzer]; gas analysis instrument
烟气分析仪 flue gas analyser [analyzer]; gas analysis instrument
烟气观测装置 smoke observation device
烟气管理条例 smoke ordinance
烟气横向冲刷束锅炉 cross baffled boiler
烟气降尘室 flue dust chamber
烟气进口 gas approach
烟气净化器 flue gas cleaner
烟气净化设备 purifier of flue gas
烟气净化装置 flue gas purifying installation
烟气扩散地段 smoke propagation zone
烟气冷却器 after-cooler
烟气粒度 particle-size of fume
烟气粒径 particle-size of fume
烟气量 amount of combustion gas; volume of smoke
烟气流程 gas path
烟气流速 velocity of flue gas
烟气灭火系统 flue gas extinguisher system
烟气浓度表 flue density meter; smoke concentration chart
烟气浓度测量 measuring of smoke density
烟气浓度计 smoke indicator
烟气排放 fume emission; smoke emission
烟气旁路装置 gas by-passing arrangement
烟气清洗 flue gas washing
烟气取样管 gas sampling probe
烟气热量的利用 flue utilization
烟气试验 smoke test
烟气室 chimney gas chamber; flue gas chamber; smoke gas chamber
烟气损害 damage by fume
烟气探测器 smoke detector
烟气探测设备 smoke detection appliance
烟气探测系统 smoke detecting system
烟气探测装置 smoke detecting plant
烟气填充防火系统 flue gas system
烟气调节 fume conditioning; gas tempering
烟气调节风门 smoke damper
烟气调质 flue gas conditioning
烟气停留时间 collecting duration
烟气褪色 atmospheric fading; fume fading
烟气脱硫 flue gas desulfurization
烟气脱硫系统 flue gas desulfurization system
烟气脱硫装置 flue gas desulfurization unit
烟气望远镜 telesmoke
烟气温度 flue-gas temperature
烟气温度探头 gas temperature probe
烟气污染 smoke pollution
烟气洗涤器 smoke washer
烟气系统 < 锅炉的 > gas circuit
烟气下沉 down washing
烟气循环风机 circulating gas fan
烟气再循环 flue gas recirculation; gas recirculation
烟气再循环风机 gas recirculating fan
烟气折流器 flue baffler

烟气自动报警设备 automatic smoke alarm service
烟取样器 smoke sampler
烟热报警器 smoke and heat detector
烟色比较图 Ringelmann chart
烟色玻璃 smoke glass
烟色玻璃窗 smoke glass window
烟色的石英 cairngorm
烟色黄玉 smoky topaz
烟色浓度标准表 Ringelmann chart
烟色石英 smoky quartz
烟射流 smoke jet
烟式石英 bull quartz
烟室 smoke compartment
烟水晶 cairngorm;smoky quartz
烟酸 niacin;nicotinic acid
烟塔窗 smoke tower window
烟炱 coom;smut
烟炱沉积 soot deposit(ion)
烟炭 soot carbon
烟通耐火拱顶 flue bridge
烟筒 chimney shaft;chimney terminal;funnel;smokestack;smoke tube;tewel
烟筒挡烟罩 smoke stack deflector;smoke stack hood
烟筒的锥形混合段 mixing cone of the chimney
烟筒的锥形扩压段 affusing cone of the chimney
烟筒顶 smoke stack top
烟筒管 flue pipe
烟筒扩压段 chimney diffuser
烟筒裙管 chimney bell
烟筒通风帽 chimney cowl
烟筒座 smoke stack base
烟突 blind pipe
烟突污染物 stack pollutant
烟团扩散 puff diffusion
烟团子模型 puff sub-model
烟污 besmoke
烟污染 smoke pollution
烟雾 fume;mist;reek;smoke;smoke cloud;smoke fog;smoke shell;smoky fog;smother;smog <烟与雾的混合物>
烟雾包围 smoke bound
烟雾报警信号 smog alert;smoke alarm;smoke alert
烟雾测风流法 smoke stick
烟雾测量器 smokemeter
烟雾层 aerosol layer;smog layer
烟雾层顶 smog horizon
烟雾带 smoke zone
烟雾的不透光性 obscuration of smoke
烟雾地带 smoke band
烟雾点 smoke point
烟雾发散 emission of smoke
烟雾发生器 aerosol producer;fog applicator;fog generator;fogger;fogging machine;smoke generator
烟雾反应 smog reaction
烟雾防护 smoke prevention;smoke protection
烟雾防治 smoke prevention
烟雾放射器 Nebelwerfer
烟雾焚烧 fume incineration
烟雾高度 smog horizon
烟雾管理条例 smoke ordinance
烟雾罐 smoke candle
烟雾化作用 aerosolization
烟雾回收 fume extraction
烟雾机 fog machine
烟雾计 nephelometer;smokemeter
烟雾剂 aerosol;smoke agent;smoke substance;smoking agent
烟雾检测器 fumes detector
烟雾警报 smog alert;smoke alarm
烟雾警告指示器 smoke warning indicator
烟雾净化 fume cleaning
烟雾控制 smog control;smoke control
烟雾控制器 smog control device
烟雾扩散 spread of smoke
烟雾扩散试验 spread of flame test
烟雾滤除效率 smoke removal efficiency
烟雾滤过器 smoke filter
烟雾弥漫的 hazy;smoke-laden;smoky
烟雾密度 smoke density
烟雾浓度 smoke concentration
烟雾浓度测定器 smokescope
烟雾浓度测量器 smokescope
烟雾浓度分级 fume rating
烟雾排出管道 fume vent line
烟雾排除设备 smoke abatement device
烟雾喷枪 fog gun
烟雾喷射机 aerosol sprayer
烟雾气溶胶 smog aerosol
烟雾(强)度系数 coefficient of haze
烟雾燃烧设备 fume combustion apparatus
烟雾色调 smoke shade
烟雾试验 aerosol test;smoke test
烟雾室 fume chamber;smog chamber
烟雾损失 fume loss
烟雾探测 smoke detection
烟雾探测器 detector of smoke;fume detector;smoke detector
烟雾天气 hazy weather
烟雾透光率 transmittance of light(ing)
烟雾透过率 transmittance of light(ing)
烟雾危害 smog injury
烟雾污染 smoke pollution
烟雾洗涤器 fume scrubber
烟雾显像密度计 smoke densitometer
烟雾消除 abatement of smoke;fume abatement
烟雾消除器 mist eliminator
烟雾信号 smoke signal
烟雾信号装置 fumes detector
烟雾形状 fume shape
烟雾性 smokiness
烟雾抑制器 fume suppressor
烟雾指示器 smokemeter
烟雾指数 smog index
烟雾质 aerosol
烟雾中毒事件 smog poisoning episode;smog poisoning incident
烟雾状的 smoggy
烟雾自动报警器 automatic smoke alarm
烟雾自动取样器 automatic smoke sampler
烟雾自动探测器 automatic smoke detector
烟下沉现象 fumigation
烟酰胺 nicotinamide;nicotinic acid amide
烟香椿 Brazilian cedar
烟箱 smoke-box;smoke chamber
烟箱T形汽管头 smoke-box tee head
烟箱鞍座 smoke-box saddle
烟箱撑杆 smoke-box brace
烟箱挡烟板 diaphragm plate
烟箱挡烟板尾板 diaphragm apron
烟箱底衬 smoke-box bottom liner
烟箱隔板台板 table plate
烟箱管板 front tube plate;front tube sheet;smoke-box tube plate;smoke-box tube sheet
烟箱过热器 smoke-box superheater
烟箱后圈 smoke-box back ring
烟箱火星网 smoke-box netting
烟箱壳 smoke-box shell
烟箱门 smoke-box door
烟箱门衬板 smoke-box door liner
烟箱门垫片 smoke-box front door washer
烟箱门夹 smoke-box door clamp
烟箱门铰链 smoke-box door hinge
烟箱门手柄 smoke-box door handle
烟箱门闩 smoke-box door catch
烟箱气流调节门 smoke-box draft regulating damper
烟箱前板 smoke-box front sheet
烟箱前端 smoke-box front
烟箱前圈 smoke-box front ring
烟箱清灰孔 smoke-box cleaning hole
烟箱清灰孔盖 smoke-box cleaning hole cover
烟箱裙管 cleaning pipe
烟箱洒水器 smoke-box sprinkler
烟箱筒 smoke-box shell
烟箱延伸圈 smoke-box extension
烟箱蒸汽管接头 smoke-box steam pipe connection
烟箱中间圈 smoke-box middle ring
烟橡 fumed oak
烟熏 besmoke;fume smoking;fumigate;smoke;smoking;smudging;soot
烟熏点 smudge pot
烟熏干燥 smoke dry(ing)
烟熏干燥法 flue curing
烟熏干燥装备 equipment for flue curing
烟熏剂 fumigant;smoke generation
烟熏胶片 smoked rubber sheet
烟熏烤干 fire-curing
烟熏器 fumigator
烟熏室 smoke house
烟熏损坏 smoke damage
烟熏褪色 gas fading
烟熏味 smoky flavour
烟熏烟 flue-cured tobacco
烟熏釉层 smoked glaze
烟熏纸 smoked paper
烟眼 smoke eye
烟叶 strip leaf;tobacco plant
烟叶干燥法 process for drying tobacco
烟叶加工 cured tobacco
烟叶质量 quality of tobacco
烟页仓库 tobacco warehouse
烟液 tobacco juice
烟羽 plume;smoke plume
烟羽不透明度 plume opacity
烟羽高度 plume height
烟羽激光探测器 plume laser detector
烟羽扩散 plume diffusion
烟羽路径 path of plume
烟羽上升 plume rise
烟羽上升高度 rising height of smoke plume
烟羽上升模型 plume rise model
烟羽湿度 plume moisture
烟羽抬升 plume rise
烟羽抬升高度 plume rise height
烟羽抬升公式 plume rise formula
烟羽抬升模式 plume rise model
烟羽污染 plume contamination
烟羽形状 plume shape
烟羽型 plume pattern
烟羽中心线 center line of plume;central line of plume
烟云 chimney cloud;cloud of smoke;smoke cloud
烟云及防撞警报 cloud and collision warning
烟云柱 cloud column
烟闸门 smoke valve
烟障 smoke shutter
烟罩 fume collector;helmet;hood for fume;petticoat pipe;smoke bell;smoke jack
烟罩座 petticoat pipe base
烟制二氧化硅 fumed silica
烟制品 tobacco product
烟钟 smoke bell
烟烛 smoke candle
烟柱 belch;column of smoke;pillaring
烟柱测距仪 hazemeter
烟柱高度 fume height
烟砖 <压缩的> negrohead
烟状雾 smoky fog
烟状云 fumulus

胭脂虫 cochineal insect

胭脂虫红 cochineal
胭脂虫栎 ground oak
胭脂虫色淀 kermes lake
胭脂红(色) Chinese lake;cochineal redcarmine;lake;rouge flame;rouge red;carmin(e)
胭脂红色淀 carmine lake;Venetian lake
胭脂红釉 rouge red glaze
胭脂树(红) annatto [aronotta]
胭脂鱼 Chinese sucker

淹地权 flowage right

淹顶丁坝 overflow groin
淹灌 artificial flooding;basin flooding;basin irrigation;flooding;flood(ing) irrigation;flush irrigation;free flooding irrigation;inundative irrigation;irrigation by flooding;level border irrigation
淹灌法 flooding irrigation method;flowing irrigation method;inundation method;submersion irrigation method
淹灌渠 inundation canal
淹灌土 flooded soil
淹灌系统 inundation irrigation system
淹过 overtopping
淹过路面 overtopping of highway
淹井 shaft submergence
淹没 covering;deluge;drown;drowning-out;flooding;flowage;inundate;inundation;oversubmergence;submerge;submersion;swallow;swamp;overwhelm <大浪洪水等>
淹没岸 drowned coast;submerged coast
淹没比 submergence ratio
淹没边界 submerged margin
淹没冰 flooded ice;inundated ice
淹没常数 submergence constant
淹没出流 submerged discharge;submerged efflux
淹没床曝气 submerged-bed aeration
淹没床曝气系统 submerged-bed aeration system
淹没(大)陆架 submerged shelf
淹没的 submerged
淹没的泵 drowned pump
淹没的海岸平原 submerged coastal plain;coastal shelf
淹没的河谷 drowned valley
淹没地 inundated land;overflowed land
淹没地区 flooded area;inundated land;inundated region;ponded area;submerged area;submersion area
淹没地区图 flood map
淹没丁坝 overflow groin;submerged spur dike [dyke]
淹没度 degree of submergence;degree of submersion;submergence
淹没范围 range of inundation
淹没盖 submerged cover

淹没管 drowned pipe; immersion pipe; submerged pipe
淹没灌溉 submerged irrigation; submersion irrigation
淹没海岸 drowned coast
淹没河 drowned stream; overflow river; overflow stream
淹没进水管 drowning pipe
淹没可能 flooding potential
淹没孔口 submerged orifice
淹没孔口式鱼道 submerged orifice fish pass
淹没矿井 inundated mine
淹没历时 duration of flooding; duration of inundation; duration of submergence; duration of submersion
淹没流 drowned flow; submerged flow
淹没滤池(法) submerged filter process
淹没滤水栅 submerged crib
淹没面积 flooded area; flooded periphery; flooding area; inundated area; ponded area; submerged area
淹没农田 inundated cultivated land
淹没喷射流 submerged jet
淹没平原土壤 flood plain soil
淹没曝气 submerged aeration
淹没期 duration of submersion
淹没桥 inundation bridge
淹没区 flooded periphery; flood(ing) area; flood-prone area; floodwater zone; inundated area; pond area; submerged area
淹没润滑 flood lubrication
淹没沙洲 submerged bar
淹没射流 submerged jet
淹没深度 depth of submergence; submerged depth
淹没时间 duration of flooding; time of submergence
淹没式 submerged type; submerged ultimate
淹没式板式膜 submerged plate type membrane
淹没式出水口 submerged outlet
淹没式垂直提升门 submergible vertical lift gate
淹没式底槛 submerged sill
淹没式电站 submergible power plant; submergible power station
淹没式丁坝 submerged dike [dyke]; submerged groin; submerged groyne; submerged jetty; submerged pier; submerged spur
淹没式定轮闸门的坝 submergible roller dam
淹没式浮标 submerged float
淹没式复合膜生物反应器 composite submerged membrane bioreactor
淹没式复合中空纤维膜生物反应器 composite submerged hollow fibre membrane bioreactor
淹没式管道 submerged conduit; submerged pipe; submerged tube
淹没式过滤 submerged filtration
淹没式涵洞 drowned culvert
淹没式弧形门 submerged tainter gate; submergible tainter gate
淹没式活性污泥中空纤维膜生物反应器 activated sludge submerged hollow fibre membrane bioreactor
淹没式间歇曝气膜生物反应器 intermittently aerated submerged membrane bioreactor
淹没式搅拌机 submerged agitator
淹没式接触曝气 submerged contact aeration
淹没式进水口 submerged intake
淹没式孔口 submerged orifice
淹没式控制器 submerged controller
淹没式滤池 submerged filter
淹没式螺旋分级机 submerged spiral type classifier
淹没式膜分离活性污泥法 submerged membrane separation activated sludge process
淹没式膜工艺 membrane-immersed process; submerged membrane process
淹没式膜生物反应器 immersed membrane bioreactor; submerged membrane bioreactor
淹没式膜生物反应器系统 submerged membrane bioreactor system
淹没式膜微滤 submerged membrane microfiltration
淹没式木笼 submerged crib
淹没式排水泵 submergible drainage pump
淹没式平板门 submergible vertical lift gate
淹没式曝气池 submerged aerated basin
淹没式曝气滤池 submerged aerated filter
淹没式曝气器 submerged paddle aerator
淹没式曝气生物膜反应器 aerated submerged biofilm reactor
淹没式曝气系统 submerged aeration system
淹没式取水口 submerged crib; submerged inlet; submerged intake
淹没式射流 submerged jet
淹没式生物接触滤池 submerged biofilter; submerged biological filter
淹没式生物接触器 submerged biological contactor
淹没式生物膜 submerged biofilm
淹没式生物膜法 submerged biofilm process
淹没式生物膜反应器 submerged biofilm reactor
淹没式水表 submerged meter
淹没式水流 submerged flow; drowned flow
淹没式水轮发电机组 submerged hydroelectric(al) unit; submergible hydrogenerator set
淹没式水软化装置系统 submerged demineralizer system
淹没式水跃 drowned-out hydraulic jump; submerged jump
淹没式索型生物膜反应器 submerged rope-type biofilm reactor
淹没式填料 submerged carrier
淹没式温度和氧量记录设备 submersible temperature and oxygen recording equipment
淹没式悬浮载体 build-in type suspended carrier
淹没式厌氧膜生物反应器 submerged anaerobic membrane bioreactor
淹没式叶轮曝气装置 immersed impeller aerator
淹没式溢洪道 submerged spillway
淹没式闸门 submerged gate
淹没式折流板 submerged deflection baffles
淹没式蒸发皿 flood(ed) evapo(u)rator
淹没式中空纤维膜反应器 submerged hollow fibre membrane reactor
淹没式中型厌氧反应器 submerged media anaerobic reactor
淹没式转子曝气 submerged rotor aeration
淹没水流 submerged flow
淹没水舌 drowned nappe; wetted nappe
淹没水深 submerged depth; water depth of submergence
淹没水跃 drowned(-out) hydraulic jump; submerged hydraulic jump
淹没损失 flood damage; flowage damage
淹没洗涤器 flooded scrubber
淹没系数 flooding factor; submergence coefficient; submergence factor; submergence ratio
淹没泄流 submerged discharge
淹没旋转子曝气 submerged rotor aeration
淹没堰 submerged weir
淹没灾害 flood hazard
淹没植物群落 hydatophytia
淹没指标 inundation index
淹没转盘式聚偏氟乙烯膜生物反应器 submerged rotating polyvinylidene fluoride membrane bioreactor
淹溺 drowning
淹水 flooding; inundation
淹水草地 water meadow
淹水草甸 water meadow
淹水草原 sunken meadow
淹水的 floodable
淹水地 submerged land
淹水地区 inundated district
淹水坡度线 submerged grade line
淹水期 hydroperiod
淹水区 flood area
淹水深度 depth of flood
淹水土壤 flooded soil
淹水屋面 submerged roof
淹水沼泽 fen
淹死 drown; drowning
淹死的 drowned

湮

湮灭算符 annihilation operator

湮没 annihilation; obliteration
湮没辐射 annihilation radiation
湮没伽马射线 annihilation γ-ray
湮没角点 obliterated corner
湮没力 obliterating power
湮没算符 annihilation operator
湮没现象 annihilation
湮没作用 annihilation

燕

燕山 Yanshan mountain

燕山联合弧形构造【地】 Yanshan conjunct arc structure
燕山期【地】 Yanshan period
燕山期地槽 Yanshanian geosyncline
燕山台褶带 Yanshan platform folded belt
燕山旋回 Yanshan cycle
燕山亚旋回 Yanshanian subcycle
燕山运动 Yanshan movement
燕山运动 1 幕 Yanshan orogeny 1st episode
燕山运动 2 幕 Yanshan orogeny 2nd episode
燕山运动 3 幕 Yanshan orogeny 3rd episode
燕山运动 4 幕 Yanshan orogeny 4th episode
燕山运动 5 幕 Yanshan orogeny 5th episode
燕山运动 A 幕 Yanshan orogeny episode A
燕山运动 B 幕 Yanshan orogeny episode B

延

延板 board drag

延爆段 delay
延爆剂 inhibitor of ignition
延爆雷管 delay(-blasting) cap
延爆器 delay detonator
延长 continue; drawn-out; dwell; extension; lengthen(ing); prolong(ation); protract(ion)
延长凹槽 elongated trough
延长摆线 superior trochoid
延长板 extension plate
延长报税期限 extension of time for filing
延长臂 adjutage
延长部分 prolongation
延长冲刷历时 prolongation of scouring duration
延长储层测试 extended storage test
延长的 prolate
延长的连发列车间隔距离 extended headway
延长的连发列车间隔时分 extended headway
延长堤线 levee extension
延长动作 overreach
延长动作时间 overreach time
延长断裂 extended fracture; extension fracture
延长墩柱概念 extended column concept
延长发盘 extend offer
延长法 decompression
延长范围 extended range
延长符号 elongation sign
延长付款期限 extension of time for payment
延长工期的权利 entitlement for extension of time
延长工期索赔 claim for extension of time
延长公里 <车站> extended kilometer [kilometre]
延长谷 extended valley
延长管 extension pipe
延长合同 extension contract
延长合同有效期 prolong a contract
延长河 extended river; extended stream
延长河谷 extended valley
延长汇点(河流) Yazoo
延长会让站 lengthening crossing loop
延长计 ductilimeter
延长记录 extension of records
延长加工周期 extension of the work cycle
延长交货时间 stretch-out
延长节理 extended [extension] joint
延长进路 extended route
延长镜头 elongated lens
延长刻度 extension scale
延长率 coefficient of extension; elongation ratio; rate elongation
延长绿灯时间 <感应信号中在初期时间以后可以延伸的绿灯时间> extendible portion
延长码 lengthened code
延长排放系统 extended discharge system
延长喷放时间 extended discharge time
延长喷射 extended discharge
延长期 extension of time
延长期限 additional period; extension of maturity
延长器 elongation pad; extender; lengthener; pad
延长区 region of elongation
延长日期 date of extension
延长软线 extension cord
延长三角形接法 extended delta connection
延长伸臂 <起重机的> extension boom
延长绳路 extension cord
延长时间 time expand
延长时距系统 prolonging time-dis-

tance system
延长时期 extended period
延长使用期保养费系数 extended-life multiplier;extended-use multiplier
延长使用寿命 extended life; extend the life;increase of service life
延长试验 protracted test
延长熟料在高温带时间 soaking
延长水线 extension water line
延长顺向河 extended consequent river;extended consequent stream
延长索赔期 extend the time for filing claims
延长索赔期限 extend the time limit for filing clause
延长统 < 下侏罗统 >【地】Yenchang series
延长筒 extension tube
延长投标有效期 extension of validity of bids
延长物 continuation
延长系数 extension coefficient
延长线 extension flex; production【数】
延长线路 elongation line;extended line; extension line;extension wire
延长线圈 extension coil
延长限度 extension limit
延长信用证有效期 extend letter of credit
延长性抑制 prolonged inhibition
延长因子 elongation factor
延长应用率 extended application rate
延长有效期 extend the expiration date;prolong the period of validity
延长支流汇点 deferred tributary junction
延长枝 elongated shoot
延长轴 spindle extension
延长装置 extension fixture
延程反射 extended range reflection
延迟 backening; defer; delay; detention;hangover;hold off;lag(ging); late;postpone
延迟板 retardation plate;retarder plate
延迟拌和 delay(ed) mixing
延迟报警继电器组 alarm delay relay set
延迟爆破 delay(ed) blast(ing)
延迟爆炸 delay(ed) blast(ing);delayed explosion
延迟闭合 retarded closing
延迟变形 delayed deformation
延迟表层流 delayed subsurface runoff
延迟补偿 delay compensation; delay equalization
延迟步重新启动 deferred step restart
延迟部件 delay unit
延迟槽 delay slot
延迟策略 delay policy
延迟沉降 delayed subsidence
延迟成熟 delayed maturity
延迟程序 delay routine
延迟程序设计 delay programming
延迟出口 deferred exit;exit
延迟处理 deferred processing; delayed time processing
延迟触发 delay trigger
延迟触发脉冲 delayed trigger
延迟触发脉冲振荡器 delayed-trigger generator
延迟触发器 delayed trigger;delay flipflop
延迟传送 delay in delivery
延迟带通滤波器 delay bandpass filter
延迟单元 delay cell;delay unit
延迟的更新过程 delayed renewal process
延迟的循环事件 delayed recurrent event
延迟地址【计】deferred address
延迟点火 delay-action firing;delayed ignition; ignition lag; late ignition; retarding ignition
延迟点火管 time delay squib
延迟点火正时 retarded spark timing
延迟点火装置 delay igniter; retard position
延迟电缆 delay cable
延迟电路 delay circuit
延迟电位 counter potential
延迟电压 delay voltage
延迟定址 deferred addressing
延迟动作 delay action; slow-operating
延迟动作的 delayed action;slow acting
延迟动作继电器 delay-action relay
延迟动作开关 delay-action switch
延迟断开 off delay;time-lag trip
延迟断裂 delayed fracture
延迟队列 deferred queue
延迟多谐振荡器 delay(ing) multivibrator
延迟发放 delay in delivery
延迟发光 delayed luminescence
延迟发火(装置) ignition lag
延迟发生器 delay generator
延迟发送 delayed delivery
延迟发送机 retard transmitter
延迟反馈 delayed feedback; lagging edge;lagging feedback
延迟反馈测听仪 delayed feedback audiometer
延迟反射 delayed reflex
延迟反应 deferred reaction; delayed reaction
延迟反应学习 delayed response learning
延迟费用 deferred charges; deferred expenses
延迟分辨率 delay resolution
延迟符合 delayed coincidence
延迟符合法 delayed coincidence method
延迟辐射效应 late radiation effect
延迟付款 delayed payment; delay in payment
延迟负责 deferred liabilities; liability for delay
延迟附加费 < 运费 > delayed surcharge
延迟跟踪 delay tracking
延迟跟踪器 delay tracker
延迟更新 deferred update
延迟故障 delay fault
延迟关闭 late release
延迟管 phantastron
延迟函数 delay function
延迟呼叫 delayed call
延迟呼叫计数器 calls waiting meter
延迟互换交易权责日 deferred swap accrual date
延迟缓解 delay release
延迟换向 lagging commutation; late commutation
延迟恢复 delayed recovery
延迟回波 delayed echo
延迟回答 delayed response;reply delay
延迟回声 deferred echo
延迟会话 delayed conversation
延迟或暂时存储 delay or temporary storage
延迟畸变 delay distortion
延迟激发 delayed initiation
延迟检验点再启动 deferred check point restart
延迟交付 delayed delivery; delay in delivery
延迟交货 back order(ing);delayed delivery
延迟交货成本 back order cost
延迟交货罚金 penalty on delayed delivery
延迟浇筑 delay in placing
延迟焦化 delayed coke
延迟焦化过程 delayed coking
延迟焦炭 delayed coke
延迟角 delay angle
延迟介质 delay medium
延迟进气 retarded admission
延迟进位 delayed carry
延迟晶体 delay crystal
延迟警报 delayed alarm
延迟警报继电器 delayed alarm relay
延迟径流 delayed run-off
延迟均衡 delay equalization
延迟均衡器 delay equalizer
延迟开发 delayed development
延迟开启 delayed opening
延迟控制 delay control;retarded control
延迟扩展 delay spread
延迟理论 endochronic theory; retardation theory
延迟力 retarding force
延迟利息 delay interest
延迟连接 delayed union
延迟量 retardation
延迟裂纹 delayed cracking;postcooling crack
延迟磷光 delayed phosphorescence
延迟滤波 delay filtering
延迟率 elongation ratio
延迟轮牧 deferred rotation grazing
延迟螺旋线 delay l`ne helix
延迟脉冲 delayed (im) pulse; delay trigger
延迟脉冲发生器 delay-pulse generator
延迟脉冲技术 delay-pulse technique
延迟脉冲振荡器 delay-pulse oscillator
延迟媒质 delay medium
延迟门脉冲发生器 delay gate generator
延迟模糊函数 delay ambiguity function
延迟模式 delay mode;delay pattern
延迟凝固 retarded coagulation
延迟偏差 delay bias;delay distortion
延迟曝气 time delay aeration
延迟期 delay period; detention period;lag phase
延迟期间 timing period
延迟启动 delayed start
延迟起爆 delayed initiation
延迟起动 delayed start
延迟器 delayer;retarder
延迟强化 delayed reinforcement
延迟切断信号 delay disconnect signal
延迟请求方式 delayed-request mode
延迟屈服 delayed yield
延迟确认时间 delayed-confirm time
延迟燃烧 retard(ed) combustion
延迟燃烧处理 fire-retardant treatment
延迟燃烧顶棚 fire-retardant ceiling
延迟燃烧天花板 fire-retardant ceiling
延迟壤中流 delayed subsurface runoff
延迟热 delayed heat
延迟热分解 retarding thermal decomposition
延迟熔断丝 time-delay fuse
延迟蠕动 retarded creep
延迟入渗 delayed infiltration
延迟扫描 delayed scan(ing);delayed sweep;ratchet time-base
延迟申请方式 delayed-request mode
延迟失真 delay distortion
延迟时间 delay(ed) time;delay period;detention period; lag time; retardation time;time of delay; time of lag;heat lag【暖通】
延迟时间分布 delay time distribution
延迟时间寄存器 delay-time register
延迟时期 detention period;period of retardation
延迟式轮牧 deferred and rotational grazing
延迟式中断卫星 delayed repeater satellite
延迟式自动音量控制 delayed automatic volume control
延迟式自动增益控制 delayed automatic gain control;quiet automatic volume control
延迟释放 time-lag trip
延迟释水 delayed drainage
延迟输出 deferred exit
延迟输出设备 delayed output equipment
延迟输入 deferred entry
延迟输入输出 deferred input/output
延迟伺服机构 time delay servo
延迟损失费 damages for delay
延迟锁定 delay lock
延迟态 deferred mode
延迟弹性 delayed elasticity;retarded elasticity
延迟弹性变形 delayed elastic deformation
延迟特性 lag characteristic
延迟条件码 deferred condition code
延迟调节 delay control;retarded control
延迟调制 delay modulation
延迟调制光波 retardation-modulated light wave
延迟调制码 delay modulation code
延迟跳闸 delayed tripping
延迟通道放大器 delayed-channel amplifier
延迟通知 notice of delay
延迟退休金 deferred retirement benefit
延迟完成 delayed finish
延迟网络 delayed network;phase lag network
延迟维持费 deferred maintenance
延迟位 delayed bit
延迟误差 delay error
延迟熄灭信号 delayed blanking signal
延迟系数 delay coefficient;delay factor
延迟系统 delay system
延迟下渗 delayed infiltration
延迟线 artificial delay line;delay cable;delay line;retarding line
延迟线存储器 delay line memory;delay line storage;delay line store
延迟线积分器 delay line integrator
延迟线计数器 delay counter
延迟线寄存器 delay line register
延迟线解码器 delay line decoder
延迟线偏转器 delay line inflector
延迟线圈 retarder
延迟线容量 delay line capacity
延迟线时钟 delay line clock
延迟线消除器 delay line canceller
延迟线引出装置 delay line ejector
延迟线振荡器 delay line oscillator
延迟响应 delayed response
延迟响应换能器系统 delayed response transducer system
延迟向量 delay vector
延迟效应 delayed effect;late effect
延迟卸货附加费 delayed surcharge
延迟信道 delayed-channel
延迟信道放大器 delay-channel ampli-

fier
延迟信号 delayed signal;inhibit(ing) signal
延迟信号放大器 delayed-channel amplifier
延迟信号装置 delayed alarm
延迟型更新 delayed transaction
延迟性 retardance
延迟修改 deferred update
延迟修理 delay repair
延迟序列 delayed sequence
延迟选通脉冲 delayed gate
延迟压实法 delayed-compaction process
延迟因数 retardation factor
延迟引起损害 delay action
延迟应变 deferred strain
延迟荧光 delayed fluorescence
延迟元件 delay component;delay element; delay unit; time delay element
延迟约束 deferred constraint
延迟运算子 delay operator
延迟再启动 deferred restart
延迟闸门脉冲 delayed gate
延迟振荡器 delay generator
延迟征聘 delay in recruitment
延迟整修处理 < 水泥混凝土路面 > delay(ed) finish
延迟支付进口货款 lags
延迟值 length of delay
延迟植入 delayed implantation
延迟指令 delay instruction
延迟指数 delay index
延迟指数响应 delayed exponential response
延迟中毒 delayed toxicity
延迟重发卫星 delayed repeater satellite
延迟重合闸 delayed reclosure
延迟周期 deference period; deferred cycle
延迟轴 retardation axis
延迟装船 delay in shipment
延迟装运罚款条款 penalty for late shipment clause
延迟装置 deferred mount;delayed action device;delay unit
延迟状态 delaying state
延迟着火 ignition lag;retarding ignition
延迟作用 delayed action;retarded action
延尺 running foot;linear foot < 沿直线英尺长 >
延度比 ductility ratio
延度计 ductilimeter; ductility tester; ductilometer;extensometer
延度试验 ductility test
延度试验机 ductibility testing machine; ductility tester
延度系数 ductility factor
延度仪 ductilimeter; ductility tester; ductility (testing) machine; ductilometer;extensometer
延发爆破 electric(al) delay blasting; delay blasting
延发导火线 retarded action fuse
延发电雷管 delayed electric(al) detonator; electric (al) delay detonator;retarded action fuse

延发雷管 delay(-action) detonator; delay (blasting) cap
延发期 delay-action stage
延发时间 delay period
延发性雪崩 delayed action avalanche
延发元件 delay element
延付 delay payment
延付催缴股款 calls in arrears
延付股息 deferred dividend
延付货款 deferred payment
延付债券 deferred bond; extended bonds
延搁 held over;hold-up
延后 postpone
延缓 delay;linger;retardation
延缓爆炸 delay blasting
延缓变形 deferred deformation; delayed deformation
延缓偿付期 moratorium
延缓沉降 delayed fallout
延缓成本 postponable cost
延缓动作报警 deferred-action alarm
延缓动作信号 deferred-action signal
延缓冻结 retardant of freezing
延缓反馈 delayed feedback
延缓反射 delayed reflex
延缓混凝土碳化作用的表面处理 carbonation-retardant concrete coating
延缓剂 retarding admixture;retarding agent
延缓减水剂 retarding and water reducing admixture; retarding and water reducing agent
延缓接续概率 probability of delay
延缓接续制 delay working
延缓利息抵押贷款 deferred interest mortgage
延缓燃烧 retarded combustion
延缓燃烧剂 fire-retardant agent
延缓生长 retarding of growing
延缓水泥强度增长 deferred losses of prestress
延缓弹簧 back-moving spring; backward spring
延缓弹性 delayed elasticity
延缓特性 delay characteristic
延缓投资 lag in investment
延缓维修 deferred maintenance
延缓系数 coefficient of retardation
延缓信号 delay signal
延缓预应力损失 deferred losses of prestress
延缓征聘 recruitment postponement
延缓滞弹性效应 delayed visco-elastic response
延缓作用 delay(ed) action
延交定货 back ordering
延米 linear meter [metre];running meter
延期 deferment;deferred;extension (of time for completion); hold-over; postpone(ment);prolong(ation);put over;respite;stand over;carry over
延期保险 extended (terms) insurance
延期爆炸性 delayed explosibility
延期补偿合同 deferred compensation contract
延期补偿款转让 assignment of deferred compensation
延期偿付 transfer moratorium
延期偿付法 moratory law
延期偿付公债 continued bond
延期偿付权 < 依法给债务人的 > moratorium
延期偿付债券 extended bonds
延期偿还 moratoria
延期偿款协议 forbearance agreement
延期成本 postponable cost
延期承保 extended cover
延期出口 deferred exit
延期处理 deferred processing
延期贷款 deferred credit; work out loan
延期点火装置 delay igniter
延期电信管 electric(al) delay fuse
延期发火 ignition lag
延期罚款 demurrage;late charges
延期返还财产托管信托书 delayed reconveyance
延期费(用) deferred charges;demurrage charges; demurrage (money) < 车辆船舶装卸超过规定时间 >; extension fee;demurrage cost
延期费账 demurrage account
延期分期偿付 extended term amortization
延期付即期信用证 deferred sight letter of credit
延期付款 business credit;defer(red) payment;delayed payment;delay in payment;pay later;payment respite
延期付款交易 deferred payment transaction
延期付款贸易 deferred payment trade
延期付款命令 moratorium
延期付款期间 moratorium
延期付款权利 deferred payment option
延期付款销货法 deferred payment sales method
延期付款信用证 deferred payment of letter of credit
延期付息股票 deferred stock
延期付息债券 deferred bond
延期负债 deferred liabilities
延期更新 deferred update
延期过账 delayed posting
延期海难抗辩书 < 船长呈交公证行的 > extended protest
延期和续定条款 extension and renewal clause
延期还本付息 deferred repayment of capital and interest
延期还债认可书 extension
延期回扣 deferred debate; deferred rebate
延期回扣制(度) deferred rebate system
延期或重订 tacit renewal
延期间接费用 deferred overhead charges
延期建设 deferred construction
延期交付 delay in delivery
延期交割 delayed delivery
延期交割交易 contango dealing
延期交割金 backwardization
延期交货 back order;delayed delivery
延期交货的截止日期 deadline for delayed delivery
延期交货费用 back order cost; delayed delivery cost
延期交货负债 deferred performance liabilities
延期交货通知单 back order memo
延期交易的交割日 prompt day
延期缴税 tax deferral
延期结果 postponed result
延期金 contango
延期拒付证书 extended protect
延期均衡器 delay equalizer
延期课税 postponement of taxation
延期宽限日 days of grace
延期雷管 delay-action detonator
延期利息 deferred interest; interest on arrears;postponed interest
延期纳税 postponement of the taxation;tax deferral
延期年金 deferred annuity
延期年金的现值 present value of a deferred annuity
延期票据 deferred sight
延期契约 extension agreement
延期日息 contango;contango price of money;price of money
延期入口 deferred entry
延期生命年金 deferred life annuity
延期十天 ten-day postponement
延期时间 defer time;demurrage < 指船舶在港期 >
延期使用码头费 penalty dockage
延期收款出口 export by deferred payment
延期收款出口信贷 export credit on deferred payment
延期收入 deferred income; deferred revenue
延期手续费 extension commission
延期输出 deferred exit
延期输入 deferred entry
延期损失不负责条款 time penalty clause
延期损失赔偿 damages for detection
延期损失赔偿额 liquidated damages for delay
延期索偿同意书 letter of licence
延期索赔 delay claim
延期停泊费 demurrage charges
延期维护 deferred maintenance
延期维修 deferred maintenance
延期维修时间 deferred maintenance time
延期未分配费用 deferred unapplied expenses
延期销售 adjourned sale
延期协议 extension agreement
延期信用 deferred credit
延期信用证 extended credit;extended letter of credit
延期一周 a week postponement
延期引爆 ignition lag
延期永续年金 deferred perpetuity
延期再启动 deferred restart
延期炸药 delay powder
延期账(户) deferred account
延期折扣 deferred rebate
延期者 deferrer
延期征聘 delay in recruitment
延期支付 deferred payment;postpone a payment;postponement of payment
延期支付标准 standard of deferred payments
延期支付销售 deferred payment sale
延期执行 compliance delay
延期滞延费 demurrage charges
延期装船 delayed shipment
延期装运 delayed shipment
延期装置 period delay
延期资产 deferred assets
延期作用 delayed action
延燃区 secondary combustion zone
延伸 continue;elongation;expansion; extension;lengthen(ing);prolongation; reach; stretching; stretching run
延伸百分率 elongation percentage
延伸板 extender plate
延伸保险范围 extended coverage
延伸比 ratio of elongation
延伸臂 extendible arm;extendible jib; extension arm
延伸边框 extension jamb
延伸变形 extensional deformation
延伸柄 extension stem
延伸波 drawn-out wave
延伸部分 continuation;extendible portion
延伸部件 extension piece
延伸材料 extender
延伸长度 development length;extension length
延伸车长 augmented car length
延伸成型 stretch forming
延伸传动轴 extension drive shaft
延伸传输驱动 transmission extension drive
延伸窗铰 easy-clean(ing) hinge

延伸大气 extended atmosphere
延伸的 continuous;drawn
延伸的地界线 extended property line
延伸的建筑红线 extended property line
延伸的用地线 extended property line
延伸电报电路 extended telegraph circuit
延伸电缆 extension cable
延伸顶系 elongated apical system
延伸度 extensibility;ductility
延伸段 stretch
延伸断裂 extension fracture
延伸范围 expanded range
延伸杆 <增加伸缩式钻塔高度的> scoop pole;extension rod
延伸杆件 lengthening bar
延伸根系 extensive root system
延伸公差带 projected tolerance zone
延伸管 extension pipe;extension tube
延伸管件 extension piece
延伸光球 extended photosphere
延伸河(流) extended river;extended stream
延伸滑槽 extension chute
延伸机 elongator
延伸机座 extension base
延伸计划 outreach program(me)
延伸接头 extension sub
延伸劲度 extensional stiffness
延伸井 extended reach well
延伸开段 augmented unblock
延伸孔 elongated hole
延伸孔型 elongation pass
延伸框架 ductible frame
延伸缆索 extension cable
延伸链动输送机 extension flight conveyer [conveyor]
延伸领海 extensive territorial waters
延伸流变学 extensional rheology
延伸流动 elongational flow
延伸率 coefficient of extension;elongation ratio;extensibility;modulus of elongation;percentage;rate of elongation;ratio of elongation;specific elongation;unit elongation
延伸(率)试验 elongation test
延伸面 extended surface
延伸模 cupping die;extension die
延伸目标 extended object
延伸黏[粘]度 elongational viscosity
延伸器 extender;stretcher
延伸圈 pull ring;shifting bolt;shifting eye
延伸散射介质 extended scattering medium
延伸声明 farther state
延伸时间 extensible portion
延伸试验负载 elongation load;elongation test load
延伸双凸透镜 prolonged lenticular
延伸顺向河 extended consequent stream
延伸速度 <钻进时> penetration feed
延伸天体 <天文摄影> extended object
延伸位错 extended dislocation
延伸误差 propagated error;propagation error
延伸系数 coefficient of elongation;coefficient of extension;elongation modulus;extension coefficient;lengthening coefficient
延伸纤维射电星云 extended filamentary radio nebula
延伸线(路) extended line;extension line
延伸斜顶式锅炉 extended wagon top boiler
延伸型的 extension-type
延伸性 ductibility;expansivity;extensibility;ductility
延伸性的 productile
延伸压力机 stretch former
延伸烟箱 extended smokebox
延伸仪 extens(i)ometer;extension meter;tens(i)ometer
延伸因子 elongation factor
延伸应变 elongation strain
延伸应力 elongation stress
延伸晕 broad halo
延伸造型 stretch forming
延伸轧机 elongating mill;elongator
延伸振源 extended shock source
延伸轴 axis of extension;extension shaft;outrigger shaft
延伸装置 extension fitting
延伸钻井 extended reach drilling
延伸钻井程序 procedure of extended reach drilling
延伸作用 progradation
延深工作 work ahead
延深井身 workover well
延生破坏 ductile failure
延绳钩 long line fishing
延绳钩法 long lining
延时 delay time;late timing;retardation time;time delay;time lag;time lapse
延时保险丝 time delay fuse
延时报警 delayed alarm
延时爆破 delayed blast(ing)
延时爆炸 delayed blast(ing)
延时闭合 time closing
延时闭合接点 time closing contact
延时操作阀 time delay valve
延时沉降 delayed settlement
延时出动 delayed response
延时处理 delayed time processing
延时磁铁 timing magnet
延时淬火 delayed quenching
延时存车费 demurrage
延时电磁继电器缓放线圈 slow releasing coil of time delayed magnetic relay
延时电雷管 electric(al) delay blasting cap
延时电流速断保护 current fast operating protection with delay
延时电路 delay circuit;time delay circuit
延时电容器 time delay condenser;time-lag condenser
延时动作 deferred action;time delayed action;time-lag action
延时断开接点 time opening contact
延时多谐振荡器 delay multivibrator
延时阀 time delay valve;timing valve
延时分析 time delay analysis
延时分析器 time delay analyser [analyzer]
延时过滤 extended filtration
延时过滤法 extended filtration process
延时计数器 delay counter
延时记录器 delay recorder
延时技术 <改善内燃机燃烧条件的技术> retarded timing
延时继电器 delay-action relay;delay(ed) relay;slow-acting relay;slugged relay;time delay relay;time delay switch;time element;time element relay;time-lag(ged) relay;time limit relay;timing relay
延时搅拌 prolonged agitation
延时接触器 timing contactor
延时解锁器 time-lag release
延时开关 delay-action circuit breaker;delay switch;extended switch;inertia switch;relay switch;time delay switch;timing contactor
延时控制 timing-control
延时雷管 delay(-blasting) cap;delay detonator
延时慢速摄影 memomotion photography
延时抹面 delayed finishing
延时凝结强力纸 <作用于表面上的> surface retarder paper
延时平衡 delay balance
延时平面位置显示器 delayed plan position indicator
延时曝气 delay aeration;extended aeration
延时曝气法 extended aeration process;prolonged aeration process
延时曝气法的活性污泥处理装置 Oxiser
延时曝气法或接触消化法的活性污泥处理装置 Oxigest
延时曝气活性污泥法 extended aeration activated sludge process
延时曝气器 extended aerator;prolonged aerator
延时曝气系统 delay aeration system
延时器 chronotron;decelerator;delayed action;delayer;delay timer;retarder (box);self-timer;time delay unit
延时熔断器 delay-action fuse
延时熔断丝 time delay fuse
延时熔丝 time-lag fuse
延时扫描 delay(ed) sweep
延时摄影 time-lapse photography
延时式平面位置指示器 delayed plane position indicator;open-centered type-P indicator
延时释放 time release
延时双稳态多谐振荡器 delay flip-flop
延时特性 time delay characteristic
延时停车费 demurrage
延时脱扣 delayed tripping
延时微调刻度 fine delay dial
延时位置计数器 delay set counter
延时系统 delayed time system
延时线 delay line
延时线前的副载波再注入 predelay-line re-insertion of subcarrier
延时响应 delayed response
延时消化效应 extended digestion effect
延时信号 delayed signal
延时性维修 deferred maintenance
延时元件 delay cell;delay unit;time delay element;time element
延时原则 <暴雨分析的> extended-duration principle
延时炸药 delay powder
延时装置 delayed device;retarding device;time delay device;time delay mechanism;timing device
延时子程序 time delay subroutine;wait subroutine
延时子程序调用 wait subroutine call
延时阻抗继电器 impedance time relay
延时作用 time-lag action
延停车数 postpone parking number
延拓 continuation
延误 aberrance [aberrancy];arrears;delay;lose time
延误比率 delay ratio
延误导致的损坏 damage on account of delay
延误的 belated
延误罚款 penalty for delay
延误费 cost of delay
延误分布 delay distribution
延误合同罚款 contract delay penalty
延误价值 cost of delay
延误检验 delayed test
延误控制模式 delay mode
延误率 delay rate
延误模型 delay model
延误时间 delay time
延误时间的价值当量 cost of delay
延误损失 loss on account of delay
延误外加时间 delay allowance
延误有效条款 continuation clause
延误运输 delayed shipment
延误支付 delayed payment
延误(滞)微分方程【数】 delay-differential equation
延误(滞)序列【数】 delayed sequence
延线 range line
延限带 range zone
延效 prolongation of effect
延性 ductileness;tensibility;tensility
延性比 ductility ratio
延性变形 ductile deformation
延性材料 ductility material
延性参数 ductility parameter
延性测定计 ductilimeter
延性-脆性转变温度 embrittlement temperature
延性的 ductile
延性度 degree of ductility
延性断口 ductile fracture
延性断裂 ductile fracture;fibrous fracture
延性反应 ductile response
延性反应谱 ductility response spectrum;spectrum of ductility
延性符号 elongation sign
延性负 length fast
延性钢 malleable steel
延性钢筋混凝土框架 ductile reinforced concrete frame
延性构件 ductile member
延性混凝土 ductile concrete
延性计 ductilimeter [ductilometer]
延性建筑物 ductile building
延性结构 ductile structure
延性金属 ductile metal
延性抗弯空间框架 ductile moment-resisting space frame
延性抗震设计 ductility seismic design
延性破坏 ductile failure;fibrous failure
延性破坏机制 ductile failure mechanism
延性破裂 ductile crack;ductile fracture
延性谱 ductility spectrum
延性屈服 ductile yield
延性设计 ductile design
延性势 ductility potential
延性试验 ductility test
延性试验机 ductilimeter;ductility tester;ductility test(ing) machine
延性试验用模 <沥青材料> ductility test mo(u)ld
延性水平 ductile level
延性特征 ductility traits
延性铁 ductile iron;forgeable iron
延性铁T形螺栓 ductile iron T-bolt
延性铁固定器 ductile iron restrainer
延性推入接头 ductile iron pipe joint restraint
延性系数 coefficient of ductility;ductility coefficient;ductility factor
延性需要量 ductility requirement
延性要求 ductility demand;ductility requirement
延性因数 ductility factor
延性正 length slow
延性支承 ductile support
延性值 ductility value
延性铸铁 Ductalloy;ductile cast iron
延性铸铁管 ductile cast iron pipe
延性铸铁铸造车间 malleable foundry
延性转变温度 ductility transition

temperature
延性状况 ductile behavio(u)r
延续 continuation;continue;last
延续拨款 continuing appropriation
延续层序 extended succession
延续进路 successive route
延续行 continuation line
延续卡片 continuation card
延续劳动力再生产 continuing the reproduction of labor power
延续时间 perdurability
延续时期模拟 extended period simulation
延续式点火器 continuous pilot
延续性 continuity;duration
延续预算 continuous budget(ing)
延英尺 <即每一英尺长,1 英尺≈0.3048米> foot run;linear foot;lin.ft;running foot
延增手续费 extension commission
延展 extension
延展承保条款 extended cover clause
延展度 extensibility
延展阶段 ductile stage
延展率 elongation
延展面 plane of flattening
延展面构造【地】planar structure
延展期限 forbearance
延展器 extensor
延展实测系列 extension of records
延展试验 ductility test
延展性 extensibility;malleability;ductility
延展性材料 ductile material
延展有效期 extend the expiration date
延展值 ductility value
延滞贷金 dead loan
延滞方程式 delay equation
延滞费 demurrage
延滞交货罚金 penalty on delayed delivery
延滞径流 delayed run-off
延滞期 delay period
延滞时间 dead time;retarding time;time of lag
延滞系统 retarding system
延滞因素 retarding factor

严冬 severe winter

严封坚固舱 sealed cabin
严格凹函数 strictly concave function
严格比较原理 strict comparison principle
严格标准 strict standard;tight standard
严格博弈 rigid game
严格不等式 strict inequality
严格操作条件 strict operating condition
严格抽样检查 tightened sampling inspection
严格初始函子 strictly initial functor
严格串行顺序 strict sequential order
严格代码 severity code
严格单侧有向图 strictly unilateral digraph
严格单调 strictly monotone
严格单调递减 strictly monotone decreasing
严格单调递减函数 strictly monotone decreasing function
严格单调递增 strictly monotone increasing
严格单调递增函数 strictly monotone increasing function
严格单调函数 strictly monotone function;strictly unimodal function
严格的规则 hard-and-fast rule;strict rules
严格的解答(法) rigid solution
严格的开根条件 strict root condition
严格的理论 rigid theory
严格的理性主义 strict rationalism
严格的使用条件 stringent condition of use
严格的弯曲理论 strict bending theory
严格的温度控制 rigid temperature control
严格递减 strictly decreasing
严格递减函数 strictly decreasing function
严格递增 strictly increasing
严格递增函数 strictly increasing function
严格递增序列 strictly increasing sequence
严格定义场合 well-defined situation
严格定义的概念 well-defined notion
严格定义的函数 well-defined function
严格对角优势 strictly diagonal dominance
严格(方)法 rigorous method
严格分离 strictly separate
严格分析 critical analysis
严格符合主义 doctrine of strict compliance
严格管理 close supervision;stringent regulation
严格归纳极限 strictly inductive limit
严格规定 hard-and-fast rule
严格规章制度 rigorously enforce rules and regulations
严格互补性 strict complementary
严格化 severization
严格寄生物 strict parasites
严格假定 stringent assumption
严格检查 close check;close examination;close inspection;meticulous inspection
严格检验 acid test;close check;close inspection;tightened inspection
严格简单排序 strict simple ordering
严格节约 rigid economy
严格解 exact solution
严格解法 rigorous solution
严格局部极值 strict local extremum
严格决定矩阵博弈 strictly determined matrix game
严格开平方条件 strict root condition
严格考验的 high test
严格可递算子代数 strictly transitive operator algebra
严格控制 close control;strict control
严格控制加班时间 keep a close control of overtime
严格拟凹性 strict quasi-concavity
严格拟凸 strictly quasiconvex
严格赔偿责任 strict liability
严格偏好 strict preference
严格平差 rigorous adjustment
严格平稳过程 strictly stationary process
严格平稳假设 hypothesis of strict stationarity
严格平稳数列 strict-sense stationary series
严格平稳性 strict stationarity
严格嵌入 strict imbedding
严格容差 stringent tolerance
严格容限 severe tolerance
严格弱有向图 strictly weak digraph
严格三角矩阵 strictly triangular matrix
严格上界 strict upper bound
严格上三角矩阵 strictly upper triangular matrix
严格上限 strict upper bound
严格试验 rigid test;severe test
严格凸对策 strictly convex game
严格凸函数 strictly convex function
严格凸空间 strictly convex space
严格伪凸 strictly pseudoconvex
严格误差界限 rigorous error limits
严格下限 strict lower bound
严格线性方程 strictly linear equation
严格线性函数 strictly linear function
严格限制 close restraint
严格限制积分器 hard limited integrator
严格性 severity
严格选择 rigorous selection
严格一致原则 strict compliance rule
严格优先 strict preference
严格有效约束 strict active constraint
严格约束 strict constraint
严格蕴涵 strict implication
严格责任(制) strict liability
严格正定算子 strictly positive definite operator
严格正泛函 strictly positive functional
严格正算子 strictly positive operator
严格正线性映射 strictly positive linear mapping
严格正向量 strictly positive vector
严格支配 dominance strict
严格执行 rigour
严格遵守规定范围 adhere to assigned limits
严格遵守合同条款 keep strictly to the terms of the contract;strict performance of contract
严寒 algidity;biting cold;bitter cold;heavy frost;inclemency;intense cold;jack frost;killing freeze;severe cold
严寒的 inclement
严寒地带 ice box
严寒地带使用的车辆 articized vehicle
严寒气候 frigid climate;harsh climate;severe climate
严寒天气 Arctic weather;subzero weather
严禁烟火 smoking and naked frames prohibited
严峻 austerity
严峻条件 stringent condition
严酷 <天气的> asperity
严酷的 crucial
严酷气候 stern climate
严酷天气 inclement weather;severe weather
严酷性 severity
严厉处罚 a stiff penalty
严厉的规则 stringent regulation
严厉条件贷款 hard loan
严密 closeness;rigour
严密的 refined;rigorous
严密的分析 refined analysis
严密的考查 closed check
严密地组织 regimentation
严密调查 closed investigation
严密度 tightness
严密封闭 tight seal
严密监督 close supervision
严密平差 rigorous adjustment
严密推理 close reasoning
严密效用函数【交】strict utility function
严密遮光物 dark blind
严密止水 tight seal
严守秘密 strictly confidential
严霜 black frost;block frost;killing freeze;killing frost;severe frost;smart frost
严丝合缝 neat seam
严重 major river bed
严重变色(木材) heavy discolo(u)ration
严重冰况 hard frost;hard ice condition
严重波动 catastrophic fluctuation
严重剥落 severe scaling
严重不足 most short
严重差错秒比 severely errored second ratio
严重沉积 damaging deposition;damaging sedimentation;heavy deposition
严重程度因子 severity factor
严重冲击载荷 extreme shock load
严重冲刷 damaging scour
严重出错码 severity code
严重出错信元块比 severely errored cell block ratio
严重错误 severe error;gross error
严重错误信息 fatal error message
严重的 momentous;rigorous
严重的天电干扰 heavy static
严重的住房短缺 acute deficiency of housing
严重电积 damaging deposition;damaging sedimentation;heavy deposition
严重淀积 damaging deposition;damaging sedimentation;heavy deposition
严重渎职 gross misconduct
严重短缺 critical shortage
严重短少 critical shortage
严重断裂的 badly faulted
严重飞溅 excessive spatter
严重风化 bad weathering;heavy weathering
严重富营养化水体 severe eutrophic water body
严重富营养化状况 severe eutrophication condition
严重干旱 severe drought
严重干扰 energetic disturbance
严重故障 catastrophic failure;serious fault
严重关头 crucial point
严重洪灾 disastrous flood
严重后果 serious consequence
严重滑坡 bad slip
严重环境健康风险 serious environmental health risk
严重毁坏 severe destruction
严重橘皮 pebble
严重亏损 heavy losses
严重裂隙化 very fissured
严重冒顶 clumb
严重磨耗 hard wear
严重磨损 heavy wear;severe wear
严重偏析 macrosegregation
严重破坏 catastrophic failure;critical damage;havoc;heavy damage;serious damage;severe damage;significant damage;wrack <车辆、机器的>
严重破坏的路面 badly tracked surface
严重破碎地带 zone of intense fracture
严重破损 serious damage
严重侵蚀 severe attack
严重全漏失 catastrophe loss;serious loss
严重缺乏 severe shortage
严重缺货时的市价 famine price
严重缺陷 critical defect
严重扰动 heavy disturbed
严重入侵 severe intrusion
严重伤人的道路交通事故 serious-injury road accident
严重烧伤 serious burn
严重渗漏点 point of gross leakage

严重失灵船 dead ship
严重失效 major failure
严重失业地区 substantial-unemployment area
严重失业时期 heavy unemployment period
严重事故 fatal accident; major accident; serious accident; serious causality
严重事故报告 serious incident report
严重事故率 severity rate
严重事件 emergency episode
严重受影响的小区 severely affected plot
严重受震区 heavily shaken area
严重衰退 slump of disastrous proportions
严重霜冻 severe frost
严重霜害 damage by severe frost
严重水污染问题 serious water pollution problem
严重损害 critical damage
严重损坏 catastrophic failure; critical damage
严重损坏车 badly damaged car
严重退化生态系统 serious degradation ecosystem
严重外表损伤 appearance failure
严重危害 severely impair
严重违背合同 material breach; serious breach of contract
严重违约 material breach; serious breach of contract
严重违章 gross violation
严重紊流 heavy turbulence
严重污染 heavy pollution; serious pollution; severe contamination
严重污染区(域) heavily polluted area
严重污染水源 heavily polluted source
严重误差 gross error
严重误码秒 severely errored second
严重性 severity
严重锈蚀 heavy rusting
严重盐碱化 heavy salinization
严重淤积 damaging deposition; damaging sedimentation; heavy deposition
严重灾害 havoc
严重胀缩地基 very severely swelling-shrinkage foundation

言

言语行为 verbal behavio(u)r

岩

岩鞍【地】phacolith
岩岸 promuntorium [复 promunturia]; rocky coast; rocky shore
岩岸群落 petrichthium
岩坝 cross cliff; rock bar
岩板尺寸 <12 英寸×24 英寸, 1 英寸≈0.0254 米> duchess slate
岩帮 country rock
岩爆 pressure bump; rock outburst; rock pressure burst; rock burst <岩石爆落现象>
岩爆岩石 popping rock
岩爆震动 rock burst shock
岩崩 avalanche of rock; detritus avalanche; rock avalanche; rock burst; rock fall; rock slide; run rockfill
岩崩坝 rock-slide dam
岩崩地震 rock-fall earthquake
岩崩沟槽 rock-fall furrow
岩崩角砾岩 avalanche breccia
岩崩石块 popping rock
岩比图 rock ratio map
岩壁 ledge; palisade
岩壁控制爆破 controlled wall blasting
岩边 cliffside
岩表坑凹 rock surface depression
岩滨港 rocky harbo(u)r
岩饼 drib(b)let
岩饼锥 drib(b)let cone hornito
岩玻屑凝灰结构 lithic-vitroclastic tuffaceous texture
岩玻屑凝灰岩 lithic-vitroclastic tuff
岩钵 mortar
岩部 petrosa; petrous portion
岩部后缘 posterior border of petrosal part
岩部尖 apex partis petrosae; petrous apex
岩层【地】stratum [复 strata]; bed terrace; earth's formation; layer; rock bedding; rock formation; rock layer; rock stratification; rock stratum [复 strata]; Strata; stratified rock; terrain; terrane
岩层边界 system boundary
岩层变形 deformation of rock
岩层表面 rock surface
岩层侧转面 over-tipped face
岩层层面 bedding surface
岩层层平面 bedding plane
岩层插座 rock socket
岩层产状 attitude of bed; attitude of rock formation; attitude of rocks
岩层产状要素 elements of attitude; formation factor
岩层产状仪 dipmeter [dipmetre]
岩层储集参数 reserved property of formation
岩层储油库 storage jug
岩层错动 strata displacement
岩层单位 lithogenetic unit
岩层的地层次序 stratigraphic(al) sequence of the beds
岩层的连续性 continuity of a bed
岩层的位态 attitude of bedded structure
岩层等斜向褶皱 isoclinal fold in rock beds
岩层底部 rock bottom
岩层底基 base of a formation
岩层顶 rock head
岩层顶层 top of rock bed
岩层断裂 fracturation; fracturation of rock
岩层对比 identification of seams
岩层反向褶皱 reverse fold in rock beds
岩层分离形成的空洞 Weber cavity
岩层风化 bed erosion
岩层封闭 rock seal
岩层观察镜 stratascope
岩层灌浆 ground cementation; ground grouting
岩层厚度 depth of rock stratum; depth of stratum; thickness of bed; thickness of rock stratum
岩层基本水准标石 rock primary benchmark
岩层加固 rock improvement
岩层间的薄黏[粘]土层 clay gouge
岩层间断 strata gap
岩层间水 intermediate water
岩层间隙 strata gap
岩层鉴别 identification of seams
岩层交互变化 alternation of beds; alternation of strata
岩层交互层 alternation of beds; alternation of strata
岩层接触关系 contact correlation of strata
岩层裂缝走向 strike of seam
岩层露头 basset; bend outcrop; outcrop(ping)
岩层锚杆 rock anchor
岩层锚杆支护 strata bolting
岩层锚栓 rock bolt
岩层面 bedding plane
岩层面标高 rock level
岩层模型 strata model
岩层内部破裂声 talking
岩层内部水 internal water
岩层内灌浆 rock grouting
岩层耙松机 rock ripper
岩层平均方向的估计 estimation of formation mean direction
岩层破裂 break of stratum
岩层剖面 strata profile
岩层普通水准标石 rock ordinary benchmark
岩层强度 strata strength
岩层倾覆 overslipping
岩层倾覆面 over-tipped face
岩层倾角 dip of strata
岩层倾向 rock tend
岩层倾斜 inclination of strata
岩层曲率计算 calculation of formation curvature
岩层缺失 omission
岩层群 group of beds
岩层上部 rock head
岩层水 native water
岩层水平状态 drift of strata
岩层素描图 rock formation sketch
岩层特性 strata behavio(u)r
岩层突裂 rock burst
岩层土 regosol
岩层稳定测试仪 seismitron
岩层污染 rock stratum pollution
岩层下沉 rock depression
岩层斜度 dip of rock layer
岩层斜向【地】slant
岩层斜向褶皱 syncline fold in rock beds
岩层性质 nature of ground; strata behavio(u)r
岩层性状 strata behavio(u)r
岩层压力 earth pressure; ground pressure; pressure of strata; rock pressure
岩层移动 earth movement; ground movement; movement of stratum; rock movement; rock stratum movement; strata displacement; strata movement; upheaval; upheave
岩层移动观测站 observation station of rock stratum movement
岩层硬度 formation hardness
岩层油气比 formation gas-oil ratio
岩层褶皱 folds in rock beds; rock fold
岩层中夹层 middleman
岩层状态 formation condition
岩层走向 direction of dip of strata; direction of strata; strike
岩层走向方位角 angle of strike
岩层走向线 strike line
岩层组构 stratofabric
岩层钻孔 rock hole
岩尘 rock dust
岩尘污害 rock-dust nuisance
岩沉积 littoral deposit
岩成土(壤) endodynamomorphic soil; regosol
岩床 bed rock; groundsel; ground sill; ledge(r) (rock); natural ground; rock bed; rock floor; rock sheet; rock sill
岩床地质层系 underlying geologic(al) formation
岩床河流 hard-bed stream
岩床深度 sill depth
岩岛海岸 skerry coast
岩底 hard ground; rock bottom
岩地沥青 <一种天然地沥青> rock asphalt
岩地沥青粉 rock asphalt powder
岩地锚 rock anchor
岩顶 caprock
岩顶标高 rock head level
岩顶锚杆支撑 rock roof bolting
岩顶凿岩机 roof drill
岩顶钻机 ceiling drilling machine
岩洞 rock cave; abra; abri; cave-in rock; cave (rn); caving; dimple; rock cavity; rock house; rock opening; rock shelter; grotto <人工开挖的>
岩洞坟墓 cave tomb
岩洞建筑 hypogeal; hypogee; hypogeum; rock architecture
岩洞落道 karst flow passage
岩洞神祠 cave chaitya hall
岩洞水 cavern water
岩洞陷落 cenote
岩洞修道院 cave monastery
岩洞制图学 speleocartography
岩堆 cliff debris; rock deposit; rock pile; talus
岩堆洞 rock stack-formed cave
岩堆滑坡 talus slide
岩堆坡 talus slope
岩堆泉 talus spring
岩堆锥 talus cone
岩粉 core borings; drill cuttings; drilling dust; drill sludge; fine dust; rock flour; inert dust; rock cuttings; rock flour; rock meal; rock powder; bug dust <空气洗井钻进时>; cuttings of boring <岩芯钻孔的>
岩粉捕集器 sludge trap; slurry trap
岩粉采样 cuttings sampling
岩粉测验仪 rock-dust testing kit
岩粉尘 drilling dust
岩粉沉淀 slime sedimentation
岩粉沉淀槽 <有隔板的> sample trough
岩粉沉淀池 slurry tank
岩粉沉淀器 setting device for sludge sample
岩粉沉淀箱 slime tank
岩粉的排除 cutting removal
岩粉分离器 dust separator
岩粉分析 sludge assay
岩粉分选器 sludge saver
岩粉管 sludge barrel; sludge bucket
岩粉管堵塞的 sludge bound
岩粉加工台 slime table
岩粉密度 sludge density
岩粉凝聚 slime flocculation
岩粉排出管 cuttings pick-up pipe; cuttings pick-up tube
岩粉撒布器 rock-dust distributor
岩粉试样 sludge sample
岩粉试样分样器 sludge splitter
岩粉收集 sludge collection
岩粉收集器 chip catcher; sample trap; sludge-catchment basin
岩粉土 alphitite
岩粉箱 sludge box
岩粉样分选工 sludge man
岩粉样品 sludge sample; dust sample <空气洗空时得到的>
岩粉样(品)钻孔 sludge hole
岩粉用量 quantity of rock powder
岩粉淤泥 slime
岩缝灌浆堵塞 rock sealing
岩缝间水 interstitial water
岩盖 laccolite; laccolith
岩盖山 laccolith mountain
岩高兰 red crowberry
岩埂 riedel
岩拱 rock arch
岩拱脚 rock abutment

岩沟 clint;gryke;lapie
岩核 rocky core
岩滑 rock slide; rock-slide in the catchment area;rock slip
岩化丝炭 petrologic(al) fusain
岩化作用【地】metamorthism;diagenesis;lithification
岩击 shock bump
岩基 abyssolith;batholite [batholith];bathylite [bathylith]; foundation in rock;ground mass;rock bed;rock foundation; rock head; rock in place;rock mantle;rocky ground
岩基壁 batholith wall
岩基觇标 rock target
岩基成矿说 batholith hypothesis of mineralization
岩基床 solid bedrock
岩基顶阶段 acrobatholithic
岩基灌浆 rock foundation grouting
岩基花岗岩 batholithic granite
岩基接触 contact of the batholith
岩基上覆土层厚度 depth of soil overlying bedrock
岩基深部带 hypobatholitic zone
岩基式的 embatholithic
岩脊 phacolite;phacolith;cleaver <突出于冰河或雪原上>
岩岬 thrum;thurm
岩架 ledge;rock(y) ledge
岩尖 rocky spur
岩槛 rock beam
岩浆 fused earth substance; igneous magma; magma; magmatic segregation; molten magma; rock grout; rock magma
岩浆爆发 magmatic explosion
岩浆标志 igneous markers
岩浆玻璃灰 magma-glass-ash
岩浆残余挥发分 residual liquid
岩浆残余物 magmatic residuum
岩浆成矿作用 magmatic ore-forming process
岩浆成因 magmatic origin
岩浆成因类型 genetic(al) type of magma
岩浆成因条件 magmatogenic condition
岩浆成因岩石 magmatogene rock
岩浆冲击成因 magma impact genesis of earthquake
岩浆储源 magma basin;magma chamber; magma pocket; magma reservoir
岩浆带 zone of magma
岩浆的 magmatic
岩浆底辟 magma diapir
岩浆地热系统 magma geothermal system
岩浆顶蚀作用 magmatic stoping
岩浆发散 emission of lava
岩浆房 macula [复 maculae]; magma chamber
岩浆房热储 magma chamber reservoir
岩浆分结作用 magmatic segregation
岩浆分凝作用 magmatic segregation; segregation in magma
岩浆分异作用 differentiation of magma;magmatic differentiation
岩浆富集 magmatic concentration
岩浆贯入矿床 magmatic injection deposit
岩浆贯入作用 magmatic emplacement;magmatic injection
岩浆过程的数学模拟 mathematic(al) simulation of magmatic process
岩浆后期的 deuteric;post-magmatic
岩浆后期矿床 deuteric ore deposit; post-magmatic ore deposit
岩浆后期蚀变 syntantexis
岩浆花岗岩 magma granite
岩浆回旋 magmatic cycle
岩浆混合作用 magmatic mixing; mixed magma
岩浆活动 magmatic activity
岩浆活动标志 magmatism marker
岩浆活动差异 difference in magmatic activity
岩浆活动次数 time number of magmatism
岩浆活动方式 magmatism pattern
岩浆活动规模 magmatism scale
岩浆活动极性 magmatism polarity
岩浆活动强度 magmatism intensity
岩浆活动时代 magmatism age
岩浆活动特征 magmatism feature
岩浆建造 magmatic formation
岩浆建造类型 type of magmatic formation
岩浆库 magma chamber;magma reservoir
岩浆矿床 magmatic (ore) deposit; pyrogenic deposit
岩浆矿田构造 magmatic orefield structure
岩浆矿物 magmatic mineral
岩浆类型 magma type
岩浆论者 magmatist
岩浆末期的 hysterogenetic
岩浆内的 intramagmatic
岩浆囊 pocket of magma
岩浆喷发 magmatic emanation;magmatic eruption
岩浆喷溢 magmatic eruption
岩浆喷溢矿床 ore deposits of magmatic eruption
岩浆期后矿床 post-magmatic mineral deposit;post-magmatic ore deposit
岩浆起源 magmatic origin
岩浆气(体) juvenile gas;magmatic gas
岩浆侵入 magma intrusion;magmatic intrusion;stoping
岩浆侵入假说 magma intrusion theory
岩浆侵入作用 intrusion of magma
岩浆圈 magmosphere;pyrosphere
岩浆泉 juvenile spring
岩浆热估算法 magmatic heat budget method
岩浆热力作用 magnetic thermogenesis
岩浆热流体 magmatic hot fluid
岩浆热液 magmatic hydrothermal solution
岩浆热液成矿说 magnetic-hydrothermal theory
岩浆热源 magmatic heat source
岩浆溶液 magmatic solution
岩浆熔蚀 magmatic corrosion
岩浆升蚀 magmatic stoping
岩浆生成论<花岗岩的> magmatism
岩浆蚀顶作用【地】foundering
岩浆水 intratelluric water; juvenile water;magmatic water;rejuvenated water;volcanic water;plutonic water
岩浆水成的 aqueo-igneous
岩浆坍顶作用 roof foundering
岩浆同化作用 magmatic assimilation; magmatic digestion
岩浆同位素成分 magma isotope composition
岩浆同源地 consanguinity
岩浆系列 magma(tic) series
岩浆玄武岩 magma basalt
岩浆旋回 magmatic cycle
岩浆学说 magma theory
岩浆岩 igneous rock;magmatic rock; magmatite
岩浆岩刺穿圈闭 igneous rock diapir trap
岩浆岩类型 magmatic rock type
岩浆岩体遮挡 igneous barrier
岩浆岩原生构造 primary structure of igneous rock
岩浆岩中天然气 natural gas in igneous
岩浆岩组合 magmatic rock association
岩浆演化 magmatic evolution
岩浆因素 magmatite factor
岩浆缘的 perimagmatic
岩浆源的 juvenile
岩浆运动 juvenile motion
岩浆杂岩 magmatic complex
岩浆蒸馏作用 magma distillation
岩浆蒸汽 magmatic steam
岩浆作用 magmatism
岩浆作用地球化学 geochemistry of magmatic processes
岩浆作用方式 magmatism way
岩浆作用迁移 magmatism migration
岩浆作用岩石组合和岩石成因类型 magmatism rock assemblage and genetic type of
岩礁 ledge (of rocks); lithoherm; rock(y) ledge;rock(y) reef
岩礁柱 rock pillar
岩角 thrum
岩脚坡 undercliff
岩晶 mountain crystal;rock crystal
岩晶屑砂状结构 lithic-crystalloclastic psammitic texture
岩精 ichor
岩颈 neck;plug;volcanic neck
岩颈丘 plug dome
岩孔 foramen petrosum;petrosal foramen
岩孔冲洗 hole flush
岩孔伸缩仪 corehole extensometer
岩窟 boss
岩窟寺 chaitya
岩窟寺院 chaitya cave
岩块 block mass;block of rock;clod; derrick stone;hunk;knocker;mass of rock; rock block; rock fragments;rock mass
岩块爆破 blockhole blasting; block holing
岩块剥落 crumbling
岩块倒落 block tipping
岩块火山角砾岩 rock block volcanic breccia
岩块集块岩 rock block agglomerate
岩块孔隙度 rock porosity
岩块砾岩 petromictic conglomerate
岩块流 block stream
岩块声波速度测试 wave-velocity test of rock
岩矿分析 rock and mineral analysis
岩矿分析送样单 sample paper of rock and mineral analysis
岩矿鉴定报告 identification report of rock and mineral
岩矿鉴定采样 sampling for mineralogic(al) and petrologic(al) determination
岩矿鉴定方法 the method for identification of minerals and rocks
岩矿鉴定取样 sampling of rock and mineral identification
岩矿鉴定仪器 identification instrument of minerals and rocks
岩矿石的复频谱类型 type of complex frequency spectrum in rock and minerals
岩矿石的各向异性特征 anisotropic(al) characteristic of rocks and minerals
岩矿石的极化类型 type of polarization of rocks and minerals
岩矿石的极化系数 polarization coefficient of rocks and minerals
岩矿石电阻率测定方法 way for measuring resistivity of rock and mineral
岩矿物理性质 physical property of minerals and rocks
岩类 clan
岩类土壤 rocky soil
岩类学 petrography
岩理学 petrology
岩沥青 kir;rock asphalt
岩沥青路面 rock asphalt pavement
岩栎 rock chestnut oak
岩粒流动变质带 anomorphic zone
岩镰 harpolith
岩裂 lithoclase
岩裂带 rock fracture zone
岩裂作用 lithofraction
岩流 rock flow (age); rock glacier; rock stream
岩流带 rock flowage zone; zone of flowage;zone of rock flowage
岩流圈 asthenosphere
岩流竖趾丘【地】steptoe
岩流岩裂带 rock flowage and fracture zone
岩瘤 boss
岩隆 build-up
岩隆泉 barrier spring
岩漏斗 ethmolith
岩脉 dike [dyke]; ledge rock; lode; rocky ledge;streaking;vein
岩脉槽探 costean
岩脉储水构造 storage structure of vein
岩脉含水带 water-bearing vein zone
岩脉脊 dike ridge
岩脉间层 astillen
岩脉控制点 control point of vein
岩脉群 dike [dyke] swarm
岩脉水 vein water
岩脉系 dike system
岩脉组 dike set
岩锚 rock anchor
岩锚的 U 形拖钩 shackle
岩棉 rock fiber [fibre];rock wool
岩棉板 rockwool board; rockwool slab
岩棉保温层 rockwool quilt
岩棉保温套管 rockwool lagging section
岩棉保温制品 rockwool insulating product
岩棉玻璃布缝毡 rockwool mat with glass cloth on one side
岩棉薄板 rockwool sheet
岩棉带 rockwool lamella
岩棉垫 rockwool mat
岩棉钢丝网缝毡 rockwool mat with wire net on one side
岩棉隔热板 rockwool insulation-grade slab
岩棉隔热材料 rockwool insulation (grade) material
岩棉隔热料 rockwool insulator
岩棉隔声板 rockwool insulation-grade slab
岩棉隔声材料 rockwool insulation (grade) material
岩棉管壳 rockwool pipe casing
岩棉管套 rockwool pipe section
岩棉建筑板 rockwool building
岩棉建筑板材 rockwool building board
岩棉卷毡 rockwool mat
岩棉绳套 rockwool cord covering
岩棉绳罩 rockwool cord covering
岩棉吸音吊顶 rockwool acoustic(al) ceiling

Y

岩棉纤维 rockwool fiber [fibre]
岩棉毡 rockwool felt; rockwool building
岩棉制品 asbestos article
岩面 rock (sur) face
岩面标高 level of base rock
岩面测标 surveying mark on rock surface
岩面加速度 acceleration of base rock
岩面开挖 ledge excavation
岩面裂缝 rock fissure
岩面清理 scaling
岩面容限 rock face tolerance
岩面突出部 rock ledge
岩面线 rock line
岩面砖 rock-faced brick
岩末 rock flour
岩漠 hammada; rocky desert
岩漠沉积 rock deposit
岩漠相 stone facies
岩盘 batholite [batholith]; bed rock; caprock; laccolite [laccolith]; rock basin; rock in place
岩盘面 rock surface
岩盘盆地 rock basin
岩配基 rockogenin
岩盆 lopolith; rock basin
岩片 rock lump; sliver
岩坡破坏的防治措施 treatment methods of rock slope
岩墙 dike [dyke]
岩墙脊 dike ridge
岩墙群 dike swarm
岩墙岩脉 vein dike [dyke]
岩桥 strata bridge; rock bridge
岩壳 rock mantle
岩芹酸 petroselinic acid
岩穹【地】dome
岩丘 lava dome
岩区 petrographic(al) province
岩群 terrain
岩壤天然热释光法 rock soil nature heat releasing light method
岩刃 akmolith
岩溶 abri; carst; karst; solution channel
岩溶暗河流量过程曲线 flow graph of underground river in karst area
岩溶暗河水系图 karst hydrographic(al) map of underground rivers
岩溶剥蚀面 karst denudation plane
岩溶残山 haystack hill
岩溶残液 residual liquor
岩溶槽 karstic channel
岩溶沉积物 karst deposit
岩溶充水矿床 mineral deposit of karst inundation
岩溶大厅 karst hall
岩溶的 karstic
岩溶地层 karstic formation
岩溶地基 karst subgrade
岩溶地基处理 treatment of karst foundation; treatment of karst subgrade
岩溶地貌 karst geomorphy; karst land feature; karst landform
岩溶地貌调查 survey of karst morphology
岩溶地貌剖面图 karst geomorphology profile
岩溶地面塌陷 karst collapse
岩溶地区 karst(ic) terrain; karst(ic) region
岩溶地区大洼地 poljie [复 polgia]
岩溶地下水 karstic groundwater
岩溶地形 karst(ic) feature; karst landform; karst topography
岩溶地质学 karstic geology
岩溶调查 karst investigation
岩溶调查数据采集 data collected in karst investigation
岩溶洞 karst cave
岩溶洞厅的面积 area of a karst hall
岩溶洞穴 karst grotto
岩溶洞穴渗漏 leakage through karst cave
岩溶洞穴特征 karst cave characteristics
岩溶堆积地貌 karst accumulation landform
岩溶堆积物 karst deposit
岩溶发育带 karst developing zone
岩溶发育密度 density of karst development
岩溶发育区范围 karst development range
岩溶发育区深度 depth of karst development region
岩溶分布规律 distribution law of karst
岩溶干宽谷 ouvala [uvala]; vala
岩溶高原 karst plateau
岩溶沟 karsten; grike
岩溶孤峰 tower karst
岩溶谷地 karst valley
岩溶管道通道 passage of karst pipe
岩溶含水层 karst aquifer
岩溶河(流) carst [karst] river; carst [karst] stream
岩溶湖 karst lake
岩溶化 carstification [karstification]
岩溶化的 karstified
岩溶化模式 karstification model
岩溶化石灰岩 karstified limestone
岩溶化岩石 karstificated rock; karstified rock
岩溶角砾岩 karst breccia
岩溶截流总合法 method of total karst cutoff
岩溶井 karst pit; karst well
岩溶景观 karst landscape
岩溶坑 aven
岩溶空隙水异常 anomaly of karst water
岩溶矿床 karst deposit
岩溶类型 karst type
岩溶类型图 karst type map
岩溶裂隙 karst fissure; solution channel
岩溶裂隙通道 passage of karst fissure
岩溶漏斗 karst funnel
岩溶漏斗直径 diameter of a doline
岩溶率 degree of karstification; factor of karst
岩溶盆地 cockpit; karst basin; karst depression; karstic basin; poljie [复 polgia]
岩溶平原 karst plain
岩溶期 karst stage
岩溶气爆 karst air-explosion
岩溶桥 karst bridge
岩溶侵蚀 karst erosion
岩溶侵蚀基准面 karst base level
岩溶区 karst; karst region
岩溶泉 carst spring; karst(ic) spring; karst source
岩溶热水 karst thermal water
岩溶石灰岩 karst limestone
岩溶竖井观测点 observation point of karst shaft
岩溶竖井深度 karst pit depth
岩溶竖井直径 karst pit diameter
岩溶水 cavern water; karst(ic) water
岩溶水动力分带 hydrodynamic(al) zonality of karst water
岩溶水文地质调查 karst hydrogeologic(al) survey
岩溶水文地质剖面图 karst hydrogeology profile
岩溶水文地质图 map of karst hydrogeology
岩溶水文地质学 karstic hydrogeology
岩溶水文学 karstic hydrology
岩溶塌陷 karstic collapse; collapsed karst
岩溶塌陷角砾岩 karstic collapse breccia
岩溶塌陷通道 passage of karst collapse
岩溶坍陷 karstic collapse; karst subsidence
岩溶特征 karstic feature
岩溶天窗通道 passage of karst louver
岩溶通道 karst corridor; karst(ic) channel; karst passage(way); karst street
岩溶洼地 karst depression; solution depression
岩溶稳定问题 problem of karst
岩溶物理地质现象 karst physical geology phenomenon
岩溶系统 karst system
岩溶现象 karst(ic) phenomenon
岩溶陷落 karst collapse
岩溶陷落柱 karst collapse-column
岩溶陷落柱通道 passage of karst collapse column
岩溶形态 karst form
岩溶形态测量 karst shape measure
岩溶形态尺寸 dimensions of karst
岩溶形态分布图 distribution map of karst form
岩溶形态类型 type of karst shape
岩溶形态特征 character of karst shape
岩溶型地基 subgrade with karst
岩溶旋回 karst cycle
岩溶学 karstology
岩溶演化 karst evolution
岩溶沼泽 karst fen
岩溶准平原 karst peneplain
岩溶作用 karstification; karst process
岩溶作用带 karstification zone
岩溶作用基底 base of karstification
岩溶作用基准面 base of karstification
岩乳 rock milk
岩塞 rock plug
岩塞爆破 chock blasting; rock plug blasting
岩塞丘 plug dome belonite
岩砂 rock sand
岩砂撒布器 salt and sand spreader
岩山脊 dike ridge
岩舌【地】tongue
岩深沟 grike
岩生植物 rock(-bearing) plant
岩生植物园 rock garden
岩石 ledge; litho; roach; rock; stone
岩石摆球硬度 ball pendulum hardness of rock
岩石搬运 casting; casting of rocks
岩石保护阶地 rock-defended terrace
岩石饱和吸附容量 saturated adsorption capacity of rocks
岩石饱水率试验 rate of saturated of rock test
岩石爆裂 rock popping
岩石爆破 rock blast(ing); rock burst
岩石爆破导孔 leading hole of blast
岩石爆破的裸露面 free face of rock blasting
岩石爆破的破裂半径 radius of rupture
岩石爆破力学 mechanics of rock blasting; mechanics of rock bursting
岩石爆破性指数 blasting index
岩石背脊丘 rock drumlin
岩石崩解性试验 slaking test of rock
岩石崩裂 rock failure
岩石崩落 rock fall
岩石崩落带 caved goaf
岩石崩碎 rock disintegration
岩石崩塌 cave-in; rock fall
岩石崩坍 rock avalanche
岩石边坡 rock(y) slope
岩石变形 deformation of rock; rock deformation; rock yield
岩石变形的影响因素 deformation influential factor of rock
岩石变形及稳定性测量 rock deformation and stability measurement
岩石变形阶段 deformation stage of rock
岩石变形试验 deformation test of rock
岩石变形图 rock deformation diagram
岩石变异 alternation of rocks
岩石变质 <因内部水分而引起的> aqueous metamorphism
岩石变质区 rock alteration zone
岩石变质作用 metamorphism
岩石标本 rock sample; rock specimen
岩石标石 rock mark
岩石表层 mantle of rock
岩石表面 rock surface
岩石表面凹窝 alveolization
岩石表面破碎 surface fragmentation of rock
岩石表面破损小比例形态 chatter mark
岩石冰川 rock glacier
岩石冰冻法 rock freezing
岩石剥落 rock disintegration
岩石剥蚀 rock disintegration
岩石薄片 rock slice
岩石不连续性 discontinuity of rock
岩石材料 rock material
岩石采样 rock sampling
岩石残余物 litho-relict
岩石残余应力 residual stress in rocks
岩石侧向约束膨胀率试验 lateral restraint swelling test of rock
岩石层 lithosphere
岩石层理 rock bedding; rock stratification
岩石层图 geologic(al) map
岩石插筋(锚固) rock pinning
岩石产品 rock product
岩石铲斗 rock bucket
岩石潮解的 air slaked
岩石沉积环境 lithotope
岩石成分 composition of rocks; petrographic(al) composition; rock constituent
岩石成矿作用 petrometallogeny
岩石成为包体 enclavement
岩石成因的 lithogenous
岩石成因分类 petrogenic classification
岩石成因格网 petrogenic grid
岩石成因类型 petrogenetic type
岩石成因论 petrogenesis
岩石成因学 lithogenesis; lithogenesy; lithogeny; petrogenesis; petrogeny
岩石成因学说 doctrines of petrogenesis
岩石承插口 <桩底> rock socket
岩石承受静荷载的强度 static rock strength
岩石齿轮钻头 rock roller bit
岩石充填 rock-fill
岩石冲击破碎 dynamic(al) breakage of rock
岩石出露面 exposed rock surface
岩石初爆 primary blasting
岩石穿孔器 gadder
岩石床再生冷却 rock bed regenerative cooling
岩石磁测 rock magnetic measurement
岩石磁导率 magnetic permeability of rock
岩石磁性 rock magnetism
岩石脆塑性分级 classification of rock

brittle-plasticity
岩石带 rock zone
岩石单位抗力系数 unit reaction coefficient of rock
岩石单位重量 rock unit weight
岩石单元 rock unit
岩石单元体 rock element
岩石单轴饱和抗压强度 uniaxial saturated compressive strength of rock
岩石单轴抗压强度试验 uniaxial compressive strength test of rock
岩石单轴压缩变形试验 uniaxial compressive deformation test of rock
岩石的 petrean; petrographic(al); petrous; rock-ribbed; rocky
岩石的包体 peterologic(al) enclave
岩石的饱水率 water saturation of rock
岩石的饱水(率)系数 water saturation coefficient of rock
岩石的变形模量 deformation modulus of rock
岩石的变形特征 deformation character of rock
岩石的泊松比 Poisson's ratio of rock
岩石的不连续(小)间隙 isolated interstice; discontinuous interstice
岩石的不整合性 unconformity of rock
岩石的单独小间隙 isolated interstice
岩石的导热性 heat conductivity of rock
岩石的动力强度 dynamic(al) rock strength
岩石的非均质性 heterogeneity of rock
岩石的分类和名称 classification and name of rocks
岩石的工程地质分类 engineering geologic(al) classification of rock
岩石的工程特性 engineering characteristics of rock
岩石的构造 rock structure
岩石的极限强度 ultimate rock strength
岩石的结构 rock texture
岩石的抗冻性 frost resistivity of rock
岩石的抗拉强度 tensile strength of rock
岩石的抗水性 water resistance of rock
岩石的抗压强度 compressive strength of rock
岩石的可变性 deformability of rock
岩石的可钻性 drilling character of rock; rock drillability
岩石的空洞 peterologic(al) cavity
岩石的空气渗透系数 permeability of rock to air
岩石的空隙和水 pore and water in rock
岩石的空隙类型 type of pore in rock; type of rock void
岩石的矿物成分 petrologic(al) mineral composition
岩石的力学性质 mechanical property of rock
岩石的摩尔数 mole fraction of rock
岩石的黏[粘]结 cohesion of rock
岩石的破碎 <在钻头作用下> rock breakdown
岩石的其他物质组分 other petrologic(al) mineral composition
岩石的强度特征 strength character of rock
岩石的强化特性 intensive characteristics of rock
岩石的热导性 heat conductivity of rock
岩石的色调 petrologic(al) tone
岩石的渗透系数 coefficient of rock permeability
岩石的渗透性分类 rock classification on permeability
岩石的水变质作用 hydrometamorphism
岩石的水平裂缝 bed joint
岩石的塑性 plasticity of rock
岩石的弹性模量 elastic modulus of rock
岩石的吸水率 absorption rate of rock; moisture content ratio of rock
岩石的线形构造 <如流线、波痕等> lineation
岩石的颜色 colo(u)r of rocks
岩石等效系数 equivalent coefficient of rock
岩石抵抗系数 coefficient of rock resistance
岩石地层 rock fabric
岩石地层带 lithostratigraphic(al) zone
岩石地层单位 geolith; lithostratigraphic(al) unit; rock-stratigraphic(al) unit; rock unit
岩石地层单位名称 name of lithostratigraphic(al) unit
岩石地层单元 geolith
岩石地层格架 lithostratigraphic(al) framework
岩石地层时 geochron
岩石地层污染 rock stratum pollution
岩石地层学 lithostratigraphy; rock stratigraphy
岩石地层学方法 method of litho-stratigraphy
岩石地基 rock foundation
岩石地沥青 rock asphalt
岩石地沥青路面 compressed asphalt; rock asphalt pavement
岩石地幔 rock mantle
岩石地锚 rock anchor
岩石地貌调查 survey of morpholithos
岩石地面 rocky ground
岩石地球化学测量 geochemical rock survey; lithogeochemical survey
岩石地球化学方法 lithogeochemical method
岩石地形图 rock chart
岩石点荷载强度试验 point load strength test
岩石点荷载强度指标 point load strength index
岩石点荷载强度指数 point load strength index
岩石点荷载强度指数仪 rock strength index log
岩石点荷载指标 point load(ing) index of rock
岩石电导率 conductivity of rock
岩石电性 rock electricity
岩石电阻 electric(al) resistance of rock
岩石雕刻 petroglyph [petrograph]
岩石雕刻术 petroglyphy
岩石吊桶 mucking bucket
岩石顶层 rock head
岩石动参数试验 dynamic(al) parameters test of rock and soil
岩石动力学 dynamic(al) rock mechanics
岩石冻失率 loss rate of rock weight by frost
岩石断裂力学性质 fracture mechanic property of rock
岩石断面 section through rock
岩石堆积 rock deposit
岩石发热 rock heat
岩石发生学 petrogenesis; petrogeny
岩石发展学 petrogenesis
岩石防蚀阶地 rock-defended terrace
岩石飞散 rock sprinkling
岩石分布图 lithologic(al) map; slid map
岩石分级 rock mass classification; rock rating
岩石分解 demorphism
岩石分类 petrographic(al) classification; rock category; rock classification; rock rating
岩石分类分析 petrographic(al) classification analysis
岩石分类体系 rock classification system
岩石分水岭 rocky ridge
岩石分析 petrographic(al) analysis; rock analysis
岩石分析样品 sample for rock analysis
岩石粉化 efflorescence
岩石粉磨机 rock pulverizer
岩石粉碎 to break up of rock
岩石粉碎机 rock pulverizer
岩石风化 rock decay; rock rot; rotting
岩石风化层 rock mantle
岩石风化程度 degree of rock decay; rate of rock decay; weathering grades of rock
岩石风化程度系数 alteration coefficient of rock; coefficient of rock decay; coefficient of weathering for rock
岩石风化作用 rock weathering
岩石缝隙 rock fissure
岩石缝隙中植物 homophyte
岩石敷面 rock facing
岩石符号 rock quality designation [RQD]
岩石腐化 rock decay
岩石覆盖层 rock cover
岩石覆盖层滑动 slough
岩石覆盖泥土层(不钻孔)爆破 mud cap
岩石刚度 rigidity of rock
岩石各向异性 aeolotropism of rock; aeolotropy of rock
岩石工程 rock engineering
岩石工程地质分类 petrographic(al) classification for engineering geology
岩石工程学 geotechnology
岩石工作面 hard heading
岩石构造 rock fabric; structure of rocks
岩石构造分析 petrofabric analysis
岩石构造学 petrotectonics
岩石构造组合 petrotectonic assemblage
岩石骨架俘获截面 matrix capture cross section
岩石骨架密度 density of matrix
岩石骨架声波时差 sonic wave interval transit time of matrix
岩石骨架视密度 apparent density of matrix
岩石骨架视中子孔隙度 apparent neutron porosity of matrix
岩石鼓丘 rocdrumlin; rock drumlin
岩石固结 rock bonding
岩石固体压缩系数 lithosolid compressibility
岩石灌浆 rock grouting; rock hardening
岩石滚落 fall of rock
岩石过爆 overbreak
岩石海岸 rocky coast
岩石海岸的潮水坑 rock pool
岩石海底 rocky bottom
岩石海底区 rocky area
岩石海湾 rocky bay
岩石含的黏[粘]土 stone gall
岩石含的杂质 stone gall
岩石耗蚀过程 process of rock wastage
岩石和土壤符号 rock and soil symbols
岩石河槽 rock channel
岩石河床 rocky (river) bed
岩石荷载 rock load
岩石荷载支承衬砌 lining for support of rock load
岩石厚度 depth of rock
岩石花纹符号 symbol of lithological pattern
岩石花园 rock garden
岩石滑动 debris slide
岩石滑落 rock slide
岩石滑坡 rock (land) slide; rock slip
岩石滑移 rock slip
岩石化学 petrochemistry
岩石化学测量 lithochemical survey
岩石化学法 rock chemistry method
岩石化学分类 chemical petrologic classification
岩石化学分析 chemical analysis of rock
岩石化学破碎 chemical fragmentation of rock
岩石化学数据分析 petrochemistry data analysis
岩石化学数据库 petrochemistry data base
岩石环境 rock environment
岩石荒漠 rock desert
岩石灰 rock ash
岩石回旋 rock cycle
岩石(回转牙轮)钻头钻进 rock bit drilling
岩石回转(钻进用)牙轮钻头 rock roller bit
岩石火成说 plutonism
岩石机械破碎 mechanical fragmentation of rock
岩石机械性碎屑化的 autoclastic
岩石机钻 machine rock drill
岩石基本性态 governing rock characteristics
岩石基础 rock base
岩石基础试验 <强度、变形等> rock foundation test
岩石基底 rock base
岩石基质压缩系数 rock matrix compressibility
岩石加工工具 rock-working tool
岩石加工骨料时产生的尘土 dust of fracture
岩石加工面 rock faced
岩石加固 rock anchorage; rock reinforcement; rock strengthening
岩石夹层 dirt band; intermediate rock
岩石架 ledge
岩石尖顶 rock pinnacle
岩石尖峰 rock pinnacle
岩石尖礁 rock pinnacle
岩石坚固系数 coefficient of rock stiffness
岩石坚固性分级 classification of rock strength
岩石坚固性分类 Protodyakonov's classification of rock
岩石坚固性系数 coefficient of rock stiffness; rock solidarity coefficient
岩石坚硬程度 hardness degree of rock
岩石间沉积物 open space deposit
岩石间航道 inner lead
岩石间隙水 interstitial water
岩石剪力试验 rock shear test
岩石检测 rock test
岩石检验 proof-testing of rock
岩石建筑种及亚种 rock construction types and its mutation
岩石鉴定 petrographic(al) examina-

tion;rock identification
岩石鉴定报告 report on rock identification
岩石鉴定样品 sample for rock determination
岩石阶地 cut terrace;esplanade;rock terrace
岩石节理 rock joint
岩石结构参数 rock structure parameter
岩石介电常数 dielectric(al) constant of rock
岩石界面位置 position of boundary of rock
岩石金属量测量 rock metallometric survey
岩石晶体 rock crystal
岩石静态负荷 lithostatic(al) loading
岩石静态压力【地】lithostatic(al) pressure
岩石静压力 lithostatic(al) pressure
岩石绝对吸水量 absolute water absorbing capacity of rock
岩石掘进 dead works;rock cut
岩石开采 rock quarry
岩石开采方案 rock-removal project
岩石开采工程 rock-removal project
岩石开掘机 rock grab
岩石开挖 ledge excavation;rock excavation
岩石开挖付价线 pay line
岩石开凿 rock cut(ting)
岩石抗冲刷系数 coefficient of scouring resistance
岩石抗冻试验 frost resistivity test of rock
岩石抗冻性 frost resistivity of rock
岩石抗拉强度 tensile strength of rock
岩石抗拉强度试验 tensile strength test of rock
岩石抗力 rock resistance
岩石抗力系数 coefficient of rock resistance; reaction coefficient of rock
岩石颗粒 rock particles
岩石颗粒的热膨胀系数 thermal expansivity of rock grain
岩石可爆性 blastability of rock
岩石可割性 rock cuttability
岩石可凿性 drillability
岩石可钻性 drillability (scale) of rock
岩石可钻性等级 drilling grade of rock
岩石可钻性分级 drillability classification of rock
岩石可钻性系数 coefficient of rock drillability;rock drillability index
岩石可钻性 drillability
岩石可钻性指数 drillability index of rock;rock drillability index
岩石坑道 rock gallery;stone drift
岩石孔 rock bore
岩石孔隙 rock pore
岩石孔隙度 rock porosity
岩石孔隙率 rock porosity
岩石孔隙仪 porosimeter
岩石控制 rock control
岩石块体顺坡滑动 gravity erosion; mass erosion
岩石矿物成分鉴定 mineral composition determination of rock
岩石扩容性 rock dilatancy
岩石类别 rock classification
岩石类土(壤) rocky soil
岩石类型 rock type
岩石棱角的爆破 chip blasting
岩石棱柱体(试件) rock prism
岩石力学 rock mechanics
岩石力学测定工具 tools of rock mechanics
岩石力学和岩石工程 <奥地利季刊> Rock Mechanics and Rock Engineering
岩石力学现场研究 in-situ rock mechanics
岩石立体图 block diagram of rock
岩石劣地 scabrock
岩石裂缝 fracture of rock;rock fissure;rock interstice
岩石裂隙 rock crevice;rock fissure; rock fracture
岩石裂隙带 rock fracture zone;zone of rock fracture
岩石裂隙化程度 fissured degree of rock
岩石裂隙性 rock fissility
岩石溜井 muck raise
岩石流 rock flowage; rock glacier; rock stream; stone-river; stone run;talus glacier
岩石流变 rock flowage
岩石流变带 zone of rock flowage
岩石流变试验 rheologic(al) teat of rock
岩石流变性 rock rheological property
岩石流动 rock flow
岩石流动带 rock flowage zone
岩石流动裂隙带 rock flowage and fracture zone
岩石露头 rock exposure; rock outcrop(ping); day stone; outcrop of rock
岩石路堤 rock-filling dike [dyke]
岩石路基 rock subgrade
岩石路堑 rock cut
岩石麓原 rock pediment
岩石落锤硬度 drop hammer hardness of rock
岩石麻花钻 rock auger
岩石锚碇装置 <维持开挖面整体稳定性的> rock anchorage;ground anchorage primary reinforcement
岩石锚杆 rock anchor; rock bolt(ing)
岩石锚杆测力计 rock-bolt dynamometer
岩石锚杆测试 rock-bolt test
岩石锚杆加固 rock bolting
岩石锚杆加固的洞室 caverns supported by rock bolts
岩石锚杆检测仪 rock boltometer
岩石锚杆结合喷混凝土支护 rock bolting and shotcrete
岩石锚杆孔 rock bolt hole
岩石锚杆理论 rock bolting theory
岩石锚杆伸长仪 rock-bolt extensimeter [extensometer]
岩石锚杆支承能力 rock-bolt load carrying capacity
岩石锚固 rock anchor
岩石锚固插筋 rock pin
岩石锚固螺栓 rock bolt
岩石锚栓 rock anchor
岩石锚栓加固 rock bolting
岩石锚栓支撑 rockbolts shore
岩石锚栓支护 rock bolting
岩石冒落 fall of ground
岩石煤 slate coal
岩石密度 density of rock;rock density
岩石棉 rock wool
岩石面层上的工程 surface construction in rock
岩石描述 designation of rock
岩石名称 name of rocks
岩石模量折减系数 reduction factor of rock modulus
岩石模式 petromodel
岩石磨碎 rock grinding
岩石莫氏硬度表 Mohs' scale
岩石莫氏硬度计 Mohs' scale
岩石内部 rock interior
岩石内全应力状态 complete state of stress in rock
岩石内围层 inlier
岩石耐磨性 abrasive resistance of rock
岩石碾碎 rock grinding
岩石凝聚力 rock cohesion
岩石凝聚强度 rock cohesion
岩石抛射距离 rock throw
岩石配置 rock arrangement
岩石膨胀压力试验 swelling pressure test of rock
岩石劈开孔 lewis hole
岩石劈理 rock cleavage
岩石劈理面 cleavage plane
岩石劈裂方向 rift
岩石劈裂(可能)性 rock rippability
岩石疲劳破碎 fatigue fragmentation of rock
岩石片理 rock schistosity
岩石平巷 hard heading; rock drift; rock heading
岩石破坏 failure of rock;rock failure
岩石破坏型式 failure type of rock
岩石破裂 pressure burst; riving; rock burst;rock disruption;rock failure; rock fracture
岩石破裂带 rock fracture zone
岩石破碎 rock breaking;rock comminution;rock destruction;rock disintegration; rock failure; rock fragmentation
岩石破碎参数 <通常以爆破压力表示> rock breaking parameter
岩石破碎带 rock breaking zone;rock fracture zone;zone of rock fracture
岩石破碎的细骨料 crushed rock fines
岩石破碎方法 method of rock fragmentation; rock fracturing technique
岩石破碎机 boulder buster; boulder cracker; crushing engine; rock breaker; rock-crushing machine; stone mill
岩石破碎机理 fragmentation mechanism of rock
岩石破碎器 boulder buster; boulder cracker; rock breaker; rock-crushing machine;stone mill
岩石潜坝 rock sill
岩石潜动 rock creep
岩石浅滩 cripple
岩石强度 resistance of rock; rock strength;strength of rock
岩石强度损失 rock strength loss
岩石强度损失率 loss rate of rock strength;strength loss ratio of rock
岩石强度系数 Protodyakonov's number
岩石强度指数 rock strength index
岩石强制位移特性 rock force-displacement performance
岩石切割力学 mechanics of rock cutting
岩石切削方法 rock-cutting process
岩石情况 rock condition
岩石屈服 rock yield
岩石取样法 rock sampling
岩石圈【地】lithosphere; geosphere; petrosphere; rock sphere; stereosphere
岩石圈板块 lithospheric plate;lithospheric slab
岩石圈次板块 lithospheric subplate
岩石圈地球化学 geochemistry of lithosphere
岩石圈地球化学异常 lithogeochemical anomaly
岩石圈地震 lithospheric earthquake
岩石圈动力学及演化计划 dynamic(al) and evolution of the lithosphere project
岩石圈断裂 lithospheric fracture
岩石圈分裂 lithospheric splitting
岩石圈均衡 lithospheric isostasy
岩石圈膨胀 lithospheric bulge
岩石圈软度 softness of lithosphere
岩石圈下板块 lithospheric subplate
岩石圈信号 lithospheric signal
岩石全分析 rock bulk analysis
岩石群落 pelrium
岩石扰动 rock disturbance
岩石热导率 thermal conductivity of rock
岩石热力钻孔 thermal drilling of stone
岩石容积压缩系数 rock-bulk compressibility
岩石溶化穿透器 rock-melting penetrator
岩石熔化器 rock melter
岩石蠕变 creep of rock
岩石蠕动 rock creep
岩石乳化炸药 rock emulsifying explosive
岩石软化系数 softening coefficient of rock
岩石软化性试验 softening test of rock
岩石软化性系数 coefficient of softening of rock
岩石三轴压力机 rock triaxial apparatus
岩石三轴压缩强度试验 triaxial compressive strength test of rock
岩石山丘 rock massif
岩石扇状地 rock fan
岩石上刻像或文字 petrography
岩石渗透性 rock permeability
岩石渗透性试验仪 Menard pressure permeameter
岩石渗透性质测定 permeability measurement of rock
岩石生成 lithogenesis;lithogenesy
岩石声波传播速度 acoustic(al) velocity in rock
岩石声阻抗 acoustic(al) constant of rock
岩石剩余磁化 rock remanent magnetization
岩石剩余磁化稳定性 rock remanent magnetization stability
岩石施工 construction in rock
岩石试件制备 petrographic(al) preparation
岩石试验 rock study;rock test
岩石试验机 rock test rig
岩石试样 rock sample
岩石试样分析 rock sample analysis
岩石室内试验研究 rock mechanics laboratory studies
岩石收缩 rock contraction
岩石受压移动 bump;crump
岩石疏浚 rock dredging
岩石水化崩解作用 hydration shattering
岩石水胶炸药 rock water gel explosive
岩石水理性质 hydrophysical property of rocks
岩石水理性质测定 measuring of hydrophysical property of rock
岩石水敏性 water sensitivity of rocks
岩石水泥 rock cement
岩石水溶性 water solubility of rocks
岩石松弛 relaxation of rock
岩石松动 overbreak
岩石塑性指数 plasticity index of rock
岩石碎块 rock fragments;fly rock <爆破后的>
岩石碎裂 rock fragmentation

岩石碎裂技术 rock fracturing technique
岩石碎裂作用 cataclase
岩石碎片 crag; rock fragments; rock slice
岩石碎屑 clast; rock debris; rock fragments; debris
岩石隧道 rock tunnel; tunnel in rock
岩石隧道掘进联合机 rock tunnel(ling) machine
岩石隧洞 rock tunnel
岩石塌落 rock fall
岩石塌陷 rock subsidence
岩石坍方 rock slide; rock slip
岩石坍落 rock fall
岩石滩 rock foreshore
岩石弹性 elasticity of rock; rock elasticity
岩石弹性模量 elastic modulus of rock
岩石弹性能 elastic energy of rock
岩石掏槽机 rock slotter
岩石特性 petrographic(al) characteristic; rock behavio(u)r; rock character; rock property
岩石特性曲线 rock response curve
岩石特性指标试验 geotechnic(al) index property test
岩石特征 petrographic(al) characteristic
岩石提升 rock hoisting; rock winding
岩石提升格 rock compartment
岩石体 lithosome
岩石体积 rock volume
岩石体积破碎 volumetric(al) fragmentation of rock
岩石体重 specific gravity of rock
岩石天然裂面 cleavage plane
岩石填料 rock-fill
岩石庭园 alpine garden; rock garden
岩石同位素成分 rock isotope composition
岩石同位素平衡 isotope equilibrium between rocks
岩石透水性 rock perviousness
岩石凸出部分 ridge of rock
岩石突出 crump; rock bump
岩石突出部分 rock protuberance
岩石土壤化学成分分析 chemical composition quality of rock and soil
岩石挖沟机 rock-trenching machine
岩石挖掘机 rock excavator; rock grapple
岩石挖掘(石方工程) rock excavation
岩石挖凿齿 rock-cutting tooth
岩石弯曲强度 rock flexural strength
岩石微晶玻璃 rock glass-ceramics
岩石为主的公园 rock garden
岩石围绕的 ironbound; rockbound
岩石温度 rock temperature
岩石温度图 rock temperature map
岩石纹理 crowfoot [复 crowfeet]; vein
岩石稳定性 rock stability
岩石稳定性分级 classification of rock stability
岩石窝孔 rock pocket
岩石物理力学性质 physical-mechanical property of rock; physio-mechanical property of rock
岩石物理破碎 physical fragmentation of rock
岩石物理学 <研究岩石孔隙空间及其特征> petrophysics
岩石物性参数 parameters of rock property
岩石物质 rock substance
岩石吸水率 rate of rock absorption
岩石吸水率试验 water absorption rate of rock test
岩石吸收性试验 water absorption test of rock
岩石系统 rock system
岩石下的地下工程 underground construction in rock
岩石纤维 rock fiber [fibre]
岩石显微镜 petrographic(al) microscope
岩石显微硬度 rock micro hardness
岩石相平衡 petrologic(al) phase equilibrium
岩石巷(道) stone drift
岩石项圈 rock necklace
岩石小爆破 boulder blasting; secondary blasting
岩石楔体 wedge of rock
岩石斜坡 inclined rock slope
岩石形变 rock deformation
岩石形变仪 rock extensimeter [extensometer]
岩石形成 rock formation
岩石形态学 petromorphology
岩石性质 lithologic(al) character; rock property
岩石徐变 rock creep
岩石学 lithology; petrography; petrology
岩石学标志 marker of lithology; petrologic(al) indication
岩石学的 lithologic(al); petrographic(al)
岩石学分类 classification of petrology; petrologic(al) classification
岩石学家 petrographer; petrologist
岩石学图件 diagrams of rocks
岩石学序言 introduction of petrology
岩石学者 petrographer; petrologist
岩石学指数 <系数/参数/公式比例> petrologic(al) index
岩石循环 rock cycle
岩石压力 rock pressure; rock thrust
岩石压力测量 measurement of rock pressure
岩石压力叠加 superposition of rock pressure
岩石压力发展过程 rock pressure development process
岩石压力迹象 rock pressure indications
岩石压力理论 rock pressure theory
岩石压裂 bump
岩石压切点 point of attack
岩石压入硬度 indentation hardness of rock
岩石压碎机 rock crusher
岩石压缩 rock compression
岩石研磨性 abrasiveness of rock; rock abrasiveness; rock abrasivity
岩石研磨性分级 classification of rock abrasiveness
岩石研磨性系数 coefficient of rock abrasiveness
岩石研磨硬度 abrasive hardness of rock
岩石颜色 petrologic(al) colo(u)r
岩石样品 rock sample
岩石样品类型 type of rock sample
岩石移动 rock movement; strata displacement; strata movement
岩石移动错位爆破 primary blasting
岩石异常 rock anomaly
岩石异常类型 type of rock anomaly
岩石异常值 anomalous value of rock
岩石应变计 rock extensimeter [extensometer]
岩石应力 rock stress
岩石应力测量 rock stress measurement
岩石应力量测 rock stress measurement
岩石应力应变特性 stress-strain characteristics of rock
岩石硬度 rock hardness
岩石硬度分级 classification of rock hardness
岩石与岩体结构 rock and rockmass structure
岩石预加应力 rock prestressing
岩石预裂爆破法 rock pre-splitting
岩石预应力 rock prestressing
岩石预应力假说 rock prestressing hypothesis
岩石园 rock garden
岩石园艺 rock gardening
岩石原生构造线性体 lineaments along original structure of rock
岩石允许流速 allowable velocity of flow of rock
岩石凿成的 rock-hewn
岩石凿子 rock chisel
岩石噪声 rock noise
岩石轧成碎石的能力 aggregate making property
岩石轧碎加工的粗骨料 crushed rock
岩石轧碎设备 crushed-stone plant
岩石炸药 quarry powder; rock explosive
岩石褶皱 rock fold
岩石震动 rock burst shock
岩石整体开挖 bulk rock excavation
岩石整体特性 mass properties of rocks
岩石正常值 normal value of rock
岩石正压力 normal rock pressure
岩石支承的 rock-supported
岩石支护 rock-support
岩石支柱 rock prop
岩石支座 rock-support
岩石直剪试验 direct shear test of rock
岩石植物 rock plant
岩石植物群落 rock vegetation
岩石制品 rock product
岩石质量 rock quality
岩石质量系数 rock quality coefficient
岩石质量指标 rock quality designation [RQD]; rock quality index
岩石质量指数 rock quality designation
岩石中的地面建筑 surface construction in rock
岩石中的泡 vesicle
岩石中锚碇孔 rock anchor hole
岩石中石油的储藏 rock storing of oil
岩石柱 rock pillar
岩石抓斗 rock-ripping bucket
岩石状断口 rock candy structure
岩石状土壤 lithosol
岩石自由膨胀率试验 free swelling test of rock
岩石族 rock family
岩石阻力 resistance of rock
岩石阻力系数 rock factor
岩石阻留阶地 rock-perched terrace
岩石组成 lithologic(al) composition; petrographic(al) composition
岩石组分 rock constituent
岩石组构 petrofabric; rock fabric
岩石组合 rock association
岩石钻 rock borer; rock drill
岩石钻机 machine rock drill; stone drill
岩石钻进冷却软化液 drilling fluid
岩石钻孔 rock boring
岩石钻孔设备 rock boring device
岩石钻探 rock boring
岩石钻探船 rock drill barge; rock drill vessel
岩石钻头 rock drill bit
岩石钻眼机 rock drill
岩石钻凿 rock machining
岩栓 bysmalith; plutonic plug; rock bolt; rock plug
岩松 stone pine
岩塌 rock fall
岩塌石块 popping rock
岩台 rock bench
岩坍 rock slide
岩滩 rock(y) beach
岩套 suite
岩梯 rock stairway
岩体 bulk rock; magmatic body; mass of rock; rock body; rock bulk; rock mass; rock matrix; terrene
岩体变形试验点数 number of deformation parameters test of rock mass
岩体变形试验方法 deformation test of rock mass
岩体变形系数 deformation coefficient of rock mass
岩体不连续产生的渗透 seepage from discontinuity
岩体不连续(节理)群的数量 number of sets of discontinuities
岩体衬砌支护 lining of rock mass
岩体承压板试验 bearing plate test of rock mass
岩体尺寸效应 size effect of rock mass
岩体出露区 terrain; terrene
岩体初始应力 initial stress of rock mass
岩体的工程地质分类 engineering geologic(al) classification of rock mass
岩体的结构类型分类 structure type classification of rock mass
岩体的埋深 buried depth of rock body
岩体的强度折减系数值 SRF [strength reduction factor] value of rock of mass
岩体的倾向 dip direction of rock body
岩体的稳定性 rock mass stability
岩体的质量分级 quality classification of rock mass
岩体的走向 strike of rock body
岩体底盘 bottom wall of intrusion
岩体顶盖 top wall of intrusion
岩体动力学 rock mass dynamics
岩体法向压力 normal rock pressure
岩体分级 classification of rock mass
岩体分类 classification of rock mass; rock mass classification
岩体风化程度 weathering degree of rock mass
岩体工程地质(力学) engineering geologic(al) of rock mass
岩体工程质量指标 engineering quality index of rock mass
岩体构造 structure of rock mass
岩体构造断裂 rock discontinuity
岩体滑动 rock mass sliding
岩体基本质量 rock mass basic quality
岩体间断(节理)之间距离 spacing of discontinuity
岩体剪断试验 rock shear test
岩体剪切试验方法 shear test of rock mass
岩体剪切试验种类 type of shear test of rock mass
岩体接触带控制点 control point of contact zone of intrusive body
岩体结构 rock texture; texture of rock mass
岩体结构类型 structural pattern of rock mass; structural type of rock mass
岩体结构面 discontinuity of rock mass; rock discontinuity structural plane; structure surface of rock mass
岩体抗力 rock mass resistibility

岩体控制 rock mass control; strata control
岩体力学 mechanics of rock masses; rock mass mechanics; rock mechanics
岩体连续性等级 continuity rank of rock mass
岩体内浅层滑动 shallow slip in rock mass
岩体喷锚支护 rockbolts and shotcrete supporting of rock mass
岩体破裂 rock breaking
岩体起源 origin of rock bodies
岩体强度 rock mass strength; strength of rock mass
岩体切割 rock cutting
岩体容矿构造 ore-containing structure of intrusion
岩体软化系数 softening coefficient of rock mass
岩体弱面 weakness of rocks
岩体深成的 massif
岩体声波速度测试 acoustic (al) wave velocity test of rock mass
岩体试验 rock formation test
岩体速度指数 velocity index of rock mass
岩体完整度 rock mass completeness
岩体完整性系数 intactness coefficient of rock mass
岩体完整性指数 intactness index of rock mass; integration coefficient of rock mass
岩体稳定问题 problem of stability of rock mass
岩体稳定性 stability of rock mass
岩体系数 <岩体模量与岩块模量之比> rock mass factor
岩体形态 shape of intrusion
岩体型斜长岩 body anorthosite
岩体压力 rock mass pressure
岩体移动 displacement of rock masses
岩体应力特性 nature of stress in rock mass
岩体应力应变曲线类型 type of stress and strain curve of rock mass
岩体应力重分布 rock mass stress redistribution
岩体原位测试 in-situ measurement of rocks
岩体原位试验 in-situ test of rock mass
岩体原位应力测试 in-situ rock stress test
岩体直剪试验 rock mass direct shear test
岩体质量指标 quality index of rock mass; rock mass quality designation; rock mass quality index
岩体中的静水压力状态 hydrostatic pressure hypothesis
岩体中应力 stress in a rock mass
岩体重 weight of rock mass
岩体自然节理 natural rock jointing
岩体钻孔 bulk drilling
岩体钻孔变形试验 drillhole deformation test of rock mass
岩体最终积累应变 final accumulated rock strain
岩筒 rock pipe; volcanic pipe
岩土边坡 rock and soil slope; earth slope and rock slope
岩土参数标准值 standard value of a geotechnical parameter
岩土成分与结构 composition and structure of rock and soil
岩土单位滞水量 specific retention
岩土的材料属性 nature of materials of rock and soil
岩土的电学性质 electric (al) property of rock and soil
岩土的电阻率 electric (al) resistance of rock and soil
岩土的动力性质 dynamic (al) property of rock and soil
岩土的放射性 radioactivity property of rock and soil
岩土的工程地质性质 engineering geologic (al) property of rock and soil
岩土的化学成分 chemical composition of soil and rock
岩土的孔隙性 porosity of rock and soil
岩土的流变性质 rheologic (al) property of rock and soil
岩土的其他工程性质 other engineering property of rock and soil
岩土的热学性质 thermal property of rock and soil
岩土的声学性质 acoustic (al) property of rock and soil
岩土的水理性质 hydraulic character for rock; rock and soil property relating water
岩土的重量 weight of rock and soil
岩土地基 rock and soil foundation; rock and soil ground
岩土 rock and soil
岩土参数标准值 standard value of geotechnical parameters
岩土调研 geotechnic (al) investigation
岩土分类 petrographic (al) classification
岩土工程 earthwork engineering
岩土工程处理方法 geotechnic (al) process
岩土工程地质分类 engineering geologic (al) classification for rock and soil
岩土工程方法 geotechnic (al) process
岩土工程分级 categorization of geotechnical projects
岩土工程分类 geotechnic (al) classification
岩土工程分析评价 geotechnic (al) analysis and evaluation
岩土工程改良 engineering modification of rock and soil
岩土工程改良方法 engineering improvement of soil and rock
岩土工程勘察 geotechnic (al) investigation; geotechnic (al) study
岩土工程勘察报告 geotechnic (al) investigation report; geotechnic (al) study report
岩土工程勘察分级 categorization of geotechnic (al) investigation
岩土工程勘察纲要 method statement for geotechnic (al) investigation works
岩土工程勘察阶段 geotechnic (al) investigation stage
岩土工程勘察任务书 specifications for geotechnic (al) investigation works
岩土工程勘探 geotechnic (al) exploration
岩土工程模型试验 geotechnic (al) model test
岩土工程评价 geotechnic (al) evaluation
岩土工程师 geotechnic (al) engineer; geotechnician
岩土工程试验标准 standard of geotechnical engineering test
岩土工程试验学报 <美国试验和材料学会季刊> Geotechnic (al) Testing Journal
岩土工程条件 engineering condition of rock and soil
岩土工程性质 geotechnic (al) property
岩土工程学 engineering rock and soil; geotechnics; geotechnology; geotechnic (al) engineering
岩土工程学报 <美国土木工程学会月刊> Journal of Geotechnical Engineering
岩土工学的预测能力 predictive capability in geotechnics
岩土荷载 geotechnical load
岩土技术 geotechnics; geotechnique
岩土技术的预测能力 predictive capability in geotechnics
岩土加固 rock and soil improvement; rock and soil reinforcement
岩土交错型地基 subgrade with soil and rock
岩土力学 geomechanics; geotechnics; geotechnique; rock soil mechanics; soil and rock mechanics
岩土力学模型 geotechnic (al) model
岩土描述术语 soil and rock descriptive term
岩土热物理测试 geothermal physical test
岩土室内试验 laboratory soil and rock test
岩土试验研究 test studies of rock and soil
岩土体 rock and soil mass (geotechnical mass); soil and rock; soil and rock mass
岩土体测斜 inclination measurement of soil and rock
岩土体位移监测 displacement monitoring of soil and rock
岩土体中的应力 stresses in soil-rock mass
岩土统一分类系统 unified classification system of soil and rock
岩土振动破坏方式 vibration failure mode of rock and soil
岩湾 rocky bay
岩温率 rock temperature gradient
岩温型 type of rock heat
岩屋 cliff dwelling (settlement); rock dwelling
岩席 drong; sheet
岩系 petrographic (al) series; rock formation; rock series; series of rocks; series of strata
岩隙 grike
岩隙植物 chomophyte
岩下荒漠群落 petrideserta
岩相【地】 petrographic (al) facies; litho-facies; magma facies; petrofacies; rock facies; rock phases
岩相变化控制点 control point of sedimentary facies change
岩相变化圈闭 lithofacies change trap
岩相变化特征 feature of lithofacies variation
岩相不整合圈闭 facies-unconformity trap
岩相层序 lithostratigraphy
岩相带划分 dividing rock fascia
岩相定量分析 petrographic (al) quantitative analysis; quantitative lithofacies analysis
岩相分析 petrofabric analysis; petrographic (al) analysis; petrographic (al) interpretation; rock facies analysis
岩相分析法 lithofacies analysis
岩相古地理条件 lithologic (al) and palaeogeographic (al) condition
岩相古地理图 lithofacies-paleotopographic (al) map; map of lithofacies and palaeogeography
岩相混合程度 entropy
岩相解释 petrographic (al) interpretation
岩相三角图 lithologic (al) triangle
岩相突变 abrupt change in facies
岩相图 lithofacies map; lithologic (al) map; facies map
岩相显微镜 petrographic (al) microscope
岩相显微镜技术 petrographic (al) microscopy
岩相学 facieology; petrography
岩相学家 petrographer
岩相研究 petrographic (al) investigation
岩相异常类型 type of lithofacies anomaly
岩相译释 petrographic (al) interpretation
岩巷 rock gangway
岩巷掘进 development works in stone; rock drivage; stone drifting; tunnel works
岩像学 microphysiography
岩楔 sphenolith
岩屑 chat; cliff debris; core borings; detritus; drill cuttings; drilling breakers; drilling cuttings; land waste; rock cuttings; rock debris; rock fragments; rock waste; rotted rock; shatter
岩屑搬运 transportation of debris
岩屑崩落 debris avalanche; debris fall
岩屑崩塌 debris avalanche; debris fall
岩屑崩坍 debris fall
岩屑比 volume per unit of depth
岩屑比重 cuttings weight
岩屑边坡 detrital slope
岩屑编号 cuttings number
岩屑冰 debris-laden ice
岩屑层 acre
岩屑长石砂岩 lithic arkose
岩屑长石砂岩质瓦克岩 lithic arkosic wacke
岩屑长石杂屑岩 lithic feldspathic graywacke
岩屑沉积 detrimental sediment
岩屑冲积扇 detrital fan
岩屑的 detrital
岩屑的排除 cutting removal
岩屑地带 detritus zone
岩屑堆 debris cone; debris dump; detrital cone; detrital dump; detritus rubbish; runoff hill stone; scree; talus cone; talus material
岩屑堆积 talus deposit
岩屑堆积物中植物 homophyte
岩屑堆徐变 talus creep
岩屑风化壳 debris residuum
岩屑厚度 cuttings thickness
岩屑滑动 debris slide
岩屑滑坡 debris slide; talus slide
岩屑流 debris (-laden) flow; debris (-laden) stream; talus glacier; talus train; tulus glacier
岩屑录井 sieve residue log
岩屑录井间距 debris log interval
岩屑凝灰熔岩 lithic tuff lava
岩屑凝灰岩 lithic tuff
岩屑平原 debris plain
岩屑坡 talus slope; debris slope
岩屑蠕变 talus creep
岩屑蠕滑 talus creep
岩屑入侵 detritus intrusion
岩屑砂 detrital sand
岩屑砂屑岩 litharenite; lithic arenite
岩屑砂岩 lithic sandstone
岩屑砂状结构 lithic psammitic texture

岩屑扇 talus fan
岩屑上返速度 cuttings returning rate
岩屑上升速度 cuttings lifting speed
岩屑升速 cuttings lifting velocity
岩屑石英砂岩 lithic quartz sandstone
岩屑石英杂砂岩 lithic quartz graywacke
岩屑试样 chip sample
岩屑体积百分比 percent of cutting volume
岩屑推移质 debris bed load
岩屑瓦克岩 lithic wacke
岩屑下滑 detritus slide;talus creep
岩屑下滑速度 cuttings sliding down speed
岩屑携带比(率) cuttings transport ratio
岩屑形状 cuttings shape
岩屑亚长石砂岩质瓦克岩 lithic subarkosic wacke
岩屑亚稳定砂屑岩 lithic sublabile arenite
岩屑岩 clasolite
岩屑移动 detritus movement;movement of detritus
岩屑抑制作用 chip hold down effect
岩屑杂砂岩 lithic graywacke
岩屑锥 acree;debris cone;detritus cone;gliders;glister [glistre];glitter;glydrs;talus cone
岩芯 bore(hole) core;bore sample;corduroy sample;core;drill core;rock core;well core
岩芯半合管 core laden split tube
岩芯半径 radius of the core
岩芯保存盒 core box
岩芯饱和度测定法 weight-saturation method
岩芯饱和率测定法 weight-saturation method
岩芯编号 core number
岩芯编录 core documentation
岩芯标本 core specimen
岩芯标记器 core marker
岩芯饼化现象 formation of disk-shaped rock cores
岩芯采取 core recovery
岩芯采取百分率 percentage core recovery
岩芯采取参数 core recovery parameter
岩芯采取工具 core recovery tool
岩芯采取率 corduroy recovery;core extraction;core recovery (percent);core run;percentage of core recovery;percentage recovery of core;recovery ratio
岩芯采收率 core extraction;core recovery percent;core run;percentage of core recovery;recovery of core
岩芯采样 core sampling
岩芯采样管 coring tube;core tube
岩芯采样率 core recovery percent
岩芯采样筒 core barrel
岩芯槽 trough core
岩芯测试 tests on cores
岩芯长度 core length;length of core
岩芯长度分类 core length classification
岩芯尺寸 core size
岩芯冲出器 blowout plug
岩芯冲蚀 core erosion;core wash
岩芯穿透深度 efficiency of core penetration
岩芯打捞筒 core basket
岩芯导向胶塞 guide core
岩芯的纯洁性 purity of core
岩芯的代表性 representativeness of core
岩芯的完整度 integrity of core
岩芯定位 core orientation
岩芯定向 core orientation
岩芯定向工作 core orientation works
岩芯定向装置 core interpreting device;core orienting device
岩芯堵塞 core blockage;core jam(ming)
岩芯断面 core cross-section;core intersection;core interval
岩芯分隔标牌 core divider
岩芯分析 corduroy analysis;core analysis
岩芯辐射取样 core radiation sampling
岩芯管 barrel;corduroy barrel;core barrel;core pipe;corer;mining core tube
岩芯管长度 length of core barrel
岩芯管导杆 core barrel rod
岩芯管堵塞 barrel blockage
岩芯管接头 corduroy barrel head;core barrel head
岩芯管丝锥 core tube tap
岩芯管稳定器 core barrel stabilizer
岩芯管自卡 barrel blockage
岩芯管组件 core assembly
岩芯管钻进速度 speed of rotation of core barrel
岩芯过早卡塞 premature block
岩芯盒 core box
岩芯盒标记 label(l)ing of core box
岩芯回采率 percentage core recovery;recovery ratio
岩芯回收百分率 percentage core recovery
岩芯回收总长度 total core recovery
岩芯获得率 core recovery;percentage core recovery;recovery ratio
岩芯机 core borer;core rig;rock borer;rock boring machine;rock drilling machine
岩芯记录 core logging;core record;drill log;drill record
岩芯记录标志的解释 explanation of symbols used on core logs
岩芯夹具 core catcher
岩芯夹具座 core catcher case;core gripper case;core spring carrier;core spring case
岩芯夹钳 core tongs
岩芯架 core rack
岩芯鉴定 core value
岩芯节段 section of core
岩芯锯 core saw
岩芯卡簧 core catcher;core spring
岩芯卡簧接头 core spring adapter
岩芯卡簧座 core catcher case;core gripper case;core spring carrier;core spring case
岩芯卡簧座接头 core-shell coupling
岩芯坑 core pit
岩芯库 core chamber;core house;core library;core storage
岩芯库房 core shack;core shanty
岩芯矿层厚度 <从岩芯看出的> core intersection
岩芯捞取器 core fisher;core picker
岩芯类型 core type
岩芯裂隙率 fissure factor of core
岩芯描述 description of core
岩芯磨损 core grinding
岩芯内管 core-laden inner tube
岩芯盘 core tray
岩芯喷出口 discharge nozzle
岩芯棚 core shack;core shanty
岩芯劈分法 core-splitting method
岩芯劈开机 core separator;core splitter
岩芯劈开器 core separator;core splitter
岩芯墙 core wall
岩芯切断器 core splitter;core cutter
岩芯切断提取器 core catcher
岩芯切割 core cutter
岩芯切取筒 core cutter
岩芯取样 boring sample;core sampling;coring
岩芯取样回收率 solid-core recovery
岩芯取样检验 core-drilling inspection
岩芯取样率 percentage of core recovery
岩芯取样器 corduroy sampler;core barrel;core catcher;core cutter;core lifter;core sampler;coring apparatus;coring device
岩芯容纳管 core container
岩芯溶蚀痕迹 dissolution trace of core
岩芯入口 <岩芯管的> core entry
岩芯上螺纹槽 rifling
岩芯上下显示矿体厚度 core intersection
岩芯示出的含矿层 cored interval
岩芯试件 core specimen
岩芯试验 corduroy analysis;corduroy test;core analysis;core test
岩芯试样 borehole sample;bore specimen;boring sample
岩芯试样强度 core strength
岩芯收获率 core recovery
岩芯收集环 core trap-ring
岩芯素描图 core sketch
岩芯损失 core loss
岩芯探管 <插入地下,抽取试样的管子> corduroy tube
岩芯提断部位 <孔内> pulling point
岩芯提断点 pulling point
岩芯提断器 corduroy cutter;core breaker;core extractor;core snatcher
岩芯提断器接头 core-lifter adapter
岩芯提断器卡簧 core breaker spring;core spring adapter
岩芯提断器外壳 core-lifter case;spring lifter case
岩芯提取器 corduroy lifter;core catcher;core cutter;core extractor;core lifter;corer;ring lifter;sample grabber;grapple
岩芯筒体 diamond core barrel
岩芯图 core graph
岩芯推出器 <液压或螺旋的> core plunger;core punch;core pusher (plunger)
岩芯退取器 core extractor
岩芯脱落 core fall off
岩芯外壳 core shell
岩芯外露长度 <介于钻头与岩芯管内末端之间的> core exposure
岩芯外皮 core shell
岩芯完整程度 degree of core completion
岩芯显示遇到的岩脉 core intersection
岩芯箱 core box;core tray
岩芯箱车刀 cove box bit
岩芯循环钻机 core reverse circulation drill
岩芯岩粉比 core-to-sludge ratio
岩芯岩溶率 karst factor of core
岩芯样(品) boring sample;core (drill) sample;corduroy sample;sample core
岩芯样(品)的制备 preparation of core samples
岩芯遗失解释 interpreting core loss
岩芯遗失控制 control of core loss
岩芯硬度 core hardness
岩芯与矿心 core and mineral core
岩芯造型 cored-up mo(u)ld
岩芯折断器 core breaker
岩芯整理 core trimming
岩芯直径 core diameter
岩芯置放台 core placing table
岩芯周长 core circumference
岩芯柱状图 graphic(al) strip log
岩芯抓长度 core catcher length
岩芯抓尺寸 core catcher size
岩芯爪 core grabber;core gripper;core snatcher
岩芯自动卡断器 automatic core-breaker
岩芯自卡被扭成螺旋形裂隙 torsion fracture
岩芯总长度 total core length
岩芯钻 corduroy drill;rock core drill
岩芯钻头 core bit
岩芯钻探 corduroy-drilling
岩芯钻车 core rig
岩芯钻机 boring rig;corduroy drill machine;corduroy drill rig;core-drilling machine;core-drilling rig;core drill machine;core drill rig;core rig;coring machine;drill for quarrying;core drill
岩芯钻机附件 core drill fittings
岩芯钻进 corduroy-drilling;core boring;core drilling;coring
岩芯钻进进尺 cored footage
岩芯钻具 core equipment
岩芯钻孔 corduroy-hole;core hole
岩芯钻孔的岩粉 cutting of boring
岩芯钻连接 core bit connection
岩芯钻取器 corer
岩芯钻探 corduroy boring;core boring;core drilling;core investigation
岩芯钻探采样 drill-core sampling
岩芯钻筒 corduroy barrel;core barrel
岩芯钻头 concave bit;core head;hollow bit;hollow drill bit;jack bit;rock bit;rock core bit;coring bit <用于回转岩芯钻进>
岩芯钻头连接 coring bit connection
岩型 lithotype
岩性 lithologic(al) character;rock character;rock characteristic
岩性比较法 method of lithologic(al) comparison
岩性边界 lithologic(al) boundary
岩性测井记录 lithologic(al) log
岩性带 lithizone [lithozone]
岩性地层带 lithostratigraphic(al) zone
岩性地层单位 lithostratic unit;lithologic(al) unit
岩性地层界线 lithostratigraphic(al) boundary
岩性地层学方法 lithostratigraphy method
岩性断面图 lithologic(al) section
岩性对比 lithologic(al) correlation
岩性分类图 classification map of lithology
岩性分析 lithology analysis
岩性分析的地层模型 earth model of lithologic(al) characters
岩性分析图 lithologic(al) analysis diagram
岩性符号 rock quality designation
岩性厚度 lithologic(al) thickness
岩性及构造简单型 simple style in lithologic(al) character and geologic(al) structure
岩性及构造较复杂型 complicated style in lithologic(al) character and geologic(al) structure
岩性较单一型 simple lithology type
岩性较复杂型 complicated lithology type
岩性阶 lithologic(al) order
岩性阶地 esplanade
岩性接触带 lithologic(al) contact zone

Y

岩性控制 lithologic(al) control
岩性流轴距 axis distance of lithologic(al) flow
岩性录井图 graphic(al) log;lithologic(al) log
岩性密度测井 lithodensity log
岩性密度测井曲线 lithodensity log curve
岩性模式 lithologic(al) pattern
岩性特征 feature of rock type;lithologic(al) character
岩性体 lithosome
岩性图 lithologic(al) map
岩性土 rocky soil
岩性相 litho-facies
岩性相等 lithologic(al) identity
岩性相对厚度 lithologic(al) relative thickness
岩性相似性 lithologic(al) similarity
岩性相组合 lithofacies association
岩性学 lithology
岩性岩相横剖面图 transverse profile of lithologic(al) phase
岩性岩相图 map of lithology and lithofacies
岩性异常类型 type of lithologic(al) anomaly
岩性因素 lithology factor
岩性油藏 lithologic(al) pool
岩性油气田 lithologic(al) oil-gas field
岩性指标 rock quality designation [RQD]
岩性转移频数 lithologic(al) transition frequency
岩性状态 lithologic(al) state
岩序 sequence of rock
岩穴 abri;grotto;rock pocket
岩穴流 cavern flow
岩压 lithostatic(al) pressure;rock pressure
岩压测定 rock pressure measurement
岩崖 ledge;rock cliff
岩盐 halite;mine salt;rock salt;salt fissile;salt rock;common salt
岩盐晶格 rock-salt lattice
岩盐棱镜 rock-salt prism
岩盐穹丘外缘坝 saline dome outer edge bar
岩盐脱水法 salt dehydration
岩盐型结构 rock-salt structure
岩盐岩溶 salt karst
岩盐柱 salt pillar
岩眼 phacolith
岩样 core sample;formation sample;muck sample;rock sample
岩样分析 formation sample analysis
岩样数 number of rock sample
岩样微剩磁测量仪 rock generator
岩榆 rock elm
岩原 rock floor
岩渣 rock waste;scoria cinder
岩渣托运 haulage of muck
岩渣锥 cinder cone
岩枕裂 fissura petro-occipitalis;petro-occipital fissure
岩支 petrous branch
岩枝 apophysis [复 apophyses];tongue;offshoot
岩质边坡 rocky slope
岩质边坡稳定性 stability of rock slope
岩质边坡稳定性计算方法 computing method of stability failure of rock slope
岩质底土 rock subsoil
岩质分类 rock classification
岩中基础 foundation in rock
岩钟 kupola [cupola]
岩钟成矿说 cupola hypothesis of mineralization
岩轴冒落 rock failure
岩株 laccolith;stock
岩柱 barrier pillar;bysmalith;pluton-ic plug;rock pillar;rock stack
岩柱支撑 rock post shotcrete
岩铸体 chonolite
岩桩 rock pile
岩状冰 ice-formed rock
岩状物质 rock-like mass;rock-like substance
岩锥 scree cone;talus cone
岩锥堆积土 talus deposit
岩渍土的地基评价 estimation for salty foundation soil
岩组【地】petrofabric
岩组分析 fabric analysis
岩组分析样品 sample for fabric analysis
岩组构造岩 R-tectonite
岩组平行面 S-plane;S-surface
岩组图 fabric diagram;petrofabric diagram
岩组学 petrofabrics

沿

沿 *X* 轴方向运动 *X*-motion
沿 *Y* 轴方向运动 *Y*-motion
沿 *Z* 轴方向运动 *Z*-motion
沿岸 inshore;littoral;seashore
沿岸凹槽 longshore trough
沿岸搬运 littoral transport;longshore transport
沿岸变化过程 littoral process
沿岸标 bankwise mark
沿岸标志 coastal aids
沿岸冰带 ice fringe
沿岸冰间航路 shore clearance;shore clearing;shore lead
沿岸冰间湖 shore polynya
沿岸冰间水道 shore clearance;shore clearing;shore lead
沿岸冰间水面 shore clearing
沿岸冰麓 pressure ice foot
沿岸冰面开裂水域 land waters
沿岸冰穴 shore polynya
沿岸波 edge wave
沿岸测量 coastwise survey
沿岸潮流 longshore current
沿岸沉积痕 strand marking
沿岸沉积(物) littoral accumulation;littoral deposit;littoral sediment;shore deposit
沿岸沉积物流 longshore drift
沿岸沉积作用 littoral sedimentation
沿岸冲刷 shore erosion
沿岸大地测量 coast and geodetic survey
沿岸带 nearshore zone
沿岸(导航)灯标 coastal light
沿岸的 alongshore;epilittoral;longshore
沿岸地磁观测站 coast and geodetic magnetic observatory
沿岸地带 littoral region;paralic zone;littoral zone
沿岸地区 shore land
沿岸地形 coastal topography;littoral landform;littoral topography
沿岸动物区系 littoral fauna
沿岸堆积 littoral accumulation
沿岸法规 riparian legislation
沿岸风 alongshore wind;littoral wind
沿岸风暴冰脚 storm ice wind
沿岸港 coastal harbo(u)r;shoreline harbo(u)r
沿岸工厂 waterfront factory;waterfront plant
沿岸工作居民 longshoreman
沿岸公园 waterfront park
沿岸沟谷 longshore trough
沿岸固定冰 landfast ice;shore floe
沿岸国(家) riparian state;coastal state;littoral state;riparian country
沿岸(海岸或环礁湖岸)泥沙埂 potrero
沿岸海床经济区域 coastal sea-bed economic area
沿岸海沟 longshore trough
沿岸海关缉私队 Preventive Service
沿岸海流 alongshore current;coastal current;littoral current;longshore current
沿岸海图 coast chart
沿岸海洋调查 inshore oceanographic(al) survey
沿岸海洋卫星监视系统 satellite-based system for coastal oceans monitoring
沿岸海洋学 coastal oceanography
沿岸海域 coastal seas;coast waters;marine belt;marginal sea
沿岸海域监视卫星 coastal ocean monitor satellite
沿岸海域监视卫星系统 coastal ocean monitor satellite system
沿岸航道 bankwise channel;coastwise channel
沿岸航线 coasting line
沿岸航行 coast;coasting;lie along the land
沿岸弧丘 morro
沿岸湖 coastal lake
沿岸环境 littoral environment;riparian environment
沿岸环流 coastal circulation
沿岸急流 sea push
沿岸监测 coastal monitoring
沿岸监视 coastal surveillance
沿岸监视系统 coastal surveillance system
沿岸建筑(物) waterfront construction;waterfront structure;waterfront building
沿岸交易 cabotage;coastal trade;coastwise trade
沿岸阶地 shore terrace
沿岸结构(物) waterfront structure
沿岸开发 development of seaside;seaside development
沿岸开发区 development of seaside area
沿岸雷达电视 shore radar television
沿岸流 longshore current;coastal stream;fringing current;nearshore current;offshore current;shore current;littoral current
沿岸流搬运有机质 organic matter by coastal current transportation
沿岸(流)冲刷 littoral erosion
沿岸流区 littoral zone
沿岸流示踪剂 littoral tracer
沿岸流障碍物 littoral barrier
沿岸码头 marginal quay;shore terminal
沿岸贸易 coasting;coast(ing) trade;coastwise trade
沿岸贸易税 coast trade duties
沿岸泥沙 littoral sediment
沿岸泥沙流 littoral mud and sand flow;littoral drift
沿岸泥沙垅 potrero
沿岸泥沙输移率 rate of littoral transport
沿岸泥沙运动 littoral sand movement
沿岸漂积物 shore drift
沿岸漂流 longshore drift
沿岸漂沙 alongshore drift;drift sand;littoral drift;littoral movement;littoral transport;longshore transport;longshore drift
沿岸漂沙变向区 nodal zone
沿岸漂沙程度 littoral drift magnitude
沿岸漂沙带 littoral drift belt;zone of littoral drift
沿岸漂沙的沉积物 littoral sediments
沿岸漂沙的上游沙滩 updrift beach
沿岸漂沙的下游 downdrift side
沿岸漂沙的下游沙滩 downdrift beach
沿岸漂沙的主导方向 downdrift
沿岸漂沙海岸 littoral drift coast
沿岸漂沙海滨 littoral drift shore
沿岸漂沙量 littoral drift quantity
沿岸漂沙率 rate of littoral transport
沿岸漂沙区 zone of littoral drift
沿岸漂砂 longshore drift
沿岸漂移 littoral drift;longshore drift
沿岸平原 coastal plain;littoral plain
沿岸浅海底的 eulittoral
沿岸浅水带 phytal zone
沿岸浅滩 shore bank
沿岸侵蚀 coastal erosion;shore erosion
沿岸区 coastal region;littoral area;littoral region;littoral zone;maritime belt;nearshore zone
沿岸群岛 coastal arching
沿岸沙坝 barrier bar;longshore bar;offshore bar
沿岸沙堤 longshore bar
沿岸沙埂 longshore bar
沿岸沙垄 fore dune
沿岸沙丘 coastal dune;fore dune;littoral dune;towan
沿岸沙质低地 machair
沿岸沙洲 barrier bar;barrier beach;longshore bar;shore bank
沿岸上升流 coastal upwelling
沿岸使用权 littoral right;riparian rights
沿岸示踪沙 littoral tracer
沿岸式码头 marginal type wharf
沿岸输沙 alongshore transport;littoral drift;littoral movement;littoral sediment transport;littoral transport
沿岸输沙的主导方向 predominant direction of transportation
沿岸输沙率 littoral transport rate
沿岸输送 longshore transport
沿岸输移 alongshore transport;littoral movement;littoral transport;longshore transport
沿岸水 littoral water
沿岸水道 intracoastal waterway
沿岸(水底)沙坝 longshore bar
沿岸水工建筑物 waterfront hydrostructure;waterfront structure
沿岸水流 alongshore current;alongshore water movement;longshore water movement
沿岸水色扫描仪 coastal zone colo(u)r scanner
沿岸水域 coastal waters;inshore waters
沿岸水域污染 polluted belts along river bank;pollution of coastal waters
沿岸台地 sea terrace;strand flat;sublittoral platform
沿岸滩地 coastal plain
沿岸滩沙埂 barrier beach
沿岸条件 littoral condition
沿岸调蓄 lateral storage
沿岸通道 shore lead
沿岸通航带 inshore navigable zone;inshore traffic zone
沿岸通航区 inshore navigable zone;inshore traffic zone
沿岸土地 littoral land;riparian land;shore land
沿岸洼槽 longshore trough

沿岸污染 coastal pollution
沿岸物质流 littoral drift;longshore drift
沿岸系 littoral system
沿岸相 littoral facies
沿岸小型航运 petty cabotage
沿岸效应 coastline effect
沿岸信号 coast signal
沿岸洋流 coastal current
沿岸移动 littoral movement
沿岸涌浪 marginal swell
沿岸用水 riparian uses;riparian water use
沿岸用水权 riparian water rights
沿岸用水权学说 doctrine of riparian rights
沿岸淤积 littoral accumulation; littoral deposit
沿岸淤积埂 potrero
沿岸淤滩 littoral nourishment
沿岸渔业 inshore fishes;longshore fishery;strand fishery
沿岸运动 littoral movement
沿岸运河 <约平行于海岸> intracoastal waterway
沿岸运输 littoral transport;longshore transport
沿岸运输许可证 transire
沿岸沼泽 coastal plain swamp
沿岸植被 riparian vegetation
沿岸植物漂流块 raft of land vegetation
沿岸资源 coastal resources
沿半径变化的螺距 radial pitch variation
沿边 along the edge;edgeways
沿边变形 edgewise distortion
沿边剥裂 edge spall
沿边擦过 side swipe
沿边穿孔卡片 verge-punched card
沿边磁铁 rim magnet
沿边的 edgewise;sideway;sidewise
沿边的横斜 sidewise
沿边光线 rim ray
沿边焊接 edgewise welding;weld edgewise
沿边剪切 edgewise shear
沿边建筑基底 edge lot
沿边建筑基地 edge lot
沿边将砖墙铺平在灰浆上 buttered
沿边节 margin knot
沿边开裂 intergranular crack
沿边铺砌 lay edgewise
沿边铺设 edgewise placing
沿边支承 supported at edges
沿边砖铺砌 edgewise brick paving
沿滨的 alongshore
沿滨漂积物 shore drift
沿侧石停车容量 curb capacity
沿层波形参数分析 wave shape parameter analysis along layer
沿层横向错动【地】 lateral movement
沿层频谱分析 spectrum analysis along layer
沿长度方向弯曲 principal curvature
沿长向斜纹理 <木材> diagonal grain
沿长轴伸展 elongate with long axis
沿程变化 streamwise variation
沿程变流量 spatially varied flow
沿程变流量输沙率 spatially varied transport rate
沿程变流速 spatially varied velocity
沿程冲刷 progressive erosion; progressive scour(ing)
沿程渐变流 gradually varied flow
沿程流速变化 streamwise variation of flow velocity; variation of flow velocity along its course
沿程流向 streamwise flow direction
沿程摩擦损失 streamwise friction loss
沿程水流剖面 streamwise section
沿程水流特性 streamwise flow condition
沿程水流条件 streamwise flow condition
沿程水流状况 streamwise flow condition
沿程水流状态 streamwise flow condition
沿程水面比降 streamwise water surface slope
沿程水头损失 friction(al) head loss; friction(al) loss of (water) head
沿程压力损失 linear pressure loss; pressure loss in flowing
沿程淤积 progressive deposition
沿尺 foot run
沿传输线电平 line level
沿垂线分布 distribution in a vertical
沿垂线平均 mean in vertical
沿垂直轴旋转的 vertical pivoted
沿道路的距离 over-the-road distance
沿道路行进 movement on roads
沿等高线开沟(耕作) contour ploughing
沿等高线修筑的雨水排除系统 contour drainage
沿等高线栽培 contour check
沿等高线岩壳 contour scaling
沿等高线种植 contour check
沿堤小路 dike path;dike way
沿底层 near-bottom layer
沿底的 demersal
沿底拖运法 bottom-pull method
沿地板面气流 floor current
沿地面集材 <用拖拉机> bobtailing
沿地质相共生 intergrown along the facies
沿顶板掏槽爆破 burn cut along tunnel roof
沿断裂带走向 along fault strike
沿浮生植物层 <河、湖上的> raft of vegetation
沿富水带走向 along the strike of abundant groundwater zone
沿腹面的外表伸展 run along the ventral surface
沿革 case history;history
沿革纪录片 history card
沿根轴伸长 elongate along the axis of the root
沿埂沟 border ditch
沿工作面 via face
沿公路的 over-the-road
沿公路的露天广告(牌) outdoor advertising along highway
沿拱方向的 archwise
沿构造方向 upstructure
沿谷道路 hollow way
沿管线敷设 stringing
沿管蒸馏 pipage
沿轨道上翼缘行走的桥架 <起重机> top-running bridge
沿轨道运行 circumrotation
沿轨误差 along-track error
沿海 nearshore;seaside;shoreface
沿海船舶 coaster
沿海船舶泊位 coaster berth
沿海船舶交通服务 coastal vessel traffic service
沿海岸单桅小船 hoy
沿海岸的 circumlittoral; coastal; intracoastal;coastwise;longshore
沿海岸的高速公路 seashore freeway
沿海岸的内陆水道 intracoastal waterway
沿海岸地区 littoral
沿海岸国 coastal state;littoral state; riparian state
沿海岸航船 coastal vessel
沿海岸航行 coasting sailing
沿海岸河流 <美国的密西西比河及其他流入大西洋、太平洋、墨西哥湾的几条河流> coastal river
沿海岸入侵 intrusion through the beach of the sea
沿海岸水域 coastal area
沿海岸铁路 coast railway
沿海保护区 coastal protected area
沿海泊位 coastal berth
沿海测量船 coastal surveying ship
沿海产卵场 inshore spawning ground
沿海潮流 littoral current
沿海沉积物 nearshore sediment
沿海城市 coastal city;coastal town
沿海船舶 coaster; coasting craft; coasting vessel
沿海船舶码头 coasting vessel quay
沿海船长 coastwise master
沿海船队 coastal fleet
沿海船(员) shoaler
沿海船只 coastwise craft
沿海大陆架 coastal shelf
沿海岛屿 offshore island
沿海的 coastal;inshore;intracoastal; littoral
沿海灯标 coasting lights
沿海低地围垦 tide-and reclamation
沿海地带 coastal area;coastal belt; coastal strip; coastland; maritime belt;oceanfront
沿海地貌 coastal feature
沿海地区 coastal belt;coastal land; coastal region;coastal zone;coasting area; coastland; downcountry; greater coasting area;littoral area; sea board;seaboard line;coastal area
沿海地区经济发展战略 economic development strategy of coastal areas
沿海地图 coastal chart
沿海地形 coastal topography;inshore bottom contour
沿海短程运输 limited coasting trade; small coasting trade
沿海防波堤 coastal breaker;coastal breakwater
沿海防洪堤 coastal flood embankment
沿海防卫工作 coast defense work
沿海干道 coast trunk road
沿海港口 coastal harbo(u)r; coastal port; coasting harbo(u)r; coasting port
沿海港口内(集装箱)喂给服务 coastal intraport feeder service
沿海高速公路 bayshore freeway
沿海个体渔业 artisanal inshore fishery
沿海各省 coastal provinces;maritime provinces
沿海各州天然气公司 <美> Coastal States Gas
沿海工程船 coastal engineering ship
沿海工地 coastal site
沿海工业 coastal industry;industry in the coastal regions; seaside-oriented industries
沿海工业地带 littoral industrial zone; seashore industrial reservations
沿海工业区 coastal industrial region; coastal industrial zone; littoral industrial area; waterfront industrial area
沿海供应船 hoy
沿海管辖权 coastal jurisdiction
沿海国的优先权利 preferential rights of coastal states
沿海国(家) coast(al) state; littoral country;littoral state;maritime country
沿海国家优先区 coastal state priority zone
沿海国家资源管辖权 coastal state resource jurisdiction
沿海海底领土 coastal submarine domain
沿海海蚀准平原 marine peneplain
沿海海图 coastal chart;coasting chart
沿海海湾 coastal bays
沿海海洋环境 coastal marine environment; inshore marine environment
沿海海域科学考察船 coastal zone research vessel
沿海航标 coastal aids;coast beacon; coast signal
沿海航船 coaster
沿海航道 coastal waterway
沿海航路 coastal waterway route
沿海航路指南 coast(ing) pilot
沿海航线 coastal service; coasting service; coasting shipping line; coastline; coastwise route; greater coasting service;coastal route
沿海航线班轮 coast(al) liner
沿海航线集装箱船 coastwise line container ship
沿海航线区 greater coasting area
沿海航线引航员 coast(al) pilot
沿海航行 coastal navigation; coasting; coasting navigation; coastwise navigation; coastwise shipping; inshore navigation
沿海航行船 coastal vessel; coaster; coasting boat
沿海航行权 cabotage
沿海航行者 coaster
沿海航行指南 coast pilot
沿海航运 coastal shipping; coasting service;cabotage;coastal traffic
沿海航运权 cabotage right
沿海河口湾 coastal inlet
沿海河流 coastal river
沿海洪水 coastal flooding
沿海护岸 coastal defence revetment
沿海环境 coastal environment; inshore environment; maritime environment
沿海汇流区 coast confluence zone
沿海豁免区 exempt coastal zone
沿海或航道待命 coast or channel for orders
沿海货船 coastal cargo vessel
沿海货运清单 coasting manifest
沿海交通 coast(wise) traffic
沿海结构物 nearshore structure
沿海经济开放区 coastal economic open zone; open coastal economic areas
沿海警告 coastal warning
沿海开发 coastal development; seaside development
沿海开发区 seaside development area
沿海开放城市 coastal city opening; coastal open city
沿海客货船 combination coaster
沿海雷达平台 offshore radar platform
沿海陆地 coastland
沿海贸易 coasting;coast(ing) trade; coastwise trade
沿海贸易船舶 coasting trade vessel
沿海贸易区 coasting trade area
沿海贸易权 cabotage
沿海贸易商船 shoaler
沿海贸易者 coaster
沿海漂积物 beach drifting
沿海漂沙 beach drifting
沿海漂移 beach drifting

沿海漂移物 < 如砂砾 > beach drifting
沿海平地 strand flat
沿海平顶方山 taboleiro
沿海平原 coastal plain
沿海平原土壤 coastal plain soils
沿海平原小丘 mendip
沿海气候 coastal climate; littoral climate
沿海汽艇 motor coaster
沿海潜洲 coastal bar
沿海区 coasting area
沿海区管理法 Coastal Zone Management Act
沿海区域 coastal region; coastal zone
沿海沙脊 chenier
沿海沙丘 coastal dune; towyn [tywyn]
沿海沙滩 coastal bar; coastal beach
沿海沙洲 coastal bar; coastal beach
沿海砂礁 coastal bar
沿海山脉 coastal range
沿海山丘 coastal mountain
沿海商船 coasting vessel
沿海生长的 littoral
沿海生态系统 coastal ecosystem
沿海生态学 coastal ecology
沿海省份 coastal provinces; maritime provinces
沿海湿地 coastal wetland
沿海市场 coastal region markets
沿海疏浚 coastal dredging
沿海水产资源 coastal resources
沿海水道 intracoastal waterway
沿海水路 coastal waterway
沿海水文测量 inshore hydrographic(al) survey
沿海水污染 coastal water pollution
沿海水下平原 coast shelf
沿海水域 coastal waters
沿海台地 coastal terrace; strand flat
沿海台站 inshore station
沿海拖运 coastal towage
沿海污染研究 coastal pollution research
沿海雾 coast fog
沿海泻湖 marginal lagoon
沿海巡船 coaster
沿海巡逻截阻艇 coastal patrol interdiction craft
沿海盐水侵入 coastal salt water intrusion
沿海引水员 coast pilot
沿海用测深锤 < 30 ~ 50 磅重 > coasting lead
沿海油船 coasting tanker
沿海油轮 coastal tanker
沿海游憩胜地 riviera
沿海渔场 coastal fishing ground; inshore fishing ground
沿海渔港 coastal fishery harbour
沿海渔业 coastal fishery; inshore fishery; inshore fishing
沿海渔业国 coastal fishery state
沿海原油浮油油膜 oil film on coastal waters
沿海运河 coastal canal; intracoastal waterway < 约平行于海岸 >
沿海运输 coastal shipping; coast(al) traffic; coastal transport(ation); coastwise carriage; coastwise transport; coastwise traffic
沿海运输的 coastwise
沿海运输业 coastwise carrier
沿海沼泽 marine swamp; paralic swamp
沿海州 coastal state
沿海自然资源 natural resources of coastal waters
沿含水层走向 along the strike of aquifer
沿航线导航 primary navigation
沿航线速度 velocity along route
沿河 along the river
沿河泊位 riverside berth
沿河产业 waterfront property
沿河长廊林 fringing forest; gallery forest
沿河城镇 riparian cities and towns
沿河冲积平原 river bottom
沿河大堤 river embankment; river levee; mainline levee
沿河道入侵 intrusion through river course
沿河低冲积地 carse
沿河地带 riverfront
沿河地区 waterfront area; waterfront region
沿河陡崖 river cliff
沿河谷渗流 thalweg
沿河建筑物 riverfront
沿河阶地 river terrace; stream terrace
沿河流放 river running
沿河流或道路未砍伐的森林带 green strip
沿河路 river side road
沿河码头 river quay; riverside wharf; riverfront
沿河排入的污染物 riverine input of pollutants
沿河泉 channel spring
沿河沙堤 hirst [hurst]
沿河省份 riparian provinces
沿河台地 river terrace
沿河土地 riparian land; riverfront
沿横斜坡挖掘 sidehill cut
沿厚度方向振动 thickness vibration
沿迹线旋转而展出的面 rabatment
沿江大堤 river embankment
沿江河建厂 riverine factories
沿江下驶 proceed down the river
沿交通干线有计划发展起来的城市 dyna(metro)polis
沿礁砂 sand apron
沿接触交代作用【地】replacement along contacts
沿接触面啮合的岩石 interlocked rock
沿街边线 front lot line
沿街大楼 storefront
沿街道路 frontage road(way); frontage street or road
沿街的前门 < 古罗马 > janua
沿街辅道 frontage road
沿街隔声(设施) frontage insulation
沿街后院 front equivalent yard
沿街建筑 abutting building; street architecture; street building
沿街建筑立面 street front
沿街宽度 frontage
沿街门面 block face
沿街(门面)长度计算(征费)法 front-foot rule
沿街面道路 service road; service street
沿街铺面 storefront
沿街墙 street wall
沿街商业区 street commercial district
沿街停车 on-street parking
沿街土地 block face
沿街屋前空地 street front
沿节理形成的洞穴 bedding cave
沿晶界破裂 intercrystalline cracking
沿晶开裂 intergranular crack
沿晶面交代作用 replacement along crystallographic(al) planes
沿径 along track
沿径槽 radial slot
沿径连接 diametric(al) connection
沿径曲线 diametral curve; diametric(al) curve
沿径向线穿云 radial penetration
沿距离射角 range aperture
沿口加强钢丝 < 轮胎 > bead wire
沿宽度方向 broadway; broadwise
沿栏杆或女儿墙过道或走廊 alur
沿量度方向 direction of measurement
沿裂隙岩溶通道入侵 intrusion through fissure-karst passage
沿流动方向 downstream
沿流向压力梯度 streamwise pressure gradient
沿露头方向 a cropping
沿路 along the road
沿路拌合料 road-mix(ture)
沿路边车道交通 flank-siding
沿路服务 < 指汽车 > on the road service
沿路拱车道行驶 crown-riding
沿路拱车道行驶的车速 crown speed
沿路搅拌机 road mixer
沿路景观 roadside landscaping
沿路开发 roadside development
沿路式电车杆 side pole
沿路用地 abutting land; adjacent land
沿路中线的开裂 centre line crack
沿脉 ore trend tunnel
沿脉勘探巷道 gopher drift
沿脉坑道 ore trend tunnel
沿脉平巷 drive
沿脉巷道 drift
沿面放电 creeping discharge
沿面放电路径 creeping distance
沿面取向 planar orientation
沿面绕组 surface winding
沿母岩侧向运移 lateral migration along source rock
沿木材年轮切向锯开 slash sawn
沿跑道滑跑 runway roll-out
沿劈理交代作用【地】replacement along cleavage
沿坡沟灌 gradient irrigation
沿坡降下 undertow
沿坡路线 supported line
沿墙格栅 wall brander
沿墙栏栅 wall brander
沿墙喷射 wall jet
沿墙喷送 < 一种隧道通风方式 > wall jet
沿墙起重机 wall crane
沿墙通道 wall passage
沿墙突部 banket(te); banquette
沿墙线照明 cornice lighting
沿倾向变化 variation along dip direction
沿倾向的下落断块 downdip block
沿倾向间距 spacing sling dip
沿倾向前进式长壁开采法 longwall advancing to the strike
沿倾向向下前进式长壁开采法 longwall advancing to the dip
沿倾斜方向 a-dipping
沿倾斜方向的工作面 dip face
沿倾斜方向向下开采 dip working
沿人行道地下室采光窗 pavement light
沿人行道停车容量 curb capacity
沿软方向定向 < 金刚石 > structurally weak orientation
沿软弱夹层面滑坡 bedding plane slip
沿山河(道) sidehill stream
沿山脊采样 ridge and spur sampling
沿山脚采样 base of slope sampling
沿山脚的 submountain
沿山(脚)铁路 submountain railway
沿山坡开挖 sidehill cut
沿视线的 radial
沿视线方向的 radial
沿水道上驶 proceed up the channel
沿水库边缘堤岸 marginal bund
沿水流方向 streamwise direction
沿水流方向变化 streamwise variation
沿丝伞属 < 拉 > Naematoloma
沿滩漂积物 beach drifting
沿滩漂沙 beach drifting
沿滩沙坝 offshore barrier
沿滩沙堤 offshore barrier
沿滩沙埂 barrier beach
沿滩沙障 offshore barrier
沿滩沙洲 offshore barrier
沿滩砂埂 beach barrier
沿踢脚板供热 skirting heating
沿踢脚板铺设的供暖管道设备 base heater
沿踢脚板铺设的管道 skirting duct
沿踢脚板铺设的管道设备 baseboard type unit
沿踢脚板铺设的暖气管 skirting board heater; skirting heater
沿踢脚板铺设的散热器 baseboard radiator
沿踢脚板铺设的通风装置 bottom register
沿踢脚板设置的散热器 skirting radiator
沿条边敲平擦亮机 welt feather bending machine
沿条裁断器 welt cutter; welt-cutting machine
沿条槽 welt groove
沿条和沿条缝擦平机 welt and seam rubbing machine
沿条厚度 welt substance
沿条加湿和阴干装置 welt-tempering machine
沿条夹持器 < 缝沿条机的 > welt holder
沿条矫直机 welt straightener
沿条浸湿 welt wetting
沿条卷 welt coil
沿条刻槽机 welt groover; welt-grooving machine
沿条宽 welt width
沿条内底 welted insole
沿条内底槽 welted insole channel
沿条内底成型 welted insole mo(u)lding
沿条内底开槽机 welted insole channeller
沿条内缝机 weit in seaming machine
沿条片边 welt feathering
沿条片边机 welt beveller; welt bevelling machine; welt-skiving machine
沿条片茬 welt butting; welt end skiving
沿条片茬机 welt butter
沿条敲平锤 welt hammer
沿条敲平机 welt beating machine
沿条切割 welt cutting
沿条清洁刷 welt brush
沿条修边 welt trimming
沿条修平机 welt-splitting machine
沿条楦 welted last
沿条压道机 welt wheeling machine
沿条压纹用轮 welt wheel
沿条样 welt pattern
沿铁路线 down line
沿艇外流动 external flow vehicle
沿同一勘探线上钻孔钻进 line drilling
沿途带送自行车及其驾驶人 bicycle bus
沿途货运 way port trade
沿途加冰站 re-icing station
沿途接送的汽车 pick-up bus
沿途零担车 less-than-carload car; local car; peddler car; road van; scrap-load car; station wagon; wayfreight car
沿途零担和摘挂列车 pick-up goods (and distribution) train
沿途零担货车 station-to-station wag-

on
沿途零担货物列车 less-than-carload freight train
沿途零担列车 less-than-carload freight train
沿途零担业务 road freight service
沿途零担摘挂列车 collection and distribution train; goods collecting train; goods way freight train; way freight train; way goods train; wayside freight train
沿途免费搭车旅行 hitchhiking
沿途条件 roadside condition
沿途停靠港 port of call
沿途摘挂车辆 detaching and attaching cars en route
沿途装卸的零担车 <俚语> padlar [pedder]
沿外轨中线测量 survey along outer-rail center line
沿外界线的天井 outer lot line court
沿纹理劈裂的 texture split
沿坞墙移动的工作平台 travel(l)ing staging
沿溪路 river side road
沿溪线【道】 valley line
沿线 trackside; wayside
沿线标地调查 line-plot survey
沿线测平 level(l)ing along the line
沿线车站 roadside station
沿线储备料 line stock
沿线存放材料 line storage
沿线地面信号机【铁】 wayside signal
沿线电压分布 line voltage distribution
沿线调车机车 road switcher
沿线分段测绘 line-plot survey
沿线各站的需要 on-line requirement
沿线各站货车 line car
沿线控制 wayside control
沿线零担仓库【铁】 less-than-carload freight store
沿线路铺管 stringing pipe
沿线路外火灾 wayside fire
沿线路装运 line haul
沿线设备 line equipment; roadside equipment; roadway equipment; wayside equipment
沿线设施 roadside facility; track-side facility
沿线时间 over-the-road time
沿线试验 over-the-road test
沿线停车场 roadside parking lot
沿线土地开发价值 land developing value along the line
沿线推进 routing
沿线无线电台 wayside radio station
沿线行车调度 line traffic control
沿线运土 line haul
沿线运行 road movement
沿线运行费用 line-haul cost
沿线站 wayside station
沿线装卸货物 station-to-station goods
沿线自动监督 automatic line supervision
沿线组织 line organization
沿巷道周边钻进 line-hole drilling
沿斜屋顶铺的顶棚 coom ceiling
沿悬索之临时狭小步道 catwalk
沿压力方向的流动 drag flow
沿岩层倾斜面向下开掘巷道 diphead
沿岩芯方向压力降 pressure drop across the core
沿翼弦方向 chordwise
沿用标志累加分类 summary tag-along sort
沿用贷款抵押品 conventional collateral for loans
沿用公式 working formula
沿原路返回的航线 retrace course
沿原路折回 backtracking
沿圆周铆接 girth rivet seam
沿直径方向的 diametral
沿直线钻孔 line drilling
沿中线切断管子 center split pipe
沿周边焊接的 welded all round
沿周缝 peripheral joint
沿轴的 direct axis
沿轴面的岩层厚度 thickness along axial plane
沿轴线布置 axial plan
沿主根排成行 arrange in rows along the taproot
沿着表面 surfacewise
沿着海岸航行 run-down a coast
沿着海岸线 along the coast
沿着某条直线的温度变化 temperature traverse
沿走向 on strike
沿走向变化 variation along strike direction
沿走向的 girtwise
沿走向规则长壁法 regular longwall along strike
沿走向间距 spacing sling strike
沿走向掘进 level drive
沿走向推进工作面 strike face
沿最陡轨迹下降 steepest descent

炎热 hot; parch heat

炎热的 torrid
炎热干旱区 hot arid zone
炎热期混凝土施工 concreting in hot weather
炎热气候 thermal climate
炎热气候下混凝土浇筑 hot-weather concreting
炎热天 dog days

研棒 pestle

研钵 mortar; mortar box; triturator
研成粉 trituration
研成粉末 triturate
研成粉末的 powdered
研成粉末状 grind into fine powder
研成粉状黏[粘]土 triturated clay
研成细末 grinding into fine powder; pulverization
研齿 gear lapping
研齿机 gear lapping machine
研杵 muller; pestle
研发部 research and development department
研粉作用 trituration
研光 milling off; polish; satin finish-(ing); silk
研光机 glazing machine; lapping machine
研光器 burnisher
研光纸 flint paper
研合 lapping-in
研合密封 grinding seal; seal by precision fit
研糊剂 mulling agent
研剂 lapping compound
研究 disquisition; investigation; research
研究班 seminary
研究班课程 seminar
研究报告 development report; dissertation; memoir; report of investigation; researches; research report
研究备忘录 research memorandum
研究薄片的制备 preparation of sections for study
研究补助金 research grant
研究部门 academic(al) body; research department
研究成本分配的费用法 expense method of apportionment of research cost
研究成果 finding; research achievement; research findings; research production
研究程度 studying degree
研究程度图 map of research level
研究处 research department
研究船 research ship
研究单位 research institution
研究的地物类型 ground material types studied
研究的目的与范围 purpose and scope of the study
研究的研究公司 research-on-research corporation
研究地区 problem area; region studied
研究对象 object of study; subject investigated
研究发展费用 research and development expenses
研究法 methodology; organon
研究范围 range of study; study area; study range; study scale; term of reference
研究方法 method of approach; method of study; research method; research technique
研究方法的进展 advanced in methodology for research
研究房 study house
研究费(用) research expenditures; research funds
研究费支出 research expenditures
研究工程师 research engineer
研究工作 development work; research effort; researches; research work
研究工作实验室 research laboratory
研究工作者 researcher; researchist; research worker
研究管理 administration of research activities
研究规划 research program(me)
研究合同 contract research
研究和发展 research and development
研究和发展成本 research and development cost
研究和发展方案 research and development project
研究和发展机能 research and development function
研究和发展人才 research and development personnel
研究和发展设备 research and development equipment
研究和发展用反应堆 research and development reactor
研究和技术集约 research and technique intension
研究和技术密集 research and technique intension
研究和开发 research and development
研究和开发成本 research and development cost
研究和开发计划 research and development plan
研究和开发投资 research and development investment
研究和开发项目 research and development project
研究和开发支出 research and development expenditures
研究和推广 research and extension
研究河段 study reach
研究环境污染的专家 environmentalist
研究环境(污染)问题的专家 environmentalist
研究会 workshop
研究活动 research activity
研究机构 development facility; research agency; research institution
研究基金 research foundation; research funds
研究级水 research grade water
研究计划 advanced project; research effort; research plan; research program(me); research project
研究阶段 research stage; study phase
研究结果 findings
研究结论 research conclusion
研究经济的方法 economic research method
研究精确度 precision of research
研究开发性试验 development test
研究课题 research opportunity; research problem; research subject; topical of development
研究力量 research effort
研究历史 history of study
研究论文 research paper
研究密集 research-intensive
研究模型 research model; study model
研究目标与试验 research target and test
研究年度 study year
研究排队现象和拥挤现象 queuing theory and waiting time problems
研究潜能 research potential
研究区 region of interest
研究区构造位置 tectonic setting of studied area
研究人员 researcher; researchist; research man; research personnel; research staff; research worker
研究人员补助金 fellowship grant
研究任务 research task
研究设备 research equipment; research facility; research instrument
研究设备耦合 research device coupler
研究设计 research design
研究生 graduate student; postgraduate
研究生毕业证/学位证 graduate diploma/graduate degree's diploma
研究生导师 supervisor
研究生教育 post-graduate education
研究生院 post-graduate research institute
研究时间 time of investigation
研究实习员 research assistant
研究实验室 research laboratory
研究实验室废物 research laboratory waste
研究市场对产品的反应 product testing
研究试验 development test; research trial
研究试验费 study and test fee
研究试制方案 research and development program(me)
研究试制工作 research and development effort
研究室 office of research; research department; research office
研究思考 research idea
研究所 academic(al) institution; academy; institute; institution; research agency; research bureau; research institute
研究所所长 director, research institute
研究讨论 brainstorm(ing)

研究题目 research topic
研究土特产的生态和地理分布类型的科学 phyochorology
研究显微镜 research microscope
研究项目 research item;research project
研究小组 research team
研究协会 Research Association
研究性的消雾系统 fog investigation dispersal operations
研究性监测 research monitoring
研究性试验 exploratory test;investigation test
研究用显微镜 research microscope
研究与发展计划 research and development program(me);research and development project
研究与发展阶段的验收 research and development acceptance test
研究与发展项目 research and development project
研究与发展用发射井 research and development silo
研究与发展中心 research and development center [centre]
研究与发展中心局 Bureau of Research and Development Center [Centre]
研究与工艺 research and technology
研究与开发 research and development
研究与开发费用 research and development cost
研究与推广 research and extension
研究域 domain of study
研究员 research fellow
研究员基金 fellowship
研究员薪金 fellowship
研究员职位 fellowship
研究原则 <科学、哲学的> organon
研究院 academy;institute;research institute;graduate school <大学里的>
研究者 investigator
研究支出预算 research budget
研究中 under study
研究中心 research center [centre];study center [centre]
研究资料 research data
研究总结报告 end-of-study report
研究总蓝图 research master blueprint
研究组 study group
研具 lapper;lap tool
研轮 muller
研磨 abrading;abrasion;abrasivity;bedding-in;combined grinding;comminution;grinding;grind on;high finish;lapping;lapping finish(ing);levigate;making good;milling;mull;pestle;pestling;reseat;rubbing;seat grinding;tripsis;trituration;whet
研磨板 lapping plate
研磨泵 grinding pump
研磨比 grinding ratio
研磨边缘 milled border
研磨表面 lapped face
研磨不正确 misgrind
研磨不足 undergrinding
研磨材料 abrading material;abrasive compound;abrasive material;abrasive substance
研磨材料储罐 abadging material bin;abrading material bin;abrasive bin
研磨材料贮仓 abadging material bin;abrading material bin;abrasive bin
研磨槽 abrasion groove
研磨测硬法 abrasion hardness test
研磨层 grinding bed
研磨层的填塞 filling of coated abrasives
研磨产品 ground product
研磨车间 grindery;grinding mill block
研磨成粉 flour
研磨程度 degree of grind(ing);grinding stage
研磨尺寸 grinding size
研磨杵 muller
研磨带 abrasive band;abrasive belt;grinding belt
研磨的 milled
研磨的表面光洁度 lapped finish
研磨的片子 lapped slice
研磨的平板玻璃 ground plate glass
研磨的瓶底 puntied base
研磨底座 puntying
研磨掉 grinding off
研磨阀面 refacing a valve
研磨阀座 reseat
研磨法 grinding method
研磨废砂 burgee;burgy;spent grinding sand
研磨废水 whet wastewater
研磨粉 abrading powder;abrasive dust;abrasive powder;crocus;grinding dust;grinding powder;lapping powder;polishing powder
研磨粉布 crocus cloth
研磨粉末 abrasive flour;ground powder
研磨膏 abrasive paste;abrasive compound;emery paste;grinding compound;grinding paste;lapping compound;lapping paste
研磨工 grinder
研磨工具 abrasive tool;lapping tool
研磨工人 mill-hand
研磨工业 abrasive industry
研磨功率 grinding power
研磨辊 grinding roll;mixing roll
研磨过的 ga(u)ged
研磨过的色浆 ground paste
研磨过度 overgrind(ing)
研磨过度的 overground
研磨过细 overgrind(ing)
研磨和抛光 grinding and polishing
研磨和抛光滚筒 grinding and polishing drum
研磨和抛光机 grinding and polishing machine
研磨痕 grinding marks
研磨烘干机 combined drying and grinding mill;grind drier [dryer]
研磨环 grinding ring
研磨混合机 putty-chaser-type mixer
研磨活塞 ground and polished piston
研磨机 abrader;abrasive machine;attrition mill;grinder;grinding machine;grinding mill;lapper;lapping lubricant;lapping machine;milling machine;mill of edge runner type;mortar mill;muller;pulping machine
研磨机挡板 patten of grind mill
研磨机的喂料器 grinder feeder
研磨机和刮板 muller and plate
研磨机件 pulverizing element
研磨机进料器 grinding mill feeder
研磨机立轴 grinder spindle
研磨机台的往复运动 table traverse
研磨机橡胶挡板 rubber patten of grind mill
研磨机械 milling equipment;grinding machine
研磨机心轴 grinder spindle
研磨机罩 grinder shell
研磨机组件 grinding element
研磨基板 grinding slab
研磨剂 abradant;abraser;abrasive;abrasive compound;abrasive filler;abrasive material;grinding agent;grinding compound;grinding medium;lapping agent;lapping compound;lapping powder;sanding agent
研磨加工 abrasive machining;honing finish;lapped finish
研磨加工留量 lapping allowance
研磨架 grinding stage
研磨浆料 ground slurry
研磨角 grinding angle;sharpening angle;sharpness angle;tooth angle;wedge angle
研磨搅拌机 Muller mixer
研磨介质 abrasive medium;attrition medium;grinding medium [复 media]
研磨介质的级配 grinding media grading;media grading
研磨介质磨耗 grinding media wear
研磨介质平均直径 average diameter of grinding media
研磨介质载量 grinding media charge
研磨精加工 lapping finish(ing)
研磨均匀 clean grind
研磨孔 grinding hole
研磨块 abrasive block
研磨粒度 grind size
研磨料 abradant;abrasive;abrasive ground;abrasiveness;gritting material
研磨裂纹 grind crack
研磨轮 edge wheel;glazer;glazing wheel;grinding miller;lapping wheel;muller
研磨密封 grinding in
研磨面 abradant surface;abrasive surface;ground surface;lapping surface
研磨敏感性 grinding sensitivity
研磨模 lap
研磨黏[粘]附作用 ground adhesion
研磨盘 abrasive disc [disk];flat plate;grinding pan;sanding disc [disk]
研磨盘砂轮 grinding disk [disc]
研磨配合 lap fit
研磨配合活塞 lap fit piston
研磨片 lapped slice
研磨平板 grinding plate
研磨平板玻璃的废砂 burgee
研磨平面 ground flat
研磨平台 grinding plate
研磨器 sanding pad
研磨强度 severity of grind
研磨切断 abrasive cutting-off
研磨球 mill ball
研磨圈 bull ring
研磨刃 wearing blade
研磨溶蚀 grinding corrosion
研磨润滑剂 lapping lubricant
研磨砂 <又称研磨沙> abrasive sand;burnish sand;grinding sand
研磨砂布 emery cloth
研磨砂粉 abrasive powder
研磨砂粒 abrasive grain
研磨砂轮 abrasive disc [disk];abrasive wheel;glaze wheel
研磨砂纸盘 sanding disc [disk]
研磨伤痕 grinding line
研磨烧伤 lapping burn
研磨设备 abrading apparatus;grinding equipment;grinding installation;grinding plant;milling equipment
研磨(升温)裂纹 grinding crack
研磨石 grinding slip
研磨石料 grinding stone;grindstone
研磨时的添加料 interground addition
研磨时间 milling time
研磨式刀片 abrasive-type blade
研磨式的 abrasive-type
研磨式混汞器 grinding amalgamator
研磨式(岩石)钻 abrasive drill
研磨试验 buffing test;grinding test
研磨水泥浆 abrasive slurry
研磨速度 grinding rate
研磨碎片 grinding chips
研磨损耗 wearing loss
研磨台 grinding stage;grinding table;revolving metal table
研磨套 grinding sleeve
研磨体 crusher ball;grinding body;grinding charge;grinding media
研磨体级配 gradation of grinding media;grinding media grading
研磨体磨耗 grinding media wear
研磨体瀑落 cascading of grinding media
研磨体容重 bulk density of grinding media
研磨体提升槽 cascading channel of grinding media
研磨体填充率 charging degree of grinding media
研磨体与熟料重量比 grinding media to clinker mass ratio
研磨体装填比例 grinding media filling ratio
研磨体装填量 load of grinding media
研磨体装载量 grinding media charge;grinding media load
研磨体自行分级 grinding media segregation
研磨添加剂 additive for grinding;interground addition
研磨条 lapping stick
研磨筒 sand cylinder
研磨头 grinding head;grinding unit
研磨头的转动部分 grinding runner
研磨(涂)层堵塞 filling of coated abrasives
研磨瓦 grinding segment
研磨系数 grindability factor
研磨细度 fineness of grind;grind fineness
研磨细度规 fineness-of-grind ga(u)ge
研磨线 grinding line
研磨消光 dull polish
研磨效率 grinding efficiency
研磨屑 lapping rejects
研磨芯轴 end-mounted spindle
研磨型树脂 grinding resin;grinding type resin
研磨性 abrasiveness;abrasive property
研磨性地层 abrasive formation
研磨性矿粉 abrasive slurry
研磨性能 grinding performance
研磨性泥浆 abrasive slurry
研磨性岩层 abrasive formation;abrasive ground
研磨循环 grinding cycle
研磨样板 grinding template
研磨样品 ground sample
研磨液 grinding fluid;grinding lubricant;lapping liquid;lapping oil
研磨液体 grinding liquid
研磨应力 grinding stress
研磨硬度 abrasion hardness;abrasive hardness;grinding hardness
研磨用凹槽 grinding relief
研磨用的 abradant
研磨用钢棒 grinding rod
研磨用磨块 emery cake
研磨用球 grinding mill ball
研磨用砂 grinding sand
研磨用砂浆 silt
研磨用砂粒 abrasive grit
研磨油 grinding lubricant;lapping oil
研磨油墨 abrasive ink
研磨与抛光装置 grinding and buffing attachment
研磨圆盘 sand disc [disk]
研磨砧板 abrasive block
研磨整形砂轮 set-up wheel

研磨纸 pouncing paper
研磨轴梗 grinding spindle
研磨(主)色浆 master grind
研磨主轴 grinding spindle
研磨助剂 dispersant;milling aids
研磨砖 abrasive brick
研磨装置 grinding unit
研磨钻孔 abrasion drilling
研磨作用 abradant action; abrasive action;grinding action
研末 levigation;pulverize
研末作用 pulverization
研墨 abrasive ink;abrasiveness of ink
研配 bedding-in;grind in;run-in
研砂机 sand crusher;sand grinding mill
研碎 bray;bruise;grind(ing);levigation;pulverization;tripsis
研碎的 chippy
研碎的粉末 trituration
研碎机 pulping machine
研碎器 pulverizer
研讨 perusal
研讨方案 idea-finding
研讨会 seminar;symposium;charrette <有各方专家参加的>
研讨会暨考察 seminar-cum-study tour
研讨曲线 learning curve
研细 porphyrize
研细白垩 prepared chalk
研细程度 grind fineness
研细的白垩 whiting
研压块皂 milled soap
研样板 buckboard;bucking iron plate
研皂机 soap mill
研制 advanced development;develop(ment);research and produce
研制费用 development cost
研制工作 development effort; development work; investigation work; research and development
研制工作的科学管理 research-on-research
研制过程 triturating
研制机 triturating machine
研制计划 development effort; development plan(ning); development program(me)
研制剂 triturate;trituration
研制阶段 development stage; engineering grade
研制开采方法 development mining system
研制期 development period
研制设备 development plan
研制生产系统 development production system
研制时间 development time
研制试验 developing test; development test;researching test
研制新产品 development of new products
研制新农具 develop new tool
研制型 development type
研制性运行 development running
研制与试验 development and test
研制中的发动机 development engine
研制周期 <产品设计至投产的时间> development cycle;lead time
研制周期补偿 lead-time offset

Y

盐巴土 salt clay

盐斑 salt cast;salting stain;salty basin
盐板结土 salt-crusted soil
盐背斜 salt anticline
盐背斜圈闭 salt anticlinal trap
盐表 salinometer
盐冰 salt ice
盐饼 salt cake
盐病 salt sick
盐剥落 salt scaling
盐槽 salt bath
盐槽焊接 salt-bath brazing
盐草潮滩 schorre
盐测流速法 salt-velocity method
盐层 salt bed;salt deposit
盐层储存 salt deposit
盐层库 salt bed storage
盐层套管 salt string
盐场 brine-pan; sal-flat; salina; salt bed; saltern; saltery; salt evaporator;salt garden;salt works
盐尘 salt dust
盐沉积 saline sediment
盐沉积物 playa deposit
盐沉清器 salt setting tank;salt settler
盐成土 halogenic soil;halomorphic soil
盐成圩区土 halogenetic polder soil
盐成圩田土壤 halogenetic polder soil
盐成作用 halomorphism
盐池 salt garden;salt lake;salt pond
盐处理石膏 aridized plaster
盐川 salt glacier
盐床 salt bed
盐刺穿 salt diapir
盐萃取的 salt-extracted
盐代谢 salt metabolism
盐淡水混合区 brackish water zone
盐的 salinows
盐的分布 distribution of salts
盐的风化 efflorescence of salt
盐的积聚 accumulation of salts
盐的耐水性 salt-water resistance; salt-water resistant
盐的起霜 efflorescence of salt
盐底辟作用【地】salt diapirism
盐地 saline land
盐电离常数 ionization constant of salt
盐店 salt store
盐斗 salt hopper
盐度 salinity;saltness
盐度变化 salinity variation
盐度表 salinity table;sali(no)meter <指仪表>
盐度表示法 expressible method of salinity
盐度测定 salinity measurement;salinity test
盐度对流 salinity convection
盐度反常 salinity anomaly
盐度范围 salinity range
盐度分布 distribution of salinity
盐度分层 salinity stratification
盐度感应器 salinity sensor
盐度环流 salinity circulation
盐度混合作用 salinity convection
盐度计 brine ga(u)ge;halometer;salimeter; salinity indicator; salinity meter [metre];salinometer
盐度控制 salinity control
盐度流 salinity current
盐度随深度增高的 katohaline
盐度梯度 salinity gradient
盐度突变层 halocline
盐度效应 effect of salinity
盐度异常 salinity anomaly
盐度跃(变)层 halocline;salinocline;transition layer of salinity
盐度指示器 salinity indicator
盐度指示器系统 salinity indicator system
盐鲕 halolite
盐分 saline concentration;saline matter; salinity; salinity content; salt content;salt graben;saltness
盐分搬运带 salt transportation zone
盐分表聚 surface accumulation of salt
盐分测定法 salinity test
盐分沉淀物 saline deposit
盐分传感器 salinity sensor
盐分带出 salt zoning
盐分电导仪 salinity bridge
盐分堆积带 salt accumulation zone
盐分法 <测量水渠流量的> salt solution method
盐分负荷 salt burden;salt load
盐分积聚 salt accumulation
盐分积累 salt accumulation
盐分来源 source of salt
盐分敏感的 salt sensing
盐分敏感性 salt sensitiveness
盐分浓度 salt concentration
盐分浓度过渡带宽度 width of transitional belt salt concentration
盐分平衡 salt balance
盐分前锋 salinity front
盐分侵入 intrusion of salt
盐分溶滤带 salt leaching zone
盐分入侵 intrusion of salt;salinity intrusion
盐分上升 uprise of salt
盐分探测器 salt detector
盐分梯度 salinity gradient
盐分调节 salt regulation
盐分析法 salimetry
盐分蓄积 salt accumulation
盐分指标 salt index
盐分状况 salt status
盐分组成 salt constituent
盐粉化 salt efflorescence
盐风 salt breeze
盐风损害 salty wind damage
盐风灾害 salty wind damage
盐肤木 Chinese sumac;Rhus chinensis;Rhus javanica;shumac;sumac(h)
盐肤木中毒 Rhus shinensis poisoning
盐浮计 brine ga(u)ge;salt ga(u)ge;salt meter
盐腐蚀 brine corrosion;salt corrosion
盐负 salt load
盐盖路面 salty overlay pavement
盐垢 salt deposit
盐硅硼钙石 serendibite
盐锅 salt pan
盐海水 sea brine
盐害 brine injury; saline contamination; saline damage; salt damage; salt injury
盐含量 salt content
盐核 salt nuclei
盐盒式建筑 saltbox
盐呼吸 salt respiration
盐湖 brine lake;saline lake;salt lake
盐湖沉积 brine lake deposit
盐湖地域 salt lake basin
盐湖矿床 deposit(e) of saline lake
盐湖盆地 salt lake basin
盐湖区 salt lake basin
盐湖相 salt lake facies
盐湖型锂矿床 salt-lake-type lithium deposit
盐湖型硼矿床 salt-lake-type boron deposit
盐湖岩溶 salt lake karst
盐花 flowers of salt
盐华 salt efflorescence
盐华状 efflorescence
盐化过程 saline process
盐化碱土 salinized alkaline soil
盐化栗土 salinized chestnut soil
盐化柠檬酸纸 citro-chloride paper
盐化热水 saliferous thermal water
盐化土 salinized soil
盐化土壤 salt soil
盐化作用 sali(ni)fication
盐基 salt base
盐基饱和 base saturation
盐基饱和度 base saturability;degree of base saturation
盐基饱和率 base-saturation percentage
盐基饱和作用【化】base saturation
盐基代换量 base exchange capacity
盐基度 basicity
盐基度稳定性 basicity stability
盐基含量 base content
盐基交换 base exchange
盐基交换络合物 base exchange complex
盐基交换容量 base exchange capacity
盐基交换软水法 base exchange softening
盐基交换物质 base exchange material
盐基交换总量 total exchangeable bases
盐基碳酸铜 copper basic carbonate;copper hydroxy carbonate
盐基脱饱和 base desaturation
盐基性铅盐稳定剂 basic lead stabilizer
盐基亚氯酸铜 copper basic chlorite;copper hydroxy chlorite
盐基乙酸铜 copper oxyacetate
盐基质 base status
盐脊 salt ridge
盐碱斑 slick spot
盐碱不毛地块 slick spot
盐碱草原 salt prairie
盐碱低地 salt bottom
盐碱地 alkali flat;alkaline earth;alkaline land; alkaline soil; salina; salinate field;saline land;salt lick;salt marsh
盐碱地群落 driumium
盐碱地造林 saline-alkali soil forestation
盐碱害 saline-alkaline damage
盐碱湖 dhand
盐碱化 salinification
盐碱化地块 slick spot
盐碱化防治 salinity control
盐碱化作用 salinization
盐碱结壳区 slick spot
盐碱情况 saline-alkali condition
盐碱使混凝土剥落 salt scaling of concrete
盐碱水 saline-alkaline water
盐碱滩 alkali flat; salina [saline];salt flat;salt marsh
盐碱滩的草 salt grass
盐碱滩地 salt marsh (land)
盐碱条件 saline-alkali condition
盐碱土 alkali saline soil;halic;saline-alkali soil; saline(-sodic) soil; salt-affected soil;szik soil;halomorphic soil
盐碱土改良 improvement of saline(-and)-alkali soil; reclamation of saline-alkali soils;reclamation of saltern;saline-sodic soil amelioration
盐碱沼泽 alkali marsh(land); salt marsh;outmarsh;salting;saltmarsh
盐碱沼泽的草 salt grass
盐碱沼泽地 lick
盐键 salt bond;sat linkage
盐结度 salt crust
盐结壳地【地】bhurland
盐结晶 salt crystallization
盐结皮 salt crust;salt incrustation
盐浸处理 brining
盐浸作用 brining
盐晶胶结 salt crystal cement
盐晶体 salt crystal
盐井 brine pit;salt pit;salt well
盐壳 pellicular salt;salt crust;salt incrustation
盐壳土 salt-crusted soil
盐坑 salt pit
盐空气热交换器 salt-air heat exchanger

盐库 salt store
盐块 salt block
盐矿 saline ore;salt mine
盐矿处置 salt mine disposal
盐矿床 salt deposit
盐渍化地表标志 ground surface symbolization of salinization
盐渍土 salinized soil
盐渍土的分布 distribution of saline soil
盐渍土分类 classification of saline soil
盐垒【地】ekzema [eczema]
盐类 salts
盐类沉淀 precipitation of salts
盐类沉积(物) halogenic deposit;saline deposit;saline lick
盐类胶结 salt cementation
盐类矿床 saline deposit
盐类矿物 saline mineral
盐类污染 pollution by salts
盐类岩 salinastone
盐类岩屑 halolite
盐类营养物 nutrient salt
盐冷的 salt-cooled
盐冷排气阀 salt filled exhaust valve
盐冷却气门 salt-cooled valve
盐粒撒布机 salt distributor
盐粒撒布器 salt spreader
盐量测定(法) sal(in)ometry
盐量计 halometer;salinimeter [salinometer];salt ga(u)ge
盐量平衡 salt balance
盐量图 salinograph
盐量仪 salinoscope
盐瘤 acromorph
盐隆 salt swell
盐漏斗 salt hopper
盐炉砖墙 mid-feather
盐卤 bittern;salt brine
盐霾 salt haze
盐帽透镜体圈闭 salt-dome cap rock lens trap
盐镁芒硝 d'Ansite
盐敏性的 salt sensitive
盐漠 salt desert;travesias
盐母 mother of salt
盐木 pegatpat
盐泥 brine sludge;salt slurry;salty mud
盐泥储罐 sludge storage tank
盐泥火山 salinelle
盐泥滩 saline mud-flat
盐泥压滤机 sludge press filter
盐浓度 saline concentration;salinity;salt content hydrology;saltness
盐浓度变化 saline variation
盐浓度测量法 salometry
盐浓度梯度 salinity gradient
盐浓度引起的异重流 salinity-driven density current
盐磐 salt pan
盐泡 salt bubble;scab;whitewash
盐喷 salt spray
盐喷腐蚀试验 salt spray corrosion test
盐喷试验 salt spray test
盐劈 salt wedge
盐劈作用 sale wedging
盐皮 saline
盐坪沉积 salt flat deposit
盐碛土 saline soil
盐墙 salt wall
盐桥【化】salt bridge
盐侵 salt cutting
盐侵蚀 salt action;salt corrosion
盐清地区 saline area
盐穹 ekzema [eczema];salt dome
盐丘 piercement dome;piercing fold;salt diapir;salt dome;salt dune
盐丘和盐背斜油气聚集带 accumulation zone of salt dome and anticline
盐丘及底辟环形体 circular features of salt domes and diapirs
盐丘角砾岩 salt-dome breccia
盐丘油藏 salt-dome oil pool
盐丘油气田 salt-dome oil-gas field
盐丘钻进 salt-dome drilling
盐泉 brine spring;bring spring;saline spring;salt spring
盐溶 salting in
盐溶角砾岩 evaporite-solution breccia;salt solution breccia
盐溶效应 salting-in effect
盐溶液 brine solution;salting liquid;salt solution;saline solution
盐溶液法 <测管道中流量用> salt solution method
盐溶液浸出 brine leaching
盐溶液浓度 salt solutions concentration
盐舌 salt tongue
盐渗析势 brine osmotic potential
盐生草甸 saline meadow;salt meadow
盐生段木贼 anabasis sp
盐生荒漠 haloeremion
盐生蓝藻类 halocyanophytes
盐生漠境 haloeremion
盐生生物 halobios
盐生土壤 halomorphic soil
盐生形态 halomorphism
盐生演替系列 halarch succession;halosere
盐生植被 halophitic vegetation
盐生植物 halophyte;halophytic vegetation
盐生植物群系 saline formation
盐石膏 marine gypsum
盐栓 salt plug
盐栓圈闭 salt plug trap
盐栓油藏 salt plug pool
盐栓遮挡-断层圈闭 trap formed by intersection of a salt plug barrier with fault
盐霜 salt efflorescence;salt rust
盐水 brine water;pickle;saline water;salt brine;salt liquor;salt water;brackish water <中等程度的>
盐水泵 brine circulator;brine pump;salt-water service pump
盐水比重计 brine ga(u)ge;salt ga(u)ge
盐水测流(速)法 cloud velocity gauging
盐水出口 brine outlet
盐水处理 brine disposal;salt-water disposal;treat with salt water
盐水淡化 saline water reclamation
盐水滴管 saline dropper
盐水底爬 salt water creep
盐水冻结方式 brine freezing system
盐水冻结系统 brine freezing system
盐水冻结装置 brine freezing system
盐水阀 brine valve
盐水法<测管道流量用> saline method
盐水分配 brine distribution
盐水格式盘管系统 brine pipe grid coil system
盐水工段 brine section
盐水供应 brine supply
盐水管 brine pipe;brine tube
盐水管线 brine line
盐水过滤器 brine strainer
盐水湖 salt lake
盐水环流 brine circulation
盐水回流箱 brine return tank
盐水回水管 brine return
盐水混合箱 brine mixing tank
盐水集管 brine header
盐水加压泵 salt pump for rising pressure
盐水浸渍试验 salt-water immersion test
盐水井 brine well;salt-water well;subsalt well
盐水冷冻 brine freezing
盐水冷却 brine cooling
盐水冷却管 brine cooler;brine pipe cooling
盐水冷却器 brine cooler;salt-water cooler
盐水冷却塔 salt-water cooling tower
盐水冷却系统 brine cooling system
盐水立管 brine riser
盐水流 saline current
盐水流量 brine flow
盐水泥浆 brine mud;salt-water mud
盐水黏[粘]土 salt-water clay
盐水浓度 brine concentration
盐水排除 salt-water disposal
盐水喷淋 brine spray
盐水喷淋系统 brine spray system
盐水喷淋制冷方式 brine spray refrigerating system
盐水喷淋制冷系统 brine spray refrigerating system
盐水喷淋制冷装置 brine spray refrigerating system
盐水喷洒系统 brine spray system
盐水喷雾 brine sparge
盐水喷雾(耐蚀)试验 salt spray test;salt fog test
盐水喷注管 brine injection pipe
盐水膨胀箱 brine balance tank;brine expansion tank;brine head tank
盐水漂浮法 saline flo(a)tation method
盐水瓶 saline bottle
盐水侵入 salt-water intrusion
盐水侵蚀 saline encroachment;salt-water corrosion
盐水泉 saline spring;salt spring
盐水溶液 saline solution;salt solution
盐水入口 brine inlet
盐水入侵 intrusion of saltwater;saline water intrusion;saline wedge intrusion;salinity intrusion;salt-water encroachment;salt-water intrusion
盐水上升锥 intrusion cone of saltwater
盐水上溯<河口的> saline wedge
盐水速度(测定)法 <测流量用> salt-velocity method
盐水体 body of salt water
盐水桶 brine drum
盐水温度 brine temperature
盐水污染 saline contamination;salt-water pollution;saline pollution
盐水稀释法 <测量水流量用> salt dilution method
盐水系统 <消防灭火用> salt-water system
盐水箱 brine tank
盐水箱保持阀 brine retaining valve
盐水楔 <潮水河道涨潮时,海水进入河道,水质变咸,称盐水楔> saline wedge;salt-water wedge
盐水楔侵入 saline wedge intrusion
盐水楔河口 salt-wedge estuary
盐水楔异重流 density current in estuary formed salt water wedge
盐水循环 brine circulation
盐水循环泵 salt-water circulating pump
盐水循环冷却 brine pipe cooling
盐水循环器 brine circulator
盐水异重流 saline density current;saline wedge
盐水预热器 brine preheater
盐水原木池 salt chuck
盐水沼泽 salt marsh
盐水针 saline infusion needle
盐水蒸发器 brine evaporator
盐水蒸馏器 evaporator
盐水中和槽 brine neutralizing tank
盐水转化 saline water conversion
盐水转化厂 saline water conversion plant
盐水总管 brine header;brine main
盐水钻井液 salt-water drilling fluid
盐酸 chlorhydric acid;hydrochloric acid;spirit (of) salt;muriatic acid <旧名>
盐酸胺类 amine hydrochloride
盐酸的 hydrochloric
盐酸分解 decomposition with HCl
盐酸副品红 pararosaniline hydrochloride
盐酸过氧化氢分解 decomposition with mixture of HCl and H_2O_2
盐酸化物 hydrochloride
盐酸氯酸钾分解 decomposition with mixture of HCl and $KClO_3$
盐酸萘乙二胺比色法 colo(u)rimetry with naphthylethylenediamine dreochloride
盐酸清洗 hydrochloric acid cleaning
盐酸生产 hydrochloric acid manufacture
盐酸酸浸废水 hydrochloric acid pickling wastewater
盐酸用量 consumption of hydrochloric acid
盐穗草 hallostachys belangeriana
盐滩 salina;salt flat
盐滩岸 saltings
盐体遮挡 salt block barrier
盐田 salt field;brine-pan;salina [saline];salt bed;saltern;salt evaporator;saltfield;salt flat;salt pan;salt pond;sump
盐土 saline land;saline soil;saltierra;solonchak;solonchak soil;white alkali soil;salt clay;reh soil <恒河印度河平原的>
盐土草原 salt meadow
盐土干草原 salt steppe
盐土沙漠 salt waste
盐土沼泽 salt marsh;salt swamp
盐土沼泽地干草 salt marsh hay
盐土沼泽群落 limnodic
盐土植被 halophytic vegetation
盐土植物 halicole;halophyte;salt plant
盐味 saline taste
盐味水 brackish water
盐味水域区 brackish water zone
盐污染 saline contamination;salt contamination;salt(y) pollution
盐误差 salt error
盐雾 salt atmosphere;salt fog;salt mist;salt haze
盐雾腐蚀速率 salt fog corrosion rate
盐雾试验 <一种加速锈蚀试验> salt fog test;salt spray test
盐雾试验箱 salt spray testing instrument
盐雾箱 salt fog spraying cabinet
盐雾演替顶级 salt spray climax
盐吸草 sheepweed
盐析 salting-out
盐析萃取法 salting-out extraction
盐析电解质 salting-out electrolyte
盐析法 salting-out method
盐析剂 salting-out agent
盐析浓度 salting-out point
盐析强度 salting-out strength
盐析色谱法 salting-out chromatography
盐析洗脱色谱 salting-out elution chromatography
盐析效应 salting-out effect
盐析蒸馏 salting-out distillation
盐析纸色谱 salting-out paper chro-

matography
盐析作用 flower of salt
盐稀释法 saline method
盐洗段 saltwash member
盐系统 salt system
盐相 saline facies
盐效校正 salt correction
盐效应 salt effect
盐楔湾 salt-wedge estuary
盐屑岩 halolite
盐形式 saline form
盐型絮凝 salt-type flocculation
盐性草沼 saline marsh;salt marsh
盐性草沼平原 salt marsh plain
盐性草沼水道 salt marsh channel
盐性林沼 salt swamp
盐性黏[粘]土 salt clay
盐性酸沼 salt bog
盐性土 salt soil
盐腌盆 pickling bath
盐岩 halogen rock;saline rock
盐液比重计 sali(ni)meter [sal(in)-ometer]
盐液测流法 salt method;salt solution method of ga(u)ging
盐液测(流)速法 salt-velocity method
盐液淡化法 salt dilution method
盐液滴定法 salt titration method
盐液法 <测流的> salt (solution) method
盐液干燥(木材) salt seasoning
盐液流速法 salt-velocity method; salt-velocity process
盐液流速计 salt-velocity meter
盐液密度计 salinometer;salt meter
盐液箱 brine solution tank
盐引起的锈蚀剥落 salt scaling
盐硬石膏 halo-anhydrite
盐釉 salt glaze;salt glazing
盐釉瓷砖 salt-glazed brick
盐釉的 salt-glazed
盐釉建筑贴面块 salt-glazed structural facing unit
盐釉面砖 salt-glazed brick; salt-glazed tile
盐釉炻器 salt-glazed stoneware
盐釉陶瓷管 salt-glazed earthenware pipe
盐釉陶罐 Bellarmine
盐釉涂面 salt-glazed finish
盐釉瓦 salt-glazed tile
盐釉质面层 hard glazed coat(ing)
盐釉砖 brown-glazed brick
盐浴(槽) salt bath
盐浴池 salt bath
盐浴除鳞 salt-bath descaling
盐浴淬火 hot salt quenching;salt-bath hardening;salt-bath quenching
盐浴淬火炉 salt-bath hardening furnace
盐浴电炉 salt-bath electric(al) furnace
盐浴回火炉 salt-bath tempering furnace
盐浴炉 salt bath;salt-bath furnace
盐浴钎焊 salt-bath brazing
盐浴热处理 salt-bath heat treatment
盐浴烧结 salt-bath sintering
盐浴渗铬 salt-bath chromizing
盐浴索氏体化处理 salt-bath patenting
盐浴铜焊 salt-bath dip brazing
盐浴退火 salt-bath annealing
盐跃层 halocline
盐泽 tidal marsh; tidal swamp; tide marsh
盐泽排水 tidal marsh drainage
盐沾污 saline contamination
盐栈 saline
盐胀 salt heaving
盐障 salt rampart
盐沼 saline;shott
盐沼地 salina;salina solonchak;saline
盐沼海滩 salt swamp beach
盐沼平原 salt marsh plain
盐沼区 salina
盐沼泽 lick;saline bog;saline marsh; salt marsh; sebkha; salitrol; salt marsh (land);tidal marshland
盐沼(泽地) salt swamp
盐枕 salt pillow
盐汁 salt brine
盐指构造 salt finger structure
盐质的 saline;salty
盐质荒漠 salt desert
盐质泥岩 halopelite
盐质适应 halophily
盐质泻湖 sansouire
盐致风化作用 salt weathering
盐致岩爆裂 salt burst
盐重计 salometer
盐株 salt stock
盐助溶 salting in
盐柱 salt pillar;salt plug
盐爪爪属 kalidium
盐砖 salt block;salt brick
盐渍 stained by salt water
盐渍草地 salt meadow
盐渍草原 salt steppe
盐渍地 salinate field;saline;salt lick
盐渍地改良 reclaiming salt area
盐渍化 salinification
盐渍化程度 degree of salinity
盐渍化的 saliferous
盐渍化调查 salinity survey
盐渍化类型 type of salinization
盐渍化形成条件 condition of saline formation
盐渍化岩石 saline rock
盐渍化作用 salinization
盐渍钠质土 saline-sodic soil
盐渍苏打土 saline-sodic soil
盐渍提液 salt extract
盐渍土的分区 partition of saline soil
盐渍土的数据采集 data collecting of saline soil
盐渍土地基危害的预防 prevention of damage to salty foundation soil
盐渍土分类 classification of salty soil
盐渍土(壤) halomorphic soil;saline soil;salinized soil;salted soil;salty soil;soil-affected soil;szik soil
盐渍污染 saline contamination

颜固定 colo(u)r fixing

颜基比 pigment binder ratio
颜料 colo(u)red pigment;colo(u)ring;colo(u)ring agent;colo(u)ring matter; dyestuff; dyeware; paint(ing);pigment (colo(u)r)
颜料比容 pigment volume ratio
颜料表面包膜剂 coating agent
颜料表面处理剂 finishing agent
颜料层 coat of colo(u)r
颜料厂废水 pigment wastewater
颜料厂废物 pigment waste
颜料沉积 pigment settlement
颜料橙 pigment orange
颜料冲洗方法 pigment flushing process
颜料的可混性 pigment miscibility
颜料的浓度 pigment concentration
颜料的散开 pigment dispersion
颜料的吸收 pigment absorption
颜料的细度 pigment fineness
颜料的相容性 pigment compatibility
颜料等级 pigment grade
颜料淀积 pigmentation
颜料发暗 pigment dulling
颜料发光能力 pigment photogenic property
颜料分散剂 pigment dispersant
颜料分散剂需量 pigment dispersant demand
颜料分散体 colo(u)rant dispersion
颜料粉 pigment powder
颜料份 pigment percentage
颜料复合包膜 colo(u)rant double coating
颜料膏 colo(u)rant;pigment paste
颜料铬黄 pigment chrome yellow
颜料工业 pigment industry
颜料刮涂检验法 pigment drawout
颜料罐 <泥土染有岩石物质的泥火山> paint-pot
颜料过度淀积(作用) over-pigmentation
颜料过滤器 paint filter
颜料含量 pigment content
颜料盒 paint box
颜料褐 bistre [bister]
颜料黑 aniline black
颜料红 pigment red
颜料红紫 pigment purple
颜料糊 pigment paste
颜料化染料 pigment dye(stuff)
颜料黄 pigment yellow
颜料混合 pigment blend
颜料火红 fire red
颜料加入量 degree of pigmentation
颜料坚牢紫 pigment fast violet
颜料浆 graining paste;pigment paste
颜料结块 agglomeration;pigment agglomerate
颜料颗粒的球度 pigment particle globularity
颜料矿物 mineral for pigment;pigment mineral
颜料粒子 colo(u)r particle;pigment grain
颜料粒子包膜层 colo(u)r particle coating
颜料粒子间的黏[粘]结力 colo(u)r particle power
颜料粒子上的反常点 anomalous site colo(u)r particle
颜料流线 paint streaming
颜料滤饼 cake press
颜料绿 pigment green
颜料密度 density of pigment
颜料磨碎机 colo(u)r mill
颜料耐久性 dye duration
颜料耐蚀试验 etch solution test
颜料捏合 pigmentation
颜料捏合性 pigmentability
颜料漆料比 pigment binder ratio
颜料漆料系统 pigment binder system
颜料染色能力 pigment binding capacity
颜料容积比 pigment volume ratio
颜料容积浓度 pigment volume concentration
颜料溶合 pigment blend
颜料润湿剂 pigment-wetting agent
颜料润湿性 pigment wettability
颜料色素 pigment colo(u)r
颜料商 colo(u)rman
颜料渗色 pigment exuding
颜料树脂片 resin colo(u)r chip
颜料刷 colo(u)ring brush
颜料双层包膜 pigment double coating
颜料酞青 monastral blue
颜料提浓物 colo(u)r concentrate
颜料体积 pigment volume
颜料体积百分比 percent pigment volume
颜料体积比 pigment volume ratio
颜料体积浓度 pigment volume concentration
颜料填充剂 loading pigment
颜料调和 pigment absorption
颜料调和法 gumption
颜料土 earth colo(u)r
颜料污点 pigment stain
颜料污染 pigment pollution
颜料吸油系数 pigment packing factor
颜料纤维 colo(u)rant fiber [fibre]
颜料研磨机 colo(u)r grinder;colo(u)r grinding machine
颜料印流 bleeding of colo(u)r
颜料油漆 colo(u)r in oil
颜料油性分散体 colo(u)r in oil;colo(u)r in varnish
颜料玉红 pigment rubine
颜料载色剂 pigment carrier medium
颜料载体 pigment carrier
颜料枣红 pigment bordeaux
颜料增量剂 pigment extender
颜料占涂料总体积的百分比 pigment volume concentration
颜料制备物 pigment preparation
颜料重量比 pigment weight percent
颜料重量浓度 pigment weight concentration
颜料着色性 pigmentability
颜料棕 pigment brown
颜隆线 facial carina
颜色 chromasia;colo(u)r;colo(u)ration;colo(u)rity;paint
颜色安定性 colo(u)r stability
颜色斑驳 variegate
颜色饱和度 colo(u)r saturation; depth of colo(u)r saturation
颜色比(例) colo(u)r ratio
颜色笔 crayon
颜色变暗 colo(u)r darkening
颜色变化 colo(u)r variation
颜色变换 colo(u)r switching
颜色辨别 colo(u)r discrimination
颜色辨认 colo(u)r identification
颜色标记 colo(u)r mark
颜色标志 colo(u)r coding;colo(u)r designation;colo(u)r mark
颜色标准 colo(u)r standard;pache
颜色表示法 colo(u)r notation
颜色缤纷 variegate
颜色波动 colo(u)r fluctuation
颜色玻璃 colo(u)r glass;pigmented glass;sun shade;tinted glass
颜色玻璃块 tinted glass block
颜色玻璃砖 tinted glass brick
颜色箔 pigment foil
颜色不规则的砖 grey stock
颜色不均 colo(u)r lacking uniformity
颜色不均匀涂层 blur
颜色不鲜明的 dirty
颜色不一致 colo(u)r variation
颜色不正 bad colo(u)r;wrong colo(u)r
颜色不正的 off colo(u)red
颜色测定 colo(u)r measurement;colo(u)r test
颜色测量 colo(u)r measurement;colo(u)r measuring
颜色测流(速)法 colo(u)r method of measuring velocity; colo(u)r-velocity ga(u)ging;colo(u)r-velocity method
颜色测试 colo(u)r definition
颜色呈现 colo(u)r rendering
颜色冲淡 desaturation
颜色重叠 colo(u)r superimposition
颜色纯度 colo(u)r purity
颜色刺激 colo(u)r stimulus
颜色搭配 colo(u)r assortment
颜色代码定向航路 colo(u)red airway
颜色带红头的 colo(u)r with red shade
颜色单位 colo(u)r unit
颜色的 chromatic
颜色的比较 comparing of colo(u)rs

颜色的范围 range of colo(u)rs
颜色的分布 range of colo(u)rs
颜色的浓淡程度 gradation colo(u)r
颜色的三属性 three-attributes of colo(u)r
颜色的深浅程度 gradation colo(u)r
颜色的吸收 colo(u)r absorption
颜色的一致性 consistency of colo(u)r
颜色的组合 colo(u)r way
颜色灯 colo(u)red lamp
颜色灯光 colo(u)red light
颜色对比 colo(u)r contrast
颜色多变化的 versicolo(u)r(ed)
颜色发飘 colo(u)r drift; colo(u)r variation
颜色法 colo(u)ring
颜色反衬 colo(u)r contrast
颜色反应 colo(u)r response
颜色范围 colo(u)r range
颜色方程 colo(u)r equation
颜色分辨力 chromatic resolving power
颜色分布 colo(u)r distribution
颜色分层 colo(u)r breakup
颜色分析 colo(u)r analysis
颜色分析器 colo(u)r analyser [analyzer]
颜色分析仪 colo(u)r analyser [analyzer]
颜色浮面釉 colo(u)r floating glaze
颜色改进剂 colo(u)r improver
颜色改良剂 colo(u)r improver
颜色高温计 colo(u)r pyrometer
颜色规格 colo(u)r specification
颜色国际标准深度 colo(u)r international standard depth
颜色很淡 very pale of colo(u)r
颜色很浅 very pale of colo(u)r
颜色环 colo(u)r circle
颜色灰暗 dingy
颜色混合 colo(u)r mixing; colo(u)r mixture
颜色混合计算机 colo(u)r mixture computer
颜色混合器 colo(u)r mixer; colo(u)r top
颜色混合曲线 colo(u)r mixture curve
颜色基调 colo(u)r motif
颜色极暗 very dark of colo(u)r
颜色极深 very dark of colo(u)r
颜色加深 saddening
颜色坚牢的 colo(u)r-fast
颜色坚牢度 colo(u)r fastness
颜色渐次变化的涂料 graded coating
颜色鉴别 colo(u)r identification
颜色键钮 colo(u)r keyed button
颜色校正过的水银灯 colo(u)r-corrected mercury lamp
颜色控制器 colo(u)r controller
颜色控制系统 colo(u)r control system
颜色控制仪 colo(u)r monitor
颜色理论 technicolo(u)r theory
颜色立体模型 colo(u)r solid
颜色亮度 colo(u)r brightness
颜色量化 colo(u)r quantization
颜色灵敏度 colo(u)r sensitivity
颜色流痕 wool-drag
颜色滤光片 colo(u)r filter; light filter
颜色轮 colo(u)r wheel
颜色密度 colo(u)r density
颜色名称 colo(u)r name
颜色明度 brightness of a colo(u)r; chroma luminance; colo(u)r brightness; lightness
颜色耐日晒牢度 colo(u)r fastness to light
颜色浓度 colo(u)r concentration; colo(u)r depth; colo(u)r strength; depth of colo(u)r; saturation
颜色浓度校正 colo(u)r concentration correction
颜色配方 pigment formation
颜色配合 colo(u)r scheme
颜色匹配 colo(u)r match(ing)
颜色品质 quality of colo(u)r
颜色铅笔 crayon
颜色浅淡 lightness
颜色强度 colo(u)r intensity; colo(u)r strength
颜色清晰度 colo(u)r definition
颜色曲线 colo(u)r curve
颜色设备 colo(u)r facility
颜色深暗的 dense
颜色生成 colo(u)ration formation; colo(u)r formation
颜色失常 losing normal colo(u)r
颜色失真 cross colo(u)r; falsification of colo(u)rs
颜色识别 colo(u)r discrimination; colo(u)r identification
颜色识别标签 colo(u)red identification label
颜色识别试验 identification of colo(u)r test
颜色视觉 colo(u)r vision
颜色适应性 colo(u)r adaptability; colo(u)r aptitude
颜色适应性试验 colo(u)r aptitude test
颜色属性 colo(u)r dimensions
颜色数标法 colo(u)r notation
颜色水泥 cement colo(u)rs
颜色随观察角度而变化的涂层 colo(u)r-flop coating
颜色特性 colo(u)r response
颜色调和 colo(u)r harmony
颜色调和手册 colo(u)r harmony manual
颜色调配 colo(u)r matching
颜色调谐 colo(u)r balance
颜色调整 colo(u)r adjusting
颜色调整剂 colo(u)r modifier
颜色完整性 colo(u)r integrity
颜色维持能力 colo(u)r retention
颜色温标 colo(u)r scale
颜色温度 colo(u)r temperature
颜色稳定剂 colo(u)r stabilizer
颜色稳定性 colo(u)r stability
颜色稳定性加速试验 accelerated test for colo(u)r stability
颜色稳定性指数 colo(u)r stability index
颜色误差 colo(u)r error
颜色鲜明的 colo(u)r vivid
颜色鲜明度 colo(u)r definition
颜色鲜艳度 colo(u)r concentration
颜色显现 colo(u)r rendering
颜色相似的 homeochromatic
颜色效应 colo(u)r effect
颜色信号 colo(u)r signal
颜色信号圆牌 colo(u)r signal disc
颜色信号圆盘 colo(u)r signal disc
颜色形成 chromogenesis
颜色性能 colo(u)ristic property
颜色选择 colo(u)r selection
颜色学 chromatics
颜色样板 colo(u)r plate
颜色样本 colo(u)r card; colo(u)r chip(ping)s; colo(u)r sample; colo(u)r swatch
颜色样卡 colo(u)r chart
颜色样品 sample of colo(u)r
颜色一致性 colo(u)r consistency; colo(u)r match
颜色釉 colo(u)red glaze; colo(u)r glaze
颜色指标 colo(u)r index
颜色指数 colo(u)r index
颜色种类 type of colo(u)r
颜色组合 colo(u)r scheme

檐 ledge; overhang; scull

檐板 brow; cant strip; cornice; roof plate; weather-board(ing)
檐板条 check rail
檐板瓦 eaves plate tile
檐板下的线脚 bed mo(u)ld(ing)
檐板下作成滴水槽 underthroating
檐板照明 cornice lighting
檐壁 frieze
檐边 eaves mo(u)lding
檐部 entablature
檐槽 cistern head; eaves channel; eaves gutter; killesse; small trough; eaves trough; rhone <苏格兰式的>
檐槽吊钩 gutter hanger; gutter hook; strap hanger
檐槽梁托架 valley beam
檐槽螺栓 gutter bolt
檐槽木板 gutter plank
檐槽(竖)板 gutter plate
檐槽挑口板 gutter board
檐槽托 bearer of gutter; gutter hook
檐槽托架 gutter bearer; gutter bracket
檐撑 eaves strut
檐椽 <古建筑> eaves rafter; verge rafter
檐椽接长木 sprocket
檐椽支撑木 sprocket
檐的下部 architrave
檐底板 planceer; plancier
檐底通风口 cornice vent; vent opening in eaves soffit
檐底托板 mutule
檐垫板 cushion board
檐垫层 eaves course
檐垫条 eaves catch
檐枋 eaves tie beam; architrave【地】
檐腹板 eaves soffit
檐高 eaves height
檐沟 chuting; condensation gutter; drainage gutter; drain gutter; eaves trough; eaves trowel; gutter; rainfall gutter; rain gutter; trough gutter; water gutter; eaves gutter
檐沟薄板 gutter sheet
檐沟出水孔 pap
檐沟导流管 eaves gutter nozzle
檐沟挂钩 gutter hanger; gutter hook
檐沟托 gutter hanger
檐沟托座 fascia bracket
檐沟雨水槽 rone
檐架 couple roof construction
檐角托饰 angle modillion
檐脚梁 eaves plate
檐口 cornice
檐口板 eaves board; eaves fascia; eaves plate; facia board
檐口板条 eaves lath
檐口衬板 starter
檐口出挑 eaves overhang; eaves projection
檐口处垫瓦 undercloak
檐口处理 cornice trim
檐口椽 verge rafter
檐口椽间木 frieze block
檐口大梁 eaves girder
檐口大样 eaves detail
檐口滴水沟 eaves gutter
檐口滴水条 chantlate
檐口滴水瓦 verge course
檐口垫底层 undercourse
檐口垫木 eaves plate
檐口垫瓦 double eaves course; doubling course; under-eaves course; under-eaves tile
檐口垫瓦条 arris fillet; doubling piece; eaves board; tilting fillet; tilting piece
檐口吊顶 cladding of roof overhang
檐口泛水 eaves flashing
檐口挂瓦条 eaves batten
檐口桁条 verge purlin(e)
檐口金属片滴水槽 drop apron
檐口卷边层 undercloak
檐口梁 eaves plate
檐口披水 eaves flashing
檐口披水板 apron eaves piece
檐口平顶 cornice soffit; eaves ceiling; flat roof fascia board
檐口铺底层 starting course; starting strip
檐口嵌条 eaves pole
檐口墙 blocking course
檐口人字木 barge couple
檐口饰 antefix(ae)
檐口双层瓦 double course; double eaves course
檐口水沟 eaves trowel
檐口通风口 eaves vent
檐口凸出物 eaves projection
檐口突出部 projection of cornice
檐口托板 eaves fascia; eaves soffit
檐口托架 eaves bearer
檐口托饰 modillion
檐口托座 cornice bracket
檐口瓦 eaves tile; starter tile; verge tile
檐口瓦层 eaves course
檐口屋面板 starting course; starting strip
檐口线脚 eaves cornice; eaves mo(u)lding; thickness mo(u)lding <檐板下>
檐口支撑 eaves strut
檐口支座 eaves bearer
檐廊 eaves gallery
檐梁 girt(h)
檐梁式建筑学 trabeated architecture
檐檩 eaves purlin(e); pole plate; raising plate; roof plate; rafter plate <支承椽下端的檩条>
檐溜石 dripstone
檐坡滴水凹圆线 underthroating
檐墙 eaves wall
檐饰线脚 eaves mo(u)lding
檐饰线条 eaves mo(u)lding
檐水 eavesdrop
檐水滴落池 eaves drop [drip]
檐头墙 blocking course
檐头石 blocking course
檐突 overhang
檐托座间 intermodillion
檐瓦 antefix tile; double roll verge tile; eaves tile; gable tile; starter tile
檐瓦垫条 eaves pole
檐瓦条木 eaves fillet
檐下齿形装饰 dentil
檐下齿形装饰间的空间 metoche
檐下廊道 eaves gallery
檐下(石)望板 cussome
檐下瓦层 under(earth)-eaves course
檐缘 draw-off
檐柱 peripheral column
檐砖 capping brick; coping brick

衍变 developing

衍彩法 <涂漆> spray and wipe painting
衍化 derivation
衍化物 derivant; derivative
衍射 inflection
衍射 P 波 diffracted P wave

衍射斑点 diffraction spot
衍射本领 diffracting power
衍射边 diffracting edge
衍射边缘 diffraction edge
衍射表 diffractometer
衍射波 diffracted wave; diffraction wave
衍射波矢量 diffracted wave vector
衍射测微计 eriometer
衍射测微器 eriometer
衍射产生的热点 diffraction-produced hot spots
衍射场 diffractional field
衍射衬度 diffraction contrast
衍射点 diffraction point
衍射点阵 diffractional lattice
衍射对比 diffraction contrast
衍射对称 diffraction symmetry
衍射法 diffraction approach; diffraction method
衍射反差 diffracted contrast
衍射反射 diffraction reflection
衍射方向 diffraction direction
衍射分光计 diffraction spectrometer
衍射分光镜 diffraction spectroscope
衍射分析 diffraction analysis
衍射峰 diffraction maximum; diffraction peak
衍射符号 diffraction mark
衍射光 diffracted light
衍射光斑半径 diffraction spot radius
衍射光环 diffraction ring
衍射光谱 diffracted spectrum; diffraction spectrum
衍射光栅 diffraction grating; grating
衍射光栅复制 diffraction-grating replica
衍射光栅干涉仪 diffraction-grating interferometer
衍射光栅光谱级 diffraction-grating spectral order
衍射光线 diffracted ray
衍射光学 diffraction optics
衍射光学平视显示器 diffraction optics head-up display
衍射花样 diffracted pattern; diffraction pattern
衍射积分 diffraction integral
衍射级 diffraction order; order of diffraction
衍射极大 diffraction maximum
衍射极限 diffracted maximum; diffraction limit; diffraction maximum; diffraction peak
衍射极限发散度 diffraction-limited divergence
衍射极限直径 diffraction-limited diameter
衍射计【物】 diffraction instrument; diffractometer
衍射间距 diffraction spacing
衍射角 angle of diffraction; diffraction angle
衍射阶 order diffraction
衍射结晶学 diffraction crystallography
衍射截面 diffraction cross-section
衍射介质 diffracting medium
衍射孔 diffracting opening; opening diffracting
衍射孔径 diffraction aperture
衍射理论 diffraction theory
衍射流 diffraction current
衍射弥散圆 diffraction blur circle
衍射模糊 diffraction blurring
衍射模糊圈 diffraction circle of confusion
衍射耦合 diffraction coupling
衍射频谱 diffraction spectrum
衍射平面 diffraction plane
衍射评价函数 diffraction-based merit function
衍射屏 diffraction screen
衍射器 diffractometer
衍射强度 diffracted intensity
衍射强度标准 diffraction intensity standard
衍射区 diffraction region
衍射曲线 diffraction curve
衍射去耦合 diffraction decoupling
衍射群 diffraction group
衍射散射 diffraction scattering
衍射射束 diffracted beam
衍射射线 diffracted ray
衍射束 diffracted beam
衍射损耗 diffraction loss
衍射损失 diffraction loss
衍射体 diffracting object
衍射条纹 diffraction fringe
衍射调制传递函数 diffraction modulation transfer function
衍射透镜 diffraction lens
衍射图(像) diffraction pattern; diffraction diagram; diffractogram
衍射图形叠加 addition of diffraction patterns
衍射图样 diffraction pattern
衍射系数 diffraction coefficient
衍射显微术 diffraction microscopy
衍射现象 phenomenon of diffraction
衍射线 diffraction line; diffraction ray
衍射限光装置 diffraction-limited optics device
衍射限镜 diffraction-limited mirror
衍射限模式 diffraction-limited mode
衍射限透镜 diffraction-limited lens
衍射像点 diffraction image point
衍射像片 diffraction picture
衍射效率 diffraction efficiency
衍射效应 diffraction effect
衍射楔 diffraction wedge
衍射学 diffractometry
衍射掩模 diffracting mask
衍射仪 diffractometer
衍射溢失 diffractive spillover
衍射应变仪 diffracto-ga(u)ge
衍射(影)像 diffraction image
衍射用 X 射线管 diffraction X-ray tube
衍射原理 diffraction theory
衍射圆面 diffraction disc [disk]
衍射晕 diffraction corona; diffraction halo
衍射增宽 diffraction broadening
衍射振荡 diffraction oscillation
衍射指数 index of diffraction
衍射锥 diffraction cone
衍射作用 diffraction
衍生 derivation; spin-off
衍生部门 derivative departments
衍生产品 derived product
衍生高聚物 derived high polymer
衍生剂 derivating agent
衍生价值 derived value
衍生结构 derivative structure; derived structure
衍生色 induced colo(u)r
衍生物 offshoot; ramification; derivant; derivative【化】
衍生(物)法 derivative method
衍生稀树干草原 derived savanna
衍生效益 derived benefit
衍生需求【交】 derived demand
衍生岩 derivative rock
衍生脂类 derived lipid
衍续经济学 sustainable economics

偃

偃伏株木 red osier dogwood
偃角 angle of underlay; hade
偃麦草 quack grass
偃松 Japanese stone pine

掩

掩蔽 bury; camouflaging; mask; masked off; masking; screening; shroud
掩蔽比 masking ratio
掩蔽部 casemate; funkhole
掩蔽处 coverture
掩蔽的阵地 masked position
掩蔽电位 masking tension; mask potential
掩蔽顶盖 canopy
掩蔽顶盖边缘 canopy lip
掩蔽法 masking method
掩蔽法脱臭 masking method deodorizing
掩蔽腐蚀 mask etch
掩蔽沟 revetted shelter trench
掩蔽级 masking level
掩蔽剂 masking agent; masking reagent; screening agent; sequestering agent
掩蔽孔径 mask aperture
掩蔽扩散 masked diffusion
掩蔽篱笆 screen fence
掩蔽良好的地方 well-sheltered place
掩蔽良好的港口 well-protected harbo(u)r; well-sheltered harbo(u)r
掩蔽露头 hidden outcrop; subterranean outcrop
掩蔽模 masking jig
掩蔽期 eclipse period
掩蔽期间 masking period
掩蔽区 <日冕仪> occulting disk
掩蔽声 masking sound
掩蔽声谱 masking acoustic spectrum
掩蔽试剂 masking reagent
掩蔽衰减器 masking pad
掩蔽隧道 shelter tunnel
掩蔽所 hide-out
掩蔽体 maskant
掩蔽贴板 butt plate
掩蔽涂装法 resist technique
掩蔽团 masked group
掩蔽脱臭装置 malodo(u)r masking equipment
掩蔽外延方法 mask epitaxial method
掩蔽物 coverture
掩蔽现象 occlusion
掩蔽消融 covered ablation
掩蔽效应 masking effect
掩蔽信号 masking signal
掩蔽烟道 blank flue
掩蔽值 masking value
掩蔽作用 masking action
掩冲大断层 overthrust fault
掩冲断层 overthrust; overthrust fault
掩冲断层背斜 overthrust anticline
掩冲岩片 overthrust slice
掩带 occultation band
掩覆 overriding
掩盖 blanket(ing); blankoff; bury; cover; shielding
掩盖层 hidden layer
掩盖功率 blanket power
掩盖力 hiding power
掩盖露头 subterraneous outcrop
掩盖式煤田 covered coalfield
掩盖物 covering
掩护 aegis; camouflage; covering; escort; haven; shelter
掩护电弧焊接 shield-arc welding; shielded metal arc welding
掩护良好的地方 well-sheltered place
掩护良好的港口 well-protected harbo(u)r; well-sheltered harbo(u)r
掩护平原 covered plain
掩护色漆 baffle paint
掩护物 shelter
掩护性投标 cover bid; cover tender
掩护支架 shield
掩码配准 mask matching
掩埋 landfill
掩埋场气体 landfill gas
掩埋处置 land disposal
掩埋谷 buried valley
掩埋管道 buried pipe line
掩埋式洪积扇 buried pluvial fan
掩摸原图 mask artwork
掩模 mask
掩模板 mask plate
掩模窗孔 mask opening
掩模对准 mask alignment
掩模对准器 mask aligner
掩模法 mask means
掩模腐蚀 mask etch
掩模复制 mask replication
掩模光刻机 mask aligner
掩模过程 mask process
掩模加工 mask processing
掩模架 mask holder
掩模检查 mask detection
掩模校准 mask alignment
掩模孔隙 mask aperture
掩模匹配 mask matching
掩模迁移 mask transfer
掩模设计 mask design; mask layout
掩模生成系统 mask generation system
掩模套准器 mask aligner
掩模图案 mask pattern
掩模外延法 mask epitaxial method
掩模位 mask bit
掩模效应 masking effect
掩模原版 master reticle
掩模罩 filter
掩模蒸镀 mask evaporation
掩模制造 mask manufacture
掩模重合 mask alignment
掩没 close-over; override
掩片 masking sheet
掩食角 occultation angle
掩食曲线 occultation curve
掩食时间 duration of eclipse; eclipse duration
掩食时刻 occultation time
掩饰 camouflage; deception; paper; varnish
掩饰的 palliative
掩饰过去 carry it off
掩饰物 cryptic mimicry
掩饰性投标 cover bid; cover tender
掩饰照明 reduced lighting
掩索 tow rope
掩体 barrier shielding; blindage; block house; breast work; bunker; dugout; pill-box; shelter
掩体后土墙 parados
掩体建筑物 buried structure
掩体墙 shadow wall
掩土住房 earth-sheltered housing
掩望镜 protectoscope
掩楔带 masking tape
掩星 obscuration; occultation
掩星法 star occultation
掩源 occultation

眼

眼斑结构【地】 ocellar texture
眼板 eye plate; flounder plate; monkey face; paɑ eye
眼保护装置 eye protector
眼鼻面罩 eye-nose mask; separate face mask
眼壁柱 eyewall chimney
眼的反差敏感度 contrast sensitivity

of eye
眼的适应 accommodation of eye
眼点 eye point
眼洞窗 oculus
眼杆 eye bar;eye rod
眼杆衬垫 eyebar packing
眼杆垫板 eyebar packing
眼杆钩 eyebar hook
眼杆填料 eyebar packing
眼杆悬索桥 eyebar suspension bridge
眼高 height of eye
眼高差 angle of dip;depression of the sea horizon;dip of horizon;dip reduction; height of eye correction; inclination
眼高差改正 dip correction height of eye correction
眼高度 eye-level
眼钩 eye hook
眼光 flair
眼光短浅 nearsighted
眼光水平 eye-level
眼护 eye goggles
眼环 collar;dead eye;eye thimble
眼环螺栓 throat bolt
眼环销螺丝卸扣 round collar pin shackle
眼基线【测】 eye base(line);eye separation
眼基线调节【测】 eyebase setting;interpupillary adjustment
眼睑式快门 eyelid shutter
眼见的 ocular
眼睫毛的 ciliate
眼界 eyeshot;eyesight;outlook;purview;scope
眼睛 eye
眼睛残留特性 eye storage characteristic
眼镜 spectacle
眼镜玻璃 spectacle glass
眼镜法 spectacle method
眼镜脚 ear piece
眼镜片玻璃球 bulb for spectacle glass
眼镜片中心距离 <60 ~65 毫米> interpupillary distance
眼距 eye base;eye separation;interocular distance
眼科 department of ophthalmology;ophthalmology
眼科室 ophthalmic room
眼孔 eyelet
眼孔螺栓连接摇枕吊 eyebolt link hanger
眼力好的 long-sighted
眼灵敏度曲线 eye sensitivity curve
眼螺栓 eye bolt
眼脑系统 eye-brain system
眼疲劳 eye strain;eyestrain
眼球 bulb of the eye
眼球片麻岩 eye gneiss
眼球形叠加褶皱 augen superposed fold
眼球状构造 augen structure; eye structure
眼球状混合片麻岩 augen amphogneiss
眼球状混合岩 augen migmatite
眼球状煤 bird's eye coal;eye coal
眼球状燧石 augen-chert
眼球状体【地】 augen
眼圈 eye thimble
眼圈接合 eye joint
眼圈螺钉 eye screw
眼神灯 eye light
眼石 eyestone
眼手机器 eye-hand machine;hand-eye machine
眼榫接头 eye joint
眼调节计 accommodometer
眼瞳孔 eye pupil
眼痛 eye-sore
眼图 eye pattern
眼窝 eye hole
眼屑状构造 eye-and-eyebrow structure
眼形窗 eye window;oculus
眼形物 oculus
眼旋(转)错觉【测】 oculo-gyral illusion;opto-gyral illusion
眼炎 ophthalmia
眼照明 <摄影用> eye light
眼罩 blind pack;eyepiece;eyecup <取景器的>
眼折射差 eye-refraction difference
眼重力错觉【测】 oculogravic illusion
眼状构造 augen structure
眼状煤 circular coal
眼状片麻岩 augen gneiss
眼状石灰斑 glazki
眼状物 eye
眼状显示图 eye pattern
眼状油气田 eye field
眼子 buttonhole
眼子板 punched-plate
眼子圈 hole circle
眼左转 levoversion

演

演变 development; differentiation; evolution

演变磁结构 evolving magnetic feature
演变规律 evolutionary regularity
演变过程 evolutionary process
演变曲线 development curve
演变中的邻里 transitional neighbourhood
演播控制室 studio control room
演播室 studio;telestudio;teletorium
演播室表演区 studio floor
演播室布景 studio decorations;studio scenery
演播室重放设备 studio display facility
演播室灯光板 studio light boards
演播室对讲 sound talkback
演播室返送电路 studio foldback circuit
演播室混合台 studio mixer
演播室监视器 studio monitor
演播室控制台 studio control console
演播室容量 studio capacity
演播室设备 studio apparatus; studio facility
演播室摄像机三脚架 studio camera pedestal
演播室摄像机通路 studio camera chain
演播室摄像机移动车 studio camera dolly
演播室摄像设备 studio camera equipment
演播室升降设备 studio crane
演播室声学 studio acoustics
演播室拾音 studio pick-up
演播室吸声 studio absorption
演播室照明 studio lights
演播中心 studio center [centre];studio complex
演出 presentation
演出公司 performance company
演出台 dais
演化 evolution;evolvement
演化程 evolutionary track
演化带 evolutionary zone
演化单位 evolutional unit
演化发生 genetic(al) inception
演化分类 evolutionary taxonomy
演化关联 genetic(al) connection
演化过程 evolutionary process
演化阶段 evolutionary phase
演化阶段名称 usable name of evolution stage
演化模式 evolutionary pattern
演化趋向 evolutionary trend
演化序列 evolutionary series
演讲 dissert(ate);lecture
演讲讲坛 pulpitum
演讲台 rostrum [复 rostra];speaker's platform;tribune
演讲坛 rostrum [复 rostra]
演讲厅 cavea;lecture hall;lecture theatre [theater];lyceum;speaking-place
演示插孔 demonstration jack
演示软件 demo-ware
演示实验 lecture experiment
演示者 demonstrator
演算 calculus [复 calculuses/calculi]
演算参数 routing parameter
演算的 computing
演算方程 routing equation
演算函数 routing function
演算节目 repertoire
演算器 exerciser
演算曲线 routing curve
演算时段 routing period
演算子程序 interpretative subroutine
演坛 dais
演替 succession
演替顶极 climax community; climax of succession
演替顶极群落 climax community
演替环境 successional habitat
演替系列 sere
演习 dry run
演习区 exercise area
演习区浮筒 practice area buoy
演绎 deduce;deduction
演绎的【数】 a priori
演绎的推论 apriorism
演绎定理 deduction theorem
演绎法 deductive approach;deductive method;method of deduction
演绎规则 deduce rule
演绎过程 deductive procedure
演绎磕 illation
演绎逻辑 deductive logic
演绎模拟 deductive simulation
演绎模拟法 deductive simulator method
演绎模型 deductive model
演绎树 deduction tree;refutation tree
演绎推理 deductive inference;deductive reasoning; modus ponens; syllogism
演绎推论 deductive inference;deductive reasoning
演绎问答程序 deductive question-answering program(me)
演绎系统完备性 completeness of deduction system
演绎型数据库 deductive database
演绎证明 deductive proof
演绎综合法 deductive generation
演员休息室 artist lounge;green room
演奏 playing
演奏节目 repertoire
演奏室 studio
演奏台 bandstand
演奏厅 concert chamber; concert hall;odeom;odeum
演奏席壳盖 orchestra shell
演奏者 executant

鼹

鼹鼠犁 mole plough

鼹鼠穴式排水沟 mole drain

厌

厌绿轮枝孢 Verticillium glancum

厌气的 anaerobic
厌气发酵 anaerobic fermentation
厌气分解 anaerobic breakdown;anaerobic decomposition
厌气接触法 anaerobic contact process
厌气接触过滤法 anaerobic contact filter process
厌气菌 anaerobe
厌气(生物)处理 anaerobic treatment
厌气塘 anaerobic pond
厌气微生物 anaerobe
厌气消化 anaerobic digestion
厌气消化池 anaerobic digester;anaerobic digestion tank
厌气性黏[粘]合剂 anaerobic adhesive
厌气性细菌 anaerobic bacteria
厌气(性细)菌腐蚀 anaerobic corrosion
厌弃鱼类 abnoxious fishes
厌食 bdelygmia
厌水性材料 hydrophobic material
厌水骨料 hydrophobic aggregate
厌恶风险 risk aversion
厌恶风险的度量 measure of risk aversion
厌恶风险函数 risk aversion function
厌恶风险假设的检验 verification of risk aversion assumption
厌恶量 dislike value
厌恶性行业 offensive trade
厌氧氨氧化 anaerobic ammonia oxidation;anammox
厌氧氨氧化反应器 anaerobic ammonia oxidation reactor
厌氧氨氧化工艺 anaerobic ammonia oxidation process
厌氧氨氧化活性 anaerobic ammonia oxidation activity
厌氧氨氧化技术 anaerobic ammonia oxidation technology
厌氧氨氧化菌 anaerobic ammonia oxidation bacteria
厌氧氨氧化速率 anaerobic ammonia oxidation rate
厌氧氨氧化体 anaerobic ammonia oxidation body;anammoxosome
厌氧氨氧化污泥 anaerobic ammonia oxidation sludge
厌氧层 anaerobic layer
厌氧产酸 anaerobic acidogenesis
厌氧超滤膜反应器 anaerobic ultrafiltration membrane reactor
厌氧沉淀条件下驯化 domestication of anaerobic-settling condition
厌氧除磷 anaerobic dephosphorization
厌氧除磷菌 anaerobic dephosphorization bacteria
厌氧除硫化物和硝酸盐 anaerobic sulfide nitrate removal
厌氧处理 anaerobic process; anaerobic treatment
厌氧处理槽 septic tank
厌氧处理法 anaerobic treatment process
厌氧代谢 anaerobic metabolism
厌氧单组分胶粘剂 anaerobic single component adhesive
厌氧挡板式反应器 anaerobic partion reactor
厌氧的 anaerobic
厌氧滴流 anaerobic trickling
厌氧滴滤池 anaerobic trickling filter
厌氧底泥 anaerobic sediment
厌氧毒性 anaerobic toxicity

厌氧毒性检测 anaerobic toxicity assay
厌氧发酵 anaerobic fermentation
厌氧法 anaerobic process
厌氧反硝化过程 anaerobic denitrification process
厌氧反应 anaerobic reaction
厌氧反应器 anaerobic reactor
厌氧分解 anaerobic breakdown; anaerobic decomposition
厌氧分解代谢 anaerobic catabolism
厌氧腐烂 anaerobic decay
厌氧附着膜膨胀床法 anaerobic attached-film expanded bed
厌氧附着微生物膜膨胀床 anaerobic attached microbial film expanded bed
厌氧复合床反应器 anaerobic composite bed reactor
厌氧杆菌 anaerobic bacillus; cillus anaerobicus
厌氧工艺 anaerobic;anaerobic process
厌氧固氮细菌 anaerobic nitrogen fixing bacteria
厌氧固定床反应器 anaerobic fixed bed reactor
厌氧固定膜反应器 anaerobic fixed film reactor
厌氧罐 anaerobic jar
厌氧过程 anaerobic process
厌氧过滤接触曝气工艺 anaerobic filtration-contact aeration process
厌氧过滤器 anaerobic filter
厌氧/好氧工艺 anaerobic/oxic process
厌氧/好氧活性污泥法 anaerobic/oxic activated sludge process
厌氧/好氧交替循环系统 anaerobic/oxic alternating cycle system
厌氧/好氧接触氧化法 anaerobic/oxic contact oxidation process
厌氧/好氧偶合强化池 anaerobic/oxic couple strengthening pond
厌氧/好氧/生物活性炭工艺 anaerobic/oxic/bioactivated carbon process
厌氧/好氧生物流化床 anaerobic/oxic biological fluidized bed
厌氧呼吸 anaerobic respiration; respiration anaerobic
厌氧呼吸带 anaerobic respiration zone
厌氧化学氧化混凝气浮法 anaerobic-chemical oxidation-coagulation-air flo(a)tation process
厌氧环境 anaerobic environment
厌氧活度 anaerobic activity
厌氧活性 anaerobic activity
厌氧活性污泥 anaerobic activated sludge
厌氧活性污泥法 anaerobic activated sludge method
厌氧兼性好氧塘 anaerobic-facultative-aerobic lagoon
厌氧兼性塘 anaerobic-facultative lagoon
厌氧降解 anaerobic decomposition; anaerobic degradation
厌氧胶粘剂 anaerobic adhesive
厌氧接触池 anaerobic contact pond
厌氧接触法 anaerobic contact process
厌氧接触反应器 anaerobic contact reactor
厌氧接触过程 anaerobic contact process
厌氧接触消化池 anaerobic contact digester
厌氧菌 anaerobe; anaerobic bacteria; anaerobic organism
厌氧颗粒污泥 anaerobic granular sludge
厌氧颗粒污泥反应器 anaerobic granular sludge reactor
厌氧流化床 anaerobic fluidized bed
厌氧流化床反应器 anaerobic fluidized bed reactor
厌氧流化床生物膜反应器 anaerobic fluidized bed biofilm reactor
厌氧滤池 anaerobic filter
厌氧膜膨胀床 anaerobic film expanded bed
厌氧膜生物反应器 anaerobic membrane bioreactor
厌氧内循环反应器 anaerobic inner loop reactor
厌氧黏[粘]合剂 anaerobic adhesive
厌氧培养 anaerobic culture; anaerobic incubation
厌氧培养基 anaerobic culture medium
厌氧膨胀床 anaerobic expanded bed
厌氧膨胀颗粒污泥床 anaerobic expanded granular sludge bed
厌氧情况 anaerobic condition
厌氧区 anaerobic zone
厌氧去除有毒有机物 anaerobic toxic organics removal
厌氧缺氧/好氧工艺 anaerobic-anoxic/oxic process
厌氧缺氧/好氧系统 anaerobic-anoxic/oxic system
厌氧/缺氧序批间歇式反应器 anaerobic-anoxic sequencing batch reactor
厌氧上层溶液 anaerobic supernatant
厌氧生化处理 anaerobic biochemical treatment
厌氧生活 anaerobiosis
厌氧生活细菌 anxybiotic bacteria
厌氧生态系统 anaerobic ecosystem
厌氧生物 anaerobe; anaerobic organism
厌氧生物处理 anaerobic biological treatment
厌氧生物处理法 anaerobic biological treatment process
厌氧生物降解能力 anaerobic biodegradation
厌氧生物降解速度 anaerobic biodegradation velocity
厌氧生物滤池 anaerobic biofilter
厌氧生物膜法 anaerobic biofilm method; anaerobic contact process
厌氧生物膜膨胀床 anaerobic biofilm expansion bed
厌氧生物絮凝 anaerobic bioflocculation
厌氧生物转化 anaerobic biotransformation
厌氧生物转盘 anaerobic biological rotating disc [disk]; anaerobic rotating biological contactor
厌氧释磷 anaerobic phosphate release
厌氧水解 anaerobic hydrolysis
厌氧水解池 anaerobic hydrolysis basin
厌氧水解高负荷生物滤池 anaerobic hydrolyzation high loading biological filter
厌氧水解生物接触氧化工艺 anaerobic hydrolyzation and biological contact oxidation process
厌氧水解酸化 anaerobic hydrolysis acidification; anaerobic hydrolytic acidification
厌氧水解酸化缺氧反硝化好氧工艺 anaerobic hydrolytic acidification-anoxic denitrification-oxic process
厌氧水解酸化氧化 anaerobic hydrolysis acidification oxidation
厌氧酸化 anaerobic acidification
厌氧酸化法 anaerobic acidification method
厌氧塘 anaerobic pond
厌氧条件 anaerobic condition
厌氧土地填埋 anaerobic landfill
厌氧脱氮法 anaerobic denitrification process
厌氧脱氮过程 anaerobic denitrification
厌氧脱色 anaerobic decolo(u)rization
厌氧微生态结构 anaerobic microecostructure
厌氧微生物 anaerobe; anaerobic microbial; anaerophyte
厌氧污泥 anaerobic sludge
厌氧污泥床反应器 anaerobic sludge bed reactor; anaerobic sludge blanket reactor
厌氧污泥消化池 anaerobic sludge digestion tank
厌氧污泥消化池作用 anaerobic
厌氧系统 anaerobic system
厌氧细菌 anaerobic bacteria
厌氧细菌降解 anaerobic bacteria degradation
厌氧细菌降解作用带 anaerobic bacteria degradation zone
厌氧消耗罐 anaerobic digestion tank
厌氧消化超滤 anaerobic digestion
厌氧消化池 anaerobic digester; anaerobic digestion tank
厌氧消化动力学 anaerobic digestion kinetics
厌氧消化法 anaerobic digested process
厌氧消化(过程) anaerobic digestion
厌氧消化污泥 anaerobically digested sludge; anaerobic digesting sludge
厌氧型 anaerobic type
厌氧性处理法 anaerobic method
厌氧性黏[粘]结剂 anaerobic adhesive
厌氧性微生物 anaerobic microorganism
厌氧性污泥消化 anaerobic sludge digestion
厌氧需氧处理 anaerobic-aerobic treatment
厌氧絮凝作用 anaerobic flocculation
厌氧悬浮床反应器 anaerobic suspended bed reactor
厌氧循环 anaerobic cycle
厌氧氧化 anaerobic oxidation
厌氧液体 anaerobic liquor
厌氧移动床生物膜反应器 anaerobic moving bed bio-film reactor
厌氧异波折复合式反应器 opposite folded plate hybrid anaerobic reactor
厌氧抑制动力学 anaerobic inhibition kinetics
厌氧有毒废水处理 anaerobic toxic wastewater treatment
厌氧预处理塘 anaerobic pretreatment pond
厌氧预发酵-间歇曝气生物膜系统 anaerobic pre-fermentation-intermittent aerated biofilm system
厌氧预反硝化 anaerobic pre-denitrification
厌氧折流板反应器 anaerobic baffled reactor
厌氧折流板污泥床反应器 anaerobic baffled sludge blanket reactor
厌氧猪粪池液 anaerobic swine lagoon liquid
厌氧状况 anaerobic condition
厌氧自养菌 anaerobic autotrophic bacteria

砚

砚屏 ink-stone screen
砚台 ink stone

宴

宴会室 banqueting room; grand chamber
宴会厅 banquet(ing) hall; convection hall; festival room; grand chamber

艳

艳黄 brilliant yellow
艳丽的 flamboyant
艳绿色 emerald
艳绿调色剂 brilliant green tone

验

验标 examination of bids
验布机 cloth inspecting machine
验残检验书 inspection certificate on damaged cargo
验残证书 certificate on damaged cargo
验舱证书 certificate on hold; certificate on tank
验槽 check of foundation subsoil; inspection of foundation subsoil
验潮 sea level observation; tidal observation; tide observation
验潮标 ga(u)ging rod; tidal pole; tidal scale; tidal staff
验潮标志 tidal bench mark
验潮杆 staff tide ga(u)ge; tidal pole; tidal scale; tidal staff; tide pole; tide staff
验潮杆基准点 tide pole reference point
验潮杆误差 staff error
验潮计 marigraph
验潮井 ga(u)ge well; ga(u)ge box; ga(u)ge chamber; ga(u)ging well; observation well; tide-ga(u)ge well
验潮井测(水位)井 ga(u)ge well
验潮器 thalasometer; tide ga(u)ge
验潮水尺 tidal observation rod; tidal scale; tide pole
验潮水准点 tidal bench mark
验潮仪 hydrologic(al) ga(u)ge; maregraph; river ga(u)ge; tidal ball; tidal ga(u)ge; tidal meter; tide ga(u) ge; tide-meter; tide register; water-level ga(u)ge
验潮仪式自记测波仪 surface ga(u)-ge type wave recorder
验潮站 ga(u)ging station; tidal ga(u)ging station; tidal monitoring station; tidal observation station; tidal observatory; tidal station; tide-ga(u)ging station; tide house; tide station
验潮站零点 zero point of tidal station
验潮站零点高程 zero elevation of tidal station
验潮站水准标准 benchmark at tidal station
验潮站有效范围 effective area of tidal station
验秤员 weigher
验船 survey of ship
验船顾问 consulting surveyor
验船师 classification surveyor; marine surveyor; ship surveyor; surveyor
验船师建议书 advice note of surveyor
验船条款 boardings
验船协会 classification society
验船证书 ship inspection certificate
验串 cement channeling proofing
验疵器 flaw detector
验磁器 magnetoscope
验错 error checking
验带器 tape certifier
验单 verification note

验电板 proof-plane
验电笔 test pencil
验电法 electroscopy
验电盘 detector; electroscope; galvanoscope; proof-plane
验电器 electroscope; phonoscope; rheoscope; galvanoscope
验电术 electrometry
验定(法) assay
验方 proved recipe
验方尺 try square
验方角尺 trying square; trial square
验方角规 trying square
验付条款 payment after inspection clause
验估人 appraiser
验关 customs clearance; customs inspection
验关仓库 clearance depot
验关地点 sites of customs inspection
验关站 clearance depot
验关证 clearance certificate
验光 refraction
验光库 appraiser's surveyor
验核基线 check base(line)
验后放行 release if in order after examination
验后概率 posterior probability
验火孔 sight hole
验货 cargo examination
验货单 particular paper; verification certificate
验货后付款 payment after inspection
验货收款台 check stand
验货销售 sale by inspection
验货员 surveyor
验检员 examiner
验卡机 card verifier
验孔 verify
验孔规 hole ga(u)ge
验流 current observation
验流器 current detector
验明 identity
验明支票收款人 identifying the payee of a check
验票 ticket check
验票机 patron operated ticket checker; validation specification
验票员 ticket inspector; chopper <俚语>
验平板 surface plate
验平石 bedding stone
验平仪 profilograph; profilometer
验气球管 gas pipet(te)
验气燃烧器 aeration test burner
验讫 check off; examined
验契税 tax for the examination of deeds
验器示硬度 scleroscope hardness
验签副本 attested copy
验签条款 attestation clause
验前风险函数 pretest risk function
验色管 chromoscope
验色剂 toner
验色器 chromatoscope; chromoscope
验声器 phonoscope
验尸 postmortem (examination)
验尸室 autopsy
验尸所 mortuary
验湿器 hygroscope
验湿仪 hygroscope
验收 acceptance (check); acceptance survey; check and accept; control reception; examination and acceptance; examine and receive; inspection and acceptance; office test; proving; receiving; receiving audit
验收报告(单) acceptance report; receiving report
验收比率 acceptance rate
验收标准 acceptable standard; acceptance criterion; acceptance level; acceptance standard; accepted standard; inspection standard; acceptable criterion
验收部门 receiving department
验收测量 acceptance survey; contract acceptance survey
验收测试 acceptable test; acceptance test; acceptance trial; approval test
验收程序 acceptance procedure
验收抽样(法) acceptance sampling
验收抽样计划 acceptance sampling plan; reception sampling plan
验收抽样试验中允许的最多不合格数 acceptance number
验收单 acceptance certificate; certificate of acceptance; receiving apron
验收的热处理工艺 acceptable heat treatment procedure
验收范围 acceptance range
验收分析 analysis on acceptance; analysis upon entry
验收盖印 acceptance stamp
验收盖章 acceptance stamp
验收工程师 inspecting engineer
验收公差 acceptance margin; acceptance off-size; accepted tolerance
验收功能试验 acceptance functional test
验收故障率 acceptable failure rate
验收规 acceptance ga(u)ge; inspection ga(u)ge; receiver ga(u)ge
验收规范 acceptability criterion; acceptance specification
验收规则 acceptance rule; regulations of inspection
验收规章 regulation of inspection
验收过程 course of receiving
验收合格 acceptance check
验收合格证(书) acceptance certificate; certificate of acceptance
验收环境 acceptable environment; acceptance environment
验收回单 acknowledge sheet
验收极限 acceptance limit
验收计划 acceptance plan
验收记录 examining and receiving record; record of acceptance
验收技术规格 inspection specification
验收检查 acceptance check; receiver inspection; receiving inspection; acceptance inspection
验收检查设备 acceptance inspection equipment
验收检验 acceptance inspection; acceptance trial; acceptance test(ing)
验收鉴定书 acceptance certificate
验收考核 acceptance test
验收模式 acceptance pattern
验收偏差 acceptance deviation
验收期 acceptance period
验收签认单 acceptance sheet
验收区域 acceptance region
验收取样方案 acceptance sampling scheme; reception sampling plan
验收取样检查 acceptance sampling inspection
验收人 accepter; payee; receiver
验收容许公差 acceptance permissible tolerance
验收容许偏差 acceptance permissible variation; acceptance tolerance
验收容许误差 acceptance permissible deviation
验收时间 reception time
验收试航 acceptance trial trip
验收试验 acceptable test; acceptance test; acceptance trial; accepting test; approval test; proof test; reception test; warranty test
验收试验程序 acceptance test procedure
验收试验手册 acceptance test manual
验收试验允许公差 <如轨道衡> acceptance test tolerance
验收数目 acceptance number
验收天数 days of examining and receiving
验收条件 acceptance condition; conditions of acceptance
验收图 acceptance drawing
验收委员会 acceptance committee
验收文件 acceptance documents
验收限度 acceptance limit
验收性能试验 acceptance functional test
验收要求 acceptance requirement
验收与测试 acceptance and checkout
验收与维护 receiving inspection and maintenance
验收与移交 acceptance and transfer
验收员 examiner; inspector
验收允许公差 acceptance permissible variation
验收允许量 acceptance tolerance
验收丈量 footage
验收者 checker and accepter
验收证(明)书 acceptance certificate; certificate of acceptance
验收值 <计量抽查时的> acceptance value
验收制度 system of check and acceptance
验收质量 acceptance property; acceptance quality
验收质量标准 acceptable quality level; acceptable quality standard
验收质量水平 acceptable quality level
验收准则 acceptance criterion
验收资料 acceptance documents
验收总结报告 acceptable summary report; acceptance summary report
验水阀 test valve
验水位塞体 ga(u)ge cock body
验水位旋塞漏斗泄水管 ga(u)ge cock drip funnel drain pipe
验水旋塞 test cock
验丝机 serimeter
验丝计 serimeter
验算 checking calculation; checking computation; checkout; chocking computation; proofing; prove; rerun the solution
验算公式 check formula
验算荷载 checking load; check load(ing)
验算截面 examined section
验算流量 checked discharge
验探槽 exploration trench
验温器 thermoscope
验线 checking of building line
验修 running repair
验压器 baroscope
验硬度器 scleroscope
验油比重计 el(a)eometer; elaiometer
验油浮计 el(a)eometer
验油机 oil tester
验油计 oleometer
验震器 seismoscope
验震器记录 seismoscope record
验震器数据 seismoscope data
验震器响应 seismoscope response
验证 probation; proofing; proving; validation; verification
验证本 certified copy
验证标记 verification mark
验证表 proof list(ing)
验证程序 procedure of verification; proving program(me); validation procedure
验证单 check list
验证方案 proof scheme
验证方法 method of verification; verification process
验证方式 verification mode
验证分析 check analysis
验证负荷 proof load
验证过程 validation process
验证荷载 proof load
验证机 proof machine
验证计量 check gauging
验证记号 proof mark; witness mark
验证技术 verification technique; validation technique【计】
验证假设 verify hypothesis
验证检查 confirmatory check
验证井 confirmation well
验证控制 proof control
验证列表 proof list(ing)
验证码 identifying code
验证设置 verification setting
验证时间 proving time
验证式 verification expression
验证试验 acceptance test; certification test; confirmatory test; proof test; retest
验证条款 attestation clause
验证性测量 confirmatory measurement
验证性试验 check off; proof test; proving test; proving trial
验证压力 proof pressure
验证压力试验 proof pressure test
验证样本 proof sample
验证样品 proof sample
验证应力 proof stress
验证用数据 verification data
验证用终端 verification terminal
验证誊本 attested copy
验证员 verifier
验证终端键盘 verification terminal keyboard
验证终端显示 verification terminal display
验证钻进 check-boring; check-drilling
验证钻孔 confirmation hole
验重器 gravi(to)meter
验准器 level trier
验资报告 capital verification report

谚

谚语 maxim; proverb

堰

堰 weir; an(n)icut; clauster; dam; spillway dam; stank; aboideau <加拿大习用>

堰岸墩 abutment sidewall
堰坝 barrage
堰坝上游阶形面 upriver stepped face
堰坝式电站 barrage power station
堰板 cut-off wall; dam baffle; gate; poling board; poling plank; weir board; weir plate
堰板缝隙 slice opening
堰板痕 slice mark
堰板浆速 slice velocity
堰板量水建筑物 weir box
堰板量水箱 weir box
堰板调节 slice setting
堰板位置 slice position
堰壁收缩 end contraction
堰舱 cofferdam
堰槽 weir notch
堰侧收缩 <水舌> end contraction
堰测(定)法 weir method
堰唇 slice lip; weir lip
堰的岸墩 abutment of weir

堰的背水面 downstream face of a weir
堰的上游 back of dam;back of weir
堰的下游面 airside face
堰堤式发电站 dam type power plant
堰堤式河漫滩 barrier flood plain
堰堤式(水力)发电厂 dam type power plant
堰底 weir floor;weir sill
堰底收缩 bottom contraction
堰顶 crest of weir;top of weir;weir crest
堰顶板 weir plate
堰顶长度 crest length
堰顶超高 crest freeboard
堰顶高程 crest elevation;crest level;invert elevation
堰顶净长 net length of crest
堰顶可倾式过水坝 tilting dam
堰顶控制 crest control
堰顶控制装置 weir crest control device;weir crest control unit
堰顶临时挡水设施的闸门 flashboard check gate
堰顶流量 crest discharge
堰顶(流下的)水帘 weir nappe
堰顶剖面(曲线) crest profile
堰顶曲线 crest curve;weir crest curve
堰顶收缩 crest contraction
堰顶水深 crest depth;weir crest depth
堰顶水头 head of water over weir;head over weir;weir head
堰顶水位计 weir ga(u)ge
堰顶挖沟形成的台道 dished channel fish pass
堰顶溢流 weir overflow
堰顶溢流水深 depth of water flowing over weir
堰顶溢水深度 depth of surcharge
堰顶圆弧包角 included angle of crest
堰顶圆弧中心 center [centre] of weir crest circle
堰顶闸门 barrage gate;crest gate
堰顶中心角 central angle at crest
堰顶自动泄水闸门 automatic crest and scour gate
堰顶自动闸门 automatic crest gate
堰端收缩 end contraction
堰段 weir section
堰墩 weir pier
堰高 height of weir;weir height
堰根 weir root
堰湖 barrier lake
堰槛 sill of weir
堰坎 sill of weir
堰孔 weir notch
堰口 crest of weir;weir
堰口测流法 weir method
堰口断面 weir section
堰口负荷 weir loading
堰口率 weir rate
堰流 flow over weir
堰流坝 weir dam
堰流公式 weir formula
堰流计 weir ga(u)ge
堰流量 weir flow
堰流系数 weir coefficient
堰门 weir gate
堰旁水池 weir pond
堰前扩流减速箱 weir box
堰前水头 barrage head
堰前缘 weir lip
堰塞 damming
堰塞湖 barrier lake;check(ed)-up lake;choked lake;dammed lake;damped lake;sag pond;imprisoned lake <用堤围成的人工湖>
堰塞湖盆 barrier basin
堰塞盆地 barrier basin
堰塞泉 barrier spring
堰上缺口 weir notch
堰上水头 weir head
堰上液层高度 crest over the weir
堰式挡板 weir-type baffle
堰式罐 weir tank
堰式流量计 weir flow meter;weir ga(u)ge
堰式施胶器 weir-type applicator
堰式压力管道 weir penstock
堰式溢洪道 barrage type spillway
堰式鱼道 weir-type fish pass;weir-type fish passway
堰水 sluice
堰水位差 afflux
堰体上游倾斜面 battered upstream face
堰下槛 weir cill [sill]
堰下游建筑部分 aft-bay
堰箱 weir box
堰溢流 weir waste
堰溢流率 weir overflow rate
堰缘 weir edge
堰闸 flash;weir and lock;weir lock
堰闸顶 weir shutter crest
堰闸门 weir gate
堰址 location of barrage
堰址测站 weir station
堰洲 barrier
堰洲坪 barrier flat
堰洲嘴 barrier spit

焰

焰白云石 flame dolomite
焰边切 flame cut edge
焰道 flue
焰底 flame base
焰锋 flame front
焰锋速度 flame front velocity
焰割边 torch cutting edge
焰管 furnace flue
焰管锅炉 multitubular boiler
焰管受热面 flue heating surface
焰黑 flame black
焰红染料 phloxine
焰弧 flame arc
焰弧灯 flame arc lamp
焰火 fire works
焰火发送器 pyrotechnic generator
焰火具 fire work
焰炬 torch
焰口 crater
焰口效应 cratering effect
焰煤 flame coal
焰桥 flame bridge
焰切机 flame cutter
焰球 flame bulb
焰热阴极 glowing cathode
焰熔法 flame fusion method
焰色 flame colo(u)ration
焰色反应 flame reaction
焰色分析 flame analysis
焰色试验 flame colo(u)ration test;flame colo(u)r test;flame test
焰烧净化法 flame cleaning
焰式窗(花)格 flamboyant tracery
焰式拱 flamboyant arch
焰式建筑 flamboyant architecture
焰式圆花窗 flamboyant rose window
焰炱发光度 soot luminosity
焰体 flame body
焰芯 flame cone;inner cone;flame heart;flame kernel
焰芯炭棒 flame-cored carbon
焰型 flame pattern
焰晕 firedamp cap;flame cap
焰轴 flame axis

雁

雁翅坝 vane dike [dyke]
雁叫声 honk
雁来红 tricolo(u)r amaranth
雁列错断 en echelon offset
雁列带长度 length of echelon array
雁列带宽度 width of echelon array
雁列断块 en echelon fault blocks
雁列角 en echelon angles
雁列节理 en echelon joints
雁列裂缝 en echelon cracks
雁列脉特征 feature of en echelon vein
雁列面产状 attitude of echelon plane
雁列丘地形 basket of eggs' relief;basket-of-eggs topography
雁列山脉 cordillera
雁列式 echelon pattern
雁列线 en echelon lines
雁列褶皱 en echelon folds
雁列注记 slopping name
雁列状 en echelon
雁列(状)断层 en echelon faults
雁列(状)构造 en echelon structures
雁行 echelon
雁行磁异常 en echelon magnetic anomaly
雁行断层 echelon faults
雁行构造 echelon structure
雁行构造形式 echelon pattern
雁行节理 en echelon joints
雁行裂缝 en echelon fissure
雁行排列 en echelon arrangement
雁行式构造 en echelon structures
雁行式裂缝 echelon fracture
雁行透镜体 en echelon lenses
雁行褶皱 echelon fold;en echelon folds
雁行褶皱作用 en echelon folding
雁行状构造 en echelon structures

燕

燕麦粉 <一种止水材料> oatmeal
燕尾【建】dovetail;fantail
燕尾扒钉 dovetail cramp
燕尾槽 dovetail groove;dovetail slot
燕尾槽拉削 broaching dovetail groove
燕尾槽锚固 dovetail slot and anchor
燕尾导轨 dovetailed way;dovetail guide
燕尾对开叠接 dovetail half-lap joint;dovetail halved joint
燕尾合页 dovetail hinge
燕尾滑板 dovetail slide
燕尾夹 dovetail cramp
燕尾键 dovetail feather;dovetail key
燕尾键槽 dovetail key way
燕尾铰 dovetail hinge
燕尾接合 dovetail groove;dovetailing;dovetail joint
燕尾锯 dovetail saw
燕尾开脚 fang;fish-tailed end
燕尾开榫机 dovetail machine
燕尾肋 dovetail rib
燕尾连接 dovetailing
燕尾龙骨 dovetail joist
燕尾律 swallow-tail law
燕尾螺栓 fish bolt;fishtail bolt;split bolt
燕尾锚(具) Danforth anchor;dovetail anchor
燕尾锚杆 dovetail anchor
燕尾刨 dovetail plane
燕尾旗 burgee
燕尾式导轨 dovetail slide bearing
燕尾式外锁闭 swallow-tailed type outside locking device
燕尾榫 combed joint;dovetail;dovetail tenon;swallow tail;undercut tenon
燕尾榫板条 dovetail margin
燕尾榫半搭接 dovetail half-lap;dovetail halved joint;dovetail halving
燕尾榫槽 dovetailed groove
燕尾榫搭角接 butterfly;butterfly wedge
燕尾榫接 dovetailed joint;laminated joint
燕尾榫接合 joint with dovetail groove
燕尾榫口 dovetail feather
燕尾铁脚 swallow-tailed ironwork
燕尾(铣)刀 dovetail cutter
燕尾楔形木条地板 dovetailed fillet
燕尾斜角接缝 dovetail mitre [miter]
燕尾形 dovetail form;swallow-tail form
燕尾形板 dovetail sheeting
燕尾形表面刻琢 dovetailing
燕尾形冲筋 splayed ground
燕尾形的 swallow-tailed
燕尾形键 splay piece
燕尾形接头 dovetail joint
燕尾形片材 dovetail sheeting
燕尾形生铁盒 anchor box
燕尾形榫眼 dovetail mortise
燕尾形斜面接合 dovetail mitre [miter]
燕尾桩 dovetail pile;pug pile
燕尾钻 dovetail bit
燕子 swallow
燕子矶【地】swallow rock

赝

赝标量 pseudo-scalar (quantity)
赝标量耦合 pseudoscalar coupling
赝布儒斯特角 pseudo-Brewster angle
赝场 pseudo-field
赝单色仪 pseudo-monochromator
赝对称 pseudo-symmetry
赝二元硫族化合物玻璃 pseudo-binary chalcogenide glass
赝光波段 <毫米波> pseudo-optic(al) band
赝光线追迹 pseudo-ray tracing
赝架 scaffold
赝晶的 pseudo-crystalline
赝晶(体) pseudo-crystal
赝晶体物质 pseudo-crystalline material
赝拉曼技术 pseudo-Raman technique
赝立体声 pseudo-stereophony
赝偶极子 pseudo-dipole
赝品 bogus;counterfeit;fake;fakement;imitant;sham;shoddy
赝平面 pseudo-flat;pseudo-plane
赝全息术 pseudo-holographic technique;pseudo-holography
赝矢量 pseudo-vector
赝随机 pseudo-random
赝锁相 pseudo-lock
赝相位共轭 pseudo-phase-conjugate
赝像 pseudo-morphism;pseudo-scopic image
赝张量 pseudo-tensor
赝正弦线 poid
赝周期 pseudo-period

央

央墙烟道 barge
央轴 central axis

秧

秧地 plant bed
秧苗 seedling;young plant
秧田 nursery bed;rice nursery;rice shoots;seeding bed
秧土层 plant bed
秧穴 plant pit

鞅 martingale

扬尘 dust development;dust emission;dust generation;escape of dust;raising of dust

扬尘点 dusty place;source of dust
扬程 delivery lift;discharge head;head of delivery;height of lift(ing);lift head;pumping head;water head;pressure head <泵的>;lifting height <指起重机升降范围>
扬程范围 head range
扬程流量曲线 head-discharge head
扬程损失 head loss
扬程特性曲线 <泵或风机的> head-capacity curve
扬德尔方程 Jander's equation
扬动流速 stirring-up velocity;velocity of lifting movement
扬斗式装载机 overloader
扬帆结 sail halyard bend;studding sail boom hitch
扬帆索 ha(u)lyard
扬返机 rereeling machine
扬返丝 rereeled silk
扬格计划 Young plan
扬灰货物 dusty cargo
扬量 capacity
扬料板 cascader;dispersing flight;lifter flight;lifting flight
扬料架 lifting shelf
扬料率 hold-up rate
扬料器 cascader
扬料叶片 lifter blade
扬料砖衬 cam lining
扬料装置 hold-up;lifting device
扬煤机 coal hoist
扬旗工【铁】flagman
扬起 raise
扬起飞尘 dust-producing
扬弃爆破 throw blasting
扬砂泵 <又称扬沙泵> sand pump;sludger
扬砂泵滑轮 sand pump pulley
扬砂管 sand lift
扬升式排水法 pumping-up drainage method
扬升水 upgrade water
扬声电话机 loud speaking telephone set
扬声喇叭筒 loudspeaker horn
扬声器 loudspeaker;radiator of sound;reproducer;sound projector;sound reproducer;speaker;talker
扬声器保护网 speaker screen
扬声器布置 loudspeaker placement
扬声器电话 speaker-phone
扬声器分频网络 loudspeaker dividing network
扬声器孔 loudspeaker hole
扬声器群 loudspeaker cluster
扬声器系统 loudspeaker system
扬声器箱 loudspeaker cabinet;loudspeaker enclosure;loudspeaker housing;speaker box;speaker cabinet;speaker enclosure
扬声器音圈 loud-speaker voice coil
扬声器障板 speaker baffle
扬声器阵 row of loudspeakers
扬声器震声 blasting
扬声器纸盒 effuser
扬声器纸盆 cone;diffuser
扬声器助音箱 box baffle
扬声器装置 speaker unit
扬声器组 loudspeaker group
扬水 lifting of water;pumping
扬水泵 lift pump
扬水带 hydraulic belt
扬水干管 rising main
扬水高程 water-raising capacity
扬水高度 delivery lift
扬水管 ascending pipe;uptake
扬水管道 pumping line
扬水灌溉 irrigation by pumping;lift irrigation;pumping irrigation
扬水机 flap wheel;water elevator;water engine;water lift;water lifting machine;water raiser;water-raising machine
扬水机械 water lifting machinery
扬水井 suction well
扬水率 pumpage rate
扬水轮 flap wheel;scoop wheel
扬水能力 pumping capacity;water-raising capacity
扬水器 water elevator
扬水曲线 lifting test curve
扬水式抽水井 absorption well
扬水式发电站 pumped-storage power station
扬水式配水系统 upfeed distribution system
扬水式水力(蓄能)发电厂 pumping-up power plant
扬水试验 aquifer test;bailing experiment;lifting test pumping;pumping test(ing)
扬水试验的抽水井 absorbing well;absorption well
扬水竖井 pumping shaft
扬水水头 pump delivery head;pumping head
扬水水头物质 pumping head matter
扬水水头扬程 pumping head
扬水系统 pumping-out system
扬水蓄能 pumped storage
扬水压力折减系数 uplift reduction coefficient
扬水站 lift station;pumped-storage station;pumping station
扬水站址工程地质勘察 engineering geologic(al) investigation of pumping station site
扬水主管 rising main
扬水作用 pumping action
扬酸器 acid elevator
扬索 halliard;halyard
扬压力 hydraulic uplift pressure;lift pressure;raised pressure;uplift pressure;upward hydraulic pressure;upward hydrostatic(al) pressure;upward pressure
扬压力计 uplift cell
扬子鳄 Alligator sinensis;Chinese alligator;Yangtze alligator
扬子期【地】Yangtze stage

羊背后凹凸地形【地】knob-and-basin topography

羊背石【地】dressed rock;roche moutonnee;sheepback rock;sheep rock
羊肠小道 meander(ing);narrow winding trail
羊齿 brake;fern
羊齿烷 eernane
羊齿形花纹 fern-like pattern
羊冲眼标记 punch mark
羊肚菌 morel
羊羔毛加工滚筒 lamb's wool cylinder
羊槛 cote
羊角 spindle
羊角臂 knuckle arm
羊角锤 adz(e)-eye hammer;carpenter's hammer;claw hammer;half hatchet;joiner's hammer;nail hammer
羊角榔头 nail hammer;claw hammer
羊角碾 club-foot roller;spiky roller
羊角图案 ram's horn figure
羊角系缆柱 cleat
羊角状花纹 ram's horn figure
羊脚捣路机 sheep-foot tamper
羊脚夯击机 sheep-foot tamper
羊脚路滚 sheep-foot roller
羊脚碾 sheep('s)-foot roller;sheep-foot tamper;tamper roller;tamping roller;tapered-foot roller
羊圈 sheepfold;sheep pen
羊栏 cot
羊毛 wool
羊毛厂废水 woolen-mill wastewater
羊毛厂废物 woolen mill waste
羊毛尘 wool dust
羊毛的 woollen;woolly
羊毛地毯 wool carpet
羊毛加工废水 wool processing wastewater
羊毛浸渍 sheep dipping
羊毛洗涤厂 wool washing plant
羊毛洗涤废水 wool scouring wastewater
羊毛纤维 wool fiber [fibre]
羊毛屑 wool shavings
羊毛油 yolk
羊毛甾烷 lanostane
羊毛毡 wool felt
羊毛毡板 wool felt board
羊毛织品 fleece-wool
羊毛脂 degras;lanolin(e);lanolinum;wool fat;wool grease
羊毛脂沥青 wool grease pitch;wool pitch
羊毛制的 woollen
羊毛状的 flocculent;woolly
羊毛状物 wool
羊皮(革) sheep skin
羊皮货 parchment
羊皮加工辊 lamb roller
羊皮纸 lambskin;membrane;parchment;sheepskin
羊皮纸文件 parchment
羊舌饰【建】lamb's tongue
羊蹄滚筒 sheep('s)-foot roller;taper(ed)-foot roller
羊蹄夯击机 sheep-foot tamper
羊蹄路碾 sheep-foot roller
羊蹄压路机 sheep('s)-foot roller;taper foot roller
羊头饰 aegricane
羊腿形的 leg of mutton
羊尾沟 barranca
羊眼钉 ring-shank nail
羊眼杆 eye bar
羊眼螺栓 eye bolt
羊眼螺丝 eye screw;screw eye
羊眼圈 screw eye
羊脂 mutton tallow
羊状岩 roche moutonnee
羊足 <碾压滚筒上的> tamping foot;sheepsfoot
羊足捣路机 sheep-foot tamper
羊足滚筒 taper foot roller
羊足夯 sheep-foot tamper
羊足碾(压机) sheep('s)-foot roller;sheep-foot tamper;tamper roller;tamping foot compactor;tamping roller;taper(ed)-foot roller
羊足碾压实 kneading compaction;sheep-foot roller compaction
羊足压路机 peg-foot roller;sheep('s)-foot roller;tamping foot compactor;tamping-type roller;taper(ed)-foot roller
羊足压实机 tamping foot compactor

阳冲模 male die

阳道砖 runner brick
阳地植物 heliophyte;sun plant
阳电 positive electricity
阳电板 electrode positive
阳电荷 positive charge
阳电荷乳胶 positex;positive latex
阳电极 positive electrode;anode <电缆防蚀装置的>
阳电溶胶 positive sol
阳电势 electropositive potential
阳电性 electropositivity
阳电性的 electropositive
阳电性元素 electropositive element
阳电性原子 electropositive atom
阳电子 anti-electron;positive electron;positron
阳电子素 positronium
阳端接头 male coupling
阳根 positive group;positive radical
阳沟 open channel;open ditch;surface ditch
阳光玻璃温室 solar house
阳光充足的房间 sun parlo(u)r
阳光充足的住房 solarium [复 solaria/solariums]
阳光充足房间 cheerful room
阳光地带 Sunbelt Area
阳光电池 solar cell
阳光反射表面 solar reflecting surface
阳光防护窗 solar shield
阳光辐射能 incident radiation
阳光辐射作用 solarization
阳光干燥 weather drying
阳光接收器 solar radiation collector
阳光控制薄膜 solar control film
阳光控制(挡)板 solar control blind
阳光控制的软百叶窗 solar controlled venetian (blind)
阳光控制膜 solar control film
阳光热量反射表面 spar finish
阳光热量反射面 solar reflecting surface
阳光入射角 solar angle
阳光栅板 solar grating
阳光室 solaria
阳光收集器 sunshine collector
阳光透入 sunlight penetration
阳光吸收率 solar absorptivity
阳光压力 sunlight pressure
阳光荫蔽 solar shading
阳光荫蔽设施 solar shading device
阳光引起的化学反应 sunlight-driven reaction
阳光照射侧 sunny side
阳辉光 positive glow
阳机压降 anode drop
阳极 anode;plate electrode;positive electrode;positive pole
阳极暗区 anode dark space
阳极靶 plate target
阳极靶茎 anode stem
阳极板 anode plate;positive plate
阳极板充电机 anode plate charging machine
阳极板静电感应除尘器 electrostatic precipitator with plate electrodes
阳极半电池 anode half-cell
阳极棒 anode bar;anode rod;anode stub
阳极饱和 anode saturation
阳极保护层 anodic coating
阳极保护(法) anodic protection
阳极保护作用 anodic protection
阳极变流器 anode converter
阳极残铜 anode scrap
阳极残头 anode crop
阳极残渣 anode mud;anodic slime
阳极沉积 anodic deposition

阳极沉积层 anode slime
阳极沉渣 anodic slime
阳极处理 anodic process;anodic treatment;anodization;anodising [anodizing]
阳极处理保护层 anode finish;anodic finish
阳极处理法 anodic treatment method
阳极导体 plate conductor
阳极导线 anode conductor
阳极的 anodal;positive;zincous
阳极电沉积 anodic electrodeposition
阳极电池组 anode battery; anodic battery;B-battery;plate battery
阳极电导 anode conductance;plate conductance
阳极电镀 anodise [anodize];anodization;anodizing
阳极电镀机 anode casting machine
阳极电感 anode inductance
阳极电解液 anode liquor;anolyte
阳极电紧张 anelectrotonus
阳极电抗器 anode reactor
阳极电流 plate current
阳极电流的直流分量 feed current
阳极电流密度 anodic current density
阳极电流起伏 anode current fluctuation
阳极电流强度 anodic current intensity
阳极电路 anode circuit;anode loop;B-circuit
阳极电抛光 anode brightening
阳极电容量 anode capacity
阳极电势降 anode drop
阳极电位 anode potential
阳极电位稳定 anode potential stabilization
阳极电压 anode voltage;B plus voltage;plate voltage
阳极电压降 anode drop
阳极电泳漆 anionic electrodeposit coating
阳极电源 anode supply;B power supply;B-source;plate supply
阳极电源整流器 B eliminator
阳极电阻 anode resistance;plate resistance
阳极电阻微调 anodic resistor trimming
阳极淀渣 anode slime
阳极镀层 anode coating;anodic coating;anodic film;anodic finish
阳极端 anode tap;positive terminal
阳极断电收缩 anodal opening contraction
阳极钝化 anode passivation
阳极钝化剂 anodic inhibitor
阳极钝态 anodic passivation; anodic passivity
阳极发光 anode light
阳极法 anodic process
阳极反馈线圈 tickler
阳极反向峰值电压 peak negative anode voltage
阳极反应 anode reaction
阳极防腐 anodize
阳极防腐板 anodic corrosion protector
阳极防腐处理铝合金线 alumin(i)um alloy anodizing wire
阳极防腐法 anodization;anodizing
阳极防腐蚀(法) anodic corrosion protection;anodic protection
阳极敷层 anodic coating
阳极腐蚀 anode corrosion;anodic attack;anodic corrosion;anodic pickling
阳极腐蚀控制 anodic corrosion control
阳极腐蚀效率 anodic corrosion efficiency
阳极腐蚀抑制剂 anodic corrosion inhibitor
阳极负荷 anode load;plate load
阳极负载 anode load;plate load
阳极附着物 anodic slime
阳极高压 anode high voltage
阳极跟随器 anode follower;plate-follower
阳极功率 plate power
阳极供电 anode supply
阳极光柱 anode column
阳极规格 size of anode
阳极耗散 anode dissipation; anodic dissipation;plate dissipation
阳极耗散功率 anode dissipation power
阳极耗散热能 radiant heat of anode
阳极糊比电阻 anode paste resistivity
阳极护罩 anodic shield
阳极化 anodic treatment;anodising;anodization
阳极化表面处理 anodised finish
阳极化处理 anodise [anodize];anodizing
阳极化处理厂 anodizing plant
阳极化处理缎光面 bright satin anodize
阳极化镀层 anodized coating
阳极化废水 anodizing waste(water)
阳极化废物 anodizing waste
阳极化过程 anodizing process
阳极化镜 anodized mirror
阳极化铝 anodized alumin(i)um
阳极化抛光 anodised [anodized] finish
阳极缓蚀剂 anodic inhibitor
阳极辉光 anode glow;anode light;positive glow
阳极回路 anode circuit;anode loop;plate circuit
阳极活接点 anode tapping point
阳极极化 anodic polarization
阳极极化铝 anodised alumin(i)um
阳极检波 anode detection;plate detection;transrectification
阳极检波器 anode detector;plate detector
阳极检波特性 transrectification characteristic
阳极键控法 anode keying
阳极降压 anodic drop pressure
阳极接地 anodic earth;plus earth
阳极接地放大器 grounded-anode amplifier;grounded plate amplifier
阳极接头 positive contact
阳极解吸 anodic stripping
阳极金属 anode metal;anodic metal
阳极浸蚀 anodic scouring
阳极静电流 anode rest current
阳极可变电感器 plate variometer
阳极空间 anode chamber;anode compartment
阳极控制 anodic control
阳极扩散电流 anodic diffusion current
阳极帘栅调幅 anode-screen modulation
阳极帽 anode button;anode cap;plate cap
阳极帽封接机 anode cap inserting machine
阳极膜 anode film
阳极母线 anode bus;positive bus
阳极泥 anode mud;anode slime
阳极泥覆盖 anode-slime blanket
阳极抛光 anodic brightening
阳极片 piece of anode [anodic] plate
阳极侵蚀 anode corrosion
阳极清洗 anodic cleaning
阳极区 anode region; anodic area; positive column; positive polarity zone
阳极去极化波 anodic depolarized wave
阳极去极剂 anodic depolarizer
阳极溶出法 anodic stripping
阳极溶出伏安法 anodic stripping voltammetry
阳极溶出极谱法 anodic stripping polarography
阳极溶解 anodic dissolution
阳极溶解波 anodic dissolution wave
阳极溶解电流 anodic dissolution current
阳极溶解抑制剂 anodic dissolution inhibitor
阳极射线 anode light;anode ray;anodic ray;canal ray;positive ion rays;positive ray
阳极射线束 anodic beam
阳极蚀刻 anodic etching
阳极室 anode compartment; anodic chamber
阳极输出电流 anode output current
阳极酸洗 anodic cleaning; anodic pickling
阳极损耗 anode loss
阳极特性 anode characteristic
阳极填充料 anodic backfill
阳极调整 plate control;power control
阳极调制 anode modulation; plate modulation
阳极铜 anode copper
阳极透明氧化被膜法 anodizing
阳极涂层 anodic coating
阳极涂料 anode coating
阳极线圈 plate turn
阳极效率 anode efficiency
阳极效应 anode effect
阳极行为 anodic behavio(u)r
阳极蓄电池 high-tension battery
阳极氧化 anode oxidizing;anodic oxidation;anodise [anodize]
阳极氧化板 anodizing sheet
阳极氧化处理 anodic treatment;anodising [anodizing]
阳极氧化电解着色 anodizing electrolytic colo(u)ring
阳极氧化(镀)层 anodic coating;anodic finish
阳极氧化法 anodic oxidation method
阳极氧化工厂 anodizing plant
阳极氧化铝 anodised alumin(i)um
阳极氧化铝加工处理法 alumilite process
阳极氧化铝加工法 alumite process
阳极氧化率 anodizing
阳极氧化膜 anode oxide film
阳极氧化膜花纹 anode oxide film streak
阳极氧化膜蓄气性发花(起泡) defect of gas accumulation of anode oxide film
阳极氧化着色 anodic colo(u)r;anodic oxidation colo(u)r
阳极氧转移反应 anodic oxygen-transfer reaction
阳极液 anode liquor
阳极阴极电催化 anodic-cathodic electrocatalysis
阳极优化 anode's optimization
阳极圆筒 anode cylinder
阳极罩 anode bag;anode cell
阳极纸 positive paper
阳极制约 anodic control
阳极柱体 anode cylinder
阳极组 positive group
阳角 external corner;outside corner
阳角壁脚弯砖 external angle to cove skirting
阳角尖嘴砖 external bird's beak
阳角接 mitre [miter]
阳角接缝做法 jointing at outside corner
阳角抹灰泥刀 arris trowel
阳角抹子 corner trowel;outside corner trowel
阳角扇形砖 external shoulder angle
阳角砖 external angle head
阳角转角 bead angle
阳刻 incised inscription
阳离子 basic ion;cation [kation];positive ion
阳离子比率 ratio of cations
阳离子表面活化剂 cationic surface active agent
阳离子表面活性消毒剂 cation surface-active disinfectant
阳离子层 cationic layer
阳离子掺和剂 cationic additive
阳离子导体 cationic conductor
阳离子地沥青乳液 cationic asphalt emulsion;cationic slurry
阳离子电荷 cationic charge
阳离子电解质 cationic polyelectrolyte
阳离子电流 positive ion current
阳离子电泳 kataphoresis [cataphoresis]
阳离子电泳速度 cataphoretic velocity
阳离子淀粉 catonic starch
阳离子发射 positive emission
阳离子分析 cationic analysis
阳离子分组 cationic grouping
阳离子附加剂 cationic additive
阳离子改性 cationic modification
阳离子改性淀粉 cationic modified starch
阳离子高分子絮凝剂 cationic polymeric flocculant
阳离子隔膜 cationic membrane
阳离子固定 cationic fixation
阳离子固化 cationic curing
阳离子含量三角图 triangular plot showing relative amount of cations
阳离子活度 cationic activity
阳离子活化 cationic activation
阳离子间最适比率 optimum ratio between the cations
阳离子交换 base exchange;cation exchange;cation interchange
阳离子交换测定 cation exchange capacity measurement
阳离子交换层析 cation exchange chromatography
阳离子交换床 cation exchange bed
阳离子交换反应器 cationic exchange filter
阳离子交换过滤器 cationic exchange filter
阳离子交换活动性 cathode exchange activity;cation exchange activity
阳离子交换剂 base exchanging compound;cation exchanger;cation exchanging compound;cationite
阳离子交换量 cathode exchange capacity
阳离子交换滤水器 cationic exchange filter
阳离子交换膜 cation exchange membrane
阳离子交换能力 cation exchange ability;cation exchange capacity
阳离子交换能力活化剂 cathode exchange capacity
阳离子交换器 base exchanger;cation cell;cation exchanger
阳离子交换容量 cation exchange capacity
阳离子交换容量试验 cation exchange capacity test

阳离子交换软水法 cation exchange method of water-softening
阳离子交换软水器 cation exchange softener
阳离子交换树脂 cation resin;positive ion-exchange resin
阳离子交换塔 cation exchange tower;cation tower
阳离子交换特性 cation exchange property
阳离子交换柱 cation exchange column
阳离子交换作用 cation exchange action
阳离子交替吸附作用 cation exchange and adsorption
阳离子净水剂 cationic clarificant
阳离子聚合 cationic polymerization
阳离子空位 cationic vacancy
阳离子沥青乳剂 cationic bituminous emulsion
阳离子沥青乳液 cationic bitumen emulsion
阳离子(沥青乳液砂土)稳定 cationic stabilization
阳离子络合物 cationic complex
阳离子密度 cationic density;positive ion density
阳离子膜电池 cationic membrane cell
阳离子黏[粘]合剂 cationic binder
阳离子黏[粘]着剂 < 改进沥青的黏[粘]着性用的 > cationic adhesion agent
阳离子排斥作用 cation exclusion
阳离子配位作用 cation coordination
阳离子平衡 cation balance
阳离子缺陷 cation defect
阳离子乳化剂 cationic emulsifier
阳离子乳化沥青 cationic emulsified bitumen
阳离子乳化液 cationic emulsion
阳离子乳剂 cation emulsion
阳离子乳胶 cationic latex
阳离子乳液 cationic emulsifier
阳离子润湿剂 cationic wetting agent
阳离子杀菌剂 cationic germicide
阳离子射线法 positive-ray method
阳离子渗入 cation penetration
阳离子试剂 cationic reagent
阳离子树脂 cationic resin;resin cation
阳离子树脂交换 resinous cation exchange
阳离子树脂交换柱 cation resin exchange column
阳离子酸 cation acid
阳离子添加剂 cationic additive
阳离子头 cationic head
阳离子物种 cationic species
阳离子吸附 cation adsorption
阳离子吸附饱和度 base-saturation percentage
阳离子吸收当量 cation equivalent
阳离子洗涤剂 cationic detergent
阳离子型表面 cathode surface active agent
阳离子型表面活化剂 cation surface active agent
阳离子型表面活性剂 cationic surface active agent;cationic surfactant
阳离子型表面活性剂胶团 cationic surface active agent micelle;cationic surfactant micelle
阳离子型化学发光单体 cationic chemiluminescent monomer
阳离子型化学发光聚合物 cationic chemiluminescent polymer
阳离子型交换树脂 cation exchange resin
阳离子型聚丙烯酰胺 cationic polyacrylamide;polycationic acrylamide
阳离子型聚电解质 cation polyelectrolyte
阳离子型去污剂 cationic detergent
阳离子型无机微粒 cationic inorganic microparticle
阳离子移变(现象) cationotropy
阳离子阴离子除盐器 cation anion demineralizer
阳离子阴离子除盐装置 cation anion demineralizer
阳离子云 positive ion cloud
阳离子皂 cationic soap
阳离子置换 cation replacement
阳离子中凝型 < 乳化沥青 > cationic medium setting
阳离子总和 total amount of cation;total cation
阳离子组成 cation composition
阳历 solar calendar
阳历年 solar year
阳历月 solar month
阳螺钉 male screw
阳螺纹 external thread;male screw;male thread;pin thread;positive thread
阳螺纹管接头 male fittings
阳螺纹配件 male fittings
阳螺旋 male screw;male thread;positive helicity
阳面 illuminated part of a surface
阳面冷藏 refrigeration with the sun
阳模 force piston;force plug;male force;male mo(u)ld;male plug;male section;male building form;mo(u)ld plunger;patrix;piston;plunger;plunger die;positive die;positive mo(u)ld;reproduction positive mo(u)ld;top force
阳模成型 positive forming
阳模冲子 punch
阳模雕刻机 punch cutting machine
阳模和阴模 patrix and matrix
阳模环 plunger retainer
阳模配合 male fittings
阳模托板 force plate
阳摩擦锥轮 male friction cone
阳坡 adret(to);sunny slope;tailo
阳起蓝闪绿帘绿泥片岩 actinolite glaucophane epidote chlorite schist
阳起石 actinolite [actynolite];stralite
阳起石板岩 actinolite slate
阳起石化 actinolitization
阳起石片岩 actinolite schist
阳起石岩 actinolite rock
阳气水分离器 three-phase separator
阳三角 < 配件砖 > external coconut piece
阳伞 parasol;sun shade
阳伞效应 umbrella effect
阳山碑材 Yangshan gravestone
阳栅 plate grid
阳射线 positive ray
阳生植物 heliad;heliophyte;oread;sun plant
阳生植物群落 heliophytia;sun plant community
阳十字接头 male cross
阳台 ante-solarium;balcony;gazebo [复 gazebo(e)s];guzebo;stoep;veranda(h);dallan < 有屋盖的波斯、印度建筑 >
阳台板 balcony plate;balcony slab
阳台表面涂层 balcony lining
阳台窗 balcony window
阳台瓷砖 balcony tile
阳台底面 balcony soffit;soffit
阳台隔墙 balcony partition (wall)
阳台横梁 pony girder
阳台栏杆 balcony balustrade;balcony rail(ing);balustrade;balconet(te) < 在建筑物正面的 >
阳台梁柱 balcony beam-column
阳台绿化 balcony greening
阳台门 balcony door;terrace door
阳台门金属附件 terrace door hardware
阳台门金属附件项目 terrace door hardware item
阳台门金属小五金 terrace door hardware
阳台门上配件 balcony door fittings
阳台门上小五金 balcony door hardware
阳台门装配附件 terrace door fitting
阳台门装置 terrace door furniture
阳台面层料 balcony facing
阳台女儿墙 balcony parapet
阳台排水 balcony drainage
阳台排水口 balcony outlet
阳台式窗栏 Balconet(te)
阳台式进水口 balcony-like intake
阳台饰面 balcony facing;balcony lining
阳台提升门 balcony lifting door
阳台通道 access balcony;terrace access
阳台外部通气管道 balcony exterior exit
阳台外廊遮篷 veranda(h) awning blind
阳台雨水口 balcony outlet;balcony rainwater outlet
阳台遮阳 terrace awning blind
阳台柱 balcony column
阳台砖 balcony tile
阳碳离子 caronium ion
阳条(瓷砖) evert corner
阳条子 positive stripe
阳凸缘 tongued flange
阳图重氮法 positive diazo process
阳纹花样 relief pattern
阳纹样本 relief pattern
阳像 positive image
阳像刻图 positive scribing
阳像制版 positive plate making
阳性的 positive
阳性反应 positive reaction
阳性胶乳 positive latex
阳性强化 positive reinforcement
阳性乳胶 positex
阳性生物 intolerant organism
阳性树 intolerant tree;light demanding tree
阳性条件反射 positive conditioned reflex
阳性预告值 positive predictive value
阳性植物 sun plant
阳肘节 male elbow
阳锥 male cone

杨

杨桉 poplar eucalyptus

杨干象甲 < 拉 > Crgptorhynchus lapathi
杨基干燥箱 Yankee drier [dryer]
杨家屯统【地】Yangchiatun series
杨柳 poplar and willow
杨木 cotton wood;poplar
杨十斑吉丁 < 拉 > melanophila decastigma
杨氏测微径计 Young's eriometer
杨氏阀动装置 Young valve gear
杨氏干涉条纹 Young's interference fringes
杨氏棱镜 Young's modulus;Young's prism
杨氏模量 modulus of elasticity;Young's modulus
杨氏模数 Young's modulus
杨氏挠曲模量 Young's modulus in flexure
杨氏全区域等积投影 Young's projection with total area true
杨氏摄谱仪 Young's spectrograph
杨氏实验 Young's experiment
杨氏双缝干涉仪 Young's two-slit interferometer
杨氏双缝实验 Young's double slit experiment
杨氏双缝隙干涉 Young's two-slit interference
杨氏双狭缝 Young's double slit
杨氏弹性模量 Young's modulus of elasticity
杨氏条纹 Young's fringes
杨氏系数 coefficient of elasticity;Young's coefficient
杨氏衍射理论 Young's diffraction theory
杨氏作图法 Young's construction
杨属 cotton wood;poplar;Populus < 拉 >
杨树 abele;poplar;populus
杨四星叶甲 < 拉 > Clytra laeviuscula
杨五星吉丁虫 < 拉 > Capnodis curiosa

疡

疡壳孢属 < 拉 > Dothichiza

洋

洋白蜡树 red ash

洋彩 foreign colo(u)r
洋葱头式圆顶 imperial cupola
洋葱形圆顶 < 教堂等的 > onion dome
洋底 fondo;ocean bottom;ocean floor;floor of the ocean
洋底变质作用 ocean-floor metamorphism
洋底测震术 ocean bottom seismometry
洋底测震仪下水深 water depth under ocean bottom seismometer
洋底沉积 fondothem
洋底的 suboceanic
洋底地壳结构 sea-bottom crust texture
洋底地图 map of ocean floor
洋底地形 fondoform;ocean bottom feature
洋底地形图 bathyorographic (al) map
洋底火山 oceanic volcano
洋底扩张 ocean-floor spread(ing)
洋底锰结核 manganese nodules on the ocean floor
洋底破裂带 oceanic-floor fracture zone
洋底壳 oceanic crust
洋底扫描声呐 ocean bottom scanning sonar
洋底山岭 mid-oceanic bottom
洋底玄武岩 ocean-floor basalt;sea-floor basalt
洋底岩层 fondothem
洋干漆 lac;shellac(k)
洋干漆片 shellac(k)
洋橄榄 common olive
洋红 aniline red;fuchsin(e)
洋红色 carmetta;carmine lake;carminette;carmin(e);magenta
洋红色花 magenta flower
洋化作用 oceanization
洋槐 acacia;black locust;false acacia;locust;robinia
洋灰搅拌机 grouting machine
洋脊 ocean(ic) ridge
洋脊背侧 back-ridge

洋脊地震带 oceanic ridge belt
洋际运河 interoceanic canal
洋梨 common pear
洋流 ocean(ic) current
洋流搬运有机质 organic matter by ocean current transportation
洋流搬运作用 transportation of ocean current
洋流剥蚀作用 ocean current denudation
洋流检测仪 ocean current monitor
洋流玫瑰图 current rose
洋流图 current diagram
洋流小浪日常 current rips
洋流要素 element of current
洋流运动特征 motion characteristic of ocean current
洋隆调查计划 rise project
洋隆拉斑玄武岩 rise tholeiite
洋面 ocean surface;offing
洋面反气旋 ocean anti-cyclone
洋内弧 intra-oceanic arc
洋内中水道 mid-oceanic channel
洋盆 oceanic basin
洋盆洋脊体系 basin-ridge system
洋器 export ware
洋壳动力学计划 oceanic crust dynamic project
洋区 oceanic province
洋区码 maritime country code;ocean region code
洋杉 red deal
洋松 Oregon pine;Puget Sound pine;Washington fir
洋铁罐 tin
洋为中用 make foreign things serve China
洋玉兰 bull bay
洋中 mid-ocean
洋中动力学研究计划 mid-oceanic dynamics experiment
洋中脊 mid-oceanic ridge
洋中裂谷 mid-oceanic rift
洋中隆 mid-oceanic rise
洋中群岛 mid-oceanic islands
洋中峡谷 mid-oceanic canyon

仰冲板块【地】 abduction plate

仰冲断层 upthrust
仰冲岩席 obducted sheet
仰冲造山带 obduction orogen belt
仰冲作用 obduction
仰打(桩) out-batter pile driving;rear raking pile driving; pile driving of backward inclination; positive batter;fore batter;forward batter
仰打的斜桩 forward raker
仰打(斜桩)导向架 out-batter pile-driving leaders
仰釜日晷 upward-looking bowl sundial
仰拱 arch invert; arch of elevation; inflected arch; invert; invert(ed) arch
仰拱闭合 invert closure
仰拱标高 inverted level
仰拱底板 inverted arch floor
仰拱高程 inverted level;invert elevation
仰拱焊接 inverted weld(ing)
仰拱基础 inverted arch foundation
仰拱模板 inverted arch form; invert form
仰拱铺砌的涵管 paved-invert pipe
仰拱砌块 invert segment
仰拱支撑 invert strut
仰灌 uphill grouting
仰光 <缅甸首都> Rangoon
仰焊缝 overhead welding seam
仰焊角焊缝 overhead fillet weld
仰焊(接) inverted weld(ing); overhead position weld(ing); overhead weld(ing); welding overhead position; twelve o' clock welding; overhead work
仰焊位置 overhead position
仰极【天文】 elevated pole
仰极高度 elevation of pole; polar altitude
仰角 altitude; altitude angle; angle of altitude; angle of elevation; angle of gradient; ascending vertical angle; degree of elevation; elevation angle; plus angle; positive altitude; positive angle of elevation; quadrant elevation; rake angle; vertical angle
仰角标度 elevation scale
仰角测位 elevation position-finding
仰角测位天线 elevation position-finding antenna
仰角差 elevation difference
仰角传动 elevation drive
仰角传动电动机 elevation drive motor
仰角传动机构 elevation drive gear
仰角的 elevation angle
仰角电平 elevation level
仰角电位计 elevation potentiometer
仰角定位 elevation setting
仰角反射特性曲线 elevation coverage diagram
仰角范围 elevation coverage
仰角方位 elevation bearing
仰角分辨能力 elevation resolution
仰角分压器 elevation potentiometer
仰角辐射图案 elevation radiation pattern
仰角跟踪 elevation tracking
仰角跟踪器 elevation tracker
仰角跟踪指示器 elevation tracking cursor
仰角焊 overhead fillet welding
仰角焊接 elevated position welding
仰角计 elevation meter
仰角精度 elevation accuracy
仰角可调天线 tiltable antenna
仰角控制电压 elevation control voltage
仰角缆包 elevation cable wrap
仰角偏转线圈 elevation yoke
仰角驱动 elevation drive
仰角驱动电动机 elevation drive motor
仰角上限 upper elevation limit
仰角视野特性曲线 elevation coverage diagram
仰角手轮 elevation handwheel
仰角伺服功率放大器 elevation servo power amplifier
仰角探测仪 elevation finder
仰角同步传动 elevation synchro gear
仰角位置测定 elevation position-finding
仰角位置雷达指示器 elevation position indicator
仰角误差 elevation angle error; elevation error
仰角误差信号 elevation error signal
仰角显示器 elevation indicator
仰角限制器 elevation stowing switch
仰角信息 elevation information
仰角旋转连接 elevation rotating joint
仰角引导单元 elevation guidance element
仰角指示器 elevation finder
仰角轴 elevation axis
仰角转换器 elevation commutator
仰角准确度 elevation accuracy
仰角自动同步传动机构 elevation selsyn drive gear
仰角自动同步传送机 altitude transmitting selsyn
仰角自动同步机 elevation selsyn
仰角自动同步机变压器 elevation selsyn transformer
仰接焊 uphead welding
仰掘机 up digging excavator
仰开滚轴平衡重 <桥梁的> rolling counterweight
仰开跨 draw spalling
仰开跨度 bascule span; draw span
仰开桥 drawbridge; leaf bridge; balance bridge <衡重式的>
仰开桥的活动桁架 bascule
仰开桥衡重臂 counterweight arm of bascule bridge
仰开桥平衡重竖井 tail pit
仰开桥桥墩 trunnion pier
仰开桥桥翼 channel arm of bascule bridge; overhanging arm of bascule bridge; river arm of bascule bridge
仰孔钻进 self-cleaning drilling; uphole boring; uphole drilling
仰孔作业 uphole work
仰面 overhead
仰坡 front slope; heading slope; overlaying slope
仰起端 rising end
仰倾高度 tipping height
仰韶彩陶 Yangshao painted pottery
仰韶文化陶器 Yangshao culture pottery
仰摄 high hat shot
仰视景观 upward landscape
仰视图 bottom view; reflected plan; up projection; upward view; worm's eye view
仰瓦屋面 concavely upward-facing tile roof
仰铣 conventional milling; up(-cut) milling
仰斜掘进 driving up the pitch
仰斜矿房 leading band
仰檐式(站台雨棚) up eaves
仰仪 scaphe
仰置虹吸管 inverted siphon [syphon]
仰置抛物线 inverted parabola

养蚕场 nursery

养蚕室 rearhouse of silkworms; silkworm nursery; silkworm (rearing) room
养橙温室 orangery
养虫室 insectarium; insectary
养分 nutrient
养分保留 retention of nutrient
养分吸收 absorption of nourishment
养蜂场 apiary; bee farm; bee house; bee yard
养蜂(业) beekeeping
养狗处 <筑有围篱的> dog run
养护 ag(e)ing; attendance; cure; curing <混凝土>; healing; maintenance
养护保温层 <混凝土的> curing blanket
养护薄膜 curing membrane
养护补给站 maintenance depot
养护不够 undercuring
养护步道 service walkway
养护步骤 curing procedure
养护材料 curing material
养护操作 maintenance manipulation
养护场 <混凝土件> curing area; curing yard
养护车 curing car; maintenance vehicle
养护成熟 mature
养护程序 maintenance process
养护池 curing tank
养护持续时间 curing duration; duration of curing
养护处治对策 maintenance treatment alternative
养护床 curing bed; curing kiln <混凝土蒸汽的>
养护催化剂 curing agent
养护定额【道】 maintenance quota
养护对策 maintenance counterproposal
养护法 curing process
养护方案比选 maintenance treatment alternative
养护方法 curing procedure; maintenance process; method of curing
养护防水纸 curing paper; waterproof paper for curing
养护费(用) cost of maintenance; cost of upkeep; maintenance cost; upkeep; cost maintenance; maintenance expenditures
养护覆盖 curing cover
养护覆盖层 curing blanket
养护覆盖物 <养护混凝土用> curing mat; curing overlay; curing blanket
养护覆盖织物 curing mat
养护覆盖纸 <混凝土> overlay paper
养护格栅 <混凝土块的> rack for curing concrete
养护工 maintainer
养护工程师 maintenance engineer
养护工队 maintenance force
养护工费用 maintenance labor cost
养护工况监测 service monitoring
养护工序准备时间 curing delay
养护工长 maintenance compound
养护工作队 maintenance crew; maintenance gang
养护工作中心 maintenance operation centre
养护规程 maintenance regulation
养护过程 <混凝土> curing process; process of curing
养护后强度 cured strength
养护后硬化的混凝土 matured concrete
养护混凝土 cure the concrete; curing concrete
养护混凝土的棉毯 cotton mat
养护混凝土覆盖物 curing blanket
养护混凝土路 cure concrete road
养护或硬化过程 <混凝土> seasoning
养护机 curing machine
养护机构 maintenance organization
养护机械 maintenance machinery
养护记录 maintenance record
养护剂 curing agent; monkey blood; curing compound <混凝土的>
养护剂撒布器 curing-compound sprayer
养护架 <混凝土块> curing rack
养护开支 maintenance expenditures
养护可靠性 maintenance reliability
养护坑 curing pit
养护空间 ambient
养护蜡涂料 <混凝土> wax concrete curing compound
养护力量 maintenance force
养护领班 maintenance compound
养护密封装置 curing seal
养护面积 maintenance area
养护棚 curing shed
养护平整机 maintainer scraper
养护铺盖 curing blanket
养护期 curing period; curing time; period of curing; curing delay <蒸汽养护混凝土的>
养护前期 precured period

养护情况 maintenance status
养护区域 maintenance area
养护人员 maintenance personnel
养护设备 curing fixture; curing rig; maintenance equipment; maintenance facility
养护湿度 curing humidity
养护时间 <混凝土> time of curing; cure time
养护时期 curing period
养护试验 curing test; maintenance test
养护室 curing chamber; curing room; fog room
养护室防潮 damp room dampproofing
养护室隔墙 damp room partition (wall)
养护室设施 damp room services
养护室照明设备 damp room light fixture; damp room luminaire fixture
养护室照明装置 damp room light fitting
养护手册 maintenance manual
养护熟化程序 curing procedure
养护水 water for curing; curing water
养护水柜 <混凝土养护用> curing tank
养护水箱 <混凝土> water curing tank
养护速率 <混凝土> curing rate; rate of cure; rate of curing
养护台座 curing bed
养护毯 <混凝土> curing blanket
养护条件 curing condition
养护维修开支 maintenance charges
养护维修预算 maintenance budget
养护温度 curing temperature
养护席 curing mat
养护系数 maintenance factor
养护箱 curing tank
养护(小)分队 maintenance division
养护性能 curing capacity
养护性维修 maintenance repair
养护性质 curing property; curing quality
养护巡路车 motor patrol
养护压办 curing pressure
养护延时 curing delay
养护窑 curing kiln
养护要求 maintenance requirement
养护液 <混凝土> curing solution
养护液喷洒器 curing compound sprayer
养护硬化后混凝土 matured concrete
养护用检测设备 maintenance test equipment
养护用棉毡 cotton mat
养护用毡 curing blanket
养护用纸 curing paper
养护优化决策 optimization decision for maintenance
养护指示器 service indicator
养护制度 curing schedule
养护质量 maintenance quality
养护质量综合值 composite value of maintenance quality; general rating of maintenance quality
养护周期 maintenance interval; maintenance period; curing cycle; curing period <混凝土>
养护状况指示仪表 service instrumentation
养鸡场 chicken farm; chicken run; hennery; pheasantry; poultry farm; fowlrun <英>
养老保险 endowment assurance; endowment insurance
养老保险单 endowment policy
养老保险制度 the old-age insurance system
养老基金 old-age pension fund; pension fund; superannuation fund
养老金 annuity; benefit; gratuity; gratuity fund; old-age benefit; old-age pension; pension; pension allowance; retirement allowance; retirement benefit
养老金的固定缴款计划 defined contribution plan; defined pension plan
养老金的固定受益计划 defined benefit plan
养老金基金准备 pension fund reserves
养老金及退休金税收优惠 superannuation and retired allowances
养老金计划 pension plan; pension schemes
养老金计划的全部受益 fully vested
养老金计划会计 pension plan accounting
养老金交费逐年计算法 annual premium costing
养老金缴款 pension contribution
养老金联营 pension pool
养老金领取权 pension rights
养老金领取者 pensioner
养老金受控筹资法 controlled funding; controlled funding pensions
养老金授予 pension vesting
养老金信托 pension trust
养老年金 endowment annuity
养老事业 old-age care
养老院 aged person's home; almshouse; asylum for the aged; geracomium [gerocomium]; home for the aged; old-age home; old folks' home
养路 maintain railways; maintain roads; maintenance of road; maintenance of way; road maintenance
养路班 section crew; section gang; section party; section team
养路班长 section boss; section foreman
养路班组 track maintenance gang
养路补充材料 maintenance patching material
养路操作 maintenance manipulation
养路测量 maintenance survey
养路道班 maintenance gang; road maintenance crew; track gang; track maintenance gang
养路道班人员 road maintenance crew
养路电话 track maintenance telephone
养路段 maintenance division; maintenance section
养路段长 maintenance superintendent
养路队 road maintenance crew; technical road patrol; track force
养路费 highway maintenance cost; road maintenance cost; road toll; toll of road maintenance
养路费员 road maintenance fee
养路分段长 roadmaster; track supervisor
养路副工长 assistant track foreman; subforeman
养路工 lengthsman; line (s) man; maintainer; maintainer scraper; track hand; track layer; trackman
养路工班 maintenance gang
养路工班长 <俚语> straw boss
养路工程 maintenance process
养路工程师 engineer of maintenance of way; engineer of way; maintenance engineer; maintenance of way engineer
养路工队 maintenance gang
养路工房 maintenance building
养路工工作队 track maintenance gang
养路工具 track maintenance tools
养路工区 maintenance division; maintenance section; track maintenance section; track section
养路工区房屋 section house
养路工区工人 trackman; trackwalker
养路工人 plate layer; road mender; section man
养路工手摇车 platelayer's troll(e)y
养路工长 maintenance foreman; roadmaster
养路工作 maintenance manipulation; maintenance work
养路工作队 maintenance; maintenance force; track maintenance force
养路工作手册 instruction book
养路管理系统 maintenance management system
养路机 maintainer; road maintainer
养路机械 maintenance machinery
养路机械和设备 maintenance machinery and equipment
养路机械化 mechanization of maintenance
养路机械修配厂 maintenance depot; repair depot
养路机械修配场 maintenance depot
养路机械作业平台 maintenance flatroof for machine; mechanized work-platform of maintenance
养路局长 roadmaster
养路领班 maintenance foreman
养路领工员 roadmaster; track supervisor
养路平路机 autopatrol grader
养路设备 maintenance equipment; maintenance facility; road maintenance equipment; service equipment
养路设备专用线 track maintenance equipment siding
养路手车 <包括马达车> section car
养路税 highway maintenance tax
养路维护设备 track maintenance equipment
养路延误工程 arrears of road maintenance
养路用平地机 patrol grader
养路用平路机 autopatrol grader
养路用汽油小车 pop car
养路主管工程师【铁】 maintenance superintendent
养路总工长 general track foreman
养路作业安全规定 limitation in maintenance operations
养马牧场 horse ranch
养牛场 cattle farm
养禽所 aviary
养生槽 curing tank
养生池养护 <混凝土预制构件的> curing by ponding
养生封面混合剂 <多用于混凝土路面> curing-sealing compound
养生混合物 curing compound
养生期间 curing period
养生室 curing room
养生温度 curing temperature
养生周期 curing period
养兔场 rabbit warren; warren
养畜场 cattle farm
养畜密度 stocking rate
养羊场地 sheep run
养羊地区 sheep country
养鱼 fish culture
养鱼场 farm pond; fish farm; fishing ground; live box; stew pond
养鱼池 farm pond; fish pond; nursery; pound
养鱼池塘 piscina
养鱼的 piscicultural
养鱼槛 crawl
养鱼设备 fish breeding installation
养鱼术 pisciculture
养鱼塘 fish pond
养鱼污水 breed fish sewage
养鱼业 pisciculture
养育 breed; nurture
养育薄膜 curing membrane
养育设施 nursing facility
养育员 breeder
养育院 house of refugee; poorhouse
养殖 aquiculture and poultry; cultivate
养殖场 hatchery; nursery
养殖泉 breed spring
养殖业 aquaculture; fish breeding and poultry raising; livestock breeding
养殖用水 aquaculture water
养殖鱼 cultured fished
养殖珍珠 cultured pearl
养珠场 pearl fishery
养猪场 hoggery; pig farm; piggery; pig unit
养猪场废水 piggery waste(water)

氧-18 漂移 oxygen-18 drift

氧饱和 oxygen saturation
氧饱和百分率 percentage of oxygen saturation
氧饱和度 oxygen saturation
氧饱和量 oxygen saturation capacity
氧饱和率 percentage of oxygen saturation; percentage of saturated oxygen
氧饱和值 oxygen saturation value
氧比 oxygen ratio
氧丙环 anprolene; ethylene oxide
氧丙酮 oxyacetone
氧丙烷切割 oxy-propane cutting
氧丙烷切割器 oxy-propane cutter
氧不足 oxygen deficit; oxygen lack
氧测井 oxygen log
氧测井曲线 oxygen log curve
氧除碳法 oxygen decarbonizing
氧传感器 oxygen sensor
氧垂曲线 dissolved oxygen sag curve; oxygen-sag curve
氧磁铁矿 oxymagnite
氧弹 oxygen bomb
氧弹老化 oxygen bomb ag(e)ing
氧弹老化试验 oxygen bomb ag(e)ing test
氧弹量热器 oxygen bomb calorimeter
氧弹试验 oxygen bomb test
氧氮硅石 sinoite
氧氮杂萘 benzoxazine
氧的 oxygenic
氧的测定 determination of oxygen
氧的纯度 oxygen purity
氧的供给 oxygen supply
氧的局部压力 partial pressure of oxygen
氧的欠缺 oxygen deficient
氧的顺磁性 paramagnetic property of oxygen
氧的吸取 uptake of oxygen
氧的吸收 oxygen absorption; uptake of oxygen

氧的消化 oxygen consumption
氧的质量摩尔浓度 molar concentration of oxygen
氧点 oxygen point
氧电极法 oxygen electrode method
氧毒性 oxygen toxicity
氧断 lancing

氧二丙腈 oxydipropionitrile
氧钒石 doloresite
氧分布 oxygen distribution
氧分压 oxygen partial pressure; partial pressure of oxygen
氧腐蚀 oxygen(-type) corrosion
氧割后退火 post-annealing
氧硅磷灰石 wilkeite
氧硅钛钠石 natisite
氧硅烷 oxosilane
氧过少 hypoxia
氧含量 oxygen content; oxygen lancing
氧焊 burnt weld; oxyacetylene welding
氧焊剂 oxyflux
氧焊接 oxygen weld(ing)
氧耗竭 oxygen depletion
氧耗量 oxygen consumption
氧合作用 oxygenation
氧黑云母 oxybiotite
氧弧焊 arc-oxygen welding; oxy-arc welding
氧弧切割 oxy-arc cutting
氧化 burning; oxidate; oxidize; oxygenate
氧化奥氏体 oxy-austenite
氧化斑痕 oxidized streak
氧化半丝质体 oxysemifusinite
氧化钡 baria; barium monoxide; barium oxide; baryta; calcined baryta; cawk
氧化钡铀 uranium barium oxide
氧化斑点 oxidation stain
氧化焙烧 oxidizing roast; roasting; roasting in air
氧化本领 oxidizing power
氧化苯乙烯 styrene oxide
氧化铋 bismuth oxide
氧化蓖麻油 blown castor oil
氧化变色 oxidation stain
氧化表层 <金属或矿物的> patina
氧化表面 oxide surface; oxidized surface
氧化丙烯 prop(yl)ene oxide
氧化材料 oxidation material
氧化槽 oxidation ditch; oxidation tank
氧化测定法 oxidimetry
氧化层 layer of oxide; oxidation layer; oxidation zone; oxide; oxide layer; oxidizing zone; zone of oxidation
氧化层窗孔 oxide window
氧化层厚度测试 measurement of oxide layer thickness
氧化产物 oxidation product
氧化厂 oxygen installation
氧化沉淀物 oxide precipitate; oxidized sediment
氧化成分 oxidizing constituent
氧化(程)度 degree of oxidation; oxidation lagoon
氧化池 oxidation pond
氧化处理 oxidation treatment; oxide treatment
氧化氚 tritium oxide
氧化促进剂 oxidation accelerator; oxidation promoter
氧化催化 oxidation catalysis
氧化催化反应器 oxidizing catalytic reactor
氧化催化过滤介质 oxidize catalytic filter media

氧化催化剂 oxycatalyst
氧化催化滤料 oxide catalytic filter media
氧化催化燃烧装置 oxide catalytic combustion equipment
氧化带 oxidation zone; oxidized zone; oxidizing zone; zone of oxidation
氧化带煤样 coal sample for determination of oxidized zone
氧化氮 nitric oxide; nitrogen oxide; oxide of nitrogen
氧化氮检定器 nitric oxide detector
氧化氮浓度 nitrogen oxide concentration
氧化当量 oxidation equivalent
氧化氘 deuterium oxide; heavy water
氧化的 oxidative; oxidized; oxidizing
氧化的涂层 oxide coating
氧化的抑制 inhibition of oxidation
氧化低铜 copper sandoz
氧化低亚铜 cuprous oxide
氧化地沥青 blown asphalt; oxide asphalt; oxidized asphalt; oxidized bitumen; oxygenated asphalt; oxidized asphaltic bitumen
氧化碘 iodine oxide
氧化电极 oxidizing electrode
氧化电位 oxidation potential; oxidizing potential
氧化淀粉 oxidation starch; oxidized starch
氧化丁烯 butylene oxide; epoxy butane
氧化顶吹转炉炼钢法 Linz-Donawitz process
氧化铥 thulia; thulium oxide
氧化动力学 oxidation kinetics
氧化镀层 oxide coating
氧化煅烧 oxidizing roasting
氧化二苯锡 diphenyl tinoxide
氧化二丁锡 dibutyl tinoxide
氧化二甲基锡 dimethyl tin oxide
氧化二烷基锡 dialkyl tin oxide
氧化发光 oxyluminescence
氧化法 hydrocarbonylation; oxidation process; oxo-process
氧化法合成醇类 oxo alcohol
氧化钒 karelianite; vanadium oxide
氧化反应 oxidation reaction; oxidizing reaction
氧化方法 oxidation process
氧化防腐剂 oxidizing anti-corrosion agent
氧化防腐膜 anti-corrosive coat(ing)
氧化防护膜 anti-corrosive coating with oxides
氧化废水 oxidized wastewater
氧化分解 oxidative decomposition; oxygenolysis
氧化釜 blowing still
氧化腐蚀 oxide etch
氧化富集带 enriched oxidation zone
氧化覆盖层 oxide coating
氧化钙 burnt lime; calcia; calcium oxide; caustic lime; dehydrated lime; lime; quicklime
氧化钙玻璃 lime glass
氧化钙含量 content of calcium oxide
氧化钙烧伤 calcium oxide burn
氧化钙熟料 calcia clinker; lime clinker
氧化干性油 oxyn
氧化干燥 drying by oxidation; oxidation drying
氧化高钴 cobaltic oxide
氧化锆 zirconia; zirconium oxide
氧化锆瓷 zirconia whiteware
氧化锆耐火材料 zirconia refractory
氧化锆搪瓷 zirconia enamel
氧化锆陶瓷 zirconia ceramics
氧化锆砖 zirconia brick
氧化镉 cadmium oxide
氧化镉光生伏打电池 cadmium oxide photovoltaic cell
氧化铬 chrome oxide; chromic oxide; oxide of chromium
氧化铬绿 chrome oxide green; chromium oxide green
氧化铬绿颜料 chromium green
氧化铬研磨膏 chromium oxide grinding grease
氧化铬颜料 chromic oxide pigment; chromium oxide pigment
氧化工艺 oxidation ditch process; oxidation technology; oxidizing process
氧化汞 mercuric oxide; mercury oxide; red precipitate; yellow precipitate
氧化汞法 mercuric oxide method
氧化沟 <净化下水> oxidation ditch; ditch oxidation
氧化沟处理 oxidation channel treatment
氧化钴 cobalt blue; cobaltous oxide; king's blue; powder blue
氧化固体 oxide solid
氧化管 oxidation tube
氧化硅 monox; silica; silicon oxide
氧化硅胶 silica gel
氧化硅胶型 silica-gel type
氧化硅四面体 silicon-oxygen tetrahedron
氧化过程 oxidation process
氧化过滤器 oxidizing filter
氧化铪 hafnium oxide
氧化合成法 oxo-process
氧化合物 oxygen compound
氧化湖 oxidation lagoon
氧化还原 oxido-reduction; reduction oxidation
氧化还原半电池反应 oxidation reduction half-cell reaction; redox half-cell reaction
氧化还原泵 redox pump
氧化还原常数 redox equilibrium constant
氧化还原沉淀 redox precipitation
氧化还原成岩变化阶段 redoxomorphic stage
氧化还原催化剂 oxidation reduction catalyst; redox catalyst
氧化还原催化剂体系 redox-catalyst system
氧化还原带 redox zone
氧化还原当量 oxido-reduction equivalent
氧化还原滴定 oxidation-reduction titration; redox titration
氧化还原滴定法 oxidimetry; reduction method; redox method; redox process
氧化还原滴定剂 redox titrant
氧化还原电池 oxidation-reduction cell; redox cell
氧化还原电对 oxidation-reduction pair
氧化还原电极 oxidation-reduction electrode
氧化还原电势 oxidation-reduction potential; redox potential
氧化还原电势突变 oxidation-reduction discontinuity; redox potential discontinuity
氧化还原电位 oxidation-reduction potential; redox potential; reduction-oxidation potential
氧化还原电位法 redox potentiometry
氧化还原电位检测器 redox potential detector
氧化还原对 redox couple; redox pairs
氧化还原法 oxidation-reduction method
氧化还原反应 oxidation-reduction reaction; redox reaction
氧化还原反应缓冲剂 poiser
氧化还原分析 oxidation-reduction analysis; redox analysis
氧化还原分析器 redox analyser [analyzer]
氧化还原过程 oxidation-reduction process; redox process
氧化还原(过程)化学 redox chemistry
氧化还原缓冲 redox buffering
氧化还原活性电对 redox-active couple
氧化还原活性介质 redox-active media
氧化还原活性乳化聚合 redox-activated emulsion polymerization
氧化还原活性物种 redox-active species
氧化还原活性系统 redox-active system
氧化还原间接脱色 redox-mediated decolo(u)rization
氧化还原阶段 redoxomorphic stage
氧化还原介体 redox mediator
氧化还原聚合 redox polymerization; reduction-oxidation polymerization
氧化还原聚合物 redox polymer
氧化还原离子交换 redox ion exchange
氧化还原离子交换树脂 redox ion exchanger
氧化还原理论 theory of redox
氧化还原率 rate of redox
氧化还原能力 oxidizing/reducing power
氧化还原偶 redox couple
氧化还原平衡 oxidation-reduction equilibrium; redox equilibrium
氧化还原热化学 redox thermochemistry
氧化还原势控制 redox control (in bleaching)
氧化还原说 redox-hypothesis
氧化还原速率常数 redox rate constant
氧化还原态 redox state
氧化还原体系 redox system
氧化还原条件 conditions of oxidation and reduction; redox condition
氧化还原系列 redox series
氧化还原系统 oxidation-reduction system; redox system
氧化还原(型)树脂 redox resin; redoxite
氧化还原引发剂 redox initiator
氧化还原引发聚合 redox initiate polymerization
氧化还原引发作用 redox initiation
氧化还原指示剂 oxidation-reduction indicator; oxido-indicator; redox indicator
氧化还原作用 oxidation reduction; redox
氧化环境 oxidizing environment
氧化混凝工艺 oxidation-coagulation process
氧化剂 oxidant; oxidation agent; oxide agent; oxidizer; oxidizing agent; oxygenant; oxidizing constituent <火箭燃料的>
氧化剂泵 oxidant pump
氧化剂泵叶轮 oxidize pump impeller; oxidizer pump
氧化剂舱 oxidizer compartment
氧化剂分配器 oxidizer distributor
氧化剂和有机过氧化物 oxide agents and organic over-oxide articles
氧化剂记录器 oxidant recorder
氧化剂模型 oxidant model
氧化剂浓度 oxidant concentration
氧化剂喷嘴 oxidize nozzle
氧化剂烟雾 oxidant smog
氧化夹杂物 dross inclusion
氧化镓 gallium oxide
氧化钾 burnt potash; kali; potassium oxide
氧化钾含量 content of potassium ox-

ide
氧化降解作用 oxidative degradation
氧化阶 oxidation degree
氧化金 auric oxide;gold oxide
氧化金属 oxidized metal
氧化金属电极 metallio-oxide electrode
氧化金属涂料 oxide paint
氧化聚合 oxidation polymer
氧化聚合油 blown stand oil
氧化聚合作用 oxidation polymerization
氧化聚明胶 oxypolygelatin
氧化钪 scandium oxide
氧化烤色 oxidizing firing
氧化矿带 oxidized ore zone
氧化矿石 oxidized ore
氧化矿物油 boleg oil
氧化铼 rhenium oxide
氧化镧 lanthanum oxide
氧化老化 oxidative ag(e)ing
氧化铑 rhodium oxide
氧化离子 oxide ion
氧化锂 lithia;lithium oxide
氧化锂瓷 lithia porcelain
氧化锂陶瓷 lithia ceramics
氧化力 oxidizing force;oxidizing power
氧化沥青 blow asphalt;blown asphaltic bitumen; blown bitumen; oxidized asphalt
氧化栗红 maroon oxide
氧化炼钢法 oxygen-steel process
氧化裂断 oxidative scission
氧化磷酸化作用 oxidative phosphorylation
氧化硫<大气污染物,主要为二氧化硫> sulfur [sulphur] oxide
氧化硫硫杆菌 thiobacillus thiooxidant
氧化炉 oxidation oven;oxidized still;oxidizing furnace
氧化镥 lutecia
氧化铝 alumina;alumin(i)um oxide;oxide of alumin(i)um;alundum
氧化铝板<吸声、隔音的> alumite tile;alutile
氧化铝厂 alumina plant;alumina producer
氧化铝瓷 alumina porcelain
氧化铝单晶 alumina single crystal;single crystal alumina
氧化铝电解槽 alumina electrolysis bath
氧化铝粉 alumina powder
氧化铝干燥剂 activated alumina
氧化铝干燥器 alumina drier [dryer]
氧化铝基催化剂 alumina base catalyst
氧化铝基片 alumina substrate
氧化铝基切削工具 alumina-based cutting tool
氧化铝结疤 alumina scale
氧化铝空心球砖 alumina bubble brick
氧化铝膜 pellumina
氧化铝膜处理法 alumite;alunite
氧化铝磨料 adamite;corundum
氧化铝耐火材料 alumina refractory
氧化铝耐火织物 alumina fiber [fibre]
氧化铝凝胶 alumina gel
氧化铝溶胶 alumina sol
氧化铝砂布 alumin(i)um oxide cloth
氧化铝砂轮 alundum wheel
氧化铝砂纸 alumin(i)um oxide paper
氧化铝石墨质耐火材料 alumina-graphite refractory
氧化铝水合物 hydrated alumina
氧化铝水化物 alumina hydrate
氧化铝水泥 alundum cement;calcium aluminate cement
氧化铝水凝胶 alumina hydrogel
氧化铝碳化硅碳质耐火材料 alumina-silicon carbide-carbon refractory
氧化铝陶瓷 alumina ceramics;alumina whiteware
氧化铝陶瓷刀具 alumina tool;alumin(i)um oxide tool
氧化铝陶瓷刀片 alumin(i)um oxide ceramic insert
氧化铝陶瓷膜 alumin(i)um oxide ceramic membrane
氧化铝陶瓷镶装刀片 alumin(i)um oxide ceramic insert chip
氧化铝涂层硬质合金 Al2O3-coated cemented carbide
氧化铝吸附 alumin(i)um oxide adsorption
氧化铝纤维 alumina fiber [fibre]
氧化铝絮凝 alumin(i)um oxide flocculation
氧化铝研磨 alumina lap
氧化铝研磨剂 alumin(i)um oxide abrasive
氧化铝载体 alumina supporter
氧化铝质耐火材料 alumina refractory
氧化铝质球磨用球 alumina balls
氧化铝-重晶石粉 alumina blanc fixe
氧化铝柱 alumina column
氧化铝砖 alumina brick
氧化率 oxidation index;oxidation rate
氧化氯 chlorine monoxide
氧化麻点 scale mark
氧化煤 oxidized coal
氧化煤柏油脂 blown coal-tar pitch;blown pitch
氧化煤焦油 oxidized coal-tar
氧化煤沥青 oxidized coal-tar
氧化镁 bitter earth;magnesia;magnesium oxide
氧化镁玻璃 magnesia glass
氧化镁瓷 magnesia ceramics
氧化镁的膨胀作用 action of magnesia;expansion tendency due to magnesia
氧化镁法烟气脱硫 flue gas desulfurization with magnesium oxide
氧化镁坩埚 magnesia crucible
氧化镁硅胶 magnesia-silica gel
氧化镁含量 content of magnesium oxide
氧化镁结合剂砂轮 oxychloride (bonded) wheel
氧化镁绝缘层<用于管线> magnesia insulation
氧化镁绝缘金属铠装电缆 magnesia-insulated metal sheathed wire
氧化镁膨胀 expansion due to magnesia;magnesia expansion
氧化镁膨胀水泥 magnesia type expansive cement
氧化镁铺面 magnesite flooring
氧化镁铺面砖 magnesite flooring tile
氧化镁清洗过程 magnesium oxide process
氧化镁石灰砖 magnesian lime brick
氧化镁试验<石灰的> oxidized magnesium test (of lime)
氧化镁水泥 magnesium oxide cement
氧化镁水泥混凝土 magnesium oxide cement concrete
氧化镁陶瓷 magnesia ceramics
氧化镁须晶 magnesia whisker
氧化镁悬浮液 magnesia magma
氧化镁引起的不安定性 magnesia unsoundness
氧化锰含量 content of manganese oxide
氧化膜 film of oxide;fire-coat;layer of oxidem;oxidation film;oxide coating;oxide film;oxide layer;oxide membrane; oxidic film fire coat;tarnish;tarnish film
氧化膜避雷器 oxide-film arrester [arrestor]
氧化膜处理 oxide-film treatment
氧化膜电容器 oxide-film capacitor
氧化膜发射体 oxide-film emitter
氧化膜铝线 alumite wire;anodised wire
氧化膜密度 film density
氧化膜色 heat tint;hot-tinting
氧化膜细孔 pore of oxide film
氧化钼 molybdena;molybdenum oxide
氧化钼矿石 oxidized Mo ore
氧化钼纤维 molybdenum oxide fibre
氧化钠 natron;sodium oxide
氧化钠含量 content of sodium oxide
氧化能力 oxidation capacity;oxidizability; oxidizing ability; oxidizing power;oxygenation capacity
氧化铌 niobium oxide
氧化镍 nickelous oxide
氧化镍矿石 oxidized nickel ore
氧化钕 neodymia
氧化偶氮化合物 azoxy compound
氧化偶氮染料 azoxy dye
氧化偶合絮凝技术 oxidation couple flocculation technology
氧化偶联作用 oxidation coupling
氧化硼 boric oxide;boron oxide
氧化硼的异常性 boric oxide anomaly
氧化硼铝 boraxal
氧化铍 berillia [beryllia]; beryllium oxide
氧化铍衬底 beryllia substrate
氧化铍瓷 beryllia porcelain
氧化铍阀座 beryllia valve seat
氧化铍耐火材料 beryllia refractory
氧化铍耐火材料制品 beryllia refractory product
氧化铍实验反应堆 experimental beryllium oxide reactor
氧化铍陶瓷 berillia[beryllia] ceramics;beryllium-oxide ceramic
氧化皮 cinder; crust of iron; dross;fire-coat;forge cinder;forge scale;furnace cinder; mill scale; oxide;oxide skin;roll scale;scale cinder
氧化皮沉淀池 scale settling tank;scale sump
氧化皮冲除箱 descaling box
氧化皮打磨机 descaling machine
氧化皮的形成 scale formation
氧化皮沟道排除系统 scale sluice system
氧化皮和聚 balling up
氧化皮壳 scale jacket
氧化皮排除系统 scale-handling system;scale removal system
氧化皮清除器 scale breaker;scale descaler
氧化皮清理机 scale breaker
氧化皮收集槽 scale trap
氧化皮损耗 scale loss
氧化皮通电去除法 dynamisator process
氧化皮黏[粘]辊 roll banding
氧化平衡仪 oxidation stabilometer
氧化瓶 gas bomb
氧化期 oxidation period;oxidizing period
氧化气氛 oxidizing atmosphere
氧化气瓶 oxygen bottle;oxygen cylinder
氧化铅 lead oxide; litharge; massicot;plumbous oxide;yellow lead (oxide)
氧化铅橙 mineral orange
氧化铅粉 lead oxide powder
氧化铅(光导)摄像管 plumbicon
氧化铅光电导层 lead oxide photoconductive layer
氧化铅光电导管 lead oxide photoconductive tube;lead oxide vidicon
氧化铅矿石 oxided lead ore
氧化铅视像管 plumbicon
氧化铅锌 lead zinc oxide
氧化铅锌矿石 oxided Pb-Zn ore
氧化强化剂 prooxygenic agent
氧化切割 oxygen-acetylene cutting
氧化侵蚀 oxidative attack
氧化青铜 oxidized bronze
氧化青铜饰面 oxidized bronze finish
氧化区 oxidation zone; oxidizing zone;zone of oxidation
氧化炔焊 oxyacetylene welding
氧化炔焊枪 oxyacetylene blowpipe;oxyacetylene torch
氧化染料 oxidation colo(u)r
氧化热 heat of oxidation
氧化溶液 oxidizing solution
氧化铷 rubidium oxide
氧化润滑剂 oxidized lubricant
氧化三丁基锡 tributyl tin oxide
氧化色 air tint;oxidation tint
氧化色斑 oxidation stain
氧化铯 cesium oxide
氧化钐 samarium oxide
氧化砷 tertiary arsine oxide
氧化石蜡 oxidized paraffin
氧化石油沥青 blown petroleum;catalytic asphalt;oxidized petroleum asphalt;byerlile [byerlyle] <贝尔勒法生产的>
氧化势 oxidation potential
氧化试验 oxidation test
氧化室 oxidizing chamber
氧化铈 ceria;cerium oxide
氧化铈夹杂物 snotter
氧化铈抛光粉 cerium rouge
氧化树脂 oxidized resin;resene
氧化树脂体 oxyresinite
氧化数 oxidation number
氧化水解(反应) oxydrolysis
氧化丝质体 degradofusinite
氧化锶 strontia;strontium oxide
氧化松节油 oxidized turpentine oil
氧化松香 oxidized rosin
氧化速率 oxidation rate;rate of oxidation
氧化损耗 oxidation loss
氧化损失 oxidational losses
氧化铊 thallium oxide
氧化塔 oxidation column; oxidation tower;oxidizing tower
氧化态 oxidation number; oxidation state
氧化钛 titanium oxide
氧化钛含量 content of titanium oxide
氧化钛陶瓷 titania ceramics;titanium oxide porcelain
氧化钛铁 iron titanium oxide
氧化钛型电焊条 high titanic type electrode
氧化钛砖 titania brick
氧化碳中毒 carbon monoxide poisoning
氧化塘 aerated lagoon;aerated pond;aeration lagoon;aeration pond;oxidation pond
氧化塘出(流)水 oxidation pond effluent
氧化塘法 lagoon process; oxidation pond process
氧化塘酚污染 phenol lagoon pollution
氧化塘工艺 lagoon process;oxidation pond process
氧化陶瓷 oxide ceramics
氧化特性 oxidation characteristic
氧化铽 terbium oxide
氧化锑 antimonic oxide;antimony ox-

ide;flower of antimony
氧化锑矿石 oxidized Sb ore
氧化体 oxidosome
氧化条件 oxidizing condition
氧化铁 ferric iron oxide;ferric oxide; iron oxide; iron sesquioxide; ironstone; mineral purple; oxide of iron;red stone
氧化铁薄膜 sull
氧化铁残渣 iron oxide residue
氧化铁橙色 Mars orange
氧化铁法 iron oxide process
氧化铁法控制硫化氢 control of hydrogen sulfide ferric oxide
氧化铁粉 ferric oxide powder
氧化铁和铝反应热焊接 iron oxide and alumin(i)um reaction thermit welding
氧化铁黑 black iron oxide; black rouge; iron oxide black; synthetic(al) magnetite
氧化铁红 ferric oxide rouge;red iron oxide
氧化铁红颜料 red iron oxide pigment;red oxide of iron pigment
氧化铁黄 yellow iron oxide
氧化铁黄颜料 iron oxide yellow pigment
氧化铁胶结物 iron oxide cement
氧化铁矿 iron oxide ore
氧化铁粒子 iron oxide particle
氧化铁硫杆菌 Thiobacillus ferro-oxidant
氧化铁煤气净化器 iron oxide purifier
氧化铁磨料 crocus abrasive
氧化铁抛光粉 crocus abrasive
氧化铁皮 iron scale;sinter
氧化铁皮斑点 kisser
氧化铁皮印痕 scale mark
氧化铁铅 lead iron oxide
氧化铁素体 oxyferrite
氧化铁铁芯 oxide core
氧化铁涂层 iron oxide coating
氧化铁涂料 iron oxide paint
氧化铁吸附剂 iron oxide adsorbent
氧化铁(细)砂布 crocus cloth
氧化铁锌 zinc iron oxide
氧化铁型 iron oxide type
氧化铁型电焊条 high iron oxide type electrode
氧化铁型焊条 iron oxide type electrode
氧化铁型膨胀水泥 ferric type expansive cement
氧化铁研磨粉 crocus
氧化铁颜料 iron oxide pigment
氧化铁氧化铝总含量 ferric and alumin(i)um oxide content
氧化铁紫 iron oxide purple
氧化铁棕 brown iron oxide;iron oxide brown
氧化同化作用 oxidative assimilation
氧化铜 aerugo;cupric oxide
氧化铜防护膜 copper oxide barrier film
氧化铜光电池 cupron cell
氧化铜光伏元件 copper oxide photovoltaic cell
氧化铜解调器 copper oxide demodulator
氧化铜矿石 oxidized copper ore
氧化铜流化床法 fluidized bed copper oxide process
氧化铜整流器 copper oxide rectifier; cupron cell;oxide rectifier;rector
氧化铜整流式电表 copper oxide rectifier instrument
氧化土 oxisol
氧化土沉积带 <近地面> ferreto zone

氧化钍 thoria;thorium oxide
氧化钍耐火材料 thoria refractory
氧化钍耐火材料产品 thoria refractory product
氧化钍凝胶 thoria gel
氧化钍凝胶化 thoria gelatination
氧化钍抛光粉 Barnesite polishing powder
氧化钍陶瓷 thoria ceramics
氧化钍微粒 thoria microsphere
氧化钍阴极 thoria cathode
氧化脱臭 oxidation deodorizing
氧化脱硫 oxidation desulfurization
氧化脱硫塔 oxidation desulfation [desulphation] tower
氧化脱色 oxidative decolo(u)rization
氧化脱碳过程 oreing
氧化微晶蜡 oxidized microcrystalline wax
氧化微粒体 degradomicrinite
氧化温度 oxidation temperature;oxidizing temperature
氧化稳定性 oxidizing stability
氧化稳定性试验 oxidation stability test
氧化污泥 oxidized sludge
氧化污泥法 oxidized sludge process
氧化污染 oxidation stain
氧化污水处理厂 oxidation sewage plant
氧化污水处理法 oxidation sewage process
氧化钨 tungsten oxide; tungstic oxide;wolframium oxide
氧化无机物 oxide mineral
氧化物 oxidant; oxidate; oxide; oxide compound;oxidizing material
氧化物斑点 oxide patch;oxide spot
氧化物半导体 oxide semiconductor
氧化物玻璃 oxide glass
氧化物薄膜 oxidation film;oxide film
氧化物层 oxide skin
氧化物沉积 oxidate
氧化物沉积侵蚀 oxide deposit attack
氧化物纯化法 oxide passivation
氧化物磁石 oxide magnet;oxide magnetic compact
氧化物磁铁 ceramic magnet;oxide magnet
氧化物磁芯 oxide core
氧化物灯丝 oxide filament
氧化物电阻 oxide resistor
氧化物粉末 oxide dust
氧化物覆膜法 oxide coating
氧化物隔离 oxide isolation
氧化物核燃料 oxide nuclear fuel
氧化物基金属陶瓷 oxide base cermet
氧化物基体 oxide matrix
氧化物间强相互作用 strong oxide-oxide interaction
氧化物结构 oxide structure
氧化物结块 oxide accretion
氧化物颗粒 oxide particle
氧化物刻蚀 oxide etching
氧化物矿石 oxide ore
氧化物矿物 oxide mineral
氧化物离解化学干扰 oxide-dissociation chemical interference
氧化物磷光体 oxide phosphor
氧化物锰矿石 Mn oxide ore
氧化物弥散强化 oxide dispersion strengthening
氧化物膜保护法 protection by oxide films
氧化物黏[粘]结的碳化硅 oxide-bonded silicon carbide
氧化物燃料堆 oxide fuel reactor
氧化物色料 oxide colo(u)r
氧化物烧结阴极 sintered oxide-coated cathode
氧化物陶瓷 oxide ceramics
氧化物(陶瓷)刀具 oxide tool
氧化物陶瓷激光器 oxide-ceramic laser
氧化物陶瓷型壳 oxide-ceramic shell mo(u)ld
氧化物添加剂 oxide addition
氧化物涂覆阴极 oxide-coated cathode
氧化物涂料 oxide coating;oxide paint
氧化物物质 oxidizing substance
氧化物物种 oxidized species
氧化物相铁建造【地】 oxide-facies iron formation
氧化物烟气 oxide fume
氧化物颜料 oxide colo(u)r
氧化物阴极 oxide(-coated) cathode; Wehnelt cathode
氧化物优先浮选 selective oxide flo(a)tation
氧化物油漆 oxide paint
氧化吸附剂 oxide adsorbent
氧化吸收法脱氮 control of NOx by absorption oxidation
氧化烯类聚 alkylene oxide polymer
氧化锡 putty powder;stannic oxide; tin ash;tin oxide
氧化锡薄膜电阻 nesacoat
氧化锡电极砖 tin oxide electrode brick
氧化锡粉 putty powder
氧化锡耐火材料 tin oxide refractory
氧化锡抛光粉 putty
氧化锡陶瓷 tin oxide ceramics
氧化锡氧化铅混合釉 <英国锡釉陶器用> calcine
氧化系底物 oxidogenic substrate
氧化系数 coefficient of oxidation
氧化系统 oxidative system
氧化纤维素 oxidized cellulose; oxycel;oxycellulose
氧化氙 xenon oxide
氧化橡胶 oxidized rubber
氧化消毒 oxidation sterilizing
氧化效应 oxidation effect
氧化锌 Chinese white; flowers of zinc;snow white;sterlingite;white zinc;zinc oxide;zinc white
氧化锌焙砂 zinc oxide calcine
氧化锌槽 zinc oxide bath
氧化锌尘雾 zinc oxide fume
氧化锌法脱硫 zinc oxide process desulfurization
氧化锌甘油(剂) glycerinum zinc oxide
氧化锌接合剂 zinc oxide cement
氧化锌晶体 zincite crystal
氧化锌矿 zinc bloom
氧化锌矿石 oxidized zinc ore
氧化锌明胶 zinc gelatin
氧化锌黏[粘]性试验 zinc oxide viscosity test
氧化锌软膏 ointment of zinc oxide; zinc oxide ointment
氧化锌碳酸钠烧结 sintering with mixture of ZnO and Na_2CO_3
氧化锌陶瓷 zinc oxide ceramics
氧化锌颜料 zinc oxide pigment
氧化锌油漆 zinc oxide paint
氧化锌-杂酚油防腐法 Card process
氧化锌增稠试验值 zinc oxide thickening value
氧化锌纸 zinc oxide paper
氧化锌中毒 zinc oxide poisoning
氧化型醇酸树脂 oxidizing type alkyd resin
氧化型邻苯二甲酸醇酸树脂 oxidizing phthalic alkyd resin
氧化型杀菌剂 oxidation-typed bactericide

氧化性 oxidizability
氧化性火焰 oxidizing flame
氧化性能 oxidation susceptibility
氧化性酸 oxidizing acid
氧化性涂料 oxide coating
氧化性质 oxidizing property
氧化絮凝工艺 oxidation-flocculation process
氧化亚氮 nitrous oxide
氧化亚铬 chromous oxide
氧化亚汞 mercurous oxide
氧化亚汞棱镜起偏(振)器 mercurous chloride prism polarizer
氧化亚麻(仁)油 linoxyn; oxidized linseed oil
氧化亚麻籽油 linoxyn
氧化亚铁 ferrous oxide; iron protoxide
氧化亚铁含量 content of ferrous oxide
氧化亚铜 copper-protoxide; copper sub-oxide; cuprous oxide; red copper oxide
氧化亚铜光电池 cuprous oxide photocell;photox cell
氧化亚铜整流器 cuprous oxide rectifier
氧化亚物 protoxide
氧化亚锡 stannous oxide
氧化延时曝气法 oxidation-ditch extended aeration process
氧化掩模 oxidation mask
氧化焰 oxidation flame; oxidizing flame;oxydizing flame;short fire
氧化页岩 oxidized shale
氧化铱 iridium oxide
氧化铱电阻器 iridium oxide resistor
氧化乙炔 oxyacetylene
氧化乙炔割断 oxyacetylene cutting
氧化乙炔割断器 oxyacetylene cutter
氧化乙炔喷焊机 oxyacetylene blowpipe
氧化乙炔切割 oxyacetylene cutting
氧化乙炔切割器 oxyacetylene cutter
氧化乙炔焰 oxyacetylene flame
氧化乙炔焰焊接 oxyacetylene welding
氧化乙烯 ethylene oxide
氧化乙酰 acetic oxide
氧化钇 yttrium oxide
氧化抑制剂 oxidation inhibitor;oxidation retarder
氧化抑制润滑脂 oxidation-inhibited grease
氧化镱 ytterbin
氧化铟 indium oxide
氧化银 argentous oxide;silver oxide
氧化银电池 silver oxide battery;silver oxide cell
氧化引起的应力 oxide-induced stress
氧化硬煤沥青 blown coal-tar pitch
氧化油 blown oil; oxidized oil; oxygenated oil
氧化铀 urania;uranium oxide
氧化铀坩埚 urania crucible
氧化有机物 oxygenated organics
氧化鱼油 degras
氧化预处理 oxidation pretreatment
氧化预防 oxidation prevention
氧化运行特性 oxidation-ditch performance
氧化再循环沟槽 oxidizing recirculation channel
氧化渣油 blown residual oil
氧化障 oxidizing barrier
氧化锗 germanic oxide
氧化针孔 oxidation pit
氧化(正)铜 copper oxide
氧化脂肪油 blown fatty oil
氧化值 oxidation number; oxidation value

氧化砖 oxidation brick
氧化转炉 oxygen converter
氧化装置 oxidizing installation
氧化状态 oxidation state;state of oxidation
氧化着色 heat tinting
氧化紫红 maroon oxide
氧化组沉积 oxidate
氧化作用 aerobic oxidation; de-electronation; oxidizing action; oxidation;oxidization [oxidisation]
氧化作用带 oxidation zone
氧还电位稳定器 redoxostat
氧缓冲 oxygen buffer
氧活度 oxygen activity
氧活化污泥系统 oxygen-activated sludge system
氧基醇 keto-alcohol
氧基化合物 oxo-compound
氧结合力 oxygen combining power
氧解离曲线 oxygen dissociation curve
氧炬切割 lance cutting
氧亏 oxygen debt;oxygen deficiency; oxygen deficit
氧亏耗 oxygen depletion
氧亏率 oxygen deficit ratio
氧亏指示器 oxygen deficiency indicator
氧扩散率 oxygen diffusion rate
氧离子 oxygen ions
氧离子迁移率 mobility of oxygen ion;oxygen ion mobility
氧离子缺位浓度 vacancy concentration of oxygen
氧利用 oxygen utilization
氧利用率 oxygen utilization efficiency
氧利用速率 oxygen utilization rate
氧利用系数 coefficient of oxygen; oxygen utilization coefficient; percentage of oxygen utility
氧量分析器 oxygen analyser [analyzer]
氧量过剩 excess oxygen
氧磷灰石 oxidapatite;oxy-apatite;voelckerite
氧硫化镎 neptunium oxide-sulfide
氧硫化碳 carbon-oxygen sulphide; carbonyl sulfide
氧硫化锑 antimony oxysulfide
氧硫化物【化】oxysulphide
氧硫化锌 zinc oxysulfide
氧硫酸镁水泥 magnesium oxysulfate cement
氧硫杂环己烷 thioxane
氧卤化物和氢氧卤化物 oxyhalides hydroxyhalides
氧氯苯磺酸 oxychlorosene
氧氯化锆 zirconium oxychloride
氧氯化物 oxychloride
氧氯化作用 oxychlorination
氧络聚合物 oxolation polymer
氧矛切割 oxygen lance cutting
氧锚(离子) oxonio
氧茂甲醇 furfuryl alcohol
氧煤气 oxycoal gas
氧煤气切割 oxy-city gas cutting
氧煤气焰 oxycoal gas flame;oxy-gas flame
氧敏感试验 oxygen tolerance test
氧钼矿 tugarinovite
氧浓度 oxygen concentration
氧漂泊 oxygen bleaching
氧平衡 oxygen balance;oxygen equilibrium
氧平衡参数评价 oxygen equilibrium parameter assessment
氧平衡模式 oxygen equilibrium model
氧平衡评价 oxygen balance evaluation;oxygen equilibrium evaluation
氧平衡曲线 oxygen equilibrium curve
氧瓶燃烧法 oxygen flask method
氧气 gaseous oxygen;gox;oxygen gas
氧气饱和量 oxygen saturation capacity
氧气表 oxygen ga(u)ge
氧气测量仪 oxygen measurement set; oxygen meter
氧气厂 oxygen installation
氧气导入管 air tube;airway
氧气电池 aeration cell;oxygen cell
氧气电弧切割 oxy-arc cutting
氧气顶吹炉炼钢 basic oxygen furnace steel making
氧气顶吹转炉 basic oxygen furnace
氧气顶吹转炉钢 basic oxygen steel
氧气煅烧 air roasting
氧气对数概率图 logistic probability figure of radon
氧气发生器 pneumatogen apparatus
氧气阀 oxygen valve
氧气法 oxygen survey
氧气防护器 oxygen protection apparatus
氧气分析仪 oxygen analyser [analyzer]
氧气割炬 oxygen lance
氧气供给器 fresh-air breathing apparatus
氧气供应管道 oxygen line
氧气罐 oxygen canister
氧气罐群体 battery of oxygen cylinders
氧气焊接 oxygen weld(ing)
氧气焊接橡胶管 oxygen welding rubber hose
氧气呼吸器 oxygen-breathing apparatus;respirometer
氧气弧切 oxy-arc cutting
氧气环境 oxygen environment
氧气汇流排 battery of oxygen cylinders
氧气活门 oxygen valve
氧气火焰切割器 oxygen lance
氧气监护控制器 oxygen monitor controller
氧气监视器 oxygen monitor
氧气炬 oxygen torch
氧气快速开关 emergency O_2 valve
氧气炼钢 oxygen blast
氧气炼钢法 oxygen steel-making process
氧气量管 oxygen measuring burette
氧气流量计 oxygen flow meter
氧气面具 oxygen mask
氧气面罩 breathing apparatus;fresh-air mask;oxygen mask
氧气(浓度)分析器 oxygen analyser [analyzer]
氧气喷枪 oxygen lance
氧气喷嘴 oxygen jet
氧气瓶 bomb; gas bottle; oxygen bomb;oxygen bottle;oxygen cylinder;oxygen receiver
氧气瓶组 oxygen cylinder unit
氧气切割 oxy-cutting; oxygen-cut(ting);oxygen lancing;oxygen machining
氧气切割机 oxygen cutter
氧气切割器 oxygen lance
氧气切割枪 oxygen lance
氧气清除剂 oxygen scavenger
氧气容器 oxygen receiver
氧气入口 oxygen inlet
氧气软管 oxygen hose
氧气烧枪 oxygen lance
氧气设备 breathing equipment; oxygen apparatus;oxygen-breathing apparatus; oxygen equipment; oxygen installation
氧气设备操纵机构 oxygen control
氧气湿化计 oxygen humidifier
氧气示流器 oxygen flow indicator
氧气随深度变化图 figure of variation with depth of oxygen
氧气随时间变化图 figure of variation with time of oxygen
氧气损耗 <水中的> oxygen depletion
氧气天然气 oxynatural
氧气天然气烧嘴 oxynatural gas burner
氧气调节器 oxygen regulator
氧气筒 oxygen bottle; oxygen cylinder;oxygen tank
氧气筒搬运车 oxygen cylinder truck
氧气筒调节器 cylinder regulator
氧气吸入器 oxygen inhalation apparatus;oxygen inhalator
氧气限量安全装置 oxygen limitation safety device;oxygen limiter
氧气需量 oxygen demand
氧气需要 oxygen demand
氧气压力表 oxygen ga(u)ge;oxygen pressure indicator
氧气压力调节器钥匙 oxygen pressure regulator key
氧气乙炔吹管 oxyacetylene blowpipe
氧气乙炔焊 oxyacetylene welding; oxygen-acetylene welding
氧气乙炔炬 oxyacetylene torch
氧气乙炔切割 oxyacetylene cutting; oxygen-acetylene cutting
氧气乙炔切割设备 oxyacetylene cutting equipment; oxygen-acetylene cutting equipment
氧气乙炔燃烧器 oxyacetylene burner
氧气乙炔软管 oxyacetylene hose
氧气乙炔焰 oxyacetylene flame
氧气与乙炔 <焊接用俚语> air and gas
氧气运输拖车 oxygen-transfer trailer
氧气运输系统 oxygen transporting system
氧气再生系统 recovery oxygen system
氧气站 oxygen station
氧气张力 oxygen tension
氧气帐 oxygen tent
氧气中毒 oxygen poisoning; oxygen toxicity
氧气转炉 oxygen converter
氧气装置 oxygen unit
氧气状况 oxygen status
氧汽油切割机 oxy-gasoline cutting machine
氧切割 oxygen cutting
氧切割阀 oxygen cutting valve
氧切割喷嘴 oxygen cutting tip
氧切割器 oxygen lance
氧氢氧化铁 iron oxyhydroxide
氧氢氧化铁固体 <即针铁矿> iron oxyhdroxide solid
氧氢氧化物 oxyhdroxide
氧圈 oxysphere
氧炔 oxyacetylene
氧炔吹管 oxyacetylene blowpipe
氧炔断割器 oxyacetylene cutter
氧炔焊吹管 oxyacetylene blowpipe
氧炔焊(接) autogenous weld(ing); oxygen-acetylene welding
氧炔焊枪 oxyacetylene blowpipe;oxyacetylene torch
氧炔喷焊器 oxyacetylene blowpipe
氧炔气焊机 oxyacetylene welding outfit
氧炔切割 acetylene cutting; gas cut(ting);oxyacetylene cutting;flame-cutting
氧炔切割/焊接设备 oxy-acetylene cutting/welding equipment
氧炔切割机 gas cutter; oxyacetylene cutter
氧炔熔化 autogenous cutting
氧炔焰 oxyacetylene flame
氧热涂层 oxythermal coating
氧容量 oxygen capacity
氧容限 oxygen tolerance
氧溶剂表面清理 powder washing
氧溶剂切割法 flux oxygen cutting method
氧溶解率 solubility of oxygen
氧熔剂电弧切割 oxyflux cutting
氧熔剂切割 chemical flux cutting;flux oxygen cutting; oxyflux cutting; powder cutting
氧熔气刨 powder scarfing; powder washing
氧上限 maximal oxygen uptake
氧输送效率 oxygen transfer efficiency
氧探测器 oxygen probe
氧探针 oxygen probe
氧碳氮共渗 oxynitrocarburizing
氧碳氮化物 oxy-carbonitride
氧碳化硅 siloxicon
氧碳化物 oxycarbide
氧碳原子比 oxygen-carbon atomic ratio
氧锑铁矿 versiliaite
氧天然气切割 oxynatural gas cutting
氧调节记录器 oxygen controller recorder
氧铁软焊 autogenous soldering
氧同位素 oxygen isotope
氧同位素比率 oxygen-isotope ratio
氧同位素地热温标 oxygen isotope geothermometer; oxygen isotopic geothermometer
氧同位素地质温度计 oxygen isotope geothermometer; oxygen isotopic geothermometer
氧同位素阶段 oxygen isotope epoch
氧同位素温标 oxygen isotope thermometer
氧微电极 oxygen microelectrode
氧吸收 oxygen absorption;oxygen intake
氧吸收率 oxygen uptake rate
氧吸收性中耳气压损伤 oxygen absorption barotrauma
氧硒矿 selenolite
氧下垂曲线 <表示水体自净过程的变化曲线> oxygen-sag curve
氧纤磷铁矿 oxykerchenite
氧相二氧化硅 aerosil
氧消耗 oxygen consumption; oxygen depletion
氧消耗量 amount of oxygen consumption
氧效应 oxygen effect
氧需要量 oxygen requirement
氧序 oxygen sequence
氧循环 cycle of oxygen;oxygen cycle
氧烟酸 oxiniacic acid
氧焰开槽设备 oxygen flame-grooving equipment
氧焰切割 oxyacetylene cutting; oxy-arc cutting;oxygen lancing
氧焰切割器 oxygen lance
氧乙炔棒枪 oxyacetylene rod gun
氧乙炔处理 oxyacetylene treatment
氧乙炔的 oxyacetylene
氧乙炔法 oxyacetylene process
氧乙炔粉体枪 ding; oxyacetylene powder gun
氧乙炔割炬 oxyacetylene cutting torch
氧乙炔焊机 oxyacetylene welder
氧乙炔焊接 oxyacetylene welding
氧乙炔焊接气焊(法) oxyacetylene welding
氧乙炔焊接器 oxyacetylene welder
氧乙炔焊炬 oxyacetylene torch
氧乙炔混合室 oxygen and acetylene

Y

mixing chamber
氧乙炔火焰 oxyacetylene torch
氧乙炔火焰喷涂 oxyacetylene flame spraying
氧乙炔火焰清理 oxyacetylene scarfing
氧乙炔火焰清理机 oxyacetylene scarfing machine
氧乙炔焦化材面防腐法 oxyacetylene process
氧乙炔炬 acetylene torch; oxyacetylene torch
氧乙炔喷焊器 oxyacetylene blowpipe; oxyacetylene torch
氧乙炔气 oxyacetylene
氧乙炔气焊机 oxyacetylene welding outfit
氧乙炔切割 autogenous fusing; oxyacetylene cutting; oxy-cutting; oxygen-acetylene cutting
氧乙炔切割程序 oxyacetylene process
氧乙炔切割器 oxyacetylene cutter
氧乙炔焰 oxyacetylene flame
氧乙炔焰焊接 oxygen weld(ing)
氧乙炔焰切割器 oxyacetylene cutter
氧逸散度 oxygen fugacity
氧茚 benzofuran; coumarone
氧茚树脂 coumarone resin
氧鎓的 oxonium
氧游离基阴离子 oxygen radical anion
氧原子 oxygen atom
氧源 oxygen source; oxygen supply
氧杂蒽 xanthene
氧杂蒽酮 xanthone
氧杂环丙烷 oxirane; oxyacyclopropane
氧杂环丁烷 oxacyclobutane
氧杂环化合物 heterocyclic(al) oxygen compound
氧杂环己烷 amylene oxide
氧杂环壬四烯 oxonin
氧杂环壬烷 oxonane
氧杂环戊二烯 oxole
氧杂环戊烷 tetrahydrofuran
氧杂环辛三烯 oxocin
氧杂环辛三烯醇 oxocinol
氧杂环辛烷 oxinane [oxocane]
氧杂茚【化】coumarone
氧债 oxygen debt
氧张力 oxygen tension
氧指数 oxygen index
氧中毒 oxygen intoxication; oxygen poisoning; oxypathy
氧转化型邻苯二甲酸醇酸树脂 oxygen convertible phthalic resin
氧转化型树脂 oxygen convertible resin
氧转移 oxygen transfer
氧转移率 oxygen transfer rate
氧转移系数 aeration coefficient
氧转移效率 oxygen transfer efficiency
氧转移指数 oxygen transfer exponent
氧阻聚作用 oxygen inhibition

痒

痒阈 < 声学术语 > threshold of tickle

样 sample

样板 control ga(u)ge; face mo(u)ld; flat ga(u)ge; former; formwork; gabarite; ga(u)ge (board); ga(u)ge plate; ga(u)ging board; guiding rule; master form; master plate; model; mother plate; mo(u)ld(ing) board; plate ga(u)ge; profile ga(u)ge; prototype; sample board; sampler; scantling ga(u)ge; screed board; screeding beam; screed plate; shaping plate; striking board; template; templet; screeding board < 刮平混凝土用 >; jigging block < 机械加工用 >; screed < 泥瓦工用 >
样板厂 demonstration plant; demo plant
样板车间 template shop
样板城市 model city
样板刀 form(ing) tool; shaping tool
样板刀成型 template forming
样板法 model method; template method
样板法规 model code
样板房 model house
样板粉光机 smoothing beam finisher
样板高级旅客列车 < 英 > prototype advanced passenger train
样板工 liner man
样板工程 prototype project; typical project
样板工具 master tool
样板刮平混凝土 screeded concrete
样板规 ligne ga(u)ge; profile board
样板和模型 templates and models
样板划线 pattern scribing
样板机 prototype board
样板机床 ga(u)ge lathe
样板夹 panel clamp
样板间 model room; prototype room
样板建筑 demonstration building
样板列车 preproduction train; prototype train
样板路段 model section
样板轮廓磨削 template grinding
样板毛坯 master blank
样板磨床 template grinding machine
样板目镜 template eyepiece
样板刨 ga(u)ge plane; mo(u)lding planer
样板曲线 master curve
样板设计 design example
样板绳 screed wire
样板式栏杆车床 spindle mo(u)lder
样板式振动器 screed board vibrator; screed vibrator
样板凸轮 model cam
样板图 drawing of sample
样板文件管理 exemplary documentation
样板铣床 template milling machine
样板下料方法 spiling
样板压机 templet shaping press
样板研磨 form grinding
样板造型 sweep mo(u)lding
样板住宅 model dwelling; model house
样本 advance(d) copy; advance sheets; brochure; example; exemplar; format; instruction sheet; literature; sample; sample book; sample copy; specification sheet; specimen; specimen copy; style-book; swatch; advance print < 指印刷 >
样本 R 阶矩 sample moment of order R
样本百分比 sample percentage
样本百分位数 sample percentiles
样本保持 sample-hold
样本本征值 sample characteristic value
样本比例 sample proportion
样本变差 sample variation
样本标准差 sample standard deviation
样本标准相关系数 sample canonical correlation coefficient
样本波动 sample fluctuation
样本波形 sample waveform
样本采集 collection of sample
样本操作系统 sample operating system
样本草图 scantling
样本产生符 sample generator
样本成功比例 sample proportion of successes
样本尺寸 sample size
样本抽选规则 sample selection rule
样本处理 handling of sample
样本大小 sample size
样本代表性 representativeness of sample
样本单位 sample unit
样本单位数 number of units in sample
样本的 sampled
样本的范围 range of the sample
样本的构形 configuration of sample
样本的最优分配 optimum allocation of sample
样本的最优设计 optimum design of sample
样本点 sample point
样本调查 sample inquiry; sample survey
样本对 sampling pair
样本多重相关系数 sample multiple correlation coefficient
样本范围 coverage of sample; range of sample; sample range
样本方案 sample plan
样本方差 sample variable; sample variance; empiric(al) variance
样本方差矩阵 sample variance matrix
样本方差协方差矩阵 sample variance-covariance matrix
样本费 cost of samples
样本分布 sample distribution
样本分配 allocation of samples
样本分数 sampling fraction
样本分位数 sample fractile
样本分析 sample analysis
样本概率测度空间 sample probability measure space
样本构成 composition of sample; structure of sample
样本构型 < 统计数学上的 > configuration of sample
样本估计值的标准误 standard error of estimate based on sample
样本规格 sample size
样本含量 sample content
样本含量估计 estimation of sample size
样本函数 sample function
样本号 catalog(ue) number
样本号码 brochure code
样本化 sampling
样本回归函数 sample regression function
样本及保存 sample and hold
样本极差 sample range
样本记录 sample record
样本监督 sample monitoring
样本检验员 sampler
样本结果的精确度 precision of sample results
样本矩 sample moment
样本均方差 sample standard deviation
样本均方或然性 sample mean square contingency
样本均方列联 sample mean square contingency
样本均值的分布 distribution of mean of sample
样本均值的精确度 precision of sample mean
样本均值分布 sample distribution of mean
样本卡片 sample card
样本可信度 sampling reliability
样本空间 sample space
样本空间的点 point in the sample space
样本扩大法 method of sample enlargement
样本离差 sample dispersion
样本量的比例分配 proportional allocation of sample size
样本量字码 sample size code letter
样本率 rate of sample; sample rate
样本偏差 sample deviation
样本偏相关系数 sample partial correlation coefficient
样本偏斜系数 sample skewness coefficient
样本平均数 sample average; sample mean; sampling mean
样本平均值 sample average; sample mean; sampling mean
样本平均值的标准误 standard error of sample mean
样本平均值的抽样分布 sampling distribution of sample mean
样本普查 sample census
样本谱函数 sample spectral function
样本确定性 sample certainty
样本容量 sample capacity; sample size; sample volume
样本容量要求 sample size requirements
样本设计 design of sample; sample design
样本生产 sample production
样本十分位数 sample decile
样本时点 sample time point
样本式栏杆车床 spindle mo(u)lder
样本试验 sampling test
样本树 sample tree
样本数 sample number
样本数据 sample(d) data
样本数据量度 sampled-data measurements
样本数据线性运算 linear operations on the sample data
样本数量 sample quantity
样本数目 number of samples
样本四分位数 sample quartile
样本似然率函数 sample likelihood function
样本算术均数 sample arithmetic(al) mean
样本特征 sample characteristic
样本特征值 sample characteristic value
样本统计量 sample statistics; sampling statistics
样本统计数 sample statistics
样本网络 network of samples
样本稳定化 sample stabilization
样本误差 sample error
样本系数 sample coefficient
样本显著性水平 sample level of significance; sample significance level
样本相关矩阵 sample correlation matrix
样本相关系数 sample correlation coefficient
样本协方差 sample covariance
样本协方差函数 sample covariance function
样本协方差矩阵 sample covariance matrix
样本协方差系数 sample covariance coefficient
样本性概差 probable errors based upon different samples
样本一致性 homogeneity of sample
样本预处理 sample pretreatment
样本值 sample(d) value
样本(值)输入 sample input
样本中位数 sample median

样本中位值 sample median
样本中心矩 sample central moment
样本众数 sample mode
样本住户收支 sample household budgets
样本资料 data sample
样本自相关函数 sample autocorrelation function
样本总数 total number of samples
样本总值 sample total
样本组数 number of samples
样饼试烧法 compressed pat kiln test
样车 reference vehicle
样冲 joint punch
样船 example ship
样带 belt transect;sample strip
样地 sample plot(ting)
样地面积 sample area
样点 sampling point
样点抽样(法) point sampling
样洞 open test pit; pit; prospect pit; trial hole; trial holing; trial pit
样方 plot; quad; quadrat
样方表 quadrat list
样方测定法 point quadrat method
样方法 quadrat method; quadrat sampling method
样方目测地 ocular estimate by plot method
样方图 quadrat chart
样规 control ga(u)ge; mitre templet
样行 guideline
样机 mock-up; model; model machine; preproduction model; prototype; prototype aeroplane; prototype instrument; prototype model; prototype unit; reference vehicle; specimen machine
样机板 prototype board
样机部件 prototyping component
样机操作系统 prototyping operating system
样机测试 prototyping testing
样机插件 prototyping card
样机插件系统 prototyping card system
样机雏形 prototype
样机的实际试验 field test of prototype
样机工厂 prototype plant
样机开发系统 prototype development system
样机开发系统模块 prototype development system module
样机逻辑操作 prototype logic(al) operation
样机试验 prototype test(ing)
样机试验工厂 prototype plant
样机试制 advanced development; advanced development of model machine
样机调试 prototype debug
样机投资 pilot machine investment
样机外传噪声的固定试验 exterior-stationary test
样机系统 prototyping system
样机质量检验 phototype qualification testing
样件 exemplar; master piece; prototype workpiece; sample piece; sample work piece
样块 sample piece
样模 shaping plate
样木 sample tree
样片 print; rush; sample wafer; swatch; coupon <供割取或加工用的>
样品 assay; dummy; exemplar; follow-on; model; muster; proof sample; prototype; sample; sample piece; scantling; specimen; swatch; test piece
样品包装方式 packing manner of sample
样品保存 sample storage
样品保存时间 sample storage time
样品杯 specimen cup
样品编号 sample number
样品标志 sample mark(ing)
样品标准偏差 sample standard deviation
样品表面积 face area of sample
样品玻璃 mother glass
样品材料 specimen material
样品采集 collection of sample
样品槽 sample chamber
样品槽钢 sample channel
样品草图 scantling
样品测试 sample testing
样品测试方法 testing method of sample
样品测试类型 testing type of sample
样品测试日期 testing data of sample
样品测试实验室 laboratory of sample testing
样品测试项目 testing term of sample
样品长度 sample length
样品陈列室 sample room
样品尺寸 sample size; specimen size
样品充填密度 loaded density of sample
样品出口 sample export
样品储藏间 store for samples
样品储存 storage of samples
样品储存器 sample container
样品储存时间 sample storage time
样品处理 handling of sample; sample treatment
样品存储器 sample container
样品大小 sample size; size of a sample
样品袋 sample container; sample sack
样品道比 sample channels ratio
样品的百分透光率 percentage transmission for sample
样品的均一性 homogeneity of sample
样品的使用或申请 application of sample
样品的室内编号 sample number in the laboratory
样品的野外编号 sample number in the field
样品的坐标 specimen coordinate
样品等分法 split sample method
样品电导率 sample conductivity
样品订单 sample order
样品断面面积 sectional area of sample
样品发票 sample invoice
样品费用 cost of samples; fee of sample; sample expenses
样品分布 sample distribution
样品分解 decomposition of the sample
样品分离 sample splitting
样品分离器 sample splitter
样品分离装置 separator of sample
样品分析 sample analysis; sampling analysis; analysis of samples
样品分析方法 analytic(al) method of sample
样品分析精度 accuracy of sample analysis
样品分析类型 analytic(al) type of sample
样品分析日期 assaying date analytic(al) date of sample
样品分析实验室 laboratory of sample analysis
样品分析项目 analytic(al) term of sample
样品分析指数 analytic(al) index for sample
样品分析质量 quality of sample analysis
样品富集 sample enrichment
样品观测值 sample observations
样品管 sample cell; sampling tube
样品罐 sample can; sampling tin
样品规格 sample dimension; sample specification
样品柜 display case
样品号 specimen number
样品合并 sample merging
样品合格率 percent of passed sample
样品厚度 thickness of sample
样品厚度测量 measurement of sample thickness
样品环管 sample loop
样品回归 sample regression
样品回路管 sample loop
样品集体 sample community
样品记录器 sample logger
样品加工 sample processing; sample treatment
样品加工公式 sample processing formula
样品加工流程 sample processing circuit
样品加工流程图 flowchart of sample treatment
样品夹 specimen holder
样品甲 Sample A
样品间 sample room
样品检查员 sample passer; sampler
样品检验 sample examination
样品检验种类 type of sample examination
样品鉴定 sample identification; sample survey
样品鉴定方法 identification method of sample
样品鉴定类型 identification type of sample
样品鉴定日期 identifying date of sample
样品鉴定实验室 laboratory of sample identification
样品鉴定项目 identification term of sample
样品交换器 sample changer
样品介质 sampling parent
样品卡 sample card; show card
样品可靠重量 reliable weight of sample
样品来源 source of sample
样品类型 type of sample
样品离差 sample dispersion
样品理论 theory of sampling
样品粒度 particle-size fraction of sample; size fraction of sample
样品零件 specimen part
样品买卖的默认条件 implied warranties in sale by sample
样品免税通关卡 commercial samples carnet
样品描述 description of sample
样品名称 name of sample
样品模 sample mold
样品目录 specimen copy
样品浓度 sample concentration
样品浓度测量 measurement of sample concentration
样品盘 sample pan
样品喷嘴 sample nozzle
样品平均数 sample mean
样品平均数差 difference of sample means
样品平均数的解释 interpretation of sample means
样品平均值 sample average; sample mean
样品瓶 sample bottle; sample bulb
样品契约 sample contract
样品强度 sample intensity; sample strength
样品情况 sample circumstances
样品取集 sampling collection
样品取向 specimen orientation
样品容积 sample content
样品容量 sample size; size of a sample
样品商 sample hunter
样品商人 sample merchant
样品烧箱 sample sagger
样品设计 sample design
样品生活力 viability of sample
样品实验编号 sample code in laboratory
样品实验室记录 sample record in laboratory
样品试池 sample cell
样品试验 pilot test; sample test; sample trial; sampling test; specimen test
样品试制 advanced development; advanced development of sample
样品试制阶段 pilot stage
样品室 sampler room
样品收集管 sample collection tube
样品收集器 sample catcher
样品输出 sample export
样品输送机 specimen-transfer mechanism
样品数 number of samples
样品数量 sample size
样品说明书 sample specification
样品送入孔 specimen-insertion port
样品酸碱度 sample pH value
样品缩分 reducing of sample; sample splitting
样品台 specimen stage
样品体积 sample volume; specimen volume
样品桶 sample bucket
样品图 sample drawing; specimen screen
样品污染 sample contamination; sample pollution
样品相关系数 sample correlation coefficient
样品箱 sales kit
样品形状 specimen shape
样品岩性 specimen lithologic(al) characters
样品研磨机 sample grinder
样品颜色 sample colo(u)r
样品氧化还原电 sample eH value
样品野外编号 field sample number
样品用皿 sample pan
样品原始编号 original code of sample
样品原始重量 initial weight of sample
样品原重 original sample weight
样品运送 sample handling; sample transport
样品沾污 contamination
样品展览会 sample fair
样品展示 information show
样品折扣 sample discount
样品正态分布 sampling normal distribution
样品正态分布曲线 sampling normal distribution curve
样品指示灯 specimen viewer attachment lamp
样品制备 preparation of sample; sample preparation; specimen preparation
样品中位数 sample median
样品中元素含量 element content in the sample
样品种类 sample material; type of samples
样品重量 sample weight
样品砖 sample brick

样品转换装置 sample changer; sample change unit
样品自动更换器 automatic sample changer
样品组数 number of samples; number of sets of sample
样品钻机 prototype drill
样品钻头 prototype bit
样品最终重量 final weight of sample
样品座 specimen holder
样勺 spoon
样式 type; style; pattern; epure (asphalt); fashion; form; manner; mode; modus
样式概念 idea of style
样式合适的构造 form-fit construction
样式摄影术 fashion photography
样书预告标志 advance copy
样体 sample work piece
样条 spline; transect
样条逼近 spline approximation
样条插值 spline interpolation
样条插值法 spline interpolation
样条法 spline finite strip
样条函数 spline function; splines
样条函数插值 interpolation of spline function
样条函数计算 evaluation of spline function
样条函数近似 spline approximation
样条函数算法 splining algorithm
样条理论 < 即样条插值法 >【数】 theory of spline
样条内插(法) spline interpolation
样条拟合 spline fit(ting)
样条拟合环 spline fit ring
样条拟合曲线 spline fit curve
样条曲线 spline curve
样图 advanced sheet; map standard; master drawing; press pull; sample copy; sample plot(ting); specimen map; specimen sheet; style sheet
样图审校 proof checking
样图照片 model photograph
样线 line transect
样线法 line-intercept method
样型 template
样页 advance sheets
样张 guideline; pull; specimen page; specimen paper; specimen sheet
样值 sample
样重比率 size-weight ratio
样子 model; specimen

漾 seiche

漾气差 atmospheric refraction

么线 unit graph; unit hydrograph

么元(素) identical element; unit element
么阵 identity matrix
么重 specific mass gravity

幺并矢 idem factor

幺模矩阵 unimodular matrix
幺模特性 unimodular property
幺模阵 unimodulus matrix
幺图 identity graph
幺旋 unitary spin; u-spin
幺元【数】 identical element; identity element
幺正八重态 unitary octet
幺正表示 unitary representation
幺正对称 unitary symmetry
幺正对称理论 unitary symmetry theory
幺正多重态 unitary multiplet
幺正矩阵 unitary matrix
幺正散射因数 unitary scattering factor
幺正十重态 unitary decuplet
幺正算符 unitary operator
幺正算子 unitary operator
幺正性 unitarity
幺重 unit weight

夭折 < 计划等 > abort

妖女饰【建】 gorgoneion; gorgoneum [复 gorgonoia/gorgonoa]; gorgonion

要求安装日期 date of installation required

要求按价支付诉讼 action to recover the price
要求办公时间外提货 application for extra work
要求报酬率 required rate of return
要求报价 request for an offer
要求标记 demand token
要求拨给特种货车 requisition of special-type wagon
要求补偿 claim for compensation
要求补偿的诉讼 action for reimbursement
要求长度 desired length
要求偿还 claim for reimbursement
要求承包 request for bid
要求承包文件 request for bid documents
要求处理 demand processing; time-sharing process(ing)
要求搭顺路车的人 thumber
要求达到的效果 desired impact
要求的活荷载 required live load
要求的进路(或径路) route claimed
要求的精确度 claimed accuracy
要求的压力 required pressure
要求递价 ask to make a bid
要求电告已否收妥的托收款项 wire fate item
要求定义 requirements definition
要求定义技术 requirements engineering
要求分布 distribution of demand
要求服务的总体 calling population
要求付给救助费 claim for salvage
要求付款 claim for payment; claim payment; claim reimbursement; demand for payment; payment on demand
要求跟踪程序 requirement tracer
要求归还原物 claim for restitution
要求厚度 desired thickness; required thickness
要求恢复 reclaim
要求恢复原状的诉讼 restitution
要求间歇使用非已有的权利 discontinuous easement
要求减价 abatement claim; claim for reduction
要求交付股款 call on shareholder
要求交货日期 required delivery date
要求接收 request to receive
要求解释 call to account
要求精(确)度 claimed accuracy; precision prescribed; required accuracy
要求均恒而适中的温度 require equable moderate temperature
要求开价 call for bids
要求列车成队运行 call for fleeting
要求流量 flow demand
要求路线图 desire line chart
要求履行义务 require performance of an obligation
要求描述语言 requirement description language
要求赔偿 appeal for compensation; claim (for) compensation; claim indemnity; demand (for) compensation; seek redress
要求赔偿的主因 subject of claim
要求赔偿评定书 adjustment letter
要求赔偿权 right of indemnity
要求赔偿书 claim letter
要求赔偿损失 claim compensation for losses; claim for damages; claim indemnity; claim of damages
要求赔偿损失的权利 right to claim for damages; right to demand compensation for losses
要求赔偿损失的诉讼 action for compensation for loss
要求赔偿违约损失的诉讼 < 或可据以进行此项诉讼的契约 > assumpsit
要求赔款 demand an indemnity
要求强度 desired strength; required strength
要求驱动的 demand-driven
要求驱动的计算机 demand-driven computer
要求取得应收价款 claim for proceeds
要求全面保护的专利 umbrella patent
要求权 claim
要求权评定 assessment of claim
要求确立产权的诉讼 petitory action
要求确认 confirmation request
要求人孔的最小直径容器 minimum diameter vessel requiring manhole
要求容量 capacity required
要求润滑的轴箱 axle-box lubricated as required
要求深度 required depth
要求时 on demand
要求矢量 requirements vector
要求使用期 demand usage time
要求收回 reclaim
要求数目 number required
要求索赔损失 claim against damages
要求提交建议 request for proposal
要求调整合同价 claim for adjustment of price
要求贴现人 applicant for a discount
要求停止某些行动的警告 caveat
要求图形 demand graph
要求退还多收费用 application of repayment of an overcharge
要求退还多收运费 claim for overcharge
要求退还原物的诉讼 action for restitution
要求退汇的金额 reexchange
要求退款 claim for refund
要求线 < 交通规划中常用的分析方法,把城市各区间的交通流量在图上以粗细的空间直线表示出来 > desire line
要求向量 requirements vector
要求性能 required function
要求宣判契约无效的诉讼 action of restitution
要求延长索赔期 request to extend time-limit of claim
要求延性分布 distribution of ductility demand
要求一览表 schedule of requirements
要求银行贷款 demand for bank funds
要求银行提交财务报表通知 bank call
要求应用的语言 application-required language
要求增加担保 claim for additional security
要求支付救助费 claim for salvage
要求支付欠款 recovery of late payment
要求作出裁决 ask for a ruling
要求作出断然抉择的 take-it or leave-it
要约 causative action of agreement; offer
要约的期限 time limit of offer
要约的有效期限 term of validity of offer
要约和承诺 offer and acceptance
要约人 offeror
要约引证 invitation for offer

腰板 expansion plate; surbase; wainscot

腰板横梁 expansion crosstie
腰部 middle; waist
腰部进水水轮 breast water wheel
腰部舷板 waist plate
腰槽 gain
腰窗 window transom
腰窗档 fanlight quadrant
腰带 belt rail
腰带补强铁 belt rail stiffener
腰带层 < 墙上沿窗台凸出的装饰层 > belt course
腰风 blowing from bosh; peripheral blowing
腰拱 rider arch
腰箍 peripheral tie beam; wall tie
腰鼓形螺纹接套 barrel nipple
腰果 acajou; cashew
腰果酚 anacardol
腰果间二酚 cardol
腰果壳酚 cardanol
腰果壳液树脂 cashew nut shell liquid resin
腰果壳油酚醛树脂 cashew nut shell oil aldehyde resin
腰果树 cashew
腰果树脂磁漆 cashew resin enamel
腰果树脂涂料 cashew resin paint
腰果树脂油灰 cashew resin putty
腰果油 apple oil; cashew nut oil
腰果油树脂 cashew resin
腰果油水性防锈底漆 cashew water coat
腰荷 intermediate load
腰横缆 waist hawser (line)
腰栏【建】 breast rail
腰梁 breast beam; breast timber; wale
腰梁垫木 kicking piece
腰墙 breast wall
腰墙板 dado panel
腰曲面 surface of striction
腰塞 < 下水泥用 > bridging plug
腰肾形料堆 kidney shaped stockpile
腰铁 iron clamping ring
腰头窗 fanlight; sublight; transom
腰头气窗 night vent; vent sash
腰窝锉刀 waist setting attachment
腰窝铣刀轴 waist cutter spindle
腰窝铣削机 waist-trimming machine
腰窝铣削铣刀 waist trimmer; waist-trimming cutter
腰线 band course; belt course; oversailing course; sailing course; string course; veneer tie; frieze < 雕有图案、花纹等 >
腰线板 frame board
腰形试件 waisted specimen
腰沿釜 waisted cauldron
腰眼 breast hole
腰圆形木盘 butler's tray
腰圆形书桌 kidney desk
腰缘鸠尾嵌接 shoulder dovetail halved joint

邀标 invitation to bid; tender invitation

邀请 bidding; invitation
邀请报告 guest lecture; invited lecture
邀请报价 invitation for offer
邀请表 invitation list
邀请参与贷款 invite subscription for a loan
邀请参与捐款 invitation for subscription
邀请参与认购 invitation for subscription
邀请出价 invitation for offer
邀请电传 invitation telex
邀请发盘 invitation for offer
邀请附有建议的投标 invitation for proposals; proposal call; call for proposals
邀请捐款 invitation for subscription
邀请开价 invitation for offer
邀请认购 invitation to subscribe
邀请视察 inspection by invitation
邀请投标 invitation to bidding; invitation for bid
邀请投标人洽谈 invitation to treat
邀请投标人商议 invitation to treat
邀请投标人谈判 invitation to treat
邀请投标人做交易 invitation to treat
邀请投标(书) bid invitation; invitation to bid; invitation to tender; inviting bid; request for bid; invite to tender
邀请投标者名单 invited bidders
邀请投标者清单 selected list of bidders
邀请信 letter of invitation
邀请延迟 invitation delay
邀请招标 <指有限范围的招标> invited tendering; selected bidding; selective bid(ding); selective tender(ing)
邀请者 inviter
邀游的 roving

尧敦方(区组)【数】 Youden square

窑斑砖 kiln-marked brick

窑壁 kiln shell
窑壁温度 kiln-shell temperature
窑变 accidental colo(u)ring; fambe; furnace transmutation
窑变绿釉 green flambe blaze
窑彩鲸鱼 whale in kiln colo(u)r
窑彩双鲸鱼 two-flying whales in kiln colo(u)r
窑彩双猫 two-cats in kiln colo(u)r
窑藏 cache
窑场 kiln factory
窑车 kiln car; kiln cart
窑车的台面 car top
窑车上耐火材料面层 desk
窑车式间歇式窑 bogie kiln
窑车式隧道窑 car tunnel kiln; railroad tunnel kiln
窑车下通风 under-car ventilation
窑车下温度 under-car temperature
窑衬 kiln liner; kiln lining; lining
窑衬砌筑支撑法 prop method
窑尺寸 kiln size
窑的额定产量 rated kiln output
窑的燃烧区 burning zone of kiln
窑的有效容积 effective capacity of kiln
窑的预热段 preheating section of kiln
窑的圆锥形外壳 hovel
窑底 furnace bottom; furnace siege; kiln bottom; siege
窑底大砖 siege block
窑底格排垛 grillage
窑底冷却 bottom cooling
窑底梁 bottom beam; grillage
窑底托梁 bottom beam
窑底砖 bottom block; bottom paving brick; furnace bottom block
窑顶 kiln crown; kiln roof; plattering
窑顶吹风 crown blast
窑顶加柴煅烧 sky firing
窑顶排气孔 blow-hole
窑顶砖 arch brick
窑洞 burrow; caved welling house; cave-house; kiln hole
窑洞建筑 cave architecture
窑洞住宅 cliff dwelling (settlement); cave dwelling
窑废热 kiln waste heat
窑缝 kiln crackle
窑干 hot-air seasoning; kiln dry
窑干贬值 kiln degrade
窑干材色变 kiln stain
窑干褐变 kiln brown stain; kiln burn
窑干木材 kiln-dried wood
窑干木料 oven-dry timber; oven-dry wood
窑干试样 kiln sample
窑干燥过程 kiln run
窑干装材车架 kiln bunk
窑干装载量 kiln charge
窑杆材 kiln-dried lumber
窑杆木材 kiln-dried
窑隔热层 kiln insulation
窑拱 arch of furnace; kiln crown
窑箍 band
窑箍立柱 anchor
窑管焊接 stove pipe weld
窑汗 kiln sweat
窑烘的 kiln-burned; kiln-fired
窑烘(干) kilning
窑烘砖 kiln brick; kiln burnt brick; kiln refractory brick
窑换 kiln transformation
窑灰 kiln dust
窑焦油 kiln tar
窑居 cave dwelling
窑具 kiln furniture
窑壳 kiln shell
窑壳温度 kiln-shell temperature
窑孔 kilneye
窑口 kilneye
窑口圈砖 discharge-end block; nose ring block
窑里层 kiln lining
窑令 furnace campaign
窑炉 furnace; kiln
窑炉操作 furnace operation
窑炉操作曲线 furnace (operation) curve
窑炉抽力 furnace draught
窑炉出口烟道 furnace offtake
窑炉的顶部 crown
窑炉的隔焰耐火罩 muffle
窑炉顶部砌接 bonded roof
窑炉隔热 furnace insulation
窑炉缓慢加热 smoking
窑炉换向控制 furnace reversal control
窑炉结构 furnace construction
窑炉烤烘 kiln roasting
窑炉控制 furnace control
窑炉炉头 firing hood
窑炉气氛 furnace atmosphere
窑炉热平衡 heat balance of kiln
窑炉热效率 heat efficiency of kiln
窑炉容量 furnace capacity
窑炉熔化性能 furnace performance
窑炉设计 furnace design
窑炉升温 furnace heating-up
窑炉石料 kiln stone
窑炉维修 furnace maintenance
窑炉效率 furnace efficiency
窑炉卸料 furnace discharge
窑炉压记录器 furnace pressure recorder
窑炉压力 furnace pressure
窑炉烟道 furnace flue
窑炉烟气 kiln gas
窑炉用油 furnace oil
窑炉转磨 kiln mill
窑炉作业 kiln operation
窑炉作业周期 furnace run
窑门 kiln entrance
窑门洞 kiln hole
窑内衬垫 kiln liner
窑内衬砌 kiln lining
窑内抽力 furnace draft
窑内飞尘 furnace dust
窑内飞料 furnace dust
窑内干燥 <木材的> kiln dry
窑内烘干 kiln drying
窑内烘干的 kiln-dried
窑内烘干木材 kiln-dry lumber
窑内空气 furnace air
窑内气氛 kiln atmosphere
窑内陶瓷坯盒 kiln furniture
窑内停留时间 retention time in kiln; time of passage through kiln
窑内通风设备 ventilation device of kiln
窑泥 pug; pugged clay
窑泥机 pugmill (pugmill mixer)
窑皮 clinker coating
窑气 kiln gas
窑器 kiln ware
窑器干燥室 hot house
窑前室 vestibule
窑墙涂料 kiln wash
窑燃烧区 burning zone of kiln
窑砂 kiln sand
窑烧 kilning
窑烧的 kiln-burned; kiln-fired
窑烧法 kiln process
窑烧石灰 lime burning
窑生盐迹 <黏[粘]土砖的> kiln white; kiln scum
窑式脱水器 kiln evapo(u)rator
窑室 kiln chamber
窑室烘干的 kiln-dried
窑霜 kiln scum
窑体保温 insulation of furnace wall
窑体钢箍结构 furnace bracing
窑体结构 kiln structure
窑填充率 degree of kiln filling
窑头灶 combustion chamber
窑尾废气 kiln gas
窑业 ceramic industry; ceramics
窑业家 ceramist
窑业作业周期 campaign
窑用燃烧设备 combustion installation for kiln
窑用输送设备 conveying device for kiln
窑釉 Bristol glaze; kiln glaze
窑渣 kiln slag
窑罩 kiln hood
窑址 kiln site
窑制耐火砖 kiln refractory brick
窑制砖 fired brick
窑中焙烧 roasting in kilns
窑中烘干 kiln dry(ing)
窑中退火 kiln annealing
窑轴线 kiln axis
窑装载量 kiln charge
窑作业 working of kiln

摇把 crank (handle); cranking bar; dodder

摇把挡 handle stop
摇把吊环 <底开门车> handle hanger link
摇把吊环安全销 handle stop safety pin
摇把吊环托 handle stop bracket
摇摆 bob; joggle; lurching; nodding; rock(ing); shimmy; sway(ing); swing(ing) motion; tremble; vacillate; vacillation; wabble; wag; waggle; wave; willow sapling; wobble; wobbling; swing
摇摆板 wobble plate; wobbler
摇摆板多缸发动机 wobble plate engine
摇摆板燃料泵 wobble plate fuel pump
摇摆泵 wobble pump
摇摆不稳定性 yaw instability
摇摆槽 rocking trough
摇摆船身脱浅 sallying the ship
摇摆的 staggering
摇摆的圆盘 buffing wheel
摇摆刚度 rocking stiffness
摇摆刮削 swing-scraping
摇摆拐臂 return motion linkage
摇摆滚锻工艺 slick process
摇摆航空摄影 aerial photography by movable camera mounting
摇摆机 wabbler; wobbler machine
摇摆机构 wig-wag mechanism; wobbler mechanism
摇摆基础 rocking foundation
摇摆架 rocking frame
摇摆角度 rocking angle; swing angle
摇摆脚手架 swinging scaffold
摇摆铰链 rocker hinge
摇摆锯 wobble saw
摇摆力 sway force
摇摆力矩 rock moment
摇摆梁 drilling beam
摇摆梁枢轴 radius beam pitman
摇摆门 double-action door
摇摆门柱 swinging post
摇摆盘式泵 wobble plate pump
摇摆炮塔 oscillating turret
摇摆曲线 rocking curve
摇摆散布器 swing diffuser
摇摆筛 oscillating sieve; rocking sieve; swinging screen
摇摆式拌和机 oscillating agitator
摇摆式车架 oscillating frame
摇摆式道口信号机【铁】 wig-wag crossing signal
摇摆式电弧炉 rocking arc furnace
摇摆式电解槽 rocking cell
摇摆式反应器 swing reactor
摇摆式杠杆 swinging lever
摇摆式沟道磨床 oscillating groove grinding machine
摇摆式活动支座 rocker bearing
摇摆式机械手 swinging mechanical hand
摇摆式进料器 wobble plate feeder
摇摆式栏木 swing barrier; swing gate
摇摆式缆索起重机 cable crane with swinging leg
摇摆式冷轧管机 rockright mill
摇摆式输送机 shaking conveyer [conveyor]
摇摆式停车标 <道口用> stop-when-swinging sign
摇摆式卸载闸门 oscillating gate
摇摆式信号 <表示列车接近> wig-wag signal
摇摆式支座 pendulum bearing; pendulum stanchion
摇摆式制粒机 oscillating granulator
摇摆输送机 swinging conveyer [conveyor]
摇摆台 tilter
摇摆调节器 oscillating treatment controller

摇摆凸轮 <完成往复运动的凸轮> rocking cam
摇摆卸车 car shakeout
摇摆卸车机 car shaker
摇摆研磨机 swinging grinding machine
摇摆摇动 swinging motion
摇摆仪 roll and pitch indicator;yawmeter
摇摆圆锯 wabble saw;drunken saw
摇摆运动 flapping; jigging motion; rocking motion; see-saw motion; weaving movement
摇摆增量 increment of roll
摇摆招牌 swinging sign
摇摆振荡 rocking oscillation
摇摆振动 rocking vibration
摇摆症 staggers
摇摆支座 oscillating foundation
摇摆止回阀 flap check valve;sewing check valve
摇摆中心 centre of oscillation
摇摆轴 pivot center [centre]
摇摆轴线 axis of oscillation
摇摆转子 pendulum roller
摇板 rocker panel
摇板式进料器 wobble plate feeder
摇板型造波机 flap type wave generator
摇臂 bell crank;rock(er) arm;rocker; rocking arm; sneezing bar; swing arm; swing jib; working beam
摇臂泵 beam pump
摇臂变焦距镜头 Zoomar lens
摇臂传动 rocker-arm drive
摇臂传动比 rocker-arm ratio
摇臂导轨 rocker-arm way
摇臂吊 bell crank
摇臂吊车 swinging crane
摇臂吊杆 derrick
摇臂复位弹簧 rocker-arm return spring
摇臂杠杆 rocker lever
摇臂杠杆系统 rocker-arm linkage
摇臂滚轮 rocker-arm roller
摇臂滚珠轴承 rocking ball bearing
摇臂横动床面 traverse bed
摇臂滑杆支架 adjust board
摇臂簧 rocker-arm spring
摇臂机构 rocker arm mechanism
摇臂机械手 swing arm mechanical hand
摇臂锯 arm saw
摇臂开关 rocker switch
摇臂连杆 rock(er) arm;rocker-arm link;rocking arm
摇臂牛头刨 rocker-arm shaper
摇臂曝气器 swing diffuser
摇臂起锚机 pump brake windlass;revolving/wing crane
摇臂起重机 jib crane;swinging crane;whipping crane
摇臂起重机架 boom derrick
摇臂墙 <门架系统的> rocker wall
摇臂润滑滤器 rocker arm lubricating filter
摇臂润滑油泵 rocker arm lubricating oil pump
摇臂润滑油柜 rocker arm lubricating tank
摇臂式接触焊机 rocker-arm resistance welding machine
摇臂式喷灌机 swing arm sprinkler
摇臂式喷灌器 circle sprinkler;swing arm sprinkler
摇臂式喷头 swing arm sprinkler
摇臂式起落架 articulated landing gear
摇臂式起重机 derrick crane
摇臂式输送机 natural frequency conveyer [conveyor]
摇臂式装载机 rocker-arm shovel
摇臂收割机 sail reaper;self-rake reaper
摇臂头 rocker head
摇臂凸轮 rocker cam
摇臂推杆 rocker-arm push-rod
摇臂万能铣床 radial universal milling machine
摇臂下保险立架 <钢绳冲击钻机> headache post
摇臂销 rocker-arm roller pin
摇臂悬置轮 lever-suspension wheel
摇臂油槽 rocker-arm oil trough
摇臂轴 inverted lever;rocker(-arm) shaft;rock(ing) shaft;wayshaft;weigh shaft
摇臂轴推力弹簧 rocker shaft spacer spring
摇臂轴油管 rocker shaft oil pipe
摇臂轴油路 oil passage to rocker shaft
摇臂轴支座 rocker bearing
摇臂轴直径 rocker shaft diameter
摇臂轴止动环 rocker shaft circlip
摇臂柱塞 rock arm plunger
摇臂柱塞簧 rocker-arm plunger spring
摇臂装料机 jib loader
摇臂钻 weigh shaft
摇臂钻床 beam drill;drilling machine with jointed arm pivotal;mounted on swinging jib machine;radial arm drill;radial drill;radial drilling machine
摇臂钻横动床面 traverse bed of radial drilling machine
摇臂钻机 radial drill(ing machine)
摇臂钻进给箱横切 head traverse of radial drilling machine
摇臂座 rocker-arm support; rocker seat
摇表 megameter;megger;megohmmeter;tramegger
摇柄操作器 rotor-type operator
摇柄木钻 short brace
摇柄钻 bit stock;center bit
摇槽 rocker vat;rocking trough
摇槽式全流取样机 rocking whole-stream cutter; rocking whole-stream sampler
摇车 track-cycle
摇车避车洞 troll(e)y shelter
摇车钟 ring;telegraph
摇尺 setting
摇尺测量 swinging staff
摇尺杆 set lever
摇尺机构 set-work
摇尺水准测量法 <司尺员前后摇动标尺,取最小读数为正确读数> waving the rod;waving rod level(l)-ing method
摇窗 pivoted casement;pivoted window;pivot light
摇床 cradle;jerking table;picking table;rocker;shaker;shaking table;table concentrator;tabling
摇床分级机 table classifier
摇床浮选 flo(a)tation-tabling;table flo(a)tation
摇床工 table man
摇床精矿 table concentrate
摇床精选 table concentration
摇床流槽选矿装置 shaking-bed sluicing device
摇床流程 table circuit
摇床面 table deck
摇床湿法洗选 wet tabling
摇床式电炉 shaker hearth electric(al) furnace
摇床尾矿 table tailings
摇床序批间歇式反应器 shaking-bed sequencing batch reactor
摇床选 table separation
摇床选矿(法) tabling
摇锤 helve hammer
摇锤式破碎机 swing hammer crusher
摇荡 shake up;toss
摇动 convulse;cranking;joggle;jolt;juggling; jutter; rocking; rock shake;shaking up;sway(ing);swaying motion;swing(ing) motion;topple;vibrate;vibration;wag;waving;wig-wag;wobbling
摇动板 cradle plate
摇动包 shaking ladle
摇动臂 swing arm
摇动槽 shaking channel
摇动床层 shaken bed
摇动阀 oscillating valve
摇动法 suspension system
摇动给料机 shaking feeder
摇动惯性力 inertia shaking force
摇动混合机 shaking mixer
摇动混合器 shaking mixer
摇动机 <测定石油中沉淀物与水分含量的离心机> shakeout machine
摇动机构 head motion;shaker mechanism;wabbler mechanism
摇动机制 censer mechanism
摇动架 jiggled frame
摇动接点式电压调整器 rocking-contact voltage regulator
摇动客车座席 rocker car seat
摇动客车座席铸架 rocker casting
摇动框架式研磨机 swing-frame grinder
摇动溜槽 bull shaker;oscillation chute;shaker chute
摇动漏斗 shaking chute
摇动炉箅 rocking grate;shaking grate
摇动炉箅加煤机 shaking grate stoker
摇动炉排 rocking grate
摇动炉条 shaking grate
摇动炉栅 rocking grate
摇动螺栓 shake-bolt
摇动螺旋面 shaken helicoids
摇动马达 oscillating motor
摇动拍溅 sloshing
摇动盘水工模型 rocking tray hydraulic model
摇动培养法 shake culture
摇动平板 saddle strap
摇动器 shaker;shaking apparatus;wobbler
摇动切片机 rocking microtome
摇动筛 bull shaker; shaker screen; shaking sieve; swing sieve; shaking screen
摇动筛倾角 shaker pitch
摇动筛条 oscillating bar
摇动筛纤维吸出装置 shaking-screen suction table
摇动栅式冷却器 shaking grate-type cooler
摇动湿度计 sling psychrometer
摇动式格筛 shaking grizzly
摇动式混汞板 shaking plate
摇动式混合机 shaking apparatus
摇动式活塞冷却法 shaker method of piston cooling
摇动式间接电弧炉 indirect arc rocking furnace
摇动式拣矸台 shaking picking table
摇动式炉 rocking furnace
摇动式筛分机 shaking grate
摇动式输送机 oscillating conveyer[conveyor]
摇动式输送器 oscillating conveyer[conveyor]
摇动式振捣器 jolt vibrator;shock vibrator
摇动试验 shake test;shaking test
摇动输送机 shaker conveyer [conveyor]
摇动套筒 shaking sleeve
摇动凸轮 oscillating cam
摇动洗净机 bogie bath machine;bogie washing machine
摇动斜槽 shaking trough
摇动卸料车 rocker-dump car;rocker-dump hand-cart
摇动者 shaker
摇动针碎机 shaking pick breaker
摇动支承 swing bearing
摇动支座 rocker bearing
摇动装置 shaking apparatus; shaking appliance;tilter
摇抖动 wobbler action
摇法 rotating manipulation
摇杆 hand hold;hinge pedestal;jiggle bar;jointed arm;main rod;operating arm;pitman;radius link;rocker;rock(er) arm;rocking bar;rocking lever;slotted lever;swing bar;trail;connecting rod <蒸汽机车的>
摇杆插销 stub key
摇杆齿轮机构 rocker gear mechanism
摇杆传动装置 rocker gear
摇杆刀架 rocker tool post
摇杆吊货法 single boom system
摇杆端 main rod end
摇杆端部 mule head
摇杆端填块 stub back block
摇杆盖 rocker cover
摇杆辊子 persuader roll
摇杆机构 end play device; rocker gear;rocker mechanism
摇杆曲拐销 driving crank pin
摇杆筛 rocker sieve
摇杆式加热炉 rocker-bar heating furnace
摇杆式模具 rocker-type die
摇杆式射孔器 swing jet perforator
摇杆铜套 main rod bearing;main rod brass;stub bushing
摇杆铜瓦 stub brass
摇杆头 main rod end;pitman head
摇杆推料炉 rocker-bar furnace
摇杆箱 rocker box
摇杆罩盖 rocker cover
摇杆支承 rocker-bar bearing
摇杆支承铰 rocker support
摇杆轴 rocker shaft
摇杆轴承 main rod bearing;pin rocker bearing;rocker bearing
摇管机构 casing tube oscillating mechanism
摇光 roller flattening
摇滚作用 <车辆运动的> rock and roll action
摇晃 dangle; labo(u)ring; swag; sway;toss;waver;weave
摇晃不定的 ramshackle
摇晃不定的船 crazy ship
摇混 shaking up
摇架 rocker casting;rocket centring
摇架式滑道 slipway with cradle
摇架铁件 rocking iron
摇角 <摇架的> cradle angle
摇绞 reeling silk
摇绞机 hank reel
摇镜头拍摄全景 pan
摇块 swing block
摇框 bracket
摇篮 cot;cradle
摇篮式继电器 cradle relay
摇连杆 rod
摇连杆铜套 rod bearing
摇连杆铜瓦 rod bearing
摇梁 rocking beam;walking beam
摇橹 scull;yuloh
摇橹划子 scull boat

摇炉杠杆 grate lever
摇炉器 grate shaker
摇炉轴 grate shaft
摇落 fetch (a)way
摇门 double-acting door
摇频 swept frequency
摇频测试 sweep check
摇频检验 sweep check
摇频信号发生器 wobbulator
摇频振荡器 frequency sweep generator; megasweep
摇频制 wobbling system
摇瓶培养 shake flask culture
摇瓶培养法 shaking culture
摇瓶培养试验法 shaking-bottle incubating test method
摇瓶试验 shake flask test
摇溶的 thixotropic(al)
摇溶剂 thixotropic(al) agent
摇溶凝胶 thixotropic(al) gel(ation)
摇溶外加剂 thixotropic(al) admixture
摇溶污泥 thixotropic(al) sludge
摇溶现象 thixotropy
摇溶性 thixotropy
摇筛 bull shaker; reciprocating grid; shaking sieve; swing(ing) screen; swing(ing) sieve; vibration screen
摇筛机 jigging sieve; mechanical sieve shaker; shaker apparatus; sieve shaker; test-sieve shaker; test sieve vibrator; vibrating grizzly; vibrating screen
摇筛器 sieve shaker
摇筛试验 shaking test
摇扇门 swinging door
摇石 log(g)an stone
摇实密度 tap density
摇式排架 rocker bent
摇式桥墩 rocking pier
摇式卸车机 rocker unloader
摇手柄 crank handle; jiggle bar; starting crank; winding handle
摇手装置 hand cranking device
摇首 yaw
摇首倾侧 yaw heel
摇闩 swing bar
摇松以使整齐 shake up
摇台箱 <伞齿切齿机的> cradle housing
摇头 hunting; panning head
摇头把杆 swivel boom
摇头窗 transom
摇头风扇 pivting fan
摇头角度 panning angle
摇头式风扇 circulating fan; oscillating fan
摇头压榨机 tilting head press
摇头振动 nosing
摇腿 oscillating mast
摇洗槽 tossing tub
摇旋振动 torsional vibration
摇檐 roof projection
摇摇晃晃 lopping
摇曳 <机车等的> nosing
摇曳力 sway force
摇移器 rocker
摇椅 rocker (car seat); rocking chair
摇匀 shake up
摇运机 oscillating conveyer [conveyor]; oscillation chute
摇针 rotating the needle; shaking the needle
摇枕 bogie bolster; bolster; bottom bolster; cradle; swing bolster; truck bolster
摇枕撑套 bolster thimble
摇枕挡 friction(al) block; truck bolster guide; column guide <拱板转向架>
摇枕吊 bolster hanger; crossbar; hanger link; link hanger; swing bar; swing hanger; swing link
摇枕吊 U 形螺栓 bolster hanger U bolt
摇枕吊耳 bolster guide carrier; bolster hanger carrier; bolster hanger lug; swing hanger carrier; swing hanger lug; swing hanger pin bearing
摇枕吊杆 bolster beam
摇枕吊下部吊销 lower swing hanger pivot
摇枕吊销 bolster hanger pin; swing hanger pin
摇枕吊眼孔螺栓 link hanger eye bolt
摇枕吊轴 bolster hanger axle; bolster hanger pivot; crossbar; mandrel pin; spring plank bearing; spring plank carrier; swing hanger axle; swing hanger crossbar; swing hanger shaft
摇枕吊轴轴颈 stub-end of swing hanger crossbar
摇枕定位 location of bolster
摇枕横向弹簧 swing bolster spring
摇枕簧座 bolster spring seat
摇枕减震器 bolster snubbing device
摇枕拉杆 bolster anchor
摇枕摩擦板 bolster chafing plate
摇枕摩擦减振器 bolster snubber
摇枕弹簧 body spring; swing motion spring
摇枕弹簧装置 bolster spring device
摇枕中心距 distance between pins
摇枕纵向挡 bolster guide
摇振试验 shaking test
摇振铁架 cradle iron
摇轴 rocker (shaft); rock(ing) shaft
摇轴链 rocker joint chain
摇轴支座 pendulum stanchion; rocker bearing; rocker support; tilting bearing
摇轴座 rocker trunnion block
摇柱式 <隔离开关的> rocking-post type
摇转 cranking
摇转拌和筒 churn
摇转吊扇 ceiling oscillating fan
摇转搅拌器 churn
摇转搅拌筒 churn
摇转马达 cranking motor
摇转竖轴窗框 swivel frame
摇转损失 cranking loss
摇钻 bit stock; brace; brace bit; bit brace
摇钻螺丝攻 bit brace tap
摇座 pendulum stanchion; rocker bearing; rocker support; rocking pier; rocking shaft; swing bearing; tilting bearing; tumbler bearing
摇座排架 rocker bent
摇座桥墩 rocking pier
摇座座板固定支座 rocker-and-masonry-plate bearing

遥

遥拜方向 <伊斯兰教徒的> kibleh
遥臂 jib arm
遥测 distance observation; distant measurement; remote indication; remote measurement; remote measuring; remote metering; remote monitoring; remote sensing; telemeasure(ment); telemetering (measurement); telereconnaissance
遥测安培表 teleammeter
遥测安培计 teleammeter
遥测报告 remote reporting; telemetry report
遥测编码 telemetry code
遥测编码器 telemetry encoder
遥测编码设备 telemetry encoder
遥测编码调制 telemetry code modulation
遥测波浪浮标 telemetering wave buoy
遥测波形图 telemetric oscillogram
遥测采集 telemetry acquisition
遥测参数 telemetry parameter
遥测操纵器 teleoperator
遥测操作的 teleoperated
遥测程序 telemetry schedule
遥测储罐液位计 remote-reading tank ga(u)ge
遥测处理 telemetry processing
遥测处理机 telemeter processor
遥测处理站 telemetry processing station
遥测传感器 telemetering pickup; telemetering sensor; remote sensor
遥测传感设备 remote sensor
遥测传送 telemeter; teletransmission
遥测传送器 teletransmitter
遥测船 telemetry ship; tracking ship <导弹靶场的>
遥测单色计 telemetering monochrometer
遥测到的信息 telemetered information
遥测地面系统 telemetry ground system
遥测地面站 telemetry ground station
遥测地震 telseis
遥测地震计 telemetering seismometer
遥测地震台网 telemetered seismic net; telemetering seismic station network
遥测地震台阵 telemetered array of seismic stations
遥测地震学 teleseismology
遥测地震仪 remote measurement seismograph; telemetering seismograph
遥测电传液面高度计 tellevel
遥测电动机 telemotor
遥测电流计 teleammeter
遥测电路 telemetric circuit
遥测电压表 televoltmeter
遥测读出 telemetry readout
遥测读数 distance reader; distant reading; remote (control) reading
遥测多路传输系统 remote multiplexing system
遥测发射机 telemetering transmitter; telemeter transmitting set; telemetry transmitter
遥测发射器 telemeter
遥测发送器 telemeasure transmitter; telemetering sender; teletransmitter
遥测发信机 telemeasure transmitter
遥测法 telemetry
遥测方法 method of telemetering; telemetering medium
遥测放大器 telemetering amplifier
遥测分光镜 telespectroscope
遥测伏特计 televoltmeter
遥测跟踪和指令站 telemetry tracking and command station
遥测跟踪系统 telemetry tracking system
遥测跟踪指令和监听站 telemetry tracking, command and monitor station
遥测功率计 telewattmeter
遥测光度表 telephotometer
遥测光度计 telephotometer
遥测光度学 telephotometry
遥测海流计 remote-reading current meter; telemetering oceanographic current meter
遥测海洋流速仪 remote-reading current meter; telemetering oceanographic current meter
遥测海洋剖面装置 telemetering ocean profiling system
遥测环境剖面装置 telemetering environment(al) profiling system
遥测回声仪 telesounder
遥测机 telega(u)ge
遥测积分器 telemetric integrator
遥测基准脉冲 telemetry reference
遥测计 telega(u)ge; telemeter; telemetering system; telemometer
遥测计数器 telemetering counter
遥测计算 telemetry computation
遥测记录 remote recording; telerecording
遥测记录地震仪 remote recording seismograph
遥测记录器 distance recorder; histogram recorder; telerecorder
遥测记录系统 distant recording system
遥测记录仪 telerecorder
遥测记录仪器 distant recording instrument
遥测技术 radio telemetry; telemetering technique; telemetry
遥测加法器 telemetering totalizer
遥测监测系统 telemetry observation system
遥测检查系统 telemetry checkout system
遥测角计 telegoniometer
遥测接收机 telemetering receiver; telemeter receiving set; telemetry receiver
遥测接收装置 telemetering receiving apparatus
遥测距离 remote measurement
遥测空气温度计 distance-type of cooling thermometer
遥测孔斜计 teleclinograph
遥测控制 telemetric control; telemetry control
遥测控制设备 telemetered control equipment
遥测控制与监听 telemetry control and monitor
遥测力平衡变换器 telemetering force balance type converter
遥测流量计 telemetering current meter; teletach(e)ometer
遥测流速仪 telemetering current meter; teletach(e)ometer
遥测滤波器 telemetry filter
遥测模拟器 telemetry simulator
遥测频带 telemetry band; telemetry frequency
遥测气象计 telemeteorograph
遥测气象学 telemeteorometry
遥测气象仪器学 telemeteorography
遥测气象站 telemetering weather station
遥测器 remote-detector; telemeter; telesounder
遥测设备 remote measuring equipment; telemeter equipment; telemetering device; telemetry equipment; telemetry unit; telesynd
遥测设备校准 telemetry equipment calibration
遥测深度仪 telemetering depth meter
遥测深水光度计 telerecording bathyphotometer
遥测湿度表 telepsychrometer
遥测湿度计 telehygrometer
遥测式应变计 remote-ready strainometer
遥测室 telemetry building

遥测收信机 telemeasuring receiver
遥测输入系统 telemetry input system
遥测术 remote measurement
遥测数据 telemeasuring data; telemetered data; telemetering data; telemetric data; telemetry data
遥测数据变换器 telemetric data converter
遥测数据处理 teledata processing; telemetered data reduction; telemetry data reduction
遥测数据处理机 telemetry processor
遥测数据处理系统 teledata processing system
遥测数据传递网 telemetry data network
遥测数据发射机 telemetric data transmitter; telemetric data transmitting set
遥测数据发射装置 telemetric data transmitting set
遥测数据分析器 telemetric data analyser [analyzer]
遥测数据记录器 telemetric data recorder
遥测数据监测仪 telemetry data monitor
遥测数据监控器 telemetric data monitor; telemetric data monitoring set
遥测数据鉴定系统 telemetry data evaluation system
遥测数据接收机 telemetering data receiving set; telemetric data receiver; telemetry data receiver
遥测数据接收与记录设备 telemetric data receiving-recording-scoring set
遥测数据门模件 telemetry data gate module
遥测数据数字转换器 telemetry data digitizer
遥测数据系统 telemetry data system
遥测数据压缩器 telemetry data compressor
遥测数据中心 telemetry data center [centre]
遥测数据转换器 telemetric data converter
遥测双目镜 telebinocular
遥测水深仪 telemetering depth meter
遥测水听器 telemetering hydrophone
遥测水位表 remote water-level indicator
遥测水位计 long-distance stage transmitter; long-distance water level recorder; long-distance water-stage recorder; remote water-level indicator
遥测水位指示器 telemetering water-level indicator
遥测水位指示仪 telemetering water-level indicator
遥测水下操纵器 remote underwater manipulator
遥测水下测音器 telemetering hydrophone
遥测水银温度计 mercury telethermometer
遥测速度计 telespeedometer
遥测台 observatory-remote; remote observatory; telemetering station
遥测台网 telemetry network; telenetwork
遥测台站 telestation
遥测探空 remote sounding
遥测天线 telemetering antenna; telemetry antenna
遥测调制系统 telemetry modulation system
遥测通道 telemetering channel
遥测网 telemetry net
遥测桅 telemetry mast
遥测温度表 distance thermometer; distant thermometer; telethermograph; telethermometer; telethermoscope
遥测温度计 distance reading thermometer; distance thermometer; distant-reading thermometer; distant thermometer; remote-reading thermometer; remote thermometer; telethermograph; telethermometer; telethermoscope
遥测温度器 telethermoscope
遥测温度自动记录器 telethermograph
遥测系换向器 telemetering commutator
遥测系统 remote measuring system; remote metering system; remote supervision system; supervising system; telemeter(ing) system; telemetry system
遥测显微器 telemicroscope
遥测线 telemetric link; telemetry link
遥测线路 telemetering link; telemetry link
遥测信号 remote signal; telemetered signal; telesignalisation [telesignalization]; telesignal(l)ing
遥测信号化 telesignal(l)isation [telesignal(l)ization]
遥测信号设备 telesignallisation [telesignallization]
遥测信息 teleinformation; telemetry information
遥测信息还原与显示系统 telemetry decommutation and display systems
遥测压力表 telemanometer
遥测压力的 telepressure
遥测压力计 telemanometer; telepressure ga(u)ge
遥测压力收受器 telepressure receiver
遥测验潮仪 tide-ga(u)ge telemetry
遥测液位 remote surveying of liquid level
遥测仪 remote-reading ga(u)ge; telega(u)ge; telemeter; telemometer
遥测仪表 distant action instrument; remote indication instrument; telega(u)ge
遥测仪器 long-distance recorder; remote-indicating instrument; telemetric instrument; distance apparatus
遥测仪器舱 module of telemetry equipment
遥测引信 telemetry fuze
遥测与指令天线 telemetry and command antenna
遥测与指令站 telemetry and command station
遥测雨量计 telemetering rain ga(u)ge
遥测载波 telemetry carrier
遥测站 telemetering station; telemetry station
遥测振荡器 telemetering oscillator
遥测振动数据 telemetered vibration data
遥测指令 telemetry command
遥测指令接收机 telecommand receiver
遥测指令系统 telemetry (and) command system
遥测指示器 distant indicator; teleindicator
遥测中断 telemetry blackout
遥测终端设备 telemetering terminal equipment
遥测终端系统 telemetry terminal system
遥测转速表 distance reading tach(e)ometer; teletach(e)ometer
遥测转速计 distance reading tach(e)ometer; distant-reading tach(e)ometer; teletach(e)ometer
遥测装置 remote measurement device; remote measuring unit; remote monitor; telemeter; telemetering device; telemetering equipment; telemetering gear; telemetering installation; telemetry unit
遥测资料 telemetry data
遥测自动处理系统 telemetry automatic processing system
遥测自动还原设备 telemetry automatic reduction equipment
遥测自动记录仪 automatic telerecorder; telerecorder
遥测自动简化系统 telemetry auto reduction system
遥测自记设备 distant recording instrument
遥导 teleguide
遥读 distance reading
遥读记录器 remote-read register
遥读水位表 remote water-level indicator
遥读系统 remote readout system
遥读转速计 remote reading tachometer
遥感 rapid reconnaissance; remote observation; remote sense; telereconnaissance
遥感标志 remote-sensing indication
遥感采样 remote sampling
遥感采样器 remote sampler
遥感测量 remote-sensing survey
遥感测量仪 remote-sensing instrument
遥感成矿地质条件分析远景评价方法 perspective evaluating method by analysis minerogenetic geologic(al) conditions based on remote sensing
遥感成像 remotely sensed imagery; remote-sensing imagery
遥感城市建筑工程地质条件评价 engineering geologic (al) condition assessment for urban architecture based on remote sensing
遥感传感器 remote sensor; sensor of remote sensing
遥感地层单位 remotely sensed stratigraphic(al) units
遥感地貌类型 remotely sensed geomorphologic(al) types
遥感地热勘查应用 remote-sensing applications for geothermal field prospecting
遥感地学多数据处理成矿统计预测 statistic(al) forecast methods for minerogenesis by processing remote sensing-geonomy data
遥感地学资料综合分析远景评价方法 perspective evaluating method by comprehensively analysing remote sensing-geonomy date
遥感地震地质调查应用 remote-sensing applications for seismogeologic(al) survey
遥感地震地质调查应用成果 results of remote sensing applications for engineering geologic(al) survey
遥感地质 remote-sensing in geology
遥感地质成矿模式 geologic(al) minerogenetic model based on remote sensing
遥感地质调查报告编写 compilation of geologic(al) investigation report based on remote sensing
遥感地质学 remote-sensing geology
遥感地质应用 remote-sensing applications in geology
遥感地质应用成果种类 category of results of remote sensing applications for geology
遥感地质制图 geologic(al) mapping based on remote sensing
遥感地质制图程序 geologic(al) mapping procedure based on remote sensing
遥感地质制图类型 kinds of remote sensing applications for geologic(al) mapping
遥感地质专题编图 geologic(al) thematic mapping based on remote sensing
遥感发动 remote start(ing)
遥感方法 remote-sensing method
遥感方式 means of remote sensing
遥感飞机 remote-sensing aircraft
遥感分类 classification of remote sensing
遥感概略地质图 geologic(al) sketch map based on remote sensing
遥感工程地质调查应用 remote-sensing applications for engineering geologic(al) survey
遥感工程地质调查应用成果 results of remote sensing applications for hydrogeologic(al) survey
遥感工程地质调查应用种类 category of remote sensing applications for engineering geologic(al) survey
遥感构造地质及地质制图应用 remote-sensing applications for structural geology and geologic (al) mapping
遥感构造地质及地质制图应用成果 results of remote sensing applications for structural geology and geologic(al) mapping
遥感构造图 structural map based on remote sensing
遥感构造要素 remotely sensed structural elements
遥感环境地质调查应用种类 category of remote sensing applications for environmental geologic(al) survey
遥感环境基础地质调查 foundational environmental geologic (al) survey based on remote sensing
遥感环境矿产资源调查 environmental mineral resources exploration based on remote sensing
遥感环境稳定性评价 environmental stability assessment based on remote sensing
遥感环境污染监测应用 remote-sensing applications for environmental pollution monitoring
遥感环境综合调查应用 remote-sensing applications for comprehensive environmental surveying
遥感环境综合调查应用成果 results of remote sensing applications for comprehensive environmental investigation
遥感火箭 remote-sensing rocket
遥感几何学 remote-sensing geometry
遥感技术 remote sensing; remote-sensing technique; remote-sensing technology; remote-transmission technique
遥感监测设备 remote supervision plant
遥感解译 remote sensing interpretation
遥感空气监测 telemetered air monitoring
遥感矿产勘查应用 remote-sensing applications for mineral exploration
遥感矿产勘查应用成果 results of remote sensing applications for min-

eral exploration
遥感矿产信息提取找矿方法 method of ore prospecting by use of extracting remotely sensed minerals product information
遥感矿产预测图 ore forecasting map based on remote sensing
遥感煤田勘查应用 remote-sensing applications for coal field prospecting
遥感能源勘查应用 remote-sensing applications for energy exploration
遥感平台 platform; platform for remote sensing; remote platform; remote-sensing platform
遥感平台编号 platform number
遥感气球 remote-sensing balloons
遥感器 remote-sensing instrument
遥感器组件 remote sensor package
遥感区域地质图 regional geologic(al) map based on remote sensing
遥感热异常图 remote thermal map
遥感扫描器 remotely sensed scanner
遥感设备 remote-sensing equipment; remote sensor
遥感设施 remote-sensing device
遥感试验 remote-sensing experiment
遥感数据 remote-sensing data
遥感水文地质调查应用 remote-sensing applications for hydrogeologic(al) survey
遥感水文地质调查应用成果 results of remote sensing applications for energy exploration
遥感水系类型 remotely sensed drainage pattern
遥感特征光谱信息找矿方法 method of ore prospecting using remotely sensed characteristic spectral information
遥感图像 remote-sensing image
遥感图像处理 remotely sensed image processing
遥感图像处理研究成果 research results of image processing
遥感图像地质解译 geologic(al) interpretation of remote sensing image
遥感图像解译 interpretation of remotely sensed images
遥感图像解译成果 remote-sensing image interpretative results
遥感图像解译法 method of remote sensing image interpretation
遥感图像水文地质解译 hydrogeological interpretation of remote-sensing images
遥感外生地质灾害调查应用成果 results of remote sensing applications for exogenic geologic(al) hazard survey
遥感温度表 telethermometer
遥感系列地质图件编制 serial geologic(al) mapping based on remote sensing
遥感系统 remote-sensing system
遥感线 remote-sensing line
遥感详细地质图 detail geologic(al) map based on remote sensing
遥感信号控制机 traffic-actuated controller
遥感信息 remote-sensing information
遥感信息处理 remote-sensing information processing
遥感信息传输 remote-sensing information transmission
遥感信息获取 remote-sensing information acquisition
遥感修编地质图 geologic(al) remapping based on remote sensing
遥感学 remote sensing
遥感岩石类型 remotely sensed lithologic(al) kinds
遥感遥测分析法 remote-sensing and measuring analysis
遥感仪器 instrument for telereconnaissance
遥感仪器设备 remote-sensing instruments and equipments
遥感应用实验室 laboratory for application of remote sensing
遥感影像标志 mark of remote-sensing image
遥感油气勘查应用 remote-sensing applications for oil-gas field prospecting
遥感与环境分析信息系统 information system of remote sensing and environmental analysis
遥感找矿 remote-sensing prospecting
遥感制图 remote-sensing cartography; remote-sensing mapping
遥感中心 remote-sensing center [centre]
遥感装置 remote control assembly; remote-sensing device
遥感资料 remote-sensing material
遥供电源 range power supply
遥截管 remote cut-off tube
遥截止 remote cut-off
遥截止管 remote cut-off tube
遥截止特性 remote cut-off characteristics
遥截止五极管 remote cut-off pentode
遥警 remote alarm
遥控 distance control; distance operation; distant control; distant operation; handle control; remote control; remote handling; remote manipulation; telearehics; telecontrol; telemechanic(al) control; telemonitor(ing); telerun
遥控 X 射线诊断机 remote-controlled X-ray unit
遥控安培表 teleammeter
遥控安全阀 remotely operated relief valve
遥控按键 remote keying
遥控靶艇 remote control target-boat
遥控板 mini-panel; operator-held control box; remote panel; telecontrol board
遥控报警键 remote alarm control key
遥控爆破车 remote-controlled demolition carrier
遥控泵 remote-controlled pump
遥控泵房 telecontrol pumping house
遥控泵头 remote head
遥控泵站 remote-controlled pump station
遥控编码 remote control coding
遥控编码器 remote control coder
遥控变电所 remote control substation; telecontrolled substation
遥控变速 remote gear control
遥控拨号器 remote control dial unit
遥控拨号装置 remote control dial unit
遥控补机 remote-controlled helper
遥控补码系统 remote patch system
遥控采煤法 push-button coal mining; remote control coal mining
遥控操舵系统 telemotor system
遥控操舵装置 remote steering gear; telemotor steering gear
遥控操纵台 remote control console
遥控操纵装置 teleoperator
遥控操作 remote(-controlled) operation; straight-forward operation
遥控操作机 remote control manipulator
遥控操作机构 remote control gear
遥控操作器 remote manipulator; teleoperator
遥控操作钳 remote handling tongs
遥控操作天平 remote-operated balance
遥控操作装置 remote control apparatus
遥控插入码系统 remote patch system
遥控车 telecar
遥控车站 remote-controlled station
遥控沉箱 remote-controlled caisson
遥控程序 controllable function
遥控处理 remote processing; teleprocessing
遥控处理计算机 remote processing computer
遥控处理系统 teleprocessing system
遥控触发 remote triggering
遥控穿孔机 telepunch
遥控传动装置 remote control of transmission; telemotor gear
遥控传感器 remotely monitored sensor
遥控船 drone ship; remote control vehicle; remote operation vehicle
遥控船舰射击靶 queen duck
遥控存取终端 remote-access terminal
遥控导阀 remote pilot valve
遥控道岔【铁】 remotely controlled points; remotely actuated points; remotely controlled switch; remotely operated points
遥控道口 remotely controlled level crossing
遥控的 distantly controlled; remotely controlled; remotely operated; robot; teledynamic
遥控的沉井挖掘机 remote control caisson excavator
遥控的监控装置 remotely controlled monitor
遥控地震法 remote seismic method
遥控点火 remote control ignition
遥控电磁测斜仪 electromagnetic teleclinometer
遥控电动机 apogee motor; telechron motor; telemotor
遥控电缆 remote control cable
遥控电路 remote control circuit
遥控电视 remote-controlled television
遥控电站 remote-controlled station; telecontrolled power station
遥控电子交换系统 remote control electronic switching system
遥控定向钻进 remote-controlled directional well drilling
遥控动力潜水器 remote-controlled motorized submersible
遥控读数 remote reading
遥控调车机车 <远程控制的调车机车> remote-controlled shunting locomotive
遥控断电 remote power-off
遥控断路器 remote control circuit breaker
遥控对光 remote control focusing
遥控对象 remotely controlled object
遥控发报 remote keying
遥控发电站 remote-controlled station; telecontrolled power station
遥控发动机 remote control engine; telemotor
遥控发射 remote launch
遥控发射指挥台 remote firing panel
遥控阀 remote-controlled valve; remote(-operated) valve
遥控阀控制 remote valve control
遥控方位角自动同步机 remote azimuth selsyn
遥控飞行器 remote vehicle
遥控分机 field equipment of conditional transfer of control
遥控分配器 remote control distributor
遥控风门 remote control door
遥控风门机构 damper with remote control device
遥控浮标 remote buoy; remotely controlled float
遥控浮子 remote float
遥控杆 remote control lever
遥控高压开关 remote control high tension switch
遥控工序 controllable function
遥控工作输入 remote job entry
遥控轨道缓行器 remote-controlled rail retarder
遥控锅炉 remote-controlled boiler
遥控海底取样 remote bed corer
遥控海底十字板剪切仪 Seavane
遥控盒 remote control box
遥控恒温器 remote bulb thermostat; remote control thermostat
遥控混凝土喷射机 robot spray
遥控机构 remote control gear; telemechanism
遥控机机构 telemotor gear
遥控机器人 radio robot; telepuppet; telerobot
遥控机械手 remote manipulator
遥控机械系统 telemechanic(al) system
遥控机械学 teleautomatics; telemechanics; telemechanism
遥控机械装置 robot mechanism; telemechanic(al) apparatus; telemechanic(al) device; telemechanic(al) installation; telemechanic(al) system; telemechanic(al) unit
遥控极谱仪 remotely operated polarograph
遥控集中联锁 remote-controlled interlocking
遥控计时钟 telechron clock
遥控记录 remote recording
遥控记录仪 remote recorder
遥控记录仪表 remote-controlled instrument; remote-indicating instrument
遥控继电器 banked relay; distance relay; remote-controlled relay; telecontrol relay
遥控加工 teleprocessing
遥控驾驶的车辆 remotely driven vehicle
遥控监测 remote monitoring; remote supervisory control; telemonitor(ing)
遥控监测仪 remote monitor
遥控监测站 remote monitoring station
遥控监视 remote surveillance
遥控监视设备 remote control equipment; remote supervisory equipment
遥控监视系统 remote control monitoring system
遥控减速器 remote-controlled rail retarder
遥控键 teleswitch
遥控键盘 remote keypad
遥控浇注 remote casting
遥控脚踏板 remote control foot pedal
遥控接管器 remote pipe connector
遥控接收机 remote-controlled receiver
遥控接收器 receiver of remote control system
遥控接收装置 remote control receiving unit

遥控接头 remote-type coupling
遥控警报器 remote alarm
遥控开窗设备 remote window control equipment
遥控开关 remote control switch; remote cut-off; remotely controlled switch; remote switch; teleswitch; trip switch
遥控开关系统 remote control switching system
遥控可变光阑 remote control iris
遥控刻图器 remote control scriber
遥控空气阀 remotely controlled air valve
遥控控制台 remote control console
遥控累积计重器 remote weight totaliser [totalizer]
遥控冷凝器 remote condenser; remote control condenser type air conditioner
遥控力矩器 teletorque
遥控力学 teleautomatics; telemechanics; telemechanism
遥控连接器 remote control connector
遥控联锁 remote-controlled interlocking
遥控联锁机 remote interlocking machine
遥控联锁架 remote interlocking frame
遥控联锁连接器 remote interlock connector
遥控灵敏装置 remote sensing
遥控罗盘 remote compass; telecompass
遥控螺线管 remote-controlled solenoid
遥控马达 telemotor
遥控面板 remote control panel
遥控命令 remote control command
遥控目标 remotely controlled object
遥控排字机 teletypesetter
遥控盘 remote control board
遥控配电站 remote control substation
遥控喷射器 remote-controlled injector; remote injector
遥控批量处理 remote batch process
遥控平交道口 remotely operated level crossing
遥控平台 remote platform
遥控启动 remote start(ing)
遥控起爆 remote-controlled priming
遥控起动 remote start(ing)
遥控起动器 remote control starter
遥控起锚机 remote-controlled anchor windlass
遥控器 remote controller; remote control manipulator
遥控潜水器 remote-operated vehicle
遥控区段 remotely controlled section
遥控取样 remote sampling
遥控取样器 remote control sampler; remote sampler
遥控容量滴定 remote volumetric titrimetry
遥控闪光继电器 remote flashing relay
遥控设备 remote control apparatus; remote control device; remotely controlled plant; remote operating equipment; telecommand equipment
遥控设施 remote control facility
遥控摄像设备 remote pickup unit
遥控式的 distance-type
遥控式干燥机 remote-controlled drier [dryer]
遥控式机器人 remote-controlled robot
遥控式摄像机 remotely controlled camera; remotely operated camera
遥控式钻机 push-button drilling machine
遥控视频显示单元 remote video display unit
遥控室 remote control office; telecontrol room
遥控释放机构 remote control releaser
遥控收发报机 remote transmitter-receiver
遥控手柄 remote control handle
遥控输入 remote input
遥控输入装置 remote input units
遥控数据处理 remote-sensing data processing
遥控数据处理中心 remote-sensing data processing center[centre]
遥控数据发射机 remote data transmitter
遥控数据收集 remote data collection
遥控数据输入 remote data input
遥控水电站 remotely supervised hydroelectric(al) station
遥控水位指示器 remote-controlled water level indicator
遥控水下机器人 remote-controlled underwater robot
遥控水下机械手 remote-controlled underwater manipulator
遥控水下推土机 remote-controlled underwater bulldozer
遥控水下挖掘机 remote-controlled underwater excavator
遥控水下运载工具 remotely controlled underwater vehicle
遥控水下作业车 remote-operated vehicle
遥控所 remote control office
遥控台 remote control board; remote station
遥控台报警器 remote station alarm
遥控探测装置 telepuppet
遥控天平 remote control balance
遥控调焦 remote control focusing; remote focus control
遥控调节器 distant regulator
遥控调谐 remote tuning
遥控调制 remote control modulation
遥控铁路火车 remotely controlled railway trains
遥控艇 distance-controlled boat
遥控通道 remote control channel
遥控同步机 telesyn
遥控同步系统 remote-synchronizing system
遥控同步装置 telesynd
遥控图像处理系统 remote information processing system
遥控图像信号源 remote video source
遥控推土机 remote control bulldozer
遥控拖拉机 robot tractor
遥控挖掘机 remote control excavator; remote-operated crane
遥控望远镜 remotely operated telescope
遥控微音器 telemicrophone
遥控维修 remote maintenance
遥控维修装置 remotely maintained plant
遥控温包 remote bulb
遥控温度计 remote-reading thermometer
遥控无人管理的工作系统 remote unmanned work system
遥控无人作业系统 remote unmanned work system
遥控无线电系统 remote control radio system
遥控吸移管 remote pipet
遥控系统 remote(-controlled) system; telechirics; telecontrol system; telemechanic(al) system
遥控系统测试 remote control system testing
遥控系统脉冲发生器 telecontrol system pulse generator
遥控显示器伺服机构 remote-indicator servomechanism
遥控显微镜 remote control microscope
遥控线路 keying line; remote line
遥控小型盾构 tele-mole
遥控信号(机) remote(-controlled) signal
遥控信号楼【铁】 remotely controlled signal box
遥控信号箱【铁】 remotely controlled signal box
遥控信号选择 remote control signals selection
遥控信号周期变更装置 remote cycle change
遥控选择器 remote selector
遥控学 telautomatics
遥控仰角自动同步机 remote elevation selsyn
遥控-遥信装置 remote operating and signal(l)ing device
遥控液压操舵装置 telemotor control steering gear; telemotor system
遥控仪表 telega(u)ge
遥控移液器 remote pipetter
遥控员 beeper
遥控运行 operate by remote control; remote-controlled operation
遥控再生 remote regeneration
遥控增益放大器 remote gain amplifier
遥控闸板 remote control gate
遥控站 remote control station
遥控真空泵加料机 remote vacuum-pump loader
遥控真空管 remote-controlled valve
遥控指令 remote control command; remote control instruction; telecommand
遥控指令接收器 commander receiver of remote control
遥控指示器 remote indicator
遥控指示压力表 remote-indicating pressure ga(u)ge
遥控制 remote control system; telecontrol system; telemechanic(al) system
遥控制导 external guidance
遥控中层拖网 remotely controlled midwater trawl
遥控中继站 remotely controlled repeater station
遥控中心 remote control center [centre]
遥控终端设备 remote control terminal
遥控周期变更装置 remote cycle change
遥控转向机构 remote control steering gear
遥控装备 remote control installation
遥控装置 control facility; remote control apparatus; remote control assembly; remote-controlled device; remote-controlled mount; remote-controlled unit; remote equipment; remotely controlled plant; remote monitor; remote unit; telecommand; teleequipment
遥控装置控制器 remote unit monitor
遥控自动测向仪 automatic direction finder remote controlled
遥控自动化采煤机 carbide miner
遥控自动化仪表板 remote control automation board
遥控自动技术 teleautomatics
遥控自动驾驶仪 fly-by-wire
遥控自动力学 teleautomatics
遥控自动同步机 magslep
遥控自动信号和联锁系统 remote-controlled automatic signal and interlocking system
遥控自动信号和联锁制 remote-controlled automatic signal and interlocking system
遥控自动装置 remote-controlled robot
遥控自封接头 remote self-sealing coupling
遥控自行装置 remote control rover
遥控自行装置学 teleautomatics
遥控作业工作臂 remotely operated work arm
遥控作用 tele-action
遥领地主 absentee owner
遥领制 absentee ownership
遥示 distant indication; remote indication; remote reading
遥示查询制 interrogation tele-signaling system
遥示管 remote-spotting tube
遥示罗经 distant-reading compass; telecompass
遥示器 remote indicator
遥示器伺服机构 remote-indicator servomechanism
遥示水尺 remote water-level indicator
遥示速率器 speed teleindicator
遥示温度计 remote-indicating thermometer
遥示旋转流量计 remote-indicating rotameter
遥示装置 remote-indicating device
遥调 distant adjustment; distant setting); remote adjustment; remote regulating; remote regulation; remote setting
遥信 remote signal(l)ing; telesignal(l)ing
遥信对象 remotely surveillance object
遥信分区 remotely surveillance subsection
遥信区段 remotely surveillance section
遥信系统 telecommunication system
遥远 far off
遥远操作 remote operation
遥远测角计 telegoniometer
遥远的 distant; off-lying; way out
遥远地区信道 remote area channel
遥远监视 telemonitor(ing)
遥远控制 remote control
遥远受控站 remote-controlled station
遥远性 remoteness
遥置能力 remote mounting capability

鳐

鳐鱼 skate

杳

杳无人迹的 uninhabited

咬

咬边 crimping; undercut <焊接缺陷>

咬茬砌合 racking bond
咬的 rodent
咬底 biting; picking up; working up; pull(ing)-up <涂料涂刷缺陷>; lifting <一种漆病>
咬封盖 omnia
咬缝管 cased tube; close joint tube

咬合 articulation; bite; clutch; coherence; holding-on; joining-up; meshing; occlusion; tooth
咬合板 bite plate
咬合不正 malocclusion
咬合采样器 bottom grab
咬合测距尺 bitegage
咬合的 meshed
咬合对辊 < 磨碎机的 > intermeshing rolls
咬合缝屋面 flat-lock seam roofing
咬合拱 bonded arch
咬合关系 occluding relation
咬合过高 supra-occlusion
咬合痕 scuffing mark
咬合计录用架 bite frame
咬合架 articulator; dental articulator; occluding frame
咬合力计 gnathodynamometer
咬合力学 gnathodynamics
咬合面 malleolar surface; occlusal surface
咬合平衡 occlusal balance
咬合砌体 bonded brick work
咬合砌砖 rat-trap bond
咬合强度 keying strength
咬合式砌肢 bonded brick work
咬合调整 occlusal adjustment; occlusal correction
咬合握裹力 mechanical bond(ing)
咬合线 lines of occlusion
咬合翼片 bite wing
咬合缘 occlusal margin
咬合运动 chewing movement; masticatory movement
咬合纸 articulating paper; occluding paper
咬合纸夹 articulating paper pliers
咬合作用 interlocking; mechanical bond(ing)
咬痕 bite mark
咬坏的 gnawed
咬颊 check biting
咬剪 double cutting snips
咬接 < 砌石的 > laid touching
咬接平瓦 flat interlocking (clay) tile
咬接砌块 interlocking block
咬接头 saddle joint
咬接瓦 interlocking tile
咬接现象 scuffing
咬紧装帧 binding
咬嚼数字 number crunching
咬口 crimping; solderless joint; hollow roll; seaming; welted nosing < 金属薄板屋面的 >
咬口接缝 hook seam; locked joint; lock seam; welted joint
咬口接合 saddle joint; seaming
咬口接头 horsed joint; saddle-back joint; saddle joint; splash lap
咬口金属带 welting strip
咬口连接 groove joint
咬口模 seaming die
咬口轧缝 tongue and groove rolling
咬口折缝 fell
咬力 biting force
咬力测验计 biting force meter
咬黏[粘] seizure
咬起 bite; picking
咬砌 stagger bond
咬入 bite
咬入角 angle of nip; entering angle; nip angle
咬伤 bite wound
咬蚀 biting
咬死 seizure
咬送辊 nip rolls
咬碎 crunch
咬住 bite; chafe; dig; galling; hitch; seize; seizing; seizure

舀取 dip

舀水 bali
舀水器 bailer; bali
舀桶 < 钻探用 > bailing bucket

药包 cartridge

药包长度 explosive cartridge length
药包点火 cartridge ignition
药包端穿孔顶杆 < 装雷管用 > prod
药包卷 explosive cartridge
药包类型 explosive cartridge type
药包临界直径 critical diameter of explosive
药包起爆方法 firing method of explosion crater
药包筒 explosive cartridge
药包直径 charge diameter; diameter of charge; diameter of explosion cartridge
药包重量 explosive cartridge weight
药包装药密度 explosive cartridge density
药标 header
药材 botanical; crude drugs; medicinal herb; medicinal materials
药厂 pharmaceutical factory
药橱 chemicals closet
药典 codex
药店 chemist's shop; drug store; medicamentarius; pharmacy
药方 prescription
药房 chemist's shop; dispensary; officina; pharmacy; potecary; drug store < 美 >
药粉 trituration
药膏 ointment; salve
药衡制 apothecaries measure < 容积单位 >; apothecaries weight < 重量单位 >; apothecaries system < 英 >
药壶爆破 chamber blasting; springing shot; squibbing blasting
药剂 medicament
药剂槽 chemical tank
药剂沉淀池 coagulation basin
药剂固定储备量 stand-by reserves
药剂喷射 chemical injection
药剂溶解箱 chemical dissolving box
药剂师 apothecary; dispenser; pharmacist; potecary; chemist < 英 >
药剂天平 dispensing balance
药剂学 pharmacy
药剂周转储备量 current reserve
药剂注入 impregnation
药酒 tincture
药卷 cartridge; cartridge explosive; powder stick
药理心理学 pharmacopsychology
药理学 pharmacology
药皮电焊条 coated arc welding rod
药皮电焊条焊接 shielded arc welding
药皮焊条 coated electrode
药皮气焊条 coated gas welding rod
药片 pellet; tablet
药品 pharmaceutical
药品沉淀 chemical precipitation; chemical sedimentation
药品房面积 area of chemical reagents
药品柜 medical cupboard; medicine chest
药品库 drug storage; drug store
药品溶解箱 chemicals dissolving tank
药品溶液池 chemical solution tank
药品箱 medicine box; medicine cabinet; medicine chest
药铺 herbal medicine shop
药腔 powder space
药商 apothecary
药室 chamber; coyote hole; gopher hole; mine chamber; pollen cells; powder chamber; powder drift
药室爆破 power drift blast
药室掘进 powder blast development
药室容量 charge capacity
药鼠李 cascara buckthorn
药水瓶 vial
药炭鼠李 glossy buckthorn
药丸 pill
药丸盒 pill-box
药物 medicament
药物分析 pharmaceutical analysis
药物化学 pharmaceutical chemistry
药物控制释放材料 drug controlled release material
药物天平 tare balance
药物污染源 pharmaceutical source
药物学 pharmacology
药线雷管 fuse cap
药箱 medicine cabinet
药芯焊条 cored electrode
药学 pharmacy
药液灌注装置 chemicals feeder
药液贮藏箱 chemicals dissolved water storage tank
药用玻璃 medicinal glass; pharmaceutical glass
药用矿物 medico mineral
药用油 medicinal oil
药用原料矿产 mineral raw material commodities for medical use
药用植物园 herb garden; medical plants garden
药柱端面 grain end
药柱炸药 pellet charge
药柱铸造 grain casting

要保人 applicant for insurance; proposer

要保书 application; proposal of insurance
要车计划表 car planned requisition list; wagon requisition plan table
要打的桩 pile for driving
要倒塌状建筑 ramshackle
要点 critical point; highlight; key feature; key link; key point; key position; main point; motif; nub(ble); outline; principal point; salient feature; subject matter; upshot
要点标签 header label
要点抄录 indicative abstract
要电线路 live line
要腐烂的(木料等) dozy
要害 strategy; vulnerability
要害部件 vitals
要害部门 key department
要害部位 critical organ; critical part; critical position; vital part
要害处 key trouble spot
要害的 strategic
要害问题 vital question
要价 asked price; asking price; charge; demand; offered price; pricing; quote; the price asked
要价过多 surcharge
要价收盘 closing asked price
要件 essential
要况报告 flash report
要领引导线 guideline
要路 helm
要目 syllabus
要拍卖的抵押不动产 distressed property
要塞 citadel; cylindric(al) keep; fort; fortification; fortress; strong hold; alcazar < 西班牙或阿拉伯的 >
要塞城 fortress-town
要塞地带 belt of fortresses; fortified zone
要塞建筑 fortification
要塞建筑技艺 art of fortification
要塞建筑艺术 art of fortification
要塞教堂 fortress-church
要塞井 fortress well
要塞礼拜堂 fortress-chapel
要塞内兵营 casern
要塞式的 fortress-like
要塞小教堂 fortress-chapel
要塞砖石建筑 fortress masonry
要素 constituent; element factor; essential; essential factor; feature; fortification; ingredient; integrant; key element; stuff
要素比价 relative factor price
要素比例 factor proportion
要素比率 factor rate
要素成本 component cost; elemental analysis; factor cost
要素成本价值 factor-cost value
要素成本会计 cost accounting by elements
要素的供给弹性 elasticity of supply of factor
要素反转 factor reversal
要素分析 component cost; factor analysis
要素服务 factor service
要素记录 feature record
要素技术 key technology
要素价格边界 factor-price frontier
要素价格差别 factor-price differential
要素价格均等化 factor-price equalization
要素价值 factor value
要素间关系 factor-factor relationship
要素检出器 feature extractor
要素利用 factor utilization
要素连接 join object
要素配置市场调节 production-element allocating market regulation
要素日 constituent day
要素生产力 factor productivity
要素市场 production elements market
要素市场均衡 factor-market equilibrium
要素收入的分配 distribution of factor income
要素替代弹性 elasticity of substitution of factor
要素投入系数 factor input coefficient
要素移动后果 effect of factor movements
要素增广的技术进步 factor augmenting technical progress
要素组合 factor combination; factor mix
要素最优使用 optimal factor employment
要素作业 elemental operation
要素作用总和 sum total of element functions
要提交的技术文件 document to be submitted
要项 staple
要项控制 key area control
要义 pith
要因分析图 cause and effect analysis chart
要因合同 causative contract
要因行为 causative action
要印修订版 recension
要员优先信号系统 very important person system
要账 demand payment of a debt; press for repayment of a loan

要旨 tenor

曜斑 yohen spot

曜变天目釉 yohen Tenmoku

耀斑 flare;flare spot;solar flare

耀斑暗晕 flare nimbus
耀斑波 flare wave
耀斑冲浪 flare surge
耀斑带 flare ribbon
耀斑光谱 flare spectrum
耀斑核 flare kernel
耀斑环 flare loop
耀斑激发 flare onset
耀斑级 flare class
耀斑级别 importance of a flare
耀斑际物质 interflare matter
耀斑喷焰 flare puff
耀斑闪光 flare-flash
耀斑指示器 flare indicator
耀度 brilliance [brilliancy]
耀目的光 dazzle
耀水晶 arenturine
耀现色球 flaring chromosphere
耀眼 dazzle;dazzling
耀眼的 dizzy
耀眼的眩光 discomfort glare
耀眼影响 glare effect
耀州窑器 Yaozhou ware
耀州窑系 Yaozhou type

椰壳炭 coco(a)nut charcoal

椰壳纤维 coco(a)nut fiber [fibre]
椰壳纤维席纹布 cocoa mat
椰枣 date palm
椰子壳 coco(a)nut
椰子壳活性炭 activated coconut charcoal
椰子木 coco wood;cocuswood
椰子奶油 coco(a)nut butter
椰子仁油 palm kernel oil
椰子(色) coco(a)nut
椰子属 <拉> Cocos
椰子树 coco(a)(palm);coco(a)nut palm (tree)
椰子油 cocoa butter;coco(a)nut oil;coco(a)nut butter
椰子油脂肪酸 coco(a)nut fatty acid;coconut oil fatty acid
椰子脂 coco(a)nut oil
椰子棕色 coco(a)nut brown
椰棕地毯 coconut matting
椰棕地毡 coco(a)nut mat(ting)
椰棕绳 coir rope
椰棕席子 coco(a)nut mat(ting)

耶茨连续性校正 Yates' correction for continuity

耶茨修正 Yates correction
耶德林带状基线尺【测】Jäderin tape
耶德林线状基线尺【测】Jäderin wire
耶尔卡雪暴 yalca
耶尔马可夫法 N. N. Yelmarkoff method
耶尔马可夫图解(法) N. N. Yelmarkoff diagram
耶尔锁 Yale lock
耶格法 Jaeger method
耶格鼓风机 Jaeger blower
耶基斯光谱分类法 Yerkes classification
耶基斯物镜 Yerkes objective
耶基斯折射望远镜 Yerkes refractor
耶克物镜 Yerkes objective
耶雷半荫仪 Jellet halfshade
耶路撒冷阿格沙清真寺 Mosque of Al-Aqsa at Jerusalem
耶路撒冷的圣墓 Holy Sepulchre at Jerusalem
耶路撒冷的圣墓教堂 church of the Holy Sepulchre at Jerusalem
耶路撒冷松 Jerusalem pine
耶拿玻璃 <用于光学、化学仪器> Jena glass
耶拿玻璃器 Jena ware
耶拿光玻璃 Jena light glass
耶拿天象馆 <1922 年设计的第一个薄壳混凝土圆屋顶> Jena planetarium
耶稣复活教堂 church of the Resurrection
耶稣复活礼拜堂 Chapel of the Resurrection
耶稣会教堂 Jesuit church
耶稣教会式 <拉丁美洲的> Jesuitical style
耶稣受难像 Calvary
耶索石膏 <南美洲刷墙用> yeso
耶西树 <表示基督的家系> Tree of Jesse

也门建筑 Yemen architecture

冶成 forming

冶金白云石砂 dead-burned dolomite grains
冶金不稳定性 metallurgic(al) instability
冶金产品 metallurgic(al) product;metallurgy product
冶金产品分析 metallurgic(al) product analysis
冶金厂 smeltery
冶金厂废气 metallurgic(al) waste gases
冶金尘雾排放 metallurgic(al) fume emission
冶金冲天炉 metallurgic(al) blast cupola
冶金粗料 metallurgic(al) crude
冶金方法 method of metallurgy
冶金废料 metallurgic(al) waste;smeltery waste
冶金废渣 metallurgic(al) slag;smeltery slag
冶金粉尘 metallurgic(al) dust
冶金工厂 metallurgic(al) plant;metallurgic(al) works
冶金工程 metallurgic(al) engineering
冶金工程师 metallurgic(al) engineer
冶金工业 metallurgic(al) industry
冶金工业部 Ministry of Metallurgical Industry
冶金工业废水 metallurgic(al) waste-water
冶金工业指标 index of metallurgic(al) process
冶金工作者 metallurgist
冶金过程 metallurgic(al) process
冶金衡算 metallurgic(al) accountability
冶金化学 metallurgic(al) chemistry
冶金回收率 smelting recovery
冶金机械 metallurgic(al) machinery
冶金计算 metallurgic(al) calculation
冶金焦炭 furnace coke;hard coke;smelter coke;metallurgic(al) coke
冶金结合 metallurgic(al) bonding
冶金菱镁矿矿石 metallurgic(al) magnesite ore
冶金流程 floe-sheet of metallurgy
冶金炉 metallurgic(al) furnace;metallurgy furnace
冶金炉排放物 metallurgic(al) furnace emission
冶金炉渣 metallurgic(al) slag
冶金煤 furnace coal
冶金镁砂 fettling magnesite grain
冶金缺陷 defect of metallurgy;metallurgic(al) imperfection
冶金燃料 metallurgic(al) fuel
冶金热处理 metallurgic(al) heat treatment
冶金热力学 metallurgic(al) thermodynamics
冶金(熔渣)水泥 metallurgic(al) cement
冶金设备 metallurgic(al) equipment
冶金试验 metallurgic(al) test
冶金术 metallurgy
冶金水泥 <掺有大量矿渣的> supermetallurgical cement
冶金特征 metallurgic(al) feature
冶金提取(率) metallurgic(al) extraction;metallurgy extraction
冶金添加剂 metallurgic(al) addition agent
冶金显微镜 metalloscope;metallurgic(al) microscope
冶金学 metallurgy
冶金学家 metallurgist
冶金学者 metallurgist
冶金烟气 metallurgic(al) fume
冶金用白云岩 dolomite for metallurgic(al) use
冶金用煤 metallurgic(al) coal
冶金用砂岩 sandstone for metallurgic(al) use
冶金助熔石 fluxstone
冶金专业证明书 metallurgic(al) certificate
冶炼 refining;smelting
冶炼比 ratio of melting
冶炼操作 smelting operation
冶炼厂 metallurgic(al) plant;metal production plant;smeltery;smelting plant
冶炼厂废水 smeltery waste;smelting waste
冶炼厂建设 metallurgy plant building
冶炼成本 metallurgy cost
冶炼工人 smelter
冶炼过程 smelting process
冶炼回收率 rate of melting recovery
冶炼技术 alloying technique
冶炼炉 smelting furnace
冶炼强度 rate of driving
冶炼设备 smelting facility
冶炼时间 duration of heat(ing)
冶铁煤 smithy coal
冶铁学 siderology

野桉 desert gum;flooded gum;Moitch eucalyptus

野餐 junket;picnic
野餐场所 halting place;picnic site
野餐地 picnic area
野餐篮 picnic basket
野餐热水瓶 picnic flask
野餐营地 picnic site
野草 quitch grass;weeds;wort
野草莓树 Strawberry tree
野地 wild land
野靛蓝素 baptisoid
野复理石 wildflysch
野橄榄树 wild olive
野葛 Rhus toxicodendron
野核桃 Chinese walnut
野花楸 checker tree
野花楸树 mountain ash
野麻 dogbane hemp
野蛮卸料 indiscriminate dumping
野蛮作业 rough handling
野猫井 wildcat well
野猫钻井者 wildcatter
野煤 wild coal
野茉莉 styrax japonica
野喷井 wild well
野漆树 Rhus succedanea;wax-tree
野漆树蜡 japan tallow;japan wax
野趣园 wild garden;wild plants botanical garden
野生 wildness
野生草场 wild pasture
野生草地 wild grassland
野生动物 wildlife
野生动物保护标志 wild animals protection sign
野生动物保护法 law of wild animals protection
野生动物保护区 sanctuary;wild animals refuge area;wildlife protection area;wildlife refuge
野生动物保护系统 wildlife protection system
野生动物(防护)栅 wildlife fence
野生动物公园 wildlife park
野生动物栖息地 wildlife habitat
野生动物区系 wild fauna
野生动物事故 wildlife accident
野生苗 seeding growth
野生生物 wildlife
野生生物保护 conservation of wildlife;wildlife conservation;wildlife preserve
野生生物保护区 conservation area of wildlife;wild animals protection area;wildlife protection area
野生生物栖息地 wildlife habitat
野生生物区系 wildlife fauna
野生生物资源 wildlife resources
野生药材资源保护区 reserve area of wild medical herbs resources
野生园 wild garden
野生植物保护法 law of wild plant protection
野生植物园 wild plants garden
野生种 wild species
野生状态 wildlife state
野外 afield
野外阿尔发卡法 field Alpha card survey
野外安装的 field-installed
野外靶场 outstation
野外编制测量 field compilation survey
野外辨认 field deciphering
野外波谱测量仪器 field spectral measurement instrument
野外补点 field completion
野外仓库 field storage
野外操作 field operation
野外草图 field map;field sketching
野外草图绘制 field mapping
野外测定 field measurement
野外测绘等高线 field contouring
野外测量 field measurement;field survey;field work;measuring tour
野外测量仓库 field survey depot
野外测量法 field surveying
野外测量基地 field survey depot
野外测量技术规定 field survey pamphlet
野外测量手簿 survey field notes
野外测量手册 field survey pamphlet
野外测量数据 field survey data
野外测量站 field survey station
野外测设的 field-established
野外测试 field assay
野外测图 field mapping;field plotting
野外测图板 field plotter

野外查勘 field reconnaissance
野外持水量 field water retaining capacity
野外持水率 field water retaining capacity
野外抽水试验 field pumping test
野外储存 field storage
野外处置 field disposal
野外磁测工作 field work of magnetic survey
野外磁秤 field balance
野外大剪试验点数 number of field shear test
野外道路 exposed road;field road
野外的 open air;outdoor;out of door
野外地形测量队 field topographic(al) unit
野外地质工作 field geologic(al) work
野外地质记录 field geologic(al) record
野外地质勘探队 geologic(al) field party
野外地质人员 field geologist
野外地质图 field geologic(al) map
野外地质学 field geology
野外调查 field edit; field examination;field investigation;field survey
野外调绘 classification survey;field annotation;field classification;field inspection
野外定位工作 field location work
野外定线 field location
野外定线工作 field location work
野外堆积 windrow composting
野外对比 field correlation
野外反射光谱仪 field reflectance spectrometer
野外方法 field method
野外放射仪 field radiation device
野外分层 field separation
野外分光光度计 field spectrophotometer
野外分局 field branch
野外分类 field classification
野外分析 field analysis
野外分析箱 field test kit
野外服务 field service
野外复查 field review
野外复核【测】 field check
野外改正 field correction
野外工作 field activity; field work; outwork
野外工作津贴 field work allowance
野外工作人员 field man;field personnel
野外工作手图 map of free hand in field
野外工作站 field station
野外估算 field evaluation
野外观测 field inspection;field observation
野外观测结果 field evidence
野外观测装置 field setup
野外观测资料 field data
野外观察 field observation
野外管道 field pipeline
野外光干涉比长器 field interference comparator
野外光谱辐射计 field spectroradiometer
野外光谱学 field spectroscopy
野外光圈 field stop
野外含水当量 field moisture equivalent
野外荷载试验 field loading test
野外绘图 field mapping
野外活动 outdoor sport
野外活动锻铁炉 field forge
野外火灾等级 class of wildfire
野外机场 field aerodrome
野外挤奶装置 field bail
野外计时 field timing
野外计算 field calculation;field computation
野外计算机系统 field computer system
野外记录 field note;field record(ing)
野外记录本 chain book;field book;field recording book
野外记录簿 chain book;field book;field recording book
野外记录卡 field recording card
野外记录数据 field recording data
野外监测 field monitoring
野外剪力仪 field shear box
野外剪切试验 field shear test
野外剪切试验方法 process of field shear test
野外剪切试验种类 type of field shear test
野外检查 field check;field edit;field examination;field-inspect;field inspection
野外检定 field calibration
野外检定基线 field comparator
野外检核 field check
野外检验 field check
野外鉴定 field identification;field recognition
野外鉴定法 field identification procedure;field identification procedure of soil
野外鉴定试验 field identification test
野外校正 field correction
野外校正装置 field adjusting device
野外经纬仪 field theodolite
野外经验 field experience
野外勘测 field work
野外勘测队 field party
野外勘察 field exploration;field investigation
野外勘探 field exploration
野外勘探系统 field exploration system
野外考察 field inspection
野外科 field section
野外科学试验场 desert establishment
野外可靠性 field reliability
野外控制 field control
野外控制点 field control point
野外控制试验 field control testing
野外立体镜 field stereoscope
野外漏斗式黏[粘]度计 field funnel
野外露头 feature of exposures
野外轮胎服务车 field tire service truck
野外弥散试验 method of dispersion test in field
野外描绘 field sketching
野外排列方式 field arrangement
野外评价 field evaluation
野外普查 field reconnaissance
野外勤务 field service
野外容许偏差 field tolerance
野外三角测量 field triangulation
野外设备 field installation
野外摄影机 field camera
野外渗透系数 field coefficient of permeability;field permeability coefficient
野外实习 field seminar;field trip
野外实验 field experiment
野外实验室 field laboratory
野外使用 field application;field usage;field use
野外使用性能 field performance
野外试验 field examination;field experiment; field experimentation; field test(ing);field trial;off road test;troop test
野外试验场 field test site
野外试验结果 field test result
野外适用性 field worthiness
野外手册 field book
野外手术车 field operation car
野外手图 freehand field map
野外数据 field data
野外数据采集 field data collection
野外数据采集系统 field digital system
野外数据单 field data sheet
野外数据的描述和分析 presentation and analysis of field data
野外数据卡 field data card
野外水文地质试验 hydrogeologic(al) test in field
野外水质自净试验 self-cleaning test of water quality in field
野外水准测量 field level(l)ing
野外素描 field sketching
野外踏勘 field reconnaissance; field trip;reconnaissance survey in field
野外体视镜 field stereoscope
野外天文点 astronomic(al) field station
野外天文观测 field astronomical observation
野外天文观测点 field astronomical point;field astronomical station
野外天文学 field astronomy
野外填图 field mapping
野外条件 field condition
野外望远镜 field glass(es)
野外维修队 field maintenance unit
野外温度计 field thermometer
野外物探人员 field geophysicist
野外显微镜 field microscope
野外修测透写图 field correction
野外修理用的半拖车 field shop repair semitrailer
野外修理站 camp shop
野外压实 field compaction
野外压实试验 field compaction trial;field compression test
野外研究 field investigation;field research;field study
野外演习 field exercise
野外演习日 field day
野外宴会 barbecue
野外验正及像片地质图编绘 field verification and image-geologic(al) mapping
野外样品 field sample
野外仪器 field instrument
野外应用 field application;field use
野外用被覆线 field cable
野外用泵 field pump
野外用电话 field telephone
野外用红外辐射源 field infrared source
野外用镜 field lens
野外用轮胎防滑链 tyre-chain for cross-country operation
野外用压缩机 field service compressor
野外油料管理员 field petroleum officer
野外油料库 field petrol depot
野外原图 field map
野外圆形剧场 amphitheater [amphithcatre]
野外运输 field transport
野外帐篷 field camp
野外照明车 field lightening truck
野外振摆仪 field pendulum
野外证据 field evidence
野外指南 field guide; field guidebook;field manual
野外制图 field mapping
野外制作 field work
野外注记 field annotation
野外装备 field equipment
野外装配 field connection
野外资料 field document
野外综合实验路 field comprehensive experimental road
野外钻探工作 field drilling operation
野外作业 field operation;field work(ing);rough terrain operation
野外作业车 field operation car
野外作业处 field operation division
野外作业计划 field program(me)
野外作业设施 field service unit
野外作业用具 field instrument
野外作业自动化 field automation
野性的 savage
野鸭 mallard
野营 camp (out);encampment
野营场地 camp ground;camping area;camping site
野营车 camper
野营会 camporee
野营区 camp site
野营拖挂车 camper trailer
野营医院 camp hospital
野战电话机 camp telephone
野战机场 field aerodrome
野战医院 ambulance;field hospital
野战炸弹库 bomb dump;bomb store
野战制印设备 field printing plant
野战制印图 locally produced map
野猪 wild boar

业

业绩管理法 results management
业绩评估 performance assessment
业绩评价 performance evaluation
业绩评价会计 performance accounting
业绩审计 performance audit
业绩指数 performance index
业经验货同意 inspected approval
业经装船 afloat;to arrive
业经装船之货 afloat goods;cargo afloat;floating cargo
业救济 dole
业权调查书 requisitions on title
业权基础 root of title
业外费用 non-operating expenses
业外营业 outside venture
业务 activity; affair; business; service;transaction;vocation;vocational work
业务报表 bordereaux
业务报酬 business consideration
业务报告 business report
业务备忘录 engagement memorandum
业务标准 service standard
业务波道 service channel
业务波段 service band
业务部 business department;business division;steward department
业务部门 business agencies;business sector;business segment
业务部门目标 business department objective;line objective
业务采购 buying for business use
业务参考 business reference
业务测试【计】 operational trial
业务策略 business game
业务差错 error of service
业务差旅费 travel(l)ing expense for business
业务成本 operating cost
业务成果 operating result
业务成果账户 operating result account
业务出行 business trip
业务储备金 operational reserves
业务处理 handling of traffic
业务代办公司 bonding company
业务代表 business agent

业务代码 service code
业务单位 operating unit
业务的 operating
业务等级 grade of service; professional grade
业务地区说明 description of territory
业务电报 service message
业务电话 business telephone; service telephone
业务电话台 service desk
业务电路 <增音站之间的> speaker circuit
业务调查 business survey
业务动态分析 dynamic(al) business analysis
业务端口功能 service port function
业务发达的商行 going concern
业务发展 trade development
业务发展过快 overtrading
业务法规 business code; business rules; code of practice
业务繁忙程度 intensity of traffic
业务繁忙的地区 region of heavy traffic
业务繁忙时间 heavy traffic period
业务繁忙时期 heavy traffic period
业务范围 business field; business scope; class of business; province of activities; scope of business
业务方法 operational approach
业务方针 business policy
业务飞行 business flying
业务费用 functional expenses; operating outlay; operational expenditures; operation cost
业务费支出 business expenditures
业务分割 business separation
业务分配 traffic distribution
业务分析 business diagnosis; traffic analysis
业务份额 business share
业务复用 service integration
业务改善 better business
业务概率 service probability
业务干扰 disturbance of service
业务工作 office work; professional work
业务工作量 magnitude of operation
业务工作时间 hours of service
业务供应品 operating supplies
业务贡献率责任额 quota of sales contribution margin
业务顾问 management consultant
业务关系 business relations
业务管理 business management; operational control; operational management
业务管理费用 job overhead
业务管理模拟训练 simulation games for business
业务管理人员 line executive
业务管理系统 service management system
业务规程 service instruction
业务规划 business planning; plan of operation
业务规章 code of practice; operating instruction; service instruction; traffic regulation; vocational rules
业务规章手册 manual of operating instructions
业务函件 official correspondence
业务合同 business contract; service contract
业务合作 business cooperation
业务和维修 service and repair
业务核算 business calculation
业务汇率 operational rate
业务会议 business conference; business meeting
业务活动 business activity; operational action; operational activity
业务活动比率 activity ratio
业务活动差异 activity variance
业务活动的评价 measurement of performance
业务活动分类 activity classification
业务活动分析 activity analysis
业务活动率 activity rate
业务或工作须知 operating instruction
业务机会 business chance
业务机械 business machinery
业务基金 operating fund
业务基金账户 business fund account
业务计划 business plan(ning); operational plan(ning); work plan(ning)
业务计划会计 accounting for management planning
业务计划体系 operation planning system
业务监查员 service observer
业务检查 service inspection; service observation
业务简介 brief description on business
业务交叉 overlapping business operations
业务交易 business transaction
业务节点 service node
业务节点接口 service node interface
业务经管人 account executive
业务经理 account executive; business manager; office manager; operation manager
业务经营表 statement of business operation
业务净利 net profit from operation
业务竞争 business game
业务决策 operating decision; operational decision making; operative decision
业务开发 operational development
业务开发费 sales expenses
业务开支 operating cost; operating expenses
业务控制 business control; operational control
业务扩充准备 reserve for expansion; reserve for extensions
业务扩张 business expansion
业务类别 class of service
业务类型 type of service
业务礼品费用 business gift expenses
业务联络线塞孔 ancillary jack
业务联系 business connection
业务联系银行 corresponding bank
业务联系员 bond(ing) agent
业务量 traffic
业务量路由 traffic route
业务量疏导 traffic grooming
业务领导机关 specialized government agencies
业务流程 operation flow; work flow
业务略语 service abbreviation
业务秘密 business secret
业务名片 business card
业务能力 professional ability; professional proficiency; professional qualification
业务评议局 operation evaluation department
业务凭证 business document; business vouchers
业务气象卫星 operational meteorological satellite
业务洽谈 business discussion
业务清淡时间 slack hour
业务情况 service condition
业务情况报告 business report of condition
业务区 service area
业务人员 business personnel; servicer
业务日志 business diary
业务容量 traffic capacity; traffic-carrying capacity; volume of business
业务设计 work design
业务审计 business audit; operational audit
业务时间 service time
业务实践 professional practice
业务实习 business practice; office practice
业务使用 operational use
业务收费 professional fee
业务收费标准 scale of professional charges
业务收入 business income
业务收益 earnings
业务收益表 earnings statement
业务守则 code of practice
业务数据处理 business data processing
业务水平 professional qualification
业务缩语 service abbreviation; service code
业务条件 professional qualification
业务通话 exchange conversation; service call
业务通信[讯]电路 service circuit
业务通信[讯]系统 service communication system
业务通信[讯]增音机 service circuit repeater
业务通知 service talk
业务统计 business statistics; operating statistics
业务图表 business graphics
业务推广奖励奖 promotional allowance discount
业务推广奖励折扣 promotional allowance discount
业务推广津贴 promotional allowance
业务完成的评价 measurement of performance
业务往来 business contact; transaction
业务往来技术计算 technical calculation of business transactions
业务位 service bit
业务系统 operation system
业务细目 service inventory
业务线 call circuit; interposition trunk; service line; servicing line
业务线电键 order wire speaking key
业务线分配器 order wire distributor
业务限制 service restriction
业务项目 operational project
业务协议 service protocol
业务协助 operational assistance
业务信道 traffic channel
业务信托机构 business trust
业务信息系统 operating information system
业务性投资 trade investment
业务性预算 business-type budget
业务须知 service instruction
业务学习 vocational study
业务循环 cycle of business operations
业务训练 on-the-job training
业务训业 on-the-job test
业务延续时间 duration of service
业务营运 operations
业务用名 business name
业务预算 operating budget
业务预算编制 operational budgeting
业务员 account executive
业务月报表 account current
业务责任 portfolio
业务增长情况 growth pattern
业务账户 account working; activity account
业务招待费 business entertainment; goodwill entertaining
业务支出 business expenses; revenue charges
业务知识 professional knowledge
业务职能 operation function
业务指标 operational indicator
业务指南 business guide
业务质量 service quality; service signal
业务中断 service interrupt(ion)
业务中断保险 business interruption insurance
业务中继线 trunk from concentrating switch
业务种类 business lines; class of business; kind of business; type of traffic
业务主任【船】 chief purser; purser
业务助理 operation assistant
业务专长 professional specialty; specialized kill
业务准备 operational reserves
业务咨询 business consulting
业务资金账户 business fund account
业务资料 operating information; service data
业务资料处理 business data processing
业务自动化 business automation
业务总部所在地 principal place of business
业务总裁 chief operating officer
业务组 traffic group
业务组织形式 form of business organization
业务最高负责人 chief executive officer
业务座席 application position; service position
业已交货 delivered
业余爱好 avocation
业余爱好工作室 hobby room
业余爱好劳作室 home workshop
业余爱好者计算机 hobby computer
业余波段 <无线电> amateur (frequency) band
业余的 amateur; amateurish; inexpert; non-professional; sparetime
业余地震预报 amateur earthquake prediction
业余工作 by-work
业余活动 amateurism
业余活动者 amateur
业余矿物学家 rock hound
业余农民 hobby farmer
业余时间 after hour; after time; off-hour; out-of-service time; sparetime
业余无线电 amateur radio
业余无线电台 amateur (radio) station; amateur set
业余性质 amateurism
业余学校 spare-time school
业余园艺家 amateur gardener
业余者 amateur
业余职业 spare-time employment
业载航程 payload range
业主 building owner; business owner; business proprietor; client; employer; householder; owner; property owner; proprietary; proprietor
业主不在位管理 absentee management
业主财产 client's property
业主财产保险 insurance of client's property
业主产权 proprietary equity
业主-承包人协议(书) owner-contractor agreement
业主承担的风险 <承包合同内列明的> accepted risk

业主代表 client's representative; owner's representative
业主代理人 owner's inspector
业主的代理人 agent for owner
业主的风险 employer's risks
业主的过失 default of employer
业主的技术人员 owner's technical personnel
业主对工程造价的限制 cost limit
业主风险 owner's risk
业主工程师 owner's engineer
业主供货 supply by owner
业主供货和工程范围 scope of works and supply by owner
业主监工员 owner's inspector
业主监理 owner's inspector
业主建造的 constructed by the owner
业主-建筑师协议书 owner-architect agreement
业主可偿还给承包商的费用 reimbursable expenses
业主控制的企业 owner-controlled firms
业主会计 proprietorship accounting
业主批准付款证书 project certificate for payment
业主弃权声明书 landlord waiver
业主权利 owner's rights
业主权益 equity ownership; proprietary equity; proprietary interest
业主权益与债务资本比率 capital gearing
业主认可 acceptance by owner
业主入住 owner occupant
业主设备 employer's equipment
业主设施 employer's facility
业主收到投标的收据 receipts of bids
业主收入 entrepreneurial income; proprietor's income
业主提供 supplied by the owner; provided by the owner
业主提款账户 drawing account
业主投资额 amount of capital invested by the owner
业主违约 default of employer; default of owner; owner default
业主协会 Home Owner's Association
业主义务 obligation of client; owner's duty; owner's obligation
业主义务保险 employer's liability insurance; owner's liability insurance
业主与承包人合同 owner-contractor agreement
业主与承包人契约 owner-contractor agreement
业主与建造师契约 owner-architect agreement
业主与建筑师合约 owner-architect agreement
业主与有关各方讨论决议的正式记录 record of decision
业主预留改变工程范围的金额 reserve for scope changes
业主愿意收回抵押 owner will carry mortgage
业主责任 liability of client
业主责任保险 owner's liability insurance
业主账户 proprietary account; proprietor account; proprietorship account
业主职员 client's personnel
业主驻现场代表 field representative; owner's representative
业主专用泊位 owner-user berth
业主专用船 owner-user ship; owner-user vessel
业主专用码头 owner-user terminal
业主咨询工程师标准服务协议书 conditions of the client/consultant model services agreement
业主资本 owner's capital; proprietary capital
业主自用的 owner-occupied
业主租约 proprietary lease

叶

叶板 acanthus leaf; blade; paddle
叶板搅拌器 paddle stirrer
叶板饰 acanthus; crop
叶瓣 leaf
叶瓣饰 feathering; foliation
叶瓣与方块花纹装饰线脚 leaf and square
叶瓣与箭头花纹装饰线脚 leaf and dart
叶背空化 back cavitation
叶泵 impeller blade
叶镖装饰线脚 leaf-and-dart mo(u)lding
叶柄 blade stem; stalk; stipe
叶柄轴承 stem bearing
叶车运输系统 demand-responsive transportation system
叶丛卷茎饰柱头 caulcole
叶丛效应 foliage effect
叶单宁 acertannin
叶的 foliar
叶底部 dorsal
叶碲金矿 nagyagite
叶碲矿 foliated tellurium; nagyagite
叶点霉属 <拉> Phyllosticta
叶顶旋涡 blade tip eddy
叶端 leaf apex
叶端喷口 tip-jet
叶端喷流透平 tip turbine
叶端损失 tip loss
叶段分布 segmental distribution
叶尔羌地块【地】Yarkant block
叶尔羌-滇东地块带【地】Yarkant-Eastern Dian block-zone
叶沸石 knolllite; zeophyllite
叶根 <螺旋桨的> blade root
叶根空化 root cavitation
叶梗饰 <考林辛式柱头的> caulcole
叶痕 leaf scar
叶环饰 civic crown
叶黄素 phytoxanthin
叶迹 leaf trace
叶尖间隙 blade tip clearance
叶尖空化 blade tip cavitation
叶尖空蚀 blade tip cavitation
叶尖汽蚀 blade tip cavitation
叶尖饰【建】finial; crop; horse
叶尖吐水 <植物> guttation
叶尖涡流 blade tip vortex
叶肩空化 blade shoulder cavitation
叶肩汽蚀 blade shoulder cavitation
叶箭饰 leaf and dart
叶桨搅拌式沥青乳化机 paddle bitumen emulsifying machine
叶桨式粉碎转子 mill rotor
叶桨吸泥机 cutter-suction dredge(r)
叶结构镜质体 phyllotelinite
叶镜质体 phyllovitrinite
叶蜡石 agalmatolite; grave wax; pagoda stone; pagodite; pyrauxite; pyrophyllite
叶蜡石高岭石岩矿石 pyrophyllite-kaolinite ore
叶蜡石矿床 pyrophyllite deposit
叶蜡石耐火砖 pyrophyllite fire brick
叶蜡石黏[粘]土 pyrophyllite clay
叶蜡石片岩 pyrophilitic schist
叶蜡石石棉 pyrophyllite asbestos
叶蜡石陶瓷 pyrophyllite ceramics
叶蜡石套 pyrophyllite sleeve
叶蜡石岩 pyrophyllitite
叶理 foliation; folium [复 folia]
叶理构造【地】foliation structure
叶理尖角转向带 kink band
叶理结构【地】foliated texture
叶理裂缝 foliation fissure
叶理面 foliation plane
叶理岩 foliated rock
叶硫砷铜石 chalcophyllite
叶硫砷银铅矿 lengenbachite
叶滤机 leaf filter
叶绿矾【地】copiapite; misylite
叶绿泥石 japanite; pennine; penninite; pouzacite; tabergite
叶绿色颜料 chloroplastic pigment
叶绿石 chlorophyllite
叶绿素 chlorophyl(l)
叶绿素体 chlorophyllinite
叶绿体 chloroplast
叶轮 blade wheel; face roller; impeller [impellor] (wheel); lobed wheel; paddle (wheel); runner; vane (wheel); propeller <泵或风机的>
叶轮板 impeller plate
叶轮包装机 impeller packer
叶轮泵 impeller [impellor] pump; peripheral pump; sickle pump; turbopump; vane pump; vane-type pump; wing pump
叶轮柄 impeller hub
叶轮充气器 simplex turbine aerator
叶轮冲击磨机 impeller impact mill
叶轮冲击式破碎机 impeller impact breaker; impeller impact crusher
叶轮出口周向速度 leaving whirl velocity
叶轮除渣器 turbo-separator
叶轮传感器 impeller sensor; paddle wheel sensor
叶轮的轮毂比 impeller hub ratio
叶轮的叶型 profile of impeller
叶轮电动泵 vane motor pump
叶轮端螺钉 impeller locking screw
叶轮发动机 vane motor
叶轮法 vane-wheel method
叶轮反击式粉磨机 impeller impact mill
叶轮反应度 impeller reaction
叶轮风扇 paddle wheel fan
叶轮风速计 rotating vane anemometer; vane anemometer
叶轮浮选 turbine flo(a)tation
叶轮给料机 revolving plow reclaimer
叶轮给料器 rotary plow feeder
叶轮毂板 impeller backplate; impeller hub disc; impeller hub plate
叶轮鼓风机 bucket wheel blower; simplex turbine aerator; turbo-blower
叶轮固定环 impeller mounting ring
叶轮后盖板 impeller hub
叶轮混合机 impeller mixer
叶轮机 turbine
叶轮机驱动泵 turbine-driven pump
叶轮激动器 impeller exciter
叶轮间隙 impeller clearance
叶轮角 vane angle
叶轮搅拌机 turbine stirrer
叶轮搅拌机轴 paddle mixer shaft
叶轮搅拌器 impeller agitator; turbine mixer
叶轮进口 impeller approach; impeller eye; impeller inlet
叶轮进口导向叶片 impeller inlet guide vane
叶轮进口面积 impeller inlet area
叶轮壳 impeller casing
叶轮扩散器间隙 <浮选机的> impeller-diffuser clearance
叶轮离合器 impeller clutch
叶轮离心式风扇 impeller centrifugal fan
叶轮离心式压气机 vane-type centrifugal compressor
叶轮流道 impeller channel; impeller passage
叶轮流量计 propeller flowmeter
叶轮轮盖 enclosing cover for impeller
叶轮轮毂 impeller hub
叶轮轮环 impeller belt
叶轮螺母 impeller nut
叶轮迷宫密封 impeller labyrinth
叶轮名义直径 impeller nominal diameter
叶轮摩擦损失 wheel friction loss
叶轮排出量 impeller output
叶轮盘 impeller disk
叶轮破碎机 impeller crusher
叶轮曝气器 paddle wheel aerator
叶轮曝气器系统 impeller-sparger system
叶轮曝气设备 simplex turbine aerator
叶轮气体吸收器 turbine gas absorber
叶轮前盖板 impeller band
叶轮腔室 wheel chamber
叶轮切割 impeller cut
叶轮驱动机构 impeller drive
叶轮入口 impeller approach; impeller eye
叶轮润滑油杯 impeller cup
叶轮筛 turbine sifter; wing screen
叶轮式 paddle wheel system
叶轮式拌和机 turbo-mixer
叶轮式泵 vane pump
叶轮式铲运机 paddle wheel scraper
叶轮式反应室 paddle reaction chamber
叶轮式粉磨机 impeller type pulverizer
叶轮式粉碎机 impeller breaker
叶轮式风速仪 impeller anemometer
叶轮式格子 turbogrid
叶轮式格子塔盘 turbogrid plate
叶轮式给料器 rotary plow feeder
叶轮式鼓风机 helical blower; impeller-blower; spiral blower
叶轮式混合器 paddle wheel aerator; turbine mixer; turbo-mixer
叶轮式计程仪 impeller type log
叶轮式减振装置 paddle dampening device
叶轮式搅拌机 paddle wheel aerator; turbine mixer; turbo-mixer
叶轮式搅拌器 turbine-type agitator; turbo-mixer
叶轮式块根切碎机 impeller type root chopper
叶轮式量水计 vane water meter
叶轮式料斗卸料车 paddle-type bunker discharge carriage
叶轮式流量传感器 impeller type flow transmitter
叶轮式流量计 vane-wheel type flowmeter
叶轮式螺旋输送机 paddle worm conveyer [conveyor]
叶轮式排气机 rotary impeller exhauster
叶轮式喷头 geared sprinkler
叶轮式破碎机 impeller type breaker; impeller type crusher
叶轮式曝气池 paddle aeration tank; paddle aerator
叶轮式汽轮机 paddle wheel steamer
叶轮式切碎机 impeller type chopper
叶轮式撒布机 cylinder paddle spreader
叶轮式栅格板 turbogrid plate
叶轮式栅格分馏塔盘 turbogrid tray
叶轮式水表 rotary-vane meter; vane (-wheel type) water meter; wing shape water ga(u)ge; wing type water meter
叶轮式水流观测器 vane wheel type water flow observer
叶轮式水轮机 impeller (type) tur-

bine
叶轮式通风机 bucket wheel blower; paddle wheel fan
叶轮式压缩机 rotary-vane compressor
叶轮式液压马达 vane hydraulic motor
叶轮式油压马达 vane oil motor
叶轮式增压器 vane-type supercharger
叶轮式轧碎机 impeller breaker
叶轮式轴流风机 vane-axial fan
叶轮式轴流鼓风机 vane-axial blower
叶轮室 propeller bowl
叶轮水表 vane water ga(u)ge
叶轮通道流速 impeller channel velocity
叶轮推动的热虹吸管 impeller assisted thermosiphon
叶轮外径 impeller outer diameter
叶轮外缘速度 impeller tip speed
叶轮无叶片的离心泵 centrifugal pump with bladeless impellers
叶轮吸气上浮法 drawn turbine flo(a)tation
叶轮吸水口 impeller suction
叶轮箱 impeller casing
叶轮效率 rotor efficiency
叶轮心 bead core
叶轮选择 impeller selection
叶轮压缩机 turbo-compressor
叶轮叶片 impeller blade; impeller vane
叶轮液压马达 vane hydraulic motor
叶轮圆周速度 circumferential speed of impeller
叶轮罩 impeller chimney
叶轮直径 diameter of impeller; impeller diameter
叶轮制动器 wing brake
叶轮中孔面积 impeller eye area
叶轮轴 impeller hub; impeller shaft; paddle shaft
叶轮轴承 impeller bearing
叶轮轴承壳 impeller bearing house
叶轮轴流式压气机 vane-type axial-flow compressor
叶轮爪 impeller finger
叶轮组 impeller assembly
叶螺旋推进器 vane-screw propeller
叶脉 nervure; vein
叶面空化 blade face cavitation; face cavitation
叶面空蚀 blade face cavitation
叶面喷洒 foliage spray
叶面汽蚀 blade face cavitation
叶面散发 foliar transpiration; stomatal transpiration
叶面蒸腾 foliar transpiration
叶面指数 leaf-area index
叶钠长石 cleavelandite
叶尼塞极性超时间带 Yanisei polarity superchronzone
叶泥炭 leaf peat
叶片 blade; blading; fan; fin; flapper; fly-bar; folium [复 folia]; leaf blade; leaf gold; lobe; paddle; runner blade; vane; runner bucket <混流式的>; throw-over blades and pick-ups <混凝土拌和机的>; baffle
叶片安放角 blade angle; blade incidence; vane setting angle
叶片安置角 blade angle; blade incidence; vane setting angle
叶片安装 rooting-in of blades
叶片安装底架 blade retainer
叶片安装工 blader
叶片安装工作 blading work
叶片安装角 blade angle; blade incidence; vane setting angle
叶片凹边 spoon of the blade
叶片坝 vane dam; vane dike [dyke]
叶片拌和器 blade mixer
叶片拌和器轴 blade mixer shaft
叶片包角 subtended angle of blade; wrapping angle of blade
叶片背面 vacuum side of blade
叶片泵 blade pump; impeller pump; paddle pump; sliding gate pump; vane pump; wing pump; rotodynamic pump
叶片边移 blade side shift
叶片标号 blade marker
叶片槽 blade groove
叶片颤动 buffeting of vane
叶片颤振 blade flutter
叶片长 blade height
叶片长度 blade length
叶片承压面 pressure side of the blade
叶片澄清机 leaf clarifier
叶片充实度 vane solidity
叶片冲角 blade incidence
叶片稠度 solidity of blades
叶片出口半径 exit radius of blade
叶片出口边 exit edge of blade
叶片出口角 <螺旋桨> blade outlet angle; exit blade angle; outlet blade angle; vane exit angle
叶片出汽边 back edge; blade trailing edge
叶片出水边 out edge of vane
叶片的 foliated
叶片的标志圆环 blade marker
叶片的扭曲度 vane twist
叶片的频谱图 interference diagram
叶片的失效部分 non-active part of vane
叶片的允许速度 bladed allowable speed
叶片的制动冲杆及凸轮轴 blade locking plunger and camshaft
叶片底叶 bottom of the blade
叶片电动机 sliding-vane motor
叶片顶部速度 tip velocity
叶片顶端旋涡 blade tip eddy
叶片顶尖 blade point
叶片端面 blade face
叶片断面汽蚀 profile cavitation
叶片阀 leaf valve; vane valve; blade valve
叶片法向厚度 normal thickness of blade
叶片反曲离心式通风机 backward type centrifugal fan
叶片分析 foliar analysis; leaf analysis
叶片分析数据 data to leaf analysis
叶片风速表 vane anemometer
叶片附根 blade shape
叶片高(度) blade height
叶片根部 root of blade
叶片工作面 pressure side of the blade
叶片骨线 center line blade profile
叶片固定法 blade fixing
叶片光合作用 leaf photosynthesis
叶片辊 paddle roller
叶片和冠层覆盖的温度 leaf and canopy temperature
叶片荷载分布图 blade loading diagram
叶片后缘 trailing edge
叶片厚度 vane thickness
叶片花饰壁缘 foliage frieze; foliated frieze
叶片花饰的尖顶 foliage cusp; foliated cusp
叶片花饰的柱顶 foliage capital
叶片花饰与带箍线条 foliage and strapwork
叶片滑石 foliated talc
叶片环 blade ring
叶片混砂机 kneader; kneader-mixer
叶片加固 blade reinforcement
叶片尖端 blade tip; blade toe; point of blade; vane tip
叶片间的 interlobe
叶片间距 blade pitch; blade spacing
叶片间流道 vane channel
叶片间隙 impeller clearance
叶片角(度) blade angle
叶片角度调整 angle setting of blade
叶片角可变的定轮 <液力变扭器的> variable-pitch stator
叶片角面 center [centre] surface of blade
叶片搅拌机 blade stirrer
叶片搅拌器 paddle agitator
叶片搅拌系统 blade paddle system
叶片搅拌轴 blade shaft
叶片接力器 blade servomotor
叶片节距 blade pitch
叶片结构 foliated texture
叶片截面 blade section
叶片紧固器 leaf fastener
叶片进口半径 entrance radius of blade
叶片进口边 centre edge of blade
叶片进口角 blade inlet angle; inlet blade angle
叶片进水边 inlet edge of vane
叶片径置叶轮 radial bladed impeller
叶片距离 impeller passage
叶片开度 blade opening; runner blade opening; runner opening
叶片开口 blade opening; runner blade opening; runner opening
叶片可调轴流风机 blade adjustable axial fan
叶片控制 vane control
叶片控制电机 vane control motor
叶片快门 blade shutter
叶片扩压器 vane(d) diffuser
叶片冷凝器 finned cooler
叶片冷却 blades cooling
叶片连接 root fixing
叶片联动装置 blade linkage
叶片流道 blade passage
叶片轮廓 blade profile
叶片轮缘 blade rim
叶片马达 sliding-vane motor
叶片面 blade face
叶片面积 blade area
叶片磨损指示器 blade wear indicator
叶片扭转 blade twist
叶片配置 blade arrangement
叶片偏移光束 blade shaft beam
叶片平均翼弦 mean blade chord
叶片气流互相干扰效应 cascade effect
叶片嵌装 inside of blades
叶片侵蚀 blade erosion
叶片侵蚀测试计 blade erosion tester
叶片倾角 blade tilt
叶片倾斜度 <水泵> blade inclination
叶片曲率 vane camber; vane curvature
叶片入口角 blade inlet angle
叶片栅距 tangential blade space
叶片伸长 blade extension
叶片式拌和机 arm mixer; pugmill; pug mixer; blade paddle mixer; paddle mixer
叶片式泵 vane-type pump
叶片式成球机 paddle-type nodulizer
叶片式出风口 vaned outlet
叶片式除粉器 paddle gummer
叶片式挡板 blade type damper
叶片式电动液压传动装置 vane-type electro-hydraulic gear
叶片式电枢 vane armature
叶片式盾构 blade shield
叶片式风动马达 vane-type rotary air motor
叶片式风扇 blade fan
叶片式风速计 vane anemometer
叶片式格栅 vane grill(e)
叶片式给料机 vane feeder
叶片式鼓风机 vane-type blower
叶片式过滤器 leaf filter
叶片式混料机 pug mixer
叶片式混砂机 blade mixer; revolving arm mixer; revolving paddle mixer; sand kneader
叶片式混砂机传动轴 blade shaft
叶片式搅拌机 arm mixer; blade paddle mixer; blade paddle stirrer; pugmill; pug mixer; paddle mixer
叶片式搅拌器 blade agitator; paddle-type agitator
叶片式卷取机 paddle-type coiler
叶片式空气制动器 fan brake
叶片式快门 bladed shutter; lamellar shutter; leaf-type shutter
叶片式流量计 vane flow meter
叶片式流速计 bucket current meter
叶片式螺旋输送器 paddle screw conveyer [conveyor]
叶片式马达 vane motor
叶片式喷嘴 blade nozzle
叶片式破碎机 blade crusher
叶片式曝气装置 bladed-surface aerator
叶片式气马达 linear motor
叶片式撒布机 vane spreader
叶片式扫雪机 blade type snow plough
叶片式输油泵 vane-type fuel transfer pump
叶片式松砂机 revivifier
叶片式弹簧 leaf spring
叶片式调节阀门 gate apparatus; guide vane apparatus
叶片式通风机 vane-type blower
叶片式洗矿机 paddle mill
叶片式压滤机 pressure leaf filter
叶片式压气机 vane-type compressor
叶片式压缩机 sliding-vane compressor; vane compressor
叶片式液压电动机 vane-type hydraulic motor
叶片式液压马达 vane-type motor; wing type hydraulic motor
叶片式液压马达旋转促动器 vane-type motor rotary actuator
叶片式仪表 vane-type instrument
叶片式油泵 vane-type oil pump
叶片式油马达 vane-type hydraulic motor
叶片式油液伺服机构 oil vane servomotor
叶片式增压器 van-type supercharger
叶片式折流板 vane-type baffle
叶片式真空泵 vane-type vacuum pump
叶片式真空过滤器 vacuum-leaf filter
叶片式轴流风扇 vane-axial fan
叶片式主弹簧 medium leaf spring
叶片式嘴组 vane-type nozzle block
叶片枢 blade pivot
叶片枢轴 blade stem
叶片数 number of leaves; vane number
叶片水热能的估计 estimating leaf water potential
叶片水势能 leaf water potential
叶片伺服马达 blade servomotor
叶片伺服器 blade servomotor
叶片损失 blade loss; bucket loss
叶片锁块 bucket locking piece
叶片弹簧 blade spring; flat spring
叶片提升机械构造 blade lift mechanism
叶片提升机械管理室 blade lift control housing

叶片提升控制小齿轮 blade lift control pinion
叶片提升连杆 blade lift link
叶片提升轴 blade lift shaft
叶片调节阀 blade regulating valve
叶片调节接力器 blade adjusting servomotor
叶片通道通流截面面积 bucket area
叶片通过频率 blade passing frequency
叶片推力 blade thrust
叶片外缘 tip of blade
叶片外缘间隙 tip clearance
叶片弯度 blade camber;blade curvature;vane camber
叶片温度 leaf temperature
叶片涡流 blade vortex
叶片涡旋 blade vortex
叶片镶嵌 blading
叶片效率 blade efficiency;efficiency of blading;vane efficiency
叶片形继电器 vane-type relay
叶片形刻刀 blade type cutter
叶片形面阻力 profile drag
叶片形皮带扫清器 blade type belt cleaner
叶片形气力振动器 vane-type air vibrator
叶片形式 blade shape
叶片形线 contour
叶片旋涡形装饰 foliage scroll
叶片压力 blade pressure
叶片液冷 liquid blade cooling
叶片液压马达 vane hydraulic motor
叶片油泵 vane oil pump
叶片有效高度 free height of blade
叶片缘 blade edge
叶片正面 pressure side of the blade
叶片制动销 blade lock
叶片轴 rubber blade trunnion
叶片轴颈 runner blade trunnion
叶片轴流式扇风机 vane-axial fan
叶片转换杆 blade shift beam
叶片转子 vane rotor
叶片装置 blade setting;blading
叶片装置角 blade angle;blade incidence;vane setting angle
叶片状 bladed;foliaceous;foliated;lamellar
叶片状的 foliate
叶片状断口 foliated fracture
叶片状隔板 blade-shaped diaphragm
叶片状构造 bladed structure;foliated structure
叶片状花岗岩 foliated granite
叶片状灰岩 laminated limestone
叶片状结构 foliated texture;leaf-like texture
叶片状结晶 leaflet crystal
叶片状晶体 bladed crystal
叶片状煤 foliated coal
叶片状习性 lath-like habit
叶片状岩(石) foliated rock
叶片组 blading;segment of blading
叶羟硅钙石 zeophyllite
叶蠕绿泥石 phyllochlorite
叶色素 chromophyll
叶砂屑岩 phyllarenite
叶栅 blade cascade; blade lattice; blade row; blading; cartridge of blade; diffuser grid; louver [louvre];vane cascade
叶栅稠度 cascade solidity
叶栅稠密度 cascade solidity; chord spacing ratio
叶栅的几何参数 geometric(al) data of cascade
叶栅节距 spalling of wing
叶栅节弦比 cascade pitch chord ratio
叶栅流动 cascade flow;lattice flow
叶栅密度 cascade solidity
叶栅式扩散器 grating diffuser
叶栅试验 cascade test
叶栅(试验)风洞 cascade wind tunnel
叶栅(试验)水洞 cascade water tunnel
叶栅外形 cascade configuration
叶栅效应 cascade effect
叶栅栅距 vane pitch
叶扇空蚀 bladed shoulder cavitation
叶梢 blade tip
叶梢空化 blade tip cavitation;tip cavitation
叶梢涡旋 blade tip vortex
叶蛇纹石 baltimorite; picrolite; schweizerite;zermattite
叶蛇纹岩 antigorite
叶式搅拌器 leafing agitator
叶饰【建】 foliage; cut foilage; fleuron; leafage; leaf-like decoration; leaf ornamentation; leafwork; rinceau
叶饰雕刻 leafwork
叶饰和箭头饰 leaf and dart
叶饰和舌饰 leaf and tongue
叶饰皇冠 leaf crown
叶饰细雕 leafwork
叶饰与针刺饰 leaf-and-dart mo(u)lding
叶饰中段柱身 caulis
叶饰柱身的柱(子) corollithic column
叶饰柱身柱 Carolithic column
叶饰柱头 foliage capital;foliated capital
叶饰柱(子) corollithic column
叶树木材 leaf wood
叶双晶石 fremontite;natromontebrasite
叶水势 leaf water potential
叶素轨迹 blade element path
叶酸 folic acid
叶铁钨华 phyllotungstit
叶纹装饰板 acanthus
叶形 leaf pattern;phylliform
叶形板 acanthus foliage;acanthus leaf
叶形桁架 lens truss;lenticular truss
叶形轮 lobe;lobed wheel
叶形耙 blade harrow
叶形饰 acanthus;foiling
叶形饰拱 foiled arch
叶形线 folium [复 folia]
叶形与卷须形饰 vinette
叶形指数 leaf index
叶形转子式流量计 lobed impeller meter
叶形装饰 acanthus; decorative foil; foil;bear's breech <古希腊科林思柱头>;angle leaf <柱脚方石板角上的>
叶形装饰的 foil decorating
叶形装饰檐壁 acanthus frieze
叶型 aerofoil; air foil; blade profile; leaf type;lobe
叶型规 contour ga(u)ge
叶型角 profile angle
叶型空化 profile cavitation
叶型泥浆搅拌机 propeller mud mixer
叶型汽蚀 profile cavitation
叶型升力 profile lift
叶型损失 profile loss
叶型样板 profile ga(u)ge
叶型阻力 profile drag
叶旋涡饰 rinceau
叶硬蛇纹石 baltimorite
叶与箭线脚装饰 heart and dart
叶与舌饰 leaf and tongue
叶展 leafing
叶展持久性 leafing retention
叶展剂 leafing agent
叶展稳定性 leafing stability
叶展型铝粉 leafing-type alumin(i)um powder
叶痣菌 <拉> Melasmia
叶状 lobate
叶状冰楔 foliated ice wedge
叶状柄 phyllode
叶状剥落 exfoliation
叶状层 folium [复 folia];leaf gold
叶状窗花格 <哥特式> leaf-shaped curve
叶状的 foliaceous;foliar;leafy
叶状地线 leafy liverwort
叶状地衣 foliose lichen
叶状拱 foliated arches
叶状沟槽铸型 frondescent furrow flute casting
叶状构造 bladed structure; foliation structure
叶状锂长石 petalite
叶状煤 leaf coal
叶状黏[粘]土 leaf clay
叶状曲线 leaf-shaped curve
叶状饰 foliation;leaf-like ornamentation
叶状饰带 leafy frieze
叶状悬崖 lobate scarp
叶状岩石 foliated rock
叶状檐壁 leafy frieze
叶状枝 leaf-like stem; phylloclade; phylloid cladode
叶状植物 thallophyte;thallose
叶状铸型 frondescent cast
叶状装饰 leaf-like decoration
叶状组织 leaf-like tissue
叶子板 fan blade; mud apron; mud guard
叶子板后视镜 fender mirror

曳

曳板 dragging-shoe;dragging-slip

曳车蟹壳桶 tractor cram-shell bucket
曳出流体 entrained fluid
曳船道 towing pass;tow way
曳锭器 retractor
曳光剂 tracer composition; tracer mixture
曳光器 tracer
曳轨器 rail puller
曳痕 dragline
曳火线 time fuse
曳力 downdrag;drag
曳力矩 drag torque
曳力系数 coefficient of drag;drag coefficient; drag factor; Faxen drag factor;hindrance factor
曳裂弧 festoon
曳木机 log haul-up
曳炮车 ordnance tractor
曳索 towing strap
曳索机架 logging frame
曳索塔 guyed tower
曳拖带 tow
曳网 ground net;trawl
曳物线 tractrix
曳物线喇叭 tractrix horn
曳下 drawdown
曳下高程 drawdown elevation
曳引标度 drafting scale; draughting scale
曳引车 towing troll(e)y;towing vehicle
曳引带去 entrainment
曳引的 tow
曳引机 road traction engine
曳引绞车 towing winch
曳引力 towing power; traction; traction power; tractive force; tractive power
曳引汽车 towing vehicle;towing winch
曳引驱动电梯 traction drive lift
曳引式拖拉机 tow tractor
曳引索 tow line
曳引体系 towing system
曳引效应 drag effect
曳阻力 drag force

页

页边说明 marginalia

页表 page table
页出口 page exit
页硅酸盐 phyllosilicate;sheet silicate
页寄存器 page register
页交换文件 page swap file
页界 page boundary
页拷贝 page copy
页框 page frame
页理【地】 lamellation
页码 pagination;dropfolio
页面 page frame
页面边界 page boundary
页泥炭 leaf peat
页片剥离山 exfoliation mountain
页片状 lamellar
页片状结构 laminated texture
页式 page mode
页式打印机 page printer;receive-only page printer
页式电传打字电报机 page teletype
页数 pagination
页数栏 folio column
页体 page body
页跳移 paper skip;paper slew
页跳移符号 paper throw character
页头 top of form
页岩 cleaving stone; shale; shale rock;shiver;torbanite;beck <其中的一种>
页岩残渣 burnt shale;oxidized shale
页岩槽 shale pit
页岩层 rammell
页岩储集层 shale reservoir
页岩底辟 shale diapir
页岩分类 classification of shale
页岩覆盖层 clod top
页岩干馏 retorting of shales;shale retorting
页岩干馏炉 retort for processing of oil shale
页岩化作用 shalification
页岩灰 shale ash
页岩混凝土板 shale concrete plank
页岩脊瓦 slate ridge
页岩夹层 batt; shale band; shale break
页岩焦油 sale tar
页岩焦油沥青 shale tar
页岩坑 shale pit
页岩控制泥浆 shale-control mud
页岩矿床 shale deposit
页岩沥青 shale tar;shale tar pitch
页岩沥青毡 batt
页岩砾砾岩 shale-pebble conglomerate
页岩垆姆 slaty loam
页岩脉理 shale stringer
页岩煤 bone coal; bony coal; slate coal
页岩煤油 photogen
页岩黏[粘]土 shale clay
页岩黏[粘]土砖 shale clay brick
页岩刨削机 shale cutter;shale planer
页岩汽油 shale gasoline
页岩壤土 slaty loam
页岩砂 shaly sand
页岩砂岩交错层 gray bed
页岩石灰 shale lime;slate lime
页岩水泥 shale cement;slate cement

页岩碎片 shale fragment
页岩太阳油 secunda oil
页岩陶粒 haydite;shale ceramisite
页岩陶粒和陶砂 coarse and fine aggregate of expanded shale
页岩筒瓦 slate roll
页岩土 white bina
页岩硬焦油沥青脂 shale tar pitch
页岩硬砂岩建造 shaly-graywacke formation
页岩油 oil-shale fuel;schist oil;shale oil;solar oil
页岩油母 oil-shale kerogen
页岩原油蒸馏时得到的最后馏分 once-run oil
页岩渣 shale ash
页岩质粗砂岩 shale grit
页岩质煤 bone coal
页岩质黏[粘]土 shale clay;slaty clay;shaly clay
页岩质砂岩 hazel
页岩质石灰岩 shaly limestone
页岩砖 brick shale;shale brick
页岩状的 shaly
页岩状构造 shaly structure
页岩状结构 shaly texture
页岩状黏[粘]土 shaly blaes;shaly clay
页岩状砂岩 shaly blaes
页状 laminated
页状剥落【地】 exfoliation;exfoliate;sheeting;sheet jointing
页状剥落节理 exfoliation joint
页状剥落丘 exfoliation dome
页状的 fissile;schistose;schistous
页状构造【地】 book structure
页状集束构造 bookhouse fabric
页状节理【地】 sheet joint
页状结构 laminated structure
页状矿物 sheet mineral
页状煤 leaf coal;paper coal
页状泥炭 paper peat
页状黏[粘]土 book clay;leaf clay
页状岩层 shaly bedding

夜班 dog shift;hoot owl;late shift;lobster shift;night duty;night gang;night shift;night watch
夜班报务员 night operation
夜班费 night duty allowance;night shift allowance
夜班工人(总称) night shift
夜班工作 graveyard shift
夜班话务员 night operation
夜班津贴 night differential;night duty allowance;night-work allowance
夜班警卫员 dark horse
夜班矿工 night pair
夜班室 night duty room
夜班守望员 jack-o-lantern
夜班守卫员 jack-o-lantern
夜班台 concentration position;night position
夜半太阳 midnight sun
夜潮 night tide
夜大学 evening college;evening university
夜盗 burglary
夜灯 night light(ing)
夜风 night wind
夜浮游生物 nyctipelagic plankton;nyctoplankton
夜工 night work
夜光 night light(ing);noctilucence
夜光标度盘 luminous diagram
夜光表 glow-watch;luminous watch;radioluminescent watch
夜光表盘 blackout dial;luminous dial
夜光罗盘 luminous compass
夜光漆 luminous paint
夜光涂料 luminous paint;phosphorescent coating;phosphorescent pigment;self-luminous paint
夜光洗片温度计 luminous thermometer
夜光颜料 luminescent pigment;luminous pigment;phosphorescent pigment
夜光余迹 noctilucent train
夜光元件 luminous element
夜光云 luminous cloud;luminous night clouds;noctilucent clouds
夜光钟 astral movement clock
夜规 night dial
夜航命令簿 night order book
夜航设备 night-flying equipment
夜航图 night-flying chart
夜航信号 night signal
夜弧 night arc
夜弧光 nocturnal arc
夜辉 airglow;night glow (emission);night sky light;night sky luminescence
夜辉光谱 night-glow spectrum
夜间 nighttime
夜间安全 night safe
夜间报警继电器 night alarm relay;night bell relay
夜间标志 night mark(ing)
夜间操作 night operation
夜间测距仪 astigmatizer
夜间测量 nocturnal measurement
夜间沉降 night setback
夜间充电 night charging
夜间臭氧垂直方向分布 night-time ozone profile;nocturnal ozone profile
夜间传话电报局 night appointed office
夜间存款窗口 night deposit
夜间存款业务 night depositing
夜间导航线 night leading line
夜间的 nocturnal
夜间低谷 night dip
夜间动力 night power
夜间发光 night illumination
夜间发光的 noctilucent
夜间发生的 nocturnal
夜间飞机救护舰 night plane guard
夜间峰值 <指电力负荷> evening peak
夜间服务 night service
夜间服务窗 night-service window
夜间辐射 nocturnal radiation
夜间负荷 night-time load(ed)
夜间负载 night load
夜间改道制 slumber-time diversion
夜间工作 night operation;night work
夜间功率 night power
夜间观测器 nightviewer
夜间观测仪 night viewing device
夜间观察 night observation
夜间航标 night signal
夜间航空摄影机 aerial night camera
夜间轰炸目标地图 night bombing target map
夜间呼叫 night call
夜间弧 night arc;nocturnal arc
夜间活动隔热装置 night insulation
夜间集中台 night-service desk
夜间计时仪 nocturnal timer
夜间加热 night charging
夜间驾驶仪 night driverscope
夜间监视 night surveillance
夜间降低 <自动控制温度控制点的> night setback
夜间降温 night setback
夜间交通 night traffic
夜间交通事故 night traffic accident
夜间接续 night-service connection
夜间截击 night interception
夜间警铃 night alarm bell
夜间净流量 net night flow
夜间俱乐部 night club
夜间看守者 night watchman
夜间可见距离 nocturnal penetration range
夜间雷暴 nocturnal thunderstorm
夜间冷却 night cooling;nocturnal cooling
夜间瞭望距离 night sighting;night-time visibility
夜间列车 night train
夜间列车运行 night service
夜间流量 night flow
夜间轮班 graveyard shift
夜间能见度 night visibility
夜间能见距 penetration range
夜间能见距离 night visual range
夜间逆温 nocturnal inversion
夜间迁移 nocturnal migration
夜间潜望镜 night periscope
夜间人口 night population;night-time population
夜间人口密度 night-time population density
夜间任务 night mission
夜间容量 night capacity
夜间散热 nocturnal cooling
夜间摄影 night photography;night shot
夜间摄影机 night camera
夜间生产能力 night capacity
夜间施工噪声 night construction noise
夜间施工增加费 construction expense added at night
夜间使用费 night-only tariff
夜间示像 night aspect
夜间视距 headlight sight distance;night sight distance
夜间视觉 night vision
夜间视宁度 night-time seeing
夜间税率 night-only tariff
夜间税则 night-only tariff
夜间送气 night charging
夜间损失 night-time loss
夜间探测器 snooperscope
夜间探向设备 night direction-finder
夜间通风孔 night vent
夜间通信[讯]距离 night range
夜间通信[讯]频率 night frequency
夜间投送电报 night letter (telegraph)
夜间望远镜 metascope
夜间误差 night error
夜间雾 night-time fog
夜间显示 night indication
夜间显示距离 night-time visibility
夜间效应 night effect;skywave trouble
夜间信号 night signal;night-time signal
夜间信号开关 night-alarm switch
夜间信号通信[讯] night signal(l)ing
夜间行车 night driving
夜间行车能见度 night-time visibility
夜间行车能见度要求 night visibility requirements
夜间修理 night repair
夜间蓄电加热器 electric(al) night storage heater
夜间蓄热式电加热器 night storage heater
夜间业务电台 station open in night
夜间医院 night hospital
夜间移栖 nocturnal migration
夜间噪声 night noise
夜间增压 night charging
夜间障碍标志 night obstruction marking
夜间照明 night illumination;night light(ing);night-time lighting
夜间照明电路 all-night circuit
夜间照明系统 night lighting system
夜间照相 night photography
夜间值 night value
夜间贮热水器 night storage heater
夜间贮水式热水器 off-peak hot water heater
夜间转换 night changeover
夜间作业 night operation;night work
夜间作用距离 night range
夜景 night scene;night view
夜景画 nocturne
夜警电路 night alarm circuit
夜空背景 night sky background
夜空辐射 night sky radiation
夜空辉光 night-glow of the sky
夜空透明度 night transparency
夜空照相机 night sky camera
夜来香【植】 tuberose
夜铃 night alarm;night bell
夜铃电键 day-and-night transfer key;night-alarm key;night bell key;night key
夜铃电路 night alarm circuit
夜铃继电器 night alarm relay;night bell relay
夜铃开关 night-alarm switch;night bell switch
夜铃装置 night alarm circuit
夜盲症 moon blindness;mooneye;night blindness;nyctatopia
夜明卡 luminous card
夜明罗经卡 luminous card
夜明盘手表 luminous dial wrist watch
夜明砂 bat's dung;bat's feces
夜明涂料 undark
夜气辉 permanent aurora
夜射程 night range
夜市 night market;night town
夜视 night viewing
夜视测距仪 see-in-dark rangefinder
夜视窗 night window
夜视航海六分仪 night vision marine sextant
夜视计划 night visibility plan
夜视镜 snooperscope
夜视距离 night-time visual range;penetration range;transmission range
夜视觉 night vision
夜视力 scotopia
夜视瞄准器 night vision sight;sniperscope
夜视器 snooperscope
夜视器件 night vision device
夜视设备 night observation device
夜视双目镜 night vision binoculars
夜视望远镜 night binocular(s);night vision telescope
夜视物镜 night vision objective
夜视系统 night-time vision system
夜视袖珍观察镜 night viewing pocket scope
夜视眼镜 night vision goggles
夜视仪(器) night-time vision device;night vision instrument;night vision viewer;snooperscope
夜视增益 gain of night vision instrument
夜视组件 night vision component
夜天 night sky
夜天背景 dark sky background
夜天辐射 night sky radiation
夜天光 airglow;light in the night sky;night airglow;night glow (emission);night sky light;night sky luminescence
夜天光度 luminosity of night sky

夜天光谱 night sky spectrum
夜天记录器 starshine recorder
夜天空 night sky
夜天亮度 night brightness
夜天摄谱仪 night sky spectrograph
夜天透明度 night transparency
夜托托儿所 child night time home
夜晚业务 night service
夜望镜 night glass(es);sniperscope;snooperscope
夜雾 night-time fog
夜销 night bolt
夜校 evening school;night school
夜星快速包裹业务 <英国铁路> Night Star parcels service
夜行列车 owl train
夜用红外传感器 night infrared sensor
夜用望远镜 night glass(es)
夜值 dog watch;middle watch
夜总会 night club;bistro;cabaret;super club;discotheque <放流行歌曲唱片供来客跳舞的>;frolic pad <俚语>;nitery <美>

液

液氨 liquid ammonia
液氨槽车 liquefied ammonia tank vehicle
液氨储存器 liquid ammonia receiver
液氨电池 liquid ammonia cell
液氨管 liquid ammonia pipe
液氨汽车槽车 tank truck for liquid ammonia
液氨贮槽 liquid ammonia storage tank
液棒运输机 gravity conveyer [conveyor]
液包膜【给】tonoplast vacuole
液包体 fluid inclusion
液泵 fluid pump;liquid pump
液泵密封 liquid pump seal
液比 liquor ratio;liquor-to-wood ratio <木材的>
液壁 liquid wall
液壁电离室 liquid wall ionization chamber
液边材 sap wood
液表面活性剂膜 liquid surfactant membrane
液波 fluid wave
液材 alburn(um);sap wood
液舱壁 tank bulkhead;tank wall
液舱舱柜构架 tank framing
液舱测量管 tank sound pipe
液舱顶 tank dome
液舱耗损量 ullage
液舱检验 tank survey
液舱吸口 tank suction
液舱压头 tank head
液舱座 tank foundation
液槽 fluid bath;fluid ditch;liquid cell
液槽压力计 cistern manometer
液层 liquid layer
液层扫描镜 teleidoscope
液层柱 liquid layer column
液称量法 liquid weigh process
液成 hydratogensis
液成沉积物 hydatogen sediment
液成的 hydatogenous
液成岩 hydatogenous rock
液成作用 hydatogenesis
液池 liquid bath
液床烘干器 <烘集料用> fluid bed drier [dryer]
液锤 liquid hammer;surging shock
液打兰 <药量单位,1 液打兰≈3.55 毫升> fluid dram
液氮车 liquefied nitrogen truck
液氮储存罐 liquid nitrogen container
液氮快速冻结装置 liquid nitrogen spray type quick freezer
液氮冷冻地基法 liquid nitrogen method of freezing ground
液氮冷阱 liquid nitrogen (cold) trap
液氮冷却 liquid nitrogen cooling
液氮冷却(超导)发电机 liquid nitrogen-cooled generator
液氮冷却系统 liquid nitrogen cooling system
液氮气化器 liquid nitrogen converter
液氮容器 liquid nitrogen vessel
液氮温度 liquid nitrogen temperature
液氮消耗量 liquified nitrogen consumption
液氮制冷 liquid nitrogen cooling
液氮制冷晶体 liquid nitrogen cooling crystal
液氮制冷探测器 liquid nitrogen-cooled detector
液道 fluid passage
液等静压成型 liquid isostatic(al) pressing
液滴 dripping;droplet;fluid drop;liquid drop
液滴捕集 entrainment trap
液滴捕集器 drop catcher
液滴法 <测定表面张力> sessile drop method
液滴分离器 demister entrainment;eliminator; entrainment separator;liquid drop separator;separator
液滴分离装置 entrainment trap
液滴计数器 drop counter
液滴加油器 drop oiler
液滴媒质 liquid drop medium
液滴模型 liquid drop model
液滴汽油 drip gasoline
液滴燃烧时间 droplet burning time
液滴润滑 drop lubrication
液滴示踪法 liquid bubble tracer
液滴指示器 drop indicator
液电破碎 hydroelectric(al) crushing
液电式推车机 hydroelectric(al) mine-car conveyer [conveyor]
液电伺服系统 hydroelectric(al) servocontrol system
液动冲击器 hydrohammer;hydropercussive tool
液动冲击钻具 hydropercussive drilling tool
液动的 fluid;liquid operated
液动阀 hydraulically operated valve;oil controlled valve
液动风扇 fluid drive fan
液动换向阀 hydraulically operated direction control valve
液动机 hydraulic actuator
液动夹板 hydraulic clamp
液动控制 hydraulic control
液动捞管器【岩】hydraulic pipe catcher
液动力矩变换器 hydrodynamic(al) torque transformer
液动螺杆钻具 hydraulic helicoids drill
液动马达 fluid motor
液动模拟 fluid flow analog(ue);hydrodynamic(al) analogy
液动牛头刨 hydraulic shaper
液动升降台 hydraulic fluid-lift platform
液动式动力传输 hydrodynamic(al) power transmission
液动式牙嵌离合器 hydromatic jaw clutch
液动伺服拖动装置 hydraulic servo-actuator
液动弹性力学 hydroelasticity
液动体冲压法 hydrodynamic(al) process
液动体润滑 hydrodynamic(al) lubrication
液动调节器 hydraulically operated controller;hydraulic controller
液动循环管路 hydraulic circuit
液动压板 hydraulic clamp
液动压力机构 hydraulic power unit
液动压润滑 hydrodynamic(al) pressure lubrication
液动闸 hydraulic brake;liquid brake
液动震击器【岩】hydraulic jar
液动阻力 hydraulic resistance
液度(估定)计 <研究土壤化学性质的装置> lysimeter
液肥混合调配器 liquid fertilizer mixer-proportioner
液肥坑 liquid mature pit
液分离器 liquid separator
液封 dip seal;fluid seal;liquid blocking;liquid packing;liquid seal;water lute
液封泵 fluid packed pump;liquid sealed pump
液封的 liquid-tight
液封点 seal point
液封放卸 liquid sealed discharge
液封管 dip pipe
液封环 lantern ring
液封提拉法 liquid encapsulated Czochralski growth
液封型真空泵 Nash pump
液封压缩机 Nash compressor
液封直拉法 liquid encapsulated Czochralski technique
液氟 liquid fluorine
液浮摆式加速度仪 liquid floated pendulous accelerometer
液浮陀螺仪 float-type gyroscope;fluid floated gyroscope;liquid floated gyroscope
液感 liquid inductance
液汞提升 liquid mercury lift
液固分离 liquid solid separation
液固混合推进剂 lithergol
液固界面 fluid solid interface;liquid solid boundary
液固色层(分离)法 liquid solid chromatography
液固色谱(法) liquid solid chromatography
液固(态)转化点 liquid solid transition point
液固吸附 liquid solid adsorption
液固吸附色谱法 liquid solid adsorption chromatography
液固系统 fluid solid system
液罐车 mobile tank;tank wagon
液罐挂车 liquid trailer;trailed tank
液罐货车 tank truck
液罐汽车 tank truck
液柜车 cistern wagon
液柜座 tank foundation
液滑现象 aquaplaning
液化 liquefaction; fluidication; liquate;liquefy(ing)
液化苯酚 liquefied phenol
液化比率 liquefaction ratio
液化变性 liquefactive degeneration
液化丙烷 liquefied propane (gas);liquid propane
液化丙烷燃料发动机 liquid propane engine
液化丙烷运输船 liquefied propane gas tanker;LPG tanker
液化补救方法 liquefaction remediation measure
液化层 fluidized bed
液化层深度 depth of liquefaction layer
液化掺和系统 liquid blending system
液化船 liquid cargo ship
液化床反应器 liquid bed reactor
液化床生物反应器 liquid bed biologic(al) reactor
液化氮 liquid nitrogen
液化氮气 liquefied nitrogen gas
液化的 diffluent;liquefied
液化等级 liquefaction category
液化地基 liquefaction foundation
液化点 flow point;liquation point;liquefaction point;liquidizing point;liquefying-point <沥青材料等>
液化度 degree of liquefaction
液化惰性气体 liquefied gas;liquefied noble gas
液化法 liquefaction process
液化概率 liquefaction probability
液化固体层滑床 bed of fluidized solid
液化过程 liquefying process
液化货船 liquid ship
液化机会图 opportunity map of liquefaction
液化剂 fluidizing agent;liquefier;liquefying agent;liquifier
液化甲烷 liquefied methane
液化甲烷气 liquefied methane gas
液化甲烷运输船 liquefied methane gas tanker;LMG [liquidized methane gas] tanker
液化焦化装置 fluid coking unit
液化抗力 liquefaction resistance
液化颗粒集合体 liquefied particles
液化可能性 liquefaction potential;possibility of liquefaction
液化空气 liquefied air;liquefying air;liquid field air;liquified air
液化沥青 asphalt cutback;road oil
液化炼厂气 liquefied refinery gas
液化灵敏度 liquefaction susceptibility
液化流 fluidized flow
液化率 fraction of gas liquefied;rate of liquefaction
液化氯气 liquefied chlorine gas
液化煤气发动机 liquid-gas engine
液化煤气贮存箱 liquefied gas storage tank
液化敏感性 susceptibility of liquefaction
液化气 blue gas
液化气泵 liquefied gas pump
液化气叉车 liquefying gas fork truck
液化气储罐站安全区 protection zone within the liquefied gas tank farm
液化气储配站 liquefied petroleum gas distribution station
液化气的储存 storage of liquefied gas
液化气低温储存 low-temperature storage of liquefied gas
液化气地下储存 in-ground storage for liquefied gas
液化气动力车 liquefied petroleum gas powered car
液化气发动机 liquefied petroleum gas engine;LPG engine
液化气管道 liquefied gas pipeline
液化气罐 liquefied natural gas tank;liquefied petroleum gas tank
液化气零售点 liquefied petroleum gas retail depot
液化气输送压缩机 liquid-gas transfer compressor
液化气体 liquefied gas;liquefying gas;liquid gas;liquified gas
液化气体槽车 liquefied gas tanker
液化气体储罐 liquefied gas tank
液化气体燃料 dry gas fuel;liquid-gas fuel;liquified fuel gas
液化气体燃料车 liquid-gas vehicle
液化气添加剂 liquefied petroleum gas

additive
液化气调节器 liquefied petroleum gas regulator
液化气雾剂 liquefied gas aerosols
液化气系统 liquid petroleum gas system
液化气运输 gas transport
液化气运输船 liquefied gas carrier; liquefied gas tanker
液化气蒸发残余试验 residue test for liquefied petroleum gas
液化气转输站 liquefied petroleum gas relaying station
液化器 liquefier; liquifier; liquilizer
液化潜热 latent heat of liquefaction
液化潜势 liquefaction potential
液化潜在能力 liquefaction potential
液化强度 liquefaction strength
液化区 liquefied region
液化热 heat of liquefaction
液化砂层 liquefying sand layer
液化砂土 liquefied soil
液化石蜡 liquefied petrolatum
液化石油气 cylinder gas; gasol; liquefied petroleum gas; liquified petroleum gas [LPG]; LP-gas
液化石油气铂重整 liquefied petroleum gas platforming
液化石油气槽车 liquefied petroleum gas tank vehicle
液化石油气储存 LPG storage
液化石油气储配站 LPG distribution station
液化石油气船 liquefied petroleum gas carrier; liquified gas carrier; LPG carrier
液化石油气发动机 liquefied petroleum gas engine; liquid-petroleum-gas engine
液化石油气钢瓶 liquefied petroleum gas bottle; liquefied petroleum gas container; liquefied petroleum gas cylinder
液化石油气高压气罐 liquefied petroleum gas high pressure holder
液化石油气公共汽车 liquefied petroleum gas bus
液化石油气供应 distribution of LPG; LPG supply
液化石油气机车 liquefied petroleum gas truck
液化石油气码头 liquefied petroleum gas terminal
液化石油气汽车 liquefied petroleum gas automobile
液化石油气汽车槽车 tank truck for liquefied petroleum gas
液化石油气燃料 liquefied petroleum gas fuel
液化石油气设备 liquefied petroleum gas equipment
液化石油气体 gasol; gas oil
液化石油气铁路槽车 liquefied petroleum gas tank-wagon
液化石油气运输 LPG transportation
液化石油气增味剂 liquefied petroleum gas odorant
液化时间 liquefying time
液化势 liquefaction potential
液化势评估 liquefaction potential assessment
液化树胶 liquefied resin
液化树脂 liquefied resin
液化松散颗粒流 liquefied cohesionless-particle flow
液化天然气 liquefied natural gas; liquid natural gas [LNG]
液化天然气槽车 LNG truck
液化天然气槽船 LNG ship navigated in river or nearshore sea
液化天然气穿梭运输气化船 shuttle and regasification vessel
液化天然气船 liquefied natural gas carrier; liquid natural gas carrier; LNG storage vessel
液化天然气返装作业 LNG ship reloading
液化天然气购销合同 LNG sale & purchase agreement
液化天然气罐 liquefied natural gas tank
液化天然气罐内泵 LNG in-tank pump
液化天然气加气机 LNG dispenser
液化天然气加气站 LNG filling station
液化天然气接收站 LNG (receiving) terminal; LNG regasification terminal
液化天然气快速相变 rapid-phase-transition of LNG
液化天然气冷能利用 LNG cryogenic energy utilization
液化天然气码头 liquid natural gas terminal; LNG jetty
液化天然气卫星站 LNG satellite station
液化天然气项目开发协议 LNG project development agreement
液化天然气卸料总管 LNG unloading header
液化天然气溢出 spillage of LNG
液化天然气运输船 LNG carrier; LNG tanker; LNG vessel
液化天然气贮槽 LNG storage vessel
液化天然气装车橇 LNG truck loading package
液化天然气装卸臂 LNG marine loading and unloading arm
液化天然气装卸臂包络范围 LNG arm envelope
液化天然气组成 LNG [liquid natural gas] composition
液化烃 blau gas
液化温度 condensing temperature; liquefying temperature
液化系数 coefficient of liquefaction
液化箱 fluidizing box; liquefying tank
液化效应 Joule-Thomson effect; liquefaction effect
液化性 liquescence
液化易燃气 liquefied inflammable gas
液化应力 liquefying stress
液化应力比 liquefaction stress ratio
液化用煤 liquefaction coal
液化沼气 liquefied methane gas
液化值 liquefaction value
液化指数 liquefaction index
液化装置 liquefaction plant; liquefying plant
液化状态 quick condition
液化作用 liquation; spontaneous liquefaction; liquefaction; liquification
液化作用破坏 liquefaction failure
液环泵 liquid rotary pump; rotary pump with liquid piston
液环式氯气压缩机 liquid circulation chlorine compressor
液环(式)压缩机 liquid piston compressor; liquid ring compressor; liquid rotary compressor
液环(式)真空泵 liquid ring vacuum pump
液回路 circuitry
液货舱容 liquid cubic
液货船 tanker
液货罐装高度 liquid cargo filling height
液货软管舱 cargo hose compartment
液货吸管 cargo suction pipe
液击 hammering; liquid hammer; liquid slugging; pipe hammer
液剂 liquor
液剂量计 liquid dosing apparatus
液加仑 liquid gallon; wine gallon
液浇塑胶屋面涂层 fluid-applied plastic roof coating
液胶 liquid glue; sol
液胶体 liquogel
液结电势 liquid junction potential
液解作用 lyolysis
液界电位差 liquid junction potential
液浸保安器 liquid fuse unit
液浸变形 liquostriction
液浸变压器 liquid-immersed transformer
液浸变阻器 liquid rheostat
液浸电抗器 liquid-immersed reactor
液浸法 immersion method
液晶 liquid crystal
液晶表 liquid crystal watch
液晶波导 liquid crystal waveguide
液晶层 liquid crystal layer
液晶成像 liquid crystal imaging
液晶存储器 liquid crystal memory
液晶发光 lyo-luminescence
液晶固定相 liquid crystal stationary phase
液晶光阀 liquid crystal light valve
液晶光学 liquid crystal optics
液晶盒 liquid crystal box; liquid crystal cell
液晶红外器件 liquid crystal infrared device
液晶混合物 liquid crystal compound
液晶聚合物 liquid crystal polymer
液晶膜 liquid film
液晶闪烁分光计 liquid scintillation spectrometer
液晶速度计 liquid crystal speed indicator; liquid crystal speedometer
液晶速度指示器 liquid crystal speed indicator
液晶态 liquid crystalline state; liquid crystal state
液晶体 liquid crystal
液晶无探伤法 liquid crystal non-destructive testing
液晶显示 liquid crystal display
液晶显示玻璃 liquid crystal display glass
液晶显示材料 liquid crystal display material
液晶显示操作 liquid crystal display operation
液晶显示 fluid crystal display; liquid crystal display
液晶显示器 liquid crystal display device
液晶显示石英手表 quartz watch with liquid crystal display
液晶显微镜 liquid crystal microscope
液晶显像 liquid crystal pictorial display
液晶相 liquid crystalline phase
液晶像转换器 liquid crystal image converter
液晶元件 liquid crystal cell
液晶注入 liquid crystal injection
液空悬挂支柱 hydropneumatic suspension strut
液控安全阀 pilot safety valve
液控传感器 fluid sensor
液控单向阀 hydraulic-operated check valve; pilot-controlled check valve; pilot-operated check valve
液控式阀 pilot-controlled type valve; pilot-operated type valve
液控溢流阀 pilot relief valve
液拉压力表 liquid manometer
液蜡 fluid wax
液冷 fluid cooling; liquid cooling
液冷变压器 liquid cooled transformer
液冷电机 fluid-cooled electrical machine
液冷定子绕组 liquid cooled stator winding
液冷发电机 liquid cooling generator
液冷钢化玻璃 liquid quench hardening glass
液冷介质 liquid cooling medium
液冷晶体管 liquid cooled transistor
液冷却系统 liquid cooling system
液冷燃气轮机叶片 liquid cooled gas-turbine blade
液冷绕组 fluid-cooled winding
液冷式的 liquid cooled
液冷式发动机 liquid cooled engine
液冷式内燃机 liquid cooled engine
液冷式喷管 liquid cooled nozzle
液冷式制动装置 liquid cooled brake
液冷转子 liquid cooled rotor
液力保险杠 hydraulic bumper
液力保险杠液力缸 hydraulic bumper cylinder
液力泵 fluid operated pump; hydraulic pump unit
液力泵喷浆机 hydraulic pump placer
液力变换器 fluid converter
液力变矩器 converter slip; hydraulic converter; hydraulic torque converter [convertor]; hydrodynamic(al) torque-converter; torque converter
液力变矩器操纵杆 torque converter lever
液力变矩器尺寸 torque converter size
液力变矩器传动 torque converter drive
液力变矩器控制阀 torque converter charging valve
液力变矩器控制压力 torque converter charging pressure
液力变矩器渗漏 torque converter leakage
液力变矩器输出安全阀 torque converter outlet relief valve
液力变矩器输出特性曲线 output characteristic curve of hydraulic torque converter
液力变矩器通用特性曲线 universal characteristic curve of hydraulic torque converter
液力变矩器循环泵 torque converter recirculation pump
液力变矩器用液 torque converter charging fluid; torque converter liquid
液力变矩器油温警告灯 torque converter oil temperature telltale
液力变矩器油箱 torque converter tank
液力变矩器油压表 torque converter oil pressure ga(u)ge
液力变矩器原动机联合特性曲线 joint characteristic curve of hydraulic torque converter and prime motor
液力变矩器主动轮 hydraulic torque converter driving wheel
液力变扭器 flow converter; hydrodynamic(al) transformer; torque converter level tell-tale
液力变扭器闭锁【机】 converter lockup
液力变扭器传动 torque converter drive; torque converter power unit; torque converter transmission
液力变扭器反应器 torque converter reactor
液力变扭器壳 torque converter housing
液力变扭器控制杆 torque converter control lever

Y

液力变扭器箱 torque converter housing
液力变扭器液温信号灯 torque converter temperature telltale
液力变扭器用液 torque converter fluid
液力变扭自动换挡 hydrodynamic(al) shift
液力变速传动 hydraulic variable speed drive
液力变速箱 hydraulic transmission box
液力薄膜测力计 hydraulic capsule
液力补偿悬挂 hydrolastic
液力采石工作 hydraulic rock cutting
液力操纵板 hydraulic panel
液力操纵的 hydraulically operated
液力操纵的侧倾卸车 hydraulically operated side-tipping wagon
液力操纵离合器 hydraulically controlled clutch
液力操作 hydrodynamic(al) operation
液力操作的 liquid operated
液力侧推力器 hydraulically powered side thrust
液力测功器 hydraulic dynamometer
液力测力环 hydraulic link
液力测力计 hydraulic cell;hydraulic dynamometer
液力测压计 hydraulic cell
液力冲击 hydraulic shock
液力传动 fluid power transmission; hydramatic transmission; hydraulic drive; hydraulic power; hydraulic transmission; hydrodynamic (al) drive; hydrokinetic transmission; hydromatic drive; hydrodynamic(al) transmission
液力传动柴油机车 diesel hydraulic transmission locomotive
液力传动机车 hydraulic drive locomotive
液力传动绞车 hydraulically powered winch
液力传动螺旋 hydraulic auger
液力传动内燃机车 diesel hydraulic transmission locomotive; diesel hydromechanical locomotive;hydraulic diesel locomotive
液力传动式割草机 hydraulically driven mower
液力传动输入输出转速差 hydraulic slip
液力传动系 fluid transmission
液力传动油 hydraulic transmission oil
液力传动中间悬挂式割草机 hydraulically driven mid-mounted mower
液力传动转差率 slip ratio of hydraulic drive; slip ratio of hydraulic transmission
液力传动装置 fluid clutch;hydraulic transmission gear;fluid drive
液力传绞车 hydraulically powered winch
液力传送器 hydrotransmitter
液力存储器 hydraulic storage
液力打滑损失 hydraulic slip loss
液力导轨钻机 hydrotrack drill
液力的 hydrodynamic(al)
液力电动机 fluid power motor
液力定速阀 hydraulic speed setting valve
液力锻压机 hydraulic forging press
液力发动机 fluid motor
液力阀 hydraulic valve
液力仿形控制 hydraulic tracing control
液力飞轮【机】fluid flywheel
液力分离器 hydroseparator
液力风扇 hydraulic fan
液力附加装置 hydraulic attachment
液力工作油热交换器 hydraulic transmission oil heat exchanger
液力供压机 hydraulic mule
液力钩 hydraulic hook
液力管接头 hydraulic coupling;hydrokinetic coupling
液力横向搂草机 hydraulic dump rake
液力缓冲底架 hydrocushion underframe
液力缓冲器 hydraulic buffer
液力换向机构 hydraulic change-over
液力回旋淘析器 hydraulic cyclone elutriator
液力机 hydraulic machine
液力机械 hydrodynamic(al) machine; hydrokinetic machine
液力机械变速器 hydromechanical transmission
液力机械操作 hydrodynamic(al) operation
液力机械传动 hydromechanical drive; hydromechanical transmission; Torqflow;hydraulic and mechanical drive
液力机械传动变速箱 Torqflow transmission
液力机械传动系统 hydromechanical drive system
液力机械传动效率 hydraulic mechanical efficiency
液力机械元件 hydrodynamic (al) mechanical element
液力夹具 hydraulic clamping device
液力间隙调整器 <柴油机> hydraulic lash adjuster
液力减速器 hydraulic retarder; hydrodynamic(al) retarder
液力剪机 hydraulic shearing machine
液力剪切式割草机 hydroclipper mower
液力绞车 hydraulic power winch;hydraulic winch
液力绞盘 hydraulic capstan
液力搅拌 hydraulic agitation
液力搅拌器 hydraulic agitator
液力静挤压 hydrostatic extrusion
液力举升器操纵 hydrojack control
液力控制 hydraulic control
液力控制阀 fluid-controlled valve
液力离合器 hydraulic clutch; liquid clutch
液力离合器总成 hydraulic clutch assay
液力力矩 hydraulic torque
液力连接器 hydraulic coupling
液力联轴节 fluid coupling; hydraulic coupling; hydrodynamic (al) coupling;hydrokinetic coupling
液力联轴器 hydraulic coupler
液力螺栓机 hydraulic bolt forcer
液力马达 fluid power motor
液力马达传动 hydraulic motor drive
液力马达驱动的绞车 hydraulic motor winch
液力铆机 hydraulic riveting machine
液力密封 liquid packing
液力摩擦制动 fluid friction brake
液力磨矿机 fluid energy mill
液力黏[粘]着 hydraulic lock
液力扭矩 hydraulic torque
液力扭矩计 hydraulic torquemeter
液力耦合离合器 fluid coupling clutch
液力耦合器 fluid clutch; fluid coupling;fluid-flywheel clutch;hydraulic clutch; hydraulic coupler; hydraulic coupling; hydrodynamic (al) coupling;hydrokinetic coupling
液力耦合器的滑移 slip in fluid coupling
液力耦合器等效特性曲线 equivalent characteristic curve of hydraulic coupling
液力耦合器效率曲线 efficiency curve of hydraulic coupling
液力耦合器原动机联合特性曲线 joint characteristic curve of hydraulic coupling and prime motor
液力耦合器原始联合特性曲线 primary characteristic curve of hydraulic coupling
液力盘式制动器 hydraulic disk brake
液力喷射 hydrojet
液力喷雾 hydraulic atomization
液力喷雾器 pump operated sprayer
液力平衡阀 hydraulically balanced valve
液力平衡悬架 hydrolastic
液力破碎法 hydrofracturing
液力起步 fluid start
液力起重缸 lifting cylinder
液力起重机 hydraulic crane; water crane
液力起重器 hydraulic jack
液力倾卸挂车 hydraulic tipping trailer
液力驱动 hydraulic propulsion; hydrostatic drive
液力驱动的 hydraulically operated; hydraulic operated
液力驱动的机械式离合器 hydraulically operated clutch
液力驱动装置 hydraulically operated equipment
液力熔断器 hydraulic fuse
液力散弹钻进方式 fluid pellet system
液力刹车油 hydraulic brake fluid;hydraulic fluid
液力升降机构 hydraulic lift mechanism
液力式联结器 hydraulic hitch
液力疏浚 hydraulic dredging
液力输送机 hydraulic conveyer[conveyor]
液力输送系统 hydraulic conveyer[conveyor]
液力双制动器 hydraulic dual brake
液力伺服马达 hydraulic servomotor
液力损失 hydraulic loss
液力提升 hydraulic lift
液力提升控制 hydraulic lift control
液力提升系统 hydraulic lifting system
液力挑顶刷帮机 hydraulic scaling rig
液力调节器 hydraulic governor
液力退刀器 hydraulic tool release
液力挖沟机 hydrotrencher
液力挖掘机 hydraulic excavator digger
液力弯板机 hydraulic plate bender
液力弯曲机 hydraulic bending machine
液力涡轮 hydraulic turbine
液力效率 hydraulic efficiency
液力斜接头 hydraulic bent sub
液力行星变速器 hydroplanetary transmission
液力蓄能器 hydraulic accumulator
液力悬挂自动调平系 hydraulically powered suspension leveling system
液力旋压 hydrospin(ning)
液力循环 fluid cycle
液力压头 hydraulic ram;hydraulic rammer
液力压弯机 hydraulic bender
液力油 hydraulic oil
液力元件 hydraulic element; hydrodynamic(al) element
液力(源) hydropower
液力源组 hydraulic power pack
液力造型机 hydraulic mo(u)lding machine
液力张紧调整器 hydraulic tension regulator
液力真空联合制动器 hydraulic vacuum brake
液力振动器 hydraulic vibrator
液力直径 hydraulic diameter
液力制动操纵阀 hydraulic brake operating valve
液力制动阀 hydrodynamic(al) brake valve
液力制动机 hydrodynamic (al) brake;hydrokinetic brake
液力制动(器) hydraulic brake; hydrodynamic(al) brake;liquid brake
液力制动油调节器 hydrodynamic(al) brake oil filling regulator
液力制动轴 hydraulic brake shaft
液力轴向推力 axial hydraulic thrust
液力助力转向 hydraulic booster steering
液力转向加力器 hydraulic steering brake
液力转向助力器 hydraulic steering booster
液力转向装置 hydraulically powered side thruster
液力转压式压榨机 hydraulic revolving-pack press
液力装载机 hydraulic loader
液力自动倾卸车 hydraulic dump truck
液力自动装置 hydraulic automatic device
液力阻塞 hydraulic lock
液粒 droplet
液量 fluid measure;fluid volume
液量盎司 fluid ounce
液量打兰 <等于 1/8 液量英两> fluidrachm;fluidram;fluid dram
液量单位 hogshead;liquid measure
液量度量 liquid measure
液量计 bubbler ga(u)ge; dosimeter; fluid volume meter
液量夸脱 liquid quart
液量指示器 telltale
液料表面处理 surface dressing
液流 flow of sap; fluid flow; fluid stream; liquid flow; liquid stream; liquor stream
液流泵 fluid flow pump
液流不稳定状态 hydraulic transient
液流洞 dog hole
液流阀 fluid valve
液流供料机 flow machine
液流横断面 cross-stream dimension
液流角 flow angle
液流开关 cock
液流理论 theory of fluid flow
液流力矩 moment of flow
液流描记器 hydrophorograph
液流模拟(法) fluid flow analogy
液流目视指示器 sight flow indicator
液流喷射器 flow gun
液流调节 flow control
液流调节孔 trim hole
液流通路 flow channel
液流图 effluogram; fluid flow pattern;flux map
液流限制 restriction of flow
液流学 rheology
液流中心 center [centre] of flow
液氯 liquid chlorine
液氯储罐 liquid chlorine storage tank
液氯工段 liquid chlorine section
液氯计量槽 liquid chlorine metering tank
液氯扩散器 liquid chlorine diffuser
液氯喷射器 liquid chlorine injector
液氯气 liquified chlorine gas
液氯消毒法 disinfection method by liquid chlorine
液满 full of liquid;hydroful

液密的 liquid-tight
液密的防水剂 liquid-tight water repeller
液密的蜡 liquid-tight wax
液密接触 fluid-tight contact
液密接头 fluid-tight joint
液密增塑剂 liquid-tight workability agent
液面 fluid level; level surface; liquid face; liquid level; liquid surface; surface of liquid; flux level <熔融玻璃>
液面标尺 ullage scale; ullage stick; ullage tape
液面标高 liquid surface level
液面表 tank ga(u)ge
液面测定 liquid level measurement
液面测定杆 dip rod
液面测杆 dipstick; fluid level ga(u)ge rod
液面测量杆 dipstick
液面池壁砖 flux-line block
液面传感器 liquid level sensor
液面传声全息法 liquid surface acoustic(al) holography
液面的气动控制 pneumatic control of liquid level
液面电位差 liquid potential
液面发送器 liquid level transmitter
液面阀 liquid elevating valve
液面浮动继电器 liquid level relay
液面负荷 liquid surface load
液面干涉仪 liquid surface interferometer
液面高程 liquid surface level
液面高度 level of liquid
液面高度计 liquid leveler; tellevel; wantage rod
液面(高度)记录仪 liquid level recorder
液面钩尺 <量液体高度的> hook ga(u)ge
液面恒定阀 constant level valve
液面滑行 hydroplaning
液面滑行速度 hydroplaning speed
液面恢复 build-up of fluid
液面回声探测仪 echo liquid level instrument
液面计 content ga(u)ge; level ga(u)ge; level meter; liquid indicator; liquid leveler; liquid level meter; liquidometer
液面计玻璃 ga(u)ge glass; ga(u)ging glass
液面计传送器 tank ga(u)ge transmitter
液面计公称长度 nominal length of liquid level ga(u)ge
液面计接管尺寸 nozzle dimensions for level ga(u)ge
液面记录控制器 liquid level recording controller
液面记录器 level recorder; liquid level recorder; recording ga(u)ge
液面记录仪 recording liquid level ga(u)ge
液面监视器 level monitor
液面检查开关 ga(u)ge cock
液面降低 depression of surface level
液面警报器 liquid level alarm
液面开关 surface cock
液面控制 level control; level(l)ing
液面控制范围 liquid level control range
液面控制开关 surface cook
液面控制器 faratron; level controller; liquid level controller
液面控制旋塞 surface cock
液面控制装置 level(l)ing device
液面落差 falling head
液面曝气器 surface aerator
液面上部气体分析 headspace analysis
液面上气相色谱分析 gas chromatographic(al) head space analysis
液面上下限警报器 level alarm high low
液面上响应 headspace response
液面升高 build-up of fluid
液面升降 fluctuation of level
液面声全息(术) liquid surface acoustic(al) holography
液面式垂直度传感器 orthogonal level transducer
液面视镜 level glass
液面收集剂 surface collection agent
液面水平测量法 liquid level method
液面调节 liquid level control
液面调节器 level controller; level regulator; liquid level controller; liquor-level regulator
液面调整器 level controller
液面吸附 liquid surface adsorption
液面下灌浆 undersurface filling
液面下指示器 undersurface indicator
液面显示玻璃 reflex glass
液面线 flux line
液面线侵蚀 cutback; flux-line attack; flux-line corrosion; glass level cut; metal line attack; metal line cut
液面信号发送器 tank sender
液面仪 liquid level ga(u)ge; liquid level indicator
液面指示变送器 liquid level indicating transmitter
液面指示玻璃管 level sight glass
液面指示计 indicating liquid level ga(u)ge; liquid level indicator
液面指示控制 level indicated control
液面指示控制器 level indicating controller
液面指示器 fluid level indicator; level ga(u)ge; level indicator; liquid indicator; liquid level ga(u)ge; liquid level indicator
液面指示水尺 fluid level ga(u)ge
液面指示仪 fluid level ga(u)ge
液面砖 flux block
液面自动记录仪 liquid level recorder
液面自动控制 automatic level control
液面自由能量 free surface energy
液面纵断面图 liquid surface profile
液膜 fluid film; liquid film; liquid membrane; liquid sheet
液膜传递系数 liquid film transfer coefficient
液膜传质系数 mass transfer coefficient
液膜萃取技术 liquid membrane extraction technique
液膜电极 liquid membrane electrode
液膜分离 liquid membrane isolation; liquid membrane separation
液膜分离废水处理 liquid membrane isolation wastewater separation
液膜分离技术 liquid membrane separation technology
液膜工艺 liquid membrane process
液膜共振 liquid film resonance
液膜厚度 film thickness
液膜控制 liquid film control(ling)
液膜密封 liquid film seal
液膜破裂 liquid sheet disintegration
液膜润滑 fluid film lubrication; liquid film lubrication
液膜式 liquid film type
液膜养护 liquid membrane curing
液膜蒸发器 film type evapo(u)rator
液膜蒸馏器 film distillation apparatus
液膜轴承 fluid film bearing
液膜阻力 liquid film resistance
液囊 liquid pocket; liquid slug
液平面 fluid level
液气比 liquid-gas ratio
液气的 hydropneumatic
液气动的 hydropneumatic
液气分布器 liquid-gas distributor
液气分界面 liquid vapo(u)r surface
液气共存 liquid vapo(u)r coexistence
液气后座系统 hydropneumatic recoil system
液气换热器 liquid-to-air heat exchanger
液气混合物 liquid vapo(u)r mixture
液气界面 liquid vapo(u)r interface
液气两相平衡区 liquid-gas equilibrium region
液气平衡线 vapo(u)rization curve
液气热交换器 liquid vapo(u)r heat exchanger
液气系统 liquid-gas system
液气悬架装置 hydropneumatic suspension
液气悬胶 ligasoid
液气装置 hydropneumatic device
液汽共存曲线 liquid-vapo(u)r-coexistence curve
液汽平衡 liquid vapo(u)r coexistence
液氢 liquid hydrogen
液氢泵 liquid hydrogen pump
液氢槽 liquid hydrogen bath
液氢发动机 hydrogen engine
液氢气泡室 liquid hydrogen bubble chamber
液氢站 liquid hydrogen plant
液区 liquid zone
液散货 bulk liquid cargo; liquid bulk; bulk liquid cargo
液栅氧化过程 wet gate oxidation process
液上气相分离 head space
液身阻塞 hydraulicking
液施 liquid application
液施法 liquid application; liquid method
液蚀 liquid corrosion
液塑限联合测定仪 combined liquid plastic limit device; liquid-plastic tester; Atterberg limits combination testing method
液碎铁粉 liquid disintegrated powder
液碎铸铁 liquid disintegrated cast iron
液态 liquid condition; liquidness; liquid state
液态氨 liquid ammonia
液态氨船 ammonia tanker; liquid ammonia carrier
液态氨分离器 liquid ammonia separator
液态包裹物 liquid inclusion
液态丙烷 liquid propane
液态丙烯 propylene liquid
液态丙烯酸聚合物系统 liquid acrylic polymer system
液态玻璃 liquid glass
液态薄膜电极 liquid membrane electrode
液态薄膜状隔膜 thin liquid film membrane
液态出渣炉 slag tap furnace
液态储存 liquid accumulation
液态单体工艺 liquid monomer process
液态单体塑胶 liquid monomer plastic
液态氮 liquid nitrogen
液态氮冻结地基法 nitrogen-method of freezing ground
液态氮发动机 liquid nitrogen engine
液态氮冷冻装置 liquid nitrogen freezing system
液态电解质电容器 liquid dielectric(al) capacitor
液态丁烷 liquefied butane
液态毒物 fluid poison
液态锻造 precast-forging
液态二氧化氮汽油炸药 anilite
液态二氧化硫 liquid sulfur dioxide
液态二氧化硫抽提过程 liquid sulfur dioxide extraction process
液态二氧化碳 liquid carbon dioxide
液态二氧化碳比率 the percentage of liquid carbon dioxide
液态二氧化碳槽车 liquefied carbon dioxide tank vehicle
液态二氧化碳洗井 well cleaning with liquid carbon dioxide
液态废料 liquid waste
液态分离作用 liquid immiscibility
液态分异作用 liquid differentiation
液态酚 liquefied phenol
液态附加剂 liquid admix(ture); liquid agent
液态汞 liquid mercury
液态氦 liquid helium
液态合金 liquid alloy
液态和膏状材料 Impervion
液态化合物 liquefied compound
液态灰浆稠化剂 liquid mortar densifier
液态混凝土 liquid concrete
液态混凝土附加剂 liquid concrete admix(ture)
液态极限 liquid limit
液态挤压钢 liquid-compressed steel
液态胶 liquid glue
液态胶乳 liquid latex
液态胶质 liquid gum
液态介质 liquid medium
液态金属 fluent metal; hot metal; liquid metal; liquor metal
液态金属盛桶 hot-metal ladle
液态金属传热系统 liquid metal heat-transfer system
液态金属磁流体发电机 liquid metal magnetohydrodynamic generator
液态金属萃取 liquid metal extraction
液态金属电刷 liquid metal brush
液态金属反应堆 liquid metal reactor
液态金属核燃料 liquid metal nuclear fuel
液态金属混流 metal turbulence
液态金属集电环 liquid metal slip-ring
液态金属冷却堆 liquid metal-cooled reactor
液态金属冷却剂 liquid metal coolant
液态金属冷却剂回路 liquid metal-coolant circuit
液态金属冷却介质 liquid metal coolant
液态金属喷雾塔 liquid metal spray column
液态金属汽轮机 liquid metal steam turbine
液态金属侵蚀 liquid-metal corrosion
液态金属燃料 liquid metal fuel
液态金属燃料电池 liquid metal fuel cell
液态金属燃料堆 liquid metal fuel reactor
液态金属热交换器 liquid metal heat exchanger
液态金属渗透 liquid metal infiltration
液态金属增殖堆 liquid metal breeder reactor
液态金属装料 liquid metal charge
液态空气 liquefied air; liquid air
液态空气阱 liquid air trap
液态空气瓶 liquid air bottle
液态空气容器 liquid air container; liquified air container
液态空气温度 liquid air temperature
液态空气循环发动机 liquid air cycle

engine
液态空气制冷探测器 liquid air-cooled detector
液态矿 liquid ore
液态(矿)渣 liquid slag
液态扩散 liquid state diffusion
液态扩张膜 liquid expanded film
液态离子交换技术 liquid ion exchange technique
液态离子交换剂 liquid ion exchanger
液态沥青 liquefied bituminous material; liquid asphalt; liquid asphaltic products; liquid bitumen
液态沥青材料 fluid bituminous material; liquid asphaltic material; liquid bituminous material
液态流动性 castability
液态硫化氢 liquid hydrogen sulfide
液态炉 wet bottom furnace
液态铝 liquid alumin(i)um; molten alumin(i)um
液态铝珠 liquid alumin(i)um globule
液态氯 liquid chlorine; liquified chlorine
液态氯化镁 liquid magnesium chloride
液态镁 liquid magnesium
液态膜分离工艺 liquid film separation process
液态膜系数 liquid film coefficient
液态奶废水 liquid milk wastewater
液态泥浆 liquid sludge
液态泥土 fluid mud
液态腻子 liquid filler
液态黏[粘]结剂 liquid bonding agent
液态凝胶 liquid gel
液态凝聚膜 liquid condensed film
液态浓缩剂 liquid densifier; liquid densifying agent
液态排渣 slag tapping
液态排渣发生炉 slagging gasifier; slagging producer
液态排渣锅炉 wet bottom boiler
液态排渣炉 slag tapping boiler
液态排渣炉膛 liquid bath furnace; wet bottom furnace; slag tap furnace
液态排渣燃烧方式 wet bottom firing system
液态起点 point of onset of fluidization
液态气体储槽 stationary tank
液态气体气化装置 liquefied gas vapo(u)rizer
液态羟燃料 hydroxygen
液态氢 liquified hydrogen
液态氢的绝热容器 liquid hydrogen Dewar
液态燃料 fluid fuel; liquid fuel
液态燃料反应堆 fluid fuelled reactor
液态燃烧 liquid phase combustion
液态热量计 liquid calorimeter
液态溶剂蜡 solvent paste wax
液态溶液 liquid solution
液态熔剂盖 liquid flux cover
液态熔渣 molten slag
液态润滑剂 liquid lubricant
液态砂浆增浓剂 liquid mortar densifier
液态生物转化工艺 liquid state bio-conversion process
液态石炭酸 liquefied carbolic acid
液态石油气运输船 liquid petroleum gas ship
液态收缩 liquid contraction; liquid shrinkage
液态树脂 liquid resin
液态霜淞 liquid rime
液态水 liquid water
液态水含量 liquid water content
液态天然气 liquefied natural gas
液态烃 liquid hydrocarbon
液态透镜 liquid lens
液态涂面料 liquid coating (material)
液态土 fluid soil
液态脱模剂 liquid bond-breaker
液态为主的地热系统 liquid dominated system
液态污泥 liquid sludge
液态污染物 fluid pollutant
液态污水污泥 liquid sewage sludge
液态污水综合处理系统 integrated system of liquid effluent treatment
液态屋面材料 liquid roofing; liquid roofing material
液态线 liquidus
液态镶嵌模型 fluid mosaic model
液态镶嵌学说 fluid mosaic theory
液态橡胶 liquid rubber
液态橡胶添加剂 liquid rubber additive
液态压缩 fluid-compressed
液态压缩气体 liquefied compressed gas; liquid pressurized gas
液态压制法 fluid compression process
液态岩浆 liquid magma; molten magma
液态养护膜 liquid membrane curing compound
液态氧 liquid oxygen; liquified oxygen; lox
液态氧储存柜 liquid oxygen storage tank
液态氧发生器 liquid oxygen generator
液态氧化二氮 liquid nitrous oxide
液态氧炸药 liquid oxygen explosive
液态乙炔 liquid acetylene; liquified acetylene
液态乙烯 liquid ethylene; liquified ethylene
液态阴离子 liquid anion
液态硬化剂 <混凝土表面处理用> liquid hardener
液态油 fluid oil
液态油脂 liquid fat
液态元素 liquid element
液态源扩散 liquid source diffusion
液态渣 liquid slag
液态炸药 liquid explosive
液态制冷剂过冷循环 liquid refrigerant subcooled cycle
液体 fluid body; influent; liquid; liquor
液体氨 ammonia liquor
液体柏油制品 liquid tar product
液体半导体 liquid semiconductor
液体包层 liquid cladding
液体包封 liquid envelope
液体包裹体 fluid inclusion; liquid inclusion
液体饱和度 fluid saturation
液体饱和蒸汽压力 saturated steam pressure of liquid
液体保安器 liquid fuse unit
液体保险丝 liquid fuse
液体报警(器) level alarm
液体泵 liquid pump; liquor pump
液体泵吸输送器 fluid pumping conveyer [conveyor]
液体比例混合器 liquid proportioner
液体比例调和器 liquid proportioner
液体比热 specific heat of liquid
液体比重 liquid specific gravity
液体比重表 hydrometric(al) table
液体比重测定法 areometry; hydrometry
液体比重测量缸 hydrometer jar
液体比重秤 areometer; hydrostatic balance
液体比重计 ar(a)eometer; densimeter; hydrometer; hydrostatic balance; picnometer [pycnometer]; specific gravity hydrometer
液体比重计分析 areometer analysis; hydrometer analysis
液体比重计瓶 hydrometer jar
液体比重计组 hydrometer set
液体比重天平测定比重 hydrostatic weighting
液体变速机 fluid power transmission
液体变速装置 fluid power transmission
液体变阻器 immersed rheostat; liquid resistance; liquid rheostat
液体表面防水剂 liquid surface waterproofing agent
液体冰点 freezing-point of liquid
液体丙烷 petrogas
液体玻璃 liquid glass
液体箔膜模型 hydrodynamic(al) lubrication model
液体薄膜 fluid film; liquid film; liquid lamella
液体薄膜养护剂 liquid-membrane curing compound
液体薄膜形成剂 <养护混凝土用> liquid membrane-forming compound
液体补充 topping up
液体捕集器 liquid trap
液体不混溶性 liquid immiscibility
液体不能透过的 liquid-tight
液体不足 fluid low
液体残渣 liquid residue
液体舱 tank
液体藏量 liquid hold-up
液体侧 hydraulic fluid side
液体测量器 liquid measure
液体测压计 liquid piezometer
液体层流阻力 viscous friction
液体层析 liquid chromatography
液体差压计 fluid differential pressure ga(u)ge
液体掺和系统 liquid blending system
液体掺碳剂 liquid cement
液体产率 liquid yield; product liquid
液体成分自动分析仪 automatic fluid analyser
液体池 liquid cell
液体充气装置 aerator
液体冲击波 hydraulic shock
液体冲蚀 fluid cut
液体抽样器 liquid sampler
液体稠度 liquid density; stiffness
液体出口 escape orifice; liquid outlet
液体除湿作用 liquid dehumidifying
液体储藏 liquor store
液体储藏器 fountain
液体传动 liquid pressure drive
液体促进剂 liquid accelerator
液体催干剂 japan; liquid drier [dryer]
液体淬火 liquid hardening; liquid quenching
液体存储器 liquid memory
液体打磨 liquid honing
液体导电 liquid conduction
液体导热性 thermal conductivity of liquid
液体的 liquid
液体的对流作用 convection of liquid
液体的含氢指数 hydrogen index of fluid
液体的焓 heat of liquid
液体的能量和压强 energy and pressure of liquid
液体的上部表面 top surface of liquid
液体的送运 fluid handling
液体的压缩性 compressibility of liquid
液体的折光率 refractive index of liquid
液体的主要性质 major property of liquid
液体的最低温度流动点 pour point
液体地沥青 <包括轻制地沥青> liquid asphalt; liquid bitumen
液体地沥青材料 fluid bituminous material; liquid asphaltic material; liquid bituminous material
液体电池 liquid cell
液体电光元件 liquid electrooptic(al) cell
液体电极 fluid electrode; liquid electrode
液体电解质 liquid electrolyte
液体电解质电容器 liquid electrolytic capacitor; wet-electrolytic capacitor
液体电介质 liquid dielectric
液体电容器 liquid capacitor; liquid condenser
液体电阻 liquid resistance
液体电阻器 liquid resistor
液体动力 fluid power; hydrodynamic(al) force
液体动力减震器 hydrodynamic(al) damper
液体动力燃料 liquid motor fuel; liquid power
液体动力学 hydrodynamics; liquid dynamics
液体动力学方程 hydrodynamic(al) equation
液体动力学研究 hydrodynamic(al) research
液体动力制动器 hydrodynamic(al) damper
液体动力轴承 hydrodynamic(al) bearing
液体动力阻尼器 hydrodynamic(al) damper
液体动压力 hydrodynamic(al) pressure
液体度量 liquid measure
液体短路 short-circuiting
液体对基底冲蚀 fluid erosion of matrix
液体对空气换热器 liquid-to-air heat exchanger
液体对空气热交换器 liquid-to-air heat exchanger
液体吨 fluid ton
液体多层料密封系统 multilayered liquid containment system
液体发泡剂 propellant [propellent]
液体阀 liquid valve
液体反射镜 liquid mirror
液体反射指数 reflective index of liquid
液体防潮作用 liquid dehumidifying
液体非弹性形变 liquid inelastic deformation
液体肥料 liquid fertilizer; liquid mature
液体废料处理 liquid waste disposal
液体废料焚化 liquid waste incineration
液体废料运输者 liquid waste hauler
液体废污 liquid waste
液体废物 liquid waste
液体废物处理 liquid waste disposal
液体废物处理技术 liquid waste disposal technique
液体废物的弃置和处理 liquid waste disposal and treatment
液体废物和固体物的弃置和处理 liquid waste and solid waste disposal and treatment
液体分尘器 liquid cyclone
液体分离 liquor separation
液体分离膜 liquid separation membrane
液体分离器 liquid separator; liquid trap; liquor separator; suction accu-

mulator;suction trap
液体分流器 liquid cyclone
液体分配器 fluid distributor
液体分配头 liquid distributor
液体分装机 liquid filling machine
液体粪肥 liquid manure
液体封闭器 fluid packed pump
液体浮力材料 liquid buoyancy material
液体辐射 fluid radiation
液体腐蚀 liquid corrosion; wet corrosion
液体负压计 liquid ga(u)ge
液体改性剂 liquid modifier
液体干料 liquid drier [dryer]
液体干涉仪 liquid interferometer
液体干燥剂 liquid desiccant; liquid drier [dryer]
液体干燥剂脱水 liquid desiccation
液体干燥作用 liquid dehumidifying
液体坩埚技术 liquid crucible technique
液体刚性 liquid rigidity
液体高热器 pressure boiler
液体工作室 fluid-filled working chamber
液体功率 liquid horsepower
液体供料器 fluid distributor
液体供应箱 fluid supply tank
液体固定相 stationary liquid phase
液体固体分离 liquid solid separation
液体固体燃料火箭发动机 liquid solid propulsor
液体观察窗 liquid indicator; liquid sight glass
液体管道 fluid pipeline
液体管(路) liquid line
液体管式压力表 liquid tubular manometer
液体管线 liquid line
液体罐车 special wagon for carriage of liquids
液体光学 optics of liquids
液体光学胶粘剂 liquid optic(al) adhesive
液体过冷却器 liquid subcooler
液体过滤 liquid filtration
液体过滤器 liquid filter
液体过滤筒 liquid filtering cartridge
液体和浆体混合器 liquid and paste mixer
液体恒温器 liquid thermostat
液体滑润轴承 flood-lubricated bearing
液体化学沉积作用 colloid chemical deposition
液体化学剂量计 liquid chemical dosimeter
液体化学品船 liquid chemicals tanker
液体化学燃料火箭 liquid chemical rocket
液体环式泵 liquid ring pump
液体缓冲挡板 liquid surge baffle
液体灰分提取法 liquid ash extraction
液体回收 dump trap liquid return
液体混合论 craseology [crasiology]
液体混合燃料 liquid combination
液体混合物 liquid admix(ture); monkey blood <喷覆在混凝土表面用以形成液膜养护的,俚语>
液体混凝土 liquid concrete
液体活塞式旋转压气机 liquid piston rotary compressor
液体活塞式压缩机 liquid piston compressor
液体火箭燃料 liquid propellant
液体火焰淬火 liquid flame hardening
液体货(物) fluid cargo; liquid cargo; wet cargo; wet goods; liquid goods
液体货物集装箱 bulk liquid container; liquid cargo container; tank container
液体货物集装箱交易机构 tank container marketing organization
液体激光器 fluid state laser; liquid dye laser; liquid laser
液体极限离心分离 liquid limiting centrifugation
液体集装箱 tank container
液体计 level meter
液体计量 liquid measure(ment)
液体计量灌注器 volumetric(al) liquid filling machine
液体计量器 liquid dosing apparatus; liquid meter; liquid metering vessel
液体计数 liquid counting
液体计数器 liquid counter
液体加气剂 liquid air-entraining agent
液体加热器 liquid heater
液体甲烷 liquid methane
液体减光器 liquid dimmer
液体减振器 liquid damper; liquid spring unit
液体溅泼 sloshing of liquid
液体浆 pulp slurry; slush pulp
液体胶 glue solution; liquid adhesive
液体焦油产物 liquid tar product
液体焦油制品 liquid tar product
液体接触电位 liquid junction potential
液体接界 liquid junction
液体接界电势 liquid junction potential
液体节温器 liquid thermostat
液体结合太阳能电池 liquid combined solar cell
液体结晶作用 crystallization form solution
液体介质 aqueous medium; fluid medium; liquid medium
液体介质电容器 liquid dielectric(al) capacitor
液体介质电阻器 liquid resistor
液体介质渗透速率 liquid media penetration rate
液体介质声延迟线 liquid medium sonic delay line
液体界面 liquid surface
液体界面指示计 liquid interface indicator
液体金属传热流体 liquid metal heat-transfer fluid
液体金属快速增殖反应堆 liquid metal fast breeder reactor
液体金属阴极 liquid metal cathode
液体紧密性 liquid tightness
液体进料泵 liquid feed pump
液体浸入法试验 liquid immersion test
液体精制法 liquid purification
液体阱 liquid trap
液体警笛 <一种超声波发生器> liquid whistle
液体净化系统 fluid purification system
液体静力的 hydrostatic
液体静力试验机 hydrostatic test(ing) machine
液体静力水准测量 hydrostatic level(l)ing
液体静力水准仪 hydrostatic level(l)ing instrument
液体静力学 hydrostatics
液体静力学方程 hydrostatic equation
液体静力轴承 hydrostatic bearing
液体静压分析 hydrostatic analysis
液体静压力 hydrostatic pressure
液体静压力计 hydrostatic ga(u)ge
液体静压润滑 hydrostatic lubrication
液体静压轴承 hydrostatic bearing
液体镜 liquid mirror
液体厩肥泵 liquid sludge pump
液体聚乙二醇 liquid macrogol
液体绝缘 fluid insulation; liquid insulation
液体绝缘材料 liquid insulating material
液体绝缘体 liquid insulator
液体栲胶 fluid extract
液体可变电阻器 liquid regulating resistor
液体可浸入的 liquid-immersible
液体可透入的 porous
液体空间 fluid compartment; fluid space
液体空间速度 liquid space velocity
液体控制阀 liquid controller; liquid control valve
液体控制器 liquid controller
液体矿产 liquid commodity
液体矿脂 liquid petrolatum; petrolax; petrosio
液体矿脂质量试验 liquid petrolatum quality test
液体扩散率 liquid diffusivity
液体扩散因子 liquid diffusion factor
液体蜡 liquid wax
液体类毒素 fluid toxoid
液体棱镜 liquid prism
液体冷却 fluid cooling; liquid cooling
液体冷却的 liquid cooled
液体冷却反应堆 liquid cooled reactor
液体冷却机器 liquid-cooled engine
液体冷却剂 liquid coolant
液体冷却剂循环系统 cooling fluid loop
液体冷却器 liquid chiller; liquid cooler
液体离合器 liquid clutch
液体离子交换 liquid ion exchange
液体离子交换剂 liquid ion exchanger
液体离子交换膜电极 liquid ion exchange membrane electrode
液体离子交换树脂 liquid resin
液体力学 fluid mechanics; hydromechanics
液体沥青 eu-bitumen; liquid (asphaltic) bitumen
液体沥青分馏试验 distillation test of liquid asphalt
液体连续搅拌机 in-line mixer; pipeline agitator
液体帘 fluid curtain
液体联轴节 liquid clutch
液体联轴器 liquid clutch
液体亮金 liquid bright gold
液体量热计 liquid calorimeter
液体流 fluid current; liquid flow
液体流出角 fluid outlet angle
液体流出物 liquid effluent
液体流动 flow of liquid; liquid flow
液体流动方程 fluid flow equation
液体流动观察玻璃 liquid flow sight glass
液体流动看窗 liquid flow sight glass
液体流动性 fluidity of liquid
液体流动指示器 liquid flow indicator
液体流化床 liquid fluidized bed
液体流量 fluid flow; liquid capacity
液体流量测量 liquid flow measurement
液体流量计 fluid flowmeter; liquid flowmeter; liquidometer; liquid quantity meter
液体流量调节器 fluid flow regulator
液体流量指示器 fluid flow indicator
液体流率 liquid flow rate
液体流束 fluid jet
液体流速分布图 fluid velocity profile
液体流速计 fluid current meter; fluid speed meter
液体硫化 liquid curing
液体漏失 fluid slippage
液体炉渣 liquid slag
液体罗经 fluid compass; liquid compass; spirit compass; wet compass
液体罗盘 liquid compass; spirit compass; wet compass
液体马力 liquid horsepower
液体慢化剂(反应)堆 liquid moderator reactor
液体门 liquid gate
液体密度表 fluid density meter
液体密度测定(法) areometry
液体密度测量计 gravimeter [gravimetre]
液体密度计 areopycnometer [areopyknometer]; densi(to)meter; fluid density meter; liquid densimeter; liquidensitometer
液体密封 hydraulic seal; liquid pack(ing); liquid sealing; sealant
液体密封的 liquid-immersible
液体密封剂 fluid sealant
液体密封料 liquid gasket
液体密封性 fluid tightness
液体密封用垫片 liquid gasket
液体灭弧熔断器 immersion liquid quenched fuse
液体灭火机 fluid fire extinguishing
液体灭火器 fluid fire extinguishing
液体名称 fluid name
液体模具 fluid die
液体模拟床 liquid simulating bed
液体膜 liquid film
液体摩擦 fluid friction; liquid friction
液体摩擦减震 by friction of liquids damping; damping by friction of liquids
液体摩擦轴承 film lubrication bearing; flood-lubricated bearing; fluid friction bearing; Morgoil bearing
液体磨机 liquid mill
液体内摩擦 internal fluid friction
液体钠 liquid sodium
液体黏[粘]度 fluid viscosity; liquid viscosity
液体黏[粘]度计 liquid viscosimeter
液体黏[粘]度理论 liquid viscosity theory
液体黏[粘]合剂 liquid adhesive; liquid binder
液体黏[粘]结剂 liquid containing binder
液体黏[粘]性 liquid viscosity
液体凝集状态 liquid aggregate state
液体凝结物 curd
液体凝聚膜 liquid condensed film
液体浓度 strength of fluid
液体浓缩混合物 liquid densifying admix(ture)
液体耦合剂 fluid couplant
液体排出 fluid discharge
液体排渣发生炉 liquid slag producer
液体抛光 liquor finish
液体抛光剂 liquid polishing agent
液体抛丸清洗 liquid blast cleaning
液体培养 fluid culture; liquid culture
液体培养基 aqueous medium; fluid medium; fluid nutrient medium; liquid medium; liquid nutrient medium
液体喷砂 liquid blasting
液体喷砂处理 liquid blasting
液体喷射 liquid jet
液体喷射药枪 liquid-propellant gun
液体喷雾器 liquid dispenser; liquid sprayer
液体膨胀温度计 liquid expansion thermometer
液体片门 liquid gate
液体漂白剂 liquid bleach
液体平衡 fluid balance; liquid balance
液体起动器 liquid starter
液体气化压力 liquid vapo(u)r pressure

液体气侵 gas cutting of fluid
液体气溶胶 liquid aerosol
液体气体 liquefied gas
液体牵引机 liquid hauler
液体强度 liquid strength
液体切力 liquid shear
液体侵入 ingress of liquids
液体氰化法 liquid cyaniding
液体区域化 liquid zoning
液体曲 liquid koji
液体取样器 fluid sampling apparatus; liquid sampler
液体取样头 liquid dividing head
液体去极化剂 liquid depolarizer
液体去湿装置 liquid absorbent dehumidifier
液体燃料 fluid fuel; liquid fuel; liquid power; liquid propellant; wet fuel
液体燃料的燃烧 fluid combustion
液体燃料的雾化 atomizing of liquid fuel
液体燃料的组成部分 reacting fluid
液体燃料堆 liquid fuel reactor
液体燃料发动机 liquid fuel motor
液体燃料矿产 liquid fuel commodities
液体燃料炉 drip furnace
液体燃料喷射系统 liquid propellant injection system
液体燃料燃烧器 liquid fuel burner
液体燃料油 liquid fuel oil
液体燃料主发动机 liquid fuel sustainer
液体燃烧剂 liquid fire
液体燃烧剂火焰 liquid fire
液体燃烧水下储存柜 sea-floor fueling station
液体热 heat of liquid; liquid heat
液体热交换器 liquid heat exchanger
液体热媒 liquid heating media
液体容积计量 liquid volume measurement
液体容量 liquid capacity
液体容器 fluid container
液体容器侧面清理方孔 clean-out box
液体容器液体乳胶 liquid emulsion
液体溶剂 liquid flux
液体溶胶 lyosol
液体溶气计 absorptiometer
液体溶液 liquid solution
液体熔断器 liquid-filled fuse unit; liquid fuse unit
液体熔丝 liquid fuse
液体熔渣 pottsco
液体入口 liquid inlet
液体入口群体温度 bulk inlet temperature
液体润滑 fluid lubrication; hydrodynamic(al) lubrication; liquid lubrication
液体润滑剂 fluid lubricant; liquid lubricant
液体润滑脂 liquid grease
液体润滑轴承 fluid lubrication bearing
液体润滑作用 fluid lubrication
液体散货 liquid bulk
液体散货码头 liquid bulk terminal
液体散货集装箱 liquid bulk container
液体色谱 liquid chromatogram
液体色谱法 liquid chromatography
液体色调剂 liquid toner
液体闪烁测量仪 liquid scintillation activity meter
液体闪烁(法) liquid scintillation
液体闪烁放射性分析 liquid scintillation radioassay
液体闪烁光谱测定 liquid scintillation spectrometry
液体闪烁计数 liquid scintillation counting
液体闪烁计数器 liquid scintillation counter
液体闪烁技术 liquid scintillation technique
液体闪烁谱 liquid scintillation spectrum; liquid scintillator
液体闪烁探测器 liquid scintillation detector
液体闪烁体 liquid scintillator
液体闪烁现象 liquid scintillation
液体烧碱 liquid caustic soda
液体射流 fluid jet stream; hydrofluidic; liquid jet
液体渗出 fluid exudation
液体渗漏 leakage of liquids
液体渗滤器 fluid percolation
液体渗碳 bath carburizing; liquid carburizing
液体渗碳处理 liquid carburizing
液体渗碳剂 liquid carburizer
液体渗透 fluid penetration
液体渗透检验(法) liquid penetrant inspection
液体渗透率 fluid permeability
液体渗透试验 liquid penetration test
液体渗透探伤 liquid penetrant examination; liquid penetrant testing
液体渗透探伤范围 extent of liquid penetrant testing
液体渗透性 permeability for liquids
液体声池 liquid sonic cell
液体石蜡 atolin(e); fluid paraffin; fluid wax; liquid paraffin; liquid wax; mineral oil <美>; paraffin(e) oil; petronol; petroxolin
液体石蜡油 saxol(ine)
液体石油膏 paroline; petralol
液体石油沥青 liquid petroleum asphaltic bitumen
液体石油脂 petronol
液体时空速度 liquid hourly space velocity
液体试剂 liquid reagent
液体试样 liquor sample; thief sample
液体试样计数管 liquid sample counter
液体收集管 liquid header
液体收集器 liquid header
液体收缩 liquid shrinkage
液体输送 fluid delivery
液体树脂 liquid rosin
液体水 liquid water
液体水平压力计 liquid level ga(u)ge
液体松脂 liquid resin
液体速凝剂 liquid acceleration
液体太阳能系统 liquid based solar heating system
液体弹簧 liquid spring
液体弹性 hydroelasticity
液体弹性形变 liquid elastic deformation
液体碳氮共渗 liquid carbonitriding
液体梯度 liquid gradient
液体提出物 liquid extract
液体提取 liquid extraction
液体体积 liquid volume
液体体积测量罐 dump tank; measuring tank
液体体积空间速度 liquid volume hourly space velocity
液体体积黏[粘]度 liquid volume viscosity
液体体积压缩模量 compression modulus of liquid volume
液体天然沥青 petroleum
液体填孔剂 liquid filler
液体填料 liquid filler
液体调节器 fluid governor; liquid regulator
液体烃 liquid hydrocarbon
液体停堆系统 liquid shut-down system
液体透镜 liquid lens
液体凸模 fluid punch
液体涂料 liquid coating (material)
液体推进剂火箭发动机 liquid propellant rocket engine
液体推进剂启动机 liquid propellant starter
液体推进剂主发动机 liquid propellant sustainer; liquid sustainer
液体退火 liquid annealing
液体脱水剂 liquid siccative
液体温度表 liquid-filled thermometer
液体温度计 liquid-filled thermometer; liquid-in-glass thermometer
液体涡动接触器 liquid vortex contactor
液体涡流 fluid vortex
液体污泥利用 utilization of liquid sludge
液体污染物 liquid pollutant
液体物理性质 physical property of liquid
液体雾化 liquid atomization
液体雾沫 liquid entrainment
液体吸附法 liquid adsorption method
液体吸附减湿器 liquid sorbent dehumidifier
液体吸气换热器 liquid suction heat interchanger
液体吸气计 absorptiometer
液体吸收剂 liquid absorbent
液体稀释了的 watered-down
液体熄弧保安器 liquid fuse
液体洗涤 liquid scrubbing
液体洗涤剂 liquid detergent
液体洗涤器 liquid scrubber
液体洗地板皂 liquid floor soap
液体显影 liquid development
液体橡胶 fluid rubber
液体效率 fluid efficiency
液体形变 liquid deformation
液体型集热器 liquid type collector
液体虚质量 virtual mass of liquid
液体悬浮培养 fluid suspension culture
液体悬浮燃料堆 liquid suspension reactor
液体悬浮液 liquid suspension
液体旋风器 liquid cyclone
液体漩涡 gurgitation
液体循环 fluid cycle; liquid cycle
液体循环温度控制 liquid circulating temperature control
液体循环系统的 hydronics
液体压舱 liquid ballast
液体压差计 fluid differential pressure ga(u)ge
液体压力 fluid pressure; liquid pressure
液体压力传感器 liquid pressure pickup
液体压力计 hydraulic pressure ga(u)ge; liquid manometer; liquid pressure ga(u)ge; manometer
液体压力计的 manometric(al)
液体压力计水头 manometric(al) delivery head; manometric(al) head
液体压力计吸水头 manometric(al) suction head; manometric(al) suction lift
液体压力计旋塞 manometer cock
液体压力计压头 manometric(al) pressure head
液体压力记录器 pressure recorder
液体压强 pressure of liquid
液体压缩 liquid squeeze
液体压缩钢 fluid-compressed steel
液体压缩系数 liquid compressibility factor; modulus compressibility of liquid
液体压头 fluid head; liquid head
液体压载 liquid ballast
液体延迟线 liquid delay line
液体研磨 hydroabrasion; liquid honing
液体研磨机 liquid honing machine
液体颜料 liquid colo(u)rant
液体掩盖技术 liquid encapsulation technique
液体仪表 meter for liquid
液体易变的 hydrolabil
液体溢流 liquid flooding
液体阴极 pool cathode
液体油脂 liquid grease
液体预膜剂 liquid prefilming agent
液体元粒子 fluid particle
液体原料 liquid charging stock
液体源镭含量 liquid source radium content
液体运输卡车 liquid hauling truck
液体载热剂 heat-exchange fluid
液体再分布器 liquid re-distributor
液体再循环 liquid recirculation
液体再循环模型 liquid recycle model
液体脏污程度 fluid dirt level
液体皂基 liquid soap base
液体增塑剂 liquid plasticizing aid
液体炸药 liquid explosive
液体炸药筒 hydraulic cartridge
液体炸药钻机 liquid explosive drill
液体张力 strength of liquid
液体蒸发换热器 liquid vapo(u)r heat exchanger
液体蒸发速度测定器 gravimetroscope
液体整流器 liquid rectifier
液体指示计 fluid ga(u)ge
液体指示器 liquid indicator
液体制冷剂减压器 liquid refrigerant pressure reducer
液体制冷剂收集器 liquid trap
液体质量 fluid mass
液体质量稳定性 liquid mass stability
液体滞留(器) liquid hold-up
液体置换 liquid displacement
液体中含油量的测定 fat-lub test
液体中矿物颗粒连续沉淀物 continuous sludge
液体种类 kind of liquid
液体重力控制 liquid gravity control
液体注射阀 liquid injection valve
液体铸塑树脂 liquid casting resin
液体转差调节器 liquid slip-regulator
液体转速计 liquid tach(e)ometer
液体状态 liquid condition
液体状污泥 liquid sludge
液体着色渗透探伤记录 liquid dye penetrant examination record
液体着色探伤 liquid penetrant inspection
液体阻尼 fluid damping; liquid damping
液体阻尼罗盘 liquid damped compass
液体阻尼器 liquid damper
液体阻尼装置 liquid damping device
液体组分 liquid component
液烃 liquid hydrocarbon
液烃燃料 liquid hydrocarbonceous fuel
液位 fluid level; liquid level
液位报警 level alarm
液位报警器 liquid level alarm
液位表 fuel level ga(u)ge; level ga(u)ge
液位玻璃管 ga(u)ge glass
液位槽 level tank
液位测量 liquid level measurement
液位差 head of liquid
液位传感器 liquid level sensor
液位高度 level of liquid
液位计 content ga(u)ge; fluid level ga(u)ge; level ga(u)ge; level indi-

cator; level meter; liquid level meter; liquidometer; tank ga(u)ge
液位计玻璃 ga(u)ge glass
液位计阀组 valves of liquid level ga(u)ge
液位记录器 liquid level recorder; liquid recorder
液位检查旋塞 liquid full try cock; liquid level inspection cock
液位开关 liquid level switch
液位控制 fluid level control; liquid level control
液位控制降低 depression of surface level
液位控制器 fluid level controller; liquid level controller
液位控制系统 < 油船 > tank level control system
液位量杆 measuring rod; metering rod
液位流量计 liquid level ga(u)ge
液位塞 level plug
液位探针 level probe
液位调节 liquid level control
液位调节器 liquid level regulator
液位突然下降 sudden drawdown
液位、温度、密度监测仪 LTD [level, temperature and density] monitoring system
液位显示管 level indication tube
液位迅速下降 rapid drawdown
液位遥测仪 liquid level transmitter
液位仪 content ga(u)ge
液位指示玻璃管 ga(u)ge glass column
液位指示管 level indication tube
液位指示计 liquid level indicator
液位指示器 fluid level ga(u)ge; fluid level indicator; level indicator; level meter; liquid level indicator
液雾灭火器 vapo(u)ring liquid extinguisher; vapo(u) rising [vapo(u)rizing] liquid extinguisher
液吸收塔 stripping liquid
液析 eliquation; liquation
液下爆破 firing under fluid
液下泵 submerged pump
液下焚烧器 submerged incinerator
液下空气吹入式浮选机 subaeration machine
液下控制阀 buried valve with key
液下屏蔽泵 submerged canned pump
液下燃烧式蒸发器 submerged-combustion evapo(u)rator
液下手轮阀 buried valve with hand wheel
液线 liquidus; liquidus line
液限 flow limit; limit of liquidity; liquid limit
液限测定 liquid limit determination
液限试验 liquid limit test
液限仪 liquid limit apparatus; liquid limit device; liquid limit machine; mechanical liquid limit device
液相 aqueous phase; fluid phase; liquid component; liquidoid; liquid phantom; liquid phase
液相百分数 percentage liquid phase
液相比率 the percentage of liquid phase
液相不混溶 liquid phase immiscibility
液相操作 liquid phase operation
液相层析 liquid chromatography
液相沉淀 liquid phase deposition
液相沉积 liquid deposition
液相成分 composition of liquid phase
液相抽提 liquid phase extraction
液相传质 mass transfer in liquid phase
液相反应 liquid phase reaction
液相分解过程 liquid phase cracking process
液相分离 liquid phase separation
液相分配色谱法 liquid partition chromatography
液相氟化 liquid phase fluorination
液相覆盖技术 liquid encapsulation technique
液相干扰 liquid phase interference
液相含量 liquid content
液相荷载量 liquid phase loading
液相活性炭处理法 liquid phase activated carbon treatment
液相胶 liquid phase gum
液相精炼 refining in the liquid phase
液相精制 liquid phase refining
液相精制过程 liquid purification process
液相聚合 liquid phase polymerization; liquid polymerization
液相聚合物 liquid polymer
液相颗粒活性炭 liquid phase granular activated carbon
液相空气氧化 liquid phase air oxidation
液相量 liquid phase volume; quantity of liquid phase
液相裂化 liquid phase cracking
液相裂化过程 liquid phase cracking process
液相流动色谱(法) hydrodynamic(al) chromatography
液相面 liquid phase surface; liquidus phase
液相面积 the area of liquid phase
液相模 liquid phase mode
液相氢化 hydrogenation; liquid phase hydrogenation
液相区 liquid phase region
液相曲线 liquidus curve; liquidus line
液相燃烧 liquid phase combustion
液相燃烧器 liquid phase burner
液相热扩散法 liquid thermal diffusion
液相热裂化 liquid phase thermal cracking
液相热裂化过程 < 其中的一种 > Jenkin cracking
液相热压 liquid phase hot pressing
液相色层(分离)法 liquid chromatography
液相色谱(法) liquid chromatogram; liquid phase chromatography
液相色谱法质谱法联用 liquid chromatography-mass spectrometry
液相色谱傅立叶红外光谱联合测定仪 liquid chromatograph/Fourier transform infrared spectrometer
液相色谱仪 liquid chromatograph
液相色谱质谱联合测定仪 liquid chromatograph/mass spectrometer computer
液相色谱质谱仪联用 liquid chromatograph/mass spectrometer
液相烧结 liquid phase sintering; liquid sintering
液相烧结机理 liquid phase sintering mechanism
液相生长 crystal growth from liquid phase; liquid growth
液相体积 the volume of liquid phase
液相外延 liquid phase epitaxy; rheotaxial; rheotaxy
液相外延法 liquid phase epitaxial method
液相外延生长 liquid phase epitaxial growth; rheotaxial growth
液相吸附法 liquid phase adsorption method
液相洗脱色谱 liquid elution chromatography
液相线 liquid curve; liquid line; liquidus < 二元系液温线图 >
液相线烧结 liquidus sintering
液相线温度 liquidus temperature
液相硝化 liquid phase nitration
液相悬浮过程 liquid phase suspension process
液相穴 crater; liquid core
液相氧化作用 liquid phase oxidation
液相柱层析 liquid column chromatogrpahy
液相柱色谱法 liquid column chromatogrpahy
液相总体积 total liquid volume
液相阻力 liquid friction
液楔 fluid wedge
液心顶弯 liquid center bending
液芯光纤 liquid core fiber [fibre]; liquid core optic(al) fiber [fibre]
液芯光纤波导 liquid core fibre-optic(al) waveguide
液芯纤维 liquid core fiber [fibre]
液性 liquidity
液性暗点 fluid sonolucent point
液性暗区 fluid sonolucent area; opaque dark area of fluid
液性极限 limit of liquidity; liquidity limit
液性界限 liquid limit
液性平段 fluid echoless segment
液性限度 liquid limit
液性限界 liquid limit
液性压力 press of fluidity
液性指数 index of liquidity; liquid index; liquidity factor; liquidity index; relative water content
液悬体 hydrosol
液压 fluid power; fluid pressure; hydrostatic pressure; line pressure; liquid pressure; oil-pressure; sap pressure
液压安全阀 hydraulic fuse; hydraulic relief valve; hydraulic safety valve
液压安全装置 fluid pressure warning device
液压拔道器 hydraulic track lining device
液压拔杆器 hydraulic puller
液压拔根机 hydraulic rooter
液压拔销器 hydraulic pin puller
液压扳手 hydraulic wrench
液压半倾倒式拖车 hydraulic semi-dump trailer
液压瓣升降杆 hydraulic valve lifter
液压保护装置 hydraulic protector
液压保险器 hydraulic fuse
液压泵 fluid operated pump; fluid pump; hydraulic (power) pump; hydraulic pump unit; liquid handling pump; oil pump; winch pump
液压泵汲设备 hydraulic pumping unit
液压泵搅拌器 pump agitator
液压泵接合杆 hydraulic pump lever
液压泵接头 hydraulic pump disconnect
液压泵扭矩 pump torque
液压泵泄放 hydraulic pump discharge
液压泵旋转斜盘 swash plate
液压泵站 hydraulic power unit
液压泵中间轴 intermediate shaft
液压泵装置 hydraulic pumping unit
液压泵组 hydraulic pump aggregate; liquid pressure pump set
液压比变换装置 hydraulic ratio changer
液压比例自控系统 automatic hydraulic proportioning system
液压比率变换器 hydraulic ratio changer
液压比重计 hydrostatic balance
液压闭路系统 closed circuit system
液压闭锁式油缸 hydrostop cylinder
液压闭锁装置 hydraulic lock
液压避震器 oleo-damper
液压臂 hydraulic arm
液压变幅 hydraulic luffing
液压变矩器 hydraulic moment variator
液压变速 hydraulic powershifting; hydraulic transmission
液压变速传动 hydraulic variable speed drive
液压变速传动装置 hydraulic variable speed gear
液压变速联轴器 hydraulic variable speed coupling
液压变速器 hydraulic transformer
液压变速箱 hydraulic variable speed gear
液压变速装置 hydraulic variable speed gear
液压变速装置传动 torque convertor transmission
液压表 hydraulic pressure ga(u)ge; hydraulic pressure indicator; hydromanometer; hydrometer
液压剥皮法 hydraulic barking
液压剥树皮机 hydraulic baker
液压补偿联轴节 hydraulic slip coupling
液压补偿调节器 hydraulic pressure compensating governor
液压步进马达 hydraulic stepping motor
液压舱 hydraulic compartment
液压操舵机 hydraulic telemotor; telemotor
液压操舵器 hydraulic telemotor; telemotor
液压操舵装置 hydraulic steering gear; hydraulic transmission gear
液压操纵 hydraulically assisted steering; hydraulic control; hydraulic operate; hydraulic operation
液压操纵铲刀 hydraulic blade
液压操纵的 hydraulically operated; hydraulic operated
液压操纵的底板 hydraulic bedplate
液压操纵的方向控制阀 hydraulically operated direction control valve
液压操纵的裂土机 hydraulic ripper
液压操纵的松土机 hydraulic ripper
液压操纵杆 hydraulic control lever
液压操纵后卸卡车 hydraulically operated rear-dump truck
液压操纵机构 hydraulic control unit
液压操纵盘状离合器 hydraulic control of disk clutch
液压操纵器 hydraulic operator; hydroman
液压操纵式离合器 hydraulically controlled clutch
液压操纵式松土器 hydraulic-operated scarifier
液压操纵式推土机 hydraulic (type bull) dozer
液压操纵手柄靠垫 hydraulic hand control cushion
液压操纵推土板 hydraulic blade
液压操纵推土机 hydraulically controlled bulldozer; hydraulic-controlled bulldozer
液压操纵挖掘机 hydraulic actuated excavator
液压操纵系统 hydraulic control system
液压操纵(显示)板 hydraulic display panel
液压操纵箱 hydraulic cabinet
液压操纵油缸 hydraulic control cylinder
液压操纵抓爪 hydraulic grab

液压操纵装置 hydraulic control apparatus;hydraulic control device
液压操作 hydraulic operate;hydraulic operation
液压操作的 hydraulically operated; oil hydraulic operated
液压操作器 hydroman
液压操作设备 hydraulically operated equipment
液压操作闸门 hydraulic operate gate;hydro-automatic gate
液压侧壁取心工具 hydraulic sidewall coring tool
液压侧臂 hydraulic side boom
液压侧倾卸车 hydraulically operated side-tipping wagon
液压测功计 hydraulic dynamometer
液压测功器 hydraulic dynamometer
液压测力盒 hydraulic load cell
液压测力计 hydraulic capsule; hydraulic cell; hydraulic dynamometer;hydraulic load cell
液压测力仪 hydraulic capsule; hydraulic dynamometer; hydraulic load cell;hydroplat
液压测试车 hydraulic test cart
液压测试器 hydroscope
液压测速仪 hydraulic tachometer
液压层合模塑 fluid pressure laminating
液压差速传动 hydraulic differential drive
液压差速器 hydraulic differential
液压铲 hydraulic shovel
液压铲斗 hydraulic bucket
液压铲斗挖掘机 hydraulic shovel
液压铲坡机 hydraulic sloper
液压铲运装载机 hydraulic shovel loader
液压超载保护装置 hydraulic overload device
液压车床 hydraulic lathe
液压车轮轮毂马达 hydraulic wheel hub motor
液压衬垫 hydrocushion
液压撑脚 hydraulic prop
液压成型 hydraulic forming; hydroform(ing)
液压成型法 hydroform method
液压成型机 hydroforming machine
液压成型压力机 hydraulic forming press
液压成型用橡胶模 hydraulically expanded rubber die
液压乘人电梯 hydraulic passenger elevator
液压乘人升降机 hydraulic passenger elevator
液压秤 <又称液压称> hydraulic scale;liquid pressure scales
液压持纤器 hydraulic steel holder
液压齿轮泵 hydraulic gear pump
液压冲采机 hydraulic excavator
液压冲床 hydraulic press; hydraulic punching machine
液压冲击 hydraulic shock; hydrodynamic(al) shock;line shock
液压冲击式凿岩机 hydraulic hammer; hydraulic percussive drill
液压冲击式钻机 hydraulic impact unit;hydraulic thrust boring machine
液压冲击振动 hydraulic shock vibration
液压冲击装置 hydraulic impact unit
液压冲孔机 hydraulic piercing press; hydraulic punching machine; hydraulic punching press
液压冲刷挖泥船 hydraulic erosion dredge(r)
液压冲钻机 hydraulic thrust boring machine
液压除震器 hydraulic shock eliminator
液压储油器 hydraulic reservoir
液压处理 hydraulic pressure treatment
液压触发器 hydraulic trigger
液压穿孔 hydraulic perforating; hydraulic piercing
液压传递 fluid transmission; hydraulic transmission; hydrostatic transmission
液压传递装置 crowding gear;hydraulic gear; hydraulic operating gear; hydrostatic transmission
液压传动 fluidrive; hydraulic action; hydraulic drive; hydraulic power drive; hydraulic power transmission; hydromatic drive; hydrostatic drive; hydrostatic transmission; liquid pressure drive; oil-operated transmission; hydrostatic transmission; hydraulic transmission
液压传动泵 fluid-driven pump; hydraulically driven pump; hydraulic power pump
液压传动操舵装置接收器 telemotor receiver
液压传动操舵装置用油 telemotor oil
液压传动铲斗 scoop with fluid drive
液压传动齿轮 fluid transmission gear
液压传动的捡拾压捆机 hydraulically driven baler
液压传动斗 scoop with fluid drive
液压传动回路 hydraulic drive circuit
液压传动机构 hydraulic gear
液压传动机械 hydraulic driving machine
液压传动控制 hydraulic steering; hydraulic transmission control
液压传动起绒机 hydronapper
液压传动起重车 hydratruck
液压传动器 fluid drive
液压传动青贮料抓斗 hydraulically powered silage grab
液压传动式拖拉机 hydraulic tractor
液压传动推土机 hydrostatic drive bulldozer
液压传动系统 hydraulic drive system;hydraulic transmission system
液压传动系统图 hydraulic circuit
液压传动液 hydraulic transmission fluid
液压传动装置 crowding gear; hydraulic actuator; hydraulic gear; hydraulic operating gear; hydraulic transmission gear; hydrodynamic(al) drive;oligear
液压传感器 hydraulic cell; hydraulic load cell; liquid pressure transducer;hydraulic capsule <囊式的>
液压传力 fluid drive
液压传力垫块 hydraulic sensing pad
液压传送 shuttle
液压传送顶车机 hydraulic pusher for transfer car
液压传送装置 crowding gear; hydraulic (operating) gear; hydrostatic transmission
液压窗开关器 hydraulic window (control) gearing
液压锤 <桩工> hydroblock; hydraulic hammer
液压促动模板 hydraulically actuated form
液压促动器 hydraulic actuator
液压搭脚手架 hydraulic scaffold(ing)
液压打包机 hydraulic press packing
液压打桩 hydraulic pile driving
液压打桩锤 hydraulic hammer
液压打桩导向架 hydraulic lead
液压打桩方法 hydraulic pile driving
液压打桩机 hydraulic pile driver
液压带唇环形密封件 hydraulic lip packing
液压单斗挖土机 hydraulic excavator; hydraulic shovel
液压导轨式凿岩机 hydraulic drifter
液压导向器 hydraulic guide
液压捣固机 hydraulic tamping machine
液压捣实 <孔内爆破时> fluid-tamping
液压道路模拟 hydraulic road simulator
液压的 hydraulic; hydraulically pressed; hydrostatic; liquid operated;oil hydraulic
液压地 hydraulically
液压点火系统 hydraulic starting system
液压电磁式制动器 hydraulic electromagnetic brake
液压电磁制动器 hydromagnetic brake
液压电(动)机 fluid pressure motor; hydraulic motor; hydromotor; oil motor
液压电机最低稳定转速 minimum steady speed of hydraulic motor
液压电机最低转速 minimum speed of hydraulic motor
液压电机最高稳定转速 maximum steady speed of hydraulic motor
液压电机最高转速 maximum speed of hydraulic motor
液压电力起落机构 hydraulic-electric(al) lift
液压电气系统 hydroelectric(al) system
液压电梯 hydraulic lift <英>; plunger elevator
液压电梯机械 hydraulic lift machine
液压垫 hydraulic die cushion
液压吊臂 hydraulic boom
液压吊货秤 hydraulic duckham
液压吊货杆 hydraulic derrick
液压顶 hydraulic ram
液压顶车机 hydraulic pusher
液压顶撑 jack shore; hydraulic shoring
液压顶撑设备 <顶拉预应力筋或支撑滑升模板> jacking device
液压顶升法顶高下沉路面板 slab jacking
液压顶升机 hydraulic elevator; hydraulic lift;plunger elevator
液压顶升支柱 jack post
液压顶重器开闭装置 hydraulic ram actuator
液压顶柱 roof jack
液压定位起吊设备 hydraulic spud lifting gear
液压定位器 hydraulic positioner
液压定位调整装置 hydraulic positioner
液压动力 hydraulic power
液压动力机构 hydraulic power unit
液压动力机械 hydrokinetic machine
液压动力机组 hydraulic power pack
液压动力破碎 hydrokinetic crushing
液压动力设备 hydraulic power unit
液压动力系统 hydraulic power system
液压动力装置 hydraulic power unit
液压动力组 hydraulic power unit
液压动组件 hydraulic power pack
液压动作 hydraulic operation
液压动作筒 hydraulic actuating cylinder;hydraulic actuator
液压斗式提升机 hydraulic scooper
液压端部装载机 hydraulic end loader
液压断裂试验 hydraulic fracture test
液压断流器 hydraulic cutout
液压锻钎机 hydraulic dressing machine
液压锻(压)机 hydraulic forging press
液压锻造 hydraulic forging; hydrostatic forging
液压锻造机 hydraulic forging press
液压堆垛机 hydraulic stacker
液压堆料机 hydraulic stacker
液压颚式破碎机 hydraulic jaw crusher
液压发动机 fluid power motor; hydraulic engine; hydraulic motor; hydromotor;oil motor
液压伐木楔 hydraulic felling wedge
液压阀 hydraulic circuitry; hydraulic valve; hydrovalve; pressure-operated valve
液压阀控制系统 hydraulic valve control system
液压阀提升器 hydraulic valve lifter
液压阀游隙调整器 hydraulic valve lash adjuster
液压法 hydraulic method
液压翻边机 hydraulic flanging machine
液压翻边压力机 hydraulic flanging press
液压翻车机 hydromechanical tipper
液压翻车系统 hydraulic tilting system
液压翻地机 hydraulic scarifier
液压翻斗 hydraulic-operated skip
液压翻斗车 hydraulic tipper
液压反铲挖掘机 hydraulic backactor
液压反铲挖泥船 hydraulic backhoe dredger
液压反铲挖土机 hydraulic backactor
液压反馈 hydraulic feedback
液压反(向)铲 hydraulic backhoe
液压方法 hydrostatic means
液压方向机 hydraulic traversing mechanism
液压方向机操纵手把 hydraulic traversing control handle
液压方向机电动机 hydraulic traversing electric(al) motor
液压方向机油泵 hydraulic traversing mechanism pump
液压方向机油箱 hydraulic traversing oil reservoir
液压方向控制阀 hydraulic directional control valve
液压防护油 hydraulic preservation oil
液压防喷器 hydraulic blow out preventer
液压防撞装置 hydraulic fender
液压仿形刀架 hydraulic copying attachment
液压仿形控制 hydraulic tracing control
液压仿形牛头刨床 hydraulic copy shaper
液压仿形铣床 hydraulic copying miller
液压放大器 hydraulic amplifier; hydraulic intensifier
液压飞轮 fluid flywheel
液压分类 hydraulic classification
液压分离器 hydraulic separator
液压分配器 pressure distributor
液压封闭 hydraulic lock;liquid seal
液压封闭器 hydraulic gate
液压扶钎器 hydraulic steel support
液压符号 hydraulic symbol
液压辅助系统 hydraulic booster system
液压辅助制动器 hydraulic retarder
液压辅助(装置) hydraulic assist(ance)
液压复轨机 hydraulic re-railer
液压改变起重臂倾角的顶杆 hydraulic derricking ram

液压改变起重臂倾角的推杆 hydraulic derricking ram
液压杆塞 hydraulic cylinder plug
液压杆组件 cylinder assembly
液压缸 fluid cylinder; hydraulic actuating cylinder; hydraulic cylinder; hydraulic ram; pressurized oil ram; ram
液压缸冲击锤 cylinder impacting type
液压缸的配置 cylinder layout
液压缸垫密片 hydraulic cylinder gasket
液压缸垫圈 cylinder packing
液压缸放出管螺钉 hydraulic cylinder bleeder screw
液压缸盖 cylinder head
液压缸给进 hydraulic cylinder feed
液压缸回动弹簧 hydraulic cylinder return spring
液压缸活动叉座 cylinder yoke
液压缸活塞 hydraulic cylinder piston
液压缸活塞杆冲程 cylinder stroke
液压缸活塞簧 < 起重机 > cylinder piston spring
液压缸活塞止动杆 hydraulic cylinder piston stop rod
液压缸式防喷器 ram-type blow-out preventer
液压缸头面积 cylinder head end-area
液压缸油封 hydraulic cylinder seal
液压缸支销 cylinder pivot pin
液压缸支座 cylinder bracket
液压缸柱塞 hydraulic ram
液压缸总成 cylinder assembly
液压缸作用力 cylinder force
液压缸座套 < 装于升降机井坑底部，用于装接液压缸 > cylinder well
液压钢筋冷镦机 hydraulic bar cold header
液压钢筋切断机 hydraulic reinforcement bar cutter
液压钢筋弯曲机 hydraulic bar bender
液压钢丝冷镦机 hydraulic wire cold header
液压钢枕 Freyssinet jack
液压钢枕法 flat jack technique
液压钢支架 hydraulic steel support
液压杠杆压砖机 hydraulic toggle press
液压高架起重机 hydraulic gantry crane
液压隔膜微量移液管 hydraulic-diaphragm micropipetter
液压给进 hydraulic feed
液压给进缸 hydraulic feed cylinder
液压给进回转器 hydraulic feed swivel head
液压给进式钻机 auger with hydraulic feed; hydraulic feed core drill; hydraulic feed equipment; hydraulic feed rig
液压给进调节器 hydraulic drilling control
液压给进系统 hydraulic feed system
液压给进装置 hydraulic feed equipment; hydrofeeder
液压耕耘机 hydraulic hoe
液压工程 oil hydraulic engineering
液压工具 hydraulic tool
液压工作液 hydraulic fluid
液压功率 fluid power; hydraulic power
液压供应 hydraulic supply
液压构件的启动压力 pilot pressure of hydraulic units
液压固结盒 hydraulic consolidation cell
液压刮管器 hydraulic pipe cutter
液压刮土机 hydraulic scraper
液压刮削器 hydraulic scraper
液压挂挡变速箱 hydroshift transmission
液压挂钩 hydraulic hook
液压关闭阀 hydraulic shutoff valve
液压管 hydraulic pipe; hydraulic tube; hydraulic tubing
液压管道 hydraulic line; hydraulic pipe
液压管道部件 hydraulic conduit sections
液压管路 fluid pressure line; hydraulic pipe line; loading line
液压管子截割机 hydraulic pipe
液压灌浆 hydraulic cementing
液压轨缝调整器 hydraulic rail gap adjuster
液压滚轴输送机 hydrostatic roller conveyor
液压焊 water-gas welding
液压焊接法 hydraulic weld process
液压夯锤水锤泵 hydraulic ram
液压荷载膜盒式装置 hydraulic load capsule set
液压盒 hydraulic cell; hydraulic pressure cell
液压横移装置 hydraulic sideshift
液压后退方法 hydraulic retraction system
液压后退缸 hydraulic retraction cylinder
液压蝴蝶阀 hydraulic butterfly valve
液压护舷 hydraulic fender
液压滑动联轴节 hydraulic slip coupling
液压滑动模板 hydraulic sliding form
液压滑模 hydraulic slipform
液压缓冲 hydraulic cushion; hydrocushion
液压缓冲护栏 hydraulic cushion guardrail
液压缓冲器 dashpot; hydraulic buffer; hydraulic bumper; hydraulic pressure snubber; oil buffer; oil-pressure fender; oil shock absorber
液压缓冲装置 hydraulic damping device; hydraulic damping mechanism
液压换挡器 hydraulic selector
液压回动装置 hydraulic reversing gear
液压回阀 hydraulic lock valve
液压回路 hydraulic circuit
液压回旋钻进 hydraulic rotary drilling
液压回转方法 hydraulic rotary method
液压回转式钻机 hydraulic rotary drill
液压回转钻进 hydraulic rotary drilling
液压混凝土压力试验机 hydraulic concrete compression testing machine
液压活动制动器 hydraulic walking chock
液压活门 hydrovalve
液压活塞 hydraulically operated piston; hydraulic oil ram; hydraulic piston
液压活塞潜水泵 free-type subsurface hydraulic pump
液压活塞取样器取样 hydraulic pressure piston coring
液压活塞式驱动器 hydraulic piston actuator
液压活塞轴 hydraulic plunger shaft
液压货梯 hydraulic freight elevator
液压机 hydraulic compressor; hydraulic machine; hydraulic press; hydraustatic press; hydropress; hydrostatic press; liquid press; oil press
液压机车 hydraulic locomotive
液压机构 hydraulic gear; hydraulic mechanism; hydraulic unit
液压机构顶盖 hydraulic top cover
液压机构工作臂 hydraulic arm
液压机件 hydraulic component
液压机具 hydropower
液压机器人 hydraulic robot
液压机室 < 船闸启闭闸门的 > cylinder pit
液压机械 hydraulic machinery
液压机械操纵 hydraulic mechanical control
液压机械控制 hydraulic mechanical control
液压机械式的 hydromechanical
液压机械式调速器 hydra-mechanical governor
液压机械式压机 hydromechanical press
液压机械手 hydraulic manipulator
液压机械停机装置 hydra-mechanical shutdown
液压机械支柱 hydromechanical leg
液压机械铸模机 hydromechanical injection mo(u)lding machine
液压机液体 hydraulic fluid
液压机用油 hydraulic (press) oil
液压积分控制器 hydraulic integral controller
液压挤压 hydraulic extrusion
液压挤压机 hydraulic extrusion press
液压计 hydraulic ga(u)ge; hydraulic pressure ga(u)ge; hydromanometer; hydrostatic ga(u)ge
液压计程仪 Pitot log
液压计算 hydraulic design
液压计算机 hydraulic computer
液压技术 hydraulic technology
液压继动阀 hydraulic relay valve
液压继动阀活塞 hydraulic relay valve piston
液压继动阀活塞皮碗 hydraulic relay valve piston cup
液压继动器 hydraulic servomotor
液压加力 hydraulic amplifier
液压加载油缸 hydraulic stroker
液压夹紧 hydraulic tightening
液压夹具 hydraulically operated fixture; hydraulic clamping device; hydroclamp
液压夹钳 hydroclamp
液压减速器 hydraulic retarder
液压减速装置 hydraulic reduction gear
液压减压器 hydraulic pressure snubber; hydraulic shock absorber
液压减振 liquid springing
液压减振防护 hydraulically cushioned protection
液压减振器 fluid damper; hydraulic buffer; hydraulic bumper; hydraulic damper; hydraulic shock absorber
液压减振设施 hydraulically cushioned protection
液压减振装置 hydraulically cushioned protection; hydrocushion
液压减振座椅 hydraulically cushioned seat; hydraulically damped seat
液压减震器 fluid damper; hydraulic damper; hydraulic pressure snubber; hydraulic shock absorber; hydraulic shock eliminator; liquid damper; oleo-buffer; oleo-cushion; oleo-damper; oleo-gear
液压减震器试验台 hydraulic damper test stand
液压减震支柱 hydraulic shock strut
液压剪床 hydraulic shear
液压剪切机 hydraulic shears
液压检查 hydrocheck
液压降低阀 fluid pressure reducing valve
液压胶管 hydraulic hose
液压绞车 hydraulic power winch
液压绞刀 hydraulic cutter
液压绞盘 hydraulic capstan
液压铰链开闭装置 hydraulic ram actuator
液压脚手架 hydraulic scaffold(ing); pump-jack scaffold
液压脚制动器 hydraulic foot brake
液压铰 adjustable fitting
液压铰接装卸臂 knuckle boom
液压接合 hydraulic joint
液压接力器 hydraulic servomotor
液压(结构的)喷水灭火系统 hydraulically designed sprinkler system
液压截流阀 hydraulic shutoff valve
液压截切机 hydraulic guillotine cutter
液压介质 hydraulic fluid; hydraulic medium; pressure medium
液压金属支架 oil-operated steel support
液压进刀 hydraulic feed
液压进给 hydraulic feed
液压进给机构 hydraulic feed unit
液压进给装置 hydrofeed(er)
液压进给钻 hydraulic power feed drill
液压进料 hydraulic feed
液压进退式滑撬机架 hydraulic retractable slide base
液压进退系统 hydraulic retraction system
液压进油管路 hydraulic oil-recharge pipeline; in-line
液压井壁刮刀 hydraulic wall scraper
液压井壁泥饼清除器 hydraulic wall scraper
液压井壁取样器 hydraulic wall sampler
液压井径仪 hydraulic hole calipers
液压静力触探仪 hydraulic static cone penetrometer
液压静力传动 hydrostatic drive; hydrostatic transmission
液压静力传输 hydrostatic transmission
液压救援设备 porto power
液压举升器 plunger elevator
液压均载器 hydraulic load equalizer
液压卡车挖掘机 hydraulic truck-mounted excavator
液压卡紧 hydraulic lock
液压卡盘 hydraulic chuck; oil chuck
液压开窗装置 hydraulic window opening device
液压开沟铲 hydraulic trench-forming shovel
液压开沟锄 hydraulic trenching
液压开沟机 hydraulic trencher
液压开沟耙 hydraulic trenching
液压开关 hydraulic selector; hydrovalve
液压开关装置 hydraulic breaker(attachment)
液压开塞机 autopour
液压客梯 hydraulic passenger elevator
液压空气锤 hydraulic pneumatic power hammer
液压空气压缩机 hydraulic air compressor
液压空气弹簧 hydropneumatic spring
液压空气悬架 oleo-pneumatic suspension
液压空气悬架系统 hydropneumatic suspension system
液压空气液压控制 fluid control; hydraulic control
液压空气制动机 hydropneumatic brake
液压孔眼测径器 hydraulic cal(l)ipers
液压控制 hydraulically assisted steering; hydraulically operated control; hydraulic control; hydrocheck
液压控制泵 hydraulic control pump

液压控制车床 hydraulic control lathe
液压控制的 hydraulic operated
液压控制动作阀 pilot-operated valve
液压控制阀 hydraulic control valve
液压控制管路 hydraulic control circuit;sense line
液压控制快速定心卡盘 fast-acting hydraulic controlled self centrical chuck
液压控制犁 hydraulically controlled plow
液压控制马达 hydraulic-controlled motor
液压控制器 hydraulically operated controller;hydraulic controller
液压控制式铲运机 hydraulically controlled scraper
液压控制式离合器 hydraulically controlled clutch
液压控制式制动器 hydraulically controlled brake
液压控制手柄 hydraulic control lever;hydraulic hand control
液压控制随动阀 modulating valve
液压控制系统 hydraulic control system
液压控制箱 hydraulic control box
液压控制旋塞 surface cook
液压控制元件 hydraulic control component
液压控制装置 hydraulic control device;hydraulic controller
液压快速移动挖泥机 hydraulic fast-travel excavator
液压扩建 hydraulic extension
液压扩孔器 hydraulic underreamer
液压扩张术 hydraulic dilatation
液压拉拔器 hydraulic puller
液压拉铲 hydraulic pull-shovel
液压垃圾切碎机 hydraulic shears pulverizer
液压拉力检定器 hydraulic pull tester
液压拉伸 hydraulic tensioning
液压拉伸矫直机 hydraulic stretcher
液压拉伸器 hydraulic elongator;tensioning jack
液压拉伸器小车 hydraulic elongator car
液压拉伸压缩疲劳试验机 pulsator
液压冷镦机 hydraulic cold header
液压冷拉机 hydraulic cold-drawing machine
液压离合器 fluid clutch;fluid coupling;hydraulic clutch;liquid clutch
液压离合器传动 oil clutch drive
液压离合器传动箱 oil clutch drive housing
液压力 hydraulic pressure
液压力反馈 hydraulic force feedback
液压立式刮软机 hydraulic universal staking machine
液压立轴钻机 hydraulic spindle core drill
液压连接器 hydraulically operated connector;hydraulic connector
液压连接装置 hydraulic hitch
液压连液压流网绞盘 fluid link hydraulic drift net capstan
液压联动机构 hydraulic power pack
液压联动装置 hydraulic lift linkage
液压联锁 hydraulic interlocking
液压联锁装置 hydraulically operated inter-locking device
液压联轴节 fluid drive coupling;hydraulic coupler;hydraulic coupling;hydraulic transmitter
液压联轴节噪声 fluid-coupling noise
液压联轴器 fluid coupling;hydraulic coupling
液压料斗装载机 hydraulic bucket loader
液压零配件 hydraulic fitting
液压流体 hydraulic fluid
液压流体罐 hydraulic lift tank
液压流网绞盘 hydraulic drift net capstan
液压滤油器 hydraulic filter
液压履带式挖掘机 hydraulic crawler excavator
液压履带式挖土机 all-hydraulic crawler-mounted excavator; hydraulic crawler excavator
液压履带式装载机 hydraulic crawler loader
液压履带式钻机 hydraulic crawler drill
液压履带张紧 hydraulic track tensioning
液压轮箍压装机 hydraulic tyre applying press
液压轮卡车 hydraulic wheel truck
液压轮牙 hydraulic cog
液压逻辑系统 hydrologic(al) system
液压螺距操纵系统 hydraulic pitch control system
液压螺母 hydraulic nut
液压螺旋管阀 hydraulic solenoid valve
液压螺旋压力机 hydraulic screw press
液压落锤 hydraulic drop hammer
液压落地千斤顶 hydraulic floor jack;roll-a-car hydraulic jack
液压马达 fluid motor;fluid pressure motor;hydraulic engine;hydraulic motor;hydraulic slave motor;hydromotor;oil hydraulic motor
液压马达车轮传动 hydrostatic wheel drive
液压马达车轮驱动 hydrostatic wheel drive
液压马达的 SAE 制动压力 hydraulic motor SAE stall pressure
液压马达动力头 top-drive hydraulic power head
液压马力 hydraulic horsepower
液压锚杆安装机 hydraulic bolting machine
液压铆钉机 hydraulic riveter
液压铆钉枪 hydraulic riveter
液压铆接 hydraulic riveting
液压铆接机 hydraulic riveter;hydraulic riveting machine
液压密封 hydraulic packing
液压密合 hydraulic pressurizing
液压模 fluid die
液压模板侧移机 hydraulic moldboard side shift
液压模板侧移器 hydraulic moldboard side shift
液压模板尖端 hydraulic moldboard side shift
液压模锻 liquid pressing
液压模拟 hydraulic analog(ue)
液压模压机 hydrodynamic(al) mo(u)lding press
液压磨矿机 fluid energy mill
液压能力 fluid power;hydraulic capacity;hydraulic competence
液压碾磨机 hydromill machine
液压碾磨造孔机 hydromilling trencher
液压牛头刨 hydraulic shaper
液压扭矩变换器 hydraulic torque converter [convertor]
液压扭力扳手 hydraulic wrench
液压耦合器扭矩曲线 torque curve of hydraulic coupling
液压排泥管 hydraulic ejector
液压盘式制动器 oil-disc brake
液压盘形制动器 hydraulic disc-type brake
液压旁承 hydraulic side bearing
液压刨床 hydraulic planer
液压喷漆 airless spraying;hydraulic spraying
液压喷涂 airless paint spraying;airless spraying;hydraulic spraying
液压喷雾器 hydraulic sprayer
液压喷嘴 hydraulic nozzle;hydraulic pressure nozzle
液压膨胀成型 hydraulic bulge forming
液压劈裂机 hydraulic splitting machine
液压劈裂器 hydraulic splitter
液压劈木机 hydraulic log splitter
液压劈石机 hydraulic splitting machine
液压平板机 hydraulic ram press
液压平衡 hydrocope;hydrocushion
液压平衡的 hydrolastic
液压平衡阀 hydraulically balanced valve
液压平衡盘 hydraulic balancing disc
液压平衡器 hydraulic leveler
液压平衡式安全阀 hydrocushion type relief valve
液压平衡式接轴 hydraulically balanced spindle
液压平衡系统 hydraulic balance system
液压平衡装置 hydrolevel(l)ing device
液压平面磨床 hydraulic surface grinding machine
液压平台升降机 aerial work platform;hydraulic light lift
液压破碎法 hydraulic splitting
液压破碎机 hydraulic crusher;hydraulic breaker
液压破碎器 hydraulic burster
液压歧管 hydraulic pressure manifold
液压启闭机 hydraulic hoist
液压启动 hydraulic starting
液压启动器 hydraulic starter
液压启门机 hydraulic-operated gate lifting device
液压起道机 hydraulic track jack
液压起动 hydraulic starting
液压起动马达 hydraulic starter
液压起阀器的油孔 hydraulic valve lifter oil passage
液压起货机 hydraulic cargo winch
液压起落机构 hydraulically operated lift;hydraulic lift mechanism;hydraulic power lift
液压起落犁 hydraulically raised plow
液压起锚机 hydraulic windlass
液压起重臂 hydraulic jack
液压起重机 hydraulic hoist;hydraulic lift;hydrocrane;hydrostatic trigger
液压起重平台 jack-up crane barge
液压起重器 hydraulic lifting jack;oil jack
液压气动的 hydropneumatic
液压气动技术 hydropneumatics
液压气动控制台 hydraulic pneumatic panel
液压气动式空气悬架 hydropneumatic suspension
液压气动式铆钉锤 hydropneumatic riveter
液压气动式铆钉机 hydropneumatic riveter
液压气动式铆钉枪 hydropneumatic riveter
液压气动式模具缓冲器 hydropneumatic die cushion
液压气动水箱 hydropneumatic tank
液压气动调平器 hydropneumatic leveler
液压气动系统 hydropneumatic system
液压气动蓄能器 hydropneumatic accumulator
液压气动学 hydropneumatics
液压气动涌浪制动器 hydropneumatic surge arrester
液压气压储能箱 surge tank
液压气压缓冲器 surge tank
液压气压联动 hydro air
液压气压联动装置 hydroair unit
液压气压蓄能器 hydropneumatic accumulator
液压汽车起重机 hydraulic autocrane;hydraulic mobile crane
液压汽车升降机 hydraulic auto-lift
液压器 hydraulic press
液压千斤顶 flat jack;fluid pressure operated jack;hydraulic jack;hydraulic operating gear;hydraulic ram;oil jack;pressure oil jack
液压千斤顶分离法 hydraulic jack separating method
液压牵引斗钩油缸 hitch cylinder
液压前端装载机 hydraulic front-end loader
液压钳 hydraulic tongs
液压嵌齿 hydraulic cog
液压强制给进机构 hydraulic pulldown
液压桥式起重机 hydraulic overhead crane
液压切断器 hydrashear cutter;hydraulic cutter
液压切断(液压倾翻)机构油缸 hydraulic tipping ram
液压切管器 hydraulic pipe cutter
液压切削支臂 hydraulic cutting boom
液压切削支臂掘进机 hydraulic cutting boom machine
液压倾斜 hydraulic tilting
液压倾斜装置 hydraulic tipper
液压倾卸 hydraulic dumping
液压倾卸机构 hydraulic tipping gear
液压倾卸系统 hydraulic tilting system
液压清洁器 hydrojet cleaner
液压清理回转筒 hydroblast barrel
液压驱动 fluid drive;fluid power drive;hydraulic drive;hydrostatic drive
液压驱动泵 hydraulically powered pump
液压驱动的 hydraulically operated
液压驱动电梯 hydraulic driven elevator
液压驱动功率输出轴 hydraulic power take-off
液压驱动横向推进器 hydraulically driven transverse propeller
液压驱动盘式制动器 hydraulically actuated disk type brake
液压驱动器 hydraulic actuator
液压驱动湿式盘式制动器 hydraulically actuated wet disk type brake
液压驱动系统 fluid power system
液压去钉器 hydraulic pin puller
液压去栓器 hydraulic pin puller
液压让压支柱 hydraulic yielding prop
液压蠕动泵 hydraulic peristaltic pump
液压软管 hydraulic hose
液压软管滚筒 hose reel
液压软管卷筒 hose reel
液压软管卷轴 hose reel
液压软管连接套 hydraulic hose connection
液压润滑器 hydrostatic lubricator
液压刹车 hydraulic brake;water brake
液压上紧(螺钉)装置 hydraulic bolting device

液压上料机 hydraulically operated loader
液压设备 hydraulic device;hydraulic equipment;hydraulic installation;hydraulic plant
液压伸轨器 <无缝线路> rail tensor
液压伸缩吊具 hydraulic telescopic

spreader
液压伸缩支腿 hydraulic outrigger
液压伸张式井壁刮刀 hydraulic expansion wall scraper
液压渗透率 hydraulic diffusibility
液压升船机 hydraulic shiplift
液压升船台 hydraulic lift dock
液压升降车 hydraulic jack;hydraulic lift(ing) truck
液压升降船坞 < 旧式的 > hydraulic dock;hydraulic lift dock
液压升降堆垛台 hydraulic lift piler
液压升降机 hydraulic elevator;hydraulic lift
液压升降平台 hydraulic lifting work platform; hydraulic working platform
液压升降塔架 hydraulic platform
液压升降台 hydraulic elevating table;hydraulic lift platform
液压升降装置 hydraulic lift
液压声能的 hydroacoustic(al)
液压绳轮升降机 hydraulic rope geared elevator
液压式拔卸器 hydraulic puller
液压式边墩 hydraulically operated bilge block
液压式测力计 hydraulic piezometer
液压式铲运机 hydraulic-controlled scraper;hydraulic scraper
液压式超重车 hydratruck
液压式打桩机 hydraulic hammer;hydraulic hammer pile driver
液压式的动臂起重机 hydraulic boom machine
液压式顶板锚栓安装机 hydraulic roof bolting machine
液压式顶板锚栓安装设备 hydraulic roof bolting equipment
液压式动力输出装置 hydraulic power take-off
液压式翻斗车 hydraulic-operated dumper
液压式反铲挖掘机 hydraulic backhoe; hydraulic backhoe excavator;hydraulic hoe
液压式防护板 hydraulic fender
液压式风窗刮水器的液力驱动器 hydraulic-powered wiper motor
液压式俯仰角调节装置 < 推土机 > hydraulically variable pitch
液压式俯仰角调整器 hydraulic tip
液压式荷载传感器 hydraulic load cell
液压式缓行器 hydraulic type of retarder
液压式缓行系统 hydraulic car retarder
液压式混凝土泵 hydraulic concrete pump
液压式挤出机 hydraulic extruder
液压式继动阀回动弹簧 hydraulic relay valve return spring
液压式夹紧装置 hydraulic clamp
液压式夹桩器 hydraulic chuck
液压式减速器 hydraulic type of retarder
液压式减压阀 hydraulic relief valve
液压式卷扬机 hydraulic winch
液压式孔隙压力计 hydraulic piezometer
液压式离合器 hydraulically actuated clutch

液压式履带松紧度调节装置 hydraulic track adjuster
液压式磨木机 hydraulic grinder
液压式磨石刮刀 hydraulic feed pulpstone doctor
液压式牛头刨 hydraulic shaper
液压式平地机 hydraulic bullgrader
液压式平衡 hydraulic balance
液压式破碎机 hydraulic breaker(attachment)
液压式起重机 hydratruck;hydraulic crane
液压式气门挺杆 hydraulic valve lifter
液压式扰流片 hydraulic spoiler
液压式渗透性试验仪 hydraulic permeability tester
液压式升船机 hydraulic shiplift
液压式隧道钻车 hydraulic tunnel-drilling rig
液压式套管塔 hydraulic telescopic tower
液压式调节偏斜角、倾斜角和俯仰角的推土机 hydraulic angle/tilt/pitch dozer
液压式调节倾斜角的推土机 hydraulic power tilt dozer
液压式调速器 hydraulic speed governor
液压式推土机 hydraulic bulldozer
液压式挖沟机 hydrotrencher
液压式挖掘机 hydraulic (actuated) excavator
液压式往复导丝控制 hydraulic traversing guide control
液压式小型翻斗车 hydraulic type small dumper
液压式悬挂器 hydraulic hanger
液压式压拔桩机 hydraulic press-extracting machine
液压式压出机 hydraulic extruder
液压式压机 hydraulic press
液压式压力试验机 hydraulic pressure tester
液压式闸门启闭机 hydraulic-operated gate lifting device
液压式振动发生器 hydraulic vibration generator
液压式振动驱动器 hydraulic type vibration actuator
液压式正铲挖掘机 hydraulic shovel
液压式制动装置 hydraulic brake
液压式转向操纵系统 hydraulic steering control system
液压式装修吊篮 hydraulic basket
液压式装修升降平台 hydraulic lifting platform
液压式装载机 hydraulic loader
液压式锥形破碎机 huydrocone crusher
液压式自动换排 hydromatic drive
液压式自卸车 hydraulic-operated dumper
液压式综合整平器 hydraulic combination screed
液压式钻车 hydraulically drive jumbo
液压试验 hydraulic pressure test;hydrostatic test(ing);water test
液压试验机 hydraulic tester;hydraulic test machine
液压试验室 hydraulic test chamber
液压试验台 hydraulic test bench;hydraulic test stand
液压试验系统 jack test
液压试验装置 water test unit
液压室 hydraulic pressure cell
液压手持式凿岩机 hand-held hydraulic drill
液压手动倾卸装置 hydraulic hand tipping gear
液压输送 hydraulic conveying;hydraulic transmission
液压输油软管 hydraulic hose
液压双动混凝土输送泵 hydraulic double-acting concrete pump
液压双动千斤顶 hydraulic double-action jack
液压双轮掘削机 hydrofraise;hydromill
液压双轮掘削造孔法 hydrofraise method
液压水流量 hydraulic flow
液压水枪 hydropressure water gun
液压水位计 pressure-actuated ga(u)ge
液压顺序阀 hydraulic sequence valve
液压伺服电动机 hydraulic servomotor
液压伺服机构 hydraulic servo;hydraulic servomechanism; hydraulic servo system
液压伺服马达 fluid servo-motor;hydraulic servomotor;hydraulic slave motor
液压伺服系统 hydraulic servo system
液压伺服执行机构 hydraulic servo-actuator;servohydraulic actuator
液压伺服执行器 hydraulic servo-actuator
液压伺服转向 hydraulic servo steering
液压伺服转子马达 servohydraulic rotary motor
液压松解的制动器 hydraulically released
液压松开 hydraulic unlock
液压松土机 hydraulic ripper;hydraulic rooter;hydraulic scarifier
液压速度控制器 hydraulic speed controller
液压随动系统 hydraulic servo
液压随动油缸 hydraulic slave cylinder
液压碎石机 hydraulic rock breaker;hydraulic rock crusher
液压隧道 hydraulic tunnel
液压隧道盾构千斤顶 hydraulic tunnel shield jack
液压隧道掘进千斤顶 hydraulic tunnel shield jack
液压隧道模板 hydraulic tunnel form
液压损失 hydraulic loss;hydraulic slip;injection drop
液压索铲 hydraulic pull-shovel
液压锁 hydraulic lock
液压锁定 hydraulic lock
液压锁定阀 hydraulic lock valve
液压锁紧器 hydraulic bias
液压锁止 hydraulic lock
液压台车 hydraulic jumbo
液压弹簧 hydraulic spring grease cup
液压镗床 hydraulic boring lathe;hydraulic boring machine
液压套管千斤顶 hydraulic casing jacking unit
液压梯度 flowing-pressure gradient;hydraulic gradient; hydraulic pressure gradient
液压梯度线 piezometric level
液压提升船坞 hydraulic lift dock
液压提升的船闸下游闸门 hydraulic lift tailgate
液压提升机 hydraulic elevator;hydraulic hoist
液压提升机构 hydraulic lift
液压提升机构盖 hydraulic lift cover
液压提升机构盖板 lift cover
液压提升能力 hydraulic lifting capacity
液压提升平台 hydraulic lifting platform
液压提升式铲运机 hydraulic lift scraper
液压提升系统 hydraulic lifting system
液压提升油缸 hydraulic lifting cylinder;hydraulic ram cylinder
液压提升装置 hydraulic pressure elevator
液压替续器 hydraulic relay
液压调节 hydraulic governing;meter in < 在压力管路中的 >
液压调节器 fluid pressure governor;hydraulic governor;hydraulic pressure regulator;hydraulic regulator
液压调平千斤顶 hydraulic leveling jack
液压调平装置 hydrolevel(l)ing device
液压调速器 fluid pressure governor;hydraulic governor;hydrodynamic(al) governor
液压跳板 hydraulically operated ramp
液压跳闸装置 hydraulic tripping device
液压挺杆 hydraulic tappet;oil tappet
液压挺液压筒 ram pot
液压头 dynamic(al) head; fluid head;hydraulic head
液压推车机 hydraulic pusher
液压推动 hydraulic actuation
液压推进 hydraulic feed
液压推进机构 hydraulic feed mechanism
液压推进系统 hydraulic propulsion system
液压推拉杆 hydraulic ram
液压腿 hydro-leg
液压退芯泵 hydraulic core pump
液压托钎器 hydraulic steel support
液压拖铲挖泥机 hydraulic drag-shovel
液压拖带铲土机 hydraulic tractor-shovel
液压拖网起网机 hydraulic trawl winch
液压脱钩装置 hydraulic tripping device
液压脱扣装置 hydraulic trip assembly
液压脱模机 hydraulic ejector
液压挖沟铲 hydraulic ditching shovel
液压挖沟机 all-hydraulic ditcher;hydraulic ditcher
液压挖壕机 hydrotrencher
液压挖掘 hydraulic excavation
液压挖掘附件 hydraulic digging attachment
液压挖掘机 hydraulic (actuated) excavator;hydraulic crawler excavator;hydraulic shovel
液压挖掘机旋转臂 off-set boom
液压挖土机 hydraulic (actuated) excavator
液压挖土装载机 hydraulic excavator-loader
液压外圆磨床 hydraulic cylindric(al) grinding machine; hydraulic traverse cylindric(al) grinder
液压弯板机 hydraulic plate bender
液压弯边机 hydraulic flanging machine
液压弯钢管机 hydraulic steel bar bender
液压弯管机 hydraulic bender; hydraulic bending machine;hydraulic pipe bender;hydraulic pipe bending machine
液压弯轨机 hydraulic rail bender
液压弯曲机 hydraulic bender; hydraulic bending machine
液压弯折机 hydraulic bender
液压万能履带式挖土机 hydraulic universal crawler type excavator
液压万能汽车挖土机 hydraulic universal mobile excavator
液压万能试验机 hydraulic universal testing machine
液压往复式刮刀 hydraulic oscillating doctor
液压微分分析器 hydraulic differential analyser [analyzer]
液压围网起网机 hydraulic purse seine winch
液压维护板 hydraulic maintenance panel
液压微型隧洞挖掘机 hydraulic mole

液压位置伺服机构 hydraulic positional servomechanism
液压温度控制 hydraulic temperature control
液压温度控制器 hydraulic temperature controller
液压稳定的 hydrolastic
液压稳定缸 hydraulic stabilizing cylinder
液压稳定器 hydraulic pressure stabilizer
液压稳定性 hydraulic stability
液压污泥斗 hydraulic drag bucket
液压屋顶支座 hydraulic roof support
液压无级调速器 variable-speed hydraulic governor
液压吸扬机 hydraulic ram
液压系泊装置 hydraulic mooring device
液压系统 hydraulic arrangement; hydraulic power line; hydraulic pressure system; hydraulics; hydraulic system; oil system
液压系统安全阀 system relief valve
液压系统保险开关 pressure safety switch
液压系统布置 hydrostatic arrangement
液压系统操纵手柄 hydraulic system control handle
液压系统操作压力 line pressure
液压系统导出孔 hydraulic tapping
液压系统的漏油 hydraulic leak
液压系统的维修 hydraulic service
液压系统地下室 hydraulic cellar
液压系统分路阀箱 hydraulic manifold
液压系统工作压力 hydraulic system working pressure
液压系统工作液体 hydraulic medium
液压系统管接头 hydraulic coupler
液压系统柜 hydraulic system tank
液压系统加长软管 hydraulic extension pipe
液压系统控制盘 hydraulic panel
液压系统日用油柜 hydraulic system supply and vent tank
液压系统容油量 hydraulic system capacity
液压系统设计 hydrostatic arrangement
液压系统图 hydraulic scheme
液压系统图解 graphic(al) diagram of hydraulic system
液压系统图形符号 symbols of hydraulic system
液压系统泄漏量 hydraulic slip
液压系统压力 fluid pressure
液压系统压力管路 fluid pressure line
液压系统液体罐 hydraulic reservoir
液压系统溢流阀 system relief valve
液压系统用油 hydraulic oil
液压系统油箱 hydraulic reservoir; hydraulic tank
液压系统重量 hydraulics weight
液压系统最大容许压力 operating pressure
液压系统作业循环时间 hydraulic cycle time
液压下套管(法) hydraulic casing
液压先导阀管道 hydraulic pilot line
液压线 hydraulic grade; hydraulic grade line; hydraulic gradient
液压箱 hydraulic fluid tank; hydraulic pillow; hydraulic tank
液压效率 hydraulic efficiency
液压楔 hydraulic wedge
液压卸除机构 hydraulic stripper
液压卸扣装置 hydraulic rod breakout kit
液压行走马达 travel(l)ing hydraulic motor
液压蓄力器【机】hydraulic reservoir
液压蓄能器 hydraulic accumulator
液压悬挂机构 hydraulic lift mechanism
液压悬挂系统 hydraulic lift system
液压悬挂系统联结装置 hydraulic lift linkage drawbar
液压悬挂系统油缸 hydraulic suspension system oil cylinder
液压悬挂装置 hydraulic lift hitch; hydraulic ram lift
液压悬挂座椅 oil-suspension seat
液压悬架装置 liquid spring unit
液压悬架座椅 hydraulic suspension seat
液压旋臂起重机 hydraulic slewing crane
液压旋转方法 hydraulic rotary method
液压旋转式钻机 hydraulic rotary drill
液压旋转式钻进 hydraulic rotary drilling
液压旋转头 fluid swivel
液压旋转钻井机 hydraulic rotary drilling machine
液压选择阀 hydraulic selector; hydraulic selector valve
液压循环澄清池 hydraulic recirculation clarifier
液压循环管路 hydraulic circuit
液压循环时间 hydraulic cycle time
液压循环系统 hydraulic circulating system
液压压拔套管机 hydraulic casing extractor
液压压力盒 hydraulic cell
液压压头 hydraulic pressure head
液压延绳钓机 hydraulic longline gurdy
液压岩石破碎机 hydraulic rock breaker
液压岩芯退取器 hydraulic core extractor
液压摇摆油缸 hydraulic oscillating cylinder
液压遥控马达 hydraulic telemotor
液压叶轮泵 hydraulic vane pump
液压叶片泵 hydraulic vane pump
液压液 hydraulic fluid
液压液速装置 hydraulic reduction gear
液压液体 pressure liquid
液压液体容器 pressure liquid tank
液压移动起重机 hydraulic mobile crane
液压移动挖掘机 hydraulic mobile excavator
液压移动装置 hydraulic shifter unit
液压用流体 hydraulic fluid
液压用途 hydraulic service
液压用液体 hydraulic fluid
液压用油 hydraulic oil
液压油 hydraulic fluid; hydraulic medium; hydraulic oil; hydraulic power oil; pressurized oil
液压油泵 hydraulic oil pump; oil-pressure pump; pressure oil pump
液压油泵传动装置壳体 hydraulic oil pump driving device housing
液压油泵托架 hydraulic cylinder bracket
液压油泵泄放管路 hydraulic oil pump discharge line
液压油低温安定性试验 low-temperature stability test for hydraulic oil
液压油发动机 hydraulic oil engine
液压油分配器 hydraulic fluid distributor
液压油封 hydraulic seal
液压油缸 hydraulic cylinder; hydraulic jack; hydraulic ram; hydrocylinder; pressure oil ram
液压油缸耳轴 cylinder trunnion
液压油缸杠杆 ram arm
液压油管道 hydraulic oil line
液压油规格 hydraulic specification
液压油柜 hydraulic oil tank
液压油冷却器 hydraulic oil cooler
液压油流量 hydraulic flow; pump supply oil flow
液压油滤清器 hydraulic filter
液压油滤网 hydraulic screen
液压油路 hydraulic circuit; hydraulic oil system
液压油路板 hydraulic manifold block
液压油路阀箱 press(ure) manifold
液压油千斤顶 hydraulic oil jack; pressurized oil jack
液压油腔 fluid chamber
液压油容量 hydraulic capacity
液压油室 fluid chamber
液压油油温传感器 hydraulic oil sender
液压油温度表 hydraulic oil temperature ga(u)ge
液压油系统 pressurized oil system
液压油箱 hydraulic oil tank
液压油箱盖 hydraulic tank cap
液压油箱挂锁 hydraulic padlock
液压油箱滤清器 hydraulic strainer; hydraulic tank filter
液压油箱滤芯 hydraulic tank element
液压油箱容积 hydraulic oil capacity
液压油油压力 hydraulic oil pressure
液压油引擎 hydraulic oil engine
液压元件 fluid pressure cell; hydraulic component; hydraulic element; hydraulic pressure cell; hydraulic unit
液压元件规格 hydraulic specification
液压元件进口压力 hydraulic unit in pressure
液压元件试验 hydraulic element test
液压圆锯床 hydraulic circular sawing machine
液压圆筒式启闭机 hydraulic cylinder; hydraulic cylinder hoist
液压圆锥破碎机 hydrocone crusher
液压圆锥旋回破碎机 hydrocone gyratory crusher
液压远距离传输 hydraulic remote transmission
液压凿井钻台 hydraulic shaft jumbo
液压凿岩 hydraulic drilling; hydraulic rock cutting
液压凿岩法液压式钻机 hydro-drilling rig
液压凿岩法液压式钻架 hydro-drilling rig
液压凿岩机 fluid power drill; fluid-powered drill; hydraulic drill; hydraulic jack hammer; hydraulic rock cutter; hydraulic rock drill
液压凿岩机腿 hydra-leg
液压凿岩台车 hydraulic drill ring
液压造塑(法) fluid pressure mo(u)lding
液压造型(法) fluid pressure mo(u)lding
液压造型机 hydraulic mo(u)lding machine
液压增力器【机】hydraulic booster
液压增强计 hydraulic intensifier
液压增强器 hydraulic intensifier
液压闸门 hydraulic gate
液压榨油机 hydraulic oil press
液压张紧装置 hydraulic tensioning equipment
液压张拉器 hydraulic puller
液压张力器 hydraulic tensor
液压张胎器 hydraulic tire expander
液压胀形试验 hydraulic bulge test
液压找包机 hydraulic pressing machine
液压找顶机 hydraulic scaling rig
液压折边机 hydraulic flange press; hydraulic flanging machine
液压折叠式舱盖 hydraulic folding hatchcover
液压真空阀 hydraulic vacuum valve
液压真空缸 hydrovacuum cylinder
液压真空制动器 hydrovac power brake; hydrovacuum brake
液压振荡式钻机 oscillatory hydraulic drilling machine
液压振动打桩机 vibro-driver
液压振动器 hydraulic vibrator
液压振动台 hydraulic shaking table
液压振动系统 hydraulic vibration system
液压整坡机 hydraulic sloper
液压正铲挖土机 hydraulic forward shovel excavator
液压正向排量泵 hydraulic positive-displacement pump
液压支臂 hydroboom
液压支臂切削掘进机 hydraulic boom machine
液压支臂设备 hydroboom equipment
液压支垛 hydraulic chock
液压支架 hydropost
液压支架设备 < 凿岩机 > hydraulic positioner
液压支架限程器 hydraulic rack limiter
液压支架装置 hydraulic setting device
液压支腿 hydraulic jack; oil-leg; hydraulic leg
液压支柱 hydraulic prop; hydropost
液压支柱补偿装置 hydraulic prop compensating gear
液压支柱推拉移动法 push-pull support system
液压执行机构 hydraulic actuator
液压执行元件 hydraulic actuator
液压直边冲压机 straight-side hydraulic press
液压止动器 hydraulic stop
液压止回阀 hydraulic back-pressure valve
液压制 hydraulic system
液压制动 hydraulic braking
液压制动操纵机构 hydraulic brake controls
液压制动车辆 hydraulic braked vehicle
液压制动缸 hydraulic braking cylinder; hydraulic checking cylinder
液压制动机 hydraulic brake
液压制动加力器 hydraulic brake booster
液压制动器 hydraulic brake; hydraulic actuator; hydraulically actuated brake; hydromatic brake; liquid brake; oil brake
液压制动软管 hydraulic brake hose
液压制动系放气软管 hydraulic brake drain tube
液压制动系统 brake fluid system; hydraulic brake system
液压制动系统加油器 hydraulic brake filler
液压制动液 hydraulic brake fluid
液压制动装置 liquid brake
液压制砌块机 hydraulic block-making machine
液压制退的炮架 hydraulic recoil mount
液压致动器 hydraulic actuator
液压致裂(法) hydraulic fracturing
液压轴承 fluid bearing
液压轴向活塞泵 hydraulic axial-piston pump
液压皱纹形成 hydraulic crowd(ing)
液压助力的 hydraulically boosted
液压助力缸 servohydraulic cylinder
液压助力器 hydraulic booster

液压助力器试验器 hydromaster tester
液压助力式旋压 power-assisted spinning
液压助力转向 hydraulic boost steering
液压助力装置 hydraulic servo
液压铸模机 hydraulic injection mo(u)lding machine
液压柱 hydraulic leg;hydraulic post;hydro-leg
液压柱塞 hydraulic plunger
液压柱塞升降机 hydraulic plunger elevator
液压柱塞式压出机 hydraulic pression extruder
液压柱塞直顶式升降机 hydraulic plunger elevator
液压抓斗 hydraulic grab; hydraulic grapple;hydroclamp
液压抓斗挖槽机 hydraulic clamshell trenching machine
液压抓岩机 hydraulic rock grab
液压爪式送料 hydraulic gripper feed
液压转换器 liquid pressure transducer
液压转盘 hydraulic swivel head
液压转速计 hydraulic tachometer
液压转向 fluid-link steering;hydrostatic power steering;hydrostatic steering
液压转向机构 hydraulic steering mechanism
液压转向加力器 hydraulic steering booster
液压转向加力装置 hydraulic steering unit
液压转向离合器 oil steering clutch
液压转向联动装置 hydraulic steering linkage
液压转向助力泵 hydraulic steering pump
液压转向助力器 hydraulic steering assist
液压转向装置 hydraulic power unit for steering;hydraulic steering;hydraulic steering device;power steering
液压转辙机 hydraulic switch machine
液压桩锤 hydraulic pile hammer;hydraulic ram
液压桩夹 hydraulic pile chuck
液压桩头破碎器 hydraulic pile head splitter
液压装车机 hydraulic loader
液压装填 hydraulic packing
液压装箱机 hydraulic casing machine
液压装岩机 hydromucker
液压装油臂 hydraulic loading arm
液压装载铲 hydraulic loading shovel
液压装置 fluid pressure supply; hydraulic arrangement; hydraulic device; hydraulic equipment; hydraulic fitting;hydraulic unit
液压装置的速度 hydraulic speed
液压装置的组装 hydraulic modulation
液压装置皮碗 hydraulic leather
液压撞锤 hydraulic oil ram;hydraulic ram
液压撞击 hammer blow
液压锥形破碎机 hydrocone breaker;hydrocone crusher
液压锥形罩 hydrocone
液压锥罩式环动破碎机 hydrocone gyratory crusher
液压自动比例控制系统 automatic liquid proportioning system
液压自动变距螺(旋)桨 hydromatic propeller
液压自动传动 hydromatic
液压自动传动系统 hydromatic system
液压自动工作(方)法 hydromatic process
液压自动焊接 hydraulic automatic weld
液压自动控制装置 Askania
液压自动式 hydramatic;hydraulically automatic
液压自动调平装置 hydraulic automatic leveling device
液压自动闸门 hydro-automatic gate
液压自动找平装置 hydraulic automatic leveling device
液压自升式突堤码头 jack-up pontoon pier
液压自卸车 hydraulic tipper
液压总管 hydraulic main
液压阻尼 hydraulic damping
液压阻尼减震座椅 hydraulic damping seat
液压组件 hydraulic package
液压钻 hydraulic drill
液压钻臂台车 hydraulic drill jumbo
液压钻车 hydraulic drill rig;hydraulic jumbo;hydro-drill rig
液压钻杆提升器 hydraulic rod extractor
液压钻机 hydraulic (feed) drill;hydro-drill
液压钻机套管 hydraulic drill collar
液压钻机套环 hydraulic drill collar
液压钻机套圈 hydraulic drill collar
液压钻架 hydro-drill rig
液压钻架台车 hydraulic drilling jumbo
液压钻架钻车 hydroboom jumbo
液压钻进 hydro-drilling
液压钻孔 hydraulic drilling
液压钻孔机 auger with hydraulic feed
液压钻孔应变计 hydraulic borehole ga(u)ge
液压钻孔应力计 hydraulic borehole stressmeter
液压钻探 hydraulic drilling
液压钻探法液压式钻机 hydro-drilling rig
液压钻探法液压式钻架 hydro-drilling rig
液压钻探机 hydraulic exploration rig
液压钻探设备 hydraulic drill rig
液压钻头功率 hydraulic bit horsepower
液压钻眼机 hydraulic drill
液压作动 hydraulic actuation
液压作动器 hydraulic actuator
液压作动湿式多片离合器 hydraulically actuated wet multiplate clutch
液压作动筒 hydraulic ram
液压作动系统 hydraulic actuation system
液压作用筒 hydraulic ram
液延生长 rheotaxial growth
液氧 liquid oxygen;lox(ygen) <液态氟气 30%,液态氧气 70%>
液氧爆破筒 liquid oxygen cartridge
液氧爆炸 liquid oxygen explosive
液氧泵 liquid oxygen pump
液氧槽车 liquefied oxygen tank vehicle
液氧的气化器 loxygen evaporator
液氧的压送 pressurization of liquid oxygen
液氧法 oxyliquid method
液氧酒精发动机 liquid oxygen-alcohol unit
液氧密封 lox seal
液氧汽油发动机 liquid oxygen-gasoline unit
液氧设备 lox unit
液氧炸药 liquid oxygen explosive;oxygen explosive;oxyliquid
液氧炸药包 oxygen cartridge
液叶片包角 scroll of blade
液液不混溶性 liquid-liquid immiscibility
液液层析 liquid-liquid chromtagraphy
液液萃取法 liquid-liquid extraction
液液萃取分离 liquid-liquid extraction separation
液液萃取气相层析 liquid-iiquid extraction gas chromatography
液液萃取气相色谱法 liquid-liquid extraction gas chromatography
液液电极 liquid-liquid electrode
液液分配色谱法 liquid-liquid partition chromatography
液液化脂肪酸 liquid oxidation fat acid
液液换热器 liquid-to-liquid heat exchanger
液液色谱法 liquid-liquid chromatography
液液提取系统 liquid-liquid extraction system
液液相分离 liquid-liquid phase separation
液液柱状层析 liquid-liquid column chromtagraphy
液液柱状色谱法 liquid-liquid column chromtagraphy
液液转移回路 liquid-liquid transfer loop
液溢 hydrorrhea;liquorrhea
液引拖拉机 haulage tractor
液印木纹 graining
液印用颜料油浆 graining paste
液油位指示管 telltale pipe
液浴 liquid bath
液匝缸总容积 total cylinder capacity
液渣底 slagging table
液汁 sap
液中油 tramp oil
液柱 column of liquid;fluid column;liquid column;liquid head
液柱高度 elevation of a liquid column;height of liquid column
液柱气压表 liquid column barometer
液柱压力 head of liquid
液柱压力表 manometer
液柱压力计 liquid column ga(u)ge;liquid column manometer;manometer
液装容量 water capacity
液状 liquidness;liquid state
液状残渣 liquid residue
液状废料焚化 liquid waste incineration
液状废料焚化炉 liquid waste incinerator
液状灰浆 wet grout
液状混凝土 liquid concrete
液状沥青材料 liquid bituminous material
液状密封剂 liquid sealant
液状凝胶 liquogel
液状石蜡 liquid paraffin;liquid petrolatum;paraffin(e) oil
液状污物 liquid waste
液(自记)气压计 aneroidograph
液阻 hydraulic lock
液阻流板 step

腋

腋脚凸榫 haunched tenon
腋(生)的【植】axillary
腋下的 axillary
腋羽 axillary

一

一把 handful
一把抓 fishing catch
一百线机键 <自动电话> one hundred point switch
一百八十度式楼梯 staircase of half-turn type
一百八十度弯头 set-off bend
一百八十度转变 somersaulting
一百八十度转角(两跑)式楼梯 staircase of half-turn type
一百磅 short hundredweight
一百磅重量 <美> hundred weight
一百多 long hundred
一百二十 <一种计数单位> great hundred
一百二十磅重量 <英> hundred weight
一百公斤 one-hundred kilogram
一百立方米 hectostere
一百米 hectometer
一百千克 one-hundred kilogram
一百三十五度弯管 half-normal bend
一百万 one million
一百五十周年纪念 sesquicentennial
一百亿 ten billion
一百英尺测链 engineer's chain
一百周年 centenary
一班制 single shift
一班制工作 single-shift work
一般爱尔朗分布 general Erlangian distribution
一般百叶窗(材料) ordinary shuttering
一般半单代数群 general semisimple algebraic group
一般包裹 regular parcel post
一般保障维护 general support maintenance
一般报价 general offer
一般背书 general endorsement [indorsement]
一般边值问题 general boundary value problem
一般标准 denominator
一般驳船队 average tow
一般步行速度 footpace
一般财产税 general property tax
一般采购广告 general procurement
一般操作程序 general operation procedure
一般操作规程 general operating specification
一般操作规定 general operating provision
一般操作条件 general operation(al) requirement
一般测试 general test
一般叉积 general crossed product
一般产权 property rights in general
一般产业统计 general industrial statistics
一般长涌 average length of swell;moderate swell
一般常数 generic constant
一般厂区 general territory
一般场地考虑 general site consideration
一般场方程 general field equations
一般超平面 general hyperplane
一般成本 general cost
一般承兑 general acceptance
一般承认的 accepted
一般程序 general procedure
一般尺寸 stock size
一般尺寸的颗粒 common-sized grain
一般冲力 general impulsive force
一般冲刷 general erosion; general scour(ing);general washout
一般储备 general reserve
一般储备的借项 charge against the general reserve
一般储备的形式 formation of supply in general
一般船运货物 <与散装货物相对应> general cargo

一般大潮 ordinary spring tide
一般大潮低潮(位) low-water of ordinary spring tide
一般大潮高潮(位) high water of ordinary spring tide
一般大潮高水位 high water of ordinary spring tide
一般大修理 general overhaul
一般代理 general agency
一般代理人 general agent
一般代数簇 general algebraic variety
一般贷款 ordinary loan
一般到达 general arrival
一般道路 ordinary road
一般道路几何 general geometry of path
一般的带有反馈的伺服机构 general feedback servomechanism
一般的制品 general-use article
一般等价物 general equivalent in value;universal equivalent
一般等价形态 universal equivalent form
一般等效点 general equivalent points
一般等效点系 general equivalent point system
一般底沙搬运 general bed load transport
一般抵押 general mortgage
一般抵押债券 general mortgage bond
一般地 universally
一般递归函数 general recursive function
一般递归性 general recursiveness
一般点 generic point
一般电路参数 general circuit parameter
一般调查 general investigation
一般对数 general logarithm
一般二次方程 general equation of second degree
一般二次曲面 general quadric surface
一般发价 general offer
一般发盘 general offer
一般反响模型 general response model
一般反应模型 general reaction model
一般方程 general equation
一般方法 general method
一般方法最优化 general method optimization
一般方针 general policy
一般放款抵押契约 general loan and collateral agreement
一般非线性 general nonlinearity
一般废物 general wastes
一般废物处置 general waste disposal
一般费用 general cost;general expenses;overhead cost
一般费用分摊 overhead allocation
一般费用再分配 general expense reappointment
一般分布【数】general distribution
一般分配法 general distribution
一般分配系统 general distribution system
一般风力 general wind force
一般风险 average risks;usual risks
一般风险分析 general risk analysis
一般服务成本 cost of common services
一般服务时间 general service-time
一般辐射 general radiation
一般腐蚀 general corrosion; normal corrosion
一般傅立叶变换 general Fourier transformation
一般概念 general concept;general idea
一般干扰 general interference
一般钢管 common steel tube
一般格式 general form; general format

一般工程研究 general engineering research
一般工龄 general length of service
一般工人 common worker
一般工业 general industry
一般工作人员 clerk general
一般公路 average highway; ordinary highway
一般公认会计原则 generally accepted accounting principles
一般公式 general formula
一般公众业务 general public service
一般供给过剩 general oversupply
一般构造 ordinary construction
一般估计 general estimate
一般固定资产 general fixed assets
一般固定资产基金 general fixed assets fund
一般固定资产投资 investment in general fixed assets
一般固定资产账类 general fixed assets group of accounts
一般观点 general points of view
一般管理 general management
一般管理费和销售费 general administrative and selling expenses
一般管理费(用) general administration cost; overhead expenditures; general administrative expenses; general cost of administration; general management expenses; general overhead charges; general overhead expenses
一般管理费用再分派 overhead reapportionment
一般管理费预算 administration expense budget; general administrative expense budget
一般管理和销售费用 general administrative and selling expenses
一般管理技术 general administration skill; general management technique
一般管理信托 general management trust
一般管制 general control
一般管制区 general control area
一般惯例 normative usage
一般灌浆 general grouting
一般光制 general finishing
一般规定 general requirement
一般规格 general requirement; general specification
一般规律 general rule; universal law
一般规则 general regulation; general rule
一般过失 ordinary negligence
一般海区界限 maritime limit in general
一般含义 general sense
一般函数 general function
一般合伙企业 general partnership
一般荷载 usual load
一般横线支票 general crossed check
一般互反律 general law of reciprocity; general reciprocity law
一般化 generalization; generalizing; universalize
一般化电机 generalized machine
一般化国别风险 generalized country risk
一般化剖面 generalized section
一般化数据操作程序 generalized data manipulation program(me)
一般化水深 soundings generalization
一般化相量 generalized phasor
一般化学分析 <区别于光谱析> wet analysis
一般划分 general partition
一般环境 general environment
一般换能 generalized transduction; general transduction

一般混凝土 general concrete
一般活动仿真程序 general activity simulation program(me)
一般货船 conventional ship; general ship
一般货物 general cargo
一般货物运费率 general cargo rate
一般积分 general integral
一般基本图 general base map
一般集合 general set
一般集论 general set theory
一般技术标准 preliminary specification
一般技术要求 general specification; preliminary specification
一般加法定理 general addition theorem
一般加法定律一般价格 price generally charged
一般家庭污水量 normal domestic sewage
一般价格水准 general level of prices
一般价值 general value
一般价值函数 general cost function
一般价值形式 general form of value; generalized form of value
一般间断模型 general discretized model
一般间接费用 general burden; general overhead; general overhead of office on cost
一般减免 general deduction
一般减税 general tax reduction
一般检查 look see
一般检查员 junior inspector
一般检验 general survey
一般见解 general considerations
一般建议 general recommendation
一般建筑 ordinary construction
一般建筑物 building in-general
一般建筑物地基勘察 foundation exploration of general construction
一般建筑商的承包建筑 contract construction by general builders
一般交叉口 regular intersection
一般交点理论 general intersection theory
一般交换环 general commutative ring
一般交易条件 general terms and conditions
一般交易条件协议书 agreement on general terms and conditions of business
一般阶段 average stage
一般结构用轧制钢材 rolled steel for general structure
一般结账 general closing
一般借款协定 general arrangement to borrow
一般紧密空间 general compact space
一般谨慎条款 general prudential rule
一般经纪人 outside broker
一般经济分析 general economic analysis
一般经济规律 general economic law
一般经济均衡 general economic equilibrium
一般经济统计 general economic statistics
一般精度 general precision
一般径流带 general runoff zone
一般竞争分析 general competitive analysis
一般竞争投标 open bid(ding); open tender
一般就业水平 general level of employment
一般均衡 general equilibrium
一般均衡分析 general equilibrium analysis
一般均衡决定因素的变动 change in determinants of general equilibrium

一般均衡(理)论 general equilibrium theory
一般均衡模型 general equilibrium model
一般均衡系统 general equilibrium system
一般均衡噪声 general equilibrium noise
一般均衡状态 general equilibrium state
一般可接受的准则 general acceptability criterion
一般客货 general cargo
一般客货码头 general cargo wharf
一般控制 common control
一般枯水年 mean dry year
一般会计 general account
一般会计原理 general accounting principles
一般会计制度 general accounting system
一般扩散方程 general diffusion equation
一般劳动 labo(u)r in general
一般劳动过程 general labor process
一般劳动力市场 market for general labor
一般理论 general theory
一般力学 general mechanics
一般利率 general rate of interest; normal interest rate
一般利润率 general profit rate; general rate of profit
一般利息折旧法 interest-in-general depreciation method
一般链环 common link
一般链锁控制 universal chain control
一般量 general quantity
一般留置权 general lien
一般滤波问题 general filtering problem
一般路基 general subgrade
一般路面研究 general pavement studies
一般绿地 ordinary open space
一般轮询 general poll
一般螺旋线 general helix
一般漫射照明 general diffuse(d) lighting
一般锚地 general anchorage
一般贸易 general trade
一般幂 general power
一般名称 generic name
一般命题 general proposition
一般模型 general model; universal model
一般模制品 general mo(u)lded goods
一般目标 general objective
一般目的程式 general purpose program(me)
一般目的系统模拟程序 general purpose system simulator
一般目镜 primitive eyepiece
一般耐火建筑 protected ordinary construction
一般能力成套测验 general aptitude test battery
一般能源 general energy
一般泥沙颗粒 average sediment particles
一般年份 average year
一般黏[粘]性土 general cohesive soil
一般欧几里得联络 general Euclidean connections
一般判定 general estimation
一般培训 general training
一般配电系统 general distribution system
一般配合力(效应) general combining ability
一般频率曲线 common-frequency

curve
一般平均值 general average
一般企业环境 general business environment
一般起吊绳索 general hoist rope
一般气动夹具 universal air-operated clamping fixture
一般气象预报 general forecast
一般砌筑工程 general masonry
一般强度混凝土 normal-strength concrete
一般勤务 general service
一般情报 general information
一般情况 general aspects; general information; general rule; general run of things
一般情况下免费 normally without charges
一般趋势 climate; general run of things; general trend
一般取芯(钻进) convectional boring
一般权利 general rights
一般人类劳动 human labor in general
一般人事费 common staff cost
一般人员 rank and filer
一般砂箱 tight flask
一般商品 general goods
一般商品交付条件 general conditions for the delivery of goods
一般上导数 general upper derivate
一般设计 general design
一般设施 general improvement
一般射影线性群 projective general linear group
一般摄动理论 general theory of perturbation
一般生产量 average production
一般失调 general detuning
一般时效期 overall limitation period; overall time limit
一般使用条件 average service conditions
一般使用要求 general operation(al) requirement
一般市场 open market
一般式 general expression
一般事故 minor accident; ordinary accident
一般事务人员职等 general service level
一般事务员 general clerk
一般试验问题 general test problem
一般适合综合学说 theory of general adaption syndrome
一般适应性综合征 general adaption syndrome
一般收入 general income
一般收入分享 general revenue sharing
一般收益 general benefit
一般属性 general property
一般数据 general data
一般数值 prevailing value
一般水平 average
一般水平以上的质量 above the average quality
一般水平以下的质量 below the average quality
一般水准 denominator
一般税率 general rate; general tariff
一般税则 general tariff
一般说明 general description; general remark
一般算法 general algorithm
一般所得税 general income tax
一般讨论 general discussion
一般特惠制 generalized scheme of preferences
一般特性 general aspects; general characteristic; general feature
一般特性指标 indices of general property
一般特征 general feature
一般提款权 ordinary drawing rights; regular drawing rights
一般天气推断 general inference
一般天气预测 general inference
一般条件 average condition; general condition; general term
一般条件下操作 moderate-duty service
一般条款 general clause; general provisions
一般投影几何 general projective geometry
一般投资惯例 normal investment practice
一般推移质搬运 general bed load transport
一般椭圆形 general elliptic type
一般拓扑空间 general topological space
一般拓扑学 general topology
一般挖方<不含大块石的> common excavation
一般外露面积 exposed area
一般外貌 general appearance
一般维护组织 general maintenance organization
一般维修 general maintenance
一般位势 general potential
一般文件 general document
一般文书工作测验 general clerical test
一般稳定判据 general stability criterion
一般稳定性准则 general stability criterion
一般问题的范围 general problem area
一般问题解决者 general problem solver
一般污染物 prevalent pollutant
一般污水<指中等浓度污水> average sewage
一般屋面坡度<37.5°>【建】ordinary pitch
一般无线电指向标 radio beacon in general
一般物价水平变动 general price level change
一般物价水平法 general price-level method
一般物价水平调整 general price-level adjustment
一般物价指数 general price index
一般物资供应基金 general supply fund
一般误差定律 general law of error
一般习惯 general custom
一般系统论 general system theory
一般系统模拟 general systems simulation
一般细菌数 general bacterial population
一般细菌总数 total general bacteria
一般下标文法 general subscript grammar
一般下导数 general lower derivate
一般显色指数 general colo(u)r rendering index
一般现金 general cash
一般现色指数 general colo(u)r rendering index
一般线性变换 general linear transformation
一般线性递归系统 general linear recurrence system
一般线性回归模型 general linear regression model
一般线性假设 general linear hypothesis
一般线性模型 general linear model
一般线性群 general linear group
一般线性微分方程 general linear differential equation
一般项【数】general term
一般项目 general mine
一般消费物价指数 general index of retail price
一般消费者 average consumer
一般销售费用 general selling expenses
一般销售税 general sales tax
一般小潮 ordinary neap tide
一般小潮低潮 low-water of ordinary neap tide
一般小潮高潮 high water of ordinary neap tide
一般效用 general utility
一般协调预算 general coordinating budget
一般信息无损耗 general information lossless
一般信用证 general credit; general letter of credit; open credit
一般行政费用 general government expenditures
一般形式 general form
一般形态 gross morphology
一般性 generality; universalism; universality
一般性拨款 general grant
一般性补充资金 general replenishments
一般性补给 general supply
一般性处理 general process
一般性的税 overhead tax
一般性技术规程 unclassified technical order
一般性检查 general inspection
一般性能飞机 average aircraft
一般性能数据 general performance number
一般性破坏 general failure
一般性热置换 general heat exchange
一般性人口变动 general population movement
一般性术语 generic term
一般性土 generally soil
一般性修理 general overhaul
一般性需要 general need
一般性引用标准 general reference to standard
一般性指标 indices of general property
一般性质 general aspects
一般修理 general repair
一般许可证 general license
一般旋轮线【数】roulette
一般询问 general inquiry
一般押汇担保函 general letter of hypothecation
一般押汇质权书 general letter of hypothecation
一般养护 general maintenance
一般要求 general requirement; modest intent
一般业务费用 general operating expenses
一般义务 general obligation; general responsibilities
一般易燃室内用品 ordinary-hazard contents
一般意外损失准备金 general contingency reserve
一般意外盈余准备 general contingency reserve
一般意义的生产劳动 productive labor in its general sense
一般因素 general factor
一般因子 general factor
一般银行业务 general banking
一般印象人员选择法 overall impression method of selection
一般应变状态 general state of strain
一般应力状态 general state of stress
一般营业费用 general expenses; general reserves
一般用单角铣刀 single angular cutter for general use
一般用法 common usage
一般用户特权级 general user privilege class
一般用途 general duty; general usage; general utility; general service
一般用途的 general purpose
一般用途的车辆 general purpose vehicle
一般用途货柜 general purpose container
一般用途信用限额 general purpose lines of overhead
一般用途抓斗 general-purpose bucket
一般优惠 average preference
一般优惠关税 general preferential duties
一般优先权 normal priority
一般有理分数 general rational fraction
一般雨量 general rainfall
一般预报 general forecasting
一般域【数】general domain
一般原理 general considerations; general principle
一般原因故障 common cause failure
一般运输方程 general transport equation
一般运输后勤车辆 general transport administrative vehicle
一般杂货 general cargo
一般责任 general responsibilities
一般责任保证 general obligation bond
一般债权人 general creditor
一般债券基金 general bonded-debt fund
一般债务债券 general obligation bond
一般涨价 general increase
一般账户收入 receipts of general accounts
一般照明 general illumination; general lighting
一般照明及电力配电 general light and power distribution
一般照明配电盘<剧院> house board
一般照明设计 general illumination design; general lighting design
一般正规方程 general normal equation
一般政策 general policy
一般政策控制 general policy control
一般政策模型 general policy model
一般知识 general knowledge
一般职工 rank and filer
一般指导原则 general guideline
一般指数修匀 general exponential smoothing
一般质量 run-of-the-mill
一般质量保护措施 average quality protection
一般质押书 general letter of hypothecation
一般注释 general comment
一般装车细则 general loading instructions
一般状况检查 general condition survey
一般着色 universal tinting colo(u)r
一般资本 capital in general
一般资产 assets in general
一般资料 general information
一般子句 general clause
一般走向 general trend
一般租船契约 gencon
一般最惠国条款 general most favo(u)red nation clause
一般最优值函数 general optimal value function
一般左递归 general left recurrence
一般作业 general job
一般作业程序 generalized operating procedure

一版清绘 individual drafting
一半 half;one-half;moiety <尤用于法律方面>
一半一半 half-and-half
一拌 <混凝土> batch
一拌干容量 dry capacity per batch
一拌混凝土 a batch of concrete
一帮 gang
一磅重的东西 pounder
一磅装(盒子) penfold
一包 a pack;a parcel;one pack
一包多件的商品包装 multipack
一包棉花 bale of cotton
一杯 drink
一倍瞄准望远镜 unit power sighting telescope
一本土壤方面的书的提纲 outline of one's book on soils
一泵站 primary pumping station
一比二 one-on-two;one-to-two
一比二坡度 one-on-two
一比一 half-and-half;same-size ratio
一比一边坡 one-to-one slope
一比一模型 full-scale model;full-sized model
一比一样板 full-scale template;full-sized template
一笔整数 a lump sum
一边 one-side
一边高一边低的 lopsided
一边禁止停车制 unilateral prohibition of waiting;unilateral waiting
一边磨耗 <钢轨> swaying defacement
一遍 single-pass
一遍操作 one-pass operation
一遍方案 one-pass scheme
一遍算法 one-pass algorithm
一遍通过 first pass
一步 one-step;single step
一步成像摄影(术) instant photography;one-step photography
一步成像照相机 instant camera;one-step camera
一步醇酸合成法 one-stage alkyd process
一步的长度 foot step
一步的距离 stride
一步迭代 single-step iteration
一步对一步地 step for step
一步发泡成型 one-step mo(u)lding
一步法氨基甲酸酯(泡沫)涂层 one-shot urethane coating
一步法树脂 one-stage resin
一步合成 one-step synthesis
一步结晶法 one-step crystallization
一步净化器 one-step purifier
一步炼钢法 direct steel process
一步摄影 one-step photography
一步洗脱法 one-step elution
一步作业 one-shot job;one-step job;one-step operation
一部分 fraction;moiety;parcel;partly;portion;subdivision
一部分的 divisional;partial
一部分切成薄片 slice
一部门增长模型 one-sector growth model
一舱进水不沉船 one-compartment ship
一舱制船 one-compartment ship
一侧测定点位 unilateral spotting
一侧观察 unilateral observation
一侧观察法 unilateral method
一侧加长并凿斜的榫眼 pulley mortise
一侧进水 water entering on one side
一侧凸起的 <有浮起装饰的> bossed (on) one side
一侧镶边 banded one side
一侧支流水系 unilateral stream
一层 parking tier <多层停车场的>;single ply <指材料如一层板片、一层油毡等>;first floor <指楼房,美国>
一层半房屋 stor(e)y-and-a-half
一层比一层高的 rampant
一层玻璃窗 single-glassed window
一层的 one-layer;one-level
一层厚厚的蜡被 a heavy bloom
一层接地平面图 grounding on ground floor
一层楼 ground stor(e)y
一层平面图 <美> first floor plan
一层浅土 spit of land
一层绳 flake
一层水 sheet of waters
一层松散物质 the loose mass of the earth's crust
一层涂层厚度 mileage
一层一层的 lit-by-lit
一铲可掘起的土的深度 <英> graft
一铲深度 spit
一铲土 spit of land
一长串公共汽车 a string of buses
一长列汽车 autocade
一长制 system of one-head leadership;system of one-man leadership
一场 scene
一场暴雨 storm event
一场雨 individual rain
一车货 wagon load
一车货物的分卸 break-bulk;break of load
一车货物分卸和整理作业 break-bulk and consolidation operation
一车载量 <泛指一车的路用材料> lorry load
一车之量 cart
一车组材 set of timber
一锄深度 spit
一触即拍的底片 point-and-shoot
一触键 one-touch key
一串 a bunch;string
一次 X 射线 primary X-ray
一次安装 one-setting
一次拌和量 batch sample
一次拌和量所需要的水 batched water
一次拌和时间 mixing cycle
一次拌和用水 <混凝土> batched water
一次保护装置 primary protection
一次保险单 named policy
一次报废式仪器 expendable instrument
一次暴雨降水量 total storm precipitation
一次曝光 single exposure
一次爆破 initial fragmentation;primary blasting;primary breaking;one-shot
一次爆破单位炸药消耗量 initial explosive ratio;primary explosive ratio
一次爆破循环进尺 depth of round
一次爆破炸药 primary blasting explosive
一次倍频 first overtone
一次被覆光纤 primary coating fiber [fibre]
一次泵 primary pump
一次泵冷水系统 chilled water system with primary pumps
一次变电所 primary substation
一次变电站 primary substation
一次变分方程 equation of first variation
一次变换 linear transformation
一次变换器 primary transducer
一次标引 primary indexing
一次拨款法 lump-sum appropriation
一次布网【测】 once establishment control network
一次采全高 full-seam mining
一次采样给料机 primary sample feeder
一次采油 primary recovery
一次采油可采储量 primary reserves
一次参数 primary parameter
一次操作 once-through operation;single job
一次测定 single determination
一次测额定电流 rated primary current
一次测额定电压 rated primary voltage;primary rated voltage
一次测管路 primary piping
一次测量仪表 primary instrument;primary measuring instrument;primary meter
一次测漏泄 primary leakage
一次测配管 primary piping
一次测熔断器 primary fuse
一次测无载抽头 primary no-load tap
一次差 first difference
一次产品 primary production
一次产业 primary industry
一次产业部门 primary industrial sector
一次尝试联想 one-trial association
一次偿付 single payment
一次偿付复利因素 single-payment compound amount factor
一次偿付复利因子 single-payment compound amount factor
一次偿付现值计算法 single-payment present worth computation
一次偿付现值因素 single-payment present worth factor
一次偿付现值因子 single-payment present worth factor
一次偿还信贷 non-instalment
一次偿清贷款 simple payment loan;single-payment loan
一次场 primary field
一次沉淀池 primary sedimentation basin
一次沉降槽 primary settling tank
一次沉降池 primary settling tank
一次沉陷 primary setting
一次衬砌 primary lining
一次成影照相机 Polaroid camera
一次冲击 one-shot;single blow
一次抽样单位 primary sampling unit
一次抽样方案 single sampling plan
一次抽样检验 single sampling inspection
一次抽样样品 single sample
一次处理 primary treatment;single treatment
一次传感器 primary transducer
一次锤击下沉量 set per blow
一次纯保费 net single premium
一次磁场 primary magnetic field
一次簇 primary cluster
一次导数 first derivative;first-order derivative
一次导数控制 anticipatory control
一次导数调节电路 rate of change circuit
一次到期债券 term bond
一次的 lineal;linear;one-dimensional
一次点火起爆 firing at one
一次点燃 single firing
一次电波 primary wave
一次电池 decomposition cell;galvanic cell;one-shot battery;primary cell;primary element
一次电池供电 primary cell supply
一次电池组 primary battery
一次电故障试验 primary fault test
一次电离 primary ionization
一次电力网 primary network
一次电流 primary current
一次电路 primary circuit
一次电路电阻 primary resistance
一次电路端子 primary terminal
一次电路原理图 primary circuit diagram
一次电路终端 primary terminal
一次电压 primary voltage
一次电源 primary power source;primary source
一次电子 incident electron;primary electron
一次冻透 straight freezing
一次断头 primary break
一次发火 one firing
一次发生的 monogenetic
一次发作时间 spell
一次法 once-through method
一次反复 one replication
一次反射 one-bounce reflection;primary reflection
一次反射脉冲 once-reflected pulse
一次反射束 once-reflected beam
一次反应 first-order reaction;primary reaction
一次反应区 primary skip zone
一次返修 repair for 1st time
一次方 first power;unit power
一次方程式 equation of the first degree;first-order equation;linear equation;simple equation
一次放射 primary emission
一次分发量 dosage
一次分离 primary separation
一次分配结算 single distribution liquidation
一次分析 batch analysis
一次风 primary air;primary air-flow;underfire air
一次风百分率 percent primary air
一次风鼓风机 firing fan;primary blower
一次风机 fan for primary air;primary air fan
一次风速 primary air velocity
一次风温度 primary air temperature
一次辐射器 primary feed;primary multiplex equipment
一次付款 compounding;lump sum;non-instalment
一次付清 lump-sum payment;pay(ment) in full
一次付清的保险费 single premium
一次付清的现金贷款 single-payment personal cash loan
一次付清年金 single-payment annuity
一次付清养老金 frozen pension
一次付清债务 compounding a debt
一次干线 primary main
一次光电流 primary photoelectric current
一次光电效应 primary photoelectric effect
一次光源 primary source
一次过渡模 master mo(u)ld
一次过闸 single lockage
一次函数 function of first degree;linear function
一次航程 stretch
一次和分 first summation
一次虹 primary rainbow
一次(化学)灌浆法 one-shot method
一次还本贷款 bullet loan;bullet maturing;bullet maturity
一次还清的信贷 non-instalment credit
一次缓解 direct release
一次换能器 primary transducer
一次回采 primary excavation;primary mining;primary stoping
一次回风 primary return air

一次回路 primary circuit; primary loop; primary system
一次回路释放系统 primary relief system
一次回路泄放系统 primary relief system
一次回收 primary return
一次回收方法 <石油> primary recovery method
一次回收率 primary recovery
一次回水 primary water
一次回转 primary revolution; rev
一次活套 primary loop
一次货物作业平均停留时间 average detention time; average time of detention per goods operation
一次计数器 start-stop counter
一次剂量 dose; dosis
一次继电器 primary relay
一次加荷重力线弹性分析 linear elastic gravity-turn-on analysis
一次加热 single heat
一次加热器 primary heater
一次加油润滑 lifetime lubrication
一次加油润滑的 lifetime lubricated
一次加载应力应变关系特性 monotonic stress-strain behaviour
一次检测元件 primary detecting element
一次检查 single sampling
一次交货量 consignment
一次浇灌量 lift
一次浇筑 one-shot placing
一次浇筑的正常厚度 <混凝土> normal lift
一次浇筑混凝土的层高 concrete lift
一次浇筑施工法 <混凝土> single-lift construction
一次胶合 primary gluing
一次搅拌的干拌物 <混凝土等> dry batch
一次搅拌的混合料干重量 dry batch weight
一次缴清保费的保险 lump-sum payment insurance
一次接触 primary exposure; single exposure
一次接触限值 primary exposure limiting value
一次结构 primary structure
一次结晶 primary crystallization
一次结清 lump-sum settlement
一次近似 approximation on the average; first approximation
一次近似值【数】 approximation of 1st degree
一次进入点 mary entry point
一次进水支管 primary feeder lateral
一次精选 primary cleaning
一次精选槽 primary cleaner
一次精选尾矿 primary cleaner tailing
一次就位 single set-up
一次局部膜应力 primary local membrane stress
一次(掘进)循环进度(长度) length of one round
一次开采方法 <石油> primary recovery method
一次开挖 primary excavation
一次颗粒 primary particle
一次空气 fresh air; primary air; underfire air
一次雷达 primary radar
一次冷床 primary cooler
一次冷镦机 single-stroke header
一次冷凝器 primary condenser
一次冷却剂 primary coolant
一次冷却剂回路 primary coolant circuit
一次冷却水 primary cooling water
一次冷却系统 primary cooling system
一次离子束 primary ion beam
一次力矩 first moment
一次连续浇捣量 placement
一次量测权 weight of one measurement
一次流过冷却 once-through cooling
一次流过冷却系统 once-through coolant system
一次流过密封 once-through seal
一次馏液 once-run distillate
一次码烧工艺 once setting in drying-firing
一次码烧隧道窑 tunnel kiln with drier [dryer]
一次幂 first power; unit power
一次敏感装置 primary sensing device
一次模 one-off pattern
一次模型材料 expendable pattern material
一次模铸造 expendable pattern casting
一次抹灰面 <两块整平板之间的墙面> floating bay
一次抹面 dinging; one-coat plastering; single-coat stucco
一次莫来石 primary mullite
一次能源 primary energy; primary energy source; prime energy
一次能源燃料 primary energy fuel
一次黏[粘]土 primary clay
一次排放 primary emission
一次配电 primary distribution; primary power distribution
一次配电干线 primary distribution main
一次配电网 primary distribution network
一次配电系统 primary distribution system
一次配矿 primary ingredient ore
一次配料工厂 one-stop batch(ing) plant
一次喷射厚度 spouting thickness of layer
一次喷涂衬里 one-shot lining
一次喷涂施用 one spray coat application
一次喷涂应用 one spray coat application
一次喷嘴 primary nozzle
一次碰撞 primary collision
一次贫化率 first dilution ratio
一次频率调节 primary frequency regulation
一次瓶 one-way bottle non-returnable
一次破碎 crushing in single pass; primary crushing; primary shredding
一次破碎工段 preliminary-crushing department
一次破碎机 preliminary crusher; primary breaker; primary crusher; primary disintegrator
一次铺筑的厚度 height of lift(ing)
一次启动 one-shot
一次起动 one push start
一次起动操作 one-shot operation
一次气孔 primary blowhole
一次气流 primary air-flow
一次迁移 <石油> primary migration
一次强化 primary reinforcement
一次清洁 lump-sum settlement
一次情报 primary communication; primary information
一次球磨机 primary ball mill
一次取样 primary sample
一次取样单 primary sampling unit
一次取样单位 primary sampling unit
一次取样检验 single sampling inspection
一次群 primary block; primary group
一次群速率 primary group rate
一次群用户接口 primary user-network interface
一次燃料循环 once-through fuel cycle
一次燃烧 primary combustion
一次燃烧区 primary combustion zone
一次燃烧室 primary zone
一次绕组 primary coil; primary winding
一次绕组电感 primary inductance
一次热空气 hot primary air
一次融透 straight thawing
一次润滑轴承 lubed-for-life bearing
一次扫视操作 one-pass operation
一次闪蒸 single flash
一次闪蒸管式釜 single-flash pipe still
一次闪蒸系统 single-flash system
一次烧成 firing process; monofired; once firing; one-step firing process; single fire; single firing
一次烧成的制品 once-fired ware; single fired ware
一次烧成工艺 single firing process
一次烧成器皿 once-fired ware
一次烧成搪瓷 single firing enamel
一次设计 once-through design; one-through design
一次设置 one-setting
一次渗碳体 primary cementite
一次石墨 primary graphite
一次石墨化 primary graphitization
一次矢函数 linear vector function
一次使用的 one-shot
一次使用的油桶 one-trip oil container
一次使用信用证 straight letter of credit
一次使用装置 expendable equipment
一次式重合闸 single-shot reclosing
一次试验 single test
一次收发的邮件 postbag
一次输电的馈电线 primary transmission feeder
一次输电线 primary transmission line
一次水循环管路 primary circulation pipe; primary flow-and-return pipe; primary flow-and-return piping
一次速率接入 primary rate access
一次酸洗薄钢板 single-pickled sheet
一次缩小 first-stage reduction
一次摊销法 once-for-all method of amortization
一次探测器 primary detector
一次搪瓷 one-coat enamel
一次提纯 primary purification
一次调风门 primary air regulator
一次调整 one-setting
一次停车标 stop once sign
一次通过 single-pass; one-shot
一次通过操作 one-pass operation
一次通过程序 blue ribbon program(me); star program(me)
一次通过的粉碎 open circuit reduction
一次通过的研磨物料量 grind per pass
一次通过非循环系统 once-through system
一次通过锅炉 single-pass boiler
一次通过过程 once-through process; one-through process
一次通过区域熔炼 single-pass zone-melting
一次通过物料平衡 once-through material balance
一次通过型蒸汽发生器 once-through type steam generator
一次同余 congruence of first degree
一次投放法 sudden injection method
一次投料压出机 batch extruder
一次投票 single ballot
一次涂层 primary coat(ing)
一次涂覆光纤 primary coating fiber [fibre]
一次脱水 primary dewatering
一次弯曲应力 primary bending stress
一次完成操作 one-pass operation
一次完成的 once-through; one-shot
一次完成法 one-shot method
一次完成作业的土壤搅拌机 one-pass soil mixer
一次完工法 completion contract method
一次危险 single risk
一次微分法 first derivative method
一次微分控制 anticipatory control; derivative control; lead control
一次文献库 primary document file
一次污染 primary pollution
一次污染物 primary pollutant
一次污染源 primary pollution source
一次误差 first-order error
一次(稀释)溶剂 primary solvent
一次系统 first-order system
一次纤维 primary fiber; primary filament
一次显示 primary display
一次显影液 one-time developer
一次线 primary line
一次线圈 primary coil
一次线匝 primary turns
一次项 first-order term
一次消费 one-time consumption
一次效果 primary effect
一次谐波 first harmonic; fundamental harmonic; primary harmonic
一次行程 one-stroke
一次形成 one-shot forming
一次型 dispensable mo(u)ld
一次性包装 disposable packaging; non-turnable container; one-way package
一次性保险丝 one-time fuse
一次性变动 once-and-for-all change
一次性裁决制度 one-time adjudication system
一次性测定 single-time measurement
一次性偿还信贷 non-instalment credit
一次性处理 single-shot process
一次性措施费用 lump-sum measure expenses
一次性调查 one-time investigation; one-time survey; single-round survey
一次性费用 non-recurring charges; non-recurring expenses; front-end fee <订约后先收的>
一次性付款 lump-sum payment; single payment
一次性高聚物 primary high polymer
一次性过程 single-shot process
一次性基础 <塔吊等机械的> expendable base
一次性集装箱 one-way container
一次性奖金 lump-sum award; one-time premium
一次性浇灌面层 one-course pavement; straight-forward surfacing
一次性浇注式施工法 single-lift construction
一次性空气过滤器 disposable air filter
一次性口袋 throwaway-type bag
一次性冷却系统 once-through cooling system
一次性滤材 disposable filter medium
一次性(滤)袋 disposable bag
一次性滤器 throwaway-type filter
一次性滤芯 disposable cartridge
一次性滤纸 throwaway-type filter
一次性灭火器 disposable fire extinguisher

一次性模 waste mo(u)ld
一次性模型 waste model
一次性起霜 plate out
一次性桥塞 expendable plug
一次性润滑系统 one-shot lubricating system;one-shot lubrication
一次性使用的餐具 disposable
一次性收入 lump-sum payment;non-recurring income
一次性收入来源 non-recurring sources of income
一次性手续费 front-end fee
一次性塑料制品 disposable plastic item
一次性损益 non-recurring profit and loss
一次性特别缴款 special contribution
一次性投资工艺 first cost technology
一次性物品 single-service items
一次性消费产品 perishable consumer goods
一次性效应 once-and-for-all effect
一次性用具 single-service article
一次性有效出入境签证 entry-exist visa valid for a single journey
一次性支出 lump-sum cost
一次性资源 fugitive resources
一次性钻头 throwaway bit
一次悬挂 primary suspension
一次选择 single selection
一次循环 primary circulation
一次循环冷却 once-through cooling
一次循环流程图 once-through flowsheet
一次压制 single-pressing
一次压制法 single-press process
一次压注模制的体积量 volume mo(u)lded per shot
一次压注容量 volume mo(u)lded per shot
一次冶成 one-shot forming
一次冶练 primary smelting
一次一步模式 chess pattern
一次仪表 primary instrument;primary meter
一次因子 first-order factor
一次印张数 paper capacity
一次应力 primary stress
一次荧光 first-order fluorescence
一次用包装 non returnable container
一次用完货品 single-use goods
一次铀燃料循环 once-through uranium-fuel cycle
一次有效的对策 one-short game
一次有效工作服 disposable overall
一次有效利用 one-time operation
一次有效装置 expendable equipment;one-shot device
一次元的 unidimensional
一次元件 primary element
一次源 primary source
一次运输 primary haulage
一次运算 once-through operation
一次运算计算机 single-shot computer
一次载热剂 primary coolant
一次载重量 turn
一次再结晶 primary recrystallization
一次再热 single reheat
一次凿岩 primary drilling
一次张拉 one-shot tensioning
一次照射 single exposure
一次遮蔽 primary shield(ing)
一次蒸发 flush distillation;single vapo(u)rization
一次蒸发管式炉 once-through pipe still
一次蒸馏 single flash
一次蒸馏的 single distilled
一次蒸汽 primary steam
一次支付补助 lump-sum subsidy
一次支付贷款 single-payment loan
一次支付公式 single-payment formula
一次支付现值系数 single-payment present worth factor
一次支付因素 single-payment factor
一次支付因子 single-payment factor
一次枝线 linear branch
一次指令 once command
一次重力加荷线性弹性分析 linear elastic gravity-turn-on analysis
一次主配电线路 primary distribution trunk line
一次助燃空气 primary air
一次注射润滑法 single-shot lubrication
一次铸模 expendable mo(u)ld
一次铸型 temporary mo(u)ld
一次转换 primary switch
一次转换时间分解 single transition time decomposition
一次转换时间分配 single transition time assignment
一次装填 one-time pad
一次装运 one shipment
一次装载时间 time per loading
一次追击机会对策 one chance pursuit game
一次自由度 single degree of freedom
一次总付 lump-sum payment
一次总付补助 lump-sum subsidy
一次总付的保险费 lump-sum premium
一次总付的方法 lump-sum basis
一次总付的(金额) lump sum
一次总付合同 lump-sum contract
一次总付金额 a lump sum
一次总付要求 blanket commitment request
一次总付制 lump-sum basis
一次总购 basket purchase;lump-sum purchase
一次总价 lump-sum price
一次总缴的人头税 lump-sum tax
一次总体薄膜应力 primary general membrane stress
一次走刀 one-pass
一次组织 primary structure
一次钻进深度 pull length
一次钻头 single-use bit
一次最高容许浓度 momentary maximum allowable concentration;primary maximum allowable concentration; single exposure maximum allowable concentration
一次作业 single job
一丛(群) cluster [clustre]
一醋精 monoacetin
一簇 cluster;tuft
一寸尺 inch plank
一打 a dozen (of)
一打锉刀 dozen file
一打卷宗 dozen file
一打装 pack in dozen
一大笔钱 a round sum
一大步 stride
一大部分 a large proportion of
一大堆 raft;wilderness
一大片 extent
一代磷酸钙 monobasic calcium phosphate;primary calcium phosphate
一代磷酸钠 monobasic sodium phosphate;sodium orthophosphate
一代人 generation
一代人再生产率 generation reproduction rate
一代盐 primary salt
一袋水泥拌料 one-sack batch
一袋水泥混凝土拌料 one-sack batch
一单位关系固定样本调查 one-unit relational panel survey
一单元 monoblock
一氮化二铁 iron nitride
一氮化钒 vanadium nitride
一氮化合物 mononitride
一氮化三锂 lithium nitride
一氮化三银 silver nitride
一氮化钛 titanium nitride
一氮化钽 tantalum nitride
一氮化铀 uranium mononitride
一挡 first gear;first speed
一挡齿轮 bottom gear; first speed gear
一挡齿轮啮合 first-gear engagement
一刀 <纸张数量单位,24 或 25 张> quire
一刀片之量 bladeful
一道成活漆 one-coat paint
一道漆 one coat
一道涂层 one coat
一的补码 complement-on-one; one's complement
一的补码电路 one's complement circuit
一的补数 complement-on-one
一等 first-order
一等舱 first-class cabin
一等舱旅客 first-class passenger;salo(o)n passenger
一等测点【测】 first-order point;first-order station;main-scheme station;primary point;primary station
一等测候站 first-order weather station
一等测量仪器 first-order instrument
一等测站 first-order station; primary station
一等大地点 main point; major controlled point;principal point
一等大地控制网 primary geodetic control network
一等大地网 first-order geodetic network
一等导线【测】 first-order traverse; primary traverse
一等导线点 first-order traverse point; first-order traverse station; primary traverse point;primary traverse station
一等导线站 first-order traverse point; first-order traverse station; primary traverse point;primary traverse station
一等点 primary control point
一等工厂 first of a kind plant
一等工人 primary worker
一等观测 primary observation
一等精度 first-order accuracy;first-order precision;primary accuracy
一等客舱甲板 salo(o)n deck
一等控制 primary control
一等控制点 primary control point
一等控制网 first-order control frame; first-order control network; primary control framework;primary control network
一等秘书 <大使馆的> chancellor; first secretary
一等品 first quality;firsts
一等三角测量点 first-order triangulation point; primary triangulation point;principal triangulation point
一等三角测量(法)【测】 first-order triangulation; main triangulation; major triangulation; primary triangulation;principal triangulation
一等三角测量三角形 primary triangle
一等三角测量网 first-order triangulation net;primary triangulation net
一等三角点 first order triangulation station;primary triangulation point
一等三角链 first-order triangulation chain
一等三角锁 first-order triangulation chain
一等三角网 first-order triangulation net;major framework;primary triangulation net
一等商业票据 first-class commercial paper
一等水文站 principal hydrometric station
一等水准(测量) first-order level(l)ing;primary level(l)ing
一等水准尺 first-order staff
一等水准点 first-order benchmark
一等水准网 first-order level network;primary level network
一等水准仪 first-order benchmark level
一等网 network of primary station
一等证书 first-class certificate
一等重力网 primary gravity network
一等砖 first quality brick
一滴 drib(b)let;blob <半流质的>
一滴一滴地混合 dribble blending
一地址 one-address
一地址码 one-address code;single address code
一地址指令 one-address instruction; single address instruction
一点流法 one-point method of stream ga(u)ging
一点的邻域 neighbo(u)rhood of a point
一点吊具 single-point suspension spreader
一点儿 vestige
一点儿也不 top or tail
一点法 <测定土的最大干密度的> one-point method
一点分布 zero one distribution
一点记录仪 single-point recorder
一点检查法 one-point detection
一点控制极 one-point gate
一点连续法 one-point continuous method
一点透视 one-point perspective
一点五产业 <介于第一和第二产业之间的产业,如食品加工业等> one-and-half industry
一点一点地 piecemeal
一点应力状态 point stress state
一碘醋酸 monoiodo-acetic acid
一碘化 monoiod(in)ate
一碘化铊 thallous iodide
一碘化铟 indium monoiodide
一碘酪氨酸 mono-iodotyrosine
一碘乙酸 monoiodo-acetic acid
一碘酯 monoiodo-ester
一吊货 draft;draught;hoist;sling
一叠卡片 card deck
一丁基锡 monobutyltin
一丁两顺(交替)砌筑法 Flemish double-stretcher bond
一丁一顺石砌体 <片石填心> emplectum
一定变异 definitive variation
一定变异性 definite variability
一定次数 designated number of times
一定的 designated
一定地区 a given area
一定发育阶段 duration of certain stages
一定级别的技术工人 technical workers with specific grades
一定距离 designated distance;measurable distance
一定量的价值 values of definite magnitude
一定量的商品 given commodity
一定量凝固的劳动时间 definite mas-

ses of congealed labor-time
一定能级的离散吸收 line absorption
一定浓度 finite concentration
一定期限内的投资限额 capital rationing
一定失利 stand to lose
一定时期 a given period
一定数量的种子 a quantity of seeds
一定体积 designated volume
一定条件 controlled condition
一定土层 designated number of layer soil
一定温度 uniform temperature
一定斜度 constant slope
一定形态的劳动量 amount of labor in a certain form
一定形式的调节作用 regulate in a particular manner
一定形状 definite shape
一定重量 designated weight
一冬冰 winter ice
一栋房屋面积 area of a building
一斗 bucketful
一度 <油漆、抹灰等> individual coat
一度(空间)的 unidimensional
一度(皮肤)烧伤 first-degree skin burn
一度漆 a coat of paint;coating
一度烧伤 first-degree burn
一度涂层 individual coat
一度重合存贮器结构 one-dimension memory organization
一端 one aspect;one end
一端搬运 end hauling
一端不通行的走廊 dead-end corridor
一端插销 banana
一端带孔的杆 eye rod
一端带弯钩的钢筋 bar hooked at one end
一端封闭的管子 closed pipe
一端封闭的滚针轴承 needle bearing with closed end
一端固定梁 beam fixed at one end; fixed beam at one end
一端固定一端自由条件 free-clamped end condition
一端加粗异径三通管接 reducing on run and outlet T
一端简支的悬臂梁 propped cantilever
一端铰接杆 rocker-bar
一端铰接杆件 rocker member
一端铰接支柱 rocker column
一端接一端 butt and butt
一端举升器 one-end lift
一端绝缘接头 one end insulated joint
一端螺纹 threaded one end
一端送电 end feeding
一端缩小异径三通管接 reducing on run only T
一端有钩的长杆 pike pole
一端有钩的杆 <立杆用> deadman
一段持续时间 spell;stretch
一段浮选 one-stage flo(a)tation
一段工作时间 spell
一段管子 individual section of pipe
一段轨道电路解锁 single track circuit release
一段楼梯 steps;grece <英国方言>
一段破碎 one-stage crushing;single-stage crushing
一段蠕变 first-stage of creep
一段钛冷器 1st titanium cooler
一段套管柱 joint of casing
一段通信[讯] one-stage communication
一段土方 <开挖渠道> chunk
一段为期两个月的时间 bimester
一段消化脱氮法 single-stage nitrification-denitrification process
一段行程 lap;stretch
一段烟道 flue belt
一段铸塑 one-stage casting
一段转化炉 one-stage converter;primary converter;primary reformer
一段转化炉管 primary reformer tubes
一段钻杆柱 joint of drill pipe
一堆 collection;crowd;lot;one pack
一堆货物单位 <码头上存放的,占地120平方英尺,1平方英尺≈0.0929平方米> lot unit
一堆木料 rick
一堆石子 hurrock
一队 gang;platoon;team
一对 couple;gemel;gymmer;pair
一对半分离房屋 double house
一对磁石 pair of magnets
一对磁铁 pair of magnets
一对电堆 pair of piles
一对多 one-to-many
一对多关系 many-one relationship
一对多线路控制系统 one-to-many route control system
一对回教尖塔 pair of minarets
一对角楼 pair of turrets
一对桥墩 pair of piles
一对双人床 twin beds
一对塔楼 pair of turrets
一对托叶 a pair of stipules
一对一 monogamy;one-for-one;one-one;piece for piece
一对一的 one-to-one
一对一翻译 one-for-one translation
一对一关系 one-one relationship
一对一交换算法 one-for-one exchange algorithm
一对一线路控制系统 one-to-one route control system
一对一映满的映射 one-to-one onto mapping
一对一映射 one-to-one mapping
一对应汇编程序 one-to-one assembler
一对中之一 doublet
一对桩 pair of piles
一吨位置 a ton by measurement
一朵斗拱 <古建筑> corbel bracket arms cluster
一垛干草 rick
一/二等水准测量 first/second-order leveling
一、二两等混合躺椅客车 two class couchette coach
一法配置 one-way layout
一方 pane;party
一方承担义务的协议 hold harmless agreement
一方当事人在场所作出的仲裁裁决 ex-parte award
一方的 unilateral
一方所提出的证据 ex-parte evidence
一方验收取样试验 one-sided acceptance sampling test
一分度 one division
一分时多路传输 <通过一对线路传送多个不同信息的技术> time division multiplexing
一分为二 one divides into two
一分钟管理 one-minute managing
一份 batch;copy
一份分配 portion
一份食品 <尤指半流质的> mess
一氟化铊 thallous fluoride
一氟三氯甲烷 trichloromonofluoromethane
一氟乙酸钠 sodium monofluoracetate
一幅图纸 a piece of drawing
一副 one set;pair
一副铅字 fo(u)nt;type font
一副字符 character font
一个半象限差 sesquiquadrate
一个半柱径式 pycnostyle
一个波长 one wavelength
一个波段的 single band
一个操作过程 cycle
一个超高频转换器 bullet transformer
一个初等量的求值 evaluation of a primary
一个点 <模数参考系统中的> modular point
一个概率分布的方差 variance of a probability distribution
一个回次 round trip
一个回合 round turn
一个设计决定 resolution on a design
一个事件 episode
一个碳单位 one carbon unit
一个往返运行 one round trip
一个循环的进度 advance per attack; advance per round
一个循环作业时间 cycle time;round time <隧洞掘进的>
一个一个地离开 drop-off
一个一组 triad
一个月内的期货汇率 short forward rate
一个整数的分解 partition of an integer
一个周期的样板(列车运行)图 cyclic(al) pattern diagram
一个字母一个字母地 letter by letter
一个自由度 one-degree-of-freedom
一个钻头进尺数 hole-per bit
一根套管 <靠近套管鞋的> starter joint
一工分做制 job-splitting
一钩车(调车) rake of wagons
一构件支护 one-piece set
一鼓一板式(护舷) one cell one board form
一关货 sling
一贯 all along;consistent
一贯单位制 coherent system of units
一贯的 unvaried
一贯的政策 consistent policy;established policy
一贯方针 consistent policy
一贯性(原则) consistency
一贯原则 doctrine of consistency
一贯运输 interior point intermodal
一贯制会计 single system accounting
一罐装 one package;single package
一罐装聚氨酯涂料 one-can urethane coating
一罐装磷化底漆 one-package wash primer
一号车底架端头 end wall car end No.1
一号船边漆 shipside paint I
一号钢丝绳 first grade steel wire rope
一河多桥 multibridge over one river
一河一桥 one bridge over one river
一户的住宅 one-family house
一户结构 one-household structure
一户住宅 one-room apartment; one-room(ed) flat
一环扣一环 closely linked
一回 turn
一回路释压凝汽箱 primary relief tank
一火轧成 roll in one heat
一机多用 a tractor serves several purposes; multiple duty; multiple function
一机拉挤多根型材 multistream profiles
一级 first-order; premium grade; single-stage
一级安全阀 primary safety valve for inner container
一级胺 primary amine
一级保养 first-class maintenance
一级备用容量 <能立即投入的> ready reserves
一级泵站 low-lift pump station
一级编址 direct addressing
一级变量 level-one variable
一级标样 primary standard; primary standard sample
一级标准 primary standard
一级标准气压表 absolute standard barometer
一级标准物质 primary standard material
一级标准样 primary standard sample
一级标准仪器 absolute instrument
一级波 first-order wave
一级采购供应站 first-level purchasing and supply station
一级测潮站 primary tide station
一级差别待遇 first-degree discrimination
一级产业 primary industry
一级潮位站 primary tide station
一级成矿远景区 the first grade of minerogenetic prospect
一级成煤远景区 the first grade of coal-forming prospect
一级程序设计语句 first-level statement
一级齿轮减速 single-stage gear reduction
一级抽样 primary sampling
一级抽样比 primary sampling fraction
一级抽样单位 primary sampling unit
一级除盐系统 primary demineralization system
一级处理 first treatment; primary treatment
一级处理沉淀池 detritus tank
一级传动装置 primary transmission device
一级醇 primary alcohol
一级存储技术 one-level store technique
一级存储器 first-level storage; one-level storage;one-level store;single level memory
一级大银行 prime bank
一级代码 code; one-level; one-level code;one-level storage
一级导数光谱 first derivative spectrum
一级道路 first-class road
一级的 first-class; first degree; one-level;single-stage
一级地址 direct address; first-level address; one-level address; single-level address
一级电池 galvanic element
一级电离 first-stage ionization
一级电离常数 first ionization constant
一级定址 direct addressing
一级动力学反应 first-order kinetic reaction
一级镀锡薄钢板 prime plate
一级颚式破碎机 primary jaw crusher
一级二阶矩法 first-order second moment method
一级发酵 primary fermentation
一级反射 first-order reflection
一级反应 first-order reaction
一级方法 one-level method
一级分叉 primary furcation
一级风 force-one wind;light air
一级风险 primary risk
一级干道 <英> trunk road
一级干线公路网 <英> primary route network (of roads)

一级高压透平 primary high-pressure turbine
一级高压涡轮机 primary high-pressure turbine
一级公路 class one highway; first-class highway; first-class road
一级公路网 primary highway system
一级公式 one-level formula
一级构造 first grade structure
一级光谱 first-order spectrum; primary spectrum
一级光学部件 top-class optic(al) component
一级和二级 firsts and seconds
一级河床 first riverbed; first bottom <正常的洪泛平原>
一级河流 first-order river; first-order stream
一级化学强化处理方法 chemically enhanced primary treatment method
一级环流 primary circulation
一级火警 first alarm
一级技术保养 primary technical maintenance
一级减速 one stage reduction; single reduction
一级减压器 primary reducer
一级检验 first-order test
一级校准法 primary calibration method
一级阶地 first-stage terrace; first terrace; second bottom
一级结构 primary formation; primary structure
一级结构面 grade one discontinuity
一级结构体 grade one texture body
一级结果 first-level outcome
一级近似 first approximation
一级近似理论 first-order theory
一级精度 extra fine grade
一级精度配合 extra-fine fit
一级精密 first-order precision
一级精确度配合 high a class fit
一级刻度 first-level calibration
一级控制点 primary control point
一级矿石 first-class ore; shipping ore
一级浪 force-one wave
一级类链 primary chain
一级连接寄存器 one-level linkage register
一级路线 <英> primary route
一级磨机 primary mill
一级母钟系统 primary master clock system
一级木材 first grade timber
一级木桩 first-class wood pile
一级内齿圈 <行星变速器的> primary annulus
一级能见度 thick fog
一级频率标准 primary frequency standard
一级品 first line; first quality ware; firsts; primes
一级品率 ratio of first-grade products
一级破碎机 coarse crusher; primary crusher; primary disintegrator
一级起步时差 single offset
一级气象站 first-order weather station
一级汽车运输公司 Class One Motor Carrier
一级潜水员 first-class diver
一级驱动 single-stage drive
一级渠道 primary canal
一级燃料 first grade fuel
一级三场编组站【铁】 first-stage/three-yard marshalling station
一级三角测量 primary triangulation
一级生化需氧量 first-stage biochemical oxygen demand
一级生活范围区 primary activity zone
一级生生物强化处理方法 bioenhancement primary treatment method
一级受益收入 primary benefit
一级水手 able-bodied seaman
一级水准点 primary benchmark
一级速度过程 first-order rate process
一级铁路 class one railroad
一级同位素效应 primary isotope effect
一级微分修正 first-order differential rectification
一级维护 first-class maintenance
一级文件 level one file
一级污染物 primary pollutant
一级污染源 primary pollution source
一级污水处理 primary sewage treatment
一级污水处理方法 method of primary treatment of sewage
一级吸附 primary adsorption
一级线纹米尺 first-order metric(al) ga(u)ge
一级相变 first-order phase transition
一级消除动力学 first-order kinetics
一级消除模型 first-order elimination model
一级消化 <指污泥> one-stage digestion
一级消息 first-level message
一级消息成员 first-level message member
一级效益 primary benefit
一级协议 first-level protocol
一级星系 first-rank galaxy
一级行星齿轮 primary planet pinion
一级行星齿轮轴座圈 primary planet ring
一级行星架 primary planet carrier
一级行政中心 town of first administrative importance
一级修正 first-order correction
一级悬浮预热器 one-stage suspension preheater
一级寻址 direct addressing
一级压缩 single-stage compression
一级验潮站 primary tide station
一级氧化处理 first-stage oxidation treatment
一级样本 primary sample
一级用户 Class I consumer
一级油气远景区 the first grade of oil-gas prospect
一级源 primary source
一级战备 No. one alert
一级证券 senior securities
一级支渠 primary canal
一级质量 first quality
一级中断 one level interrupt
一级中断处理程序 first-level interrupt handler
一级主题词 primary term
一级铸件 primary quality casting
一级砖 first-class brick
一级子程序 first-order subroutine; one-level subroutine
一级子例行程序 one-level subroutine
一级租船人 first-class charter
一技之长 professional skill
一加一地址 one-over-one address; one-plus-one address
一加一地址指令 one-plus-one address instruction
一加一地址指令格式 one-plus-one address instruction format
一家的房地产 homestall
一家独用住房 one-family detached dwelling
一甲胺 monomethylamine
一甲基胺 monomethylamine
一价 monovalence; univalence [univalency]
一价的 monatomic; monovalent; univalent
一价的对策 univalent game
一价定律 law of one price
一价基的 monadic(al)
一价碱 monoacid base; monovalent base
一价金的 aurous
一价离子 monovalent ion
一价铊的 thallous
一价铜的 cuprous
一价物的 monadic(al)
一价阴离子 single valence anion
一价银的 argentous
一价元素的 monadic(al)
一碱价的 monobasic
一碱(价)酸 monobasic acid
一碱磷酸钠 monobasic sodium phosphate
一件 a piece (of); article; piece
一件工具 come-along
一件设备 <俚语> come-along
一件一件地 piece by piece; piecemeal
一角商店 <卖便宜货的商店> dime store
一角银币 <美国、加拿大> dime
一绞 hank; skein
一阶 first-order
一阶必要条件 first-order necessary condition
一阶表达式 first-order expression
一阶差(分) <两相邻值之差> first-order difference
一阶差分变换 first difference transformation
一阶差分法 first difference method
一阶差分方程 first-order difference equation
一阶差分数列 first difference series
一阶导数 first derivative; first-order derivative
一阶导数分光光度法 first-order derivative spectrophotometry
一阶的 single order
一阶段抽样 one-stage sampling
一阶段设计 one-phase design; one-stage design; one-step design
一阶段制管工艺 single-stage process
一阶法 first order method
一阶反馈系列 first-order feedback system
一阶反应 first-order reaction
一阶方程 first-order equation
一阶方程组 systems of first order equation
一阶函数值 first-order functional value
一阶环路 first-order loop
一阶矩 first moment
一阶矩矩阵 first moment matrix
一阶理论 elementary theory; first-order theory
一阶力 first-order force
一阶零点 simple zero of order 1
一阶逻辑 first-order logic
一阶马尔可夫过程 first-order Markov process
一阶马尔可夫型 first-order Markov scheme
一阶马尔可夫序列 first-order Markov sequence
一阶面积矩 first moment of area
一阶面积矩法 first moment area method
一阶平稳 first-order stationary
一阶齐次非平稳过程 first-order homogeneous nonstationary process
一阶上同调运算 primary cohomology operation
一阶设计 first-order design
一阶摄动 first-order perturbation
一阶条件 first-order condition
一阶通路 one-level channel
一阶微分方程 differential equation of first order; first-order differential equation
一阶微分灵敏度 first-order differential sensitivity
一阶谓词 first-order predicate
一阶谓词逻辑 first-order predicate logic
一阶谓词演算 first-order predicate calculus
一阶稳定体系 system stable of first order
一阶误差 first-order error
一阶吸附动力学方程 one-order adsorption kinetic equation
一阶系统 first-order system
一阶线性函数 first-order linear function
一阶线性微分方程 linear first-order differential equation
一阶效应 first-order effects
一阶修正量 first-order correction
一阶序列相关 first-order serial correlation
一阶衍射斑点 first-order diffraction spot
一阶因素分析 first-order factor analysis
一阶有限差 first difference
一阶约束品性 first-order constraint qualification
一阶子程序 first-remove subroutine
一阶自回归方程 first-order autoregressive equation
一阶自回归型式 first-order autoregressive scheme
一阶自相关系数 first-order autocorrelation coefficient
一节点内的场 field within a node
一节货车满载量 carload
一节链 shot of a cable
一节杆 rod section
一节烟道 flue belt
一景二像结合的主体印象 <航空摄影术语> stereopair
一九八五年国家高程基准 National Height Datum 1985; National Vertical Datum 1985
一九零七海牙国际法会议 Hague Conference 1907
一九六六年美国公路安全法令 Highway Safety Act of 1966 (US)
一九七一年国际重力基准网 International Gravity Standardization Net 1971
一九五六年黄海高程系统 Huanghai Vertical Datum 1956
一九五四年北京坐标系 Beijing Geodetic Coordinate System 1954
一聚硅盐酸 polysilicate
一卷 bolt; hank
一卷绳索 fake
一卡通 octopus card; smart card; special pass
一开 <木料锯成两半> half sawing
一开间 severy
一开始 first instant
一刻钟 quarter hour
一刻钟制度 quartering system
一孔轻载 one-span light load
一孔重载 one-span heavy load
一控双达标政策 policy of one control; Two-Goals
一口价 one price

Y

一口井的总产量 total capacity of a well
一口体积 bitesize
一跨(的宽度) stride
一块 a piece (of);lot;parcel
一块板道路 single carriageway;single carriageway road
一块半宽的瓦 slate-and-a-half slate
一块地 splat
一块地板 < 常指可掀开的 > floor board
一块地皮 lot of land
一块地(皮的)面积 plottage
一块好林木 a good plot of tree
一块平地 parterre
一块土地 a piece of land; parcel of land
一块园地 a garden patch
一捆 a bundle;bolt;budget;bunch;bundle;one pack;packet < 邮件等的 >
一捆板条 a bundle of laths
一捆匹头货 one bundle of piece goods
一篮子贷款 basket loan
一篮子货币 a basket of currencies
一览表 check list; compendium [复 compendiums/compendia];comprehensive list;conspectus;data sheet; descriptive schedule; directory; glance tabulation; list; schedule; specification; synopsis; table; table look-up; summary; systematic inventory;catalog(ue)
一览图 chorographic(al) map;general chart; general drawing; general map
一揽子 bouquet;package;wholesale
一揽子安排 package arrangement
一揽子保险单 blanket insurance policy;blanket policy
一揽子采购 basket purchase; lump-sum purchase
一揽子承包 package(d) deal;packed deal
一揽子承包的 turnkey
一揽子承包合同 package deal contract
一揽子筹资计划 financing package
一揽子贷款 basket loan;credit package;package loan
一揽子担保 blanket bond
一揽子订购单 blanket order
一揽子定单 blanket order
一揽子发盘 lump offer
一揽子方案 package arrangement
一揽子浮动贷款制度 pool based variable-rate tending system
一揽子购买 basket purchase
一揽子合同 blanket contract; contract package; lump-sum contract; package contract
一揽子汇率 basket of exchange rate
一揽子货币 basket of currencies
一揽子集资方案 financing package
一揽子计划 package plan; package program(me)
一揽子价格 blanket price;flat price; lump price;package price
一揽子建议 package proposal
一揽子交易 package(d) deal;packed deal
一揽子交易合同 package deal contract
一揽子解决办法 overall solution; package solution
一揽子经济体制改革 economy ought to be restructured in a package deal
一揽子决策 decision package
一揽子均衡守则 < 关贸总协定术语 > a balanced package
一揽子免责 catch-all exceptions
一揽子权利要求 blanket claim
一揽子索赔 compound claim
一揽子调整方法 adjustable basket technique
一揽子投标 bid package
一揽子投资 package(d) investment
一揽子项目 all-in-one program(me); package project; umbrella project
一揽子协议 bouquet agreement; umbrella agreement
一揽子信托契据 blanket deed of trust
一揽子信誉保证书 blanket fidelity bond
一揽子许可证 blanket license [licence]
一揽子要价 package bid
一揽子易货 package barter
一揽子运价率 blanket rate
一揽子赠款 block grant
一揽子中心汇率 basket central rate
一揽子租赁 master lease
一类边界外节点 external nodal point of first kind boundary
一类抵押债券 first mortgage bond
一类建筑物 type I important construction
一类容器 first category vessel
一类商品 first category of commodities
一类物资 first category of goods;means of production in category 1; Supplies under Category 1
一立方厘米液体 fluigram(me)
一粒传法 single-seed descent
一链长 cable length
一梁一柱式支架 half frame
一辆接一辆 bumper to bumper
一列 procession;string
一列壁柱 pilastrade
一列舱口 one-row hatch
一列的 in-line
一列料车 string of cars
一列式的 in-line
一列式全息图 in-line hologram
一列式拖带驳船队 barge train;pull-tows in series
一列式拖带法 in-line barge transportation
一列式组装机 in-line assembly machine
一列围墙 stockade
一列栅栏 stockade
一磷化二钴 cobalt phosphide
一令 < 纸张计数单位,一般为 500 张 > ream
一流的 first-class
一流商业票据 prime commercial paper
一硫代磷酸钠 sodium monothiophosphate
一硫二氧化钼 molybdenum dioxysulfide
一硫二氧化物 dioxysulfide
一硫化钡 barium monosulfide
一硫化二铊 thallous sulfide
一硫化二铟 indium monosulfide
一硫化汞 cinnabar(ite)
一硫化锰 manganese monosulfide; manganous sulfide
一硫化铁 iron monosulfide
一硫化物 monosulfide;monosulphide
一硫化铀 uranium monosulfide
一六五矿油 Oleoparaphene
一楼 < 英 > ground floor
一炉 batch
一卤代苯酚 monohalogenated phenol
一律 uniformity;without exception
一氯胺 monochloroamine
一氯苯 monochlorobenzene; monochlorobenzol
一氯丙酮 monochloroacetone
一氯醋酸 monochloroacetic acid
一氯代苯 phenyl monochloride
一氯代乙酸 chloroacetic acid
一氯二氟甲烷 < 制冷剂 R22 > monochlorodifluoro; chlorodifluoromethane
一氯酚 chlorphenol;monochlorphenol
一氯化 monochlor(in)ate
一氯化碘 iodine monochloride
一氯化汞 mercurous chloride
一氯化硫 sulfur chloride
一氯化铊 thallium monochloride; thallous chloride
一氯化物 monochloride
一氯化铟 indium monochloride
一氯甲硅烷 monochlorosilane
一氯三氟甲烷 < 制冷剂 R13 > monochlorotrifluoro; chlorotrifluoromethane
一氯五氟乙烷 chloropentafluoroethane
一氯乙酸 monochloroacetic acid
一氯乙酸盐 monochloroacetate
一氯乙烷 monochlorethane
一轮 round;turn
一满仓 binful
一满匙 spoonful
一满斗 binful;scoopful;shovelful
一门古老的科学 a very old science
一米程差法 method of path difference of one meter
一米处每小时伦琴 roentgen per hour at one meter
一米轨距铁路 meter ga(u)ge railway
一米深温度 one meter depth temperature
一面凹一面平的 concavo-plane
一面凹一面凸的 concave-convex; concavo-convex
一面板暴露 one-side panel exposure
一面光 good one side
一面光硬质纤维板 screen-back hardboard
一面落水(俗称) one-way slope
一面排水路拱 one-way crown
一面平一面凹的 piano-concave
一面平一面凸的 piano-convex
一面坡排水 single drainage
一面坡屋 lean-to
一面坡屋顶 half-span roof
一面倾斜的 single slope
一面数一面分开 count out
一面凸的 anti-clastic
一面凸一面凹的 convexo-concave
一面凸一面平的 convexo-plane
一面楔形砖 feather end
一面有锯齿形凹凸花纹(折光)的玻璃 prismatic(al) glass
一面圆边(釉面)砖 round edge tile
一秒(级)经纬仪 one-second theodolite;one-second transit
一模多铸铸型 family mo(u)ld
一目布尔算符 unary Boolean operator
一目布尔运算 monadic Boolean operation
一目减四元组 unary minus quadruple
一目算子 monadic operator
一目运算 monadic operation; unary operation
一目运算符 unary operator
一幕 scene
一逆 one mistake made in treatment
一年第 30 个最大小时交通量 < 公路设计中,取指定年度中按小时运量的第三十个最高数值作为设计依据 > thirtieth highest annual hourly volume
一年间的 yearly
一年间第 20 个最大小时交通量 twentieth highest annual hourly volume
一年间第 10 个最大小时交通量 tenth highest annual hourly volume
一年两茬连作 double cropping
一年两次的 biannual; semi-annual; semi-yearly
一年两次循环湖 dimictic lake
一年两熟 double cropping
一年内两次以上连作 < 但不间作 > sequential cropping
一年期的长期债务 current maturity of long-term debt
一年期的定期存款 deposits of one-year maturity
一年期的公债 certificate of indebtness
一年三茬连作 triple cropping
一年三熟 triple cropping
一年生苗 yearling
一年生牧草 annual grass
一年生越冬植物 hibernal annual plant
一年生杂草 annual weed
一年生植物 annual plant;fructus industriales; therophyte; yearling; yearly plant
一年生作物 annual crop
一年四茬连作 quadruple cropping
一年四熟 quadruple cropping
一年通过能力 annual throughput
一年一次 yearly
一年一度的 annual
一年一年的 year-to-year
一年一续的保险 one-year renewable term
一年枝 one-year growth twig
一排 procession;windrow
一排板桩 sheet piling
一排半露方柱 pilastrade
一排壁橱 bank of closets
一排壁柱 pilastrade
一排电梯 bank of elevators
一排房屋 tier of blocks
一排房子 range of building
一排建筑物 row of buildings
一排库房 row of stores
一排楼房 block
一排锚链 range of cable
一排汽车库 battery garages
一排商店 row of stores
一排筒仓 battery of silos
一排卧室壁橱 bank of bedroom closets
一排小便槽 urinal range
一排烟囱 chimney tun
一排支座 row of seats
一排桩 row of piles
一排钻孔 coring row
一排座椅 row of chairs;row of seats
一盘 hank
一盘混凝土 batch of concrete
一盘绳索 a fake
一盘文件带 file reel
一盘纸带 coil
一跑楼梯 single-flight stair(case)
一跑楼梯的长度 going of the flight
一批 batch;collection;lot;panel;parcel
一批备用药品 pharmacy
一批次货 bad batch
一批顾客 constituency
一批货(物) a unit of cargo;consignment;one consignment of goods
一批交货合同 single delivery contract

一批料量 lot size
一批零担货物 part-load consignment
一批特别托运的货物 special consignment
一批托运的货物 consignment
一批托运货物 consignment of goods
一批圆木 bunkload
一批制品量 lot size
一皮丁砖 heading course
一皮砖 a course of bricks
一匹布 bolt
一片 pane;piece;wisp;single ply <如胶合板薄层>
一片海藻 field of seaweed
一片积水 piece of water
一片巨大的冰川体 ice field
一片水 sheet of waters
一片土地 parcel of land
一片相同的房屋 tract house
一片一片的 piecemeal
一票货 parcel cargo;parcel lot
一品红 tyraline
一平方米 cent(i)are;one square metre
一平方英里 < =640 英亩> section of land
一期工程 first phase of construction;first-stage construction;first-stage project
一期灌浆孔 primary grout hole
一期混凝土 first-stage concrete
一期修补术 primary repair
一齐跌价 all-round decline
一齐投入 blank operation
一齐涨价 all-round advance
一起发生 conjunction
一起工作的人 fellow worker
一起使用 be used together
一千个字节 kilobyte
一千年 chiliad;millennium
一千万 ten million
一千万亿 <等于 1×10^{15}> quadrillion
一千位 kilobit
一千小时条款 one thousand hour clause
一前一后的 in tandem
一前一后排列 tandem;tandem position
一羟醇 monohydric alcohol
一羟(基)的 monohydric
一羟(基)酚 monohydric phenol
一羟甲基脲 monomethylolurea
一切 bodily;top and tail
一切财产留置权 general lien
一切费用在内价格 all prices
一切海损不保 free of all average
一切海损及救助费均不负责 free of all average and salvage charges
一切海损均不赔偿 free from all average
一切海洋运输及战争险 all marine and war risks
一切海洋运输险 all marine risks
一切其他风险 all other perils
一切顺利 all goes well
一切条款与免除条款 conditions and exceptions
一切通常险 all usual risks
一切为用户服务 customer first
一切险 all risks;all risks insurance
一切险保单 all-risk cover;all risks covers;all risks policy
一切险保险 all risks insurance
一切险保险单 all-risk policy
一切险条款(保险) all risks clauses
一圈 complete turn;one coil;round
一圈长度 length of round
一瘸一拐地走 scuff
一群 cluster;cohort;gang;parcel;squadron
一群人 gang;platoon
一人操作的振动器 one-man vibrator
一人操作(的作业) one-man operation
一人公司 one-man business
一人景象 one-man picture
一人可搬动的石块 one-man (size) stone
一人控制 one-man control
一人能完成的作业 one-man operation
一人手推刮土机 one-man scraper
一人一年的工作量 man-year
一人一年(劳动)当量 man-year equivalent
一人一票 one-man one vote
一人一小时的工作量 man-hour
一人在一班中所做的工作量 manshift
一人住户 one-person household
一日 <到午夜 12 点为止> legal day
一日八小时劳动制 eight-hour's day
一日必需量 minimal daily requirement
一日潮 single day tide
一日贷款 overnight loan
一日的变化幅度 daily variation
一日对冲 day-to-day swaps
一日剂量 daily dose
一日量 daily dose
一日内测潮所得的海平面 daily sea level
一日生化需氧量 one-day biochemical oxygen demand
一日游 day trip
一日装车数 number of cars loaded per day
一色 of the same colo(u)r;plain shade
一色的 self-colo(u)r
一闪信号 glitch
一砷化铀 monoarsenide
一生 life time
一时 for a time
一时性繁荣 boom and bust
一时性黑蒙 amaurosis fugax;blackout
一时性收入 transitory income
一世纪一次的 secular
一式 uniformity
一式八份 in octuplicate
一式九份 in nonuplicate
一式两份 bipartite;duplicate;in duplicate
一式六份 in sextuplicate
一式七份 in septuplicate
一式三份 in triplicate;triplicate;triplicate copy
一式十份 in decuplicate
一式四份 in quadruplicate;quadruplicate
一式四份的文件 quadruplicate
一式四份中的一份 quadruplicate
一式五份 in quintuplicate
一室户 one-room(ed) flat;single room apartment
一室户公寓 efficiency apartment or unit
一室户住宅 one-room dwelling;one-room house
一室体系 one-room system
一室住宅 one-room(ed) flat
一手持砖一手持镘刀的砌砖法 pick and dip
一手可拿的砌块 one-hand block
一手可拿的砖瓦 one-hand tile
一束 a bunch of;budget;bunch;hank;skein
一束熟铁 fag(g)ot
一双 dyad;gemel;pair
一水【船】 able bodied sailor quarter
一水化合物 monohydrate
一水兰铜矾 posnjakite
一水铝石型铝土矿 monohydrate bauxite ore
一水软铝石 boehmite
一水碳酸钠 monohydrated sodium carbonate;thermonatrite
一水铁矾 szomolnokite
一水型铝土矿矿石 boehmite-diaspore ore
一水杨酸一缩二丙二醇酯 dipropytene glycol monosahcylate
一水硬铝石 diaspore
一顺一丁砌式 cross bond
一顺一丁砌砖法 Flemish bond;out-and-in bond;quarter bond
一瞬 <3/100 秒> wink
一瞬间 blink;breathing
一送多受 single feeding and multiple receiving
一送多受轨道电路 single feeding and multiple receiving track circuit
一艘船的航行道 manoeuvring lane;shipping lane
一酸(价)碱 monoacid base
一岁的 yearling
一缩二丙二醇 dipropylene glycol
一缩二甘油 diglycerin
一台机车行车凭证 one engine in steam token
一台机车行车制 <在短支线上> one engine in steam system
一台机器运转一小时工作量 machine-hour
一碳化锆 zirconium carbide
一碳化铌 columbium carbide
一碳化三镍 nickel carbide
一碳化物 monocarbide
一碳化铀 uranium monocarbide
一套 a set;a suit;battery;gang;one set;pair
一套宝石饰物 parure
一套表格 chain-track tractor
一套财务报表 report package
一套穿孔卡片【计】 card deck
一套窗框 sash gang
一套房间 suite;tenement <公寓的>;apartment
一套服装 suit
一套工具 kit
一套公寓房间 <由个人占用的> condominium
一套公寓房间占两层楼 duplex apartment
一套机架 a suite of racks
一套家具 set of furniture
一套拉出器 puller set
一套模板 battery of formwork;battery of shutters
一套铅字 type font
一套筛 battery of screens;battery of sieves
一套筛子 a set of sieves;bank of sieves;nest of screens;nest of sieves
一套相似物 nest
一套小型公寓房间 <有工作室的公寓> studio apartment
一套颜料 palette
一套专用设备 dedicated set of equipment
一套总钥匙 grandmaster-keyed series
一套钻机 set of drills
一套钻头 drill set
一体化城市交通系统 integrated urban transport system
一体化的 integrative
一体化发展 integrated development
一体化反应堆 integral reactor
一体化公共交通系统 integrated public transport system
一体化环境设计 integrated environmental design
一体化经济 integrated economy
一体化区域合作 integrated regional cooperation
一体化直接销售 direct marketing by integration
一体式复合式生物膜反应器 integrated hybrid biofilm reactor
一体式固定化生物活性炭生物反应器 integrative immobilized biological activated carbon bioreactor
一体式固着膜活性污泥法 integrated fixed-film activated sludge process
一体式化学生物絮凝悬浮介质床工艺 integrated chemical biological flocculation-suspended media blanket process
一体式六箱活性污泥反应器 integrative six tanks activated sludge reactor
一体式膜分离 integrated type membrane separation
一体式膜复合生物反应器 integrated membrane-composite bioreactor
一体式膜工艺 integrated membrane process
一体式膜生物反应器 integrated membrane bioreactor
一体式膜生物反应器系统 integrated membrane bioreactor system
一体式膜系统 integrated membrane system
一体式膜序批间歇式生物反应器 integrated membrane-sequencing batch bioreactor
一体式曝气过滤反应器 integrated aeration-filtration reactor
一体式生化反应器 integrated biochemical reactor
一体式生化污水处理装置 integrated biology-chemistry equipment for sewage treatment
一体式生物塘系统 integrated biological pond system
一体式生物质分离膜生物反应器 integrated biomass separation membrane bioreactor
一体式水解酸化好氧生物滤池 integrated hydrolytic acidification-biological aerobic filter
一体式微波/紫外光照法 integrated microwave/ultraviolet-illumination method
一体式污水膜生物反应器 integrated sewage membrane bioreactor
一体式厌氧好氧曝气生物滤池 anaerobic-aerobic integrated biological aerated filter
一体式氧化沟 integrative oxidation ditch
一体式中空纤维膜生物反应器 integrated hollow fiber [fibre] membrane bioreactor
一天贷款 day loan
一天的工作 darg
一天中的工作时间 working day
一条 swath(e);wisp
一条鞭法 single tax in silver
一条龙 a connected sequence;a coordinated process
一条龙协作 integral coordination
一条履带 <制动转向时被驱的> turning track
一条面包 loaf
一条铁路 <美国通常指一条短铁路> pike
一条线 <模量参考系统中的> modular line

一条纵长形的狭窄裂缝 a narrow fissure running lengthwise
一帖 < 四张纸对折成八张一叠 > quire
一桶 bucketful;cask;barrel bulk < 松散物料体积计量单位约合 0.14 立方米 >
一桶的量 pail
一头不通的车库 dead-ended shed
一头不通的(飞)机库 dead-end hangar
一头不通的滑行道 bulkhead taxiway;stub-end taxiway
一头不通的棚 dead-ended shed
一头不通行的走廊 dead-end corridor
一头沉桌 heavy-at-one-end desk;single-pedestal desk
一头和分叉的 45 度 Y 形管 forty-five degree lateral reducing on one run and branch
一头和分叉缩径分叉管 lateral reducing on one run and branch
一头接一头 butt and butt
一头梅花形、一头开口式双头扳手 double-ended multitype wrench with one end open;one end double hexagon
一头缩径分叉管 lateral reducing on one run
一头缩小三通 tee reducing on one run
一头圆针 strongylote
一头胀大的钢管 one end swelled steel pipe
一图形中的顶点 vertex in a graph
一团 tuft
一团糟 mull
一万亿分之几 parts per trillion
一望无际的海洋 unbounded ocean
一维 linear dimension;one dimension
一维表示法 one-dimensional representation
一维波 one-dimensional wave
一维波谱 one-dimensional wave spectrum
一维布朗运动 one-dimensional Brownian motion
一维畴壁 one-dimensional domain wall
一维传递函数 one-dimensional transfer function
一维传热 one-dimensional heat transfer
一维代码 one-dimensional code
一维单形 one-dimensional simplex
一维单元 one-dimensional element
一维单元表示法 one-dimensional element representation
一维的 one-dimensional; unidimensional
一维等离子体 one-dimensional plasma
一维点 one-dimensional point
一维点阵 one-dimensional lattice
一维二次漂移 one-dimensional quadratic drift
一维二相流 one-dimensional two-phase flow
一维放大 one-dimensional magnification;unidimensional magnification
一维非线性波传播 one-dimensional quasi-non-linear wave propagation
一维非线性波传播理论 one-dimensional quasi-non-linear wave propagation theory
一维傅立叶变换 one-dimensional Fourier transformation
一维概率分布 one-dimensional probability distribution
一维固结 one-dimensional consolidation;unidimensional consolidation
一维光栅 one-dimensional grating
一维河口水质模型 one-dimensional estuary water quality model
一维基本形 one-dimensional fundamental form
一维激光器 one-dimensional laser
一维几何形状 one-dimensional fundamental form
一维计算 one-dimensional manipulation
一维剪切波传播 one-dimensional shear propagation
一维晶体 one-dimensional crystal
一维空间 one-dimensional space;unidimensional space
一维空间带宽乘积 one-dimensional space-bandwidth product
一维孔径 one-dimensional aperture
一维孔径综合 one-dimensional aperture synthesis
一维连续性方程 one-dimensional continuity equation
一维邻域 one-dimensional neighborhood
一维流 one-dimensional current; one dimension flow; unidimensional flow
一维流动 one-dimension(al) flow
一维流动理论 streamline theory
一维流形 one-dimensional manifold
一维滤波 one-dimensional filtering
一维模型 one-dimensional model
一维挠群 one-dimensional torsion group
一维碰撞 one dimension collision
一维偏转 one-dimensional deflection
一维偏转调制显示器 one-dimensional deflection-modulated display
一维球面波 one-dimensional spheric(al) wave
一维趋势分析 one-dimensional trend analysis
一维全息图 one-dimensional hologram;unidimensional hologram
一维数学模型 one-dimensional mathematical model
一维数组 one-dimensional array
一维衰坏 one-dimensional failure
一维搜索 linear search; one-dimensional search
一维铁电体 one dimensional ferroelectrics
一维通路 one-dimensional path
一维统计量 one-dimensional statistic
一维问题 one-dimensional problem
一维无序 one-dimensional disorder
一维系统 system with one degree of freedom
一维相关 one-dimensional correlation
一维寻查 unidimensional search
一维循环 one-dimensional cycle
一维一次漂移 one-dimensional linear drift
一维移动加权平均 one-dimensional weighted moving average
一维应力 one-dimensional stress
一维元素 one-dimensional element
一维源 one-dimensional source
一维运动 motion in one dimension
一维运用 one-dimensional operation
一维阵列 one-dimensional array
一维准非线性波传播理论 one-dimensional quasi-non-linear wave propagation theory
一维总线 unidirectional bus
一维最优化 one-dimensional optimization
一位 one bit;single bit;single order
一位处理机 one-bit processor
一位错误校正 single-bit-error correction
一位代码 unitary code
一位计算机 bit machine;one-bit machine
一位加法器 one-column adder; single-digit adder
一位减法器 single-order subtracter
一位群接线器 one-digit group connector
一位时间 bit time
一位数 one-digit number;unidigit
一位数的位置 digit place;digit position
一位数延迟 one-digit delay
一位数字 number with one digit
一位随机表 one-digit random number
一位位移 single-place shift
一卧室单元 one bedroom unit
一屋子的人 roomful
一〇五九 < 一种杀虫、杀螨剂 > demeton
一系列 a series of;round-robin
一系列壁柱 pilastrade
一系列车轮荷载 < 测试桥梁用 > train of loading
一系列的转卖 series of resales
一线的 unilinear
一线多用 joint use
一线检修 first line servicing
一线库场 transit shed and yard
一线前进式 way of straight going
一相毒性 primary toxicity
一箱多批运输 container transport by loading more than one consignments of goods in one container
一向分类 one-way classification
一向分组 one-way classification
一向固结 one-dimensional consolidation
一向延伸 one-dimensional stretch
一硝基苯 mononitro-benzene
一硝基苯胺 mononitraniline
一硝基酚 mononitrophenol
一硝基化 mononitration
一硝基化合物 mononitro-compound
一硝基甲苯 mononitrotoluene
一硝基甲烷 < 喷气燃料 > mononitromethane
一硝基萘 < 油料的去荧光剂 > mononitronaphthalene
一硝基衍生物 mononitro-derivative
一硝酸盐 mononitrate
一小撮 a handful of
一小份 modicum
一小块地 a small area of ground
一小批 packet
一小时定额 one-hour rating
一小时额定出力 one-hour rating
一小时(防火)门 one-hour door
一小时(防火)墙 one-hour wall
一小时功率 one-hour rating
一小时空闲时间的价格 price of an hour's leisure
一小时平均编解车数 average number of cars sorted per hour
一小时值 hourly value;one-hour value
一些表现型 several phenotypes
一些热带土壤 some tropical soil
一些引入材料 some of the introduction
一泻千里 flow down vigorously
一心一意的 single-minded
一星期客票 weekly season ticket
一行 string;swath(e);windrow
一行代码 line code
一行人 party
一型多模铸造法 pattern grouping
一型和二型概率 type I and II probabilities
一型计数模型 counter model type one;type one counter model
一型四十五度弯头 type forty-five degree bend
一溴 monobromo
一溴代苯 monobromo-benzene
一溴代酯 monobromo-ester
一溴酚 monobromophenol
一溴化 monobromination
一溴化萘浸液系统 monobromonaphthalene immersion system
一溴化铊 thallous bromide
一溴化物 monobromide
一溴化硒 selenium bromide
一溴化铟 indium monobromide
一溴醚 monobromo-ether
一溴三氟甲烷 bromotrifluoromethane
一溴乙酰苯胺 monobromacetanilid
一溴樟脑 monobromcamphor
一序列车辆 caravan of vehicles
一学期 semester
一旬 < 10 天 > dekad
一氧化氮 nitric oxide;nitrogen oxide
一氧化氮测定 determination of nitric oxide
一氧化碲 tellurium monoxide
一氧化二氮 nitrogen monoxide; nitrous oxide;oxydum nitrosum
一氧化二钠 sodium monoxide
一氧化二铅 lead suboxide
一氧化二铅定性试验 qualitative test of lead monoxide
一氧化二铅防锈漆 lead suboxide anti-corrosive paint
一氧化二铊 thallous oxide
一氧化二铜 cuprous oxide
一氧化铬 chromous oxide
一氧化钴 cobalt black
一氧化硅 silicon monoxide
一氧化铑 rhodium monoxide
一氧化硫 sulfur monoxide
一氧化卤素 hypohalous
一氧化锰 manganese monoxide; manganous oxide
一氧化钼 molybdenum monoxide
一氧化镍 nickel monoxide; nickel protoxide
一氧化铅 lead monoxide;lead oxide; lead protoxide; litharge; lithargite; massicot; plumb monoxidum; plumbous oxide;yellow lead (oxide)
一氧化铅混合剂 litharge stock
一氧化铅与甘油灰泥 litharge-glycerine mortar
一氧化铅中毒 lithargysmus
一氧化铯 cesium monoxide
一氧化铊 thallium monoxide;thallium oxide;thallous oxide
一氧化碳 carbonic oxide; white damp;carbon monoxide
一氧化碳报警系统 CO-warning system
一氧化碳变换 shift conversion of carbon monoxide
一氧化碳测定 determination of carbon monoxide
一氧化碳测定仪 CO meter
一氧化碳电池 carbonic oxide cell
一氧化碳定量分析 quantitative analysis of CO gas
一氧化碳发生量 carbon monoxide emission
一氧化碳分解 carbon monoxide disintegration
一氧化碳分析仪 carbon monoxide analyser [analyzer]
一氧化碳过滤器 carbon monoxide filter
一氧化碳记录器 carbon monoxide recorder
一氧化碳记录仪 carbon monoxide recorder
一氧化碳监测仪 carbon monoxide

Y

meter;carbon monoxide monitor
一氧化碳检测器 carbon monoxide detector
一氧化碳检测仪 carbon monoxide detector
一氧化碳控制器 carbon monoxide controller;co-controls
一氧化碳连续监测仪 continuous analyzer for carbon monoxide
一氧化碳弥散量 carbon monoxide diffusion capacity
一氧化碳浓度 concentration of carbon monoxide
一氧化碳排放率 carbon monoxide emission rate
一氧化碳容许浓度 permissible concentration of carbon monoxide
一氧化碳污染 carbon monoxide pollution
一氧化碳吸收剂 absorbents of carbon monoxide
一氧化碳消毒罐 carbon monoxide canister
一氧化碳源 source of carbon monoxide
一氧化碳允许浓度 allowable concentration of carbon monoxide;carbon monoxide allowable concentration
一氧化碳指数 carbon monoxide index
一氧化碳中毒 carbon monoxide poisoning
一氧化碳转化器 carbon monoxide converter
一氧化碳自动监测仪 continuous analyzer for carbon monoxide
一氧化碳自救器 carbon monoxide self-rescuer;gas mask
一氧化铁 ferrous oxide
一氧化铜 copper monoxide; copper oxide
一氧化物 mon(o)oxide
一氧化铀 uranium monoxide
一氧化锗 germanium monoxide;germanous oxide
一叶片之量 bladeful
一夜工夫 overnight
一夜之间的 overnight
一一表示模 faithful representation module
一一的函子 faithful functor
一一的模 faithful module
一一对应 monogamy; one-for-one; one-to-one correspondence;one-to-one mapping; one-to-one transformation
一一对应的关系 one-to-one correspondence
一一对应翻译程序 one-to-one translator
一一对应函数 one-to-one function
一一反对应 reciprocal one-to-one correspondence
一一平坦射 faithfully flat morphism
一一全函子 faithfully full functor
一一线性表示 faithful linear representation
一一映射 one-one mapping; one-to-one mapping
一一置换表示 faithful permutation representation
一乙醇胺 monoethanol amine
一乙烯(基)乙炔 monovinyl acetylene
一亿分之几 parts per hundred million
一应俱全 all kinds kept in stock
一英尺长的尺 <1 英尺≈0.3048 米> foot ruler
一英寸板 <1 英寸≈0.0254 米> inch board;inch plank
一英寸比例尺 one-inch scale
一英寸地图 one-inch map
一英寸号钉 twopenny nail
一英寸厚的木板 inch stuff
一英寸木板 one-inch plank
一英寸直径管孔 24 小时的放水量 <约 14 立方米> water-inch
一英里射电望远镜 one-mile telescope
一英里预告标志 one-mile advance sign
一油二硬脂 oleodistearin
一油二棕榈脂 oleodipalmitin
一油清 monoolein
一油酸 monooleate
一铀酸盐 monouranate
一有机会就装船 shipment by first opportunity
一与半选比 one-to-partial select ratio
一与零编码 one and zero code
一与零(输出)比 one-to-zero ratio
一浴法 one-bath method
一遇机会即装运 shipment by first opportunity
一元波谱 one-dimensional wave spectrum
一元伯醇 isobutyl alcohol
一元布尔运算 monadic Boolean operation
一元布尔运算符 monadic Boolean operator
一元材积表 single-entry volume table
一元操作 unary operation
一元操作数 monadic operand
一元产权形式 form of a monistic property right
一元醇 monohydric alcohol
一元的 monadic(al);monobasic;one-dimensional;unary;Unitarian
一元的本利和 amount of one Yuan
一元的拉伸 unitary extension
一元方程 equation in one unknown; simple equation
一元非线性回归方程 non-linear regression equation of one variable
一元分布 univariate distribution
一元分析法 one-dimensional analysis
一元酚 monohydric phenol
一元复利本利和 compound amount of 1 Yuan
一元关系 unary
一元函数 function of one variable; one-variable function
一元化的 unified
一元化管理 centralized management
一元化制度 unitary system
一元回归分析 simple regression analysis
一元继电器 single-element relay
一元假说 unitarian hypothesis
一元碱 monoacidic base
一元经济结构 monistic economic structure
一元流(动) one-dimensional flow
一元论 monism; monophyletic theory;unitarian hypothesis
一元论学说 unitarian theory
一元论者 monophyletist
一元配置 one-way layout
一元式气体热叠合过程 Unitary thermal polymerization
一元数组 unary array
一元素 unicomponent system
一元酸 monoacid; monobasic acid; monohydric acid
一元酸酯 monobasic acid ester
一元酮 monoketone
一元统计学 univariate statistics
一元系 unary system; unicomponent system
一元系统 one-component system;single system;unary system
一元现值 present value of 1 Yuan
一元现值定期付款 periodic(al) payment with present value of one Yuan
一元现值系数 present worth of one factor
一元线性回归方程 linear regression equation of one variable
一元线性回归模型 unary linear regression model
一元相图 one component phase diagram
一元形式 unary form
一元样条函数 univariate spline
一元一次方程式 simple equation
一元应力 one-dimensional stress
一元运算 monadic operation; one-place operation;unary operation
一元运算对象 monadic operand
一元(运)算符 monadic operator; unary operator
一元正态分布 monistic normal distribution
一元指示 monadic indication
一元指示符 monadic indicant
一元子查询 unary subquery
一月解冻 January thaw
一月两次的 bimonthly;semi-monthly
一再 time and again
一闸水 lockfull of water
一站枢纽 one station junction terminal;railway terminal with one station
一站直达货物列车 solid train
一站直达空车列车 solid train of empties
一站直达整装零担车 solid car;straight car
一张 piece
一针 stitch
一阵 slatch
一阵风 slant
一阵寒潮 spell of cold weather
一阵好天气 slatch;spell of fine weather
一阵狂风 flaw
一阵烈风 flaw
一阵顺风 slant of wind
一阵雪丸 graupel shower
一阵阵的 spasmodic
一阵阵地吹 whiffle
一阵阵冒出 belch
一阵子 <工作等的> snatch
一整车 carload
一整套筹资措施 financing package
一整套措施 package
一整套方法 methodology
一整套技术 technologic(al) package
一支车队 road train
一直线钻孔 line drilling
一直向前 go ahead
一直向前推动 push straight on
一直有效 good through
一纸空文 scrap of paper
一指宽 digit
一指(之)长 finger
一指(之)阔 finger
一致 accordance; agreement; all-in-one;coincidence;concurrence;conformation; conformity; correspondence match; fall-in; fall in with; hand in hand;unification;uniformity; union; unite; unity; consecutive consensus <意见等的>
一致逼近 uniform approximation
一致逼空间 uniformly approximated
一致遍历定理 uniform ergodic theorem
一致标度 uniform scale
一致标准 consensus standard
一致表决 unanimous vote
一致裁定 unanimous verdict
一致程度 degree of uniformity
一致程序 consensus proceeding
一致抽样比 uniform sampling fraction
一致创新 congruent innovation
一致存储系统 unified storage system
一致殆周期函数 uniformly almost periodic function
一致单调映射 uniformly monotone mapping
一致的 concordant;congruous;unified;uniform;united
一致的方法 unified method
一致的分配 conforming imputation
一致的证明 certificate of compliance
一致等价函数 uniformly equivalent function
一致等价距离 uniformly equivalent distance
一致等连续集 uniformly equicontinuous set
一致定子空间 uniformly definite subspace
一致法 consensus method
一致范数 uniform norm
一致分布 uniform distribution
一致分数检验 uniform scores test
一致共振 uniform resonance
一致估计 consistent estimate
一致估计量 consistent estimator
一致和满意 accord and satisfaction
一致化算子 unifier
一致基准 uniform reference
一致集合 unification set
一致几何刚度 consistent geometric(al) stiffness
一致校验方程 parity check equation
一致校验矩阵 parity matrix
一致校验数字 parity check digit
一致较大功效检验 uniformly more powerful test
一致较好决策函数 uniformly better decision function
一致接口 uniform interface
一致节点荷载 consistent nodal load
一致结果 consistent result
一致近似法 uniform approximation
一致局部紧空间 uniformly locally compact space
一致局部连通空间 uniformly locally connected space
一致开映射 uniformly open mapping
一致可行方向 uniformly feasible direction
一致空间 uniform space
一致连续性 uniform continuity
一致满意条款 accord and satisfaction clause
一致门 coincidence gate
一致年龄 concordant ages
一致奇偶校验 parity check
一致曲线法 Concordia method
一致曲线上交点 upper Concordia intercept
一致曲线下交点 down Concordia intercept
一致熔融 congruent melting
一致熔融点 congruent melting point
一致熔融化合物 congruent melting compound
一致收敛 uniform convergence
一致数组【计】 uniform array
一致随机变量 uniform random variable
一致弹性 unitary elasticity
一致条件 uniform condition
一致通过 assent and consent; unanimous vote
一致同意 by common consent; unanimity;unanimous agreement
一致同意的文本 agreed text

一致统计量 consistent statistic
一致凸空间 uniformly convex space
一致推算子 consistent estimator
一致拓扑 uniform topology
一致稳定性 uniform stability
一致线 Concordia
一致线图 Concordia diagram
一致限 union bound
一致销售法 Uniform Sales Laws
一致行动 concerted action
一致性 cohesion; compatibility; concordance; conformability; uniformity; consistency [consistence]
一致性比率 consistency ratio
一致性参数 parameter of consistency
一致性操作 consistency operation
一致性程序 consistency routine
一致性错误 consistency error
一致性的约束 consistency constraint
一致性法则 unanimity rule
一致性关系 consistency relation
一致性检查 consistency check; parallel shot
一致性检验 consistency check; consistency test
一致性校核 coincidence checking
一致性校验 consistency check
一致性条件 condition of consistency; consistency condition
一致性系数 coefficient of agreement; coefficient of concordance; consistency coefficient
一致性指标 consistency index
一致性指数 index of conformity
一致性子句 conformity clause
一致序 consistent order
一致延限带 coincident-range-zone
一致有界 uniform bound; uniformly boundness
一致有界变差 uniformly bounded variation
一致有界原理 principle of uniform boundness; uniform boundedness principle
一致原始递归函数 uniformly primitive recursive function
一致原则 consistency principle
一致指标 coincidence indicator; coincident indicator
一致质量矩阵 consistent mass matrix
一致置换 unifier
一致置换符 unifier
一致置换合成 unifying composition
一致转动 coherent rotation
一致装置 consistent unit
一致最大功效不变检验 uniformly most powerful invariant test
一致最大功效检验 uniformly most powerful test
一致最大功效无偏检定 uniformly most powerful unbias(s)ed tests
一致最大功效置信区域 uniformly most powerful confidence region
一致最佳不变风险估计量 uniformly best constant risk estimator
一致最佳检验 uniformly most powerful test
一致最佳距离功效检验 uniformly best distance power test
一致最小方差 uniformly minimum variance
一致最小风险 uniformly minimum risk
一致最优方案 uniformly optimal plan
一致最优方案的不足 lack of uniformly optimal plan
一致最优方案的计算 computation of uniformly optimal plan
一种测样省时技术法 a timesaving technique for measuring
一种滴灌系统 a trickle irrigation system
一种钙化物 a calcium compound
一种概率分布的均值 average value of a probability distribute
一种概率分布的期望值 expected value of a probability distribution
一种干黏[粘]土石 paretta
一种高塑性黏[粘]土 botting clay
一种红外线探测器 pyroscan
一种化学元素 a chemical element
一种快速法 a rapid method
一种快速原型 a rapid prototype
一种立体悬挑的蜂窝状装饰 honeycomb vault
一种沥青和铺路材料混合物 colasmix
一种耐火黏[粘]土水泥 pyruma
一种情况接情况的基础 a case-by-case basis
一种三步骤过程 a three-step process
一种试验方法 a testing program(me)
一种酸性溶液 an acid solution
一种同步驱动发电机 motor-torque generator
一种铜制扁圆形挠性活塞环 oblate ring
一种文字图例 monolingual legend
一种新式化学方法 a new chemical method
一种絮凝剂 magnafloc
一种淹水装置 a water bath system
一种最好的方式 one best-way
一周 hebdomad; one week
一周长度 length of round
一周的总工时 workweek
一周工作计划 schedule of work to be performed during a week
一周年割面 yearling face
一周岁至两周岁之间的动物 yearling
一轴的 uniaxial
一轴负光性 uniaxial negative character
一轴晶 optic(al) uniaxial crystal; uniaxial crystal
一轴晶干涉图 monoaxial interference figure; uniaxial interference figure
一轴晶光率体 indicatrix of optic(al) mono-axial crystal
一轴伸长一轴缩短型应变椭圆 one axis extend and another compressed type
一轴正光性 uniaxial positive character
一肘之地 elbow room
一昼夜的 intradiel
一昼夜的限额 days quota
一昼夜的消耗 day's expenditures
一砖半厚墙 brick-and-a-half wall; one-and-half brick wall [37cm brickwall]
一砖半墙 one-and-half brick wall [37cm brickwall]
一砖(厚的)墙 heading bond; one-brick (masonry) wall; whole-brick wall
一转 complete turn
一字不改 word for word
一字(槽)螺钉 slotted head screw
一字槽螺丝帽 slot head; straight slot head
一字槽头螺钉 minus screw
一字尺 parallel ruler
一字锚泊 ordinary moor
一字头(螺杆泵) manifold
一字线 line shaped track
一字形冲击式钻头 chisel bit
一字形冲击钻头 spade-type bit
一字形的 single-edged
一字形钢板桩 straight-web steel sheet pile
一字形钎头 bull bit; single chisel bit; straight bit; two-point bit
一字形桥台 head wall abutment; straight abutment
一字形整体钎 chisel-shaped steel
一字形整体钎杆 chisel rod
一字形钻头 bull bit; single chisel bit; straight bit; two-point bit
一字闸门 single revolving gate
一自由度系统 one-degree-of-freedom system; system with one degree of freedom
一宗 a parcel
一组 battery; cluster; gang; one group; one pack; one set; panel; platoon; squadron; suite; team; volume
一组存储器 storage stack; store stack
一组独立式房子 free-standing block
一组服务台 storage
一组搁板 shelving
一组锅炉 battery of boilers
一组荷载 group of loads
一组缓冲 pool of buffer
一组货车 rake of wagons
一组卡片 card deck
一组码头 quayage
一组炮眼 hole set; round of holes
一组炮眼布置 layout of round
一组曲线 curve family; series of curves; set of curves
一组试验筛 stack of test(ing) sieves
一组梯形座排 <古剧场> cuneus
一组问题 questionnaire
一组细筛 nest of sieves
一组线脚 battery mo(u)ld
一组线脚的最后一道线脚 thickness mo(u)lding
一组仪器 kit
一组运输工具 battery of hauling equipment
一组转辙器 pair of switches
一组装饰线条 battery mo(u)lding
一组钻孔 slabbing round

伊

伊奥尼亚海 Ionian Sea
伊达矿 idaite
伊登坩埚 crucible
伊登阶【地】Edenian
伊碲镍矿 imgreite
伊丁石 iddingsite; oroseite
伊豆-小笠原海沟 Izu-Ogasawara trench
伊尔 <非洲> Zaire
伊尔布兰德减摩合金 Eel Brand antifriction metal
伊尔弗拉库姆层 Ilfracombe bed
伊尔格纳型飞轮 Ilgner flywheel
伊尔加恩陆核 Yilgran nucleus
伊尔加因古陆 Yilgran old land
伊尔科维奇方程 Ilkovič equation
伊尔库茨克极性超时 Irkutsk polarity superchron
伊尔库茨克极性超时间 Irkutsk polarity superchronzone
伊尔库茨克极性超带 Irkutsk polarity superzone
伊尔铝合金 Earlumin; Earlumin alumin(i)um alloy
伊冯光度计 Yvon photometer
伊盖杜尔铝合金 Igedur
伊高尔特水准仪 <法国制造> Egault level
伊格尔-皮切尔型起泡试验箱 Eagle-Picher blister box
伊格尔装置 Eagle mounting
伊格尼尔定律 Egnell's law
伊格塔洛伊钨钴硬质合金 Igatalloy
伊管 itron
伊硅钙石 istisuite
伊硅钠钛石 ilmajokite
伊红 eosin
伊红染液 eosin stain
伊加斯玛琋脂 <一种沥青玛琋脂止水材料的商品名> Igas
伊奎帕特雨 equiparte
伊拉克国家石油公司 Iraq National Oil Company
伊拉克石油公司 Iraq Petroleum Company
伊拉克中央银行 Central Bank of Iraq
伊兰合金 electron metal
伊朗国际财团 Iranian Consortium
伊朗国家石油公司 National Iranian Oil Company
伊朗国家银行 Bank Markazi Iran; National Bank of Iran
伊朗建筑 Iranian architecture
伊朗梅里银行 Bank Melli Iran
伊朗石油合资者有限公司 Iranian Oil Participants Ltd
伊朗式脚手架 ballic pole
伊雷巴任公式 Irabarren formula
伊里诺冰期 Illinoian
伊里亚阶 <中泥盆世>【地】Erian (stage)
伊里亚统【地】Erian
伊里亚运动 <中泥盆世>【地】Erian movement
伊里亚造山运动 Hibernian orogeny; Erian orogeny
伊丽莎白港 <南非> Port Elizabeth
伊丽莎白建筑形式 Elizabethan style
伊丽莎白式建筑 Elizabethan style
伊利姆合金 Illium
伊利黏[粘]土 grundite
伊利诺斯 <美国州名> Illinois
伊利诺斯大陆公司 <美> Continental Illinois Corporation
伊利诺斯大陆银行 <美> Continental Illinois National Bank
伊利诺斯中央海湾铁路公司 <美> Illinois Central Gulf Railroad Company
伊利色斯神庙 Temple on the Ilissus
伊利石 glimmerton; grundite; hydromuscovite; ledikite; shiny clay; illite
伊利石结构 illite structure
伊利石黏[粘]土 illite clay
伊利(石)水云母 illite hydromica
伊利铁路公司 <美> Erie Railroad Company
伊罗科木 iroko
伊洛瓦底江 Irrawaddy River
伊梅迪姆砂滤池 Immedium sand filter; upward-flow sand filter
伊梅冈镜头 Imagon lens
伊瑞克提翁神庙 <古希腊> Erechtheion
伊舍伍德框架 Isherwood frame
伊舍伍德体系 Isherwood system
伊士托循环接头 Easto circulating sub
伊斯顿红 Easton red
伊斯兰堡 <巴基斯坦首都> Islamabad
伊斯兰大教堂 jami
伊斯兰教建筑 Islamic architecture; Mohammedan [Muhammadan] architecture; Muslim architecture
伊斯兰教经学院 Islamic academy
伊斯兰教历 Mohammedan calendar; Moslem calendar
伊斯兰教寺院 masjid; mosque
伊斯兰教寺院光塔 minaret
伊斯兰教寺院中为妇女设的席位 tecassir
伊斯兰教中要求祈祷壁龛朝向麦加的方位 qibla
伊斯兰开发银行 Islamic Development Bank
伊斯兰寺院天井 sah
伊斯门测斜仪 Eastman survey instrument

伊斯门造斜涡轮 Eastman whipstock turbine
伊斯坦布尔的蓝清真寺 Blue Mosque at Istanbul
伊斯特尔坟墓 Easter sepulchre
伊索克石 isokite
伊特鲁斯坎风格 Etruscan style
伊特鲁斯坎建筑 <古意大利> Etruscan architecture
伊特鲁斯坎建筑的柱 Etruscan column
伊特鲁斯坎庙宇 Etruscan temple
伊藤石 itoite
伊万诺夫柔性路面设计法 Ivanof's flexible pavement design method
伊万-萨葛抗弯强度表达式 <对弯扭联合作用下铰的> Evant-and-Sarkar expression
伊万斯-伯尔单位 Evans-Burr unit
伊万斯带钢轧机 Evans mill
伊万斯分级机 Evans classifier
伊文式人力合抱取土螺钻 manual soil auger of lwan pattern
伊蚊 aedes
伊希斯神庙 Temple of Isis
伊辛模型 Ising model
伊辛耦合 Ising coupling
伊泽特非时效钢 <锅炉用软钢> Izett steel
伊兹桥 <1867～1874 年美国密西西比河上第一座用悬臂法建造的钢拱桥> Eads Bridge

衣

衣阿华水利研究所 <美> Iowa Institute of Hydraulic Research

衣氨酸 isamic acid
衣橱 clothes closet; dressing locker; wardrobe
衣橱挂衣钩 closet bar
衣橱间 dressing-locker room
衣橱门执手 closet knob
衣橱门执手心轴 closet spindle
衣橱式卧室壁橱 bedroom closet bank
衣服 clothing; habiliments
衣服干燥室 clothes drying cabinet
衣服烘干机 clothes dryer [drier]
衣服沾染监测器 clothing monitor
衣钩 coat hook
衣冠冢 cenotaph
衣柜 armoire; chest of drawers; commode; garderobe; kas; wardrobe
衣夹 peg
衣架 coat hanger; hanger
衣架货柜 hanging container; hanging garment container
衣架模 coat hanger die
衣酒石酸 itatartaric acid
衣康酸 itaconic(al) acid
衣康酸二甲酯 dimethyl itaconate
衣康酸酐 itaconic(al) anhydride
衣康酸钠 sodium itaconate
衣兰油 anona oil; ylan-ylan oil
衣料段 end
衣料麻纱 dress linen
衣帽壁橱 coat closet; wardrobe closet
衣帽壁橱墙板 wardrobe type closet bank
衣帽钩 coat-and-hat hook; hat-and-coat hook
衣帽柜 dressing locker; locker
衣帽(寄放)间 check room
衣帽架 cloak stand; clothes stand; clothes tree; hall stand; hall tree; hat tree; pin rail
衣帽间 cloakroom; coat room; locker room
衣帽存放间 check/cloak room
衣帽间锁柜 cloakroom locker
衣帽间围屏 cloak screen
衣帽室围屏 cloak screen
衣坯 garment blank
衣片 garment piece
衣苹酸 itamalic acid
衣散油 Isano oil; ongueka oil
衣上垂片 tab
衣室 wardrobe
衣物袋止挡 <铁路卧铺上的> rack catch
衣物柜环 locker ring
衣物间 coat space; locker room
衣下烧伤 under clothing burn
衣箱 suitcase; trunk
衣藻属 chlamydomonas
衣着 apparel
衣着费 clothing expenses

医

医疗 medical treatment

医疗保健 medical care
医疗保险 hospitalization insurance; insurance for medical care; medical coverage; medical insurance
医疗保险费 hospitalization premium; medical insurance premiums
医疗保险基金 medical benefits fund
医疗保险制度 system of medical care insurance
医疗保障 medical security
医疗保障方案 Medicare plan
医疗补贴 medical subsidies
医疗补助计划 Medicaid
医疗车 medical vehicle
医疗船 floating hospital; hospital ship; medical boat
医疗单位标志 caducei symbol
医疗地质学 medical geology
医疗队 medical corps; medical team
医疗费保险 medical expense insurance
医疗费(用) hospitalization cost; medical and dental expenses; medical service; medical payment <美国汽车保险项目之一>
医疗服务 medical service
医疗管理 medical control
医疗光学仪器 medical optic(al) instrument
医疗合同 hospital service contract
医疗护理 medical assistance; medical attention
医疗化学 therapeutic chemistry
医疗机构 medical institution
医疗记录 medical record
医疗监督 medical control
医疗建筑 medical building
医疗津贴 medical benefits
医疗救护站 clearing station
医疗矿泉 healing spring; medicinal spring
医疗矿泉水标准 quality standard of mineral spring water for medical treatment
医疗密封舱 <沉箱病防治用> medical lock
医疗气闸 medical air lock
医疗器材 medical appliance
医疗器械 medical apparatus; medical apparatus and instruments; medical appliance
医疗器械厂 medical apparatus and instruments factory
医疗区段 physics block
医疗泉 medical spring; medicinal spring
医疗设备(施) medical facility; medical installation
医疗设施系统 hospital service system
医疗体育室 therapeutic gymnastic room
医疗图表柜 medical chart cabinet
医疗卫生 health and medical community
医疗卫生工作 medical and health work
医疗卫生建筑 medical treatment and sanitation building
医疗卫生人员技术服务 technical service by medical and health workers
医疗卫生系统 health system
医疗休养环境 medical recuperate environment
医疗仪器及器械研究所 institute of medical instruments and apparatus
医疗椅 medichair
医疗用壁灯 wall hospital luminaire fixture
医疗用气压舱 medical air lock
医疗用气压室 medical air lock
医疗用照明装置 wall hospital luminaire fixture
医疗援助 medical assistance
医疗闸 hospital lock; medical lock <沉箱的>
医疗站 health center [centre]; medical station
医疗照射 medical exposure
医疗诊断书 medical certificate
医疗中心 medical center [centre]
医生 doctor; physician
医生证明书 medical certificate
医师 physician
医师办公室 doctor's office
医士学校 medical assistant's school
医务部 department of medical administration
医务部门 medical department
医务部主任 head of the department of medical administration
医务车 hospital car
医务处 health service; medical department
医务船 hospital ship
医务的 medical
医务劳动鉴定委员会 committee on medical appraisal of labor fitness
医务室 doctor's office; medical room
医务所 clinic; dispensary; firmary; infirmary
医务艇 ambulance/clinic boat; medical launch
医务助理人员 paramedic
医学地质学 medical geology
医学机构 medical establishment
医学检查 medical check-up; medical examination
医学评价 medical assessment
医学数据库 medical data base
医学院 medical faculty
医药 physic
医药的 medical
医药费 medical charges; medical cost; medical expenses
医药柜 medical locker
医药化学 medicinal chemistry
医药卫生补助金 medical and health subsidy
医药卫生费 medical and dental expenses
医药用品商店 drug and hospital equipment store
医用包 medical bag
医用绷带 medical bandage
医用橱(柜) medical cabinet
医用气闭室 <用于治疗减压病的特殊房间> medical lock
医用手套 medical gloves
医用温度计 clinical thermometer
医用洗手盆 surgical lavatory
医用显微镜 medical microscope
医用箱 medical case
医用遥测计 medical telemeter
医用原子反应堆 medical atomic reactor
医用真空泵 medical vacuum pump
医院 health building; hospital
医院壁灯 wall-mounted hospital luminaire fixture
医院壁装式照明装置 wall-mounted hospital luminaire fixture
医院病床 hospital bed
医院病床提升器 hospital bed lift
医院病房 hospital ward; ward
医院船 hospital ship
医院的带臂门拉手 hospital arm pull
医院废水 hospital wastewater
医院妇产科 mother and baby unit
医院供暖 hospital heating
医院固体废物 hospital solid waste
医院计算机系统 hospital computer system
医院家具 hospital furniture
医院建筑 hospital architecture
医院洁净室 hospital clean room
医院空气调节 hospital air conditioning
医院门 hospital door
医院门诊部 hospital clinic
医院手术室 hospital operating room
医院输送舰 hospital transport
医院污水 hospital wastewater
医院污水处理 hospital sewage treatment; hospital wastewater disposal
医院污水污染源 source of hospital wastewater
医院血库 hospital blood bank
医院银行 hospital bank
医院用床 hospital bed
医院院长 director of hospital; warden <英>
医院诊所 hospital clinic
医院肘臂执手 hospital arm pull

依

依阿华法 <一种采用低坍落度密实混凝土罩面的方法> Iowa method

依比例尺空中基线距离 scale distance of air base
依比例符号 symbol by proportional point
依比重分层 gravity segregation
依测度收敛 convergence in measure
依常例顺序 in regular turn
依此法顺延 be prolonged accordingly; be prolonged in a similar manner
依次传送制 sequential system
依次淬火 progressive hardening
依次的 ordinal; round-robin
依次滴定 stepwise titration
依次叠进 telescope
依次灌水 irrigation in their given order
依次进行 succession
依次排列的 in tandem
依次取样检验 sequential sampling inspection
依次缩小行间距离的挂瓦法 diminishing course work
依次弹射 sequence ejection
依次先后排列施工 tandem work
依次相连的 consecutive
依次溢出 progressive overflow
依次张拉的应力损失 sequence-stressing loss
依次转接电路 sequential switching circuit
依次装货 loading in turn
依从关系 dependence
依存度 degree of dependence

依存分化 correlative differentiation
依存关系 dependence relationship
依存效应 dependence effect
依存因子 dependent factor
依存于位置的 location-dependent
依地酸二钠钙 disodium calcium ethylene diamine
依地酸钴 cobalt edetate
依地形布置的道路网 topographic(al) street system
依地形低空飞行 contour-chasing
依法 ipso jure
依法惩处 impose punishment in accordance with the law
依法登记的名称 legal name
依法登记的住所 legal residence
依法减税 tax cut by law
依法律观点 in the eye of the law
依法取消抵押品赎回权 statutory foreclosure
依法无效 null and void
依法献地 statutory dedication
依法行政 administration according to law
依法占有 seizure
依法占有的财产 seizin
依法占有土地 seisin;seizen
依法执行 prosecute
依法终止合同 ipso jure termination of a contract
依法仲裁 arbitration in law
依法仲裁员 arbiter juris; arbitrator de jure
依法准许 duly admitted
依分布收敛 convergence in distribution
依附 cling;dangle
依附点 adherent point
依附方案 dependent alternative
依附关系 relations of dependence
依附结构 dependence structure
依附力 adherence
依附图 adjoint
依概率收敛 convergence in probability
依格达胶 <光弹模型材料> igdantin
依格泡 <表面活性剂> igepons
依工作种类授权 delegation of authority by type of work
依国际海事组织 international maritime organization
依航标航行 point-to-point
依结果管理 management by results
依据标志【测】 reference mark
依据的法律 governing law
依据点 reference point
依据概率预算 probabilistic budget
依据合同或契约的诉讼 action ex contract
依据验货付款 as inspected; as maintained
依据轴 reference axis
依据主要文件名称 name of main documents
依据资历提升 promotion by seniority
依据资料 background information
依靠工资为生者 payroller; wage earner
依靠计算装置的进行顺序 computer-aided approach sequencing
依靠进口 depend on import
依靠空气和水 depending air and water
依靠农业为生的人口 population dependent on agriculture
依靠人的因素 reliance on the human element
依靠投机或高利贷发财的人 money spinner
依靠阻力分离 drag-separating
依来铬黑 eriochrome black
依来铬红 eriochrome red
依来铬蓝黑 eriochrome blue black
依赖保持 dependency preservation
依赖的等价 equivalence of dependency
依赖关系 dependency relationship
依赖关系图 dependence graph
依赖进口 dependence on import
依赖外国船的进出口贸易 passive commerce
依赖性 dependence
依赖性公理 axiom for dependency
依赖性试验 dependence test
依赖性需求存货系统 inventory system for dependent demand
依赖于机器的特性 machine-dependent feature
依赖于密度的因素 density-dependent factor
依赖于震级的峰值加速度 magnitude-dependent peak acceleration
依赖于状态的服务 state-dependent service
依赖域 domain of dependence
依曼劳硬瓷 <德国> Ilmenau porcelain
依毛利计算 calculated on gross profit
依平均值收敛 convergence in mean
依期付款 payment in due course
依墙柱 engaged column
依时风应力 time-dependent wind stress
依时海流 time-dependent ocean current
依时效而取得的权利 prescriptive rights
依时运动 time-dependent motion
依氏烧瓶 Erlenmeyer flask
依数法 dependence method
依数效应 colligative effect
依数性 colligative property
依数性质 colligative property
依斯特瓷器 <软瓷,意大利> Este porcelain
依斯特曼彩色胶片 Eastman colo(u)r
依托服务城市 depending on the city and serving the city
依瓦尼姆铝合金 Ivanium
依限付款 pay on due date
依限提出 due presentment
依序整理 collating
依约废止 denunciation
依约照付 pay as may be paid
依照票面价格 at par
依照入息支付 pay-as-you-earn
依照通知 as per advice
依照习惯做法 conform to conventional practice
依照先例 stare decisis
依照协定 as agreed
依指数定律衰变 exponential decay

铱

铱铂合金 irid(i)oplatinum; platinoiridium
铱的 iridic
铱锇合金 iridosmine; osmiridium
铱锇矿 iridosmine; osmiridium
铱黑 iridium black
铱金 iraurite
铱矿 iridium ores
铱丝 iridium wire

仪

仪表 apparatus; appliance; device; instrument; instrumentation; measurement device; measuring appliance; measuring device; meter; telltale device
仪表百叶箱 instrument shelter
仪表板 board; console; control cluster; dash board; dash panel; dial plate; disposition board; facia panel; fascia [复 fa(s)ciae/fa(s)cias]; fascia board; ga(u)ge board; ga(u)ge plate; ga(u)ging board; instrumentation console; instrument board; instrument cluster; meter board; meter panel; operation cluster; panel; panel board
仪表板玻璃字盘 instrument-board dial glass
仪表板衬 instrument-panel gasket
仪表板灯 dashboard light; instrument lamp; dashlight; instrument board lamp; instrument cluster lamp; instrument-panel light; panel light
仪表板时钟 instrument-board clock
仪表板托架夹 instrument-board bracket clamp
仪表板小灯 panel lamp
仪表板信号灯 instrument-panel warning light
仪表板应急照明电门 emergency panel lights switch
仪表板照明灯 dashboard lamp; instrument-panel lamp; panel light; dashlamp
仪表板照明指示灯 instrument light
仪表板振动器 panel vibrator
仪表板支架 instrument-panel bracket; instrument-panel support
仪表板装置 dash unit
仪表板总成 panel assembly
仪表包装 meter enclosure
仪表笔 instrument pen
仪表壁龛 meter niche
仪表编号 meter number
仪表变压器 instrument transformer
仪表标定 meter calibration
仪表标度盘 instrument dial
仪表玻璃 apparatus glass; instrument glass
仪表玻璃标度盘 instrument glass dial
仪表布线 electric(al) wiring
仪表布置 instrumentation
仪表舱 guidance section; instrument bay
仪表操纵板 ga(u)ge plate; instrument panel
仪表操作 instrumental operation
仪表操作人员 instrument man
仪表操作台 control console; instrumentation operation station
仪表测量 instrument survey
仪表测量系统 instrumentation system
仪表测试按钮 ga(u)ge test button
仪表测试机械 meter testing machine
仪表测试设备 instrumentation
仪表插头 instrument plug
仪表长度 meter run
仪表常数 apparatus constant; instrumental constant; instrument constant; meter constant
仪表厂 instrument and meter plant
仪表车 instrument truck
仪表车床 instrument lathe
仪表程序 instrument program(me)
仪表程序控制性 instrument programmability
仪表齿轮 instrument gear
仪表储藏室 meter niche
仪表导航 blind navigation
仪表导降系统滑道 instrument landing system glide path
仪表导线 instrument lead
仪表的 instrumental
仪表的测量机构 meter movement
仪表的可调支架 adjustable instrument mounting
仪表的滞后 instrument lag
仪表灯 ga(u)ge lamp; instrument light; panel light
仪表地窖 meter pit
仪表电动机 instrument motor
仪表电路 appliance circuit; instrument circuit; metering circuit
仪表电绳 instrument cord
仪表电阻器 instrument resistor
仪表读数 dial reading; ga(u)ge reading; instrument reading; meter reading; reading
仪表读数偏差 variation of instrument reading
仪表读数校正 index correction
仪表读数自动记录 data logging
仪表读数自动记录装置 logger
仪表端子 instrument terminal
仪表惰性 instrument lag
仪表飞行规则 instrument flight rules
仪表飞行气象条件 instrument meteorological condition
仪表飞行条件 instrument flight rules condition
仪表飞行图 instrument chart; instrument flying chart
仪表分接点 ga(u)ge-tapping point
仪表分流器 instrument shunt
仪表封装 meter enclosure
仪表服务部 instrument service division
仪表附加电阻 instrument multiplier
仪表高度 indicated altitude
仪表工程学 apparatus engineering; instrumentation engineering
仪表工地外露情况 field exposure condition
仪表工业 instrumentation; instrument industry
仪表工作范围 working range
仪表工作状态 instrument mode
仪表观测 instrumentation; instrument observation
仪表管路 instrument piping
仪表柜 meter compartment
仪表过量程防护 overrange protection
仪表盒 instrument case; meter box
仪表盒子 meter enclosure
仪表滑行道 instrument runway
仪表化(观测) instrumentation
仪表化钻进 monitored drilling
仪表活接头 ga(u)ge union
仪表机械工 instrument mechanician
仪表及其控制方法 instruments and controls
仪表记录 instrument record
仪表记录摄影机 instrument recording camera
仪表记录照相 instrument recording photography
仪表技术 instrumental technique
仪表技术员 instrument technician
仪表继电器 instrument relay
仪表架 dial holder; instrument stand; meter rack
仪表间 instrument and meter workshop; instrument cubicle; instrument room; meter compartment; meter room
仪表监测 instrumentation
仪表鉴别器 instrument discriminator
仪表鉴定 instrument identification
仪表校正 adjustment of instrument; instrument calibration; regulating
仪表校正程序 instrument calibration procedure
仪表校正因数 meter correcting factor
仪表校准程序 instrument calibration procedure
仪表接地 instrumentation grounding

仪表接头 instrument adapter [adaptor]
仪表接线端子 instrument terminal
仪表接线柱 instrument terminal
仪表截止阀 instrument block valve
仪表进场导航图 instrument approach procedure chart
仪表进场着陆 instrument approach
仪表精密度 precision of instrument
仪表精(确)度 accuracy of instrument;instrument precision
仪表开关 instrument switch; meter switch
仪表壳 meter housing
仪表刻度 meter [metre] scale; scale of a meter [metre]
仪表刻度盘 dial of meter; ga(u)ge dial; indicating head; meterman; meter [metre] dial
仪表刻度校正 index correction
仪表孔 instrument port
仪表控制 instrument control
仪表控制板 instrument control panel
仪表控制架 instrumentation control rack
仪表控制器 instrument controller
仪表扩程器 instrument multiplier;voltage multiplier;voltage range multiplier
仪表廊 instrument gallery
仪表量测廊道 metering gallery
仪表量程 instrument range; range of (an) instrument
仪表量程倍增器 meter multiplier
仪表灵敏度 ga(u)ge factor; meter sensitivity; sensitivity of instrument
仪表灵敏系数 ga(u)ge factor
仪表零点 ga(u)ge zero
仪表零位调整 zero setting
仪表领航 blind navigation
仪表领航图 blind navigation chart
仪表满标值 end scale value
仪表面 dial
仪表面板 dial needle;instrument face panel;meter panel;panel
仪表目镜十字线 instrument cross hair
仪表盘 dash board; ga(u)ge board; ga(u)ge panel; instrument board; instrument panel; meter board; metering switch board; meter panel; operation cluster
仪表盘灯 instrument light;panel lamp
仪表盘电线束 instrument panel harness
仪表盘盖 dash cover
仪表盘装置 ga(u)ge unit
仪表跑道 instrument runway
仪表配备 instrumentation
仪表配置 instrument array
仪表皮革 <无漏孔性> meter leather
仪表屏 instrument panel;meter panel
仪表歧管 ga(u)ge manifold
仪表气源 instrument air
仪表器具用涂料 appliance finish
仪表前盖 bezel
仪表区段 instrument range
仪表容器 instrument capsule
仪表塞门 ga(u)ge cock
仪表设备 instrumentation
仪表设施 metering device
仪表示值读数 readout
仪表式继电器 instrument-type relay; meter-type relay
仪表饰带卡环 instrument fascia bezel
仪表饰框 instrument luster
仪表室 instrument cubicle;instrument room; ga(u)ge chamber; meter cabinet; meter compartment; recorder house;control room

仪表数字技术 instrument digital technique
仪表伺服机构 instrument servo mechanism
仪表伺服系统 instrument servo
仪表速度 speed of instrument
仪表速率 rate of instrument
仪表台 instrumentation console; instrument desk;instruments and meters table
仪表天气 instrument weather
仪表调节装置 adjustable instrument mounting
仪表调整者 meterman
仪表图 meter diagram
仪表外壳 metre case
仪表网络 instrument network; network of instruments
仪表维修规程 instrument maintenance code
仪表误差 instrument(al) error; meter error;site error
仪表系统 instrumentation system
仪表系统图 instrumentation diagram
仪表弦控指针 string pointer
仪表显示 panel display
仪表显示屏 meterman
仪表显示损耗 meter loss
仪表线路 appliance circuit
仪表线路测量 instrument route survey
仪表线路示意图 block diagram
仪表线圈 meter coil
仪表箱 instrument box; instrument case;meter box;meter compartment; meter enclosure
仪表信号灯 instrument light
仪表型继电器 meter-type relay
仪表修理间 instrument workshop
仪表修整速度 <飞机> calibrated air speed;rectified air speed
仪表学 instrumentation
仪表引线槽板 instrument duct
仪表应用 instrumentation
仪表用保护电路 meter protection circuit
仪表用变流器 instrument converter
仪表用变压器 meter transformer; potential transformer
仪表用电路 appliance circuit
仪表用风 instrument air
仪表用互感器 instrument transformer
仪表用计算机 instrument computer
仪表用空气系统 instrument air system
仪表用微计算机 instrument microcomputer
仪表用线路 appliance circuit
仪表用自耦变压器 instrument autotransformer
仪表用综合互感器 instrument transformer for all purpose
仪表油 instrument oil;meter oil
仪表有效荷载 instrument payload
仪表噪声 meter noise
仪表站 meter station
仪表照明 dial illumination
仪表罩 instrument mask;meter enclosure
仪表支架 implement carrier
仪表支线 instrument branch; meter branch
仪表脂 instrument grease
仪表指示 meter display
仪表指示的温度 probe temperature
仪表指示灯 instrument lamp; meter lamp
仪表指示器 meter indicator
仪表指示数 reading
仪表指针 ga(u)ge needle; ga(u)ge pointer;meter hand;meter needle
仪表指针跳动 dancing
仪表指针校正 index correction
仪表制造学 instrumentation
仪表制造业 instrument industry
仪表滞后 instrumental lag
仪表钟 appliance clock
仪表重量 weight of instrument
仪表专业人员 instrumentalist
仪表转换开关 meter switch
仪表装配 payload build-up
仪表装置 instrumentation;metering device
仪表装置传感器 instrumentation sensor
仪表准确度 accuracy of an instrument
仪表着陆 blind landing; instrument landing
仪表着陆无线电台 instrument landing station
仪表着陆系统 instrument landing system
仪表组 cluster ga(u)ge; instrument cluster
仪表组合 combination of instruments
仪表座板 instrument mounting plate
仪测浮标 instrumentation buoy
仪导滑行道 instrument runway
仪器 apparatus; equipment; instrument;tooling
仪器安排 meter arrangement
仪器安置高度 apparatus-fixed height
仪器安装 set-up of instrument
仪器百叶箱 instrument screen; thermometer screen;thermometer shelter; thermoscreen; instrument shelter
仪器板 instrument board
仪器本底 instrument background
仪器变量 instrumental variable
仪器变量法 instrumental variable method
仪器标尺 instrument dial;setting dial
仪器标记 instrument identification
仪器表 meter board
仪器表面板 instrument panel
仪器玻璃 apparatus glass; laboratory glass
仪器玻璃标度盘 instrument glass dial
仪器不稳定性 instrumental instability
仪器布置 meter arrangement; set-up instrument
仪器参数 instrument(al) parameter
仪器舱 instrument bay; instrument cabinet; instrument compartment; instrument room; instrument section
仪器操纵飞行 instrument flight
仪器测定的音节 instrumental segment
仪器测定的震中 instrumental epicenter [epicenter]
仪器测定(法) instrumental measurement
仪器测定烈度表 instrument intensity scale
仪器测定震级 instrument magnitude
仪器测量 instrumental measurement; instrument(al) survey
仪器测量法 instrument-survey method
仪器测量范围 instrument range
仪器测制 mapping by instrument
仪器常数 apparatus constant; constant of instrument;instrument(al) constant
仪器厂 instrument plant
仪器车 dog house;instrument truck
仪器储柜 equipment cupboard

仪器垂直旋转轴中心 center [centre] of instrument
仪器带电粒子活化分析 instrumental charged-particle activation analysis
仪器导向 instrument direction
仪器的 instrumental
仪器的鉴别力 instrumental resolution
仪器的抗偏差性 freedom from bias of the instrument
仪器的可靠性 reliability of the instrument
仪器的稳定性 stability of the instrument
仪器的重复性 repeatability of the instrument
仪器地震学 instrumental seismology
仪器递降 instrumental degradation
仪器电源 electric(al) supply of instrument; instrument feeding; instrument power supply
仪器订正 instrument correction
仪器读出时间 instrument reading time
仪器读数 instrument reading; meter reading
仪器读数灵敏度 sensitiveness of reading
仪器对心 centering of instrument
仪器对中 centering of instrument
仪器墩 instrument stand
仪器法 instrumental method; instrumental survey
仪器反褶积 instrument deconvolution
仪器方位(角) instrumental azimuth; instrumental bearing
仪器方位校正 correction for instrument azimuth
仪器放大器 instrumentation amplifier
仪器飞行起落跑道 instrument runway
仪器分辨率 instrumental resolution
仪器分析 instrumental analysis; instrumentation analysis
仪器分析法 instrumental method
仪器符号 instrument symbol
仪器改正 instrumental correction
仪器高 elevation of sight
仪器高程 <测量时的> elevation of sight; instrument elevation; instrument level;level of instrument;elevation of instrument
仪器高度 height of instrument;instrumental height
仪器工作 instrument operation
仪器功能 instrumental function
仪器供电 instrument feeding
仪器观测室 ga(u)ge house
仪器光学 instrument(al) optics
仪器光学率定 optic(al) calibration
仪器光子活化分析 instrumental photon activation analysis
仪器柜 instrument case
仪器函数 apparatus function
仪器和人力需求 instruments and manpower requirements
仪器回收 recovery of instruments
仪器绘图 instrumental drawing
仪器绘图桌 tracing table alongside the apparatus
仪器活化分析 instrumental activation analysis
仪器积分法 mechanical integration
仪器基座 instrument base; machine base
仪器计量的挖泥量【疏】measurement by instrument
仪器记录 instrument record
仪器架 instrument stand
仪器监测 instrument monitoring
仪器检测垂直度 instrument plumbing

仪器检测修理实验室 instrument test repair laboratory
仪器检出限 instrument detection limit
仪器角 instrumental angle
仪器校正 instrument adjustment; adjustment of instrument; instrument(al) correction
仪器校正场 adjusting yard
仪器解译方法 instrumental interpretation method
仪器经度 instrumental longitude
仪器精确度 accuracy of instrument
仪器开关 instrumental switch
仪器刻度范围 scale limitation
仪器刻度盘 instrument dial; meter dial
仪器空中三角测量 instrumental aerial triangulation
仪器控制 instrument control
仪器类型 instrument type
仪器连测法 instrumental bridging procedure
仪器量测 instrumental measurement
仪器量程 full-scale range; instrument range; range of measurement
仪器量距 range of measurement
仪器灵敏度 ga(u)ge factor; instrumental sensitivity
仪器零点漂移 instrumental drift; zero creep
仪器零点校正 instrument zero correction
仪器零位置 instrument zero
仪器轮廓 instrumental contour
仪器埋设 internal instrument installation
仪器埋置深度 depth of instrument burial
仪器面板 board
仪器名(称) instrument name; name of instrument
仪器内大气折射 instrumental refraction
仪器盘 facia panel
仪器配备 instrumentation
仪器配色 instrument-matching
仪器配色测示法 instrumental match prediction
仪器配置 location of apparatus
仪器偏差 instrumental bias
仪器偏心(率) eccentricity of instrument
仪器漂移效应 instrument drift effect
仪器频率响应 instrument frequency response
仪器平台 instrument platform
仪器评定 objective ranking
仪器谱 apparatus spectrum
仪器谱线轮廓 instrumental line profile
仪器歧离 instrument straggling
仪器气象状况 instrument meteorological condition
仪器缺陷 instrument(al) defect
仪器热中子活化分析 instrumental thermal neutron activation analysis
仪器入水深度 ga(u)ge depth
仪器上输入数据 instrument setting; machine setting
仪器设备层 apparatus floor
仪器设备工程学 apparatus engineering
仪器设备通风管 appliance ventilation duct
仪器实测等高线 instrumental contour
仪器使用 instrument use
仪器使用范围 instrument(al) limit; instrument(al) range
仪器使用图 instrumentation diagram
仪器式人体模型 <耐燃试验用> instrumental mannequin
仪器视差 instrumental error; instrumental parallax
仪器室 apparatus room; ga(u)ge house; instrument bungalow; instrument room
仪器手册 instrument manual
仪器台 instrument stand
仪器调节 meter set
仪器调节系统 modular instrument system
仪器调零 adjustment to zero; zeroing of instrument
仪器调零点装置 zero adjuster
仪器调零钮 zero adjusting screw
仪器调整 instrument adjustment
仪器图 instrument drawing
仪器外形 instrument contour
仪器维护编号 instrument maintenance code
仪器纬度 instrumental latitude
仪器误差 apparatus error; instrument(al) error; instrument(al) straggling
仪器误差改正量 instrument correction
仪器系统 instrument system
仪器下沉 settlement of instrument
仪器显示 instrument indication
仪器线 instrument line
仪器箱 casing; ga(u)ge house; instrument box; instrument cabinet; instrument case; instrument container
仪器响应 instrumental response
仪器像片三角测量 instrument photo triangulation
仪器效能 instrumental effect
仪器型号 instrument model; mode of instrument
仪器性能数据 characteristic data of equipment; characteristic data of instrument
仪器修正 instrument correction
仪器延迟 instrument lag
仪器仪表 instruments and meters; instrumetation
仪器仪表制造业 instrument industry
仪器因素 equipment factor
仪器用润滑剂 instrument lubricant
仪器用油 instrument oil
仪器游标盘 vernier circle
仪器有效荷载 instrumentation payload
仪器站 instrument station
仪器照片三角测量 instrument photo triangulation
仪器罩 instrument shelter
仪器折射 instrumental refraction
仪器指针 needle of instrument
仪器制图 instrumental drawing
仪器制造 making instrument
仪器制造工业 instrument-making industry
仪器制造学 instrumentation
仪器致宽 instrumental broadening
仪器置平 level(l)ing of instrument; levelness of instrument
仪器中心 center [centre] of instrument
仪器中子活化分析 instrumental neutron activation analysis
仪器重量 instruments weight
仪器轴 instrumental shaft; instrument axis
仪器装备 instrumentation
仪器装配 instrument assembly
仪器装置 instrument device
仪器桌 instrument table
仪器资料 instrument data
仪器综合作用 integration of instrument
仪器组 instrument unit
仪器组合 equipment combination; instrument combination
仪器坐标 instrument coordinates
仪器座 instrument stand
仪容 appearance of staffs
仪式 ceremony; memorial; Ritual
仪式的 ceremonial
仪式化 ritualization

夷

夷低谷 degrading valley
夷低河 degrading stream
夷平 raze (to the ground)
夷平毁坏 level(l)ing
夷平面 planation surface; plane surface
夷平剖面 <河流的> profile of equilibrium
夷平作用 deplanation; level(l)ing; planation

沂

沂蒙矿 yimengite

宜

宜拆毁的 demolishable
宜耕荒地 reclaimable wasteland
宜和兰红 neolan red
宜和兰蓝 neolan blue
宜和兰绿 neolan green
宜和兰染料 neolan colo(u)r
宜和兰桃红 neolan pink
宜喷射稠度 gun grade
宜取变换 admissible transformation
宜取策略 admissible strategy
宜取形势 admissible situation
宜人的草原 amenity grass land
宜涂漆木制品 paint grade
宜兴陶瓷 boccaro
宜于拆毁的 fit for demolition; fit for wrecking
宜于加固的 fortifiable
宜于设防的 fortifiable
宜于做把手或手柄的一段木材 handle blank

移

移变 shift variant
移驳绞车 barge warping winch
移泊 berth shifting
移泊费 shifting charges
移泊拖轮 berthing tug (boat); dock tug
移泊许可证 shifting permit
移舱 hatch shifting
移舱不停止装卸作业 switch holds without interrupting loading
移测法 traverse method
移测卡规 transfer cal(l)ipers
移测卡钳 transfer cal(l)ipers
移测显微镜 travel(l)ing microscope
移层电子 metastasic electron
移车器 cherry picker
移车台【铁】 travel(l)ing platform; car shifting platform; sliding bridge; transfer table; traverse table
移车台车 traverse carriage
移出 shift out
移出常数 removal constant
移出断面 removed section
移出符号 shift out character
移出国 country of emigration
移出截面 removal cross-section
移出扩散法 removal diffusion method
移出脉冲 shift out pulse
移出剖面 removed section
移出剖视 removed section
移出通量 removal flux
移出物 explant
移船车 <船台的> transfer car
移船台 transfer platform; transfer table <船台上的>; transfer carriage <升船机的>
移船轨道 ship railway
移船绞车 pontoon manoeuvring winch
移船锚 kedge
移带叉 belt fork; belt strike; blinker
移带器 belt shifter; belt striker; shifter
移档 berth shifting
移道机 rail shifting machine; track shifter
移底式加热炉 reheating furnace with movable hearth
移点器【测】 plumbing arm; plumbing fork; centring bracket
移电性 electrotaxis
移定桩延迟系数【疏】 spud moving delay factor
移锭器 retracter
移锭装置连接器 ingot adapter
移动 dislodge; dislodging; displacement; drift; floating; forward motion; jockey; locomote; locomotion; migration; motion; movement; progressive motion; removal; remove; removing; shift(ing); transfer; transference; translation【数】; travel; upfold
移动板块 active plate
移动办公室 mobile office; movable office
移动暴雨 moving rainstorm
移动泵 portable pump
移动笔 moving-pen
移动闭塞 movable block
移动闭塞系统 moving block system
移动闭塞原理 moving block principle
移动篦式冷却器 travel(l)ing grate cooler
移动篦式热交换器 travel(l)ing grate heat exchanger
移动篦式碳化炉 travel(l)ing grate retort
移动篦式窑 kiln with travel(l)ing grate
移动篦式预热器 travel(l)ing grate preheater
移动边界问题 moving boundary problem
移动变电站 mobile substation; movable substation
移动标尺 moving scale
移动标灯 portable beacon
移动标线 moving wire
移动冰 running ice
移动冰碛 active moraine; moving moraine
移动波 advanced wave; translation wave; travel(l)ing wave
移动波光参量振荡 travel(l)ing-wave optic(al) parameter oscillation
移动播音室 mobile studio
移动槽 shifting chute
移动测量 traverse measurement
移动叉轴 shifting fork shaft
移动插塞 shift plunger
移动差距图 the moving-range chart
移动长度 travel length
移动场 ling field; moving field; shifting field
移动车 mobile vehicle
移动车驾驶员 travel(l)ing man
移动车架 travel(l)er
移动称料斗 travel(l)ing weigh hopper
移动齿轮 carrier wheel; travel(l)ing gear
移动船舶站 mobile ship station

移动床 movable bed
移动床层 mobile bed; moving bed
移动床(超)吸附器 moving bed adsorber
移动床重整 moving bed reforming
移动床法 moving bed process; moving bed system
移动床过程 moving bed process
移动床过滤 moving bed filtration
移动床过滤器 moving bed filter
移动床活性炭接触器 moving bed activated carbon contactor
移动床接触 moving bed contacting
移动床离子交换 moving bed ion-exchange
移动床离子交换设备 moving bed ion-exchange equipment
移动床气化 moving bed gasification
移动床沙滤池 moving bed sand filter
移动床生物膜反应器 moving bed biofilm reactor
移动床吸附剂 moving bed adsorption
移动床吸附器 travel(l)ing bed adsorber
移动床吸附柱 moving bed adsorption column
移动打磨机 portable grinder
移动带 dislocation area
移动带式传送机 moving band conveyer [conveyor]
移动带式输送机 moving band conveyer [conveyor]
移动挡板 moving stop
移动刀架式牛头刨 travel(l)ing head shaper
移动导管(装置) pipe guide
移动导数 translatory derivative
移动的 bias(s)ed; erratic; migratory; motive; travel(l)ing; moving
移动的道岔 moving points
移动的关闭方木或钢轨 <必要时横放在钢轨上> movable scotch block
移动的路拌设备 road pug travel-mix plant
移动的沙丘 marching dune
移动的砂 shifting sand
移动灯 portable lamp; portable light
移动地球站 mobile earth station
移动点荷载 moving point load
移动电动机 travel motor
移动电话 <俗称大哥大> cellular phone; mobile (tele) phone
移动电话交换局 mobile telephone switching office
移动电极 travel(l)ing electrode
移动电缆 travel(l)ing cable
移动电势 migration potential
移动电刷架 brush-rocker
移动电刷式推斥串激单相牵引电动机 brush shifting repulsion series single phase traction motor
移动电台通信[讯] mobile service
移动电位 migration potential
移动吊车 jenny; yard crane
移动吊机 jenny
移动吊架 travel(l)ing bridge
移动叠影 travel ghost
移动定差 shifted gradient
移动定位 <航空或航海> running fix
移动陡波 abrupt translatory wave
移动断裂模型 moving dislocation model
移动轭销 shifting yoke pin
移动发电机组地面电缆 trailing cable
移动阀 movement of valve
移动法兰 rotary flange; running flange
移动范围 moving range; range of movement; successive range
移动分量 translational component
移动负载 moving load
移动杆 motion lever; setting lever; shifting bar; shipper rod; working lever
移动杆弹簧 shifting lever spring
移动感应圈 travel(l)ing induction coil
移动钢桁架式膺架 centering Rogla
移动杠杆 live lever
移动杠杆拉杆 live lever connecting rod
移动杠杆拉杆托 live lever connecting rod guide
移动杠杆上滑轮 live lever top connecting rod guide
移动杠杆上拉杆 live lever top connecting rod guide
移动杠杆上拉托 live lever top connecting rod guide
移动杠杆支点 live lever guide
移动杠杆支托 live lever guide
移动钢滚柱 dolly roller
移动隔墙 sliding partition (wall)
移动工作架 finish jumbo
移动工作台 travel(l)ing table; traverse table
移动刮板集泥器 travel(l)ing flights sludge collector
移动观察技术 moving observer technique
移动观察钟 <潜水装置> mobile observation bell
移动光标 travel(l)ing light spot
移动光点 travel(l)ing light spot
移动光缆 portable optic(al) fibre cable
移动轨道 track orbit; track slewing
移动滚轮 travel(l)ing roller
移动滚式运输机 portable roller conveyer [conveyor]
移动滚柱式传送机 roller conveyer [conveyor]
移动滚子 travel(l)ing roller
移动海滩 drifting beach; travel(l)ing beach
移动函数 move function
移动河床 movable bed; movable river bed
移动河湾 migrating meander
移动荷载 movable load; moving load; travel(l)ing load
移动荷载试验 moving-load test
移动滑车 movable block; moving block; removable tackle; running block; travel(l)ing block
移动滑轨 shifting sledge
移动滑坡 downhill creep
移动环 shift(ing) eye; shift(ing) ring
移动换向漏斗 movable divertor hopper
移动黄道 movable ecliptic
移动汇率 movable rate
移动活塞 shifting piston
移动货物的倒装 reloading of displaced consignment
移动机构 travel mechanism
移动基金 mobile fund
移动基期 shifting base period
移动极限角 travel(l)ing limit angle
移动几何平均法 method of moving geometric(al) average
移动计算网络 mobile switching network
移动季节变动 moving seasonal variation
移动继续年金 portable pension
移动加权平均法 moving weighted average method
移动加权平均预测法 moving weighted average forecasting
移动加热法 travel(l)ing heater method
移动加热器 travel(l)ing heater
移动加速度 acceleration of translation
移动键 shifting bonds; shifting links
移动键控 shift keying
移动交换中心 mobile switching center [centre]
移动角 angle of draw; dislocation angle; moving angle; travel(l)ing angle
移动绞车 crab winch; movable winch; travel(l)ing crab; travel(l)ing winch
移动脚手架 French scaffold; moving scaffold; travel(l)ing scaffold(ing)
移动搅拌机 travel(l)ing mixer
移动接头 shift joint
移动界面 <指牵伸变速点> moving boundary
移动界面电泳 moving boundary electrophoresis
移动界面法 moving boundary method
移动锯木机 portable saw mill
移动均衡 moving equilibrium
移动坑线 moving ramp
移动空气污染源 mobile source of air pollution
移动控制 translational control
移动拦阻射击 predicted barrage
移动链篦碳化炉 moving chain grate retort
移动梁 travel(l)ing girder
移动量 amount of movement; travel motion
移动料仓 batch truck
移动列车通信[讯]系统 mobile train communication system
移动列车通信[讯]制 mobile train communication system
移动临界水深 critical water depth for sand movement
移动流 slump flow
移动流化床光致反应器 moving fluidized bed photoreactor
移动楼梯 travel(l)ing stair(case)
移动漏斗的导轨 guide rail for shifting the hopper
移动漏斗的滑轮 sliding wheel for shifting hopper
移动炉法 reciprocating furnace process
移动炉排 travel(l)ing grate stoker
移动炉栅烧结 travel(l)ing grate sintering
移动路面 moving pavement
移动率 flow rate; mobility; rate of flow
移动螺母 travel(l)ing nut
移动螺栓 shifting bolt
移动门 shifter gate
移动门架式悬臂起重机 travel(l)ing portal jib crane
移动门(式起重)机 gantry travel(l)er
移动门座起重机 travel(l)ing portal crane
移动面 displaced surface
移动模板 moving die plate; moving form(work); moving plate; travel(l)ing formwork; peg mold <线脚抹灰用>
移动模板浇筑的混凝土 walled concrete
移动模架 form on travel(l)er
移动模型 moving model
移动模铸造法 moving mo(u)ld casting process
移动目标探测仪 motion detection equipment
移动能 migration energy
移动能量系统 <如铁路机车> mobile energy system
移动泥沙 moving sediment
移动耙料机 travel(l)ing plough
移动泡沫混合发生器 mobile foam generator
移动配电站 mobile substation; movable substation
移动配重式测力计 running weight type dynamometer
移动皮带 floating belt
移动皮带杆 shifter pole
移动频率 travel frequency
移动平衡重块 moving poise
移动平均比率法 ratio-to-moving-average method
移动平均成本法 moving average cost method
移动平均法 moving average; moving average method; moving average process
移动平均过程 moving average process
移动平均(数) moving average; progressive average; progressive mean; running mean; rolling average
移动平均数法【数】 method of moving averages
移动平均数模型 moving average model
移动平均调整 moving average adjustment
移动平均预测法 moving average forecast method
移动平均值 moving average; moving mean; removed mean value
移动平均值法 moving average method
移动平均值图 moving average value figure
移动屏 moving screen
移动屏(测流速)法 travel(l)ing screen method
移动起重机 crane hoist; travel(l)ing crane; travel(l)ing hoist; yard crane
移动起重机台架 gantry travel(l)er
移动器械 mover
移动桥 travel(l)ing bridge
移动桥架式起重机 moving bridge crane
移动桥式吊车 gantry travel(l)er; travel(l)ing bridge crane; travel(l)ing crane
移动桥式刮泥机 travel(l)ing bridge scraper
移动桥式刮土机 travel(l)ing bridge scraper
移动桥式集泥器 travel(l)ing bridge sludge collector
移动桥式起重机 gantry travel(l)er; travel(l)ing bridge crane
移动区 turnover zone
移动曲线 displacement curve
移动溶剂法 travel(l)ing solvent method
移动洒油车 travel(l)ing distributor
移动沙波 migrating sand wave
移动沙垄 drifting dune
移动沙丘 blowing dune; drifting dune; migrating dune; moving dune; shifting dune; temporary silo; travel(l)ing dune; wandering dune
移动沙纹交错纹理构造 moving-ripple cross lamellar structure
移动沙闸 moving sand debris
移动沙洲 migrating bar; travel(l)ing bar
移动筛 travel(l)ing screen
移动筛脱器 moving screen concentrator
移动沙丘 migrating dune; migratory

Y

dune;wandering dune
移动烧火区 moving-fire zone
移动设备 mobile equipment;movable plant
移动设备成本 cost of removing equipment
移动摄影 follow-scene;travel(l)ing shot;travel shot
移动摄影车 dolly
移动升降机 movable lift
移动升降台 cherry picker
移动十字丝 movable retic(u)le
移动时间模型 travel-time model
移动实验室 mobile laboratory
移动式 move mode;portable type;slip-on type
移动式X射线机 mobile X-ray machine;mobile X-ray unit
移动式暗室 transportable darkroom
移动式拌和机 mixer-lorry;mobile mixing plant;portable agitator;portable mixer;removable agitator;travel(l)ing agitator;travel(ling) mixer
移动式拌和设备 travel(l)ing mixing plant;mobile batching plant
移动式保养设备 mobile maintenance plant
移动式备用方式【电】moving spare system
移动式泵 portable pump
移动式臂架起重机 travel(l)ing jib crane
移动式边墩 hauling bilge block
移动式变电所 portable mining substation
移动式变电站 portable station
移动式变压器 mobile transformer;transportable transformer
移动式标志 portable sign;stanchion sign
移动式波浪浮标 disposable wave buoy
移动式薄膜厩液泵 mobile diaphragm liquid manure pump
移动式采矿钻车 mobile mine drill
移动式餐馆 mobile restaurant
移动式侧面提升吊车 mobile side-lift hoist
移动式测高仪 transportable height finder
移动式柴油发电机 portable diesel electric(al) generator
移动式柴油发电机组 portable diesel generating set
移动式铲斗挖泥船 walking scoop dredge(r)
移动式铲斗挖泥机 walking scoop dredge(r)
移动式长臂输送机 portable boom conveyer [conveyor]
移动式厂拌 travel(l)ing plant mixing
移动式车顶工作平台 movable roof working platform
移动式车架 traversing carriage
移动式车体支座 movable car body support
移动式称量装置 portable batch plant
移动式称料斗 travel(l)ing weigh hopper
移动式冲击破碎机 portable impact breaker
移动式抽风管 portable duct
移动式初级破碎厂 mobile primary crushing plant
移动式储罐 skid tank
移动式处理厂 portable plant
移动式传送带 mobile conveying belt
移动式船台 traversing slipway
移动式磁粉探伤机 movable magnetic powder detector
移动式打磨 portable grinding
移动式打桩机 travel(l)ing pile driver;walking driver
移动式带式输送机 mobile belt conveyer [conveyor];portable belt conveyer [conveyor]
移动式挡墙 removable stopping
移动式车间 mobile workshop
移动式船边提升机 portable shipside elevator
移动式导口 shifting type entry guide
移动式捣固车 travel(l)ing tamper
移动式捣碎机 mobile hammer mill
移动式道岔 portable shunt
移动式的 in powder form;movable;non-stationary;walking
移动式底板推移 sliding floor ejection
移动式地沥青拌和设备 portable asphalt plant;mobile asphalt plant
移动式地沥青混合料拌和厂 portable asphalt plant;travel(l)ing bituminous mixing plant
移动式地沥青混合料拌和设备 portable asphalt plant;travel(l)ing bituminous mixing plant
移动式地面信标 transportable ground beacon
移动式地面站 mobile earth station;transportable earth station
移动式地热电站 mobile geothermal-power station
移动式点焊机 portable spot welder
移动式电动吊车 electric(al) travel(l)ing crane
移动式电动葫芦 runway hoist
移动式电动机 travel(ling) motor;propelling motor <自行钻机的>
移动式电动起重机 electric(al) travel(l)ing crane;mobile electric(al) crane;travel(l)ing motor crab;travel(l)ing motor hoist
移动式电动切砖机 portable electric(al) saw
移动式电动索铲挖土机 electric(al) walking dragline
移动式电焊设备 portable electric(al) welding equipment
移动式电葫芦 runaway hoist
移动式电话 mobile phone;portable phone
移动式电话系统 mobile telephone system
移动式电极 mobile electrode;movable electrode
移动式电锯 portable electric(al) saw
移动式电磨机 portable electric(al) grinder
移动式电台 mobile station;transportable station
移动式电子修理所 transportable electronic shop
移动式吊车 creeper crane;mobile crane;mobile hoist;movable crane;travel lift;walking crane;working carriage
移动式吊管机 jenny
移动式吊机 jenny;mobile derrick
移动式吊篮 travel(l)ing cradle
移动式吊桥 travel(l)ing bridge crane
移动式调车转辙器 sliding floor
移动式顶层支架 continuous roof support
移动式顶管机【给】dolly mounted jack
移动式动力装置 mobile power unit
移动式斗带升运机 mobile belt trough elevator
移动式斗带升运器 mobile belt trough elevator
移动式斗式挖泥机 walking scoop dredge(r)
移动式堆料机 travel(l)ing stacker;mobile stacker
移动式对流层散射通信[讯]设备 transportable tropospheric scatter communication equipment
移动式对流加热器 mobile space heater
移动式多路通信[讯]中心 transportable multiplex communication center [centre]
移动式颚板 walking jaw
移动式颚式破碎机 portable jaw crusher
移动式发电机 mobile generator;portable engine;portable generator
移动式发电机组 mobile generator set;portable generating set
移动式发电设备 mobile generating equipment;mobile generating set;portable electric(al) power plant
移动式发电站 portable electric(al) power plant
移动式发射机 mobile transmitter;portable transmitter
移动式伐树机 mobile harvester
移动式反应堆 mobile reactor
移动式防波堤 movable breakwater
移动式防雪栏 removable snow fence;removable snow fencing
移动式防雪栅 removable snow fencing
移动式防御雷达 transportable defense radar
移动式放大器 portable amplifier
移动式分料斗 movable divertor hopper
移动式分盘拌和设备 portable batcher plant
移动式分配器 movable distributor
移动式分批搅拌设备 portable batcher plant
移动式分批干燥机 mobile batch drier
移动式分批搅拌机 mobile batch-mixer
移动式分条整经机 mobile warping frame
移动式分析器 mobile analyser [analyzer]
移动式分线盒 movable distributor
移动式粉碎机 mobile crusher
移动式风机 portable fan
移动式封套窑 moving-envelope kiln
移动式辐照器 mobile irradiator
移动式辐照装置 portable irrdadiation facility
移动式俯仰起重机 travel(l)ing luffing crane
移动式伽马辐照器 mobile gamma irradiator
移动式干燥机 travel(l)ing drier [dryer]
移动式钢轨探伤设备 rail detection by portable ultrasonic equipment
移动式高架工作平台 mobile elevating work platform
移动式高架起重机 movable gantry;overhead travel(l)ing crane;travel(l)ing overhead crane
移动式高架卸船机 travel(l)ing unloading tower
移动式高压空气供应台 mobile high-pressure air stand
移动式格筛 movable grizzly;portable grizzly
移动式格筛输送机 travel(l)ing grizzly feeder
移动式格栅 travel(l)ing screen
移动式隔断 movable partition(wall)
移动式隔墙 rolling partition(wall);movable partition
移动式给料机 travel(l)ing feeder
移动式跟踪 mobile tracking
移动式跟踪扫描器 mobile tracking scanner
移动式跟踪装置 mobile tracking unit
移动式工厂 mobile plant
移动式工程机械 portable equipment
移动式工地生活设施 transportable site accommodation unit
移动式工房 mobile site cabin
移动式工业搬运设备 mobile industrial handling equipment
移动式工业装卸设备 mobile industrial handling equipment
移动式工作平台 mobile work platform;travel(l)ing stage
移动式工作台 moving bolster;travel(l)ing stage
移动式供冷机组 portable cooling unit
移动式供水系统 portable water system
移动式拱架 travel(l)er(arch) center[centre];travel(l)ing(arch) centering[centring]
移动式沟槽支撑 trench box;trench shield
移动式构架 travel(l)ing framework;travel(l)ing gantry
移动式构架桥 portable frame bridge
移动式构造 travel(l)ing configuration
移动式刮板机 travel(l)ing scraper
移动式刮土机 slushier
移动式观测装置 mobile observation unit;mobile observing unit
移动式管垫【给】travel(l)ing pipe shield
移动式罐 portable tank
移动式罐装设备 movable filling arrangement
移动式轨道 sectional track
移动式滚筒 travel(l)ing roller
移动式锅炉 portable boiler
移动式焊接机 portable welder
移动式焊接机组 mobile welding set
移动式焊接设备 mobile welding equipment
移动式夯击压实机 portable impactor
移动式夯具 mobile compactor
移动式核子反应堆 portable nuclear reactor
移动式烘炉 portable drier [dryer];portable mo(u)ld drier [dryer]
移动式滑车 sliding crown-block;travel(l)ing hoist
移动式滑轮 travel(l)ing sheave
移动式灰浆搅拌机 movable mortar mixer;portable mortar mixer
移动式回转调幅起重机 portable derrick crane
移动式混合砂处理装置 portable sand conditioning unit
移动式混凝土拌和车 mixer-lorry
移动式混凝土拌和站 movable concrete mixing plant
移动式混凝土泵 portable concrete pump;mobile concrete pump
移动式混凝土浇注机 mobile concrete placer
移动式混凝土浇筑机 mobile concrete placer
移动式混凝土搅拌机 portable concrete mixer;travel(l)ing concrete mixer
移动式混凝土摊铺机 mobile concrete placer
移动式混砂机 mobile sand mill;portable muller
移动式活塞夯实机 walking ram
移动式火箭燃料净化装置 transportable rocket fuel purification system

Y

移动式机动铲 mobile mechanical shovel
移动式机器 travel(l)ing plant
移动式机器人 mobile robot
移动式机体 walking mechanism
移动式机械 mobile unit;travel plant
移动式机械起重机 mobile mechanical crane
移动式级配控制的称量设备 portable batch plant with gradation control
移动式集尘装置 movable dust collector
移动式集料加工设备 portable aggregate plant
移动式集料制备车间 portable aggregate plant
移动式集装箱吊运车 transtainer
移动式计算机 movable computer
移动式加速器 mobile accelerator
移动式驾车机 mobile lifting jacks
移动式架空装置 mobile unit
移动式间歇干燥机 mobile batch drier
移动式剪式升降工作平台 scissor lift mobile elevating work platform
移动式桨叶拌和机 travel(l)ing paddle mixer
移动式交汇控制 moving merge control
移动式交通信号 portable traffic signal
移动式胶带筛分装载机 portable screen belt loader
移动式胶带输送机 movable belt conveyer [conveyor]; portable belt conveyer [conveyor]
移动式绞车 mobile hoist; portable hoist;travel(l)ing crab;travel(l)ing winch
移动式铰接臂可升降工作平台 articulated boom mobile elevating work platform
移动式脚手构架 travel(l)ing framework
移动式脚手架 mobile scaffold;movable falsework; movable scaffold(ing); portable mobile scaffold; transfer platform;travel(l)er; travel(l)ing centering [centring]; travel(l)ing frame; travel(l)ing scaffold(ing); French scaffold
移动式搅拌厂 mix-in-travel plant; travel(l)ing mixing plant
移动式搅拌机 mixer-lorry; movable mixer; portable agitator; portable mixer; removable agitator; transit mixer;travel(l)ing mixer;travel(l)ing mixing machine; travel(-plant) mixer
移动式搅拌楼 travel-plant mixer
移动式搅拌器 travel(l)ing agitator
移动式搅拌设备 portable mixer;portable mixing plant
移动式搅拌站 travel(l)ing mixing plant
移动式搅拌装置 movable batch plant
移动式接触焊机 portable resistance welder
移动式金刚石刀架 diamond held trailing
移动式紧急提升设备 mobile emergency winding equipment
移动式近海钻井平台 offshore mobile drilling rig
移动式近海钻探平台 offshore mobile drilling rig
移动式矩阵标志 portable matrix sign
移动式巨型起重机 goliath(crane)
移动式锯 ripsnorter; travel(l)ing saw
移动式锯木机 portable saw mill
移动式聚光灯 portable spotlight
移动式卷扬机 mobile hoist;travel(l)ing hoist;hoisting truck
移动式开关柜 mobile switching cabinet
移动式砍伐树枝机器 delimber
移动式可伸缩卷扬机 ant queen
移动式可伸缩装置 ant queen
移动式可升降工作平台 mobile elevating work platform
移动式坑柱 walking pit prop
移动式空气压缩机 mobile air compressor; movable air compressor; portable (air) compressor
移动式空压机 mobile compressor
移动式空中工作平台 aerial platform
移动式空中工作升降机 aerial lift
移动式控制 moving mode control
移动式控制台 remove control panel
移动式矿井支柱 walking pit prop
移动式框架 travel(l)ing frame
移动式垃圾储藏箱 bulk waste storage container
移动式拉铲表土剥除机 walking dragline stripper
移动式拉索脱模机 walking dragline stripper
移动式缆道 travel(l)ing cableway
移动式缆索起重机 travel(l)ing cableway crane
移动式冷拌和厂 travel(l)ing coldmix plant
移动式冷藏间 portable cold room
移动式冷间 portable cold room
移动式冷却装置 portable cooling unit
移动式沥青拌和厂 travel(l)ing bituminous mixing plant
移动式沥青拌和机 travel(l)ing bituminous mixing plant
移动式沥青拌和设备 mobile bitumen mixing plant
移动式沥青釜 <养路用> patrol kettle
移动式沥青工厂 portable asphalt plant
移动式沥青混合料拌和设备 mobile bituminous mixing plant
移动式沥青混凝土搅拌机 movable asphalt mixer
移动式沥青混凝土搅拌设备 movable asphalt mixer;movable asphalt mixing plant; mobile asphalt mixing plant
移动式沥青炉 <养路用> patrol kettle;portable kettle
移动式沥青熔化装置 travel(l)ing asphalt melting unit
移动式沥青制备设备 portable asphalt plant
移动式沥青筑路设备 mobile bitumen(ite) road-making plant; mobile bitumen roadmaking plant
移动式联合混砂机 sand cutter
移动式料仓 ambulatory bin; batch truck;portable bin
移动式料斗 portable bin; portable hopper;travel(l)ing crab
移动式龙门吊 travel(l)ing bridge crane;travel(l)ing gantry crane
移动式龙门架 travel(l)er gantry; travel(l)ing gantry;mobile gantry
移动式龙门起重机 travel(l)ing gantry crane; gantry travel(l)er; portal travel(l)ing crane;travel(l)ing bridge crane; travel(l)ing derrick crane; mobile gantry crane; mobile portal crane
移动式楼板推移 sliding floor ejection
移动式楼梯 travel(l)ing stair(case)
移动式漏斗 portable hopper
移动式炉 portable furnace
移动式炉箅 travel(l)ing grate
移动式炉排 travel(l)ing grate
移动式轮胎充气装置 mobile tire inflation unit
移动式码垛机 portable stacker
移动式码头起重机 mobile dock crane
移动式门(窗)框 mobile frame
移动式门架 travel(l)er;travel(l)ing gantry
移动式门(架)式起重机 travel(l)ing gantry crane
移动式模板 travel(l)ing forms;travel(l)ing shutter;travel(l)ing shuttering (work)
移动式模板系统 form traveler
移动式模架 form on travel(l)er
移动式模具 travel(l)ing forms
移动式模壳 moving form(work); travel(l)ing forms
移动式磨(砂)轮 portable grinding wheel
移动式木材储料场架空索道搬运机 mobile yarder
移动式目标指示器 moving target indicator
移动式耐压试验台 movable voltage testing stand
移动式暖房 rolling greenhouse
移动式泡沫挡帘 travel(l)ing foam plug
移动式泡沫炮 portable foam cannon; portable foam monitor
移动式配电器 movable distributor
移动式配水器 movable distributor; travel(l)ing water distributor
移动式喷灌机 mobile sprinkler;movable sprinkler; portable irrigation sprinkler; travel(l)ing sprinkler; walking sprinkler
移动式喷灌系统 portable sprinkle irrigation system; portable sprinkler system
移动式喷淋器 walking sprinkler
移动式喷洒机 movable distributor
移动式喷洒器 travel(l)ing distributor
移动式喷雾室 travel(l)ing spray booth
移动式喷油车 travel(l)ing distributor
移动式棚屋 portable cabana
移动式皮带传送带 carryable belt conveyer [conveyor]
移动式皮带输送机 mobile belt conveyer [conveyor]; portable belt conveyer [conveyor]
移动式皮带运输机 portable belt conveyer [conveyor]
移动式皮带装载机 mobile belt loader
移动式平带升运机 mobile flat belt elevator
移动式平均 running average;running mean
移动式平均法 method of moving averages
移动式平均分析 moving average analysis
移动式平台 mobile platform; mobile stage;slip stage
移动式平台模板 travel(l)ing table mould
移动式坡道 portable ramp
移动式破碎厂 mobile crushing plant; portable crushing plant
移动式破碎机 mobile breaker;mobile crusher; movable crusher; portable breaker; portable crusher; portable crushing plant; walking crushing plant
移动式破碎机组 mobile crushing plant
移动式破碎及筛分设备 movable crusher and screen classifier
移动式破碎筛分厂 mobile screening and crushing plant
移动式破碎筛分装置 rock ranger
移动式破碎设备 mobile crushing plant;portable crushing plant
移动式铺路机 travel(l)ing form paver
移动式启闭机 platform hoist
移动式起重吊架 <悬挂在导轨上的> carter's cradle
移动式起重机 crab derrick; lift about hoist; mobile crane; mobile hoist; movable crane; moving crane; platform hoist;self-propelled crane;transit crane; travel(l)er; travel(l)ing crab;travel(l)ing crane;travel(l)ing hoist; walking crane; jenny; portable crane;runabout crane <小型的>
移动式起重架 mobile derrick; travel(l)er
移动式起重器 railroad jack
移动式汽车库 portable garage
移动式汽油计量器 portable gasoline measuring
移动式千斤顶 mobile jack
移动式牵引机 donkey
移动式前炉 receiver ladle
移动式桥式起重机 moving bridge crane
移动式轻便钻机 portable trailer-mounted drill rig; travel(l)ing portable rig
移动式倾卸装置 travel(l)ing tripper
移动式清垢器 travel(l)ing cleaner
移动式清管平台 movable go-devil platform; movable pig platform; movable scraper platform
移动式清选分级机 mobile cleaner-grader
移动式取料机 travel(l)ing reclaimer
移动式取暖器 portable heater
移动式燃气轮机 mobile gas turbine
移动式热拌挂车 portable hot mix trailer
移动式人字起重机 portable derrick crane;travel(l)ing derrick;travel(l)ing derrick crane;travel(l)ing type derrick
移动式容器 movable vessel
移动式润滑设备 mobile lubrication equipment
移动式润滑油加注器 oil bucket pump
移动式撒布机 travel(l)ing distributor
移动式散射通信[讯]设备 mobile scatter communication equipment
移动式筛分机 portable screening machine
移动式筛分破碎装置 portable crushing and screening plant
移动式上向运输机 mobile elevating conveyer [conveyor]
移动式设备 mobile equipment;portable installation; portable plant; transportable equipment;transportable set; travel(l)ing plant;travel plant
移动式设备拌和法 travel-plant method
移动式设备拌制 travel-plant mixing
移动式设备拌制(混凝土)法 travel-plant method
移动式伸缩臂升降工作平台 telescopic boom mobile elevating work platform
移动式升降机 portable elevator
移动式升降平台 movable lifting platform [table]

移动式升降梯 free elevator
移动式实验室 laboratory vehicle; mobile laboratory; mobile laboratory
移动式试验台架 mobile test stand
移动式收发站 transportable transmit and receive station
移动式收集系统 portable collection system
移动式售票台 <市郊列车> portable ticket booth
移动式输煤带 mobile coal belt
移动式输送带 moving conveyer [conveyor]
移动式输送机 portable conveyer [conveyor]; travel(l)ing belt; trimmer conveyer [conveyor]
移动式输送螺旋 transportable auger
移动式数字计算机 mobile digital computer
移动式双颚式轧碎机 double movable jaw crusher
移动式水泵机组 portable pumping unit
移动式水处理设备 portable water treatment equipment
移动式水管灌溉 portable-pipe irrigation
移动式水力起重机 mobile hydraulic jack
移动式水力起重器 mobile hydraulic jack
移动式水力千斤顶 mobile hydraulic jack
移动式水泥泵 portable cement pump
移动式水泥筒仓 portasilo
移动式水枪 jet cutting car
移动式水位计 portable level recorder
移动式水污染源 mobile source of water pollution
移动式水箱 portable tank
移动式水跃 moving hydraulic jump
移动式碎石机 portable crushing machine
移动式塔吊 transportable tower crane; travel(l)ing tower crane
移动式塔吊基座 travel base
移动式塔架 <缆道> travel(l)ing tower
移动式塔式起重机 mobile tower crane; transportable tower crane; travel(l)ing tower crane
移动式台座 mobile mounting unit
移动式摊铺机 travel(l)ing form paver
移动式探测雷达 transportable detection radar
移动式镗孔岛 mobile boring island
移动式套管起拔机 portable pulling machine
移动式梯子 portable ladder
移动式提管机 portable pulling machine
移动式提升机 mobile winder
移动式天线 mobile antenna; movable antenna; portable antenna
移动式跳板 moving ramp
移动式铁水沟 rocking runner
移动式通信[讯]地面站 transportable ground communication station
移动式通信[讯]系统 transportable communication system
移动式脱粒机 portable thresher; travel(l)ing thresher
移动式挖掘机 mobile excavator; walking excavator
移动式挖泥船 walking dredge(r)
移动式挖泥机 walking dredge(r)
移动式挖土机 mobile excavator; walking excavator; walking scoop dredge(r)
移动式外海建筑物 mobile offshore structure
移动式外海平台 mobile offshore platform
移动式弯沉仪 travel(l)ing deflectometer
移动式微波无线电通信[讯]设备 transportable MW radio communication
移动式微型港口 flexiport
移动式桅杆 movable mast; travel(l)ing mast
移动式维修保养设备 mobile maintenance and support equipment
移动式尾水堰 movable tail weir
移动式卫星地球站 transportable satellite earth station
移动式卫星通信[讯]线路终端机 transportable satellite communications link terminal
移动式卫星通信[讯]终端设备 transportable satellite communication terminal
移动式围堰 movable cofferdam
移动式稳定土厂拌设备 mobile stabilized soil mixing plant; movable stabilized soil mixing plant
移动式稳船架 movable bilge arm
移动式涡浆搅拌机 travel(l)ing paddle mixer
移动式坞壁悬臂脚手架 travel(l)ing dock arm
移动式污水处理装置 package treatment plant
移动式屋顶 rolling roof
移动式无线电传通信[讯]电台 mobile radio teletype station
移动式无线电话专用线路 mobile radiotelephone private line
移动式无线电机用电源 power plant for mobile radio apparatus
移动式无线电台 mobile radio unit
移动式无线电通信[讯] mobile radio
移动式无线电通信[讯]系统 <列车调度员与列车联络用> mobile radio-communication (Railcom)
移动式舞台 travel(l)ing stage
移动式物料提升机 portable material elevator
移动式系数 coefficient of movement
移动式箱格导柱【船】 movable cell guide post
移动式消防泵 trailer pump
移动式消防栓 portable hydrant
移动式小冲模 removable subdie
移动式小型起重机 cherry picker
移动式卸货机 travel(l)ing unloader
移动式卸货塔 travel(l)ing unloading tower
移动式卸料(箕)斗 travel(l)ing skip
移动式卸料器 travel(l)ing tripper
移动式卸料装置 travel(l)ing tripper
移动式信号灯 <建筑道路时用的> portable road beacon (for highway construction)
移动式行车 travel(l)ing gantry
移动式修配间 mobile workshop
移动式悬 adjustable boom
移动式悬臂吊车 travel(l)ing portable jib crane
移动式悬臂起重机 travel(l)ing jib crane
移动式悬挂定重称料机 travel(l)ing type suspended weigh-batcher
移动式旋臂堆料机 travel(l)ing slewing stacker
移动式旋转电刨 portable electric(al) rotary planer
移动式旋转起重机 mobile jib crane; mobile rotary crane; mobile slewing crane
移动式旋转悬臂起重机 non-slewing jib mobile crane
移动式压板 movable platen
移动式压风机 mobile compressor
移动式压实工具 mobile compactor
移动式压实机(械) mobile compactor
移动式压缩机 movable compressor
移动式压延台 roller tray
移动式厌氧污泥床反应器 anaerobic migrating sludge blanket reactor
移动式摇架 travel(l)ing cradle
移动式遥控装置 mobile robot; mobot
移动式叶轮给料器 travel(l)ing rotary plow feeder
移动式液化气罐 skid tank
移动式液压锤 mobile hydraulic hammer
移动式液压吊车 mobile hydraulic crane
移动式液压起重机 mobile hydraulic crane
移动式液压千斤顶 mobile hydraulic jack
移动式液压校正器 portable power
移动式鹰架 travel(l)ing centering
移动式引桥 moving ramp
移动式油箱 kinetic tank
移动式浴缸 movable bathtub; moving bathtub
移动式浴盆 movable bathtub; moving bathtub
移动式圆盘锯 chop saw
移动式运料车 working carriage
移动式运输机 portable conveyer [conveyor]
移动式凿岩机支架 portable rig
移动式轧石厂发电厂 portable crushing plant
移动式轧石厂发电机 portable crushing plant
移动式轧石机 movable crusher
移动式轧石设备 portable crushing plant
移动式轧碎机 mobile crusher; movable crusher
移动式闸门启闭机 travel(l)ing gate hoist
移动式栅栏 movable barrier
移动式帐篷 transportable tent
移动式照明 portable lighting
移动式照明平台 travel(l)ing lighting gallery
移动式照明通道 travel(l)ing lighting gallery
移动式真空除尘器 mobile vacuum cleaner; portable vacuum cleaner
移动式真空吸尘器 portable vacuum cleaner
移动式蒸汽机 donkey engine; transportable steam engine
移动式整流器 portable rectifier
移动式支撑 walking support
移动式支架 removable support; walking support
移动式只收站 transportable receive-only station
移动式制材厂 mobile mill; mobile saw-mill
移动式中间工作平台 movable working table
移动式中间门 removable intermediate bulkhead
移动式重锤 floating weight
移动式轴承 shifting bearing
移动式轴流通风机组 mobile axial-flow fan unit
移动式贮料箱 portable bin
移动式柱座标志 portable sign
移动式抓斗起重机 travel(l)ing bucket crane; travel(l)ing clamshell crane
移动式转转臂起重机 mobile jib crane
移动式转筒干燥机 mobile rotary drum drier
移动式转子冲击破碎机 mobile impeller impact breaker
移动式装船机 travel(l)ing shiploader; moving loader
移动式装袋机 mobile packager
移动式装料嘴 mobile loading spout
移动式装填车 mobile loader vehicle
移动式装卸机械【机】 mobile handling machinery
移动式装卸桥 travel(l)ing gantry; bridge-type travel(l)ing shiploader; travel(l)ing gantry bridge
移动式装卸台 moving platform
移动式装修吊篮 travel(l)ing basket
移动式装载机 mobile loader; portable loader; portable stacker; port stacker
移动式装置 movable device; portable installation; travel(l)ing plant; travel plant
移动式自动电传打字机 automatic mobile teleprinter
移动式自动落纱机 travel(l)ing autodoffer
移动式自升平台 walking jack-up platform
移动式自支承升降机 mobile self-supporting hoist
移动式钻车 mobile drill; mobile jumbo
移动式钻床 portable drill
移动式钻机 mobile drilling machine; mobile drill(ing) rig; portable rig; wagon drill
移动式钻机台 jumbo
移动式钻井机 mobile rig
移动式钻孔岛 mobile drilling island
移动式钻孔机 portable borer
移动式钻孔平台 mobile boring island
移动式钻(孔)台 mobile drilling platform
移动式钻探设备 portable drilling rig
移动式钻头 removable drill bit; slip-on bit
移动数值 translation value
移动水波 wave of translation
移动丝 movable hair; movable retic(u)le; moving hair; moving thread; moving wire; travel(l)ing hair; travel(l)ing thread; travel(l)ing wire
移动速度 migration velocity; rate of travel; ratio of travel; translational speed
移动速率 rate of migration; rate of movement; rate of travel; travel rate
移动算术平均法 method of moving arithmetic(al) average
移动所需时间 shifting hour
移动塔 <缆道的> travel(l)ing tower
移动塔架 travel(l)ing tower
移动塔式装卸桥 travel(l)ing unloading tower
移动台 travel(l)ing carriage; mobile station <电波设备>
移动台板式文件复印机 moving platen document copying machine
移动台架 transfer platform
移动台座 travel(l)ing platform
移动探针 travel(l)ing probe
移动梯 travel(l)ing stairway
移动调节环 pusher collar
移动停车标 portable stop sign
移动停车标志牌 movable stop sign board
移动停车牌 portable stop board

移动挺杆起重机 travel(l)ing jib crane
移动通信[讯] mobile communication
移动通信[讯]网 mobile communication network
移动头磁盘 moving head disc [disk]
移动脱轨辙叉 movable derailing frog
移动弯沉仪 travel(l)ing deflectometer
移动网络计算机参考规格 mobile network computer reference; mobile network computer reference specification
移动卫星 mobile satellite
移动卫星业务 mobile satellite service
移动位置 shift position
移动污染源模式 mobile pollution source model
移动无线电 mobile radio
移动无线电导航台 radio navigation mobile station
移动无线电电台 mobile radio station
移动无线电通信[讯] mobile radio
移动无线电业务 mobile radio service
移动物标探测系统 moving target detection system
移动洗涤床 moving bed scrubber
移动狭缝 travel(l)ing slit
移动咸淡水界面近似方程 approximation equation for migrating interface between salt and fresh water
移动线 portable cord
移动线路 shiftable haulage line
移动相 mobile phase
移动相关控制 travel-dependent control
移动相位 travel(l)ing phase
移动相位离子色谱法 mobile phase ion chromatography
移动向量 translation vector
移动小车 travel(l)ing car; travel(l)ing guide carriage
移动效应假设 shift effect hypothesis
移动卸料器 travel(l)ing tripper
移动信号 movable signal
移动信号机【铁】 movable signaller; portable signal
移动信号员 mobile signalman
移动信号圆牌 portable signal disc
移动信号圆盘 portable signal disc
移动行程 shift motion; travel(l)ing motion
移动性 mobility; portability; transferability
移动性城市污染源 moving urban pollution source
移动性低气压 travel(l)ing depression(low); travel(l)ing low depression
移动性反气旋 migratory anticyclone; mobile anticyclone
移动性高气压 migratory anticyclone; mobile anticyclone; travel(l)ing anticyclone; travel(l)ing high
移动性活性炭床 activated carbon moving bed
移动性控制断面 shifting control section
移动性能 locomotiveness
移动性农业 ladang
移动性气旋 travel(l)ing cyclone
移动性沙洲 travel(l)ing bar
移动性污染源 mobile pollution source; mobile source of pollution
移动性污染源模型 mobile source model
移动性污染源排污 mobile source emission
移动压板 moving platen
移动烟幕 rolling smoke curtain
移动养路工班 migrating crew
移动遥测台 mobile telemetering station
移动业务 mobile service
移动抑制试验 migration inhibition test
移动抑制因子 migration inhibition factor
移动翼式堆料机 travel(l)ing winged stacker
移动用户 mobile subscriber
移动邮政局 travel(l)ing post office
移动油缸冲程 retracting cylinder stroke
移动预备林 shifting reserves
移动预算 moving budget
移动圆盘信号 hand disc signal
移动圆盘信号机【铁】 hand disc signal
移动载荷 travel(l)ing load
移动站台 travel(l)ing platform
移动站台滚子 travel(l)ing platform runners
移动罩反冲滤池 movable hood backwashing filter
移动罩滤池 mobile cover filter; movable hood filter
移动罩式喷漆室喷雾净化装置 movable purifying device for paint sprayer room
移动真空吸尘器 portable vacuumatic cleaner
移动支架逐跨施工法 span-by-span method
移动支柱 floating post
移动制材厂 portable mill
移动中心架 follow rest
移动终端 mobile termination
移动轴 shifting axle; slide shaft
移动轴承 shifting bearing
移动轴锁球弹簧 shifter shaft lock ball spring
移动轴支架 shifter shaft support
移动主梁门式起重机 gantry crane with movable girder
移动住房 mobile residence
移动住房停车场 mobile residence park
移动柱座标志 stanchion sign
移动铸造起重机 travel(l)ing foundry crane
移动抓取设备 floating grab(bing) installation
移动转臂起重机 travel(l)ing derrick crane
移动装车斜坡台 mobile loading ramp
移动装船机 travel(l)ing shiploader
移动装配法 moving assembly method
移动装置 mobile device; movable installation; running gear; shift device; shifter; shifting unit; travel(l)ing equipment
移动状态 translational state
移动总量 moving total
移动钻杆 rod walk
移动钻机 moving in a rig
移动钻孔平台 mobile drilling platform
移堆机 cocklifter
移峰效应 peak shifting effect
移幅键控 amplitude shift keying
移附污染 migratory stain
移管法 transfer pipe method
移管机 pipe-mover
移轨 removed rail
移轨保护 removed rail protection
移轨机 track shifter
移轨器 rail(road) slewer; track shifter
移行【计】 line advance; new line
移后 setback; zoom back
移后角 angle of lag
移后扣减 carry-over
移花 paper transfer
移画印花法 decal; decal comania method
移画印花图案 decal comania
移换 transvection; transferring of rail <钢轨的>
移积 allochthonous deposit; removal
移积土 allochthonous soil; transported soil
移交 devolution; farm out; hand-over; taking-over; transfer; turnover
移交财产或管理权给受托人 trustee
移交车 loaded cars to be delivered at junction station
移交车周转时间 average turnaround time of loaded wagons to be delivered
移交的 handing over
移交服务 deliverable service
移交邻路的货物吨数 tonnage delivered to connecting carriers
移交路 transferor railway
移交前使用 use before taking over
移交前最后检验 final inspection
移交试验 service-type test
移交条件 transfer condition
移交协定 devolution agreement
移交协议 devolution of agreement
移交证书 handing-over certificate; take-over certificate; taking-over certificate
移交重车 loaded wagons delivered at junction station
移交重车保有量计划 plan of number of loaded wagons for delivery at junction station to be kept
移进 shift-in
移进波 translational wave; translatory wave
移进符号 shift in character
移居 transmigrate; migration <尤指移居外国>
移居的 migratory; transmigrant
移居法 <美> Homestead Act
移居国外 emigrate; external migration; emigration
移居国外的人 emigrant (out)
移居国外者 expatriate
移居入境 immigrate; immigration
移居者 emigrant (out); immigrant in; migrant; migrator; resettler; settler; transmigrant; transmigrator
移距 displacement; shift; travel distance
移距系数 coefficient of offset; extension coefficient
移开 off bear
移来 in-migration
移来的 immigrant
移离 moving apart
移离目标物 <指电视摄像机> zoom out
移列部件 column-shift unit
移码 frame shift
移锚 anchor shifting; shifting anchor
移锚不停车【疏】 shifting anchor without engine stop
移锚船 anchor barge
移锚吊杆 anchor boom
移锚绞车 anchor handling winch; anchor hoisting winch
移锚位 anchorage shifting; shift anchorage
移锚延迟系数 anchor moving delay factor
移苗栽培 cultivation by shifting seeding
移民 emigrant (out); immigrant; immigration; migration; resettler; settler; transmigrator; transmigrant <尤指中转移民>
移民安居 resettlement
移民安置 resettlement
移民安置补偿费 resettlement cost
移民安置策略 resettlement strategy
移民安置村 squatter village
移民安置范围 extent of resettlement
移民安置费补偿 resettlement compensation cost
移民安置费估算 resettlement cost estimate
移民安置费预算 resettlement budget
移民安置规划 resettlement planning
移民安置经费 resettlement expenditures
移民安置可行性 resettlement feasibility
移民安置区 host area; resettlement area
移民安置实施方案 resettlement scenario
移民安置条例 resettlement regulation
移民安置政策 resettlement policy
移民安置总方针 general resettlement policy
移民出境 emigration
移民船 emigration ship <从国外移入>; emigrant ship <向国外移居的>
移民点 settlement
移民动态 dynamics of migration
移民队 colony
移民法 immigration law; resettlement regulation
移民工人 migrant labo(u)r; migratory labo(u)r
移民工作 resettlement effort
移民官员 immigration officer
移民汇款 immigrant remittance
移民计划 resettlement effort
移民局 immigration and naturalization office; immigration office; resettlement bureau
移民局官员 immigration officer
移民垦荒 moving people to other places for land reclamation
移民锚地 <从国外移入> emigration anchorage
移民区 migration area; zone of migration
移民手续 immigration procedure
移民水位 resettlement level
移民投资预算 resettlement budget
移民团体 immigrant community
移民政策 emigration policy; policy on resettlement
移膜涂饰(法) transfer finishing; transfer coating
移挪补空 kiting
移频 frequency shift; shift frequency
移频编码 frequency shift coding
移频变换器 frequency shift converter; shift converter
移频传输 frequency shift transmission
移频单音频 frequency shifted audio tone
移频叠加 frequency shift overlay
移频发射机监控装置 frequency shift transmitter monitor
移频轨道电路 audio-frequency shift modulated track circuit
移频键控 frequency shift keying
移频键控电报 frequency shift keying telegraph
移频键控器 frequency shift keyer
移频键控射频信道 frequency shift keyed radio channel
移频键控通信[讯] frequency shift keying working
移频键控信号音 frequency shifted keying tone

移频接收器 frequency shift receiver
移频数据 frequency shift data
移频调幅变换器 FS/AM [frequency shift/amplitude modulation] converter
移频调制 frequency shift modulation
移频调制音频电报制 frequency shift modulated voice-frequency telegraph system
移频系统 frequency shift system
移频制 frequency shift system
移频制通信[讯] frequency shift communication
移频子空间迭代法 subspace iteration method with shifts
移频自动闭塞 automatic block with audio frequency shift modulated track circuits; frequency shift automatic block
移栖 migration
移栖的 migratory
移栖动物 migrant
移栖(物)种 migratory species
移气吸管 gas pipet(te)
移前 set forward;zoom in
移前角 angle of lead
移前扣减 carry back
移情作用 < 美学 > empathy
移去 dislodge
移去荷载 cut loose
移去类别 removal class
移去能力 bleed-off
移圈 barring-on
移圈式调压器 shifting coil voltage regulator
移入 shift-in
移入物 engraft
移入移出的差额 balance of migration
移入者 immigrant
移数网络 shift network
移刷环 brush-rocker ring
移刷型电动机 brush shifting motor
移送 deport
移送机 chain-and-ducking dog mechanism;drag-over unit;pull-over gear;skid;transfer bed;transfer gear
移送机构 transfer mechanism
移送模块 evoke module
移送台架 transfer bank
移酸滴管 portacid
移头器 shifter
移图印花法 transfer process
移图印花装饰 transfer decoration
移挖充填 cutting and fill
移挖作填 cut-and-fill (excavation);cut-fill transition
移位 dislodge;displacement;relocate;relocation; replacing; shift (ing); translocate; transpose; transposition;travel motion;translation【地】
移位变化 metastasis [复 metastases]
移位补偿 bit shift compensation
移位不变性 position-invariance;shift invariant
移位操作 shifting function;shift operation shifting
移位除数 shift(ed) divisor
移位触发器 carry flip-flop;displacement contact
移位次数 shift count
移位电路 shifting circuit
移位电子 metastasic electron
移位定理 shift(ing) theorem
移位动作 shift motion
移位多数元组 shift multibyte
移位罚函数 shifted penalty function
移位服务 deliverable service
移位符号 shift character
移位感觉 referred sensation
移位公式 shift formula
移位功能变换器 displacement transducer
移位过程 shifting process
移位函数 shifting function
移位激励器 shift driver
移位及轮转 shift and rotate
移位及轮转指令 shift and rotate instructions
移位计数器 shifting counter;stepping counter
移位计数值 shift count
移位计算 shift operation
移位记发码 shift register code
移位寄存器 shift(ing) register
移位寄存器发生器 shift register generator
移位键控 shift keying
移位接触器 displacement contact
移位矩阵 shift(ed) matrix
移位控制 shift control
移位矿体 displaced ore body
移位累加器 shifting accumulator
移位链 shift chain
移位裂缝 rent of displacement
移位露头 misplaced outcrop
移位滤波 shift filtering
移位逻辑 logic(al) with shift
移位逻辑指令 shift logic(al) instruction
移位码 shift code
移位脉冲 shift pulse
移位门 shift gate
移位器 shifter
移位器装置 shifter;shift unit
移位缺陷 shift defect
移位绕组 shift winding
移位输入端 carry input
移位数 carry number
移位数字 carry digit
移位算符 shifting operator
移位算子 shifting operator
移位锁定 shift lock
移位网络 shift network
移位位置 shift position
移位系统 shifts system
移位效应 displacement effect
移位型十进位计数器 shift-type decade counter
移位岩体 displaced mass
移位运算 shift operation
移位栈 shift stack
移位折叠 shift fold
移位植物 colonist
移位指令 shift(ing) order;shift instruction
移位指数分布 < 一种连续型概率分布 >【数】 displaced exponential distribution;shifted exponential distribution
移位制 transposition system
移位终止 end of shift;shift end
移位作业方式 non-stationary way of operation
移位作用 translocation
移坞装置 dock moving equipment
移线 transferred position line; shipping line transfer【船】
移线船位 running fix
移线定位 running fix
移线位置线 replaced position line
移相 dephasing;outphasing;phase shift
移相变换器 phase shifting transformer;shift converter
移相变位 phase deviation
移相变压器 phase shifting transformer;phasing transformer
移相触发器 phase shift trigger
移相电流 dephased current
移相电路 phase setting circuit;phase shift(ing) circuit;shifting circuit
移相电容器 phase shifting condenser
移相定理 shifting theorem
移相段 jayrator
移相键控 phase shift keying
移相角(度) phase shifting angle
移相开关 phase switcher
移相控制 phase shifting control
移相量度 dephasing measure
移相滤波器 < 使信号延迟但波形不失真的滤波器 > all-pass filter;phase shift filter
移相器 phase adapter;phase changer;phaser;phase shifter;phase shifting device;phase switcher;shifter
移相示功图 shifted indicator diagram
移相式振荡器 phase shifter; phase shift oscillator
移相调制 phase shifting modulation
移相调制方式 out phasing modulation system
移相用自耦变压器 phase shifting autotransformer
移箱输送机方式 portveyor system
移向目标 zooming
移向目标物 < 指电视摄像机 > zoom in
移项【数】 transpose (term); rearrangement;shift;transposition
移项的 transpositive
移项器 transposer
移像光电摄像管 ariscope;iconotron;image shift iconoscope;photicon
移像光电稳定摄像管 photo-electron-stabilized-photicon
移像式光电摄像管 image iconoscope;super-iconoscope
移像正析摄像管 image vericon
移行式装吊车 creeper crane
移雪器 deflector
移液管 decantation tube; suction pipette;volumetric(al) pipet(te)
移液管法 pipet(te) method
移液管法颗粒分析 pipet(te) particle analysis
移液吸(移)管 pipet(te);transfer pipet(te);volumetric(al) pipet(te)
移液细管 transfer pipet(te)
移用的材料 borrowed material
移运 deport
移栽前处理 preplant treatment
移栽前土壤熏蒸剂 replanting soil fumigant
移载 load shifting
移载法 < 船舶脱浅法 > shifting method
移站定线法 < 在不能设站的两固定点之间的直线上设中间点 >【测】 wiggling-in
移植 explant;explantation;graft;plant out; transplantation; transplant(ing)
移植半熟树 transplanting semi-mature tree
移植法 implantation;transfer method
移植工具 migration tool
移植幼苗 sprigging
移置 allochthonous deposit;displacement;remove;transportation;transpose;transposition
移置暴雨 transposed storm
移置大陆边缘 translation continental margin
移置的 allochthonous
移置地下水 allochthonous groundwater
移置法 transposition method
移置风暴 transposed storm
移置煤 allochthonous coal
移置体【地】 allochthon(e)
移置体的【地】 allochthonous
移置调整 transposition adjustment
移置线路信号机【铁】 relocated wayside signal
移置限制 transposition limit
移置学说 < 煤形成的 > allochthonous theory
移置作用 displacement
移重工具 heaver
移重机 transfer crane
移轴机构 swing and tilt mechanism
移注 decant;decantation;transfuse
移注阀 transfer valve
移注液 transfer passage
移柱立式镗床 travel(l)ing column vertical boring machine
移转 aversion
移转价值 transfer value
移准距点透镜 anallatic lens
移走泥炭 removal of peat
移走型砖 < 由制砖黏[粘]土场顶面层处移去型砖 > encallowing
移坐标轴 displacement of coordinate axis

遗

遗产 bequest; decedents' estates; estate;heritage;inheritance;legacy

遗产承受人 legatee
遗产管理人 estate administrator
遗产管理人的契约 administrator's deed
遗产管理委任状 letters of administration
遗产继承人 legatee of inheritance
遗产继承中的选择权 elective share
遗产税 death duty;estate duty;estate tax; legacy duty; settlement estate duty
遗产税法 inheritance tax law
遗产税投资信托公司 Estate Duties Investment Trust
遗产转让 transfer at death
遗传 entail;heredity;inheritance
遗传的 descendant
遗传的特征 inheritance
遗传工程 genetic (al) engineering;hereditary engineering
遗传河 epigenetic river;superimposed river
遗传环境协方差 covariance between inheritance and environment
遗传密码 genetic(al) code
遗传平衡 genetic(al) equilibrium
遗传趋势 inherited tendency
遗传生态学 genecology
遗传算法 genetic(al) algorithm
遗传物质 genetic(al) material
遗传性格学 hereditary characterology
遗传学 genetics
遗传学的 genetic(al)
遗憾临界值 regret critical value
遗迹 historic (al) remains; ichnite;ichnolite;monument;old ruins;relic; remainder; remains; rudiment;rudimentum;traces;vestige
遗迹河 indapted river; misfit river;underfit river
遗迹化石 ichnofossil;trace fossil
遗迹(化石)相 ichnofacies
遗迹化石学 ichnology
遗迹类型 type of seismic remains
遗觉像 eidetic image
遗留 leave behind
遗留负载 remaining load
遗留工程 remaining works
遗留谷 inherited valley
遗留河 inherited river;inherited stream
遗留水系 inherited drainage;superim-

posed drainage
遗留体制 hold-over system
遗留物 remnant
遗留误差 inherited error
遗漏 missing;omission;escape【计】
遗漏不在此限 omission excepted
遗漏错误 errors of omission;missing error
遗漏划线 missing line
遗漏码 missing code
遗漏设计 missing design
遗漏误差 error of omission;missing error
遗漏系数 omission factor
遗漏值 missing value
遗漏中断处理程序 missing interruption handler
遗弃 dereliction
遗弃人 abandoner
遗失 missing
遗失的托运货物(或行李包裹) missing consignment
遗失的行李 missing luggage
遗失的支票 lost check
遗失后补缴全部保险费 full premium if lost
遗失或损坏保险 insurance against loss or damage
遗失货物 missing cargo;missing goods
遗失货物报告 return of missing goods
遗失货物目录清单 missing list
遗失误差 dropout error
遗失险 risk of non-delivery
遗失信息 dropout information
遗失行李 missing baggage
遗忘 amnesia
遗忘因子 forgetting factor
遗物 relic;remainder;remains
遗物箱 reliquary
遗误除外 error and omission excepted
遗益享受保险单 income benefit insurance policy
遗赠 bequest;causa donation;causa mortis;demise;devise;legacy
遗赠财产 real property
遗赠动产 bequeath
遗赠人 bequeather;devisor
遗赠受益人 devisee
遗赠物 <一般指动产> legacy
遗赠者 bequeather
遗址 archeologic(al) site;legacy;remainder;ruins
遗嘱 testament;will
遗嘱更改 codicil
遗嘱检验法庭 probate court
遗嘱人 estator
遗嘱上的地产 estate at will
遗嘱信托财产 testamentary trust
遗嘱修改附录 codicil
遗族恤金 survivor's benefit

颐和园佛香阁 <北京> pavilion of the Fragrance of Buddha

疑存 <指图示水下障碍物> existence doubtful

疑符定位器 erasure locator
疑惑区间 zone of uncertainty
疑难的 problematic(al)
疑难工程 problem job
疑难榴石 griphite

乙胺 ethyl amine

乙苯 phenylethane
乙苯的溶解度 solution ethylbenzene
乙苄纤维素 ethyl benzyl cellulose
乙丙二烯单体 ethylene propylene diene monomer
乙丙共聚物 ethylene propylene copolymer
乙丙胶垫 ethylene-propylene packing
乙丙烷碳同位素值间距 separation of ethane propane carbon isotope value
乙丙橡胶 ethylene-propylene rubber
乙草酸 ethoxal
乙撑氯醇 ethylene chlorohydrin
乙撑亚胺 aziridine;ethyleneimine
乙橙 ethyl orange
乙川【化】 acet
乙醇 ethyl alcohol;grain alcohol;spirit of wine
乙醇胺 cholamine;ethanolamine
乙醇胺法 ethanol amine process
乙醇苯沥青 A 含量 ethyl alcohol-benzene bitumen A content
乙醇不溶物 alcohol insoluble matter
乙醇萃取物 ethanolic extract
乙醇化物 alcoholate
乙醇胶试验 ethanol gel test
乙醇镁 magnesium ethoxide;magnesium ethylate
乙醇钠 sodium alcoholate;sodium ethylate
乙醇醛 glycolaldehyde
乙醇醛二聚物 glycolaldehyde dimer
乙醇溶液 <金相试剂> nital
乙醇酸 glycollic acid
乙代己酸盐 two-ethyl hexoate
乙底酸 <即 EDTA,乙二胺四醋酸> editic acid
乙电 B-source
乙电池 plate battery
乙电池组 anode battery;B-battery
乙电路 B-circuit
乙电源 B power supply
乙二铵四乙酸铬 chromium ethylenediamine tetraacetic acid
乙二铵四乙酸钴 cobalt ethylenediamine tetraacetic acid
乙二胺 ethylenediamine
乙二胺醋酸 ethylene diamine tetraacetic acid
乙二胺四乙酸 ethylene-diamine tetracetic acid
乙二胺四乙酸铵盐速测法 EDTA-ammonium salt method
乙二胺四乙酸滴定法 EDTA titration
乙二胺四乙酸盐 Edentate
乙二醇 ethylene glycol;glycol
乙二醇-苯醚 phenyl cellosolve
乙二醇-丙醚 propyl cellosolve
乙二醇单丁醚 butyl cellosolve;butyl glycol
乙二醇单甲醚乙酸酯 ethylene glycol monomethyl ether acetate
乙二醇单乙醚月桂酸酯 ethylene glycol monoethyl ether laurate
乙二醇-丁醚 ethylene glycol butyl ether
乙二醇丁酸酯 ethyleric glycol butyrate
乙二醇二醋酸酯 glycol diacetate
乙二醇二甲基丙烯酸酯 ethylene glycol dimethacrylate
乙二醇二甲醚 ethylene glycol dimethyl ether
乙二醇二硝酸酯 nitroglycol
乙二醇二乙酸酯 glycol diacetate
乙二醇二月桂酸酯 glycol dilaurate
乙二醇防冻剂 <其中的一种> prestone
乙二醇-甲醚 ethylene glycol monomethyl ether
乙二醇-甲醚乙酸酯 methyl cellosolve acetate
乙二醇醚 glycol ether
乙二醇醚乙酸酉目 glycol ether acetate
乙二醇醚酯(类) glycol ether ester
乙二醇水溶液防冻剂 glysantine
乙二醇顺丁烯二酸酯 glycol maleate
乙二醇缩乙醛 glycol ethylidene-acetal
乙二醇吸附 glycol adsorption
乙二醇吸附法 alcohol adsorption method
乙二醇硝酸酯 glycol nitrate
乙二醇盐 glycolate
乙二醇乙醚 glycol monoethyl ether
乙二醇乙醚蓖麻酸酯 ethoxyethyl ricinoleate
乙二醇乙醚乙酸酯 cellosolve acetate;ethoxy ethanol acetate
乙二醇乙酸酯 ethylene glycol monoacetate;glycol monoacetate
乙二醇异丙醚 glycol isopropyl ether
乙二醇酯 glycol ester
乙二磺酸 ethionic acid
乙二磺酸盐 ethanedisulphonate
乙二腈 cyanogen;oxalonitrile
乙二醛 ethylene dialdehyde;glyoxal
乙二酸 ethanedioic acid;oxalic acid
乙二酸钡 barium oxalate
乙二酸盐 oxalate
乙二酰氯 ethanedioyl chloride;oxalyl chloride
乙方 party B;second party;the second party
乙方检验 second party inspection
乙-呋喃甲醛 furol
乙钢 Z-iron;Z-steel
乙硅酮树脂 ethyl silicone resin
乙硅烷 disilane;silicoethane
乙硅烷硫基 disilanylthio
乙硅烷氧基 disilanoxy
乙焊机 acetylene welder
乙(换)硫酸 sulphovinic acid
乙基 ethyl
乙基苯 ethyl benzene;phenylethane benzene
乙基苄基醚 ethyl benzyl ether
乙基苄基纤维素 ethyl benzyl cellulose
乙基橙 ethyl orange
乙基碘 ethyl iodide;hydroiodic ether;iodoethane
乙基硅酮树脂 ethyl silicone resin
乙基硅油 ethyl silicon oil
乙基红 ethyl red
乙基化 leading
乙基化作用 ethylation
乙基环己烷 ethylcyclohexane
乙基环戊烷 ethylcyclopentane
乙基己醇 ethyl hexanol
乙基甲基醚 ethyl methyl ether
乙基磷酸汞剂 <木材防腐> timsan
乙基硫酸 ethyl hydrogen sulfate;ethyl sulfuric acid
乙基醚 ethyl ether
乙基汽油 <含四乙铅的> ethyl gasoline;ethyl petrol
乙基羟乙基纤维素 ethyl hydroxyethyl cellulose
乙基燃料 ethylized fuel
乙基溶纤剂 ethyl cellosolve
乙基噻吩 ethylthiophene
乙基三氟甲硅烷 ethyl trifluorosilane
乙基三甲基甲硅烷 ethyl trimethyl silane
乙基四氢化萘 ethyltetrahydronaphthalene
乙基缩水甘油醚 glycidyl ethyl ether
乙基戊基甲酮 ethyl amyl ketone
乙基纤维(类)塑料 ethyl cellulose plastics
乙基纤维素 ethyl cellulose
乙基纤维素漆 ethyl cellulose lacquer
乙基橡胶 ethyl rubber
乙基液 ethyl fluid;ethyl liquor;motor mix
乙基乙炔 ethyl acetylene
乙基乙烯基醚 ethyl vinyl ether
乙基异丁基甲酮 ethyl mobutyl ketone
乙基正丁基甲酮 ethyl n-butyl ketone
乙级 <指货品> B grade
乙级道路 <联邦资助次要干道> B system
乙级检验 class B inspection
乙级木材 B-grade wood
乙交换机 B-position
乙交换台 B board;B switchboard
乙阶段 B-stage
乙阶(段)树脂 B-stage resin
乙阶酚醛树脂 resitol;resolite
乙腈 acetonitrile;ethanenitrile;methyl cyanide
乙类放大 Class B amplification
乙类钢 B-type steel
乙硫醇 ethanethiol
乙硫磷 ethion
乙硫橡胶 thiocol [thiokol]
乙硫橡胶基材料 thiocol[thiokol] based material
乙硫橡胶镶尖 thiokol tip
乙纶绳 polythene rope
乙醚 ether;ethyl ether;ethyl oxide
乙醚辅助起动装置 ether starting
乙醚胶囊 ether capsule
乙醚排水装置 ether discharger
乙醚喷注冷起动 ether injection starting
乙醛 acetaldehyde;aldehyde;ethanal
乙醛二乙酸酯 ethyudene diacetate
乙醛聚合物 acetaldehyde polymer
乙醛酶 acetaldehydase
乙醛树脂 acetaldehyde resin
乙醛酸 ethanol acid;glyoxalic acid;oxaldehydic acid
乙醛缩二乙醇 acetal;ethylidene ether
乙醛肟 acetaldoxime
乙炔 acetylene;acetylene gas
乙炔苯 ethynylbenzene
乙炔苯酚树脂 acetylene phenol resin
乙炔测定 determination of acetylene
乙炔吹管 acetylene torch
乙炔灯 acetylene burner;acetylene lamp
乙炔灯(光)浮标 acetylene light buoy
乙炔灯照明的浮标 acetylene gas lighted buoy
乙炔灯照明的救生圈 acetylene gas lighted buoy
乙炔发生器 acetylene generator;acetylene producer;carbide-feed generator
乙炔发生站 acetylene generating station
乙炔管 acetylene pipe
乙炔罐 acetylene cylinder
乙炔焊 acetylene weld(ing);air acetylene welding;autogenous weld(ing);gas welding
乙炔焊接 acetylene weld(ing);oxyacetylene welding;oxy welding
乙炔焊(接)吹管 acetylene welding torch
乙炔焊接橡胶管 acetylene welding rubber hose
乙炔焊炬 acetylene burner;acetylene welding torch
乙炔焊枪 acetylene welding torch
乙炔化(合)物 acetylide
乙炔化三氯 acetylene trichloride;trichloroethylene
乙炔化亚铜 cuprous acetylide

乙炔化作用 ethinylation
乙炔还原方法 acetylene reduction method
乙炔火焰 acetylene flame
乙炔基 ethynyl
乙炔基苯 acetylenyl benzene
乙炔基甲醇 acetylenyl carbinol; ethynyl carbinol
乙炔基金属 metal acetylide
乙炔基氯 acetylene chloride
乙炔计数器 acetylene filled counter
乙炔(加热)解除应力 acetylene removal of stress
乙炔炬 acetylene burner; acetylene torch
乙炔钠 sodium acetylene
乙炔喷灯 acetylene torch
乙炔瓶阀 acetylene cylinder valve
乙炔起动机 acetylene starter
乙炔气割 acetylene cutting; autogenous cutting
乙炔气管道 acetylene gas pipe
乙炔气焊工 acetylene welder
乙炔气焊机 acetylene welder
乙炔气焊枪 acetylene welder
乙炔气焊设备 acetylene apparatus; acetylene welder
乙炔气汇流排 battery of acetylene cylinders
乙炔气瓶 acetylene bottle; acetylene cylinder
乙炔气瓶组 acetylene cylinder unit
乙炔气闪光器 acetylene flasher
乙炔气(体) acetylene gas
乙炔气调节器 acetylene regulator
乙炔气筒 acetylene cylinder
乙炔汽油 acetyl gasoline
乙炔前灯 acetylene headlight
乙炔切割 acetylene cutting
乙炔切割器 acetylene cutter
乙炔燃烧器 acetylene burner
乙炔熔化 autogenous cutting
乙炔闪光灯 acetylene flashing lamp; acetylene flashing light
乙炔闪光器 acetylene flasher
乙炔烧割器 acetylene cutter
乙炔设备 acetylene apparatus
乙炔炭黑 acetylene carbon black
乙炔锌 zinc acetylide
乙炔压力自动调整装置 acetylene regulator
乙炔焰 acetylene flame; acetylene torch
乙炔焰渗碳 acetylene flame carburizing
乙炔银 silver acetylide
乙炔渣 acetylene residue; acetylene sludge
乙炔站 acetylene station
乙炔照明信号前灯 acetylene headlight
乙炔照明信号头灯 acetylene headlight
乙胂化硫 ethyl arsine sulfide
乙酸 acetic acid
乙酸铵 ammonium acetate
乙酸苯汞 phenylmercuric acetate
乙酸苄酯 benzyl acetate; phenyl methyl acetate
乙酸丙酸纤维素 acetopropion cellulose; cellulose acetate propionate; cellulose acetopropionate
乙酸丙酮镍 nickel acetylacetonate
乙酸丙酯 propyl acetate
乙酸淀粉 amylose acetate
乙酸丁酸纤维素 acetobutyryl cellulose; acetyl butyryl cellulose; cellulose acetate butyrate; cellulose acetobutyrate
乙酸丁酸纤维素喷漆 cellulose acetate butyrate lacquer
乙酸丁酸纤维素塑料 cellulose acetate butyrate plastics
乙酸丁酯 butyl acetate
乙酸法铅铬黄 acetate chrome
乙酸法铅铬绿 acetate green
乙酸钙 calcium acetate
乙酸钙镁 calcium and magnesium acetate
乙酸甘油酯 acetoglyceride; glycerol acetate
乙酸酐 acetic anhydride
乙酸高镍 nickelic acetate
乙酸镉 cadmium acetate
乙酸根离子 acetate ion
乙酸庚酯 heptyl acetate
乙酸癸酯 decyl acetate
乙酸环己酯 adrollal; cyclohexyl acetate
乙酸基 acetate
乙酸己酯 hexyl acetate
乙酸甲基环己酯 heptaline acetate
乙酸甲(基)戊酯 methyl amyl acetate
乙酸甲基异丁基甲酯 methyl isobutyl carbinol acetate
乙酸甲氧基丁酯 methoxy butyl acetate
乙酸甲酯 methyl acetate
乙酸铝 alumin(i)um acetate
乙酸镁 magnesium acetate
乙酸锰 manganese acetate
乙酸钠 sodium acetate
乙酸镍 nickel acetate; nickelous acetate
乙酸钕 neodymium acetate
乙酸铅 acetate of lead; lead acetate; lead sugar; sugar of lead
乙酸铅法 lead acetate method
乙酸铅试验 lead-acetate test
乙酸壬酯 nonyl acetate
乙酸溶纤剂 cellosolve acetate
乙酸三丁基锡 tributyl tin acetate
乙酸叔丁酯 tertlary butyl acetate
乙酸双氧铀 uranium acetate; uranyl acetate
乙酸锶 strontium acetate
乙酸铊 thallium acetate
乙酸铜 copper acetate; crystal aerugo; crystals of Venus; verdigris
乙酸戊脂灯 amyl acetate lamp
乙酸戊酯 banana oil; isoamyl acetate; oil of pears
乙酸烯丙酯 allyl acetate
乙酸细菌 acetic acid bacteria
乙酸纤维素 acetyl cellulose; cellulose acetate
乙酸纤维(素)蒙皮(清)漆 acetate clear dope
乙酸纤维素漆 acetyl cellulose lacquer
乙酸纤维酯蒙皮漆 acetate dope
乙酸纤维酯涂布漆 acetate dope
乙酸锌 zinc acetate
乙酸亚汞 mercurous acetate; mercury acetate; mercury protoacetate
乙酸亚砷酸铜 copper acetate senite
乙酸亚铁 ferrous acetate; iron acetate
乙酸亚铁溶液 printer's liquor
乙酸盐 acetate
乙酸盐雾试验 acetic acid salt spray test
乙酸乙汞 ethyl mercury acetate
乙酸乙烯-苯乙烯共聚物 vinyl acetate-styrene copolymer
乙酸乙烯苯酯 vinylphenyl acetate
乙酸乙烯-丙烯酸乳胶 vinyl acetate acrylic latex
乙酸乙烯-丙烯酸酯共聚 vinyl acetate-acrylic ester
乙酸乙烯-顺丁烯二酸酐共聚物 vinyl acetate-maleic anhydride copolymer
乙酸乙烯-乙烯共聚物 vinyl acetate-ethylene copolymer
乙酸乙烯(酯) vinyl acetate
乙酸乙烯酯乳化漆 vinyl acetate emulsion paint
乙酸乙烯(酯)树脂 vinyl acetate resin
乙酸乙烯酯树脂涂料 vinyl acetate resin coating
乙酸乙酯 ethyl acetate
乙酸乙酯共聚体乳液黏[粘]合剂 vinrez
乙酸乙酯水解 hydrolysis of ethyl acetate
乙酸异丙酯 isoprol acetate
乙酸异丁酯 isobutyl acetate
乙酸异己酯 two-ethyl butyl acetate
乙酸异辛酸酯 two-ethyl hexyl acetate
乙酸银 silver acetate
乙酸正丙酯 n-propyl acetate
乙酸正己酯 n-hexyl acetate
乙酸正壬酯 n-nonyl acetate
乙酸(正)铜 cupric acetate
乙酸正戊酯 amyl acetate
乙酸酯 acetate; acetic ester
乙酸酯法 acetin method
乙酸酯涂层聚乙烯织物 acetate coated polyethylene fabric
乙酸酯油墨 acetate ink
乙酸仲丁酯 secondary butyl acetate
乙酸仲戊酯 secondary amyl acetate
乙缩醛 acetal
乙缩醛共聚物 acetal copolymer
乙缩醛树脂 acetal resin
乙台 incoming junction position; incoming position; inward position; B switchboard; B station <电话交换台>
乙台话务员 B-operator
乙替甲酰胺 ethyl-formamide
乙替乙酰替苯 n-ethyl acetanilide
乙酮醇 ketol
乙烷 ethane; ethyl hydride
乙烷的溶解度 solubility of ethane
乙烷二磺酸 ethane disulfonic acid
乙烷二甲酸 ethane dicarboxylic acid
乙烷四甲酸 ethane tetracarboxylic acid
乙烷碳稳定同位素组成 stable carbon isotopic composition of ethane
乙戊酮 ethyl amyl ketone
乙烯 ethylene; ethyl vinyl acetate
乙烯丙烯二烯三元共聚物 ethylene propylene diene tripolymer
乙烯丙烯二烯系单体 ethylene propylene diene monomer [EPDM]
乙烯丙烯二烯烃合成橡胶 ethylene propylene synthetic rubber [EPDM]
乙烯丙烯二烯系共聚橡胶 EPDM rubber
乙烯丙烯共聚物 ethylene propylene copolymer
乙烯丙烯酸 ethylene acrylic acid
乙烯丙烯酸乙酯共聚物 ethylene-ethyl acrylate copolymer
乙烯丙烯酸酯共聚物 ethylene acrylate copolymer
乙烯丙烯弹性体 ethylene propylene elastomer
乙烯丙烯橡胶 ethylene-propylene rubber
乙烯箔 cello foils; vinyl foils
乙烯薄板 vinyloid sheet; vinyl sheet(ing)
乙烯薄膜 vinyl film; vinyl membrane
乙烯布 vinoleum
乙烯叉 vinylidene
乙烯叉二氯 vinylidene chloride
乙烯掺和剂 vinyl blend
乙烯产品 ethylene product
乙烯厂 ethylene plant
乙烯撑 vinylene
乙烯船 ethylene tanker
乙烯醇 ethenol
乙烯醇缩乙醛 vinyl acetal
乙烯醇缩乙醛树脂 vinyl acetal resin
乙烯醋酸乙烯共聚物 ethylene-vinyl acetate copolymer
乙烯地板 ethyl flooring
乙烯地板面层 sheet vinyl floor finish
乙烯地毯 vinoleum
乙烯丁二烯共聚物 ethylene butadiene copolymer
乙烯丁醛树脂 vinyl butyral resin
乙烯丁烯共聚物 ethylene-butylene copolymer
乙烯砜染料 vinylsulfone dyes; vinyl sulfone dyestuff
乙烯复合钢板 vinyl covered steel plate
乙烯工程 ethylene project
乙烯工业废水 ethylene manufacture wastewater
乙烯共聚沥青 ethylene copolymer bitumen
乙烯共聚物 ethylene copolymer
乙烯管 ethylene tube
乙烯护墙面 vinyl wall facing
乙烯化作用 vinylation
乙烯环己烷 vinyl cyclohexane
乙烯磺酸 vinyl sulfonic acid
乙烯基 vinyl
乙烯基板 vinyl tile
乙烯基包裹线 <用盐化乙烯基树脂包皮电线> vinyl covered cord
乙烯基苯 vinylbenzene
乙烯基苯酚 vinylphenol
乙烯基吡咯烷酮 vinyl pyrrolidone
乙烯基薄膜 vinyl film
乙烯基衬里 vinyl liner
乙烯基次乙基 vinyl ethylene
乙烯基醋酸 vinylacetic acid
乙烯基底漆 vinyl lacquer
乙烯基地板覆盖层 vinyl flooring cover(ing)
乙烯基地板终饰 vinyl flooring finish
乙烯基碘 vinyl iodide
乙烯基丁酸酯 vinyl butyrate
乙烯基氟 vinyl fluoride
乙烯基改性醇酸树脂 vinylated alkyd resin
乙烯基改性油 vinylated oil
乙烯基共聚物 vinyl copolymer
乙烯基硅橡胶 vinylsiloxane rubber
乙烯基和丙烯酸胶粘剂 vinyl and acrylic acid adhesive
乙烯基化物 vinyl compound
乙烯基甲苯 methyl styrene; vinyltoluene
乙烯基甲醚顺丁二酸酐共聚物 polyvinyl methyl ether-maleic anhydride copolymer
乙烯基甲酸 vinyl formic acid
乙烯基甲酮类 vinyl ketones
乙烯基胶片 vinyl film
乙烯基聚合物 vinyl polymer
乙烯基卷材地面 vinyl sheet flooring
乙烯基均聚物 Lustrex
乙烯基咔唑 vinyl carbazole
乙烯基立方体 vinyl cube
乙烯基楼板覆盖层 vinyl flooring cover(ing)
乙烯基楼面覆盖层 vinyl composition tile
乙烯基楼板终饰 vinyl flooring finish
乙烯基氯 vinyl chloride
乙烯基醚树脂 vinyl ether resin
乙烯基密封剂 vinyl-base sealant
乙烯基黏[粘]合剂 vinyl adhesive; vinyl bonding adhesive; vinyl cementing agent

乙烯基泡沫衬垫 vinyl foam cushioning
乙烯基皮革 vinyl leather
乙烯基铺地砖 vinyl flooring tile;vinyl tile
乙烯基墙纸 vinyl wall paper
乙烯基氰 vinyl cyanide
乙烯基壬基醚 vinyl nonyl ether
乙烯基溶胶美术漆 vinyl sol pattern paint
乙烯基溶胶涂料 vinyl sol paint
乙烯基溶液 vinyl solution
乙烯基溶液型涂料 vinyl solution coating
乙烯基软管 vinyl hose
乙烯基三氯硅烷 vinyl trichlorosilane
乙烯基石棉板 vinyl-asbestos tile
乙烯基石棉薄板 vinyl-asbestos sheet
乙烯基石棉材料 vinyl-asbestos material
乙烯基石棉地板覆盖层 vinyl-asbestos floor cover(ing)
乙烯基石棉地板饰面 vinyl-asbestos floor(ing) finish
乙烯基石棉化合物 vinyl-asbestos compound
乙烯基石棉楼板覆盖层 vinyl-asbestos floor cover(ing)
乙烯基石棉楼板饰面 vinyl-asbestos floor(ing) finish
乙烯基石棉团块 vinyl-asbestos mass
乙烯基石棉组成 vinyl-asbestos composition
乙烯基饰面钢板 vinyl steel plate
乙烯基树脂 vinyl(it) resin
乙烯基树脂板 vinyl tile
乙烯基树脂复合面板 vinyl composition tile
乙烯基树脂(覆面)石膏板 vinyl-surfaced gypsum
乙烯基树脂糊 vinyl paste
乙烯基树脂楼面板 vinyl flooring
乙烯基树脂墙壁覆盖物 vinyl wall covering
乙烯基树脂软片 vinylit sheet
乙烯基(树脂)石棉瓦乙烯基石棉砖 vinyl-asbestos tile
乙烯基树脂涂料 vinyl paint
乙烯基树脂瓦(管) vinyl tile
乙烯基树脂型防潮层 vinyl-type vapo(u)r barrier
乙烯基树脂型汽隔 vinyl-type vapo(u)r barrier
乙烯基树脂油漆 vinyl paint; vinyl resin paint
乙烯基塑料 vinyl plastics
乙烯基塑料地板覆盖层 vinyl plastic floor cover(ing)
乙烯基塑料地板终饰 vinyl plastic floor(ing) finish
乙烯基塑料覆面薄(钢)板 vinyl-coated sheet
乙烯基塑料管 vinyl pipe
乙烯基塑料楼板覆盖层 vinyl plastic floor cover(ing)
乙烯基塑料楼板终饰 vinyl plastic floor(ing) finish
乙烯基塑料片 vinyl plastic sheet
乙烯基(塑料)铺地料 vinyl flooring
乙烯基(塑料通风)管道 vinyl duct
乙烯基踢脚板 vinyl base;vinyl floor base
乙烯基涂料 vinyl coating
乙烯基涂(面)层 vinyl overlay;vinyl coating
乙烯基(瓦)板 vinyl tile
乙烯基乙醇 allyl carbinol;vinyl-ethyl alcohol
乙烯基乙醇酸 vinylglycollic acid
乙烯基乙醚 vinyl ethyl ether
乙烯基乙炔 vinylacetylene
乙烯基乙酸酯 vinyl acetate
乙烯基酯 vinyl ester
乙烯基转移作用 trans-vinylation
乙烯甲醚 vinyl methylether
乙烯胶乳基 vinyl latex base
乙烯聚合物 ethylene polymer
乙烯聚合油 ethylene polymerised oil
乙烯聚合作用 vinyl polymerization
乙烯绝缘软性电缆 vinyl cabtyre cable
乙烯类聚合物 polyvinyl
乙烯类塑料 vinyl group of plastics
乙烯醚类 vinyl ethers
乙烯密封衬垫 vinyl liner
乙烯嵌花(装饰)品 inlaid vinyl goods
乙烯墙衬 vinyl wall lining
乙烯乳液 vinyl emulsion
乙烯树脂车辆 vinyl resin vehicle
乙烯树脂覆面石膏墙板 vinyl-covered gypsum wallboard
乙烯树脂介质 vinyl medium;vinyl resin medium
乙烯树脂黏[粘]合剂 vinyl adhesive
乙烯树脂调和漆 vinilex mixed paint
乙烯树脂载体 vinyl resin vehicle
乙烯四氟乙烯共聚物 ethylene-tetrafluoethylene copolymer
乙烯塑料 ethylene plastics
乙烯塑料壁纸 vinyl wall cladding
乙烯塑料地板布 vinyl plastic covering
乙烯塑料盘 vinyl disc
乙烯塑料溶胶 vinyl plastisol
乙烯塑料罩 vinyl cover
乙烯缩醛树脂胶 vinyl glue
乙烯烃 vinyl group
乙烯烃塑料 vinyl plastics
乙烯酮 ethenone;keten(e)
乙烯酮灯 ketene lamp
乙烯酮法 ketene process
乙烯系单体 vinyl monomer
乙烯系共聚物 vinyl copolymer
乙烯(系)化合物 vinyl compound
乙烯系墙布 vinyl wall covering
乙烯(系)乳化漆 vinyl emulsion paint
乙烯系弹料 vinyl elastomer
乙烯系涂料 vinyl paint
乙烯纤维 vinyl fiber
乙烯纤维素 vinyl cellulose
乙烯镶板 vinyl panel
乙烯镶嵌物 vinyl insert(ion)
乙烯型基 ethenoid group
乙烯型聚合物 ethenoid polymer
乙烯(型)树脂 ethenoid resin
乙烯型塑料 ethenoid plastics
乙烯压出品 vinyl extrusion
乙烯压缩机 ethylene compressor
乙烯乙酸乙烯热熔胶 ethylene-vinyl acetate copolymer hot-melt adhesive
乙烯乙酸乙烯酯 ethylene-vinyl acetate
乙烯乙酸乙烯(酯)共聚物 ethylene-vinyl acetate copolymer
乙烯乙酸乙烯酯橡胶 ethylene-vinyl acetate rubber
乙烯原 ethylogen
乙烯载体 vinyl vehicle
乙烯酯类 vinyl esters
乙烯装置 ethylene unit
乙酰 acet(yl);ethanoyl (acetye)
乙酰氨基 acetamino
乙酰胺 acetamide;ethanamide
乙酰苯 acetophenone
乙酰苯酚 acetylphenol
乙酰苯间二酚 acetyl-resorcin
乙酰苯偶姻 benzoin acetate
乙酰吡啶 acetyl pyridine
乙酰吡嗪 acetyl pyrazine
乙酰蓖麻酸甲氧基乙酯 ethylene glycol monomethyl ether acetyl ricinoleate
乙酰蓖麻酸乙二醇单甲醚酯 ethylene glycol monomethyl ether acetyl ridnolcatelene
乙酰蓖麻酸乙酯 ethyl acetyl ricinoleate ether
乙酰丙 levulinic
乙酰丙醛 levulinic aldehyde
乙酰丙酸 levulinic acid
乙酰丙酸钙 calcium laevulinate;neocalcium
乙酰丙酸盐 levulinate
乙酰丙酮 acetyl acetone;diacetone
乙酰丙酮钛 titanium acetylacetonate
乙酰丙酮铀 uranium acetyl acetonate
乙酰纯黄 acetyl pure yellow
乙酰醋酸盐 acetoacetate
乙酰达玛树脂 acetylated dammar
乙酰碘 acetyl iodide
乙酰靛红 n-acetyl isatin
乙酰丁醇 acetobutyl alcohol
乙酰丁酸 acetobutyl acid
乙酰呋喃 acetyl furan
乙酰过氧化苯甲酰 acetyl benzoyl peroxide
乙酰化材 acetylated wood
乙酰化剂 acetylating agent
乙酰化棉 acetylated cotton
乙酰化木材 acetylated wood; acetyl wood
乙酰化器 acetylator
乙酰化羟乙基纤维素 acetylated hydroxyethyl cellulose
乙酰化烧瓶 acetylization flask
乙酰化作用 acetylation
乙酰基 acetyl base;acetyl group
乙酰(基)蓖麻酸甲酯 methyl acetyl ricinoleate
乙酰基蓖麻油酸酯 acetyl ricinoleate
乙酰基丙酮 pentanedione
乙酰基缩二脲 acetyl biuret
乙酰基纤维素塑料 acetyl cellulose plastics
乙酰基乙醇酸乙酯 ethyl acetyl glycola(e)
乙酰甲胺磷 acephate
乙酰甲醇 acetol;oxyacetone
乙酰解作用 acetolysis
乙酰亮蓝 acetyl brilliant blue
乙酰磷酸 acetyl phosphate
乙酰氯 acetyl chloride
乙酰汽油 acetyl gasoline
乙酰染料 acetyl colo(u)r
乙酰(替)苯胺 acetanilid(e)
乙酰猩红 acetyl scarlet
乙酰亚砷酸铜 copper aceto-arsenite
乙酰氧基三甲基硅烷 acetoxytrimethylsilane
乙酰氧基硬脂酸甲氧基乙酯 methoxy ethyl acetoxystearate
乙酰乙酸 acetoacetic acid; etheric acid
乙酰乙酸丙酯 propyl acetoacetate
乙酰乙酸丁酯 butyl acetoacetate
乙酰乙酸甲酯 methyl acetoacetate
乙酰乙酸钕 neodymium acetyl acetonate
乙酰乙酸盐 acetoacetate
乙酰乙酸乙酯 acetoacetic ester; diacetic ether;ethyl acetoacetate
乙酰乙酸酯 acetoacetate
乙酰乙酰芳胺黄 diarylide yellow
乙酰皂化值 acetyl saponification number
乙酰值 acetyl number;acetyl value
乙酰转移作用 transacetylation
乙酰唑胺 acetazolamide
乙型肝炎 hepatitis B
乙型水准标石 model B benchmark
乙亚胺 ethyleneimine;ethyliminum
乙氧基苯 ethoxy benzene
乙氧基苯胺 ethoxyaniline
乙氧基苯甲酸 ethoxy benzoic acid
乙氧基苯偶姻 ethoxy-benzoin
乙氧基蓖麻油 ethoxy castor oil
乙氧基丙酸乙酯 ethyl ethoxyl propionate
乙氧基丙烷 ethoxy propane
乙氧基测定 ethoxy determination
乙氧基二硫代甲酸 xanthic acid
乙氧基化物 ethoxylate
乙氧基己烷 ethoxy hexane
乙氧基甲烷 ethoxy methane
乙氧基金属 ethoxide
乙氧基钠 sodium alcoholate
乙氧基三甲基甲硅 ethoxytrimethylsilane
乙氧基三乙基硅烷 ethoxy triethyl silane
乙氧基钛 titanium ethoxide
乙氧基戊烷 ethoxy pentane
乙氧基纤维素 ethoxy cellulose
乙氧基辛烷 ethoxy octane
乙氧基乙醇 ethoxy-ethanol
乙氧基乙醇乙酸酯 ethoxy ethanol acetate
乙氧基乙基邻苯二酸盐 ethoxyethyl phthalate
乙氧基乙酸 ethoxyacetic acid
乙氧基乙烯 ethyl vinyl ether
乙氧钠 sodium ethylate
乙酯 ethyl ester
乙种比重计 ethyl gravitometer
乙种粒子 beta particle
乙种射线 beta ray
乙种水密门 B watertight door
乙字管 offset;offset pipe;pipe offset
乙字头 offset
乙字形存水弯 double-return siphon; double-return trap
乙字形(存水)弯管 double-return siphon
乙字形连接管 pipe offset
乙座席 B-position

已安装的工厂生产能力 installed plant capacity

已安装能力 installed capacity
已安装配件 installed fittings
已按每月保比例保险 held covered
已扳转的握柄 reversed lever
已保存系统 saved system
已保险 assured
已保险失业率 insured unemployment rate
已保险银行存款 insured bank deposits
已报废的固定资产 abandoned property
已报关货物 declared goods
已爆破岩石 shot rock
已背书票据 backed note
已背书债券 backed bond
已被拒付的应收票据 notes receivable protested
已被认可的 approved
已闭塞线路 blocked line
已编号卡 numbered card
已编排阵列 organized array
已编型号的整机设备 nomenclatured equipment set
已贬值的美元 cheap dollar
已变更的条件 changed condition
已变化成冰川的 glaciated
已变址地址 indexed address
已拨款的贷款 loan disbursed
已拨款项 allotment issued
已不存在的公司 defunct company

已不通电流的电线 dead wire
已布设三角网地区 triangulated area
已采地区 bare ground
已测的 surveyed
已测点 take-off spot
已测路中线 chained centre line of road
已测煤气量 measured gas
已测日工资率 measured day rate
已测压力 measured pressure
已查传票 audited voucher
已查核账目 certified account
已查核资产负债表 certified balance sheet
已查证账目 certified account
已查证资产负债表 certified balance sheet
已拆除房屋的场地 area cleared of buildings
已拆毁铁路 dismantled railroad; dismantled railway
已拆散的 demounted
已产生的正当权利 accrued rights
已偿还遗债 debts of the deceased paid
已超过服务年限 past service life
已潮的石灰 stale lime
已沉陷的混凝土板抬高 raising of sunken concrete slabs
已陈化换向器 seasoned commutator
已称重的 weighted
已成图区 covered surface; mapped surface
已承担间接费用 absorbed burden
已承兑 hono(u)red
已承兑的 accepted
已承兑的票据 accepted bill (draft)
已承兑汇票 accepted bill (draft); accepted draft
已承兑信用证 accepted letter of credit
已承兑债券 accepted bond
已承诺费用 committed cost
已充电的 charged
已冲洗胶片 processed film
已除去汽油烃的干气 stripped gas
已处理的 seasoned
已处理的种子 treated seeds
已处理过的毛毡 treated felt
已处理木材 treated timber
已处理水 treated water
已处理污水 treated sewage
已穿孔卡片 punched card
已存数据 canned data
已打内涂层 primed
已担保的背书 endorsement guaranteed
已到偿还期 fall-in
已到汇票 arrival draft
已到价合约 in the money
已到期保险金 earned premium
已得纯利 net profit earned
已得利润 profit earned
已登记支付书 registered warrant
已登录词汇 entry vocabulary
已登录词条 entry term
已抵补的美元 covered dollars
已抵偿的 balanced
已抵除增加额 balanced addition
已抵押出去的地产 estate encumbered with mortgages
已抵押债券 backed bond
已抵押资产 pledged assets
已电离的气体 ionized gas
已订舱货物 booked cargo
已订购 on order
已订购物资 material on order
已定的 made-up
已定购材料表 schedule of materials on order
已定界 determinate(d) boundary
已定义变量 defined variable
已定义项 defined item
已定账目 account stated
已兑现支票 cashed check
已发出的信号 distinguished symbol
已发出通融票据 accommodation bills issued
已发放的贷款 outstanding on loan
已发功率 power developed
已发生成本账户 cost incurred account
已发生损失 incurred losses
已发生债务 obligation incurred
已发送的 off the stock
已发现储量 discovered reserve
已发行公司债的现行价值 carrying value of a bond issued
已发行公债 outstanding bond
已发(行)股本 capital stock issued
已发行股票 issued shares; issued stock; outstanding shares; outstanding stock
已发行债券 already-issued bond; bonds issued
已发行证券 outstanding securities; securities issued
已发育的洪泛平原 developed floodplain
已发在外流通股本 capital stock outstanding
已发展地区 built-up area; improved land
已发展技术的 state-of-the-art
已发展区域 built-up area
已发证券 securities issued
已放弃产权财产 abandoned property
已废弃的 antiquated; obsolete
已废信息 obsolete information
已分解货车 detached wagon
已分解客车 detached vehicle
已分类文件 sorted file
已分配材料管理费用 applied material handling expenses
已分配成本 absorbed cost; allocated cost; applied cost; distributed cost
已分配的存储器 allocated storage
已分配的费用 distributed expenses
已分配的要素收入 distributed factor income
已分配费用 absorbed expenses; applied expenses
已分配工厂间接费用 applied factory overhead expenses
已分配管理费 applied administrative expenses
已分配间接费 applied burden
已分配间接费成本 absorbed overhead (cost)
已分配间接费用 absorbed burden; absorbed overhead (cost); applied overhead
已分配间接制造费用账户 applied manufacturing overhead account
已分配利润 distributed profit
已分配制造费用 absorbed manufacturing expenses; applied manufacturing expenses
已分配制造费用账(户) burden-credit account
已分摊差滞 distributed lag
已分摊的成本 allocated cost
已分摊跌价差额 absorbed declination
已分摊制造费用汇总表 summary of cost of manufacturing expenses applied
已风化侵蚀 erosional exposed surface
已付定金 down payment paid
已付费交货地点 point of free delivery
已付分包款金额 amount paid to subcontract
已付股本 paid-up capital
已付股利 dividend paid; paid dividends
已付股息 paid dividends
已付(关)税 duty paid
已付合同保证金 contract deposit paid
已付合同存款 contract deposits paid
已付价持票人 holder for value
已付金额 disbursements to date
已付款 account paid
已付款的保险 paid-up insurance
已付款凭证 paid voucher
已付款收货通知单 notice of delivery paid
已付款项 scot
已付利息 interest paid
已付赔款 losses paid
已付凭单档案 paid voucher file
已付其他费用 paid other expenses
已付讫票据 retired bill; take-up bill
已付现金股息 cash dividend paid
已付薪金 paid salaries
已付印 gone to press
已付邮资 post paid
已付杂项利息 sundry interest paid
已付杂项手续费 sundry commission paid
已付账款 account paid
已付支票 cancelled check; paid check; paid-up cheque
已付资本 paid-up capital
已付租金 paid rent
已负担的费用 absorbed expenses
已负担间接成本 absorbed burden
已改良的排放 improved discharge
已改良的土壤 reclaimed soil
已盖章合同 sealed contract
已告"线路出清" line clear sent
已告"线路开通" line clear sent
已购人寿保险年金 purchased life annuity
已刮灰木材 plugged lumber
已关闭的外汇市场 foreign exchange market closed
已关门企业的股票 obsolete securities
已观测方向 observed direction
已观测目标 observed object
已观测值 observed value
已灌溉的 irrigated
已灌浆的 grouted-in
已过的时间 elapsed film
已过滤废水 filtered wastewater
已过年数 elapsed years
已过时效的债权 barred claim
已过账的 posted
已航摄区 covered surface
已耗成本 expired cost
已耗费用 experimentation expenses; expired expenses
已耗效用 expired utility
已耗专利权价值 expired patent value
已核准发票 vouchered invoice
已化验试样 dump sample
已还清的贷款 paid-up loan
已回收债券 retired bond
已回填砂石场 filled tope
已婚者住宅 married quarters
已获得的计时利息比率 time interest earned ratio
已获得良好记录的模型 well documented model
已获利息 earned interest
已获收入 earned income; earned revenue
已获盈利 profit earned
已获盈余 earned surplus; surplus from profit
已激励的 energized
已记录的最低水平面 lowest recorded level
已记录的最低水位 lowest recorded stage
已加工表面 processed surface
已加工材 timber
已加工材料的试验 processed material test
已加工工件 finished work
已加工件 wrought stuff
已加工面 finished surface; machined surface
已加工木材 worked lumber
已加工木料 treated timber; worked lumber
已加工(原)材料 fabricated material; processed material
已加热的 heated
已检波的信号 detected signal
已检查 inspected
已检验的存货 certificated stock
已建的 existent; existing
已建的建筑 existing building
已建造的 constructed
已降低的价格 reduced price
已降低额定值的 derated
已降落 landed
已交付的 delivered
已交数量 quantity delivered
已交税的 duty paid
已搅拌混凝土 ready-mix concrete
已缴 paid in; paid-up
已缴付的保费 paid-up insurance premium
已缴股本 call-up capital; capital stock paid-up; paid-up capital
已缴股份 fully paid stock; paid-up share; paid-up stock
已缴清保费 paid-up insurance premium
已缴许可证费 paid-up licence[license] fee
已缴印花税的保单 stamped policy
已缴运输进款 paid-up transport revenue
已缴资本 paid-in capital; paid-up capital
已校对样本 censored sample
已校正格网摄影法 calibrated reseau photography
已校正像片 annotated photograph
已校准的电阻 calibrated resistance
已校(准)仪表 calibrated meter
已接受的 accepted
已接受订货总数 backlog of orders
已接受而尚未运出的定货总数 backlog of orders
已接通的电路 established connection
已结关 cleared
已结平账款 closed account
已结清账户 closed account; closed accounting
已结清账款 settled account
已结清账目 settled account
已结账户 closed account
已结账目 settled account
已解调的信号 demodulated signal
已解约 cancelled
已进行的工程 project already undertaken
已进行过的调查工作 investigation work
已经废除的 overpassed
已经刮草的地块 scraped plot
已经关闭信号的闭塞区段 controlled block
已经关闭信号的区间 controlled block
已经加工表面 finished surface
已经韧化的 now-toughened
已经探伤部分 ultrasonically tested area
已拒付的票据 note dishonored
已拒付的应收票据 note receivable

discounted; note receivable protested
已具备施工的地块 manufactured lot
已浚挖的航道 dredged channel
已开发的地热田 developed geothermal field
已开发的洪泛平原 developed floodplain
已开发地区 developed quarter; developed site
已开发探明储量 proved developed reserves
已开发土地 developed area; developed land
已开航 sailed
已开拓矿体 ore developed
已勘探矿区 proved field
已砍伐的林地 denuded area
已垦殖的洪泛平原 developed floodplain
已扣缴预提税款 tax withheld
已枯竭的井 exhausted well
已亏损后 after loss realization
已老化换向器 seasoned commutator
已累缴利润 cumulative profit
已累缴利税总额 cumulative profit and tax
已累缴税额 cumulative tax
已利用河段 utilized river reach; utilized river section; utilized river stretch
已励磁的 energized
已连接导线【测】connected traverse
已列入近期规划 in the near future programme
已录音磁带 prerecorded tape
已滤波全息图 filtered hologram
已履行的对价 executed consideration
已满期的订单 order expired
已灭瀑布 extinguished waterfall
已磨浆料 refined stock
已抹底灰的砖砌体 rendered brickwork
已没收担保品的销售 foreclosure sale
已没收担保品价值 foreclosure value
已没收货物 confiscated goods
已募得的总金额 subscription realized
已纳税 assessment paid; duty proof
已纳税的 duty paid
已纳税货物 clear goods
已纳税款扣除法 tax deduction method
已凝结的 concretionary
已排水的 drained
已判定的共振 resolved resonance
已刨木纹 torn grain
已批准的 approved
已批准的贷款 loan approved
已批准的贷款项目表 statement of loans approved
已批准的但尚未发行的债券 bonds authorized and unissued
已批准的工程项目 authorized project
已批准发行但尚未实际发行的债券 bonds authorized and unissued
已平差点位 adjusted position; adjuster position
已平整的地面以上 above-grade
已评定的收入 measured income
已破波 broken wave
已破坏的乳化液 broken emulsion
已破浪作用力 force of broken wave
已砌的拱 sprung arch
已签名盖章的文据 sealed instrument
已签署的 signed
已切割木材 worked lumber
已清偿的债务 liquidated obligation
已清偿(了结)的损失赔偿金 liquidated damages
已清偿票据 bills retired
已清偿损失额 liquidated damages
已清偿债务 liquidated debt
已清偿债务的破产 discharged bankrupt
已清缴保费 paid-up insurance premium
已清理的公司 liquidated corporation
已清算账户 liquidated account
已清洗气 washed gas
已驱动的火烟监测器 smoke detector actuated
已取得的利息 interest earned; interest received
已取得的折扣 discount taken
已取得盈余滚存 acquired surplus
已取芯样的混凝土梁 cored beam
已全部付清 fully paid-up
已全部缴款股份 fully paid stock
已全部缴款股票 fully paid share
已全部折旧的 fully depreciated
已全部折旧资产 fully depreciated assets
已全穿孔带 chadded tape
已确认的背书 endorsement confirmed
已燃气 burned gas; burnt gas; combustion gas
已热处理表面 heat-treated surface
已热处理的 heat-treated
已认付的 accepted
已认购的普通股 common stock subscribed
已认购股本 subscribed capital stock
已认股本 paid-in capital
已熔金属 molten metal
已扫雪的跑道 open runway
已上底漆 primed
已上税物品搬运证 removal permit for duty-paid goods
已上涨的价格 advanced price
已申请(专利) claims priority
已审查过的净销售额 audited net sales
已审查凭单 audited voucher
已审定的 approved
已审定账目 audited accounts
已审核储量 reserve awaiting ratification
已审批储量 ratified reserves
已审凭单 audited voucher
已生锈表面 rusted surface
已失时效的债务 outlawed debt
已实现的速度 speed made good
已实现利润 realized profit
已实现收入 realized gains; realized revenue
已实现损益 realized gains or losses
已实现投资 realized investment
已实现盈余 realized surplus
已实现增殖 realized appreciation
已实现折旧 realized depreciation
已实现资产持有利得 realized holding gain
已使用财产 used property
已使用固定资产的费用 cost of used fixed assets
已使用过的固定资产 used fixed assets
已使用年限 in the years already spent
已适应的 seasoned
已收保证金 earnest money received
已收存款 deposit(e) received
已收股本 paid-in stock
已收合同保证金 contract deposits received
已收合同存款 contract deposits received
已收回股份 retired stock
已收回债券 retired bond
已收回注销的股票 retired stock
已收入金额 amount received
已收贴现票据 bill received discounted
已收杂项利息 sundry interest received
已收账目 receipted account
已受损 in damaged condition
已售完 out of stock
已疏浚航道 dredged channel
已赎回债券 retired bond
已水化水泥 hydrated cement paste
已税产品 taxed product
已税品 taxed article
已税商品 taxed commodity
已说明方差 explained variance
已碎波 broken wave
已损害混凝土 loosened concrete
已损坏混凝土 loosened concrete
已损坏图形 damaged drawing
已损毁货车 damaged wagon
已损毁客车 damaged vehicle
已锁闭道岔 locked switch
已摊还的借款 amortized loan
已摊税金 apportioned tax
已摊销成本 amortized cost
已探明藏量 known economic reserves in places; proved reserves
已探明地区 proven territory
已提出拒绝证书后承兑 acceptance for honour supra protest
已提高账面价值 written-up
已提供押款金额 sum lent on a mortgage
已提货的提单 accomplished bill of lading
已调波放大器 modulated wave amplifier
已调等幅波 modulated continuous wave
已调电压 modulated voltage
已调放大器 modulated amplifier
已调幅波 amplitude-modulated wave
已调幅脉冲 amplitude-modulated pulse
已调幅信号 amplitude-modulated signal
已调绘像片 annotated photograph
已调节的空气 conditioned air
已调连续波 modulated continuous wave
已调配好的 well modulated
已调频的 frequency-modulated
已调载波 modulated (wave) carrier
已调载频 modulated (wave) carrier
已调整基准 adjusted basis
已调制波 modulated wave
已调制电流 modulated current
已调制信号 modulated signal
已调制载波电压 modulated carrier voltage
已调制正弦波 modulated sinusoid
已贴现票据 bill discounted; discounted bill; discounted note
已贴现应收客户款 customer's account discounted
已贴现应收票据 notes and bills receivable discounted
已贴现账款登记簿 discounted account register
已停业公司 defunct company
已停业务的会计处理 accounting for discontinued operations
已通电的 energised [energized]
已通知还本的公债 called bond
已通知信用额度 advised line of credit
已投产的产品 production version
已投成本 sunk cost
已投股本 contributed capital
已投资本 vested capital
已推算时间 dead-reckoning time
已退税款 tax refunded
已退原材料报表 materials returned report
已退原材料报告 materials returned report
已脱气的泥条 vacuumed clay
已挖区 dredged area
已完成的 off the stock
已完成的绘线图形 finished line drawing
已完成的销售 executed sale
已完成订单 completed order
已完成工程 finished construction
已完成工程成本 cost applicable to construction revenue
已完成工程量成本 cost of work performed
已完成工作量 completed work amount; executed amount; executed work amount; workdone
已完成工作量成本 cost of work performed
已完成会计事项 completed transaction
已完成销售合同 executed contract of sale
已完工程 construction finished; projects completed
已完工程成本 cost applicable to construction revenue
已完工合约的收益 income from completed contract
已完工批号日记账 completed jobs journal
已完工作 finished work
已完税 assessment paid; duty paid; duty proof
已完税货物 duty-paid goods
已稳定滑坡 stabilized landslide
已吸收成本 absorbed cost
已吸收跌价 absorbed declination
已吸收费用 absorbed expenses
已吸收证券 digested security
已吸收制造费 absorbed manufacturing expenses
已熄灭的灯标 extinguished light
已熄灭灯 extinguished lamp
已洗蒸汽 washed steam
已下水船 launched ship
已显像 developed image
已消失的水热活动区 extinct hydrothermal area
已消逝成本 expired cost
已消逝费用 expired expenses
已消逝专利权价值 expired patent value
已消亡的地热田 extinct geothermal field
已销售货物 goods sold
已卸车辆 wagon discharged
已卸货物 cargo unloaded
已形成的洪泛平原 developed floodplain
已形成的通路连接 built-up connection
已修改的设计 revised design
已修整的跑道 patchy runway
已修整的坡度 finished grade
已修整(好的)表面 finished surface
已修正的结果 corrected result
已修琢成的建筑天然石料 natural stone dressed ready for building
已锈表面 rusted surface
已宣布分配的股息 declared dividend
已宣布股息 dividend declared
已宣布利润 declared profit
已渲染设计 rendu
已硬化的混凝土 hardened concrete
已硬化的水泥浆体 paste matrix
已硬化的外缘 hardened outside verge
已用拨款 expended appropriation
已用时间 elapsed time

已用提单 spent bill of lading;used up bill of lading
已有词条 entry term
已有的 available;propriate
已有调查报告名称 name of investigation work
已有调查工作比例尺 scale of investigation work
已有调查工作单位 unit of investigation work
已有调查工作阶段 stage of investigation work
已有调查工作日期 date of investigation work
已有调查工作性质 type of investigation work
已有港口 established harbo(u)r;existing harbo(u)r
已有工程 existing work
已有裂缝 preexisting crack
已有排水系统 antecedent drainage
已有数据 canned data
已有资料 available data;available information
已逾时效的规定 statute-barred
已预定包房 reserved compartment
已预定座席 reserved seat
已预付 prepaid
已预缩的 preshrunk
已在仓 in position
已占空间 excluded volume
已占容积 excluded volume
已占态 occupied state
已占用资本 locked-in capital
已折成本 depreciated cost
已(折)耗成本 depleted cost
已折旧价值 depreciated value
已振捣混凝土 vibrated concrete
已整电流 rectified current
已整治河流 regulated river;regulated stream
已证明的技术 proven technology
已支付的 paid
已知边 known side
已知场地 given site
已知储量 known reserves
已知错误条件 known error condition
已知大小的容器 container of known size
已知的 given
已知的损失 known loss
已知地址 known address
已知点 given point;known point
已知段 known segment
已知段表 known segment table
已知范围 prescribed limit
已知方向 known direction
已知负债 known liability
已知函数 known function
已知基线 known base line
已知价格 given price
已知角 known angle
已知节点 datum node
已知矿床号 number of known deposits
已知量 known quantity
已知流量边界 boundary of known flow
已知频率 given frequency
已知破损 known damage
已知溶液 known solution

已知三角点 trigonometric(al) control point;trigonometric(al) fixed point
已知试样 known sample
已知数【数】 known number;datum;given number;known (quantity);prescribed value
已知数据 given data
已知水位边界 boundary of known water level
已知条件 data;known condition
已知物质 known substance
已知误差 conscious error;known error
已知样品 known sample
已知整体 known universe
已知值 given value;known value
已知重量的干土样 dry soil sample of known weight
已知重量的土样 soil sample of known weight
已知状态 known state
已知组分 known component
已知最大流量 highest ever-known discharge
已知坐标的地面点 located station
已知坐标点 coordinated point
已执行的订单 order executed
已执行完毕的合同 executed contract
已指定用途的款项 money appropriated
已置平模型 flattened model
已注册的 registered
已注册公司 registered corporation
已注册股本 capital stock registered
已注册设计 registered design
已注册资本 registered capital
已注释像片 annotated photograph
已注水泥的套管柱 cemented casing
已注销订单 order cancelled
已注销债券 cancelled bond
已注销支票 cancelled check
已转嫁的税收 shifted tax
已转让利润 profit transferred
已转让应收账款 accounts receivable assigned
已装备车辆 equipped vehicle
已装船 on board;shipped on board
已装船背书 on-board endorsement
已装船背书提单 on-board endorsement bill of lading
已装船海洋清洁提单 ocean clean on board bill of lading
已装船货提单 shipped bill
已装船货物 afloat cargo
已装船批注 on-board notation
已装船清洁提单 clean on board bill of lading
已装船提单 on-board bill of lading;shipped bill of lading
已装货物 afloat goods;cargo loaded
已装框架的 enframed
已装料的堆芯 loaded core
已装片暗盒 loaded magazine
已装修的楼梯斜梁 finished string
已准出舱 free delivered
已走距离 covered distance
已组织的数组 organized array
已钻地区 drilled area
已作废合同 voided contract
已作废支票 voided check
已做的功 workdone

以20 英尺长作为 1 个集装箱的换算单位 twenty feet equivalent units

以百分比表示的误差 error given in percent
以百分比计的调制深度 modulation percentage
以磅计算的牵引力 pound-drawbar pull capacity
以保险单作抵押的贷款 cash policy loan
以本国船只运输的贸易 active commerce
以本国货币表示的价格 prices in local currencies
以本机振荡器控制的激励 adjustable local-oscillator drive
以泵向岸排泥 pump ashore discharge
以便继续浇筑混凝土 slush grouting
以便做凸缝或嵌缝 raked out
以波形瓦盖屋顶 pantiling
以补充点为基准的订货方式 replenishment-based ordering system
以不计营利为基础的核算方程式 non-profit accounting equation
以不同方法或不同资料进行核对 cross-check
以参数量表示 parametrized
以拆卸状态装车 knocked down in carload
以产定销 basing sales on production;sales determined by products
以产品偿还的计划 product-pay-back scheme
以产品为对象的车间布局 plant layout by products
以产业供债权人分配者 cessionary bankrupt
以长期贷款易取短期贷款的投资者 take-out investor
以成本发生经常与否为基础 base of regularity of occurrence
以成本回收为基础 cost recovery basis
以成本与产品间的关系为基础 base of relationship to the cost of unit
以成批方式 in batch mode
以承包方式 on contract terms
以城市为中心的 urban centered
以传感器为基础的设备 sensor-based
以船舶装运 shipment by steamer
以船运费最廉的航线装运 shipment by cheapest route
以船作抵押的贷款 bottomry
以船作抵押的借款 bottomry
以床为寝具的居住方式【建】 bed system
以次充好 taking substandard products as fine products
以存款作担保 deposit(e) as collateral
以存款作抵押的贷款 deposit(e) loan;passbook loan
以存支贷 using the money from the deposits for loans
以打计 by dozen
以大信号激励电子管 drive the tube hard
以代理人的身份 by procuration
以代理人资格签署 sign per pro
以代收银行为受货人 consigned to collecting bank
以代收银行为受货人 unto collecting bank
以贷代款 in kind
以到岸价计 by coast, insurance and freight
以道路为主 on road
以低于平均价格买进 average down
以地球为中心的 geocentric
以地要房 exchanging land for housing
以第三者为托运人 third party shipper
以第三者为装船人的提单 third party bill of lading
以点为基础的线性 point-based linearity
以垫舱货垫舱 flooring off
以队组为中心的目标管理模型 team-oriented management by objective model
以吨计的二氧化碳当量 tonnage CO2 equivalent
以(二进制)编码表示的十进制记数法 coded decimal notation
以法令规定 decree
以房地产收入作抵押的贷款 income property loan
以非现金购得子公司股本 non-cash acquisition of subsidiary stock
以废治废 treatment of a waste with another waste
以费挤税 replace taxes by charges
以分期付款方式 on easy terms
以丰补歉 make-up for possible shortages with surpluses
以弗所的古神殿 <古希腊城市的> Archaic temple at Ephesus
以弗所古庙 <小亚细亚> Archaic temple at Ephesus
以付款凭单核对证明 vouching
以付款凭单(核对)证明的账户 vouching account
以钢为纲 taking steel as the key link
以高速度 at full speed
以高于平均价格卖出 average up
以各种方式维持(商品)价格 valorize
以工代干 allow a worker act as a cadre;using workers as cadres
以工代账 work relief
以工代赈 provide work as a form of relief;public work form of relief
以工代租 rent service
以工业为基地的学员 industry-based student
以工资为生者 wages-earner
以工作为依据的参与 task-based participation
以工作为依据的评价 task-based appraisal
以公里里程计算的轮胎寿命 life of tire in kilometers
以公债充作基金 public fund
以供应市场为目的的菜园 market garden
以购货人为受货人 consigned to buyer;unto buyer
以股票付股息 stock dividend
以股票支付的股息 dividend payable in capital stock
以股票作红利 share bonus
以股息方式摊还资本 capital returned to stockholders in dividends
以股易股 split-off
以顾客为中心 customer-oriented
以管理是否能控制为基础 base of administrative control
以广招徕 in order to promote patronage or sales
以壕围绕 entrench
以核心为基础的系统 kernel-based system
以后的租船运费 ulterior chartered freight
以化学法粉碎 disagglomeration
以环境为基础的学习 environment-based learning
以环氧树脂为底的叠合接面 decopolymer flooring
以黄金偿付的债券 gold bond
以黄金支付的外国汇票 foreign bills payable in gold
以黄金作抵押的贷款 gold-secured loans
以汇票为担保的证券 bill on deposit
以汇票为抵押证券 bill deposited as collateral security;bill on deposit
以活动为中心的领导方式训练 action-centered leadership
以火灭火 <一种以火切断火路的方法> back-fire
以货币表示的 monetary term
以货币购物 pecuniary exchange
以货币支付的利息 explicit interest
以货抵债 cession bonorum
以货易货交易 barter deal
以货易货制 barter system
以获有舱位为条件 subject to shipping space available

以极低温度冷藏 deep freeze
以计算机为基础的 computer-oriented
以计算机为基础的仿真模型 computer-based simulation model
以记录为基础的系统 record-based system
以记账支付的支票 check payable in account
以箭头代表作业 activity on arrow
以交钥匙的形式 on turn-key terms
以角度表示的斜度 angular pitch
以角相接 cornering
以……校核 check with
以较高级职工代替低级职工 back-tracking;bumping
以金钱为准的工资 nominal wages
以进口地报关的受货人 consigned to a foreign custom house broker
以进养出 import for expanding export
以经济建设为中心 make economic development our central task
以旧物折价换取新物 trade-in
以就业工人人数而论的工厂规模 size of establishment by employment
以举债方式筹措资金 debt financing
以巨人与神的战斗为主题的作品 gigantomachy
以卷形为主题的装饰 scroll work
以卡或焦耳值表示 heating value
以开证银行为受货人 consigned to issuing bank
以客车/小时计的基本交通量 base capacity in passenger vehicle s/hr
以空间为基地的次系统 space-based subsystem
以会计期间为基础 base of accounting period
以离岸价格计 by free on board
以礼貌规劝的交通警 courtesy cop
以里程计的寿命 mileage life
以里程计算的差旅费 mileage allowance
以立柱作支承的梁式桥 leg bridge
以利润再投资 plow back
以利息化为资本 capitalization of interests
以链计量 chain measure
以链形花纹装饰的 catenated
以量词限定(命题等) quantify
以邻为壑的进口壁垒 beggar-my-neighbo(u)r import barriers
以邻为壑政策 beggar-my-neighbo(u)r policy
以零为基数的计划编制核预算 zero base planning and budgeting
以六十进制元的系统 sexagesimal system
以履带装置为底盘的 crawler mounted
以螺旋坡道连接的室内多层停车场 spiral car park
以买方检验为最后依据 Buyer's inspection to be final
以满期缺口补进部位 covered position with a maturity gap
以毛重作净重 gross for net
以毛作净 gross for net
以每股股息额表示 dividend-per-share presentation
以美国轮船载运的立法 ship in American bottoms legislation
以美国为权数的购买力比价 US weighted purchasing power parity
以美国为权数的几何平均指数 geometric(al) mean of U.S. weighted index
以美国支出为权数 US expenditures weight
以美元支付的进口 dollar import
以秒计的弧度 second of arc
以模拟装置为基础的培训 simulator-based training
以木标划界 peg(ging out)
以目标利润率定价 target-rate-of-return pricing
以目标为中心 goal-orientation
以内侧尺寸计算 in the clear
以内宽计算 in the clear
以耐火黏[粘]土为主成分的耐火砂浆 fireclay-base refractory mortar
以年为基础换算 converted to an annual basis
以黏[粘]土润滑 puddling
以抛物线速度再入 parabolic(al) reentry;parabolic(al) velocity reentry
以批准为条件 subject to approval
以期定量 fixing production on a periodic basis
以期换现 against actual
以期货换现货 exchange of future for cash
以起岸重量为准 landed weight final
以气候学方法预报 climatological forecast
以汽车为主的运输 truck-dominated transportation
以铅锤测 plumb
以铅丹为基底的胶粘水泥 minium-based building mastic
以铅丹为基底的胶粘涂料 minium-based building mastic
以签订契据为条件 subject to contract
以前的 preceding
以前有代表性的时期 a previous representative period
以人民币计 by Renminbi
以人为本的设计 ergonomic design
以容积计(费) measurement basis
以容量为标准的道路 <用于人口密集区> capacity road
以闪光信号表示 blink
以上发送收讫 above transmitted as receive
以生产(成绩)为中心的 production-centered[centred]
以十计的 decuple
以十为底的对数 denary logarithm
以石膏为基料的 gypsum based
以石灰为基底的抹灰用料 plaster mixture based on lime
以石楔咬紧的 keyed
以时间为基础的程序 time-based program(me)
以使用者为基础 user basis
以市中心区为终点的交通 downtown terminal traffic
以收定支 expenditure is determined by revenue
以收支抵支 use of revenue to finance expenses
以手控代自控时节流阀调整 override throttle setting
以手推车计量 measurement by wheelbarrow
以手指沾墨在纸上抹样 tap-out
以数据为基础的微指令 data-based microinstruction
以数据为基础的微指令周期 data-based microinstruction cycle
以数量表示 quantification
以水表计算水费 water metering
以松节油为基料的金属蚀剂 mordant based on turpentine
以台架存放物品的仓库 rack warehouse
以太波 ether wave
以太计算机网 Ethernet
以太漂移 ether drift
以太网(络) Ethernet
以太曳引 ether drag
以碳素测定年代 carbon-date
以体积计 by volume; measuring by volume
以体积计量 cubing
以体积计量的产品 cubic(al) product
以体积计算 cubing
以贴补形式维持物价 revalorization
以铁路为主 on rail
以投标形式 by-tender
以投资为目标的项目 investment oriented project
以图表说明 represent graphically
以图像表示的 pictorial
以土地使用权入股 contribute the land usage right as parts of its investment
以(土壤)干重百分比计的含水量 water content in percent of dry weight
以(土壤)干重百分比计的含水率 water content in percent of dry weight
以外币发行的公司信用债券 debenture in foreign currency
以往的 bygone
以往地震活动性 preceding seismic activity
以微程序装备 microprogram(me)
以维持团体关系为主(M形)的(领导方式) maintenance-directed
以五乘之 quintuple
以物衬砌 lining-up
以物品担保的贷款 loan secured by things
以物为证 take something as a pledge
以物易物 barter trade;truck
以锡为主的锡锑铜轴承合金 adamant metal
以险情为基础设计 risk-based design
以现金计算的价值 cash value
以现金交易的顾客 cash customer
以像素为基础 pixel-by-pixel basis
以销售为目标的生产 market-oriented production
以销售为目的果菜园 market garden
以小时计 hourly basis
以效用差为基础的公理 axiom based on utility difference
以信用担保的贷款 loan secured by credit
以压力保持阀缓解 retaining valve release
以颜色分类 colo(u)r-key
以氧化合 oxygenate;oxygenation
以谣诼压价收买房地产 block busing
以液化天然气为燃料的车辆 liquefied natural gas vehicle
以乙炔为能源的 acetylene gas-powered
以应收票据担保 notes receivable as collateral
以应收账款为抵押的(信贷) pledging of accounts receivable
以英币计算的汇率 sterling exchange rate
以英尺表示的长度 footage
以英寸为单位的水银柱绝对压力 in Hg abs [inch of mercury absolute]
以英寻表示的水深 depth in fathoms
以用户检验为最后依据 user's inspection to be final
以油漆房屋为职业的人 huse
以有舱位为准 subject to shipping space available
以与常规方向相反的方向卷绕所得的品质或状态 heterostrophy
以远权 beyond right
以运输进款作为运输收入 take transport income as transport revenue
以暂时股票付红利 scrip dividend
以债券偿付利息 funding debenture
以针选择卡片 needle
以整数计 in round number
以证券为担保的贷款者 taker-in
以证券为担保的借款者 giver-on
以支票提取的活期存款 checking deposit
以执行任务为主的(领导方式) performance-directed
以职员优先股计划筹集资金 capital raised under share incentive scheme
以指令为基础的结构 instruction-based architecture
以中心轴分室的多室料仓 central cone silo
以重计量 measurement by weight
以重量计 by weight
以烛光表示的发光强度 candle power
以桩支护 stake out
以字体增添装饰 letter enrichment
以自己为收款人的汇票发票人 drawer-paper
以租入财产担保的债券 bond secured by leased property
以租养房 maintenance with rent
以足践踏 stamp
以最好价格订购 order at the best

钇

钇硅磷灰石 britholite-(Y);abukumalite

钇褐帘石 muromontite
钇镓石榴石 yttrium gallium garnet
钇矿 yttrium ores
钇榴石 yttrogarnet
钇铝石 yttroalumite
钇铝石榴石 <激光用料> yttrium alumin(i)um garnet
钇铌钽石 hjelmite
钇铌铁矿 yttrocolumbite
钇球墨铸铁 yttrium nodular iron
钇石榴石 yttrium garnet
钇钛烧绿石 titanbruchevite
钇钽矿 loranskite
钇钽铁矿 yttrotantalite
钇铁石榴石 ferrite garnet;yttrium iron garnet
钇铁石榴石材料 yttrium iron garnet material
钇铁石榴石滤波器 yig filter
钇铁氧体 yttrium ferrite
钇钨华 yttrotungstite
钇钨矿 yttrotungstite
钇榍石 keilhauite;yttrian titanite
钇易解石 aeschynite-(Y);aeschynity;blomstrandin(it)e;priorite
钇萤石 yttrofluorite
钇铀矿 cleveite;nivenite
钇铀烧绿石 obruchevite; yttropyrochlore
钇杂铌矿 nuolaite

舣

舣装 fitting out;outfitting

舣装吨位 equipment tonnage
舣装港池 wet slip
舣装码头 equipment and repair quay; equipment quay; fitting-out berth; fitting-out pier; fitting-out wharf; gantry berth;outfitting pier
舣装品 equipment
舣装起重机 fitting-out crane
舣装数 equipment number;fitting-out number

蚁

蚁巢 ant nest; formicarium; formicary

蚁垤 formicarium;formicary

蚁害 ant attack;ant injury
蚁科 <拉> Formicidae
蚁客共生 myrmecoclepty
蚁类研究 myrmecology
蚁侵蚀 ant attack
蚁醛 formaldehyde
蚁醛试验 formaldehyde test
蚁山 formicary
蚁酸 formic acid
蚁酸铜 ant salt of copper
蚁酸盐 formate
蚁穴 ant net;formicary
蚁学 myrmecology
蚁学家 myrmecologist
蚁咬病 formiciasis
蚁贼 thievery
蚁冢 anthill;termite hill
蚁冢动物 myrmecophile

倚焊 gravity type arc welding;gravity type welding;gravity weld

倚在 abutting

椅背的一块竖板 splat

椅背套 anti-macassar
椅背突板 misericord(e)
椅背硬套 anti-macassar
椅背中部纵板 splat
椅车 wheel chair
椅带 chair tie
椅式 chair form
椅调整 seat adjustment
椅突板 patience;subsellium
椅形字符 <光学字符读取机中用的一种字形> chair
椅状劈裂 barber chair
椅子 chair
椅子车工 bodger
椅子架 seat stand
椅子靠背板 miserere

义捐物品拍卖 rummager sale

义卖 charity sale; sale of goods for charity
义卖集市 <美国慈善性质的> kermess
义卖市场 baza(a)r;charity bazar
义务 burden; commitment; duty; incumbency; liability; obligation; onus; responsibility
义务承担费 commitment charges
义务船 burdened vessel; give-way vessel;obliged vessel
义务的 compulsory;obligatory
义务护林员 fire co-operator
义务建筑学校 free school of architecture
义务教育 compulsory education
义务年限 obligatory term
义务人 obliger [obligor]

亿 a hundred million

亿吨 hundred million tonne
亿分率 parts per hundred million
亿立方米 hundred million stere
亿万分之一 <10^{-12}> part per trillion
亿万年 aeon

艺术 <包括诗歌、绘画、雕塑、建筑、音乐等> fine arts

艺术爱好者 art-lover
艺术标准 aesthetic criterion; artistic criterion;canons of art
艺术表现 artistic expression
艺术表现形式 artistic formation; artistic form of expression
艺术玻璃 art glass;artistic glass;amberina <一种美产的>
艺术玻璃器皿 art glassware; fancy glass ware
艺术玻璃制品 artistic glassware
艺术博物馆 art museum
艺术才能 artistry
艺术处理 aesthetic treatment; artistic treatment
艺术创造 artistic creation
艺术创作 artistic creation
艺术瓷 artistic porcelain; art porcelain
艺术瓷窑 ground-hog kiln
艺术大理石 art marble
艺术的 artistic
艺术的爱好者 dilettante
艺术的独特风格 individual form of art
艺术的个人形式 individual form of art
艺术范围 artistic circle
艺术风格 artistic style
艺术感觉 aesthetic sense
艺术感染 art infection
艺术感受力 aesthetic feeling
艺术宫 Palace of the Arts
艺术顾问 artistic adviser
艺术挂毯 artistic tapestry
艺术馆中心 artistic circle
艺术花纹 art weave
艺术混凝土 architectural concrete
艺术纪念牌 artistic monument
艺术技巧 artistry
艺术加工 embellishment
艺术家 artist
艺术家的 artistic;virtuosic
艺术家工匠 artist-craftsman
艺术家工作室 artists' studio
艺术价值 artistic value
艺术剪纸 artistic engraving of papers
艺术鉴定家 art connoisseur;art-lover
艺术鉴赏 appreciation of arts
艺术鉴赏家 art connoisseur;art-lover
艺术鉴赏力 aesthetic feeling
艺术教育 art education
艺术解剖 art anatomy
艺术解剖学 artistic anatomy
艺术界 artistic circle
艺术金属物 art metal
艺术剧院 artistic theatre;art theatre
艺术夸张 artistic exaggeration
艺术领域 realm of art
艺术论 technics
艺术美 beauty of art
艺术木雕 xyloglyphy
艺术喷泉 architectural fountain
艺术批评 art criticism
艺术批评家 art critic
艺术品 artwork;virtu;work of art
艺术品爱好者 art connoisseur
艺术品复制体 duotone
艺术品复制图 rendering
艺术品鉴赏家 art connoisseur; virtuoso
艺术器皿 artistic ware
艺术情调 aesthetic feeling
艺术上成熟 artistic maturity
艺术设计 artistic design;art layout
艺术炻器 art stoneware
艺术收藏(品) art collection
艺术手法 artistry
艺术水平 state of art
艺术塑造品 artistic modelling
艺术陶(瓷) art ceramics;art pottery
艺术陶器 studio pottery
艺术特征 artistic characteristics
艺术厅 arts center [centre]
艺术吸引力 aesthetic charm
艺术系 arts faculty
艺术效果 aesthetic effect; art effect; artistic effect;artistry
艺术心理学 psychology in art; psychology of art
艺术形式 art form; artistic form; forms of art
艺术型 artistic type
艺术性 artistic quality;artistry
艺术学 technics
艺术学校 art school
艺术学院 art college;arts faculty
艺术釉 artistic glaze;artware glaze
艺术造(成)形 artistic shaping
艺术造型 artistic expression; artistic formation; ornamental mo(u)ld(ing)
艺术造型设计 artistic design
艺术造诣 artistic attainments
艺术哲学 philosophy of art
艺术珍品 art treasure;museum piece
艺术直观 artistic intuition
艺术指导(者) art director
艺术制品 artware
艺术中心 art creation;arts center[centre]
艺术装修 architectural treatment and finish
艺术作品 art creation; artifact; artistic creation;artwork
艺术作品特色 motif
艺徒培训 apprentice training

议案 bill;proposal

议标 bid negotiation; discuss bids; negotiated bidding; negotiated contract; negotiated tendering; negotiating bids; negotiating tenders; negotiation of bids; negotiation tendering
议标合同 negotiated contract; negotiation contract
议标阶段 negotiation phase
议长 chairperson
议程 agendum [复 agenda]; order of business
议程项目 agenda item
议定 come to terms
议定贷款 originate loan
议定的费率 agreed rate
议定的价格 agreed price
议定发包合同 negotiated contract
议定付款 agreed payment
议定关税 agreed duty
议定积载因素 agreed stowage factor
议定金价 gold fix
议定贸易量 agreed quantity of trade
议定书 protocol
议定书草案 draught protocol
议定数额 agreed amount
议定文本 agreed text
议付金额 negotiated amount
议付票据 negotiated bills
议付期限 negotiating date
议付手续费 negotiation commission
议付外国汇票 negotiation of foreign bills
议付信用证 negotiation credit
议付业务 negotiating transaction
议付(银)行 negotiating bank
议付正本 negotiable original copy
议购 purchase on negotiation
议会 legislature;parliament
议会大楼 congress building; congress house; parliament building; parliament house
议会大厦 congress building; congress house; parliament building; parliament house;capitol
议会厅 <罗马政府的> curia
议价 bargain(ing); chaffer; drive a bargain; negotiated price; negotiation price;price arbitration
议价单位 bargaining unit
议价购买 buy at a bargain price
议价过程 bargaining process
议价合同 negotiated contract
议价交易 bargaining transaction
议价能力 ability to bargain; bargaining power
议价人 chafferer
议价市场 negotiated market
议价收购 negotiate purchase
议价条件 negotiating condition
议价者 bargainer
议决分派的股利 dividend declared
议决小区试验方案 pass a plot program(me)
议事规程 standing order
议事规则 standing order
议事机构 deliberative organ
议事录 transaction
议事日程 agendum [复 agenda]; docket;order of the day
议事室 <中世纪教堂置于门廊或圣器室局部的> salutatorium
议事摘要 synopsis of proceedings
议事桌 tapis
议题范围 theme-circle
议题小组 theme-group
议销价格 sale at negotiated price
议员 assemblyman;council(l)or
议员豁免权 parliamentary immunity

屹立 soar

异艾氏剂 <杀虫剂> Isodrin

异苯二甲酸 kophthalic acid
异苯二甲酸二甲酯 dimethyl isophthalate
异苯二甲酸烯丙酯 diallyl isophthalate
异壁厚管 butted tube
异变态 tetometamorphism
异变组构 heterotactic fabric
异冰片 isoborneol
异冰片烯 isobornylene
异冰片酯 isobornyl thiocyanoacetate
异丙胺 isopropyl amine
异丙苯的溶解度 solution isopropyl benzene
异丙苯过氧化氢 cumyl hydroperoxide; isopropyl benzene hydroperoxide
异丙苯氧化装置 isopropyl benzene oxidation unit
异丙醇 isopropanol;isopropyl alcohol
异丙醇铝 alumin(i)um isopropoxide
异丙醇试验 isopropanol test
异丙(基)苯 cumene; cumol; isopropyl benzene
异丙醚 diisopropyl ether;isopropyl ether
异丙氧化物 isopropoxide
异丙氧基苯 isopropoxy benzene
异丙氧基钛 titanium isopropoxide
异丙氧基铀 uranium isopropoxide
异剥钙榴岩 rodingite
异剥橄榄石 wehrlite
异剥辉石橄榄岩 wehrlite
异剥辉石岩 diallagite
异剥石 diallage

异补堆积结构 heteracumulate texture
异步 asynchronization;asynchrony
异步变频器 asynchronous frequency changer
异步并行算法 asynchronous parallel algorithm
异步补偿机 asynchronous condenser
异步操作 asynchronous operation;asynchronous working
异步测功机 induction dynamometer
异步测量 asynchronous measurement
异步出口 asynchronous exit
异步出口例行程序 asynchronous exit routine
异步处理 asynchronous processing
异步传输 asynchronous transmission
异步传输方式 asynchronous transfer mode
异步传输模式 asynchronous transfer mode;asynchronous transmit mode
异步传送 asynchronous communication
异步传送输出 asynchronous transmission output
异步串行数据 asynchronous serial data
异步存储器 asynchronous memory
异步的 non-synchronous;asynchronous
异步电动机 asynchronous dynamo;asynchronous motor;induction motor;non-synchronous motor
异步电动机的转差率 slip of induction motor
异步电花隙 asynchronous spark gap
异步电机 asynchronous dynamo;asynchronous machine
异步电路 asynchronous circuit
异步电容器 asynchronous condenser
异步定时器 non-synchronous timer
异步断开方式 asynchronous disconnected mode
异步多路转换器 asynchronous multiplexer
异步遏制 asynchronous quenching
异步发电机 asynchronous generator;induction generator;inductive generator;inductor generator
异步发送 asynchronous transmission
异步发送方式 asynchronous sending mode
异步方式 asynchronous system
异步分相机 asynchronous phase splitter
异步(感应)电动机机车 asynchronous (induction) motor locomotive
异步工作 asynchronous operation;asynchronous working
异步化同步电动机 asynchronized synchronous motor
异步回波滤除 defruit(ing)
异步机 asynchronous machine
异步激发 asynchronous excitation
异步计时器 asynchronous timer
异步计数 asynchronous counting
异步计算机 asynchronous computer;non-synchronous computer
异步记录操作 asynchronous record operation
异步接口 asynchronous interface
异步结构 asynchronous structure
异步(径)流 asynchronous flow
异步均衡方式 asynchronous balanced mode
异步控制 asynchronous control
异步链路 asynchronous link
异步启闭式六颚板抓具 asynchronous opening and closing 6-jaw grab
异步牵引电动机动力机车 locomotive with single phase/three phase converter set
异步请求 asynchronous request
异步入口点 asynchronous entry point
异步时分多路方式 asynchronous time division multiplexing
异步时分多路器 asynchronous time division multiplexer
异步时分复用 asynchronous time division multiplexing
异步时序网络 asynchronous sequential network
异步事件 asynchronous event
异步输出 asynchronous output;asynchronous transmission output
异步输入 asynchronous input;deferred entry
异步鼠笼感应牵引电动机 asynchronous squirrel cage induction traction motor
异步数据 asynchronous data
异步数据传输 asynchronous data transmission
异步数据传输通道 non-synchronous data transmission channel
异步数据传送 asynchronous data transfer
异步数据流 asynchronous flow
异步数据收集 asynchronous data collection
异步数据输出 asynchronous output
异步数据信道 asynchronous data channel
异步伺服电动机 asynchronous servomotor
异步速率 asynchronous speed
异步算子 asynchronous operator
异步调相机 asynchronous condenser;asynchronous phase modulator
异步调制解调控制器 asynchronous modem controller
异步调制解调器 asynchronous modem
异步通信[讯] asynchronous communication
异步通信[讯]接口 asynchronous communication interface
异步通信[讯]接口适配器 asynchronous communication interface adapter
异步通信[讯]控制附件 asynchronous communication control attachment
异步通信[讯]控制器 asynchronous communication controller
异步通信[讯]控制适配器 asynchronous communication control adapter
异步通信[讯]卫星 asynchronous communication satellite
异步退出 asynchronous exit
异步网(络) asynchronous network
异步系统 asynchronous system
异步显示 asynchronous display
异步信号 asynchronous signal
异步信号发送 asynchronous signal(l)ing
异步行波 asynchronous travel(l)ing wave
异步性 antisynchronism;asynchronism
异步要求 asynchronous request
异步移位寄存器 asynchronous shift register
异步抑制 asynchronous quenching
异步应答方式 asynchronous response mode
异步有限自动机 asynchronous finite state machine
异步运行 asynchronous operation
异步振动 asynchronous oscillation
异步振子 non-synchronous vibrator
异步整相器 asynchronous phase modifier
异步中断请求 asynchronous interrupt request
异步终端 asynchronous terminal
异步终端支持 asynchronous terminal support
异步终结 asynchronous termination
异步转矩 induction torque
异步转速 asynchronous speed
异步装置 asynchronous device
异步字符 asynchronous character
异步总线接口 asynchronous bus interface
异步总线系统 asynchronous bus system
异步阻抗 asynchronous impedance
异材板梁 hybrid plate girder
异操作 anti-coincidence operation;non-equivalence operation
异侧岔尾相对道岔 sides of the original track and with the tails of frogs facing each other
异侧对向道岔 two-sets of turnouts facing each other and laid on opposites sides of the original track
异侧顺向道岔 two-sets of turnouts facing each other and laid on opposites sides of the original track
异常 anomaly;disorder;out-of-the-way;unusualness
异常包裹体 extraordinary inclusion
异常保险 abnormal risk
异常报告 exception report(ing)
异常报文 exception message
异常本征函数 improper eigenfunction
异常标准离差 anomaly standard deviation
异常表征 abnormal attribute
异常波 extraordinary wave;freak wave
异常波型 abnormal wave pattern
异常材 abnormal wood
异常测定器 anomaly finder
异常差错 abnormal error
异常场 anomalous field
异常潮 abnormal tide;anomalistic(al) tide
异常潮水位 anomalous tide level
异常潮位 abnormal sea level;anomalous sea level;anomalous tide level
异常车行条件 abnormal operating condition
异常衬度 anomaly contrast
异常成本 abnormal cost
异常成分 abnormal component
异常尺码的 bastard
异常冲刷 abnormal erosion;abnormal scour;extraordinary scour
异常臭气 abnormal odo(u)r
异常出口 exception exit
异常出售 abnormal sale
异常处理 exception handling
异常处理程序 exception handler
异常传播 anomalous propagation
异常传输 transmission anomaly
异常磁差 abnormal magnetic variation
异常磁场 abnormal field;anomalous field
异常磁化 anomalous magnetization
异常次生加厚 anomalous secondary thickening
异常大的 monster
异常大小 anomaly size
异常带 anomalous zone
异常的 abnormal;above normal;anomalistic(al);anomalous;colossal;deviant;extraordinary;inordinate;remarkable;singular;transnormal
异常的分类 anomaly classification
异常的交通运输 extraordinary traffic
异常的圈定 anomaly delineation
异常等值线 anomaly contour
异常等值线图 anomaly map
异常地层压力 abnormal formation pressure
异常地震 anomalous earthquake
异常地震烈度 abnormal seismic intensity
异常电场 anomaly electric(al) field
异常电离层 abnormal E layer
异常电流 abnormal current;off-state current
异常电压 abnormal voltage
异常调度(例行)程序 exception scheduling routine
异常断层 abnormal fault
异常发酵 abnormal fermentation
异常返回 abnormal return
异常返回地址 abnormal return address
异常分化 abnormal differentiation
异常分派程序 exception dispatcher
异常分散模式 anomaly dispersion pattern
异常分水岭 anomalous watershed
异常丰度 anomalous abundance
异常丰富的水资源 exceedingly rich to water supply
异常风险 abnormal risk
异常峰类型 anomaly peak type
异常峰值 anomaly peak value;peak value of anomaly
异常服务例行程序 exception service routine
异常干旱 unusual drought
异常干涉色 abnormal interference colo(u)r;anomalous interference colo(u)r
异常高潮 extraordinary high tide;tidal wave
异常高潮位 anomalous high water-level;extraordinary high tide level;tidal wave crest
异常高水位 abnormal high water-level;anomalous high water-level;extraordinary high tide level
异常工况 unusual service condition
异常工作条件 abnormal operating condition
异常构造 anomalous structure
异常光线 extraordinary light ray
异常规模 dimension of anomaly
异常过电压 abnormal overvoltage
异常过载 abnormal overload
异常海流 abnormal current;abnormal sea current
异常函数 abnormal function
异常荷载 abnormal load
异常洪水位 exceptional flood level
异常化学岩【地】 allochemical rock
异常辉光放电 abnormal glow discharge
异常回波 angel
异常回答 exception response
异常积分 improper integral
异常级别 anomaly grade
异常极大值 maximum value of anomaly
异常季节性变 anomaly seasonal variation
异常检查 anomaly inspection;inspection of anomaly;reexamination of anomalies
异常检查报告 examination report of anomaly
异常检查平面图 inspection map of anomalies
异常接触 abnormal contact
异常结束 abend;abnormal end;ab-

normal end of task;abort
异常结束包 abort packet
异常结束控制表 abnormal end control table
异常截止点 anomaly cutoff
异常解释 anomaly explanation
异常解释平面图 plane figure of anomaly interpretation
异常界线划定 identification of anomaly boundary
异常金属元素 anomalous metallic element
异常晶粒生长 abnormal grain growth; exaggerated grain growth
异常聚声 acoustic(al) dazzle
异常均匀性 anomaly homogeneity
异常宽度 anomaly width;width of anomaly
异常烈度 abnormal intensity;anomalous intensity
异常林倾向 dipping of anomaly body
异常码 exception code
异常蒙气差 abnormal refraction
异常密度 abnormal density
异常磨损 abnormal wear; inordinate wear
异常木材 rotholz
异常逆矩阵 transnormal inverse matrix
异常浓度分带 anomaly concentration zoning
异常偏差 abnormal deviation
异常贫化 impoverishment of anomaly
异常平均值 anomaly mean value
异常评价 anomaly evaluation
异常评序 anomaly ranking
异常剖面 abnormal profile
异常期 anomalistic(al) period
异常气候 abnormal climate;abnormal weather
异常气味 abnormal odo(u)r; abnormal smell
异常铅 anomalous lead
异常强度 anomaly intensity
异常强化 enhancement of anomaly
异常侵蚀 abnormal erosion
异常情况 abnormal circumstance; abnormality; abnormal situation; uncommon condition
异常情况处理 exception handling
异常情况单 exception list
异常请求 exception request
异常区 anomalous zone
异常曲线 abnormal curve
异常任务终结 abnormal task termination
异常色觉 anomalous colo(u)r vision
异常筛选 sieving of anomaly
异常上地幔 anomalous upper mantle
异常升温 overheating
异常生长 abnormal growth
异常收益 abnormal return
异常属性 abnormal attribute
异常衰减模式 anomaly decay pattern
异常衰减曲线 anomaly attenuating curve
异常水位 abnormal water level; exceptional water level
异常水系 abnormal drainage
异常水晕 abnormal water halo
异常算子 exclusive operator
异常态 anomalous mode
异常特征 anomaly characteristics
异常特征分析法 anomaly characteristic analysis method
异常体长度 length of anomaly body
异常体宽度 width of anomaly body
异常体埋深 buried depth of anomaly body
异常体走向 strike of anomaly body
异常天气 extraordinary weather; unusual weather
异常条件 exceptional condition
异常条件中断 abnormal condition interrupt
异常停机 cancel closedown
异常图 anomaly map
异常土(壤) abnormal soil; extra-normal soil
异常推断 anomaly deduction
异常弯曲 abnormal curvature; gryposis
异常危险 abnormal risk
异常位 anomalous potential
异常物质 anomalous material
异常析出 hillock
异常系数 anomalous coefficient
异常下限 threshold
异常现象 abnormal phenomenon
异常现象出现层位 layer location of anomaly emerged
异常现象出现的孔深 depth of anomaly emerged
异常响应 exception response
异常向量 exception vector
异常项 exception item
异常项编码 exception-item encoding
异常项目 abnormal item
异常像 extraordinary image
异常信息 abnormal information
异常行为 aberant behavio(u)r; abnormal behavio(u)r
异常形成 malformation
异常形状 anomaly shape
异常旋转 improper rotation
异常寻址【计】 abnormal addressing
异常压力 abnormal pressure
异常压力下形成的岩层 abnormal pressure formation
异常岩浆 anomalous magma
异常应答 exception response
异常应力 abnormal stress
异常涌浪 abnormal swell
异常语句 abnormal statement
异常原因 abnormal cause
异常原则系统 exception principle system
异常允许 exception enable
异常运行 abnormal operation; misoperation
异常噪声 abnormal noise
异常增加 unusual increase
异常照射 unusual exposure
异常折光 abnormal refraction
异常折射 anomalous refraction
异常褶皱【地】 abnormal fold
异常真测试 abnormal true test
异常振动 abnormal shock
异常震区 anomalous felt zone
异常值 abnormal value; anomalous value
异常指示 abnormal indication
异常滞后 anomalous lag
异常中止 abort
异常中止卸出 abend dump
异常终结 abnormal termination
异常终止 abend; abnormal end; abnormal termination
异常终止程序 abort program(me)
异常终止命令 abort command
异常终止序列 abort sequence
异常终止作业 aborting job
异常种类 kind of anomaly; type of geochemical anomaly
异常重要 exceptional importance
异常转储 abnormal dump
异常转动 anomalistic(al) revolution
异常转矩 accidental torque
异常状态 abnormal condition; abnormal state; error state
异常着色 chromatism
异常姿态机动 unusual attitude maneuver
异常综合图 general anomaly map
异常走向 strike of anomaly
异常组分分带 anomaly composition zoning
异成分熔融 incongruent melting
异程回水式系统 direct return system
异程式布置 direct return scheme
异程式系统 direct return system
异臭 abnormal odo(u)r; foreign odo(u)r; taste and odor
异臭氧化物 isozonide
异地成煤说 drift theory
异地堆积 allochthony
异地灰岩 allochthonous limestone
异地角砾岩 allochthonous breccia
异地泥晶灰岩 allomicrite
异地砂屑 allodapic
异地生成煤 allochthonous coal; allochthony coal
异地岩块 allochthonous block
异地浊积灰岩 allodapic limestone
异地浊积岩 allodapic turbidite
异点 difference; dissimilarity; singular point
异电路 anti-coincidence circuit
异靛 isoindigo
异靛蓝 bioxindol
异丁 isobutyric
异丁氨基甲酸乙酯 isobutyl urethane
异丁醇 isobutanol; isobutyl alcohol
异丁二酸 isosuccinic acid; methylmalonic acid
异丁基溶纤剂 isopropyl cellosolve
异丁基橡皮胶 butyl rubber (building) mastic
异丁基溴 isobutyl bromide
异丁基乙炔 isobutyl acetylene
异丁酸 isobutyric acid
异丁酸异酯 isobutyl isobutyrate
异丁烷的溶解度 solubility of isobutane
异丁烷碳稳定同位素组成 stable carbon isotopic composition of isobutane
异丁烯 isobutene; isobutylene
异丁烯基溴 isocrotyl bromide
异丁烯腈 methacrylonitrile
异丁烯聚合物 isobutene polymer
异丁烯醛 methacrolein
异丁烯酸 methacrylate; methacrylic acid
异丁烯酸甲酯 methyl methacrylate
异丁烯酸异丁酯 <加强混凝土用的> isobutyl methacrylate
异丁烯酸酯 methacrylate; methacrylic ester
异丁烯橡胶 butyl rubber; isobutene rubber; isobutylene-isoprene
异丁烯橡胶带 butyl rubber tape
异丁烯橡胶嵌缝膏 butyl rubber sealant
异丁烯橡胶墙角护条 butyl rubber beading
异丁烯-异戊间二烯橡胶 isobutylene-isoprene rubber
异丁橡胶 buttress rubber; isobutene rubber; isobutylene-isoprene copolymer; isobutylene-isoprene rubber
异丁橡胶基层底板 butyl rubber base
异丁橡胶泡沫 butyl rubber foam
异丁橡胶嵌缝膏 butyl rubber sealant
异丁橡皮胶 butyl rubber (building) mastic
异端分配 heretical imputation
异端集 heretical set
异端连接 heterojunction
异二溴丁二酸 isodibromosuccinic acid
异发演替 allogenic succession
异方差性 heteroscedasticity
异分同晶性 allomerism
异分同晶质【化】 allomer
异分子聚合物 copolymer; interpolymer
异分子聚合作用 copolymerization
异佛尔酮 isophorone
异佛尔酮二异氰酸酯 isophorone diisocyanate
异构化催化剂 isomerization catalyst
异构化度 degree of isomerization
异构化反应 isomerization reaction
异构化干性油 isomerized drying oil
异构化过程 isomerization process
异构化合物 isocompound; isomeric compound
异构化聚合 isomerization polymerization
异构化橡胶 isomerized rubber; isorubber
异构化亚麻籽油 isomerized linseed oil
异构化油 isomerized oil
异构化组分 iso-component
异构化作用 isomerization
异构聚合物 isomeric polymer
异构裂化 isocracking
异构裂化物 isocrackate
异构硫 isosulf
异构热 isomerization heat
异构体 isomer; isomeride
异构烃 isohydrocarbon; isomeric hydrocarbon
异构烷烃 isoparaffin
异构烷烃的溶解度 solubility of isoalkane hydrocarbon
异构物 isomeric compound
异构现象 isomerism
异构型多处理机 heterogeneous multiprocessing
异国风尚 exotic fashions
异国情调 exotic atmosphere; exoti(ci)sm
异号 contrary sign; jack per station; opposite sign
异号电荷 unlike charge
异号相关 unlike-signed correlation
异乎寻常的 unconventional
异化 alienation; dissimilation
异化变质作用 allochemical metamorphism
异化产物 catabolin; catabolite
异化分裂 heterokinesis
异化过程 dissimilatory process
异化粒 allochem
异化粒灰岩 allochemical limestone
异化系数 dissimilarity coefficient
异化岩 allochemical rock
异化作用 catabolism; disassimilation; dissimilation
异辉锑铅银矿 diaphorite; ultrabasite
异或 anti-coincidence unit
异极 heteropole
异极场磁铁 heteropolar field magnet
异极的 heteropolar
异极点阵 heteropolar lattice
异极电机 heteropolar dynamo; heteropolar machine
异极对称型 heteropolar class
异极发电机 heteropolar generator
异极附着能 heteropolar attachment energy
异极感应子发电机 heteropolar inductor generator
异极化合物 heteropolar compound
异极键 heteropolar bond
异极交流发电机 heteropolar alternator
异极矿 calamine; galmei; hemimorphite; lapis cal aminaris

异极石 calamine
异极像 < 晶体的 > hemi-morphism; hemi-morph
异极形 hemi-morphic form
异极型轴对称 heteropolar type of axial symmetry
异极性 hemi-morphism; heteropolarity; opposite pole; reversed polarity
异极性液体 heteropolar liquid
异极性噪声 opposite-polarity noise
异己酮 hexone
异价 unequal valent
异教徒的 ethenic
异教徒居住区 pagan dom
异金属 < 如钛、锆、铪等 > exotic metal
异晶结构 heterostructure
异晶体 xenolith
异径 T 形管接管 reduced tee pipe coupling
异径 T 形接头 reducing tee [T]
异径承插管 reducing socket
异径法兰 reduced flange; reducing flange
异径缸 step-bore cylinder
异径管 convergent pipe; convergent tube; converging pipe; converging tube; reductor; stepped taper tube; transition pipe
异径管承插接头 socket reducer
异径管箍 reducing pipe coupling
异径管件 reducing fittings
异径管接 increaser; pipe reducer; reducer
异径管接头 reducing joint; adapter; reducing coupling; reducing socket; tapered pipe
异径管接头配件 reducing fittings
异径管节 reducing coupling; reducing pipe joint; reducing socket
异径管配件 reducing pipe fitting
异径活管接 reducing union
异径接箍 combination collar
异径接管 reducing piece
异径接头 bell nipple; bump joint; change-over sub; eccentric reducer; increaser; reducing joint; reducing piece; reducing union; sub; subcoupling; substitute sub; swedged nipple; screw-to-rod adapter < 螺旋立轴接与钻杆连接的 >
异径井身 tube construction with different diameter
异径连接管 adapting pipe
异径连接器 adapter coupling
异径连接套 reduction sleeve
异径螺纹管接头 reducing nipple
异径螺纹接套 reduced nipple; reducing nipple
异径内外螺纹管接头 reducing extension piece
异径喷嘴 reducing nozzle
异径三通管接 reducing T
异径三通(接头) reducing tee [T]; bull-head(ed) tee; reduced tee [T]
异径三通双弯接头 reducing twin bend; reducing twin elbow
异径三通双弯头 reducing twin elbow; reducing twin ell
异径十字(形接)头 reducing cross
异径四通(接头) reducing (pitcher) cross(ing)
异径管弯道接头 reducing elbow
异径弯管 street elbow; reducing elbow
异径弯头 adapter bend; reducing bend; reducing elbow; reducing ell; reducing pipe elbow; street elbow
异径支管 reducing branch
异径直管接头 reducing pipe coupling
异径管直线接头 straight reducer
异径止水 water sealing in different diameter
异径肘管 street elbow
异孔菌属 < 拉 > Heteroporus
异喹啉 isoquinoline
异类 heterogeneity
异类结构 heterogeneous structure
异类组 heterogeneous group
异离法 schlieren
异离体 schlieren
异粒子【物】 exotic particles
异亮氨酸 isoleucine
异量异序元素 heterobaric heterotope
异量元素 heterobar
异磷铁矿 heterosite
异磷铁锰矿 heterosite; neopurpurite
异鳞云杉 redtwig dragon spruce
异流 training
异流涵洞 by-pass tunnel
异流片 guide blade; guide plate
异硫氰酸 isothiocyanic acid
异硫氰酸盐 isothiocyanate
异硫氰酸酯 isothiocyanate
异硫锑铅矿 heteromorphite
异卤层 halocline
异氯甲桥萘 lsodrin
异码变化不归零制【计】 non-return-to-zero change
异门【计】 difference gate; distance gate; diversity gate; exclusive OR gate; anti-coincidence element; exclusive OR modulo-two sum gate; exjunction gate; non-equality gate; non-equivalence gate
异面力 non-coplanar forces
异面弯曲 non-uniplanar bending
异面异弹性 orthotropy
异名关系 synonymy
异平面力 non-coplanar forces
异谱同色 metameric colo(u)r
异倾斜的 heteroclinal
异氰化物 carbylamine; isocyanide; isonitroso-antipyrine
异氰脲酸 isocyanuric acid; tricarbimide
异氰脲酸三缩水甘油酯 triglycidyl isocyanurate
异氰脲酸酯泡沫 isocyanurate foam
异氰脲酰亚胺 melanuric acid
异氰酸 carbimide; isocyanic acid
异氰酸苯甲酰酯 benzoyl isocyanate
异氰酸苯酯 phenyl isocyanate
异氰酸丙酯 propyl isocyanate
异氰酸汞 mercuric isocyanate
异氰酸基含量 NCO content
异氰酸基指数 NCO index
异氰酸甲酯 methyl isocyanate
异氰酸树脂 isocyanate resin
异氰酸盐 isocyanate
异氰酸盐清漆 isocyanate varnish
异氰酸盐树脂 isocyanate resin
异氰酸盐塑料 isocyanate plastics
异氰酸乙酯 ethyl isocyanate
异氰酸酯 isocyanate
异氰酸酯黏[粘]合剂 isocyanate adhesive
异氰酸酯清漆 isocyanate varnish
异巯基丁酸 isomercaptobutyric acid
异曲表面 non-conformal surface
异燃料 exotic fuel
异染物 metachromasia metachromate
异染性 metachromasy; metachromatism
异溶化作用 heterolysis; heterophagy
异溶源性 allolysogeny
异三聚氰泡沫 isocyanurate foam
异色刺激 heterochromatic stimuli
异色的 heterochromatic; heterochromous
异色光度学 heterochromatic photometry
异色路面 contrasting pavement
异色柿 Philippine ebony
异色异构结晶 merchrome
异色异构物 chromo-isomer
异色异构(现象) chromogen isomerism; chromo-isomerism; chromotropism; chromotropy
异色异构颜色变化 chromotropism; chromotropy
异声 abnormal sound
异十九烷 pristane
异时的 heterchronous
异时发生 heterochrony
异时观测 different time observation; non-simultaneous observation; observation at interval
异时节律 heterochronic rhythm
异时趋同 heterochronous convergence
异时生成的 heterochronous
异时性 heterochronia; heterochrony; heterochronism
异势差连接法 heterostatic(al) method
异数 heteromerism; heteromery
异水胆矾 heterobrochantite
异水菱镁矿 giorgiosite
异速传动 friction(al) drive
异速生长 allometric growth; allometry
异态 anomaly; differential mode
异态动力学 heterodynamics
异态现象 heteromorphism
异态性 heteromorphism; heteromorphy
异酞腈 isophthalonitrile
异体 foreign body; variant; variant character
异体熔化 incongruent melting
异体设计 variantional design
异位差 correction for reduction to the zenith
异位差的 heterostatic(al)
异位差连接法 heterostatic(al) method
异味 off-flavo(u)r; off odo(u)r; off taste
异温动物 heterothermic animal
异无熔化 incongruent melting
异戊二烯 isoprene
异戊基环戊烷 isopentylcyclopentane
异戊间二烯 isoprene
异戊间二烯橡胶 isoprene rubber
异戊间二烯型烷烃 isoprenoid alkane
异戊烷 isopentane
异戊烷碳稳定同位素组成 stable carbon isotopic composition of isopentane
异戊烯 isoamylene; isopentene
异物 foreign body; foreign material; foreign matter
异物探测器 foreign body locator
异物探寻器 foreign body finder
异物同名 homonym
异物同形 homeomorph; homeomorphism
异物同形现象 homeomorphism
异物凿 foreign body gouge
异烯(烃)【化】 iso-olefine
异细胞 idioblast
异线 wrong track
异响 unusual sound
异相变质作用 allophase metamorphism
异相薄膜 heterofilm
异相沉积 heteropic(al) deposit
异相成分 out-of-phase component
异相成核 heterogeneous nucleation
异相的 out phase
异相电流 out-of-phase current
异相电压 out-of-phase voltage
异相分量 out-phase component
异相交叉相关 out-of-phase cross correlation
异相聚合 heteropolymerization
异相膜电极 heterogeneous membrane electrode
异相能谱 out-of-phase energy spectrum
异相燃烧 heterogeneous combustion
异相调制 Chireix modulation; outphasing modulation
异相(位) out-of-phase
异相信号 out-of-phase signal
异相振动 out-of-phase vibration
异相制 out-phase system
异向 contrary flexure
异向连接 perikinetic conjunction
异向模具 inverted die
异向捻 ordinary lay
异向凝结作用 perikinetic coagulation
异向三开道岔 three-throw turnout of contrary flexures
异向双开岔道【铁】 unsymmetric(al) double curve turnout
异向拖褶皱 incongruent drag fold
异向弯曲 contrary flexure
异向斜板沉淀池【给】 tube plate settler
异向信号 out-of-phase signal
异向性 anisotropy
异向性的 anisotropic(al)
异向旋转式 heterodromy
异向运动 perikinetic motion
异像 heteromorphism
异像岩【地】 heteromorphic rocks
异硝基 aci-nitro
异辛醇 ethyl hexanol
异辛酸盐 two-ethyl hexoate
异辛酸盐催干剂 ethyl hexoate drier
异辛酸酯 two-ethyl hexoate
异辛烷 isooctane
异形 abnormity; heteromorphosis; heterotype; allotype
异形板 contoured sheet; irregular slab; profile sheet(ing); shouldered sole plate; sketch plate
异形板坯 sketch blank
异形棒钢 deformed (reinforcing) bar
异形边 shaped bevel
异形扁钢 deformed flat steel
异形波导 irregular waveguide
异形玻璃研磨机 shaped glass grinding machine
异形槽钢 deformed channel steel
异形铲斗 irregular-shaped bucket
异形车辆 non-conventional type vehicle
异形沉箱 deformed caisson; special-shaped caisson
异形初轧坯 shaped block
异形带钢 profile strip; section strip
异形导线 shaped conductor
异形的 deformed; odd-shaped
异形垫板 compromising pad
异形垫圈 profile gasket
异形方块码头 modified cube quay wall
异形钢 fashioned iron; figured steel; rolled section steel; figured bar iron; profiled bar; profile iron; shaped iron; special rolled steel bar
异形钢板 deformed metal plate
异形钢板桩 deformed sheet pile
异形钢棒 profiled bar
异形钢材 irregularly shaped steel; special section; profiled bar; shaped bar; shaped steel; special-shaped steel; angle splice bar < 制造连接板用的 >; rolled section < 不包括圆、方、六角等简单断面的型材 >
异形钢带 profile(d) strip

异形钢轨 <两端连接不同型钢轨的> taper rail; compromise rail; compromising rail; transition rail
异形钢筋 deformed reinforcing bar; deformed (steel) bar; figured bar; high-bond bar
异形钢坯 <轧制工字梁用的> shaped block; beam blank
异形钢丝 profiled wire; section wire; shaped wire; deformed wire; deformed reinforcement
异形钢丝编成的长孔筛子 profiled wire slotted screen
异形钢索 deformed reinforcement
异形格栅 shaped joist
异形拱桥 variant arch bridge
异形构件 deformation member; non-standard member; profiled element
异形股钢丝绳 flat-strand rope
异形管 adapting pipe; deformed pipe; deformed tube; pipe for special service; profiled pipe; profiled tube; special pipe; specials; special section tube; taper pipe; taper tube
异形管件 special piece
异形管片 tapered segment
异形轨 compromise joint
异形焊 shape welding
异形护面块体 <防波堤等的> special-shaped armo(u)r block; special-shaped armo(u)r unit
异形黄铜条 rolled section brass
异形混凝土沉箱 deformed concrete caisson
异形混凝土块体 irregular concrete block; irregular concrete unit; shape-designed concrete block
异形挤出 contour extrusion
异形加工机 pantograph
异形夹板 compromise joint; compromising joint
异形夹板接头 <连接不同断面钢轨的> compromise joint
异形浇筑混凝土镶板 profiled cast concrete panel
异形接头 compromise joint; transition(al) joint; special-shaped coupling
异形接头夹板 compromise joint bar; compromising joint bar
异形截面玻璃 profile glass
异形金属板 deformed metal plate; stamped plate
异形绝热件 shaped insulation
异形绝缘子 shaped insulator
异形刻花 shaped engraving
异形孔挤压模 spider die
异形孔形 section groove
异形拉杆 deformed tie bar
异形冷拔模 heterogeneous die
异形联会 heterosynapsis
异形六边形 deformed hexagon
异形六角形 deformed hexagon
异形路肩石 profiled curb; profiled kerb
异形路签 strange staff
异形路缘石 profiled curb; profiled kerb
异形铝板 profiled alumin(i)um panel
异形铝片屋面材料 profiled alumin(i)um roof(ing) sheet
异形镘刀 sleeker; smoother
异形木板 profiled board
异形耐火砖 quarl(e)
异形配件 fashion parts; special-shaped parts
异形拼接缝板 special splice plate
异形砌块 special-shaped block
异形腔流体压力计 shaped-chamber manometer
异形墙砖 trimmer
异形射线 heterogeneous ray
异形双反射面 shaped dual reflector
异形条钢 special rolled steel bar
异形网状 heterobrochate
异形现象 heteromorphism
异形镶板 profiled panel
异形信号电极 difference type pickup electrode
异形形成 heteromorphosis
异形型材 bar stock
异形型钢 bar steel; shape steel rolled stock
异形性 heteromorphism
异形压实板 shaped pressure squeeze board
异形岩 heteromorphous rock
异形异链 allelomorphism
异形鱼尾板 compromising joint bar; step joint bar; transition fishplate; various shaped fishplate; cranked fishplate
异形预应力钢丝 deformed prestressing steel wire
异形圆钢筋 round deformed bar
异形再生 heteromorphosis; homeosis
异形珍珠 baroque pearl
异形织物 shaped fabric
异形植物 heterophyte
异形砖 angle brick; quaint brick; shaped brick; specially mo(u)lded brick; special-shaped brick; squint brick; brick shape; complicated shape brick; figurate stone; special form brick
异形作用 heterotypic(a) effect
异性 isomerism
异性电 opposite electricity
异性交替 heteromorphic alternation
异性石 eucolite; eudialite [eudialyte]
异性条件 diversity condition
异性霞石正长岩 lujavrite
异性岩 eudialite [eudialyte]
异性质 heterogeneity
异序素【化】isobar
异序同晶系 eutropic series
异序同晶现象 eutropy
异序元素 heterotope
异序作业 job shop
异旋双折射 allogyric birefringence
异压岩层 abnormal pressure formation
异亚丙基丙酮 mesityl oxide
异亚硝基 hydroxyimino; isonitroso
异亚硝基樟脑 isonitrosocamphor
异岩性不整合 nonconformity
异养好氧生物膜 heterotrophic aerobic biofilm
异养菌平板计数 heterotrophic bacterial plate count
异养菌总数 total heterotrophic bacteria count; total number of heterotrophic bacteria
异养生物 heterotroph; heterotrophic organism
异养微浮游生物 heterotrophic nanoplankton
异养微生物 heterotrophic microbe; heterotrophic microorganism
异养细菌 heterotopic organisms; heterotrophic bacteria
异养硝化 heterotrophic nitrification
异养性 heterotrophism
异养循环 heterotrophic cycle
异养植物 heterophyte; heterotrophic plant
异样 diversity
异样色痕 ferrite ghost; ghost
异腰梯形车场 trapezium yard
异叶果玉茶 heteroleaf false panax
异叶铁杉 Western coast hemlock; western hemlock
异乙炔 isoacetylene
异议 demurer; disagreement
异议报告 exception report(ing)
异议和索赔条款 discrepancy and claim clause
异议书 protest
异吲哚 isoindole
异吲哚金属络合物颜料 isoindole metal complex pigment
异吲哚啉酮系颜料 isoindolinone pigment
异硬脂酸 isostearic acid
异油酸 isooleic acid; vaccenic acid
异域物种 allopatric species
异元件 anti-coincidence element; non-equivalent-to-element
异元母质层 paralichic contact
异元熔点 incongruent melting point
异原子序元素 heterotope
异源包体 enclave enallogene
异源地体 exotic terrane
异源发生 heteroblastic
异源过程 allogenic process
异源火山角砾岩 allogenic volcanic breccia
异源集块熔岩 allogenic agglomerate lava
异源集块岩 allogenic agglomerate
异源角砾熔岩 allogenic breccia lava
异源凝灰熔岩 allogenic tuff lava
异源凝灰岩 allogenic tuff
异源熔结集块岩 allogenic welded agglomerate
异源熔结角砾岩 allogenic welded breccia
异源熔结凝灰岩 allogenic welded tuff
异源融合 heteromixis
异源生物群落 allobiocenosis
异源转化 allogenic transformation
异运算 modulo-two-sum; non-equivalence operation
异樟脑 isocamphor
异质 alloplasm; heterogeny
异质包体 enclave antilogene
异质的 heterogeneous
异质功能多处理机 heterogeneous multiprocessor
异质货物 heterogeneous cargo
异质结材料 heterojunction material
异质结二极管 heerodiode
异质结构 heterojunction structure; heterostructure
异质结光电池 heterojunction photovoltaic cell
异质结光电阴极 heterojunction photocathode
异质结红外探测器 heterojunction infrared detector
异质结太阳电池 heterojunction solar battery
异质晶体 heterocrystal
异质凝结 heterocoagulation
异质同晶共聚树脂 polyallomer resin
异质同晶共聚物 polyallomer
异质同晶聚合物 polyallomer
异质同晶(现象) allomerism; homeomorphy; homomorphism
异质同晶性 allomerism
异质外延 heteroepitaxy
异质外延光波导 heteroepitaxial optic(al) waveguide
异质外延膜 heteroepitaxial film
异质纤维 hetero fiber [fibre]
异质相加 heterogeneous summation
异质性 heterogeneity
异质性的方差 heterogeneous variance
异质性群体 heterogeneous population
异质氧化物 heterogeneous oxide
异种处理机 dissimilar processor
异种的 foreign
异种化感 allelopathy
异种晶体 foreign crystal; heterocrystal
异种路面 contrasting pavement
异种凝集 heteroagglutination
异种凝集素 heteroagglutinin
异种溶解 heterolysis
异重层 pycnocline
异重流 current of higher density; densimetric flow; density current; density stratified flow; stratified current; stratified flow; turbidity current
异重流沉积作用 turbidite sedimentation
异重流底层 density current bed
异重流含沙量比尺 scale of sediment content of density current
异重流模型 density current model; density flow model
异重流潜流 density underflow
异重水层 density water stratification
异重水流 density flow; interfluent flow
异重效应 density effect
异轴性 desaxe
异轴增生 epitaxial overgrowth
异棕檬酸 isocitric acid
异组分体 allobar

役龄 enlistment age

役畜 draft animal

抑爆峰值 explosion suppressing peak value

抑爆剂 knock suppressor
抑爆控制器 explosion suppression control unit
抑爆浓度 deflagration suppression concentration
抑尘 dust suppression
抑尘的 dust allaying
抑尘剂 dust preventive
抑低增益 reduced gain
抑浮剂 depressant
抑弧 arc suppression
抑景 blocking view; obstructive scenery
抑菌剂 bacteriostatic(al) agent; fungistat; germifuge; bacteriostat
抑菌圈 bacteriostasis diameter
抑菌素 ablastin
抑菌性增塑剂 bacteriostatic(al) plasticizer
抑菌作用 bacteriostasis; bacteriostatic action
抑霉菌剂 fungistat; fungistatic agent
抑泡剂 foam depressant; foam inhibiting agent; foam inhibitor; foam reducing composition; foam suppressing agent; foam suppressor
抑染剂 suds suppressor
抑限制 curbing
抑谐网络 harmonic suppression network
抑音器 damper
抑油圈 oil deflector
抑噪器 noise remover
抑真菌剂 fungistat; fungistatic agent
抑蒸发剂 anti-evapo(u)rant
抑殖素 ablastin
抑止 hold-back; suppression
抑止电极 retarding electrode; retarding potential
抑止杆 check rod

抑止剂 retardant
抑止(频)带 stop band
抑止频带宽度 barrage width
抑止频率 blanketing frequency
抑止器 killer
抑止设备 stopper
抑止谐波的扼流圈 harmonic choke
抑止载波 quiescent carrier
抑止载波调制 quiescent-carrier modulation
抑制 bate; constraint; containment; damping; dam up; depress; depression; elimination; gulp; jamming; killing; nullification; repress; repression; restrain; retard; stifle; stopping; trapping; withhold
抑制比 rejection ratio
抑制比率 inhibition ratio
抑制边带 suppressed sideband
抑制长度指示器 suppress length indicator
抑制场 limiting field
抑制场制 suppressed field system
抑制尘埃系统 dust suppression system
抑制程度 containment procedure
抑制冲击器 shock suppresser [suppressor]
抑制窗孔 rejection iris
抑制措施 disincentive
抑制带 inhibition zone
抑制的 disincentive
抑制点 rejection point
抑制电极 retarded electrode; retarding electrode; suppressor electrode
抑制电流 suppression current
抑制电路 inhibit circuit; suppressed circuit; suppression circuit
抑制电阻 suppression resistance
抑制电阻器 suppression resistor
抑制阀 inhibitor valve; suppression valve
抑制反射 inhibitory reflex
抑制分解阶段 arrested decay
抑制敷料滚筒 damper applicator roller
抑制腐蚀 inhibition of corrosion; retarding corrosion
抑制干扰 barrage jamming
抑制干扰电阻(器) suppression resistor; suppressor resistor
抑制工作时间 suppression operate time
抑制购买价格 prohibitive price
抑制故障率 reject failure rate
抑制管 killer tube
抑制过压器 suppressor overvoltage
抑制海浪 kill the sea
抑制焊接 restraint welding
抑制红色灵敏度 suppressed red response
抑制恢复 suppression recovery
抑制挥发剂 devolatilizer
抑制活动 inhibited activity
抑制火花装置 spark arrester [arrestor]
抑制级 inhibit stage; killer stage
抑制极 suppresser electrode
抑制集中 concentration of inhibition
抑制剂 depressant; depressor; inhibiting agent; inhibitor; localizator; localizer; poison; restrainer; suppressor; inhibitory substance
抑制剂感应 inhibitor response; inhibitor susceptibility
抑制剂染料 inhibitor dye
抑制剂脱臭法 inhibitor sweetening
抑制剂脱硫 inhibitor desulfurization
抑制剂效力 inhibitor effectiveness
抑制碱集料膨胀 inhibiting alkali-aggregate expansion
抑制降解 inhibiting degradation
抑制截面 stopping cross-section
抑制解除 disinhibition
抑制控制 suppression control
抑制扩散 irradiation of inhibition
抑制类似物 inhibitory analogue
抑制联系 inhibitory connection
抑制邻近频率信号 reject signals at nearby frequencies
抑制滤波器 suppression filter
抑制脉冲 killer pulse; suppression pulse; suppressor impulse; suppressor pulse
抑制膜 inhibiting film
抑制浓度 concentration of inhibition
抑制频带 rejection band; suppressed frequency band; stop band
抑制频率 rejection frequency
抑制期 inhibition period; inhibitory stage
抑制器 allayer; eliminator; inhibitor; killer; rejector; suppresser [suppressor]
抑制器电路 killer circuit; rejector circuit
抑制器检波放大器 killer detector-amplifier
抑制侵蚀的 erosion deterrent
抑制区 inhibitory area; reject region
抑制栅 cathode grid; suppression grid; suppressor; suppressor grid
抑制栅电压 suppressor grid voltage
抑制栅电子管 suppressor grid tube
抑制栅极 earth(ed) grid; gauze; grounded grid; guard net; suppressor (electrode); suppressor grid
抑制栅极电位 suppressor potential
抑制栅极调变 suppressor grid modulation
抑制栅极调幅 suppressor grid amplitude modulation
抑制栅极调制 suppressor grid modulation; suppressor modulation
抑制栅孔 suppressor holes
抑制栅调制电路 suppressor grid modulation circuit
抑制栅调制放大器 suppressor-modulated amplifier
抑制栅透镜 cathode grid lens
抑制栅真空计 suppressor grid ga(u)ge
抑制生产作用 disincentive
抑制时间 inhibition time; suppressed time
抑制时间延迟 suppressed time delay
抑制试验 inhibition test; suppression test
抑制输入 inhibitory input
抑制损耗 suppression loss; suppressor loss
抑制天线 suppressing antenna
抑制添加剂 suppressant additive
抑制图像 rejection image
抑制网 suppressor mesh
抑制无线电干扰的成套装置 suppressor kit
抑制物 inhibitor
抑制相 inhibitory phase
抑制效能 inhibitor effectiveness
抑制效应 depression effect; inhibiting action retarding effect; inhibitory effect; stopping effect
抑制信号 inhibit signal
抑制行动平衡 inhibition-action balance
抑制型 inhibitory type
抑制型电离真空计 suppressor ion ga(u)ge
抑制性毒性 inhibitory toxicity
抑制性突触后势能 inhibitory postsynaptic potential
抑制延迟时间 suppression hangover time
抑制氧化 inhibited oxidation
抑制因数 rejection factor
抑制因子 inhibiting factor; inhibitor; suppressor factor
抑制油 inhibited oil
抑制元件 straining element
抑制载波 <在无调制周期内> quiescent carrier; suppressed carrier
抑制载波传输 suppressed carrier transmission
抑制载波的双边带调制 double-sideband suppressed-carrier modulation
抑制载波电话 quiescent-carrier telephone
抑制载波发射机 suppressed carrier transmitter
抑制载波体制 suppressed carrier system
抑制载频光学调制 suppressed carrier optic(al) modulation
抑制噪声 noise abatement; noise elimination; noise suppression; silencing
抑制噪声方法 silencing method
抑制噪音 squelch
抑制真菌的 fungistatic
抑制蒸发 evaporation reduction; evaporation suppression
抑制值 inhibiting value; threshold value <防腐剂的>
抑制(周界)灌浆 containment grouting
抑制柱 suppressor column
抑制装置 suppression device; suppressor
抑制状态 inhibitory state
抑制作用 effect of restraint; inhibited effect; inhibit function; inhibitory action; retarding effect; inhibition

译

译本 translation; version
译电室 coding office; coding room
译码室 decoding room
译青表 transliteration key
译审 professor of translation

易

易暗示性 suggestibility
易搬空的 evacuable
易保养涂饰剂 easy care finish
易爆材料 explosive material
易爆的 detonable
易爆发性建筑材料 volcanic building material; volcanic construction(al) material
易爆货物锚地 explosive anchorage
易爆及危险货物锚地 explosives and dangerous goods anchorage
易爆可燃尘埃 combustible dust
易爆气(体) explosion hazard gas; explosive gas
易爆气体环境 explosive atmosphere
易爆燃料 explosive fuel
易爆危险品旗 powder flag
易爆物 explosion hazard
易爆物料 explosive material
易爆性 explosion hazard
易爆炸品 explosive goods; explosives
易爆炸品锚地 explosive anchorage area
易北河 Elbe River
易被剪切的 shear-susceptible
易泵唧性 pumpability
易泵性 <指砂浆、混凝土> pumpability
易变 mutability; unsteadiness; versatility
易变半导体 liquid semiconductor
易变的 fluid; labile; uncertain; unstable; unsteady; versatile; volatile
易变钙钒石 sherwoodite
易变硅钙石 tacharanite
易变河段 variable reach
易变荷载 mobile load
易变辉石 pigeonite
易变拦门沙 mouth-bar liable to change
易变平衡 mobile equilibrium
易变情况下的决策 decision-making under uncertainty
易变区 labile region
易变色 allochroism
易变天气 variable weather
易变文件 volatile file
易变物 variable
易变现金的财产 liquid assets
易变形板 lithe board
易变形的 deformable; yielding
易变形梁 slender beam
易变性 unsteadiness; volatility
易变振荡器 labile oscillator
易变质货物 goods subject to early deterioration
易变质军需品 deteriorating supplies
易变(状)态 labile state
易辨认点 identifiable point
易玻化坯体 vitrifiable body
易玻璃化黏[粘]土 vitrifiable clay
易剥离的 fissile
易剥裂页岩 fissile shale
易剥落的 flaky
易剥落性 friability
易采的岩石 kindly
易操纵的步行式机械 steerable walking mechanism
易操纵的滚筒 steerable roll(er wheel)
易操纵的履带拖拉机 steerable crawler
易操纵的压路机 steerable roll(er wheel)
易操纵性 handiness; manageability
易拆 easy off
易拆除的模板 collapsible formwork
易拆式管接头 breakaway connector
易拆式吸管 easy disconnected suction tube
易拆卸的 quick detachable
易拆卸顶棚 removable ceiling
易掺混的 easy-to-blend
易颤期 vulnerable period
易潮解的 deliquescent
易潮解物质 deliquescent material; deliquescent matter
易潮湿的 retentive
易潮石 trudellite
易潮物质 deliquescent matter
易沉降的 free settling
易沉降固体 settleable solid
易沉陷(的软)土 yielding ground
易沉陷地面 yielding ground
易沉陷基础 yielding foundation
易沉陷路基 yielding foundation
易成粉末状 powdery form
易成型性 formability
易冲蚀的河岸 erodible
易冲蚀的河岸质 easily eroded bank material
易冲蚀土 erodible soil
易冲刷材料 very erodable material
易冲刷的河流 erosive river
易冲刷河槽 erodible bed channel
易冲刷河床 erodible bed
易出故障处 trouble spot
易出事故 accident proneness
易出事故的驾驶员 accident-prone driv-

er
易出事故地带 troublesome zone
易出事故地点 accident exposure
易处理的 disposable;tractable <材料等>
易处理性 amenability of treatment
易磁化方向 direction of easy axis;preferred direction;preferred direction of magnetization
易磁化轴 easy axis;easy magnetizing axis;preferred axis of magnetization
易脆顶板 short ground
易带出的元素 more active element
易地处理 ex-situ treatment
易点燃性 ease of ignition
易冻的 frost-prone;frost susceptible
易冻坏性 liability to frost damage
易冻货物 perishable freight;perishable goods
易冻货物预冷 precooling of perishable freight;precooling of perishable goods
易冻结性 susceptibility to frost action
易冻性 frost susceptibility;liability to frost damage
易冻胀的 frost susceptible
易冻胀颗粒 frost-susceptible particle
易冻(胀)土 frost-susceptible soil;frost-active soil
易读的 easy-to-read;legible;readable;self-reading
易读地图 decipherable map
易读性 legibility;readability
易堵塞的 cloggy
易断接头 dry joint;dry weld
易发生的 incidental
易发生事故的 accident-prone
易发震区 earthquake prone area;earthquake prone region
易翻的 cranky
易分解化合物 unstable compound
易分解有机质 labile protobitumen
易分解原沥青 labile protobitumen
易分离耦合 frangible coupling
易分散颜料 shear-sensitive pigment
易粉化的 efflorescent
易粉化型颜料 chalking pigment
易风化成碎屑的 eugeogenous
易风化性 liability to weathering
易风化岩 eugeogenous rock
易浮的 buoyant
易腐败的 perishable
易腐败的东西 perishable
易腐补给品 perishable supplies
易腐材料 perishable material
易腐货快运 meat run
易腐货物 perishable cargo;perishable goods;perishable product;perishables
易腐烂的 putrescible
易腐烂货物 perishable cargo;putrescible goods;perishable freight
易腐烂商品 perishable commodities;putrescible goods
易腐烂物质 perishables;putrescible matter
易腐烂性 perishability
易腐品<尤指食物> perishables
易腐蚀性 corruptibility
易腐食品 perishable items
易腐物储存 perishable inventory
易腐物品 perishable goods
易腐物质 perishables
易感动的 susceptible
易感光的<如软片> sensitive
易感受期 susceptible period
易感性 liability;susceptibility;susceptivity
易感应的 sensitive
易感者 susceptible
易感状态 sensitization
易感状态增强机制 sensitization-invigoration mechanism
易割冒口 knock-off head;neck-down riser;washburn riser
易割冒口圈 riser pad
易割冒口芯片 knock-off core;wafer core
易汞齐化金 free-milling gold
易管理性 manageability
易灌注性 remo(u)lding effort
易焊接性 solderability
易焊锰钢<其中的一种> Vanity steel
易耗的包装 expendable packing
易耗件 consumable item
易耗品 consumable goods
易耗托盘<不必回送的> expendable pallet
易耗资源 fugitive resources
易糊钻地层 balling formation
易滑的 slippy
易滑动的积雪 fast snow
易滑土 slip clay
易滑脱的 labile
易滑性 lability
易滑移面 plane of easy slip
易化 facilitation
易化扩散 facilitated diffusion
易化区 facilitated area
易换速变速器 easy-change transmission
易挥发成分 more volatile component
易挥发的 highly volatile
易挥发货物 volatile cargo
易挥发味 fugitive flavor
易挥发物 volatile matter
易挥发性 effumability
易毁的 vulnerable
易混槽法炭黑 easy processing channel black
易货 barter
易货合同 barter contract
易货汇兑 barter exchange
易货交易 barter deal;barter transaction
易货贸易 barter;barter trade;barter versus;goods exchanging trade
易货协定 barter agreement;barter arrangement
易货协议 barter agreement
易货者 trucker
易货制度 barter system
易积水的地质构造 structural trap
易激晶体 crystal of high activity
易加工的钢材 easy-machining steel
易加工时间 workable time
易加工树脂 easy processing resin
易加工性 workability
易见渗入剂 visible penetrant
易溅出泥浆的 sloppy
易降解洗涤剂 ease degradation organic matter
易浇注性 remo(u)lding effort
易接近的 accessible;approachable
易接近性 accessibility
易洁性 cleanability
易结渣煤 clinkering coal
易解开的绳结 granny knot
易解绳结 granny knot
易解石 aeschynite;aesehynite;eschynite
易开白铁皮罐 easy-open tin can
易开包装 pull tab or button drown
易开塞子 quick-release stopper
易刻性 scratchability
易控变量 manageable variable
易控制性 manageability
易拉长的 ductile
易涝洼地 land liable to flood
易利用水 readily available water
易利用水分 readily available moisture
易利用水分含量 management allowed deficiency
易裂的 fissile
易裂性 fissility;liability to cracking
易流动 easy flow
易流动液体 mobile liquid
易流阀 free flow valve
易流混凝土 fluid concrete
易流渣 free running slag
易漏 leakiness
易冒落顶板 easily falling roof
易密性 compactability
易磨光的 polish-susceptible
易磨光集料<对磨光敏感的集料> polish-susceptible aggregate
易磨损材料 breakout material
易磨系数 grindability factor;grind factor
易磨性 easiness to milling;grindability
易磨性试验 grindability test
易磨性系数 grindability coefficient;grindability factor
易磨性指数 grindability index
易挠性 limberness
易凝结的 clotty
易弄干净的 easily cleanable
易排屑型 chip flow type
易膨化的黏[粘]土 swelling clay
易劈向 grain
易劈岩 freestone
易劈页岩 fissile shale
易疲劳性 easy fatigability
易破坏的 damage vulnerable
易破坏性 damageability
易破碎的细粒砂岩 packsand
易破损的 damageable
易起反应 responsiveness
易迁移的元素 elements of easy migration
易切不锈钢的小孔 pitting of free-machining stainless steel
易切的砂石 freestone
易切(砂)岩 freestone
易切石料 freestone
易切削的 free cutting
易切削钢 automatic steel;free cutting steel;free-machining steel
易切削钢丝 free cutting steel wire
易切削黄铜 free cutting brass
易切削性 free cutting machinability
易切削铸铁 machinable cast-iron;soft cast iron;soft iron
易侵蚀的 erodible
易侵蚀河槽 erodible bed channel
易侵蚀土 erodible soil
易侵蚀性 erodibility
易倾斜的 crank(y);tender【船】
易倾斜的船 crank vessel;tender ship;tender vessel
易倾性 burnability;tenderness
易曲的 limp
易燃材料 combustible material;flammable material;inflammability material
易燃材料仓库 inflammable store
易燃成分 inflammable constituent
易燃的 burnable;combustive;deflagrable;fire-susceptible;flammable;free-burning;inflammable
易燃等级 hazard classification
易燃废物堆 fire trap
易燃粉尘 combustible dust
易燃干木料 light wood
易燃构造 combustible construction
易燃固体 flammable solid;inflammable solid;spontaneous combustion articles and combustion in humidity articles
易燃挥发液体 inflammable volatile liquid
易燃货品 inflammable cargo;inflammable freight;inflammable goods
易燃货物 inflammable cargo;inflammable freight;inflammable goods
易燃剂 inflammable compound
易燃建筑物 combustible construction;tinderbox
易燃焦 fast-burning coke
易燃结构 combustible construction
易燃空气油气混合物 flammable air-vapo(u)r mixture
易燃框架结构 combustible frame construction
易燃垃圾 combustible waste
易燃料 inflammable substance
易燃煤 free-burning coal
易燃木材 light wood
易燃品 flammable material;inflammable material;inflammable substance
易燃品储藏库 inflammable store
易燃品库 inflammable material storage
易燃气煤 gas flame coal
易燃气体 fire gas;inflammable gas;inflammable vapo(u)r
易燃气体爆炸 fire-gas-explosion
易燃石油产品 dangerous oil
易燃瓦斯 combustible gas
易燃危险 fire hazard
易燃围篱 combustible fence
易燃物 combustibles;fire hazard;inflammables;inflammable substance
易燃物仓库建筑 combustible storage building
易燃物品 combustible material
易燃物品红标签 red label
易燃纤维 combustible fiber [fibre]
易燃限度 inflammable limit
易燃性 combustibility;fire hazard;flammability;ignitability;inflammability
易燃性材料 inflammable material
易燃性固体废物 inflammable solid waste
易燃性试验 inflammability test
易燃压缩气体 inflammable compressed gas
易燃液体 flammable liquid;inflammable liquid
易燃液体泊位 inflammable liquid berth
易燃栅栏 combustible fence
易燃蒸气 inflammable vapo(u)r
易燃制冷剂 flammable refrigerant
易溶的 diffluent
易溶解 readily dissociate
易溶聚合物 hemi-polymer
易溶黏[粘]土 slip clay
易溶物质 rapidly soluble material
易溶性化合物 easy-soluble compound
易溶性岩石 strongly soluble rock
易溶盐 lyso-soluble salt
易溶盐含量 content of lyso-soluble salt;content of readily soluble salts
易溶盐含量百分比 percent proportion content of soluble salt
易溶盐含量试验 strongly soluble salt content test
易溶盐试验 easy-soluble salt test;strongly soluble salt test
易溶阳离子组 soluble anion group;soluble cation group
易溶元素 soluble elements
易熔安全塞 safety fusible plug
易熔玻璃 fusible glass;soft glass
易熔部件 fusible member
易熔瓷土 fusible clay;vitrifying clay

易熔的 colliquable;fusible;readily fusible
易熔点 eutectic point;eutectoid point;vitrifying point
易熔铬铁 chromsol
易熔锅炉塞合金 boiler plug alloy
易熔焊接 eutectic bonding
易熔焊料 fusible solder;quick solder
易熔焊条 eutecrod
易熔合金 bend alloy;boiler plug alloy;eutectic alloy;fuse alloy;fusible alloy;lower melting alloy;low-melting alloy;low-melting point alloy
易熔合金定温火灾探测器 fixed temperature detector using fusible alloy
易熔化黏[粘]土 vitrified clay
易熔灰(分) fusible ash;easy-fusible ash
易熔辉石 eulite
易熔接件<安全防火用> fusible link
易熔金属 fuse metal;fusible metal
易熔连杆 fusible link
易熔连接 fusible link
易熔黏[粘]土 fusible clay;slip clay;vitrifying clay
易熔熔丝 quick-break fuse
易熔塞 drop plug;fusible plug
易熔塞变形温度 yield temperature
易熔塞合金 boiler plug alloy
易熔色料 hot melt
易熔生料 easy fluxing mix
易熔栓 fusible plug
易熔体结构 eutectic structure
易熔性 fusibility
易熔釉 fusible glaze
易熔渣 fusible slag
易熔质 eutectic
易软性 limberness
易润湿颜料 wetting colo(u)r
易散粉尘 fugitive dust
易散性排放 fugitive emission
易散性污染源 fugitive source
易散性烟尘排放 fugitive dust emission
易扫除干净的 easily cleanable
易伤道路使用者 vulnerable road users
易烧结粉末 sinterable powder
易烧性 burnability
易烧性系数 burnability factor
易烧性指标 burnability scale
易烧性指数 burnability index
易摄养分 available nutrient
易生物降解物质 biodegradable substance;readily biodegradable organics
易失存储器 volatile store
易失火的建筑物 fire trap
易失性 volatility
易失性测试 volatility test
易失性存储器 volatile memory;volatile storage
易蚀土壤 erodible soil
易使水污染的农药 water-pollution-prone agricultural chemicals
易受 exposed to
易受潮的 susceptible to moisture
易受冲蚀的土壤 erodible soil
易受冲刷的 erodible
易受腐蚀的 erodible
易受攻击 liable to attack
易受洪水泛滥区 flood-prone area
易受洪灾地区 flood-prone area
易受机械损伤 subject to mechanical injury
易受侵蚀材料 erodible material
易受侵蚀的 erodible
易受侵蚀的土壤 erodible soil
易受伤的 vulnerable
易受损害的 damageable;predispose to damage
易受污染损害的 vulnerable to pollution
易受影响的 accessible;susceptible
易受灾区域 disaster-prone region
易树脂化 resinophore
易刷性 brushability
易水化黏[粘]土 easily hydrated clay
易碎白云岩 friable dolomite
易碎玻璃 brittle glass
易碎玻璃密封 breakable glass seal
易碎材料 brittle material;fragile material;friable material
易碎成粉末的 powdery
易碎大理石 friable marble
易碎的 brashy<木材等>;breakable;cracky;crisp;crumbling;crumbly;fluffy;fragile;frail;frangible;friable;pulverulent
易碎褐煤 moor coal
易碎货物 fine cargo;fragile cargo;fragile goods
易碎件 quick-wear parts
易碎晶格 brittle lattice
易碎颗粒 friable particle
易碎块煤 lively coal
易碎裂的 splintery
易碎煤 friable coal;weak coal
易碎黏[粘]土 friable clay
易碎品 breakables;fragile;fragile article
易碎石 spallable rock
易碎物品 fragile
易碎性 brashness;breakage susceptibility;brittleness;fragileness;fragility;frangibility;friability
易碎岩石 friable rock
易碎岩芯 friable core
易碎沼煤 moor coal
易碎纸 egg-shell paper
易损 rapid wear
易损部分 vulnerable area
易损的 vulnerable
易损附件 vulnerable accessories
易损坏的 brittle;damageable;short-lived
易损坏性 damageability;vulnerability
易损货物 fragile cargo;vulnerable cargo
易损件 high-wear item;vulnerable parts;wearing parts
易损零件 easily-worn parts;quick-wear parts;vulnerable parts
易损期 vulnerable period;vulnerable phase
易损伤性 vulnerability to accidents
易损系统 vulnerable system
易损性 susceptibility to failures
易损性分析 vulnerability analysis
易损性函数 vulnerability function
易塌的积雪 fast snow
易塌页岩 heaving shale
易坍的 insecure
易坍塌的 free caving
易坍塌的粗状岩层<须用锤击钻进> collapsible granular stratum
易坍塌地层 cavey formation;caving formation;caving ground
易坍塌井 caving hole
易坍塌孔 caving hole
易坍塌岩层 caving formation
易坍塌页岩 caving shale
易炭化物 readily carbonizable substance
易探测圈闭 easily detectable trap
易通过的 readily accessible
易褪色的染料 fugitive dye
易褪色的色料 fugitive colo(u)r
易褪色的颜色 fugitive colo(u)r
易褪着色 fugitive tinting
易脱模板条 drip strip
易脱水污泥 drainable sludge
易脱楔块 easing the wedges;easing wedge
易脱性彩粉浆 colo(u)r wash
易脱性彩色涂料 colo(u)r wash
易弯的 flexible;pliable
易弯曲的 lithe;lithesome;supple;whippy
易弯(屈)性 flexibility;flexibleness
易为变现金的(财产等) liquid
易位 displacement;relocation;repositioning;translocation;transposition
易位求积分法 integration by transposition
易位损伤 translocated injury
易位作用 metathesis [复 metatheses]
易误修复 error-prone repair
易洗的 easily cleanable
易陷地面 yielding ground
易消失的 fugitive
易消失的湖 evanescent lake
易销货物 marketable goods
易效水 readily available water
易效水分<土壤内的> readily available water;readily available moisture
易泄漏的 leaky
易卸包 readily removable pack
易卸接头 frangible coupling
易卸压缩机<现场用> accessible compressor;field service compressor
易新价值 trade-in value
易行车 free runner;free running car;good runner;easy rolling car;good rolling car【铁】
易行线 easy rolling track;easy running track;easy track;good rolling track;good running track
易修整性 finishability
易朽材料 perishable material
易朽性 perishability;perishableness
易选 easy washing
易选的 free-milling
易选矿石 free-milling ore
易淹地区 land liable to flood
易延展材料 ductile material
易氧化材料 oxidizable material
易逸度 fugacity
易逸性 fugacity
易逸性平衡模型 fugacity equilibrium model
易引起火灾的废物堆 fire trap
易引起误解的标记 misleading indications
易淤浅性 liability to shoaling
易于被糟蹋的环境 environment subjected to misuse
易于操纵的 easy-to-handle
易于操作 ease of operation
易于操作的 maneuverable
易于拆卸的半成品 semi-knockdown
易于拆卸的东西 knock down
易于出错的 error-prone
易于发生地震的 quake-prone
易于发生事故 liable to accident
易于干旱区域 drought-prone region
易于感到的 perceptible
易于加工的玻璃 sweet glass
易于进入的 easily accessible
易于筛选材料 easy-to-screen material
易于使用的 easy-to-use
易于受到污染的 pollution-prone
易于受热带气旋侵袭的地区 tropic(al) cyclone-prone area
易于疏的矿床 ore deposits of easy drainage
易于水污染的农药 water-pollution-prone pesticide
易于通达的位置 easy reach
易于维修 easy maintenance;service-friendly
易于卸货的 evacuable
易于引起传染病的物质 substance liable to cause infection
易于转换的 easy-to-change
易于自燃物质 substance liable to spontaneous combustion
易与水混合的 water miscible
易运输性 transportability
易遭受危害 be much more likely to suffer injury
易展性 laminable
易蒸发的 evaporable
易致事故的 accident-prone
易主 change hand
易转成现金的 fluid
易装 easy-on
易装锁 rigid lock
易着火 ease of ignition;fire hazardous
易自溃堤段 fuse plug levee
易自燃 easily self combustion
易钻地层 easy-to-drill formation

驿

驿车 stage carriage;stage coach
驿车旅行 staging
驿车业 staging
驿道<中国> ancient post road;courier route
驿馆<旧时的> post house
驿路 post road
驿站<旧时的> courier station;roadside station;post house
驿站站长 postmaster

疫

疫病 pest
疫霉属<拉> Phytophthora
疫苗接种程序<用来探测计算机病毒,并用触发警报打断操作等办法来防止破坏> vacination program
疫区 pest hole;plague spot
疫区处理 anti-epidemic measures in a focus
疫源地 epidemic focus;infections focus
疫沼 palus epidemiarum

益

益本比 benefit-cost ratio
益本分析 cost-effectiveness analysis;cost-benefit analysis
益处 benefit

翌

翌年 ensuring year

逸

逸出 breakaway;diverging;emit(ting);escape;liberate;overshoot;overswing
逸出常轨 swerve
逸出的 outgoing
逸出点 exit point;point of egress
逸出电子 outcoming electron;outgoing electron
逸出阀 escape valve
逸出概率 escape probability
逸出功 work function
逸出流速 exit velocity
逸出坡降 escape gradient
逸出气分析法 evolved gas analysis

method
逸出气检测法 evolved gas detection
逸出气曲线 evolved gas curve
逸出气体 emergent gas
逸出深度 escape depth
逸出时间 exit time;time of liberation
逸出速度 exit speed;velocity of escape
逸出值 outlier
逸出字符 escape character
逸度 fugacity
逸度系数 fugacity coefficient
逸度逸度图 fugacity-fugacity diagram
逸航 rambling
逸离轨道 escape trajectory
逸漏时间 time for escape
逸气 boil-off gas
逸气管 freeing pipe
逸气口 escape vent
逸去常轨 swerve
逸散 dissipation
逸散损失 walk-off-loss
逸水门【给】escape
逸性 fugacity
逸转 race rotation

意表符号 ideogram;ideograph

意大利阿格里真托的赫拉拉西尼亚神庙 Temple of Hera Lacinia at Agrigentum
意大利北部建筑 Lombard architecture
意大利产的白花和绿花大理石 cipoline
意大利产的白绿花纹的大理石 cipol(l)in(o)
意大利船黄 Turner's yellow
意大利船级社 Registro Italiano Navale
意大利大教堂 duomo
意大利的新艺术复兴 <一九四五年后> neo-liberty
意大利方形泥刀 Italian square trowel
意大利高(凸)浮雕 alto-relievo
意大利哥特式 Italian Gothic
意大利硅铝合金 italsi
意大利红 Italian red
意大利结构模型研究所 <意大利> Instituto Sperimentale Modellie Structure
意大利罗马风格 Italian Romanesque
意大利罗马式 Italian Romanesque
意大利面砖 Italian tile
意大利铅黄 Verona yellow
意大利嵌镶物 Italian mosaic
意大利柔性路面设计法 Italian flexible pavement design method
意大利石棉 Italian asbestos
意大利式簿记 bookkeeping by Italian method
意大利式窗 Italian window
意大利式建筑 Italian architecture
意大利式卷帘 Italian blind
意大利式开挖法 Italian excavation method
意大利式马赛克 Italian mosaic
意大利式掏槽 Italian cut
意大利式庭园 Italian Renaissance style garden
意大利式瓦 Italian roof tile;pan and roll tile;Italian tile
意大利式屋顶 Italianized roofing
意大利式屋面盖瓦 pan and roll roofing tile
意大利式柱式 Italian order
意大利式柱型 Italian order
意大利丝带 Italian ferret
意大利贴砖法 <彩色瓷砖贴法> Italian tiling
意大利筒瓦 Italian tile;roll tile
意大利瓦 Italian tiling
意大利瓦屋面 Italian tiling
意大利晚花杨 black Italian poplar
意大利文艺复兴 Italian Renaissance
意大利文艺复兴时期建筑 Cinquecento architecture
意大利屋顶 Italian roof
意大利屋瓦 <同时使用平、弧两种瓦> Italian roof tile
意大利五针松 Italian stone pine;pinea;stone pine
意大利现代式 Italian modern style
意大利镶嵌花砖 Italian mosaic
意大利新艺术形式的复兴 liberation of heat
意大利新艺术运动派 Italian modern style
意大利形式 Italianare
意大利艺术 <16 世纪的> cinquecento
意大利艺术时期 trecento
意大利硬币钢 acmonital
意大利园林 Italian gardens
意大利圆形泥刀 Italian round trowel
意大利早期文艺复兴时期 <15 世纪> vestigially period
意大利赭色 Italian ochre
意大利赭石 Italian ochre
意大利柱型 Italic order
意见登记簿 record of complaints
意见调查 complaint investigation;opinion sounding
意见调查表 questionnaire
意见调查员 ombudsman
意见交换制度 real-time information system
意见库 idea bank
意见书 opinion book;position paper;prospectus
意见听取会 public hearing
意见箱 suggestion box
意匠 drafting
意匠图 plan of weave
意境 artistic conception;poetic imagery
意料 on the supposition that
意料结果 expected conclusion;foregone conclusion
意水硼钠石 ezcurrite
意思自主 autonomy of the will
意图 purpose
意图传播 intent propagation
意图模型法 view modeling approach
意图模型化【计】view modeling
意图模型技术 view modeling technique
意图声明 statement of intent
意图综合 view integration
意图综合过程 view integration process
意外 emergency
意外保险 accidence insurance;accident(al) insurance;casualty insurance
意外爆炸 unplanned explosion
意外备用费 contingence reserve
意外变动 accidental fluctuation;after-clap
意外出现 drop in
意外错误 graunch
意外的 accidental;contingent;extraordinary;fortuitous;unexpected;unforeseen
意外的挫折 jolt
意外的利润 accidental profit
意外的赔偿数额 random claim amount
意外的损坏 unexpected failure
意外地形 accidental form
意外发生 supervene;supervention
意外费用 accidental cost;contingence;contingence [contingency] sum;contingency cost;contingency fund;contingency reserve
意外风险 excepted risk;unexpected risks
意外辐射 accidental exposure
意外副产品 fallout
意外故障 chance failure
意外荷载 contingency loading;accident(al) load(ing)
意外洪水流量 emergency flood flow
意外及伤亡保险 accident and death benefit insurance
意外开支 unexpected expenses;unforeseen expenses
意外开支拨款 contingency allocation
意外开支权 contingency authority
意外利润 windfall profit
意外利益 windfall profit
意外漏油 accidental oil spill;accidental spillage
意外排放 incidental discharge;incidental emission;incidental release
意外喷出物 accidental ejecta
意外碰撞 accidental collision
意外气泡 accidental air;entrapped air
意外牵伸 misdraft
意外倾卸 incidental emission
意外情况 unforeseen circumstance
意外伤 accidental wound
意外伤害 accident injury
意外事故 accident(al) hazard;casualty;chance failure;contingence [contingency];fortuitous accident;fortuitous event;fortuity;misadventure;unexpected accident;unforeseen incident
意外事故保险 contingency insurance;special hazards insurance
意外事故储备金 accident fund
意外事故基金 accident fund
意外事故赔偿 accident compensation
意外事故赔偿损失 accident and indemnity
意外事故频率 accident frequency
意外事故死亡 accident death
意外事故损坏 accident damage
意外事故折让 contingency allowance
意外事故准备 contingent reserve
意外事件 contingency;contingent event;supervention;unobserved event
意外事项补偿 contingency allowance
意外事项准备金 contingent reserve
意外释放 incidental release
意外死亡 accidental death
意外死亡保险 accident death insurance
意外损害 accidental damage
意外损坏 accidental damage
意外损伤 accidental damage
意外损失 accidental damage;accidental loss;uninsured loss
意外停泊 unscheduled call
意外停机 hang-up;unexpected halt
意外停机预防 hang-up prevention
意外停靠 unscheduled call
意外停歇 disorderly closedown
意外脱开 inadvertent disconnection
意外污染 accidental contamination;accidental pollution
意外消耗 accidental consumption
意外泄放 accidental discharge
意外泄漏 accidental release
意外泄油 accidental discharge of oil
意外性破坏 accidental destruction
意外遗漏 accidental omission
意外逸出 accidental spillage
意外因素 contingency factor;surprise factor
意外灾害保险 casualty insurance
意外障碍 snag
意外支出 extraordinary payment;unexpected expenses;unexpected pay
意外制动 unexpected braking
意外中断 involuntary interrupt
意外准备金 contingency reserve
意外准备金账户 contingency account
意味深长的 meaningful
意向表 intention list
意向论 intentionalism
意向书 comfort letter;intention agreement;intention letter;letter of interest;letter of intent(ion) <签订合同的>
意向协议书 agreement of intent;intention agreement
意向心理学 act psychology
意向性规划 conceptual planning
意向性协定 intention agreement
意向性协议 agreement of intent
意向运动 intention movement
意象 imago
意象型 imagery type
意义含糊 amphiboly
意义信息 semantic information
意译 paraphrase
意译法 free translation
意志法 voluntary law
意志力 will power

溢岸流 overbank flow

溢岸水位 overbank stage
溢出 effluence;egress;extravasation;flushoff;kick back;overfill;overflow;overrun;overshoot;run over;slop-over;spillage;spilling;spillover;superfusion
溢出报警位 overflow alarm bit
溢出杯 flow cup
溢出标记 overflow flag
溢出表 overflow table
溢出捕俘 overflow trap
溢出操作 overflow operation
溢出处理 overflow handling
溢出磁带 overflow tape
溢出错误 overflow error
溢出的顺序存取方法 overflow sequential access method
溢出的油 oil spill
溢出登记项 overflow entry
溢出点 spill point
溢出点埋深 buried depth of spill point
溢出阀 spill valve
溢出概率 overflow probability
溢出高度 spillover level
溢出埂 flash ridge
溢出沟渠 spill channel
溢出管 discharger
溢出管道 spill channel
溢出过程 overflow process
溢出呼叫处理 overflow call treatment
溢出回流 flood back
溢出记录 overflow record
溢出寄存器 overflow register
溢出检测器 overflow detector
溢出检测系统 overflow detection system
溢出检查 overflow check
溢出检查指示器 overflow-check indicator
溢出校验 exceed capacity check
溢出接点 overflow contact
溢出接卡 overflow stacking
溢出卷 spill volume
溢出控制 spillage control
溢出口 overfall;overflow port
溢出块 overflow block
溢出浪 spilling wave

溢出类型 overflow type
溢出连接 overflow connection
溢出量 spillage
溢出漏 overflow drain
溢出脉冲 overflow pulse
溢出盆地 spill-basin
溢出区 overflow area;spillover zone
溢出泉 boundary spring; descending spring; overflowing spring; spilling spring
溢出溶液 spillage solution
溢出散列 overflow hash
溢出石油 spilled oil
溢出式(铸)塑模 cut-off mo(u)ld; flash mo(u)ld
溢出事故 accidental spill
溢出水 spilling water
溢出水槽 spill-basin
溢出水流 spillway stream
溢出碎波 spilling breaker
溢出条件 overflow condition
溢出条件测试 overflow condition test
溢出通道 spill channel
溢出桶 overflow bucket
溢出位 overflow bit
溢出位置 overflow position;spillover position
溢出问题 overflow problem
溢出物 ex(s)udation;overspill;spill; spilth
溢出误差 overflow error
溢出项 overflow entry
溢出信号 spillover
溢出型塑模 flash
溢出堰 effluent weir
溢出油 spillage oil
溢出渣 runoff slag
溢出指示 overflow indication
溢出指示器 overflow-check indicator;overflow indicator
溢出指示仪 overflow indicator
溢出中断 overflow trap
溢出状态 overflow status
溢出状态组 overflow status set
溢出自陷 overflow trap
溢道 spillway
溢顶 overtopping
溢短残损单 overload and damage list
溢/短卸货报告单 over/short landed cargo report
溢额部分 surplus share
溢额分保合同 surplus reinsurance treaty
溢额再保险 surplus reinsurance
溢光灯 photoflood lamp
溢光灯光 kliegshine
溢过堰顶薄层水流 nappe
溢洪坝 spillway dam; water notch dam
溢洪道 by-channel;by lead;by-wash; conduit spillway; flood discharge outlet; flood gate; flood spillway; lasher; overfall spillway; overflow channel; overflow flood spillway; overflow spillway; spillway; waste weir; wastewater <英 spillway 的旧称>
溢洪道鼻坎 bucket lip
溢洪道表面 spillway surface
溢洪道出口底槛 spillway outlet sill
溢洪道唇 spillway lip
溢洪道导水墙 spillway training wall
溢洪道的前池 absorption pond
溢洪道地质测绘 geology mapping of spillway
溢洪道顶(部) crest of spillway;overfall crest; overflow crest; spillway crest
溢洪道顶控制 crest control
溢洪道顶以上蓄洪 storage above spillway crest
溢洪道顶最低溢流水位 top water level
溢洪道陡槽 spillway chute
溢洪道墩间跨径 spillway bay
溢洪道墩间跨距 spillway bay
溢洪道反弧段曲线 bucket curve of spillway
溢洪道工程地质勘查 engineering geologic(al) investigation of spillway
溢洪道构筑物 spillway structure
溢洪道海墁<即溢洪道防冲铺砌或护坦> spillway apron
溢洪道护坦 spillway apron
溢洪道建筑物 spillway structure
溢洪道截面 overflow spillway section;spillway section
溢洪道静水池 spillway basin
溢洪道控制(流量)设施 spillway control device
溢洪道流量控制设备 spillway control device
溢洪道流量控制装置 spillway control device
溢洪道排水量 spillway capacity
溢洪道砌面 spillway face
溢洪道前缘 spillway lip
溢洪道桥 spillway bridge
溢洪道容量 capacity of spillway;spillway capacity
溢洪道设计洪水 spillway design flood
溢洪道水渠 spillway channel
溢洪道水舌形堰顶 nappe-shaped crest of spillway
溢洪道挑流戽斗 deflecting bucket
溢洪道下消能 energy dissipation below spillway
溢洪道下泄水舌 plunging nappe for the spillway
溢洪道消力池 spillway basin
溢洪道消力戽 spillway bucket
溢洪道消力戽曲线 bucket curve of spillway
溢洪道泄洪能力 flood-releasing capacity of spillway;spillway capacity
溢洪道泄量 spillway discharge
溢洪道泄水槽 spillway channel
溢洪道泄水底槛 spillway outlet sill
溢洪道蓄水曲线 storage curve of spillway
溢洪道堰 spillway weir
溢洪道堰顶高程 spillway crest level
溢洪道堰孔 spillway bay
溢洪道溢流 overtopping of spillway; spillway overflow
溢洪道运行特性 spillway performance
溢洪道闸墩 spillway pier
溢洪道闸孔 spillway bay
溢洪道闸孔径 spillway bay
溢洪道闸(门) crest gate; spillway gate
溢洪陡槽 spillway chute
溢洪河道 flood canal
溢洪能力<溢洪道的> spillway capacity;capacity of spillway
溢洪桥 spillway bridge
溢洪缺口 pilot channel
溢洪式水表 overflow type water meter
溢洪水道 spillway channel
溢洪速度 overflow velocity
溢洪隧洞 tunnel spillway
溢洪堰顶 overfall crest
溢洪堰顶控制 crest control
溢洪堰关系曲线 rating curve of spillway
溢洪闸 spillway gate
溢洪闸门 gate for spillway
溢呼 overflow;overflow call
溢呼表 overflow meter
溢呼处理 overflow call treatment
溢呼次数计 overflow call meter
溢呼计次器 overflow register
溢呼控制继电器 overflow control relay
溢呼路由 overflow route
溢呼业务量 overflow traffic
溢剂 bleeding
溢价 premium;premium price
溢胶 bleeding
溢晶石 tachydrite;tachyhydrite
溢开账款 overcharge and account
溢料 flash
溢料缝 flash groove
溢料间隙 flash clearance
溢料空隙 spew relief
溢料口砖 flow block
溢料漏斗 spillage hopper
溢料门 door trip
溢料面 flash ridge
溢流 overflow(ing) (current);overrun;crest flow;effluence;effluent; effusion;flooding;flow over;overfall;run-out;spill current;spilling water; spillover; spillwater; weir waste;flow off
溢流坝 anicut;crest control weir;gole;overfall dam;overflow dam;overtopped dam;spillway;spillway dam; spillweir dam; supercharged dam; supercharger dam; waste dam;weir dam
溢流坝顶桥 spillway bridge
溢流坝段 spillway section
溢流坝面 roller-way face
溢流板 deck spillway;overflow plate;overflow slab;spillway slab
溢流板延伸 spill plate extension
溢流泵 flood pump
溢流比率 overflowing rate
溢流边缘 flood-level rim; overflow rim
溢流标高 spillover level
溢流部分 overflow section;spillway section
溢流仓 overflow bin;overflow bunker;overflow silo
溢流槽 overflow launder; overflow tank; overflow trough; spillage tank; spillway channel; spillway tank;spillway trough
溢流侧堰 baffled side weir
溢流测定法 weir method
溢流岔道 overflowing bypass
溢流(超饱和)排队长度 overflow queue length
溢流(超饱和)停车率 overflow stop rate
溢流池 overflow basin;runoff pit;spillage tank;spill pit;spillway tank
溢流床洗涤器 flooded bed scrubber
溢流次数 overflowing frequency
溢流淬火 flush quenching
溢流挡板 overflow baffle;overpass
溢流挡墙 overflow stand
溢流道 by-pass canal; by-pass conduit; drainage trunk; drain trunk; overfall; overflow passage; spillway;spillway channel;wasteway
溢流道护坦 apron slab of overflow; apron slab of spillway
溢流道截面 overflow spillway section
溢流道排水量 overflow capacity; spillway capacity
溢流的 flooded
溢流等效电力 overflow electric(al) power
溢流堤 overfall dike [dyke];overflow dike [dyke];overflow embankment
溢流堤泄水槽 dam spillway channel
溢流地 overfall land;overflow land
溢流地段 overflow section
溢流地役权<水在他人土地流过之权> flowage easement
溢流点 flooding point
溢流丁坝 overfall dike [dyke];overflow dike [dyke];overflow groin
溢流顶峰 crest of overflow
溢流短管 overflow nipples
溢流段 overflow section; spillway section
溢流段截面 overfall section
溢流断面 overflow section
溢流堆石坝 rock-fill spilling dam
溢流阀定位 relief valve setting
溢流阀(门) overflow valve;bleed-off valve;by-pass valve;easing valve; escape valve; excess flow valve; flooded valve; flood(ing) valve; overfall valve;overflow cock;relief valve; relieving valve; spillover valve;spill valve;surplus valve
溢流阀式喷油泵 spill valve type fuel injection pump
溢流法 overflow process;press over system <鞣液的>
溢流峰值 overflow crest
溢流扶壁 overflow buttress
溢流扶垛 overflow buttress
溢流负荷 overflow load
溢流负荷量 weir loading
溢流高程 overflowing level
溢流高度 height of overflow; overflow height
溢流高架水箱 overflow overhead tank
溢流隔板 weir divider
溢流拱坝 overflow arch dam
溢流拱顶 overflow crown
溢流沟 overfall
溢流管 by-pass pipe;cross-over connection; down corner; downflow pipe; downspout conductor; flooding pipe; overfall pipe; overflow conduit;overflow pipe [piping]; overflow tube [tubing]; run-down pipe [piping]; run-down tube [tubing]; spillover; spill pipe [piping]; spill tube [tubing]; spillway; trickle drain; upflow tube; warning conduit;warning pipe
溢流管线 overflow line
溢流管嘴 overflow cock
溢流灌溉 flooding irrigation
溢流罐 overflow tank;weir tank
溢流柜 overflow tank
溢流涵洞 spillway culvert
溢流河槽 bypass(ed) channel
溢流河道 overflow channel;overflow stream
溢流虹吸管 overflow siphon
溢流话务 overflow traffic
溢流计 weir meter
溢流剂 flooding agent
溢流建筑物 overflow structure;spillway structure
溢流浇注 overcasting
溢流绞刀输送机 overflow screw conveyer [conveyor]
溢流结构 overflow structure
溢流截面 overflow section
溢流井 overflow chamber;overflowing weir;overflow well
溢流阱 overflow trap
溢流警报器 overflow alarm
溢流开关 flood cock;relief cock
溢流空气 spill air
溢流孔 bleeder hole;overflow hole; overflow orifice;overflow port;re-

lief hole; spout hole; trim hole
溢流控制设备 overflow control device; spillway control device
溢流控制装置 overflow control device; spillway control device
溢流口 outgate; overfall gap; overflow (edge); overflow outlet; overflow port; overflow section; relief opening; rollway; weir waste
溢流量 overflow discharge; quantity of overtopping; spillway discharge; weir flow
溢流量调节 overflow governing
溢流流量 excess flow; overflow discharge
溢流龙头 overflow tap
溢流滤池 overflow filter
溢流率 overflow rate; overflow rating; weir loading
溢流脉冲 overflow pulse
溢流冒口 pop-off
溢流门 overflow gate
溢流面板 overflow panel; overflow slab; spillway slab < 大头坝的 >
溢流模 flash mo(u)ld
溢流能力 overflow capacity
溢流排队 overflow queue
溢流排队长度 overflow components of queue length
溢流排洪道 overflow spillway
溢流排水管 relief pipe; relief piping; relief sewer; relief tube; relief tubing
溢流排水口 overflow gutter
溢流排污管 overflow sewer; relief sewer; relieving sewer
溢流盘 seal cup
溢流喷发 effusive eruption
溢流撇油装置 weir skimmer
溢流频率 frequency of overfall; frequency of overflow
溢流曝气生物滤池 biologic(al) aerated flooded filter
溢流汽化器 flooded carburet(t)or
溢流器具 flooded fixture; overflow fixture
溢流墙 overflow wall
溢流桥 relief bridge
溢流丘 lava flow dome
溢流渠槽 spillway channel
溢流渠(道) overflow channel; spillway channel; waste canal; waste channel; by-pass channel; channel spillway
溢流泉 overfall spring; overflow(ing) spring
溢流容器 overflow vessel
溢流润滑 flooded lubrication; flooding system lubrication
溢流塞 flooding plug; relief fitting
溢流三角洲 spillover delta; washover delta
溢流设备 overflowing facility; overflow installation
溢流设施 overflow installation
溢流深度 overflow depth
溢流时间 overflowing time
溢流式厂房 overflow type power house
溢流式大气冷凝器 bleeder type condenser
溢流式大气冷凝装置 bleeder type condensing plant
溢流式堤 overflow dike [dyke]
溢流式电站 overflow type power plant
溢流式丁坝 overflow groin
溢流式虹吸管 overflow siphon [syphon]
溢流式均衡室 spilling surge chamber
溢流式冷凝器 bleeder type condenser
溢流式量水堰 measuring dam
溢流式球磨机 overflow ball mill
溢流式溶解器 flooded dissolver
溢流式水电站 overflow station; overflow type power plant; tailwater power plant
溢流式水库 overflow reservoir
溢流式调压井 spillway surge tank
溢流式调压室 spilling surge chamber
溢流式土堤 overflow earth dike[dyke]
溢流式溢洪道 overfall spillway; overflow type spillway
溢流式溢水口 overflow type spillway
溢流式鱼类通道 overfall type fish pass
溢流式闸门 overflow gate
溢流式蒸发器 flooded evapo(u)rator
溢流室 overflow chamber; spilling chamber
溢流收集管 overflow gallery pipe
溢流受器 overflow
溢流竖管 overflow standpipe
溢流栓 overflow cock; overflow tap
溢流水 overfalling water; overflow(ing) water; overpouring water; sheet drift water
溢流水槽 overflow trough
溢流水层 overflowing sheet
溢流水道 overfall channel; overflow channel
溢流水加热器 overflow water heater
溢流水库 overflow reservoir
溢流水量 amount of overflowing
溢流水舌 free nappe; inclined nappe; nappe; nappe over spillway; overflow(ing) nappe
溢流水舌截面 nappe profile; overflow nappe profile
溢流水深 overflow depth
溢流水头 overflow head
溢流水位 flood level; overflow(ing) level; spillover level
溢流水箱 overflow chamber
溢流速度 flooding velocity
溢流速率 flooding rate; overflow rate
溢流隧道 overflow tunnel; spillway tunnel
溢流隧洞 overflow tunnel; spillway tunnel
溢流损失 overflow loss; spillage loss; spill loss
溢流损失电力 overflow electric(al) power
溢流塔 overflow tower
溢流潭 spillover pool
溢流调节 bleed-off
溢流调节环 adjustable weir ring
溢流调整器 over flow regulator
溢流跳汰机 overflow jig
溢流停车率 overflow components of stop rate
溢流停止 overflow connection
溢流通路 overflow passage
溢流桶 overflow ladle
溢流土堤 overflow earth dike [dyke]
溢流土石坝 overflow earth-rock fill(ed) dam
溢流污染【疏】 overflowing spoil
溢流污水管 relief sewer
溢流系数 coefficient of overflow
溢流系统 overflow system
溢流下料槽 overflow sluice
溢流下水道 relief sewer
溢流下泄速度 downwash velocity
溢流线 overflow line
溢流箱 runoff box; spill box
溢流效率 spillover efficiency
溢流斜槽 overchute; overflow trough
溢流卸料孔 overflow discharge opening
溢流形成的片状水流 overflowing sheet of water
溢流旋塞 flooding cock
溢流压力 hydraulic relief pressure
溢流延误 overflow components of delay; overflow delay
溢流岩 effusive rock
溢流堰 barrage type spillway; clear overflow weir; crest control weir; diverting weir; downflow weir; effluent weir; flush weir; gole; lasher; overfall; overfall dam; overfall spillway; overfall weir; overflow dam; overflow(ing) weir; overflow spillway; overtopped dam; rolling dam; sluice weir; spillway dam; spillweir; surface weir; waste weir; weir
溢流堰坝的水幕 nappe
溢流堰槽 weir trough
溢流堰导流堤 < 导流堤有一段设置溢流堰，以便旁道输沙 > weir jetty
溢流堰顶 crest of overfall; crest of overflow; crown of overfall; crown of overflow; overflow crest; spillway crest
溢流堰顶高程 overflow weir level
溢流堰(顶)护坦 apron slab of spillway
溢流堰顶控制设备 crest control equipment
溢流堰顶面不平度 straightness of weir plate top
溢流堰顶水头 head-on the spillway
溢流堰顶缘 overflow (chute) edge
溢流堰顶闸门 spillway crest gate; weir penstock
溢流堰段 weir section
溢流堰反弧鼻坎 upcurved spillway bucket
溢流堰负荷 weir loading
溢流堰负荷率 weir loading rate
溢流堰高度 height of weir plate
溢流堰进水口 weir inlet
溢流堰流量系数 weir discharge coefficient
溢流堰流量与堰顶水深关系曲线 rating curve of spillway
溢流堰排料 weir discharge
溢流堰水平 weir level
溢流堰水头 depth of water flowing over spillway; depth of water flowing over weir
溢流堰箱 weir box
溢流堰消能鼻坎 upcurved spillway bucket
溢流堰消能挑坎部分曲线 bucket curve of spillway
溢流堰消能挑流坝曲线 bucket curve of spillway
溢流堰堰缘 spillway lip
溢流堰闸门 weir gate
溢流堰闸室 overflow chamber
溢流堰至容器内壁的距离 distance of weir to inside surface of tank
溢流液出口 overflow box outlet
溢流引水 weir inlet spillway
溢流影响 impact of overflows
溢流油管道 bleed line
溢流闸门 by-pass damper; overflow gate; spillway gate
溢流闸室 overflow chamber
溢流支管 spill pipe
溢流止回阀 excess flow check valve
溢流指示器 overflow-check indicator
溢流重力坝 overflow gravity dam
溢流铸模 overflow mo(u)ld
溢流装置 overflow mechanism
溢流总管 overflow main line; return main
溢流嘴 overflow lip
溢漏 overflow drain
溢漏处理承包人 spill disposal contractor
溢漏风险 spill hazard
溢漏事故 spill incident
溢漏事故现场 spill site
溢漏事故应急技术 spill technology
溢漏事故应急用品 spill response product
溢漏危险 spill hazard
溢漏污染物的清除活动 spill cleanup activities
溢漏应急计划 pre-spill planning
溢满信号器 overflow alarm
溢满指示器 overflow indicator
溢漫 spill
溢漫岸 spill-bank
溢漫河谷 spill-hollow
溢漫坡 spill-slope
溢气阀 snifting valve
溢气泉 bubbling spring
溢弃水 weir waste
溢泉 depressing spring; depression spring
溢缺额账户 over-and-short account
溢失功率 spillover power
溢式模具 flash mo(u)ld
溢式压塑模 flash mo(u)ld
溢收货 appropriate goods
溢水 overfall; overtopping; spilling water; spillwater; surplus water
溢水坝 overfall dam; overflow spillway dam; overtopped dam; supercharged dam
溢水槽 by-pass channel; by-pass conduit; leak-off chute; overflow trough
溢水池 overflow tank; spillway basin
溢水出口结构 spillway outlet structure
溢水道 by-channel; bye channel; byewash; diversion cut; flood-level rim; gole; overfall; overflow channel; overflow passage; sluice; spillway; spillweir; water escape; weir waste
溢水陡槽 overflow chute
溢水阀 release valve
溢水沟 overflow chute; waste canal
溢水构筑物 overflow structure
溢水管 conduit spillway; escape pipe; flooding pipe; flow pipe; flush pipe; overflow pipe; overflow tube; warning conduit; warning pipe; warning tube
溢水集水道 spillway gallery
溢水接管 spud
溢水结构 overflow structure
溢水井 overflow well
溢水口 cut-out overflow; overflow; riser; spill; spillway outlet
溢水口渡槽 spillway flume
溢水流量 discharge of overflow
溢水滤筛 overflow strainer
溢水期 overtopping stage
溢水渠 by-pass canal; waste canal
溢水设备 overflow installation
溢水式沉降计 hydraulic overflow settlement cell; hydraulic overflow settlement ga(u)ge; hydraulic settlement cell; water overflow settlement ga(u)ge
溢水式沉降仪 hydraulic overflow settlement cell; hydraulic overflow settlement ga(u)ge; water overflow settlement ga(u)ge
溢水竖管 standing waste

溢水塔 overflow tank
溢水通道 overflow passage
溢水通路 overflow passage
溢水显示管 telltale pipe
溢水箱 dump tank;overflow tank
溢水堰 flush weir; overfall weir; overflow weir
溢水闸 spillway gate
溢水装置 overflow installation
溢土量 overflow;spill
溢位 overflow
溢卸货 overlanded cargo
溢液 discharge;exudation
溢液池 impounding basin; retention pool;retention pound
溢液阀 spill valve
溢液隔墙 spill wall
溢涌槽 surge bin
溢用 override
溢油 oil spilling; oil spills; overflow oil;spillage oil
溢油处理剂 oil-spillage cleaning agent;oil-spill chemicals
溢油阀 spill valve
溢油分散剂 oil-spill dispersant
溢油风险分析 oil-spill risk analysis
溢油警报系统 oil-spill warning system
溢油孔 spill port
溢油量 oil spillage
溢油清除船 spill clearance vessel; spill combating vessel
溢油清除法 oil-spill clean-up method
溢油收集船 oil spill recovery vessel
溢油围栅 oil barrier
溢油形成油膜 oil slick
溢油性能响应评估法 spill performance response assessment technique
溢油应急计划 responsive plans for unexpectedly oil spilling
溢油预防、控制和防范措施 oil-spill prevention, control and countermeasure
溢油灾难模型 oil-spill fates model
溢渣 slag overflow
溢装两件 over shipped tow packages

翳影 veiling

翼 < 风车、轮机等的 > vane

翼坝 spur dike [dyke]; vane dike [dyke];wing dam
翼板 alar plate; deck slab; wing panel;wing plate
翼板和腹板连接处 root of flange
翼瓣 petal;wing
翼瓣安全阀 wing flap relief valve
翼瓣控制阀 wing flap control valve
翼瓣止回阀 wing clack check valve
翼壁 alar wall
翼部 alar part
翼部地层 strata in fold limbs
翼部挠曲 flexure limb
翼舱 wing tank
翼舱壁 wing bulkhead
翼侧声径 sound flanking path
翼长 wing length;wing span (of aircraft)
翼窗 wing light
翼次射信号 salvo signal
翼刀 fence
翼导 flyer lead
翼堤 vane dike [dyke];wing dike [dyke];wing levee
翼点 pterion
翼点发生器 wing spot generator
翼锭 flyer spindle
翼锭络筒机 flyer winding frame
翼动角 flapping angle
翼端 wing tip
翼端喷嘴 wing-tip nozzle
翼段 wing panel
翼墩 wing abutment
翼帆 wingsail
翼缝 wing slot
翼缝扰流板 interceptor plate
翼杆 wing bar;wing beam
翼隔翼展比 gap span ratio
翼构件 wing member
翼轨 wing rail
翼轨动程 < 辙叉 > opening of wing rails
翼轨升高 < 辙叉 > wing wheel riser
翼柜 side tank
翼航状态 foil-borne
翼荷载 wing load
翼滑 wing slip
翼簧 leaf spring
翼尖 wing tip
翼尖浮筒 wing-tip float
翼尖航行灯 wing navigational light
翼尖梢 wing tip
翼尖斜削 wing-tip rake
翼尖硬架航电仪 wing-tip rigid boom air-borne electromagnetic system
翼间角 interlimb angle
翼间距 rib interval
翼角钢 flange angle
翼角铁 flange angle iron
翼铰链 wing hinge
翼结构 wing structure
翼肋 profile rib
翼犁 wing plough
翼篱 side fence
翼力 wing power
翼梁 longeron;spar;wing beam;wing spar
翼梁腹板 spar web
翼梁荷载 spar loading
翼梁铣床 spar miller
翼轮 alar wheel;wing wheel
翼轮风速计 swinging-vane anemometer
翼轮风速仪 vane anemometer
翼轮廓 wing contour
翼轮廓线 wing curve
翼轮式水表 wheel type water-meter
翼轮推进器 vane-screw propeller; vane wheel
翼罗盘 wing compasses
翼蒙布 wing covering fabric
翼面 air foil; airfoil surface; wing plane;wing surface
翼面积 blade area
翼面散热器 wing radiator;wing surface radiator
翼挠曲 wing-warp
翼片 air foil;fin;tab;wing panel
翼片安装 tab assembly
翼片换热器 finned tube exchanger; fin tube heat exchanger
翼片继电器 vane relay
翼片加热管 gilled heating pipe
翼片加热器 gilled heater
翼片取暖器 gilled heater
翼片热交换器 fin tube heat exchanger
翼片热面 finned surface
翼片砂轮 flap wheel
翼片式 propeller-type
翼片试验 tab test
翼片斜板沉淀池 alar sloping settling tank
翼片组合 tab assembly
翼剖面 air foil;airfoil section;wing section
翼剖面阻力 profile resistance
翼前羽 wing-front
翼前缘 nose of wing
翼墙 abutment wall;aliform;flare wall; guide wall; ramp wall; return wall; side fence; turnout wall; wing abutment;wing masonry wall;wing wall; aileron < 教堂侧廊、通道、耳堂等 >
翼墙的张开部分 wingwall flare
翼墙上的防浪墙 approach parapet
翼墙式桥台 abutment with wing wall;wingwall abutment
翼倾角 wing incidence
翼区 pterion
翼栅 latticed wing
翼上走道 wing-walk
翼梢 wing tip
翼梢板 wing-tip plate
翼梢半径 wing-tip radius
翼梢灯 wing-tip light
翼梢发射架 wing-tip launcher
翼梢方向舵 wing-tip rudder
翼梢缝 wing-tip slot
翼梢浮筒 wing float;wing-supporting float;wing-tip pontoon
翼梢副翼 wing-tip aileron
翼梢构架 wing-tip truss
翼梢滑橇 wing skid
翼梢炮塔 wing-tip turret
翼梢伞 wing-tip parachute
翼梢损失 wing-tip loss
翼梢天线 wing-tip antenna
翼梢外挂物 wing-tip store
翼梢涡流 wing-tip vortex; wing tip vortices
翼梢弦 wing-tip chord
翼梢旋涡 tip vortex;wing-tip vortex
翼梢着陆闪光灯 wing-tip flare;wing-tip landing flare
翼梢阻流方向舵 wing-tip drag rudder
翼狮像 griffin
翼狮像壁画 fresco of griffins
翼式泵 wing pump
翼式除尘机 fan duster
翼式除尘器 wing duster
翼式风速表 vane anemometer; wing type anemometer
翼式风速计 vane anemometer
翼式封盖 wing type seal cap
翼式拦污栅 wing screen
翼式桥台 wing abutment
翼式散热管 wing heating tube
翼式水表 wing type meter
翼式天线 wing antenna
翼式托盘 wing pallet
翼式雪犁 wing snow plough
翼式重力桥台 gravity wing
翼式轴流风扇 vane-axial fan
翼塔 flanking tower
翼尾颤振 buffeting
翼尾排列 wing-tail arrangement
翼尾修整装置 trimming gear
翼隙出口 exit gap
翼隙角 wing clearance angle
翼隙进口 entry gap
翼下 underwing
翼弦 chord of airfoil;wing chord
翼弦安放角 chord angle
翼弦线 chord line
翼弦斜度 wing taper
翼弦轴 chord axis
翼斜削度 wing taper
翼形 aerofoil; aerofoil contour; air foil; airfoil profile; airfoil section; airfoil shape; wing profile; wing type;wing section < 指托盘两端吊翼 >;wing form;wing shape
翼形的 alar(y);aliform;thumb
翼形堤 wing dam;wing dike [dyke]; wing levee
翼形阀 flutter valve;wing valve
翼形风速计 Biram's wind meter
翼形高背软垫椅 wing chair
翼形拱座 splayed abutment
翼形管 finned tube
翼形滚筒 wing type pulley
翼形混砂机 wing mixer
翼形基础 wing footing
翼形集装架 wing type pallet
翼形继电器 vane-type relay
翼形件 thumb piece
翼形截面 air foil
翼形量规 wing dividers
翼形螺钉 thumb screw; wing (ed) screw
翼形螺帽 ear nut;fly nut;thumb nut; thumb screw;wing nut
翼形螺母 butterfly nut;eared nut;ear nut;fly nut;thumb nut;wing nut
翼形螺栓 wing bolt;wing-headed bolt
翼形螺丝 wing screw
翼形帽螺栓 bolt with winged nut
翼形培土器 wing cover
翼形坡脚 wing toe
翼形剖面 aerofoil profile
翼形桥台 splayed abutment; wing abutment
翼形曲面 aerofoil camber
翼形绕流 profile flow
翼形散热器 radiator with thin fin; wing radiator
翼形散射 wing scattering
翼形升力 profile lift
翼形水轮 flutter wheel
翼形托盘装饰 < 古建筑 > wing disc [disk]
翼形系列 profile set
翼形系数 section coefficient
翼形压力分布 profile pressure distribution
翼形叶片 airfoil fan
翼形叶片风扇 airfoil fan
翼形叶栅 wing cascade
翼形栅栏 wing fence
翼圆盘 winged disc
翼圆饰【建】flying dish
翼缘 flange;flange girth
翼缘板 boom plate;chord plate;cover plate;flange(d) plate;flange slab
翼缘板厚度 flange thickness
翼缘端(边) wing tip
翼缘断面 flange section
翼缘杆 flange rod
翼缘刚度 flange stiffness
翼缘钢 flange steel
翼缘焊接 edge flange weld
翼缘横断面 flange cross-section
翼缘横截面 flange cross-section
翼缘加固 flange strengthening
翼缘加劲板 flange plate
翼缘角钢 boom angle; chord angle; flange angle
翼缘紧固件 flange brace
翼缘卷曲 flange curling
翼缘宽度 chord width;flange width
翼缘连系构件 flange brace
翼缘联结 flange coupling
翼缘铆钉 flange rivet
翼缘面积 flange area
翼缘面积法 flange area method
翼缘拼接板 flange splice
翼缘平行梁 beam of constant depth
翼缘墙 fin wall
翼缘翘曲 flange warping
翼缘切割 flange cut
翼缘切口 flange cut
翼缘屈曲 flange buckling
翼缘镶板 flange splice; flange splice plate
翼缘效应 edge effect
翼缘应力 flange stress
翼缘有效宽度 effective width of flange
翼缘与腹板联结 flange-to-web-joint

翼缘增强 flange strengthening
翼展 over wings diameter;wing span <飞机>;wing spread;span【机】
翼展方向 spanwise
翼展中央断面 mid-span section
翼罩 wing covering
翼震动 wing flutter
翼支柱 cabane strut;wing strut
翼重 wingheaviness
翼重的 wing heavy
翼轴线 wing axis
翼柱 cabane; flanking column; wing mast
翼桩 wing pile
翼状 wing form
翼状薄壁组织 aliform parenchyma
翼状导管阀 wing guided valve
翼状的 alar(y);aliform
翼状横撑支柱 wing stull
翼状扩孔钻头 paddle reaming bit; wing reaming bit
翼状螺钉 wing screw
翼状木栓 winged cork
翼状穹顶肋 wing groin
翼状物 wing
翼状钥匙 wing key
翼状钻头 <回转钻进用的> bladed bit; drag bit; drag-type bit; wing bit;winged scraping bit
翼子板灯 fender light
翼子板托架 wing stay
翼子板支架 fender support
翼足类软泥【地】pteropod ooze
翼阻力 wing drag;wing resistance
翼组结构【地】wing cellule
翼座 wing abutment

臆

臆测确实性 validity of assumptions
臆拟暴雨 hypothetical storm
臆拟单坡 hypothetical single slope
臆拟洪水 hypothetical flood

癔

癔病性视觉障碍 hysterical vision disturbance

镱

镱玻璃激光器 ytterbium glass laser
镱镓石榴石 ytterbium gallium garnet
镱矿 ytterbium ores
镱兴安石 higganite-(Yb)

因

因阿铝合金 Inalium
因罢工关闭的 struck
因变化而产生的 variational
因变量 dependent variable
因变数【数】dependent variable; dependent number
因长期使用而获得的通行权 prescriptive easement
因传动机构误差留在工件上的痕迹 gear mark
因次 dimension
因次单位 dimensional unit
因次的 dimensional
因次法 method of dimensions;dimensional method
因次方程 dimensional equation
因次分析 dimensional analysis
因次分析法 dimensional method
因次公式 dimensional formula
因次关系 dimension(al) relation
因……的作用 under the action of
因低潮不能进出港的 tide bound
因地制宜 according to local condition; adaptation to local condition; to adopt measures suiting local condition;to suit measures to local condition;treatment in accordance with local condition
因地制宜从而发挥优势 suiting local needs and making full use of favo-(u)rable condition
因钉孔而产生渗漏的 nail sick
因钉孔太多而漏水的船 nailsick boat
因钉孔太多而失去强度的木板 nail-sick board
因多次钉钉而不结实的 nail sick
因风雨褪色的 weather stained
因钢 invar
因钢带尺 invar tape
因钢精密水准尺 invar precise levelling staff
因钢丝 invar wire
因钢线尺 invar wire
因格松图解(法) Ingson diagram
因工人无经验造成故障 weevils got him eat up
因工业化人口外迁 industrialization emigration
因公 on business
因公出差 business trip
因公出差旅行 travel on official business
因公负伤 work-related injury
因公死亡 death in line of duty
因购房而致贫的业主 house-poor homeowner
因购置和使用固定资产的费用 expenditure for acquisition and use of fixed assets
因故停潜时间 diving downtime for some reason
因故暂停 out-of-service
因故障停工时间 downtime
因光异色现象 metachromasia; metachromatism
因果 causal and effect; cause and effect
因果分析 causal analysis;causality analysis
因果分析法 causal method
因果分析图 cause and effect diagram
因果分析预报法 causal forecasting method
因果关系 causal relation;causal relationship;cause and effect relation; cause-and-effect relationship
因果关系的认知 perception of social causality
因果关系概率 causality probability
因果关系推定 inference of causation
因果关系研究 causal research
因果关系预测模式 causal forecasting model
因果假设 causal hypothesis
因果联系 causal association;causal nexus
因果链 causal chain; chain of causation
因果链图 cause and effect chain diagram
因果链系统图 causes and effects chain system diagram
因果律 determinism; law of causality;law of causation
因果(律)条件 causality condition
因果模型 causal model
因果生物地层学 causal biostratigraphy
因果顺序 causal sequence
因果图 cause and effect diagram; cause-consequence chart
因果图法 cause-effect graphing
因果途径分析 causal path analysis
因果系统 causal system; nonanticipatory system;physical system
因果系统图 cause and effect system diagram
因果性 causality
因果性网络 causal network
因果循环 causal circle
因果预报 cause-and-effect forecast-(ing);deterministic forecast
因果预测 causal forecasting
因果预测模型 casual forecasting model
因果转化 interchange of cause and effect
因经受风雨日晒而褪色 weather stain
因卡法 <浅层曝气法> Inka process
因卡曝气(法)【给】Inka aeration
因卡式曝气系统 Inka aeration system
因康镍合金 inconel
因考虑环境问题而带来的其他问题 environmental spillover
因科洛伊 <一种耐高温的镍铬铁合金> Incoloy
因科奈尔镍合金料罐 Inconel charge can
因科镍 Inco nickel
因科镍合金 Incochrome nickel
因硫碲铋矿 ingodite
因陋就简 make do with the meagre means at hand
因破产而拍卖的货物 bankruptcy stock
因情况而异的方法 contingent approach
因缺乏耐性所致 resulting from lack of tolerance
因日晒雨淋而斑驳变色的 weather stained
因施工段封闭段 section closed due to construction
因施工封闭区间 section closed due to construction
因时效丧失权益 discharge of the right of action by lapse of time
因时制宜 treatment in accordance with seasonal condition
因式 factor
因式分解(法) factorization;factor analysis; factoring; factorize; rank factorization
因收入增减引起的需求变动 demand shifts from income
因数 coefficient; facient; factor; module; modulus [复 moduli]; multiple digit;multiplier digit;submultiple
因数的 factorial
因数定理 factor theorem
因数法 factoring method
因数分解【数】resolution into factors
因数分解法【数】factorization
因数分析 factor analysis
因数计算法 factoring process
因数重量 factor weight
因素 efficient;element;factor
因素比较 factor comparison
因素比较法 factor comparison system
因素比例 factor proportion
因素成本 factor cost
因素代替 factor substitution
因素分析(法)【数】factor analysis
因素论者 factorist
因素水平 factor level
因素型实验 factorial type experiment
因特殊需要调整工作时间制度 <英> spread-over system
因特网 <一种国际互联网>【计】Internet
因特网电话 IP-phone
因特网服务提供商 internet service provider
因特网内容提供商 internet content provider
因特网全书 interpedia
因特网协议 internet protocol [IP]
因天气受阻的 weather-bound
因投资收入而发生的费用 expense attributable to the investment income
因瓦 <又称殷瓦> invar
因瓦线尺 invar wire
因为场地受到影响的限制 site-specific constraint
因违约终止合同 termination for default
因无能被解职 discharged for inefficiency
因雾受阻 fog bound
因袭 stylization
因袭管理法 conventional management;rule-of-thumb
因循守旧 lockstep
因循守旧过程 lockstep procedure
因延迟给付而发出的传票 default summons
因盐度差异引起的海流 salinity current
因征用引起房地产向货币的强制转换 involuntary conversion
因骤冷引起的漆膜失光或起雾 quenching
因子 efficient;facient;factor;multiplier;divisor【数】
因子变换 factor transformation
因子表示 factor representation
因子补偿作用 factor compensation
因子代数 factor algebra
因子得分 factor score
因子的 factorial
因子定理 factor theorem
因子对 factor pair
因子分解 factor analysis; factoring; factorization;rank factorization
因子分解(方)法 factorization method
因子分解技巧 factorization technique
因子分析 factor(ial) analysis
因子分析法 factor analysis method
因子分析模型 factor analysis model
因子负荷量 factor loading
因子函数 saturation
因子互换检验 factor reversal test
因子集 factor set
因子价格界限 factor-price frontier
因子结构 factor structure
因子矩阵 factor matrix
因子模 factor modulus
因子模型 factor model;factor pattern
因子排列 factorial arrangement
因子判别式分析 factorial discriminant analysis
因子平面图 factor plane map
因子群 factor group
因子群分析 factor group analysis
因子设计 factorial design
因子试验 factorial experiment;factorial test;factorial type experiment
因子数字 multiplier digit
因子套设计 nested design of experiment
因子显影 factorial development
因子相互作用 factor interaction
因子效应 factorial effect
因子选择 factoring
因子载荷 factor loading
因子指标 level of factor
因子总计 factor total
因子组 factor set;factor system
因钻孔缩径而不能继续钻进 come out of a well

阴

阴暗 cloudiness;obscuration
阴暗的 adumbral; dull; fuliginous; lu-

rid
阴暗色调照明 low-key lighting
阴暗天空 gloom
阴暗线 hidden line
阴版 negative plate
阴沉的 dark
阴沉的坏天气 heavy weather
阴沉天空 heavy overcast
阴沉天气 dull weather;ugly sky;ugly weather
阴处 shade
阴丹士林 indanthrene
阴丹士林暗蓝 indanthrene dark blue
阴丹士林橄榄绿 indanthrene olive
阴丹士林红紫 indanthrene red violet
阴丹士林黄 flavanthrene;indanthrene yellow
阴丹士林金橙 indanthrene golden orange
阴丹士林金黄 pyranthrone
阴丹士林蓝 indanthrene blue;indanthrone
阴丹士林蓝绿 indanthrene blue-green
阴丹士林亮橙 indanthrene brilliant orange
阴丹士林亮蓝 indanthrene brilliant blue
阴丹士林亮绿 indanthrene brilliant green
阴丹士林亮玫瑰红 indanthrene brilliant rose
阴丹士林染料 indanthrene dyes
阴丹士林瓮染色 indanthrene vat colo(u)r
阴丹士林印染蓝 indanthrene printing blue
阴丹士林枣红 indanthrene bordeaux
阴丹士林蒸箱 indanthrene steamer
阴丹士林直接黑 indanthrene direct black
阴丹酮 indanthrone
阴的 female
阴地栅地放大器 grounded-cathode grounded-grid amplifier
阴地生的 umbraticolous
阴地植物 dryad;sciophyte;shade serum
阴电 negative electricity
阴电荷 negative charge
阴电荷胶体 negative colloid
阴电荷溶胶 negative sol
阴电极 electrode negative;negative electrode
阴电势 electronegative potential;negative potential
阴电性 electronegativity
阴电性的 electronegative
阴电性根 electronegative radical
阴电性离子 electronegative ion
阴电性凝胶 electronegative gel
阴电性元素 electronegative element
阴电原子价效应 electronegative valency effect
阴电子 negative electron;negatron
阴雕 intaglio
阴钉 edge nailing
阴法兰 female flange
阴干 dried in shade
阴干材 shed drying stock
阴沟 blind drain;cloaca;conduit;culvert;drain;foul drain;foul sewer;foul water sewer;kennel;sewer;sink;underdrain
阴沟标记 drain sentinel
阴沟标志 drain sentinel
阴沟沉泥井 sewer catch basin
阴沟出口 outlet culvert
阴沟出口河流 outfall
阴沟存水弯 gull(e)y trap
阴沟方木 sewer plank
阴沟分布图 sewer layout plan
阴沟盖 manhole cover
阴沟盖板 steel grating
阴沟隔断 caponier;sewer cutoff
阴沟管模板 dod(d)
阴沟集泥井 sewer catch basin
阴沟进口 sewer inlet
阴沟口 sink hole
阴沟口滤污器 catch basin
阴沟排水 blind drainage;valley drainage
阴沟气中毒 sewer gas poisoning
阴沟疏通杆 drainage rod
阴沟水 black water;sewage
阴沟头 gull(e)y trap
阴沟污水 sullage water
阴沟污物 sullage
阴沟系统 system of sewers
阴沟用楔形砖 wedge sewer brick
阴沟中的浮渣 drier scum
阴沟砖 blue brick;sewer brick
阴河 groundwater stream
阴晦天气 gloomy weather
阴极 cathode;kathode;negative electrode;negative pole
阴极暗电流 cathode dark current
阴极暗区 cathode dark space;Crookes dark space
阴极斑点 cathode spot
阴极板 cathode plate;minus plate;negative plate
阴极棒 cathode bar
阴极保护 cathode protection;cathodic protection;electrolytic protection
阴极保护底漆 cathodic protective primer
阴极保护电流 impressed current
阴极保护法 cathode protection method;cathodic protection
阴极保护腐蚀 cathodic protection corrosion;sacrificial corrosion
阴极保护膜 cathodic protection coating
阴极保护设备 cathodic protection equipment
阴极保护涂层 cathodic protection coating
阴极崩解 cathode disintegration
阴极臂 cathode leg
阴极边框 cathode border
阴极变压器 cathode transformer
阴极波 cathodic wave
阴极剥落 cathodic peeling
阴极剥落物 crop of cathodes
阴极补偿 cathode compensation
阴极材料 cathode material
阴极材料喷涂机 cathode-spray machine
阴极参量 cathode parameter
阴极层电弧技术 cathode layer arc technique
阴极层富集 cathode layer enrichment
阴极层弧光法 cathode layer method
阴极层效应 cathode layer effect
阴极插入孔 cathode aperture
阴极产物接收器 cathode product receiver
阴极沉积过程 cathode run
阴极沉积(物) cathode deposit;cathodic deposit(ion)
阴极炽点 cathode spot
阴极打火 cathode sparking
阴极导电棒 cathode collector bar
阴极导电母线 cathode busbar
阴极导向器 cathode phase inverter
阴极的 cathodic
阴极的机械谐振 mechanical cathode resonance
阴极灯 cathode lamp
阴极电沉积 cathodic electrodeposition
阴极电池 cathodic cell
阴极电感 cathode inductance
阴极电辉 cathode glow
阴极电解液 cathode liquor;catholyte
阴极电解质 catholyte
阴极电紧张 catelectrotonus
阴极电流 cathode current;cathodic current
阴极电流产生的偏压 cathode bias
阴极电流强度 cathode current intensity;cathodic current intensity
阴极电流效率 cathode efficiency
阴极电流最大值 peak cathode current
阴极电路 cathode circuit
阴极电路调谐振荡器 tuned-cathode oscillator
阴极电容 cathode capacitance
阴极电势 cathode potential
阴极电位 cathode potential
阴极电位调节器 cathode potential regulator
阴极电位稳定 cathode potential stabilization
阴极电压 cathode voltage
阴极电压降 cathode drop;cathode fall
阴极电压调制 cathode voltage modulation
阴极电泳 cathodic electrodeposition
阴极电源接头 cathode power connection
阴极电源引线 cathode power lead
阴极电子放大器 thermionic amplifier
阴极电子激发光 cathodoluminescence
阴极电子射线束 cathode beam
阴极电子束 cathode-ray beam
阴极电子束靶 cathode-beam target
阴极电阻 cathode resistor
阴极电阻器 cathodic resistor
阴极淀积 cathode deposit
阴极镀层 cathodic coating
阴极端子 cathode terminal
阴极断裂 cathode breakage
阴极发光 cathode light
阴极发光图像 cathode luminescence image
阴极发射 cathode emission
阴极发射电流密度 cathode emission current density
阴极发射率 cathode emissivity
阴极发射体 cathode emitter
阴极发射效率 cathode emission efficiency
阴极反馈 cathode feedback
阴极反馈放大器 cathode feedback amplifier
阴极反应 cathodic reaction
阴极防腐(法) cathodic corrosion protection
阴极防腐剂 cathodic protector
阴极防蚀 cathodic protection
阴极防蚀法 cathode non-corrosive method;cathodic corrosion protection
阴极放大器 cathamplifier
阴极放电管 cathode arrester;glim lamp
阴极分析 cathode analysis
阴极敷层 cathode coating
阴极辐射灵敏度 cathode radiant sensitivity
阴极腐蚀 cathode corrosion;cathodic corrosion
阴极负反馈 cathode degeneration;cathode degradation
阴极负反馈电阻 cathode degeneration resistance
阴极负反馈电阻器 cathode degeneration resistor
阴极负载 cathode load
阴极负载的 cathode-loaded
阴极负载二极管 bootstrap diode
阴极负载辅助调制器 bootstrap driver
阴极高频补偿 cathode peaking
阴极(高频)峰化 cathode peaking
阴极隔膜 cathodic membrane
阴极镉 cathode cadmium
阴极跟随器 cathamplifier;cathode follower
阴极汞放电灯 hot cathode mercury discharge lamp
阴极挂车 cathode trailer
阴极光电流 cathode photocurrent
阴极光照灵敏度 cathode luminous sensitivity
阴极(护)罩 cathode shield
阴极还原 cathode reduction;cathodic reduction
阴极环 cathode loop
阴极缓蚀剂 cathodic inhibitor
阴极辉光 cathode luminance;cathode luminescence;cathodic light;negative glow
阴极辉光电 cathode glow
阴极辉光区 cathode glow space
阴极回轰 cathode back-bombardment
阴极回路 cathode loop
阴极回授 cathode feedback
阴极活性 cathode activity
阴极击穿 cathode breakdown
阴极基金属 cathode base metal
阴极(激)发光 cathode luminescence
阴极激活 activation of filament;cathode activation
阴极激励 cathode drive
阴极极化 cathodic polarization
阴极加热器 cathode heater
阴极溅镀 cathode sputtering
阴极溅射(镀膜) cathodic spattering;cathode sputtering
阴极溅射镀膜法 cathode sputtering process
阴极溅射法 cathodic spattering method
阴极键控(法) cathode keying
阴极接地 cathodic ground;minus earth
阴极接地的 earthed cathode
阴极接地放大器 cathamplifier;cathode-base amplifier;grounded-cathode amplifier
阴极接地三极管 earth-cathode triode
阴极接头 negative contact
阴极介层 cathode interlayer
阴极界面间电容 cathode interface capacitance
阴极界面阻抗 cathode interface impedance;layer impedance
阴极绝缘反相电路 floating paraphase circuit
阴极开关 cathode switch
阴极空间 cathode chamber
阴极空腔谐振器 cathode cavity
阴极控制 cathodic control
阴极馈电线 negative feeder
阴极冷端效应 cathode cold end effect
阴极量子效率 cathode quantum efficiency
阴极灵敏度 cathode sensitivity
阴极漏泄 cathodic leakage
阴极脉冲调制的 cathode pulsed
阴极锰片 manganese cathode chip
阴极密度 cathode density
阴极膜 cathode film
阴极母线 cathode bus;negative bus;negative busbar
阴极耦合 cathode coupling

阴极耦合的 cathode-coupled
阴极耦合电路 cathode-coupled circuit
阴极耦合多谐振荡器 flopover circuit
阴极耦合放大器 cathode-coupled amplifier
阴极耦合器 cathode-coupled stage; cathode coupler; cathode follower
阴极耦合式振荡器 cathode-coupled oscillator
阴极耦合限幅器 cathode-coupled clipper
阴极喷镀 cathodic spattering
阴极偏压 cathodic bias
阴极偏压电阻器 cathode bias resistor
阴极屏幕上的点状(接收)记录 dot recording on cathode ray screen
阴极起始发射 initial cathode emission
阴极气流 cathode flame
阴极清洗 cathode cleaning
阴极区 cathode column; cathode space; cathodic area; negative polarity zone
阴极区富集法 cathode layer enrichment
阴极燃烧器 negative burner
阴极溶出伏安法 cathodic stripping voltammetry
阴极栅 cathode grid
阴极烧坏 cathode breakdown
阴极射线 cathode beam; cathode ray; negative ion ray
阴极射线测向仪 cathode-ray direction finder
阴极射线储存器 cathode-ray memory tube
阴极射线处理 cathode-ray treatment
阴极射线存储管 graphec(h)on
阴极射线的 electron-beam
阴极射线电子枪 cathode-ray gun
阴极射线发光 cathodic luminescence; electroluminescence
阴极射线发光法 cathode-ray luminescent method
阴极射线伏特计 cathode-ray voltmeter
阴极射线管 cathode [cathode]-ray tube; Braun tube; crookes; display tube; Farnsworth oscillight; scope; triniscope
阴极射线管表示器 cathode-ray tube indicator
阴极射线管存储器 cathode-ray tube storage
阴极射线管封接机 cathode-ray tube sealing machine
阴极射线管光点扫描器 cathode-ray tube spot scanner
阴极射线管绘图仪 cathode-ray tube plotter
阴极射线管记录 cathode-ray tube record
阴极射线管胶片印刷机 cathode-ray tube and film printer
阴极射线管图形显示终端 cathode-ray tube graphics terminal
阴极射线管显示 cathode-ray tube presentation
阴极射线管显示控制器 cathode-ray tube display controller
阴极射线管显示器 cathode-ray tube display
阴极射线管显示器终端机 cathode-ray tube terminal unit
阴极射线管型功率计 cathode-ray tube type powermeter
阴极射线管荧光屏 cathode ray tube screen
阴极射线管折射地震仪 cathode-ray tube refraction seismograph
阴极射线管指示器 cathode-ray tube indicator
阴极射线管终端 cathode-ray tube terminal
阴极射线光谱辐射计 cathode-ray spectroradiometer
阴极射线极谱法 cathode-ray polarography
阴极射线极谱仪 cathode-ray polarograph
阴极射线记录仪 cathode-ray oscillograph
阴极射线加速 cathode-ray acceleration
阴极射线加速器 cathode-ray accelerator
阴极射线接收管 cathode-ray receiving tube
阴极射线磷光 cathodophosphorescence
阴极射线炉 cathode-ray furnace
阴极射线录波器 cathode-ray oscillograph
阴极射线屏幕 cathode-ray -screen
阴极射线屏幕显示 cathode-ray-screen display
阴极射线曲线图示仪 cathode-ray curve tracer
阴极射线扫描显示(器) cathode-ray scan display
阴极射线摄影机 cathode-ray camera
阴极射线示波管 cathode-ray oscilloscope; oscillotron
阴极射线示波记录仪 cathode-ray oscillograph
阴极射线示波器 cathode-ray oscillograph; cathode-ray oscilloscope
阴极射线示波术 cathode-ray oscillograph
阴极射线示波图 cathodogram
阴极射线式存储管 memotron
阴极射线式存储器 memorytron
阴极射线输出 cathode-ray output
阴极射线束 cathode-ray beam; cathode-ray pencil
阴极射线特性曲线描记器 cathode-ray curve tracer
阴极射线显像管 cathode-ray picture tube
阴极射线仪器 cathode-ray apparatus; cathode-ray instrument
阴极射线荧光 cathodofluorescence
阴极射线致发光 cathode luminescence; cathodoluminescence
阴极射线致色 cathodochromism
阴极射线致色暗迹管 cathodochromic dark-trace tube
阴极射线致色的 cathodochromic
阴极蚀刻 cathodic etching
阴极势降 cathode fall
阴极室 cathode compartment
阴极寿命 cathode life
阴极输出放大器 bootstrap amplifier; cathode follower amplifier
阴极输出激励器 bootstrap driver
阴极输出级 cathode follower stage
阴极输出检波器 cathode follower detector
阴极输出器 cathode follower; grounded-anode amplifier; grounded plate amplifier
阴极输出器符合线路 cathode gate
阴极输出器检波器 cathode follower detector
阴极输入 cathode injection
阴极输入放大器 cathode-input amplifier
阴极丝 cathode filament
阴极送话器 cathode phone; cathodophone; ionophone
阴极碳化 cathode carbonization
阴极(碳)块 cathode block
阴极调制 cathode modulation
阴极铜 cathode copper; tough cathode
阴极透镜 cathode lens; first electron lens
阴极涂层 cathode coating; cathodic coating
阴极涂层粒子 cathode particle
阴极涂覆法 cathodic coating
阴极涂料 cathode coating; cathodic coating
阴极脱黏[粘] cathodic disbonding
阴极稳压 cathode regulation
阴极污染 cathode contamination
阴极析出 cathodic stripping
阴极显示仪 cathodic-ray display
阴极线路 cathode circuit
阴极陷波电路 cathode trap
阴极箱 cathode compartment
阴极效应 cathode effect
阴极芯 cathode base
阴极抑制剂 cathodic inhibitor
阴极引出端 cathode end
阴极引出线 cathode lead
阴极引线 cathode leg
阴极引线电感 cathode lead inductance
阴极荧光像 cathode fluorescence image
阴极有效系数 cathode active coefficient
阴极预热时间 cathode preheating time
阴极圆筒 cathode cylinder
阴极罩 cathode cell
阴极真空喷镀 sputtering
阴极蒸发 cathode evaporation
阴极支架 cathode anchor
阴极支路 cathode lead
阴极种板 cathode blank
阴极注入 cathode injection
阴极柱体 cathode cylinder
阴极铸造机 cathode-casting machine
阴极蚀法 cathiodic protection
阴极组 cathode assembly
阴角 inside corner; internal corner; reentrant corner
阴角处线条 base shoe corner
阴角接缝做法 jointing at inside corner
阴角抹子 angle trowel; feather-edge; inside-angle tool; inside corner trowel; margin trowel
阴角扇形砖 internal shoulder angle
阴角压条 inside corner mo(u)lding
阴角砖 internal angle bead; internal angle to cove skirting
阴井篦盖 area grating
阴冷的 rheumy
阴离子 anion; negative ion; negatively charged ion; negion; posion
阴离子半透膜 anionic semipermeable membrane
阴离子表面活化剂 anionic surface active agent
阴离子表面活化剂胶团 anionic surfactant micelle
阴离子表面活化剂溶液 anionic surfactant solution
阴离子部位 anionic site
阴离子沉淀剂 anion precipitant
阴离子催化聚合作用 anionic catalytic polymerization; anionoid recombination
阴离子导体 anionic conductor
阴离子的 anionic; anionoid
阴离子地沥青乳液 anionic asphalt emulsion
阴离子电沉积漆 anionic position coating
阴离子电荷 anionic charge
阴离子电迁移率 anionic electric(al) mobility
阴离子电泳 anaphoresis
阴离子发生器 anion generator
阴离子分析 anion analysis
阴离子分组 anion grouping
阴离子高分子电解质 anionic polyelectrolyte
阴离子含量三角图 triangular plot showing relative amount of anions
阴离子合成表面活性污染物 anionic synthetic(al) surface-active pollutant
阴离子合成洗涤剂 anionic synthetic(al) detergent; negative synthetic(al) detergent
阴离子环化聚合作用 anionic cyclopolymerization
阴离子活化剂 anion activator; anion active agent
阴离子活性洗涤剂 anion active detergent
阴离子交换 anion(ic) exchange
阴离子交换材料 anion(ic) exchange material
阴离子交换床 anion exchange bed
阴离子交换过滤器 anion(ic) exchange filter
阴离子交换剂 anion exchange material; anion exchanger; anionite
阴离子交换滤池 anion(ic) exchange filter
阴离子交换膜 amberplex anion; anion exchange membrane; permutite A
阴离子交换能力 anion exchange capacity
阴离子交换器 anion exchanger
阴离子交换(容)量 anion exchange capacity
阴离子交换色谱(法) anion exchange chromatography
阴离子交换树脂 anion exchange resin
阴离子交换塔 anion(ic) exchange tower
阴离子交换物质 anion exchange material
阴离子交换系数 anion exchange coefficient
阴离子交换柱 anion exchange column
阴离子接枝 anionic grafting
阴离子界面 anionic surface
阴离子聚丙烯酰胺 anionic polyacrylamide
阴离子聚合电解质 anionic polyelectrolyte
阴离子聚合物 anionic polymer
阴离子聚合作用 anionic polymerization
阴离子空位 anion(ic) vacancy
阴离子空穴 anion(ic) vacancy
阴离子亏损 anion defect
阴离子沥青乳液 anionic bitumen emulsion
阴离子络合物 anionic complex
阴离子膜 anionic membrane
阴离子黏[粘]土 anionic clay
阴离子黏[粘]土吸附 anionic clay adsorption
阴离子配位聚合作用 anionic coordinate polymerization
阴离子迁移数 anionic transference number
阴离子去垢剂 anionic detergent
阴离子缺陷 anion defect
阴离子染料 anionic dye
阴离子乳化剂 anionic emulsifier
阴离子乳化沥青 anionic asphalt emulsion; anionic emulsification bitumen; anionic emulsified bitumen
阴离子乳液 anionic emulsion

阴离子润湿剂 anionic wetting agent
阴离子渗入作用 anion penetration
阴离子渗透膜 anionic permeable membrane
阴离子树脂 resin anion
阴离子树脂交换 resinous anion exchange
阴离子树脂交换能力 anion resin capacity
阴离子树脂交换柱 anion resin exchange column
阴离子透膜 anionic permeable membrane
阴离子团 anionic radical
阴离子吸持作用 anionic retention
阴离子吸附 anion(ic) adsorption
阴离子洗涤剂 anion(ic) detergent
阴离子纤维素三酯 anionic cellulose triester
阴离子消去试法 elimination test of anion
阴离子型 anionic
阴离子(型)表面活化剂 anionic surfactant
阴离子(型)表面活性剂 anion surface active agent
阴离子型表面活性剂混合物 <其中的一种> Adoform
阴离子型表面片性剂 anionic surfactant
阴离子型活性剂 anionic active agent;negative ion active agent
阴离子型聚表面活性剂 anionic type surface active agent
阴离子型聚丙烯酰胺 anionic type polyacrylamide
阴离子型聚丙烯酰胺絮凝剂 anionic type polyacrylamide flocculant
阴离子型聚电解质 anionic type polyelectrolyte
阴离子移变(现象) anionotropy
阴离子引发作用 anionic initiation
阴离子有机污染物 anionic organic contaminant
阴离子再生柜 anionic regeneratant tank
阴离子中凝型 <乳化沥青> anion medium setting
阴离子总和 total amount of anion
阴历 lunar calendar [calender]
阴历年 lunar year
阴历时 lunar time
阴历月份 lunar month;moon month
阴螺钉 female screw;internal screw
阴螺栓拉杆 she bolt
阴螺丝 female screw
阴螺纹 female screw;female thread;internal screw thread;negative thread
阴螺纹管接头 female fitting
阴螺纹规 plug screw ga(u)ge
阴螺纹接头 female coupling
阴螺旋 female screw
阴霾天气 thick weather
阴面 cold side
阴模 bed die; bottom force; cavity die;die cavity;dies cavity;female;female die;female mo(u)ld;matrix [复 matrixes/matrices]; mo(u)ld cavity; negative die; negative mo(u)ld
阴模板 female templet
阴模成型 negative forming
阴模块 cavity block
阴模模槽 female cavity
阴模模穴 female cavity
阴模式阴极 matrix cathode
阴模旋坯成型 jolleying
阴模压铸混合机 cavity transfer mixer
阴膜 cavity block
阴片 negative plate;opaque copy
阴坡 opaco;saylo;shady slope
阴戗天沟 valley gutter
阴窍 lower orifice
阴晴不定的天气 broken weather
阴燃 smo(u)lder(ing)
阴日 overcast day
阴三角 internal coconut piece
阴三角砖 internal coconut piece
阴山石 yinshanite
阴射线 negative ray
阴湿的 dank;rheumy
阴湿地 damp sand
阴蚀 erosion of vulva
阴碳离子 carbanion
阴碳离子加成作用 carbanion addition
阴天 cloudy day; cloudy weather; gloomy weather;heavy weather;overcast;overcast day;overcast sky;thick weather;heavy overcast
阴天的 lowering
阴天率 cloudy day factor
阴天微隙 breaks in overcast
阴条 <瓷砖的> invert corner
阴条角尖嘴砖 internal bird's beak
阴条三角 internal bird's beak
阴条砖 <釉面砖的配件砖> internal angle bead
阴凸缘 female flange
阴图版 negative engraving
阴图重氮法 negative diazo process
阴线 line of shade
阴像 negative image
阴像地图 negative map
阴像刻图【测】 negative engraving;negative cutting
阴像刻图膜 opaque coating
阴像刻图员 negative cutter
阴像制版 negative plate making
阴性反应 negative reaction
阴性胶体 negative colloid
阴性强化 negative reinforcement
阴性树 shade-bearer;shade-bearing tree
阴性树种 shade species
阴性条件反射 negative conditioned reflex
阴性元素 negative element
阴性植物 heliophobous plant
阴阳插头 hermaphroditic connector
阴阳极比 cathode/anode ratio
阴阳极间距离 anode-cathode separation
阴阳离子 zwitterion
阴阳离子半径比值 cation-anion radius ratio
阴阳离子平衡 anion-cation balance
阴阳历 lunisolar calendar
阴阳面 male and female face
阴阳年 lunisolar year
阴阳色 two-tone colo(u)r
阴阳榫 mortise and tenon;tenon and mortise
阴阳榫接缝 offset
阴阳条织物 shadow stripe
阴阳瓦 mission tile
阴阳屋顶瓦 mission roofing tile
阴影 blight;shade;shadow(ing)
阴影边界 shadow boundary
阴影部分 dash area;shaded portion
阴影的 shaded
阴影的分布 distribution of shadow
阴影地带 shadow zone
阴影地区 shaded area
阴影段 shadow section
阴影法 hill shading;radio wave penetration method;shadowing method
阴影混合岩 nebulitic migmatite;shady migmatite
阴影角 shadow angle
阴影校正 shadow correction
阴影晶体 ghost crystal; phantom crystal
阴影滤波器 shadow filter
阴影率 eclipse factor
阴影面 shadow surface
阴影面积 blighted area; dash area; shaded area;shielded area
阴影模式分析 shadow pattern analysis
阴影莫阿试验 shadow-Moire test
阴影屏蔽 shadow shield
阴影区衰减 shadow attenuation
阴影区(域) shadow region; shaded region;shadow area;shadow zone; umbra area; umbra [复 umbrae/umbras]
阴影日照时间 shadow-sunlight time
阴影散射 shadow scattering
阴影色调 shade;shadow tone
阴影色调效果 shaded effect
阴影栅 shadow grid
阴影扇形区 dark sector
阴影式调谐指示器 shadow-tuning indicator
阴影衰减 shadow attenuation
阴影损耗 shadow loss
阴影条纹 shadow fringe
阴影条纹试验 shadow fringe test
阴影投影显微镜 shadow projection microscope
阴影图 shadowgraph;shadow pattern
阴影图解 shadow graphing
阴影图形 shading pattern
阴影微绕射 shadow microdiffraction
阴影误差 shade error
阴影系数 shadow(ing) factor
阴影显微镜 shadow microscope
阴影显微术 shadow microscopy
阴影线 dash(ed line);hatch(ing line);hatchure;line of shadow;shade line;hatching;cross hatching
阴影线标 shadow line label
阴影线部分 dashed area
阴影像 shadow image
阴影效果 hatching effect
阴影效应 shadow effect
阴影学 shades and shadow
阴影掩模 shadow mask
阴影掩模屏 shadow-mask screen
阴影页表 shadow page table
阴影仪 direct-shadow set-up; shadow-meter;shadow system
阴影因子 shadow factor
阴影圆柱 shadow cylinder
阴影云纹法 shadow-Moire method
阴影照片 shadow photograph; shadow picture
阴影照相法 shadowgraphy
阴影照相机 shadowgraph camera
阴影照相系统 shadowgraph system
阴影周期 dark period
阴影状构造 shady structure
阴影状混合花岗闪长岩 nobular migmatitic granodiorite
阴影状混合花岗岩 nobular migmatitic granite
阴影锥 shadow cone
阴影锥形 shadow pyramid
阴影资源 shadow resource
阴与影 shade and shadow
阴雨的 rainy
阴雨期 rain spell
阴郁 drabness
阴圆角 rounded internal angle
阴障隐蔽 barrier shielding
阴罩管 shadow-mask tube
阴质子 negative proton
阴转环 female swivel
阴锥 female cone

洇色 bleeding

荫蔽 overshadow

荫蔽的 shaded
荫蔽地 ground cover;natural cover
荫蔽地区 closed country; enclosed country; enclosed ground; shadow ground
荫蔽环 shading coil;shading ring
荫蔽三色管 shadow-mask tricolo(u)r tube
荫蔽因数 shadow factor
荫地植物 Sciad
荫度 arboricity
荫屏作用 shadowing
荫栅 aperture grill
荫生林 coppice (wood);copse wood
荫生植物 shade plant
荫性树 shade tree
荫性植物 shade plant
荫影 shading
荫影角 shadow angle
荫影校正 shading correcting
荫影散射 shadow scattering
荫影显微绕射 shadow microdiffraction
荫罩 apertured mask;apertured shadow mask
荫罩板 planar mask
荫罩密度检测仪 mask inspection densimeter
荫罩屏 shadow-mask screen
荫罩式彩色显像管 colo(u)rtron

音爆 sonic boom

音标 phonetic writing
音标铅字 phonotype
音波 acoustic(al) oscillation;acoustic(al) wave; sonic wave; sound wave
音波的 sonic
音波发光机 phonophote
音波计 audiometer [audiometre]
音波面【物】 wavefront
音波(深海)散射层 deep scattering layer;sonic scattering layer
音波探查 acoustic(al) prospecting
音波振动描记器 tonoscope
音叉 diapason;tuning fork
音叉的股 prong
音叉放大器 fork amplifier
音叉换能器 tuning-fork transducer
音叉频率 fork frequency
音叉频率控制 tuning-fork frequency control
音叉式 tuning-fork type
音叉式安培计 tuning-fork type ammeter
音叉式棒条筛 tuning-fork grizzly
音叉音 fork tone
音叉音调制 fork-tone modulation
音叉振荡器 fork generator; tuning-fork oscillator
音差 beat
音程 interval;musical interval;pitch interval;pitch of sound;volume range
音程的 intervallic
音触终端 voice-actuated terminal
音道 tone channel
音调 musical note;tonality
音调补偿音量控制 tone-compensated volume control
音调测量学 tonometry
音调的 tonal
音调改变 dodging
音调高度 pitch of sound
音调计 tonometer
音调校正 tone correction

音调接收机 tone receiver
音调结构 acoustically treated construction
音调均衡器 tone equalizer
音调析离 tone separation
音调频率 tone frequency
音调调整 tonal response adjustment; tone control; tone correction
音调调整补偿器 pitch trim compensator
音调修正 pitch correction
音调遥测术 tone telemetry
音调再现 tone reproduction
音度 loudness of sound
音符 musical note
音符式橡胶止水 music-note type rubber seal
音杆 tuning rod
音感 tone sense
音高镜 tonoscope
音弧 musical arc
音簧(钟) gong
音级 sound level
音级计 sound level meter
音级仪 sound level meter
音阶 gamut; musical scale
音节 syllable
音节明晰度 percentage syllable articulation
音控防鸣器 voice-operated antising-ing device [VODAS]
音控防鸣式短波电话终端机 VODAS radio telephone terminal
音控门限电平 threshold level of voice operated circuit
音乐播音室 music studio
音乐餐厅 cabaret; cafe chantant
音乐池 orchestra pit
音乐回声 musical echo
音乐会场 music hall
音乐酒吧间 <有自动电唱机的> juke joint
音乐频率 musical frequency
音乐室 music room; music studio
音乐台 <装有壳状反射板的> band shell; bandstand
音乐厅 auditorium [复 auditoria]; concert hall; music hall; philharmonic hall
音乐图书馆 music library
音乐心理学 psychology of music
音乐学院 academy of music; college of music; conservatory of music
音乐演奏厅 ode(i)on
音乐钟 musical clock
音量 loudness; sound volume; speech volume; volume of sound
音量表 volume indicator
音量补偿控制器 compensated volume control
音量单位 voice unit; volume unit
音量单位表 volume unit meter
音量单位警系统 voice unit
音量等响线 equal loudness contours
音量范围 volume range
音量范围控制 volume range control
音量符号 quantity mark
音量级 loudness level
音量计 audiometer [audiometre]; audio-volume indicator
音量开关 volume switch
音量控制 loudness control; volume control
音量控制开关 volume control switch
音量控制器 fader; volume controller; volume expander
音量控制器具 volume control device
音量控制旋钮 volume control knob
音量扩大 volume expansion
音量扩展 volumetric(al) expansion
音量扩展器 volume expander
音量衰耗当量 <传输系统> volume equivalent
音量衰减调整 fader control
音量衰减器 volume attenuator
音量调节器 volume adjuster; volume control device
音量调整旋钮 volume control knob
音量压缩 volume compression
音量压缩扩展器 volume compander
音量压缩器 volume compressor
音量指示器 volume indicator
音码测距法 tone code ranging
音帕克托镍锰钼钢 Impacto
音频 <30 赫 ~ 20 千赫> acoustic(al) frequency; audio frequency
音频保真度 audio fidelity
音频报警设备 audible alarm
音频逼真度 audio fidelity
音频逼真度控制 audio-fidelity control
音频编码器 audio-frequency coder
音频变化 sound variation
音频变压器 audio-frequency transformer; audio transformer
音频拨号 tone dial(l)ing; voice-frequency dial(l)ing; voice frequency keying
音频拨号制 voice frequency selecting system
音频波 audio-frequency wave
音频波段 audio band
音频波段电路 voice grade channel
音频布线找寻器 tone locator
音频差频振荡器 audio interpolation oscillator
音频传送 tonic transmission
音频单音解码器 audio tone decoder
音频导航 audio navigation; audio piloting
音频电报 acoustic(al) telegraph; tone telegraphy; voice frequency telegraph
音频电报法 audio-frequency telegraphy
音频电报术 audio-frequency telegraphy
音频电分割轨道电路 audio-frequency track circuit with electric(al) disconnecting joints
音频电话线 voice frequency telephone line
音频电缆 audio cable; voice-frequency cable
音频电流 audio-frequency current; voice-frequency current
音频电路 audio circuit; tone circuit; voice-frequency circuit
音频电码 tone code
音频电子轨道电路 audio-frequency electronic track circuit
音频调度电话分机 voice frequency dispatching tele-phone subset
音频调度电话总机 voice frequency dispatching tele-phone control board
音频叠加电路 audio-frequency overlay circuit
音频叠加系统 audio-frequency overlay system
音频叠加制 audio-frequency overlay system
音频扼流圈 audio-frequency choke
音频发生器 audio-frequency generator; audio generator; tone generator
音频反射 sound reflecting
音频范围 audio-frequency range; audio-frequency region
音频放大 note magnification
音频放大器 audio amplifier; audio-frequency amplifier; sound amplifier
音频峰值限制器 audio peak chopper
音频浮标 acoustic(al) buoy
音频副载波 audio subcarrier
音频隔离线 audio-channel wire
音频各站电话 voice frequency interstation telephone
音频各站(养路)电话分机 voice frequency interstation/track maintenance telephone subset
音频各站(养路)电话总机 voice frequency interstation/track maintenance telephone control board
音频功率 audio-frequency power
音频轨道电路 audio-frequency track circuit
音频海流计 audio current meter
音频混频管 audio mixer
音频级 audio-frequency stage
音频级线路 voice grade line
音频继电器 voice frequency relay
音频加重电路 accentuator
音频加重器 accentuator
音频检波器 audio detector; aural detector
音频键控 <长途通信[讯]> voice frequency keying
音频晶体管列车调度电话 audio-frequency transistor train dispatching telephone
音频控制 audio control
音频控制的 voice-operated
音频控制设备 voice-operated device
音频流速仪 audio current meter
音频滤波器 tone filter
音频脉冲测听计 pulse-tone audiometer
音频脉冲发生 voice-frequency pulsing
音频脉冲遥感 remote-sensing with audio-frequency pulse
音频命令点 audio-frequency command spot
音频内插振荡器 audio interpolation oscillator
音频偏移调制 audio-frequency shift modulation
音频频段 audio band; audio-frequency range; tonal range
音频频率计 audio-frequency meter
音频频谱 audible spectrum; audio spectrum; sound spectrum
音频频谱计 audio-frequency spectrometer
音频频谱仪 audio-frequency spectrometer; sound spectrograph
音频前置放大器 audio preamplifier; speech-input amplifier
音频设备 audio-frequency apparatus
音频输出限制器 audio output limiter
音频输入系统 audio input system
音频数据通信[讯] voice data communication
音频调谐 audio tuning
音频调整 voice control
音频调制 audio modulation; buzzer modulation; tone modulation
音频调制解调器 acoustic(al) coupler
音频通信[讯]电路 audio-frequency circuit
音频通信[讯]设备 audio-communication equipment
音频通信[讯]系统 audio-communication system
音频通信[讯]线路 audio-communication line; sound line
音频无缝轨道电路 audio-frequency jointless track circuit
音频误差振荡器 audio interpolation oscillator
音频限幅器 audio peak limiter
音频响应 audio response
音频信道时隙 voice channel time slot
音频信号 audible alarm; audio signal; tonic train signal(l)ing; voice frequency signal(l)ing
音频信号的谐波畸变 audio-frequency harmonic distortion
音频信号发生器 audio signal generator; tone signal generator
音频信号混合器 sound mixer
音频信号设备 voice frequency signal(l)ing
音频信号试验 audible test
音频信号指示器 audio signal detector
音频信扰比 audio-frequency signal-to-interference ratio
音频选号调度电话 audio-frequency selective ringing traffic control telephone
音频选号呼叫 audio-frequency selective ringing
音频选叫 voice frequency selective calling
音频选择制 voice frequency selecting system
音频扬声器 audio tweeter
音频应答器 audio response unit
音频载波 audio carrier; sound carrier; voice carrier; voice-frequency carrier
音频噪声 audio-frequency noise
音频增益 audio gain
音频增音机 <调度、各站分机用> voice frequency repeater
音频增音站 voice frequency repeater office
音频照准系统 audio-frequency pointing system
音频振荡器 audio-circuit or frequency oscillator; audio-frequency oscillator; audio generator; note oscillator; tone generator
音频振动沉桩机 sonic pile driver
音频振幅比较式定位器 tone localizer
音频振铃 voice frequency ringing
音频振铃电流 tone ringing current
音频振铃设备 voice frequency ringing set; voice frequency signaling set
音频直流转换器 VF/DC [voice frequency/direct current] converter
音频指示器 audio indicator
音频终端 audio terminal; voice frequency terminal
音频终端架 voice frequency terminating equipment bay
音频终端设备 voice frequency terminal equipment
音频终端装置 audio-frequency terminating set
音频转接段 audio-frequency section
音品 timbre; tone colo(u)r; tone quality
音平 tone level
音强 acoustic(al) intensity; sound intensity
音强度计 phon(o)meter
音圈 voice coil
音圈电动机 voice coil motor
音色 musical quality; timbre; tone colo(u)r
音色失真 audio-frequency harmonic distortion
音哨 tone whistle
音室 tone chamber
音素 phoneme
音素学 phonemics
音速 acoustic(al) speed; acoustic-

(al) velocity; sonic speed; sonic velocity; sound speed; sound velocity; speed of sound
音速冲响 sonic bang
音速喷管 sonic nozzle
音速线 sonic line
音速压力 sonic pressure
音速仪 <水中的> sound velocimeter
音调制 tone modulation
音调制器 tone modulator
音纹噪声 surface noise
音响 sounding
音响板 sounding board
音响报警 acoustic(al) warning; audible alarm
音响报警浮标 sound warning buoy
音响报警设备 audible alarm
音响报警系统 voice warning system
音响报警信号 audible warning; aural warning; sound warning
音响报警装置 audible warning; aural warning; sound warning
音响壁板 sounding board
音响标度 sone scale
音响表示器 audible indicator; sound indicator
音响不连续性 acoustic(al) discontinuity
音响材料 acoustic(al) material
音响测高计 sound ranging altimeter
音响测距 sound ranging
音响测距仪 sound ranger
音响测量 acoustic(al) investigation; acoustic(al) survey(ing); acoustic(ing); echo measurement; sound measurement
音响测深法 echo sounding
音响测深机 echometer; echo sounder
音响测深图 echogram
音响测深自动记录仪 echograph
音响测向仪 aural direction finder
音响车辆探测器 sound-sensitive vehicle detector
音响冲击吸收器 acoustic(al) shock absorber
音响处理 acoustic(al) treatment
音响传感器 acoustic(al) type strain ga(u)ge
音响传输 sonic propagation; sound propagation
音响磁带盒 audio cassette
音响磁性水雷 acoustic(al) magnetic mine
音响道口信号 sound crossing-signal
音响的 acoustic(al)
音响灯(浮)标 lighted sound buoy; light whistle buoy
音响电报 acoustic(al) telegraph
音响定位参照系统 acoustic(al) position reference system
音响定位器 sonar [sound navigation and ranging]
音响定位仪 acoustic(al) locating device
音响断路开关 audio cut off switch
音响发射性 acoustic(al) reflectivity
音响反射板 orchestra shell
音响反射性 acoustic(al) reflectivity
音响方位 sonic bearing
音响浮标 acoustic(al) buoy; sonobuoy; sound buoy; trumpet buoy
音响告警 aural warning
音响共鸣 acoustic(al) resonance
音响海流计 acoustic(al) (water) current meter
音响航标 audible aids; audible navigation signal; audible signal; sound signal
音响航向 aural course
音响呼叫 audible call
音响呼叫信号 phonic call signal
音响呼救开关盒 voice distress switch box
音响环境 acoustic(al) environment
音响计时机构 acoustic(al) timing machine
音响记录 sonic record
音响降低系数 acoustic(al) reduction factor
音响检测器 acoustic(al) detector
音响检漏器 audible leak detector
音响检验 sonic inspection
音响降低系数 sound reduction factor
音响接收 reception by sounder
音响警报 audible alarm; aural warning
音响警报浮标 sound warning buoy
音响警报器 acoustic(al) alarm (unit); audible alarm unit
音响警报装置 audible acoustic(al) alarm unit
音响警告信号 acoustic(al) warning signal
音响绝缘板 acoustic(al) insulation board
音响控制 sound control
音响控制室 sound-control booth
音响控制台 sound-control console
音响流量计 sonic flowmeter
音响流速仪 acoustic(al) (water) current meter
音响榴弹 sound signal shell
音响脉冲 sound pulse
音响脉冲收发两用机 acoustic(al) transponder
音响脉冲转发器 acoustic(al) transponder
音响模拟 sound simulation
音响莫斯电报机 sounder Morse instrument
音响内部通信[讯]系统 audible intercommunication system
音响疲劳 sonic fatigue
音响起伏 acoustic(al) scintillation
音响器 box-sounding relay; sounder; sounding relay; tapper
音响器电键 sounder key
音响器共鸣箱 sounder resonator
音响强度 sound intensity
音响式检漏仪 sonic leak detection device
音响式流量计 sonic flowmeter
音响试验 acoustic(al) test; sonic test
音响收报 sound reading
音响收发装置 sound transceiver
音响数据 acoustic(al) data
音响衰减 sound damping
音响水雷 acoustic(al) mine
音响损失系数 sound-reduction factor
音响探测器 acoustic(al) detector; aural detector; sonic finder
音响特性 acoustic(al) characteristic; acoustic(al) property <建筑物的>
音响通知 audible annunciation
音响(透过)损失系数 sound-reduction factor
音响无线电信标 aural beacon; aural radio beacon; talking radio beacon
音响雾号 acoustic(al) fog signal
音响吸收的 sound absorbing
音响系统 sound system
音响效果 acoustic(al) effect; sound effect
音响效果传声器 effects microphone
音响效果电路 effects circuit
音响效果发生器 acoustic(al) effects generator
音响效果混合器 acoustic(al) effects mixer
音响信标 aural beacon; pinger
音响信号 acoustic(al) signal; audible signal; aural signal; sonic signal; sound signal
音响信号编码器 audio coder
音响信号定向能力 orientability of a sound signal
音响信号盘 annunciator panel
音响信号系统 audible signal system; audible signage
音响信号装置 acoustic(al) signal device; audible signalling; sound signal(l)ing
音响修正 acoustic(al) homing; acoustic(al) modification
音响学 acoustics
音响仪 <探测混凝土强度和裂缝深度用的> soniscope [sonoscope]
音响鱼雷 acoustic(al) torpedo
音响指向标 aural beacon
音响制导 sound guidance
音响装置 acoustics
音响资料库 sound archives
音像同步装置 moviola
音信 tidings
音信不通的 silent
音型 sound type
音选调度电话分机 dispatching telephone set with VF selective system; dispatching telephone subset with VF selective calling
音选调度电话总机 dispatching telephone control board with (voice frequency) selective system
音选分机测试仪 tester for selective calling subset; tester for VF-telephone subset
音选各站(养路)电话分机 interstation/track maintenance telephone subset with VF selective calling
音选各站(养路)电话总机 interstation/track maintenance telephone control board with selective system; voice frequency interstation/track maintenance telephone control board
音选双向增音机 two-way repeater for VF selective calling
音选同线电话分机 party-line telephone subset for VF selective system; party-line telephone subset with VF selective calling
音选同线电话总机 party-line telephone control board with voice frequency selective system
音压 sound pressure
音译 transliteration
音域 diapason
音源标定 sound measurement
音源基准 power level
音障 sonic barrier
音值 allophone; phonetic value
音质 acoustic(al) fidelity; acoustics; musical quality; quality of sounds; timbre; tone quality
音质校正 acoustic(al) correction
音质控制 tone control
音质评价标准 acoustic(al) criterion; psychoacoustic criterion; psycho-acoustic(al) criterion
音质设计 acoustic(al) design; acoustic(al) quality design
音质失真 quality distortion
音质研究者 audiophile

殷

殷钢 invar metal; invar steel
殷钢尺 invar (plotting) scale; invar tap
殷钢尺引伸仪 invar tape extensometer
殷钢带尺 invar ribbon; invar tape
殷钢横基尺 invar subtense bar
殷钢绘图尺 invar (plotting) scale
殷钢卷尺 invar tape
殷钢水准标尺 invar level(l)ing rod
殷钢水准尺 invar level(l)ing staff
殷钢丝伸缩仪 invar extensometer
殷钢线尺 invar wire
殷霍夫沉淀漏斗 Imhoff sedimentation funnel
殷霍夫固体 Imhoff solid
殷霍夫井 Imhoff pit
殷霍夫(式沉淀)池 Imhoff tank
殷霍夫式化粪池 Imhoff tank
殷霍夫式锥形管 Imhoff cone
殷霍夫圆锥管试验 <污水沉淀> Imhoff cone test
殷霍夫锥形杯 <污水沉淀试验用的> Imhoff cone
殷霍夫锥形测定 Imhoff cone test
殷珀兹 <一种不溶于松节油的天然沥青> Imposite
殷实的 solid
殷实铺保 substantial firm for security
殷瓦摆 <又称因瓦摆> invar pendulum
殷瓦标尺【测】 invar rod
殷瓦尺伸长计 invar tape extensometer; invar wire extensometer
殷瓦尺引伸仪 invar tape extensometer; invar wire extensometer
殷瓦尺丈量 invar measurement
殷瓦带尺 invar ribbon
殷瓦杆 invar rod
殷瓦横基尺 invar subtense bar
殷瓦绘图尺 invar (plotting) scale
殷瓦基线尺 invar baseline wire
殷瓦基线尺丈量 invar base tape measurement
殷瓦劳合金钢 invaro
殷瓦轮 invar wheel
殷瓦横测尺 invar subtense bar
殷瓦视距尺 invar stadia rod
殷瓦水准尺 invar level(l)ing rod; invar level(l)ing staff
殷瓦线尺 invar wire

铟

铟钢丝引伸仪 invar wire extensometer
铟黄锡矿 sakuraiite
铟矿 indium ores
铟矿床 indium deposit
铟银焊料 <含铟 90%、银 10% 的合金> indalloy

银

银白的 argent
银白色 argent; silvery-white
银白色的 silver
银白色页状方解石 argentine
银白颜料 silver white
银白杨 abele; white poplar
银白珍珠 Australian pearl
银板 silver plate
银板硫锑铅矿 rayite
银本位 silver standard
银币合金 coin silver
银币饰 bezant
银箔 silver foil; silver leaf; silver paper
银薄片 silver lame
银衬套 silver bushing
银触点 silver contact
银触头继电器 silverstat relay
银触型电压调整器 silverstat regulator
银枞松 silver fir
银达理木 red tulip oak; silver stone-

wood
银弹 silver shot
银道 galactic equator
银道面 galactic plane
银道圈 galactic center [centre]
银道坐标 galactic coordinate
银道坐标系 galactic coordinate system;galactic system
银的 argentic
银的回收 silver recovery
银的凝固点 silver point
银滴定电量计 silver titration coulometer
银点 silver spot
银电极 silver electrode
银电解式电量计 silver voltameter
银垫圈 silver gasket
银锭榫 dovetail cramp;slate cramp <石或金属制的>;double dovetail key;dovetail feather;hammerhead key <硬木制的>
银镀膜 silver coating
银对银接点 silver-to-silver contact
银耳属 <拉> Tremella
银方铅矿 argentiferous galena;silver lead ore
银沸石 silver zeolite
银粉 powdered silver;silver powder;alumin(i)um powder
银粉底漆 alumin(i)um primer
银粉浆 alumin(i)um paste
银粉漆 alumin(i)um enamel;alumin(i)um paint;alumin(i)um powder paint;silver paint
银粉漆涂层 alumin(i)um paint coating
银粉颜料 alumin(i)um pigment
银粉(油)墨 alumin(i)um ink
银膏 silver paste
银镉电池 silver-cadmium cell
银镉焊料 silver-cadmium flux
银镉蓄电池 silver-cadmium battery;silver-cadmium storage battery
银根 liquidity
银根紧缩 currency deflation;tight money
银根松 ease of money;loose money
银根松的 easy money
银汞膏 argental;argental mercury
银汞合金 silver amalgam
银汞矿 moschellandsbergite
银汞齐 arquerite
银汞齐合金 arquerite alloy
银光彩料 silver luster [lustre]
银光风暴 silver storm
银光花纹 silver grain
银光面漆 argentine finish
银光闪闪的 clinquant
银光霜 silver frost;silver thaw
银光纹理 felt grain;silver figure;silver grain
银光纹木材 felt wood;silver grain wood
银光皂 silver soap
银光泽彩 silver luster [lustre]
银果胡颓子 silverberry
银焊 silver brazing;silver soldering
银焊料 mattisolda;silver solder
银焊料合金 easy-flo
银行 banking house
银行保函 banker's letter of guarantee;bank guarantee
银行保证 bank guarantee
银行保证书 bank guarantee
银行背书 bank endorsement;bank stamp
银行本票 bank's order;cashier's check
银行拨款单 bank money order
银行长期汇票 banker's long draft
银行偿付保证书 bank refund guarantee;bank refundment guarantee
银行偿付信用 bank's disposal credit
银行承兑 banker's acceptance
银行承兑的票据 bank note;bank paper
银行承兑限额 acceptance line
银行持股公司 bank holding company
银行出具的履约保证书 bank guarantee for performance bond
银行出具的投标保证书 bank guarantee for bid bond
银行出具的资信证书 certificate of credit standing issued by bank
银行储备金 bank reserves
银行存户 bank account
银行存款 bank deposit;cash on bank;deposit(e) at bank;cash in bank
银行存款保证额 banker's margin
银行存款解冻 release of bank accounts
银行存款日记账 bank deposit journal
银行存款调节表 bank deposit reconciliation statement
银行存款余额 bank balance
银行存折 bank book;bank passbook;bank savings account;pass book
银行大厅 bank(ing) hall
银行大厅地面 bank hall floor
银行大厦 bank block
银行贷款 bank accommodation;bank advance;bank credit;bank loan
银行担保 bank credit;bank guarantee
银行担保(支票)有效 <美> certify
银行等级选择法 banker's rate selection method
银行抵押业务 mortgage banking
银行电汇 bank telegraphic(al) transfer
银行垫款 bank advance
银行对账单 bank copy
银行兑换单 bank exchange memo;bank slip
银行兑换券 bank bill;bank note
银行发票 banker's invoice
银行法 bank law
银行房屋 bank building
银行分支行 branch bank;money shop
银行购买汇率 bank buying rate
银行股本 bank stock
银行股份 bank stock
银行股利专户 dividend bank account
银行柜台装置 bank fittings
银行贵重物品保险库 bank strongroom
银行划拨 bank transference
银行划转 bank transfer
银行汇款 bank transfer
银行汇款手续费 bank remittance fee
银行汇率 banker's rate
银行汇票 bank's draft;bank's bill;bank money order;bank post bill;bill of exchange;order on a bank
银行即期汇票 bank's demand draft
银行家 banker
银行间划拨清算 bank clearing
银行间汇率 interbank rate
银行间利率 interbank rate
银行间票据交换 bank clearing
银行间转账 bank transfer;interbank transfers
银行监督 banking supervision
银行交易 banking transaction
银行结存 bank balance
银行结存的证明 certificate of bank balance
银行结余 balance at the bank
银行结账单 bank statement
银行开户 open account
银行可承兑的 <证券等> bankable
银行肯担保项目 bankable project
银行控股公司 bank holding company
银行会计 bank accounting
银行来往账户 bank advance;nostro account
银行来往账余额 bank balance
银行利率 banking interest;bank rate
银行利息 bank interest
银行联合 <美> money pool
银行留的底本 bank copy
银行留置权 banker's lien
银行卖出汇票 bank selling rate
银行票据 bank bill;bank draft;banker's bill;bank money
银行票据交换 bank clearing
银行破产 bank failure
银行欠款 due from banks
银行取款单 counter-cheque
银行券 bank bill;bank note
银行确认的信用证 confirmed letter of credit
银行审计 bank audit
银行事务 banking
银行收付款 banking
银行收益 bank return
银行手续费 bank charges;bank commission;bank service charges
银行授信 credit standing
银行送金单 bank receipt
银行特许条例 bank charter act
银行贴现 bank discount
银行贴现率 bank discount rate;bank rate;bank rate of discount
银行同业汇价 interbank rate
银行投资 bank financing;bank investment
银行投资能力 banking power
银行透支 bank overdraft
银行托收 bill of exchange;collection through the bank
银行往来账 bank account
银行危机 bank crisis
银行现金保证函 letter of credit [L/C]
银行协议书 banking agreement
银行信贷 bank credit
银行信贷基金 bank credit fund
银行信贷资金 bank credit capital
银行信托基金 bank trust fund
银行信用卡 bank(er's) card
银行信用证 bank credit;bill of credit;letter of credit [L/C]
银行信用状 banker's letter of credit
银行学 banking
银行业 banking
银行业务 banking;banking business;banking practice
银行业者 banker
银行营业所 banker's shop
银行营业资金 banking fund
银行佣金 bank commission
银行优惠利率 bank favo(u)rable rate
银行有价证券 bank portfolios
银行再贴现率 bank rate of rediscount
银行账户 bank account
银行账目核对 reconciliation of bank accounts
银行证明书 bank reference
银行支票 bank cheque;bank draft;bank(er's) check
银行支票户头 bank checking account
银行支票账款 checking account
银行纸币 bank bill
银行纸币发行额 circulation of a bank
银行转账清算(制度) giro
银行资产流动性 bank liquidity
银行资信书 bank reference
银行总行 head office
银行最优惠利率 bank prime rate
银合欢【植】Leucaena glauca;silver wattle;bedge acacia;hedge acacia
银合金 silver alloy
银合金焊 silver-alloy brazing
银合金焊料 silver solder
银合金接触头 silver alloyed contact
银河【天】galaxy;Via Lactea;Milky Way
银河系 Galactic system;galaxy;Milky Way system;star system
银河系的 galactic
银河(星系射电)噪声 galactic noise
银河中心 galactic center [centre]
银河坐标系 <用银经和银纬来确定天体位置> Galactic system of coordinates
银华树 silky oak
银化合物 argentine compound;silver compound
银桦 silky oak
银桦属 <拉> Grevillea
银桦木 silver birch
银还原剂 silver redactor
银黄 silver yellow
银黄扩散着色 silver staining
银黄色光泽彩 cantharides lustre
银黄锡矿 hocaritite
银黄着色 silver stain
银灰 silvery grey
银灰木 <印度产的黑白两种硬木> greywood
银灰漆 alumin(i)um paint;alumin(i)um paste;silvery grey paint
银灰色 grey silver;silver gray [grey]
银灰色的 metallic grey
银辉铅铋矿 alaskaite
银货两讫 both sides clear;cash on delivery
银基填充合金 silver base filler alloy
银激活盖革计数器 silver-activated Geiger counter
银极 galactic pole;silver plate
银加厚液 sliver intensifier
银匠式建筑风格 <十六世纪西班牙> plateresque style
银胶菊 guayule
银接触 silver contact
银接触式构件 silverstat
银接点 silver contact
银金矿 electrum
银金矿石 Ag-Au ore
银金木 <印度和东印度产的坚硬的深褐色硬木> ingyin
银经 galactic longitude
银镜反应 silver mirror reaction
银镜试验 silver mirror test;silver staining test
银菊橡胶树 guayule rubber
银聚度 galactic concentration
银康热电偶 silver-constantan thermocouple
银矿床 silver deposit
银矿(石) silver ore
银矿指示植物 indicator plant of silver
银离子毒性 silver ion toxicity
银离子聚集 accumulation of silver
银亮钢 bohler;silver steel
银亮钢丝 silver steel wire
银量法 argentometry
银楼 silverware shop
银绿色 silver green
银-氯化银参比电极 silver-silver chloride reference electrode
银氯化银电极 silver-silver chloride electrode
银毛矿 silver jamesonite
银镁镍合金 silver-magnesium-nickel alloys
银锰铝特种磁性合金 silmanal
银冕 coronal of the Galaxy;galactic corona
银面触点 silver-faced contact

银面触头 silver-faced contact
银膜 Argentea;silver film
银膜过滤器 silver membrane filter
银墨 silver ink
银钼合金 silver-molybdenum
银幕 film screen; motion-picture screen; projection screen; silver screen;viewing screen;wall screen
银幕合成 screen process
银幕架 screen frame (unit)
银幕框架 screen frame (unit)
银幕倾角 angle of screen
银幕斜度 screen inclination
银镍黄铁矿 argentopentlandite
银凝胶 silver gel
银盘 galactic disk
银盘日射强度表 silver disk pyrheliometer
银盘日射强度计 silver disk pyrheliometer
银平价 parity of silver
银槭 silver maple;soft maple;white maple
银企关系 relation between bank and enterprise
银企合作 cooperation between banks and enterprises
银器 silverware
银器钢 silver steel
银器喷漆 silver lacquer
银迁移 silver migration
银钎焊 silver brazing
银钎焊料 silver solder
银钎焊片 silver brazing tape
银钎料 silver brazing alloy;silver solder
银铅 argentalium;argentiferous lead
银铅焊条合金 silver-lead solder alloy
银铅合金 stannum
银氰化钾 silver potassium cyanide
银球 silver shot
银鹊树 Chinese falsepistache
银染色法 silver staining
银熔点 silver point
银熔线 silver fuse
银色 silver
银色的 argentine
银色光泽 silvering
银色金属 argentine
银色片状玻璃 feathered washboard
银色漆 alumin(i)um bronze paint; silver paint
银色锡锑合金 Argentine metal
银色页状方解石 argentine
银砂 cinnabar(ite);silver sand
银山毛榉 silver falsebeech
银山楂 silver hawthorn
银石墨电刷 silver-graphite brush
银石墨制品 silver-graphite
银饰品 silver-work
银树 silver tree
银水 bright silver
银丝 filamentary silver;silver wire
银丝玻璃 filigree glass
银丝嵌饰 filigree enrichment
银丝嵌饰件 filigree decorative fixture
银丝网 silver gauze
银松 Scotch fir
银锁砌合 <丁顺交替,顺砖立式砌合> silver-lock bond
银碳接点 silver-impregnated carbon contact
银锑合金 silver-antimony alloy
银条 silver bar; silver bullion; silver ingot
银铁矾 argentojarosite
银铜 silver-bearing copper
银铜合金 silver bronze;silver-copper alloy
银铜矿石 Ag-bearing copper ore
银铜氯铅矿 boleite
银团 bank consortium;syndicate
银团贷款 syndicated loan
银纬 galactic latitude
银纹 craze;crazing
银钨电接触器材 silver-tungsten contact material
银钨合金 silver tungsten
银钨双金属带 silver-tungsten bimetallic band
银锡合金汞齐 silver-tin amalgam
银锡软焊料 plumbsol
银纤焊 silver-alloy brazing
银线 silver wire
银线系 silverline system
银像 silver image
银心 galactic center [centre]
银心轨道 galactic orbit
银心区 galactic center region
银心区源 galactic center source
银锌电池 silver-zinc battery; silver-zinc cell
银锌壳 zinc crust
银锌蓄电池 Andre-Venner accumulator; silver-zinc accumulator; silver-zinc storage battery
银星石 lasionite; wavellite; zepharovichite
银杏 gingko(biloba);ginkgo
银杏目 Ginkgoales
银杏树 maidenhair tree
银盐 argentic salt;silver salt
银盐定量计 argentometer
银盐复制法 silver halide process;silver process
银盐感光材料 silver sensitized material
银盐感光照片 silverprint
银盐工艺 silver process
银盐全息干板 silver holographic plate
银盐溶液槽 silver bath
银氧铯光阴极 Ag-O-Cs photocathode
银叶树 looking glass tree
银液滴定法 argentimetry
银黝铜矿 freibergite; silver fahlore; silver tetrahedrite
银黝银矿 leucargyrite
银浴器 silver bath
银云母 cat silver
银云杉 silver spruce;sitka spruce
银晕 galactic halo
银晕辐射 galactic halo emission
银枣 silverberry
银值 silver number
银纸沉着病 argyrism
银制品 silverware
银质尘着病 argyria
银质的 argental
银质接触环 coin-silver ring
银中毒 argyria; argyrism; silver poisoning
银舟 silver boat
银朱 red mercuric sulphide;vermil(l)ion
银朱涂料 vermil(l)ion paint

霪雨 excessive rain(fall of long duration);steady rain

霪雨期 rain spell

引板 back slab; corbel back slab; runoff tab

引爆 detonate;detonation;firing;shot-firing;shot firing
引爆按钮 firing button
引爆导线 blasting fuse
引爆电键 firing key
引爆电缆 explosive cable
引爆电路 detonating circuit; igniting circuit;shot-firing circuit
引爆电桥 bridge
引爆电线 ignition cable
引爆方式 firing pattern
引爆管 blasting cap;blasting primer
引爆火药 detonating powder
引爆激光器 initiating laser
引爆剂 <爆破工程> flashing compound; booster; detonating agent; detonator;flashing composition; igniter
引爆雷管 detonating primer;fuse cap; knocker;knocker
引爆能力 detonating capacity;detonation power
引爆器 blasting machine; blasting unit;igniting primer
引爆器钥匙 firing key
引爆人 powderman
引爆式投放器 pyrotechnic dispenser
引爆顺序 firing order
引爆速率 detonating rate; ignition rate
引爆索 primacord;primacord fuse
引爆危险 risk of igniting explosive
引爆物 trigger
引爆线 detonating cord; detonating fuse; detonating string; exploding wire;lead wire
引爆药 amorce;detonating charge;initiator
引爆药包 igniter [ignitor] pad
引爆药卷 detonating cartridge
引爆炸药 ignition charge; primacord explosive
引爆装置 detonating equipment; explosive device;igniter [ignitor]
引泵 pump priming
引槽 approach(ing) channel;channel of approach
引长杆 projecting lever
引潮力 tidal force;tidal generation force; tide-forming force; tide-generating force;tide-producing force; tide raising force
引潮渠 tidal opening
引潮时间 tidal opening
引潮势 tide-generating potential;tide-producing potential
引潮位 tide-generating potential;tide-producing potential
引虫剂 insect attractant
引出 abstraction; draw-off; elicitation;emerge;lead(ing)-out
引出板 end tab; runoff plate; runoff tab
引出次数 out-degree
引出点 exit point;tapping point <管道的>
引出电缆 outgoing cable
引出电流 projected current
引出电位 extraction potential
引出端(子) carry-out terminal;lead-out terminal;outlet terminal;output terminal; pinout; eduction end; exit; leading-out end; lead-out; pigtail;outlet <地下电线、空气管道等>
引出端(子)出线端 leading-out terminal
引出杆 leading-off rods
引出管 fair leader;pigtail
引出辊 pull roll
引出过程 spill process
引出盒 outlet box
引出井 exit well
引出口 outlet
引出连接线 outconnector
引出率 rate of draft
引出曲线 exit curve
引出刃 exit lip
引出设备 <通常用于限制进出的道路> exit facility
引出束 ejected beam; exit beam; extracted beam
引出水流 diverted flow
引出速率 rate of withdraw
引出纹槽 lead-out groove; run-out groove
引出物 derivative
引出线 expansion line;extension line; leader line; leading-off; lead(ing)-out (wire); lead through; lead wire; outgoing line; outlet; outlet line;output lead
引出线电缆 lead(-covered) cable
引出线罩 boot
引出箱 outlet box;outlet case
引出指令 exit instruction
引出轴 take-up reel
引出装置 ejector; extractor; leading-off;lead-out;take-away belt (conveyor);haul off <螺杆的>
引出组 outgoing group
引船绞车 hauling winch
引船进坞 haul in
引船进坞设备 haul in equipment; leading-in equipment
引船进坞小车 hauling-in-trolley
引船进坞装置 leading-in gear
引船小车 hauling carriage
引船小车轨道 ship hauling trolley track;hauling carriage track
引词 index term
引带 belt;gubernaculum;leading tape; tape leader
引带接头 belt lacing
引导 aim(ing); call on; guidance; guide;homing;leader pilot;leading; manuduction; marshal; pilot; steer(ing);vector(ing)
引导按钮 calling on button
引导臂板 calling-on arm
引导边缘 guide edge;guide margin
引导波束 lead beam
引导部分 leader
引导操作 pilot operation
引导场 guide field
引导程序【计】 bootstrap program(me); bootstrap routine; director; vectoring procedure
引导传播 guided propagation; trapping
引导代码 guidance code
引导带 pilot tape
引导灯(桩) leading light;pilot lamp; pilot light
引导电极 leading electrode;recording electrode
引导电缆 pilot cable;leader cable
引导电缆式自动导向系统 lead cable automatic guidance system
引导电路 steering circuit
引导端 leading end
引导阀 pilot relay valve;pilot valve
引导飞机着陆方向 take in an aircraft
引导符号 aiming symbol
引导缸筒 guide cylinder
引导高度 steering level
引导跟踪方式 lead pursuit
引导工作 pilotage
引导光束 lead beam
引导缓冲器 index buffer
引导货车 leading wagon
引导机 vectoring aircraft
引导机车 pilot engine

引导记录 home record;leader record
引导(降落)伞 drogue parachute
引导胶片 leader film
引导角 pilot angle
引导阶段 vectoring phase
引导精度 guidance accuracy
引导卡片 guide card
引导客车 leading coach
引导块 bootstrap block
引导棱镜 guiding prism
引导例行程序 bootstrap routine
引导连杆 guide link(age)
引导零 leading zero
引导轮 idler;track idler
引导轮传动小齿轮 idler pinion
引导轮挡圈 idler collar
引导轮导向板 idler guide
引导轮护扳 idler guard
引导脉冲 pilot pulse
引导模型 pilot model
引导目镜 guiding ocular
引导跑道 pilot lane
引导气流 steering current; steering flow
引导驱动器 bootstrap driver
引导绕纱 jack off
引导人员 pathfinder
引导塞规 pilot ga(u)ge
引导哨 vectoring post
引导示像 calling-on aspect
引导式地面运输 <如铁路气垫列车> guided ground transport
引导式塞规 pilot ga(u)ge
引导手信号 calling-on hand signal
引导输入程序 bootstrap input program(me)
引导数据 vectoring information
引导数据计算机 vectoring computer
引导伺服阀 pilot servomotor
引导通风冷却塔 induced-draft cooling tower
引导投资方向 give guidance to the direction of investment
引导图形符 leading graphic
引导误差 vectoring error
引导系统 guidance system;pilot system
引导显示 calling-on aspect
引导显微镜 guiding microscope
引导线补偿 lead wire compensation
引导消费 guide consumption
引导信号 call-to-enter signal; pilot signal;calling-on signal
引导信号臂板 calling-on signal blade
引导信号机【铁】 calling-on signal; call-to-enter signal
引导信号控制继电器 call-on signal control relay
引导星 guiding star
引导行车制 pilot system
引导序列 homing sequence
引导压力启动阀 pilot-operated valve
引导余量 guide margin
引导域 aiming field
引导员 pilotman
引导员引导列车运行 train working by pilotman
引导原则 vectoring doctrine
引导圆 aiming circle
引导站 vectoring station
引导指令【计】 bootstrap driver;bootstraping;key instruction
引导指令程序 bootstrap program(me)
引导指令方法 bootstrap instructor technique;bootstrap technique
引导指令记录 bootstrap record
引导指示 vectoring instruction
引导中心 direction center[centre]
引导轴 leading axle
引导装入程序 bootstrap loader;bootstrap loading routine;output lead
引导装入例行程序 bootstrap loading routine
引导装置 guide arrangement;guiding;guiding arrangement; pathfinding apparatus
引导组织 transmitting tissue
引道 access;access connection;access road;approach;approach bank;approach road;approach way;guide passage;rampway;slip road;slope ramp
引道板 approach slab
引道便道 access lane
引道标志 approach sign
引道垫层 approach cushion
引道高架桥 approach viaduct
引道接线 approach alignment
引道结构 approach structure
引道尽头 approach end
引道尽头处理 approach end treatment
引道开挖 approach cutting
引道(路)堤 approach embankment;approach fill
引道路肩 approach shoulder
引道排架栈桥 approach trestle
引道坡 approach ramp; ramp approach
引道坡度 approach grade; approach gradient
引道区 approach zone
引道区域 approach zone district
引道填筑 approach fill
引道突堤 approach jetty
引道挖方 approach cutting
引道斜坡段 approach ramp
引道展线 approach development
引道栈架 approach trestle
引道栈桥 access viaduct;approach trestle
引道折板 <移动式桥的> approach flap
引得号数 reference number
引堤 approach bank; approach embankment
引点 derived point
引电弧 striking arc
引定理 lemma [复 lemmata/lemmas]
引锭 dummy ingot
引锭杆 dummy bar
引锭坯 dummy sketch
引动模式 priming model
引杜林(染料) induline
引杜林色基 induline bases
引渡 deliver;deport;extradition;tradition
引渡法 extradition law
引渡时间 delivery time
引发 priming
引发电弧 starting arc
引发电路 detonator circuit
引发机构 initiating mechanism
引发机理 triggering mechanism
引发剂 booster;ignition primer;initiator;reaction initiator;trigger
引发开关 proximity switch
引发曲线 <强度发挥曲线> mobilization curve
引发添加剂 initiating additive
引发温度 kick-off temperature
引发污染源 pollution-creating source
引发物 trigger
引发物连锁反应 trigger chain reaction
引发语句 raise statement
引发作用 initiating ability;initiation
引放温度 releasing temperature
引风 induced air;induced draft;natural draft
引风道引风平巷 fan drift
引风锅炉 induced-draft boiler
引风机 draft fan; draught fan; exhauster;induced-draft fan;induced draught fan;induced fan
引风强度 draft intensity
引风式凉水塔 induced-draft cooling tower
引风水冷却塔 induced-draft water-cooling tower
引盖 drive cap
引割求全法 completion by cuts
引轨 approach rail
引辊 pick-up roll
引航 pilot service
引航标志 approach aid(s)
引航船 pilot boat;pilot cutter;pilot vessel
引航道 access channel;adjacent navigation channel;approach channel;dock entrance;navigation approach channel
引航道灯 channel indicating lights
引航道旁边泊船水域 approach basin
引航电缆 pilot cable
引航费 pilot dues;pilot fee;pilotage
引航费率 pilotage rate
引航工会 <英> Trinity House
引航工作 pilot service
引航管理 pilot management
引航管理当局 pilot authority
引航管理机构 pilot authority
引航管理机关 pilot authority
引航技术 pilotage;piloting
引航价率 pilotage rate
引航快艇 pilot cutter
引航旗 pilot flag
引航区 pilotage district;pilotage waters;piloting district;piloting waters
引航水道 entrance channel
引航水域 pilotage waters; piloting waters;pilot waters
引航伺服机 pilot servomotor
引航图 pilot chart
引航无线电台 pilot radiostation
引航信号 pilot signal
引航员 pilot
引航员离船 pilot left ship
引航员留置权 pilot's lien
引航员上下船地段 pilotage ground
引航员疏忽责任 default of pilot
引航员梯 pilot ladder
引航员巡航区 pilot's cruising ground
引航员执照 pilot licence [license]
引航站 pilot station
引航证 pilotage certificate
引号 quotation mark
引河 pilot channel;pilot cut;pilot cut-off;channel of approach
引洪灌溉 spate irrigation
引洪漫地 irrigation with flood water
引洪漫灌 irrigation with flood water
引洪淹灌 irrigation with flood water
引洪淤灌 irrigation with flood water
引弧 generating of arc;starting arc;striking the arc <电焊的>
引弧板 end tab;end tab assembly;run-on plate;run-on tab
引弧棒 starting rod
引弧电压 striking voltage
引弧端 arc end
引弧环 arc ring;short-proof ring
引辉孔 glow priming hole
引火 fire fighting;fire-leading;pilot light;priming
引火不良 misfiring
引火材料 matches
引火柴 kindling wood
引火点 flash(ing) point
引火点范围 firing range
引火放电 priming discharge
引火粉 pyrophoric powder
引火管 capped fuse;flash tube
引火管小火 central pilot
引火合金 pyrophoric alloy;pyrophoric metal
引火具 portfire
引火料 priming mixture
引火木(柴) kindling wood; touch-wood
引火器 lighter; match; pilot; pilot burner;priming apparatus
引火绳 hauling line;heaving line;messenger
引火试验 inflammation test
引火铁 pyrophoric iron
引火物 fire lighter;pyrophore;pyrophorus;spunk;tinder;kindling <美>
引火线 firing cable;ignition harness;shot-firing cable
引火药 ignition charge;priming
引火用风扇 starting fan
引火针 spark pin
引火着火 piloted ignition
引火嘴调风板 pilot air shutter
引接线 lead wire
引接线电阻 lead wire resistance
引进的 adventive;alien
引进灯 <机场> lead in light
引进方 licensee
引进国外技术 introducing of foreign technology;introduction of foreign technology
引进国外投资 introducing of foreign investment
引进国外资本 introducing of foreign capital
引进建筑物的给水干管 building main
引进刃 entrance lip
引进设备 imported equipment
引进树种 exotic tree
引进水源 import water
引进外国投资 introduction of foreign investment
引进外资 introduce foreign capital;introduction of foreign capital
引进先进技术 import of advanced technology;introduction of advanced technology
引进植物 adventive plant;introduced plant
引进作物 alien crop
引晶技术 seeding
引晶作用 adductive crystallization
引开 entrainment
引跨 approach space;approach span
引缆绳 hauling line;heaving line;messenger
引缆索 hauling line;messenger;heaving line
引离辊 pick-off roll
引力 attraction force;attractive force;force of attraction;force of gravitation;gravitation;gravitational force;gravitative attraction;gravity;gravity force
引力扁率 gravitational flattening
引力波 gravitational wave
引力波天文学 gravitational astronomy
引力波源 gravitational wave source
引力不稳定性 gravitational instability
引力层 gravisphere
引力常数 constant of gravitation;gravitational constant
引力场 gravitational field; gravity field
引力场理论 gravitational field theory

引力场强度 intensity of gravitational field
引力潮 gravitational tide;gravity tide
引力成团(说) gravitational clustering
引力带 girdle
引力弹弓模型 slingshot model
引力的 gravitational
引力范围 gravisphere
引力防波堤 attracting groyne
引力分界 gravipause
引力辐射 gravitational radiation
引力荷 gravitational charge
引力红移 gravitational red shift
引力昏暗 gravitational darkening
引力计 attraction meter
引力加速度 acceleration due to attraction;gravitational acceleration
引力流体 gravitating fluid
引力模式 gravity model
引力能 gravitational energy
引力凝聚 gravitational condensation
引力盘 gravitating disk
引力碰撞 gravitational encounter
引力偏移 gravitational shift
引力平衡 gravitational equilibrium
引力球谐函数展开式 expansion of gravitational into spheric(al) harmonics
引力扰动 gravitational disturbance
引力摄动 gravitational perturbation
引力势 gravitational potential
引力势阱 gravitational potential well
引力收缩 gravitational contraction
引力坍缩 gravitational collapse
引力梯度 gravitational gradient
引力通量 gravitational flux
引力同步加速辐射 gravitational synchrotron radiation
引力透镜 gravitational lens
引力透镜效应 gravitational-lens effect
引力弯曲 gravitational deflection
引力位 attractive potential; gravitational potential
引力位函数 gravitational potential function
引力位能 gravitational potential energy;potential energy of gravitation
引力位移 gravitational displacement; gravitational flux density
引力吸引 gravitational attraction
引力向量 gravitational vector
引力效应 gravitational effect
引力谐函数 gravitational harmonic function
引力佯谬 gravitational paradox
引力张量 gravitational tensor
引力正常位 normal potential of gravitation
引力质量 gravitational mass
引力中心 barycenter [barycentre]; center [centre] of attraction
引力子 graviton
引力最小值 gravitational low
引力作用 attractive effect; gravitational interaction
引流 conduction;diversion;tapping; water spreading
引流槽 pilot channel
引流丁坝 attracting groin
引流管 draft tube; drainage tube; draught tube
引流墙 cutoff wall
引流渠 diversion ditch
引流室 diversion chamber
引流术 drainage
引流竖管 uptake tube
引流竖管曝气器 simple(x) aerator
引流水坝 attracting groin
引流物 drain
引流型放大元件 induction fluid amplifier
引流叶片 suction vane
引流叶片齿轮 suction vane gear
引路 approach road
引路(标志)钉 road stud;traffic stud
引路速度 velocity of approach
引帽 drive cap
引黏[粘]土止水 clay sealing
引喷温度 <自动消防器的> releasing temperature
引频感测器 pilot sensor
引坡 road approach
引坡匝道 approach ramp
引起 create; evocation; fallout; originate;procure;promote;provoke
引起变形的力 transmutative force
引起的弯矩 ensuring moment
引起地貌改变的活动 land-disturbing activity
引起反感的 obnoxious
引起腐烂的 saprogenic;saprogenous
引起改变的 alterative
引起共鸣的 resonant
引起火灾的废料堆 fire trap
引起火灾的废物堆 fire trap
引起火灾的垃圾堆 fire trap
引起污染的 pollutive
引起应力 set up stress
引起应力的外力 stress-producing force
引起张力的 tensive
引起重分布弯矩 moment-inducing redistribution
引起注意 attention; catch attention; command attention
引起注意的信号 attention
引起注意符 attention symbol
引起注意键 attention key
引起注意列表 attention list
引起注意信号 attracting attention signal
引起注意中断 attention interrupt(ion)
引起注意装置 attention device
引气 air-entrainment; entrained air; entrainment of air
引气促流井 gas lift flowing well
引气硅酸盐水泥 air-entraining Portland cement
引气混凝土 air-entrained concrete
引气剂 air entrainer;air-entraining admixture;air-entraining agent
引气减水剂 air-entraining and water reducing admixture
引气量 air-entrained content
引气量测定仪 air-entrainment meter; entrainment meter
引气量试验 air-entrainment test
引气砂浆 air-entrained mortar
引气水泥 air-entrained cement
引气作用 air-entraining
引前电刷 leading brush
引前相供电臂 leading phase feeding section
引桥 access bridge;access space;access span;approach bank;approach bridge; approach space; approach to a bridge;landing bridge;ramp bridge; bridge approach;approach span;approach trestle
引桥跨 approach spalling
引桥跨度 approach span
引桥路 approach of bridge;road approach
引桥路堤 approach embankment;bridge approach fill
引桥坡道 approach ramp
引桥式码头 approach trestle pier;pier with approach bridge;pier with approach trestle;T-head pier
引桥填方 approach fill
引桥挖方 approach cutting
引桥挖方量 approach cutting quantity
引擎 engine;subframe
引擎安装面 engine mounting face
引擎发电机 engine-generator
引擎分析器 engine analyser [analyzer]
引擎基座 engine base
引擎空转试验 racing test
引擎起动油泵 engine primer
引擎散热器 engine radiator
引擎调速器 engine governor
引擎外套 engine jacket
引擎消声器 engine muffler
引擎消音器 engine muffler
引擎油 motor oil
引擎罩 bonnet;hood
引渠 access canal;approach channel; channel of approach
引取水量 offtaking
引燃 firing;inflame;pilot ignition
引燃点火喷射 ignition injection
引燃点火室 igniter chamber
引燃电极 igniter;ignitor
引燃电缆 firing cable;shot-firing cable
引燃电路 firing circuit; ignition circuit;shot-firing circuit
引燃电位 firing potential
引燃电压 ignition voltage; priming potential;striking voltage
引燃管 ignition rectifier; ignitron; trigatron
引燃管机车 ignitron locomotive
引燃管机车设备 ignitron locomotive equipment
引燃管机车装置 ignitron locomotive equipment
引燃管接触器 ignitron contactor
引燃管控制 ignitron control
引燃管引线 ignitron leads
引燃管整流器 glass tank rectifier
引燃火花 ignition spark; incendiary spark
引燃火焰 pilot flame
引燃极 igniter
引燃极振荡 ignitor oscillation
引燃剂 detonator;ignition powder
引燃角 firing angle
引燃开关 glow switch
引燃脉冲 firing pulse
引燃能量 fire ignition energy
引燃喷射 pilot injection
引燃喷嘴 pilot burner
引燃器 igniter;lighter
引燃热点 hot spot
引燃失败 firing failure
引燃式 ignition type
引燃线 ignition wire;ignitor cord
引燃阳极 ignition anode
引燃药 igniting primer
引燃油束 pilot jetting
引燃炸药 fire the charge
引人注目的冲击 dramatic impact
引入 indraft; indraught; intake; introduce; lead(ing)-in; lead through; pull-in
引入板 lead in plate
引入变量法 method of leading variable
引入槽 lead-in groove;lead screw
引入车道 access lane
引入衬管 inlet bush
引入导线 lead-in conductor
引入道路 approach road
引入的植物 exotic plant
引入地址 call address
引入电杆 office pole
引入电缆 entrance cable;lead(ing)-in cable;service cable
引入电刷 lead-in brush
引入端 entrance; lead-in; leading-in end
引入端子 leading-in terminal
引入方向 incoming direction
引入符号 created symbol
引入杆 leading-in pole
引入管 induction pipe;inlet pipe;service pipe
引入辊 draw roll;lead in roll
引入呼叫 call entry
引入滑轨 leading ramp
引入回灌量 <含水层的> intake recharge
引入架 derrick;leading-in bracket; lead in rack
引入尖脉冲 peaking
引入角 angle of indraught
引入绝缘管 leading-in insulator;leading-in tube
引入绝缘子 window insulator
引入开关 insertion switch
引入空气 draw-in air; entrain; introduction of air
引入空气量的指示器 <在混凝土或砂浆中> entrained air indicator
引入连接 service connection
引入量 intake
引入率 introductory rate
引入螺钉 lead in screw
引入螺线 lead-in groove
引入频率 pull-in frequency
引入品种 introduced variety
引入歧管 entrance branch
引入曲线 entrance curve
引入人孔 <电缆管道的> intake chamber
引入式通风 draw-in ventilation; supply ventilation
引入试验架 incoming test raceway; lead-in (and) test rack
引入(竖)井 service shaft
引入数(符号) call number
引入羧基 carbonation
引入套管 lead-in bushing
引入套筒 inlet bush
引入体 lead-in
引入通道 <气体的> approach channel
引入同步 pulling-in step
引入线 connecting track; drop line; incoming line;inlead;lead-in (conductor); leading-in line; leading-in wire;lead track;service conductor; service entrance conductor;service line;service wire
引入线瓷管 lead in porcelain tube
引入线短接管 lead in nipple
引入线夹 lead-in clamp
引入线绝缘套 entrance bushing
引入线绝缘子 lead in insulator
引入线室 intake chamber
引入线套管 leading-in tube
引入线箱 service box
引入箱 draw-in box; leading-in box; pull-in box
引入序列 calling sequence
引入窄脉冲 peaking
引入植物 introduced plant
引入指令 calling indicator; call instruction
引入指示 calling indicator
引入转换架 <调度电话> lead-in and change-over
引入子程序 calling subroutine
引入字 call word
引入组 incoming group
引上玻璃原板宽度 ribbon width
引上池 drawing basin
引上电缆 leading up cable; lead-out

cable
引上法 pulling method
引上工 drawing worker
引上工段 drawing section
引上机 drawing machine
引上机利用率 efficiency of drawing machine
引上率 drawing rate
引上速度 drawing speed
引上通路 connecting canal; drawing channel
引上线 leading down wire
引上线用管 lifting pipe
引上窑室 drawing chamber; drawing kiln; drawing pit; drawing tank
引射 injection
引射泵 jet pump
引射虹吸器 jet siphon
引射器 eductor; gas orifice; injector; jet exhauster
引射式风洞 induction tunnel
引射式空气增压机 ejector type air compressor
引射式喷燃器 nozzle mixing burner
引射式燃烧器 injection burner; injector burner; nozzle mixing burner
引射式热水提升器 injector hot-water lifter
引射式真空泵 vacuum ejector
引射推力增强器 bleed-off augmenter
引伸 explicate; prolong
引伸产品 augmented product
引伸冲断层 stretch fault; stretch thrust
引伸法 projective method; derivative method
引伸计 extensometer
引伸接头 extension piece
引伸压力机 cupping press; drawing press
引伸延期 prolong
引伸仪 extensometer; tens(i)ometer
引渗池 spreading basin
引渗井 absorbing well; leakage well
引示波 pilot; reference
引示灯 pilot lamp
引示电池 pilot cell
引示电缆 guide cable; leader cable
引示电缆套管 pilot cable duct
引示铃 pilot bell
引示射束 index beam
引示替续器 pilot relay
引示线 pilot wire
引示信号 pilot
引示制 pilot block system
引束孔道 beam hole; glory hole
引数 argument
引水 admission of water; diverting water; water diversion; water abstraction; pilotage【船】
引水坝 diversion barrage; diversion dam; diverting dam; intake dam
引水板 apron flashing
引水槽 approach flume; diversion flume; gutter; head race; headrace channel; headwater channel
引水船 pilot boat
引水道 channel of approach; conduit; diversion waterway; mill stream; penstock
引水(道)式电站 diversion canal type of power plant; diversion conduit type hydropower station
引水堤 diversion dike [dyke]; diverting dike [dyke]
引水点 heading; water point
引水渡槽 diversion aqueduct
引水阀 diversion valve
引水方法 method of diversion
引水费 pilotage charges
引水干渠 water main
引水工程 abstraction works; collecting works; diversion project; diversion works; headworks; intake works
引水沟 diversion ditch; feed(ing) ditch; field ridge
引水构筑物 diversion construction; diversion structure; diversion works; intake structure; intake works; water-diverting structure; water intake structure
引水管 aqueduct; water conduit
引水管道 conveyance conduit; pipe penstock
引水管灌溉 pipe irrigation
引水海图 pilot chart
引水航路 pilot fairway
引水户 abstractor
引水建筑物 diversion construction; diversion structure; diversion works; intake structure; intake works; water-diverting structure; water intake structure
引水角 diversion angle
引水结构 water-diverting structure; diversion structure
引水井 diversion chamber
引水口 diversion intake
引水量 abstraction volume; quantity of water intake
引水龙头 curb cock
引水率 diversion ratio
引水盲沟 diversion blind drain
引水锚地 pilot anchorage
引水喷头 water fountain
引水前土壤含水量 antecedent soil water
引水区域 pilotage ground
引水区域界线外附近海面 offing
引水渠 approach channel; channel-type head race; delivery canal; diversion canal; diversion channel; feed(er) canal; headrace channel; headwater channel; intake canal; rhine; channel of approach; mill stream; water race <工矿的>
引水渠槽 head race
引水渠道 head race
引水渠坑道 headrace gallery
引水渠式发电站 canal system power station
引水渠首 intake head(ing)
引水渠首工程 diversion headwork
引水入大田 carry water across field
引水设施 abstraction works; diversion works
引水渗沟 diversion blind drain(age)
引水式(发)电站 diversion power plant
引水式工程 diversion canal development
引水式水电站 diversion type (hydroelectric) power plant
引水式水力发电厂 river-run plant
引水枢纽 water-diversion project
引水隧道 diversion tunnel; headrace tunnel; intake tunnel; seepage tunnel; water-diversion tunnel; water gallery; water tunnel
引水隧洞 diversion tunnel; headrace tunnel; intake gallery; intake tunnel; seepage tunnel; water-diversion tunnel; water gallery
引水隧洞式电站 diversion tunnel power plant
引水填充 open hydraulic fill
引水土沟 earth feed ditch
引水线路 route of water transfer
引水需要量 diversion requirement
引水堰 diversion weir
引水业务 lode
引水员 pilot
引水闸 water intake sluice; diversion gate
引水闸灌溉率 head-gate duty of water
引水闸门 crown gate; head gate; intake gate
引水自流灌溉地 commanded land
引缩大陆海 attracted continental sea
引索 inhaul; straw line
引铁 <继电器的> armature iron
引铁间隙 armature gap
引头 leader
引头子 making-up; mudding up
引退 fall back
引文 quotation
引文索引 citation indexing
引物 primary matter
引下线 down lead
引下线阻抗 down lead impedance
引先限位开关 lead limit switch
引线 blaster fuse; feed-through; lead(er); leader line; lead(ing) wire; pigtail; R-wire; terminal; wire lead
引线保护装置 pilot protection
引线标志 approach sign; lead identification
引线补偿 lead wire compensation
引线插头 terminal pin
引线插头装置 pin-and-plug assembly
引线差动继电路 pilot differential relay
引线导板 lead bushing
引线导管 entrance bushing; lead bushing
引线点火器 fuse lighter
引线电感 lead in inductance
引线电缆 leader cable; leading-in cable
引线电容 lead capacitance
引线电阻 lead resistance
引线端子 end block; lead terminal
引线分配 pin assignment
引线封焊 lead seal
引线高架桥 approach viaduct
引线管 fairlead; fair leader
引线焊接 lead attachment; lead bonding; lead sealing
引线盒 drawing-in box; pull box <消防用>
引线环 feed-through collar
引线夹 lead clamp
引线夹头 wire clamp
引线接合 lead bonding
引线接合器 wire bonder
引线卷轴 wire spool
引线孔 fairlead; fair leader; pin-hole; pulling eye
引线孔板 lead bushing
引线框式键合 lead frame bond
引线拉力 lead strain
引线连接(法) lead attachment; lead connection; pin connection
引线路堤 approach embankment; approach fill
引线扭力 lead torque
引线疲劳试验 lead fatigue test
引线坡度 approach grade
引线强度试验 terminal strength test
引线识别 lead identification
引线隧道 approach tunnel
引线头 lead riser
引线图案 lead pattern
引线修齐 lead trimming
引线悬挂的片子 lead-suspended chip
引线曳放器 lead splitter
引线展延装置 lead spreader
引线阻抗 lead impedance
引向单元 director element
引向反射天线 director-reflector antenna; Yagi antenna
引向目标 vector
引向器 director; sender
引向振子 director dipole
引鞋 guide shoe; shoe guide
引信 blasting cap; blasting fuse; detonator; exploder; fuse(e); fuse mechanism; fuze(e); igniter fuse
引信保险隔板 interrupter
引信定秒器 fuel setting key
引信定时器 fuze setter
引信复雷管 fuse cap
引信固定环 setting ring
引信检验器 exploder tester
引信开始动作 fuse action; fuze action
引信头部 fuse head
引言 epigraph; introduction
引药线 lead in wire
引液泵 priming pump
引用 quotation; quoting; refer to reference
引用变量 reference to variable
引用标准 quoted standard; reference to standard
引用表 reference list
引用表达式 refer expression
引用表指令 table look-up instruction
引用补给半径 recharge radius
引用层 reference level
引用程序 reference program(me)
引用程序表 reference program(me) table
引用串 quoted string
引用的精确性 precision of reference
引用的强度 strength of reference
引用调用 call by reference
引用符号 quotation mark; reference to symbol
引用关键字 key of reference
引用过程 invocation of procedure
引用计数 reference count
引用计数器 reference counter
引用记录 reference record
引用结束 unquote
引用矩阵 citation matrix
引用库的程序设计 library reference programming
引用宽度 quoted width
引用流 reference stream
引用流量 substitute discharge
引用水头值 substitute water head value
引用私人 nepotism
引用索引 index of reference
引用文献 literature cited
引用性出现 use occurrence
引用影响半径 substitute influence radius
引用终止标志 ending quotation mark
引油环 lubricating ring
引诱 temptation
引诱剂 attractant
引诱利率 teaser rate of interest
引诱物 attractant
引语结束 <电报、听写等中用语> unquote
引原水的隧道 raw water tunnel
引炸药 initiating explosive; primary explosive
引张破裂 tension rupture
引张器 stretcher
引张线 floating wire; straining wire; stretched wire; tension wire; tension wire alignment
引张线法 method of tension wire alignment
引张线观测 tension wire alignment observation
引证 citation; quotation; quote

引证串 quoted string
引证符号 quote symbol
引证索引 citation index(ing)
引证映像 quote image
引证证据 adduce evidence
引纸辊 paper carrying roll
引种造林 reforestation with exotic species
引种植物 adventitious plant
引注器件 attention device
引砖 draw(ing) bar;submerged draw bar <用于无槽法>
引砖(产生的)气泡 drawbar bubble
引砖加强筋 drawbar rib
引砖刻面 drawbar contour;drawbar profile
引砖深度 depth of the draw bar;immersion of the draw bar
引走热量 heat elimination
引座员 usher
引座员室 ushers room

吲哚【化】 indol(e);benzazole

吲哚基醋酸 indoleacetic acid
吲哚试验 indole test
吲哚乙酸 indoleacetic acid
吲唑 benzdiazole;indazole

饮 drink(ing)

饮茶室 tea room
饮酒间 room for drinks
饮酒司机 drinking driver
饮酒中毒 acute alcoholism
饮疗 drinking-cure
饮料 beverage;drink;victuals
饮料杯 goblet
饮料储藏室 still room
饮料店 beverage store
饮料废水 drink wastewater
饮料工业废水 beverage industry wastewater
饮料间 beverage room
饮料可饮的 potable
饮料矿泉水标准 quality standard of mineral spring water for drinking
饮料配制器 beverage dispenser
饮料瓶 beverage bottle;carafe
饮料水 sweet water
饮料亭 drive-in restaurant
饮料纸杯 beverage cup;Dixie cup
饮泣墙 <耶路撒冷犹太人会堂的残壁> Wailing Wall
饮食 diet;fare;food and drink;repast
饮食标准 dietary standard
饮食店 canteen store;eating and drinking establishment;eating house;refreshment saloon
饮食店柜台 canteen counter
饮食店建筑 canteen building
饮食工业 food and beverage industry
饮食习惯 food habit
饮水 potable water
饮水杯 dispenser
饮水舱用无味防腐黑漆 tank black
饮水槽 <畜生的> water(ing) trough
饮水池 drinking water pool;watering-place
饮水处 watering-place
饮水传染 water-borne infection
饮水缸 drinking jar;scuttle butt
饮水供给 drinking water supply;potable water supply
饮水供区 drinking-place
饮水供应 potable water supply
饮水柜 drinking tank
饮水柜台 scuttle butt
饮水加氟 fluoridation of drinking water
饮水口 bubbler
饮水龙头 drinking fountain;scuttle butt;water spout
饮水喷头 drinking fountain;scuttle butt;water fountain
饮水器 drinker;drinking bowl;drinking cup;drinking fountain;drinking trough;waterbowl;water dispenser;waterer;watering device
饮水器阀门 water bowl paddle
饮水台 drinking fountain stand
饮水系统 potable water system
饮水中的氟化物 fluoride in drinking water
饮畜池 livestock pool;livestock reservoir
饮畜塘 earth tank
饮用玻璃杯 drinking glass
饮用的 drinking;potable
饮用喷泉 drinking fountain
饮用泉 drinking spring
饮用水 drinking water;feed water;potable water;sweet water;tap water
饮用水保护 protection of drinking water
饮用水泵 drinking water pump
饮用水标准 standard for potable water;standard of potable water;drinking water standard
饮用水不足 shortage of drinking water
饮用水槽 drinking water tank
饮用水出售器 drinking water vendor
饮用水除氟 drinking water defluoridation
饮用水处理 drinking water treatment
饮用水处理厂 drink(ing) water treatment plant
饮用水处理系统 drinking water treatment system
饮用水的处理 processing of drinking water
饮用水的回收 recovery of potable water
饮用水地下水源 underground source of drinking water
饮用水短缺 drinking water shortage
饮用水分区保护法令 Drinking Water Conservation Zoning Regulation
饮用水氟化 drinking water fluoridation
饮用水供给 potable water supply
饮用水供水 drinking water supply
饮用水供水管 drinking water supply pipe;potable water supply pipe
饮用水供水水源 drinking water supply source
饮用水供水水质 drinking water supply quality
饮用水供水系统 drinking water supply system
饮用水供应 drinking water supply
饮用水供应范围 potable water supply area
饮用水管理者 drinking water administrator
饮用水管网 drinking water network;potable water network
饮用水管线 drinking water pipeline
饮用水罐 drinking water tank
饮用水过滤器 drinking water filter
饮用水含污染物容许量 drinking water tolerance
饮用水基本标准 primary drinking water standard
饮用水基本规程 primary drinking water regulation
饮用水基准 drinking water criterion
饮用水加氟 drinking water fluoridation;fluoridation of drinking water
饮用水加氯消毒 chlorination of drinking water;chlorination of potable water
饮用水加热器 drinking water heater
饮用水监测 drinking water surveillance
饮用水监测程序 drinking water surveillance program(me)
饮用水检测 drinking water test
饮用水检测法 drinking water testing method
饮用水检查法 drinking water testing method
饮用水检验 test of drinking water
饮用水井 drinking water well;well for drinking
饮用水净化 purification of drinking water;drinking water purification
饮用水库 drinking water reservoir
饮用水冷却器 drinking water cooler;water dispenser
饮用水氯化处理 chlorination of drinking water
饮用水配水系统 drinking water distribution system
饮用水喷出器 drinking fountain
饮用水取水口 potable water intake
饮用水生物稳定性 drinking water biostability
饮用水水库 drinking water reservoir
饮用水水箱 drinking water tank;potable water tank
饮用水水源 drinking water source
饮用水水源区 drinking water source area
饮用水水源区污染防治管理条例 Regulations on Administration of Pollution Prevention and Control in Drinking Water Source Area
饮用水水质 drinking water quality
饮用水水质标准 drinking water quality standards;water quality standards for drinking water
饮用水水质恶化 degradation of drinking water quality
饮用水水质基准标准 drinking water criteria standard
饮用水水质监测 drinking water quality monitoring
饮用水水质检验法 water quality testing method
饮用水水质评估 drinking water quality assessment
饮用水水质容限 drinking water quality limit
饮用水水质浊度指标 turbidity index of drinking water quality
饮用水所需化学性质 required chemical characteristics of drinking water
饮用水桶 scuttle butt
饮用水外观 appearance of drinking water;appearance of potable water
饮用水微生物污染 drinking water microbial contamination
饮用水卫生标准 drinking water sanitary standard;sanitary standard for drinking water
饮用水问题 drinking water issue
饮用水污染物 drinking water contaminant
饮用水系统 potable water system
饮用水箱 drinking water tank
饮用水消毒 drinking water disinfection
饮用水消毒富产物 drinking water disinfection byproduct
饮用水需求量 drinking water demand
饮用水蓄水池 drinking water reservoir;drinking water storage tank;potable water reservoir
饮用水压力管 drinking water pressure pipe;potable water pressure pipe
饮用水源 drinking water source
饮用水制备 drinking water preparation
饮用水质量 drinking water quality
饮用水质量标准 drinking water quality standard
饮用水质量评价 drinking water quality assessment
饮用水中污染物容许量 drinking tolerance
饮用杂用双水系统 dual water supply system
饮用纸杯 Dixie cup

隐靶 vanishing target

隐孢子内壁体 cryptointosporinite
隐孢子外壁体 cryptoexosporinite
隐爆火山口 cryptoexplosion crater
隐爆角砾岩 underground explosion breccia
隐爆炸构造 cryptoexplosive structure
隐蔽 camouflage;concealment;privacy;secrecy;shading;shelter;stifle;wrap-up
隐蔽暗梁 back lintel
隐蔽爆炸装置 booby trap
隐蔽壁龛 blind niche
隐蔽不易发现的缺陷 latent defect
隐蔽不整合 blind unconformity
隐蔽布线 buried wiring;concealed wiring
隐蔽财产 hidden wealth
隐蔽层 blind layer;blind zone;hidden layer
隐蔽场所 sheltered site;sheltered situation
隐蔽沉箱结构 caisson perdu(e) system
隐蔽沉箱系统 caisson perdu(e) system
隐蔽成本 hidden cost
隐蔽处 blind;concealment;hide-away;mews;nook;refuge;shelter
隐蔽窗口 blind window
隐蔽的 blind;concealed;private;recessed;secret;subterranean;subterraneous;underground
隐蔽的导线 hidden conductor
隐蔽的道路 covered way
隐蔽的电缆 secret cable
隐蔽的对称性 concealed symmetry
隐蔽的煤气管道 concealed gas piping
隐蔽的配件 rough hardware
隐蔽的照相机 <侦察汽车用的> hidden camera
隐蔽灯光 dim-out
隐蔽地 shelter
隐蔽地区 enclosed ground
隐蔽电线 concealed wring
隐蔽钉子 concealed nailing
隐蔽丢失 <货物包装完整但实际已发现有丢失> concealed loss
隐蔽阀 hidden valve
隐蔽反应 masked reaction
隐蔽返航路线 return route concealment
隐蔽费用 hidden cost
隐蔽峰 occult peak
隐蔽缝合 cryptic suture
隐蔽扶壁 blind abut(ment)
隐蔽港湾 closed harbo(u)r;sheltered harbo(u)r

隐蔽阁楼楼梯 stair disappearing
隐蔽工厂 shadow factory
隐蔽工程 concealed works; embedded construction; embedded works; sheltered works
隐蔽工程记录 record of concealed works
隐蔽工程验收 acceptance of concealed works; acceptance of hidden subsurface works; hidden work acceptance; subsurface work acceptance
隐蔽工程验收记录 record of acceptance of concealed works
隐蔽工事 cover-up
隐蔽构造 closed construction
隐蔽故障 latent defect
隐蔽过梁 back lintel
隐蔽海岸线 sheltered coastline; sheltered shoreline
隐蔽海湾 closed bay; enclosed bay; sheltered arm of the sea
隐蔽基岩 concealed bedrock; sheltered bedrock
隐蔽剂 sequestering agent
隐蔽加热设施 concealed heating
隐蔽角落 nook
隐蔽铰链 invisible hinge
隐蔽接头 housing joint
隐蔽结构 closed construction; embedded construction
隐蔽结构墙 blind wall
隐蔽结构物 buried structure
隐蔽鸠尾榫 concealed dovetail
隐蔽决定簇 hidden determinant
隐蔽空间 concealed space
隐蔽空气温度 shade air temperature
隐蔽空隙 void in shotcrete
隐蔽矿物 occult mineral
隐蔽来楼梯 cut-string stair(case)
隐蔽梁 concealed girder
隐蔽楼梯 hidden stair(case); stair disappearing
隐蔽露头 concealed outcrop; hidden outcrop; suboutcrop
隐蔽锚地 blind anchorage; sheltered anchorage
隐蔽锚碇 blind anchor
隐蔽门 secrete door
隐蔽密丛 covert
隐蔽面 hidden face; underside
隐蔽木板支撑 closed sheeting
隐蔽期 eclipse phase
隐蔽墙 barrier shield
隐蔽桥台 blind abut(ment); secret abutment
隐蔽倾销 hidden dumping
隐蔽取暖器 recessed heater
隐蔽缺陷 hidden weakness
隐蔽色 cryptic colo(u)r(ation)
隐蔽色油漆 concealment paint
隐蔽设备 concealed installation; privacy device
隐蔽伸缩缝 enclosed expansion joint
隐蔽式安装 secret fixing
隐蔽式壁架 closed shelving; concealed shelving
隐蔽式道牙 flush curb
隐蔽式对流器 concealed convector
隐蔽式供暖 concealed heating; panel heating
隐蔽式供暖器 background heater
隐蔽式固定 secret fixing
隐蔽式固定装置 secret fixing
隐蔽式铰链 secret hinge
隐蔽式扣件 blind fastener; concealed fastener
隐蔽式拦沙障 concealed fencing
隐蔽式淋浴器 concealed shower
隐蔽式锚固 concealed fastening
隐蔽式暖气片 concealed radiator
隐蔽式桥台 dead abutment; secret abutment
隐蔽式散热片 concealed radiator
隐蔽式手纸架 concealed toilet paper holder
隐蔽式天沟 concealed gutter; secret gutter
隐蔽式浴槽阶梯 concealed scale
隐蔽式照明 built-in lighting
隐蔽式支座 concealed bearing seat
隐蔽式装配 concealed fixing; secret fixing
隐蔽事故 latent defect
隐蔽水域 sheltered area; sheltered waters
隐蔽损坏 concealed damage
隐蔽损伤 concealed defect
隐蔽通风式采暖炉 vented recessed heater
隐蔽围墙 ha-ha; haw-haw
隐蔽物 concealment; defilade
隐蔽系数 hiding factor
隐蔽显示 concealed display
隐蔽线 concealed line; hidden line
隐蔽像 blind image
隐蔽效应 masking effect
隐蔽选择操作 hide option
隐蔽因子 hiding factor
隐蔽照明 concealed lighting; cove lighting
隐蔽支座 blind abut(ment)
隐蔽轴线系统 concealed system of axes
隐蔽装置 hidden installation; secret installation
隐蔽准备 hidden reserve
隐蔽作用 hiding effect; hinding effect; sequestration; sheltering effect
隐壁床 <不用时折叠在壁橱内的> Murphy bed
隐避处 shelter
隐变量 hidden variable; implicit variable
隐变数 hidden variable
隐参数 implicit parameter
隐藏 concealment; disappearance; hiding; occultation
隐藏策略 hiding strategy
隐藏策略的值 value of a hiding strategy
隐藏成本 hidden cost
隐藏的 latent; non-salient; snug
隐藏的接缝 joint hiding
隐藏的霉菌 fungus subterraneous
隐藏的缺陷 concealed defect
隐藏灯 concealed lamp; recessed light
隐藏灯光 recessed light
隐藏钉 concealed nail(ing)
隐藏洞穴 pool wall hole
隐藏角 hidden corner
隐藏接缝 concealed joint
隐藏面 hidden surface
隐藏面算法 hidden surface algorithm
隐藏商誉 implied goodwill
隐藏式灯 built-in lamp
隐藏式环形天线 recessed loop aerial
隐藏式淋浴器 concealed shower
隐藏式散热器 built-in radiator; covered radiator
隐藏式阴极 recessed cathode
隐藏式照明装置 recessed fixture
隐藏所 shelter
隐藏线 hidden line
隐藏线算法 hidden line algorithm
隐藏线消除 hidden line elimination; hidden line removal
隐藏信息 information hiding
隐藏与寻找 hide and seek
隐藏在树林中 embower
隐藏照明 built-in lighting
隐成的 cryptogenic
隐(磁)极 non-salient pole
隐刺穿褶皱 blind piercement fold
隐带 zone of avoidance
隐灯 concealed lamp
隐灯信号 concealed-lamp sign
隐点霉属 <拉> Cryptostictics
隐雕 slightly-embossed carving
隐钉 edge toe nailing; secret nail
隐定义 implicit definition
隐定义阵列 implicitly defined array
隐鲕状 cryptoolitic
隐二次方程 implicit quadratic equation
隐二次函数 implicit quadratic function
隐方式 implicit mode
隐伏捕虏体 blind xenolith
隐伏冲断层 blind thrust
隐伏地裂缝 non-outcropping ground fissure
隐伏断层 blind fault; hidden fault
隐伏缝合线 cryptic suture line
隐伏矿床 blind deposit
隐伏矿体 concealed ore body
隐伏矿体异常 hidden ore body anomaly
隐伏露头 buried outcrop; suboutcrop
隐伏煤田 concealed coal field
隐伏热源 buried heat source
隐伏水系的 cryptor(h)eic
隐伏异常 hidden anomaly
隐腐熟腐殖质 crypto mull
隐拱 blind arch
隐构造 cryptostructure
隐谷 hock; hoek
隐轨手推车 floor insert trolley
隐过程 implicit procedure
隐过程参数 implicit procedure parameter
隐含 implication imply
隐含保证 implied warranty
隐含表 implication table
隐含操作 implication function
隐含成本 implicit cost
隐含程序 implication routine
隐含存储器 cache
隐含的契约 implied contract
隐含的条件 implied condition
隐含迭代法 implicit iterative method
隐含法 implicit method
隐含关系 implication relation
隐含函数 implicit function
隐含解 implicit solution
隐含可定义性 implicit definability
隐含枚举法 implicit enumeration
隐含门 conditional implication gate
隐含式 implication
隐含数 implicant
隐含同意 implied consent
隐含图 implication graph
隐含小数点 implied decimal point
隐含性 implicity
隐含寻址 implied addressing
隐含运算 if-then operation
隐含租金价值 implicit rental value
隐函时间差分法 implicit time-difference method
隐函数变换 implicit function generation
隐函数定理 implicit function theorem
隐函数法计算 implicit computation
隐函数法则 implicit function rule
隐函数反向差分法 implicit backward finite difference
隐函数方程 implicit function equation
隐函数解 implicit solution
隐弧焊 smothered arc welding
隐花湿生植物 cryptogamic hygrophyte
隐花植物 agamous plant; cryptogams
隐花植物纲 Cryptogamia
隐化池 <即殷霍夫池> Inhoff tank; digestion tank
隐患 concealed damage; hazardous condition; hidden danger; hidden peril; hidden trouble; internal hazard; latent danger; snag; hidden defect <堤防>
隐患处理 deflect treatment
隐火 hidden fire
隐火山构造 cryptovolcanic structure
隐火山型 cryptovolcanic type
隐或门电路 implied OR circuit
隐极机 non-salient pole machine
隐极式转子 distributed polar rotor
隐极同步发电机 non-salient pole alternator; non-salient pole synchronous generator
隐极转子 rotor with non-salient poles
隐钾锰矿 cryptomelane
隐胶质镜质体 cryptogelocilinite
隐接头 blind joint
隐节理 blind joint
隐结构 implicit structure
隐结构镜质体 cryptotelinite
隐晶 cryptocrystal
隐晶的 cryptocrystalline
隐晶花岗状 cryptogranitic
隐晶基斑状 aphanophyric
隐晶结构 aphanocrystalline texture; aphanitic texture
隐晶粒状 cryptograined
隐晶流纹岩 lithoidite
隐晶石墨 cryptocrystalline graphite
隐晶文象结构 cryptographic texture
隐晶岩 aphanite; cornean; cryotomere; cryptocrystalline rock; cryptomere [kryptomere]; cryptomerous rock
隐晶岩的 kryptomerous
隐晶荧光 cryptofluorescence
隐晶质 cryptocrystalline; extremely finely crystalline
隐晶质的 aphanitic; cryptomerous
隐晶质灰岩 aphanitic limestone
隐晶质胶结物结构 cryptocrystalline cement texture
隐晶质结构 aphanitic texture; cryptocrystalline texture
隐晶质石墨 cryptocrystalline graphite ore
隐晶质土状石墨矿床 aphanitic earthy graphite deposit
隐镜屑体 cryptovitrodetrinite
隐居 privacy
隐孔菌属 <拉> Cryptoporus
隐孔隙度 closed porosity
隐孔隙率 closed porosity
隐框玻璃幕墙 hidden framing glass curtain wall
隐篱 hah-hah fence
隐粒的 cryptograined
隐粒结构 cryptograined texture
隐磷铝石 bolivarite
隐流的 cryptor(h)eic
隐庐 hermitage
隐轮滑车 clew line block; secret block; strap-bored block
隐瞒的 underlying
隐瞒事实真相 concealment of fact; subreption
隐瞒性失业 disguised unemployment; hidden unemployment; latent unemployment
隐瞒资产 hidden assets
隐没的 dormant
隐没河 sinking creek
隐门 gib door; vanishing door <隐入墙内的拉门>; jib door <与墙齐平

的>
隐秘圈闭 subtle trap
隐密 privacy
隐密玻璃 privacy glass
隐名股东 sleeping partner
隐名合伙合同 contract of dormant partnership
隐名合伙人 dormant partner; silent partner; sleeping partner
隐名委托人 unnamed principal
隐匿 veil
隐匿处 hide-out; nook
隐匿剂 masking agent
隐匿面 buried surface
隐匿气泡 <塑料层内的> boil bubble
隐匿微孔 microvoid hiding
隐匿位 hidden bit
隐判定 implicit decision
隐棚 blind; hide
隐皮菌属 Cryptoderma
隐桥台 dead abutment; secret abutment
隐燃 smo(u)ldering
隐入河 drowned river; drowned stream; losing river; losing stream; lost river; lost stream; sunken river; sunken stream
隐入墙内的拉门 vanishing door
隐色还原染料 leuco vat dye
隐色亚甲基蓝 leucomethylene blue
隐色颜料 leucopigment
隐栅(场效应晶体)管 gridistor
隐伤 latent defect; rind gall
隐身材料 stealth material
隐生的 cryptogenic
隐生物岩 cryptobiolith
隐式 implicit expression
隐式板撑 closed sheathing
隐式编址 implied addressing
隐式辨识 implicit identification
隐式标识 implicit identification
隐式标识技术 implicit identification technology
隐式差分方程 implicit difference equation
隐式差分方法 implicit difference method
隐式差分格式 implicit difference scheme
隐式存储管理 implicit storage management
隐式打开 implicit opening
隐式地震 kryptoseismology
隐式地震的 kryptoseismic
隐式地址 implicit address
隐式迭代法 implicit iterative method
隐式法 implicit scheme
隐式方法 implicit method
隐式访问 implied addressing
隐式格式 implicit scheme
隐式公式 implicit formula
隐式估计法 implicit estimation technique
隐式积分 implicit integration
隐式积分公式 implicit integration formula
隐式计算 implicit computation
隐式计算机 hidden computer
隐式交替方向 alternative direction implicit
隐式接合 concealed fixing
隐式结合 implied association
隐式控制流 implicit control flow
隐式控制模式 implicit control pattern
隐式控制算子 implicit control operator
隐式类型 implicit type
隐式类型结合 implicit type association
隐式类型转换 implicit type conversion
隐式链 disappearing chain
隐式楼梯 disappearing stair(case)
隐式逻辑 hidden logic
隐式木板支撑 closed sheeting
隐式排水 buried drain
隐式平滑 implicit smoothing
隐式数据 hidden data
隐式数学模型 implicit mathematical model
隐式属性 implicit attribute
隐式说明 implicit declaration
隐式踏脚板 concealed running board; concealed safety step
隐式网格 network for the implicit method
隐式微分(法) implicit differentiation
隐式系统 implicit system
隐式线性约束 implied linear constraint
隐式限定 implicit qualification
隐式寻址 implied addressing
隐式引用 implicit reference
隐式有限差分 implicit finite difference
隐式域 hidden domain
隐式装置(法) concealed installation
隐式自校正器 implicit self-tuner
隐事 cover-up
隐输出和诱导输出 implied and induced output
隐数法 implicit enumeration
隐丝高温计 disappearing filament pyrometer
隐丝光学高温计 disappearing filament optic(al) pyrometer
隐丝式光测高温计 disappearing filament pyrometer
隐梯 disappearing stair (case); loft ladder
隐同步信号 implicit synchronizing signal
隐头喷泉 secret fountain
隐图编码卡片 cryptographically coded card
隐团块镜质体 cryptocorpocollinite
隐退 privacy; reclusion
隐微分法 implicit differentiation
隐微粒 cryptomere
隐微粒现象 cryptomerism
隐温矿床 cryptothermal deposit
隐纹长石 cryptoperthite
隐窝 crypt
隐显灯 occulting light
隐显墨水 invisible writing ink
隐显目标 bobtail target; disappearing target
隐显油墨 secret ink; sympathetic(al) ink
隐现 glint
隐现凹凸 photographing
隐现花斑 ghosting
隐线 invisible line
隐线绘图 hidden line plot
隐线浓淡处理 hidden line rendering
隐线算法 hidden line algorithm
隐像 latent image
隐斜视【物】heterophoria
隐屑的 cryptoclastic
隐屑结构 cryptoclastic texture
隐屑岩 cryptoclastic rock
隐形玻璃 invisible glass
隐形材料 stealth material
隐形的 cryptomorphic
隐形垫圈 <用于墙板的> concealed washer
隐形式 implicit form
隐形收入 invisible gain
隐性的 recessive
隐性经济 hidden economy
隐性品质 recessive trait
隐性特征 recessive trait
隐选择 default option
隐循环 implied circulation
隐循环表 circulation implied list
隐芽植物 cryptophytes
隐岩浆矿床 cryptomagmatic deposit; cryptomagmatic mineral deposit; kryptomagmatic deposits
隐与门电路 implied AND circuit
隐语 cryptology
隐域土 interzonal soil; intrazonal soil
隐域性的 intrazonal
隐域植物群落 intrazonal community
隐域植物群系 intrazonal formation
隐喻 metaphor
隐约看见陆地 looming of the land
隐置设备 concealed installation
隐轴 hidden axis; invisible axis

印

印澳界【地】Indo-Australian realm
印版 forme; plate; printing forms
印版版台 carriage
印版的图文部分 printing elements
印版滚筒磨损 cylinder wear
印版晃动 plate slap
印版接口 plate nip
印版空白部分 non-photographic(al) base
印版墨辊 forme rollers
印版石灰岩 lithographic(al) limestone
印版直径 printing diameter
印版钻孔机 block drill
印边机 selvedge stamping machine
印标志 franking
印成电阻器 printed resistor
印成线路 printed wiring
印冲带 obduction zone
印出 <以打印方式表示的计算机计算结果> printout
印出任务 printout task
印戳 overprint
印错 misprint
印第安红齿菌 Indian paint fungus
印第安花纹手织毛毯 Indian blanket
印第安纳 <美国州名> Indiana
印第安纳灯 <测定烟点用> Indiana lamp
印第安纳鲕状灰岩 Indiana limestone
印第安纳高级细布 <经防水处理可用作橡胶布> Indiana; Indiana cloth
印第安纳氧化法 Indiana oxidation test
印第安纳异戊烷法 Indiana isopentane process
印第安人建筑 American-Indian architecture; Amerindian architecture
印第安石 Indialite
印第安夏 Indian summer
印第安烟管 Indian pipe
印度、巴基斯坦-日本航线 India-Pakistan/Japan route
印度白檀 Indian sandalwood
印度板块 India plate
印度半岛陆核 Indian peninsula nucleus
印度标准学会 Indian Standards Institution
印度表柱 lat
印度铲 Indian shovel
印度赤铁树 illipe
印度船级社 Indian Register
印度次大陆 Indian subcontinent
印度大潮潮面 India spring water; India tide plan
印度大潮低潮面 Indian spring water; Indian tide plane
印度大陆与冈瓦纳大陆分离 India-Gondwana separation
印度大陆与欧亚大陆碰撞 India-Eurasia collision
印度大麻 hasheesh; India hemp
印度道路会议 Indian Road Congress
印度地盾 Indian shield
印度地台 Indian platform
印度-地中海双壳类地理大区 Indo-Mediterranean bivalve region
印度动力与河谷开发学报 <印度双月刊> Indian Journal of Power and River Valley Development
印度方尖庙 vimana
印度方尖寺院 vimana
印度佛教建筑 Indian Buddist architecture
印度古陆 pal(a)eo-India
印度合欢 ko(k)ko; siris
印度河 Indus River
印度河盆地 Indus river basin
印度红 Indian red; iron saffron
印度红木 Indian mahogany; Indian redwood
印度红颜料 Indian red pigment
印度红赭石 Indian red ochre
印度黄 Indian yellow; pioury; piuri; purree
印度黄檀(木) Indian rosewood; sissoo
印度混凝土杂志 <期刊> Indian Concrete Journal
印度吉纳树胶 Indian kino
印度纪念塔 minar
印度纪元 Indian era
印度季风【气】Indian monsoon
印度建筑 Hindoo architecture
印度胶 ghatti; ghatti gum
印度教庙宇 mandir; vihara
印度栲 Indian chestnut
印度栗木 Indian chestnut
印度楝 margosa
印度庙宇 Hindoo temple; Hindu temple
印度庙宇山门 gopura(m)
印度苜蓿 Indian alfalfa
印度苜蓿和红三叶草的混合料 mixture of Indian alfalfa and red clover
印度尼西亚强干南风 Selatan
印度尼西亚地槽 Indonesia geosyncline
印度尼西亚-日本航线 Indonesia/Japan route
印度榕 India fig (tree)
印度榕树 banian; banyan (tree); pagoda-tree
印度三角洲 Indian delta
印度桑 Indian mulberry
印度神龛塔 stupa
印度石 Indialite; Indian stone
印度式 Indian style
印度式堆石坝 Indian rock-fill dam
印度式堆石堰 loose rock Indian type weir
印度式建筑 Indian architecture
印度式建筑的瓶形顶部装饰 kalasa
印度式引水堰 Indian type diversion weir
印度树胶 Indian gum
印度寺庙前廊亭 mantapam
印度寺院的高塔 shikkara
印度松木 Indian cedar
印度塔 stupa; tope-mound
印度太平洋箭石地理大区【地】Indo-Pacific belemnite region
印度太平洋界 Indo-Pacific realm
印度太平洋三叶虫地理区系 Indo-Pacific trilobite realm
印度太平洋双壳类地理大区 Indo-Pacific bivalve region
印度桃花心木 Indian mahogany
印度乌木 Indian ebony
印度西太平洋有孔虫地理区系 Indo-

West Pacific foraminiferal realm
印度犀 Rhinoceros unicornis
印度橡胶 Indian caoutchouc tree;Indian rubber;Indian rubber fig
印度橡胶电缆 India-rubber cable;India-rubber wire
印度橡胶软管 India-rubber hose
印度橡胶树 Assam rubber; caoutchouc;Indian rubber tree
印度亚麻平布 Indian linen
印度亚区 Indian subregion
印度烟草 Nicotiana bigelovi
印度岩土工程学报 <印度季刊> Indian Geotechnical Journal
印度洋 Indian Ocean;the Indian Ocean
印度洋板块 Indian Ocean plate
印度洋-北美洲波系 Indian Oceanic-Canada crustal-wave system
印度洋赤道洋流 Indian Ocean equatorial current
印度洋大潮低潮面 Indian spring low water
印度洋地区 Indian Ocean area;Indian Ocean Region
印度洋海岭地震构造带 Indian ridge seismotectonic zone
印度洋航海警告 navigational warning for Indian Ocean
印度洋季风海流 Indian monsoon current
印度洋季节风 monsoon
印度洋麻 Indian okra
印度洋区域 Indian Ocean region
印度洋区域的覆盖范围 Indian Ocean coverage
印度洋热带性低压 Mauritius hurricane
印度洋卫星 Indian Ocean Satellite
印度洋夏季吃水线 Indian summer loadline
印度洋中脊 mid-Indian ocean ridge
印度洋中央海岭 mid-Indian ocean ridge
印度遥感卫星 Indian remote sensing satellite
印度银灰木 Indian silver greywood;white chuglum
印度硬木 chuglum
印度油石 <修表工磨雕刻刀用> India oil stone
印度柚木 bastard teak
印度玉米 Indian corn
印度原始建筑 Indian architecture
印度圆顶塔 tope
印度月桂树 Indian laurel
印度月桂油 Indian laurel oil
印度早期坟墓 dagobas
印度樟木 Indian camphor
印度赭红 Indian ochre
印度赭石 Indian ochre
印度纸 Indian paper
印度中央道路研究所 Central Road Research Institute
印度种植园 tope
印度紫红色 Indian purple
印杆 print bar
印格 grid printing
印工消耗 printer waste
印辊 cylinder
印果油 marking nut oil
印号锤 die hammer
印号机 die hammer
印号码 branding
印痕 dent;dint;ichnite;ichnolite;impression;imprint;marking;moulage;print
印痕测定器 impression block
印痕法 imprint method
印痕控制器 type impression control
印痕面积 area of indentation
印痕深度 <布氏硬度试验> depth of impression;depth of indentation
印痕硬度 indentation hardness
印花 calico-printing;printing;tax stamp
印花板 stencil
印花玻璃 configurated glass; painted glass
印花布 <美> calico [复 calico(e)s]
印花厂 printwork;printery <棉布的>
印花窗帘布 printed casement
印花大理岩 calico marble
印花地毯 printed carpet
印花粉 pounce
印花粉齿轮 pounce wheel
印花粉印图 pouncing
印花格子布 print gingham
印花辊 graining roll
印花机 decorating machine; printwork
印花间 printing room
印花胶合板 printed plywood
印花棉布 chintz
印花漆布 printed linoleum
印花墙纸 printed wallpaper
印花色浆 all over colo(u)r
印花税 documentary tax;stamp duty;stamp tax
印花税戳 documentary tax stamps
印花税法 stamp act;stamp tax act
印花税票 bill stamp;revenue stamp
印花税完税凭证 documentary tax stamps
印花税务局 stamp-office
印花瓦 printed picture tile
印花岩 calico rock
印花油地毯 stamped lino(leum)
印花油地毡 printed linoleum
印花油漆布 stamped lino(leum)
印花云斑 cloud print
印花纸坯 paper substrate
印花砖 printed picture tile
印花装饰 stenciled decoration; filigree <硬木或夹板木镶板上戳印装饰>
印花装饰布 cretonne
印迹记录 ink mist recording
印迹作用 imprinting
印记 brand(ing);chop;impression;imprint;marking;mintage;signet;stamp
印记冲模 letter punch;numbering die
印记收集器 impression packer
印加建筑 Inca architecture
印尖霍丝试验 Ingenhausz's experiment
印件分堆器 kicker
印鉴 broad seal;signature
印鉴样本 facsimile signature
印鉴证明书 letter of indication
印校样的人 proofer
印金 bronze printing
印金用清漆 bronze printing varnish
印金油墨 bronze printing ink
印块 <钻孔吊捞用> impression block
印流 bleeding
印码电报机 recorder
印码机油墨记录器 pen recorder
印码轮 inking wheel
印模 crooked drill pipe;die (plate);impression;moulage;stamp
印模材料 moulage
印模法 impression method
印模化石 mo(u)ld fossil
印模孔隙 mo(u)ldic porosity
印模盘 impression-cup;impression-tray
印模术 moulage
印木 eng
印尼船级社 Indonesian Register
印泥 ink pad;red ink paste used for seals
印片用胶片 print film
印票机 ticket printing machine
印茄属 ifil;ipil(e);Intsia <拉>
印染 print dyeing;printing and dyeing
印染厂 printing and dyeing mill
印染厂废水 dye-house wastewater
印染废水 dy(e)ing wastewater;printing and dyeing wastewater
印染工业 printing and dyeing industry;textile mill
印染工业废水 printing and dyeing industry wastewater
印染工业废水处理 wastewater treatment in textile-mills
印染机 dy(e)ing machine
印染浆 printing paste
印染染料 printing dye
印染设计 printed design
印染水污染 water pollution by dyeing
印染台 printing table
印染污泥 printing and dyeing sludge
印染助剂 printing and dyeing assistant
印色盒 pad
印商标 badging
印时器 <电话> time check
印试样 pull a proof
印数 impression; printing number;printing quota;run
印数说明 printing notes
印刷 press;print(ing)
印刷版 forme;impression;printed panel;printed plate;printing plate
印刷版边部插头 printed circuit broad connector
印刷版插头 edge connector
印刷版插座 printed circuit board jack
印刷版的分离通风 winded
印刷版电路 plated circuit; plated printed circuit
印刷版堵(油)量 blinding of lithographic(al) plates
印刷版面 printing surface
印刷本 printed book
印刷本主题索引 printed subject indexes
印刷标记 typographic mark
印刷标志 printed mark
印刷布线 printed wiring
印刷部件 print member
印刷材料 printing material
印刷厂 press;printing house;printing plant;printshop
印刷车间 pressroom;printing room
印刷衬底 printed substrate
印刷程序 print routine
印刷穿孔机 printing punch
印刷传输【计】 print through;print transfer
印刷次数 printed times
印刷错误 letter error
印刷大楼 press building
印刷单位 printed unit
印刷导线 printed circuit cable;printed conductor; printed electrical conductor
印刷的 typographic
印刷的风格 printed pattern
印刷的设计 printed pattern
印刷的图文部分 printing areas
印刷电感开关 printed inductance switch
印刷电感线圈 printed inductor
印刷电缆 printed cable
印刷电路 copper-surfaced circuit;electronic print;plated circuit;prefabricated circuit;printed circuit;printed electronic circuit;printed wiring;printing circuit
印刷电路板 etched circuit card;printed circuit board;printed circuit card;printed wiring plate
印刷电路板引出端 edge board contact
印刷电路板座 card socket
印刷电路产生器 printed circuit generator
印刷电路底板 printed chassis
印刷电路基板 tellite
印刷电路技术 printed circuit technique
印刷电路开关 printed circuit switch
印刷电路图形 circuit pattern
印刷电路学 printed circuitry
印刷电路装配 printed wiring assembly
印刷电路组件 printed circuit assembly
印刷电容器 printed capacitor
印刷电阻器 printed resistor
印刷读出器 printing reader
印刷反差比 print contrast ratio
印刷反差信号 print contrast signal
印刷方法 printing process
印刷复穿孔机 printing reperforator
印刷干燥器 print drier [dryer]
印刷杆 print bar
印刷杠子 gear mark
印刷格式 typography
印刷工 <英> machineman
印刷工人 pressman;printer;typographer
印刷工业 printing industry
印刷工艺学 printing technology
印刷构件 print member
印刷海流计 printing current meter
印刷换算装置 scaler-printer
印刷机 printer;printing machine;printing press;printing unit;typer
印刷机房 pressroom; printing machine room
印刷机构 printing mechanism
印刷机器 imprinter;printing mechanism
印刷机色带 printer tape
印刷机械 printing mechanism
印刷机压板 platen
印刷记录 hard copy;printer record
印刷技工 printing house craftsman
印刷间 pressroom;printing room
印刷间隔表格 printer spacing chart;spacing chart
印刷接触 printed contact
印刷控制符号 print control character
印刷量 number to be printed
印刷密度 printing density
印刷面 printing surface
印刷面光泽 ink gloss
印刷名 print name
印刷命令 print command
印刷模糊 slur
印刷模写纸 decal comania paper
印刷墨 printer's ink;printing ink
印刷母墨 base ink
印刷目录卡 printed card
印刷能力 press capacity
印刷品 black and white;literature;printed matter;printed output
印刷品配页设备 copy preparation
印刷品邮件 book post
印刷平滑度 printing smoothness
印刷漆 printer's varnish
印刷器 (im)printer
印刷器的轻质夹纸框 frisket
印刷清样 press proof
印刷日期 date of printing
印刷容量 printing capacity
印刷设备 printing equipment;printout equipment
印刷设计 printed design
印刷式电动机 printed motor

印刷式键控穿孔机 printing keypunch
印刷适性 printability; printing ability
印刷输出设备 printout equipment
印刷术 printing art
印刷数字 press figure; press number
印刷说明 printing notes
印刷速度 print speed printing speed
印刷所 printery; printing office; printing shop; printshop
印刷套色顺序 colo(u)r sequence
印刷体 block letter; print hand
印刷体字 lettering
印刷体字母 block letter
印刷天线 printed antenna
印刷筒 printing cylinder
印刷头 print head
印刷图 fair chart
印刷图案 printed pattern
印刷线路 printed wiring
印刷线路基板 printed wiring substitution
印刷用房屋 printing building
印刷用纸 printings
印刷油墨 printer's ink; printing ink; stamp-pad ink
印刷釉 screen printing glaze
印刷元件 printed parts
印刷原稿 printed original
印刷原墨 base ink
印刷原图 printed original map
印刷云斑 cloud print
印刷周期 print cycle
印刷装饰 printed decoration
印刷装置 typer
印刷字母 printed character
印刷字体 letter form; printing
印刷作业 presswork
印条 print bar
印铁 decorative printing; tin-printing
印铁法 iron printing process
印铁漆 tin-decorating finish
印铁清漆 overprinting varnish
印铁线 printing line
印铁像 iron printing process
印铁油墨 tinplate ink
印铁罩光漆 overprinting lacquer
印铁罩光涂层 overprinting coating
印透 print through
印图 impression
印图机 graphic(al) printer
印图样 pull a print
印图纸 atlas paper; enamel(led) paper; printing paper
印纹炻器 impressed stoneware
印纹陶 press marked pottery
印纹陶文化 Stamped Pottery Culture
印纹轧光机 embosser
印纹轧压机 gauffer calender
印纹纸 plating paper
印相 printing
印相机 contact printer; printer
印相密度 printing density
印相设备 printing apparatus
印相纸 photographic (al) paper; printout paper; printing-out paper
印象 neurogram
印象派 impressionism
印像速度 printing speed
印像图 image map
印压 coining; indentation; stamps
印压机 platen press
印牙齿用烧石膏 impression plaster
印样 <地图> advanced copy
印样绘制 proof plotting
印样控制 proof control
印油 cold oil; squeegee oil; stamp ink
印有浮雕图的器皿 sprigged ware
印在空白区的标记 overprinting
印在文件面上 enface
印轧 print pad

印张 printed sheet
印张容量 printed sheet capacity
印章 seal
印章管理 control of stamping
印章印文检验 verification of seals and stamps
印支地块 Indochina massif
印支古陆 pal(a)eo-Indo-China
印支期【地】 Indo-Chinese epoch; Indo-Sinian period
印支期地槽 Indochina geosyncline
印支旋回 Indo-Sinian cycle
印支亚旋回【地】 Indochina subcycle
印支运动【地】 Indo-Chian movement; Indochina orogeny; Indo-Sinian movement
印支造山运动 Indo-Sinian orogeny
印制 printing
印制板 printed plate
印制板插头 edge connector
印制标签的传热装置 heat-transfer labeling equipment
印制导线 printed conductor
印制导线衬底 printed wiring substrate
印制电路板插座 printed circuit (edge) connector
印制电路底板 printed circuit backplane
印制电路环氧树脂层 printed circuit epoxy-type coating
印制矩阵布线 printed matrix wiring
印制卡片格式 printed card format
印制设备 printed apparatus; printing apparatus; printout equipment
印制线 track
印制线路模板 production master
印制线圈 printed coil
印制元件 printed component; printed element
印字部 platen
印字带 ink ribbon
印字电报 printergram; printing telegraph
印字电报机 printing telegraph; telegraph printer; type printer; type printing machine
印字复凿孔机 printing reperforator
印字机 character printer; inker; ink writer; marking machine; printer; type printer

茚 indene

窨

窨井 back drop; detritus chamber; detritus chamber or pit; detritus pit; detritus tank; manhole; intercepting chamber <分开排水系统与阴沟>

窨井盖 gull(e)y cover; manhole cover
窨井检查盖板 access gulley
窨井进人孔 drop manhole
窨井井身 stub riser for the access hatch
窨井前存水弯 chamber interceptor
窨井圈 manhole ring
窨室 surface box

应

应按防爆设计 to be designed as explosion proof

应保险 insurable
应报废而勉强航行的船 crazy ship
应被审判的 judicable
应变 strain(ing); deformation
应变百分率 strain percentage
应变棒 strain bar
应变爆裂 strain burst
应变比法 <用于沉降计算中压缩层的确定> strain ratio method
应变标志 strain markers
应变标志测量数目 number of measurements
应变波 strain wave
应变波速 strain wave velocity
应变不变量 strain invariant
应变不连续性 strain discontinuity
应变部署 emergency arrangement
应变部署表 muster list; muster roll; muster station; station bill; station list
应变测定 extensometry; strain measurement
应变测定棒 strain bar
应变测定仪 strain-ga(u)ge logger
应变测力仪 strain-ga(u)ge dynamometer
应变测量 strain ga(u)ging; strain measurement
应变测量点位置 localities of strain measurement point
应变测量点至参考线距离 distance from locality of strain measurement to reference line
应变测量电桥 strain measuring bridge
应变测量方法 techniques of strain measurement
应变测量技术 strain-ga(u)ge technique; strain measurement technique
应变测量设备 strain measuring apparatus; strain measuring device
应变测量数据 data of strain measurement
应变测量仪 strain measuring instrument
应变测量装置 strain measuring device
应变测验仪 strain viewer
应变场 strain field
应变秤 extensometer balance
应变稠化 strain thickening
应变传递 strain transfer
应变传感器 strain pickup; strain transducer
应变淬火 strain quenching
应变大小 strain level; strain size
应变的 strained
应变的分解 resolution of strain
应变地震计 strain seismometer
应变地震图 strain seismogram
应变地震仪 strain seismograph; strain seismometer
应变点 strain point
应变电桥 extensometer bridge; strain bridge
应变电阻花 strain rosette
应变电阻丝 ga(u)ge line
应变动测 deformation dynamic inspection
应变二次曲面 strain quadric
应变二次式 strain quadric
应变发电站 reserve power station
应变范围 range of strain; strain range
应变方向 direction of strain
应变放大器 strain amplifier
应变分辨率 strain resolution; strain sensitivity
应变分布 strain distance; strain distribution
应变分解 resolution of strain; strain resolution
应变分离 strain release
应变分量 component of strain; strain component
应变分析 analysis of strain; strain analysis
应变峰值 peak strain
应变浮升 escalation
应变幅(度) strain amplitude; strain range
应变干涉仪 interferometer strain ga(u)ge
应变杆 telltale (rod); strain rod <桩工>
应变感应系数 influence factor of strain
应变刚度 strain rigidity
应变岗位 alarm post
应变公差 strain tolerance
应变功 deformation work; strain work
应变功率 power of straining
应变固体潮 strain tide
应变关税 contingent duty
应变观察器 strain viewer
应变管 strain tube
应变管理 contingency management
应变光性方程 strain-optic(al) equation
应变光学灵敏度 strain-optic(al) sensitivity
应变光学松弛 strain-optic(al) relaxation
应变光学系数 strain-optic(al) coefficient
应变规 strain sensor; strain transducer
应变轨迹 trajectory of strain
应变过程 strain path
应变和应力 strain and stress
应变盒 strain cell
应变痕 strain mark
应变花 strain-ga(u)ge rosette; strain rosette
应变滑动 strain-slip
应变滑动劈理 strain slip cleavage
应变滑动褶皱 strain slip folding
应变滑劈理 strain slip cleavage
应变环 strain hoop
应变恢复 strain recovery
应变恢复率 strain recovery ratio
应变回复曲线 strain recovery curve
应变回复特性 strain recovery characteristic
应变回弹 strain rebound
应变回跳 strain rebound
应变积累 accumulation of strain; strain accumulation; strain build-up
应变集中 strain concentration
应变集中系数 strain concentration factor
应变集中因数 strain concentration factor
应变集中因子 strain concentration factor
应变计 deformeter; extensometer; strain cell; strain ga(u)ge; strain-(ing) meter; strainometer; tautness meter
应变计传感器 strain-ga(u)ge transducer
应变计法 strain-ga(u)ge method
应变计荷载盒 strain-ga(u)ge load cell
应变计划 contingency plan
应变计玫瑰图 strain rosette
应变计扫描器 strain-ga(u)ge scanner
应变计式多点记录器 strain-ga(u)ge scanner recorder
应变计式加速度计 strain accelerometer
应变计式位移计 strain-ga(u)ge type displacement meter
应变计算 strain calculation
应变计压力盒 strain-ga(u)ge pressure cell
应变计针架 strain-ga(u)ge sting
应变记录 strain recording

应变记录器 strain-ga(u)ge logger
应变记录仪 recording strain ga(u)-ge; strain-recording equipment; strain-recording instrument
应变加速度 strain acceleration
应变架 straining frame
应变间断 strain discontinuity
应变检查仪 strain finder; strain viewer
应变检流计 strain galvanometer
应变检验仪 strain detector
应变阶 strain step
应变结晶 strain crystallization
应变截距 strain-offset
应变矩阵 strain matrix
应变开裂 strain crack(ing)
应变控制 strain control
应变控制荷载试验 strain-controlled load test
应变控制剪力仪 shear apparatus equipped with strain control; strain control shear apparatus
应变控制式直剪设备 strain controlled direct shear apparatus
应变控制式直(接)剪力仪 strain controlled direct shear apparatus
应变控制试验 controlled strain test; strain controlled test
应变老化 strain ag(e)ing
应变类型 strain type
应变理论 strain theory
应变力 strain force
应变力学效应法<路面设计的> method of mechanical effects of strain
应变历时曲线 strain-time curve
应变历史 strain history
应变量 dependent variable; magnitude of strain; strain capacity; strain magnitude
应变量测计 strain ga(u)ge
应变量值 strain quantity
应变裂缝 strain crack(ing); strain slip cleavage
应变裂纹 strain crack(ing)
应变灵敏度 strain sensibility; strain sensitivity
应变灵敏度系数 strain factor; strain sensibility factor
应变路径 strain path
应变率 rate of strain(ing); strain rate; strain ratio
应变率方程 strain rate equation
应变率敏感性 strain rate sensitivity
应变率矢量 strain rate vector
应变率张量 strain rate tensor
应变螺线管 extensometer solenoid
应变脉冲 strain-pulse
应变面 deformation plane; strain plane
应变敏感性 strain sensitivity
应变模式 strain pattern
应变莫尔圆 Mohr's circle of strain
应变能 strain energy; work of deformation
应变能存储 strain-energy storage
应变能法<计算超静定结构的> strain-energy method
应变能函数 strain-energy function
应变能力 adaptability to changing; strain capacity
应变能量 energy of deformation; energy of strain; power of straining; strain energy
应变能量系数 strain energy factor
应变能密度 strain-energy density
应变能释放率 strain-energy rate; strain-energy release rate
应变能振幅 strain-energy amplitude
应变扭转仪 strain-ga(u)ge torquemeter
应变盘 strain disk
应变疲劳 strain fatigue
应变片 foil ga(u)ge; strain foil; strain ga(u)ge
应变片传感器 strain-ga(u)ge transducer
应变片丛 rosette ga(u)ge; strain-ga(u)ge rosette; strain rosette
应变片电阻 ga(u)ge resistance
应变片花<为同时测得不同方向应变值而特制的一种组合的应变丝片> strain(-ga(u)ge) rosette
应变片技术 strain-ga(u)ge technique
应变片桥 strain-ga(u)ge bridge
应变片式测试仪 strain-ga(u)ge tester
应变片系数 ga(u)ge factor
应变片转矩计 strain-ga(u)ge torque; strain-ga(u)ge torquemeter
应变偏差 strain deviation
应变偏差器 strain deviator
应变偏量 deviator(ic) of strain; strain deviation; strain deviator
应变偏斜张量 strain deviator
应变偏移 strain deviation; strain shift
应变偏张量 deviatoric tensor of strain
应变破坏 strain failure
应变破裂 strain break
应变谱 strain spectrum
应变起伏法 strain relief method
应变器 effector; strainer
应变强度 strain intensity; strain strength
应变强度因子 strain intensity factor
应变强化 strain-hardening; strain-strengthening
应变桥加速度计 strain bridge accelerometer
应变区 strained zone; strain region
应变曲线 strain curve
应变蠕变 strain creep
应变软化 strain softening
应变软化材料 strain softening material
应变软化型 strain soften form
应变弱化 strain softening
应变三角形单元体 strain triangular element
应变伸长测定 measurement of elongation by strain
应变时间曲线 strain-time curve
应变时效 strain age; strain ag(e)ing
应变时效脆性 strain-age(d) brittleness; strain-age(d) embrittlement
应变时效脆性试验 strain-age-embrittlement test
应变时效的 strain-aged
应变时效钢 strain-aged steel
应变时效裂纹 strain-aged cracking
应变时效硬化 strain-aged hardening
应变拾音器 strain pickup
应变蚀刻法 strain etch method
应变矢量 strain vector
应变式波向仪 strain-ga(u)ge type wave direction meter
应变式测波仪 strain-ga(u)ge type wave meter
应变式传感器 strain sensor
应变式加速度计 strain accelerometer
应变式压力传感器 strain pressure transducer
应变式张力计 strain ga(u)ge tensiometer
应变式直剪仪 direct strain apparatus
应变势能 potential energy of strain
应变试验 strain-ga(u)ge test
应变释放 strain release; strain relief
应变释放法 strain relief method
应变释放方式 release types
应变释放加速度 strain-release acceleration
应变释放曲线 release curve of strain
应变数 dependent variable
应变双折射 strain birefringence
应变水平 strain level; stress level
应变松弛 strain relaxation
应变速度 rate of strain(ing); strain velocity
应变速率 rate of strain(ing); strain-(ing) rate
应变速率系数 strain rate factor
应变速率效应 effect of rate of strain
应变速率影响 strain rate effect
应变探测器 strain detector
应变梯度 strain gradient
应变条款 contingency clause
应变突击队 emergency squad
应变图 diagram of strains; flow figure; strain diagram; strain figure; strain sheet
应变图像 strain map; strain pattern
应变图形 strain figure; strain pattern
应变途径 strain path
应变推迟 strain retardation
应变退火 strain annealing
应变椭球(体) deformation ellipsoid; strain ellipsoid; ellipsoid of strain
应变椭球圆截面 circular section of strain ellipsoid
应变椭球圆锥面 elliptic(al) one of strain ellipsoid
应变椭圆 ellipse of strain; strain ellipse
应变微细裂纹 strain line
应变位移方程 strain-displacement equation
应变位移关系 strain-displacement relation
应变位移矩阵 strain-displacement matrix
应变稀化 strain thinning
应变系数 strain coefficient; ga(u)ge factor
应变系统 strain system
应变弦 strain wire
应变显示 strain indicating
应变线 strain line
应变相容性 strain compatibility
应变相(协)调性 compatibility of strain
应变向量 strain vector; vector of strain
应变橡胶 strained rubber
应变消除 strain relief
应变消失点 strain-release point
应变小组 emergency squad
应变协调 strain compatibility
应变协调方程 equations of compatibility of strains
应变协调因子 strain coordination factor
应变信号 strain signal
应变型传感器 strain-ga(u)ge type sensor
应变巡检箱 strain scanning unit
应变循环 strain cycling
应变岩石突出 strain burst
应变遥测技术 telemetry for strain measurement
应变仪 deformability meter; deformeter; extensometer; strain ga(u)ge; strain indicator; straining meter; strain instrument; strain measuring instrument; strain meter; strainometer; tens(i)ometer; tensotast
应变仪测力计 strain-ga(u)ge dynamometer
应变仪测量 strain-ga(u)ge measurement
应变仪传感器 strain-ga(u)ge pickup
应变仪读数 rosette ga(u)ge reading; strain-ga(u)ge reading
应变仪法 strain-ga(u)ge method
应变仪放大器 strain-ga(u)ge amplifier
应变仪负荷传感器 strain-ga(u)ge load cell
应变仪记录器 strain-ga(u)ge recorder
应变仪设备 strain-ga(u)ge installation; strain-ga(u)ge instrumentation
应变仪式加速度仪计 strain-ga(u)ge accelerometer
应变仪式压力表 strain-ga(u)ge pressure indicator
应变仪系统 strain-ga(u)ge system
应变仪型传感器 strain-ga(u)ge type pickup
应变仪叶片 strain-ga(u)ge rosette
应变仪支架 strain-ga(u)ge sting
应变仪指示器 strain-ga(u)ge indicator
应变阴影 strain shadow
应变应力关系 strain-stress relation
应变应力环线 strain-stress loop
应变应力曲线 strain-stress curve
应变应力图解 strain-stress diagram
应变影 strain shadow
应变影响系数 strain influence factor
应变影响因数 strain influence factor
应变硬度 strain-hardness
应变硬化 strain-hardening
应变硬化材料 strain-hardening material
应变硬化的 strain-hardened
应变硬化钢 strain-hardening steel
应变硬化规律 strain-hardening law
应变硬化速率 rate of strain hardening
应变硬化系数 coefficient of strain-hardening
应变硬化效应 strain-hardening effect
应变硬化性 strain hardenability
应变硬化指数 strain-hardening exponent
应变诱导的分子取向 strain-induced molecular orientation
应变诱导断裂 strain-induced fracture
应变诱导开裂 strain-induced cracking
应变诱导析出 strain-induced precipitation
应变诱发的各向异性 strain-induced anisotropy
应变余能 complementary strain energy; strain complementary energy
应变阈值 strain threshold
应变再结晶作用 strain recrystallization
应变增量 strain increment
应变张力计 strain ga(u)ge tensiometer
应变张量 strain tensor
应变值 strain value
应变指示计 strain indicator
应变指示漆 strain indicating lacquer
应变指示器 strain indicating instrument; strain indicator
应变指示仪 strain indicating instrument; strain indicator
应变指数 strain exponent
应变中心 strain center [centre]
应变轴 axis of strain; strain axis
应变主轴 principal axis of strain; strain axis
应变转向装置 strain deviator
应变状态 state of strain; strained condition; strain regime; strain state
应变状态和性质 state and property of strain

应变准静态分量 quasi-static component of strain
应变组合片 rosette
应拆除的居住区 clearance area
应拆除的住宅区 clearance area
应出勤工日数 calculated man-days in attendance
应出勤人日数 calculated man-days in attendance
应当到达时间 due time of arrival
应到未到的列车数 trains due
应得报酬的投资 investment worthy of compensation
应得部分 quotient
应得的权利 entitlement
应得价款的请求 claim for proceeds
应得价款索取 claim for proceeds
应得收益率 required rate of return
应电性 electrotropism
应电作用 electrotaxis
应舵时间 delay of swing-response
应舵时间间隔 answer interval
应发未发电量 expected energy not served
应罚款的 fin(e)able;pecuniary;picuniary
应罚款违法行为 pecuniary offence
应付 cope with;due to;manipulation
应付承兑汇票 acceptance payable
应付承兑票据 acceptance payable
应付出现意外灾害时的计划 contingency plan
应付代收外局款 foreign bureau incurred charge payable
应付贷款 loan payable
应付的 callable;payable
应付电费 liability for light and power
应付费的 chargeable
应付费用 expense payable
应付费用项目 accrued items of expenses
应付分期账款 installed account payable
应付福利费审计 welfare payable audit
应付工薪税 liability for payroll tax
应付工资 accrual wage;accrued payroll;wage payable
应付工资审计 wage payable audit
应付股息 dividend payable
应付顾主款项 account due to custom
应付关税 liable to customs duty
应付关税的 dutiable
应付国际联运清算款 international account settlement payable
应付海关费用 customs due
应付红利 bonus payable
应付汇价 giving quotation
应付汇票及电汇 draft and telegraphic(al) transfer payable
应付货款 trade accounts payable
应付货款周转期 trade payable turnover period
应付寄销账款 due to consignor
应付奖金 bonus payable
应付借款 loans payable
应付金额 amount payable
应付开支 accrued expenditures
应付款 accrue payable;amount due;due;money due
应付款明细分类账 account payable subsidiary ledger
应付款日期 due date
应付款项 accrued payable
应付利润审计 profit payable audit
应付利息 imputed interest;interest in red;interest payable
应付利息准备 interest payable reserve
应付贸易货款周转期 trade payable turnover period
应付贸易账款 trade accounts payable; trade accounts receivable; trade payable
应付票据 bill payable;note payable
应付票据审计 bill payable audit
应付票据贴现 discount on note payable
应付凭单 voucher payable
应付凭单登记簿 voucher payable register;voucher register
应付凭单账户 voucher payable account
应付凭单制 voucher system
应付期间 payable period
应付期票 note payable
应付期票账 notes payable account
应付其他基金款 due to other funds
应付上级单位账款 accounts receivable from superior units
应付收付款凭单 warrant payable
应付手续费 commission payable
应付水运运杂费 sundry charges for water transport payable
应付税 liable to duty;tax payable
应付税捐 tax payable
应付税款账户 tax payable account
应付所得税 income tax payable
应付外汇汇价 paying quotation
应付外债 external debt-service obligation
应付未付费用 expense accrued;outstanding expenses
应付未付票据 bills payable
应付未付账 account rendered
应付未付账款 outstanding account payable
应付息券 coupon's payable
应付下级单位账款 accounts receivable from subordinate units
应付现金凭证 document payable in cash
应付薪金 salaries payable
应付押款 mortgage payment
应付养恤金 benefit payable
应付运营款 operating account payable
应付债款 debt payable;debt payment
应付债款账 debt payable account
应付债券 bonds payable
应付债券审计 bonds payable audit
应付债券息票 coupon payable
应付债券折价 discount on bonds payable
应付账款 accounts payable; payable account
应付账款登记簿 accounts payable register
应付账款分类账 accounts payable ledger
应付账款款项 inter-fund balance payable
应付账款明细账 account payable ledger
应付账款内部控制制度审计 internal control system audit of accounts payable
应付账款清单 statement from creditor
应付账款审计 accounts payable audit
应付账款增多 accounts payable stretching
应付账款账户 accounts payable account
应付账款真实性和合法性审计 accounts payable authenticity and legality audit
应付账款周转率 turnover of payment;turnover of ratio of accounts payable
应付账明细表 schedule of accounts payable (receivable)
应付账目 account payable
应付租金 rent payable
应负责任的 responsible
应该做的 competent
应改正的错误 corrigendum
应归成本 imputed cost
应归利息 imputed interest
应归属折价的收益 gains contribution to depreciation
应核销投资性费用审计 audit of investment expenses that should be cancelled after verification
应激子 stresser
应计不动产税(金) accrued real estate taxes
应计成本 should-be cost
应计的 accrual;accrued <资产或负债,收入或支出>
应计抵押利息 accrued interest on mortgage
应计费用 accrued charges; accrued expenses
应计负债 accrued liability
应计工资 accrued wages
应计股利 accrued dividend
应计股息 accrued dividend
应计股息账 accrued dividend account
应计奖金 accrued bonus
应计净租金 imputed rent
应计开支 accrued expenditures
应计开支项目 accrued items of expenses
应计累计股利 accrued cumulative dividends
应计利润 accrued profit
应计利息 accrual of interest;accrued interest;interest accrued
应计年假 accrued annual leave
应计收入 accrued income
应计收益 accrued income; accrued revenue
应计税金 accrued taxes;tax accrued
应计税捐 accrued taxes
应计税款 accrued taxes;tax accrual
应计所得税款 accrued tax on income
应计投资利息 accrued interest on investment
应计未付不动产税 accrued real
应计未付费用 accrued expense payable;accrued expenses
应计未付工资 accrued payroll
应计未付借款利息 accrued interest on loan
应计未付期票利息 accrued interest on notes payable
应计未付售出债券利息 accrued interest on bond sold
应计未付税金 accrued taxes payable
应计未付税捐 accrued taxes
应计未付所得税款 accrued income tax
应计未付项目 accrued payable
应计未付薪金 accrued salaries payable
应计未付债券利息 accrued interest on bonds
应计未付租金 accrued rent payable
应计未收股利 accrued cumulative dividends
应计未收款项 accrued receivable
应计未收利息 accrued interest receivable
应计未收收益 accrued income receivable
应计未收佣金 accrued commission receivable
应计未收债券利息 accrued bond interest receivable
应计未收租金 accrued rent receivable
应计项目 accrual;accrued item
应计薪金 accrued salaries
应计养恤金 pensionable remuneration
应计营业收入 accrued revenue
应计佣金 accrued commission
应计增值税款 accrued tax on value added
应计债券利息 accrued bond interest
应计债务 accrued liability
应计账户 accrued account
应计折旧 accrued depreciation
应计折旧的 depreciable
应计折旧的固定资产总值 total value of fixed assets to be depreciated
应计折旧资产 depreciable assets
应计折旧总额 total depreciation accrued
应计支出制 accrued-expenditures basis
应计资产 accrued assets
应计租金 accrued rent(al)
应减去的 subtractive
应交会费 compulsory contribution
应交纳货物税的 excisable
应交纳执照税的 excisable
应交税金 tax payable
应交税金审计 tax payable audit
应交税所得 taxable income
应交债券 bond payable
应缴税金 accrued taxes;taxes due;taxes payable
应缴税物料 dutiable store
应缴税物品 dutiable article
应缴铁路建设基金 railway construction funds payable
应缴运输进款 transport income payable
应接板 adapter [adaptor]
应接管 adapter [adaptor]
应接室 general public room
应解汇款 drafts and telegraphic transfers payable
应具报的 notifiable
应具备的能力和技艺 reasonable care and skill
应课税的 chargeable
应课税收入净额 net chargeable income
应扣缴预提税收入 income subject to withholding
应拉木 <一种阔叶材缺陷> tension wood
应力 internal force;stress
应力安全范围 safe range of stress
应力安全系数 stress factor of safety
应力斑 stress pattern
应力斑痕 stress mark
应力板 stress(ed) plate
应力包络线 stress envelope
应力包络线图 stress-endurance diagram
应力包线 stress envelope
应力饱和 stress repletion
应力比法 <用于沉降计算中压缩层的确定> stress ratio method
应力比(率) stress ratio
应力边界条件 stress boundary condition
应力变动 stress fluctuation
应力变幅 range of stress; stress amplitude
应力变化 alternation of stress; stress change; stress variation; variation of stresses
应力变化范围 range of stress
应力变化历程 stress history
应力变换 stress alternation
应力变向 stress reversal

应力变向速度 speed of stress alternation
应力变形 stress deformation
应力变形关系 stress-displacement relation
应力变形特性 stress-deformation characteristic
应力变形图 stress-deformation diagram
应力标记 stress mark
应力表 stress sheet
应力表层 monocoque
应力并矢式 stress dyadic
应力并向量 stress dyadic
应力波 stress wave
应力波传播 stress wave propagation
应力波动 stress fluctuation
应力波发射 stress wave emission
应力波方程法 stress wave equation method
应力波理论 stress wave theory
应力波速度 stress wave velocity
应力波形 stress wave
应力补偿 balancing of stresses; stress compensation
应力不变量 stress invariant
应力不连续 stress discontinuity
应力不足 distress; understress(ing)
应力不足的 understressed
应力不足的混凝土 understressed concrete
应力部位 stress area
应力测量 stress measurement
应力测量技术 stress measuring technique
应力测量装置 stress measuring apparatus; stress measuring device; stress measuring unit
应力测试 stress measurement
应力测试技术 stress measuring technique
应力测试装置 stress measuring apparatus; stress measuring device; stress measuring unit
应力测验仪 strain viewer
应力层 stress layer
应力层积木桥面 stress-laminated wood deck
应力差 stress deviation; stress difference
应力差线 stress-difference line
应力产生 stress generation
应力产生在一起 stress together
应力铲试验 stress shovel test; total pressure cell test
应力场 stress field; stress pattern
应力(场)强度系数 stress intensity factor
应力超限 over-stressing
应力超限的 over-stressed
应力超限的岩体 over-stressed rock mass
应力超越 stress overshoot
应力程序 stress program(me)
应力弛豫 relaxation of stress; stress relaxation
应力重分布 redistribution of stress; stress redistribution
应力重分布区 stress redistribution zone
应力重分布系数 coefficient of redistribution of stress; coefficient of stress redistribution
应力重复 stress repetition
应力重复循环 repeated stress cycle
应力重取向 stress reorientation
应力初始状态 initial state of stress
应力传播 stress propagation
应力传播波浪理论 wave theory of the propagation of stress
应力传播能力 stress dispersion capacity
应力传递 propagation of stress; stress at transfer; stress transfer; stress transmission
应力传递的波浪理论 wave theory of the propagation of stress
应力传递系数 stress transmission coefficient
应力传感器 force transducer
应力传送路线 stress transmission path
应力次数曲线 <疲劳试验中的> S-N diagram; stress-number curve
应力脆性涂料法 stress brittle coating
应力大小 magnitude of stresses; stress intensity
应力带 stress zone
应力单位 unit of stress
应力单元 stress element
应力导 stress guide
应力倒向 inversion of stress
应力的边界条件 boundary state (of stress)
应力的法向分量 normal component of stresses
应力的非均质状态 non-homogeneous state of stress
应力的含混性 ambiguity of stresses
应力的极限范围 limiting range of stress
应力的膜态 membrane state of stress
应力的确定 stress determination
应力的下降 decline of stress
应力的线性分布 linear stress distribution
应力的线性状态 linear state of stress
应力的消除 backing-off
应力的行径 path of stressing forces
应力的主平面 principal plane of stress
应力等级 stress grade; stress level
应力等值线 stress contour; stress isopleth
应力点 stress point
应力电流模拟 stress-current analogy
应力电压模拟 stress-voltage analogy
应力叠加 superimposed stress; superposition of stress
应力叠加原理 principle of superimposed stresses
应力冻结 <光弹试验> stress freezing
应力冻结法 stress-freezing method; stress frozen method
应力冻结烘箱 stress-freezing oven
应力冻结效应 stress-freezing effect
应力断裂 stress crack(ing)
应力断裂曲线 stress rupture curve
应力多边形 S-polygon
应力二次曲面 stress quadric
应力二次式 stress quadric
应力法 stress method
应力反复 stress repetition; stress reversal
应力反复图 <材料疲劳试验的> stress-endurance diagram
应力反复周期 cycle of stress reversal; repeated stress cycle
应力反射系数 stress reflection coefficient
应力反向 reversal of stress; stress reversal
应力反应 reversal of stress
应力范围 stress range
应力方程 stress equation
应力方向线 stress-director line
应力放散 stress-relief <焊接长钢轨>; de-stressing <无缝线路>
应力放松钢线 stress-relieved wire
应力分布 distribution of stress; force condition; stress distance; stress distribution
应力分布近似法 stress distribution approximation
应力分布能力 stress distribution property
应力分布图 stress distribution diagram; stress envelope
应力分布型式 stress pattern
应力分布性质 stress distribution property
应力分担 <指钢筋与混凝土在钢筋混凝土中分担应力> stress division
应力分级 stress grading
应力分级法 stress system
应力分级木材 stress-graded lumber; stress-graded timber
应力分解 resolution of stress
应力分量 stress component
应力分量不变式 invariant of stress
应力分配 stress division
应力分析 stress analysis; structural analysis
应力峰(值) peak stress; stress maximum; stress peak
应力幅度 amplitude of stress; stress amplitude; range of stress; stress range <用于疲劳强度>
应力腐蚀 stress corrosion; stress etching
应力腐蚀断裂 stress corrosion fracture
应力腐蚀界限强度因子 stress erosion limiting intensity factor
应力腐蚀开裂 stress corrosion cracking
应力腐蚀裂缝 season cracking; stress corrosion cracking
应力腐蚀裂纹 stress corrosion cracking
应力腐蚀裂纹试验 stress corrosion cracking test
应力腐蚀破坏 stress corrosion failure
应力腐蚀破裂 stress corrosion cracking
应力腐蚀试验 stress corrosion test
应力感应图 influence diagram of stress
应力钢丝 stress wire
应力各向异性 stress anisotropy
应力工作台 bed stressing
应力功 stress work
应力估计 assumed stress approach; stress estimation
应力贯入曲线 stress-penetration curve
应力光图 stress-optic(al) pattern
应力光学常数 stress-optic(al) constant
应力光学定律 stress-optic(al) law
应力规律 stress pattern
应力轨迹 stress trajectory; trajectory of stresses
应力轨迹图 stress pattern
应力轨迹线 stress path
应力轨线 <平行或垂直于主应力方向的轨线> stress trajectory
应力过程 stress path
应力过大的 over-stressed
应力函数 stress function
应力函数单元 stress function element
应力合成 stress resultant
应力合力 stress resultant
应力和应变场 stress and strain fields
应力环 proving ring; stress ring
应力(环)架 proving frame
应力缓冲 relaxation of stress; stress relaxation
应力缓冲薄膜 stress alleviating membrane
应力缓冲薄膜中间层 stress alleviating membrane interlayer
应力缓冲器 stress buffer
应力换向循环 reversed cycle of stress
应力恢复 stress restoration
应力恢复法 stress recovery method; stress restoration method
应力恢复法原位应力测试 test for in-situ stress recovery method
应力回火 stress tempering
应力汇总表 stress summary sheet
应力活化流动 stress-activated flow
应力活化模型 stress activated model
应力迹线 trajectory of stresses
应力积聚 accumulation of stress; stress accumulation
应力积累 accumulation of stress; stress accumulation
应力积蓄 accumulation of stress; stress accumulation
应力及腐蚀引起的开裂 stress corrosion cracking
应力级 stress level
应力级位 stress level
应力极限 endurance range; limiting range of stress; limit of stress; stress limit
应力集中 accumulation of stress; bulb pressure; concentration of stress; stress accumulation; stress concentration
应力集中部位 stress concentration area
应力集中处 <如刻槽等> stress raiser
应力集中点 concentration point of stress; focal point of stress; stress concentration point
应力集中区 stress concentration area; stress raiser
应力集中区范围 location of stress concentration zone
应力集中系数 coefficient of stress concentration; factor of stress concentration; fatigue-notch factor; stress concentration factor
应力集中效应 notch effect; stress concentration effect
应力集中因数 stress concentration factor; stress raiser
应力集中因素 stress concentration factor
应力集中因子 fatigue strength reduction factor; stress concentration factor
应力集中源 stress raiser
应力计 stress detector; stress ga(u)ge; stress meter; stressometer; taseometer
应力计数曲线 stress-number curve
应力计算 stress calculation
应力计算表 stress sheet
应力记录仪 stress recorder
应力加厚 stress thickening
应力加热 stress heating
应力间断 stress discontinuity
应力减除 stress-relief
应力减除穴 stress-relieving cavity
应力减低 stress reduction
应力减少 stress reduction
应力减少系数 stress reduction factor
应力减小 drop of stress; stress decrease
应力剪胀理论 stress shear dilatancy theory
应力检测系统 stress measuring system
应力检流计 stress galvanometer
应力检验计算 stress check calculation
应力检验片 stress section
应力交变 reversal of stress; stress alternation; stress reversal

应力交变速度 speed of stress alternation
应力交变速率 speed of stress alternation
应力交变周期 cycle of stress reversal
应力交替 alternation of stress;stress alternation
应力交替变化速度 speed of stress alternation
应力校核 stress check
应力阶段 stage of stress
应力阶跃 stress step
应力结晶 stress crystallization
应力解除 stress release;stress-relief;stress-relieving
应力解除爆破 relieving shot;stress-relief blast
应力解除槽 <岩石试验的> stress-release channel
应力解除处理 relief treatment of stress
应力解除的高抗拉强度钢丝 stress-relieved high-tensile-strength wire
应力解除法 stress-relief method
应力解除法原位应力测试 test for in-situ stress-relief method
应力解除炉 stress-relieving furnace
应力解除切槽 stress-relief slot
应力解除切口 stress releasing cut
应力解除热处理 stress-relief heat treatment; stress-relieving heat treatment
应力解除岩芯 stress-relief core
应力解除钻进 overcore
应力解除钻孔 stress releasing bore-hole
应力筋 stress reinforcement;stress wire
应力近似分布 stress approximation distribution
应力矩阵 stress matrix
应力聚合物 stressed polymer
应力聚集 stress build up
应力均匀化 adequation of stress;stress equalizing
应力开裂 stress crack(ing)
应力空间 stress space
应力空心墙板 spancrete panel
应力控制 control of stress; stress control
应力控制的 stress-controlled
应力控制荷载试验 controlled-stress load test;stress-controlled load test
应力控制剪力仪 stress control shear apparatus
应力控制器 stress controller
应力控制式直(接)剪力仪 stress-controlled direct shear apparatus
应力控制试验 controlled-stress test
应力控制张拉 prestressing under stress control
应力矿物 stress-mineral
应力扩散 stress dispersion
应力扩散角 angle of stress dispersion
应力扩散能力 stress dispersion capacity
应力累积 accumulation of stress;stress accumulation
应力累积故障 repeated stress failure
应力理论 stress theory
应力力矩 stress moment
应力历程 stress history
应力(历)史 stress history
应力历史与归一化土工参数 <用于土体沉降计算的> stress history and normalized soil engineering parameter
应力历史与归一化土工参数法 Shansep approach
应力立体单元 stress solid element
应力利用系数 factor of stress utilization
应力裂缝 stress crack(ing)
应力裂纹 stress crack(ing)
应力灵敏 stress sensitive
应力流 stream of stress;stress flow
应力流线 flow of stresses
应力路径 stress path
应力路径沉降分析 stress-path settlement analysis
应力路径法 stress path method
应力路径模型 stress path model
应力路线 stress path
应力率 stress rate
应力脉冲 stress impulse
应力脉动 stress fluctuation
应力密度 level of stress
应力密度系数 stress-inducing factor;stress intensity factor
应力密度因子 stress intensity factor
应力面 stress plane
应力面积 stress area
应力模量 stress modulus
应力模拟 stress analogy;stress modelling
应力模式 stress pattern
应力模型 stress model
应力木 reaction wood
应力挠度曲线图 stress deflection diagram
应力能 stress energy
应力逆转 reversal of stress
应力偶 stress couple
应力泡 bulb of pressure;stress bulk
应力疲劳 stress fatigue
应力片集流器 electric(al) resistance wire current collector
应力偏差 stress deviation
应力偏量 stress deviator
应力偏张量 deviatoric tensor of stress
应力平衡 balancing of stresses;stress balancing;stress equilibrium
应力平衡状态 equilibrium of stress state
应力平面 stress plane
应力评价 stress evaluation
应力破断 stress rupture
应力破坏试验 stress rupture test
应力破坏系数 stress damage coefficient
应力破裂 stress rupture
应力破裂强度 stress rupture strength
应力剖面图 stress profile
应力谱 stress spectrum
应力奇点 stress singularity
应力起伏 stress fluctuation
应力强度 intensity of stress;stress intensity;stress strength
应力强度比 stress strength ratio
应力强度范围 stress intensity range
应力强度系数 stress-inducing factor;factor of stress intensity
应力强度因子 factor of stress intensity;stress intensity factor
应力强度因子范围 stress intensity factor range
应力强度值 stress intensity value
应力强化因子 stress intensity factor
应力切变膨胀理论 stress dilatancy theory
应力侵蚀 stress corrosion
应力侵蚀开裂 stress corrosion cracking
应力情况 stress case;stress pattern
应力区 stress block; stressed zone;stressful region
应力区间 stress interval
应力屈服限 proportional limit
应力曲线 stress curve
应力全部释放技术 <岩石力学的> complete stress relief technique
应力全解除技术 <岩石力学的> complete stress relief technique
应力全息干涉测量 stress holo-interferometry
应力热光系数 stress thermo-optic(al) coefficient
应力软化 stress-softening
应力软化效应 stress-softening effect
应力三角形 stress triangle
应力散失 stress dissipation
应力散失区 stress-free zone
应力上升 stress raising
应力上升处 stress raiser
应力伸长比 stress-elongation ratio
应力伸长曲线 stress-elongation curve
应力升高值 lift-off stress value
应力剩余状态 residual state of stresses
应力施加速率 rate of stress application
应力时间叠加作用 stress time superposition
应力时效 stress ag(e)ing
应力史应变史关系式 stress-strain-history relation
应力矢量 stress vector
应力式直剪仪 direct shear apparatus
应力势能 potential stress energy
应力试验 stress test
应力释放 relaxation; stress release; release of stress; relief of stress; stress-relief;stress-relieving
应力释放初始应变 stress-free initial strain
应力释放带 de-stressed zone; distressed zone
应力释放区 de-stressed zone
应力释放岩芯 stress-relief core
应力受控流变仪 controlled-stress rheometer
应力树 utility tree
应力衰减 stress decay
应力双折射 stress birefringence;stress-induced birefringence
应力水平 stress level;level of stress
应力松弛 relaxation of stress;stress decay; stress relaxation; stress slackening
应力松弛反应 stress relaxation response
应力松弛计 relaxometer;stress relaxometer
应力松弛试验机 stress relaxometer
应力松弛仪 relaxometer
应力速度 stress rate
应力损失 loss of stress;stress loss
应力探测计 stress detector
应力特性 stress behavio(u)r;stress response
应力梯度 gradient of stress; stress gradient;stress riser
应力梯级 stress riser
应力体系 stress(ed) system
应力条纹 stress cord
应力调节 stress adjustment
应力调整 stress adjuster; stress adjustment;stress rearrangement
应力调质 stress tempering
应力跳动 stress jump
应力凸缘 tension flange
应力图 diagram of stresses;stress diagram; stress map; stress sheet; stress trajectory
应力图示法 diagrammatic representation of stresses
应力图形 stress map;stress pattern
应力图样 stress pattern
应力涂层 stress coat
应力涂料 <测定应力和应变使用的一种脆性涂漆,不同标号的涂漆在相应的应力下脆裂> stress coat
应力涂料法 stress coating method
应力途径 stress path
应力途径法 stress path method
应力退火 permanent stress annealing
应力椭球(面) ellipsoid of stress;stress ellipsoid
应力椭圆 ellipse of stress;stress ellipse
应力椭圆面 ellipsoid of stress
应力椭圆体 ellipsoid of stress;stress ellipsoid
应力外皮结构 stressed-skin construction
应力位移矢量 stress-displacement vector
应力温度特性 stress-temperature characteristic
应力稳定性 stress stability
应力问题 stress problem
应力吸收薄膜 <常用沥青橡胶材料> stress absorbing membrane
应力吸收薄膜中间层 <常用沥青橡胶材料> stress absorbing membrane interlayer
应力吸收能力 stress absorbability
应力系数 stress factor
应力系统 stress system
应力下降 stress decline;stress drop
应力下降因数 stress reduction factor
应力线 stress line
应力线图 sheet;stress sheet
应力限程 range of stress
应力限度 limit of stress
应力相关模量 stress-dependent modulus [复 moduli]
应力相似 stress analog(ue);stress similitude
应力响应 stress response
应力向量 stress vector
应力消除 equalization; stress releasing; stress-relief; stress-relieving; stress removal
应力消除层 <为减少对应裂缝而设的> stress-relieving interlayer
应力消除加热 stress-relief heating
应力消除裂纹 stress-relief crack(ing)
应力消除热处理 stress-relief treatment;stress-relieving
应力消除试验 stress-relief test
应力消减 <设计细梁和长柱时降低容许应力的系数> stress reduction
应力消减层 <路面> stress-relief course;stress-relieving interlayer
应力消失 stress-relieved
应力效应 stress effect
应力形变 stress deformation
应力形变曲线 stress-deformation curve
应力形状 stress behavio(u)r
应力型式 stress pattern
应力锈蚀 stress corrosion
应力锈蚀裂缝 stress corrosion cracking
应力循环 stress alternation;stress circulation;stress cycle;stress cycling
应力循环次数图 stress-number diagram of stress [S-N diagram of stress]
应力循环脉动系数 stress amplitude ratio
应力循环图 stress-cycle diagram
应力岩体破坏方式 failure mode of stress rock mass
应力验算 stress check
应力依从性质 stress-dependent behavio(u)r
应力依存性 stress-dependence
应力依赖性 stress-dependence
应力仪 stress detector;stress ga(u)ge;stress meter;stressometer

Y

应力移除 stress removal
应力银纹 stress crazing
应力引起的各向异性 stress-induced anisotropy
应力应变 stress-strain
应力应变比 stress-strain ratio; stress to strain ratio
应力应变测量 stress-strain measurement
应力应变成正比假说 straight-line theory
应力应变方程 stress-strain equation
应力应变关系 stress-strain relation(ship)
应力应变关系曲线 stress-strain relation curve
应力应变函数 stress-strain function
应力应变回线 stress-strain loop
应力应变矩阵 stress-strain matrix
应力应变模量 stress-strain modulus
应力应变曲线 stress-strain curve; stress-strain diagram
应力应变曲线的塑性阶段 plastic stage
应力应变曲线回环 loop of strain stress curve
应力应变曲线上超过屈服点的平直部分 ductility plateau
应力应变时间反应 stress-strain time response
应力应变时间关系 stress-strain time relation
应力应变时间性状 stress-strain time behavio(u)r
应力应变试验 stress-strain tester
应力应变试样 stress-strain sample
应力应变特性 stress-strain characteristic; stress-strain property
应力应变体积变化 stress-strain-volume change
应力应变图 stress-strain graph; load-deformation curve; stress-deformation diagram; stress-strain curve; stress-strain diagram; stress-strain plot
应力应变图解 stress-strain diagram
应力应变温度空间 stress-strain-temperature space
应力应变相容分析 stress-strain compatibility
应力应变性能 stress-strain behavio(u)r
应力应变性质 stress-strain property
应力应变性状 stress-strain behavio(u)r
应力应变仪 stress-strain ga(u)ge
应力应变圆 stress-strain circle
应力应变直线比例法 straight-line method of stress-strain
应力应变状态 stress-strain behavio(u)r; stress-strain condition; stress-strain state
应力影响 stress effect
应力影响函数 stress influence function
应力影响模量 stress-dependent modulus [复 moduli]
应力影响系数 stress influence factor
应力硬化 stress hardening
应力诱导 stress-induce
应力诱导的各向异性 stress-induced anisotropy
应力诱导取向 stress-induced orientation
应力诱导生长 stress-induced growth
应力诱导系数 <路面板的> intensity factor; stress-inducing factor
应力诱导相变 stress-induced phase transformation
应力诱导有序位错固定 stress-induced order locking of dislocation
应力诱发结晶 stress-induced crystallization
应力与变形关系特性 stress-deformation characteristic
应力与强度比 stress to strength ratio
应力与温度关系特性 stress-temperature characteristic
应力与应变 stress and strain
应力阈值 stress threshold
应力圆 circle of stress; stress circle
应力再分布 stress redistribution
应力再分配 stress redistribution
应力增长 stress increase
应力增大 build-up of stress
应力增加 stress increase
应力增宽 stress broadening
应力增量 stress increment
应力张量 stress tensor; tensor of stress
应力张量计 stress tensor ga(u)ge
应力张量主值 principal value of stress tensor
应力张量主轴 principal axis of stress tensor
应力折减系数 stress reduction factor
应力折旧的固定资产 depreciable fixed assets
应力(振)幅 <最大应力与最小应力之差> stress amplitude
应力之下的回弹力 resilience under stress
应力直线变化 linear progression of stresses
应力值 magnitude of stresses; stress value
应力指示涂层 stress indicating coating
应力指示线 stress-director line
应力指数 stress exponent
应力指数法 stress index method
应力中断期间 stress interval
应力中断器 stress-breaker
应力中心 center [centre] of stress
应力周期 cycle of stress; stress cycle
应力周期变化 cyclic(al) variation of stresses
应力轴方位 orientation of stress axes
应力轴(线) stress axis; axis of stress
应力主平面 principal plane of stress
应力主轴 principal axis of stress; principal stress axis
应力转换 inversion of stress
应力桩 tension pile
应力状态 states of stress; stress condition; stress regime; stress state; stress system
应力状态变量 stress state variable
应力状态分析 stressed state analysis
应力状态与变形 state of stress and deformation
应力自记仪 stress recorder
应力自由度 stress degree of freedom
应力综合 stress resultant
应力组合 combination of stresses
应力最大值 stress maximum
应列贷方科目 accounts to be credited
应列借方科目 accounts to be debited
应履行债务日期 date on which the claim becomes due
应募公司债簿 subscription book
应募股票人 applicant for shares
应募入伍 enlistment
应纳税 taxability
应纳税财产 ratable property
应纳税单位 taxable unit
应纳税的 taxable
应纳税的纯入息 net chargeable income
应纳税的房地产 ratable estate
应纳税的入息 chargeable income
应纳税货品 dutiable goods
应纳税货物 dutiable goods
应纳税金 tax liability
应纳税进口商品报单 entry for dutiable goods
应纳税商品 tax article
应纳税资产净值 net taxable assets
应评税的收入 assessable income
应升值资产 appreciable assets
应驶距离 distance to go
应氏航海表 Inman's Nautical Tables
应收承兑汇票 acceptance receivable
应收承兑票据 acceptance receivable
应收筹资账款 accounts receivable financing
应收贷款 loans receivable
应收到期账款 accounts due
应收抵押 mortgage receivable
应收分期账款 instalment account receivable
应收股东票据 notes receivable from stockholder
应收股利 dividends receivable
应收股息 dividend receivable
应收汇价 receiving quotation
应收货款 trade accounts receivable
应收及预付款项的审计 receivable and advanced payment audit
应收寄销人款 due from consignor
应收军运后付运费 military transport charges receivable
应收款 due accounts
应收款让售 factoring of receivables
应收款项 due from; receivable
应收款项账户 accrued debit account
应收款周转率 receivable turn over
应收款作担保的借款 loan secured by account receivable
应收利息 interest in block; interest receivable
应收票据 bill receivable; note receivable; receivable
应收票据审计 bill receivable audit
应收票据贴现 notes receivable discounted
应收票据与承兑汇票的贴现 discounting notes and acceptance receivable
应收票据账户 bill receivable account
应收期票 note receivable
应收期票账 notes receivable account
应收取劳务费 chargeable labo(u)r
应收认缴股款 capital stock subscriptions receivable; due from subscribers
应收日期 accrual date
应收入金额 amount receivable
应收收益 accrued revenue
应收手续费 commission receivable
应收水运运杂费 sundry charges for water transport receivable
应收税款 tax receivable
应收外局代收款 foreign bureau incurred charge receivable
应收未收客户款 collection receivable for customer
应收未收款项 accrued assets
应收未收票据 bills receivable
应收未收资产 accrued assets
应收下级款 receivable from subordinary unit
应收银行到期存款 deposit(e) due from bank's account
应收银行款 due from banks
应收应付会计制 accrual accounting
应收应付制 accounting on the accrual basis; accrual basis
应收邮运运费 postal transport charge receivable
应收逾期账款 overdue account receivable
应收与实收的现金 cash vs. accrual
应收运营款 operating account receivable
应收债券利息 bond interest receivable
应收账净额 net account receivable
应收账款 accounts receivable; debt receivable; outstanding account; receipt account; receivable account
应收账款保险 account receivable insurance
应收账款筹资 account receivable financing
应收账款的分期 ag(e)ing accounts receivable
应收账款的挪后补前 leading and lading
应收账款登记簿 account receivable register
应收账款分类账 account receivable ledger; debtor's ledger
应收账款坏账处理 bad debt audit
应收账款净额 net account receivable
应收账款控制 account receivable control
应收账款明细账 account receivable ledger
应收账款内部控制制度审计 account receivable internal control system audit
应收账款平均收款期 average collection period of receivables
应收账款平均收现日数 number of days' sales in receivables
应收账款平均收账期 average collection period of receivables
应收账款平均账龄 average age of receivables
应收账款审计 audit of account receivable
应收账款贴现 account receivable discounted; discounted of account receivable; discounting of account receivable
应收账款贴现佣金支出 commission paid on discounted accounts
应收账款折现佣金支出 commission paid on discounted account receivable
应收账款周转率 turnover ratio of receivable account
应收账明细表 schedule of accounts payable (receivable)
应收账目 outstanding account; receipted account
应收政府定额补贴 quoted government subsidies receivable
应收滞纳税款 tax receivable delinquent
应收转账款项 inter-fund balance receivable
应受惩处的犯法行为 punishable offence
应受……处分 on penalty of; under penalty of
应受法院审理的争端 justiciable dispute
应受谴责的 obnoxious
应税基数 taxable base
应税金额 taxable amount
应税金额的扣除 tax deduction
应税利润 taxable profit
应税年度 taxable year
应缩应力 stress under compression
应提偿债基金 debt service fund requirements
应提折旧 accrued depreciation
应显示旗信号而不显示 short flagging
应信信号 countersign

Y

应需线路 on-call circuit
应压木 compression wood
应用 application;utility
应用安全性 safety of use
应用标准 application standard
应用冰川学 applied glaciology
应用不当 misapplication
应用参考手册 application reference manual
应用操作例行程序 application function routine
应用层 application layer
应用层模型 models of application layer
应用衬里 applied linings
应用成本比例计算期末存货 applying cost ratio to ending inventory
应用程序 application routine; utility program(me); utility routine
应用程序包 application package; utility package
应用程序标识(符) application program(me) identification
应用程序出口例行程序 application program(me) exit routine
应用程序的构造程序 application builder
应用程序管理程序 application program(me) manager
应用程序接口 application program(me) interface; application programming interface
应用程序均衡装入 application load balancing
应用程序控制代码 application control code
应用程序库 application library
应用程序设计 application programming
应用程序设计员 application programmer
应用程序输出范围 application program(me) output limit
应用程序映象 application program(me) image
应用程序员 application programmer; utility programmer
应用程序站 closed shop
应用程序主节点 application program(me) major node
应用程序装入表 application load list
应用程序子例程集 application program(me) stub
应用处理操作 application processing function
应用到应用的服务程序 application-to-application service
应用的 practical
应用的出现 applied occurrence
应用的详细规范 specification for applications
应用的详细说明 specification for applications
应用的详细细则 specification for applications
应用地层学【地】applied stratigraphy
应用地点 point of use
应用地理学 applied geography
应用地貌图 applied geomorphologic(al) map
应用地球物理学 applied geophysics
应用地热学 applied geothermal
应用地震学 applied seismology
应用地质学 applied geology; practical geology
应用定义记录 application definition record
应用范围 application area; application field; applied range; area of application; field of application; range of application; range of use; scope of application
应用方法 application method
应用方式 application mode
应用分析 applied analysis
应用服务 application services
应用服务程序 application program(me)
应用服务器【计】application server
应用辐射学 radiology
应用辐射学的 radiological
应用负荷 applied load
应用负载均衡 application load balancing
应用概率 applied probability
应用概率论 applied probability theory
应用功能例行程序 application function routine
应用构造地球化学 applied tectonogeochemistry
应用管理程序 application management program(me)
应用光谱学 applied spectroscopy
应用光学 applied optics
应用过程 application process
应用过程管理应用过程 application management application process
应用过程组 application process group
应用海洋学 applied oceanography
应用函数例行程序 application function routine
应用号 application number
应用化学 applied chemistry; practical chemistry
应用环境地球化学 applied environmental geochemistry
应用绘图程序 application plot program(me)
应用基 as fires basis
应用基低位发热量分级 as fired basis met calorific value graduation
应用级协议 application level protocol
应用集成模块 application integrated module
应用计量学 applied metrology
应用计算机 appliance computer
应用计算机科学 applied computer science
应用技术卫星 application technology satellite
应用价格 applied cost
应用监督程序 application monitor
应用件 application ware
应用建立程序 application builder
应用键 utility key
应用接口模块 application interface module
应用进程 application process
应用经济学 applied economics
应用静力学 applied statics
应用开发工具 application development tools
应用开发系统 application development system
应用开发周期 application development cycle
应用科学 applied science
应用科学实验室 applied science laboratory
应用空气动力学 applied aerodynamics
应用控制表 application control table
应用控制代码 application control code
应用控制块 application control block
应用控制语言 application control language
应用控制语言指令 application control language instruction
应用会计 applied accounting
应用快速制动 rapid brake application
应用矿物学 applied mineralogy
应用力学 applied mechanics
应用力学评论 <期刊> Applied Mechanics Reviews
应用力学实验室 applied mechanics laboratory
应用力学学报 <季刊> Journal of Applied Mechanics
应用力学杂志 Journal of Applied Mechanics
应用例行程序 application routine
应用领域 application area
应用流体动力学 applied fluid dynamics
应用流体力学 applied fluid mechanics
应用率 application rate; utility ratio
应用美术 applied art
应用目标的形态 application image
应用喷射技术 spray application technique
应用气候学 applied climatology
应用气象学 applied meteorology
应用权 application right
应用热力学 applied thermodynamics
应用人类学 applied anthropology
应用日渐广泛 ever-growing use
应用容量 practical capacity
应用软件 application-oriented software; application software
应用软件包 application software package
应用软件工程 application software engineering
应用软件语言 application software language
应用生态系统分析 applied ecosystem analysis
应用生态学 applied ecology
应用实例 examples of application
应用实体地址 application entity address
应用手册 application manual; application workbook
应用树 utility tree
应用数据 field data
应用数据库 application data base; applied database
应用数据研究 applied data research
应用数学 applied mathematics
应用数学家 applied mathematician
应用水分 applied moisture
应用水力学 applied hydraulics
应用水文学 applied hydrology
应用弹性力学 applied elasticity
应用调试(程序) utility debug
应用调整 application tuning
应用统计理论【数】applied theory of statistics
应用统计学【数】applied statistics
应用图 application drawing; utility graph
应用图像 application image
应用图学 applied graphics
应用土壤生物学 applied soil biology
应用土壤学 edaphology
应用网关【计】application gateway
应用微生物生态学 applied microbial ecology
应用微生物学 applied microbiology
应用温度范围 temperature limit
应用文件 application file
应用问题 rider
应用物理学 applied physics
应用系数 application factor
应用系统 utility system
应用系统分析 applied systems analysis
应用协议数据单元 application protocol data unit
应用心理学 applied psychology
应用新技术 use of the new technology
应用信息服务 application information services
应用信息服务程序 application information services
应用信息管理系统 applied information management system
应用研究 application study; applied research; exploratory development
应用要求 applicable requirement
应用要求语言 application-required language
应用艺术学院 college of applied art
应用艺术展览 exhibition of applied art
应用因素 application factor
应用因子 application factor
应用于不规则的砖砌墙面 hatching and grinning
应用语言 applicative language
应用预测 applied forecasting
应用源程序 application source program(me)
应用云物理学 applied cloud physics
应用造船学 applied naval architecture
应用造型 applied mo(u)lding
应用支持 application support
应用支援 application support
应用支援程序包 application support package
应用制造研究 applied manufacturing research
应用终端 application terminal
应用注释 application note
应用装入表 application load list
应用资源图像开发系统 applied resource image exploitation system
应用子系统 application subsystem
应用组名 application group name
应有的奖励 due reward
应有注意 due care; due diligence
应折旧财产 depreciable property
应折旧成本 depreciable cost
应折旧价值 depreciable cost
应折旧年限 depreciable life
应折旧资产 depreciable assets
应争保险开关 intervention switch
应征税的 dutiable; ratable; rateable
应征税的货物 dutiable goods
应征税价格 dutiable price
应征税价值 dutiable value
应征税商品 dutiable article
应支付的 due to be paid

英

英安斑岩 dacite porphyry
英安流纹岩 dacite liparite; dellenite
英安凝灰岩 dacite tuff
英安岩 dacite; quartz andesite
英安岩类 andesite group
英安质凝灰熔岩 dacitic tuff lava
英安质凝灰岩 dacitic tuff
英安质熔结集块岩 dacitic welded agglomerate
英安质熔结角砾岩 dacitic welded breccia
英安质熔结凝灰岩 dacitic welded tuff
英版潮汐表 Admiralty tide tables
英版灯标雾号表 Admiralty List of Lights and Fog Signals
英版海图 admiralty chart; British Admiralty Chart
英版海图零点 Admiralty Chart Datum
英版海图图夹节本 Abridged Admiralty Chart Folio

英版海图图夹索引图 List and Index of Admiralty Chart Folio
英版航海通告 Admiralty Notices to Mariners
英版无线电航标表 Admiralty List of Radio Signals
英镑 <1 英镑 = 453.603 克> English pound;pound (sterling)
英镑承兑票据 dollar acceptance;sterling acceptance
英镑汇率 sterling exchange rate
英镑集团 pound bloc;sterling bloc
英镑结存 sterling balance
英镑区 sterling area
英镑远期价格 sterling forward price
英镑证券 sterling bonds
英币 sterling
英币常数 sterling constant
英币图像 sterling picture
英才教育 meritocracy
英池式水箱 Intze tank
英尺 foot
英尺磅 <英制功的单位> foot-pound
英尺磅秒 <英国单位制> foot-pound-second
英尺磅秒单位制 foot-pound-second system
英尺磅达 foot-poundal
英尺朗伯 <亮度单位每平方英尺的面积上辐射或反射一个流明的亮度> foot-Lambert
英尺秒 foot-second
英尺标准集装箱 twenty equivalent unit
英尺测链 Gunter's chain
英尺程 foot run
英尺/分 feet per minute
英尺换算单位 <集装箱> forty-foot equivalent unit
英尺换算箱 <集装箱> forty-foot equivalent unit
英尺计量 foot measure
英尺米双面标尺 foot-meter rod
英尺/秒 feet per second
英尺数 footage
英尺/天 feet per day
英尺/小时 feet per hour
英尺压差 feet head
英尺烛光 <照度单位> foot-candle
英尺烛光计 foot-candle meter
英寸磅 inch-pound [in-lb]
英寸 <1 英寸 = 2.54 厘米> English fathom;inch
英寸板 inch board;inch plank
英寸材 inch stuff
英寸点数 dots per inch
英寸吨 tons per inch
英寸吨曲线 curve of tons per inch immersion
英寸汞柱 inches of mercury
英寸强度 inch strength
英寸水 <水头压力单位,水深单位> inch of water
英寸水柱 inches of water
英寸以内的长度 fraction-of-an-inch fraction
英寸英尺消防水费 inch-foot charge for water service
英达洛依焊料 indalloy
英担 centum weight;hundred weight <1 英担 = 1/20 吨,英国为 112 磅,美国为 100 磅>;quintal <质量单位,英制 1 公担 = 50.8029 千克,美制 1 公担 = 45.3597 千克>
英淡歪细晶岩 quartz-bostonite
英吨 <英制质量单位,1 英吨 = 1101.6 千克> gross ton;British ton;English ton;long ton
英分 <1 英分 = 1.295 克> scruple
英戈尔德刀具 Ingold cutter
英格兰砌筑 Saxon masonry work
英格兰山字型构造体系 England Epsilon structural system
英格索尔光泽测定仪 Ingersoll glarimeter
英格索尔(偏光)光泽计 Ingersoll gloss meter
英国纽林海平面基准点 Newlyn datum
英国白 English white
英国摆式抗滑试验仪 British pendulum
英国半木式建筑 English half-timbered architecture
英国半透明瓷 English translucent china
英国保险协会 British Insurance Association
英国标准 British Standard
英国标准磅 British imperial pound
英国标准尺度 British standard scale
英国标准粗牙螺纹 British standard coarse thread
英国标准单位 British association unit;British Standard unit
英国标准感光度 British Standard speed
英国标准管道 British Standard Pipe
英国标准管螺纹 British Standard Pipe Thread
英国标准管用螺纹 British Standard Pipe Thread
英国标准规范 British standard specification
英国标准规格(书) British standard specification
英国标准(规格)协会 British Standard Institution
英国标准惠氏螺纹 British Standard Whitworth Thread
英国标准技术规范 British standard specification
英国标准加仑 <1 加仑 = 4.546 升> imperial gallon
英国标准局 British Standard Board
英国标准开业守则 British Standard Code of Practice
英国标准零点 British Ordinance Datum
英国标准螺纹 English thread
英国标准铅字合金 electrotype metal
英国标准筛制 British standard screen scale;British Standard Sieve
英国标准时 British standard time
英国标准水准点 British Benchmark
英国标准网目 British Standard Sieve
英国标准细牙螺纹 British standard fine thread
英国标准线规 British Standard Wire Ga(u)ge;English standard wire ga(u)ge
英国标准协会 British Association of Standard
英国标准型钢 British standard section;British standard section iron
英国标准学会 British Standard Institution
英国标准烛光 British standard candle
英国财政部 Board of Exchequer
英国测绘局 British Survey;Ordnance Survey
英国常衡制 avoirdupois weight
英国出口局 British Export Board
英国储蓄公债 British savings bonds
英国船舶研究会 British Ship Research Association
英国船级社 British Corporation Register
英国达比铁路中心 British Rail;Derby Railway Technical Centre
英国大伦敦道路与交通处 Greater London Roads & Traffic Division
英国大水槽 Yankee gutter
英国单位制 British system of units
英国道路联合会 British Road Federation
英国道路指南 <全称为新建道路路面结构设计指南> British Road Note
英国电气工程师学会 British Institute of Electrical Engineers
英国冬季时间 British winter time
英国都铎式建筑 Anglo-Tudor architecture;English Tudor architecture
英国度量标准 British standard dimension
英国法定度量衡制的 imperial
英国法定海里 admiralty knot
英国法定蒲式耳 Imperial bushel
英国法蜡熔点 English melting point
英国风景式庭园 English landscape style garden
英国钢铁研究协会 British Iron and Steel Research Association
英国港口协会 British Ports Association
英国哥特式建筑 English Gothic architecture
英国哥特式建筑风格之后期垂直线建筑风格 late pointed style
英国工程标准协会 British Engineering Standards Association
英国工程单位制 British engineering units
英国工程科学数据库 Engineering Science Database of United Kingdom
英国工联 Federation of British Industries
英国古典建筑 Anglo-classic architecture
英国管螺纹标准 British Standard Pipe Thread
英国广播公司 British Broadcasting Corp.;British Broadcasting Corporation
英国国际贸易促进会 British Council for the Promotion of International Trade
英国国家高程基准面 <英国陆军测量局地图中采用的基准> Ordnance Datum
英国国家煤炭部袖珍岩石强度锥探仪 NCB cone indenter
英国国家平衡点 national balance point of UK
英国国教 Episcopal church
英国国教的宫殿 Episcopal palace
英国海军部 admiralty
英国海军潮汐表 Admiralty tide tables
英国海军打捞组织 Admiralty Salvage Organization
英国海军单位 Admiralty unit
英国海军型系泊浮筒 admiralty type mooring buoy
英国海平面基准点【测】 Ordnance Datum
英国海事法规 admiralty law
英国海事审判管辖权 admiralty jurisdiction
英国海外航空公司 British Overseas Airways Corporation
英国海外贸易局 British Overseas Trade Board
英国海运总局 British National Maritime Board
英国焊接学报 British Welding Journal
英国焊接研究会 British Welding Research Association
英国焊接研究协会 British Welding Research Association
英国航空公司 British Airways
英国合金 English metal
英国合众社 British United press
英国河道局 British Waterways Board
英国褐栎属 English brown oak
英国红 English red
英国华林福水力学研究所 Hydraulics Research Limited Wallingford, UK
英国化学情报处 United Kingdom Chemical Information Service
英国环境部 Environment Ministry of United Kingdom
英国环境法 Environmental Law of Great Britain
英国皇家建筑师学会 Royal Institute of British Architects
英国皇家特许测量员学会 Royal Institution of Chartered Surveyors
英国皇家学会会刊 Proceedings of Royal Society
英国皇家植物园 kew garden
英国黄 English yellow
英国惠氏标准螺纹 British Standard Whitworth Thread
英国货币 sterling
英国货运码头局 British Transport Docks Board
英国集料建材工业公司 British Aggregate Construction Materials Industries
英国给水工程学会会刊 <月刊> British Waterworks Association Journal
英国给水协会 British Waterworks Association
英国给水协会会刊 British Waterworks Association Journal
英国建筑 British Architecture;English architecture
英国建筑法规 London Building Acts
英国降雨组织 British Rainfall Organization
英国金雀花王朝式(建筑) Plantagenet style
英国绝对单位制 British absolute system of units
英国军用坐标网 British Military Grid
英国坎德拉 British candle
英国抗滑试验仪 British skid resistance tester
英国科茨窝德式建筑 English Cotswold architecture
英国科学发展协会 British Association for the Advancement of Science
英国会计师与审计师协会 British Association of Accountants and Auditors
英国劳埃船级社 British Lloyd
英国老式砌砖法 old English bond
英国雷恩文艺复兴时期 English Renaissance
英国栎 common oak;English oak;Quercus robur
英国联邦农业科技情报研究所 Commonwealth Agricultural Bureau
英国联合电气工业公司 Associated Electrical Industries
英国流体力学研究会通报 Bulletin of British Hydromechanics Research Association
英国流体力学研究协会 British Hydromechanics Research Association
英国陆地测量部 Ordnance Survey
英国陆地测量部地图 Ordnance Survey Datum
英国陆军测量局水准点 Ordnance Bench mark
英国陆军地形测量 Ordnance Survey
英国路面摆式抗滑指数 British portable skid-resistance number
英国路用柏油协会 British Road Tar Association

英国路用焦油沥青协会 British Road Tar Association
英国罗马式（建筑） English Romanesque
英国罗马式建筑艺术 English Romanesque
英国螺纹学会 British Association of Threads
英国贸易部 Board of Trade
英国黏[粘]度单位 British viscosity unit
英国纽林标准基准面 ordnance datum Newlyn
英国欧洲航空公司 British European Airways
英国普通水泥标准 British Portland cement Standard
英国桥梁荷载标准 <指一种常用的等代荷载> HA loading
英国桥梁荷载标准之一 <指一种特殊轴重，大小由有关当局指定，在英国最大为每轴重 45 吨，共 4 轴> HB loading
英国壳牌石油公司路面设计手册 Shell Pavement Design Manual
英国热量单位 British thermal unit
英国人 Briton
英国砂砾协会 Sand and Gravel Association of Great British
英国砂石 English dinas
英国商船法案 Merchant shipping act
英国商品混凝土协会 British Ready Mixed Concrete Association
英国十八世纪安娜女王时代的建筑家具式样 Queen Anne style
英国十字缝砌砖法 <即荷兰式砌法> English bond
英国（石料）磨光值 British polishing number
英国石油公司 British Petroleum Company
英国式房屋底层 English basement
英国式桁架 <三角桁架>【建】English truss
英国式建筑 English architecture
英国式砌合 English（cross）bond
英国式砌砖法 <顶砖层与顺砖层交错> English bond
英国式隧道支撑法 English method of timbering
英国式庭园墙砌砖法 English garden-wall bond
英国式屋架 English roof truss
英国式屋面瓦 English roof(ing) tile; English shingle
英国式园林 English style garden
英国数字资料自动化系统 Ada; British Action Data Automation System
英国双曲线无线电近程导航系统 G-system
英国水泥标准 Portland cement British standard specification
英国水准集点【测】Ordnance Bench mark
英国丝绒 English velvet
英国条件平安险 free of particular average English condition
英国铁路 British Railways
英国铁路局 British Railway Board
英国铁路旅游乘车证 BritRail pass
英国铁路旅游国际公司 British Rail Travel International Incorporation
英国铁路条例手册 British Rail Rulebook
英国铁路运煤巡回列车系统 BR's merry-go-round system
英国涂料研究协会 British Paint Research Association
英国土的分类体系 British Soil Classification System
英国土壤学会 British Society of Soil Science
英国晚期哥特式建筑 perpendicular style
英国文艺复兴建筑 English Renaissance architecture
英国细牙螺纹标准 British Standard Fine Thread
英国协会 British Association
英国协会标准单位 British Association Unit
英国协会螺纹 British Association Thread
英国新建道路柔性和刚性路面结构设计指南 a guide to the structural design of flexible and rigid pavement for new roads
英国学术 British Association
英国压缩空气协会 British Compressed Air Society
英国雅可布式风格 <1603～1625 年> Good King James's Gothic
英国岩土工程学会 British Geotechnical Society
英国养路学会 Permanent Way Institute
英国伊利莎白式建筑 English Elizabethan
英国银朱 English vermilion
英国邮电局式继电器 Post Office relay
英国油漆制造商与相关行业协会 British Paint Manufactures and Allied Trades Association
英国油脂与颜料化学家协会 British Oil and Colo(u)r Chemists Association
英国榆 Ulmus campestris
英国雨情 <英国年刊> British rainfall
英国园墙式砌合法 English garden-wall bond
英国原子能管理局 United Kingdom Atomic Energy Authority
英国运输博物馆协会 Association of British Transport Museums
英国运输码头局 British Transport Docks Board
英国造船工程师学会 Royal Institute of Naval Architects
英国植物学会 Botanical Society of the British Isles
英国中央电力研究所 Central Electricity Research Laboratories, UK
英国中央电气管理局 British Central Electricity Authority
英国烛光 English candle; international candle
英国住宅式 Queen Anne style
英国铸铁研究协会 British cast Iron Research Association
英国专利 British Patent
英辉闪长岩 andendiorite
英辉正长（斑）岩 akerite
英吉利海 Chunnel
英吉利海峡 the English Channel
英吉利海峡渡船 cross channel vessel
英吉利海峡隧道 chunnel
英碱正长岩 nordmarkite
英金镑 pound sterling
英卡石 chrysoprase
英夸（脱） imperial quart
英里 <1 英里等于 5280 英尺或 1760 码，合 1.609 公里> mile; statute mile; English mile
英里标 mile post; milestone
英里程标 mileage point; mileage post
英里程标石 milestone
英里程点 <运价表计算运费的> mileage point
英里程计 mileage counter; mileage sensor; mileometer
英里程记录 mileage recording
英里程记录仪 mileage recorder
英里程客票 <按每次乘车英里程计算> mileage ticket
英里里程 mileage
英里里程津贴 <铁路对私有车辆支付的> mileage allowance
英里数 mileage
英里数计 mile meter
英里/小时 mile/hour; miles per hour
英联邦特惠关税 British preferential tariff
英联邦特惠制 Commonwealth Preference System
英两 ounce
英码 yard
英美编目条例 Anglo-American Cataloguing Rules
英美法系 common law system
英美学派 British-American schools
英美烟草公司 British-American Tobacco
英美制 British and America system
英亩 <1 英亩 = 6.07 亩> acre(age)
英亩英尺 acre-foot
英亩英寸 acre-inch
英亩英寸日 <每日每英亩面积上水深以英寸计> acre-inch day
英亩产量 acre yield
英亩数 acre(age)
英女皇维多利亚时代建筑形式 <1837～1901 年> Victorian style
英普逊焦沥青 impsonite
英钱 <衡量名，常量 = 1.772 克 = 1/16 英两，药量 = 3.888 克 = 1/8 英两> dram
英闪玢岩 esterellite
英闪细晶岩 yukonite
英石 <英国重量名，1 英石 = 14 磅> stone
英石岩 quartzfels; silexite
英氏光泽计 Ingersoll glarimeter
英式地下室 English basement
英式仿古机制花边 English antique lace
英式敷设管道方法 English tubing
英式滚边 English edging
英式活动板手 English spanner
英式卷纸机 English reel
英式罗纹组织 English rib
英式磨光轮试验 British Polishing Wheel test
英式砌法 English bond
英式擒纵叉 English lever
英式湿捻 English system of wet twisting
英式十字缝砌砖法 English cross bond
英式（十字）砌法 English cross bond; St. Andrew's cross bond
英式屋面瓦 English tile
英式无线电信号 English time signals
英式有结织网机 English knot net making machine
英式预制弧形块井壁 English-type tubing
英斯特朗电子强力测试仪 instron tensile (strength) tester
英斯特朗电子强力机 Instron apparatus
英斯特朗毛细管流变仪 Instron capillary rheometer
英特尔 <芯片商标> Intel
英特尔移动模块 Intel mobile module
英特克石 mountkeihite
英驼洛克斯鞍形填料 Intolox saddle (packing)
英王（查理一世）宫廷 Court of the Lions
英文品名 name of commodity in English
英伍德系数 <用于计算年金的系数> Inwood factor
英希拉米月谷 Vallis inghirami
英雄墓舍 heroum
英寻 fathom
英译 English translation
英译本 English version
英云红柱角岩 proteolite
英云角岩 kerilite
英云角页岩 keralite
英云片麻岩 quartz-mica gneiss
英云闪长斑岩 tonalite porphyry
英云闪长岩 tonalite
英制 British system; imperial scale; English system
英制比例 inch scale
英制比例尺 jointed rule; scale of feet
英制标准 Inch-standard
英制标准电线规格 imperial wire
英制标准螺纹 English standard thread; Whitworth screw thread
英制尺 foot rule; inch rule
英制尺寸 imperial sizing; inch size
英制的量尺 foot ruler
英制度量衡单位 British system of units
英制发电单位热耗 <1 英制发电单位热耗 = 252 卡路里/千瓦时> BTU/kwh sendout
英制钢丝网 B.R.C. fabric; British steel wire
英制工程单位 English engineering unit
英制海里 <1 英制海里 = 6080 英尺，折合 1853.24 米> admiralty measured mile; admiralty mile
英制机号 English ga(u)ge
英制计量单位 English calibration; English unit
英制加仑 <1 英制加仑约等于 1.2 美国加仑，等于 4.546 升，容量为 277.42 立方英寸> British imperial gallon; UK gallon; imperial gallon
英制刻度 graduated scale
英制里 <即英制海里，1 英制海里 = 6080 英尺，折合 1853.24 米> English sea mile
英制里程碑 mile post
英制螺距规 English screw pitch ga(u)ge
英制螺纹 English screw; English thread; inch screw thread; inch system screw thread; inch thread
英制螺纹板牙 inch screw die
英制螺纹丝锥 inch screw tap
英制马力 <功率单位，1 英制马力 = 0.746千瓦> British horse power; watt's horse power
英制模数 inch module
英制捻系数 English twist multiplier
英制欧姆 British Association ohm
英制千分表 dial ga(u)ge
英制热量单位 British thermal unit
英制石油产品量桶 UK barrel
英制硬度 Clark degree; English degree; English hardness <水的>
英制支数 English count
英制直尺 foot ruler
英制烛光 British candle; English candle

婴儿车 baby car; pram
婴儿秤计 pedometer
婴儿的 infantile
婴儿室 baby room
婴幼儿载具 infant carrier

瑛珞柏【植】common juniper

缨状胶束 fringed micelle

缨子 tassel

罂粟 eschshltzia mexicana

罂粟花 poppy
罂粟花饰 poppy; poppy head
罂粟油 poppy oil
罂粟子油 poppy (seed) oil
罂酸 linoleic acid
罂子桐 Aleurites cordata (steud); Japan woodoil tree
罂子桐油 Aleurites ardata oil; Japanese tung oil

樱草黄 primrose (chrome) yellow

樱草灵黄 primuline yellow
樱草绿 primrose green
樱草色 primrose (chrome) yellow
樱草素 primin; primuline
樱草园 primula garden
樱红色 cherry
樱红色桃花心木 cherry mahogany
樱花 Japanese flowering cherry; oriental cherry; Yoshino cherry
樱花树 flowering cherry
樱煤 cherry coal
樱木 cherry
樱色 cerise
樱桃核 cherry stone; Nux Pseudocerasi
樱桃红木 cherry mahogany
樱桃红热 cherry-red heat
樱桃红(色) cherry red; bright cherry-red
樱桃花红色 cherry blossom
樱桃加工 cherry processing
樱桃(木) cherry
樱桃蔷薇色 cherry rose
樱桃色 cerise; cherry
樱桃树 cerasus; cherry; cherry tree

鹦鹉蓝 parrot blue

鹦鹉绿 parrot green
鹦鹉螺【动】nautilus
鹦鹉螺动物地理区 nautiloid faunal province

膺窗【建】Yingch'uang

膺架 centering scaffold; falsework; scaffold
膺架垫块 camber block
膺架浇筑法 cast on scaffolding method
膺架式架设法 erection with scaffolding method

鹰雕 hawk eagle

鹰鼎 eagle shaped din
鹰钩 dog
鹰钩板 dog board
鹰壶 eagle ewer
鹰徽 eagle
鹰架 trestle works
鹰铃 hawkbell
鹰狮头雕像 hieracosphinx
鹰石 eagle stone
鹰头怪兽壁画 griffin fresco
鹰头狮身像壁画 fresco of griffins
鹰头柱头 eagle capital
鹰眼石 hawk's eye
鹰嘴 olecranon
鹰嘴剪 hawkbill snips
鹰嘴钳 eagle nose pliers; hawbill snips; hawkbill pliers

迎宾处 greeting arriving

迎冰坡 stoss side
迎波面坡 apron slope
迎春(花)【植】winter jasmine
迎风 aweather; wind ahead
迎风岸 weather shore; windward bank; lee whore
迎风侧 weather side; windward
迎风潮 weather tide; windward flood; windward tide
迎风的 upwind; weather side; windward
迎风海岸 windward shore; exposed shore
迎风航驶 close-hauled ply; upon a wind
迎风航行 weather; wind run
迎风进气口 ram intake
迎风冷泪 irritated epiphora with cold tear
迎风流 windward set
迎风面 luvside; stoss face; stoss side; stoss surface; weather face; weather side; windward face; windward side
迎风(面)桁架 windward truss
迎风面积 core area; face area; front face area; windward area
迎风抛八字锚 moor with an open hawse
迎风漂流 windward drift
迎风坡 stoss slope; windward side; windward slope
迎风气流 relative wind
迎风停泊 lay by(e); lying to
迎风位置 upwind position
迎风舷 weather-board(ing)
迎风斜驶【航海】beat about
迎风行驶的 close-hauled
迎风一侧的起落架 upwind gear
迎风涨潮 windward flood
迎合 cater
迎火 <把草原或森林中的一块地带先纵火烧光> back-fire
迎击航向引导 collision course homing
迎角 angle of attack; angle of incidence; attack; attack angle; incidence (angle) yaw; rigging-angle of incidence
迎角传感器 angle of attack detector
迎角探测器 angle of attack detector
迎角指示器 angle of attack indicator
迎浪 head sea; imaging sea; oncoming sea
迎浪风 opposing wind
迎浪角 angle of wave approach
迎浪行驶 heading the sea
迎流 incident flow
迎面 bow-on; face side; front side
迎面车流 oncoming traffic
迎面冲撞 head-on collision
迎面打火 back-fire
迎面大风雨 leeward squall
迎面导航方式 lead pursuit
迎面而来的(车辆) oncoming
迎面风 adverse wind; dead-wind; head-on wind
迎面风速 face velocity; head wind velocity
迎面辐射 head-on radiation
迎面航线 head-on course
迎面火 back-fire
迎面解锁 facing release
迎面来的船 meeter
迎面力面积 drag area
迎面流 counter-flow; free stream flow; incident flow
迎面碰撞 central collision; head-on collision; knock-on collision
迎面平行航向双机编队通过 sequenced crossover
迎面坡 apron slope
迎面气流 approach flow
迎面气流速度 air speed; approach stream velocity
迎面强风 muzzler; nose ender
迎面送换乘 kiss-and-ride
迎面涌浪 head swell
迎面阻力 drag force; head resistance; leading-end resistance
迎面阻力张线 drag wire
迎收电路 acceptor circuit
迎水侧截流戗堤 <围堰> outer closure dike [dyke]
迎水堤面 floodside of dam
迎水路堤 upstream fill
迎水面 stoss face; stoss side; upstream face; waterside face; water face <坝的>; leading face <叶片>
迎水面承压管 impact tube
迎水面护堤 outer banquette; outer berm(e)
迎水面马道 outer banquette; outer berm(e)
迎水面坡度 upstream batter; upstream slope
迎水面坡脚 heel; upstream heel; upstream toe
迎水面斜坡 outer slope
迎水面堰顶曲线 upstream radius of crest
迎水坡 front slope; outer slope; outside slope; riverside slope; upstream batter; upstream slope; waterside bank; waterside slope; wet slope; water slope <堤坝的>
迎水坡阶梯式护坡工程 stepwork
迎水坡墙 wave wall
迎水坡水下护坦 lower apron
迎送换乘 <公共交通换乘的一种方式,乘客由别人驾驶来车送到公共交通车站> kiss-and-ride
迎头对遇 head on; stem on stem(ship)
迎头法 frontal method
迎头航线 head-on course
迎头浪 head-on sea
迎头碰撞 head-on impact
迎头撞击 head-on collision
迎铣 conventional milling; out-milling; up(-cut) milling
迎着船头 head-on bow

盈反比【数】reciprocal rate

盈非空的【数】non-empty
盈功 gain(ed) work
盈亏 gain and loss
盈亏报表 profit and loss statement
盈亏拨补账 profit and loss appropriation account
盈亏点 break-even point
盈亏分界图表 break-even chart
盈亏分析 break-even analysis
盈亏分析图表 break-even chart
盈亏计算 profit and loss account
盈亏两抵 average out
盈亏临界点 break-even point
盈亏临界点分析 balance analysis of profit and loss; break-even point analysis
盈亏平衡 break-even
盈亏平衡点 break-even point
盈亏平衡点法 break-even point method
盈亏平衡分析 break-even analysis
盈亏平衡图 break-even chart
盈亏通知书 difference account
盈亏相抵 break-even point
盈亏一览结算线 bottom line
盈利 earnings; free of charges; in the black; payoff; profit
盈利比率 profit ratio
盈利对股息比率 dividend cover
盈利对利息的倍数 interest coverage ratio; times-interest earned
盈利分析 profitability analysis
盈利机构 revenue-producing activities
盈利可能率 profitability
盈利率 payout ratio; profit rate; rate of return
盈利敏感率 profit sensitivity ratio
盈利能力 earning power; profitability
盈利能力测验 test of profitability
盈利能力的资本化 capitalization of earning power
盈利能力指数 profitability index
盈利企业 paying concern; profitable firm
盈利区 profit area
盈利实况 earning performance
盈利税 business profit tax
盈利税申报书 business profit tax return
盈利投资 lucrative investment
盈利性的 profit-making
盈利佣金 profit commission
盈式干粉碎机 dry pass
盈水河 gaining stream
盈水溪 gaining stream
盈损比较表 comparative profit and loss statement
盈悬链曲面 catenoid
盈余 override; rest; surplus
盈余表 statement of surplus
盈余拨抵折旧法 depreciation-appropriation method
盈余电子 excess electron
盈余电子的 excess electronic
盈余公积金审计 surplus fund audit
盈余公积金使用的合法性审计 audit of legality of using surplus funds
盈余价格比率方法 earnings/rice ratio method
盈余明细表 statement of surplus
盈余调节 surplus reconcilement
盈余准备 surplus reserves
盈余资金 surplus fund
盈余总额 total net earnings; total net profit
盈月 waxing moon

荧光 fluorescence light; luminescence; phosphorescence; photoluminescence; reflected colo(u)r

荧光 X 射线 fluorescent X-ray
荧光 X 射线光谱分析 fluorescent X-ray spectrometry; X-ray fluorescence spectrometric analysis
荧光 X 射线光谱学 fluorescent X-ray spectroscopy
荧光 X 射线计数 fluorescence X-ray counting
荧光 X 射线照相术 fluororoentgenography
荧光安全出口标志系统 photoluminescent exit marking system
荧光胺 fluorime

荧光板 fluorescent plate; fluorescent screen
荧光板技术 fluorescent plate technique
荧光标记细菌 fluorescently labeled bacteria
荧光标识 X 射线 fluorescent characteristic X-ray
荧光标识处理剂 fluorescence-labeled water treatment agent
荧光标志 fluorescent sign
荧光标准物 fluorescence standard substance
荧光表盘 fluorescent scale
荧光表盘照明灯 luminous dial lighting
荧光玻璃 fire-fly glass; fluorescent glass; luminescent glass
荧光玻璃探测器 fluorescent glass detector
荧光薄层 fluorescence thin layer
荧光薄膜技术 fluorescent film technique
荧光材料 fluorescent material
荧光测定 fluorescence determination; fluorometric assay
荧光测定法 fluoremetry
荧光测定术 fluoremetry
荧光测量方法 fluorescence survey
荧光层 fluorescence coating; fluorescent layer
荧光产额 fluorescence yield; fluorescent yield
荧光产物 fluorescence causing substance
荧光抽运 fluorescent pumping
荧光穿透试验 fluorescent-penetrate test
荧光床料示踪剂 fluorescent bed-material tracer
荧光磁粉 fluorescent magnetic particle (powder)
荧光磁粉探伤 fluorescent magnetic powder detection
荧光磁粉探伤机 fluorescent magnetic powder detector
荧光磁性检测 fluorescent magnetic inspection
荧光促进剂 fluorescence improver
荧光淬火 fluorescent quenching
荧光淬灭法 fluorescence quenching method
荧光代谢物 fluorescent metabolite
荧光带 fluorescence band; fluorescent strip
荧光单体 fluorescent monomer
荧光的 fluorescent
荧光灯 fluorescence discharge lamp; fluorescent lamp; fluorescent light(ing); fluorescent lighting fixture; luminescent lamp; phosphorescent light; daylight fluorescent; black light
荧光灯带 continuous row fluorescent fixture
荧光灯封接机 fluorescent lamp sealing machine
荧光灯管 cold cathode lamp; fluorescent tube
荧光灯广告牌 video sign
荧光灯具 fluorescent fixture; fluorescent light illuminator
荧光灯启动器 glow starter
荧光灯启辉器 fluorescent lamp starter
荧光灯稳压器 fluorescent lamp stabilizer
荧光灯照明 fluorescent lighting
荧光灯照明设备 fluorescent lighting fixture
荧光灯镇流器 fluorescent lamp ballast
荧光灯装置 fluorescent fittings
荧光滴定 fluorescence titration; fluorometric titration
荧光滴定法 fluorescence titrimetric method
荧光底物 fluorogenic substrate
荧光地图 fluorescent map; map with fluorescent colo(u)ring
荧光点 phosphor dot
荧光点滴法 fluorescence spot-out method; fluorescence titrimetric method
荧光点阵 phosphor dot array
荧光电解液 fluorescent electrolyte
荧光电影摄影 cinefluorography
荧光发射 fluorescence emission
荧光发射光谱 fluorescence emission spectrum
荧光发生反应 fluorigenic reaction
荧光法 fluorescence method; fluorescent method; fluorimetric method
荧光反光灯 fluorescent reflector lamp
荧光反应 fluorescence reaction
荧光放电 fluorescence discharge
荧光放射增强 fluorescence stimulation
荧光分光光度测定分析 spectrofluorometric assay
荧光分光光度法 fluorescence spectrophotometry; fluorescent spectrometry
荧光分光光度计 fluorescence spectrophotometer; fluospectrophotometer; spectrophotofluorometer
荧光分光光度纸层析 paper chromatography spectrophotofluorometry
荧光分光光谱检测器 spectrophotofluorometric detector
荧光分光计 fluorescence spectrometer
荧光分析 fluorescent analysis; fluorimetric analysis
荧光分析法 fluorescence analysis; fluorescent analysis; fluorimetry
荧光分析光谱 fluorescent analysis spectrum
荧光分支比 fluorescence branching ratio
荧光粉 fluorescent powder; phosphor; phosphor powder
荧光粉饱和度 phosphor saturation
荧光粉彩色响应 phosphor colo(u)r response
荧光粉粒的反射 reflections from phosphor particles
荧光粉烧伤 phosphor burning
荧光粉余辉 phosphor persistence
荧光峰 fluorescence peak
荧光敷层 fluorescent coating
荧光辐射 fluorescence radiation; fluorescent radiation
荧光汞灯 fluorescent mercury lamp
荧光汞放电灯 fluorescent mercury discharge lamp
荧光管 luminescent tubing; phosphor-coated tube; fluorescent tube
荧光管灯 tubular fluorescent lamp
荧光光度 fluorophotometric
荧光光度测定法 fluorophotometry
荧光光度滴定 fluorophotometric titration
荧光光度分析 fluorophotometric analysis
荧光光度计 fluorescent photometer; fluorimeter; fluorophotometer
荧光光密度分析法 fluorodensitometry
荧光光谱 fluorescence spectrum; fluorescent spectrum; luminescence spectrum
荧光光谱法 fluorescence spectrography; fluorescent spectrometry; fluorescent spectroscopy
荧光光谱分析 fluorescence analysis; spectrofluorimetry
荧光光谱分析测定法 spectrofluorimetry
荧光光学 fluorescent optics
荧光化合物 fluorescent compound
荧光黄 eosin yellow; fluorescein(e); fluorescent yellow; uranin yellow
荧光混合物 fluorescent composition
荧光激发光谱 fluorescence excitation spectrum
荧光计 fluorimeter[fluorometer]; photofluorometer; spelled fluorimeter
荧光技术 fluorescent technique
荧光剂 fluorescer; phosphor
荧光剂量棒 fluorod
荧光检测器 fluorescence detector
荧光检查 flouroscopy; fluorescent inspection
荧光检查法 fluoroscopy
荧光检查器 fluoroscope
荧光检定仪 fluoroscope
荧光检验 fluorescent inspection; fluoroscopic inspection
荧光晶体 fluorescent crystal
荧光镜检法 fluorescence microscopy
荧光镜(屏)<量测烟气密度用> fluoroscope
荧光聚合物 fluorescent polymer
荧光菌落 fluorescent colony
荧光蓝 eosin blue
荧光沥青分析 fluorometric bituminological analysis
荧光粒子大气示踪剂 fluorescent particle atmospheric tracer
荧光粒子技术 fluorescent particle technique
荧光量的变化 changes of fluorescence
荧光量子计数器 fluorescence quantum counter
荧光量子效率 fluorescence quantum efficiency; fluorescent quantum efficiency
荧光流量表 fluorometer
荧光密度测定 fluorodensitometry
荧光面 fluorescent face
荧光膜 fluorescent film
荧光能量 fluorescent energy
荧光偏振 fluorescence polarization; fluorescent polarization
荧光偏振显微镜 fluorescence polarization microscope
荧光漂白 fluorescence bleach; fluorescent bleach
荧光漂白剂 optic(al) bleaching agent
荧光屏 fluorescence screen; fluorescent screen; fluoroscopic screen; luminescent screen; phosphor-coated screen; phosphorescent screen; phosphor plate; photofluoroscope; salt screen; telescreen; viewing screen; window of tube; face-plate <阴极射线管>; bezel
荧光屏持久性 screen persistence
荧光屏淀积 screen condensation
荧光屏发光效率 screen actinic efficiency
荧光屏可轴向移动的阴极射线管 peritron
荧光屏亮度 screen bright; screen intensity
荧光屏面 face of the screen
荧光屏上黑白扫描线 spoke
荧光屏烧坏 screen disintegration
荧光屏烧毁 screen burning
荧光屏摄影机 photofluoroscope
荧光屏摄影术 fluorography; photofluorography
荧光屏深度 depth of screen
荧光屏图像摄影 photofluorography
荧光屏图像照片 photofluorogram
荧光屏显示 present on a screen
荧光屏颜色 screen colo(u)r
荧光屏阴影区 blind sector
荧光屏有效直径 screen effective diameter
荧光屏余辉 perpetual of screen; screen afterglow; screen persistence
荧光屏余辉时间 screen storage time
荧光屏余辉特性 decay characteristic phosphor
荧光屏噪声 screen noise
荧光屏照片 fluorogram
荧光屏照相机 screen camera
荧光屏字符显示器 alphanumeric fluorescent screen display equipment
荧光谱图 fluorogram
荧光谱线 fluorescent line
荧光漆 fluorescent paint(ing)
荧光强度 fluorescence intensity; fluorescence strength
荧光染料 fluorescent dye; fluorochrome; luminescent dye
荧光染色 fluorescent staining
荧光染色法 fluorescent staining
荧光染色剂 fluorescent dye
荧光染色砂 fluorescent dyed sand
荧光散射 fluorescent scattering
荧光扫描技术 fluorescent scanning technique
荧光色 fluorescence colo(u)r
荧光色材 fluorescent colo(u)r
荧光色谱法 fluorescence chromatography; ultra-chromatography
荧光色谱图 fluorescence chromatogram
荧光色散 epipolic dispersion
荧光沙 <又称荧光砂> fluorescent sand
荧光射线摄影 photoroentgenography
荧光摄影 fluorescent photography
荧光渗透剂 fluorescent penetrant
荧光生 fluorescin
荧光生成标记技术 fluorigenic labeling technique
荧光石英玻璃 luminescent silica glass
荧光示踪 fluorescent tracing
荧光示踪技术 fluorescent tracer technique
荧光示踪剂 fluorescent tracer
荧光示踪物试验 fluorescent tracer experiment
荧光示踪装置 fluorescent tracing unit
荧光试剂 fluorescent reagent
荧光试验 fluoroscopy
荧光寿命 fluorescent lifetime
荧光数码管 fluorescent digital display tube
荧光衰变 fluorescence alteration
荧光水处理剂 fluorescent water treatment polymer
荧光水溶性聚合物 fluorescent water soluble polymer
荧光水银灯 fluorescent mercury lamp
荧光素 fluorescein(e); luciferin
荧光素钠 uranin yellow
荧光素配位剂 fluorexone
荧光素染料 fluorescein(e) dye
荧光素试验 fluorescein(e) test
荧光素试纸 fluorescein(e) paper
荧光塑料 fluorescent plastics
荧光态 fluorescence state
荧光探测法 fluorescent penetrant meth-

od
荧光探测器 fluorescent probe
荧光探伤 fluorescent detection; fluorescent inspection; fluorescent test
荧光探伤法 fluorescent penetrant inspection; fluorescent penetrant method; Zyglo penetrant method; zyglo inspection
荧光探伤器 zyglo
荧光探伤仪 fluorescent fault detector
荧光探头 fluorescent probe
荧光探针技术 fluorescent probe technique
荧光碳 fluorocarbon
荧光搪瓷 fluorescent enamel; luminescent enamel
荧光特性曲线 fluorescent characteristic curve
荧光特征曲线 fluorescent characteristic curve
荧光体 fluorescent substance; fluorinate; fluorophor; self-luminescent material
荧光体余辉 phosphor persistence
荧光添加剂 fluorescent additive
荧光透视 fluoroscopic viewing
荧光透视检查 fluoroscopic examination
荧光图 fluorograph
荧光图案 phosphor pattern
荧光图像摄影 fluorograph
荧光图像增强器 fluoroscopic image intensifier
荧光图照相术 fluorography
荧光涂层 fluorescent coating; lumiphor coat
荧光涂料 fluorescent paint(ing); luminous colo(u)r
荧光团 fluorogen; fluorophore
荧光微波双共振法 fluorescence microwave double resonance
荧光物质 fluorescent material; fluorescent substance; phosphor
荧光物质沉积法 screen settling
荧光细菌 fluorescent bacteria
荧光显示 fluorescent display; fluorescent show
荧光显示器件 fluorescent display device
荧光显微法 fluorescence microscopy
荧光显微(技)术 fluorescence microscopy
荧光显微镜 fluorescence microscope; fluorescent microscope
荧光现象 phenomenon of fluorescence; phosphorescence
荧光线宽 fluorescence line width; fluorescent line width
荧光消抑 killing of fluorescence
荧光小体 fluorescent body
荧光效率 fluorescence efficiency
荧光效应 fluorescence effect; fluorescence yield; fluorescent effect
荧光效应修正 fluorescent effect correction
荧光性 fluorescence
荧光性的 epipolic
荧光性辐射 fluorescent radiation
荧光性(忽)绿(忽)紫的颜色 iridescent
荧光学 fluorescopy [fluoroscopy]
荧光循环 fluorescence cycle
荧光颜料 daylight fluorescent pigment; fluorescent pigment
荧光颜色 fluorescence colo(u)r
荧光阳极 fluorescent anode
荧光液检裂法 hyglo
荧光仪 luminoscope
荧光仪检查 fluorographic study
荧光油 fluorescent oil
荧光油墨 daylight fluorescent ink; fluorescence ink; fluorescent ink
荧光诱蛾灯 fluorescent light trap
荧光余辉 phosphorescence afterglow
荧光余辉时间 fluorescence decay time
荧光玉红类颜料 fluorubine pigment
荧光阈 luminescence threshold
荧光跃迁 fluorescent transition
荧光再放射 allochromy
荧光噪声发生器 fluorescent noise generator
荧光增白剂 fluorescent brightening agent; fluorescent whitener; fluorescent whitening agent; fluorescer; optic(al) bleaching agent; optic(al) brightener; optic(al) whitener
荧光增白染料 fluorescent whitening dye
荧光增感屏 fluorescent intensifying screen
荧光增进剂 fluorescence improver
荧光增强 stimulation
荧光照明 fluorescent illumination; radioactive lighting
荧光照明光源 fluorescent lighting source
荧光照明设备 fluorescent light fixture
荧光照明装置 fluorescent fixture
荧光照相 photofluorography
荧光照相术 fluorescence photography
荧光指示分析 fluorescence indicator analysis
荧光指示计 fluorescent indicator
荧光指示剂 fluorescence indicator; fluorescent indicator
荧光指示剂吸附法 fluorescence indicator adsorption method; fluorescent indicator adsorption method
荧光指示剂吸附分析 fluorescent indicator adsorption analysis
荧光指示剂吸附术 fluorescent indicator adsorption te hnique
荧光指示器 fluorescent flag; fluorescent indicator
荧光质点示踪研究 fluorescent particle tracer study
荧光质衰退 dark burn
荧光中心 fluorescence centre; fluorescent center [centre]
荧光转换效率 fluorescence conversion efficiency
荧光着色技术 fluorescent staining technique
荧光紫外线冷凝光照水淋耐候仪 fluorescent ultraviolet condensation light and water-exposure apparatus
荧红环 fluorubin
荧火管 lighting tube
荧色物 fluorochrome
荧石 fluor(phosphor)
荧石花岗岩 fluorite granite
荧烷 fluorane

萤石 cand; cann; blue john; bruiachite; calcium fluoride; derbyshire spar; derby spar; fluorbaryt; fluorite; fluorspar; kann; liparite; ratofkite

萤石光学 fluorite optics
萤石含量 fluorite
萤石花岗岩 fluorite granite
萤石化 fluoritization
萤石矿床 fluorite deposit
萤石矿石 fluorite ore
萤石棱镜 fluorite-lens; fluorite-prism
萤石石英矿石 fluorite quartz ore
萤石石英云英岩 fluorite quartz greisen
萤石物镜 fluorite objective
萤石型结构 fluorite structure
萤石云母云英岩 fluorite muscovite greisen
萤石云英岩 fluorite greisen

营巢区 nesting site

营地 camp(ing) ground; camp(ing) site
营地拖车 tent trailer
营房 barrack; encampment
营房船 house boat
营房建筑 barracks block; building of barracks
营火 bonfire; camp fire
营火处 fireplace
营建管理 construction administration
营救 rescue; save
营救橙 rescue orange
营救船 rescue ship; wrecker
营救船舶 salvage ship
营救当局 rescue authority
营救队 salvage crew
营救人员 salvage crew
营救任务 rescue mission
营救设备 salvage appliance
营救协调中心 rescue co-ordination center [centre]
营救用具 salvage appliance
营救者 rescue
营救作业 rescue operation
营力 agent
营力活动带范围 range of active seress zone
营林区 range
营林区长 divisional officer
营林所长 district officer
营林员 forest ranger; ranger
营幕 marquee
营区 encampment
营私舞弊 jobbery; malpractice
营销费用 marketing expenses
营养 alimentation
营养比 nutritive ratio
营养比例 nutritive proportion
营养钵 culture pan; nutrition pot
营养钵育苗 seeding in nutritive cup
营养钵栽植 pot planting
营养补给条件 condition of nourishment
营养补助费 foodstuff subsidization
营养不良 alimentary deficiency; dystrophy; ill-nourished; malnutrition; poor nutritional state
营养不足 abiotrophy; alimentary deficiency; atrepsy; hypoalimentation
营养层 nutritive layer
营养层次 trophic level
营养成分 nutritional ingredient
营养厨房 diet kitchen; invalid kitchen
营养的 nutritive
营养分解层 tropholytic zone
营养分析 trophic analysis
营养丰富 eutrophy; nutrient enrichment
营养物丰富的湖泊 eutrophic lake
营养广温性 vegetative eurythermy
营养广盐性 vegetative euryhaline
营养规格 nutrient specification
营养过度 hypernutrition; hypertrophy
营养级 tropic(al) level
营养期 nutrition period; trophophase; vegetation period
营养情况 nourishment
营养缺乏 alimentary deficiency
营养室 nutritive chamber
营养水平 trophic level
营养土钵压制机 plant balling machine
营养土钵压制器 plant baller
营养土块 nutritive cube; turf muck block
营养卫生研究所 research institute of nutrition
营养物 nurture; nutrient
营养物富集 nutrient enrichment
营养物循环 nutrient cycle
营养狭温性 vegetative stenothermy
营养狭盐性 vegetative stenohaline
营养学家 dietitian; nutritionist
营养液 nutrient solution; nutritive medium
营养周期 vegetative cycle
营养砖 planting brick
营养状况 nutrient status; nutritional status; nutriture
营养状况预测 trophic state prediction
营养状况指数 trophic state index
营养作用 nutrition
营业 prosecute; revenue service
营业报表 operating statement; statement of operation
营业报告 account of business; business report; operating statement
营业比率 <即营业支出对营业收入的比> operating ratio; operation ratio
营业边际 operating margin
营业表 operating statement
营业不振 business slump
营业部 business office; business department; business division; division of business; sales department; sales office
营业部经理 marketing manager; business manager
营业部门 business sector
营业仓库 business store
营业车访问调查 commercial vehicle interview survey
营业车辆 commercial vehicle
营业车辆运行系统 commercial vehicle operation system
营业成本 cost in business; operating cost
营业处 business establishment; business place
营业处长 chief traffic manager; traffic manager
营业处所 seat of business
营业处主管 terminal manager
营业存款 business deposit
营业大厅 business hall
营业贷款 business loan
营业道德 business ethics; business mortality
营业的 operating
营业登记费 business registration fee
营业登记证 business registration certificate
营业地点 place of business
营业额 business volume; turnover; volume of business
营业发达的企业 going concern
营业范围 line of business
营业费(用) business expenses; trade expenses; operating cost; operating expenses
营业费用表 statement of operating expenses
营业费用预算 operating expense budget
营业负债 working liability
营业杠杆率 operating leverage
营业公里数 working kilometrage
营业估价 appraisal of business
营业股权公司 operating holding company
营业股权总账 operating ledger
营业后勤 business logistics
营业环境风险信息 business environment risk information
营业货票 revenue waybill

营业机构 business agency
营业基金 business fund
营业计划 operating plan
营业价值 going concern value
营业驾驶员执照 commercial driver license
营业轿车 taxi
营业结果 operating result
营业经理 business manager; circulation manager; sales manager
营业净利(润) net operating profit; net profit from operation
营业净收入 net revenue account
营业净收益 net income from operations
营业决策 operating decision
营业开支 management and general expense; operating cost; operating expenses; overhead
营业科目 account of business
营业控股公司 operating holding company
营业会计 accounting of business
营业亏损 business loss; operating deficit
营业里程 revenue mile
营业利润 business profit; opening profit; operating profit; operation profit; trading profit
营业利润税退税 business profit tax return
营业量 operational capacity
营业流程 flow of operations
营业毛利 operating margin
营业面分配 assignment of space
营业名称 fictitious name
营业目录 catalog(ue) of business
营业能力 business ability
营业年度 business year; fiscal accounting year
营业票据 business paper
营业评估 business assessment; business estimate
营业期 business period
营业晴雨表 business barometer
营业区 business district; business quarter; business zone
营业人员 operating personnel; operative
营业日 business day
营业日记 business diary
营业设备 business equipment
营业设备维修保养费 office equipment repairs and maintenance expenses
营业时间 business hours; office hours; shophours
营业实绩 operating performance
营业实绩比率 operating performance ratio; operating ratio
营业实绩收益表 operating performance income statement
营业实体 business entity
营业室 parlo(u)r
营业事务 business affairs
营业收款凭证 vouchers for business receipts
营业收入 business income; business receipt; operating income; operating receipt; operating revenue
营业收入表 statement of operating revenue
营业收入的履约确定法 completed contract method of revenue recognition
营业收入的收付实现确定 cash basis of revenue recognition
营业收入调整 adjustment of business income
营业收入债券 income bond
营业收入折旧法 depreciation revenue method; revenue depreciation method
营业收入总额 total revenue
营业收益 operating earnings; operating income
营业收益税 business profit tax
营业收支 operating income and expenditures
营业收支比率 operating ratio
营业收支分类账 operating ledger
营业收支汇总表 summary of operating revenue and expenses
营业税 business tax; excise tax; receipt tax; sales tax; taxes on business; turnover tax
营业税法 act of business taxes
营业速度 service speed
营业损失 operating loss
营业损益表 operating statement
营业所 abode; place of business
营业特许 business patent
营业体制 structure of business unit
营业条件 operating condition
营业厅 business hall; selling area; shopping hall
营业外的 no-operating
营业外费用 non-operating expenses; other charges
营业外经营 outside venture
营业外净支出 net disbursement besides business
营业外利润 non-operating profit
营业外收入 non-business income; non-operating revenue; non-revenue receipt
营业外收入审计 non-operating income audit
营业外收益 non-operating earnings; non-operating income; unrelated business income
营业外收支 non-operating revenue and expenditures
营业外收支净额 net value of non-operating revenue and expenditures
营业外损益 non-operating profit and loss
营业外损益表 non-operating profit and loss statement
营业外损益净额 net amount of non-operating profit and loss
营业外项目 non-operating item
营业外支出表 non-business expenditure statement; non-operating expense audit
营业外支出(费用) cost outside business; non-business expenditures; non-operating outlay
营业外支出审计 non-business expenditure audit; non-operating expense audit
营业危机 business crisis
营业下降 business dip; business downturn
营业现金流动 business cash flow; cash flow from operation
营业现金余额 operating cash balance
营业线路公里 kilometers of road operated
营业线路公里数 kilometers of track operated
营业项目 business item; item of business
营业信托 business trust
营业信誉 business reputation
营业行为 business behavio(u)r
营业性驳船 commercial franchised barge
营业性饭店 commercial hotel
营业性旅店 commercial hotel
营业性停车库 commercial garage
营业性戏院 legitimate house
营业性游乐场 commercial recreation
营业循环 earning cycle; operating cycle
营业盈余 earned surplus; earning surplus; operating surplus
营业盈余表 retained earnings statement
营业盈余净额 net earned surplus
营业用财产 property used in business
营业用款 operating expenditures; running expenditures; working expenditures
营业用器具账 fixture and furniture account
营业(有收入的)货物装车数 revenue freight carloadings
营业(有收入的)整车货 revenue carload
营业预算 operating budget
营业员 assistant; clerk; salesman; saleswoman
营业运行 revenue running
营业(运营)车辆安全联盟 commercial vehicle safety alliance
营业站 operating station
营业账(户) operating account; trading account
营业者 undertaker
营业证 business licence [license]; certificate of business
营业证书 commercial instrument
营业支出 charge to income; nominal expenditures; operating expenses; revenue charges; revenue expenditures
营业执照 business licence [license]; certificate of business; operating license [licence]
营业执照税 tax on licenses
营业指南 business guide
营业中断保险 business interruption insurance
营业中断收益损失保险 loss of profit insurance
营业中心 seat of business
营业周期 operating cycle
营业主任 business manager; circulation manager; sales manager
营业专利权 business patent
营业转让支付费 business transfer payment
营业状况 business picture; business status
营业准备金 operating reserves
营业资本 operating capital; trade capital
营业资本收益率 operating earnings rate
营业资本周转率 operating capital turnover ratio
营业资产 business assets; operating assets; trade assets; working assets
营业资产周转率 operating assets turnover
营业资金 business fund; operating fund
营业自动化 business automation
营业自主权 business autonomy
营业总额 business sales
营运报告 operating statement
营运差额补贴 <美> operating differential subsidy
营运车辆调度管理系统 commercial vehicle operation/fleet management
营运成本 operating cost; operating expenses; operation cost; operation expenses; running charges; running cost; running expenses
营运成本表 operating cost statement
营运吨位 earning capacity
营运发电机 service generator
营运方针 operational policy
营运费(用) operating expenses; operation cost; operation expenses; running charges; running cost; running expenses; operating cost
营运费账目 operating expense accounts
营运辅助系统 operational assistance system
营运功能 operational function
营运故障 operational trouble
营运航速 sea speed
营运荷载 service load
营运计划 operating program(me); operation scheme; plan of operation; schedule of operations
营运交通量 service volume
营运决定 operative decision
营运里程 <新铁路线修筑期间,分段开放交通> operating kilometer [kilometre]
营运利润 operating profit
营运区保证 trading warranty
营运商船队 active trading fleet
营运设施 operation facility
营运试验 operational test
营运收入 operating income; operation income
营运数据 operational data
营运速度 service speed; operating speed
营运速率 speed of service
营运条件 operating condition; operational condition
营运效率 operating efficiency
营运性能 behavio(u)r in service; running ability
营运运输能力 paying load; payload
营运章程 operating rule
营运证 operation certificate
营运支出 operational expenditures
营运职能 operational function
营运指标 service index
营运重量 payload
营运状况 service condition
营运资本 operational capital; working capital
营运资产 working assets
营运资产净额 net working assets
营运资金 circulating capital; operation funds; working capital
营运租赁 operating lease
营造 building operation; construction; plant forest
营造厂 building yard; construction company; contracting business
营造厂商 contract builder
营造工房 construction shed
营造工长 masterbuilder
营造商 builder; building contractor; merchant builder
营造商的保证书 builder's warranty
营造商支付给分包商的总价 prime cost sum
营造师 masterbuilder
营造物 building
营造学 science of building; tectonics
营造业主 building owner
营帐 encampment

楹间 ai(s)le

楹联 couplet on pillar

蝇拍 flyflap; fly-swat(ten)

蝇眼电子光学 fly's eye electron op-

tics
蝇眼透镜 fly's-eye lens;fly lens

赢得 beam off

赢得(对策)矩阵 payoff (game) matrix
赢亏拨补账 surplus appropriation account
赢利股 bonus dividend;bonus stock
赢水河 effluent river;effluent stream

影蔽因数 shadow factor

影壁 screen wall
影测量法 shadow measurement
影带 shadow bands;shadow zone
影道 shadow channel
影格板 groove mask
影痕 ghost;phantom
影集 photograph album
影剧院建筑 cinema-theatre building
影剧院预售票处 play-guide
影孔板 apertured shadow mask;shadow mask
影孔板彩色管 shadow-mask tube
影孔板式荧光屏 shadow-mask screen
影片 anaglyph;film
影片翻印 dub(bing)
影片库 film storage;film vault
影片框 film loop
影片乳剂 film emulsion
影频干扰 image interference
影青瓷 shadowy blue glaze porcelain;shadowy blue ware
影区 shadow region;shadow zone;silent zone
影区风化 shadow weathering
影色畸变 pattern distortion
影射商标 counterfeit trademark;imitation brand
影视客车 cinema coach
影视音像中心 audio-video center [centre]
影纹布 jaspé cloth
影纹布的长度 length of jaspe sheet
影纹地毯 jaspe carpet
影显微镜 shadow microscope
影线 hachure (line);line of shadow;shade line
影线图 hachure map;hatching map
影相术 shadowgraphy
影响 influence;repercussion
影响半径 radius of influence
影响半径计算方法 calculating method of influence radius
影响报告 impact statement
影响层 affected layer
影响场 influence field
影响程度 degree of incidence
影响大的 far-reaching
影响带 influence zone
影响带宽度 width of influence zone
影响的研究 impact study
影响地域分类 classification on the influence
影响范围 area of coverage;area of influence;circle of influence;influence basin;influence range;range of influence;sphere of influence;zone of influence
影响费独立评估计算法 independent impact fee assessment calculation
影响分析 impact analysis
影响幅度 magnitude of the effect
影响辐射效应气体 radiatively active gas
影响函数 circle of influence;coverage;influence function;sphere of influence
影响及后果研究 < 工程项目的 > impact study
影响集中点 < 采矿 > focal point of subsidence
影响径流的因素 factor affecting runoff
影响矩阵 influence matrix
影响力 leaven
影响量 influence quantity
影响漏斗 cone of depression;cone of influence;crater of depression < 井内抽水形成的 >;cone of recharge < 向地下注水形成的 >
影响面 influence surface
影响面积 area of extraction;area of getting;area of influence;area of working;influence area;area of mining < 抽水时的 >
影响盆地 influence basin
影响皮肤的 cutaneous
影响评定 impact assessment
影响评价 effect evaluation;impact evaluation
影响气候的活动 climate-sensitive activity
影响区(域) area of coverage;area of infection;area of influence;contributing region;influence basin;influence zone;zone of influence;range of influence
影响圈 influence circle;sphere of influence
影响圈半径 radius of influence circle
影响全身的 systemic
影响人口 affected people;affected population
影响深度 depth of influence
影响生态系统的土壤因素 edaphic factor
影响试验反应的各种因素 factors affecting response of testing
影响数 influence number
影响水头 affected head
影响特性 influence characteristic
影响图 influence chart;influence diagram
影响土壤的通气 influence aeration of the soil
影响土壤水分的供应 to influence the water supply of the soil
影响外力地质作用的因素 factors influencing exogenic process
影响问题工作组 working group on impacts
影响系数 coefficient of influence;contribution factor;influence coefficient;response coefficient
影响系数法 influence coefficient method
影响系数值 influence coefficient value
影响线 influence line;line of influence
影响线分析 influence line analysis
影响线面积 area of influence line
影响线图 influence line diagram
影响向量 influence vector
影响(行为)的工具 channel of influence
影响因素 affecting factor;contributory factor;factor affecting;influencing factor
影响因子 factor of influence
影响域 domain of influence;region of influence
影响圆 < 水降曲线的 > circle of influence;influence circle
影响圆半径 radius of influence circle
影响者 influent
影响值 influence value
影响重力场因素 the effect on gravity field
影响资源形势的因素 effect factors of resources situation
影响纵坐标 influence ordinate
影像 video;figure;image;image picture;optic(al) impression;phantom;presentation;display;screenage
影像斑点 smudges
影像保留幕板 image retaining panel
影像比 aspect ratio
影像编码 image coding
影像表 shadow table
影像差 image difference
影像成帧 framing of images
影像重合 image registration
影像重建 image reconstruction
影像传感 image sensing
影像传感器 image sensor
影像地貌图 image geomorphologic(al) map
影像地热显示标志 image criteria of geothermal feature
影像地图 image format map;image map;image photomap;photographic(al) map;photomap
影像地图制图学 image-cartography
影像地下水露头类型 groundwater outcrop type on image
影像地质图 geologic(al) photomap
影像地质学 photogeology
影像电流 image current
影像对称 image symmetry
影像反差 image contrast
影像反差大 great contrast
影像反差适中 middle contrast
影像反差小 small contrast
影像反转过程 image inverse process
影像放大 image enlargement;zoom
影像放大率 image magnification
影像分辨力 image resolution
影像分辨率 resolving power of image
影像分光系统 image splitting optic(al) system
影像分离 image separation;picture separation
影像分析 image analysing computer;image analysis
影像分析法 image analysis method
影像扶正 image erection
影像副本 image copy
影像更新数据库 image refresh database
影像光度计 shadow photometer
影像环形体 image circular features
影像环形体图 image circular feature map
影像畸变 image distortion
影像几何(性质) image geometry
影像计 eiconometer
影像减薄 photographic(al) reduction
影像结构 image texture
影像解译 image interpretation
影像矩阵的奇异值分解 singular value decomposition of image matrix
影像矩阵酉变换 unitary transform of image matrix
影像聚焦电极 image focusing electrode
影像拉长 stretch
影像雷达 imaging radar
影像联机扫描 on-line scanning of images
影像亮度 brightness of image
影像亮度控制 image brightness control
影像录制 image record(ing);image transcription
影像轮廓线法 < 根据航空照片上的影像轮廓获得线划 > photo-line production
影像密度控制 photographic(al) density control
影像明晰度 tonal value
影像模糊 image fog
影像目标 silhouette target
影像凝合 fusion of image
影像判读 image interpretation
影像判读性 image interpretability
影像判译 image interpretation
影像配准 image registration
影像匹配 image match(ing)
影像剖视仪 photographic(al) image
影像清晰度 definition of image;image sharpness
影像清晰性 acutance
影像全息图 image hologram
影像锐度 image acutance
影像色调 image tone
影像深浅度 image density
影像声呐 imaging sonar
影像失真 image fault
影像蚀变类型图 image alteration pattern map
影像视差 image parallax
影像视觉清晰度 apparent image sharpness
影像数据采集 image data acquisition
影像数据系统 viewdata
影像衰减 image attenuation
影像衰减常数 image attenuation constant
影像水系图 image drainage network map
影像损失 image-loss
影像同步 image synchronization
影像图 picture;shadowgraph;striograph
影像微密度计 photo microdensitometer
影像位移的补偿 motion-compensation
影像文件 image file
影像污点 image smear
影像误差 image error
影像细部 image detail;photographic(al) detail
影像显示不良地质现象 harmful geologic(al) phenomenon displayed on image
影像线性体 image lineaments
影像线性体图 image lineament map
影像相干性 coherence of image formation
影像相关 correlation of image;image correlation
影像相减 image subtraction
影像相位常数 image phase constant
影像镶嵌图 image mosaic
影像信息 video information
影像信息预处理 image information preprocessing
影像旋转 image rotation
影像学 iconography
影像延迟 image lag
影像移动 image motion;image movement
影像移动补偿法 < 用于航测中 > image motion compensation
影像移动速度 travel(l)ing velocity of photo
影像油气显示标志 image criteria of oil-gas indication
影像预处理 image preprocessing
影像再现装置 picture reproducer
影像暂留 image persistance
影像增强 image enhancement
影像质量 image quality

影像中心 image center [centre]
影像转换 image transfer
影像转绘 image transfer
影像转移常数 image transfer constant
影像转制 image transfer
影像阻抗 image impedance
影像作图法 image construction
影移法 shadow-Moire method
影印 autotype;photocopy;photographic(al) print(ing);photomechanical printing; photo-offset copy; photoprint;xerox
影印凹版 photogravure
影印版 photoengraving;photoetch(ing)
影印本 facsimile;photo-offset copy;photostat copy
影印法 photolithography;photostat
影印副本 photocopy
影印机 photoprinter;photorepeater
影印件 photocopy
影印石版 photolithograph
影印术 autotype
影源 eikonogen
影锥 shadow cone
影锥角 shadow angle
影锥椭圆 ellipse of shadow cone
影子 shadow
影子工程法 shadow project method
影子工资 shadow wage
影子工资率 shadow wage rate
影子公司 dummy corporation;shadow corporation
影子汇率 shadow exchange rate
影子极小 shadow minimum
影子价格 shadow price
影子价值向量 shadow value vector
影子利率 shadow interest rate
影子内存 shadow RAM
影子内阁 shadow cabinet
影子盘 shadow disc [disk]
影子外汇率 shadow foreign exchange rate
影子效应 shadow effect
影子照片 shadowgraph

应答 acknowledge; acknowledgement; answer back; answering; feedback; respond; response; talk-back

应答标 responsive tender
应答标题 response header
应答表 answer list;answer sheet
应答部件 response unit
应答插头 answering plug
应答单 answer sheet
应答单元 response unit
应答的 responsive
应答灯 answer lamp
应答灯光信号 return light
应答电话 answerphone
应答电键 talking key
应答动作 responder action
应答队列元素 reply queue element
应答发射机 transmitter responder;transponder
应答方式 response mode
应答分析程序 response analysis program(me)
应答浮标 transponder buoy
应答机构 answer-back mechanism;answer-back unit; answering back mechanism
应答机控制部分 transponder control unit
应答机(器) responder;responsor
应答机信标 transponder beacon
应答监视灯 answering supervisory lamp
应答监视继电器 answering supervisory relay
应答检测码型 answering detection pattern
应答键 answering key
应答交替程序通信[讯]块 response alternate program(me) communication block
应答进程 answering process
应答雷达 secondary radar
应答率 response rate
应答码 answer-back code
应答脉冲 answering pulse;reply pulse
应答脉冲比 countdown
应答脉冲间隔 reply pulse spacing
应答密码 reply code
应答频率编码 reply-frequency coding
应答器 responder;responsor;secondary radar;transponder (set)
应答器编码 responder coding
应答器的应答效率 transponder reply efficiency
应答器寂静时间 transponder dead time
应答器式雷达信标 transponder radar beacon
应答器损耗 responder loss
应答器天线 transponder antenna
应答器效率 transponder reply efficiency
应答器信标 responder beacon;transponder beacon
应答请求 reply request
应答曲线 response curve
应答塞 inquiry plug
应答塞孔 answering jack;answering sleeve;listening jack;local jack
应答塞绳 answering cord
应答塞头 listening plug
应答塞子 answering plug
应答设备 answering equipment
应答时间 answering time;response time
应答式监视雷达 responder assisted surveillance radar
应答式雷达信标 transponding radar beacon
应答式无线电信标 responder beacon;transponder beacon
应答式无线电指向标 responder beacon;transponder beacon
应答首部 response header
应答速度 speed of answer
应答算法 responsive algorithm
应答台 answering board
应答头 response header
应答线路 responder link
应答消息 response message
应答效率 reply efficiency
应答信标 respond beacon;transponder transponding beacon
应答信道 responder link
应答信号 answering signal;reply signal;response message
应答信号灯 answer lamp
应答信息 response message
应答行为 respondent behavio(u)r
应答延迟 answering delay; beacon delay
应答音 answering tone
应答指示灯 answer lamp
应答中断 acknowledge interrupt
应答周期 acknowledge cycle
应答装置 answer-back device
应急 emergency;intervention;preemergency
应急安全开关 emergency safety switch
应急安全坑 escape gallery
应急安全坑道 emergency gallery
应急安全评价 emergency safety evaluation
应急安全装置 emergency release
应急按钮 emergency button; emergency key; intervention button; panic button;scram button
应急把手 abort handle
应急包 emergency kit
应急保险电路 emergency safety circuit
应急保险电门 emergency safe switch
应急保险开关 intervention button
应急报告系统 emergency reporting system
应急报警电路 emergency warning circuit
应急备料 emergency material
应急备用电源 emergency standby power
应急备用电站 emergency reserve station
应急备用发电机 emergency generator
应急备用基金 emergency reserve fund
应急备用锚 emergency anchor
应急备用设备 emergency apparatus;emergency plant;on-premise stand by equipment
应急备用闸门 safety lock gate
应急泵 expedition pump;jury pump;stand-by pump
应急舭水泵 emergency bilge pump
应急壁龛设施 <消防、电话等> emergency niche equipment
应急避难所 emergency shelter
应急补给品与装备 readiness reserves
应急部署 contingency planning
应急部署表 station bill;station list
应急材料储备 stock of urgently needed material
应急参考水平 emergency reference level
应急舱 emergency hatch
应急舱壁 temporary bulkhead
应急舱壁灯 box lamp
应急舱底水泵 emergency bilge pump
应急舱口 escape hatch;escape hole
应急操纵 emergency control;emergency maneuver
应急操作 emergency operation
应急操作步骤 emergency procedure
应急操作规程 emergency procedure
应急操纵台 emergency console
应急操作系统 emergency operations system
应急超控电门 emergency override switch
应急车 emergency bus
应急车辆 emergency service vehicle
应急撤离程序 emergency evacuation
应急成本 crash cost
应急程序 emergency procedure
应急池 emergency lagoon
应急齿轮 emergency gear
应急出口 emergency exit; escape hatch;escape scuttle;escape vent;escape way
应急出口标牌 emergency exit placard
应急出口舱盖 emergency exit hatch
应急出口挡板 emergency exit spoiler
应急出口建筑物 emergency exit structure
应急出口孔洞 emergency egress opening
应急出口梯 emergency exit ladder
应急出口围壁 escape trunk
应急出口围阱 escape trunk
应急出水面 <潜水时> emergency surfacing
应急储备 emergency capacity;emergency stock;emergency storage
应急储备金 contingency reserves
应急处理 emergency treatment;emergent treatment
应急处理程序 emergency program(me)
应急处置 emergency disposal
应急传号线 emergency call wire
应急船 emergency boat
应急窗 emergency window
应急措施 contingency measures; crash program(me); emergency action;emergency management;emergency measure
应急措施设备 emergency facility
应急担架 emergency stretcher
应急的 emergence [emergency];jury;panic
应急灯 emergency lamp;emergency light
应急灯具 emergency luminaire
应急电话 emergency phone;emergency telephone
应急电话电缆 emergency phone cable
应急电话系统 emergency phone system;emergency telephone system
应急电话选择器 emergency telephone selector
应急电键 emergency key
应急电缆 breakdown cable;emergency cable;interruption cable;jumper cable;relief cable
应急电力装置 emergency power installation
应急电流 emergency electric(al) current
应急电台 contingency station
应急电源 emergency pack;emergency power;emergency power source;emergency supply
应急电源控制台 emergency power controlling set
应急电源手柄 emergency power handle
应急电源组 emergency power package
应急电站 emergency (reserve) power station
应急吊杆索具 rechange
应急叠梁闸门 emergency stoplog
应急定位发射机 emergency locator transmitter
应急定位信标 crash-locator beacon
应急定位应答器 emergency location transponder
应急动阀 emergency valve
应急动力 emergency power
应急动力电源 emergency power supply
应急断电(开关) emergency-off
应急断开 emergency disconnection
应急断开电门 emergency disengage switch
应急舵 jury rudder;relieving rudder;substitute rudder
应急舵柄 emergency tiller;relieving tiller
应急舵缓冲绞辘 rudder tackle
应急舵机 emergency steering gear
应急舵链 rudder chain;rudder vang
应急舵链挂环 rudder horn
应急舵链索 boggin line;rudder pendant
应急发电机 emergency generator;abort engine; reserve generator;stand-by generator

应急发电机室 emergency generator room
应急发电机组 emergency generating set
应急发电装置 emergency generator
应急发动机 emergency engine
应急发射机 emergency transmitter
应急发射指向标 emergency transmitter beacon
应急发送机 emergency transmitter
应急阀(门) emergency closing valve;stand-by valve
应急帆 jury rig;jury sail
应急反应 emergency reaction
应急返航发动机 take-home engine
应急方案 emergency plan
应急防喷器 emergency blowout preventer
应急放电 emergency discharge
应急放起落架手柄 emergency extension lever
应急放下 emergency extension
应急放下系统 emergency extension system
应急放油 emergency discharge
应急放油泵 fuel jettison pump
应急放油装置 fuel jettison
应急费(用) contingence [contingency] cost;contingency fund;contingency reserve;contingency sum;contingent charges;crash cost;emergency fund
应急分离火箭 escape rocket
应急服务(机构) emergency service
应急服务线 emergency service line
应急辅助消防水管 standpipe-first-aid
应急复印图 tide-over run
应急干粮 emergency ration
应急高频备用(干)线 emergency high frequency back-up trunk
应急隔墙 emergency stopping
应急工程 emergency works
应急工程项目 emergency project
应急工具 emergency kit
应急工作 emergency operation;emergency works
应急工作状态 stand-by mode of operation
应急供电 emergency service
应急供水 emergency water supply
应急供氧设备 emergency oxygen equipment
应急供应 emergency accommodation
应急构件 jury
应急鼓风机 emergency blower
应急关闭装置 emergency closing device;emergency closure gate
应急关断 emergency cut-out
应急关断开关 emergency stop cock
应急管理系统 emergency management system
应急管闩 emergency stop cock
应急管栓 emergency stop cock
应急管线 emergency pipe line
应急轨 emergency rail
应急轨夹 emergency rail clamp;emergency strap
应急过程 contingency procedure
应急互接 emergency interconnection
应急话音报警通信[讯]系统 emergency voice alarm communication system
应急话音通信[讯]系统 emergency voice communication system
应急活动计划 emergency action plan
应急基地 emergency centre
应急基金 contingent fund;emergency fund
应急计划 contingency plan;crash program(me);crash project;emergency plan;emergency program(me);project crashing
应急记录标绘 emergency recorder plot
应急剂量 emergency dose
应急加速度计 crash accelerometer
应急监测 emergency monitor(ing)
应急减灾措施 emergency relief measure
应急检测系统 emergency detection system
应急交换 emergency exchange
应急教堂 emergency church
应急接地 emergency grounding
应急接收机 emergency receiver
应急接头夹板 emergency joint bar
应急接线 emergency connection
应急经费 contingency appropriation
应急井 emergency shaft
应急警铃信号 emergency alarm signal
应急救生浮标 emergency buoy
应急救生艇 accident boat;emergency lifeboat
应急救险修理 emergency repair
应急救援操作 emergency rescue operation
应急救援设备 emergency and rescue equipment
应急救助气垫船 air cushion crash rescue vehicle
应急居住计划 emergency housing scheme
应急开关 emergency cock;emergency stop cock;emergency switch;intervention switch;panic button
应急靠泊设备 emergency docking facility
应急空气系统 emergency air system
应急空气压缩机 emergency air compressor
应急控制站 emergency control station
应急快速解脱装置 emergency quick-release device
应急冷却剂再循环 emergency coolant recirculation
应急冷却剂注入 emergency coolant injection
应急冷却器 emergency cooler
应急冷却系统 emergency coolant system;emergency cooling system
应急离机舱口 escape hatch opening
应急离机操纵 emergency escape control
应急离机程序 emergency escape procedure;emergency procedure;escape sequence
应急离机出口 escape exit
应急离机救生包 <内装地图> escape kit
应急离机设备 emergency escape equipment;emergency escape provision;escape equipment
应急离机事故 escape emergency
应急离机手柄 escape handle
应急离机手段 means of escape
应急离机系统 emergency escape system
应急林 emergency district
应急临时建筑 emergency temporary construction
应急临时住宅 emergency house
应急路由 emergency route
应急罗盘 emergency compass
应急锚 jury anchor;kelleg
应急锚地 emergency anchorage
应急门 emergency door;maintenance gate
应急排风 emergency exhaust
应急排气 emergency exhaust
应急排气阀 emergency exhaust valve
应急排气风扇 emergency exhaust fan
应急排气孔 emergency vent
应急排气口 emergency vent
应急排汽 emergency exhaust
应急排水泵 salvage pump
应急抛载装置 emergency ejecting device
应急配电板 emergency switchboard
应急配电盘 emergency switchboard
应急频率 emergency frequency
应急评估法 contigent valuation method
应急起落架 emergency landing gear
应急起重机 breakdown crane
应急气瓶 emergency cylinder
应急汽油箱 emergency gasoline tank
应急器材 emergency duty
应急器材柜 damage control locker
应急桥 freeway bridge;treadway bridge
应急切断 emergency switching-off
应急切断装置 emergency cut-off
应急情况分析 emergency analysis
应急取料斗 emergency reclaim hopper
应急权宜措施 expedient measure to meet an emergency
应急燃料 emergency back-up fuel;emergency fuel
应急燃料切断开关 emergency fuel cut-off switch
应急燃料调节器 emergency fuel regulator
应急燃料箱 emergency fuel tank
应急人员 emergency personnel
应急日程表 crash schedule
应急润滑 emergency lubrication
应急刹车 emergency brake
应急设备 emergency accommodation;emergency equipment;emergency installation;emergency outfit;emergency plant;emergency set;panic equipment
应急设备壁龛 niches for emergency equipment
应急设施 emergency facility;emergency service
应急生命保障系统 emergency life support system
应急绳索 emergency rope
应急时间 crash time
应急识别信号 emergency identification signal
应急食品 emergency food
应急释放开关 emergency release push
应急收报机 emergency receiver
应急收信机 emergency receiver
应急手泵连接器 emergency hand pump connector
应急手动装置 emergency hand-drive
应急水阀门 emergency water valve
应急水龙带 first-aid hose
应急水下通话 emergency through-water speech
应急水下通信[讯] emergency through-water communication
应急水下作业 emergency under-water work
应急说 emergency theory
应急送硼探测器 emergency feed boron detector
应急速闭阀 quick-closing emergency valve
应急速度 emergency speed
应急索具 jury rig
应急弹射 emergency ejection
应急弹射座椅 emergency ejector seat
应急胎链 emergency tire [tyre] chain
应急太平门建筑物 emergency exit structure
应急梯 emergency escape ladder;emergency stair(case);escape ladder;fire escape ladder
应急梯子 emergency ladder
应急天线 emergency antenna
应急调速器 emergency governor
应急跳闸机构 emergency trip mechanism
应急停车场 emergency parking
应急停车车道 emergency stopping lane
应急停电 emergency power shut-off
应急停堆系统 last moment emergency shut-down
应急通道 escape hatch
应急通风 emergency vent;emergency ventilation
应急通风口 emergency venting
应急通路 emergency route
应急通信[讯] crisis communication;critical communication;emergency communication
应急通信[讯]网 emergency communication net
应急通信[讯]业务 emergency service
应急通信[讯]终端站 emergency communication terminal
应急图 provisional chart
应急拖缆 emergency tow rope;insurance hawser;salvage hawser
应急脱扣 emergency trip
应急脱离舱 emergency escape capsule
应急脱离索 emergency escape cable
应急脱离塔 escape tower
应急脱险舱 emergency escape chamber
应急桅 jury mast
应急维修 emergency maintenance
应急维修时间 emergency maintenance time
应急卫星通信系统 emergency satellite communications system
应急喂料机 emergency feeder
应急涡轮发电机 emergency turbine generator
应急污水泵 emergency bilge pump
应急屋顶走道 <防火灾用> emergency roof walkway
应急无线电设备 emergency radio
应急无线电示位标 emergency position-indicating radio beacon
应急无线电装置 radio emergency installation
应急物资 critical material
应急系统 emergency system;stand-by system
应急响应计划 emergency response plan
应急消毒措施 emergency decontamination measure
应急消毒设备 emergency disinfection apparatus
应急消防泵 emergency fire pump
应急(消防)立管 first-aid standpipe
应急泄洪口 emergency outlet
应急信号灯 emergency signal lamp
应急信息处理 emergency message handling
应急行动 emergency action
应急修理 emergency maintenance;jury repair
应急修理车 emergency repair truck;hurry-up wagon
应急修理船 emergency repair ship
应急需要 emergency requirement
应急旋塞 emergency cock
应急旋转换流机 emergency rotary converter
应急掩蔽部 emergency shelter
应急演习 emergency drill

应急堰 emergency weir
应急氧气 emergency oxygen
应急氧气设备 emergency oxygen apparatus
应急遥测术 emergency telemetry
应急移动通信 emergency mobile communication
应急溢洪道 emergency spillway
应急溢流 emergency overflow
应急溢流管 emergency overflow pipe
应急因素 contingency factor
应急用备料 emergency material
应急用泵 emergency pump
应急用电池 emergency battery; emergency cell
应急用电台 emergency set
应急用广播系统 emergency broadcast system
应急用户电报路由 emergency telex route
应急用料储备 stock of urgently needed material
应急用路线 emergency route
应急用桥 emergency bridge
应急用无线电信道 emergency radio channel
应急用装置 emergency feature
应急油箱 emergency tank
应急预案 emergency program(me)
应急钥匙 emergency key
应急(运行)方式 emergency mode
应急运转 emergency operation
应急运转特性 emergency-run characteristic
应急再启动 emergency restart
应急闸 emergency lock
应急闸门 emergency closure gate; emergency gate; emergency guard gate; guard gate; safety lockage; bulkhead gate
应急账户 contingency account
应急照明 emergency illuminance; emergency illumination; emergency lighting
应急照明电池 emergency lighting battery
应急照明电机 emergency lighting set
应急照明电门 emergency light switch
应急照射 emergency exposure
应急支柱 jury strut
应急(指令)序列 intervention sequence
应急制动控制继电器 emergency braking control relay
应急制动装置 emergency brake gear
应急住房 emergency housing
应急住所 emergency accommodation; emergency dwelling unit
应急住宅 emergency dwelling; temporary dwelling
应急注浆材料 grouting material in urgent need
应急贮液器 emergency receiver
应急转储 panic dump
应急转储程序 panic dump routine
应急装置 emergency apparatus; emergency device; emergency installation; emergency set; emergency unit
应急状态 emergency condition
应急准备金 contingent reserve
应急着陆场 emergency landing ground
应急着陆程序 emergency landing procedure
应急着陆地带 emergency landing strip
应急自溃堤段 fuse plug
应急走廊 < 防火灾用 > emergency corridor
应急组织 emergency organization
应聘者 candidate
应求系统 on-demand system; on-request system
应时的 in season
应时色 season colo(u)r
应试教育 examination-oriented education system
应试者 candidate
应征人 applicant
应征人员 applicant

映

映成 mapping of a set onto another
映出轮廓 silhouette
映光 illumination
映光器 illuminator
映画器 delineascope; magic lantern; projecting lantern; projection lantern
映绘台 light table
映满的映射 onto mapping
映谱仪 spectral projector
映入 map into; mapping of a set in another
映入的映射 into mapping
映入自同构 meromorphism
映入自同构的 meromorphic
映山红 micranthum
映射【数】 mapping
映射变换 mapping transformation
映射次数 number of image
映射的合成 composition of mappings
映射的线性化扩充 linearized extension of map
映射的限制 restriction of mapping
映射的值域 range of a mapping
映射点类型 type of image point
映射抖动 mapping jitter
映射度 degree of mapping; mapping degree
映射法 image method; reflection method
映射干涉仪 projection interferometer
映射管 projection tube
映射函数 mapping function
映射缓冲区 mapped buffer
映射集 mapping ensemble
映射井 image well
映射空间 mapping space
映射扩张 mapping extension
映射类 mapping class
映射问题 mapping problem
映射误差 mapping error
映射序列 sequence of mapping
映射要素 image factor
映射载 mapping truck
映射轴平面图 mapping axis map
映射柱 mapping cylinder
映射装置 mapping device
映射锥 mapping cone
映象 image; map(ping); reflex
映象保真度 eye fidelity
映象编码 coded image
映象表 map table
映象变换信息 mapping
映象程序 map program(me); program(me) image
映象处理 image process(ing)
映象处理程序 image processor
映象存储器 mapped memory; mapped storage
映象存储区 image store
映象的多重性 multiplicity of image
映象度 mapping degree
映象段 image section
映象段描述符 image section descriptor
映象法 method of images
映象方式 image mode
映象分析 image analysis
映象函数 mapping function
映象缓冲器 mapped buffer
映象激活程序 image activator
映象记忆 iconic memory
映象空间 mapping space
映象名 image name
映象内存显示 memory-mapped video
映象平面 image plane
映象屏 projection screen
映象器 projection instrument
映象图形学 image graphics
映象系统 image system
映象信号 image signal
映象旋量 image screw
映象语言 mapping language
映象装置 mapping device

硬

硬安全帽 hard hat
硬拔线 hard(-drawn) wire
硬白垩 hard chalk
硬柏油 tar pitch
硬板 hard board; rigid plate
硬背装订 drawn on
硬焙烘 hard cure
硬币 coin; effective coin; hard cash; hard currency; hard money; metallic currency; mintage; real money; base coin < 英 >
硬币边纹 milling
硬币等的正面 obverse
硬币反面 verso
硬币划痕试验 coin scratch test
硬币控制 coin control
硬币上的印记 mintage
硬币识别器 coin validator
硬币试验 coin test
硬币样品箱 pyx
硬币准备 specie reserve
硬币自动机 slot-machine
硬边 hard-edge
硬变 cirrhosis
硬表层 crust
硬表面 crust; hard (sur)face
硬表面板 hard surface plate
硬表面的 hard-faced
硬玻璃 Bohemian glass
硬布线逻辑 hard-wire logic
硬部分子 hard parton
硬材 broadleaf (wood); broad-leaved wood; hard wood; leaf wood
硬材板 hardwood board
硬材扶杆 hardwood rail
硬材干馏 hardwood distillation
硬材焦油 hardwood tar
硬材沥青 hardwood tar pitch
硬材料垫层 hard core bed
硬材木炭 hardwood charcoal
硬材扦插 hardwood cutting
硬材树种 < 一些豆科的 > locust
硬材枕木 hardwood sleeper
硬彩 hard colo(u)r; strong colo(u)r
硬草草本群落 duriherbosa
硬草草甸 duriprate
硬层 hard pan; rind
硬层锈菌属 < 拉 > Stereostratum
硬层钻进 hard drilling
硬插条 hardwood cutting
硬掺和料 < 提高沥青稠度 > hard flux
硬衬布 buckram
硬衬毛毡 bratisheen
硬成分 hard component
硬齿多用锯片 hard tooth combination saw blade
硬虫胶 hard lac (resin)
硬稠混凝土 stiff concrete
硬稠性 stiff consistency [consistence]; stiff scale of index
硬触点 hard contact
硬瓷 hard paste; hard porcelain
硬磁材料 magnetically hard material
硬磁合金 magnetically hard alloy
硬磁盘机 hard disk drive; rigid disk drive
硬磁铁 rigid magnet
硬磁铁氧体 hard ferrite; permanent-magnetic ferrite
硬磁性材料 hard magnetic material; magnetically hard material; retentive material
硬磁性合金 retentive alloy
硬粗纱 hard roving
硬簇射 hard shower
硬脆地层 friable hard formation; hard broken ground
硬脆蜡 ceresin(e)
硬脆岩石 hard brittle rock
硬搓绳 hard laid; short laid
硬错误 hard error
硬错误率 hard error rate
硬错状态 hard error status
硬带材 hard strip
硬贷款 hard loan
硬导管 fibre conduit
硬道路面 hard road
硬道面滑水现象 rigid surface planning
硬的 douke; petrosal; robust; stony; unyielding
硬的金刚石类材料 hard diamond like material
硬等待 hard wait
硬底 hard bottom; hard ground
硬底层 hard core
硬底构造 hard ground structure
硬底海滩 hard beach
硬底土 hard subsoil
硬底质 hard-bed material
硬地板漆 hard floor paint
硬地板砖和砌砖 < 由黏[粘]土烧结的 > attoc
硬地层 hard formation; hard layer; hard subsoil
硬地(基) hard ground; solid ground
硬地基土 hard subsoil
硬地蜡 boryslowite
硬地沥青混合料 < 采用硬地沥青粉作黏[粘]结料的混凝土混合料 > hard asphalt mix
硬地沥青 < 针入度 10 以下的 > hard asphalt
硬地面 firm ground; hard stand
硬地面肥育地 paved feedlot
硬地面通过机动性 hard surface mobility
硬地面通过特性 hard surface mobility
硬地停车场 hard parking-place
硬地网球场 hard court
硬垫 hard packing
硬垫层 hard core
硬垫土 pan soil
硬凋的 contrasty
硬调 high contrast
硬调透镜 hard focus lens
硬调图像 hard image
硬顶 hard top
硬顶式集装箱 < 一种顶开式集装箱 > hard top container
硬顶小客车 hard-top minibus
硬冻胶砂强度试验法 earth-dry mortar strength test
硬冻土 hard-frozen soil
硬度 degree of hardness; hardness; hardness solidness; rigidity; solidity; stiffness

硬度比 hardness ratio;solidity ratio
硬度变化量 hardness change
硬度标 hardness ga(u)ge;hardness point;hardness scale (degree)
硬度标度 hardness degree; hardness scale (degree);hard scale;scale of hardness
硬度标准 scale of hardness
硬度表 hardness scale (degree); scale of hardness
硬度测定(法) hardness determination; determination of hardness; hardness test(ing)
硬度测定计 durometer
硬度测定器 durometer
硬度测量 hardness measurement
硬度测量仪 < X 射线 > qualimeter
硬度测试 hardness test
硬度测试机 hardness testing machine
硬度测试仪 hardness tester;hardness testing device;hardness testing machine
硬度测试自动化 hardness testing automation
硬度尺寸 hardness scale (degree)
硬度等级 class of hardness;hardness scale (degree);scale of hardness; hardness class
硬度法 < 非破损性混凝土强度测定法 > hardness method
硬度范围 hardness range; range of hardness
硬度分类 class of hardness
硬度分析 hardness analysis
硬度分析仪 hardness analyser [analyzer];sclerometer;scleroscope
硬度划分 hardness demarcation
硬度划痕试验 scratch test
硬度换算 hardness conversion
硬度机 indentation machine
硬度极限 hardness limit;hardness range
硬度计 hardness ga(u)ge;hardness meter; hardness testing device; hardness testing machine;hardometer; hard scale; sclerometer; hardness scale (degree)
硬度计表盘刻度 hardness scale (degree)
硬度计测定的硬度 sclerometric hardness
硬度计球 indenting ball
硬度计压头 indenter
硬度计针入度 indentation of durometer
硬度计锥形压头 conic(al) indenter of durometer
硬度结垢 scale of hardness
硬度控制 hardness control
硬度老化 hardness ag(e)ing
硬度裂缝 hardness crack
硬度判定 hardness determination
硬度器 < 一种装有金刚石尖头的标准锤 > scleroscope
硬度曲线 hardness curve
硬度蠕变 hardness creep
硬度软化剂 hardness reducer
硬度深度曲线 hardness-depth curve
硬度实验 hardness experiment
硬度实验机 hardness experiment machine
硬度矢量 hardness vector
硬度试验 hardness test(ing);scratch test
硬度试验荷载 stiffness test load
硬度试验机 hardness tester;hardness testing machine
硬度试验计 hardness tester;hardness testing machine
硬度试验球凹 ball impression of hardness test
硬度试验压入深度 depth penetration of hardness test
硬度试验压头 penetrator
硬度试验仪 durometer
硬度适当 moderately robust
硬度衰减 hardness loss
硬度损失 hardness loss;loss of hardness
硬度系数 coefficient of hardness;coefficient of stiffness;stiffness coefficient
硬度显示器 scleroscope
硬度相纸 contrast grade
硬度消失 hardness loss
硬度泄漏 hardness breakthrough
硬度值 hardness factor; hardness number;hardness value
硬度值转换 hardness number conversion
硬度指标 hardness number
硬度指数 hardness index;indentation hardness number
硬度坠落试验 shatter test
硬断面 rigid section
硬锻带 hardenability band
硬而有光泽 to be hard and shiny
硬珐琅质 hard gloss
硬珐琅质涂层 hard glazed coat(ing)
硬反馈 follow-up;rigid feedback
硬反馈控制器 rigid feedback controller
硬防蚀剂 hard ground
硬沸石 bavenite;duplexite
硬费用 hard cost
硬分段 hard sectoring
硬酚醛树脂 hard phenolic resin
硬风景 hard landscape
硬枫木 hard maple
硬腐 < 初期腐朽 > hard rot
硬腐泥 saprocol
硬副本 hard copy
硬盖 carapace;hard coat
硬干 hard dry
硬干时间 hard drying time
硬杆扫测【疏】 bar sweeping
硬刚纸轴承 hard fiber bearing
硬钢 < 即高碳钢 > high-speed steel; high steel;rigid steel;hard steel
硬钢杆 hard-grade billet
硬钢管 rigid steel conduit
硬钢筋 hard-grade bar
硬钢坯段料 hard-grade billet
硬钢丝 hard-wire
硬钢丝绳 coarse wire rope;inflexible steel rope;inflexible steel wire
硬革 hard leather
硬格式 hard format
硬铬 hard chrome;hard chromium
硬铬电镀 hard chromium plating
硬铬镀层 hard chromium plating
硬铬尖晶石 chrompicotite
硬垢水 hard filth water
硬垢系数 coefficient of hard filth
硬垢总量 total weight of hard filth
硬骨料 high stone
硬固性条件 condition of stiffness
硬刮槽工具 latterkin
硬管 rigid conduit
硬管脉冲发生器 hard-tube pulser
硬管调制器 hard-tube modulator
硬光漆 hardening gloss paint; hard gloss paint
硬光泽 hard gloss paint
硬硅钙石 eakleite;xonotlite
硬果 < 如胡桃、栗等 > nut
硬果松 bristle cone pine
硬海绵 hard-head sponge
硬焊 braze; braze welding; brazing; hard brazing
硬焊的 brazed;hard soldering
硬焊合金 brazing alloy
硬焊剂 hard soldering
硬焊接 brize weld(ing)
硬焊接的 hard soldered
硬焊接缝 hard soldered joint
硬焊接合 hard soldered joint
硬焊料 brazing filler metal;hard solder;ironstone solder;spelter solder
硬焊料焊接 hard solder joint
硬焊药 hard solder
硬合金补强 set with hard alloy
硬合金钢心辙叉 hard-centered frog
硬合金钻头 hard metal bit;tungsten-carbide bit
硬核(心) hard core
硬横跨 portal structure
硬横梁 portal beam
硬红木 beef wood
硬红木着色用土红(颜料) abraum
硬滑石 indurated talc
硬滑脂 hard grease
硬化 cementation; feeding-up; harden; indurate; metallization; petrify; sclerotization; setting up; vulcanize;maturing < 混凝土的 >;sclerosis < 木材细胞壁的 >;live ring < 指涂料油漆的变质 >
硬化饱和脂 hard saturated fat
硬化保护层 hardening rule
硬化本领 hardening capacity
硬化变形 curing strain
硬化玻璃 hard glass
硬化薄壁组织 sclerotic parenchyma
硬化不足 under-ag(e)ing;undercuring
硬化材料 curing material; hardened material
硬化层 reinforcing coat
硬化层积材 < 用酚脂胶合 1/16 ~ 1/30英寸单板 > amber wood
硬化层积木材 hardened layer wood
硬化层压木板 densified laminated wood
硬化成分 hardener
硬化程度 degree of hardness;degree of induration; hardenability; hardening degree
硬化处理 carburising;carburizing; hardening treatment
硬化催化剂 hardening catalytic agent
硬化带 hardenability band
硬化的 hardened;indurated;petrean; petrified;petrous;vulcanized
硬化的水泥 matured cement
硬化的岩石 indurated rock
硬化点 condensation point; hard(ening) point
硬化电路 hardened circuit
硬化度 hardness penetration
硬化法 hardening process
硬化反应 reaction of hardening;sclerous reaction
硬化放射虫软土 radiolarite
硬化粉剂 hardener dust
硬化钢 chilled steel; hardened steel; quenched steel
硬化钢环箍 hardened steel tire [tyre]
硬化工 hardener
硬化规则 hardening rule
硬化滚压 < 轴颈圆角等的 > hard rolling
硬化过程 hardening behavio(u)r; hardening process; process of curing; process of hardening; process of setting;setting process
硬化灰浆 hardened mortar; mortar stone
硬化混凝土 < 反复荷载后消除残余变形,呈稳定弹性的混凝土 > work-hardened concrete; hardened concrete;matured concrete
硬化机 hardener
硬化机理 hardening mechanism
硬化基部 sclerobase
硬化剂 accelerator;catalyst;curing agent; hardener; hardening admix(ture); hardening agent; rigidizing agent; setting-up agent; stiffener; stiffening agent;vulcanizing agent
硬化加速剂 hardening accelerating admixture; hardening accelerator (admixture)
硬化金属 hardening metal
硬化进程 progress of hardening
硬化浸填体 sclerotic tylosis
硬化矩阵 hardening matrix
硬化聚氯乙烯 hardened PVC
硬化绝缘板 vulcanized fiber [fibre]
硬化理论 theory of hardening
硬化沥青 hardening asphalt; hardening bitumen;manjak;munjack
硬化量 hardening capacity
硬化龄期 age of hardening
硬化炉 hardening furnace
硬化铝合金 hardening alumin(i)um alloy
硬化率 cementation index; hardening rate;hardening ratio
硬化面 hardened face
硬化面层 hardening coat
硬化模量 hardening modulus
硬化能 energy of hardening
硬化能力 ability to harden;hardening capacity;hardening energy
硬化能量 hardening energy
硬化泥浆 hardened mortar; mortar stone
硬化黏[粘]土 clunch;indurated clay
硬化黏[粘]土岩 skleropelite
硬化凝胶 stiff gel
硬化谱 hardened spectrum
硬化谱分布 hardened distribution
硬化期 hardening age;hardening period;period of hardening;setting period
硬化钳 hardening tongs
硬化强度 setting strength
硬化侵填体 sclerotic tylosis
硬化区 hardening zone;hard zone
硬化曲线 curve of hardening;hardening curve
硬化屈服值 hardening yield value
硬化热 hardening heat; heat of hardening;setting heat
硬化乳化层 hardened emulsion
硬化砂浆 hardened mortar; mortar stone
硬化砂浆找平层 hard screed
硬化烧石膏 set plaster
硬化深度 < 钢材的 > hardness penetration;depth of hardening
硬化时间 hardening time
硬化时效 hardening age;hardening ag(e)ing;hardness ag(e)ing
硬化试验 hardening test
硬化室 hardening chamber;hardening room
硬化水 petrifying water;water of setting < 水泥的 >
硬化水泥 hardened cement;solidified cement
硬化水泥浆(体) hardened cement paste
硬化水泥石中的硫铝酸钙结晶 Candlot's salt
硬化松香 hardened rosin
硬化速度 hardening rate;rate of hardening;setting rate;velocity of hardening
硬化速率 hardening rate;rate of cure; rate of hardening

硬化塑料 rigid plastics
硬化碳 hardening carbon
硬化特性 hardenability; hardening characteristic; property of setting
硬化特征 hardening characteristic
硬化体 hardenite
硬化填充料 hardener filler
硬化填料 hardened dust
硬化温度 hardening temperature; hard-temperature; stiffening temperature
硬化细胞 sclereid
硬化纤维板 chemboard; vulcanized fiber[fibre]
硬化箱 hardening cabinet
硬化橡胶 ebonite; hardened rubber; rigidified rubber; vulcanite
硬化效应 hardening effect
硬化锌合金 binding metal
硬化型双曲线性系统 hardening bilinear system
硬化性能 ability to harden; hardenability; hardening capacity
硬化性质 property of setting
硬化窑 hardening kiln
硬化液 hardening bath; hardening liquid; petrifying liquid
硬化应变 hardening strain
硬化应力 hardening stress
硬化硬煤沥青 sulphurized coal-tar pitch
硬化油 fixed oil; hardened oil; hardening oil
硬化与研磨 harden and grind
硬化浴 hardening bath
硬化原 sclerogen
硬化炸药 hardened explosive
硬化脂 hardened fat
硬化纸板 fibreboard; parchmentized fibre; vulcanized fiber [fibre]; vulcanized paper
硬化纸板锤 fibre hammer
硬化指数 cementation index; hardenability value
硬化中的混凝土 till cured concrete
硬化作用 hardening action < 混凝土 >; vulcanicity; vulcanization; induration
硬黄熟期 hard-dough stage
硬黄铜 hard brass; high brass
硬黄质结晶 sclerocrystalline
硬灰煤 hard ash coal
硬灰石 rag stone
硬灰质板岩 slater
硬辉沥青 harbolite
硬回火薄板 < 布氏 1 号硬度 > hard temper sheet
硬回火的薄钢板 hard temper sheet
硬击穿 hard breakdown
硬基 solid bed
硬基土 hard subsoil
硬基岩 hard-base rock
硬极限 hard limiting
硬挤出机 stiff extrusion press
硬给水 hard feed water
硬加法器 hard adder
硬碱 hard base
硬件 hardware
硬件表示(法) hardware representation
硬件部件 hardware component; hardware subassembly
硬件层 hardware layer
硬件成分 hardware component
硬件错误恢复管理系统 hardware error recovery management system
硬件的估算 evaluation of hardware
硬件堆栈 hardware stack
硬件仿真程序 hardware emulator
硬件封锁 hardware lockout
硬件复位 hardware reset
硬件构造 hardware construction
硬件固件软件权衡 hardware firmware software made-off; hardware firmware software trade-off
硬件故障 hardware fault; hardware malfunction
硬件关联 hardware context
硬件管理 hardware management
硬件互锁寄存器 hardware interlock register
硬件汇编程序 hardware assembler
硬件汇编器 hardware assembler
硬件基地址 hardware base address
硬件计时器 hardware timer
硬件计算机 bare computer; bare machine
硬件技术 hardware technology
硬件兼容性 hardware compatibility
硬件监督程序 hardware monitor
硬件监视 hardware monitoring
硬件监视器 hardware monitor
硬件检验 hardware check; wire-in check
硬件接口 hardware interface
硬件进程控制块 hardware process control block
硬件开发 hardware development
硬件刻图膜 hard texture
硬件控制 hardware control
硬件逻辑 hardware logic
硬件描述语言 hardware description language
硬件模块 hardware module
硬件模块化 hardware modularity
硬件模拟程序 hardware simulator
硬件内核 hardware core
硬件判优器 hardware arbiter
硬件配置 hardware configuration
硬件缺陷 hardware deficiency
硬件冗余(法) hardware redundance [redundancy]
硬件软件对接 hardware-software interface
硬件软件接口 hardware-software interface
硬件软件支持 hardware-software support
硬件软件支援 hardware-software support
硬件设计 hardware design
硬件数字处理 hardware digital processing
硬件体系结构 hardware architecture
硬件填充序列 hardware padding sequence
硬件通 hacker
硬件同步机制 hardware synchronization mechanism
硬件图形发生器 hardware pattern generator
硬件网络 hardware net
硬件系统 hardware system
硬件系统可靠性 hardware system reliability
硬件系统可用性 hardware system availability
硬件项目 hardware item
硬件选择标准 hardware selection criteria
硬件选择器 hardware selector
硬件遥测 hardware telemetry
硬件优先(权)中断 hardware priority interrupt
硬件支持 hardware support
硬件支援 hardware support
硬件指令 hardware instruction
硬件中断级 hardware interrupt level
硬件中断设备 hardware interrupt facility
硬件资源 hardware resource
硬件资源分配 allocation of hardware resource
硬件资源控制 hardware resource control
硬浆 hard stock
硬胶 sinew glue
硬胶板 hard rubber sheet
硬胶蔽电池极板 ebonite clad plate
硬胶皮螺丝帽 ebonite stud
硬胶丝 hard gummed skein
硬胶套管 ebonite bush
硬焦炭 hard coke
硬焦油沥青 asphalt-tar pitch; hard pitch
硬焦油脂 hard pitch; pitch; tar pitch
硬焦油脂灌浆 pitch grout(ing)
硬焦油脂碎石路 pitch-grouted macadam
硬焦油脂填缝料 pitch filler
硬礁岩 klintite
硬角叉心 acute frog
硬接蜡 solid wax
硬接线 hard-wired
硬接线逻辑 hard-wired logic
硬节 < 木材等的 > firm knot; knurl; sound knot
硬结 concretion; consolidate; durinode; hard-drying; hardening; hard-settling
硬结表(土)层 dry crust
硬结的水泥 solidified cement
硬结地层 concretionary horizon
硬结构 hard structure
硬结混凝土 hardened concrete
硬结期【地】 locomorphic phase; locomorphic stage
硬结热 heat of setting
硬结填料 hardening filler
硬结土(壤) hard cemented soil; indurated soil
硬结作用 induration
硬金属病 hard metal disease
硬金属管 rigid metal conduit
硬金属铠装电缆 hard metal sheathed cable
硬金属胎体 hard matrix
硬金属套电缆 hard metal sheathed cable
硬金云母 pholidolite
硬紧黏[粘]土 gumboti
硬浸焊 dip brazing
硬晶岩石 hard crystalline rock
硬静止起动 hard still enable
硬聚合物 hard polymer
硬聚氯乙烯 rigid PVC; unplasticised polyvinyl chloride
硬聚氯乙烯管 hard polvinyl choride pipe; rigid PVC conduit
硬卷装 hard yarn package
硬拷贝 hard copy; page copy
硬拷贝材料 hard-copy material
硬拷贝记录 hard-copy log
硬拷贝输出 hard-copy output
硬拷贝图板 hard-copy plotting board
硬拷贝外围设备 hard-copy peripherals
硬拷贝终端 hard-copy terminal
硬科学 hard science
硬科学家 hard scientist
硬颗粒 hard particle; rigid granule
硬壳 duricrust; encrusting matter; incrustation; rind; scale
硬壳坝 shell dam
硬壳蚌 hard shell clam
硬壳的 hard-shelled
硬壳蛤 hard shell clam
硬壳化 encrustation
硬壳机身 monocoque body
硬壳结构的 monocoque
硬壳帽 hard hat
硬壳煤气表 cast case meter; hard case meter
硬壳球体模型 hard shell sphere model
硬壳式机身 monocoque
硬壳式结构 monocoque; monocoque construction
硬孔菌属 < 拉 > Rigidoporus
硬库巴树脂 hard copal
硬块 compact mass; hard lump; solid mass
硬块材料 hardpan material
硬块沥青 pitch block
硬矿块 burr
硬矿石 hard ore
硬矿渣 blast-furnace slag
硬拉 hard draw
硬拉的 hard drawing; hard drawn
硬拉钢丝 hard-drawn steel wire
硬拉铜接触线线夹 hard-drawn copper contact wire clip
硬拉铜线 hard-drawn copper wire
硬拉线 hard-drawn wire
硬蜡 hard wax; geocerite
硬蜡熟 hard dough
硬冷拔 hard draw
硬沥青 base-tar; candle pitch; mineral rubber; solid asphalt; uinta(h)ite
硬沥青熬煮锅 pitch boiler
硬沥青地沥青混合料 pitch-bitumen mixture
硬沥青地沥青结合料 pitch-bitumen binder
硬沥青灌浆 pitch grout(ing)
硬沥青检定 hard asphalt determination
硬沥青块 pitch grit
硬沥青玛瑞脂 hard mastic asphalt
硬沥青玛瑞脂地面 pitch mastic flooring
硬沥青玛瑞脂面层 pitch mastic flooring
硬沥青泥炭 pitch peat
硬沥青砂胶 pitch mastic
硬沥青填料 pitch filler
硬沥青毡 pitch felt
硬砾岩 conglomerate
硬粒 hard grain; hard seeds; shot; solids
硬连接 hard-wired
硬连接数控装置 hard-wired numerical control
硬连线 hard-wired
硬连线电路 hard-wired circuit
硬连线互连 hard-wired interconnection
硬连线控制器 hard-wired controller
硬连线逻辑 hard-wired logic
硬连线数控器 hard-wired numerical control
硬连线系统 hard-wired system
硬连线系统工程师 hard-wire-oriented engineer
硬连线指令 hard-wired instruction
硬练 < 混凝土或灰浆的 > dry consistency[consistence]; stiff consistence
硬练灰浆 dry mortar
硬练砂浆 dry mix; dry mortar
硬练砂浆强度试验(法) early-dry mortar strength test
硬练砂胶强度试验(法) earth-dry mortar strength test

硬练试验 test by dry mortar
硬链段 hard segment
硬量子 hard gamma
硬裂纹黏[粘]土 stiff fissured clay
硬裂隙黏[粘]土 stiff fissured clay
硬硫铋铅铜矿 lindstroemite
硬路基 hard core
硬路肩 hardened shoulder; hardened

verge;hard strip
硬路面 hard standing;rigid material;rigid pavement;road metal
硬路面的公路 hard-topped highway
硬铝 dural;duralumin;duralumin(i)-um;hard alumin(i)um
硬铝板 dural plate;duralumin plate
硬铝铆钉 duralumin(u)n rivet
硬铝石黏[粘]土 diaspore clay
硬铝线 hard-drawn alumin(i)um wire
硬率 hardness scale (degree)
硬绿泥石 chlorite spar;chloritoid;ottrelite;phyllite
硬绿泥石板岩 ottrelite-slate
硬绿泥(石)片岩 chloritoid schist
硬绿泥石千枚岩 chloritoid phyllite
硬绿黏[粘]土 stiff blue clay
硬绿蛇纹石 bowenite
硬麻布 <装订用> buckram
硬麻点 hard spot
硬镘光表面 hard-troweled surface
硬毛 bristle;hard hair
硬毛浮游生物 chaetoplankton
硬毛辊 bristle roller
硬毛漆刷 bristle brush
硬毛刷 bristle brush; dabber; scrubbing brush;stiff-bristled brush
硬锚 fixed termination
硬煤 anthracite; anthracite coal; bastard coal; coal stone; hards; hard coal <即白煤或无烟煤>
硬煤层刨煤机 hard coal plough
硬煤国际分类 hard coal international classification
硬煤沥青 coal pitch;coal-tar pitch
硬煤沥青地沥青混合料 pitch-bitumen
硬煤沥青悬浮液 coal-tar pitch dispersion
硬煤煤饼 hard coal briquet(te)
硬锰矿 black iron ore;psilomelane <氧化锰的混合物>
硬锰矿矿石 psilomelane ore
硬冕玻璃 hard crown
硬面 hard surface
硬面层 hard standing
硬面场地 hardstand(ing) apron
硬面处理 hard-facing
硬面道路 hard surface road
硬面合金 hard-facing alloy
硬面灰膏 hard-finished plaster
硬面层胶合板 plyron
硬面犁铧 hard-faced share
硬面路 hard surfaced road
硬面停车场 hard standing
硬面停机坪 hard standing
硬模 die
硬模件 hardware module
硬模浇铸 permanent casting
硬模铝合金 alumin(i)um permanent mo(u)ld alloy
硬模铸造 die casting;metal mo(u)ld casting;permanent mo(u)ld casting
硬模铸造铝合金 alumin(i)um permanent mo(u)ld casting alloy
硬膜 hard coat;hard film
硬膜防锈剂 hardening film preventives
硬磨石 hard grinding stone
硬母线 rigid busbar
硬木 deciduous wood;iron bark;ironwood;lignum vitae;tough wood
硬木板 hardwood board; hardwood plank;gerwood
硬木板条 hardwood strip
硬木板条地板 hardwood strip floor
硬木表饰 hardwood finish
硬木材 hard wood
硬木插 hardwood cutting
硬木插条法 hardwood cutting method
硬木产区 <美国南部> hammock
硬木导块 hardwood guide
硬木地板 hardwood floor(ing)
硬木刮槽工具 <撬起窗上铅条的硬木工具> latterkin
硬木轨枕 hardwood sleeper; hardwood tie
硬木护(舷)木 hardwood fender
硬木家具 hardwood furniture
硬木剪力键 hardwood shear key
硬木键 hardwood dowel
硬木胶合板 hardwood plywood
硬木焦油沥青 hardwood tar
硬木焦油脂 hardwood tar pitch
硬木节 sound knot
硬木块楼地面 hardwood block flooring
硬木块镶嵌地板 hard mosaic floor; hardwood mosaic floor
硬木料 firm wood
硬木毛地板 hardwood rough board flooring
硬木拼花地板 hardwood parquet flooring; overlay flooring; parquet flooring
硬木拼花企口板 hardwood tongue-groove parquet floor strips
硬木群落 hammock
硬木树 broad-leaved tree
硬木条形(企口)地板 hardwood strip flooring
硬木坞墩 hardwood timber block
硬木席纹地板 hardwood basket weave parquet flooring
硬木镶边条 railing
硬木镶块地板 hardwood mosaic floor
硬木楔 hardwood wedge
硬木旋钉 <管工用> turnpin
硬木旋塞 turning pin
硬木圆棒 mandril
硬木杂酚油 hardwood creosote
硬木直尺 hardwood straight edge
硬木轴承 hardwood bearing
硬木装修 hardwood finish
硬木座圈 <便桶用的> fiton pad
硬目标 hard target
硬耐火土 cat
硬泥 stiff mud
硬泥成型法 stiff-mud process
硬泥法 stiff-mud process
硬泥灰岩 marlite
硬泥挤坯 stiff extrusion
硬泥浆 stiff mud
硬泥岩 clift
硬泥造砖机 stiff-mud brick machine
硬泥制砖法 stiff-mud process
硬泥砖 stiff-mud brick
硬拈 hard twist
硬黏[粘]土 hard clay;firm clay;stiff clay;flint clay;leck;bass;batt;blue bind;blue stone
硬黏[粘]土层 clay pan;iron pan
硬黏[粘]土质页岩 irestone
硬捻 hard lay;hard twist
硬镍马氏体白口铸铁 Ni-hard martensitic white cast-iron
硬凝过程 hardening process
硬盘 hard pan;hard disk [disc];rigid disc【计】
硬盘叠 hard disk pack
硬盘盒 hard disk cartridge
硬盘接口 hard disk controller
硬盘驱动器 hard disk drive
硬盘土 hard-packed clay; orterde; ortstein;pan;pan soil
硬盘(土)层 <由铁质或钙质胶结的> duripan;hard pan
硬盘系统 hard disk system
硬盘柱(面) hard disk cylinder
硬磐 duripan;hard pan
硬磐泉 hardpan spring
硬磐土 indurated soil
硬泡 hard bubble;super-bubble
硬泡沫隔热板 rigid foam insulant
硬泡沫隔热衬料 rigid foam insulation furring
硬泡沫隔声板 rigid foam insulant
硬泡沫夹心的钢构件 steel rigid foam sandwich element
硬泡沫聚酯 rigid polyester foam
硬泡沫速率胶合剂 rigid foam bonding adhesive
硬泡沫塑料 rigid expanded plastics; rigid foam
硬泡沫塑料板 rigid foam board;rigid foam plastic board
硬泡沫塑料夹层板 rigid foam laminate(d) board
硬泡沫塑料黏[粘]结剂 rigid foam adhesive
硬泡沫塑料填料 rigid foam fill(ing)
硬泡沫塑性芯墙板 rigid foam core(d) wall pane
硬泡抑制 hard bubble suppression
硬喷漆 hard lacquer
硬硼钙石 colemanite
硬膨罐 hard swell
硬膨罐腐败 hard swell spoilage
硬坯 hard paste
硬皮 file-hard shell;hard bark;incrustate;late bark;sclerodermite
硬皮革齿轮 rawhide gear
硬皮书 hardback
硬片 plate negative
硬片暗盒 plate carrier
硬片递换 plate change
硬片晾匣 plate holder
硬片量测 plate measurement
硬片洗印槽 plate tank
硬片像对 pair of plates
硬片岩 hard schist
硬片照相机 plate camera
硬片转换盒 plate adapter
硬槭木 hard maple
硬漆 hard varnish
硬钎焊 <用铜锌合金焊接> brazing; hard soldering;solder brazing
硬钎焊连接 brazed joint
硬钎焊态 as-brazed
硬钎接合 brazed joint
硬钎料 brazing alloy; brazing filler metal;hard solder
硬铅 antimonial lead;hard lead
硬铅管(道) hard lead pipe;antimonial lead pipe
硬铅合金 hard lead alloy
硬铅皮 hard lead sheet
硬墙纸板 hardwall board
硬切刀 slab cutter
硬切换 direct cut operation
硬青黏[粘]土 stiff blue clay
硬青铜 hard bronze
硬球 hard sphere
硬球模型 hard-sphere model
硬球形流体 hard-sphere fluid
硬区 hard area
硬乳酪状黏[粘]土 cheese-hard clay
硬朊【化】 albuminoid
硬软件协调 hardware-software harmony
硬/软交错施工相互割切现场灌注桩挡土墙 hard/soft secant pile wall
硬软酸碱原理 hard-soft acid-base principle
硬润滑脂 hard grease
硬扫帚 stiff-bristled broom
硬砂砾 steel grit
硬砂岩 clasmoschist;graywacke[greywacke];hard sandstone;whine-rock
硬砂岩砂石 graywacke sandstone
硬山 flush gable roof
硬山搁檩屋顶 purlin(e) roof
硬山屋顶 Chinese gabled roof
硬珊瑚灰岩 coral rag
硬扇区格式 hard sectored format
硬伤 bruising;mechanical damage
硬烧 dead burn(ing);hard burning; hard burnt
硬烧白云石 shrunk dolomite
硬烧的 hard-burned
硬烧镁氧 hard magnesia
硬烧耐火器材 hard-burned refractory ware
硬烧耐火制品 hard-burned refractory ware
硬烧石膏灰泥 hard-burned plaster
硬烧陶瓷 hard-fired ware
硬烧氧化镁 hard-burned magnesia
硬烧游离石灰 hard-burned free lime
硬蛇纹石 antigorite; baltimorite; hampdenite;picrolite
硬蛇纹石石棉 picrolite asbestos
硬设备 equipment unit;hardware
硬设备校验 hard machine check
硬设施 hard facility
硬射流 hard shower
硬绳 coarse laid wire rope;hard laid wire rope
硬剩磁磁化 hard remanent magnetization
硬失效 hard failure
硬石 hard stone; rag stone; rock hard rock;adamant <原指金刚石和刚玉>
硬石层 solid bedrock
硬石产品 hard rock product
硬石粉末 hard rock flour
硬石膏 anhedritite; anhydrite; anhydrous calcium sulfate; anhydrous gypsum; cube spar; dead-burned gypsum; dead-burnt gypsum; gypsum anhydrite;karstenite;muriacite
硬石膏板 anhydrite board;anhydrite sheet
硬石膏绷带 anhydrite bander
硬石膏地带 anhydrite zone
硬石膏灰浆 anhydrite plaster; anhydrous gypsum plaster
硬石膏灰浆泥抹 anhydrite screed
硬石膏灰泥 <用硬石膏与催化剂制成的> anhydrite plaster; anhydrite cement;anhydrous gypsum plaster
硬石膏胶结料 anhydrite bander; anhydrite cement;Keene's cement;Martin's cement <含碳酸钾外加剂的>
硬石膏胶结物 anhydrite cement
硬石膏矿床 anhydrite deposit
硬石膏矿石 anhydrite ore
硬石膏凝结 concretion of anhydrite
硬石膏砌块 anhydrite block
硬石膏砂浆 anhydrite mortar
硬石膏-石膏矿石 anhydrite-gypsum ore
硬石膏石灰砂浆 anhydrite lime mortar
硬石膏水泥 anhydrite cement
硬石膏瓦 anhydrite tile
硬石膏无缝地板 anhydrite jointless floor(ing);anhydrite screed
硬石膏无缝面层 anhydrite jointless floor(ing)
硬石膏岩 anhydrock
硬石膏蒸发盐 anhydrite evaporite
硬石膏砖 anhydrite block;anhydrite tile
硬石膏砖块隔墙 anhydrite block partition
硬石灰 hard lime
硬石灰板 hard calcareous slate

硬石灰膏 adamant plaster
硬石灰石 ashburton marble
硬石灰质岩石 rag stone
硬石混凝土 hard rock concrete
硬石开采(场) hard rock quarry
硬石蜡 hard paraffin;ozokerite pitch;paraffin(e) wax;paraffinum durum
硬石石板 hard rock slab
硬石岩 hard rock
硬石岩骨料混凝土 hard rock concrete
硬石岩集料混凝土 hard rock concrete
硬石钻头 rock bit
硬矢面 hard vector surface
硬式操纵 rigid control
硬式传动机构 push-pull rod linkage;rigid control linkage
硬式飞艇 rigid airship;rigid dirigible
硬式浮舟 rigid boat
硬式结构 hard structure
硬式离合器板 rigid type clutch plate
硬式扫床【疏】 rigid bed-sweeping
硬式扫海 bar sweep
硬式巡逻飞艇 rigid patrol airship
硬式训练飞艇 rigid training airship
硬势 hard potential
硬饰面 hard finish
硬饰面石膏 hard finish plaster
硬书皮装订的 hardbound
硬树胶 copal;hard gum
硬树脂 animi resin;cured resin;hardened resin;rigid resin
硬数据 hard data
硬刷刷毛 <混凝土路面用> bristle brooming
硬刷(子) stiff broom;stiff brush
硬水 earth water;earth water scale-producing water;earthy water;hard water;scale-production water
硬水肥皂 hard water soap
硬水井 hard water well
硬水铝矿 diaspore
硬水铝石 diaspore;diasporite
硬水铝石高岭石岩矿芒 diaspore-kaolinite ore
硬水入口 hard water inlet
硬水软化 softening of water;water softening
硬水软化厂 water softening plant
硬水软化法 Clark's process;water softening method
硬水软化剂 water softening agent
硬水软化器 water softener
硬水软化设备 water softening plant;water softening unit
硬水软化装置 water softening equipment;water softening unit
硬伺服 hard servo
硬松 hard pine
硬塑料 plastomer
硬塑料板材 rigid plastic sheets
硬塑料成型法 stiff-plastic making;stiff-plastic process
硬塑料穿孔管 rigid plastic perforated pipe
硬塑料管 hard plastic tube;rigid plastic tube
硬塑料夹心板 rigid plastic sandwich panel
硬塑料模压 stiff-plastic mo(u)lding
硬塑料压制 stiff-plastic mo(u)lding
硬塑料造型(法) stiff-plastic mo(u)lding
硬塑料制模 stiff-plastic mo(u)lding
硬塑料制造 stiff-plastic making
硬塑压制法 stiff-plastic process
硬酸 hard acid
硬碎地层 broken hard rock
硬碎石 hard rock chip(ping)s
硬缩 sclerostenosis
硬炭 hard carbon
硬炭薄膜 hard carbon film
硬陶 terra-cotta clay
硬弹簧 hard spring
硬弹性区域 hard elastic region
硬弹性纤维 hard elastic fibre
硬天幕 cyclorama
硬填缝料 hard filler
硬铁 hard iron
硬铁核壳 ferricrust
硬停 hard stop
硬挺整理 hard finishing
硬通货 hard cash;hard currency;hard dollars;hard money;scarce currency;strong currency
硬铜 crimped copper;hard copper
硬铜绞合线 hard-drawn copper stranded conductor
硬铜绞线 hard-drawn copper strand wire
硬铜母线 rigid copper bus
硬铜线 hard-drawn copper wire
硬头 hard head
硬头榔头 hard-headed hammer
硬图像 hard picture;high contrast image
硬涂层系统 hard coat system
硬土 firm bottom;firm earth;firm ground;firm soil;hard soil;pan soil;stiff soil;stony soil;terra firma
硬土操作工人 hard ground man
硬土层 firm soil stratum;iron pan;orterde;ortstein;pan (formation);hard pan;fragipan
硬土层破碎机 pan breaker
硬土场地 firm site
硬土底盘 fragipan
硬土地基 firm ground
硬土堆 hard lump
硬土块 hard clod
硬土稳固地基 terra firma
硬弯 kinky
硬丸衬垫 shot backing
硬微丝煤 hard fusite
硬污泥 hard sludge
硬钨合金 Elwotite
硬物 adamant
硬物震动声 rattle
硬硒钻矿 trogtalite
硬锡合金 hard tin alloy
硬席车厢 <美> day coach
硬席的 semi-cushioned
硬席卧车 carriage with semi-cushioned berthes
硬席卧铺 semi-cushioned berth
硬席座车 carriage with semi-cushioned seats
硬系统 hard system
硬细石 <镶嵌细工用> piercing saw;pietra dura
硬下层土 hard subsoil
硬纤蛇纹石 metaxite
硬纤维 indurated fiber [fibre]
硬纤维板 hard board;masonite
硬纤维垫圈 hard fiber washer
硬纤维模板 fiber pan
硬纤维塞 hard fiber plug
硬纤维芯 hard fiber core
硬线逻辑 hard-wire logic
硬线逻辑系统 hard wired logic(al) system
硬限幅 hard limiting
硬限幅转发器 hard limiting transponder
硬限量积分器 hard limited integrator
硬限制 hard limiting
硬橡胶 ebonite;hardness rubber;hard rubber;solid rubber;vulcanite;vulcanized rubber
硬橡胶板 ebonite board;hard rubber board
硬橡胶薄板 hard rubber sheet
硬橡胶撑轮圈 hard rubber bead
硬橡胶垫 hard rubber base
硬橡胶方向盘 hard rubber steering wheel
硬橡胶管 hard rubber pipe;hard rubber tube
硬橡胶滚轮 hard rubber roller
硬橡胶护套 cabtire sheath(ing);cabtyre sheath(ing)
硬橡胶介质 solid rubber dielectric
硬橡胶绝缘电缆 vulcanized rubber insulated cable
硬橡胶壳 hard rubber box
硬橡胶轮胎 hard rubber tire [tyre]
硬橡胶片 hard rubber sheet
硬橡胶套管 cabtire sheath(ing)
硬橡胶贴接面 hard rubber meeting faces
硬橡胶屋面 vulcanite roofing
硬橡胶蓄电池壳 hard rubber battery box
硬橡胶圆辊 hard rubber circular rod;hard rubber round rod
硬橡胶轴衬 ebonite bushing
硬橡皮 ebonite;hard rubber;vulcanite
硬橡皮擦 hard eraser
硬橡皮衬套 ebonite bush
硬橡皮电缆护套 cab-tyre sheath(ing)
硬橡皮扣 ebonite knot
硬橡皮座 ebonite socket
硬芯 hard core
硬芯箍缩 hard core pinch
硬芯会切 hard core cusp
硬芯胶合板 stout heart plyboard
硬芯喷嘴 hard center nozzle
硬芯势 hard core potential
硬锌 hard zinc
硬型 gravity die
硬型界面活性剂 hard type surfactant
硬型铸造 gravity die-casting
硬性 hardness;inflexibility;solidness;stiffness
硬性(X)射线 hard X-ray
硬性表面活性剂 hard type surfactant
硬性残留物 hard residue
硬性底片 hard film
硬性电子管 hard tube
硬性辐射 hard radiation
硬性感光纸 hard paper
硬性光学玻璃 crown glass
硬性规定 hard-and-fast rule;hard rule
硬性规则 hard-and-fast rule
硬性灰浆 hard mortar
硬性接头 hard joint
硬性结构面 rigid structural plane
硬性金属 hard metal
硬性聚乙烯 hard polyethylene
硬性离子 hardness ions
硬性联轴节 fast coupling
硬性射线 hard ray
硬性手感 boardy feel
硬性受限积分器 hard limited integrator
硬性树脂 hard resin
硬性洗涤剂 hard detergent
硬性显影液 hard bath
硬性限幅器 hard limiter
硬性像纸 contrast paper
硬性阴影 hard shadow
硬性釉 stiff glaze
硬性转动 rigid rotation
硬烟末 hard black
硬岩(采)矿工 hard rock miner
硬岩采石场 hard rock quarry
硬岩测试 testing in hard rock
硬岩层 hard formation;hard layer;rock head
硬岩层钻头 hard formation cutting head
硬岩打进 hard driving
硬岩地质学 hard rock geology
硬岩颚式碎石机 hard rock jaw crusher
硬岩反击式破碎机 hard rock reaction crusher
硬岩粉末 hard rock flour
硬岩礁 klint [复 klintar]
硬岩阶地 rock terrace
硬岩金刚石钻头 diamond rock drill crown
硬岩掘进工人 hard heading man
硬岩掘进工作 hard heading work
硬岩掘进机 hard rock mole
硬岩开采 hard rock mining
硬岩开采工人 hard ground man
硬岩开采机 hard rock miner
硬岩开巷机 hard rock tunneling machine
硬岩块 bastard
硬岩块砂浆 bastard mortar
硬岩矿 hard rock mine
硬岩磷酸盐 hard rock phosphate
硬岩盘 iron pan
硬岩盘泉 hardpan spring
硬岩磐 hard pan
硬岩平巷掘进机 hard rock drift machine
硬岩平巷掘进机钻进法 hard rock tunnel boring
硬岩(石) hard rock
硬岩石板 hard rock slab
硬岩石掘进机联合机 hard rock tunneling machine
硬岩隧道掘进机 hard rock tunneling machine
硬岩隧道施工 hard rock tunneling
硬岩掏槽 hard cutting
硬岩天井钻进机 hard rock raise boring machine
硬岩钻工 hard rock driller
硬岩钻进 cut rock;driving in stone;hard rock boring;hard rock drilling;rough drilling
硬岩钻探程序 hard rock boring program(me)
硬盐 hartsalz
硬叶 sclerophyll
硬叶常绿灌木 sclerophyllous shrub
硬叶常绿灌木群落 durifruticeta
硬叶常绿林 sclerophyllous forest
硬叶灌木群落 durifruticeta
硬叶乔木群落 durishilvae
硬叶树林 durilignosa
硬叶树木 sclerophyllous woodland
硬叶榆 cedar elm
硬叶植物 sclerophyllous plant;sclerophyte;stiff leaved plant
硬页岩 raw shale
硬乙烯膜剂量计 rigid vinyl film dosimeter
硬/硬交错施工相互割切现场灌注桩挡土墙 hard/hard secant pile wall
硬油 hard oil
硬油清漆 gold size;short-oil varnish
硬(淤)渣 hard sludge
硬宇宙射线 hard cosmic ray
硬玉 axe stone;chalchewete;jadeite;jadestone
硬玉蓝闪片岩 jadeite glaucophane schist
硬玉透辉石 jadeite-diopside
硬玉岩 jadeitite
硬缘锤 hard edged hammer
硬杂木 hard wood
硬皂 hard soap
硬渣 black cinder;grit;solid slag

硬胀罐 <用手指压不回去> hard swell can
硬罩 hard coat
硬折痕 hard fold
硬支承座 solid bearing
硬脂 hard fat;stearin(e);tristearin
硬脂醇 stearyl alcohol
硬脂的 stearic
硬脂腈 stearonitrile
硬脂精 stearin(e);tristearin
硬脂沥青 stearin(e) pitch
硬脂内酯 stearolactone
硬脂醛 stearaldehyde; stearic aldehyde
硬脂炔酸 stearolic acid
硬脂酸 canoic acid;stearic acid;stearin(e)
硬脂酸铵 ammonium stearate
硬脂酸胺 stearin(e) amine
硬脂酸钡 barium stearate
硬脂酸丁酯 butyl octadecanoate;butyl stearate
硬脂酸钙 calcium stearate
硬脂酸铬 chromium stearate
硬脂酸铬络化物 stearate chromium complex
硬脂酸汞 mercuric stearate;mercury stearate
硬脂酸化物 stearate
硬脂酸环己酯 cyclohexyl stearate
硬脂酸甲酯 methyl stearate
硬脂酸焦油 candle pitch
硬脂酸精 sterin
硬脂酸锂 lithium stearate
硬脂酸沥青 candle pitch
硬脂酸铝 alumin(i)um stearate;alumin(i)um stearic acid
硬脂酸氯化铬 stearatochromic chloride
硬脂酸镁 magnesium stearate
硬脂酸膜 stearic acid film
硬脂酸钠 sodium stearate
硬脂酸铅 lead stearate
硬脂酸山梨醇酯 sorbitol stearate
硬脂酸铈 cerium stearate
硬脂酸铁 ferric stearate;iron stearate
硬脂酸戊酯 amyl stearate
硬脂酸锌 zinc stearate
硬脂酸盐 stearate
硬脂酸酯 stearate
硬脂酮 stearone
硬脂酰胺 octadecanamide;stearamide;stearic amide
硬脂芯焊丝 stearin(e)-cored solder wire
硬脂芯焊条 stearin(e)-cored solder wire
硬脂油 hard grease;stearin(e) oil
硬脂制造业 stearinery
硬纸板 box board; cardboard; fibreboard; hard board; mill board; pasteboard; sheet paper; strawboard;tar-board
硬纸板包装 cardboard package;customer-size package
硬纸板厂废水 hardboard mill wastewater
硬纸板衬垫 pulpboard liner
硬纸板导管 fiber conduit
硬纸板灯芯 cardboard wick
硬纸板管 fiber [fibre] tube
硬纸板机 hardboard machine
硬纸板纤维 cardboard fiber
硬纸导管 fiber duct;fibre conduit
硬纸垫密片 fiber gasket
硬纸管 cardboard tube
硬纸管状模具 fiber duct
硬纸卡 pasteboard
硬纸模板 paper form;paper mo(u)ld;stiff paper form

硬纸套管 cardboard bushing
硬纸套筒 cardboard sleeve
硬质氨基甲酸乙酯泡沫 rigid foam urethane
硬质巴比合金 hardening Babbitt; hard head
硬质白垩 clunch
硬质白灰罩面 hard white coat
硬质板 hard board;rigid board
硬质板地面 hardboard floor(ing)
硬质板门 hardboard door
硬质板铺面 hardboard finish
硬质保温材料 hard insulating material;hard thermal insulation material
硬质边界 hard boundary
硬质玻璃 hard glass
硬质玻璃灯壳 hard glass bulb
硬质玻璃瓶 hard bottle;Pyrex bulb
硬质材料 hard material;mechanically resistant material
硬质长纤维 hard line fibre
硬质衬砌 hard lining
硬质抽水管 hard suction
硬质瓷 <旧用名词> hard paste
硬质瓷器 hard porcelain
硬质粗面贴面 rugged facing
硬质粗石贴面 ragstone facing
硬质的夹心 hard centre
硬质地层 hard pan
硬质地面 hard-surface
硬质地面层 hard floor finishes
硬质地面的空地或道路 hard top
硬质点夹杂 hard inclusion
硬质镀铬工具 hard chromium plated tool
硬质翻砂石膏 hard mo(u)lding plaster for foundry industry
硬质酚醛塑料板 rigid phenolic slab
硬质复合材料 rigid composite material
硬质钢 converted steel
硬质高岭土矿床 hard kaolin deposit
硬质高强度合金 hard metal alloy
硬质高强精陶 ironstone china
硬质隔热 rigid insulation
硬质隔热材料 rigid insulation(grade) material
硬质隔热泡沫(材料) rigid insulating foam
硬质隔声 rigid insulation
硬质隔声材料 rigid insulation(grade) material
硬质隔声泡沫(材料) rigid insulating foam
硬质骨料 hard aggregate
硬质管壳 rigid pipe section
硬质光学纤维棒 multifiber rod
硬质海底 hard bottom
硬质焊敷层 hard-facing;hard surfacing
硬质焊料 hard solder
硬质合金 carbide alloy;carbide tungsten; cemented carbide; cermet; hard metal; kentanium; sintered carbide;hard alloy
硬质合金侧铣 carbide side cutter
硬质合金齿尖 cemented carbide tip
硬质合金冲模 carbide punch
硬质合金冲压模 carbide press die
硬质合金刀具 carbide cutter; carbide-tipped tool;hard alloy cutter; hard metal tool;inserted tool
硬质合金刀片 carbide blade;carbide chip
硬质合金刀刃 carbide-tipped cutter bit
硬质合金刀头 carbide bit; carbide tip;carbide-tipped cutter bit; cermet bit
硬质合金的 carbide-tipped
硬质合金电焊 electric(al) brazing

硬质合金顶尖 carbide-tipped center [centre]
硬质合金端铣刀 carbide end mill
硬质合金堆焊 hard surfacing
硬质合金粉末 cemented carbide powder
硬质合金覆面 hard surfacing
硬质合金钢 hard alloy steel
硬质合金工具 carbide tool;sintered-carbide tool
硬质合金工具磨床 carbide-tip tool grinding machine
硬质合金工业 hard carbide industry
硬质合金刮刀 carbide scraper
硬质合金化合物 hard metal compound
硬质合金铰刀 carbide-tipped reamer
硬质合金锯条 carbide-tipped saw blade
硬质合金空心钻 carbide-tipped core drill
硬质合金块式钻头 carbide type bit
硬质合金拉模 carbide die
硬质合金拉延环 carbide draw ring
硬质合金零件 cemented carbide part
硬质合金模具 sintered-carbide die
硬质合金磨块 hard alloy grinding block
硬质合金喷嘴 carboloy nozzle
硬质合金片 carbide tip;hard metal tip
硬质合金钎头 carbide bit
硬质合金切割工具 carbide tool
硬质合金切削具 carbide cutting element
硬质合金球 sintered-carbide ball
硬质合金球齿牙轮钻头 slug-type bit
硬质合金取芯钻头 carbide-insert core bit
硬质合金去毛刺装置 carbide bur
硬质合金熔焊 hard-welding
硬质合金三面刃铣刀 carbide side cutter
硬质合金十字钻头切割边 carbide cross bit cutting
硬质合金胎体的钻头 carboloy-set bit
硬质合金凸模 carbide punch
硬质合金托板 carbide blade
硬质合金铣刀 carbide-tipped milling cutter
硬质合金(镶)的钻头 carbide-tipped bit
硬质合金镶钢钎 carbide-tipped steel
硬质合金镶卡瓦 carbide-tipped slip
硬质合金镶块 carbide insert
硬质合金镶片式活钎头 detachable insert bit
硬质合金镶片式可卸式钎头 detachable insert bit
硬质合金镶片钻头 tipped bit
硬质合金镶嵌物 hard metal insert
硬质合金银片钎头 carbide insert bit
硬质合金圆柱平面铣刀 carbide cylindric(al) surface cutter
硬质合金铡刀 hard alloy guillotine blade
硬质合金制品 hard metal article; hard metal product
硬质合金钻进 hard metal drilling; tungsten-carbide coring drilling; tungsten-carbide drilling; tungsten-coring drilling
硬质合金钻具的风动冲击钻进 carbide percussion drilling
硬质合金钻头 carbide drill; carbide bit; carbide-tipped drill; cemented carbide bit; hard alloy bit; hard metal alloy;hard metal tipped bit; inserted drill; carbide-insert/tipped bit
硬质合金钻头切割边 carbide drill bit cutting edge
硬质河床 hard bed;rigid bed;rigid riverbed
硬质河底 hard bottom
硬质灰膏 hard plaster
硬质胶 ebonite
硬质胶屑 hard rubber scraps
硬质胶制品 hard rubber article
硬质金属弯管 adamant bend
硬质精陶 earthenware;hard fine pottery
硬质景观 hard landscape
硬质聚氨酯泡沫 rigid polyurethane foam;rigid urethane foam
硬质聚合物 rigid polymer
硬质聚氯乙烯 rigid polyvinyl chloride [PVC];unplasticized polyvinyl chloride
硬质聚氯乙烯板材 rigid PVC sheet
硬质聚氯乙烯底面 rigid PVC soffit
硬质聚氯乙烯管 rigid PVC pipe
硬质聚氯乙烯护墙板 rigid PVC siding
硬质聚氯乙烯化合物的 rigid polyvinyl
硬质聚氯乙烯内异形挤出型材 rigid PVC interior-profile extrusion
硬质聚氯乙烯泡沫塑料 foamed rigid polyvinyl chloride
硬质聚乙烯 rigid polyethylene; rigid polythene
硬质聚异氰酸酯泡沫塑料 rigid polyisocyanurate foam
硬质绝热板 rigid insulation board
硬质绝缘 rigid insulation
硬质矿棉 rigid mineral wool
硬质拉丝筒 rigid forming tube
硬质沥青 hard asphalt;hard (grades of) bitumen; pitch; solid asphalt; gilsonite
硬质路肩 hard shoulder
硬质路面 hard pavement;hard surface; hard surfacing;hard top;road metal
硬质路面的道路 hard surface road
硬质路缘带 hard strip;marginal strip
硬质氯乙烯 hard vinyl chloride
硬质面层 hard (sur)facing
硬质面层海滩 hard beach
硬质面层胶合板 hardboard-faced plywood
硬质面粉 hard flour
硬质模具石膏 hard mo(u)lding plaster for ceramic industry
硬质母线 solid bus
硬质木材 hard lumber;hard timber; hard wood
硬质木屑混凝土 hard-textured chipped wood concrete
硬质耐火黏[粘]土 dunstone;flint fireclay;stiff fireclay
硬质泥球 armo(u)red mud ball
硬质腻子 iron filler
硬质黏[粘]结剂 hard adhesive
硬质黏[粘]土 flint clay; tough clay; hard clay
硬质黏[粘]土矿石 hard clay ore
硬质镍合金衬板 Ni-hard lining
硬质泡沫 rigid foam
硬质泡沫聚苯乙烯 rigid cellular polystyrene
硬质泡沫聚苯乙烯绝热 rigid cellular polystyrene thermal insulation
硬质泡沫聚氯乙烯 foamed rigid polyvinyl chloride
硬质泡沫塑料 rigid foam plastics;rigid plastic foam
硬质硼硅玻璃 hard borosilicate glass
硬质青石灰岩 hard blue limestone
硬质砂胶沥青 hard mastic asphalt
硬质砂岩 bing brick
硬质石灰层 hard lime

硬质石灰岩 hard limestone
硬质石灰岩石板 hard calcareous slate
硬质石屑 hard rock chip(ping)s
硬质石油沥青 oil pitch
硬质塑胶管 plastic conduit
硬质塑料 duroplastics;rigid plastics
硬质塑料管 plastic conduit
硬质塑料溶胶 rigisol
硬质碎料 hard core
硬质炭黑 hard carbon black
硬质陶器 ironstone china;ironstone ware
硬质陶土 hard china clay
硬质填充料 hard stopping
硬质填(塞)料 hard stopping; hard stopper
硬质土(壤) hard soil
硬质吸水管 hard suction
硬质细胞团 grit
硬质纤维 hard fiber [fibre]
硬质纤维板 chemboard;fiberboard[fibreboard] panel;hardboard (panel) <一种由废木屑压成的夹板>;hard fiber[fibre] board;solid fiberboard [fibreboard];stiff fiber [fibre] board
硬质纤维板板壁 hardboard siding
硬质纤维板垫板 hardboard underlayment
硬质纤维板垫层 hardboard underlayer;hardboard underlayment
硬质纤维板护墙板 hardboard siding
硬质纤维板贴面(胶合)板 hardboard-faced plywood
硬质相 hard phase
硬质橡胶 ebonite;hard solid rubber; hard rubber
硬质橡胶贴接面 hard rubber meeting face
硬质岩毛石工程 rag rubble
硬质岩石 competent rock;hard rock; rag;rag stone;strong rock
硬质阳极氧化膜 hard anodic oxidation coating
硬质氧化铝膜处理法 alumilite process;hardas process
硬质医用模具石膏 hard mo(u)lding plaster for medical use
硬质乙烯树脂 rigid vinyl
硬质油灰 body coat;hard putty
硬质油性腻子 hard oil putty; hard putty
硬质杂粒 hard sundry particle
硬质罩面 hard finish
硬柱蓝闪石英片岩 lawsonite glaucophane quartz schist
硬柱石 lawsonite
硬柱石蓝闪石钠长片岩 lawsonite glaucophane albite schist
硬柱石绿泥石绿帘片岩 lawsonite chlorite epidote schist
硬铸钢 hard cast steel
硬铸件 hard casting
硬铸铁 hard casting;hard cast iron
硬铸铁滚筒 hard-grained roll
硬砖 hard brick; hard-burned brick; well-burnt brick
硬砖压制机 tile press
硬装修地面 hard finish floor
硬着陆 hard landing;rough landing
硬着陆宇宙飞船 impactor
硬紫胶树脂 hard lac resin
硬自激 hard self excitation
硬自激振荡 hard self-excited oscillation
硬自励 hardness self-excitation; hard self excitation
硬棕毛刷 stiff bass brush
硬钻 hard drill
硬钻头 solid bit
硬座 rigid chair;semi-cushioned seat
硬座标记定员 marked number of semi-cushioned seats on passenger train
硬座超成定员 overload number of semi-cushioned seats on passenger train
硬座车 carriage with semi-cushioned seats
硬座客车车厢<美> day coach
硬座实际定员 real number of semi-cushioned seats on passenger train

佣

佣工 hireling

佣金 brokerage;commission;commission charges; commission fee; factorage; middleman's fee; percentage;premium;rack off;rake-off
佣金代理人 commission agent
佣金抵押借款 commission mortgage loan
佣金费用 brokerage charges;brokerage expenses
佣金经纪人 commission broker
佣金商 commission agent
佣金账目 commission account
佣金支出 commissioning expenditures

拥

拥包 upheaval

拥壁 retaining wall
拥护 advocate;stand for
拥护者 advocate;supporter
拥挤 crowd(ing)
拥挤不堪的 sardine-fit
拥挤车流 heavy traffic stream
拥挤沉淀 crowed sedimentation;hindered sedimentation
拥挤程度检测 congestion detection
拥挤处 close quarter
拥挤的 aclutter
拥挤的地区 warren
拥挤的交通 bumper-to-bumper traffic;congested traffic
拥挤地区 congested area
拥挤定价【交】congestion pricing
拥挤度<交通的> congestion degree
拥挤房屋 warren
拥挤管理系统【交】congest management system
拥挤交通 jam;congestion
拥挤交通流 congested flow
拥挤街道 congested street
拥挤控制 congestion control
拥挤控制算法 congestion control algorithm
拥挤密度 crowd density
拥挤期间 rush period
拥挤时间 rush hours
拥挤时刻高峰小时 rush hours
拥挤市区 congested urban area
拥挤系数 volume-capacity ratio
拥挤效应 crowding effect
拥挤住宅 rabbit warren
拥挤总分英里<一拥挤分钟×拥挤路段长度> total minutes miles of congestion
拥塞 bottleneck;congestion;engorgement
拥塞程度型 congestion mode
拥塞检测 congestion detection
拥塞交叉口控制 congested intersection control
拥塞交叉口控制联同排队控制【交】congested intersection control coupled with queue management control
拥塞时间 congestion time
拥塞系数 volume-capacity ratio
拥升动作<船舶> heaving motion
拥有半数以上股权 majority holding
拥有成本及工作成本<设备的> O & O cost
拥有出口优势 hold the trump-card in the export of
拥有费用 owning cost
拥有过半数股权的附属机构 majority-owned subsidiary
拥有期 ownership period
拥有权<财产等> holding
拥有设备的成本 owning cost
拥有使用期 ownership usage
拥有土地的 landed
拥有者 owner
拥有者进程 owner process
拥有资源所居位数 number of the ore reserve to arrange in order
拥有资源所占百分比 the ore reserves at percentage

庸

庸俗的 vulgar

庸俗建筑装饰 gingerbread work
庸俗装饰 pastiche [postiche]

壅

壅冰水位预报 forecasting of ice jam stage; forecast of ice-jam stage

壅高 damming;dam up;heading up
壅高水 pent-up water
壅高水位 affluent level;backup water level;banked-up water level;raised water level;raising of water level
壅高水位线 line of raised water level
壅塞喉道 chocked throat
壅塞湖 obstruction lake
壅塞效应<风洞> choking effect
壅水 backwater; banked-up water; dammed water; damming; dam-up water;heading up
壅水坝 flood(ing) dam
壅水板<小水道中代替堰的> retention board
壅水波 surge wave
壅水长度 damming length
壅水(船)闸 damming lock
壅水堤 backwater levee
壅水丁坝 retention dike [dyke]; retention groin;retention groyne
壅水范围 damming limit;damming range
壅水幅度 amplitude of setup
壅水高程 backwater level; retention elevation
壅水高(度) backwater height;height of backwater;height of swell;backwater jump;height of damming
壅水过高 over-dammed
壅水函数 backwater function
壅水涵洞 submerged culvert
壅水极限 damming limit
壅水建筑物 water-controlling structure
壅水结构 backwater structure; damming structure
壅水界限 damming limit; limit of backwater
壅水距离 backwater distance
壅水库容 backwater storage
壅水区 dammed region;damming area;ponding area
壅水曲线 back curve;backwater curve; backwater profile; rising surface curve
壅水水平 banked-up water level
壅水水位 banked-up water level; retention level;retention water level
壅水梯级 damming step
壅水位 pool level
壅水线 backwater curve
壅水效应 backwater effect
壅水蓄量 backwater storage

鳙

鳙 big-head carp

永

永备发射点 pill-box

永备防御工事 permanent defense [defence]
永备工事 permanent works
永闭式 non-open type
永磁步进电动机 permanent-magnet stepper motor
永磁材料 permanent-magnet material
永磁场 permanent-magnetic field
永磁磁导率 permeability of permanent magnet
永磁的 permanent-magnet
永磁电动机 permanent-magnet motor
永磁电动式扬声器 permanent dynamic(al) speaker
永磁电话 magnetotelephone
永磁电机 magnetoelectric (al) machine;magnetoinductor;permanent-magnet machine
永磁电铃 magnetobell
永磁电阻表 magnetoohmmeter
永磁动圈式扬声器 permanent dynamic(al) speaker
永磁动圈式仪表 permanent-magnet moving coil instrument
永磁动铁式仪表 permanent-magnet moving-iron instrument
永磁发电机 dynamo magneto; magnetic generator; magnetodynamo; magnetoelectric (al) generator; magneto-generator; permanent-magnet alternator;permanent-magnet field generator;permanent-magnet generator;magneto
永磁发电机电枢 magnetoarmature
永磁感应器 permanent-magnetic inductor
永磁钢 permanent-magnet steel
永磁合金 permanent-magnet alloy
永磁极 permanent-magnet pole
永磁交流发电机 magnetoalternator; permanent-magnet alternator
永磁聚焦 permanent-magnet focusing
永磁抹音头 permanent-magnet-erasing head
永磁偏心 permanent-magnet off-centering
永磁起重器 permanent-magnet lifting device
永磁绕组 permanent-magnet winding
永磁拾音器 permanent magnet pick-up
永磁式 static type
永磁式安培计 permanent-magnet ammeter
永磁式的 magneto
永磁式继电器 static type relay
永磁式欧姆表 magnetoohmmeter
永磁式扬声器 permanent-magnet type loudspeaker
永磁式仪表 permanent-magnet instrument;permanent-magnet meter
永磁受话 permanent magnet receiver
永磁体 permanent-magnet
永磁铁 permanent-magnet
永磁同步的 permasyn
永磁同步电动机 permasyn motor
永磁同步发电机 magnetoalternator
永磁透镜 permanent-magnetic lens
永磁显微镜 permanent-magnetic mi-

croscope
永磁消音头 permanent-magnet-erasing head
永磁性 permanent-magnetism
永磁性质 permanent-magnetic property
永磁扬声器 permanent-magnet speaker
永磁直流发电机组 magdynamo;magdyno
永磁转子 p-m rotor
永磁装置 permanent magnet assembly
永存 perpetuity
永存惯性力 ever present inertia force
永存物 perpetuity
永存物使复杂化 perpetuity
永存预应力 effective prestress;permanent prestress
永电体 electret
永定流 permanent current
永动机 perpetual motion machine
永冻 permafrost
永冻层 congelisol;ever-frost layer;ever frozen;ever-frozen layer;permafrost frozen layer;permafrost horizon;permafrost layer;eternal frost
永冻层层面线 permafrost table
永冻层内的 intrapermafrost
永冻层融化(过程) depergelation;permafrost degradation
永冻层融解(过程) depergelation
永冻层上的土层 superpermafrost
永冻层下水 subpermafrost water
永冻层消退 degradation of permafrost
永冻带 permafrost zone
永冻带的 intrapermafrost
永冻的 thawless
永冻地 eternal frozen ground;permafrost
永冻地区 permafrost area
永冻覆盖层 frozen overburden
永冻过程 pergelation
永冻湖 amictic lake;perennially frozen lake
永冻基土 ever-frozen subsoil
永冻降级 permafrost degradation
永冻降解 permafrost degradation
永冻气候 eternal frost climate;ice-cap climate;perpetual frost climate
永冻区水文学 permafrost hydrology
永冻圈 cryosphere
永冻土 permafrost soil;congelisol;eternal frozen ground;ever-frozen soil;frozen ground;perennial frost soil;pergelisol;permafrozen ground;permanently frozen ground;permanently frozen soil;perpetually frozen soil;stable frozen ground;ever frost
永冻土层内水 intrapermafrost water
永冻(土)层深度线 permafrost table
永冻土层深度线以上区域 suprageli-sol
永冻土层下不冻区 subgelisol
永冻土的加积 aggradation of permafrost
永冻土活动层 suprageliso l;suprapermafrost layer
永冻土面 pergelisol table;permafrost table
永冻土壤 perpetually frozen soil
永冻土上层 supragelisol
永冻土上层水 superpermafrost water
永冻土上的水 suprapermafrost water
永冻土升高 permafrost aggradation
永冻土中的 intrapermafrost
永冻土钻孔 permafrost drilling
永冻退化 permafrost degradation
永冻线之上的土层 suprapermafrost
永固白 permanent white
永固绿 Pelletier's green
永固银朱 chlorinated para red
永固油墨 permanent ink
永恒的 timeless
永恒的关系 permanent relationship
永恒方程式 eternal equation
永恒接地 static ground
永恒理论 permanence theory
永恒平衡 secular equilibrium
永恒润滑 lifetime lubrication
永恒性 timelessness
永恒运动 perpetual motion
永假式 contradiction
永久 perpetuity
永久坝 permanent dam
永久白 permanent white
永久变位 permanent deflection
永久变形 irreversible deformation;non-reversible deformation;permanent deformation;permanent distortion;permanent strain;residual deformation;set deformation
永久变形测定法 set method
永久变形率 rate of permanent set
永久变形应力测定法 offset stress method
永久变形有效温度 permanent deformation effective temperature
永久变余应力 permanent set stress
永久标石 permanent monument;permanent mark
永久标志 permanent mark
永久标桩 permanent peg
永久标准地 permanent sample plot
永久冰壁 permanent ice foot
永久冰冻 ever frost;perennial frost
永久冰雪 firn;neve (snow)
永久冰雪区 firn area
永久拨款 permanent appropriation
永久波 permanent wave
永久不变的 immutable
永久残余应力 permanent set stress
永久草地 permanent pasture
永久草甸 permanent meadow
永久测站 permanent station
永久层化 <湖中水体的> meromixis
永久产权 freehold
永久潮湿 permanent moisture
永久沉淀 permanent precipitation
永久沉陷 permanent settlement
永久衬里 permanent liner
永久衬砌 secondary lining
永久撑杆 permanent bracing
永久撑木 permanent shore
永久成员 permanent member
永久橙 permanent orange
永久储存器 non-volatile memory
永久储量 permanent reserves
永久处置 permanent disposal
永久船磁性 ship permanent magnetism
永久船闸 permanent navigation lock
永久磁棒 constant bar;permanent bar
永久磁钢 permanent-magnet steel
永久磁化 permanent-magnetization
永久磁铁 magnet keeper;polarized magnet;ferroxdure
永久磁铁装置 permanent magnet assembly
永久磁性 permanent-magnet;permanent-magnetism
永久磁性材料 permanent-magnet material
永久磁性吸盘 permanent-magnet chuck
永久存储器 non-leak memory;permanent memory;permanent storage;readout memory
永久错误 permanent error
永久大地觇标 permanent geodetic beacon
永久贷款 permanent loan;perpetual loan
永久的 aeon;permanent
永久地不结硬 permanently nonhardening
永久地塑性 permanently plastic
永久地弹性 permanently elastic
永久地网 permanent earth
永久地下水面 permanent water table
永久地下水位 permanent water table
永久电容器式电动机 permanent-capacitor motor
永久凋零点 ultimate wilting point
永久凋萎 permanent wilting;ultimate wilting
永久凋萎点 permanent wilting point
永久凋萎含水率 permanent wilting percentage
永久凋萎系数 permanent wilting coefficient
永久顶撑 permanent shore
永久冻土 constant congelation soil;constantly frozen ground;perennial frost;perennially frozen ground;pergelisol;permafrost;permafrost soil;permanent frost soil;perpetually frozen soil;stable frozen ground
永久冻土区 permanent frozen earth
永久冻土生态系统 permafrost ecosystem
永久冻土之上的 superpermafrost
永久段 permanent segment
永久防滑保护 permanent non-slip protection
永久分段区域 permanent segment area
永久封冻湖 amictic lake;permanent amictic lake
永久封固 permanent mounting
永久封严轴承 sealed-for-life bearing
永久浮力舱 permanent buoyant tank
永久符号 permanent symbol
永久符号表 permanent symbol table
永久干涸湖 permanently extinct lake
永久隔墙 permanent partition (wall)
永久工作舱 permanent working enclosure
永久关键码 permanent key
永久河床 permanent bed;permanent river bed
永久河(流) permanent stream;permanent river
永久荷载 perpetual load;quiescent load
永久红 permanent red
永久红色原 permanent red tone
永久红调色剂 permanent red tone
永久红颜料 permanent red pigment
永久花坛 permanent garden
永久环流 permanent circulation
永久黄 permanent yellow
永久黄颜料 permanent yellow pigment
永久机构 self-perpetuating bureaucracy
永久积雪 band of firn;firn cover;firn snow;neve snow;permanent snow-cover;perpetual snow
永久积雪冰 neve ice
永久积雪测量 perpetual snow measurement
永久积雪带 band of firn
永久积雪界限 firn limit;firn line
永久积雪区 firn-line zone;firn zone;region of perpetual snow
永久积雪线 firn edge;firn limit;firn line;line of firn;neve line
永久积雪形成作用 firnification
永久积雪原 firn field;neve field
永久基点 permanent mark
永久极化的电介质 electret
永久记忆 permanent memory
永久坚牢紫 permanent fast violet
永久接地 continuous earth
永久接合 closed coupling;permanent joint
永久接头 permanent coupling
永久结冰 perpetual ice
永久界碑 permanent monument
永久界桩 monumented boundary peg
永久居所原则 principle of permanent residence
永久卷曲 permanent curl
永久可重新签订的租约 perpetually renewable lease
永久空地 permanent open space
永久孔雀蓝 permanent peacock blue
永久控制(点) permanent control
永久拉伸 permanent extension;permanent stretch(ing);permanent elongation
永久蓝 permanent blue
永久蓝色原 permanent blue toner
永久蓝调色剂 permanent blue toner
永久蓝颜料 permanent blue pigment
永久连接 permanent connection
永久漏口 permanent chute
永久绿 Brunswick green;permanent green
永久绿色原 permanent green toner
永久绿调色剂 permanent green toner
永久绿颜料 permanent green pigment
永久螺栓 permanent bolt
永久密封的衬套 sealed-for-life bush
永久密封的销套 sealed-for-life bush
永久密封叶片泵 sealed-for-life vane pump
永久苗圃 permanent nursery
永久模板 stay-in-place form(work)
永久模壳 permanent shuttering
永久目标 permanent object
永久挠度 permanent deflection
永久挠曲 permanent deflection
永久凝固 permanent set
永久扭转 permanent twist
永久偶极定向 permanent dipole orientation
永久偶极子 permanent dipole
永久盘存 perpetual inventory
永久盘存法 perpetual inventory method
永久膨胀 permanent expansion
永久皮质 permanent cortex
永久偏移 permanent deviation;permanent drift
永久屏蔽室 permanent cave
永久气体 permanent gas
永久潜水面 permanent water table
永久强度 permanent strength
永久翘曲 permanent deflection
永久群落 permanent community
永久日历 perpetual calendar
永久润滑的 lifetime lubricated;lubricated lifetime
永久伤残 permanent disability
永久商人 permanent merchant
永久设备 permanent plant
永久伸长 permanent elongation;permanent extension;permanent stretch(ing)
永久失效 permanent failure
永久式衬砌 permanent lining
永久式路面 permanent pavement;Permapave <商品名>
永久式模板 permanent shuttering
永久式钻井岛 permanent drilling island
永久收缩 permanent shrinkage
永久收益 permanent income
永久束(筋) permanent tendon
永久数据组 non-temporary data set
永久霜花 crater bloom

永久水尺 permanent ga(u)ge
永久水位 permanent water level
永久水位站 permanent ga(u)ge
永久水文站 permanent hydrometric station
永久水源 permanent water source
永久水准(基)点 permanent bench mark; permanent standard point; monumented benchmark
永久损耗 permanent loss
永久所有权 perpetuity
永久锁定信件 permanently locked envelope
永久弹性 permanent elasticity
永久特性 permanent character
永久提升机 permanent hoist; stationary hoist
永久通风的大厅 permanently ventilated lobby
永久通风的天窗 permavent
永久通航建筑物 permanent navigation structure
永久通航设施 permanent navigation facility
永久通信[讯]地址 permanent mailing address
永久投资 permanent investment
永久弯曲 permanent deflection
永久完井封隔器 permanent-completion packer
永久位移 irreversible displacement
永久文件 permanent documents; permanent file
永久稳定性 permanent stability
永久吸附水 permanently absorbed water
永久吸收水 permanently absorbed water
永久线路中线橛 permanent line stake
永久线路中线桩 permanent monument
永久线路桩 permanent line stake; permanent monument
永久巷道 permanent openings
永久信号 permanent glow signal
永久形变 permanent deformation
永久形式 permanent form
永久型检验器 permanent type detector
永久性 permanence [permanency]; perpetuity
永久性保存 permanent stores
永久性保留 call hold
永久性变形 permanent deformation; permanent set
永久性标记 permanent marker; permanent sign
永久性标签 permanent label
永久性标志 permanent label; permanent marker; permanent sign
永久性兵营 permanent post
永久性财产 permanent property
永久性草地 permanent grassland
永久性草原 permanent meadow
永久性测量觇标 permanent geodetic beacon
永久性缠扎用油麻细绳 seizing line
永久性常驻卷 permanently resident volume
永久性沉箱 permanent caisson
永久性衬砌 permanent lining
永久性衬砌材料 permanent lining material
永久性衬砌支撑 permanent lining support
永久性成分 permanent member
永久性成员 mandatory member
永久性充填 permanent filling
永久性抽屉存储器 permanent drawer store
永久性筹资 permanent financing
永久性船闸 permanent shiplock
永久性磁铁 permanent-magnet
永久性存储计算机 permanent-memory computer
永久性存储器 non-volatile storage; permanent memory; permanent storage
永久性道路 permanent way
永久性的 life time
永久性的活动房屋位置 permanent mobile home space
永久性的伪装 fixed camouflage
永久性的桩材(防护) permanent piling
永久性抵押 <抵押十年以上> permanent mortgage
永久性地锚 permanent ground anchorage
永久性地图记录 permanent map record
永久性地下径流 permanent groundwater runoff
永久性地下上层滞水 permanent perched groundwater
永久性调动 permanent change of station
永久性冻土 permanently frozen soil
永久性毒物 permanent poison
永久性对象 permanent object
永久性对心 permanent alignment
永久性对准 permanent alignment
永久性多种用途猪舍 permanent multi-use house
永久性防御工程 permanent defense [defence]
永久性防御工事 permanent fortification
永久性防御设施 infrastructure
永久性附属装置 permanently attached equipment
永久性干涸 permanent extinction
永久性工程 permanent project; permanent works
永久性供水沟渠 permanent supply ditch
永久性构筑物 permanent structure
永久性故障 permanent fault
永久性观测站 permanent counting station
永久性管 eternit pipe
永久性灌溉 permanent irrigation
永久性河流改道 permanent canal diversion
永久性荷载 permanent load(ing)
永久性护岸 permanent bank protection
永久性回粘 permanent tack
永久性基础 permanent foundation
永久性基地 infrastructure
永久性加荷 permanent load(ing)
永久性建筑(物) permanent building; permanent construction; permanent structure
永久性结构 permanent construction; permanent structure
永久性局部伤残 permanent partial disability
永久性控制 permanent control
永久性库容 permanent pool; permanent storage
永久性拉伸 permanent extension
永久性雷达基地 radar infrastructure
永久性冷铺罩面材料 permanent cold-lay surfacing material
永久性连接 permanent coupling
永久性路线 permanent route; permanent way
永久性锚泊系统 permanent mooring system
永久性密封 sealed-for-life
永久性命令 standing order
永久性模板 permanent form(work); permanent shuttering
永久性破坏 permanent damage
永久性破坏极限状态 limit state of permanent damage
永久性起爆线 permanent blasting wire
永久性欠固化 permanent undercure
永久性桥(梁) permanent bridge
永久性上油(润滑) greased-for-life
永久性设备 lasting equipment
永久性设施 permanent feature
永久性渗漏 permanency leakage
永久性收缩 permanent contraction
永久性数据集 permanent data set
永久性数据文件 permanent data file
永久性水尺 permanent water ga(u)ge
永久性损坏 <不能修复的损坏> irremediable defect
永久性梯级船闸 permanent flight locks
永久性调整 permanent adjustment
永久性铁路 permanent way
永久性听觉丧失 permanent hearing loss
永久性听觉损伤 permanent hearing defect
永久性听阈位移 permanent threshold shift
永久性通告 permanent notice
永久性弯曲 permanent twist
永久性网络结点 permanent network juncture
永久性未熔生料毯 permanent blanket of unfused batch
永久性屋面 permanent roofing
永久性物资 permanent property
永久性误差 permanent error
永久性吸收相位条带 permanent absorbing phase strip
永久性限制 permanent limitation
永久性协定 permanent agreement
永久性泄水孔 permanent sluice
永久性修补 permanent patch
永久性修理 permanent repair
永久性悬挂式脚手架 permanent suspended scaffold
永久性压密 permanent compaction
永久性压实 permanent compaction
永久性整理 permanent finish
永久性整治建筑物 permanent regulation structure
永久性支撑 permanent shore
永久性支护系统 permanent support-ing system
永久性植被 permanent vegetation
永久性止水 permanent water sealing
永久性轴向应变 permanent axial strain
永久性住宅 permanent housing
永久性铸模 permanent mo(u)ld
永久性资产 permanent assets
永久修理 permanent repair
永久虚拟电路 permanent virtual circuit
永久雪线 firn-line zone; permanent snow-limit; permanent snow-line
永久雪原 neve field; permanent snow-field
永久压力 permanent compression
永久压缩变形 permanent compression set
永久淹水舱 permanent flooded tank
永久延伸 permanent extension
永久阳极 permanent anode
永久页面存储器 permanent page store
永久仪差 secular error
永久阴影面 <建筑物常年没有日照部分> area of perpetual shadow
永久应变 eternal strain; permanent deformation; permanent strain
永久应力 permanent stress
永久硬度 non-carbonate hardness; permanent hardness
永久硬水 permanent hard water
永久预加应力 permanent prestressing
永久预应力 permanent prestress
永久运动 perpetual motion
永久杂合体 permanent heterozygote
永久载重 permanent load(ing); perpetual load
永久载重线区 permanent zone
永久债券 perpetual debenture
永久栈道 permanent trestle
永久支撑 permanent bracing; permanent shore; permanent sustain
永久支护 permanent support
永久支架 permanent support
永久支柱 permanent shore
永久支座 permanent bearing
永久植被 permanent vegetation cover
永久滞水 perched permanent ground water
永久重量 permanent weight
永久贮存量 permanent storage
永久驻留前台 permanent resident foreground
永久铸型 permanent mo(u)ld
永久桩 permanent monument
永久资本 permanent capital
永久紫 permanent violet
永久租户 life tenant
永久租借地 <美> manor
永久租借权 perpetual lease
永久租用权 perpetual lease
永久钻井 permanent drilling well
永久钻井岛 permanent drilling island
永久作用 permanent action
永亮灯 perpetual lamp
永留模板 permanent shuttering
永膨胀 permanent expanding
永气体 permanent gas
永世 aeon
永态反馈 permanent feedback
永态调差率 permanent speed drop
永态转差系数 steady-state speed drop
永态转差系数图 speed droop graph
永态转速降 permanent drop
永停滴定法 dead stop titration
永停终点法 dead stop end point method
永外磁棒 constant bar; permanent bar
永续变异 permanent modification
永续年金 perpetual annuity; perpetuity
永续盘存 continuous inventory; perpetual inventory
永续盘存明细记录 detailed perpetual inventory records
永续盘存账户 perpetual inventory account
永续盘存制 balance of stock; perpetual inventory system
永续实地盘存制度 continuous system of physical inventory
永续收获 sustained yield
永续收获经营 sustained yield management
永续预算 perpetual budget
永压树脂 durable-press resin; permanent-press resin
永真公式【数】valid formula
永真式 identically true formula; universally valid formula
永真问题 validity problem
永真性 validity
永真蕴涵 tautological implication
永滞层 <指湖水或海水下部> monimolimnion
永租地权 perpetual lease
永租权 lease in perpetuity

甬道 close;corridor

甬道门 stack door

泳道 <池的> swim lane

泳动电势 streaming potential
泳透力 throwing power
泳透性 throwing power

勇敢者游戏场 adventure playground

涌波 bore;surge;surging

涌波传播 propagation of surges
涌波高度 surge height
涌波抑制器 surge suppressor
涌潮 breaking surge;eager [eagre];eigre;mascaret;sea bore;surge wave;tidal bore;tide bore;bore <河口的>
涌出 gush;spillage;upwelling;springing out;well out;well up <水、油>
涌出的水 outgoing water;rushing-out water
涌出水 gushing water
涌出水流 water efflux flow
涌出油 outgush oil
涌幅 amplitude of swell
涌高 height of swell
涌毁 fountain failure
涌级 swell scale
涌浪 long rolling sea;surging sea;swell(ing)(wave);swelling of sea;swelling sea;upsurge;bore <明渠的>
涌浪泵 surge pump
涌浪标志 surge sentinel
涌浪波速 surge celerity
涌浪波型 surging profile
涌浪冲击 surging shock
涌浪传播 propagation of surges
涌浪挡板 surge plate
涌浪等级 scale for swell;scale of swell;swell scale
涌浪叠加 stacking of surges
涌浪陡度 slope of surge
涌浪方向 swell direction
涌浪分级标准 standard of swell scale
涌浪高度 height of swell;surge height
涌浪缓冲滑闸 surge plunger
涌浪级别 character of swell;classification of swell
涌浪记录装置 sea-swell wave recording system;sea wave analyzing system
涌浪减退 swell abatement
涌浪界限 limit of uprush
涌浪力 surging force
涌浪前沿 surge front
涌浪区 splash zone
涌浪衰减 swell attenuation;swell decay
涌浪速度 surge celerity;surge velocity
涌浪调整槽 surge tank
涌浪系数 surge coefficient;surge factor
涌浪线 surge line
涌浪消减设备 surge suppressor
涌浪效应 surge effect
涌浪型海滩剖面 swell beach profile
涌浪抑制 surge suppression
涌浪抑制器 surge suppressor
涌浪预报 swell forecast
涌浪振幅 surge amplitude
涌浪制动设备 surge arrester equipment
涌浪周期 period of swell
涌浪最高水位 top surge water level
涌流 flashy flow;inrush current;outgush;surge current
涌流的水 inrush water
涌流分流器 surge diverter
涌流风缸 surge reservoir
涌流接地 surge arrester [arrestor]
涌流孔 gushing water hole
涌流区 upwelling area;upwelling region
涌流式灌溉 surge flow irrigation
涌流式灌溉控制器 surge irrigation controller
涌泥 mud surge
涌起 run-up;surging
涌泉 artesian spring;blow-well;boiling spring;estavel;gushing spring;jet fountain;rising spring;water spring;sprout fountain <庭园中的>
涌泉井 artesian well;free flowing bore
涌入 blow in
涌入水 inrush water
涌入速度 influx rate
涌塞 bottleneck
涌沙 <又称涌砂> boiling of sand;heaving sand;sand boil
涌上 run-up
涌升流 ascending current;upwelling;upwelling current
涌水 gushing water;inrush of water;spring water;water inflow
涌水层 water producing formation
涌水层部位法 <沿井筒测量液体电阻以确定孔内的> strataflow process
涌水池 artesian spring;artesian spring tank
涌水点 point of water inrush
涌水量 groundwater discharge;inflow of water;inflow rate;make of water;specific yield;water-make
涌水量参数计算方法 calculating method of water yield parameter
涌水量水位降深曲线方程 equation discharge-drawdown curve
涌水量水位降深曲线类型 type of discharge-drawdown curve
涌水率 inflow rate
涌水区 headwater
涌水事故 water troubles
涌水速度 issuing velocity
涌土 mud boil;soil inrush
涌现出 open up

蛹 pupa

用 U 形钉钉住 staple

用安全锚头竖钻塔 cathead a derrick up
用安置值纠正 rectification with predetermined settings
用暗沟排水 underdrain
用暗销等接合 dowel
用凹槽试件进行冲击试验 test with notched test piece
用坝堵高水位 dam up
用百乘 centuple
用百分率方式 percentagewise
用百分数方式 percentagewise
用柏油铺路 petrolize
用摆动测量(砂等) vibrate
用板铺砌 slab paving
用板条钉住 batten(ing)
用板条钉住照明装置 batten lighting fixture
用板条箱装 crate
用板围住 board up
用板遮住 board up
用板组成的预制木屋 homansote
用半圆壁龛支撑的 apse-buttressed
用半圆规描绘 protract
用拌浆锄拌石灰砂浆 larried up
用拌泥机拌和的砂浆 mill-run mortar
用棒捣实 rodding
用棒干捣的 dry-rodded
用棒干捣实的骨料 dry-rodded aggregate
用棒通沟 rodding
用……包上(或涂上) coated with
用保证书或许可证证明 certify
用堡垒保护 bulwark
用爆炸法填土 settlement by blasting
用锛子加工 adzing
用本国货币支付的费用 expense in local currency
用苯酚磺酸光电比色法 using a phenolsulfuric acid colo(u)rimetric method
用绷带缚上 bandage
用泵补偿运河需水 canal feeding by pumping
用泵抽水 pump up
用泵抽水灌溉 irrigation by pump
用泵抽送 stirring
用泵的人 pumper
用泵灌注混凝土 concrete pump placing
用泵加速循环 accelerated circulation by pump
用泵排出 pump out
用泵生产强迫循环 forced circulation by pump
用泵输送 pumpability
用泵卸车 discharge by pump
用比例尺画图 scale drawing
用比例尺制图 protract
用壁龛支撑的 niche-buttressed
用编枝排列 line with wicker-work
用扁斧加工 adz(e)
用扁斧加工的 adz(e)-hewn
用变矩器驱动的 converter-driven
用变扭器驱动的 converter-driven
用杓取出 scoop
用标准构件组装 modular construction
用表格表示 tabulate
用表面振捣器捣实混凝土 compacting concrete by surface vibrator
用冰冷却的 iced
用波纹钢板及钢构架制造的半圆形活动房屋 Quonset hut
用波形花纹装饰的 damascened
用玻璃和钢架建造的建筑 crystal palace
用驳船运液体污泥 barging liquid sludge
用薄板覆盖 laminate
用不同方法(计算)所得结果 cross check
用不同方法校核 cross check
用不正当手段得到的 ill-gotten
用不正当手段取得的 misgotten
用不正当手段预先安排的 put up
用布覆盖的 draped
用材高 timber height
用材林 commercial forest;economic forest;economic trees;timber forest
用材林改进法 timber stand improvement
用材木 <直径 3 英寸以上,1 英寸≈0.0254 米> timber wood
用材形数 timber form factor
用材中央干围 mid-timber girth
用材中央直径 mid-timber diameter
用彩色涂料粉刷 colo(u)r wash
用餐地方 dining corner;dining place
用槽贮藏(法) tankage
用草覆盖 laydown
用草皮或植物覆盖 mulch
用测杆测地下水位 dowsing
用测井曲线划分的地层 elector-bed
用测链测 chain off
用测链丈量过的路中线 chained centre line of road
用测平杆测得的管沟斜度 boning fall
用测平杆测定孔洞的标高 borning
用测深法 by sounding
用测斜仪测定的高度 clinometric height
用叉分选 sorting by forks
用叉杆探水 water diving
用柴油机发动的挖掘机 diesel power shovel
用柴油发电机驱动的 diesel-electric
用铲插捣 <混凝土的> spading
用铲除草等 spud
用铲刀作业 blading operation
用铲开挖竖直面 face excavated by shovel
用铲取样 shovel sampling
用长索拖动重物 snaking
用场地交通影响分析 site traffic impact analysis
用车搬运 van(ning)
用车床加工 lathe
用车辆运载的 vehicular
用车送到 drive home
用车完毕 finished with engines
用车运出 vehicular disposal
用车运输 go wagon;chariot
用车运送 bowl;cart(ing)
用车装运 cart
用沉淀法分取 decant
用撑架突出 corbel out
用撑架托住 corbel(ling)
用成组吊钩吊起 hooking up
用程序规划 programming
用匙取出 spoon
用匙舀 spoon
用尺测深 dipstick gauging;dipstick metering
用齿条退回 rackback
用翅托挑出 corbel outwards
用冲子打标号 dab
用抽查法证实 verification by test and scrutiny
用抽风机吸尘 fanning
用臭氧处理【给】 ozonize
用杵捣 pestle;pestling
用传力杆接合 dowel joint;dowel(l)ing
用传票传唤 subpoena
用传票索取 subpoena
用船渡运 ferry
用船体反复冲击海水 bucking
用船运到 ship off to
用船装运 by boat
用垂球测定垂直度 plumb perpendicular
用锤打 chap
用锤打实 driving fit
用锤钉上 hammer down
用锤分块的 hammer blocked
用锤击碎石料 knobbling
用锤加工花岗石表面 nudging
用锤尖锤击 penning
用锤尖打 peening
用锤尖加工 peening
用锤尖敲打 peening
用锤尖敲击 peening
用锤链翻动的圆牌 hinged disc
用锤敲入、用螺丝刀起出的螺钉 screwnail
用锤修琢 dabbing
用锤凿面的 hammer faced
用锤整修 hammer dressing

用词 wordage;wording
用瓷夹板固定电线 cleat wiring
用次级电子机构组成的等离子体 multipacting plasma
用打孔枪穿孔机 gun-perforate
用打字机打字 typewrite
用大板构造的房屋 block of large slabs
用大车搬运 dray
用大钉和套箍固定的檐沟装置 spike-and-ferrule installation
用大理石贴面 marmoration
用大卵石铺的 cobbly
用大铁锤敲 sledge
用大字体书写 engross
用代号【测】encipher
用代赭石涂 reddle
用带穿边 lace
用带传送 taper transport
用带导向杆的钻头扩孔 stinger ream
用带系紧 girt(h)
用带捆扎 strap;taping <加固装载>
用带抛光表面 finishing by belt
用带修饰表面 finishing by belt
用带子捆扎 strap
用贷款投资 financial leverage
用单搭板对接 butt joint with single strap
用氮气扫罐 nitrogen purging of tank
用刀具切割加工 tooling
用刀片拌和 blade mixing
用导管灌注(的水下)混凝土 tremie concrete
用导管灌筑(的水下)混凝土 tremie concrete
用导轨支承的 rail-borne
用导接线导通 bonding
用堤坝防护 protection by dykes [dikes]
用堤围住的 endyked
用地 land take;right of way
用地标 land mark
用地补偿 indemnity for area loss
用地测量 land-use survey; right-of-way survey
用地磁电动仪测海流 geking
用地的取得 right-of-way acquisition
用地递减率 rate of land reduction
用地分类 classification of land use
用地封闭 land closure
用地规则 right-of-way rule
用地计划 land-use plan
用地界 right-of-way boundary
用地界碑 land monument
用地界石 right-of-way monument
用地界线 land line
用地宽度 right-of-way width; width of right-of-way
用地面积 plot area
用地平衡 land-use balance
用地区划强化 down zoning
用地申请 land-use claim
用地调整 land-use adjustment
用地图 land-use map;right-of-way map
用地围墙 land closure
用地线 property line; property line right of way line
用地信息系统 land information system
用地栅栏 right-of-way fence
用地资料 land-use data
用点虚线表示 dotting
用碘处理 iodate
用电 electricity utilization
用电车等运输 tram
用电池工作的记录器 battery operated recorder
用电池供电的 battery powered
用电池启动器 battery-driven
用电的地板采暖(系统) floor heating by electricity
用电费 energy charges
用电户 power consumer
用电键发送器直流拨号 direct current keying
用电解法把金属熔融沉积到另一金属表面(增强硬度) metallide
用电解法处理 electrolyze
用电量递减收费制 step meter rate
用电量指针 demand pointer
用电脑登记 log in
用电瓶发动的 battery powered
用电设备 utilization equipment
用电视收看 televiewing
用电视遥控的 television-directed
用电指标 electric(al) characteristic
用电子计算机处理 computerise;computerization
用电子计算机控制 cybernation
用电子计算机模拟 computerized simulation
用电子计算机评定的 computer-rated
用电子束烧穿 electronic burn
用电阻调整 resistive padding
用垫环焊接 backing weld
用垫片调整 shimming; skim adjustment
用垫圈或球珠绝缘 beaded insulation
用雕像装饰 decorated with status; statue
用吊门关闭 portcul(l)is
用吊桶打水的井 sweep well
用钉冲孔 nail set
用钉钉住 nail
用钉固定 fixing by nails
用定量计加料 graduated hopper-charging
用定位槽板换挡 gate gear shift
用定位焊接的非结构性连接 non-structural connection by tack welding
用定位桩插孔 spudding
用动力夯夯 leapfrog
用动力调节铲刀偏斜角的推土机 power angle blade dozer
用动力调节铲刀偏斜角和倾斜角的推土机 power angle and tilt blade dozer
用动力调节铲刀倾斜角的推土机 power tilt blade dozer
用动力推动的 powered
用斗出料 bucket discharge
用斗式提升机加料的筒仓 silo fed by bucket elevator
用短线接地 short ground return
用短桩固定 pegging
用断续高压作电缆击穿试验 flashing test of cables
用盾构法开挖隧道 shield tunnel(l)ing
用盾构掘进法开挖隧道 shield-driven tunnel(l)ing
用盾构施工法掘进隧道 shield-driven tunnel(l)ing
用多轮断面仪测定的平整度质量值 <按每英里英寸计> Q-value by multi-wheel profilometer
用多种材料粗砌的墙 <古罗马> maceria
用舵操纵 rudder control
用舵点 wheel over point
用舵完毕 finished with the wheel
用二次爆破来破碎岩石 pop shotting
用二甲苯脱蜡 removing paraffin with xylol
用二氧化碳冷冻 carbon freezing
用发动机驱动的动力 motoring power
用发动机驱动的(控制位)槽 motoring notch
用发动机作动力的 engine-powered
用阀门调节 valving
用法 directions
用法说明 directions for use; directions of use;explanatory notes
用法线切断的 subnormal
用法指导 directions of use
用翻斗车卸料 dumping with tilting bucket
用反射光观测 observation by reflected light
用泛光灯照亮 flood light
用方木材构筑的脚手架 builder's scaffold
用防腐法 preservatize
用防腐剂处理木材 <一种专利方法> Haskinizing
用防火分隔铜条装配玻璃 copper light glazing
用房屋抵押 house mortgage
用肥皂处理过的 soap-treated
用肥皂洗 soap
用沸水清洗 scald
用沸水烫 scald
用分级碎石填方 classified rubble fill
用分块法求逆矩阵 matrix inversion by partitioning
用分期付款办法购买 buy on the installment plan
用分数表示的坡度 fractional slope
用酚消毒 phenicate
用粉煤灰制成的轻质骨料 lytag
用粉饰 application of plaster
用粉状炭接触 contacting with powdered carbon
用风镐开采 picker work
用风缆拉紧的钢烟囱 guyed steel stack
用风冷却 wind chill
用风枪吹除 lance
用封泥封 lute
用氟利昂升华冷却 evaporation cooling with Freon
用浮标指示(礁、水道等) buoy
用浮点编码的数据精简法 floating-point coding compaction
用浮桥渡河 pontoon
用浮石粉抛光 float stone grinding
用浮石摩擦 pumicing
用浮石磨 pumicate;pumice
用浮子操作的 float-operated
用浮子控制的 float-operated
用符号表记的货物列车 symbol freight train
用符号表示 symbolically; symbolization
用辐射灭菌(或消毒) radiosterilize
用斧修整石料表面 regrating
用斧斩削的 adz(e)-hewn
用斧者 axman
用改装方法控制噪声 retrofit noise control
用盖料覆盖路面 mulch
用干材料筑成的隔墙 drywall partition
用干浇注法制成的管 dry cast pipe
用杆旁推调车 poling of cars
用杆清通排水沟 rodding
用钢笔过账 pen-and-ink posting
用钢齿饰石细工 dragged work
用钢带裹着的光面轮胎 beadless tire [tyre]
用钢箍加强的柱 <美> tieback
用钢筋网铰接的混凝土块 <护岸工程用的> articulated concrete block
用钢梁组成的格床 girder grillage
用钢丝扫床【疏】wire sweep
用钢丝扫过的海 swept by wire
用钢丝刷(进行)表面抛光 wire brush finish
用钢丝系缚 wiring
用钢条捆扎 strapping
用高度机械化方法修筑的 super-mechanical
用高频扫描快速读出 radio-frequency reading
用镐的矿工 pike man
用格言表达 proverb
用隔板分开 compartmentalize
用各种手段 try every means
用铬酸盐处理 chromate
用工 recruit and use work force
用工厂制造的模数化部件建筑的住屋 modular housing
用工程完工百分数来核算财务的方法 percentage-of-completion method of accounting
用工具修琢面层 <石料> tooled finish
用工量 amount of labo(u)r used
用公共汽车接送 bus(s)ing
用公式表示 formulate;formulize
用公制 in metric
用公制表示 metricate;metricize
用汞处理 mercurialize
用拱顶石连住的 keyed
用拱顶石连住的拱腹 hacked soffit
用拱顶石支承 keystone
用拱跨越 imbow
用钩抓住 dog
用构架和拉线加固的电杆 truss-guyed pole
用固定联轴器连接的泵 close-coupled pump
用刮路机及拖斗车平整路面 blading and dragging
用刮路刷刷路 drag brooming
用挂毡装饰 tapestry
用管道输出 pipe away
用管冷却 pipe cooling
用管连接的 pipe-connected
用管钳拧上管子 tong pipe up
用管线连接管系机 pipeline machine
用管子连接的 pipe-connected
用管子排除(混凝土)屋面积水 pipe floating the surface
用管子取样 pipe sampling
用管子输送 pipage
用贯入法测定(土的)密实度 needle density
用贯入法的结构层 penetration course
用贯入法铺筑的面层 penetration surface course
用贯入法筑路 penetration construction
用灌浆方法固结的冲积层 grouted alluvium
用灌木覆盖的 b(r)ushy
用光电池的 photronic
用光面带拖平 <混凝土路面> belting
用光学仪器检查 optic(al) test
用光学仪器接近 optic(al) approach
用规检验 ga(u)ging
用硅胶干燥冷却空气法 silica-gel system
用轨滑动舞台 slip stage
用滚轮滚动的活动桥 rolling drawbridge;roller bridge
用棍棒打 club
用过的 exhausted;second-hand
用过的催化剂 used catalyst
用过的核燃料 spent fuel
用过的旧设备 second-hand plant
用过的砌筑材料 second-hand masonry materials
用过的热 used beat
用过的溶剂 spent solvent
用过的润滑脂 used grease; worked grease
用过的石灰 used lime
用过的水 second-hand water; spent

wash;wasted water
用过的酸 spent acid
用过的液体 used liquefaction; used liquid
用过的油 used oil;waste oil
用过的针入度 worked penetration
用过金刚石 used diamond;used stone
用过润滑脂的针入度 worked grease penetration
用海绵揩拭 sponge
用海绵润湿 sponge
用壕沟围绕 entrench
用合成树脂黏[粘]结 bonded with synthetic(al) resinous material
用合成橡胶改性的酚醛树脂黏[粘]合剂 plastilock
用合同形式制定减价运价 contract rate
用核子密度计在现场测得的容重 in-place nuclear density
用鹤嘴斧掘 pick ax(e)
用桁构支承 truss
用桁架加劲的吊桥 truss-stiffened suspension bridge
用横撑木支持 needle
用横缆锚地紧固 fastening with transverse cables anchored to ground
用红色刷背景 rubrication
用红色涂写 rubrication
用红字写的账户 rubricated account
用红字印的账户 rubricated account
用虹吸管供水 water syphoning
用户 audience; consumer; consumer buyer;customer;customer's subscriber;end-user;occupier;subscriber;ultimate consumer;ultimate purchaser;user
用户安全检验程序 user security verification routine
用户安装 customer set-up
用户安装产品 customer set-up product
用户安装程序 installed user program(me)
用户包装 consumer packaging
用户保护装置 subscriber's protector
用户闭路 subscriber's loop
用户编码的虚存 user-coded virtual memory
用户便览 user handbook
用户标号 user label
用户标号出口程序 user label exit routine
用户标号处理 user label handling
用户标号信息 user label information
用户标记 user flag
用户标识 identification of user
用户标识符 user identifier
用户标志器 subscriber's marker
用户标准 user's specification
用户表 customer list; service meter <自来水、电、煤气表等>
用户拨号脉冲继电器 line (impulse) control relay
用户拨号盘 subscriber's dial switch
用户操作 user operation
用户插座 subscriber's jack
用户长途拨号 <美> subscriber trunk dial(l)ing
用户长途直拨 subscriber toll dialing
用户成本 user cost
用户程序 user program(me)
用户程序表 user program(me) list
用户程序错误 user program(me) error
用户程序库 user library
用户程序区 user program(me) area
用户出口程序 user exit routine
用户处理机 line processor;user processor
用户传真 telefax
用户存储器 line memory
用户错误处理过程 user error procedure
用户打字电报机 telex
用户代号 subscriber number;user number
用户代理 user agent
用户到用户 house-to-house
用户到用户线路 user-to-user circuit
用户的程序设计 user programming
用户的热容量 heat capacity demand
用户的最终验收 customer's final acceptance check
用户等待时间 period of reservation of number
用户等效距离误差 user equivalent range error
用户第二级寻线机 <自动电话> local secondary line switch
用户第一主义 consumerism
用户电报 subscriber's telegraph;telegraph exchange; telex; house telegraph
用户电报挂号 booking of telex calls
用户电报挂号的有效期限 validity of telex booking
用户电报回路 telex circuit
用户电报机 subscriber's telegraph
用户电报交换机 teleprinter exchange
用户电报通报费 call charges
用户电报网 telex network
用户电报站号码 number of telex station
用户电表 supply meter
用户电话机 subscriber's apparatus; subscriber's telephone set;subset
用户电话机拨号盘 subscriber's dial
用户电话交换机 private branch exchange
用户电缆 house cable;internal cable;service cable;subscriber cable
用户电路 subscriber's line circuit
用户电视电报 teletext
用户吊牌 subscriber's drop
用户调查 user survey
用户调访 canvass customer opinions
用户定向产品 user-oriented product
用户定义函数 user defined function
用户定义子系统 user-defined subsystem
用户端 user side
用户端口功能 user port function
用户对用户通信[讯] user-to-user communication
用户对用户业务 user-to-user service
用户访问级别【计】 user access level
用户费(用) user charges;user cost;user fee
用户分布图 user profile
用户分机 subscriber's extension (station)
用户分类 classification of service
用户分区 user partition
用户分时 user time-sharing
用户分线 subscriber branch line
用户分线箱 house connecting box
用户服务 user service
用户服务处 service department
用户服务设施 consumer's service
用户服务台 teller console
用户服务系统 teller system
用户辅助设施 ancillary customer
用户复式塞孔 subscriber's multiple jack;subscriber multiple
用户干线 service main
用户(给水)旁管 service lateral
用户(给水)支线 service lateral
用户工作区【计】 user working area
用户公用网络 common user network
用户供电变压器 bulk supply transformer
用户供电控制(装置) consumer's supply control
用户供电站 consumer's power station;consumer's terminal
用户供水 service water supply; user water supply
用户管 domestic mains;house dead end line;house lateral;service line;service pipe
用户管鞍座和接头 service saddle and connector
用户管穿管 service entry
用户管道 installation pipe
用户管阀 shut-off cock; curb shut-off; curb valve; curb stop; pavement service valve <设在路边的>
用户管管夹 service clip;service saddle
用户管接头及配件 service couplings and fittings
用户管总阀 <一般在户外> service stop;service valve
用户号改变音 changed-number tone
用户号码簿 subscriber list
用户号码核对电路 number checking circuit
用户号码排列 directory listing
用户号码识别装置 number identifier
用户和记录合并台 combined line and recording position
用户呼叫方式 user calling pattern
用户呼叫业务 custom calling service
用户花费 user effort
用户话机 customer station set; subscriber's station;substation
用户话机保安器 subscriber's protector
用户话务员 subscriber's operator
用户环路 local loop;subscriber loop
用户换能器 user transducer
用户回路 user loop
用户回线 local loop;subscriber's loop
用户活动连接器 subscriber connector
用户机 subscriber's machine
用户级 subscriber stage;user class
用户级模块 line module
用户级协议 user level protocol
用户集团 user group
用户集线网 line concentrator
用户计费表 subscriber's meter
用户记发器 subscriber's register; subscriber's sender
用户记录器 user writer
用户记事 user account
用户继电器 calling party's line relay
用户加热器 domestic water heater
用户间的连接 user-to-user connection
用户检查 customer inspection
用户交换机 private exchange
用户交换台 private branch switchboard;subscriber's exchange
用户接地阻抗 consumer's earth resistance
用户接管 communicating pipe;house connection;service connection
用户接口 user interface
用户接入 customer access
用户接线 terminal block
用户结点接口 user node interface
用户解密钥匙 user decryption key
用户界面 user interface
用户界面管理系统 user interface management system
用户进户线 service wire
用户进水管 service pipe
用户距离误差 user range error
用户卷宗 user volume
用户决定形式 user-determined form
用户均衡【交】 user's equilibrium
用户开关 subscriber switch
用户可编程序数据采集系统 user programmable data acquisition system
用户可编程序 user program(me)
用户控制查找 <即交互式查找>【计】 user controlled searching
用户控制通路 user controlled path
用户馈电电缆 feeder cable
用户馈电和检视装置 line feed and supervision modifier
用户扩充工作编码 user own coding
用户垃圾费 user collection charges
用户类别 class of line;class of service
用户类别标记 class-of-line detection
用户类别路 subscriber class circuit
用户类服务 user class of services
用户立管 vertical service pipe;vertical service run
用户连接管 service connection
用户连接设备 subscriber connecting equipment
用户联机检索资料 customized online search
用户链 user chain
用户链标识连接器 line-link marker connector
用户链路驱动器 line-link switch controller
用户链路网 line-link network
用户密度 consumer density;customer density;user density
用户命令 user command
用户内部通话装置 subscriber's intercommunication installation
用户排水管 service drain(age)
用户排水支管 building connection sewer
用户培训 customer's training
用户区(域) user area
用户驱动系统【计】 user-driven system
用户确定的程序库 user-defined library
用户确定的关税 user-defined tariff
用户群 user group
用户热力站 consumer thermal substation
用户热水器 domestic water heater
用户认可 customer's approval
用户任务安排 user task scheduling
用户任选项 user option
用户入口线 subscriber access line
用户塞孔 subscriber's jack
用户扫描 line scan
用户扫描器 line scanner; subscriber scanner
用户设备 consumer's installation; consumer's plant; customer's installation;customer's plant;subscriber's apparatus;subscriber equipment;user equipment;user facility
用户设备传输指标 subscriber plant transmission index
用户设计命令 user-design command
用户设施 user facility
用户时间 user time
用户识别符 user identification
用户使用说明 instructions to the user
用户试验 user test
用户室内安装的煤气管 gas service pipes
用户手册 user's manual
用户首标 user header label
用户授权文件 user authorization file
用户属性数据集 user attribute data set
用户数据协议 user datagram protocol
用户数据组 user data set
用户水表 domestic meter;domestic water meter;service meter

用户水管 service pipe
用户水塔 service tank
用户水箱 service tank
用户说明 instruction manual
用户台 subscriber's board;subscriber's switchboard
用户特权 user privileges
用户特许文件 user authorization file
用户调压器 service governor;service regulator
用户通话计次器 line register
用户通话计数器 subscriber's meter;subscriber's register
用户通信[讯]接口 user communication interface
用户图 user map
用户图形界面 graphic(al) user interface
用户图形文件 user graphic file
用户网络接口 subscriber network interface;user network interface
用户微程序编制加工器 user microprogrammer processor
用户微程序编制器 user microprogrammer
用户微程序控制性 user microprogrammability
用户微程序设计 user micro-programming
用户尾部标记 user trailer label
用户文本【计】user version
用户文件 user file
用户文件编制 user documentation
用户文件目录 user file directory
用户污水 house sewage
用户污水管 collection line;house sewer
用户屋内设备 installation pipework
用户席 line desk
用户系统 custom system
用户线 central office line; exchange line; subscriber's line; subscriber loop
用户线电路 line circuit
用户线继电器 line relay
用户线空闲 subscriber line free
用户线路 individual line;subscriber's line;subscriber's loop
用户线路接口电路 subscriber's line interface circuit
用户线路设备 subscriber's line equipment
用户线模块 subscriber's module
用户线占线信号 subscriber's line busy signal
用户线自动测试器 automatic subscriber line testing equipment
用户小交换机 private branch exchange; telephone private branch exchange
用户小交换机中继线代号 private branch exchange hunting number; private branch exchange trunk number
用户效益 user benefit
用户协会 user group
用户写入程序 user writer
用户信号灯 called subscriber held lamp
用户信息控制系统 customer information control system
用户需求 user's needs;user's requirement;user's demand
用户需求规格书 user requirement specification
用户需要 user's needs
用户旋转式选择器 subscriber's uniselector
用户选定油漆 custom paint
用户选择器 subscriber's selector
用户训练 customer training
用户要求 customer requirement;user requirement
用户意见 consumer complaint
用户意图数据 user view data
用户引入线 branch to a building; drop wire;service drop conductor; service entrance; service entrance conductor
用户引水龙头 service cock
用户引线管 service pipe
用户友好的 user-friendly
用户友好人机界面 user man-machine interface
用户友善性【计】user friendliness
用户友谊 user friendship
用户与系统交互作用【计】user/system interaction
用户载波终端机 subscriber carrier terminal
用户占线信号 subscriber-busy signal
用户栈 user stack
用户栈指示器 user stack pointer
用户站 subscriber station
用户账单 subscriber's account
用户支管 house sewer; service branch
用户直接呼叫指示的人工电话 non-coded call display working
用户直接用计算机网络检索 customized online search
用户直通传真电报 telex facsimile
用户止水阀铁箱 curb cock box
用户指定的 user-defined
用户指南 userguide;user's guide
用户指示灯 user lamp
用户终端 consumer's terminal;user terminal
用户终端设备 subscriber terminal equipment
用户终端系统 client terminal system; user terminal system
用户终端业务 teleservice
用户主机 main station; subscriber's main station
用户驻地设备 customer premises equipment
用户驻地网 customer premises network; subscriber's premises network
用户专用线 subscriber's line
用户状态 user mode
用户状态表 user status table
用户状态监视电路 subscriber supervisory circuit
用户自动计费 private automatic message accounting
用户自动交换机 private automatic exchange
用户自动交换机中的值班分机 night attendant
用户总出口程序 user totaling exit routine
用户总机 subscriber's main station
用户总开关 consumer main switch
用户租赁服务 custom tenant service
用户组 group of customers; user group
用户最优【交】user optimum
用户最优交通分配模型【交】optimum traffic assignment model;user optimum traffic assignment model
用花键接合的柄 splined hub
用花键接合的衬套 splined hub
用花键接合的独立轴 splined independent axle
用花键接合的钢衬套 splined steel hub
用花键接合的钢轮毂 splined steel hub
用花键接合的钢轴 splined steel hub
用花键接合的轮毂 splined hub
用花键接合的轴 splined hub;splined shaft
用花装饰的窗子 flower window
用滑槽运料 chuting
用滑槽运送混凝土 chuting concrete
用滑车吊起 bouse
用滑车举起 pulley
用滑模浇筑的混凝土板 slipformed slab
用滑模修筑成的边缘 slipformed curb
用滑模修筑成的路缘 slipformed curb
用滑石处理 talc(um)
用化学方法去湿 chemical dehumidification
用化学溶剂清除发动机沉积物 tune-up
用化学药物再生活性炭法 reactivation of carbon by chemical method
用划线器划线 scribe
用划针描样板轮廓 scribe
用画笔画 pencil
用坏 stale;wear out
用环氧树脂补裂缝 epoxy repair of crack
用环氧树脂黏[粘]合的 epoxy-bonded
用环装饰的 ringed
用环做记号的 annulated
用黄铜镶(饰) braze
用黄铜制造 braze
用灰浆法配合混凝土成分 proportioning on a cement-paste basis
用灰泥粉刷 coated with stucco
用灰泥填塞屋顶板或瓦下面的空隙 torching
用灰砂制的砖块 brick made from sand-lime
用挥发性漆加深颜色的印刷品 urushi-ye
用混凝土泵灌注 concrete pump placing
用混凝土泵浇注 concrete pump placing
用混凝土层加固的柱子 buttress lift
用混凝土浇入 concrete in
用混凝土铺料机建筑运河坡 canal slope concrete paver
用混凝土嵌缝 filling with concrete
用混凝土墙锚碇 concrete wall anchorage
用混凝土套接的木桩 concrete-spliced wood pile
用混凝土填充 filling with concrete
用混凝土修补 concrete patching
用混凝土修路面 concrete patching
用活动扳手操作的水龙头 bib and spanner
用火安全要求 fire safety requirement
用火加热的加热器 direct heating apparatus
用火炉烘干 stove
用火炉加温 stove
用火焰喷射器开挖隧道 flame-jet tunnel(l)ing
用货盘装运 palletised[palletized] load
用货盘装运货物 palletised cargo
用机床加工的 machined
用机器混合抹灰 machine applied mixed plaster
用机器人施工 robot-controlled construction
用机械探寻(地下水源、矿脉等)桩 dowse
用机械探寻(水源)的专业人员 dowsing practitioner
用机械筑路 road machining
用积分原理的脉冲选择器 integrating divider
用激光控制方向 laser-type directional control
用集装箱运的货物 palletised load
用集装箱运输 containerization
用几何图形表示 geometrize
用挤水法生产的颜料 flushing process pigment
用挤压法制成的管子 extruded pipe
用计算机的应力标绘 stress plotting by computer
用计算机仿真 computerized simulation
用计算机检验 counter control
用计算机(进行)设计 computer-aided design [CAD]
用计算机进行探测 computer-aided detection
用计算机控制的交通信号通控制 computerized traffic signal control
用计算机设计信息软件 computer-aided design information software
用计算机诊断 computer-aided diagnosis
用计算机作上市登记 computer listing
用加热法处理(玻璃料) frit
用夹板夹 splint
用夹具加工 jig
用夹子连接软管 clamp hose connection
用架支撑屋顶 cradling roof
用架支住 cradle
用尖锄掘 picking
用尖锤琢石 nigging
用间隔隔开 interspace
用碱性当量溶液滴定 alkalimetry
用键固定 key on
用键接合的 keyed
用键控穿孔机 key punch
用桨划 paddle
用降落伞降落 parachute
用糨糊封 impaste
用交叉撑加劲力 single bridging
用交叉线画成阴影的 cross-hatched; cross-sectioned
用胶结材料处理的碎石路 oil macadam
用胶结法压装轮对的轻型轴 light axle attached by gluing
用胶囊包 encapsulate;encapsulation
用胶囊包起来的 encapsulated
用胶泥堵塞 puddle
用胶水和蛋黄调绘 al secoo painting
用胶体乳化剂乳化沥青 asphaltic emulsion with colloidal emulsifier
用胶质乳化剂乳化沥青 bitumen emulsion with colloidal emulsifier
用角尺分步画线 stepping-off
用角度表示经度 longitude in arc
用绞车提取 take a hitch on
用脚手架进行安装 erection by staging
用脚座安装 foot mounting
用铰刀铰孔 realm
用接合板连接 fish
用接近井径钻铤钻进技术<一种防井斜技术措施> packed hole technique
用结合环的加深梁 deepened beam with connectors
用截断机截断 guillotine
用金箔装饰 tinsel
用金刚石雕刻的玻璃表面图饰 diamond-point engraving
用金刚石钻孔 diamond drilling
用金刚石钻取岩芯 core-drilling with diamonds
用金刚石钻头钻探 diamond boring
用金属板作悬挂屋顶材料 metal plate suspension roof(ing)
用金属包镀 metal
用金属包覆 metal(l)ing
用金属衬里 bush
用金属处理 metallize
用金属敷镀 metal(l)ing

Y

用金属件固件固定的陶瓷面砖 anchored-type ceramic veneer
用金属丝加固的 wired
用金属丝捆缚的 wired
用金属丝系 wired
用金属条接地 <电缆铠甲或铅壳上用的> bonding strip
用禁运办法限制装车 restriction of loading by embargo
用晶体稳定的 crystal checked
用井筒托换基础 underpinning with cylinders
用鸠尾榫结合 dovetail
用旧 stale
用旧的 worn-out
用锯加工成的镶面板 sawn veneer
用具 apparatus; appliance; equipage; furniture; implement; tool; utensil <尤指厨房用>
用具储藏室 kit locker
用具进气管 drop line
用具设备通风管 appliance ventilation duct
用具箱 kit;work box
用具旋塞 drop valve
用卷尺测量 tape measure;taping
用卷扬机绞起 windlass
用绝缘板隔热 slab insulant
用卡钳量 caliper
用开水洗或烫 scald
用可变钥匙开启的锁 key changing lock
用可动心轨辙叉的单式交分道岔 single slip switch with movable point-frog
用可动辙叉的复式交分道岔 double slip switch with movable point frog
用空气吹冷的 air-cooled
用空气净化 purifying with air
用空气冷却的 air-cooled
用空气驱除油罐气 removing tank gas by air
用空头支票进行交易 kiting transaction
用空心铆钉接合板材 eyeletting
用控制点纠正 rectification with control points
用控制图分析资料 control chart for data analysis
用快镜摄影 snap
用款计划 expenditure plan
用框架固定 frame mounting
用捆绑连接的系杆 tie bar joint with rope lashing
用拉杆固定的堵头 tied bulkhead
用拉杆固定的闷头 tied bulkhead
用拉杆加固 grapple
用拉杆加固的板桩岸壁 tied bulkhead
用拉索拉紧的铁烟囱 guy steel stack
用蜡保护层 application of wax resist
用蜡笔画 crayon
用蜡密封的 wax-sealed
用蜡纸印 stencil
用来擦光瓷砖边缘的金刚石 rubbing stone
用来捆扎的东西 tie-up
用来弥补赤字的公债 deficit-covering bond
用蓝矾(硫酸铜)浸渍 boucherize
用蓝铅笔校改 blue-pencil
用蓝色铅笔做记号 blue-pencil
用缆绳悬吊的屋顶 rope cantilevering roof
用缆索操纵的 cable-operated
用缆索操纵的挖掘机 cable-operated excavator
用缆索系固 cabled
用捞砂筒清理钻孔 bail down;bailing up
用烙铁熨平 iron out
用雷诺数表示的消色力 Reynold's reducing power
用肋材加强的钢板 ribbed plate
用肋加固 finning
用肋加劲的 rib-strengthened
用肋加强的 rib-strengthened
用棱镜底部的水准仪 prismatic (al) bottom level
用冷凝油墨印刷 cold-set printing
用离心浇注法预制的 centrifugally prefabricated
用篱笆排列 line with wicker-work
用力的 forcible
用力刮平 forced screening
用力过度 overexertion;over-strain
用力过猛 overpressure
用力划 stretch-out
用力拉 rouse up
用力敲 rap
用力投(掷) hurl
用立柱加固 underprop
用沥青拌制的骨料 asphalt-coated aggregate
用沥青补路 bituminous patching
用沥青材料黏[粘]结 bonded with asphaltic material
用沥青处理 bituminization
用沥青处理的 bituminized
用沥青处理过的表面 treated bituminous surface
用沥青的人 asphalt man
用沥青滴敷设屋面石棉板 drop point
用沥青粉拌和的冷铺沥青混合料 Westphalt
用沥青封丝扣 <为防止漏水> pitching the threads
用沥青覆盖的 pitchy
用沥青灌注的 bitumen-impregnated
用沥青浸渍的 bitumen-impregnated
用沥青料修补 bituminous patching
用沥青铺路面 blacktopping
用沥青油修补 oil pitching
用砾石填充的集水沟 gravel-type gully
用砾石填充的进水井 gravel-type gully
用粒料加固的 granular-stabilized
用粒状软木覆盖 covered with granulated cork
用连续施工法铺筑的道路 <每次铺半边宽度> road laid in continuous construction (half width at a little)
用连字符连接 hyphen
用镰刀割 scythe
用链丈量 chaining
用凉亭遮蔽 embower
用梁加劲的 girder-stiffened
用梁托支出 corbel outwards
用两耳的 binaural
用两个字母标的尺寸 double letter dimension
用两脚规截取距离 compasses setting;compass setting
用两种语言出版的书 diglot
用量 addition;dosage;use level
用量分配 distribution of demand
用量高峰 consumption peak
用量规量 tramming
用量计量仪 demand register
用量记录计 demand register
用量投药配量 dosage
用料表 cutting list
用料成本预算 material cost budget
用料(清)单 bill of materials
用料取样 as-used sample
用料预计单 bill of materials
用流水作业法铺筑路面 straight-forward surfacing
用流网设计排水沟 designing drains with flow net
用硫酸酸解 sulphatising
用柳条或竹笼装的细口瓶 carboy
用柳条笼装的细口瓶 demijohn
用六柱的 hectastyle
用镂花模板复制(图案) stencil work
用芦苇和木材制成的柴排 Dutch mattress
用路拌堆土法稳定土 mulch method
用路油修补路面 oil patching
用辘轳拉上来 tackle
用履带式拖拉机牵引 caterpillar traction
用氯化汞浸渍(木材) kyanize
用氯量 chlorine dosage
用氯灭菌 sterilization by chlorine
用氯消毒 sterilization by chlorine
用卵石铺的路 bouldering
用卵石铺砌 paving with pebbles
用卵石铺设的 cobbled
用卵形掘土机建成的排水渠 mole drain
用乱石砌筑的 snecked
用轮带运输 belting
用螺钉拧紧 screw
用螺钉拼接 screwing
用螺栓固定 fasten with bolt; fixed with bolt
用螺栓固定的接合板 bolt-up fishplate
用螺栓固定的鱼尾板 bolt-up fishplate
用螺栓扣紧 fasten with bolt; fixed with bolt
用螺栓连接 bolting;pin together
用螺栓连接的节点 bolt-connected joint
用螺栓支护 bolting
用螺丝接合 joining by screw
用螺旋钢筋的 spirally reinforced
用裸线连接 bare wiring
用麻布作为填充物封隔套管环状间隙 shirt-tail packer
用麻绳放线 string lining
用麻绳放样 string lining
用马歇尔法测定的(沥青混凝土)性质 Marshall property
用码测量的长度 <英制> yardage
用镘板粉光 smoothing board finish
用镘刀刮出一定深度的灰缝 trowel-(l)ed joint
用镘刀抹灰浆 buttering
用镘刀涂抹 trowel;trowel off
用镘刀修整 float finish
用镘涂灰浆 buttering
用盲板堵管 blanking
用茅草盖屋顶 thatch
用锚测深 anchor as a lead
用锚将船固定住 grapple
用煤供暖 coal heating
用煤加工生产的煤气 producer gas
用煤加热 coal heating
用煤油轻制的沥青 kerosene cutback
用煤油助溶的 kerosene-fluxed
用美元结算 settle accounts in US dollar
用美元支付的汇票 dollar draft
用米制表示 metricize
用密封垫镶玻璃 gasket glazing
用幂公式逼近 approximation by power
用模板的 formed
用模板印成的花样 stencil
用模板印刷 stencil
用模型压印 stamp
用磨料进行喷砂处理 abrasive blasting
用抹子修饰 trowel finish
用木板铺屋面 shingle roofing
用木板支撑 planking
用木材撑牢的 well-timbered
用木材加固的 well-timbered
用木材支护的 timber-lined
用木钉钉定 pegging
用木钉钉牢 nogging;peg down
用木块垫起出轨车辆 packing
用木排铺成的 corduroy
用木栅栏围住的 encircled by a palisade
用木杖探寻(地下水源、矿脉等) dowse
用木支撑 timbering
用木制造 timbering
用木质纤维毡压制的绝缘板 celotex; celotex board
用木柱支撑 nogging
用木桩标线 peg(ged) out
用木桩定位的沉排 pegged mattress
用木桩围上 picket
用目镜读数 eyepiece reading
用幕覆盖 draping
用幕隔开 curtain
用内撑框架的钢板桩围堰 steel sheet-piling cofferdam with internal bracing frame
用内木框架支撑的板桩围堰 timber sheet piling cofferdam with internal frame bracing
用内折流板折流 internal baffling
用耐火材料保护的 fire-retarded
用泥封固 lute
用泥浆泵钻土 bailer-boring
用泥浆钻进 bore with slurry; drill with slurry
用泥浆钻井 bore with slurry
用黏[粘]泥封闭接合处 luting
用黏[粘]泥涂 slime
用黏[粘]土粘合的 clay-bonded
用黏[粘]土夯筑围堤 stock ramming
用黏[粘]土加固渠道底槽 lining a canal bottom with clay
用碾压实 compaction by rolling
用尿素脱蜡 dewaxing with urea
用耙翻起 spike up
用耙子装载 hoeing
用排水控制 control by drainage
用排水砂井加速固结 acceleration consolidation with sand drains
用抛掷机进行装填 mechanical stowing with slingers
用配重平衡 counterweigh
用喷枪吹洗 lancing
用喷洒法施肥 fertilizer application by sprinkling
用喷砂或喷丸清理过的表面 abraded surface by blasting
用棚布遮盖 overcanopy
用棚架支撑 trellis
用膨胀水泥灌浆混合物 expansive cement grouting compound
用皮革包盖 American leather
用品箱 <运输设备的> cellar
用品箱下垫木 cellar bolster
用平版印 planograph
用平地机及刮路机整平 blading and dragging
用平地机将材料堆成长条堆 blade into a single windrow
用平地机挖沟 ditching by grader
用平地机修筑路堤边坡 fill sloping with grader
用平地机在工地拌和材料 blade mixing
用平地机整平 blade grading
用平地机整平(道)路 road blading
用平地机重新整型 reblade
用平衡重开启的吊桥 counterpoise bridge
用平均法求经验公式 empiric (al) formula by method of average
用平路机平路 blading
用平路机修筑斜坡 sloping with grader
用平头钉钉住 tack
用屏保护的 screen-protected
用期 period
用漆刷窄边涂漆 <俚语> ride the

brush
用旗(灯)信号显示保护列车的规则 flagging rule
用旗(灯)信号指示列车停止或开动 train flagging
用旗信号保护列车的办法 flag protection
用旗作信号 flagging
用起动燃料点火 starting-fluid ignition
用起重机搬运 crane
用起重机吊装 crane loading
用起重机起吊 derricking; lift with crane
用起重桅装(货) steeve
用气袋堵管 bagging-off
用气定额 gas consumption quota
用气堵(塞)住 air lock
用气分类 class of service gas
用气分离 air separation
用气炬烧 torching
用气量 gas consumption
用气量指标 gas consumption quota
用气普及率 customer penetration;gas customer penetration
用气塞堵住 air lock
用气体盖覆 blanketing with gas
用气体净化管道 gas cleaning pipe
用气体燃烧的 gas-fired
用气体再生活性炭法 reactivation of carbon by gas method
用气相色谱(法)分析 gas chromatograph
用气压力 standard service pressure
用气运送 convey by air
用汽车拖带的活动住房 trailer house
用汽力开动 steam
用汽水钻孔 drilling with aerated water
用汽油润湿 moistened with gasoline
用契约约束 hold
用器械探矿 dowse
用器械探水 dowse
用千斤顶 jacking
用千斤顶沉放 sinking by jacking
用千斤顶抵桩 jacking of pile
用千斤顶顶拉 jacking
用千斤顶顶起 jacking;jack(ing) up
用千斤顶顶托 jacking
用千斤顶起重 jacking
用千斤顶下沉 sinking by jacking
用千斤顶张拉(钢筋)<预应力混凝土中的> jacking;tensioning jack
用铅笔写 pencil
用铅笔重绘 repencil
用铅锤检查垂直度 plumbing
用铅搭叠 lead lap
用铅条密封 drawn lead trap
用铅重叠 lead lap
用钱大方的 freehanded
用钳子夹紧 cramp
用前滩物质修复砾石海滩顶部 beach banking;beach bumping
用枪打入的紧固件 powder-set fasteners
用枪打入的锚固件 powder-set fasteners
用墙承重的 wall-bearing
用墙堵死 walling up
用墙堵住 walling up
用墙围起 circumvallate;enclose
用墙围住 enclosing by masonry wall; wall
用锹填塞小碎石 shovel packing
用切碎机切碎 devil
用轻石擦 pumice
用轻型钢制作的钢构件 light-ga(u)ge steel member
用球磨机磨碎 ball mill
用曲柄连接 crank
用全力 with might and main
用全息照相术拍摄 holograph
用燃料冷却的 fuel-cooled
用热户 heat consumer
用热要求 heat demand
用人工方法引来的水 new water
用人物装饰的 historiated
用溶剂打磨到无光 rub to dull
用溶剂稀释的 solvent-thinned
用入射光观测 observation by incident light
用软管浇水 hose
用软管洗涤 hose down
用软木板护壁 cork lagging
用软木板护壁的 cork-lagged
用软木为骨料的混凝土 concrete with cork aggregate
用软木作骨料的喷射混凝土 concrete-spraying with cork aggregate
用软皮摩擦 buff
用润滑剂 with lubricator
用塞子封堵<管道、通道等的> blankoff
用三角测量法测量高度的仪器 hypsometer
用三片锲块锚住一根钢丝为一单元 Gifford-Udall anchorage
用扫帚扫 broom
用色惯例 colo(u)r convention
用纱团擦涂 fading
用砂打磨 sanding
用砂打磨机具 sanding apparatus
用砂袋筑坝(壅水) damming by sand bags
用砂堵塞 sand up
用砂浆修补 slush
用砂砾喷射的 grit blasted
用砂量 sand content
用砂密封 sanding sealer
用砂轻磨 touch sanding
用砂修整 sand finish
用砂纸打磨 sanding;sandpapering
用筛分级 screen sizing
用栅隔开 fence off
用栅栏防护 fence
用栅栏围住 stockade
用栅围进 fence in
用栅围拢 fence in
用栅围绕 palisade
用栅围住 pale;picket
用膳者 diner
用商品抵押的预付款 advance against goods
用射水方式冲洗 flush out
用射线照相时 when radiographed
用升汞浸渍木材 kyanize
用绳穿过 reeve
用绳捆 rope
用绳拉住 guy
用绳索或铁索绷紧的 funicular
用绳索捆扎保护 keckle
用绳索起下油泵 wire-line oil pump
用绳索系牢 tether
用绳索运转的 funicular
用绳拖曳 tow
用绳系 tether
用石板盖屋面 helying
用石板铺路 flagstone paving
用石板铺砌 slab paving
用石板瓦盖屋顶 slate board(ing)
用石板瓦盖屋面 slate
用石板瓦铺屋顶 slating
用石灰处理 lime
用石灰结合的 lime-bound
用石灰(石)处理细料 limit material
用石灰刷白<石材的临时保护> lime washing
用石灰水粉刷 lime whitewash
用石蜡处理 paraffin(e)
用石油处理 petrolize
用石油点燃 petrolize
用时间单位表示经度 longitude in time
用实物缴税 taxes in kind;taxes paid in kind
用式子表示 formulation
用视线高度测水平法 collimation system
用试坑勘测地基 foundation reconnaissance by test pits
用试坑研究地基 foundation investigation by test pits
用试坑作地基勘探 foundation exploration by test pits
用试配法配合混凝土成分 proportioning by trial method
用试算法解题 solve by trial
用试验法配合混凝土成分 concrete proportioning by trial method
用试验进行开发 test development
用饰面板处理的房屋正面 veneered facade
用手 by hand
用手柄转向 hand lever steering
用手操纵的 hand steer(ing)
用手操作的 manually operated
用手操作的研磨石 handstone
用手操作检查 hand-on inspection
用手车运料 barrow
用手充分拧紧螺母后的状态 finger tight position
用手导回原位 hand-restoring
用手动 manually
用手工做圆木 thurm
用手供料 hand feeding
用手驾驶的 hand steered;hand steer(ing)
用手进料 hand feeding
用手控制的 hand-controlled
用手拉紧 hand taut
用手拧紧 back-out by hand
用手拍发 hand-typing
用手切割者 hand cutter
用手填充 hand filling
用手填实 hand filling
用手调整 hand set(ting)
用手推车运料 barrow
用手摇发电机和蜂鸣器的磁石式电话机 local battery with magneto and buzzer calling telephone exchange
用手摇计算机计算 hand computation; manual computation
用手引导的 hand-guided
用手掌握的 hand-guided
用手钻钻孔 gimlet
用梳刀刻螺纹 chase
用树脂材料黏[粘]结 bound with resinous material
用树脂处理 resin;resinification
用树脂浸渍的木料 bakelite-impregnated wood
用数量表示 quantification;quantify
用数字表达<数据等> digitalize
用数字表示 figuring
用数字表示的 numeric(al)
用刷上色 brush over
用刷涂的填料 filler for brush application
用刷涂敷的 brush-applied
用刷子清除 brush clearing
用刷子刷(油漆涂料) application by brushing
用刷子涂上 apply by brushing
用刷子完成的工作 brush work
用双锚固定的系锚 span mooring; two-arm mooring;two-leg mooring
用双重元件保证可靠性 reliability by duplication
用水 water usage
用水安全 water security
用水饱和<计算砂容积的> inundation
用水比 water-use ratio
用水标准 standard of water use
用水部门 department of water use
用水掺和 blunge
用水冲的 jetted
用水冲法下沉混凝土桩 concrete pile jetting
用水冲击岩芯 flushing core
用水冲洗 water flushing
用水冲洗的 water-flushed
用水淬火 water hardening
用水的 watered
用水点处理 point-of-use treatment
用水点处理装置 point-of-use treatment device
用水点工业 point-of-use industry
用水点装置 point-of-use device;point-of-use system
用水点装置工业 point-of-use equipment industry
用水调度规则 water operating rule
用水定额 water consumption norm; water consumption quota; water consumption rate;water duty
用水份额 water allotment
用水隔油 line clearance with water
用水供水阀 stop ferrule
用水管理 management of water use; water management
用水灌溉了的 watered
用水灌满 flood
用水规章(制度) water operating rule
用水果装饰成弧形垂莲花彩 fruit work
用水户 water consumer;water user
用水户协会 water users association
用水灰比法配合混凝土成分 proportioning by water-cement ratio;concrete proportioning by water-cement method
用水回收率 recovery rate of water
用水计量 water metering
用水间<厨房、浴室等的> water section
用水搅拌 blunge
用水冷却 hydrocooling
用水力学方法演算洪水 hydraulic routing
用水量 service discharge;water consumption;water duty
用水量标准 rate of water demand; rating of water demand
用水量大的工业 water-using industry
用水量累积图 accumulation diagram of water demand
用水量曲线 curve of water consumption
用水龙带冲洗 hose down
用水龙浇水 hose
用水率 duty of water;rating of water consumption; water application rate;water duty;water rate;water-rating
用水磨细 wet reduction
用水泥固定套管 set casing
用水泥灌浆法稳定土壤 ground stabilization by cement grouting
用水泥灌入 cement in
用水泥浆加固 consolidation grouting
用水泥浇筑 cast in cement
用水泥抹光 float with cement
用水泥抹平 float with cement
用水泥砂浆重衬里管道 pipe relined with cement mortar
用水泥填充 fill with cement;grout in with cement
用水强度 intensity of water
用水权 right of water; water privilege;water rights
用水设备 plumbing system appliance

用水填充 filling with water
用水文学方法演算洪水 hydrologic(al) routing
用水系数 water use coefficient
用水下冲法沉混凝土 concrete pile jetting
用水效率 application water efficiency;water use efficiency
用水许可制度 licences system of using water
用水优先权 prior appropriation;priority of water use;priority use of water
用水再循环系统 water recycle system
用水蒸气和肥皂清洁石块的方法 Roxor
用水制度 system of water consumption
用水专利权 water appropriation
用水专有权 appropriated right
用水装满 filling with water
用松土机松土 rotovation
用松香擦 rosin
用塑像装饰 decorated with status
用酸拌和的 acid pugged
用酸处理 acidize
用酸侵蚀测定钻孔倾斜法 acid dip survey
用算盘计算 click-clack
用碎石加固的土 crushed-stone soil
用碎石铺路 metal(ling)
用碎石铺砌 set in broken stones
用碎石嵌实 set in broken stones
用碎瓦片填高屋脊瓦片的位置 galleting
用隧道缩短线路长度 tunnel cut-off
用榫钉合缝 dowel(l)ing
用榫钉连接 dowel(l)ing
用榫接合 tenon jointing
用榫眼接合 mortise
用索缚紧 frap
用台车凿岩 drilling by drill carriage
用钛稳定的钢 steel stabilized with titanium
用弹簧加压的 spring-loaded
用弹簧卡子紧固轮箍 fastening of the tyre by spring clip
用弹簧圈紧固轮箍 fastening of the tyre by spring ring
用探杖探矿 dowse
用探杖探水 dowse
用碳还原 reduction by carbon;reduction with carbon
用碳粒送话器的电话机 carbon telephone
用套管扶正钻杆 collaring
用套加热 jacketing
用套冷却 jacketing
用套色法印刷 chromatograph
用特殊字体印的标题 rubricus
用梯段法开采 stope
用梯攀登 escalade
用梯子爬墙 escalade
用梯子上达 ladder access
用天篷遮覆 canopy
用天然砂土筑路 topsoiling
用天然页岩作室内装修 beaulate
用填砂筑路 topsoiling
用挑砖砌的墙压顶 verge course
用条状垫板的焊接 weld with backing strip
用调色剂染色 stain
用贴墙光源照亮墙面 wall-washing
用铁(沉淀)置换 <以析出某种金属物质> cement with iron
用铁链拴系的浮体防波堤 tethered breakwater
用铁路运输 by rail
用铁盘烘 griddle
用铁锹挖土 soil cut by spade
用通风器吸尘 fanning
用同样步调 step for step
用统计方法控制质量 statistic(al) quality control
用桶装运 bucket;carrying by means of barrels
用投标报价 offer by tender
用透明薄织物做的 peek-a-boo
用透视光观测 observation by transmitted light
用突出宽带作装饰的半露柱体 banded pilaster
用图表表示 graphing
用图表示 diagrammatic (re) presentation;diagrammatization;diagram(m)ing;figuring
用图表说明 illustration;illustration with figures
用图画表示的 pictorial
用图积分法【数】graphic(al) integration
用图解法表示 diagrammatize;diagram(m)ing
用图解法求经验公式 empiric(al) formula by plotting
用途 purpose
用途费用报表 objective statement
用途分区 use zoning
用途区划 use zoning
用土覆盖 heal
用土权 easement
用土填满 land up
用推进装置减速 propulsion-system braking
用推土机清除 doze;dozing
用推土机推成行 blade into a single windrow
用推土机推平 doze
用推土机推土 bulldozing
用拖把涂沥青 mopping
用拖拉机底盘的挖掘机 tractor-excavator
用拖拉机牵引的 tractor-drawn;tractor-hauled
用拖拉机驱动的 tractor-driven
用拖轮牵引 tug
用挖泥材料填筑的路堤 filling of dredged material
用瓦盖屋面 helying
用外钢套拼接钢管桩 steel pipe pile splicing by tight steel outside sleeve
用外皮包裹 incrustation
用外皮覆盖 crust
用弯爪起钉的拔钉器 nail puller with bent claw
用完 exhausting;outwear;run-out;use up
用完的蓄电池 discharged battery
用完时间符号 pad character
用网捞取 dredger
用网眼刺绣镶边的 peek-a-boo
用围栅围绕 palisade [palisado]
用帷装饰的 draped
用温度计测得的 thermometric
用文字表示的信息 graphic(al) information
用文字的 verbal
用蜗杆调节 worm adjustment
用污水灌溉的农田 sewage farm
用屋顶光照明 lighting by roof-light(s)
用屋面卷材覆盖的屋顶 roof cladding with roll roofing;roof covering with roll roofing
用无线电广播 air-cast
用雾化作用进行的喷淋 spray by atomizing
用雾化作用进行的喷洒 spray by atomizing
用吸尘箱防护的钻机 drill guard with dust bag
用吸扬式挖泥船挖泥 hydraulic dredging
用系列小爆破形成大爆破硐室 chambering;springing
用细钢丝的预应力混凝土 prestressed concrete with fine wires
用细灰泥抹光表面 hard finish
用隙灰比法配合混凝土成分 concrete proportioning by void-cement method;proportioning by void-cement ratio
用隙灰比法配料 <混凝土混合料> proportioning by void-cement ratio
用纤牵曳 warp
用现货 upon the spot
用现金彻底购买 buying outright
用现金付出的费用 out-of-pocket expenses
用现金买 buy for cash
用现金支付 pay down
用现款 on-the-line
用现值法计算的效益成本比率 discounted benefit/cost ratio
用线钉(书) stab
用线划分的 lined
用线划(分)开 line off
用线性电动机对溜放车辆加减速度的控制 acceleration and deceleration speed control by linear motor
用线性电动机推进 linear-motor propulsion
用线性电动机推进的车辆 linear-motor powered vehicle
用线性感应(异步)电动机推进的车辆 linear induction motor propulsion vehicle
用线性排量关系计算测流堰 measuring weir with linear discharge
用线性同步电动机推进的车辆 linear synchronous motor propulsion vehicle
用线支承的 line supported
用相对显微镜的 phase contrast
用象牙制的 eburnean
用橡胶处理的涂料 rubberized paint
用橡胶刮板刮拭 squeegee;squeeze
用橡胶辊滚压 pneumatic rolling
用橡胶辊碾压 pneumatic rolling
用橡胶液浸渍的帆布 rubberized canvas
用橡木雕刻扇形的装饰术 nulling
用硝酸处理【化】nitrate
用销钉连接的键接 dowel(l)ed connection
用销连接 pegging
用销子固定的 pin-locking
用小车运走 buggy away
用小方块镶嵌物嵌成的 tessellar
用小块石砌筑的墙或砌体 rag work
用小连杆与转向架构架相连的轴箱 axle-box fastened to the bogie frame by means of small articulated rods
用小梁支撑基础 needle beam underpinning
用小石块填缝 sneck
用小石填筑的 snecked
用小数表示的分数值 decimal equivalent
用小炸药包爆破 squib shooting
用小桩标出的测线 staking a line
用楔锤破地 beatway
用楔固定 fixing by wedges
用楔块塞瓷 hand wedging
用楔劈 wedging
用楔止住 chock
用楔子固牢 cleat
用楔子支持 quoin
用斜槽进料 chute
用斜铲推土机推 angle-dozing
用斜(滑)槽运输混凝土 chuting concrete
用心做成 elaboration
用锌版印 zincograph
用锌包 zinc
用锌处理的紧固件 zincked fastenings
用信号灯调节或控制的电位器 lamp check potentiometer
用信号铃的呼叫系统 call-bell system
用信号通知 signalize
用信号通知的速度 signal(l)ed speed
用信号寻找线路故障 signal tracing
用信件报盘及接受 offer and accept by post
用信件方式预约公共汽车 mail a bus
用星号标明的 starred
用星装饰 star
用星装饰的 starred
用型铁做 swage
用溴处理 bromize
用畜队运 team
用选速按钮操纵的缓行器 retarder operated by push-button selection of speeds
用选择溶剂的共沸蒸馏 selective azeotropic distillation
用雪覆盖 snow
用雪橇运输 sled(ge)
用循环水冷却 cooling by means of circulating water
用压痕或打孔来装饰的 pounced
用压力盒测出的压力 cell pressure
用压力计量的 manometric(al)
用压路机碾压 steam-roll
用压缩空气扬水 air lift
用亚麻籽油启动 priming with linseed oil
用言辞的 verbal
用岩石锚碇 rock anchor
用盐处理 salt treatment
用盐处理的 salted
用盐水处理 brine;brining;salination
用眼评定 eye assay
用羊皮纸所写的文件 parchment
用羊足碾压实塑性土壤 kneading compaction
用养护罩保护新拌混凝土 protection of green concrete by curing overlay
用氧饱和 oxygenate
用氧气烧枪在混凝土上打孔 oxygen lance concrete drilling
用样板刮平 screed
用样板拉线脚 running a mo(u)lding
用样板找平 screed
用药量 dosage;therapy dose
用药浓度 dose rate
用叶片拌和 blade mixing
用液压千斤顶稳定 stabilizing by hydraulic jacks
用一袋水泥的混凝土拌和料 one-sack batch
用一个字母标注的尺寸 single letter dimension
用仪器测量 instrument measurement
用仪器观测 instrumental observation
用仪器记录 instrumental record
用仪器进行竖井铅锤测量 instrumental shaft plumbing
用乙醚萃取 ether extraction
用以薄涂的涂料 scumble
用以代替 serve as a substitute
用以伸长的渗水件 short extension
用以识别心线的有色线 <电缆> colo(u)red tracer thread
用以调节的 accommodating
用异形碎石、卵石、晶石等镶嵌的图案 rock rash
用益权 <在不损害产业的条件下使用他人产业并享受其收益的权利> usufruct;usufructuary right

用殷钢尺丈量 invar measurement
用印刷法涂荧光屏 phosphor printing
用影线表示 hachure
用影印石版印刷 photolithograph
用硬橡皮做的电缆护套 cab-tyre sheath(ing)
用硬渣喷射的 grit blasted
用硬质合金敷焊牙轮的齿 fused carbide tipping
用硬质弹性沥青片铺盖屋面 vulcanite roofing
用优势法简化 reduction by dominance
用邮船运送 packet
用油萃取 oil extraction
用油灰安装玻璃 putty glazing
用油灰接合 putty
用油灰填平 butter with mastic
用油灰填塞 putty
用油灰填塞裂隙 puttied split
用油灰镶嵌 putty
用油灰粘牢 butter with mastic
用油回火【机】 oil tempering
用油浸渍的 oil-impregnated
用油率 rate of application
用油墨冷模压 cold stamping
用油漆盖死 paint out
用油漆涂抹 paint out
用油设备 oil-using units
用油石修整锯齿侧边 side jointing
用油洗涤的空气滤清器 oil-washing air cleaner
用游标调整 vernier adjustment
用有色材料对大理石的天然孔进行补洞 waxing
用右手的 dext(e)rous;dextral;right-handed
用于半挂车的牵引车 motor tractor for semitrailer
用于拆除建筑物的铲斗 bucket for demolition
用于穿孔螺塞部分 for trepanning plug sections
用于对钎焊容器的连接开孔 openings for connections to brazed vessels
用于多机牵引设备的后楼梯电路 back-stairs circuit for multiple-unit equipment
用于恶劣环境的机器人 robot for poor environment
用于粉刷的木刮板 timber screed for plastering
用于辐射三角测量的菱形锁 chain of lozenge
用于高工作面的钻车 high face jumbo
用于公路的沥青乳化液 bitumen highway emulsion
用于管子连接的开孔 openings for pipe connections
用于海水环境的黏[粘]结剂 marine adhesive
用于海水环境的胶合板 marine plywood
用于集会或会议的一组建筑物 conference block
用于计算机的信息处理标准 information processing standards for computers
用于空调系统的组装式消声器 packaged sound attenuator for air-conditioning system
用于马赛克的砖石 abaciscus
用于排放的开孔 openings for drainage
用于铺屋面油毡的铁钉 nail for prepared roofpaper
用于山墙的边缘砖 marginal tile for gables
用于生产的有形资产 <土地除外> capital goods
用于提染料的硬木 dye woods
用于涂油前的 underglaze
用于外压容器 for vessels under external pressure
用于镶嵌的小块 mosaic piece
用于消费的产品 consumer's product
用于畜舍的铺地缸砖 flooring quarry (tile) for animal shelters
用于运输工程的地理信息系统 transportation geographic information system
用于再投资的利润 plow back profits
用于织物表面的塑料 alabascote
用于自然界的对策 game against nature
用鱼雷等破坏 torpedo
用语 parlance;wording
用语索引 concordance
用育草土嵌填的接缝 soil filled joint
用预算管制 control through budget
用预应力结合的预制件 jointing cast units by prestressing
用预制构件建造 preengineered
用预制件制造的 preengineered
用预制件装配的门框 knocked down frame
用原子反应器加热的锅炉 A-boiler
用圆盘研磨 lap
用圆凿打眼 gouge
用远近法缩小绘画 foreshortening
用钥匙锁闭属具与闭塞机联锁的调车凭证 shunting token interlocked with block instrument by means of a key lock attachment
用越仓法摊铺的混凝土 concrete laid in alternate bay
用运输带配料 batching by conveyor belt
用运输带作衡器的秤 measuring conveyor belt scale
用运输皮带测量(运输量) measuring by conveyer [conveyor] belt
用杂酚油防腐处理过的木材 creosoted timber
用杂酚油防腐处理过的木桩 creosoted pile
用杂酚油灌注 impregnate with creosote
用再生橡胶补轮胎 top cap
用凿刀修琢的石工 tooled work
用凿雕刻的木工 gouge work
用凿开采块石 gadding
用凿毛锤琢面 bush hammer finish
用凿劈开 chip out
用凿石锤整修的 bush-hammered
用凿石锥加工 tooled by Crandall
用凿削切 chiseling
用帐篷遮盖 overcanopy
用招贴宣传 poster
用照相凹版印刷 photogravure
用针选取的穿孔卡片 needle-operated punched card
用珍珠装饰的 pearled
用真迹石版术复制 autograph
用真空吸尘机打扫 Hoover
用振动台振实 table vibration
用蒸汽机航行 under steam
用蒸汽开动 steaming
用蒸汽清洗 steam cleaning
用蒸汽驱除油罐气 tank gas freeing by steam
用蒸汽挖掘机开挖 steam shovel excavation
用蒸压器蒸 autoclave
用整除部分的乘法 multiplication by aliquot parts
用整平板压实整平 screed
用正方形制的分隔墙 quarter red partition
用之不尽 inexhaustibility
用支撑顶住 shore up
用支撑防止大变形 supports to prevent distortion
用支索拉住 guy
用支柱撑住 shore;shore up
用支柱撑住的 shored-up
用支柱加固 propping
用支柱支撑 propping
用枝条编织的 wattled
用织物纤维增强抹灰层 scrimble
用直尺刮平 straight-edging
用直尺检查平面的平整度 straight-edging
用直升机救援 helicopter rescue
用直线围着的 rectilinear
用直爪起钉的拔钉器 nail puller with straight claw
用纸 form;paper-used
用纸裱糊 papering
用纸重铺 repaper
用纸重贴 repaper
用纸重新包装 repaper
用纸作底层的 paper-backed
用指甲切试法试硬度 finger-nail test
用指模弄 fingering
用制动器给进 feed by the brake
用制动(装置)操纵 steering by brakes
用制链器制链 bowse to a chain
用雉堞墙装饰 decorated with battlements
用中间不绝缘钢管制作的轨距尺 non-insulated pipe center track ga(u)ge
用中间绝缘钢管制作的轨距尺 insulated pipe center track ga(u)ge
用种草来稳定边坡 slope stabilization by seeding
用肘关的水龙头 <手术室用> elbow operated tap
用助铲机助铲 <铲土机> push-loading
用柱垫石的基础 post stone foundation
用柱支持 pillar
用柱装饰 pillar
用柱子支撑 shoring up
用抓斗式挖泥船挖泥 clamshell dredging
用抓斗疏浚 grab bucket dredge(r)
用爪爬 claw
用爪止住 pawl
用爪抓 crab
用专用工具勾灰浆缝 tooled joint
用砖砌面 facing in brick
用砖砌入木框架 nogging
用砖塞孔 bricking-up;plugged with brick
用砖石墙围住 enclose by masonry wall
用砖填充墙架 nogging
用砖填塞 bricking-up;building-up
用砖填塞门窗 build-up
用砖镶面 veneered with brick
用砖筑墙 brick walling
用桩标出 stake
用桩标出岸线 staking out the bank line
用桩标出的测线 stake line
用桩标出堤线 staking out the bank line
用桩定出 stake out
用桩加固的柴排 fag(g)ot with pile
用桩加固地基 palification
用桩加固土壤 palification
用"桩头朝下"法打桩 drive the piles "butt down"
用桩围住 picket;stake up
用装料台卸载 off-loading ramp
用装饰性肋的穹顶 lierne vaulting
用锥角环带法 <解析摄影测量> cone angle banding
用锥子锥 gimlet
用琢石砌面 ashlaring
用紫外线辐照 irradiation with ultra-violet light
用字符填充 character fill
用自动驾驶仪操纵 gyrorudder control
用自卸车尾部倾斜卸料 end tipping (from the shore)
用钻孔法沉井 shaft-sinking by drilling
用钻头数 numbers of bit used
用最小二乘方的近似法 approximation by least squares
用左手的 sinistral
用作边界、路标和纪念牌的石头堆 ahu
用作边界、路标和纪念牌的土丘 ahu
用作仓库的废船 hulk
用作家具和内部细工的红木板材 timbo
用作壳体的管子 pipe used for shells
用作门厅的 vestibular
用作拼接的螺栓 splice bolt
用作砌体面层 quarry-faced
用作石膏硬化剂的氟硅酸盐 fluosilicate hardener for gypsum
用作石膏硬化剂的氟化物 fluate hardener for gypsum
用作树篱的树木 quickset
用作支撑的 portative

优

优比 ratio of greater inequality
优策略 dominant strategy
优超性 superiority
优待 preference; preferential treatment
优待价 privileged price
优待减价票 <铁路职工待遇> privilege ticket
优等 premium grade
优等的 high class;high grade
优等金刚石 excellent diamond
优等路 excellent level highway
优等品 adequate quality;best quality; high class product;premium grade
优等投资 prime investment
优等延性油 ductile base oil
优等原木 prime log
优地槽【地】 eugeosycline
优地槽型沉积建造 eugeosyncline type formation
优地斜 eugeocline
优点 advantage; excellence; merit; strong point
优点和缺点 merits and demerits
优点夸大化效应 leniency effect
优点评价 merits rating
优度 goodness
优分 optimal sorting
优抚安置制度 the special care and placement
优共轭弓形 major conjugate segment
优共轭弧 major conjugate arc
优共轭角 major conjugate angle
优函数 major function
优点和缺点 pros and cons
优弧 major arc;super-arc
优弧角 reflex angle
优化 majorization;optimization;optimize
优化遍数 optimization pass
优化参数 optimal parameter; optimum parameter
优化程序 optimiser [optimizer]
优化稠度 optimized dispatch
优化调度 optimizing regulation
优化法 optimum seeking method
优化翻译程序 optimizing translator
优化方法 optic(al) method

优化方法的应用 application of optimization
优化分析 optimality analysis
优化功能 optimizational function
优化规范 optimat
优化函数 majorized function
优化后分析 post-optimality analysis
优化级数 majorizing series
优化集 majorized set
优化计算 optimization calculation
优化计算机 optimizing computer
优化技术的应用 application of optimization
优化阶段 optimizing phase
优化抗震设计 optimum seismic design
优化可能性 optimization possibility
优化控制 optimal control; optimization control
优化控制动作 optimizing control action
优化控制理论方法 optimal control theory
优化理论 theory of optimization
优化模式 optimization mode
优化模型 optimization model
优化剖面 optimum profile
优化器 optimizer
优化容量 optimizing capacity
优化设计 optimal design; optimum design; refined design
优化设计波高 optimum design wave height
优化时差 optimising offset
优化适合性 optimization suitability
优化水管理系统 optimization of water management system
优化四元组 optimized quadruple
优化特性 optimizing characteristic
优化外形 optimum configuration
优化问题 optimization problem
优化线性映射 majorizing linear mapping
优化序列 majorizing sequence
优化原理 principle of optimality
优化匝道限流系统【交】 optimal ramp metering system
优化支出结构 optimize the structure of expenditures
优化装卸程序 loading sequence optimum
优化准则 optimal criterion
优化资源配置 optimize the allocation of resources
优化子波处理 optimum wavelet analysis
优化子集 majorized subset
优化钻进 optimized drilling
优惠 concession; favo(u)r; immunity; preference; privilege; sufferance
优惠办法 preferential measure
优惠保险费 deviated rate
优惠差额 margin of preference
优惠差幅 preferential margin
优惠存车 preferential parking
优惠存款 preferential interest
优惠代表价 central preference price
优惠待遇 preferential treatment
优惠贷款 concessional loan; concessionary loan; loan on favo(u)rable terms; preferential loan; soft loan
优惠贷款和赠款的净值 net concessionary loan and grant
优惠贷款利率 prime interest rate
优惠的 concessional; favo(u)rable
优惠方位 privileged direction
优惠放款利率 prime lending rate
优惠幅度 margin of preference; preferential margin
优惠附加注册费 preferential additional registration fee
优惠关税 preferential duties; preferential tariff
优惠关税率 preferential tariff
优惠国 favo(u)red nation
优惠汇率 preferential rate; preferential rate of exchange
优惠机构 preferential shop
优惠价(格) concessional rate; favo(u)rable price; preferential price
优惠减税额 preferential tariff cut
优惠利率 bank prime rate; deviated rate; fine rate; preferential interest rate; prime interest rate; prime rate; prime rate of interest
优惠率 rate of concession
优惠码头 sufferance quay; sufferance wharf
优惠期 days of grace; grace period; period of grace
优惠权 preferential rights
优惠税率 benefit tariff
优惠条件 concessional term; concessionary term; favo(u)rable condition; favo(u)rable term; soft term
优惠条件特批 concessionary grant
优惠条款 favo(u)rable term; preferential clause
优惠贴现率 fine rate
优惠通信[讯]电路 preference circuit
优惠投标法 preferential tendering
优惠项目最低税额 minimum tax on reference item
优惠销售 concessionary sale
优惠性援助 concessional aid
优惠性资金流入 flow of concessionary fund
优惠运价 preferential rate
优惠运价表 preferential tariff
优惠政策 preferential policy
优惠支付 ex gratia payment
优惠自行车道 < 允许自行车停在车队前面并先行 > fast cycle lane; fat cycle lane
优级纯 guarantee reagent
优角 reflex angle
优角的 < 大于 180 度小于 160 度 > reflex
优晶质的 eucrystalline
优境学 euthenics
优聚合物 eupolymer
优良船艺惯例 observance of good seamanship
优良的商品 best possible merchandise
优良和易性 high workability
优良河段 excellent river stretch
优良河段模拟法 excellent-stretch simulation method
优良结构 well structure
优良能见度 excellent visibility; very-good visibility
优良品 non-defective unit
优良清晰度 fine details
优良设计标志 < 工业制品的 > good design mark
优良因素 factor of merit
优良指数 figure of merit
优良种子 quality seed
优劣比 ratio of greater inequality; ratio of less inequality
优劣顺序 order of quality
优硫胺 alinamin; propyldisulfide thiamine
优美 grace
优美的 delicate
优美的弧度 rondure
优缺点 advantages and disadvantages; merits and demerits; merits and drawbacks; merits and faults; relative merit; strong and weak points; virtues and defects
优若藜 eurotia ceratoides
优生学 eugenetics; eugenics
优胜法 grand slam
优胜者 champion; superior
优势 advantage; ascendance [ascendancy]; dominate; goodness; odds; predominance; superiority; upper hand; vantage; whip hand; dominance [dominancy]
优势比 odds ratio
优势波 predominant wave
优势-持续-强度指数 prevalence-duration-intensity index
优势的 superior
优势动机 prepotent motive
优势方向 predominant direction
优势分布 complementarity
优势构象 preferential conformation
优势积分 dominating integral
优势级数 dominating series
优势金属物种 dominant metal species
优势类型 dominant species
优势流 predominant current
优势率 odds ratio
优势木 dominant tree
优势偏向 directional preponderance
优势频率 dominant frequency
优势迁移 removal of dominance
优势区图 predominance area diagram
优势弱酸 dominant weak acid
优势生物 dominant organism
优势树 dominant tree
优势物种 dominant species
优势氧化态 dominant oxidation state
优势藻类物种 dominant algae species
优势种 dominant social; dominant species
优势种群 dominant population
优势转移 removal of dominance
优势作物 dominant crop
优素尼雅住宅 < 建筑大师赖特设计的一种纯美国风格住宅 > Usonian Houses
优系 major clique
优先 precede; precedence [precedency]; preempt; priority
优先编码 priority encoding
优先编码器 priority encoder
优先标记 priority flag
优先标识 priority symbol
优先参数 priority parameter
优先操作 priority service
优先车道 priority lane; preferential lane
优先车组 block of priority wagons
优先沉淀 preferential precipitation; selective precipitation
优先沉积 preferential deposition
优先程序 priority routine
优先程序触发器 priority program(me) flip-flop
优先程序状态 privileged program(me) state
优先处理 foreground processing; preferential treatment; priority processing
优先传递 priority in transmission
优先次序 precedence; priority; priority order
优先次序安排 priority programming
优先次序的 priority-ranked
优先次序规划 priority programming
优先次序号码 precedence number
优先次序开关 priority switch
优先存取 priority memory access
优先错误转储 priority error dump
优先贷款信托文契 first trust deed
优先的 preferential; preferred
优先等级 priority level; priority rating; superior class
优先等级符号 priority designator
优先等级列车 < 行车规则及时刻表规定 > train of superior class
优先抵押权 first mortgage
优先电路 priority circuit
优先电源 preferred power supply
优先调度 priority scheduling
优先调度程序 priority scheduler
优先调度系统 priority scheduling system
优先定位 preferred orientation
优先定向 preferred direction
优先定义 priority definition
优先发送 preference sending
优先发展区 priority development area
优先法 precedence method; priority method
优先法则 sequence rule
优先方案 preferred plan; preferred scheme
优先方式 mode of priority; priority mode
优先方向 privileged direction; superior direction
优先方向列车 < 行车规则及时刻表规定 > train of superior direction
优先放行线【交】 priority route
优先分级单位 priority ranking unit
优先分级系统 priority ranking system
优先分析 prior analysis
优先服务 priority service
优先浮选 differential flo(a)tation; preferential flo(a)tation; preferred flo(a)tation; selective flo(a)tation
优先复合 preferential recombination
优先供应单位 priority supplier
优先购买 preempt
优先购买合同 option to purchase contract
优先购买权 buy the refusal of; option; preemption; refusal; right of preemption
优先购买权协议 agreement on buying option
优先股 preferred share; senior shares
优先股本 capital stock preferred
优先股成本 cost of preferred stock
优先股和普通股 both preference and ordinary shares
优先股利息 dividend for preference share holders
优先股票 priority stock; referential stock; preference stock; preferred stock
优先关系 precedence relation
优先规定 precedence [precedency]; priority discipline
优先规则 priority rule
优先函数 precedence function
优先号 priority number
优先呼叫 preference call; priority call
优先化学品 priority chemical
优先回铃音 precedence ring-back tone
优先级 priority level; priority sequence
优先级控制 priority control
优先级通信多路传输器 priority communication multiplexer
优先级相位 priority phase
优先级仲裁线路 priority-arbitration circuit
优先技术 precedence technique; precedence technology
优先寄存器 priority register
优先监测 priority monitoring
优先监测污染物 priority monitoring pollutant
优先拣选 selective sorting
优先检查 priority check
优先检定 priority rating
优先建设 priority construction

优先建筑 priority construction
优先结构 priority structure
优先结晶 preferential crystallization
优先进路 preferential route;preferred route;priority route
优先进入线路的信号联动系统 <各道口信号灯周期及其灯式变换时间，随优先进入线路的车列长短与到达道口停车线的时间而变换> Plident system;platoon identification;scheme system
优先径路 preferential route;preferred route;priority route
优先矩阵 precedence matrix
优先决定器 priority resolver
优先决定权 first refusal (right)
优先开发地区 development priority area
优先开发区域 <城市规划中的> action area
优先开工率 preferred operating rate
优先考虑的材料 priority material
优先靠码头 free stem
优先控制 priority control
优先控制部件 priority control unit
优先控制方案 priority control scheme
优先控制污染物 priority considered pollutant
优先理论 Preference theory
优先联线(道) priority link
优先列车 <与次要列车相对应> superior train
优先留置权 prior(ity) lien
优先律 law of priority
优先率调整 rate action control
优先逻辑 priority logic
优先模块 privileged module
优先目标 priority target
优先内函数 priority built-in function
优先排队 priority queue
优先排队规则 priority queue discipline
优先破碎 differential disintegration;selective crushing
优先取舍权 first refusal (right);first step of refusal;refusal
优先权 preemption;precedence;preference; preferential; preferential rights;priority;right of priority;superiority;superior right
优先权列车 <列车命令规定> train of superior right
优先权文件 priority documents
优先权相关索引 priority concordance;priority concordance index
优先权指示符 priority indicator
优先燃烧 preferential combustion
优先认购权 preemptive right
优先溶解 preferential solution;selective solution
优先润湿 preferential wetting;selective wetting
优先设备 priority facility
优先生态系统 priority ecosystem
优先时差 priority phase offset
优先使用的到发线 <各方向到发列车按照一种优先的规则选择到发线> preference siding for arrival and departure
优先使用的跑道 preferential runway
优先受偿的抵押 tacking mortgage
优先输入 priority-in
优先数 preferred number; priority; priority number
优先数据库 underlying data base
优先数系 preferred number
优先数字 preferred number; preferred value
优先水解 selective hydrolysis
优先水流 <通过土壤中大孔洞及裂隙的水流> preferential flow
优先顺序 order of precedence;order of priority
优先顺序表 priority sequence table
优先顺序电路 priority circuit
优先算法 priority algorithm
优先索赔权 prior claim
优先替换使用的合作系数 preferred alternative index
优先条款 priority clause
优先停车 priority parking
优先通行交叉口 <主干道上包括停车标志和让路标志交叉口> priority intersection
优先通行街道 <在交叉口可以直通穿过，无须停车> preferential street
优先通行路 <在交叉口可以直通穿过，无须停车> preference road;preferential road
优先通行权 priority of starting
优先图解法 precedence diagram
优先网络 priority network
优先伪变量 priority pseudo-variable
优先文法 precedence grammar
优先污染权制度 priority pollution rights system
优先污染物 priority pollutant
优先吸附 selective absorption
优先吸收 preferential absorption
优先系数 preferred number
优先系统 priority system
优先显示度 <交通标志设计质量指标之一> priority level;priority value
优先信号 priority signal
优先(信号)时差 preferential offset;priority offset
优先行动 dominant action
优先行驶权 right of way
优先需求目标单 priority requirement objective list
优先选用的 preferred;preferred to be selected;with the priority to be selected
优先选择 preference; priority selection
优先选择路线 preferred route
优先选择中断 priority selection interrupt
优先研磨 differential grinding
优先掩护的设施 priority installation
优先要求权 prior claim
优先业务 priority facility
优先用户插入 priority extension override
优先用气 premium uses of gas
优先用水权原则 prior appropriation doctrine
优先有机污染物排放 organic priority pollutant compound
优先有序中断 priority ordered interrupt
优先于 <权利、索赔等> underlie
优先语言 precedence language
优先预算法 budgeting to priorities
优先运输 priority transport
优先运输货物 preferential freight
优先运行 priority in running
优先运行权 <列车> precedence in running
优先再结晶 selective recrystallization
优先债权人 senior creditor
优先振铃 priority ring
优先证券 senior securities
优先值 preferred value
优先指令 advantage instruction
优先指示符 priority indicator
优先指示器【计】priority indicator
优先指示信息 priority indicator
优先中断 priority interrupt
优先中断表 priority interrupt table
优先中断方案 priority interrupt scheme
优先中断禁止开关 priority interruption inhibit switch
优先中断控制 priority interrupt control
优先中断控制模件 priority interrupt control module
优先中断控制器 priority interrupt controller;priority interrupt control unit
优先中断通道 priority interrupt channel
优先中断制 priority interrupt basis
优先专用权原则 prior appropriation doctrine
优先装运 shipment by first opportunity
优先准则 prior criterion
优先作业 priority job
优秀的 top flight
优秀能见度 excellent visibility
优秀设计 excellent design;outstanding design
优选 selective preference
优选参数值 optimizing parameter value
优选打分制 merit point system
优选法 optimized method; optimum seeking method;optimization
优选方案 preferred schematization;preferred scheme
优选方位 preferred orientation
优选方位图 preferred orientations diagram
优选方位形式 pattern of preferred orientation
优选分析 release analysis
优选规则 preferred plan
优选技术 optimal search technique;optimization technique
优选径节 preferred diametral pitch
优选理论 decision theory;theory of decision
优选木材 select lumber
优选切割装置 cutting optimizer
优选设计 decision design
优选水解 selective hydrolysis
优选研磨粒度 break
优选值 preferred value
优选组构 preferred fabric
优异能见度 excellent visibility
优异数据 champion data
优于 overpass;surpass;in preference to
优于失去平衡 outbalance
优越 best;vantage
优越坝址 attractive site
优越工程地址 attractive site
优越化比例 optimum ratio
优越解 solution by dominance
优越情结 superiority complex
优越性 precedence [precedency];superiority;predominance
优越性能 high performance
优越者 superior
优择方式 selective preferential type
优值 figure of merit;Q-value
优值计 Q-meter
优质 best quality;first quality;goodness;irreproachable quality;premium;sound quality;superior quality
优质板垛箱 prime piling box
优质丙纶缆 super-poly propylene rope
优质薄板 prime sheet
优质薄钢板 prime sheet
优质材料 sound material
优质产品 high grade products;high-quality products;products of quality;quality goods;quality products
优质出水 high-quality effluent
优质处理 polishing process
优质窗玻璃 selected glazing quality
优质次氯化物 high test hypochloride
优质次氯酸盐 high test hypochlorite
优质大理石 <一般指意大利白大理石> statuary marble
优质的 high class; high test; rich; well-behaved; high grade; high quality
优质等级 high-quality grade
优质地沥青混凝土路面 fine asphalt surfacing
优质锻件 quality forging
优质发动机油 premium motor oil
优质方坯 rerolling quality blooms
优质钢 extra-fine steel;fine(d) steel;high grade steel;high-quality steel;quality steel
优质高速钢 super-high speed steel
优质工程 high-quality project
优质工程率 rate of best quality engineering
优质光学玻璃 crown glass
优质函数 merit function
优质焊缝 high quality welded joint
优质黄铜 high brass
优质混凝土 quality concrete
优质混凝土生产 quality concrete production
优质集料 quality aggregate
优质结构(木)材 select structural wood
优质金刚石 gem grade diamond;high diamond;tool stones
优质金属 high test metal
优质精铜 best selected copper
优质绝缘 premium insulation
优质冷轧板 cold-rolled prime
优质沥青 fine asphalt
优质沥青混凝土 fine asphalt concrete
优质沥青煤 bright coal
优质沥青砖 fine asphalt tile
优质淋浴器 de luxe shower
优质芦苇 best reed
优质率 factor of merit
优质镁铝锰合金 peraluman
优质木板 high grade timber board
优质木材 bright wood; high grade timber; selected merchantable; clean stuff <无节疤等>
优质耐火材料 high-duty refractory
优质泥浆 premium mud
优质屏蔽的 supershielded
优质普鲁士蓝 high quality Prussia blue
优质砌块 special-quality block
优质浅色阿拉伯树胶 hashab
优质燃料 premium fuel
优质润滑剂 quality lubricant;superior lubricant
优质赛马场草皮 certified sod
优质商品 selected merchantable
优质商品冷轧薄板 cold-rolled commercial quality sheet
优质烧透砖 body brick
优质设计 quality design
优质深冲钢 extra deep-drawing steel
优质生产 quality production
优质生产厂 quality producer
优质生铁 high-duty pigiron
优质石材 <取自矿中最佳部位的> liver rock
优质石料 quality stone
优质石棉 fine asbestos
优质手工制法 quality-crafted
优质熟铁 merchant iron
优质水 excellent water;good-quality water;high-quality water
优质水量 quantities of high quality water

优质水泥 high grade cement; high-quality cement; high test cement; premium cement; sound cement
优质水硬石灰 eminently hydraulic lime
优质饲料 heavy feeder
优质弹性橡胶 snappy rubber
优质弹性橡皮 snappy rubber
优质碳钢 quality carbon steel
优质碳钢管 quality carbon steel pipe
优质碳素工具钢 quality carbon tool steel
优质碳素结构钢 carbon construction(al) quality steel; carbon structural quality steel
优质填料 crown filler
优质投入 better input
优质图像 excellent picture; qualitative picture
优质橡胶 top-quality rubber
优质橡木 prime-grade oak
优质心(部木)材 sound heartwood
优质羊毛 wool of fine staple
优质因数 figure of merit
优质饮用水 high-quality drinking water
优质饮用水源 quality drinking water resource
优质油 branded oil
优质油渣 neutral bottoms
优质原油 good oil
优质再生水 high-quality renovated water
优质再用水 high-quality reuse water
优质纸 <绘图用> Bristol paper; rag paper
优质指数 figure of merit; index number of merit
优质制绳钢丝 plough steel wire
优质种子 quality seed
优质主镜 high-quality primary mirror
优质铸件 high-quality cast iron; premium casting
优质砖 best quality brick; special-quality brick; well-burned brick
优质锥形锚具 oriental cone anchorage

忧

忧郁症 melancholia

幽

幽暗 dusk

幽闭恐惧症 claustrophobia
幽谷 dell; dingle; glen
幽径 by-walk
幽深沉积 euxinic deposit(ion)
幽影 looming

悠

悠哒窗 <弧形转动窗> pivoted casement

尤

尤蒂卡页岩 Utica shale

尤耳明铝合金 ulminium
尤格风 Youg
尤卡坦海流 Yucatan Current
尤卡坦海峡 Yucatan channel
尤拉奎洛风 euraquilo; euroa-quilo; euroclydon
尤拉纽斯镍络合金钢 Uranus
尤里康岩层 Uriconian rocks
尤马尔铝合金 Ulmal
尤钠钙矾 eugsterite
尤尼杰尔炸药 Unigel
尤尼洛伊镍铬钢 Uniloy
尤尼马特 <一种机器人> Unimate

由

由安全阀和泄压阀泄放 discharge from safety and relief valves

由岸上搬上船 fetch off
由岸向水推进的施工方法 end tipping (from the shore)
由岸向水中推进的尾卸式施工方法 end-dumping
由被叫用户付费的通话 collect message
由泵独立传动的液压系统 live hydraulic system
由标准程序的说明 interpretation by standard procedure
由薄板组成的 laminar
由不均匀沉降导致的不稳定性 lability due to uneven settlement
由部件装配的 built-in sections
由测斜仪测定的高程 clinometric height
由车辆交换铁路交来 delivery by interchange railroad
由车辆运到 roll up
由沉船造成的损害 <水底隧道要防止的> damage by a foundering ship
由沉陷引起的裂缝 settlement crack
由齿条齿轮液压驱动的系统(机构) rack and pinion hydraulically driven system
由臭氧引起的表面裂纹 corona
由船东负责 at ship owner's risk
由船东选择 at carrier's option
由垂直图切水平面图 section cut vertical map
由大而小排列 decreasing order of magnitude; size down
由大而小顺序 descending order
由带到卡片 tape-to-card
由单一弯曲断层形成的圈闭 trap formed by a single curved fault
由等离子体产生的冷凝气溶胶 plasma generated condensation aerosol
由地震产生的震动 seismic shock
由第三者保存 escrow
由第三者执行的让售委托书 sales escrow instructions
由电缆上的深度记号确定深度 depth determined from cable mark
由顶向下法 top-down method
由定价估计价值 value-put-in-place estimate; value put in value place estimate
由东南流来的 southeast flowing
由断层与构造鼻交切形成的圈闭 trap formed by intersection of structural "nose" with a fault
由断层与褶皱交切断层形成的圈闭 trap formed by intersection of fault with fold
由多家承包商承包的项目合同 multiple prime contract
由多种不稳固卵石构成的砾岩 polymict
由尔康铜铅系重载轴承合金 Ullcony metal
由二交切断层形成的圈闭 trap formed by two intersecting faults
由二马牵引的古代双轮战车 <常用于雕塑> biga
由发货点算起 ex point of origin
由发货人承担风险 on shipper's risk
由发货人负责 at sender's risk
由附墙柱和檐头框起的拱 arch order
由钢板和角钢拼制的梁 plate girder
由高势面和非渗透遮挡共同封闭 enclosed jointly by higher potential plane and impermeable barrier
由高势区向低势区 toward low potential from high potential
由高压干管流向低压的支管 take-off line
由公司免费运输 on company service
由汞引起的 mercurial
由共鸣加强的 resonant
由购买者在货到时付运费 freight in
由管子制造 made from pipe
由海关经营的港口 customs port
由海向陆的风 inshore wind
由合并土地所增加的价值 plottage value
由红线后退进去的房屋 setback building
由后推 boost
由后推进力 boosting power
由湖面吹向岸边的微风 lake breeze
由花岗岩做成的 granitic
由滑移导致的不稳定性 lability due to sliding
由灰构成的 cinereour
由火成岩栓侵入造成的背斜圈闭 anticlinal trap by igneous plug
由或代表船东签发 by or on behalf of the ship's owner
由货主担负一切责任 at owner's risks and responsibilities
由积云形成的 cumulous
由极大到极小 peak-to-peak
由几部分组成的 multipart
由计算机自动控制操纵的缓行器 retarder operated through automatic control from computers
由加工板边去除金属 metal removal from edges of plates
由绞车放钢绳 lengthen drilling line
由结果追溯到原因的 a posterior
由经销商安装的机械 dealer-installed equipment
由经销商安装的任选装置 dealer-installed optional equipment
由客户直接退回查账员的对账回单 direct verification
由空载到满载的减速变化 speed droop
由空中运输 air freight
由孔泄出 vent
由库存股本交易所得资本 capital from treasury stock transactions
由垃圾产生的 refuse-derived
由浪花冻结的岸冰 storm ice foot
由雷达测出的物标距离 radar range
由砾石组成的 gravelly
由联邦或州政府赠予或转让给个人的土地 <美> government patent
由联邦政府发给的土地证书 land certificate
由两部分构成的教堂 bipartite church
由炉箅下的少量煤 riddlings
由买方负担 for buyers account
由买方制造 fabricated by the buyer
由苗圃供应的成批树苗 nursery stock
由摩擦所致的 friction(al)
由木焦油蒸馏得到的强溶剂 tar spirit
由内向外 ento-ectad
由南流来的 south-flowing
由泥底群造成的背斜圈闭 anticlinal trap by mud diapir
由扭矩产生的不稳定性 torsional instability
由扭矩引起的临界荷载 torsional buckling load
由扭矩引起的破坏 torsional failure
由棚屋建成的 hutted
由偏转电路获得的高压电路 flyback circuit
由企业开支的招待费 entertainment on the expense account
由起爆火药驱动的工具 powder-actuated tool
由汽车拉的活动房屋 mobile home
由戗 inverted V-shaped brace
由曲线出岔的道岔 turnout from curved track
由曲线经校正后确定深度 depth determined from corrected curves
由曲线同向出岔的道岔 turnout from curve of similar flexures
由曲线异向出岔的道岔 turnout from curve of contrary flexure
由燃烧引起的 pyric
由热引起的 pyrogenic; pyrogenous
由人造卫星带入轨道后放出的物体 subsatellite
由若干交切断层形成的圈闭 trap formed by several intersecting faults
由三部形成的 triatic
由上到下 from the top downward
由上而下分段 descending stage
由上而下分段灌浆 descending stage grouting; staged grouting proceeding from the top down
由深度转换轮确定深度 depth determined from the roll of depth inversion
由湿冷引起的弯曲 cold-laid moist bending
由十个组成的 decadal
由始至终的管理 cradle to grave management
由市电供电 current supply from public mains
由收货人自负责任 at consignee's risk
由水库提取的水量 reservoir draft
由水龙头控制 faucet control
由水陆形成的 terraqueous
由水路 by-water
由水运转到铁路的 water-to-rail
由四部分形成的 quadripartite
由四部分组成的 quad
由四个部分组成的 quadruple
由四根钻杆组成的立根高度塔板 fourble board
由四人组成的 quadripartite
由塑性发生的变化 plastic modified
由碎屑形成的 detrital
由他人代理缺席业主进行经营管理 absentee management
由碳得到的 carbonic
由天体高度测的方位角 attitude azimuth
由调整资本而产生的资本 capital arising from re-capitalization
由铁路运输 rail
由桶中移装瓶内 bottle off
由投资引起得消费增加 induced consumption
由托运人选择 a shipper's option
由托运人自行承担危险的货物 adventure
由外墙及防火楼层包围部分房屋 horizontal compartment
由外向内 ecto-entad
由涡流侵蚀形成的盆地 basin due to erosion
由五部分组成的 quinquepartite
由稀释剂(引起)的发白 bluishing caused by thinner
由下向上支护井筒 underpin
由下向上钻进 <在坑道内> drill upward
由下向上做支撑 underpin
由销售者付运费 freight out
由小河来的水 branch water
由小粒形成的 granulons
由斜井 VSP 数据取得孔隙度横向变化 lateral porosity variations from deviated well VSP data
由心集中荷载 central concentrated load
由心配电盘 central distribution switch board

由心主惯性矩 central principal moment of inertia
由行车钢轨和护轮轨组合的线路交叉 two rail crossing
由(行车轨、护轮轨和补强轨)三种钢轨组合的线路交叉 three-rail crossing
由压缩空气操作的管道 < 输送设备 > carrier air tube system
由演播室演播 direct pick-up
由已知终值和现值求算未知 known future value and present value-compute unknown
由盈余增加的资产 addition to property through surplus
由于冲刷的破坏 failure due to scouring
由于电车和横向交通的同时阻滞 synchronizing delays due to trams and cross-traffic
由于都市化建设而丧失自然景色的地区 subtopia
由于断层作用而陷落 downfolding
由于发展引起的人口迁移 development-induced displacement
由于呼吸作用的 respiratory
由于机器过度摩擦而烧黑加工的木材 machine burn
由于矿藏量衰竭而减税 depletion deduction
由于轮缘爬上钢轨而出轨 derailment due to mounting of wheel-flange on rail
由于轮缘跳上钢轨而出轨 derailment due to jumping of wheel-flange on rail
由于起始条件的输出 output due to initial condition
由于侵蚀所需要的裕度 allowance for erosion
由于热轴或其他不良状态而摘下停放在岔线上的车辆"set-out" car
由于溶剂滞留而引起的起泡现象 solvent pop
由于疏忽冒进危险信号 inadvertent passing of signal at danger
由于输入作用的输出 output due to input
由于信号前沿不足引起的失真 underthrow distortion
由于月、日引力的 lunisolar
由原因推及结果的前提 a priori premise
由真菌引起的 fungous
由政府机关拆屋 wrecking by government authorities
由酯组成的合成油 ester oil
由中介持有地契的时期 holding escrow
由中心向四方扩散的 quaquaversal
由钟表示的时间 time of day
由铸石空心砌块建成的房屋 Mac-Girling
由资产重估所得资本 capital from appraisal adjustment
由自动闭塞信号机间隔开的列车 train spaced by automatic block signals

犹

犹他 < 美国州名 > Utah

犹他式挤奶站 Utah-type milking pool
犹太教堂 synagogue
犹太教堂中的讲坛 Jewish bema
犹太沥青 bitumen Judaism;Dead Sea asphalt
犹太人居住区 ghetto
犹太人区 ghetto
犹太石层 Jew stone
犹太式建筑 Jewish architecture

邮

邮包 parcel;postal package

邮包钩 mailbag hook;pouch hook
邮包架 mailbag rack;pouch rack
邮包通知书 advice note
邮包邮务处 parcel post
邮包运输保险 parcel post insurance
邮包装卸机 mail catcher
邮保风险 parcel post risks
邮车 mail car(t);mail-coach;mall car;postal car
邮传电报 mailgram
邮船 liner;mail boat;mail carrier;mail liner;mail ship;passenger liner
邮船定期 mail and passenger steamer
邮船定线 mail and passenger steamer
邮船公司 liner company
邮船开航日 packet day
邮戳 indicia;postage stamp;post mark
邮戳日期 date of postmark
邮袋 mailbag;mail pouch;postbag;pouch
邮袋帆布 mailbag duck
邮递电报 < 发往邮局,再由邮局送交收报人 > mailgram
邮递业务 mail service
邮递员 mail man;messenger;postman
邮电部 Ministry of Posts and Tele Communication
邮电大楼 post-telecommunication building
邮电费 post and telecommunication expenses
邮电工作人员 signalman
邮电局 post and tele-communications office;telecommunication post bureau
邮电所 post and tele-communications office;post-telegram office
邮费 postage
邮费表 postal tariff
邮费付讫 postage paid
邮费已付 postage paid
邮封 mail cover
邮购 mail order
邮购公司 mail-order firm
邮购商店 mail-order house
邮购业务 mail-order trade
邮航机 mail plane
邮汇 mail transfer
邮汇汇票 postal order;post office order
邮寄包裹 parcel post
邮寄订购商店 mail-order house
邮寄笼 shipping cage
邮寄设备 mailing facility
邮件 mail;postal matter
邮件编码号 postal number
邮件存局待领 general delivery
邮件的存局候领处 general delivery
邮件递送 mail carrying
邮件调查 mail survey;questionnaire mail
邮件分拣室 mail sorting room
邮件服务器 mail server
邮件过滤器 mail filter
邮件滑送槽 mail chute
邮件间 mail apartment
邮件截止日 mail day;packet day
邮件快艇 mail cutter
邮件列表 mailing list
邮件门环 postal knocker
邮件旗 mail flag;mail pennant;mail signal
邮件声报单 mail declaration
邮件室 mail room
邮件收发机 mailing machine
邮件收发时间 post-time
邮件投递口 < 墙上或门上的 > mail slot
邮件网关 mail gateway
邮件运输 mail traffic
邮件炸弹 mail bomb
邮件转运 post transportation
邮局 post office
邮局办事员 mail clerk
邮局包裹说明书 parcel post statement
邮局的 postal
邮局电话线 post office telephone cable
邮局红 post office red
邮局汇票 money order
邮局式(电阻箱)电桥 post office bridge
邮路 mail route;postal route
邮路地图 mail route map;postal route map
邮轮 mail steamer;cruise liner;mail boat;packet boat
邮区编号 < 美 > zip code
邮亭 postal kiosk
邮筒 mailbox;mail drop;pillar-box < 英 >;postbox;self-mailer
邮筒式消防栓 post hydrant
邮务船 post boat;stage boat
邮务的 postal
邮务艇 post boat;stage boat
邮箱 letter drop-off;mailbox;postbox
邮箱号码 box number
邮运收入 postal service revenue
邮运运价 mail tariff
邮政 mail;penny post;post
邮政包间 compartment reserved for mail
邮政编码 zip code < 美 >;post(al) code < 英国、加拿大 >
邮政便船 mail boat
邮政部门 post office department
邮政车 mail bus;mail truck;mail wagon;postal truck;postal van;postal wagon;post office car;mail van
邮政行李车收益 headend revenue
邮政船 packet boat
邮政的 postal
邮政地图 local map;postal map
邮政飞机 mail plane
邮政管理局 postal administration
邮政轨道车 mail railcar
邮政汇款 postal remittance
邮政汇票 postal order
邮政机器人 mailbot
邮政局 post office
邮政局长 postmaster
邮政快递 emergency mail service
邮政路线 postal route
邮政汽车 mail car(t);mail motor truck
邮政所 post house
邮政信箱 PO Box;post office box
邮政业务 mail service;postal service
邮政运输 post transport
邮政运输收入 mail revenue
邮政专用间 mail compartment
邮政总局 general post office;head post office;post central office
邮政总局电信塔楼 General Post Office tower
邮政总局通信[讯]电缆 General Post Office cable
邮政总局通信[讯]线路 General Post Office line
邮资 postage

油

油桉树 lemon scented gum

油白 oil white
油斑 fat spot;grease marks;oil mark;oil patch;oil print;oil slick;oil spot;oil stain
油斑腐蚀 oil stain corrosion
油斑光度计 grease spot photometer
油板 oiled-plate method
油拌 < 即拌以沥青材料 > oil mix(ture)
油拌混合料 oil mixed type surface;oil mix(ture)
油拌式面层 oil mixed type surface;oil mix(ture)
油包 oil box;oil pocket;pancake
油包水 water-in-oil
油包水乳剂 water-in-oil emulsion
油包水型 water-in-oil type
油包水(型)沥青乳液 inverted asphalt emulsion
油包水型乳化沥青 inverted asphalt emulsion
油包水(型)乳液 water-in-oil emulsion
油包酸乳液 acid-in-oil emulsion
油杯 grease chamber;grease cup;oil bowl;oil cistern;oil cup;oiler;oil-holder;oil lubricator;wiper cup;lubricating cup;plain oil cup
油杯放洁塞 bowl drain plug
油杯盖 oil cup cap;oiler cap
油杯盖簧 oil cup cover spring
油杯盖螺钉 oil cup cover screw
油杯润滑 cascade oiling
油杯润滑脂 cup grease;tank car;tank wagon
油杯润滑轴承 bearing axle-box lubricated as required
油泵 oil pump;fill pump;lubricating press;lubropump;oil keeper;dump pump
油泵安全阀 oil pump relief valve
油泵安全阀弹簧 oil pump relief valve spring
油泵保护凸缘 pump protection
油泵保险阀 oil pump safety valve
油泵齿轮 oil pump gear
油泵齿条杆 pump rack bar
油泵出油阀 oil pump outlet valve
油泵出油管 oil pump outlet line;oil pump outlet tube
油泵传动轴 oil pump drive spindle
油泵从动齿轮 oil pump driven gear
油泵从动齿轮轴 oil pump driven wheel spindle
油泵导管 fuel pressure manifold
油泵电动机 fuel oil pump motor;oil pump motor
油泵垫圈 fuel pump gasket;oil pump washer
油泵发动机 oil pump motor
油泵房 oil pump house
油泵浮筒滤网 oil pump float screen
油泵盖 oil pump cover
油泵盖密封垫 oil pump body cover gasket
油泵给油杆 fuel pump priming lever
油泵工作员 oil pumper
油泵供油 oil pump feed
油泵过滤器 oil pump screen
油泵护片 oil pump shield
油泵簧 oil pump spring
油泵回油器 scavenge gears
油泵集流腔 fuel pressure manifold
油泵进油阀 oil pump inlet valve
油泵净油器 oil pump purifier
油泵壳(体) oil pump casing
油泵空转齿轮 oil pump idler gear
油泵空转齿轮轴 oil pump idler gear shaft
油泵扣紧簧 oil pump retaining spring
油泵拉杆 oil pump rod
油泵联结 oil pump coupling
油泵滤(清)器 oil pump cleaner;oil pump strainer
油泵滤网 oil pump screen

油泵滤网盖 oil pump screen cover
油泵滤网护圈 oil pump screen retainer
油泵密封垫 oil pump gasket
油泵排油道 pump drain line
油泵旁通阀 oil pump bypass valve
油泵喷油器摇杆 rocker of injector-pump
油泵膨胀塞 oil pump expansion plug
油泵偏心轮 oil pump eccentric wheel
油泵驱动 fuel pump drive;oil pump drive
油泵式手术台 oil hydraulic operating table
油泵试验装置 oil pump testing device
油泵室 oil pump room
油泵输出管 scavenge delivery
油泵随轮【机】 oil pump following gear
油泵体 oil pump body
油泵体衬垫 oil pump body gasket
油泵体衬套 pump body bushing
油泵调节阀 oil pump regulating valve
油泵调节器 fuel pump governor
油泵筒 oil pump cylinder
油泵推杆 fuel pump push rod
油泵外壳 oil pump housing
油泵吸油管 oil pump suction tube
油泵吸油管密封片 oil pump suction pipe gasket
油泵心轴 oil pump spindle
油泵心轴止推轴承 oil pump spindle thrust bearing
油泵心轴轴承壳 oil pump spindle bearing housing
油泵压力安全球阀 oil pump pressure relief valve ball
油泵摇臂 fuel pump rocker arm
油泵叶轮 fuel pump impeller
油泵罩 oil pump shield
油泵支架 oil pump bracket
油泵支架螺栓 oil pump bracket bolt
油泵钟形罩 oil suction bell
油泵轴【机】 oil pump shaft
油泵轴齿轮 oil pump shaft gear
油泵轴导管 oil pump shaft guide
油泵主动齿轮 oil pump drive gear;oil pump driving wheel;oil pump pinion
油泵主动齿轮套 oil pump drive gear sleeve
油泵主动轴 oil pump drive shaft
油泵主动轴齿轮 oil pump drive shaft gear
油泵主动轴支架 oil pump drive shaft support
油泵柱塞 fuel feed pump plunger;fuel pump plunger;oil pump plunger
油比重测定计 oil ga(u)ge
油比重计 el(a)eometer;elaiometer;oil hydrometer;oleometer
油笔 oil pike
油篦子 oil grating
油变性 oil reactivity
油变性酚醛树脂 oil-reactive phenolic resin
油标 oil level indicator
油标表 oil ga(u)ge;oil leveler;oil pointer;oil scale
油标钢皮卷尺 oil ga(u)ging tape
油标志 oil indication
油表 oil ga(u)ge;oil scale
油表操纵开关 fuel ga(u)ge control switch
油表浮子 fuel ga(u)ge float
油表管接头 fuel ga(u)ge adapter
油表开关 fuel ga(u)ge breaker
油表旋塞 ga(u)ge cock
油饼 oil cake;pancake <灌浇沥青不均匀而形成的>
油饼碎裂机 oil-cake breaker
油驳 fuel oil barge;oil barge;oil storage barge
油捕 oil trap
油捕集器 trap for oil
油不足 shortage of oil
油布 American leather;leather cloth;oil cloth;oil-coat;oiled linen;oil-skin;tarp(aulin);wax cloth
油布干燥器 oil cloth drier [dryer]
油布清漆 lino(leum) oil cloth varnish
油布雨衣 oil-coat;oil-skin;slicker
油彩 grease paint;oil paste
油彩地 <罗马建筑的彩色路面> asarotum [复 asaroria]
油彩地面 asarotum [复 asaroria]
油菜籽 rape weed
油舱 fuel oil tank;oil bunker;oil-holder;oil tank
油舱标尺 tank ga(u)ge;tank scale
油舱底部水量测定器 thief sampler;water finder
油舱加热管系 tank heating line
油舱口 oil tank hatch;tank hatch
油舱量尺 float ga(u)ge
油舱漆 oil tank paint
油舱清底 tank stripping
油舱清洗 tank cleaning
油舱清洗船 tank-cleaning vessel
油舱疏气阀 vapo(u)r cock
油舱梯 oil tank ladder
油舱通风机 tank ventilation fan
油舱通风系统 tank ventilation system
油舱通气管 vapo(u)r pipe;vent
油舱熏洗系统 tank steaming-out and cleaning system
油舱液面计 tank level indicator
油舱液面控制 tank level control
油舱液位控制 tank level control
油舱液位指示系统 tank level indicator system
油舱装卸观测反光镜 tank mirror
油藏 oil accumulation;oil pool
油藏等渗透率图 oil isoperm
油藏等效半径 equivalence radius of pool
油藏地质模型 geologic(al) model of pool
油藏动力学 reservoir dynamics
油藏开始注水时压力 pool pressure at starting influx
油藏平面图 planimetric(al) map of oil pool
油藏剖面图 sectional drawing of oil pool
油藏驱动机制 oil-well drive;reservoir drive mechanism
油藏生态学 ecology reservoir
油藏数学模型 mathematic(al) models of pool
油藏循环法 reservoir cycling
油藏研究和评价 research and evaluation of pool
油藏贮量 oil in place
油藏综合压缩系数 complex compressibility of pool
油藏总压降 total pressure drop of pool
油槽 bunker;bunker boat;channel for oiling;grease groove;oil bath;oil channel;oil groove;oil pan;oil pit;oil pocket;oil receiver;oil receptacle;oil sink;oil slot;oil space;oil sump;oil tank(er);oil trough;oil well;petroleum tank;sump;tanker
油槽驳船 tank barge
油槽车 cistern;mobile tank;motor tank trunk;oil car;oil tank car;oil tank truck;oil truck;road tank car;road transport tanker;tank car;tanker;tank wagon;trough truck
油槽车冲洗 oil tanker washing
油槽车床 oil grooving lathe
油槽车吹扫机 tank car blower
油槽车底壳 tank car bottom shell
油槽车顶 tank car head
油槽车加热器 tank car heater
油槽车检查工 tank tester
油槽车穹顶 tank car dome head
油槽车穹室罐车圆顶 tank car dome
油槽车容量 capacity of tank car;shell capacity
油槽车上交货价格 tank wagon price
油槽车身 tank body
油槽车梯子 tank car ladder
油槽车卸载 tank unloading
油槽车装卸台 tank car loading rack
油槽衬 tank car lining
油槽船 tanker;tanker ship
油槽船长度 tanker length
油槽船船舱 tanker oil compartment
油槽船船身 tanker hull
油槽船的设备 tanker facility
油槽船队 tanker fleet
油槽船吨位 tanker tonnage
油槽船横隔框 tanker transverse bulkhead
油槽船梁 tanker beam
油槽船排出 tanker discharging
油槽船上油舱与锅炉间的隔墙 tanker cofferdam
油槽船运输 tanker transportation
油槽船中线隔框 tanker centre line bulkhead
油槽船周转过程 tanker turnaround
油槽船装油码头 tanker loading terminal
油槽船装载 tanker loading
油槽断路器 oil-trough circuit-breaker
油槽管线 oil enclosed pipe line
油槽罐车 tank body truck
油槽横梁 tank cross member
油槽加热器 oil tank heater
油槽壳皮 tank car shell sheet
油槽可替换运油车 relay tank truck
油槽冷却器 oil tank cooler
油槽列车 tank train
油槽汽车 road tank(er);tank body truck;tank lorry;tank truck;truck tank
油槽汽车泵 tank truck pump
油槽汽车负载架 tank truck loading rack
油槽汽车拖车 tank truck trailer
油槽入口接受过滤器 oil tank hopper
油槽润滑 oil bath lubrication
油槽式充油断路器 oil bath type oil circuit breaker
油槽托 anchor plate
油槽拖车 tank-mounted trailer;tank trailer
油槽铣刀 oil groove milling cutter
油层 oil-bearing bed coat;oil bed;oil-coat;oil sheet;oily layer;oily sheet
油层边界测试 reservoir limit test
油层采收率 reservoir recovery
油层的压裂处理 fracture treatment
油层顶部 top of oil horizon
油层顶构造等深图 isobaths of structure in the top of oil layer
油层工程 reservoir engineering
油层工程师 reservoir engineer
油层含油范围 areal limits of oil sand
油层厚度 oil reservoir thickness;thickness of strata with oil
油层检测装置 bed detector
油层内流体分布 fluid distribution in an oil reservoir
油层平衡压力 equilibrium reservoir pressure
油层平均渗透率 average permeability of oil bed
油层剖面 section of reservoir
油层上部气 gas above oil reservoir
油层深层气 deep gas below oil reservoir
油层深度 oil-bearing depth
油层水 oil reservoir water
油层水力压裂 reservoir fracturing;well fracturing
油层套管 oil-bearing formation casing;oil string
油层套管深 oil casing depth
油层温度 reservoir temperature
油层物性参数 physical property parameter of reservoir
油层压力 reservoir pressure
油层压力保持法 pressure maintenance process
油层油气比 reservoir gas-oil ratio
油层有效厚度 effective thickness of oil bed
油层原油 crude oil in reservoir
油茶 oil tea;oiltea camellia;tea oil tree
油长 oil length
油潮气 oily moisture
油沉淀池 oil sump
油沉淀池过滤器 oil sump strainer
油沉淀柜 oil settling tank
油承 oil keeper
油橙 oil orange
油池 oil basin;oil bath;oil pool;oil sump;oil trough;sump
油池泵 sump pump
油池放油塞 drain plug for oil sump
油池漏斗 coning in oil reservoir
油池式折皱滤清元件 reservoir element
油池水 pool fire
油匙 dipper;oil dipper;oil skip;oil spoon
油尺 dipstick;metering rod;oil dipstick
油齿轮 oil gear
油齿轮压缩机 oil gear compressor
油充橡胶 oil-extended rubber
油冲洗 oil flush
油绸 oil(ed) silk
油储藏 oil stock
油储藏损失的控制 oil stock loss control
油处理 oil treatment
油处理剂 oil-treatment agent
油处理室 oil purification room
油传动 oil drive
油传动刹车 oleo-pneumatic brake
油船 cargo tank;fuel bunker;oil barge;oil carrier;oil carrying ship;oiler;oil tanker;petrol carrier;petroleum ship;ship tank;tanker;tank ship;tank vessel
油船吃水 tanker draft
油船船队 tanker fleet
油船吨位委员会 tanker tonnage committee
油船锅炉舱 tanker stokehold
油船码头 oil tanker jetty;tanker terminal
油船锚地 petroleum anchorage
油船排水量 tanker displacement
油船上管理泵的船员 pumpman
油船深度 tanker depth
油船失事事件 oil tanker accident incident
油船艉 tanker aft
油船污染 pollution of oil ship
油船坞 oil dock
油船系泊 tanker mooring
油船用阻火器 flame arrester for oil-tanker

油船栈桥 tanker terminal
油船自动装卸压载系统 LOGISTRIP automatic system
油船纵向舱壁 tanker longitudinal bulkhead
油窗 oil window
油床 oil accumulation; oil pool; oil reservoir
油锤作用 oil hammer
油醇 oleic alcohol; oleyl alcohol
油醇二硫酸盐 oleylalcohol disulfate
油醇酸树脂漆 oil alkyd paint
油催干剂 oil drier
油淬 oil quenching
油淬断流器 oil-quenched cut-out
油淬钢 oil-hardened steel; oil-hardening steel
油淬火 oil hardening; oil-quench(ing)
油淬火的 oil hardened
油淬火钢 oil-quenching steel
油淬强化 oil hardening
油淬硬的 oil-quenched
油淬硬化 oil hardening
油淬硬化钢 oil-hardened steel
油带 bank of oil; oil belt; oil zone
油袋 oil bag
油挡 oil catch
油导管 oil piping layout
油倒空 oil emptying
油道 oil canal; oil gallery; oil groove; oil lead; oil line; vittae
油道塞 oil channel plug
油的 oleic
油的包装 oil packaging
油的本色 oil true colo(u)r
油的变质 breakdown of an oil
油的表面温度 oil surface temperature
油的表面张力 oil surface tension
油的澄清 breakdown of oil; breaking of oil
油的冲淡 oil thinning
油的出口温度 oil outlet temperature
油的储藏 oil storage
油的储藏损失 oil stock loss
油的处理 oil handling
油的处理接受器 oil handling receptacle
油的触变行为 oil thixotropic behavio(u)r
油的代用品 oil substitute
油的防磨损性质 oil wearing quality
油的分解 oil decomposition
油的分离 oil separating; separating of oil
油的工作特性 oil performance
油的供应 oil supplying
油的规格 oil requirement
油的挥发性 oil volatility
油的回收 oil recovery; recovery of oil
油的混合性 oil compatibility
油的加温回收法 thermal methods of oil recovery
油的兼溶性 oil compatibility
油的鉴定 oil identification
油的搅乳(乳化) churning
油的结膜性质 film-forming property of oil
油的净浮力 net buoyancy of oil
油的净化装置 oil purification plant
油的聚合 bodying of oil; oil bodying
油的馏分 oil fraction
油的摩擦阻力 oil drag
油的黏[粘]度 oil viscosity
油的黏[粘]度号 oil grade
油的黏[粘]结 caking of oil
油的牌号 oil grade
油的喷淋 oil spray(ing); oil throwing
油的喷射 oil injection
油的气化 oil gasification
油的氢化 hydrogenation of oil
油的情况 oil condition
油的热聚合 heat bodying of oil
油的热聚合作用 thermal polymerization of oil
油的闪蒸 oil flash
油的石脑油溶液 oil-naphtha solution
油的碳化指数 carbonization index of oil
油的提取 extraction of oil
油的体度 oil body
油的体积黏[粘]度 oil mass viscosity
油的体积热膨胀系数 thermal expansivity of oil volume
油的添加剂 oil additive
油的脱水装置 oil dehydration plant
油的脱吸 oil denuding
油的吸收 oil absorption
油的稀释 oil dilution
油的稀释槽 oil dilution tank
油的相容性 oil compatibility
油的行驶里程 oil mileage
油的性质 oil property
油的需要量 oil requirement
油的延展长度 length of oil
油的氧化稳定性 oil oxidation stability
油的要求 oil requirement
油的荧光 oil bloom
油的预热 preheating of oil
油的皂化值 oil saponification value
油的蒸馏物组分 oil distillate fraction
油的主要成分 main body of oil
油的转化过程 oil conversion process
油的组成 oil composition
油灯 oil burner; oil lamp; oil lantern
油灯玻璃罩 chimney
油灯工 lampman
油灯升降机 <点油灯的臂板信号机用> lamp hoist; lamp winch
油灯信号 oil-lit signal
油灯信号机【铁】oil-lit signal
油灯照明 oil lighting
油滴 oil drippage; oil drippings; oil droplet
油滴接斗 oil catcher
油滴颗粒 olesome
油滴盘 oil drip pan
油滴收集盘 drip pan
油滴污垢 oil drippings
油滴釉 oil-spot glaze
油底槽桶板 oil pan
油底壳 drain pan; oil pan; oil sump tank; sump; sump oil reserve
油底壳容量 oil sump capacity
油底壳温度 sump temperature
油底壳预热器 oil pan heater
油底盘 oil pan; oil sump
油底盘桶板 oil pan
油底盘装甲 oil pan armor
油底漆 oil-base(d) paint
油砥石 hone; oil stone
油地毯 lino(leum); sheet lino(leum)
油地毡 lino(leum)
油地毡地面 lino(leum) surface
油地毡胶粘剂 lino(leum) (bonding) adhesive; lino(leum) cement
油地毡焦油 lino(leum) tar
油地毡块 lino(leum) felt; lino(leum) tile
油地毡楼地面 lino(leum) flooring
油地毡面砖 lino(leum) tile
油地毡铺地面 lino(leum) flooring
油地毡用油 lino(leum) oil
油点 oil drop; oil spot
油点火器 oil ignitor; oil lighter
油电晕 oil corona
油垫 oil mat; oil pad
油顶起轴承 oil lift bearing
油定色颜料 oil-fixative pigment
油动泵 oil-driven pump
油动插入式振捣器 gas internal vibrator
油动阀 oil controlled valve
油动换向装置 oil-operated reverse gear
油动机部件 oil power cylinder block
油堵螺母 oil nut
油度 <漆料的油和树脂的比例> oil length
油断路开关 oil-break switch; oil-filled type circuit breaker
油断路器 oil-break circuit-breaker; oil circuit breaker; oil-filled type circuit breaker; oil switch
油发动机台 engine platform
油发泡 oil aeration
油阀 fuel tap; oil valve
油法 oil process
油帆布 canvas tarpaulin; oiled canvas
油反射炉 oil-fired air-furnace
油反应度 oil reactivity
油反应性 oil reactivity
油反应性树脂 oil-reactive resin
油分 cut; oil content <指油的含量>
油分监测装置 oil content monitoring arrangement
油分解 oil decomposition
油分离槽 oil separate chamber
油分离器 oil eliminator; oil extractor; oil separator
油分浓度 oil concentration
油分浓度计 oil concentration analyser [analyzer]; oil content meter
油分配器 oil dispenser
油分配器的管道 oil distributor pipe
油分配器润滑 lubrication by means of oil distributor
油分散剂 oil dispersant
油分散剂混合液 oil dispersant mixture
油分散性 oil-dispersing property
油分水机 oil purifier
油分析 oil analysis
油粉 oil-bound distemper; oil-bound water paint
油封 conservation; grease packing; grease seal; liquid seal; oil retainer (thrower); oil seal; oil seal snap ring
油封板 oil seal plate
油封泵 oil-sealed pump
油封槽 oil seal groove; seal with annular groove
油封拆卸工具 oil seal remover
油封齿轮箱 oil-enclosed power
油封挡圈 oil seal collar
油封的 oil-tight
油封垫 oil seal gasket
油封垫圈 oil seal washer
油封防护圈 backing ring
油封封面圈 gland follower
油封盖 oil seal cover
油封革 oil seal leather
油封更换 oil seal replacement
油封管 oil seal tube
油封护挡 oil seal guard
油封护圈 oil seal retainer
油封件的唇部 oil seal lip
油封金属内圈 inner case (of seal)
油封金属外圈 outer case (of seal)
油封壳 oil seal shell
油封毛毡 oil seal felt
油封摩擦速度 oil seal rubbing speed
油封期限 preservation life
油封圈 oil seal; oil seal ring; oil seal washer
油封设备 oil lock
油封弹簧 oil seal spring
油封套 oil seal sleeve
油封填料 oil-sealed stuffer; packing seal
油封填料盖 oil gland; oil-sealing gland
油封填料箱 oil seal housing
油封毡 oil seal felt
油封毡护圈 oil seal felt retainer
油封装置 oil lock; oil-sealing arrangement
油封阻力矩 oil seal torque
油浮 oil float
油浮筒 oil float support
油浮选 oil-buoyancy flo(a)tation; oil flo(a)tation
油辐射器 oil radiator
油腐蚀 oil corrosion
油腐蚀试验 oil corrosion test
油改性醇酸树脂 oil-modified alkyd; oil-modified alkyd resin
油改性聚氨酯 oil-modified urethane
油改性聚氨酯清漆 oil-modified polyurethane varnish
油改性聚酯树脂 oil-modified alkyd resin
油改性树脂 oil-modified resin
油改性顺(丁烯二酸)酐醇酸树脂 oil-modified maleic alkyd
油干料 oil drier
油干燥 oil drying
油感 oiliness
油橄榄 common olive; olive
油缸 actuator; hydraulic cylinder; hydrocylinder; oil cylinder
油缸布置 cylinder arrangement
油缸操纵杆 ram controls
油缸促动器 cylinder actuator
油缸盖 cylinder cap
油缸活塞杆 cylinder rod
油缸活塞杆刮垢器 cylinder wiper
油缸集装箱 tank container
油缸联结销 ram anchor
油缸套盖 cylinder jacket
油钢化 oil tempering
油港 fuel oil bunkering port; offshore oil terminal; oil harbo(u)r; oil port; oil terminal; petroleum port
油膏 ca(u)lk; factice; factis; filling compound; oil paste; ointment; salve; unction; unguent
油膏保油性 oil retention of caulk
油膏低温柔性 low-temperature flexibility of ca(u)lk
油膏缸 ointment jar
油膏挥发率 ca(u)lk volatility
油膏剂 salve
油膏耐热度 heat-resistance of ca(u)lk
油膏黏[粘]接性 ca(u)lk adhesion
油膏黏[粘]结性 ca(u)lk adhesion
油膏碾磨器 ointment roller
油膏施工性 ca(u)lk workability
油膏填料 grease seal
油膏涂料 grease paint
油膏选矿 grease-surface separation
油工 oil painter
油供应 oil supply
油沟 oil duct; oil gallery; oil (ing) groove; oil passageway
油垢 clingage; oil crust; oil dirt
油管 fuel pipe; fuel tube; oil conduit; oil lead; oil line; oil passageway; oil pipe; oil tube [tubing]; oil-well tubing
油管安全接头 safety tubing joint
油管安装 oil piping installation
油管打捞矛 tubing dog; tubing spear
油管打捞筒 tubing socket
油管大钩 tubing hook
油管道 line of oils; oil piping
油管的布置 oil piping layout
油管垫 oil pipe grommet
油管吊卡 tubing elevator

油管吊钳 tubing tongs
油管吊绳 tubing line
油管堵塞 oil-line plugging
油管堵塞器 tubing plug
油管封隔器 tubing packer
油管干线 trunk line
油管钢级 grade of tubing steel
油管工作嘴类型 type of landing nipple
油管工作筒 handing nipple in tubing
油管挂 mandrel hanger
油管和抽油杆打捞筒 tubing and sucker rod socket
油管滑车 tubing block
油管夹 clip tube;oil pipe clamp;oil pipe clip
油管夹持器 tubing catcher;tubing hanger
油管接长 oil-line extension
油管接箍 tubing coupling;tubing joint
油管接环 tubing connecting links
油管接头 fuel pipe union;oil connection
油管卡 tubing clamps
油管卡盘 tubing spider
油管卡套 tubing ring with wedges
油管控制系统 oil pipe control system
油管扣型 tubing thread form
油管路 oil line
油管路接头 lubricator adapter
油管锚 tubing anchor
油管内成渣 sludging in oil line
油管内径 oil hose inner diameter
油管配件 oil pipe fitting
油管清扫 oil-line scavenge
油管深 oil pipe depth
油管丝扣 tubing thread
油管通称直径 nominal diameter of tubing
油管头 landing head;tubing head
油管头法兰 tubing-head adapter flange
油管托架 oil tube bracket
油管完成 tubing for casing completion
油管下入深度 tubing depth
油管线 oil pipeline
油管泄放阀 tubing bleeder
油管旋吊头 tubing swivel
油管压力 oil tubing pressure
油管栈桥 oil pipe access trestle;oil pipe way;pipe way
油管蒸汽吹扫 oil-line steaming
油管柱 flow column;oil string
油管转动器 tubing rotator
油管钻头 oil tube drill
油罐 oil basin;oil can;oil reservoir;oil storage facility;oil tank;petroleum tank;storage tank;store holder
油罐安全阀 oil pot relief valve
油罐安装 tankage installation;tank setting
油罐鞍座 tank saddle
油罐鞍座垫板 slabbing;tank slabbing
油罐保养 tank maintenance
油罐边缘 tank rim
油罐标尺 tank ga(u)ge
油罐表面修整 tank surface reconditioning
油罐驳 oil storage barge;tank barge
油罐驳船 tank lighter
油罐舱 tank compartment
油罐操作 tank operation
油罐场 tank battery;tank farm
油罐场储存油料 tank farm stock
油罐场管线 tank farm pipe line
油罐超限警报 overfill alarm
油罐车 fuel(l)ing vehicle;fuel tanker car;fuel tanker truck;oil car;oil tank car;oil tanker;oil tank wagon;oil truck;road tank car;tank car;tank lorry;transportation tank
油罐车内梯支撑铁 manhole ladder brace
油罐车洗刷所【铁】 oil tank washing plant
油罐车油泵 tank truck pump
油罐车运输燃料 tank truck pickup
油罐车装载架 tank truck loading rack
油罐沉积槽 tank sump
油罐衬里 tank lining
油罐充水预压 preloading of oil tank
油罐出油口 nozzle;tank nozzle;tank outlet
油罐出油口盖 tank outlet cap
油罐出油口缩径管 tank outlet reducer
油罐储存设施 tank storage facility
油罐的固定 tank anchoring
油罐的清扫口 tank clean-out opening
油罐的通气隔膜 tank breathing diaphragm
油罐的修理 rehabilitating of tank
油罐的周围长度 tank circumference
油罐的锥形部分 tapered ring of tank
油罐底 tank base
油罐底凹处 tank pocket
油罐底残渣 bushwash
油罐底缘 tank bottom chime
油罐顶 tank deck
油罐顶浮子 tank deck float
油罐顶盖 tank top
油罐顶环 tank ring
油罐顶栏杆 tank railing
油罐顶上的开口 roof openings
油罐顶小门 tank batch
油罐顶檐 tank roof cave
油罐端板 tank head
油罐端板外隆 convex outward
油罐放出口 tank drain
油罐放气阀 tank relief valve;tank vent valve
油罐分隔室 tank compartment
油罐分类 tank classification
油罐缝试验 tank seam testing
油罐浮动 tank floating
油罐浮水移动 tank water riding
油罐附件 tank accessories
油罐盖 tank cap
油罐钢板 tank plate
油罐高度 tank elevation
油罐隔板 tank spacer
油罐挂车 tank trailer
油罐过滤阀 tank filter valve
油罐过滤器 tank filter
油罐焊接 tank welding
油罐横梁 tank cross member
油罐呼吸阀 tank breathing valve
油罐呼吸隔膜 tank breathing diaphragm
油罐呼吸管 tank breather tube
油罐换气 tank ventilation
油罐混凝土基座 tank concrete pad
油罐货柜 tank container
油罐火灾 tank fire
油罐基础 tank foundation
油罐加热管线 tank heating line
油罐加热盘管 oil tank coil heater
油罐加热器 tank oil heater
油罐加热旋管 tank heating coil
油罐加温设备 bulk heater
油罐加油孔盖帽 tank filler cap
油罐加油口 tank filler
油罐加油口盖 tank filler cap
油罐架 tank rack
油罐间距 tank spacing
油罐检查工 tank tester
油罐建造 tank erection
油罐建造工人 tankies
油罐绞车 tank winch
油罐脚蹬 tank step
油罐搅拌器 tank agitator
油罐校正 tank calibration
油罐校正表 tank table
油罐接地 ground the tanks;tank grounding
油罐接缝 tank joint
油罐结构零件 tank details
油罐绝缘 tank insulation
油罐卡车 tank(er) truck
油罐卡带 tank band
油罐卡带座 tank band fastener
油罐壳厚度 tank shell thickness
油罐壳体 tank shell
油罐空气 tank air
油罐空气包 tank dome
油罐空气驱除器 tank air-mover
油罐冷却器 tank cooler
油罐连接管 tank connection
油罐量器 tank content ga(u)ge
油罐列车装油栈桥 tank train filling rack
油罐铆合 tank riveting
油罐内部零件 internal accessories of tank
油罐内部设备 internal fittings of tank
油罐内部压力 pressure in the tank
油罐内梯 inside ladder;manhole ladder
油罐内体积 internal measurement of tank
油罐内稀释 tank dilution
油罐内液面遥控显示器 remote tank lever indicator
油罐清洗 tank cleaning
油罐区 tank farm;tank park;petroleum tank farm;storage tank farm
油罐区面积 oil tank area
油罐区围埂 bund wall
油罐人孔 tank manhole
油罐人孔盖 tank hood
油罐容积图表 tank volume charts
油罐容量 tank storage capacity
油罐容量计算 tank sizing
油罐设计 tank design
油罐失火 tank fire
油罐失火时溅出物 tank fire boil-overs
油罐使用期 tank life
油罐式变动 oil-canning
油罐探测管 tank sound pipe
油罐田 tank field
油罐通风孔 tank draft hole
油罐通气 tank venting
油罐通气管 tank vent pipe
油罐通气口 oil tank vent
油罐桶数 stock tank barrels
油罐透视灯 tankoscope
油罐涂料层 tank coating
油罐涂料贮罐涂料 tank coating
油罐托 anchor plate
油罐微生物 tank microflora
油罐位置 tank site
油罐洗出物 tank cleanings
油罐信号发送装置 tank unit
油罐旋管 tank coil
油罐旋管公式 tank-coil formula
油罐选择阀 tank selector valve
油罐压力计 tank pressure ga(u)ge
油罐压力洒油车 tank pressure distributor
油罐油面上部的空间 ullage
油罐与集装箱清洗 tank and container clean
油罐圆顶 tank dome
油罐远距离测量 remote gauging of tanks
油罐远距离计量 remote gauging of tanks
油罐站 tank station
油罐蒸气 tank vapo(u)r
油罐蒸气回收 tank vapo(u)r recovery
油罐蒸气空间 tank vapo(u)r space
油罐整形 tank truing
油罐支管 tank manifold
油罐支管阀 tank manifold valve
油罐支架系统 tank support system
油罐中混合气 tank air
油罐中损耗 tank outage
油罐中压力 tank pressure
油罐周围的沟 tank ditch
油罐转证价 tank cession price
油罐装油设备 tank loading facility
油罐装油系统 tank filling system
油罐装置 tank installation;tank work
油罐总容量 total tankage
油罐组 tank battery
油罐组驳船 tank battery barge
油罐座 tank pad
油光暗淡的琥珀 oily-looking dim amber
油光薄板 slick sheet
油光锉 barrette file;dead smooth file
油光调和漆 oil gloss paint
油光涂料 oil gloss paint
油光泽 oil gloss
油光纸 parchamyn paper
油规 fuel oil meter;oil ga(u)ge
油柜 oil tank;tank
油柜漆 tank paint
油柜清洁器 tank cleaning machine
油柜数 number of tanks
油滚柱 oil roller
油锅辊 grease-pot rollers
油果松 nut pine
油过滤器 oil cleaner;oil filter
油过滤清器 oil cleaner
油耗 fuel consumption;oil consumption;oil wear
油耗量 oil consumption
油耗曲线 fuel consumption curve
油和煤气 oil and gas
油和树脂比 <清漆的> oil/resin ratio
油和脂肪 fat and oil
油盒 keep of axle box;oil box;oil cellar;cellar
油盒楔 oil cellar wedge
油鹤 oil crane
油黑 oil black
油黑印码器 inker
油恒温器 oil thermostat
油红 oil red
油虹吸管 oil siphon
油壶 dashpot;lubricator;oil can;oil cup;oiler;oil feeder;oil jack;oil lubricator;oil pot
油壶的自动盖头 oil-can ratchet autolock
油壶的自锁棘轮 oil-can auto-lock ratchet
油壶盖 can top;oil-can lid
油壶继电器 dashpot relay
油壶润滑 oil-can lubrication
油糊白铅 paste-in-oil white lead
油糊碱性碳酸铅 paste-in-oil basic lead carbonate
油花 oil bloom;oil slick
油花分离器 oil slick separator
油滑道路 greasy road
油滑的 greasy;slick
油滑地段 oil slick portion
油滑剂 slip agent
油滑组织 oil lubricating tissue
油化矿石 oiled ore
油画 canvas painting;oil paint(ing);painting (in oil);paintwork
油画板 academy board
油画底色 dead colo(u)r
油画蓝 academy blue
油画漆 artist paint
油画颜料 oil colo(u)r;oil paint

油画载色剂 megilp
油画载色体 megilp
油环 oil control ring;oil(-retaining) ring;revolving ring;wiper ring
油环挡圈 oil-ring retainer
油环导槽 oil-ring guide
油环高度 height of oil ring
油环卡住 oil-ring plugging
油环黏[粘]结 oil-ring sticking
油环润滑 oil-ring lubrication;ring-oil(ing);ring lubrication
油环润滑的 ring-oiled
油环润滑法 oil-pad lubrication
油环润滑器 ring lubricator
油环润滑式滑动轴承 ring-oiled sleeve bearing
油环润滑轴承 oil-ring bearing;ring lubrication bearing;ring-oiled bearing
油环润滑轴颈箱 ring-oiling journal box
油环式轴承 oil-ring bearing
油环轴承 ring oiling bearing
油环注油器 ring oiler
油环自动润滑 oil-ring self-lubrication
油缓冲罐 oil dashpot
油缓冲壶 oil dashpot
油缓冲器 oil buffer;oil bumper;oil dashpot;oleo-gear
油缓冲器行程 oil buffer stroke
油缓冲支柱 oleostrut
油灰 sealing compound;badigeon;chinking;clairecolle;clearcole;glazing putty;lute;luting;oil putty;painter's putty;putty;sal ammoniac;smoothing cement;wood filling;slush <白铅石灰>
油灰操作性 putty finishability
油灰槽口 rebate for putty
油灰刀 dovetail cutter;glazier's chisel;stopping knife;putty knife
油灰底层<玻璃的> bed putty
油灰粉 putty powder
油灰缝 putty joint
油灰附着力 putty adhesive force
油灰工 puttier
油灰勾缝 pointing with putty;putty pointing
油灰刮刀 hacking(-out) knife;putty chaser;putty scraper;putty knife
油灰龟裂试验 putty chap cracking test
油灰唧筒 putty pump
油灰麻绳填料环 grommet ring
油灰密封 putty seal(ing)
油灰磨 putty chaser
油灰抹刀 putty knife
油灰嵌缝 putty seal(ing)
油灰清除 putty removal
油灰填缝料 joint filler
油灰涂层 slush coat;stopper coat
油灰土 putty soil
油灰土路面 bituminous clay-lime pavement
油灰镶玻璃法 glazing with putty;putty glazing
油灰样可移动的底层 bedding putty
油灰油 putty oil
油灰整 wood filling
油灰装玻璃 putty glazing
油回火 oil temper(ing)
油回火钢丝 oil temper wire
油回流管 oil return pipe
油回流管道 oil return passage
油回收 oil reclamation;oil recovery;withdrawal of oil
油回收器 oil saver
油回收系统 oil recovery system
油混合物 oil mix(ture)
油活化性 oil reactivity
油活节的 articulated
油击穿 oil puncture
油击穿试验 oil puncture test
油机 oil engine
油机发电机 engine-driven generator
油迹 grease marks;oil stain;oil trace;oil wake
油迹滞留指数 oil residence index
油基 base stock;body of oil;oil base;petroleum base
油基薄涂法使用的瓷釉 oil-based scumble glaze
油基薄涂釉面 oil scumble glaze
油基建筑用胶粘剂 oil-based building mastic
油基建筑用玛琋脂 oil-based building mastic
油基胶粘剂 oil-based mastic
油基介质 oil-based medium
油基绝缘漆布 empire cloth
油基玛琋脂接缝密封垫 oil-based mastic joint sealer
油基媒液 oil vehicle
油基泥浆 oil-based mud;oil mud
油基泥浆钻取的岩芯 oil-based core
油基腻子 oil putty
油基漆 oil-based coating;oil-based paint
油基漆稀料 thinner for oleoresinous paint
油基嵌缝料 oil-based ca(u)lking compound
油基清漆 oil-based varnish
油基溶剂 oil-based vehicle
油基润滑剂 oil lubricant
油基色斑 oil-based stain
油基树脂的 oleo-resinous
油基树脂清漆 oleo-resinous varnish
油基污点 oil-based stain
油基印墨 oil-based ink
油基油灰 oil-based putty
油基油漆 oil-based paint
油基油漆涂层 oil-based paint coating
油基油漆涂膜 oil-based paint coating
油基脂胶 mastic based on oil;oil-based mastic
油极 oil pole
油急冷(处理) oil quenching
油计量 oil measurement
油计量系统 oil-metering system
油剂 daubing;oil;oil-soluble concentrate;oil solution
油加氢 hydrogenation of oil
油加热隔阻器 heating oil barrier
油加热管 oil heater
油加热截流器 heating oil interceptor
油加热器 oil heater
油加热栅栏 heating oil barrier
油加热装置 oil heating unit
油夹套 oil lagging
油煎噪声 fry
油减振器 oil bumper;oil dashpot
油减震 oil damping
油减震器 oil damper;oil dashpot;oil shock absorber
油减震器底盘 oleo undercarriage
油减震式起落架 oleo undercarriage
油减震柱 oleostrut
油检查员 oil inspector
油浆沉降器 slurry settler
油浆法 slurry process
油浆系统 slurry service
油降解菌 oil-degrading bacteria
油胶 factice
油胶树脂 oleo-gum-resin
油焦 oil coke
油焦质 carboids
油焦置换比 oil-to-coke replacement ratio
油脚 bottom oil;foots;foots oil;oil foot(ing)
油脚不清装油法 load on top
油脚不清装油制度 load on top system
油脚和水 bottom sludge and water
油接触器 oil contactor
油接收器 oil receiver
油截止阀 oil cut-off valve
油介质的 oil-medium
油金 shell gold
油浸 oil bath;oil impregnation;oil-immersion
油浸白云岩 oil-impregnated dolomite
油浸板 oil-treating board
油浸变压器 bath transformer;immersed transformer;oil transformer;liquid-immersed transformer;oil-filled transformer;oil-immersed transformer
油浸变阻器 oil-immersed rheostat
油浸槽 oil bath
油浸池 oil(-immersion) bath
油浸的 oil-immersed
油浸电缆 oil compression cable;oil-filled cable;oil-impregnated cable
油浸电缆油 oil-filled cable oil
油浸电容器 oil capacitor;oil condenser;oil-filled condenser;oil-impregnated condenser
油浸电容器式套管 oil-filled condenser type bushing
油浸断路器 oil(-immersed) circuit breaker
油浸法 immersion oil method;immersion technique;oil-immersion method
油浸反射率 oil-immersion reflectivity
油浸防腐木材 creosoted wood;creosoted timber
油浸防腐木桩 creosoted-wood pile
油浸风冷式 oil-immersed forced-air cool
油浸固体绝缘体 oil-impregnated solid insulator
油浸罐头 oiled can
油浸锅 penetrating oil pot
油浸过的木材 oil-impregnated wood
油浸灰岩 oil-impregnated limestone
油浸剂 infused oil;oleo-infusion;oleol
油浸镜头 oil-immersion lens
油浸镜头组 immersion series
油浸开关 oil(-immersed) switch
油浸离合器 wet clutch
油浸砾岩 oil-impregnated conglomerate
油浸木材圆筒 creosoting cylinder
油浸木桩 creosoted pile
油浸泥灰质泥岩 oil marly clay
油浸泥岩 oil-impregnated clay
油浸起动器 oil-immersed starter
油浸起动装置 oil-immersed starter
油浸强迫风冷 oil-immersed air-blast cooling
油浸熔燃断路器 oil-fuse cutout
油浸熔丝保险器 oil-quenched cut-out
油浸砂岩 oil-impregnated sandstone
油浸生物灰岩 oil-impregnated biolithite
油浸绳 tarred rope
油浸石灰岩 oil-impregnated limestone
油浸式变压器 oil-filled transformer
油浸式点火线圈 oil-filled ignition coil
油浸式电弧控制设备 oil-immersed arc control device
油浸式电弧控制装置 oil-immersed arc control device
油浸式电力变压器 oil-immersed power transformer
油浸式电流互感器 oil-immersed current transformer
油浸式断路器 oil-immersed breaker
油浸式空气滤清器 oil moistened air filter
油浸式离合器 oil-type clutch
油浸式起动器 oil-immersed starter
油浸式轴承 oil-immersed type bearing
油浸试验 oil-immersion test
油浸水冷却 oil-immersed water cooling
油浸水冷式 oil-immersed water-cool
油浸套管 oil-filled bushing
油浸式调整变压器 oil-immersed regulating transformer
油浸物镜 oil-immersion objective
油浸系<显微镜的物镜> immersion system
油浸显微镜 oil-immersion microscope
油浸显物镜 oil-immersion microscope objective
油浸压碎法 oil-immersion crushing method
油浸岩芯 oil wet core;weeping core
油浸衍射仪 oil-immersion diffractometer
油浸浴 oil bath
油浸折射计 immersion refractometer
油浸枕木 impregnated sleeper
油浸纸 oil paper
油浸纸条 oiled paper strip
油浸纸质电容器 oil-filled paper condenser
油浸制冷 oil-immersion cooling
油浸桩(处理)<用杂酚油防腐处理过的木桩> creosoted pile
油浸自然冷却 oil-immersed natural cooling
油浸自然冷却式 oil-immersed self cool
油浸渍的木材 oil-impregnated wood
油浸渍的木料 oil-impregnated wood
油浸渍的纸 oil-impregnated paper
油浸渍过的 oil-impregnated
油精 olein
油精馏器 oil rectifier
油井 oil borehole;oil well
油井爆破 oil-well blasting;shooting of oil wells
油井爆破器 go-devil
油井爆破药筒 bullet;torpedo
油井产量测定 well ga(u)ging
油井产油率 rating of well
油井衬管 oil-well screen
油井的汲油半径 radius of drainage
油井的排水半径 drainage radius of an oil well
油井的排油半径 radius of drainage
油井底 well face
油井动态 well performance
油井堵塞 oil-well packing
油井额定生产能力 well rating
油井废水 oil-well wastewater
油井封隔 oil-well packing
油井管 oil-well pipe
油井管滤器 well tube filter
油井灌浆 oil field cementing;oil-well grouting
油井灌浆水泥 oil-well grouting cement
油井过滤筛 oil-well screen
油井计量 ga(u)ging of oil wells
油井记录 oil log
油井架 oil derrick
油井架供应船 oil rig supply vessel
油井架腿 oil derrick legs
油井井架 oil-well derrick
油井井喷 oil well blowing
油井枯竭 decline of well
油井滤管 oil-well screen
油井喷油 oil well blowing
油井平台 well platform
油井平台混凝土 oil-well platform

concrete
油井起套管 well pulling
油井气 casing-head gas
油井群 line of wells
油井射孔 oil-well shooting
油井生产率 performance of a well; well flow index
油井生产能力 productive capacity of well
油井衰退 decline of well
油井水泥 <一种耐高温水泥> oil-well cement
油井水泥稠化时间测定仪 thickening time tester for oil well cement
油井水泥分类 oil-well cement classification
油井水泥实验仪器 oil-well cement experiment instrument
油井水泥熟料性质 oil-well cement clinker property
油井水泥物理性质 oil-well cement physical property
油井塔架 oil derrick
油井套管 oil field casing; oil-well casing
油井温度 oil-well temperature
油井线 line of wells
油井性能试验 well performance testing
油井旋转钻头 oil field rotary bit
油井压力 oil-well pressure
油井延深钻进 old-well deeper drilling
油井盐水污染 brine pollution from oil well
油井用水泥 oil-well cement
油井增产措施 stimulation of wells
油井支撑物 oil-well proppant
油井注水 flushing
油井注水泥 oil-well cementating; oil-well cementation
油井柱塞泵 oil-well plunger pump
油井转盘 oil field rotary
油井准备 well conditioning
油井钻工 well sinker
油井钻进 oil-well drilling
油井钻探船 spudding drill
油井钻探设备 oil derrick; oil rig
油井钻头 oil field bit; oil-well bit
油净化 oil purging; oil purification
油净化处理 oil purification
油净化处理器 oil treater
油净化器 oil cleaner; oil purifier
油净化室 oil purification room
油净化系统 oil cleaning system
油净化箱 oil cleaning tank
油净化装置 reconditioner
油镜 immersion objective
油聚结器 oil coalescer
油卷 roller-towel
油绝缘 oil insulation
油绝缘变压器 bath transformer; immersed transformer; oil transformer
油绝缘纸 oiled paper
油开关 circuit breaker; oil-break switch; oil (circuit) breaker; oil switch
油开关脱扣 oil switch trip
油勘探 exploration activity
油可溶性 oil solubility
油空气热交换器 oil-to-air heat exchanger
油孔 oil hole; oillet(te); oil orifice; oil passage; oil pit
油孔螺钉 oil-hole screw
油孔配置标记 oil-hole location marks
油孔塞 oil-hole plug; Welch plug
油孔钻(头) oil-hole drill
油控的 oil controlled
油口接头 port connection; port fitting

油库 bulk oil; bulk plant; fuel depot; oil bunker; oil depot; oilery; oil house; oil reservoir; oil storage; oil store; oil tankage; petroleum storage depot; tank farm
油库管线装设 storage piping installation
油库建筑 combustible storage building
油库区 oil depot
油库设备 terminal facility
油库设施【港】 marine bunker facility
油块 oil clot; oil in mass
油矿 oil mineral
油矿泵 jack pump
油矿两用船 ore-cum-oil carrier; ore/oil carrier
油矿气井 gasser
油矿油罐 lease tank
油扩散泵 oil diffusion pump
油垃圾焚烧炉 oil refuse incinerator
油栏 oil barrier; oil fence
油蓝 oil blue
油廊 transept
油类 oil and grease; oils
油类比重计 acrometer
油类薄涂法使用的瓷釉 oil-type scumble glaze
油类捕集器 oil interceptor
油类防腐剂 preservative oil
油类副产品 oil byproducts
油类记录簿 oil record book
油类精制的过滤器 oil refiner pack
油类精制器 oil reclaimer
油类流出口 oil drain(age)
油类流动性 oil mobility
油类流动性的倒数 reciprocal of oil mobility
油类黏[粘]结剂 oil-based binder
油类喷雾剂 oil spray
油类燃料 oil fuel
油类软化剂 oil-softener
油类润滑剂 oil lubricant
油类杀虫剂 oil insecticide
油类收集器 oil scoop
油类污斑 oil-type stain
油类污染 oil material contamination; oil pollution
油类污染物 oil pollutant
油类物质保护 oil-borne preservative
油类盐剂混合 <电杆处理> oil-salt combination
油类硬化 hardening of oils
油类与油脂去除 oil and grease removal
油类蒸发器 oil vapo(u)rizer
油冷变压器 bath transformer; immersed transformer; oil transformer
油冷的 oil-cooled
油冷电抗器 oil-immersed reactor
油冷电子管 <发射用电子管> oil-cooled tube
油冷定子绕组 oil-cooled stator winding
油冷法 oil-cooled system
油冷汽轮发电机 oil-cooled turbogenerator
油冷器管道 oil cooler line
油冷却 oil cooling
油冷却变压器 oil-cooled transformer; tank transformer
油冷却的 oil-cooled; oil cooling
油冷却法 oil cooling method
油冷却管 oil cooling pipe
油冷却器 oil cooler
油冷却器百叶窗 oil cooler shutter
油冷却试验 oil cold test
油冷式的 oil-cooled
油冷式离合器 oil-cooled clutch
油冷式制动器 oil-cooled brake

油冷真空管 oil-cooled valve
油冷装置操作盘 control board for cooling unit
油离心机 oil centrifuge
油沥青 oil asphalt
油沥青质的 bituminiferous
油粒 elaioleucite; elaioplast
油粒综合体 oil-solid system
油连接管 oil connecting pipe
油炼法 creosoting process
油链润滑轴瓦 chain-oiled bearing
油量 oil mass; oil quantity
油量表 fuel ga(u)ge; fuel level ga(u)ge; fuel quantity meter; oil meter; tank ga(u)ge
油量差 outage
油量分配阀 dividing valve
油量分配器 fuel flow divider; oil distributor
油量计 fuel content ga(u)ge; fuel ga(u)ge; fuel meter; oil flowmeter; oil ga(u)ge; oil measurer; oil meter; oil quality ga(u)ge; oil quality indicator; oleometer; oil quantity ga(u)ge
油量计玻璃 fuel ga(u)ge glass
油量节流阀 oil volume restrictor
油量控制针 jet needle
油量热计 oil calorimeter
油量调节 fuel flow control
油量调节阀 oil regulator
油量调节器 oil regulator
油量调整 oil volume adjustment
油量限流器 oil volume restrictor
油量指示器 oil index; oil quantity indicator
油料变稠结块 live ring
油料薄膜层爆轰 film detonation
油料沉积物 oil deposit
油料成本 oil fuel cost
油料澄清器 oil clarifier
油料抽提机 oil extractor
油料稠化 oil thickening
油料储存 oil storage; oil storing
油料的沉淀 oil dregs
油料的压力输送 positive delivery of oil
油料费 oil fuel cost
油料分布器 oil header
油料分析试验设备 oil-testing equipment
油料过滤器 oil strainer
油料技术规格 oil specification
油料加入器 oil filler
油料加入器管 oil filler pipe
油料林木 oil-yielding shrubs
油料炉黑 oil furnace black
油料滤清器 oil filter
油料色粉涂饰 oil-bound distemper
油料事故 oil hazard
油料试验机 oil-testing machine
油料脱水器 oil dehydrator
油料稀释 oil dilution
油料消耗 oil consumption
油料用亚麻籽油 linseed oil for paints
油料注入器 oil feed injector
油料装运 oil shipment
油料籽 oil seed
油料作物 oil-bearing crop; oil plant; oil(seed) crop
油料作物和森林害虫 oil plant and forest pests
油裂化器 oil cracker
油裂解 oil-breaking
油裂解气量 amount of cracking genetic gas of oil
油流 oil stream
油流出管道 oil-out line
油流继电器 oil flow relay
油流检定继电器 oil flow detection relay

油流量 oil flow
油流量继电器 oil flow relay
油流图 oil flow picture
油流系数 oil flow coefficient
油漏斗 oil funnel
油炉 oil(-fired) furnace; oil-stove
油滤壳体 filter housing
油滤纸 oil filter paper
油路 channel for oiling; oil channel; oil circuit; oil duct; oiled road; oil groove; oil passage(way); oil pipeline; oil-way
油路分配盘 port plate
油路开关 oil circuit breaker
油轮 ocean tanker; oil carrier; oiler; oil ship; oil tanker; tanker (vessel); crude oil carrier; tank ship
油轮安全与防污条约 Tanker Safety and Pollution Prevention
油轮泊位 oil tanker berth; tanker berth; oil berth; petroleum berth
油轮舱口装置 tank coaming and lid
油轮船东关于油污自愿协定 Tanker Owners Voluntary Agreement Concerning Liability for Oil Pollution
油轮船东污染责任协议 Pollution Liability Agreement Among Tanker Owners
油轮的水线长度 waterline length of tanker
油轮干舷 tanker freeboard
油轮管线 tanker line
油轮航次租 tanker voyage charter
油轮货预热系统 tank heating system
油轮接地电缆 bonded cable
油轮控制系统 tanker control system
油轮码头 oil tanker jetty; tanker berth; tanker terminal
油轮码头及炼油作业区 oil refinery and tanker terminal
油轮期租 tanker time charter
油轮清洗驳 tank-cleaning barge
油轮清洗系统 tank-cleaning system
油轮洗舱清洁法 Butterworth tank cleaning system
油轮卸油管线 tanker unloading pipeline
油轮卸油设备 tanker discharging facility
油轮压舱水 tanker ballast water
油轮主油管从油泵室到甲板出口的一段 deck delivery
油轮自动化洗舱装置 automatic cleaning equipment
油轮自动装卸压载系统 automatic logistrip
油麻绳 marlin(e); tarred rope
油麻绳包裹的钢索 fibre clad rope
油麻丝 oakum
油马达 hydraulic motor; oil motor
油码头 oil berth; oil dock; oil jetty; oil terminal; oil wharf; petroleum berth
油毛毡 asphalt felt; asphalt mattress; asphalt sheet; bitumen sheet(ing); bituminous sheet(ing); building paper; felt; hair felt; malthoid; rag felt; repellent building paper; saturated felt; sheathing paper; tar(red) felt; tar roofing felt; ruberoid <覆盖屋顶用的>; roofer felt <铺屋面的>; Stuko-bak <用于外粉刷背面的>
油毛毡衬层 <置于屋顶或瓦的下面> sarking felt
油毛毡垫层支座 felt pad bearing
油毛毡胶粘剂 felt adhesive; felt cement
油毛毡卷筒 dry felt web
油毛毡卷轴 roll roofing winding mandrel

油毛毡喷灯焊接 felt torching
油毛毡嵌入物 felt insert
油毛毡填充料 plug of felt
油毛毡条 taping strip
油毛毡屋顶 malthoid roofing
油毛毡屋顶材料 tar felt roofing
油毛毡屋顶覆盖层 tar felt roof cladding
油毛毡屋面 asphalt felt roof;malthoid roofing
油毛毡用钉 clout nail
油毛毡纸夹 asphalt felt paper clamp
油毛毡制造 dry felt manufacture
油媒污渍 oil-borne stains
油煤气焦油 oil-gas tar
油煤气焦油沥青 oil-gas tar
油煤气焦油脂 oil-gas tar pitch
油煤气硬沥青 oil-gas tar pitch
油门 accelerator; butterfly throttle valve;petcock;throttle
油门操纵杆 accelerator lever
油门操纵杆弧座 throttle quadrant
油门操纵机构 throttle control
油门杆 throttle lever
油门关闭 gasoline shut-off
油门回位缓冲器 throttle return damper
油门开合度 throttle opening
油门控制杆系 throttle linkage
油门手柄 throttle control handle
油门锁 throttle lock
油门踏板 accelerating pedal
油门调节 throttling
油门自动关闭按钮 autothrottle cut-out button
油门自动控制 automatic throttle control
油密舱壁 oil-tight bulkhead
油密舱盖 oil-tight hatchcover
油密舱口 oil-tight hatch
油密的 fuel oiltight;oil-tight
油密滴油盘 oil-tight drip tray
油密封 oil-tight
油密封垫 oil pad
油密封接头 oil-tight joint
油密封控制装置 seal oil control equipment
油密封试验 oil-tight test
油密封套 oil seal housing
油密封系统 seal oil system
油密盖 oil-tight cover
油密工作 oil-tight work
油密横舱壁 oil-tight transverse bulkhead
油密接头 oil-tight joint
油密肋板 oil-tight floor
油密铆钉 oil-tight riveting
油密铆接 oil-tight riveting
油密试验 oil-pressure test for tightness;oil-tight test
油密性 oil-tightness
油棉布剂 sindon oleatanae
油面 fuel level;grease (sur) face;oil level;oil surface
油面标志 oil level mark
油面表 fuel level ga(u)ge
油面测杆 oil level dipstick
油面测量计栓 oil level plug
油面高度 fuel head;oil level
油面观察玻璃 oil sight glass
油面管 oil level pipe
油面计 oil ga(u)ge
油面计开关 oil level cock
油面检查孔螺塞 oil level plug
油面控制机构 oil level control
油面路 oil surface road
油面螺塞 oil level screw
油面视镜 oil sight glass
油面信号灯 oil level tell-tale
油面指示管 oil ga(u)ge pipe
油面指示器 fuel level indicator; oil level ga(u)ge;oil level indicator; oil pointer
油苗【地】 oil seepage;oil show;seepage
油苗学 seapology
油灭弧 oil blast
油灭弧断路器 oil-blast circuit breaker
油灭弧开关 oil-blast switch
油敏感的 oil-susceptible
油膜 oil(-bound) film; oil-coat; oil paint;oil slick;slick
油膜表面 surface of film
油膜范围 oil film extent
油膜刚度 oil film rigidity
油膜光阀 oil deformation valve
油膜光阀投影 oil film light modulator projection
油膜滚珠轴承理论 oil ball bearing theory
油膜过滤器 oil film filter
油膜厚度 oil film thickness;thickness of oil slick
油膜剪应力 friction(al) stress
油膜摩擦损失 oil drag loss
油膜黏[粘]度 film viscosity; oil film viscosity
油膜破坏 oil film breakdown
油膜破裂 rupture of oil film
油膜强度 film strength;oil film strength
油膜强度试验 oil film test
油膜强度添加剂 film strength agent
油膜润滑 film lubrication
油膜式投影电视 eidophor
油膜损耗 oil film loss
油膜涡动 oil whirl
油膜下沉 sinking of oil slicks
油膜旋涡 oil whirl
油膜预报 forecasting oil slick
油膜振荡 oil whip(ping)
油膜振动 oil film vibration;oil whipping
油膜轴承 film lubrication bearing; flood-lubricated bearing;kelmat;oil film bearing;oil flooded bearing
油膜轴承合金 kelmat
油磨 oil abrasion
油磨刀石 oil stone
油磨法 oil-mull technique method
油磨光 oil polishing
油磨石 oilstone slip
油墨 printing ink;printing oil
油墨半成品 ink intermediate
油墨变质 decomposition of ink
油墨擦掉或弄污 ink crocking
油墨层 ink lay
油墨层厚度 ink-film thickness
油墨的肝化 flocculation of the vehicle
油墨的固着 setting of ink
油墨干后变暗 ink dry back
油墨干燥 drying of ink
油墨缸 ink well
油墨刮板 ink knife;ink slice
油墨刮刀 ink knife
油墨刮涂膜 pulldown
油墨光泽 ink gloss
油墨辊 doctor;inker
油墨记录 ink recording
油墨记录器 ink recorder
油墨胶凝 gelation of ink
油墨胶体分解 breakdown of ink colloid
油墨颗粒 ink particle
油墨拉力计 inkometer
油墨蓝 ink blue
油墨轮 inking wheel
油墨膜 skin
油墨膜断裂 ink-film breakdown
油墨黏[粘]性计 inkometer
油墨黏[粘]脏 set-off of ink
油墨黏[粘]纸 sticking
油墨浓度 ink grade
油墨浓度电子控制器 inkatron
油墨浓黏[粘]度调节器 idotron
油墨喷射打印机 ink-jet printer
油墨热致变色 colo(u)r burn-out
油墨乳化 ink emulsification
油墨色浆 ink base
油墨渗透 ink penetration
油墨渗透率 rate of ink penetration
油墨台 ink-up table
油墨添加剂 easer
油墨涂饰 blackout
油墨污垢 colo(u)r soiling
油墨吸收能力 ink receptivity
油墨移印 retroussage
油墨印件 inking print
油墨印码器 ink writer
油墨用三辊磨 ink roll mill
油墨展开面积 ink mileage
油墨纸张关系 paper ink relation
油墨转移 ink transfer
油母岩对比 oil and source rock correlation
油母岩质 kerabitumen; kerogen; petrologen
油母页岩 oil-forming shale; kerogen shale
油木 oiled wood
油幕 oil curtain
油囊体 oleocyst;oleocystidium
油泥 dirt;fatlute;grease;grease filth; greasy filth; oil sediment; oil sludge;plastocene;sludging;slush <厨房废弃油脂>
油泥沉积 sludge deposition
油泥刀 glazier's chisel
油泥过滤 sludge filtration
油泥浆 oil mud
油泥心 oil-core
油腻 grease
油腻的 greasy;unctuousness
油腻料 gunk
油腻水 greasy water
油腻性 greasiness;oiliness;unction
油腻子 oil loam;oil putty
油黏[粘]度计 fluidimeter [fluid(o)-meter]
油黏[粘]度品级 oil grade
油捻式润滑器 wick lubricator
油凝胶 oleogel
油盘 oil catcher; oil disc [disk]; oil pan;oil receiver;oil tray;sump
油盘孢属 <拉> Elaeodema
油盘衬垫 oil pan gasket
油盘垫片 oil pan gasket
油盘放油塞【机】 oil pan drain plug
油盘放油旋塞 oil pan drain cock
油盘加油口盖 oil pan filler cap
油盘间隙 oil pan clearance
油盘紧固件 oil pan fastener
油盘进油喇叭口 suction bell
油盘密封垫 oil pan packing
油盘深度计 oil pan depth ga(u)ge
油盘油封座 oil pan filler block
油盘制动器 oil-disc brake
油泡 oil vacuole
油喷淋器 oil sprayer
油喷射 oil spurts
油喷射泵 oil ejector pump
油喷射冷却 oil injection cooling
油喷射器 oil ejector
油喷雾器 oil atomizer;oil sprayer
油盆 oiler can
油漂白 oil bleaching
油撇取器 oil skimmer
油品 oil type
油品安定性 oil stability
油品包装 oil packaging
油品泵送 oil pumping
油品泵送系统 oil pumping system
油品比重计 oil hydrometer
油品变稠结块 liver
油品变黑 oil darkening
油品变质 oil deterioration
油品槽车 oil tank vehicle
油品储运 storage and shipping of oil products
油品的基 oil base
油品的牌号 oil brand
油品的破乳化性 oil demulsibility
油品的寿命 oil life
油品的脱色 oil decolorization
油品分类 oil classification
油品腐蚀 oil corrosion
油品规格 oil specification
油品净化 oil purification
油品乳化液 oil emulsion
油品色调计 oil tintometer
油品输送 oil delivery
油品损耗 oil loss
油品添加剂 oil dope
油品脱除胶质 oil degumming
油品脱水 oil dehydrating
油品脱水装置 oil dehydrating plant
油品在发动机内工作特性 oil performance in engine
油品装卸设备 oil-loading facility
油品装运 oil shipment
油平台楼梯 platform stair(case)
油漆 beaded paint;coating;oil paint; paint(ing);pek;keystone <一种用于新水泥工程的>
油漆泵 paint pump
油漆比较试验板 paint patch panel
油漆笔 painting pen
油漆变稠 fatten(ing)
油漆变质 paint degradation
油漆标线 painted road stripe
油漆表面裂纹 bitty;orange peel
油漆表皮的污点 cessing
油漆表皮的小孔 cessing
油漆剥落 paint peeling
油漆布 lino(leum);oil-varnished cambric
油漆残渣混合物 smudge
油漆仓库 paint store
油漆操作 painting practice
油漆层 skin of paint
油漆长加仑 oil varnish length-gallons
油漆厂 lacquer factory
油漆车间 finishing room; painter's shop;paint plant; paint shop; paint workshop
油漆沉淀物 paint residue
油漆成分 paintingredient
油漆承包人 painting contractor
油漆稠度 body of paint;paint consistency
油漆疵病 paint defect
油漆次序 painting system
油漆催干剂 paint drier [dryer];siccative
油漆打底 paint base
油漆打底层 primer;priming coat
油漆打粉底 chalking
油漆刀 spattle;spatula
油漆的含油率 oil length of varnish
油漆的金属板 painted sheet metal
油漆的伪装层 camouflage coat of paint
油漆的颜色 painting colo(u)r
油漆的主题 painted motif
油漆底层 paint filler;rough coat
油漆底涂层 first coat
油漆底子 primary;paint base
油漆发白 blushing
油漆方法 paint staining
油漆废水 paint wastewater

油漆干燥剂 paint drier [dryer]
油漆干燥期 drying period
油漆膏 paint paste
油漆工 brother of the brush;colo(u)-rer; japanner; lacquerer; painter; paintwork
油漆工厂 japanning works
油漆工程 painter's work;paintwork
油漆工工具 painter's tool
油漆工脚手架 ladder jack scaffold
油漆工具 painter's tool; painters' tool;painter
油漆工喷灯 painter's torch
油漆工人 varnisher
油漆工序 paint system
油漆工业 paint industry
油漆工艺 paint technology
油漆工艺学会联合会 <美> Federation of Societies for Coatings Technology
油漆工作 painting;painting work;paintwork
油漆刮刀 painter's spattle; paint scraper;point scraper;spatula
油漆罐 paint bottle; paint-pot; paint tank
油漆辊 paint roller
油漆滚筒 paint roller
油漆滚子 paint roller
油漆过程 paint system
油漆过程废水 painting process wastewater; paint process wastewater
油漆过滤器 paint filter
油漆过黏[粘] after-tack;residual tack
油漆行 paint shop
油漆行业 painter trade
油漆和清漆洗除剂 paint and varnish remover
油漆烘干(法) infrared drying
油漆厚度 thickness of coating
油漆化学家 paint chemist
油漆混浊膜 blushing
油漆或清漆干而且失去黏[粘]性 tack-free dry
油漆机 painting machine
油漆机械 painting machinery
油漆基层工艺 bodying
油漆基料 paint base
油漆及涂料等的耐热温度 paint resist temperature
油漆计划 painting scheme
油漆技师 paint technician
油漆间 japanning room;paint locker; paint shop;paint store
油漆溅出 paint spillage
油漆匠用胶水 painter's glue
油漆匠用刷子 painter's brush
油漆搅拌器 paint agitator;paint mixer;paint shaker
油漆结晶纹饰 crystallization finish; crystallized finish
油漆结皮 <油漆疵病> skin drying; surface drying
油漆介质 paint medium
油漆局部光泽工艺 flashing
油漆开裂 cracking in paint; chip crack
油漆抗议 paint-in
油漆颗粒 paint particle
油漆库 paint storage
油漆块渣 caking
油漆快干段 <装配线> flash-off area
油漆裂缝 paint checking
油漆裂隙缺陷 catface
油漆流淌 bleed-off point
油漆漏涂 holiday
油漆麻点 pin-holing;pock marking
油漆媒液 paint vehicle
油漆密封层 sealing coat of paint
油漆面斑点 spotting
油漆面变暗 saddening-down
油漆面层 overpaint;paint-coat;paint skin
油漆面缺陷的修理 teasing
油漆膜 paint membrane
油漆木料前用以填塞裂缝的塑性材料 stopping
油漆腻子 painter's putty;putty
油漆黏[粘]土 paint clay
油漆配方 paint formulation
油漆配色 paint scheme
油漆喷枪 compressed-air painting gun;paint air gun;paint gun;paint-spray gun;paint spraying gun
油漆喷涂泵 paint spray pump
油漆喷涂法 paint spraying process
油漆喷涂幅度 band of paint
油漆喷涂器 compressed-air painting apparatus
油漆喷涂室 paint spray room
油漆喷雾器 paint atomizer
油漆破坏剂 paint-destroying agency
油漆起泡 bubbling
油漆起皮 skinning
油漆清除剂 paint remover
油漆缺点 painting defect
油漆燃除 burning off paint
油漆溶剂 lacquer solvent; paint solvent
油漆溶剂油 paint and varnish naphtha;paint naphtha;paint oil
油漆软管 paint hose
油漆色彩变更 flash
油漆上底 precoating
油漆施工系统 paint system
油漆饰面 painted enrichment;painted (ornamental) finish(ing)
油漆稀释剂 glaze medium;paint thinner;petroleum spirit <石油提炼的>
油漆刷 flat(paint) brush;paint brush
油漆台位 painting position
油漆提桶 paint bucket
油漆填充料 asbestine
油漆填料 <油漆前填塞木孔等用的混合物,如油灰> paint filler; painter's putty
油漆调稠 box the paint
油漆调和料 varnish extender;extender pigment
油漆调和颜料 extender pigment
油漆调色工 toner
油漆桶 paint kettle;paint pail;paint-pot
油漆桶凹边 chimb;chime
油漆涂层 oil paint coat;paint coating
油漆涂底 clearcole
油漆涂绘手法 paint display
油漆褪光 mattness
油漆褪色 blushing
油漆未干 <警告用语> wet paint
油漆污泥 paint sludge
油漆雾浊 blushing
油漆稀料 lacquer diluent;thinner
油漆稀释剂 paint thinner
油漆箱 paint dipping tank
油漆小工 painter's labo(u)rer
油漆循环系统 paint circulating system
油漆迅速干燥法 flash drying
油漆颜料 colo(u)ring pigment;paint pigment
油漆咬底 picking up
油漆溢出 paint spillage
油漆硬结 caking
油漆用脚手架或吊架 paint stage
油漆用矿物质溶剂 mineral solvent for paints
油漆罩面 finish coating;paint finish
油漆之前灰浆底层 gesso
油漆制造厂 paint manufacturer
油漆制造工业 paint manufacturing industry
油漆制造和施工用石脑油 varnish maker's and painter's naphtha
油漆中间涂层 flat coat
油漆皱皮 orange peel(ing)
油漆装饰 painted decoration
油漆装饰工 painter and decorator
油漆装饰面层 painted decorative finish
油漆作坊 paint shop
油漆作业 paintwork
油漆做法 paint system
油歧管 oil manifold
油歧管阀 oil manifold valve
油起动器 oil starter
油气 oil gas;oil vapo(u)r
油气柏油 oil-gas tar
油气爆炸物 fuel-air explosive
油气泵 oil air pump
油气比(率) gas-oil ratio;fuel-air ratio;gas factor;gas-to-oil ratio
油气藏 oil and gas pool
油气藏底界埋深 burying depth of pool base
油气藏顶界埋藏深度 burying depth of pool top
油气藏分布区 distribution area of oil and gas pool
油气藏分类 classification of pool
油气藏平面图 planimetric(al) map of oil and gas pool
油气藏破坏和油气再分布 destruction of pool and oil-gas redistribution
油气藏剖面图 sectional drawing of oil and gas pool
油气藏图示 pool diagram
油气藏图示及参数 pool diagram and its parameter
油气层等级 grade of oil and gas bed
油气层容积 volume of oil and gas bed
油气储集层类型 type of reservoir of petroleum and gas
油气储量 oil and gas reserves
油气储量计算 computation of oil and gas reserve
油气处理储存船 gas and oil process and storage tanker
油气处理设备 gas and oil pressing equipment
油气地球化学对比 geochemical correlation of oil and gas
油气动式压力盒 oil-pneumatic pressure cell
油气发生器 oil-gas generator
油气分界面 gas-oil interface
油气分离罐 catch pot
油气分离器 air-oil separator;gas-oil separator;gas trap;oil and gas separator;oil field separator; oil separator
油气分离站 gas and oil separating plant
油气分离装置 gas and oil separating plant
油气过渡带厚度 thickness of oil-gas transition zone
油气合用的 airdraulic
油气化 oil gasification
油气化法 oil-gas process
油气化炉 oil gasifier
油气化探 geochemical exploration for oil and gas
油气换热器 oil-to-air heat exchanger
油气混合比指示器 fuel-air ratio indicator
油气(混合)井 oil and gas well
油气混输泵 oil-gas transfer pump
油气混用燃烧器 combination firing burner
油气减震器 oleo-pneumatic shock absorber
油气检测平面图 plan of hydrocarbon detection
油气焦油 oil-gas tar
油气焦油的沥青 oil-gas tar pitch
油气焦油脂 oil-gas tar pitch
油气校正 hydrocarbon correction
油气接触(面) gas-oil contact;oil-gas contact
油气界面 gas-oil surface; oil gas interface
油气界面张力 boundary tension between oil-gas contact
油气井产物 well effluents
油气聚集大带 big zone of oil and gas accumulation
油气聚集带和趋向带 oil-gas accumulation and trend zone
油气聚集和油气藏 oil-gas accumulation and pool
油气聚集作用 accumulation process of oil and gas
油气勘探 oil-gas exploration; oil-gas prospecting
油气扩散真空泵 oil diffusion pump
油气联合燃烧器 combination gas and oil burner
油气两用机 alternative-fuel engine
油气两用燃烧器 alternative multi-fuel burner;dual-fuel burner
油气盆地 oil-gas basin
油气侵 oil and gas cutting;oil and gas inflow
油气驱除设备 tank gas-freeing installation
油气省 oil and gas province
油气水混层 oil-gas-water mixed bed
油气水物性参数 physical property parameter of oil/gas and water
油气体干燥法(木材) oil vapo(u)r process
油气田 oil and gas field
油气田边缘租地 edge lease
油气田的储集层岩性分布 distribution of oil and gas field by lithology of reservoir
油气田的深度分布 distribution of oil and gas field by depth
油气田的时代分布 distribution of oil and gas field by geologic(al) age
油气田放射晕 radio-haloes of combination gas field
油气田分布 distribution of oil and gas field
油气田分类 classification of oil and gas field
油气田规模 size of oil and gas field
油气田开发地质 development geology of oil-gas field
油气田水 oil-gas field water
油气田油气聚集带及其他油气聚集单元 oil-gas field accumulation and other unit
油气田在盆地中位置的分布 distribution of oil and gas field by position in basin
油气同层 oil-gas bed
油气显示 showing of oil and gas
油气显示和固体沥青 oil-gas show and solid bitumen
油气悬挂装置 oil-pneumatic suspension
油气悬架【机】 oleo-pneumatic suspension
油气硬沥青 oil-gas tar pitch
油气源对比 oil-gas and source rock correlation
油气远景评价方法 evaluating method

for oil-gas-bearing prospect
油气运移 oil-gas migration
油气转化率 transformation ratio
油气钻井 drilling
油汽空间 oil-vapo(u)r space
油铅油漆 lead-and-oil paint
油枪 grease pressure gun;hand lubricator;lubricating press;oil gun;oil pistol;oil syringe
油枪口 external burner lip
油枪润滑 shot lubrication
油枪式黏[粘]度计 grease gun viscometer
油枪嘴 gun adapter
油腔 oil reservoir;oil sump
油强化 oil quenching
油清漆 oil varnish
油球 globule of oil
油球体 <沥青乳液中的> oil drippings;oil globule
油驱动电梯 hydraulic elevator
油圈 scraper ring
油圈闭 oil trap;trap
油泉 oil-spring
油燃料 oil fuel
油燃料泵 oil fuel pump
油燃料仓库 oil fuel depot
油燃料罐 oil fuel tank
油燃料输送泵 oil fuel transfer pump
油燃烧 oil combustion
油燃烧器 oil burner;oil-fired burner;oil firing burner
油燃烧器的点火装置 ignition equipment of oil burner
油热处理硬质(纤维)板 oil-tempered hardboard
油热处理油质(纤维)板 oil-tempered hardboard
油热法 oil heating
油热炼过度 over cooking of oil
油热设备 oil heating installation
油热式沥青贮仓 hot oil heating asphalt storage
油热水器 oil-water heater
油容量 oil capacity;oil volume
油容器 oil vessel
油溶的 oil-soluble
油溶黑 nigrosine
油溶红 oil red
油溶剂 oil solvent
油溶胶 oleosol
油溶解度 oil solubility
油溶气 dissolved gas in oil
油溶染料 fat colo(u)r;oil-soluble dyes
油溶树脂 oil-soluble resin
油溶性 oil solubility
油溶性表面活化剂 oil-soluble surfactant
油溶性防腐剂 oil-borne preservative;oil-solvent preservative
油溶性酚醛树脂 oil-soluble phenotic resin
油溶性抗氧化剂 oil-soluble inhibitor
油溶性清净分散剂 oil-soluble detergents
油溶性染料 oil colo(u)r;oil-soluble dye stuff
油溶性溶剂 oil-dissolving solvent
油溶性树脂 oil-soluble resin
油溶性颜料 oil colo(u)r
油溶胀 oil swell
油熔断器 liquid fuse unit;oil-break fuse
油熔丝断路器 oil-fuse cutout
油鞣 oil tanning
油鞣废油 moellon
油鞣革 chamois leather;oil tannage;shammy;wash leather
油鞣余物 degras;moellen
油乳化泥浆 oil emulsion mud
油乳化液 oil emulsion
油乳化佐剂 oil emulsion adjuvant
油乳剂 oil emulsion
油乳胶 oil emulsion
油乳胶浆 oil emulsion mud
油乳胶组成 oil emulsion composition
油乳水泥 oil emulsion cement
油乳液 fat liquor
油乳浊液 oil emulsion
油乳浊液浸渍处理 oil dipping
油乳佐剂 oil emulsion adjuvant
油入口 oil-in;oil inlet
油润 lubrication
油润滑 oiling;oil lubrication
油润滑管路 oil lubricating piping
油润滑器 oil lubricator
油润滑轴承 oil-lubricated bearing
油润湿的磨石 oil-wetted sharpening stone
油润湿性能 oil wettability
油塞 oil plug
油塞孔 oil plug holing
油塞移动增压油枪 grease gun with movable plunger
油散货混装船 combination carrier
油散货矿石三用船 oil-bulk-ore carrier
油散矿兼用船 ore/bulk/oil carrier
油散热器 oil radiator
油刹车 hydraulic brake;oil-pressure brake
油砂 <拌沥青的砂> oil-bonded sand;oil(ed) sand
油砂层 pay sand
油砂加工厂 oil sand processing plant
油砂芯 oil-bonded core;oil-core
油筛堵塞 oil-screen clogging
油杉属 <拉> Keteleeria
油栅 oil fence
油栅管 oil boom tubing
油商 oilman
油勺 oil dipper;oil scoop;oil skipper
油渗出 oil seepage
油渗透率 oil permeability
油生成泡沫 oil foaming
油绳 oil-saturated hemp rope;oil string;oil wick;wick
油绳虹吸 wick siphon [syphon]
油绳接头 oil wick adapter
油绳润滑 oil wick lubrication;wick lubrication
油绳润滑器 wick lubricator;wick oiler
油湿式离合器 oil-type clutch
油湿式制动器 oil-type brake
油石 abrasive stick;buhrstone;gouge slip;grinding slip;hone;Indian stone;oil coated stone;oilslip;oil stone;oilstone slip;sharpener;strickle;whetstone
油石比【道】 asphalt-aggregate ratio;bitumen-aggregate ratio
油石比自动控制装置 asphalt-aggregate ratio auto-device
油石粉 oil stone dust;oil stone powder
油石珩磨机 honing machine
油石磨孔机 honing machine
油使用期限 life-span of oil
油示踪法 oil tagging system
油势 oil potential
油势梯度 oil potential gradient
油室 grease chamber;oil cavity;oil chamber
油室口【机】 aperture of oil chamber
油收集器 oil collector;oil interceptor;oil scoop;oil trap
油输送 oil-transferring
油树油 yuju oil
油树脂 oleoresin;wood oil
油树脂黏[粘]合剂 oleo-resinous adhesive
油树脂清漆 oleoresin varnish
油树脂体 oil resinite
油刷 oil brush
油刷润滑的 brush-lubricated
油栓子 oil-embolus
油霜 bloom
油水比 oil-water ratio;water factor
油水边界 oil-water boundary
油水边界异常值 edge value
油水舱布置 tankage arrangement
油水舱管系布置 tank piping
油水沉渣 oil-water sludge
油水淬火槽 oil-and-water quenching tank
油水分界面 oil-water surface
油水分离 oil-water separation
油水分离槽 oil-separating tank
油水分离剂 oil-water separation agent
油水分离器 oil-and-moisture trap;oil-and-water separator;oil-and-water trap;oil catcher;water (-and)-oil separator;deoiler
油水分离设施 oil and water separation device
油水分离系统 oil-water separation system
油水分离装置 oil-and-water separating equipment
油水柜存量表 sounding table
油水柜存量图 sounding diagram
油水柜内实际油水量的测量 innage
油水过渡带厚度 thickness of oil-water transition zone
油水耗量 oil-water consumption
油水换热器 oil-to-water exchanger
油水混合物 oil-water mixture
油水积储缸 sump reservoir
油水接触面 oil-water boundary;oil-water surface
油水界面 oil-water boundary;oil-water interface;water-oil interface
油水界面测定器 interface detector;oil-water interface detector
油水界面探测仪 interface detector;oil-water interface detector
油水界面张力 boundary tension between oil-water contact
油水阱 oil-and-water trap
油水两相渗流模型 oil-water two-phase flow model
油水面 oil-water contact;oil-water surface
油水浓度计 oil concentration analyser [analyzer]
油水容量 fluid capacity
油水溶胶 oil hydrosol
油水乳化液 water-oil emulsion
油水收集器 oil-and-water trap
油水同层 oil-water bed
油水同层厚度 thickness of oil-water layer
油水系统接触角 contact angle of oil-water system
油水系统界面张力 interfacial tension of oil-water system
油水相乳化液 <油为内相,水为连续外相> oil-in-water emulsion
油松 Chinese pine;Pinus Sinensis
油酸 olei(ni)c acid;red oil
油酸铵皂化的石蜡油 petrox(olin)
油酸钡 barium oleate
油酸苯汞 phenylmercuric oleate
油酸铋 bismuth oleate;oleo-bi
油酸捕收剂 oleic collector
油酸的 oleic
油酸丁酯 butyl oleate
油酸钙 calcium oleate
油酸根【化】 oleate
油酸汞 mercuric oleate;mercury oleate
油酸甲酯 methyl oleate
油酸铝 alumin(i)um oleate
油酸镁 magnesium oleate
油酸锰 manganese oleate
油酸木溜油 oleocreosote
油酸钠 natrium oleinicum;sodium oleate
油酸钠凝胶 sodium oleate gel
油酸钠皂 sodium oleate soap
油酸铅 lead oleate;plumbic oleate
油酸铅硬膏 lead oleate plaster
油酸三乙醇胺酯 triethanolamine oleate
油酸同系物 oleic acid series
油酸铜 copper oleate
油酸锌 zinc oleate
油酸盐 oleate
油酸乙酯 ethyl oleate
油酸酯 oleate;olein
油酸制剂 oleate;oleatum
油台 tank deck
油台浮子 tank deck float
油态脱模剂 oil phased mo(u)ld cream
油炭 parrot coal
油炭质残渣 oil-carbon sludge
油碳 oil carbon
油套 oil jacket;oil sleeve
油提灯 oil lantern
油提取器 oil extractor
油提升管 oil riser
油体 oil body
油田 oil deposit;oil field
油田泵 oil field pump
油田边缘的油井 outstep boring
油田成本 exoil field cost
油田出水来源 source of water troubles
油田储藏量 oil reserves
油田的开发 development of a well
油田地下水 oil-field water
油田废水 oil field waste(water)
油田废水处理 oil field wastewater treatment
油田构造图 dip-arrow map
油田含油废水 oil field oil-bearing wastewater
油田和气田开发 oil and gas field exploitation
油田开采权图 map of oil field concession
油田开发初期 early field life
油田卤水 oil field brine
油田每日产量 field potential
油田模型 oil field model
油田气 gas field;oil field gas
油田区域 oil patch
油田乳化油 oil field emulsion
油田设备 oil field equipment
油田水 oil field brine;oil-field water
油田水按盐度分类 classification of oil field water according its salinity
油田水比重 specific gravity of oil field water
油田水产状 occurrence of oil field water
油田水常规分析数据 usage analytic(al) data of oil field water
油田水导电率 specific electric(al) conductance of oil field water
油田水分析数据 analytic(al) data of oil field water
油田水化学剖面图 hydrochemistry profile for oil field
油田水化学图 hydrochemical map for oil field

油田水化学组成 chemical composition of oil field water
油田水化学组成板式图解 stiff diagram shoeing chemical composition of soil field water
油田水化学组成直方图 bar graph shoeing chemical composition of oil field water
油田水净化剂 purification agent
油田水来源 source of oil field water
油田水类型 type of oil field water
油田水黏[粘]度 viscosity of oil field water
油田水其他分类 other classification of oil field water
油田水酸碱度 pH value of oil field water
油田水同位素组成 isotopic composition of oil field water
油田水文地质调查 hydrogeologic(al) survey of oil field
油田水文地质图 hydrogeologic(al) map for oil field
油田水物理参数 physical parameter of oil field water
油田水物理性质 physical property of oil field water
油田水专项分析数据 special analysis data of oil field water
油田特性 field performance
油田图 field map
油田压力 field pressure
油田盐水 oil field brine
油田用泵 field pump
油田用管材设备 oil country tubular goods
油田贮罐 oil field tank
油田租出协议 farm out agreement
油田作业自动化 field automation; oil field automation
油填料 oil packing
油调着色剂 oil stain
油桐 Aleurites fordii; tung (oil) tree; wood oil tree
油桐属 <拉> Aleurites
油桐树 Chinese wood-oil tree
油桐籽 bancoul nuts
油桶 barrel; oil barrel; oil drum; oil vessel
油桶泵 barrel oil pump; drum pump
油桶仓库 barrel-house
油桶打油泵 oil barrel pump
油桶箍条 tank strap
油桶架 drum cradle; oil drum stand
油桶口 bunghole
油桶手摇泵 barrel oil pump
油桶提闩 tank toggle
油桶再装油 refilling of oil tank
油桶照明灯 drum lighting lamp
油头 oil injector; sprayer
油涂料 oil paint
油土 <用铺路油类或沥青处治过的土> oiled earth
油土基层 <混凝土路面的> oiled clay base
油土路 oiled earth road; oiling earth road
油土路面 oiled earth surface
油土面层 <用沥青或铺路油处治过> oiled earth surface
油土筑路法 oiled earth construction
油团 oil puddle
油团浮选 bulk-oil flo(a)tation
油腿 oil-leg
油脱蜡 oil dewaxing
油脱色 oil bleaching
油脱水装置 oil dehydrating plant
油网过滤装置 viscous screen filter
油纬 oiled pick
油位 oil level
油位报警试验 oil level alarm test
油位表 fuel (level) ga(u)ge; oil hole; oil level(l)er; oil (level) ga(u)ge
油位表玻璃 oil level ga(u)ge glass
油位玻璃管 ga(u)ge glass; oil ga(u)ge glass
油位玻璃托 ga(u)ge glass bracket
油位测量杆 oil level rod
油位测量计 oil-depth ga(u)ge
油位观察玻璃 oil sight glass
油位观察孔 oil sight glass
油位管 oil level pipe
油位计 oil ga(u)ge; oil level ga(u)ge
油位计玻璃 oil level ga(u)ge glass
油位监视标志 level filler
油位塞 fuel level line plug
油位调节器 oil level regulator
油位指示器 fuel indicator; fuel level ga(u)ge; oil level indicator; oil-measuring ga(u)ge
油温保护 oil temperature protection
油温表 oil temperature ga(u)ge; oil thermometer
油温操作的 oil temperature operated
油温度 oil temperature
油温度计 oil temperature ga(u)ge
油温计 oil thermometer
油温切断器 oil temperature cut-out
油温上升解脱器 oil temperature high trip device
油温调节器 oil temperature regulator
油温自动调节器 automatic oil temperature regulator
油纹 oil stripping
油窝 oil socket
油污 dirt; greasy dirt; gunge; oil contamination; oil failure; oil slick; oil soil; oil stain
油污斑渍 oil discolo(u)ration
油污带 sleek field
油污的 greasy; oil polluting; stained with oil
油污垢 oil foulant
油污柜 oil drain tank
油污货物 dirty cargo
油污迹 oil stain
油污染 oil contamination
油污染残渣 oil pollution residue
油污染的 oil polluted
油污染地表水 oil-contaminated surface water
油污染防范措施 countermeasures for oil pollution
油污染防治 oil pollution control
油污染公约 convention on oil pollution
油污染海滨 oil polluted seashore
油污染海鸟 oiled birds
油污染环境治理 oil pollution environmental control
油污染监视系统 oil pollution surveillance system
油污染检测 oil pollution detection
油污染控制 oil pollution control
油污染试验 oil pollution experiment
油污染水体 oil polluted waters
油污染水域 oil polluted waters
油污染损害 oil pollution damage
油污染物测定 detection of oil pollution
油污染应急计划 oil pollution emergency plan
油污染指示剂 oil pollution indicator
油污染致命性指数 oil vulnerability index
油污收集消防船 oil skimming and fire boat
油污水汇集柜 oil bilge collecting tank
油污条痕 oil streak
油污消除剂 sludge breaker
油污自动平衡刮集装置 self-level(l)ing unit for removing pollution
油雾 oil fog; oil mist; oily moist
油雾捕集 oil mist trapping
油雾发生器 oil fog generator
油雾化 oil atomization; oil fogging
油雾化器 atomizer; oil atomizer; oil fogger
油雾监测 oil mist monitoring
油雾检测 oil mist detection
油雾检测器 oil mist detector
油雾滤清器 oil mist filter
油雾喷射器 oil sprayer
油雾润滑 atomizer lubrication; oil fog lubrication; oil mist lubrication
油雾润滑器 oil fog lubrication
油雾润滑系统 oil mist system
油雾探测 oil mist detection
油雾探测器 oil mist detector
油雾系统 oil mist system
油吸试验 oil absorption test
油吸收法 oil absorption process
油吸收剂 oil absorber
油吸收塔 oil absorption tower
油吸值 oil absorption (value)
油吸作用 oil absorption
油烯基 oleyl
油烯基硫酸酯 oleyl sulfate
油烯基羧酸盐 oleyl carboxylate
油洗 oil wash
油洗涤器 oil scrubber
油洗装置 oil-washing apparatus
油系统 distributed system
油隙 oil clearance
油显剂 <浮选用> emulsol
油显示 oil show
油线 oil string
油腺 oil droplet
油香树脂合剂 mistura oleobalsamica
油箱 fluid reserve; fuel tank; oil basin; oil box; oil case; oil container; oil reservoir; oil tank; sump; breather <电缆用>
油箱壁接头 tank-wall connector
油箱残渣 tank sludge
油箱舱 tank bay; tank space
油箱车 oil tank wagon
油箱沉积物 oil cell bottom
油箱抽气器 oil tank gas extractor
油箱出口 fuel tank outlet
油箱储油量指示器 tank ga(u)ge
油箱底 reservoir bottom; tank bottom
油箱底渣 oil cell bottom
油箱底座 oil tank base support
油箱吊架 fuel tank hanger
油箱盖 filler cap; fuel tank cap; tank cup
油箱管 tank line
油箱呼吸器 oil tank breather
油箱护板 fuel tank guard
油箱回油 tank drainback
油箱加油口 fuel tank filler
油箱接线头 <接地用> tank unit terminal
油箱壳体 tank envelope
油箱口 reservoir port
油箱冷却器 tank colder; tank cooler
油箱龙筋 oil box type spindle rail
油箱漏斗 water coning in oil reservoirs
油箱螺盖 tank screw-cap
油箱内剩余燃料 outage
油箱气孔通道 tank air vent access
油箱容积 fuel tank capacity
油箱容量 fuel tankage; tankage; tank capacity
油箱润滑 tank lubrication
油箱润滑器 tank lubricator
油箱润滑油稀释 tank dilution
油箱散热器 oil tank radiator
油箱室 tank room
油箱调整 tank truing
油箱通风孔 fuel tank vent
油箱通气管 oil tank breather; tank breather pipe
油箱压扁 tank collapsing
油箱压力 tank pressure
油箱已满指示 tank full indication
油箱英里程 tank mileage
油箱英里数 tank miles
油箱用滤油器 reservoir filter
油箱油量表 tank fuel indicator
油箱油量指示器 tank contents indicator
油箱油面传感器 tank-level sensor
油箱油位 fuel level in tank
油箱增压 fuel tank pressurization
油箱支架 fuel tank bracket; tank hanger
油箱中扩散管 diffuser tube in tank
油箱总成 full oil tank assembly
油相 oil wet phase
油相对渗透率 oil relative permeability
油相消失带 oil phase-out zone
油消光 oil delustering
油楔 oil film wedge; oil wedge
油楔作用 oil wedge action
油芯 oil-core; oil wick; wick
油芯管 oil core tube
油型 oil type
油型防腐剂 oil-type preservative
油型防锈剂 oil-type rust preventive
油型分析 oil-type analysis
油性 oiliness; unctuosity
油性瓷漆 oil enamel
油性打底漆 oil primer
油性的 unctuous
油性底层涂料 bed oil paint; bench oil paint; oil primer
油性底漆 oil(-based) primer
油性底漆和二道漆两用漆 oil primer surfacer
油性底漆涂层 oil primer coat
油性堵(塞)缝 oil ca(u)lking
油性度 oiliness degree
油性二道浆 oil surfacer
油性二道漆 oil surfacer
油性防腐剂 oil(-borne) preservative
油性废水 oily wastewater
油性光漆 oil gloss paint
油性过滤器 oil filter
油性剂 oiliness agent
油性沥青漆 black japan
油性膜 oiliness film
油性腻子 oil putty; oil stopper
油性黏[粘]土 fat clay; soapy clay; unctuous clay
油性墙粉 oil-bound distemper
油性清漆 long oil varnish; oil varnish
油性染色剂 oil stain
油性润滑剂 oily lubricant
油性色粉涂饰 oil-bound distemper
油性色剂 <用于木料着色> oil stain
油性色浆 colo(u)r in oil; paste in oil; pigment oil stain
油性色料 oil colo(u)r
油性色漆 oil paint
油性试验 oiliness test
油性试验器 oiliness tester
油性树脂漆 oleo-resinous paint
油性树脂清漆 oleo-resinous varnish
油性树脂色漆 oleo-resinous paint
油性树脂涂料 oleo-resinous coating
油性水彩颜料 oil-bound water paint
油性添加剂 oiliness additive; oiliness agent; oiliness improver
油性填充剂 oil-extender
油性填缝料 oil filler
油性填孔剂 oil filler

油性填料 oil stopper
油性调和漆 oil-prepared paint
油性涂料 oil paint
油性污水 oily sewage
油性污物 oily pollutant
油性载体 oiliness carrier
油性遮盖力 oiled hiding power
油性指标 oiliness index
油性中层漆 oil surfacer
油性着色剂 oil stain
油性着色料 oil stain
油旋管 oil coil
油旋塞 oil cock
油旋转泵 oil rotary pump
油选溜槽 greased sluice
油选台 greased table
油循环 oil circulation
油循环反应器 oil circulating reactor
油循环过程 oil circulation process
油循环加热 oil circulation heating
油循环冷却 oil circulation cooling
油循环润滑(法) lubrication by oil circulation; oil circulating lubrication
油循环润滑系统 oil circulation lubricating system
油循环系统 oil circulating system; oil circulation system
油压安全阀 oil-pressure relief valve
油压安全阀盖 oil-pressure relief valve cap
油压安全阀簧 oil-pressure relief valve spring
油压安全阀螺母 oil-pressure relief valve nut
油压安全阀密封垫 oil-pressure relief valve plug gasket
油压安全阀塞 oil-pressure relief valve plug
油压安全阀体 oil-pressure relief valve body
油压安全系统 oil-pressure safety system
油压保持性 oil-pressure retention
油压保护开关 oil protection switch
油压保护装置 oil-pressure cut-out protection
油压报警器 oil-pressure warning unit
油压泵 oil hydraulic pump
油压表管路 oil-pressure ga(u)ge pipe
油压表进油管 oil-pressure ga(u)ge feed pipe
油压表连接管 oil-pressure ga(u)ge connection
油压表油管 oil ga(u)ge pipe
油压操舵机调整 telemotor adjustment
油压操舵机用油 telemotor oil
油压操舵机油管路 telemotor circuit
油压操舵系统 telemotor system
油压操作的 oil-pressure operated
油压操作电压调整器 oil-operated voltage regulator
油压传动 hydraulic transmission; oil-operated transmission
油压传动装置 oil gear
油压的 oil hydraulic
油压的调节螺丝 oil-pressure adjusting screw
油压低 low oil pressure
油压电缆 oil-ostatic(al) cable
油压顶起装置 <向轴承注高压油的泵> jacking oil pump
油压动力室 hydraulic pressure engine room
油压动力系统 oil hydraulic power system
油压断缆器 oil-pressure cutter
油压断流阀 oil-trip valve
油压阀 fuel pressure valve; oil-pressure valve
油压放松阀 oil-pressure release valve
油压缸 oil cylinder; oil hydraulic cylinder; oil-pressure reservoir
油压工程 oil hydraulic engineering
油压管 oil-pressure pipe; oil-pressure tube
油压缓冲叉 oleo fork
油压缓冲器 liquid damper; oil buffer; oil-pressure fender; oleo-buffer; oleo-cushion; oleo-damper
油压回路 oil hydraulic circuit
油压机 hydraulic press; hydraustatic press; liquid press; oil hydraulic press; oil press; oleodynamic (al) press
油压挤塑机 hydraulic extruder
油压计 fuel pressure ga(u)ge; oil manometer; oil-pressure ga(u)ge
油压继电器 oil-pressure relay
油压减压阀 oil-pressure reducing valve
油压减震器 hydraulic damper; hydraulic shock absorber; oil shock absorber; oil (-type) buffer; oleo-damper; oleo-gear; oil-pressure damper
油压减震器外罩 oil damper casing
油压减震柱 oleo leg; oleostrut
油压降低解脱器 oil-pressure low trip device
油压警告信号 oil-pressure warning signal
油压卡盘 oil chuck
油压开关 oil-pressure cut-out; oil-pressure shut-off switch; oil-pressure switch
油压开关螺丝 oil-pressure switch screw
油压开式调速器系统 <涡轮机的> open-governor oil pressure system
油压控制 fuel pressure control; oil control; oil-pressure control
油压控制继电器 oil-pressure control relay
油压控制系统 oil-pressure control system
油压离合器 oil clutch
油压力 hydraulic pressure; oil-pressure
油压力泵 oil-pressure pump
油压力表 oil-pressure ga(u)ge
油压力计 oil manometer
油压马达 hydraulic motor; oil hydraulic motor
油压起重器 pressure oil lift
油压气动补偿器 oleo-pneumatic compensator
油压气动悬挂装置 oleo-pneumatic suspension system
油压千斤顶 hydraulic jack; oil jack
油压千斤顶张拉 hydraulic jack prestressing
油压强度试验 oil-pressure test for strength
油压切断器 oil failure switch; oil-pressure cut-out
油压燃烧器 oil-pressure burner
油压刹车 oil brake
油压烧嘴 oil-pressure burner
油压设备 oil-pressure supply system; oil-pressure unit
油压式防舷材 oil jack fender
油压式滚筒定位 hydraulic roll balancing
油压式缓行器 oil-pressure unit type car retarder
油压式伺服电动机 oil-type servomotor
油压式自动换挡 hydromatic drive
油压事故继电器 oil-pressure fault relay
油压试验 hydraulic pressure test
油压室 hydraulic pressure cell
油压调节 oil-pressure adjustment
油压调节阀 oil-pressure adjusting valve
油压调节螺钉 oil-pressure adjusting screw
油压调节器 oil-pressure regulator
油压调节系统 oil-pressure governing system
油压调整装置 oil-pressure control unit
油压推杆式操舵装置 telemotor steering gear
油压稳定器 oil-pressure stabilizer
油压雾化喷燃器 oil-pressure atomizing burner
油压系统 oil hydraulic system; oil pipe control system; oil-pressure system; pressurized oil system
油压楔 pressure wedge
油压泄放阀 oil-pressure release valve
油压型雾化器 oil-pressure type atomizer
油压选择器 oil-pressure selector
油压压紧装置 oil gag
油压油泵 oil booster pump
油压闸 oil-pressure brake
油压遮断器 oil-pressure shut-off switch
油压真空 hydrovacuum
油压真空制动器 hydrovacuum brake; hydro-vac
油压支承 oil lift
油压止回阀 oil-pressure check valve
油压指示灯 oil-pressure indicator lamp
油压指示器 oil-pressure indicator
油压制动 oil brake; pressurized oil braking; oil check
油压制动缸 oil hydraulic braking cylinder
油压制动器 oil-pressure brake
油压制动装置 oil brake
油压轴承 oil-pressure bearing
油压转盘钻机 rotary drills with hydraulic
油压转向装置 hydraguide
油压装置 oil-pressure installation; oil-pressure supply system; oil-pressure unit; pressure oil device
油压自动控制传动装置 hydramatic transmission
油压自动式 hydramatic
油压阻尼器 oil shock absorber
油压组式减速器 oil-pressure unit type car retarder
油压钻机 hydraulic feed drill
油烟 chimney soot; lamp black; oil fume; oil smoke; vegetable black
油岩 oil rock
油研磨 oil grinding
油眼 oillet(te); oil well; resin pocket; pitch pocket <木材缺陷>
油焰 oil flame
油氧化后生成的有害物 objectionable oil oxidation products
油样 oil sample
油样分离 separation of oil sample
油样化验报告 report of oil analysis
油样收集器 drip cup
油窑 oil-fired furnace
油页岩 asphalt bearing shale; bitumenite; bituminous rock; bituminous shale; coal stone; combustible shale; dunnet shale; kerosene coal; kim coal; oil-bearing shale; oil shaker; oil shale; petroleum shale; pil shale; pyroshale; resinous shale; tasmanite; wax shale; kuchersite <产于波罗的海>
油页岩尘肺 bituminite pneumoconiosis
油页岩干馏 oil-shale retorting; oil-shale distillation
油页岩干馏厂 oil-shale retorting plant
油页岩干馏釜 oil-shale retort
油页岩胶凝材料 oil-shale cement
油页岩焦油 oil-shale tar
油页岩矿渣 oil-shale slag
油页岩炼制过程 oil-shale eduction process
油页岩群 oil-shale group
油页岩烧制集料 <一种拌制水泥混凝土用的轻集料> burned oil-shale aggregate; burnt oil-shale aggregate
油页岩石蜡 shale paraffin(e)
油页岩水泥 oil-shale cement
油页岩渣 oil shale waste
油液 dubbin; dubbing
油液减振装置 dynamic(al) oil-damper; oil damper
油液降解 oil degradation
油液净化装置 oil cleaning device; oil reconditioner
油液空气减震支柱 oleo-pneumatic strut
油液空气减震柱 oleostrut
油液流点 pour point
油液罗经 oil compass
油液黏[粘]度 viscosity grade
油液散热器 oil radiator
油液弹簧减震柱 oleo spring shock (absorber) strut
油液吸振装置 dynamic(al) oil-damper
油液橡皮减震柱 oleo rubber shock (absorber) strut
油液支柱 oildraulic strut
油易染的 oil-susceptible
油溢测量 oil-spill detection
油溢流阀 oil-overflow valve; oil relief valve
油溢消除 oil-spill remover
油引聚合 bodying of oil
油印 mimeograph; printing by stencil; stencil
油印标记 grease pit
油印池 copygraph
油印机 duplicator; mimeograph; mimeograph machine; multigraph
油印机室 mimeograph room
油印蜡纸 stencil; stencil paper
油印品 mimeo(graph)
油印纸 multigraph paper
油硬脂 beef stearin; oleostearin(e)
油硬脂酸盐 oleostearate
油用温度计 oil dual thermometer
油油对比 oil and oil correlation
油油苗对比 oil and show correlation
油/油脂废水 oil/grease wastewater
油淤 oil sludge
油与白土的混合物 oil-clay mixture
油与氢气的混合物 oil-hydrogen mixture
油与溶剂的混合物 oil-solvent blend
油与水的界面 oil-water boundary
油与水之间的交界面 oil-water interface
油浴齿轮传动 oil bath gear
油浴电磁线圈 oil-immersed solenoid
油浴电阻炉 oil bath resistance furnace
油浴法 oil bath oiling
油浴(锅) oil bath

油浴过滤器 oil bath filter
油浴回火 oil tempering
油浴技术 oil bath technique
油浴空气过滤器 oil-type air cleaner
油浴灭菌法 sterilization by oil bath
油浴盘式停车闸 oil disc parking brake
油浴盘式停车制动器 oil disc parking brake
油浴盘式制动器 oil-disc brake
油浴器 oil bath
油浴润滑 bath lubrication; oil bath; oil bath lubrication
油浴润滑式轴承 oil bath type bearing
油浴式车闸 oil-type brake
油浴式齿轮箱 oil bath gearbox
油浴式多片盘式制动器 oil-immersed multiplate disk brake
油浴式空气净化器 oil bath type air cleaner
油浴式空气滤清器 oil bath air cleaner; oil bath filter; oil-wetted air cleaner; on-bath type air cleaner
油浴式离合器 oil bath clutch; oil(-disc) clutch; wet clutch
油浴式滤清器通气管 oil bath air breathing
油浴式制动器 oil-type brake
油浴制动器 oil-immersed brake
油源 oil seepage; oil show; oil sources
油源丰度 richness of oil source
油源区距离 distance from source area
油源岩 oil source rock
油再生 oil recovery
油再生法 pan process
油再生性 oil reactivity
油在水中弥散 oil-in-water dispersion
油皂杀虫剂 oil and soap insecticide
油渣 foots; grease residue; oil ballast; oil deposit; oil foot(ing); oil refuse; oil residue
油渣饼碎片 oil meal
油渣促进剂 sludge promoter
油渣和水 bottom sludge and water
油渣形成 sludge occurrence
油渣抑制剂 sludge inhibitor
油闸 <发电厂及输配电线用> oil switch
油炸(煎)锅 fryer
油毡 asphaltic belt; asphaltic felt; bitumen felt; bituminous felt; malthoid; pitched felt; roofing felt; roof paper; saturated felt; sorbing felt; tarred felt
油毡板 board of felt(ed fabric); lincout
油毡不透水性 water impermeability of malthoid
油毡衬 sarking felt
油毡衬背 felt back(ing)
油毡衬层 underslating felt
油毡衬底 felt underlining; felt undersarking
油毡衬垫 bitumen felt packing; felt washer pad; sarking
油毡衬里 lino(leum) lining
油毡床面 lino(leum) deck
油毡大气稳定性 atmosphere resistance of malthoid
油毡刀 lino(leum) knife
油毡底层 lino(leum) base
油毡垫 oil felt pad
油毡调节装置 membrane regulating device
油毡钉 clout nail; felt nail; felt washer; roofing nail; tack
油毡豆石屋面 felt and gravel roof(ing)
油毡断裂 rupture of felt
油毡防潮层 felt damp-proof course
油毡防霉性 anti-fungal property of malthoid
油毡防水层 waterproof asphaltic-felt
油毡防雨层 fabric flashing
油毡覆盖层 felt carpet
油毡覆盖面层 felt carpet
油毡覆盖屋面 felt carpet roof
油毡规格 lino(leum) ga(u)ge
油毡护面 lino(leum) cover
油毡基布 roofing fabric; woven cloth for asphalt felt
油毡级滑石 asphalt-felt-grade talc
油毡甲板 lino(leum) deck
油毡浸涂总量 saturated and coated bitumen amount of malthoid
油毡抗水性 water-resistance of malthoid
油毡拉力 tensile strength of malthoid
油毡沥青 oil felt pad asphalt
油毡绿豆砂屋面 gravel roof(ing)
油毡面板条 paper-backed lath
油毡耐热度 heat-resistance of malthoid
油毡黏[粘]结剂 lino(leum)(bonding) adhesive; lino(leum) cement
油毡铺地 lino(leum) floor(ing)
油毡铺地面层 lino(leum) floor cover(ing)
油毡铺砌 felting
油毡铺贴 felting
油毡起鼓 blistering of felt; felt blistering
油毡清漆 lino(leum) varnish
油毡热稳定性 heat stability of malthoid
油毡柔度 flexibility of malthoid
油毡润滑 oil-pad lubrication
油毡撒绿豆砂屋顶 felt and gravel roof(ing)
油毡砂砾屋面 felt and gravel roof(ing)
油毡石子屋顶 felt and gravel roof(ing)
油毡收口压条 closure strip; closure strip of felt
油毡条 felt strip; stripping felt
油毡停留机 finished product looper
油毡瓦垫层 shingle underlayment
油毡屋面 asphalt roll(ed-strip) roofing; bitumen felt roof(ing); malthoid roofing
油毡吸水性 water absorption of malthoid
油毡印刷漆 lino(leum) print paint
油毡折边 nib
油毡织物片材 sheet of felt(ed fabric)
油毡纸 asphalt paper; felt paper; lincrusta
油站 oil depot
油折射计 oleorefractometer
油针 needle type valve
油真空泵 oil vacuum pump
油枕 expansion tank; oil breather; oil conservator; treated timber sleeper; treated wooden sleeper
油蒸馏器 oil still
油蒸馏塔 oil rectifier
油蒸气 oil vapo(u)r
油蒸汽泵 oil-vapo(u)r pump
油蒸汽扩散泵 oil-vapo(u)r diffusion pump
油整面涂料 oil surfacer
油症 oil induced symptom
油脂 celvacenie grease; fat; fatty oil; grease; grease and flats; lipa; lubricant; oil and grease; tallow
油脂杯 grease cup
油脂焙烧的 grease burning
油脂泵 grease pump
油脂捕集器 grease trap
油脂槽 grease tank
油脂测定器 grease tester
油脂澄清离心机 oil polisher
油脂带处理 grease-belt treatment
油脂带分选 grease-belt separation
油脂垫 grease pad
油脂废水 grease waste(water); rendering waste(water)
油脂分解 fat splitting
油脂分解菌 grease decomposition bacteria
油脂分离 grease separation
油脂分离及收集装置 grease trap
油脂分离器 fat collector separator; grease interceptor [intercepter]; grease separator; grease trap
油脂分配槽 grease distributing groove
油脂浮选池 grease flo(a)tation tank
油脂隔离器 grease trap
油脂供应 supply of lubricants
油脂罐 grease tank
油脂光泽 greasy luster [lustre]; oily luster
油脂过滤器 grease filter
油脂含量 grease content
油脂化工厂 grease chemical industry
油脂环 grease collar
油脂回收 recovery of grease
油脂剂 stearol
油脂加工厂 fatty oil processing factory
油脂截留器 grease interceptor [intercepter]
油脂截留设施 grease interceptor [intercepter]
油脂井 grease pit
油脂抗氧化剂 oil antioxidant
油脂孔 grease hole
油脂密封圈 grease retainer seal
油脂膜 greasiness
油脂喷壶 grease nipple
油脂喷射 grease shot
油脂皮带机 grease belt
油脂枪 grease compressor; grease lubricator
油脂枪喷嘴 grease nipple
油脂清除池 grease removal tank
油脂球 grease ball
油脂去除 grease removal
油脂去除器 grease remover; grease stripper
油脂染料 fat dye
油脂溶剂 fat solvent
油脂润滑剂 oiliness agent
油脂润滑器 grease lubricator
油脂生产废水 edible oil production wastewater
油脂石灰水涂刷 lime tallow wash
油脂实验室 oil laboratory
油脂试验机 grease testing machine
油脂水解车间 oil splitting plant
油脂酸 fatty acid
油脂通路 grease passage
油脂桶 grease tank
油脂涂层 grease coating; grease paint
油脂污染的 grease polluted
油脂匣 fat box
油脂性 greasiness; unction
油脂性粉刺 oil acne
油脂性基质 oleaginous base
油脂选矿 grease-surface separation
油脂选矿法 greased-surface concentration
油脂选矿机 greased-surface concentrator
油脂压力枪 grease pressure gun
油脂焰 greasy flame
油脂摇床选 grease tabling
油脂皂化 saponification of fats
油脂状冰 grease ice; ice flat; lard ice
油脂状物 grease
油脂嘴 grease nipple
油止回阀 oil check valve
油纸 building paper; pitched paper; rozin paper; saturating paper; saturation paper; oiled paper <绝缘用>
油纸皮电缆 paper cable
油指示剂 oil indicator
油指示器 oil indicator
油制动器 oil brake
油制气 oil gas
油制气废水 manufacturing gas with oil-wastewater
油质白铅 white lead in oil paste
油质的 oily; oleaginous
油质底层涂料 bed oil paint; oil primer
油质底漆 oil primer
油质废水排放 oily discharge
油质废液 oily waste liquor
油质胶结料 oil mastic
油质玛琋脂 oil mastic
油质(木材)防腐剂 oil-type preservative
油质腻子 oily putty
油质黏[粘]合剂 oil binder
油质清漆 oil varnish
油质树脂漆 oleo-resinous paint
油质树脂清漆 oleo-resinous varnish
油质体 elaiosome; oleosome
油质涂料 oil paint; unctuous paint
油质污泥 oily sludge
油质污染物 oily pollutant
油质污水 oily sewage; oily wastewater
油质无光油漆 flat oil paint
油质颜料 oil colo(u)r
油质整面涂料 oil surfacer
油滞性 oleostasis property
油中沉淀物 oil sediment
油中淬火 oil hardening; oil quenching
油中含水量测定 water-in-oil test
油中和剂 oil neutralizing agent
油中回火 blazing-off
油中水 water-in-oil
油中水型乳液 water-in-oil emulsion
油中研细(涂料) ground in oil
油中杂质 well-cuts
油中杂质引起的磨损 oil borne abrasion
油盅 lubricant plug
油盅盖 oil cup cap
油朱红 oil vermilion
油贮 oil reservoir; petroleum reservoir
油柱 column of oil
油柱高度 height of oil column
油状的 oily
油状废水 oily waste
油状乳液 oil emulsion
油状石英 greasy quartz
油状液体 oily liquid
油着色 oil-staining
油着色剂 oil stain agent
油籽 oil-bearing seeds
油渍 grease smudge; grease spot; grease stain; oil mark; oil stain; smear; oil-stained by oil
油渍险 risk of oil; risk of oil damage
油渍岩石 oil-stained rock
油棕 oil palm
油棕榈酸盐 oleopalmitate
油阻流器 oil trap
油阻尼 oil damping
油阻尼器 fluid damper; oil damper
油嘴 bean; discharge nozzle; nozzle tip; venting nipple
油嘴垫片 oil nipple gasket
油嘴管汇 choke manifold
油嘴集管 choke manifold

油嘴下流压力 pressure at downstream of choke flow line
油嘴柱 oil nipple bolt

柚木 Burma teak; common teak; Rangoon teak; teak(wood); Tectona grandis

柚木板 teak plank
柚木窗 teak window
柚木代替材料 samar
柚木甲板铺板 teakwood decking
柚木属 Tectona
柚木踏板 teak tread
柚木踏梯 teak tread
柚木梯级 teak tread
柚木装修 teak trimming
柚(子) pomelo; shaddock

疣皮桦 Swedish birch

疣齐墩果 warty olive
疣双孢锈菌属 <拉> Tranzschelia

铀234 法 234U method

铀 234 年代测定法 uranium-234 age method
铀 234 铀 238 法 234U/238U method
铀 234-铀 238 年代测定法 uranium-234 to U238 age method
铀 235-镤 231 法 uranium-235/protactinium-231 method
铀 235 铅 207 等时线 235U-207Pb isochron
铀 238 的热产率 heat productivity of U-238
铀 238 铅 206 等时线 238U-206Pb isochron
铀棒 uranium bar; uranium rod; uranium slug
铀棒栅格 uranium-rod lattice
铀玻璃 uranium glass
铀补给 uranium makeup
铀储备 uranium reserves
铀储量 uranium reserves
铀道窗宽 uranium channel window width
铀道灵敏度 sensitivity of U-channel
铀的回收 uranium recovery
铀的热产率 heat productivity of uranium
铀电解槽 electrolysis cell
铀氡镭水 uranium-radon-radium water
铀氡水 uranium-radon water
铀氡异常对比图 contrast figure of U/Rn anomaly
铀堆 uranium-fuelled reactor; uranium pile; uranium reactor; pile
铀矾 uranopilite
铀反应堆 uranium reactor
铀方钍矿 uranothorianite
铀覆盖层 uranium coating
铀钙石 liebigite
铀钙铜矿 uranochalcite
铀汞齐 uranium amalgam
铀共振 uranium resonance
铀含量 uranium content
铀含量等值图 contour map of uranium content
铀含量计 uranium content meter
铀含量平剖图 profile on plane of uranium content
铀含量异常 uranium content anomaly
铀红 uranium red
铀后 transuranium
铀后废物 transuranium waste
铀后金属 transuranium metal
铀后元素 transuranium element; tranuranic element
铀华 uranium ocher [ochre]
铀化合物 uranium compound
铀还原法 uranium reducing method
铀黄 uranium yellow
铀基燃料 uranium-base fuel
铀钾比值 ratio of U/K
铀钾比值平剖图 profile on plane of U/K ratio
铀钾等值图 contour map of U/K ratio
铀钾异常 uranium/potassium anomaly
铀精矿 uranium concentrate
铀精炼厂 uranium refinery
铀块 uranium button
铀矿 uranium mine; uranium ore
铀矿床 uranium deposit
铀矿化学 geochemical exploration for uranium
铀矿加工厂 uranium ore processing plant
铀矿开采 uranium mining
铀矿类 uranite
铀矿脉 uranium vein
铀矿普查 uranium reconnaissance
铀矿石 uranium mineral; uranium ore
铀矿物 uranium mineral
铀矿异常 anomaly of uranium ore
铀矿渣 uranium mine waste residue
铀镭衰变系 uranium-radium decay series
铀量测定 uranometric survey
铀量测量 uranometric survey
铀裂变 uranium fission
铀磷灰石 uran-apatite
铀硫酸盐 uranocher
铀络合物 uranium complex
铀钼矿 sedovite
铀铌钽铁矿 toddite
铀年龄 uranium age
铀浓缩 uranium enrichment
铀浓缩工厂 uranium enriching plant
铀浓缩物 uranium concentrate
铀铅 uranogenic lead
铀铅测年 uranium-lead dating
铀铅二阶段模式 uranium-lead two-stage model
铀铅年代测定法 uranium-lead age method
铀燃料棒 uranium fuel rod
铀燃料循环 uranium fuel cycle
铀燃料元件 uranium fuel element
铀锐钛矿 uranoanatase
铀栅格 uranium lattice
铀烧绿石 betafite; hatchetttolite; uranpyrochlore
铀石 coffinite
铀石墨堆 uranium-graphite reactor
铀石墨栅格 uranium-graphite lattice
铀试剂 uranol
铀水 uranium water
铀水栅格 uranium water lattice
铀水钍石 uranohydrotorite
铀酸 uranic acid
铀酸钕 neodymium uranate
铀酸盐 uranate
铀钛磁铁矿 davidite
铀碳钙石 uranospinite
铀碳化物 uranium carbide
铀锑催化剂 uranium antimony catalyst
铀提纯废水 uranium purification wastes
铀铁 ferrouranium
铀同位素分离器 calutron
铀同位素年代 uranium-isotope age
铀铜矾 johannite; uranvitriol
铀铜矿 uranolepidite
铀钍比值 ratio of U/Th
铀-钍比值平剖图 profile on plane of U/Th ratio
铀-钍等值图 contour map of U/Th ratio
铀-钍混合异常 U-Th mix anomaly
铀-钍-铅测年法 the U Th-Pb dating method
铀-钍-铅法 uranium-thorium-lead method
铀-钍-铅年代测定法 uranium-thorium-lead age method
铀-钍异常 uranium-thorium anomaly
铀蜕变系 uranium decay series
铀外壳 uranium jacket
铀系 uranium family; uranium (-radium) series
铀系法 uranium series method
铀系元素 uranide(s); uranium series element; uranoid
铀细晶石 uranmicrolite
铀霞石 studtite
铀酰氯 uranyl chloride
铀酰手 uranyl
铀酰盐类 uranyl salts
铀线储量外检误差 external examining errors of uranium linear reserves
铀消耗 uranium consumption
铀循环 uranium cycle
铀盐 uranium salt
铀盐加厚法 uranium intensification
铀盐调色法 uranium toning
铀氧矿物 uranium-oxygen mineral
铀易解石 uranoaeschynite
铀有利性指标限 uranium advantage marked limit
铀有利性指标直方图 histogram of beneficial mark of uranium
铀远景区指标 uranium perspective area marked limit
铀云母 uran-mica; uranite
铀中毒 uranium poisoning
铀重水反应堆 uranium-and-heavy-water reactor
铀重水栅格 uranium-deuterium lattice

游标 slider; slipper; vernier; cursor <计算尺的>

游标闭合计 vernier closure meter
游标臂 vernier arm
游标标度 vernier division
游标测尺 Sopwith staff
游标测高规 vernier depth ga(u)ge
游标测径规 vernier cal(l)ipers
游标测径器 vernier cal(l)ipers
游标测微计 micrometer with vernier
游标测微器 vernier micrometer
游标尺 moving scale; vernier (cursor); nonius; vernier scale; vernier ga(u)ge
游标尺每格读数值 reading value per division of vernier
游标齿厚规 vernier gear tooth ga(u)ge
游标齿厚仪 gear tooth vernier ga(u)ge
游标电位计 vernier potentiometer
游标读数 vernier reading
游标读数闭合差 <游标第一次和最后一次读数之差> vernier closure
游标读数器 vernier-read instrument
游标读数压力计 vernier reading manometer
游标度盘 slide rule diagram; vernier dial
游标对准视觉锐度 vernier acuity
游标发动机 vernier engine
游标发动机关车 vernier cutoff
游标分度规 vernier protractor
游标分度器 vernier protractor
游标分划 vernier division; vernier scale
游标分划尺 nominal scale
游标符号 cursor mark
游标高度尺 vernier height ga(u)ge
游标高度规 vernier height ga(u)ge
游标规 cal(l)iper square; vernier ga(u)ge
游标滑动卡尺 vernier slide cal(l)ipers
游标活动量角器 vernier bevel protractor
游标经纬仪 vernier theodolite; vernier transit
游标精度 vernier accuracy
游标镜 optic(al) vernier
游标卡尺 caliper; slide cal(l)ipers; slide ga(u)ge; sliding cal(l)ipers; square cal(l)pers; trammel; vernier cal(l)ipers; vernier ga(u)ge
游标卡尺的量具测径规 vernier cal(l)ipers ga(u)ge
游标卡规 vernier ga(u)ge
游标卡钳 vernier cal(l)ipers
游标刻度 vernier division; vernier graduation
游标刻度盘 vernier dial; vernier scale
游标量角器 vernier protractor
游标零点 nonius zero; vernier slide zero; vernier zero
游标零分划 minus zero; nonius zero; vernier zero; zero of vernier
游标罗盘(仪) vernier compass
游标名 cursor name
游标目镜 vernier eyepiece
游标盘 upper circle; upper plate; vernier circle; vernier plate
游标盘最小读数 least count of vernier
游标千分尺 micrometer with vernier; vernier micrometer
游标扇形(刻)度盘 vernier quadrant
游标深度尺 vernier depth ga(u)ge
游标式联轴器 vernier coupling
游标式六分仪 vernier sextant
游标式迷宫汽封 vernier labyrinth gland
游标视距仪 vernier tach(e)ometer; vernier tachymeter
游标水准管 vernier level
游标水准器 vernier level
游标调节 vernier control
游标调谐 vernier tuning
游标调整 vernier adjustment
游标微调 vernier tuning
游标误差 vernier error
游标显微镜 vernier microscope; vernier reading microscope
游标线 vernier line
游标斜分度规 vernier bevel protractor
游标元件 vernier element
游标钟 vernier clock
游标重合误差 vernier alignment error
游标装置 vernier arrangement; vernier device
游标准距计 vernier tachymeter
游彩 vagrant colo(u)rs
游车 <长大货物超过一辆车长度时的加挂车> idle(r) car; idler; runner wagon; travel(l)ing block
游尘 dust
游程编码 run-length encoding
游程长度 length of run; run length
游程长度受限码 run-length limited code
游程个数 number of runs
游程检验 run test
游程距 length of run
游尺 nonius; sliding ga(u)ge

游尺调节 vernier adjustment
游船 pleasure boat
游船码头 marina
游船区 boating spot
游荡 interlace; interlacing; meandering; migration; wander(ing)
游荡波 migrating wave; migratory wave
游荡焊接工序 wandering sequence
游荡河槽 interlaced channel
游荡河道 meandering channel; twisting channel
游荡河段 migrating reach
游荡河弯 migration of meander
游荡沙坝 braid bar
游荡水系 interlacing drainage; interlacing drainage pattern
游荡速率 migration rate
游荡型河道 braided channel
游荡型河流 braided river; braided stream; meandering river; meandering stream; wandering river; wandering stream; walker river
游荡性河槽 mobile channel; shifting channel
游荡性河床 mobile bed; shifting bed
游荡性河段 braided reach; fluctuating reach; wandering reach
游荡性河流 meandering river; meandering stream; mobile bed stream; movable-bed stream; walker river; walker stream; wandering river; wandering stream
游荡性河湾 migrating meander; random meander
游荡性曲流 migrating meander; random meander
游荡性沙洲 alternate bar; random bar; shifting bar; travel(l)ing bar
游荡性水道 meander channel; twisting channel; wandering channel
游荡性心滩 interlaced island
游荡性洲滩 random bar
游荡运动 meandering movement
游钓鱼类 game fishes
游动 excursion; ply; wandering
游动变量 running variable
游动的 travel(l)ing
游动吊卡 travel(l)ing spider
游动度盘 vernier dial
游动阀 travel(l)ing valve
游动阀门 travel(l)ing valve
游动凡尔 travel(l)ing valve
游动钢绳 rotary drill line; rotary line
游动杠杆 brake equalizer; floating lever
游动杠杆连接杆 floating connecting rod
游动杠杆托 floating lever bracket
游动杠杆托架 floating lever hanger
游动杠杆支点连杆 floating lever fulcrum connecting rod
游动光标 <计算机屏幕上的> cursor
游动滑车 travel(l)ing sheave
游动滑车护罩 travel(l)ing block guard
游动滑车组 rotary block; travel(l)ing block; troll(e)y block
游动滑轮 travel(l)ing sheave; free wheel
游动杠杆和固定杠杆的连接 floating lever and fixed lever connection
游动量 lost motion
游动能力 ability to swim; swimming ability
游动偏心轮 loose-eccentric wheel
游动偏心器 loose eccentric
游动沙丘 migrating dune; migratory dune; moving dune; wandering dune
游动式平台 mobile platform
游动式织物密度镜 travel(l)ing counting glass
游动输送带 shuttle conveyer belt
游动输送机 shuttle conveyer [conveyor]
游动系统效率 efficient of travel(l)ing system
游动下标 running subscript
游动效应 <汽车行驶或飞机起降时的> wandering effect
游动芯棒 floating plug
游动引线 wandering lead
游动指针 cursor
游动轴承 non-locating bearing
游管 play pipe
游滑轮 dead pulley; loose wheel; uncoupled wheel; loose pulley
游滑轮部件 freewheel unit
游滑皮带轮 loose pulley
游间过量 excessive play
游客 sightseer; tourist; visitor
游客容量 tourist capacity
游客休息室 visitor room
游客涌进(到) influx of visitors
游客用车 visitor's car
游框 slider
游览 pleasure trip; sightsee(ing); tour
游览车 park sightseeing bus; phaeton; pleasure traffic; recreational vehicle; sightseeing bus; tourer; touring car; tourist coach; rubberneck bus; rubberneck car; rubberneck wagon <美>
游览车厢 observation car
游览城市 city for sight-seeing; resort city; sightseeing city
游览船 excursion boat; pleasure cruise boat; pleasure steamer; recreational craft; cruise vessel
游览道路 sightseeing road; tourist road
游览地 tourist site
游览渡船 excursion ferry
游览港 sightseeing harbo(u)r
游览公共汽车 excursion bus; observation coach; sightseeing bus; touring bus
游览公路 scenic highway
游览公园 pleasure-garden
游览观赏慢车道 <汽车可以时驶时停> promenade lane
游览广场 grand sweet
游览航班服务台 sightseeing flight counter
游览环境 recreational environment
游览机 joy-riding plane; touring plane; tourist plane
游览交通 pleasure traffic
游览客票 excursion ticket
游览列车 excursion train
游览路 promenade lane; sightseeing road; tourist road
游览路线 route of journey; scenic route; touring route; tourist circuit
游览旅馆 resort hotel
游览轮 excursion steamer
游览汽艇 pleasure steamer
游览区 esthetic area; excursion area; leisure area; open-to-public area; point of interest; recreational area; sightseeing resort; touristic zone
游览胜地 excursion center [centre]; place for sightseeing; resort
游览图 hiking map; touring map; tourist map; travel map
游览团体 excursion
游览拖车 travel trailer
游览用水 recreational water
游览优待票 tourist ticket
游览者 sightseer; tourer; tourist
游览指南 tope guide
游廊 cernada(h); lanai; stoep; veranda(h); prothyron <古希腊建筑门前>
游廊步道 cloister walk
游乐场 amusement garden; pleasance; recreation ground
游乐场所 recreational area
游乐车 fun about
游乐公园 pleasure-garden
游乐湖 recreational lake
游乐汽车 Arvee
游乐区 recreational precinct
游乐水域 recreational waters
游乐性车辆 recreational vehicle
游乐性出行 recreational trip
游乐用小汽车 fun about
游乐园 amusement garden; pleasance
游离 dissociate; extricate; free(ing); liberation
游离氨和盐氨 free ammonia and saline ammonia
游离柏油 free tar
游离卟啉 free porphyrin
游离层 ionosphere
游离的 unbound
游离地沥青 free asphalt
游离电解质 free electrolyte
游离度 freeness
游离端 free end
游离二氧化硅 free silica; free silicon dioxide
游离二氧化硅含量 content of free silica
游离二氧化碳 free carbon dioxide
游离腐殖酸 freely humic acid
游离硅酸 free silicic acid
游离灰 free ash
游离基 free radical
游离基共聚合 radical copolymerization
游离基机制 free radical mechanism
游离基加成 free radical addition
游离基聚合 radical polymerization
游离基清除剂 radical scavenger
游离基取代 free radical substitution
游离基型调聚反应 radical telomerization
游离基引发聚合 free radical polymerization
游离碱 free alkali; free base; free soda
游离焦虑 free floating anxiety
游离焦油(沥青) free tar
游离金 free gold
游离金属 free metal
游离金属离子 free metal ion
游离类脂物 free lipid
游离离子 free ion
游离氯 free chloride
游离面 free surface
游离漆膜 free film
游离气 free gas
游离气顶 free gas cap
游离氢 free hydrogen
游离氢氧化钙 free calcium hydroxide; uncombined lime
游离氰(化物) free cyanide
游离三氧化二铝含量 content of free alumina
游离三氧化二铁含量 content of free ferric oxide
游离色 variegate
游离渗碳体 free cementite
游离剩余氯 free residual chlorine
游离石灰 free calcium oxide; free lime; uncombined lime
游离石灰膨胀 expansion due to free lime
游离石墨 free graphite
游离室体积 volume of dissocial chamber
游离水 free water; mobile water; uncombined water
游离水分 free moisture; mobile moisture
游离水(不包括骨料吸附水)与水泥之比 detrital ratio
游离酸 free acid
游离酸度 free acidity
游离碳 free carbon; uncombined carbon
游离碳含量 free carbon content
游离碳化物 free carbide
游离体 corpus liberum
游离铁 free iron
游离铁素体 free ferrite
游离同位素 free isotope
游离瓦斯 free gas
游离系泊装置 free mooring arrangement
游离纤维 wandering fibre
游离型空腔 cavity detachment
游离性氯 free chlorine
游离性有效氯 free available chlorine
游离性有效余氯 free available residual chlorine
游离性余氯 free residual chloride
游离性余氯化 free residual chlorination
游离盐酸 free hydrochloric acid
游离盐酸测定 free hydrochloric acid determination
游离氧化钙 <水泥中> free calcium oxide
游离氧化钙含量 free calcium oxide content; free lime content
游离氧化钙总量 total free lime
游离氧化铝 free alumina
游离氧化镁 free magnesia
游离氧化物试验 free oxides test
游离余氯 free residual chlorine
游离脂肪酸 free fatty acid; non-esterified fatty acid
游离脂酸结晶 free fatty acid crystal
游离状态 free state; set free
游离资本 freed capital; released capital
游离作用【化】 dissociation
游历 peregrination
游历的 travel(l)ing
游梁抽油井 beam well
游梁吊架 beam hanger
游梁头 beam head
游梁托架 beam hanger
游梁悬支 beam hanger
游梁柱 beam post
游轮 belt tightening pulley; idle pulley
游码 rider <精密天平上的>; sliding poise
游码标尺 rider bar
游码钩 rider carrier; rider hook
游牧生活 nomadic life; nomadism
游牧业 nomadic herding
游泥沉淀 sludge settling
游憩地带 leisure zone
游憩设施 recreational facility
游憩设施的贷款 recreation facility loans
游憩中心 recreational center [centre]; recreation center [centre]
游钎套 rotation chuck
游禽 natatorial bird
游禽类 natatores
游清器傍流保护 by-pass protection
游人小道 tourist path
游散地 esplanade
游赏庭园 tour garden
游丝 balance spring; filament; hair(line); hair spring; hair wire
游丝测微器 filar micrometer
游丝除垢剂 hairspring cleaner

游丝定长 hairspring setting
游丝分规 hair-spring divider
游丝刚度 rigidity of hairspring
游丝夹 hairspring holder
游丝架 hairspring support
游丝校平器 hairspring lever
游丝螺钉 hairspring screw
游丝内桩 hairspring inside stud
游丝内桩拆除器 hairspring collet removers
游丝内桩工具 hairspring collet tool
游丝镊子 hairspring tweezer
游丝调整杆 hairspring set lever
游丝外桩 hairspring stud
游丝外桩螺钉 hairspring stud screw
游丝销 hairspring pin
游艇 cruising craft; excursion boat; pleasure boat; pleasure craft; recreational boat; yacht
游艇保险 yacht insurance
游艇发动机 recreational boat engine
游艇港 yacht harbo(u)r; marina; yacht marina
游艇港池 marina; yacht basin
游艇甲板流水沟 plank sheer
游艇驾驶人员 yachter
游艇驾驶术 yachting; yachtsmanship
游艇俱乐部 yacht club
游艇旅客馆 boatel
游艇旅社 boathouse
游艇码头 yacht landing stage; yacht wharf
游艇锚泊地 yacht marina
游艇停泊场 boathouse
游艇停泊区 yacht landing area
游艇系缆绳 yacht rope
游艇用图 yacht chart
游艇之家 boathouse
游艇制造厂 yacht building yard
游息公园 recreation(al) park
游息林 recreational forest
游戏 playing; sport
游戏操纵杆 joy stick
游戏场 game court; play(ing) field; play(ing) yard
游戏场地 playground; play space; playing field
游戏雕刻 play sculpture
游戏机 game machine
游戏街道 play street
游戏结构物 play structure
游戏室 playroom
游隙 back lash; back play; clearance; lash; windage
游隙可调的滚针轴承 needle roller bearing with variation
游星齿轮 pinion; pinion gear
游行 parade
游檐木 tassel [torsel]
游移 back lash; divagation channel
游移底栖生物 vagrant benthos
游移函数 oscillating function
游移河 divagative river; divagative stream
游移河槽 divagative river channel; shift(ing) channel
游移河床 divagation channel
游移河弯 migrating meander
游移湖 wandering lake
游移块焊接 wandering block welding
游移群落 migratory community
游移沙丘 migrating dune; wandering dune
游移沙洲 migrating bar
游移水流 shift current
游艺室 fun house
游泳 swim
游泳场 bathing beach; swimming beach; swimming place
游泳池 bathing-pool; swim(ming) pool
游泳池池边 swimming pool nosing
游泳池大厅 swimming pool hall
游泳池底泳道线 marking
游泳池防滑面板 pool deck
游泳池覆盖棚 swimming pool cover(ing)
游泳池盖 swimming pool cover(ing)
游泳池过滤器 swimming pool filter
游泳池加热系统 swimming pool heating system
游泳池甲板 lido deck
游泳池棚 swimming pool shelter
游泳池式反应堆 swimming pool reactor
游泳(池)水 swimming pool water
游泳池水的处理 processing of swimming pool water
游泳池涂料 swimming pool paint
游泳池卫生 sanitation of swimming pool
游泳池污染 pollution of bathing place
游泳池型反应堆 swimming pool reactor
游泳池用漆 swimming pool paint
游泳动物 nekton; nekton organism; swimmer
游泳馆 natatorium; pool hall; swimming hall
游泳和跳水池 swimming and diving pool
游泳建筑物 aquatic building(s)
游泳区 swimming area
游泳人数 swimming population
游泳生物 necton
游泳训练池 training pool
游泳训练馆 training pool hall
游泳者 swimmer
游泳足 swimmeret
游勇 free-lance
游园会 garden party
游轴 blind axle
游资 disposable capital; dormant capital; floating capital; floating money; free capital; free resources; idle fund; idle money; loose fund; unemployed capital; volatile money; hot money <为获得高利或保障币值由一国转移至另一国的流动资金>
游资移动 disposable capital; hot money volatile money

友

友好城市 sister city
友好合作 friendly cooperation
友好解决 mutual consultation
友好界面 friendly interface
友好调解条款 amiable composition clause
友好通商条约 treaty of amity and commerce
友好协商解决 <业主和承包商之间争端解决途径之一的> amicable settlement
友矩阵 companion matrix
友谊厅 social hall
友谊医院 friendship hospital

有

有V形肋的钢丝网 rib lath
有碍健康的 unsanitary
有碍能见度的颗粒 visibility-impairing particle
有安全装备的 fail safe
有安全装置的 foolproof
有鞍点对策 saddle-point game
有案可查事项 matter of record
有暗销的舌槽接缝 dowel(l)ed tongue and groove joint
有凹槽的 fluted; grooved
有凹槽的瓷砖 fluted tile
有凹槽的钢管桩 fluted steel tube pile
有凹槽的模板 fluted formwork
有凹槽混凝土块 shadowed block
有凹槽门框 buck frame
有凹槽圆头螺钉 fillister head screw
有凹槽柱身 fluted column
有凹凸的 irregular
有凹凸榫的管子 rebated pipe
有凹凸榫的金属板 rebated sheet metal sheath
有凹陷的盖苫篷布 hollow sheeting
有凹圆线脚的钢门樘 cove-mo(u)ld frame
有八面的 octahedral
有疤的 scabbed
有把的大杯 mug
有把水罐 jug
有把握的 confident
有把握的方法 foolproof method
有把握地 confidently
有坝河段 diked reach
有坝式 reservoir with dam
有坝引水 dammed intaking
有白点的 flaky
有百叶窗的通风口 outlet ventilator
有百柱建筑 hecatonstylon
有斑 spottiness
有斑点的 dappled; granulated; mottled; speckled; spotted; spotty
有斑点的表面 mottled surface
有斑痕的木材 spave
有斑马纹的 zebra
有斑纹的 brinded; brindle; mottled; patchy
有版权的作品 work in private domain
有版权、有专利权状态 private domain
有半角锥体撑住的塔尖 broached spire
有瓣的 valvular
有瓣膜的壳体 shell with valve
有帮助 subservience
有帮助的相互作用 benevolent interaction
有包覆膜的(颜料)粒子 particle having a finish
有包装货物 package freight; packed goods
有保持力的 retentive
有保护层的路肩 protected shoulder
有保护层的路面 mat-covered pavement
有保护的电弧焊 shielded arc welding
有保护的蒸汽供热 protective steam heating
有保护电弧焊 shielded bridge welding
有保护面层的抛石堤 armoured rubble mound
有保护网的风扇 cased type fan
有保留的同意 qualified approval
有保留接受 acceptance with reservation
有保留验收 acceptance with reservation
有保险的贷款 insured loan
有保证的委员会决议 certified board resolution
有保证市场投资 guaranteed market investment
有报酬的 remunerative; rewardful
有报酬的职业 gainful occupation
有暴洪的河流 flashy stream
有爆炸危险 possible explosive risk
有爆炸危险的环境 explosive atmosphere
有爆炸危险的空气 explosive atmosphere
有背架的镶木地板 plate parquet
有比例地图 scale map
有彼此成直角之裂缝的 orthoclastic
有壁龛墙单元 wall unit with recess
有边带 tape with selvage
有边界线的目标 edge target
有边框隔墙 framed partition (wall)
有边框墙的系梁人字木屋架 collar beam roof with jamb walls
有边框墙的系梁三角屋架 collar tie roof with jamb walls
有边梁的板 slab with edge beam
有边条 edge banding
有边筒管 spool
有边筒子 spool
有边细孔 bordered pit
有鞭状枝 flagellate
有便溺器的盥洗室 commode-type toilet
有变化的 varied
有变酸倾向的 acidoid
有标称排水孔的金刚石钻头 diaphragm with conventional waterways bit
有标号公用块 label(l)ed common block
有标号语句 label(l)ed statement
有标名的列车 named train
有标志路线 marked route
有表决权股票 voting stock
有柄插入式六角套筒扳手 hexagon socket spinner wrench
有柄插入式振捣器 immersion vibrator with handle
有柄插入式振动器 immersion vibrator with handle
有柄插销 bolt-up with handle
有柄瓷埚 casserole
有柄瓷皿 casserole
有柄瓷蒸发皿 porcelain casserole
有柄的 ansate
有柄浸入式振捣器 immersion vibrator with handle
有柄浸入式振动器 immersion vibrator with handle
有柄曲拐 winch
有柄旋塞 turncock
有柄蒸发皿 casserole
有病的 sick
有病毒的 virose
有拨号盘的电话机 post-selector
有波斯结的地毯 carpet with Persian knots
有波纹的 fluted; moire; watered; wavy
有波纹的水流 undulating flow
有波形舷部的船 ship with corrugated side
有波形烟道的兰开夏锅炉 Lancashire boiler with corrugated flues
有波折的 checkered; chequered
有波状下缘的线脚 nebule mo(u)lding
有玻璃窗的阳台 gazebo [复 gazebo(e)]
有玻璃窗和纱窗的门 storm and screen door
有玻璃格条的窗子 glazing bar window
有玻璃天窗的摄影室 glasshouse
有玻璃砖填充的钢筋混凝土 reinforced concrete with glass tiled fillers
有薄层基础的 thin-bedded
有薄片的 lamellate
有薄雾的 hazy
有补偿的 balanced
有补偿破坏 compensatory damage
有补格 complemented lattice
有不变水平推力的拱形桁架 arched

girder with invariable horizontal thrust
有不法行为的船员 barrater [barrator]
有擦痕 with striations
有擦痕的表面 torn surface
有擦痕的基岩 striated bedrock
有擦痕的黏[粘]土 slickensided clay
有擦痕的岩基 striated bedrock
有采暖装置的货车 wagon with heating apparatus
有彩斑大理石 variegated marble;compound marble
有残留气体的电子管 soft tube
有操纵轮的阀门 valve with handwheel
有糙面 present shagreen surface
有槽半圆头平端螺钉 slotted halfround head flat point screw
有槽半圆头自攻木螺钉 slotted halfround head selftapping wood screw
有槽导轨 grooved rail
有槽导子 gorget;grooved director
有槽的 furrowed;reeded;riffled
有槽的钢丝绳卷筒 grooved drum
有槽的混凝土方块 grooved cube
有槽的门樘 rabbeted door jamb
有槽口的抹子 notched trowel
有槽的平台 notched stage
有槽电枢 grooved armature;slotted aperture
有槽阀杆 grooved valve stem
有槽钢轨 grooved rail
有槽工具磨边 edge with a groove
有槽管式取样器 <取散装水泥等的> slotted tube sampler
有槽滚筒 grooved cylinder
有槽滚轴 grooved roller
有槽滑轮 grooved pulley
有槽活塞环 grooved piston ring
有槽卷筒 grooved barrel
有槽空心砖 scored tile
有槽六角头螺丝 slotted hexagon head screw
有槽螺母 groove nut;slotted nut
有槽门窗 door frame with rebate
有槽门框 rebated door
有槽盘形凸轮 grooved disk cam
有槽拼模块 slot piece
有槽铅条 <花格窗上的> lead;came
有槽球顶 slotted ball top
有槽球座 grooved ball seat
有槽双端螺栓 grooved stud
有槽弹簧杆 grooved spring bar
有槽弹簧用圆钢 bar for grooved springs
有槽铁芯 slotted core
有槽凸(圆)头螺钉 fillister head screw
有槽凸缘 notched flange
有槽突缘 notched flange
有槽瓦板 grooved tile
有槽纹的 fluted
有槽心形凸轮 grooved heart cam
有槽引上法 <平板玻璃> debiteuse method
有槽引针 gorget;grooved probe
有槽圆顶 slotted round top
有槽圆盘 grooved disk
有槽圆头螺钉 fillister head screw
有槽圆头螺母 slotted round nut
有槽轧辊 grooved roll
有槽整流子 slotted commutator
有槽(止动)铁柱 grooved casting
有槽轴 fluted shaft
有槽柱 notched column
有槽铸铁轧辊 grooved iron roll
有槽转子 grooved rotor;grooving rotor
有槽锥形销 grooved taper pin
有草的开阔高地 <英格兰南部的> downs
有草皮的 turf bound
有侧翻斗的货车 wagon with side-tipping bucket
有侧房的 aisled
有侧面梯级的梯 side-step ladder
有侧限的抗压试验 laterally confined compression test
有侧限抗压强度 confined compressive strength
有侧限盆式支座 confined pot bearing
有侧限膨胀 confined swelling
有侧限试样 laterally confined specimen
有侧限速度 confined velocity
有侧限压缩试验 confined compressive test
有侧音电话机 sidetone telephone set
有测绘资料地区 covered surface
有测距设备的全方向无线电信标 omnibearing distance facility
有层车(船) decker
有层次的 stratified
有叉齿的起重电磁铁 electric(al) lifting magnet with tines
有插图的 pictorial
有插图的出版物 pictorial edition
有插图的书 illustrated book
有插图书中的正文 letter press
有差别 make a difference
有掺和料水泥 additive cement
有产权的 proprietary
有产权的环境保护技术 proprietary environmental protection technology
有产权的配方 proprietary formulation
有产权的情报 proprietary information
有长槽的板 slotted plate
有长穿孔眼的砖 brick with horizontal perforations
有长秒针的钟表 sweep hand;sweep-second
有长坡度车顶的汽车 <美> fast-back
有长曲线 rectifiable curve
有偿贷款 onerous loan
有偿调出 payment allocated-out
有偿付能力 loan worthy;solvent
有偿付能力声明 declaration of solvency
有偿契约 <负有法律责任的> onerous contract;considerable contract
有偿援助 assistance on a refundable basis
有偿转拨 aid transmission;paid transmission
有偿转让技术 transfer of technique with compensation
有唱诗班席位的大厅 <教堂> hall quire
有唱诗班席位的多角形大厅 polygonal hall-choir;polygonal hall-quire
有唱诗班席位的多角形会堂 polygonal hall-choir;polygonal hall-quire
有超高的轨道 track with cant
有超高的弯道 super-elevated turn
有潮泊位 tidal berth
有潮差港 tidal harbo(u)r
有潮港池 open basin;open dock;tidal basin;tidal dock
有潮港(口) open tidal harbo(u)r;open tidal port;tide harbo(u)r
有潮海湾 tidal bay
有潮河 tidal river
有潮河段 tidal reach
有潮河口(湾) tidal estuary
有潮码头 tidal dock;tidal quay;tidal wharf
有潮区 tidal compartment
有潮渠 wave channel
有潮水域 tidal waters
有潮小海湾 tidal channel;tidal inlet
有车标志 positive train identification
有车的轨道电路 occupied track circuit
有车家庭 car owning household
有车区间 occupied section
有车厢的货车 box freight car
有车辙道路 rut(ted) road
有车辙的 rutted;rutty
有车者术速度 vehicle available
有尘空气 dust-laden air
有尘空气排空系统 airlift evacuation system
有沉砂室的竖井 collecting manhole
有衬层的明沟 lined ditch
有衬垫接合 gasketed joint
有衬里的 lined
有衬里的贮水器 copper-lined cistern
有衬砌的倒拱底板 paved invert
有衬砌的地下排水沟 mole-channel with liner
有衬砌的反弧底板 paved invert
有衬砌的渠道 lined canal
有衬砌沟渠 lined ditch
有衬砌路面 packing course
有衬砌跑道 packing course
有撑架的板桩 framing sheet pile
有承插接头的上釉陶管 plumbing tile (pipe)
有(承受水平荷载)支撑的梁 braced beam
有(承受水平荷载)支撑的柱 braced column
有承口的 belled;bell-mounted
有城垛的 embattled
有城垛的桥梁栏杆 battlemented bridge parapet
有城墙的城镇 walled town
有尺寸图纸 dimensional drawing;dimensioned drawing
有尺度的 dimensional
有尺度图 dimensional drawing
有齿的 dentary
有齿飞轮 cogged flywheel
有齿槛的静水池 dented stilling basin
有齿槛的消力池 dented stilling basin
有齿槛消能池 dented stilling basin
有齿接环 toothed ring dowel
有齿拉钩 toothed retractor
有齿链 toothed chain
有齿镊 pincers;toothed forceps
有齿皮带 cogged belt
有齿石 rhyncholite
有齿饰的檐板 denticulation corona
有齿套筒 toothed sleeve
有齿形边线的线脚 nebule mo(u)lding
有齿形起拱线的拱顶 vault with dentate springing lines
有齿形装饰的线脚 indented mo(u)lding
有翅盘管 finned coil
有充分保障的 fully secured
有充分保证 fully ensured
有充分理由 with reason
有充气车胎的手推车 wheel-barrow with inflatable tyre [tire]
有充填物的不连续面 filled discontinuity
有充填物的软弱结构面 filled discontinuity
有冲击式活塞的工具 percussive piston tool
有冲突相位 <交通信号中允许左、右转车辆同对向车流或行人有冲突的相位> permitted phase
有冲突转弯 <信号交叉口上允许同对向车流或行人有冲突的左、右转弯> permitted turn;unprotected turn
有冲洗液钻进 watering drilling
有重新当选资格的 reeligible
有抽水单元 pumping element
有抽屉的办公桌 <英> bureau [复 bureau/bureaus]
有抽屉的柜 bachelor chest
有抽头的次级线圈 split secondary
有抽吸现象的混凝土板 pumper
有酬荷载 payload
有酬劳动 paid labo(u)r
有臭味的 odo(u)riferous
有臭氧的 ozoniferous
有出口湖 lake with outlet
有出口湖出流 lake with outlet
有初变形的圆柱面壳 almost cylindric(al) shell
有厨房卫生设备的公寓套房 <一至二居室> efficiency apartment
有触觉的工业机器人 industrial robot with tactile sensing
有传导性的砖瓦 conductive tile
有传动装置的雷达信标 hayrack
有传力杆的接缝 dowelled joint
有传力杆的企口 dowel(l)ed tongue and groove joint
有传力杆的企口接缝 dowel(l)ed tongue and groove joint
有传力销的企口接缝 dowel(l)ed tongue and groove joint
有船即装 shipment by available steamer
有船闸的堰 weir with lock
有船闸的运河 canal with locks
有窗的 fenestrated
有窗格条的窗子 glazing bar window
有窗孔的浇制板 cast panel with window opening
有窗孔的墙板 panel with window opening
有创造力的 creative;originative;pregnant
有垂直边梁的壳体 shell with edge vertical beams
有垂直摄影资料地区 vertical photograph coverage
有槌状头部的凿子 hammerheaded
有锤痕的玻璃 hammered glass
有次序状态 ordered state
有刺短灌木丛 monte community
有刺钢丝 barbed wire;dog wire;steel barbed wire;wire of irregular shape
有刺激性的 pungent
有刺落叶灌木丛 garide
有刺铁丝 barbed wire;dog wire
有刺铁丝网 entanglement;wire entanglement
有丛毛的 comose
有粗点的 asperses
有粗线道的玻璃 glass with heavy cords
有存款的银行账户 active bank account
有存款账户的主顾 account customer
有错文电 error message
有错误的 vicious
有答案的计算程序测试 benchmark test
有大量藻类水 algal water
有代表性的 representative
有代表性的曲线 symptomatic curve
有带的 bandy
有待核准 ad referendum
有待开放的地段 open land area
有单独采暖设备的房间 zoned unit
有单独引出线的阳极 separated anode
有单柱和斜支撑的檩支屋顶 purlin(e) roof with king post and slanting studs

有担保的贷款 guaranteed loan;loan guaranteed
有担保的信贷 guaranteed credit
有担保的债权人 secured creditor
有担保的债券 backed bond;collateral bond
有担保定期贷款 fixed loan secured
有担保公债 debt funded
有担保债券 endorsed bond
有挡板的护栏 apron rail
有挡板的围栏 apron rail
有挡链 stud chain;stud link chain
有挡链环 stud link
有挡土板的开挖 sheeted excavation
有挡土结构的开挖 supported excavation
有刀锋的锉刀 knife file
有导径的深孔麻花钻 guide twist drill
有导向肋骨的 wing guided
有倒钩的接合销 < 木工用 > barbed dowel pin
有灯槽的反光灯 concealed light
有等高线的地形图 chart with contour lines
有等高线的平面图 plan with contour lines
有凳门廊 exedra
有堤防河道 leveed river
有堤河段 embanked reach
有堤渠道 leveed channel
有堤水道 leveed channel
有滴滤池的净水场 purification plant with trickling filters
有底的蛋形圆筒 egg-shaped barrel with base
有底盘安装的喷雪机 carriage mounted snowblower
有底座的管道 pipe with base
有抵抗力的 resistant
有抵押保证的证券 mortgage-backed securities
有抵押的票据 bill with collateral securities
有抵押品的公司债券 mortgage debenture
有抵押权担保的债权 claim secured by mortgage
有地表均匀渗入 uniform infiltration from surface
有地界的地段 land plat
有地下室房屋 basement house
有地震感知的 seismic-conscious
有地震意识的 seismic-conscious
有地址的 addressed
有点儿 to a measure
有点外行味道的 amateurish
有点温热的 lukewarm
有点线的 dotted
有电部(零)件 live part
有电导线 live conductor
有电电路 live circuit
有电电线 hot wire
有电顶锻余量 current-on upset allowance
有电感线圈起动器 impedance starter
有电木面层的墙板 Apco
有电梯的多层住宅 elevator multiple dwelling;lift multiple dwelling
有电梯的公寓大楼 elevator apartment house
有电梯的公寓房子 lift apartment house
有电梯的居住房屋 elevator residence house;elevator residential house
有电梯的居住房子 lift dwelling house
有电梯的居住建筑 lift residential building
有电梯的住房街坊 lift dwelling block
有电梯的住宅大楼 elevator residence block;elevator residential block
有电梯的住宅房屋 elevator dwelling house
有电梯的住宅建筑 lift residence building
有电梯的住宅建筑街坊 elevator dwelling block
有电梯的住宅街坊 lift residence block
有电梯的住宅区段 lift residence block
有电线 live wire
有电压电路 alive circuit;live circuit;active circuit
有雕塑的壁龛门道 blocked doorway
有雕像点缀的大马路 statued avenue
有吊顶的耐火楼板 fire-proof floor with suspended ceiling
有吊平顶的不易燃楼层 non-combustible floor with suspended ceiling
有钉齿的瓦 studded tile
有钉横梁 < 舞台拴住布景绳用的 > pin rail
有钉木砖 anchor brick
有顶撑的开挖 strutted excavation
有顶的通道 covered shaft
有顶的走廊 veranda(h)
有顶的走廊商场 covered mall
有顶的走廊商店建筑 covered mall building
有顶洞 shelter cave
有顶盖的长廊 pawn
有顶盖的货车 closed-top van
有顶盖的通道 pawn
有顶盖市场 covered market (place)
有顶购物中心 covered mall center [centre]
有顶棚的人行道 arcade sidewalk
有顶天井 closed shaft
有顶通道 dogtrot
有顶走廊 < 由教堂至修道院的 > slype
有定额的 normed
有定价的 valued
有定位心轴的模 horn die
有定向转向架的移动式消防炮 portable director
有定型产品的 off the peg
有动力的 powered
有动力装置的 powered
有动物浮雕的壁缘 frieze with animal reliefs
有毒 virulence
有毒材料 toxic gas;toxic material
有毒的 deleterious;nocuous;noisome;noxious;pernicious;poisonous
有毒废气 noxious exhaust gas
有毒废水 deleterious effluent;noxious wastewater;poisonous wastewater;toxic discharge;toxic wastewater
有毒废物 deleterious waste;poisonous waste;toxic waste
有毒废物处置 toxic waste disposal
有毒废物的倾弃 dumping of toxic waste
有毒蜂蜜中毒 toxic honey poisoning
有毒工业 offensive industry;toxic industry
有毒工业废物 toxic industry waste
有毒害的 deleterious
有毒害的事物 pestilence
有毒害性 noxiousness
有毒化学废水 toxic chemical wastewater
有毒化学废物 toxic chemical waste
有毒化学品 toxic chemicals
有毒化学品登记 register of toxic chemicals
有毒化学物质污染 pollution by toxic chemicals;toxic chemical substances pollution
有毒环境 toxic environment
有毒货物 toxic cargo
有毒金属 poisonous metal;toxic metal
有毒垃圾 toxic waste
有毒力的 virulent
有毒栎 poison oak
有毒排放物 toxic discharge
有毒气的 mephitic(al)
有毒气体 nocuous gas;pernicious gas;poisonous gas;toxic gas
有毒溶剂 toxic solvent
有毒水 harmful water;poisonous water;toxic water
有毒水生植被 noxious aquatic vegetation
有毒水引起的疾病 disease caused by poisonous water
有毒酸 noxious acid
有毒涂料 toxic paint
有毒危险 danger toxic hazard
有毒微粒 hazardous particle
有毒污染 toxic pollution
有毒污染物 toxic pollutant;toxic contaminant
有毒污染物浓度 toxic pollutant concentration
有毒污染物水平 level of toxic pollutant
有毒污水 poisonous wastewater
有毒物 deleterious agent;noxious product;poison material;toxicant
有毒物质 poison(ous) material;poisonous substance;toxicant;toxic material;toxic substance
有毒物质的进入环境 entry into the environment of a toxic material
有毒物质的协同效应 synergistic effects of toxic substance
有毒物质点污染源排放 toxics point source discharge
有毒物质控制法令 < 美 > Toxic Substance Control Act
有毒物质生物循环 biologic(al) cycle of toxic substance
有毒物质污染 toxic substance pollution
有毒性的 toxic(ant);virose
有毒性反应的最低浓度 lowest observable adverse effect level
有毒烟气 noxious fume
有毒烟雾 killer smog;noxious fume;toxic smog
有毒盐类 < 土壤的 > toxic salts
有毒颜料 poisonous colo(u)r
有毒阳离子 toxic cation
有毒液体 noxious liquid;poisonous liquid;toxic liquid
有毒油漆 toxic paint
有毒有机化合物 toxic organic compound
有毒有机污染物 toxic organic contaminant;toxic organic pollutant
有毒有机物 toxic organic material;toxic organics;toxic organic substance
有毒釉 poisonous glaze
有毒鱼 ichthyotoxic fishes;poisonous fish
有毒元素 toxic element
有毒藻类 toxic algae
有毒蒸气指示剂 toxic vapo(u)r indicator
有毒植物 poisonous plant
有毒作用 toxic action
有渡线连接的线路 cross-over road;cross-over track
有渡线联结的线路 cross-over line
有镀层的钢丝 steel wire with plated cover
有端部装饰的椽子 rafter with ornamental end
有对角线通风的砖坯 scintled
有钝头的凿子 hammerheaded
有多边形状的拱形桁架 arched girder with polygonal outlines
有多层外皮的 multiwall
有多层外皮的水泥袋 multiwall cement bag
有多方面才能的人 generalist
有多方面知识的人 generalist
有多个独立居住单位的建筑物 multiple dwelling building
有多输出通路互不相扰的电子管放大器 trap valve amplifier
有多种解释 polysemy
有多种用途的 polychrestic
有恶臭的 noisome
有恶臭货物 malodo(u)rous cargo
有耳的 auricled
有耳房的 aisled
有耳螺母 eared nut
有耳砖 lug brick
有发酵力的 fermentative
有发射的接收机 blooper
有发生故障的迹象 conk
有阀冲击器 valve air hammer
有阀的 valvular
有阀门的 valvular
有法定时间限制的息票 statute barred coupon
有法定资格的 competent
有法律效力的协议 legally binding agreement
有法律约束力的合同 binding contract
有反馈作用的伺服机构 reaction servomechanism
有反应的 responsive
有反应器的蒸馏塔 integrated tower
有方位度盘的测角仪 azimuth circle instrument
有方向的动作 directional action
有方向性 aeolotropy
有方向性的 directional
有防波堤的储罐 mounded tank
有防潮矮墙的丁坝 groyne with short training wall
有防潮矮墙的防波堤 groyne with short training wall
有防潮闸的港池 tidal basin
有防尘网的进气口 screened air intake
有防护道口 guarded crossing;protected crossing;protected level crossing;watched crossing
有防护的水域 sheltered area of waters
有防护与导向设施的高速专用道 high-speed guideway
有防滑短的瓷砖 tile with non-slip pattern of short ribs
有防渗墙的坝 < 土石坝 > diaphragm dam
有防渗心墙的坝 water-tight diaphragm dam
有防锈层的铁板 stainless-clad plate
有房产未来指定权的人 remainderman
有纺土工织物 woven geofabric
有纺型土工织物 woven geotextile
有放大器的控制 servo-operated control
有飞边的旧钢轨 worn rail with overlapping fin
有飞边模锻 closed-die forging
有非结构填料的平板梁 slab-and-beam with non-structural fillers

Y

有分车带的公路 divided-lane highway
有分寸的 measured
有分隔带的公路 divided highway
有分(系)杆的桁架 truss with subties
有分支的电线干线 cable-tap system
有风洞 cave with wind
有风格的 mannered
有风味的 racy
有风险的投资 risky investment
有风险的运费 freight at risk
有峰的 peaky
有峰曲线 peaky curve
有缝 T 形接合 open tee [T] joint
有缝 T 形接头 open tee [T] joint
有缝薄膜 slit-film
有缝的 seamed;seamy
有缝的泵 split-casing pump
有缝的倾卸门 split-bottom dump
有缝的小屋 splitting shanty
有缝钢管 slotted pipe
有缝钢轨 slotted rail
有缝管 seamed pipe;seamed tube;slit tube;welded tube
有缝轨道 jointed track
有缝夹套 collet;split collet
有缝料斗 slit hopper
有缝试件抗裂试验 slit type cracking test
有缝套爪 collet
有缝线路 jointed track
有缝岩石 seamy rock
有扶垛拱的 < 如护岸墙用 > counter-arched
有扶垛支撑的支柱 buttress bracing strut
有服务员的停车场 < 代客停车,取车 > attendant parking
有浮雕的金属小梁 chase lintel
有浮雕陶制小装饰品 cameo ware
有浮雕装饰的 bossed
有浮力的 buoyant
有浮力装置隔水管 riser with flo(a)tation modules
有符号常数 signed constant
有符号二进制数 signed binary
有符号数 directed number
有辐轮毂 spoked assembly
有辅助能源的闭环控制 closed loop power-assisted control
有腐蚀力的 corrodent
有腐蚀性的 corrosive
有腐蚀性的水 active water;aggressive water
有付款条件的合同 deed money escrow
有负荷的零件 loaded parts
有负荷感觉的伺服机构 reaction servomechanism
有负载的 on-load(ing)
有负载时的 Q 值 loaded Q (value)
有负载线 terminated line
有附件的票据 bill with attached documents
有附文的 provisory
有附着力的 tenacious
有附着性的 cohesive
有复本保有权者 copyholder
有复式塞孔盘的交换机 coupled positions
有复验性的结果 repeatable result
有复杂因素而可猜寻的问题(优选法) stochastic problem

有副产品的炼焦法 by-product coke process
有副产品回收的煤气发生炉 by-product recovery gas product
有覆盖层的中等密度胶合板 medium density overlay
有覆盖的走廊 < 教堂中的 > slype
有覆盖平原 covered plain
有覆面的材料 cladded material
有盖舱口驳 covered barge
有盖层的地热含水层 caprock-type geothermal aquifer
有盖敞车 covered gondola
有盖敞口驳 covered hopper barge
有盖大篮 hamper
有盖的小天窗 scuttle
有盖坩埚 closed pot
有盖沟槽 closure channel
有盖货车 box car;goods van;van vehicle;van < 铁路 >
有盖集装箱 covered container
有盖铰链 capped butt
有盖可开关的天窗 scuttle
有盖漏斗车 covered hopper car
有盖马桶 close-stool
有盖(清)水池 covered reservoir
有盖汤盘 soup tureen
有盖天窗 scuttle
有盖箱式托盘 covered box-pallet
有盖小舱口 cap scuttle
有盖烟灰盒 lidded ash
有干扰变量补偿的比例控制器 proportional controller with disturbance-variable compensation
有杆泵 sucker rod pump
有杆电钻 rod electrodrill
有杆电钻钻进 rod electrodrilling
有杆锚 admiralty (pattern) anchor;common anchor;stock(ed) anchor
有杆锚孔上下的杆箍 nuts of an anchor
有杆转爪锚 close stowing anchor
有感地震 felt earthquake;sensible shock
有感地震范围 earthquake-felt area
有感电阻 inductive resistance
有感电阻器 inductive resistor
有感负载 inductive load
有感区 area of perceptibility;felt area
有感染力的 catching;contagious
有感线圈 inductive coil
有刚性腹杆的梁式桁架 braced girder
有刚性轴的插入式振动器 stiff-shaft internal vibrator
有刚性轴的内部振动器 stiff-shaft internal vibrator
有刚性轴的振动器 stiff-shaft vibrator
有钢轮缘的轮子 steel-rimmed wheel
有钢丝绳装备滑车系统 stringing the wire line
有钢套的 steel-encased
有钢套管的扩底桩 cored pedestal pile
有高程注记的地形剖面图 profile in elevation
有高级知识的 A-level
有高起缘石的分车岛 curbed separator
有疙瘩的玻璃 knotty glass
有鸽舍的墙 pigeon-holed masonry wall
有歌舞厅餐厅 caboret
有格栅的下水道进口 yard gull(e)y
有格式读语句 formatted read statement
有格子细工的筒形圆屋顶 barrel vaulted with trelliswork
有隔担子菌类 < 拉 > Phragmobasidiomycetes
有隔挡的平车 flat car with bulkhead
有隔墙的坝 diaphragm dam
有隔墙的开口沉井 drop shaft with cross walls
有隔热包层的墙板 thermally insulated cladding panel
有隔栅的水落管 < 平屋顶 > roof outlet
有各种地图资料的地区 heterogeneous map coverage
有根据的 authentic;well-founded
有根据的值 codified value
有根树 rooted tree
有根水生植物 hydrophyte radicanta
有根图 rooted graph
有根有向图 rooted directed graph
有梗花栎 common oak
有工伤危险的故障 critical failure
有工业价值的油井 commercial well
有公度的 commensurable
有公隔墙的两毗连房屋 semi-detached house
有公共隔墙的两间毗连房屋 pair of semi-detached houses
有公用事业综合设施的顶棚 combined service(d) ceiling
有功部分 active component;active constituent;energy component;power component;real component;watt component
有功的 wattful
有功电流 active component;active current;effective current;effort current;energy current;virtual current;watt current
有功电路网络方程 active circuit mesh equation
有功电能量 active energy
有功电压 active voltage
有功分量 active component;energy component;power component;real component;watt component
有功负荷 active load
有功负载 real load;resistive load
有功功率 active power;real power;wattful power
有功功率成组调节装置 group active power regulating device
有功功率继电器 active power relay
有功率输入的稳定 powered stabilization
有功输出 active output;useful output
有功损耗 active loss
有功损失 active loss
有功效的 powerful
有功效的东西 medicine
有功效率 available power efficiency
有拱顶的长方形基督教堂 vault christian basilica
有拱顶的长方形建筑物 vault basilica
有拱顶的长方形教堂 vault basilica church
有拱顶的长廊 piazza
有拱顶的大厅 vaulted hall
有拱顶的大厦 vaulted edifice
有拱顶的房屋 vaulted building
有拱顶的教堂 vaulted church
有拱顶的内厅 vaulted interior
有拱顶的室 vault
有拱顶的通廊 gallery arcade
有拱顶的小室 vaulted chamber
有拱顶的走道 < 两旁设商铺 > arcade;vaulted walk
有拱顶的走廊 vault corridor
有拱廊的人行道 arcaded sidewalk
有拱廊街道 arcade
有沟槽的玻璃 broad reeded
有沟的 plicate;reeded
有沟壳 grooved shell
有沟探针 grooved probe
有沟探子 grooved probe
有钩扳手 hooked spanner
有钩长臂 < 启闭门上的 > long arm
有钩的 hooked
有钩的长杆 bait
有钩的下滑轮 hook bottom block
有钩吊板 hoisting pad
有购买权的租约 lease with option to purchase
有估计能力的 estimative
有固定轴线的开启桥 balance bridge with fixed axis
有故障的 defective
有故障的电路 faulty circuit
有故障的接头 defective joint
有故障的绝缘节 defective joint
有故障的装备 malfunction(ing) equipment
有故障电路 faulted circuit
有故障飞机 ailing aircraft;sick aircraft
有关本行的谈话 shop talk
有关波长 relevant wavelength
有关部门 appropriate body;related department;the department concerned
有关参数 pertinent parameter
有关产品 related product
有关成本 related cost;relevant cost
有关船级社遗留问题 outstanding recommendation against class
有关单位 related parties
有关当局 the authorities concerned;the proper authorities
有关当事人 client concerned;interested party;party concerned
有关的 concerned;related
有关的各运输单位 interested carrier
有关的价值中心 centrality of related values
有关的细节 pertinent detail
有关的小时 relevant hour
有关方(面) interested parties;relevant parties;the parties concerned
有关费用 relative cost
有关各界 interested circle
有关工种 related trades
有关哈托尔女神及其头像柱的 Hathoric
有关海上打捞爆炸物的说明 instruction regarding explosive picked up at sea
有关航行方面的问题 navigational aspects
有关合同双方对于索赔、付款争议等的和解协议 accord and satisfaction
有关河流的信息库 river database
有关恒载 contributory dead load
有关环境的后续行动 environmental follow-up
有关环境的资料 environmentally-related measurements
有关机构 appropriate body
有关机关 appropriate body
有关建筑的 architectural
有关来文 related communications
有关劳务的统计 returns of labo(u)r
有关旅游客运的 cruise-oriented
有关情况 pertinent condition
有关区域 domain of dependence
有关数据 pertinent data;relevant data
有关数据文件 relational data file
有关水下资料 information concerning submarine
有关税收的规定 tax-related provisions
有关条文 relevant clauses
有关系的 pertaining
有关系的一方 interested party
有关行业 related trades
有关一方要求赔偿的放弃 waiver by interested party
有关因数 pertinency[pertinency] factor
有关应提供数据的标准 standard on data to be provided
有关章节 pertinent passage
有关证件 appertaining documents
有关支付的 pay

有关仲裁的 arbitrational
有关主管机关 interested administration
有关主题 related topics
有关注意事项 matters needing attention
有关资料 interrelated data;pertinent data
有冠飞轮 flywheel crowned for flat belt
有管的 tubular;tubulate;tubulose
有管接头的水龙头 union cock
有管辖权的机关 authority having jurisdiction
有管造成的 tubular
有贯通风管的车辆 through pipe car
有光白 gloss white
有光布 lustre cloths
有光彩的铅质玻璃 strass
有光瓷漆 gloss enamel
有光化性的 actinic
有光蜡 self-polishing wax
有光面漆 gloss paint finish
有光漆 gloss paint
有光清漆 gloss varnish
有光乳化漆 gloss emulsion
有光乳胶漆 gloss emulsion
有光色料 gloss colo(u)r
有光涂料 gloss paint
有光油墨 gloss ink
有光泽 glossiness
有光泽表面 glossy surface
有光泽的 non-matting;shiny;sleek;watered
有光泽的劈裂面 cleavage fracture
有光泽的釉面瓦 satin-glazed tile
有光泽的釉面砖 satin-glazed tile
有光泽清漆 glossy varnish
有光泽釉料 glossy glazing
有光整理 lustrous finish
有光纸照片 glossy print
有规度【化】tacticity
有规反式构形 trans-tactic
有规划的地段发展 planned unit development
有规划的地段建设 planned unit development
有规划的住宅地段建设 residential planned community
有规聚合物 regulated polymer
有规立构的 stereoregular;stereospecific
有规立构聚合物 tactic polymer
有规律变化 rhythmic(al) change;rhythmic(al) variation
有规律的检查 methodical inspection
有规律的循环运动 rhythm
有规律移动 rhythmic(al) movement
有规溶胶 tactosol
有规则 orderly;regularity
有规则穿孔的 regularly perforated
有规则的 measured;systematic
有规则的层理 rhythmic stratification
有规则的沉积 rhythmic sedimentation
有规则的分层 rhythmic layering
有规则的累积 <海滩或海底的> rhythmic accumulations
有规则的横向沉陷 regularly-spaced transverse depression
有规则进程 regular procession;regular progression
有轨车道 railroad;railway
有轨车辆 tracked vehicle
有轨打桩机 railway pile driver
有轨导向车辆 rail-guided vehicle
有轨道电路的闭塞区段 track-circuited block section
有轨道电路的闭塞区间 track-circuited block section
有轨道电路容许闭塞 track circuit permissive block
有轨道结构 ballasted track structure
有轨的 rail-mounted;trackbound
有轨电车 railroad car;rattler;street car;tram(car);tramway motor-car;troll(e)y;electric(al) troll(e)y <美>
有轨电车(车)道 car track lane;street railroad;street railway
有轨电车车库 tram depot
有轨电车道交叉 tramway crossing
有轨电车轨道 tram rail;twin-cable ropeway;twin-cable tramway
有轨电车架空线 troll(e)y line;troll(e)y wire
有轨电车交通 tram service
有轨电车路 tram road
有轨电车路线 car track line;tramcar;tramway
有轨电车内噪声 noise inside the railroad car
有轨电车探测器 street-car detector
有轨电车停车场 tram depot
有轨电车停车站 tram depot
有轨电车线(路) street railway;tramline;tramway track;street-car line <美>
有轨电车转换器 tramway switch
有轨电车转辙器 tramway switch
有轨高速交通 rapid rail transit
有轨滑模 railed slipform
有轨机动车 rail motor car
有轨箕斗 rail skip
有轨交通 rail transit
有轨交通方式 rail transit mode
有轨快速交通 rail rapid transit
有轨快速交通(路)线 rapid transit line
有轨快速客运系统 fixed guide way transit system
有轨链斗挖掘机 rail bucket ladder excavator
有轨马车 horse tram-car
有轨马车道 horse car line;horse tramway
有轨起重器 railroad jack
有轨气垫车 tracked air-cushion vehicle
有轨气垫式车辆 tracked air cushion vehicle
有轨气垫式快车 tracked air cushion vehicle
有轨汽车 railway motor car
有轨设备 rail-mounted equipment
有轨挖土机 rail shovel
有轨巷道堆垛工作级别 classification group of S/R [storage/retrieval] machine
有轨巷道堆垛起重机 S/R machine
有轨巷道堆垛设计重量 design mass of S/R machine
有轨巷道堆垛水平运行 travel(l)ing of S/R machine
有轨巷道堆垛稳定性 S/R machine stability
有轨巷道堆垛重量 total mass of S/R machine
有轨小车 rail bogie;rail buggy
有轨斜坡的垫板 canted tie plate
有轨移动式破碎机组 rail-mounted mobile crushing plant
有轨运输 rail transit;rail travel;track haulage;tramming;rail haulage
有轨运输坡度 grades for rail haulage
有轨运送拌和车 transit-mixing rail-car
有轨凿岩车 track-mounted jumbo
有轨抓斗起重机 rail grabbing crane
有轨(自动)车 rail car
有轨钻车 jumbo for rail service;track-mounted jumbo;track-type jumbo
有国际意义的生态系统 internationally significant ecosystem
有过穿堂风的 draughty
有过梁的 linteled
有过失的 blameworthy;delinquent
有害 disadvantage
有害边缘水 harmful marginal water
有害变化 detrimental change
有害变形 injurious deformation
有害变异 harmful variation
有害材料 deleterious material;hazardous material
有害残渣 harmful residue
有害产品 noxious product
有害沉降 detrimental settlement
有害成分 harmful ingredient;injurious ingredient;noxious constituent;objectionable constituent;pernicious ingredient
有害刺激(物) noxious stimulus
有害的 deleterious;detrimental;epinosic;injurious;nocuous;noisome;noxious
有害的大气环境 hazardous atmosphere
有害的大气污染物 hazardous air pollutant
有害的东西 detrimental
有害的海绵状结构(能引起沉降危险的土结构) detrimental sponginess
有害的空气污染物 hazardous air pollutant
有害的毛细管升高 detrimental capillary rise
有害的水 deleterious water;harmful water
有害的炭沉积 harmful carbon deposition
有害的土地使用 adverse land use
有害的中子吸收剂 poison
有害反射气 detrimental reflection sound
有害反应 adverse reaction;deleterious reaction
有害废料 hazardous waste
有害废料处理 hazardous waste management;treatment of hazardous waste
有害废料焚烧处理 hazardous waste incineration
有害废料管理单元 hazardous waste management unit
有害废料管理设施 hazardous waste management facility
有害废料控制系统 hazardous waste control system
有害废水 deleterious effluent;noxious wastewater;pernicious effluent
有害废物 deleterious waste;harmful waste;hazardous waste
有害废物处理 hazardous waste disposal
有害废物管理 hazardous waste management
有害废物海洋焚化 ocean incineration of toxic waste
有害废液 hazardous waste
有害粉尘 harmful dust
有害辐射 harmful radiation
有害辐射防护学 health physics
有害干扰 harmful interference
有害工业 noxious industry;offensive industry
有害含量 injurious amount
有害化合物 hazardous compound
有害化学品 hazardous chemical
有害化学物 hazardous chemical
有害环境 hostile environment
有害环境区域 hazardous area
有害混杂物 harmful impurity
有害货物 noxious goods
有害健康的 unwholesome
有害健康的事物 health hazard
有害健康物质的控制 control of substance hazardous to health
有害菌 harmful bacteria
有害空间 noxious space;gap in the frog;open throat <辙叉的>
有害昆虫 harmful insect
有害昆虫侵袭 insect infestation
有害垃圾 harmful refuse;hazardous waste
有害排放物 noxious emissions
有害配料 harmful ingredient
有害膨胀 detrimental expansion
有害坡道 severe gradient
有害坡度 severe gradient
有害坡段 harmful district
有害气流 bad air
有害气体 deleterious gas;harmful gas;harmful gas and vapo(u)r;nocuous gas;noxious gas;pernicious gas;toxic fume;toxic gas
有害气体容许浓度 allowable density of harmful gas
有害气体中毒 harmful gas poisoning
有害气味 objectionable odo(u)r
有害缺陷 injurious defect
有害溶剂 harmful agent
有害三废处理 hazardous material disposal
有害色素 noxious colo(u)ring matter
有害闪烁 objectionable flicker
有害生长 obnoxious growth
有害生物 pest
有害生物综合治理 integrated pest management
有害树脂 harmful pitch
有害土壤 detrimental soil
有害微粒 hazardous particle
有害微粒元素 harmful microelement
有害污染(物) detrimental contamination
有害污染物质 harmful pollutant
有害物的输送 movement of hazardous material
有害物扩散 diffusion of noxious substance
有害物指标 nuisance value
有害物质 deleterious matter;deleterious substance;deleterious waste;harmful material;harmful matter;harmful substance;harmful waste;hazardous material;injurious material;nuisance;harmful agent;poisonous material
有害物质积累 accumulation of harmful material
有害物质间接作用 indirect action of noxious substance
有害物质浓度 concentration of harmful substance
有害物质容许限值 permissible limit of harmful substance
有害吸收 unwanted absorption
有害细菌 harmful bacteria;malignant bacteria
有害效应 harmful effect;ill-effect
有害性 harmfulness
有害性物质 hazardous substance
有害烟尘 mineral dust
有害烟雾 noxious fume
有害盐类 deleterious salts
有害溢流的临时容器 containment box
有害因素 adverse factor
有害影响 harmful effect

有害有机化合物 hazardous organic chemicals
有害于健康 health hazard; health risk
有害元素 harmful element; hazardous element; toxic element
有害运动 destructive movement
有害杂草 injurious weed
有害杂质 detrimental impurity; harmful admixture; harmful impurity; inimical impurity; objectionable impurity; troublesome impurity
有害杂质平均允许含量 admissible average content of harmful impurity
有害藻类水华 harmful algal bloom
有害种 obnoxious species
有害阻力 detrimental resistance; passive drag
有害组分 harmful component; harmful constituent
有害组分平均允许含量 average allowable amount of harmful components
有害作用 detrimental effect; harmful effect
有害作用阈 threshold of adverse effect
有焊缝的屋面 seam roofing
有航差像片 crabbed photograph
有航摄资料地区 photographic(al) coverage
有耗电介质 lossy dielectric
有耗网络 dissipative net(work)
有耗终端负载 lossy termination
有合法利益的人际关系 privity
有和易性混凝土 workable concrete
有河流的 watered
有核火山泥球构造 nuclear volcanic mud ball structure
有荷因数 duty cycle
有荷载的木结构 timber loaded structure
有荷载膨胀 loaded swelling
有荷载膨胀率 loaded swelling rate
有黑色条纹的深褐色硬木 <产于新西兰> black maire
有横肋的 cross-ribbed
有横楣分隔的窗 transom window
有横木踏板的绳梯 accommodation net
有横墙的格仑 diaphragm cell
有横墙的开口沉箱 caisson with digging wells
有横向钢筋的侧限区 confined region
有红条纹的红灰色硬木 <中美洲产> viraru
有厚橡皮套的 cabtire
有呼叫指示器的电话局 call indicator exchange
有湖灰岩洼地 lake polje
有湖坡立谷 lake polje
有虎斑的 brinded
有护板的 lined
有护壁的鼠道 mole-channel with liner
有护道的料堆 bermed pile
有护顶的机棚 protecting roof
有护轨的线路 check rail track
有护栏的铁路平面交叉 guarded railway crossing
有护栏楼梯 stairway enclosure
有护面的河岸 paving bank
有护面的河床 armo(u)red bed
有护套的 jacketed
有护套室的空调房间 jacketed air conditioning room
有护网风扇 cased type fan
有护罩的汽缸 shield-carrying cylinder
有花斑纹的砖瓦 mottled tile
有花洞的墙 pierced wall
有花岗石纹的器皿 granite ware
有花饰的拱 lobed arch
有花饰的面柱 column-figure portal
有花纹的 checkered; dappled; figured
有华盖的祭坛 baldachin altar
有华盖的墓 baldachin tomb
有滑动触点调节的感应线圈 inductance with sliding-contact adjustment
有滑溜危险的 skid-inducing
有滑轮的卸扣 roller shackle
有滑移危险的 skid-inducing
有滑油腔的轴承 pocketed bearing
有划痕的底涂 scratch coat
有环境意识的 environmentally conscious
有环纹的 annulated
有环纹的柱 annulated column
有环氧树脂涂层的钢筋 epoxy-coated rebar
有环轴颈 journal with collar
有环组成的 annulated
有缓冲器的终端 buffered terminal
有缓和段的圆曲线 circular curve with transition
有缓行器的编组场【铁】retarder classification yard; retarder-equipped yard
有缓行器装备的驼峰【铁】retarder-equipped hump
有簧塞孔 spring jack
有灰斑的 grizzled
有灰斑(缺陷)镀锡薄板 dry plate
有回弹力的 resilient
有回弹力的材料 resilient material
有回叫装置的电话机 call-back telephone set
有回扣销售 sale on commission
有回声的 resound(ing)
有混凝土块填心的毛石砌体 opus incertum
有混凝土填心的块石工程 opus incertum
有混凝土填心的块石砌体 opus incertum
有混凝土填芯的块石工 opus antiquum
有活动翻板的桌子 drop-leaf table
有活动河床的河流 river with shifting bed
有活动开合板的汽车顶 sun roof
有活动座椅的客车【铁】chair car
有活节的 articulated
有活塞止回阀的提升泵 lifting pump with bucket valve piston
有火灾征兆 fire-breeding
有火蒸汽机车 live engine
有或无毒作用 all-or-none toxic effect
有或无继电器 all-or-nothing relay
有机氨肥 ammoniate
有机半导体 organic semiconductor
有机包膜剂 organic coating agent
有机变质程度 level of organic metamorphism
有机变质作用 organic metamorphism
有机表面活性剂 organic surface active agent
有机玻璃 acryl glass; lucite; methyl methacrylate; organ(ic) glass; perspex; plastic glazing; Plexiglass; polymethyl methacrylate
有机玻璃板 perspex sheet
有机玻璃调刀 perspex spatula
有机玻璃定向 orientation of organic glass
有机玻璃模型 Plexiglass model
有机玻璃水槽 Plexiglass trough
有机玻璃瓦楞天窗 perspex
有机薄膜电容器 organic film condenser
有机补强剂 organic reinforcing agent
有机部分 organic fraction
有机材料 organ(ic) material
有机材料摩擦衬片 organic lining
有机残渣 organic detritus; organic slime
有机差热分析 organic differential thermal analysis
有机常数 organic constant
有机超导波导 organic superconducting waveguide
有机超导共振器 organic superconducting resonator
有机超导天线 organic superconducting antenna
有机超声探头 organic ultrasonic probe
有机车设备的车厢 dummy car
有机尘埃 organic dust
有机沉淀物需氧量 benthal demand
有机沉淀效应 organic sedimentation effect
有机沉积海岸 organic deposition coast
有机沉积矿床 organic deposit
有机沉积(物) cumulose deposit; organic deposit; organic precipitation; organic sediment; organogenous sediment; organogenic precipitation
有机沉积需氧量 benthal demand
有机成分 organic composition; organic constituent
有机成因 organic origin
有机成因的 organogenic
有机成因气 organic genetic gas
有机成因说 organic origin theory
有机冲洗 organic irrigation
有机除草剂 organic herbicide
有机促进剂 organic accelerator
有机淬灭计数管 organic quenched counter tube
有机萃取 organic extraction
有机氮 organ(ic) nitrogen; organonitrogen
有机氮化合物 organic nitrogen compound; organonitrogen compound
有机氮农药 organonitrogen pesticide
有机氮农药污染 pollution by organonitrogen pesticide
有机氮杀虫剂 organic nitrogen insecticide; organonitrogen insecticide
有机氮杀虫药 organic nitrogen pesticide; organonitrogen pesticide
有机氮源 organic nitrogen source
有机的 organic
有机底泥 organic bottom sludge
有机底泥释放的生物需氧量 biochemical oxygen demand releasing from organic bottom sludge
有机地球化学 organic geochemistry
有机碘 organic iodine
有机碘化合物 organic iodine compound
有机电化学 organic electrochemistry
有机电极过程 organic electrode process
有机电解液 organic electrolyte
有机电解液电池 organic electrolyte cell
有机电解质 organic bath
有机定量分析 organic quantitative analysis
有机毒剂 organic toxicant
有机毒物 organic poison; toxic organic material
有机毒物指数 organic toxic index
有机多硫化物 organic polysulfide
有机二硫化物 organic disulfide
有机发光体 organic luminophor
有机反应 organic reaction
有机反应动力学 organic reaction kinetics
有机反应机理 organic reaction mechanism
有机反应物 organic reactant
有机反应型增充剂 organic reactive extender
有机防腐剂 organic anticorrosion agent; organic corrosion inhibitor
有机放射性汞制剂 organoradiomercurial
有机非金属化合物 organo non-metallic compound
有机非线性光学材料 organic non-linear optic(al) material
有机肥耕作 biologic(al) husbandry
有机肥料 ammine; organic fertilizer
有机废料 organic refuse; organic waste(water)
有机废料回收设施 organic refuse recycling facility
有机废气处理 control of organic waste gas
有机废(弃)物 organic waste(water)
有机废水 organic effluent; organic waste(water)
有机废水处理 organic wastewater treatment
有机废液 organic waste liquor
有机沸石 organic zeolite
有机分析 organic analysis
有机分析标准 organic analytic(al) standard
有机分析试剂 organic analytic(al) reagent
有机分子 organic molecule
有机酚型添加剂 organic phenol type additive
有机粉尘 organic dust; particulate organic matter
有机粉土 organic silt
有机粪肥 organic manure
有机风化 organic weathering
有机氟残留 organofluorine residue
有机氟硅烷 organofluorosilane
有机氟化合物 organic fluro-compound
有机氟化学 organ fluorine chemistry
有机氟农药 organic fluorine pesticide
有机氟杀虫剂 organic fluorine insecticide
有机氟中毒 organofluorine poisoning
有机符号 organic symbol
有机腐败物 organic decay product
有机腐朽 organic decay
有机负荷 organic loading
有机负荷率 organic loading rate
有机富锌底漆 organic zinc-rich primer
有机覆盖物 organic mat
有机改性膨润 organic modified bentonite
有机高分子絮凝剂 organic polymeric flocculant
有机高分子重金属捕集絮凝剂 organic macromolecule heavy metal trapping flocculant
有机高分子助凝剂 organic polymer coagulant-aid
有机铬处理剂 organo-chromium finish
有机铬试剂 organo-chromium reagent
有机工业废水 organic industrial wastewater
有机汞 organic mercury
有机汞测定 organic mercury determination
有机汞除草剂 organomercurial herbicide
有机汞防霉剂 organomercurous fungicide
有机汞化合物 <木材防腐剂> organic mercurials; organomercury compound
有机汞化合物中毒 organo-mercuric

compounds poisoning
有机汞灭真菌剂 organomercury fungicide
有机汞农药 organomercury pesticide
有机汞农药中毒 organomercury pesticides poisoning
有机汞杀虫剂 organic mercury pesticide
有机汞杀菌剂 organomercurous fungicide
有机汞杀菌剂中毒 organomercury germicide poisoning
有机汞制剂 organic mercurials
有机汞中毒 organic mercury poisoning
有机汞转化作用 organomercurial transformation
有机骨料 organic aggregate
有机骨料混凝土 organic aggregate concrete
有机固体 organic solid
有机光学纤维 organic fiber [fibre]
有机硅 organic silicon; organo-silicone
有机硅醇酸树脂 silicone alkyd
有机硅的 organosilyl
有机硅废水 organo-silicone wastewater
有机硅呋喃树脂 organic silicon furan resin
有机硅化合物 organic silicon compound; organosilicon compound; organo-silicone compound; silicoorganic compound
有机硅胶 silastic
有机硅聚合物 organo-silicone polymer; organosilicon polymer; polysilicone; silicone polymer
有机硅绝缘漆 organic silicon insulating varnish
有机硅泡沫塑料 organic silicon foamed plastics
有机硅润滑剂 organo-silicic oil
有机硅树脂 organic silicon resin; silicone; silicone resin
有机硅树脂漆 silicone lacquer; silicone paint
有机硅树脂涂料 silicone resin coating
有机硅树脂橡胶(加)热电缆 silicone rubber heating cable
有机硅树脂液体 silicone liquid
有机硅树脂油 organic silicone oil
有机硅塑料 organic silicon plastics; silicone plastics
有机硅酸盐 organosilicate
有机硅酸盐涂层 organosilicate coating
有机硅涂层 silicone coat
有机硅涂料 organic fluorescent paint; silicone paint
有机硅烷 organosilan(e)
有机硅橡胶 organic silicon rubber; organo-silicone rubber
有机硅消泡剂 silicone defoamer
有机硅氧烷 organo-silicone; organosiloxane
有机硅氧烷聚合物 organic siloxane polymer
有机过渡金属 organotransition metal
有机过渡金属化合物 organo-transition metal compound
有机过酸 organic peracid
有机过氧化氢 organic hydroperoxide
有机过氧化物 organic over-oxide articles; organic peroxide; organoperoxide
有机含磷化合物 organophosphorus compound
有机含氯润滑油 organic chlorine oil
有机合成 organic synthesis
有机合成材料容器 polymeric material vessel
有机合成化学 organic synthetic(al) chemistry
有机合成农药 organic pesticide
有机合成颜料 organic synthetic(al) pigment
有机合成阳离子型絮凝剂 organic synthesis cationic flocculant
有机合成装置 organic synthesis plant
有机红颜料 <由煤焦油制得的> toluidine red
有机化合物 organ(ic) compound
有机化合物去除 organic compound removal
有机化合物中毒 organic compound poisoning
有机化金属 organ metal
有机化学 organ(ic) chemistry
有机化学家 organic chemist
有机化学可生物降解性 organic chemical biodegradability
有机化学实验室 organic chemistry laboratory
有机化学试剂 organic chemical reagent
有机化学污染物 organic chemical pollutant
有机化学物 organic chemicals
有机化学物生物降解 biodegradation of organic chemicals
有机化学纤维 organic chemicals fiber [fibre]
有机化学药品 organic chemicals
有机环境 organic environment
有机环境因素 biotic environment factor
有机环状化合物 organic ring compound
有机缓蚀剂 organic inhibitor
有机黄(颜料) organic yellow
有机灰尘 organic dust
有机混凝剂 organic coagulant
有机基 organic group; organic matter basis; organic radical
有机基体 organic matrix
有机基质 organic substance
有机集料 organic aggregate
有机集料混凝土 organic aggregate concrete
有机计数管 organic counter tube
有机剂 organic substance
有机碱 organic base
有机碱交换料 organolite
有机建筑 organic architecture
有机建筑理论 organic agriculture
有机建筑物 organic building
有机降解 organic degradation
有机交联膨润土 organic cross-linked bentonite
有机胶结料 organic cement
有机胶泥 organic mastic
有机胶态铁 organic colloidal iron
有机胶体 organic colloid
有机胶体分散剂 organic colloidal dispersant
有机胶体物质 organic colloidal matter
有机胶粘剂 organic adhesive
有机礁 organic reef
有机礁圈闭 organic reef trap
有机结合剂 organic bond
有机结合料 organ(ic) binder; organ(ic) binding agent
有机介电功能材料 organic dielectric(al) functional material
有机介质 organic medium
有机界 organic sphere; organic world
有机界的合理性 organic fitness
有机界进化 organic evolution
有机金属 organic metal
有机金属催化剂 organometallic catalyst
有机金属单体 organometallic monomer
有机金属的 organometallic
有机金属合成 organometallic synthesis
有机金属化合 metailorganic compound
有机金属化合物 organic metal compound; organometal; organometallic compound; pentamethide
有机金属(化合物)抗爆剂 organometallic anti-knock
有机金属化学 organometallic chemistry
有机金属聚合物 organometallic polymer
有机金属络合物 organometallic complex
有机金属取代作用 organometallic substitution
有机金属杀菌剂 organometallic fungicide
有机金属稳定剂 organometallic stabilizer
有机晶体 organic crystal
有机聚合物 organ(ic) polymer
有机绝缘(材料) organic insulation
有机颗粒 organic particle
有机颗粒磷 organic particulate phosphorus
有机颗粒物 organic particulate matter
有机空间 organic space
有机矿泥 organic slime
有机矿物 organic mineral
有机矿质超滤膜 organic mineral ultrafiltration membrane
有机矿质土(壤) organic mineral soil
有机矿质团粒 organic mineral aggregate
有机垃圾 organic refuse
有机来源 organic origin
有机冷却剂 organic coolant
有机冷却型原子反应堆 organic moderated reactor
有机理论 organic theory
有机锂化合物 organolithium compound
有机磷 organic phosphorus; organophosphor
有机磷残留 organophosphorus residue
有机磷残渣 organophosphorus residue
有机磷除草剂 organophosphorus herbicide
有机磷毒物 organophosphorus poison
有机磷光体 organic phosphor
有机磷化合物 organic phosphorous compound; organophosphorus compound
有机磷化学物 organophosphorus chemicals
有机磷农药废水 organics phosphorus pesticide wastewater; organophosphorus pesticide wastewater
有机磷农药检测 organic phosphorus pesticide determination; organophosphorus determination
有机磷农药污染 pollution by organophosphorus pesticide
有机磷农药中毒 organophosphorus intoxication
有机磷杀虫剂 organic phosphorous insecticide; organophosphorus insecticide; systox
有机磷杀虫剂中毒 organophosphorus insecticide poisoning
有机磷试剂 organophosphorus reagent
有机磷酸盐 organic phosphate; organophosphate
有机磷酸酯 organophosphate; organophosphorus ester
有机磷制剂 organophosphorus agent
有机磷中毒 organophosphorus poisoning
有机流出物 organic effluent
有机流体 organic fluid
有机硫 organic sulfur [sulphur]; organosulfur
有机硫化合物 organosulfur compound
有机硫化物 organic sulfide
有机硫黄化合物 organic sulfur compound
有机硫农药 organosulfur pesticide
有机硫排放物 organosulfur emission
有机硫杀菌剂中毒 organosulfur germicide poisoning
有机硫杀真菌剂 organic sulfur fungicide
有机硫酸盐 organic sulfate
有机硫酸酯 organic sulfate; organosulfate
有机馏分 organic fraction
有机卤化物 organic halide
有机卤素 organo-halogen
有机卤(素)化学物 organic halogen compound
有机氯 organochlorine
有机氯胺 organic chloramines
有机氯残渣 organochlorine residue
有机氯化合物 organic chlorine compound; organochlorine compound
有机氯化物 chloro-organic compound; organic chloride
有机氯农药 organochlorine pesticide
有机氯农药残留 organochlorine pesticide residue
有机氯农药污染 organochlorine pesticide pollution; pollution by organochlorine pesticide
有机氯农药中毒 organochlorine pesticide poisoning
有机氯杀虫剂 endosulfan; organochlorine insecticide; organochlorine pesticide
有机氯杀虫剂中毒 organochlorine insecticide poisoning
有机氯污染 organochlorine contamination
有机络合剂 organic complexing agent
有机络合物 organic complex
有机慢化剂 organic moderator
有机慢化冷却堆 organic moderated and cooled reactor
有机媒质 organic medium
有机镁化合物 organo-magnesium compound
有机镁卤化合物 organo-magnesium halide
有机锰化合物 organo-manganese compound
有机面 organic surface
有机膜 organic membrane
有机摩擦材料 organic(-type) friction material
有机耐磨材料 organic friction material
有机耐热涂层 organic thermal control coating
有机黏[粘]合法 organic adhesive method
有机黏[粘]合剂 organic cement
有机黏[粘]胶 organic cement
有机黏[粘]结剂 organic adhesive; organic binder; organic binder bond; organic binding agent
有机黏[粘]土 organ clay
有机凝胶 organogel
有机凝聚剂 organic coagulant
有机农药 organic pesticide

有机农用化学物 organic agricultural chemicals
有机泡沫塑料 organic foam plastic
有机硼 organoborane
有机硼化 organoboration
有机硼化合物 organoboron compound
有机硼烷 organic-borane
有机膨润土 organic bentonite; organobentonite
有机平面波导 organic planar waveguide
有机屏蔽层 organic shield
有机漆 organ(ic) varnish
有机气体 organic gas
有机气体计数管 organic vapo(u)r counter tube
有机铅化合物 organo-lead compound
有机铅抗爆剂 organometallic lead antiknock compound
有机氰化物 organic cyanide
有机氰化物废水 organic cyanide wastewater
有机燃烧 organic combustion
有机染料 organic dye; organic dyestuff
有机染料溶液激光器 organic dye solution laser
有机染料示踪剂 organic dye-tracer
有机溶剂 organic solvent
有机溶剂防腐剂 organic solvent preservative
有机溶剂木素 organosol lignin
有机溶剂涂布 organosol coating
有机溶剂脱脂(法) organic solvent degreasing
有机溶剂型防腐剂 organic solvent type preservative
有机溶剂中毒 organic solvent poisoning
有机溶胶 organisol [organosol]
有机溶解组分 organic dissolved component
有机溶媒 organic solvent
有机溶液 organic fluid; organic solution
有机溶液防腐剂 organic solvent preservative
有机溶质 organic solute
有机润滑剂 organic lubricant
有机润滑脂 organic grease
有机色料 organic colo(u)ring matter
有机杀虫剂 organic insecticide
有机杀菌剂 organic bactericide
有机杀真菌剂 organic fungicide
有机闪烁溶液 organic scintillating solution
有机闪烁探测器 organic scintillation detector
有机闪烁体 organic scintillator
有机砷化学物 organoarsenic chemical
有机砷农药 organoarsenic pesticide
有机胂 organic arsine
有机生长物质 organic growth substance
有机生成沉积物 organogenous sediment
有机生成的 organogenic
有机生油物质 organic source material
有机石灰岩 organic limestone
有机试剂 organic reagent
有机疏散 organic decentralization
有机疏水性 organo-phobicity
有机树脂 organic resin
有机衰减 organic decay
有机水化学 organic hydrochemistry
有机水污染物 organic water pollutant
有机塑料 organic plastics
有机酸 organ(ic) acid
有机酸废水 organic acid wastewater
有机酸利用试验 organic acid utilization test
有机酸溶性磷 organic acid soluble phosphorus
有机酸性 organic acidity
有机酸盐 salt of organic acid
有机碎屑 organic bits; organic debris; organic fragment
有机钛酸酯 organic titanate
有机碳 organic carbon
有机碳法 organic carbon method
有机碳富集作用 organic carbon concentration
有机碳含量 organic carbon content
有机碳门阳值 threshold value of organic carbon
有机碳黏[粘]土 kibusi clay
有机碳平均值 average value of organic carbon
有机碳下限含量 lower limit of organic carbon
有机碳总量 total organic carbon
有机碳最大的最小值 max-min value of organic carbon
有机碳最大值 maximum value of organic carbon
有机碳最小的最小值 min-min value of organic carbon
有机碳最小含量 minimum quantity of organic carbon
有机碳最小值 minimum value of organic carbon
有机体 organic body; organism; organization
有机体的化学组成 chemical composition of organism
有机体发光的 photic
有机体类型和来源 type and source of organism
有机体指数 organosomatic index
有机体主要类型 principal type of organism
有机添加剂 organic addition agent; organic additive; organo-additive
有机填充物 organic filler
有机填料 organic filler
有机调色剂 organic toner; toner
有机铁化合物 organo-iron compound
有机涂层 organic coating
有机涂层钢板 organic coating steel sheet
有机涂盖层 organic coating
有机涂料 organic coating
有机土 histosol
有机脱臭剂 organic deodorizer
有机完整性 organic completeness
有机微污染物 organic microcontaminant; organic micropollutant
有机尾矿 organic debris
有机稳定剂 organic stabilizer
有机稳定性二氧化氯 organic stabilized chlorine dioxide
有机污泥 organic sludge
有机污染 organic contamination
有机污染标准 organic pollution criterion
有机污染生物指数 biotic index of organic pollution
有机污染物 organic contaminant; organic pollutant
有机污染物标准 organic pollutant criterion
有机污染物分布自记仪 distribution register of organic pollutant
有机污染物分析 organic pollutant analysis
有机污染物负荷 organic pollutant load
有机污染物降解 organic pollutant degradation
有机污染物累积 organic pollutant accumulation
有机污染因子 organic pollution factor
有机污染指数 organic pollution index
有机污水 organic sewage; organic waste(water)
有机污着底泥 organic fouling deposition
有机物 organism; organic matter
有机物残体 organic debris
有机物残渣 debris
有机物沉淀效应 organic sedimentation effect
有机物沉积 organic deposit
有机物成分 organic component
有机物底泥指数 organism-sediment index
有机物堆积 organic accumulation
有机物多相光催化降解反应 organism multiphase photo catalysis degradation reaction
有机物废料 organic waste
有机物分离效率 organic matter separation efficiency
有机物分子量分布 molecular weight distribution of organics
有机物封填双层玻璃窗 organic-seal double-glazed unit
有机物风化 organic weathering
有机物干重 organic dry weight
有机物含量 organic content
有机(物)合成化 organ synthesis
有机物降解菌 organism degradation bacteria
有机物降解作用 degradation of organic matter; organic matter degradation
有机物结合耐火材料 organic bonded refractory
有机物垃圾转换热能 thermal conversion
有机物冷却堆 organic-cooled reactor
有机物内浸出 intra-organic matter diffusion
有机物浓度 organics concentration
有机物汽化分析 organic vapo(u)r analysis
有机物去除 organic matter removal
有机物容量 volume of organic matter
有机物如植物(树木)苔藓等 organic materials like plant/wood moss
有机物生物降解 organic matter biodegradation
有机物湿度计 organic hygrometer
有机物水污染 water pollution by organic substance
有机物填充的电位器 organic solid composition potentiometer
有机物外形 organic profile
有机物污染 organic contamination; organic pollution
有机物循环 cycles of organic matter
有机物元素 organogen
有机物杂质 organic impurity
有机物黏[粘]合 organic bond
有机物蒸气 organic vapo(u)r
有机物指标 organic index
有机物指数 index organism; organic index
有机物质 organ(ic) matter; organ(ic) substance; organic material
有机物质分解 decay of organic matter
有机物质沥青毡 organic felt
有机物质自然腐化 natural decay of organic materials
有机物种 organic species
有机物转化 conversion of organic matter
有机物组分 organic component; organic constituent
有机吸附剂 organic adsorbent
有机稀释剂 organic thinner
有机锡 organotin
有机锡化合物 organotion compound
有机锡化合物污染 organotin compounds contamination
有机锡杀菌剂 organotin-fungicide; organotin germicide
有机锡杀菌剂中毒 organotin germicide poisoning
有机锡羧酸盐 organotin-carboxylate
有机锡稳定剂 organotin stabilizer
有机锡中毒 organotin poisoning
有机纤维 organic fiber [fibre]
有机显微组分 organic micro-constituent
有机相 organic facies; organic phase
有机相连续系统 organic-continuous system
有机相水相界面 organic-aqueous interface
有机硝基化合物 organic nitro-compound
有机形态 organic form
有机性粉尘 organic dust
有机性需要 organic need
有机溴 organic bromine
有机溴化合物 organic bromine compound
有机需要量 organic demand
有机絮凝剂 organic coagulant; organic flocculant
有机絮凝作用 organic flocculation
有机蓄积 organic accumulation
有机悬沙 organic suspended sediment
有机选择 organic selection
有机循环 organic cycle
有机亚磷酸酯 organic phosphite; organophosphite
有机岩 organic rock; organolite
有机岩屑 organ(ic) debris; organ(ic) detritus
有机盐 organic salt
有机颜料 organic pigment; pigmentary dyestuff
有机衍生物 organic derivatives
有机阳离子 organ(ic) cation
有机阳离子化合物 organic cationic compound
有机阳离子稳定(土壤)剂 organic cationic stabilizer
有机氧化 organic oxidation
有机液体 organic liquid
有机液体激光器 organic liquid laser
有机液体胶体 organic liquid gel
有机液体冷却剂 organic liquid coolant
有机乙酰氧 organoacetoxysilane
有机抑制剂 organic inhibitor
有机阴离子 organic anion
有机荧光漆 organic fluorescent paint
有机荧光体 organic fluorescent
有机营养 organic nutrition; organotrophy
有机油扩散泵 organic oil-vapour pump
有机油类 organic oil
有机油漆溶剂 organic lacquer solvent
有机油毡 organic coating; organic felt
有机铀化合物 organo-uranium compound
有机铀矿 urano-organic ore
有机铀络合物 organo-uranium complex
有机有毒物质 organic toxic substance

有机有害废物 organic hazardous waste
有机淤泥 organic silt; organic slime
有机元素 organic element
有机元素分析 organic element analysis
有机杂质 organic impurity
有机杂质含量检验 <骨料的> organic impurity test
有机杂质检验 organic impurity test
有机杂质试验 organic impurity test
有机皂土 bentonite
有机增稠剂 organic thickener
有机脂胶黏[粘]合剂 organic mastic adhesive
有机脂黏[粘]合剂 organic mastic adhesive
有机酯 organic ester
有机质 organic material; organic matter; soil-ulmin; terreau
有机质层 dirt bed; organic horizon
有机质沉积环境 depositional environment of organic matter
有机质成熟作用演化阶段 maturation evolution stage of organic matter
有机质成烃作用演化阶段 generation hydrocarbon evolution stage of organic matter
有机质成岩作用演化阶段 digenesis evolution stage of organic matter
有机质代谢作用带 metabolism zone of organic mater
有机质的沉积和保存 deposition and conservation of organic matter
有机质的降解作用 degradation of organic matter
有机质的量和产率 quantity and productivity of organic matter
有机质底泥 organic bottom mud
有机质堆积速率 accumulation rate of organic matter
有机质粉土 organic silt
有机质丰度 abundance of organic matter
有机质丰度指标 abundance indicator of organic matter
有机质高岭石高岭土矿石 kaolin ore of organic-kaolinite type
有机质含量 content of organic matters; organ(ic) content; organ(ic) matter content
有机质含量高的土 highly organic soil
有机质含量检验 organic (matter content) test
有机质含量试验 organic (matter content) test
有机质和有机常数 organic matter and constant
有机质活化能 activation energy of organic matter
有机质积累 accumulation of organic substance
有机质降解作用带 degradation zone of organic mater
有机质结构 organic texture
有机质来源 source of organic matter
有机质类型指标 indicator of subdividing organic matter type
有机质泥沙 organic silt
有机质泥砂 organic clay sand
有机质黏[粘]土 organic clay
有机质膨润土矿石 organic bentonite ore
有机质破坏 destruction of organic material
有机质谱法 organic mass spectrometry; organic mass spectroscopy
有机质谱分析 organic mass spectrometry; organic mass spectroscopy
有机质溶解成分 organic dissolved component
有机质生物化学作用带 biochemistry zone of organic mater
有机质试验 organic matter test
有机质熟化作用 maturation of organic matter
有机质土(壤) organic soil
有机质污染 organic pollution
有机质演化 evolution of organic matter
有机质演化阶段 evolution stage of organic matter
有机质组成 anabolism
有机致癌物 organic carcinogen
有机中间体 organic intermediate
有机重组 organic reconstruction
有机助催化剂 organic promoter
有机助凝剂 organic coagulant aid
有机准金属化合物 organometalloidal compound
有机着色剂 dyes and dyestuffs
有机阻蚀剂 organic corrosion inhibitor
有机组成 organic composition
有机组分 organic composition; organic constituent
有机组分和溶解气 organic constituent and dissolved gas
有机组合 organic constituent
有机组合直线空间 organic linear space
有机作用 organic effect
有基础的 well grounded
有基聚硅烷 organo-poly-siloxane
有基台模型 models with a sill
有基座的管道 cradle-invert pipe
有激浪日 surf day
有级变速 step speed change
有级钎肩式钎尾 staved collar drill shank
有极电路 polarised circuit
有极分子 polar molecule
有极分子吸附 absorption of polar molecule
有极继电器 polarised [polarized] relay
有极继电器接点 polarized relay contact
有极加强继电器 retained neutral polarized relay with heavy-duty contacts
有极键 heteropolar bond
有极接点 polar contact
有极音响机 polarized sound
有急弯的木工制品 quick sweep
有几分 to a measure
有几何图形的罗马式马赛克 opus signinum
有脊丛山 ridge mountains
有脊屋顶 ridge(d) roof
有计划怠工 absenteeism
有计划的 systematic
有计划的焚烧 prescribed burning
有计划的周期 planned period
有计划商品经济 planned commodity economy
有计量连接 metered connection
有记忆力的 retentive
有记忆信道 channel with memory
有纪念意义的建筑物立面 monumental facade
有技术根据的检查 engineering demonstrated inspection
有技艺的 workmanlike
有加力钢筋的板角 <混凝土> protected corner
有加强筋的 corrugated; finned
有甲套电缆 sheathed cable
有价成分 valuable constituent
有价成分含量 valuable content
有价约因 valuable consideration
有价债券 documentary securities
有价证券 documents of values; negotiable note; portfolio; valuable papers; value papers; value securities; active security <在市场流通使用的>
有价证券经纪人 <股票等> note broker
有价证券税 negotiable note tax
有价证券投资 portfolio investment
有价值的 valuable; valued
有价值的发现 valuable discovery
有价值的情报 valuable information
有价值的设备 equipment in value
有价值开采的矿层 pay formation
有价值物品 articles of value
有价资产 admitted assets
有驾驶台的挂车 driving trailer
有驾驶台的自动客车 driving coach
有尖叉的起重电磁铁 lifting magnet with tines
有尖齿的起重电磁铁 lifting magnet with tines
有尖顶的 steepled
有尖顶的纵断面 cusped profile
有尖角的砂 sharp sand
有尖塔的穹顶 dome with spire
有坚硬路面的道路 hard surface road
有间隔带 gapped tape
有间隙根部 open root
有间隙焊接 space welding
有间隙接头 open joint
有间柱的建筑物 stud work
有肩螺钉 shoulder screw
有肩钎杆 shanked drill rod
有监控的喷水灭火系统 supervised sprinkler system
有检查孔的管道 pipe with access eye
有减载平台的码头岸壁 platform wharf wall
有鉴别性的振铃(呼叫)系统 discriminating call system; discriminating ring system
有奖储蓄 bonus bearing certificate
有奖储蓄券 bonus bearing certificate
有奖励(条款)合同 incentive contract
有奖设计竞赛 prize competition design
有交叉槽的凸轮 cross grooved cam
有交叉撑杆的(桩基)突堤码头 braced pier
有交叉的四坡屋顶 hip-and-valley roof
有交叉间壁的开口沉箱 open caisson with cross walls
有交叉线的 retiform
有交错层理的 cross-bedded
有交错层理的岩石 cross-bedded rock
有交通管理的交叉口 controlled intersection
有交通信号的交叉口 signalized intersection
有交通信号控制的交叉口 signal-controlled intersection
有交折射率 effective index
有胶土填心的堤坝 puddle dyke [dike]
有角的 angulate
有角度规定的连接法 angular attachment
有角砾石 torpedo gravel
有角楼屋顶 turreted roof
有角螺母 horned screw nut
有角性 angularity
有角柱的楼梯 newel stair(case)
有脚大杯 beaker
有脚的架子 horse
有脚支架 horse
有铰 T 形刚构 T-frame with hinge
有铰拱 hinged arch
有铰接的翻斗车 articulated dump truck
有铰链窗 casement light
有铰链的门 flapper
有搅拌器的球磨机 agitator ball mill
有搅拌装置的(混凝土)运输车 agitator truck
有较高级视听设备和器材进行社会教育的机构 audio-visual center [centre]
有较好混响装置的播音室 live studio
有阶段的 scalary
有阶段缓解的制动机 brake with graduated release
有阶段缓解作用的分配阀 distributing valve with graduated release
有接点部件 contact component
有接点元件 contact component; contact element
有接缝的 jointed; seamy
有接缝配筋的混凝土路面 jointed reinforced concrete pavement
有接坡的交叉口 ramp crossing
有节的 nodal; nodose
有节点荷载的拉杆 loaded chord
有节理的岩层 jointed rock
有节木料 knot wood
有节乳液管 articulate latex duct
有节植物 arthrophyte
有节制的 low-key(ed); temperate
有节制的照明 low-key lighting
有节制溢流堰 controlled weir
有节制闸的桥 bridge-cum-regulator
有节奏涌波 rhythmic(al) surge
有节奏涌浪 rhythmic(al) surge
有洁净作用的 abstergent; abstersive
有结合力的 cohesive
有结节的 nodular; nodulated
有截口的梁 kerfed beam
有截沙井的进水口 drop inlet
有解 solution existence
有界变差 bounded variation
有界变差函数 function of bounded variation
有界变分 bounded variation
有界变更 limited variation
有界变量 bounded variable
有界变数 bounded variable
有界点集 bounded set of points
有界分数 limited fraction
有界函数 bounded fraction; bounded function; limited function
有界集(合)【数】 bounded set
有界矩阵 bounded matrix
有界区域 enclosed area
有界收敛 bounded convergence
有界输入有界输出 bounded-input-bounded-output
有界输入有界输出的稳定性 BIBO [boundary input and boundary output] stability
有界数集 bounded set of numbers
有界算子 bounded operator
有界投影 bounded projection
有界线性变换 bounded linear transformation
有界序列 bounded sequence
有界域 bounded region
有金属包层的 metal-clad
有金属光泽的 specular
有金属框的窗 metal framed window
有金属轮圈的轮子 metal-to-metal wheel
有金属味 metallic taste
有津贴的房屋 subsidized housing
有筋的 ribbed
有尽的 finite
有尽解【数】 finite solution
有尽连分数 terminating continued fraction
有尽小数 terminating decimal; finite

decimal
有经济价值的矿物 economic mineral
有经济效益的 cost-effective
有经济效益的荷载 economic load
有经验承包商 experienced contractor
有经验工程师 experienced engineer
有经验者 expert
有茎植物 anthophyta
有晶体管放大器的唱机 transistorized phonograph
有景观室内 landscaped interior
有径流时候 runoff event
有竞争能力的费率 competitive rates
有竞争能力的价格 competitive price
有竞争性投标 competitive bid
有镜面 with slickenside
有镜面光泽的 specular
有居民的 inhabited
有具体偿还期债券 dated stock
有距离的 distant
有锯齿状的 jaggy
有锯痕的背面 sawed back;sawn back
有决定性的 pacing
有决心的 decisive
有绝缘柄的钳子 insulated pliers
有绝缘节的轨道电路 insulated joint track circuit
有开采价值的矿藏 promising deposit
有开采价值的矿层 commercial bed
有铠装承载面的输送机 armo(u)red face conveyer [conveyor]
有看护的坝 dam with attendance
有抗风支撑的吊杆 wind-braced boom
有抗力的 resisting
有抗倾斜交叉撑的框架 braced frame
有颗粒的 seedy
有壳绳状熔岩 shelly pahoehoe
有壳水轮机 encased turbine
有可见呼叫指示的人工电话制 panel call indicator operation
有可开合后盖的小汽车 landau
有可能 most likely;within the bounds of possibility
有可逆式卸货的圆筒混合机 drum mixer with reversing discharge
有可以右转弯标志的直行车道 right-turning lane
有可以左转弯标志的直行车道 left-turning lane
有刻度觇标 scaled target
有刻度的 calibrated;scaled
有刻度的控制器 indicating controller
有刻度倾斜螺旋 gradienter screw
有刻痕的冲击试件 keyhole specimen
有刻痕的冲击试块 keyhole specimen
有客厅楼层 piano nobile
有坑槽的路面 pitted surface
有坑洞的混凝土路面 potted concrete surface
有坑洞的路面 potted surface
有坑洼的 pitted
有空洞的 porous
有空气隙铁心 gapped core
有空气制动机的棚车 covered air-brake van
有空位的电子层 vacant shell
有空心肋的 hollow-ribbed
有空心柱塞的提升泵 lifting pump with hollow plunger
有空中交通管制雷达的环境 radar environment
有空重车的制动系统 empty-loaded braking system
有孔板<陈列杯瓶用的> abacus [复 abaci/abacuses]
有孔薄膜 slit-film
有孔材 broad-leaved wood; pored wood;porous wood
有孔虫 foraminifer
有孔虫补偿深度 foraminiferal compensation depth
有孔虫动物地理区 foraminiferal faunal province
有孔虫目<拉> Foraminifera
有孔虫溶跃层 foram lysocline
有孔虫软泥 foraminiferal ooze
有孔虫(石)灰岩 foraminiteral limestone
有孔虫岩 foraminite
有孔的 holey;meshed;pervious;pierced
有孔的砌块 hole block
有孔的砖块 hole block
有孔洞或缺口的基础 footing with holes or notches;foundation with holes or notches
有孔端 eye end
有孔防波堤 porous breakwater
有孔封头 perforated head
有孔盖板 perforated cover plate
有孔杆 eye bar
有孔钢板拼装的跑道 steel pierced-plank runway
有孔给水管 perforated feed pipe
有孔管 perforated pipe
有孔滚筒 perforated roller
有孔集水管 perforated collector pipe
有孔胶片 perforated film
有孔井管 perforated casing
有孔巨型浇钢用砖 tundish brick
有孔抗剪墙 pierced shear wall
有孔路面 porous pavement
有孔螺母 capstan nut
有孔黏[粘]土 keramite
有孔平面 punctured plane
有孔墙 perforated wall
有孔套管 perforated pipe casing
有孔镶板墙 pierced panel wall
有孔小珠 bead
有孔屑纸带 chadless tape
有孔眼的套管 perforated casing
有孔(硬)纤维板 ornamental hardboard;perforated hardboard
有孔渣 inflated slag
有孔中心板 female center plate
有孔砖 air brick;perforated brick
有控水库(洪水)演算 routing in controlled reservoir
有控制的供水 controlled water supply
有控制的垃圾填埋 controlled landfilling
有控制的倾度 controlled dumping
有控制的人行横道 controlled pedestrian crossing
有控制的(分层压实)填土 controllable fill
有控制的混凝土配合比设计 controllable design concrete
有控制的间隙密封 controllable-gap seal
有控制的接近海底的拖曳法 controllable above-bottom pull method
有控制的水下接近海面的浮运法 controllable underwater floatation method
有控制的蓄水 controlled storage
有控制的溢洪道 controlled spillway
有控制的溢流堰 controlled weir
有控制的制动 control braking
有控制点镶嵌图 checked mosaic;controlled mosaic
有控制交通 restricted traffic
有控制螺丝刀 clutch-type screwdriver
有控制设施的溢洪道 controlled spillway
有控制镶嵌图 controlled photomosaic
有库存的 in stock
有库河流 impounded river
有库容的电站 plant with storage
有跨线桥的环形立交 flyover roundabout
有跨线桥的环形立体交叉 flying roundabout
有宽敞空间的 roomy
有矿物钻孔 ore hole
有阔孔的 wide-bore
有拉撑和支撑的表面 braced and stayed surface
有拉杆的岸壁 anchored bulkhead
有拉杆的岸墙 tied wall
有拉杆的板桩 anchored sheet piling
有拉杆的板桩岸壁 anchored sheet pile bulkhead
有拉杆的船坞墙 tied wall
有拉杆的挡土墙 tied retaining wall
有拉杆的拱 arched girder without horizontal thrust
有拉杆的拱式构架 tied-arch frame
有拉杆的拱形大梁 arched girder with tie
有拉杆的肋拱 tied rib arch
有拉杆的码头岸壁 anchored bulkhead quay; anchored bulkhead wharf
有拉杆的墙 anchored wall
有拉杆的人字屋 close-couple roof
有拉杆的悬臂结构 tied cantilever construction
有拉杆的悬臂桥 tied cantilever bridge
有拉杆的柱 tied column
有拉条拱桥 bowstring arch bridge
有拉线的电杆 guyed pole
有拦道<有拦路横木的场内小路> barway
有拦门沙的港口 bar port
有拦门沙的河口 bared river mouth
有拦污栅的进水口 screened intake
有栏杆的床 box bed
有栏杆的小床 crib
有栏杆的阳台 parapeted terrace
有栏路横木的场内小路 bar way
有栏栅扶手的楼梯 railing stair(case)
有浪花的 surfy
有浪水流 chopping flow;choppy flow
有老虎窗的屋顶 dormer(ed) roof
有肋(薄)壳 ribbed shell
有肋的 ribbed
有肋的输送机胶带 ribbed conveyer belt
有肋拱 ribbed arch
有肋拱顶 ribbed vault;rib vaulting
有肋结构 ribbed construction
有肋框架 ribbed frame;ribbed framing
有肋梁筏片基础 floating foundation costae girder
有肋楼板 ribbed slab
有肋面 extended surface
有肋片的 finned
有肋片和凸纹的毛玻璃 reeded glass
有肋汽缸 ribbed cylinder
有棱断面 bluff-section
有棱角的粗骨料 angular coarse aggregate
有棱角的骨料 sharp aggregate
有棱角的集料 angular aggregate
有棱角的砂 angular sand;sharp sand
有棱角骨料 angular aggregate
有棱角混凝土滑料 angular concrete aggregate
有棱角颗粒 angular fragment;angular grain;angular particle
有棱角砂 harsh sand
有棱角石料 angular stone
有棱角碎片 angular fragment
有棱锥形(羊)足的压路机 pyramid feet roller
有棱锥形(羊)足的压路碾 pyramid feet roller
有冷藏设施的 cool
有冷气设备的 air-conditioned
有冷却套压缩机 jacket-cooled compressor
有里衬的永久性模板 permanent shuttering with inner lining
有理逼近 rational approximation
有理变换 rational transformation
有理变换群 rational transformation group
有理标准矩阵 rational canonical matrix
有理标准型 rational canonical form
有理表示 rational representation
有理不变式 rational invariant
有理插值函数 rational interpolating function
有理磁面 rational magnetic surface
有理簇 rational variety
有理代数簇 rational algebraic variety
有理代数分数 rational algebraic fraction
有理代数式 rational algebraic form
有理单位系统 rationalized system
有理单位制 rational system of units
有理的 rational;unicursal
有理等价类群 rational equivalence class group
有理点 rational point
有理对合 rational involution
有理对合性代数 rational involutorial algebra
有理对应 rational correspondence
有理多项式 rational polynomial
有理方程式 rational equation
有理分式 rational formula; rational fraction
有理分式插值 rational fraction interpolation
有理分数 rational fraction
有理分数函数 rational fractional function
有理分析 rational analysis
有理根【数】 rational root
有理函数 rational function
有理函数逼近 rational function approximation
有理函数近似 rational function approximation
有理函数域 rational function field
有理化 rationalization;rationalize
有理化表示法 rational representation
有理化单位系统 rationalized system of units
有理化分母 rationalizing denominator
有理化径流公式 rational runoff formula
有理化式 rational formula
有理化因子 rationalizing factor
有理化运算 rational operation
有理回转型曲面 rational swinging surface
有理活度系数 rational activity coefficient
有理积分表示式 rational integral expression
有理积分函数 rational integral function
有理级数 rational series
有理交截定律 law of rationality
有理近似 rational approximation
有理矩阵的规范型 canonic(al) form of rational matrix
有理可除代数 rational division algebra
有理空间 rational space

有理零点 rational zero points
有理蒙皮曲面 rational skinning surface
有理内射性 rational injectivity
有理区间 rational interval
有理区间函数 rational interval function
有理区域 register of rationality
有理曲面 rational surface
有理曲线 rational curve; unicursal curve
有理曲型简化 rational canonical reduction
有理实数 rational real number
有理式 rational expression; rational form; rational formula
有理数 rational; rational number
有理数域 rational number field; region of rationality
有理素数 rational prime number
有理特征标 rational character
有理同态 rational homomorphism
有理微分方程 rational differential equation
有理微分组 rational differential system
有理型的 meromorphic
有理性 rationality
有理循环 rational cycle
有理映射 rational mapping
有理由的 well-founded
有理元素 rational element
有理真分式 rational proper fraction
有理整函数 rational entire function
有理整数 rational integer
有理整数环 ring of rational integers
有理整数解 rational integral solution
有理值巢 rational-valued nests
有理指数 rational index
有理指数定律 law of rational indices
有理秩 rational rank
有理子群 rational subgroup
有理作用 rational action
有力搬运的 portative
有力的 convincing; powerful; predominant; strong; vigorous; weighty
有力的握手 hand grip
有力矩理论 shell theory with moments
有力证明 eloquent proof
有历史意义的建筑 building of historic interests
有立体感的彩色壁画 stereochrome
有立体感的地形图 hypsography
有立柱车体的卡车 stake body truck
有利坝址 viable site
有利比 advantage ratio
有利变异 favo(u)rable variation
有利标志状态 useful index state
有利程度 profitability
有利大气条件 favo(u)rable atmosphere condition
有利的 advantageous; favo(u)r; favo(u)rable; lucrative; payable; profitable; remunerative
有利的购买 comparative advantage
有利的坡度 favo(u)rable grade
有利的前景 favo(u)rable prospect
有利的生产 comparative advantage
有利的投资 good investment; lucrative investment
有利的影响 benefit effects
有利地点 favo(u)rable location
有利地位 favo(u)rable position; inside track; perch; vantage
有利断面 <隧道> pay-section
有利发射时机 launch window
有利方法 rational method
有利方面 positive aspect
有利工程地址 viable site
有利观测条件 favo(u)rable observation condition
有利害关系的人 privy
有利汇兑 favo(u)rable exchange
有利几何条件 favo(u)rable geometry
有利价格 remunerative price
有利结果 favo(u)rable result
有利可图 revenue-earning
有利(可图)的投资 profitable investment
有利跨河地点 favo(u)rable stream crossing
有利目标 favo(u)rable target
有利情形 favo(u)rable case
有利示像 favo(u)rable aspect
有利事件 favo(u)rable event
有利台阵场地 favo(u)rable array location
有利天气 favo(u)rable weather
有利条件 advantage; favo(u)rable condition; favo(u)rable term
有利外部条件 external economics
有利显示 favo(u)rable indication
有利相位 favo(u)rable phase
有利性 profitability; stead
有利性调查 profitability investigation
有利岩层 favo(u)rable beds
有利岩层与断裂交接地段 intersection of favo(u)rable bed with fractures
有利岩层与侵入体交接地段 intersection of favo(u)rable bed with intrusion
有利因素 plus factor
有利于 profit
有利跃迁 favo(u)red transition
有利状态 favo(u)red state
有沥青盖层的碎石路 coated macadam
有连拱廊的街道 arcade
有连廊的列车 vestibule train
有联锁设备的车站 interlocking station
有链条炉排的层燃炉 chain grate stoker
有良好耕性的土壤 soil with good tilth
有凉台的平房 bungalow
有两分隔间的箱式大梁 box girder with two compartments
有两个泊位的码头 dual berth facility
有两个房间的房屋的内室 ben
有两个牛腿的柱子 column with two brackets
有两个平衡架的筛子 screen with two counterbalanced frames
有两个托臂的柱子 column with two brackets
有两个翼部的寺庙 dipteros
有两文字的地图 bilingual map
有两翼的建筑物 double-wing(ed) building; two-winged building
有两支的 biramous
有两种功能的催化剂 duofunctional catalyst
有量纲量 dimensional quantity
有瞭望顶棚的守车 cupola car
有瞭望圆顶的客车 dome car
有裂缝的 chinky; flawy; seamy
有裂缝岩石 fissured rock; seamy rock
有裂纹的玻璃 crackled glass
有裂纹的水泥混凝土板 cracked slab
有林地 tree site
有林溪谷 dene
有林小陡谷 dean; dene
有鳞的 squamose
有鳞片的 scaled
有菱形图案的 diamond-patterned
有菱形图案的玻璃 diamond-patterned glass
有菱形辙叉的双开道岔 double turnout with diamond crossing
有龄期的 seasoned
有流体排出的泉 discharging spring
有流通性的 <货币、报刊等> circulative
有流线型尾部的 boat-tailed
有六边形柄的钻 drill with hexagonal shank
有六个的 sextuple
有六个驱动轮的运货汽车 six-by-six
有六根柱子的建筑物 <古代寺庙> hexastylos
有六角刀架的冲床 revolving head punch
有六角的 hexangular
有六柱的 hectastyle
有龙头过滤器 faucet-mounted filter
有楼梯斜梁的楼梯 string stair(case)
有漏窗的墙 perforated wall; porous wall
有漏斗状容器的卡车(或拖车) <运输混凝土的> gondola
有漏孔的 leaky
有漏元件 leaker
有路拱的 crowned
有路拱的(车行道)横断面 crowned section
有路拱的道路 crowned road
有路拱的面层 crowned surface
有路面的道路 surfaced road
有路面的路线 pavement line
有路面的码头前沿地带 paved apron
有路缘石的 curbed
有路缘石的分车岛 curb separator
有路缘石的横断面 curbed section
有路缘石的路段 curbed section
有绿化布置的办公室 landscaped office room
有孪生殿的庙宇 temple with twin sanctuaries
有轮布景 stage wagon
有轮布景架 wagon stage
有轮车 wheeler
有轮舞台 wagon stage
有轮子的(车身)底盘 wheeled chassis
有螺杆和螺母可用作螺栓的钩子 bolt hook
有落底式进水口涵洞 drop-inlet culvert
有麻点的 pocky
有麻窝的 pocky
有码计数器 mark counter
有脉动信号的伺服机构 pulse servo mechanism
有毛病 out-of-order; out-of-true; out of truth
有毛病的 awkward; cranky
有毛病的钻孔 faulty drilling
有毛的 comate
有锚单排板桩墙结构 anchored single-wall sheet-pile structure
有锚碇墙支撑的板桩堤岸 sheet-pile bulkhead with anchor-wall support
有锚格形板桩结构 anchored cellular sheet-pile structure
有锚链孔的船首部分 hawse
有帽管 capped pipe
有帽(紧固)螺钉 cap(ped) screw
有霉斑的 pecky
有霉斑木材 pecky timber; peggy timber
有门互通房间 communicating rooms
有门廊的 porched
有门楣的门框 door frame with transom
有门厅的 vestiluled
有名的 well-known
有名公用区 named common area
有名目标 named destination
有名无实的 titular
有名无实的审判 mock trial
有名无实条款 nominal term
有名义而无酬的 unsalaried
有模板的 formed
有模板浇灌 form placing
有磨蚀作用的 abrasive
有墨迹的 inky
有木框的镜子 wood-framed mirror
有木裂纹的板的质量标准 quality standard shake
有木屋顶的会议厅 wooden-roofed basilica
有木支撑的工作面 <隧洞开挖的> timbered stope
有目标的 object-oriented
有目的的使用 telesis
有目的的行为 purposive behavio(u)r
有目的地 purposefully
有目的的建造的 purpose-built
有内部热源的热交换器 heat-exchanger with internal heat source
有内套的 inner-cased
有耐火炉衬的 refractory faced
有能力的 competent; energetic
有泥舱的挖泥船 hopper dredge(r)
有年轮树 exogen(ous) tree
有年限的地产 estate for years
有黏[粘]结的 cementitious
有黏[粘]结力的 cohesive
有黏[粘]结力的预应力 prestress with bond
有黏[粘]结体内预应力桥 bonded internally prestressed bridge
有黏[粘]结性的混合物 cementitious blend
有黏[粘]结性集料 cementitious aggregate
有黏[粘]聚性的 cohesive
有黏[粘]着力的 adhesional
有鸟眼纹理的煤 bird's eye coal
有凝固力的 concretive
有凝结水管热网 heating network with condensating pipe
有凝聚力的 cohesive
有牛腿的柱 column with bracket
有排泥设备的挖泥船 combined dredge(r)
有排水通道的玻璃窗 drained glazing
有牌照出租车的驾驶员 medallion
有旁侧踏步的梯 side-step ladder
有旁证的 marginal
有抛物线型弦杆的拱形桁架 arched girder with parabolic chord
有泡钢锭 bleb ingot
有泡沫的 frothy
有疱的 blotchy
有配线分界点 intermediate train distance point with siding; train spacing point with distribution tracks
有棚起重机 capping crane
有棚停车处 sheltered car place
有棚运货车 wagon truck
有棚站台 roofed platform; sheltered refuge
有篷布遮盖的运料车 sheeted lorry
有篷货车 box car
有篷货摊 booth
有篷料车 cover wagon
有篷轮椅 bath chair
有篷运货车 wagon truck
有篷运货汽车 van
有篷载重汽车 wagon truck
有批准权的权威单位 approving authority
有皮带的手控铲运车 hand scraper with belt
有偏抽样 bias(s)ed sampling
有偏的 bias(s)ed
有偏估计 biased estimation
有偏估计量 bias(s)ed estimator
有偏估值 biased estimator
有偏见统计 biased statistics

有偏结果 biased result
有偏统计量 bias(s)ed statistic
有偏误差 bias error
有偏预测 biased forecast
有漂砂的沿岸地带 littoral drift coast
有平衡器的自动阀门 automatic counterweight gate
有平衡重的百叶窗 counterweight shutter
有平衡重的起重机 balance crane
有平面控制的地形测图摄影 <指雷达定位的同时进行摄影> horizontally controlled photography
有平台的坟墓 bench tomb
有平行防波堤的海港 harbo(u)r with parallel jetties
有平行码头的海港 harbo(u)r with parallel jetties
有平行深槽的 sulcate
有平行铁杆的横梁 page
有屏蔽的 shielded
有屏障泊位 sheltered berth
有屏障港池 sheltered basin;sheltered dock
有屏障港湾 protected harbo(u)r; sheltered harbo(u)r
有屏障海岸线 sheltered coastline
有屏障海湾 enclosed bay; protected harbo(u)r; sheltered arm of the sea
有屏障锚地 sheltered anchorage
有屏障水域 sheltered area;sheltered waters
有坡度的 acclive;graded;shelving
有坡度的楼面 grade floor
有坡轨道 inclined track
有坡口焊缝 groove weld(ing)
有坡口焊接接头 groove-weld joint
有坡线路 inclined track
有铺面的路 paved road
有铺面的码头前沿 paved apron
有铺位的游艇 house boat
有铺筑道面机场 surfaced airfield
有七根柱子的建筑物 <如古典式寺庙> heotastylos
有期转移限制的继承权 fee simple subject to an executory limitation
有鳍的 finned
有鳍动物 finny creature
有起伏的(建筑物)正面 undulating facade
有起伏沙丘的河床 dune covered riverbed
有起重机的驳船 crane lighter
有气动套筒的铆机 jam rivet(t)er
有气孔的 porous
有气门的窗 valved window
有气泡的 alveolate;vesicular
有气室水准器 chambered level tube
有气味的 odo(u)riferous; odo(u)rous;ordo(u)rant
有气味的化合物 odo(u)rous compound
有气味的货物 smelled cargo
有气隙的磁路 aeroferric circuit; aeromagnetic circuit
有砌石护岸的岸坡 beached bank
有钎电钻 electric(al) drill with drilling rod
有铅箔油毡 rag felt with lead foil
有铅垫板的沥青防水层 ledbit
有铅管的白铁管 tin pipe with lead jacket
有签署的合同 sealed contract
有前提的 hypothetical
有前厅的 vestibular
有前厅房间 <古希腊或罗马的> procoeton
有嵌板的门框 door frame with panels
有戗脊的老虎窗 hip dormer

有强制通风的加热炉 furnace with forced draft
有墙帐篷 wall tent
有切管轮的切管机 pipe cutter with cutting wheels
有亲笔签名的文件 subscription
有侵蚀性 erodible
有侵蚀性的水质 corrosive water
有轻雾的 misty
有倾口的烧杯 beaker
有倾斜摄影资料地区 oblique coverage
有倾斜支撑的檩支屋顶 purlin(e) roof with slope studs
有倾斜柱的半抛物线大梁 semi-parabolic girder with sloping end posts
有倾卸斗的手推车 hopper
有清偿能力 solvency
有穷差 finite difference
有穷的 finite
有穷格局 finite configuration
有穷观点 finite standpoint
有穷过程 finite process
有穷函数 finite function
有穷级数【数】 finite progression; finite series
有穷集 finite aggregate;finite set
有穷几何 finite geometry
有穷论的空间 finitistic space
有穷论者 finitist
有穷数学 finite mathematics
有穷图 finite graph
有穹隆的教堂 domed church
有穹隆的神庙 domed temple
有求即应 on-demand basis
有求即应时间分配制 on-demand time sharing; on-request time sharing
有求即应制 on-demand system; on-request system
有区别的 differentiated
有曲形齿的弧形锉刀 circular-cut file with curved teeth
有全缓解作用的分配阀 distributing valve with direct release
有全权的 omnicompetent;plenipotentiary
有权 entitle;in power
有权持有的 titular
有权持有的财产 titular possession
有权力的 powerful
有权支出 authorized outlay
有泉华包壳的陡坎 sinter-encrusted scarplet
有泉华包壳的泉盆 sinter-encrusted basin
有缺点的 defective; flawy; vicious; vulnerable
有缺点的工程 defective work
有缺口 resources reserve gap
有缺口的 jagged;jaggy
有缺损圆木 waney log
有缺陷材料 defective materials
有缺陷的 brack;unsound
有缺陷的电镀制品 mender
有缺陷的路面 defective pavement
有缺陷的人 defective
有缺陷的施工质量 imperfect workmanship
有缺陷工程 defective works
有缺陷焊缝 poor weld
有缺陷混凝土 defective concrete
有缺陷胶合 defective gluing
有缺陷钎焊 defective brazing
有缺陷室内管工 defective plumbing
有缺陷涂层 defective coating
有缺陷卫生设备系统 defective plumbing
有缺陷原料 defective materials
有缺陷铸件 misrun casting

有确实可靠的根据 on good authority
有人操纵的 manned
有人段(载波电话) attended repeater section
有人观测站 attended station
有人管理灯标 watched light
有人管理站 attended station
有人看守道口 guarded crossing; manned level crossing; staffed level crossing;watched crossing
有人看守的平交道口 manned level crossing
有人落水 man overboard
"有人使用"显示牌 occupancy indicator
有人无人钮 indicator button
有人无人牌 occupancy indicator
有人无人锁 indicator bolt
有人烟绿洲 inhabited oasis
有人增音机 attended repeater
有人增音站 attended repeater station
有人站 <配备有工作人员的车站> manned station;staffed station
有人值班操作 attended operation
有人值班的非自航驳 manned non-self propelled barge
有人值班的浮驳 manned pontoon
有人值班的增音站 attended repeater
有人值守的 attended
有人值守的电站 attended station
有人值守的支局 attended substation
有人住的房屋 occupied house
有刃的工具 edge tool
有刃镊 pince-ciseaux
有绒毛的 fuzzy
有溶解力的 dissolvent;resolvent;solvent
有溶解能力的 dissolvent
有乳头突起状的片材 nippled sheet
有乳油色斑点的石灰石 <产于英国约克郡> Roche Abley
有软百叶窗的 Venetianed
有若干座位的汽车或飞机 seater
有洒水设备的横向管道 lateral (pipe)line containing sprinklers
有塞试管 stoppered test tube
有三边的 trilateral
有三部分 triform(ed)
有三叉的 trifurcate
有三个半圆形室的 tri-apsidal
有三个坑位的厕所 three-holder
有三角格栅的结构框架 triangulated grid framework
有三角形断面的 triquetrous
有三角形花样的 triangulate
有三面的 trihedral
有三套住房的房屋 triplex building
有三叶的 trifoliate
有三种本质的 triform(ed)
有三种形式的 triform(ed)
有伞形顶的看台 umbrella stand
有散热片的 finned
有散热片的管 finned pipe
有散热片的汽缸 finned cylinder
有散热套汽缸 jacketed cylinder
有色标记 colo(u)r marking
有色表示镜 colo(u)red roundel
有色玻璃 carbon amber glass; colo(u)red glass;flash glass;pigmented glass; pot metal; pot-metal glass; stained glass; sun glass; tinted glass;tinting shade
有色玻璃窗 stained-glass window; tinted glass window
有色玻璃滤光片 colo(u)red glass filter;colo(u)r glass filter
有色玻璃墙 tinted glass wall
有色玻璃砖块 colo(u)red glass block
有色玻璃装配 colo(u)red glazing
有色衬底涂料 colo(u)red background coating

有色粗轧玻璃 tinted rough rolled glass
有色的 chromatic;colo(u)red
有色的装修 colo(u)red finish
有色(灯)光 colo(u)red light
有色地沥青封层 asphalt colo(u)r coat
有色镀锌板 colo(u)r tutanaga
有色防护眼镜 colo(u)red safety lens
有色废水 colo(u)red wastewater
有色钢化玻璃 tinted tempered glass
有色工业废水 non-ferrous metallurgical industry wastewater
有色光学玻璃 colo(u)red optic(al) glass
有色合金 non-ferrous alloy
有色化合物 colo(u)red compound
有色灰雾 colo(u)red fog
有色混凝土 colo(u)ring concrete
有色混凝土路(面) colo(u)red concrete road
有色介质 colo(u)r medium
有色金属 brass and bronze; non-ferrous metal
有色金属材料 non-ferrous material
有色金属的 non-ferrous
有色金属废水 non-ferrous metal waste water
有色金属分析 non-ferrous metal analysis
有色金属工业 non-ferrous industry
有色金属焊条 non-ferrous electrode
有色金属合金 non-ferrous alloy
有色金属矿产 non-ferrous metal commodity
有色金属容器 non-ferrous vessel
有色金属冶金工业 non-ferrous metal refinery industry
有色金属冶金学 non-ferrous metallurgy
有色金属冶炼 non-ferrous metallurgy
有色金属冶炼厂 non-ferrous metal smelting work
有色金属冶炼工业 non-ferrous metal refinery
有色金属渣 non-ferrous metallic slag
有色金属轧机 non-ferrous metal rolling mill
有色金属铸件 non-ferrous casting
有色晶体 colo(u)red crystal
有色离子 colo(u)red ion
有色沥青封层 asphalt colo(u)r coat
有色沥青玛琋脂 colo(u)red mastic asphalt
有色粒(体) chromoplast(id)
有色毛玻璃 tinted rough rolled glass
有色墨水 colo(u)red ink
有色黏[粘]土 bole
有色清漆 varnish stain
有色人造石铺石的 tinted granolithic
有色溶解有机物 chromophoric dissolved organic matter
有色纱线 colo(u)red thread
有色石英岩 Vermont slate
有色树脂粉末 pigmented resinous powder
有色树脂试剂 colo(u)red resin reagent
有色水 colo(u)red water
有色水泥 colo(u)r(ed) cement
有色体 chromoplast(id)
有色透镜 colo(u)r lens
有色涂料 colo(u)red paint; tinted paint
有色物质 colo(u)ring agent
有色雾翳 colo(u)red fog
有色性 colo(u)redness
有色絮凝体 colo(u)r floc
有色悬浮物 colo(u)red suspended matter
有色烟幕 colo(u)red smoke
有色眼镜 tinted spectacles
有色眼镜玻璃 colo(u)red ophthal-

mic glass
有色眼镜片 colo(u)red ophthalmic glass
有色冶金工业废水 wastewater from non-ferrous metallurgical industry
有色冶金学 non-ferrous metallurgy
有色阴影 colo(u)red shadow
有色油灰 tinted putty
有色油漆 tinted paint
有色有机物 colo(u)red organics
有色釉料涂层 colo(u)red glazed coat(ing)
有色噪声 colo(u)red noise
有色噪音 colo(u)red noise
有森林的 wooded
有沙滩的 beachy
有沙纹河床 ripple-covered bed
有沙洲河口 bar-build estuary
有砂的 gritty
有砂岩及页岩的冰碛平原 chenango
有筛网的底阀 foot valve with strainer
有山墙的 gabled;pedimented
有山墙的房屋 gabled house
有山墙的屋顶天窗 gabled roof dormer
有扇状叶脉的 diadromous
有伤痕的 seamy
有商标产品 branded product
有商务关系的人 correspondent
有少量裂缝 a little fissures
有设施用地 improved land; serviced land
有射孔套管 perforated casing
有伸缩刀刃的刀子 knife with retractable blades
有伸缩式罩套的货车 telescopic(al) hood wagon
有深远影响的 far-reaching
有升降翻斗的货车 wagon with lifting and tipping bucket
有升降台板的舞台 stage with trap
有生产力的 creative;productive
有生力量 effective force
有生气 verdure
有声的 vibrant
有声电影 moving picture;phonofilm; talking picture;talker <俚语>
有声电影机 kinetophone
有声电影系统 <利用唱片录放音的> vitaphone
有声电影制片厂 sound film studio
有声号志 audible signal
有声器材 noise equipment
有声区 insonified zone
有声热轴检测器 talking hotbox detector
有声望的 prestigious
有声无声显示 aural-null present
有声显示 aural present
有声信号 audible signal
有声应答器 acoustic(al) transponder
有声影厂 sound film
有声影片 talkie
有声运转发电机 noisy generator
有石拱顶的走廊 stone arcade
有石铺走廊的史前墓地 gallery grave
有时间记录的像片 time-correlated photograph
有时可感 sometimes felt
有时刻表的线路容量 <在一条有已定线路运营时刻表的线路上,每小时每一方向能运送的客位数> scheduled line capacity
有时刻表的线路容量利用系数 scheduled line capacity utilization coefficient
有实测资料河流 ga(u)ged river;ga(u)ged stream
有实测资料流域 ga(u)ge drainage area;ga(u)ged watershed
有实际经验的 practically minded
有实际经验者 practical operator
有实践经验的 experienced; with practical experience
有实力的房地产投资者或开发商 heavy hitter
有蚀痕的 eroded
有史时期 historic(al) time
有史以来的 on record
有使用价值的 employable
有事业心的 enterprising
有势力的 powerful
有势涡 potential vortex
有饰面层的 veneered
有适应性的 adaptative
有收入的车辆公里 revenue car-kilometers
有收入的吨公里 revenue ton-kilometers
有收入的货票 revenue waybill
有收入的货物吨公里 freight revenue ton-kilometers
有收入的列车运行 revenue service
有收入的业务 revenue service
有收缩力的 contractive
有收缩性的 contractive
有收益资产 earning assets
有手柄的振动筛 poker vibrator with handle
有手法的 mannered
有手艺的 workmanlike
有手艺的油漆工 brush hand
有枢栓的铰 snibill
有枢轴窗 casement light
有输送能力的 <美> burdensome
有树胶的 gummiferous
有树林的 wooded
有树林的高地 hurst
有树瘤的木材 burr wood
有树木的场地 wooded site
有树丘 herst
有竖框窗 mullion window
有数种不同显示器的雷达系统 complex display
有刷痕涂层 ropy finish
有衰减水平推力的拱形桁架 arched girder with diminished horizontal thrust
有闩的 barred
有栓塞活门的热水加热器 insulated single point water heater
有栓旋塞 plug cock
有双歧接头的歧管 two-connector manifold
有水厕 flush toilet
有水的 watered
有水的壕沟 wet fence
有水、电、暖、煤气设备 have electricity, gas, running water and steam heating system
有水库的水电站 reservoir power plant
有水力梯度的隧洞 grade tunnel
有水连接 <水管的> wet connection
有水平齿的弧形锉刀 circular-cut file with straight teeth
有水平带形线脚的拱基 banded impost
有水隧洞 water tunnel
有水外流的湖 open lake
有水源的适耕地 irrigable land
有水钻孔 water hole
有税的 dutiable
有税品 dutiable goods
有说服力的 persuasive
有司机棚的机车单元 cab unit
有(私)利的 expedient
有四个正面的 quadrifrontal
有四面的 tetrahedral
有四叶的 quadrifoliate
有四足的 quadruped
有塑料成分的灰泥 plastic mortar
有酸味的 acetous;acidulous
有髓心木材 timber with pith
有损耗材料 lossy material
有损耗的 lossy
有损耗电介质 lossy dielectric
有损耗电缆 lossy cable
有损耗介质 lossy medium
有损耗媒质 lossy medium
有损耗衰减器 lossy attenuator
有损耗同轴电缆 lossy coaxial cable
有损探伤法 destructive method
有损压缩 lossy compression
有榫槽的 tongue-and-grooved
有榫钉的企口缝 dowel(l)ed tongue and groove joint
有榫腋脚的榫接头 haunched mortise-and-tenon joint
有所发现 discover
有锁的柜 locker
有锁键的按钮 latching button
有塔的 towered
有塔的火神庙 towered fire-temple
有塔架的桥跨 <起重机> tower span
有塔楼屋顶 turreted roof
有踏步的混凝土通道 stepped concrete path
有台基上部线脚的 surbased
有台阶的 stepped
有台口的剧院 proscenium theatre [theater]
有台口的舞台 proscenium stage
有滩陡岸 bank barrier
有弹簧的塞绳 spiral-conductor flexible cord
有弹力的 buoyant;springy
有弹性的 elastic;resilient;whippy
有弹性的卡箍 resilient clip
有弹性地 elastically
有套泵 jacketed pump
有套的柱 cased post
有套管的混凝土桩 sleeved concrete pile;shelled concrete pile
有套管绳索冲抓钻机 casing rope percussion-grab drill
有套管钻孔 cased bore hole;lined borehole
有套气缸 jack(eted) cylinder
有套汽缸 jacket cylinder
有套筒的压注管 sleeved injection tube
有特权的 prerogative
有特色的 distinctive
有特色的窗 feature window
有特殊抗力的 heavy-duty;high duty
有特殊科学价值的地区 site of special scientific interest
有梯级平台 stage platform
有梯卡车 ladder truck; ladder troll(e)y <修理电线用>
有提升平台的卡车 lift platform truck
有提升平台的装卸车 lift platform truck
有体式钻头 bodied bit
有天沟的屋顶 valley roof
有天然入流的水库 reservoir with natural inflow
有填充墙的两柱檩支屋顶 purlin(e) roof with two posts and wind filling
有填充墙的三柱檩支屋顶 purlin(e) roof with three posts and wind filling
有填充物的树脂 <一种绝缘化合物> electrose
有填充物的天然树胶 electose
有填缝的砌石护坡 stone pitching with filled joints
有条不紊的方法 systematic approach
有条件承兑 conditional acceptance; qualified acceptance; special acceptance
有条件贷款 conditional loan
有条件的 conditional
有条件的不动产销售 conditional sales of real property
有条件的承担 conditional commitment
有条件的承认 conditional recognition
有条件的筹资 conditional financing
有条件的处理 conditional disposition
有条件的多数 qualified majority
有条件的合同 conditional contract
有条件的买主 qualified buyer
有条件的判决 conditional sentence
有条件的批准 conditional ratification;qualified approval
有条件的平衡 conditional equilibrium
有条件的契约 conditional contract; escrow
有条件的条款 conditional clause
有条件的销售 conditional sale
有条件的转让契约 escrow agreement
有条件禁运(制)品 conditional contraband
有条件期付款项 bill and accounts payable with terms
有条件支付的汇票 draft payable with terms
有条件转让协议 escrow agreement
有条理的 methodical
有条纹玻璃 cordy glass;striated glass
有条纹的冰川黏[粘]土 banded glacial clay
有条纹的河床质运动 striped bed load movement
有条纹的推移质运动 striped bed load movement
有调节池的径流式电站 run-of-river plant with pondage
有调整气温设备的 air-conditioned
有调中转车 transit car with resorting
有调中转车停留时间 detention time of car in transit with resorting
有调中转列车技术作业过程 operating procedure of transit train with resorting
有贴脸的门洞 cased opening;trimmed opening
有铁蒺藜的围栏 barbwire fence
有铁丝网的 wired
有停车场的路旁旅馆 motor inn
有停车管制的交叉口 stop-controlled intersection
有停车库的公寓 garage apartment
有通风(槽的)电枢 armature with ventilation
有通风孔砌块 vent block
有通过台的客车 vestibule type coach
有通气孔的蓄电池 vented battery
有同等高度的头部 <浅浮雕中> isocephalic
有同类地图资料地区 uniform coverage
有同量的 commensurable
有同轴短线的半偶极子 half-dipole with coaxial stub
有头键 gib key
有头螺钉 cap screw
有头螺栓 cap screw;tap bolt
有投票权者 voter
有透光孔的画廊 loopholed gallery
有凸出花纹的墙纸 anaglyphic wallpaper
有突板的翼式托盘 wing pallet with projecting deck
有突出部分的板 bellied
有突堤的港口 jetty harbo(u)r
有突片的 finned
有突起圆边的玻璃 hammered cathedral
有图案玻璃 figured glass
有图案的 figured;historiated
有图案轧制的暗色玻璃 amazon

有图瓷砖 picture tile
有图地区 map coverage; mapped area; mapped surface
有图护封 picture jacket
有涂层的水管 coating pipe
有涂层铸铁管 coated cast-iron pipe
有土堤的运动场 earth embankment stadium
有土耳其结的地毯 carpet with Turkish knots
有团粒土壤 aggregated soil
有腿货箱 tote bin; tote box
有退还抽样 sampling with replacements
有托板的胶带给料机 plate-supported belt feeder
有托板飞檐 mutule cornice
有托块的飞檐 mutule cornice
有挖泥和排泥设备的(自航式)挖泥船 combined hopper dredge(r)
有挖填方的截面 grade section
有外层壁的窗框 window recess with slipping walls
有外壳的 crustacean
有外壳(的钢)结构 encased structure
有外壳的计数管 skirt counter
有外壳的磨床 housing grinder
有外壳的压力桩 shell-pressure pile
有外壳的研磨机 housing grinder
有外壳的肘形弯管 encased elbow bend
有外壳的桩 encased pile; encased sheet pile; box section sheet pile
有外螺纹的短管 male nipple
有外螺纹的管接头 male nipple
有外向槽的门框 <供外开门用> giblet check
有弯钩的钢筋 hooked bar
有弯钩的箍筋 hooking stirrup
有弯曲线条的 vermiculate
有完善信息的存货量期望值 expected value with perfect information
有万能钥匙的系列 series of master-keyed locks
有网状饰的线脚 reticulated mo(u)lding
有网状小孔的 openwork
有望的 prospective
有危险的空气 hazardous atmosphere
有危险性的 dangerous
有威信的 prestigious
有微波的水域 broken waters
有微动刻度的电容器 vernier-control capacitor
有围壁室的船 trunk vessel
有围壁室的甲板船 trunk deck vessel
有围护的房屋 closed building
有围护的楼梯 stairway enclosure
有围栏床 box bed
有围墙的 walled
有围墙的宫殿 walled palace
有围墙的花园 walled garden
有围墙的通道 walled passage
有围墙的修道院 walled monastery
有围墙内院 peribolos
有围墙寺院 peribolos
有尾窗的 hatchback
有尾窗的汽车 hatchback
有尾槛的戽斗式静水池 bucket basin with sill
有尾槛的戽斗式消力池 bucket basin with sill
有卫星城的城市 conurbation
有卫星城镇 conurbation
有位置指示器的控制开关 indicating control switch
有味物质 odo(u)rant; odo(u)rous material
有喂料滚筒的斗式铺砂机 hopper type gritter with feed roll
有文字雕刻的壁缘 frieze with inscription
有纹理的 veined
有纹理的板 textured board
有纹理的大理石 veined marble
有纹理的丝或毛织品 rep(p)
有纹饰的 apsilate
有纹铸铁管 veined pipe
有问题的 problematic(al)
有蜗壳的水轮机 spiral-cased turbine
有握裹力预应力 prestress with bond
有污点的 spotted; spotty; tarnished; tarnishing
有污染资源 pollution-carrying resource
有屋顶的 covered; roofed
有屋顶的观众席 roofed spectator's stand
有屋顶的堰 roof weir
有屋顶式闸门堰 bear-trap weir
有屋盖的阳台 dalan
有屋面的木桥 covered timber bridge
有无车辆检查法 car presence detection
有无车辆检查器 car presence detector
有无对比法 with-and-without approach
有无线电台的吉普车 radio jeep
有五部分 quinquepartite
有误差的 erroneous; inaccurate
有误差的地图数据 error map
有雾的 foggy; misty
有雾日雾天 foggy day
有吸收力的 absorptive
有吸收能力的 absorbent; absorptive
有吸收性的 absorbent; spongy
有吸引力的 attractive
有吸引力的承租人 key tenant
有吸引力的价格 charm(ing) price
有希望得标的投标者 prospective bidder
有希望的 promising; prospective
有息存款 interest bearing deposit
有稀薄空气的孔穴 cavity with rarefied air
有溪流(通过)的峡谷 flume
有系杆的板梁结构 stayed girder construction
有系杆的板梁桥 stayed girder bridge
有系杆的拱梁 arched girder with tieback
有系杆的结构 tied structure
有系梁椽 close rafter
有系梁的人字木屋(顶) collar rafter roof
有系梁的人字木屋架 collar beam roof truss
有系绳气球 tethered balloon
有系统的 methodical; organic; systematic
有系统的技术进步奖 scientific
有系统应用 systematically use
有细胞的 cellulose
有细扣方钻杆 threaded kelly
有细扣钻杆 threaded drill pipe
有细粒的 granular
有隙黏[粘]合层 open binder
有狭孔门 loophole door
有下铰的凸出窗 bottom hung projecting window
有先见的 far-seeing
有纤毛的 ciliate
有现款 to be in cash
有线传输 cable communication; wire transmission
有线传输的电视系统 wired television system
有线传真 wire photo
有线传真系统 wire facsimile system
有线传真照片 wire photo
有线的 wired
有线电报 line telegraphy
有线电传机 wire teletype
有线电话 cable phone; line circuit; line telephone; wired call; wire telephone
有线电话设备分排 wire telephone installation section
有线电路 line circuit; metallic channel; metallic circuit
有线电视 cable television [cable TV]; closed circuit television; community television; piped television; wired television [wired TV]
有线电视放大器 wired TV amplifier
有线电视广播节目 piped program(me)
有线电视机 cable box
有线电视接收机 wired television receiver
有线电视文字传输系统 wired teletext system
有线电视系统 cable television system; jeep; wired television system
有线电信 line telecommunications
有线广播 carrier frequency wire broadcast(ing); electrophone; line broadcast(ing); public address system; rediffusing broadcast(ing); rediffusion on wire; wire(d) broadcast(ing); wired radio
有线广播接收机 public address receiver
有线广播节目 piped program(me)
有线广播系统 public address broadcasting system; public address system
有线广播站 rediffusion station; wired broadcast station
有线脚的石砌层 curstable
有线脚前缘的实心踏步 solid step with profiled noising
有线控制 wire-control
有线联系 wire-link
有线路图的配电板 graphic(al) panel
有线探空仪 wire sonde
有线调制解调器 wire-line MODEM
有线通信[讯] cable communication; line communication; wire communication; wire telecommunication
有线通信[讯]电路 wire communication circuit
有线通信[讯]工具 means of line communication
有线通信[讯]设备 means of line communication; wire communication facility
有线通信[讯]线路 wire communication line; wire-link
有线同步方式 wire synchronization system
有线/无线转接器 wire/wireless switcher
有线系统 wired system
有线线路 wire circuit; wire line
有线性流量关系的堰坝 weir with linear discharge relation
有线遥测术 hard-wire telemetry; wire-link telemetry
有线遥控 wire-control; wired remote control
有线印刷装置 on-line printer
有线远距离联合通信[讯] wired telecommunication
有线远距离通信[讯] wired telecommunication
有线载波广播 carrier frequency wire broadcast(ing)
有线载波通信[讯] line radio (communication); wire(d) radio
有线直通扬声器 wire-connected direct calling loudspeaker communication
有线制导 cable guidance; capture guidance; wire guidance; wire guide; wire-link guidance
有线中继 line relay; wired trunking; wire relaying
有线中继设备 line relaying equipment
有线中继制 line relay system
有线终端设备 line termination unit
有线转播 wire relay broadcasting
有限 finitude
有限阿贝耳群 finite Abelian group
有限安全 limited safe
有限板状体的综合参数 synthetic(al) parameter of finite sheet
有限棒束 finite rod bundle
有限闭合含水层 finite closed aquifer
有限闭区间 finite closed interval
有限闭子范畴 finitely closed subcategory
有限变化 limiting variety
有限变形 finite deformation; limited deformation
有限表 finite table
有限表观流动 limited apparent flow
有限表示 finite (re)presentation
有限波列 finite wave train
有限博弈 finite game
有限薄片 finite thin sheet
有限不连续性 finite discontinuity
有限步数 finite number of steps
有限部分 finite part
有限参加股 partially participating stock
有限测度 finite measure
有限测度空间 finite measure space
有限差(分)【数】 finite difference
有限差分逼近 finite difference approximation
有限差分表示式 finite difference expression
有限差分法 analysis by finite differences; calculus of finite difference; finite difference calculus; method of finite difference; finite difference method
有限差分方程 finite difference equation
有限差分格式 finite difference scheme; finite differencing scheme
有限差分公式 finite difference formula
有限差分公式模型 finite difference equation model
有限差分合成记录 finite difference synthesized record
有限差分解 finite difference solution
有限差(分)近似(法) finite difference approximation
有限差分理论 finite difference theory
有限差分模拟(法) finite difference modelling
有限差分能量法 finite difference energy method
有限差分偏叠加 finite difference migration
有限差分深度偏移 finite difference depth migration
有限差分式系数 coefficient of finite-difference-equation
有限差分水质模型 finite difference water quality model
有限差分算子 finite difference operator
有限差分析分法 analysis by finite differences
有限差数 finite difference
有限长锭料 finite ingot

有限长度 finite length
有限长度串 finite length string
有限长度的发射装置 finite length launcher
有限长度楔 finite wedge
有限长断层 finite length fault
有限长函数 finite length function
有限长梁 finite beam
有限长效应 finite length effect
有限超越次数 finite transcendence degree
有限乘法群 finite multiplicative group
有限乘积 finite product
有限乘数 finite multiplier
有限程序 finite process
有限持续时间 finite duration
有限尺寸板 finite-size panel
有限充液恒温器 gas charged thermostat; limited liquid charge thermostat
有限冲激响应滤波 filtering finite impulse response filtering
有限抽样行为 finite sample behavio(u)r
有限出现代数 finitely presented algebra
有限出现函子 finitely presented functor
有限出现群 finitely presented group
有限纯不可分扩张 finite purely inseparable extension
有限次映射 limited image number
有限存储滤波器 finite memory filter
有限存储器 finite memory
有限存储自动机 finite memory automaton
有限错动模型 finite dislocation model
有限代数 finite algebra
有限代数扩张 finite algebraic extension
有限代数数域 finite algebraic number field
有限带宽白噪声 band-limited white noise; bandwidth-limiting white noise
有限带宽的 band-limited
有限带宽放大器 bandwidth-limiting amplifier
有限带宽函数 band-limited function
有限带宽频谱 band-limited spectrum
有限带宽随机过程 band-limited random process
有限带宽信道 band-limited channel
有限带宽信号 band-limited signal
有限单纯复形 finite simplicial complex
有限单群 finite simple group
有限单元 elements of finite order; finite element
有限单元逼近 finite element approach
有限(单)元法 <对一般工程结构物,可看作是很多单元体在无数节点上互相联结的一个组合,当个别单元体的应力和变形关系为已知时,借助结构分析,可求出整个结构物的性质和作用> finite element method; method of finite element; finite element technique
有限单元法源程序 source program(me) of finite element method
有限单元分析 finite element analysis
有限单元公式 finite element formula(tion)
有限单元基岩 finite element matrix
有限单元技术 finite element technic; finite element technique
有限单元结构 finite element framework
有限单元解法 finite element solution
有限单元近似解 finite element approximation
有限单元理想化 finite element idealization
有限单元力法 finite element force method
有限单元梁格 finite element grid
有限单元数值模拟 finite element numerical modeling
有限单元体模式 <按有限单元法的计算模式> finite element model
有限单元体模型 finite element model
有限单元体特征值的直接公式化 direct formulation of finite element characteristics
有限单元网络 finite element network; finite element mesh
有限单元系统 finite element system
有限单元域 finite element domain
有限单元子空间 finite element subspace
有限的 finite
有限的公共面积 limited common area
有限的公开招标行动 limited submission
有限的过去 finite past
有限的筛孔尺寸 limit screen size
有限的筛网尺寸 limit screen size
有限的越野机动性 limited cross-country mobility
有限的资源 depletable resources
有限等价集 finitely equivalent set
有限递归法 limited recursion
有限点 finite point
有限点阵 finite lattice
有限迭代 finite iteration
有限定义 finitely defined
有限堆 finite reactor
有限对策 finite game
有限多面体 finite polyhedron
有限翻译机 finite transducer
有限反馈 limiting feedback
有限反时限继电器 inverse time-lag relay with definite minimum release
有限范围 limited range
有限方差 finite variance
有限方向射限 limited traverse
有限访问数据 limited access data
有限非空集 finite nonempty set
有限非空子集 finite nonempty subset
有限分辨率 finite resolving power
有限分段 finite segmenting
有限分解 finite decomposition
有限分析法 finite analysis method
有限分枝 finite branch
有限风区 fetch-limited
有限傅立叶变换 finite Fourier transform
有限傅立叶级数 finite Fourier series
有限覆盖 finite covering
有限伽罗瓦扩张 finite Galois extension
有限概率 finite probability
有限概型 finite scheme
有限公司 company limited; (in) corporation; limited company
有限共完全范畴 finitely complete category
有限构型相互作用 limited configuration interaction
有限股东 limited partner
有限固溶体 limit solid solution
有限观测时间 finite observation time
有限管束 finite tube bundle
有限管辖权 qualified jurisdiction
有限归纳法 finite induction
有限轨道不稳定性 finite-orbit instability
有限国际招标 limited international bidding
有限含水层 finite aquifer
有限函数 finite function
有限航区 restricted service
有限合伙 limited partnership
有限合伙公司 limited partnership
有限合伙人 limited partner
有限和 finite sum(mation)
有限核 finite nucleus
有限核大小效应 finite nuclear size effect
有限亨克尔变换 finite Hankel transform
有限后域 limited converse domain
有限厚度 finite thickness
有限厚度含水层 aquifer with limited thickness
有限糊精 limit dextrin
有限互溶性 limited mutual solubility
有限滑动栓接缝 limited slip bolted joint
有限划分 finite partition
有限环 finite ring
有限环道 finite circuit
有限回流 finite reflux
有限回路 finite circuit
有限活化期 limited pot life
有限活化寿命 limited pot life
有限迹 finite trace
有限积分变换 finite integral transform
有限积分器 limited integrator
有限基 finite basis
有限基底定理 finite basis theorem
有限基数 finite cardinal number
有限级数 finite progression; finite series
有限极宽磁铁 finite-pole-width magnet
有限集(合)【数】 finite collection; finite set; finite aggregate
有限计算自动机 finite counting automat
有限记忆滤波器 finite memory filter
有限记忆信道 finite memory channel
有限加性测度 finitely additive measure
有限加性集函数 finitely additive set function
有限加性类 finitely additive class
有限价 finite order
有限间断 finite discontinuity
有限交换群 finitely Abelian group
有限交换域 finite commutative field
有限交性 finite intersection property
有限角 finite angle
有限阶元素 elements of finite order
有限节点数 finite number of nodal point
有限解【数】 finite solution
有限介质 bounded media; bounded medium; finite medium
有限进口 conditional import
有限精度数 finite precision number
有限精确度 finite precision
有限净空高 limited headway
有限径向滑动轴承 finite journal bearing
有限竞争性招标 limited competitive bidding
有限矩定理 finite moment theorem
有限矩阵 finite matrix
有限均值 finite time average
有限开加细 finite open refinement
有限开区间 finite open interval
有限可测函数 finite measurable function
有限可分扩张 finite separable extension
有限可加性 finite additivity
有限可解群 finite solvable group
有限可靠性 fail soft
有限可控天线 limited steerable antenna
有限空间 confined space; finite space; restricted space
有限宽度 finite width
有限宽度隙 finite gap
有限宽效应 finite-width effect
有限扩展 restrained expansion
有限扩张 finite extension
有限拉莫尔半径稳定法 finite Larmor radius stabilization
有限类质同象【地】 restricted isomorphism
有限棱柱法【数】 finite prism method
有限离散系统 finite discrete system
有限利用度 limited availability
有限连分数 finite continued fraction
有限链 finite chain
有限梁 finite beam
有限量 finite quantity
有限量物料 finite lot
有限裂缝 limited cracking
有限流动 limited flow
有限流动变形 <与液化现象有关的土体变形> limited flow deformation
有限流动应变 <与液化现象有关的土体应变> limited flow strain
有限流体流动 limited fluid flow
有限滤子 finite filtration
有限率 finite rate
有限论 finitism
有限马尔可夫链 finite Markov chains
有限脉冲响应 finite pulse response
有限贸易 conditional trade
有限幂零群 finite nilpotent group
有限面积喷淋系统 limited area sprinkler system
有限描述 finite description
有限模 finite module
有限目标距离 finite object distance
有限耐久性 finite life
有限挠度 finite deflection
有限能见度 limited visibility
有限能量分辨率 finite energy resolution
有限能量修正 finite energy correction
有限能量值 finite energy correction
有限年金 temporary annuity
有限扭转 finite twisting
有限泡胀 limited swelling
有限频带波道 band-limited channel
有限频率范围 limited frequency range
有限平面 finite plane
有限屏蔽层积累因数 finite shield build-up factor
有限屏幕 finite baffle
有限期的 terminable
有限期的产权 terminable interests
有限期的合同 terminable contract
有限期性 terminability
有限前馈 limiting feedforward
有限区间 finite interval
有限区污染 pollution in limited area
有限区域 finite region
有限全域 finite universe
有限群 finite group
有限扰动 finite disturbance
有限热导率不稳定性 finite heat conductivity instability
有限日期 date of expiry
有限容积法 finite volume method
有限三角剖分 finite triangulation
有限散发 limited distribution
有限扫掠点波束 limited-scan spot beams
有限筛选 limiting screen
有限上链 finite cochain
有限射 finite morphism
有限射程范围 limited coverage range

有限射流 finite length jet
有限射束源 finite beam source
有限射影 finite projection
有限射影几何 finite projective geometry
有限射影几何码 finite projective geometry code
有限射影平面 finite projective plane
有限深度 finite depth
有限生产 limited production
有限生长 determinate growth
有限生成对象 finitely generated object
有限生成环 finitely generated ring
有限生成扩张 finitely generated extension
有限生成模 finitely generated module
有限生成群 finitely generated group
有限生成射影模 finitely generated projective modules
有限生成系 finite system of generators
有限生成域 finitely generated field
有限生成自由模 finitely generated free modules
有限时间 finite time
有限时间滤波 finite time filtering
有限寿命 finite lifetime
有限寿命部件 limited life component
有限寿命疲劳强度 fatigue strength for finite life
有限寿命区 finite life region
有限寿命设计 finite life design
有限输入源 finite input source
有限数据传递及显示装置 limited data transfer and display equipment
有限数学归纳法 finite induction
有限数值 finite value
有限水深 finite water depth; restricted water depth
有限水深的码头 finite water depth terminal
有限水深水域 restricted depth water area
有限说明 finite specification
有限素因子 finite prime divisor
有限塑性范围 finite plastic domain
有限塑性流动 limited plastic flow
有限算法【数】 finite algorithm
有限态 finite state
有限态表 finite state table
有限态技术 finite state technique
有限态识别程序 finite state recognizer
有限态算法 finite state algorithm
有限态随机对策 finite state stochastic games
有限态语言 finite state language
有限态装置 finite state device
有限弹黏[粘]塑性 finite elastoviscoplasticity
有限弹塑性理论 finite elastic-plastic theory
有限弹性层 finite elastic layer
有限特征 finite character
有限特征条件 condition of finite character
有限体积 finite volume
有限体积格式 finite volume scheme
有限体积近海模型 finite volume coast and ocean model
有限条分法 finite slice method; finite strip(e) method
有限条件 finite condition
有限跳跃 finite jump
有限通行屋顶 limited access roof
有限投标 closed bid
有限透镜长度 finite lens length
有限图 finite graph
有限土层 finite layer of soil
有限外推 definite extrapolation
有限外自同构群 finite group of outer automorphism
有限弯曲 finite bending
有限微扰理论 finite perturbation theory
有限维代数 finite dimensional algebra
有限维的 finite dimensional
有限维分布 finite dimensional distribution
有限维复流形 finite dimensional complex manifold
有限维空间 finite dimensional space
有限维控制 finite dimensional control
有限维流形 finite dimensional manifold
有限维欧几里得空间 finite dimensional Euclidean space
有限维射影几何 finite dimensional projective geometry
有限维数模 finite dimensional module
有限维数映射 finite dimensional mapping
有限维线性空间 finite dimensional linear space
有限维向量空间 finite dimensional vector space
有限维子空间 finite dimensional subspace
有限位移 finite displacement
有限位移(理)论 finite displacement theory; limited displacement theory
有限稳定性 limited stability
有限问题 finiteness problem
有限线宽(度) finite linewidth
有限线性黏[粘]弹性 finite linear viscoelasiticity
有限限度 finite limit
有限限幅 finite clipping
有限限制 finite restriction
有限消息机 finite message machine
有限形变 finite deformation; limited deformation
有限形变弹性理论 finite strain theory of elasticity
有限型 finite type
有限性 finite; finiteness
有限性破碎(法) arrested crushing
有限性条件 finiteness condition
有限性问题 finiteness problem
有限修正 finite correction
有限序列 finite sequence
有限序数 finite ordinal number
有限旋转群 finite rotation group
有限询价 limited inquiry
有限循环群 finite cyclic(al) group
有限压缩层 finite compressible layer; finite compressive stratum
有限延迟时间 finite delay time
有限样本生物测定 limited sample bioassay
有限叶栅 finite cascade
有限移动源模型 finite movement source model
有限应变 finite strain
有限应变椭圆长轴与 X 轴夹角 angle between long axis of finite strain ellipse and X
有限应变椭圆轴率 axial-ratio of finite strain ellipse
有限应力 finite stress
有限映射 finite mapping
有限优化 limited optimization
有限余弦变换 finite cosine transform
有限预(加)应力 limited prestressing
有限预应力混凝土 limited prestressed concrete
有限预应力混凝土桥梁 limited prestressing concrete bridge
有限域 finite domain; finite extent; finite field; Galois field
有限域模式 limited area model
有限域细网格模式 limited area fine mesh model
有限元逼近 finite element approximation
有限元程序设计 finite element programming
有限元的单元体 finite element unit
有限元反应谱分析 limited element response spectrum
有限元分析(法) finite element analysis
有限元(件) finite element
有限元离散化 finite element discretization
有限元模拟 finite element model(l)ing; finite element simulation
有限元模型化 finite element model(l)ing
有限元数学 finite mathematics
有限元水质模型 finite volume water quality model
有限元素法 finite element method
有限元体 finite element
有限元网络 finite element grid; finite element network
有限圆柱体 finite cylinder
有限圆柱形堆 finite cylindric(al) reactor
有限远距离 finite distance
有限远物点 finite object point
有限运动 finite motion
有限运动源 finite moving source
有限责任 finite liability; incorporated liability; limited liability
有限责任的合伙股东 limited partners
有限责任公司 company with limited liability; limited company; limited liability company <英>
有限责任股东 limited partner
有限增长率 finite rate of increase
有限增量 finite increment
有限增量法 method of finite increment
有限增益 finite gain
有限增益放大器 limited gain amplifier
有限招标 invited tendering; limited tendering; selected bidding
有限振动 restricted vibration
有限振幅 finite amplitude
有限振幅波 finite amplitude wave
有限振幅测深声呐 finite amplitude depth sonar
有限正比区(域) region of limited proportionality; limited proportionality region
有限正熵 finite positive entropy
有限正弦变换 finite sine transform
有限直线 finite straight line
有限值 finite value
有限值函数 finitely valued function
有限指数 finite index
有限制的 conditional; limited
有限制劲度 conditioned stiffness
有限制条件的房地产继承权 qualified fee estate
有限制条件的继承权 fee simple determinable
有限制条件的世袭家产 conditional fee estate
有限制条件的销售合同 conditional sales contract
有限制责任运输货物 released freight
有限秩 finite rank
有限置换 finite substitution
有限置换群 finite permutation group
有限种类 limiting variety
有限转动液力促动器 limited-rotation hydraulic actuator
有限转换器 finite transducer
有限桩 finite pile
有限状态机 finite state machine
有限状态图 finite state diagram
有限状态信道 finite state channel
有限资金 limited resources
有限资源 exhaustible resources
有限子复形 finite subcomplex
有限子集 finite subset
有限子可加性 finite subadditivity
有限子群 finite subgroup
有限子族 finite subfamily
有限自动机 finite automat(ion); finite state automation; finite state machine
有限自动机推断 finite automat inference
有限自同构群 finite group of automorphism
有限自旋 finite spin
有限自由度 finite degrees of freedom
有限自由度机器人 limited-degree-of-freedom robot
有限总体 finite population
有限阻力的接地线 earth wire of limited resistance
有相邻边界的 limitrophe
有相(位)差的 dephased
有香气的 odo(u)riferous
有香味的 aromatics
有厢房教堂 aisled church
有箱造 box mo(u)ld
有镶饰材料的墙体 masonry veneer
有向边 directed edge
有向边序列 directed-edge train
有向回路 directed circuit
有向集 directed set
有向角 directed angle
有向进入边 directed forward edge
有向距离 directed distance; oriented direction
有向离开边 directed away edge
有向路 directed-path
有向路长度 length of directed-path
有向欧拉线 directed Euler line
有向偶图 bidiagraph
有向平面 oriented plane
有向曲线 directing curve; directing graph
有向树 directed tree; oriented tree
有向树矩阵 directed-tree matrix
有向树枚举 oriented tree enumeration
有向树图 directed-tree graph
有向体积 oriented volume
有向通信[讯]网 oriented communication network
有向图 digraph; oriented graph; directed graph
有向图顶点 vertices of diagraph
有向线 vectorial line
有向线段 directed line segment
有向线素 lineal element
有向性 directed property
有向性试样 oriented sample
有向圆 oriented circle
有向直线 directed line
有项目 with project
有象征意义的图案 allegory
有像差光学系统 aberrated optics
有橡胶衬垫的 rubber sealed
有橡胶衬里纤维水带 fabric rubber lined hose
有橡胶特性的 elastomeric
有橡皮轮胎的推土机 bulldozer fitted to wheel tractor
有销铰接 pinned joint
有销孔的硬度质纤维板 pegboard
有销路的 merchantable
有销路的经济作物 marketable indus-

trial crop
有销路市场 promising market
有小齿的 denticulate
有小厨房及卫生间设备的居住单元 efficiency unit
有小孔的 alveolate;foraminate(d)<拉>
有小门的大门 stable door
有小泡的 vesicular
有小气泡玻璃 seedy glass
有小塔的 turriculate
有小窝的 alveolate
有效 avail;usefulness
有效f数 effective f-number;effective relative aperture
有效安培 effective ampere
有效安匝数 virtual ampere turn
有效八面体应力 effective octahedral stress
有效靶面积 effective target area
有效摆动角<转子发动机的> effective angle of obliquity
有效斑点尺寸 effective spot size
有效板高 effective depth of slab;effective plate height
有效板厚 effective depth;effective depth of slab
有效板宽 effective width of ribbon;effective width of slab
有效办公空间 available office space
有效半径 action radius;effective radius
有效半排出期 effective half-life
有效半衰期 effective half-life;effective half period
有效伴流 effective wake
有效饱和度 effective saturation
有效保险 amount in force;insurance in force
有效曝光范围 useful exposure range
有效曝光时间 effective exposure time
有效爆破 effective fragmentation
有效爆炸深度 effective shot depth
有效倍率 useful magnification
有效倍增常数 effective multiplication constant
有效倍增系数 effective multiplication factor
有效比 availability ratio;effective ratio
有效比表面积 effective specific surface area
有效比冲 effective specific impulse
有效比例限度 effective proportional limit
有效比特 significant bit
有效比重 effective specific gravity
有效臂长 effective rake
有效编码 actual coding;efficient coding
有效变换群 effective transformation group
有效变形<强夯> effective deformation
有效标记 significant notation
有效标书 bona fide bid
有效标准偏差 effective standard deviation
有效表面 active surface;available surface;effective surface
有效表面积 available surface area;effective surface area
有效表面空隙室 effective superficial porosity
有效表面孔隙率 effective superficial porosity
有效并联阻抗 effective shunt impedance
有效波 effective wave;significant wave
有效波长 effective wave length;significant wave length
有效波尔磁子 effective Bohr magneton
有效波高 effective wave height;significant wave height
有效波浪 effective wave
有效波浪推算图 significant wave prediction chart
有效波斜度 effective wave slope
有效波周期 significant wave period
有效波阻抗 effective wave impedance
有效部分 active component;available part;effective portion;energy component;power component;real component
有效参数 valid parameter
有效残留阻力 effective residual drag
有效舱容 availability space;available capacity
有效操作 valid operation
有效操作期【计】up-time
有效操作数地址 effective operand address
有效侧向压力 effective lateral pressure
有效测程 effective range
有效测定范围 useful range
有效测量 actual measurement
有效测量范围 effective range;useful range
有效测试 validity test
有效缠结密度 effective entanglement density
有效产层 effective pay
有效产量 effective output;useful output
有效长(度) active length;available length;effective length;serviceable length;usable length;useful length;working length;action length
有效长度系数 effective length factor
有效长宽比 effective aspect ratio
有效常数 effective constant;real constant
有效场 effective field
有效场强 effective field intensity
有效超载功率 effective overload output
有效超载压力 effective overburden pressure
有效沉淀量 effective precipitation
有效沉降量<强夯的> effective deformation
有效成本 effective cost
有效成分 active constituent;available component;effective constituent;effective ingredient;worth
有效承载力 payload capacity
有效乘员 efficient crew
有效程度 significant degree
有效程度达…… effective up to…
有效程序 effective procedure
有效持水量 available moisture capacity;effective moisture capacity;effective water-holding capacity
有效持水能力 available water-holding capacity;effective water-holding capacity
有效持续时间 effective duration
有效尺寸 available size;effective dimension;effective size
有效齿高 working depth
有效齿廓 active profile
有效充填 effective placement
有效冲程 effective stroke
有效冲洗 effective cleaning
有效重叠期 effective overlap period
有效抽力 effective draft
有效抽速 effective pumping speed
有效出力 available power;effective output;effective power;useful output
有效出水量 net output
有效初压力 effective initial pressure
有效储备 active storage;live storage
有效储存量 effective storage
有效储量 available reserves;available storage (capacity);net amount
有效储料仓 live storage bin
有效储料库 live storage
有效传播速度 effective propagation velocity
有效传递特性 effective transfer characteristic
有效传动力 effective driving force
有效传热系数 effective thermal transmittance
有效传输 effective transmission
有效传输当量 effective transmission equivalent
有效传输等效值 effective transmission equivalent
有效传输定额 effective transmission rating
有效传输率 effective transmission rate
有效传输速度 effective transmission speed
有效传输增益 effective transmission gain
有效传真频带 effective facsimile band
有效磁场比 effective field ratio
有效磁导率 effective permeability
有效磁化率 effective susceptibility
有效磁化强度 effective magnetization
有效磁化倾角 effective magnetized inclination
有效磁通(量) working flux;useful flux
有效磁子数 effective magneton number
有效次数 effective degree
有效粗糙度 effective roughness
有效存储量 active storage capacity
有效存储容量 active storage capacity
有效存取 valid memory access
有效措施 effective measure
有效大地辐射 effective terrestrial radiation
有效大气层 effective atmosphere
有效大气压(力) effective atmosphere
有效带 effective zone
有效带的水井公式 water-well formula in effective zone
有效带宽 effective bandwidth
有效单极(子)辐射功率 effective monopole radiated power
有效单位重量 effective unit weight
有效导磁率 effective permeability
有效导纳 effective admittance
有效导体 active conductor
有效道路路线 efficiency road route;efficient road route
有效的 acting;active;actual;approved;available;effective;effectual;efficient;formal;operative;useful
有效的层次存取时间 effective hierarchy access time
有效的分级筛孔尺寸 effective separating size
有效的拱跨 effective arch span
有效的化学品 chemicals that are useful
有效的跨距/长度比 effective span/length ratio
有效的契约 valid contract
有效的所有权 good title
有效的烟囱高度 effective chimney height
有效的专利 patent in force
有效地层压降 effective formation pressure drop
有效地面储蓄量 effective surface retention
有效地面辐射 effective terrestrial radiation
有效地面滞留量 effective surface retention
有效地排除岩粉 efficient removal of cuttings
有效地平线 effective horizon
有效地球半径 effective earth radius;effective radius of the earth
有效地球辐射 effective terrestrial radiation
有效地势 available relief
有效地下水 available ground water;available groundwater
有效地震力 effective earthquake force
有效地址 actual address;valid address;effective address
有效递减率 effective decline rate
有效点 available point;efficient point;significant point
有效电导 effective conductance
有效电导率 effective conductivity
有效电动势 effective electromotive force
有效电感 effective inductance
有效电荷 effective charge
有效电极 active electrode
有效电抗 effective reactance
有效电离截面 effective cross section of ionization
有效电量 available capacity
有效电流 active current;effective current (flow);effort current;energy current;root mean square current;virtual current;watt current
有效电流极限 limits of effective current range
有效电路 effective circuit;efficient circuit
有效电平 significant level
有效电平变换 active level transformation
有效电容 effective capacitance;useful capacitor
有效电势 action potential
有效电位 action potential
有效电压 active voltage;effective voltage;useful voltage;virtual voltage
有效电压梯度 effective voltage gradient
有效电阻 effective resistance;virtual resistance
有效电阻率 effective resistivity
有效垫片宽度 effective gasket width
有效动磁导率 effective dynamic(al) permeability
有效动力 actual power;effective power
有效动态模量 effective dynamic(al) modulus
有效冻结时间 effective freezing time
有效陡度 effective steepness
有效毒性 effective toxicity
有效度 level of efficiency;significance;validity
有效度比 availability ratio;effective ratio
有效端 live end
有效断面 active cross-section;effective (cross-) section;useful area;working section
有效断面面积 effective cross-section(al) area

有效堆放高度 available height for storage
有效吨公里 ton-kilometers available
有效吨数 tons available
有效发电量 available energy output
有效发电水头 effective power head; power head
有效发射率 effective emissivity
有效发射线 effective launcher line
有效发射信号 effective signal radiated
有效法向应力 effective normal stress
有效反射 effective reflection; usable reflection
有效反射率 effective reflectivity
有效反射面 effective reflecting surface
有效反射面积 specular cross section
有效反射系数 effective reflectance
有效反应 active reaction
有效反映 usable reflection
有效反作用 active reaction
有效范围 effective range; effective region; effective zone; efficient range; range of action; range of effectiveness; range of validity; service area; useful range
有效范围等达…… effective up to…
有效方法 available approach; available method
有效方式 effective ways
有效防护 effective protection
有效放大 effective amplification
有效放大倍率 useful enlargement
有效放大系数 effective gain; effective multiplication factor
有效分布系数 effective distribution coefficient
有效分度 significant graduation
有效分耕 valid till
有效分离 effective segregation
有效分离密度 effective separating density
有效分离系数 effective separation factor
有效分量 active component; active constituent; effective constituent
有效分权 efficient departmentalization
有效分散参数 effective dispersion parameter
有效分析 available analysis; effective analysis
有效分选半径 effective separation radius
有效分选比重 effective separating density
有效分支因素 effective branching factor
有效分子量 effective molecular weight
有效风区长度 effective fetch length
有效风载 effective wind load
有效风阻面积 effective wind blocking area
有效峰数 effective peak number
有效峰值加速度 effective peak acceleration
有效缝隙 effective slit
有效缝隙宽度 effective gap
有效肤深 penetration of current
有效伏安 active volt-ampere; effective volt-ampere
有效俘获功率 effective radiation power
有效浮力 available buoyancy; positive buoyancy
有效幅面 effective format; effective picture size
有效辐射 effective emittance; effective radiation
有效辐射功率 effective radiated power
有效辐射通量 effective radiant flux
有效辐射温度 effective radiation temperature; resulting radiation temperature
有效辐射系数 effective radiation constant
有效辐照度 effective irradiance
有效负荷 actual load(ing); effective load; live load; payload; real load; usage load; useful load(ing)
有效负荷传感器 payload sensor
有效负荷范围系数 efficiency load-range factor
有效负荷监测装置 payload monitoring
有效负荷量 payload capacity
有效负荷曲线 efficiency load curve
有效负荷容量 effective carrying capacity
有效负载 active load; actual load(ing); available load; effective load; live load; payload; real load; useful load(ing)
有效负载能力 payload capability
有效负载体积 payload volume
有效负载通信[讯]系统 payload communications system
有效负载质量 payload mass
有效复合截面 effective section recombination
有效复合速度 effective recombination velocity
有效复合系数 effective recombination coefficient
有效富裕 effective margin
有效赋值 effective valuation
有效覆盖层厚度 effective overburden depth
有效覆盖层压力 effective overburden pressure
有效覆盖范围 effective coverage
有效干扰 favo(u)rable interference
有效干扰面积 effective confusion area
有效干燥面积 effective drying surface
有效感觉噪声分贝 effective perceived noise decibel
有效感觉噪声级 effective perceived noise level
有效感应 actual induction
有效刚度 effective stiffness
有效刚性 effective stiffness
有效钢筋 effective reinforcement
有效钢筋面积 effective area of reinforcement
有效杠杆比 effective lever arm
有效高度 apparent height; effective height; operational altitude; significant height; usable height; useful height; virtual height; working height; radiation height <天线>; effective depth <指钢筋混凝土拉力钢筋中心到混凝土受压边缘的距离>; operating altitude
有效高度达…… effective up to…
有效隔距长度 <强力测试的> effective ga(u)ge length
有效隔绝 active isolation
有效隔离 active isolation
有效隔振 active isolation
有效镉截止值 effective cadmium cutoff
有效各向异性 effective anisotropy
有效给水量 available water supply; effective water supply
有效根层深度 effective rooting depth
有效更新率 effective up-date rate
有效工程 effective engineering
有效工时 effective production hour; effective working hour; productive time; work time
有效工业废水排放指标 effective index of industrial wastewater emission
有效工资率 effective pay rate
有效工作 available work; effective work; useful work
有效工作半径 <起重机> clear outreach
有效工作长度 effective travel length
有效工作面积 effective working area; useful working area
有效工作区 effective service area; service area
有效工作时间 effective working time; available machine time <机器的>; effective time
有效工作系数 net efficiency
有效工作系统 availability system
有效工作小时 effective working hour
有效公差 effective tolerance
有效公斤 effective kilogram
有效公式 valid formula
有效功 effective work; useful work
有效功率 active power; actual (horse) power; available power; effective capacity; effective horsepower; effective output; effective power; gross effect; mechanical effect; net horsepower; real power; true power; useful efficiency; useful (horse) power; useful output; water horsepower
有效功率分析 real power analysis
有效功率级 available power level
有效功率损耗 active power loss
有效功率效率 available power efficiency
有效功率因数 effective power factor
有效功能 effective efficiency
有效供给 effective supply
有效供水 available water supply; efficient delivery
有效供水量 available (water) supply
有效供水水源 active water supply source
有效供油行程 effective delivery stroke
有效拱 active arch
有效共振积分 effective resonance integral
有效共振截面 effective cross-section for resonance
有效估计量 effective estimator; efficiency estimator; efficiency estimate; efficient estimator; efficiency estimation
有效鼓风 working blast
有效观察距离 effective observation distance
有效官能度 effective functionality
有效管径 effective diameter of pipe
有效管理 effective management; value engineering
有效管子长度 effective tube length
有效惯量 effective inertia mass
有效惯性矩 effective moment of inertia
有效惯性力 effective inertia force
有效灌溉 fertile irrigation
有效灌水率 <根系层的> application efficiency
有效罐存期限 effective pot life
有效光点大小 effective spot size
有效光阑 effective diaphragm; effective stop
有效光强 effective intensity of light
有效光圈 effective f-number
有效光通 useful luminous flux
有效滚动半径 effective rolling radius
有效滚柱长度 <轴承的> effective roller length
有效过滤压 effective filtration pressure
有效过水断面 active cross-section; available water carrying section
有效过水(断面)面积 obstruction free flow area
有效含水饱和度 effective water saturation
有效含水量 available moisture capacity; available moisture content; available water-holding capacity
有效含盐量 effective salinity
有效含氧量 available oxygen
有效焓 effective enthalpy
有效焊缝厚度 effective throat depth
有效夯实深度 effective compacted depth
有效行 active line
有效合同 enforceable contract; valid contract
有效合同价格 effective contract price
有效荷载 actual service load; effective load; final load; payload; real load; service load; useful load; virtual load
有效荷载比 payload ratio
有效荷载舱 payload capsule
有效荷载传感器 payload sensor
有效荷载能力 payload capacity
有效荷载(容)量 payload capacity
有效荷载与车辆总重比 payload-to-vehicle weight ratio
有效荷载载体 payload carrier
有效荷载展开和恢复系统 payload deployment and retrieval system
有效荷载质量比 payload-mass ratio
有效荷载重量 payload weight
有效横断面 effective cross-section; useful cross-section
有效横流 effective cross current
有效红灯时间 effective red time
有效宏观截面 effective macroscopic cross-section
有效喉道截面 effective throat
有效厚度 effective depth; effective thickness; effective throat <焊缝>
有效呼叫计数器 effective-call meter
有效呼叫 effective call
有效滑动速度 effective slip velocity
有效滑距 effective slip
有效滑距比 effective slip ratio
有效滑移系数目 number of effective slip system
有效化 effectuation
有效化合价 active valence
有效环境 effective environment
有效缓发 effective delayed neutron fraction
有效换气量 effective ventilation
有效换热面积 effective heat exchange area
有效簧圈 active coil; actuating coil
有效灰度系数 effective gamma
有效挥发度 effective volatility
有效挥发性 effective volatility
有效回波区【水文】effective echo area
有效绘图面积 active drawing area; effective drawing area
有效活荷载 service live load
有效霍耳参数 effective Hall Parameter
有效机电耦合系数 effective electromechanical couple factor
有效机能 useful function

有效积分 effective integral
有效积温 effective accumulated temperature
有效基金 bank roll
有效基线 effective base(line);virtual base(line)
有效激励时间 effective actuation time
有效级 level of significance
有效极面 active polar surface
有效极限 validity limit
有效集电极 effective collector
有效集总参数质量 effective lumped-parameter mass
有效记录长 effective record length
有效记录纸宽度 effective chart width
有效剂量 effective dose
有效剂量当量 effective dose equivalent
有效剂量中值 median effective dose
有效加强面积 area available for reinforcement
有效加热表面 effective heating surface
有效加热面积 effective heating surface
有效加热时间 effective heating time
有效加速度 effective acceleration
有效价格 effective price
有效价值 standing value
有效间隔 significant interval
有效间隔理论持续时间 theoretic(al) duration of a significant interval
有效间距 effective spacing
有效间隙 effective clearance
有效监测 effective monitoring
有效减速力 <汽车的> effective retarding force
有效剪力 effective shear
有效剪切面积 effective shear area
有效剪切应变能准则 effective shear strain energy criterion
有效剪应变 effective shear strain
有效剪应力 effective shear stress
有效检索宽度 effective sweep width
有效检验 test of significance
有效碱 available base;effective alkali
有效碱度 effective alkalinity
有效碱性物 available base
有效降深 effective drawdown
有效降水(量) available precipitation;effective precipitation;effective rainfall
有效降水入渗系数 efficient infiltration coefficient of precipitation
有效降雨 effective rain
有效降雨量 effective precipitation;effective rainfall;net rainfall
有效降雨量指数 precipitation effectiveness index
有效降雨强度 intensity of effective rainfall
有效降雨损失量 <总雨量与净雨量之差> effective abstraction;effective abstractions of precipitation
有效交叉系数 effective crossing coefficient
有效交互作用 effective interaction
有效交换 effective crossing over
有效交混长度 effective mixing length
有效焦距 effective focal length
有效角 effective angle
有效接触 effective exposure
有效接触半径 effective contact radius
有效接触面积 effective contact area
有效接触面积比 effective contact area ratio
有效接触压力 effective contact pressure;effective contact stress
有效接触应力 effective contact stress
有效接地 effective grounding;effectively grounded;useful earthing
有效接地面积 effective ground contact area
有效接地线 effective grounded line
有效接地压力 effective ground pressure
有效接收时间间隔 effective acceptance time interval
有效节距 effective pitch
有效结论 effective conclusion;valid conclusion
有效截面 effective crossover;effective cross-section;effective section;net section;useful cross-section;clear opening <涵管等>
有效截面惯性矩 effective moment of inertia
有效截面积 clear area;free area
有效截面面积 effective cross-section(al) area;effective sectional area;net sectional area;useful cross-sectional area
有效截止频率 effective cut-off frequency
有效截止日期 expiry date
有效解 <计算机的> non-trivial solution
有效介电常数 effective dielectric(al) constant
有效界膜 effective film
有效界限 effective margin
有效金额 amount in force
有效筋 effective reinforcement
有效劲度 effective stiffness
有效劲度比 effective stiffness ratio
有效劲度常数 effective stiffness constant
有效进尺 effective footage
有效进气 effective admission
有效浸润线 effective saturation line
有效经济使用期 useful economic life
有效精度 available accuracy;effective accuracy;operational accuracy
有效井径 effective well radius
有效颈厚 <焊缝最小厚度> effective throat thickness
有效净容限 effective net margin
有效净水头 effective net head
有效竞争 effective competition
有效静区 effective shadow
有效距离 coverage;effective distance;effective range;operating range;radius of action;significant space
有效距离达…… effective up to…
有效距离截止 range-cutoff function
有效聚束角 effective bunching angle
有效卷曲半径 effective crimping radius
有效卷曲数 effective wave number
有效菌株 effective strain
有效开关场 effective switching field
有效开孔面积 effective opening
有效抗剪刚度 effective shearing rigidity
有效抗剪强度参数 effective shear strength parameter
有效抗张强度 effective tensile strength
有效靠泊能量 effective berthing energy
有效颗粒度 effective size of grain
有效颗粒密度 effective particle density
有效可降水分 effective precipitable water
有效空间 effective dimension;stowage space;useful space
有效空泡数 effective cavitation number
有效空隙(度) active porosity
有效空隙率 dynamic(al) porosity;effective porosity
有效空载时间 effective dead time
有效空中路线 effective air path
有效孔径 effective aperture;effective diameter;effective opening;effective pore radius;efficacious opening;real aperture;usable aperture;useful aperture
有效孔径系数 coefficient of effective aperture
有效孔隙 available pore space;effective pore space
有效孔隙度 effective porosity;practical porosity;active porosity
有效孔隙率 active porosity;drainable porosity;effective drainage porosity;effective porosity
有效孔隙容积 effective pore volume
有效孔隙体积 effective pore volume
有效口径 effective aperture
有效库容 active (reservoir) capacity;active storage (capacity);available capacity;available storage;conservation storage;effective capacity;effective storage;live capacity;live storage;live storage capacity;operating storage;reservoir live storage;usable capacity;usable storage;useful storage;useable storage capacity
有效跨度 effective space;effective spacing;effective span
有效宽度 effective width;usable width;useful width
有效宽度系数 effective width coefficient
有效亏数 effective deficiency
有效扩散 effective diffusion
有效扩散半径 effective diffuse radius
有效扩散参数 effective dispersion parameter
有效扩散常数 effective diffusion constant
有效扩散率 effective diffusivity
有效扩散速度 effective diffusion velocity
有效扩散系数 effective diffusion coefficient
有效拉力 effective pull;effective tension
有效劳动 effective labo(u)r
有效劳动量 effective workload
有效雷诺数 effective Reynolds number
有效镭含量 effective radium content
有效类目 significant link
有效冷量 useful refrigeration effect
有效冷却面积 active cooling surface
有效离子电荷 effective ionic charge
有效离子淌度 effective ionic mobility
有效理论塔板等效高度 height equivalent to an effective theoretical plate
有效力 active force;effective force
有效力矩 final moment;net torque
有效力向量 effective force vector
有效历时 effective duration
有效利用 effective utilization
有效利用率 availability
有效利用系数 availability factor
有效粒度 effective size of grain
有效粒间压力 effective intergranular pressure
有效粒径 effective diameter of grain;effective (grain) diameter;effective (grain) size
有效粒子速度 effective particle velocity
有效粒子直径 effective particle diameter
有效链 active chain
有效链路 active link
有效梁高 effective depth;effective depth of beam
有效量 effective dose;effective quantity;root mean square quantity;useful quantity
有效量程 effective range
有效裂纹尺寸 effective flaw size
有效裂隙 effective flaw
有效磷 available phosphorus
有效灵敏度 effective sensitivity
有效流动参数 effective flow parameter
有效流动带 active flow zone
有效流动截面 effective obstruction free flow area;obstruction free flow area
有效流动区 active flow zone
有效流距 effective flow pitch
有效流量 available discharge;available flow;effective discharge;effective flow;usable flow;useful discharge;useful flow;virtual flow
有效流速 effective velocity;field velocity
有效流速与船速比 <螺旋桨处> effective velocity ratio
有效流体压力 active fluid pressure;effective fluid pressure
有效流转压头 effective circulating head
有效流阻 effective flow resistance
有效硫 available sulfur
有效漏失面积 effective leakage path
有效炉箅面积 useful grate area
有效炉膛 furnace useful tank
有效滤过压 effective filtration pressure
有效滤率 effective filtration rate
有效路肩 usable shoulder;useful shoulder
有效率 effective percentage;effective rate;virtual rating
有效率的 efficient
有效率系数 coefficient of efficiency
有效绿灯时间 available green time;effective green time
有效绿灯时间取代长度 effective green displacement
有效氯 active chlorine;available chlorine
有效氯法 available-chlorine method
有效氯浓度 effective chlorine concentration
有效轮压力 effective wheel pressure
有效论证 valid argument
有效螺距 effective pitch
有效螺距比 effective pitch ratio
有效螺距角 effective pitch angle
有效螺纹 effective thread
有效螺旋角 effective helix angle
有效落差 effective head;net head;net pressure head;useful head
有效马力 actual horsepower;duty horsepower;effective horsepower;net horsepower;root mean square horsepower;useful horsepower
有效脉冲 available impulse;effective impulse
有效满功率天数 effective full power hours
有效锚定长度 effective wheel pressure
有效铆钉中距 effective pitch
有效密度 density effect;effective density
有效密封材料 effective sealing material
有效密集消防射流 effective solid fire stream

有效面 significant surface；useful surface
有效面积 active area；active face；actual area；clear area；clear opening；effective area；effective cross-section；free area；operating area；significant area；usable area；useful area；working area
有效面积系数 effective area coefficient
有效描述集合论 effective descriptive set theory
有效灭火区 effective extinguishing zone
有效名 basonym；valid name
有效模量 effective modulus
有效模态质量 effective modal mass
有效模型 effective model；gross model；neat model
有效膜面积 effective membrane area
有效摩擦角 effective angle of friction；virtual angle of friction
有效母岩顶面埋深 buried depth of top of effective source rock
有效母岩厚度 effective thickness of source rock
有效母岩平均厚度 average effective thickness of source rock
有效母岩最大厚度 maximum effective thickness of source rock
有效母岩最小厚度 minimum effective thickness of source rock
有效目标长度 effective target length
有效目标面积 effective target area
有效内聚力 effective cohesion
有效内摩擦角 effective angle of internal friction
有效能 availability energy；available energy；useful energy
有效能灯时间 available energy
有效能力 available capacity；effective capacity
有效能量 available energy；effective energy
有效能态密度 effective density of state
有效年限 effective age
有效黏[粘]度 effective viscosity；virtual viscosity
有效黏[粘]合力 effective cohesion
有效黏[粘]聚力 effective cohesion；effective cohesion intercept
有效黏[粘]土 active clay
有效黏[粘]（滞）性 virtual viscosity；effective viscosity
有效碾压深度 effective rolling depth
有效凝聚力 effective cohesion
有效扭矩 effective torque；effective torsional moment
有效浓度 effective concentration
有效偶极矩 effective dipole moment
有效爬高值【水文】 significant turn-up
有效排放高度 effective emission height；effective height of emission
有效排放量 effective discharge
有效排气速度 effective exhaust velocity
有效排水孔隙度 effective drainage porosity
有效排水面积 active drainage area
有效炮孔线 effective gun bore line
有效炮眼 effective perforation
有效（配）筋 effective reinforcement
有效喷射时间 effective discharge time
有效膨润土 effective bentonite
有效碰撞 effective collision
有效碰撞截面 effective collision (cross-) section；effective cross-section of collision
有效票 valid ballot
有效频带 effective band
有效频带宽度 effective bandwidth
有效平均功率 effective mean power
有效平均时间 effective averaging time
有效平均温差 effective mean temperature difference
有效平均压力 effective mean pressure
有效平均值 actual mean
有效平均指示压力 effective indicated mean pressure
有效评价 effective evaluation
有效屏幕面积 useful screen area
有效屏频常数 effective shielding constant
有效坡道制动 effective grade braking
有效坡度 effective gradient；effective pitch
有效破裂带 effective fractured zone
有效破裂带深度 depth of effective fractured zone
有效破片 effective fragment
有效剖面 effective section
有效剖面模数 effective section modulus
有效谱线宽度 effective line width
有效期 available period；effective date；effective period；indate；period of validity；usable life；validity period
有效期的延长 extension of validity
有效期间 duration of validity；effective duration；period of availability；time of effect
有效期满 exhaustion of effect；expiration of effect
有效期满日期 date of expiry of validity
有效期未满 unexpired term
有效期限 activity duration；can stability；date of availability；date of expiration；life span；time of efficiency；valid period；term of validity；useful life；storage life
有效期限调度 effective dead-line scheduling
有效期（限）截止 expiration of limitation period
有效期（限）终了 expiration of limitation period
有效期（限）终止 expiration of limitation period
有效期延长 extension of availability
有效期已满的专利 expired patent；lapsed patent
有效奇偶校验 valid parity check
有效起伏 available relief
有效气孔率 effective porosity
有效气压 effective atmosphere
有效汽耗 useful steam consumption
有效砌合 effective bond
有效迁移率 effective mobility
有效迁移速度 effective migration velocity
有效牵引功率 effective tractive power
有效牵引力 effective traction；effective tractive effort；effective tractive force；effective tractive power；net tractive effort
有效强度 active strength；effective intensity；effective strength
有效强度参数 effective strength parameter
有效墙厚 <用于计算细长比> effective thickness of a wall
有效墙宽 <用于计算细长比> effective thickness of a wall
有效氢 available hydrogen
有效清洁技术 active cleaning technique
有效情报 rapid information
有效区 coverage
有效区间 section of validity
有效区域 effective coverage
有效驱进速度 effective migration velocity of particle
有效取样面积 effective sampling area
有效圈 <弹簧的> effective coil
有效圈数 effective turn
有效群体大小 effective population size
有效热 available heat；net heat；useful heat
有效热负荷 effective heat duty
有效热价 effective heat price
有效热降 useful heat drop
有效热截面 effective thermal cross-section
有效热量 available heat；useful heat
有效热特性曲线 service thermal characteristic curve
有效热效率 effective thermal efficiency
有效热值 available heating value；net heating value
有效热中子截面 effective thermal neutron cross-section
有效热阻 effective thermal resistance；thermal resistance
有效日期 date of expiration；date of expiry；date of validity；effective date
有效日数 days when available
有效容积 active volume；available volume；dischargeable capacity；effective volume；effective volume capacity；live volume；net volume；usable capacity；useful capacity；useful volume
有效容积利用系数 coefficient of utilization of useful capacity
有效容量 active volume；actual capacity；available capacity；available storage (capacity)；capacity payload；effective capacity；effective content；effective unit weight；effective volume；effective volume capacity；effective volume content；useful capacity
有效容量曲线 service capacity curve
有效容重 effective unit weight；submerged unit weight；useful capacity
有效溶剂 active solvent
有效融雪量 effective snowmelt
有效散热面 active cooling surface
有效散射截面 effective scattering cross-section
有效散射质量 effective scattering mass
有效扫除容积 effective swept volume
有效杀伤程度 useful casualty levels
有效筛孔 effective opening；effective screen aperture
有效筛孔度 effective sieve aperture size
有效筛孔面积 open area；open screening area
有效筛面积 effective screen area
有效筛眼孔径 effective sieve aperture size
有效上覆压力 effective overburden pressure
有效射程 effective distance；effective firing range；effective range；effective reach；effective range of jet；effective throw <水枪的>
有效射界 effective field of fire
有效射距 effective jetting distance
有效射速 effective rate of fire
有效伸长 effective elongation
有效伸长率 effective aspect ratio parameter
有效砷 available arsenic
有效深度 available depth；effective depth；significant depth
有效渗透（率） effective permeability
有效渗透系数 effective coefficient of permeability；effective permeability
有效渗透性 effective permeability
有效升力 disposal lift；effective lift；useful lift
有效生产量 effective output
有效生产时间 actual productive time
有效生化需氧量 workload of biochemical oxygen demand
有效生物剂量 effective biological dose
有效生物膜去除法 efficient biofilm removal method
有效生物膜载体 efficient biofilm carrier
有效生油岩 effective source rock
有效声功率 effective acoustic (al) power
有效声速 effective sound velocity
有效声压 effective acoustic(al) pressure；effective sound pressure
有效声压级差 effective sound pressure level difference
有效声影区 effective shadow zone
有效声源中心 effective acoustic (al) center [centre]
有效失真 effective distortion
有效湿度 available moisture；effective humidity；effective moisture
有效石灰 available lime
有效时间 active session；available time；duration of validity；effective duration；operational use time；significant interval
有效时间常数 effective time constant
有效时期 period of availability
有效使用率 availability
有效使用年限 <建筑物> effective age
有效使用期 effective age
有效使用期限 effective life (time)
有效使用时间 effective use time；operational use time
有效使用寿命 acceptable life
有效视场 available field of view
有效视场角【测】 effective angular field
有效收入 effective yield
有效收缩量 effective shrinkage
有效寿命 active life；actual life；available life；effective life (time)；real-life；useful life；useful life span
有效寿期 effective life (time)
有效受风面积 effective frontal area
有效受热面 effective heating surface
有效疏松度 effective porosity
有效输出 actual output；useful output
有效输出长度 effective length of delivery
有效输出导纳 effective output admittance
有效输出额 effective output
有效输出功率 effective output power
有效输出能量 available energy output
有效输出阻抗 effective output impedance
有效输入 effective input
有效输入导纳 effective input admittance
有效输入电阻 effective input resistance
有效输入功率 effective power input；useful power input
有效输移率 effective transport rate
有效数 effective number；significance；weight
有效数据 effective data

雨水污迹 rainwater stain; weathering stain
雨水污染 precipitation pollution
雨水污水 storm sewage
雨水污水沉淀池 sedimentation/storm-sewage tank
雨水污水池 storm-sewage tank
雨水污水分流系统 separate system
雨水污水分流堰 storm-sewage diversion weir
雨水污水合流系统 storm-sewage system
雨水洗涤 rainwash
雨水洗刷 rainwash
雨水系统 storm-water system
雨水下水道 rainwater sewer; storm sewer
雨水下水道沉沙池 grit chamber
雨水下水道工程 storm sewerage works
雨水下水道系统 storm-water system
雨水险 pluvious insurance
雨水箱 rainwater cistern; rainwater tank
雨水泄水管 rain outlet
雨水性质 quality of rainfall
雨水蓄池 rainwater cistern
雨水溢流 storm-water overflow
雨水溢流井 storm overflow chamber
雨水溢流装置 storm-water overflow device
雨水溢水沟 storm(-sewage) overflow
雨水溢水管 storm overflow sewer
雨水溢水设施 storm overflow
雨水阴沟 surface water sewer
雨水与污水的混合处理 dual disposal
雨水与污水分流系统 separate system
雨水源河 rain-fed river; rain-fed stream
雨水沾污的 rain spotting
雨水制品 drainage goods; drainage product
雨水滞留能力 absorptive capacity for rain water
雨水滞留容量 absorptive capacity for rain water
雨水中污染物质 pollutants in rainfall
雨水贮池 rain cistern
雨水作用而成的 pluvial
雨凇 <下降时呈液态,着地后就冰冻> freezing drizzle; freezing rain; glaze; glazed frost; nebelfrost; rain ice; silver thaw; sleet; glaze(d) ice; silver frost
雨凇暴 glaze(d) storm
雨凇冰 glazed frost; glaze(d) ice; verglas
雨凇指示仪 glazed frost indicator
雨套 rain hood
雨天 rain(fall) day; rain-weather; wet weather
雨天备用处理设备 wet-weather treatment plant
雨天高峰流量 wet weather peak flow
雨天工资 wet time
雨天交通事故 wet traffic accident
雨天流量 wet weather flow
雨天流污水调蓄池 storage reservoir for wet weather flow; storage tank for wet weather flow
雨天流污水中的污染物质 pollutants in wet weather flow
雨天气候 rainy weather
雨天顺延条款 weather permitting clause
雨天天气 rainy weather
雨天污水量 wet weather flow
雨天作业 rain work
雨蛙 hylid; tree frog
雨温比 pluviothermic ratio
雨污合流下水道系统 combined sewage system
雨污水分流堰 separating weir
雨污水分流制 separate system
雨污水合流排水管 storm drain combined with sanitary drainage
雨污水合流系统 combined sewage system; combined sewerage system; storm-sewage system; storm-sewerage system
雨污水合流下水道系统 combined sewerage system
雨污水合流下水(管)道 building combined drain; building combined drain sewer; building combined sewer; combined sewer
雨污水合流制 combined sew(er)age system
雨污水合流制下水 water-borne sewerage
雨雾衰减 rain fog attenuation
雨雾凇 glime
雨洗理论 rainfall purification theory; theory of washout by rain
雨向计 vectopluviometer
雨形成机理 mechanism of rain formation
雨型【气】rainfall pattern
雨靴 water boot
雨学 hyetology; ombrology
雨雪 rain with snow
雨雪测量器 rain and snow ga(u)ge
雨雪干扰抑制 rain clutter suppression
雨雪计 snow-rain ga(u)ge
雨雪静电干扰 precipitation static interference
雨雪量 moisture condition
雨雪量器 rain and snow ga(u)ge
雨雪杂波 precipitation clutter
雨雪噪声 precipitation noise
雨檐 window eaves
雨燕 swifts
雨衣 cravenette; mack(intosh); raincoat; rainproof; rain wear; waterproof; raingear <施工人员穿用的>
雨衣雨裤 wet weathers
雨影 rain shadow
雨影区 rain shade
雨影区荒漠 rain-shadow desert
雨育农业 rain-fed agriculture
雨育种植 rain-fed cropping
雨源河床 ephemeral channel
雨云 nimbus [复 nimbi/nimbuses]; rain cloud
雨云数据处理系统 nimbus data handling system
雨云卫星 nimbus satellite
雨云阴影 rain shadow
雨运物 rainwash
雨罩 appentice; canopy; marquee; rain cap; rainfall hood
雨致蠕动 rainwash
雨致衰减 rain attenuation
雨中工作 rain work
雨中浇捣混凝土 concreting in rain
雨中浇注混凝土 concreting in rain
雨中污染物 pollutant in rain
雨中作业 rain work
雨珠饰 gutta [复 guttae]
雨柱 rain pillar
雨状回波 rain echo

语控防鸣器 vodas [voice-operated antisinging device]; voice-operated device antisinging

语控装置 voice-operated device
语言频率 <300 ~ 3400Hz> speech frequency
语音插空系统 <利用语言间歇的电路时间交错分割多路通信[讯]制> time assignment speech interpolation system
语音系统 phonetic system

玉滴石 hyalite

玉雕 jade carving; jade sculpture
玉符山石 American jade; californite
玉工车床 jewelers lathe
玉红 rubine; rubine red
玉红玻璃 red glass; ruby glass
玉红磨料 rubine grain
玉红色 rubine; ruby
玉兰 yulan magnolia
玉绿色 jade green
玉米 maize; corn <美>
玉米板 <一种绝缘保温材料> Maizewood
玉米产品废物 corn product waste
玉米带气候 corn belt climate
玉米蛋白纤维 zein fiber [fibre]
玉米的小穗 <美> nubbin
玉米淀粉加工厂 corn starch processing mill
玉米淀粉生产废水 corn starch production wastewater
玉米粉碎机 corn grinder; maize grinder
玉米黄 maize yellow
玉米灰岩 cornstone
玉米螟(虫) maize borer
玉米碾碎机 corn crusher
玉米片状铝粉 corn flakes
玉米清选机 corn cleaner; shelled corn cleaner
玉米湿加工面粉厂 corn wet mill
玉米纤维 zein fiber [fibre]
玉米油 corn oil maize oil
玉米油皂 maize oil soap
玉米油脂肪酸 corn oil fatty acid
玉米雨季 <东非 2 ~ 5 月间的一种大雨> maize rains
玉器 jade article; jade object; jadeware
玉器工厂 jade workshop
玉色 jade green; light bluish green
玉色玻璃 jade glass
玉石 jade(stone); jadete
玉石雕刻术 lapidary
玉石粉红色 cameo pink
玉石工 lapidary
玉石黄色 cameo yellow
玉石矿床 jade deposit
玉石蓝色 cameo blue
玉石绿色 cameo green
玉石棕色 cameo brown
玉蜀黍 corn; maise [maize]; zea moys
玉蜀黍地带 corn belt
玉蜀黍油 oil of maize
玉蜀黍属 maize
玉髓 <石英的变种> chalcedonite; c(h)alcedony
玉髓蛋白石 chalcedony-opal
玉髓的卵形结石 thunder egg
玉髓骨料 chalcedony aggregate
玉髓胶结物 chalcedony cement
玉髓绿色 chrysoprase green
玉髓燧石 beekite
玉质透辉石 diopside-jadeite

育空河 Yukon River

育空石 yukonite
育空-坦那纳高地地体 Yukin-Tanana upland terrane
育林学 silviculture
育龄组 child boarding group
育苗 nursery sock growing
育苗场 nursery
育苗区 nursery garden
育苗室 phytotron(e)
育土层 gley
育婴室 nursery
育婴堂 baby-care unit; founding hospital
育婴院 baby farm
育种 breed(ing)
育种场 breeding ground; breeding place; breeding station
育种人员 breeder

郁闭 crown closure; crown-contact

郁闭度 canopy density; crown density; shade density
郁闭林 close stand; dense stand
郁闭群落 closed community
郁闭疏开 moderate opening of canopy
郁金香栎木 red tulip oak
郁金香木 tulipwood
郁金香形的 tulip-shaped
郁李 dwarf flowering cherry

狱长 <英> governor

狱吏 warder
狱门桥 <美国纽约> Hell Gate Bridge
狱门锁 hasp lock

峪槽 bath; sanitary tub

浴槽炉 tank furnace

浴厕隔板 stall
浴场 bathing place; outdoor bathing place; thermal-bath
浴场更衣室 <古罗马、古希腊的> apodyterium
浴场建筑物 aquatic building(s)
浴场式窗 thermal window
浴场污染 pollution of bathing place
浴池 common bathing pool; plunge bath; plunge pool; piscine; alveus <古罗马时嵌入楼层的>
浴池附件 bath accessory
浴池水龙头 bath faucet
浴池油身室 <古罗马> alipterion
浴负荷 <公共浴室的> bathing load
浴缸 bathtub; tub
浴缸边缘 bathtub ledge
浴缸瓷漆 bath enamel
浴缸高度 bathtub height
浴缸或淋浴装置 bathtub/shower installation
浴缸龙头 bathtub faucet
浴缸排水塞子 bathtub plug
浴缸铅隔汽具 bathtub lead trap
浴缸石 monk's park
浴缸式卡车 <运水泥混凝土用> bathtub truck
浴缸手握把 bathtub hand grip
浴缸水龙头 globe tap
浴缸凸边缘 bathtub rim
浴缸突出部分 bathtub ledge
浴缸形曲线 bathtub curve
浴巾架 bath towel holder
浴疗室 unctuarium
浴炉 liquid furnace
浴炉隔墙 bridge wall
浴炉加热 heating in salt bath furnace
浴盘式卡车 <送混凝土水泥用的> bathtub-type truck
浴盆 bathing tub; bathtub

Y

浴盆存水弯 bath trap
浴盆龙头 tub cock
浴盆塞字 bathtub cock
浴盆上抓手 bath grip
浴盆式卡车 <运水泥混凝土用> bath-tub-type truck
浴盆式曲线 bathtub curve
浴盆弯管 bath trap
浴盆下水管存水弯 tub lead trap
浴器 bath
浴室 bath (house); bathroom; hot house; shower room; toilet; bagnio <意大利、土耳其式的>
浴室壁橱 bathroom closet
浴室布置 bathroom layout
浴室窗帘 bathroom curtain
浴室的人体涂油室 <希腊、罗马> ceroma
浴室附属件 bathroom accessories
浴室附属设备 bathroom accessories
浴室供暖器 bathroom heater
浴室供热设备 bathroom heating appliance
浴室挂钩 bathroom hook
浴室及厕所 bath and W.C.
浴室集水沟 trapped bathroom gull(e)y
浴室门 bathroom door
浴室内热水 califont
浴室内塑料进排水口 plastic bathroom inlet gull(e)y
浴室内塑料排水口 plastic bathroom outlet
浴室内装置 bathroom fixtures
浴室排水沟 bathroom gull(e)y; trapped bathroom gull(e)y
浴室配件 bathroom fittings
浴室汽水分隔器 bathroom trap
浴室汽水分离器 bath trap
浴室取暖 bathroom heating; bathroom warning
浴室设备 bathroom equipment; bathroom installation
浴室卫生 sanitation of bath-room
浴室用具 bathroom fittings
浴室用品 bathroom product
浴室用品箱 bathroom cabinet
浴室用热水器 geyser
浴室用物品 bathroom article
浴室照明装置 bathroom lighting fixture
浴室装饰小五金件 bathroom fitments
浴室装置 bathroom fixtures
浴水快热器 califont; geyser
浴堂 bagnio
浴洗废水管 waste pipe
浴箱 bath cabinet
浴用管子 bath tube
浴用龙头 bath-cock

预安排 precondition

预安装 pre-installation
预案 prearranged planning
预白化 prewhitening
预白噪声化滤波器 prewhitening filter
预办 presetting
预办闭塞 preworking a block
预办控制 presetting control
预拌 preblend; precoat; premix(ing)

预拌材料 premixed material; premixed stuff; ready-mixed stuff
预拌车间 premixing plant; ready-mix plant
预拌处理 premix treatment
预拌的 coated; ready-mixed
预拌法 premixing method
预拌粉刷 ready-mixed plaster
预拌工厂 premixing plant; ready-mix plant
预拌骨料 premixed aggregate
预拌好的石灰砂浆 ready-mixed lime-sand mortar
预拌和黏[粘]结剂 ready-mixed compound
预拌灰浆 premixed plaster
预拌灰泥 premixed plaster
预拌混合料 preblend; premix (ed mixture); ready-mix
预拌混凝土 premixed concrete; preshrunk; ready-mix(ed) concrete
预拌混凝土拌运车 premixed truck; ready-mixed truck
预拌混凝土操作者 ready-mixed concrete operator
预拌混凝土(工)厂 ready-mixed concrete plant
预拌混凝土楼 ready-mixed concrete tower
预拌混凝土穹顶 precast concrete cupola
预拌混凝土生产者 ready-mixed concrete operator
预拌混凝土输送车 ready-mixed concrete lorry
预拌混凝土摊铺机 premixed facility; ready-mixed concrete spreader; ready-mixed distribution facility
预拌混凝土运送车 premixed truck; ready-mix(ed) truck
预拌混凝土站 ready-mixed concrete plant; ready-mixed concrete tower
预拌集料 precoated aggregate; premixed aggregate
预拌搅拌机 premixer
预拌搅拌器 premixer
预拌沥青混合料路面 premixed bituminous surface
预拌沥青混合料面层 premixed bituminous surface
预拌沥青路面 premixed bituminous surface
预拌沥青面层 premixed bituminous surface
预拌料 mill mixture
预拌路面材料 premixed surfacing
预拌嵌缝混合料 ready-mixed joint compound
预拌轻质石膏粉饰 premixed light (weight) gypsum plaster
预拌弱混凝土 ready-mixed lean concrete
预拌砂浆 ready-mixed mortar
预拌设备 premixing plant
预拌湿的浆状物 premixed wet paste
预拌湿糨糊 ready-mixed wet paste
预拌石膏材料 premixed gypsum stuff
预拌石膏灰 gypsum ready-mixed plaster; ready-mixed plaster
预拌石膏灰浆 premixed gypsum plaster
预拌石膏混凝土 mill-mixed gypsum concrete
预拌石膏砂浆 gypsum ready-mixed plaster
预拌石膏石灰浆 premixed gypsum-lime plaster; ready-mixed gypsum-lime plaster
预拌石屑 coated chippings; premixed chip(ping)s
预拌式表面处治 premix type of treatment
预拌水结碎石路 premixed water-bound macadam
预拌水泥输送车 ready-mix cement truck
预拌碎石 coated macadam; premixed macadam
预拌填料 ready-mixed stuff
预拌涂料 ready-mixed stuff
预拌油基涂料 ready-mixed oil-base(d) paint
预拌着色的拉毛粉饰 premixed colo(u)red stucco
预拌着色的外墙灰浆 premixed colo(u)red exterior plaster
预拌着色的外墙抹灰 premixed colo(u)red rendering
预包络 preenvelop
预饱和 presaturation
预饱和器 presaturator
预饱和柱 presaturation column
预保温 preincubation
预报 advance information; forecast(ing); forerunner; prediction; prognosis [复 prognoses]; prognostication
预报变量 predication variable
预报潮(汐) forecast tide; predicted tide
预报订正 amendment to forecast; forecast amendment
预报法 method of prediction
预报方程 forecast equation; prognostic equation
预报方法 forecast method; forecasting; forecasting technique
预报仿真 simulation in forecasting
预报风险的影响 predicting the effects of risk
预报改正 amendment to forecast; forecast amendment
预报概率 forecast probability
预报高水位 predicted high water
预报告 predictor
预报工作 forecasting service
预报公告 forecast bulletin
预报公式 prognostic formula
预报估测 forecast estimation
预报估计 forecast estimation
预报过高 overprediction
预报海洋学 synoptic(al) oceanography
预报函数 forecasting function; prediction function
预报机构 forecasting service
预报技术 forecasting technique
预报检验 forecast verification
预报校正法 predictor-corrector method; predictor-corrector procedure
预报校正器 predictor-corrector
预报阶段 warning stage
预报结果 predictor result
预报精度 accuracy of forecasting; forecast accuracy
预报警状态 prealarm state
预报径流 forecasted flow
预报可靠性 reliability of forecast
预报控制 predictive control
预报理论 prediction theory
预报量 predictand
预报量控制 predictor control
预报流量 forecasted flow; forecasting flow
预报模式 forecast model; prediction model
预报模型 forecast model; prediction model
预报判据 forecasting criterion
预报期 forecast period
预报期间 forecasting period
预报器 precursor; predictor
预报区(域) forecast area; forecast district; forecast zone
预报曲线 forecast curve
预报时段 prediction interval
预报时效 earliness of forecast; period of forecasting; period of validity
预报术语 forecast terminology
预报水文过程线 forecasted hydrograph
预报算子 predictor
预报天气图 predicted chart; prognostic chart
预报天气者 weatherman
预报通知 advance report
预报通知单 advance report notice
预报图(表) forecast chart; prognostic chart
预报误差 forecast error
预报系数 prediction coefficient
预报信号系统 annunciator system
预报验证 verification of forecast
预报因子 forecaster; predictor
预报(预)见期 earliness of forecast
预报员 forecaster; predictor
预报站 forecasting station
预报者 predictor
预报正反检验 forecast-reversal test
预报值 forecast value; predictand; predicted value; predictor
预报中心 central forecasting office; forecasting center [centre]
预报周期 forecast cycle; forecast period
预报装置 predictor
预报准确度 correctness of forecast; forecast accuracy; predictability
预报准确率 correctness of forecast; forecast accuracy; predictability
预报准确性 correctness of forecast; forecast accuracy; predictability
预曝光 preexposure; prefogging
预曝光调节 preexposure setting
预爆 preshoot
预爆震 preknock
预备 make-ready; preparation; preparative
预备班 preparatory course
预备报告 preliminary report
预备步骤 preliminary step
预备操作 preliminary operation
预备测试 preliminary checkout; range finding test; screening test
预备程序 preliminary program(me)
预备处理 conditioning treatment; preparatory treatment
预备的 preliminary; preparatory
预备调查 preliminary review
预备定理 preparation theorem
预备伐 advanced feeling; preparatory cutting; preparatory stage
预备费 allowance; contingency cost; contingency fund; provisional sum; reserve funds; stand-by fund
预备分配色谱法 preparative partition chromatography
预备浮力 reserved buoyancy; safety buoyancy
预备副滑车 watch tackle
预备工程 preliminary works
预备工作成本 preparatory cost
预备功室 stand-by power
预备贯入 seating drive
预备号 preparative
预备会议 preliminary conference
预备间隔 preparatory interval
预备接点 preliminary contact
预备金 budget allowance; reverse funds
预备卷绕 preliminary winding
预备轮 extra wheel
预备轮胎 stepney
预备锚 spare anchor
预备门 bulkhead gate
预备苗床 preparatory nursery
预备期 preparatory period; probation; probationary period
预备区 zone of preparation
预备人员 spare hand

预备容量 reserve capacity
预备设计 preliminary design
预备时间 free time;readiness time
预备实验 range finding experiment
预备式 ready mode
预备室 preparation room
预备数据 preliminary data
预备信号 preparatory signal
预备性拨款 preparatory allocation
预备性鉴定试验 preevaluation test
预备性交接试验 preliminary acceptance trial
预备性研究 preliminary study
预备针梳机 gill preparer
预焙 prebake
预焙烧 preparatory roasting;preroast
预焙烧炉 preroast furnace
预焙烧阳极 baked anode
预焙炭板 prebaked carbon slab
预焙阳极电解槽 prebaked-anode type cell
预焙阳极铝电解槽 prebaked cell for alumin(i)um-reduction
预苯酸 prephenic acid
预编 preliminary cataloging
预编程序 preprogram(me)
预编辑 preedit
预编辑程序 preedit program(me)
预编辑检查程序 preedit check program(me)
预编解释系统 preedited interpretive system
预编译程序 precompiler;precompiler program(me)
预编译时间 precompile time
预变形 <用于筒形容器,加高压进行预应力处理,使生产一定量的永久变形> autofrettage;predeformation
预变形钢 prestressing steel
预变形器 preformer
预变形曲率 allowance for camber
预变形轧辊 <锻压机的> maxirolling
预标度计数器 prescaler counter
预标准 prestandard
预并条 predrawing
预拨经费 advance appropriation
预拨资金 advance funding
预剥离 prestripping
预剥离加速 prestripping acceleration
预补偿 precompensation;preemphasis
预操作助剂 preprocessing aid
预测 anticipation; calculate; calculation; forecast(ing); foreshadow(ing);predicting;prediction;preestimate; preliminary survey; presetting; preview; prognosis [复 prognoses]; prognosticate; prognostication
预测比例尺 predicting scale
预测编码 predictive coding
预测变量 forecast variable
预测波通信[讯] predicted-wave signal(l)ing
预测不满百分比 predicted percentage dissatisfied
预测财务报表 projected financial statement
预测差异 forecast variance
预测长度 predict length
预测沉降 forecasting settlement
预测成本 future cost;managed cost;predicted cost;prediction cost
预测成果 predicted conclusion
预测程序 predictor
预测储层特性参数 prediction of reservoir characteristic parameter
预测储量 predictive reserves
预测单位 predicting organization
预测单元 predictive cell
预测的层位 predict strata
预测的方法 prediction method
预测的构造单元 predict structure unit
预测的时间间隔 period of forecast;time span of prediction
预测的数学模型 mathematic(al) models of prediction
预测的正确性 validity of prediction
预测地层状态 predicted strata performance
预测地区 predicted district
预测地震 forecast earthquake
预测点 future position;predicted point;predicted position
预测电路 predictor circuit
预测额 amount forecasted;amount of forecast
预测法 predicative technique;predicted method
预测反褶积 predictive deconvolution
预测范围 predicted range
预测方程 predictor equation
预测方法 method of prediction
预测非点污染源风险 predicted nonpoint source risk
预测分析法 predictive analysis
预测风险 forecasting risk
预测概率 anticipated probability
预测公式 predictor formula;prognosis formula;prognostic formula
预测估计 prediction estimation
预测固定成本 managed fixed cost
预测过程 forecasting process
预测过滤器 prediction filter
预测耗水量 projecting consumption
预测货运量 forecast cargo traffic;predicted freight traffic volume
预测机 predicting machine
预测技术 forecasting technique
预测价格 forecast price
预测价值 forecast price
预测间隔 predicting interval
预测间接成本 forecasted overhead
预测间接费用 forecasted overhead
预测减法编码 prediction-subtraction coding
预测降深值 prediction of drawdown value
预测校正法 predictor-corrector method;predictor-corrector procedure
预测校正器 predictor-corrector
预测校正图式 predictor-corrector scheme
预测径流量 forecasting runoff
预测距离 predictive distance
预测开采量 prediction of mining quantity
预测开始和完成时间 predicted starting and finishing times
预测可计算性函数 predictably computable function
预测可靠性 prediction reliability
预测控制 prediction control;predictive control;predictor control
预测矿藏量 ore expectant
预测矿产资源量 forecasting mineral resources
预测矿床号 number of predicted deposits
预测矿种 predicted mineral commodity
预测理论 prediction theory
预测量化系统 predictive quantizing system
预测滤波 predictive filtering
预测脉冲形状网络 predicted-pulse-shape network
预测面积 area of predicted district
预测模式 prediction model
预测模型 forecast model; predictive model(ling);prognostic model
预测目的 prediction aim
预测碰撞探测器 predictive crash sensor
预测偏低 underprediction
预测漂移区 anticipated area of drift
预测平均热感觉 predictive mean vote
预测破裂 projecting breaks
预测期 forecast period
预测器 anticipator;predicting apparatus;predictor
预测区 prediction region
预测区间 prediction interval
预测区面积 area of prediction region
预测区域 estimation range
预测区元素丰度 element abundance of prediction region
预测人口 anticipated population
预测日期 predicting date
预测深度 predicted depth
预测使用年限 anticipated service life;life expectancy;probable life
预测使用期限 anticipated service life;life expectancy;probable life
预测式 predictive form
预测式控制系统 predictor-type controller
预测式预算 forecast-type budget
预测数的变化范围 prediction interval
预测数据融合 predictive data fusion
预测衰减量 predicted attenuation value
预测水泥强度 predicting cement strength
预测顺序 forecasting sequence
预测算法 prediction algorithm;prognostication algorithm [algorism]
预测算子 prediction operator
预测体积 predicted volume
预测条件 predicting condition
预测图 map for prediction;prognostic map
预测危险区 predicted area of danger
预测问题 prediction problem
预测污染物负荷 predicting pollutant loading
预测误差 forecasting error;prediction error
预测误差反褶积 prediction error deconvolution
预测误差滤波 prediction error filtering
预测系数 prediction coefficient
预测型用户最优【交】 predictive user optimum
预测性风险评价 predictive risk assessment
预测性能 estimated performance
预测性试验 predictivity test
预测需水量 forecasting demand
预测序列 forecasting sequence
预测学 science of forecast
预测因子 predictor
预测原理 prediction principle
预测远景区 predicted prospective area
预测运转成果 predicted performance
预测执行 speculative execution
预测值 predictand; predicted value; predictor;prognostic value
预测指标 predictor
预测指数 predictive index
预测资产负债表 forecast balance sheet
预测资源量 predictive amount of resources
预测坐标 predicted data;prediction data
预层【数】 presheaf
预察工作 forecasting service
预掺和作用 preblending action
预掺混料 preblend(ing)
预掺料灰泥 premixed plaster
预掺气水跃 preentrained jump
预掺砂石膏灰 gypsum ready-sanded plaster
预沉池 preliminary sedimentation tank;presedimentation tank; presettling tank
预沉淀 preformed precipitate;preliminary precipitation;preliminary sedimentation; preprecipitation; presedimentation;presettling
预沉淀池 preliminary settling tank;presedimentation tank; presettling tank;primary settling tank
预沉积 presedimentation
预沉降槽 preliminary sedimentation tank
预沉降作用 preliminary sedimentation;presedimentation
预沉井 presettling chamber
预沉作用 preliminary sedimentation
预衬垫(密封)带 preshimmed tape
预衬垫密封膏 preshimmed sealant
预称量斗仓 prebatching bin
预称量料斗 prebatching bin
预撑力 setting pressure
预成泡沫 preformed foam
预成说者 preformationist
预成型 perform(ation); preformed shape;preforming;preliminary shaping
预成型的黏[粘]结水泥 preformed mastic
预成型电极 preform electrode
预成型对模模塑法 preform die mo(u)lding
预成型钢丝绳 preformed wire rope
预成型工艺 preforming technique
预成型机 preformer; preform(ing) machine
预成型绝缘材料 preformed insulation
预成型绝缘层 preformed insulation
预成型模塑 preform mo(u)lding
预成型模塑法 preformed matched die
预成型模头 preforming die
预成型模压法 preforming
预成型绳 <钢丝绳经过处理,使各股预先成为扭曲形,在扭结成绳后不致松散> preformed rope
预成型制品 preformed shaped
预澄清 predefecation;presettling
预充 precharge;precharging
预充电 precharge;precharging
预充电脉冲 precharge pulse
预充电栅 precharge gate
预充电时间 precharge time
预充气 preliminary charge
预充水阀 prefill valve
预充填柱 prepacked column
预充压力 precharge pressure
预冲孔 subpunching
预冲孔冲头 passing punch
预冲洗 preflush;prewash(ing)
预冲洗筛分机 preliminary rinsing screen
预冲洗筛子 preliminary rinsing screen
预冲洗液 preflush fluid
预抽 forepumping
预抽泵 prepump
预抽压力 forepressure
预抽真空泵 backing pump;fore-vacuum pump; forvacuum pump; vacuum booster pump;forepump
预抽真空室 preevacuated chamber
预筹地下铁道 <一种轻型有轨交通系统,设计成具有易于转换为快速有轨交通(地铁)的条件> premetro
预臭氧化 pre-ozonization

预初步试验 trial test
预初试验 primary test(ing)
预除节 preknotting
预储备 prestocking
预处理 first treatment;initial handling;precondition(ing);preliminary treatment;preparation;preparatory treatment;preprocess(ing);preservice;pretreat(ment)
预处理槽 pretreatment tank
预处理程序 preprocessing program(me);preprocessor
预处理程序变量 preprocessor variable
预处理程序过程 preprocessor procedure
预处理池 pretreatment tank
预处理的平底船 pretreatment pontoon
预处理方法 pretreatment procedure
预处理费用 pretreatment cost
预处理氟硅酸盐 pretreatment fluosilicate
预处理工艺 pretreatment process
预处理过的地表水 pretreated surface water
预处理过的河水 pretreatment river water
预处理过的矿石 pretreated ore
预处理过的水 pretreated water
预处理过的岩石 pretreated rock
预处理机 preprocessing unit;preprocessor
预处理剂 pretreatment agent
预处理器 preprocessor;pretreater
预处理水 pretreated water
预处理塔 pretreater
预处理涂底料 pretreatment primer
预处理涂料 pretreatment coating
预处理文本 preprocessed text
预处理装置 pretreatment equipment;pretreatment unit
预处理子程序 preprocessing subprogram(me)
预处置 pretreatment
预触发 pretrigger
预触发脉冲 pretriggering pulse
预触发器 pretrigger
预触发时间选择器 pretrigger time selector
预穿孔 prepunched hole;prepunching
预穿孔卡片 prepunched card
预传导 preconduction
预传导电流 preconduction current
预吹 fore-blow
预垂弯传送机 presag conveyer [conveyor]
预纯化 prepurification
预磁化 premagnetization
预磁化式变流器 premagnetization type current transformer
预次序 preorder
预促进聚酯树脂 preaccelerated polyester
预催化聚酯树脂 precatalyzed polyester
预淬火 prequenching
预淬火模具钢 prehardened mo(u)ld steel
预存(储) prestore;prestoring
预存队列 prestored queue
预存费总数 amount of deposit
预搓丝股的钢索 preformed lay wire;preformed wire rope;tru-lay rope
预打 seating drive
预打样 preproof
预导电流 preconduction current
预捣固 preliminary compaction
预捣实粗集料 prepacked coarse aggregate
预涤气装置 prescrubber
预底涂 preprimed
预抵期 estimated time of arrival
预地图 predictive map;prognostic map
预点火阶段持续时间 preliminary-stage duration
预点火室 preignition chamber
预电解 preelectrolysis
预垫的 imprest
预淀积 predeposition
预顶极 preclima
预订 advance order;booking;subscription
预订舱位 booking space
预订单 advance order;booking list
预订的货 bespeak
预订费 subscription
预订和发售客票(系统)制度 system for making reservations and sale of tickets
预订时间 lead time
预订座席 seat reservation
预订座席计划 seat reservation plan
预定 prearrange(ment);preconcert;precontract;predetermination;predetermine;subscribe
预定保险 general insurance policy;open policy;reinsurance open cover
预定保险通知书 declaration under open policy
预定标器 prescaler
预定表皮 presumptive epidermis
预定参数 preset parameter
预定产量 preset output
预定成本 predetermined cost
预定成本计算 predetermined cost accounting
预定程序 preprogrammed schedule;preset program(me);preset sequence
预定程序表 preset program(me)
预定程序电路 preprogrammed chain
预定程序控制系统 preprogrammed control system
预定程序自动驾驶仪 preprogrammed autopilot
预定尺寸 predetermined dimension;preliminary dimension
预定尺寸标准 predetermined dimensional standards
预定处理(过程) predefined process
预定船位 prearranged position
预定的 preestablished;scheduled;preset
预定的采伐 regular cutting
预定的处理符号 predefined process symbol
预定的轨迹 projected path
预定的过程 predefined procedure
预定的极限 preestablished limit
预定的申请 preplanned request
预定调节参数 preselect controls
预定动作时间 predetermined motion time
预定方位 preset bearing
预定房间 bespeak
预定放大 preset enlargement
预定飞行高度 prearranged altitude
预定格式记录 preformatted record
预定工程 scheduled works
预定工程用时间 scheduled engineering time
预定工作量 projected workload
预定公式 predetermined formula
预定拱度 predetermined camber
预定估计制度 predetermined estimate system
预定航次 intended voyage
预定航迹 preset flight path
预定航线 predetermined course;preset course;scheduled course-line
预定航向 prearranged heading;preset course
预定航行计划 sailing schedule
预定好的顺序 predefined procedure
预定核实 book confirmation
预定及购买协议 subscription and purchase agreement
预定极限 preset limit
预定计划 preset program(me)
预定计划工作 schedule work
预定计数 preset count
预定价格 target price
预定价格法 estimated price method
预定间接费用分配率 predetermined overhead rate
预定间歇 prepause
预定角度 predetermined angle
预定结束日期奖金 bonus target date
预定经营标准 predetermined operations standard
预定距离 preset distance
预定开航时间 scheduled sailing time
预定(开始或结束)日期 target date
预定客票 booking of ticket
预定利率 assumed interest rate
预定利润率 target profit rate
预定流量 prescribed flow
预定路线 projected route
预定命运图 fate map
预定模式 preassigned pattern
预定目标 destination;projected goal
预定目的港 port of intended destination
预定能级 preset level
预定扭矩 preset torque
预定扭矩扳手 preset torque-wrench
预定浓度 predetermined concentration
预定频率 preset frequency
预定期间 scheduled period
预定期限 stipulated date
预定起拱 predetermined camber
预定气流调节 timed flow control
预定气压充气装置 preset air meter
预定强度 desired strength
预定区域 presumptive area
预定取样点 preselected site
预定任务 preplanned mission
预定任务之申请 preplanned mission request
预定深度 preselected depth
预定时间 predetermined time;preset time;scheduled time;set time
预定时间标准 predetermined time standard;scheduled time standard
预定时间标准法 predetermined time standard method;scheduled time standard method
预定时间付款汇票 bill payable at a definite time
预定时间划分 predetermined time-sharing
预定时间区分 predetermined time-sharing
预定时间研究 predetermined time study
预定实验 designed experiment
预定输出控制 schedule control
预定数据 expected data;tentation data;tentative data
预定数字 predetermined figure;target figure
预定速度 preset velocity
预定提前时间 procurement lead time
预定条件 postulated condition
预定完成时间 estimated completion time;time for completion
预定完工标准时间 allowed time
预定完工日期 scheduled completion date
预定违约金 liquidated damages
预定维修 scheduled maintenance
预定维修计划 predetermined maintenance project
预定位控制 preoperative control;preselector control
预定位控制机构 preselector mechanism
预定位置 precalculated position;preset position;target position
预定温度 preset temperature;target temperature
预定温度控制 preselected temperature control
预定误差 presumptive error
预定先定 predesign
预定线路 projected route
预定信号 prearranged signals
预定行动开始时间 zero hour
预定形收缩 preset shrinkage
预定形整理 precure finish
预定性能 predetermined characteristic
预定压力 predeterminated pressure;preset pressure
预定一空 be booked up
预定义 predefine
预定义的文本表示法 predefined text representation
预定义的析线表示法 predefined polyline representation
预定义的颜色表示法 predefined colo(u)r representation
预定意义 prospective significance
预定用途 destination
预定造价 projected cost
预定照度法 interflectance method
预定震级 preestablished magnitude of earthquake
预定政策 predetermined policy
预定支付 grace payment
预定值 predetermined value;preset value
预定址方式 preaddressed mode
预定指数 provisional index
预定周期 predetermined cycle;predetermined period
预定周期式交通信号 pretimed signal
预定装配时间 closed assembly time
预定装置 preset device
预定座位计算机化设备 computerised seat reservation facility
预动触点 preliminary contact
预冻结 prefreezing
预镀 preplate
预断 prejudge
预断公式 predictive equation
预煅烧 precalcination;previous calcining
预锻 blocking;dummying;heading;preforging
预锻模膛 blocker;blocking impression
预堆边焊 buttering (run)
预堆熔焊 buttering
预对偶 predual
预镦粗 preupset
预发承付款项 forward commitment
预发件 advance copy
预发泡 prefrothing
预发泡的 preexpanded
预发泡剂 prefoaming material
预发泡沫 preformed foam
预反硝化法 pre-denitrification process
预反应 prereacting;prereaction
预反应料 prereacted raw batch
预反应镁铬砖 prereacted magnetite-chrome brick
预防 anticipation;precaution;preclu-

sion;prevention;provide against
预防白点退火 hydrogen relief annealing
预防办法 precaution
预防保健措施 prophylactic health measure
预防保修 preventive servicing
预防层 preventive stratum
预防成本 prevention cost
预防处理 preventive treatment
预防措施 cautionary measures;counterplan; counterplot; measures of prevention; precaution; precautionary measure; precautioning measure(ment);preventive action;preventive treatment
预防的 precautionary; preservative; prevenient; preventive; prophylactic;protective
预防地震措施 pre-earthquake measure
预防法 preventive; preventive treatment; prophylactic; prophylactic treatment;prophylaxis
预防费用 preventive cost
预防功能 prophylactic function
预防故障的 trouble-proof-saving;trouble-saving; trouble-saving trouble-free
预防剂 preven(ta)tive;preventer; prophylactic
预防检修 preventive maintenance; prophylactic repair
预防接种 vaccination
预防开裂 precaution against cracking
预防疗法 preventive therapy
预防裂纹 precaution against cracking
预防(林)带 preventive belt
预防碰撞的雷达仪 radameter
预防破坏 prevention of damage
预防器 anticipator
预防溶液 preventive solution
预防设备 prevention equipment
预防设施 preventive device
预防损坏 prevention of damage
预防损失 prevention of losses
预防危急安全器 anticipatory gear
预防为主 prevention first
预防为主、防治结合的政策 policy of prevention in the first place and integrating prevention with control
预防维护合同 preventive maintenance contract
预防维护时间 preventive maintenance time
预防维修 preventive maintenance
预防污染 prevention of pollution
预防污染收费 pollution prevention pays
预防线圈 preventive coil
预防性安排 preventive maintenance
预防性安排时间 preventive maintenance time
预防性安全措施 preventive safety
预防性保养的检查和修理 preventive maintenance checks and services
预防性保养计划 preventive maintenance program(me)
预防性保养时间 preventive maintenance time
预防性补救 preventative remedy
预防性措施 precautionary measure; preventive (safety) measure
预防性大修 preventive maintenance overhaul
预防性的 preventative
预防性的木材防护 preventive wood protection
预防性后方保养 preventive depot maintenance
预防性技术保养 preventive maintenance
预防性技术保养检查 preventive maintenance inspection
预防性检查 preventive inspection
预防性检修 preventive maintenance; prophylactic measure
预防性砍树处理 preventive snagging
预防性清障 preventive snagging
预防性托换 precautionary underpinning
预防性维护 preventative maintenance
预防性维护检查 preventive maintenance inspection
预防性维护时间 preventive maintenance time
预防性维修 condition-based maintenance; preventive maintenance; programmed maintenance; protective maintenance
预防性维修计划 preventive maintenance schedule
预防性维修检查 preventive maintenance inspection
预防性维修时间 preventive maintenance time
预防性卫生监督 preventive sanitary supervision
预防性消毒 prophylactic disinfection
预防性修补 preventative remedy
预防性修理 preventive overhaul;preventive repair(ing)
预防性养护 prevention maintenance; preventive maintenance; remedial maintenance
预防性养护计划 preventive maintenance program(me)
预防性医疗 preventative remedy
预防性着陆航线 precautionary landing pattern;precautionary pattern
预防修理 preventive overhaul
预防修理日 off-day
预防药 preservative;preventive;prophylactic
预防医学 preventive medicine
预防质量降低条例 <空气质量标准中的> anti-degradation clause
预防注射 inoculation
预放大(率) premagnification
预放电 prearcing; predischarge; presparking
预放气 pregassing
预分多址卫星系统 preassigned multiple access satellite system
预分解 predecomposition
预分解技术 precalcining technology
预分解窑系统 precalciner kiln system
预分类 pre-sort(ing)
预分离 predissociation; pre-separation
预分馏塔 prefractionator;preliminary fractionator
预分配 prearranged assignment
预分配多址 preassigned multiple access
预分配系统 preassignment system
预分配制 preassignment system
预分散 predispersing
预分散颜料 pigment preparation
预分析 preanalysis
预分选 pre-sort(ing)
预粉磨 preliminary grinding; prepulverize
预粉磨室 pregrinding chamber
预粉碎 precomminution
预粉碎的破碎机 prebreaker
预敷层 precoat
预浮处理 prefloat treatment
预浮流程 prefloat circuit
预浮选 prefloat
预腐蚀 preetching
预付 advancement; advances; in advance;prepay
预付保险费 advance call;insurance prepaid; prepaid insurance; deposit (e) premium
预付保证金 advance payment bond; deposit(e) premium
预付备料款 advances on materials
预付部分现款 partial cash advanced
预付材料款 advance payment for materials
预付的 prepaid
预付定金 advance money on a contract;down payment
预付费用 advanced charges; charge prepaid; expense paid in advance; prepaid expenses;prepaid materials
预付费用表计 prepayment meter
预付费用账 prepaid expense account
预付工料款项 advance for work in process and construction materials
预付工资 advanced wage;dead horse;wage advance plan
预付关税 duty forward
预付话费挂号 call with indication of charge
预付回报费 prepaid reply
预付回报凭单的款项 value of the reply paid voucher
预付汇款 advance remittance
预付货款 advance to the supplies; cash in advance; cash with order; paying in advance; payment forward
预付价格 advanced price; price for forward delivery
预付建筑工程款 advance on construction;advance payment on construction
预付金 advance fee;advance payment bond;prepayment money
预付金额 prepaid amount
预付经费 advance fund
预付款 advance charges;advance(d) payment; advance load for mobilization; advances; ante; deferred charges;down payment;hand money; imprest; make advance; money paid in advance; pay down; pay in advance; payment in advance; prematurity payment;prepayment
预付款保函 advance payment guarantee;prepayment guarantee
预付款不足 insufficiently prepaid
预付款偿还 repayment of advance
预付款持有人 advance holder
预付款净额 net advance
预付款煤气表 slot meter
预付款器材订单 advance material order
预付款审计 advanced payment audit
预付款收费制 prepayment tariff
预付款项 advance allocation;advance payment;imprest account;payment in advance
预付款仪表 prepayment meter
预付款账户 advance account
预付款支付 payment of advance
预付利息 interest advance; interest prepaid;prepaid interest
预付年度保险 annuity in advance
预付年金 prepaid annuity
预付年金保险 annuity in advance
预付式 prepaid
预付式电度表 prepayment meter
预付式煤气表 prepayment gas meter
预付收费系统 prepaid fare system
预付税款 tax paid in advance;tax repayment
预付现金 advance in cash; cash advance;cash deposit in advance
预付项目 prepaid item;prepayment
预付营业资金 prepaid operating capital
预付优惠 prepayment privilege
预付约金 advance money on a contract
预付运费 advance(d) freight;carriage prepaid; freight paid in advance; freight prepaid; prepaid freight
预付站 pay station;prepay station
预付账户 account of advances
预付账款 prepayment
预付转承(包)者款 advances to subcontractors
预付资本 advanced capital
预付总额 lump-sum in advance; lump-sum prepaid
预付租金 advance rental payment; prepaid rent;rent prepayment
预负荷 preload
预富集 preconcentration;preenrichment
预赋值 preassignment
预干期 predrying period
预干室 predrier [predryer]
预干燥 preliminary drying;predry(ing)
预干燥处理 predrying treatment
预感 forefeel;premonition;presage; presentiment
预感的 aural
预感光板 precoated plate
预感器 anticipator
预钢化处理 prestrain(ing)
预告 advance(d) notice; announce in advance;forenotice;foreshadow;give warning; preacquaint; preannounce; preliminary announcement; preliminary notice;premonition;prognostication;warn(ing notice)
预告按钮 advance button
预告闭塞 advance block
预告比塞信号 advance block signal
预告闭塞信号机【铁】 distant block signal
预告臂板 fishtail end (semaphore) blade; fishtail type arm; notched semaphore blade
预告标 <列车进站> advance(d) announcing post;distant post
预告标志 advance sign;warning sign
预告的 warning
预告复示式显示制度 approach indication system
预告接近示像 advance approach aspect
预告接近中速示像 advance approach medium aspect
预告铃 advance bell
预告目录 prospective bibliography
预告牌 approach board;distant warning board;warning board
预告片 prevue
预告示像 warning aspect
预告式显示系统 approach indication system
预告式显示制 approach indication system
预告书 follow-up letter
预告位置 warning position
预告显示 warning position
预告显示制度 approach indication system
预告信号 advance warning signal;anticipating signal;offer signal;prefix signal;presignal(l)ing;warning signal
预告信号机【铁】 advance warning signal(l)er;distant signal(l)er
预告信号机构 distant signal mecha-

nism
预告信号继电器 distant signal relay
预告信号铃 preliminary signal bell
预告信号示像 distant signal aspect
预告信号装置 prediction signal(l)ing system
预告音 pre-information tone
预告值 predictive value
预告周期 warning period
预告装置 distant warning device
预隔滤 prestrain(ing)
预给器 presetting mechanism
预根式 preradical
预更换 advance replacement
预拱度 camber (of ceiling); counter-camber; precamber
预拱法 pre-arch method; pre-cutting method
预拱机 cambering machine
预购 buy for future delivery; forward purchase; purchase in advance
预购定金 advance payments for future purchase
预购合同 forward purchasing contract
预估 prejudge
预估变量 predication variable
预估差异沉降 estimate differential settlement
预估冲刷深度 estimated erosion depth
预估的要求 forecast demand
预估地面速度 predicted ground speed
预估公式 predictor formula
预估使费 pro forma disbursement
预估需水量 forecasting demand
预估值 discreet value
预估制造费用 predetermined factory overhead
预估重现期 interval of expectancy
预固化 precure; prehardening; procuring
预固化期 procured period
预固结 preconsolidation; preliminary consolidation
预观 preview
预观察 previewing
预贯入击数 seating drive
预灌溉 advance irrigation
预灌浆 pregrouting
预灌浆钻孔 precementation borehole
预光室 viewing room
预规格化 prenormalization; prenormalize
预滚动 preroll
预裹沥青的石屑 precoated chip(ping)s
预过滤 prefiltration
预过滤器 prefilter
预过热 presuperheating
预过热器 presuperheater
预焊 preweld(ing); tack weld
预烘 predry(ing); preliminary drying
预烘干 prebake
预烘干仓 predrying compartment
预烘干机 predryer
预烘干室 predrying chamber; preliminary drying chamber
预烘机 predrier
预烘燥装置 predrying unit
预后 prognosis [复 prognoses]
预后指数 prognostic index
预还原 prereducing
预还原处理 reducing pretreatment
预还原法 prereduction method
预还原炉料 prereduced burden material
预缓解 prerelease
预灰槽 preliming tank
预汇编时间 preassemble time
预混 precompounding
预混合 preblending; premixing
预混合的铺层 premixed carpet
预混合法 premixing method
预混合料 premix
预混合模塑成型 premix mo(u)lding
预混合模塑料 premix mo(u)lding compound
预混合气 premixed gas
预混合器 premixer
预混合烧嘴 premix burner
预混合室 premixing cavity; premixing chamber
预混合物 premixture
预混合型燃烧器 premix type burner
预混合作用 preblending action
预混火焰 premixed flame
预混剂 precoating agent
预混(空气)烧器 aerated burner; premixed burner
预混料 gunk; precompound stock; premix compound; premix (material)
预混料模塑 premix mo(u)lding
预混模制 premix mo(u)lding
预混凝超滤 pre-coagulation-ultrafiltration
预混泡沫溶液 premixed foam solution
预混气燃烧器 premixed air and gas burner
预混气体燃烧器 premixed gas burner
预混燃烧 premixed combustion
预混燃烧器 premixed burner
预混溶液 premixed solution
预混室 premixer
预混素材 premixed elements
预混焰 premixed flame
预混柱塞式注模机 premix-plunger machine
预击穿 prebreakdown
预击穿态 prebreakdown state
预击穿特性 prebreakdown characteristic
预畸变 predistortion
预激波 preexcitation wave
预激励 predrive
预级 prestage
预极化材料 prepolarized material
预极化场 prepolarizing field
预计 estimate; estimation; outlook; prejudge
预计编制 budgeting
预计财务报表 forward financial statement; predetermined financial statement; projected financial statement
预计财务报告 forward financial statement; predetermined financial statement; projected financial statement
预计财务状况变动表 expected financial conditions changing sheet
预计差异沉降量 anticipated differential settlement
预计产品寿命 life expectancy
预计沉降 prediction of settlement
预计沉降量 predicted settlement; settlement expectancy
预计沉降速率 estimated settlement rate
预计沉陷量 anticipated settlement
预计成本 anticipated cost; predetermined cost; projected cost
预计成本制 predetermined cost system
预计储量 expected reserves
预计存量 anticipated stock
预计到达日期 expected data of arrival
预计到达时间 estimated time of arrival; expected time of arrivals
预计到完工时的成本 estimate to complete
预计的 estimated; expected
预计抵港时间 estimated time of arrival
预计地下水产量 anticipated ground water yield
预计定点 projected point
预计法 prediction method
预计返回时间 estimated time of return; expected time of return
预计费用 estimated disbursement; estimated expenses; projected expenditures
预计分批成本制度 predetermined job-cost system
预计分摊率 estimated burden rate
预计风险 anticipated risk; calculated risk
预计负债 estimated liabilities
预计改变工程的询价单 <发给承包商的> contemplated change notice
预计工资率 estimated wage rate
预计供应日 estimated supply day; R day
预计航迹 desired track; intended track; prospective path
预计航线角 precomputed course
预计航向 estimated course
预计合同范围 estimated contract coverage
预计滑动距离 expected distance of sliding
预计计时工作 predetermined time work
预计价格 anticipate(d) price
预计交船期 expected time of delivery
预计交货期 expected time of delivery
预计交货时间 estimated time of delivery
预计交通量 anticipated volume of traffic; predicted volume of traffic
预计交通时 future traffic volume
预计接近时间 expected approach time
预计开采资源 predicted exploitation resources
预计开航时间 estimated time of departure; expected sailing time; expected time of departure
预计靠岸时间 estimated time of berthing
预计可用年限 life expectancy
预计离岸时间 estimated time of departure; estimated time of unberthing
预计离港时间 estimated time of departure
预计利润 anticipated profit
预计粒度曲线 expected gradation curve
预计流量 anticipated discharge
预计满顶流量 expected rate of overtopping
预计模式 prediction model
预计目标 scheduled target
预计能耗 scheduled energy
预计能力 <水体纳污等的> predictive ability
预计能量 scheduled energy
预计年限总数折旧法 depreciation-sum of expected life method
预计启航时间 estimated time of departure; expected sailing time; expected time of departure
预计人口增长 anticipated population increase
预计渗流量 anticipated seepage
预计渗漏量 anticipated seepage
预计渗透量 anticipated seepage
预计渗涌量 anticipated seepage
预计施工期限 expected working period
预计时间 scheduled time
预计使用年限 anticipated service life; expected life; life expectancy; probable life
预计使用期限 anticipated service life; life expectancy; probable life; life expectance [expectancy]
预计收益表 budgeted income statement; projected income statement
预计寿命 expected life
预计寿命信息系统 advanced life information system
预计数据 predicted data; prediction data; scheduled data
预计损失 estimated loss
预计损失费 initial loss estimated
预计损益表 expected profit and loss sheet
预计途中时间 estimated time of enroute
预计完成情况 budgeted performance
预计完成时间 estimated completion time; time for completion
预计完工前成本 forecast to complete
预计完工前费用 forecast to complete
预计完工日期 estimated complete date; estimated date of completion; expected date of completion; predicted finishing date
预计完工时的造价 estimate at completion
预计完工时间 estimated completion time; expected completion time; estimated time of completion
预计误差 determinate error; predicted error; prediction error; premeditated mistake
预计现金需要量 estimated cash requirement
预计卸毕时间 estimated time of finishing discharging; expected time of finishing discharging
预计卸完时间 estimated time of finishing discharging; expected time of finishing discharging
预计性能 predicted performance
预计修理量 repair expectancy
预计需要量 anticipated requirement
预计应力 estimated stress; intended stress
预计油耗量 fuel consumption prediction
预计造价 estimated cost of construction; expected cost of building; projected cost
预计增加经费 estimated additional requirements
预计折扣 anticipated discount
预计支出 estimated disbursement
预计执行时间 expected time
预计值 expected value; predicted value
预计装货清单 pro forma invoice
预计装货时间 anticipated shipping date
预计装完时间 estimated time of finishing loading; expected time of finishing loading
预计装运日期 estimated shipping date
预计资产负债表 estimated balance sheet; expected balance sheet
预计总额 estimate total amount
预计最大灾害 damage potential
预计最早完成时间 earliest event occurrence time
预计最终原位应力 anticipated final field stress
预记录 prerecord
预记录操作 preregister operation
预记信号 prerecorded signal
预剂量 predose

预加成聚合物 preadduct
预加臭氧 pre-ozonization
预加的应力荷载 stressing load
预加反向变形 prestrain(ing)
预加负荷 preloading
预加负荷的 preloaded
预加工 conditioning; predigestion; preelaboration; prefabrication; prefinish(ing); preparation; preparative treatment; preparatory cutting; preparatory treatment; preprocess(ing); pretreat(ing); pretreatment; prior operation
预加工程序 preprocessor
预加荷载 initial load; precompressed load; preload(ing)
预加荷载固结 consolidation by preloading
预加荷载压实 preload compaction
预加灰 prelime; preliming
预加拉力 pretension
预加力 preapplied force
预加氯(处理) prechlorination
预加气 precharge; precharging
预加钎料钎焊 preplaced brazing
预加热 preheat(ing)
预加热器 primary heater
预加润滑剂的轴承 prelubricated bearing; prepacked bearing
预加润滑脂的轴承 prepacked bearing
预加石灰处理 lime pretreatment
预加水成球技术 prewatering granulating technology; prewatering nodulizing technique
预加水成球系统 preapplied water nodulizing system
预加水分解 prehydrolysis
预加速 pre-acceleration
预加弯力 preflex(ion)
预加温 prewarming
预加性范畴 preadditive category
预加压 precharge; precharging; prepressurization
预加压力 precompress
预加压式压力容器 prepressurized diaphragm expansion tank
预加应力 initial stress; prestress(ing); preload; prestressing force; prestretching
预加应力壁柱 prestressing lesene
预加应力长度 stressing distance
预加应力场地 prestressing yard
预加应力程序 sequence of prestressing
预加应力的 prestressed; under-prestressed
预加应力的步骤 stressing process
预加应力的方法 stressing process
预加应力的钢筋束 stretching tendon
预加应力的钢丝束 stretching tendon
预加应力的距离 prestressing distance
预加应力的距离增量 prestressing distance increment
预加应力的力矩 stressing moment
预加应力的力量 stressing force
预加应力的面积 prestressing area
预加应力的隧道 prestressing tunnel
预加应力的弹性损失 elastic loss of prestress
预加应力的(通风)管道 prestressing duct
预加应力端 stressing end
预加应力法 prestress(ing method)
预加应力工厂 prestressing factory
预加应力工程 prestressing engineering
预加应力工艺步骤 prestressing process
预加应力工艺过程 prestressing process
预加应力工作队 prestressing gang
预加应力工作线 stressing line
预加应力构件 prestressing element
预加应力过程 process of prestressing
预加应力混凝土设计体系 stressed concrete design system
预加应力距离 stressing distance
预加应力块体 prestressing block
预加应力力矩 prestressing moment
预加应力模具 prestressing mo(u)ld
预加应力区 prestressing zone
预加应力顺序 stressing order
预加应力台 prestressing bed
预加应力台座 prestressing bed
预加应力用千斤顶 prestressing jack
预加应力用楔块 prestressing wedge
预加应力于 prestress
预加应力元件 tensioning element
预加应力张拉线 prestressing line
预加应力值 prestressing value
预加应力装置 <钢筋> take-up set
预加应力座 prestressing bed
预加永久变形 preset
预加载 preloading; pretension
预加载荷 preload
预加张力 pretension
预加重 preaccentuation; preemphasis; preload
预加重电路 preemphasis circuit
预加重滤波器 preemphasis filter
预加重网络 emphasizer; preemphasis network
预加重信号 preemphasized signal
预坚膜 preharden
预剪 preshear
预剪法 preshearing
预检 dry run; precheck; preexamination; preinspection check; pretest(ing); preventative maintenance; preview
预检波带宽 predetection bandwidth
预检波器 predetector
预检方式 service mode
预检监视器 preview monitor
预检接通 preview switching
预检时间 servicing time
预检修 preventive maintenance
预检验 precheck
预见 foresee; foresight; precognition; preconceive; prediction; preview; previse
预见到 look-ahead for
预见的 prospective
预见的能力 foresight
预见期 forecasting period
预见条件 predicted condition
预见性 foreseeability; predictability
预建的 prebuilt
预浇 precoat
预浇的 precoated
预浇电缆坑道 precast cable tunnel
预浇钢筋混凝土车棚 precast reinforced concrete garage
预浇钢筋混凝土门窗过梁 precast reinforced concrete lintel
预浇钢筋混凝土门窗楣 precast reinforced concrete lintel
预浇钢筋混凝土汽车库 precast reinforced concrete garage
预浇钢筋混凝土探查孔 precast reinforced concrete manhole
预浇钢筋混凝土探井 precast reinforced concrete manhole
预浇工业 precast industry
预浇拱桥 precast arch bridge
预浇混凝土电灯杆 precast concrete lighting column
预浇混凝土工作 precast concrete work
预浇混凝土涵洞 precast concrete culvert
预浇混凝土浇筑构件 precast architectural concrete member
预浇混凝土结构件 precast structural concrete member
预浇混凝土框架(工作) precast concrete frame(work)
预浇混凝土楼面板 precast concrete floor unit
预浇混凝土墙板单元 precast concrete wall unit
预浇混凝土墙板结构 precast concrete wall structure
预浇混凝土桥面板 precast concrete floor unit
预浇混凝土梯级 precast concrete flight
预浇混凝土贴面构件 precast concrete face member
预浇混凝土巷道筒壁弧形砌块 precast concrete tube segment
预浇混凝土照明杆 precast concrete lighting column
预浇路缘石 precast kerb
预浇面层 precoating
预浇模型 mo(u)ld for precast work
预浇石膏建筑单元 gypsum cast building unit
预浇隧道衬管(节) precast tunnel tube
预浇外露骨料板 precast exposed aggregate panel
预浇预应力钢筋混凝土工场 precast prestressed concrete manufacturing yard
预浇预应力混凝土构件 precast prestressed concrete member
预浇制造业 precast industry
预浇铸的 precast
预胶凝 pregelatinization
预绞式护线条 preformed armo(u)r rod
预绞先张法 prehinged pretensioning method
预矫 preequalization
预矫正 predistortion
预矫正网络 predistorting network
预铰的 prehinge
预搅拌作用 preblending action
预搅动 preagitation
预缴保险费 advance payment of premium
预缴公司税 advance corporation tax
预缴外汇证明书 certificate for advance surrender of export exchange
预缴押金 advance deposit
预校正 prealignment; precorrection
预校正电路 preemphasis circuit
预校正器 presetter
预解方程 resolvent equation
预解核 resolvent kernel
预解集 resolvent set
预解(矩阵)恒等式 resolvent identity
预解式 resolvent
预解算子 resolvent operator
预借 borrowing in advance; imprest
预借提单 advanced bill of lading
预金属络合的染料 premetallised dye
预紧 precompact; preloading; pretighten
预紧的滚动轴承 preloaded bearing
预紧的锚固件 prestressing anchorage fixture
预紧的锚固装置 prestressing anchorage fixture
预紧集 precompact set; totally bounded set
预紧力 initial tension; pretension; pretightening force; pretightening load; pretightening up force
预紧密封比压 pretightening unit sealing load
预紧器 preloader
预紧橡胶衬套 preloaded rubber bushing
预紧轴承 preloaded bearing
预进气 preadmission
预进位 precarry
预进站戗堤 pretipped banks
预浸 predip; preimpregnation; preliminary dip; preliminary impregnation; preliminary steeping; presoak(ing)
预浸材料 prepreg
预浸出 preleach
预浸处理 prepreg
预浸工艺 preimpregnated process
预浸骨料 presoaked aggregate
预浸集料 presoaked aggregate
预浸胶机 prepreg machine
预浸料 prepreg
预浸料成型 prepreg mo(u)lding
预浸料坯 prepreg
预浸木材 preextracted wood
预浸水 prewetting
预浸透 presaturation
预浸洗剂 presoak(ing)
预浸增强塑料 preimpregnated reinforced plastic
预浸轧 prepadding
预浸渍 predip; preimpregnation; preliminary dip
预浸渍带 prepreg tape
预浸渍的 preimpregnated
预浸渍法 prepreg method
预浸渍工艺 preimpregnated process
预浸渍绝缘 preimpregnating insulation
预浸渍漆 preimpregnated varnish
预浸渍体 prepreg
预浸渍线圈 preimpregnated coil
预浸渍毡 prepreg mat
预浸渍制品 prepreg
预精炼 preliminary refining; prerefining
预精练剂 prescouring agent
预精选 preconcentration
预精轧机座 pony stand
预警 early warning; premonition; prewarning
预警技术 early-warning technique
预警接触器 prealarm contactor
预警界限 warning limit
预警雷达 early-warning radar
预警模块 prealarm module
预警器 precaution device
预警设备 source of early warning
预警无线电网 warning net
预警系统 forecasting and warning system
预警信号 early-warning signal
预警装置 prealarm; prealarm module
预净除尘器 pre-collector
预净化 preliminary cleaning
预净化器 pre-purifier
预局部化子范畴 prelocalizing subcategory
预具资格 prequalification
预聚合 prepolymerization
预聚合物 preformed polymer; prepolymer
预聚焦 prefocus(ing)
预聚焦灯 prefocus lamp
预聚焦灯头 prefocus base
预聚焦镜头 prefocusing lens
预聚焦束 prefocused beam
预聚焦透镜 prefocus lens
预聚焦烛台卡口灯头 bayonet candelabra prefocusing collar base
预聚束 prebunch(ing)
预聚束器 prebuncher
预聚物 preformed polymer; prepolymer
预聚物模塑 prepolymer mo(u)lding
预聚物凝胶 prepolymer gel

预均化 preblending;prehomogenizing
预均化技术 prehomogeneous technique
预均化料堆 preblending pile
预均化效率 prehomogenizing efficiency
预开材料 precut lumber;precut timber
预开槽 pre-slot;preslotting
预开沟槽法 pre-slit method
预开螺栓 puller bolt
预开木材 precut;precut lumber
预开木料 precut lumber
预开坡口 prebeveling
预开起动阀 advanced starting valve
预开挖 preexcavation
预开钥匙系统 preparatory key system
预开账单 advance billing
预看混合 preview matrix
预看监视 preview monitoring
预看开关 preview switch
预看控制 preview control
预看选择器 preview selector
预看增益 preview gain
预科 preparatory course
预科生<大学的> preppy
预可行性分析 prefeasibility study
预可行性勘探 prefeasibility exploration
预可行性研究 preliminary feasibility; preliminary feasibility study; prefeasibility study
预控热 heat precontrol
预控制 precontrol
预扣赋税 withholding tax
预扣利息放款者 discounter
预扣税款证书 tax withholding certificate
预扣所得税 pay-as-you-earn
预馈脉冲 prepulsing
预扩散 prediffusion
预扩散工艺 prediffused technique
预扩展器 preexpander
预拉 predraw; prestretching; pretensioning
预拉钢(绞)索 prestretched strand
预拉钢筋 pretensioned steel
预拉钢丝 prestretched wire; pretensioned wire
预拉力 pretension
预拉伸 predraft; predrawing; preliminary elongation; prestress (ing); prestretching; pretension; pretensioned wire; Hoyer method of prestressing
预拉伸的 prestressed
预拉伸过程 predrawing process; prestretching process
预拉系统 pretensioned system
预老化 preag(e)ing;preconditioning
预垒混凝土 prepacked concrete
预冷 fore-cooling;precool
预冷篦板 precooling grate
预冷藏 pre-refrigeration
预冷淬火 delayed quenching
预冷的 precold
预冷骨料 precooled aggregate
预冷剂 precoolant
预冷间 precooling room
预冷凝器 precondenser
预冷凝作用 precondensation
预冷器 fore cooler;precooler;recooler

预冷却带 precooling zone
预冷却(法)<骨料、集料的> initial cooling;precooling
预冷却混凝土材料 precooling concrete material; precooling room concrete material
预冷却盘管 precooling coil
预冷却器 fore cooler;precooler
预冷时间 precooling time
预冷室 fore-cooling room
预冷装置 precooling apparatus; precooling plant
预冷作硬化 prepeening
预离解作用 predissociation
预离期 estimated time of departure
预离析 pre-separation
预料 anticipation;contemplate;expectation; in prospect; preconceive; prediction;presupposition
预料的 prospective
预料准许进场时间 expected approach clearance time
预裂 pre-split(ing)
预裂爆破 preshearing blasting; presplitting blasting; stress-relieving blasting
预裂爆破法 preshearing blasting;presplitting blasting; presplitting technique; stress-relieving blasting; presplit blast(ing);presplitting
预裂法 precracking;presplitting (method)
预裂技术 presplitting technique
预裂空隙 pre-split void
预裂孔 presplit hole
预裂面 pre-split face
预裂炮孔 preshear hole
预裂掏槽 presplit cut
预裂隙理论 theory of induced cleavage
预令 preliminary command; preliminary order;preparatory command
预留按钮 available button
预留备淤深度 extra depth for siltation
预留变形量 allowable for deformation;reserved deformation
预留不可预见经费 contingency fund
预留槽 form duct;preformed groove; provided groove;reserved groove
预留长度 reserve length
预留沉降量 reserve settlement;settlement allowance
预留沉落量 reserve settlement
预留地 reservation area
预留第二线 prepared double-line;reserved second line
预留洞 hole to be provided
预留缝 ready-made joints
预留钢筋 stub rod
预留拱度 counter-camber
预留观测孔 reserved observation borehole
预留管 sleeve pipe
预留混凝土孔 block joint
预留活载发展系数 developing coefficient of live load reservation; preserved live load increasing factor
预留活载提高系数 preserved live load increasing factor
预留机组 future unit;skeleton unit
预留机组段<厂房的> skeleton bay
预留间隙 intended gap
预留阶梯形齿缝 racking back
预留阶梯形接头 racking back
预留接缝的无筋混凝土路面 jointed plain concrete pavement
预留金额 provisional sum
预留开采孔 reserved mining borehole
预留空间 head space;outage
预留空洞模板<永久性埋设在混凝土内的> void form
预留空隙 reserve gap
预留空余 kick-up
预留孔 break-in; foreseen hole; preformed cavity;preformed hole;prepared hole; preremained hole; preserved hole;provided hole
预留孔道 tendon profile
预留孔洞 blockout; tendon profile; preformed cavity
预留孔洞图 recess drawing
预留孔堵块<浇混凝土后移去> block out
预留孔模 box out
预留孔模板 core form
预留孔平面图 plan of provision of holes
预留梁洞 pocket;wall pocket
预留量 reserve capacity
预留锚固孔 cast-in socket
预留膨胀缝 expansion allowance
预留强度 reserved strength
预留曲度 bewel
预留容积<容器内的防液体膨胀> outage
预留上拱度 preformed camber
预留时间 contingency allowance
预留收缩长度 shrinkage allowance
预留收缩量 shrinkage allowance
预留踏步接头 raking back
预留土地 land for future extension; reserved land
预留弯(拱)度 counter-camber
预留锈蚀富余量 allowance for metal loss
预留锈蚀量 corrosion allowance
预留余地 make allowance for
预留预应力钢索管道的塑料管或金属管 duct former
预留在线管中拉电线的铁丝 wire-puller
预留砖孔 block joint
预留砖石孔 block joint
预硫化 precure
预硫化胶乳 prevulcanized latex
预录的电视节目 prerecorded television program(me)
预滤 prefiltration;pre-flock;prefiltering
预滤池 prefilter; preliminary filter; roughing filter
预滤过性水 prefiltered water
预滤器 catch tank
预氯化 prechlorination
预氯化池 prechlorination tank
预氯化处理 prechlorinating
预氯化消毒 prechlorinating
预埋 U 形环 prebuilt in U ring
预埋的 pre-embedded
预埋的穿引电线管道 trench header duct
预埋地脚螺栓 pre-embedded foot bolt
预埋吊件 built-in hanger; embedded hanger
预埋附固件 sticker
预埋钢板 pre-embedded steel plate
预埋钢筋 embedded bar; embedded reinforcement
预埋工程<管子或电线的> carcass work
预埋工事<管道电缆等> carcase works
预埋构件 building in;embedded element;unit strut
预埋骨料<后用水泥浆灌注成混凝土> prepacked aggregate
预埋骨料灌浆 advanced slope grouting;advancing slope grouting
预埋管线工程 carcase works;carcassing
预埋集料 grouted prepacked aggregate;prepacked aggregate
预埋件 building in fitting; embed parts;insert;built-in fitting;embedded inserts;embedded parts;inserted piece;prebuilt-in parts;pre-embedded parts;pre-embedded pieces <管子或电线等>;handling reinforcement<混凝土构件的>
预埋开关 embedded switch
预埋连接钢筋 stub bar
预埋联结盒 anchorage point
预埋螺杆 embedded bolt; inserted bolt
预埋螺母管 nut socket
预埋螺母和螺栓 prebuilt in anchor and nut
预埋螺栓 embedded bolt;insert(ed) bolt
预埋锚杆 recessed anchor
预埋木块 block out
预埋木砖 built-in wooden brick;embedded wood brick
预埋砌块 embedded block; insert block
预埋墙系筋 prefabricated tie
预埋式温度传感器 embedded temperature detector
预埋式温度计 embedded temperature detector
预埋式温度探测器 embedded temperature detector
预埋式蜗壳 embedded spiral case
预埋受钉木块 fixing block
预埋套管 buried sleeve; service sleeve
预埋条状物 embedded strap
预埋铁箍 cast-in iron band
预埋铁件 pre-embedded iron member
预埋网 embedded mesh
预埋系统 embedded system
预埋芯 superimposed core
预埋型芯 kiss core
预埋药包 sleeper charge
预埋有管道或空心墙、空心楼板的建筑 cavity construction
预埋在混凝土墙或楼板内的套筒 pipe sleeve
预埋在混凝土中 embedded in concrete
预埋在混凝土中的充气管 ductube
预铆铆钉 shop rivet
预描绘 prepaint(ing)
预瞄 lead sight
预敏化 presensitizing
预敏化抗蚀剂 presensitized resist
预模塑 premo(u)lding
预模压加热蠕变成型 prestress forming
预模制套口接头 premo(u)lded gasket joint
预模制填缝条 premo(u)lded filler
预膜<保护金属表面的> prefilming
预膜过滤器 precoated filter
预膜剂 prefilming agent;pretreatment filming agent
预磨 pregrinding
预磨的 pregrinding
预磨机 pregrinder
预磨粒状渣 preground granulated slag
预磨削用研磨膏 pregrinding paste
预磨渣 ground slag
预匿影 pre-blanking
预捏合作用 preblending action
预凝 presetting
预凝胶(体) pregel
预凝结 precoagulation;precondensation
预凝(结)时期 presetting period
预凝期 preset period
预凝器 precondenser
预凝时间<混凝土> presetting period
预浓缩 pre-concentration;prethickening
预浓缩器 preconcentrator
预欧姆 preohmic
预欧姆腐蚀 preohmic etch
预欧姆接触 preohmic contact
预排 preset(ting)
预排程序的 preprogrammed
预排电路 presetting circuit

预排工作 walk-through
预排进路 route presetting
预排控制 presetting control
预排气 predischarge;preevacuate
预排水 predraining
预排水法 predrainage
预排演 dry run
预派的 preassigned
预抛光 prepolish(ing)
预抛拦石坝 <截流用的> dumping rock-blocking sill in advance
预培训中心 advance education center [centre]
预配 preassembly
预配线的 prewired
预配线架 prewired rack
预配选择 prewired option
预喷漆 prepaint(ing)
预偏心桥墩 pre-excentric pier
预偏振 prepolarization
预偏置 prebias
预漂 prebleaching
预拼装 preframe
预拼装的 preframed
预频率校正电路 preaccentuator
预平衡成分 pre-equilibrium composition
预评 pre-evolution
预评价试验 prevaluation test
预破碎 precrushing
预破碎粉磨系统 precrushing and grinding system
预破碎室 precrushing chamber
预铺骨料灌浆混凝土 preplaced-aggregate concrete
预曝气 preaeration
预曝气池 preaeration tank;preaerator
预曝气箱 preaeration tank;preaerator
预期 anticipation; bargain for; contemplation; expectance [expectancy];look-ahead for;prospection
预期报酬 expected reward
预期报酬率 expected rate of return
预期爆心投影点 desired ground zero
预期变形 anticipated deformation
预期产额 expected yield
预期产量 expected yield
预期成本 anticipated cost;expected cost
预期成本效益法 expected benefit-cost method
预期成效 planned performance
预期持续时间 expected duration
预期的 anticipant;contemplated;prevenient;prospective
预期等候时间 expected waiting time
预期地震 anticipated earthquake;expected earthquake
预期地震烈度图 seismic intensity expectancy map
预期电流 prospective current
预期电路 anticipation network
预期费用 anticipated cost
预期分析 predictive analysis
预期峰值 prospective peak value
预期负荷 anticipated load(ing)
预期轨道 nominal orbit
预期荷载 anticipated load(ing);predicted load
预期环境政策 anticipated environmental policy
预期患病率 expected morbidity rate
预期价格 anticipated price; expected price;price expectation
预期检修 planned preventive repair
预期交通量 anticipated traffic;anticipated traffic volume; anticipated volume of traffic
预期结果 expected result
预期精(确)度 anticipated accuracy;expected accuracy
预期开发费 anticipated use or development expenditures
预期开支 anticipated expenditures
预期会计 forward accounting
预期利润 anticipated profit;expected profit;projected profit
预期利润点 desired-profit point
预期利润率 expected rate of profit
预期利润损失保险 loss of advanced profits insurance
预期利益 anticipated profit;imaginary profit
预期流量 anticipated discharge
预期路面性能 expected road surface performance
预期目标反应 anticipatory goal response
预期疲劳寿命 expected fatigue life
预期品质水准 expected quality level
预期平均寿命 expected average life
预期平均值 predicted average;predicted mean
预期坡度 expected gradient of river bed
预期前方点 set-forward point
预期潜像 latent preimage
预期强度 desired strength; target strength
预期缺货 expected shortage
预期缺货量 expected value of stock-out
预期使用年限 life expectancy
预期使用期 expected life time; intended performance
预期使用期限 life expectance [expectancy]
预期使用寿命 expected service life; life expectance [expectancy]; life expectancy
预期收入 anticipated revenue;expected income; prospective earnings; prospective yield
预期收益 anticipated effect; desired result; expected earnings; expected return
预期寿命 expectation of life;expected life;life of expectance
预期输出值 desired output
预期数量短缺 expected quantities short
预期死亡 expectation of death
预期死亡率 expected death rate;expected mortality
预期损失 expected loss
预期投资回收率 expected rate of return on investment
预期完成 desired performance
预期效果 anticipated effect; desired effect; desired result; expectation effect
预期效果的可达到程度 valuable measure of effectiveness
预期效益 expected benefit
预期效用 expected utility;prospective utility
预期行动 anticipated action
预期性能 expected performance
预期性指标 anticipatory index
预期影响 anticipated effect; desired effect
预期有石油的地面积 prospective acreage
预期运费 anticipated freight
预期运行结构使用期限 anticipated structural life
预期运行事件 anticipated operation(al) event
预期造价 expected cost
预期责任 due care
预期折扣 anticipated discount
预期值 desired value; due value; expectation value; expected value; predicted mean value; predicted value; prospective value; required value
预期值理论值 due value
预期质量水平 expected quality level
预期资产价值 expected wealth value
预期纵断面 anticipated profile
预期最大负荷 maximum demand
预期最大水量 maximum demand
预期最低收益率 minimum desired rate of return
预漆工作 prepainting work
预启阀 equilibrated valve; preinlet valve
预汽蒸 presteaming
预汽蒸室 presteaming chamber;presteaming vessel
预砌 prelaying
预牵伸 preliminary draft
预切刀 stocking cutter
预切割 precut
预清机 precleaner
预清机的防护罩 precleaner guard
预清洁 preleasing
预清理 pretreatment
预清洗 preleasing;prerinse
预球化 prespheroidizing
预驱动 predrive
预取 look ahead;prefetch
预取数 prefetched operand
预取微指令 prefetch microinstruction
预取向度 preorientation degree
预燃 preburning; preflame; preignition; preliminary combustion; primary combustion
预燃灯 simmering lamp
预燃反应 preflame reaction
预燃期 preignition period
预燃器 premix burner
预燃曲线 prearc curve;preburn curve;prespark curve
预燃烧 precombustion
预燃烧反应 precombustion reaction
预燃时间 prearc period;preburn(ing) time;prespark period
预燃式柴油机 prechamber diesel engine
预燃室 energy cell; miniature chamber; prechamber; precombustion chamber; premixing cavity; antechamber <内燃机的>
预燃室点火装置 precombustion chamber igniter
预燃室喷嘴 precombustion chamber nozzle
预燃室式柴油机 antechamber diesel; diesel engine with antechamber; prechamber diesel; precombustion diesel;precombustion engine;diesel engine with precombustion chamber; precombustion chamber type diesel engine
预燃氧化 preflame oxidation
预染污 presoil
预热 heat addition; initial heating; prefiring;preheat;preliminary heating; prewarming; reversible key; warming;warming up
预热薄膜 preheated film
预热部分 regenerator section
预热层 preliminary heating zone
预热沉积 preheating deposit
预热处理 preheating treatment
预热带 preheating zone;zone of preparation
预热的 preheated
预热的热阴极灯 preheat hot-cathode lamp
预热的要求 preheat requirement
预热的钻进用水 hot drilling water
预热电弧焊接 preheated electric(al) arc welding
预热电流 preheat(ing) current
预热电炉 electric(al) preheating furnace
预热电路 preheat circuit
预热段 preheating section
预热法 preheating method;preheating process
预热风扇 preheat fan
预热骨料 preheated aggregate; preheating aggregate
预热管 economizer bank; preheating pipe;preheating tube
预热辊 preheat roll
预热锅 preheating pot
预热锅炉 preboiler;preheating boiler
预热回路 preheating circuit
预热火焰 preheat(ing) flame
预热集料 preheated aggregate
预热加工 preheat treatment machining
预热阶段 preheater section
预热空气 preheated air;tempered air
预热空气用蛇管 coil for preheating air
预热孔 preheating gate
预热炼 prebodying
预热料仓 preheating bin
预热料斗 preheating bin
预热留量 preheating allowance
预热炉 mill furnace; preheating furnace;reheating furnace
预热路面磨平机 hot planing machine
预热铆钉 hot driven rivet
预热黏[粘]合剂 hot glue
预热盘管 preheat(ing) coil
预热期 warm-up period
预热起动 heat-start;preheat starting
预热起动器 preheat starter
预热器 fore heater;fore warmer;preheater; primary heater; regenerator; reheat; temperature booster; heat booster
预热器调节阀 preheater adjusting valve
预热器级 preheater stage
预热器旁路系统 preheater bypass system
预热器室 forehearth
预热器塔 cyclone tower
预热器塔架 precalciner tower; preheater tower
预热器系列 preheater string
预热器旋风筒 preheater cyclone
预热器窑 preheater kiln
预热区 preheating range; preheating region; preheating section; preheating zone; preliminary heating zone;zone of preparation
预热燃烧器 preheat burner
预热燃烧室式柴油机 diesel engine with precombustion chamber
预热塞 glow plug;heater plug
预热塞线路接头 wiring harness for glow plugs
预热设备 preheating equipment
预热时间 preheating period; preheating time; warm-up period; warm-up time
预热式热阴极灯 preheat hot-cathode lamp
预热室 prechamber;preheating chamber
预热送风 blast heating
预热隧道 preheating tunnel
预热温度 preheating temperature
预热线圈 preheat coil
预热压力焊 hot-pressure welding

预热压缩空气 preheated compressed air
预热压延机 warming calender
预热焰 preheating flame
预热氧 preheat oxygen
预热液态金属 preheating liquid metals
预热荧光灯 preheat fluorescent lamp
预热与局部加热系统 preheat and hot spot system
预热裕度 preheat margin
预热蒸发器 preheating evaporator
预热装置 preheating unit;primary heater unit
预溶化 prethawing
预溶解 predissolve
预熔 fritting;premelting
预熔共晶 prefused eutectic
预熔化 premelt
预熔聚合物 prefluxed polymer
预熔炼 presmelting
预熔漏板 premelter bushing
预熔器 foremelter
预熔融 prefusion
预熔融转变 premelting transition
预鞣 pretan(ning)
预鞣剂 pretanning agent
预润滑 preoiling
预润滑泵 pregreasing pump;prelubed pump
预润滑封闭式球轴承 prelubricated sealed ball bearing
预润滑密封轴承 prelubricated sealed bearing
预润滑球 prelubricated ball
预润滑装置 preoiler
预色散单色仪 predispersing monochromator
预色散光栅 predispersing grating
预色散器 predisperser
预筛分 preliminary screening;prescreening
预筛分机 prescreener
预筛分器 prescreener
预筛分装置 prescreener
预筛筛分机 presizing screen
预闪 preflashing
预闪蒸 preflash
预商 preconcert
预上胶 presizing
预上黏[粘]胶瓷砖 pregrouted tile
预烧 burn-in;fore-fire;preburning;precalcine;prefire;prefiring
预烧法 preburning
预烧结 pre-sinter(ing)
预烧结棒 presintered bar
预烧坯 presintered compact
预设 preinstall
预设比值 default ratio
预设标志 premarking
预设标志点 premarked point
预设点 default point
预设计的 predesigned;preengineered
预设计师 preengineer
预伸索缆 prestretching cable
预审 precanvass;precensor
预审法官 examining magistrate
预生产车 preproduction vehicle
预生产型 preproduction version
预生铜绿 pre-patinate
预生铜锈 pre-patinate
预生氧化膜的合金 prefilmed alloy
预失真 predistortion
预失真滤波器 predistortion filter
预湿 premoistening <混凝土>;prewet(ting) <湿陷性土>
预湿剂 prewetting agent
预湿筛 prewet screen
预时效 preag(e)ing

预实完全一致性 preread complete uniformity
预示 adumbrate;adumbration;forebode;foreshadow;foretaste;foretell;prediction;presage;prophecy [prophesy];signify
预示保证 earnest
预示变量 predictor
预示方式 predictive mode
预示校正法 predictor-corrector method
预示识别算法 predictive recognizer
预示算子 predictor
预示压力 pilot pressure
预视 preview(ing)
预试 preexamination;prerun;pretest(ing);trial test
预试打支撑桩 preset underpinning;pretest underpinning
预试反射波 pre-echo
预试基础 pretesting footing
预试期 preliminary trial period
预试托换基础 pretest underpinning
预试验 preliminary examination;preliminary test(ing);range finding test;screening test;trial test
预试验的 preproved
预试载 preload
预试柱 pretest column
预试桩 premo(u)lded pile;pretest pile <托换工程的>
预试桩托换 pretest pile underpinning
预饰胶合板 prefinished plywood
预饰面墙板 prefinishing wall board
预饰面纸面石膏板 prefinishing gypsum plasterboard
预饰刨花板 prefinished particle board
预收 receive in advance
预收保证金 margin money
预收尘器 dust precleaner;dust precollector
预收催缴资本 calls in advance
预收费用 advance charges
预收工料款 advances on materials
预收合同款 advance received on contract
预收货款 advance on sale;deposit(e) received on sale
预收货款账户 advance collection account
预收集 precollection
预收款 deferred liabilities;deposit(e) received;money received in advance
预收款项 advance collection;items received in advance
预收收益 income in advance
预收税款 taxes collected in advance
预收缩 preshrinking;preshrunk
预收缩混凝土 preshrunk concrete
预收缩搅拌 preshrink mixing
预收项目 items received in advance
预收账款审计 audit of accounts received in advance
预受力的 prestressed
预售 presale
预熟成 preripening
预熟化 precure
预竖立坡度控制金属丝 pre-erected grade wire
预竖立坡度线 pre-erected grade line
预水冲刷理论 theory of washout by rain
预水合 prehydration
预水合膨润土 prehydrated bentonite
预水化 prehydration
预水解 prehydrolysis
预塑 premo(u)ld
预塑的 premo(u)lded
预塑地沥青板 premo(u)lded asphalt plank

预塑缝 preformed joint;premo(u)lded joint
预塑钢丝绳 preformed wire rope
预塑化 precure;preplasticizing
预塑化装置 preplasticizer
预塑结构防水垫层 preformed structure gasket
预塑沥青板 premo(u)lded asphalt plank
预塑沥青嵌缝板 premo(u)lded asphalt sealing strip
预塑炼 preplastication
预塑嵌缝板 precast joint filler;preformed asphalt joint filler;premo(u)lded joint filler
预塑嵌缝式沥青嵌缝板 premo(u)lded asphalt sealing strip
预塑嵌缝式沥青嵌缝条 premo(u)lded asphalt sealing strip
预塑嵌缝条 precast joint filler;preformed joint filler;premo(u)lded joint filler
预塑式沥青嵌缝板 premo(u)lded asphalt sealing strip
预塑式嵌缝板 preformed joint filler
预塑填缝料 preformed seal
预塑桩 premo(u)lded pile
预算 account valuation;budget;budget estimate;costing;estimate;estimate of cost;estimation;precalculation;preestimate;prefigure
预算安排 budget layout
预算百分比(曲线) fee curve
预算办法 budget practice
预算保证 budgetary support
预算报告 budget report
预算本体 main budget
预算编排 budget layout
预算编制 budget(ary) making;budget(ary) planning
预算编制程序 budget(ary) procedure;budget(ary) process
预算编制方法 budgeting technique
预算编制方式 budget present;presentation of budget
预算编制格式 budget presentation
预算编制人 budgeter
预算(编制)委员会 budget committee
预算编制者 budgeter
预算标准 budget level
预算表 budget;budget sheet;budget statement;estimate sheet
预算拨款 budget allocation;budget allotment;budget(ary) appropriation;budget provision
预算拨款明细账 subsidiary appropriation ledger
预算不敷垫款 deficiency advance
预算部门 budget department
预算财务报表 budgeted financial statement
预算草案 budget layout;budget proposal;drafted budget
预算超额 excess budget
预算成本 budgetary cost;budget(ed) cost;estimated capital;estimated cost
预算程序 budget procedure;budget process;estimating procedure
预算赤字 budget(ary) deficit;budget gap
预算赤字贷款 budget deficit loan
预算储备金 budget reserve
预算处 budget division
预算单 estimate sheet
预算到港时间 estimated time of arrival
预算的 budgetary;precomputed
预算的汇总 estimate summary
预算的收入净额 net flow of revenue in budget

预算的延期 budget deferral
预算抵押贷款 <贷款人每月付税费和保险费> budget mortgage
预算定额 budget norm;detailed estimate norm;norm for detailed estimates
预算定额套用的审计 audit of using budget quota
预算额 allowance;estimating
预算法 budget law
预算方案 budget program(me)
预算费用 budgeted expenses;estimated cost
预算分类 budget classification
预算分类账 budget ledger
预算分配 allocation of budget;budget allocation
预算分配比例 estimate burden rate
预算分配表 allotment ledger
预算分配额 budget allotment
预算份额 budget share
预算付款 budgetary payments
预算概念 budgetary concept
预算高度 precomputed altitude;precomputed height
预算格式 budget present;form of budget
预算工程量 estimated amount
预算工程师 budgetary engineer;estimating engineer
预算工作 budgeting
预算估计 budget estimate
预算观点 budgetary view-point
预算管理 administration of budget;budget control;budget management
预算规划项目 projections
预算核查 budget audit(ing)
预算汇总 budget summary
预算汇总表 budget procedure
预算活动 budget activity
预算基数 budget base
预算计划 budgetary planning
预算价(格) budget(ary) price;budgetary cost;extended price;estimated cost
预算价值 budgetary value
预算简表 budget summary
预算建议书 budget proposal
预算节余 budgetary saving
预算结余 budgetary saving;budgetary surplus
预算金额 budgeted amount
预算紧缩 budget retrench
预算经费 budgetary resource
预算经费短缺 scarcity of budgetary resources
预算经营费限额户 normed deposit account of budgetary expenditures
预算局 budget bureau;budget office;bureau of budget
预算开航时间 estimated time of departure
预算控制计划 budgetary control plan
预算控制(数) budget(ary) control
预算会计 budgetary accounting
预算离港时间 estimated time of departure
预算利润 estimated profit
预算两年期 budgetary biennium
预算明细比较表 green sheet
预算明细表 budget schedule
预算内开支 budgeted expenses
预算内资金 budgetary fund;capital in budget
预算能力 budgeted capacity
预算年度 budget year
预算平衡 budgetary equilibrium
预算平衡表 budgetary balance sheet
预算期 budget period
预算期限的长度 length of budget pe-

riod
预算权限 budget authorization
预算申请书 budget message; budget statement
预算审查 budgetary security; budget audit(ing); budget review
预算审定 examination of budget
预算师 quantity surveyor
预算实况 budgetary position
预算实施 budget execution
预算收入 budgetary revenue; budget receipt
预算收入账户 budgetary receipts account
预算书 budget documents; budget statement; estimate documentation; written estimate
预算书的检查 checking of estimate
预算数据 budget data
预算数量 estimated amount
预算说明书 budget statement
预算损益表 pro-form income statement
预算调整 adjustment of budget
预算通知 budget notice
预算图表 budget chart
预算外 extra-budgetary; off-budget
预算外计划 extra-budgetary program(me)
预算外经费 extra-budgetary fund
预算外开支 expenditure out of budget
预算外开支的 out-of-pocket
预算外来源 extra-budgetary source
预算外收入与支出 receipts and expenditures out of budget
预算外收支 receipt out of budget
预算外投资 investment outside the budget
预算外支出 expenditure out of budget; out-budget expenses
预算外资金 capital out of budget; extra-budgetary fund
预算外资源 extra-budgetary source
预算文件 budget documents; documents of estimate; estimate documentation
预算系统 budgeting system
预算限度 budget limitation
预算限额 budgetary allowance
预算限制 budget(ary) constraint; budget restraint; budget restriction
预算项目 budget account; budget items; terms of estimate
预算削减 budgetary reduction
预算效果 budgetary effect
预算修正案 budget amendment; budget revision
预算需要 budgetary requirement
预算循环 budgetary cycle
预算一览表 budget calendar
预算盈余 budget surplus
预算用表 estimating forms
预算余绌 surplus or deficit of budget
预算与决算制度 budget and financial statement system
预算与实际开支的差额 budget variance
预算与实际损益比较表 comparative budget and actual income sheet
预算预报 budgetary forecast
预算员 budgeter; budget officer; estimate clerk; estimator
预算造价 cost of detailed estimation; estimated amount; estimated cost
预算账户 budget(ary) account
预算政策 budgetary policy
预算支出 budgetary expenditures; budgetary outlay; budget outlay
预算支出限额 existing amount of budgetary expenditures
预算支出账户 budgetary expenditures account
预算执行 budget enforcement; budget execution; budget implementation; execution of the budget
预算执行部门 budget executive
预算执行情况 budget(ary) performance; performance of budget
预算执行者 budgeteer
预算直接费 flat cost
预算值 budgetary value; estimated value
预算职能结构 functional frame for budget
预算指南 budget manual
预算制度 budget system
预算制造费用 budgeted overhead
预算周期 budget cycle
预算主管当局 budget authority
预算主管人 budget director
预算专业 budgetary discipline
预算准备金 budgetary reserve
预算咨文 budget message
预算资本 budgeted capital
预算资产负债表 budgeted balance sheet
预算资金 budgetary fund; budgetary resource
预算资料 data of estimate
预算资源 budget source
预算总表 estimate summary
预算总(金)额 budget amount; overall budget level
预算总造价 overall estimated cost
预碎机 prebreaker
预缩 <纺织物> preshrunk
预缩工序 shrinking processes
预缩合物 precondensate
预缩合作用 precondensation
预缩聚物 prepolycondensate
预缩整理机 sanforizer
预探井 exploratory test well
预提 withholding
预提费用 prevision for expenses
预提费用审计 projected expenditure audit
预提坏账准备 doubtful debts provision
预提取 predraw; preextraction
预提税 withholding tax
预提税抵免 withholding tax credit
预提税率 withholding tax rate
预提所得税 withholding income tax
预提折旧 provision for depreciation
预填 prepack; ramup
预填充载体 prepacked support
预填充柱 prepacked column
预填粗骨料 prepacked coarse aggregate
预填粗集料 prepacked coarse aggregate
预填骨料 prepacked aggregate
预填骨料法 prepakt method
预填骨料灌浆法 colcrete method; prepacked method
预填骨料灌浆混凝土 prepacked aggregate concrete; colcrete; grouted aggregate concrete; preplaced aggregate concrete
预填骨料混凝土 colloidal concrete; grouted aggregate concrete; intruded aggregate concrete; intrusion concrete; packaged concrete; prepackaged concrete; prepacked aggregate concrete; prepakt concrete; preplaced-aggregate concrete
预填骨料结构 precast construction
预填骨料压灌砂浆 colgrout
预填骨料压浆混凝土 colcrete; prepacked concrete
预填骨料专用砂浆 colgrout
预填混合料 prepackaged mix(ture)
预填混凝土 prepacked aggregate concrete
预填集料 prepacked aggregate
预填集料灌浆法 prepacked method
预填集料灌浆混凝土 prepacked aggregate concrete; prepakt
预填集料混凝土 grouted aggregate concrete; grouted concrete; intrusion concrete; prepacked concrete
预填集料压力灌浆法 prepacked method
预填砾石过筛筒 prepacked gravel sleeve
预填料混凝土 prepacked concrete
预填黏[粘]土砖 prepacked clay brick
预填日期 date forward; foredated; postdate
预填入 preload
预填物管子 <填有沙子> loading pipe
预调 pre-adjusted; preliminary adjustment; preset regulation
预调彩虹光栏系统 preselector iris system
预调参数控制 preselect parameter controls
预调初始荷载 preset yield load
预调刀具 preset tool
预调的 preset
预调灯光衰减器 preset fader
预调电路 preconditioning circuit; presetting circuit
预调电容器 preset capacitor
预调电位计 preset potentiometer
预调定时器托架 preset timing bracket
预调定压力 preset pressure
预调干扰机 preset jammer
预调光阑 preset stop
预调和油漆 ready-mix paint
预调机床 presetting machine
预调机构 preset mechanism
预调计数 preset count
预调计数器 predetermined counter
预调焦灯座 prefocus lamp base
预调节 preadjustment; preconditioning; prefading; presetting
预调节器 preconditioner; preregulator
预调控制机构 pretuning control
预调控制钮 preset control
预调棱镜 preset prism
预调频率 preset frequency
预调设备 preset apparatus
预调时间 preset time
预调式仪器 preset measuring instrument
预调试 precommissioning
预调衰减器 preset attenuator
预调水温 presetting of water temperature
预调透镜 preset lens
预调位置 preset position
预调物镜 preset lens
预调谐 preset adjustment; preset tuning
预调谐电路 presetting circuit
预调卸载高度 pre-set dump height
预调旋钮 preset knob
预调压阀 surge anticipator
预调振荡器 preset oscillator
预调整 preset; preset adjustment
预调整装置 anticipatory control
预调直的 pre-straightened
预调制 premodulation; anticipating control
预调准 preset adjustment
预同步 presynchronization
预投准确率 forecast accuracy
预图 forecast chart; forecast map; prognostic map
预图案寄存器 prepattern register
预图案模板匹配 prepattern template matching
预涂布 precoat
预涂材料 precoating compound
预涂层 precoat(ing); precoated sheet; precoating layer
预涂层的黏[粘]合料 binder for precoating
预涂层过滤 precoat filtration
预涂层过滤器 precoated filter
预涂层沥青碎石 precoated chip(ping)s
预涂的 precoated
预涂底漆 blast primer; precoating primer; prefabrication primer; shop primer
预涂钢板 precoated steel plate
预涂骨料 precoated aggregate
预涂护墙板 prefinished wall panel
预涂混合剂 precoating composition
预涂基层 precoated base
预涂集料 precoated aggregate
预涂剂 precoating agent
预涂胶水 presizing
预涂胶纸 pregummed paper
预涂金属 precoating metal
预涂金属板 precoated metal plate; prefinished metal plate
预涂刨花板 prefinished particle board; prefinished wallboard panel
预涂漆 prepaint(ing)
预涂熔剂 prefluxing
预涂色彩层 prefinished colo(u)r coating
预涂砂粒 precoated grit
预涂石屑 precoated chip(ping)s
预涂铜绿的混凝土屋面瓦 pre-patinated concrete roof(ing) tile
预涂铜绿的黏[粘]土屋面瓦 pre-patinated clay roof(ing) tile
预涂铜绿的油漆 pre-patinating paint
预涂铜绿剂 pre-patinating agent
预涂助滤剂的过滤机 precoated filter
预推(法)施工 push-out construction
预脱氮法 pre-denitrification process
预外差旁通技术 preheterodyne bypassing technique
预弯 preflex(ion); prespring(ing)
预弯大梁 preflex girder
预弯的 pre-curved
预弯法预应力 prestressing without tendon by pre-bending
预弯钢轨 precurved rail; precurving rail
预弯辊 cambering roll
预弯焊件 prespringing
预弯梁 prebending girder; preflex beam
预弯梁桥 prebending girder bridge; preflex beam bridge; preflex girder bridge
预弯曲 prebend(ing); prebuckling
预温罩 preheat shroud
预吸取 predraw
预吸收 preabsorption
预稀释的 prediluted
预熄灭 pre-blanking
预洗 prewash(ing)
预洗槽 prewash tank
预洗涤 prescrub
预洗涤装置 prescrubber
预洗机 prewasher; prewashing machine
预洗矿筒 preliminary washing drum
预洗炼 prescouring
预洗期 prewash phase
预先 beforehand; previously
预先安排 prearrange(ment); preconcert
预先安排计划 prearrange; prescheduling
预先安装 preassembling; preset

预先安装的 preestablished
预先拌好的混合料 premix
预先拌和 premix
预先拌和的 premixed
预先拌和水泥 pre-mixed cement
预先包装 prepack(age);prepacking
预先报关 preentry
预先报警系统 advanced warning system
预先报知 ample warning
预先爆破 preliminary demolition;preshoot
预先焙烧 preroast
预先焙烧过的 prefired
预先编辑 preedit(ing);preedition
预先编码 predictive coding
预先编制程序 preprogramming
预先变形钢丝绳 preformed rope
预先变质处理 premodification
预先补偿 precompensation
预先部位分离作用 preliminary fractionation
预先采购 advance buying
预先偿还 advance repayment
预先沉淀 presedimentation
预先成型 preshaping
预先成型的绝热材料 sectional insulation
预先成型的绝缘材料 sectional insulation
预先承兑 anticipated acceptance
预先充电 precharge
预先冲洗流通式计数管 preflush flow counter
预先抽水法 < 如用井点降水等 > predraining method
预先筹划的 preengineered
预先除气 predegassing
预先处理 anticipation;prepreparation;pretreat(ment)
预先触发 pretriggering
预先触发脉冲时间选择器 pretrigger time selector
预先穿孔 prepunch
预先磁化 premagnetization
预先存储 prestoring;prestore
预先存在裂缝 preexisting crack
预先的条件 precondition
预先点火 prefire
预先调度法 forward scheduling
预先定位传力杆 prefixed dowel bar
预先定向 predetermined orientation
预先定义 preliminary definition
预先读出 preliminary reading
预先读数头 preread head
预先对准 prealignment
预先放好 preposition
预先放置的物资 prepositioned stock
预先分布 predistribution
预先分级 preliminary classification
预先分节 presegmentation
预先分配 predistribution;preplanned allocation
预先分配的 preassigned
预先分批配料的 prebatched
预先分析 preanalysis
预先付款 prepaid;prepay
预先付款的物资分发 prepayment issue
预先负载 preloading
预先富集 preconcentration
预先概率 prior probability
预先干燥 predry(ing)
预先格式 preformat
预先给定的数位格式 preset digit layout
预先给定的数位配置 preset digit layout
预先构成 precompose
预先购买 anticipated buying
预先估计 preliminary evaluation
预先估(计)值 advance estimate
预先固结 preconsolidate;preconsolidation;previous consolidation
预先灌浆法 pre-injection
预先规定 preordain
预先规定的 predefined
预先规定方法 predefined process
预先规定过程 predefined process
预先规定合同价格 flat fee contract
预先规划 preplan(ning)
预先过滤 prefiltering
预先航路点 prebriefed point
预先核准 prior approval
预先烘干 prebaking
预先呼叫 advance calling
预先还原 prereduction
预先混合 preblend;preliminary mixing;premix
预先混合喷射器 premix injector
预先混合器 premixer
预先火法精炼 preliminary fire refining
预先集装的托盘化货物 prepalletized cargo
预先计划 forward planning;premeditation;preplan(ning)
预先计划好的 designed
预先计划搜索 preplanned search
预先计算 precomputation
预先计算的 precomputed
预先计算的高度曲线 precomputed altitude curve
预先计算好的 precalculated
预先计算曲线 precomputed curve
预先加工 preparation;preprocess(ing)
预先加力 preforce
预先加料 preload(ing)
预先加热 hot spotting;preheat
预先加压用油泵 pressurizing pump
预先加油 preoiling
预先加油器 preoiler
预先假定 presuppose
预先检查 preexamine;prior examination
预先检定 precalibration
预先检验 precheck;pretest(ing)
预先搅拌 premixing
预先校验 precheck
预先校正 precorrection;preediting
预先接线的 prewired
预先解决 foreclose
预先解锁 prerelease
预先进入阵地 preposition
预先浸湿 prewatering;prewetting
预先浸湿的 presoaked
预先浸渍 bitumen preimpregnating;bitumen preimpregnation
预先浸渍的 preimpregnated
预先精炼 prerefining
预先警报系统 early-warning system
预先警告 forenotice;forewarn;premonition;prewarning
预先警告范围 precautionary range
预先警告牌 distant warning board
预先警告信号 advance warning signal
预先警告装置 distant warning device
预先具有资格 prequalify
预先聚焦 prefocusing
预先决定论 predetermination
预先开挖 < 打桩前的 > preexcavation
预先考虑 anticipation;preconsideration
预先控制 precontrol
预先拉伸 prestretching
预先拉伸应力 prestressing stress
预先老化 preliminary ag(e)ing
预先连接的小口径水带 preconnect handline
预先联系 preliminary correspondence
预先硫化 prevulcanization
预先录音 prescoring
预先锚碇的 preanchored
预先蒙导 preliminary mentor
预先瞄准 minus sight
预先模制 premo(u)ld
预先模制的 premo(u)lded
预先磨矿 preliminary grinding
预先磨碎 foregrinding
预先捏和作用 premastication
预先排气 preexhaust;prerelease
预先排水 predraining
预先判断 prejudge;prejudication
预先泡水 prehydration
预先配料的 prebatched
预先配线的 prewired
预先起的拱度 predetermined camber
预先起动 preliminary start
预先起拱 predetermined camber
预先汽化进料 previously vaporized charge
预先清偿 advance repayment
预先清理 preleasing
预先确定的值 predetermined value
预先润滑 prelubrication;preoiling
预先润滑群接触器 pregreasing group contactor
预先润湿 prewet(ting)
预先润脂继电器 pregreasing relay
预先筛分 prescreening;presizing;probable sizing
预先烧结的玻璃配合料 prereacted glass batch
预先设计 preliminary design
预先设立 preestablish
预先设立的 preestablished
预先审查 precognition
预先施测的地面控制 preplaced ground control
预先湿润 prewatering
预先试车 pretest
预先试验 pretest(ing);previous test
预先释放 premature release;prerelease
预先疏干 advance dewatering
预先水化 prehydration
预先水洗 prewash(ing)
预先算出的 precomputed
预先锁闭 advance locking;preliminary locking
预先碳化 precarburization
预先碳化处理 pre-carbonation (treatment)
预先填充 prefill
预先填上日期的 foredated
预先填上日期的支票 < 票上日期早于实际出票日 > foredated check
预先调谐 pretuning
预先调整 pre-adjusted;preadjustment;preset(ting);anticipating control
预先通知 advance(d) notice;pervious notice;preacquaint;previse;warn
预先通知期限 period of advance notice
预先同意 preconcert
预先投资 upfront investment
预先退火 preannealing
预先脱水 primary dewatering
预先弯曲 prebend(ing)
预先稀释的润滑油 prediluted oil
预先舾装 advance fitting-out
预先洗涤 prewash(ing)
预先消化 predigest
预先卸载 predischarge
预先形成 preformation;preshaping
预先形成的 preformed;preforming
预先形成的沉淀物 preformed precipitate
预先形成的泡沫 < 制造蜂窝状混凝土的 > preformed foam
预先型的 preforming
预先修改 premodification
预先修正 premodification
预先选择 preselection
预先压碎机 preliminary breaker
预先压缩 precompression
预先印好的 preprinted
预先硬化反应 prehardening reaction
预先油量 preliminary fueling
预先油漆 prepaint(ing)
预先与水泵连接的水带 preconnected line
预先约定 by appointment
预先约定的机动动作 prearranged aircraft maneuver;prearranged maneuver
预先掌握 premastery
预先整平的 prelevelled
预先指定 preassign
预先指定的 preassigned;previously designated
预先制定 preestablish
预先注水 preliminary infusion
预先铸模 premo(u)lding
预先装定的引信 precut fuze
预先装配 preassemble
预先装配安装法 preassembly method
预先装配部件 prefabricated component
预先装配的 preassembled
预先装填 prepack(age)
预先装置 preset
预先装置调节 preset regulation
预先准备 advance preparation
预先准备的 ready-made
预先组装 preassemble;preassembling;preassembly
预先组装的 preconstructional
预先钻孔 advancing boring;preboring
预想 preconceive;preconception;premeditation
预消化 predigestion
预消化器 preslaker
预消化污泥 previously digested sludge
预消隐脉冲电平 prepedestal level
预效应 preliminary effect
预斜网络 emphasis network;preemphasis network;pre-equalizer
预泄 < 洪水 > advance release
预信号 presignal
预信号操纵 preregister operation
预行程 pretravel
预行分级 forecasting grade
预行警告 previse
预行润滑 pregreasing
预修 preven(ta)tive maintenance
预修和维护 preventive maintenance overhaul
预修项目表 preventive maintenance roster
预修正电路 predistorter
预絮凝粒 prefloc
预絮凝体 preformed floc
预旋 prerotation
预旋器 prerotator;prewhirler
预旋叶片 prerotation vane;prewhirl vane
预旋转 < 气流的 > prewhirl
预选 forward selection;prechoose;preconcentration;preelect;preliminary concentration;preselect
预选变速器 preselective gearbox
预选操纵 preoptive control;preselective control
预选的 preselective
预选阀 preselector valve
预选工程 preselected project
预选光栏 preselected stop
预选换挡变速器 preselective gearbox
预选机 preselector
预选级 preselection stage;preselector stage;stage of preselection

预选开关 preselection switch
预选控制 preselective control; preselector control
预选器 line switch; preselector; subscriber's selector
预选器架 line switch board; preselector rack
预选式变速箱 preselector gearbox
预选式齿轮变速器 preselector gearbox
预选式的 preselective
预选送 preselection
预选通脉冲延迟 prestrobe delay
预选尾矿 preliminary rejection
预选性 preselection
预选寻线机 preselecting line switch; preselector line switch
预选择 forward selection; preselection
预选值 preselected value
预选制 presetting system
预选质量数 preselected masses
预选装置 preselector
预压 preliminary compaction; preliminary pressure; prepressing; preload(ing)
预压波 precompression wave
预压的 precompressed
预压法 preloading method; preload system
预压浮沉模 prepressing-die-float
预压浮模控制杆 prepressing-die-float control
预压骨料法 prepakt method
预压骨料混凝土 intruded aggregate concrete; prepackaged concrete; prepakt concrete
预压固结 preconsolidation; preloading consolidation; surcharge preloading consolidation; consolidation by preloading
预压固结荷载 preconsolidated [preconsolidation] load
预压固结两用仪 preload consolidation apparatus
预压固结压力 preconsolidated pressure
预压合片材 mangle
预压痕弯曲 pinch bending
预压花 precreping
预压机 preformer; preforming press
预压集料混凝土 prepakt concrete
预压加固 preload compaction; preload consolidation
预压加固法 preloading method; prepressing method
预压紧 preliminary compaction
预压空气 precompressed air
预压力 precompression; ramming pressure
预压力区 precompressed zone
预压木材 precompressed wood
预压黏[粘]土 precompressed clay
预压排水固结法 consolidation by preloading method
预压坯 preformed compact
预压器 precompressor
预压腔 spacer
预压区 <预应力锚头的> precompressed zone
预压时间 <电阻焊时> squeeze time
预压实 <在碾压前由其他机械附带压实> precompaction; preconsolidation; preliminary compaction
预压实黏[粘]土砖 prepacked clay brick
预压实土 preconsolidated soil
预压式接触焊机 press-type resistance welder
预压室 precompression chamber
预压受拉区 precompressed tensile zone
预压水箱 preload tank
预压缩 precompression
预压弹簧 set-up spring
预压填土 preconstruction fill
预压条机 preplodder
预压应力 compressive prestress
预轧孔型 intermediate pass
预言 foretell; presage; prognosticate; prophecy [prophesy]
预言能力 prophecy [prophesy]
预言在未来发生的 futuristic
预研 advanced research; pre-implementation study
预演 walk-through
预阳极化 preanodize
预养护 precure; precuring
预养护期 precured period; preset period; presteaming period
预养护时期 precured period
预养期 <混凝土蒸汽养护前的> presteaming period; presetting period
预氧化 pre-oxidation
预液泛 preflooding
预抑制 initial-suppression; presuppression
预译器 pretranslator
预引比例因子 prescaling
预印本 advance copy
预印符号 preprinted symbol
预印图案 preprinting
预应变 autofrettage; prestrain(ing)
预应变焊接 shrink welding
预应介质 prestressed medium
预应拉力 prestressed tension
预应力 inherent stress; preload stress; prestress; prestressing force; pretensioned force
预应力 H 形截面梁柱 prestressed H beam column
预应力 JM-12 型锚具 JM-12 anchorage (device)
预应力坝 prestressed dam
预应力班组 stressing gang
预应力板 prestressing plate
预应力板桩 prestressed sheet pile
预应力背拉杆 prestressed diagonal
预应力壁柱板条 prestressing pilaster strip
预应力玻璃 prestressed glass
预应力薄底板 <装配式整体结构用> slab soffit
预应力薄壳 prestressed shell
预应力薄膜 prestressed membrane
预应力材料 prestressing material
预应力槽瓦 load-bearing structural tile
预应力层 prestressed layer
预应力缠丝 <用以制造管材> prestressed wire winding
预应力厂 stretching factory
预应力超限的 over-prestressed
预应力成型 prestress forming
预应力承载 bearing preloading
预应力重分布 prestressed redistribution; prestress redistribution
预应力传递 prestress transfer; transfer of prestresses
预应力传递长度 transfer length of prestress; transmission length of prestress
预应力次序 prestressing order
预应力倒 T 形钢 inverted prestressed T bar
预应力道路 prestressed road
预应力的 prestressed
预应力的传递 transfer of prestresses
预应力地锚 <代替后座顶墙> prestressed ground anchors
预应力度 degree of prestress
预应力对接拼装 prestressed abutted assemblies
预应力反拱 prestressing camber
预应力分布 distribution of prestress; prestress distribution
预应力杆 prestressing rod
预应力钢材 stressing strand; prestressing steel
预应力钢杆 prestressing rod
预应力钢绞索 stretching cable; prestressing strand
预应力钢绞线 prestressed stranded steel wire; stressing strand
预应力钢筋 prestressed steel; prestressing bar; prestressing reinforcement; prestressing steel; prestressing (steel) tendon; stressing bar; stressing reinforcement; stressing rod; stressing steel; stretching reinforcement; prestressed reinforcement
预应力钢筋层张拉端头 jacking end of a bed
预应力钢筋的应力损失 anchorage loss
预应力钢筋法 tendon method
预应力钢筋弧线 prestressed steel profile; steel profile
预应力钢筋混凝土 cylindric(al) prestressed concrete shell; prestressed reinforced concrete
预应力钢筋混凝土 T 形刚构架 prestressed concrete T-frame bridge
预应力钢筋混凝土板桩 prestressed reinforced concrete sheet pile
预应力钢筋混凝土大梁 prestressed reinforced concrete girder
预应力钢筋混凝土浮坞门 prestressed concrete floating dock gate
预应力钢筋混凝土刚架 rigid prestressed concrete frame
预应力钢筋混凝土钢筒管 prestressed concrete cylinder pipe
预应力钢筋混凝土管 prestressed reinforced concrete pipe
预应力钢筋混凝土轨枕 prestressed reinforced concrete sleeper; prestressed reinforced concrete tie
预应力钢筋混凝土连续梁桥 continuous prestressed concrete bridge
预应力钢筋混凝土梁 prestressed reinforced concrete beam
预应力钢筋混凝土锚碇板 anchor plate in prestressed concrete
预应力钢筋混凝土桥 prestressed reinforced concrete bridge
预应力钢筋混凝土系材 prestressed reinforced concrete tie
预应力钢筋混凝土桩 prestressed reinforced concrete pile
预应力钢筋腱 <包括其组成部分,钢丝、钢绞线、钢筋、筋束以及锚具等> tendon; prestressing tendon
预应力钢筋截面面积 area of prestressed steel
预应力钢筋孔道 conduit for prestressing steel arrangement
预应力钢筋束 prestressed tendon; steel tendon; stressed tendon; prestressing tendon
预应力钢筋束锚具 strand grip
预应力钢筋松弛损失 loss due to tendon relaxation
预应力钢筋张拉机 prestressed steel bar tensioning machine
预应力钢筋张拉器 jack for prestressed concrete
预应力钢筋张拉千斤顶 prestressed steel bar drawing jack
预应力钢筋张拉设备 prestressed steel bar tensioning equipment
预应力钢缆 stressing cable
预应力钢缆法 internal cable method
预应力钢缆线 prestressing strand
预应力钢束滑移 tendon slip
预应力钢丝 compressor wire; prestressed concrete wire; prestressed steel wire; prestressed wire; prestressing wire; pretensioned wire; reinforcement wire; reinforcing wire; string wire
预应力钢丝混凝土 prestressed concrete wire; prestressed concrete with wires
预应力钢丝绳 prestressed cable; prestressed strand; prestressing cable
预应力钢丝束 male cone; prestressed tendon; prestressing tendon; pretensioned tendon; steel tendon; stressed tendon
预应力钢丝束轮廓线 cable profile
预应力钢丝束套管 sheath of tendon
预应力钢丝索 pretensioned tendon
预应力钢丝网混凝土船 prestressed-ferro-concrete ship
预应力钢丝网水泥 prestressed ferro cement
预应力钢丝网水泥船 prestressed-ferro-cement ship
预应力钢索 prestressed cable; prestressed strand; prestressing cable; tendon
预应力钢索的锚固部分 anchor tendon
预应力钢索护套 tendon sheathing
预应力钢索联结器 sheath coupler
预应力钢索轮廓线 tendon profile
预应力钢索锚具 strand grip
预应力钢索弯曲段摩擦 curvature friction
预应力钢索吻合线形 concordant tendon profile
预应力钢索与混凝土黏(粘)结的构件 bonded member
预应力钢弦混凝土 prestressed wire concrete
预应力钢桩 prestressed pile
预应力高强度钢丝 shaft plumbing wire
预应力工场 stressing yard
预应力工程 prestressed construction
预应力工具 prestressing tool
预应力工艺 prestressing technique
预应力构件 prestressed component; prestressed member; prestressed unit; stressing element; stretching element
预应力股绳 stressing strand
预应力管道 stressing duct
预应力管柱 prestressed concrete column; prestressed concrete drilled caisson
预应力管桩 prestressed concrete pipe pile
预应力轨迹线 trajectory of prestressing force
预应力过张的 over-prestressed
预应力合力轨迹线 path of prestressing force
预应力后张 post-tensioned
预应力后张现场记录 post-tensioning field log
预应力厚板 prestressed plank
预应力厚壁管 prestressed thick-wall pipe
预应力灰浆 prestressed mortar
预应力混凝土 preload concrete; prestressed concrete; pretensioned concrete
预应力混凝土 T 形刚构加吊梁桥 prestressed concrete T-frame bridge with suspended beam

预应力混凝土T形刚构桥 prestressed concrete T-frame bridge
预应力混凝土T形梁 prestressed concrete T(ee)-beam
预应力混凝土坝 prestressed concrete dam
预应力混凝土板 preload concrete slab;prestressed concrete slab
预应力混凝土板构件 prestressed concrete slab unit
预应力混凝土板基 prestressed concrete plate base
预应力混凝土板桥 prestressed concrete slab bridge
预应力混凝土板桩 prestressed concrete sheet pile
预应力混凝土板桩墙 prestressed concrete sheet pile wall
预应力混凝土壁板 prestressed concrete wall panel
预应力混凝土薄壳 prestressed concrete shell
预应力混凝土薄壳屋顶 prestressed concrete shell roof
预应力混凝土槽形梁 prestressed concrete through girder
预应力混凝土插入件 prestressed concrete insert
预应力混凝土插入物 prestressed concrete insertion
预应力混凝土长跨距板 prestressed concrete long-span slab
预应力混凝土长线张拉法 long line prestressed concrete process
预应力混凝土车间 prestressed concrete plant
预应力混凝土沉箱 prestressed concrete caisson
预应力混凝土衬里钢(筒)芯 prestressed concrete lined cylinder
预应力混凝土衬里管柱 prestressed concrete lined cylinder
预应力混凝土衬里桩 prestressed concrete lined pile
预应力混凝土储液器 prestressed concrete reservoir
预应力混凝土串联梁 prestressed concrete series beam
预应力混凝土窗过梁 prestressed concrete window lintel
预应力混凝土带大梁楼板 prestressed concrete girder floor
预应力混凝土带上下桁条的梁 prestressed concrete Vierendeel
预应力混凝土道路 prestressed concrete road
预应力混凝土的分类 classification of prestressed concrete
预应力混凝土地板梁 prestressed concrete floor beam; prestressed concrete joist
预应力混凝土电力杆 prestressed concrete power pole
预应力混凝土顶棚格栅 prestressed concrete ceiling joist
预应力混凝土顶升楼板构造 prestressed concrete lift-slab construction
预应力混凝土多层停车场结构 prestressed concrete parking structure
预应力混凝土反应堆容器 prestressed concrete reactor vessel
预应力混凝土方格天花板 prestressed concrete coffer plate
预应力混凝土方桩 square prestressed concrete pile
预应力混凝土房屋 prestressed concrete building
预应力混凝土非钢筒心管 prestressed concrete non-cylinder pipe
预应力混凝土废水管 prestressed concrete refuse water pipe
预应力混凝土分块拼装梁 prestressed block-beam
预应力混凝土佛伦第尔大梁 prestressed concrete Vierendeel
预应力混凝土附加钢筋 auxiliary reinforcement
预应力混凝土复合构件 prestressed concrete compound unit
预应力混凝土杆(件) prestressed concrete bar
预应力混凝土刚性构架 prestressed concrete rigid frame
预应力混凝土钢筋 prestressed concrete bar; prestressed concrete reinforcement bar; steel prestressed concrete
预应力混凝土钢筋锚具应力损失 anchorage deformation
预应力混凝土钢丝 prestressed concrete wire
预应力混凝土高架道路 prestressed concrete elevated road
预应力混凝土高压管 prestressed concrete high-pressure pipe
预应力混凝土格栅 prestressed concrete joist
预应力混凝土隔板 prestressed concrete cross wall
预应力混凝土给水管龙头 prestressed concrete pipe penstock
预应力混凝土给水管栓 prestressed concrete pipe penstock
预应力混凝土工厂 prestressed concrete plant
预应力混凝土工程 prestressed concrete engineering; prestressed concrete work
预应力混凝土工字梁 prestressed concrete I-beam
预应力混凝土工作 prestressed concrete work
预应力混凝土弓弦 prestressed concrete bowstring
预应力混凝土公路 prestressed concrete highway
预应力混凝土公路桥 prestressed concrete road bridge
预应力混凝土公路桥梁 prestressed concrete highway bridge
预应力混凝土构架 prestressed concrete skeleton
预应力混凝土构件 prestressed concrete element; prestressed concrete section; prestressed concrete unit; pretensioning concrete member
预应力混凝土构造 prestressed concrete construction
预应力混凝土骨架 prestressed concrete skeleton; prestressed concrete supporting skeleton
预应力混凝土管 prestressed concrete pipe; prestressed concrete piping; prestressed concrete tube; prestressed concrete tubing
预应力混凝土管道 prestressed concrete pipeline
预应力混凝土管件 prestressed concrete segment
预应力混凝土管线 prestressed concrete pipeline
预应力混凝土管柱 prestressed concrete cylinder
预应力混凝土管柱沉箱 prestressed concrete drilled caisson
预应力混凝土管桩 prestresses tubular concrete pile
预应力混凝土储罐 prestressed concrete container
预应力混凝土轨枕 prestressed concrete tie; prestressed reinforced-concrete sleeper
预应力混凝土柜 preload tank
预应力混凝土合成梁 prestressed concrete composite beam
预应力混凝土桁架 prestressed concrete truss
预应力混凝土桁架梁 prestressed concrete truss bridge lining segment
预应力混凝土桁架桥 prestressed concrete truss bridge
预应力混凝土桁条 prestressed concrete purlin(e); prestressed concrete string(er)
预应力混凝土横梁 prestressed concrete header
预应力混凝土横墙 prestressed concrete cross wall
预应力混凝土护墙板 prestressed concrete wall panel
预应力混凝土花格大梁 prestressed concrete lattice girder
预应力混凝土环梁 prestressed concrete ring beam
预应力混凝土混成梁 prestressed concrete composite beam
预应力混凝土机场跑道面 prestressed airport concrete pavement
预应力混凝土架空地板 prestressed concrete suspended floor
预应力混凝土架空悬桥面 prestressed concrete suspended floor
预应力混凝土简支梁 prestressed concrete simple supported beam
预应力混凝土建筑 prestressed concrete building; prestressed concrete construction
预应力混凝土浇筑 prestressed concrete cast(ing)
预应力混凝土铰接结构 Freyssinet concrete hinge
预应力混凝土铰索 prestressed concrete bowstring
预应力混凝土结构 prestressed concrete structure
预应力混凝土结构体系 prestressed concrete structure system
预应力混凝土结构用钢绞线 prestressed concrete steel wire strand
预应力混凝土结构用刻痕钢丝 deformed prestressed concrete steel wire
预应力混凝土局部加粗桩 prestressed concrete pile with blisters
预应力混凝土矩形网络板 prestressed concrete rectangular grid slab
预应力混凝土壳体 prestressed concrete shell
预应力混凝土壳体构造 prestressed concrete shell roof
预应力混凝土空腹梁 prestressed concrete Vierendeel
预应力混凝土空心板桥 prestressed concrete voided slab bridge
预应力混凝土空心梁地板 prestressed concrete hollow beam floor
预应力混凝土空心梁楼板 prestressed concrete hollow beam floor
预应力混凝土空心楼板 hollow cast prestressed concrete floor; prestressed concrete hollow-core slab floor
预应力混凝土空心毛地板 prestressed concrete hollow-core plank floor
预应力混凝土宽枕【铁】 prestressed concrete broad sleeper
预应力混凝土框架 prestressed concrete frame
预应力混凝土肋板 prestressed concrete ribbed slab
预应力混凝土棱柱壳体屋顶 prestressed concrete prismatic(al) shell
预应力混凝土离心浇制管 prestressed concrete spun pipe
预应力混凝土离心浇注管 prestressed concrete centrifugally cast pipe
预应力混凝土连续梁 prestressed concrete continuous beam
预应力混凝土连续梁桥 prestressed concrete continuous girder bridge
预应力混凝土梁 prestressed concrete beam
预应力混凝土梁板结构 prestressed concrete slab-and-beam
预应力混凝土梁格栅 prestressed concrete binder joist
预应力混凝土梁构件 prestressed concrete bridge member
预应力混凝土梁紧邻锚具的应力分布区 primary distribution zone
预应力混凝土梁楼板 prestressed concrete floor
预应力混凝土梁桥用预制分段平衡旋臂施工 prestressed concrete beam bridge by precast balanced cantilever segmental construction
预应力混凝土梁式桥 prestressed concrete beam bridge
预应力混凝土檩条 prestressed concrete purlin(e)
预应力混凝土龙门架 prestressed concrete portal frame
预应力混凝土楼板梁 prestressed concrete floor beam
预应力混凝土楼面板构件 prestressed concrete floor slab unit
预应力混凝土楼梯斜梁 prestressed concrete string(er)
预应力混凝土路 prestressed concrete road
预应力混凝土路面 prestressed concrete (road) pavement
预应力混凝土埋藏式钢筒芯管 prestressed concrete embedded cylinder pipe
预应力混凝土锚固长度 anchorage distance
预应力混凝土锚固件区域 anchorage zone
预应力混凝土锚固区 <先张法或后张法> anchorage zone
预应力混凝土锚固装置 anchorage device
预应力混凝土锚具 concrete anchor; stressing head
预应力混凝土门过梁 prestressed concrete door lintel
预应力混凝土门式支架 prestressed concrete portal frame
预应力混凝土挠曲试验梁 prestressed concrete flexure test beam
预应力混凝土牛腿 prestressed concrete bracket
预应力混凝土跑道 prestressed concrete pavement for airport; prestressed concrete runway
预应力混凝土平板基础 prestressed concrete plate foundation; prestressed concrete slab foundation
预应力混凝土平腹梁 prestressed concrete plain web(bed) beam
预应力混凝土平台 prestressed concrete platform
预应力混凝土平行桁架 prestressed concrete parallel truss
预应力混凝土嵌空心砖梁 prestressed concrete block beam
预应力混凝土嵌入物 prestressed concrete insertion

预应力混凝土墙 prestressed concrete wall
预应力混凝土桥 prestressed concrete bridge
预应力混凝土桥施工方法 <原联邦德国 Polensky 和 Zoller 公司创造的,亦称导梁施工法> PZ method
预应力混凝土桥墙 prestressed concrete cross wall
预应力混凝土穹隆 prestressed concrete dome
预应力混凝土圈梁 prestressed concrete ring beam
预应力混凝土容器 prestressed concrete vessel
预应力混凝土三角孔桁架 prestressed concrete triangular truss
预应力混凝土三角形构架 prestressed concrete triangular truss
预应力混凝土伞形结构 prestressed concrete parachute
预应力混凝土伞形物 prestressed concrete umbrella
预应力混凝土设备 prestressed concrete equipment
预应力混凝土施工 prestressed concrete construction
预应力混凝土实体大梁 prestressed concrete solid girder
预应力混凝土实心腹板梁 prestressed concrete solid web(bed) beam
预应力混凝土实心桩 solid prestressed concrete pile
预应力混凝土受压管 prestressed concrete pressure pipe
预应力混凝土双向格子板 prestressed concrete waffle slab
预应力混凝土双向密肋板 prestressed concrete waffle slab
预应力混凝土水池 preload tank
预应力混凝土水平拉杆 prestressed concrete horizontal tie back
预应力混凝土水塔 prestressed concrete water tower
预应力混凝土塔 prestressed concrete tower
预应力混凝土铁道轨枕 prestressed concrete railroad sleeper; prestressed concrete railway sleeper
预应力混凝土铁路轨枕 prestressed concrete railroad sleeper; prestressed concrete railway sleeper
预应力混凝土停车甲板 prestressed concrete parking deck
预应力混凝土停车楼层 prestressed concrete parking deck
预应力混凝土筒仓 prestressed concrete silo
预应力混凝土筒形拱顶 prestressed concrete wagon vault
预应力混凝土筒形拱壳体 prestressed concrete barrel vault
预应力混凝土凸边板 prestressed concrete panel slab
预应力混凝土土木工程结构 prestressed concrete for civil engineering structure
预应力混凝土托架 prestressed concrete bracket
预应力混凝土托座 prestressed concrete bearer
预应力混凝土污水管 prestressed concrete foul water pipe; prestressed concrete sewage pipe
预应力混凝土屋顶板 prestressed concrete roofing slab
预应力混凝土屋顶大梁 prestressed concrete roof-girder
预应力混凝土屋顶梁 prestressed concrete roof beam
预应力混凝土屋顶望板 prestressed concrete roof sheathing
预应力混凝土屋面 prestressed concrete roof
预应力混凝土系杆 prestressed concrete tie-rod
预应力混凝土系统 <采用高拉力钢丝的> magnet-blaton; prestressed concrete system
预应力混凝土匣形梁 prestressed concrete box beam
预应力混凝土箱板 prestressed concrete cassette plate
预应力混凝土箱形大梁 prestressed concrete box girder
预应力混凝土箱形(截面)梁 prestressed concrete box beam
预应力混凝土镶板 prestressed concrete wall panel
预应力混凝土协会 <美> Prestressed Concrete Association; Prestressed Concrete Institute
预应力混凝土斜板式屋顶 prestressed concrete tilted-slab roof
预应力混凝土斜拉桥 prestressed concrete cable-stayed bridge
预应力混凝土芯棒 prestressed concrete bar
预应力混凝土芯管 prestressed concrete central tube
预应力混凝土蓄水池 prestressed concrete reservoir
预应力混凝土蓄液池 prestressed concrete tank
预应力混凝土悬挂式屋顶 prestressed concrete suspended roof
预应力混凝土旋转浇制管 prestressed concrete spun pipe
预应力混凝土学会会刊 <美国双月刊> Journal of the Prestressed Concrete Institute
预应力混凝土学会志 <美国双月刊> Journal of the Prestressed Concrete Institute
预应力混凝土压力管 prestressed concrete pressure pipe
预应力混凝土压力容器 prestressed concrete pressure vessel
预应力混凝土研究所 <美> Prestressed Concrete Institute
预应力混凝土异形钢丝 deformed prestressed concrete steel wire
预应力混凝土用钢丝 steel wire for prestressed concrete; stretching wire; wire for prestressed concrete
预应力混凝土用钢索 cable for prestressed concrete
预应力混凝土用刻痕钢丝 deformed prestressed concrete steel wire; indented prestressed concrete steel wire
预应力混凝土有肋楼板 prestressed concrete ribbed slab
预应力混凝土预制板 prestressed precast concrete slab
预应力混凝土预制构件 prestressed precast concrete member
预应力混凝土预制空心楼板 hollow plank
预应力混凝土圆筒拱顶 prestressed concrete tunnel vault
预应力混凝土圆筒形壳体 prestressed concrete cylindrical shell
预应力混凝土圆筒形桩 cylindrical prestressed concrete pile
预应力混凝土圆屋顶 prestressed concrete cupola
预应力混凝土栅栏 prestressed concrete fence
预应力混凝土张拉设备 stretching device
预应力混凝土照明灯柱 prestressed concrete lighting column
预应力混凝土照明桅杆 prestressed concrete lighting mast
预应力混凝土折板结构 prestressed concrete folded-plate structure
预应力混凝土折板屋顶 prestressed concrete hipped plate roof
预应力混凝土枕(木) prestressed concrete sleeper
预应力混凝土支撑 prestressed concrete support
预应力混凝土支承骨架 prestressed concrete bearing skeleton
预应力混凝土支架 prestressed concrete support
预应力混凝土支座 prestressed concrete bearer; prestressed concrete support
预应力混凝土(制件)销售工程师 prestressed concrete sales engineer
预应力混凝土中的锚头 circular stressing head
预应力混凝土中间楼面 prestressed concrete intermediate floor
预应力混凝土中心螺旋筋 prestressed concrete center spiral
预应力混凝土贮罐 prestressed concrete storage tank
预应力混凝土贮水箱 prestressed concrete storage tank
预应力混凝土贮液槽 prestressed concrete storage tank
预应力混凝土贮液池 prestressed concrete tank
预应力混凝土柱 prestressed concrete column
预应力混凝土桩 prestressed concrete pile; prestressed concrete piling
预应力混凝土装饰板 prestressed concrete trimmer plank
预应力活塞环 prestressed steel piston ring
预应力机构 prestressing mechanism
预应力机架 prestressed stand
预应力技术 prestressing technique
预应力加气混凝土 prestressed aerated concrete
预应力夹板 prestressed yoke assembly
预应力浇注工场 prestress casting yard
预应力浇注混凝土构件 prestressed cast-concrete member
预应力浇筑(混凝土)件 prestressed casting
预应力绞合线 prestressed strand
预应力绞合线夹具 compression grip
预应力接合 stressed connection
预应力接头 stretched connection
预应力结构 prestressed structure
预应力筋 tendon
预应力筋的松弛 relaxation of prestressing steel
预应力筋腱 prestressing reinforcement; prestressing tendon
预应力筋锚固长度 <混凝土> transmission length
预应力筋松弛 relaxation of prestress tendons
预应力筋套管 sheath of tendon
预应力筋楔锚器 wedge anchor(age)
预应力井架 prestressed tower
预应力空心板 prestressed hollow core slab
预应力空心梁 prestressed hollow beam
预应力控制张拉 prestressing under stress control
预应力块梁 block beam
预应力连接 prestressed connection
预应力连续梁 prestressed continuous beam
预应力梁 prestressed beam; prestressed girder
预应力梁张拉台 beam bed
预应力锚 prestressed anchor
预应力锚杯 prestressed anchor cone
预应力锚杆 prestressed anchor; prestressed bolt; prestressed rock bolt; prestressing anchor
预应力锚固 anchoring of prestress
预应力锚固滑动损失 prestressing loss due to slip at anchorage
预应力锚固筋 bonded member
预应力锚塞 male cone
预应力锚索 prestressed cable; prestressed wire
预应力面层屋面板 stressed-skin roof deck
预应力模具镶块 prestressed die insert
预应力摩擦损失 prestressing loss due to friction
预应力黏[粘]土 prestressed clay; Stahlton
预应力黏[粘]土板 plank in prestressed clay; Stahlton plank
预应力黏[粘]土厂 Stahlton plant
预应力黏[粘]土窗过梁 prestressed clay window lintel; Stahlton prestressed window lintel; window lintel in prestressed clay
预应力黏[粘]土地板 Stahlton prestressed floor
预应力黏[粘]土地面 floor in prestressed clay
预应力黏[粘]土工厂 prestressed clay factory; prestressed clay plant; Stahlton factory
预应力黏[粘]土过梁 prestressed clay lintel; Stahlton prestressed lintel
预应力黏[粘]土厚板 prestressed clay plank
预应力黏[粘]土梁 prestressed clay beam; Stahlton prestressed beam
预应力黏[粘]土门过梁 Stahlton prestressed door lintel
预应力黏[粘]土墙板 Stahlton prestressed wall panel; Stahlton prestressed wall slab; wall-panel in prestressed clay
预应力黏[粘]土墙面板 prestressed clay wall panel
预应力黏[粘]土墙体单元 Stahlton wall unit
预应力黏[粘]土屋顶 prestressed clay roof; Stahlton prestressed roof
预应力黏[粘]土屋顶板 prestressed clay roof slab; roof slab in Stablton
预应力黏[粘]土屋面 roof in Stahlton
预应力黏[粘]土屋面板 Stahlton prestressed roof slab
预应力黏[粘]土砖板 Stahlton prestressed clay plank
预应力黏[粘]土砖窗过梁 Stahlton plank window lintel
预应力黏[粘]土砖楼板 Stahlton floor
预应力黏[粘]土砖墙板 wall slab in prestressed clay
预应力黏[粘]土砖屋顶 Stahlton roof
预应力黏[粘]土砖屋面板 Stahlton roof slab
预应力偏心钢筋 <一般指曲线预应力筋> eccentric tendon
预应力偏心钢丝束 eccentric tendon
预应力偏心距 prestress eccentricity
预应力拼装 multielement prestressing
预应力拼装构件 segmental member
预应力普通混凝土 prestressed normal concrete

预应力砌体 prestressed masonry
预应力砌体工程 prestressed masonry works
预应力千斤顶 prestressing jack
预应力牵拉缆索 prestressed stay
预应力墙 prestressed wall
预应力轻质混凝土 prestressed lightweight concrete
预应力轻质混凝土屋面板 prestressed lightweight concrete roof slab
预应力区 stressing zone
预应力圈 <凹模的> shrink ring
预应力全消失 total loss of prestress
预应力砂浆 prestressed mortar
预应力设备 prestressing equipment; stressing device
预应力设计 prestress design
预应力施工 prestressed construction
预应力式拉杆 prestressed tie rod
预应力收缩损失 prestressing loss due to shrinkage
预应力束 stressing tendon
预应力双 T 形构件 prestressed double tee
预应力丝孔道 conduit
预应力损失 loss of prestress; prestress(ed) loss; prestressing loss; stretching loss
预应力索偏心距 cable eccentricity
预应力索线 prestressing wire
预应力索张拉程序先后引起的应力损失 sequence-stressing loss
预应力台座 prestressing bed
预应力弹性胶管伸缩缝 prestressed elastic neoprene tube expansion joint
预应力陶瓷 prestressing ceramics
预应力弯曲钢筋 deflected tendon
预应力圬工水池子 prestressed masonry tank
预应力圬工水罐 prestressed masonry tank
预应力圬工水箱 prestressed masonry tank
预应力屋面 prestressed roof skin
预应力系统 prestressing system
预应力先张法 pretensioning system
预应力箱梁 prestressed box girder
预应力橡胶管伸缝 prestressed elastic neoprene tube expansion joint
预应力楔块 stressing wedge
预应力楔片 stressing wedge
预应力斜(拉)杆 prestressed diagonal
预应力形钢 prestressing steel
预应力徐变损失 prestressing loss due to creep
预应力蓄液池 prestressed concrete tank
预应力悬臂梁桥 prestressed cantilever bridge
预应力悬索屋盖 prestressed cable roof
预应力岩锚块 prestressed rock anchor block
预应力岩石 prestressed rock
预应力岩石独股锚杆 prestressed mono-strand rock anchor
预应力岩石锚栓 prestressed rock anchor
预应力预浇混凝土建筑 prestressed precast concrete construction
预应力预浇混凝土结构 prestressed precast concrete construction
预应力预浇混凝土装配式建筑 prestressed precast concrete system building
预应力预浇梁 prestressed precast beam
预应力预制装配式混凝土地板 prestressed prefabricated concrete floor
预应力预制装配式混凝土楼板 prestressed prefabricated concrete floor
预应力张拉 initial stress tensioning
预应力张拉超过设计规定 over-stretching
预应力张拉端 jacking end
预应力张拉力 jacking force
预应力张拉设备 prestressing equipment; jacking device
预应力张拉时最大应力 jacking stress
预应力张拉台 prestressing bed; pretensioning bed; prestressing table
预应力折板 prestressed folded plate
预应力整体混凝土墙 prestressed monolithic concrete wall
预应力直线强拉 linear prestressing
预应力值 amount of prestress; prestress value; stressing value
预应力制作者 prestress fabricator
预应力重力坝 prestressed gravity dam
预应力砖工 prestressed brickwork
预应力桩 prestressed pile
预应力自动喷浆机 prestressed and automated shotcrete machinery
预应力组合水池 prestressed composite tank
预映 preview
预硬传送带 precuring conveyer [conveyor]
预硬化 prehardening
预硬化期 prehardening period
预约 engage; forward contract; precontract; reservation; subscribe; subscription
预约保险 open policy insurance
预约保险单 open cover(age); open policy
预约承运期间 booking period
预约初保单 open ship
预约初保凭证 open ship
预约处 booking hall
预约登记簿 <美> appointment book
预约电话 sequency call
预约付款权力 forward commitment authority
预约公共汽车 subscription bus
预约挂号 call on hand; carried forward call
预约呼叫 advance calling
预约卷宗 reserved volume
预约码头泊位 accommodation berth
预约式页面调度 prepaging
预运转 elimination run
预载 preload
预载垫圈 preloading washer
预载机器 preload machine
预载离子交换填料床吸附反应器 preloaded ion exchange packed bed adsorption reactor
预载压实 preloading compaction
预载弹簧 preloaded spring
预早购置 early acquisition
预增白 prebrightening
预增频 accentuation
预增强器 preaccentuator
预增压水 prepressurized water
预扎吊索 presling
预轧材 first shape
预轧道次 previous pass
预轧机 pony rougher
预轧机座 pony-roughing stand of rolls
预展 private view
预占 camp-on
预占线转接 camp-on switching
预占业务 preemption service
预占制接续 delay basis operation
预张绑线 <电机> pretensioned binding
预张法混凝土管道 pretensioned concrete pipe
预张法预应力混凝土 pretensioning prestressed concrete
预张法预应力混凝土管道 pretensioned pipe
预张钢筋 pretensioned steel
预张紧的拉杆 prestressed tieback
预张紧的拉条 prestressed tieback
预张紧的牵索 prestressed tieback
预张紧的岩石锚碇 prestressed rock anchor
预张紧荷载 prestressing load
预张紧力 prestressing load
预张拉的预应力混凝土大梁 pretensioned prestressed concrete girder
预张拉钢筋 pretensioned bar; pretensioned reinforcement; pretensioned steel
预张拉钢丝束 pretensioned tendon
预张拉混凝土钢心管 pretensioned concrete cylinder pipe
预张拉损失 loss of pretension
预张力 pretension; stressing force
预张力锚杆 pretensioned bolt
预张台 pretensioning bed; pretensioning bench
预张系统 prestressing system
预兆 forerunner; foreshadowing; omen; precursor; premonition; premonitoring symptom; presage
预兆价格 shadow price
预兆性天空【气】 emissary sky
预照射 preirradiation
预真空 fore-vacuum
预真空容器 fore-vacuum vessel
预真空样品瓶 prevacuated sampling tube
预真空组件 fore-vacuum subassembly
预振贫混凝土 previbrated lean concrete
预震 preparatory shock
预整定装置 presetting apparatus
预整理 preliminary finish
预整理壁纸 pretreatment wallpaper; pretrimmed wallpaper
预整形 preform
预整修的 prefinished
预正则线性算子 preregular linear operator
预支 advance (payment); anticipation; cash in advance; draw on the future; make advance; receive in advance
预支差旅费 travel advance
预支的承兑汇票 anticipated acceptance
预支工资 payment in advance
预支掘进 forepoling; spilling
预支款 advanced payment
预支款登记簿 advances register
预支款申请书 application of advance
预支款账户 advance account
预支条款 red clause
预支通知单 advance note
预支未到期存款 break a deposit
预支运费 anticipated freight
预支账户 account of advances
预知 foreknowledge; foresee; foresight; precognition; prognosis [复 prognoses]
预值 predictor
预指定 preassignment
预制 precasting; precut preforming; prefabricate; premold; shop fabrication
预制 L 形码头岸壁 precast L-wall quay; precast L-wall wharf
预制安装 prefabrication
预制安装构件 prefabricated parts
预制安装建筑 prefabricated building
预制安装平台 prefabricated platform
预制安装组合单元 prefabricated unit
预制百分率 factory content; factory fraction; factory percentage
预制柏油薄型屋顶材料纸 tar ready sheet roofing paper
预制板 precast(ing) slab; prefabricated slab; prefabricated tiling
预制板材 prefabricated panel
预制板材隔墙 prefabricated tile partition-(wall)
预制板材构造 precast panel construction
预制板材结构 precast panel construction
预制板材施工(法) panel construction; prefabricated panel construction
预制板隔墙 slab partition (wall)
预制板工人 <升运未经养护的> vacuum man
预制板结构 precast panel construction
预制板楼盖 precast slab floor
预制板面上的覆盖纸 paper overlay
预制板式房屋 panel type house; precast panel house
预制板式轨道 precast slab track
预制板式结构 precast slab type of construction
预制板砖 panel brick
预制棒 blank
预制包衬 ladle liner
预制保温用加热管段 preformed pipe insulation section for heat protection
预制保温罩 lagging section
预制壁板式地下墙(地下连续墙) prefabricated wall with identical panel
预制标准化运输机 preengineered conveyer [conveyor]
预制标准间的灭火系统 pre-engineered suppression system
预制薄壳构件 thin-shell precast
预制薄纸板屋面材料 prepared sheet roofing (paper)
预制部分构造 sectional construction
预制部分集合构造 section construction
预制部件 prefabricated component; prefabricated subassembly; ready-made unit
预制部件构成房屋 sectional building
预制部件装配式建筑 industrialized building; system building
预制槽形楼板 precast trough shaped floor unit
预制侧石 precast curb
预制层 preply
预制厂 <混凝土构件> factory for prefabrication; prefabricated factory; prefabricated plant; precast factory; precasting plant
预制场 casting bay; manufacturing yard
预制场地 casting bed
预制车行道铺面板 <桥梁> roadway casting
预制沉放管段 precast immersed tunnel section
预制衬里 preformed liner
预制衬砌 precast lining
预制成型炮泥 preformed stemming
预制承重板建筑 beam panel system; bear panel system
预制承重墙板式建筑 bearing panel system
预制冲模 preformed die
预制处理厂 prefabricate treatment plant

预制窗间墙板 prefabricated window-wall unit
预制床 casting bed
预制大板构造 large panel construction
预制大跨石膏板 long-span precast gypsum slab
预制大理石块 precast marble tile
预制大砌块 blockwork
预制单元 precast unit; preconstructional unit; prefabricated element; prefabricated unit
预制导波系统 preplumbed system
预制的 factory-made; off-the-shelf; prebuilt; precast(ing); preconstructional; prefabricated; preformed; premade; premo(u)lded; ready-made; shop-fabricated; preforming
预制的地板格栅 benfix
预制的防空掩体 precast bomb shelter
预制的钢筋混凝土单元 reinforced cast(ing)
预制的钢筋混凝土楼面 reinforced cast floor
预制的混凝土和玻璃板组合 translucent concrete
预制地沥青填缝料 preformed asphalt joint sealer
预制顶推箱拼装节段 precast thrust section
预制短跨石膏板 short-span slabs for precast gypsum
预制段 precast segment
预制多单元构件 multielement member
预制法 preassembly method
预制法施工 precast construction
预制方法 precast measure
预制方块 precast block
预制防冻保护层 lagging section for cold protection
预制防空庇护掩体 precast air-raid sheltering bunker
预制房屋 braithwaite; prefab; shamahduplex <钢结构和塑性墙板组成的>; alcrete <预制房屋中的一种>
预制房屋商 home manufacturer
预制放热保护层 lagging section for heat protection
预制分段船体的装配 prefabrication
预制分段施工 precast segmental construction
预制缝 preformed joint
预制浮石混凝土 precast pumice concrete; prefabricated pumice concrete
预制感光板 presensitized plate
预制钢筋骨架 prefabricated reinforcement; prefabricated reinforcing cage
预制钢筋混凝土 precast reinforced concrete
预制钢筋混凝土板材 precast reinforced concrete panel
预制钢筋混凝土板桥 precast reinforced concrete slab bridge
预制钢筋混凝土板桩 precast reinforced concrete sheet pile
预制钢筋混凝土挡块 precast reinforced concrete retaining unit
预制钢筋混凝土房屋构件 precast reinforced concrete member
预制钢筋混凝土格笼式挡土墙 crib retaining wall built with precast reinforced concrete member
预制钢筋混凝土构架 reinforced precast frame
预制钢筋混凝土构件 reinforced precast element
预制钢筋混凝土管 precast reinforced concrete pipe
预制钢筋混凝土结构 precast reinforced concrete construction; precast reinforced concrete structure
预制钢筋混凝土空心梁 hollow beam
预制钢筋混凝土块砌筑的岸壁 quay wall of precast reinforced concrete units
预制钢筋混凝土块砌筑的岸墙 quay wall of precast reinforced concrete units
预制钢筋混凝土框架 precast reinforced concrete frame
预制钢筋混凝土肋 precast reinforced concrete rib
预制钢筋混凝土楼板 reinforced precast floor
预制钢筋混凝土墙 reinforced precast wall; precast reinforced concrete wall
预制钢筋混凝土桥梁 precast reinforced concrete bridge
预制钢筋混凝土石笼挡土墙 crib retaining wall built with precast reinforced concrete member
预制钢筋混凝土支撑 precast reinforced concrete support
预制钢筋混凝土柱 precast reinforced concrete column
预制钢筋混凝土桩 precast reinforced concrete pile
预制钢筋网 <埋置在空斗墙内的> prefabricated tie
预制钢筋网衬砌 fabricated steel liner
预制钢束法 <简称 PPWS 法> prefabricated parallel wire strands method
预制钢型 preformed die
预制港口结构 Mulberry harbo(u)r
预制高层公寓 high industrial block (of flats); precast residence tower; precast tall flats
预制高层公寓建筑 precast high(rise) block of flats
预制高层建筑楼板 precast high(rise) floor
预制高架桥 prefabricated viaduct
预制隔墙 prefabricated partition wall(ing); ready-made partition
预制隔热板 prefabricated panel
预制工场 block yard
预制工程 precast construction
预制工艺 precast; prefabrication technique
预制工作台 fabricated platform
预制弓形块 precast-segmental
预制拱 precast concrete vault
预制沟管段 precast segment; precast-segmental sewer
预制沟渠管节 precast-segmental sewer
预制构架 precast framework
预制构架建筑 prefabricated construction
预制构件 assembly unit; building block; finished building fabric; precast; precast component; precast element; precast member; precast section; precast structural member; precast unit; prefabricated component; prefabricated element; prefabricated member; prefabricated unit; ready-made unit; knocked down <易于拆卸的>
预制构件厂 factory for prefabrication; prefabricated-component factory
预制构件场 casting yard
预制构件场地 casting bed
预制构件工艺工程 process of precast member
预制构件建筑 prefabricated construction
预制构件建筑物 manufactured building
预制构件立面 industrialized facade
预制构件模板 formwork for precast work
预制构件模型 mo(u)ld for precast work
预制构件拼装结构 section(al) construction
预制构件桥 precast bridge
预制构件上预留的安装孔 shearkey hole
预制构件屋顶 roof of prefabricated elements
预制构件窄桥 strait bridge
预制构件装配施工法 tilt-up construction method
预制构造(物) precast construction
预制骨料灌浆混凝土 guncreting
预制管 prefabricated pipe
预制管道隔热块 lagging section
预制管道系统 prefabricated (pipe) conduit system; prefabricated pipe system
预制管段 prestressing tube; preformed section
预制管段沉放法 construct by immersion method
预制管段的沉放 sinking of precast tube
预制管涵 preformed culvert
预制管水底隧道 precast subaqueous tunnel; prefabricated subaqueous pile
预制管子连接环 preformed joint ring
预制硅藻土管段 <保温用> preformed Tripoli-powder section
预制轨道 prefabricated track
预制过梁 precast lintel
预制涵管 precast pipe unit
预制焊模 prefabricated welding mo(u)ld
预制合金 prealloy
预制合金锭料 prealloyed ingot
预制护岸混凝土块 precast concrete revetment block
预制花格板 prefabricated lattice panel
预制混凝土 precast(ed) concrete; prefabricated concrete; prepacked concrete
预制混凝土板 precast concrete plank; precast concrete slab
预制混凝土板面层 precast concrete slab pavement
预制混凝土板铺面 precast concrete slab pavement
预制混凝土板桩 precast concrete sheet pile; prefabricated concrete sheet pile
预制混凝土背板 precast concrete lagging
预制混凝土壁柱砖 cast concrete pilaster tile
预制混凝土表面处理 cast concrete cladding
预制混凝土薄衬砌环 precast thin-wall concrete ring
预制混凝土薄壳 thin-shell precast
预制混凝土薄壳构件 thin-shell precast concrete members
预制混凝土厂 precast concrete plant
预制混凝土车间 precast concrete plant
预制混凝土沉箱 precast concrete caisson
预制混凝土衬砌 precast concrete lining
预制混凝土衬砌管片 precast concrete liner segment
预制混凝土衬砌隧道 precast concrete lined tunnel
预制混凝土成型板 cast concrete profile(d) panel
预制混凝土成组模板 gang mould
预制混凝土承重骨架 precast concrete weight carrying skeleton
预制混凝土窗槛 cast concrete cill[sill]; precast concrete cill [sill]
预制混凝土窗台 cast concrete cill[sill]; precast concrete cill [sill]
预制混凝土从入模到通蒸汽升温的时间 holding period
预制混凝土从入模到通蒸汽升温养护的时间 precuring period; preset period; presteaming period
预制混凝土单元 precast concrete unit
预制混凝土倒虹吸 precast concrete inverted siphon
预制混凝土地板单元 concrete building floor unit
预制混凝土顶板 precast concrete ceiling
预制混凝土端块 precast concrete end-block
预制混凝土方块墙 precast concrete block wall
预制混凝土方块围堰 precast concrete blockwork cofferdam
预制混凝土防空洞 cast concrete bomb shelter
预制混凝土防空掩蔽所 cast concrete shelter
预制混凝土房屋 orlet; precast concrete house
预制混凝土房屋的工厂 factory for precast concrete buildings
预制混凝土房屋构件 concrete building unit
预制混凝土房屋框架 structural precast concrete building frame
预制混凝土盖板 precast concrete cover
预制混凝土高层住宅 precast residence tower
预制混凝土隔水墙 precast concrete umbrella
预制混凝土工 precast concrete worker
预制混凝土工厂 precast concrete factory
预制混凝土构件 blockwork; cast member; concrete building unit; precast concrete element; precast concrete member; precast concrete part; precast concrete unit; structural precast concrete; precast concrete component
预制混凝土构件场 block yard
预制混凝土构件房屋 precast concrete house
预制混凝土构件中埋件 concrete anchor
预制混凝土骨架 precast concrete skeleton
预制混凝土管 precast concrete pipe
预制混凝土管道 precast concrete flue
预制混凝土管片 precast concrete segment
预制混凝土过梁 cast lintel; precast concrete lintel
预制混凝土弧形块 precast concrete segment
预制混凝土护面块体 precast concrete armo(u)r unit
预制混凝土护墙 precast concrete curtain wall
预制混凝土化粪池 precast concrete septic tank

预制混凝土环状板 precast concrete stave
预制混凝土基础 precast concrete foundation
预制混凝土检查井 precast concrete manhole
预制混凝土检查井井筒 precast concrete manhole shaft
预制混凝土节点详图 precast concrete connection details
预制混凝土结构 system-built concrete structure
预制混凝土结合梁块 bond beam block
预制混凝土进水管 precast concrete penstock
预制混凝土井圈 precast concrete shaft ring
预制混凝土看台 precast concrete grandstand
预制混凝土(空心)块墙体 blockwork
预制混凝土空心砌块 precast hollow concrete block
预制混凝土块 blockwork
预制混凝土块(铺)面层 concrete block paving
预制混凝土块砌体 blockwork;precast concrete blockwork
预制混凝土块体 precast concrete block
预制混凝土块烟道 precast concrete block flue
预制混凝土框格式围堰 precast concrete crib cofferdam
预制混凝土廊道 precast concrete gallery
预制混凝土立面填充板 precast concrete filler slab
预制混凝土梁 precast concrete beam;precast concrete girder
预制混凝土梁底模板 precast concrete soffit form
预制混凝土梁式楼板 cast concrete beam floor
预制混凝土檩条 precast concrete purlin(e);prefabricated concrete purlin(e)
预制混凝土楼板 precast concrete floor
预制混凝土楼板肋条 cast concrete floor rib
预制混凝土楼层高墙板 cast concrete stor(e)y height wall panel
预制混凝土楼盖 precast concrete floor
预制混凝土楼梯踏板 cast concrete tread
预制混凝土楼梯斜梁 precast concrete string
预制混凝土路边石 precast concrete kerb
预制混凝土路面 precast concrete pavement
预制混凝土路面元件 precast concrete pavement unit
预制混凝土路缘石 precast concrete curb;precast concrete kerb
预制混凝土锚碇 precast concrete anchor
预制混凝土门槛 prefabricated concrete threshold
预制混凝土面板 precast concrete facing panel
预制混凝土面板构件 precast concrete deck unit
预制混凝土铺面块 precast concrete paver
预制混凝土模板 precast concrete facing form;precast concrete form
预制混凝土木笼围堰 precast concrete crib cofferdam
预制混凝土排水管 formed drain
预制混凝土铺板 precast slab;prefabricated concrete slab
预制混凝土铺块 precast concrete block;precast concrete paving block
预制混凝土铺面 precast concrete pavement
预制混凝土砌块 precast concrete block;concrete masonry unit
预制混凝土墙 prefabricated concrete wall
预制混凝土墙板 precast concrete wall panel
预制混凝土墙垛块材 precast concrete pilaster tile; prefabricated concrete pilaster tile
预制混凝土墙砌体 prefabricated concrete walling
预制混凝土桥 precast concrete bridge
预制混凝土穹隆 precast concrete dome
预制混凝土人行道石板 precast concrete flag
预制混凝土三面棱柱 precast concrete T Pees
预制混凝土山墙 prefabricated concrete gable
预制混凝土山墙梁 precast concrete gable beam
预制混凝土实心砌块 cast concrete block
预制混凝土实心砖 cast concrete solid tile
预制混凝土实心桩 solid precast concrete pile
预制混凝土饰面 precast concrete cladding;precast concrete facing
预制混凝土饰面构件 fair-faced precast concrete component;fair-faced precast concrete member
预制混凝土饰面件 fair-faced precast concrete unit
预制混凝土水槽 precast concrete flume
预制混凝土梯段 precast concrete flight of stair(case);stair unit
预制混凝土梯斜梁 prefabricated concrete string
预制混凝土天沟 cast concrete gutter
预制混凝土天沟槽形梁 cast concrete valley beam
预制混凝土填板 cast concrete infill(ing) slab
预制混凝土通风管道 cast concrete vent(tilation) duct
预制混凝土桶板 precast concrete stave
预制混凝土筒仓 cast concrete silo;precast concrete silo
预制混凝土外挂板 precast concrete cladding
预制混凝土外围框架 precast concrete perimeter frame
预制混凝土围护板 precast concrete curtain
预制混凝土屋架 precast concrete roof truss
预制混凝土屋面槽形大梁 cast concrete valley girder
预制混凝土屋面天沟 cast concrete valley gutter
预制混凝土镶边 precast concrete edging
预制混凝土镶面 cast concrete cladding
预制混凝土斜撑 precast concrete raking strut
预制混凝土新砖 cast concrete green tile
预制混凝土学术讨论会 cast concrete symposium
预制混凝土压力水管 precast concrete penstock
预制混凝土烟囱 precast concrete chimney
预制混凝土烟囱顶罩 cast concrete umbrella
预制混凝土烟道 precast concrete flue
预制混凝土檐槽 cast concrete gutter
预制混凝土异形板 cast concrete profile(d) panel
预制混凝土制品 precast concrete goods
预制混凝土周边框 prefabricated concrete perimeter frame
预制混凝土住宅 precast concrete home
预制混凝土柱 prefabricated concrete column
预制混凝土砖坯 cast concrete green tile
预制混凝土桩 precast concrete pile; prefabricated concrete pile; premo(u)lded pile
预制混凝土桩吊装点 pick-up points for precast concrete piles
预制混凝土桩起吊点 pick-up points for precast concrete piles
预制混凝土组合单元 precast concrete compound unit
预制积木式(模数制)单位 prefabricated modular units
预制及预成型件 prefabricated or preformed parts
预制剂 preformulation;preparation
预制加合物 preformed adduct
预制加筋石膏板 precast reinforced gypsum slab
预制加气混凝土 precast aerated concrete
预制加气混凝土建筑构件 precast gas concrete building unit
预制检修井的混凝土 manhole block
预制件 precast product;precast unit; preconstructional unit; prefabricated parts; prefabricated unit; prefabricated member; preformed unit;prefabricated element
预制件平衡悬拼法 precast balancing cantilever method
预制件装配 shop-fabricated member
预制建造 component construction
预制建造法 systems building
预制建筑 prefabricated architecture
预制建筑薄板 prefabricated building sheet
预制建筑构件生产 <混凝土> architectural modelling
预制建筑混凝土制品 architectural precast concrete product
预制建筑密封条 preformed architectural strip seal
预制建筑物 preformed building
预制建筑压缩密封料 preformed architectural compression seal
预制建筑组件 prefabricated building assembly
预制铰接混凝土沉排 articulating precast concrete mat
预制接缝 precast joint
预制接缝密封料 preformed joint sealant
预制接缝填料 preformed joint seal
预制接缝止水 preformed sealant
预制(节)段拼装施工法 segmental construction
预制结构 precast construction; prefab;prefabricated structure;prefabricated construction;sectional construction;system-built structure
预制结构钢拱架 prefabricated structural steel centering
预制结构构件 prefabricated structural element
预制结构体系 precast construction system
预制结合梁 precast composite beam
预制晶种 preseed
预制绝热材料 prefabricated heat insulation;sectional insulation
预制绝缘材料 sectional insulation
预制抗震结构体系 precast seismic structural system
预制可受钉混凝土厚板 precast nailable concrete plank
预制空心混凝土块 precast hollow concrete block
预制空心块 blockwork
预制空心块组成的梁 beam made of precast hollow blocks
预制空心楼板 concrete plank;precast hollow concrete plank;precast hollow floor unit
预制空心桩 hollow preformed pile
预制块 precast(ing) block
预制块段(梁) precast segment
预制块及过梁构造 precast block and lintels
预制块砌体 blockwork
预制块体 precast block
预制框架 framer;precast frame;prefabricated frame
预制拉杆 prefabricated tie
预制拉结网片 <在砌墙内的> prefabricated tie
预制肋形板 precast ribbed slab
预制肋形拱顶 prefabricated rib vault
预制立面 system-built facade
预制沥青板 prefabricated asphalt plank;premo(u)lded asphalt panel;premo(u)lded bituminous panel
预制沥青粗麻布面 bituminized jute hessian cloth
预制沥青粗麻布铺面 prefabricated bituminized hessian surfacing
预制沥青混凝土板路面 prefabricated bituminous surfacing
预制沥青混凝土板面层 prefabricated bituminous surfacing
预制沥青混凝土铺面 prefabricated bituminous surfacing
预制沥青路面 prefabricated asphaltic bitumen surfacing
预制沥青片 prefabricated asphalt sheet(ing)
预制沥青铺盖 prefabricated asphaltic blanket
预制沥青铺面 prefabricated asphaltic bitumen surfacing
预制沥青嵌缝条 preformed asphalt joint filler;premo(u)lded asphalt joint filler;premo(u)lded bituminous joint filler
预制沥青填缝料 preformed asphalt joint filler
预制沥青条 premo(u)lded bitumen strip; premo(u)lded bituminous strip
预制沥青屋面板 asphalt prepared roofing;asphalt ready roofing
预制沥青屋面材料 bitumen ready roofing
预制沥青屋面覆盖料 bitumen prepared roofing
预制沥青屋面料 bitumen wool felt
预制沥青屋面油毛毡 bitumen rag-felt
预制梁 precast beam;precast girder; prefabricated beam; prefabricated

girder
预制梁构件 precast beam unit
预制楼板 precast floor slab;prefabricated floor
预制楼板构件 precast floor segment; precast floor unit; prefabricated floor member
预制楼板组合件 <一种注册的建筑材料产品> eagle
预制楼梯 precast stair(way)
预制(路)缘石 precast curb
预制铝材洁净室 alumin(i)um prefabricate clean room
预制氯丁橡胶垫片 neoprene preformed gasket
预制卵形沟管 precast-segmental sewer
预制码头 Mulberry harbo(u)r
预制锚固铁件 precast anchor
预制美术磨石地面 precast artistic terrazzo flooring
预制门 prefinished door
预制门窗贴脸 factory-made trim;package trim
预制密封垫 preformed gasket
预制密封胶 preformed sealant
预制密封胶带 preformed tape
预制密封料 preformed sealant
预制密封圈 preformed gasket
预制面板 front panel curtain wall; precast cladding panel; precast cladding slab
预制面层 prefabricated lining
预制面层块体 paving unit
预制模板 prefab-form(work);prefabricated form(work);prefabricated shuttering unit; prefabrication form; sectional formwork; formwork panel
预制模板系统 prefabricated form system
预制模件住宅 modular home
预制模具 preformed die
预制木材工字桁条 prefabricated wood-I-joist
预制木构件 millwork
预制木桁架 prefabricated wood truss
预制木混凝土块 precast wood concrete tile
预制木建筑 wooden prefabricated construction
预制木结构 timber system construction
预制木结构房屋 ready cut house
预制木零件 prefabricated timber section
预制木屋 Ibo
预制木纤维混凝土砌块 precast wood concrete block
预制木制件 prefabricated timber section
预制泥浆 ready-made mud
预制牛皮纸 Ibeco
预制暖气道 se-duct
预制排水管 prefabricated drain
预制泡沫 preformed foam
预制泡沫法 prefoaming method
预制泡沫混凝土 precast foam concrete
预制泡沫混凝土构件 prefabricated expanded concrete member
预制泡沫混凝土建筑构件 precast expanded concrete building component; precast gas concrete building unit
预制配件者 prefabricator
预制配筋混凝土墙 prefabricated reinforced concrete wall
预制配筋墙 prefabricated reinforced wall
预制配筋桩 prefabricated reinforced pile
预制拼装 dress up
预制拼装桩 precast sectional pile
预制品 precast product;prefab;prefabricate;prefabrication
预制品绕组 preformed winding
预制平台 prefabricated platform
预制平行钢丝束 prefabricated parallel wire strand
预制平行钢丝束股法 <悬索桥主缆施工的> prefabricated parallel wire strands method [PPWS method]
预制铺面混凝土块 patio block
预制砌块 blockwork; precast block(work)
预制砌块衬砌 prefabricated lining
预制砌块墙板 masonry panel
预制砌体工程墙板 masonry panel
预制砌体墙板 prefabricated masonry panel
预制砌筑墙板 masonry panel
预制嵌缝料 preformed joint sealant
预制嵌缝条 premo(u)lded filler strip
预制墙 system-built wall
预制墙板 component panel; panel wall unit; precast panel; precast wall panel; prefabricated panel; prefabricated wall panel; walling slab
预制墙板结构 precast wall panel structure;precast wall structure
预制墙板式房屋 panel type house
预制墙板式结构 panel type construction
预制墙壁 industrialized wall
预制墙构件 precast wall unit
预制墙箍 prefabricated tie
预制墙砌体 prefabricated walling
预制墙体砌块 walling unit
预制墙砖板 panel brick
预制轻便式房屋 portable prefabricated building
预制轻混凝土板 precast light concrete slab
预制轻混凝土构件 precast lightweight concrete unit
预制轻混凝土块 light block; lightweight precast concrete block
预制轻质混凝土槽形板 featherweight precast concrete channel slab;lightweight precast concrete block; lightweight precast concrete channel slab; lightweight precast featherweight concrete channel slab
预制人字梁 cast gable beam
预制乳状液 preemulsion
预制塞缝条 premo(u)lded filler strip
预制伞形结构 precast parachute
预制山墙 precast gable
预制烧结衬里 preformed sintered liner
预制伸缩缝填料 premo(u)lded expansion joint filler
预制施工法 system construction method
预制石膏构件 gypsum precast building unit
预制石膏灰泥外墙板 gypsum sheathing
预制石膏灰泥组合墙板 gypsum wallboard composite
预制石膏制品 precast gypsum product;prefabricated gypsum product
预制石灰岩面板 limestone-faced precast panel
预制饰面 prefabricated surfacing
预制饰面混凝土组合件 fair-faced precast compound unit
预制双面沥青涂层的沥青屋面板 asphalt ready roofing with asphalt coat(ing) on both
预制水底隧道 prefabricated subaqueous tunnel
预制水底隧洞 prefabricated subaqueous tunnel
预制水管 prefabricated pipe
预制水磨石 fabricated terrazzo; factory-made terrazzo;precast terrazzo;prefabricated terrazzo
预制水磨石板 precast terrazzo slab; precast terrazzo tile
预制水磨石地面 precast terrazzo flooring
预制水磨石铺贴 precast terrazzo tiling
预制水磨石墙板 precast terrazzo
预制水磨石踢脚 precast terrazzo skirting
预制水泥构件 precast concrete component
预制水泥混合物 <加水即可使用> cemixene
预制水泥混凝土块 cement block
预制水泥空心砌块 precast cement hollow block
预制水泥块体 cement block
预制水泥型块 cement block
预制水刷石混凝土板 precast concrete facade
预制水下沉埋管式隧道 submerged prefabricated tunnel (tube)
预制四角方块 precast tetrapod block
预制四角空心方块 precast tetrapod hollow block
预制塔式公寓 precast apartment tower
预制塔式住宅 precast dwelling tower;precast residence tower
预制台 precasting table
预制台座 casting bed
预制陶瓷 preceramics
预制陶瓷贴面 prefabricated ceramic tiling
预制陶瓷芯 preformed ceramic core
预制陶粒混凝土 precast ceramisite concrete;precast haydite concrete
预制体系 precast system
预制填充材料条 strip of pre-formed filling material
预制填充物 pretested packing
预制填缝板 precast jointing plate; precast jointing slab; preformed joint filler
预制填缝材料 precast joint filler;preformed joint filler; premo(u)lded joint filler
预制填缝料 prefabricated joint filler
预制填缝条 prefabricated joint filler
预制铁片屋面建筑 scan roof construction
预制投加料系统 prefabricated feed system
预制突堤码头构件 prefabricated pier component
预制涂料 ready coating
预制涂膜片 precoated sheet
预制弯头 factory bend;factory ell
预制卫生间 precast bathroom and toilet
预制圬工墙板 prefabricated masonry panel;prefabricated masonry slab
预制污水管渠 precast-segmental sewer
预制屋顶板 precast roof slab; roof deck(ing)
预制屋顶的制造 prepared roofing manufacture
预制屋顶构件 precast roof unit
预制屋顶木瓦 prepared roofing shingle
预制屋面 prefabricated roofing; prepared roofing
预制屋面覆盖层 prepared roof covering
预制屋面料 prepared roofing; ready roofing
预制屋面油毛毡 ready sheet roofing paper
预制下水道管段 precast-segmental sewer
预制下水管 precast-segmental sewer
预制线 prewire
预制线圈 prefabricated coil
预制镶嵌地沥青 mosaic asphalt
预制小梁空心砖楼盖 precast beam and hollow-tile floor
预制斜撑杆 fabricated single-post shore
预制型材 preformed section
预制型窗密封垫 window preformed gasket
预制型锚杆嵌岩桩 precast concrete pile or steel pipe pile anchored in rock
预制型芯柱嵌岩桩 precast concrete pile or steel pipe pile with plug socketed in rock
预制型植入嵌岩桩 precast concrete pile or steel pipe pile socketed in rock
预制烟囱 prefabricated stack
预制烟道 prefabricated flue
预制移动式脚手架 prefabricated mobile scaffold
预制永久性的轻质混凝土模板 precast permanent light(weight) concrete formwork
预制预应力的 precast prestressed
预制预应力混凝土 precast prestressed concrete
预制预应力混凝土T形梁桥 precast prestressed concrete T-beam bridge
预制预应力混凝土板 spancrete
预制预应力混凝土构件 precast prestressed concrete unit
预制预应力混凝土空心板 hollow core precast prestressed concrete slab
预制预应力混凝土空心板桥 precast prestressed concrete voided slab bridge
预制预应力混凝土模板 prestressed precast concrete form
预制预应力混凝土箱梁桥 precast prestressed concrete box girder bridge
预制预应力梁 precast prestressed beam
预制预应力轻质骨料混凝土 precast prestressed light(weight) aggregate concrete
预制造 preproduction
预制整体橡皮垫 bonded rubber cushion(ing)
预制正面 system-built facade
预制制品 precast articles
预制蛭石混凝土 precast vermiculite concrete
预制柱构件 precast column unit
预制砖板 brick panel
预制砖板建筑单元 brick panel(building) unit
预制砖墙 panel brick
预制桩 precast pile;prefabricated pile; preformed pile;premo(u)lded pile
预制桩基础 prefabricate piling foundation
预制装配 prefabrication
预制装配采光井墙 prefabricated area wall
预制装配的 prefabricated
预制装配法 preassembly method
预制装配房屋 panelized house
预制装配构件 open parts
预制装配构造 package-type construction
预制装配木结构 wooden prefabricated construction
预制装配桥 precast assembled bridge

预制装配施工法 prefabricated construction method
预制装配式 open system
预制装配式板材衣橱 prefabricated tiling cubicle for clothes
预制装配式保温管道 precasting insulating pie
预制装配式采光井墙 prefabricated area wall
预制装配式房屋 prefab;prefabricated building;seco;prefabricated house
预制装配式沟管 precast-segmental sewer
预制装配式混凝土板材围墙 prefabricated concrete panel fence
预制装配式混凝土房屋 prefabricated concrete building
预制装配式混凝土建筑构件 prefabricated concrete building member
预制装配式混凝土外围护墙板 prefabricated concrete cladding panel
预制装配式混凝土园艺建筑单元 prefabricated concrete garden building unit
预制装配式建筑 prefabricated block
预制装配式建筑物 precast construction;prefabricated building
预制装配式结构 package-type construction; prefabricated construction;prefabricated structure
预制装配式楼板 <由墙、柱或梁支承的> floor panel
预制装配式墙 prefabricated wall
预制装配式线脚 laid-on mo(u)lding
预制装配式浴室单元 prefabricated bathroom unit
预制装配式整体结构 prefab monolithic structure
预制装配式住宅 prefabricated dwelling house
预制装配式住宅构件 prefabricated house unit
预制装配式砖石建筑 brick(work) system building
预制装配烟道 prefabricated flue
预制装配造船法 precast panel shipbuilding
预制装配整体结构 prefab monolithic structure
预制组合构件 prefabricated compound unit
预制组装件 prefabricated assembly
预置 initialization;initialize;preset
预置参数 preset parameter
预置操作 initialize operation
预置程序 initialize program(me);initialize routine
预置触发器 pretrigger
预置导行系统【交】 preset guidance system
预置电平 predetermined level
预置定时器 preset timer
预置方式 preset mode
预置分流器 prevent shunt of a track circuit
预置钢筋网 preset reinforcement mat
预置格式 initialize format
预置功能 preset function
预置骨料灌浆混凝土 grouted aggregate concrete; grouting pre-placed aggregate
预置骨料横向压力灌浆 advance slope grouting
预置骨料混凝土 preplaced-aggregate concrete
预置光阑 preset diaphragm
预置级 preset level
预置集料横向压力灌浆 advance slope grouting
预置计数器 predetermined counter; preset decimal counter
预置角 preset angle
预置校验 initialize verification
预置控制 preset(ting) control;reset control
预置控制器 preset controller
预置梁 encastre
预置码 preset code
预置时间 set time
预置实用程序 initialize utility
预置式自动均衡器 preset automatic equalizer
预置数据 initialize data
预置位 presetting bit
预置信号 preset signal
预置型光圈 preset iris
预置型开关 preset type switcher
预置于地面的管子 projected pipe
预置语句 initialization statement
预置值 present level;preset value;prevalue
预置制导 preset guidance
预置制导系统【交】 preset guidance system
预置桩 preliminary pile
预煮 preliminary cook
预注 prefill
预注浆 pregrouting;pre-injection
预注浆防水设计 pre-injection technique for waterproofing
预注浆扫孔 pre-injection clearing hole
预注浆设计 pre-injection design
预注入器 preinjector
预注装置 primer
预贮存池 preliminary storage tank
预柱 pre-column
预筑的地下铁道 premetro
预铸 precasting;premo(u)ld
预铸孔 precast hole
预转 prerotation
预转定序器 preroll sequencer
预转移动 preroll shift
预桩 pre-piling
预装 preloading
预装部件 preassembled parts;preassembled unit
预装单元 preassembled unit
预装的 prepositioned
预装法 prepack method
预装构件 preassembled member;preassembled section
预装好管道的 prepiped
预装件 preassembled parts
预装角 preset angle
预装门 prehung door
预装配 first fixings;preassemble;preassembling; pre-fitting; preframe; preassembly
预装配板 preframed panel
预装配的 preassembled;preframed
预装配间 prefabrication shed
预装配桥 preassembled bridge; precast bridge
预装配式房屋 prefab; prefabricated house
预装配式结构 prefabricated structure
预装配整套浴室 instant bathroom (unit)
预装配整体造船法 prefabricated-monolithic shipbuilding
预装配装置 preassembled system
预装入 preload
预装润滑脂 prepacked with grease
预装饰隔热、隔音板 predecorated insulating board
预装饰混凝土砌块 prefaced concrete masonry unit
预装饰胶合板 prefinished plywood
预装饰墙板 predecorated wallboard
预装饰石膏板 predecorated gypsum board
预装饰石膏灰泥墙板 predecorated gypsum wallboard
预装饰纤维板 predecorated insulating board
预装饰镶面墙板 predecorated wallboard panel; prefinished wallboard panel;prefinished wall panel
预装饰硬质纤维板 prefinished hardboard;predecorated hardboard
预装饰预制板 prefinished panel
预装提花轮 prefilled pattern wheel
预装填 prefill(ing);prepacking
预装修 prefinish(ing)
预装修的 prefinished
预装修石膏板 predecorated gypsum board
预装药 sleeper charge
预着色 precolo(u)ring
预着色混合料 precolo(u)red compound
预着色树脂 precolo(u)red resin
预组装 preassembly;prepack(age)
预组装车间 preassembly shop
预组装单元 precast unit;preconstructional unit;prefabricated unit
预组装件 prefabricated parts
预组装式止动器 preassembled lock
预组装锁 preassembled lock
预钻 preaugering; preboring; predrilling
预钻插桩法 socketing pile in prebored hole
预钻的 pre-augered
预钻钉眼 preboring; preboring for nails
预钻孔 preaugering; prebored hole; preboring; precoring; predrilled hole;predrilling <桩工>
预钻孔的 predrilled
预钻孔注桩法 prebored pile driving method
预钻孔桩 piles in pre-bored holes
预钻量 subdrilling
预钻式旁压试验 preboring pressuremeter test
预作用 preact
预作用阀 preaction valve
预作用灭火系统 preaction system
预作用喷水灭火系统 preaction sprinkler system
预做饰面 prefinish
预做饰面的 prefinished;prefinishing

域 domain;field

域标识符 realm identifier
域长 field length
域当前指示符 realm currency indicator
域的代数闭包 algebraic(al) closure of a field
域的代数扩张 algebraic(al) extension of a field
域的扩张 extension of a field
域扩张 field extension
域流分配法 stream allocation procedure
域论 field theory;theory of fields
域描述项 realm description entry
域名 realm name;domain name
域名服务器 domain name server
域名系统 domain name system
域式劈理 domainal cleavage
域套 nested domain
域外评价法 extrapolation
域像差 zonal aberration
域组构结构 texture of domainal cleavage fabric

阈 波长 threshold wavelength

阈波数 threshold wave number
阈场 threshold field
阈抽运 threshold pumping
阈臭 threshold odo(u)r
阈处理 threshold treatment
阈电流 threshold current
阈电路 threshold circuit
阈电压 starting voltage; threshold voltage
阈定律 threshold law
阈恶臭 threshold odo(u)r
阈反应 threshold reaction
阈管 threshold tube
阈管理员 domain administrator
阈含水量 threshold moisture content
阈函数 threshold function
阈极限 threshold limit
阈极限值 threshold limit value
阈剂量 threshold dosage;threshold dose
阈加速度 threshold acceleration
阈加速率 threshold acceleration
阈监测 threshold detection
阈结构 threshold structure
阈解码 threshold decoding
阈亮度差 threshold luminance difference
阈量 threshold dosage;threshold dose
阈裂变物质 threshold fissioning material
阈灵敏度 threshold sensitivity
阈流速 threshold speed;threshold velocity
阈逻辑【数】 threshold logic
阈逻辑网络 threshold-logic(al) network
阈能 threshold energy
阈浓度 threshold concentration
阈漂移 threshold drift
阈频率 threshold frequency
阈评价 threshold evaluation
阈气味浓度 threshold odo(u)r concentration
阈气味值 threshold odo(u)r number
阈前 prethreshold
阈前的现象 prethreshold phenomenon
阈墙 jamb wall
阈区 threshold region
阈上刺激 supraliminal stimulus
阈上级 level above threshold; sensation level
阈上水平 level above threshold; sensation level
阈收缩 threshold contraction
阈探测器 threshold detector
阈调节 threshold modulation
阈调整 threshold adjustment
阈温 threshold temperature
阈下的 subliminal
阈下剂量 subthreshold dose
阈下知觉 subliminal perception
阈限 threshold
阈限对比 threshold contrast
阈限分析 threshold analysis
阈限级 threshold level
阈限摩擦 threshold friction
阈限浓度 threshold concentration
阈限坡降 threshold gradient
阈限深度 threshold depth
阈限通行 <在主要车流停止时挤进通过> filtering
阈限旋转试验 threshold shift test
阈限压力 threshold pressure
阈限增量 threshold increment
阈限值 limit value; threshold level; threshold value
阈限值上限 threshold limit valve ceiling

阈信号干扰比 threshold signal-to-interference ratio
阈移 threshold shift
阈移位 threshold shift
阈因素 threshold element
阈因子 threshold factor
阈域 threshold region
阈域上升 raising of threshold
阈域增长 increase of threshold
阈展解调器 threshold extension demodulator
阈照度 threshold illuminance; threshold illumination
阈振动 threshold vibration
阈值 threshold(ing); threshold quantity; threshold value; value of threshold
阈值持续时间 threshold duration
阈值带滤波 threshold zonal filtering
阈值电平 threshold level
阈值电位 threshold potential
阈值电压 threshold voltage
阈值对比度 liminal contrast; threshold contrast
阈值反衬 contrast threshold; liminal contrast; threshold contrast
阈值反转 threshold inversion
阈值方式 threshold mode
阈值放大器 threshold amplifier
阈值功率 threshold power
阈值功率密度测量 threshold power density measurement
阈值函数 threshold function
阈值函数表 threshold function table
阈值函数的构造 structure of threshold function
阈值后工作状态 afterthreshold behavio(u)r
阈值极限 threshold value limit
阈值剂量计 go-no-go dosimeter
阈值减低 threshold reduction
阈值检测 threshold detection
阈值检测器 threshold detector
阈值简并 threshold degeneracy
阈值开关 threshold switch
阈值抗扰度 threshold immunity
阈值控制 threshold control
阈值流量密度 threshold flow density
阈值逻辑 voting logic
阈值逻辑电路 threshold logic(al) circuit
阈值迁移 threshold shift
阈值声强 threshold sound intensity
阈值探测器 threshold detector
阈值条件 threshold condition
阈值调整 threshold setting
阈值通量密度 threshold flow density
阈值效应 threshold effect
阈值信号 threshold signal
阈值信噪比 threshold signal-to-noise ratio
阈值性状 threshold character
阈值雅可比法 threshold Jacobi's method
阈值有效低值 threshold effect low value
阈值有效极限 threshold effect limit
阈值元件 threshold element
阈值运算器 threshold operator
阈值增益 gain for threshold value; threshold gain
阈质 threshold substance
阈状态 threshold condition
阈作用 threshold effect

寓

寓所 abiding place; boziga; digs; dwelling; family dwelling unit; lodg(e)ment; tenement

御

御谷 pearl millet

御坍棚 protecting roof

裕

裕度 allowance; clearance; margin; overmeasure; tolerance

裕量 overmeasure; surplus capacity

遇

遇变层 <岩土工程勘察术语> at every identifiable change of strata whichever is met earlier

遇到 come actress; encounter
遇到危险 encounter danger
遇故障时自动打开的 fail open
遇故障时自动关闭的 fail closed
遇见 foresight
遇忙自动回叫 automatic call back-busy
遇难 distress; wreckage <船只、火车、飞机等的>
遇难报警 distress altering
遇难船 castaway; doomed vessel; shipwreck; vessel in distress
遇难船残骸 wreckage
遇难船货物 distress cargo
遇难船只 ship in distress; wreck
遇难船只的飘浮残骸(或其货物) flotsam
遇难灯号 distress light
遇难飞机 wreck
遇难呼号 distress call
遇难呼救信号 rescue message; distress signal
遇难火车 wreck
遇难火号 distress light
遇难火箭 distress rocket
遇难阶段 distress phase
遇难频率 distress frequency
遇难旗 distress flag
遇难求救程序 distress procedure
遇难确认 distress acknowledgment
遇难特殊信号 special distress signal
遇难无线电呼叫系统 distress radio call system
遇难无线电话操作程序 radio telephone distress procedure
遇难信号 signal of distress
遇难信号波 distress wave
遇难优先度申请电文 distress priority request message
遇难转播 distress relay
遇湿易燃物品 combustion-in-humidity articles
遇水剥落 <沥青与集料遇水分离> water stripping
遇水燃烧物 combustion-in-water article; dangerous when wet
遇险 distress
遇险报警 distress altering
遇险船 doomed vessel; shipwreck; vessel in distress
遇险船残骸 wreckage
遇险船只 ship in distress
遇险灯号 distress light
遇险电话 distress call
遇险呼号 distress call
遇险呼救信号 rescue message; distress signal
遇险火号 distress light
遇险火箭 distress rocket
遇险阶段 distress phase
遇险频率 distress frequency
遇险旗 distress flag
遇险求救程序 distress procedure
遇险确认 distress acknowledgment
遇险特殊信号 special distress signal
遇险无线电呼叫系统 distress radio call system
遇险无线电话操作程序 radio telephone distress procedure
遇险信号 distress call; emergency signal; signal of distress
遇险信号控制板 emergency keyer panel
遇险信息 emergency message
遇险优先度申请电文 distress priority request message
遇险转播 distress relay
遇有列车或障碍物等能立即停车 stop short of train, obstruction etc.
遇阻沉积 eoposition
遇阻堆积 encroachment

愈

愈创木 guaiacin; pockwood; guaiaci lignum; Guaiacum officinale; guaiac-wood

愈创木酚 guaiacol
愈创木树脂 gualac(um)
愈创木属 lignum vitae
愈创木油 guaiac wood oil
愈创木脂 gumacum
愈合 coalescence; healing
愈合材 calluswood
愈合节 <木材的> intergrown knot
愈合节理 healed joint
愈合组织 callus

誉

誉清的航海日记 smooth log

誉写机 duplicating machine

鸢

鸢式浮子 kite float

鸢式转角斜踏步 kite winder
鸢尾花、百合花饰 fleur-de-lis
鸢尾花、百合花形纹章 fleur-de-lis

鸳

鸳鸯 mandarin duck

鸳鸯插头 hermaphroditic connector

元

元胺【化】 diamine

元胞自动机模型 cellular automat model
元宝车 depressed center flat car; low loader; well car; well wagon
元宝钢筋 bend bar
元宝筋 bend bar
元宝螺钉 butterfly screw; eared screw
元宝螺帽 thumb nut; wing nut
元宝螺母 butterfly nut; thumb nut; thumb screw; wing nut
元宝螺栓 bolt with winged nut; thumb bolt
元宝螺丝 thumb screw; wing screw
元宝槭 purpleblow maple; truncate-leaved maple
元宝榫 nipping
元宝铁 V-block
元宝瓦 hogsback (tile)
元编译程序 metacompiler
元波 elementary wave
元部件损坏 element failure
元程序 metaprogram(me)
元代码 metacode
元电流 elementary current
元定理 metatheorem
元对 element pair
元符号 metasymbol
元钢筋的 plain
元古代【地】 Agnotozoic age; Proterozoic (era)
元古代的 Algonkian
元古代基底 Proterozoic basement
元古界【地】 Agnotozoic erathem; Proterozoic erathem
元古宇【地】 Proterozoic Eonothem
元古宙【地】 Proterozoic Eon
元股 primer strand
元光束 elementary beam
元弧 arc element
元汇编程序 metaassembler
元激发 elementary excitation
元羁烷 friedelane
元件 cell; component; component element; component parts; elements; organ; parts; units
元件包壳 element jacket
元件泵 dual element pump
元件表 component chart
元件测试装置 component test set
元件的检测 detection of elements
元件的清洗 cleaning of element
元件故障率 element error rate
元件盒 subassembly wrapper
元件间距 element spacing
元件控制 element control
元件密度 component density
元件描述符 element descriptor
元件破损检测器 burst detector
元件设计 element design
元件失效 component failure
元件识别名 component identification
元件寿命 component life
元件损伤 element failure
元件套 element housing
元件误差 component error; element error
元件运输机 element conveyor
元件自动装配 automatic component assembly
元件组 component group
元件组合建筑 unit construction
元件组合桥 unit construction bridge
元件组合桥系统 unit construction bridge system
元结合 member aggregate
元理论 metatheory
元力 elementary force
元粒子 elementary particle
元逻辑 metalogic
元面积 elemental area; elementary area
元偶极子 elementary dipole
元排列 element array; identical permutation
元色学说 primary colo(u)r theory
元数据 metadata
元数学 metamathematics
元素 element; organ
元素半导体 elemental semiconductor
元素比值 element ratio
元素比值图 diagram showing the ratio of elements
元素变化趋向图 diagram showing the variation trend of element
元素表 list of element
元素产额 element yield
元素成分 elemental composition; elementary composition
元素垂直分布图 diagram showing the vertical distribution of element
元素存在形式 manner of occurrence of element; mode of element occurrence
元素大小 element size
元素带 strip of element
元素的 K 系列临界激发电压 K-system critical excitation voltage of elements
元素的 L 系列临界激发电压 L-system critical excitation voltage of ele-

ments
元素的 M 系列临界激发电压 M-system critical excitation voltage of elements
元素的地球化学分布 geochemical distribution of element
元素的地球化学分配 geochemical partition of element
元素的丰度 abundance of element
元素的摩尔数 elemental mole fraction
元素的迁移 elements transportation
元素的宇宙丰度 cosmic(al) abundance of element
元素地方分布 local distribution of elements
元素地球化学 element geochemistry
元素地球化学行为 geochemical behavio(u)r of elements
元素地球化学循环图 diagram showing the geochemical cycle of elements
元素地球化学演变图 diagram showing the geochemical evolution of elements
元素递变 transmission of element
元素电荷 elemental charge
元素定量分析 quantitative elementary analysis
元素定性分析 qualitative elementary analysis
元素对 pair of elements
元素反应 elementary reaction
元素分布 distribution of element
元素分布频率 distribution frequency of element
元素分布频率直方图 frequency histogram showing the distribution of elements
元素分布曲线图 distribution curves of element
元素分布图 diagram showing the distribution of elements
元素分散作用 disperse of elements
元素分析 elemental analysis; elementary analysis; ultimate analysis
元素分析硫 ultimate sulfur
元素丰度 elemental abundance
元素丰度变化图 diagram showing the variation in abundance of elements
元素丰度模式 element abundance patterns
元素符号 symbol of element
元素富集因子 element enrichment factor
元素富集作用 enrichment of elements
元素光谱特征线条 raies ultimes
元素过程 elementary process
元素过渡系数 transfer coefficient of element
元素过剩 excess of elements
元素含量 element content
元素化学分带模式图 diagram showing the chemically zonation model of elements
元素活动性 mobility of elements
元素交换作用 exchange of elements
元素绝对活动性 absolute mobility of element
元素磷 elemental phosphorus
元素硫 elemental sulfur [sulphur]
元素氯 elemental chlorine
元素浓度 concentration of elements
元素浓度相对变化图 diagram showing the relative change in the concentration of elements
元素配分 partition of elements
元素迁移 element migration; element transfer
元素迁移能力 ability of migration of elements; element transfer capacity
元素迁移实验 element migration experiment
元素迁移系列 sequence of migration of elements
元素迁移系数 element transfer coefficient
元素迁移序列 sequence of element migration
元素迁移作用 migration of elements
元素区域分布 regional distribution of elements
元素全球分布 global distribution of elements
元素缺乏 deficiency of elements
元素色谱法 elemental chromatography
元素碳 elemental carbon
元素铁 elemental iron
元素/同位素丰度 element/isotopic abundance
元素误差 elemental error; elementary error
元素细长比 element aspect ratio
元素相对活动性 relative mobility of element
元素相互干扰 inter-element interference
元素效应 element effect
元素形成 elementary form
元素有机分析 elementary organic analysis
元素有机高聚物 elementary organic polymer
元素有机化合物 elementary organic compound; organic element compound
元素域 element field
元素周期 period of element
元素周期表 periodic(al) table; periodic(al) table of elements
元素周期律 periodic(al) law
元素周期系 periodic(al) system of elements
元素主要地球化学作用及途径图 diagram showing the main geochemical processes and their pathway of elements
元素转变 element transformation
元素组 element group
元素组成 elemental composition; elementary composition; ultimate composition
元素组分 elemental constituent; elementary constituent
元素组合特征 element association characteristics
元铁 rod iron
元信令 mega signal(l)ing
元质点 elementary material particle; elementary particle
元组 tuple

员工 employee; worker

员工补偿金 employee compensation
员工酬劳金 employee bonus
员工福利 employee welfare
员工福利服务设施安排 employee comfort and services in layout
员工福利基金 employee benefit fund
员工福利设施 employee welfare facility
员工福利咨询委员会 works council
员工津贴 employee benefit
员工救济金 employee relief fund
员工培训 personnel training
员工票 employee pass
员工退休津贴 employee retirement allowance
员工薪金册 employee payroll

园标 mark and garden

园标工程 garden engineering
园道 parkway
园凳 garden bench
园地 field; garden plot; scope
园地划分 land subdivision
园丁 gardener; nurseryman
园浮雕墙 wall rosace
园工 gardener
园花饰瓷器 rose medallion
园景树 specimen tree
园景系统 gardening system
园篱 garden fence; garden hedge
园林 garden and park; gardens; grove; landscape; park
园林保护 landscape conservation
园林匾额 garden tablet
园林驳岸 pool embankment in garden
园林布局 garden layout
园林城市 landscape city
园林大道 pkwy [parkway]
园林道路设计 garden path design
园林地形改造 topographic(al) remo(u)ld of garden
园林雕塑 gardens sculpture
园林分区规划 garden block planning
园林工程 landscaping landing
园林工程师 landscape engineer
园林工程学 landscape engineering
园林工人 gardener; topper <剪去树木顶端的>
园林管理规划 management plan of gardens
园林管理机构 garden public administration
园林(管理)局 landscape bureau
园林灌溉 treegarden irrigation
园林规划 garden planning
园林化 landscaping
园林化郊区 garden suburb(an)
园林机械 garden machine
园林建设 landscape development
园林建筑 building in landscape; ornament architecture
园林建筑家 garden architect
园林建筑师 landscape architect
园林建筑学 garden architecture; landscape architecture; park architecture
园林建筑艺术 park architecture
园林教育 education of landscape architecture
园林(经营)管理 garden management
园林景区 park
园林空间 garden space
园林理水 water treatment in garden
园林露天剧场 open garden theater[theatre]
园林路 green way
园林绿化 gardening and greening; gardens afforestation; landscaping
园林绿化与旅游 garden and tourism
园林美学 garden aesthetics
园林平面布置 landscape plan
园林区划 garden area division
园林色彩艺术 art of garden colo(u)rs
园林设计 garden design; landscape design; landscape treatment; landscaping
园林设计师 designer of formal gardens and parks; landscape architect; landscapist
园林施工 garden layout
园林式 landscape style
园林系统 park system
园林小品 a piece of scene; small garden ornaments
园林小气候 garden microclimate
园林型 landscape type
园林学 landscape architecture
园林艺术 garden art; landscape art
园林艺术布局 artistic layout of garden
园林意境 artistic conception in gardening
园林楹联 garden couplet
园林造景大理石 landscape marble
园林造景镶板壁 landscape panel
园林植物 garden plant
园林总体规划 garden master planning
园路工程 garden path work; garden walks and pavement
园圃播种机 garden seeder
园圃操作 garden operation
园圃耕作 garden tillage
园圃纸 mulsh paper
园墙 garden wall
园墙交替砌合 mixed garden-wall bond
园墙砌法 garden bond
园墙砌合 garden bond; garden wall bond
园桥 garden bridge
园田壤土 garden loam
园亭 garden house
园土 garden mo(u)ld
园岩油 garden rocket oil
园椅 garden chair; garden seat
园艺 horticulture
园艺玻璃 garden glass
园艺播种机 garden drill
园艺场 farm garden; garden spot
园艺方案 horticultural scheme
园艺房 horticultural building
园艺工作 garden work
园艺工作者 horticulturist
园艺规划 horticultural scheme
园艺规则 horticultural scheme
园艺机器 horticultural machine
园艺机械 horticultural machinery
园艺技能 green fingers; green thumb
园艺技师 gardening engineer
园艺家 gardener; horticulturist; landscape architect; landscape gardener
园艺建筑物 horticultural structure
园艺建筑学 landscape architecture
园艺美化家 landscape gardener
园艺美化师 landscape gardener
园艺喷雾用油 horticultural spray
园艺品种 garden variety
园艺拖拉机 horticultural tractor; midget tractor
园艺污染 pollution in horticulture
园艺学 floriculture; gardening; landscape gardening; horticulture
园艺用泵 garden engine
园艺用玻璃 horticultural glass; plain rolled glass; roofing glass
园艺用喷灌机 garden sprinkler
园艺用平板玻璃 horticultural sheet glass
园艺用拖拉机 garden tractor
园艺用镇压器 garden roller
园艺植物 ornamental plant
园艺作物 garden crop; horticultural crop; horticultural plant
园用铲 garden shovel
园用镐 garden mattock
园用工具 garden tool
园址测量图 garden site survey map
园中小径 allee
园主 nurseryman

垣 <黄土高原的一种地形> flat ridge

原岸砾石 run-of-bank gravel

原白色 colo(u)r prime white
原白蚁科 <拉> Termopsidae
原白云岩 proto-dolomite
原柏油 primary tar;raw asphalt
原板 parent plate;raw sheet
原板废品率 reject rate of raw sheet
原板木架 raw sheet horse
原板破损率 breakage rate of raw sheet
原版 master mo(u)ld;master printing;original edition;shell mo(u)ld
原版磁带 first generation tape;magnetic original
原版海图 original chart
原版胶片 production acetate
原版卡片 master card
原版正片 master film positive
原包装 original package
原保险费 original premium
原报价 original quotation
原北极 proto-arctic
原贝壳杉酸 noragathenic acid
原本 master copy;original copy;originality;script;source version
原本的副本 duplicate original
原本价值 inherent value
原苯酚钠 sodium ortho-phenylphenate
原比电离 primary specific ionization
原比(例)尺 natural scale;original scale
原边 primary side;secondary side
原边漏抗 primary leakage reactance
原边线圈 primary coil;primary winding;secondary winding
原边阻抗 primary impedance
原标书 original bid
原标准 primary standard
原标准声源 primary standard sound source
原表皮层 protoderm
原波形 original wave
原玻璃 parent glass;bare glass
原玻璃棉 pouring wool
原玻璃纤维 virgin glass fiber
原步法 natural pace
原材 rough stock
原材料 base material;basic material;feed stock;parent material;primary material;raw material;raw stock;rough material;rough stock;source material;starting material;stuff;virgin material;crude material
原材料标准 material standard
原材料采购 material purchase
原材料仓库 raw material store
原材料成本 cost of material;material cost
原材料成本差异 material cost variance
原材料成本法 material cost method
原材料储备 material stock
原材料储藏 raw materials storage
原材料处理费用 material handling expenses
原材料存货指数 inventory index of raw materials
原材料搭配差异 material mix variance
原材料单价 material unit price
原材料订购 ordering of raw materials
原材料费用计算 material cost accounting
原材料分配 material distribution
原材料分配单 material distribution sheet
原材料工业 raw and semi-finished materials industry
原材料供应 material supply
原材料供应和产品销售的一体化 backward and forward integration of material
原材料供应计划 material supply plan
原材料购买存储预算 materials purchasing and stock budget
原材料管理 material control
原材料耗用 material consumption
原材料耗用定额 material consumption norm
原材料耗用率 materials rate
原材料混合 raw mix
原材料积压 excess of materials
原材料及加工中物料的资金定额 funds norm for raw and processed materials
原材料加工的废物 primary manufacturing residue
原材料价格差异 material price variance
原材料接收 material receipt
原材料卡 store card
原材料库存价值 value of materials in stock
原材料明细表 material specification
原材料明细分类账 material ledger
原材料审计 material audit
原材料生产率 materials productivity
原材料(使用)效率差异 material yield variance
原材料退库 materials returned in the stockroom
原材料消耗指数 consumption index of raw material
原材料需求计划 material requirement planning
原材料需要量 material requirement
原材料延误 delay of material
原材料研磨 raw material grinding
原材料用量差异 material quantity variance
原材料预算 material budget
原材料周转率 material turnover rate
原采出的 pit-run
原采石料 run-of-pit
原拆原建 compensatory replacement of demolished housing;demolishing and rebuilding
原产 origin
原产的动植物 native
原产地 country of origin;native habitat;origin;place of origin;raw product
原产地标志 mark of origin
原产地证明书 certificate of originals
原产国 country of origin
原潮汐 primary tide
原车过轨 transfer of wagon directly from one country railway to another
原沉积顶阶地 filltop terrace
原成本 original cost
原成胶 precollagen
原承包人 original contractor
原承包商 original contractor
原尺 standard;standard measure;typical measure
原尺寸 full size;life size;whole size
原尺寸大样 full-size detail
原尺寸的 full-scale;lifesize(d)
原尺截面图 true section
原尺寸模型 full-scale model
原尺寸模型试验 full-scale model test
原尺寸全径钻头 fun-sized bit
原尺寸图 natural-size drawing
原尺寸图案 epure (asphalt)
原尺寸之半 half-size
原尺寸值 full-scale value
原尺寸钻头 full-sized bit
原尺度 natural scale
原虫胶 stick lac
原初的 primeval
原初级污泥 raw primary sludge
原初作用过程 primary effect process
原储藏量 original store
原(船)队过闸 fleet lockage
原创办人 original founder
原吹填围堰面 face of existing reclamation bund
原磁带 raw tape
原磁铁 primary magnet
原刺激 primary stimuli
原粗结构 crude structure
原大的 full-scale;full-size(d);life-size(d)
原大西洋 proto-Atlantic ocean
原大小 life size;natural size
原等高线 eoisohypse
原底 original form
原底片 master negative;original negative
原底噪声 ground noise
原地 in-situ
原地表面 existing ground surface
原地槽 primary geosyncline;protogeosyncline
原地槽阶段 protogeosyncline stage
原地沉淀 in-situ precipitation
原地沉淀法 in-situ precipitation method
原地沉积 autochthonous deposit;deposit(e) in-situ;sedentary deposit
原地成岩变化阶段 locomorphic phase;locomorphic stage
原地处理 in-service treatment
原地的 autochthonous
原地堆积 autochthony
原地堆积物 sedentary deposit
原地分析 in-situ analysis
原地风化物质 in situ weathered material
原地花岗岩 autochthonous granite
原地回收 in-situ recovery
原地混合岩化方式 ectexis way
原地混凝土 cast-in-situ concrete;in-situ concrete
原地激振试验 in-situ impulse test
原地剪切强度 intact shear strength
原地建造【地】katachthonous formation
原地交付 delivery on field
原地交货 delivery on field
原地浇筑混凝土 cast-in-situ concrete;cast-in-place concrete
原地角砾岩 autochthonous breccia
原地浸出 in-place leaching;in-situ leaching
原地聚合 in-situ polymerization
原地砾石 in-situ gravel
原地铝土矿 autochthonous bauxite
原地煤 autochthonous coal
原地密度 <不采样测定的密度> in-situ density
原地面 existing ground;natural ground;original ground
原地面标高 existing ground level;level of original bed;natural ground level
原地面高程 existing ground level;level of original bed;natural ground level;natural surface level
原地面坡度 natural slope
原地面线 existing ground line;ground line;line of existing ground;natural surface line;natural ground
原地黏[粘]土 intact clay
原地生成的 autochthonous
原地生成的地下水 autochthonous groundwater
原地生成煤 autochthonous coal
原地生成说 situ theory
原地生长说 in-situ theory
原地石灰岩 autochthonous limestone
原地试验 in-place testing
原地水 native water
原地台 primary platform
原地台阶段 protoplatform stage
原地土(壤) indigenous soil;in-situ soil;original soil;sedentary soil;soil in-situ;autochthonous soil
原地土壤含水当量 field moisture equivalent
原地土壤试验 in-situ soil test
原地系统地层 strata of autochthon
原地显色染料 in-situ developed dye
原地形 initial landform
原地形成 in-situ formation
原地压力计 in-situ pressiometer;in-situ pressure meter
原地岩 autochthon(e) [复 autochthon(e)s];solid rock
原地岩石 rock in place
原地岩石应力 in-situ rock stress
原地岩体 aborigines
原地野外测量 in-situ field measurement
原地淤积 autochthonous deposit;deposit(e) in-situ;sedentary deposit
原地再生 in-place regeneration;regenerate in situs
原地褶皱 sedentary fold(ing)
原地转弯 pivot turn;spot turn
原地转向 pivot steering;spin turn
原点 base point;initial point;origin;original point;point of origin;punch mark;zero;zero point
原点导纳 direct mobility
原点改变 change of origin
原点矩 moment about the origin
原点偏移(距) origin(al) offset
原点向量 origin vector
原点移动 shift of origin
原点转移 shift of origin
原电池 battery cell;decomposition cell;galvanic battery;galvanic cell;galvanic element;primary cell;primary element;voltaic battery
原电池产生的电 galvanism
原电池腐蚀 galvanic corrosion
原电池组 galvanic battery;primary battery
原电离 primary ionization
原电流 primary current
原电流调制 primary current modulation
原电路 original circuit;primary circuit;primary wire
原电压 primary voltage
原电子 primary electron
原垫底 original copy
原定到期时间 original expiration date
原定费率 original premium rate
原定或延长偿还期 original or extended maturity
原东德工业标准 German Industry Norms
原动的 motive
原动机 drive;driving machine;motive power;mover;primary mover;prime motor;prime mover;prime power;priming mover;self-propelled engine
原动机的调速器 prime-mover governor
原动机的循环 prime-mover cycle
原动力 dynamic(al);generative power;impetus;motive force;motive power;motivity;motor;moving

force; moving power; prime mover; prime power
原动力船 atomic-powered ship; atomic-propelled ship
原动质【化】 actor
原动轴 primary shaft
原断裂 primary fault(ing)
原堆 initial pile
原队过闸 barge train lockage
原队过闸船闸 barge train lock
原发票 original invoice
原发时间 time of origin
原发性污染 primary pollution
原发性污染物 primary pollutant
原发性污染源 primary pollution source
原钒酸 ortho-vanadic acid
原钒酸盐 orthovanadate
原方 <土或岩石开挖前的> bank yards
原方案 original design
原方解石 protocalcite
原方向 zero direction
原废负荷 raw waste load
原废水 crude foul water; crude refuse water; crude wastewater; fresh wastewater; raw effluent; raw wastewater
原废水灌溉 raw wastewater irrigation
原废物 crude refuse
原分生组织 promeristem
原封面 original cover
原封皮 original board; original cloth
原氟 protofluorine
原幅 opening width; original format
原辐板 primary radial
原辐射 primary radiation
原辐射器 primary radiator
原腐殖质 prohumic substance
原钢 raw steel
原高铅酸钙 calcium orthoplumbate
原稿 manuscript; original; original copy; original manuscript; original pattern
原稿架 copyholder
原稿像片 second copy original
原隔距长度 original ga(u)ge length
原铬酸 orthochromic acid
原给矿 original feed
原根 primitive root
原沟 primitive groove
原购入价 original purchase money
原古的 primeval
原鼓室 primary tympanic cavity
原固溶体 primary solid solution
原惯性力 primary inertia force
原光电流 primary photoelectric current
原光电效应 primary photoelectric effect
原光谱标准 primary spectroscopic standard
原光束 elementary beam
原光源 primary light source
原规划 initial plan(ning); original planning
原硅酸 ortho-silicic acid
原硅酸钡 barium orthosilicate
原硅酸钙 calcium orthosilicate
原硅酸镉 cadmium orthosilicate
原硅酸甲酯 methyl ortho-silicate
原硅酸钠 sodium orthosilicate
原硅酸四乙酯 tetraethyl orthosilicate
原硅酸盐 orthosilicate
原硅酸脂 orthosilicate
原海底深度 existing seabed depth; original seabed depth
原函数 object function; primary function; primitive function
原合同 original contract; prime contract
原合同期满 prime contract termination
原合同终止 prime contract termination
原河床 original bed; original river bed
原河流 donor stream
原迹线 original trace
原积土 residual soil
原级粒子 elementary granule
原籍 domicile; domicile of origin
原籍国 country of origin
原籍类别 membership class
原计划 original plan(ning)
原记分 raw score
原记述 protolog
原寄局 office of origin
原钾霞石 kalsilite
原价 cost of price; cost price; first cost; initial cost; original cost; original price; original value; prime cost
原价工具书 desk edition
原价购买 full-price offer
原价税 tax on cost
原件 master copy; original copy; original production; original unit
原件第一代 first generation
原建筑图 releve
原浆 protoplasm
原浆勾缝 high-joint pointing; jointing
原降落水头 raw fall head
原浇面 as-cast finish
原胶片 raw film
原胶原 procollagen
原焦 raw coke
原焦油 heavy tar; primary tar; rock tar
原焦油沥青 raw tar
原结 primitive knot; primitive node
原结点 original node
原结构 original structure
原金 native gold
原酒精 raw spirit
原居村落 indigenous village
原矩阵 original matrix
原开石 quarry-pitched stone; rough stone; rubble stone
原开石面 quarry face of stone; quarry-faced stone
原坑石料 run-of-pit
原空气 primary air
原口 blastopore
原矿 crude; green ore; mine run; original ore; rough ore; rude ore; run-of-mine; run-of-mine ore; unbeneficiated ore; unscreened ore
原矿仓 crude-ore bin; primary bin; run-of-mine bin
原矿储量 source of crude
原矿的 mine run
原矿块 ore bloom
原矿年产量 annual output of ores
原矿年产值 annual values of ores
原矿石 crude ore; green ore; pit-run ore; raw ore; run-of-mine type ore
原矿石配矿 natural ore ingredient
原矿试样 head sample
原矿体爆破 blasting-off the solid
原矿样 primary sample
原矿摇动筛 run-of-mine shaker
原矿总产量 total ores
原来尺寸 natural size; original size
原来处理 original treatment
原来的 aboriginal; undisturbed
原来号数 original number
原来流水号 original sequence number
原来强度 original strength
原来位置 home position; homing position
原来问题 primal problem
原来系统 primal system
原来形式 original pattern
原来形状 initial form; original form
原来预算 original budget
原来状态 original state
原棱镜 protoprism
原理 axiom; principle; rationale
原理草图 key diagram
原理电路 schematic circuit
原理(方)框图 functional block diagram
原理接线图 principle connection diagram
原理图 basic circuit; block diagram; diagrammatic layout; illustrative diagram; principle drawing; schematic circuit; schematic diagram; schematic drawing; simplified schematic; skeleton drawing; synoptic(al) diagram
原理性电路图 <标明所有电气设备> elementary diagram
原力矩 initial moment
原沥青 primary tar
原粒体 elementary body
原料 basic material; feed material; input material; parent material; producer's stock; raw material; raw stuff; rough material; staple; stock; stuff; feed stock <送入机器或加工厂的>
原料仓 feed bin; raw material bin
原料产地国 source nation
原料场 stock yard
原料车间 raw material shop
原料成本 material cost
原料成球 raw pelletizing
原料储备堆放 stockpile (re-) handling
原料储备管理 stockpile (re-) handling
原料储备输送 stockpile (re-) handling
原料储备移动 stockpile (re-) handling
原料储藏设备 raw material storing facility
原料储存堆放 stockpile (re-) handling
原料储存管理 stockpile (re-) handling
原料储存量 raw material deposit
原料储存输送 stockpile (re-) handling
原料储存移动 stockpile (re-) handling
原料储库 raw material storage
原料处理 material handling
原料传送带 staple conveyer [conveyor]
原料的 raw
原料的性质 raw material property
原料的转化 raw material transformation
原料堆(场) raw storage pile
原料粉磨 raw material grinding
原料粉磨控制 raw material grinding control
原料高聚物 raw polymer
原料罐 head tank
原料耗费 material consumption
原料烘干 raw material drying
原料回收坑道 stockpile tunnel; stock reclaiming tunnel
原料混合物 raw mixture
原料基地 raw material base
原料间 stock yard
原料胶 initial rubber
原料金属 feed metal
原料均化 homogenization of raw material
原料均化控制 control of raw material homogenizing
原料库 raw material storage; stock room
原料库房 raw material storage building
原料来源 source of feed
原料离心分析机 centrifuge
原料粒度分级 raw material sizing
原料流 feed stream
原料磨 raw mill
原料配料计算机控制 computer control of raw material mixing
原料棚 raw material storage shed
原料皮 raw hide
原料气 feed gas; raw gas
原料气脱酸装置 acid gas removal unit
原料气预处理 feed gas pretreatment
原料气体 unstripped gas
原料入口 feed inlet; raw material inlet
原料生产国 primary producing country
原料石棉 crude asbestos
原料输出国 primary exporting country
原料危机 raw material crises
原料锡 raw tin
原料细磨 finish raw grinding
原料细磨机 finish raw mill
原料消耗 raw material consumption
原料油 raw oil
原料预热器 feed preheater
原料预热室 dog house
原料原码 true code; true form
原料制备 raw material preparation
原料制备车间 raw department
原料周转 material flow
原料准备 raw material preparation
原料着色 mass colo(u)ring
原料组分 raw constituent
原硫酸 ortho-sulfuric acid
原硫酸盐 ortho-sulfate
原铝酸 ortho-aluminic acid
原铝酸盐 ortho-aluminate
原绿色 primitive green
原氯苯胺 orthochloroaniline
原氯苯酚 orthochlorophenol
原码 true code
原码加 true add
原码形式 true form
原煤 altogether coal; brat; cleck coal; coal as mined; mine (run) coal; raw coal; rough coal; run-of-mine coal
原煤仓 raw coal silo
原煤给煤机 raw-fuel feeder
原煤合计 raw coal amounts to
原煤灰分产率 ash yield of raw coal
原煤挥发分产率 volatile matter yield of raw coal
原煤进口 raw coal inlet
原煤气 crude gas; rough gas
原煤水分 raw coal moisture
原煤脱硫 raw coal desulfurization
原煤油 primary oil
原蒙脱石 protomontronite
原密度范围 original density range
原棉 bulk wool; cotton wool; loose wool; raw cotton; raw wool; rude cotton; seed cotton
原棉毡 unbonded felt; unbonded mat
原瞄准线 original line of sight
原模 grand master pattern; master form; master mo(u)ld
原模图 protograph
原墨 base ink
原母色 orthochrome
原木 log; new wood; raw log; root timber; round timber; timber; un-

dressed lumber; undressed timber; unmanufactured wood; unsawn timber; roundwood
原木搬运设备 logging equipment
原木搬运装置 logging equipment
原木板英尺材积表 log rule
原木变色 log stain
原木标记漆 log marking paint
原木膘皮 slab
原木表面干燥变硬 <内层没有干燥> case-harden(ing)
原木材积 log volume
原木材积表 log rule; log scale; log volume table
原木材积记录 log tally
原木采运 shortwood logging
原木测杆 <有板英尺刻度的> log rule
原木测杖 log rule
原木场 log storage
原木车 logging car
原木池 log pond
原木处理 log treatment
原木打捞机 catamaran
原木大头 stub end
原木带锯 log band sawing machine
原木导堤 log training wall
原木导墙 <护岸用> log training wall
原木等级 log class; log grade
原木等级曲线 log grading curve
原木底侧剥皮 <便于滑行> Rossing
原木堵塞 log jam
原木堵塞水路 logjam
原木堆场 log yard
原木垛 log pile
原木盖的桑拿浴室小屋 log walled sauna but
原木刮路器 log drag
原木号印 log brand; log mark
原木护岸 log revetment
原木滑道 log chute
原木集运设备 logging equipment
原木集运装置 logging equipment
原木计数 log tally
原木夹叉 log grapple
原木铗具 logging scissors
原木检尺 scaling of logs; log scale; log scaling
原木检量尺 mill scale
原木结构 log construction
原木径截 quarter-sawn conversion (of log)
原木锯 head saw; log saw
原木锯制量 log run; run-of-the log
原木库存量 log survey
原木框锯 log frame saw
原木蓝变 log blue
原木楞台 log deck
原木溜槽 log chute
原木流放渠 log sluice
原木流送槽 log flume
原木木头 bottom end
原木平行下锯法 through-and-through sawing
原木起运装置 log haul-up
原木汽蒸锅 log digester
原木去膘 slabbing
原木生境 log habitat
原木四分下锯 quarter sawing; quarter-sawn
原木四开的 quarter-sawed
原木四开锯法 <使年轮与板面交叉大于 45 度> slash-grained; radial cut; rift-grained
原木素 protolignin
原木提升机 log elevator
原木提升设备 log hoisting apparatus
原木推送机 log kicker
原木拖曳架 logging frame
原木拖曳阻力 log skidder resistance
原木拖运臂 logging arm
原木销 tip end
原木小端 tip end
原木形式 in the log
原木堰 log weir
原木圆锯 log circular saw
原木运输车 log truck
原木运输车厢 log body
原木运输船 log-ship
原木运转车 log carriage
原木止水 log seal
原木制材量 log run
原木制品 rustic woodwork
原木中心板 Wainscot plank
原木抓取器 wood grabber
原木抓扬机 log grapple
原木装车机 logger
原木装饰品 rustic woodwork
原木装卸滑道 skidway
原木装载机 log loader
原能 proper energy
原浓度 original concentration
原排放空气污染物 primary emission
原配件 genuine part
原硼砂 tincal
原硼酸 ortho-boric acid
原硼酸盐 ortho-borate
原坯 green compact
原皮质 dermatosomen
原片 original film
原片密度 original film density
原坡高 initial gradient
原桥模型 prototype bridge model
原切模数 initial tangent modulus
原秦岭古陆 proto-Qinling old land
原曲线 parent curve
原燃料 crude fuel
原燃料电池 primary fuel cell
原绕组 primary winding
原绕组电压 primary winding voltage
原溶液 original solution
原色 colo(u)r prime white; elementary colo(u)rs; matrix [复 matrixes/matrices]; primary colo(u)r; primitive colo(u)r
原色版 original block
原色哔叽 Beige
原色边材 bright sapwood
原色的 non-colo(u)r
原色调 self-tone
原色调涂料 full-tone coating
原色激励方式 primary colo(u)r signal drive
原色滤色镜 primary colo(u)r filter; primary filter
原色母 orthochrome
原色三角形 colo(u)r(ing) triangle (original)
原色素 primary pigment; protochrome
原色颜料的混合方法 mixing of primary pigment colo(u)rs
原色颜料的混合过程 mixing of primary pigment colo(u)rs
原色重氮盐蓝 variamine blue salt B
原纱染色的 ingrain
原砂 bank-run sand; crude sand; dead sand; raw sand; roughing sand
原设备制造厂家 original equipment manufacturer
原设计 initial design; original design
原设计能力 designing capacity
原设计水位 designing water level
原射线 primary ray
原射线束 primary beam
原砷酸 ortho-arsenic acid
原砷酸银 silver orthoarsenate
原生 protogenesis
原生白云岩 primary dolomite
原生半丝质体 primary semifusinite
原生包裹体 primary inclusion
原生变应素 primary allergen
原生冰铜 primary matte
原生材 virgin growth
原生层理【地】 direct stratification; primary stratification; original bedding
原生沉积【地】 conjunction deposit; connate deposit; primary deposit
原生沉积带 gone of primary deposit; zone of primary deposit
原生沉积构造 directional structure; primary sedimentary structure
原生代【地】 Proterozoic (era)
原生带 original zone; primary zone
原生的 genetic(al); primordial; protogen(et)ic; sedentary
原生地槽 primary geosyncline
原生地层圈闭 primary stratigraphic trap
原生地沥青 original asphalt
原生地幔 juvenile mantle
原生地下水 native groundwater; juvenile groundwater
原生动物 protozoa
原生动物杀虫剂 protozoan insecticide
原生冻原 primary tundra
原生放射性元素 primordial radioelement
原生分散 primary dispersion
原生分散模式 primary dispersion pattern
原生分散晕 primary dispersion halo
原生腐殖酸 primary humic acid
原生各向异性 initial anisotropy
原生构造【地】 primary structure
原生构造类型和性质【地】 type and property of primary structure
原生构造岩 primary tectonite
原生谷 original valley
原生固溶体 primary solid solution
原生海岸 primary coast
原生含金硫化物 primary gold-bearing sulfide
原生河(流) original river; original stream
原生化石 primary fossil
原生化学沉积 orthochem
原生环境 original environment; primary environment; primitive environment
原生环境保护区 wilderness area
原生环境水文地质问题 hydrogeologic(al) problem of primary environment
原生黄土 primary loess
原生灰分 inherent ash
原生活废水 raw sanitary waste
原生活污水 raw sanitary sewage; raw sanitary wastewater
原生间隙 original interstice; primary interstice
原生胶体物质 primary colloidal material
原生节理 original joint; primary joint
原生结构面 primary discontinuity
原生解理【地】 protoclase
原生金属 primary metal; virgin metal
原生空隙 original interstice
原生孔洞 primary opening
原生孔隙【地】 primary opening; original pore; primary pore
原生孔隙度 primary porosity
原生矿 primary ore
原生矿床 primary deposit
原生矿带 primary ore zone
原生矿泥 primary slime
原生矿石 primary ore
原生矿物 genetic(al) mineral; original mineral; primary mineral; sedentary mineral
原生矿物质 primary mineral mater
原生沥青 coarse asphalt; raw pitch
原生裂隙 primary fracture; primary joint
原生裂隙水 original fissure water
原生林 original forest; primary forest; primeval forest; primitive forest; virgin forest
原生流动构造 primary flow structure
原生铝 primary alumin(i)um
原生糜棱岩【地】 protomylonite
原生面滑坡 original plane landslide
原生木素 protolignin
原生木质部 protoxylem
原生黏[粘]土 primary clay
原生劈理【地】 original cleavage
原生片麻理 primary gneissosity
原生片麻岩 granite-gneiss; primary gneiss; protogene gneiss
原生片麻状条带 primary gneissic banding
原生偏析 primary segregation
原生平缓节理 L joint
原生气顶气 original gas cap
原生铅 primary lead; primordial lead
原生倾斜 original dip
原生曲流 primary meander
原生群落 primary community
原生砂 virgin sand
原生珊瑚体 founder polyp
原生渗流洞 primary seepage cave
原生渗透性 primary permeability
原生生物 protist
原生生物界 protista
原生剩余磁化 primary remanent magnetization
原生剩余磁化强度 primary remanent magnetization strength
原生石油 protopetroleum
原生石油沥青 crude asphalt
原生树标 mark with trees
原生双晶 primary twin
原生水 conjunction water; connate water; fossil ground water; fossil water; intratelluric water; juvenile water; primary water; primitive water
原生水平层平原 plain of formation
原生水压 primary pressure
原生顺向湖 original consequent lake
原生丝质体 primary fusinite
原生塑料 virgin plastics
原生碎屑构造 primary structure; protoclastic structure
原生碎屑结构 protoclastic texture
原生他形【地】 autallotriomorphic
原生天然火山灰 raw natural pozzolan
原生铁 base iron
原生同位素 parent isotope; primeval isotope
原生土层 azonal soil
原生土地 aboriginal land
原生土结构 primary soil structure
原生土(壤) genetic(al) soil; indigenous soil; natural soil; original soil; primary soil; sedentary soil; soil in-situ; undisturbed ground; virgin soil; autochthonous soil; parent soil
原生土砖 primitive brick
原生围岩 original host rock
原生污泥 raw sludge
原生污染 primary pollution
原生污染物 primary contaminant; primary pollutant
原生污染影响 primary pollution effect
原生污染源 primary pollution source
原生污水 primary effluent

原生污水过滤 primary effluent filtration
原生污着膜 primary fouling film
原生物 protobiont
原生显微组分 original maceral
原生型 prototype
原生玄武岩 primary basalt
原生岩浆 primary magma
原生岩类 primary rocks;primitive rocks
原生岩(石) Archean rock; rock in place; virgin rock; original rock; primitive rock; native rock; parent material; parent rock; primary rock; protogenic rock; protogine; source bed;source rock
原生盐土 primary saline soil
原生盐渍化 primary salinization;primary salinized soil
原生演替 primary succession
原生演替系列 primary sere
原生要素 primary element
原生叶理 primary foliation
原生异常 primary anomaly
原生铀矿 primary uranium ore
原生有机质 primary organic matter
原生晕 primary halo
原生造山作用 primary orogeny
原生植被 native vegetation; original vegetation;primary vegetation
原生植物 protophyte
原生质【生】 sarcode;biomolecule;bioplasm; plasma; plasmogen; protoplasm
原生质素 bioplasmin
原生质体 protoplast
原生自由面【地】 primary free face (of rock surface)
原生棕壤 primary brown earth
原声录音带 master
原石 raw stone;run-of-pit-quarry
原石灰 run-of-kiln lime
原石库 raw stone store
原石蜡 protoparaffin
原石水分 quarry damp
原石英岩 protoquartzite; quartzose subgraywacke
原石油 petroleum crude
原石油质 protobitumen
原史时代 protohistoric age
原史学 protohistory
原始爱奥尼克柱头 proto-Ionic capital
原始巴洛克建筑 proto-Baroque architecture
原始饱和度 initial saturation
原始饱和压力 primary saturation pressure
原始被子植物 protangiospermae
原始编录 basic documentation
原始变态式建筑 proto-Baroque architecture
原始变种 primitive variety
原始标记 original marking
原始标准 original standard;primary standard
原始标准海水 original standard seawater;primary standard sea-water
原始表 original table;primary table
原始表列 primary tabulation
原始波 original wave;primary wave
原始波群 original wave group
原始玻璃 primitive glass
原始玻璃(状)体 primary vitreous
原始部分 initial portion
原始材料 original material; raw file; source material
原始参考标高 original levels of reference
原始操作 primitive operation
原始侧 primary side
原始测井曲线 raw log
原始层理【地】 primary stratification; direct stratification
原始产品 initial production
原始产物 primary product
原始沉积构造 primary sedimentary structure
原始衬度 primary contrast
原始成本 aboriginal cost; historic(al) cost;initial cost; initial expenses; initial investment; old cost; original cost;prime cost;first cost <未计利息>
原始成本定率折旧法 depreciation-percentage of original cost method
原始成本法 original cost method
原始成分 primitive component
原始承办人 originator
原始承销人 originator
原始程序 original program(me); source module
原始程序卡片组 source deck
原始吃水 original draught
原始尺寸 original dimension;original size
原始齿形 standard basic rack tooth profile
原始冲量 primary signal
原始抽样 initial sample
原始储量 primary reserves
原始瓷器 proto-porcelain
原始磁带 grandfather magnetic tape
原始粗觉 protopathic sensitivity
原始大气 protoatmosphere
原始大洋 proto-ocean
原始带 original tape
原始单据 basic document; original documents;source documents
原始氦 primordial helium
原始的 aboriginal; base line; coarse; original; primeval; primordial; pristine; protopathetic; raw; underived; unwrought; virgin; first hand
原始登记日 date of original register
原始地表 original ground surface
原始地层 primary formation; prime stratum
原始地层学 prostratigraphy
原始地层压力 initial formation pressure; original formation pressure; virgin formation pressure
原始地面 original ground; original ground surface;unmade ground
原始地面标高 original ground level
原始地面高程 original ground level
原始地区 original lot;primitive area; primitive region
原始地图 primitive map
原始地形 initial (land) form;original (land) form
原始地形模型 initial relief model
原始地质编录 initial geologic(al) logging
原始递归 primitive recursion
原始递归函数 primitive recursive function
原始递归性 primitive recursiveness
原始递归余数函数 primitive recursive remainder function
原始电动机 primary motor
原始调查 primary investigation
原始洞穴 primary excavation
原始读数 primary reading
原始堆积孔隙 constructional-void porosity
原始对偶算法 primal algorithm
原始多项式 primitive polynomial
原始发报人 originator
原始反射 primary reflection
原始反射光 primary reflection light
原始方案 original scheme;original version
原始方程 original equation;primitive equation
原始方程模式 primitive equation model
原始方位 reference azimuth
原始放射性同位素 primary radioisotope
原始飞机 prototype aircraft
原始费用 baseline cost;original cost
原始分解 primary decomposition
原始分类 original classification
原始分录 original entry
原始分配 original distribution
原始分子 primary molecule
原始粉末 starting powder
原始缝隙 original interstice
原始符号 original symbol
原始辐角 primary amplitude
原始辐射率 raw radiance
原始干缩(率) initial dry(ing) shrinkage
原始感觉 sentience
原始刚度矩阵 original stiffness matrix
原始高程 original elevation;unchecked spot elevation
原始高程数据 manuscript level data
原始格子 primitive lattice
原始各向等压固结曲线 virgin isotropic consolidation curve;virgin isotropic consolidation line
原始根源 primary source
原始耕作 infant farming
原始构造形式 baseline configuration
原始固结 primary consolidation
原始故障 primary fault(ing)
原始观测方程 original observation equation
原始光源 primary light source
原始光栅 original grating;original grid
原始轨道 original orbit
原始过程 original procedure
原始海岸剖面 initial coastal profile
原始海洋 primitive ocean
原始海洋纪 oceanic period
原始海洋时代 oceanic era
原始含气饱和度 initial gas saturation
原始含水饱和度 initial water saturation
原始含水量 initial moisture (content); original water content
原始含油饱和度 initial oil saturation
原始合金 virgin alloy
原始合作 proto-cooperation
原始核 pronucleus
原始横断面坡度 initial cross-sectional gradient
原始横隔【医】 septum transversum
原始宏观模型 prototype macro-model
原始化合物 parent compound
原始环境 primal environment;primitive environment
原始荒漠景观 primitive desert landscape
原始黄土 original loess
原始混合料 base mix
原始混合物 base mix;original stock
原始火山的 eovolcanic
原始机械装置 original machine unit
原始积累 original accumulation; primary accumulation;primitive accumulation
原始基准点 original point
原始基准线 original line
原始计算机 <为其他计算机输入的一种计算机> source computer; primitive computer
原始记录 first record; historic(al) records;home record;mastering recording; original entry; original record(ing);primary accounting;primary record; protocol; source recording
原始记录账簿 book of original entry
原始记录制作 source recording
原始技术条件 original specification
原始加速度记录图 original accelerogram
原始价值 aboriginal cost;aboriginal value;initial value;original value
原始假设 initial hypothesis
原始剪冲断层 initial shear thrust
原始件 original
原始建筑 pioneer architecture
原始交通量 raw traffic capacity
原始阶段 preliminary stage
原始接收 primary reception
原始结构 primary construction; primary formation
原始结晶年龄 original crystallization age
原始截面面积 original cross-sectional area
原始解 primitive solution
原始进料量 initial input
原始晶格 primitive lattice
原始晶体 parent crystal
原始净费率 original net rate
原始距离 initial range
原始聚合物 base polymer;raw polymer
原始卡片组 source deck
原始抗滑力 initial skid resistance
原始孔隙比 initial void ratio;original void ratio
原始孔隙度 in-situ porosity;primary porosity
原始孔隙率 in-situ porosity
原始孔隙压力 original pore pressure
原始矿藏 primary ore deposit
原始矿浆流 primary pulp stream
原始垃圾 raw refuse
原始蜡盘 wax master;wax original
原始类型 initial form
原始立方 primitive cube
原始粒子 direct particle; initiating particle;primary particle
原始联络矩阵 primitive connection matrix
原始裂片 primary fission fragment
原始裂纹 <混凝土骨料周围的> initial flaw
原始林 old growth; original forest; primitive wood; virgin growth; virgin timber;virgin wood;wild wood
原始林区 primitive area;virgin area
原始流 primary flow
原始流程表 primitive flow table
原始流与回流 primary flow-and-return
原始陆块 initial landmass
原始陆生动物 primitive land animal
原始陆体 initial landmass
原始论文 original paper;primary paper
原始脉冲 original pulse
原始美 pristine beauty
原始密度 field density;original density
原始描述 original description
原始命令 original directive
原始模型 archetype; master mo(u)ld; master pattern; model mo(u)ld;original model;prototype model
原始母板 original mother plate
原始目标函数 primal objective function
原始内聚力 origin cohesion
原始能 prime energy

Y

原始能量 primary energy
原始黏[粘]度 initial viscosity
原始黏[粘]聚力 origin cohesion
原始凝聚力 original cohesion
原始农业 infant farming
原始浓度 primary concentration
原始频率标准 primary frequency standard
原始凭单 original voucher
原始凭证 original evidence; original voucher; source documents; underlying documents
原始凭证汇总表 summary of original vouchers
原始坡度 depositional gradient
原始谱 original spectrum
原始气动力特性 primary aerodynamic characteristic
原始迁移 primary migration
原始铅 primeval lead
原始强度 pristine strength; virgin strength
原始切线模量 initial tangent modulus
原始倾斜 initial dip; original dip; primary dip
原始倾斜岩层 initial inclined bed
原始情绪 protopathetic emotion
原始区 natural area; region of initiation
原始曲线 primary curve; primitive curve; virgin curve
原始取得 original acquisition
原始群体 initial population
原始任务 ancestral task
原始日志 raw log
原始森林 first growth; indigenous forest; primary forest; primeval forest; primitive forest; timber forest; virgin forest; wild wood
原始森林大部分已伐除的土地 cut-over land
原始森林气候 climate of Taiga
原始上覆压力 virginal overburden pressure
原始设计 original design; preliminary design; primary design
原始射线 initial ray
原始射影 primitive projection
原始生态系统 primordial ecosystem
原始生物 primitive organism
原始生物的 eobiontic
原始生物化学 protobiochemistry
原始生物期 initial life-stage
原始湿度 virginal humidity
原始石器 eolith
原始石器时代 Eolithic Age
原始时代 rude times
原始试样 primary sample
原始收受站 crude terminal
原始收缩裂缝 initial contraction crack
原始束 original bunch
原始束流 primary beam
原始数据 base data; basic data; crude data; historic (al) data; incoming data; initial condition; initial (izing) data; input data; manuscript data; original data; preliminary data; primary data; primary information; raw data; raw digital data; raw information; raw manuscript data; zero initial data
原始数据报表 preliminary data report
原始数据存储器 raw storage
原始数据格式 raw data form
原始数据矩阵 primary data matrix
原始数据来源 original data source
原始数据图 raw data map
原始数据组 grandfather
原始数值 raw value
原始水 primary water; primitive water
原始水灰比 initial water ratio; original water-cement ratio
原始水平 original horizontality
原始水平地面 original level ground surface
原始水生植物 primary aquatic plant
原始水文资料 basic hydrologic (al) data
原始水线 original water line
原始水准点 original bench mark
原始速度记录 original velocity recording
原始弹性极限 primitive elastic limit
原始陶瓷器 primitive pottery
原始陶立克柱式 proto-Doric
原始陶立克(柱)型 proto-Doric column
原始梯度 constructional gradient
原始体积 initial volume
原始条件 initial condition; original condition; rest condition
原始条件方程 original condition equation
原始统计 primary statistics
原始投资 original capital cost
原始图件 original map
原始土 raw soil
原始土层厚度 initial thickness of soil stratum; original thickness of soil stratum
原始土路 primitive road
原始土壤 initial soil; original soil
原始微观模型 prototype micro model
原始维数 primary dimensionality
原始位置 home position; original position; reference position
原始温度 original temperature
原始温度场 original temperature field
原始文件 original documents; primary documents; primary file; source documents; source file
原始文献 original documents; source documents
原始文献出版物 primary publication
原始文艺复兴时期建筑 proto-Renaissance architecture
原始文章传送 original article delivery
原始文字材料 original literal material
原始污泥 primary sludge
原始污染 primary pollution
原始污染源 primary pollution source
原始物料 initial materials; staring materials
原始物体 primary body
原始误差 initial error; original error
原始纤维 pristine fiber
原始现场 original scene
原始项 primitive term
原始信号 original signal; primary signal
原始信息 original information; primary information; raw information
原始信用状 master credit; prime credit
原始形成层 procambium [复 procambia]
原始形状 initial form
原始性比率 primary sex ratio
原始性驱力 primary drive
原始性质 primitive character
原始雪 wild snow
原始压力 reset pressure
原始压力下原油体积系数 volume factor of crude at the primary pressure
原始压密 virgin compaction; virgin consolidation
原始压缩 virgin compression
原始压缩曲线 virgin compression curve
原始压缩线 virgin compression line
原始氩 primordial argon
原始岩 primitive rock
原始岩代 lithic era
原始岩浆 original magma; primary magma; protomagma
原始岩类 primitive rocks
原始样本 early sample; initial sample; intact sample
原始样号 sample original number
原始样品 early sample; initial sample; intact sample
原始一批 original lot
原始异地生成煤 primary allochthonous coal
原始因子 primitive factor
原始应力 virgin stress
原始应力状态 virgin state of stress
原始营地 <无现代设备的> primitive campground
原始营养细菌 prototrophic bacteria
原始营养源 primary sources of nutrient
原始影响 original influence
原始硬材 primary hardwoods
原始硬度 initial hardness
原始油层平均压力 original average reservoir pressure
原始油层压力 initial reservoir pressure
原始油气比 primary gas-oil ratio
原始有机碳 primary carbon
原始有机质 primary organic matter
原始宇宙丰度 primordial cosmic (al) abundance
原始域元素 primitive field element
原始责任 original responsibility
原始账簿 book of original entry
原始振荡器 primary oscillator
原始证件 original certificate
原始证据 original evidence
原始直径 green diameter
原始值 original value
原始植被 primeval vegetation
原始植物 primordial plant
原始指令 presumptive instruction
原始中心 archicenter; primary center [centre]
原始种 initial species; original species
原始重量 initial weight; original weight
原始周期 primitive period
原始周期平行四边形 primitive period parallelogram
原始主盘 original master
原始主应力 principal virgin stress
原始铸锭 starting ingot
原始状态 aboriginality; initial state; nature; original condition; raw state; reset condition
原始状态土地 raw land
原始资本 seed capital
原始资金 seed money
原始资料 base data; base line; basic data; crude data; first-hand data; first-hand information; ground material; initial data; original data; original information; original material; primary information; raw data; raw information; source book; source documents; source information; source material; starting material; primary data
原始资料收集 original data collection
原始自然环境保护区 wilderness preservation area
原始总收益率 original gross rate
原始纵坐标 initial ordinate
原始组分 original composition
原始钻孔记录表 preliminary borehole log
原始坐标 origin coordinate
原试样 true sample
原收款单位 original payee
原收款人 original payee
原收原交 discharged as loaded
原树脂 primary resin; unmodified resin
原水 crude water; natural water; raw water; untreated water <未经处理过的水>
原水泵 raw water pump
原水泵站 raw water pumping station
原水供水 raw water supply
原水供应 raw water supply
原水进水管 raw water inlet
原水来水管 raw water influent
原水氯化 raw water chlorination
原水氯消毒 raw water chlorination
原水平 present level
原水平层理 original horizontal stratification
原水水网 raw water network
原水水源 raw water source
原水水质 quality of raw water; raw water quality
原水头 raw head
原水箱 raw water tank
原水性质 quality of raw water
原水样本 raw water sample
原水贮存池 raw water storage basin
原水浊度 raw water turbidity
原丝筒处理 conditioning of cake
原丝筒回潮 conditioning of cake
原丝筒调节 conditioning of cake
原松浆油 crude tall oil
原速调节 <从定子侧调速> primary speed control
原速调整 primary speed control
原宿营地 primitive campground
原酸 ortho acid; raw acid
原酸澄清槽 raw acid settling tank
原酸化 ortho acid
原酸酯 orthoester
原太阳 protosun
原态河道水流 free flowing streamflow
原态河流 <未受人工措施影响的> free flowing stream; free flowing river
原钛酸 titanic hydroxide
原糖 crude sugar; raw sugar
原糖液 crude sugar solution
原体 elementary body; protomer; prototype; subunit
原体爆破 primary blasting
原体测量 prototype measurement
原体观测 prototype measurement
原体节 primary segment; primitive body segment
原体试验 prototype test (ing)
原体试验数据 prototype data
原条 pole with top; prestreak; primitive streak; stemwood; timber stripe
原条装载机 tree length log loader
原铁 native iron
原铁水 base iron
原投资额 original investment
原图 artwork; base design; basic design; master manuscript; master map; master sheet; original; original artwork; original chart; original drawing; original map; original pattern; original picture; original plan; original plot; parent map; primitive map; protracting map

原图板 copy board
原图编辑 original map editing
原图拼贴 mount a map; mount the copies
原图系统 artwork system

原图显示 manuscript display
原图信息带 artwork tape
原图纸 parent map
原土 original soil saw soil;raw soil
原土地基 raw subgrade
原土夯实 original soil rammed
原土夯实地基 original soil rammed foundation
原土回填 cut-and-cover
原土开挖 primary excavation
原土路基 original subgrade;raw subgrade
原土石层开挖 primary excavation
原土挖掘 primary excavation
原土状态 bank state
原顽火辉石 <人造的> protoenstatite
原维管束 provascular strand
原位 field; home position; in-situ; original position
原位安装 on-site installation
原位测量 in-situ measurement
原位测试 in-situ measurement;in-situ measuring;in-situ test(ing)
原位承载试验 field loading test
原位抽水渗透性试验 field pumping test;pumping test in-situ
原位处理 in-situ treatment
原位单轴抗压试验 in-situ uniaxial compressive test
原位地层图 dip-corrected map
原位定量 in-situ quantitation
原位动力工程性质 dynamic(al) engineering property in-situ
原位动力压实法 in-situ dynamic(al) compaction procedure
原位发动机 engine in situ
原位分析 in-situ analysis
原位复合材料 in-situ composite
原位固结时间 field consolidation time
原位贯击数校正 correction of field blow count
原位含水量 in-situ water content
原位化学处理 in-situ chemical treatment
原位击实试验 Proctor compaction test
原位激振试验 in-situ impulse test
原位计量工程数量 in-place quantity measurement
原位加固土桩 in-situ stabilized column
原位加州承载比试验 in-situ California bearing ratio test;in-situ CBR test
原位监测 in-situ monitor(ing)
原位键 home key
原位孔隙比 field void ratio;in-situ void ratio
原位密度 field density;in-place density;in-situ density
原位模塑 in-place mo(u)lding
原位强度 field strength;in-situ strength
原位容重 field unit weight;in-situ unit weight
原位上覆压力 in-situ overburden pressure
原位渗入 field infiltration
原位渗透系数 field coefficient of permeability
原位渗透性试验 field permeability testing;in-situ permeability test
原位生物处理 in-situ biological treatment
原位生物膜 in-situ biofilm
原位生物修复 in-situ bioremediation
原位实时观察法 real-time and in-situ observation
原位示踪剂 in-situ tracer
原位试验 field test(ing);in-situ test(ing)
原位试验土 in situ testing soil
原位水当量 field moisture equivalent
原位土工试验 field soil test;in-situ soil test
原位土密度 density of soil in place;filed density
原位土(壤) in-situ soil;field soil;soil in-situ
原位推裂试验 in-situ thrust test
原位推压试验 in-place push test
原位挖方数量 bank measure
原位挖方体积 bank cubic yard
原位压实力 field compactive effort
原位应力 in-situ stress
原位有效上覆压力 in-situ effective overburden pressure
原位有效压力 field effective pressure;in-situ effective pressure
原位直(接)剪(切)试验 in-situ direct shear test
原位置 in-situ
原文 master sheet;original reading;original text
原稳定流量 initial steady discharge
原稳定水位 initial steady level
原窝 primitive pit
原污泥 crude sludge; fresh sludge; primary sludge
原污染物 parent pollutant
原污水 crude refuse water;crude foul water;crude sewage;raw effluent;raw sewage;raw wastewater
原污水泵 crude sewage pump
原污水分析 raw sewage analysis
原污水过滤(法) raw-sewage screening
原污水浓度 raw sewage concentration
原污水排放 raw sewage discharge
原污水道污泥 raw sewage sludge
原物 original;original object
原物尺寸 life size;natural size
原吸收量 initial absorption
原下水道污泥 primary sewage sludge
原先吨位 builder's tonnage
原纤化作用 fibrillation
原纤维 bare glass fiber; basic fiber [fibre];protofibre
原显微形态 preexisting micro-feature
原线 elementary line; original line; primitive streak
原线圈 primary coil;primary winding
原线圈线头 out-primary
原线圈阻抗 primary impedance
原相图 phase diagram
原像 inverse image;original image;primary image
原像负片 original picture negative
原橡胶绝缘电缆 cabtire cable;cabtyre
原硝酸 orthonitric acid
原形 primary form
原形光栅 original grid
原形质【生】 plasma
原型 antetype;original;original form;original mo(u)ld
原型变量 prototype variable
原型操作系统 prototyping operating system
原型测量 prototype measurement
原型车 <最初研制出来的车辆> prototype car;prototype vehicle
原型成本限额 prototype cost limits
原型尺寸 original size
原型处理 prototype treatment
原型单元 prototype unit
原型的 archetype; full size; prototype;original;full-scale
原型电机 prototype machine
原型堆芯 prototype core
原型反应堆 prototype reactor
原型观测 field observation;prototype measurement; prototype observation
原型光栅 original grating
原型荷载试验 full-scale load test
原型化方法 prototype approach
原型环境浮标 prototype environment(al) buoy
原型机车 prototype locomotive
原型机研制 prototype development
原型机研制阶段 prototype development phase
原型机制造 prototyping
原型基础试验 prototype foundation test
原型基金 prototype carbon fund
原型级配 prototype gradation
原型集装箱 prototype container
原型监测 prototype monitoring
原型阶段 prototype stage
原型结构 full-scale structure; prototype structure
原型快速反应堆 prototype fast reactor
原型量测 full-scale measurement
原型列车 preproduction train; prototype train
原型滤波器 prototype filter
原型模式 prototype pattern
原型模型 full-scale model
原型配位体 prototypic(al) ligand
原型汽轮机 prototype turbine
原型桥 proto-bridge;prototype bridge
原型桥试验 test on prototype bridge
原型侵入岩 archaeomorphic rock
原型软件 prototype software
原型设备 prototype equipment
原型试验 full-scale test;prototype experiment;prototype test(ing)
原型试验研究 full-scale investigation
原型试验装置 prototype installation
原型数据 prototype data
原型水轮机 actual turbine;full-scale turbine;prototype turbine
原型图 prototype drawing
原型系统 prototype system
原型验证 prototype verification
原型样机 prototype
原型样品 prototype hardware
原型液体 prototype liquid
原型运输机 prototype transport
原型载荷试验 prototype load testing
原型装置 prototype plant
原型资料 prototype data; prototype information
原型自动机床 original machine
原序 first-order
原亚砷酸 ortho-arsenous acid
原亚砷酸钠 sodium orthoarsenite
原亚砷酸铜 copper orthoarsenite
原亚砷酸盐 ortho-arsenite
原亚砷酸银 silver orthoarsenite
原亚锑酸 ortho-antimonous acid
原亚锑酸盐 ortho-antimonite
原烟层 smoke pall
原岩 country rock; primitive rock; rock in place
原岩层 primitive rock stratum
原岩建造 primary rock formation
原岩浆 original magma
原岩年龄 protolith ages
原岩体 in-situ rock mass
原岩温度 original rock temperature
原岩应力 in-situ stress; original rock stress;stress of primary rock
原盐 crude salt
原盐水 crude brine
原盐效应 primary salt effect
原养生物 prototroph
原养型细菌 prototrophic bacteria
原氧钒石 protodoloresite
原氧化物 primary oxide
原样 original sample;original shape
原样的 undisturbed
原样实验 full-scale experiment
原样未动的 intact
原样重量 original sample weight
原野 champaign;champion;rough terrain
原野的 moory
原页岩油 crude shale oil
原液 raw liquor;stock solution
原液柜 <泡沫灭火剂> concentrate tank
原液染色 dope dyeing; solution dyeing
原液体 primary liquid
原液着色 mass colo(u)ring
原乙酸 ortho-acetic acid
原乙酸乙酯 ethyl orthacetate
原因 cause reason;sake
原因不明的 agnogenic
原因不明的动作 doubtful operation
原因不明的故障 passive defect
原因不明时间 debatable time
原因调查 causal investigation
原因论 aetiology;etiology
原饮用水 raw drinking water
原印海图 base chart
原应力 initial stress;original stress
原营养型微生物 prototroph
原硬度 natural hardness
原永冻层 passive permafrost
原用名 protonym
原油 base oil;crude oil;crude petroleum; earth oil; heavy oil; mother oil; naphthalene base; petroleum crude;primary oil;protopetroleum;raw oil;raw petroleum
原油拔顶气 tops from crude distillation
原油泵 crude oil pump; petroleum pump
原油泵房 crude oil pumping station
原油比重计 hydrometer
原油(薄)膜 crude oil film
原油产量 crude production
原油产率 oil production rate
原油产品 crude oil products
原油初馏 primary distillation of crude-oil
原油储藏 crude storage
原油储存能力 crude storage capacity
原油储罐 crude oil storage tank
原油船 crude oil carrier
原油船舶输送 shipment of crude
原油的含氢指数 hydrogen index of oil
原油的加工 processing of crude oil
原油的炼油装置 crude oil processing plant
原油的密度 oil density
原油的轻馏分 light fraction of oil
原油的相对渗透率 relative permeability to oil
原油的有效渗透率 effective permeability to oil
原油的重馏分 heavy fraction of oil
原油发动机 crude oil engine
原油分析 crude analysis;oil analysis
原油高压物性试验设备 petroleum voltage test equipment
原油管线 petroleum pipeline
原油含水量 moisture content in crude oil
原油含盐量 salt content of crude oil
原油计量站 metering station
原油加工 crude processing
原油加热器 raw juice heater
原油鉴定法 crude assay

原油接收站 crude tar
原油库容 crude storage capacity
原油裂化 crude oil cracking
原油裂解 crude oil cracking
原油码头 crude oil terminal
原油汽车 crude oil automobile
原油输送 crude oil transportation
原油输送管线 crude line
原油输油管 crude pipe line
原油体积换算系数 volume conversion factor of crude oil
原油脱硫 oil desulfurization
原油脱水器 oil dehydrator
原油脱盐 desalting of crude oil
原油稳定塔 crude stabilizer
原油洗舱 crude oil washing
原油样瓶 oil sampling bottle
原油一次加工 one-step refining of crude oil
原油(油)库 crude storage
原油油轮 crude carrier;dirty tanker
原油蒸馏 crude distillation
原油蒸馏锅 crude still
原油蒸馏装置 crude oil unit
原油中的中馏分 medium fraction of oil
原油重力能 gravity energy of crude
原油组分分析设备 oil composition analysis equipment
原有波动 initial oscillation
原有不稳定性 intrinsic(al) instability
原有沉降 existing settlement;inherent settlement;initial settlement
原有城市 parent city
原有村社 host farming community
原有贷款 underlying financing
原有的根系 original root system
原有的零件 original detail
原有的质量 initial quality
原有垫褥 original bedding
原有断层 preexisting fault
原有对偶算法 primal dual algorithm
原有房屋 existing building;existing house;original building
原有港池 established harbo(u)r basin
原有港口 established harbo(u)r
原有港区 established sector
原有公司 constituent company
原有公众 host community
原有航道 established channel
原有合同 original contract
原有河道 original river
原有河底 existing riverbed
原有基层 original bedding
原有空隙 original interstice
原有孔隙 preexistent pore
原有裂缝 preexisting crack;preexisting fracture
原有漏损 original leakage
原有挠度 initial deflection
原有破裂带 preexisting fractural zone
原有人口 host population
原有设备 existing equipment;existing facility
原有湿度 antecedent wetness;inherent moisture
原有湿度条件 antecedent moisture condition
原有湿度状况 antecedent moisture condition
原有湿气 inherent moisture
原有树木 existing tree
原有水道 existing water-course;existing waterway
原有水深 initial depth
原有特性 primary characteristic
原有天然气 native gas
原有投资 original investment
原有温度 original temperature
原有稳定性 inherent stability
原有物 predecessor
原有误差 inherited error
原有形状 initial form
原有研究程度 studied precision
原有盈余 old surplus
原有约束 primal constraint
原有资本 invested capital
原宇宙辐射 primary cosmic radiation
原语 primitive
原语接口过程 primitive interface procedure
原元素 newtonium
原圆 primitive circle
原云 primitive nebula
原杂基 protomatrix
原载波 primary carrier
原在腐蚀 existent corrosion
原在胶 potential gum
原则 axiom;principle;tenet
原则(方向)比较线 directional comparative lines
原则上的 of principle
原则性布置 diagrammatic arrangement
原则性问题 matter of principle
原张力 initial tension
原褶 primitive fold
原蒸汽 primary steam
原正片 original positive
原正弦波 primary sinusoid
原值 original cost;original value
原植体植物 thallophyte
原纸 base paper;organic felt
原纸吸油量 kerosine absorption of felt
原纸吸油速度 oil absorption velocity of felt
原制造厂 original manufacturer
原质机油 non-fluid oil
原质量 proper mass
原质区 <未受热影响区> unaffected area
原蛭石 protovermiculite
原置成本 original cost
原置成本法 original cost method
原中隔孔 ostium primum
原种 mother seed;original breed;protospecies;registered seed
原种培养 stock culture
原种圃 original seed farm
原种植区 stock plots
原种种子 original seed
原重 original weight
原轴 primitive axis
原竹 bamboo
原住地 country of origin; country where you live
原驻地 proper station
原柱期 primordial shaft
原著 original;original work
原著者 original author
原装 original binding
原装漆 packaged paint
原装设备 original equipment
原装设备制造业 original equipment manufacture
原装原卸 discharged as loaded
原装主机制造业 original equipment manufacture
原状 as dug condition;in place;raw condition
原状标本 undisturbed sample
原状沉淀 undisturbed settling
原状的 raw;rudimentary;undisturbed;untouched
原状碟形黏[粘]土 undisturbed disk-shaped clay
原状废水 crude wastewater
原状骨料 raw aggregate;as-dug aggregate
原状固结【地】diagenesis
原状含水量 field moisture capacity;field moisture content
原状黄土 original loess
原状基岩 in-place bedrock;undisturbed bedrock;undisturbed rock
原状集料 raw aggregate
原状砾石 as-dug gravel
原状路基 original subgrade
原状密(实)度 field density;in-place density;in-situ density
原状黏[粘]土 intact clay;undisturbed clay
原状强度 in-situ strength;undisturbed strength
原状取土器 undisturbed soil sampler
原状砂 undisturbed sand
原状砂样 undisturbed sand sample
原状试件 intact specimen
原状试件强度 undisturbed (testing piece) strength
原状试块 block sample
原状试样 undisturbed sample
原状态 raw state;undisturbed state
原状碳氢化合物 raw hydrocarbons
原状土 undisturbed soil;parent soil
原状土测渗计 monolithic lysimeter
原状土剪力试验 in-place soil-shearing test
原状土剪切试验 field soil-shearing test;in-place soil-shearing test;in-situ soil-shearing test
原状土开挖 primary excavation
原状土开挖数量 cubical yard bank measurement
原状土密实度 in-place density
原状土曲线 undisturbed soil curve
原状土取样 undisturbed sample boring;undisturbed sampling
原状土取样器 sampler for undisturbed samples
原状土壤标本 undisturbed sample of soil
原状土试件 undisturbed soil sample
原状岩石爆破 primary blasting
原状土样 monolith;original soil sample;undisturbed (soil) sample
原状土样的溶度测定仪 monolith lysimeter
原状土样个数 number of in situ soil samples
原状土样钻取 undisturbed sample boring
原状土柱试样 soil column
原状土柱样品 soil column
原状污泥 raw sludge
原状岩芯 undisturbed core
原状样品 undisturbed sample
原子半径 atomic radius
原子爆破开挖 atomic blast excavation
原子爆炸 atomic blast
原子爆炸烟云 atomic blast cloud
原子标度 atomic scale
原子参数 atomic parameter
原子磁化率 atomic susceptibility
原子簇化合物 cluster compound
原子弹 atomic bomb
原子弹掩蔽室 atomic bomb-proof shelter;atomic shelter
原子的 atomic
原子的电子结构 electron structure of atom
原子的基本粒子 <如质子、中子等> atomic sub-particles
原子的偶数项 even term of atom
原子灯 atomic lamp
原子地质年代表 atomic time scale
原子电池 radioisotope battery
原子电荷 atomic charge
原子动力 atomic power
原子动力船 atomic-powered ship;nuclear powered vessel
原子动力的 atomic-powered
原子动力反应堆 atomic power reactor
原子动力航空母舰 atomic-powered air-craft carrier
原子堆 atomic pile
原子发动机 atom engine
原子发射 atomic emission
原子发射光谱(分析)法 atomic emission spectrometry
原子发射谱线 atomic emission line
原子反应 atomic reaction
原子反应堆 atomic furnace;atomic (reactor) pile;chain-reacting pile
原子反应堆构筑物 atomic reactor containment structure
原子反应堆建筑物 atomic reactor containment structure
原子防护 anti-atomic defence;atomic defence
原子防御 anti-atomic defence;atomic defence
原子分裂 atomic fission;atomic fissure
原子丰度 abundance by atom
原子符号 atomic symbol
原子辐射 atomic radiation
原子复合 atomic recombination
原子工业废料处理 treatment of atomic waste
原子共振 atomic resonance
原子共振谱线 atomic resonance line
原子光电效应 atomic photoelectric(al) effect
原子光谱 atomic spectrum
原子光谱选择定则 selection rules for atomic spectra
原子光谱学 atomic spectroscopy
原子轨道 atomic orbit(al)
原子轨道函数线性组合 linear combination of atomic orbitals
原子轨函数 atomic orbital
原子锅炉 atomic boiler
原子锅炉舱 reactor room
原子核 atomic kernel;atomic nucleus;centron;nucleus [复 nuclei/nucleuses];nucleus of atom
原子核的起源 nucleogenesis
原子核的中子激发 neutron excitation of the nucleus
原子核动力 nuclear power
原子核堆 nuclear pile
原子核反应 nuclear reaction
原子核反应堆 nuclear reactor
原子核分裂作用 nuclear fission
原子核辐射 nuclear radiation
原子核工程学 nucleonics
原子核构造 nuclear structure
原子核化学 atom chemistry;nuclear chemistry
原子核类 nucleid
原子核裂变 atomic fission;atomic nucleus fission;nuclear fission
原子核裂度 atomic nucleus fission
原子核嬗变 transmutation
原子核实验所 hot lab(oratory)
原子核损伤 nuclear damage;nuclear injury
原子核物理学 nuclear physics
原子核转变 nuclear event
原子核组成 nuclear contribution
原子恒量 atomic constant
原子化 atomisation [atomization];atomize
原子化效率 atomization efficiency
原子化学 atomic chemistry
原子环【化】ring
原子火箭动力装置 atomic rocket power plant

原子火箭发动机 atomic rocket power plant
原子机车 A-locomotive;atomic locomotive
原子基态 atomic ground state;atomic unexcited state
原子基准 atomic standard
原子激光器 atomic laser
原子极化 atom polarization
原子极化率 atomic polarizability
原子极化强度 atomic polarization
原子加速器 atomic accelerator
原子价 atomicity;quantivalence[quantivalency];valence[valency]
原子价态 valence state
原子假说 atomic hypothesis
原子间干涉 interatomic interference
原子间距离 atomic separation;interatomic distance;interatomic spacing
原子间抗磁性电流 interatomic diamagnetic current
原子间力 interatomic force
原子键 atomic binding;atomic bond
原子键力 interatomic bonding force
原子结构 atomic structure
原子结构化学 metachemistry
原子结合力 atomic binding forces
原子截面 atom section
原子晶格 atomic crystal lattice;atom(ic) lattice
原子壳层 atomic shell
原子扩散 atomic diffusion
原子力显微镜 atomic force microscopy
原子粒子 atomic particle
原子链 atomic link
原子链式反应 atomic chain reaction
原子量 atomic mass;atomic weight;isotopic weight
原子量标度 atomic weight scale
原子量单位 atomic weight unit
原子论 atomics;atomism
原子逻辑式 atom logic(al) expression
原子慢化比 atomic moderation ratio
原子密度 atomic density
原子面 atomic plane
原子模格 atomic modular lattice
原子模型 atomic model
原子内的 subatomic
原子能 A-energy;atomic energy;nuclear energy
原子能采暖 nuclear heating
原子能船 nuclear ship
原子能道岔表示灯 atomic switch lamp
原子能的 atomic;nuclear(y)
原子能电池 atomic battery;atomic energy battery
原子能电站 atomic power plant;atomic power station;nuclear power plant
原子能电站网 nuclear grid
原子能动力装置 atomic power plant
原子能发电 nuclear electric(al) power generation;nuclear power generation
原子能发电厂 atomic energy plant;atomic power plant;nuclear power plant
原子能发电站 A station;atomic electric(al) plant;atomic power station;atomic station;nuclear power station
原子能反应堆 atomic reactor
原子能反应堆结构 nuclear reactor structure
原子能反应堆设备 nuclear reactor facility
原子能废物 atomic waste
原子能工业 atomic energy industry
原子能工业废料 nuclear waste
原子能工业废水 atomic wastewater
原子能工业废物 atomic waste;nuclear waste;waste from atomic energy industry
原子能工业企业 atomic energy installation
原子能级 atomic energy level
原子能控制 A-control
原子能联合委员会 Joint Commission on Atomic Energy
原子能驱动 nuclear propulsion
原子能委员会 Atomic Energy Commission
原子能研究 atomic research
原子浓度 atomic concentration
原子偶极矩 atomic dipole moment
原子频率标准 atomic frequency standard
原子破冰船 atomic ice breaker
原子破裂 atomic disruption
原子谱线 atom line
原子潜艇 atom powered submarine
原子氢焊机 atomic hydrogen welding apparatus
原子氢焊(接) atomic arc welding;atomic hydrogen welding
原子氢焰 atomic hydrogen torch
原子取向 atomic orientation
原子缺陷吸收 atomic defect absorption
原子燃料 atomic fuel
原子热(容) atomic heat
原子热损伤 nuclear heat damage
原子容积 atomic volume
原子散射 atomic scattering
原子散射本领 atomic scattering power
原子散射因子 atomic scattering factor
原子射线 atomic ray
原子湿度仪 atomic moisture meter
原子时 atomic time
原子时标 atomic time scale
原子时代 atomic age
原子式 atomic formula
原子束 atomic beam
原子数 atomicity
原子衰变 atomic disintegration
原子水分密度测量仪 nuclear density moisture ga(u)ge
原子顺磁性 atomic paramagnetism
原子态氧 nascent oxygen
原子体积 atomic volume
原子团 aggregate;atomic cluster;atomic group
原子蜕变 atomic disintegration
原子陀螺仪 atomic gyroscope
原子位移 <晶格中的> discomposition
原子位移效应 discomposition effect
原子位置 atom site
原子武器 atomic weapon;nuclear weapon
原子物理学 atomic physics
原子吸收 atomic absorption
原子吸收法 atomic absorption method
原子吸收分光光度法 atomic absorption spectrophotometry
原子吸收分光光度计 atomic absorption spectrophotometer
原子吸收分光光谱法 atomic absorption spectrophotometry
原子吸收分光计 atomic absorption spectrometer
原子吸收光测法 atomic absorption photometry
原子吸收光度法 atomic absorption photometry
原子吸收光度计 atomic absorption photometer
原子吸收光谱 atomic absorption spectrum
原子吸收光谱测定 atomic absorption spectrometry
原子吸收光谱法 atomic absorption spectrography;atomic absorption spectrophotometry
原子吸收光谱分析 atomic absorption spectrographic analysis
原子吸收光谱分析法 atomic absorption spectrometry
原子吸收光谱学 atomic absorption spectroscopy
原子吸收火焰光谱仪 atomic absorption flame spectrometer
原子吸收型测汞仪 atom-absorption mercury analyser[analyzer]
原子性 atomicity
原子序 atomic number;ordination number
原子序数 atomic number;charge number;ordination number
原子序数修正 atomic number correction
原子学 atomics;atomology
原子学说 atomism
原子氧 atomic oxygen;elemental oxygen
原子氧层 atomic oxygen layer
原子移变作用 prototropy
原子荧光光度计 atomic spectrophotofluorometer
原子荧光光谱法 atomic fluorescence spectrophotometry
原子荧光光谱分析法 atomic fluorescence spectrometry
原子荧光型测汞仪 atom-fluorescence mercury analyser [analyzer]
原子有序化 atomic ordering
原子云 atomic cloud
原子折射度 atomic refraction
原子振荡 atomic oscillation
原子振动 atomic vibration
原子质量 atomic mass
原子质量测定 atomic mass determination
原子质量单位 atomic mass unit
原子钟 atomic clock;hydrogen clock
原紫胶 stick lac
原作 original articles;original work

圆鞍形填料 Berl saddle packing

圆凹饰 bead and quirk
圆凹线脚 hollow chamfer
圆凹状盆地 cup
圆把 bundle of circles
圆把手门锁 knob door lock
圆把手配件 knob fittings
圆柏 Sabina chinensis
圆板 circular flat-plate;disc
圆板过滤器 American filter
圆板缓冲器 plate buffer
圆板耙路机 disc harrow
圆板信号机【铁】 target
圆板形玻璃 crown glass
圆板牙 circular die;circular screwing die;round die
圆蚌线 circular conchoid
圆棒 bar of circular section;circular bar;cylindric(al) rod;mandrel;round bar
圆棒材 round bar
圆棒抗裂试验 bar type crack(ing) test;round bar crack test
圆棒索梯 round-rung ladder
圆棒形线脚 ressaut
圆棒张力试验 rod-tension test
圆棒直槽滚压 bar rolling
圆堡 <防御海岸用> Martello tower;roundel
圆背座椅 bucket seat
圆鼻车刀 round-nose tool
圆鼻的 roundness
圆鼻錾 hollow chisel
圆鼻凿 round-nose chisel
圆舭 round bilge
圆舭船 round bilge ship
圆舭型船体 round bilge hull
圆币饰 bezant
圆壁龛 roundel
圆边 <平板玻璃磨成的> bulb edge;edge rounding
圆边扁钢 round-edged(steel) flat
圆边扁钢丝 round-edged flat steel wire
圆边的 round-edged
圆边滴水槽 bottle-nose curb;bottle-nose drip
圆边混凝土 edge rounding concrete
圆边击平锤 holding-up hammer;round set-hammer;snap hammer
圆边加工 pencil edging
圆边铰链锉 round edge joint file
圆边轮圈 round-edged tyre [tire]
圆边刨 beading plane;beading tool
圆边踏步级 bottle-nose curb;bottle-nose step;ottler-nose step
圆边瓦 rounded edge tile
圆边瓦装修 round edge tile fitting
圆边外角砖 tile with round edge external corner
圆边圆角砖 jamb brick
圆边砖 round edge tile
圆编 round sennit
圆冰丘 moutonnee hummock
圆柄钥匙 round key
圆波导 circular waveguide
圆波导过渡器 circular waveguide taper
圆波峰 round crest
圆玻璃 roundel
圆玻璃窗 bull's eye
圆玻璃切刀 circle glass cutter
圆材 bar of circular section;billetwood;log;round timber;round wood
圆材的四分割 quarter-cut
圆材结 timber hitch
圆材结加半结 kill(i)ck hitch;timber and half hitch
圆材锯 head saw;log saw
圆材燃管 <煤气暖炉的> gas log
圆槽 circular groove;circular slot;rough groove
圆槽方钻杆 round spline kelly
圆槽抗裂试验 circular groove crack test
圆槽释放法 trepanning method
圆槽头螺栓 round slotted head bolt
圆层磷灰石 naurite
圆觇板 sighting disc
圆铲 round-pointed shovel;round shovel;round spade
圆池 circular pond;circular tank
圆齿 knuckle tooth
圆齿齿轮(装置) knuckle gear(ing)
圆齿状板 round-toothed plate
圆冲头 round punch
圆船尾 round stern
圆窗 eye;fenestra rotunda;oculus;oeil-de-boeuf window;round window
圆窗膜破裂 rupture of round window membrane
圆锤 round set-hammer;top fuller
圆锤头 ball peen

圆唇钻头 round-nose bit
圆磁场 circular field
圆锉 circular file;hollowing file;rat tail(ed) file;round file
圆锉刀 circular-cut file
圆带线条 <柱头或柱脚凸起的> bead mo(u)ld(ing)
圆挡木 retaining log
圆刀刀架 round cutter tool holder
圆导轨 round guide
圆导线 round wire
圆道钉 round spike
圆的 conglobate;orbicular;orbiculate;spheric(al)
圆的包迹 envelop to circles
圆的标准偏差 circle standard error
圆的第二渐伸线 second involute of the circle
圆的极点 pole of a circle
圆的平头钉 round plain head nail
圆的周长 perimeter of a circle
圆灯笼 jack-o-lantern
圆灯罩 lamp globe
圆凳 <无扶手和靠背柱形的> tabo(u)ret
圆底铲斗 radius bucket;round bottom bucket
圆底船体 round bottom hull
圆底唇钻头 round-face bit
圆底的扩展基础 spread footing with circular base
圆底离心管 round bottom centrifuge tube
圆底模 fuller block
圆底刨刀 round sole plane
圆底烧瓶 round bottom flask
圆底线脚 torus [复 tori]
圆底形钢制沉箱 <用作桥墩基础,俚语> cookie cutter
圆底有刻度离心管 graduated round bottom centrifuge tube
圆底圆筒 round-ended cylinder
圆点 circular point;dot;round spot
圆点花纹 polka dot
圆点花纹多彩涂料 polka-dot paint
圆点曲线【数】circular curve
圆点饰面 cribbled
圆点图 dot chart
圆电缆 round cable
圆电刷 round brush
圆电线 round wire
圆垫圈 circular washer
圆雕 detached statuary;full relief;round sculpture
圆雕饰 medallion
圆雕饰板 circular carved panel;medallion
圆雕饰线脚 medallion mo(u)lding
圆吊灯 <教堂的> corona
圆钉 nail
圆顶 calotte;cupola;globe-roof;round crest;spheric(al) calotte
圆顶包卷接头 <金属屋面的> round-topped roll
圆顶边缘 dome edge
圆顶表面 dome surface
圆顶长方形教堂 domed basilica church
圆顶齿 kunckle tooth
圆顶储仓 dome storage
圆顶窗 round-head(ed) window
圆顶大厅 domed hall
圆顶地下式石灰窑 dome kiln
圆顶顶点 dome apex
圆顶顶端 dome top
圆顶顶棚区段 savory
圆顶顶棚一跨 savory
圆顶端 domed end
圆顶房屋 <干垒毛石的> trullo
圆顶(分)片 dome segment
圆顶拱顶 dome crown
圆顶拱基 dome impost
圆顶鼓风炉 cupola
圆顶环 dome ring
圆顶活塞 dome-head piston
圆顶基础 dome foundation
圆顶集装箱 igloo
圆顶夹层 intercupola;interdome
圆顶建筑 igloo
圆顶建筑拱 rotunda arch
圆顶建筑物 rotunda
圆顶建筑艺术 art of vaulting
圆顶结构 domed structure
圆顶居所 dome housing
圆顶扩散器 dome diffuser
圆顶量水堰 rounded crest measuring weir
圆顶炉 dome-shaped roof
圆顶路缘石 drum curb
圆顶螺帽 acorn nut;ring nut
圆顶螺母 acorn nut
圆顶清真寺 domed mosque
圆顶穹隆 arched dome;circular vault;pendant vault
圆顶丘 coupole;dome
圆顶人孔 dome manhole
圆顶山 round-top mountain
圆顶十字形教堂 domed-cruciform church
圆顶石屋 <史前撒丁岛独有的> nuraghi
圆顶式建筑 domical architecture
圆顶塔 tope(-mound)
圆顶塔导游人 tope guide
圆顶塔基础 tope-mound base
圆顶塔基座 tope-mound base
圆顶塔神龛 tope shrine
圆顶天窗 dome skylight;round-head(ed) dormer window
圆顶天窗圈梁 lantern ring
圆顶挑檐 dome cornice
圆顶亭 domed pavilion
圆顶小丘 cop
圆顶小塔楼 domed diminutive tower;domed turret
圆顶斜拉条 diagonal tie of a dome
圆顶形坝 cupola dam
圆顶形界限法 arch-bound method
圆顶形式 dome form
圆顶形天窗 circular headed skylight
圆顶漩涡状装饰 round-topped
圆顶堰 round crested weir;rounded crest weir
圆顶窑 dome kiln
圆顶伊斯兰教寺院 domed mosque
圆顶应力 dome stress
圆顶帐篷 yurt(a)
圆顶支承结构 tholobate
圆顶中的空间 hollow space in cupola
圆顶柱头螺钉 fillister head screw
圆顶装料机 cupola loader
圆顶状冰山 dome-shaped iceberg
圆顶状的 domal;dome like
圆顶子午肋 meridian rib of a dome
圆洞 pot-hole
圆洞口 circular orifice
圆洞片拱桥 plate ribbed arch bridge with circular openings;thin ribbed cellular arch bridge
圆肚窗 bow window;compass window
圆度 <表示集料等表面起棱角的程度> roundness;circular degree;degree of roundness;perimeter;sphericity
圆度比 circularity ratio;convexity ratio;roundness ratio
圆度槽 round bottom slot
圆度公差 roundness tolerance
圆度盘 circular dial;divided circle
圆度系数 coefficient of circularity;coefficient of roughness
圆度仪 roundness ga(u)ge;roundness measuring equipment
圆度指数 roundness index
圆端 round nose;round ends <两端可转动>;nose circle <轴的>
圆端臂板 <信号机> round-end (semaphore) blade
圆端丁砖 bull(-nose) header
圆端面钻头 round-face bit
圆端顺砖 bull(-nose) stretcher
圆端踏步 bull-nose step;rounded step;round-ended step
圆端形 round-ended shape
圆端形桥墩 round-ended pier
圆断面 annular section
圆断面唇部钻头 full-radius crown;full round nose bit
圆断面蜗壳 round section spiral casing
圆墩式突式码头 cylinder jetty
圆耳柳 rounder willow
圆二色散 circular dichroism
圆阀(门) drum gate;round valve
圆法兰 circular lip;round flange
圆分度 circular division
圆分度器 circular protractor
圆分度头 circular index
圆风阀 circular damper
圆缝 circular slot
圆浮雕 rosace;tondo
圆盖 bonnet;dome
圆盖板 closing disc
圆盖螺母 dome nut
圆盖屋顶 dome-shaped roof
圆盖形屋顶 dome roof
圆杆 bar of circular section;circular shaft;round rod
圆杆材 <胸高直径 5~10.9 英寸,1 英寸≈0.0254 米> pole-timber
圆杆构成的网状构件 cylindric(al) rod web
圆杆锚固件 round-stock anchor
圆杆门拉手 round bar handle
圆杆网件 cylindric(al) rod web
圆冈 hummock
圆钢 rod iron;round bar;round rod;round steel;smooth bar;plain bar
圆钢材 rod stock;round stock;steel round
圆钢叉子 round steel fork
圆钢钉 wire nail
圆钢箍筋 round bar web
圆钢剪切机 bar-cropping machine
圆钢矫直机 round straightener
圆钢筋 circular bar;circular reinforcement bar;cylindric(al) reinforcement bar;cylindric(al) reinforcement rod;reinforced steel bar round;reinforcing round steel;rod steel;steel round;round steel bar;round re-bar;round reinforcement bar
圆钢筋束 <混凝土> bundle of round reinforcing bars
圆钢筋条组成的网 bar fabric
圆钢筋柱上用棘轮 tension wheel assembly for tubular steel pole
圆钢捆 rod bundle
圆钢拉杆 trussed rod
圆钢拉条 bar tie rod
圆钢锚 round steel anchor
圆钢丝 round steel wire
圆钢丝刷 wire wheel brush
圆钢丝网筛布 round wire screen cloth
圆钢条 rod stock;round bar steel;round iron
圆钢斜撑 trussed rod
圆钢柱 cylindric(al) steel column
圆钢柱的杯形基座 cup base
圆格构图 <海上勘探定位用> circle lattice chart
圆格型板桩结构 circular cells sheet-piled structure
圆格型围堰 circular type cellular cofferdam
圆隔板 circular orifice;disk spacer;spacer disc[disk]
圆隔盘 spacer disc [disk]
圆隔片 circular orifice
圆根切刀 recessing tool
圆工咨询 employee counseling
圆弓形 circular segment
圆拱 bull's eye arch;obtuse arch;one-centered arch;round arch
圆拱边沿 cupola edge
圆拱边缘 cupola edge
圆拱的 rounded
圆拱顶 annular vault;circular vault;cul-de-four;cupola crown
圆拱顶盖 vault head
圆拱墩 cupola pier
圆拱墩柱 arch abutment
圆拱结构 birdcage
圆拱式横断面 rounded crown section;rounded cross-section
圆拱挑檐 round-arched corbel-table
圆拱形 cupola
圆拱形城堞 round-arched merlon
圆拱形的 round-arched
圆拱形拱顶 round-arched barrel vault;round-arched wagon vault
圆拱形开口 round-arched opening
圆拱檐口 cupola
圆拱缘饰 cupola
圆钩 choker hook
圆箍 circular hoop
圆箍筋 round stirrup
圆箍失落 rings missing
圆箍脱落 rings off
圆箍线 annulet
圆箍线脚 gradetto
圆谷冰川 circus glacier
圆股钢丝 round strand
圆股钢丝绳 round strand cable;round strand rope;round strand steel wire rope;round strand wire rope
圆股绳 round strand rope
圆股提升钢丝绳 round strand hoisting rope
圆鼓干燥骨架 skeleton drum
圆鼓屋顶 drum;drum of dome
圆鼓形薄壳 drum shell
圆刮刀 round scraper
圆挂钩 round hanger
圆管 circular duct;circular pipe;circular tube;cylindric(al) tube;round pipe;round tube
圆管涵 pipe culvert
圆管桁桥 tubular arch bridge;tubular bridge
圆管结构 tubular frame
圆管框架 tubular frame
圆管钳 dolly wrench
圆管桥 tubular arch bridge;tubular bridge
圆管式电集尘器 wire and pipe precipitator
圆管式下水道 circular conduit-type sewer
圆管铁壳水银温度计 round tube mercury thermometer with iron casing
圆罐 round tank
圆光罩 round-opening openwork screen
圆规 circling attachment;compasses;pair of compasses;pencil compasses

圆规扳子 compasses key
圆规尺 radial bar
圆规脚 leg of compasses
圆规接腿 compass lengthening bar
圆规延伸杆 compass lengthening bar
圆轨旋转式桥吊 overhead slewing crane
圆辊 roller drum
圆辊坝 roller dam
圆辊错齿线脚 round billet mo(u)lding
圆辊防撞装置 roll fender
圆辊活动坝 roller weir
圆辊门运料斗 bucket with a roller gate
圆辊堰 roller weir
圆辊闸门 roller drum gate;roller gate;rolling gate
圆辊闸门挡板 roller gate shield;roller shield
圆辊闸门堰 roller drum gate weir
圆滚摆 cycloidal pendulum
圆滚线【数】cycloid
圆滚线拱 cycloidal arch
圆函数 circular function
圆涵洞 circular culvert
圆夯 circular hammer; circular rammer
圆盒水准器 circular level
圆弧 arc of circumference; circular arc;curve
圆弧凹形嵌边饰 conge
圆弧测定器 cyclometer
圆弧尺 circular arch rule; dial type scale
圆弧齿 circle-arc tooth
圆弧齿轮 circular tooth gear
圆弧齿圆柱蜗杆减速机 circular tooth and cylindric(al) worm reducer
圆弧导航 arc navigation
圆弧点啮合齿轮 arc-point-mesh gear
圆弧顶面齿 knuckle tooth
圆弧段拱的门 segment head
圆弧断面履带板 circular arc shoe
圆弧法 circular arc method
圆弧分析法 circular arc analysis
圆弧拱 coved arch; one-centered arch;circular arch;skene arch <相对角小于 180 度>
圆弧拱梁 circular arched girder
圆弧拱桥 circular arch bridge
圆弧构成的椭圆 false ellipse
圆弧规 arcograph; bow compasses; cyclograph
圆弧滑动 circular slide; circular slip; rotational slide
圆弧滑动分析 circular-slide analysis; slip circle analysis
圆弧滑动分析法 method of slip circle analysis
圆弧滑动面 circular sliding surface; circular slip surface
圆弧滑动面分析法 soil friction circle method
圆弧机车库 round house
圆弧锯 keyhole saw
圆弧棱边 half-round edge
圆弧量测 circular measure
圆弧面 crown face
圆弧(面)高度 crown height
圆弧面密集补强 concentrated reinforcement with curved surface

圆弧磨石 gouge slip
圆弧内插法 circular interpolation
圆弧刨 capping plane;circular plane; compass plane;round shave
圆弧频率 circular frequency
圆弧破坏分析 circular failure analysis
圆弧切口 circular lance
圆弧锁面 astragal front
圆弧天花板 coved ceiling
圆弧椭圆 false ellipse
圆弧瓦 segmental tile
圆弧形格体 arc cell
圆弧形拱柱坝 dam with segmental-headed counterforts
圆弧形滑坡 rotational landslide; circular slide
圆弧形金属水槽 circular metal flume
圆弧形楼梯 geometric (al) stair-(case)
圆弧形门楣 segment head
圆弧形木板条水槽 circular wood-stave flume
圆弧形实心砖 compass solid brick
圆弧形土层滑塌 rotational earth slump
圆弧形土层滑移 rotational earth slump
圆弧形屋顶 compass roof
圆弧形硬砖 compass hard brick
圆弧形转弯的底座线脚 bent shoe
圆弧形撞击坑 semi-circular impact pits
圆弧样板 radius former
圆弧叶型 circular arc profile
圆弧应力分析法 soil circle stress method
圆弧钻头 gouge bit
圆花窗 Catherine wheel;marigold window;rosace;rose (tte); rose window;rugosa;wheel window
圆花钉 rose nail
圆花饰 paterage;rosace;rosette
圆花装饰 rose trim
圆花钻 rose drill
圆滑过渡 rounding-off
圆滑褶皱 rounded fold
圆画 tondo
圆环 annulus; ba (s) ton; circular ring;cirque;disc;ring
圆环猜想 annulus conjecture
圆环衬垫 ring gasket
圆环的 circumferential
圆环电弧焊 cyc-arc welding
圆环阀 ring gate;ring valve
圆环格形围堰 circular type cellular cofferdam
圆环格状围堰 circular type cellular cofferdam
圆环回跳现象 ring resilience
圆环架 ring stand
圆环拉强试验 <水泥及混凝土> ring test
圆环链 circle chain;round links
圆环密封 O-ring seal
圆环面 torus [复 tori]
圆环喷嘴 ring nozzle
圆环圈式法兰 ring-type joint flange
圆环式密封装置 ring-type seals
圆环式破碎机 rolling ring type crusher
圆环(试验)法 ring test
圆环锁 roller lock
圆环体钉 ring-shank nail
圆环图 doughnut
圆环形 circular ring
圆环形接头 O-ring joint
圆环形凸线脚 tore
圆环形匣钵 ringer
圆环形烛灯圈 <教堂用> corona lucis
圆环窑 ring kiln
圆环域 annulus [复 annuli/annuluses]
圆环轴套 packing sleeve
圆环柱心 toroidal core
圆环状平板 circular ring (shaped) plate
圆环坐标 toroidal coordinates
圆缓点 curve to spiral point
圆簧止销 spring pin
圆簧组 nest spring
圆汇流线 bus rod
圆混凝土柱 cylindric (al) concrete column
圆火道 circular flue
圆火花 ring fire
圆或椭圆形壳 toroidal shell
圆或椭圆形盘旋楼梯 winding stair (case);winding stairs
圆机壳电动机 <无突出轴承> round-body motor;round frame motor
圆基础 circular footing;circular foundation
圆极化波的旋转 rotation of circular polarized waves
圆脊 breakover
圆脊角钢 round backed angle
圆脊檩 ridge rod;ridge roll
圆计算尺 circular slide rule
圆尖半径 radius of curvature
圆尖嘴钳 long round nose pliers
圆尖石凿 round-pointed stone chisel
圆剪 circle sheet;circular snips
圆礁丘 reef knoll
圆角 arris;circular bead;fillet(ing); round(ed) angle;round(ed) corner;rounding;round-off corner
圆角半径 radius of corner;radius of rounded angle; radius of rounded corner;round radius
圆角边距 bull-nose trim
圆角边刨 fillet plane
圆角边坡 rounded slope
圆角边缘 buffet
圆角扁钢 bulb flat steel
圆角槽铁 D-iron
圆角侧石 rounded-lip curb
圆角大砖 bullnose block
圆角的 well-rounded
圆角丁砖 bull header
圆角方钢 quarter octagon steel
圆角方料 round-corner square
圆角方坯 round-corner square billet
圆角方形衬板 square plate with round corner
圆角方形螺旋分级衬板 round angled square screwed type classifying lining
圆角骨料 rounded aggregate
圆角光子 round sleek
圆角规 fillet ga(u)ge;radius ga(u)ge
圆角滚轧 fillet rolling
圆角焊缝 fillet bead
圆角焊肉厚 leg of fillet weld
圆角和倒角 corners and fillets
圆角混凝土块体 bull-nose block
圆角机翼 round-off wing
圆角集料 rounded aggregate
圆角铰刀 bull trowel
圆角块体 bull-nose block;bull-nose unit
圆角路缘 roll curb
圆角帽梁 bull-nose coping
圆角面 hollows
圆角磨削装置 radius attachment
圆角抹子 arches;bull trowel
圆角木材 wane lumber
圆角木扶手 <栏杆上> capping plane
圆角泥刀 bull trowel
圆角刨 astragal plane;bead plane
圆角砌块 bull-nose block
圆角嵌条 corner filler;corner fillet
圆角切刀 radial tool;radius rod segment; radius segment; radius segment tool;radius tool;segment tool
圆角刃口 round edge
圆角势阱 round edge well
圆角顺砖 bull-nose stretcher
圆角踏步 bull-nose step; commode step;rounded step
圆角踏步板 bull stretcher
圆角铣刀 corner rounding (milling) cutter
圆角斜踏步 <楼梯的> bull-nose winder
圆角样板 fillet ga(u)ge;radius former
圆角缘石 rounded-lip curb
圆角凿 fillet chisel
圆角砖 bull-nose unit
圆节 round knot
圆结核体岩石 nablock
圆截面垫片 round cross section gasket
圆截面缝 concave tooled joint;convex joint
圆截面弯梁 circular curved beam
圆截面橡胶垫圈 round section rubber packing ring
圆截面橡胶盘条 round section rubber strip
圆截面型材 circular profile
圆截面凿岩机 fuller hammer
圆截盘 circular jib
圆金柑 round kumquat
圆井 derrick cellar;round well;well cellar
圆井抓斗 well-digging clamshell
圆颈螺母 round neck nut
圆径概率误差 circle of equal probability;circle of probable error;circular probable error
圆径千分尺 annular micrometer
圆剧场 arena
圆剧场兽室 carcer
圆锯 annular bit; annular saw; buzz saw;circuit saw;circular buzz saw; compass saw; dapper; locksaw; rim saw;ring saw;swage saw;trepan;trephine
圆锯齿顶修整 topping
圆锯齿间间隙 saw gullet
圆锯锉 gulleting saw file
圆锯护罩 crown cover
圆锯机 buzz saw;circular sawing machine;disc saw;circular saw
圆锯夹盘 collar
圆锯架 circular saw bench
圆锯口松弛 loose
圆锯摩擦切割 circular saw friction cutting
圆锯片 circular saw blade
圆锯刃磨机 circular saw sharpener;circular saw sharpening grinding machine
圆锯台架 saw bench
圆锯座架的轴 saw arbor
圆颗粒 round(ed) grain
圆刻度 circle graduation
圆刻度盘温度计 dial thermometer
圆空刨 round and hollow plane
圆孔 annular opening;circuit aperture; circular hole; oculus; round hole; round orifice
圆孔板 round-hole plate
圆孔电锯 electric(al) hole saw
圆孔电锯条 electric(al) hole saw blade
圆孔垫圈 round-hole washer
圆孔骨架 round-hole framework
圆孔筋板 perforated metal screen
圆孔锯 crown saw;cylinder saw
圆孔卡片 round-hole card
圆孔壳属 <拉> Amphisphaeria
圆孔空心混凝土板 circular voided concrete slab
圆孔空心楼板 tube floor (slab)
圆孔口 circular orifice
圆孔拉刀 round broach
圆孔排种盘 round-hole plate
圆孔切割器 circle cutter
圆孔筛 circular hole sieve;round-hole mesh; round-hole sieve; round-meshed screen;circular screen

圆孔筛板 circular screen;round-hole punched plate; round-punched sheet;round-hole screen
圆孔试验筛 round-hole test sieve
圆孔试验网筛 round mesh test screw
圆孔衍射 diffraction by a circular aperture
圆孔凿 round-nose chisel
圆孔直径 round-hole diameter
圆口铲土机 shovel with round mouth
圆口灯 beak
圆口钉 beak
圆口剪 scroll pivoted snips
圆口流量计 rounded-entrance flow meter
圆口钳 round mouth tongs
圆口錾 gouge chisel
圆扣接箍 round thread coupling
圆扣套管 round thread casing
圆扣套管接箍丝扣抗拉强度 round casing coupling thread tensile strength
圆扣套管丝扣抗拉强度 round casing thread tensile strength
圆库卸料系统 silo-discharge system
圆库卸料装置 silo-discharge device
圆库卸料锥体 silo-discharge cone
圆库中间仓 circular silo compartment
圆块 nahlock;cob <煤、石头等的>
圆块煤 cobble
圆括号 curves;parenthesis [复 parentheses]; round brackets; round parenthesis
圆括号的嵌套深度 depth of parenthesis nesting
圆括弧 circular bracket;round brackets;round parenthesis
圆拉刀 circular broach;round broach
圆拉条 circle brace
圆喇叭 circular horn
圆喇叭形进口 circular bell mouth entrance
圆喇叭形进水口 circular bell mouth entrance
圆栏杆形饰 <爱奥尼亚柱头侧面的> pulvinus
圆肋钢筋 bulb bar
圆棱 diminishing stop bevel
圆棱边 round arris edge
圆楞条 cymbia
圆犁刀 rolling colter
圆砾 gravel
圆砾结构 conglomeratic texture
圆砾石 pudding stone;round gravel
圆砾岩 pudding rock;pudding stone
圆粒 rounded grain
圆粒金刚石 bort(z)
圆粒砂 rounded grain
圆粒砂子 round sand
圆链法 circle-chain method
圆梁 circular beam
圆量尺 cyrtometer
圆料金刚石 boart
圆列柱廊 circular peristyle; spacer peristyle
圆流 circular jet
圆楼层 round stor(e)y
圆楼梯踏步 round(ed) step
圆炉箅 circular grate
圆卵石 pebble roundstone
圆轮闸门 roller gate
圆罗盘 circular compass
圆螺帽钩头扳手 hooked tommy
圆螺母 circular nut;round nut
圆螺母扳手 hook wrench
圆螺丝板牙 round die
圆螺纹 Edison screw thread; knuckle screw thread; rope thread; round thread
圆螺纹螺丝 round-threaded screw
圆马蹄形拱 round horseshoe arch
圆满 completion;roundness
圆满成功 consummation
圆满竣工 final completion
圆墁杆 club tool
圆镘刀 round trowel
圆锚 mushroom anchor
圆锚筋 tubular dowel
圆铆钉头 round rivet head
圆帽 bonnet
圆煤块 cobcoal
圆密尔(英制面积单位) circular rail
圆密耳 circular mil
圆面包状节理【地】loaf-like jointing
圆面粗车刀 round-faced roughing tool
圆面滑动 rotational slip
圆面滑坍 rotational failure
圆面积 area of circle
圆面积计算法 circular area method
圆面刨 rounds
圆明园遗址 Relics of Yuan Ming Yuan
圆模 round die
圆磨 cone crusher
圆抹角 cymbia;rounded corner;round-off corner
圆木 gum pole;log;log wood;raddle;round-log; round timber; round wood;whole beam
圆木暗销 disk dowel
圆木搬运 logging
圆木搬运保护网 logging sweep and protective screen
圆木搬运叉 log(ging) fork
圆木搬运架 logging arch
圆木材 round timber
圆木采伐 logging
圆木撑柱 round timber prop
圆木锉 round rasp
圆木道 <沼泽地区铺路的> cord read
圆木叠框 log-crib
圆木堆 log raft
圆木扶手 mopstick
圆木杆 mope pole
圆木隔墙 round timber bulkhead
圆木钩 cant hook
圆木构造 round-log construction
圆木构筑成墙 round-log construction
圆木刮路器 log drag
圆木轨枕 pole tie
圆木涵洞 log culvert
圆木滑道底木 base log
圆木滑道两侧的顶木 top log
圆木建筑 cobwork;construction with logs
圆木建筑柱 round timber construction pole
圆木脚手架 round timber falsework
圆木结构 log construction
圆木锯成板 breaking down
圆木锯方 conversion of timber
圆木锯解 conversion of timber
圆木料 ricker
圆木路 <沼泽地上的> ground bridge; corduroy road
圆木路刮 log drag
圆木码垛机 log yarder
圆木耙 log harrow
圆木铺成的路 <沼泽地上> ground bridge
圆木墙小屋 log cabin
圆木桥 log bridge
圆木清扫锯 <扫除圆木杂质> barking saw;rock saw
圆木梢径 top diameter
圆木十字形锯开的木材 quarter-sawed lumber
圆木输送机 log conveyer [conveyor]
圆木四开锯法 quarter-cut
圆木榫 disk dowel
圆木梯 pole ladder
圆木提升机 log elevator
圆木头螺钉 cup head wood screw
圆木拖运机 log skidder
圆木围长法 quarter-girth rule
圆(木)桅杆 round mast
圆木屋的外墙披叠板 log cabin siding
圆木匣 round chip box
圆木小屋 log hut
圆木堰 log weir
圆木折合立方数法则 quarter-girth rule
圆木折合立方数量 quarter-girth measurement
圆木支撑 round timber support
圆木支承 round timber support
圆木支架 round timber support
圆木制材 stuff
圆木柱 log column
圆木抓具 wood grab;wood grapple
圆木抓扬机 log grapple
圆木桩 log(ged) pile;round timber pole
圆木桩栅栏 round fence picket
圆木装卸叉 log(ging) fork
圆木装卸机 flipper
圆木装卸钳 log handling tongs
圆木装运车 log wagon
圆木装载机 log loader
圆木作 circular work
圆内接 inscribed in a circle
圆内旋轮线 hypocycloid
圆内旋转线【数】hypercycloid
圆泥刀 circle trowel
圆钮定位法 buttoning
圆钮开关 switching knob
圆排号志 disc signal
圆牌杆 rod of disc
圆盘 collar plate;disc [disk]
圆盘凹度 disk dish
圆盘拌和机 circular pan mixer
圆盘比色计 disk colo(u)rimeter
圆盘除根机 undercutter with disc
圆盘除脱器 roller remover
圆盘传送带 car(r)ousel
圆盘传送器 car(r)ousel
圆盘打磨机 disc [disk] sander
圆盘刀 circular cutting disc [disk]; colter disk;cutting disc [disk];knife disk;rotary knife
圆盘刀片的心轴 knife arbor
圆盘导翼阀 wing valve
圆盘的 discoid
圆盘电枢 built-up rotor; disc armature
圆盘叠片式机油滤清器 disk oil filter
圆盘钉 disk nail;impulse pallet;roller pin
圆盘(对径)压缩试验 disc test
圆盘伐树刀 peripheral cutter
圆盘阀 disc [disk] valve;poppet valve
圆盘翻土机 clay buster
圆盘翻转双向犁 reversible disk plow
圆盘风扇 disk fan
圆盘浮标 disc float
圆盘干燥器 disk drier [dryer]
圆盘给料机 circular disk feeder;disk feeder;plate feeder;revolving table feeder;rotary plough feeder;table feeder
圆盘给水器 table feeder
圆盘刮泥机 disk scraper
圆盘规 disk ga(u)ge;roller jewel ga(u)ge
圆盘辊 spool roll
圆盘滚刀 disk hob
圆盘(滚轴)筛 disk grizzly;revolving disc grizzly
圆盘花饰 bezantee;disc frieze;patera [复 paterae]
圆盘回动装置 circle reverse
圆盘计量器 disk meter
圆盘计算尺 disk slide rule
圆盘记录器 circular recorder; disk recorder
圆盘加料机 disc feeder
圆盘剪床 circular shears
圆盘剪旁的带卷台 slitter ramp
圆盘剪(切机) circle shears
圆盘搅拌机 disk mixer
圆盘接地器 circular disk grounding device
圆盘精研机 disk refiner
圆盘锯 buzz saw;circular saw;rotary saw;sawing disk;slasher saw;table saw;disc [disk] saw
圆盘锯台架 saw bench
圆盘开沟器 disc [disk] boot
圆盘快门 disc shutter
圆盘离合器 disc [disk] clutch; plate clutch
圆盘犁 disc harrow; disc [disk] plough;grubbing harrow
圆盘犁刀 rolling coulter
圆盘联轴节 disc [disk] coupling;lamination coupling
圆盘路犁 disk plough
圆盘镘平板 disc type power float
圆盘铆钉 disc [disk] rivet
圆盘面积仪 disk planimeter
圆盘灭茬犁 disk paring plow
圆盘摩擦 disc [disk] friction
圆盘摩擦锤 friction-board hammer
圆盘摩擦离合器 disk friction clutch
圆盘摩擦损耗 disc friction loss;dish friction loss
圆盘摩擦损失 disc friction loss;dish friction loss
圆盘磨 attrition mill
圆盘磨床 disc [disk] grinder
圆盘磨光机 disc [disk] sander; disk emery cloth
圆盘磨光器 disc [disk] sander
圆盘耐磨硬度试验机 wear hardness machine
圆盘碾压机 disk roller
圆盘耙 disc [disk] harrow; soil pulverizer
圆盘耙耙片 disc
圆盘耙组方轴 arbor bolt
圆盘抛光机 disk polisher
圆盘片 disc
圆盘平地机 disk planer
圆盘平直涡轮式搅拌器 flat disk turbine agitator
圆盘破碎机 disc crusher
圆盘求积仪 disk planimeter
圆盘刹车 disk brake
圆盘筛 disc grizzly;disc [disk] screen
圆盘上料机 circular disk feeder
圆盘式变阻器 dial type rheostat
圆盘式冲沙闸门 dished sluice gate
圆盘式除草机 disk weeder
圆盘式放电器 disc discharger
圆盘式飞轮 disc [disk] flywheel
圆盘式分离器 disc separator
圆盘式分选器 disc separator
圆盘式粉碎机 disc mill
圆盘式干燥器 disk drier [dryer]
圆盘式缸径规 dial type cylinder ga(u)ge
圆盘式给料器 disc [disk] feeder
圆盘式挂钟 plate clock
圆盘式滚刀 disc cutter
圆盘式滚刀刃口 disc cutting edge
圆盘式滚筒 disc roller
圆盘式过滤器 disk filter;leaf-type filter
圆盘式划线 disk ruling
圆盘式划行器 disk-type marker

圆盘式挤压机 disc extruder
圆盘式记录纸 circular chart
圆盘式加料机 disk feeder
圆盘式加料器 disk feeder
圆盘式加湿器 disk humidifier
圆盘式搅拌机 circular pan mixer
圆盘式金刚石锯 diamond saw (splitter)
圆盘式块根切碎机 disk-type root cutter
圆盘式离心粉碎机 Kek mill
圆盘式料浆过滤器 disc filter
圆盘式路机 disc harrow
圆盘式路犁 disc plough
圆盘式路耙 disc [disk] harrow
圆盘式磨削机 disc grinding machine
圆盘式判读样片 disk key
圆盘式抛料机 disk material thrower
圆盘式破碎机 disk crusher
圆盘式铺砂机 rotary disk type gritter
圆盘式启闭机 operating machinery with disk rotor
圆盘式切边剪 rotary side trimming shears; rotary trimming shears
圆盘式切割机 disc [disk] cutter
圆盘式切削工具 disc cutter
圆盘式绕组 disc winding
圆盘式撒布机 disc spreader
圆盘式刹车 disc brake
圆盘式输送机 disc [disk] conveyer [conveyor]
圆盘式水表 disk meter; disk-type water-meter
圆盘式碎土机 disk cultivator
圆盘式挖掘机 dish digger
圆盘式喂料机 circular feeder
圆盘式污水滤网 disc sewage screen
圆盘式泄水闸门 dished sluice gate
圆盘式旋转流量计 nutating-disk meter
圆盘式压路机 disc roller
圆盘式研磨机 disk grinder
圆盘式张力器 disc tensioner
圆盘式转子 disc type rotor
圆盘式钻架 push-pull jacking rig
圆盘试验 disc test
圆盘水表 disk water meter
圆盘松土器 disk ripper
圆盘碎石机 disk cultivator
圆盘碎土机 disk crusher
圆盘损失 disc friction loss
圆盘天线 disk antenna
圆盘铁栅筛 disk-grizzly screen
圆盘图 pie chart
圆盘挖掘器 digging wheel
圆盘喂料秤 dosing rotor weigher
圆盘喂料机 disc feeder; round plate feeder; table feeder
圆盘衔铁 disc [disk] armature
圆盘信号 disc signal
圆盘信号机【铁】 disc signal; signal disc [disk]
圆盘信号机的开放位置 off-position of disc signal
圆盘形扁锥 rotary disk bit
圆盘形扁钻 rotary disk bit
圆盘形表示器 disc type indicator
圆盘形当量缺陷 circular disc defect
圆盘形浮雕 tondino
圆盘形复示器 disc type indicator
圆盘形回转钻 rotary disk bit
圆盘形记录仪表 round chart instrument
圆盘形切割器 disc [disk] cutter
圆盘形绕组 disc winding
圆盘形水表 disk water meter
圆盘形弹簧 disc spring
圆盘形透水石 perforated disc
圆盘形图 tondino
圆盘形装饰 besant [bezant]; byzant

圆盘型机电选别器 electromechanical spindle slot
圆盘旋(楼)梯 circular geometrical stair(case)
圆盘选别器 rotary electric(al) slot
圆盘压裂试验 disc test
圆盘压实机 disc-compactor
圆盘压碎机 disk crusher
圆盘叶轮风扇 disc fan
圆盘硬度计 disc hardness ga(u)ge
圆盘原理 theory of discs
圆盘月形孔抛光工具 roller crescent tool
圆盘轧碎机 disk crusher
圆盘闸 disk brake
圆盘振动器 disc vibrator
圆盘指针式天平 dial balance
圆盘制板法 crown process
圆盘制动器 disk brake
圆盘中耕机 disk harrow
圆盘柱 disc [disk] pile
圆盘装车机 disk car loader
圆盘状的 discoid; disklike
圆盘状电极 circular electrode
圆盘状辐射体 disk-shaped radiator
圆盘钻(头) disc [disk] bit
圆盘作用 disc [disk] action; sheet action
圆刨(床) buzz plane; circular plane; heel plane; round(ing) plane
圆喷口 round nozzle
圆喷嘴 round nozzle
圆片 disc [disk]; discus; pellet; planchet; push penny < 电线导管口上遮挡异物掉入的 >
圆片车刀 disc cutter
圆片窗玻璃 rondel
圆片法 disc method
圆片划线器 wafer scriber
圆片截煤机 disc cutter
圆片离合器 plate clutch
圆片式锯轨机 circular blade type rail sawing machine
圆片式气锯 circuit saw
圆片弹簧 disk spring
圆片铣刀 disc cutter
圆片形开槽锯 circular slitting saw
圆片形缺陷 disc type flaw
圆片原理 theory of discs
圆偏差 circular deflection
圆偏光 rotatory polarization
圆偏光器 circular polariscope
圆偏极光仪 circular polariscope
圆偏振 circular polarization
圆偏振波 circularly polarized wave
圆偏振光 circularly polarized light
圆偏振环形 V 天线 circular polarized loop vee[V]-antenna
圆偏振双折射 allogyric birefringence
圆频率 cyclic(al) frequency; circular frequency
圆平板 circular flat-plate
圆平头钉 round flat-headed nail
圆剖面小线脚 < 柱头或柱脚的 > astragal
圆漆刷 round paint brush
圆气泡水准器 circular bubble
圆牵引加万向接头 circle draft frame ball socket
圆钳 round pliers
圆钳子 round pliers
圆嵌条 cymbia
圆墙上的拱顶 cupola
圆锹 round shovel
圆鞘管 round sheath
圆切割 round cut
圆切面 circular section
圆穹 mamelon
圆穹顶 circular dome; dome; spheric(al) dome

圆穹顶坝 dome dam
圆穹顶肋 rib of dome
圆穹隆 domical vault
圆丘 carn; dome; hammock; hump(y); knob; knoll; mamelon
圆丘礁 knoll reef
圆丘泉 knoll spring
圆丘形礁 reef knoll
圆丘状的 dome-shaped
圆求方问题 quadrature of the circle
圆球半径 crown radius
圆球冲击试验 ball punch impact test
圆球灯 spheric(al) lamp
圆球灯罩 globe; light globe
圆球度 sphericity
圆球脚 bun foot
圆球筛 ball sieve
圆球饰 ball-flower
圆球体 spheroplast; spherosome
圆球形 spheric(al) shape
圆球形气球 rotary balloon
圆曲管 circle bend
圆曲线 circular curve; simple curve
圆曲线半切线型尖轨 semi-tangential form of circular curve of switch rail
圆曲线测设 circular curve location; layout of circular curves; setting-out of circular curves
圆曲线放样 layout of circular curves
圆曲线起点 beginning of circular curve
圆曲线型导轨 circular lead rail
圆曲线要素 circular curve data; elements of circular curve
圆曲线终点 end of circular curve
圆曲线主点 principal points of circular curve
圆圈 conic(al) ring
圆圈百分分度 centesimal(circle) graduation
圆圈测天仪 astrolabe; mariner's ring
圆圈等分法 circle bisect line method
圆圈地震 circular earthquake
圆圈法管网分析法 circle method of pipe-grid analysis
圆圈飞行中心 orbit point
圆刃螺丝起子 round blade screwdriver
圆刃楔形锤 fuller
圆蠕虫 round worm
圆筛 circular sieve; rotary strainer; round screen
圆筛条 round bar
圆山头饰 round pediment
圆扇形 circular sector
圆上方差 circle variance
圆上极差 circle range
圆上平均偏差 circle average deviation
圆上平均散度 circle average dispersion
圆射流 circular jet
圆石 boulder; cobble (boulder); cobblestone; field stone; pebble stone; round stone
圆石基层 boulder base
圆石脊 slate ridge; slate roll
圆石砾岩 conglomerate
圆石铺砌 cobbling
圆石铺砌路面 cobble pavement; cobblestone pavement
圆试样 round specimen
圆室 chevet
圆束 pencil of circles
圆树节 round knot
圆双曲线系统 circle-hyperbolic system
圆双折射 circular birefringence
圆水落管 round downpipe
圆水平仪 circular level; circular spirit level; spirit circular level

圆水准校正螺旋 circular level adjustment screw
圆水准器 ball fluid level; ball spirit level; box bubble; box level; bull's-eye fluid level; bull's eye level; bull's-eye spirit level; circular bubble; circular l fluid level; circular spirit level; spheric(al) fluid level; spheric(al) level vial; spheric(al) spirit level
圆水准器中心圆 limiting circle
圆水准仪 circular level; round spirit level
圆丝板 round die
圆算图 < 列线图 > circular nomogram [nomograph]
圆碎屑【地】 spheroclast
圆塔 circular tower; round tower
圆踏步 round step
圆台 round table; truncated cone
圆台平面磨床 rotary grinder
圆弹簧 coil spring
圆弹簧排水管 coil spring drain
圆体针 round bodied needle
圆天窗 bull's eye
圆条 round bar; round iron; round rod
圆条钢 round bar
圆条母线 bus rod
圆跳动 circular runout
圆铁 rod iron; round bar (iron)
圆铁钉 wire nail
圆铁辊 round iron bar
圆铁锚具 round iron anchorage
圆铁片 < 保持钢筋间隔用 > doughnut
圆铁条 round iron
圆铁桶 keg
圆厅 rotunda
圆亭 circular pavilion; rotunda; round kiosk
圆桶 barrel
圆桶搬运机 barrel handler
圆桶式冷乳液直接喷射装置 direct from drum cold emulsion sprayer
圆桶式直接喷射装置 direct from drum type sprayer
圆桶桶壁块 cylinder segment
圆桶形沉淀池 cylindric(al) sedimentation tank
圆筒 barrel; cylinder; drum
圆筒扳手 cylinder wrench
圆筒杯形砂轮 cylindric(al) cup wheel
圆筒仓 round silo; silo; silo bin; tower silo
圆筒仓室 silocell
圆筒插销 barrel bolt
圆筒搀和机 blunder
圆筒铲 round-pointed shovel; round-pointed spade
圆筒储存器 drum storage
圆筒(带槽)凸轮 barrel cam; drum cam
圆筒的 bullet-headed
圆筒丁字钢 bulb tee
圆筒顶 barrel camber
圆筒墩 cylinder pier
圆筒阀 ring gate; spool valve; cylindrical valve; cylinder gate < 水轮机的 >
圆筒法平板玻璃 cylinder glass
圆筒分级机 classifying drum
圆筒分选筛 separating cylinder
圆筒浮动式护舷 cylindric(al) floating fender
圆筒钢轨 bull-head(ed) rail
圆筒给水管 cylinder feed pipe
圆筒拱顶 tunnel vault
圆筒管 cylindric(al) pipe
圆筒柜 round tank
圆筒滚柱轴承 cylindric(al) roller

bearing
圆筒过滤布 bag filter fabric
圆筒混合机 drum mixer
圆筒混合器 drum mixer
圆筒加热器 cartridge heater
圆筒检验法 cylinder method
圆筒件锻造 ring forging
圆筒搅拌干燥器 agitated cylinder drier [dryer]
圆筒搅拌机 blunder
圆筒结构 cylindric(al) structure
圆筒锯 crown saw; cylinder saw; drum saw
圆筒卷筒绞车 cylindric(al) drum hoist
圆筒拉制法 <平板玻璃> cylinder drawing process
圆筒理论 barrel theory; theory of barrels; cylinder theory <拱坝设计>
圆筒粮仓 grain silo; tower silo
圆筒炉 cylindric(al) furnace; drum furnace
圆筒螺钉 button-headed screw
圆筒螺旋管 tube solenoid
圆筒门 cylinder gate
圆筒磨 cylindric(al) mill
圆筒内径测定器 cylinder ga(u)ge
圆筒内陷法 dipping cylinder method
圆筒碾磨机 cylinder mill; cylindric(al) mill
圆筒喷灌器 rotor sprinkler
圆筒碰锁 cylinder night latch
圆筒片拱桥 plate ribbed arch bridge with circular openings
圆筒屏蔽 roller shield
圆筒气瓶阀 cylinder valve
圆筒球磨机 cylindric(al) ball mill
圆筒筛 cylinder sieve; cylindric(al) screen; cylindric(al) sieve; drum screen; drum sieve; drum sifter; perforated cylinder; riddle drum; rotary drum screen; rotary screen
圆筒升降机 barrel elevator; barrel hoist
圆筒式丁坝 cylinder-type jetty
圆筒式丁堤 cylinder-type jetty
圆筒式分级机 cylinder sorter
圆筒式干燥炉 circular kiln
圆筒式拱 cylindric(al) arch
圆筒式过滤器 cylindric(al) filter
圆筒式结晶器 drum cylinder crystallizer
圆筒式掘孔 <隧道爆破的> cylinder cut
圆筒式粮舱 cylindric(al) granary
圆筒式码头 cylinder-type jetty; cellular type wharf
圆筒式千斤顶 column jack; cylindric(al) jack
圆筒式球磨机 drum mill
圆筒式热交换器 cylindric(al) heat exchanger; drum-type heat exchanger
圆筒式水下检波器 cylindric(al) hydrophone
圆筒式弹簧碰锁 cylinder (rim) night latch
圆筒式调压室 cylinder surge chamber
圆筒式突堤 cylinder-type jetty
圆筒式橡胶囊液压机 rubber pad hydraulic press of tunnel type
圆筒式闸坝 cylindric(al) barrage
圆筒式(真空)过滤器 drum vacuum filter; rotary drum filter
圆筒式轴流通风机 tube axial-flow fan
圆筒式装料斗 barrel hopper
圆筒刷布机 cylinder brushing machine
圆筒提升机 barrel elevator; barrel hoist
圆筒体 hollow cylinder
圆筒体滚光 barrel-burnishing
圆筒体抛光 barrel-burnishing
圆筒桅杆 pipe mast
圆筒喂料机 cylindric(al) screen feeder
圆筒洗涤器 cylindric(al) washer
圆筒洗矿机 drum washer
圆筒隙缝天线 pylon antenna
圆筒下落式黏[粘]度计 falling cylinder viscometer
圆筒销子锁 cylinder head lock; cylinder lock; night lock
圆筒形安全锁 safety cylinder lock
圆筒形坝 cylindric(al) barrage
圆筒形拌和盘 cylindric(al) mixing pan
圆筒形保险锁 safety cylinder lock
圆筒形被动筛 cylindric(al) passive screen
圆筒形壁灯 cylinder wall lamp
圆筒形薄膜存储器 cylinder-type magnetic thin film memory
圆筒形沉淀池 cylindric(al) sedimentation tank
圆筒形沉箱 cylinder caisson; cylindric(al) caisson
圆筒形冲洗筛 cylindric(al) washing screen
圆筒形储气罐 cylindric(al) gas holder
圆筒形窗 bared-shaped window
圆筒形挡水板 cylinder shield
圆筒形的 barrel-shaped; cylindric(al); wagon-headed
圆筒形吊斗 circular skip
圆筒形顶棚 cylindric(al) ceiling
圆筒形端铣刀 shell end mill
圆筒形发电机 barrel-shaped generator
圆筒形封闭式胶带运输机 cylindric(al) type conveyer [conveyor]
圆筒形浮标 cylindric(al) buoy
圆筒形浮筒 cylindric(al) buoy
圆筒形干燥器 cylindric(al) drier [dryer]
圆筒形钢筋混凝土薄壳 cylindric(al) reinforced concrete shell
圆筒形格筛 cylindric(al) sieve
圆筒形格子球磨机 Marcy mill
圆筒形拱 cylindric(al) arch
圆筒形拱坝 cylindric(al) arch dam
圆筒形拱顶 circular barrel vault
圆筒形拱顶的 tunnel-vaulted
圆筒形罐 bullet type tank
圆筒形虹吸管 circular barrel siphon
圆筒形滑动 <土坡的> cylindric(al) slide
圆筒形混凝土沉井 concrete cylinder
圆筒形加热坑 barrel heating pocket
圆筒形加热器 barrel heater
圆筒形交流分级机 Hardinge countercurrent classifier
圆筒形绞筒提升 parallel drum winding
圆筒形搅拌机 blunder; cylindric(al) mixer
圆筒形结构系统 tubular structure system
圆筒形掘进盾构 drum digger shield
圆筒形掘进铠框 drum digger shield
圆筒形开口沉井 cylinder caisson
圆筒形壳 circular cylindric(al) shell; cylindric(al) shell
圆筒形空气分布器 cylindric(al) air distributor
圆筒形冷凝器 cylindric(al) condenser
圆筒形滤器 barrel screen filter; cylindric(al) type filter
圆筒形码头 cylinder pier
圆筒形门锁 cylindric(al) lock
圆筒形木工旋风除尘器 cylindric(al) carpenter's cyclone dust collector
圆筒形排气泵 barrel exhausting pump
圆筒形排水泵 barrel exhausting pump
圆筒形喷管 nozzle with a parallel shroud
圆筒形千斤顶 cylindric(al) jack
圆筒形墙壁的倾斜度 <支撑圆屋顶的> drumminess
圆筒形穹顶 circular barrel vault; circular wagon vault
圆筒形裙座 cylindric(al) skirt support
圆筒形燃烧室 parallel-sided combustion chamber; throatless chamber
圆筒形容器 cylindric(al) vessel; tub
圆筒形散热器 barrel radiator
圆筒形(砂层)取样器 sampling cylinder for sand layer
圆筒形石笼 wire-cylinder
圆筒形水池 cylindric(al) pond; cylindric(al) tank
圆筒形水塔 standpipe
圆筒形水箱 cylindric(al) tank
圆筒形掏槽法 cylinder cut
圆筒形掏心法 cylinder cut
圆筒形天线阵 cylindric(al) array
圆筒形线圈 solenoid; solenoid coil
圆筒形橡胶护舷 cylindric(al) rubber fender
圆筒形橡胶缓冲器 <船尾/船头顶入泊位的> cylindrical rubber end-on
圆筒形消能器 cylindric(al) energy absorber
圆筒形泄水闸门 cylindric(al) sluice gate
圆筒形闸门 drum gate
圆筒形闸门坝 drum dam
圆筒形抓斗 barrel grab
圆筒形钻 cylindric(al) drill
圆筒型汽缸 barrel-type casing
圆筒旋涡 cylindric(al) vortex
圆筒选矿机 drum cobber
圆筒堰 cylinder weir; cylindric(al) weir; drum weir
圆筒闸门 roller drum gate; cylinder gate <水轮机的>
圆筒闸门进水口 cylinder gate intake
圆筒真空过滤机 drum vacuum filter
圆筒制粒机 drum pelletizer
圆筒轴承 barrel bearing; cylindric(al) bearing
圆筒柱 circular cylinder
圆筒转动筛 cylindric(al) screen
圆筒桩 cylinder pile; cylindric(al) pile
圆筒状拱顶 circular wagon vault
圆筒状屏 cylindric(al) screen
圆筒状旋涡 cylindric(al) vortex
圆筒状自由旋涡 cylindric(al) free vortex
圆头 pommel [pummel]; rounded front; round nose; snap head; ball peen <锤的>; round head <大头坝的>; fillister head <螺栓或螺钉>
圆头T形玻璃窗心条 bulb tee patent glazing bar
圆头T形玻璃格条 bulb tee patent glazing bar
圆头T形不用油灰的玻璃格条 bulb tee puttyless glazing bar
圆头槽铣刀 ball end mill
圆头铲 rounded tip spade; round-pointed shovel; round spade
圆头长钉 button-head spike
圆头车刀 round-nose tool; round-nose turning tool
圆头锤 ball hammer; ball peen hammer; round-face hammer; stamper
圆头锉 burr(-drill)
圆头大木槌 raising hammer
圆头带槽螺栓 stove bolt
圆头捣棒 bullet-nosed rod
圆头道钉 button-head spike
圆头的 blunt; bullet-headed; bullet-nosed; obtuse; round-pointed
圆头雕刻铣刀 ball end mill
圆头丁砖 bull head(er)
圆头丁字钢 bulb rail; bulb rail steel; bulb tee
圆头丁字铁 bulb tee
圆头钉 bullen; bull pin; button-head screw; cup head screw; French nail with round head; round-head(ed) nail; wire nail
圆头方颈螺栓 round-head(ed) square neck bolt; round top square neck bolt
圆头扶垛 round-head buttress
圆头杆 ball-head rod; bullet-nosed rod
圆头钢轨 bull-head(ed) rail
圆头钢条 bulb steel
圆头缓冲器 buffer disc [disk]
圆头活塞 round-head(ed) piston
圆头键 round-end key
圆头角材 bulb angle section
圆头角钢 bulb angle iron; bulb edge
圆头角料 bulb angle-bar
圆头角铁 bulb angle; bulb angle iron
圆头开关 button switch; push-button switch
圆头空心铆钉 snap head rivet with hollow shank
圆头窥孔 round-head(ed) loophole
圆头扩孔器 round reamer
圆头漏洞 round-head(ed) loophole
圆头螺钉 button-head (cap) screw; cheese (head) screw; cup head wood screw; round-head (carriage) bolt; round screw; round-head(ed) screw
圆头螺母 dome nut
圆头螺栓 button-head bolt; cape bolt; case bolt; round head(ed) bolt; snap(ped) head bolt
圆头螺丝 button-head screw; cheese head screw; cheese screw
圆头螺丝钉 blunt screw
圆头镘刀 bead tool
圆头锚碇 button-head anchorage
圆头铆钉 bull-head(ed) rivet; button head rivet; button rivet; cup-headed rivet; full head rivet; round-head(ed) rivet; snap head(er); snap head rivet; snapped rivet; snap rivet; snap rivet head
圆头帽钩头扳手 hooked tommy
圆头木螺钉 round-head(ed) wood screw; round wood screw
圆头木螺丝 cup head wood screw; round-head(ed) wood screw
圆头木铆钉 cup head wood screw
圆头平面锤 rounded flatter
圆头钳 round-nose(d) pliers
圆头锹 round-pointed shovel
圆头切刀 round-nose cutting tool
圆头式张拉锚具系统 button-head tensioning anchorage system
圆头式支墩坝 round-head(ed) type buttressed dam
圆头手锤 ball peen hammer
圆头铁 bulb iron
圆头铁条 bulb bar; bulb iron
圆头屋面螺钉 spring-head roofing nail
圆头销钉 French nail with round head; pin with round head
圆头形 bulb edge

圆头形窗 round-head(ed) window
圆头形挑出的面层 <由一片牛腿支承的> round-head(ed) corbel-table
圆头型 brachycephalic
圆头型材 bulb section
圆头旋压工具 round-nose spinning tool
圆头压板 round end clamp
圆头叶片 round-nosed blade
圆头(油漆)刷 ground brush
圆头錾 prick punch
圆头凿 mallet-headed chisel; round-nose chipping tool
圆头支墩 round-head buttress
圆头支墩坝 round-head buttress dam
圆头砖 bull-nose brick; round-nosed brick
圆头桩 round-head(ed) stake
圆头桩基 button bottom
圆凸缝 beaded joint
圆凸勾缝【建】bead joint
圆凸接缝 bead-jointed
圆凸卷边接缝 torus roll
圆凸面 circular face
圆秃秃的山顶 scalp
圆突形线脚装饰 ovolo
圆图 pie chart
圆图表 circular graphical chart
圆瓦当 round eaves-tile ornament
圆外旋轮线 epicycloid
圆弯(波导管) corner beveling
圆弯管 circle bend
圆弯头 round elbow
圆碗状物 circular bowl
圆网 cylinder screen; rotary screen
圆网成型机 cylinder forming machine
圆网辊筒印花机 rotary screen roller printing machine
圆网烘燥机 rotary screen dryer
圆网孔筛 round mesh screen; round mesh sieve
圆网屏 round screen
圆网印刷法 rotary screen printing
圆围栏断面 round guard section
圆尾 rounded tail
圆纹曲面 cyclic(al) surface
圆涡旋 circular vortex
圆屋顶 cupola; dome; domed roof; round roof; vaulted roof; tholos [复 tholoi] <希腊古典风格>
圆屋顶采光 ceiling dome light
圆屋顶窗 eye of dome
圆屋顶的 domical; domy
圆屋顶的中间隔层 interdome
圆屋顶构架托座 dished-out
圆屋顶观察窗 dome eye
圆屋顶间 interdome
圆屋顶上的水平压力 belt stress
圆屋顶式拱 arched dome
圆屋顶天窗 round roof-light
圆屋顶预均化堆场 preblend dome
圆屋顶支承环 thrust ring
圆屋顶座 tholobate
圆屋顶座盘 dome-drum
圆屋脊 round ridge
圆屋窑 dome kiln
圆屋中间隔层 interdome
圆物 roundel
圆铣 circular milling
圆铣刀 circular-cut file
圆线板条工作 bead-and-batten work
圆线材 cylindric(al) wire
圆线脚 ba(s)ton; round; round mo(u)lding
圆线脚平头 <板接缝的> bead and butt
圆线脚平头接 bead butt
圆线刨 astragal plane; beading plane; beading tool
圆线条 bead; bead mo(u)ld(ing)
圆线条带饰 bead seat band
圆线条刨 beading plane; beading tool; bead plane
圆线条(装)饰 gadroon
圆线型飞檐 astragal cornice
圆镶边 bull-nose trim
圆(向量)图 circle diagram
圆向预应力 circular prestress
圆橡皮条 round rubber stripe
圆销 round pin
圆小杆 boltel
圆楔块 round wedge
圆斜脊瓦屋面 round-hipped-plate roof
圆心 center [centre] of a circle; peripheral core
圆心轨迹【数】deferent
圆心角 angle at center [centre]; center [centre] angle; radius angle; central angle
圆行程的振动器 vibrating screen with circular movement
圆形 cabochon; circle; circular shape; rondure; roundness
圆形按钮 knurl
圆形暗销 round dowel pin
圆形凹地 amphitheater [amphithcatre]
圆形疤 popper
圆形把手 knob [knop]
圆形百叶式启动阀 circular louvred starting damper
圆形板 circular plate; cycloidal plate; dome slab; round plate; round slab
圆形板调风阀 circular damper
圆形板窝 round cella
圆形办公室 round office block
圆形半圆顶室 round apse
圆形堡垒 round bastion
圆形边 rounded edge
圆形边导器 circular edge guide
圆形边围 circular enclosure
圆形边缘 round edge
圆形标准差 circular standard deviation
圆形冰原岛峰 rognon
圆形冰原石山 rognon
圆形波 circular wave; cycloidal wave
圆形波导管 circular waveguide; round waveguide
圆形玻璃塔楼 round glass tower
圆形薄膜 circular membrane
圆形材 bar stock
圆形材料 rounded material
圆形槽 circular recess; encircling groove; ring channel; round-shaped slot
圆形测量仪表 roundness measuring instrument
圆形层 circular layer
圆形插销 barrel bolt
圆形插座 round socket
圆形觇板 sighting disc
圆形场 cirque
圆形车刀 circular bit
圆形车顶 arched roof
圆形沉淀池 circular sedimentation tank; cylindric(al) sedimentation tank
圆形沉箱 circular caisson
圆形衬砌 circular lining
圆形城堡主塔 circular donjon; circular dungeon
圆形城墙 <古希腊> cyclopean
圆形澄清池 circular clarifier
圆形池 orbital basin
圆形齿根轮廓 rounded root contour
圆形齿环 O-type dowel
圆形出口锥体 circular outlet cone
圆形窗 circular window; oeil de boeuf
圆形磁化 circular magnetization
圆形磁极 round pole
圆形粗齿锯 circular rip saw; ripping circular saw
圆形打捞母锥 <钢绳冲击钻的一种> round spud
圆形大剧场 amphitheater [amphithcatre]
圆形大梁 circular girder; round girder
圆形大露天剧场 <古罗马> Flavian architecture
圆形大厅 circular auditorium; circular hall; rotunda
圆形大珍珠 paragon
圆形带连接弧的格型钢板桩结构 steel-sheet pile cellular structure of circular cells with inter connecting arcs
圆形单元 circular cell
圆形导边 rounded leading edge
圆形导轨 circular guideway; circular lead rail
圆形导线 round conductor
圆形的 circular; circular shaped; cycloid; disciform; rotund
圆形的红木工件 circular work
圆形等代缺陷 circular disc defect
圆形堤头 round-head construction
圆形底 rounded bottom
圆形底脚 round footing
圆形抵座 <夹卡筒形工件用> cup dolly
圆形地脚 circular footing
圆形地牢 circular dungeon; round dungeon
圆形地下消防栓 round underground fire hydrant
圆形电缆 round cable
圆形电流计 circular galvanometer
圆形雕饰 medallion
圆形吊艇柱 bar davit
圆形丁字钢 T-bulb steel
圆形顶端饰物 pommel
圆形顶盖 circular cover
圆形顶棚窗 vaulted deck window
圆形定位销 round dowel pin
圆形洞室 circular openings
圆形斗 circular bowl
圆形度 circularity
圆形端头 button-head
圆形断面 bulb section; circular section
圆形断面排水管 circular conduit-type sewer
圆形断面隧洞 circular section tunnel
圆形锻模 dolly
圆形堆场 circular pile
圆形对称光纤 circularly symmetric fiber
圆形墩身 circular-shaft pier
圆形盾构 circular shaped shield
圆形多色性 circular pleochroism
圆形多室料仓 circular multi-compartment bin; circular multi-compartment silo
圆形剁刀 round gutter cutter
圆形发生炉 annular producer
圆形法 orbital process
圆形反射器 circular reflector
圆形房间 circular chamber
圆形房屋 tholos [复 tholoi]
圆形分布 circular distribution
圆形分度工作台 circular indexing table
圆形风管 circular duct
圆形风管防火阀 fire-resisting valves in round ducts
圆形风管调节板式送回风口 round duct air supply and return openings with damper
圆形风帽 round cowl
圆形封头 circular head
圆形扶手 mopstick; mopstick handrail; rounded handrail
圆形浮雕 medallion; tondo
圆形浮雕装饰品 pellet ornament
圆形钢 bar steel
圆形钢垫片 steel plate circular washer
圆形钢筋笼 circular reinforcing cage
圆形钢丝绳 circular wire brush
圆形钢条 round wire rod
圆形钢柱 round steel column
圆形格筛 cylindric(al) sieve
圆形格体 circular cell
圆形格箱板桩墙 circular cell
圆形格箱板桩围堰 circular cellular cofferdam
圆形格型钢板桩 circular cell
圆形格型结构 circular cell
圆形格栅 circular grate
圆形工件 round piece; round work
圆形拱 cycloidal arch
圆形拱窗 round-arched window
圆形拱或壳体 circular arch or shell
圆形共用淋浴设备 Bradley fountain
圆形钩 round hook
圆形构件 round element
圆形谷 circus; corrie
圆形骨料 rounded aggregate
圆形管道 circular conduit
圆形光圈 circular iris
圆形光圈启动阀 circular diaphragm starting damper
圆形广场 circus
圆形轨道 circular orbit
圆形轨道速度 circular orbital velocity
圆形滚柱粉碎机 annular roller mill
圆形涵洞 circular culvert
圆形核孔 nuclear pore
圆形核心 round core
圆形荷载 circular load(ing)
圆形荷载面积 circular loaded area
圆形盒底 round keep
圆形盒式磁盘数据存储装置 disk cartridge data storage unit
圆形桁架大梁 rounds-type trussed girder
圆形横断面 circular cross-section
圆形喉通道 circle throat clearance
圆形厚浆池 circular thickener
圆形花格窗 circular traceried window; round traceried window
圆形花坛 round bed
圆形滑动面 circular sliding plane; circular slip-plane; circular slip surface
圆形滑动面法 circular sliding plane method
圆形滑坡分析 circular-slide analysis
圆形环路 circular circuit
圆形回复机 circular purl machine
圆形回转试验水池 circular turning basin
圆形混凝土骨料 rounded concrete aggregate
圆形混凝土集料 rounded concrete aggregate
圆形混凝土铰支座 circular type concrete hinge-bearing
圆形混凝土烟道砌块 round concrete flue block
圆形或椭圆形窗 bull window
圆形机车车库 round house
圆形机车库 circular shed; circular engine house; round (locomotive) house【铁】
圆形基础 circular foundation; round footing; round foundation; circular footing
圆形基础柱 circular foundation pier

圆形基脚 circular footing
圆形极地波 circularly polarized wave
圆形极化波 circular polarized wave
圆形集尘器 circular dust collector
圆形记录计 circular recorder
圆形继电器 round type relay
圆形加固笼 circular reinforcing cage
圆形加载板 circular load plate
圆形监狱 panopticon
圆形检查车 circular manhole
圆形检查孔 round manhole; round manway
圆形建筑(物) circular building; cylindric(al) building; rotunda; round building; tholos [复 tholoi]; hunch
圆形键 round key
圆形浇注器 round distributor
圆形角材 bulb angle section
圆形教堂 church-in-the-round; round church
圆形接杆 round extension rod
圆形截面 circular (cross-) section
圆形截面钉 wire nail
圆形截面隧洞 circular canal tunnel
圆形金属饰 bullion
圆形晶体 rod-like crystal
圆形精选台 Sperry buddle
圆形井筒 round shaft
圆形井筒凿 round shaft chisel
圆形竞技场 amphitheater [amphithcatre]; arena; circus
圆形剧场 arena type stage
圆形剧场式舞台 arena type stage
圆形剧场中表演区 arena
圆形均布 circular area with fl uniform distribution
圆形均布荷载 circular uniform load
圆形均热炉 circular pit
圆形颗粒 rounded particle
圆形壳 toroidal shell
圆形刻度盘 clock-type dial
圆形坑 crater
圆形空气分布器 circular air distributor
圆形空心梁 circular box beam
圆形孔 ox-eye
圆形孔口 rounded orifice
圆形孔型 round pass
圆形控制堰 circular control weir
圆形扩孔器 round type reamer
圆形扩散器 round air diffuser; round diffuser
圆形廊 circular gallery
圆形肋 circular fin
圆形棱堡 round bastion
圆形冷锯 circular cold saw
圆形离合器组件 circular clutch pack
圆形礼拜堂 chapel-in-the-round
圆形立面 round facade
圆形砾石 subrounded pebble
圆形练马场 lunge
圆形链锁管 circular interlocking pipe
圆形梁 round beam
圆形列柱式庙宇 circular perimeter temple
圆形列柱式神殿 circular perimeter temple
圆形陵墓 tomb tholus
圆形流法 circular flow method
圆形隆起物 hunch
圆形楼面 round floor
圆形楼梯 circular stair(case); round stair(case)
圆形楼梯井 round stair(case) well
圆形露天剧场 < 古罗马 > amphitheater [amphithcatre]; coliseum
圆形露天运动场 coliseum
圆形滤叶加压叶滤机 pressure filter with cycloid filter leaves
圆形罗纹机 circular rib machine
圆形螺帽扳手 lug wrench
圆形螺纹 round thread
圆形螺纹梳刀 circular chaser
圆形马灯 bull's eye lantern
圆形码头 circular pierhead
圆形茅草独室房屋 < 非洲 > toukul
圆形茅屋 rondavel
圆形煤(基) circular coal
圆形门把手 knop
圆形门捏手 round knob
圆形门捏手装置 round knob door furniture
圆形门梃 < 双向门的 > rounded forend; rounded front
圆形面 rounded front
圆形庙宇 round temple
圆形膜盘 membranous disk
圆形木笼岸墩 log-crib abutment
圆形木榫 round dowel
圆形木条板 quirk bead
圆形木楔 pellet
圆形木柱头 wood roll
圆形内插法 circular interpolation
圆形内殿 circular cella; round cella
圆形浓缩机 circular thickener
圆形诺模图 circular nomograph
圆形排锤 round beater
圆形刨削 circular planing
圆形炮塔 round turret
圆形喷灌机 circular sprinkler
圆形喷水式洗手器 circular wash fountain
圆形喷雾器 round diffuser
圆形喷嘴 ring nozzle
圆形棚 round shed
圆形棚屋 circular hut
圆形披屋 circular shed
圆形偏极光 circularly polarized light; circulating polarized light
圆形拼合板牙 round split die
圆形平板 flat-sheet circle
圆形平面 round plan
圆形平面反光镜 round plane mirror
圆形平行板电离室 parallel circular plate chamber
圆形破坏模式 < 梅耶霍夫关于混凝土路面破坏形式的一种设想 > circular rupture model
圆形铺面 circular paving
圆形漆刷 sash tool
圆形砌块 round block
圆形墙内的水平钢筋 radius bent
圆形切割导向器 circle-cutting guide
圆形穹顶 annular vault; domical vault; arched dome
圆形丘 hummock
圆形囚牢 circular keep
圆形曲梁 circularly curved beam
圆形圈梁 circular ring girder
圆形缺口 circumferential notch
圆形热态锯切机 circular hot sawing machine
圆形人孔 round manhole; round manway
圆形塞子 circular plug
圆形扫描 circular scan(ning); scan-round
圆形扫描显示器 circular trace indicator
圆形扫描指示器 circular trace indicator
圆形筛孔 circular screen perforation
圆形(山)谷 cirque
圆形射流 round jet
圆形神庙 circular temple
圆形渗透计 cylinder permeameter
圆形石塔 < 奥克尼群岛，苏格兰本土的一种史前建筑 > broch
圆形视镜 circular sight glass
圆形输水隧道 circular canal tunnel
圆形输水隧洞 circular canal tunnel
圆形竖井 circular shaft
圆形竖流式沉淀池 circular vertical-flow settling tank
圆形竖炉 vertical cylindric(al) furnace
圆形水槽 full flush
圆形水池 ring-shaped basin
圆形水落管 circular down pipe
圆形水渠 circular conduit
圆形水准器 universal level
圆形隧道 circular tunnel
圆形隧道衬砌 circular tunnel lining
圆形隧道拱顶 circular tunnel vault
圆形隧道掘进机 rotary excavator
圆形索环眼 round thimble
圆形塔 round tower
圆形塔楼 circular turret; round turret
圆形台锯 circular bench saw
圆形钛铜复合棒 circular titanium cladded copper bar
圆形提升台 < 水泥试验用 > circular raised table
圆形天线 circular antenna
圆形天线阵 circular array
圆形调风器 circular register
圆形跳汰机 circular jig
圆形通风栅 circular air grid
圆形统计图 pie chart
圆形筒仓 round silo
圆形筒壳 circular cylindric(al) shell
圆形筒柱 cycloidal cylinder
圆形头 bell end; rounded nose
圆形凸出的门廊 cyrtostyle
圆形凸模 round punch
圆形图 circle chart; circle diagram; circle graph; circular chart; pie diagram; pie graph
圆形图记录器 round chart recorder
圆形外柱廊式 monopteral
圆形外柱廊式建筑 monopteral architecture
圆形弯管机 round bend
圆形网格法 scribed circle method
圆形网目板 circular screen
圆形微粒 round particle
圆形围栏外形 round guard rail profile
圆形围堰 circular cofferdam
圆形围柱式庙宇 round peripteral temple
圆形紊动射流 circular turbulent jet
圆形蜗壳 round section spiral casing
圆形污水管 circular sewer
圆形屋顶 arched roof; circular dome; rounded roof
圆形屋顶的柱间墙 tambour
圆形无头钉 round lost head nail
圆形物 rondure; rotundity; roundel; round stone
圆形雾化器 round air diffuser
圆形吸边器 circular edge guide
圆形吸附色谱 circular adsorption chromatogram
圆形吸附色谱法 circular adsorption chromatography
圆形铣切装置 circular milling attachment
圆形细孔(筛网) round mesh
圆形下水道 circular sewer
圆形舷窗 ordinary sidelight
圆形舷缘 rounded gunwale
圆形显示 round indication
圆形线材 round wire rod
圆形线荷载 circular linear load
圆形线脚 circular mo(u)lding; roll mo(u)lding; rounded mo(u)lding
圆形镶块抗裂试验 circular-patch crack(ing) test
圆形像 circular image
圆形销钉 round dowel
圆形小房 round cella
圆形小教堂 round chapel
圆形小室 circular cella
圆形小塔楼 circular diminutive tower; round diminutive tower
圆形斜板屋顶 round tilted-slab roof
圆形斜脊端 spheric(al) hipped end
圆形泄水孔 circular sluice
圆形心轴 < 车床及铸造用的 > mandrel
圆形芯墙 round core
圆形芯体 circular core
圆形芯线 round core wire
圆形蓄水池 circular reservoir
圆形悬索屋顶 wheel-type hanging roof
圆形牙顶 rounded crest
圆形牙钻 round bur
圆形烟囱 circular chimney
圆形烟道 circular flue
圆形堰 circular weir
圆形堰顶 circular sill
圆形阳台 circular balcony
圆形翼梢 round wing tip
圆形荧光灯 fluorescent O-lamp
圆形峪【地】 cwm
圆形预均化堆场 circular preblend stockpile
圆形预均化库 circulating preblending silo
圆形源 circular source
圆形照明器 circular luminaire
圆形折板屋顶 circular folded slab roof; circular hipped-plate roof
圆形折板屋面 round folded plate roof
圆形针织物 circular web
圆形振动板 cylindric(al) vibration plate
圆形振动(撞击)台 bumping table
圆形支承 circle support
圆形支承板 circular bearing plate
圆形纸色谱法 circular paper chromatography
圆形钟楼 circular campanile; round campanile
圆形钟塔 circular campanile
圆形周边流动沉砂池 circular perimeter flow grit settling tank
圆形轴封 round-body packing ring
圆形住房群 circus
圆形贮仓 circular silo
圆形柱顶板 circular abacus
圆形柱钢模 round column steel form
圆形柱(廊)建筑 monopteral; monopteron[复 monoptera]
圆形柱头 round capital
圆形柱头托板 circular raised table
圆形铸锭机 casting wheel
圆形砖窗 round brick window
圆形桩 circular pile
圆形桩码头 circular pile terminal
圆形装饰 medallion; round decoration; cording < 家具等 >
圆形锥体 cuneoid
圆形着陆航线 landing circle
圆形自转挤奶台 rotary dairy
圆形组织图表 circular organization chart
圆形钻 < 俚语 > mouse hole
圆形钻杆 round drill rod
圆形钻头 circular bit
圆形作业室 circular drawing chamber
圆悬杆 round hanger
圆压板 round clamp
圆烟囱砌合 circular chimney bond
圆阳角 arris rounded
圆窑 circular kiln; round kiln
圆翼形散热器 fin and tube type radiator; finned radiator
圆阴角 coving; fillet
圆英寸 circular inch
圆油刷 round paint brush
圆圆定位系统 range positioning system

圆圆航法 rho-rho system navigation; range-range navigation
圆缘 rounded edge
圆缘孔 rounded orifice
圆缘梁 bulb beam
圆缘螺帽 collar nut; flange nut
圆缘螺母 collar nut; flange nut
圆缘刨 astragal plane; fillet plane
圆缘凿石 backsetting
圆攒尖 round pavilion roof
圆攒尖顶 conic(al) roof
圆凿 scalper
圆凿钳 gouge forceps; gouge-nippers
圆凿子 circular chisel; gouge; paring gouge; turning gouge
圆闸门 drum gate
圆罩 bonnet; spheric(al) calotte
圆罩式曝气器 dome aerator
圆针 round needle
圆枕 torus [复 tori]
圆枕木锅炉座 saddle
圆振动 circular vibration
圆(振)二(向)色性 circular dichroism
圆整 rounding
圆整数 round number
圆正态分布 circular normal distribution
圆正形投影 circular orthomorphic projection
圆支墩 round pillar
圆直点 curve to tangent point; point of tangent
圆直面接合 circular mitre [miter]
圆纸色谱 paper-disk chromatography
圆趾角钢 angle bulb; bulb angle(-bar)
圆趾铁 bulb iron
圆趾桩 bulb pile
圆周 circle; circum; periphery
圆周半径 radius of a circle
圆周爆破 perimeter blasting
圆周闭合差 closure error of horizon; error of closure of horizon
圆周闭合条件 condition for closing the horizon
圆周标度 circular scale
圆周波纹 circumferential corrugation
圆周槽口 circumferential notch
圆周长(度) peripheral length; circumference length
圆周齿距 circular pitch
圆周的 circumferential
圆周的四分之一 quadrant
圆周定位螺丝 circumferential register screw
圆周对接焊 circular butt welding
圆周法 three-figure method
圆周反应 peripheral reaction
圆周分速度 peripheral component of velocity
圆周规 circumference ga(u)ge
圆周焊缝的允许偏移 allowable offset in circumferential joints
圆周焊接 circular butt welding; circular seam welding; circumferential welding
圆周滑动卡尺 circumference slide calipers
圆周积分 circulation integral
圆周加速度 circular acceleration
圆周间隙 circumferential clearance; circle clearance <转车盘>
圆周剪切机 circle-cutting shears
圆周角 angle in a circular segment; angle of circumference; circumferential angle; perigon; round angle
圆周角闭合条件 conditions for closing the horizon
圆周角条件 condition for closing the horizon
圆周接缝 circumferential joint; circumferential seam; transverse seam
圆周节距 circular pitch; circumferential pitch
圆周进给 rotary feed
圆周拉应力 circumferential tensile stress; tensile circumferential stress
圆周累计齿节 cumulative circular pitch
圆周力 circumference force; circumferential force; force of periphery; peripheral force
圆周流 circumferential flow
圆周率 circular constant; circumference ratio; ratio of the circumference of a circle to its diameter; pi【数】
圆周轮磨 circular grinding accessory
圆周罗盘 circumferentor [circumferenter]
圆周铆钉法 circumferential riveting
圆周磨削 peripheral grinding
圆周喷射 peripheral jet
圆周频率 circular frequency
圆周平均法 method of circle average
圆周切割装置 circle-cutting attachment
圆周切线 tangent to periphery
圆周切线荷载 circumferential line load
圆周绕流 peripheral flow
圆周扫描 circular scanning
圆周扫描声呐接收器 sonar receiver-scanner
圆周扫描调制盘 circular-scanning reticle
圆周升降机 peripheral hoist
圆周受拉状态 state of all-round tension
圆周搜索 round search
圆周速度 circular speed; circumference velocity; circumferential speed; circumferential velocity; peripheral speed; peripheral velocity; peripheric velocity; rim speed; circular velocity; surface feet per minute <英尺/分>
圆周速率 peripheral speed; peripheral velocity
圆周凸轮 peripheral cam
圆周误差 circular error
圆周铣削 peripheral milling
圆周线焊机 circular seam welder
圆周线速度 peripheral speed
圆周形 circle pattern
圆周压力 circumference pressure; circumferential pressure
圆周堰 peripheral weir
圆周应变 circumferential strain
圆周应力 circumference stress; circumferential stress; hoop stress; peripheral stress; ring stress
圆周预加应力(法) circular prestressing
圆周运动 circling motion; circumferential motion; circular motion
圆周运行轨道 <波浪运动中水质点的> circular orbit
圆周振动 circle vibration
圆周支承 supported at circumference
圆轴 circular shaft
圆轴车床 shaft turning lathe
圆肘管 round elbow
圆珠笔 ball pen; ball point pen; biro <可以吸墨水的>
圆竹节钢筋 deformed round steel bar
圆住头铆钉 cylindric(al) head rivet
圆柱 circular column; cylinder; cylindric(al) post; round column; slug
圆柱棒 <粉碎材料用> cylpeb
圆柱表面 periphery
圆柱波 cylindric(al) wave
圆柱测径规 cylindric(al) plug ga(u)ge
圆柱插头 pin plug
圆柱齿轮 cylindric(al) gear
圆柱大厅 hall of columns; stoa <有屋顶的拜占庭建筑的>
圆柱的阻力系数值 drag coefficient value for circular cylinder
圆柱等积投影 cylindric(al) equal-area projection
圆柱顶 circular capital
圆柱顶板 <其上搁柱顶过梁> abacus [复 abaci/abacuses]
圆柱顶板花纹 abacus flower
圆柱顶冠 abacus [复 abaci/abacuses]
圆柱度 cylindricity; roundness
圆柱段 cylindric(al) section
圆柱墩 circular-shaft pier
圆柱阀 spool
圆柱方程 equation of cylinder
圆柱浮标 spar buoy
圆柱(钢)棒 cylpeb
圆柱拱坝 cylindric(al) arch dam
圆柱箍圈 parallel
圆柱滚动轴承 cylindric(al) roller bearing
圆柱滚子轴承 roller bearing
圆柱函数 cylindric(al) function
圆柱环规 cylinder ga(u)ge; cylindric(al) ga(u)ge
圆柱基部的虎爪形装饰 griffe
圆柱集 cylinder set
圆柱铰刀 cylindric(al) reamer
圆柱径 caliber [calibre] size
圆柱壳 barrel vault(ing)
圆柱孔 cylindric(al) hole
圆柱控制器 spatial carpenter inverter
圆柱立式钻床 column upright drill machine
圆柱螺母 barrel nut
圆柱螺线 cylindric(al) spiral
圆柱螺旋弹簧 cylindric(al) coiled spring
圆柱螺旋线 circular helix; cylindric(al) helix
圆柱码头 round pier
圆柱面 cylinder surface; cylindric(al) surface
圆柱面断层 cylindric(al) fault
圆柱面活塞环 straight face piston ring
圆柱面极坐标图 circle cylindric(al) polar coordinate plot
圆柱面刨削装置 circular planing attachment
圆柱面数据 cylinder data
圆柱面荧光屏 cylindric(al) face
圆柱磨头 cylinder grinding head
圆柱木芯块 mandrel
圆柱泥门 cylindric(al) hopper door
圆柱扭曲 cylinder twist
圆柱平尖 flat cylindrical point
圆柱强度 cylinder crushing strength
圆柱桥墩 round pier
圆柱穹隆 cylindric(al) vault
圆柱塞规 plug ga(u)ge; round plug ga(u)ge
圆柱筛 cup screen
圆柱渗透计法 <测定导水率的> cylinder permeameter method
圆柱石 assize
圆柱式泵 plunger pump; plunger type pump
圆柱式风暴信号 storm drum
圆柱式封闭蓄水池 standpipe
圆柱式码头 cylinder-type jetty
圆柱式燃油泵 plunger fuel pump
圆柱式碎石机 cylinder-type crusher
圆柱式栈桥码头 braced cylinder type wharf
圆柱试件 cylindric(al) specimen
圆柱试体 test core; test cylinder; push-out cylinder <现场从未硬化混凝土中取出的>
圆柱收分线 entasis [复 entases]
圆柱锁 bored lock
圆柱弹簧 cylindric(al) spring
圆柱体 column; cylindric(al) object; pipe barrel
圆柱体部分 segment of cylinder
圆柱体长度 length of cylinder
圆柱(体)的 cylindric(al)
圆柱体抗压强度 cylinder compression strength; cylinder crushing strength; cylindric(al) compressive strength
圆柱体抗压强度试验 cylinder test
圆柱体量规 cylinder ga(u)ge
圆柱体劈裂试验 split-cylinder test
圆柱体劈裂试验抗拉强度 split-cylinder tensile strength
圆柱体强度 cylinder strength
圆柱体强度试验 cylinder strength test
圆柱体石块 assize
圆柱体式浮护舷 cylindrical floating fender
圆柱体试件 <混凝土> cylinder specimen
圆柱体(试件)开裂试验 cylinder splitting test
圆柱体试件抗压强度 <混凝土> cylinder compressive strength; cylinder crushing strength
圆柱体试件强度 cylindric(al) specimen strength
圆柱体试件试验 core test
圆柱体(试件)试验机 <混凝土> cylinder tester; cylinder testing machine
圆柱体试块 test cylinder
圆柱体(试块)试验 cylinder test
圆柱体试样 cylinder sampler; cylindric(al) sample
圆柱体(试样的)压缩试验 cylindric(al) compression test
圆柱体压碎强度 cylinder crushing strength
圆柱体压缩 cylindric(al) compression
圆柱体岩芯 cylinder core
圆柱体支承 cylinder support
圆柱体支座 cylinder support
圆柱调和函数 cylindric(al) harmonics
圆柱筒壳 circular cylindric(al) shell
圆柱头 circular capital
圆柱头顶板 tailloir
圆柱头内六角螺钉 socket cap screw
圆柱投影 <测绘地图的一种画法> conic(al) projection; cylindric(al) (map) projection
圆柱投影反求法 inverse solution of cylindric(al) projection
圆柱透视(法) cylindric(al) perspective
圆柱凸轮 axial cam
圆柱尾物体 cylindrically based body
圆柱蜗杆 cylindric(al) worm; straight worm
圆柱锡矿 cylindrite
圆柱锡石 cylindrite
圆柱销 cylinder pin; standard pin; straight pin
圆柱谐函数 cylindric(al) harmonic function
圆柱形 cannon; circular cylinder
圆柱形玻璃塔 cylindric(al) glass tower
圆柱(形薄)壳 cylindric(al) shell
圆柱形薄壳屋顶 cylindric(al) shell

roof
圆柱形插锁 cylinder mortice [mortise] lock
圆柱形插头 cannon plug
圆柱形沉降槽 cylindric(al) settler
圆柱形沉井 large-diameter cylinder
圆柱形沉箱 cylinder caisson;cylindric(al) caisson
圆柱形城堡主塔 cylindric(al) donjon;cylindric(al) keep <中世纪>
圆柱形齿轮滚刀 parallel hobbing cutter
圆柱形船头 cylindric(al) bow
圆柱形导杆 cylindric(al) guide
圆柱形的 columniform;cylindric(al)
圆柱形断层 cylindric(al) fault
圆柱形对称 cylindric(al) symmetry
圆柱形墩 cylindric(al) pier
圆柱形焚化炉 wigwam;wigwam-type incinerator
圆柱形浮动式(径向受力)护舷 cylindric (al) floating diametrically loaded fender
圆柱形钢板滚柱 cylindric(al) plate-steel roller
圆柱形钢管插针 cylindric(al) steel needle
圆柱形钢筋 cylindric(al) bar
圆柱形钢筋混凝土薄壳结构 cylindric(al) concrete shell structure
圆柱形鼓筒 cylindric(al) drum
圆柱形光栅 cylindric(al) lenticulation
圆柱形滚筒 rotating cylindric(al) drum
圆柱形滚柱 <输送机的> cylindric(al) roller
圆柱形滚柱推力轴承 cylindric(al) roller thrust bearing
圆柱形滑动 cylindrical slide
圆柱形混凝土人行巷道 cylindric(al) concrete manway
圆柱形混凝土试件 cylindric(al) concrete test specimen;cylinder specimen;cylindric(al) specimen
圆柱形混凝土试样 cylindric(al) concrete test specimen
圆柱形火药柱 cylinder grain
圆柱形货物 cylindric(al) goods
圆柱形建筑(物) cylindric(al) building
圆柱形铰接支座 cylindric(al) knuckle bearing
圆柱形铰链 cylindric(al) hinge
圆柱形卷筒 cylindric(al) drum
圆柱形卷装 cylindric(al) build package
圆柱形壳 circular cylindric(al) shell
圆柱形壳顶 cylindric(al) shell roof
圆柱形壳体 cylindric(al) shell
圆柱形壳体结构 cylindric(al) shell structure
圆柱形空腔谐振器 cylindric(al) cavity
圆柱形孔道 cylindric(al) vent
圆柱形立罐 cylindric(al) tank
圆柱形螺母 cylindric(al) nut
圆柱形螺纹 cylindric(al) thread;straight thread
圆柱形门 cylindric(al) gate
圆柱形面 cylindroid
圆柱形摩擦卷筒 cylindric(al) friction drum
圆柱形磨碎凿尖 cylindric(al) grinding point
圆柱形千斤顶 cylindric(al) jack
圆柱形桥墩 cylindric(al) pier;cylinder pier
圆柱形切片机 cylinder microtome
圆柱形燃烧室 cylindric(al) combustion chamber
圆柱形人孔 cylindric(al) manhole
圆柱形人行巷道 cylindric(al) manway
圆柱形容器 cylindric(al) vessel
圆柱形锐孔 hat orifice
圆柱形筛面 cylindric(al) screening surface
圆柱形石块 assize
圆柱形石柱 drum
圆柱形试件抗压试验 <混凝土> cylinder test
圆柱形试块 <混凝土> test cylinder
圆柱形(试样的)压缩试验 cylindric(al) compression test
圆柱形锁 barrel lock;cylindric(al) lock
圆柱形陶槽 cylindric(al) cut
圆柱形天线 cylindric(al) antenna;hoop antenna;sausage antenna
圆柱形(投影)海图 cylindric(al) chart
圆柱形透镜 cylindric(al) lens
圆柱形推进器 cylindric(al) impeller
圆柱形托辊 <输送机的> cylindric(al) roller
圆柱形围堰 cylindric(al) cofferdam
圆柱形涡流层 cylindric(al) vortex sheet
圆柱形屋脊 cylindric(al) piend
圆柱形铣刀 cylindric(al) face milling cutter;plain spiral milling cutter
圆柱形系留气球 sausage balloon
圆柱形系数 cylindric(al) coefficient;longitudinal coefficient;prismatic(al) coefficient
圆柱形线弧 link frame field
圆柱形线圈 <变压器的> cross-over coil
圆柱形橡胶防撞圈 cylindric(al) rubber fender ring
圆柱形橡胶防撞装置 cylindric(al) rubber fender
圆柱形橡胶护舷 cylindric(al) rubber fender
圆柱形橡胶碰垫 cylindric(al) rubber fender
圆柱形心轴弯曲柔韧性 flexibility cylindric(al) mandrel
圆柱形型芯 cylindric(al) core
圆柱形穴 churn hole
圆柱形烟囱 cylindric(al) chimney
圆柱形岩石 chimney rock
圆柱形叶片 blade cylinder
圆柱形阴极 cylindric(al) cathode
圆柱形油石 cylindric(al) oil stone
圆柱形有机玻璃壳 plexiglass cylindrical shell
圆柱形钥匙 cylinder key
圆柱形闸门 cylinder gate
圆柱形褶皱 cylindric(al) fold
圆柱形支承轭架 cylinder support yoke
圆柱形支承轭状物 cylinder support yoke
圆柱形直角器 cylindric(al) cross staff
圆柱形钟楼 cylindric(al) campanile
圆柱形桩 cylinder pile;cylindric(al) pile
圆柱形准直器 column collimator
圆柱形钻 cylindric(al) drill
圆柱形钻头 cylindric(al) bit;cylindric(al) drill
圆柱压缩仪 cylinder compressometer
圆柱牙轮扩孔器 roller reamer
圆柱域 cylindric(al) domain
圆柱闸门 cylinder gate
圆柱闸门进水口 cylinder gate intake
圆柱正齿轮 cylindric(al) spur gear
圆柱正形投影 conformal cylindric(al) projection
圆柱直角器 cylindric(al) cross staff
圆柱中微凸线样板 entasis reverse
圆柱中心投影 central cylindric(al) projection
圆柱轴 cylinder axis;spindle column <多倍仪 Γ 形柱的一部分>
圆柱轴颈 cylindric(al) journal
圆柱注水泥法 cylinder method
圆柱状的 columnar
圆柱状断层 cylindric(al) fault
圆柱状管型 cylindric(al) cast
圆柱状盒 tubular case
圆柱状孔道 cylindric(al) vent
圆柱状排列的盥洗盆 column lavatory basin
圆柱状排列的清洗盆 column wash basin
圆柱状排列的水洗槽 column wash bowl
圆柱状排列的洗脸盆 column wash basin
圆柱状石拱顶 stone cylindrical vault
圆柱状探测器组合 detector assemblage with cylindric(al) shape
圆柱状体 cylindroid;cylinder
圆柱自升式平台 column jack-up
圆柱钻 cylinder bit
圆柱坐标 circular cylindric(al) coordinate;cylindric(al) coordinates
圆柱坐标机器人 cylindric(al) coordinates robot
圆柱坐标式机器人 cylindric(al) coordinates robot
圆柱坐标式机械手 cylindric(al) coordinate manipulator
圆柱坐标系(统) cylindric(al) coordinate system
圆炷菌属 <拉> Gyrophana
圆转盖板梳棉板 revolving flat card
圆装饰线脚 caulis
圆状 rounded
圆锥 conus
圆锥半角 half-angle of projection
圆锥棒型联结件 taper rod type tie
圆锥杯突试验 conic(al) cup test
圆锥表面 <保障飞机场周围航行安全> conic(al) surface
圆锥柄轴 conic(al) arbor
圆锥波道聚光器 cone-channel condenser optics
圆锥薄壳 cone shell
圆锥测定法 cone test method
圆锥常数 constant of the cone
圆锥车床 taper turning lathe
圆锥沉淀槽 settling cone
圆锥承载力值 cone bearing value
圆锥承载试验 cone bearing test
圆锥承重试验 cone bearing test
圆锥齿轮 bevel gear wheel;cone gear;conic(al) gear
圆锥齿轮轴 bevel pinion shaft
圆锥触探 cone penetration
圆锥触探器 cone penetrometer;conic(al) penetrometer
圆锥触探试验 cone penetration test
圆锥触探(试验)仪 cone penetrometer
圆锥锉 taper file
圆锥的高 cone height
圆锥底 taper base
圆锥顶点 conic(al) tip
圆锥顶销 taper knock
圆锥动力触探试验 dynamic(al) penetration test
圆锥度 conicity;coning;degree of coning
圆锥墩形顶 motte-top castle
圆锥二次曲线 conic(al) section
圆锥法 <测定混凝土坍落度的> cone method
圆锥法试验 <测定混凝土坍落度或砂的容重> cone test
圆锥分级机 cone classifier
圆锥分选机 separating cone
圆锥辐射计 cone radiometer
圆锥高丘式样 sugar-loaf fashion
圆锥鼓轮 <起重机的> scroll drum
圆锥管 conic(al) pipe;conic(al) tube
圆锥贯入度 cone penetration
圆锥贯入度试验 cone penetration test;deep penetration test
圆锥贯入度仪 conic(al) penetrometer
圆锥贯入器 cone penetrometer
圆锥贯入试验 cone bearing test;cone penetration test
圆锥贯入仪 cone penetrometer
圆锥贯入指数 cone penetration index
圆锥贯入阻力 cone resistance
圆锥贯入阻力值 cone-resistance value
圆锥辊 tapered roll
圆锥辊颈轧辊 taper-neck roll
圆锥滚柱轴承 roller bearing;tapered roller bearing
圆锥滚子轴承 conic(al) roller bearing
圆锥号型 conic(al) shape
圆锥花序 panicle
圆锥交叉拱穹棱 conic(al) groin
圆锥角 coning;coning angle
圆锥角环带法 cone angle banding
圆锥绞筒提升机 conic(al) drum hoist
圆锥接头 cone joint
圆锥截体 frustum of a cone
圆锥颈桶 cylindric(al) taper drum
圆锥卷筒绞车 conic(al) drum hoist
圆锥壳 bowl
圆锥壳体 conoid shell
圆锥孔 conic(al) bore
圆锥离合器连接 clutch bevel
圆锥连接配合 conic(al) fit
圆锥连接锅筒 conic(al) connection
圆锥流限仪 cone penetrometer
圆锥螺纹 tapered thread
圆锥锚具 Freyssinet cone anchorage
圆锥铆钉 cone neck rivet;conic(al) head rivet;tapered neck rivet
圆锥面 circular conic(al) surface;conic(al) surface
圆锥磨法 cone grinding
圆锥磨轮 coned grinding wheel
圆锥磨凿尖 conic(al) grinding point
圆锥浓缩机 thickening cone
圆锥破碎机 cone crusher;conic(al) breaker;conic(al) crusher;gyratory;rotary cone crusher;taper crusher
圆锥破碎机的破碎面 crushing concave
圆锥破碎机的腔部 concave
圆锥曲面 conicoid
圆锥曲线 conic(al) curve;conicoid
圆锥扫描自动跟踪系 cone scan auto-tracking
圆锥式采样器 conic(al) sampler
圆锥式破碎机 cone-type crusher
圆锥式碎石机 cone crusher
圆锥试验 cone test
圆锥试验法 cone test method
圆锥饰 gutta [复 guttae];medallion gutta;treenail
圆锥四分(取样)法 cone quartering
圆锥碎石机 gyratory crusher;rotary crusher
圆锥台 frustum cone;frustum of pyramid
圆锥探头阻力 cone (penetration) resistance

圆锥体 circular cone; cone; penetrator; circular conoid; conoid
圆锥体冲洗式触探头 conic(al) wash bit
圆锥体触探头 conic(al)-shaped point
圆锥体的 conic(al)
圆锥体尖顶 pinnacle terminating in conic(al) form
圆锥体破坏 conic(al) failure
圆锥体屋顶 conic(al) roof; conoid roof
圆锥筒试验 cone method
圆锥头的 cone-headed
圆锥头螺栓 cone-headed bolt; conic(al) head bolt
圆锥头铆钉 conic(al) head rivet
圆锥投影地图 conic(al) projection chart; conic(al) projection map
圆锥投影(法) conic(al) projection
圆锥投影反求法 inverse solution of conic(al) projection
圆锥投影面展开 development of a conic(al) projection surface
圆锥外形 < 喷射混凝土形成的 > gunning pattern
圆锥网 brailer
圆锥屋脊 conic(al) hip
圆锥销 tapered bolt; tapered cotter
圆锥形 cone; conoid; loaf; sugar-loaf fashion; taper
圆锥形拌和机 conic(al) mixer
圆锥形侧倾式搅拌机 conic(al) tilting mixer
圆锥形澄清池 conic(al) type clarifier
圆锥形刀具 < 加工砂轮的 > conic(al) cutter
圆锥形的 cone-shape(d); conic(al); coniform; coning; cylindroconic(al)
圆锥形顶棚散流器 cone-type ceiling diffuser
圆锥形堆垛 coneply stacking
圆锥形风暴信号 storm cone
圆锥形风标 air cock; wind sock; wind cone
圆锥形风箱 conic(al) bellows
圆锥形浮标 conic(al) buoy; nun buoy
圆锥形钢轨导接线塞钉 conic(al) rail bond pin
圆锥形钢滚子 tapered steel roller
圆锥形高丘 sugar loaf
圆锥形固定拱 tapered circular fixed arch
圆锥形管嘴 conic(al) nozzle
圆锥形滚筒 scroll drum
圆锥形锅炉筒 conic(al) shell course
圆锥形锅筒 gusset course
圆锥形红外整流罩 conic(al) irdome
圆锥形滑车 cone pulley
圆锥形搅拌机 conic(al) mixer
圆锥形截面 conic(al) section
圆锥形晶状体 lenticonus
圆锥形静区 cone of silence
圆锥形壳 conic(al) shell
圆锥形可移动交通标志 traffic cone
圆锥形喇叭 conic(al) horn
圆锥形漏斗管 petticoat pipe
圆锥形螺母 tapered nut
圆锥形螺栓 conic(al) head bolt
圆锥形螺旋 tapered auger
圆锥形锚塞 < 预应力混凝土的 > conic(al) plug
圆锥形泥门【疏】 conic(al) hopper door
圆锥形浓缩机 cone thickener
圆锥形平面轮 conic(al) face wheel
圆锥形破碎机 cone crusher
圆锥形穹顶 conic(al) dome
圆锥形丘 conic(al) loaf; sugar loaf
圆锥形球磨机 cylindroconic(al) ball mill
圆锥形裙座 conic(al) skirt support
圆锥形燃烧室 conic(al) combustion chamber
圆锥形塞 conic(al) plug; male cone < 预应力用的 >
圆锥形扫描 conic(al) scan
圆锥形石堆 cairn
圆锥形石建筑 nuraghi
圆锥形水管 conic(al) water pipe
圆锥形水轮机 conic(al) turbine
圆柱形弹性销 cylindric(al) resilient key
圆锥形调压室 conic(al) surge chamber
圆锥形投影 conic(al) projection
圆锥形尾水管 conic(al) draft tube
圆锥形屋顶 conic(al) roof
圆锥形纤维 conic(al) fiber [fibre]
圆锥形线圈 cop
圆锥形小山 sugar loaf
圆锥形扬声器 cone-type loudspeaker
圆锥形轧碎机 cone crusher
圆锥形褶皱 conic(al) fold
圆锥形正轴位置 normal position of the cone
圆锥形重介质选矿机 conic(al) heavy-medium separator
圆锥形轴棒 conic(al) mandrel
圆锥形轴尖 conic(al) pivot
圆锥形柱 batter post
圆锥形转塞阀 conic(al) plug valve
圆锥形转筒筛 conic(al) trommel
圆锥形桩靴 conic(al) pile point
圆锥选矿机 separating cone; separator cone
圆锥牙 conic(al) tooth
圆锥造球机 cone pelletizer
圆锥针入度 deep penetration
圆锥针入度仪 conic(al) penetrometer
圆锥振动筛 cone shaker
圆锥正割投影 conic(al) projection with two standard parallels; secant conic(al) projection
圆锥指数 cone index
圆锥轴承 cone bearing
圆锥转子式选粉机 conic(al) rotor type air separator
圆锥状堆积 conic(al) accumulation
圆锥状褶皱 conic(al) fold
圆坠砣 circular weight
圆桌 round table
圆桌会议 round-table conference; round-table meeting
圆紫铜棒 round copper bar
圆族 family of circles
圆钻 circular bit; round bur
圆钻杆 circular rod
圆钻头 blank bit
圆嘴钳 round pliers
圆坐标 circle coordinates

援

援救潜水员 assisting diver; rescue diver

援款限制 aid tying
援外的 aid-foreign
援外运输 aid-foreign traffic
援助 aid; bolster; stand-by; succo(u)r; support
援助的附带条件 aid tying
援助额 aid disbursement
援助方式 aid modality
援助国 donor country
援助合作协议 aid cooperation agreement
援助机构 aid agency
援助基金 aid fund
援助设备 aid equipment
援助项目 aid project
援助协定 aid agreement
援助协议 aid agreement
援助一体化 integration of assistance
援助者 booster; reliever; supporter

缘

缘 rand; rim

缘斑 marginal spot
缘板 marginal plate; marginal scute
缘比 peripheral ratio
缘边 edge; lip
缘边焊 flanged edge weld
缘层 marginal layer
缘成型术 marginoplasty
缘承吊桥 rim-bearing draw
缘齿菌属 < 拉 > Acia
缘垂 marginal lappet
缘丛 marginal plexus
缘带 marginal zone
缘带线 marginal fasciole
缘端 acies
缘垛 festoon
缘杆 edge bar
缘故 sake; the whys and wherefores
缘嵴 marginal crest
缘间沟 intermarginal sulcus
缘间线 intermarginal line
缘结节 marginal tubercle
缘晶形器 marginal lens-like organ
缘孔 marginal pore
缘孔纹 pit border
缘口防护装置 rim guard
缘墙板 curb plate
缘墙的山墙盖瓦 curb gable tile
缘石 border; kerb curb
缘石标志 curb mark(ing)
缘石街沟 curb and gutter
缘石排铺成行 curb arrangement
缘石坡道 curb ramp
缘石修面 curbstone finish
缘饰【建】 border; fringe; narrow fabric skirt
缘饰胶合板 mo(u)lded plywood
缘饰托梁 border joist
缘饰镶板 mo(u)lded panel
缘速比 peripheral ratio
缘条 fillet
缘褶 marginal fold
缘趾 < 轮胎的 > toe of bead
缘趾线环 toe bead wire
缘周板 contour plate

源

源板 source plate

源包 source packet
源保护 source protection
源倍增法 source multiplication method
源编辑程序 source editor
源编辑应用程序 source edit utility option
源编码 source code
源标准化 source calibration
源表示 source-representation
源部件 source block
源采样 source sampling
源参量 source parameter
源参数 source parameter
源层成分 composition of source diapir
源层厚度 thickness of source layer
源层埋深 depth of source layer
源程序 source program(me); source routine; subject program(me)
源程序变量 source program(me) variable
源程序带 source program(me) tape
源程序符号 source program(me) symbol
源程序卡片套 source deck
源程序库 source library; source program(me) library
源程序模块 source module
源程序设计 source programming
源程序信息 source image
源程序形式 source program(me) form
源程序运算 source program(me) operation
源程序正文 source text
源程序指示字 source program(me) pointer
源程序字 source program(me) word
源程序字符 source program(me) character
源程序最佳化程序 source program(me) optimizer
源出于地球的 telluric
源磁泡 source bubble
源大小 source size
源代码 source code
源代码指令 source code instruction
源带 source tape
源带交叉汇编程序 source tape cross-assembler
源带准备 source tape preparation
源导纳 source admittance
源的工作条件 source condition
源的相互校准 source intercalibration
源等离子体 source plasma
源地 source
源地区 source area; source region
源地震 source earthquake
源地址 source address
源地址指令 source address instruction
源地址字段 source address field
源点 source; source point
源点函数 source function
源电导 source conductivity
源电极 source electrode
源电流 source current
源电路 source circuit
源电平 source level
源电压 source voltage
源电阻 source bulk resistance; source series resistance
源端子 source terminal
源发射度 source emittance
源分布 source distribution
源分布函数 source distribution function
源赋值指令 source designation instruction
源功率 source power
源管 source capsule; source tube
源宏定义 source macro-definition
源汇法 source-sink method
源机器 source machine
源激励 source forcing
源极 source electrode; source pole
源极端 source terminal
源计算机 source computer
源记录 source record
源寄存器字段 source register field
源检测干扰 source detected interference
源接点 source contact
源节点 source node
源距 source spacing; source detector separation < 放射性勘探 >
源卡片叠 source deck
源刻度实验室 source-calibration laboratory
源控制 control of source
源库存程序 source library program-

(me)
源类别 source category
源例行程序 source routine
源联锁装置 source interlock
源流 head of a river; head stream; source flow
源流控制 source control
源漏间电压 source-drain voltage
源漏特性 source-drain characteristics
源炉温 source oven temperature
源密度 source density
源名 parent name
源模件 source module
源模块 source module
源目的地指令 source-destination instruction
源盘 source tray
源偏压效应 source bias effect
源气体 source gas
源强(度) source intensity; source strength
源强度测定 source strength measurement
源强度归一化 source strength normalization
源区 source region; source zone
源区段 source range
源取样 source sampling
源泉 fountain; head spring; well head
源容器 source container
源栅电极 source grid electrode
源熵 entropy of the source
源实用程序 source utility
源势流 source potential flow
源室 source housing
源受体关系 source-receptor relation
源输出电路 source follower circuit
源输出器探头 source follower probe
源输出效率 source output efficiency
源数据 source data
源数据采集 source data acquisition
源数据结构 source data structure
源数据项 source data entry; source data item
源数据自动化 source data acquisition
源速度 source speed
源体 source body
源调制频率 source modulation rate
源调制器 modulator of source
源头 fountain head; head stream; source; spring head
源头坝 <小流域治理过程> watershed dam
源头保水性 headwater retention
源头沉积 <属冰碛沉积> head deposit
源头高程 altitude of source
源头河保护区 headwater protection area
源头控制 headwater control
源头区 headwater region
源头水渠 headwater channel
源头污染预防 pollution prevention at the source
源头削减 source reduction
源头与壑 source and sink
源位置 source position
源位置控制 source positioning
源温 source temperature
源文件 source file
源文件编辑程序 source file editor
源线 source line
源线中心 center [centre] of source line
源项 source item
源像 source image
源像法 method of images
源效率 source efficiency
源芯 <放射性> source core
源信息 source information
源岩 mother rock; source rock
源岩层 source bed
源样距 distance between source and sample
源移动 source movement
源引出线 source terminal
源于公路的污染 highway source
源于人类活动的物质 substance of anthropogenic origin
源于生物的氯 biogenic chlorine
源于生物的气体 biogenic gas
源语句 source statement
源语言 original language; source language
源站 source station
源支线 source branch
源种类 source category
源主机 source host
源准备 source preparation
源自宇宙射线的 cosmogenic
源阻抗 source impedance
源组件 source module
源作业 subject job
源座 source-holder

辕 thill

辕杆 hitch pole
辕杆垫板 beam block
辕马 wheelhorse

远

远岸沉积 infradeposit; infra-littoral deposit
远岸的 outshore
远岸漂移 longshore drift
远岸浅海底带 ellitoral zone
远岸浅海底的 ellitoral; sublittoral
远岸浅海地带 ellitoral zone
远岸钻探 offshore boring; offshore drilling
远螯聚合物 telechelic polymer
远滨 offshore
远滨沉积 offshore deposit
远冰川沉积 extraglacial deposit
远不是这样 far out of
远部 distal part
远操连接器 remote connector
远侧 offside
远侧部 distal portion
远侧通行优先规则 off-side priority rule
远测的 distal
远测温度计 distance thermometer
远场 far field; distant field
远场表示式 far-field expressions
远场测试 far-field testing
远场地震(地面)运动 far-field earthquake motion
远场地震信号 far-field seismic signal
远场分布 far-field distribution
远场分析器 far-field analyser
远场干涉图 far-field interference pattern
远场极辐射 far-field polar radiation
远场近似 far-field approximation
远场卡塞格伦天线 far-field Cassegrainian antenna
远场灵敏度 far-field sensitivity
远场谱 far-field spectrum
远场区域 far-field region
远场全息术 far-field holography
远场水流模型 far-field flow model
远场条纹 far-field fringes
远场图形 far-field pattern
远场位移 far-field displacement
远场应力 far-field stress
远场噪声 far-field noise
远超重核 far-super heavy nuclei
远成热液矿床 telethermal deposit
远成熟液作用 telethermal process
远程 global range
远程包集中器 remote packet concentrator
远程报警 remote alarm
远程备份系统 remote backup system
远程操纵 remote control; remote operation
远程操纵变速器机构 remote gearbox control
远程操作 remote operation
远程操作面板 remote operator panel
远程操作台 remote console
远程操作系统 remote operating system
远程测试 remote debugging; remote testing
远程长波导航设备 Dectra
远程成批操作 remote batch operation
远程成批处理 remote batch; remote batch processing
远程成批处理终端 remote batch terminal
远程成批处理终端系统 remote batch terminal system
远程成批计算 remote batch computing
远程成批输入 remote batch entry
远程成批站 remote batch station
远程程序翻译 remote program(me) translation
远程程序控制 teleprocessing
远程程序输入 remote program(me) entry
远程程序装入器 remote program loader
远程传输 remote transmission
远程传送 remote transmission; teletransmission
远程传送方式 remote mode
远程从属显示器 remote slave display
远程存取 remote access
远程存取和控制 remote-access and control
远程存取数据处理网络 remote-access data processing network
远程存取终端 remote-access terminal
远程打印机 remote printer
远程单元接口 remote unit interface
远程导航 long-range navigation [loran]
远程导航区 long-range navigation zone
远程导航设备 long-range navigation aids
远程导航图 aeronautical Loran charts; long-range air-navigation chart
远程导航系统 long-range navigation system
远程的 distance-type; far-ranging; long range
远程登录【计】remote login; telnet
远程电动旅客列车 long-distance motor-coach train
远程电信会议 teleconference
远程定位岸台 long-range radio location land station
远程定位系统 long-range position determination system
远程读出器 remote reader
远程读数 remote reading; remote readout
远程读数液位计 remote-reading tank ga(u)ge
远程读数转数计 remote-reading tach(e)ometer
远程多路转换器 remote multiplexer
远程发射机 distant transmitter
远程访问服务 remote-access service
远程访问计算系统 remote-access computing system
远程工作端 remote work station
远程供暖 distant heating
远程共存程序 tele-symbionts
远程观测系统 telemetry observation system
远程航图 long-range chart
远程航行 long-range navigation [loran]
远程激光雷达 long-range laser radar
远程集线器 remote concentrator
远程集中器 remote concentrator
远程计算机 remote computer
远程计算机连网 remote computer networking
远程计算机系统语言 remote computing system language
远程计算监督系统 remote computing monitor system
远程计算器 remote calculator
远程计算系统 remote computing system
远程计算系统的一致性误差 remote computing system consistency error
远程计算系统记录表 remote computing system log
远程计算系统监督程序 remote computing system monitor
远程计算系统交换机 remote computing system exchange device
远程计算系统交换器 remote computing system exchange
远程记录仪 remote recording apparatus
远程记录仪器 long-distance recorder
远程监控 telemonitor(ing)
远程监控机 remote supervisory and control(ling) equipment
远程监控设备 remote supervisory and control(ling) equipment
远程监控制 remote-watch-and-control system
远程监视与控制设备 remote supervisory and control system
远程交互计算 remote interactive computing
远程精度 long-range accuracy
远程警戒雷达 early-warning radar
远程开关 teleswitch
远程可见显示器 remote visual display unit
远程控制 distant control; long-range control; remote control; telecontrol
远程控制操作 remote-controlled operation
远程控制的 remotely operated
远程控制计量仪表 remote control meter
远程控制台 remote console
远程控制信号 remote control signal
远程控制学 telautomatics
远程控制站 remote-controlled station
远程馈电 remote-fed
远程雷达 long-range radar
远程雷达导航(系统) long-range radar navigation [loran]
远程雷电 sferics [spherics]
远程雷电探测仪 spherics sounder
远程离阀门控制 remote valve control
远程连接审核 authority of remote connection
远程轮询 remote polling
远程耦合常数 remote coupling constant
远程软件 remote software
远程设备 remote device
远程设置 remote setup
远程声传播 long-distance sound propagation
远程声呐 long-range sonar

远程识别符 remote identifier

远程实时分支控制器 remote real-time branch controller

远程实时终端 remote real time terminal

远程拾波 remote pickup

远程拾波中继 remote pickup relay

远程示值读数 remote readout

远程式 distance-type

远程试验设备 remote test equipment

远程输入 long-range input

远程输入输出站 remote input-output station

远程输入网络 remote entry network

远程输入装置 remote entry unit

远程输送 long-range transmission; remote transmission

远程数据 long-range data

远程数据处理 remote data processing

远程数据集中 remote data concentration

远程数据集中器 remote data concentrator

远程数据库 remote data base

远程数据收集 remote data capture; remote data collection

远程数据输入 remote data entry

远程数据站 remote data station

远程数据终端 long-range data terminal; remote data terminal

远程双曲线导航系统 loran [long-range navigation]

远程水表读数计 remote meter reader

远程探测 remote probe

远程探询 remote polling

远程调试 remote debugging

远程调整 remote regulation

远程调制解调器 remote modem

远程通信[讯] telecommunication

远程通信[讯]存取法 telecommunication access method

远程通信[讯]控制台 remote communications console

远程通信[讯]时间 telecommunication time

远程同步遥控装置 telesynd

远程拖网船 long-range trawler

远程挖土运输机 long-range excavator conveyer [conveyor]

远程网 telenet

远程网络 remote network

远程网络服务系统 remote networking system

远程卫星机 remote satellite computer

远程位置指示器 remote position indicator

远程温度选择器 remote temperature selector

远程文件系统 Telefile

远程无线电导航系统 navaglobe; nar-arno; long range radio-navigation system

远程无线电导航系统航标站 navaglobe beacon station

远程无线电导航制 navaglobe system

远程无线电台 radio station for long distance

远程无线电信标 long-range radio beacon; remote radio beacon

远程误差读出 remote error sensing

远程系统 remote system

远程先导控制阀 remote pilot control valve

远程显示 <列车运行和调车场控制楼> remote indication

远程显示器 remote display

远程显示系统 remote display system

远程线路集中器 remote line concentrator

远程相互作用 long-range interaction

远程信号 distant signal; remote signal

远程信息 remote information; teleinformatic

远程信息处理 remote message processing; teleprocessing

远程信息处理系统 teleprocessing system

远程信息服务 teleinformatic services

远程型 long-range version

远程询问 remote inquiry

远程询问功能 remote inquiry function

远程岩浆煤化作用 telemagmatic coalification

远程岩浆热变质作用 telemagmatic metamorphic

远程氧等离子体反应器 remote oxygen plasma generator

远程遥测 long haul telemetry; long-range telemetering

远程遥测浮标 long-range telemetering buoy

远程遥信分区 relayed surveillance subsection

远程遥信网络 relayed surveillance network

远程液面指示器 remote indication system

远程预警线操作员 radician

远程预警线通信[讯]系统 distance early warning line communication system

远程越界空气污染公约 Convention on Long-range Transboundary Air Pollution

远程运送 long haul(age)

远程再启动 remote restart

远程再启动过程 remote restart process

远程诊断 remote diagnosis

远程诊断系统 remote diagnosis system

远程帧 remote frame

远程指示 distant indication

远程指示系统 remote data processing

远程指示元素 long-range indicator element

远程终端 remote console; remote terminal

远程终端处理 remote terminal processing

远程终端辅助设备 remote terminal support

远程终端绘图仪 remote terminal plotter

远程终端控制台 remote terminal console

远程终端类型 remote terminal type

远程终端设备 remote terminal support; remote terminal unit

远程终端显示器 remote terminal display

远程终端装置 remote terminal unit

远程助航设备 long-range aids

远程转场外挂油箱 travel pod

远程装入 remote loading

远程自动控制学 telautomatics

远程作业的收发 remote job receiving and dissemination

远程作业进入 remote job entry

远程作业输出 remote job output

远程作业输入 remote job entry

远程作业输入网络 remote job entry network

远程作业输入系统 remote job entry system

远程作业输入终端系统 remote job entry terminal system

远程作用 remote action

远程作用系统 remote-action system

远处激励源 source of remote excitation

远传水位计 long-distance stage transmitter

远传送测量仪 remote transmitting ga(u)ge

远传信号灯 telelight

远传自计水位计 long-distance stage transmitter

远传自记水位计 long-distance water level recorder

远传自记仪器 long-distance recorder

远大的 far-reaching

远道位置 position of far traces

远的 distant

远的一边 offside

远的一面 far side

远堤破浪 broken wave

远地点 < 月球或任何行星轨道上距离地球最远的点 > apog(ee)

远地点潮 apogean tide; apogee tide

远地点潮差 apogean (tidal) range

远地点潮流 apogean current

远地点火箭 apogee rocket

远地交换局集中器 remote exchange concentrator

远地震 teleseism

远点 far point; distant point【测】

远点方位角 azimuth of distant point

远东的 oriental

远东植物地理大区 far east floral region

远动发送机 telemechanic(al) transmitter

远动化 telemechanisation [telemechanization]

远动接收器 telemechanic(al) receiver

远动控制 telemechanic(al) control

远动跳闸控制 remote trip control

远动系统 telemechanic(al) system

远动学 telemechanics

远动用频率 telecontrol frequency

远动装置 telemechanic (al) apparatus; telemechanic (al) device; telemechanic (al) installation; telemechanic(al) unit

远读测斜器 distant-reading inclinometer

远读记录器 remote recorder

远读罗盘 remote-indicating compass

远读水位计 remote water-level indicator

远端 distal end; distance end; far end; remote end; remote terminal; terminal

远端操作的终端回波抑制器 far-end operated terminal echo suppressor

远端操作服务单元 remote operation service element

远端场 far-end field

远端串话 far-end crosstalk

远端串扰 far-end crosstalk

远端串音 far-end crosstalk; output-to-output crosstalk; receiving-end crosstalk

远端串音防卫度 far-end crosstalk ratio

远端串音衰减 far-end crosstalk attenuation

远端的 distal

远端电话局 distant exchange; distant office

远端风暴沉积 distal storm deposit

远端干扰 far-end interference

远端交换模块 remote switch unit

远端节点 remote node

远端局 distant station

远端漏话 far-end crosstalk

远端模块 remote subscriber

远端目标 remote target

远端失效指示 remote failure indication

远端蚀积岩 distal turbidite

远端数据站 remote data station

远端台 distant station

远端用户单元 remote subscriber unit

远端用户单元接口模块 remote subscriber unit interface module

远端用户集中器 remote subscriber concentrator

远端用户模块 remote line module

远方 outlying

远方观察 televiewing

远方机场 remote airport

远方集中器 remote concentrator

远方监控水电站 remotely supervised hydroelectric(al) station

远方马赫数 remote Mach number

远方跳闸保护(装置) remote trip protection

远方跳闸装置 remote tripping device

远方脱扣装置 remote tripping device

远方雾区 jack's land

远方效益 off-site benefit

远方终端 remote terminal unit

远拂肘管 long-sweep ell

远感色 receding colo(u)r

远隔联想 remote association

远隔锚地 remote anchorage

远隔作用 remote effect

远供系统 remote power feeding system

远拱点 apoapsis

远古 high antiquity; hoary antiquity

远古的 dateless

远光 distance light; driving beam; far beam; long-distance light; upper head lamp beam <汽车大灯>

远光灯 high beam

远光灯丝 headlight high beam filament

远光头灯 far-reaching headlamp

远海 deep sea; off-lying sea

远海测量 pelagic(al) survey

远海沉积物 globigerina ooze

远海的 pelagic(al)

远海国家 distant water state

远海海水 ocean water

远海拖网渔轮 deep-sea trawler

远海相 pelagic(al) facies

远海渔业 deep-sea fishery

远航程平均价格 long-run average cost

远航船长 extramaster

远航的 ocean-going

远航计算机 long-range navigation computer

远航食品 food for long voyage

远核点 aphelion [复 aphelia/ aphelions]

远核圈 outer loop

远红光 far-red light

远红外 far infrared

远红外波段 far infrared band

远红外成像 far infrared image(ry)

远红外窗口 far infrared window

远红外的 far red

远红外干涉 far infrared interference

远红外干涉仪 far infrared interferometer

远红外干燥 far infrared drying

远红外光电导 far infrared photoconductivity

远红外光谱 far infrared spectrum

远红外加热器 far infrared heater

远红外控制器 far infrared controller; ircon

远红外区 far infrared region; far infrared band

远红外扫描 far infrared scanning

远红外探测器 far infrared detector; far infrared photoconductor

远红外通信[讯] far infrared communication
远红外透射滤光片 far infrared transmission filter
远红外吸收光谱 far infrared absorption spectrum
远红外线 far infrared ray
远红外(线)辐射 far infrared radiation
远红外线干燥 far infrared drying
远红外线激光器 far infrared laser
远红外线加热式沥青贮仓 far infrared rays heating asphalt storage
远红外仪器 far infrared gear
远红外余辉带光谱 far infrared reststrahlen spectrum
远幻日 mock sun;paranthelion
远回波 distance echo
远及的 far-reaching
远极孔 ulcer
远间距平行跑道 far parallel runways
远见 farsightedness
远见力 far-sight
远郊 outer suburbs
远郊城镇 subtopia
远郊富裕阶层居住区 exurb
远郊区 exurb; outer suburban district;outer urban region
远郊区列车运行 outer suburban service
远焦点 overfocus
远焦透镜 afocal lens
远焦系统 afocal system
远近比例 perspective scale
远近结合 far-and-near combination
远近镜 split-field filter
远近视差 near and far visual difference
远近适应反射 accommodatory reflex
远近适应力 accommodatory ability
远景 distance scene;distant scene;distant view;far future;long-range perspective;long shot;lookout;offing;perspective;vista
远景储量 future reserves;possible ore;prospective ore;prospective reserves
远景单元 perspective cell
远景的 far-seeing;far-sighted;prospective
远景电流 prospective current
远景调查 prospective study
远景发展 long-term development
远景发展地段 future development area;future expansion area
远景发展规划 advanced development objective; long-range development plan;long-term development plan
远景负荷需要量 future load demand
远景规划 advanced planning;advanced project; far-reaching plan; far-seeing plan; long-range plan(ning); long-term planning;planning horizon;prospective plan;secular plan
远景规划纲要 long-term outline plan
远景货运量 future traffic volume
远景价值 prospective value
远景交通量 future traffic volume; prospect(ive) traffic volume
远景结构 good structure
远景开发计划 long-term development plan
远景勘探线 prospective investigation line
远景矿产分类图 classification map showing prospective mineralization area
远景矿产种类 type of prospective ore deposits
远景扩建 future enlargement
远景目标 long-term goal;long-term objective
远景平均日交通量 future average daily traffic
远景评价 prospect evaluation
远景区定位 localization of perspective region
远景区级别 grade of prospective district
远景区面积 area of prospective district
远景区位置 location of prospective district
远景区资源量估计 perspective region resources estimation
远景设计 far-reaching design
远景透视 distance vision
远景图 perspective view(ing)
远景显示 perspective representation
远景线网 long-range line network
远景研究 advanced research; prospecting study
远景研究部 advanced research division
远景研究规划 advanced research planning
远景运量 future traffic; prospective traffic
远景资源量 prospective resource
远距 X 线照片 teleroentgenogram
远距 X 线照相术 teleradiography;teleroentgenography
远距标志 distant sign
远距波导通信[讯] long-distance guided communication
远距采样探头 remote sampling probe
远距操纵机构 remote control gear; remote controls
远距操作天平 remote balance
远距测定法 telemetry
远距测定装置 telemetric apparatus
远距测量 remote measurement
远距测速计 teletach(e)ometer
远距长途通信[讯]电路 long haul toll circuit
远距传动器 autosyn
远距传感器 remote transducer
远距传送压力表 distant-reading manometer
远距传物 teleportation
远距垂直陀螺仪 remote vertical gyroscope
远距导航台 outer locator
远距道岔 outlying points; outlying switch
远距道岔闭锁器 outlying switch lock
远距道岔防护信号机【铁】distant switch signal
远距道岔锁闭器 outlying switch lock
远距道岔锁闭器操纵握柄 outlying switch lock lever
远距动力控制系统 remote power control system
远距读数指示器 remote-reading indicator
远距端机装置 remote station
远距防护 distance protection
远距分光镜 telespectroscope
远距感应 remote sensing
远距给料泵 remote head pump
远距跟踪方式 long-range tracking mode
远距光度计 telephotometer
远距光谱辐射计 telespectroradiometer
远距机动化 telemechanization
远距记录 remote recording
远距监视 remote inspection; remote monitoring;telemonitor(ing)
远距控制 distance control
远距控制运载工具 remotely controlled vehicle
远距窥视系统 remote-viewing system
远距喇曼光谱法 remote Raman spectroscopy
远距离 long distance
远距离保护 distance protection
远距离报警装置 distant warning device
远距离表盘流量计 remote dial flowmeter
远距离操纵 distance control; handle change; remote control operation; remote handling;remote operation; telemechanization;telerun
远距离操纵的 pilot-operated;telechiric;teleoperated
远距离操纵阀 remote-controlled valve
远距离操纵盘 remote control panel
远距离操纵器 remote control manipulator;teleoperator
远距离操纵桥 remote control bridge
远距离操纵设备 remote handling tool
远距离操纵手 telechiric
远距离操纵手柄 joy stick
远距离操作 distant operation;remote (-controlled) operation;remote manipulation;remote servicing
远距离操作的 remote-operated
远距离操作活接头 remote disconnect
远距离操作设备 remote handling equipment; remote manipulating equipment; remote operating equipment
远距离操作手柄 joy stick
远距离操作系统 remote handling system
远距离操作装置 remote handling device;remote handling gear
远距离测量 remote measurement;remote measuring; remote metering; telemeasurement; telemetering; telemetry
远距离测量系统 remote measuring system
远距离测量仪(表) telega(u)ge
远距离测量仪器 distance apparatus
远距离测量元件 remote measuring element
远距离测量装置 telemetering medium
远距离插索雷达 early-warning radar
远距离查阅目录法 telereference
远距离铲 long-range shovel (for stripping)
远距离程序控制 remote programming control;teleprocessing
远距离处理 remote processing
远距离处理式计算机 teleprocessing type computer
远距离传播 long-range propagation
远距离传感器 remote sensor
远距离传输线 remote line
远距离传送 long-range transmission
远距离(大冲击)碰撞 distant collision
远距离导航 remote navigation
远距离导航系统 long-distance navigation system; long-range navigation system
远距离的 distant;stand-off
远距离低损耗传输 long-distance low loss transmission
远距离地震 distant earthquake
远距离递减的运费率 tapering freight rates
远距离电测装置 telemeter
远距离电源 remote source
远距离读出 remote readout; remote sensing
远距离读数 telereading
远距离读数仪器 remote-reading instrument
远距离发射机 remote transmitter
远距离发射控制 remote launch control
远距离发信号 remote signal(l)ing
远距离辐照 teleirradiation
远距离给定值调整 remote set point adjustment
远距离跟踪 following in range
远距离供电 long-distance supply
远距离供能 long-distance supply
远距离供水 long-distance water supply
远距离供水管 long-distance water pipeline
远距离观测 remote observation; distance observation
远距离观察 distance view(ing); distant surveillance;distant viewing
远距离观察设备 remote-viewing equipment
远距离轨道车 long-distance rail-car
远距离过程控制 process remote control
远距离海岸 outshore
远距离航行船 deep-going vessel
远距离合闸 distant switching-in
远距离后备保护(装置) remote backup
远距离呼叫 long-distance call
远距离换挡杆系 remote shift linkage
远距离机械化 telemechanisation[telemechanization]
远距离计数器 remote recorder
远距离计算机 telecounter
远距离记录 remote recording
远距离记录仪 long-distance recorder
远距离驾驶杆 joy stick
远距离监测 remote monitoring
远距离监测器 remote monitor
远距离监控系统 remote monitoring system
远距离监视 remote monitoring; remote surveillance
远距离监视和控制系统 remote-watch-and-control system
远距离监视控制 remote supervisory control
远距离监视器 remote monitor
远距离监视设备 remote monitoring equipment
远距离交通 long-distance traffic
远距离校准 remote calibration
远距离接收 distance reception; distant reception
远距离节流控制 remote throttle control
远距离进给控制 remote feed control
远距离警报装置 distant warning device
远距离静电印刷 long-distance xerography
远距离开关控制 remote switching control
远距离开关控制钢绳 <井场发动机的> telegraph cord
远距离可见显示 remote visual display
远距离可调准直器 remotely adjustable collimator
远距离控制 distant control; handle change;kinegraphic(al) control; remote control;remote handing;teleautomatics;telecontrol;teletype control
远距离控制操作 remotely controlled operation
远距离控制的泵站 remote-controlled piping stations
远距离控制的泵装置 remote pumping unit
远距离控制阀 remote-controlled valve
远距离控制方式 telemeter system
远距离控制起重机 remote control rack

远距离控制器 remote(-operated) controller
远距离控制设备 remote control apparatus
远距离控制台 remote console; remote control panel
远距离控制装置 kinegraphic(al) control unit
远距离联系 telecommunication; teleconnection; teleconnexion
远距离罗盘 telecompass
远距离瞄准 remote aiming
远距离目标 distant object
远距离能源 remote source
远距离起爆 induced detonation
远距离气动控制 pneumatic remote control
远距离气压表 telebarometer
远距离气压计 telebarometer
远距离倾斜摄影 <用长焦距窄视角透镜摄影机的摄影> long range oblique photography; lorop photography
远距离驱动 distant drive; remote drive
远距离取样 remote sampling
远距离热源 long-distance heat
远距离热源引进 long-distance heat intake
远距离散射 long-distance scatter
远距离设备 remote equipment
远距离设定点调整器 remote set point unit
远距离设定值调整 remote set point adjustment
远距离摄片 teleoroentgenography
远距离摄影 long-distance shot
远距离摄影镜头 telephoto lens
远距离摄影照片 telephotography
远距离拾波 remote pickup
远距离拾波中继 remote pickup relay
远距离输电 long-range transmission
远距离输送 long-distance conveyance; long-distance transport; long-range transmission
远距离数据 remote data
远距离数据处理 remote data processing
远距离数据传递部件 remote data box
远距离数据传输 remote data transmission
远距离数据指示器 remote data indicator
远距离数字流量计 remote dial flowmeter
远距离双曲线低频导航系统 radux
远距离水平控制 remote level control
远距离水位计 long-distance water-stage recorder
远距离水位指示计 remote water-level indicator
远距离水下操纵 remote underwater manipulation
远距离水银温度计 mercury distant-reading thermometer
远距离探测技术 remote probing techniques
远距离调节 remote regulating
远距离通信[讯] haul communication; long-distance communication; range communication; remote communication
远距离通信[讯]会议 teleconference
远距离通信[讯]全套设备 remote communications complex
远距离通信[讯]系统 telecommunication system
远距离同步 distant synchronization
远距离维护 remote servicing
远距离位置 remote location
远距离位置指示器 distant position indicator; remote position indicator
远距离温度表 distant thermometer
远距离温度选择器 remote temperature selector
远距离无线电导航系统 loran [long-range navigation]
远距离无线电站 long-distance radio station
远距离误差传感 remote error sensing
远距离显示 distant indication; remote sensing
远距离显示器 remote display unit
远距离显示水位表 remote-reading water level indicator
远距离线路 remote line
远距离信号 distance [distant] signal; distant indication; long-distance signaling; remote signal; telesignalisation [telesignalization]
远距离信号控制 long-distance signal operation
远距离信号设备 telesignal(l)isation [telesignal(l)ization]
远距离信号装置 remote signal(l)ing plant
远距离信息输入输出 remote message input/output; message input/output
远距离选择器 remote selector
远距离压力计 remote indication manometer
远距离遥测记录 remote recording
远距离液位计 remote-reading tank ga(u)ge
远距离液位指示器 distant level indicator
远距离液位指示系统 remote level indicating system
远距离用户 remote subscriber
远距离增益控制 remote gain control
远距离增益控制度盘 remote gain control dial
远距离站 remote station
远距离照相的 telephoto
远距离照准目标 remote target
远距离指标值调整 remote index value adjustment
远距离指示 remote indication
远距离指示罗经 magnesyn compass
远距离指示器 distant indicator; teleindicator
远距离指示系统 teleindicating system
远距离指示压力计 remote indication manometer
远距离指示遥测计 remote indication telemeter
远距离指示仪表 distant-indicating instrument
远距离指示仪器 remote-indicating instrument
远距离指向 remote sensing
远距离智能终端 remote intelligent terminal
远距离终端 remote terminal
远距离周率继电器 distance frequency relay
远距离转换开关 remote plug
远距离装料 remote loading
远距离自动校准系统 remote automatic calibration system
远距离自动学 teleautomatics
远距离自动遥测设备 remote automatic telemetry equipment
远距离自动装置 teleautomatics
远距离自记设备 distant recording instrument
远距离自记仪器 distant recording instrument
远距离钻井 stepout well
远距离作用 action at a distance; telekinesy
远距抛掷(爆破) widespread throw
远距气象测定学 telemeteorography
远距气象计 telemeteorograph
远距射线照相术 teleradiography
远距摄影术 telephotography
远距双星 wide pair
远距水面指示器 remote water-level indicator
远距水位记录仪 long-distance water level recorder
远距台 remote station
远距调节 distance regulation
远距调整 teleadjusting
远距卫星 remote satellite
远距位移调整器 long-shift control
远距温度计 telethermometer
远距显示 remote readout
远距线性调整器 long-linearity control
远距信号 remote signal(l)ing
远距信号装置 remote signal(l)ing
远距液面指示器 remote level indicator
远距移液 remote pipetting
远距荧光屏检查 telefluoroscopy
远距用户 distant subscriber
远距照相镜头 telephoto
远距诊断 telognosis
远距指点标 outer mark; outer marker beacon
远距指示器 teleindicator
远距中继方式 distance relaying system
远距中继设备 remote pickup equipment
远距转速调节 remote speed adjustment
远控泵站 remote-controlled station
远控吸移管 remote control pipet(te)
远控吸移装置 remote control pipetting device
远控信息存取 remote batch access
远控制门 remote gate
远离岸边 offing
远离岸的 infra-littoral
远离城市的原野 boonies
远离城镇的地方 hinterland
远离大都市的乡镇 hicktown
远离的 off-center [centre]; outlying
远离地表水体 far from surface water
远离锅炉 stow away from boiler
远离海岸的岛 off-lying island
远离海滨 outshore
远离海洋的 midland
远离扩音器 off-mike
远离热源 stow away from heat; keep away from heat
远离物标 off an object
远离中心的 outlying
远路 roundabout
远虑 foresight; forethought
远锚锚泊 ride a long peak
远凝聚接触 remote aggregation contact
远泡点道号 far shot-point trace number
远破波 broken wave
远破波荷载 broken wave load
远期 far-future stage; prospective period
远期贷款承诺 take-out commitment
远期的 long-dated; long range; long-term
远期辐射效应 late radiation effect
远期付款 payable at usance
远期付款交单 distance of payment after date; document against payment-after-sight
远期规划 long-term planning
远期合同 long(er)-term forward contract
远期后效 remote aftereffect
远期环境效应 long epoch environmental effect; long-term environmental effect
远期环境影响 long epoch environmental effect; long-term environmental effect
远期环境终端 long-term environmental effect; long-term environmental impact
远期汇率 forward exchange; forward rate
远期汇票 time draft; usance bill(draft)
远期汇票买入价 buying rate for usance bill
远期汇票贴现率 usance bill rate
远期建议 far-reaching proposal
远期交货 forward delivery; future delivery; futures
远期利息 interest on arrears
远期列车计算长度 train calculation length for long term
远期目标 long-term goal
远期票据 bill at long sight; date bill; long bill; time bill
远期票据贴现 time bill discount
远期平均成本 long-run average cost
远期平均日交通量 future average daily traffic
远期期票 long(-dated) bill
远期天气预报 long-range forecast
远期天气展望 further outlook
远期外汇 forward exchange
远期效应 late effect; remote effect
远期信用证 acceptance letter of credit; usance letter of credit
远期影响 late effect; remote effect
远期预测 long-range prediction; long-term prediction
远期预付 future advance
远期运转维修费 future operation and maintenance
远期装运 forward shipment
远气化的 telepneumatolytic
远区 far field; far zone
远区地形改正精度 accuracy of terrain correction farther distance
远区红外线 far infrared
远区水流模型 far-field flow model
远区条件 far zone condition
远驱水深 <水跃的> sweep-out depth
远驱水跃 remote jump
远日点【天】 aphelion [复 aphelia/aphelions]
远日点距离 aphelion distance
远沙坝 distal bar; far bar
远沙坝沉积 distal bar deposit
远射程喷灌机 long-range sprinkler
远射程喷灌器 rain gun
远射程喷嘴 gun jet nozzle
远射程人工降雨机 circular spray sprinkler
远摄镜头 long-distance lens; telephoto lens
远摄物镜 telephoto lens
远摄像片 long-range photograph
远摄照片 long-range photograph; telephotograph
远深海沉积物 pelagic(al) abyssal sediment
远生矿床 telescoped ore deposit
远示图 perspective representation
远示温度计 remote-indicating thermometer
远示压力表 remote indication manometer
远视 distant view; farsightedness

远视的 far-sighted;long-sighted;telescopic(al)
远视面 far face
远视眼 presbyopia
远途运费递减制 tapering distance rate system
远外堤岸 far bank
远心沉淀 centrifugalization
远心的 telecentric
远心点 apocenter [apocentre];apofocus
远心端 distal end
远心光阑 telecentric stop
远心光路系统 telecentric optic(al) system
远心光学系统 telecentric optic(al) system
远心镜 telecentric mirror
远心距 apothem
远心照明 telecentric light
远星点 apastron
远行外存储器 transaction file
远洋 high sea;mid-ocean
远洋班轮 ocean liner
远洋波 ocean wave
远洋驳船 ocean barge
远洋捕鱼作业 offshore fishing
远洋不定期货船 ocean tramp
远洋不定期货轮 ocean tramp
远洋舱区 ocean-going area
远洋沉积(物) eupelagic sediment;ocean deposit;pelagic(al) deposit;pelagic(al) sediment;eupelagic deposit
远洋船(舶) ocean carrier;ocean(-going) vessel;ocean(-going) ship;ocean range vessel;transoceanic vessel
远洋船队 ocean-going fleet
远洋岛 pelagic(al) island
远洋的 deep-sea;deep water;eupelagic;ocean-going;pelagic(al);transoceanic
远洋顶推驳船队 ocean-going pusher-barge combination
远洋动物区系 pelagic(al) fauna
远洋废物处理 ocean disposal
远洋浮游生物 eupelagic plankton
远洋港口 ocean port
远洋航程 deep-water voyage
远洋航船上甲板 ocean-going platform
远洋航路 ocean lane;ocean route
远洋航轮 ocean-going vessel
远洋航区 ocean-going area
远洋航线 ocean line;ocean route;ocean routing;ocean service;deep-sea route
远洋航行 ocean navigation;deep-sea navigation
远洋航行的 ocean-going;sea going
远洋航运 ocean-going shipping;ocean navigation
远洋豪华客轮 ocean palace
远洋环境 pelagic(al) environment
远洋灰岩 pelagic(al) limestone
远洋洄游的 oceanodromous
远洋货轮 ocean-going freighter;sea-going freighter
远洋货物 ocean cargo;ocean-going goods
远洋货运 ocean freight
远洋集装箱 deep-sea container;transcontainer
远洋舰艇 sea-going ship
远洋交通 transoceanic traffic
远洋交通船 ocean boarding vessel
远洋救助船 ocean salvage vessel
远洋客轮 motor passenger ship liner;ocean-going liner;passenger liner
远洋客轮班轮 ocean liner
远洋轮船 ocean liner;ocean-going vessel;sea-going vessel;transoceanic steamer;ocean-going ship
远洋贸易 deep-sea trade;long trade;ocean-going trade
远洋黏[粘]土 eupelagic clay
远洋区 ocean-going area;pelagic(al) division;pelagic(al) realm;pelagic(al) zone
远洋区域 pelagic(al) region
远洋驱逐舰 sea-going destroyer
远洋群岛 mid-ocean arching
远洋软泥 pelagic(al) ooze
远洋商船 ocean-going vessel;oceanic vessel
远洋深海带 abyssal pelagic zone
远洋深海动物区系 abyssal pelagic fauna
远洋深海动物群 abyssal pelagic fauna;abyssopelagic fauna
远洋深海生态学 abyssal pelagic ecology;abyssopelagic ecology
远洋深海生物 abyssal pelagic organism;abyssopelagic organism
远洋深水 deep sea
远洋生物 pelagic(al) organism
远洋生物群 pelagic(al) community
远洋食物链 pelagic(al) food chain
远洋水险 ocean marine insurance
远洋特快班轮 ocean greyhound
远洋通信[讯] transoceanic communication
远洋拖带 ocean towing
远洋拖航 ocean towing
远洋拖轮 deep-sea tug;ocean-going tug;ocean tug-boat;sea-going tug;ocean-going towage
远洋拖网渔船 sea-going trawler
远洋污染 pelagic(al) pollution
远洋相 eupelagic facies
远洋性黏[粘]土 pelagic(al) clay
远洋悬浮 pelagic(al) suspension
远洋油船 transoceanic tanker
远洋油轮 ocean-going tanker
远洋渔业 deep-sea fishery;distant water fishery;pelagic(al) fishery;sea-going fishery
远洋域 pelagic(al) realm
远洋运价 ocean freightage
远洋运输 ocean carriage;ocean freight;ocean traffic;ocean shipping;ocean trade;ocean transport(ation)
远洋运输舱单 transportation manifest
远洋运输代理公司 ocean shipping agency
远洋运输业 ocean shipping trade;oversea shipping trade
远洋种 pelagic(al) species
远洋资源 pelagic(al) resources
远洋作业图 oceanic plotting sheet
远因 causa remota;incipient cause;remote cause
远缘杂种 distant hybrid
远源地震 distant earthquake;earthquake of distant origin
远源气化方式 telepneumatolitic way
远远离开陆地 lower the land
远月潮 apogean tide;apogee tide
远月潮潮幅 apogean (tidal) range
远月点 apocynthion;apolune;aposelene
远运距 long haul(age)
远运距材料 long haul material
远震 distant earthquake;distant shock;earthquake of distant origin;far earthquake
远震波 teleseismic wave
远震到时 teleseismic arrival
远震记录 teleseismic record
远震距离 teleseismic distance
远震事件 teleseismic event
远震台网 teleseismic network
远震台站 teleseismic station
远震探测 teleseismic detection
远震微震噪声 teleseismic microseismic noise
远震信号 teleseismic signal
远震仪 distant earthquake instrument
远震噪声 teleseismic noise
远震震级 teleseismic magnitude
远震震中定位 teleseismic epicenter location
远震资料 teleseismic data
远置重量平衡配重 remote mass-balance weight
远轴傍管薄壁组织 abaxial paratracheal parenchyma
远轴的 abaxial
远主焦点 apocenter [apocentre];apofocus
远紫外辐射 extreme ultraviolet radiation;far ultraviolet radiation
远紫外激光器 uvaser
远紫外区 extreme ultraviolet region;far ultraviolet region
远紫外线 extreme ultraviolet ray;far ultraviolet ray
远足 excursion;tramp
远足者 tripper

院

院部办公室 general office of the hospital

院长 master;provost;rector
院长办公室 deanery
院长宅邸 deanery
院廊 alure
院落 <古建筑中> aula
院门 gate
院内管网 inside courtyard pipe network
院内蓄水池 <古罗马> compluvium;impluvium
院墙 courtyard wall
院士 academician
院系大楼 faculty block
院中花园 courtyard garden
院中清真寺 courtyard mosque
院子 courtyard;curtilage;hypaethral;hypaethron;yard

垸

垸田 diked marsh

愿

愿付原则 willingness to pay principle

愿望线 desire line;wishing line

约

约旦标准型【数】Jordan canonical form;Jordan normal form

约当产量 equivalent unit
约当代数的根 radical of a Jordan algebra
约当弧【数】Jordan arc
约当矩阵 Jordan matrix
约当曲线 Jordan curve
约定 appoint;commitment;make an appointment;make an arrangement;promise;stipulate;with the understanding that
约定保险价值 agreed insured value
约定的总价 stipulated sum;stipulation sum
约定地租 contract rent
约定定义 stipulative definition
约定付款 make commitments
约定付款会计 encumbrance accounting
约定付款数 purchase commitment
约定骨料 convectional aggregate
约定呼叫 appointment call
约定汇价 given quotation
约定集料 convectional aggregate
约定剂量当量 committed dose equivalent
约定价格 committed cost;stipulated price
约定交货期 period stipulated for delivery
约定利率 contract rate of interest
约定利息 contract interest
约定留置权 contract lien
约定频率 agreed frequency
约定书 agreement
约定税率 conventional tariff
约定速度 contract speed
约定条件 stipulation
约定效率 conventional efficiency
约定信号 previously arranged signal
约定邮件运输 contract mail service
约定者 stipulator
约定支付的 promissory
约定值 default value
约访 appointment call
约分【数】abbreviation;reduction of a fraction;reduce a fraction to its lowest terms
约翰晶体几何学 Johann crystal geometry
约翰色粉 <一种粉刷墙壁用的可洗的含油色粉> John bull
约翰森法 <丹麦工程师约翰森的屈服线理论,用于钢筋混凝土板的极限荷载计算> Johansen's method
约翰逊阀 <高落差水轮机阀> Johnson valve
约翰逊规块 Johnson ga(u)ge (block)
约翰逊计数器 Johnson counter
约翰逊精选机 Johnson concentrator
约翰逊空间中心 Johnson space center [centre]
约翰逊块规 Johansson block
约翰逊式调压塔 Johnson regulator
约翰逊效应 <热噪效应> Johnson effect
约翰逊噪声 Johnson noise
约翰逊噪声电压 Johnson noise voltage
约翰逊轴承青铜 Johnson's bronze
约化阿贝耳群 reduced Abelian group
约化表示 reduced representation
约化残差方程 reduced residual equation
约化残余方程 reduced residual equation
约化代数 reductive algebra
约化代数概型 reduced algebraic scheme
约化德布罗意波长 reduced de Broglie wavelength
约化二次形式 reduced quadratic form
约化二次型 reduced quadratic form
约化法方程式 reduced normal equation
约化方程 reduced equation
约化分布函数 reduced distribution function
约化分宽度 reduced partial width
约化复形 reduced complex
约化格 reduced lattice
约化格式 reduced scheme
约化关联矩阵 reduced incidence matrix
约化积空间 reduced product space
约化极值距离 reduced extremal distance

约化矩阵元 reduced matrix element
约化空间 reduced space
约化宽度 reduced width
约化理论 reduction theory
约化连接 reduced joint
约化量 reduced quantity
约化律 reduction law
约化密度效应 reduced density effect
约化偶 reductive pair
约化频率 reduced frequency
约化普朗克常数 reduced Planck constant
约化齐性空间 reductive homogeneous space
约化剩余系 reduced residue system
约化双角锥 reduced suspension
约化算子 reductive operator
约化态 reduced state
约化特征标 reduced character
约化条件方程式 reduced condition equation
约化同调群 reduced homology group
约化图 reduced graph
约化位能 reduced potential energy
约化相关矩阵 reduced correlation matrix
约化应力 reduced stress
约化映射锥 reduced mapping cone
约化有序对群 reduced ordered pair group
约化约当代数 reduced Jordan algebra
约化跃迁几率 reduced transition probability
约化张量 reduced rank tensor
约化正规方程 reduced normal equation
约化正则方程 reduced normal equation
约化质量 reduced mass
约化质量效应 reduced mass effect
约化重力 reduced gravity
约化锥 reduced cone
约计 approximate; in round number; rough estimate
约计成本 approximate cost
约计重量 approximate weight
约减变量 reduced variable
约简表 reduced unitized table
约简长度 reduced length
约简方程式 reduced equation
约简矩阵 reduced matrix
约卡尔布劳牌合金 <一种制管用铝铜合金> Yorcalbro
约克窗 Yorkshire light
约克怀特牌金属 <一种制管用白色金属> Yorcwyte
约克阶 <晚石炭世>【地】Yorkian
约克郡砂岩 York stone; Howley park <英>
约量 submultiple
约硫砷铅矿 jordanite
约略的估计 rule-of-thumb
约略的衡量 rule-of-thumb
约略估计 rough estimate; rough guess
约略数量 approximate quantity
约略调整 approximate adjustment
约去 cancellation; divide out
约瑟夫式混凝土分隔栏 Jersey barrier
约瑟夫松 Jersey pine
约瑟夫逊效应 Josephson effect
约瑟夫主撞栏 Jersey barrier
约森米克牌屋面卷材 <商品名> Yosenmite
约氏量块【机】Jo blocks
约束 bind(ing); bondage; comfinement; constrain(ing); constraint; restriction
约束板 restrained slab
约束爆炸 confined explosion
约束变量 apparent variable; bound variable
约束变项 apparent variable; bound variable
约束变形 restraint deformation
约束步长法 restricted step method
约束层 restraint layer
约束长(度) confined length; constraint length
约束成本 committed cost
约束程度 degree of restraint
约束大梁 constrained girder
约束导流 confinement of flow
约束的 obligatory; restrictive
约束的钢筋混凝土楼板 restrained reinforcement concrete slab
约束的设备 restraint equipment
约束地下水 fixed groundwater
约束点 obligatory point; obliged point
约束端 restrained end
约束反力 constraint reacting force; constraint reaction
约束反作用力 constraint reacting force; constraint reaction
约束方程式 constraint equation; equation of constraint
约束(方)法 constraint method
约束钢筋 confined reinforcement
约束拱 constrained arch
约束固定成本 committed fixed cost
约束固定装置 restraint fixing
约束规范 constraint qualification
约束规格 constraint qualification
约束缓和曲线 constraint transition curve
约束混凝土 confined concrete
约束极值 constrained extremum
约束集 constraint set
约束降温试验 confined cooling test
约束节点 restraint joint
约束结构 restrained structure; restraining structure
约束劲度 restraint stiffness
约束矩阵 constraint matrix
约束开裂 restraint crack
约束力 binding force; constrained force; constraining force; restraining force; restraint force; restrict power
约束力矩 constraining moment; force of constraint; restraining moment; restraint moment
约束利润 constraint profit
约束梁 constrained beam; restrained beam
约束裂缝 restraint crack
约束流动 bonded flow
约束锚索 restraining anchor cable
约束面积 confined area
约束模量 constrained modulus; restraint modulus
约束模态 constraint mode
约束扭转 constraint torsion
约束扭转变形 constrained twisting deformation
约束膨胀 restrained expansion
约束平差 constrained adjustment
约束曲线 constraint curve
约束伸胀 restrained expansion
约束式自适应控制 adaptive control of constraint
约束视距 restrictive sight distance
约束收缩 restrained shrinkage
约束水 bound water
约束税率 bound rate
约束条件 condition of constraint; constrained condition; constraint condition; restrained factor; restraint condition
约束条件式 constraint equation
约束条件限定 definition of constraints
约束条件与变数 constraint and variable
约束推理 constraint reasoning
约束弯矩 constraining moment; locking moment; restraining moment; restraint moment
约束涡流 bound vortex
约束系数 constraint factor; restraint coefficient
约束系统 restraint system
约束(下)焊接 restraint welding
约束线性系统 constrained linear system
约束项【数】bound term
约束效应 restraining effect
约束信号 seizing signal
约束性 binding character
约束性合同 firm contract
约束性条款 mandatory provision; obligatory term
约束性通(运)行【交】unoperation
约束性运行【交】constrained operation
约束压力 restraining pressure
约束因数 restraint factor
约束因素 constrained factor
约束应力 confined stress; confinement stress; reaction stress
约束优化(数) constrained optimization
约束运动 constrained motion; constrained movement
约束运行 constrained operation
约束整数 rounded number
约束支承 restrained support
约束柱 restrained post; restraint column
约束转动力矩 holding moment
约束桩 restrained pile
约束自由面 constricted free face
约束阻抗 blocked impedance
约束最大化问题【数】constrained maximization problem
约束最小化问题 restrained minimization problem; constrained minimization problem
约束最优化 constrained optimization
约束作用 effect of contraction; effect of restraint
约束坐标 constrained coordinates
约数 approximate; divisor; submultiple
约四分之一块砖 ordinary closer
约特尼统 <前寒武纪> Jothian series
约简线性微分方程 reduced linear differential equation
约相关矩阵 reduced correlation matrix
约一半 moiety
约整数 rounded number; rounding-off number
约整误差 round-off error

月

·台【岩】driller-month

月暗期 dark of the moon
月白灯 lunar white light
月白灯部件 lunar unit
月白灯单元 lunar unit
月白灯光 lunar white light
月白色 lunar white colo(u)r
月白色玻璃 lunar white glass
月白色闪光 flashed lunar white
月白色透镜 lunar white lens
月半径 lunar radius
月报 monthly magazine; monthly report; monthly settlement
月报表 monthly returns; monthly settlement; monthly statement; monthly tables
月变幅 monthly amplitude
月变化 monthly variation
月不均系数 monthly unbalance factor
月财务报表 monthly financial statement
月产量 current yield
月长 length of the month
月长石 hecatolite; moonstone
月长石状玻璃 moonstone glass
月潮 lunar tide
月潮的 lunitidal
月潮低潮间隔 low-water interval; low-water lunitidal interval
月潮低潮间隙 low-water lunitidal interval
月潮高潮间隔 high water lunitidal interval
月潮高潮间隙 high water lunitidal interval
月潮间隔时间 lunitidal interval
月潮间隙 lagging of tide; lunar tide interval; lunitidal interval; tidal tide interval
月潮流间隙 lunicurrent interval
月潮引潮力 gravitational tidal force of the moon; lunar tide generating force; tidal force of the moon; tide-forming force of the moon
月尘 lunar dust
月池 new moon pool
月冲 opposition of moon
月出 moon rise
月初 beginning of month
月大气潮 lunar atmospheric tide
月堤 circle levee; counter dike [dyke]; hooping dike [dyke]; ring embankment; ring levee
月底 end of month
月地间运输系统 moon-earth transportation system
月地空间 cislunar space
月洞门 <中国建筑中的> moon gate
月度报告 monthly report; monthly settlement
月度变化 monthly variation
月度货物运输计划 monthly freight transport plan; monthly goods transport plan
月度计划 monthly plan
月度计算 monthly statement
月度检查 monthly inspection
月度结算 monthly statement
月度旅游卡 monthly travelcard
月度生产计划 manufacturing program(me) of a month
月度要车计划表 monthly wagon requisition form
月度运输计划 monthly transport plan
月度证明 monthly certificate
月份 month
月份的 monthly
月份结算报告 monthly settlement report
月份结账 accounting by month
月份派款 monthly allotments
月峰荷 monthly peak load
月斧 round ax(e)
月付款申请 monthly statement
月付款证书 monthly payment certificate
月负荷历时曲线 monthly load duration curve
月负荷因数 monthly load factor
月负载率 monthly load factor
月负载曲线 monthly load curve
月工资 monthly wages
月光 moonlight; moonshine
月光回照仪 <一种反射月光的仪器,

Y

用于长距离观测 > selenotrope
月光汽油 moonlight gasoline
月光束 moonbeam
月光下的活动 moonlighting
月桂 laurelor;laurus;laurus nobilis
月桂苯酮 laurophenone
月桂醇 lauryl alcohol
月桂二酸 dodecanedioic acid
月桂粉红 < 淡红色 > laurel pink
月桂果油 oil of bayberry
月桂果脂 bayberry oil
月桂蜡 laurel wax
月桂木 laurel wood
月桂色 bay
月桂属 laurel
月桂树 baytree;laurel;myrtle burl
月桂树蔽藏道 baytree
月桂酸 dodecanoic acid;lauric acid
月桂酸丙二醇酯 propylene glycol laurate
月桂酸醇酸树脂 lauric alkyd resin
月桂酸丁氧基乙酯 butoxyethyl laurate
月桂酸甘油酯 glyceryl laurate ester
月桂酸马来酸二丁锡 dibutyltin laurate-maleate
月桂酸钠 sodium laurate
月桂酸戊酯 amyl laurate
月桂酸盐 laurate
月桂酸乙氧基乙酯 ethoxyethyl laurate
月桂酮 laurone
月桂酮酸 lauronic acid
月桂烷 laurane;myrceane
月桂烯 laurene;myrcene
月桂烯酸 lauroleic acid
月桂烯酮 myrcenone
月桂酰 lauroyl
月桂叶 laurel leaf; myrcia; bay leaf < 用于半圆形线脚装饰 >
月桂叶花环 laurel-leaf swag
月桂叶饰品 laurel-leaf swag
月桂叶油 bay oil;laurel
月桂油 myrica oil;nikkel oil
月桂脂 bayberry tallow; laurel oil; laurel tallow
月桂子油 bayberry oil;bay-tree oil
月耗电量 monthly consumption
月耗水量 monthly consumption use
月和日的 lunisolar
月核 lunar core
月洪水 monthly flood
月虹 lunar (rain)bow;moonbow
月弧销 woodruff key
月湖 lunar lacus
月华 corona;lunar corona
月极 lunar pole
月计数器 month counter
月计账户 abstract account
月际变化 intermonthly variation
月际变异 intermonthly variation
月际变异率 intermonthly variability
月季【植】Chinese rose;China rose
月季花 ever flowering rose; monthly rose;China rose < 又名月月红 >
月季旅客 < 持有月季票的旅客 > season-ticket holder
月季票价率 season-ticket rate
月季票旅客 commuter
月降水量 monthly precipitation
月降水量分布 monthly distribution of precipitation
月降雨量 amount of monthly rainfall
月交通量变化 monthly variation
月交通量变化图 diagram of monthly traffic variation
月角【天】cusp
月角差 anomalistic(al) inequality;parallactic equation;parallactic inequality
月较差 monthly amplitude
月结算 monthly settlement
月结算报告 monthly settlement report
月进尺 month footage
月进度 monthly progress
月径流 monthly runoff
月径流量 monthly water discharge
月均值 monthly mean value
月开挖进度 monthly progression of excavation
月壳 lunar crust
月离 moon's motion
月历表 ephemeris of the moon
月利息 monthly interest
月梁 crescent beam
月亮出没时间表 moonrise and moonset tables
月亮航用表 lunar table
月亮角距 lunar distance
月亮金字塔 < 位于墨西哥特奥蒂瓦坎 > Pyramid of the Moon
月亮罗盘定向 moon compass orientation
月亮门 moon gate
月亮视差 lunar parallax
月亮销 woodruff key
月亮中天 moon culmination
月龄 age of the moon;moon's age
月龄周期 lunar cycle;Metonic cycle
月龄装置 lunar work;moon work
月流量过程线 monthly distance hydrograph;monthly hydrograph
月流量历时曲线 monthly flow duration curve
月轮 moon's disc
月落 moon down;setting of moon
月幔 lunar mantle
月面 lunar surface
月面测量控制 selenodetic control
月面测量学 selenodesy
月面断层 selenofault
月面拱形结构 domes
月面谷 rill(e)
月面弧 lunar arc
月面环形山 lunar crater
月面降落 lunar-impact camera
月面经度 selenographic longitude
月面景色 mooscape
月面曲率 lunar curvature
月面摄影机 lunar camera
月面投影 selenographic projection
月面图 lunar chart;lunar map;moon map;selenograph
月面图集 lunar atlas;moon atlas
月面图制图学 lunar cartography
月面土 lunar soil
月面网 selenodetic network
月面纬度 selenographic latitude
月面学 selenography
月面坐标 selenographic coordinate
月明细表 monthly detailed schedule
月末报告书 monthly report
月末交付 month-end delivery
月末盘存制 system of taking inventory at end of month
月末支付 month-end payment
月末资金 month-end fund
月没 moon set
月奶石 moon milke
月盆 lunar basin
月偏蚀 partial lunar eclipse
月偏食 lunar partial eclipse
月票 commutation ticket; monthly ticket;season-ticket
月票乘客 commuter
月票(乘客)出行 commuter trip
月票乘客集中地区 commuter zone
月票乘客居住区 commuter land
月票乘客率 commuting ratio
月票乘客平均行程距离 average length of commuters' journey
月票乘客行程 commuter journey
月票居民区 commuter;villa
月票客居住区 bedroom town
月票客流 commuter movement
月平均 monthly average
月平均大气压力值 monthly mean value of atmospheric pressure
月平均等值线 < 特指气温的 > isomenal
月平均海平面 monthly mean sea level
月平均海水面 monthly sea level
月平均含沙量 monthly mean sediment concentration
月平均降雨量 average monthly rainfall; monthly average rainfall; monthly mean rainfall
月平均交通量 monthly average traffic
月平均径流量 average monthly runoff;monthly mean runoff
月平均绝对湿度 monthly mean absolute humidity
月平均流量 mean monthly discharge; monthly average discharge; monthly mean discharge
月平均气温 mean monthly air temperature;monthly mean air temperature;monthly mean temperature
月平均日交通量 monthly average daily traffic
月平均湿度 mean monthly humidity
月平均疏浚量 average monthly dredging quantity
月平均输沙量 monthly mean sediment discharge
月平均数 monthly average
月平均数量 average monthly quantity
月平均水位 mean monthly stage; monthly average stage; monthly mean stage
月平均温度 mean monthly temperature;monthly average temperature; monthly mean temperature
月平均温度线 isomenal
月平均相对湿度 monthly mean relative humidity
月平均雨量 average monthly rainfall
月平均(值) monthly average;monthly mean
月平均最低气温 mean monthly minimum temperature
月平均最低水位 mean monthly lowest water level
月平均最低温度 mean monthly minimum temperature
月平均最高潮位 mean monthly highest tidal level
月平均最高气温 mean monthly maximum temperature
月平均最高水位 mean monthly highest water level
月平均最高温度 mean monthly maximum temperature
月球测图 moon mapping
月球潮汐 moon tides
月球大地测量学 selenodesy
月全食 total lunar eclipse
月热 selenothermy
月色白垩 moon-chalk
月神庙 < 古希腊 > Artemission
月生产能力 monthly capacity
月石 borax
月时差 < 月球在格林尼治子午圈与地方子午圈中天时刻之差 > lunar interval
月时间隔 lunar interval
月蚀 eclipse;lunar eclipse
月食 eclipse of the moon; lunar eclipse
月视差不等 parallax inequality
月台 landing; platform; railway platform; side platform; station platform;stoop
月台地道 platform tunnel
月台顶棚 platform roofing
月台空间 platform dimensions
月台棚 platform roof;railway roof
月台起重机 platform crane
月台隧道 platform tunnel
月调节水库 monthly regulating reservoir
月统计表 monthly statement
月投资计划 monthly investment plan
月凸轮 month cam
月土 lunar regolith;lunar soil
月吞吐量不平衡系数 monthly unbalance coefficient of cargo handled at the port
月外空间 < 月球轨道外的空间 > translunar space
月弯 long radius elbow;sweep fitting
月误差 monthly error
月雾色 moonmist
月息 interest per mensem; interest per month
月下点 sublunar point
月下位置 sublunar position
月相 lunar phase;moon phase;phase of the moon
月相不等 lunar phase inequality
月相不等潮龄 age of phase inequality;age of tide;phase age
月相关 monthly correlation
月心轨迹 selenocentric trajectory
月心坐标 selenocentric coordinate
月心坐标系 selenocentric coordinate system
月薪 monthly pay;monthly salary; monthly wages
月星轮 month star-wheel
月行差 < 地球运动 > lunar inequality
月形的 luniform
月形物 moon-shaped object;selene
月修 monthly maintenance
月牙凹 crater
月牙板 swing link
月牙板吊杆 link lifter
月牙板吊杆座 link saddle
月牙板滑块 die block;link block
月牙板滑块销 link block pin
月牙板夹板 link plate
月牙槽 crescent;passing hollow
月牙刀 moon knife
月牙堤 ring embankment
月牙儿形的 crescent-shaped
月牙杆 crescent-shaped lever
月牙桁架 crescent truss
月牙键 semi-circular key; woodruff key
月牙卡铁 crescent
月牙老虎窗 barrel light
月牙肋 crescent rib
月牙肋岔管 crescent rib bifurcation
月牙筛 crescent screen
月牙式屋架 crescent roof truss
月牙饰 crescent;meniscus [复 menisci/meniscuses]
月牙纹变形钢筋 semi-lunar deformed bar
月牙形 crescent;lune
月牙形齿轮泵 crescent pump
月牙形的 crescent;semi-lunar
月牙形隔板 crescent(-shaped) separator
月牙形拱 crescent(-type) arch
月牙形桁架 crescent truss
月牙形桁架拱桥 crescent-type truss arch bridge

月牙形间隙 <内啮合齿轮泵中的> crescent-shaped land
月牙形开裂 <推挤裂纹> crescent cracking
月牙形梁 crescent beam
月牙形曲率 meniscus curvature
月牙形水压力 crescent water pressure
月牙形弯曲 meniscus curvature
月牙形屋盖 crescent roof
月牙形屋盖桁架 crescent roof truss
月岩 lunabase;lunar rock
月岩矿物 lunar mineral
月岩球 lunar nodule
月岩学 lunar petrology
月样圆面容 moon-face;moon-shaped face
月影 lunar shadow
月预算 monthly estimate
月圆 <苏格兰> broch
月运 lunar system
月运动不等性 variational inequality
月晕 aureola [aureole];halo;lunar aureole;lunar halo
月债务与收入的比率 debt-to-income ratio
月站 nakshatra
月折旧回收 monthly depreciation reserves
月振幅 monthly amplitude
月震 moonquake
月震计 lunar seismometer
月震检收器 moonquake monitor
月震学 lunar seismology
月震仪 moon seismograph
月蒸发量 monthly evaporation discharge
月支付证书 certificate of monthly payment
月指数 monthly index
月志 menology
月志学 lunar topology
月中付款证书 monthly interim payment certificate
月中结算 mid-month settlement
月中天 moon culmination;moon transit
月中心角 angle at center [centre];central angle
月钟 moon clock
月周期性 lunar periodicity
月柱 moon pillar
月状沟 lunate sulcus
月资产负债表 monthly balance sheet
月子 dinking die
月总额 monthly total
月最大 monthly maximum
月最大(洪峰)流量 monthly flood
月最大洪水(流)量 monthly flood
月最低温度 monthly minimum temperature
月最高负载 monthly maximum load
月最小 monthly minimum

岳

岳麓山爱晚亭 Love Dusk pavilion on Yuelu Mountain

钥

钥匙 unlocking key

钥匙安全点火 safety key ignition
钥匙壁箱 key locker
钥匙标签 key tag
钥匙槽 key slot
钥匙齿 bit
钥匙齿槽 ward
钥匙齿形变化 key change
钥匙齿型 key change
钥匙定位钉 broach
钥匙阀门 key valve
钥匙柜 key cabinet
钥匙号码 key number
钥匙夹套 keytainer
钥匙架 key rack
钥匙开关 key switch
钥匙孔 key hole
钥匙孔板 escutcheon;key escutcheon
钥匙孔盖 escutcheon;keyhole cover;scutcheon
钥匙孔盖板 drop key;keyhole plate
钥匙孔盖片 key drop
钥匙孔盖销 escutcheon pin
钥匙孔(金属护)板 key plate
钥匙孔式领圈 keyhole neck
钥匙孔形喷孔 keyhole shaped orifice
钥匙孔形切口试样 keyhole specimen
钥匙孔形缺口冲击试样 Charpy key hole specimen
钥匙孔罩 keyhole escutcheon
钥匙控制器 key controller
钥匙控制系统 key control system
钥匙联锁 key interlocking
钥匙链 key chain
钥匙路牌 key staff;key tablet;staff with the key;tablet with the key
钥匙路签 key staff;staff with the key
钥匙路签器 staff key container
钥匙牌 key tag
钥匙凭证 key token
钥匙凭证机 key token instrument
钥匙凭证系统 key token system
钥匙凭证制 key token system
钥匙式槽口 keyhole type notch
钥匙锁 bit keyed lock
钥匙锁闭功能 key shut-out feature
钥匙锁闭机构 key lock mechanism
钥匙锁闭器 Annett's lock;key lock
钥匙锁闭器的钥匙 Annett's key
钥匙锁闭装置 key lock apparatus
钥匙锁定开关 key lock switch
钥匙箱 key box
钥匙销轴 key pin
钥匙形片 blank key
钥匙形缺口冲击试样 keyhole Charpy impact test specimen
钥匙占用系统 key occupancy system
钥匙占用制 key occupancy system
钥匙中相应的凹凸部 ward

悦

悦目 eye-pleasing

悦目色 eye-rest colo(u)r

钺

钺石 axe stone

阅

阅报室 newspaper room

阅兵 parade
阅兵场 parade ground;reviewing ground
阅读工具 reading medium
阅读机 reader;reading machine
阅读器 reader;reader unit;reading device;reading machine
阅读台 reading station
阅览隔间 carol(le);carrel(l);carrel(l)cubic(al);carrel(l)stall
阅览设备 reading off device
阅览室 athen(a)eum;reading room
阅览室藏书 reading room library
阅览厅 reading hall
阅览证(情) admission card
阅书架 bookrest

跃

跃变 abrupt change;jumping

跃变层 thermal barrier
跃变函数 jump function
跃波 green water
跃步 galloping
跃步 1 和 0 测试 <存储器测试法> galloping 1's and 0's
跃步图 galloping pattern
跃层 skip floor
跃层式公寓 skip-floor apartment (house)
跃长 length of jump
跃动 jerking motion
跃幅 saltus
跃后水深 sequent depth
跃阶恢复 step recovery
跃进 leap forward;quantum jump
跃距 skip distance
跃迁 transition
跃迁比 transition ratio
跃迁概率 transition probability
跃迁几率 transition probability
跃迁矩 transition moment
跃迁率 transition rate
跃迁频率 transition frequency
跃迁时间 transition time
跃迁系数 transition coefficient
跃迁信号 transition signal
跃迁型变化 variation of twinkling type
跃升 inshot;zooming
跃升阀 inshot valve
跃升进入横滚 pull-up into a roll
跃升压力 inshot pressure
跃升止阀 quick rise check valve
跃升装置 inshot
跃式拱顶 <一种具有西里西亚地方特色的建筑形式> jumping vault
跃水分水堰 leaping weir
跃水式溢洪道 straight drop spillway
跃移 movement by saltation;transportation by leaps and bounds
跃移颗粒 saltating grain;saltating particle
跃移泥沙 saltation load
跃移速度 <泥沙的> saltating soil velocity;saltation velocity
跃移土粒 saltating soil particle
跃移质 <泥沙等> saltation load
跃移质输送 saltating transport;saltation transport
跃移质输移 saltating transport;saltation transport
跃移质输移量 saltation load discharge
跃障复位装置 stump-jump device
跃障机构 stump-jump mechanism
跃障式圆盘犁 stump-jump disk plow

越

越岸沉积 overbank deposit

越边掩蔽 cross masking
越波【港】 overtopping
越车 over-take
越车道 passing track
越车动作 passing maneuver
越车规则 overtaking rule
越车辆 overtaken vehicle
越城交通 through city traffic
越城隧道 crosstown tunnel
越程波段 skip band
越带分离器 overband separator
越带指令 tape skip
越堤冲岸浪 overwash
越顶波浪 overtopping wave
越顶(洪水) overtopping
越顶浪及渗漏海水的排水沟 delph ditch
越顶水量 overtopping water
越冬 hibernation;overwinter(ing);overyear(ing);winter(ing)
越冬场所 hibernaculum;wintering ground;hibernacle【动】
越冬地 wintering ground
越冬谷物 winter cereals
越冬牧草 overwintering grass
越冬停滞期 winter stagnation period
越冬性 winter hardiness
越冬植物 hibernal plant;overwintering plant
越冬状态 hibernating state
越冬作物 hibernal plant
越负荷 extra duty
越轨 aberrance [aberrancy]
越轨式道岔 <通向安全线> run over type turnout
越过 breakover;come over;crossing;jump over;outreach;overreach;overrunning;surmount
越过轨道 trespassing
越过轨道行人 trespasser
越过建筑物的高速公路 overbuilding freeway
越过交叉口的公共汽车站 far-side bus stop
越过目标的时间 time over target
越过其他道路或结构物的道路 fly-over
越过其他管道或障碍物的弯道 pass-over bend
越过梯阶 negotiating stair(case)
越过障碍 bridge over
越过作业线的边坡 sidehill
越海的 transmarine
越河水准测量 overriver level(l)ing
越级跳闸 over level tripping
越级组构 hiatal fabric
越脊沙流 <指沙流悬布坡上如同冰川> sand glacier
越建筑物(高架)高速干道 overbuilding freeway
越江设施 river-crossing facility
越江隧道 subaqueous tunnel
越界 off-normal;transgress
越界调查 external study
越界空气污染 trans-atmospheric pollution;transboundary air pollution
越界旅行 outtravel
越界水域 transboundary waters
越境 border crossing
越境流域管理 transboundary river basin management
越境污染 trans-frontier pollution
越境污染物 transboundary pollutant
越境污染物输移 transboundary transport of pollutant
越境转移 transboundary movement
越橘 <北美灌木> huckleberry
越距 skip distance
越空探测 transosonde
越浪 overtopping;wave overtopping
越浪堆积 storm delta;washover;wave delta
越浪量 overtopping volume
越岭隘口 mountain crossing
越岭河段 summit level;summit reach;summit reach canal
越岭渠道 summit canal
越岭取直 summit cut-off
越岭水库 divide cut reservoir
越岭隧道 summit tunnel;watershed tunnel
越岭线 mountain line;ridge crossing line
越岭选线 location of line in mountain;location of mountain line;location over mountain
越岭垭口 mountain crossing
越岭运河 divide cut canal;summit canal
越流 transfluence;leakage
越流补给 leakage recharge
越流补给量 quantity of leaky re-

charge
越流补给条件 condition of leaky recharge
越流层 leakage layer
越流层垂向渗透系数 vertical permeability of leaky aquifer
越流层厚度 thickness of leaky aquifer
越流负荷 weirload
越流理论 leaky theory
越流排泄 leaking discharge
越流区面积 area of leaky region
越流渗透 upward seepage
越流渗透距离 distance of leaky percolation
越流水 leakage water
越流系数 leakage coefficient
越流系统 leaky system
越流系统井函数 well function of leaky system
越流性含水层 leaky aquifer
越流因数 leakage factor
越流因素 leakage factor
越南古陆 Vietnam old land
越南山地 Vietnam mountains
越南榆 Ulmus tonkinensis
越年 overwintering;overyearing
越年生牧草 overwintering grass
越浅水波 extra shallow wave
越区 overzoning
越区供电 over-zone feeding
越区回游的 diadromous
越渠渡槽 overflume
越渠飞槽 overflume
越权 excess of authority (power);over-step
越权存取 unauthorized access
越权的 unauthorized
越权行为 ultra-vires action
越山管线 trans-mountain line;trans-mountain pipeline
越位的 offside
越限 out of bounds
越限应力 overtension
越行 overtaking
越行车辆 passing vehicle
越行点 overtaking point
越行线 passing loop;passing siding;passing track;refuge siting;side pass by;side-track;liebye <英>
越行站 overtaking station;passing point;passing station
越洋电缆 transoceanic cable
越洋电缆系统 transoceanic cable system
越洋飞行 transoceanic flight
越洋海底电缆 transoceanic submarine cable
越洋航行 transoceanic navigation
越洋距离 transoceanic distance
越洋探空仪 transoceanic sonde;transonde
越洋线路 transoceanic link
越野 across country
越野搬运车 site handler
越野叉车 rough terrain forklift;rough terrain truck
越野车 go-anywhere vehicle;off-road vehicle
越野车底盘 cross-country chassis
越野车队 off-road train
越野车辆 cross-country vehicle;off-highway;off-(the-) road vehicle;off vehicle highway vehicle
越野车辆用的轮胎 off-road tire;off-the-road tire [tyre]
越野车辆用轮胎 off-road ground tyre [tire]
越野的 cross country;off-highway;off-road;rough terrain
越野地(带) cross country
越野地方<高低不平、无道路的地方> cross country
越野吊车 cross-country crane
越野工程机械 off-road work machine
越野机动性 off-highway manoeuvrability;off-road mobility
越野卡车 cross-country truck;off-highway truck;off-road truck;rough terrain truck
越野路线<试验汽车越野性能的路线> cross-country track
越野轮式起重机 off-highway wheel crane;rough terrain wheeled crane
越野轮胎 cross-country tyre [tire];off-highway tyre [tire];off-road tire;off-the-road tire [tyre]
越野面包车 cross-country minibus
越野摩托车 cross-country motorcycle;off-road motorcycle
越野能力 cross-country power
越野起重机 rough terrain crane
越野汽车 cross-country car;cross-country vehicle;off-highway vehicle;off-the-road vehicle
越野汽车运输装备 off-road motor transport equipment
越野牵引 off-road traction
越野牵引车 cross-country tractor;off-highway hauler;off-highway tractor
越野塞车 autocross
越野设备 off-road equipment
越野设备用轮胎 off-road equipment tyre [tire]
越野式 cross country
越野式车辆 off(-the)-road
越野式工程机械 off-highway construction equipment
越野式建筑机械 off-highway construction equipment
越野式林业机械 off-highway forestry-machine
越野式起重机<允许在公路上行驶的> all-terrain highway crane
越野式起重机底盘 rough terrain carrier
越野式原木运输车 off-highway logging truck
越野试验 off road test
越野铁道 interurban railroad [railway]
越野土方机械 off-highway earthmoving machinery
越野小型汽车 cross-country minibus
越野行动 cross-country operation;cross-country performance
越野行驶 cross-country run(ning);off-the-road locomotion
越野行走轮 flo(a)tation wheel
越野(型)叉车 rough terrain forklift-truck
越野(型)后卸汽车 off-highway end dump truck
越野性能 cross-country mobility;cross-country performance
越野性试验 off-the-road test
越野移动式起重机 rough terrain mobile crane
越野运输 all-terrain transportation;off-road haulage;roadless transport
越野运输车 off-highway hauler;off-highway vehicle
越野运输工具 off-road transporter
越野运输牵引车 all-terrain hauler tractor
越野运输拖拉机 all-terrain hauler tractor
越野运土车 hauler
越野运行 cross-country service
越野载货车 cross-country cargo carrier
越野载重(汽)车 cross-country cargo carrier;cross-country truck;off-highway truck;off-highway lorry;off-road truck;all-terrain vehicle
越野作业 off-road operation
越占(左侧)车道 encroach on left lane
越站干扰 over-reach interference
越障测量 obstacle detouring;obstacle detouring survey
越障概率 probability of obstacle-clearing
越障性能 obstacle performance
越组约化 block elimination

云

云白辉长岩 puglianite
云斑<图像上> chilns;cloud;cloud-(ing) point;cloudy patch
云斑闪长岩 antisohite
云斑天牛<拉> Batocera horsfielde
云斑印痕 cloud print
云斑油地毡 Moiré linoleum
云豹 clouded leopard
云层 cloud cover;cloudland;cloud layer;condensation layer;cover of cloud;veil of cloud
云层播雨 cloud seeding
云层分析 nephanalysis
云层分析图 nephanalysis;neph chart
云层衰减 cloud attenuation
云催化剂 cloud seeding agent
云带 cloud band;cloud bar
云的催化 cloud seeding
云的反照率 cloud albedo
云的分类 cloud classification
云的回波 cloud echo
云的破裂 cloud break
云的人工影响 cloud modification
云的形成 cloud formation
云堤 cloud bank;bank of clouds
云滴 cloud particle
云滴谱 cloud droplet spectrum
云滴取样器 cloud-drop sampler
云底 ceiling;cloud base
云底高(度) ceiling;height of cloud base
云底记录仪 cloud-base recorder
云底亮度图 sky map
云底能见度 ceiling visibility
云地间放电 cloud-to-ground discharge
云点 cloudy patch
云顶 cloud top
云顶高度 cloud-top height
云顶温度 cloud-top temperature
云度 cloudiness;degree of cloudiness;nebulosity
云粉红色 cloud pink
云符号 cloud symbol
云辐射相互作用 cloud-radiation feedback
云覆盖 cloud cover
云盖 cloud deck;cloud veil;veil of cloud
云橄粗安岩 macedonite
云橄黄煌岩 modliboyite
云高 cloud height;cloud level
云高计 ceil(l)ometer
云高指示器 ceilometer;cloud height indicator
云冠 cloud crest;sansan
云核 cloud nuclei
云虹 cloud box
云厚 cloud thickness
云煌霏细岩 minette-felsite
云煌石 minette
云煌岩 minette
云辉斑岩 cuselite
云辉黄煌岩 holmite
云辉蓝方黄煌岩 luhite
云级 cloud scale
云级图 atlas of cloud
云间放电 cloud-to-cloud discharge;intercloud discharge
云街 cloud street
云卷花纹 spider legs
云卷饰 curled clouds
云块 cloud mass
云蓝色 cloud blue
云粒 cloud particle
云量 cloudage;cloud amount;cloud cover;cloudiness;cover of cloud;sky cover
云量计 nephelometer
云量系数 cloud cover factor
云列 cloud street
云流 cloudy flow;cloudy stream
云龙 cloud-dragon
云幔 veil
云母 anthrophylite;cat gold;diatom-earth;glimmer;glist;isinglass;isinglass stone;katzengold;katzensilver;mica;sheep silver;specular stone
云母安山玢岩 mica granular porphyrite
云母安山岩 mica andesite
云母白云(石)碳酸岩 mica rauhaugite
云母斑岩 micaphyre;mica porphyry
云母斑岩状紫色彩饰 micaceous porphyry
云母板 mica in sheet;mica plate;micarta;plate mica
云母板套管 micanite sleeve
云母板岩 mica slate
云母板状岩 fake
云母玻璃 micalex [mycalex]
云母箔 mica cloth;mica foil;micafolium
云母薄片 mica flake;mica in sheet
云母薄片绝缘 mica-flake insulation
云母布 mica cloth
云母插片 mica insert
云母尘肺 mica pneumoconiosis
云母赤铁矿 micaceous iron ore
云母窗 mica window
云母大理石 cipol(l)in(o)
云母大理岩 cipol(l)in(o)
云母带 mica tape
云母带加热器 mica band heater
云母带纸 paper for mica tape
云母电气石云英岩 muscovite tourmaline greisen
云母电容器 foil mica capacitor;mica capacitor;mica condenser;micadon
云母电阻 mica resistance
云母垫 mica mat
云母垫片 mica spacer
云母垫圈 mica washer
云母雕刻 talc plastics
云母端窗计数管 mica end-window counter
云母方解石碳酸岩 mica alvikite
云母霏细岩 niva felsite
云母分隔片 mica partition
云母分裂 mica splitting
云母分选筛 mica separating screen
云母玢岩 mica-porphyrite
云母粉末 mica dust
云母粉(状物质) mica flour;mica powder;powdered mica
云母橄榄岩 mica peridotite
云母构造 micaceous structure
云母管 micanite pipe;mica tube
云母含量 mica content
云母黑云碳酸岩 mica sovite
云母花岗岩 micaceous granite
云母滑石 mica talc
云母化作用 micatization

云母环 mica ring
云母黄玉云英岩 muscovite topaz greisen
云母煌斑岩 mica lamprophyre; mica trap
云母辉石岩 mica pyroxenite
云母火花塞 mica spark(ing) plug
云母加热单元 mica heating unit
云母间蒙脱石 tarasovite
云母检查板 mica test-plate
云母胶合板 micanite
云母金伯利岩 mica kimberlite
云母晶块 mica pig
云母绝缘 mica insulation
云母绝缘夹 mica clip
云母绝缘片修整刀 mica undercutter
云母绝缘子 mica insulator
云母颗粒 mica particle
云母苦橄岩 mica picrite
云母块 mycalex
云母矿床 mica deposit
云母类 mica group
云母裂缝 mica cleavage
云母律 mica law
云母镁云碳酸岩 mica beforsite
云母模制绝缘物 mica mo(u)lded insulator
云母膜片 mica;mica diaphragm
云母黏[粘]土 mica clay
云母盘 mica disc [disk]
云母劈理(面) mica cleavage
云母片 laminated mica; mica book; mica cloth; mica flake; mica lamination; mica segment; mica sheet; sheet mica; shell mica
云母片比表面积 specific surface area of mice sheet
云母片截门 mica flap valve
云母片绝缘 mica-sheet insulation
云母片轮廓面积 outline area of mice sheet
云母片麻斑岩 mica schist porphyry
云母片麻岩 mica(ceous) gneiss
云母片面积 area of mice sheet
云母片型号面积 ranked area of mice sheet
云母片岩 mica(ceous) schist; micacite; mica slate; schistose mica
云母片有效面积 effective area of mice sheet
云母圈 mica collar
云母砂 micaceous sand
云母砂岩 micaceous sandstone
云母石 mica alba; micalex
云母石英岩 micaceous quartzite
云母石英云英岩 muscovite quartz greisen
云母塑胶板 micanite
云母塑料板 micanite
云母碎片 mica flake; mica fragments
云母碎屑 mica dust
云母探测器 mica detector
云母条 mica strip
云母铁矿 micaceous iron ore; micaceous iron oxide
云母铁矿涂料 micaceous iron oxide paint
云母铁矿油漆 micaceous iron oxide paint
云母铜矿 chalcophyllite; copper mica
云母土 glimmerton; micaceous soil
云母瓦 micaceous tile
云母矽肺 mica silicosis
云母箱 mica box
云母小体 mica body
云母屑 scrap mica
云母玄武岩 mica-basalt
云母岩 glimmerite
云母颜料 mica pigment
云母氧化铁 micaceous iron oxide
云母叶片 mica flap
云母页片 mica flap
云母页片阀 mica flap valve
云母页岩 micaceous shale; mica flap; mica shale
云母英化作用 greisenization
云母萤石云英岩 muscovite fluorite greisen
云母铀矿 uranite
云母云英岩 muscovite greisen
云母正长岩 mica-syenite
云母纸 mica(nite) paper
云母质的 micaceous
云母质砂岩 fake
云母珠光颜料 mother-of-pearl pigment
云母状斑点 mica specks
云母(状)赤铁矿 micaceous hematite
云母状的 micaceous
云母状铁锈屑 iron mica
云幕 ceiling; streamer; cloud ceiling
云幕灯 ceiling lamp; ceiling light; cloud searchlight; ceiling projector <测云高度的射光器>
云幕底面 cloud base ceiling
云幕分类 ceiling classification
云幕高度 ceiling height
云幕高度探照灯 ceiling height indicator
云幕高度指示仪 ceiling height indicator
云幕气球 ceiling balloon
云幕仪 ceilograph; ceilometer
云南红杉 sikkim larch
云南柳 salix cavaleriei
云南山字型构造体系【地】 Yunnan epsilon tectonic system
云南松 Yunnan pine
云南铁杉 Yunnan hemlock
云南樟 Indian camphor
云凝结核 cloud condensation nuclei
云凝聚核 cloud condensation nuclei
云盆 cloud basin
云气候学 cloud climatology
云区 cloud field; cloudland
云雀资源火箭 skylark resource rocket
云色玻璃 dolomite glass
云杉 aerial ladder; black hill spruce; dragon spruce; fir; spruce
云杉大黑天牛 < 拉 > Monochamus urussovi
云杉木 sitka
云杉松木 Northern pine
云杉属 < 拉 > picea
云杉屋面盖板 spruce pine
云杉小蠹 Seolytus sinopiceus
云闪花岗岩 andengranite
云深 cloud depth
云石 granular limestone; marble
云石板铺面 marble flag pavement
云石边 marbled edge
云石粉 marble dust
云石灰浆 marble plaster
云石灰泥 marble plaster
云石砂 marble sand
云石纹刷 mottler
云石镶嵌 marble inlay
云石镶嵌装饰 marble intarsia
云石形玻璃 marbled glass
云石纸 marble paper
云室 expansion chamber; fog chamber; nepheloscope
云速计 nephoscope
云梯 aerial ladder; scaling ladder
云梯操纵手 pole man
云梯顶部 fly ladder
云梯救火车 ladder truck
云梯消防车 ladder truck
云梯消防队 ladder company
云铁防锈底漆 micaceous iron oxide anticorrosive primer
云图 cloud atlas; cloud chart; cloud picture; nephogram
云团 cloud cluster
云纹 cloudiness
云纹玻璃 clouded glass
云纹绸 moire
云纹(方)法 Moiré method
云纹工艺 clouding
云纹条纹 Moiré fringe
云纹图像 Moiré pattern
云纹印涂 marbling print
云物理学 cloud physics
云雾径迹 cloud track
云雾林 mist forest
云(雾)室【物】 cloud chamber
云雾天气 soupy weather
云雾物理探测器 sounding instrument for cloud and fog physics
云雾状态 cloud behavio(u)r
云雾状污泥 cloud form sludge
云雾状装饰线脚 nebule mo(u)lding; nebulous mo(u)lding
云吸收 cloud absorption
云系 cloud system; nephsystem
云系分界线 nephcurve
云霞黄长岩 mica nepheline melilitite; turjaite
云霞正长岩 miascite
云霞正长岩型霞石正长岩 miaskitic nepheline syenite
云下层 subcloud layer
云下的 sub-cloud
云下洗脱 washout of subcloud layer
云相图 cloud-phase chart
云消散 burn-off
云斜煌岩 kersantite
云形板 drawing curve; flexible curve; French curve; irregular curve
云形成过程 cloud process
云形规 French curve
云形截锯 scroll saw
云型 type of cloud
云学 nephology
云翳 cloudiness; mistiness; nephelium
云英花岗岩 esmeraldite
云英岩【地】 greisen
云英岩化(作用)【地】 greisenization
云英岩型锡矿床 greisec-type tin deposit
云英岩型锡石矿石 cassiterite ore of greisen type
云英岩异常 anomaly of greisen
云影 cloud shadow
云幛 cloud shield
云砧 plume
云芝 rainbow conk
云芝属 < 拉 > Polystictus
云中放电 cloud discharge; cloud flash; intracloud discharge
云周围环境系统 cloud environment system
云柱 cloud column
云状 cloud form
云状的 cloudy
云状非金属夹杂 slag clouds
云状粉尘 cloud dust
云状花纹 chilling; clouding
云状花纹表面 curtains
云状空化 cloud cavitation
云状饰 nebulous mo(u)lding
云状挑台 nebuly corbel table
云状物 cloud
云状线脚 nebulous mo(u)lding

匀

匀斑岩 skedophyre
匀变分布负荷 uniformly varying load
匀变速运动 uniformly variable motion
匀布 equipartition
匀称的 symmetric(al); well-balanced
匀称屋顶 homogeneous roof
匀称照明 proportional illumination
匀磁线 unifluxor
匀调延迟线 continuously variable delay line
匀度 evenness
匀度系数 uniform coefficient; uniformity factor
匀镀能力 throwing power
匀隔 equipartition
匀光仪器 dodging instrument
匀光装置 amplifier for dodging
匀衡计算区等级 classification of balance calculation area
匀衡计算区面积 area of balance calculation
匀衡重力异常图 isostatic(al) gravity anomaly map
匀厚平板 uniform-thickness slab
匀化 homogenize
匀化处理 homogenizing treatment
匀化器 homogenizer
匀化作用 homogenization
匀货舱口 trimming hatch(way); trimming hole
匀加速度 constant acceleration; uniform acceleration
匀加速流体 steadily accelerated fluid
匀加速运动 uniformly accelerated motion
匀减速运动 uniformly retarded motion
匀浆 homogenate; refining
匀浆辊 doctor roll
匀浆器 homogenizer
匀浆填充(法) slurry packing
匀晶 uniform grain
匀料 refining
匀料辊 distributing roller
匀料机 refiner
匀料筒 revolving sleeve
匀流功率 even flowing power
匀流坎 spreader
匀墨辊 mouse roller; rider; rider roller; riding roller; wavers
匀泥尺 dirt screed; screed
匀排光谱 normal spectrum
匀配 equipartition
匀平状态 in trim
匀染 level dyeing
匀砂 sand brooming
匀饰性 flow level(l)ing
匀速 uniform speed; uniform velocity
匀速卸料技术 uniform discharging technique
匀速旋转流体 uniformly rotating fluid
匀速圆周运动 uniform circular motion
匀速运动 motion of uniform velocity; steady motion; uniform motion
匀速直线平移 uniform rectilinear translation
匀速直线运动 uniform rectilinear translation
匀涂 equalizing
匀涂合成剂 level(l)ing compound
匀细度 uniformity
匀相乳油 homogeneity of concentrate
匀行裕度 justify margin
匀整坡度 boning in
匀质半无限体 isotropic(al) semi-infinite solid
匀质材料 isotropic(al) material
匀质混合物 homogeneous mixture
匀质混凝土 homogeneous concrete; uniform concrete
匀质砂浆 uniform mortar
匀质土 homogeneous soil; isotropic

soil
匀质土坝 homogeneous earth dam
匀质屋顶 homogeneous roof
匀质系数 coefficient of uniformity;uniformity coefficient; uniformity factor
匀质系统 homogeneous system
匀质纤维板 building board
匀质性 homogeneity;homogenization
匀质岩 monogenic rock
匀质因数 uniformity factor

芸香烯 terebene

耘土机 spring-tooth cultivator

允变变化 allowable variation

允差 acceptable tolerance;allowance;franchise
允给折扣 discount offered
允诺程度 state of compliance
允写环 write enable ring
允许保护 guard enable
允许暴露时间 permitted exposure time
允许闭合差 allowable closure
允许变分 permissible variation
允许变形量 fairness limit
允许波况 acceptable wave condition
允许不圆度 permissible out of roundness
允许采伐量 allowable cut
允许插入 intromission
允许拆除 wrecking permit
允许产量 allowance production
允许超车 free passing
允许超宽 allowance over-width
允许超深 allowable over-depth;overdepth allowance
允许超挖线 B-line
允许车辆连挂的冲击速度 tolerated speed of impact
允许承载力 <桩的> point-bearing capacity
允许程序存储 program(me) store enable
允许吃水 admissible draught;allowance draft
允许尺寸偏差 allowable dimension variation
允许冲击荷载 impact allowance load
允许抽水降深 allowed pumping drawdown depth
允许抽水量 allowed pumping amount
允许抽水时间 allowed pumping tie
允许当量裂纹尺寸 allowed equivalent crack length
允许的 approved
允许的辐射级 tolerance level
允许的粒径范围 grading limit
允许的最高水位 allowable maximum water level
允许电流 permissible current
允许电平 enable level
允许电压升降 permissible voltage fluctuation
允许调头的中央车道 reversible center [centre] path
允许掉头的交叉口 turn crossing
允许定时中断 timer interrupt enable
允许放射性 allowable radioactivity
允许飞机通过 clear an aircraft to pass
允许废品 permissive waste
允许腐蚀度 corrosion allowance
允许腐蚀厚度 corrosion allowance
允许负差 negative allowance
允许负荷 permissible load
允许(工厂)排污量 discharge allowance
允许工作负荷 safe working load(ing)
允许工作压力 safe working pressure
允许公差 acceptable tolerance;acceptance tolerance;allowable deviation; allowance tolerance; tolerance
允许故障 mission failure
允许故障率 acceptable failure error;acceptable failure rate; admissible error; allowable error; permissible error
允许过载的持续额定(功率)值 continuous rating permitting overload
允许含尘量 permissible dustiness
允许和尺寸下限的误差 lower permitted deviation
允许荷载 allowable load;proof load
允许荷载量 carrying capacity
允许极限 allowable limit;permissible limit
允许剂量级 tolerance level
允许间隙 permissible clearance;safety clearance
允许接近速度 permissible approach speed
允许解除 permissive release
允许进场 approach clearance
允许进入 admission
允许进位信号 carry clear signal
允许静水压力值 allowable value of static(al) water pressure
允许距离保护 <闭锁式的> permissive distance protection
允许绝对偶然误差 admissible absolute random error
允许开采量 allowable mining yield;allowance production
允许开采模数法 permissible mining modulus method
允许开孔直径 allowable opening diameter
允许离差 permissible deviation
允许力矩 moment allowance
允许量 allowance
允许量分配模型 tolerance distribution model
允许流量 flow capacity
允许流速 permissible velocity of flow
允许硫含量 permissible sulfur
允许免税限度 exempt allowable
允许模式 permissive mode
允许磨耗 tear-and-wear allowance
允许磨蚀度 corrosion allowance
允许磨损量 permissible wear
允许挠曲 allowable deflection
允许能带 allowed band
允许浓度 allowable concentration
允许偶然误差 admissible random error
允许排放速率 allowable emission rate
允许配带最大功率 allowable mating maximum power
允许偏差 admissible variation;allowable deviation;allowable variation;allowance variation;permissible deviation;tolerance deviation
允许偏心 allowable offset
允许频率 allowed frequency; tolerance frequency
允许平均声强度 specific normal acoustic(al) admittance
允许起飞 permission to take off
允许强度 allowance strength; proof strength
允许区截 permissive block
允许趋近极限 acceptable close limit
允许缺陷标准 acceptable defect level
允许热损失 permitted heat loss
允许上偏差 upper permitted deviation
允许渗透的排水管线 drain pipe line allowing infiltration
允许渗透量 permitted leak
允许试验量 permissible sample size
允许试验压力 allowable test pressure
允许试样量 permissible sample size
允许收缩量 shrinkage allowance
允许输出 output enable
允许数值 permissible level
允许数字 admissible mark
允许闩锁 latch enable
允许水力梯度 permissible hydraulic gradient
允许速度 permissible speed;permissible velocity
允许算术俘获 arithmetic(al) trap enable
允许算术自陷 arithmetic(al) trap enable
允许坍塌的护脚棱体 falling apron
允许条件 enabled condition
允许通车间隔 allowance gap
允许通行车间隔 <用于感应信号的延长绿灯时间> allowable gap
允许土壤流失量 allowable soil loss
允许外部中断 external interrupt enable
允许温差 allowable temperature difference
允许温度 allowance temperature; working temperature
允许文件输入 file enable input
允许污染极限 permissible contaminant limit
允许污染物浓度 admissible pollutant concentration
允许污水排放量 allowable discharge of waste water
允许污水排放浓度 allowable concentration of waste water disposal
允许误差 allowance error;error excepted; leeway; permissible error; permission error
允许吸上真空高度 allowable suction lift;permitted vacuum height of intake
允许下偏差 lower permitted deviation
允许显示 favo(u)rable indication
允许限度 consent limit
允许限度级 tolerance level
允许相对偶然误差 admissible relative random error
允许(小车)超车 free passing of cars
允许写入 write enable
允许信号 enable signal;enabling signal;signal-enabling
允许行动联络装置 permissive action link device
允许形式 permissible type
允许型式 type permitted
允许压降 allowable pressure drop
允许压力 authorised[authorized] pressure
允许烟雾浓度 allowance smoke concentration
允许淹没深度 allowable flooding depth
允许延期付款 indulge
允许摇摆程度 lurching allowance
允许摇摆度 <铁路桥梁上车辆的> lurching allowance
允许液面 level allowance
允许应变法 allowable-strain method
允许应答 acknowledge enable
允许应力 allowance stress;permissible stress;stress allowance
允许有一定增减的条款 plus or minus clause
允许预推信号 allow to approach the humping signal
允许载货吃水线 <船舶> deep load-line
允许在道路上行驶 roadable
允许在公路上行驶的卡车 on-highway truck
允许噪声级 acceptable noise level
允许折旧 allowed depreciation
允许振动 allowance vibration; permissible vibration
允许证 permit
允许支承力 allowable bearer
允许值 accepted value;allowable value
允许质量指标 acceptable quality level;acceptance quality level
允许中断 enabled interruption;interrupt enable
允许中断标志 interrupt enable flag
允许中断触发器 interrupt enable flip-flop
允许(中断)模块 enable module
允许状态 enabled state
允许状态信号 state enable signal
允许着陆 permission to land
允许字 enable word
允许最大计算挠度 maximum allowable computed deflection
允许最低发热量 admissible minimum of caloricity
允许最低回收率 cut-off rate
允许最高转速 maximum allowable speed
允许作用 permissive action

陨玻长石 maskelynite

陨石 aerolite;asiderite;ceraunite;falling stone;meteoric stone;meteorite;meteorolite;stony meteorite;uranolite
陨石分类 minerals of meteorites
陨石坑 crater;meteor crater
陨石坑湖 meteoritic crater lake
陨石矿物 classification of meteorites;meteoric minerals
陨石年龄 meteorite ages
陨石撞击构造 meteorite-impact structure
陨碳铁 cohenite
陨碳铁矿 haxonite
陨铁 aerosiderite; chalybite; cosmic(al) iron;iron meteorite;meteoric iron;siderite
陨铁大隅石 merrihueite
陨铜硫铬矿 gentnerite
陨星坑 meteorite crater
陨致地震 meteoric seism

孕穗期 boot stage

孕镶金刚石 flush-set diamond
孕镶金刚石的碳化钨刮刀 diamond-impregnated tungsten carbide blade
孕镶金刚石扩孔器 impregnated diamond reaming shell
孕镶金刚石钻头 diamond-impregnated bit;impregnated diamond bit
孕镶式钻头 <瑞典> diaborit
孕镶胎体 impregnated matrix
孕镶套管靴 impregnated casing shoe
孕镶套管钻头 impregnated casing bit
孕镶岩芯钻头 impregnated core bit
孕镶钻头 impregnated bit
孕育铸铁 impregnated cast iron;inoculated cast iron

孕甾烷 pregnane
孕震断裂 earthquake pregnant fault
孕震构造 seismic structure; seismogenic structure
孕震构造体系 earthquake pregnant tectonic systems

运

运搬工作 handling work
运拌混凝土 transit-mixed concrete
运冰滑道 ice chute
运兵船 troopship
运驳母船 barge carrier; barge on board ship
运材 transportation of wood
运材车 timber wagon
运材道路 haul road
运草捆车 bale carrier
运程 haul distance; length of haul; length of run; shipping distance
运程距 length of run
运程牵引 long pull
运程无线导航设备 long-distance aid
运筹的 operational
运筹分析【数】 operations analysis
运筹符 operator
运筹学 operational analysis; operation-(al) research; opsearch
运筹学方法 operation research techniques
运筹学模型 operation research model
运筹研究 operation research
运出 carting; hauling away; shipped
运出港 harbo(u)r of export; port of export
运出货物 freight outward; outbound freight
运出货物登记簿 register for goods forwarded
运出汽车装车区 haul-away loading area
运船小车 cradle and roller
运袋车 sack barrow
运袋卡车 bag truck; sack truck
运袋器 sack transporter
运单 bill of freight
运单副本 duplicate of way bill
运单上的目的地 waybill destination
运到 carting
运到船上交货的价格 price FOB [free on board]
运到时间 time of arrival
运到收费 freight to be collected
运电杆拖车 pole trailer
运锭块的车 ingot transfer car
运动 locomotion; motion; movement
运动包络线 kinematic(al) envelops
运动边界 moving boundary
运动边界条件 kinematic(al) boundary condition
运动冰川 active glacier
运动波(浪) kinematic(al) wave
运动波浪说 kinematic(al) wave theory
运动波理论 kinematic(al) wave theory
运动波演算 kinematic(al) routing
运动波涌 kinematic(al) surge
运动补偿器的敏感系统 sensing system for motion compensator
运动不确定性 kinematic(al) indeterminacy
运动部分 moving element
运动(部)件 moving parts
运动参数 kinematic(al) parameter
运动参数记录器 motion recorder
运动残像测定仪 movement after image measuring instrument
运动测量器 cinesimeter; kinesimeter
运动常数 constant of motion
运动场 athletic field; game court; playground; playing field; promenade; sport area; sports field; sports ground; stadium [复 stadia]
运动场的后阻篱障 athletic back-stop fence
运动场地 sport area; sport ground
运动场滚筒 sportsfield roller
运动场看台入口 vomitorium
运动场入口 vomitory
运动场设备 playground equipment item
运动场设备项目 playground equipment item
运动场所 <古希腊、古罗马建筑中的> ephebeion
运动场压路机 sportsfield roller
运动场用的串联式压路机 sports ground type tandem roller
运动场用的双轮压路机 sports ground type tandem roller
运动场用垫 gym mat
运动场帐篷 sports ground pavilion
运动传递 transmission of motion
运动传动机构 motion work
运动传感器 movement sensor
运动单位 motor unit
运动的 kinetic; locomotive; motive
运动的返回 return of movement
运动定律 law of motion; motion law
运动对偶 kinematic(al) pair
运动方程 equation governing the motion; equation of motion; kinetic equation; motion equation
运动方程式 kinematic(al) equation
运动方式 mode of motion
运动方式分类 classification on the movement mode
运动方向 direction of motion; direction of movement
运动方向标 traffic direction sign
运动方向分类 classification on the movement direction
运动房 sports building; sports hall
运动分析器 motion analyzer
运动副 pair
运动公园 playfield park
运动关系 kinematic(al) relation
运动规律 characteristics of motion
运动海洋学 kinematic(al) oceanography
运动后效 after-effect of motion
运动机构 motion; running gear
运动机件 moving parts; motion work
运动技巧 acrobatics and tumblings
运动甲板 sports deck; sun deck
运动检测器 motion detector
运动结果分类 classification on the movement result
运动静力学 kinetostatics
运动矩阵 kinematic(al) matrix
运动空间 motion space
运动空间向量【物】 degree of freedom
运动控制系统 kinetic control system
运动控制中心 telecontrol center [centre]
运动力计 motor meter
运动力矩 motoring torque
运动链 kinematic(al) link; kinematic(al) chain
运动链换向 inversion of kinematic(al) chain
运动链系 kinematic(al) chain; motion link
运动量 amount of exercise; magnitude of motion
运动灵巧性 motor dexterity
运动流动性 kinematic(al) fluidity
运动流度 coefficient of mobility; kinematic(al) fluidity
运动毛管率 kinematic(al) capillarity
运动密度测定法 kinedensigraphy
运动面积取样 moving-area sampling
运动模拟 motion simulation
运动模拟器 motion simulator
运动模式 motor pattern
运动模型 motion model
运动摩擦 friction of motion
运动目标电子探测器 petoscope
运动目标自动指示器 automatic moving target indicator
运动能力试验 motor activity test
运动能量 energy of motion
运动泥沙 moving sediment
运动黏[粘](滞)度 kinematic(al) viscosity
运动黏[粘](滞)度系数 coefficient of kinematic viscosity; kinemati(al) viscosity coefficient
运动黏[粘](滞)度系数比例 scale of coefficient of kinematic viscosity
运动黏[粘]滞率 kinematic(al) viscosity
运动黏[粘]滞系数 coefficient of kinematic viscosity; kinematic(al) viscosity coefficient; kinematic-viscous coefficient
运动黏[粘]滞性 kinematic(al) viscosity
运动黏[粘](滞)性系数 coefficient of kinematic viscosity
运动平衡 equilibrium of motion
运动谱 motion spectrum
运动器材室 sports equipment closet
运动器官 motorium
运动器械 exerciser
运动前区 premotor area
运动潜水 sport diving
运动强度分类 classification on the movement intensity
运动筛 moving screen
运动设施 sports facility
运动神经 motor nerve
运动时间 run duration
运动时间分析 motion time analysis
运动势 kinetic potential
运动水流因数 Froude number; kinetic flow factor
运动速度 celerity; kinematic(al) velocity; pace; velocity of movement
运动速度分类 classification on the movement velocity
运动速率线 <车身的> dashing lines
运动算子 kinematic(al) operator
运动弹性 kinematic(al) elasticity
运动特点分类 classification on the movement character
运动特性 kinetic characteristic; motion characteristic; state of motion
运动特性曲线 kinetic characteristic curve
运动提前量 kinetic lead
运动体 mobile; vehicle
运动体航向 heading of moving vehicle
运动图像 movement picture
运动图像专家组 motion picture experts group; movement picture experts group
运动图形 movement picture
运动微分方程 differential equation of motion
运动系数 kinematic(al) coefficient
运动系统 kinematic(al) scheme
运动线法 arrowhead method
运动相似的 kinematically similar
运动相似性 kinematic(al) similarity; kinematic(al) similitude; kinetic analogy
运动像差 motion aberration; movement aberration
运动型小客车 sporting sedan
运动学 dynamics; kinematics
运动学分析 kinematic(al) analysis
运动学设计 kinematic(al) design
运动循环 cycle of motion; motion cycle
运动延性 kinematic(al) ductility
运动硬化 kinematic(al) hardening
运动用的 sport
运动油质污染物输移模型 kenematic oily pollutant transport model
运动渔业 sport fishery
运动员村 athletes' village
运动员风度 sportsmanship
运动员更衣室 clubhouse
运动员外流 brawn drain
运动原理模拟 analogy of movement principle
运动增强 motion divergence
运动知觉 motion perception; movement perception
运动中称重 dynamic(al) weigher; motion weigher
运动中枢 motor center[centre]; motorium
运动中心 center [centre] of motion; center [centre] of movement
运动周期 period of motion
运动状态 motion state; state of motion
运动自由度 freedom of motion; freedom of movement
运动阻力 motional resistance; resistance of motion
运动阻力的统计评价 motion resistance statistic(al) evaluation
运动阻力特性 motion-resistance-force characteristic
运动坐标 coordinates of motion
运动坐标系统 moving coordinate system
运垛机 cocklifter
运筏渠 canal for rafting wood
运法 haulage; Merchant shipping act
运费 carriage (freight); cartage; cost of freight; freightage; freight charges; freight cost; haulage; hauling charges; port(er)age; toll; transportation cost; transportation expenses; wagonage
运费包干租赁 lump-sum charter
运费保险 freight insurance
运费保险单 freight policy
运费、保险费付至 CIP [carriage and insurance paid to]
运费标准 freight base
运费表 scale of rates; table of freight
运费补贴 primage
运费从价标准 price basis
运费存款 deposit(e) for carriage
运费待付 carriage forward
运费待收 freight to be collected; freight to collect
运费单 freight bill
运费到付 carriage charges to pay; carriage to pay; cost to be paid; cost to collect; freight forward; freight payable at destination; freight to be paid at destination; freight to collect
运费等级 classification rating
运费低廉的交通线 differential route
运费吨 freight ton(nage)
运费额 freight amount
运费发票 freight bill
运费罚金 freight penalty
运费费率表 freight tariff
运费分摊 split of total freight
运费付讫 carriage free; carriage paid;

freight paid
运费付至…… free on carriage paid to;carriage paid to
运费付至目的交货地 carriage paid to the named point of destination
运费和保险费付至 carriage, insurance paid to;freight, insurance paid to
运费和延滞费 freight and demurrage
运费回扣 charges rebate;freight rebate;primage
运费回扣制 fidelity rebate system;rebate system
运费货物收讫交毕 collected and delivered
运费计算 calculation of freight
运费加佣金(价) cost, insurance, freight and commission
运费价格申报 declared value for carriage
运费价目表 rate tariff
运费交货时付 carriage forward
运费交货照付 carriage forward
运费经纪人 freight broker
运费净数 net freight charges
运费留置权 lien for freight
运费率 rate of freight;volume rate;rate of passage
运费率一览表 schedule of freight rates
运费免付 carriage free;carriage paid
运费免收 freight free
运费明细表 freight list
运费平均估计 average freight rate assessment
运费清单 freight account;freight bill;freight note
运费收据 freight receipt
运费损失 loss of freight
运费条款 freight clause
运费退款 charges refund
运费托收通知单 notice of freight collection
运费未付 carriage forward
运费未收 carriage forward
运费先付 freight prepaid
运费协定 freight agreement; rate agreement;tariff agreement
运费已付 carriage charges paid;carriage free;carriage paid;free of carriage charges;freight paid
运费已付登记簿 carriage paid record book
运费以外的小额酬金 primage
运费由收货人付 carriage forward
运费由提货人负担 freight collect;freight forward
运费由提货人支付 freight collect;freight forward
运费余额 balance of freight
运费预付 freight prepaid;prepayment of freight
运费在内价 cost and freight
运费账单 freight account;freight note
运费支出 transportation outlays
运费支付条款 freight clause;hire and payment clause
运费制 freight basis
运费准免 franco
运钢轨车 rail wagon
运钢坯车 billet car
运管叉车 pipe lift truck
运管理系统 intermodalism management system
运管拖车 pipe hauling trailer
运轨车 rail-carrying wagon
运氦车 helium car
运焊条 weave
运河 artificial navigable waterway;canal; canal river; communicating canal;navigation canal;shipping canal;water course
运河岸(坡) canal bank
运河泵站 canal pumping station
运河比降 canal gradient
运河边 canal side
运河边坡 canal side slope; channel bank;channel slope;slope of canal
运河边坡衬砌 canal slope lining
运河边墙 canal wall
运河驳船 canal barge
运河长度 canal length
运河衬砌 canal lining;canal liner
运河储水处 overfall
运河船 canal boat;canaler
运河船舶 canal boat;canal craft
运河船坞 canal dock
运河船员 canal(l)er
运河船闸 canal (navigation) lock;lock of canal
运河的从属权利 canal appurtenance
运河的横断面 canal cross-section
运河的水 canal water
运河堤 canal embankment
运河堤岸 canal bank
运河底 canal bottom
运河底宽 canal base width;canal bottom width
运河地带 calzone
运河地区 canal zone
运河定线 canal alignment
运河渡口 canal crossing
运河段 canal reach;canal section
运河断面 canal cross-section
运河舵 canal rudder
运河防渗 canal seepage prevention
运河分析 <利用高速摄影> motion analysis
运河改道 canal diversion
运河干河 arterial canal
运河港 canal harbo(u)r;canal port
运河港埠 canal port
运河工程 canal construction; canal engineering
运河工程河底及边坡整平机 canal trimmer
运河供水 water supply for canal
运河过水截面 wet cross section of canal
运河航标 aids-to-navigation on canal
运河航行 canal navigation
运河航运 canal navigation;canal traffic
运河河槽 canal channel;canal prism <棱柱形的>
运河河床 canal bed
运河河堤 canal bank
运河河段 canal reach
运河河况 canal regime
运河横截面 canal (cross) section
运河护岸 canal bank protection;canal slope protection
运河护坡 canal bank;canal slope protection
运河滑模 canal slipform
运河化 canalization
运河化河流 canalization river
运河(计费)吨位 canal tonnage
运河加宽 canal widening
运河加深 canal deepening
运河建造 canal construction
运河建筑物 canal construction
运河交汇口 canal junction
运河交通 canal traffic
运河口 canal mouth
运河宽度 canal width;width of a canal
运河阔幅段 lock bay
运河类型 kind of canal
运河立交渠 canal aqueduct
运河路线 channel way
运河轮船 canal steamer
运河轮廓 canal profile
运河码头 canal dock
运河目标 moving object
运河(内陆)船闸 canal inland lock
运河坡保护 canal slope protection
运河坡度 canal slope
运河起点构筑物 canal head
运河起点建筑物 canal head
运河戗道 canal berm
运河桥 canal aqueduct;canal bridge;over canal bridge
运河区 canal reach;canal zone
运河区法规 Canal Zone Code
运河曲线段 canal curve
运河入口 canal entrance
运河上港口 canal port
运河上航运 canal navigation
运河上游河段 upper canal reach
运河渗漏 canal seepage
运河升船机 canal lift
运河式船坞 canalled dock
运河收费 canal charges;canal tolls
运河枢纽 hydrojunction of navigation canal
运河水级 canal lift
运河水力特性 characteristics of canal
运河水力学 canal hydraulics
运河水量补给 canal feeding
运河水量损失 canal wastage
运河水深 canal (water) depth
运河水位 canal level
运河水位控制设备 canal check
运河水位升降(装置) canal lift
运河水闸 canal lock
运河水闸门 head gate
运河税 canal dues;canal fee
运河隧道 canal tunnel
运河隧洞 canal tunnel
运河梯级 canal ramps
运河庭园 <文艺复兴时期的一种几何式花园> canal garden
运河通过能力 canal capacity;carrying capacity of canal
运河通过时间 canal transit time
运河通航水位控制 canal navigable stage control;navigable stage control of canal
运河通航隧洞 canal tunnel
运河通行费 canal dues
运河通行税 canal tolls
运河挖泥船 canal dredge(r)
运河挖泥机 canal dredge(r)
运河弯道段曲率 canal curvature
运河弯段 canal bend
运河弯曲线 canal curve
运河网 canal network
运河系统 canal system
运河下闸门 tail gate
运河斜面升船机 canal incline lift
运河斜坡道 canal incline
运河需水量 water demand of canal;water requirement of canal
运河引航员 canal pilot
运河淤积 canal silting
运河运输 canal transport
运河运输能力 canal capacity;carrying capacity of canal
运河闸(门) canal lock;lock of canal
运河支汊 lateral canal;arm of canal
运河支流 canal branch
运河支流施工 canal branching
运河支线 spur canal
运河中驳船转头加宽段 winding hole
运河中间消涌池 intermediate pool
运河中水闸 draw gate
运河轴线 canal axis
运河专用的 canal-built
运河纵断面 canal profile
运缓列车 out-of-course train;out-of-schedule train
运灰车 ash wagon
运灰机 ash conveyer [conveyor]
运灰浆车 grout car(t)
运回 carry back
运混凝土卡车 concrete bowl lorry
运混凝土倾卸车 concrete tip wagon
运混凝土小车 concrete cart
运活鱼渔船 live fish carrier
运货 ship
运货班轮 cargo liner
运货舱 cargo module
运货车 block truck; cargo carrier; cargo truck; cartload; conveyer car; delivery truck; gantry truck; goods-carrying vehicle; lorry; track line; transport truck; van vehicle; wagon(age);sling cart <车轴上有吊链的>
运货车底盘 goods chassis
运货车辆 wagonage
运货车轮 truck wheel
运货车转台 truck turntable
运货船 freight boat;freight ship
运货大马车 wain
运货代理商 freight agent;shipping agent
运货单 bill of lading;bill of loading; consignment invoice; freight bill; shipping order; shipping ticket; waybill
运货的马 cart-horse
运货电梯 freight elevator;freight lift; goods elevator;goods lift
运货吊桥 goods lift bridge
运货吊梯 goods lift
运货洞 fall way
运货飞机 air truck; cargo airplane; flying lorry;freight carrier
运货浮动平台 floating cargo landing stage
运货工具 cargo carrier
运货航次 freighting voyage
运货合同 contract for carriage of goods; contract for delivery; contract of carriage
运货机 aerovan
运货机车 freight train locomotive
运货机具 cargo carrier
运货价目表 freight tariff
运货绞车 cargo winch
运货经纪人 freight broker
运货卡车 deliver(ing) truck;freight car; freight truck; goods vehicle; lorry;motor truck
运货列车 goods train; wagon train; freight train;rattler <美>
运货马车 cart;dray;horsedrawn van; horsedrawn vehicle; horse wagon; wain
运货马车费 drayage
运货汽车 automobile wagon;automotive truck;autotruck;cargo truck; commercial vehicle; freight car; freight carrying vehicle;goods-carrying vehicle;motor lorry;van;cargo vehicle; freight vehicle; goods car;motor truck;truck <美>;lorry <英>
运货汽车车道 truck lane
运货汽车车身 truck body
运货汽车驾驶员 gear jammer
运货汽车轮链 truck track
运货汽车轮辋 lorry rim
运货汽车入口 truck entrance
运货汽车输送垃圾 transportation of refuse by motor-lorry
运货汽车终点站 lorry terminal;truck terminal
运货人 freight forwarder

运货人自装自算 < 未经铁路检查 > shipper's load and count
运货上储料堆的输送机 inloading conveyor
运货升降机 freight elevator;goods elevator;goods lift
运货升降机门 dumbwaiter door
运货竖道 fall way
运货索赔 freight claim
运货通道 passage for freight handling
运货同盟 conference
运货拖拉机 cargo tractor
运货文件 shipping documents
运货证书 bill of lading
运货装置 shipper
运积矿床 transported deposit
运积黏[粘]土 transported clay
运积速度 transporting velocity
运积土 carried soil;transported soil;travel(l)ed soil
运积物 transported deposit
运积物异常 transported overburden anomaly
运家禽时在车站围栏使用权 yardage
运价 tariff;tonnage rate for hauling;transportation rate
运价比价 tariff parity
运价变更通知 notice of change in rates
运价表 freight list;freight tariff;miscellaneous charge book;schedule of freight rates;tariff;rate table < 车站至车站间 >
运价表补充表 tariff supplement
运价表档案 tariff file
运价表的施行 application of tariff
运价表的一节 section of tariff
运价表的重新编制 recasting of tariff
运价表附录 appendixes to a tariff
运价表规定的条件 tariff condition
运价表基础 basis of tariff
运价表内的货物分等 tariff classification of goods
运价表通告 tariff circular
运价表文件 tariff file
运价表项目 items in tariff
运价的合理性 reasonableness of rates
运价的双费率结构 dual charge structure of tariff
运价等级基数 basis of scale
运价公会 freight conference;shipping conference;steamship freight conference
运价公会决定的费率 conference rate
运价鼓励 tariff incentive
运价规章 tariff regulations
运价合同 < 美国铁路与大货主间签订 > rate contract
运价和价格政策 tariff and price policy
运价和票价的非标准化 de-standardization of rates and fares
运价核算员 rates clerk;tariff clerk
运价汇编 rate compilation
运价计算 rate calculation
运价加、减成计费 charge by adding or subtracting a ratio
运价降低 tariff rate;tariff reduction
运价结构 construction of rates
运价里程 kilometrage for charging rates;rate-making distance;tariff kilometerage [kilometrage];tariff mileage
运价理论 the theory of railway tariff
运价联盟 tariff union
运价率 tariff (rate);freight rate
运价率表 scale of rates
运价率的剪刀差 < 即最高率和最低率的差别 > fork of rates
运价率基数 rate base
运价(率)一览表 schedule of freight rates
运价区 < 根据运输密度和运营条件划分 > rate zone
运价手册 miscellaneous charge book
运价水平 rate level
运价特许 tariff concession
运价体制改革与运输企业定价 the reform of the railway tariff system and the pricing of the transport industry
运价调整 rate adjustment
运价削减 rate cut(ting)
运价协定 rate agreement
运价协调 tariff coordination
运价协议 freight agreement;rate agreement;tariff agreement
运价协议会 shipping conference
运价修订 tariff revision
运价增加 tariff increase
运价战 rate war
运价政策 tariff policy
运价指数 freight index;tariff index
运价制定 rate-making
运价制定权 rate-making power
运价总簿 rate book
运架一体机 transporting-erecting machine
运交不合格商品 bad delivery
运金属的车辆 ingot buggy
运金属的卡车 ingot buggy
运金属屑用拖车 salvage trailer
运进货物登记簿 register for goods received
运进加工 inward processing
运距 haul(ing) distance;haul (length);length of haul;load distance;shipping distance
运客工具 people mover
运客装置系统 people mover system
运款列车 cash train
运矿车 mine car;ore transfer car
运矿船 ore carrier
运矿石车 mine car
运矿石船 ore-carrying vessel
运垃圾驳(船) sludge barge
运来的 shipped in
运砾石手推车 gravel barrow
运梁车 beam transportation car
运粮船 grain carrier;grain ship;victual(l)er
运粮大驳船 provision hoy
运粮拖车 grain carrier;grain cart
运量 amount of traffic;freight traffic;freight volume;traffics;traffic volume
运量波动系数 freight volume fluctuation coefficient
运量单位 traffic unit
运量的正常增长 normal traffic growth
运量调查 transport volume survey
运量繁忙的线路 heavily trafficked track
运量分配 division of traffic;traffic assignment;traffic distribution
运量观测 traffic survey;transportation survey
运量观测记录 traffic recorder;traffic records
运量观测数据 traffic figures
运量观测站 traffic count station
运量密度 density of traffic
运量适应图 transport capacity vs traffic-volume curve
运量特征 character of traffic
运量吸引面积 traffic area
运量预测 traffic estimate;traffic volume forecast;traffic volume prediction
运量增长 traffic growth
运料车 batch cart;carryall;lorry;material mover;push car < 铁路 >
运料车轨道 tram rail
运料道路 haul(ing) road
运料吊车 material handling crane
运料斗 conveyer [conveyor] bucket
运料罐笼 material cage
运料监督 traffic supervisor
运料路 hauling road
运料路线 path of haul
运料平板升降机 material platform hoist
运料起重机 material handling crane
运料气闸 material lock
运料升降机 material hoist
运料手推车提升机 barrow-hoist
运料桶 carrying ladle;shipping cask;transfer cask
运料用闸口 muck lock
运料闸门 material lock
运磷酸盐车 phosphate car
运流 convection current
运流电流【电】 convection current
运流漂移 convection drift
运率差 differential
运率平均运价 average freight rate assessment
运马车棚 horse box
运马棚车 horse box
运锚船 anchor hoy
运锚小艇 mooring craft
运煤 coal preparation
运煤驳(船) coal lighter;haulabout
运煤驳船队 coal tows;coal towing
运煤长列车 < 俚语 > black snake
运煤车 bunker car;coaler
运煤车辆 coaler
运煤船 coal carrier;coaler;coaling ship;collier
运煤船船员 collier
运煤吊车 coaling crane
运煤方驳 coal compartment boat
运煤工 headman
运煤火车 coal car
运煤机 coal conveyer [conveyor];conveyer [conveyor]
运煤坑道 coal conveyer tunnel
运煤/矿石船 coal-ore-carrier
运煤廊 enclosed coal (conveyor) trestle
运煤列车 unit train
运煤桥 coal bridge
运煤手车 coal barrow
运煤铁路 coaler;coal railway
运煤栈桥 coal conveyer trestle
运煤专用线 coaler
运木材车 log wagon;timber truck
运木材机车 logging locomotive
运木材卡车 logging truck;lumber body
运木车 straddle carrier;timber cart
运木船 log (and lumber) carrier;log-boat;lumber carrier;timber carrier;timber ship;wood carrier;wooden ship
运木船干舷 lumber freeboard
运木港 lumber port;timber port
运木拱架 log arch
运木滑板 jumbo
运木机 log conveyer [conveyor]
运木夹具 claw for lumbering;claw for timbering
运木链条 logging chain
运木斜梯 jack ladder
运泥驳船 dump barge;sludge barge
运泥车 muck car;sludge tank truck
运泥船 barge;dump scow;hopper barge;mud lighter;mud scow
运泥工具 earthmover equipment
运坯架 stilliard
运坯手推车 hack barrow
运期合同 forward contract
运期契约 forward contract
运期援助计划 forward aid program(me)
运汽车带架平车 automobile rack flat car
运汽车的货车 automobile car
运汽油车 gasoline tank truck
运人车辆 staff vehicle
运人吊篮 mammy chair
运入 carry-in
运入的 shipped in
运入货物 freight inward
运入运费 inward freight and cartage;transportation-in
运入重量 intake weight
运砂车 sand transportation truck
运砂船 sand carrier
运砂机 sand drag
运生铁手车 pig-iron barrow
运生土 allochthonous soil
运牲口的火车 stock car
运尸体门 luch gate;scallage
运石半挂车 rock hauling semi-trailer
运石驳 rock delivery barge
运石车 go-devil
运石开底驳 stone carrying hopper barge
运石平底船 stone boat
运石平底橇 stone boat
运石雪橇 stone boat
运石用的窄轨机车 dink(e)y locomotive
运式挖掘机 elevator digger
运输 carriage;carting;convey(ing);freight;handlage;haul;portage;teaming;traffic;tramming;transport;transportation;transporter-erector-launcher;transporting;transport movement
运输安全 safety in transportation;transportation safety
运输安全管理 safety management of traffic
运输安全监察 safety supervision of traffic
运输安全检查 safety inspection of traffic
运输安全评估 safety evaluation of traffic
运输安全系统工程 safety system engineering of traffic
运输拌和机 on the road mixer
运输包装 transport package;transport packaging
运输保安 safety of traffic
运输保险 transportation insurance
运输报表 traffic returns
运输报告和控制系统 traffic reporting and control system;transportation report and control system
运输变更 < 在货物到达到达站前和后 > reconsignment;diversion;alterations in goods transport
运输变更地点 point of diversion
运输标志 shipping mark
运输波动 fluctuation in traffic;surge in traffic
运输驳 transport barge
运输补贴 transport subsidy
运输不均衡系数 imbalance factor of traffic
运输不平衡系数 coefficient of freight (traffic) unbalance
运输部 Ministry of Transport < 英 >;Department of Transportation < 美 >
运输部标志系统 department of transportation system of labeling
运输部部长 minister of Transport

运输部门 transport operator; transport sector
运输部门车场 traffic sector yard
运输部门支出 transport sector expenses
运输槽 conveying trough; transport pan; troughing belt
运输产品 <以旅客公里和吨公里计算> transport product
运输车 carrier vehicle; chariot; forwarder; haulage car; highway tractor; transfer car; transport car(t); transporter; transport troll(e)y
运输车场 transportation pool; transport park
运输车队 haulage fleet
运输车架 carrier frame
运输车辆 carrier vehicle; haulage vehicle; hauling unit; rolling stock; transport(er) vehicle; wagon
运输车辆停放场 transport vehicle park
运输车辆用轮胎 transport tire
运输车流 transport stream
运输车数 total lorries
运输车用户 transport user
运输成本 carrying cost; freight cost; transport(ation) cost
运输成本计划 transportation cost plan
运输成本计算表 calculation schedule of transportation costs
运输成本理论 cost-of-service theory
运输成本利润率 ratio of traffic profit to transportation cost
运输成本指标 index of transportation cost; target of transportation cost
运输成绩 traffic performance; transport performance
运输承包人 hauling contractor; transport contractor
运输承包商 hauling contractor; transport contractor
运输承包者 hauling contractor; transport contractor
运输承载能力 transport carrying capacity
运输尺寸 transportation clearance; transport dimension
运输充塞 transportation congestion
运输处 freight department; traffic operating department
运输处长 chief traffic manager; chief traffic superintendent; superintendent of transportation; traffic manager
运输船 carrier; transporter; transport ship
运输船舶 shipping
运输船队 fleet
运输船军医 transport surgeon
运输船上运输部门的负责军官 transportation officer
运输船上运输部门负责军官代理人 transportation agent
运输次数 times of transportation
运输大队 transportation group
运输代办人 forwarding agent; forwarding broker; freight forwarder
运输代办行 forwarding agent; shipping agent
运输代理 transport agency
运输代理人 forwarding agent; forwarding broker; freight forwarder; transportation broker
运输代理人提单 house bill of laden
运输代理商 shipping agent
运输代理行 freight forwarder; landing agent
运输带 conveyer [conveyor]; conveying belt; conveyor belt(ing); transport band; travel(l)ing apron; travel(l)ing belt; weigh belt
运输带的托辊 apron roll
运输带的窝球节 conveyer ball joint
运输带给矿机 travel(l)ing belt feeder
运输带给料机 travel(l)ing belt feeder
运输带鼓轮机 conveyor rolling machine
运输带回摆控制 swing control
运输带配料用的秤 batching conveyor belt scale
运输带筛 travel(l)ing belt screen
运输带支承滚轴 idler
运输带最大坡度 maximum belt slope
运输带最大张力 maximum belt tension
运输单据 transport documents
运输单位 traffic unit; traffic volume unit
运输淡期 light traffic period
运输导洞 haulage heading
运输道(路) haul road; haulage way; roller way
运输道坡度 haulage grade
运输的 transportational
运输的路途费用 line-haul cost for transport
运输的提供 supply of transport
运输的无规律性 irregular nature of traffic
运输的站场费用 terminal cost for transport
运输等级指快 class of traffic
运输等级指慢 class of traffic
运输底层结构 transport infrastructure
运输地理信息系统 transportation geographic information system
运输电机车 electric(al) haulage locomotive
运输调查 transportation survey
运输调查研究局 <设在美国纽约> Transportation Research Board
运输调度 traffic control; traffic dispatch; traffic movement control; transportation control; transport movement control
运输调度员 traffic operator; transport controller
运输调度站 regulating station; transportation control point
运输调度中心 transport control center [centre]; transport movement control center [centre]
运输定额 transport quota
运输动力消耗 power consumption of transport
运输动脉 transport artery
运输斗 tote box
运输堵塞 congested with traffic; traffic block
运输队 transport and supply column; transportation column; transportation control; transport corps; uniting convoy
运输吨数 tons transported; transportation tons
运输发展 traffic evolution
运输发展理论 transport development theory
运输发展趋向 traffic trend
运输法 transport law
运输法案 transport acts
运输繁忙干线 heavy-density trunk line
运输繁忙期间 heavy traffic period
运输繁忙铁路 heavy-traffic railroad
运输方案 traffic program(me); transport program(me)
运输方法 mode of transportation; transportation resource; way of transportation; transportation method
运输方式 haulage; means of transport(ation); mode of transportation; transportation medium; transport(ation) mode; transportation system; form of transport
运输方式的分化 modal split
运输方式的选择 modal optimum
运输方式内的 intramodal
运输方向 shipping direction
运输方向间的竞争 competition of directions
运输方针 transportation policy
运输飞机 air carrier; cargo airplane; transport plane; transshipment plane
运输非常状态 transportation emergency
运输费率 carry rates; freight rate
运输费(用) conveying cost; cost of transportation; cost of truck and transportation; delivery expenses; forward expenses; operating expenses; transportation charges; transformation cost; carriage charges; transportation cost; transport(ation) expenses; fare
运输费用指数 index of transportation cost
运输分析办公室 <美国州际商务委员会 ICC 管辖> Office of Transportation Analysis
运输风险 risk of carriage
运输服务 transportation services
运输符号 transport symbol
运输辅助线 feeder line
运输附属服务设施 transportation service accessory
运输改善计划 transportation improvement program(me)
运输干道 traffic artery
运输干扰 traffic disturbance
运输干线 traffic artery
运输干燥机 conveying drier [dryer]
运输钢丝绳 haulage rope; hauling wire rope; tram rope
运输高峰 traffic peak
运输工 haulage-man; haul(i)er
运输工程师 traffic engineer; transit engineer; transportation engineer
运输工程师学会 <英> Institution of Transport Engineers
运输工程学 transportation engineering; transshipment engineering
运输工具 conveyancer; delivery vehicle; haulage means; hauling equipment; hauling operation; hauling unit; means of transit; means of transport(ation); mode; rig; transportation; transportation equipment; transportation means; transportation medium; transporter vehicle
运输工具分担模型 modal share model
运输工具分配 mode split
运输工具角色 conveyance case
运输工具涂料 transportation finish
运输工具注册费 carrier registration expenses
运输工具转运式客机坪 transporter type apron
运输工人工薪 delivery workers' salaries
运输工时消耗 man hours consumption of transport
运输工作 hauling operation
运输工作班计划 shift traffic working plan
运输工作成绩 conveyance performance
运输工作技术计划 plan of technical indices for freight traffic
运输工作量 amount of transport
运输工作人员工资 wage of traffic operation staff
运输工作日常计划 day-to-day traffic working plan
运输工作日计划 daily traffic working plan
运输工作通行证 transport work ticket
运输公司 carrier; common carrier; forwarding agent; transport(ation) company; transportation firm
运输功 <运输单元和它们通过距离的乘积> transportation work
运输功能 function of transportation
运输供给 transportation supply
运输供给模式 transportation supply mode
运输供求关系 transportation demand and supply
运输共同体 transport community
运输钩 hold-up hook
运输固定设备 transport fixed installations; transshipment fixed installations
运输固定设备结构 transshipment infrastructure
运输关系 traffic relation
运输管道 transport pipeline
运输管道的卡车 pipe truck
运输管理 traffic management; transportation administration; transportation management; transportation movement
运输管理工 transport attendant
运输管理机构 transit authority; transportation operating agencies
运输管理信息系统 transportation management information system
运输管理员 transport officer
运输管制 transport regulation
运输规程 traffic regulation
运输规划 transport(ation) planning; transportation program(me); transportation programming
运输规划程序 transportation planning process
运输规划改进系统 transportation improvements planning system
运输过程 transportation process
运输过程中拌和 mixed-in-transit
运输过程中搅拌 in-transit mixing
运输行 common carrier; forwarder; freight agent; haul(i)er
运输行收货凭证 forwarding agent's certificate of receipt
运输行业 carrier's trade
运输航空机场 air carrier airport
运输号码 transport No.
运输合理化 rationalization of transport(ation); traffic rationalization
运输合同 carriage contract; contract for carriage; contract of carriage; contract of transportation; traffic agreement
运输和出口 <进口货物到达进口港后通过本国直达其他国家的某一地点> transportation and exportation
运输和装卸过程 transportation and handling procedure
运输滑道 slideway
运输环境工程学 transport ergonomics
运输混凝土的卡车 concrete delivery truck
运输货车牵引车 truck tractor
运输货物保险 floater

运输货物标准代号 standard transportation commodity code
运输货物标准符号 standard transportation commodity code
运输货箱 transportainer
运输机 aerotransport; air freight; belt conveyer [conveyor]; cargo-transport plane; conveyer [conveyor]; haul(i)er; hauling machine; transport aircraft; transport airplane; transporter; transport machine; transport plane; transport vehicle; sky truck <大型的>
运输机壁 conveyer jib
运输机分输系统 conveyer distributing system
运输机钩杆 <使货物卸入斜槽内> tripper
运输机构 agency of transportation; transshipment machinery
运输机化 conveyerisation [conveyerization]
运输机回程皮带惰轮 return idle
运输机具 cargo carrier
运输机坑道 convey tunnel
运输机皮带 conveyer belt(ing); travel(l)ing apron; travel(l)ing belt
运输机皮带滚轴 apron roll
运输机平巷 convey tunnel
运输机桥 transporter bridge
运输机停机坪 transport apron
运输机械 transport(ation) machinery
运输机械折旧储备金 depreciation reserve of delivery equipment
运输机用跑道 transport runway
运输机装车溜槽 <俚语> yo-yo
运输机装料工 conveyer filler
运输机装料漏斗 conveyer charging hopper; conveyer loading hopper
运输机装卸漏斗 conveyer charging hopper; conveyer loading hopper
运输基本设备结构 transport infrastructure
运输基本设施结构 transshipment infrastructure
运输基础结构 transport infrastructure
运输基础设施 transport infrastructure
运输及工业部 Department of Transport and Industry
运输及竖起安装车 transporter-erector
运输及通风平巷 haulage and ventilation drift
运输及通信[讯]委员会 Transport and Communication Commission
运输极限 transportation limitation
运输集散线 feeder and distribution line
运输计划 long period transport plan; movement plan; shipping plan; traffic plan; transport(ation) plan
运输计划图表的编制 transport scheduling
运输计重 shipping weight
运输技术 transportation technology
运输技术经济指标 technical-economic index of transport
运输夹具 transportation tongs
运输驾驶台 transport station
运输监护 <液化气> custody transfer
运输监控系统 transport monitor system
运输舰 troopship
运输交换 interchange of traffic
运输交通调度组 transport traffic regulating group
运输交易市场 freight market
运输角度 haulage angle
运输绞车 haulage winch
运输搅拌 mixed-in-transit
运输搅拌混凝土 transit-mixed concrete; transit mix(ture); truck-mixed concrete
运输阶段 haulage stage
运输结构 traffic structure; transport structure
运输结构物 transport structure
运输结算 closing transport account
运输界 transport sector
运输界协会 <包括工、农、商、运各界> Shipping Association
运输津贴 transportation subsidy
运输进款 transport income
运输进款挂钩收入 transport earning linked revenue
运输进款清算办法 settlement method of transportation income
运输进款收支报告 receipt and payment report of transportation income
运输进款月报 monthly report of transport revenue
运输进款占用天数计算表 railway transport revenue impropriate days calculation sheet
运输经费 transportation spending
运输经纪行 freight forwarder
运输经纪人 freight broker; freight forwarder; shipping broker; warding broker
运输经济范围 sphere of transport economy
运输经济距离 economic transport distance
运输经济平衡表 balance sheet of transport economy
运输经济学 traffic economics; transport(ation) economics; transshipment economics
运输经营者 transport operator
运输径路 traffic route
运输径路间的竞争 competition of routes
运输竞争 transport competition
运输局 traffic department
运输距离 transportation range; transport distance
运输卡车 delivery lorry; delivery truck; haulage truck; hauling truck
运输科 traffic department
运输坑道 <连接地下开挖的出入坑道> haulage way
运输控制编号 transport(ation) control number
运输控制仓库 transportation control depot
运输控制程序 shipping control procedure
运输控制卡片 transportation control card
运输控制系统 shipment control system; transport control system
运输控制中心 transport control center [centre]
运输宽度 shipping width
运输会计代码 transportation account code
运输垃圾的大型集装箱 pull-on container
运输类型 type of traffic
运输里程 carriage kilometerage; carriage mileage; transport mileage; transshipment mileage
运输力 transporting force
运输力的补给 supply of transport
运输力弱 transport fatigue
运输利润 traffic returns
运输连接点 traffic node
运输联合会 transport Association
运输联合企业 transport combine
运输联锁系统 gate interlock
运输联营集合体 <德国> transshipment co-ordination and integration (in FRG)
运输联营组织 traffic pool
运输链 conveying chain; transport chain
运输量 amount of traffic; discharge; earning capacity of traffic; haul; operational throughput; quantity to be conveyed; quantity to be hauled; traffic; traffic-carrying capacity; traffic performance; transportation burden; transportation volume; transported quantity; transport performance; volume of transport; haulage
运输量的分配额 part of traffic; proportion of traffic; share of traffic
运输量调查 transportation investigation; transportation survey
运输量过磅站 traffic weighing station
运输量流向 current of traffic
运输流程图 transport flowchart; transshipment flowchart
运输流体 conveyance fluid
运输流向的最优合理化 optimum rationalization of transport flows
运输路径 transportation route
运输路线 haulage way; haulway; traffic route; transit route; transportation route
运输旅客和行李 processing passengers and baggage
运输密度 traffic density
运输命令 transportation order
运输模式 mode of transport
运输能力 carrying capacity; hauling capacity; load-transfer capacity; movement capacity; shipping capacity; traffic capacity; transportability; transportation capacity
运输能力的调查 research of transport capacity
运输排土桥 transport-and-dumping bridge
运输皮带 travel(l)ing belt
运输皮带廊道 conveyer gallery
运输票据 carriage document; transport documents
运输平洞 gangway
运输平硐 haulage tunnel
运输平盘 haulage berm
运输平巷 gate road; haulage drift; haulage entry; haulage heading; haulage tunnel; haulage way; hauling entry; hauling gallery; transport drift
运输平巷弯道 haulage curve
运输凭单 transportation voucher
运输凭证 carriage voucher
运输期限的计算 calculation of transit period
运输企业成本利润率 transport enterprise cost profit rate
运输企业资产增值率 transport enterprise assets added value rate
运输企业资金利税率 transport enterprise capital profit-tax rate
运输起重机 transport crane
运输气垫车 transport air cushion vehicle
运输汽车 carrier (vehicle); transport automobile
运输汽车基地 transportation motor pool
运输汽车连 troop truck company
运输汽车枢纽 <铁路> automobile transport terminal
运输汽车用半拖车 semi-trailer for car
运输契约 contract for carriage; contract of affreightment; contract of carriage
运输契约的签订 conclusion of the contract of carriage
运输契约的执行 execution of the contract of carriage
运输契约规定外的里程 overrun
运输契约规定外的运程 extension of route
运输器 carrier; conveyer [conveyor]; transport vehicle
运输牵引车 cargo tractor; transporter truck-tractor
运输前检查 pre-trip inspection
运输潜力 traffic potential
运输潜水艇 transport submarine
运输强度 handling strength; intensity of traffic
运输桥 conveyor bridge; overburden conveying bridge <露天开挖系统的>
运输清淡期间 slack period
运输清淡时间 slack hour
运输清算收入 transport liquidated revenue
运输情报 transportation intelligence
运输情况 traffic condition
运输区 traffic zone; transportation zone
运输扰乱 disturbance of traffic
运输人 carrier
运输容器 haulage vehicle; hauling container; shipping cask; shipping unit; transfer cask; transport basket; transport container
运输散装货物的有漏斗状容器的卡车 gondola
运输筛分机 conveyer screen
运输设备 handling device; facility for transportation; handling equipment; haulage plant; hauling equipment; hauling plant; hauling unit; transportation device; transportation plant; transporting equipment; transport vehicle; transport(ation) facility
运输设备利用率 utilized efficiency of transport equipment
运输设备一条龙 battery of hauling equipment
运输设施 transportation facility; transshipment facility
运输申请 transportation request
运输生产 transportation production
运输生产定额 transport production quota
运输生产率 carrying productivity
运输生态学 transportation ecology
运输石门 haulage cross-cut
运输石油产品的橡胶管 rubber cargo hose
运输时的高度 transport height
运输时间 <船的> shipping time; haul(age) time; time of haul; time of transport
运输时外形尺寸 transporting outer size
运输食品列车 train carrying foodstuffs
运输市场 transportation marketing; transport market
运输市场及相关经济理论 transport market and economic theories related
运输式桥 transporter bridge
运输式燃气轮机 vehicular gas turbine
运输式透平 vehicle turbine
运输事故 shipping accident

运输事务所 shipping office
运输事业 haulage business;transport undertaking
运输试验中心 <美> Transportation Test Center
运输适温 suitable appropriate transport temperature
运输收入 income of transportation;operation revenue;traffic receipt;transportation revenue
运输收入管理的审计 transportation revenue management audit
运输收入会计核算审计 transportation revenue accounting audit
运输收入计划完成情况审计 transportation revenue plan accomplishment audit
运输收入交款单 transport revenue payment bill
运输收入进款计划完成情况表 statement about transport revenue plan finished
运输收入进款计划完成情况附表 auxiliary statement about transport revenue plan finished
运输收入进款收支报表 the income and expense report of transport revenue
运输收入进款资产负债表 general sheet of transport revenue
运输收入率 rate of traffic revenue
运输收入票据 receipt of transport income
运输收入审计 transportation revenue audit
运输收入往来 account current transport income
运输收入责任制 responsibility system of traffic revenue
运输收入专户 special account for traffic revenue;specified account for traffic revenue
运输收入子系统 subsystem of transport revenue
运输收入总表 balance sheet of transport revenue
运输手段 transportation medium
运输枢纽 transportation junction
运输数据系统 transportation data system
运输水平 haulage horizon;haulage level;tramming level
运输税 tax on transport;transportation tax
运输税金及附加 transport tax and addition
运输司机室 transport station
运输隧道 conveyance tunnel;haulage tunnel;transport tunnel
运输隧洞 conveyance tunnel;haulage tunnel;transport tunnel
运输损耗 conveyance loss
运输损伤 shipping damage
运输损失 conveyance loss
运输所 transport point
运输索 carrying cable;track rope
运输塔 transporter tower
运输体系 transport system
运输条件 condition of transport(ation);traffic condition;transport condition
运输条款 transit clause
运输调整 regulation of traffic;traffic adjustment;traffic regulation
运输调整计算机 traffic regulation computer
运输停顿 hold-up
运输通报系统 traffic annunciator system
运输通知书 traffic advice
运输统计 traffic statistics;transport statistics
运输统计报表 traffic returns
运输统计报告(表) traffic returns
运输统计原始凭证 original evidence of railway transport statistics
运输桶 carrying ladle;transport cask
运输投入产出调查 transport input and output survey
运输投资 transportation investment
运输途中货损 goods damage in transport
运输途中受震 shocks received during transit
运输途中损坏 damage in transit
运输挖方法 haul-away
运输外部性 externality of transportation
运输外的 non-traffic
运输外的收入 nontraffic receipt
运输网 transport(ation) network;transshipment network
运输网路理论 transportation networks theory
运输网络 transportation network
运输网生产能力 throughput capacity of the transport network
运输危险 service danger
运输委托书 forwarding order
运输委员会 Transport Committee
运输位置 carry position
运输问题 transportation problem
运输吸引地区 catching area;catchment area
运输吸引范围 traffic catchment area
运输吸引区 traffic catchment area
运输系数 ratio of freight traffic
运输系统 conveying system;conveyance system;transit system;transportation chain;transportation system;transporting system
运输系统的取舍 transportation system alternative
运输系统的组成部分 components of a transportation system
运输现代化 traffic modernization
运输线 <铁路、公路、水运、航空等> transport line;artery;traffic line;transportation lifeline;transportation routine
运输线路 traffic line;transport route
运输线系统 transport chain
运输限度 transportation limit
运输限制 transport limitation
运输箱 shipping case;transport case;transport container;travel(l)ing box
运输巷道 belt heading;haulage drift;haulage heading;haulage (road) way;haulage tunnel
运输巷道装车溜井口 haulage box
运输小车 push-away buggy;travel(l)ing bogie
运输小泡 transport vesicle
运输效率 conveying efficiency;transport efficiency
运输协作 transport coordination
运输卸载区域 transport area
运输性能 traffic performance;transportability;transport performance
运输需求 transportation demand
运输需求分析 transportation demand analysis
运输需求管理 transportation demand management
运输需求理论 transportation demand theory
运输需要 traffic requirement;transportation need
运输需要量 demand for transport;traffic demand
运输许可证颁发 transport licensing
运输学会 Institution of Transport <英>;Institute of Transport
运输学院 transportation institute
运输循环 haul cycle
运输咽喉 traffic bottleneck
运输延误 delay in traffic;transit delay
运输研究 traffic study;transportation research
运输研究讨论会 <美> Transportation Research Forum
运输研究学会 <美> Transportation Research Institute
运输掩护分队 movement protection detachment
运输演变 traffic evolution
运输要道 traffic artery
运输要口 traffic gateway
运输业 carrying trade;transport(ation) industry;transport service;transport trade
运输业固定资产 fixed assets of transportation
运输业规模经济与范围经济 economics of scale and scope in transport industry
运输业网络经济特性 carrying trade network economy characteristic
运输业务 forwarding transport service;shipping;transport operation
运输业务办事员 shipping clerk
运输业务促进 traffic promotion
运输业者 carrier;forwarder;forwarding agent
运输业者的扣押权 carrier's lien
运输业者的留置权 carrier's lien
运输业者间的竞争 intercarrier competition
运输意外事故 transportation contingence
运输英里程 traffic mil(e)age
运输用半挂车 transport semitrailer
运输用的起重机 transport crane
运输用户 transport user
运输用集装箱 shipping container
运输用铰链机构 transport linkage
运输用履带拖拉机 transport caterpillar tractor
运输用燃料 transport fuel
运输用拖拉机 transport tractor
运输优先权 transportation priority
运输鱼的缆车或空中索道 fish trap and transportation
运输与安装设备 transportation and erection equipment
运输与道路研究实验室 <英> Transport and Road Research Laboratory
运输与电信情报 transportation and telecommunication intelligence
运输预测 traffic forecast
运输预测系统 travel forecasting systems
运输原木小吊车 timber bob for hauling
运输(运送)特权 transit privilege
运输载料斗 conveyer loading hopper
运输噪声 transportation-produced noise
运输责任 obligation to carry
运输增加值 transport added value
运输战略 transportation strategy
运输栈桥 haulage gantry
运输站 transportation depot;transportation station
运输账户代码 transportation account code
运输者 carrier;transportee;transporter
运输者的责任 liability of carrier
运输者签署承认接运的货物完好的提货单 clean bill of lading
运输者所负责任(数量) amount of the carrier's liability
运输证明 certification of carriage
运输证明书 certification of carriage
运输政策 transportation policy
运输支出 transport expenditures;transport expenses
运输支出补偿的清算收入 compensate transport expense liquidated revenue
运输支出审计 transport expense audit
运输直升机 transport rotorcraft
运输指挥部 transport command
运输指示 traffic instruction
运输指数 transport index
运输质量 transportation quality
运输中的漏损 transportation leakage
运输中断 traffic suspension;transportation disruption
运输中心 transportation center [centre]
运输中心枢纽 corduroy transportation center[centre];core transportation center [centre]
运输终点站 transportation terminal;transport terminal activity
运输种类 category of traffic
运输重量 hauled weight;shipping weight
运输周期 haul-cycle (time);hauling cycle;shipping cycle
运输周转量 traffic turnover
运输主任 chief of transportation
运输专题讨论 transportation workshop
运输砖的车子 brick carrier
运输装运条款 shipping clause
运输装载车 transport and loading vehicle
运输装载机 conveyer loader;payloader
运输装置 conveyer [conveyor];conveying arrangement;transportation plant;transporter
运输状态 transport position;travel(l)ing position;travel position
运输状态全宽 overall shipping width
运输咨询委员会 Transport Advisory Committee
运输资料(数据)协调委员会 <美> Transportation Data Coordinating Committee
运输综合发展 comprehensive development of transport
运输总产值 transport gross output
运输总吨数 tonnage
运输总支出 general cost of transportation
运输走廊 transportation corridor
运输阻滞 traffic interruption
运输最佳化技术 transportation optimization technique
运输作业 handling operation;hauling operation;traffic operation
运输作用 translocation
运水驳船 water barge
运水车 carriage water;tank car;water(ing) cart;water wagon
运水船 water carrier
运水卡车 water truck
运水泥专用船 cement barge
运水小车 water barrow;water cart
运饲料车厢 forage box
运送 carry;consignation;conveyance;convey(ing);deport;forwarding;freight;hauling;shipment;shipping;transport
运送拌和补坑【机】 transit-mix patch
运送拌和机 <混凝土> transit mixer

运送材料 delivery of material
运送材料的闸门 material lock
运送材料双车卷扬机 barrow-hoist
运送车 transfer cart;transit truck
运送车拌和混凝土 transit-mixed concrete
运送成本 shipment and delivery cost; transformation cost
运送带 moving belt
运送单一物品的火车 unit train
运送的 carrying
运送吊车 transfer crane
运送吨数 tons carried;tons conveyed
运送方向 shipping direction
运送费 convey(ing) cost;transport(ation) expenses; transshipment expenses
运送负载 carrying load
运送工具 means of conveyance
运送骨料铁路 aggregate railroad [railway]
运送固体材料泵 solids handling pump
运送轨道 forwarding track
运送合约终止条款 termination of contract of carriage clause
运送混凝土车辆 car gondola
运送混凝土吊斗 concrete skip
运送混凝土吊罐 concrete bucket
运送混凝土料斗 concrete skip
运送混凝土溜槽 concrete chute
运送混凝土小车 concrete buggy
运送火箭燃料的油罐车 propellant trailer
运送货车 haul-away wagon
运送货物吨数 number of tons of freight carried
运送货物或旅客时制动机调整器 goods-passenger brake change-over device
运送机 transporter;transveyer
运送搅拌 <混凝土> mixed-in-transit
运送搅拌车 <混凝土> transit mixer
运送搅拌机 transit mixer
运送距离 distance carried; distance covered
运送空白 travel blank
运送快件的列车 express truck
运送沥青混凝土车辆 material transfer vehicle
运送量 volume of movement
运送流 transporting stream
运送旅客人数 number of passengers carried;passengers carried
运送能力 transportability; transport competency;transporting capacity; transporting power
运送品中途截留 stoppage in transit
运送凭证 certificate of conveyance
运送期间终止 suspension of transit period
运送期限 transit period
运送期限的计算 calculation of transit period
运送器 transveyer
运送热沥青铺路面的供应车 material tender
运送人 carrier
运送人员小车 man car
运送容器 hauling container
运送时间 shipping time;time in transit; time of haul; time of transit; transit time
运送式桥 transported bridge; transporter bridge
运送收据 shipping receipt
运送速度 transporting speed; transporting velocity;travel speed
运送条件 conditions of carriage;term of shipment
运送圬工材料的小车 brick buggy
运送应力 transport stress
运送油品的车辆 oil delivery truck
运送有收入的货物吨数 revenue tons came;revenue tons carried
运送者 conveyer [conveyor]
运送中拌和的混凝土 transit-mix(ed) concrete
运送中货物 goods in transit
运送砖头的手推车 barmac
运送装置 vehicle
运送子 transporton
运酸船 acid tanker
运算 calculate;calculation;counting operation;operate;operation;tally
运算变换对 operation transform pair
运算标识符 arithmetic(al) identifier
运算标志 operation token
运算部分 arithmetic(al) portion
运算部件 arithmetic(al) facility; arithmetic(al) unit; computation unit
运算参数 operational parameter
运算差分放大器 operational differential amplifier
运算乘法器 operational multiplier
运算程序 operating procedure;operating program(me); operational process;operational sequence;operator routine;running program(me)
运算程序控制 operation sequence control
运算程序块 operating block
运算代码 operation part
运算单位 arithmetic(al) unit;computation unit
运算单元 arithmetic(al) element; arithmetic(al) unit
运算的 operating; operational; operative
运算地址寄存器 operation-address register
运算电路 arithmetic(al) circuit(ry); operation(al) circuit
运算对策 operational gaming
运算对象 operand
运算法则 algorithm
运算方案 interpretative version
运算方程式 operational equation;machine equation
运算方法 arithmetic(al) mode;compute mode;operational method
运算方式 arithmetic(al) mode;operating mode; operation mode; compute mode;operate mode
运算方式选择按钮 operation mode selection button
运算放大器 computing amplifier;operational amplifier
运算放大器部件 operational amplifier block
运算分类单位 operational taxonomic unit
运算分析 operational analysis
运算符 operator
运算符号 arithmetic(al) sign;operational character; operation (al) symbol; sign of operation; symbol of operation
运算符文法 operator grammar
运算符优先权 priority of operator
运算符栈 operator stack
运算工具 operational tool
运算工序 operational process
运算公式 operational formula
运算故障模型 arithmetic(al) fault mode
运算函数 operating function
运算换算 reduction of operation
运算机 processing unit
运算积分器 operational integrater
运算记号 sign of operation
运算寄存器 A register;arithmetic(al) register
运算校验 arithmetic(al) check
运算结果指示符 resulting indicator
运算精度 operational precision
运算可靠性 operation(al) reliability
运算控制 operating control;operation control
运算控制部分 arithmetic(al) control section
运算控制部件 arithmetic(al) control unit
运算控制键 operation control key
运算控制器 arithmetic(al) and control unit;arithmetic(al) controller; operation controller
运算控制设备 operation control unit
运算控制系统 computing control system;operation control system
运算控制装置 arithmetic(al) and control
运算例外中断 arithmetic(al) exception interrupt
运算流程图 operational flowchart
运算流水线 arithmetic(al) pipeline
运算律 operational rule
运算逻辑 arithmetic(al) logic
运算逻辑部件 arithmetic(al) (and) logic(al) unit
运算逻辑单元 arithmetic(al) and logic(al) unit
运算逻辑器 arithmetic(al) and logic(al) unit
运算码 operation(al) code
运算模式 arithmetic(al) mode
运算能力 arithmetic(al) capability; operational capability
运算器 arithmetic(al) and logic(al) unit;arithmetic(al) device;arithmetic(al) logic unit;arithmetic(al) organ;arithmetic(al) unit; processing unit
运算器板件 wafer
运算球度 operational sphericity
运算软件 arithmetic(al) software
运算设备 arithmetic(al) facility
运算设计与分析 operational design analysis
运算时间 operating time;operation time
运算式 arithmetic (al) expression; working equation
运算数 operand;operator
运算数长度 operand length
运算数据 arithmetic(al) data;operational data
运算数学 operational mathematics
运算数页面 operand page
运算数有效地址 operand effective address
运算水文学 operating hydrology
运算顺序控制 operation sequence control
运算说明 calculation specification; operation declaration
运算速度 arithmetic(al) speed;operating speed; operational speed; operation time speed;speed of operation
运算速率 arithmetic(al) speed;operating rate
运算特征 operating characteristic
运算特征函数 operating characteristic function
运算体制模式 operational systems model
运算条件 calculation condition
运算调试 arithmetic(al) debug
运算通式 general form
运算图 arithmograph
运算微积 operational analysis;operational calculus
运算微积法 operational calculus method
运算微积分 operational calculus
运算位置 work location; work position
运算问题 operational problem
运算误差 error in operation
运算系统 arithmetic(al) system;operating system;operation system
运算系统矩阵 operational system matrix
运算(行)率【计】operation factor[ratio]
运算型式 arithmetic(al) mode;operational form
运算性能 operational performance
运算学 opsearch
运算延迟 operating delay
运算移位 arithmetic(al) shift
运算异常代码 operator exception code
运算译码器 operation(al) decoder
运算溢出 arithmetic(al) overflow
运算因子 operating factor
运算与逻辑 arithmetic(al) and logic(al) unit
运算语句 arithmetic(al) statement
运算域 operand
运算元件 arithmetic(al) element;operational unit
运算元素 arithmetic(al) element
运算指令 arithmetic(al) instruction; operational command; operational order
运算中断 interruption of operation
运算中断码 operator interruption code
运算装置 arithmetic(al) device; arithmetic(al) section
运算状态 arithmetic(al) mode;compute mode
运算准则 operational criterion
运算子程序 arithmetic(al) subroutine
运算子的分解 resolution of operator
运算字段 arithmetic(al) field
运算阻抗 computing impedance; operational impedance
运套管车 casing wagon
运条 arc manipulation
运条途径 path of electrode
运艇车 boat car
运途中受损 damage in transit
运土 earth hauling;earthmoving;movement of earth; movement of earth mass;muck haulage
运土铲斗 soil shovel
运土铲土机 earthmoving scraper
运土车 hauling unit
运土车辆 earthmoving vehicle
运土车厢 earth body
运土承包公司 earthmoving contracting firm
运土传动装置 earthmoving gear
运土船 barge
运土的 earthmoving
运土斗车 muck car
运土方 haul volume
运土方数 haul yardage
运土工 earth loader
运土工序 earthmoving operation
运土工作 earthmoving operation
运土刮土机 earthmoving scraper; hauling scraper
运土机轮胎 earthmover tire [tyre]
运土机(械) dirt mover; earth mov-

er; earthmoving machine; earthmoving plant; earthmoving equipment; hauling unit; earth loader; earth mover;loader
运土机械设备 earthmoving plant
运土计划 schematization of haul; scheme of haul
运土筐 earth carrier
运土量 haul yardage; quantity to be hauled
运土列车 earthmoving train
运土设备 earthmoving equipment; earthmoving plant
运土手推车 navvy barrow
运土用的窄轨机车 dink(e)y locomotive
运土作业 earthmoving operation
运务经理 managing operator
运物升降机 material handling elevator
运销 marketing
运行 circulation; functionate; motion; operate; operation; operation running; performance; running; set in operation; wheeling
运行安全 operational safety; safety of operation; safety of traffic
运行安全地 operation-safe ground motion
运行安全系统 running protective system
运行安全性 safety in operation
运行饱和度试验 running saturation test
运行保持设备 run-on facility
运行保证考核 performance guarantee test
运行报告 operating report
运行备件 operational equipment
运行备用容量 operating reserve capacity
运行泵 process pump
运行标志 traffic sign
运行标准 performance standard
运行表 operating schedule; run chart
运行表示灯 traffic indication lamp
运行不可靠的 unserviceable
运行不可靠性 unserviceability
运行不稳定 fluctuation of service
运行不正常 malfunction
运行步骤 operational sequence; operation schedule
运行参数 operating parameter; operating variable; operational factor; operational parameter; working parameter
运行操作规范 operation instruction
运行操作位置 run location
运行测试 performance test
运行长度编码器 run-length coder
运行场所 operational site
运行超高 operational free board
运行车辆称重铁路轨道衡 motion weighing railway track scale
运行车辆电子轨道衡 electronic weigh-in-motion scale
运行车辆轨道衡 weighing-in-motion scale
运行车速 operating speed; operation speed; running speed
运行成本 running cost
运行程序 operating procedure; operational procedure; operational sequence; running program(me); working procedure
运行程序文件 run program(me) file
运行程序语言 running program(me) language
运行持续期间 run duration
运行持续时间 duration of operation; operating period
运行出力 service capacity
运行储备 operating storage; operational storage
运行储备器 working stock
运行储存分配 operating storage allocation
运行磁场与运动方向相同的制动 plugging brake
运行存储分配 runtime storage allocation
运行大楼 operations building
运行单位 run unit
运行导轨 travel(l)ing guiding runways
运行导向轮装置【机】 travel(l)ing guiding wheels device
运行的 travel(l)ing
运行的规律性 regularity of movement
运行的可靠性 reliability of operation
运行的平稳性 travel(l)ing comfort
运行等级 running class
运行点 operating point
运行电流 running current
运行电压 operating voltage; running voltage; working voltage
运行调度 run scheduling
运行调度子系统 run scheduling subsystem
运行定时 run timing
运行吨公里 moved ton-kilometer; ton-kilometers hauled; ton-kilometers of transportation; ton-kilometers operated; ton-kilometers transported
运行范围 range of operation; travel(l)ing range
运行方法 operational gaming
运行方面 run phase
运行方式 method of operation; operating duty; operating mode; run mode; mode of operation
运行方式试验 operating duty test
运行方式选择按钮 operation mode selection button
运行方向 traffic direction; travel(l)ing direction
运行方向按钮 direction button
运行方向保持继电器 traffic stick relay
运行方向表示灯 traffic indication light
运行方向表示灯光 traffic indication light
运行方向继电器 traffic direction relay
运行方向上的优先权 superiority by direction
运行方向锁闭 traffic locking
运行方向照明表示器 illuminated traffic indicator
运行方向转接器 receiver selector
运行方向自闭继电器 traffic stick relay
运行费(用) cost of operation; operating charges; operating cost; operating expenses; running cost; working expenses; operation cost; running expenses
运行分析 operational analysis
运行负荷 operating load; performance load
运行负载 operating load
运行改进 operational development
运行钢轨 running rail
运行工程师 operating engineer; plant engineer
运行工程学 <人机关系> human engineering
运行工况 condition at operation; running condition
运行工况试验 operating duty test
运行工作方式试验 operating duty test
运行功率 service rating
运行功能 performance function
运行故障 active defect; operational defect; operational trouble
运行管理 operational guidance
运行管理中心 operation control center [centre]
运行管理自动化 computer-aided traffic control
运行管线 operating (pipe) line
运行规程 operating instruction; operating rule; operating standard
运行规定 operating provision
运行规划 operating scheme
运行规则 operating rule; operation rule; service regulation
运行规章条例 operating regulation
运行轨道 orbit; travel(l)ing runways【机】
运行轨道高程 travel(l)ing runways level
运行轨迹控制装置 track control device
运行过程 operational process; run procedure
运行合同 operational contract
运行和维护 operation and maintenance
运行和维护成本 operating and maintenance cost
运行荷载 operating load; operation load(ing); performance load
运行机动性 operational flexibility
运行机构 operating unit; operational mechanism; running gear
运行基本地震 operating basis earthquake
运行基准地震 operating basis earthquake
运行级地震 operating level earthquake
运行计划 operating schedule; operational plan(ning); traffic schedule
运行计时器 running time meter
运行记录 log(out); operating record; operational log; operation(al) record; service record
运行技术 operational technology
运行技术标准 performance specification
运行监测 operational monitoring
运行检查 operational check; running check
运行检修 operating maintenance
运行鉴定 operational rating
运行阶段 operating phase; operational stage; run phase; target phase
运行结构 running structure
运行结果 operating result
运行结束(例行)程序 end-of-run routine
运行进路 traffic route
运行经验 operating experience; operational experience
运行径路卡 routing card
运行径路牌 routing card
运行距离 rangeability
运行决策 operating decision
运行开关 sail switch
运行开始 commencement of operation
运行考核 post-installation review
运行可靠性 functional reliability; operating reliability; operational reliability; safety in operation
运行控制 operating control; operational check; operational control; traffic control; movement control
运行控制中心 operation control center [centre]; traffic control center [centre]
运行库存 operational stock; operational store
运行块长度 block run length
运行李电梯 baggage elevator; luggage elevator
运行李升降机 baggage elevator; luggage elevator
运行里程 operation kilometrage; operation mileage
运行哩油量 operational mile
运行灵活性 operational flexibility
运行流 run stream
运行流程 operational scheme; operational sheet
运行路程 distance travelled
运行路线 walking thread; walk route
运行率 operating ratio; operation ratio; service factor; external operation ratio
运行命令 action command; run command
运行模式 operating mode; travel pattern
运行目标 motive destination
运行能力 running ability
运行频率 operating frequency
运行平稳性试验 ride index test
运行平稳性指标 index of running stability
运行评价 operational evaluation
运行期 operating period; service period
运行期检修 operating repair
运行期弃水 <灌溉的> operation(al) waste
运行期限 period allowed for conveyance [conveyancy]
运行器材 operational aid
运行前操作 precommissioning operation
运行前试验 preoperation(al) test
运行情况 operating situation; operational aspect; operational situation; running condition; serviceable condition
运行区 operation(al) area; region of operation
运行曲线 operating rule curve; operational curve; performance curve
运行曲线图 operation (curve) diagram; performance diagram
运行人员 attendant; operating crew; operating personnel
运行日期 commissioning date
运行设备 operational equipment
运行失效 operational failure
运行时程序库 runtime library
运行时存储管理 runtime storage management
运行时管理程序 runtime administration routine
运行时环境 runtime environment
运行时间 operating time; operation time; period of duty; running time; runtime; time of operation; working hours; working time
运行时间百分率 <设备的> percentage of operating time rate
运行时间表 operating schedule; operation schedule
运行时间计数器 running time counter
运行时间记录器 service recorder
运行时间库 runtime library
运行时间统计 runtime statistics
运行时刻表 running schedule
运行时累加器 runtime accumulator
运行时数 service hours

运行时数据区 runtime data area
运行时限 < 商品混凝土搅拌车 > time of haul
运行时栈顶 runtime stack top
运行时诊断 runtime diagnosis
运行时组织 runtime organization
运行实验 operating test
运行事故 operational trouble
运行事故率 operating accident rate
运行试验 field survey; operating test; operation test(ing); performance test(ing); run(ning) test; service experience; service test; service trial
运行试验设备 operational testing equipment
运行适合性 suitability for running
运行手册 run book
运行寿命 operating life; service life
运行受阻 interruption of operation
运行输出功率 service output
运行数据 operating data; performance data; service data
运行水头 operating head
运行水位 operating level; operating water level
运行水温 operating water temperature
运行顺序 operating sequence
运行说明书 run(ning) book
运行速 travel speed
运行速度 operating speed; operating velocity; operation speed; running speed; running velocity; service velocity; speed of service; travel(l)ing speed; service speed; speed of operation
运行速率 maneuvering speed; operating rate
运行索 travel(l)ing cable; travel(l)ing rope
运行特性 operational characteristic; performance characteristic; running characteristic; running performance; operating characteristic
运行特性曲线 operating characteristic curve; performance characteristic curve; performance curve
运行条件 operating condition; operational condition; running condition
运行条件的模拟试验 environmental test(ing)
运行调节 operational conditioning; operation regulation
运行调整 regulation of traffic
运行通量 running flux
运行图 performance diagram; performance record; run diagram; running chart; service diagram; traffic diagram
运行图表 operating chart; running program(me)
运行图不成对系数 coefficient of number of trains in the predominant traffic direction to that in non-predominant traffic direction
运行图横轴 horizontal axis of graph
运行图记录仪【铁】 train diagram recorder
运行图描绘仪【铁】 train diagram recorder
运行图天窗 gap in the train diagram; skylight in the train diagram
运行图周期 period in the train(working) diagram
运行维护 operating maintenance; running attention
运行维护费 operation and maintenance cost
运行维护和管理 operation, maintenance and management
运行维修 operating maintenance
运行位置 operational site; run location; run(ning) position
运行温度 operating temperature; operative temperature; running temperature
运行稳定性 riding stability; running stability
运行系数 availability factor; efficiency factor; operating factor; performance factor
运行系统 operating system; operation(al) system; production system
运行线 operating line; travel line
运行线路 operative route; working line
运行线路的对焊 in-track buttweld
运行线圈 operating coil
运行限位开关 travel limit switch
运行限制开关 limit switch for travel motion
运行限制值 operational limit
运行小时 running hours
运行效率 operating efficiency; operational efficiency
运行效益 operational efficiency
运行协定 operating agreement; operation agreement
运行信号灯 operating signal lamp
运行性能 operating characteristic; operating performance; running performance; performance characteristic
运行性压力波 travel(l)ing pressure wave
运行序列 operational sequence
运行循环 operating cycle
运行压力 operating pressure
运行延误 delay in movement
运行研究 operation research
运行要求 operating requirement
运行依据地震 operating basis earthquake
运行优先方向 traffic direction preference
运行有效载重 operational payload
运行有准备 ready for operation
运行预测 operational forecasting
运行裕量 operating margin
运行障碍 interruption of service
运行振动 operational shock
运行蒸汽 operating steam
运行政策 operating policy
运行值 runtime value
运行指令 active instruction; operating instruction
运行指示器 operation indicator; run indicator
运行制动 travel brake
运行中 in service
运行中的程序 active program(me)
运行中的作业 active job
运行中断 interruption of operation; outage of running
运行中管道 charged main
运行中换挡 on-the-go shift
运行中检查 in-service inspection
运行中调节 on-the-go adjustment
运行中心 operational center [centre]; operations center [centre]
运行终点 end of run
运行终端限速器 travel end limiter
运行终端限制器 travel(l)ing speed end limiter
运行终了 end of run
运行重量 operating weight
运行周期 cycle of operation; operating cycle; operational period; operation cycle; performance period; running cycle; running period
运行贮存容 operating storage
运行转速 running speed; speed of operation
运行装置 running gear
运行状况 condition at operation; operating condition
运行状态 condition at operation; operating mode; running condition; running position; running state; running status; run phase
运行准备 readiness for operation
运行准备就绪 ready for operation
运行资料 operational data; run book
运行子程序 runtime subroutine
运行自由度 freedom of motion
运行阻力 resistance due to running; resistance to movement; running resistance
运行作业费 operating expenditures
运盐船 salt carrier
运洋集装箱货轮 sea freight liner
运洋轮船运输单位 deep-sea carrier
运洋轮船运输公司 deep-sea carrier
运液体汽车 tank truck
运移 migration
运移动力 power of migration
运移方向和通道 direction and pathway of migration
运移机理 migration mechanism
运移阶段和过程 stage and progress of migration
运移距离 distance of migration
运移距离时间和深度 migration distance/time and depth
运移模式 migration model
运移时间 time of migration
运移通道 escape route
运移相态 phase state of migration
运移相态和油气溶解度 phase state of migration and solubility of oil and gas
运营 operating; operation
运营比率 < 即运营支出对运营收入的比率 > rate of working
运营补贴 subsidization in operation
运营策划室 operation planning office
运营成本 operating cost; operation(al) cost; running cost
运营成本控制 controlling of operating cost
运营成绩 operating performance
运营处 operating department; operation department
运营处长 chief operating manager; chief operating superintendent
运营单位资金周转额 assets turnover of operating unit
运营的 operational
运营的房地产 operative property
运营段 operating division
运营段段长 division superintendent
运营吨公里 ton-kilometers operated
运营方案 operational concept
运营方法 operational know-how
运营方式 operation mode
运营费(用) operating cost; operating expenses; operation expenses; running cost; operating charges; operating expenditures; running expenditures; working expenditures
运营辅助系统筹体系 operational assistance system
运营工作量 operating workload
运营公里里程 operating kilometrage
运营功率 operating power
运营固定资盘 operating (fixed) assets
运营管理 operation management
运营管理中心 maintenance control center [centre]
运营轨道 operated track
运营轨道公里数 kilometers of track operated
运营货车 carrier lorry
运营计划 operating plan
运营计划和执行体系 operation planning and execution system
运营结果 operating result
运营进款收支报告 receipts and payments report of operating income
运营经济学 operating economics
运营卡车 carrier truck
运营控制中心 operation control center [centre]
运营控制中心设备 operation control center equipment
运营亏损 operating deficit; operating loss
运营利润 operating profit
运营模式 mode of operation; operation mode
运营情况 operating condition
运营人员 operating personnel
运营设备检修 repair of operating equipment
运营时间 operating time
运营收入 operating revenue
运营数据 operating data; operational data
运营速度 operating speed
运营条件 operating condition
运营铁路 operating railway; railway in operation; railway in service
运营铁路线 line operated; operated line
运营停车点 operational stop(ping) point
运营通风 operation ventilation
运营系数 < 运营支出对运营收入的比 > operating coefficient
运营现金收入 cash provided by operations
运营线路 line operated to traffic
运营线路长度 length of line in operation; length of line open to traffic; length of line operated; length of line operation; line of road open to traffic
运营线路公里 kilometrage of track operated
运营线路广播系统 broadcasting equipment for the service line
运营效率 operating efficiency
运营因数 operational factor
运营用品 operating supplies
运营运行 service running
运营支出总量 quantum of operating expenditure
运营职工 operational staff
运营制度 operating system; operation system; system of working
运营中心终点站 hub terminal
运营周期 operation period; running period
运营主管人 operating officer
运营资本变动表 statement of changes in working capital
运营资本总额及净额周转率 turnover ratio of gross and net working capital
运营资产总额周转率 turnover ratio of total operating assets
运营资金 working fund
运营资金周转 working capital turnover
运营总段 < 主管总段内运、机、工、电等工作 > general division
运营总段长 < 主管总段内运、机、工、电等工作 > general superintendent

运营总管 <主管全局运、机、工、电等工作> general manager
运营组织 operation(al) organization
运用 apply;put into service;take advantage of;utilization;utilize
运用参数 operating parameter;operational parameter
运用车 applied train;car for traffic use;cars open to traffic;serviceable car;wagon for traffic use
运用车保有量 daily stock of serviceable cars;number of serviceable cars held kept;number of serviceable cars to be kept
运用车动载量 number of dynamic load of serviceable wagon
运用车工作量 number of serviceable cars turnround
运用车轨道衡 weigh-in-motion scale
运用车日产量 daily output per serviceable car;serviceable work done per car per day
运用车数 quantity of using trains
运用程序 operating procedure;operating schedule
运用导入保护涂层 application importing of protective coating
运用的管道 duct for service
运用法 modus operandi
运用机车 locomotive in operation;operating locomotive;serviceable locomotive
运用机车调整 regulation of the number of locomotive in services
运用机务段 locomotive running depot
运用客车需要数 number of serviceable carriages needed
运用库 running shed
运用困难 operational difficulties
运用率 operational availability
运用频率 frequency of operation;operational frequency
运用期 servicing period
运用期弃水 operational waste
运用刹车 application of brakes
运用试验 test-use
运用水头 operating head
运用损失 operating loss;operating waste
运用性能 serviceability;service behavio(u)r
运用研究 operation(al) research
运用预算 working budget
运用制动器 brake
运用状态 behavio(u)r in service
运用资本 working capital
运用资产 working assets
运用资产净额 net working assets
运油车 oil can;oiler;oil tank car;oil tanker;tank truck
运油车的终点站 bulk-oil terminal
运油车辆 oiler
运油船 dirty ship;oil tanker;ship-tanker
运油和散(装)货船 oil and bulk carrier
运油业务 oil transport
运鱼船 fish carrier;fish transport ship
运鱼设备 fish handling facility
运杂费 freight and miscellaneous charges
运杂费计算和核收的审查 check of calculating and collecting petty expenses in transport
运杂费收据 receipt of freight and miscellaneous charges
运杂费收据整理报告 adjustment report of the receipts of freight and miscellaneous charges
运载 carriage;carry;reshipment
运载车 carrier loader
运载传递 transport transmission
运载单元 travel unit
运载的 carrying
运载范围 operating range
运载方法 means of delivery
运载飞灰的车斗底板 fly ash bed
运载飞机 mother aircraft;parent aircraft
运载飞行阶段 carrier phase
运载负荷 carrying load
运载工具 booster;carrier;delivery vehicle;haul vehicle;launch vehicle;travel unit;vehicle
运载挂车 trailer carrier
运载火箭 carrier booster;carrier rocket;carrier vehicle;freight rocket;launch(ing) vehicle;transport rocket
运载机动船的挂车 utility power-boat trailer
运载机构 transport mechanism
运载机械 hauling machine
运载空气 carrier air
运载链斗的梯形链条 ladder chain carrying bucket
运载能力 carrying capacity;diode current capacity;hauling capacity;load-carrying capacity;load-transfer capacity
运载器 vehicle
运载器架 carrier bearer
运载牵引力 tractive effort on haul
运载桥 transporter
运载体 carrier;carrier vehicle;transporting species;vehicle
运载托盘升降机 pallet elevator
运载拖车 trailer carrier
运载卫星 vehicle satellite
运载装备 carrier assay;carrier assembly
运载装置 delivery system;toter
运渣 muck haulage
运渣车 jumbo;spoil wagon
运渣方式 haulage system
运渣方数 haul yardage
运渣机 excavating leader;slag conveyer [conveyor];slag-conveying machine
运渣机车 ballast engine
运渣卡车 muck truck
运渣列车 ballast train;muck train
运渣桶 slag truck ladle
运渣装置 slag-conveying machinery
运纸浆车 pulpwood car
运砖板 brick pallet
运砖车 brick truck
运砖的轻便车 brick cart
运砖架 hack barrow
运砖框 bed of brick
运砖手推车 hack barrow
运转 operation running;running;set in operation;turn round
运转安全度 operational dependability;reliability of operation;working safety
运转保证周期 operational proof cycle
运转备用 spinning reserves
运转备用容量 spinning reserve capacity;spinning reserve content
运转变扭器 running torque converter
运转标志 operation notice
运转标准 standard of performance
运转标准化 operational standardization
运转不规律 <发动机> run harshly
运转不平稳 <发动机> galloping;run rough
运转不稳的发动机 rough engine
运转不稳定 fluctuation of service;gallop
运转材料 running material
运转操作 running operation
运转测试 operation test(ing);performance test
运转差动机构 running differential mechanism
运转场 train operating yard
运转场地 manoeuvering area
运转车长 conductor;train conductor;train guard
运转成本 operating cost
运转程序 running program(me)
运转迟滞 operational delay
运转初期 start of run
运转次序 running order
运转的 locomotive
运转的可靠性 maintainability
运转的质量 running quality
运转点 running point
运转电动机 run motor
运转电脑化 computerization of operation
运转电压 running voltage
运转发热 running warm
运转范围 operating range;travel(l)ing range
运转方法 method of operation
运转废水 operation waste
运转费用 cost of operation;operating cost;running cost;operation cost;operating charges;operating expenses
运转服务条件 service-simulated condition
运转负荷 operating load
运转钢索 track line
运转工作的联合 <铁路合并的一种方式> working union
运转功率 running power
运转功能 function of motion
运转故障 operating disturbance;operational failure;operational trouble
运转辊距 running nip
运转过程 duration of runs
运转和维护费用 operation and maintenance cost
运转计 operameter
运转计划 operation scheme
运转计时机 countermeasure machine
运转记录 operation record;service record
运转记录器 running recorder
运转寄存器 motion register
运转间隙 running clearance
运转检查 operational checkout
运转检验周期 service proofing cycle
运转经验 running experience
运转卷扬机 rotahoist
运转可靠性 authority for operation;operational reliability
运转历时 duration of runs
运转链 propelling chain
运转量 volume of traffic
运转流动 running flow
运转率 operating factor;operational factor;running rate
运转轮系 going train
运转模 operational mode
运转磨损 service wear
运转末期温度 end-of-run temperature
运转能力 operational capability;operational capacity;outperform;running ability
运转年份 years of operation
运转黏[粘]度 <泥浆> running viscosity
运转泡 transport vesicle
运转频率 operating frequency
运转平稳性 smoothness of operation
运转期 on-stream period
运转弃水 operation waste
运转钳 carrying tongs
运转情况 operating situation;operational situation;service behavio(u)r;service performance;working condition;working order
运转曲线 <供列车运转计划用> run(ning) curve
运转日期记录 log sheet
运转筛机 starting-up a screen
运转筛组 starting-up a screen
运转设施 handling device
运转失灵 glitch
运转时间 duration of runs;length of run;machine hours;operating time;operation time;running period;running time;run-on time;runtime;time of operation;time of run;working hours
运转时间比 operating time ratio
运转时数 running hours
运转事故报告 operational hazard report
运转试验 duration test;field investigation;maneuvering trial;operation(al) test(ing);running test;running trial;service test
运转室 operation office for train receiving-departure;traffic operation office;train operating room【铁】;operation office;shunting cabin
运转适应性试验 operational suitability test
运转寿命 running life;operational life
运转数据 service data
运转水文学 operational hydrology
运转说明书 operating manual;operational manual
运转速度 operational speed;running speed;operating speed;work speed
运转速度范围 operating speed range
运转(所需的)物料 running material
运转特性(曲线) operating characteristic;performance characteristic
运转体 runner
运转天数 days of operation
运转调整器 running adjuster
运转维护 during-operation service;running attention;operational maintenance
运转位 driving position;running position
运转位置 run(ning) position
运转温度 running temperature
运转系数 operating factor;stream factor
运转小时 operating hours;running hours
运转小时指示器 running hours indicator
运转效率 running efficiency
运转性能 behavio(u)r in service;operational performance
运转需求量 operation demand
运转要求 operation requirement;running requirement
运转依据地震 operating basis earthquake
运转用发热 run hot
运转预报 running warn
运转原理 principle of operation
运转噪声 running noise
运转噪音 running noise
运转者 runner
运转正常 run faultlessly;well running
运转指示器 running indicator
运转指数 running index
运转质量能 running quality
运转中 in service
运转中的部件 live part

运转中的价值 going value
运转中断 outage
运转中计算机 living computer
运转中检查 on-stream inspection
运转中清洗 on-stream method
运转中维修 on-stream maintenance
运转中修理 current running repairs
运转终了 end of run
运转重量 operating weight
运转周期 operating cycle; operation cycle; run duration; running period
运转轴 operating shaft
运转轴臂 operating shaft arm
运转装置 running gear
运转状况 behavio(u)r in service; operational behavio(u)r
运转状态 operating condition; operative condition; run position; service condition; working order; working state
运转准备试验 operational readiness test
运转总成本 overall operating cost
运转阻力 running resistance
运转作业报告系统 operations reporting system
运转作业革新 operational innovation
运走 cart away
运走损坏工程的废料 debris removal
运作办公室 operating office
运作地区 operational area
运作工程师 operating engineer
运作中心 operation center [centre]
运作装置 operating apparatus
运作总重量 total operating weight
运作租约 operating lease

晕边 corona

晕彩 iridescence
晕长石 peristerite
晕车(病) car sickness; motion sickness
晕车的 trainsick
晕车晕船 car sickness and sea sickness
晕船 motion sickness; naupathia; nausea; sea sickness
晕船的 sea-sick
晕(的现象)【气】 halo phenomenon
晕点法 stipple
晕电荷 halation effect
晕动 motion sickness
晕光蒙片 halomask
晕光作用【物】 halation
晕化电极 discharge electrode
晕环异常 halo-type anomaly
晕机 airsickness
晕抗 corona resistance
晕流 corona current
晕轮 halo
晕轮效应 halo effect
晕切弧【气】 tangent arc
晕圈 halo
晕圈环形体 halo ring circular features
晕圈效应 halo effect
晕圈状光 flare spot
晕染 gradation dyeing
晕色 iridescence; muted colo(u)rs; scumbling
晕色条纹 muted stripe
晕色用罩光漆 <上光剂> scumble glaze
晕色用着色剂 scumble stain
晕色釉 shaded glaze
晕水 halo water
晕嗡 hairy caterpillar
晕嗡法 hachuring
晕嗡图 hachured map
晕嗡线 hachure (line)
晕滃 hatching; hill shading
晕滃线 linear hachure
晕线【测】 set of hachures; ruling
晕线加密 crowding
晕线面积 dashed area
晕线描绘器 hatching apparatus
晕线仪 dead beat
晕眩 dizziness
晕眩的 dizzy
晕渲 <画木线的阴影> hill shading
晕渲程序 shading program(me)
晕渲地貌 shaded relief; wash-off relief
晕渲地形图 shaded-relief map
晕渲法 <表示地形> brush-shading (method); hachure method; hill-shading method
晕渲面 shaded picture
晕渲数据 shading data
晕翳表示法 hachuring
晕翳线 hachure
晕影 halation
晕映图像 vignette
晕转态 yrast state

酝酿 brewing; deliberation; ferment; incubation

酝酿计划 gestation
酝酿阶段 embryonic stage
酝酿区 incubation zone
酝酿设想 gestation

韵律 rhythm

韵律层【地】 cyclothem; rhythmic(al) unit; rhythmite; zebra layering
韵律层理 rhythmic(al) bedding
韵律层序 rhythmic(al) succession
韵律沉积作用 rhythmic(al) sedimentation
韵律成层 rhythmic(al) layering
韵律感 rhythmic(al) image
韵律厚度图 map of rhythmic(al) thickness
韵律混乱 rhythmic(al) disturbance
韵律结晶 rhythmic(al) crystallization
韵律粒级层理构造 rhythmic(al) graded bedding structure
韵律图 rhythmic(al) map
韵律型火山泥球构造 rhythmic(al) volcanic mud ball structure
韵律性 rhythmicity
韵律性变化 rhythmic(al) variation
韵律学 rhythmics

熨斗 flat iron; smoothing iron

熨斗式运煤船 flat iron collier
熨斗形山 flat iron
熨斗状外形 flat iron shape
熨化器 ironing machine
熨沥青路面机 asphalt road roller
熨平 ironing; plating; pressing
熨平板 screed
熨平混凝土路面 floating of concrete pavement
熨平机 ironer; ironing machine
熨平装置 screed unit
熨烫板 ironing board
熨烫台 ironing board
熨烫样板 ironing-screed
熨压 ironing
熨压器 ironing machine
熨衣机 ironing machine
熨衣室 ironing room
熨衣台 ironing board; ironing table

蕴藏量 deposit(e); reserve

蕴含 implicant; implication; implicit; inclusion; if-then <一种逻辑算符>
蕴含关系 implication relation
蕴含门 inclusion gate
蕴含式 implicant
蕴含选址 implied addressing
蕴含运算 if-then operation

Z

匝 turn; coil; convolution; loop

匝比调节器 ratio adjuster
匝道 ramp; slip connection; slip road; approach ramp <进出高速公路的>
匝道车流控制设施 ramp metering system
匝道车流量 ramp flow; ramp volume
匝道车流调节 ramp metering
匝道端点 ramp nose
匝道高速干道交接段 ramp-freeway junction
匝道辊 gandy stick
匝道集成系统控制 ramp integrated system control
匝道交接点 ramp junction
匝道交通量 loop traffic volume; ramp flow; ramp traffic volume
匝道交通调节 ramp metering
匝道交织段 ramp-weave section
匝道街道交接点 ramp-street junction
匝道进口 ramp entrance; ramp entry
匝道控制 ramp control
匝道控制前置预告标志 advance ramp control warning sign
匝道口车辆汇流 ramp merge
匝道连接处 ramp junction
匝道前端 ramp nose
匝道桥 ramp bridge
匝道让车道 ramp turnout
匝道容车量 ramp volume
匝道设计车速 ramp design speed
匝道驶出段 off ramp
匝道枢纽 ramp terminal
匝道通过能力 ramp capacity
匝道线形设计 alignment design of ramps
匝道限流 ramp metering
匝道限流控制信号 ramp metering signal
匝道限流率【道】ramp rate
匝道需求容量控制 ramp demand-capacity control
匝道终点 ramp terminal
匝道专用车道 ramp meter bypass lanes
匝的平均长度 length of mean turn; mean length of turn
匝间 interturn; turn-to-turn
匝间保护 interturn protection
匝间电场强度 field intensity between turns
匝间电容 turn-to-turn capacitance
匝间垫条 turn separator
匝间短路 interturn short circuit; turn-to-turn fault; turn-to-turn short (circuit)
匝间短路保护 interturn short circuit protection
匝间故障 interturn fault; turn-to-turn fault
匝间间隔片 turn separator
匝间绝缘 interturn insulation; turn(s)-insulating; turn-to-turn insulation
匝间旁路电容量 turn-to-turn shunt capacity
匝距 pitch of turn
匝连 linkage
匝数 number of turns; number of windings; turn number
匝数比 rate of turn; ratio of winding; turns ratio
匝数计 turn indicator
匝数降压比 turns step down ratio
匝效应 turn effect

杂白钙沸石 cerinite

杂斑的 parti-colo(u)red
杂斑模纹 mottle
杂变柱石 racewinite
杂波 clutter; noise wave
杂波滤波器 clutter filter
杂波输出 noise output
杂波输出功率 spurious-wave output power
杂波调制干扰机 noise-modulated jammer
杂波调制器 noise modulator
杂波响应 squrious response
杂波形式 noisy mode
杂波抑制 anti-clutter; clutter rejection; clutter suppression
杂波抑制器 black spotter; clutter suppressor; interference inverter; noise limiter
杂波噪声 clutter noise; noise from clutter
杂操作 miscellaneous operation
杂草 hog weed; natural grass; weeds
杂草采取权 right to grass
杂草残茬搂耙 trash rake
杂草草原 mixed prairie
杂草策略 ruderal strategy
杂草丛生 heavy weed growth; invasion of grass; invasion of weeds; overrunning weed; rank growth
杂草丛生的 weed-covered
杂草丛生河流 weed-filled river
杂草丛生湖 over-grown lake; weed-filled lake
杂草的 ruderal
杂草防除法 weed-control method
杂草防治 control of weed; weed control; weed prevention
杂草分离器 weed separator
杂草焚 acidluoren sodium
杂草控制 weed control
杂草切除机 weed cutter
杂草切碎器 trash cutter
杂草侵入 invasion by grass
杂草清除链 trash chain
杂草清除器 weed cleaner
杂草筛 weed remover
杂草似的 weedy
杂草为害 weed encroachment; weed injury
杂草型草地 weedy grassland
杂草与残茬抛撒器 trash remover
杂草植物 ruderal plant
杂层 stray pay; symmicton; diamicton【地】; stray
杂赤铁土 plinthite
杂醇油 fusel oil
杂醇油沥青 fusel-oil tar
杂凑法 hashing
杂存石块的 calculous
杂的 sundry
杂碲金银矿 muthmannite
杂电 stray
杂多化合物 heteropoly compound
杂多蓝 heteropoly blue
杂多钼酸 heteropoly molybdic acid
杂多酸 heteropoly acid
杂芳族化合物 heteroaromatics
杂废排水 non-fecal drainage
杂废水 non-fecal drainage
杂费 current expenses; fittage; incidental charges; incidental cost; incidental expenses; incidental fee; incidentals; miscellaneous expenses; on cost; petty expenditures; petty expenses; sundry charges; sundry expenses
杂费及未预计费用 miscellaneous and contingencies
杂费及预备费 miscellaneous and contingencies
杂费已付 charge paid
杂费账 petty expenses account
杂费支付通知 miscellaneous charge order
杂分散 heterodisperse
杂分散性 heterodispersity
杂酚浸过的木材 creosoted timber
杂酚浸过的木桩 creosoted-wood pile
杂酚浸过的桩 creosoted pile
杂酚漂白油 creosote bleaching oil
杂酚油 <木材防腐油> creosote[kreosote]; creosote oil; heavy oil
杂酚油防腐处理 creosoting
杂酚油防腐处理法 creosoting process
杂酚油灌木 creosote bush
杂酚油灌注 to impregnate with creosote
杂酚油加压浸渍 pressure creosoting
杂酚油浸处理 creosote treatment
杂酚油浸防腐木桩 creosoted-wood pile
杂酚油浸渍(防腐) impregnate with creosote; creosoting
杂酚油煤焦油混合物 creosote-coal-tar mixture
杂酚油煤焦油溶液 creosote-coal-tar solution
杂酚油黏[粘]层 creosote primer
杂酚油乳剂 creosote emulsion
杂酚油石油溶液 creosote-petroleum solution
杂酚油污染水 creosote-contaminated water
杂酚油型防腐剂 creosote-type preservative
杂酚油压浸木材 pressure creosoted timber
杂酚油油酸酯 oleocreosote
杂酚油注入 creosoting
杂酚皂液 lysol
杂辐射 heterogeneous radiation
杂工 backman; builder's handyman; common labo(u)r; do-all; factotum; handy man; jobber; odd-job worker; plasterer's labo(u)rer; supernumerary; various jobs
杂工队 floating gang
杂工工头 hook tender; pennydog
杂共轭 hetero conjugation
杂谷 miscellaneous cereal
杂光 flare spot; flare veiling glare; parasitic(al) light; stray light; veiling glare
杂规聚合物 heterotactic polymer
杂硅氧烷 heterosiloxane
杂号电荷 heterocharge
杂合 heterozygosis
杂合的 heterozygous
杂褐铁矿 esmeraldaite
杂化 hybrid; hybridization
杂化轨道 hybridized orbital; hybrid orbital
杂化合物 heterocompound
杂化键 hybrid bond
杂化流程 hybrid process
杂化模型 hybrid model
杂化配合物 hybrid complex
杂化作用 hybridism; hybridization
杂环 heteroatomic ring; heterocycle; heterocyclic(al) ring
杂环氮化物 heterocyclic(al) nitrogen compounds
杂环的 heterocyclic(al)
杂环的系统 heterocyclic(al) system
杂环化合物 oxazolel; heterocyclic(al) compound; heterogeneous ring compound
杂环结构染料 heterocyclic(al) configuration dye
杂灰色沙 salt and pepper sand
杂辉锑银铅矿 fizelyite
杂烩 hotch-pot(ch)
杂混规则 scramble rule
杂混网络 scramble network
杂活 char(e)
杂货 combination cargo; fancy goods; general cargo; general goods; general merchandise; grocery; miscellaneous cargo; miscellaneous goods; mixed cargo; mixed goods; sundries; sundry goods
杂货班轮 break-bulk liner
杂货泊位 break-bulk berth
杂货车 merchandise car
杂货船 break-bulk carrier; break-bulk ship; break-bulk vessel; general cargo ship
杂货店 chandler; drug store; general store; grocery; notion-store; variety shop; variety store
杂货费率 general cargo rate
杂货货柜 break-bulk container ship
杂货集装箱 dry cargo container
杂货集装箱船 break-bulk container ship
杂货集装箱两用船 convertible container ship
杂货零售商 chandler
杂货零售商店 miscellaneous retail store
杂货码头 general(cargo) terminal; general(cargo) wharf
杂货批发商 general merchandise wholesaler
杂货铺 general shop
杂货商 chandler; grocer; provider
杂货商人 chandler
杂货市集 fancy fair
杂货业 chandler
杂货运货率 general cargo rate
杂货运输 break-bulk transport; general cargo traffic
杂货作业区 general terminal
杂基支撑 matrix-supported
杂集 miscellanea
杂技场 circus; gaff; hippodrome; variety hall
杂技与戏剧结合演出剧院 acrotheater [acrotheatre]
杂件 miscellaneous accessories; miscellaneous parts; sundries
杂键的 heterodesmic
杂交模拟模型 hybrid analog(ue) model
杂交模型 hybrid model
杂交元 hybrid element
杂礁灰岩 reef complex
杂居群聚 sympolyandria
杂聚合物 heteropolymer
杂聚合作用 heteropolymerization
杂块 gob; pell-mell block
杂块花岗石 taspinite
杂块砌的建筑 pell-mell construction
杂矿石 matrices
杂类草 forb
杂类储集层 miscellaneous reservoir
杂类船货 break-bulk cargo
杂离子 hetero-ion
杂粒 detritus; foreign particle; grit
杂粒沉沙池洗涤 <污水处理厂操作> grit washing

杂粒池 grit chamber
杂粒冲洗机械 grit washing mechanisms
杂粒斗 grit hopper
杂粒量 quantity of grit
杂粒排放坑 grit discharge pocket
杂粒运动轨迹 trajectory of grit particles
杂链 heterochain
杂链化合物 heterogeneous chain compound
杂链聚合物 heterocatenary polymer; heterochain polymer
杂粮 miscellaneous cereal
杂菱银矿 selbite
杂硫铋铜矿 dognacskaote
杂硫碲铋矿 rubiesite
杂硫铁砷钴矿 alloclasite
杂硫银铜矿 cocinerite
杂卤石 polyhalite
杂卤石岩 polyhalite rock
杂绿脱高岭石 faratsihite
杂乱 amorphism; litter
杂乱的 unsystematic(al)
杂乱的东西 litter
杂乱电流 random electron current
杂乱分布 random distribution
杂乱回波 clutter; wave clutter
杂乱间隔 random interval
杂乱裂纹 random crack(ing)
杂乱脉冲干扰 random pulse jamming; hash <显示器屏幕上的>
杂乱毛方石砌体 snecked rubble(masonry)
杂乱毛石 snecked rubble
杂乱弄乱 pie
杂乱起伏数据 random fluctuating data
杂乱缺陷 random defect
杂乱散射 incoherent scatter(ing)
杂乱色 random colo(u)r
杂乱闪光 random flashes
杂乱石料 <料坑的> quarry run
杂乱时间分布 random time distribution
杂乱土层 erratic subsoil
杂乱无章 criss-cross
杂乱信号 hash; random signal
杂乱形状技术 random geometry technique
杂乱影像 scrambled image
杂乱噪声 fuzz; hash; hash noise; random noise
杂乱噪声消除 random noise correction
杂乱噪音 random noise
杂络物 heterocomplex
杂芒硝 tychite
杂镁榴石 ransatite
杂木 miscellaneous tress; weedtree
杂木林 spinn(e)y
杂木围墙 brushwood fence
杂钠钾盐 natrikalite
杂铌矿 wiikite
杂凝胶 heterogel
杂凝聚 heterocoagulation
杂拼物 medley
杂品存储室 utility room
杂品堆放室 utility room
杂品铸铁 mottle cast-iron
杂青金石 lapis lazuli; lazuli
杂糅构造 pell-mell structure
杂散变量 stray parameter
杂散波 stray waves
杂散参数 stray parameter
杂散场 stray field
杂散场效应 stray-field effect
杂散磁场 stray magnetic field
杂散磁通 stray flux; stray magnetic flux
杂散磁通量 stray flux
杂散的【物】 stray
杂散地电流 stray earth current
杂散电场 stray electric(al) field
杂散电磁场 stray electromagnetic field
杂散电动势 stray electromotive force
杂散电感 leakage inductance; stray inductance; stray induction
杂散电抗 stray reactance
杂散电流 stray electric(al) current; terrestrial current; vagabond current
杂散电流测定仪 stray current measuring instrument
杂散电流电解 stray current electrolysis
杂散电流腐蚀 stray current corrosion
杂散电流危险 <引起提前爆炸> stray current hazard
杂散电容 spurious capacitance; stray capacitance
杂散电容耦合 stray-capacity coupling
杂散电容效应 stray-capacity effect
杂散电子 stray electron
杂散电阻 stray resistance
杂散发射 spurious emission; stray emission
杂散反射 <微波测距> stray reflection
杂散辐射 spurious radiation; stray radiation
杂散感应 stray induction
杂散钢块 tramp iron
杂散功率 stray power
杂散功率法 stray power method
杂散功率损耗 stray power loss
杂散光辐射 stray light emission
杂散光校正 flare correction
杂散光滤光片 optic(al) light filter
杂散光滤器 stray light filter
杂散光(线) stray light
杂散光照 stray illumination
杂散行情 split quotation
杂散荷载 stray load
杂散荷载损耗 stray load loss
杂散回波 angel echo
杂散耦合 spurious coupling; stray coupling
杂散束 stray beam
杂散损耗 stray loss
杂散损耗系数 stray-loss factor
杂散天空 scattered
杂散铁块 tramp iron
杂散通量 stray flux
杂散铜损 stray copper loss
杂散系统 polydisperse system
杂散谐振 stray resonance
杂散元件 stray element
杂散噪声拾波 stray pick-up
杂散振荡 spurious oscillation
杂散阻抗 spurious impedance
杂色 mottle; mottled effect; parti-colo(u)r; variegated colo(u)r
杂色布 motley
杂色大理石 compound marble; variegated marble
杂色的 motley; mottled; multicolo(u)-red; party colo(u)red; pied; roan; variantly colo(u)red; varicolo(u)red; varied; variegated; versicolo(u)r(ed)
杂色浮标 assorted colo(u)red buoy
杂色光 heterogeneous light
杂色琥珀 romanite
杂色棉布 jaconet
杂色木料 mottled wood
杂色呢 motley
杂色黏[粘]土 mottled clay; multicolo(u)red clay; varicolo(u)red clay
杂色砂岩 bunter sandstone; mottled sandstone; variegated sandstone; graywacke; bunter <常成为丰沛的含水层>
杂色麝香石竹 carnation
杂色石灰石 cornstone
杂色刷子 mottling brush
杂色铁 mottled(pig) iron
杂色亚黏[粘]土 mottled subclay; multicolo(u) red subclay; varicolo(u) red subclay
杂色釉 miscellaneous waste glaze
杂砂砾的 petromict
杂砂片岩 graywacke [greywacke] schist
杂砂石板 graywacke slate
杂砂石灰岩 graywacke limestone
杂砂岩 apogrit(e); greywacke; whine-rock
杂砂岩储集层 greywacke reservoir
杂蛇纹石 verd(e) antique
杂砷铜矿 mohaw-algodonite
杂砷银矿 huntilite
杂生的植物 <如野草> mat-forming plant
杂声计 psophometer
杂石 <填筑用> gob; rock matrix
杂石膏重晶石 dreelite
杂石滩 boulder stone; gravel and shingle foreshore
杂食性动物(总称) omnivore
杂树丛 copse wood
杂树林 copse
杂耍剧场 variety theatre; vaudeville theater
杂水榴石 hydrogrossular; plazolite
杂税 irregular tax; miscellaneous tax
杂酸 heteroacid
杂碎 chop-suey; chow chow
杂碎骨 garbage bone
杂碎油脂 garbage grease
杂缩合 heterocondensation
杂填土 landfill waste in; made ground; miscellaneous fill; mixed soil; random fill; rubblish fill
杂填土滑坡 miscellaneous fill landslide
杂贴广告的临时棚栏 advertisement boarding
杂铁 tramp iron
杂铁捕集 metal trap
杂同(立构)聚合物 heterotactic polymer
杂铜 assorted brass; composition brass
杂顽火无球粒陨石 whitleyite
杂纹方石工程 broken ashlar; random work
杂污染源 miscellaneous sources of pollution
杂务 chore; general work; miscellaneous business
杂务工 bottle washer
杂务间 utility room
杂务梯 service stair(case)
杂务院 service court; service yard; utility yard
杂物 adulterant; foreign body; raffle; sundries
杂物传播 transmission by formite
杂物袋 holdall
杂物柜 glory hole
杂物及油污 impurity and greasy dirt
杂物间 cubby; hovel; litter receptacle; lumber room
杂物容器 catch-all
杂物室 glory hole
杂物箱 catch-all; litter receptacle; tidy
杂锡砂 squat
杂系参数 hybrid parameter
杂项 miscellaneous; miscellaneous items; sundries; sundry items
杂项补贴 miscellaneous subsidy
杂项补助 miscellaneous subsidy
杂项存款 sundry deposit
杂项的 miscellaneous
杂项费用 overhead charges; overhead cost; overhead expenses; sundry charges
杂项敷线 miscellaneous wiring
杂项工程 general work
杂项工具 miscellaneous tool
杂项管理费 miscellaneous general expense; sundry overhead
杂项规定 miscellaneous provisions
杂项耗用品 sundry supplies
杂项荷载 miscellaneous load
杂项及意外费用 miscellaneous and contingencies
杂项架 miscellaneous bay
杂项津贴 miscellaneous subsidy
杂项开支 miscellaneous expenses; overhead charges; overhead(cost)
杂项利润 miscellaneous profit
杂项收入 miscellaneous revenue; sundry receipt
杂项说明 miscellaneous declaration
杂项损失 miscellaneous loss
杂项铁件 miscellaneous iron
杂项项目 sundry items
杂项型钢 miscellaneous shape
杂项营业收入 miscellaneous revenue
杂项用电 miscellaneous power
杂项账户 miscellaneous account
杂项支付 miscellaneous payments
杂项制品 miscellaneous products
杂项准备 sundry reserves
杂项资产 miscellaneous assets; sundry assets
杂形斗 accidental whorl
杂形矿 heteromorphite
杂薰衣草油 Lavandine oil
杂岩 complex; diamictite; mixtite
杂样灯 mixed light lamp
杂伊利石 santorine
杂役僧侣食堂 refectory for lay brethren
杂役僧侣斋堂 frater lay brethren
杂音 bloop; complex tone; noise; stray noise
杂银星绿松石 callainite
杂用泵 general service pump; service pump
杂用船 miscellaneous vessel; service craft; yawl
杂用机械 miscellaneous machinery
杂用建筑 <如仓库、厨房、马厩或附属在田园的等> agricultural service building
杂用品 chattels
杂用时间 incidental time
杂用室 utility room
杂用水 non-potable water; service water; water for miscellaneous use
杂用水泵 service water pump
杂用水泥 non-constructive cement
杂用水箱 station service water tank
杂用艇 jolly boat
杂用小绳 fox
杂油液 miscella
杂有石块的 calculous; rocky
杂鱼 rough fish; waste fish
杂鱼类 miscellaneous fishes
杂原子链 heteroatomic bond
杂院 multihousehold compound
杂支 overhead
杂支费 miscellaneous expenditures
杂脂 heterolipid
杂志架 magazine rack
杂志室 periodic(al) room
杂志阅览室 magazine room
杂质 impurity; contaminant; dirt; dirt

and foreign matter; dross; external material; extraneous matter; fancy; foreign material; foreign matter; foreign substance
杂质半导体 extrinsic semiconductor
杂质泵 solid pump; trash pump
杂质补偿 impurity compensation
杂质沉淀 contamination precipitation
杂质带传导 impurity-band conduction
杂质带传导理论 impurity-band conduction theory
杂质带迁移率 impurity-band mobility
杂质导电 impurity conduction
杂质导电区 extrinsic range
杂质电离能 impurity ionization energy
杂质多的矿石 halvans; hanaways
杂质分布 impurity distribution
杂质分布测量 impurity distribution measurement
杂质分布剖面图 impurity distribution
杂质分离器 trash eliminator
杂质分凝 impurity segregation
杂质粉粒 foreign particle
杂质钙沸石 cerinite
杂质光电导 impurity photoconductivity
杂质光电导体 impurity photoconductor
杂质含量 impurity content
杂质琥珀 bastard amber
杂质活性 impurity activity
杂质积聚 accumulation of impurities
杂质激活 impurity activation
杂质激活能 impurity activation energy
杂质夹层 dirt parting
杂质检测极限 limit of detection of impurities
杂质金属 contaminating metal
杂质控制 impurity control
杂质矿物 impurity mineral
杂质扩散 impurity diffusion
杂质离子 contaminant ion; foreign ion
杂质离子不稳定性 impurity ion instability
杂质粒子 foreign particle
杂质量 impurity level
杂质煤 attrital coal; attritus
杂质能带 impurity band
杂质能级 impurity energy level
杂质浓度 impurity concentration
杂质迁移率 impurity mobility
杂质缺陷 impurity defect; interstitial defect
杂质散射 impurity scattering
杂质渗入 impurity penetration
杂质试验极限 limit of detection of impurities
杂质损伤 impurity damage
杂质态 impurity state
杂质梯度 impurity gradient
杂质条纹 impurity striation
杂质吸收 absorption by impurities; impurity absorption
杂质徙动 impurity migration
杂质限度 impurity limitation; limit of impurities
杂质选择 impurity selection
杂质元素 impurity element
杂质原子 foreign atom
杂质源 impurity source
杂质沾污 impurity contamination
杂质中心 impurity center[centre]
杂质着色的玻璃 glass tinted by impurities
杂作坊 service yard

砸

砸道镐 beater pick; tamping pick
砸道棍【铁】tamping rod; tamping bar
砸道机【铁】tamping machine
砸开 crack
砸伤 injury accident from falling
砸碎 beat in; crush
砸铁道棍【铁】tamping bar
砸弯 <钉尖> clinching[clenching]

灾

灾变 cataclysm; catastrophe; convulsion
灾变带 casuzone
灾变假说 catastrophic hypothesis
灾变理论 catastrophe theory
灾变论 convulsionism
灾变事件 catastrophic event
灾变说【地】catastrophism; cataclysm theory
灾变学说【地】convulsionism
灾度 disaster magnitude
灾害 calamity; catastrophe; disaster; havoc; hazard rate; maelstrom; pest; plague
灾害保险 casualty insurance
灾害的 disastrous
灾害等级界限 hazard level limitation
灾害地质学 disaster geology; hazard geology
灾害多发区 disaster-prone area; disaster-prone region
灾害防备和救济 disasters preparedness and relief
灾害管理 disaster management
灾害救助制度 the natural disaster relief system
灾害链 disaster chain
灾害模式 disaster mode
灾害潜势 disaster potential
灾害清理工作 disaster clean-up operation
灾害伤害 hazard-induced injury
灾害生态学 calamity ecology
灾害受害者的复原 rehabilitation of victims of a disaster
灾害统计 damage statistics
灾害危险区 calamity danger district
灾害危险性 disaster potential
灾害消除 hazard elimination
灾害性暴雨 catastrophic cloudburst
灾害性变化 catastrophic change; catastrophic variation
灾害性倒毁 catastrophic collapse
灾害性倒塌 catastrophic collapse
灾害性地震 catastrophic earthquake
灾害性反应 catastrophic reaction
灾害性洪水 catastrophic flood; damaging flood; disastrous flood
灾害性泥石流 calamitous mudflow; catastrophic mudflow
灾害性破坏 catastrophic damage; catastrophic failure
灾害性切断电源 disaster dump
灾害性印出 disaster dump
灾害性转储 disaster dump
灾害学 disaster science
灾害易发区 disaster-prone area
灾害预报 hazard forecast; hazard prediction
灾害预测 calamity foreknowledge
灾害预防 prevention of disasters
灾害造成的漂集堆 catastrophic drift
灾后恢复 post-disaster recovery
灾后阶段 post-disaster phase
灾后救济 post-disaster relief
灾后气体 after-gas
灾后损害调查 post-disaster damage survey
灾后损失保险 consequential loss insurance
灾后重建 post-disaster reconstruction
灾后综合征 post-disaster syndrome
灾荒 calamity
灾祸 calamity; catastrophe; disaster; misadventure; mishap; peril
灾级 disaster magnitude
灾难 calamity; catastrophe; disaster; fatality; peril
灾难的边缘 precipice
灾难分析 disaster analysis
灾难风速 catastrophic wind speed
灾难救援 disaster relief
灾难清理作业 disaster clean-up operation
灾难现象 catastrophic phenomenon
灾难性暴雨 catastrophic cloudburst
灾难性错误 catastrophic error
灾难性的 catastrophic; disastrous
灾难性洪水 disastrous flood
灾难性后果 disastrous consequence
灾难性碰撞 catastrophic collision
灾难性破坏 catastrophic failure
灾难性山崩 catastrophic landslide
灾难性事故 disastrous accident
灾难性事件 catastrophic event
灾难性死亡 catastrophic mortality
灾难性坍塌 catastrophe collapse
灾难性转储 disaster dump
灾难性转贮 disaster dump
灾年 year of famine
灾前规划和准备工作 pre-disaster planning and preparedness
灾前活动 pre-disaster activity
灾前资料 prefire information
灾情 condition of disaster
灾情贷款 disaster loan
灾情调查 survey of disasters
灾区 devastated land; disaster area; distress area; emergency area

甾

甾醇(类) <旧名固醇>【化】sterol
甾类化合物 steroid
甾烷 sterane
甾烷的分子参数 molecular parameter of steranes
甾烷含量 steranes content
甾烯 sterene
甾质【化】steroid

栽

栽 cultivation technique
栽成一丛树 clump
栽苗机 plant setter; plant setting machine
栽培 cropping; cultivate; culture; grow; planting
栽培材料 planting material
栽培草地 cultivated grassland
栽培地 milpa
栽培法 cultivation method; training
栽培化 cultivation
栽培技术 cultural practice; culture technics
栽培类型 cultivated type
栽培面积 cultivated area
栽培面积的类型 the type of area cultivated
栽培面积逐步增加 gradually increased the cultivation area
栽培牧草 tame forage grass
栽培品系 cultivar
栽培品种 cultivated variety
栽培区(域) cultural area
栽培群落 agrium; culture community
栽培生态学 hemeroecology
栽培试验 experiment in cultivation
栽培树木 the cultivation of trees
栽培苋 love-lies-bleeding
栽培者 grower
栽培植被型 cultivated vegetation type
栽培植物 cultivated plant; hemerophyte
栽培植物用框架 garden frame
栽培种 cultigen; cultivated species
栽培作物 raise crops
栽绒地毯 knotted pile carpet
栽树 plant trees
栽丝 <焊前的> studding
栽体 carrying agent
栽植 field planting; plantation; planting out; transplanting
栽植标桩 planting stick
栽植材料 planting stock
栽植铲 spud
栽植锄 planting hatchet; planting mattock
栽植带 planting belt
栽植地 planting site
栽植点 planting point
栽植方式 planting system
栽植幅度 transplanting width
栽植镐 planting mattock
栽植沟 planting furrow; planting trench
栽植机 planter; planting machine; transplanter
栽植距离 planting distance
栽植开沟器 supply colter
栽植配置 planting composition
栽植平茬苗 stump plant
栽植器 planting apparatus
栽植锹 planting bar
栽植绳 planting cord
栽植桶 plant tub
栽植挖坑机 planting borer
栽植挖穴机 dibbler; planting borer; planting hole machine
栽植穴 planting hole
栽植于下的 underplanted
栽植与中耕附加装置 planting-and-cultivation attachment
栽植造林 forest planting
栽植砖 planting brick
栽植装置 plant feed unit; plant setting mechanism
栽种 grow; plant
栽种花木的盆 plant tub
栽种花木的桶 plant tub
栽种计划 planting plan
栽种绿肥的休耕地 green fallow
栽种品种 commercial variety
栽种深度 planting depth

宰

宰里拜圈地 <位于东北非> Zariba

载

载包车 bale truck
载病体 disease carrier
载波 carrier wave; signal carrier; carrier
载波摆值 carrier swing
载波包络 carrier envelope
载波闭锁方向保护(装置) directional protection with carrier current blocking
载波闭锁距离保护(装置) distance protection with carrier current blocking
载波不定波 floating carrier wave
载波差拍 carrier beat; intercarrier
载波差柏哼声 intercarrier hum
载波差拍噪声抑制器 intercarrier noise suppressor
载波差拍制 intercarrier system
载波颤动 carrier flutter
载波传声器 ticker[tikker]
载波传输 carrier transmission

载波传输滤波器 carrier transfer filter
载波传输系统 carrier transmission system
载波导频系统 carrier pilot system
载波的分布位置 carrier position
载波的频谱 spectrum of carrier
载波电缆 carrier wire
载波通信[讯]网 carrier network
载波通信[讯]系统 carrier system; carrier communication system
载波制遥测 remote measurement by carrier system
载波终端设备 carrier frequency terminal equipment
载玻片 ground slide; microscopic slide; microslide; slide; slide micrometer
载玻片培养 slide culture
载驳舱 barge stowage hold
载驳船 barge carrier ship; barge-carrying ship
载驳船的船 barge of lighter aboard ship system
载驳船系统 lighter aboard ship system[LASH]
载驳货船 barge carrier; barge-carrying ship; barge on board ship; lighter aboard ship; sea-barge carrier
载驳货船的子驳 shipborne barge
载驳货船码头 barge carrier terminal; LASH dock; ship dock
载驳货船系统 < 浮船坞式 > barge clipper system
载驳货轮及子驳共用码头 combination ship and barge dock
载驳/集装箱船 lighter/container ship
载驳快船 sea-barge clipper
载驳快船方式 barge clipper system
载驳双体船 barge aboard catamaran vessel; barge above catamaran; barge-carrying catamaran
载驳运输 barge carrier transport
载驳运输方式 float on/float off system
载驳子母船 lighter aboard ship
载铂石棉 platinized asbestos
载车船 car-carrier
载车渡船 vehicle ferry
载车列车 vehicle-train system
载车气垫轮渡 car ferry hovercraft
载带系统 loading system
载电的 current-carrying
载电轨 live rail
载电能力 current carrying capacity
载电线 live wire
载额估计 loading estimate
载割阻力系数 coefficient of coal cutting resistance
载供 carrier supply
载供系统 carrier supply system
载轨车 carrier car
载荷 weight
载荷半径 load radius
载荷比 load ratio
载荷变形 deformation under load
载荷变形测定 load deformation measurement
载荷变形图 load-strain diagram
载荷补偿 load compensating
载荷补偿制动机 load compensating brake
载荷长度 loaded length
载荷吃水 load draft
载荷处理 cargo handling
载荷传递 load transfer
载荷传感变扭器 load sensing torque converter
载荷传感调节器 load responsible control
载荷的精确控制 precise load control
载荷的偏心度 eccentricity of loading
载荷阀 loaded valve
载荷反应制动器 load-reaction brake
载荷分布 load spread
载荷分布图 loading distribution chart
载荷分布因素 load distribution factor
载荷分配 load spread
载荷分配因数 factor of load distribution
载荷感应阀 load respondent valve
载荷隔离器 load isolator
载荷估计 loading estimate
载荷极限 load limit
载荷计 load meter
载荷夹钳 load grippers
载荷率 loadability; load factor
载荷能力 loadability
载荷能力分布特性 load-spreading property
载荷判据 loading criterion
载荷强度 density of load; intensity of load(ing)
载荷时间 load time
载荷损失 loss of load
载荷调整器 load adjuster
载荷推出器 load ejector
载荷托架 load bracket
载荷系数 loading factor
载荷弦 < 桁架的 > loaded chord
载荷箱 load box
载荷因数 load factor
载荷增长 load disproportionate
载荷增量 increment of load; load increment
载荷者 carrier
载荷重量 weight of load
载荷重心 center of gravity of load
载回弹弯沉 static rebound deflection
载货 lading; shipment
载货驳船的停泊处 quay for goods-carrying barges
载货驳船的停泊码头 quay for goods-carrying barges
载货驳货码头 quay for goods-carrying barges
载货舱位 cargo space
载货吃水标志 Plimsoll's mark
载货吃水线 load line; Plimsoll's line; Plimsoll's mark
载货船只 carrying ship; carrying vessel
载货电梯 freight elevator; lift
载货定额 cargo capacity
载货吨 cargo(deadweight) ton; shipping ton【船】; revenue ton
载货吨位 cargo carrying capacity; cargo tonnage; useful capacity; useful deadweight; freight tonnage; cargo deadweight
载货吨位数 loaded tonnage
载货返航 back haul
载货飞机 lorry
载货浮力 working buoyancy
载货港(口) loading port; port of loading; port of shipment
载货挂车 truck trailer
载货过多 over-freight
载货货柜 loaded container
载货集装箱 loaded container
载货甲板 cargo deck
载货监视系统 load monitoring system
载货卡车 camion; cargo vehicle; commercial vehicle; goods car; lorry
载货量 cargo carrying capacity; carrying capacity; freight ton(nage); cargo deadweight
载货马车 dray
载货面积 loading area
载货能力 cargo capacity; cargo carrying ability
载货气球列车 freight carrying train of air-balloons
载货汽车 automotive truck; mechanized lorry; motor truck
载货汽车磅秤 truck scale
载货汽车底盘 truck chassis
载货汽车绞盘 lorry winch
载货汽车轮胎 truck tire[tyre]
载货汽车轮胎拆卸工具 truck-tire [tyre] remover
载货汽车停车场 truck stop
载货汽车蒸发器 truck evapo(u)rator
载货汽车制造厂 truck plant
载货清单 manifest
载货容积 bale cubic capacity; cargo capacity; cargo space; payload space
载货容积吨数 measurement tonnage; tonnage of load volume
载货容量 bale cubic capacity; cargo capacity; cargo space; freight-hold capacity; payload space
载货上船 load a ship with cargo
载货升降机 goods lift
载货隧道 cargo tunnel
载货台 carriage
载货提升机 goods hoist
载货体积单位 measurement ton(nage)
载货系数 stowage factor
载货舷门 side port
载货限定外形尺寸 clearance loading ga(u)ge
载货行驶里程 cargo-mile
载货运清单 freight manifest
载货重量 cargo capacity; cargo deadweight; payload
载货重量吨 deadweight cargo tonnage
载货总量 total amount on board; total on board
载胶体 emulsion carrier
载客 passenger loader
载客舱位 passenger space
载客车辆 passenger accommodation; passenger vehicle
载客电梯 passenger elevator; passenger lift
载客电梯门 passenger elevator door; passenger lift door
载客电梯厢 passenger elevator car
载客吊笼 cab(in); cable car
载客定额 passenger carrying capacity
载客定员 seating capacity
载客渡船 passage boat; passenger ferry
载客渡轮 passage boat; passenger ferry
载客飞机 passenger carrier
载客客机 passenger carrier
载客量 passenger carrying capacity; passenger loader; people carrying capacity
载客面积 passenger capacity area
载客能量 passenger carrying capacity
载客汽车 passenger automobile; passenger service vehicle; carryall < 两边座位相对 >
载客舢板 wherry
载客升降机 passenger lift
载客输送机 passenger conveyer[conveyor]
载客位置 loading position
载客运输带 man conveyer[conveyor] belt
载客运输皮带 passenger conveyer belt
载控继动阀 load dependent relay valve
载冷剂 refrigerating medium; secondary refrigerant
载量 capacity
载料空间 loading space
载流倍数 interception ratio
载流部件 current-carrying part
载流承力索 current-carrying carrier
载流导体 current-carrying conductor
载流的 current-carrying
载流等离子体 current-carrying plasmas
载流法 carrier method
载流量 carrying capacity; current capacity; current-carrying conductor; ampacity < 以安培表示 >
载流能力 current-carrying capacity; current-carrying conductor
载流容量 carrying capacity; current-carrying capacity
载流通道 carrying channel
载流系数 current-carrying factor
载流线 current-carrying conductor; elementary stream
载流轴 current-carrying shaft
载流状态 current flow condition
载流子 carrier; charge carrier; current carrier
载满 laden
载煤量 coaling capacity
载明保证 express warranty
载模板 backing board; backing plate
载膜 film carrier
载木台 travel(l)ing carriage
载能旋涡 energy containing eddy
载泥船 mud boat
载片 slide glass
载频 resting frequency
载频摆幅 carrier swing
载频放大 carrier frequency amplification
载频放大器 carrier amplifier; carrier frequency amplifier
载频分配器 carrier frequency distributor
载频供给架 carrier frequency supply equipment bay
载频和导频供给架 channel carrier and pilot supply bay
载频缓冲器 carrier buffer
载频精确偏置 carrier precision offset
载频控制的干扰抑制器 codan
载频脉动 carrier ripple
载频偏移 carrier deviation; carrier shift
载频偏移测试器 carrier deviation meter
载频偏移计 frequency carrier deviation meter
载频偏置 carrier frequency offset
载频漂移 carrier shift
载频平衡 carrier balance
载频驱动继电器 carrier-actuated relay
载频全息图 carrier frequency hologram
载频同步 carrier frequency synchronization
载频信号对干扰信号比 ratio of carrier-to-interfering signal
载频振荡器 carrier frequency oscillator; carrier generator
载频中继 carrier current relaying
载频重置 carrier reinsertion
载气 carrier gas; supporting gas
载气泵 gas ballast pump
载气流速 flow rate of carrier gas
载气逆流 carrier gas inversion
载气平均流速 average flow rate of carrier gas
载汽车电梯 vehicular elevator; vehicular lift
载热固体 Thermofor
载热剂 coolant; heat carrying agent; heating agency; heat-transfer a-

gent; heat-transfer material; heat-transfer medium
载热减速剂 coolant moderator
载热介质 heat carrier; heat carrying medium
载热量 heat load
载热流体 heat carrying fluid; heat-conducting fluid; heat-exchange fluid; heat-transfer fluid; heat transport(ing) fluid
载热能力 heat carrying capacity
载热体 heat carrier; heat carrying agent; heating agency; heating medium; heat-transfer material; Thermofor; thermophore
载热体入口群体温度 bulk inlet temperature
载热油 heat-transfer oil
载热质 coolant; heating agency
载人的 inhabited; manned
载人电梯 manlift; passenger elevator; passenger lift
载人吊运车 man trolley
载人渡船 foot boat
载人飞船 manned spacecraft; manned vehicle
载人飞行 manned flight
载人飞行器 manned craft; satelloid
载人罐笼 man cage
载人轨道航天站 manned orbital space station
载人轨道空间站 manned orbital space station
载人轨道实验室 manned orbiting laboratory
载人海底站 manned undersea station; manned underwater station
载人海底作业 manned underwater work
载人航海日志 entry in log
载人航天器 manned spacecraft
载人卷扬机 passenger hoist
载人可潜器 manned submersible
载人缆道 <水文测验用的> manned cableway
载人气球 manned ballon
载人潜水器 manned submersible; manned vehicle
载人日志簿 log
载人升高的工作平台 manlift
载人升降机 personnel-carrying hoist
载人水下工作系统 manned underwater work system
载人水下工作站 manned underwater station
载人水下结构 manned underwater structure
载人水下实验室 manned underwater laboratory
载人水下作业 manned underwater work
载人水下作业系统 manned underwater work system
载人踏板 <电动走道的> treadway
载人提升机 manlift
载人提升机的踏步板 manlift step platform
载人提升机的行程 manlift travel
载人微型潜水器 manned microsubmersible
载人卫星 inhabited satellite; manned satellite
载人系缆式微型潜水器 manned tethered microsubmersible
载人宇宙飞船 manned spacecraft; manned spaceship
载人宇宙飞船中心 manned spacecraft center[centre]
载容量 passenger capacity
载入平台 <起重机的> platform
载色剂 binding agent; colo(u)r carrier; vehicle
载色体 chromatophore; vehicle
载色体的 chromatophorous
载沙液 sand carrier
载上 load on
载湿 moisture carry-over
载水剂 water-carrying agent
载水式压路机 water ballast type roller
载水手推路碾 water ballast hand roller
载水箱拖车 water tank trailer
载台 microscope carrier; microscope stage
载体 bearer; carrier(compound); carrier material; isotopic carrier; supporter; supporting medium; vehicle
载体产生 carrier generation
载体大小 carrier size
载体的沉淀 carrier precipitation
载体的钝化 deactivation of support
载体电泳图法 electropherography; ionography
载体分离器 carrier separator
载体分馏 carrier distillation
载体分馏法 method of carrier distillation
载体胶粘剂 supported adhesive
载体介质 carrier medium
载体金属 carrier metal
载体矿物 host mineral
载体密度 support density
载体浓度 carrier concentration
载体添加法 carrier addition method
载体同位素 carrier radioisotope
载体涂层开管柱 support-coated open tubular column
载体吸收 absorption of vehicle
载体效应 carrier effect; support effect
载体信道 information bearer channel
载体性转运 carrier-mediated transport
载体振荡器 carrier source
载体蒸馏 carrier distillation
载体置换法 carrier displacement
载桶拖车 barrel carrying trailer
载图记录仪 chart recorder
载拖式牵引车 saddle tractor
载物玻璃 <显微镜> object carrier
载物玻片 glass slide
载物玻片橱 slide cabinet
载物玻片盒 slide box
载物玻片染色机 slide staining machine
载物薄片 slide
载物片 microslide; slide
载物片加温器 slide warmer
载物片架 glass rack; slide rack
载物片清洗器 slide cleaner
载物台 objective table; object stage; stage
载物台测微计 stage micrometer
载物台粗调螺旋 coarse adjustment knob
载物台横向运动螺旋 lateral movement knob
载物台架 object stage
载物台前后运动螺旋 forward and backward movement knob
载物台细调螺旋 fine adjustment knob
载物体 object carrier
载像 image-bearing
载型板 bottom board
载畜过多的草场 over-stooked pasture
载畜量 carrying capacity; carrying rate; grazing capacity; grazing load; rate of stocking; stock capacity; stocking density; stocking rate
载畜率 degree of grazing; rate of stocking; stocking rate
载岩车厢 <双层底板的> rock body
载氧体 oxygen carrier
载液 carrier liquid; supporting liquid
载液胶片 carrier liquid film; Bimat film
载油舱 cargo tank
载有生物的人造地球卫星 biosatellite
载有装运集装箱的拖车队 trailvan train
载运 carriage; carrier transport
载运杓 carrying ladle
载运斗 travel(l)ing skip
载运管 carrier pipe
载运货重 carrying load
载运机 load-and-carry equipment
载运距离 load distance
载运量 carrying capacity
载运能力 transport competency
载运气体 carrier gas
载运索 <架空索道的> track cable; carrying cable
载运违禁品 carriage of contraband
载运因素 conveyance factor
载噪比 carrier(-to)-noise ratio
载噪声 air-borne sound
载质 charge material
载重 burden(ing); carrying capacity; lading; load; loading ga(u)ge; loading weight; payload; weight; weighted high-duty
载重把持法 load holding device
载重板车 dray
载重保持装置 load holding device
载重崩溃 load to collapse
载重比试验 bearing ratio test
载重标尺 deadweight scale; displacement scale; draft scale; immersion scale
载重标志 load mark
载重标准 loading ga(u)ge
载重表尺 deadweight scale
载重车 cross-country truck; load-carrying vehicle; loader; load vehicle; truck
载重车磅秤 scale for load vehicles
载重车车身 truck body
载重车队 train
载重车混入率 commercial vehicle rate
载重车辆 heavy car; load-carrying vehicle
载重车前挂钩 breast hook
载重车式混凝土搅拌机 concrete delivery mixer
载重车箱 truck box
载重车用柴油机 truck diesel
载重沉箱法 sinking of caisson by loading
载重吃水 load draught
载重吃水标尺 loading ga(u)ge
载重吃水标记 loadline mark; load water mark
载重吃水标志 loadline mark; load water mark
载重吃水线 load line; load waterline; loadline
载重吃水线标志 Plimsoll's mark
载重船 general cargo
载重带 felloe band
载重的 high duty; load carrying; loaded; weight-carrying; weighted
载重等级标志 structure-classification symbol
载重定额 loading instruction
载重吨 burden; freight ton(nage); tons deadweight; deadweight ton
载重吨位 burden; deadweight capacity; deadweight tonnage
载重翻斗车 athey wagon
载重峰值 load peak
载重钢缆 load cable
载重钢丝绳 carrier cable; holding line
载重拱 relieving arch
载重构架 weight-carrying frame
载重骨架 weight-carrying skeleton
载重骨架构件 weight-carrying skeleton member
载重骨架结构 weight-carrying skeleton construction; weight-carrying skeleton structure
载重规定 loading instruction; regulation of load
载重滑车 troll(e)y
载重滑车架 load pulley block
载重滑车组 load pulley block
载重机 weight carrier
载重机铲斗堆装容量 load bucket capacity
载重机构 weight-carrying mechanism
载重机制 weight-carrying mechanism
载重计 load ga(u)ge
载重计算法 tonnage noting
载重检测器 weight detector
载重结构 weight-carrying structure
载重卡车 cargo truck; cargo vehicle; motor truck
载重卡车运输 road delivery
载重框架 weight-carrying frame
载重肋 weight-carrying rib
载重力 lifting power
载重力矩限制器 load moment limiter
载重链 load chain
载重量 bearing value; capacity tonnage; carrying capacity; carrying power; dead load; deadweight carrying capacity; deadweight tonnage; loadage; load-carrying ability; load-carrying capacity; loading capacity; magnitude of load; weight capacity
载重量标度表 deadweight scale
载重量利用率 tonnage utilization ratio
载重列车英里 loaded train-miles
载重率 rate of loading; tonnage rating (of traction)
载重轮胎 band tire[tyre]; high-capacity tire[tyre]; lorry tire[tyre]; truck tire[tyre]
载重马车 <车身低的> dray
载重面 weight-carrying plane
载重面积 loading area
载重木结构 timber load-carrying structure
载重能力 cargo capacity; deadweight capacity; lifting capacity; loading capacity; weight capacity; weight-carrying power
载重排水量 load displacement
载重排水量系数 deadweight displacement coefficient; deadweight ratio
载重平板车 heavy-duty bogie wagon
载重平底船 canal barge
载重普通腹板结构 weight-carrying structure of plain web girders
载重汽车 automotive truck; autotruck; bogie bolster; cargo truck; cargo vehicle; motor lorry; single-unit truck; truck; straight job <俚语,无拖车的>; motor truck <美>
载重汽车保有量 truck population
载重汽车补给 motor-truck supply
载重汽车不足 truck shortage
载重汽车车道 truck lane
载重汽车车架 truck frame
载重汽车车轮 truck wheel
载重汽车冲洗 truck wash(down)
载重汽车出租公司 truck leasing firm
载重汽车底盘 truck chassis
载重汽车队 truck convoy

载重汽车吨位 truck ton
载重汽车附起重机 truck with crane
载重汽车规格 truck size
载重汽车和铁路货车 tarvan; truck and rail van
载重汽车荷载 truck load(ing)
载重汽车驾驶室 truck cab
载重汽车驾驶员 trucker
载重汽车轮胎外胎 truck-tire[tyre] casing
载重汽车一览表 truck schedule
载重汽车用轮胎 lorry tire[tyre]
载重汽车运输费 truckage
载重汽车载重量 truck capacity
载重汽车站 motor-truck terminal
载重汽车装载清单 lorry loading list
载重牵引车 prime-mover truck; truck tractor
载重牵引车半挂车 truck tractor semitrailer
载重墙 weight-carrying wall
载重墙结构 weight-carrying wall construction; weight-carrying wall structure
载重桥 loading bridge
载重情报 weight information
载重曲线 curve of loads; load curve
载重曲线图 load chart
载重时间挠曲线 load-time deflection curve
载重实心腹板梁结构 weight-carrying structure of solid web girders
载重试验 load-bearing test; load-carrying test; load(ing) test(ing)
载重手推车 skid platform
载重水线 load water-line
载重索 carrier cable; load cable; main cable
载重调节器 weight-loaded governor
载重拖车 athey wagon; truck tractor; truck trailer
载重稳定性 stability under load; stability under working conditions
载重圬工砌筑工作 weight-carrying masonry work
载重圬工墙 weight-carrying masonry wall
载重系数 bearing capacity factor
载重系统 weight-carrying system
载重下降速度 load-lowering speed
载重线标圈 loading disk; loadline disc [disk]; Plimsoll's disk
载重线标志 freeboard mark; loadline mark; load water-line; Plimsoll's line; Plimsoll's mark
载重线导向块 loadline guide block
载重线导向轮 loadline guide wheel
载重线定期检验 periodic(al) load line inspection
载重线公约 loadline convention
载重线规则 loading regulation; loadline regulation
载重线规则、地带、区域及季节期 loadline rules, zone, areas and seasonal periods
载重线合格证定期更新 periodic(al) load line renewal
载重线检验 loadline survey
载重线勘定 assignment of loadline; loadline assignment
载重线证书 certificate of approval of the marking; certificate of load line; freeboard certificate; loadline certificate
载重限度 loading limit
载重限制 loading limitation
载重小车式起重机 crab trolley type bridge crane
载重信息 weight information
载重性能 weight-carrying ability
载重因数 bearing capacity factor
载重增加 increase of loading; load increment
载重增量 increment of load
载重指示器 load indicator
载重柱 load column
载重砖 weight-carrying brick
载重转移装置 load transfer device
载重状态 load condition
载重/自重比 load/tare rate; load to empty weight rate; payload-weight ratio
载装码头 loading quay; loading terminal; loading wharf

再搬运 rehandling
再版 future edition; reprint edition; republication; republish; second edition
再版本 republication
再拌和 reblending; remixing; retempering
再拌混凝土 remixed concrete
再拌机 remixer
再拌重铺方法 remixing process
再包土的 balled
再包土栽植 balled planting
再饱和器 resaturator
再饱和作用 resaturation
再保险 reinsurance
再保险单 reinsurance policy
再保险分出人 cedant
再保险附加保费 additional premium-reinsurance
再保险人 reinsurer
再保险条款 reinsurance clause
再保险协定 reinsurance agreement
再保险佣金 reinsurance commission
再保证 reassurance
再报价 repeat offer
再爆破 secondary breakage
再闭合 reclosing
再闭合时间 reclosing time
再编辑 re-editing
再编制 reorganization
再标定 re-proving
再补法 recomplementation
再补给 replenish; resupply
再补给单 resupply voucher
再参加 rejoin
再参加考试 <英> resit
再测 regauging
再测量 re-measuring; resurvey
再测信度 retest reliability
再插发射机 re-insert transmitter
再查验 reinspection
再掺气作用 reaeration
再阐述 reformulate
再车削 reface
再沉淀 resedimentation; resettling
再沉淀作用 after precipitation; redeposition; reprecipitation
再沉积 redeposit; reworking
再沉积动力学 dynamics of resedimentation
再沉积黄土 redeposited loess
再沉积凝灰岩 reworked tuff
再沉积岩 resedimented rock
再沉积作用 redeposition; resedimentation
再沉降 reprecipitation; resedimentation
再成型 post forming; reshape
再成岩【地】 deuterogenous; deuterogenic; deuterosomatic
再城市化 reurbanization
再炽热 recalesce; reglowing
再炽热点 recalescence point
再充电 additional charge; booster charge; boosting charge; recharge; recharging; replenishing
再充电电流 recharge current
再充电时间 recharging time
再充二氧化碳 recarbonation
再充灌水池 recharging basin
再充满 refill(ing)
再充气 backfill(ing); reaeration; recharge; recharging; refilling
再充水 refill(ing)
再充水过程 refilling process
再充填 refilling
再充氧 reoxygenation
再抽汲 repumping
再抽空 re-evacuate; re-evacuation
再抽提 reextraction
再筹资金 refinancing
再出口 re-exportation
再出售 <进口商无力付款提货,只好在进口地点减价出售> resale
再出租 release; relet
再除尘器 reduster
再处理 after-treatment; rehandle; rehandling; reprocessing; rerun; retreat(ing); retreatment; rework
再处理铲运机 rehandling scraper
再处理机械 rehandling machinery
再处理设备 rehandling unit
再处理设施 rehandler
再处理系统 retreating system
再处理油 rerun oil
再触发 restrike
再触发器 retrigger
再传 repeated transmission
再吹 reblowing
再纯化 repurification; repurify
再纯器 repurifier
再磁化 remagnetization; remagnetize
再次拨款 reappropriation
再次插入 reinsertion
再次点火 reignite; relight(ing)
再次堆焊 resurfacing by welding
再次分级 reclassification
再次分选 reclamation
再次粉磨 regrinding
再次粉磨的 reground
再次粉碎 repulverize
再次干燥 after-drying
再次构造 secondary structural elements
再次回火 retemper(ing)
再次混合 reblending
再次加冰 re-icing
再次加工 reworking
再次甲基化 remethylation
再次检测 revision survey
再次检查 reinspection
再次校测 revision survey
再次接入 reclosure; re-engagement
再次接通 reclose
再次精选 secondary cleaning
再次浚挖 redredge
再次刻纹 regrooving
再次刻纹机 regroover
再次肯定的 reaffirmed
再次冷却 aftercooling
再次离心 recentrifuge
再次离子化 reignition
再次硫化 recure; revulcanization
再次蔓延 reinfestation
再次培养 subculture
再次起动 <又称再次启动> restart-(ing); re-engagement; relight(ing)
再次起动能力 restarting capability
再次强调 reemphasizing
再次侵染 secondary infection
再次确认 reaffirming
再次燃烧 after-combustion
再次筛分 re-screen(ing)
再次声明 restate
再次试飞 reflight
再次挑顶卧底 rebrushing
再次投标 rebid
再次研磨 regrind(ing)
再次氧化 reoxidation
再次应答 secondary response
再次增稠 rethickening
再次涨潮 turn of the tide
再次褶皱 posthumous fold(ing)
再次转让许可证 sublicense[sublicence]
再次装船 reship
再次装料 recharge
再存取 reaccess
再打混凝土桩 spud pile
再打桩 repiling
再贷款 re-lending
再导杆 reconducting lever
再登记 re-entry
再抵押 re-hypothecate; re-hypothecation; submortgage
再点火 relight(ing); restrike; restrike of arc
再电解 re-electrolysis
再电离 reionize
再调查 recheck
再钉 renail
再订购点 reordering point
再订货 duplicate order; reorder
再订货点 reorder point
再订货量 reorder level
再定比例 reproportion
再定比率 reproportion
再定时 retiming
再定位 refixation; relocatability; relocation; reorientation; repositioning; reregister; reregistration
再定位二进制程序 relocation binary program(me)
再定位装配程序 relocating loader
再定位字典 relocation dictionary
再定线 relocation
再定义 redefine; redefinition
再定中心标定系统 recentering calibration system
再定中心机构 recentering mechanism
再冻 refreeze; regelation
再冻结 refreezing
再度沉淀 reprecipitation
再度分解 secondary decomposition
再度过滤的油 refiltered oil
再度熔化 remolten
再度盐(碱)化 re-salinization
再度硬化条纹 rehardened streak
再度运行 rerunning
再度蒸馏 rerun
再对位射极 reregistered emitter
再对准 reregister
再发 reappear; recurrence; relapse
再发的 recurrent
再发反应 recurrent reaction
再发率 recurrence rate
再发盘 repeat offer
再发射 re-emission
再发射能力 refire capability
再发生 recur; recurrence; reprise
再发危险率 recurrence risk rate
再发现 rediscover(y)
再发行 republish
再发性 recidivity
再发展 redevelopment
再翻挖 rescarify
再反射 reecho
再返 re-entering
再返的 reentrant
再犯 recommit
再放大 re-amplification; re-amplify-(ing)
再放电 reload(ing)
再放养 repopulation; restocking

再放映 replay
再沸 reboil
再沸比 reboil ratio
再沸炉 reboiler furnace
再沸器 reboiler
再沸器循环泵 reboiler circulation pump
再分 subdivide; subdividing; subdivision
再分保 retrocession
再分布 redistribute
再分布挡板 redistribution baffle
再分布桁架 subdivided truss
再分布器 redistributor
再分布效应 redistribution effect
再分布作用 redistribution
再分导线 subconductor
再分等 reclassification
再分杆 intermediate diagonal; subdivided member
再分格桁架 truss with sub-divided panels
再分隔 repartition
再分隔断 subdividing partition
再分隔墙 subdividing partition
再分桁格 subdivided panel
再分桁格桁架 subdivided panel truss
再分桁架系 inverted triangular truss
再分化 redifferentiation
再分级 regrade
再分节间 subdivided panel
再分节间桁架 truss with sub-divided panels
再分界 subdivision
再分类 re-assort; reclassification; regrading; resorting
再分离 reselection
再分派 reassign
再分配 reapportion; reassignment; redistribute; redistribution; redivide; relocation; repartition
再分配记录 redistribution writing
再分普蒂桁架 Pettit truss
再分区 resubdivision
再分区段 subdividing partition
再分三角桁架 subdivided triangular truss
再分散法 redispersion
再分散性 redispersibility
再分散作用 redispersion
再分式桁架 double decomposition web member truss; subdivided truss
再分式桁架桥 subdivided truss bridge; truss bridge with subdivided panels
再分式华伦桁架 subdivided Warren truss
再粉碎程度 regrinding degree
再封(死) reseal(ing)
再敷面剂 resurfacing agent
再俘获 recapture
再浮选 cleaner flo(a)tation; cleaning flo(a)tation; reflo(a)tation; secondary flo(a)tation
再浮选精矿 refloated concentrate
再辐射 reradiate; reradiation
再辐射功率 reradiation power
再辐射面 reradiating surface
再辐射误差 reradiation error
再辐射系数 reradiation factor
再腐化室 secondary digestion chamber
再附聚 reagglomeration
再附着 reattachment
再附着流 reattachment flow
再复性 reproducibility
再复以 resheet
再复制 reduplicate
再复制能力 reproducibility
再赋值 reassignment
再钙化时间试验 recalcification time test
再钙化作用 recalcification
再割机 rebreaker
再割区 rebreaking zone
再耕 secondary tillage
再供给 refurnish
再供给煤 recoal
再巩固 reconsolidate
再勾缝 tuck pointing
再构成技术 restructuring technique
再构造 reframe
再购回 repurchase
再估计 restudy
再估价 reappraisal; reappraise; reassess(ment); revaluate; revaluation; revalue
再固化 recuring
再固结 reconsolidation
再关闭 reclosing
再灌充 refill
再灌注 replenish
再归一化 renormalization
再归一化常数 renormalization constant
再规定 reenact
再过滤 refilter; refiltration
再过热 resuperheat(ing)
再过热器 resuperheater
再过筛 rescreen
再夯实 retamp
再合并 reunion
再合成 resynthesis
再核对 recheck
再核算 recalculation
再烘干 redry
再呼叫信号 rering signal
再化合 recombination
再划分 redivide
再环流(润滑)油 recirculating oil
再恢复模拟图像 reconverted analog picture
再挥发 resublimation
再辉 recalesce; recalescence; reglowing
再辉点 point of recalescence; recalescence point
再回火 retemper
再回声 reecho
再回收 recycle
再回音 reecho
再汇编封锁 reassembly lock-up
再汇合支流 <指重新汇入主流的江河支流> anabranch
再会聚的 reconvergent
再绘制 redesign
再混合 decomposite; decompound; remix; remixing
再混合池 recycling tub
再活化 reactivate; revivify
再活化管路 reactivation line
再活化作用 reactivation; remobilization
再击穿电压上升率 rate of rise restriking voltage
再积黄土 redeposited loess; secondary loess
再激活 reactivate; reactivation
再汲出 re-evacuation
再汲取 repumping
再集中 recentralizing
再计算 recalculation
再计算视极移曲线 recalculated A.P.W.[apparent polar wave] curve
再记入 re-entry
再加倍 redouble
再加冰 re-icing
再加工 remachine; remachining; reprocessing; retooling; retreat(ing); retreatment; rework; secondary conversion
再加工宝石 reconstituted stone
再加工材 shop timber
再加工厂 reprocessing plant
再加工车间 mill
再加工成本 cost of further processing; reprocessing cost
再加工木材 remanufactured lumber
再加工塑料 rework plastics
再加工系统 retreating system
再加工循环 reprocessing cycle
再加工装置 refabrication plant
再加荷模量 reload(ing) modulus
再加荷模型 reload model
再加荷曲线 reloading curve
再加价 additional mark-up
再加料 refill; replenish
再加强 reconsolidate
再加热 reheat(ing); resuperheating; saddening
再加热拉伸型拉吹模塑机 reheat stretch-type stretch-blow mo(u)lding machine
再加热炉 reheating furnace
再加热膨胀 reheating expansion
再加热收缩 after-contraction; reheating shrinkage
再加热调节器 reheating regulator
再加热系统 reheat(ing) system
再加入 rejoin
再加速 reaccelerate; reacceleration
再加压 repress(ing); repressuring
再加油 regas; reoil
再加载 reload(ing)
再加载循环 reload cycle
再夹带 reentrainment
再夹紧 rechuck
再夹(住) reclamp
再剪切 re-shearing
再检查 recheck; reexamine; reinspection
再检查油位 recheck oil level
再检校 recontrol
再检验 recheck; reexamine; reinspection
再检阅 reinspection
再碱化作用 realkalization
再建 rebuild; reconstruct; reconstruction
再建房屋 rehousing
再建价 replacement value
再鉴定 reappraisal; reassessment
再浆化槽 repulper
再浆机 repulpator
再交换 reexchange
再浇 repour
再浇封层 reseal(ing)
再浇油 reoiling
再胶结岩石 recomposed rock
再胶溶 repeptization
再校准 recalibrate; recalibration
再接合 rejoin; rejointing; relatching
再接通 reclosing
再结冰 refreeze
再结合 reunion
再结合云母 re-integrated mica
再结晶 recrystallize
再结晶边界 recrystallization boundary
再结晶层 recrystallized layer
再结晶的 recrystallized
再结晶钢 reacting steel
再结晶构造 recrystallized structure
再结晶焊接 recrystallization welding
再结晶结构 recrystallization texture
再结晶晶粒 recrystal grain
再结晶孪晶 recrystallization twin
再结晶区 recrystallization section; recrystallization zone; recrystallized zone; refined zone
再结晶热处理 recrystallizing heat treatment
再结晶石灰岩 recrystallized limestone
再结晶石墨 recrystallized graphite
再结晶石英岩 recrystallized quartzite
再结晶退火 full annealing; process annealing; recrystallization annealing
再结晶钨丝 recrystallized tungsten wire
再结晶性质 recrystallization behavio(u)r
再结晶雪 recrystallized snow
再结晶阈值 recrystallized threshold
再结晶织构 recrystallization texture
再结晶组织 recrystallization texture
再结晶作用 recrystallization
再结块 reagglomeration
再结雪壳 sun crust
再进口(货物) re-import(ation)
再进入 reenter; re-entry
再进入的 reentrant
再精选 recleaner flo(a)tation; recleaning; reconcentration; reselection
再精制 rerefining; retreat(ing); retreatment
再净化 repurification
再就业服务中心 re-employment service center[centre]
再就业工程 the project of making some people unemployed obtain employment again
再锯 resaw
再锯材 resawn lumber
再锯车间 resaw mill
再锯机 resaw
再聚集剂 reaggregation agent
再聚焦旋转光束 refocused revolving beam
再卷装置 rewinder
再均夷作用 regradation
再开动 rerun
再开发 redevelopment
再开发区 redevelopment district
再开始 re-open; resumption
再勘测 resurvey
再勘察 resurvey
再看 review
再刻石 reburring; reconditioning of stone
再矿化作用 remineralization
再扩散 rescatter
再拉 redraw
再拉拔 redrawing
再拉条 redraw rod
再拉延试验 redrawing test
再老化 reag(e)ing
再冷 aftercooling
再冷冻 refreezing
再冷凝器 after-condenser; subcooling condenser; recondenser
再冷却 aftercooling; recool(ing)
再冷却器 after-cooler; subcooler
再冷设备 recooling plant
再冷液体 subcooling liquid
再冷轧 double cold reduction
再冷轧用轧机 double cold reduction mill
再冷装置 recooling plant
再立 re-erect
再利用 recycle; reutilization
再利用水 reuse(d) water
再连接 reconnection; relinking
再联锁 re-engagement; re-interlocking
再裂化 recracking
再流平 reflow
再流平涂料 reflow coating
再流入主流的支流 anabranch
再流行 resurrection
再硫化 resulfurize; resulphurize; re-

Z

vulcanize
再卤化 rehalogen(iz)ation
再氯化作用 rechlorination
再镘 resmoothing
再弥散 redispersion
再密封 reseal(ing)
再密封存水弯 resealing trap
再密封压力 resealing pressure
再命名 redenomination
再磨 regrind(ing)
再磨部分 regrinding section
再磨光 repolish
再磨合 reburnishing
再磨回路 regrinding circuit
再磨机 regrinding unit; regrinding mill;retreatment mill
再磨快 resharpen
再磨流程 regrinding circuit
再磨锐 resharpen
再磨物料 reground material
再逆 another mistake(in treatment)
再啮合 re-engagement
再凝 regelation
再凝结 after precipitation; resolidification
再浓缩 reconcentration;reenrichment
再耦合 recoupling
再排序点 restart sorting point
再抛光 repolish
再配合 deuterogamy
再配平 retrimming
再配准 reregister
再喷入 reinjection
再膨胀 re-expand;re-expansion
再匹配 rematch
再平层 relevel(l)ing
再平衡 redress;retrim
再平展 resetter-out
再平整 relevel(l)ing
再评价 reassess
再破碎 recrushing
再铺钢轨 relayed rail
再铺砂 resanding
再曝气 re-aerate;reaeration
再曝气常数 reaeration constant
再曝气池 reaeration tank
再曝气率 reaeration rate
再曝气试验 reaeration test
再曝气污泥 reaeration sludge
再曝气系数 reoxygenation coefficient
再崎岖作用 recragging
再启动 relight(ing);rerun
再启动程序 restarting procedure; restarting routine
再启动地址 restart address
再启动点 restart point
再启动过程 restart procedure
再启动检查点 restart checkpoint
再启动控制 restart control
再启动命令 restart instruction
再启动能力 restart capability
再启动时钟 delta clock
再启动条件 restart condition
再启动指令 restart instruction
再起 resurrection
再起动 restart
再起动按钮 reset button
再起动点 restart point
再起动能力 refire capability; restart capability
再起弧 reignition of arc
再起弧电压 reignition voltage; restriking voltage
再起用 resurrect
再气化 regasification;regasify
再汽化 re-evapo(u)ration; re-vapo(u)rization
再迁移 remigration
再签署 resign
再切削 re-cut
再求补 recomplementation
再区分 redivide
再取回 resumption
再取样 resampling
再取样法 resample method
再确订(班机座位等)reconfirm
再确认 reassurance;reconfirmation
再燃 exacerbation; reburn(ing); recrudescence
再燃烧 secondary combustion
再燃烧室 reburning chamber
再绕 recoil(ing);rewind
再绕机 respooling machine
再热 reheat
再热备用气门 reheat emergency valve
再热槽 reheating bath
再热锅炉 reheat boiler
再热回热式装置 reheater-regenerative plant
再热回热循环 reheating regenerative cycle
再热回热蒸汽循环 reheat regenerative steam cycle
再热级 reheating stage
再热截流阀 reheating interceptor valve
再热联合气门 combined reheat stop and intercept valve; combined reheat valve
再热裂纹 reheating crack(ing); stress-relief crack(ing)
再热炉 reheater; reheating oven; reheating furnace
再热凝汽式汽轮机 reheat condensing turbine
再热盘管 reheat(ing)coil
再热气门 reheat stop valve
再热气循环 reheating vapo(u)r cycle
再热汽门 reheat emergency valve
再热汽压 reheat pressure
再热器 interheater; intermediate superheater;reheater
再热燃烧室 reheat combustion chamber
再热式发动机 reheat engine
再热式空气调节系统 reheat air conditioning system
再热式汽轮机 reheat turbine
再热式燃气轮机 reheat gas turbine
再热试验 reheating test
再热收缩 reheat shrinkage
再热调节阀 reheat control valve
再热系数 reheat factor
再热型 reheat-type
再热循环 reheat(ing)cycle
再热循环燃气轮机 reheat cycle gas turbine
再热压 hot repress(ing)
再热再生循环 reheating and regenerative cycle
再热蒸汽 reheat(ing)steam
再热蒸汽引入点 reheat-return point
再热装置 reheat machine; reheat plant
再溶解 redissolution;redissolve
再溶解接着法 resoluble method
再熔法 remelt process
再熔坩埚 remelt crucible
再熔工艺 remelting technology
再熔合金 remelted alloy; remelting alloy;secondary alloy
再熔化 refusion;remelt(ing)
再熔技术 remelt technique
再熔结 remelt junction
再熔炉 remelter;remelting furnace
再熔器 remelter
再熔铁 remelt(ed)iron
再熔锌 remelt zinc
再熔作用 anatexis
再乳化的 reemulsified
再乳化作用 reemulsification
再入 re-entry
再入大气层试验设备 re-entry test facility
再入点 re-entry point
再入定位 re-entry positioning
再入辅助设备 re-entry aid
再入观察预报与试验 re-entry observable prediction and experiments
再入轨道 re-entry trajectory
再入弧长引信 re-entry arc length fuze
再入环境 re-entry environment
再入环境与系统工艺 re-entry environment and system technology
再入回 re-entry recovery system
再入级 reclassification
再入假目标 re-entry decoy
再入角 re-entry angle
再入控制 re-entry control
再入烧蚀 re-entry ablation
再入式回路 reentrant loop
再入式气体冷却 reentrant gas cooling
再入式谐振腔 reentrant cavity
再入速度 re-entry velocity
再入体 re-entry body
再入条件 re-entry condition
再入通过区 re-entry window
再入头锥 re-entry nose cone
再入系统环境防护 re-entry system environmental protection
再入钻探 re-entry drilling
再润滑 re-lubrication
再润湿 rewetting
再润湿剂 rewetting agent
再润湿能力 rewetting ability
再散列过程 rehashing procedure
再筛 re-screener
再筛分 re-screen(ing)
再闪击 restriking
再闪击电压 restriking voltage
再上油 reoil
再烧 after burning;reburn
再烧过程 reburning
再烧结 post sinter
再设计 redesign;re-engineering
再摄取 reuptake
再摄影 retake
再审定(资格)requalification
再渗滤作用 repercolation
再渗入<地下水>repercolation
再渗碳 recarbur(iz)ation; skin recovery
再渗透 re-impregnating;repercolation
再升华 resublimation
再升华碘 resublimed iodine
再生 after-growth; breed; reactivate; recuperation; refining; reforming; rejuvenate; renaissance; reprocessing; reproduction; reshaping; resurgence; retroaction; rewrite; resource recovery
再生白土 sweet clay
再生帮电机 regenerative repeater
再生保养 preventive maintenance; refurnishment
再生泵 regenerative pump; turbine pump
再生比 regenerative level
再生冰川 recemented glacier; regenerated glacier
再生部分 regenerating section
再生材料 recycled material; breeder material; reclaimed material; regrown material; reprocessed material; secondary material; twice-laid stuff
再生草 after-grass;aftermath;silage after math
再生草地 aftermath pasture
再生层控矿床 palingenetic strata bound deposit
再生产 reproducibility;reproduction
再生产成本 cost of reproduction; reproduction cost
再生产的 reproductive
再生产费用 cost of reproduction; expense of reproduction; reproduction cost
再生产过程 process of reproduction; reproduction process
再生产基金 reproduction fund
再生产价值 reproducible value; reproduction value
再生产率 rate of reproduction; reproduction rate
再生产能力 reproducibility
再生产投资 reproductive investment of capital
再生产系数 coefficient of reproduction; reproduction coefficient
再生产循环 cycle of reproduction
再生产周期 cycle of reproduction
再生长 regrowth
再生常数 reproduction constant
再生成矿说 palingenetic ore-forming theory
再生程度 regeneration level
再生程序 reproducer;rerun routine
再生池 reaeration tank
再生处理 reconditioning; regeneration treatment
再生触发电路 regenerative trigger circuit
再生床净化器 regenerative bed demineralizer
再生磁道 regenerative track
再生淬火 regenerative quenching
再生存储器 regeneration memory; regenerative memory; regenerative storage;regenerative store
再生蛋白质纤维 azlon; regenerated protein fiber[fibre]
再生的 recurrent; regenerant; regenerate;regenerative
再生的橡胶溶液 reclaimed rubber solution
再生地槽 revive geosyncline
再生地槽阶段 rejuvenation stage of geosyncline
再生地沥青路面 reclaimed asphalt pavement; recycled asphalt pavement
再生点 regeneration point
再生电池 regenerated cell; regenerative cell
再生电路 regenerating circuit; regeneration circuit; regenerative circuit; regenerative feedback loop; regenerative loop;regenerator
再生电能 regenerative power energy
再生电能吸收装置 devices for absorbing regenerative electric(al) energy
再生电位 potential of regeneration
再生电压 regenerative voltage
再生电阻制动 rheostatic braking
再生顶板 mat
再生段 regenerator section
再生段开销 regenerator section overhead
再生段终端 regenerator section termination
再生断层崖 rejuvenated fault scarp
再生发动机 regenerative motor
再生阀 regenerating valve
再生反馈 regenerative feedback
再生反射器 regenerative reflector

再生反应堆 regenerative reactor
再生范围 reproducibility range
再生方法 reclaiming procedure; renovation process
再生放大 regenerative amplification
再生放大器 reproducing amplifier
再生分频器 regenerative divider
再生复激电动机 regenerative compound motor
再生干扰 regenerative impediment
再生革 reclaimed leather
再生跟踪 regenerative tracking
再生工厂 recovery plant
再生工艺 regeneration technology
再生骨料混凝土 recycled-aggregate concrete
再生锅炉 recovery boiler
再生过程 reactivation process; regeneration
再生函数 renewal function
再生合金 secondary alloy
再生河 palingenetic river; palingenetic stream
再生花岗岩 reconstructed granite
再生环路 regenerative loop
再生换热器 regenerator
再生黄土 secondary loess
再生回热式燃气轮机 regenerative cycle gas turbine engine
再生回授 regenerative feedback
再生回用 regeneration reuse
再生混合料 recycled mixture
再生混凝土 recycled concrete
再生活性炭 regenerating activated carbon
再生激光放大器 regenerative laser amplifier
再生级联 stripping cascade
再生集料 reclaimed aggregate material
再生集料混凝土 recycled-aggregate concrete
再生挤出系统 reclaim extrusion system
再生计数器 regeneration counter
再生技术 regenerative technology
再生剂 reclamite; recycling agent; regenerant; regenerative agent; regenerator; rejuvenating agent; rejuvenator
再生剂耗量 chemical consumption; regenerant consumption
再生剂计量 chemical measurement; regenerant measurement; regenerant metering
再生剂量 regenerant level
再生加工用水 recreational water
再生加热 regenerative heating
再生检波 regenerative detection
再生检波器 regenerative detector
再生碱土 regraded alkali soil
再生胶 reclaim; recuperated rubber; regenerated rubber; rejuvenated rubber; restored rubber
再生胶改性沥青 reclaimed rubber modified asphalt
再生胶胶料 reclaim mix
再生胶沥青防水涂料 asphalt reclaimed rubber waterproofing paint
再生胶油毡 asphalt reclaimed rubber roofing
再生接收机 retroactive receiver
再生结节 regenerated nodule
再生介质 regenerating medium
再生金属 secondary metal
再生晶体 regenerated crustal
再生聚合物 regenerated polymer
再生控制 regeneration control
再生控制极 generative gate
再生矿床 regenerated deposit; rejuvenated ore deposit
再生冷却 regenerative cooling
再生冷却燃烧室 regenerative cooled combustion chamber
再生冷却系统 regenerative cooling system
再生利用 recycling; reutilization
再生利用率 rate of reutilization; recycling level
再生利用水平 recycling level
再生沥青混合料 reclaimed asphalt mixture; recycling bituminous mixture
再生沥青路面 reclaimed bitumen pavement; reclaimed bituminous pavement; recycled bituminous pavement
再生沥青路面材料 reclaimed asphalt pavement material
再生沥青铺面 reclaimed bitumen pavement
再生连接 regeneration connection
再生联锁阀 regenerative interlock valve
再生链路 regeneration link
再生林 reforestation; restocking; secondary forest; secondary forest growth; secondary woody growth
再生炉 regenerating furnace; regeneration furnace; regenerative furnace; regenerator; regenerator kiln
再生炉中燃烧 combustion in the regeneration furnace
再生路面 reclaimed bituminous pavement
再生铝 secondary alumin(i)um
再生律 regeneration law
再生率 reproducibility
再生脉冲 regeneration pulse
再生模拟汇合控制器 second-generation analog merging controller
再生木材 recycled wood
再生能力 power of regeneration; regeneration capacity; regenerative capacity
再生能量 regenerated energy
再生能量储存 regenerative energy storage
再生能量储存系统 regenerative energy storage system
再生能源 renewable energy resources; renewable sources of energy
再生泥浆 reconditioned mud
再生年龄 resetting age
再生农业 regenerative agriculture
再生耦合 regenerative coupling
再生喷射器 restarting injector
再生偏转 regenerative deflection
再生偏转器 regenerative deflector
再生频率 recurring frequency
再生曝气(法) regeneration aeration
再生曝气试验 reaeration test
再生气 resurgent gas
再生气泡 blister after refining; reboil bubble; reboiling
再生气泡倾向 tendency to reboil
再生气泡效应 reboil effect
再生气体 regeneration gas
再生器 actifier; reactivator; regenerator; rejuvenator; reproducer; revivifier
再生器定时发生器 regenerator timing generator
再生铅 reviver; secondary lead
再生区 blanket; regeneration field; regeneration zone; regenerative zone; regrown region
再生区功率 blanket power
再生区热交换器 blanket heat exchanger
再生区转换比 blanker conversion ration
再生燃料 generative fuel; regenerated fuel; regenerative fuel
再生燃料电池 regenerative fuel cell
再生燃料电池系统 regenerative fuel cell system
再生热拌混合料 recycling hot mix
再生热拌沥青 recycled hot mix asphalt
再生容量 regenerated capacity
再生溶液 actified solution; regenerative solution; rework solution
再生乳状液 reconstituted emulsion
再生润滑油 reclaimed lubricating oil; rerefined oil; recovered oil
再生砂 reclaimed sand; reclamation sand; reconditioned (moulding) sand; system sand
再生栅极检波电路 regenerative grid detection circuit
再生设备 recreational facility; regeneration facility
再生设施 recreational facility; regeneration facility
再生绳 rhumbow line; twice-laid rope
再生石膏 reclaimed gypsum
再生时的速度 speed in regeneration
再生时间 recovery time; regeneration time
再生式发动机 regenerative engine
再生式方波放大器 regenerative squaring amplifier
再生式放大器 regenerative amplifier
再生式分频器 regenerative frequency divider
再生式过程 regenerative process
再生式后燃器 regenerative afterburner
再生式滑动转向 regenerative skid steering
再生式换热器 regenerative heat exchanger
再生式加热 recuperative heating
再生式检波器 regenerative detector
再生式交流换热器 regenerative regenerator
再生式接收法 regenerative reception; retroactive reception
再生式接收机 regenerative receiver; regenerative receiving set
再生式开关 regenerative switch
再生式空气预热器 regenerative air heater; regenerative air preheater
再生式链路 regenerative link
再生式滤材 renewable filter media
再生式面罩 rebreather-type mask
再生式汽缸 regenerative cylinder
再生式燃气轮机 regenerative gas turbine
再生式热风机 recuperative heater; regenerative air heater
再生式热交换器 recuperative heat exchanger; regenerately heat exchanger; regenerative heat exchanger
再生式生命保障系统 regenerative life support system
再生式调制器 regenerative modulator
再生式图像增强器 regenerative image intensifier
再生式涡轮螺桨发动机 regenerative turboprop engine
再生式氧气设备 regenerating oxygen apparatus
再生式氧气系统 rebreather
再生式液体燃料火箭 regenerative liquid-fuel rocket
再生式中继机 regenerative repeater
再生试验 regeneration test
再生手柄 regeneration handle
再生水 reclaimed water; regenerated water; regenerative water; rejuvenated water; renovated water; resurgent water; reuse(d) water
再生水回用 reclaimed water reuse
再生水泥混凝土 reclaimed concrete
再生水平 regeneration level
再生水系 palingenetic drainage; regenerated drainage
再生松香 reclaimed rosin
再生速率 regeneration rate
再生塑料 reprocessed plastics
再生酸 recovered acid; restored acid
再生酸储罐 regenerated acid storage tank
再生塔 actifier column; regenerating column; regeneration column; regeneration tower
再生塔回流泵 regenerator reflux pump
再生塔回流槽 regenerator reflux drum
再生炭 regenerated carbon
再生碳粒 granular regenerable carbon
再生陶土 regenerated clay
再生特征 palingenetic character
再生添加剂 recycling additive; regenerating additive; regenerating agent
再生添加物 recycling additive
再生同位素 fertile isotope; regenerative isotopes
再生铜 reclaimed copper; secondary copper
再生温度 reactivating temperature; regeneration temperature
再生涡轮泵 regenerative-turbine pump
再生污泥 regeneration sludge
再生污染质 secondary pollutant
再生污水 reclaimed effluent; reclaimed sewage; reclaimed wastewater; regeneration wastewater
再生物 regenerant
再生物质 regenerated material
再生系数 conversion factor; efficiency of regeneration; gain factor
再生系统 recycling system; regeneration system; regenerative system
再生纤维 regenerated fibre[fiber]
再生纤维素 regenerated cellulose
再生纤维素纤维 regenerated cellulose fibre[fiber]
再生现象 regeneration phenomenon
再生线圈 regenerative coil; tickler
再生限幅器 regenerative clipper
再生橡胶 regenerated rubber; reclaimed rubber
再生橡胶粉 reclaimed rubber dust
再生橡胶轮胎 reclaimed rubber tire [tyre]
再生消色器 regenerative colo(u)r killer
再生效果 regeneration effect
再生效率 effectiveness of regenerator; efficiency of regeneration; regeneration effect; regeneration efficiency; regenerative efficiency
再生信号 regenerated signal; regeneration signal
再生型断层 rejuvenated fault
再生型山区 rejuvenated mountain land
再生性 reactivity; reproducibility
再生修理 preventive maintenance; refurnishment
再生循环 cycle of regeneration; regeneration cycle; regenerative cycle; reprocessing cycle
再生岩 deiterosomatic rock; regenerated rock
再生盐土 regraded saline soil
再生页理 refoliation

再生液置换 displacement slow rinse; rinse displacement
再生异常 regenerated anomaly
再生因子 regeneration factor
再生引出 regenerative extraction
再生硬度 tempering hardness
再生油 reclaimed oil;refiltered oil
再生有色金属 secondary nonferrous metal
再生鱼尾板 reformed bar
再生圆偏振 circular repolarization
再生运算放大器 regenerative operational amplifier
再生杂草 weed re-growth
再生造纸废水 reproduced paper-making wastewater
再生噪声 regenerated noise; regenerative noise
再生振荡器 regeneration oscillator
再生织物 recycled fabric
再生植被 revegetation
再生殖能力 reproducibility
再生纸板 chipboard
再生纸浆废水 wastewater from paper pulping
再生制动 recuperative brake
再生制动控制系统 regenerative braking control system
再生制动器 regeneration brake; regenerative brake
再生制动器切断开关 regenerative brake cut-off switch
再生制动时的速度 speed in regeneration
再生制动性能 regenerative braking performance
再生制动作用 over-synchronous braking;regenerative braking
再生中继 regeneration and repetition
再生中继段 regenerative section
再生中继器 regenerator
再生重介质 recovered dense medium;regenerated dense medium
再生周期 regenerate cycle; regeneration period
再生助剂 reclaiming aid
再生转发器 regenerative repeater
再生转换堆 regenerative converter
再生装置 reclaimer;reconditioner;regeneration unit; regenerative apparatus; regenerative device; regenerator
再生资源 renewable resources
再生资源保护 conservation of recreation resources
再生自然资源发展研究所 Institute for the Development of Renewable Natural Resources
再生作用 actification; reclaiming action; rectification; regenerative action; regeneration; revivification; palingenesis【地】
再生作用方式 palingenesis way
再湿性胶粘剂 remoistening adhesive
再湿性黏[粘]合剂 remoistening adhesive
再使用 reapply;reuse
再使用的 re-operational
再使用工业废水 reusing industrial effluent
再试 retry
再试验 re-experiment; retest; retrial; revision test
再收敛节点 reconvergent node
再收敛扇出通路 reconverging fan-out path
再收入 resorb
再收缩 after-contraction;re-shrinkage
再售价格 resale price
再输出 reexport;re-exportation
再输入 reimportation
再输入免税单 bill of stores
再输注 reinfusion
再数 recount;renumber
再刷新 refurbish
再水合作用 rehydration
再水化 rehydration
再顺次成河 resequent-subsequent stream
再顺的 resequent
再顺断线崖 resequent fault line scarp
再顺谷 resequent valley
再顺河 resequent stream
再顺后成河 resequent-subsequent river
再顺水系 reconsequent drainage; resequent drainage
再顺向河 reversional consequent; resequent river;resequent stream
再塑造 reclaiming by plastification
再索赔 re-hypothecation
再坍缩 recollapse
再摊派 reassess
再碳化 recarbonization; recarbonize; recarbur(iz)ation
再碳化剂 recarburizer
再碳酸化料液 recarbonation feed
再碳酸化塔 recarbonation tower
再碳酸化作用 recarbonation
再镗 reboring
再镗孔 rebore
再提出(议案等)recommit
再提炼系统 retreating system
再填 refill
再填充 backfilling;refilling
再填缝 rejoint(ing)
再调附件 reset attachment
再调和 reblending
再调节 reaccommodation
再调节水库 reregulating reservoir
再调平 retrim(ming)
再调速度 speed of reset
再调整 overcorrection; over-travel; readjust(ment);retrim
再调制 remodulation
再调制器 converter;remodulator
再贴现 rediscount
再贴现率 rediscount rate
再贴现票据 bills rediscounted; rediscounted bill
再通过 rethread
再同步 resynchronization
再投入 reclosing; re-engage; re-engagement
再投资 reinvest;reinvestment
再投资的收益 reinvested earning
再投资利润 plow back earnings; reinvested profit
再投资利润的特殊免税 special exemption of reinvested profits
再投资率 reinvestment rate
再投资用的利润 plough back profit
再涂 recoat;repaste
再涂附着力 recoat adhesion
再涂干时间 dry-to-recoat time
再涂抗力 recoatability
再涂时间 recoat time
再涂试验 recoating test
再涂性 overcoatability;recoatability
再退火 reanneal
再脱位 redislocation;reluxation
再挖 re-cut
再委派 reassignment
再委托 recommit
再稳定作用 restabilization
再污染 re-contamination; re-pollution;resoiling
再吸附 readsorption
再吸气 re-inspiration
再吸入 resorption
再吸收 reabsorption; resorb; resorption
再吸收成本 reabsorbed cost
再吸收率 reabsorption rate
再吸收式制冷系统 resorption refrigerating system
再吸收系统 resorption system
再吸作用 resorption
再稀释 redilution
再洗槽 re-wash launder
再洗跳汰机 re-wash box; secondary washbox
再洗循环 re-wash recirculation
再洗溢流 re-wash overflow
再显危险率 recurrence risk
再显影 redevelopment
再现 playback; reappear; reconstruction; recurrence; rendition; repetition; representation; reproduction; resurgence
再现波前 reconstructed wave front
再现当量 repetition equivalent
再现的 reproducing
再现定律 reproductive law
再现度 degree of reproducibility
再现范围 reproducibility range
再现光束 reconstructing beam; reconstruction beam
再现交通事故 reconstructing traffic accident
再现精度 playback accuracy
再现亮度 reconstruction brightness
再现率 recall factor
再现器 reconstructor
再现设备 reproducer
再现时间 recovery time;return period
再现水 resurgent water
再现速度 playback speed
再现特性 reproducing characteristic
再现图像 reproduced image
再现系统 reconstruction system
再现性 duplicability; reduceability; repeatability;reproducibility
再现性精密度 multilaboratory precision
再现耀斑 recurrent flare
再现装置 reproducer;transcriber
再现状态 playback mode
再消化室 secondary digestion chamber
再新设备 re-equipment
再行称量 re-weigh
再行衡量 re-weigh
再行增订 additional subscription
再修机车 locomotive under repair
再修饰 refinishing
再修整 refinish
再絮凝 reflocculation
再悬 resuspension
再悬浮 resuspending
再悬浮作用 resuspension
再悬移作用 resuspension
再选 reconcentration
再选浮选槽 retreatment cell; secondary cell
再选给料 feed for retreatment
再选机 recleaner
再选水力旋流器 secondary hydrocyclone
再选跳汰机 re-wash jig; secondary washbox
再选重介质旋流器系统 secondary heavy medium cyclone circuit
再寻址 readdressing
再驯化 reacclimatization
再循环 recirculate; recirculating; recirculation; recirculatory; recycle; recycling
再循环泵 recirculating pump
再循环比 recirculation ratio; recycle ratio
再循环标准 recycling criterion
再循环程序 recycling program(me)
再循环程序区 recirculating loop
再循环池 recirculating pool
再循环的有机质 recycled organic matter
再循环阀 recycle valve;recycling valve
再循环风 recirculated air;reduced air
再循环风道 recirculating air duct
再循环风机 recirculating fan
再循环供暖机组 draw-through heater
再循环管 recirculating pipe; recirculating tube; recirculation pipe; recirculator;return pipe;return tube
再循环管路 recirculating line
再循环管线 recirculating(pipe)line
再循环空气 recirculated air; recirculating air;recirculation air
再循环冷却器 recycle cooler
再循环冷却水 recirculating cooling water;recirculation cooling water
再循环冷却系统 recirculated cooling system
再循环流量 recirculating mass
再循环暖气供暖 heating with recirculated air
再循环气体混合法 recycled-gas mixing method
再循环器 recirculator
再循环燃料 recycled fuel
再循环(润滑)油 recirculating oil
再循环上升管 return riser
再循环水 recirculating water; recirculation water;recycled water
再循环水分配 recirculating water portion; recirculation water portion;recycled water portion
再循环物料 recycle stock
再循环洗脱 recycling elution
再循环系统 recirculating system; recycle system;recycling system
再循环烟气挡板 recirculating gas damper
再循环养分 recycled nutrient
再压 repress(ing);second pressing
再压力 recompression pressure
再压滤器的进料罐 repulp filter feed tank
再压密 recompact
再压实 recompact; recompaction; reconsideration;reconsolidation
再压缩 recompression
再压缩带 zone of recompression
再压缩模量 recompression modulus
再压缩曲线 recompression curve
再压缩指数 recompression index
再压制 recompaction
再淹没 reinundation
再氧化作用 reoxidation; reoxygenation
再液化 reliquefaction
再议价准备 renegotiation reserves
再引弧 restarting;restarting a weld
再引弧电压 reignition voltage
再引种 repopulation;restocking
再印刷 republication
再应力下锈蚀 restress corrosion
再硬化 reharden(ing);reinduration
再硬结 reinduration
再用 resurrection;reuse;reutilization
再用保险丝 renewable fuse
再用材料 secondary material
再用的旧轨 second-hand relaying rail
再用轨 relaying rail;second-hand rail
再用回花 reworkable waste
再用金属 secondary metal
再用硫处理 resulfurize
再用率 reuse rate
再用试件 reusing sample

Z

再用试样 reusing sample
再用水 backwater
再用水的输送 conveyance of reuse water
再用性 reusability
再油漆 repaint(ing)
再育 proliferation;prolification
再运行 rerun
再运行点 rerun point
再运行检验点 rerun check point
再运移 remigration
再再保险 retrocession
再造 reconstruction; re-establishment; restoration
再造沉积物 reworked deposit
再造林 reafforestation;reforestation
再造铝红壤 lateritite
再造木板 reconstructed wood panel
再造木材 reconstituted wood
再造石 cast stone; precast stone; reconstituted stone; reconstructed stone
再造石材分裂机 splitter for reconstructed stone
再造熟铁棒材 merchant bar
再造图 palinspastic map
再造想象 reproductive imagination
再造岩 authineomorphic rock
再造油 reconstituted oil
再造作用 rebuilding;reworking
再增长 regain
再增溶作用 resolubilization
再增碳剂 recarburizer
再增压 repressurize
再轧 reroll
再轧钢轨 rerolling rail
再轧机 rerolling mill
再轧棉 reginned cotton
再轧坯 rerolling feed
再长 <饲料作物割后的> aftermath; after-growth
再招标 rebid;re-tender
再折叠 redoubling
再折扣 rediscount
再者 postscript
再振捣 revibration
再振动 revibration
再征收 reassessment
再蒸发 re-evapo(u)ration; regasification; re-vapo(u)rization; re-vapo(u)rize
再蒸发器 re-evaporator
再蒸馏 after-fractionating; doubling; redistill;redistillation;rerunning
再蒸馏产率 rerun yield
再蒸馏釜 rerun still
再蒸馏锅 reboiler;rerun(ning) still
再蒸馏后的残油 rerun bottom
再蒸馏设备 rerunning plant
再蒸馏水 redistilled water
再蒸馏塔 rerunning tower
再蒸馏塔塔底产物 rerun tower bottoms
再蒸馏油 rerun oil
再蒸馏装置 rerunning unit
再整顿 rehandling
再整合 reintegration
再证实 reconfirm
再植(被) re-cultivation
再植林木 reforest
再指定 reassignment
再酯化作用 reesterification
再制(备) refabricate; refabrication; reproduce;rework
再制定 reenact
再制动 braking after release; reapplication(of brake); reapply
再制木材 reconstituted wood
再制生铁 refined pig iron; synthetic(al) pig iron
再制试件 reconstituted specimen; reconstituted specimen
再制试样 reconstituted specimen; reconstructed specimen
再制铁 synthetic(al) pig iron
再制造 refabrication
再种 replanting
再重复 reduplication
再周转备料 recycling stock
再周转原料 recycling stock
再主张 reassert
再煮 reboil
再煮旋管 reboiler section
再注入分配 distribution of reinjection
再转变 reconvert
再转换 reconversion
再转置 retransposing
再装 refill;reload
再装船 redelivery;reshipment
再装货 reload
再装料 recharging
再装满 refill(ing);replenish
再装配 reassembling; reassembly; repack(age)
再装填 recharge
再装填器 recharger
再装药周期 reload cycle
再装运 re-expedition
再装载 recharge;reload
再组成 recomposition;reconstitution
再组合 recombination; reconfiguration
再组合车轮 retired[retyred] wheel
再组合花岗岩 recomposed granite
再组合系统 reconfiguration system
再组织 reframe
再钻 re-drill
再作 refashion
再作用 reworking
再作用面层理构造 reactivation surface structure
再做 repeat

在

在A音阶上分贝音量 decibel of sound on an A-scale

在105℃干燥炉中干燥24小时 bone dry
在安装安全阀时 in setting safety valves
在岸金融市场 onshore financial market
在岸上 ashore
在岸上的 onshore
在岸上预制、外海安装的大型部件 offshore module
在岸下风航行 run under the lee of the shore
在暗处抹灰 backplastering
在把柄上装有墨槽或油漆槽的划线刷或漆刷 fountain brush
在搬运中 in hauling
在板平面内受荷的平板 slab loaded in its own plane
在板上可见 to be visible on plate
在板条上抹底层灰泥 pricking up
在办公室值班 <房地产经纪人事务所> floor duty
在半路 midway
在半湿润至干旱范围内的变化 to vary from subhumid to arid
在保险单末尾签署者 underwriter
在背风岸边航行 run under the lee of the shore
在背后 posterior
在背上 piggyback
在壁画上乱涂 graffiti
在边缘地区的 downcountry
在编保养人员 organization maintenance personnel
在编人员 in-service personnel; permanent staff
在变动负载下运行 varying duty
在变形中 undergoing deformation
在表面应力条件下的结构体系 structure system in surface stress condition
在表面展开 surface spreading
在别处未加说明 not elsewhere specified
在波浪上 awave
在玻璃上吹砂 sanded sheet
在泊位区挖泥 dredging in berth area
在泊位上 berthing;on berth
在薄的材料边上加狭条 <起加强作用> lining-up
在不确定情况下制订决策 decision-making under uncertainty
在布拉格赫拉德察尼山上的圣乔治罗马风格巴西利卡 Romanesque basilica of St. George on the Hradcany Hill at Prague
在擦净点上涂底漆 prime in the spots
在材料中的缺陷 defect in materials; discontinuities in materials
在舱内 in hold
在舱外 out of hold
在操作中 in operation;on-the-run
在草地上放牧 run on grassland
在册车数 car registration
在册船舶 listed ship;vessels in service
在册地址 registered land
在册人员 staff on the rolls
在册土地 lots of record
在侧面加固 flank(ing)
在产工人 labo(u)r in process
在产品 goods in process; product in process
在产品分类账 work-in-process ledger
在产品盘存 product (work) in process inventory
在长度方向卷边 bowing in the length dimension
在场经营 business presence
在潮湿冲积层中发现 be found in alluvium
在潮湿的环境中 in a humid environment
在潮湿地区的土壤上 on soil of humid area
在车间涂底漆 shop priming
在车辆中操作 in-vehicle operation
在车辆中控制 in-vehicle operation
在车上 on board
在车站平面层之上(或下)的轨道 track level above(or below) station floor level
在车站自动摘下的客车 <直通运行中的列车> drop coach
在沉船下穿引千斤索 pass lifting slings under a wreck
在承受外压的容器 in vessels subjected to external pressure
在冲刷循环中的青年期海岸 adolescent coast
在出售中 under offer
在初轧机上轧制 roll cogging
在储层内侧向运移 lateral migration in the reservoir
在储层内向上运移 vertical migration in the reservoir
在穿墙裸线外加上套管及绝缘包皮 knob-and-tube wiring
在船舶中段 amidship
在船的正横以后 abaft the beam
在船模前面装置的促进乱流的金属卷线 trip wire
在船内 aboard; ashipboard; inboard; shipboard
在船上 aboard; afloat; on board; on-board ship; on shipboard; shipboard
在船上的锚链节 inboard shot
在船实习契约 indenture
在船首舷处 on the bow
在船头 fore
在船外 overboard
在船尾 abaft;aft;astern;on the quarter
在船位 on the berth
在椽子端再加一个角 bird's-mouth
在垂直面的角变位 derricking
在此阶段的末期 at the end of this period
在次页 overleaf
在大门挑出的防卫阳台 mushrabiya
在大气中停留时间 atmospheric residence time
在大气中运动 atmospheric movement
在大油门下工作 operate at full throttle
在单位时间里 in unit time
在单一条件下的结构体系 structure system in single stress condition
在氮气层之下 under nitrogen blanket
在到达站交付前的费用已付 delivered free to destination
在道岔和站台上撒盐(融雪) salt spread in switches and on platforms
在道岔区进退两难 cornering at points
在道路或山坡上挖掘的水沟 thank-you-madam
在道路上行驶 over-the-road
在等高线上分散 dispersal on contour lines
在等深吃水处 at even keel
在低湿条件下 under low moisture conditions
在地壳集中的元素 lithophile element
在地面标高以上 above ground level
在地面上的 overground
在地面下的截水墙 positive cut-off
在地面以上 above-ground
在地面以下 below day;below ground
在地球上循环 circulation on the earth
在地上 above-ground
在地下 underground
在第二试验点 at site two
在第一层阶地以下的洪泛平原 first bottom
在第一次取样时 at the first sampling
在颠波中的锚泊 hawse full; riding hawse fallen;riding hawse full
在电话电路上叠加电报电路 superimposing of telegraph circuits on telephone circuits
在调查或研究中 in question
在调查中 under investigation
在调车前放出列车制动管中的余气 bleeding of air in the train line
在东北 northeastward
在动的人 mover
在陡峭的地层上施工 steep working
在端梁间的车辆长度 length of car over end sills
在锻件中的缺陷 discontinuities(defect) in forgings
在对流层停留时间 tropospheric lifetime;tropospheric residence time
在多数品种中 inmost of varieties
在恶劣条件下从事繁重劳动 sweat
在恶劣条件下试验 severe duty test
在二硫化碳中的溶解度 solubility in carbon disulphide
在发动机上方的司机室 cab over engine
在发动机之后的司机室 cab-behind-engine

在法定时间内未提出要求 non-claim
在反位的握柄 reversed lever
在方格纸上作图 <缩放用> graticulation
在房地产所有权以外的室外广告牌 off-premises sign
在房地产所有权之内的室外广告牌 on-premises sign
在房屋修建中从事困难工作的酬金 boot money
在飞机上 on board
在峰顶爬上车顶的制动员 car rider
在浮标黏[粘]度计内测定的黏[粘]度 float viscosity
在浮船上钻探 drilling from floating vessels
在负荷下 on-load(ing)
在附近 in the vicinity
在附近的 cephalic
在覆层及所用衬里 in cladding and applied linings
在干处保管 keep in a dry place
在干旱期间 during a drought
在干燥的情况下操作 run dry
在干燥状态下形成的 formed in dry state
在岗(人员) on-the-job
在钢化玻璃上的应力斑 mottled patterns in tempered glass plates
在港耗量 consumption in port
在港及出航时期的保险 at and from insurance
在港记录 port log
在港困守 port bound
在港时间 time-in port; port time
在港外锚地抛锚 anchor in roadstead
在高处 on high
在高精度范围内 within close tolerance
在高水位以上 above high water mark
在各区发生的 allopatric
在各种潮位溯河航行 navigate upstream at all tides
在给定时间间隔内的计数 preset-time counting
在工厂安装的 factory-mounted
在工厂交货 at factory
在工厂上的涂层 factory-applied coating
在工厂修理好的 works-reconditioned
在工厂验收 acceptance at works
在工地的 on-job
在工地铆的铆钉 field rivet(ing)
在工地上 on(construction) site
在工地上作业的机械产品 field product
在工作岗位报到 report at post
在工作工程中 on-stream
在工作台边工作的人 table man
在工作条件下 under operation condition
在工作现场交付 jobsite delivery
在工作中 on-the-line
在工作中的 on-the-job
在工作状态 in running order; in working order
在公差范围内的尺寸 acceptable size
在公共水域或陆地回填材料数量 discharge of fill material
在公共水域排泥量 discharge of dredged material
在公路、航道竖立公里里程碑 kilometrage
在公路以外的 off highway
在公路以外地区使用的起重机 rough-terrain crane
在公文等上附加提要 docket
在构筑物上挖洞 potholing
在古代 in ancient times
在股道中的仪器 in-line meter

在固定时间内播送的 across-the-board
在故障状态的 non-serviceable
在关闭位置 in off position
在关键路线法计划表中先于某个项目的所有作业最早的完成时间 earliest event occurrence time
在关键路线法中最早可能开工日期 early start date
在关键路线法中最早能够完工日期 early finish date
在关栈货物 goods in bond
在观测者眼角处的张角 angle at eye of observer
在观察中 under investigation; under observation
在管道中掺和 line-blending
在管道中的仪器 in-line meter
在管道中进行掺和 in-line blending
在管理上不加以控制的 out of control
在管线内 in-line
在灌浆期 in the watery stage
在规定层上 on-level
在规定的制度下 in regime
在规定地点停车 positioned stop
在规定地点停车装置 prescribed point stopping device
在规定高度上卸载距离 reach at specified height
在规定期限 at staked periods
在规定时间之外 overtime
在轨道尽头的汽车搬运站台 trucking platform around the stub end of track
在轨道上 on the track
在轨道上滑动的道岔 <施工中隧道用> slide points
在轨上滑动的 underhung
在轨枕下填实道砟 packing
在滚子上移动 use rollers
在国际范围 at the international level
在国内的 onshore
在国内范围 at the national level
在国外 abroad
在海岸工作的 longshore
在海面以下的 undersea
在海上 afloat; at sea
在海外 beyond the sea
在海外的 ultramarine
在海下面进行的 undersea
在海中的 undersea
在海中填筑的陆地 inning
在焊缝边上 at weld edge
在焊缝上 in welded joints
在焊缝中的缺陷 defect in welds; discontinuities in welds
在焊接容器上 on welded vessels
在航 underway
在航采样斗 scoopfish
在航采样戽 scoopfish
在航船 vessel underway
在航船总数 all the shipping afloat
在航吨位 active tonnage
在航海途中 at sea; half seas over; out at sea
在航线上 on course
在合同上签字 execute a contract
在河道中间 amid-river; amid-stream
在河口段溯河航行 navigate an estuary upstream
在河口处 downriver
在荷载下 on-load(ing)
在荷载作用之下 stand under load
在恒压下经过电阻充电 modified constant-voltage charge
在后 posteriority
在后背 rearward
在后部 posterior
在后的 hinder; posterior

在后方 rearward
在后面的船 ship astern
在后期阶段 during the late stages of development
在后台的 backstage
在户外的 alfresco
在滑板上 ski
在环境中稳定的 environmentally stable
在灰泥表面刻痕以便铺筑新层 deviling
在回转架上递送行李包裹 merry-go-round for baggage
在混凝土梁板增加柱或支撑以支承超过设计的荷载 repropping
在活塞上切环槽 fraising
在活性炭上的浸渍 impregnated on activated carbon
在火车上 aboard
在货栈交货 at godown
在获准开工并完成土方后 housing start
在机车里信号再显示 signal repetition on locomotive
在机车前端连挂 nose on
在机械应力下 under mechanical stress
在基本建设投资决策中的非货币因素 non-monetary consideration in capital investment decisions
在基质内 in the substrate
在基质中的传播速度 <声波或地震波> matrix velocity
在计划阶段 in the planning stage
在计划外的 unplanned
在计划中 in the planning stage
在加工中 work-in-process
在家购物网络 home-shopping network
在甲板集合 turn up
在甲板上 on deck
在假期中 on leave
在肩上 piggyback
在检查中 under review
在碱性土壤中 in alkaline soils
在建改良工程 improvement in process
在建工程 construction in process; construction in progress; construction work-in-process; job in process; project under construction; under construction; assets under construction
在建工程合同 construction contract in process
在建工程检验 inspection during construction
在建铁路 railway under construction
在建项目 item under construction; project under construction
在建造中 under construction
在建筑物一旁的运输设施 transportation facility on one side of building
在建筑中 under construction
在键齿侧配合 tooth sides fit of spline
在交换的基础上 on exchange basis
在交涉商量中 be in terms
在交涉谈判中 be in terms
在交易所出售 sale at an exchange
在交易所登记 listing on the exchange
在交易所挂牌 listing on the exchange
在焦点上的 focal
在较低的山底 on the low hills
在较高的放牧场上 on the higher grazing
在今后的二十年内 within the next two decades
在金属导体中电流的透入深度 penetration of current
在紧急的情况下 in case of emergency

在近处 on hand
在近郊 uptown
在近旁 close by; near at hand
在进化过程中 during evolution
在进行中 afoot; on the way to; underway
在进行中的生产活动 carrying-on-activity
在禁止超车区划出中线 centerlining of no-passing zone
在井架上安装设备 rig the derrick
在九小时的光周期下 under a 9-hours photoperiod
在酒精灯上将载玻片加热 heat the slide over an alcohol burner
在旧房间隙处填建房舍 infilling
在旧沥青路面上先铺一层沥青找平 leveling course
在开舱卸货支付 collect before breaking bulk
在考查中 under examination
在考虑中 under consideration; under review
在可剥离的涂层上涂刷 brushed on strippable coating
在可能范围内 within the bounds of possibility
在空间 in-space
在空气中扩散 dispersion in the air
在空气中散播 dissemination in the air
在空气中硬化的 hardened in air
在空中的 midair
在孔底加温液体的装置 volcanic burner
在孔口中形成唇 eyeletting
在控制内的 in control
在控制条件下 under controlled conditions
在控制温度下运送的货物运输 traffic carried under controlled temperature
在控制下停机 controlled stop
在库房内结块的水泥 stockhouse set
在跨中的弯矩 bending moment at mid-span
在冷处保管 keep in a cool place
在离岸不远处 offshore
在离岸不远处的抛锚 offing
在离岸方向 off the land
在立体(中) in-space
在连接处板边的偏差 offset of edges of plates at joints
在良好的制片中 in good preparations
在凉爽气候条件下 in a cool climate
在两方向进行仲裁 arbitrate between two parties
在两个平面中受弯 bending in two planes
在两个停车站间的平均速度 average speed between stops
在列车上发售 on-train issue
在列车头部挂钩 coupling of locomotive at head of train
在列车运行中扳动道岔 reversing of points under a moving train
在另行通知以前 until further notice
在另一方面 per contra
在流程线上调节 in-line equalization
在流程中 on-stream; stream on
在流传中 afloat
在龙骨下面 under-keel
在楼上 above stair(case)
在楼下 downstair(case)
在露天 in the open
在陆地上 onshore
在陆上 ashore
在路上(运行的) <指机械设备和动态> on highway
在伦敦的圣殿骑士教堂 Templar's Church at London

在码头交货 ex dock
在锚地锚泊的船 roadster at mooring
在萌发前施用的除草剂 preemergent herbicides
在面板后面的装配 back of panel mounting;rear of panel mounting
在模拟的商品生产条件下 under simulated commercial conditions
在磨石上磨 hone
在末端 endway;endwise
在末端的 distal
在末尾的 rearward
在某点上 in a way
在某港湾靠 touch at a port
在某些方面 in some respects
在抹灰中嵌入砾石、卵石等骨料 dry dash
在木板框架中的砖或混凝土基础 box sill
在木板上打腻子 knotting
在木头上破坏性刻划 whittle
在目前十年内 in the present decade
在内部起作用 inward
在内舱 inboard
在内地 up-country
在内脚手架上砌砖作业 overhand work
在南北回归线之间的 intertropic(al)
在南的 southern
在南方的 meridianal
在皮革上涂油 dub
在漂石上钻孔 blockhole
在票据上背书 backing a bill
在平板封头上的开孔 opening in flat head
在平板上作模拟试验 breadboard
在平面刚度 in-plane rigidity
在平面荷载 in-plane loading
在平面剪裂缝 in-plane shear crack
在平面力 in-plane force
在平面内 planar
在平坦的黏[粘]壤 on flat clay
在平稳状况下 at even keel
在漆膜半硬时将物件压出永久印痕 printing
在其背风面产生的低气压区 downdraft
在其他条件都是一样的情况下 all other things being equal
在启运港凭单据付现 cash against documents at port of shipment
在气候潮湿的地区 in humid climates
在气象部门的协助下 supported by the meteorology development
在汽缸里不爆发 cylinder cutout
在前的 fore;former;prevenient
在前方 bear forward
在钳工台上安装 bench assembly
在轻质松散的土壤上 on light friable soil
在倾斜的位置上的 astoop
在倾斜地上钻孔 boring on the rake
在秋季 in the fall
在区域范围 at the regional level
在曲线航道段航行 navigation in curved fairway
在曲线上的尖锐噪声 squealing on curve
在渠道下游终点附近 near the lower end of the canal
在全国范围内 nationwide
在全宇宙中 universally
在缺水情况下 in-water-stressed conditions
在确定情况下制订决策 decision-making under certainty
在燃火 going fire
在燃烧时发出蓝焰 to burn with a blue flame
在人工气候室 in the controlled environment chamber
在人体组织中积累 accumulation in body tissues
在任何气候条件下的稳定性 all-weather stability
在任何情况下 through thick and thin
在任何一个发育阶段 at any stage of development
在日光灯下 under fluorescent light
在日光下晒干 sun cured
在溶剂中的最大溶解度 solvent tolerance
在溶解环境中 in solution environment
在若干种岩石中找到 be found in several kinds of rocks
在色彩上 colo(u)risically
在砂浆层上刻出 V 形槽 furrowing
在山场 at stump
在山的阳面下部 below the face of a mountain
在山脚下的 submontane
在山脚下附近 near the foot of a mountain
在山麓的 piedmont
在上风 in the wind;windward
在上风正横方向 on the weather beam
在上面的 overlying
在上面做成拱形 overarch
在设计进程中的成本控制 cost planning
在设计组合梁时只考虑翼缘抗弯矩的方法 chord method
在申请专利中 patent pending
在深水中的生长力 elongation in deep water
在审议中 under consideration;under discussion
在审阅中 under review
在升压之前 prior pressure application
在生产中 on-stream
在生产中的 in production;on-stream
在施工现场 at the construction field;at the construction site;in-site;in-situ
在施工现场制造的 job's housekeeping made
在施工中 construction in process;in course of construction;under construction
在石灰灰泥板上抹灰 plaster on plasterboard
在石面刻对角线墙 broached work
在石头中开挖的运河 canal tunnel(l)ed in rock
在时间流上的价值函数 value function over time stream
在实地 afield
在实际火情中 in actual fire conditions
在实际司机室(地板上)培训 foot-plate training
在实验地设置的量雨表 a rain ga(u)ge installed at the experimental site
在实验室的条件下 under laboratory conditions
在使用的面积 occupied area
在使用时受到中等限制的土壤 soils subject to moderate limitations in use
在使用中 in service
在使用中试验 fleet testing
在试管内 in test tube;in vitro
在试验中 on trial;underproof;under test
在试用期间 during probation
在室内 indoors
在室温中 at room temperature
在适当的潮湿情况下 under proper moisture conditions
在适当条件下 under the proper condition
在手边 near at hand
在首舷方向 on the bow
在受力时 understress
在售价上加的假设运费 phantom freight
在枢轴上转动的窗框 centre-hung sash
在数量上超过 outnumber
在数值上相等 numerically equal
在水池中处理(废水)tank treatment
在水面航行 navigating on water surface
在水面静止 dead in water condition
在水面以上 above water;clear of water
在水上安装和连接 assembly and jointing afloat
在水上的 over water
在水下的 under water
在水线以上 clear of the waterline
在水中 in aqueous
在水中分散的黏[粘]土 dispersive clay
在司机室后面的 behind cab
在四氯化碳中的溶解度 solubility in carbon tetrachloride
在速度上超过 outtravel
在所有条件下 under all conditions
在太阳正下面的 subsolar
在特定条件下 under given conditions;under specified conditions
在天然岩层中的材料 in-situ material
在天然状态的土开挖路堑 cut bank
在条例中引用标准 reference to standard in regulations
在铁道平车上 piggyback
在铁路上 on rail
在停泊中 on berth
在同块地上 on the same land
在同样时间内的同一劳动 same labo(u)r exercised during equal periods of time
在同一平面的 flush with
在同一平面的力 coplanar force
在同一平面上 at-grade
在同一直线上的【数】collinear
在同重干样中介形虫个致分布图 distribution map of number of the ostracoda in the dry sample with the same weight
在同重干样中有孔虫个致分布图 distribution map of number of the foraminifera in the dry sample with the same weight
在统计上是显著的 though statistically significant
在头上的 cephalic
在图上打方格 graticulation
在图纸规定公差范围 within the tolerance specified in the drawing
在途材料 material in transit;store in transit;stores in transit
在途货物 goods en route;goods in transit;goods on the way;in the channel;stock in transit
在途货物时间价值 value of goods on the way
在途搅拌混合料 transit mix(ture)
在途列车的位置 location of trains on line
在途损耗 loss during transit
在途天数 days in transit
在途物品 items in transit
在途现金 cash in transit
在途账 account in transit
在途中停留的溜放车 car standing in midway
在土壤和空气中得到的原料 raw materials took from the soil and air
在土壤中施加石灰 given lime in soil
在土壤中向下移动 downward transport in the soil
在外表面之间 between outside faces
在外舷 outboard
在外压作用下 under external pressure
在完工后看不见混凝土面的模板 back form
在望 in prospect;in sight
在尾舷方向 on the quarter
在未来数十年内 in decades to come in
在位安装 on-site installation
在位处理 on-site processing
在位支承 on-site support
在文件下方或末尾签署者 underwriter
在污染源减少 source reduction
在屋面瓦下嵌灰泥 underdrawing
在无路面或无道路条件下使用的轮胎 off-the-road tire[tyre]
在无外荷载时重力对构件产生的应力 gravity prestress
在物标方向 off an object
在峡湾形成的冬冰 fiord ice
在狭窄海峡内的海区 closed sea
在下层面 downstair(case)
在下的 underlying
在下风 leeward
在下风的 leeward;under the lee;under the wind
在下风头 down the wind
在下面 underlie
在下面的 subjacent
在下面划线的 under scored
在下面签名 undersign
在下面签名的 undersigned
在下游 downriver
在下游的 downriver
在先关系图 precedence diagram
在舷内 inboard
在舷外 outboard;over the side
在现场 at site;in the field;on-site;on situ;on-the-spot
在现有用户上补装 retrofitting existing service
在线操作 on-line data reduction;on-line operation;on-line working
在线操作模型 on-line model
在线测试 engaged test;on-line test(ing)
在线查询打字机 on-line inquiry typewriter
在线超声膜生物反应器 on-line ultrasound-assisted membrane bioreactor
在线程序 in-line procedure
在线处理 in-line processing;on-line processing
在线传感元件 on-line sensing cell
在线存储器 on-line storage
在线(的)on-line;in-line
在线等待 <完工后可重新开始> make good
在线迭代算法 on-line iterative algorithm
在线读出 in-line readout
在线服务 on-line service
在线固相萃取 on-line solid phase extraction
在线宏指令 in-line macro
在线化学清洗 on-line chemical cleaning
在线混凝 in-line coagulation
在线混凝超滤工艺 in-line coagulation-ultra-filtration process
在线计数 on-line counting
在线计算 on-line computation
在线计算机 in-line computer;on-line computer
在线监控 on-line monitoring and control

在线检测 on-line monitoring
在线检测技术 on-line monitoring technology
在线可换单元 on-line replaceable unit
在线控制 in-line control;on-line control
在线模拟量输出 on-line analog(ue) output
在线清洗 on-line cleaning
在线容量 capacity online
在线设备 on-line equipment
在线实时检测 on-line real-time detection
在线实时识别 on-line and real-time identification
在线试验 on-line test(ing)
在线数据处理 on-line data process(ing);in-line data process(ing)
在线数据立即处理 on-line data reduction
在线水检测 on-line water monitoring
在线调节 on-line equalization
在线调试 debugging on-line;on-line debug(ging)
在线涂装 on-line coating
在线涂装涂料 original equipment manufacture coating
在线微波消解 on-line microwave digestion
在线维修 on-line maintenance;on-line repair
在线系统 on-line system
在线修改 on-line modification
在线一致 in-line
在线仪表 in-line meter;on-line meter
在线运算 on-line data reduction;on-line operation
在线蒸馏 on-line distillation
在线装置 on-line equipment
在线子程序 in-line subroutine;on-line subroutine
在线自动监测系统 on-line automatic monitoring system
在线自检功能 on-line self-check function
在相当时候 in due course
在小潮溯河航行 navigate upstream at neap tide
在斜坡上 aslope
在斜坡上作业 sidehill-operation
在行动中 underway
在修船只代号 job number
在修理中 under repair
在修饰 refinishing
在许多方面 in a number of ways
在许多世纪前沉积下来的 to be deposited many centuries ago
在悬架上运转 gantry running
在压气条件下工作的工人 <俚语> sand hog
在压缩条件下 under compression
在岩石或混凝土表面涂水泥(砂)浆 slush grouting
在岩芯盒中保存 preservation in core box
在沿海地区的 downcountry
在沿途操作 over-the-road operation
在沿途运行时间 <铁路货物> on-rail transit time
在沿途运作 over-the-road operation
在沿线使用 over-the-road use
在研究中 under consideration;under investigation;under review;under study
在窑内干燥 kiln dry
在窑中焙烧 kilning
在窑中干燥 kilning
在药包内掏雷管窝 corkscrewing
在野外 afield;in the field
在业 in work
在液面扩展 surface spreading
在一百拓等深线内航行 on soundings
在一百拓等深线外航行 off soundings
在一般等级以上 above-grade
在一般水平之上 above average
在一方缺席时作出的仲裁裁决 avoidance rendered by default
在一个地区内居住单元的总数 housing stock
在一个站点钻一束钻孔 from one setup
在一国内的外国领土 enclave
在一条直线上 point-blank
在移动中卸车 unload in-motion
在遗嘱检验法庭监督下出售财产 probate sale
在已建墙上砌合新墙 keying in
在易破碎岩层中钻进 easy drilling
在应力状态下 on-load(ing);understress
在营运吨位 active tonnage
在用材料 secondary material
在用票价表 active fare table
在用燃具总数 population of installed appliances
在有测量水深的地区 in soundings area
在有压管上穿孔 tapping pressure pipe
在右边的 rightward
在右舷 astarboard
在宇宙空间的 spaceborne
在预定的航线上 lie the interred course
在预定时间前 ahead of schedule
在预算内的 on budget
在阈值以下运转 below-threshold operation
在原地 in-situ
在原地处理 in situ treatment
在原位 in place;in-situ
在源头减少污染 reduction at source
在远离沿岸海面 offshore;off the coast
在月光下从事活动 moonlight
在运输过程中拌和 <混凝土> mix en route;mixing-in-transit;mixed-in-transit
在运输过程中拌和的 mixed-in-transit
在运输途中 in transit;on passage
在运输途中拌和 mixed-en-route;mixed in route;mix-in-transit
在运输途中搅拌 mixed-en-route;mixed-in-transit
在运输中 on passage;on route
在运输中的车辆 cars in transit
在运送途中 in transit
在运行列车上办理出境手续 frontier formality in moving train
在运行情况下的荷载 load in running order
在运行中安装 in-service installation
在运行中换挡 on-the-go
在运移路线上的圈闭 trap on the migration line
在运用中 in service
在运转期 on-stream
在运转中 on-stream;on-the-run
在载的 on-carriage
在站报告 on-station report
在站列车位置报告 on train location report
在站台规定地点停车 platform spotting
在折返段停留时间 time of layover at transfer depot
在真空内 in vacuo
在真空条件下保持性能稳定的 vacuum stable
在整体钢轮上打标记 branding solid steel wheel
在整型时压实 blading compaction
在正横方向 bear abeam
在正后方【船】bear astern
在政府公地上定居以图获得所有权的人 squatter
在支撑后开挖 poling back
在支架上的起重能力 lifting capacity on outriggers
在支路驾车者 shunpiker
在(支票等)背面签名 endorsing
在职博士生 on-job doctorate
在职的 on-the-job
在职环境教育 environmental education on the job
在职教育 in-service education
在职培训 in-service training;on-service training;on(-the)-job training;training on the job
在职培训计划 on-service training program(me)
在职人员 in-service staff;on-the-job
在职人员名单 payroll
在职学习 in-service training
在职训练 in-service training;on-service training;on-the-job test;on(-the)-job training;training on the job
在职职工 active staff;staff on active duty
在职专业人员培训 on-the-job professional training
在职专业训练 on-the-job professional training
在指定地点称重 spot weigher
在指定地点集合 rendezvous
在制材料 material in process
在制订货 in-process order
在制定规定下 in regime
在制费用账(户) burden-in-process account
在制品 in-process material;work-in-process
在制品费用 manufacturing expense in process
在制品估价 valuation work in process
在制品间接费用 burden-in-process
在制品明细表 schedule of work in process
在制品盘存 work-in-process inventory
在制品数量 goods-in-process inventory
在制品周转 turnover of goods in process
在制人工 labo(u)r in process;labo(u)r in progress
在制物料 material in process
在制造期间检查 inspection during fabrication
在制造时 during fabrication
在制造中的货品 work-in-process
在制账户 manufacturing account
在制制造费用 burden-in-process;manufacturing expenses-in-process
在中流 <船舶停泊不能卸货> in stream;amid-stream
在中途 midway
在中纵线上 amidship;midships
在重力作用下 by gravity
在周围 about;circa
在周围回填 backfill around
在轴承内的润滑油流动 oil flow in bearing
在轴端的 overhanging
在轴线上 en axe
在轴向荷载下 under axial loading
在住宅或其他建筑顶上的平屋顶或平台 azotea
在助铲下铲装 <铲土机> pushing loading
在贮水池中蓄水 impound
在柱身刻出凹槽或凸筋 striation
在砖、石上再做上一层贴面 reclad
在转运时货车的运行轨道 interchange track
在自力的影响下 under the influence natural forces
在自然条件下 under field conditions;under field factors
在钻孔内高压提升钻具 snub out
在钻孔中插入的桩 cored pile
在钻探报表上记进尺 pad the log book
在钻头上开槽 groove bit nose
在最大高度上卸载距离 <挖掘机> reach at maximum dump height
在最短期限内 at the earliest possible date
在左边的 leftward
在左舷 aport
在左舷船侧后半部 on the port quarter
在左舷船道 on the port bows
在座用(快)餐 cartering

簪

簪梁 <托换基础用> needle beam

攒

攒 set of bracket

攒尖【建】pyramidal roof
攒尖顶【建】pavilion roof;polygonal roof;conic(al) broach roof;pyramid roof
攒尖式屋顶 pavilion roof
攒尖饰 finial;finial pinnacle;pinnacle

暂

暂保单 covernote;risk note

暂保收据 binding receipt
暂不能利用储量 transient unusable reserves;usable in future reserves
暂储区 temporary storage area
暂存储区 working area
暂存带 scratch tape
暂存带分类 tape scratch sort
暂存段 working storage section
暂存工作控制 stacked job control
暂存节 working storage section
暂存盘组 scratch pack
暂存器 temporary memory;temporary register;temporary storage;working memory;working storage
暂存区 working storage section;working storage area
暂存剩余磁化强度 temporary magnetization
暂存阵列 temporary storage array
暂存资金 suspense fund
暂等 camp-on
暂钉 <暂时钉住> spike fastening;working fastening
暂定 provisional
暂定按金 indicated deposit;initial margin
暂定坝址 tentative site
暂定标准 temporary standard
暂定厂址 tentative site
暂定成本 provisional cost
暂定船名 ambiguous name of ship
暂定船位 tentative booking
暂定措施 interim measure
暂定的 provisory;tentative
暂定的发货单 pro forma invoice
暂定发票 pro forma invoice
暂定方法 provisional method
暂定港址 tentative site
暂定工程 provisional works
暂定工程数量 <工程量清单中的> provisional quantity
暂定工程项目金额 provisional sum

Z

暂定工作及材料供应和服务项目总和 provisional sum
暂定股利 interim dividend
暂定合同 ad referendum contract
暂定金额 provisional sum
暂定金额支付 payment against provision
暂定契约 ad referendum contract
暂定权 provisional weight
暂定数据 tentation data
暂定项目 provisional item
暂定项目表 preliminary list of items
暂定项目金额 <承包合同中的> provisional sum
暂定项目预留金额 cash allowance
暂定指标 preliminary specification
暂定资源 conditional resource
暂定总金额 provision sum
暂冻土 briefly frozen soil
暂付 <作为部分付款> payment on account; on account
暂付款 interim payment; on-account payment; suspense debitor; suspense debt; suspense payment
暂付款项 suspense debit; temporary payment
暂付款账户 suspense payment account
暂搁 abeyance
暂估 interim estimate
暂管 escrow
暂焊 positional weld(ing); positioned weld(ing); tack weld(ing)
暂缓 abeyance; defer; postpone
暂缓勘探 transient stop exploration
暂缓执行令状 supersedeas
暂记 suspense
暂记待结转账户 clearing account
暂记贷项 suspense credit
暂记借项 suspense debit
暂记欠款 temporary advance and sundry debtor
暂记时间 memory time
暂记预支 temporary advance
暂记账 <即待分配的费用账户> pool account
暂记账户 clearing account; suspense account
暂记资本账户 temporary proprietorship account
暂交第三者保管 sequestration
暂借款 day-to-day money; money on call; overnight money
暂静 lull
暂居地段 accommodation area
暂留点 holding point
暂留人口 transient population
暂签 signature ad referendum
暂设房屋 temporary building
暂设工程 temporary construction; temporary works
暂设金额 provisional sum
暂设数据 tentation data; tentative data
暂设住宅 temporary dwelling
暂生境 temporary habitat
暂时 for a certain period; for a time; interim
暂时搬迁 temporary relocation
暂时变稠 transient thickening
暂时变化 transitory variation
暂时变形 temporary deformation
暂时表形显示 temporary tabular display
暂时不收保费 nominal premium
暂时程序库 temporary library
暂时稠化 transient thickening
暂时出口 temporary export
暂时储藏 interim storage
暂时储存区 transient accumulation area
暂时辍学 <以从事其他活动的大学生> stop out
暂时磁化 temporary magnetization
暂时磁体 soft magnet; temporary magnet
暂时磁铁 temporary magnet
暂时磁铁措施 temporary magnet
暂时磁性 temporary magnetism
暂时存储 intermediate memory
暂时存储单元 temporaries
暂时存储器 scratched memory; temporary memory; temporary storage
暂时存储区域 working area; working space
暂时存款 sundry credit; temporary deposit
暂时措施 interim measure
暂时错误 temporary error
暂时代用品 makeshift
暂时贷款 cash credit
暂时单元 temporary location
暂时的变化 provisional variation; temporal variation
暂时地下水面 temporary ground water-table
暂时电流 contraflow; extra current
暂时凋萎 temporary wilting
暂时钉住 sprigging
暂时顶极群落 temporary climax
暂时毒物 temporary poison
暂时堵塞 temporary plug(ging)
暂时堆积区 transient accumulation area
暂时阀 temporary valve
暂时繁荣 temporary boom; boomlet <美>
暂时分解 transitory decomposition
暂时分配 temporary assignment
暂时缝上 tacking
暂时干涸 temporary extinction
暂时干枯 temporary wilting
暂时割让 temporary cession
暂时搁置 abeyance
暂时工作区 transit working area
暂时估计 interim estimate; provisional estimate; provisional evaluation
暂时估价 interim estimate; provisional evaluation; provisional estimate
暂时估算 interim estimate; provisional estimate; provisional evaluation
暂时固定 fir fixed
暂时关闭 temporary close
暂时规定 transitory provision
暂时含水层 temporary water-bearing layer
暂时合伙 temporary partnership
暂时湖 temporary lake
暂时基准面 temporary base level
暂时价格 interim price
暂时接地 swinging earth
暂时结合 temporary joint
暂时解雇率 lay-off rate
暂时解雇期 lay-off
暂时借款 temporary loan
暂时进口 temporary import
暂时禁猎区 temporary hunting prohibited area
暂时均衡 temporary equilibrium
暂时扣下的财物 hold-back
暂时库容 temporary storage
暂时冷凝 temporary condensation
暂时连接 tacking
暂时联系 temporary connection
暂时逻辑转出 logic(al) roll-out
暂时贸易 temporary trade
暂时黏[粘]合 temporary adhesion
暂时逆差 temporary deficit
暂时凝固 temporary set
暂时膨胀 temporary expansion
暂时批准 interim authorization
暂时拼装 temporary installation
暂时平衡 transient equilibrium
暂时平价 temporary par of exchange
暂时铺设 temporary laying
暂时侵占 <堆放施工材料> temporary encroachment
暂时区 temporary area
暂时趋同 temporary convergence
暂时渠 temporary canal
暂时泉 temporary spring
暂时缺货 temporarily out of stock
暂时燃料 temporary fuel
暂时融通资金 day-to-day accommodation
暂时丧失工作能力 temporary disability
暂时生境 temporary habitat
暂时生物 transient biont
暂时失效 temporary failure
暂时适应 temporary adaptation
暂时收入 temporary income; transitory income
暂时数据集 temporary data set
暂时双折射 temporary birefringence
暂时水位 temporary water table
暂时替用的 temporarily replaced; band-aid <美>
暂时条款 temporary provision
暂时调整 temporary adjustment
暂时停车 stopper
暂时停顿 hold-back
暂时停工的费用 cost of suspension
暂时停工命令 order to suspend
暂时停工情况下的支付 payment in event of suspension
暂时停机 hesitation
暂时停用文件 scratch file
暂时停钻的井 shut-down well
暂时图像显示 temporary pictorial display
暂时弯沉 <路面在行车通过时的> transient deflection
暂时萎蔫 temporary wilting
暂时文件 temporary file
暂时稳定 metastable; temporary stability
暂时稳定平衡 metastable equilibrium
暂时无法满足的订货 back order
暂时吸收水 temporally absorbed water
暂时熄火 temporarily extinguished
暂时现象 transient phenomenon
暂时限速 temporary speed restriction
暂时项 transient term
暂时消融区 transient ablation area
暂时性 impermanence[impermanency]; temporality; transience
暂时性补给 temporary recharge
暂时性衬砌 emergency lining support
暂时性充填 temporary filling
暂时性的 fugacious; transient; provisional; temporary
暂时性毒剂 transient agent
暂时性浮游生物 meroplankton; temporary plankton
暂时性感应 temporary induction
暂时性洪流 temporary torrent
暂时性流水 temporary current; temporary water flow
暂时性旅馆 transient hotel
暂时性缺氧现象 oxygen debt
暂时性渗漏 temporary leakage
暂时性水系 ephemeral system
暂时性听力缺损【救】 temporary hearing defect
暂时性停车处 reservoir parking space
暂时性阈移 temporary threshold shift
暂时性止水 temporary water sealing
暂时悬着地下水 perched temporary ground water
暂时雪线 temporary firn-limit; temporary snow-line
暂时压力 transit press(ing); transit pressure
暂时压载物 temporary ballast
暂时演替顶极 temporary climax
暂时页面存储器 temporary page store
暂时液化 temporary liquefaction
暂时议程 tentative agenda
暂时因素 short-term factor
暂时应变 temporary strain
暂时应力 temporary stress
暂时影响 temporary influence
暂时硬度 carbonate hardness; temporary hardness; transient hardness
暂时硬水 temporary hard water
暂时预支 temporary advance
暂时指示字变量 temporary pointer variable
暂时滞水 perched temporary ground water
暂时贮存 interim storage
暂时装置 stopgap unit
暂时装置品 temporary fixture
暂时状态 interim status; momentary state; transition condition
暂时自动记录器 transient self-recorder
暂时组合法 interim combination technique
暂收款 advance received; suspense credit; suspense receipt; temporary credit; temporary receipt
暂收款账户 advance received account; suspense receipt account
暂态 transient condition; transient state; transition condition
暂态 X 射线源 transient X-ray source
暂态波动 transient swing
暂态的 transient
暂态电抗 transient reactance
暂态电流 transient current
暂态电压 transient voltage
暂态发电机 transient generator
暂态法 transient method
暂态反馈 temporary feedback; transient feedback
暂态分析 transient analysis
暂态分析仪 transient analyser[analyzer]
暂态过程曲线 recovery curve
暂态调差率 temporary speed drop
暂态特性 transient behavio(u)r
暂态温度 transient state temperature
暂态稳定储备系数 transient stability margin
暂态稳定储备余量 transient stability margin
暂态稳定性 transient stability
暂态误差 transient error
暂态响应 transitional response
暂态效应 transient effect
暂态性故障 transient-cause forced outage
暂态运动 transient motion
暂态振动 transient vibration
暂态转差率 temporary droop
暂态转差系数 momentary speed droop; temporary droop
暂态转速变化 momentary speed variation
暂态转速变化增量 incremental momentary speed variation
暂态转速降 temporary droop
暂态转速降调速器 temporary-droop governor
暂态转速下降 incremental momentary speed droop
暂停按钮 pause button
暂停辩论 adjournment of debate

暂停补实的职位 post put in escrow
暂停车辆 <不是等候信号而因故暂停在路上的汽车> standing vehicle
暂停处理的案件 suspension case
暂停的 suspensive
暂停的交通 standing traffic
暂停服务 out-of-service
暂停付款 payment withheld
暂停工作 break off
暂停工作状态 pause mode
暂停活动 moratorium;stand-down
暂停驾驶执照 suspension of driver's license[licence]
暂停交易 trading suspended
暂停开关 halt switch
暂停控制 pause control
暂停施工 suspension of work
暂停时间 suspension period
暂停使用 laid up
暂停输入/输出 pause input/output
暂停营业 temporary cessation of business
暂停支付 cessation of payment;suspend payment
暂停指令 pause instruction
暂停状态 halted state;pause status;suspended state
暂停字 pause word
暂停钻井 suspending a well
暂未满足的需求 pent-up demand
暂息 <暴风雨等的> lull
暂现X射线源 sporadic X-ray source;transient X-ray source
暂现事件 transient event
暂行办法 interim procedure;make-shift method;temporal method
暂行标准 interim criterion;tentative standard
暂行程序 tentative program(me);tentative routine
暂行初级饮用水规程 interim primary drinking water regulations
暂行处理指南 interim treatment guide
暂行措施 interim means
暂行的 tentative
暂行法规 interim primary regulation
暂行方法 temporary method;tentative method
暂行规程 tentative specification
暂行规定 interim provision;temporary provision;tentative provision;tentative specification
暂行规范 interim specification;temporary standard;tentative code
暂行规格 tentative specification
暂行合约 conditional contract
暂行基本规则 interim primary regulation
暂行技术规程 tentative technical specification;tentative specification
暂行技术规范 tentative technical specification
暂行每日容许摄入量 temporary acceptable daily intake
暂行契约 conditional contract
暂行设计 tentative design
暂行食品最大允许含量 temporary food tolerance
暂行条款 conditional clause
暂行条例 interim regulations;provisional regulation;tentative provisions
暂行协定 modus vivendi
暂行一周容许摄取量 provisional weekly tolerance intake
暂行议事规则 provisional rules of procedure
暂行状态 suspended state
暂行准则 interim criterion;tentative criterion
暂行最大允许含量 temporary tolerance
暂休 time-out
暂养池 storage pond
暂隐与再现 status change
暂用程序 transient program(me)
暂用方法 temporary method
暂用联轴节 dummy coupling
暂用跑道 temporary runway
暂用设备 temporary arrangement
暂用数据 tentation data
暂用罩冠 temporary crown
暂用中继线 intercept trunk
暂载率 <焊机的> arcing time factor;duty cycle;utilization factor
暂住 harbo(u)r
暂住人口 transient population
暂住性旅馆 transient hotel
暂驻程序库 transient library
暂驻存储区 transient area;transient storage area
暂驻目录 transient directory
暂准建筑物 tolerated structure

赞

赞比西河 Zambezi River

赞比亚陆核 Zambia nucleus
赞卡冰阶【地】Zyrianka stage
赞助公司 sponsoring firm

錾

錾 gad;moil point;track chisel

錾锉砧 file cutting anvil
錾锉座 filing block
錾刀 burin;cold set
錾光石面 chiselled stone face
錾花 chase
錾紧铆钉 ca(u)lked rivet
錾平 chipping
錾平锤 chipping hammer;scaling hammer
錾石锤 chipping hammer
錾形的 chisel-like;chisel-shaped
錾形钻头 chisel-shaped bit
錾凿 chipper;chisel(ing)
錾子 bull prick;chisel;chipper
錾座 file cutting anvil

赃

赃物 booty;pickings

脏

脏版 greasing;scumming

脏冰 dirty ice
脏臭液 foul solution
脏的 putrid
脏空气 contaminated air
脏煤 dirty coal
脏砂层 dirty sand
脏刷子引起漆膜斑点 seedy
脏水 befouled water;contaminated water;dirty water;filthy water;foul water
脏水盆 slop basin
脏污道床 contaminated ballast bed;fouled ballast bed
脏污费 dirty money
脏污集料 dirty aggregate
脏物混入 dirt entry
脏雪球模型 dirty snowball model
脏衣物溜槽 laundry chute;clothes chute
脏衣洗涤槽 soiled linen chute
脏蒸汽 dirty steam
脏渍 dirt-stained

葬

葬 bury

葬礼寺庙 funerary temple
葬礼仪式 mortuary cult
葬礼艺术 funeral art
葬礼用长方形建筑 funeral basilica

藏

藏红 safranine

藏红花 crocus;saffron crocus
藏红花的 crociate
藏红花色 croceous
藏红花柱头 saffron
藏红色 saffron
藏角 hidden corner
藏晶体 crystal receptacle
藏木香 Radix Inulae Racemosae;Tibet inula root
藏纳接头 housed joint;housing joint
藏品编目 cataloging of objects
藏瓶地窖 bottle cellar
藏器 Zang ware
藏器系数 organ coefficient;ratio of organ to body weight
藏青色 dark blue;navy;navy blue
藏式接头 housed joint;housing joint
藏式喇嘛庙 Tibetan-style lamaist temple
藏式塔 Tibetan pagoda
藏原羚 goa
藏族建筑 architecture of the Zang nationality

遭

遭岔枕木 turnout sleeper

遭火灾的房屋 gutted structure
遭难 wreck
遭难地区 distressed area
遭受虫害 insect infestation
遭受到 exposed to
遭受地震地带 nervous earth
遭受洪水的危害 subject to overflow damage
遭受扣押的财产 property subject to attachment
遭受危险 jeopardise[jeopardize]
遭受危险海岸 coastal hazard
遭受严重破坏 subject to severe risks of damage
遭遇 confrontation;undergo
遭遇频率 encounter frequency
遭灾 bet hit by a natural calamity

糟

糟化辉长岩 euphotide

糟化辉绿岩 leucophyre
糟绿帘石 zorsite
糟粕 draff;spent wash;trash
糟朽 decay;rot
糟朽木材 decayed timber;rotted wood

凿

凿 backing-off;chisel;fang

凿边 chisel(l)ing
凿冰鹤嘴锄 ice pick
凿冰器 ice auger
凿柄体 back head
凿糙墙面 stab
凿槽 chisel groove;form a flute;gouge carving;gouge(out);raggle;raglan[raglin]
凿槽刀 cope chisel
凿槽工具 croze
凿槽机 mortising slot machine
凿成特殊形状的穿孔卡 load card
凿齿锯 chisel(l)ed tooth saw
凿齿松土铲 chisel-tooth shovel
凿出 cutting out;gouge out
凿船虫 borer;pin-hole borer;ship borer;teredo;teredo navalis
凿船虫属 teredo
凿锤石膏 bush-hammered plaster
凿刀 chisel knife
凿刀钢 chisel steel
凿的刃角 bezel
凿点 chisel(l)ing point
凿吊楔孔的工具 lewising tool
凿掉 chiseling off
凿钉 chisel-pointed nail
凿断工 cleanser
凿断凿 fettling chisel
凿锋 chisel edge
凿缝 chisel(l)ing;staking
凿缝凹模 staking die
凿缝凸模 staking punch
凿斧导体 pick axe conduit
凿镐 chipper
凿沟 channeling;channel(l)ing;rifle;riffling
凿沟机 channel(l)er;channel(l)ing machine
凿焊龟裂 chisel-bond cracking
凿划石面 stroked
凿环 shearing bushing
凿环切割陶管法 back edging
凿混凝土机 concrete breaker
凿击 batting
凿尖 chisel edge;chisel point;chopping bit;moil point;nose of chisel
凿尖钉 chisel-pointed nail
凿尖角 chisel edge angle
凿紧 ca(u)lking
凿进机 thrust borer
凿井 bore well;cable tool well;drilled well;drilling well;misering;sink a shaft;sinking a well;sinking of borehole;sink of bore hole;well boring;well drilling;well lowering
凿井安全盖板 sinking bogie
凿井班 sinking shift
凿井包工 sinking contractor
凿井爆破 sinking blast(ing)
凿井沉箱 Boston caisson;caisson;caisson pile
凿井吊灯 sinking lamp
凿井吊盘 sinking platform
凿井吊盘安全门 safety sinking door
凿井吊桶 shaft-sinking pump;sinking barrel;sinking bucket;sinking kibble
凿井吊桶的导向架 sinking rider
凿井吊桶之间的联系 communicator for sinking-kibble
凿井法 shaft sinking;sinking technique
凿井方法 sinking method;sinking system
凿井格 sinking compartment
凿井工 shaft sinker;well borer;well sinker
凿井工程 sinking operation
凿井工作 drilling
凿井罐道 sinking guide
凿井机 miser;sinking machine;trepan;well borer;well digger
凿井机器 well-rig
凿井机组 cable tool rig
凿井箕斗 sinking skip
凿井加工留量 shaft allowance
凿井绞车 sinking winch;sinking winder
凿井绞盘 sinking winch
凿井进度 sinking advance
凿井井壁 sinking lining
凿井井架 sinking headframe;sinking tipple
凿井取得的 phreatic
凿井设备 sinking equipment;sinking

plant
凿井式沉箱 sinking caisson; Gow caisson
凿井水泵 sinking pump
凿井速度 sinking advance; sinking rate
凿井提升机 sinking hoist
凿井用泵 sinking pump
凿井用大钻机 shaft-sinking drill
凿井用吊泵 sinking pump
凿井用吊桶 hoppit;kibble
凿井用多钻式凿岩机 multiple drill shaft sinker
凿井用手持式凿岩机 hammer sinker
凿井用凿岩机 sinker drill
凿井用凿岩机的气腿 sinker leg
凿井用凿岩机气压腿 sinker leg
凿井装岩 shaft mucking
凿井装岩机 shaft mucker
凿井钻 well drill
凿井钻车 shaft drill jumbo; sinking drill jumbo;sinking jumbo
凿井钻机 shaft drill jumbo
凿井钻台 shaft drill jumbo;shaft jumbo
凿井作业 shaft service;sinking
凿净 peel
凿净铸件 peeling
凿开 beat away;chiseling;rip
凿孔 drilling; gouge; perforate; punching
凿孔打字机 perforating typewriter
凿孔的 punctured
凿孔法 holing
凿孔机 gadder; perforator; punch card machine;puncher
凿孔机构 punching mechanism
凿孔机头 perforating head
凿孔卡 punched card
凿孔卡片栏 card field
凿孔卡设备 punched-card equipment
凿孔空心钻头 jagged core bit
凿孔器 perforator;punch
凿孔深度 drilling depth; hole depth (by percussive drilling)
凿孔速度 penetration velocity
凿孔速率 perforating speed; penetration rate
凿孔条倒退键 backspace key
凿孔停止 punch off
凿孔纸带 perforated slip; perforated tape;punched tape
凿孔纸带记录器 perforated tape recorder
凿孔纸条 perforated slip
凿孔纸条转发 perforated tape retransmission
凿毛 chipping;hacking;roughen(ing) (by picking);green cutting <混凝土浇筑层层面>; bush-hammering <混凝土>
凿毛层 roughening course
凿毛锤 bush hammer;boucharde <石面>
凿毛的表面 roughened surface
凿毛机 roughening machine
凿毛面 bush-faced;pitch-faced
凿毛石(路)面 bush-hammering
凿毛圬工 punch-dressed masonry
凿密 fuller(ing);ca(u)lking
凿密成套工具 ca(u)lking set
凿密锤 ca(u)lking hammer
凿密法 ca(u)lking
凿密封 ca(u)lk
凿密缝 ca(u)lked joint; ca(u)lked seam
凿密工具 ca(u)lking tool; fullering tool
凿密具 ca(u)lking set
凿面 pitch face;tooling
凿面的 <石料> pitch-faced
凿面方石 tooled ashlar
凿面块石砌体 bush faced masonry
凿面块石圬工 bush faced masonry
凿面石板 chisel(l)ed slate
凿面石建筑 pitch-faced masonry
凿面石料 pitch-faced stone
凿面圬工 punch-dressed masonry
凿木虫 wood borer
凿平 chipping;chiseling;picked dressing;scaling;wasting
凿平的(墙面)方石 broached
凿平的石面 plain work
凿平修饰 picked dressing
凿钎 chisel bar
凿钳 chisel tong
凿墙面钢筋 stab bar
凿、敲试验 <简易测定水泥硬度> pick and click test (for soil cement)
凿去混凝土表面砂浆露出石头的机械 concrete scabbler
凿切割 cold cut
凿刃 bit wing; chisel edge; chisel point
凿山劈岭 tunnel through mountains and cut across ridge
凿山隧道 tunnel a hill
凿石 axed work; boasting; dressed stone;stone cutting;stugging
凿石船 rock cutter vessel
凿石锤 boucharde; bush hammer(ing); knapping hammer; roughening tool
凿石锤粗加工的 hammer dressed
凿石锤土路 bush hammer
凿石锤修饰混凝土或石头表面 bush-hammer finish
凿石锤整修 bush-hammering
凿石锤整修的灰泥涂层 bush-hammered plaster
凿石锤整修的墁灰 bush-hammered plaster
凿石工 block chopper;stone dresser; stone drifter
凿石工场的标志 stone dresser's sign
凿石工场的牌号 stone dresser's sign
凿石工程 rock-cut work;rock works; stonework
凿石工具 dressing tool; stone surfacer;stone tool
凿石工人 knapper
凿石工作 rock drilling;rock-cut job; rock works;stone-dressing work
凿石机 knapping machine;rock cutter;stone cutter
凿石块 joint block
凿石面 batting; quarry-faced stone; tooled finish;tooled surface
凿石使平 drove
凿石镶面 cut-stone veneer
凿石楔 spalling wedge
凿石斜面纹理 rock face
凿石锥 crandall
凿式 dental formula
凿式犁 chisel
凿竖井 shaft sinking
凿竖井用 rotary drill for shaft sinking
凿碎机 mechanical pick
凿榫 morticing[mortising]
凿榫工 mortiser
凿榫机 mortiser; mortising (slot) machine
凿榫眼 chase mortising;mortising
凿榫眼机 mortise machine;mortiser
凿榫钻孔两用机 mortising and boring machine
凿通 cut-through
凿通孔 hole punching
凿头 chisel bit;point tool
凿头钩 bit hook
凿头刃口 chisel-type bit cutting edge
凿挖的隧道 bored tunnel
凿挖门梃的锁 cut lock
凿挖岩石以建造地基 rock cutting in formation
凿挖岩石以建造路基 rock cutting in formation
凿纹 daubing
凿纹方石 tooled ashlar
凿削 paring
凿形播种机 chisel planter
凿形焊接 chisel bond
凿形犁铧 wedge-shaped share
凿形犁松土 chiseling
凿形犁松土种植法 chisel planting
凿形钻头 bull bit;chisel bit
凿岩 drilling off; gadder; rock cutting;rock lifting
凿岩爆破 drilling and blasting; rock blast; rock drilling and blasting; waste blasting;waste firing
凿岩爆破参数 parameter of drilling and blasting
凿岩爆破作业 drilling and blasting operation
凿岩车 drill mobile
凿岩成本 drilling cost
凿岩船 floating rock crusher; rock breaking vessel; rock (cutter) dredge(r)
凿岩锤 jack hammer
凿岩地点 drill site
凿岩电钻 electric(al) rock drill
凿岩峒室 drill chamber
凿岩而成的 rock-hewn
凿岩方法 drilling method
凿岩方式 drilling pattern
凿岩辅助设备 drilling auxiliary equipment;rock drilling auxiliary
凿岩钢钎头 steel-rib bit
凿岩钢钻头 steel-rib bit
凿岩工 borer;drill man;drill runner; holer
凿岩工班效率 drilling performance per driller and shift
凿岩工长 drillmaster
凿岩工地 drilling site
凿岩工具 drill tool
凿岩工艺 drilling technology
凿岩工作 stone work
凿岩工作平台 drilling platform
凿岩工作作业 drilling job
凿岩机 anvil type percussion drill; banjo; blast hole drill; bore hammer;borer;drifter;drifting machine; drill; drill hammer; drilling machine;drill plugger;drill rig;excavator; hammer drill; hammer rock drill;jack drill;plugger;pulsator; pulsator jig; pulsator rig; rock boring machine; rock breaker; rock cutter; rock drill drifter; rock drilling machine; rock excavator; rock hammer drill; stone drifter; stone drill; stoper; thrust borer; trepan; rock drill
凿岩机操作员 drill runner
凿岩机车 jambo
凿岩机齿尖头 rock point tip
凿岩机冲程 drilling stroke
凿岩机冲击行程 drilling stroke
凿岩机导向套 drill chuck bare
凿岩机导向套筒 chuck bare
凿岩机顶柱 jackshaft
凿岩机动力推进装置 drill power feed
凿岩机阀门箱 drill valve chest
凿岩机风管 drill hose
凿岩机干式捕尘器 rock drill dust exhauster
凿岩机供水管 drill water hose
凿岩机活塞 drill piston;hammer piston
凿岩机活塞冲程 drill piston stroke
凿岩机活塞杆 drill piston bars
凿岩机夹钎爪 drill chuck jaw
凿岩机架 drill rests;drill stand
凿岩机节流阀 drill throttle valve
凿岩机节流阀手柄 drill throttle valve handle
凿岩机结构 rock drill design
凿岩机连接管接头 drill hose connection
凿岩机排气(管的)冻结 freezing of drill exhaust
凿岩机排气装置 drill exhaust
凿岩机气管接头 drill hose connection
凿岩机气腿 rock drill column
凿岩机汽缸 drill cylinder
凿岩机钎尾套筒 drill chuck bushing
凿岩机钎子 jack bit
凿岩机润滑油 rock drill oil
凿岩机水阀门 drill water valve
凿岩机台车上部结构 drilling superstructure
凿岩机梯架 ladder
凿岩机头 drill front head
凿岩机推进器 drill feed
凿岩机腿架 pusher leg
凿岩机外壳 drill casing
凿岩机械 drilling equipment
凿岩机修理车间 drill repair shop; rock drill shop
凿岩机修理工 doctor drill
凿岩机修配车间 drill repair shop; rock drill shop
凿岩机旋转棘轮机构 drill rotation ratchet
凿岩机液压腿 hydro-leg
凿岩机硬质合金衬片 rock drill insert
凿岩机支架 drifter bar
凿岩机支架垫板 foot board
凿岩机柱齿 button
凿岩机装置 drill setup
凿岩机撞锤 drill anvil block
凿岩机自动推进 drill power feed
凿岩机钻机架 rock drill mounting
凿岩机钻孔 hammer-drill hole
凿岩机钻头 drifter drill;rock drill bit
凿岩技术 drilling technique
凿岩进尺 drill(ing) footage
凿岩巨型台空 drilling jumbo
凿岩螺旋钻 rock auger
凿岩落锤 bore hammer
凿岩能力 drilling capacity
凿岩拧钎杆机 drill-screwdriver
凿岩平巷 drill drift
凿岩器 gadder
凿岩钎 jumper bar
凿岩钎钢 rock drill steel
凿岩钎头 rock bit;stone chisel bit
凿岩钎子 rock drill jumper
凿岩设备 drilling equipment
凿岩设备折旧摊销及大修费 drill equipment depreciation apportion and overhaul charges
凿岩生产率 drilling effect
凿岩试验 rock drill test
凿岩台班 drill shift
凿岩台班效率 drilling performance per drill and shift
凿岩台车 boom-mounted drifter;drill(ing) carriage;drill(ing) jumbo;rigmounted drill; rock breaking jumbo; rock drilling jumbo; truck-mounted drill (ing rig); wagon-mounted drill
凿岩梯段 drilling bench
凿岩位置 drilling site
凿岩现场 drilling site
凿岩效率 drilling performance; rock penetration performance

Z

凿岩延米数 drillmeter
凿岩英尺 drill(ing) footage
凿岩装车联合机 jumbo loader
凿岩钻 well auger
凿岩钻车 drill(ing) jumbo; drill(ing) rig; drill wagon; jumbo drill; wagon drill
凿岩钻钢 rock drill steel
凿岩钻头 cone rock bit; rock bit
凿岩钻头镶嵌物 tock bit insert
凿岩钻头用硬质合金片 rock bit insert
凿岩作业 drilling operation
凿眼 mortise preparation
凿眼爆破 hole drilling and blasting
凿眼垫墩 < 帆布冲眼用具 > cutting block
凿眼机 mortiser; mortising machine
凿油井机 oil-well drilling machinery
凿有节疤的粗面石头 opus rusticum
凿整锭面机 deseaming machine
凿琢 picking
凿琢面 batted surface
凿子 bit; chipper; chisel; cold chisel; gad picker; nicker; plow bit
凿子钢 chisel steel
凿钻头 bull bit

早

早阿尔卑斯期地槽 early Alpine geosyncline
早奥陶世【地】Lower Ordovician
早白垩世【地】Lower Cretaceous
早白垩世气候分带 Early Cretaceous climatic zonation
早班 fore shift; mooring watch; morning shift
早爆 preloading firing; premature blast; premature explosion
早宾夕法尼亚世【地】Lower Pennsylvanian
早冰川寒冷期 early glacial epoch
早材 early wood; spring wood
早材层 zone of early wood
早材带 zone of early wood
早材管胞 spring tracheid
早餐室 breakfast room; morning room
早餐座 breakfast nook
早插 early transplanting
早潮 morning tide
早晨高峰交通量 morning peak traffic
早晨高峰期 morning peak hours
早晨星组 < 用于纬度测量 > morning group
早成 precocity
早成年期谷 early mature valley
早成组构 epigenetic fabric
早春 early spring
早春的 prevernal; vernal
早春灌溉 early spring irrigation
早春季相 prevernal aspect
早代测验 early generation test
早稻 early(season) rice
早得利亚斯冰阶 oldest Dryas stade
早第三纪【地】Eogene(period); Old Tertiary; Pal(a)eogene period
早第三系【地】Eogene system
早断 early cut-off
早二叠世【地】Lower Permian
早二叠世海浸【地】early Permian transgression
早二叠世海退【地】early Permian regression
早高峰 morning peak hours; morning rush hours
早高峰时间 AM peak hour
早高强 high-early strength
早高强混凝土 high-early strength concrete
早更新世 lower Pleistocene
早更新统【地】Eopleistocene
早古生代 Eopaleozoic
早光谱型 early spectral type
早海西期地槽 early Hercynian geosyncline
早寒武世【地】Lower Cambrian
早寒武世海浸【地】Early Cambrian transgression
早花 early blooming
早加里东期地槽【地】Early Caledonian geosyncline
早检测 early detection
早交货 prompt delivery of goods
早开的投标 prematurely opened bid
早落的 caducous
早密西西比世【地】Lower Mississippian
早泥盆世【地】Lower Devonian
早泥盆世海浸 early Devonian transgression
早凝 early set(ting); premature setting; premature stiffening < 水泥混凝土的 >
早凝水泥 early setting cement; quickly taking cement; quick-setting cement; rapid setting cement
早期 earlier period; early stage; spring time
早期安全水压值 value of safe water pressure in the early days
早期巴洛克式建筑艺术风格 early Baroque
早期保持压力方案 early pressure maintenance scheme
早期报警 early warning
早期报警装置 early-warning device
早期爆破 initial blasting
早期变态式(装饰过分的)建筑 early Baroque
早期变质的【地】cometamorphic
早期变质作用 cometamorphism
早期玻璃 prior glass
早期不稳定阶段 early transient regime
早期测验 early test
早期沉降物 early fallout
早期成岩带【地】eogenetic zone
早期成岩的 eogenetic
早期成岩作用 early diagenesis; syndiagenesis; syngenesis
早期城市 < 古希腊的城市国家 > Polis
早期冲蚀 early erosion; incipient erosion
早期抽苔 premature bolting
早期出现 anticipation
早期储水系数 storativity in early period
早期处理 early-time treatment
早期的 lower
早期低强度水泥 low-early-strength cement
早期段 early-time portion
早期发生型 early period prevalence
早期发展 early development
早期法国哥特式建筑风格 early French Gothic(style)
早期反射声 early reflection
早期反应 early reaction
早期防砂 early sand control
早期放射性沉降物 early fallout
早期佛罗伦萨文艺复兴风格 early Florentine Renaissance(style)
早期辐射反应 early radiation effect
早期腐朽 advanced decay
早期高稳定混凝土 concrete with high early stability
早期告警设备 early-warning equipment
早期告警系统 early-warning system
早期哥特式【建】primary Gothic; early Gothic
早期哥特式建筑 early Gothic architecture
早期哥特式教堂 early Gothic style church
早期固化 premature cure
早期故障 early failure period; incipient failure; initial failure
早期故障检测 early failure detection
早期灌溉 advance irrigation
早期海上勘探 early offshore exploration
早期海上钻井 early offshore drilling
早期核辐射 initial nuclear radiation
早期火烧 early burning
早期火灾探测 early fire detection
早期基督教的建筑物 early Christian structures
早期基督教(会)建筑 early Christian architecture
早期基督教堂 basilica
早期基督教堂式建筑 early Christian church architecture
早期加载 early loading
早期见水 early water breakthrough
早期鉴定 early evaluation
早期胶凝 premature gelation
早期结合 early bond
早期解离 deconjugation
早期金雀王朝式建筑 early Plantagenet style
早期警报 early warning
早期警报监测系统 early warning monitoring system
早期警报试验 < 工程结构的 > early-warning test
早期警报系统 early-warning system
早期聚合 premature polymerization
早期开采系统 early production system
早期开发 early development
早期开发装置 early production installations
早期开裂 early cracking; immediate crazing; preloading cracking
早期抗压强度 early compressive strength
早期枯叶病 early leaf blight
早期历史 early history
早期裂缝 early crack; incipient crack; incipient fracture; initial crack; preloading cracking
早期裂纹 early-age cracking
早期硫化 premature vulcanization; prevulcanize
早期硫化橡胶 scorched rubber
早期罗马风格 pre-Romanesque style
早期罗马风格的雕刻 pre-Romanesque sculpture
早期罗马式建筑 early Romanesque style
早期灭火快速反应 early suppression fast-response
早期灭火快速反应洒水喷头 early suppression fast-response sprinkler
早期模拟 early simulation
早期磨损 early wear; initial wear
早期黏[粘]合 early bond
早期破坏 premature destroy; premature failure
早期破损 premature failure
早期强度 early strength; initial strength
早期强度增长 early strength gain
早期乔治式【建】early Georgian style
早期三叠系砂岩 Lower Triassic sandstone
早期烧结 premature sintering
早期生产的产品 early production
早期生产系统 early production system
早期生油说 early origin theory of petroleum
早期失效 earlier failure; early failure; initial failure
早期失效期 debugging period; earlier failure period; early failure period
早期施工的钻孔 early hole
早期试验 previous test
早期收缩 early contraction; early shrinkage; initial shrinkage; premature contraction
早期数据 early-time data
早期死亡 early death
早期塑性流动 incipient plastic flow
早期损坏 early damage
早期损坏率 infant mortality
早期损坏率周期 infant-mortality period
早期探测 < 火灾发生时的探测 > early detection
早期碳酸盐化作用 eogenetic carbonatization
早期突破 early breakthrough
早期褪色 fugitive; fugitive dye
早期温度裂缝 early-age thermal crack
早期温度收缩 early thermal contraction
早期温度收缩裂缝 early thermal shrinkage crack
早期文艺复兴时代建筑式 early Renaissance(style)
早期文艺复兴时的半球形屋顶 early Renaissance cupola
早期文艺复兴时期圆屋顶 early Renaissance dome
早期文艺复兴式【建】early Renaissance(style)
早期稳定性 early stability
早期希腊建筑 < 有一个全遮蔽的屋顶 > cleithral; clithral
早期效应 early effect
早期斜面接触 premature contact of incline plane
早期新华夏构造体系【地】Early Neocathaysian system
早期形成的裂缝 early stage cracking
早期休闲 early fallow
早期修剪 early pruning
早期压力 early compression
早期压缩 early compression
早期岩浆分凝作用 early magmatic segregation
早期岩浆矿床 early magmatic ore deposit
早期养护 precure; precuring
早期养护阶段 early curing period
早期养护周期 early curing period
早期因素 early cause
早期英国尖顶建筑 early Pointed
早期英国尖拱式建筑 early Pointed
早期英国式教堂 early English church
早期英国天主教堂式建筑 early English cathedral style
早期英式 early English style
早期英式建筑 early English architecture
早期英式建筑窗 early English style window
早期英式建筑柱基 early English style base
早期硬化 early hardening; prehardening; premature hardening
早期用途 early uses
早期预警试验 < 工程结构的 > early-warning test
早期支承 early bearing
早期注水 initial filling

早期阻力 early resistance
早启绿灯 leading green
早强波特兰水泥 high-early(strength) Portland cement;high initial(strength) Portland cement
早强掺和剂 early strength admixture
早强的 high-early strength
早强硅酸盐水泥 high initial strength Portland cement;high-early strength Portland cement; rapid hardening Portland cement
早强混凝土 concrete of high early strength; concrete with high early strength; early strength concrete; high-early(strength) concrete
早强剂 early strength;early strength agent; hardening accelerator (admixture); high-early strength agent;early strength component(of concrete) < 水泥中的 > ; strength accelerating admixture
早强减水剂 hardening accelerating and water-reducing admixture
早强锚杆 early strength rockbolt
早强水泥 early strength cement;high-early cement; high-speed cement; quick-hardening cement; rapid hardening cement
早强添加剂 early strength addition
早强外加剂 early strength admixture
早强型混凝土 high-early strength concrete
早强型水泥 high-early strength cement
早强油井水泥 early strengthened oil well cement
早强组分 < 水泥中的 > early strength component(of concrete)
早燃 preignition; premature combustion
早三叠世【地】Lower Triassic
早膳厨房 breakfast kitchen
早生植物 xeromorphic vegetation;xerophyte
早石炭世海浸【地】Early Carboniferous transgression
早世代材料 early generation material
早释 premature disconnection
早熟 earliness;precocity;precox;prematurity
早熟的 premature
早熟禾 annual meadow grass; blue grass
早熟品种 early maturing variety
早衰效应 devitalizing effect
早霜 early frost
早填日期 antedate
早维斯康辛 early Wisconsin
早魏克塞尔冰期【地】early Weichselian glacial epoch
早燕山期地槽【地】early Yanshanian geosyncline
早燕山亚旋回【地】early Yanshanian subcycle
早已成熟的 overdue
早已到期 long past due
早已过期 long past due
早已准备好的 cut and dried;cut-and-dry
早硬结 early stiffening
早元古代冰期【地】early proterozoic glacial stage
早志留世【地】Lower Silurian
早中三叠世海浸【地】early-mid-Triassic transgression
早侏罗世【地】Lower Jurassic
早侏罗世气候分带 early Jurassic climatic zonation
早壮年期 submature

枣红色 auburn;bordeaux;claret;purplish red

枣红松节油 Bordeaux turpentine
枣树 jujube;jujube tree
枣椰树 date palm
枣属 jujube

蚤缀属 sandwort

澡盆 balaneion;bathtub

澡盆龙头 bath-cock;tub cock;tub tap
澡盆旋塞 bath-cock;tub cock
澡堂 bathhouse;hamman

藻白云岩 algal dolomite

藻饼 algal biscuit
藻层 algal layer; algal zone; gonidial layer
藻沉淀池 algae settling pond
藻的 algal
藻的繁殖 algae growth
藻叠层石硅质岩 algal stromatolitic siliceous rock
藻豆粒 algal pisolite
藻堆 alga glomerules
藻鲕 algal ooid
藻褐素 fucoxanthin
藻华 algae bloom;algal bloom
藻华控制 algal bloom control
藻华现象 algal bloom phenomenon
藻环 algal rim
藻灰结核 oncolite
藻灰岩 algal limestone
藻脊 algal ridge
藻胶 algin
藻胶体 phycocolloid
藻礁 algal reef
藻结构 algal structure
藻结沙坪 algal bound sand flat
藻井 caisson < 天花板的 > ; ceiling caisson;coffer;sunk panel;trave
藻井顶板 lacunar
藻井顶棚 cassette ceiling; coffered ceiling
藻井平顶【建】caisson ceiling;coffer(ed) ceiling;coffering;lacunar
藻井天花板 caisson ceiling; coffered ceiling;coffering;lacunar
藻井形态 caisson pattern
藻菌共生系统 algae-bacterial symbiotic system
藻菌型 algal-bacteria type
藻菌(植物) phycomycete
藻坑 algal pit
藻类 alga[复 algae]
藻类产量 algal productivity
藻类除去法 algae control
藻类处理废水 wastewater treatment by algae growth
藻类混合深度 algal mixing depth
藻类结构体 alginite-telinite
藻类控制 algae control;algal control
藻类离心分离 centrifuging of algae
藻类密度 algal density
藻类泥炭净化床 algae turf scrubber
藻类曝气氧化塘 algae aerated lagoon
藻类群落 algal community
藻类群落结构 algal community structure
藻类群体 algae population;algal population
藻类溶源菌 algae-lysing bacterium
藻类生长 algae growth
藻类生长潜力 algae growth potential
藻类生长曲线 algae growth curve
藻类生长势 algae growth potential; algal growth potential
藻类生长势测试 algal growth potential test
藻类生长势试验 algae growth potential test
藻类生物量 algal biomass
藻类生物指数 algal biological index
藻类水华 algal bloom
藻类水面增殖 algae bloom
藻类丝状体 algal filament
藻类体 alginite
藻类无结构体 alginite-collinite
藻类细菌 algae bacteria
藻类细菌黏[粘]土废水处理系统 algae-bacterial-clay wastewater treatment system
藻类絮凝体 algae floc
藻类学 algology;phycology
藻类氧化塘 algal oxidation pond
藻类遗骸泥 awja
藻类营养物 algae nutrient; algal nutrient
藻类真菌 algal gungi
藻类植被 algal vegetation
藻类植物 phycophyta
藻沥青 balkacchite
藻砾 maerl
藻粒硅质岩 algal pelletic siliceous rock
藻磷块岩 algal phosphorite
藻锚石 algal anchor stone
藻煤 algal coal;bog head coal;gelosic coal; sapromixite[sapromyxite]; tomite
藻膜体 phycoplast
藻凝块 algal lump
藻青素颗粒 cyanophycin granule
藻球 algal ball
藻球粒 algal pellet
藻群 algae mats;algal mat
藻溶胶 algosol
藻朊酸 alignic acid
藻(朊)酸盐 alginate
藻丝 algae mats;trichome
藻酸 alginic acid
藻酸铵 ammonium alginate
藻酸钙 calcium alginate
藻酸钠 sodium alginate;sodium polymanuronate
藻酸盐增稠剂 alginate thickener
藻滩相 algal bank facies
藻塘 algae pond
藻体锥 soredium
藻田构造 coalfield structure
藻团粒 maerl
藻污 weed fouling
藻席 algal mat
藻油页岩 boghead cannel; boghead shale
藻渊 algal pit
藻缘 algal rim
藻质体 alg(a)inite
藻质型 algal type
藻烛煤 boghead cannel; torbanite; boghead(ite) coal
藻状迹 fucoid
藻状菌纲 < 拉 > Phycomycetes

灶 cooker;cooking stove

灶具烟道 appliance flue
灶面板 hot plate
灶面燃烧器 boiling burner; cooker burner;top burner
灶旁热水锅炉 range boiler
灶神 household god
灶体 frame work
灶体构架 body frame
灶体构造 body frame

皂厂杂油 niger oil

皂碟 soap tray
皂碟砖 soap dish tile;soap tray tile
皂矾黑颜料 copperas black
皂粉 powdered laundry soap; soap powder
皂垢 soap scum
皂含量 soap content
皂糊 soap paste
皂滑性 soapiness
皂化 saponify(ing)
皂化槽 soap tank
皂化程度 saponification degree
皂化处理 soap treatment
皂化处理的 soap-treated
皂化醋酯纤维 saponified acetate
皂化当量 saponification equivalent
皂化的甘油三油酸酯 saponified oleine
皂化锅 soap kettle
皂化剂 saponification agent;saponifier;saponifying agent
皂化价 saponification number;saponification value
皂化率 saponification degree;saponification ratio
皂化木质树脂 saponified wood resin
皂化瓶 saponification flask
皂化器 saponifier
皂化乳化剂 saponified emulsifier
皂化石油 saponated petroleum
皂化试验 saponification test
皂化(速)率 saponification rate
皂化塔 saponification column
皂化危险 saponification risk
皂化系数 saponification number
皂化性 saponifiability
皂化油 saponifiable oil;saponified oil
皂化值 saponification number;saponification value
皂化阻碍 saponification resistance
皂化作用 saponification
皂基 soap base
皂荚 Chinese honeylocust
皂碱液 soap lye;soap solution
皂脚 niger
皂类 < 有机酸与金属的化合物 > soap
皂类防水剂 saponaceous waterproofing agent
皂粒 neat soap
皂模 soap dye;soap mo(u)ld
皂膜 soap bubble; soap film; soap membrane
皂膜比拟法 soap film analog(y); soap-bubble analog(y)
皂膜法 soap film method
皂膜计 soap film meter
皂膜流量计 soap film flowmeter
皂膜模拟 soap film analog(y)
皂膜气量计 soap film gas meter
皂膜张力 soap film tensor
皂沫法 soapsuds method
皂沫试验 soapsuds test
皂莫属 honey locust
皂泡 soap bubble
皂泡法 < 测定火焰基本速度的 > soap-bubble method
皂泡检漏 soap-bubble leak detection
皂泡流量计 soap-bubble flowmeter
皂泡试验 soap-bubble test
皂坯 soap base
皂坯洗涤粉 soap-based powder
皂皮 soap bark
皂片 chip soap;soap chip
皂片干燥机 soap chip drier
皂溶液 soap solution

皂色谱 soap chromatography
皂石 bowlingite; mountain soap; piotine; saponite; smegmatite; soap earth; soaprock; soapstone; zebedassite
皂水膜 soap water film
皂水器 <洗手用> liquid soap dispenser
皂水试验 soap test
皂素废水 saponin wastewater
皂碳氢化合物胶体 soap-hydrocarbon gel
皂体 soap body
皂条 soap bar
皂土 bentonite(clay); soapy clay; wilkinite
皂土絮凝试验 bentonite flocculation test
皂丸 soap tablet
皂洗 soaping
皂洗机 soaper
皂相构造 soap-phase structure
皂型乳化剂 soap-type emulsifier
皂性 nature of soap
皂液 soap lye;soap solution
皂液甘油 soap lye glycerine
皂液瓶 soap holder
皂因数 soap factor
皂用杀菌剂 soap germicide
皂油 soap oil
皂油分散体 soap-oil dispersion
皂油系统 soap-in-oil system
皂质乳液 <用肥皂作乳化剂的乳液> soap emulsion
皂质油乳化液 soap-oil emulsion
皂砖 soap brick
皂渍 soap-mark;soap spot

造

造岸 bank building

造坝混凝土 dam concrete
造币厂 mint
造壁 wall(ing)up
造壁作用 <泥浆的> plastering action
造标 construction of signal;erect of signal;signal erection;tower building
造标组 building party
造表 billing
造滨现象 beach-building phenomenon
造波 wave generation
造波板 wave blade;wave paddle
造波槽 wave channel;wave tank
造波抵抗 wave drag
造波机 <船模试验池的> wave machine;wave generator;wave maker
造波器 wave generating apparatus; wave generator;wave maker
造波设备 wave generating equipment;wave-making apparatus
造波水槽 wave flume;wave tank
造波水池 wave basin;wave tank
造波装置 wave machine
造波阻力 wave-making resistance
造材 buck(ing); cross-cut(ting); roughhew
造材搬钩 log jack
造材工 bucker
造材检量 laying off
造册 tab;table;tabling;tabulation
造车槽钢 wagon building channel
造车工匠 cart wright
造成 create
造成惨重损失的 disastrous
造成沉陷的地震 subsidence earthquake
造成的损害 damage done
造成弓形 camber
造成很大破坏 to do a lot of damage
造成结构出现裂缝、损坏的沉降 detrimental settlement
造成亏损 come to deficit
造成隆隆声的路面装置片 <用来警告减速或警醒驾驶员之用> rumble strip
造成缺勤的工伤 disabling injury
造成缺勤的工伤发生率 disabling injury frequency rate
造成缺勤的工伤指数 disabling injury index
造成伤害的最低限值 threshold for injury
造成事故 cause an accident
造成压力 build-up of pressure
造成意外事件的疏忽 contributory negligence
造成中心黑影 dark hole
造船 boatbuilding;naval construction
造船保单 builder's policy; construction policy
造船编号 yard number
造船泊位 <英> shipbuilding berth
造船材 shipbuilding material
造船差价补贴 construction differential subsidy
造船厂 boat builder; boatbuilding yard; building yard; dock yard; naval yard; shipbuilding works; shipbuilding yard; ship-plant; shipyard
造船厂船坞起重机 naval yard crane
造船厂工段长 quarterman
造船场 shipbuilding basin; shipbuilding yard
造船船台 shipbuilding berth; building slipway;building berth
造船船坞 building basin; building dock;shipbuilding dock
造船的 naval
造船调度员 coordinator of shipbuilding
造船干坞内下水 launching dock
造船钢 hull steel
造船工 boatwright
造船工程 marine engineering; naval architecture
造船工程师 marine architect; marine engineer; naval architect; naval engineer;shipbuilding engineer
造船工程师协会 Institution of Naval Architects
造船工程学 naval architecture; naval engineering
造船工人 shipwright
造船工业 shipbuilding industry; shipping industry
造船工作者 shipbuilder
造船公司 shipbuilder; shipbuilding corporation
造船滑道 ground way; shipbuilding slip; shipbuilding way; shipway; slipway; building berth; building way;launching way;launchway
造船滑道的基础部分 <美> slipway
造船滑道上的轨道 launchway
造船滑台 building slip;slipway
造船划线工 ship fitter
造船机械 shipyard machinery; shipyard machine tool
造船及轮机工程师试协会 the Society of Naval Architects and Marine Engineers
造船几何画法 laying down;laying off
造船技师 naval architect; naval technician
造船技术 marine technique
造船胶合板 consuta plywood
造船界 shipbuilding circles
造船留置权 shipbuilding lien
造船龙骨垫 building block
造船门式起重机 shipbuilding gantry crane
造船木材 ship timber
造船木工 shipwright
造船棚 ship-house
造船起重机 dock crane; shipbuilding crane;ship crane
造船契约图 contract plan
造船曲线板 ship('s)curve
造船渠 building basin
造船设备 shipbuilding facility
造船设计 boatbuilding design; ship calculation;ship design
造船师 ship constructor
造船湿坞 ship wet dock;slip dock
造船所 dock yard;shipbuilding yard
造船台 builder berth; building berth; building slip;shipway
造船(台)位 building berth
造船坞 building(dry)dock; construction dock;shipbuilding dock
造船学 naval architecture; shipbuilding;ship construction
造船学会 Society of Naval Architect
造船业 shipbuilding
造船业者 shipbuilding circles; shipbuilding industry;shipbuilding trade
造船用材 shipbuilding timber
造船用浮式起重机 shipyard floating crane
造船用钢板 ship plate
造船与轮机工程师学会 Society of Naval Architects and Marine Engineers
造船者 shipbuilder
造船支架 poppet
造船装配工 ship fitter
造船总工程师 chief constructor
造床过程 bed-building process; bed-forming process; channel forming process;fluvial process
造床流量 bed-building discharge; bed-forming discharge; bed-generating flow;channel forming discharge;discharge for river bed function; dominant discharge; dominant formative discharge; formative discharge; regime(n) discharge; generating flow; bed generative discharge <输移粗泥沙的最大一级流量>
造床泥沙 bed-building material; bed-building sediment;bed-forming material
造床水位 bed-building(water)stage; bed-built water level;bed-built water stage;bed-forming stage
造床水位计 bed-built water ga(u)ge
造袋术 marsupialization
造地工程 land forming
造防波堤 groining
造拱 turning an arch
造海的 thalassogenic
造海运动 thalassogenic movement
造海作用 thalassogenesis
造洪暴雨 flood producing storm
造价 building cost; calculative cost; capital cost; capital charges; capital expenditures; construction cost; cost of construction; cost of manufacture; fabricated cost; fabricating cost; fabrication cost; first cost; manufacturing cost
造价参考指南 cost reference guide
造价动态管理 cost management dynamic
造价分配 cost allocation
造价分析 analysis of prices; cost analysis;cost study
造价浮动 cost fluctuation
造价工程 cost engineering
造价工程师 value engineer
造价估算 cost estimate; cost estimating;cost estimation
造价规划 cost plan(ning); cost programming
造价合理计定 cost estimate and determine reasonably
造价结算 cost estimation
造价控制 cost control
造价口径分析 cost-aperture analysis
造价限额 cost limit
造价影响因素 cost factor
造价优化模型 cost optimization model
造价有效控制 cost control effectively
造价预算 account transaction; account valuation
造价指标 cost index
造价最低结构 minimum-cost structure
造假账 cook the accounts; falsify accounts;wangle accounts
造假账者 fiddler
造建晶格 lattice-building
造浆量 make slurry volume
造浆率 mud fluid yield; rate of making mud
造浆黏[粘]土 mud-forming clay
造浆黏[粘]土粉 Baroco
造浆岩层 <指在钻进过程中可以形成泥浆的岩层> mud caking formation;mud making formation
造礁珊瑚 bermatypic coral; hermatype;reef-building coral
造礁藻类 reef-forming algae
造井机具 penetration drilling of well completion
造景工程 landscape engineering
造景设计 landscape design
造景学 landscape architecture
造景园艺师 landscape gardener
造景栽植 landscape planting
造酒厂 distillery
造孔 pore-creating;pore-forming
造扣长度 length of thread ring
造块 agglomeration
造矿废水 wastewater from ore dressing
造矿元素 mineralizer; ore-forming element
造粒 granulation; palletizing; pelleting; pelletization[pelletisation]; prill(ing);size enlargement
造粒板 marume plate
造粒工艺 prilling
造粒机 comminutor; granulator; palletizer; pelleting machine; tablet(t)ing machine;tablets press
造粒机头 pellet head
造粒剂 nodulizer
造粒喷头 prilling spry
造粒塔 granulation column; prilling tower
造粒装置 granulators plant
造粒作用 pelletizing
造林 afforest(ation); culture; forest culture; foresting; plantation; planting;planting out;silviculture
造林百分比 percentage afforestation
造林播种机 tree-seed drill
造林地 reproducing area; site of afforestation
造林法 forestation
造林费 cost of formation; cost of planting
造林工程 afforestation project
造林计划 afforest(ation); cultural plan
造林密度 density of plantation
造林面积 afforestation area; afforested area
造林区 afforestation area; afforested area

造林区划 planting series
造林设计 project planning for machine planting
造林失败地 failed area;failed place
造林学 arboriculture;silviculture
造林学原理 silvics
造林者 forest grower
造林治沙(砂) afforestation for sand control
造林专家 silviculturist
造流降雨 discharge-producing rain
造陆 land reclamation
造陆沉积作用【地】 epeirogenic sedimentation
造陆的 epeirogenetic;epirogenic
造陆地裂运动【地】 epeirogenic taphrogenesis
造陆隆起 epeirogenetic uplift
造陆期 epeirocratic period
造陆上升 epeirogenic uplift
造陆相 epeirogenic facies;epeirogenic phase
造陆优势期 epeirocratic condition
造陆运动【地】 epeirogenesis; continent making movement; epeirogenetic movement;epeirogeny
造陆运动的 epeirogenic
造陆褶皱 epeirogenic fold
造陆作用 epeirogenesis;epeirogenetic movement;ep(e)irogeny
造沫能力 foam-producing ability
造盆运动【地】 basin building
造皮机 skin-making machine
造气 gas making
造气剂 gas-forming agent; gas-forming constituent;volatile constituent
造桥 bridge building;bridging
造桥人 bridge builder
造球机 pelletizer
造球盘 balling disc[disk]
造球性 < 矿料的 > ballability
造球作用 balling
造山变质作用【地】 orogenic metamorphism
造山不整合 orogenic unconformity
造山沉积 orogenic sedimentation
造山成因 orogeny
造山带 fold belt; orogene; orogenic belt;orogenic zone
造山带内凹 embayment
造山的 acogenic; mountain-building; mountain-making; orogenetic; orogenic
造山地槽 orogeosyncline
造山地带 orogen
造山过程 orogenetic process
造山后的 apotectonic
造山后煤化 postorogenic coalification
造山后期 postorogenic phase
造山阶段 orogenic stage
造山力 mountain-making force
造山脉动期 pulsation
造山幕 orogenic phase
造山期 mountain building period;orogenic period; orogenic phase; period of orogenesis
造山期后 postorogenic
造山期后的 post-tectonic; postkinematic;postorogenic
造山期后的裂谷 postorogenic rift
造山期后煤化作用 postorogenic coalification
造山期煤化作用 synorogenic coalification
造山前的 preorogenic
造山前期 preorogenic phase
造山前期煤化作用 preorogenic coalification
造山区 orogenic area;orogenic region
造山省 orogenic province
造山事件 orogenic event
造山旋回 geotectonic cycle;orogenic cycle
造山循环 orogenic cycle;tectonic cycle
造山运动【地】 tectogenesis;concentrated earth movement;earth movement; mountain-building;mountain folding; mountain-making; mountain-making movement; mountain orogenesis; orogenetic disturbance; orogenetic movement; orogensis; orogeny; tectonic cycle
造山运动过程 mountain-making process
造山运动前的 preorogenic
造山周期 orogenic period
造山作用 concentrated earth movement; mountain-making; orogenesis;orogeny
造水机 fresh water generator
造水装置 distilling plant
造滩波浪 constructive wave; beach building wave < 向岸输送泥沙的波浪 >
造滩的 beach building
造滩泥沙 beach-building material
造梯田 terracing
造田 empolder;garden making
造田坝 check dam for building farmland
造田工程 land forming
造艇场 boatyard
造艇工厂 boatyard
造土 soil building
造涡 churning
造窝 dimpling
造小尖塔 pinnacle
造斜【岩】 deflect(ing);side-tracking
造斜点 kick-off point; whipstock point
造斜工具 deflecting tools; deflection tool;kickover tool
造斜力 deflecting force
造斜率 deflection rate
造斜器 deflector;whipstock
造斜塞 deflecting plug; deflection plug
造斜楔 deflecting wedge; deflection wedge
造斜楔定位环 deflector wedge ring
造斜钻进 defection drilling
造斜钻孔法 deflection method
造斜钻头 deflecting bit;deflection bit
造芯混合料 core sand
造形的 plastic
造形术 neoplasty
造型 build;mo(u)ld(ing);mo(u)ld-making;shaping
造型白石膏 white mo(u)lding plaster
造型板 mo(u)ld board
造型部件 mo(u)lding unit
造型材 sand strength
造型材料 creative design material; mo(u)lding material; mo(u)lding medium;plasticine
造型操作 mo(u)lding operation
造型车间 make-up department; mo(u)ldery;mo(u)lding shop
造型尺寸 mo(u)lded dimension
造型锤 mo(u)lder's hammer
造型底板 follow board; mo(u)lding board;mo(u)lding plate
造型地坑 mo(u)ld(ing) pit
造型地貌 imaginative geomorphologic figuration
造型垫 forming pad
造型顶棚 shaped ceiling
造型法 mo(u)lding method; mo(u)lding practice
造型法变形 mo(u)lding texturing
造型感情 plastic sentiment
造型工 mo(u)lder
造型工地 foundry floor
造型工段 mo(u)lding department; mo(u)lding departure;mo(u)lding room;mo(u)lding shop
造型工具 mo(u)lder tool; rigging < 铸工的 >;gagger
造型工艺 formative technology
造型构思 plastic thinking
造型刮板 mo(u)lding scraper;strickle board;sweep template
造型刮托 strickle board support
造型光 model(l)ing light
造型混合料 mo(u)lding compound; mo(u)lding mixture
造型机 mo(u)lder; mo(u)lding machine
造型机组 mo(u)lding unit
造型及设计原理 theory of form and design
造型技术 sculpt technology
造型空间 model space
造型冷压 cold-molding
造型模板 mo(u)ld
造型黏[粘]土 bonding clay; clay bond;foundry loam;ladle clay;mo(u)lding loam
造型配色 plastic colo(u)r
造型平板 mo(u)lding board;turning over board
造型气眼针 pricker
造型润滑剂 mo(u)ld lubricant
造型润滑油 mo(u)lding grease
造型砂 casting sand
造型砂铲 mo(u)lding spade
造型砂箱 flask
造型设备 mo(u)lding apparatus;mo(u)lding equipment
造型设计 constructive design;figurative design
造型设计系统 constructive design system
造型生产线 mo(u)lding production line
造型石膏 gypsum mo(u)lding plaster;mo(u)lding plaster
造型术(构模) model(l)ing
造型台 banker
造型特性 plastic character
造型铁铲子 mo(u)lder's peel
造型通气针 mo(u)lder's brad
造型涂层 mo(u)lding plaster
造型涂料 blacking; black wash; mo(u)ld wash;slip
造型挖泥 configuration dredging
造型细木工艺 shaped work
造型线 mo(u)lding track
造型效果 plastic effect
造型性能 mo(u)ldability
造型修刀 parting-off tool
造型学 plasticism
造型压力 mo(u)lding pressure
造型艺术 figurative art; formative arts;plastic arts
造型艺术作品 figurative work of art
造型用灰泥 model(l)ing plaster
造型用煤气喷灯 gas torch for mo(u)lding;model(l)ing burner
造型用涂料 mo(u)lding ink
造型用小工作台 snap bench
造型油脂 mo(u)lding grease
造型语言 plastic language;plastic sign
造型者 mo(u)lder
造型震实机 foundry jolter
造型周期 mo(u)lding cycle
造型转台 mo(u)lding turntable
造性 creativeness
造雪机 snow maker
造崖层 cliff-maker
造岩 lithogenesis;lithogenesy
造岩的 lithogenous;petrogenic;rock-forming
造岩过程 lithogeneous process
造岩矿物 petrogenic mineral; rock-forming mineral;rock mineral
造岩元素 petrogenetic element;petrogenic element; rock-forming element
造岩组分 lithogenous component
造岩作用 lithogeneous process
造油体 oleoplast
造釉器 amelogenic organ;enamel organ
造雨 cloud seeding;rain producing
造雨过程 precipitation producing process
造雨云 rain-bearing cloud; rain producing cloud
造渊运动 bathygenesis
造园 garden construction; garden marking;landscape design
造园工程 landscape engineering;recreation engineering
造园工程师 landscape engineer
造园技术 technology of garden construction
造园家 landscape architect;landscape gardener;landscaper
造园林 landscaping
造园师 landscape gardener
造园术 garden craft
造园学 garden making;landscape architecture;landscape gardening
造园艺术 art of garden-making
造云器 meteotron
造渣 building of slag;flux(ing);formation of slag; scorify; slag; slag formation; slag-forming; slagging, slag-making;slag-off
造渣过程 slagging process
造渣剂 fluxing medium;slag former; slag-forming constituent;slagging agent;slagging constituent
造渣能力 slaggability
造渣期 slag-formation period
造渣试验 slag(ging) test
造渣型焊条 flux-coated electrode
造渣作用 fluxing effect
造者 creator
造纸 paper-making; paper manufacture
造纸材 paper wood;pulpwood
造纸材料 paper stock
造纸厂 paper factory;paper mill
造纸厂废水 paper mill waste; paper mill wastewater; wastewater from paper mill; wastewater of paper mill
造纸厂废水处理 paper-making wastewater treatment
造纸厂废液 black liquor
造纸厂废纸 broke
造纸厂泡沫油 paper mill(foam) oil
造纸厂弃纸 broke
造纸车间 paper machine room
造纸帆布 paper felt duck;paper-making canvas
造纸废料 paper-making waste
造纸废水 paper-making wastewater; paper waste(water)
造纸废物 paper-making waste
造纸废液稳定土(工艺或技术) < 利用含木质素的硫酸盐等的 > lignin stabilization
造纸工 paper maker;paper worker
造纸工业 paper industry;paper manufacturing industry
造纸工业废水 pulp and paper mill wastewater;wastewater from pulp

and paper mills
造纸工业废水处理 wastewater treatment of paper mills
造纸工业水污染物排放标准 discharge standard of water pollutant for paper industry
造纸工业污水 paper industry sewage
造纸工艺学 paper technology
造纸和纸浆废水 paper and pulp wastewater
造纸和纸浆工业 paper and pulp industry
造纸和纸浆工业废水 paper and pulp industry wastewater
造纸黑液 black paper-making wastewater
造纸机 paper machine; paper-making machinery
造纸机器废水 wastewater from paper machine
造纸机前筛浆机 paper machine screen
造纸机械 paper manufacturing machinery
造纸明矾 paper makers' alum
造纸木材 paper bolts; paper wood; pulpwood
造纸污水净化 purification of paper-making effluent
造纸纤维 <指适宜造纸的纤维> paper fiber[fibre]
造纸业 paper-making
造纸用陶土 paper clay
造纸用压光机 roll presses for paper industry
造纸原料 paper-making raw material
造纸纸浆工业 paper and pulp industry

噪暴 noise storm; storm burst

噪暴辐射 storm radiation
噪度 noisiness; perceived noisiness
噪度级 noise level
噪扰 nuisance; pigeons
噪扰带 noise fringe
噪哨 noise whistler
噪声 noise; noisiness; rushing sound; undesired sound; unpitched sound; gas noise <放电气体游离所产生的>
噪声白化滤波器 noise whitening filter
噪声斑点 noise speckle; noise spot
噪声暴露 noise exposure
噪声暴露计 noise exposure meter
噪声暴露监控计 noise exposure monitor
噪声暴露评价 noise exposure rating
噪声暴露时间 noise exposure time
噪声暴露限度 noise exposure limit
噪声暴露预测 noise exposure forecast
噪声背景 noise background
噪声本底 noise background
噪声比 noise signal ratio; speech-noise ratio
噪声比率 noise ratio
噪声边带 noise sideband
噪声边限 noise margin
噪声标限 noise objective
噪声标准 noise standard; noise criterion
噪声标准曲线 noise-criterion curve
噪声标准值 noise-criterion value
噪声表 psophometer; sound meter
噪声病 noise disease
噪声波形图 noise audiogram
噪声补偿法 noise compensation method
噪声补偿(费) noise compensation
噪声不定性 noise ambiguity
噪声槽测试频率 noise slot measuring frequency
噪声测定 noise measurement; noise-measuring
噪声测定技术 noise-measurement technique
噪声测距 noise ranging
噪声测距声呐 noise ranging sonar
噪声测量 measurement of noise; noise measurement; noise-measuring; noise survey
噪声测量浮标 noise-measuring buoy
噪声测量计 noise dosimeter
噪声测量设备 noise-measuring equipment
噪声测量仪 noise-measuring set; psophometer
噪声测量仪器 noise-measuring instrument
噪声测试 noise test(ing)
噪声测试计 noise meter; noise tester
噪声测试器 noise-measuring meter; noise meter; noise tester
噪声测试装置 noise-measuring system
噪声测温法 noise thermometry
噪声测向声呐 direct listening sonar
噪声查验 noise investigation
噪声产生 noise development
噪声常数 noise constant
噪声场 noise field
噪声场强 noise-field intensity
噪声场强测量仪 noise-field intensity measurement equipment
噪声程度 sound level
噪声冲击 noise impact
噪声冲击指数 noise impact index
噪声重发 noisy reproduction
噪声处理电路 noise processing circuit
噪声传播 noise emission; noise propagation; noise transmission
噪声传递 noise transmission
噪声传输 noise transmission
噪声传输损害 noise transmission impairment
噪声丛 noise burst
噪声带宽 noise bandwidth
噪声单位 noise unit
噪声当量 noise equivalent
噪声当量角 noise equivalent angle
噪声导生振动 noise-inducing vibration
噪声倒换 noise switching
噪声的标定方法 noise label
噪声的产生 noise generation
噪声的隔绝 isolation of noise
噪声的衰减 noise alleviation
噪声的削减 noise abatement
噪声的有效电压 root mean square noise voltage
噪声灯 noise lamp
噪声灯引燃器 noise lamp ignitor
噪声等级 noise grade; noise level
噪声等级评价数 noise rating number
噪声等效带宽 noise equivalent bandwidth
噪声等效电路 noise equivalent circuit
噪声等效电压 noise equivalent voltage
噪声等效电阻 noise equivalent resistance; noise resistance
噪声等效辐照度 noise equivalent irradiance
噪声等效功率 noise equivalent; noise equivalent power
噪声等效功率密度 noise equivalent power density
噪声等效目标温差 noise equivalent target temperature difference
噪声等效强度 noise equivalent intensity
噪声等效声压 equivalent noise
噪声等效输入 noise equivalent input
噪声等效通带 noise equivalent passband
噪声等效通量 noise equivalent flux
噪声等效通量密度 noise equivalent flux density
噪声等效温度 noise equivalent temperature
噪声等效源 noise equivalent source
噪声等效中子源 noise equivalent neutron source
噪声等值 noise equivalent
噪声等值线 contours of noise; noise contour; noise equivalent line; noise isoline
噪声等值信号 noise equivalent signal
噪声巅值 noise peak
噪声巅值限制器 noise peak clipper; noise peak limiter
噪声电动势 psophometric electromotive force
噪声电流 noise current
噪声电流发生器 noise current generator
噪声电平 level of noise; noise level; snow level
噪声电平表 noise level meter
噪声电平测量 noise level measurement
噪声电平记录 recording of the noise level
噪声电平监测器 noise level monitor
噪声电平调整 shading adjustment
噪声电平限幅器 amplitude noise control gate
噪声电平指示器 noise level indicator
噪声电压 hum voltage; noise voltage; psophometric voltage
噪声电压发生器 noise voltage generator
噪声电压计 psophometric voltage
噪声电压平均值 average noise value
噪声电阻 noise resistance
噪声调查 noise survey
噪声定位声呐 passive sonar; listening sonar; listen only sonar
噪声定向站 sound bearing station
噪声抖动 noise dither
噪声度 noise level
噪声对人的影响 effect of noise to man
噪声额定值 noise rating number
噪声发射 noise emission
噪声发射测试 noise emission test
噪声发射极限 limiting noise emission
噪声发生器 flatter generator; noise generator; noisemaker
噪声发生源 noise-producing source
噪声发生装置 noise-producing equipment
噪声发送器 noise transmitter
噪声发展 noise development
噪声罚款 noise compensation
噪声法校准 noise method calibration
噪声反馈编码器 noise-feedback coder
噪声反射面 noise reflecting surface
噪声范围 noise margin; noise range
噪声方式 noisy mode
噪声防护 noise protection
噪声防护器 noise protector
噪声防止 noise prevention
噪声防治工作者 noise crusader
噪声放大 noise amplification
噪声放大电路 noise amplifier circuit
噪声放大器 noise amplifier
噪声放电管 noise discharge tube
噪声分布 noise distribution
噪声分级 <以分贝计> sound level; sound level of equipment; noise rating; sound level difference
噪声分级曲线 noise rating curve
噪声分级仪 sound level recorder
噪声分级值 noise rating number
噪声分类 noise classification
噪声分量 noise component
噪声分配 noise assignment
噪声分区 noise zoning
噪声分析 noise analysis
噪声分析器 noise analyser[analyzer]; noise sound analyser [analyzer]; sonic noise analyser[analyzer]
噪声风洞 noise wind tunnel
噪声封闭器 <桩工> noise cover; sound-proof enclosure
噪声峰值 noise peak
噪声幅度 amplitude of noise; range of noises
噪声辐射 noise radiation
噪声辐射图 noise radiation pattern
噪声负荷比 noise load ratio
噪声负载 noise loading
噪声负载比 noise loading ratio
噪声负载测试设备 noise loading test set
噪声负载法 noise loading method
噪声负载试验 noise loading test
噪声改进因数 noise improvement factor
噪声改善系数 improvement factor; noise improvement factor; signal-to-noise improvement factor
噪声概率密度 noise probability density
噪声干扰 acoustic(al) radiation; noise disturbance; noise hindrance; noise interference; noise jamming; noise nuisance
噪声干扰机 noise jammer
噪声干扰灵敏度 noise interference sensitivity
噪声感应 noise induction
噪声隔绝套 noise shelter
噪声隔离 noise insulation
噪声隔离器 noise isolation unit
噪声隔离因素 noise insulation factor
噪声隔声等级 noise insulation class
噪声隔声分类 noise insulation class
噪声隔声率 noise insulation factor
噪声跟踪 noise tracking
噪声公害 noise nuisance
噪声公害测定仪 noise pollution(level) meter
噪声功率 interference power; noise power
噪声功率比 noise power ratio
噪声功率比较仪 noise power comparator
噪声功率测量 noise power measurement
噪声功率级 noise power level
噪声功率谱 noise power spectrum
噪声功率水平 sound power level
噪声估定系统 noise-rated system
噪声估计试验 noise evaluation test
噪声管 noise tube
噪声管理 noise management
噪声管理条例 noise control regulations
噪声管制法 noise control law
噪声管制立法 noise control legislation
噪声光电学 noise photoelectronics
噪声光电子 noise photoelectron
噪声光脉冲 noise spike
噪声轨迹 noise track
噪声过滤 noise filtering
噪声函数 noise function
噪声衡消电路 noise balancing circuit
噪声衡消制 noise balancing system
噪声环境 acoustic(al) noise environ-

ment;noise circumstance;noise environment
噪声基准 noise criterion
噪声基准曲线 noise-criterion curve
噪声畸变 noise distortion
噪声激发 noise-excitation
噪声及次数指标 noise and number index
噪声及振动评估 noise and vibration assessment
噪声级(别) noise level;level of noise
噪声级测量法 method for measurement of noise level
噪声级记录纸 recording paper of sound level
噪声级试验 noise level test
噪声级数 noise rating number
噪声级验收试验 noise level acceptance test
噪声极限 noise margin
噪声集中区 sound foci
噪声计 acousimeter;audio noise meter;noise event meter;noise level meter; noise-measuring meter; noise meter; psophometer; sound level meter;sound meter
噪声计权曲线 weighted curve for noise measurement
噪声计数 noise count
噪声记录 noise record(ing)
噪声记录仪 recording noise meter
噪声剂量 noise dose
噪声加权 noise weighting
噪声假信号 noise glitch
噪声尖峰 noise spike
噪声监测 noise monitoring;noise survey
噪声监测点 noise monitoring site
噪声监测接头 noise monitor junction
噪声监测仪 noise monitoring equipment
噪声监界区 noise critical area
噪声监控 noise monitoring;noise monitoring and control
噪声监控部门 noise monitoring unit
噪声减除 noise abatement
噪声减低 noise reduction
噪声减低设施 noise abatement device
噪声减少 noise reduction
噪声检测 noise measurement
噪声检测电路 noise detecting circuit
噪声检测器 noise monitor
噪声检定值 noise rating number
噪声建议容许级 suggested permissible level of noise
噪声降低 noise abatement;noise reduction
噪声降低测定 noise reduction rating
噪声降低系数 noise reduction coefficient
噪声矫正仪 noise rectifier
噪声校正 noise compensation
噪声接触 noise exposure
噪声接触预报 noise exposure forecast
噪声结构传输路径 structure-borne path
噪声界限 noise margin
噪声金属导线的 noise-metallic
噪声空气传输路径 air-borne path
噪声空气污染 noise-air pollution
噪声控制 noise abatement;noise control;sound control
噪声控制标准 criteria for noise control
噪声控制法 noise abatement act; noise control law
噪声控制规划 noise control program(me)
噪声控制技术 noise technique
噪声控制立法 noise control legislation
噪声控制评价标准 criterion for noise control
噪声控制系统 noise control system
噪声控制学 noise control acoustics
噪声控制要求 criterion for noise control
噪声控制准则 noise control criterion
噪声扩展 noise stretching
噪声雷达 noise radar
噪声离散 noise variance
噪声利用 noise use
噪声量 noise dose;noise level;noisiness
噪声量指数 noise and number index
噪声灵敏度 noise sensitivity
噪声流体传输路径 fluid-borne path
噪声录音机 applauseograph
噪声滤波器 noise filter
噪声码字 noise word
噪声脉冲 noise count;noise impulse
噪声脉冲串 noise burst
噪声脉冲宽度 noise pulse width
噪声门限 noise gate
噪声门限式自动增益控制 noise-gated automatic gain controller
噪声密度 noise density
噪声敏感地区 noise-sensitive area
噪声敏感性 noise susceptibility
噪声模拟器 noise simulator
噪声模拟装置 noise simulator
噪声模型 noise model
噪声耐量 noise tolerance
噪声能量 noise energy
噪声扭变 noise distortion
噪声排除 sound exclusion
噪声判据 criterion for noise; noise criterion
噪声疲劳 acoustic(al) fatigue
噪声频程 frequency division of noise
噪声频带 noise band
噪声频率响应 noise-frequency response
噪声频谱 noise spectrum
噪声频谱成型网络 noise shaping network
噪声平衡抑制装置 noise balancing system
噪声评级 noise rating
噪声评级曲线 noise rating curve
噪声评级数 noise rating number
噪声评价 noise assessment;noise rating
噪声评价曲线 noise rating curve
噪声评价首曲线 preferred noise criterion curve
噪声评价标准 criterion for noise
噪声评价参数 noise rating parameter
噪声评价曲线 noise-criterion curve
噪声评价试验 noise rating number
噪声评价数 noise rating number
噪声评价系数 psophometric weight
噪声评价值 noise-criterion number
噪声评价准则 criterion for noise
噪声屏蔽 noise screen(ing); noise shielding
噪声屏障 noise insulation;noise barrier
噪声屏障因子 noise barrier factor
噪声谱 noise spectrum
噪声起伏 noise bounce;noise fluctuation
噪声气势 noise climate
噪声器 noise meter
噪声强度 intensity of noise; noise density;noise dose;noise intensity;noise level;noisiness
噪声强度标准 noise strength standard
噪声强度曲线 noise-intensity contour
噪声强度试验 noise level test
噪声区 noise range;noise zone
噪声曲线 noise curve
噪声容限 noise margin
噪声容许标准 noise standard
噪声容许标准曲线 noise-criterion allowance curve
噪声容许极限 noise criterion
噪声容许量 noise tolerance
噪声散射 noise emission
噪声骚扰 noise pollution
噪声骚扰级 noise pollution level
噪声闪烁 noise flashes
噪声生理效应 physiologic(al) effect of noise
噪声声级的中间值 median of sound level
噪声声级计 noise level meter
噪声声垒 noise barrier
噪声声压级试验 noise pressure level test
噪声失真 noise distortion
噪声识别 noise identification
噪声拾波线圈 noise pickup coil
噪声拾取 noise pickup
噪声矢量 noise vector
噪声使听觉丧失 noise-induced hearing loss
噪声势 noise potential
噪声试验 noise test
噪声试验仪 noise testing instrument
噪声室 noise room
噪声输出量 noise output
噪声数位 noise digit;noisy digit
噪声数字 noisy digit
噪声衰减 attenuation of noise;noise attenuation; noise degeneration; noise reduction
噪声衰减器 noise muffler
噪声衰减系数 noise reduction coefficient
噪声水平 noise level;sound level
噪声水平标准 noise rule
噪声速度定律 noise-velocity law
噪声速调管 noise klystron
噪声损害 noise nuisance
噪声损伤 noise injury
噪声态 noise mode
噪声探测器 noise detector
噪声探测仪 relative ionospheric opacity meter;riometer
噪声探头 noise sensor
噪声特性 noise characteristic;noisiness;spurious response
噪声调幅干扰 noise AM jamming
噪声调相干扰 noise phase modulation jamming
噪声调制 noise modulation
噪声调制的 noise-modulated
噪声调制干扰 noise-modulated jamming
噪声跳动 noise bounce
噪声听力图 noise audiogram
噪声通路 noise channel;noisy channel
噪声图 noise figure;noise pattern
噪声图测 noise survey
噪声网络 noise network
噪声危害 noise hazard; noise nuisance
噪声围蔽 enclosure of noise
噪声维护部件 noise maintenance component
噪声温度 noise temperature
噪声温度比 noise-temperature ratio
噪声温度计 noise thermometer
噪声闻阈图 noise audiogram
噪声稳定度 noise stability
噪声稳定率 noise stability
噪声问题 noise problem
噪声污染 acoustic(al) pollution;noise emission;noise pollution;noise pollution level;sound pollution
噪声污染级 noise pollution level
噪声污染监测 noise pollution monitoring
噪声污染预测 noise pollution forecasting
噪声污染源 source of noise pollution
噪声污染综合防治 integrated control of noise pollution
噪声物理学 noise physics
噪声吸收 noise-absorbing; noise absorption
噪声吸收电路 noise-absorbing circuit
噪声吸收器 noise absorber;noise killer
噪声熄灭装置 noise blanker
噪声系数 figure(of) noise;noise coefficient;noise factor;noise figure
噪声系数测定 noise-factor measurement
噪声系数测量 noise-factor measurement
噪声系数列线图 noise-factor nomogram;noise-factor nomograph
噪声系数指示器 noise-figure indicator
噪声(细)条 grass
噪声下限 noise lower limit
噪声限度 noise limit
噪声限度情况 noise-limited condition
噪声限止 clipping of noise
噪声限止器 noise clipper
噪声限制 noise control;noise limitation
噪声限制程序 noise abatement procedure
噪声限制电路 noise limiter circuit
噪声限制范围 noise-limited range
噪声限制规则 noise regulations
噪声限制检测器 noise-limited detector
噪声限制接收机 noise-limited receiver
噪声限制器 interference inverter; noise clipper;noise limiter;noise silencer;noise killer
噪声限制上升 noise abatement climb
噪声相关系统 noise correlation system
噪声响应 noise response
噪声向量 noise vector
噪声消除 noise cancellation; noise distribution;noise elimination;noise suppression;silencing of noise
噪声消除放大器 noise-cancelling amplifier
噪声消除开关 noise cancel switch
噪声消除器 noise canceller;noise eliminator; noise killer; noise quencher; noise silencer; noise squelch
噪声消除调整 noise cancel adjustment
噪声消除系数 noise reduction coefficient
噪声消除装置 noise killer;noise suppressor
噪声消散 dissipation of noise
噪声消声器 noise silencer
噪声消隐 noise blanking
噪声效应 effect of noise
噪声信道 noisy channel;noisy communication channel
噪声信道定理 noise channel theorem;noisy channel theorem
噪声信号 noise signal
噪声信号比 hum-to-signal ratio;jar-to-signal ratio
噪声信号补偿 shading adjustment
噪声形式 noise development
噪声性耳聋 noise-induced deafness
噪声性永久性阈移 noise-induced temporary threshold shift

噪声性质 noiseness
噪声选通同步分离器 noise-gated sync separator
噪声选择 noise selection
噪声研究 noise research
噪声掩蔽 masking by noise
噪声掩蔽级 noise masking level
噪声样函数 noise-like function
噪声仪 acoustimeter
噪声抑止器 noise remover
噪声抑制 muting; noise abatement; noise elimination; noise inversion; noise limiting; noise rejection; noise stopping; noise suppressing; noise suppression
噪声抑制程序 noise abatement procedure
噪声抑制单稳态触发电路 noise inhibit monostable
噪声抑制电路 muting circuit; noise limiting circuit; noise suicide circuit; noise suppression circuit; squelch circuit
噪声抑制电容器 noise suppression capacitor; suppression condenser
噪声抑制电阻 noise suppression resistor
噪声抑制管 noise-rejected tube
噪声抑制控制 noise suppression control
噪声抑制灵敏度 noise-limited sensitivity; noise quieting sensitivity
噪声抑制器 anti-hum; noise eliminator; noise inverter; noise killer; noise reducer; noise silencer; noise squelch; noise stopping machine; noise suppresser[suppressor]; noise trap; sound damper; sourdine
噪声抑制器效应 noise suppressor effect
噪声抑制设备 noise suppression device
噪声抑制塔 noise suppression tower
噪声抑制网络 noise suppression network
噪声抑制装置 noise-suppressing device; noise suppressing system; noise suppression arrangement; noise reduction
噪声因数 noise factor
噪声因数特性 noise-figure characteristics
噪声因数指示器 noise-figure indicator
噪声因素 noise factor
噪声音响时间 noise exposure time
噪声引起的斑点 noise spot
噪声印迹 noise footprint
噪声影响 noise effect
噪声与振动控制 noise and vibration control
噪声预防 noise precaution
噪声阈 noise threshold
噪声源 noisemaker; noise origin; noise source; origin of noise; source of noise
噪声源测量 measurement of noise source
噪声源功率测量 noise source power measurement
噪声源规律测量 noise source power measurement
噪声源鉴别 identification of noise source
噪声源识别 noise source identification
噪声源系数 noise source coefficient
噪声运算 noise operation
噪声造成的传输质量降低量 noise transmission impairment
噪声闸流管 tacitron
噪声障 noise barrier
噪声照射 noise immission
噪声振幅 noise amplitude
噪声整流器 noise rectifier
噪声值 level of noise; noise figure
噪声值数 noise rating number
噪声指标 noise index
噪声指示计 indicated noise meter
噪声指数 figure (of) noise; index of noise; isopsophic index; noise factor; noise figure; noise index
噪声指数测量 noise-factor measurement; noise-figure measurement
噪声指数计 noise-figure meter
噪声指数列线图 noise-factor nomogram
噪声致害 noise damage
噪声中值 median of sound level
噪声种类 noise type
噪声主观评价 subjective assessment of noise
噪声注入接收机 noise-adding receiver
噪声自动限制器 automatic noise level(l)er
噪声自灭电路 noise suicide circuit
噪声自然衰减量 natural attenuation quantity of noise
噪声自相关 noise autocorrelation
噪声阻力 noise resistance
噪温比 noise-temperature ratio
噪压计 psophometer
噪音 noise; undesired sound
噪音测定表 noise measuring meter
噪音测温法 noise thermometry
噪音钢轨 roaring rail
噪音计 sound meter
噪音计权曲线 weighted curve for noise measurement
噪音绝缘 noise insulation
噪音强度 intensity of noise
噪音水平 noise level
噪音污染控制 noise pollution control
噪音消除 noise cancellation
噪音消除器 killer
噪音消减 noise abatement; noise reduction
噪音抑制 mute; noise suppression
噪音抑制器 sourdine
噪音闸流管 tacitron
噪音致害 noise damage

燥漆 paste drier[dryer]; patent drier[dryer]; patent drier in paste

燥石膏 drierite
燥液 liquid drier[dryer]
燥油 drying fatty

躁郁性气质 cyclothymia

择多 majority

择多解译码算法 majority-rule decoding algorithm
择多逻辑 majority logic
择多判定元素 majority decision element
择伐 selecting cutting; selection cutting; selection felling
择伐矮林作业 selection coppice system
择伐更新 regeneration under selection system
择伐更新法 selection system of natural regeneration
择伐林 all-aged forest; culled forest; selection forest
择伐式矮林 shelter-wood coppice
择伐式间伐 selection cutting
择伐式疏伐 selection thinning
择伐作业 selection method; selection system
择机代价 alternative cost
择期 to select a good time
择其最优 alternative optima
择试解法 trial-and-error solution
择一成本 <亦称机会成本> alternative cost; opportunity cost
择一活动 alternation activity
择一假设 alternative hypothesis
择一假说 alternative hypothesis
择一文件属性 alternative file attributes
择一属性 alternative attribute
择一最优值 alternative optimum
择优的 preferred
择优定向 preferred orientation
择优复合 preferential recombination
择优关系 preferential relation
择优浸蚀 preferential etching
择优勘探 optional exploration
择优理论 theory of optimization
择优侵蚀 preferential attack
择优取向 preferred orientation
择优蚀刻 preferential etching
择优选购 purchase goods on a selective basis
择优氧化 preferential oxidation

泽边苔藓 bog moss

泽地的 moory
泽地土壤 moor soil
泽尔科锌铝合金 Zelco
泽康铝合金 Zirkonal
泽普天线 Zepp antenna
泽托尼阿铅锑锡合金 Zetonia
泽乌斯伊柳塞里沃斯的拱顶柱廊 Stoa of Zeus Eleutherios

责令关闭 ordered issued to shutdown an enterprise

责令停业 order enforced to suspend operation
责令限期治理 restraining order to control pollution within prescribed time
责任 accountability; burden; incumbency; liability; limitation clauses; obligation; onus; responsibility
责任保险 liability insurance
责任保险承保人 liability insurer
责任编辑 executive editor
责任成本 responsibility cost
责任成本制度 accountability cost system
责任成本中心 accountability cost center[centre]; cost accountability center[centre]
责任代替 substitution of liability
责任单位会计 segment accounting
责任的分担 apportionment of liability
责任的限制 restriction of liability
责任的终结 termination of liability
责任范围 extent of liability; range of responsibility; scope of cover
责任分担 allocation of responsibility; division of responsibility
责任工程师 engineering officer
责任话务员座席 responsible operator position
责任会计 activity accounting; responsibility accountant; responsibility accounting
责任会计制度 responsibility accounting system
责任积累 accumulation of risks
责任解除 release of liability
责任救助 liability salvage
责任局 responsible administration
责任开始 commencement of liability
责任内容 responsibility content
责任能力 responsibility
责任期限 duration of liability; period of responsibility
责任区 first response district; responsibility zone
责任审计 responsibility audit
责任事故 accident due to negligence; human element accident; liability accident; negligent accident; responsible accident
责任条款 liability clause
责任险 liability insurance
责任限额 limit of liability
责任限制 limitation of liability; limit of liability
责任预算 responsibility budget
责任原则 obligation principle
责任者项 responsibility area
责任证明 accountability verification
责任支付命令 <指未经事前审核的> accountable warrants
责任制度 liability system; responsibility system; system of job responsibility; system of personal responsibility
责任终止 ending of liability
责任终止条款 cesser clause
责任准备金 liability reserve

增安型电动机 increased-safety motor

增白 whitening
增白剂 brightener; brightening agent
增白力 brightening power
增爆剂 booster
增拨款项 additional provision; make additional appropriations; make additional provision
增补 append; appendix[复 appendices/ appendixes]; enlargement; supplement
增补本 enlarged edition
增补的通用合同条件 supplemental general conditions
增补服务 <顾问工程师的> supplemental services
增补供水管道 supply conduit supplement
增补详图 clarification drawing
增补信息 supplemental information
增补专利 patent of addition
增产 increase in yield; increasing in yield
增产百分率 percentage increase
增产不增人力 increase production without increasing the work force
增产灌溉 fertile irrigation
增产奖励工资 incentive wage
增产节约 increase production and practice economy
增产量 increased unit
增产油量 extra-oil production
增产指标 target for increased production
增充剂 expander; extender
增充树脂 extender resin
增稠 bodiness; dewater; fatten(ing); setting up; thicken(ing)
增稠的 thickened
增稠过程 thickening; thickening process
增稠剂 plasticizing admixture; bod-

ying agent; puffing agent; settler; stiffening agent; thickening agent; thickening material; thickener; flocculating agent <用于泵送混凝土>
增稠期 stiffening time
增稠器 thickener
增稠时间 thickening time
增稠速度 thickening rate
增稠污泥 thickened sludge
增稠油 thickened oil
增大 augmentation; boosting; ekeing; enlarge; enlargement; heighten(ing); increment; magnification; magnifying power; maximize; multiplying; swell
增大冲洗液流 rising current
增大船的稳性 stiffen
增大的 built-up
增大的安全系数 increased factors of safety
增大的波导 elevated duct
增大的巡航航程 increased cruising range
增大功率 gain power
增大机翼安装角 raising the wing
增大金刚石钻探侧刃 extend ga(u)ge
增大矩阵【数】augmented matrix
增大漏磁的变压器 transformer with additional leakage flux
增大牵引力的措施 measures for increasing tractive effort
增大前进波 increasing forward wave
增大速度 push the speed
增大速率 gathering speed
增大推力 augmented thrust
增大系数 augmenting factor; enhancement coefficient; multiplication factor
增大型芯头 augmented core print
增大压力 build-up pressure
增大增长速度 growthmanship
增大转速 rev up
增大最暗部分的黑度 expand extreme dark
增大作用 accretion
增导注入 seeing
增到四倍 quadruplication
增得的净效益 incremental net benefit
增电子 electronation
增订版 added edition; amplified and revised edition
增订本 adapted edition; enlarged edition
增订条款 additional article
增毒作用 enhancing toxic action; potentiation
增多 proliferation
增多的设备 enlarged equipment
增发射安全玻璃 enhanced reflecting safety glazing material
增幅【电】amplification; amplify
增幅波 growing wave; increasing wave
增幅管 amplitron
增幅率 amplification factor
增幅偏航 diverging yaw
增幅器 amplifier
增幅深切曲流 ingrown meander
增幅系数 amplification
增幅因子 amplification factor
增幅振荡 divergent oscillation; diverging oscillation; increasing oscillation
增辐振动 divergent oscillation
增改 touch in
增感箔 intensifying foil
增感剂 sensitizer; sensitizing agent
增感屏 intensifying screen
增感染料 sensitizing dye
增感效应 enhancement effect
增感型胶片 screen-type film
增感作用 sensibilization; sensitizing action
增高 heighten(ing)
增高标价 mark-up
增高部分 raised part
增高电压 boosted voltage; boost voltage
增高灵敏度的冲压器 enhanced sensitivity manostat
增高倾卸装置 high dumper
增高时间 attack time
增高水位 raising of water level
增高卸料铲斗 high-dump bucket
增股筹资 equity financing
增固架 reinforced frame
增固梁 reinforced girder
增固填料 reinforcement filler
增光电极 intensifier electrode
增光电路 intensifier circuit
增光电位 intensifier potential
增光环 intensifying ring
增光剂 glossing agent; sensitiser[sensitizer]
增光屏 intensifying screen
增广 augmented
增广矩阵 augmented matrix
增广拉格朗日函数 augmented Lagrangian function
增广最小二乘法 extended least squares
增过程 increasing process
增和易性剂 workability admixture
增衡器 stabilizer
增厚 thicken; thickening
增厚地层 expanding bed
增厚混凝土 haunching concrete
增厚剂 intensifier
增滑剂 slipping agent
增滑涂层 anti-friction coating
增滑性 anti-friction property
增辉 unblanking
增辉电路 intensifier
增辉力 brightening power
增辉脉冲 intensifying pulse
增辉器 intensifier
增辉望远镜 scotoscope
增加混凝土或砂浆流动性的外加剂 fluidifier
增积永冻层 aggradation permafrost
增记 write up
增加 addition; addition extension; add to; aggrandizement; ekeing; fortify; heighten(ing); jack; magnification; multiply; proliferation; rev; upgrade
增加班次 extension of service; increase of number of runs
增加保费 increase of premium
增加材料 enrichment
增加采样 additional sampling
增加产量 raise the output of products
增加产量的一种品质 the quality of producing a greater yield
增加(成长)法 <按束强度的增加观察共振的方法> flop-in method
增加承保范围 additional coverage
增加粗(糙)度的构件 roughness elements
增加到最大限度 maximize
增加到最大值 maximize
增加的价值 increased value
增加的领料单 additional requisition
增加的屈服值 increased yield value
增加的蠕变 incremental creep
增加的现金流动 incremental cash flow
增加的支付 additional payment
增加抵抗力系统 increase resistant system
增加额 additional amount; increase in amounts
增加(发动机的)马力或效率 soup
增加方 increase side
增加份额 additional provision
增加风险 increase of risk
增加辅助面积 adding additional area; adding additional space
增加感光度的 super-sensitive
增加感应 super-induction
增加工资 increase wage
增加供给 increase of supply
增加供水 additional water supply
增加供水源 additional water supply source
增加供应 reinforce provision
增加轨道 track addition
增加荷载 additional load(ing)
增加环球辐射反射 increase global radiation reflection
增加积累 increase accumulation
增加级数 multistaging
增加降水 precipitation enhancement
增加降雨量计划 precipitation enhancement project
增加金额 increase amount
增加军需储备 military reserve increase
增加开支 jack up expenditures
增加空间 gain in space
增加库存 increase of stocks
增加利润 increasing return
增加量 recruitment
增加了轻馏分的原油 enriched oil
增加流通渠道 multiply the channels of circulation
增加率 augmentation; increment rate
增加贸易作用 trade creation effect
增加密度 increased density
增加品种 diversify
增加气压的 air boosted
增加器 supercharger
增加器填料 booster packing
增加强度 gain in strength
增加曲线 logistic curve
增加三倍 quadruple; quadrupling
增加色彩 chrome
增加设备 enlarged equipment
增加射出长波辐射 increased outgoing long wave radiation
增加射入长波辐射反射 increased incoming long wave radiation reflection
增加生产 increase of production
增加收成 yield improvement
增加收入估计数 estimated additional income
增加水分 moisturize
增加水分利用率 increased water use efficiency
增加水量 water gain
增加速度 speed of increase; speed-up
增加速率 advance the speed; changing up
增加体积 bodying
增加填土密度 densification of fill
增加土壤保水力 increase in the water-holding capacity of soil
增加土壤酸性 increasing soil acidity
增加推力 assist
增加危险 increase of risk
增(加位)差设备 fall increaser
增加物 accrual; augmentation; increment; accession
增加系统 add-on system
增加线路 extension of service; increase of number of runs
增加项 increase entry
增加效率 increase of efficiency
增加需求 increase of demand
增加压缩空气 air-feed(ing)
增加一倍 doubling
增加应力 restressing
增加硬度 hardening
增加预算 increase of budget
增加折旧 accrued depreciation
增加者 multiplicator; multiplier
增加值 accrual; increment
增加重量 add the weight
增加资本 capital increase
增加资金 additional fund
增加钻孔杠杆作用的工具 old man
增减 fluctuation
增减变质 allochromatic metamorphism
增减光度的闪光 flickering light
增减记账法 addition-subtraction bookkeeping method
增减继电器 add-and-subtract relay
增减率 coefficient of increase and decrease; gradient
增减手段 escalator
增减条款 increase or decrease clause
增减因数 excess multiplication factor
增减音器 swell
增减音器踏板 swell pedal
增建第二线【铁】complement with a second track
增建工程 additional construction
增建矿山概率 probability of new mining
增建新矿山 development new mines
增阶码 biased exponent
增接轴 jackshaft
增洁剂 builder
增进 enhance
增进抵抗力系统 ergotropic system
增进剂 improver
增进健康 salubrity
增进健康的 salubrious
增进期 stadium augment; stadium increment
增进土壤作用 improve soil structure
增矩器 torque amplifier
增菌法 enrichment
增菌培养基 enriched medium
增刊 extra edition; supplement
增孔剂 prefoaming material
增宽度 broaden(ing)
增宽河段 lay by(e)
增力 reinforcement
增力泵部件 booster pump assembly
增力储量完成率 rate of increasing reserves accomplishment
增力阀 power valve
增力加浓阀 power enrichment valve
增力梁架 reinforced beam
增力喷口阀 power jet valve
增力器垫片 booster gasket
增力器总成 booster assembly
增力柱 backup post
增力装置 step-up system
增链剂 chain extender
增链聚合作用 popcorn polymerization
增亮 brighten(ing); fade up
增亮剂 brightener
增亮图像 intensified image
增亮涂层 lustering coating
增量 augmenter; gain; incremental quantity; incrementation; override
增量百分率 increment percent
增量备份 increment pack up
增量比 incremental ratio; increment(ary) ratio; quotient of difference
增量闭合差改正数 incremental discrepancy correction
增量编码器 incremental coder; incremental encoder
增量编译程序 incremental compiler
增量变化 incremental change
增量表 increment list
增量表示法 incremental representa-

Z

tion
增量部分 incremental portion
增量参数 incrementation parameter
增量成本 incremental cost
增量程序 increment routine
增量处理 incrementation processing
增量磁导率 incremental permeability
增量磁感应强度 incremental induction
增量磁化率 incremental magnetic susceptibility
增量磁滞回线 incremental hysteresis loop
增量导磁率 incremental permeability
增量地址 incremental address
增量电导 incremental conductance
增量电感 incremental inductance; incremental induction
增量电感调谐器 incremental inductance tuner
增量电荷极谱法 incremental-charge polarography
增量电阻 incremental resistance
增量电阻率测量 incremental resistivity measurement
增量法 increasing technique; incremental method; incremental technique; method of addition; reinforcement method【计】
增量方式 incremental mode
增量分配 incremental assignment; secondary allocation
增量分析 incremental analysis
增量辐射线 incremental radial line
增量感应 incremental induction
增量刚度 incremental rigidity; incremental stiffness
增量函数 increasing function
增量荷载法 incremental load method
增量荷载方法 increment load procedure
增量荷重技术 <有限单元法的一种演算法> incremental load technique
增量后援程序 incremental back-up procedure
增量绘图机 incremental plotter
增量绘图器 incremental plotter
增量绘图速率 incremental plotting rate
增量绘图仪 incremental plotter
增量绘图仪控制 incremental plotter control
增量积分器 incremental integrator; saturating integrator
增量计算 incremental calculation; incremental computation
增量计算机 incremental computer
增量记录 incremental recording
增量记录器 incremental recorder
增量剂 bulk filler; bulking agent; diluent; extender
增量剂型颜料 extender pigment
增量寄存器 increment register
增量加载 increment adding
增量劲度 incremental stiffness
增量精简数据法 incremental compaction
增量聚合物 extender polymer
增量控制 incremental control
增量控制器 incremental controller
增量理论 incremental theory; incremental strain theory <锻压的>
增量利润 incremental benefit; incremental profit
增量灵敏度 incremental sensitivity
增量率 incremental rate
增量脉冲 incremental pulse
增量模型 incremental model
增量频率控制 incremental frequency control
增量频移 incremental frequency shift
增量平衡方程 incremental equilibrium equation
增量破坏 incremental collapse
增量起动器 increment start start device
增量取样法 increment sampling
增量三元表示法 incremental ternary representation
增量深度 incremental depth
增量失稳 incremental collapse
增量式部件 incremental unit
增量式传感器 incremental transducer
增量式磁带机 incremental unit
增量式数字记录器 incremental digital recorder
增量式伺服传动 incremental servo-drive
增量式调谐器 incremental tuner
增量式显示 increment mode display
增量试验法 test dosing
增量数据 incremental data
增量数值解 incremental numeric(al) solution
增量数字化器 incremental digitizer
增量数字计算机 incremental digital computer
增量衰减 incremental attenuation
增量速度 increment velocity
增量速度调整 incremental speed regulation
增量塑性理论 incremental plastic theory
增量算子 growth operator
增量损耗 incremental losses
增量弹-塑性本构矩阵 incremental elastic-plastic constitutive matrix
增量调节 incremental regulation
增量调谐 delta tune
增量调整 incremental regulation
增量调制 delta modulation
增量调制编码 delta modulation coding
增量调制系统 delta modulation system
增量铁损 incremental iron loss
增量位移 incremental displacement
增量位移灵敏度 incremental displacement sensitivity
增量位置 incremental position
增量系统 incremental system
增量显示 incremental display
增量向量 incremental vector
增量效益费用比 incremental benefit-cost ratio
增量型积分器 incremental integrator
增量修正 augmentation correction
增量寻址 increment addressing
增量压力 incremental pressure
增量压缩 incremental compaction
增量益本比 incremental benefit-cost ratio
增量应变 incremental strain
增量应力 incremental stress
增量应力比 incremental stress ratio
增量与改正量 increment and correction
增量与改正量表 increment and correction table
增量预算法 increment budgeting
增量运动 incremental motion; incremental movement
增量运动方程 increment equation of motion
增量运算 increment operation
增量噪声 incremental noise; incremental permeability
增量增塑剂 extender plasticizer
增量中断 increment interrupt
增量转储 incremental dump; increment dump
增量装入 increment load
增量坐标 incremental coordinates
增列项目 additional item
增磷 phosphorization
增流计算 increased current metering
增流剂 flow promoter[promotor]
增流双工 incremental duplex
增硫 resulphurization
增密 compaction; density
增密测记仪 compaction recorder
增密工艺 densification process; thickening technology
增密剂 <水泥> densifying agent
增密炉 densifier
增密器 densifier
增棉系数 fiber-increasing coefficient
增面燃烧 progressive burning; progressive combustion
增面性燃烧火药柱 progressive burning charge
增敏 enhanced sensitivity
增磨剂 anti-slip agent
增能 energization
增能剂 energizer[energizor]
增能密封元件 energized packing element
增能器 booster; energizer[energizor]
增能酸 energized acid
增能液 energized liquid
增黏[粘]剂 adhesion agent; adhesion promoter; anchoring agent; keying agent; viscosifier; viscosity increaser; tackifier; tack producer; tack-producing agent
增黏[粘]涂层 bond coat
增浓 densification; densify; enrich; thicken(ing)
增浓剂 densifier
增浓效应 effect of increased concentration
增暖期 warm-up period
增抛串联锚 back(ing) an anchor
增批地段 extended lot
增频变频 up-conversion
增频转换 up-conversion
增强 accentuation; amplify; build-up; densified impregnated wood; enhance(ment); heighten(ing); intensification; intensify (ing); strengthen(ing); reinforcing
增强靶 intensifier target
增强板 reinforced board; reinforcement plate
增强玻璃 reinforced glass; strengthened glass
增强玻璃板 glass-reinforced panel
增强薄板 reinforcing sheet
增强薄膜用网布 scrim for reinforcing films
增强材料 reinforced material; reinforcement; reinforcement material; reinforcer; reinforcing material
增强材料几何形状 reinforcement geometry
增强层 reinforcing coat
增强产品的竞争能力 make its goods more competitive
增强场 enhanced field; rising field
增强衬圈 gudgeon
增强充电 intensified charging
增强出口计划 export enhancement program(me)
增强刺激 reinforcement stimulus
增强带 reinforcing band
增强的 armo(u)red; fortifying; intensive; reinforced
增强的隔板 reinforced partition
增强的隔墙 reinforced partition wall
增强电极 intensifier electrode; intensifying electrode
增强电路 intensifier circuit
增强度 intension
增强发射 enhanced emission
增强(发射)镜 reinforcing mirror
增强反差 enhanced contrast; increased contrast
增强反应 intensified response
增强方法 reinforcing
增强放大器 booster
增强辐射 ation; enhanced rad
增强复合材料 matrix material; reinforced composite
增强钢筋 bar reinforcement; steel reinforcement
增强光谱 enhanced spectrum
增强光谱线 enhanced spectral line
增强红外图像 enhanced IR imagery
增强环 intensifier ring
增强环氧树脂气体压力管 reinforced epoxy resin gas pressure pipe
增强环氧树脂气体压力管件 reinforced epoxy resin gas pressure fitting
增强混凝土用玻璃纤维增强塑料模板 glass-reinforced plastic forms for reinforced concrete
增强级 booster stage
增强剂 accentuator; amplifier; enhancer; fortifier; intensifier; reinforcer; reinforcing agent; stiffening additive
增强胶 fortified glue
增强接缝 reinforced joint
增强结点板 reinforcing gusset
增强筋 reinforcing bar
增强聚苯乙烯塑料 reinforced polystyrene
增强聚丙烯塑料 reinforced polypropylene
增强聚四氟乙烯 reinforced polytetrafluoro ethylene
增强聚酯层压材料 reinforced-polyester laminate
增强矿床勘探工作 ore deposit exploration strengthen
增强沥青涂层 reinforced bituminous coat
增强沥青油毡 reinforced bitumen felt
增强流量 increasing runoff
增强脉冲 intensifier pulse
增强模塑料 reinforced mo(u)lding compound
增强膜 reinforcing diaphragm; strengthening film
增强能力 increasing strength
增强黏[粘]性的底涂层 key coat
增强刨花板 reinforced woodwool
增强谱线 enhanced spectral line
增强器 accentuator; amplifier; augmenter[augmentor]; booster; enchancer; intensifier
增强器类型 intensifier type
增强器有效增益 intensifier effective gain
增强器增益 intensifier gain
增强群呼【无】enhanced group call
增强热固性树脂 reinforced thermosetting resin
增强热固性塑料 reinforced thermosetting plastics
增强热塑性片材 reinforced thermoplastic sheet
增强热塑性塑料 reinforced plastics; reinforced thermoplastics
增强砂轮布 grinding wheel cloth; woven cloth for grinding wheel
增强砂轮网布 scrim for reinforcing grinding wheel
增强石棉橡胶板 reinforced asbestos rubber sheet
增强石墨板材 reinforced graphite sheet

Z

增强束 enhanced beam
增强树脂 reinforced resin
增强塑料 reinforced plastics
增强塑料敷层方法 layup procedure
增强塑料排水板 reinforced plastic drain sheet
增强塑料艇 reinforced plastic boat
增强塑料用玻璃织物 glass fabric for reinforcing plastics
增强弹簧 reinforcing spiral
增强填充料 reinforcing filler
增强填料 reinforcer
增强图像 enhanced image
增强物 reinforce;reinforcement
增强系数 amplification coefficient
增强纤维与聚合物 reinforced fiber and polymer
增强线圈 intensifier coil
增强橡胶基布 woven cloth for reinforcing rubber
增强橡胶用玻璃织物 glass fabric for reinforcing rubber
增强小型设备接口【计】enhanced small device interface
增强效应 enhancement effect;synergistic effect
增强型 enhanced type;enhancement mode;enhancement type;enlargement mode
增强型不归零 enhanced non-return-to zero
增强型场效应晶体管 enhancement mode field effect transistor
增强型非硫化聚合物片材 reinforced non-vulcanized polymeric sheet
增强型负载 enhancement load
增强型沟道 enhancement type channel
增强型集成传动电子线路 the enhanced integrated drive electronics
增强型浸润剂 fiber size for reinforcement
增强(型模)式 enhancement mode
增强型曲线 enhancement type curve
增强型数据输出【计】enhanced data out
增强型图形适配器 super-video graphic(al)array
增强性颜料 reinforcing pigment
增强压器 augmenter[augmentor]
增强液体涂料包封产品 reinforced liquid coating encapsulation product
增强因子 enhancement factor;enhancer
增强应力 reinforced stress
增强用玻璃纤维 reinforcing glass fiber
增强用网布 reinforcing screen
增强用织物 reinforcing fabric
增强载波解调 enhanced carrier demodulation
增强毡 reinforcing mat
增强正像 positive intensified image
增强纸带 reinforced paper tape
增强中的风 freshening wind
增强砖砌体 reinforced brick work
增强作用 enhancement action;potentiation
增热 gain of heat;heat boosting;heat gain
增热量 heat gain
增热器 booster heater;heat booster
增韧 plasticization;plastification;toughening
增韧剂 elasticizer;flexibility agent;flexibilizer;plasticiser[plasticizer];plasticizing agent;softener;toughener;toughening agent
增韧陶瓷 toughened ceramics
增韧橡胶 plasticized rubber
增容 upgrading
增容剂 extender
增容颜料 extended pigment
增溶化 solubilization
增溶机理 solubilizing mechanism
增溶剂 solubilizer;solubilizing agent
增溶溶解 enhanced solubility
增溶色谱法 solubilization chromatography
增溶相图 solubilizing phase diagram
增溶作用 solubilization
增柔剂 flexibility agent
增色的 hyperchromic
增色剂 toner
增色剂微粒 toner particle
增色晶体 colo(u)red crystal
增色团 hyperchrome
增色效应 hyperchromic effect;hyperchromicity
增色性 hyperchromicity
增沙(海)岸 rich coast
增设 superimposing
增设电路 extension circuit
增设住宅辅助设施 re-equipping a house
增设装置 extension set
增生 accretion;proliferation
增生板块 accreting plate
增生板块边缘 accreting plate boundary
增生边界 accretionary boundary
增生材料 bred material
增生常数 growth constant
增生大陆 accreting continent
增生带 accreting zone
增生地体 accreted terrane
增生晶体 accretive crystallization
增生聚合物 proliferous polymer
增生聚敛板块边界 accreting convergent plate boundary
增生矿脉 accretion vein
增生(棱)柱体 accretionary prism
增生盆地 accretionary basin
增生线 incremental line
增生楔 accretionary wedge
增生性边缘 constructive margin
增生性聚合物 proliferous polymer
增湿 humidifying
增湿车间 humidifying plant
增湿机 dampening machine;humidifying machine;moistening machine
增湿剂 humidifier;humidizer;moistening agent;wetter
增湿加热 heating with humidifying
增湿减湿法 humidification-dehumidification
增湿减湿技术 humidification-dehumidification technique
增湿器 humidifier;humidifying unit;humidizer;moistening apparatus
增湿强度 wet strengthening
增湿燃烧 humidified combustion
增湿设备 humidifying equipment
增湿塔 conditioning tower;evaporation cooler;humidifier tower
增湿系统 humidification system
增湿装置 humidifying plant;moistening installation
增湿作用 humidification;moistening
增收保险费 additional call;supplementary call
增水【水文】anstau;pile-up;set-up
增水幅度 amplitude of setup
增水量 amount of setup
增水倾斜度 wind denivellation
增送 presentation
增速 gear up;increase in speed;increase of speed;speed increase;speed-up
增速按钮 speed-up press-button
增速比 speed increasing ratio
增速齿轮 increasing gear;over-drive gear;speed increase gear;speed increasing gear;speed(ing)-up gear;step-up gear
增速传动 gearing-up;overdrive;step-up drive
增速传动装置 multiplying gear;overdrive system;step-up gear;increase gear
增速传送 step-up transmission
增速的 geared-up;speed up
增速滑车(轮)组 pull block for a gain in speed;pulley block for a gain in force;pulley block for a gain in speed
增速滑轮组 speeding-up pulley block
增速机 speed increaser
增速剂 rate accelerating material
增速交流发电机 booster alternator
增速轮 multiplying wheel
增速排气管 augment tube
增速器 over-drive gear;speeder;speed increaser;speed increasing gear;speeding-up gear;step-up gear
增速特性曲线 progressive performance curve
增速效应 speed-up effect
增速装置 increasing gear;speeder;speed increase unit
增塑 plastic(is)ation;plasticise[plasticize];plasticizing [plasticizing];plastification;plastify;soften(ing)
增塑材料 plasticized pitch
增塑的 plasticized
增塑度 degree of plasticization
增塑防冻剂 plasticising[plasticizing] frost protection agent
增塑防霜剂 plasticising[plasticizing] frost protection agent
增塑糊 plastipaste;plastisol
增塑混合物 plasticizing admixture
增塑剂 co-plasticizer;elasticizer;flexibilizer;fluidizing agent;plasticising agent;plasticity agent;plasticizer;plasticizer additive;plasticizing agent;plastifier;plastifying agent;softener;workability admixture;workability agent;workability aid;arochlor <密封料中使用的>;plasticizing admixture
增塑剂废水 plasticizer wastewater
增塑剂迁移 migration of plasticizer;plasticizer migration
增塑剂渗移 migration of plasticizer
增塑剂污染 pollution by plasticizer
增塑加气剂 combined plasticizer and air entraining agent
增塑聚氯乙烯片材 plasticized polyvinyl chloride sheet
增塑黏[粘]合剂 plasticizer adhesive
增塑凝胶 plastigel[plastogel]
增塑溶胶 plastisol
增塑速率 plasticating rate;plasticizing rate
增塑橡胶 plasto-rubber
增塑性 plasticity
增塑用聚酯 paraplex
增塑油 plasticising[plasticizing] oil
增塑指数 plasticity index
增塑组分 plasticizing component
增塑作用 plasticising[plasticizing] action;plasticization
增酸作用 acylation
增碳 carbonization;carbon pick-up;carburat(t)ing;carburet;carburetion;fill-up carbon;recarburizer;cementation【冶】
增碳剂 carbon raiser;carburant;carburetant;carburizing reagent;recarburizer;recarburizing agent
增碳空气 carburet(t)ed air
增碳期 carbonizing period
增碳器 carburet(t)er[carburet(t)or]
增碳水煤气 carburet(t)ed water gas
增碳作用 carburet(t)ion;recarburizing;recarburation
增添 addition
增添活力作用 vitalization
增添技术 add-on technology
增添量 recruitment
增添燃料 refuel
增添污染 additional pollution
增添项 addition item
增添运算码 augmented operation code
增添装饰 enrichment
增透膜 anti-reflection coating;anti-reflection film;reflection reducing coating
增透膜层 anti-reflective coating
增透涂层 anti-reflecting coating;anti-reflective coating
增推力燃烧 progressive burning
增味剂 odo(u)rant
增温电热器 booster heater
增温度/日 heating degree-day
增温期 anathermal
增温深度 geothermic depth
增温时间 heating time;heat-up time;warm-up time
增稳装置 tranquil(l)izer
增息贷款 interest-extra loan
增隙带 crushing strip
增消过程 birth and death process
增效 synergia;synergy
增效的 synergistic
增效环 piperonyl cyclonene
增效剂 potentiating agent;synergism;synergist;synergistic agent;synergizing agent
增效醚 butoxide
增效组分 building component
增效作用 potentiation;synergism;synergized action;synergy
增斜钻具 angle gaining(building-up) tool assembly
增信码 augmented code
增锈剂 corrosion promoter
增序列 increasing sequence
增选(新成员)co-opt
增压 blowdown;boosting;build-up;charging;forced;pressure boost;pressure charging;pressurize;pressurizing;supercharge;supercharging boosting;pressure charger <内燃机>
增压泵 backing pump;blower pump;booster(ing)pump;compressing pump;compression pump;force lift pump;force pump;forcing pump;inflator;press pump;pressure pump;relay pump;supercharged pump;surcharge pump
增压泵动作 pumping action
增压泵站 booster pump(ing)station;booster station
增压比 boost ratio;pressure ratio;rate of supercharging;supercharging ratio
增压变压器 booster transformer;boosting transformer
增压表 boost ga(u)ge;pressure ga(u)ge
增压舱 normal-air cabin;pressure cabin;pressure capsule;pressurized bay;pressurized cabin;sealed cabin
增压柴油发动机 pressurized oil engine
增压柴油机 pressure oil engine;supercharged diesel;supercharged diesel engine;supercharged engine
增压厂 booster plant
增压程度 degree of boost
增压齿轮 <油泵的> pressure gear

增压稠度仪 pressurized consistometer
增压袋 pressurization bag
增压单向阀门 pressurizing non-return valve
增压的 air boosted;pressurized
增压电压 built-up voltage
增压发动机 engine with supercharger;forced induction engine
增压阀 blower valve;boost(er) valve;delivery clack;forcing valve;pressure increasing valve;pressure valve
增压法 supercharge method
增压防爆型电动机 pressurized-enclosure motor
增压风机 blowdown fan;blower fan
增压风冷 ram cooling
增压风扇 forced(draft) fan;force fan
增压服 pressure suit
增压供气风机 pressurization supply fan
增压供水 boosted water supply
增压供氧调节器 pressure breathing oxygen regulator
增压鼓风机 booster blower;booster fan;positive blower;super-blower;supercharging blower
增压管 ascending pipe;booster tube;pressure inlet;pressure pipe;ram pipe
增压管道 boost line
增压管路 pressure piping(-line);pumping main
增压罐 pressurizing vessel
增压锅炉 supercharged boiler;supercharged steam generator
增压过程 pressurization
增压和通风设备 pressurizing and ventilation equipment
增压环境 pressurized environment
增压活门 delivery civilization;delivery clack
增压机 air blower;blower;booster supercharger;charge compressor;charging compressor;positive booster;supercharger
增压机压力计 supercharger pressure ga(u)ge
增压机组 booster set
增压集气管 discharge header
增压计 boost ga(u)ge
增压计量 positive battery metering
增压剂 pressure intensifier
增压加油 reservoir priming
增压检验台 pressurizing stand
增压交流发电机 booster alternator
增压结构 pressurized construction
增压进气 supercharge
增压空气 supercharged air
增压空气管 booster air pipe;booster air tube
增压空气冷却器 charger-air cooler
增压空压机 booster compressor
增压控制 boost control
增压控制器 booster pressure controller
增压块 multistage anvil
增压扩散泵 booster diffusion pump
增压楼梯间 pressurized stair(case)
增压炉膛 supercharged furnace
增压螺旋泵 booster propeller pump
增压能量 pressurization energy
增压平台 booster platform
增压气瓶 pressurized gas cylinder
增压器 augmenter;blower;blowing machine;booster;booster pressurizer;charger;intensifier;pressure amplifier;pressure blower;pressure booster;pressure intensifier;pressure multiplier;pressure unit;pressurizing unit;supercharged engine;supercharger;turbocharger
增压器传动液压机 intensifier driven hydraulic press
增压器分液环 blower slinger
增压器节流阀 supercharger blast gate
增压器软管 booster hose;forcing hose
增压器调节器 supercharger regulator
增压器叶轮 supercharger impeller
增压区 anallobar
增压曲轴箱 pressurized crankcase
增压燃烧蒸汽锅炉 supercharged furnace-fired steam generator
增压容器 pressurized container
增压扫气系统 pressure charging/scavenging system
增压设备 pressure unit;supercharging equipment
增压时间 pressurization time
增压式 pressure-charged
增压式发动机 pressure-charged engine;supercharged engine
增压式锅炉 pressure combustion boiler
增压式呼吸设备 pressure breathing system
增压式火箭 balloon-type rocket
增压式空压机 air-boost compressor
增压式扩散泵 booster-type diffusion pump
增压式内燃机 supercharged engine
增压式涡轮螺旋桨发动机 supercharged turboprop
增压式压气机 air-boost compressor;boost compressor
增压式主油缸 booster-type master cylinder
增压室 plenum chamber;pressurized area;pumping chamber
增压输送管路 force-feed main
增压水泥罐车 pressurized cement lorry
增压水射流 pressurized jet of water
增压素 hypertensin
增压速率 rate of pressurization
增压特性 supercharging performance
增压调节 boost control;supercharger control
增压调节器 boost controller;booster adjuster
增压通风机 force(d) fan
增压桶 pressure-feed container
增压涡轮机组 charging-turbine set
增压系统 pressure charging system;pressurization system;pressurizing system
增压系统导管 pressurizing duct
增压下辊锻 roll forging under increasing compression
增压线 boost line
增压限制器 debooster
增压箱 pressurizing tank
增压辛烷值 supercharge octane number
增压形式 type of pressure charging
增压压力 boost(er) pressure;loading pressure;pressurization pressure;supercharged press(ing);supercharged pressure
增压压力表 boost ga(u)ge;boost pressure ga(u)ge
增压压气机 pressure charge compressor;supercharging compressor
增压压缩机 booster compressor;pipeline compressor;supercharging compressor
增压养护箱 pressurized curing chamber
增压油泵 booster oil pump;oil booster pump;oil supercharging pump;pressure oil pump;pressurized oil pump
增压油扩散真空泵 booster oil diffusion pump
增压运行 pressure operation
增压运行的 supercharged
增压站 booster installation;booster plant;booster station
增压蒸汽锅炉 supercharged steam generator
增压直流锅炉 forced-flow once-through steam generator
增压制动器 booster brake
增压重油机 pressure oil engine
增压装置 booster;charging set;pressure device;pressure unit;pressurizer;supercharging device
增压总管 pumping main
增压作用 supercharging
增压座舱 positive pressure cabin;sealed cabin
增压座舱盖 pressurized canopy
增艳剂 brightener;brightening agent
增氧燃烧 oxygen-enriched combustion
增益 gain(ing);amplification;increment;magnification;transmission gain
增益棒 booster rod
增益比 ratio of gain
增益变化 gain variation
增益变化范围模拟 gain ranging analog
增益变换器 gain changer
增益标么值 per-unit gain
增益波动 gain fluctuation
增益参数 gain parameter
增益测量 gain measurement
增益测量器 gain measuring set
增益测量装置 gain measuring device
增益差 gain inequality
增益常数 gain constant
增益程序化放大器 gain-programmed amplifier
增益程序控制 programmed gain control
增益窜渡频率 gain crossover frequency
增益窜渡 gain crossover
增益带宽 gain bandwidth
增益带宽乘积因数 gain bandwidth factor
增益带宽积 gain bandwidth product
增益带宽因数 GB factor
增益带宽指标 gain-band merit
增益导纳 gain admittance
增益电平 gain level
增益反转 gain inversion
增益范围 gain margin
增益方程 gain equation
增益放大 gain amplification
增益放大器 gain amplifier
增益分贝数 decibel gain
增益函数 gain function
增益恢复 gain recovery
增益级 gain level;gain stage
增益极限 gain margin
增益渐近线 gain asymptote
增益降低 gain turn-down
增益交叉 gain crossover
增益交叉频率 gain crossover frequency
增益校正 gain calibration
增益校正调整 spotting gain control
增益校准 gain calibration
增益矩阵 gain matrix
增益矩阵控制 matrix gain control
增益开关 gain switch
增益控制 amplification control;gain control
增益控制范围 gain control range
增益控制方式 gain control mode
增益控制放大器 gain-control(led) amplifier
增益控制键 gain control key
增益控制器 attenuater[attenuator];gain controller
增益灵敏度控制 gain sensitivity control
增益匹配 gain matching
增益漂移 gain drift
增益频率关系 gain frequency relationship
增益频率特性(曲线) gain frequency characteristic
增益平衡 gain balance
增益平稳度 gain flatness
增益起伏 gain fluctuation
增益器 multiplicator;multiplier
增益曲线 gain curve;gain trace
增益容限 gain margin
增益时间 gain time
增益时间控制 gain time control
增益时间控制器 gain time controller
增益衰减 gain reduction
增益衰减指示器 gain reduction indicator
增益台阶精度 gain step accuracy
增益台阶数 gain step number
增益特性(曲线) gain characteristic
增益天线 gain antenna
增益条件 gain condition
增益调节 gain adjustment;gain control
增益调节器 volume controller;volume regulator
增益调整 amplification control;gain adjustment;gain control;gain setting
增益调整器 attenuator;fader;gain controller;gain control set
增益调制 gain modulation
增益通带宽度指标 gain-band merit
增益同噪声温度比 gain to noise temperature ratio
增益稳定 gain stabilization
增益稳定性 gain stability
增益稳定余量 gain stability margin
增益误差【计】 gain error
增益系数 gain coefficient;gain factor
增益下限 gain floor
增益显示 gain display
增益限幅器 gain clipper
增益相位分析 gain phase analysis
增益相位特性 gain phase characteristic
增益斜率 gain slope
增益选择 gain select
增益选择开关 gain selector switch
增益压缩 gain compression
增益压缩比 compression rating
增益压制器 amplitude delimiter
增益因数 gain factor
增益因子 gain factor
增益有限接收机 gain limited receiver
增益余量 gain margin
增益与截止频率的关系 gain cut-off relation
增益与频率的关系特性 gain frequency characteristic
增益裕度 gain margin
增益元件 booster element
增益自动下降 gain turn-down
增音段 repeater section
增音器 amplifier;gain set;repeater;sound intensifier;swell
增音系统 repeated line system;repeater system
增音选择器 selector-repeater
增音站 relaying station;repeater office;repeater station;repeating station

增音站距离 repeater station interval
增印 new print
增硬剂 hardener;integral hardener
增硬退火 three-quarter hard annealing
增援 follow-up; reinforce; reinforcement
增援处理机 attached support processor
增载 increment of loading
增载添加剂 <一种能提高润滑油承载能力的添加剂> load-carrying additive
增泽过滤器 polishing filter
增长 build-up; grow; growth; increment;propagation;rise
增长板块【地】accreting plate
增长波 increasing wave
增长部分 incremental portion
增长常数 growth constant
增长次序 ascending order
增长的 rising
增长的百分率 increased percentage
增长的百分数 percentage increase
增长的极限 the limits to growth
增长的绝对数 absolute figure increased;increase in absolute figures
增长的限度 limit to growth
增长反应 reaction of propagation
增长分析 incremental analysis
增长感应 growth response
增长管理 managing growth
增长海滩 accretion beach
增长机 increaser
增长极论 growth pole theory
增长交通量 increasing traffic volume
增长接头 lengthening joint
增长类推法 growth analogy
增长量 increasing amount;increment
增长率 growth rate; increment rate; percentage increase; rate of accretion; rate of growth; rate of increase; rate of increment; accumulation rate <冰川>; rate of accumulation <冰川>
增长率法 growth rate method
增长期 accretionary phase
增长潜力 growth potential
增长区 growth area
增长曲线 build(ing)-up curve; cumulative curve; growth curve; logistic curve
增长沙坝 accretionary bar
增长时间 build(ing)-up time; rise time; rising period; time of rise; built-up time
增长时期 growth stage
增长速度 increment speed; rate of rise; rising rate; speed of growth; velocity of increase; rate of growth
增长速率 rate of rise
增长特性 rising characteristic
增长停滞 growth retardation
增长系数 extension coefficient; growth factor; quotient of increase; modulus of growth
增长系数法 growth factor method; present pattern method
增长性自动机 growing automaton
增长序 growth order
增长因数 shade factor
增长因素法 growth factor method
增长因素模型 growth factor model
增长因子 growth factor
增长指数法 <交通预测> growth factor method
增长指数律 exponential law of growth
增长滞后 build-up lag
增长中心 growth center[centre]
增长周期 rising period
增长主河道 enmaining
增长作用 accretion;generated effect
增涨作用 accretion
增丈 torque multiplication
增支成本 incremental cost; marginal cost
增支成本节约额 incremental cost saving
增支费用 incremental expenses
增支固定成本 incremental fixed cost
增值 added value; appreciation; increase of value; increasing value; increment (of value); increment value; multiplication; self-expansion;value-added
增值资产 accrued assets
增殖 breed(ing); multiplication; proliferation;reproduction
增殖比速 specific growth rate
增殖材料 breeder material
增殖常数 growth constant
增殖的 multiplicative
增殖堆电站 breeder plant
增殖反应 breeder reaction
增殖反应堆 breeder; breeder converter; breeder reactor; internal breeder
增殖范围 multiplication nursery
增殖过程【数】birth process
增殖减退期 retardation phase
增殖阶段 multiplicative stage
增殖率 coefficient of multiplication; reproducibility;reproduction rate
增殖率倒数 reciprocal multiplication
增殖潜力 multiplication potentiality; reproduction potentiality
增殖速率 multiplication rate
增殖体 vegetation
增殖系数 growth coefficient; K factor;multiplication constant
增殖系数黏[粘]度值 K-value
增殖效应 multiplier effect
增殖抑制因子 proliferation inhibiting factor
增殖因素 multiplication factor; reproduction factor
增殖铀 bred uranium
增殖转换反应堆 breeder converter
增殖作用 multiplier effect
增至三倍 triplication
增重 ballasting; gaining in weight; liveweight gain; liveweight increase; liveweight increment; weight gain;weight increase
增重的丝绸 weighted silk
增重分批掺和机 gain in-weight batch blender
增重环 <旋冲钻杆的> drive clamp
增重剂 weight gaining agent
增重经济效益 economy of gain
增重净能 net energy for gain
增重快 fast gaining
增重量 gain in weight;weight gain
增重率 percentage of liveweight growth; rate of body weight gain; rate of live weight growth
增重能力 gained ability;gaining ability
增重效率 efficiency of gain;gaining ability
增资 capital additions;capital increase
增阻力的 late-bearing
增组 reduplication

憎 恶 allergy

憎寒 loathing of cold
憎溶剂的 solvophobic
憎色的 chromophobic
憎色性 chromophobe
憎水 anti-hydro
憎水白云石 stabilized dolomite
憎水玻璃细珠 hydrophobic glass beads
憎水材料 hydrophobic material; water-repellent material
憎水处理 water-repellent treatment
憎水的 hydrophobic; moisture-repellent; water-fearing; water repellent
憎水骨料 water-repellent aggregate
憎水硅酸盐水泥 hydrophobic Portland cement
憎水化 hydropbobisation[hydrophobization]
憎水化合物 water-repellent compound
憎水混合剂 water-repellent admixture
憎水基 hydrophobic group
憎水剂 anti-hydro; hydrophober; hydrophobic admixture; water-repellent admixture; water-repelling agent; water repellant compound
憎水剂处理木材 water repellent preservative
憎水胶体 hydrophobic colloid
憎水膨胀水泥 hydrophobic expansive cement; waterproof expansive cement
憎水膨胀珍珠岩 hydrophobic expanded pearlite
憎水溶胶 hydrophobic sol
憎水砂浆 water-repellent mortar
憎水水泥 hydrophobic cement; hydrophobic Portland cement; water repellent cement
憎水添加剂 water-repellent admixture
憎水物质 hydrophobic substance
憎水性 hydrophobicity; hydrophobic nature
憎水性防腐剂 water-repellent preservative
憎水性骨料 hydrophobic aggregate
憎水性混凝土 water-repellent concrete
憎水性集料 hydrophobic aggregate
憎水性能 hydrophobic property; water-repellent property
憎水性水泥 water-repellent cement
憎水性填缝材料 hydrophobic sealant
憎水性外加剂 hydrophobic admixture; water-repellent additive; water-repellent admixture; water-repellent agent
憎水纸 water repellent paper
憎液的 lyophobic
憎液胶体 lyophobe colloid; lyophobic colloid
憎液溶胶 lyophobic sol
憎液性 lyophobic

甑 retort

甑馏法 retort process
甑馏炼焦 retort coking
甑式焚化炉 retort-type incinerator
甑碳 retort carbon

赠 刊索阅单 gift request

赠款 grant;outright grant
赠款援助 assistance on grant terms
赠与证书 deed of gift
赠援 assistance on grant terms
赠阅书刊 supplied free issue

渣 boiler slag;bottom ash;muck

渣岸 slag bank
渣坝 slag dam
渣板 slag plate
渣包 cinder ladle; slag ladle; slag pot; sludge ladle
渣保护 flux shielding; slag coverage; slag shielding
渣泵 pulp pump
渣比 slag ratio
渣壁过渡 flux-wall guided transfer
渣饼 bedder; blob of slag; slag cake; slag pancake
渣槽 slag chute;slag runner;slag trough
渣层 slag blanket
渣场 slag dump;slag tip
渣车 slag car;slag wagon
渣尘 crushed dust
渣沉积 slag deposit
渣成分调整 slag conditioning
渣池 slag basin;slag bath;slag pool
渣池深度 depth of slag bath
渣床 slag-bed
渣床沉降 settlement of ballast
渣道车 troll(e)y
渣的 slaggy
渣的成分调整 slag conditioning
渣底 slag fill;slagging hearth
渣堆 escorial; scoria[复 scoriae]; slag; slag heap; slag muck; slag stockpile;slag tip
渣堆输送机 slag heap conveyer[conveyor]
渣分离器 pulp separator
渣粉 ground slag
渣风口 slag tuyere
渣沟 slag runner
渣沟流嘴 cinder spout
渣罐 cinder ladle; slag basin; slag pot; slag receiver
渣罐车 dump cinder car; slag buggy; slag ladle and carriage; thimble slag car
渣痕 skim;slag streak
渣化 flux; fluxing; scorify; slag; slag formation;slagging
渣化层 slagged surface
渣化法 scorification
渣化面 slagged surface
渣化皿 scorifier
渣化能力 fluxing power
渣灰印迹法 slag print
渣肩 shoulder of ballast
渣坎 clinker dam;fuel dam
渣壳 crust; flux cover; flux envelope; incrustation; scull; scurf; skull; slag crust;slag skull
渣壳块 crust block
渣壳破碎机 skull cracker
渣壳熔炼炉 skull melting furnace
渣坑 cinder fall; cinder pit; hunch pit; slag dump;slag pit;slag pocket
渣孔 cinder hole; flushing hole; scum spot; slag blowhole; slag eye; slag hole;slag inclusion;slag pin hole
渣口 cinder notch; flushing hole; monkey;slag notch aperture;slag pin hole
渣口堵塞器 slag notch stopper
渣口工 cinder snapper;teaser
渣口冷却器 jumbo;monkey cooler
渣口冷却套 cinder cooler; monkey jacket;slag notch cooler
渣口喷火 blowing on the monkey
渣口塞 slag bot(t)
渣口塞杆 stopping bar; tap-hole plugstick
渣块 clinker; slag block; slag lump; slag patch
渣块水泥 grappier cement; grappler cement
渣粒 slag particle;sliver

渣粒灰浆 slag plaster
渣量 level of residue;quantity of slag
渣料 slag charge
渣流道 slag runner
渣留地沥青 <自石油蒸馏残渣取得> residual oil asphalt
渣瘤 adhesion of slag; clinker adhesion;slag blob;slag nodule
渣炉 slag hearth
渣煤 dross coal
渣绵 slag cotton
渣棉 cinder wool;slag wool
渣面 slag(ging) level;top of slag
渣盘 slag pan
渣皮 slag crust
渣坡 slag bank;slag grade
渣球 shot;slug
渣球百分比含量 percent shot
渣球含量 shot content
渣绒 <隔音、隔热、防火材料> mineral wool;slag wool
渣熔点 slag melting point
渣筛 ballast screen
渣蚀 slagging
渣室 grit chamber
渣水比 slag water ratio
渣水混合物 <水力输渣的> slag water slurry
渣水浴 slag slurry bath
渣铁比 slag ratio
渣铜浇铸 cast slag copper
渣桶 cinder pot;slag basin;slag ladle;slag pot;sludge ladle
渣桶车 slag ladle car
渣污染 slag pollution
渣锡 prillion
渣洗 wash heat
渣洗法 wash heating
渣洗平炉底 swealing
渣洗熟铁 shotting
渣系 slag system
渣线 slag line
渣箱 dross box
渣屑桶 slack barrel
渣烟化法 slag fuming process
渣眼 slag blister
渣样 slag specimen
渣液流度 slag fluidity
渣印法 slag print
渣油 non-distillate oil; residual fuel; residual oil; residue; residuum [复 residua]
渣油加氢裂化 residuum hydrocracking
渣油加氢脱硫 residuum hydrodesulfurization
渣油加氢转化 residuum hydroconversion
渣油焦化 residuum coking
渣油焦油 residuum tar
渣油裂化过程 residuum cracking process
渣油路 residual oil road
渣油路面 residual oil pavement
渣油稳定砂路面 residual oil stabilized sand pavement
渣油转化过程 residuum conversion process
渣砖 cinder brick
渣状 scoriaceous;scoriform
渣状的 drossy;slaggy
渣状火成碎屑岩 slag
渣状熔岩 scoriaceous lava
渣状熔岩流 aa flow
渣子 dregs
渣滓 culls;dregs;dross;dross coal;feces; fecula; feculence [feculency]; foots; leavings; lees; offal; offscourings; recrement; refuse; residue; scum; sediment; slag; spent material;sullage;tailing
渣滓形成 dregs formation
渣滓油 foots oil

扎 bundle;dub;sheaf

扎绑脚手架 scaffold lashing
扎柴排 wattle
扎出气孔 mo(u)ld venting
扎除 removal by ligature
扎尔德赛德牌杀虫剂 <一种木材杀虫剂> Zaldecide
扎缚 astringe
扎钢 rolling
扎钢箍铁丝 stirrup wire
扎钢筋 wire tie
扎钢筋工 iron fighter; iron worker; rod fixer
扎钢筋铅丝 <混凝土中的> concrete reinforcement wire
扎钢筋网 netting
扎格罗斯地槽 Zargros geosyncline
扎格罗斯山前盆地 Prezagros basin
扎根介质 rooting medium
扎根深的 trench rooted
扎根深度 working depth
扎钩 dog
扎钩板 dog board
扎接 butting
扎筋夹 patent binder clip
扎紧 tighten
扎克唐手拉车 Jak-tung truck
扎筘杆 reed baulk
扎筘机 reed-making machine
扎捆机 bundler;tying machine
扎捆台 binding board
扎牢 fasten(ing)
扎马克压铸锌合金 Zamak
扎帽形截面 top hat section
扎密阿姆镍铬合金 Zamium
扎帕塔式三脚结构 <钻井船> Zapata tripod structure
扎帕塔式蝎尾型结构 <钻井台架> Zapata scorpion; Zapata scorpion structure
扎气眼 venting
扎圈 frapping turns
扎入角 <碎石机的> angle of nip
扎绳头 sailor whipping
扎石砂 crusher sand
扎束用物 tie-up
扎丝 binding wire;tie wire
扎锁木 <木排的> lock down
扎特牌防水乳状液 <商品名> Zat
扎铁 reinforcement tying
扎铁工 iron fighter
扎铁丝 wire tie
扎瓦里茨基法 Zavaritsky method
扎梳梯绳 ratlin(e)down
扎线 belting; binding wire; cotton binder;girt(h);strand wire bond; tie line;tie wire
扎依尔地穹列 Zhayier geodome series
扎用钢丝 fastening wire
扎在土中的 earthbound
扎住的刀头 frozen bit
扎住的井管 frozen casing
扎住的钻头 frozen bit
扎住钻井管 frozen drill pipe
扎锥 <木工或皮革工用具> awl

轧

轧凹凸机 blanking and embossing press
轧疤 fins;back edge;backfin
轧板厂 slab yard
轧板机 mangle;plate mill;sheeter
轧边 edge finish(ing);edge milling; edging;rolling on edge
轧边扁坯轧机 slabber-edger
轧边辊 edger roll
轧边机 edger;edger mill;edging mill; rolling edger
轧边金属板 deformed metal plate
轧边孔型 edging pass
轧边输入辊道 edger approach table
轧边压机 edging press
轧薄机 thinning machine
轧布轮 mangle gear
轧材 mill bar;rolled metal;rolled stock
轧槽 groove;pass
轧槽边斜角 groove angle
轧槽宽度 width of groove
轧槽轮廓 contour of groove
轧槽斜度 taper of groove
轧槽中线 axis of groove
轧成带材厚度指示仪 readout
轧成的骨料 crushed stone
轧成的块状炉渣 crushed lump slag
轧成钢坯 bloom
轧成骨料的性能 aggregate making property
轧成碎石的能力 <岩石> aggregate making property
轧刀 breaking knife;guillotine
轧道 mill train;pass
轧道垫板 tie pad
轧斗 jaw bucket
轧锻机 forging rolls
轧飞雷管 blown primer
轧粉机 pulverator
轧钢 steel rolling
轧钢厂 rolling mill;steel mill
轧钢厂废水 rolling steel mill wastewater
轧钢车间 blooming mill
轧钢车间铲斗 steel mill bucket
轧钢车间料斗 steel mill bucket
轧钢电动机 rolling mill motor
轧钢废水 rolling effluent; rolling wastewater
轧钢工 mill operator;mill roller;roller;roller man
轧钢工人 millman
轧钢轨设备 rail mill
轧钢机 mill; rolling machine; rolling mill;steel rolling machine
轧钢机架 roll stand
轧钢机列 train
轧钢机台 rolling platform
轧钢机台架 hot bed
轧钢机用电动机 screw-down motor
轧钢机组 mill train
轧钢鳞皮 mill scale
轧钢排出液 rolling effluent
轧钢切头机 end shears
轧钢设备 mill
轧钢条厂 bar mill
轧钢屑 mill furnace cinder
轧钢芯机 pressure engraving machine
轧钢自动化 automation of rolling
轧管厂 pipe mill;tube mill
轧管机 pipe and tube mill;pipe mill; tube mill
轧光 mill finish;press polish
轧光机 glaz(i)er;glazing calender; planishing mill
轧光印花织物 chintz
轧轨机 rail mill
轧辊 mill roll;padder;press roll;roll(er)
轧辊表面凸度控制 crown control
轧辊部件 roll assembly
轧辊车床 roll turning lathe
轧辊的弹簧式平衡 spring roll balance
轧辊的型缝 groove pass
轧辊的轴向调整 lateral adjustment of roll
轧辊电热炉 electric(al)roll beater
轧辊垫圈 rolling gasket
轧辊垫圈接头 rolling O-ring joint
轧辊端头 roll end
轧辊锻压机 roll forging machine
轧辊堆放架 holster
轧辊刚性 roll stiffness
轧辊工作表面 rolling face
轧辊工作表面冷却液 roll coolant solution
轧辊箍 crushing roll segment
轧辊辊身 roll barrel
轧辊辊套 roll sleeve
轧辊环 collar
轧辊机列 train of rollers
轧辊机座 roller cage
轧辊间距离 opening between rolls
轧辊间隙 roll seam
轧辊矫直法 roll straightening
轧辊开度 mill opening; roll(gap) opening
轧辊开度测量仪 ga(u)ge meter
轧辊开口度 roll gap
轧辊刻纹 riffling
轧辊孔型 caliber;roll calibre
轧辊孔型设计 grooving of roll
轧辊孔型中线 roll parting line
轧辊冷却乳液 roll coolant
轧辊冷却系统 roll coolant system
轧辊冷却装置 roll cooling device
轧辊梅花头 roll wobbler
轧辊磨床 roll grinder; roll grinding machine
轧辊磨光 roller burnish
轧辊挠度 roll bending;rolling deflection
轧辊偏心的修正 roll eccentricity correction
轧辊平衡锤 roll balancing counter weight
轧辊平衡装置 roll balance gear
轧辊强度 roll strength
轧辊润滑乳化液 rolling oil
轧辊升程 roll stroke
轧辊式干草压扁机 roller crusher conditioner
轧辊寿命 roll campaign;roll life
轧辊速度 speed of rolls
轧辊弹起度 roll spring
轧辊弹性压扁 roll flattening
轧辊调整机构 roll-adjusting gears; roll-separating mechanism
轧辊调整速度 roll movement
轧辊调整装置 roll-separating mechanism
轧辊筒 roll barrel
轧辊凸面 crown;roll crown
轧辊弯曲压力机 roll bender
轧辊箱 roll-box
轧辊型缝 caliber [calibre]; roll calibre; roll ga(u)ge; roll pass; roll path
轧辊型缝校准 roll calibrating
轧辊压力补偿 roller pressure compensation
轧辊压下装置 roller press-down apparatus;roll flattening;screw-down gear
轧辊压制 roll compacting
轧辊咬入轧件 biting;roll bite
轧辊液压平衡系统 hydraulic roll-balance system
轧辊预热器 roll preheating device
轧辊圆周速度 roll speed
轧辊正压力 normal roll pressure
轧辊重车系数 redressing coefficient of rolls
轧辊轴承 roll bearing;roll neck bearing
轧辊轴承座 chock;roll chock
轧辊轴承座导板 chock slide

轧辊轴线 roll axis
轧辊轴线偏斜 crossing of the rolls
轧焊 roller welding
轧痕 roll mark;roll pick-up
轧痕钢丝 indented wire
轧后净棉机 post-ginning cleaner
轧花垫圈 knurled collar
轧花机 cotton gin;embosser
轧花机的气吸式集棉筒 vacuum dropper
轧花机底部梳齿板 gin saw
轧花机落棉 gin fall
轧花墙纸 embossed paper
轧花压包机 gin compress
轧机 mill;roll;rolling mill
轧机布置 mill layout
轧机部件内的摩擦 mill friction
轧机参数 mill data
轧机操纵室 mill pulpit
轧机操作人员 mill personnel
轧机导板 rolling mill guide
轧机的传动装置 rolling mill drive apparatus
轧机的附属设备 mill auxiliaries
轧机的压下装置 rolling mill screw down
轧机电机 stand motor
轧机吊索工具 mill tackle
轧机负荷 mill load(ing)
轧机辊道 mill table
轧机后辊道 back mill table
轧机机架 mill housing;roll housing
轧机机架间张力 mill interstand tension
轧机机组 mill train
轧机机座 mill stand
轧机机座的弹跳 mill spring
轧机机座盖 mill stand cap
轧机进口侧操纵台 ingoing control post
轧机控制 rolling mill control
轧机跨 mill aisle;mill bay
轧机联接轴 mill spindle
轧机牌坊窗口 mill housing window
轧机牌坊下横梁 housing rocker plate
轧机前的推床导板 front side guard
轧机设备的总布置 mill general layout
轧机生产率 rolling rate
轧机生产能力 mill production
轧机输出辊道 mill delivery table
轧机输入辊道 mill approach table
轧机弹跳 mill spring(ing)
轧机调整 mill setting
轧机铁鳞 mill tap
轧机压下螺钉 mill screw
轧机轧座 mill shoe
轧机主传动电机 main mill drive motor
轧机装置 rolling mill assembly
轧尖 tagging
轧尖机 pointing machine
轧剪 double cutting snips
轧件 feed;mill bar;rolled piece;stock;workpiece
轧件表面上的结疤 dog's ears
轧件缠辊 collar(ing)
轧件的鱼尾端 fishtail end
轧件对轧辊的单位压力 resistance to deformation
轧件横向厚度波动值 lateral gauge variation
轧件横向移送台架 transfer bank
轧件前端下弯 turn down
轧件上弯 overdraft;overdraught
轧件下弯 underdraft
轧浆辊 quetsch
轧浆细度 degree of grinding;fineness of grind
轧焦机 coke breaker
轧节螺栓 fish bolt
轧紧 handling tight
轧孔 piercing
轧孔钳 punch pliers
轧砾机 crushed gravel
轧砾石厂 gravel-breaking plant
轧铝机 alumin(i)um rolling plant
轧麻荚机 boll crushers
轧煤 crushing
轧棉厂 cotton ginning factory;gin house
轧棉机 cotton gin
轧棉毛机 linter
轧棉子机 gin
轧面粗糙度 scallop
轧膜 rolled film
轧膜成 roll forming
轧膜机 roller film machine
轧墨机 ink grinding mill
轧坯 strip plate
轧片机 flaking machine
轧票 punch
轧平 set-off
轧平点 break-even point
轧平机 flattening mill
轧齐曲线 curve of rolling neat step shaft;cut-off curve
轧前板坯 incoming slab
轧前厚度 ingoing ga(u)ge
轧切机 punching machine
轧去 roll-off
轧染 pad dy(e)ing
轧染机 padding mangle
轧入 roll bite
轧入角 angle of rolling
轧入氧化皮 rolled-in scale
轧色浆 colo(u)ring
轧砂 sand manufacture
轧砂厂 sand mill
轧砂机 sand crusher
轧砂设备 sand plant
轧石 crushed rock;crushed stone;rock breaking
轧石残渣 crushed sand;crusher dust;crusher rock dust
轧石厂 crushed-stone plant;crushing mill;crushing plant;rock-crushing plant;stone crushing plant
轧石场弃石 run-of-crusher stone
轧石机 aggregate producer;chippings breaker;chippings crusher;crusher;crushing engine;crushing machine;crushing plant;disintegrator;rock crusher;stone crushing machine
轧石机的颚板 crushing plate
轧石机筛余的 throughs
轧石机筛余屑 run-of-crusher stone
轧石机受料口 crusher head
轧石机受料器 crusher head
轧石砂 crusher sand;stone sand
轧石筛分厂 crushing and screening plant
轧石设备 crushed-stone plant
轧水机 padder extractor
轧丝锚 anchorage with rolled screw thread;anchorage with screw rod;cold rolled thread anchorage
轧丝锚具 cold extruded thread anchorage
轧碎 breaking;buck;crush;crushing;granulate;size reduction
轧碎板 crusher plate
轧碎板岩 broken slate
轧碎棒 broken stick
轧碎玻璃 finely crushed glass;granulated glass
轧碎材料 broken material;granulated material
轧碎产出量 crushing output
轧碎产品 broken product
轧碎成粒 granulate
轧碎程度 degree of breaking
轧碎大理石 broken marble
轧碎的 crushed
轧碎高炉泡沫溶渣 broken foamed (blast furnace) slag
轧碎工作 granulating
轧碎骨料 broken aggregate;broken stone;crushed aggregate
轧碎辊 broken stick
轧碎滚筒 crushing roll
轧碎过程 crushing process
轧碎花岗岩 broken granite
轧碎混凝土形成的砂 broken concreting sand
轧碎机 breaker;chippings breaker;chippings crusher;crusher;crusher chute;crushing engine;crushing machine;granulator;impact breaker;knapper
轧碎机出料口 crusher opening
轧碎机传动齿轮 crusher drive gear
轧碎机传动小齿轮 crusher drive pinion
轧碎机的供料门斗 crusher feed hopper
轧碎机颚板 crusher jaw
轧碎机钢球 crusher ball
轧碎机滚筒 breaker roll
轧碎机进料口 crusher mouth;crusher opening
轧碎机轮壁 crusher wall
轧碎机主动轴 crusher drive shaft
轧碎机锥体 crushing cone
轧碎机组 crusher unit
轧碎集料 broken aggregate;broken stone;crushed aggregate
轧碎坚硬岩石 broken hard rock
轧碎块状矿渣 broken lump slag
轧碎矿渣 broken slag;crushed slag
轧碎砾石 crushed gravel
轧碎粒状炉渣 broken granulated cinder
轧碎炉渣砂 broken cinder sand
轧碎卵石 crushed gravel
轧碎能力 crushing capacity
轧碎黏[粘]土砖 broken clay brick
轧碎破坏 crushing failure
轧碎熔岩 broken lava
轧碎砂 <轧碎石料时所得的砂> crushed sand;manufactured sand
轧碎设备 crushing equipment
轧碎石膏 crushed gypsum
轧碎石灰石 broken limestone
轧碎石灰石砂 broken limestone sand
轧碎石料 crushed stone
轧碎石料供料器 crusher feeder
轧碎细骨料 broken fine aggregate
轧碎细集料 broken fine aggregate
轧碎细炉渣 slag sand
轧缩辊 reduction roll
轧体前端的分层 crocodile
轧条 kicker
轧条机 bar mill
轧铜厂 copper mill
轧铜机 brass mill machine
轧头 breaking head;dog
轧头螺钉 dog screw
轧瓦楞机 crimping machine
轧纹钢筋 ribbed rebar
轧纹钢条 ribbed rebar
轧细砂 crushed sand
轧屑 cinder;mill cinder;mill scale
轧压 roll compacting
轧压成型 compacting by rolling
轧压成型法 rolling process
轧压机 roller press
轧压坯块 rolled compact
轧压碎石 rolled stone
轧液机 extracting machine;mangle
轧印钳 forming pliers;swage pliers
轧用钢丝 fastening wire
轧用油 roll oil
轧圆型材的轧钢机 rolling mill for circular shapes
轧直的账户 account balanced
轧制 milling;roll;roll down;roller milling;rolling
轧制板 milled plate;milled sheet
轧制板材 rolled plate;rolled sheet
轧制板钢 rolled sheet metal
轧制半径 rolling radius
轧制棒材 rolled bar;rolled rod
轧制包覆 cladding by rolling
轧制包覆 roll cladding
轧制边 rolled edge
轧制边缘 running edge
轧制玻璃 rolled glass
轧制玻璃板 rolled plate glass
轧制薄板 rolled sheet metal
轧制部件 rolled parts
轧制材 rolled product;rolled stock
轧制材料 rolling material;rolling stock
轧制侧向压下 indirect rolling action
轧制长度 mill length
轧制车轴 rolled axle
轧制成品 milled product;mill product;rolled article
轧制成型 roll forming
轧制程序 rolling program(me)
轧制淬火 rolled hardening
轧制带肋型材 extruded rib section
轧制道次 geat;git;pass;rolling pass
轧制的 rolled;rolled-on
轧制的U形钢 rolled channel (section)
轧制的槽钢 rolled steel channel
轧制的产品 rolled product
轧制的非标准钢板 universal mill plate
轧制的钢材 rolled steel
轧制的钢槽 rolled channel(section)
轧制的混凝土用砂 crushed concreting sand
轧制的金属 rolled metal
轧制的粒状高炉渣 crushed granulated blast-furnace slag
轧制的粒状矿渣 crushed granulated slag
轧制的粒状炉渣 crushed granulated slag
轧制的丝材 rolled wire
轧制的细骨料 crushed fine aggregate
轧制的细集料 crushed fine aggregate
轧制的有色金属 rolled nonferrous metal
轧制低碳型钢 rolled mild steel section
轧制发纹 rolled flaw
轧制法兰 rolled flange;rolled-on flanges
轧制方向 rolling direction
轧制粉坯的烧结 sintering of rolled powder
轧制负荷 rolling load;rolling loading
轧制钢 rolled iron;rolled steel
轧制钢板 roiled sheet steel;rolled plate;rolled sheet iron;rolled sheet material;rolled sheet steel
轧制钢材 rolled steel
轧制钢格栅 rolled steel joist(beam)
轧制钢管 wrought-steel pipe
轧制钢横梁 rolled cross girder
轧制钢筋 rolled steel bar
轧制钢梁 rolled steel beam
轧制钢梁桥 roiled beam bridge
轧制钢龙骨 rolled joist;rolled steel joist(beam)
轧制钢丝 rolled wire
轧制钢条 rolled steel bar
轧制钢托梁 rolled steel joist(beam)
轧制工艺润滑剂 rolling lubricant
轧制工艺润滑冷却液 rolling solution
轧制工字梁 rolled steel joist(beam)

轧制公差 mill limit;rolling tolerance
轧制骨料 crushed aggregate
轧制管 rolled pipe;rolled tube
轧制光(洁)度 mill finish
轧制规范 mill condition;pass schedule
轧制过程 operation of rolling;rolling process
轧制机台 hot rack
轧制加工 mill processing
轧制剪切机 mill shears
轧制件 rolled parts
轧制件形状均匀性 rolled stock shape consistency
轧制角钢 rolled angle
轧制结构 rolled structure
轧制结构钢 rolled structural steel; structural rolled steel
轧制结构型材 rolled structural shape
轧制金箔 rolled gold
轧制金属 wrought metal
轧制金属板 rolled sheet metal
轧制金属门 rolled sheet metal door
轧制空心钢材 rolled hollow section
轧制力矩 roll torque
轧制梁 rolled beam;rolled girder
轧制裂缝 rolling crack
轧制裂纹 rolling crack
轧制鳞皮 rolling mill scale
轧制鳞片 mill scale
轧制铝合金 alumin(i)um wrought alloy;wrought alloy of alumin(i)-um;wrought alumin(i)um alloy
轧制螺纹 rolled-on thread; rolled thread
轧制梅花头 wobbler
轧制摩擦 rolling friction
轧制泡疤表面 process blister
轧制品的粒状表面 pebble
轧制平面 rolling plane
轧制器筋的孔型 deforming groove
轧制铅皮 rolled lead
轧制缺陷 rolling defect
轧制润滑系统 rolling oil system
轧制砂 broken sand; crushed sand; crusher screenings
轧制设备 rolling mill equipment;rolling mill machinery
轧制石砂 stone screenings
轧制实践 mill rolling practice
轧制速度 mill speed;rolling speed; speed of travel
轧制特种型材 rolled special shape
轧制条钢 rolled bar iron
轧制条痕 mill streaks
轧制铁鳞 mill roll scale;mill tap;roll scale;mill scale
轧制铁皮 rolled sheet iron
轧制图表 rolling program(me)
轧制围盘 repeater
轧制温度 rolling temperature
轧制无缝钢管 seamless rolled steel pipe;seamless rolled steel tube
轧制无缝管 seamless rolled pipe; seamless rolled tube
轧制无缝紫铜管 wrought seamless copper tube
轧制细砂 fine-crushed sand
轧制线 axis of rolling;pass line;rolling line;roll line
轧制形变热处理 ausrolling
轧制型材 rolled bar; rolled profile; rolled shape
轧制型钢 rolled bar iron;rolled section;rolled shape;rolled steel section;rolling section
轧制压力 draught pressure; pressure of rolling;rolling pressure
轧制压缩比 ratio of roiling reduction
轧制异型钢 rolled steel shape
轧制硬化 rolled hardening
轧制用钢锭 rolling ingot
轧制用钢坯 rolling billet
轧制(余热)淬火 mill hardening;mill quenching
轧制裕量 rolling margin
轧制支承钢板 rolled steel bearing plate
轧制织构 rolling texture
轧制制度 rolling pattern; rolling schedule
轧制周期 rolling cycle
轧制专业用语 mill language
轧制状态 as-rolled

札

札哈罗夫石 zakharovite

闸

闸 anchor gate;brake;gating;lock

闸坝 gate dam; lock and dam; lock and weir
闸坝并列式 juxtaposition type of dam and lock
闸坝的下游 tail water
闸坝法 <改善航道> lock-and-dam method
闸坝分离式 separated type of dam and lock
闸坝改善航运法 lock-and-dam method
闸坝下游水位 tail elevation
闸板 baffle plate;bib(b);brake ring; braking rim;cut-off plate;damper; damper plate;gate leaf;gate plate; paddle; restrictor; shutter; slide; slide damper; stop plank; stop plate; valve gate; gate; flashboard <坝顶调节水位的>;pipe ram <防喷器的>
闸板边墩 wicket girder
闸板操纵缸 slide actuating cylinder
闸板槽 gate slot
闸板导向构架 paddle frame
闸板阀 paddle valve;slide valve;gate valve
闸板给料斗 gate feed hopper
闸板开度 opening of damper
闸板门 lock-gate hatch
闸板升降机 gate lift
闸板式挡渣浇包 dam type pouring ladle
闸板式防喷器 ram preventer
闸板式浇口杯 dam type pouring basin
闸板式节制闸门 flashboard check gate
闸板行程 gate travel
闸板堰 flush-board weir;sliding timber weir
闸板砖 damper block;damper tile
闸板转换阀 plate-type changeover valve
闸板组件 gate assembly;valve assembly
闸柄 trigger(piece)
闸波脉冲 pulse gate
闸补整杆 brake equalizer
闸操纵杠杆 brake operating lever
闸操作杆 brake operating lever;brake operating rod
闸操作销 brake operating pin
闸槽 gate groove
闸测热效率 brake thermal efficiency
闸衬 braking surface
闸衬光面器 brake lining refacer
闸衬面 brake lining
闸衬片 brake bush; brake facing; brake lining
闸衬片材料 brake lining material
闸程 lockage;lock lift
闸池 lockage basin;lockage bay;lock bay
闸带 brake band;brake strap;strap of the brake
闸带张力 strap tension
闸刀 disconnecting link; male contact;plug-in strip;switch blade
闸刀保险销 blade latch
闸刀衬垫 knife-edge liner
闸刀盒 blade
闸刀开关 chopper switch; closing switch; contact breaker; floor push; knife-break switch; knife-edge switch; leaf actuator; plug-in strip;knife switch
闸刀开关铜片 switch blade; switch tongue
闸刀切纸机 ream indexer
闸刀式剪切机 guillotine plate shears; guillotine shears
闸刀式接触 knife-edge contact
闸刀式接点 knife-edge contact
闸刀式开关 knife-edge contact
闸刀式切纸机 guillotine cutter; guillotine cutting machine;ream cutter
闸刀式叶片型风机 guillotine blade-type damper
闸道流量计 head flow chamber
闸的滑动 slipping of brake
闸底 pocket floor;chamber floor <船闸的>;lock bottom <船闸的>
闸底板 sluice board
闸底板高程 elevation of lock bottom; elevation of lock floor; invert elevation
闸底灌水系统 bottom filling system
闸底涵洞 culvert under floor; floor culvert;underfloor culvert
闸底涵管 culvert under floor
闸底廊道 galley under floor
闸底平面图 ground plan
闸底输水涵洞 floor culvert
闸底系船柱 platform bollard
闸底泄水管 bottom emptying gallery
闸底纵干管灌水系统 <船闸的> bottom longitudinal distribution system
闸电流 brake current
闸电流强度 strength of brake current
闸吊 brake lifter
闸顶面 esplanade
闸顶系船栓 <船闸的> cope bollard
闸顶系船柱 platform bollard
闸段 <河运两闸门之间> lockage chamber
闸墩 pier
闸墩分水尖 upstream nose
闸墩分水头 upstream nose;upstream nosing
闸墩尖端 pier nose
闸墩前端 pier nose
闸墩上游鼻端 <护墩防冲、破冰等用> upstream nosing
闸墩式电站 pier head power station
闸墩收缩系数 pier contraction coefficient
闸墩首部 upstream nose
闸墩首部墩尖 pier nose
闸墩头(部) pier head
闸墩尾部 downstream pier nosing
闸墩下游端 downstream pier nosing
闸阀 brake valve;discharge gate;gate valve; gate shutoff valve; plate valve; screw-down valve; shut-off valve;sliding gate;sluice valve
闸阀间 intervalve
闸阀室 gate valve chamber
闸阀钥匙 gate valve key
闸防尘罩 brake dust shield
闸房 gate house
闸杆 brake bar; brake lever; brake mast;braking lever;throwing lever
闸杆安全链眼 brake beam safety chain eye
闸杆支架 brake lever fulcrum
闸缸 brake cylinder;braking cylinder
闸缸杠杆 cylinder lever guide
闸缸杠杆轴臂 brake cylinder lever shaft arm
闸缸管 brake cylinder pipe
闸缸金属垫料 press disc
闸缸推杆头 eye plate
闸杠杆止动器 brake lever stop
闸沟 gull(e)y plugging;sluiceway
闸鼓车床 brake drum lathe
闸管 brake tube;gate tube
闸管丁字管节 brake tube tee
闸管关断旋塞 brake tube shutoff cock
闸管滤气管 brake pipe strainer
闸管托架 brake tube bracket
闸辊 braking club
闸函数 barrier function
闸盒 brake chamber;cut-out box;gate trap;switch box【电】
闸盒推杆 brake chamber push rod
闸盒推杆弹簧 brake chamber push rod spring
闸后护坦 rear apron
闸机端挡 end gate stanchion
闸棘轮 brake ratchet wheel
闸颊板 brake check
闸架 damper frame
闸间航道 intervening channel
闸间河段 pool;pool reach
闸间水库 intermediate pool
闸间隙 brake clearance
闸槛式区洪道堰顶 obstructed crest of spillway
闸槛 clap(ping)sill;dam sill;lock sill
闸槛高程 invert elevation
闸槛上水深 depth over the sill
闸槛水深 draft on sill;sill depth;water depth on sill
闸槛斜度 rise of sill
闸臼 <坞闸转动的座> hollow quoin
闸龛 gate trap
闸孔 sluice opening
闸孔出流 flow under sluice gate
闸孔放水冲刷 sluicing and scouring
闸孔开度 gate open(ing)
闸控港池 wet dock
闸控开关 gate-controlled switch; knife switch
闸控溢洪道 gated spillway; obstructed spillway
闸控整流器 gate-controlled rectifier
闸控制器 brake controller
闸口 lid; lock entrance; plug socket; sluice;sluiceway
闸块 block
闸块式制动器 block brake
闸况等级 grade of lock condition
闸里 braking surface
闸里有效面积 effective lining area
闸连杆 brake link;connection rod
闸联动装置 brake linkage
闸联杆 brake connecting rod
闸梁 stop log
闸梁安全链眼铁 brake beam safety chain eye bar
闸梁调整器 brake beam adjuster
闸梁调整器导承 brake chamber push rod guide
闸梁系带环 brake beam strap link
闸流管 discharge tube; gas control tube; gas-filled triode; gas relay; gastriode;gastriode relay;grid-controlled rectifier; hot cathode gas-filled valve;thyratron;thyristor
闸流管变流器 thyratron inverter
闸流管电动机 thymotor; thyratron

motor
闸流管电动机传动 thyratron-motor drive
闸流管电动机控制 thymotrol
闸流管换流器 <交流变直流> thyratron inverter
闸流管计算器 thyratron counter
闸流管继电器 thyratron relay
闸流管接触器 thyratron contactor
闸流管开关 thyratron switch
闸流管控压比 control ratio
闸流管控制发电机 thyratron controlled generator
闸流管控制器 thyratron control
闸流管控制特征曲线 thyratron control characteristic curve
闸流管门 thyratron gate
闸流管逆变器 thyratron inverter
闸流管频闪观测器 thyratron stroboscope
闸流管驱动 thyratron drive
闸流管驱动系统 thyratron-rectifier drive system
闸流管伺服系统 thyratron servo
闸流管振动检测器 thyratron chatter detector
闸流管整流 thyratron commutation
闸流管整流器 thyratron rectifier
闸流管自动电压调整器 thyratron automatic voltage regulator
闸流(晶体)管 thyristor
闸流控制接通时间 gate-controlled thyristor turn-on time; gate-controlled turn-on time
闸楼 gate house; operating control point【铁】
闸轮 brake drum; brake pulley; brake wheel; braking wheel; ratchet; ratchet wheel
闸轮防尘罩 brake drum dust shield
闸轮机构 clickwork
闸轮连接 anti-kick device
闸轮联轴节 ratchet coupling
闸轮式钻床 brake drum drilling machine
闸轮制动鼓 brake drum
闸门 anchor gate; apron plate; caisson; clough; draw gate; flap; gate closure; geat; gole; hatch; hatchway; lockage gate; lock chamber gate; lock check gate; lock gate; paddle; paddle bar; paddle door; paddle gate; penstock; shutter; strobe; valve; waste gate; water gate; water sluice gate; wicket
闸门凹座 gate recess
闸门坝 penstock dam; shutter dam
闸门板 gate sheet; sluice panel
闸门板面 skin plate
闸门边槽 side recess
闸门边框 side frame of gate(leaf)
闸门边木 side recess
闸门边柱 side frame of gate(leaf)
闸门表面钢板 gate leaf
闸门布置 gate position
闸门操纵室 gate operating chamber
闸门操作便桥 gate operating deck
闸门操作机构 gate operating mechanism
闸门操作机械 gate operating machinery; gate operating mechanism
闸门操作间 gate house
闸门操作平台 gate operating deck; gate operating platform
闸门操作桥 gate operation platform
闸门操作室 gate house
闸门槽 gate recess; gate slot; gate well; gate chamber
闸门侧滚轮 side roller
闸门侧立柱 heel post
闸门挡板 seal apron
闸门导槽 guide groove of gate
闸门导轨 gate guide
闸门导向槽 gate guide; gate groove
闸门的顶部构架 top frame of gate
闸门的钢轧头 steel dog
闸门底板 gate ground
闸门底部放水管 bottom header pipe
闸门底脚 gate footstep
闸门底枢 lower gudgeon; lower pintel; lower pivot
闸门底缘 bottom edge of gate
闸门电路 gate circuit; gating circuit; strobotron circuit
闸门电路断开 gate turnoff
闸门吊架 gate hanger
闸门顶 top of gate
闸门顶部围绕闸门边柱的锚环 collar strap and anchor
闸门顶枢 upper gudgeon; upper pintle; upper pivot
闸门段 gate bay
闸门对口端柱 miter post
闸门墩 gate block; gate monolith; gate pier; hollow quoin
闸门墩头部 gate pier nose
闸门墩铸钢垫座 quoin bar; quoin plate; quoin reaction casting
闸门耳轴 gate gudgeon
闸门阀 sluice cock; sluice valve; gate valve
闸门放大器 gated amplifier
闸门放松位置 release position
闸门分水设备 <调节水量用> gate diversion works
闸门杆 gate stem
闸门钢轧头 steel dog
闸门港 dock basin
闸门隔层装置 door seals
闸门工作平台 gate operating deck; gate operation platform
闸门工作桥 gate operating deck; gate operation platform
闸门钩 gate dog
闸门关闭 gate closure
闸门关闭器 gate closer
闸门关断可控硅整流器 gate turnoff thyristor
闸门管 gate tube
闸门灌水损失 lockfull lost
闸门涵洞 underfloor culvert
闸门河段 intermediate pool
闸门横梁 gate beam
闸门活页 band and gudgeon
闸门机构 gate mechanism
闸门机制 gate mechanism
闸门价格 <欧洲共同市场规定的最低进口价> sluice-gate price
闸门间 gate house
闸门建筑物 closing structure
闸门降落机构 gate lowering mechanism
闸门角柱 quoin post
闸门铰链 shutter hinge
闸门铰座 articulated gate shoe
闸门进料 gate feed
闸门阱 <又称闸门井> gate chamber; gate shaft; gate trap; gate well
闸门开闭装置 operating gear
闸门开度 gatage; gateage
闸门开度指示器 gatage indicator
闸门开启度 gate open(ing)
闸门开启钩 gate dog
闸门龛 gate recess
闸门槛 lock(gate) sill; tripsill; gate sill
闸门坑 gate well
闸门孔 gate open(ing)
闸门孔全开 full gate opening
闸门控制的废水道 gate-controlled wasteway
闸门控制的废水口 gate-controlled wasteway
闸门控制的堰 gated weir
闸门控制的溢洪道 gate-controlled wasteway
闸门控制的溢洪口 gate-controlled wasteway
闸门控制室 gate control house; headhouse <渠首等>; gatehouse
闸门控制学说 gate control theory
闸门控制溢洪道 gated spillway
闸门库 gate chamber; gate niche; gate recess
闸门框墙 jamb wall
闸门立轴的凹形支承 pintle pot
闸门立柱 meeting post
闸门流量 discharge of sluice
闸门漏水量 gate leakage
闸门漏水试验 gate leakage test
闸门脉冲发生器 gate pulse generator
闸门锚具 gate anchorage
闸门门槛 gate seat; gate sill
闸门门扇 gate leaf
闸门门叶 gate flap; gate leaf
闸门(门叶)的边柱 side frame of gate (leaf)
闸门门座 gate seat
闸门面板 skin plate; skin plating; gate skin plate
闸门面积 gate area
闸门木梁 sluice timber
闸门排料 gate-and-dam discharge
闸门排料式跳汰机 gate-and-dam jig
闸门排水 penstock drain
闸门启闭 gate handling
闸门启闭操纵 gate operating control
闸门启闭机 gate hoist; gate lifter; gate lifting device; gate lowering mechanism
闸门启闭机构 hoisting gear
闸门启闭装置 gate lift; gate lifting device
闸门千斤顶 gate jack
闸门前面 lock front
闸门桥 water gate bridge
闸门全开度运行 full gate operation
闸门闪光 stroke flash
闸门上的输水口 gate paddle
闸门升降机 gate lift
闸门升降机构 gate-towering mechanism
闸门式坝 gate dam
闸门式给料器 gate feeder
闸门式节流交叉口 gating intersection
闸门式节流信号 <公交车优先的一种方法> gating signal
闸门式快门 drawer type shutter; guillotine shutter
闸门式水坝 <有水平转轴的> curtain dam
闸门式送料装置 gate feeder
闸门式消防栓 gate type hydrant
闸门式消火栓 gate type hydrant
闸门式泄水系统 curtain drain
闸门式堰 gate dam
闸门式振荡计数器 gated oscillator counter
闸门式振动给料器 gate type feeder
闸门室 gate house; gate vault; gate bay <船闸的>; gate chamber; caisson chamber
闸门室底板 gate platform
闸门枢轴 apron pivot; gate pivot; gate pintle
闸门枢轴锚定设备 gate anchorage
闸门输水口 paddle hole
闸门竖井 gate shaft
闸门锁定 gate lock
闸门锁定设备 gate latching device; gate logging device
闸门锁定装置 gate dogging device; gate latching device; gate logging device
闸门提升机 gate hoist
闸门提升设备 gate lifting device; gate lift
闸门提升装置 gate lifting device
闸门贴接面 meeting face
闸门推拉杆 gate strut
闸门位置 closure position
闸门销子座 apron bracket
闸门泄水口 gated outlet
闸门信号 gate signal
闸门行程 gate stroking
闸门堰顶 <设有调节水位装置的坝顶> controlled crest
闸门液压推拉杆 hydraulic gate rain
闸门罩 gate trap
闸门振动 gate vibration; shutter vibration
闸门支臂 gate arm
闸门支枢 gate pivot
闸门支枢锚定螺栓 pivot anchor bolt
闸门止水 gate seal
闸门制动器 articulated gate shoe
闸门轴 penstock shaft
闸门轴柱 quoin post
闸门作用 gate action
闸面 brake surface
闸木 sluice timber; timber stop
闸内船只 locked vessel
闸内水量 lockage water
闸内蓄水 lockage water
闸能 braking energy
闸盘 brake disc[disk]
闸片 brake lining
闸气管 brake line
闸前护坦 fore apron; front apron
闸墙 chamber wall; lock head wall; lock wall
闸墙凹槽内的系船柱 recessed bollard
闸墙超高 freeboard of lock wall
闸墙出水口 side port
闸墙顶部 lock cope
闸墙顶拦墙 coping wall
闸墙顶面 lock cope
闸墙顶面高程 height of lock cope
闸墙孔口系统 side port system
闸墙进水口 side port
闸墙内输水涵洞 lock wall culvert
闸墙输水涵洞 lock wall culvert
闸墙输水廊道 lock wall culvert
闸墙消能设施 wall baffle
闸桥 lock bridge; sluice(-gate) bridge
闸软管 brake hose; brake hose pipe
闸软管及接头 brake hose and coupling
闸山沟 dam up ravine
闸栅式安全保护装置 gate guard
闸扇枢轴端 quoin end
闸式测功器 brake
闸式充水管道 chamber filling conduit
闸式阀 gate valve
闸式阀门 seal leg
闸式水力测功器 hydraulic brake
闸式鱼道 Borland type fish pass; lock type fish pass
闸式真空液压制动器缸 brake hydrovac cylinder
闸室 sluice chamber; chamber; lock basin; lock bay; lock chamber; navigation chamber <船闸>; brake chamber; lock chamber wall
闸室岸壁 lock chamber wall
闸室边墙 lock side wall
闸室充水管 chamber filling conduit
闸室充水管道 chamber filling conduit
闸室充水廊道 chamber filling conduit
闸室底 floor of lock chamber

闸室底板 chamber floor; chamber slab; gate chamber floor; lock floor
闸室底部 invert of lock chamber
闸室顶板 gate chamber roof
闸室槛水深 water depth above sill
闸室进水总管 intake main of lock chamber
闸室宽度 width between lock walls
闸室内阶梯 lock ladder
闸室内水量 lockage water
闸室排水道 discharge manifold of lock chamber
闸室排水总管 discharge manifold of lock chamber
闸室墙 chamber wall; gate chamber wall; lift wall; lock chamber wall; lock side wall
闸室墙板 chamber slab
闸室容水量 lock bay
闸室式船闸 chamber navigation lock
闸室式鱼道 lock chamber type fishway
闸室水深 chamber depth
闸室有效长度 < 船闸 > useful chamber length; useful length of the chamber
闸室有效宽度 < 船闸 > useful width of the chamber
闸室闸门 chamber gate
闸室中心消能设施 centre baffle
闸首 < 船闸的 > lock gate; lock head
闸首充水管道 end filling conduit
闸首充水管道系统 end filling conduit system
闸首充水系统 end filling system
闸首充泄水系统 end filling and emptying system
闸首底板 sill block; sill plate
闸首底部放水管 bottom header pipe
闸首墩 gate block; gate monolith; gate pier
闸首灌水管道 end filling conduit
闸首灌水管道系统 end filling conduit system
闸首灌水系统 end filling system
闸首灌泄水系统 end filling and emptying system
闸首墙 lift wall; lock head wall
闸首区 head reach
闸闩 lock
闸踏板 brake foot plate; brake step; brake treadle
闸调器 brake adjuster; brake adjusting device
闸调整工具 brake adjusting tool
闸头 sluice head
闸凸轮 brake cam
闸凸轮杆 brake cam lever
闸凸轮杠杆 brake cam lever
闸凸轮轴杆 brake cam shaft lever
闸凸轮轴环 brake camshaft collar
闸凸缘 brake flange
闸瓦 brake block; brake drum; brake lining; brake-shoe; hem shoe; shoe plate; skid; skid shoe; slipper
闸瓦插销 brake-shoe key
闸瓦衬带磨光器 brake lining grinder
闸瓦导(定位)销 brake-shoe guide pin
闸瓦导簧 brake-shoe guide spring
闸瓦导柱 brake-shoe guide
闸瓦吊 brake block hanger; brake-shoe hanger
闸瓦钢背 brake-shoe back
闸瓦固定板 brake plate anchor
闸瓦固定夹 shoe retaining clips
闸瓦固定销 brake-shoe anchor pin
闸瓦规 brake-shoe ga(u)ge
闸瓦间隙 brake-shoe clearance
闸瓦间隙调整器 brake-rigging regulator; brake slack adjuster
闸瓦间隙自动调节器 automatic slack adjuster
闸瓦减振器 anti-shoe rattler
闸瓦减震减声器 anti-shoe rattler
闸瓦卡规 brake calliper
闸瓦扩张器 brake-shoe expander
闸瓦离合器 block clutch
闸瓦联杆销 brake-shoe link pin
闸瓦磨床 brake-shoe grinder
闸瓦片带摩擦片 brake-shoe lining
闸瓦上的作用力 brake-shoe force
闸瓦试验器 brake-shoe tester
闸瓦调节器 brake-shoe adjuster
闸瓦调整垫片 brake block insert
闸瓦调整器 brake-shoe adjuster
闸瓦调整凸轮 brake-shoe adjusting cam
闸瓦托 brake block carrier; brake block holder; brake head; brake-shoe holder; shoe holder; truck side bearing
闸瓦托板 brake block holding plate
闸瓦托吊 brake hanger; brake head hanger
闸瓦托吊耳 brake head hanger lug
闸瓦托吊拉力 hanger pull
闸瓦托吊螺栓 brake hanger bolt
闸瓦托吊托 brake hanger carrier
闸瓦托吊销 brake hanger pin; brake head hanger lug
闸瓦托吊座 brake hanger bracket
闸瓦托弹簧 balance spring; brake head spring
闸瓦托调整弹簧 brake head adjusting spring; head adjusting spring
闸瓦销 brake block key; brake-shoe key
闸瓦悬置 suspension of brake shoe
闸瓦压杆 brake block holder
闸瓦压力 brake block pressure; brake-shoe pressure
闸瓦支持销 brake anchor
闸瓦止动销 brake-shoe stop pin
闸瓦制动器 block brake; shoe brake
闸瓦制动装置 block brake unit
闸瓦轴承 brake-shoe bearing
闸坞工程 dock engineering; dock work
闸下冲刷 sluice scour(ing)
闸下出流 sluice flow
闸下游建筑部分 aft-bay; afterbay
闸压床 brake-press
闸压力 brake pressure
闸堰 sluice weir
闸用软管 brake hose
闸隅槽 hollow quoin
闸隅涵洞 quoin culvert
闸隅柱 quoin post
闸站【铁】 block post with distribution tracks; lock station
闸障碍物 barrage
闸支板 brake support shield
闸址 location of barrage; lock location < 船闸 >; lock site < 船闸 >
闸址工程地质勘察 engineering geologic(al) investigation of sluice gate site
闸轴杆 brake shaft lever
闸轴架 brake shaft carrier
闸砖 stopper
闸装置 brake-gear
闸锥 brake cone
闸阻 brake drag
闸阻设备 dragging brake equipment

铡

铡草机 chaff cutter; chaff slicer; crop cutter; forage cutter; hay cutter; straw breaker; straw shredder
铡除刀 guillotine
铡床 slotter; slotting machine
铡刀 cutting knife; fodder chopper; hand hay cutter; straw chopper
铡刀剪切机 gate shears
铡刀式钢材剪切机 guillotine bar shear
铡刀式闸门 knife gate; shear gate
铡合 rivet
铡料台 cutting bench
铡刨 slotting
铡碎粗料 shredded fodder
铡碎坚果的粉碎机 nut cracker
铡碎小麦干草 wheaten chaff
铡砧 anvil

眨

眨眼一瞥 blink

乍

乍得 < 非洲 > Chad
乍冷 cold snap

诈

诈病 malingering

栅

栅栏 barricade; barrier; boom; brandreth; catch frame; corral; fence; fencing; hedge; hurdle; impalement; pale fencing; pale; palisade; panel; rail fence; stake rack; stockade; Wild fence < 魏尔德雨量器四周围用 >; pile screen
栅栏材料 fence material
栅栏侧墙 slatted side
栅栏层 palisade layer
栅栏顶板 ledger board
栅栏附属物 fencing accessories
栅栏杆 rack stake
栅栏高度 height of fence
栅栏横杆 rails
栅栏尖板条 paling
栅栏看守房 gateman's lodge
栅栏篱笆 palisade fence
栅栏立柱 fencing stake
栅栏门 barrier; braided door
栅栏木板 fence board
栅栏筛 bar sieve
栅栏施工与保养 construction and maintenance of fences
栅栏式板夹机 plate-folding machine
栅栏式搅拌器 picket-fence stirrer
栅栏式拖车 rack-type wagon
栅栏式污泥浓缩器 picket-fence type sludge thickener
栅栏式污泥耙 picket-fence sludge rake
栅栏式折页机 buckle folding machine; pocket folder
栅栏挑水坝 hurdle dike[dyke]
栅栏透水堤 hurdle dike[dyke]
栅栏透水丁坝 hurdle groin; hurdle groyne
栅栏图 fence diagram
栅栏网 boom net
栅栏线 fence line
栅栏效应 barrier effect; fence effect
栅栏圆柱 paling
栅栏柱 barrier pillar; barrier post; fence post; fence stake; fencing post
栅栏桩 fence picket; fence stake
栅栏装置 safing
栅子地板底梁 floor sill

炸

炸包 bales burst
炸边 crack edge
炸槽 exploded channel
炸层厚度 thickness of blasting layer
炸穿深度 bomb penetration
炸弹 bomb
炸弹库 explosives magazine
炸弹量热计 bomb calorimeter
炸弹箱 cassette
炸弹型车身 cow; cowling
炸弹炸穿深度 bomb penetration
炸弹贮藏处 bomb dump; bomb store
炸弹贮藏面积 bomb store area
炸弹状储气瓶 bomb
炸点距离偏差观测 range spotting
炸点平均高度 mean height of burst
炸高 height of burst
炸后残眼 dead hole
炸后废墟 bombed site
炸毁 blow-up; demolition(blast); dynamite
炸胶 blasting gelatin(e); gelignite
炸礁 reef blasting; reef explosion
炸礁船 reef blasting ship; underwater drilling and blasting ship
炸开导坑填筑路堤法 < 沼泽地上 > trench-shooting
炸开孔穴填筑路堤法 < 沼泽地上 > underfill method
炸开钻孔含水层的裂隙 < 以扩大出水量 > shotfiring
炸孔 shothole
炸孔压力 borehole pressure
炸裂 blow; break open; bursting; explosive disruption
炸裂花纹 crackled
炸裂面积 area of explosion
"炸面饼"圈式的城市发展 doughnut phenomenon
炸面圈 < 美 > doughnut
炸泥 mud blasting
炸砰 scrap
炸破层 bomb breaking layer
炸散 explosive disruption
炸碎 shatter
炸纹废砖 chuff; shuff
炸药 blasting agent; blasting material; blasting powder; burster; burster charge; bursting charge; bursting explosive; bursting powder; dynamite; explosive; explosive agent; explosive compound; explosive filler; explosive material; explosive powder; explosive substance; percussion powder; powder; powder charge; soup; stick powder; dooly < 俚语 >; lignose < 一种含有硝化甘油和木质纤维的 >
炸药包 blasting charge; cartridge(dynamite); detonator cartridge; powder bag; satchel charge; explosive charge
炸药爆破 dynamiting; explosive explosion
炸药爆破锚固锚杆法 explosively anchored; rock bolt
炸药爆破系数 explosive factor
炸药爆速 explosion velocity of explosive
炸药残孔 misfired hole
炸药厂废水 explosive plant waste
炸药车 transporting car for explosive
炸药冲击感度 sensitiveness to impact
炸药储藏室 explosive store
炸药储存安全距离表 quantity distance tables
炸药存贮量安全距离表 quantity distance tables
炸药袋 powder pocket; powder sack
炸药的分布 dispersion of explosives
炸药的氧化剂 explosive oxidizer

炸药发放员 powder monkey
炸药费 explosive charge
炸药工艺学 powder technology
炸药管理员 powderman
炸药罐 can(n)ister;explosive can(n)ister
炸药混合浆 slurry
炸药混合器 explosive mixer
炸药检验 testing of explosives
炸药警告旗 powder warning flag
炸药卷 explosive cartridge;stick of explosive;stick powder
炸药可用性 dynamites available
炸药孔 breech opening
炸药孔炮泥 stemming
炸药孔填塞 stemming
炸药库 blasting agent storage;dynamite magazine;explosives magazine;powder house;powder magazine
炸药雷管 powder fuse
炸药类型 type of explosive
炸药量 amount of explosive;quantity of explosive charge
炸药帽 percussion cap
炸药猛度 explosive brisance
炸药猛度测定 explosive grading
炸药密度 density of explosive
炸药敏感度 powder sensibility
炸药名称 name of explosive
炸药摩擦感度 friction(al)sensitiveness
炸药捻线 twist-on explosives
炸药起爆 priming of explosive
炸药腔 powder space
炸药驱动装置 explosive-actuated device
炸药热感度 heat sensitivity of explosive
炸药生产 production of explosives
炸药式震源 explosive source
炸药室 bursting chamber;powder chamber
炸药特性阻抗 characterisation impedance of explosive;characteristic impedance of explosive
炸药体(或筒)强度 cartridge strength
炸药筒 cartridge dynamite;cartridge explosive;charged cartridge;detonator cartridge;stick powder
炸药威力 brisance;explosive power;explosive strength
炸药箱 powder box
炸药消耗量 explosive consumption
炸药消耗率 explosive factor;power factor
炸药氧平衡 oxygen balance of explosive
炸药药芯 explosive core
炸药引信 powder fuse
炸药用量 amount of used explosive
炸药(有效)系数 <单位体积爆破松散岩石的炸药重量数> powder factor
炸药运输车 explosive carriage
炸药震力 brisance
炸药震力指数 brisance index
炸药震源 dynamite source;explosive energy source
炸药之间的砂夹层 sand-stemming between the charge of explosives
炸药装量 dynamite charge
炸药装填 dynamite charge
炸药装填量 charge of an explosive;explosive charge
炸药装填系数 load coefficient of explosive charged
炸药装置 blasting charge
炸药撞杆 powder lance
炸药撞杆切断 powder lancing
炸药撞针 powder lance
炸油 blasting oil
炸釉 glaze craze

蚱

蚱蜢 <小型起重机的俚语> grass hopper

榨 wring

榨板条 strap
榨出 expression;squeeze-out
榨干机 drying press;wringer
榨棉机 cotton-press
榨取 squeeze
榨取机 squeezer
榨取者 squeezer
榨水机 wringer
榨油 pressing
榨油厂 oil mill;oil press;oil pressing mill
榨油大豆 soybean for oil
榨油机 oil expeller;oil mill;oil press
榨汁机 juicer

摘 pluck

摘车 detaching;detachment
摘车临修【铁】 axle-box lubrication check-up of vehicles detached from train
摘车修 repair of vehicles detached from train
摘车轴检 causal repair of vehicle detached from train
摘出器 enucleator;extractor
摘除 extirpate;extirpation
摘除器 enucleator;extirpator
摘掉钢丝绳 throw-off the rope
摘锭传动轴 spindle drive shaft
摘锭罩 spindle shield
摘钩 decoupling;disconnecting hook;release catch;releasing the coupling;throw of hook;uncouple;unhitch;unhook;unlink;unlocking;Janney <俚语>
摘钩工 flatter
摘钩机构的系牢 <跨装长大货物时> securing of uncoupling mechanism
摘钩设备 tripping device
摘钩员 pin puller
摘钩装置 tripping device
摘挂 rolling off
摘挂调车 detaching and attaching of cars
摘挂钩工人 crane man;slinger
摘挂列车 pick-up and drop train
摘挂列车甩挂车作业计划 district local trains pickup plan
摘花 deflower
摘环 disconnecting link;disconnecting stirrup
摘机 off-hook
摘机车 <从列车上> uncouple the engine
摘机位置 off-hook position
摘解器 <架空索道> uncoupling device
摘开 hook off;unhitched;unhook
摘开齿轮 throw out of gear
摘开离合器 disconnect a clutch;disengage a clutch
摘录者 abstractor
摘门机 door extractor
摘取 picking
摘取力 tearout
摘取物 picking
摘取装置 plucker;stripper unit
摘去 pick-off
摘下 drop-off;uncouple;set-off <自列车摘下车辆>
摘下的客车 <旅客列车在直通运行中> slip-carriage
摘要表 abstract summary;recapitulation statement;summary table
摘要说明书 abstract statement
摘要账 book of abstract

宅 tort

宅边水域 home waters
宅邸 country mansion;hotel
宅地 building lot;curtilage;toft
宅地地租率 dwelling land rate
宅地面积 homestead area
宅第 big house;family mansion;manor;manor house
宅基 home stead;homestead;toft;tort
宅基地 family land plot;private lot
宅基豁免法 homestead exemption law
宅基界线 party line
宅基免税 homestead tax exemption
宅门 house door
宅旁地 croft
宅旁果园 home-yard orchard
宅旁游憩屋 plaisance
宅旁杂草 chomophyte;ruderal
宅前花园 front garden
宅园 home garden
宅院 messuage
宅院铺面 yard pavement
宅院渗水井 yard catch basin
宅院圬工墙 yard masonry wall

翟

翟混淆现象【数】 aliasing

窄

窄板 <宽度在 12 厘米以下> strap;narrow board

窄板材 ribbon tap;ribbon tape
窄板条 slat;stripite;strip lath
窄边工字钢 I-steel
窄扁钢 flat bar iron;strip bar
窄波段宽度 narrow band width
窄波段滤光片 narrow band filter
窄波段星等 narrow band magnitude
窄波束 narrow channel
窄波束雷达 narrow beam radar
窄槽 slot
窄槽木刨 rebate plane
窄槽平缝接合的 flat-joint jointed
窄槽平缝接合勾缝 flat-joint jointed pointing
窄槽稳定器 gutter stabilizer
窄槽型急滩 rapids of narrow channel pattern;rapids of narrow channel type
窄长窗 gap window
窄车体的 narrow-bodied
窄车体电车 narrow-bodied tramcar
窄尺 narrow rule;narrow scale
窄尺寸 narrow dimension
窄处铆接机 jam rivet(t)er
窄窗座 banket(te);banquette
窄带 narrow band
窄带逼近滤波器 narrow band pass filter
窄带闭塞滤波器 narrow band blocking filter
窄带材 narrow strip;ribbon tape
窄带材围盘 strip repeater
窄带材用围盘 strip repeater
窄带测光 narrow band photometry
窄带长波全向无线电信标 narrow band-long-wave omnidirectional range
窄带传输 narrow band transmission
窄带传输线 narrow band transmission line
窄带电平 narrow band level
窄带发射 narrow line emission
窄带放大器 narrow band amplifier
窄带分析器 narrow band analyser[analyzer]
窄带干扰 narrow band interference;selective interference
窄带干涉 narrow band interference filter
窄带钢 ribbon iron;ribbon steel;riffled iron;strip iron
窄带高温计 narrow band pyrometer;spectral pyrometer
窄带过程 narrow band process
窄带接收 narrow band reception
窄带接收机 narrow band receiver
窄带截止滤波器 narrow cut filter
窄带晶体滤波器 narrow band crystal filter
窄带锯 narrow band saw
窄带卷 slit coil
窄带宽内插 narrow band width interpolation
窄带宽内插法 narrow band width interpolation
窄带(宽)体系 narrow band system
窄带滤波片 narrow band filter;short pass filter
窄带滤波器 low-bandwidth filter;narrow band filter;short pass filter;spike filter
窄带滤光片 narrow band filter;short pass filter
窄带滤光器 narrow band filter;short pass filter
窄带滤色片 narrow cut filter
窄带脉冲信号 narrow band ping
窄带频率 narrow band frequency
窄带频率响应 narrow band frequency response
窄带谱 narrow band spectrum
窄带色差解调器 narrow band colo(u)r-difference demodulator
窄带色度信号 coarse chrominance primary
窄带射频频道 narrow band radio frequency channel
窄带数字滤波 narrow band digital filtering
窄带随机过程 narrow band random process
窄带随机响应 narrow band random response
窄带随机振动 narrow band random vibration
窄带随机振动波 narrow band random vibration wave
窄带特征 narrow band characteristic
窄带天线 <用于检测单车道上的交通> narrow band antenna
窄带调频 narrow band frequency modulation;narrow frequency modulation
窄带调制 narrow band frequency modulation
窄带通滤波器 narrow band pass filter
窄带通滤光片 narrow band pass filter
窄带通路 narrow band path
窄带通信[讯]系统 narrow band communication system
窄带通信[讯]线路 narrow band link
窄带吸收 narrow band absorption
窄带信道 narrow band channel
窄带选择接收机 narrow band-selective receiver
窄带选择性伏特计 narrow band-selective voltmeter
窄带遥测 narrow band telemetry

Z

窄带用户终端 narrow band subscriber terminal
窄带噪声 narrow band noise
窄带噪声测量 narrow band noise measurement
窄带噪声因数 narrow band noise factor
窄带振动 narrow band vibration
窄带直接印字 narrow band direct printing
窄带制 narrow band system
窄带综合业务数字网 narrow band integrated serviceable digital network
窄带阻塞干扰 narrow band barrage jamming
窄导轨 narrow guide
窄道用搬运车 narrow aisle truck
窄的齿端 toe
窄的高墙 slender wall
窄底阶地 narrow base terrace
窄点 pinch-point
窄电闸电路 narrow gate circuit
窄端头楔形砖 feather end on edge
窄范围比例调节 narrow band proportional control
窄范围控制器 narrow band controller
窄方头铲 sharp-shooter
窄防波堤 narrow breakwater
窄防波堤口门 narrow breakwater opening
窄沸程产品 narrow boiling range product
窄沸点混合物 close-boiling mixture
窄风道口 slot outlet
窄缝 narrow fissure; narrow slit; narrow slot; small clearance space
窄缝槽模拟 analogy of narrow aperture trough
窄缝光门 sound slit
窄缝螺旋线 narrow gap helix
窄缝式挑流鼻坎 slit type flip bucket
窄缝镶嵌玻璃 needle glazing
窄幅地毯 narrow carpet
窄幅织品 small wares
窄幅织物 narrow fabric
窄工作面 narrow face; narrow place; narrow workings
窄工作面装煤机 shortwall loader
窄公路 minimotorway
窄共振 narrow resonance
窄共振近似 narrow resonance approximation
窄钩尺 narrow hook rule
窄光束 narrow beam; sharp beam
窄规胶片 narrow-ga(u)ge film
窄轨(道) narrow-ga(u)ge track; narrow track
窄轨道系统 narrow-ga(u)ge track system
窄轨的 decauville
窄轨斗车 decauville truck; decauville tub; decauville wagon
窄轨废料车 narrow-ga(u)ge dump car
窄轨火车头 narrow locomotive
窄轨机车 narrow locomotive; dink(e)y locomotive <运土、石用的>
窄轨距 <1.067 米及以下的>【铁】 narrow-ga(u)ge; three-feet and six inches ga(u)ge
窄轨距轨道 narrow-ga(u)ge railroad [railway]
窄轨矿车 narrow rail bogie
窄轨料车 decauville truck
窄轨内燃机车 diesel narrow-ga(u)ge loco(motive)
窄轨平板自卸车 narrow-ga(u)ge flat dump car
窄轨轻便铁道 decauville portable railway; decauville railway
窄轨轻便铁路 decauville railway
窄轨轻便自卸车辆 tipping wagon
窄轨倾斜车 narrow-ga(u)ge dump car
窄轨四轮转向架 narrow rail bogie
窄轨台车 narrow rail bogie
窄轨铁道 <轨距在 1.435 米以下> narrow-ga(u)ge railway; narrow railway
窄轨铁路 decauville; decauville railroad; decauville railway; decauville track; light-ga(u)ge railroad; light-ga(u)ge railway; light-ga(u)ge track; narrow-ga(u)ge line; narrow-ga(u)ge railroad; narrow-ga(u)ge railway
窄轨铁路电动牵引机 decauville motor tractor
窄轨铁路及设备 decauville plant
窄轨铁路器材 light railway material
窄轨铁路桥 narrow-ga(u)ge railway bridge
窄轨铁路系统 jubilee track system; light railroad system; light railway system
窄轨铁路线 narrow railway track
窄轨小机车 dink(e)y
窄轨蒸汽机车 narrow-ga(u)ge steam loco(motive); narrow steam locomotive
窄轨自卸小车 tipping wagon
窄焊道 string(er) bead(ing)
窄焊道焊机 stringer bead welder
窄焊道焊接 bead welding; stringer bead welding
窄焊条 <焊条不横摆> stringer bead
窄河槽 narrow channel; neck channel
窄厚扁钢 bar of flat
窄化 narrowing
窄火箱 narrow fire box
窄基槽 <埋设管线等的> narrow trench
窄基础 narrow base
窄基底 narrow base
窄基底杆塔 narrow base tower
窄级配 narrow gradation
窄极限 close limit
窄急海流 stream current
窄尖大脉冲 giant pulse
窄间隔法 close-spaced method
窄间隙单面焊 narrow gap one side welding
窄间隙单面自动电弧焊 narrow gap one-side automatic welding
窄间隙焊接 narrow gap welding
窄胶片 narrow-ga(u)ge
窄角镜头 narrow angle lens
窄角配位仪 narrow angle coordinator
窄角摄影机 narrow angle camera
窄角搜索 narrow angle acquisition
窄角透镜 narrow angle lens
窄角坐标方位仪 narrow angle coordinator
窄街道 narrow street
窄截槽 <采矿> thin kerf
窄颈测流槽 throated flume
窄颈段 neck reach; throat
窄锯条机锯 jig saw
窄可倾炉 Detroit rocking furnace
窄孔道 narrow passage
窄孔柱 narrow-bore column
窄口 slot
窄口的 narrow-mouth(ed)
窄口轮缘 substandard rim
窄矿房 narrow stall
窄矿柱 rance
窄框屏幕 narrow frame screen
窄框纱窗 narrow frame screen
窄粒级产品 short-range product
窄量程电桥 limit bridge
窄流道叶轮 narrow path impeller
窄馏分 narrow boiling cut; narrow boiling range fraction; narrow fraction; narrow range cut
窄路 narrow pass; neck; throat
窄履带距 narrow tread
窄履带距的推土机 narrow-ga(u)ge bulldozer
窄履带式拖拉机 narrow track tractor
窄轮距 narrow tread
窄轮距起落架 narrow track landing gear
窄轮距拖拉机 narrow track tractor
窄脉冲 barrow pulse; narrow pulse; spike pulse
窄脉冲多谐振荡管 sanatron
窄脉冲发生器 narrow pulse generator
窄脉冲取样 narrow pulse sampling
窄门道 <中世纪建筑> transyte; tresaunce; trisantia
窄门电路 narrow gate; narrow gate circuit
窄门多谐振荡器 narrow gate multivibrator
窄门梃 diminished stile; diminishing stile
窄面 small face
窄面法兰 narrow face flange
窄面环 narrow face ring
窄面掘进 driving with narrow face
窄年轮的 slow-grown
窄年轮的木材 narrow-ringed
窄刨 rifler
窄频带 narrow band; narrow frequency band
窄频带波分析仪 narrow bandwave analyser
窄频带传输 narrow band transmission
窄频带电路 narrow band circuit
窄频带电视 narrow band television
窄频带放大器 narrow band amplifier
窄频带特性观测器 narrow band charactascope
窄频带特性观测设备 narrow band charactascope
窄频带特征观测设备 narrow band charactascope
窄频带体系 narrow band system
窄频带通信[讯]方式 narrow band communication system
窄频带信道 narrow band channel
窄频带信号 narrow band signal
窄频道信道 narrow band channel
窄频滤波器 exclusive filter
窄频偏调频发射机 narrow deviation frequency modulated transmitter
窄平底自动倾覆车 narrow flat dump car
窄平钳 narrow flat pliers
窄平行光栅 raster
窄剖面 slit
窄谱线轮廓 barrow line profile
窄齐头平锉 narrow pillar file
窄浅航道 restricted channel
窄浅河段 restricted reach
窄浅水道 restricted waterway; shallow and narrow channel
窄浅水域 restricted waters
窄桥标志 narrow bridge sign
窄切犁铧 narrow cut share
窄扫描 narrow scan
窄射束 narrow beam
窄声束 narrow sound beam
窄式车架 inswept frame
窄式刮刀 plough
窄式火箱 narrow fire box
窄式开沟铲 scooter
窄视角透镜 lens covering a small angle of field
窄束灯光设备 narrow-ga(u)ge lighting
窄束几何 narrow beam geometry
窄束减弱 narrow beam attenuation
窄束声波 narrow beam sound wave
窄束声回声测深 narrow beam echo sounding
窄束声回声测深仪 narrow beam echo sounder
窄束声源 narrow beam sound source
窄束水深测量器 narrow beam bathymeter
窄束天线 narrow beam antenna
窄束吸收 narrow beam absorption
窄束吸收系数 narrow beam absorption coefficient
窄水道 narrow channel
窄松土铲 bull tongue
窄条 slip; strip
窄条饱和油砂层 shoestring sand
窄条金属网板 stripite; strip lath
窄条锯 scroll saw
窄条梁 narrow slip beam
窄条排锯 narrow blade frame saw
窄条手锯 coping saw
窄条形基础 narrow strip foundations
窄通带轴 narrow band axis
窄凸缘 strait flange
窄突堤 finger pier
窄雾锥喷嘴 solid-stream nozzle
窄系列滚动轴承 narrow type bearing
窄隙防爆式电机 lamina explosion proof machine
窄隙火花室 narrow gap spark chamber
窄峡 gut
窄峡激流 rost
窄狭水域 narrow waters
窄线光谱 narrow line spectrum
窄线宽 narrow linewidth
窄线组合 narrow line array
窄巷 pipe alley
窄巷道 narrow opening
窄小木工刨 thumb plane
窄小巷道 monkey heading
窄心墙坝 thin membrane dam
窄行播种 close dill planting; close drill sowing; narrow row drilling; narrow row sowing
窄行播种机 narrow row seeder
窄行的 narrow row
窄行条播 close planting; sowing in close dill; sowing in narrow dill
窄型拖拉机 narrow width tractor
窄选脉冲放大器 narrow gate amplifier
窄选通电路 narrow gate circuit
窄选通脉冲 narrow gate pulse
窄选通脉冲电路 narrow gate circuit
窄选通脉冲多谐振荡器 narrow gate multivibrator
窄选通脉冲门 narrow gate
窄牙管钳 narrow jaw pipe wrench
窄烟云模式 narrow plume model
窄移频系统 narrow shift system
窄移频制 narrow shift system
窄凿 cross-cutting chisel
窄闸门 narrow gate
窄闸瓦 narrow shoe
窄周期 narrow period band
窄轴式多腔焙烧炉 Herreshoff furnace
窄柱 diminished column
窄柱式 <间距为柱径的二倍> systylos
窄走廊 <中世纪建筑> transite; trisantia

债 debt

债的担保 security of obligation
债的转移 transfer of obligation
债方条款 terms of credit
债户 borrower
债款 debt
债票 bill of debt
债权 claim; credit; creditor's rights; encumbrance <如在不动产上设定的抵押权等>; financial claim; money claim; obligatory right; right of action; right of claim; right of credit; right of creditor
债权差额 balance of indebtedness
债权抵押 pledge of obligation
债权法 law of obligation
债权国 credit country; creditor country; creditor nation; lending country
债权人 claimant; creditor; debtee; loaner; obligee
债权人产权 creditors equity
债权人的债权人 creditor's creditor
债权人分户账 creditor's ledger
债权人开出票据 creditor's bill
债权人破产申请书 creditor's petition
债权受益人 creditor beneficiary
债权索赔清单 creditor's bill
债权替代条款 subrogation clause
债权投资人 creditor investor
债权银行 creditor bank
债权证书 documents of obligation
债权转让 assignment of a claim
债权转让书 subrogation form
债权转站 assignment of debts
债券 bond; bond certificate; debenture; debenture certificate; instrumentality; issues; loan bond
债券交易所 security exchange
债券资本 debenture capital
债台高筑 be involved in debt; debts are rolling up; deep in debt; heavily involved in debts; involved in debt
债务 arrearage; debitum; debt; financial obligation; indebtedness; liability; obligation
债务担保 security for debt
债务到期日 date of expiration of the obligation; date of maturity of the obligation; date on which the claim becomes due
债务转移 assignment of debts
债务总额 outstanding amount; outstanding amount of debt
债息 dividend
债主 creditor; debtee; obligee
债主权益 creditor's equity

寨墙 bulwark

寨墙堡垒 bulwark
寨主 castellan
寨子 stockaded village

沾满泥的 miry

沾染 besmirch; contamination; impurity
沾染的 contaminative
沾染剂量 contamination dose
沾染监测器 contamination monitor
沾染伤害 impurity damage
沾染物 contaminant; contaminator
沾色 staining colo(u)r; transfer of colo(u)r
沾湿 moisten; wetting
沾湿表面分馏塔 wetted surface column
沾湿部件 wetted parts
沾湿面积 wetted area
沾污 begrime; blemish; blur; contamination; defile; maculate; maculation; pollute; pollution; soiling; staining; tarnish
沾污测量计 contamination meter
沾污程度 fouling rate
沾污的 speckled; spotted
沾污的催化剂 contaminated catalyst
沾污了的 contaminated
沾污事故 contamination accident
沾污速率 fouling rate
沾污损害 taint damage
沾污物 contaminant; pollutant
沾污系数 fouling factor
沾污险 risk of contamination
沾污指数 fouling index
沾吸作用 blotting

毡 blanket; felt

毡包 felt camp; yurt(a)
毡笔 felt pen
毡布 felt cloth
毡层 blanket; blanket coat; blanket course; carpet; carpet coat; carpet veneer; felt layer
毡层黏[粘]结剂 carpet adhesive
毡衬 felted fabric base; packing felt
毡衬垫 felt gasket; felt washer【机】
毡衬黄麻 felted fabric jute
毡衬里 felt back(ing)
毡衬洗矿槽 blanket strake
毡带 felt strip
毡底地毯 felt-base rug
毡底油地毡 felt-backed lino(leum)
毡垫 felted fabric base; felted fabric pad; felt layer; felt pad; packing felt
毡垫层 felted fabric mat
毡垫底 felt back(ing); felted fabric underlay
毡垫片 felted fabric insert
毡垫圈 felt gasket; feltless ring; felt washer
毡钉 clout nail; felt nail
毡幅 felted fabric web
毡辊 felted fabric roller; felt roll
毡合织物 milled cloth
毡护圈 felt(less) retainer
毡化 felting
毡环 felt collar; felt cylinder; felt ring
毡环密封 felt-ring seal
毡基 felted fabric base
毡绝缘条 felted fabric insulating strip
毡滤器 blanket filter; felt filter
毡轮 felt wheel
毡轮抛光 mop polishing
毡密封 felted fabric seal; felt seal
毡片 felted fabric board; felted fabric sheet
毡片成型 mat formation; mat forming
毡球壳属 <拉> Neopeckia
毡圈 felt collar; felt cylinder; felted fabric cylinder; felt ring <阻尘或防油用>
毡圈密封 felt-ring seal
毡刷 felt finger; felt wiper
毡条 felted fabric strip; felt strip; mat stripping
毡贴面木抹子 carpet float
毡席圈闭 blanket trap
毡油封 felt oil seal; felt seal
毡预浸机 mat impregnator
毡纸 felt(ing) paper
毡制品 felting product
毡质阻 felt ring
毡状的 felty
毡状结构 felty texture
毡状泥炭 blanket peat
毡状酸沼 blanket bog
毡状体 felty body; fibrous body
毡子 blanket

粘层 tacky surface

粘插页机 tip-on machine
粘尘剂 dust bond
粘尘铺路油 dust-binding oil
粘稠 thickness
粘稠度测定 toughness test(ing)
粘稠度指数 toughness index
粘稠水 tough water
粘稠性 toughness
粘带涂层法 tape coating system
粘的 tenacious
粘底拉裂 drag
粘度 degree of viscosity
粘度减小 drop of viscosity
粘度温度曲线 temperature-viscosity curve
粘度温度曲线图 temperature-viscosity chart
粘度系数 twist multiplier
粘封带 tape sealant
粘附插页 tip-in
粘附剂 tacky producer
粘附抗拉强度 tensile adhesion
粘干状态 tack dry
粘合的 tacky
粘合剂 tiradaet; binding agent
粘合面积 bond area
粘胶 dope
粘结层 tack coat(ing); tack course; tie coat; bond course <砌体水平接缝间的>
粘结层之间粘结力 ply adhesion
粘结灰泥 bonding plaster
粘结剂 tiradaet; bonding admixture; jointing compound
粘结类型 type of bond
粘结力 <混凝土/砂浆与钢筋之间的> bond stress
粘结力强的材料 tenacious materials
粘结料 cementitious material
粘结耐久性 bond durability
粘结破坏 destruction of bond
粘结强度 bond strength
粘结式薄层罩面 thin bonded overlay
粘结水 adhesive water; pellicular water
粘结涂层 tack coat; bond coat
粘结物质 bonding compound
粘结系数 coefficient of cohesion
粘结应力 bond stress
粘聚度 degree of cohesiveness
粘粒组 clay fraction
粘扣 thread sticking
粘磐旱成土 durargid
粘磐土 terras soil
粘韧性的 tenacious
粘砂圆盘磨光机 disk sander
粘束孢属 <拉> Graphium
粘弹性理论体系 theory of viscoelastic system
粘贴墙面砖的快凝灰泥 casting plaster
粘性 ductility; tackiness; tacky; tenacity
粘性耗散函数 dissipation function
粘性计 tackmeter
粘性减震器 viscous damper
粘性(空气)过滤器 viscous filter
粘性流动 viscous flow
粘性膨胀系数 dilatation coefficient of viscosity
粘性铺路油 ductile base oil
粘性期 tack range
粘性纱布 tack rag
粘脏器皿 dirty ware
粘展剂 spreader-sticker
粘质砂岩 <含有矿脉的> dauk; dawk; dabbing
粘质页岩 dawk
粘滞度 viscosity; treacliness
粘滞流 viscous flow
粘滞系数 coefficient of viscosity
粘滞阻尼 viscous damping
粘滞性 treacliness
粘滞阻尼系数 tenacity damping factor
粘着的水泥 sticky cement
粘着点 tack point
粘着剂 tackifier
粘着力 tack strength
粘着力的去除 decohesion
粘着能力 tackability
粘着牵引力 tractive force by adhesion
粘着强度 adhesion strength; bond strength
粘着性 tackiness
粘着应力 bond stress
粘着状态 tacky state

詹布法 <边坡稳定分析用> Janbu method of slope stability analysis

詹金斯过滤器 Jenkin's filter
詹姆斯粉 Jame's powder
詹姆斯型摇床 James table
詹姆斯摇床 Jame's concentrator
詹内科坐标 Janecke coordinate
詹尼式车钩 Janney coupler
詹尼型浮选槽 Janney flo(a)tation cell
詹尼型压气机械搅拌浮选机 Janney mechanical-air machine
詹森不等式 Jenssen's inequality
詹森方程 Janssen's equation
詹森废水洗涤器 Jenssen's exhaust scrubber
詹森分类法 Jenssen's classification
詹森-海斯计算需水量法 Jensen-Haise method
詹森喷灯 Janssen's burner
詹森系统 Janssen's system

瞻性 crumbliness

斩拌机 cutmixer

斩波 chop; chopped wave
斩波电流 current chopper
斩波电路 chopping circuit
斩波放大器 chopper amplifier
斩波分光计 chopper spectrometer
斩波开关 chopper switch
斩波控制 chopper control
斩波控制的(电力)机车 chopper-controlled(electric) locomotive
斩波控制器 chopper controller
斩波控制设备 chopper control equipment
斩波频率 chopping frequency
斩波器 chopper; clipper; interrupter; lopper
斩波器电源 chopper supply
斩波器供电电动机 chopper motor
斩波器控制 chopper control
斩波器控制的车辆 chopper-controlled car
斩波器控制的列车 chopper-controlled train
斩波器稳定放大器 chopper-stabilized amplifier
斩波调制主镜 chopping primary mirror
斩波系统 chopper system
斩波再生制动器 chopper regenerative brake

斩波作用 chopping-off action
斩除树根的斧头 grubbing axe; grubbing hoe
斩断波 chopping
斩断法 process of chopping
斩根斧 grubbing axe
斩光器 chopper
斩假石 artificial stone; axed artificial stone; stone plaster
斩假石面 depretor; hammered granolithic finish
斩假石墙面 artificial stone coating
斩角砖 bevel(l)ed closer
斩面方石 hammer dressed ashlar
斩面条石 hammer dressed ashlar
斩平的拱腹面 keyed soffit
斩切光 chopped light
斩石 pick-dressed ashlar
斩首 behead
斩砖 cutting brick
斩砖泥刀 masonry cutting blade
斩琢工 adzed work

展布 spreading

展布力 spreading power
展布频谱 spread spectrum
展布频谱传输 spread spectrum transmission
展布频谱技术 spread spectrum technique
展布器 spreader
展布系数 spreading coefficient
展长 extension; lengthen(ing)
展长的径路 extended route
展长轧辊型缝 straight part
展成部分分数 expansion in partial fractions
展成齿廓 generated profile
展成刀具 generating tool
展成法 gear generator
展成法齿轮 generated gear
展成副 generated pair
展成傅立叶级数 Fourier analysis
展成鼓轮 generating cam
展成机构 generating mechanism
展成级数 expansion in series; series development
展成渐开线 generated involute
展成扇形 fan
展成行程 generating stroke
展成有向路 expansion in directed path
展成圆 rolling circle
展成运动 generating motion
展出 exhibit; protend
展出单位 exhibitor
展出地 show place
展出间隔 display booth
展出者 exhibiter; shower
展带接收机 bandspread receiver
展点 machine plotting; mark a point; plot point; plotting; plotting of points; pricking
展点图 point plot
展点误差 error of plotting points
展点仪 plotting machine
展度计 ductilimeter; ductilometer
展锻 stretching
展帆 cast off the sail; spread the sail
展反射 extensor reflex
展幅机 expanding roller; stentering machine; stretcher; tenter; tentering machine
展痕 <薄板表面人字形裂缝> spreader mark
展绘 cartometric scaling; plot; plotting
展绘导线 plot a traverse
展绘格网 construct on projection
展绘控制点 plot the control points
展绘碎部 plot the data; plot the details
展绘误差 plottable error; plotting error
展绘坐标网 plot a grid
展接电路 extension circuit
展接装置 <电杆> extension fixture
展卷机 payoff reel
展开 development; dispread; fan out; open out; open up; splay; spread; uncoil; unfold; unfurl; unroll; unwind
展开槽 developing tank; separation chamber
展开槽饱和 chamber saturation
展开长度 developed length; length of run
展开成级数 expansion in series
展开程序 unwind(ing)
展开尺寸 developed dimensions
展开的 unfolded
展开定理 expansion theorem
展开法 expansion method; method of development
展开反射排列 expanding reflection spread
展开方程式 expansion equation
展开方形搜索 expanding square search; square search
展开公式 expansion formula
展开罐 developing tank; development tank
展开画法 aligned view
展开剂 developer; developing agent
展开剂前沿 solvent front
展开角 angle of divergence; angle of spread; flare angle; reaming angle
展开截面 developed section
展开卡记录 spread card record
展开宽度 developed width
展开立面 developed elevation
展开面 developable surface; developed surface; extended surface
展开面积 developed area; development area; total floor area
展开目标 extended target
展开排列 expanding spread
展开排列垂直回路技术 expanding spread vertical-loop technique
展开排列剖面 expanding spread profile
展开平面形 developed platform
展开求积分法 integration by expansion
展开曲线 developed curve
展开时间 development time; duration of runs; set-up time
展开式 expanded form; expanded formula; expansion equation
展开式模板 expandable form
展开式扫描 compensated scan(ning)
展开式原理图 principle evolutionary diagram
展开(透)视图 expanded view
展开图 developed drawing; developed pattern; developed view; development drawing; development representation; expanded view; expansion plan; flat pattern; stretched-out view; unfolding drawing
展开图生成 flat pattern generation
展开图示法 developed representation
展开为泰勒级数 expansion in Taylor series
展开系数 expansion coefficient
展开线 evolute
展开信号旗 unfold the flag
展开行列式 expansion of determinants
展开叶片面积 developed blade area
展开圆筒 opening of the cylinder
展开轴 axis of dilatation
展宽 broaden(ing); expand; splay; spreading; stretch
展宽背景 extended background
展宽的 splay
展宽电子束 extended electronic beam
展宽端跨 spread span
展宽格式 wide format
展宽机 widener
展宽开挖 splay cut
展宽露边侧砖 bull stretcher
展宽轮廓 extrusion contour
展宽器 stretcher
展宽式交叉(口) enlarged intersection
展宽式交叉路口 enlarged intersection; flared crossing; flared intersection; splayed intersection
展宽式接缝 splayed joint(ing)
展宽数据 extrusion data
展宽调节 spread control
展宽系数 coefficient of spread
展宽效应 peak broadening effect
展扩 spread
展览 display; expose; parade
展览场地 exhibition ground; exposition ground
展览场所 exhibition place; show place
展览车 exhibition car
展览橱窗 display window; show case
展览船 exhibition vessel; sample fair ship
展览大楼 exhibition block
展览大厅 exposition hall
展览大厦 exhibition palace
展览大帐篷 exhibition pavilion
展览的图片 exhibition drawing
展览宫 exhibition palace; exposition palace
展览馆 exhibition building; exhibition center[centre]; exhibition hall; exposition pavilion
展览花园 exhibition garden
展览会 expo; exposition; show; exhibition
展览会场 show ground; show-ring
展览会大楼 exposition block
展览会花园 exposition garden
展览会建筑 exposition architecture
展览会开幕 opening exhibition
展览会面积 exposition area
展览货物 exhibition goods
展览架 display shelf
展览建筑 exhibition architecture
展览看台 exhibition stand; exposition stand
展览廊 display gallery; exhibition gallery
展览模型 exhibition model
展览品 displaying items; exhibit; exhibition goods; items on display; show piece
展览区 display area; exhibition area
展览区段 exhibition block
展览日 field day
展览室 exhibition room; show room
展览台 display stand
展览厅 exhibition hall; fair building
展览图(片) display drawing; exhibition print; show drawing
展览物 spectacle
展览物的 spectacular
展览型 show type
展览中心 exhibition center[centre]
展棉机 silk spreader
展频通信[讯]技术【电】 spread spectrum communication technique
展品 show piece
展平 expansion; flatten; unbend; flattening out <指洪水波扩散>
展平材料 flattening material
展平锤 planishing hammer; spreader
展平反褶积 level deconvolution
展平辊 nip rolls
展平滤波器 choke filter; smoother
展平区 flattened region
展平区半径 flattening radius
展平曲线 faired curve
展平石块 flattening stone
展平试验 flattening test
展平凿 smoothing chisel; span chisel
展铺砂浆 stringing mortar
展期 extension; postpone; prolongation
展期贷款 loan extended; roll-over credit
展期费 extension fee
展期合同 renewal a contract
展期交易 continuation exchange
展期利率 continuation rate
展期条款 continuation clause
展期信用证 extended credit; extend letter of credit
展期债务 deferred liabilities
展全帆 clean full
展色剂 vehicle
展色料 vehicle of stain
展时光谱图 time-resolved spectrum
展示板组合 group stand
展示窗玻璃 display window
展示盖篷 commercial canopy
展示会上 show business
展示机 layout machine
展示图 display drawing; exploded view
展室交货价 ex-showroom
展缩比 expansion ratio
展缩器 compander[compandor]
展胎器 tire spreader
展毯辊 spread roll
展望 envisagement; forecast; outlook; perspective; preview; prospect; vista
展望孔 peep hole
展弦比 aspect ratio <水翼的>; fineness ratio; span-chord ratio
展现 reveal; revelation
展现出来 open out
展线 development of line; extension of line; line development; route development
展线曲线 curve of development
展线系数 coefficient of developed line; coefficient of extension line; coefficient of line development
展像镜 axicon
展销 exhibition sale; sales exhibition; selling at exposition
展销产品 demo
展销店 show shop
展销会 fair
展销会场 exhibition park
展销交货(价格) ex-showroom
展销室 salesroom; show room
展销台 show stand
展性 malleableness
展性锻铁 malleability wrought iron; malleable wrought iron
展性钢 malleable steel
展性合金 ductile alloy
展性活动链 malleable detachable chain
展性镍 malleable nickel
展性销链 malleable pintle chain
展性铁 malleable iron
展性铸件 malleable casting
展性铸铁 malleable cast-iron; malleable(pig) iron
展性铸铁箍 malleable iron clip
展性铸铁管 malleable iron pipe

展延保险 extended insurance
展延保险费 renewal premium
展延保险批单 extended coverage endorsement
展延短期系列 extending of short-term records
展延破坏 failure by spreading
展延实测系列 extension of records
展延水位-流量关系曲线 extending of rating curve;extrapolation of rating curve
展业费 acquisition cost
展源 extended source
展直 straightening
展直线法 spread out straight method
展纸车 <混凝土道路工程的> paper cart
展着剂 speader

崭 新的 brand-new; fire-new; in mint condition;unprecedented

辗 锻(辗轧)用心轴 becking bar

辗钢轮 rolled steel wheel; wrought-steel wheel
辗宽的轮缘 spread rim
辗料间 mill room
辗泥机 mud mixer
辗平 ironing
辗实土坝 roller-compacted earth dam
辗碎机 mill;mulling machine
辗压遍数 roll pass
辗压辊 compression roll
辗压滚轮 tamping drum
辗压机 rivet bucker;roller mill
辗压角 angle of roll
辗压路堤 rolled embankment
辗压总宽度 rolling total width
辗轧 roll-off
辗轧玻璃 rolled glass
辗轧机(轧管) elongator
辗转相除法 algorithm of division;division algorithm

占 本级产率百分数 occupy ones fractions rate per cent

占卜神使庙宇 oracular temple
占地 land take
占地费用 space expenses
占地面积 floor space;plot area;floor area <设备的>
占地运动 enclosure movement
占机维护时间 engineering time
占机信号 seizing signal
占机证实信号 seizing acknowledgement signal
占积率 coefficient of admission
占据 colonization;seize;take possession of;tenancy
占据或生活于一个地区 occupancy [occupance]
占据机动起始位置 position for the maneuver
占据密度 population density
占据者 tenant
占空 duty
占空比 duty cycle; duty ratio; pulse duration ratio;pulse time ratio
占空度 duty cycle
占空间城市 spatial town
占空率 duty ratio
占空系数 duty cycle;duty factor;duty ratio;occupation coefficient;occupation efficiency; pulse duration ratio;pulse time ratio;space factor; coefficient of charge <线圈的>;activity coefficient <线圈的>
占空因数 duty cycle;duty factor;duty ratio;space factor
占空因数计 duty cyclometer
占空因数作用 duty factor action
占领 occupancy[occupance];occupation
占领地 occupied territory
占领区 occupied territory
占领区货运保单 occupational freight warrant
占领市场 capture market
占领学说 occupation theory
占领原则 occupation principle
占领者 occupant
占全样百分数 occupy total sample per cent
占全样产率百分数 occupy total sample rate per cent
占位继电器 position busy relay
占先能力 preemption capability
占线闪光信号 busy-flash signal
占线信号灯 engaged lamp; hold lamp;visual busy lamp;visual engaged lamp
占线指示灯 busy lamp;hold lamp
占线指示器 busy indicator
占用 appropriation;employ;occupancy[occupance];occupation;occupy
占用表示灯 occupancy lamp;occupancy light;occupation lamp
占用表示灯光 occupancy light
占用场地费用 space charges
占用的闭塞区段(或区间) blocked section
占用的闭塞区间 occupied block
占用的轨道电路 occupied track circuit
占用的区间 block occupied
占用费用 fee charged for the use of; fee for possession and use of;occupancy expenses
占用概率 acquisition probability
占用股道 occupied track;track occupied
占用轨道 track occupied
占用轨道电路 holding track circuit
占用荷载 occupancy load
占用计数器 demand meter
占用率 occupancy factor;occupancy rate
占用面积 occupied area
占用频带宽度 occupied bandwidth
占用区 get area;used area
占用区段 occupied block
占用区间 occupied section
占用人 occupant
占用人保护 protection of occupant
占用人行道 sidewalk encroachment
占用扫描 busy scan
占用时间 elapsed time;holding time; occupancy time; occupation period;time of occupation
占用时间记录器 holding time recorder
占用室 occupied room
占用现场 possession of site
占用线 occupied line;occupied track
占用线监视继电器 called line supervisory relay
占用线路 busy line
占用许可证 occupancy permit
占用钥匙 occupancy key;occupation key
占用者 occupant;tenant
占用状态 busy condition
占用资金付费原则 principle of payment for the use of funds
占优势 dominate;override;predominate
占优势超驰控制 override
占优势的 dominant;overriding;overwhelming;predominant;prevailing
占有 co-opt;occupancy[occupance]; occupation; occupy; possession; seize;take possession of;tenure <占有财产、职位等>
占有比 occupation ratio
占有标记 hold mark
占有的实现 realization of appropriation
占有地 occupation
占有动产 choses in possession
占有概率 occupation probability
占有轨道 occupied orbital
占有几率 occupation probability
占有空间的 space occupied
占有留置权 possessory lien
占有率 occupancy[occupance];occupancy rate
占有率控制 occupancy control
占有面积 occupied area
占有能带 occupied band
占有能级 occupied level
占有频带 occupied bandwidth
占有频带宽度 occupied frequency bandwidth
占有期 tenure
占有权 occupation;tenure
占有权契约 covenant of seisin
占有人 occupant;possessor
占有时间 occupation period
占有数 occupation number; population
占有态 occupied state
占有条件 tenure
占有统治地位的公司 dominant firm
占有物 thing possessed
占有性的留置权 possessory lien
占有许可证 certificate of occupancy
占有一切或一无所有 all or nothing
占有因数 occupancy factor
占有者 holder;occupant
占整层楼面的公寓 floor-through
占支配地位的周期 predominant period
占住空间 excluded space
占住者权利 squatter's rights

战 备 strategic stock

战备机场 redeployment airfield
战备桥 emergency bridge
战备区 readiness area
战备物资 war reserves
战备状态 readiness posture
战场 battle field; seat of war; war theater[theatre]
战车 chariot
战刀形电动手锯 saber[sabre] saw
战地掩蔽沟道 dugout
战斗 army operation
战斗陈列馆 fighting gallery
战斗船 fighting ship
战斗机 fighter
战斗机基地 fighter station
战斗力 fighting strength;tactical efficiency
战斗列车 combat train
战斗射击场 field firing range
战斗艇 combatant craft
战斗桅楼 fighting top
战斗装备 fighting equipment
战俘收容所 collecting post
战壕 trench
战后的 postwar
战后宏观经济政策 postwar macroeconomic policy
战后建的房屋 postwar house
战后建造的房子 postwar block
战后建筑 postwar architecture;postwar building
战后剩余的 war surplus
战后剩余物资 war surplus
战后衰退(不景气) postwar slump
战后住房 postwar housing
战绩标志 scoresheet
战舰 battle ship;ship of war
战舰上承桥楼面 battle deck(bridge) floor
战况图 situation record
战利品 booty;capture;prize
战利品雕饰 trophy
战栗 tremor
战列舰 battle ship
战略部署 strategic deployment
战略材料 strategic material
战略城市 strategic city
战略储备 strategic reserve
战略地理学 geostrategy
战略地区 strategic area
战略地图 strategic map
战略分析 strategic analysis
战略格网 strategic grid
战略观点 strategic point of view
战略管理 strategic management
战略规划 strategic plan(ning)
战略规划模型 strategic planning model
战略航空图 aeronautical planning chart
战略环境评价 strategic environmental assessment
战略环境评价系统 strategic environmental assessment system
战略环境影响评价 strategic environmental impact assessment
战略基地 strategic base
战略计划模型 strategic planning model
战略计划书 strategic plan(ning)
战略警报声响系统 strategic alert sound system
战略警报信号灯 strategic alert light
战略据点 key point
战略决策 strategic decision
战略均势 strategic balance
战略空军基地 strategic air base
战略空军司令部 Strategic Air Command
战略控制 strategic control
战略矿物 strategic mineral;war mineral
战略模型 strategic model
战略目标 strategic objective;strategic target
战略目标标示航图 strategic outline chart
战略数据规划 strategic data planning
战略水规划 strategic water planning
战略思想 strategic concept
战略铁路 strategic railway
战略托管区 strategic trust area
战略武器发射控制站 strategic weapon firing post
战略武器库 strategic arsenal
战略物资 strategic material
战略性防御工事 strategic fortifications
战略性公路研究计划 <美> strategic highway research program(me)
战略性规划 strategic plan(ning)
战略要点 strategic point
战略优势 strategic advantage
战略与稀有重要原料 strategic and critical raw material
战略与战术地图 strategic-tactical map
战略原料 strategic raw material
战略贮备 strategic stockpiling
战前的 prewar

战前公寓 pre-war housing
战前住房 pre-war housing
战区 tactical locality; theater of operations; theater of war; war theater [theatre]
战区仓库 theater depot
战区海军司令部 theater navy headquarters
战区空军物资区 theater air material area
战区司令部 theater headquarters
战区战备物资 theater war reserves
战区作战设施 theater operational facility
战时保险 war risk insurance
战时出入口 wartime passageway
战时封堵 wartime blockage
战时锚地 war anchorage
战时送风口 wartime air supply duct
战时隧道 wartime tunnel
战时体制 war establishment
战时通信[讯] war communications
战时伪装油漆 war paint
战时限制配给的材料 critical material
战时应急管线 war emergency pipe line
战时征用权 angary
战时住房 war housing
战术地图 tactical map
战术航程 tactical range
战术航空图 aeronautical plotting chart; aircraft position chart; air navigation plotting chart; operation navigation chart; tactical pilotage chart
战术计算机 tactical computer
战术决策 tactical decision
战术空中导航设备 tactical air navigation
战术空中导航系统 Tacan; tactical air navigation system
战术空中管制中心 tactical air control center[centre]
战术控制 tactical control
战术模型 tactical model
战术目标 tactical target
战术目标图 supplement supporting graphics; tactical target illustration
战术目标资料图表 tactical target materials program(me)
战术情况显示器 tactical situation display
战术数据通信[讯]中心 tactical data communication center[centre]
战术态势图 marked map
战术通信[讯]卫星 tactical communication satellite
战术通信[讯]系统 tactical communications system
战术作业砂盘 eggcrate model tactical-planning model
战台 <要塞大门或通道上建筑的> assommoir
战争、罢工及暴动条款 war, strike and riot clause
战争保险 insurance against war risk
战争爆发 conflagration
战争储备 strategic stock
战争储备物资 war reserve stock
战争(风)险 insurance for war risks; war perils; war risk
战争风险附加费 additional expense war risks
战争纪念碑 warm memorial
战争破坏地区 war-damaged area; war-devastated area; war-ravaged area
战争绕航条款 war deviation clause
战争险(保)险 war risk insurance
战争险保险单 war risk policy
战争险费率 war rate; war risk rate
战争险水面协定 water-borne agreement
战争险条款 war clause
战争险协定 war risk agreement
战争险注销条款 war cancellation clause
战争行为 act of hostility; act of war
战争遗迹 material remnants of war
战争遗留物 material remnants of war
战争状态 warfare

栈 repertory; shed; stack

栈变换 stack transformation
栈标志 stack marker
栈操作 stack manipulation; stack operation
栈处理 stack manipulation; stack processing
栈处理机 stack handler
栈处理技术 stack processing technique
栈存储点 stack memory
栈存储器 stack memory
栈单 delivery order; dock warrant; godown warrant; warehouse certificate; warehouse receipt; warehouse warrant; warrant
栈单元 stack cell
栈道 berm(e); elevated road; plank way; trestle; trestle road along cliff; viaduct; ancient plank way (built along the face of a cliff) <中国>
栈道高架铁路 elevated railway on trestle work
栈的非终极符 stack nonterminal
栈底 bottom of stack
栈地址 stack address
栈地址控制 stack address control
栈顶 stack top
栈顶单元 stack top location
栈顶寄存器 top register
栈顶算符 top operator
栈段 stack segment
栈房 godown; store; store house; warehouse
栈访问 stack addressing
栈费 warehouse rent
栈符号 stack symbol
栈构架 stack frame
栈归约 stack reduction
栈基址寄存器 stack base register
栈寄存器 stack register
栈架 trestle; trestle stand
栈架工程 trestle works
栈架结构 trestle structure
栈架式柱 trestle type column
栈结构 stack architecture
栈界限存储器 stack limit register
栈梁结构 trestle works
栈内优先数 in-stack priority
栈能力 stack capability
栈桥 access trestle; conveyor bridge; gantry; jetty; landing bridge; landing pier; landing stage; open jetty; pier; scaffold bridge; transshipment gallery; trestle; trestle bridge; trestle stand; trestle works; viaduct; overburden conveying bridge <露天开挖系统>
栈桥钢塔 steel tower for high viaduct
栈桥高架线 railway on trestle
栈桥工程 trestle works
栈桥构架 trestle frame
栈桥架 trestle stand
栈桥结构 trestle structure
栈桥跨度 trestle bay
栈桥跨距 trestle bay
栈桥梁 gantry beam
栈桥码头 berth jetty; landing pier; shore bridge
栈桥帽木 trestle cap
栈桥排架 trestle bent
栈桥平堵截流 trestle horizontal closure
栈桥桥跨 trestle bay; trestle bent span
栈桥式坝 trestle dam
栈桥式渡槽 trestle flume
栈桥式进水口 water intake on scaffold bridge
栈桥式码头 jetty pier; piled jetty; trestle and pier; trestle type wharf
栈桥式水槽 trestle flume
栈桥式引桥 access trestle; approach trestle
栈桥塔座 trestle tower
栈桥头 jetty head
栈桥卸车线 trestle unloading siding
栈桥引道 trestle approach
栈桥支墩 trestle tower
栈桥支柱 trestle pier
栈桥纵梁 trestle stringer
栈桥钻探 pier drilling
栈扫描 stack scan
栈深度 stack level
栈式处理 stacked job
栈式存储器 stacked memory
栈式作业 stacked job
栈式作业处理 stacked job processing
栈式作业控制 stacked job control
栈算子 stack operator
栈台 scaffold; trestle works
栈向量 stack vector
栈选择区 stack option
栈页面 stack page
栈溢出 stack overflow
栈溢出中断 stack interrupt
栈指令 stack instruction; stack order
栈指示寄存器 stack point register
栈指示器 stack indicator; stack pointer
栈指示字定义 stack pointer definition
栈中断 stack interrupt
栈自动化装置 stack automaton
栈租 godown rent; warehouse rent
栈作业 stacked job
栈作业控制 stacked job control

站

站别卸车计划 scheme of unloading by stations; unloading plan by stations
站别装车计划 loading plan by stations; scheme of loading by stations
站长 station agent; station master
站长室 station master's room
站场 station yard; terminal yard
站场布置 station layout
站场布置电路技术 geographic(al) circuit technique
站场布置关系 geographic(al) relationship
站场车辆【铁】 yard vehicle
站场费用 terminal cost
站场扩音对讲系统 talk-back
站场模拟图 mimic diagram
站场模型图 mimic diagram
站场排水 yard drainage
站场排水系统 water drainage system for yard
站场前动员 <俚语> pin lifter
站场通信[讯] station and yard communication
站场无线电话 station and yard radio telephone
站场无线电话系统 radio telephone system in station and yard
站场现状图 present situation map of station
站场型网络 geographic(al) circuitry
站场咽喉区 throat of yard
站场制动员【铁】 fielder; field man; ground-hog
站场作业 yard operation; yard work
站车环境舒适度 comfortableness of environment at station and on train
站车预报【铁】 advance report on available seating accommodation on trains by stations and train conductors
站到场 <集装箱> container freight station to container yard
站到门 container freight station to door
站到站 container freight station to container freight station
站到站运输 station-to-station transport
站得住脚的 tenable
站点对接 end-to-end join
站调楼【铁】 switching tower; yard controller's tower
站段 depot
站对站 station-to-station
站房 passenger building
站房地坪【铁】 station building site
站杆 snag; stub
站岗 point-duty
站号 station number
站后折返 reverse after moving forward out of the station; reverse behind station according to the proceeding direction
站计数器 station counting unit
站际合作 interstation cooperation
站间 interstation
站间电话 interstation telephone
站间话频多路调制器 intersite VF multiplexer
站间货物运输密度 cargo transportation density among stations
站间间距 distance between stations
站间距离 distance between center [centre]-lines of stations; interval of stations; spacing of passing sidings; station spacing; distance between stations
站间联系 interstation connection
站间联系电路 connecting circuit between stations
站间区间 section between stations
站间线 station line
站间线路 intersite line
站间相关 interstation correlation
站间行车电话【铁】 interstation train operation telephone
站间运输 <即不办理取送业务的运输> station-to-station traffic
站间运转时分 station-to-station time
站界 station limit
站界标 station boundary sign; station limit post; station limit sign
站界内搬运 goods transport within a station
站控微机 station computer control system
站控制器 station controller
站口鞍座 terminal saddle
站立便餐柜台 stand-up counter
站立乘客 standee
站立地位 <市郊和市区列车客车上> standing room
站立进食的酒吧 stand bar
站立空间 standing room
站立人数 number of standup
站码 station code
站名表 station nomenclature
站名牌 name board; station mark; station sign

站名索引 indexing of stations;station index
站内 intra-station
站内闭塞<进站信号机至出发信号机间> station block
站内闭塞机 station block apparatus; station block instrument; station block unit
站内道口联系电路 connecting circuit of highway crossings within the station
站内电话 station communication system
站内电缆 in-station cable
站内设备 station equipment
站内通信[讯]系统 station communication system
站内运输 transport within a station
站内中继线 sender link
站内装货马车 internal drag
站内自动化系统 local automation system
站年法<水文分析的> station year method
站牌 destination board
站票观座区域 standee area
站票看客<戏院中的> standee
站票席 standing space
站票席位<戏院的> standing room
站平台 platform
站坪【铁】station site
站坪长度 length of station site
站坪坡度 grade of station site;gradient of station site
站起来 rise to one's feet
站前服务系统<机场办理登记手续的> frontal system(apron)
站前广场 station front;station place; station plaza;station square;terminal square;station forecourt<火车站>
站前坡度 access gradient
站前折返 reverse after moving back out of the station;reverse in front of the station according to the proceeding direction
站前中间广场 nave
站群集器 station cluster
站容 appearance of terminal
站上无车时间 bus-stop clearance time
站识别 station identification
站式 standing posture
站饰 ornamental door
站台 dock;platform;station platform; loading zone<公共汽车和电车乘客上下>
站台搬运车 platform lorry
站台边缘 edge of platform;edge to platform
站台边装卸线 platform road
站台标 platform post
站台表示器 platform indicator
站台层 platform level
站台层回排风机控制箱 control box for platform air return/exhaust fan
站台长度 platform length
站台挡土墙 platform wall
站台地道 platform subway;platform tunnel;platform underpass;station platform tunnel
站台顶棚 station roof
站台端部设备用房 equipment room at station end
站台防雨棚 platform roofing
站台高度 platform height
站台工人 dock worker
站台股道占用表示握柄 platform lever
站台广播 platform announcement point
站台轨道 station platform track
站台缓冲垫 dock bumper
站台间的地下通道 inter-platform subway
站台间地道 subway leading to platforms
站台间人行道 foot crossing between platform
站台结构 platform structure
站台紧急按钮箱 emergency stop button box
站台进口 access to platforms
站台宽度 platform width
站台扩音器系统 platform loudspeaker system
站台扩音系统 station platform loudspeaker system
站台楼梯 station platform stair(case)
站台门 dock door
站台棚 platform roof;platform shed; railway roof
站台票 platform ticket
站台屏蔽门 platform screen door
站台衔接板 dock plate
站台装卸区 dock spot
站台坡道 station platform ramp
站台坡度 platform grade
站台设备 loading facility
站台升降机 station platform lift
站台式 platform type
站台式屋顶 station roof
站台手推车 platform car
站台隧道 station platform tunnel;station tube
站台铁路线 loading platform track
站台位置<候机室的> gate position
站台屋顶 station platform roof
站台屋面 station roof
站台线 platform road;platform line; platform track
站台信号 platform signal
站台信号机【铁】platform signal
站台形式 platform form
站台巡察员 platform inspector
站台有效长度 effective length of platform
站台雨棚 platform awning;platform canopy;platform roofing;platform shed;platform shelter
站台栅栏 platform barrier
站台栅门 platform gate
站台钟 platform gong
站厅 concourse
站厅层 concourse level
站厅换乘 concourse transfer
站厅回排风机控制箱 control box for concourse air return/exhaust fan
站厅信号灯 hall lantern
站停止 station deactivation
站外储存 off-depot storage
站外传输单元 outstation transmission unit
站外换乘 out of station transfer
站外设备 outside-the-station equipment
站网 network;network of stations; observation network
站网规划 network planning
站网密度 density of station network; network density
站网设计 network design
站位 erect position;standing place; station location
站务 station operation
站务及列车费用 station management and train expenses
站务室 station operation office
站席<戏院的> standing room
站席定员<客车、动车的> standing capacity
站线 sections;siding;square stations; vertical stations;house track;station track;yard track;terminal and trunkline【港】
站线布置 arrangement of station tracks
站线的分段(使用)sectioning of station tracks
站线瓶颈<站线汇集狭窄处> bottleneck
站线全长 full length of station track
站线延展长度 extended length of station track
站线有效长 effective length of station track
站心 topocenter[topocentre]
站心赤道坐标 topocentric equatorial coordinates
站心赤经 topocentric right ascension
站心赤纬 topocentric declination
站心地面坐标系 terrestrial topocentric coordinate system;topocentric terrestrial coordinate system
站心观测量 observed topocentric quantity
站心光行差 topocentric aberration
站心距离 topocentric range
站心球面坐标 topocentric coordinates
站心速度 topocentric velocity
站心天平动 topocentric libration
站心天球坐标系 topocentric celestial coordinate system
站心原点 topocentric origin
站心直角坐标 topocentric Cartesian coordinate
站心坐标 topocentric coordinates
站心坐标系 terrestrial topocentric system;topocentric coordinate system
站修所【铁】freight car repairing depot
站修线 repair track at station;vehicle repair track at station
站译码电路 station decoding circuit
站运行间隔时间 time interval between following movements in a station
站在旁边 stand-by
站址 station address;station location
站址环境 environment at station site
站址勘察 investigation of station site
站址识别 station identification
站址选择 selection of plant site
站坐标 station coordinates

绽

绽裂 fray
绽线 rip

蘸

蘸漆量 picking-up of paint
蘸塑 dip forming;dip mo(u)lding
蘸液折射计 dipping refractometer

张

张 sheet<薄钢板>;plat<厚钢板>

张臂式 overhang
张臂式支架 sting support
张臂式支柱 cantilever leg
张弛 relaxation
张弛测量器 relaxometer
张弛长度 relaxation length
张弛电路 relaxation circuit
张弛法 relaxation method
张弛频率 relaxation frequency
张弛曲线 relaxation curve
张弛时间 relaxation time
张弛振荡器 relaxation generator;relaxation oscillator;relaxor;relaxor oscillator
张弛振荡器报警器 relaxation-oscillator alarm
张弛震荡 relaxation oscillation
张弛震荡器 relaxation oscillator
张弛周期 relaxation period
张弛时间效应 effect of relaxation time
张弛试验 relaxation test
张德勒周期 Chandler period
张灯结彩 decorating with colo(u)rful light
张断层 tensile fault;tension fault
张断裂 open fracture;tensile crack
张断面 tensile fracture
张帆 bear sail;set sail
张帆桁 boom
张帆汽船 steam-sailing ship
张帆索 bowline
张杆 tension rod
张剪边界 tenso-shear boundary
张剪性节理 tenso-shear joint
张剪性深断裂 tenso-sheer deep fracture
张角 divergence angle;field angle; flare angle;tensor angle;angular aperture<天线>
张角偏转仪 Thompson's deflector
张节理【地】tension(al)joint
张紧 take-up;tension
张紧带 tension band
张紧单缆系泊 taut single line mooring
张紧垫圈 tension washer
张紧度 rate of tension;strain;tenseness;tensity
张紧钢丝 tensional steel wire
张紧工具 tightener
张紧工作绳的工具 jerk shoe
张紧辊 idler roller
张紧辊装置 tension unit
张紧滚轮 tightener sheave
张紧滚筒 tension drum
张紧过度 overtension
张紧滑车组 heel tackle
张紧滑轨 stretching slide;tightening rail
张紧滑轮 heel sheave;tension pulley block
张紧环 spudding ring
张紧机构 strainer;tensioning device adjusting screw
张紧夹具 stress accommodation
张紧卷筒 drawing capstan
张紧拉力调节装置 tension adjusting gear
张紧缆索式锚泊 taut-line mooring
张紧力 tensioning force
张紧链 tension chain
张紧链轮 tensioner sprocket;tightener sprocket
张紧轮 belt idler;dead idler;holddown roller;idler pulley;idler tumbler;jockey;snub pulley;stretching pulley;take-up tumbler;tensioner; tension pulley;tension sheave; tightening pulley
张紧螺母 tensioning nut
张紧蒙皮板 stressed-skin panel
张紧膜片 stretched diaphragm
张紧扭矩 tightening torque
张紧配重 tension weight
张紧皮带 take-up the belt;tightening belt;tighten the belt
张紧皮带轮 idler
张紧器 strainer;tensioner;tightener
张紧器刻度盘 tightening scale dial
张紧千斤顶 tension jack
张紧鞣法 tension tannage
张紧绳 tension rope
张紧式索道 tight sky line
张紧索 taut-wire

张紧索系统 taut-wire system
张紧弹簧 tension spring
张紧托滚 tightening idler
张紧系泊 taut moor(ing)
张紧系泊式平台 taut moored platform
张紧线 span wire
张紧液压缸 tension ram
张紧应变 take-up strain
张紧指示器 tightening indicator
张紧轴 tension shaft
张紧柱 tension pole
张紧装置 stretching device;stretching unit; take-up; take-up assembly; take-up device; take-up mechanism;take-up unit;tension bridle; tension device;tension equipment; tensioner;tightener
张紧座 backstand
张开 gape;open up;splay
张开的 opening
张开的或打褶的装饰边 frill
张开的阔度 gape
张开断层 gapping fault;open fault
张开花被卷迭式 open aestivation
张开及滑移型 open and slip mode
张开角 <桥墩间> open-angle
张开节理 open joint
张开口 flaring
张开裂缝 gapping fissure; gapping joint-fissure;open tension fissure
张开器 speculum[复 specula]
张开式开展 mode of tension
张开式锚碇 spread anchorage
张开式移动 opening movement
张开天幕 spread an awning
张开销钉 expanding dowel
张开斜角 open bevel
张开型裂纹 <I 型,用于断裂力学> opening mode(I) of crack;mode of tension
张开帐篷 tent
张口 dehisce
张口裂缝 gapping crack
张口裂隙 gapping crack
张口破裂 gash fracture
张拉 jacking;staying;stretching;tensioning
张拉场 stretching area
张拉成型法 stretch forming
张拉成型机 stretch-draw forming machine
张拉成型模 stretch-draw die
张拉成型压力机 stretch-draw press
张拉程序 stretching order;tensioning procedure
张拉次序 order of stressing;order of stretching;stressing order;stressing processor;stressing sequence
张拉端 jacking end; stressing bed; stressing end;tensioning end
张拉端锚具 stressing anchorage
张拉端锚头 stressing anchorage
张拉断裂 tension cut-off;tension rupture
张拉法 stretching method
张拉方法 stretching method
张拉杆 stretching bar
张拉缸回程 tension cylinder return
张拉钢筋 stretching reinforcement; stretching rod;stretching steel
张拉钢筋油泵 prestressed steel bar drawing oil pump
张拉钢束 stretching strand
张拉钢丝 <预应力用的> stretching wire;straining wire;tensioned wire
张拉钢丝法 wire drawing;wire stretching
张拉钢丝线型 tension wire alignment
张拉工厂 stretching factory; stretching plant
张拉工场 stretching yard
张拉构件 tensile component
张拉固定 tensile fixation
张拉过程 process of tensioning;stressing process;stretching process
张拉过度 overdraw(ing)
张拉荷载 tensioning load
张拉滑车 straining pulley
张拉滑轮 stretching block
张拉环 tension ring
张拉机 tensioning machine
张拉夹具 tension grip
张拉夹具套 stressing head
张拉结构 tensile structure; tension structure
张拉截面 tension cut-off
张拉开裂 tensile capability; tensile cracking
张拉控制应力 control stress for prestressing;jacking control stress
张拉缆索 stretching cable
张拉力 jacking force;tensioning force
张拉连接 stretched connection
张拉两根钢丝的千斤顶 two-wire jack
张拉裂缝 tension crack; tension fissure
张拉螺杆 stretching screw;tensioning screw rod
张拉螺母 tension nut
张拉锚碇 jacking anchorage; tension anchorage
张拉锚具 <预应力钢筋的> jacking anchorage
张拉面积 stretching area;tension area
张拉模具 stressing mo(u)ld
张拉模量 extension modulus; tensile modulus
张拉疲劳 fatigue in tension
张拉破裂 tensile crack;tension crack
张拉器 tensioner;tensor
张拉器储存室 stretcher store
张拉千斤顶 stressing jack;stretching jack;tensioning jack
张拉强度 tensile strength
张拉强度试验 tension strength test
张拉强度试验机 tension strength tester
张拉强度试样 tension strength test sample
张拉桥 cable-stayed bridge
张拉区(域) stretching zone; stressing area;tensile zone;tension zone
张拉设备 jacking assembly; straining apparatus; straining device; straining equipment; straining plant; straining unit; stressing device; stretching apparatus; stretching device; stretching equipment; stretching plant; stretching unit; tension apparatus
张拉十二根钢丝的预应力千斤顶 twelve-wire jack
张拉式压力盒 tensile type load cell
张拉试验 pulling test;tensile test
张拉顺序 stressing order
张拉速度 tensioning speed
张拉损失 stretching loss
张拉索 <缆索铁路和架空索道> stretching rope
张拉索接 tensile-socket
张拉台 bench abutment; stretching bed <预应力混凝土的>
张拉台座 stretching bed
张拉体系 stretching system
张拉头 tensioning head
张拉屋盖结构 tension roof structure
张拉细丝 stretching wire
张拉楔块 stretching wedge
张拉楔片 stretching wedge
张拉斜缆接合 jacking accessible cable connections
张拉行程 tension(ing) stroke
张拉岩石锚杆 rock bolt
张拉液压缸 drawing hydraulic(al) cylinder;puller cylinder
张拉液压缸行程 puller cylinder range
张拉应变 tension strain
张拉应力 jacking stress; tensioning stress
张拉用钢丝 stretching wire
张拉用液压千斤顶 tensioning hydraulic(al) jack
张拉预应力钢筋千斤顶 prestressed steel bar drawing jack
张拉元件 stretching element
张拉值 stretching value
张拉装置 tension device; tensioner; tension gear
张拉状态 state of tension
张拉组件 tensile component
张拉作用 tensioning
张拉座 bench abutment;stretching bed
张力 pull-up;stretching force;tensile force;tension;tension force
张力保持器 tensiostats
张力杯 tension cup
张力臂 tension arm
张力变动方法 tension variation method;T-method
张力变形法矫形 stretch straightening
张力表 tens(i)ometer
张力波 capillary wave;tensile wave; tension wave
张力波前沿 tension wave front
张力补偿 strength tensioning
张力补偿架空线 counterweight for tightening line; overhead contact line
张力补偿器 tension compensator
张力补偿调节 tension compensation regulation
张力补偿装置 automatically tensioned equipment; balance weight device; balance weight termination
张力测定法 tonometry
张力测定方法 strain measuring system
张力测定仪 tension measuring device
张力测功计 tension dynamometer
张力测量法 tensometry
张力测试仪 tension tester
张力测验 tensiometric measurement
张力层 tensile layer
张力场梁 tension-field beam
张力场效应 tension field action
张力冲击 tensile impact
张力冲击试验 tensile shock test;tension impact test
张力传感器 tension pick-up
张力锤 straining weight;tension weight
张力导纱器 tension finger; tension guide
张力的 tensile;tensive
张力地带 tension zone
张力电磁铁驱动 tension solenoid drive
张力垫圈 tension washer
张力动力计 tension dynamometer
张力断层 tension fault
张力断裂 tension rift
张力惰轮 gallows pulley;jockey pulley;tension brake-wheel
张力分布 strain distance
张力分析器 tension analyser[analyzer]
张力封隔器 tension packer
张力负荷 tensile load;tension load
张力干燥器 tension dryer
张力杆 tension rail
张力感受器 tonic receptor
张力功率计 dynamometer for stretching the tape;tension dynamometer
张力辊 jockey pulley;jockey roller; strain roll;stretch roll;tension roll(er)
张力辊系统 tension roller system
张力滚柱 tension reel; tension roll(er)
张力含水量曲线 tension-moisture curve
张力滑板 tension slide;tightening slide
张力环连接 tensile loop joint
张力换能器 tonotransducer
张力及强度不等的 anisotonic
张力棘爪 tension click
张力棘爪簧 tension click spring
张力计 extensometer;pull tension ga(u)ge;strainometer;tensile ga(u)ge;tens(i)ometer;tension ga(u)ge; tension indicator; tension meter;tonometer
张力计测量 tensiometric measurement
张力架 tension bracket;tension tower
张力架线 tension stringing
张力减低 hypotension
张力减径 <钢管的> stretch reducing
张力减径机 stretch reducer; tension reducing mill
张力检测器 tension detector
张力检流计 strain galvanometer
张力结构 tension structure
张力紧板 tension cable plate
张力紧轮 tensioning roller
张力卷取机 pull reel;winder
张力卷筒 tension reel
张力绝缘器 strain insulator
张力开关 tension switch
张力控制 tension control
张力控制器 tensor controller
张力控制装置 tenslator
张力拉伸关系式 tensile force-extension relation
张力冷轧带钢机 coiler tension rolling mill
张力裂缝 tension crack;tension fracture
张力轮 straining pulley;tension block; tension pulley;tightening pulley
张力螺钉 strain screw
张力描记器 tonograph
张力模量 tension modulus
张力摩擦 tenso-friction
张力内弹簧 tension inner spring
张力盘 tension pulley
张力盘簧 helical tension spring
张力疲劳试验 fatigue tension test; tensile fatigue test
张力平整 tension level(l)ing
张力平整机 tension level(l)er
张力器 pull tension ga(u)ge;tens(i)ometer; tensioner; tension meter; tensor
张力千斤顶 tensioning jack
张力牵伸 tension draft
张力区 tensile region
张力曲线 tension curve
张力圈 tension ring
张力入渗计 tension infiltrometer
张力绳索 tensioning rope
张力时间指数 tension time index
张力式卷取机 pull-type recoiler
张力试验 pulling test;tension test
张力试验机 tensile machine; tensile testing machine; tensioning machine;tension tester
张力释放 release of tension
张力数 tensile figure
张力水头 tension head
张力伺服 tension servo

张力松弛 tension recovery
张力速度曲线 tension-velocity curve
张力损失 loss of tension; tensioning loss
张力弹簧 tension spring
张力探测器 tension feeler
张力套筒 screw shackle
张力调节臂 tension regulator arm
张力调节辊 dancer roll
张力调节器 tension adjuster; tension link
张力调节器限位片 tension regulator stopper
张力调节装置 tension adjusting gear
张力调整臂 waffle arm
张力调整装置 tensioning equipment
张力梯度 tension gradient
张力图 tonogram
张力腿(式)平台 tension leg platform; vertically anchored platform
张力腿式生产平台 tension leg platform
张力托架 tensioning bracket
张力系数 coefficient of tension; tension coefficient; tension factor
张力细丝 tonofilament
张力线 tension line; tension wire
张力型油管挂 tension-type hanger
张力性空洞 tension cavity
张力修正 tension correction
张力旋压 spinning with tension
张力学 tesiometry
张力学说 strain theory; tension theory
张力-压缩应变波 tension-compression wave
张力仪 strain ga(u)ge; strain meter; tensotast
张力原丝 tonofilament
张力载重 tensile load(ing)
张力增量 tension increment
张力轧机 all-pull mill
张力障碍 dystonia
张力值 tensioning value
张力指示器 tension indicator
张力重 tension weight
张力重块 tensioning weight
张力重力自动补偿装置 automatically tensioned catenary
张力轴 tension axis
张力桩 tension pile
张力状态 state of tension
张力自动补偿装置 automatic tensioning device
张链器 chain tensioner
张量 tension quantity; tensor
张量变换 tensor transformation; transformation of tensor
张量表示 tensor representation
张量场 tension field; tensor field
张量(乘)积 tensor product
张量乘积法 tensor-product method
张量磁化率 tensor susceptibility
张量丛 tensor bundle
张量代数 tensor algebra
张量导磁率 tensor permeability
张量的分量 component of tensor
张量的阶 order of a tensor; rank of tensor
张量的内积 inner product of tensors
张量的权 weight of tensor
张量等式 tensor equality
张量电导率 tensor conductivity
张量定理 tensor theorem
张量短缩 tensor contraction
张量二次曲面 tensor quadric
张量发散 tensor-divergence
张量发散定理 tensor-divergence theorem
张量方程 tensor equation
张量分解 resolution of tensor
张量分析 tensor analysis; tensor calculus
张量赋值函数 tensor-valued function
张量函数 tensor function
张量和 tensor sum
张量积表示 tensor product representation
张量积代数 tensor product algebra
张量积码 tensor product code
张量极化 tensor polarization
张量集 tensorial set
张量几何 tensor geometry
张量计算 tensor calculus
张量空间 tensor space
张量力 tensor force
张量力相互作用 tensor force interaction
张量密度 tensor density
张量黏[粘]性 tensor viscosity
张量切变 tenso-shear
张量势 tensor potential
张量势阱 tensor well
张量算符 tensor operator
张量缩并 contraction of tensor
张量特征 tensor property
张量椭球 tensor ellipsoid
张量微分 tensor differentiation
张量相互作用 tensor interaction
张量形式 tensor form; tensorial form
张量性质 tensor nature
张量演算 calculus of tensors
张量约缩 tensor contraction
张量运算 tensor operation
张量-张量效应 tensor-tensor effect
张量主不变量 principal invariant of tensor
张量足标 tensor subscript
张裂 tension crack
张裂带 extensional fracture belt
张裂缝 gash fracture; tensile crack; tension fissure; gull <地滑造成的>
张裂拉力 splitting tension
张裂面 plane of tensile fracture
张裂区 cracked tension zone; extensional fracture region
张裂隙 separation fracture; tensile crack
张裂型石香肠 tension-type boudin
张满 belly
张满全风帆 pack on all sails
张幕索 awning side stops
张扭性的 tension-shearing
张扭性断层 tension-shear fault
张扭性结构面【地】 structural plane of tenso-shearing origin; tension-shear structural plane; tenso-shear structural plane
张扭性盆地 tension-shear basin
张破裂 subsidiary fracture; tensile fracture
张刃式钻头 paddy bit
张设天幕的小艇 tilt-boat
张绳 <测海上距离用的> taut-wire
张绳测距法 <海上测距用> taut-wire traverse
张丝 tensile spring; tensioned wire
张算 tensor calculus
张缩导套 magic guide bush
张缩接合 expansion joint
张缩接头 expansion joint
张所有的帆 every stitch set
张索 guy(rope); guy wire; vang
张索绞车 guy winch
张索输油法 span-wire method
张贴传单 bill-posting; bill-sticking
张贴广告人 sticker
张贴建筑许可证 <在工地上> posting of building permit
张贴式列车时刻表 timetable poster
张贴物 sticker
张帖 <学术会议的> poster
张网捕鱼 netting
张线 bracing wire; tie wire
张线法 stretched-wire method
张线滑轮 cord tension pulley
张线接合 stretched-wire joint
张线式传感器 strain ga(u)ge
张线式高温传感器 high-temperature strain ga(u)ge
张线式悬挂模型 wire-supported model
张线准直法 tension wire alignment method
张陷 taphrogeny
张楔式制动器 expanding-wedge brake
张性断层【地】 extension(al) fault
张性断裂 extensional faulting; tension fracture
张性构造 extensional structure
张性构造岩 tensile tectonite
张性节理 extension joint
张性结构面 tensile structural plane; tensional structural plane
张性破裂 extension fracture; tension fracture
张性深断裂 tensional deep fracture
张性追踪构造 tension track structure
张应变 tensile strain
张应力 tensile stress; tension(al) stress
张褶皱 tension fold
张致电阻效应 tensoresistance

章

章程 article; by-law; regulation; statute

章程的 constitutional
章动 <地轴的微动> nutation
章动常数 nutational constant
章动齿轮液压马达 nutating gear hydraulic motor
章动传动 nutation drive
章动传动装置 nutation drive units
章动轮 nutation wheel
章动器 nutator
章动椭圆 nutational ellipse
章动周期 nutation period
章动轴 nutation shaft
章度 <色度学的>【物】 saturation
章伐耳式水轮机 <平行或轴向水流水轮机> Jonval type turbine
章伐耳水轮 Jonval wheel
章回接排 run-on chapters
章节 chapters and sections
章节标题 dropped head
章氏硼镁石 hungchaoite
章印 nutation
章鱼系统【航海】 Octopus system
彰色 saturated colo(u)r

獐

獐耳细辛属植物 hepatica

樟

樟丹 minim

樟红天生 <拉> Pyrestes haematicus
樟科 <拉> Lauraceae
樟木 camphor wood; common pine; dark pine
樟木板 cassia lignea
樟脑 camphor
樟脑玻璃 <呈浑浊白色，其表面形状与块装樟脑相似> camphor glass
樟脑火棉浆 camphoid
樟脑球 camphor ball
樟脑油 camphor oil
樟树 camphor laurel; camphor tree; cinnamomum camphora
樟树油 ohba-gusu oil
樟烷 camphane
樟形天线阵 curtain array
樟油 ocotea oil
樟属 camphor; cinnamon
樟属大树 <产于圭亚那> green heart
樟属绿心硬木 <圭亚那所产，用于造船> green heart
樟属植物 cinnamon

蟑

蟑螂 blackbeetle; cockroach; roach

长

长草河流 grassed water course

长得比较高大 bigger on the hoof
长得高大的植株 plant growing quite tall
长得过大或过快 overgrow
长得宽罩衫 smock frock
长满 overgrow
长满草皮的水沟 turf drain
长满灌木之地 scrub land
长满青苔的 moss-covered
长霉试验 mo(u)ld growth test
长新梢 extending shoot
长在墙上的植物 wall plant
长在小溪旁的 rivalis
长者 senior

涨

涨潮 egre; flood; flowing tide; flux; incoming tide; rise of tide; rising flood; rising tide; swelling tide; tidal flood; tidal lift; tide; tide epoch; tide flood

涨潮标志 flood mark
涨潮不足 cut tide
涨潮潮差 flood tide range
涨潮潮幅 flood tide range
涨潮持续时间 duration of flood; duration of flood current; duration of rise
涨潮初期 young flood
涨潮船闸 tidal lock
涨潮岛 <退潮时是半岛> bridge islet
涨潮低降 flood tide depression
涨潮点 tidemark
涨潮顶点 top of the flood
涨潮海峡 flood channel
涨潮航道 flood channel; flood tide channel
涨潮和落潮 rise-and-fall
涨潮河滨 tidal creek
涨潮间隔 flood interval; tidal flood interval
涨潮界 tidal limit; tide head limit
涨潮力 tide raising force
涨潮历时 duration of flood tide; duration of tidal lift; duration of flood; duration of rise; flood tide duration
涨潮量 flood tide volume; volume of flood tide
涨潮流 flood current; flood stream; flood tide current; flood tide stream; ingoing stream; tidal flood current
涨潮流河道 flood channel
涨潮流历时 duration of flood current
涨潮流量 volume of water
涨潮流速 flood speed; flood strength; strength of flood
涨潮流月潮间隙 flood interval; tidal flood interval
涨潮流轴(线) flood axis
涨潮落潮 flux and reflux
涨潮末 ending of flood
涨潮平流 slack-water on the flood
涨潮期间 period of flushing; period of

rising tide;period of increasing tide
涨潮起点 young flood
涨潮憩流 slack-water on the flood
涨潮强度 tidal flood strength
涨潮三角洲 flood tidal delta
涨潮三角洲沉积 flood tidal deltaic deposit
涨潮时的静止水位 slack-water on the flood
涨潮时段 flood to peak interval
涨潮时关闭的船闸闸门 flood gate
涨潮时间 period of flushing;tide epoch;time of rise;time of setting in
涨潮时形成的小(海)湾 pil;pyll
涨潮水道 flood channel; flood tide channel
涨潮水量 volume of water entering on the flood tide;volume of flood
涨潮水位计 flood meter
涨潮死航道 blind flood channel
涨潮坞门 flood tide gate
涨潮线 tide line
涨潮线与落潮线之间的地带 shore
涨潮行波 travel(l)ing tidal wave
涨潮淹没地带 littoral
涨潮与落潮 flood and ebb
涨潮闸门 flood tide gate
涨潮主槽 flood-predominating channel
涨潮总量 total volume of flood tide
涨出地 inning
涨大水 spate
涨方(木材)bulkage
涨风 upward trend of price
涨轨【铁】track buckling;buckling of rail
涨轨力 buckling force
涨轨跑道 buckling of track
涨洪 flood rise
涨洪段 <过程线的> concentration curve; concentration limb; rising limb
涨洪开始期 beginning of rise
涨洪历时 duration of rise
涨洪期 period of rise
涨洪起点 young flood
涨洪时段 flood to peak interval
涨洪时刻 time of rise
涨急 maximum flood
涨价 advance(in)price;appreciation; cost escalation;hike in prices;inflation of prices; mark-up; price advance; price upswing; rise in prices;rising price;run-up
涨价趋势 upward price trend
涨价盈余 appreciation surplus
涨价准备 reserve for appreciation
涨价总额 gross appreciation
涨壳式锚杆 expansion shell bolt
涨落 fluctuation; rise-and-fall; seesaw
涨落差 range
涨落潮 ebb and flood;ebb and flow; ebb tide and flood tide
涨落潮差 range of flood and ebb
涨落潮更替 turn of the tide
涨落潮历时差 time difference of flood and ebb
涨落潮流构造 ebb and flow structure
涨落潮流路 flow path of flood and ebb
涨落潮图 current diagram
涨落潮形成的地层结构 ebb-and-flow structure
涨落潮淤积物 tidal mud deposit
涨落潮造成的海冰裂缝 tide crack
涨落潮闸门 ebb and flow gate
涨落潮周期 period of flood and ebb
涨落持续时间 duration of flushing; duration of rise
涨落范围 range of fluctuation
涨落校正因素 fluctuation correction factor
涨落井 ebbing well
涨落理论 fluctuation theory
涨落历时 duration
涨落谱 fluctuation spectrum
涨落期 period of ebb and flow
涨落泉 ebbing and flowing spring
涨落循环 cycle of fluctuation;cycle of flushing
涨落沼泽生物控制 helophytia
涨落周期 cycle of fluctuation
涨满 overflow;suffuse
涨末 end of flood
涨起 rising
涨前段 <水位过程线起涨前部分> approach curve;approach segment; approach limb
涨圈 ca(u)lking ring
涨圈槽 piston ring groove;piston ring slot;ring groove
涨圈槽嵌圈 ring groove insert
涨圈搭口 scarf
涨圈簧 plate enlargement ring
涨水 rising water
涨水段 rising limb
涨水高度 height of rising flood; height of swell
涨水阶段 rising flood stage
涨水历时 duration of rise
涨水期 rising flood period; rising flood phase;rising(flood)stage
涨水期间 rinsing flood time
涨水线段 <水位过程线的上升段,即曲线的上升段> rising limb
涨水涌浪 flood surge
涨缩盒 sylphon
涨缩屋顶 breather roof
涨滩 <潮涨则淹、潮退则现的岸> foreshore
涨停板 <交易所> limit up
涨压 turgor pressure

掌

部双层帆布手套 double palm canvas mittens

掌撑 palm stay
掌尺 <宽约 4 英寸,长约 8 英寸,1 英寸 = 0.0254 米> palm
掌舵 helm;steerage;steering
掌舵帆 steering sail
掌舵柜 coxswain's box
掌舵形撑 palm stay
掌钩 palm hook
掌管 in charge
掌上电脑 palmsize personal computer
掌式撑条 palm stay
掌式桩 palm pile
掌握 mastery;prehension;wield
掌指状水系型 digitate drainage pattern
掌状撑条 palm stay
掌状的 digitate;palmate
掌状节 splay knot
掌状深裂的 palmately parted
掌状水系 digitate drainage pattern
掌状物 palm
掌状系杆 throat brace
掌状羽状的 digitated pinnate;digitatopinnate;palmate-pinnate
掌状圆裂的 palmately lobed
掌子面 active face; coal face; drift face;driving face;face of work area;heading;tunnel face;tunnel heading; working face; working place; heading face <隧洞掘进的>; mixed face <土石方同时进行的>
掌子面采样 handing sampling
掌子面防塌撑木 breast timber
掌子面防塌护板 breast board
掌子面工效 face capacity
掌子面上的炮眼数 holes per face
掌子面通风 heading ventilation

丈

尺 yardstick

丈量 admeasurement;chaining;measure;measurement
丈量标 measuring mark
丈量标准 basis measurement
丈量船货 measurement cargo;measurement goods
丈量吨(位)measurement ton(nage)
丈量法 measured method;measuring method;method of measurement
丈量公式 formula of measurement
丈量换算表 measurement conversion table
丈量截面 tonnage section
丈量链条 measuring chain
丈量疏伐 stick thinning
丈量误差 chaining error

帐

shroud

帐杆 tringle
帐帘 dossal;dosser;drapery
帐幔 valance
帐幕 veil
帐幕般的 pavilion-like
帐幕干燥窑 curtain kiln
帐幕式教室 pavilion-type classroom
帐幕式结构 pavilion-type structure
帐幕式棚架木杆 cruck
帐幕式学校 pavilion school
帐篷 camp;tabernacle;tent(age);velum <罗马剧场观众席的>
帐篷薄膜 tent membrane
帐篷顶横杆 ridge pole
帐篷盖 tent fly
帐篷构造 camp structure;tepee structure;tent structure
帐篷结构 tent structure;tent texture
帐篷绳 tent line
帐篷绳钩 tent string hook
帐篷式顶棚 camp ceiling
帐篷式堆场 tent roofed shed
帐篷式房屋 cruck house
帐篷式干燥的 tent-dried
帐篷式拱 tented arch
帐篷式料堆 tent-shaped pile
帐篷式熟料堆场 tent-shaped clinker store
帐篷式熟料贮库 tent-shaped clinker store
帐篷式天花 camp ceiling
帐篷式卧床 tent bed
帐篷式屋顶 tilt roof
帐篷式住宅 bayt
帐篷索 tent-guy
帐篷拖车 tent trailer
帐篷屋 cabana
帐篷屋顶 tent roof
帐篷屋子 wicky
帐篷系列 tent system
帐篷形料堆 coned tent pile
帐篷桩子 tent peg
帐篷状物 tent
帐台放钱的抽屉 till
帐形纹 tented arch

杖

cane;wand

杖责发生会计 accounting on accrual basis

胀

杯 <传力杆活动端的帽套> expansion cup

胀崩现象 popped out
胀槽 expansion slot
胀差指示计 differential expansion indicator
胀尺 expansion rule
胀大 bulking;expansion;swelling
胀大系数 coefficient of bulking
胀带式离合器 expanding band clutch
胀带闸 expanding band brake
胀度 expansibility
胀方 earth swell
胀缝 expansion joint
胀缝传力杆套 <混凝土路面的> expansion cap
胀缝带 expansion joint tape
胀缝拱起 arising of expansion joint
胀缝器 expansion joint device
胀缝填料 expansion joint filler
胀缝隙 expansion spacing
胀杆 expander roll;expanding arbor
胀毂式锚杆 expansion shell bolt
胀管 expander tube;expand tube
胀管槽 expanded tube hole groove
胀管度 expanding grade
胀管机 pipe expander
胀管接头 expanded tube joint
胀管口 expand
胀管连接 expanded tube joint(with tubesheet)
胀管螺栓 expansion bolt
胀管器 casing roller; casing swage; tube expander
胀管器滚子 expander roll
胀管试验 bulge test;expand(ing)test
胀轨 buckled track
胀过的管子 rolled tube
胀焊 expanded and welded tube joint
胀环 cuff;tensioner ring
胀环离合器 expanding ring clutch
胀簧 expansion spring
胀接 expansion expanded tube joint; flared joint;rolled joint
胀接式点焊枪 expansion gun
胀紧螺栓 cinch bolt
胀紧套筒 expansion sleeve
胀开式衬砌 expanded lining
胀开式封隔器 expanding plug
胀开式弹簧筒夹 expanding chuck
胀开式心轴 expanding mandrel; expanding arbor;expansion mandrel
胀开式心轴法 expanding mandrel method
胀壳式锚杆 expansion shell bolt;expansion sleeve bolt;expansion-type rock bolt
胀壳式岩石锚杆 expansion rock bolt
胀口 bulged finish
胀口工具 swaging tool
胀力 expansive force
胀连 expanded joint
胀量 bulk
胀裂 bursting crack;expansion cracking;spalling【地】
胀裂脊 pressure ridge
胀裂强度 bursting strength
胀裂丘 pressure dome
胀流型分散体 dilatant dispersion
胀流型流动 dilatant flow;shear thickening
胀流型流体 dilatant fluid
胀流型体 dilatant
胀流性 shear thickening
胀膨黏[粘]土混凝土砌块 bloating clay concrete solid block
胀破强度 burst strength
胀破试验 burst tearing test

胀破压力 bursting pressure
胀起 blow-up; upfold; upheaval; upheave
胀气 flatulence
胀圈 cup ring; expander ring; expansion loop; expansion ring; packaging; packing ring; piston ring
胀圈槽 ring slot
胀圈套管灌浆法 Soletanche method
胀沙 heaving sand; swell of a mo(u)ld
胀砂 buckle
胀式锚杆 expansion bolt
胀式圆锥 expanding cone
胀式制动器 expanding brake
胀缩 movement
胀缩百分率 percent swell and shrinkage
胀缩变量 <木材随含水率变化而引起的膨胀或收缩> working
胀缩波 compressionally dilatational wave; compression-dila(ta)tion wave
胀缩补偿装置 lazy jack
胀缩鼓轮 expansion drum
胀缩间隙 clearance for expansion; expansion clearance
胀缩件 compensator
胀缩接头 expansion joint
胀缩卷筒式卷取机 expanding drum coiler
胀缩卷筒式开卷机 expanding drum uncoiler
胀缩可逆性 convertibility of expansion and shrinkage
胀缩联轴节 expansion coupling
胀缩联轴器 expansion coupling
胀缩率 percentage of swell and shrinkage; percent of swell and shrinkage
胀缩敏感性 shrink-swell potential
胀缩器 expansion bend; expansion loop
胀缩试验 reduction-in-expansion test
胀缩土(壤) swell-shrinking soil
胀缩弯管 expansion bend
胀缩弯曲膜盒 expansion and flexure bellows
胀缩弯头 bend type expansion joint; expansion bend
胀缩性 swell-shrinkage behavio(u)r; swell-shrink characteristic
胀缩性土 swell-shrinkage soil
胀缩总率 total linear swelling and shrinkage; total swelling and shrinkage rate
胀胎 expander tube
胀凸 <金属皮> cupping
胀凸试验 <检测金属皮延性的> cupping test
胀隙 clearance for expansion
胀限 swelling limit
胀箱 exudation; swell; swelling
胀箱力 swelling power
胀销螺栓 expansion shell bolt
胀形 bulge; bulging; expand
胀形模 bulging die; expanding die
胀形内爪 expanding inner jaw
胀形塞 expanding plug
胀形系数 bulge coefficient
胀性流动 dilatant flow
胀性(黏[粘])土 <一种拌制混凝土用的轻集料> bloating clay
胀压硬度 indentation hardness
胀闸 expanding brake; expansion braker
胀轴 expanding arbor

账本底线 bottom line

账簿 account(ing) books; book of accounts; financial book; ledger; reckoning book
账簿记录 book entry
账簿纸 account book paper; ledger paper
账簿制度 book system
账册 account books
账册用纸 account book paper
账存数量 book quantity
账单 account bill; account note; bill; check; statement of account; tab
账单插入件 bill insert
账单登记卡 bill register card
账单付讫 account paid
账房 bookkeeper's departure; casher's department
账房先生 accountant and secretary; bookkeeper
账号 account(ing); charge number
账户 account; bank account
账户编号 account number; symbolization of accounts
账户标记法 symbolization of accounts
账户簿 accounting form
账户的对应关系 correspondence of accounts
账户分类 account category; account classification; classification of accounts
账户分类说明 classification manual of accounts
账户分析 account analysis
账户格式 account form
账户滚存余额结单 running balance statement of account
账户计划 plan of accounts
账户结构 account structure; structure of accounts
账户结平 account balanced
账户结清借方余额 closing debit balance in account
账户结清余额 closing balance in account
账户结余 account balance
账户结转 bring down on account
账户例行程序 accounting routine
账户流程图 account flow chart
账户名称 account heading; account name; account title; name of account; title of account
账户明细表 detailed statement of accounts
账户式收益表 account form of income sheet; account form of income statement
账户式损益计算表 account form of profit-loss statement
账户式资产负债表 account form of balance sheet
账户体系 system of accounts
账户项目 account item
账户性质 nature of account
账户一览表 account chart
账户余额 account balance; balance of accounts
账户转换 account transfer
账款 funds on account; value in account
账款抵消 offset against accounts
账款付讫 account paid
账款付清 account paid
账款回收率 collection rate
账款结清 account settled
账款清单 settlement of account
账款贴现簿 discount account register
账龄分类 age distribution
账龄分析表 ag(e)ing schedule
账面差额 book balance
账面成本 book cost; recorded cost
账面赤字 deficit on the books
账面存货 book inventory
账面负债 stated liabilities
账面汇率 accounting rate
账面记录 book entry
账面价格 accounting price; book price; original book value
账面价值 book value; recorded valuation
账面价值的有意减低 write down
账面价值股票 book-value shares
账面价值减少 decrease in book value
账面结存价值 carrying value
账面金额 book amount
账面净值 net book value
账面亏损 book loss
账面利率 book rate
账面利润 accounting profit; book profit; paper profit
账面利益 paper profit
账面盘存 book inventory
账面盘存表 book inventory sheet
账面盘存法 book inventory method
账面盘存记 book inventory record
账面欠债 book debt
账面上的余额 balance of accounts
账面审计 book audit
账面收益 accounting income
账面收益率 book rate of return
账面税金差额 book tax difference
账面损失 book loss; paper loss
账面信贷 book credit
账面信用 book credit
账面盈余 book surplus
账面原值 gross book value
账面债券 book debt
账面债务 book debt
账面折扣 book depreciation
账面资产 ledger assets
账面总价值 gross book value
账目 account; tab
账目表 statement of account
账目不清 account not in order; accounts are not in order
账目不实 false itemization of accounts
账目单 account note
账目分析 account analysis
账目管理 administration of accounts
账目核对 verification of accounts
账目混乱 confused account
账目结余 amount of balance
账目清算 clearance of accounts
账目设账 keep additional accounts in addition to the authorized ones
账目事务 account transaction
账目摘要 abstract of account; extract of an account
账棚木杆 cruck
账上货币 money on account
账外财产 concealed property; off-the-book property
账外负债 liability out of book
账外借款 off-the-book loan
账外物资 hidden assets
账外资产 assets out of account books; assets out of accounts; hidden assets; invisible assets; non-ledger assets; off-the-book property; unlisted assets
账务分类 account classification
账务机构识别码 accounting authority identification code
账务科 account section
账务调整 accounting adjustment
账务员 account clerk
账项 account(ing) item
账项付清 account paid
账页参考 folio reference

障碍 barrage; ba(u)lk; clog; drawback; handicap; hindrance; hitch; hurdle; mar; mischance; obstruct; obstruction; pravity; blockage <美>

障碍板 balk board
障碍保持 fault holding
障碍报警 obstruction warning
障碍表示器 obstacle indicator
障碍产生的水头损失 loss of dynamic(al) head due to obstructing object
障碍船具 hamper
障碍的 obstructive
障碍灯标 obstruction light; obstacle light
障碍地带 obstacle belt; obstacle zone
障碍定位仪 fault locator; trouble locator
障碍段水头损失 loss of head due to obstructions
障碍蜂音信号 out of order tone
障碍服务台 complaint desk; repair clerk's desk
障碍浮筒 obstruction buoy
障碍感觉 obstacle sense
障碍高度与车轮直径比 obstacle-height to wheel-diameter ratio
障碍规 obstruction ga(u)ge
障碍函数 barrier function
障碍函数反换式 barrier function inversion
障碍痕 obstacle scour marking
障碍记录 fault recording; report of irregularity
障碍记录器 fault recorder
障碍记录台 fault clerk's desk
障碍间隙 obstacle clearance
障碍间隙框 obstacle clearance box
障碍检测系统 obstacle detection system
障碍检测装置 obstacle detecting device
障碍检查器 fault finder
障碍勘测 fault location
障碍控制台 fault clerk's desk
障碍理论 obstruction theory
障碍流 obstacle flow
障碍轮廓尺寸 obstacle profile
障碍模拟 obstacle approximation
障碍桥 obstructive bridge
障碍区 diked area
障碍区间 fault section
障碍绕射 obstacle diffraction
障碍绕射损耗 obstacle diffraction loss
障碍赛跑跑道 steeplechase course
障碍设置 obstacle construction
障碍设置计划 obstacle plan
障碍试验线 plugging-up line
障碍衰落 obstruction fading
障碍缩尺模型研究 scale model study of obstacles
障碍台 fault complaint service; test board desk
障碍探测 obstacle detection
障碍停车 handicapped parking
障碍突破舰船 barrier breaker
障碍图 obstruction chart
障碍位置(测定) fault location
障碍位置测定器 fault localizer
障碍位置测定仪 fault finder
障碍物 abates; abat(t)is; barricade; blockage; clog; clogging; dike [dyke]; drawback; encumbrance; entanglement; gorge; hindrance; impediment; interrupter; interruption; obstruction; remora; stumbling block; barrier; defilade; obstacle; fraise <铁丝网或木桩>

Z

障碍物标记 obstacle marking
障碍物标示 obstruction mark(er)[mark(ing)]
障碍物标示旗 obstacle flag
障碍物标志 obstacle marking;obstruction mark(er);obstruction mark(ing)
障碍物标志灯 obstruction lamp;obstruction light
障碍物标桩 obstruction beacon
障碍物灯标 obstruction light
障碍物浮标 obstruction buoy
障碍物回波反射 back echo reflection
障碍物间距 obstruction spacing
障碍物间隙 obstruction clearance
障碍物净空 obstacle clearance requirement
障碍物探测 obstacle detection;obstruction sounding
障碍物限界 obstruction ga(u)ge limit
障碍物限制重量 obstacle-limited weight
障碍物信标 hazard beacon;obstruction beacon
障碍物引起的动水头损失 loss on head due to obstruction
障碍线收容继电器群 hospital relay group
障碍信号塞孔 trouble jack
障碍学说 <关于气旋发生的一种学说> barrier theory
障碍音塞孔 out of order tone jack
障碍影 obstacle shadow
障碍增益 obstacle gain
障碍指示器 obstacle indicator
障碍作用 obstructive action
障板 baffle;baffle block;easel mask;mask;shadow mask
障板式彩色显像管 colo(u)rtron
障板室 baffle chamber
障板稳定火焰 bluff body flame
障板稳焰 bluff body flameholding
障板稳焰器 baffle flame holder
障板扬声器 baffle loudspeaker
障蔽之物 blind
障壁 baffler;barrier
障壁坝 barrier bar
障壁岛 barrier island
障壁岛沉积 barrier island deposit
障壁岛复合体【地】back barrier complex
障壁岛相 barrier island facies
障壁岛泻湖层序 barrier-lagoon sequence
障壁岛泻湖沉积 barrier-lagoon deposit
障壁岛泻湖环境 barrier-lagoon environment
障壁岛泻湖体系 barrier-lagoon system
障壁堤 barrier
障壁复合体 barrier complex
障壁后复合体 back barrier complex
障壁后退毯状沉积 barrier-retreat carpet
障壁礁油气藏趋向带 barrier reef of pool trend
障壁沙坝 <又称障壁砂坝> barrier bar
障壁沙坝沉积 barrier bar deposit
障壁沙坝圈闭 barrier bar trap
障壁沙坝相 barrier bar facies
障壁沙嘴 barrier spit
障壁台地 barrier platform
障壁滩 barrier beach
障壁滩沉积 barrier beach deposit
障壁滩相 barrier beach facies
障壁泻湖 barrier lagoon
障冰 barrier ice
障层电池 barrier-layer cell
障层光电管 rectifier photocell
障风装置 abat-vent
障积岩 bafflestone
障景 blocking view;obstructive scenery
障景种植 screen planting
障块 baffle block
障膜整流器 barrier film rectifier;barrier-layer rectifier
障热板 dead plate
障塞器 obturator
障栅 aperture grill;barrier grid
障眼物 blinder
障子 shoji screen
障阻 barrier
障阻沼泽 barrier marsh

嶂谷 narrow gorge

幛形天线 curtain antenna

瘴毒 miasma

瘴疟 malignant malaria
瘴气 malaria;pestilential pathogen;miasma [复 miasmata/miasmas] <由沼泽地产生的>

招爆性 blastability

招标 bid call;bid invitation;bid wanted;calling for tenders;invitation for bid;invitation for offer;invitation for tender;invite bids;invite for tender;invite tenders;invite to tender;tendering;call for tenders
招标采购 open tender;purchasing by invitation to bid
招标程序 tendering procedure
招标出售 sales by tender
招标单 invitation for tender
招标底价 bottom bidding price
招标抵押承包 contract with competitive bidding and secured by mortgage
招标发行 issue by tender
招标法 fixed price tendering
招标公告 advertisement for bids;advertising for bid;announcement for open tender;announcement of tender;bidding announcement;bid notice
招标广告 advertisement for bids;bidding advertisement
招标后 post tender
招标介绍 tender notice
招标落选者 unsuccessful bidder
招标内容 bid form
招标评议委员会 adjudicating panel
招标期 bidding period
招标期限 bid date;bid time
招标前 pretender
招标前会议 pre-bid meeting
招标人 tenderer
招标设计 tender design
招标审计 inviting bids audit
招标书 form of tender;invitation for bidding;invitation for tendering;invitation to bid;invitation to tender;proposal request <建筑师、工程师发出的>
招标说明 instruction to bidders;instruction to tenderers
招标条件 terms of tender
招标条款 bid form
招标通告 bidding notification;invitation for bid;invitation for bidding;invitation for tendering;invitation to bid;invitation to tender;call for bids
招标通知 announcement of tender;bidding advice;bidding notification;bid invitation;invitation to bid;invitation to tender;tender notice
招标通知书 notice to bidders
招标投标法 the bid and tender law
招标投标制 the public bidding system
招标文件 bid(ding) documents;invitation for bidding;invitation for tendering;invitation to bid;invitation to tender;tender documents;tendering documents
招标文件的澄清 clarification of tendering documents
招标项目 list of bidding items
招标形式 form of tender
招标须知 instruction for bidding
招标有效期 tender validity;term of tender validity
招标制 competitive bidding system;public bidding system
招潮属 uca
招待 servicing;serving
招待费 entertaining expenses;entertainment expenses;entertainment fee;hospitality expenses;hospitality requirement;table money
招待会 entertainment;recept(ion)
招待券 complementary ticket
招待所 boarding house;guest house;hospice;hostel;rest house;personnel accommodation
招待所住客 hostel occupant
招待员 steward;usher;floorwalker <美>
招风斗 air scoop;wind scoop
招工 hands wanted;hiring of labo(u)r;recruitment;recruit worker
招股 call for capital;raise capital by floating shares
招雇 wanted
招呼乘坐 captive rider
招呼停车站 <公共汽车的> flag stop
招呼站 halt;request stop <公共交通的>
招唤装置 annunciator
招回 call-back
招来 call in
招徕顾客 tout
招徕广告人 <俚语> adman[复 admen]
招徕生意的广告 ballyhoo
招揽 canvassing
招揽生意 bring in business
招揽途径 <指促使房地产成交过程> procuring cause
招揽运输 canvassing of traffic;traffic solicitation
招募 recruitment
招牌 facia;fascia[复 fa(s)ciae/fa(s)cias];name board;placard;sign board
招牌广告 employment advertisement
招牌可见范围 sign area
招牌上电器设备设计及安装技师 sign-master
招牌上装电器的人 sign-journeyman
招牌用清漆 signboard varnish
招聘 invite applications for a job;position vacant;wanted
招请引航员旗号 pilot flag;pilot jack
招请引水员旗号 pilot flag
招人领取公债券金额 call a bond
招商局 China Merchants Steamship Company
招商局集团 Merchants Group
招商局(中国)轮船公司 China Merchants Steamship Navigation Company
招收 recruit
招手停车站 <公共汽车的> flag stop
招贴 bill-posting;placard;poster;show bill;sticker
招贴画 poster
招贴画颜料 poster colo(u)r
招贴墙 sign board
招贴纸 poster paper
招投标程序 bid procedure;tender procedure
招投标委员会 tendering committee;adjugate panel
招致 procure
招租 house-to-let

朝阳工业 sun-rise industry

啁啾干扰 <两相邻波道边带干涉所生的干扰> monkey-chatter interference

啁啾声 chirp;monkey chatter <两相邻波道边带差频所引起的干扰>
啁哳声 chirping

着火 ablaze;burst into flame;catch fire;ignite;ignition;inflammation;kindling;on fire;take fire

着火的 ablaze;inflame
着火点 burning point;fire point;firing point;flare point;flash point;ignition point;ignition temperature;kindling point;kindling temperature;point of ignition
着火点测定 fire test
着火点试验器 ignition point tester
着火电位 flashing potential;priming potential;striking potential
着火范围 fire extend;firing range;ignition range
着火极限 inflammability limit
着火极限浓度 combustible limit
着火加速剂 <柴油的> ignition accelerator
着火落后时期 ignition delay period
着火浓度极限 combustible limit;explosive limit;flammable limit
着火浓度下限 lower limit of flammability
着火热 ignition heat
着火栅压 starting grip voltage
着火时间 ignition time
着火(时)转速 firing speed
着火时状况 behavio(u)r under fire
着火试验 flammability test
着火顺序 fire ignition sequence
着火速率 <炸药> ignition rate
着火提前 ignition advance
着火危害 ignition hazard
着火危险 danger of ignition
着火危险性 exposure hazard
着火温度 catch fire temperature;charge temperature;fire temperature;firing temperature;ignition temperature;kindling temperature
着火温度试验 fire test
着火性 ignitability
着火性能 flammability
着火延迟 ignition lag
着火滞后 ignition lag
着蓝 bluing
着力点 acting point;action point;pick-up point;point of application;point of exertion;point of strength

找补 touching up

找补腻子 putty retouch;re-putty

找出 grub
找带装置 tape search unit
找到 hit(up)on hit
找到故障 isolation;isolation of blunders
找地下水 tracing groundwater
找顶 scaling;trimming
找顶工作 <隧洞> scaling operation
找方 square up
找工作的人 job hunter
找回 retrieve
找矿 mineral prospecting;ore-search
找矿标志 indicator;prospecting indications and guides
找矿矿物学 ore-finding mineralogy
找零备用金 change fund
找零钱 change;to give change
找煤 look for coal
找平 level finding【测】;level up【测】;align;alignment;dubbed off;level(l)ing;level(l)ing off;level(l)ing work;screeding;spreading and level(l)ing;take elevation;take the level
找平板 level(l)ing board;level(l)ing plank;level(l)ing screed(material)
找平层 browning;dressing course;grade course;level(l)ing blanket;level(l)ing coat;level(l)ing course;level(l)ing layer;level(l)ing underlay;screed;screed-coat;trowel(l)ing course;underlayment
找平层促凝剂 screed accelerating agent
找平尺 screed
找平的混凝土层 blinding concrete course
找平垫层 racking course
找平工具 level(l)ing tool
找平刮板改进器 screed improver
找平合成物 feathering compound
找平灰浆 level(l)ing mortar
找平混凝土 oversite concrete
找平混凝土层 level(l)ing concrete
找平机 cal(l)iper machine
找平靠尺 screed rail
找平密封层 screed seal(ing)
找平砌块 <楼板下的> soffit block
找平器 screeder
找平千斤顶 level(l)ing jack
找平砂垫层用的平底船 screed barge
找平砂浆 level(l)ing mortar
找平砂浆表面 screed surface
找平砂浆层 level(l)ing binder course;level(l)ing screed(material)
找平砂浆改良剂 screed modifier
找平砂浆接缝 screed joint
找平碎石 leveling stone
找平填充料 level(l)ing mass
找平修整 screed finish
找平样板 screed guide;screed rail
找平砖 irregular brick
找平作业 level(l)ing operation
找水仪 water finder;water witch
找线器 line finder
找正 alignment;centering;spotting
找正测微仪 shaft alignment ga(u)ge
找中心 centering[centring]
找中钻头 balanced bit
找准 capturing

沼地 bog;fen bog;fenland;hag;marsh land;pan;slew;swamp;swamping land

沼地挡潮闸 aboideau;aboiteau
沼地道路 swamp road
沼地的 moory
沼地河流 muskeg stream
沼地螺旋推进水陆两用车 marsh screw amphibian
沼地排水 leam
沼地上圆木(铺成的)路 ground bridge
沼地土 fen soil
沼地性 quagginess
沼灰泥 lake marl
沼灰土 bog lime;lake marl
沼津灰壤 slough podzol
沼津灰土壤 bluff podzol
沼矿 bog mine ore;bog ore
沼煤 bog coal;boghead(ite)coal;moor coal;moor peat;peat bog
沼锰矿 bog manganese
沼气 biogas;biologic(al)gas;digester gas;filty;firedamp;marsh gas;methane;methane gas;sewage gas;sludge gas
沼气包 methane pocket
沼气保护 firedamp protection
沼气报警铃 methane phone
沼气爆炸 firedamp explosion
沼气测定 methane determination
沼气测定器 methane tester
沼气产量 gas yield
沼气池 biomass pool;marsh gas tank;methane-generating pit;methane tank
沼气重整法 firedamp reforming process
沼气储藏罐 sludge gas holder
沼气点火 ignition of gas
沼气点燃 firedamp ignition
沼气发电 gas-based power generation;gas-fired power generation;marsh gas power generation
沼气发电机 gas engine;gas motor
沼气发动机 methane fuelled engine
沼气发酵法 biogas fermentation process
沼气发生 marsh gas generation;marsh generation
沼气发生池 methane-generating pit
沼气放出 methane liberation
沼气肥 firedamp fertilizer
沼气格栅 methane grid
沼气锅炉 gas boiler
沼气含量 firedamp content
沼气化 methanation
沼气检测仪 firedamp detector
沼气检查员 fire trier;fire viewer
沼气检定器 firedamp detector;firedamp indicator;methanometer
沼气警报 firedamp alarm
沼气聚积层 fire layer
沼气开发 gas development
沼气利用 gas utilization;sludge gas utilization
沼气排放 firedamp drainage;methane drainage
沼气排放管 methane drainage pipe
沼气喷出 methane outburst
沼气燃烧器 sewage gas burner
沼气热值 heating value of digester gas
沼气栅网 methane grid
沼气生产井 producing gas well
沼气示意器 firedamp indicating detector
沼气式电机 firedamp proof machine
沼气室 gas dome
沼气探测器 firedamp detector
沼气泄出 methane emission
沼气形成 gas formation
沼气终端站 methane terminal
沼气贮气罐 sludge gas holder
沼气自动记录探测计 recording firedamp indicating detector
沼生 paludose
沼生草本群落 aquiherbosa
沼生栎 basket oak;cow oak;pin oak;swamp oak;white oak
沼生生物 helobios
沼生植物 helophyte
沼松 swamp pine
沼铁矿 bog iron ore;bog mine ore;bog ore;lake ore;marsh ore;meadow ore;morass ore;swamp ore
沼土 cripple;peat mo(u)ld;peat soil
沼穴 bog hole;pot-hole;rotten spot
沼油 <爱尔兰泥炭地所生的油脂状碳化氢> bog butter
沼油页岩 boghead cannel;boghead cannel shale;boghead shale
沼泽 aqua marsh;backswamp;bent;bog;carr;cienega;dismal;fell;marsh;marshal;mire;moor;morfa;mose;moss;muskeg;palus[复 pali];quagmire;slack;slew;slough;strode;swale;swamp;turfary;turf moor;fens <指英格兰东部沼泽地区>
沼泽桉 swamp mahogany
沼泽爆炸法 swamp shooting method
沼泽边缘物质 bog margin material
沼泽草本群落 emersiherbosa
沼泽草本相 marsh herbaceous facies
沼泽草地 marshy meadow;marshy pasture;everglade <用复数形式,尤指佛罗里达州南部的大沼泽>
沼泽草地海岸 marsh-grass coast
沼泽草甸 swamp meadow
沼泽草原 grass moor
沼泽层 marsh formation
沼泽车 swamp buggy
沼泽沉积矿床 swamp sedimentary deposit
沼泽沉积上的滩脊 Chenier
沼泽沉积物 marsh deposit;marsh sediment
沼泽沉积作用 marsh sedimentation
沼泽城堡 <爱尔兰在湖边建的> crannog
沼泽船 swamp boat
沼泽丛林 swamp muck;swamp woodland
沼泽岛 marsh island
沼泽的 callow;logged;paludal;paludine;swampy
沼泽的生成 formation of bog;formation of marsh
沼泽低地 slash;swale
沼泽地 beach swale;bog ground;bog land;curragh;everglade;fen(land);glade;hag;marshy area;marshy ground;marshy land;moorland;moss-land;muckland;organic terrain;overwet land;quag;quagmire;slump;swamp(land);wetland;morass;swampy ground;slue <美>
沼泽地爆破 blasting of peat
沼泽地爆破挤淤筑路法 swamp shooting
沼泽地表层 top moor layer
沼泽地菜园 muck garden
沼泽地带 carr;cripple;marsh land
沼泽地带使用的车辆 swamp buggy
沼泽地道路 swamp road
沼泽地的 marshy
沼泽地的封闭小湖 tarn
沼泽地的疏干沟 leam
沼泽地改良 bog reclamation
沼泽地垦殖 bog reclamation
沼泽地犁 marsh plough
沼泽地履带拖拉机 swampy caterpillar tractor
沼泽地泥灰岩 bog marl
沼泽地泥炭 fen peat
沼泽地排水 bog drainage;swamp drainage
沼泽地排水沟 <英> leam
沼泽地气味 swampy odor
沼泽地区 boggy country;marsh land;marshy district;swampy district
沼泽地渠道 bog canal
沼泽地生的 helobious
沼泽地(似)的 founderous
沼泽地土壤 fen soil;moorland soil
沼泽地推土机 bulldozer swamp;swamp bulldozer
沼泽地挖掘机 marsh excavator
沼泽地围垦 empoldering of marshland
沼泽地小船 swamp boat
沼泽地用车辆 swamp buggy
沼泽地用开沟器 fen colter
沼泽地用轮胎 marsh tire[tyre]
沼泽地用汽车 marsh buggy;swamp buggy
沼泽地炸开孔穴填筑路堤法 underfill method
沼泽地质作用 geologic(al)process of marsh
沼泽堆积 swamp deposit
沼泽鳄 mugger
沼泽浮游生物 heloplankton
沼泽腐泥(土) swamp muck;bog muck
沼泽腐殖土 bog muck;marsh muck;swamp muck;swamp woodland
沼泽革木 leatherwood
沼泽沟 swamp ditch
沼泽灌丛 swamp thicket
沼泽过程 swamp process
沼泽河 <其河漫滩为沼泽> swamp river;swamp stream
沼泽湖 bog lake;dystrophic lake;swamp lake
沼泽化 bogging;paludification;swamping
沼泽化产物 swamping product
沼泽化的 boggy;sloughy
沼泽化地 swampy land
沼泽化泥炭地 drag turf
沼泽化泻湖相 swamping lagoon facies
沼泽环境 swamp environment
沼泽荒地 curragh;marshy waste land
沼泽荒漠 travesias
沼泽灰泥 bog marl
沼泽灰壤 marshy podzol
沼泽混交栎林 swamp sugrudok
沼泽径迹测量 track survey on bog
沼泽类型 swamp type
沼泽林 bagon;sund(a)ri;swamp forest
沼泽隆起 bog burst
沼泽履带拖拉机 swampy caterpillar tractor
沼泽木质相 swamp woody facies
沼泽泥灰土砖 marsh marl
沼泽泥灰岩 marsh marl
沼泽泥流 bog flow
沼泽泥煤 marsh peat
沼泽泥滩 darg
沼泽泥炭 bog muck;marsh peat
沼泽泥炭土 fen peat;bog peat
沼泽泥土 muskeg
沼泽排水 bogaz;bog drainage
沼泽排水沟 swamp ditch
沼泽平原 muck flat
沼泽区 swamp(y)area
沼泽群落 fen;helic;helium;limnodium;swamp community
沼泽热 swamp fever
沼泽森林泥炭 swamp-forest peat
沼泽森林群落 helophylium
沼泽沙堆 Chenier
沼泽沙丘 chenier
沼泽生态型 swamp ecotype

沼泽生态学 marsh ecology
沼泽生物 helobios
沼泽湿地 bog ground; marsh land; swamp land
沼泽湿地蒸发 evaporation from swamp and wetted area
沼泽石灰 bog lime
沼泽式(成土作用)bog type
沼泽疏干 bogaz;bog drainage
沼泽疏林地 helorgadium
沼泽疏林区 helodium
沼泽水 boggy water; bog water; swamp water
沼泽水的补给类型 charge type of swamp water
沼泽水文调查 hydrologic(al) survey of swamp
沼泽松林 swamp pine forest
沼泽土化 bogging
沼泽土(壤)bog earth;bog soil;histosol;moorland soil;moor soil;fen soil; marsh soil; marshy ground; swamp(y) soil
沼泽洼地 < 沙丘之间的 > niaye
沼泽味 bog taste
沼泽物质 bog material
沼泽物质样品 swamp material sample
沼泽相 swamp facies
沼泽小岛 hammock
沼泽形成 marsh forming process; swamp formation
沼泽形成作用 swamp formation
沼泽性 bogginess
沼泽性土(壤) boggy soil; marshy ground;marshy soil
沼泽淤泥 bog lime; boglime lake marl;freshwater marl
沼泽园 bog and marsh garden
沼泽原生有机质 swamp autochthonous organic matter
沼泽植被 marsh vegetation;mire vegetation;swamp vegetation
沼泽植丛群落 helodric
沼泽植物 bog plant; ericelal; helad; limnocryptophyte;marsh plant;pelophyte
沼泽植物群落 hydrophytium
沼泽中的高地 hummock
沼泽中的硬地 hag
沼泽中干地 hope
沼泽种类 limnophilus
沼泽状的 boggy

召

召回通知书 notice of recall

召集 convene;convocation
召集会议 convoke
召集人员 call persons
召开会议的函件 note of convocation

兆

兆安 mega-ampere;megampere

兆巴 mega bar
兆比特 megabit
兆比特/秒 megabits/second
兆泊 < 黏[粘]滞度单位 > megapoise
兆达因 < 力的单位,1 兆达因 = 100 万达因 >【物】megadyne;megavolt
兆电子伏特 mega electron-volt; million electron-volt
兆吨 megatonnage
兆吨(百万吨)当量 equivalent megatonnage
兆伏 megavolt
兆伏安 megavolt ampere
兆伏(特)计 megavoltmeter
兆赫 megacycles per second; megahertz
兆亨(利)megahenry
兆加仑/日 mega gallons per day
兆焦(耳) < 1 兆焦(耳) = 100 万焦(耳) > megajoule
兆居里 megacurie
兆卡 < 热量单位,1 兆卡 = 10^6 卡 = 4.184×10^6 绝对焦耳 = 3967 英热单位 > thermie;megacalorie
兆克 Mg[megagram]
兆克力 megagramme force
兆克力米 megagramme force meter
兆拉德 megarad
兆力线 < 磁通单位 > megaline
兆立方米 mega cubic(al) meter;megastere
兆米 megameter
兆牛顿 meganewton
兆欧表 megameter; megohmmeter; tramegger;megger
兆欧计 earth(o) meter; megameter; megger
兆欧(姆)megaohm
兆千瓦小时 gigawatt hour
兆升 megaliter[megalitre]
兆升/日 megaliters [megalitres] per day
兆瓦 megawatt
兆瓦日 megawatt day
兆瓦时 megawatt hour
兆位 megabit memory; megabit storage
兆位存储器 megabit memory; megabit storage
兆兆 tera
兆兆位 terabit
兆周 megacycle
兆周每秒 megahertz
兆周/秒 megacycles per second; megahertz
兆字节 megabyte

赵

赵石墨 chaoite

赵州桥 < 又名安济桥 > Zhaozhou Bridge

照

照比例 draw to scale

照壁 screen wall
照标 sight vane
照查锁闭【铁】 check locking; traffic locking
照查锁闭继电器 traffic locking relay
照查锁闭握柄 check lock lever
照查条件 check requisition
照查握柄 check lever
照常规办理单独海损免赔额 average accustomed;customary average
照尺度翻样 draw to scale
照尺度放样 draw to scale
照尺度制图 draw to scale
照尺员 rodman
照单发货 shipment on order
照地灯 dipper
照地址指令格式 single address instruction format
照度 illuminating power; illumination; illumination intensity; intensity of illumination;intensity of light;light intensity; luminosity; luminous flux density
照度标准 illumination standard;lighting standard; standard of illumination
照度表 lux(o) meter
照度测定 illumination measurement; illumination photometry
照度单位 unit of illumination
照度等级 level of illumination
照度分布 aperture illumination
照度分配图 isolux diagram
照度横向分布 lateral light distribution
照度级 illuminative level;lighting level;illumination level
照度计 illumination meter;illuminometer; light meter; lumen meter; lumeter;luminance meter;luminometer;lux ga(u)ge;lux(o) meter
照度计算 illumination calculation
照度均匀度 illuminance uniformity; uniformity of illuminance
照度均匀系数 coefficient of even lighting
照度平衡 candle balance
照度曲线 illumination curve
照度衰减 fall-off in illumination
照度水平分布 lateral light distribution
照度系数 illumination factor; luminance factor
照度仪 illuminometer
照度阈值 threshold of illuminance; threshold of illumination
照付 hono(u)r draft;meet draft;pay draft;protect draft
照付不议 take or pay[TOP]
照付薪资的产假 paid maternity leave
照付薪资的休假 paid vacation
照管房屋的工友 janitor
照管人员 attending personnel
照惯例的 customary
照光玻璃 illuminating glass
照光检查 candling egg
照光频谱 irradiation spectral
照级配曲线配料 proportioning by grading charts
照价付款 pay according to the arranged price
照价赔偿 compensation according to cost
照旧的 unaltered
照来样制成样品 counter sample
照例 as customary;by usage
照亮 illuminate; illuminating; illumine;kindle;lighten;light up
照亮半球 illuminated hemisphere
照亮的目标 illuminated target
照亮面积 illuminated area
照亮圆面 illuminated disk
照料 attendance;oversee;take care about
照料旅客 attendance to passengers
照料人员 minder
照秒发亮 illuminate second flashing
照明 brightening;illuminating;lighten(ing);lighting
照明本领 illuminating power
照明比 light ratio
照明壁 diffusing wall
照明臂板 illuminated blade
照明变压器 lighting transformer
照明标志牌 light sign board
照明标准 illumination standard;lighting standard
照明标准电气设备 electric(al) standard equipment for lighting
照明表盘 illuminated dial
照明表示盘 illuminated diagram
照明玻璃 illuminating glass;illumination glass;lighting glass
照明玻璃制品 illuminating glassware; lighting glass ware
照明布线 lighting wiring
照明布置图 lighting arrangement
照明材料 light material
照明测量 illumination measurement
照明插座 light point
照明产热 heat generated by lighting
照明长方形阅读放大镜 illuminated rectangular reader
照明车 < 工地用 > mobilite
照明出线口 lighting outlet
照明窗 illuminating window; lighting aperture
照明带 light band;strip lamp
照明单位 lighting unit;unit of illumination
照明弹 flare; flare bomb; illuminated rocket;illuminating flare;illuminating projectile; light shell; missile; star shell; target indicator; marker < 英 >
照明弹架 flare adapter
照明的 illuminant
照明的均匀性 uniformity of illumination
照明灯 exciter lamp;head lamp;headlight;illuminating lamp;illumination light; illuminator; lighting burner; ring lighting
照明灯管 < 带屏蔽框和屏蔽膜的 > valance lighting
照明灯具 lighting device
照明灯聚光镜 light condenser
照明灯泡 lighting bulb
照明灯艇 gas boat
照明灯油 illuminating kerosene;middle distillate
照明灯罩 lighting cover
照明灯柱 lighting standard
照明等级 level of illumination; lighting level
照明点 lighting point
照明电度表 lighting meter
照明电费表 lighting tariff
照明电价表 lighting tariff
照明电缆 cable for lighting; lighting cable
照明电缆管道 light conduit
照明电缆盘 lighting cable drum
照明电流 illumination current; lighting current
照明电流发动机 lighting current motor
照明电路 lamp circuit;light(ing) circuit
照明电路图 illuminated circuit diagram
照明电线 lighting wire
照明电源 illuminating power;lighting mains;lighting supply source;mains lighting supply
照明电源插座 lighting outlet
照明电源箱 lighting power box
照明电源引出线 lighting outlet
照明吊棒 illuminated batten
照明吊顶 luminous ceiling
照明顶棚 diffusing ceiling;illuminated ceiling;luminous ceiling
照明度 illuminated scale; optic(al) exposure; illuminance; illuminating value
照明度盘式安培计 illuminated dial ammeter
照明对比 lighting contrast
照明对比度 lighting contrast ratio
照明发电机 lighting dynamo; service dynamo
照明法 illumination
照明范围 illumination zone
照明方法 means of illumination; means of lighting
照明方式 lighting system;mode of illumination
照明方向 direction of illumination
照明方向图 primary pattern
照明放大镜 illuminated magnifying

glass
照明费率 lighting tariff
照明费(用) cost of light; lighting cost; lighting expenses
照明分量 illumination component
照明分配 illumination distribution
照明峰荷 lighting peak
照明符号 illuminated symbol
照明负荷 electric(al) lighting load; illumination load; light(ing) load(ing)
照明负载 lighting load(ing)
照明干线 lighting mains
照明干线系统图 diagram for lighting feeder
照明钢柱 steel lighting column
照明杠杆开关 lighting lever switch
照明高度 illumination height; illumination level; illuminator level
照明格板 lighting panel
照明工程师 illumination engineer; lighting engineer
照明工程学 illuminating engineering; illumination engineering; lighting engineering
照明工程学会 <美> illuminating Engineering Society
照明工作人员 light man
照明功率 illuminating power; illumination power; lighting power
照明供暖 lighting and heating
照明构成部分 lighting component
照明光度学 illumination photometry
照明光束 illumination bundle
照明光线 illuminating ray
照明光源 illumination source; source of illumination
照明轨道模型 illuminated track model
照明过渡段 lighting transition section
照明过压保险装置 lightning arrester
照明函数 grading function
照明耗量 lighting consumption
照明护柱 illuminated guard-post
照明环 ring lighting
照明环境 light condition
照明火箭 illuminating rocket; light rocket
照明机 flare-aircraft
照明计 luminometer
照明计算 illumination calculation; lighting calculation
照明计算法 method of calculating lighting
照明技术 lighting engineering; lighting technique
照明剂 illuminant
照明尖峰 lighting peak
照明监督 light(ing) director
照明渐减调整 fader control
照明渐弱调整 fader control
照明角度 angle of illumination; angle of lighting
照明接地线 ground terminal
照明景象 lighting display
照明镜 illuminated mirror; illumination mirror
照明均匀度 uniformity of illumination
照明开关 illumination switch; light(ing) switch
照明开关导线 light connector cable
照明开关手柄 lighting switch handle
照明开关箱 lighting switchboard
照明开关肘节 light switch toggle
照明开关装置 lighting switchgear
照明开始 light on
照明控制 illumination control; lighting control
照明控制板 illuminated graphic(al) panel
照明控制带 light control tape
照明控制继电器 lighting control relay
照明控制开关 bull switch
照明控制盘 lighting control panel; visual control panel
照明控制器 lighting controller
照明控制室 illumination control room; lighting control room
照明控制特性 lighting control characteristic
照明控制箱 lighting control box
照明口镜 illuminating mouth mirror
照明馈路 lighting feeder
照明冷负荷 cooling load of light
照明率 light illuminating factor
照明脉冲 brightening pulse; intensifier pulse; intensifying gate; sensitizing pulse
照明煤气 blue gas; illuminating gas
照明模拟图 illuminated mimic diagram
照明能力 illuminating power; lighting strength
照明排灯 lighting row
照明盘 illuminated panel; illuminated screen
照明跑道 <使飞机在夜间安全降落> flare path
照明配电 light distribution; lighting sub-distribution
照明配电盘 distribution board for lighting; illuminated diagram switch board; lighting distribution panel; lighting panel
照明配电室 lighting distribution room
照明配电箱 distribution box for lighting mains; lighting distribution box; lighting power distribution panel
照明配件 light(ing) fitting
照明配线板 lighting panel
照明配线盘 lighting panel
照明平面图 layout of lighting
照明屏 illuminated screen
照明气 lighting gas; luminous gas
照明器 illuminating apparatus; illuminator; lamp; lighter; lighting unit; luminaire; luminary
照明器材 lighting fixture
照明器分级 luminaire classification
照明器具 lighting apparatus
照明器皿 illuminating glassware
照明器效率 luminaire efficiency
照明强度 illumination intensity; intensity of illumination; lighting intensity; strength of illumination; luminous intensity
照明强度差异比 diversity ratio of illumination
照明墙 illuminated wall; illuminating wall
照明切换箱 lighting changeover box
照明区(域) illuminated area; light area; light sector
照明曲线 illumination curve
照明散热量 heat gain from lighting
照明设备 illuminating equipment; illumination; illumination device; illumination equipment; illumination installation; lighting equipment; lighting facility; lighting fitting; lighting fixture; lighting installation; lighting instrument; lighting plant; luminaries; means of illumination; means of lighting; set light; luminaire <包括灯泡、反射器及附属装置>
照明设备折旧摊销及大修费 lighting plant depreciation apportion and overhaul charges
照明设备制造商协会 <英> Lighting Equipment Manufacturer's Association
照明设计 illumination design; lighting design; lighting layout; lighting set-up
照明设施 lighting facility
照明时间 lighting hour; lighting time; lighting-up time
照明时期 lighting period
照明式标志 <用外部光源照明> illuminated sign
照明式轨道显示图 illuminated track diagram
照明式镜面推拉窗 illuminated mirror slide
照明式区域位置示意图 illuminated sectional directory
照明式示向标志 illuminated indicator
照明式室内喷泉 illuminated indoor fountain
照明式天花板 illuminated ceiling
照明式围栏 illuminated barrier
照明视镜 lighting sight glass
照明适应段 lighting adaptation section
照明收费表 illumination tariff
照明束 primary beam
照明双针气压计 illuminated duplex air ga(u)ge
照明水平 level of illumination
照明损耗 light loss
照明塔 illumination tower; lighting mast; lighting pylon; lighting tower
照明炭棒 illuminating carbon
照明体 illuminant; illuminator
照明天花板 diffusing ceiling
照明天平 illuminated ceiling
照明天桥 catwalk; lighting bridge
照明天线 primary antenna
照明条件 illumination condition; lighting condition
照明调节 light conditioning; light control
照明调节器 light(ing) regulator
照明停车标 illuminated stop sign
照明通风两用筒 lighting and ventilating shaft
照明透镜 illuminating lens
照明图 illuminated diagram
照明图表 light chart; light plot
照明图示板 illuminated graphic(al) panel
照明网络 lighting mains; lighting network
照明网络稳压器 lighting voltage regulator
照明网天线 mains antenna
照明桅 lighting mast
照明卫生要求 sanitary requirement of illumination
照明系缆柱 illuminated bollard
照明系数 illumination factor; light illuminating factor; luminaire efficiency; luminosity factor
照明系统 illuminated system; illuminating system; illumination system; light(ing) system
照明系统防爆固定件 explosion-proof fixture
照明显示器 luminous display
照明线 illuminating line; lighting line
照明线路 illuminating circuit; illumination circuit; lighting circuit; lighting cable
照明效果 illuminating effect
照明效率 illuminating efficiency; illumination efficiency; lighting efficiency; luminous efficiency
照明效应 illuminating effect; lighting effect
照明信号 illumination sign
照明信号标志 illuminated signal sign
照明信号圆牌 illuminated signal disc
照明信号圆盘 illuminated signal disc
照明需量 lighting demand
照明学 illumination
照明验布机 illuminated inspection machine
照明仪器 illuminating glassware
照明艺术 lighting art
照明因数 illumination factor
照明用导线 lighting lead
照明用灯 illuminator lamp
照明用电动发电机组 lighting motor-generator set
照明用电缆 illumination cable
照明用电线 cable for lighting
照明用发电机 lighting dynamo; lighting generator
照明用分支电路 lighting branch circuit
照明用户 lighting consumer
照明用户电度表 lighting meter
照明用煤气灯 luminous flame burner; luminous gas burner
照明用油 illuminating oil
照明用支路 lighting branch circuit
照明油 signal oil
照明元件 illumination component
照明员 lighting man
照明源 light source
照明值 illumination value
照明质量 illuminating quality; quality of lighting
照明中断 blackout
照明柱 lamp post
照明装置 equipment for illumination; illuminating device; illumination installation; illuminator; intensifier; lighter; light fitting; lighting apparatus; lighting attachment; lighting device; lighting facility; lighting fixture; lighting installation; lighting plant; lighting set; lighting unit; light installation; set light
照明装置电缆管道 lighting fixture raceway
照明总开关 central lamp switch
照明组件 illumination component
照目标利润定价 target-rate-of-return pricing
照片 camera record; finished print; image picture; paper print; photo(graph); portrait; vectograph <用偏光眼镜看的>; Kodak <用小型照相机拍的>
照片比例(尺) photographic(al) scale; scale of the photograph; contact scale; image scale; picture scale
照片档 photograph collection
照片导线 photo-polygonometric(al) traverse; photo polygon
照片地平线 ground trace; horizon trace
照片地图 photomap
照片放大器 projection printer
照片分辨力 image resolution; picture resolution
照片分色 colo(u)r separation of photograph
照片分析 photo analysis; photographic(al) analysis
照片迹线 picture trace
照片基准面 photographic(al) datum plane
照片记录法 photographic(al) record
照片记时仪 photochronograph
照片解释 photointerpretation
照片纠正仪 photo transformer
照片控制点 supplemental control point; supplementary control point
照片控制点定位法 photographic(al)

control
照片连测点 picture control point
照片略图 composite photograph
照片拍摄范围 photographic(al) coverage
照片盘 photo carriage; photo carrier plate; picture carrier; pix carrier; vision carrier
照片判读 photointerpretation; photoreading
照片判读仪 photopret; Topopret < 德国蔡司制造 >
照片判读员 photointerpreter
照片拼接 photomosaic
照片平面图 photoplan
照片倾角 photo tilt; picture tilt
照片三角测量 phototriangulation
照片三角测量航线 triangulated strip; triangulation strip
照片三角测量仪 phototriangulator
照片上的方位点 picture point
照片上的控制点 picture point
照片上色剂 toner
照片收藏室 photograph library
照片索引 print reference
照片天空部分 sky portion
照片镶嵌图 composite photograph; photomontage
照片野外控制点 picture control point
照片印晒过度 overprint
照片印刷 photographic(al) printing
照片制图 photocharting
照片质量 image quality; photographic(al) quality
照片重叠调节器 photographic(al) coverage regulator; photographic(al) overlap regulator
照片转绘仪 sketch master; Rectoplanograph < 商品名 >
照片锥形法 < 测定像片倾斜的一种解析方法 > photograph pyramid
照片资料馆 picture archive
照片组 block of photographs
照片坐标系统 plate coordinate system
照片坐标轴 photograph axis
照山红【植】micranthum
照射 bombardment; illumination; irradiation; radiate; radioactive bomb; raying
照射标记 exposure label(l)ing
照射不足 under-exposure
照射场 exposure field
照射处理 radiation treatment
照射到 beat-down on
照射递减效率 illumination taper efficiency
照射点 point of irradiation
照射反照率 exposure albedo
照射范围 range of exposures; range of irradiation
照射分解 radiation-induced decomposition
照射供暖 beam heating
照射固化涂层 irradiated coating
照射管 exposure tube
照射光束 illuminating beam; illumination beam
照射光源 radiation source
照射过的 irradiated
照射盒 irradiation capsule
照射盒开盒装置 irradiation capsule opener
照射积累因子 exposure buildup factor
照射剂量 expose dose; exposure dose; irradiation dosage; irradiation dose
照射剂量率 exposure dose rate
照射角 illumination angle
照射校正 irradiation correction
照射聚乙烯 irradiated polyethylene
照射孔道 exposure hole; irradiation hole
照射量 exposure
照射量计 exposure meter
照射量率 exposure rate
照射量率常数 exposure rate constant
照射量率计 exposure ratemeter
照射量因子 exposure factor
照射量指示器 exposure indicator
照射灵敏度 illumination sensitivity
照射路径 exposure pathway
照射面 illumination surface
照射面积 irradiated area
照射面积剂量监视器 areal dose monitor
照射评价 assessment of exposure
照射器 irradiator
照射腔 exposure cavity
照射强度 exposure intensity
照射容器 exposure cage; exposure container; irradiation container
照射杀菌 irradiation
照射栅格 exposure cage
照射伤害 irradiation injury
照射时间 exposure time; irradiation time; time of irradiation
照射时间表 exposure chart; exposure time table
照射时间范围 exposure time range
照射室 exposure cell
照射衰减器 exposure dimmer
照射塑料 irradiated plastics
照射损伤 irradiation damage
照射通量密度 radiant emittance
照射透入 penetration of radiation
照射吸收 absorption of radiation
照射效应 illumination effect; irradiation effect; radiation effect
照射野 irradiation field; radiation field
照射业务 irradiation service
照射引发 radiation-initiated
照射诱导 radiation-induced
照射装置 irradiation unit
照射作用 radiation
照时显示 illuminate time display
照市价 at the market
照先例 follow suit
照相 photographing
照相凹版油墨 gravure ink
照相凹版 conventional gravure
照相凹版雕刻 gravure engraving
照相凹版法 photogravure
照相凹版胶印机 gravure offset printing press
照相凹版轮转印刷 rotogravure
照相凹版式涂敷 gravure coating
照相凹版印刷 gravure printing; photogravure; photogravure printing
照相凹版印刷机 gravure machine
照相凹版印刷术 gravure
照相暗室 photographic(al) dark room; photographic(al) laboratory
照相凹版制版法 gravure process
照相凹版转轮印刷 rotogravure printing
照相凹版自动腐蚀机 automatic gravure etching machine
照相版 photocopy; process plate; photoetch(ing)
照相版雕刻法 autogravure
照相版印刷纸 heliographic(al) paper
照相报告 photographic(al) report
照相比例尺 photo scale
照相壁画 photomural
照相材料 photographic(al) material
照相测光 photographic(al) photometry
照相测光仪 photodetector
照相测量 photographic(al) measurement
照相测量大气折射 photogrammetric(al) refraction
照相测量方法 photogrammetric(al) method
照相测量术 photogrammetry
照相测量学 photogrammetry
照相沉淀物分析 photosedimentation analysis
照相处理设备 photographic(al) processing equipment
照相磁像仪 photographic(al) magnetograph
照相存储 photographic(al) store
照相存储器 photographic(al) memory; photographic(al) storage
照相的 photographic(al)
照相底板 backing
照相底板用玻璃 glass for photographic(al) plate; photographic(al) glass
照相底版 photographic(al) plate
照相底片 negative; negative matrix; negative plate; photographic(al) film; photographic(al) negative; photographic(al) plate; photoplate
照相底片读数 photographic(al) plate reading
照相底片感光速度 speed of photographic(al) plate
照相底片漏过光的 light-struck
照相底片密度 photographic(al) density
照相底片探测 photographic(al) plate detection
照相地形测量 phototopography
照相地形图 photopography
照相雕刻版 photoglyph
照相定时器 photogrammetric(al) intervalometer
照相法 photographic(al) process
照相翻拍法 photocopying process
照相反照率 photographic(al) albedo
照相泛光灯 photoflood
照相放大 blow-up; photographic(al) amplification; photographic(al) enlargement
照相放大组 photographic(al) enlargement section
照相废液 photographic(al) waste
照相废银回收 photographic(al) silver recovery
照相分辨率 photographic(al) resolution
照相分光光度测量 photographic(al) spectrophotometry
照相分光光度法 photographic(al) spectrophotometry
照相分光镜 photographic(al) spectroscope
照相辐射点 photographic(al) radiant
照相负片 photographic(al) negative
照相复印 photographic(al) copy; photographic(al) reprint
照相复印机 Listomatic camera; photocopy machine; photorepeater
照相复制 photocopy; photo-duplicating; photographic(al) copying; photographic(al) facsimile; photostat(print)
照相复制本 photocopy; photoduplicate
照相复制法 photo-duplication
照相复制机 photocopier
照相复制品 photoduplicate; photoprint
照相复制设备 photo-reproduction means
照相复制图 photodraft
照相复制重印本 photographically reproduced reprint
照相感光 sensitization
照相感光材料 sensitive photographic(al) material
照相感光范围 photographic(al) spectrum
照相感光制版 photoengraving
照相高度 photographic(al) altitude
照相跟踪 photographic(al) tracking
照相工业废水 photographic(al) industry wastewater
照相工艺的 photomechanical
照相观测 camera observation; photographic(al) observation
照相馆 photographer studio; photostudio; studio
照相光变曲线 photographic(al) light curve
照相光度测量 photographic(al) photometry
照相光度计 photographic(al) photometer
照相光度术 photographic(al) photometry
照相光度学 photographic(al) photometry
照相航线 photographic(al) flight-line
照相化学 photographic(al) chemistry
照相化学腐蚀制造法 photofabrication
照相绘图机 photographic(al) plotter
照相机 camera; photographic(al) camera
照相机背盖 camera back
照相机机构 camera section
照相机机身 camera body
照相机架 camera-mount
照相机镜头 camera gun; camera lens
照相机聚焦镜 focusing camera mirror
照相机快门 camera shutter
照相机皮腔 camera bellows
照相机三脚架 camera crane
照相机扫描图形 camera-scan(ning) pattern
照相机视场 camera coverage
照相机视界 camera coverage
照相机外套 camera container; camera housing
照相机作用范围 camera coverage
照相计时器 phototimer
照相记录 photographic(al) recording
照相记录表 record table of photograph
照相记录法 photographic record
照相记录加速度仪 photographic(al) recording acceleration; photographic(al) recording accelerograph
照相记录器 photographic(al) recorder; photorecorder
照相记录仪 photographic(al) recording instrument
照相记录装置 photographic(al) sensory unit
照相记时仪 photochronograph
照相技术中队 photographic(al) technical squadron
照相剂量测定法 photographic(al) dosimetry
照相剂量学 photodosimetry
照相加厚 redevelopment
照相架次 photographic(al) sortie
照相减薄剂 photographic(al) reducer
照相检测法 photodetection
照相胶版 heliotype
照相胶版印刷 lithoprint
照相胶卷 photographic(al) film; roll film
照相胶片 pellicle; photographic(al) film

照相胶片存储器 film memory;film storage
照相胶片记录器 film recorder
照相胶片剂量计 photographic(al) film dosimeter
照相胶片佩章 photographic(al) film badge
照相胶印法 photo-offset
照相经纬仪 photo-theodolite;phototheodolite camera;phototransit
照相晶粒 photographic(al) grain
照相径迹 photographic(al) track
照相镜头 photographic(al) lens
照相镜头视角 camera angle
照相累积 photographic(al) integration
照相亮度 photographic(al) brightness
照相灵敏度 photographic(al) sensitivity
照相录声机 selenophone
照相滤器 photographic(al) filter
照相密度 photographic(al) density
照相密度计 photographic(al) densitometer
照相密度控制 photographic(al) density control
照相敏化剂 photographic(al) sensitizer
照相木版印刷术 photoxylography
照相排版 photoset;photosetting;phototypesetting
照相排版术 filmsetting
照相排字机 filmsetter;photocomposer;phototypesetter
照相判读 photographic(al) reading
照相判读报告 photographic(al) interpretation report
照相判读资料 photographic(al) interpretation data
照相片感光速率 speed of photographic(al) plate
照相偏振测量 photopolarimetry
照相拼版 photomontage
照相平版 aquatone;photolitho plate
照相平版印刷品 photolithograph
照相平版印刷术 photolithography
照相平印术 photolithography
照相器材 photographic(al) apparatus;photographic(al) equipment
照相潜像 photographic(al) latent image
照相枪 gun camera;photographic(al) gun
照相强化剂 photographic(al) intensifier
照相情报 photographic(al) intelligence
照相情报资料 photographic(al) intelligence data
照相全景 photographic(al) panorama
照相乳剂 photographic(al) emulsion
照相乳剂检测器 photographic(al) emulsion detector
照相乳剂粒度 photographic(al) graininess
照相乳胶 photoemulsion;photographic(al) emulsion
照相乳胶技术 photographic(al) emulsion technique
照相乳胶液 photographic(al) emulsion
照相软片 film
照相三角测量 photographic(al) triangulation
照相散光灯 overrun lamp;photoflood lamp
照相闪光灯 photoflash
照相设备 camera installation
照相设备舱 photographic(al) compartment
照相石版 photolithograph
照相蚀刻法 photoengraving
照相蚀刻术 photohyalography
照相式极谱仪 photographic(al) recording polarograph
照相室 photographic(al) studio;photography room;photostudio;studio
照相输出设备 photographic(al) output device
照相术 photography
照相数据存储器 photograph data memory;photographic(al) data memory
照相丝网印刷法 artogravure
照相速度 photographic(al) speed
照相缩版 photoreduction
照相缩小 photographic(al) reduction;photomechanical reduction
照相探测器 photographic(al) detector
照相天顶筒 photographic(al) zenith tube
照相天体测量学 photographic(al) astrometry
照相天体光度学 photographic(al) astrophotometry
照相天体光谱学 photographic(al) astrospectroscopy
照相天文学 photographic(al) astronomy
照相铜版 half-tone
照相铜版印刷油墨 half-tone ink
照相凸版 photoengraving;photoetch(ing)
照相凸版黑墨 half-tone black ink
照相凸版(术) phototype
照相凸版印刷 photoengraving
照相图像 picture image
照相望远镜 photographic(al) telescope;phototelescope
照相文献 photographic(al) documents
照相物镜 photographic(al) objective
照相吸收系统 photographic(al) absorption coefficient
照相洗印室 photographic(al) laboratory
照相显微分析 photomicrographic analysis
照相显微管 photomicroscope
照相显微镜 photomicroscope
照相显影两用机 processor-camera
照相显影液 soup
照相镶嵌图 photomosaic
照相消色差 photo-achromatism
照相效应 photographic(al) effect
照相锌版 photozincograph
照相锌版印刷品 photozincograph
照相星表 astrographic(al) catalogue;photographic(al) star catalog(ue)
照相星等 photographic(al) magnitude
照相星历表 photographic(al) ephemeris
照相星图 astrographic(al) chart
照相旋转技术 photographic(al) rotation technique
照相掩模 photographic(al) mask
照相药品 photographic(al) chemical
照相仪 photoinstrumentation
照相印花 photographic(al) printing
照相印刷 photographic(al) printing
照相印刷设备 photographic(al) printing unit
照相印刷术 autotype;phototypography
照相用闪光灯 photoflash lamp
照相原图 artwork master
照相噪声 photographic(al) noise
照相增感剂 photosensitizer
照相炸弹 photoflash bomb
照相正片 photographic(al) positive;positive matrix
照相织造 photographic(al) weaving
照相纸 gravure tissue;photographic(al) paper
照相纸曝光量范围 exposure scale
照相纸记录 photographic(al) paper recording
照相纸无光泽表面 mat surface
照相制凹版法 photogravure
照相制版 photochemigraphy;photoengraving;photographic(al) stencil;photomechanical plate making;photomechanical process;process engraving
照相制版厂 photomechanical plant
照相制版法 heliography;photomechanical
照相制版复制 photomechanical copying
照相制版工艺 photomechanical process;photomechanics
照相制版机 phototype machine
照相制版镜头 printer lens;process lens;reproduction lens
照相制版美术品 photomechanical art publication
照相制版网屏 half-tone screen
照相制版印刷版 photomechanical plate
照相制版印刷品 photomechanical print
照相制版印刷术 photomechanics
照相制版纸 photomechanical paper
照相装饰法 photographic(al) decoration
照相装置 camera device
照像的 photographic(al)
照样 follow suit
照耀的 radiant
照原保险单赔偿 pay as may be paid thereon
照原价偿还 refund the price
照准 aim(ing);collimate;collimation;ranging into line;sight(ing);take a shot
照准板 sight(ing) board;sight(ing) vane
照准标 alidade target;mire;sighting mark
照准标志【测】 sighting target
照准部偏心 eccentricity of alidade
照准部偏心差 eccentric error of alidade
照准部水准器 lower bubble;plate bubble;plate level;transversal bubble
照准部水准器格值 level scale of alidade
照准部水准仪 plate level
照准部微动螺旋 horizontal fine motion drive
照准部旋转 rotation of plate
照准部制动螺旋 alidade clamp
照准差 collimation error
照准觇标 take in transit;target in transit
照准尺 alidade rule;sight rule
照准灯标 stand on light beacon
照准地线 alignment wire;ground wire;screed wire
照准点 aiming point;collimated point;collimating point;laying point;pivot point;sighting point;target point;pivot station <两点间的>
照准点高程校正 reduction for the height of sighting point
照准点归心 reduction to target center[centre];sighting centering[centring]
照准点归心校正 reduction to the center[centre] of signal
照准杆 collimating pole
照准光线 collimated light
照准归心改正 diopter reduction
照准规 alidade rule
照准轨 sight rail
照准架 alidade
照准精度 pointing accuracy
照准目标相位差 phase error of sighting object
照准偏心 eccentricity of signal
照准器 alidade;diopter[dioptre];peep-sight;sighting gear;sighting vane;vane <罗盘等的>
照准设备 sighting device
照准十字丝 sighting wire;sight reticule;hairline;target wire
照准视线 collimation;collineation
照准丝 sighting wire;sight reticule
照准头 aiming head;sighting head
照准误差 error of sighting;pointing error;sighting error
照准线 aiming line;collimation axis;sighting line;sight line;transit line
照准线法 <一种矿井垂直投影法> target-line-method
照准线距离测量 measured of line of sight distance
照准仪 collimator;cross staff;diopter[dioptre];sight alidade;sight rule;alidade;surveyor's alidade
照准仪度盘 alidade circle
照准仪分度器 alidade protractor
照准仪分划 diopter scale
照准仪罗盘 compass with diopter
照准仪偏心 eccentricity of alidade
照准仪视距法 alidade stadia method
照准圆 aiming circle
照准圆筒 sighting cylinder
照准圆柱(觇标) cut-off cylinder
照准轴 aiming axis;collimation axis;axis of sight
照准主点 collimated principal point
照准装置 sighting device

罩

罩 boot;calotte;canopy;clothing;cowing;dome;housing;inclosure;jacket;lantern

罩杯式感应电动机 drag-cup motor
罩杯式转速发电机 drag-cup tachogenerator
罩布 casing;cover;cover lid;hood;jacket;tidy
罩布模具 drape mo(u)ld
罩灯壳 cowl lamp case
罩灯泡 cowl lamp bulb
罩垫 hood pad
罩盖 mantle;shrouding;canopy <圣坛、神龛上的>;lookum <行车或起重滑轮上的>
罩盖程度 hood wrap
罩盖法试验 hood test
罩盖披檐 <圣坛、神龛上的> canopy lip
罩冠 veneer crown
罩冠固位体 veneer retainer
罩光 glazing
罩光喷漆 overlacquer
罩光漆 final coating;finish coat;varnish
罩光清漆 clear top(coating);coating varnish;flatting varnish
罩光涂层 overcoating;glaze coat
罩光油 overprinting varnish
罩环 shroud ring
罩极 consequent pole;shaded pole;shading-pole
罩极法 shaded pole method

罩极环 shading ring
罩极绕组 consequent poles winding
罩极式电动机 shaded pole motor
罩极线圈 shading coil
罩极原理 shaded pole principle
罩金属板条 metal cover strip
罩金属带状物 metal cover strip
罩壳 cover piece;encloser;lagging
罩壳心材 boxed heart
罩口风速 face velocity
罩框部分 mask part
罩帽 cup
罩面 blanket; coating; covering; facing; finish coat; ga(u)ged skim coat; mat coat; overlay; skin covering of the surface; surface coating; surmount; veneer on; overlay of pavement <路面>; finishing layer <路面>
罩面白灰 white coat(ing); white finish coat
罩面板 covering plate; skin plate
罩面材料 covering material; facing material; masking material; skin material
罩面层 outside coat; setting stuff; skim(ming) coat; skin(coat); wash coat; finishing coat; setting coat
罩面层(用的)石膏灰浆 ga(u)ging plaster
罩面骨料 cover aggregate
罩面光漆 French polish
罩面烘漆 stovewood varnish
罩面灰浆 thin-coat plaster; thin-wall(ed) plaster
罩面基层灰 brown coat
罩面胶(泥) finish cement; finish compound
罩面抗性试验 finish coat resistance test
罩面抹灰层 finishing coat
罩面抹灰料 finishing compound
罩面漆 finish; finish coat paint; finishing paint; outside coat of paint
罩面清漆 finishing varnish; glaze coat
罩面石膏(灰)浆 ga(u)ging plaster; ga(u)ged stuff
罩面石料 cover stone; cover-up
罩面石屑 cover chip(ping)s; cover screenings
罩面使用年限 overlay life
罩面使用寿命 overlay life
罩面涂层基层 undercoater
罩面屋 setting
罩面用树脂 overcoating resin
罩面终饰层【建】 finishing coat
罩盘 rosette
罩棚 awning
罩棚月台 sheltered platform
罩篷 dorsal
罩漆 clear lacquer; screening varnish
罩漆试验 recoating test
罩铅板 lead-sealed sheets
罩嵌条钩扣 hood mo(u)lding fastener
罩衫 blouse; overall; smock
罩式对流器 cabinet convector
罩式干燥器 apron drier[dryer]
罩式机车 bonnet type locomotive
罩式炉 bell furnace; bell-type furnace; cap cover furnace; cover furnace; portable cover furnace; tilting furnace; top hat furnace
罩式炉退火 cover annealing
罩式滤池 mantle filter
罩式滤器 mantle filter
罩式配电盘室 switchgear room
罩式退火炉 hood-type annealing furnace; top hat annealing furnace
罩锁钩 hood lock hook
罩胎轮 tire[tyre] wheel
罩套 envelope of hood
罩套构件 mantle piece
罩瓦 bonnet hip tile; bonnet tile
罩衣 slop
罩印 cover printing
罩在光亮的底层漆或金属表面上的光亮清漆或有色透明漆 flamboyant finish coating
罩子 bell cot; bonnet; encasing
罩子钩扣 hood fastener
罩子挂钩 hood fastener

肇

肇事者 culprit

遮

遮暗 blind
遮暗设备 darkening plant
遮板 blinder; casing; cover; curtain board; screen board; shingle; shrouding
遮被土壤 shading soil
遮蔽 covering; cover shadow; obscuration; occultation; overshadow; screening; shade; shading; shadowing; shield
遮蔽板 baffle
遮蔽舱壁 screen bulkhead
遮蔽的 obscure; housed
遮蔽的钻井架 enclosed(drilling) derrick
遮蔽灯 obscured light
遮蔽地 dead land; ground cover; natural cover; sheltered ground
遮蔽构型 eclipsed conformation
遮蔽技术 masking technique
遮蔽剂 covering reagent; masking agent; masking compound
遮蔽甲板 shelter deck
遮蔽甲板船 shelter decker; shelter deck ship; shelter deck vessel
遮蔽甲板空间 shelter deck sheerstrake; shelter deck space
遮蔽胶带 masking tape
遮蔽角 blind angle; masking angle
遮蔽孔 masking aperture
遮蔽孔隙 shelter pore
遮蔽框架 masking frame
遮蔽露头 incrop
遮蔽面积 blighted area; dead area
遮蔽膜片 masking film
遮蔽能力 screening capacity
遮蔽屏 shadow mask
遮蔽器 eclipser
遮蔽扇形 obscured sector
遮蔽式泊位 covered berth
遮蔽式船台 covered shipbuilding berth
遮蔽式救生艇 housed-in lifeboat
遮蔽式码头 covered terminal
遮蔽式桥梁 closed bridge
遮蔽试验 sheltering test
遮蔽水域 sheltered waters
遮蔽位置 dead position
遮蔽物 cover-device; housing; hovel; hulk; veil; shelter
遮蔽系数 sheltering coefficient; sheltering factor; shielded factor
遮蔽线圈 shielding coil
遮蔽效应 blackout effect; capture effect; masking effect; screening effect
遮蔽信号 masked signal
遮蔽信号发生器 mask signal generator
遮蔽栽植 screen planting
遮蔽阵地 position defilade; sheltered position
遮蔽指示剂 screened indicator
遮蔽组灯 group occulting light
遮蔽作用<指裂缝> blighted effect; bridging effect
遮槽 drip cap
遮窗 jalousie
遮弹层 bomb protection layer ignition; layer against bomb
遮挡百叶 sight proof louver; vision-proof louver
遮挡板 edging board
遮挡的 shaded
遮挡灯光架的横幕 concert border
遮挡类型 type of barrier
遮挡式主机 screened host
遮挡涂漆 cutting-in
遮掉 dropout
遮顶<船、车、帐篷、地摊等> tilt roof
遮断 barrage; blocking out; breaking; interrupt; interruption; obstruction; occlude; rupture
遮断操作 rupturing operation
遮断电流 breaking current
遮断阀 cut-off value; cut-out valve; intercepter valve; shut-off valve
遮断功率 interrupting capacity; rupturing capacity
遮断功率试验 rupturing capacity test
遮断角 angle of obstruction
遮断控制指示器 cut-off control indicator
遮断膜片 blowout diaphragm
遮断能力 breaking capacity; closing capacity
遮断器 breaker; interceptor
遮断容量 breaking capacity; closing capacity; load-break rating; rupturing capacity
遮断容量试验 rupturing capacity test
遮断塞门 cut-off cock; cut-out cock; isolating cock; shut-off cock; shut-out cock
遮断塞门手把 cut-out cock handle
遮断射流传感器 interruptible jet sensor
遮断式传感器 interruptible sensor
遮断位置 derailing position; lap position
遮断物 intercepter[interceptor]; interruption
遮断信号按钮 monoindication obstruction signal button
遮断信号机【铁】 monoindication obstruction signal
遮断纸 barrier paper
遮断装置 cut-off; tripping device
遮风 bonnet
遮风屏 wind shield
遮封能力 holdout
遮缝带 joint masking tape
遮盖 deception; hide; mask; overspread; under cover
遮盖半圆饰 overlapping astragal
遮盖本领 hiding power
遮盖不足 underlap
遮盖材料 covering material
遮盖的楼梯梁 closed string stair(case)
遮盖的楼梯斜梁 closed string(er)
遮盖顶棚 covered ceiling
遮盖防护 canopy-protected
遮盖格栅 cover grate
遮盖经济性 hiding economy
遮盖力 capacity of coverage; coating ability <油漆等的>; coverage power; dry capacity; masking power; hiding power <油漆等的>; opacity hiding power <油漆的>
遮盖力测定计 coverimeter; hidimeter; cryptome(te)r
遮盖力测定仪 cryptome(te)r
遮盖力黑白格板 hiding power black and white checker board
遮盖力计 criptome(te)r
遮盖力弱 lack of hiding
遮盖力试验纸 brushout cards; hiding power chart
遮盖力值 hiding power value
遮盖列板 covering strake
遮盖炉箅 cover grate
遮盖木节 knotted over
遮盖能力 covering capacity; healing power; masking power; covering power
遮盖漆 concealment paint; masking paint; screening varnish
遮盖天花板 covered ceiling
遮盖物 cope; housing; overcover; pall; cover
遮盖颜料 prime pigment
遮盖养护 shading cure
遮盖罩 covering cap
遮盖(纸)带 masking tape
遮盖着的 covered
遮盖作物 cover crop
遮光 anti-glare; cut-off for lighting; dim-out; dodging; shading
遮光百叶窗 dark jalousie; light-proof louver
遮光板 anti-dazzle device; baffle; bezel; blinder; diaphragm; dowser[douser]; gobo; kitchen hood; light screen; light slide; nigger; occluder; orifice; shade; spill shield; sun louver[louvre]; sun shield; sun visor [viser]; hood
遮光板支撑 shutter dog
遮光保护 glare protection
遮光玻璃 anti-dazzle glass; anti-sun glass; front glass; muffled glass; shade glass; solar control glass; sun glass
遮光彩色负片 masked colo(u)r negative film
遮光窗 dark window; light-tight window
遮光窗窗钩 shutter dog
遮光窗的支撑 shutter stay
遮光窗活络板 shutter flap
遮光窗铰链 shutter flap
遮光窗帘 blackout blind; shade; window blind; window shade blind
遮光挡板 blackout blind
遮光的 light-proof; light-tight
遮光固定百叶窗 light-tight jalousie
遮光黑布 gobo
遮光滑板 dark slide
遮光剂 opacifier; opalizer; sun screen drug
遮光角 angle of obstruction; cut-off angle; lamp shielding angle; shielding angle
遮光卷帘 light-proof blind
遮光控制钮 dimmer knob
遮光框 framing mask; masking frame
遮光帘 anti-glare shading; light shade; shades; window shade
遮光滤光器 barrier filter
遮光铝百叶 alumin(i)um slatted blind
遮光率 shading coefficient
遮光门 dark door
遮光模 photomask
遮光膜<贴在玻璃上的> solar control film
遮光幕 light intercepting curtain
遮光盘 blanking disc[disk]
遮光棚 skylight; skylight visors; visors
遮光篷 shade
遮光片 anti-dazzle screen; anti-dazzling screen; gobo
遮光频率 chopper frequency
遮光屏 anti-glare shading; bezel; blanking screen; eclipser; shading screen; spill shield

遮光漆 blackout paint;shade paint
遮光器 dimmer;dimmer device;light chopper;light dimmer;photochopper;tinting shade
遮光墙 sun blind wall;sun screen wall
遮光墙基础 footing of sun screen wall
遮光栅 skylight window;spill shield
遮光栅格 louver[louvre]
遮光栅格天棚 louvered ceiling
遮光栅终点 end of sun screen
遮光摄影 fading
遮光提灯 dark lantern
遮光体 occulter
遮光筒 shade tube
遮光涂层 anti-sun coating;sun shielding layer
遮光涂料 opaque;opaque paint
遮光物 blind;light shade;shade
遮光系数 opacity factor;shading coefficient;vignetting factor
遮光系统 light-tight system
遮光效果 light-proofness;obscuration;shaded effect
遮光性 light-proofness
遮光眼罩 eyeshade
遮光叶片 cover wing;cutting blade;cutting wing;masking blade;cover blade;cutting blade
遮光栽植 anti-glare planting;light screen planting
遮光罩 compendium[复 compendiums/compendia];glare shield;high hat;hood;lens hood(case);lens shade;lens shield;light protective cone;light shield;louver[louvre];sun shade;rayshade <仪器的>
遮护 screen(ing);shielding;visor
遮护板 baffle plate;deflecting plate;deflector;escutcheon;shield
遮护滑板 screening slide
遮护物 baffle
遮灰护目镜 blinkers
遮火罩 flame shield
遮拦法 masking
遮拦孔径 obstructed aperture
遮帘 blind;sun blind
遮帘匣 blind box
遮帘作用 <桩工程> barrier effect;screen effect
遮没 blanking;cover(ing)over
遮没的平顶 covered ceiling
遮幕 cloak screen;masking;masking piece
遮泥板 dasher
遮黏[粘]带 adhesive masking tape
遮棚 penthouse
遮棚撑梁 canopy brace
遮篷 awning;canopy;penthouse;shade;sun shade
遮篷用织物 awning fabric
遮嵌效果 <混凝土表面用金属栅> filigree effect
遮去角 angle of cut-off
遮热板 cut-off;protective blanket;thermal baffle
遮日罩 sun shade
遮声障板 gobo
遮视区 blind area
遮水板 dash board;flashing;splash panel;dasher <船只用>
遮水楣 brow;eye brow;rigol;watershed;wriggle
遮腿挡板 <桌前或台前的> modesty panel
遮线 shade line
遮眩板 glare screen
遮檐 coping;dorse;hood
遮檐板 apron eaves piece;apron flashing;apron piece;hood
遮掩 embosom;hold-back;occultation;overlay
遮掩面 occulting disk
遮掩物 veil
遮掩现象 veil phenomenon
遮掩性反射 veiling reflection
遮掩性亮度 veiling brightness
遮掩噪声的音响 acoustic(al) perfume
遮阳 abatjour;adumbration;awning;shading;sun blind;sun shade;sun shading;tilting
遮阳百叶 louvered awning blind;louvered overhang;sun-shading louver
遮阳百叶板 sunscreen;sun shade
遮阳百叶窗 awning blind
遮阳板 ante-venna;anti-dazzling screen;apron eaves piece;solar control blind;solar screen;sun breaker;sun louver[louvre];sun-shading board;sun shield;sun visor;visors;window visor
遮阳玻璃 sun glass;sun shielding glass
遮阳布棚 shadow shield
遮阳布篷 blind curtain;sun curtain
遮阳步行道 shaded walk
遮阳窗 awning window;Venetian shutters
遮阳窗帘 window shade
遮阳窗纱 solar screening
遮阳带 shadeband
遮阳的 adumbral
遮阳顶篷 canopy
遮阳盖 sun visor;visor
遮阳光板 shadow shield
遮阳光器 anti-dazzle screen
遮阳花格 sunshade grill(e)
遮阳花格墙 solar screen
遮阳甲板 shade deck;sports deck;sun deck
遮阳甲板船 awning deck vessel;shade deck vessel
遮阳角度 shading angle
遮阳帘 shade screen;sun blind;sunscreen;window blind
遮阳幔 side screen
遮阳幕 ante-venna;blind curtain
遮阳棚 louver[louvre];shade roof;sunscreen;sun visor
遮阳篷 awning;louvered awning;shade screen;solar screen;sun blind
遮阳屏 anti-dazzling screen;solar screen
遮阳曲线 sun shadow curve
遮阳栅 sunscreen
遮阳设备 shading device
遮阳设施 sun shade;sun shield
遮阳挑檐 anti-sun cantilever roof;overhang for sun-shading;sun-shading overhang
遮阳物 sun shade
遮阳系数 shading coefficient
遮阳效果 shaded effect
遮阳信号 visiting signal;visor signal
遮阳罩 sun guard;sun shade
遮阳装置 sun breaker;sun-shading device
遮阴度 shade density
遮阴棚 sun shade
遮阴棚的支柱 posts of shade
遮荫设备 shading facility
遮阴天棚 sun shade
遮阴影区 shadow-tree zone
遮油板 grease baffle;grease shield
遮油物 grease baffle;grease shield
遮有天篷的 canopied
遮雨板 dash board;rain visor
遮雨罩 rain shade
遮罩 shade;shadow mask
遮住 curtain
遮阻甲板 awning deck
遮阻温度计 shelter thermometer

折

岸浪回流时冲成的沟 rill(e)

折百叶窗 folding shutter
折板 flap;folded plate;folded slab;table flap <折叠式桌面的>
折板混凝土屋顶 folded concrete;folded plate concrete roof
折板机 press brake
折板建筑 folded plate construction;folded plate structure
折板结构 folded plate construction;folded slab structure;folded plate structure;hipped-plate construction;hipped-plate structure
折板壳屋顶 folded plate shell roof;polygonal shell roof
折板理论 folded plate theory
折板面积 folded area
折板穹隆 folded plate dome;folded slab dome
折板式的 Hy-rib
折板式钢筋混凝土楼面 hy-rib reinforcement floor
折板式钢片 hy-fib steel sheet;hy-rib steel sheet
折板式结构 <连续板两端有横隔板以构成自承式箱形结构> folded plate structure
折板式金属网 <商品名> Hy-rib
折板式圆屋顶 folded plate cupola;folded slab cupola
折板屋顶 folded plate roof;folded slab roof
折板效应作用 folded plate effect
折板形分离器 zigzag separator
折板型除沫器 deflector-type separator
折板作用 folded plate action
折半插入 binary insertion
折半插值 interpolation to halves
折半查找 binary search
折半的 dimidiate
折半轮班 dogging the watch
折半内插 interpolating to halves
折半信度 split-half reliability
折壁(屏) folding partition(wall)
折臂式高架起重机 folding-jib gantry
折臂式塔式起重机 gooseneck jib tower crane
折边 crimp;edge fold;flanging;folding(ing);straight flange;transition knuckle;welted edge
折边板 grip plate
折边的 flanged
折边的金属丝网布 gauze wire cloth with folded edges
折边的锥形封头 flanged conic(al) head;toriconic(al) head
折边钢板网 hy-rib steel sheet
折边工具 bordering tool
折边护角 dog ear
折边机 cramp folding machine;creasing machine;crimping machine;flanging machine;folding machine;folding press;seaming machine
折边金属丝布 woven-wire cloth with folded up edges
折边金属网混凝土楼面 hy-rib reinforcement floor
折边梁 flanged beam
折边试验 <管口边与管轴成垂直> flanging test
折边收口 dog-ear(ed) fold;dog's ear
折边双反折 double reverse bend
折边屋顶盖面 welted seam roof cladding
折边无梁板结构 hipped-plate structure
折边无梁木板屋顶 timber hipped-plate roof
折边压床 flanging machine
折边压机 flanging press
折边压力机 hemming press
折边肘板 flanged bracket;flanged knee
折边装配 flanging arrangement
折变坡度 break grade
折标 backstep;backstep marks
折标尺【测】 folding staff
折波机 crimper
折玻璃窗 folding casement
折布 cuttling;plaiting
折布机 cloth plaiting machine;cuttler;doubling and plaiting machine;folder;plaiter
折测线 meander line
折插包 in-and-in
折成美元 convert into dollars
折尺 angle rule;folding pocket measure;folding ruler;folding scale;jointed rule;pocket rule;zigzag rule
折尺接头 rule joint
折冲水流 deflected stream
折除工程 demolition
折窗框 folding sash
折窗扇 folding sash
折床 Murphy
折倒天线 reclining radio antenna
折凳 folding stool
折点 break point;salient point
折点加氯 chlorination breakpoint;chlorine breakpoint
折点氯化法 break-point chlorination
折点试验 break-point test
折垫 pleated cushion
折迭偶极子反射器 folded dipole reflector
折迭频率 folding frequency
折叠 aliasing;collapse;crease;doubling;enfold;fold;foldover;furl;jointing;lap;overlap;plait;ply;pucker;wrinkle;pinchers <因耳子造成的条钢缺陷>
折叠百叶窗 boxing
折叠百叶门 corded door
折叠便座 <用于客车、剧院等> subsellia
折叠标尺 <测水准用> folding staff
折叠表面互相穿插的结构体系 structure system through interpenetration of folded surface
折叠玻璃窗 folded casement;folding casement
折叠布轮抛光 folded buff
折叠车顶 folding car roof
折叠尺 folding rule
折叠传送带 folding conveyer belt
折叠窗 folding window
折叠窗扇 folding casement;back fold
折叠床 fold-down bed;folding bed;recess bed <折入凹墙或壁柜内的>
折叠大门 folding gate;collapsible gate
折叠刀式 clasp knife
折叠刀式门 jackknife door
折叠的 accordion;back-folding;folded;folding
折叠的大张插页 pull out
折叠滴水线 welted drip
折叠底板 bottom flap
折叠对称偶极天线 folded doublet
折叠帆布椅 yacht chair
折叠方案 folding scheme
折叠放大器 folding amplifier
折叠杆 folding rod
折叠隔断 folding partition(wall)
折叠隔屏 accordion partition;folding

Z

partition(wall)
折叠隔墙 folding wall;folding partition
折叠弓形板 folded plate segment;folded slab segment
折叠罐 pillowcase tank
折叠光束 folded light beam
折叠过滤器 pleated filter
折叠和槽式接缝 folded and grooved seam
折叠画架 folding easel
折叠机 doubler;folder;folding machine;folding press
折叠家具 folding furniture
折叠架 folding leg
折叠胶黏[粘]机 folder gluer
折叠铰链 folding-hanged
折叠接缝 fold joint
折叠金属带 welting strip
折叠空腔 folded cavity
折叠栏木 folding barrier;folding gate
折叠链长度 folding length
折叠裂缝 fold crack
折叠裂纹 fold crevice
折叠门 accordion door;doubled-up door;folding gate;folding(shutter) door;multifolding door
折叠门窗 folding casement
折叠门家具 accordion door furniture
折叠门配件 folding door fitting;folding door hardware
折叠门五金 folding door hardware
折叠门箱 folding door box
折叠门装置 folding door furniture
折叠偶极(天线) folded dipole;folded dipole antenna
折叠偶极子 folded dipole
折叠偶极子反射器 folded dipole reflector
折叠片状结构 pleated sheet structure
折叠频率 folding frequency;Nyquist frequency
折叠屏风 folding screen
折叠曝气生物滤池 two-double biological aerated filter
折叠起来 fold up(wards)
折叠起重臂 folding boom
折叠气泡 lap blister
折叠器 folder
折叠钳 folding tongs
折叠区域 fold domain
折叠扇形板 folded plate segment;folded slab segment
折叠伸臂 folding boom
折叠失真 aliasing distortion
折叠式百叶 boxing shutter;folding shutter
折叠式百叶窗 folding shutter
折叠式板壁 folding partition(wall)
折叠式标尺 folding staff
折叠式布料杆 folding placing boom
折叠式舱(口)盖 folding hatch cover;hinged folding hatch cover
折叠式插页 folded leaf;fold-out;gate fold
折叠式产床 folding obstetric table
折叠式车顶 falling top;folding head;folding top;ragtop
折叠式车顶的货车 wagon with folding roof
折叠式车门 folding door
折叠式车篷 folding till(top)
折叠式担架 folding litter
折叠式的 accordion;collapsible;turn down;folding
折叠式登车阶梯 folding step
折叠式地图 accordion;accordion map;folding map
折叠式吊杆 folding boom
折叠式发动机罩 folding engine bonnet
折叠式反光立体镜 folding mirror stereoscope
折叠式反射镜 folding mirror
折叠式放大镜 folding magnifier
折叠式风挡 folding windshield
折叠式浮标 accordion buoy
折叠式附加轮箍抓地齿 folding tractor strakes
折叠式钢护舷 collapsible steel fender
折叠式钢模板 steel-ply form
折叠式搁脚板 folding foot rest
折叠式隔断 accordion partition;folding partition(wall)
折叠式隔墙 concertina partition;folding partition(wall)
折叠式公司 <一种投机性质的房地产公司> collapsible cooperation
折叠式共振腔 folded cavity
折叠式辊道输送机 accordion roller conveyer[conveyor]
折叠式过滤器 folded filter;folder filter
折叠式护板 folding shield
折叠式花格大门 folding lattice gate
折叠式滑门 folding sliding door
折叠式滑移五叶门 five-leaf sliding folding shutter door
折叠式混凝土 folded concrete
折叠式混凝土模板 folding concrete form
折叠式活动百叶窗 folding sliding shutter door
折叠式活动百叶门 folding sliding shutter door
折叠式活动格栅 folding sliding grille
折叠式活动隔断 accordion shades
折叠式集装箱 collapsible container;folding container
折叠式间(隔)壁 folding partition(wall)
折叠式胶片照相机 folding roll-film camera
折叠式井架 cantilever derrick
折叠式救生艇 collapsible lifeboat;dinge;dingey cockboat;dinghy;folding(life) boat
折叠式靠背 folding back
折叠式空气过滤器 fold media-type air filter
折叠式拉门 folding sliding door
折叠式喇叭 reentrant horn
折叠式里脚手 folding trestle
折叠式连接件 <集装箱> folding lash plate
折叠式连续环 folded continuous loop
折叠式楼板 folded flooring
折叠式炉栅 hinged grid
折叠式滤芯袋 folded septum envelope
折叠式门 concertina door;folding door <客车>
折叠式门装配 concertina door fitting
折叠式模板 folding mo(u)ld
折叠式木质斜屋顶 timber tilted-slab roof
折叠式篷顶的弧拱 folding top bow
折叠式篷顶架 folding top structure
折叠式篷顶汽车 <美> ragtop
折叠式平台 folding platform
折叠式坡道 folding ramp
折叠式千斤顶 swing jack
折叠式桥 jackknife bridge
折叠式桥节 folding bridge bay
折叠式轻便钻塔 collapsing mast
折叠式热电偶 folded thermocouple
折叠式容器 collapsible container;folding container
折叠式三脚架 folding tripod
折叠式射钉枪 combinative nail gun
折叠式升运器 folding elevator
折叠式手柄 folding handle
折叠式输送机 accordion conveyer[conveyor]
折叠式踏板 folding step
折叠式踏步 folding step
折叠式躺椅 deck chair
折叠式凸缘成型模 return flanging die
折叠式图板 accordion plate
折叠式桅杆 folding mast;jackknife drilling mast <钻采用>
折叠式尾翅 folding fin
折叠式屋顶 accordion plate roof
折叠式洗脸台 folding basin
折叠式线圈 accordion coil
折叠式箱形托盘 folding box pallet
折叠式镶板门 folding panel
折叠式小艇 collapsible boat;folding boat
折叠式小桌 folding table
折叠式写字台 roll-top desk
折叠式袖珍放大镜 folding pocket magnifier
折叠式悬臂 folding jib
折叠式液压划行器【道】 hydraulic folding marker
折叠式椅背 folding back
折叠式椅子 joint stool
折叠式硬片照相机 folding plate camera
折叠式油罐 flexible rolled up tank
折叠式圆柱面 folded cylindrical surface
折叠式照相机 folding camera
折叠式褶页 accordion fold
折叠式装订本 accordion binding
折叠式桌 gate legged table
折叠式桌子 gate table
折叠式钻杆 collapsible drilling mast;folding drilling mast
折叠式钻机 jackknife rig
折叠式钻塔 collapsible derrick;jackknife derrick;stacked derrick
折叠式座位 flap-up seat;jump seat
折叠式座席 folding seat
折叠式座椅 tilting seat;tumbler seat
折叠试验 crease-flexible over test;double-over test;doubling-over test;folding test
折叠躺椅 steamer chair
折叠梯 disappearing ladder;double ladder;folding ladder;folding stair(case)
折叠天线 folded antenna
折叠条纹 compression cord
折叠艇 collapsible boat
折叠突缝 welted nosing
折叠图纸 folding of drawings
折叠腿 gate leg
折叠桅杆 folding mast
折叠线 fold line
折叠线材 slivery wire
折叠形 folded form
折叠形钢楼板 cellular steel floor
折叠形记录纸 folding type recording chart
折叠形空气过滤器 folded media type air filter
折叠形压片机 laminator
折叠悬臂式触点簧片 folded cantilever contact
折叠悬臂式桅杆 jackknife cantilever mast
折叠亚麻布饰面镶板 linenfold
折叠椅 faldstool;flap chair;folding chair;folding seat
折叠圆面 throwing circle
折叠圆柱面的交互贯(穿) interpenetration of folded cylindrical surfaces
折叠圆柱天线开关管 folded cylinder TR tube
折叠钥匙 folding key
折叠噪声 aliasing noise
折叠帐篷拖车 folding tent trailer
折叠者 folder
折叠转角 dog ear;dog-ear(ed) fold
折叠状 rugosity
折叠状花饰 plexiform
折叠状褶皱 accordion like wrinkles
折叠桌 card table;collapsible table;folding table;yip-top table
折叠桌折板 table flap
折叠组织 folded tissue
折叠座位 misericord(e)
折钉 dog nail;rag nail
折顶 folding roof
折顶车身 folding roof body
折顶天线 folded top antenna
折断 breakage;breakdown;break off;break short;disconnect;disjunction;failure;fluting;fracture;knockout;rupture;wreckage
折断点 point of break
折断荷载 breaking load;buckling load;crippling load;load of breakage
折断面 plane of fracture;plane of rupture
折断模量 modulus of rupture
折断模数 rupture modulus
折断频率 <频率特征曲线的折断点> break frequency
折断试验 breakdown test
折断线 break line
折断销装置 breaking pin device
折断延伸率 breaking elongation
折断叶子 break the leaf
折断钻具焊接点 worry box
折反射望远镜 catadioptric telescope
折反射物镜 catadioptric objective
折反射系统 mirror lens system
折返 turn-back
折返段 back-turning section;engine terminal
折返轨 reversing track
折返环线 reverse loop
折返间隔时间 reverse interval
折返路段 turn-back section
折返能力 reverse capacity;turn-back capability
折返坡线 zigzag ramp
折返三角线 reserving triangle;triangle track;triangular track;turn-around wye;wye track;Y-track
折返设备 turn-back facility
折返时间 reverse time
折返无人机 unattended repeater with ground temperature compensation and power-feed loop back
折返线 reverse track;turnaround line;turn-back track
折返线有效长度 effective length of turn-back track
折返站 reversing station;servicing terminal;switchback station;switching depot;transfer depot;turn-around depot;turnaround station;turn-back station
折方向线 broken ray
折缝 crease;crease line;creasing;crimple
折缝机 crimper
折缝间距 <贴边的> welted seam spacing
折缝接合 <金属板的> grooved seam
折幅缝筒机 doubling and tacking machine
折干计算 calculation on dry basis;dry basis
折杆锚 folding-stock anchor

折隔屏 accordion partition; folding partition(wall)
折隔扇 accordion partition
折管目镜【测】broken eyepiece
折光 refraction
折光本领 dioptric power
折光标 diopter scale
折光表 refraction table
折光玻璃 refracting glass
折光部分 light deflecting part
折光差 astronomic(al) refraction; refraction
折光差改正 correction for refraction; refraction correction
折光差改正项 refraction term
折光成像 refracted image
折光的 dioptric
折光的色彩 overtone
折光灯 dioptric lighting
折光度 diopter[dioptre]; refraction
折光度计 dioptrometer
折光法 refractometry
折光检测器 refractive index detector
折光镜 enoscope
折光棱镜 <地下室采光用> maximum daylight
折光棱镜支架 prism bracket
折光力 refracting power
折光率 index of refraction; index test; refractive index; refringence
折光率表 refractive index table
折光率法 refraction method
折光率检验器 refractive index detector
折光器 dioptric apparatus
折光室 index of refraction
折光天线 dioptric antenna
折光透镜 dioptric lens
折光望远镜 refracting telescope
折光物体 prism; pavement prism <地下室采光用>
折光系数 refraction coefficient; refraction factor
折光系统 dioptric system
折光性 refractivity
折光学 dioptrics
折光仪 reflectometer; refractometer
折光元件 dioptric element
折光正常 emmetropia; normal refraction; stigmatism
折光指数 index of refraction; refraction index; refractive exponent; refractive index
折光指数表 refractive index table
折光指数增值 refractive index increment
折航表 traverse table
折航指示钟 zigzag clock
折耗 depletion
折耗成本 depletion cost
折耗费用 depletion expenses
折耗会计 depletion accounting
折耗准备 reserve for depletion
折合 back-folding; commute; contain; equivalent to; matrixing; reduce
折合摆长 reduced pendulum length
折合板 folded plate
折合比 reduced ratio
折合表尺板 folding leaf
折合参数 reduced parameter
折合掺杂浓度 reduced doping concentration
折合长度 foreshortened length; reduced length
折合处裂缝 fold crack
折合单位 reduced unit
折合到真空 reduced to vacuum
折合高度 reduced height
折合公制 metric(al) value
折合箍 folding hoop
折合光程 reduced optic(al) length
折合积分 convolution integral
折合架 folding leg
折合角 fold angle
折合距离 scaled distance
折合康普顿波长 reduced Compton wavelength
折合宽度 reduced width
折合量 reduced quantity
折合率 reduction coefficient
折合密度 reduced density
折合模数 reduced modulus
折合能量 reduced energy; scaled energy
折合黏[粘]度 reduced viscosity
折合偶极子 folded dipole
折合频率 reduced frequency
折合柔量 reduced compliance
折合市制 Chinese equivalent
折合式 faltung
折合式侧板 folding side-wall
折合式侧壁 folding side-wall
折合式敞篷汽车 convertible car
折合式搅拌机梁架 folding mixer frame support
折合式耙斗 folding scraper
折合式整体水泥筒仓 folding integrated cement silo
折合数量 converted amount
折合水准 reduced level
折合速度 reduced velocity
折合为海平面值 reduction to sea level
折合系数 conversion factor; reduced factor; reduction coefficient; referring factor
折合性质 reduced property
折合压力 reduced pressure
折合因数 reducing factor
折合因子 conversion factor; reduced factor
折合硬度 reduced hardness
折合有效质量 reduced effective mass
折合振子天线 folded dipole antenna
折合值 scaled value
折合质量 reduced mass
折合质量流量 reduced mass-flow
折合重力 reduced gravity
折合走时 reduced travel time
折合阻抗 reduced impedance
折痕 buckle; crease; fold; rack marks
折回 backtrack; doubling; recurvature; recurve; reflect; reflex; replicate; retrace; retroflection; retroflexion; return; run-round; turn-back
折回反射镜 retroreflector
折回效应 re-entry effect
折回原路 untread
折毁 destruction; knockout
折击变形 panting strain
折积 convolution; faltung
折积法 volumenometry
折基线 broken baseline
折价 depletion of value; reduced value
折价积累 accumulation of discount
折价价值 trade-in value
折价累积 accumulation of discount
折价轮胎 trade tire[tyre]
折价模型 trade-in model
折价时间 trade-in time
折价物 trade-ins
折价物价格 trade-in
折价债务 discount liability
折减 deduct; make a deduction
折减比 reduction ratio
折减长度 reduced length
折减地面运动 reduced ground motion
折减公式 reduction formula
折减固有频率 reduced natural frequency
折减加速度谱 reduced acceleration spectrum
折减流量 reduced discharge
折减率 discount rate
折减模量 reduced modulus
折减坡度 compensating grade; compensation grade; reducing grade; slope of compensation
折减水头 reduced head
折减弹性模量 reduced elastic modulus
折减系数 coefficient of compensation; coefficient of reduction; deamplification factor; factor of reduction; reduction coefficient; reduction factor
折减因数 factor of reduction; reduction factor
折减因子 factor of reduction; reduction factor
折角 break angle; breat angle; fold angle
折角车流 angular wagon flow
折角船尾 knuckle stern
折角阀 angle valve
折角塞门 angle cock; coupling cock
折角塞门手把 angle cock handle
折角塞门托 angle cock holder
折角塞门心 angle cock key
折角饰 lancet
折角条 lancet; strickle
折角线 knuckle line
折角运输 <在调车场需要重复分解的货车> angular traffic
折接 folded over joint; joggle; joggle(d) joint
折接板 joggle plating
折接肋骨 joggle frame; joggle timber
折井口 well-head disassembling
折旧 amortization; amortize; amortized depreciation; depreciate; depreciation; obsolescence allowance; write down
折旧按原值百分计算法 depreciation-percentage of original cost method
折旧备抵 allowance for depreciation
折旧表 depreciation table
折旧不足 under-depreciation
折旧残值 depreciation salvage value
折旧产量法 depreciation-service output method
折旧偿债基金法 depreciation-sinking fund method; sinking fund method of depreciation
折旧超额扣除 excessive depreciation deduction
折旧成本 cost of depreciation
折旧储备比率 depreciation reserve ratio
折旧单位成本法 depreciation-unit cost method
折旧的 depreciated
折旧的延期养护理论 deferred maintenance theory of depreciation
折旧定额 amortization quota; depreciation quota
折旧范围额度 allowed depreciation
折旧方程式 mortality equation
折旧方法 depreciation method; method of depreciation
折旧费 amortization cost; amortized cost; committed capacity cost; cost of depreciation; depreciation allowance; depreciation fund; redemption fund
折旧费计算方法 method of calculating depreciation charges
折旧费率 rate of depreciation charges
折旧费用 depreciation charge; depreciation cost; depreciation expenses
折旧分期递减法 depreciation reducing instalment method
折旧分期定额法 depreciation-fixed instalment method
折旧分摊递减法 diminishing balance depreciation
折旧复利法 compound interest method of depreciation
折旧根据修缮费用计算法 depreciation-maintenance method
折旧工作小时法 depreciation-working hour method
折旧估价 depreciation appraisal
折旧估价法 appraisal method of depreciation; depreciation-appraisal method
折旧估算 assessment of depreciation
折旧过多 over depreciated
折旧核对 reconciliation of depreciation
折旧后净利润 net profit after depreciation
折旧后净收入 net income after depreciation
折旧回收 depreciation recapture; recapture of depreciation
折旧会计 depreciation accounting
折旧基础 basis for depreciation
折旧基点 depreciation base
折旧基价 depreciation base
折旧基金 depreciation fund; funding depreciation; redemption fund
折旧基数 depreciable basis; depreciation base
折旧及摊提 depreciation and amortization
折旧计算方法 method of calculating depreciation
折旧价 depreciation price
折旧价值 depreciable value; depreciation value
折旧减免额 depreciation deduction
折旧降低纳税系数 depreciation factor
折旧金 depreciation allowance
折旧金提成 depreciation allowance
折旧累计法 depreciation table
折旧率 allowance for depreciation; amortization factor; depreciation charge; depreciation factor; depreciation rate; percentage of wear and tear; rate of depreciation; depreciation accrual rate
折旧明细表 schedule of depreciation
折旧年金法 depreciation-annuity method
折旧年限 depreciable life
折旧期 depreciation period; period of amortization; period of depreciation
折旧前净利润 net profit before depreciation
折旧前净收入 net income before depreciation
折旧全年总收入法 depreciation-gross earning method
折旧任意决定法 depreciation arbitrary method
折旧税收抵免 depreciation credit
折旧特别计算法 special methods of computing depreciation
折旧提成 allowance for depreciation; amortization charges; capital allowance; depreciation deduction
折旧提存 allowance for depreciation
折旧五成法 depreciation-fifty percent method
折旧系数 depreciation factor
折旧效能 efficiency of depreciation
折旧研究 depreciation study
折旧因子 depreciation factor
折旧余额递减法 declining balance method of depreciation
折旧余值 depreciated value

Z

折旧与保养合并法 combined depreciation and upkeep method
折旧预储保险金法 depreciation-insurance method
折旧预算 depreciation budget
折旧原始成本定率法 depreciation-percentage of original cost method
折旧原因 cause of depreciation
折旧原值减残值百分法 depreciation-charging percentage of cost less scrap method
折旧增值率 depreciation accrual rate
折旧账户 depreciation account
折旧照递减值定率百分法 depreciation-fixed percentage of diminishing value method
折旧直线法 depreciation-straight-line method
折旧重置成本法 depreciation-replacement method
折旧准备 provision for depreciation; reserve for depreciation; depreciation reserve
折旧准备金 depreciation reserve
折旧准备率 depreciation reserve ratio
折旧资产 depreciation assets
折旧资金 depreciation fund
折旧综合平均寿命法 depreciation-composite life method
折旧总额 total depreciation
折卷布机 doubling and rolling machine
折卷机 feed reel; unwinding reel
折扣 agio; depreciation charge; discount; rebate; reduction
折扣备抵 allowance for discounts available
折扣额 deduction; discount
折扣发行 issue bonds at discount
折扣费用 discounted cost
折扣和让价 discount and allowance
折扣回扣 rebate
折扣计费率 rebate tariff
折扣价格 discount cost; discounted price; prices at a discount; reduced price
折扣量 quantity discount
折扣率 discount factor; discount rate; rate of discount
折扣期限 discount period
折扣商店 discount store
折扣失效 discount lapsed; missed discount
折扣收入 discount earned
折扣售房 buydown
折扣损失 discount loss
折扣习惯 coustomary discount
折扣系数 discount factor
折扣现金流量 discounted cash flow
折扣因素 discount factor
折扣值 discounted value
折垒电缆 accordion cable
折力作用 <机车> jackknifing action
折裂 fold crack
折裂点 break-off point
折裂面 surface of fracture
折裂强度 strength of rupture
折流 baffling; deflection; deflexion; flow deflection; turn-back flow
折流坝 groin(e); groyne
折流坝方向 direction of groyne
折流坝工程 groyne work
折流板 baffle; baffle board; baffle plate; cross baffle; deflecting baffle; deflecting damper; deflector; deflector plate; transverse baffle; turtle back; air baffle <烟道>; flue baffler <烟道>
折流板弓形缺口 baffle cut; window cut of baffle
折流板间距 baffle spacing; pitch of baffle
折流板排列法 baffle arrangement
折流板曝气生物滤池 baffled biologic(al) aerated filter
折流板切口分数 percent of baffle cut
折流板式鱼道 vane fish ladder
折流板旋转填料床 baffled rotating packed bed
折流挡板 hydraulic barrier
折流丁坝 separation groin
折流分离 baffle separation
折流分离器 deflection separator
折流河 diverter
折流桨叶 baffle paddle
折流器 baffler; deflector; jet deflector
折流墙 baffle wall
折流清洗器 baffle washer
折流热交换器 baffled exchanger
折流设施 stream deflector
折流(式)洗涤器 baffle type scrubber; baffle scrubber; baffle washer
折流叶片 baffle blade; deflecting vane
折流蒸发器 baffled evapo(u)rator
折门 accordion curtain; accordion door; bellow-framed door; collapsible gate; folding door; hinged door; jackknife door; leaf door; telescopic-(al) door; quadrifores <古建筑>
折门零件 bellow-framed door fitting; double door hardware
折门零配件 double door fittings
折门设备 double door furniture
折门五金 double door hardware
折门装配件 bellow-framed door furniture
折面桌 butterfly table; drop-leaf table
折磨 rack
折棚 accordion; diaphragm; vestibule diaphragm
折棚顶平衡器弹簧 vestibule overhead equalizer spring
折棚盖 diaphragm cover
折棚(盖)安装铁 diaphragm cover seat
折棚横顶罩 <客车> accordion hood
折棚架 bellow frame
折棚帘 vestibule curtain
折棚帘钩 vestibule curtain hook
折棚帘摩擦板 vestibule curtain box liner
折棚帘手把 vestibule curtain handle
折棚帘箱盖 vestibule curtain box cover
折棚帘罩 vestibule curtain shield
折棚司机室 vestibuled cab
折棚弹簧 diaphragm type spring; upper buffer spring
折棚弹簧箍 diaphragm spring band
折棚弹簧卡子 diaphragm spring band
折棚铁板 diaphragm face plate; faceplate; vestibule face plate
折棚柱 diaphragm post; vestibule diaphragm post
折棚柱撑木 diaphragm post brace
折棚柱上横铁 diaphragm post connecting beam
折屏风 folding screen
折屏障 accordion curtain
折坡 knick point
折坡断块图 kantographic(al) block diagram
折坡梯田 zig bench terrace
折起 turn up
折起来 fold back
折曲 bending; crimple; crippling; joggle; warpage; warping; flex【地】
折曲光程 folded optic(al) path
折曲机 joggler; joggling machine
折曲模 joggle die
折曲试验 joggling test
折曲应变 crippling strain
折曲应力 crippling resiliency; crippling resistance; crippling stress
折曲准则 criterion of buckling
折让 allowance; discounts and allowances
折让储备金 allowance reserve
折让及回扣 allowance and rebate
折让准备 allowance reserve
折入 invagination
折入边 tuck-in selvage
折山头屋顶 clipped gable roof
折扇隔壁 accordion partition
折扇隔断 accordion partition
折扇隔墙 accordion partition
折扇门 bellows-framed door
折扇状的 plicate
折射 anacampsis; refract; swerve
折射爆破 refraction shooting
折射本领 refractive power; refractivity
折射比 ratio of refraction
折射波 Mintrop wave; refracted wave; refraction wave; refractive wave; transmitted wave
折射波对比 refracted wave correlation
折射波法 refraction method
折射波勘探法 refraction wave exploration method
折射波速 refraction velocity
折射玻璃 refracting glass
折射参数 refraction parameter
折射测定仪 refractoscope
折射层 refracted layer; refracting layer; refractors
折射常数 refraction constant
折射成像 dioptric imaging
折射冲击波 refraction shock wave
折射的 dioptric; refractive
折射灯(光) dioptric lighting
折射地震勘探 refraction seismic prospecting
折射地震学 refracted seismology; refraction seismology
折射地震仪 refraction seismograph
折射定律 law of refraction; refracted law; refractive law; Snell's law
折射度 refraction; refrangibility; specific refraction; specific refractivity
折射度测定 refraction determination
折射度的退火增加 annealing increment of refraction
折射法 refraction process
折射法地震勘探 refraction seismic prospecting
折射法调查 seismic refraction survey
折射法勘探 refraction profiling; refraction survey; refractor survey
折射反射波 refraction-reflected wave
折射分析 refractometric analysis
折射分析法 refractometry
折射改正 refraction correction
折射各向异性 refraction anisotropy
折射功效 refractive quality
折射光 dioptric light; refracted light; refraction light
折射光灯塔 dioptric lighthouse
折射光线 refraction line; refraction ray
折射光学 dioptrics; refractive optics
折射光学系统 dioptric system
折射计 refracting meter; refractometer
折射计法 refraction analysis; refractometry
折射技术 refractive technique
折射校正 correction for refraction; refraction correction
折射介质 refracting medium; refractive medium
折射镜 refractor
折射勘探法 refracted prospecting; refraction prospection; refraction survey
折射(棱)角 refracting angle; angle of refraction; refracted angle; refraction angle
折射棱镜 refracting prism; refraction prism
折射力 refracting power
折射量技术 refractometry
折射滤光片 refraction filter
折射率 index of refraction; power of refraction; refracting power; refraction exponent; refraction index; refractive exponent; refractive index; refractive power; refractivity; specific refractive power; specific refractory power
折射率测定 refractive index determination
折射率测量 refractometry
折射率滴定 refractometric titration
折射率分布参数 profile parameter
折射率分布曲线 dispersion curve
折射率分布色散 profile dispersion
折射率分布系数 refractive index profile coefficient
折射率分布指数 profile exponent
折射率校正 refractive index correction
折射率界面 refractive index interface
折射率可变介质 medium of varying refractive index
折射率匹配 index matching
折射率匹配晶体 index matched crystal
折射率色散 refractive index dispersion
折射率椭球 refractive index ellipsoid
折射率椭圆体 indicatrix
折射率温度系数 temperature coefficient of refractive index; thermal refractive index coefficient
折射率液 index liquid
折射率仪 refractometer
折射脉冲 refracted pulse
折射面 plane of refraction; refracted plane; refracting surface; refraction surface
折射模式 refraction pattern
折射模数 refractive modulus
折射能力 refracting ability; refractive power; refractivity; refrangibility
折射劈理 refraction cleavage
折射器 baffler; refractor
折射球体 refracting sphere
折射散焦线 diacaustic
折射式地震仪 refraction seismograph
折射式投影机 refractive projector
折射视差【测】 parallactic refraction
折射试验 refraction test
折射损耗 refraction loss
折射损失 refraction loss
折射特性 property of refraction; refraction property
折射体系 <只包括透镜> dioptric system
折射透镜 hot spot; hot spot lens; refractor
折射图 refracted drawing; refraction diagram
折射望远镜 refractor
折射微差 differential refraction
折射物镜 refraction objective
折射误差 refraction error
折射系数 coefficient of refraction; index of refraction; refraction coefficient; refractivity; refringency [refringence]; specific refraction
折射现象 refraction effect

Z

折射线 ray of refraction;refracted ray
折射效应 effect of refraction;refraction effect
折射楔 refracting wedge
折射型望远镜 refracting telescope
折射性 refractivity;refrangibility
折射性能 refractive property
折射仪 refractometer
折射指数 refraction index;refractive exponent; refractive index; refractometer
折射轴 refracted axis
折射纵波垂直时距曲线 vertical hodograph of refracted p-wave
折射作用 interception effect;refraction; refraction action; refraction effect
折声学 diacoustics
折实单价 parity unit
折实单位 parity unit
折式地图 accordion
折式盥洗台 folding lavatory
折死弯 dead folding
折算 conversion; convert; reduce; transform
折算标高 reduced level
折算表 commutation table
折算参数 reduced parameter
折算长度 converted length; reduced length
折算的 reduced
折算电流 reduced current
折算电压 reduced voltage
折算电阻 reduced resistance
折算断面 converted section;reduced section;transformed section
折算风险暴露 exposure to translation risk;translation exposure
折算高程 reduced level
折算高度 reduced height; reduced level
折算公式 reduction formula
折算功率 reduced power
折算固有频率 reduced natural frequency
折算惯量 referred inertia
折算惯性矩 reduced moment of inertia
折算价值 trade-in value
折算截面法 method of transformed section
折算金额 monetary equivalent
折算局部阻力系数 effective coefficient of local resistance
折算宽度 converted width; reduced width
折算流量 reduced discharge
折算率 conversion rate
折算密度 reduced density
折算铺设长度 reduced laying length
折算热负荷 reduced heat input
折算深度 reduced depth
折算渗透压 reduced osmotic pressure
折算水头 reduced head
折算速度 specific speed
折算损益 translation gain or loss
折算弹性模量 reduced modulus of elasticity
折算体积 reduced volume
折算调整 translation adjustment
折算投资占用 converted investment employed
折算为现值 discount to present value
折算温度 reduced temperature
折算系数 conversion factor; reduction coefficient
折算压力 reduced pressure
折算应力 equivalent stress; reduced stress
折算值 discounted value; discounting worth;reduced value;referred value
折算质量 reduced mass
折算自振频率 reduced natural frequency
折算阻抗 referring impedance
折碎 fleding;scrap
折损 derating
折损应力 crippling stress
折腾 calendar cut
折梯 folding ladder; folding stair (case);loft stair(case);stepladder
折贴的 commutable
折贴现边际收益 discounted marginal revenue
折瓦 flap tile
折弯 bend;kink;knee bend;rebend
折弯成型机 bending and forming machine
折弯的 folding
折弯机 bender; bending device; bending machine;folding machine
折弯模 bender;brake die
折弯切割两用机 bender and cutter
折弯设备 bending former
折弯试验 folding test
折弯压力机 bending press
折弯应力 folding stress
折纹 fluting
折弦 curved chord
折弦拱形大梁 polygonal bowstring girder
折弦拱形桁架 polygonal bowstring truss
折弦桁架桥 curved chord truss bridge
折弦角桁架 Pennsylvania truss
折弦(式)桁架 curved chord truss; polygonal truss
折弦再分桁架 Pennsylvania truss
折现 discount to present value
折现率 discount(ing) rate
折现偏好 discount preference
折现投资盈利率 discounted rate of return
折现系数 discount factor
折现因子 discount factor
折线 break line; broken line; crease line; dogleg; fold line; irregular curve;polygonal line;polyline
折线编码律 segmented encoding law
折线变坡滑道【船】 broken-sloped slipway;knuckling line slipway
折线表示法 polyline representation
折线的 unconformable
折线断面 three-level section
折线法 broken-line method; piecewise linear approximation; polygon method
折线分析法 broken-line analysis
折线弓弦桁架 polygonal bowstring truss
折线轨迹 dog-leg path
折线函数 polygonal function
折线函数发生器 function fitter
折线扫描 zigzag scanning
折线上弦(梁) broken top chord
折线式(横)断面 split plane cross-section;split-section
折线式屋顶 French roof
折线式养护窑 multiangular tunnel curing chamber
折线隧洞 dogleg tunnel
折线索引 polyline index
折线索引表 list of polyline indices
折线屋面 curb roof;knee roof
折线下弦 raised chord
折线线形 polygonal shape
折线形(布置)外部(预应力)筋 external polygonal tendon
折线形的 dog-legged;mansard
折线形桁架 camel-back truss; mansard truss
折线形金字塔 bent pyramid
折线形井 dog leg well
折线形棱锥(体) bent pyramid
折线形路拱 broken line crown
折线形平屋顶 mansard flat floor
折线形上弦杆 polygonal top-chord
折线形屋顶 double pitch(ed) roof; mansard(roof);mansard truss roof
折线形屋顶窗 mansard dormer window
折线形屋顶飞檐线条 mansard cornice mo(u)lding
折线形屋顶桁架 mansard roof truss
折线形屋架 segmental roof truss
折线形斜坡屋顶 <英> hipped mansard roof
折线形(预应力钢筋)束 draped tendon
折线形边坡 broken line slope
折线形滑动面 zigzag slip surface
折线颜色索引 polyline colo(u)r index
折线堰顶 broken crest; broken weir crest
折线窑 zigzag kiln
折线凿 dog leg chisel
折线状的 broken linear
折线总计法 trapezoidal rule
折向板 deflecting plate; deflection plate;deflector plate
折向防护屏 <使失控汽车转折方向的> deflective screen
折向节点 knuckle joint
折向流 refracted flow
折向喷嘴 deflecting coating
折向器 jet deflector;water deflector
折卸搬迁费用 debugging expenses
折卸工具 puller
折卸式刮刀钻头 drag rotary detachable bit
折形基线 broken base
折烟板 smoke deflector
折焰板 splasher
折焰器 flame diverter
折腰 articulation
折腰轮式装载机 articulated wheeled loader
折腰式车辆 articulated vehicle
折腰式拖拉机 hinged-frame tractor
折腰式装载机 articulated loader
折腰式自卸车 articulated dump truck
折腰转向式装载机 articulated steer loader
折叶 hinge
折叶板撑 fly rail
折叶式悬臂 hinged boom
折叶圆桌 butterfly table
折页器 folder
折页三角板夹纸辊 bending roller
折页式舱盖 hinged hatch cover
折页销 hinged pin;link pin
折页销插 hinge pin bracket
折页桌 drop-leaf table; envelope table;flap table
折移式轻便铁路 constructor's railway
折椅 collapsible seat; flap-seat; folding chair;misericord(e)
折翼 flap
折翼祭坛 folding altar
折翼缘 flanging
折印 <薄板叠轧缺陷> pinch;pinchers
折余价值 depreciated value; salvage value;sound value
折缘 flanging
折缘板 flanged sheet
折缘管节 flange union
折缘机 hemming machine
折楂 <原木根端突出部> sloven
折照灯光 reduced lighting
折遮 accordion door
折摺式触点 bellows contact
折摺式簧片 bellows contact
折中 trade-off
折中办法 half measure
折中的 give-and-take
折中观点 middle ground
折中使用办法 <设备损坏后的> fail soft
折中压力 mean pressure
折衷 give-and-take
折衷办法 half measure; mean method;trade off
折衷(粗调)平衡网络 compromise balancing network
折衷方案 compromise alternative;compromise proposal;compromise;made-off;trade-off
折衷分类 compromise sort
折衷式建筑 eclectic architecture; eclectic structure
折衷式庭园 eclectic-garden
折衷网络 compromise network
折衷选择 trade-off
折衷研究 trade-off study
折衷原则 principle of mediocrity
折衷主义 eclecticism
折衷主义建筑 eclectic architecture; eclectic structure
折轴反光望远镜 elbow refractor
折轴反射望远镜 Coudé reflector
折轴光谱学 Coudé spectroscopy
折轴焦距比数 Coudé f-number
折轴经纬仪 broken-back transit; prism transit
折轴(镜)焦点 Coudé focus
折轴目镜 diagonal eye(piece)
折轴摄谱仪 Coudé spectrograph
折轴望远镜 broken type telescope; Coudé telescope;elbow telescope
折轴折射望远镜 Coudé refractor; elbow refractor
折轴中星仪 broken transit instrument;prismatic (al) transit instrument
折轴装置 Coudé mounting
折肘 turned knee
折皱 corrugation; crease; crumple; pinch;puckering;pinch bar <带钢缺陷>
折皱变形 scuffing
折皱标样 wrinkle pattern
折皱的 corrugated
折皱回复度 crease recovery;recovery from creasing;wrinkle recovery
折皱回复度试验仪 wrinkle-recovery tester
折皱加热弯管 hot elbowing with wrinkle
折皱锐度 sharpness of creasing
折皱试验 folding test
折皱涂层 pinched coating
折皱弯管 creased pipe bend
折爪锚 folding anchor
折转 deflection;deflexion;turn about
折转板 abat-vent
折转的 reflex
折转角 turning angle
折转器 deflector
折状线圈 accordion coil
折桌 folding table
折子地图 accordion plate
折子式褶页 accordion fold
折子式装订本 accordion binding
折嘴锚 <用于固定的锚碇> fluke turned anchor

哲

哲学博士 <拉> Philosophiae Doctor

蛰

蛰伏 slumber;torpor
蛰伏的 dormant

摺 turn down

摺板 flap
摺边 flange;welted drip
摺边试验 < 边与管轴成垂直 > flanging test
摺叠 pucker
摺叠(窗)扇 back fold
摺叠的 plaited
摺叠式钢模板 steel-ply form
摺缝刀 ceasing stick
摺痕 crease
摺门 accordion door;leaf door
摺扇门 bellows-framed door
摺式烟筒 lowering chimney
摺皱缝 pucker

辙 wake

辙叉 crossing;frog
辙叉部分轨距 ga(u)ge at frog
辙叉长度 frog length
辙叉长心轨 point rail
辙叉导距【铁】lead of crossing
辙叉的左槽 left groove of the crossing
辙叉垫板 frog plate
辙叉吊钳 frog tongs
辙叉端趾开口距 toe spread
辙叉短心轨 short point rail of frog
辙叉返回原位 return of points
辙叉跟 frog heel;heel of blade;heel of crossing;heel of frog
辙叉跟长度 heel length;heel length of frog
辙叉跟垫板 twin tie plates
辙叉跟端 frog toe;toe end of frog
辙叉跟端接合螺栓 heel joint bolt
辙叉跟端开口 frog heel spread
辙叉跟端鱼尾板 heel fishplate
辙叉跟高间隔铁 heel riser(block)
辙叉跟高间铁块 heel riser(block)
辙叉跟间铁块 heel block
辙叉跟距 heel distance
辙叉跟开口距 heel spread of frog;spread at heel
辙叉跟宽 frog heel spread;heel spread of frog
辙叉跟理论长度 theoretic(al) heel length
辙叉跟延长(轨) heel extension
辙叉号数 frog number
辙叉喉 < 两翼轨之间的最小距离处 > throat of frog
辙叉喉间隙 neck gap
辙叉喉宽度 width of throat
辙叉后垫板 rear bearing plate of frog
辙叉护轨 frog guard rail
辙叉尖(端) frog point;point of switch
辙叉尖端和理论尖端间的距离 distance between nose and fine nose
辙叉交点 < 铁路的 > theoretic(al) point of frog
辙叉角 angle of crossing;crossing angle;frog angle
辙叉理论尖端 theoretic(al) nose of crossing;theoretic(al) point of frog
辙叉理论中心 theoretic(al) center[centre] of frog
辙叉两工作边的查照间距 back-to-back spacing in a crossing
辙叉(轮缘)槽 frog flangeway
辙叉螺栓 frog bolt
辙叉前垫板 front bearing plate of frog
辙叉人字尖 V-piece of crossing
辙叉实际尖点 actual frog point;nose of frog;point of frog
辙叉实际尖端 actual nose of crossing;actual point of frog;nose of crossing;nose of frog;point of frog
辙叉尾端 frog heel;heel of crossing;heel of frog
辙叉心 frog center[centre]
辙叉心轨 nose rail;point rail of frog
辙叉心轨尖端 actual point of frog;crossing nose;nose of crossing;nose of frog
辙叉心轨理论尖端 theoretic(al) frog point
辙叉心轨实际尖端 actual frog point
辙叉咽喉 < 两翼轨之间的最小距离处 > frog throat;throat of frog
辙叉翼 frog wing;wing
辙叉翼轨 rail wing
辙叉翼轨加高 frog wing riser
辙叉翼轨斜展 flare of frog
辙叉有害空间 < 在辙叉喉和实际尖端间的距离 > gap in gauge line;gap in the frog;open throat;unguarded flange way
辙叉有害空间长度 guideless length;length of unguided running section
辙叉趾 frog toe;toe of frog
辙叉趾长度 toe length;toe length of frog
辙叉趾端 toe end of frog
辙叉趾端开口 frog the spread
辙叉趾端开口距 spread at toe
辙叉趾端连接螺栓 heel joint bolt
辙叉趾端前插直线 tangent adjacent to toe of frog
辙叉趾间距铁 toe block
辙叉趾理论长度 theoretic(al) toe length
辙叉趾连接螺栓 toe joint bolt
辙叉趾延长(轨) toe extension
辙查杆 depression bar
辙查锁闭 detector locking
辙跟垫板 heel plate
辙跟间隔铁 heel block
辙跟头部加强锁闭杆 reinforcing head block bar
辙轨 < 活动交叉 > point rail;nose rail;tongue
辙后双垫板 twin tie plates
辙后支距垫板 turnout plate
辙迹消除器 eraser
辙夹指示器 point indicator
辙尖 point;point of switch(tongue);switch blade;switch tongue
辙尖隔离块 point separator
辙尖轨 point rail
辙尖卡铁 switch clamp
辙尖扣夹 switch clip
辙尖拉杆 bridle rod
辙尖理论尖端 vertex of switch
辙尖锁闭器钥匙 key for point lock
辙尖斜切 chamfer cut
辙尖移动滚轮 beam path roller
辙尖枕木 head block;head block tie
辙尖指示器 point indicator
辙器 goat
辙握柄座 switch box
辙轧 switch point;switch rail
辙枕 chair

锗 germanium

锗半导体 germanium semiconductor
锗半导体三极管 germanium semiconductor triode
锗磁铁矿 brunogeierite
锗低频大功率晶体三极管 germanium low frequency high power triode
锗二极管 germanium diode;germanium rectifier
锗光电管 germanium photocell
锗光敏电阻 germanium photo sensitive resistance
锗化物 germanide
锗晶体 germanium crystal
锗晶体管 germanium transistor
锗晶体三极管 germanium triode
锗矿 germanium ores
锗矿床 germanium deposit
锗三级管 germanium triode
锗石 argutite;germanite
锗整流器 germanium rectifier

赭 褐(色)的 ochre brown

赭黄土 yellow earth
赭绿 green ocher
赭色 brown sienna;chocolate;ocher[ochre];ruddle;salmon;sienna
赭色赤铁矿 ochre;red ocher[ochre]
赭色的 auburn;umber;dark red;ochreous
赭色颜料 sienna
赭砂 ferruginous sand
赭石 ocher[ochre]
赭石橙色 Spanish ocher
赭石红 Pompeian red
赭石黄 yellow ocher[ochre]
赭石泥 ochre puree
赭石突变 ochre mutation
赭石污染 ochre pollution
赭土 iron clay;iron minium;yellow ocher[ochre];sienna < 一种矿物颜料 >
赭土的 ochreous
赭土混凝土 Ocrate concrete
赭土色 umber
赭土颜料 umber
赭叶杜鹃花 rusty-leaved

褶 plait;ruckle

褶板 folded area
褶板结构 folded plate structure
褶边 frilling;ruffle
褶边机 creasing machine
褶边装置 bluffing
褶布饰镶板 drapery panel;linen-fold panel
褶层(胶合板) pleat
褶点 plait point
褶叠层 folding layers;pleat
褶叠频率 fold frequency
褶缝 tuck
褶合编码 convolution code
褶合(成) faltung
褶合定理 faltung theorem
褶合法 convolution method
褶合积分 faltung integral
褶合式变换 convolution transform
褶痕 crease
褶积 convolution;faltung
褶积变换 convolution transform
褶积定理 convolution theorem
褶积积分 convolution integral
褶积结果 convolution result
褶积矩阵 convolution matrix
褶积滤波 convolution filtering
褶积滤波法 convolution filtering method
褶积运算 convolution operation
褶集 convolution
褶集器 convolver
褶挤角砾岩 riebungsbreccia
褶裥窗帘 cafe curtain
褶裥花纹 < 木板的 > quilted figure
褶隆区 culmination
褶劈理 crenulation cleavage
褶曲 fold
褶曲层理 crinkled bedding
褶曲构造地貌 landform of fold structure
褶曲山 fold mountain
褶曲系 fold system
褶曲形态特征 feature of fold shape
褶曲翼 limb of fold
褶曲指数 index of fold
褶升区 culmination
褶纹 foliation
褶纹线理 crenulation lineation
褶形管 creased pipe
褶形花边的 scalloped
褶轴顶点 culmination
褶皱 accident of the ground;corrugation;fold;plication;reeding;wrinkle;wrinkling
褶皱板 folded plate
褶皱伴生构造 associated structures of fold
褶皱鼻 nose of fold
褶皱闭合度 closure of fold
褶皱编号 number of fold
褶皱编号和名称 number and name of fold
褶皱波长 fold wavelength
褶皱波幅 fold amplitude
褶皱层【地】contorted bed;folding stratum;plicated layer
褶皱长度 length of fold
褶皱沉降海岸 cheiragratic coast
褶皱尺度 fold scale
褶皱代号 code of fold
褶皱带【地】belt of folded strata;fold(ing) belt;fold(ing) zone;mobile zone;zone of belt;zone of fold(ing)
褶皱倒向【地】fold vergence
褶皱的 folded;folding
褶皱的地层 folded stratum[strata]
褶皱的位置 location of fold
褶皱的圆柱面 folded cylindrical surface
褶皱底【地】lower apex of fold
褶皱地垒 folding horst
褶皱地震 folding earthquake
褶皱地质年代 chronology of folding
褶皱顶 fold apex;upper apex of fold
褶皱顶点 apex[复 apices/apexes]
褶皱断层 fold(ed) fault
褶皱断层山 fold-fault mountain
褶皱分布位置 location of fold distribution
褶皱构造 folded structure
褶皱构造的数学模拟 mathematic(al) simulation of folded structure
褶皱构造地球化学 fold tectono-geochemistry
褶皱构造倾伏地段 plunge area of fold
褶皱规模 scale of fold
褶皱海岸 fold coast
褶皱和断层线 trend line
褶皱核【地】fold kern;fold core
褶皱弧 arc of folding;fold arc
褶皱基底 folded basement
褶皱级别 orders of fold
褶皱挤离的核心 detached core
褶皱检验 fold test
褶皱角砾岩 fold breccia
褶皱卷入地层 involved strata in fold
褶皱宽度 width of fold
褶皱类型和性质【地】type and property of folds
褶皱连锁【地】linkage of fold
褶皱链 fold train
褶皱裂缝 folding fissure
褶皱面 folded area;plane of fold
褶皱面向 fold facing
褶皱名称 folding name;name of fold
褶皱幕 folding phase

褶皱盆地 fold basin
褶皱期【地】folding period;period of folding
褶皱器 crimper
褶皱强度 fold intensity
褶皱强裂的构造 highly folded structure
褶皱倾伏角【地】plunge of fold
褶皱区 folded region;zone of fold(ing)
褶皱区岩溶 folded region karst
褶皱圈闭 fold trap
褶皱群【地】bundle of folds
褶皱容矿构造 ore-containing structure of fold
褶皱山 folded mountain
褶皱山脉 folded chain;fold mountain
褶皱石灰岩 folded limestone
褶皱时代确定依据 basis for dating fold
褶皱世代 fold generation
褶皱式窗棱 fold-mullion
褶皱枢纽控制点 control point of fold-hinge
褶皱束 bunchy folds;bundle of folds
褶皱推复 folding nappe
褶皱推覆体 folding nappe
褶皱外缘坝 folded outer edge bar
褶皱弯管 creased pipe bend
褶皱系 fold system
褶皱形成时间 time of fold formation
褶皱型石香肠 fold-type boudin
褶皱岩层性质 character of folded beds
褶皱要素 fold elements
褶皱要素数据 data of fold elements
褶皱翼 fold limb;slope
褶皱运动 folding movement
褶皱展布范围 distribution range of fold
褶皱展布面积 extent of fold
褶皱轴【地】axis of folding;fold axis
褶皱轴带 axial belt of folding
褶皱轴迹经过地点 locations passed by axial-trace
褶皱轴迹弯曲地段 bending area of fold axial trace
褶皱轴面 axial plane of fold; axial surface of a fold
褶皱轴位置 location of fold axle
褶皱轴线 axial line of fold;axis of fold
褶皱轴向 trend of fold axis
褶皱装置 draper
褶皱状混合岩 folded migmatite
褶皱作用【地】similar fold(ing)<半径近似一致的>;folding;plication
褶皱作用深度 depth of folding
褶皱作用形成的山 mountain formed by folding

这

这也适用于 the same holds true for

柘

柘榴虫胶 garnet lac
柘榴石 garnet

蔗

蔗蜡 cerosin(e)
蔗糖 cane-sugar;sucrose
蔗糖八乙酸酯 sucrose octa-acetate
蔗渣 bagasse
蔗渣板 cane fiber[fibre] board
蔗渣沥青纤维板 celotex board
蔗渣纤维板 bagasse fibreboard

针

针 needle;pike
针坝 needle dam;Poir(e)dam
针板<用以刺穿薄膜纤维使之透水> barbed needle
针板坝 needle dam
针板螺杆 faller screw
针板式电集尘器 point plane precipitator
针背 needle back
针标 pin mark
针标地图 pin map
针表面 wire surface
针冰 acicular ice;frazil ice frazil;needle ice
针柄 needle handle
针布钢丝 card wire
针布环套 ring clothing
针槽 needle tray;needle trough
针槽壁 needle land;needle wall
针测密实度 needle density
针插 pincushion
针场电离室 needle chamber
针尺 needle ga(u)ge
针齿轮 lantern gear
针齿轮刨齿机 pin gear shaper
针齿盘 wheel spider
针齿式圆盘耙 spiker
针赤铁矿 raphisiderite
针雏晶 belonite;spiculite
针穿孔记录系统 pin punching record system
针穿孔寄存器 pin punching register
针穿硬度计 penetrometer
针床 needle bar;needle bed
针床扳手 needle bar wrench
针床处理系统 reed bed treatment system
针床水平运动轴 pressing shaft
针床凸轮杆 needle bar cam lever
针床脱水处理系统 reed bed dewatering and treatment system
针瓷肩架 cantilever for pin insulator
针刺 acupuncture;needle prick;needle punching;needling;prod
针刺玻璃纤维毡 needled glass fiber mat
针刺地毯 punched carpet
针刺法 needle point method;needle punched process
针刺机 needle machine;needling machine
针刺检查(电杆)prod(d)ing
针刺角度 angle of needle insertion
针刺轮 spiked cylinder
针刺轮式拔取器 spiked-wheel lifter
针刺土工布<无纺布的一种> needle punched geotextile
针刺毡 needled mat;punched carpet
针刺毡机 needled mat machine
针枞 picea;spruce;tiger-tail spruce
针锉 needle file
针带式输送器 pinned conveyer[conveyor]
针导法 belonospasis
针导管 wire guide
针道 needle passage;needle track
针碲铋矿 wehrlite
针碲金铜矿 kostovite
针碲金银矿 goldschmidtite;graphic(al)tellurium;sylvanite;white tellurium;yellow tellurium
针碲矿 nobilite
针点矩阵 pin dot matrix
针电极 catwhisker;needle electrode
针端柔性 needle point compliance
针对 against;contraposition
针对型干扰发射机 point jammer
针阀 leak valve;needle control valve;needle-plug valve;needle valve;pin valve;reed valve;spear valve;drip valve<调节液压给进的>;diaphragm valve<给水调整阀>
针阀冲程 needle valve stroke;spear stroke
针阀导阀 spear pilot valve
针阀副 nozzle tip
针阀杆 needle stem
针阀簧 needle valve spring
针阀尖 spear tip
针阀间隙 needle clearance
针阀减振器 needle valve shock absorber
针阀接力器 needle servometer;spear servomotor
针阀开启压力 needle opening pressure
针阀控制杆 needle control rod
针阀控制堰 needle controlled weir
针阀偶件 pintle nozzle matching parts
针阀升程 needle lift
针阀式喷油器 needle type injector
针阀伺服机构 needle servomotor
针阀伺服马达 needle servomotor
针阀调节喷嘴 measuring jet;needle jet
针阀座 needle seating;needle valve seat
针阀座角 needle seat angle
针阀座气密度试验器 needle valve seat hermetic test device
针钒钙石 hewettite
针钒钠锰矿 santafeite
针沸石 mazzite
针锋相对 blow for blow;tit for tat
针钙镁铀矿 rabbittite
针杆 needle stem
针杆拦线板 needle bar thread guard
针杆曲柄 needle bar crank
针杆行程 needle bar stroke
针杆座 chuck liner
针根 needle root;root of needle
针钩 needle beard;needle hook
针钩集圈 tuck-in the hook
针刮样板 scratch template
针管 needle tube[tubing]
针硅钙铅矿 margarosanite
针硅钙铅石 margarosanite
针硅钙石 hillebrandite
针硅磷灰石 steadite
针硅铀矿 uranosite
针辊 needle roll;porcupine roller
针辊处理 pin-roller treatment
针划涂层黏[粘]结力试验 scratch drawing test
针黄铜矿 chalmersite;cubanite
针辉铋铅矿 giessenite
针火花隙 needle spark gap
针迹 needle tracking
针夹具 pin vice[vise]
针尖 needle point; needle tip; pinpoint;point;tip of the needle
针尖反向 pointing backwards
针尖放电避雷器 pinpoint arrester;pinpoint lightning arrestor
针尖火花隙 needle point spark-gap
针尖镊 pinpointed forceps
针尖同向 pointing forwards
针尖研磨 needle point grinding
针尖状气孔 pin-hole
针尖状絮凝体 pinpoint floc
针碱钙石 yuksporite
针脚 stitch
针脚点焊法 stitch welding
针脚式接合 stitch bond
针脚式接合单块片 stitch-bonded monolithic chip
针脚式接合法 stitch bonding
针脚形点焊 stitch welding
针接头 adapter
针节<直径小于 1/4 英寸,1 英寸 = 0.0254米> pin knot
针灸 acupuncture
针灸科 department of acupuncture and moxibustion
针灸室 acupuncture treatment room
针具 needling instrument
针壳贫属<拉> Irenopsis
针刻 needle etching
针刻腐蚀凹版 drypoint etching
针孔 blacking hole;needle hole;orifice;pinhead blister;pin-hole;pinholing
针孔板计算机 pinboard machine
针孔成像 pin-hole imaging
针孔冲 prick punch
针孔传输 pin feed
针孔传输形式 pinfeed form
针孔电子透镜 aperture lens
针孔阀 needle control valve;needle type valve
针孔检测仪 holiday detector
针孔馈送表格纸 pinfeed form
针孔馈送压纸卷轴 pinfeed platen
针孔裂缝 pin-hole leak
针孔缺陷<铸件> pin-hole
针孔润滑法 needle lubrication
针孔润滑器 needle lubricator
针孔筛目 needle mesh;needle number
针孔摄影机 pin-hole camera
针孔式触点 pin contact;pin-socket contact
针孔试验<分散性土的试验> needle hole test;needle test;pin-hole test
针孔透镜 aperture lens;pin-hole lens;single-aperture lens
针孔透镜比 lens aperture ratio
针孔型浇注口 pinpoint gate
针孔油枪 needle lubricator
针孔元件 punctured element
针孔照相机 pin-hole camera
针孔状疏松 pin-hole porosity
针孔准直 pin-hole collimation
针控制杆 needle control lever
针阔叶混交林 mixed needle-broadleaf forest
针蜡【地】geomyricin;geomyricite
针烙 cautery therapy with heated needle
针栎 pin oak
针梁<托换基础用> needle beam
针梁法 needle beam method
针梁脚手架 needle beam scaffold
针磷铝铀矿 upalite
针磷铁矿 koninckite
针磷钇矿 weinschenkite
针硫铋铅矿 aikinite;needle ore
针硫铋铅铜矿 needle ore
针硫铋铜铅矿 neyite
针硫铅铜矿 betekhtinite
针六方石 nocerine;nocerite
针绿矾 coquimbite;white copperas
针轮 pin gear;pin wheel
针轮传动 lantern-wheel gearing
针轮传动装置 pin-wheel gear
针轮大齿轮 lantern gear
针轮啮合 pin gearing
针轮平面图 pinwheel ground plan
针轮小齿轮 lantern pinion
针茅 spear grass
针茅蜡 reed wax
针钠钙石 osmelite;pectolite;ratholite
针钠锰石 serandite
针钠铁矾 ferrinatrite
针镍矿 capillary pyrite;millerite
针排堰 needle weir
针盘 dial;limb
针盘传动滑轮 dog
针盘读数 clock reading
针盘式测厚规 dial thickness ga(u)ge
针盘式测厚计 dial thickness ga(u)ge
针盘式伸长计 dial extensimeter[ex-

Z

tensometer]
针盘式形变计 dial extensometer
针盘式压力计 dial ga(u)ge
针盘式应变计 dial indicator strain ga(u)ge
针盘指针 index
针铅铋银矿 matildite;schapbachite
针钳 needle pliers;pin vice[vise]
针枪 needle gun
针青铜 needle bronze
针入度 penetration
针入度保温蝶 penetration dish
针入度比 penetration ratio
针入度(测定)值 <沥青的> penetration number;penetrometer number
针入度碟 penetration dish
针入度级 penetration grade
针入度计 penetrameter [penetrometer];penetrator
针入度计圆锥体 penetrometer cone
针入度黏[粘]度数 penetration viscosity number
针入度试验 <沥青的> penetration resistance test; penetration test; penetration testing
针入度-温度敏感性(系数) penetration temperature susceptibility
针入度-温度指数 penetration temperature index
针入度仪 penetration test apparatus; penetrator;penetrometer
针入度指数 penetration index
针入法 probe method
针入法(硬度)试验 needle penetration test
针入密实度 <用贯入法测得土的密实度> needle density
针入深度 depth of penetration
针入式黏[粘]度计 penetro-visco(si)meter
针入仪 <检验水泥凝结时间用> needle penetrometer
针入硬度计 penetrometer
针塞 bullet;needle plug;needle type valve
针舌阀 annular valve
针蚀 streak rust
针示电报机 needle telegraph
针示刻度盘 indicator
针式除锈器 needle scaler
针式打印 dot matrix printer
针式打印机 dot printer;matrix printer;wire printer
针式电位计 needle electrometer
针式浮子 nail float
针式贯入法试验 needle penetration test
针式贯入器 needle penetrometer
针式贯入仪 needle penetrometer
针式绝缘子 needle insulator;pin insulator
针式喷管 needle nozzle
针式喷嘴 needle nozzle;pintle nozzle
针式拾音器 needle pick-up
针式调节喷嘴 needle regulating nozzle
针式轴承 needle bearing
针输送纸卷筒 pinfeed drum;pinfeed platen
针束除锈器 needle scaler
针束式气动除锈器 pneumatic needle scaler
针栓 pintle
针栓喷雾器 pintle atomizer
针栓式燃油喷嘴 pintle type fuel nozzle
针水砷钙石 vladimirite
针碎机 pick breaker; pin breaker; spike mill
针碳钠钙石 gaylussite
针铁矿 goethite; needle ironstone; rubinglimmer; ruby mica; xanthsiderite
针铁矿骨料 goethite aggregate
针铁矿集料 goethite aggregate
针铁矿岩 goethite rock
针铁绿泥石 mackensite
针铁石 needle ironstone
针头 pinhead
针头水滴 pinhead-size water droplet
针头钻 pin drill
针托 needle holder
针尾座 adapter bracket
针隙 needle gap
针隙避雷器 needle gap lightning
针线盒 work box
针线铺 haberdasher's shop; haberdashery
针形冰 crystals ice
针形测压计 <气体或蒸汽的> needle ga(u)ge
针形插孔 pin jack
针形导阀 spear pilot valve
针形的 aciform;acicular
针形电极 needle electrode;pin-shaped electrode
针形阀 annular valve;needle control valve; needle nozzle; needle type valve;needle valve;pintle valve
针形阀塞 needle plug
针形阀自动关闭压力 automatically shutting pressure of needle valve
针形管 needle tube[tubing]
针形光电管 pin-photo diode
针形滚筒 spiked cylinder;spiked drum
针形换热器 needle recuperator
针形活塞 needle piston
针形活塞导管 needle piston guide
针形极 pin-shaped electrode
针形结晶 needle crystal
针形绝缘子 pin insulator;pin type insulator
针形梁 needle beam
针形喷雾器 pintle atomizer
针形喷嘴 needle nozzle
针形热电偶 needle thermocouple
针形水尺 needle ga(u)ge;point water ga(u)ge
针形水位计 point ga(u)ge
针形套筒 needle sleeve
针形调节喷嘴 needle-regulating nozzle
针形突起 nadel
针形图案 needle pattern
针形纤维 raphioid fiber
针形阴极 needle-shaped cathode
针形闸门 needle gate
针形振捣器 needle vibrator
针形支柱绝缘子 pin type support insulator
针形轴承 nail bearing;needle bearing
针形柱塞 needle plug
针形桩 needle pile
针形状电极 pin-shaped electrode
针型除锈器 needle scaler
针型阀 pin valve
针型润滑器 needle lubricator
针绣的 stitched
针选穿孔卡片 needle-operated punched card
针压 stylus pressure
针压力 needle force
针眼 pin-hole;pin-holing;spit-out
针眼渗漏 pin-hole leak
针堰 needle weir
针焰点 pinpoint flame
针叶材板材 softwood board
针叶材板条 softwood strip
针叶材的厚板 softwood plank
针叶材胶合板 softwood plywood
针叶常绿林 needle-leaved evergreen forest
针叶灌木林 conifruticeta
针叶灌木群落 aciculifruticeta; conifruticeta
针叶林 coniferous forest;forest of conifer species;needle-leaved forest
针叶林带 coniferous forest region
针叶林群落 aciculignosa;conophorium
针叶木 needle bush; needle-leaved wood
针叶木本林 conilignosa
针叶木本群落 aciculignosa;conilignosa
针叶木材 dealwood
针叶乔木林 conisilvae
针叶乔木群落 aciculisilvae;conisilvae
针叶树 acerous tree; acicular-leaved tree; cone bearer; cone bearing tree;coniferous tree;needle-leaved tree;non-porous timber;conifer
针叶树材 coniferous timber; coniferous wood; non-pored timber; softwood
针叶树材焦油 softwood tar
针叶树材焦油沥青 softwood tar pitch
针叶树的 acerose;coniferous
针叶树胶合板 softwood plywood
针叶树林 aciculisilvae; coniferous forest;softwood forest
针叶树林和灌木 aciculignosa
针叶树木材 wood of coniferous tree
针叶树上的含油树脂 oleoresin from conifers
针叶树树脂 conifer balsam
针叶相思树 needle acacia
针叶云母 siderophyllite
针叶植被 aciculignosa
针叶植物 conifer; coniferals; coniferophyte;needle-leaved plant
针叶装饰 <科林思柱> acanthus spinose
针叶状的 acerose
针用黄铜 pin metal
针扎地毯 needled carpet
针织 knit
针织厂 hosiery;knit goods mill;knitting mill
针织的绒头地毯 stitched-on pile carpet
针织地毯 knitted carpet
针织机架 knitting frame
针织品 hosiery; knit article; knit goods;knitwear
针织手套 knitted glove
针织土工布 knitted geotextile
针织物 knit goods;knitted fabric
针柱石 dipyre;dipyrite;mizzonite
针柱状变晶结构 acicular-prismatic blastic texture
针状 acerous
针状贝氏体 needle-like bainite
针状变晶结构 acicular blastic texture
针状冰 acicular ice;candle ice;needle ice;spicular ice
针状测试点 needle test point
针状虫胶 needle lac
针状磁粉 needle-shaped particle
针状的 acerate;acerose;acicular;aciform;needle-like;needle-shaped
针状电极 pin-shaped electrode
针状断口 needle fracture
针状阀【机】 needle valve;needle type valve
针状粉粒 needle
针状粉末 acicular powder; needle powder
针状峰 aiguille
针状构造【地】 needle-shaped structure;acicular structure
针状骨料 elongated aggregate
针状滚柱轴承 needle roller bearing
针状活门座 needle valve seat
针状火焰燃烧器 needle flame burner;pin-hole burner;rat-tail burner
针状计数管 needle counter
针状尖顶 needle spire
针状焦 needle-like coke
针状结构 acicular texture; needle-shaped texture
针状结晶 aciculate crystal; crystal needle
针状结晶体 acicular crystal
针状结晶岩石 needlestone
针状晶束 raphide
针状晶体 acicular; needle; needle crystal
针状晶体结构 needle-shaped crystal structure
针状晶体锡 needle-tin
针状颗粒 elongated particle; elongated piece;needle-shaped particle
针状蜡 needle wax
针状蜡结晶 needle wax crystals
针状马氏体 acicular martensite; needle type martensite
针状煤 needle coal
针状盘 spider
针状喷浴 needle bath
针状喷嘴 nozzle with spear
针状润滑器 needle lubricator
针状筛孔 needle mesh
针状石油焦 acicular coke; needle crystal
针状霜 needle frost
针状丝质体 fusinite needles
针状碳化物钻头 pin type carbide bit
针状体 needle;spicule
针状体簇 cluster of needles
针状体束 rosette of needles
针状铁素体 acicular ferrite
针状突起 nadel
针状物 needle;solar spicule;spicule
针状物际区 interspicular region
针状纤维 raphidine
针状锌白 needle zinc white
针状颜料 acicular pigment; needle-like pigment
针状氧化锌 acicular type zinc oxide; aciculate type zinc oxide
针状叶 needle leaf
针状硬合金钻头 needle tungsten carbide bit;pin type carbide bit
针状渣孔 slag pin hole
针状振捣器 needle vibrator
针状支座 needle support
针状轴 needle shaft
针状铸铁 acicular iron
针状装饰 styliform ornament
针状紫胶 needle lac
针状组织 acicular constituent; acicular structure
针状组织钢 needle steel
针撞打桩锤 buttering ram
针纵 spruce
针座 handle file; needle file; needle stand
针座环 needle ring
针座三角 needle lock cam
针座轴 needle ingot shaft

侦 测飓风气球 hurricoon

侦查实验 investigative test
侦察 reconnoiter[reconnoitre];scout
侦察船 reconnaissance ship; scout ship;surveillance ship
侦察飞船 reconnaissance flying boat
侦察飞行 reconnaissance flight
侦察观测水上飞机 scout-observation seaplane
侦察活动中心 reconnaissance operations center[centre]

侦察机 scout
侦察基准点 reconnaissance reference point
侦察记录装置 reconnaissance recorder
侦察舰 scout
侦察接收机 ether scanner;melodeon;panoramic receiver
侦察雷达 recce radar;reconnaissance radar
侦察目标 reconnaissance objective;spot
侦察设备舱 recce pack
侦察设备吊舱 recce pod;reconnaissance pod
侦察摄影 intelligence photography;surveillance photography
侦察摄影机 detective camera
侦察声呐 reconnaissance sonar
侦察探孔 scout hole
侦察艇 reconnoitring boat
侦察通信[讯]船 aviso
侦察图 reconnaissance diagram
侦察卫星 photoreconnaissance satellite;radar mapping satellite;reconnaissance satellite;spy-in-the-sky;spy satellite
侦察行动图 escape and evasion graph
侦察性测量 detection survey
侦察巡航艇 scout cruiser
侦察仪器舱 reconnaissance module
侦察仪器箱 reconnaissance pallet
侦察用轨道式稳定平台 reconnaissance orbiting stabilized platform
侦察鱼群的网 look on net;tave
侦察与水下爆破队 reconnaissance and underwater demolition group
侦察员 investigator;scout
侦察站 intelligence platform;listening station
侦察者 reconnoiterer
侦车器 vehicle detecting equipment;vehicle detector
侦车设备 vehicle detecting equipment
侦疵光电装置 aniseikon
侦声器 horn-hunter
侦听站 intercept station

珍宝 gem;valuable

珍宝珐琅 jeweler's enamel
珍宝业 gemstone and jewel(l)ery industry
珍本 curiosa;unique copy
珍本书 rare book
珍本书库 rare book storeroom
珍本图书馆 rare book library
珍藏品橱 trophy case
珍藏品匣 trophy case
珍贵的 rare;scarce
珍贵动物 precious animal
珍贵树种 precious tree;rate tree
珍贵物品 articles of value
珍贵物种 precious tree
珍品 curio;curiosa;curiosity;gem
珍品展览室 panopticon
珍视的 valued
珍稀动物 rate animal
珍稀动物保护 conservation of rate animal
珍稀物种 rare species
珍珠 pearl
珍珠白 pearl white
珍珠斑岩 pearlite porphyry
珍珠层 nacre
珍珠灯 pearl lamp
珍珠粉 pearl powder
珍珠构造【地】perlitic structure
珍珠光玻璃 satin finish glass;velvet-finish glass
珍珠光清漆 nacreous varnish
珍珠光泽 nacreous luster[lustre];pearly luster[lustre]
珍珠灰 pearl ash
珍珠结构 nacrite;perlitic texture;shelly structure
珍珠克拉 <珍珠的重量单位> pearl grain
珍珠母 mother;mother-of-pearl
珍珠泉风景区 pearl spring scenic area
珍珠色 pearl
珍珠石 nacrite
珍珠素 pearl essence
珍珠陶土 nacrite[nakrite]
珍珠岩 pearlite;pearlstone;pearlyte;perlite
珍珠岩爆裂 pearlite popping
珍珠岩堵漏粉 controlite
珍珠岩粉 ground perlite
珍珠岩隔热混凝土 pearlite insulating concrete
珍珠岩隔音混凝土 pearlite insulating concrete
珍珠岩骨料 pearlite aggregate;perlite aggregate
珍珠岩灰浆 pearlite plaster;perlite plaster
珍珠岩灰泥 pearlite plaster;perlite plaster
珍珠岩混凝土 pearlite concrete;peripheral concrete;perlite concrete
珍珠岩混凝土衬背墙 pearlite concrete backup wall
珍珠岩混凝土屋顶 pearlite concrete roof
珍珠岩混凝土屋顶板 pearlite concrete roof slab
珍珠岩混凝土屋面板 pearlite concrete roof slab
珍珠岩集料 perlite aggregate
珍珠岩绝热材料 pearlite insulating material
珍珠岩绝缘 pearlite insulation
珍珠岩矿 pearlite ore
珍珠岩矿床 perlite deposit
珍珠岩沥青混凝土 pearlite asphalt concrete
珍珠岩砾石 pearlite gravel
珍珠岩黏[粘]结剂 pearlite binder
珍珠岩墙 pearlite wall
珍珠岩轻质混凝土 pearlite lightweight concrete
珍珠岩砂浆 pearlite plaster
珍珠岩石膏吊顶板 pearlite-gypsum ceiling board
珍珠岩石膏灰泥 gypsum perlite plaster
珍珠岩松散填充绝热材料 pearlite loose fill insulation
珍珠岩松散填充料 pearlite loose fill
珍珠岩松填 pearlite loose fill
珍珠岩松填隔热 pearlite loose fill insulation
珍珠岩吸声平顶 pearlite absorbent ceiling;pearlite acoustic(al) ceiling;pearlite sound absorbent ceiling
珍珠养殖 pearl culture
珍珠养殖场 pearl farm
珍珠釉 pearl glaze
珍珠云母 lime mica;margarite
珍珠质 nacre
珍珠蛭石 dudleyite
珍珠状物 pearl

真斑岩 true porphyry

真半径 true semidiameter
真报表 true statement
真北 geographic(al) north;true north;true north heading
真北测量 true north survey
真北方位角 true north azimuth
真北方向 true north direction
真奔赴点 true vertex
真本原二次型 properly primitive quadratic form
真本原二元二次型 properly primitive binary quadratic form
真比例尺 real scale;true scale
真比重 intrinsic(al) density;real specific gravity;true density;true specific gravity;true specific weight
真扁率 real flattening
真变形 true strain
真标准差 true standard deviation
真表达 true statement
真表达式 true expression
真表示 true representation
真并行 actual pairing
真补数 true complement
真不二价 never quote two price
真材实料 real stuff;real thing
真彩色航片 colo(u)r airphoto
真彩色图像 true colo(u)r image
真草甸 true meadow
真草原 true steppe
真侧壁摩擦力 true side-wall friction
真侧壁摩阻力 true side-wall friction resistance
真差 actual error;true error
真长度 proper length
真常绿林 true evergreen forest
真常数 true constant
真潮 actual tide;natural tide
真陈述 true statement
真诚的 single hearted;single-minded
真程序 proper program(me)
真赤道 true equator
真出口 true exit
真出没 <太阳、月亮> true rising and setting
真船位 true position
真垂线 true perpendicular;true vertical
真垂直圈 true vertical
真春分点 true equinox;true vernal equinox
真纯度 true purity
真磁荷 true magnetic charge
真枞 true firs
真当地时间 local true time
真底板 true bottom
真底岩 ledge rock
真底岩爆破 chip blasting
真地平经圈 true vertical
真地平面 true horizon plane;true horizontal plane
真地平(圈) celestial horizon;rational horizon;real horizon;true horizon;astronomic(al) horizon
真地平线 celestial horizon;true horizon
真地速测速仪 true ground speed meter
真地下水流速 true groundwater velocity
真点突变 true point mutation
真电阻 true resistance
真顶点 true vertex
真顶距 true zenith distance
真定向 <按真子午线定向> true orientation
真东西圈 true prime vertical
真断距【地】true slip
真二次曲面 proper quadric;proper quadric surface
真返航向 true homing
真方位 true amplitude;true direction
真方位角【测】true azimuth;true bearing
真方位显示装置 true bearing unit
真方位线 true line of bearing
真方位指示装置 true bearing adaptor;true bearing unit
真方向 true direction;true heading;true track
真方向电子计算机 true heading computer set
真方向角 true bearing
真芳基硝基化合物 true aryl nitro-compound
真沸点-气液色谱法 true-boiling point gas liquid chromatography
真分数 proper fraction;true fraction
真丰度 true abundance
真风 true wind
真风速 true air speed
真风向 true wind direction
真缝 true joint
真浮游 euplankton
真幅 <相对于真东或真西的摆幅> true amplitude
真辐合 true convergence
真俯角 <真地平线与摄影轴之间的交角> true depression angle
真高度 corrected sextant altitude;true altitude
真光层 euphotic layer
真光轴角 true axial angle
真航高 real flying height
真航迹向 true track
真航迹指示器 time tracking indicator
真航向 made course;true bearing;true course;true heading
真核微型浮游生物 eukarytic
真黑耀岩 true obsidian
真恒星时 true sidereal time
真厚度 true thickness;true width
真黄土 true loess
真黄炸药 true dynamite
真回归面 true regression plane
真活化能 true activation energy
真机雷诺数 prototype Reynolds number
真机水轮机 actual turbine;full-scale turbine
真机水头试验 real-head test rig
真迹版 autography
真迹石印版 autograph
真极 true pole
真极大 proper maximum
真家菌 house fungus
真胶体 eucolloid
真角距 true angular distance
真近点角 true anomaly
真经度 true longitude
真距离 proper distance;true distance
真均值 true mean
真菌 fungus[复 funguses/fungi]
真菌孢子 fungal spore
真菌孢子病 sporomycosis
真菌病 mycosis
真菌的 fungous
真菌冬孢子 teleutospore
真菌腐烂 fungal decay
真菌纲 <拉> eumycetes;Fungi
真菌降解 fungal degradation
真菌菌质体 fungosclerotinite
真菌(菌种)培养 fungi cultivation
真菌类 eumycetes
真菌侵蚀 fungus attack
真菌杀虫剂 fungal insecticide
真菌生长 fungi growth
真菌生物质吸附剂 fungal biomass sorbent
真菌体 funginite
真菌性褐变 fungous brown stain
真菌絮凝剂 fungus flocculant
真菌学 mycology
真菌芽孢 gemma[复 gemmae]
真菌质结构镜质体 fungotelinite
真菌状的 fungiform;fungoid

真柯巴树脂 true copal
真空 air-vacuum; depression; empty space; negative pressure; underpressure; vacuo; vacuum
真空 X 射线分光计 vacuum X-ray spectrometer
真空安全阀 antivoid valve; pressure vacuum relief valve; vacuum relief valve; vacuum safety valve
真空白炽灯 vacuum filament lamp; vacuum incandescent lamp; vacuum lamp
真空白炽灯泡 vacuum incandescent bulb
真空板 vacuum plate
真空包埋机 vacuum investing machine
真空包装 vacuumize; vacuum package; vacuum packaging
真空包装的 vacuum packed
真空饱和的 vacuum saturated
真空饱和法 vacuum saturating; vacuum saturation
真空保藏法 vacuum preservation
真空保干器 vacuum desiccator
真空保压阀 vacuum relief valve
真空报警器 vacuum horn
真空曝光架 vacuum exposure frame
真空爆裂气体测量法 vacuum decrepitation gas measurement method
真空泵 ejector; exhauster; pick-up pump; sucking pump; suction pump; vac pump; vacuum fan; vacuum pump
真空泵抽空压气机 vacuum pump compressor
真空泵抽气 exhaust of vacuum pump
真空泵电源 vacuum pump power supply
真空泵高速阀 high-speed valve of exhauster
真空泵管道 vacuum pump line
真空泵接头 vacuum pump adapter
真空泵调节阀 exhauster regulating valve
真空泵调压器 exhauster governor
真空泵扬程 vacuum pump lift
真空泵摇臂 vacuum pump rocker arm
真空泵油 pumping fluid; vacuum pump oil
真空泵运输机 vacuum pump transporter
真空比重计 < 测定沥青混合料最大比重的一种试验仪器 > vacuum pycnometer
真空笔 vacuum pencil
真空避雷器 vacuum arrester; vacuum lightning arrester; vacuum lightning protector; vacuum type lightning protector
真空便池 vacuum toilet
真空变速 vacuum gear shift(ing)
真空表 vacuum ga(u)ge; vacuum manometer; vacuum meter; vacuum watch
真空波长 vacuum wavelength
真空波强度 expansion strength
真空玻璃 vacuum window glazing
真空补充制动器 vacuum servo brake
真空捕尘凿岩机 vacujet
真空采暖 vacuum heating
真空采暖系统 vacuum heating system; vacuum type steam heating system
真空采气法 vacuum sampling method
真空采芯管 vacuum corer
真空采样器 vacuum corer; vacuum core sampler
真空操纵的 vacuum operated
真空操纵离合器【机】vacuum operated clutch
真空操作的热水加热器调节阀 vacuum operated hot water heater control valve
真空操作模板 vacuum operation mat
真空槽 vacuum tank
真空侧 inlet side
真空侧管 vacuum bypass
真空测定计 vacuum ga(u)ge; vacu(um)ometer
真空测辐射热计 vacuum bolometer
真空测量 vacuum measurement
真空测量仪表 vacuum measuring instrument
真空测压计 < 量测低于周围气压的压力用 > vacuum meter; vacuum manometer
真空层合 vacuum lamination
真空常压法 vacuum normal pressure method
真空厂房 vacuum building
真空场 vacuum field
真空超载预压加固法 consolidation by vacuum and surcharge preloading method
真空车长阀 vacuum conductor's valve; vacuum emergency valve
真空沉积 vacuum deposition
真空沉积的 vacuum deposited
真空沉积金属膜 vacuum metallizing
真空衬护 vacuum lining
真空成长晶体 vacuum grown crystal
真空成型的 vacuum formed
真空成型法 vacuum forming
真空成型机 vacuum forming machine
真空成型立体模型 vacuum drum model
真空成型立体图 vacuum mo(u)lded relief map
真空成型设备 vacuum forming equipment
真空澄清器 vacuum clarifier
真空池 vacuum tank
真空充填法 vacuum fill
真空充填机 vacuum stuffer
真空充填器械包 vacuum plugging kit
真空冲击电流计 vacuum quantometer
真空冲洗便池系统 vacuum flush toilet system
真空冲洗大便器 vacuum flush toilet
真空冲洗式大便器 vacuum flush toilet system
真空抽风机 vacuum blower
真空抽气泵 vacuum air pump; vacuum exhaust pump
真空抽气机 vacuum air pump; vacuum fan; vacuum pump
真空抽气机组 vacuum group
真空抽气设备 vacuum pumping equipment
真空抽气系统 vacuum pumping system
真空抽提 vacuum extraction
真空抽吸 vacuum draw
真空抽吸成型 vacuum suction process
真空除尘 vacuum cleaning
真空除尘法 vacuum arresting method
真空除尘管 vacuum cleaning line
真空除尘管线 vacuum cleaning line
真空除尘集合管 vacuum manifold for dust removal
真空除尘歧管 vacuum manifold for dust removal
真空除尘器 suction sweeper; vacuum cleaner; vacuum dust catcher; vacuum manifold for dust removal; vacuum sweeper
真空除气 vacuum deaeration; vacuum degasing
真空除气器 suction deaerator; vacuum deaerator
真空除氧器 suction deaerator; vacuum deaerator
真空储藏 vacuum storage
真空处理 vacuum handling; vacuuming; vacuumization; vacuum treating; vacuum treatment
真空处理的 vacuum treated
真空处理的混凝土 vacuum treated concrete; vacuum-processed concrete
真空处理法 vacuum process; Billnear method < 混凝土 >
真空处理钢 vacuum treated steel
真空处理混凝土 air-free concrete; vacuum processed concrete
真空处理混凝土管 vacuum concrete pipe
真空传感阀 vacuum sensitive valve
真空磁控溅射涂膜玻璃 vacuum magnetron sputtering coating glass
真空带式过滤器 vacuum belt filter
真空袋 vacuum bag
真空袋模法 vacuum bag process
真空袋模塑 vacuum bag technique
真空袋模塑成型 vacuum bag mo(u)lding
真空袋模制法 vacuum bag mo(u)lding
真空导磁率 space permeability
真空导杆 vacuum guide
真空导管 vacuum pipe(line); vacuum piping
真空导向器 vacuum tape guide
真空导向器伺服 vacuum guide servo
真空的 air-free; vacuous
真空灯 vacuum lamp
真空灯泡 vacuum bulb
真空低温恒温器 vacuum cryostat
真空地带 no man's land
真空点火提前机构 vacuum spark advance mechanism
真空点火提前控制阀 vacuum advance control valve
真空点火提前装置 vacuum advancer; vacuum ignition advancer; vacuum spark advancer
真空点火正时 vacuum controlled sparking timing
真空电磁阀 vacuum solenoid
真空电镀 vacuum plating
真空电弧炉 vacuum arc furnace
真空电弧熔化法 vacuum arc melting process
真空电弧熔化炉 vacuum arc furnace melting
真空电弧熔炼 vacuum arc melting
真空电接触器材 vacuum contact
真空电绝缘 vacuum electric(al) insulation
真空电离测定器 vacuum ionization detector
真空电离传感器 vacuum ionization sensor
真空电离计 vacuum ionization ga(u)ge
真空电离检测器 vacuum ionization detector
真空电离室 vacuum ionization chamber
真空电流铸造机 vacuum casting
真空电炉 vacuum electric(al) furnace; vacuum oven
真空电容率 permittivity of vacuum; space permittivity
真空电容器 vacuum capacitor
真空电子束焊 electron bombardment welding; electronic bombardment welding; vacuum electron beam welding
真空电阻炉 vacuum resistance furnace
真空垫 vacuum pad; vacuum mat < 从混凝土中吸出水分时用 >
真空淀积 vacuum deposition
真空淀积电路 vacuum deposited circuit
真空吊车 vacuum crane
真空吊货装置 vacuum lifting gear
真空动力汽缸 vacuum power cylinder
真空冻结 vacuum freezing
真空度 degree of vacuum; subatmospheric(al) pressure; vacuity; vacuum degree; vacuum level; vacuum pressure; vacuum tightness
真空度降低 vacuum down
真空度升高 vacuum up
真空度系数 tenuity factor
真空度英寸数 inch of vacuum
真空度指示器 vacuum indicator
真空镀敷法 vacuum deposited coating
真空镀敷金属 vacuum metallizing
真空镀积法 vacuum vapo(u)r deposition method
真空镀金 vacuum metallizing
真空镀(金属)膜涂料 vacuum metallizing coating
真空镀镁 vacuum deposition of magnesium
真空镀膜 evaporation coating; sputtering; vacuum deposition; vacuum evapo(u)rated film; vacuuming metalling; vacuum plating; vapo(u)r deposition; vapo(u)r plating
真空镀膜法 vacuum coating
真空镀膜烫金法 metallizing
真空镀膜涂装法 vacuum metallizing
真空断流器 vacuum breaker
真空断路器 vacuum breaker; vacuum circuit breaker; vacuum interrupter
真空多级蒸发器 vacuum multistage evapo(u)rator
真空发动机 vacuum engine
真空发生器 vacuum generator
真空阀 dropper; vacuum valve
真空法测镭含量 radium content determined by vacuum method
真空法抽水井点 vacuum method well point
真空法处理 vacuum process
真空法混凝土 vacuum concrete; vacuum processed concrete
真空法井点排水 vacuum process of wellpointing
真空法精制地沥青 vacuum-reduced asphalt; vacuum-refined asphalt
真空法试验砂 vacuum method of testing sands
真空反吸成型 vacuum snap-back forming
真空反应 vacuum response
真空放电 discharge in vacuum
真空分光计 vacuum spectrometer
真空分离 vacuum separation
真空分离器 vacuum separator
真空分馏 vacuum topping
真空粉末 evacuated powder
真空粉末绝热 vacuum power insulation
真空风挡刮水摆臂 vacuum windshield wiper swing arm
真空风挡刮水器 vacuum windshield wiper
真空风挡刮水旋轴 vacuum windshield wiper pivot kicker
真空风缸 vacuum chamber
真空风扇 vacuum fan
真空风速表 vacuum anemometer
真空风速计 vacuum anemometer
真空封罐机 vacuum seamer
真空封接 vacuum seal
真空封蜡 picein wax

真空封帽 vacuum end cap
真空封泥 vacuum cement
真空封瓶 vacuum capping
真空封装 vacuum encapsulation
真空浮动式 vacuum suspended type
真空浮升法 method of vacuum floating
真空浮选 vacuum flo(a)tation
真空浮选法 vacuum flo(a)tation process
真空浮渣池 vacuator
真空釜 vacuum kettle
真空辅助成型 vacuum assisted mo-(u)lding
真空辅助树脂注射工艺 vacuum assisted resin injection process
真空辅助污泥脱水床 vacuum assisted sludge dewatering bed
真空负压 vacuum negative pressure
真空干碾机 vacuum pug mill
真空干涉仪 vacuum interferometer
真空干燥的 vacuum drying
真空干燥法 boulton process;vacuum drying;vacuum seasoning
真空干燥罐 vacuum drying pan
真空干燥柜 vacuum shelf dryer[drier]
真空干燥烘箱 vacuum drying oven
真空干燥炉 vacuum drying oven
真空干燥器 vacuum desiccator;vacuum drier[dryer];vacuum drying apparatus
真空干燥箱 vacuum drying chamber
真空干燥装置 Minton dryer;vacuum drier[dryer]
真空感应 vacuum induction
真空感应电炉 vacuum induction furnace
真空感应电熔炉 vacuum induction melting furnace
真空感应炉 vacuum inductance furnace
真空感应炉重熔法 vacuum induction remelt(ing)
真空感应熔化 vacuum induction melting
真空感应铸造 vacuum induction casting
真空缸 vacuum cylinder
真空缸活塞弹簧 vacuum cylinder piston spring
真空缸及柱塞总成 vacuum cylinder and plunger assembly
真空缸推杆导管 vacuum cylinder push rod guide
真空钢 vacuum steel
真空高温计 suction pyrometer
真空隔断器 vacuum breaker
真空隔离开关 vacuum interrupter
真空隔气具 vacuum trap
真空隔热 evacuated insulation;vacuum insulation
真空给水泵 vacuum feed(-water)pump
真空给油箱 vacuum feed tank
真空工程 vacuum engineering
真空工艺 vacuum technology
真空供给 vacuum feed
真空供暖系统 vacuum heating system
真空供油 vacuum fuel feed
真空(供油)装置 autovac
真空鼓式干燥机 vacuum drum drier [dryer]
真空固化 vacuum solidification
真空固结法 vacuum consolidation;vacuum method;vacuum preloading
真空刮水器 vacuum windshield wiper
真空管 electron(ic) tube;evacuated tube;vacuum tube[tubing];vacuum valve
真空管避雷器 vacuum-tube arrester
真空管玻璃 vacuum-tube glass
真空管残余气体测量仪 omegatron
真空管测试器 vacuum-tube test set
真空管道 vacuum pipe(line);vacuum piping
真空管道操作 vacuum line technique
真空管道输送机 vacuum pipeline conveyer[conveyor]
真空管滴定计 sectrometer
真空管电键器 vacuum-tube keyer
真空管电路 vacuum valve circuit
真空管电势计 valve potentiometer
真空管电压表 vacuum-tube voltmeter
真空管电压调整器 vacuum-tube voltage regulator
真空管发送器 vacuum-tube transmitter
真空管放大器 vacuum-tube amplifier
真空管敷层 vacuum-tube coating
真空管换能器 vacuum-tube transducer
真空管集热器 evacuated tube collector
真空管计算机 first generation computer
真空管计算器 vacuum tube counter
真空管继电器 vacuum-tube relay
真空管加速计 vacuum-tube accelerometer
真空管间耦合 vacuum intervalve coupling
真空管检波器 vacuum-tube detector
真空管键控法 vacuum-tube keying
真空管键控器 vacuum-tube keyer
真空管脚 valve pin
真空管静电计 vacuum-tube electrometer
真空管开关 vacuum-tube switch
真空管漏气放电现象 spot softening phenomenon
真空管时效 vacuum-tube ag(e)ing
真空管特性 vacuum-tube characteristic
真空管调制器 hard-type modulator
真空管线 vacuum line
真空管噪声 vacuum-tube noise
真空管振荡器 vacuum-tube generator;vacuum-tube oscillator;vacuum valve oscillator;valve generator
真空管整流器 vacuum-tube rectifier
真空管助听器 vacuum-tube hearing aid
真空罐 autovac;vacuum chamber;vacuum reserve;vacuum tank
真空罐式洒布机 vacuum tank spreader
真空罐式液肥喷洒机 vacuum tank spreader
真空光电池 vacuum photocell
真空光电辐射元件 vacuum photoemissive cell
真空光电管 vacuum photocell;vacuum phototube
真空光电元件 vacuum phototube
真空光具座 vacuum optic(al)bench
真空光谱法 vacuum spectrography
真空光速 speed of light in vacuum;vacuum light speed;vacuum light velocity
真空规管 vacuum ga(u)ge
真空柜 vacuum tank
真空辊 suction roll
真空锅 vacuum kettle;vacuum pan;vacuum pot
真空锅水 vacuum pan water
真空锅制盐 vacuum pan salt
真空过滤产水量 vacuum filtration yield
真空过滤法 vacuum filtration process
真空过滤过程 vacuum filtration process
真空过滤机 vacuum filter
真空过滤机脱水 dewatering of vacuum filter
真空过滤器 suction filter;vacuum filter
真空过滤作用 vacuum filtration
真空焊接 vacuum brazing;vacuum welding
真空和压力安全装置 vacuum and pressure relief assembly
真空黑漆 glyptal
真空黑橡胶板 Viton sheet
真空烘焙法 vacuum baking
真空后缩聚 vacuum finishing
真空呼吸阀 pressure vacuum vent valve
真空还原 vacuum reduction
真空环境 vacuum environment
真空缓冲筒 vacuum dash pot
真空换挡 vacuum gear shift(ing);vacuum power shift
真空换移 vacuum power shift
真空黄酸盐混合机 vacuum xanthate mixer
真空挥发 vacuum volatilization
真空回灌 vacuum recharge
真空回火装置 vacuum tempering apparatus
真空回水采暖系统 vacuum return line heating system
真空回水供暖系统 vacuum return line heating system
真空回水管 vacuum return pipe
真空回水系统 vacuum return line system
真空回水装置 vacuum return line device
真空回扬 vacuum back-pumping
真空回转过滤器 vacuum rotary filter
真空混合机 vacuum mixer
真空混凝土 vacuum concrete
真空混凝土法 vacuum concrete process
真空活塞 vacuum cock;vacuum piston
真空活塞板 vacuum piston plate
真空火花法 vacuum spark technique
真空火花放电 vacuum spark discharge
真空火花离子源 vacuum spark ion source
真空击穿 vacuum breakdown
真空机动闸 vacuum power brake
真空机械 vacuum machine
真空及燃油泵输出压力测试仪 vacuum fuel tester
真空级 vacuum stage
真空极化 vacuum polarization
真空集尘器 vacuum collector
真空集装箱 vacuum container
真空挤泥机 de-airing extruder;vacuum pug
真空挤压成型 extrusion under vacuum
真空挤压机 vacuum extruder;vacuum extrusion press
真空计 suction ga(u)ge;vacuometer;vacuscope;vacustat;vacuum ga(u)ge;vacuum indicator;vacuum meter
真空计管 vacuum ga(u)ge pipe
真空计控制电路 vacuum ga(u)ge control circuit
真空计线路箱 cabinet ga(u)ge control unit
真空计压力 subatmospheric(al)pressure;under-pressure;vacuum ga(u)ge pressure
真空技术 vacuum technique
真空继电器 low vacuum load-tripping device;vacuum relay
真空继动阀 vacuum relay valve
真空加工的 vacuum processed
真空加固法 consolidation by the vacuum method
真空加力的 vacuum assisted
真空加力式制动器 vacuum boost brake
真空加力液压制动器 hydrovac brake
真空加力制动 booster brake
真空加氯机 vacuum chlorinator
真空加热 vacuo heating;vacuum heating
真空加热巴氏杀菌器 vacu-therm pasteurizer
真空加热干燥 vacuum and heating drying
真空加热器 vacuum heater
真空加热提炼 vacuum hot extraction
真空加热直接杀菌器 vacu-therm instant sterilizer
真空加湿器 vacuum tempering apparatus
真空加压法 vacuum compression method;vacuum pressure process
真空夹管机 vacuum pipe clamp
真空夹盘 vacuum chuck
真空夹头 vacuum chuck
真空夹砖器 vacuum pad
真空减速机构 vacuum retarding mechanism
真空检漏 vacuum leak hunting
真空检漏器 sniffer;vacuum leak detector
真空检漏仪 vacuum leak detector
真空检验 vacuum test
真空降低跳闸试验 low vacuum trip test
真空降水预压法 prepressing process vacuum drawdown
真空浇注 vacuum assisted pouring
真空浇铸 vacuum casting
真空焦油 vacuum tar
真空焦油沥青 vacuum tar
真空搅拌 vacuum mixing
真空搅拌机 de-airing mixer;vacuum mixer;vacuum pug mill
真空搅土机 de-airing pug mill
真空校正 vacuum correction
真空校准浴 vacuum calibration bath
真空接受器 vacuum receiver
真空节制阀 vacuum moderating valve
真空结合 vacuum bond
真空结晶 vacuum crystallization
真空结晶器 vacuum crystallizer
真空截止阀 vacuum cut valve
真空解除阀 vacuum release valve;vacuum relief valve
真空解吸 vacuum desorption
真空介电常数 space permittivity
真空金属涂层 vacuum metallizing
真空紧急制动阀 vacuum conductor's valve;vacuum emergency valve
真空紧密的 vacuum tight
真空浸蜡包埋法 vacuum paraffin embedding method
真空浸渗设备 mogullizer
真空浸提 vacuum diffusion
真空浸透 vacuum impregnation
真空浸涂 vacuum press-in coating
真空浸渍 vacuum impregnating;vacuum impregnation;vacuum potting
真空浸渍的 vacuum impregnated
真空浸渍设备 vacuum impregnation equipment
真空精炼 vacuum refining
真空精炼炉 vacuum refining furnace
真空精馏 rectification under vacuum;vacuum recrystallization
真空精馏装置 vacuum rectifying apparatus
真空精蒸馏塔 vacuum fractionating distilling column;vacuum fractionator
真空精制 vacuum refining
真空井 vacuum well
真空井点 vacuum well point
真空井排水系统 vacuum well dewatering system

Z

真空净化 vacuum clear; vacuum purification
真空净化器 vacuum cleaner
真空净化器用合金 vacuum cleaner alloy
真空净化设备 vacuum cleaning apparatus; vacuum cleaning installation
真空距离 vacuum distance
真空聚爆式震源 flexichoc
真空聚合 vacuum polymerization
真空卷边器 vacuum crimper
真空绝热 vacuum insulation; vacuum thermal insulation
真空绝热的 vacuum tight
真空绝热罐 vacuum thermal insulation tank
真空绝热箱 vacuum insulated tank
真空绝缘 vacuum insulation
真空绝缘的 vacuum-insulated
真空开关 vacuum switch
真空开关管 vacuum switch tube
真空开关装置 vacuum switchgear
真空靠背 vacuum back
真空壳汞整流器 dynectron
真空可变电容器 vacuum variable capacitor
真空空间 vacuum space
真空空气两用制动机 double brake
真空孔隙 evacuated pore
真空控制 vacuum control
真空控制的 vacuum controlled
真空控制点火定时装置 vacuum operated spark control
真空控制点火提前机构 vacuum spark control
真空控制电磁阀 vacuum control modulator valve
真空控制阀 vacuum control valve
真空控制风挡刮水器 vacuum windshield wiper control
真空控制换挡 vacuum controlled gearshift
真空控制活塞 vacuum control piston
真空控制开关 vacuum switch
真空控制离合器 vacuum control clutch
真空控制式齿轮变速 vacuum controlled gear shifting
真空控制式点火提前 vacuum controlled ignition advance
真空控制式离合器 vacuum controlled clutch
真空控制提前点火 vacuum-controlled advance
真空控制系统 vacuum controlled system
真空控制液压 vacuum modulator pressure
真空控制装置 vacuum control unit
真空快速 vacuum flash
真空快速蒸发器 vacuum flash vaporizer
真空框 vacuum box
真空扩散 vacuum diffusion
真空扩散泵 vacuum diffusion pump
真空扩散焊接 vacuum diffusion welding
真空垃圾车 vacuum garbage truck
真空垃圾清扫车 vacuum loader
真空冷冻 vacuum refrigeration
真空冷冻法 vacuum-freezing process; vacuum refrigerating process
真空冷冻干燥 vacuum freeze drying
真空冷冻干燥器 vacuum freeze dryer [drier]
真空冷冻机 vacuum refrigerating machine
真空冷冻喷射吸附法 vacuum-freezing ejector adsorption method
真空冷冻蒸汽压缩法 vacuum-freezing vapo(u)r compression
真空冷阱 vacuum cold trap; vacuum trap
真空冷凝 vacuum condensation
真空冷凝点 vacuum condensing point
真空冷凝器 vacuum condenser
真空冷却 vacuum chilling; vacuum cooling
真空离心泵 vacuum centrifugal pump
真空练泥机 de-airing pug mill; vacuum deairing machine; vacuum pug mill
真空量 vacuum capacity
真空料道 vacuum track
真空列车管 vacuum train line
真空漏气速率 vacuum leak rate
真空漏气指示器 vacuum leak detector
真空漏泄 vacuum leak
真空炉 vacuum firing furnace; vacuum furnace; vacuum oven
真空炉测湿法 vacuum oven method
真空炉加热 vacuum furnacing
真空炉容器 vacuum furnace container
真空炉烧结 vacuum furnacing
真空炉冶炼 vacuum furnacing
真空滤池 suction filter; vacuum filter
真空滤油机 vacuum oil strainer
真空螺旋泵 aspirating propeller pump
真空马达 vacuum motor
真空毛细管黏[粘] vacuum capillary viscometer
真空密闭 vacuum tightness
真空密闭罐 vacuum tight retort
真空密闭室 vacuum tight chamber
真空密封 hermetic seal; leak tight; vacuseal; vacuum seal(tight); vacuum tightness
真空密封窗 vacuum tight window
真空密封的 vacuum tight
真空密封接头 hermatical seal
真空密封蜡 vacuum seal wax
真空密封炉盖 vacuum tight furnace cover
真空密封容器 vacuum tight container
真空密封外壳 vacuum envelope; vacuum tight sheath
真空密实法 vacuum process
真空面 suction side; suction surface
真空模板 vacuum form(work); vacuum shutter(ing); vacuum mo(u)ld(ing); vacuum panel <真空作业用的>
真空模板衬 absorptive form lining
真空模板混凝土 vacuum form concrete; deaerated concrete
真空模塑法 vacuum mo(u)ld(ing)
真空模铸 vacu-forming; vacuum casting
真空模子 vacuum tight mo(u)ld
真空膜盒 aneroid capsule; aneroid chamber; bellows; capsule; evacuated bellows
真空膜盒压力计 bellows ga(u)ge
真空膜盒元件 evacuated aneroid element
真空膜匣 aneroid chamber
真空能级 vacuum level
真空泥料搅拌混合器 vacuum pug mixer
真空黏[粘]度计 vacuum visco(si)meter
真空黏[粘]合 vacuum bonding
真空黏[粘]结剂 vacuum cement
真空捏合机 vacuum kneader
真空捏土机 de-airing pug mill
真空镊子 vacuum tweezer
真空凝气瓣 vacuum trap
真空浓缩 vacuum concentration
真空浓缩机 vacuum decker
真空排放 vacuum discharge
真空排粉 vacuum pick-up
真空排气 vacuum deaeration; vacuum exhaust; vacuum pumping
真空排气法 vacuum deaeration; vacuum removal method
真空排气扇 vacuum fan
真空排水 drainage by suction; suction method drainage; vacuum dewatering
真空排水法 vacuum dewatering method; vacuum drainage method; vacuum method of drainage; vacuum drainage
真空排水管道系统 vacuum drainage pipe system
真空排水井系统 vacuum well system
真空排水系统 vacuum sewerage system
真空排移 vacuum power shift
真空盘 vacuum pan
真空盘架干燥器 vacuum shelf tray dryer[drier]
真空泡 vacuole; vacuum bubble; vacuum envelope
真空配件 vacuum fittings
真空喷镀 vacuum evapo(u)rating; vacuum evapo(u)ration
真空喷枪 airless spray gun
真空喷砂 vacuum blast
真空喷砂法 vacuum blasting
真空喷砂清理 airless blast cleaning
真空喷射泵 vacuum jet
真空喷射器 vacuum ejector
真空喷射式柴油发动机 compressorless injection diesel(engine)
真空喷射装置 vacuum jet device; vacuum jet package
真空喷涂 airless spray; vacuum spraying
真空喷涂法 vacuum deposition; vacuum spray painting method
真空喷涂金属设备 vacuum metallizing equipment
真空喷雾器 vacuum dissipator
真空喷雾蒸发器 vacuum spray evapo(u)rator
真空膨胀计 vacuum dilatometer
真空膨胀器 vacuum expander
真空皮管 suction hose
真空漂浮法 vacuum flo(a)tation
真空漂移 vacuum fluctuation
真空瓶 Dewar flask; vacuum bottle; vacuum flask; vacuum jacketed container; vacuum jar; vacuum vessel; Dewar bottle <独瓦所发明的镀银的双层玻璃瓶，夹层中抽去空气，如热水瓶胆>
真空破坏 vacuum breaking
真空破坏阀 air admission valve; vacuum break(ing) valve
真空破坏管 vacuum-breaker line
真空破坏器 vacuum breaker
真空破碎机 vacuum breaker
真空歧管 vacuum manifold
真空启动 priming by vacuum
真空起吊 vacuum lifting
真空起动 priming by vacuum
真空起重 vacuum lifting
真空气泵 vacuum air pump
真空气浮法 vacuum flo(a)tation process
真空气缸垫密片 vacuum cylinder gasket
真空气缸加强板 vacuum cylinder reinforcing plate
真空气缸油 vacuum cylinder oil
真空气流干燥器 pneu-vac drier
真空气密水泥 vacuum cement
真空气密性 vacuum tightness
真空气球 vacuum balloon
真空汽缸 vacuum cylinder
真空汽化 vacuum vapo(u)rization
真空汽化器 vacuum carburet(t)or
真空器 vacuum apparatus
真空钎焊 vacuum brazing
真空钎焊换热器 braised heat exchanger
真空牵拉装置 vacuum take-down
真空侵入 vacuum infiltration
真空青贮 vacuum silage
真空清舱系统 vacuum stripping system
真空清除法 vacuum cleaning
真空清洁车 vacuum car
真空清洁机具 vacuum cleaning plant
真空清洁器 vacuum cleaner
真空清洁设备 vacuum cleaning plant
真空清洁装置 vacuum cleaning plant
真空清扫器 vacuum sweeper
真空球 vacuum sphere
真空球磨法 vacuum grinding method
真空区 empty space; region of no pressure
真空区域 vacuum area
真空曲线 vacuum curve
真空驱动式风窗刮水器 vacuum windshield cleaner
真空驱动式毛剪 vacuum operated clipper
真空驱动系统 vacuum actuated system
真空驱气 vacuum purge
真空取土器 vacuum corer
真空取样法 vacuum sampling method
真空取样管 vacuum corer; vacuum sampler
真空取样器 vacuum sampler; vacuum sampling tube
真空去气混凝土 vacuum deairing concrete
真空燃料进给【机】 vacuum fuel feed
真空燃烧 vacuum combustion
真空染色 vacuum dyeing
真空热爆浸取法 vacuum decrepitation extraction method
真空热处理 vacuum heat treatment
真空热传输 vacuum condition heat transfer
真空热电堆 vacuum thermopile
真空热电偶 evacuated thermocouple; vacuo-junction; vacuum junction; vacuum thermocouple; vacuum thermoelement
真空热还原 vacuum thermal reduction
真空热还原法 vacuum thermal method
真空热精炼 vacuum thermal refining
真空热天平 vacuum thermobalance
真空热压成型 vacuum hot pressing
真空容积 vacuum space
真空容器 chamber under vacuum; evacuated chamber; vacuum tank; vacuum vessel
真空熔化 vacuum fusion
真空熔化法 vacuum fusion method
真空熔化气体分析 vacuum fusion gas analysis
真空熔炼 vacuum melting
真空熔炼法 vacuum melting method
真空熔融 vacuum fusion; vacuum melting
真空熔融分析 vacuum fusion analysis
真空熔融技术 vacuum fusion technique
真空熔融色谱法 vacuum fusion(gas) chromatography
真空熔制 vacuum melt
真空融熔气相色谱 vacuum fusion gas chromatography
真空软管 vacuum hose
真空润滑油 vacuum grease
真空三极管 vacuum-tube triode

真空三轴试验 vacuum triaxial test
真空三轴试验仪 vacuum triaxial test apparatus
真空三轴试验装置 vacuum triaxial test apparatus;vacuum triaxial test device;vacuum triaxial test unit
真空扫地机 vacuum sweeper
真空杀菌 vacreation
真空杀菌器 vacreator
真空刹车 vacuum brake
真空砂试验方法 vacuum method of testing sand
真空晒版机 pneumatic printing frame;vacuum printing frame
真空晒版架 vacuum frame
真空晒相架 vacuum copy-holder
真空闪蒸 flash vapo(u)rization;vacuum flash distillation; vacuum flashing;vacuum flash vaporization
真空闪蒸器 vacuum flash vaporizer
真空闪蒸塔 vacuum flasher
真空上浮法 vacuum flo(a)tation process
真空烧成 vacuum firing
真空烧结 vacuum sintering
真空烧结炉 vacuum sintering furnace
真空设备 vacuum apparatus;vacuum device;vacuum equipment;vacuum system
真空设备接头 vacuum connection
真空设备视镜 sight glass for vacuum equipment
真空设备用附件 vacuum accessories
真空摄谱仪 vacuum spectrograph
真空伸长计 vacuum extens(i)ometer
真空深层抽拔 vacuum deep drawing
真空渗镀 vacuum coating
真空渗碳 vacuum carburization;vacuum carburizing
真空渗碳法 vacuum carburizing
真空升华 vacuum sublimation
真空升降机缓冲器 vacuum lifter pad
真空升降装置 vacuum lifting gear
真空省油器 vacuum controlled economizer
真空失蜡浇铸 vacuum investment casting
真空湿润 vacuum wetting
真空式布滤除尘器 suction type cloth filter dust collector
真空式布滤集尘器 suction type cloth filter dust collector
真空式(窗玻璃)刮水器 <汽车> vacuum wiper
真空式大便器 vacuum latrine
真空式干燥机 vacuum drier[dryer]
真空式给纸器 suction feeder
真空式供暖 vacuum heating
真空式供暖系统 vacuum heating system
真空式捡拾器 vacuum pick-up
真空式冷却器 vacuum cooler
真空式剖面 <溢流坝的> vacuum profile
真空式气动测微计 vacuum type pneumatic micrometer
真空式气体取样器 vacuum gas sampler
真空式汽泵 steam vacuum pump
真空式热绝缘 vacuum type insulation
真空式试验 leak(age) test
真空式输送装置 suction conveyer [conveyor];suction pneumatic conveyer[conveyor]
真空式水溶性胶体搅拌机 vacuum hydrosolver
真空式调节器 vacuum modulator
真空式通风系统 vacuum ventilation system
真空式吸送器 vacuum blower
真空式卸载机 vacuum unloader
真空式溢流道 aerated spillway
真空式溢流堰 aerated weir
真空式张力计 vacuum type tensiometer
真空式蒸汽供暖 vacuum steam heating
真空式蒸汽系统 steam vacuum system
真空式自动提前点火装置 vacuum automatic spark advance device
真空式自动转换阀门 vacuum relay valve
真空式自动装瓶机 vacuum type automatic bottling machine
真空试验 vacuum test;vacuum testing
真空试验台 vacuum rig
真空试验仪 vacuum test machine
真空室 evacuated chamber;vacuum case; vacuum chamber; vacuum tank;vacuum vessel
真空室壁 vacuum chamber wall
真空室尺寸 vacuum chamber dimensions
真空室度计 vacuum camera
真空室孔径 vacuum chamber aperture
真空室压强 pressure in vacuum tank
真空释放阀 vacuum relief valve
真空收尘装置 vacuum dust-collector
真空收集器 vacuum collector
真空输送 vacuum transfer
真空输送管 vacuum conveyer tube
真空输送机 suction conveyer[conveyor]; vacuum conveyer[conveyor]
真空栓 vacuum lock
真空水平 vacuum level
真空丝网印花机 vacuum screen printing machine
真空伺服机构 vacuum servo
真空伺服液压制动系统 vacuum servo hydraulic brake system
真空伺服制动器 vacuum servo brake
真空速 true air speed
真空速表 true airspeed indicator
真空速计算器 true airspeed calculator;true airspeed computer
真空塑料金属喷涂 plastic plating
真空随动泵 vacuum follower pump
真空随动装置 vacuum follower instrument
真空碎污设施 <给水管道上用的> vacuum breaker
真空缩合 vacuum condensation
真空锁 vacuum lock
真空塔顶产物 vacuum overhead
真空塔顶馏出物 vacuum overhead
真空塔式望远镜 vacuum tower telescope
真空太阳望远镜 vacuum solar telescope
真空碳化 vacuum carburization
真空套 vacuum jacket
真空套瓶 vacuum jacketed flask
真空套塔 vacuum jacketed column
真空提前控制 vacuum-controlled advance
真空提前器 vacuum advancer
真空提取器 vacuum extractor
真空提升 vacuum lift(ing)
真空提升的吸力垫板 vacuum pad
真空提升梁 vacuum lifting beam
真空提升润滑系统 vacuum lift system of lubrication
真空天平 vacuum balance
真空填充 vacuum filling
真空填充法 vacuum packing method
真空填充机 vacuum filling machine
真空填充器 vacuum filler
真空填充物 vacuum packing
真空填料器 vacuum sizer
真空挑料机 suction gatherer
真空调节阀 vacuum breaker
真空调节器 vacuum governor;vacuum modulator;vacuum regulator
真空调速器 vacuum governor
真空调整 vacuum settle
真空调整器 vacuum corrector
真空通风 vacuum ventilation
真空通路 vacuum passage
真空铜焊 vacuum brazing
真空筒 vacuum cylinder
真空头 vacuum head
真空涂漆 effect lacquer
真空退火 vacuum annealing;vacuum firing
真空脱泡 vacuum defoamation
真空脱气 vacuum deairing; vacuum degasing
真空脱气罐 vacuum receiver
真空脱气机 vacuum exhauster
真空脱气器 vacuum degasifier
真空脱水 vacuum dehydration;vacuum dewatering; vacuum hydro-extracting
真空脱水工艺 vacuum dewatering technique
真空脱水机 vacuum hydroextractor
真空脱水技术 vacuum dewatering technique
真空陀螺仪 vacuum drive gyroscope
真空瓦斯油 vacuum gas oil
真空外壳 vacuum casting
真空外延 vacuum epitaxy
真空望远镜 vacuum telescope
真空温差电堆 vacuum thermopile
真空温度计 vacuum thermometer
真空温台显微镜 vacuum hot stage microscope
真空稳定性 vacuum stability
真空污水处理车 cesspoolage truck
真空污水系统 vacuum sewerage
真空吸板 vacuum suction board
真空吸杯 vacuum cup
真空吸尘 vacuuming
真空吸尘车 vacuum truck
真空吸尘抽风机 vacuum cleaner fan
真空吸尘法 vacuum cleaning
真空吸尘机 vacuum cleaner
真空吸尘器 dust-collecting fan;Hoover;vacuum cleaner;vacuum dust-collector;vacuum sweeper
真空吸尘设备 cleaning vacuum plant;vacuum cleaning plant;vacuum equipment
真空吸尘系统 cleaning vacuum system;vacuum cleaning system
真空吸尘装置 cleaning vacuum device;cleaning vacuum plant;vacuum cleaning installation
真空吸出高度 net positive suction head
真空吸出器 vacuum extract(ion) still
真空吸带缓冲器 vacuum absorbing tape buffer
真空吸垫 vacuum mat
真空吸粪车 vacuum sewer cleaner
真空吸附 vac-sorb
真空吸附泵 vac-sorb pump
真空吸附提升法 vacuum lifting
真空吸辊 vacuum suction roll
真空吸空装置 cleaning vacuum
真空吸力 pull of vacuum
真空吸力极限 limit of vacuum pull
真空吸力式板料送进装置 suction strip feed
真空吸料 vacuum intake
真空吸料吹制 vacuum blowing
真空吸料吹制法 vacuum and blow process
真空吸料法 suction feeding
真空吸滤 vacuum suction filter
真空(吸)滤器 vacuum filter
真空吸泥机 vacuum-cleaner dredge(r)
真空吸盘 vacuum chuck;vacuum cup;vacuum grip device;vacuum lift
真空吸盘的垛板机 vacuum operated sheet piler
真空吸盘升降台 vacuum cup lifter
真空吸盘式升降机 vacuum cup crane
真空吸盘提升设备 vacuum lifting equipment
真空吸气板 vacuum suction plate
真空吸气剂 getter
真空吸入喷砂清理 vacuum blast
真空吸入式输送器 vacuum conveyer [conveyor]
真空吸水板 vacuum mat
真空吸水处理混凝土 vacuum treated concrete;vacuum concrete
真空吸水机械 vacuum water sucker
真空吸水排水车 vacuum suction drainer vehicle
真空吸水设备 vacuum water sucker
真空吸水式覆盖 vacuum mat
真空吸水箱 suction box
真空吸污车 vacuum sewer cleaner
真空吸铸 suction casting; suction pouring;vacuum casting
真空吸嘴 vacuum slot
真空洗涤机 suction washer
真空系数 vacuum coefficient
真空系统 vacuum and sub-atmospheric system;vacuum system
真空系统的管道连接 vacuum connection
真空系统试验 vacuum system testing
真空系统压强 pressure in vacuum system
真空下降 breaking of vacuum
真空下降负荷限制器 vacuum operated load reducer
真空限制阀 vacuum relief valve
真空限制器 vacuum limiter
真空箱 autovac; depression tank; evacuated container; vacuum box; vacuum case; vacuum chamber; vacuum column; vacuum tank; pneumatic water barrel <沉井排水用的>
真空泄漏阀 vacuum release valve; vacuum relief valve
真空泄漏检测仪 vacuum leak detector
真空卸荷阀 vacuum relief valve
真空卸料 vacuum unloading
真空卸料系统 vacuum unloading system
真空卸载机 vacuum unload machine
真空信号输送管 vacuum sensing line
真空行为 vacuum behavio(u)r
真空性能 vacuum performance
真空熏蒸 vacuum fumigation
真空循环式 vacuum circulation system
真空循环脱氧法 circulation degassing process
真空压花 vacuum embossing
真空压捆 vacuum baling
真空压力泵 combined vacuum pressure pump;vacuum pressure pump
真空压力 subatmospheric(al) pressure;vacuum pressure
真空压力表 compound ga(u)ge;vacuum manometer
真空压力阀 vacuum pressure valve
真空压力防腐法 <木材的> boulton process
真空压力活塞式震源 seismovac
真空压力计 pressure vacuum ga(u)ge; vacuum manometer;vacuum pressure ga(u)ge;vacuum ga(u)ge
真空压力浸漆 vacuum pressure im-

Z

pregnation
真空压力浸涂法 vacuum pressure impregnation
真空压力浸渍 vacuum pressure impregnation
真空压力两用表 vacuum pressure ga(u)ge
真空压力两用计 compound ga(u)ge
真空压力通风系统 vacuum plenum
真空压力指示灯 vacuum pressure indicator lamp
真空压力铸造 vacuum die casting
真空压平板【测】vacuum back
真空压塑 vacuum mo(u)ld(ing)
真空(压塑)成型 vacuum mo(u)ld(ing)
真空压缩机 vacuum compressor
真空压条机 vacuum plodder
真空压榨 suction press
真空压制 vacuum pressing
真空压制成型 vacuum press
真空压铸 evacuated die-casting process; vacuum pressure die casting
真空岩芯提取器 vacuum corer
真空养护法 <混凝土> Billnear method
真空氧化精炼 vor refining
真空冶金学 vacuum metallurgy
真空冶炼技术 vacuum melting technique
真空叶滤机 vacuum-leaf filter
真空叶滤器 leaf-type vacuum filter
真空仪 vacuscope
真空因数 vacuum factor
真空荧光灯 vacuum fluorescent lamp
真空硬化法 vacuum hardening
真空油 vacuum oil
真空油槽车 vacuum truck
真空油浸 vacuum impregnating
真空预浓缩器 vacuum preconcentrator
真空预压 vacuum preloading
真空预压法 <软基加固> combined vacuum electroosmotic surcharge preloading method; preloading by vacuum; vacuum method; vacuum method of preloading; vacuum preloading method; atmospheric pressure method; vacuum preconsolidation method
真空预压加固 consolidation by vacuum preloading
真空预压排水 vacuum pre-loading and drainage
真空预压排水法 vacuum preloading drainage method; vacuum drain method
真空元件 vacuum unit
真空圆筒滤器 Oliver filter
真空圆网抄纸机 rotoformer
真空再蒸馏 vacuum redistillation; vacuum rerun
真空造型 vacu(o)-forming
真空造型阴模 die box mo(u)ld
真空增加制动闸 vacuum brake
真空增力制动器 vacuum power brake unit
真空增强器 vacuum augmenter; vacuum intensifier
真空轧染 vacu-pad
真空轧制 vacuum rolling
真空闸 easamatic power brake; vacuum lock
真空罩 vacuum collector; vacuum hood; vacuum tight housing
真空遮断装置 vacuum trip device
真空蒸镀 vacuum evapo(u)rating; vacuum evapo(u)ration coating
真空蒸发镀金属 vacuum evapo(u)ration coating
真空蒸发镀膜 vacuum vapo(u)r plating
真空蒸发工艺 vacuum evapo(u)ration technology
真空蒸发积淀 vacuum evapo(u)ration deposition
真空蒸发技术 vacuum evapo(u)ration technique
真空蒸发器 cold boiler; vacuum evapo(u)rator
真空蒸发涂敷金属 vacuum evapo(u)ration coating
真空蒸发作用 vacuum evapo(u)ration
真空蒸馏 reduced pressure distillation; vacuum distillation; vacuum distilling; vacuum topping
真空蒸馏釜 vacuum still
真空蒸馏过程 vacuum distillation process
真空蒸馏瓶 vacuum distilling flask
真空蒸馏器 vacuum distillator; vacuum distilling apparatus; vacuum still
真空蒸馏设备 vacuum distillation plant
真空蒸馏塔 vacuum column; vacuum distilling column; vacuum tower
真空蒸馏塔部分冷凝器 vacuum partial condenser
真空蒸馏提取器 vacuum extract(ion) still
真空蒸馏甑 vacuum retort
真空蒸馏制沥青 vacuum asphalt
真空蒸馏装置 vacuum distillation plant
真空蒸汽蒸馏 vacuum steam distillation; wet vacuum distillation
真空整流管 vacuum rectifier tube
真空整流器 vacuum rectifier
真空支承 vacuum support
真空支管 vacuum manifold
真空脂 vacuum grease
真空止回阀 vacuum check valve; vacuum control check-valve
真空纸板机 vacuum board machine
真空制备的试样 vacuum prepared sample
真空制动车辆 vacuum braked stock
真空制动缸 vacuum brake cylinder
真空制动缸活塞 vacuum cylinder piston
真空制动管 vacuum brake pipe
真空制动机 vacuum brake
真空制动器 train brake; vacuum brake
真空制动器软管 vacuum brake hose
真空制动软管 vacuum brake flexible hose
真空制动系统 vacuum braking system
真空制动转换旋钮 air-vacuum brake change over rotary knob
真空制冷法 vacuum refrigeration
真空制冷机 vacuum refrigerating machine
真空钟罩 vacuum bell jar
真空重量 weight in vacuo; weight in vacuum
真空轴承 vacuum bearing
真空助力 vacuum servo
真空助力泵 vacuum booster pump
真空助力换挡 vacuum power gear change; vacuum power gear shift
真空助力器 vacuum booster
真空助力式前轮盘式制动器 vacuum assisted front disc
真空助力液压制动系统 vacuum assisted hydraulic brake system
真空助力制动器 vacuum brake; vacuum servo brake
真空注浆成型 suction casting; vacuum casting
真空注射 vacuum injection
真空铸罐 vacuum ladle
真空铸造 suction casting; suction mo(u)ld; vacuum casting; vacuum pressing and casting
真空转鼓过滤机 vacuum type drum filter
真空转鼓烘燥机 vacuum drum drier [dryer]
真空转筒式烘燥器 vacuum tumbler dryer
真空转移钳 vacuum type transfer tongs
真空转移容器 vacuum transfer vessel
真空装货机 vacuum loader
真空装料 vacuum loading
真空装载机 vacuum loader
真空装置 autovac; vacuum apparatus; vacuum installation; vacuum plant; vacuum unit
真空状态 state of vacuum
真空锥形干燥器 vacuum cone dryer
真空紫外辐射 vacuum ultraviolet radiation
真空紫外光谱学 vacuum ultraviolet spectroscopy
真空紫外光谱仪 vacuum ultraviolet spectrometer
真空紫外光子 vacuum ultraviolet photon
真空紫外线 vacuum ultraviolet
真空紫外线区域 vacuum ultraviolet region
真空自动泵 vacamatic pump
真空自动阀 vacamatic valve
真空自动加料机 vacuum autoloader
真空自动控制 vacuum servo
真空自动控制变速箱 vacamatic transmission
真空自动式 vacamatic
真空自耗炉 consumable electrode vacuum furnace
真空自蒸发装置 vacuum flash unit
真空钻进 vacuum boring; vacuum drilling
真空作业 vacuum process
真空作业混凝土 vacuum concrete; vacuum processed concrete; vacuum treated concrete
真空作用离合器 vacuum operated clutch
真孔隙率 <孔隙体积和整体体积的比率,以%计> true porosity
真宽度 true width
真矿脉 true lode
真矿泥 true slime
真亏数 effective deficiency
真肋 true rib
真类 proper class
真棱锥体 true pyramid
真离解度 true degree of dissociation
真理 truth
真连续吸收 true continuous absorption
真亮度 real bright; real brightness
真临界剪应力 true critical shear stress
真临界温度 true critical temperature
真临界性质 true critical property
真临界压力 true critical pressure
真零点 true zero
真流动性 true fluidity
真路 proper path
真卯酉圈 prime vertical
真密度 actual density; intrinsic(al) density; real density
真面积 true area
真命题 true proposition
真摩擦力 true frictional resistance
真摩擦系数 true coefficient of friction
真摩阻力 true frictional resistance
真木 true-wood
真内摩擦 true internal friction
真内摩擦角 angle of true internal friction; intrinsic(al) internal angle of friction; true angle of internal friction
真南 true south
真南北线 true north-and-south line
真逆流流动 true counter-current flow
真年轮 true annual ring
真黏[粘]度 true viscosity
真黏[粘]结力 true cohesion
真黏[粘]聚力 true cohesion
真黏[粘]着力 true cohesion
真凝聚力 true cohesion
真欧(姆) Rayleigh ohm
真劈理 true cleavage
真皮 cuff; derma; dermis; true skin
真平均 true mean
真平均温度 true mean temperature
真平位置 true mean place
真漆 lacker; lacquer
真旗流形 proper flag manifold
真气测量 vacuum measurement
真气孔率 true porosity
真气温 true air temperature
真气相过程 true vapo(u)r process
真前角 true rake angle
真前缀 proper prefix
真浅海的 eulittoral
真强度参数 <土的> Hvorslev parameter
真倾度 true dip
真倾角 full dip; full dip angle; true dip; true dip angle
真倾斜 true dip
真秋分点 true September equinox
真屈服点 true yield point
真热膨胀系数 true coefficient of thermal expansion
真热容量 true heat capacity
真日时组 true date-time group
真溶剂 true solvent
真溶(解)度 real solubility
真溶液 simple solution; solutide; true solution
真溶液化学沉积矿床 chemical sedimentary deposit by true solution
真溶液化学沉积作用 solution chemical deposition
真润滑 true lubrication
真三轴剪力仪 true triaxial shear apparatus
真三轴试验 cubic(al) triaxial test; true triaxial(shear) test
真三轴试验仪 cubic(al) triaxial test apparatus
真三轴仪 true triaxial apparatus
真色度 true colo(u)r
真色度单位 true colo(u)r unit
真闪长岩 true diorite
真射【数】proper morphism
真射程 true range
真深度 true depth
真生长率 true growth rate
真声带 true vocal cord
真剩余群 proper factor group
真时 true time
真实比重 actual specific gravity
真实变形 true strain
真实表示 faithful representation
真实差数 true difference
真实常数 true constant
真实成本 bona fide cost; real cost; true cost
真实程度 <指路面设计> degree of reliability
真实持有人 bona fide holder
真实尺寸 actual size; full size; true

(to) size
真实储量 <指地下水、矿藏等> positive reserves
真实的表现 true representation
真实的判定 realistic decision
真实地震 real earthquake
真实地震加速度图 accelerogram of real earthquake
真实电位 real potential
真实定义 real definition
真实读数 true plot
真实度 degree of similitude
真实断裂应力 actual breaking stress
真实断面图 true section
真实发光率 true luminance
真实发光率信息 true luminance information
真实发票 definite invoice
真实反应性 true reactivity
真实方位变化率 true bearing rate
真实飞行航道 actual flight path
真实负载法 actual loading method
真实刚度 true stiffener
真实高度 true height
真实工业污水 actual industrial effluent
真实工作条件下疲劳试验 fatigue testing under actual service condition
真实拱顶 true vault
真实故障 true fault
真实航程指示 true range indication
真实河流 real river;real stream
真实滑脱比 true slip ratio
真实滑移 true slip
真实环境 true environment
真实机理 actual mechanism
真实级 live stage
真实极大 true maximum
真实计数对偶然计数比 real-to-random ratio
真实接触 true contact
真实结构 actual structure
真实进动 real precession; true precession
真实进位 true carry
真实晶体 imperfect crystal
真实精度 realistic accuracy
真实孔径 real aperture
真实孔径雷达 real aperture radar
真实孔径天线 real aperture antenna
真实利率 real interest rate
真实亮度 true brightness; true luminance
真实灵敏度 true sensitivity
真实流动 actual flow
真实流体 real fluid
真实流体流 real fluid flow
真实美元价值 real-dollar value
真实模型 truth model
真实年龄 true age
真实黏[粘]度 true viscosity
真实黏[粘]着力 true adhesion
真实偏差 true deviation
真实票据 real bill
真实平衡 true equilibrium
真实平均数 true average;true mean
真实气体 actual gas;real gas
真实牵引力 true tractive force
真实倾角读数图 direct dip-reading chart
真实色度单位 true colo(u)rity unit
真实商域 true quotient field
真实生活污水后生物处理 real domestic wastewaters post-biotreatment
真实时间 actual time
真实世界 real-world
真实寿命 microscopic lifetime; true lifetime
真实水平 true horizon
真实速度 true velocity
真实酸度 true acidity
真实体积 true volume
真实条件 full-scale condition; natural condition
真实条件试验 full-scale test
真实图 true view
真实图表 true plot
真实图像 true picture
真实图形 actual plot
真实外景 original location
真实温度 actual temperature; true temperature
真实显示 realistic display
真实相关 true correlation
真实信道 real channel
真实形态 true form
真实形状的 true-to-shape
真实性 authenticity; trueness; truth; validity
真实性检验 validity check
真实性能 actual performance
真实徐变 basic creep
真实样品 authentic sample
真实液体 actual liquid
真实应变 logarithmic strain;true strain
真实应力 true stress
真实值 true value
真实中位数 true median
真实重力 actual gravity
真实重量 actual weight
真实自我 real self
真矢量 true vector
真适温的 euthermophilous
真适盐种 euhalobion
真树脂 true resin
真数 anti-logarithm
真双星 physical double star
真水面下漂浮生物 euhyponeuston
真水平 true horizon
真水平的 true horizontal
真水平线 rational horizon
真丝 natural silk
真速度 true velocity
真速度向量 true speed vector
真酸类 true acids
真太阳 real sun;true sun
真太阳日 true solar day
真太阳时 apparent solar time; true solar time
真体积 <孔隙不计的体积> true volume
真天顶 true zenith
真天顶距 true zenith distance
真天平动 true libration
真条件转移 branch on true
真同伦【数】proper homotopy
真椭圆性 proper ellipticity
真伪判定问题 decision problem
真伪试验 true-false test
真伪值 truth value
真尾 proper tail
真位置 true place;true position
真温度降 true temperature drop
真稳态 true steady state
真无线电方位 true radio bearing
真午 true noon
真午时 mean time of true noon
真午夜 true midnight
真误差 real error; resultant error; true error;true fault
真吸出高程试验 net positive suction head test
真吸收 true absorption
真吸收系数 true absorption coefficient
真系数 true coefficient
真系统 real system
真细菌 eubacterium[复 eubacteria]; true bacteria
真相 actual state of affairs; naked truth; real situation; rock bottom; truth
真相关 real correlation
真象限角 true bearing
真硝甘炸药 true dynamite
真信号 sure signal
真星型 euaster
真虚假实 deficiency in reality with pseudo-excess symptoms
真旋转 proper rotation
真雪杉 true cedar
真亚纯映射 proper meromorphic mapping
真岩层电阻率 true formation resistivity
真沿岸的 eulittoral
真液体 true liquid
真异步式 truly asynchronous
真因子分析 true factor analysis
真应变 true strain
真应力 true stress
真应力-应变曲线 true stress-strain curve
真映射 proper mapping
真硬度 true hardness
真圆 proper circle
真圆锥射影 true conic(al) projection
真月平均降水量 true monthly mean precipitation
真运动 true motion;true tracking
真运动雷达 true motion radar; true tracking radar
真运动显示 true motion display
真运动显示器 true motion display; true motion indicator;true tracking indicator
真运动指示 true motion indication
真载试验 actual loading test
真褶皱 true folding
真振幅保持叠加 real amplitude preservation stack
真振幅恢复 true amplitude recovery
真振幅剖面 true amplitude section
真蒸汽压力 true vapo(u)r pressure
真(正)比重 absolute specific gravity
真正成本 bona fide cost;true cost
真正的 intrinsic(al);substantial
真正的拱桥 genuine arch bridge
真正的霜害 true frost injury
真正地平【气】celestial horizon
真正电解质 real electrolyte
真正滚动 true rolling
真正价格 true price
真正价值 sterling worth;true value
真正浸出物含量 true extract
真正径尺寸 absolute bore size
真正镜质体 genuine vitrinite
真正利润 true profit
真正平衡价格 true equilibrium price
真正平均散布 true mean dispersion
真正收益 true earning
真正所得 true income
真正通货膨胀 true inflation
真正无面值股票 actual no-par value stock
真正午 true noon
真正休眠 true dormancy
真正债权人 bona fides creditor
真正(真诚)买户 bone fide purchaser
真正真菌素 true fungi
真值 actual value; real value; sterling worth;true value;truth value
真值表 Boolean operation table; matrix[复 matrixes/matrices]; true table;truth table
真值表产生试验 truth table generation testing
真值表法 truth table method
真值表计算机 truth table computer
真值函数 truth function; truth value function
真值集合 truth set
真值矩阵 truth value matrix
真值命题 truth value of proposition
真中分纬度 middle latitude;true middle latitude
真重校表仪 deadweight tester
真珠贝 nacre
真转动 proper rotation
真锥尖阻力 true cone tip resistance
真锥头阻力 true cone tip resistance
真子集 proper subset
真子空间 proper subspace
真子午线 true meridian
真子午线方向 true north heading
真子午线观测 observation for true meridian
真自动引导 true homing
真自转周 true rotation
真总温度 true total temperature
真坐标 true coordinates

砧 bick-iron;hammering block

砧板 chopping block;cutting and chopping block
砧锤重量比 anvil ratio
砧角 anvil beak;anvil horn;beakiron; beck iron;bickern;bick-iron;horn
砧块 anvil block
砧面 anvil face
砧面垫面 anvil plate
砧面垫片 anvil pallet;anvil plate
砧面托 intermediate block
砧木 rootstock; stock; tree stock; understock
砧木影响 stock influence
砧钳 anvil vise
砧台 anvil block
砧台虎钳 anvil vice
砧铁型模 anvil swage
砧用圆刃楔形锤 anvil fuller
砧用圆刃楔形套柄锤 bottom fuller
砧凿 anvil chisel;anvil cutter
砧枕 anvil cap;anvil cushion
砧状积雨云 cumulonimbus incus
砧状云 anvil cloud; incus; thunderhead
砧子 smithy
砧嘴 anvil beak
砧座 anvil;anvil base;anvil bed;anvil block;anvil seat;anvil stand
砧座震击 anvil jolt
砧座柱 anvil pillar

祯幅 frame size

祯面高 image height;picture height

甄别苗圃 screening nursery

甄别器 discriminator
甄别试验 screen(ing) test(ing)
甄别阈 discriminating threshold

榛树枝条篱栏 hazel wattle hurdle

榛属 filbert;hazel;Corylus <拉>
榛子树 Chinese hazel
榛子油 hazelnut oil

帧 frame;vertical

帧边缘 frame border
帧编号 frame number
帧差错 frame error
帧差滤除 frame difference filtration

帧场 frame field
帧传递 frame transfer
帧传输 frame transmission
帧磁滞 frame hysteresis
帧存储 frame storage
帧存储器 frame memory;frame store
帧存取功能 frame access function
帧大小 frame sign
帧代码 frame code
帧的纵横尺寸比 picture ratio
帧定位 frame alignment
帧定位恢复时间 frame alignment recovery time
帧定位时间隙 frame alignment time slot
帧定位信号 frame alignment signal
帧定相 field phasing;vertical phasing
帧丢失 loss of frame
帧放大器 frame amplifier
帧分频器 frame divider
帧幅 frame amplitude
帧幅限制 frame limiting
帧幅直线性 vertical linearity
帧格式 frame format;picture format
帧滚动 frame roll
帧号码 frame number
帧滑动 frame slip
帧缓冲存储器 frame buffer memory
帧缓冲器 frame buffer
帧回程 frame flyback
帧回描 field retrace; frame flyback; frame retrace; image flyback; picture flyback;vertical retrace
帧回描脉冲 vertical flyback voltage
帧回描时间 frame retrace time;vertical flyback period
帧回描熄灭 frame blanking; frame flyback suppression
帧回描熄灭周期 frame-suppression period
帧回描消隐 frame blanketing
帧回扫消隐 vertical blanking
帧基本功能 frame basic function
帧畸变 frame distortion
帧级功能 frame level function
帧级过程 frame level procedure
帧级接口 frame level interface
帧级数据 frame level data
帧计时 frame timing
帧计数 frame count
帧间 interframe
帧间编码 frame-to-frame coding; interframe coding;interframe encode
帧间变化 frame-to-frame variation
帧间差 frame-to-frame difference
帧间差值电路 frame difference circuit
帧间阶距 frame-to-frame step response
帧间脉冲 interframe pulse
帧间时间填充 interframe time filling
帧间跳动 frame-to-frame jump
帧间跳动响应 frame-to-frame step response
帧间线性预测 frame-to-frame linear prediction
帧间相关性 frame-to-frame correlation
帧间响应 frame-to-frame response
帧监视管 frame monitoring tube
帧检测器 frame detector
帧校验 frame check
帧校验序列 frame check sequence
帧接收器 frame grabber
帧结构 frame structure
帧结构控制 frame structure control
帧距 frame pitch
帧控制程序 frame handler
帧控制字段 frame control field
帧宽度调节 picture width control
帧理论 frame theory
帧脉冲 frame pulse
帧脉冲定相 field phasing
帧脉冲同步 frame pulse synchronization
帧脉冲同步码 frame pulse pattern
帧面 pattern;picture;picture plane
帧面高度 frame height
帧面积 frame area;picture area
帧面宽度 picture width
帧面宽高比 picture aspect ratio
帧面色 pattern colo(u)r
帧模式 frame pattern
帧内编码 intraframe coding
帧内图像编码器 intraframe image coder
帧逆程 picture flyback
帧逆程时间 frame retrace time; picture flyback time
帧偏移 vertical shift
帧偏转 field deflection;frame deflection;vertical deflection
帧偏转线圈 frame deflector coil
帧频 frame frequency; frame rate; frame repetition rate; image frequency;picture repetition frequency;picture scanning frequency; repeated frequency; repetition frequency;vertical frequency
帧频倒脉冲 inverted frame pulse
帧频锯齿波补偿信号 frame tilt
帧频控制 field frequency control
帧频起动脉冲 frame drive pulse
帧频调整 picture-frequency adjustment;picture-frequency setting
帧频微调 vertical hold
帧起始 frame start
帧倾斜 frame tilt
帧取样 frame grab;frame sample
帧扫描 field scan;frame scan(ning); picture timebase;frame time base
帧扫描电路 vertical deflection circuit
帧扫描电压 picture sweep voltage
帧扫描频率控制 field frequency control
帧扫描速度 frame-scanning speed
帧扫描振荡器 vertical scanning generator
帧扫描振幅调整 frame amplitude control
帧扫描直线性调整 frame linearity control
帧扫描周期 frame-scan(ning) period
帧扫描装置 frame sweep unit
帧失步 out-of-frame
帧失步秒 out-of-frame second
帧失步时间 out-of-frame alignment time
帧失步调整时间 out-of-frame alignment time
帧失调 out-of-frame
帧梯形失真 frame key-stone
帧调节器 framer
帧调整 framing control
帧跳动 frame-to-frame jitter
帧同步 frame alignment; frame synchronization; picture synchronization;vertical hold
帧同步电路 frame alignment circuit
帧同步恢复时间 frame alignment recovery time
帧同步控制 frame synchronization control; picture synchronization control
帧同步逻辑 frame synchronization logic
帧同步码 frame synchronization code
帧同步脉冲 picture synchronizing impulse
帧同步脉冲分离器 frame synchronizing pulse separator
帧同步器 frame synchronizer
帧同步调整 vertical-hold control
帧同步信号 frame alignment signal; frame synchronizing signal
帧同时制 picture simultaneous system
帧头 preamble
帧图像变形 frame bend
帧尾 postamble
帧熄灭放大器 frame-blanking amplifier
帧线性 vertical linearity
帧像周期 frame time
帧消隐放大器 frame-blanking amplifier
帧信号波形 frame waveform
帧信号放大器 frame amplifier; vertical amplifier
帧信号输出管 frame output valve
帧型 picture format
帧型计算机 pictorial computer
帧型调整 framing
帧序制 frame sequential system
帧序制的 field-sequential
帧页面 frame page
帧语句 frame statement
帧直线性 frame linearity
帧指示 framing indication
帧指示器 framing bit
帧中继 frame relay
帧重复频率 frame repetition frequency
帧周期 frame duration;frame period; image duration;picture period
帧转移 frame transfer

诊

诊察台 bed table

诊断表层土 diagnostic surface horizon
诊断程序 diagnosis program(me);diagnostic program(me); diagnostic routine; diagnotor; error search program(me)
诊(疗)室 consulting room;infirmary
诊(疗)所 clinic;cottage hospital;dispensary;infirmary

枕 pillow

枕板 <路面接缝下的> sleeper plate
枕藏电话机 pillowphone
枕底清筛机 sleeper bed sieve machine
枕垫 bolster
枕垫雕饰带 cushioned frieze
枕垫形的 pillowed
枕垫形凸块表面装饰处理 pillowwork
枕垫形柱头 pillow capital
枕垫座 <闸门的> receiving base
枕端捣固机 sleeper end consolidator
枕轨隙挖掘斗 railroad cribbing bucket
枕簧 body spring;bolster spring
枕基爬松器 tie-bed scarifier
枕基耙土器 tie-bed scarifier
枕间捣固机 crib consolidator; sleeper-crib tamping machine
枕间道砟清筛机 crib-ballast cleaner
枕距规 distance ga(u)ge for sleepers
枕距护木 bond timber
枕块 pillow;plumber block
枕梁 body bolster; bolster; sleeper beam
枕梁补强铁 body bolster stiffener
枕梁垫板 body bolster cover
枕梁腹板 body bolster diaphragm; body bolster web
枕梁盖板 body bolster cover
枕梁旁承座补强板 bolster stiffener
枕梁上盖板 body bolster top cover plate
枕梁下盖板 body bolster compression bar;body bolster tie(bottom cover)plate
枕料总厂 general store
枕木 bed timber; dormant tree; floor bar; ground beam; ground brace; rail-tie;sleeper;timber sleeper;timber tie; transverse sleeper; wooden sleeper;wooden tie;cross tie <美>
枕木背板 sleeper lagging
枕木插入机 sleeper inserter; tie inserter
枕木衬板 sleeper lagging
枕木抽换机 sleeper bailer
枕木捣固 tie tamping
枕木捣固机 ballast tamper;tie tamper
枕木捣固器 ballast tamper;tie tamper
枕木垛 cogs; crib chock; saddle; timber blocking
枕木防腐 sleeper impregnation yard
枕木干燥 sleepers seasoning
枕木夹紧板 sleeper clamping plate
枕木接头 sleeper joint
枕木紧线器 wooden sleeper binding wire tight
枕木锯 sleeper saw
枕木开槽机 tie scoring machine
枕木连接 sleeper joint
枕木梁 timber blocking
枕木面 tied plan
枕木木塞 <修理枕木钉孔或改变钉孔用的> tie plug
枕木铺设机 sleeper laying machine
枕木钳 tie tongs
枕木下垫块 shim
枕木削平机 sleeper planing machine; tie adzer
枕木削整 tie adzing
枕木挟窄器 wooden sleeper compactor
枕木支承板 bottom plate
枕木重砍 re-adzing of tie
枕木状砌筑 bolster work
枕木钻孔机 sleeper drill;sleeper drilling machine; tie border; tie-boring machine
枕木钻孔样板 boring template;boring templet
枕套及褥罩织物 ticking
枕头箱 <卧车> pillow box
枕形变形 pillow deformation
枕形磁场 pincushion field
枕形畸变 cushion-shaped distortion; negative distortion; pillow distortion;pincushion distortion
枕形畸变校正 pillow distortion equalizing; pincushion distortion equalizing
枕形校正器 pincushion corrector
枕形抗裂试验 pillow test
枕形气密试验 pillow test
枕形失真 negative distortion; pillow distortion;pincushion distortion
枕形失真校正电路 pincushion correction circuit
枕形失真调整磁铁 pincushion magnet
枕形致偏磁场 pincushion field
枕状构造【地】 pillow structure
枕状节理【地】 pillow joint
枕状熔岩 ellipsoidal lava;pillow lava
枕状熔岩流 pillow lava flow
枕资 sleeper block

阵

阵飑 guzzle

阵地 position
阵地的 positional

阵地地域草图 position area sketch
阵地工事 bastion
阵点 lattice point
阵点间距 interlattice point distance
阵发 spasm
阵发性 gustiness
阵发性大雨 rain gush
阵发性客流 intermittent passenger traffic
阵发性流 intermittent passenger traffic
阵发压力 popping pressure
阵风 blast of wind;flaw;flurry;gust;gust of wind;puff of wind;wind blast;wind gust
阵风波 gusty wave
阵风分量 gustiness components;intensity of turbulence
阵风风速 gustiness
阵风荷载 gust load
阵风频数 gust frequency
阵风谱 gust spectrum
阵风时间 gust duration
阵风速度 gust speed;gust velocity;wind gust speed
阵风探空仪 gustsonde
阵风污染 pollution under gusty condition
阵风系数 gusset factor;gust factor;gustiness factor
阵风性 gustiness
阵风最大风速 gust peak speed
阵阶 order of matrix
阵帘 array curtain
阵列 array;installing;parade
阵列操作 array operation
阵列测试器 array tester
阵列乘法器 array multiplier
阵列橱窗 cabinet bulb
阵列处理 array processing
阵列处理机 array processor;array unit
阵列处理器 array processor
阵列传感器 sensor array
阵列打印机 array printer
阵列单元 array element;array unit
阵列断面 cross-section of array
阵列计算机 array computer;array-type computer
阵列简化分析电路 array reduction analysis circuit
阵列结构 array architecture
阵列逻辑【计】array logic
阵列模块 array module
阵列匹配 matrix matching
阵列声呐 array sonar
阵列式车辆-路面压力传感器 vehicle-road surface pressure transducer array
阵列式存储器 matrix storage
阵列探测器 array detector
阵列天线 array antenna
阵列运算 array operation
阵列组件 array component
阵码 horizontal and vertical parity check code
阵容 line up
阵线 front
阵性风【气】gusty wind
阵性降水 shower
阵选管 matricon
阵雪 showery snow;snow flurry;snow shower
阵硬 age hardness
阵雨 brash;gust;gust of rain;rain shower;scud;shower;showery rain
阵雨般的 showery
阵雨的 brashy
阵雨型 spotting rainstorm pattern;spotty rainstorm pattern
阵雨云 shower cloud

振

振摆 run-out;thrash[thresh]
振摆式重力仪 vibrating-pendulum gravimeter
振摆仪 physical pendulum
振板式黏[粘]度计 vibratory plate visco(si)meter
振波 vibration wave
振波探漏仪 vibroscope
振颤空间 oscillation space
振颤片 trembler
振颤容积 oscillation volume
振沉桩 vibrator sunk pile
振冲 vibro-compaction;vibroflo(a)tation;vibroimpacting;vibro-jet;vibroshock
振冲法<地基处理方法之一>vibro-compaction;vibroplacement;vibrating replacement process;vibroflo(a)tation method;vibroflotating process
振冲法托换 underpinning by vibroflo(a)tation
振冲挤密 vibro-compaction
振冲器<地基振浮压实用>vibroflot;vibrating impacter
振冲碎石桩 vibro-replacement stone column
振冲碎石桩法 vibro-replacement stone column method
振冲旋压钻进 vibro-percussion rotary-thrust boring
振冲旋转施压<一种钻进方式>vibro-percussion rotary-thrust
振冲压密 vibro-compaction
振冲压密法<处理松砂地基加速沉降的一种方法>vibroflotation
振冲置换 vibro-replacement
振锤测定 vibrating hammer test;vibratory hammer test
振锤测定土壤含水量法 vibrating hammer method
振锤试验 vibrating hammer test;vibratory hammer test
振打冒灰 rapping puff
振打器 shaker;vibrator
振打强度 rapping intensity
振打清灰 cleaning by shaking;dedust by rapping
振打系统 rapping system
振打装置<清灰>rapping equipment[ash cleaning]
振荡 oscillation;oscillating
振荡板 oscillating plate
振荡板式柱 oscillating plate column
振荡倍频器 oscillator doubler
振荡闭锁装置 surge guard
振荡变压器 oscillating transformer;oscillation transformer
振荡波 oscillating wave;oscillation wave;oscillatory wave;wave of oscillation
振荡波传播 normal mode propagation
振荡波腹 oscillating loop;oscillation antinode;oscillation loop
振荡波痕 oscillation ripple;oscillation ripple mark
振荡波节 node of oscillation
振荡薄膜 oscillating membrane
振荡补偿 oscillation compensation
振荡不稳定性 oscillatory instability
振荡不足 undervibration
振荡部分 oscillating part
振荡彩色顺序 oscillating colo(u)r sequency
振荡槽 oscillating trough;oscillation trough;shaking chute;shaking trough
振荡槽路 oscillation tank circuit
振荡槽路品质因数 circuit quality factor
振荡侧向传递 oscillation flanking transmission
振荡叉 oscillation fork
振荡常数 oscillation constant
振荡场 oscillating field
振荡持续时间 hunting time
振荡冲击 impulsing
振荡触点 oscillating contact
振荡传递 oscillation transmission
振荡传感器 osciducer;oscillation pickup
振荡传感系 oscillating pickup system
振荡串 train of oscillations
振荡锤 vibrating spear
振荡磁场 oscillating magnetic field
振荡磁体 oscillating magnet
振荡磁吸收 oscillatory magneto-absorption
振荡磁阻 oscillatory magneto-resistance
振荡次数 number of oscillations
振荡单元 oscillating unit;oscillator unit
振荡的 oscillating;oscillatory
振荡的冲击激励 impulsing
振荡的初始值 oscillatory initial value
振荡的水体 opposing water mass
振荡滴 oscillating drop
振荡点 oscillating point;oscillation point
振荡电场 oscillating electric(al) field
振荡电动势 oscillating electromotive force
振荡电荷 oscillating charge
振荡电弧 oscillating arc
振荡电流 oscillating current;oscillatory current;periodic(al) current
振荡电路 oscillating circuit;oscillating contour;oscillation circuit;oscillator circuit;oscillatory circuit
振荡电路电容器 tank condenser
振荡电路线圈 tank coil
振荡电平 oscillation level
振荡电势 oscillatory potential
振荡电位 vibration potential
振荡电位差 oscillating potential difference
振荡电压 oscillating voltage
振荡电子矩 oscillating electron moment
振荡堆垒 oscillation pile-up
振荡对流 oscillatory convection
振荡发散的 oscillating divergent
振荡法 oscillation method
振荡反射镜扫描系统 oscillating mirror scan system
振荡反应 oscillating reactions
振荡反阻尼 oscillation antidamping
振荡范围 hunting range;oscillating range;oscillating region;oscillation limit
振荡方程 oscillation equation
振荡方式 mode of oscillation
振荡放大器 monofier;oscillator amplifier;oscillator amplifier unit
振荡放电 oscillating discharge;oscillatory discharge
振荡放电离子源 oscillating discharge source
振荡分类 oscillating sort
振荡分类法 oscillating sort
振荡分频器 oscillator divider
振荡分选机 oscillating separator
振荡幅度 amplitude of oscillation;oscillation amplitude
振荡负荷 oscillating load
振荡负载 oscillating load
振荡干扰 oscillatory response
振荡杆 rocking beam
振荡杆簧 rocking bar spring
振荡杆轮 rocking bar wheel
振荡杆桩 rocking bar stud
振荡给料器 oscillating feeder
振荡功率 fluctuating power;hunting power;oscillatory power
振荡功率测试仪 oscillation power tester
振荡管 oscillating tube;oscillation tube;oscillator tube
振荡光阑 vibrating diaphragm
振荡规律性 regularity of oscillations
振荡过程 oscillatory process
振荡过头 overvibration
振荡函数 oscillating function;oscillation function
振荡荷载 oscillating load;oscillatory load
振荡互补偿晶体 bimorph crystal
振荡回路 oscillation circuit
振荡回路常数 resonance constant
振荡回路电感 tank circuit inductance;tank inductance
振荡回路电容 tank capacitance
振荡回路线圈 tank coil
振荡混频第一检波器 oscillator-mixer-first detector
振荡活塞 oscillating piston
振荡基型 dominant mode
振荡激发 oscillatory excitation
振荡激励 vibratory stimulation
振荡级 oscillator stage
振荡级数 oscillating sequence;oscillating series;oscillatory series
振荡极限 oscillation limit
振荡计时器 oscillator timer
振荡计数器 oscillation counter
振荡加料器 oscillating feeder
振荡加速度 oscillating acceleration;oscillatory acceleration
振荡剪切 oscillatory shear
振荡剪切流 oscillatory shear flow
振荡检测器 oscillating detector
振荡键 oscillating link;oscillating linkage
振荡角 hunting angle
振荡搅拌机 oscillating agitator;oscillating mixer
振荡搅拌器 oscillating agitator;oscillatory mixer
振荡节点 node of oscillation
振荡界限 oscillation boundary;oscillation limit
振荡进料器 oscillating feeder
振荡晶体 oscillating crystal;vibrating crystal
振荡控制的随动系统 oscillating control servomechanism
振荡控制面 oscillatory control surface
振荡控制器 oscillating controller
振荡控制伺服机构 oscillating control servomechanism
振荡类型 type of oscillation
振荡冷却 shaker cooling
振荡离心机 vibratory centrifuge
振荡理论 oscillation theory
振荡力矩 oscillatory torque
振荡连接 oscillating joint
振荡量 oscillating quantity
振荡量子数 oscillation quantum number
振荡列 train of oscillations
振荡流 oscillatory current
振荡流动 oscillating flow
振荡炉箅 oscillating grate
振荡炉排片 oscillating bar
振荡孪生 oscillatory twinning
振荡螺旋 oscillatory spin
振荡脉冲 oscillating impulse;oscillating pulse;oscillation impulse;oscillatory impulse;oscillatory pulse

Z

振荡面 oscillating plane
振荡鸣声 singing of oscillation
振荡模式 oscillating mode
振荡磨 vibration mill;vibrator mill
振荡能量 oscillation energy
振荡黏[粘]弹计 oscillatory viscoelastometer
振荡耙 oscillating rake
振荡喷嘴 oscillating nozzle
振荡碰撞 generating collision
振荡片 oscillator plate
振荡频率 frequency of oscillation; generated frequency; hunting frequency;oscillating frequency;oscillation frequency
振荡频率倍增器 oscillatory-frequency multiplier
振荡频率漂移 oscillator drift
振荡频率选择器 oscillator
振荡平板夯 vibratory plate compactor
振荡破坏荷载 shakedown load
振荡谱 oscillation spectrum
振荡器 agitator; alternator; generator; oscillating guide; oscillator; Poulsen-arc converter; producer; shaker
振荡器部分 oscillator section
振荡器触点 vibrator contact
振荡器触发 oscillator priming
振荡器的固有运动 motion of the oscillator
振荡器的谐振模 resonator mode of oscillator
振荡器电路 Pierce circuit
振荡器电压 oscillator voltage
振荡器电源 oscillator supply
振荡器动作电压 oscillator actuating voltage
振荡器负载 oscillator load
振荡器工作区域 oscillator operating region
振荡器功率 oscillator power
振荡器管 oscillion
振荡器柜 oscillator box; oscillator house
振荡器基准相位 reference oscillator phase
振荡器晶体 oscillator crystal
振荡器滤波器 oscillator filter
振荡器频率 oscillator frequency
振荡器频率容限 oscillator frequency tolerance
振荡器石英 oscillator quartz
振荡器式波长计 oscillator wavemeter
振荡器输出 oscillator out
振荡器停振 blocking of oscillator
振荡器同步 synchronization of oscillator
振荡器统调 oscillator tracking
振荡前电流 preoscillation current
振荡强度 intensity of oscillation;oscillation intensity
振荡区域 limit of oscillation;oscillating region;oscillatory region
振荡曲线 oscillating curve
振荡燃烧 oscillation burning;oscillatory combustion;rough burning
振荡热沉淀器 oscillating thermal precipitator
振荡热离子真空管 oscillating thermionic valve
振荡筛 oscillating sieve; oscillation screen; shaking screen; oscillating screen
振荡筛分机 oscillating sieving machine
振荡(栅)极 oscillator grid
振荡射流法 <测动表面张力> oscillating jet technique; vibrating jet method
振荡石英晶体 vibrating quartz crystal
振荡时间 oscillating time;time of oscillation
振荡矢量 oscillating vector
振荡式 oscillatory type
振荡式除杂机 shake willey
振荡式给料机 oscillating feeder
振荡式刮软机 vibration staking machine
振荡式刮土板 oscillating scraper
振荡式拉软 vibration staking
振荡式流槽 oscillating launder
振荡式磨光机 oscillating sander
振荡式取样器 oscillating sampler
振荡式筛网 oscillating screen
振荡式输送机 oscillator
振荡式调节器 on-off regulator;oscillating regulator
振荡式投料器 oscillating feeder
振荡式土铲 oscillating subsoiler
振荡式压缩机 oscillating compressor
振荡式仪表 oscillating meter
振荡式轧机 oscillating rolling mill
振荡式镇压器 oscillating roller
振荡试验 oscillation test;shake test
振荡适应控制 oscillating adaptive control
振荡受力疲劳试验 flexural loading test
振荡输出 oscillator output
振荡输入 oscillator input
振荡输送机 oscillating conveyer[conveyor];shuttle
振荡衰减量 decrement of oscillation
振荡水体 oscillating water mass
振荡水跃 oscillating hydraulic jump
振荡水质 oscillating water mass
振荡速度 hunting speed
振荡碎粒机 oscillating granulator
振荡损失 oscillation loss
振荡特性 oscillating characteristic
振荡条件 oscillating condition
振荡同步 oscillator sync
振荡筒式黏[粘]度计 oscillating cylinder viscometer
振荡突起 hard start of oscillation
振荡图 oscillogram
振荡土压力 oscillating earth pressure
振荡砣 oscillating weight
振荡完全停止 absolute damping
振荡微分方程 oscillatory differential equation
振荡卫星 oscillating satellite
振荡稳定性 stability against oscillation
振荡吸附法 oscillating adsorption method
振荡洗涤机 Vibrotex machine;vibro washer
振荡系列 train of oscillations
振荡系数 coefficient of oscillation
振荡系统 oscillating system; oscillation system;oscillatory system
振荡现象 oscillation phenomenon;oscillatory occurrence
振荡线路 oscillator circuit
振荡线圈 oscillating coil; oscillator coil
振荡限度 oscillation limit
振荡相 oscillating phase
振荡相位 oscillation phase
振荡效应 oscillation effect
振荡信号 oscillating beacon;oscillator signal
振荡形式 oscillation form
振荡型 oscillating mode
振荡型分离 mode frequency separation
振荡型滤波器 mode filter
振荡型频差 mode separation
振荡型式 oscillatory formation
振荡型同步 mode lock
振荡型图 mode pattern
振荡型转移 mode shift
振荡型自动稳压器 Isenthal automatic voltage regulator
振荡性冲击 oscillatory surge
振荡性电涌 oscillatory surge
振荡性衰减 oscillatory damped
振荡性位移 oscillatory displacement
振荡性响应 oscillatory response
振荡选择器 oscillector
振荡压路机【机】 oscillatory shaking roller
振荡研磨机 oscillating mill
振荡仪 shake apparatus
振荡抑制 oscillation suppression
振荡抑制器 oscillation suppressor
振荡应力 oscillation stress; oscillatory stress
振荡圆盘式黏[粘]度计 oscillating disk visco(si)meter
振荡圆筒式黏[粘]度计 oscillating cylinder viscometer
振荡运动 oscillating motion; oscillating movement; oscillatory movement; oscillatory motion; undulatory motive <水质点的>
振荡照相法 oscillation photography
振荡正应力 oscillatory normal stress
振荡指示器 cymoscope; cymoscope wave detector;oscillation indicator
振荡制粒器 oscillating granulator
振荡滞后 oscillating hysteresis;oscillation hysteresis
振荡中断区 sink region
振荡中心 center[centre] of oscillation;oscillating center[centre];oscillation center[centre]
振荡中性(点)oscillating neutral
振荡周期 complete time of oscillation;hunting period;oscillating period; oscillation period; oscillatory period;period of oscillation
振荡主模 dominant mode;fundamental mode
振荡柱 gantry post
振荡转向轴 oscillating steering axle
振荡装置 rocking equipment
振荡状态 oscillatory regime
振荡子 oscillator
振荡自由度 oscillatory degree of freedom
振荡总振幅 peak-to-peak
振荡阻抗 oscillation impedance
振荡阻尼 oscillation dampening; oscillation damping
振荡钻头 oscillating bit
振荡作用 oscillation
振捣 compaction by vibration (and compression); jolt ramming; vibrating; vibration puddling; vibro; vibrocast; vibroramming; vibrotamping
振捣斑伤 vibrating poker burn
振捣板 float vibrator; vibrating board; vibrating tamper; vibratory tamper;vibrorammer
振捣棒 immersion-type vibrator; immersion(-type) vibrator; vibrating needle;vibrating spear
振捣不够 undervibration
振捣不实的混凝土 loosely spread concrete
振捣不足 <混凝土> undervibration
振捣不足的混凝土 loosely spread concrete
振捣常规混凝土 vibrated conventional concrete
振捣大体积混凝土 vibrated bulk concrete;vibrated mass concrete
振捣电锤 electric(al) hammer
振捣堆石 vibrated rockfill
振捣刮平机 tamping-level(l)ing finisher
振捣过的混凝土 vibrated concrete
振捣过的新浇混凝土 vibrated fresh concrete
振捣过度 overtamping
振捣夯 vibrorammer
振捣混凝土 shock concrete;vibrocast concrete
振捣混凝土管 vibrated concrete pipe
振捣梁 <混凝土摊铺机的> vibratory beam; tamping beam; vibrating beam
振捣器 <振捣混凝土用> air pick(er); vibrator; vibratory unit; vibrorammer
振捣器镘面板 vibrating screed
振捣器组 gang vibrators
振捣钎杆 immersion-type vibrator
振捣设备 compaction equipment
振捣压路机 tamping-type roller
振捣置换法 vibro-replacement
振捣置换碎石桩 vibro-replacement stone column
振捣最佳值 vibrated optimum value
振底炉 shock bottom furnace;vibrating plate furnace
振底式炉 shaker hearth furnace
振动 vibrate;vibration movement;vibratory motion; vibratory movement
振动凹凸杆 <用于混凝土路面刻槽防滑> vibrating(relief) bar
振动拔桩机 vibrating extractor; vibratory extractor;vibratory pile extractor;vibro-extractor
振动靶 vibrating target
振动板 plate vibrator;platform vibrator; vibrating board; vibrating plate; vibrobatten; vibroplate; vibratory plate
振动板捣实器 vibrating plate compactor
振动板夯实机 vibrating board compactor;vibrating plate compactor
振动板盘式振动器 vibratory plate
振动板式萃取器 vibrating plate extractor
振动板压实器 vibrating board compactor;vibrating plate compactor
振动拌和机 vibrating mixer
振动拌面器 vibrating finisher
振动棒 poker vibrator;spud vibrator; spur vibrator;vibrating head;vibrating needle; vibrating spear; vibrator beam
振动暴露时间 vibration exposure time
振动爆破 shock blasting
振动爆炸 vibrating explosion;vibratory explosion
振动杯进料器 vibratory bowl feeder
振动泵 sonic pump;vibration pump
振动篦式冷却器 shaking grate cooler;vibrating grate cooler
振动臂 shaker arm
振动变换器 chopper; vibratory converter
振动变流器 vibrator inverter; vibratory converter;vibropack
振动变形 dynamic(al) deflection
振动病 vibration disease
振动波 vibration wave
振动波节 vibration node
振动铂电极 vibrated platinum

振动铂微电极 vibrating platinum microelectrode
振动不稳定性 vibrational instability
振动不足 undervibration
振动部分 oscillating component
振动擦洗机 vibrating scrubber
振动材料 vibration damping material
振动采取岩芯 vibro-coring
振动参数 vibration parameter
振动槽 rocking trough;vibra shoot;vibrating trough;vibration chute
振动槽式分批箱 vibrating trough batcher
振动槽式输送机 jog-trough conveyer[conveyor]; vibrating trough conveyer[conveyor]
振动槽式输送器 jog-trough conveyer[conveyor]
振动槽式装载机 shaking pan loader
振动侧向传输 vibration flanking transmission
振动测定 vibrating measuring;vibration measurement
振动测定仪 vibration measuring apparatus
振动测量 vibrating measuring;vibration measurement; vibration survey;vibrographing
振动测量计 vibration meter;vibrometer
振动测量器 vibration measuring device
振动测量设备 vibration measuring equipment
振动测量系统 vibration measuring system;vibration survey system
振动测量仪 vibration measuring set
振动测深杆 vibro-sounding rod
振动测试仪 vibration measurement instrument
振动测试仪器 measuring instrument for vibration
振动叉 vibrating tine
振动铲 vibrospade
振动长镘刀 vibratory bullfloat
振动长抹子 vibratory bullfloat
振动常数 vibration constant
振动车 <物探和地震勘探用> vibrator vehicle
振动沉拔桩机 vibration pile-driver extractor; vibrating pile-driver extractor
振动沉拔桩架 vibratory pile driving and extracting frame
振动沉管灌注桩 British steel piling vibro piles
振动沉柱法 pile driving by vibration;vibrosinking
振动沉桩 pile vibrosinking;vibration piling; vibrohammer pile driving;vibrosinking pile
振动沉桩法 pile vibrosinking method;piling by vibration;vibrosinking
振动沉桩机 vibratory pile driver
振动成型 jolt(ing) mo(u)lding;vibration mo(u)lding;vibration shaping; vibratory compaction; vibromo(u)lding
振动成型机 jolt(ing) mo(u)lding machine
振动成型砂浆立方试件试验 vibrated mortar cube test
振动成型砂浆立方试块试验 vibrated mortar cube test
振动乘法器 vibration multiplier
振动乘坐环境 vibration ride environments
振动持续时间 duration of shaking
振动冲程 vibrating stroke
振动冲击 vibratory impact;vibroimpact
振动冲击沉桩机 percussion-type vibratory pile driver
振动冲击成孔机 vibrating-impacting boring machine
振动冲击夯 vibratory rammer;vibratory tamper
振动冲击荷载 vibratory shock load
振动冲击试验 jerk test
振动冲击压路机 vibrating-impacting roller
振动冲击钻孔机 vibrating-impacting boring machine
振动出芯机 core vibrating unit;core vibrator
振动除尘器 vibrating filter
振动锄 vibro-hoe
振动触点 oscillating contact; vibrating contact
振动触点式电压调整器 vibrating contact voltage regulator
振动穿孔芯棒 vibrating void-forming mandrel
振动传播 propagation of vibration;vibration transmission
振动传递 vibration transmission
振动传递率 vibration transmissibility
振动传感器 vibrating sensor; vibration detector; vibration pickup; vibration transducer;vibro-pickup
振动床 vibrated bed
振动锤 vibration hammer; vibratory hammer;vibrating hammer
振动磁针 vibrating needle
振动粗筛 vibrocribble
振动打/拔桩机 vibratory pile driver/extractor;vibro-driver extractor
振动打炉衬 vibrated lining
振动打磨机 vibratory sander
振动打桩 pile driving by vibration;vibratory pile driving;vibropiling
振动打桩锤 vibrating pile hammer; vibrating piling hammer; vibro-pile hammer;vibratory pile hammer
振动打桩锤或拔桩的夹具 pile clamp
振动打桩机 vibrating pendulum; vibrating pile driver; vibration piledriver; vibrator; vibratory driver; vibratory driving machine; vibratory hammer; vibratory pile driver; vibratory pile hammer; vibro-driver;vibro-pile driver
振动大镘刀 vibratory bullfloat
振动大抹子 vibratory bullfloat
振动带 shuttle belt;vibration band
振动刀 vibrating blade
振动捣打机 jolt-ramming machine
振动捣实 compaction by vibration (and compression); consolidation by vibration;jolt ramming;mechanical puddling;vibration puddling
振动捣实混凝土 compacting concrete by vibration
振动捣实梁 vibrating compaction beam
振动捣实器 plate compactor;vibrator(y plate) compactor;vibrorammer
振动捣实型 tamping-vibrating design
振动的 vibrant; vibratile; vibrating; vibrational;vibrative;vibratory
振动的螺旋体 shaken helicoids
振动等级 vibration level
振动等响曲线 equal vibration feeling contour
振动底板 vibrating base-plate
振动底片 vibrating base-plate
振动地震 vibroseis
振动地震量测 vibroseismic survey
振动点阵 vibrational lattice
振动电动机 vibrating motor
振动电弧堆焊 percussion arc pile-up welding;vibratory arc surfacing
振动电弧焊接 percussion welding; percussive welding
振动电机 vibratory motor
振动电键 vibroplex
振动电开关 vibroswitch
振动电流计 vibrating galvanometer
振动电路 vibrating circuit
振动电枢 vibrating armature
振动电枢式电动机 oscillating armature motor
振动调查 vibration investigation
振动动能 vibration kinetic energy
振动断裂 vibratory fracturing
振动断续器 vibrating break
振动钝化 vibrational deactivation
振动发生器 vibration generator
振动发生装置 vibration generator system;vibrator generator system
振动法 lash method; vibrating method; vibration method; vibratory method; vibratory drilling method 【岩】
振动法压实 vibration compaction
振动翻车机 rotary vibrating tippler
振动烦扰 vibration nuisance
振动反向 reversal of vibration
振动范围 range of oscillation
振动方程 equation of oscillation
振动方法 vibrational method; vibration measure
振动方孔筛 square mesh vibrating screen
振动方式 mode of vibration; vibration mode
振动方式密度 vibration model density
振动方向 direction of shock; vibration direction
振动防护开关 vibration-and-shock safety switch
振动分级 vibration separation
振动分级机 vibro-classifier
振动分离筛 shaker separator
振动分配器 oscillating distributor
振动分析 vibration analysis
振动分析和偏差概念 vibration analysis and deviation concept
振动分析器 vibration analyser[analyzer]
振动分析仪 vibration analyser[analyzer]
振动粉光抹子 vibrosmoothing trowel
振动粉磨 vibratory grinding
振动粉碎机 vibrating mill; vibrating pulverizer
振动粉碎作用 vibro-pulverization
振动浮选机 vibroflo(a)tation machine
振动幅度 amplitude of vibration
振动幅值 amplitude of vibration
振动附加荷载 dynamic(al) loading
振动杆 vibrating arm; vibrating needle;vibration rod
振动杆加油器 vibrating rod oiler
振动感受器 vibration-sensitive receptor;vibratory-sensitive receptor
振动缸 vibrating cylinder
振动钢路滚 vibratory steel roller
振动钢路碾 vibratory steel roller
振动钢轮压路机 vibratory steel roller
振动钢片式孔隙水压力计 vibrating strip piezometer
振动钢丝应变仪 vibrating wire strain ga(u)ge
振动钢弦 vibratory string
振动钢弦式伸缩计 vibrating string extensometer
振动钢弦式应变计 vibrating string extensometer
振动格筛 vibrating bar grizzly;vibration grizzly;vibrating grizzly
振动格栅进料器 vibrating grizzly feeder
振动隔绝 vibration insulation
振动给进 vibration-actuated feed
振动给料斗 vibratory feed unit
振动给料机 shaker feeder; shaking feeder; vibrofeeder; vibrating feeder;vibratory feeder
振动给料器 shaker feeder;vibrofeeder
振动给药机 vibrating reagent feeder
振动工具 vibrating tool
振动工作台 shaking table
振动公害 vibration hazard for citizen
振动公式安全系数 <提升钢丝绳> spring formula factor of safety
振动供料 vibrator supply
振动共鸣法 vibration resonance method
振动鼓形碾 vibrating drum-type roller
振动固结法 consolidation by vibration
振动故障 vibration hazard
振动刮板 vibrating screed
振动灌浆法 vibro-cem process
振动灌浆混凝土 vibro-casting concrete;vibro-grouted aggregate concrete
振动灌注桩 vibrex pile;vibro pile
振动光辊压路机 vibratory smooth drum roller
振动光谱 vibronic spectrum; vibrational spectrum
振动辊 vibrating drum
振动辊压刮平器 vibratory roller screed
振动辊压机 vibrating roller;vibratory roller
振动滚机 vibrorolling
振动滚路机 vibrating roller;vibration roller
振动滚筒碾 vibrating drum-type roller
振动滚筒式压路机 vibrating drum-type roller
振动滚压 vibratory rolling;vibrorolling
振动滚轧 vibrorolling
振动过的砖墙板 vibrated brick panel
振动过度 <混凝土> overvibration
振动过滤器 vibrating filter
振动夯 vibrating rammer; vibrating tamper;vibrorammer;vibrotamper
振动夯板 ramming foot;ramming shoe
振动夯击 jolt ramming
振动夯击机 jolt-ramming machine
振动夯实 jolt ramming;vibrating compaction
振动夯实混凝土 concrete compacted by jolting; vibrated and tamped concrete
振动夯实机 jolt-ramming machine;vibratory compacting roller
振动夯土机 vibro-tamper
振动和冲击试验 vibration-and-shock testing
振动荷载 racking load; vibrating load; vibration(al) load; vibration force;vibratory load
振动弧 arc of oscillation
振动弧度频率 radian frequency of vibration
振动环境 vibration environment
振动环形管道可逆输送(系统) reversing vibrating circular pipeline
振动换流器 chopper
振动换流器供电 vibrator power supply
振动簧片式频率计 reed frequency meter
振动回火 vibration tempering
振动回转钻进 vibro-rotary drilling

Z

振动混合器 vibromixer
振动混凝土柱桩 concrete vibro-column pile
振动混凝土桩 concrete vibro-column pile;vibro-concrete pile
振动活塞取样管 vibratory piston corer;vibratory piston sampler
振动机 bobbing machine; mechanical shaker; shaker; vibration machine; vibrator;rocker <冲洗石屑用>
振动机构 vibrating mechanism;vibration programmer
振动机械 vibrating machinery
振动机械整流器 <低压直流变高压直流用> vibrating mechanical rectifier
振动积累效应 cumulative effect of shaking
振动基础 vibration foundation
振动基础厂房地基勘察 exploration of factory foundation and vibrating ground
振动激励 vibrational excitation
振动激励器 vibration exciter[excitor]
振动激烈性 vibration severity
振动级计 vibration level meter
振动极限 vibration limit
振动挤出法 vibro-extrusion method
振动挤入锚 vibro-driven anchor
振动挤压混凝土管 vibrohydropressed concrete pipe
振动挤压机 jolt-ramming machine
振动计 vibrameter; vibrating meter; vibration measurer; vibratormeter; vibrograph;vibroscope
振动计算 vibration calculation
振动计重量秤 vibratory weigh scale
振动记录 vibro-record
振动记录图 vibrorecord drawing
振动记录仪 vibration recorder;vibration recording equipment; vibrograph
振动技术 vibrotechnique
振动继电器 vibrating relay
振动加固 consolidation by vibrating; stabilization by vibroflo(a)tation
振动加料机 oscillating batch charger
振动加料器 vibrating feeder
振动加密 vibration compaction
振动加密区 vibro-compaction zone
振动加速度 vibrational acceleration
振动加速级 vibration acceleration level
振动加压 vibration plus pressure;vibratory pressing
振动架 vibrating grid
振动监测 vibration monitoring
振动监测器 vibration monitor
振动监测仪 vibration monitor
振动减低器 vibrating dampener
振动剪 nibbling shear
振动检波器 vibrating detector
振动检测器 vibration detector;vibration monitor
振动检验计算 vibration check calculation
振动键 oscillating bond
振动浇注 vibrating casting; vibrocasting
振动浇筑混凝土 vibrocast concrete
振动角速度传感器 vibratory rate gyroscope
振动角速度陀螺仪 vibratory rate gyroscope
振动搅拌机 vibratory mixer
振动搅拌器 oscillating agitator
振动接触式电压调整器 vibrating contact voltage regulator
振动接触式速度调节器 rocking-contact speed regulator
振动接点 vibrating contact
振动节点 nodal point of vibration;vibration nodal point;vibration node
振动节面 nodal plane
振动节型 nodal pattern
振动结 vibration node
振动结构 vibrational structure;vibrating structure
振动界限 vibration limit
振动金属丝应变计 vibrating wire strain ga(u)ge
振动紧实 vibration ramming
振动进给器 vibrofeeder
振动浸取器 vibratory leacher
振动经受时间 vibration exposure time
振动静压复合压路机 combined roller
振动镜偏转器 vibrating mirror deflector
振动锯缝机 vibratory joint cutter
振动绝缘 vibration insulation;vibration isolation
振动绝缘体 vibration indicator; vibration insulator
振动觉 vibration sensation
振动开沟机 vibratory plow
振动可变装置 variable device of vibratory element
振动可传性 transmissibility of vibration
振动空气分选机 vibrating air separator
振动控制 vibration control
振动控制器 vibrating controller
振动控制系统 vibration control system
振动块料机器 block vibrating machine
振动框(架) vibrating frame
振动类比试验 vibration analogic test
振动类型 type of vibration
振动累积效应 cumulative effect of shaking
振动冷却器 vibratory cooler
振动离心机 vibratory centrifuge
振动犁铧 oscillant share;shaker share
振动理论 theory of vibration;vibration theory
振动力 dynamic(al) force; vibrating force; vibration force; vibratory force;vibratory power
振动力学 vibration mechanics
振动砾磨机 vibrating pebble mill
振动连接 wobble joint
振动梁 vibrating beam;walking beam
振动量 vibratory output
振动量的全行程 total excursion of an oscillating quantity
振动量子数 rotational quantum number;vibrational quantum number
振动料道 vibratory track
振动料斗送料器 vibratory hopper feeder
振动裂缝 vibration crack
振动铃 vibrating bell
振动溜槽 jigging chute;oscillating deliver; shaking launder; shaking trough; vibrating chute; vibratory sluice
振动溜管 <浇混凝土用的> vibrating tube;vibrating pipe
振动流槽 vibrating chute
振动流动 vibration flow
振动流度法 vibration-flow method
振动流化床 vibrated fluidized bed
振动炉箅 jigging grate;vibrating grate
振动炉排 vibrating grate
振动滤袋法 bag swinging
振动路滚 vibrating roller; vibration roller;vibratory roller
振动路机 vibrating compactor
振动率 flutter rate
振动轮激振力测定 vibration drum exciting force measurement
振动轮频率测定 drum-vibration frequency measurement
振动轮振幅测定 drum vibration amplitude measurement
振动落料栅架 knocking-out grid
振动落砂机 shakeout equipment; shakeout machine;vibrating shakeout machine; vibratory shakeout machine
振动落砂架 shaking grid; vibrating shake-out grid
振动马达 vibrating motor
振动埋入锚 vibro-driven anchor
振动脉冲 vibratory impulse
振动铆钉机 vibrating riveter
振动密实 compaction by jolting; vibration compaction
振动密实成型 vibro-casting
振动密实法 compaction by vibration (and compression)
振动密实性 <混凝土混合料> vibrability
振动面 plane of vibration; vibration plane
振动模板 vibrating mo(u)ld
振动模具 vibrating mo(u)ld
振动模拟器 vibration simulator
振动模式 mode of vibration; vibrational mode
振动模型 mode of vibration; vibrational model
振动模制 jolt(ing)mo(u)lding
振动膜 diaphragm;jockey;tympanum [复 tympana/ tympanums]; vibrating membrane; vibration diaphragm
振动膜片 vibrating diaphragm
振动磨 pulverator
振动磨光机 vibratory sander
振动磨机 vibrating mill;vibratom;vibromill
振动磨矿法 vibration milling; vibrogrinding
振动磨碎机 vibrating mill;vibromill
振动抹灰板 vibrating concrete float; vibrating float
振动抹浆 vibrating coating
振动抹面机 vibrating and finishing machine;vibrating finisher
振动抹平镘板 vibrosmoothing trowel
振动能 vibration energy
振动能分级器 vibroenergy separator
振动能级 vibrational energy level
振动能力 vibration ability
振动能量 energy of vibration;vibrational energy
振动黏[粘]度计 vibration visco(si)meter
振动碾 vibratory roller
振动碾机 vibrating roller;vibroroller
振动碾压法 vibroroller compaction; compaction by vibrating roller; vibroroller compacting;vibrorolling
振动凝结试验 jar tests for coagulation
振动耦合 vibrational coupling;vibronic coupling
振动拍摄 tonguing
振动排放 shock discharge
振动盘 pan vibrator; vibrating disc [disk];vibrating frame
振动盘法 oscillating disk method
振动盘反应器 vibrating tray reactor
振动盘黏[粘]度计 oscillating disk visco(si)meter
振动盘式给料机 vibrating tray feeder
振动盘式压实机 vibrating pan compactor
振动抛光机 vibrating smoother; vibratory finishing machine; vibratory smoother
振动配料单元 vibrating batching unit
振动疲劳 vibration fatigue
振动疲劳强度 fatigue strength under oscillation stresses
振动疲劳试验 vibration fatigue test
振动疲劳试验机 vibration fatigue machine
振动片 trembler; vibrating plate; vibrating reed; vibrational resonant disc[disk]
振动片黏[粘]度计 vibrating plate visco(si)meter
振动片式 vibrating-reed type
振动频率 frequency of oscillation; frequency of shock; frequency of vibration;vibration frequency
振动频率表 vibration frequency meter
振动频率的动力阻尼 dynamic(al) impedance against frequency
振动频率分析器 vibration frequency analyser[analyzer]
振动频率分析仪 vibration frequency analyser[analyzer]
振动频率计 vibration frequency meter
振动频率指示器 reed frequency detector
振动频谱 rumble spectrum
振动频谱分析器 vibration spectrum analyser[analyzer]
振动品质 quality of vibration
振动平板 swing plate
振动平板夯装置 vibrating plate tamper
振动平板压实机 vibrating base-plate compactor
振动平衡 vibration balancing
振动平衡器 vibration neutralizer
振动平面 plane of vibration
振动平碾 smooth drum vibrating roller
振动平筛机 scalper
振动平台 shaking platform
振动平整机 vibrofinisher
振动评价标准 vibration criterion
振动屏蔽 screening of vibration
振动坡法 <测混凝土和易性> vibrating slope method
振动破坏 flutter failure;vibration damage
振动破裂 vibration fracturing
振动破碎试验 vibrating crushing test
振动谱带 vibrational band
振动谱线 vibrational line
振动气泡 oscillating bubble
振动汽缸 vibratory cylinder
振动汽油发动机 vibrating petrol engine
振动器 driver;rapper;rocker;rocking frame; shaker; shaking device; shaking machine; shocker; stirring gear;ticker;trembler;vibrating machinery;vibrator
振动器离合器操纵杆 vibrator clutch lever
振动器板 vibrator plate
振动器插入棒 vibration head
振动器齿轮 shaker gear
振动器触点 vibrator contact
振动器电动机 vibrator motor
振动器电源 vibrator power supply
振动器极 vibratode;vibrolode
振动器校准 vibration machines calibration
振动器接触头 vibrator contact
振动器控制杆 vibrator lever
振动器套筒 vibrator cylinder
振动器膝盖阀 vibrator knee valve

振动器型反向变流器 vibrator type inverter
振动器振实 vibrator compaction
振动器轴 vibrator shaft
振动器组 gang vibrators;vibrator unit
振动钎杆 immersion-type vibrator
振动强度 intensity of shaking;intensity of vibration;vibration amplitude;vibration intensity;vibration strength;vibratory strength
振动强度的抵消 cancellation of intensities
振动强度和土质条件 shaking intensity and soil condition
振动(强)烈度 vibration severity
振动敲平锤 vibrating jigger
振动切缝刀 vibrating knife
振动切缝机 vibrating joint cutter
振动切入板 <切缝机的> vibrating plate
振动切碎器 rock mincer
振动清灰 vibration cleaning
振动清理 vibratory cleaning
振动求和定则 vibrational sum rule
振动球磨 vibratory milling
振动球磨机 vibrating ball mill;vibrating mill;vibration ball mill;vibratory mill;vibromill;vibratom
振动曲线 vibration curve
振动趋性 vibrotaxis
振动取土器 vibro-corer
振动取(岩)芯器 vibro-corer
振动取样 vibro-coring
振动取样管 vibrating coring tube
振动取样器 vibro-corer
振动容限 vibration tolerance
振动熔接 vibration welding
振动三轴试验 dynamic(al) triaxial test
振动三轴仪 vibration triaxial apparatus
振动筛 hummer screen;impact screen;impact type screen;jigging screen;jigging sieve;pulsating screen;pulsator;riddler;rocker sieve;screen shaker;shaker screen;shaking bar grizzly;shaking screen;shaking sieve;slugger screen;swinging screen;swinging sieve;vibrating frame;vibrating grizzly;vibrating mesh;vibrating sieve;vibrating screen;vibration screen;vibration sieve;vibratory screen;vibrosieve;wobbler screen
振动筛的激振器 mechanical sieve shaker
振动筛分机 laboratory sifter;sieve shaker;vibrating screen classifier
振动筛机 vibrating scalper
振动筛离心机 vibrating screen centrifuge
振动筛连杆 shaker shoe pitman
振动筛盘 shaker pan
振动筛溶解器 vibrating screen dissolver
振动筛筛分 bolting
振动筛式挖掘机 shaking-sieve digger
振动筛条 vibrator bar
振动筛组 deck screen
振动栅 vibrating grid
振动熵 vibrational entropy
振动烧结 vibratory sintering
振动烧瓶 shaking flask
振动设备 vibration drill;vibratory equipment;vibrodrill
振动设备的有效作用范围 effective range of vibrating equipment
振动渗滤器 shock diffuser
振动升华干燥器 vibration freeze-drier[dryer]
振动生理效应 vibration physiological effect
振动声 chatter
振动施荷器 vibrating loader
振动时间 time of vibration;vibration time
振动式板压机 vibrating plate compactor
振动式拌和机 vibratory mixer
振动式棒条筛 vibrating bar grizzly
振动式布袋除尘器 vibrating bag filter
振动式材料输送设备 vibrating materials handling equipment
振动式超薄切片机 oscillating blade microtome
振动式传送机 shaker conveyer[conveyor];vibration conveyer[conveyor]
振动式打拔桩机 vibratory pile driver
振动式打桩机 vibrating pile driver
振动式单轮压路机 single vibratory drum roller
振动式捣固机 vibrating tamper
振动式捣固器 vibrating tamper
振动式电缆埋设机 vibrating cable-layer
振动式电缆铺设机 vibrating cable-layer
振动式电压调节器 autopulse
振动式电压调整器 vibrator voltage regulator
振动式断路器 vibrating type circuit breaker
振动式断续器 vibrating breaker
振动式风力装载机 vibrating pneumatic loader
振动式干燥机 shaker drier[dryer]
振动式钢轮路碾 vibratory steel roller
振动式钢轮压实机 vibratory soil roller
振动式格筛进料器 vibrating grizzly feeder
振动式给矿机 shaking feeder
振动式给料机 vibrating feeder;vibratory feeder
振动式给料器 cushion feeder;vibration feeder;shaking feeder;vibratory feeder
振动式功率计 vibration power meter
振动式供给器 vibrafeeder
振动式供料机 vibrating feeder
振动式供料器 cushion feeder;shaking feeder;vibration feeder
振动式沟堑压实机 vibrating trench compactor;vibrator trench compactor
振动式刮路机 vibrating smoother;vibratory smoother
振动式管磨机 vibrating tube mill;vibratory tube mill
振动式灌注桩 vibro pile
振动式光面辊 vibratory smoothing wheeled roller
振动式轨道夯实机 vibrating type track tamper
振动式滚压机 vibration roller;vibratory smoothing wheeled roller
振动式过滤机 vibrating filter
振动式过滤器 vibrating filter
振动式夯板器 vibrating plate rammer
振动式夯实锤 vibratory impact rammer
振动式滑盘压机 vibrating sliding table press;vibratory sliding table press
振动式混凝土搅拌机 vibratory concrete mixer
振动式混凝土切缝机 vibrating joint cutter
振动式混凝土整平板 concrete vibrating screed
振动式计量器 vibrameter;vibratormeter
振动式继电器 flutter relay
振动式加力床 vibrating stressing bed;vibratory stressing bed
振动式加力台 vibrating stressing bed;vibratory stressing bed
振动式检流计 vibration galvanometer
振动式搅拌机 oscillating agitator
振动式进料器 vibrating feeder
振动式拉杆夯实机 vibratory tie tamper
振动式离心机 vibrating centrifuge
振动式励磁机 vibration exciter[exciter]
振动式溜槽 shaking chute
振动式炉箅 jigging grate;shaker grate
振动式炉排 dumping grate
振动式路碾 impact roller;vibrating roller;vibroroller;vibratory roller
振动式乱石粉碎机 vibrating pebble mill
振动式螺旋管提升机 vibratory spiral pipe elevator
振动式螺旋提升机 vibrating spiral elevator
振动式埋管机 vibrating plough
振动式埋缆犁 vibrating plough
振动式铆钉枪 vibrator
振动式磨矿机 vibration mill
振动式木片平筛 shaker-type(flat) chip screen
振动式逆变器 vibrator inverter
振动式黏[粘]土机 vibrating earth borer
振动式碾压机 impact roller;vibratory roller
振动式皮带清扫器 vibratory belt cleaner
振动式频率计 resonance frequency indicator;vibration-type frequency meter;sonometer
振动式平板夯 vibrating plate compactor;vibroplate;vibratory plate compactor;vibrating plate <切缝机的>
振动式平板压实机 vibratory plate compactor
振动式砌块制造机 vibratory block making machine
振动式清洗机 shaker washer
振动式取样器取样 sampling by vibro-sampler
振动式砂浆拌和机 vibratory mortar mixer
振动式筛滤 vibrating fine screen
振动式手扶压路机 vibrating hand-roller
振动式输送机 vibratory conveying machine;vibrating conveyer[conveyor];vibratory conveyer[conveyor];vibrating conveying machine
振动式输送器 vibratory conveyer[conveyor]
振动式双辊压路机 vibratory double drum roller
振动式双轮压路机 vibrating tandem roller;vibratory tandem roller
振动式送料器 vibratory feeder
振动式淘汰盘 vibrating frame
振动式调节器 vibrating type regulator
振动式土壤夯具 vibratory soil compactor
振动式土壤压实板 vibratory soil compacting plate
振动式土壤压实器 vibratory soil compactor
振动式土钻 vibratory earth borer
振动式推进装置 vibration-actuated feed
振动式推土机 vibrating dozer
振动式脱水机 vibrating dehydrator
振动式洗矿筛 shaking-screen washer
振动式斜槽 jigging conveyer[conveyor];oscillating chute
振动式循环管线 vibrating circular pipeline
振动式压沟机 vibratory trench roller
振动式压力机 vibratory press
振动式压路机 vibrating roller;vibration roller;vibroroller
振动式压路碾 vibroroller
振动式压实机 vibratory compactor
振动式压缩机 vibration type compressor
振动式研磨机 vibrating grinding mill;vibratory grinding mill
振动式羊脚滚筒 vibrating sheepsfoot roller;vibratory sheepsfoot roller
振动式羊脚压路机 vibrating sheepsfoot roller;vibratory sheepsfoot roller
振动式羊足碾 vibratory pad-foot roller
振动式羊足压路机 vibratory sheepsfoot roller
振动式运输机 shaker-type conveyer[conveyor];vibrating conveyer[conveyor]
振动式运输机槽 shaking conveyor pan
振动式运输机给料车 shaking conveyor supply truck
振动式运输机滚珠支架 shaking conveyor ball frame carriage
振动式运输机回转座 shaking conveyor swivel
振动式运输机连杆 shaking conveyor rod
振动式运输机螺旋顶杆 shaking conveyor screw jack
振动式运输机支架 shaking conveyor carriage
振动式砸道机 vibratory tie tamper
振动式轧管机 vibratory tube mill
振动式振捣器 jolt vibrator
振动式镇压器 vibrating roller
振动式整流器 pendulum rectifier;vibrating type rectifier;vibrator type rectifier
振动式整平板 vibrating finishing screed;vibrating screed;vibratory smoothing screed
振动式整平尺 vibratory finishing beam
振动式整平机 vibrating smoother;vibratory finishing machine;vibratory smoother
振动式整平器 vibratory finishing beam;vibratory float
振动式直线剪切机 vibratory linear shear apparatus
振动式中耕机 vibrator cultivator
振动式重力仪 vibration gravimeter
振动式转速计 vibration tachometer
振动式转筒筛分机 revolving vibrating screen
振动式桩锤 vibrator pile hammer
振动式装载机 vibrating loader
振动式自动电压调整器 fuss type automatic voltage regulator;vibrating type automatic voltage regulator
振动试验 jolt-and-jumble test;shake-down test;shake test;shaking test;swing test;vibrating test(ing);vibration test(ing);vibratory test(ing)
振动试验机 vibration rig;vibration testing machine;vibratory testing machine
振动试验器 shaker
振动试验设备 shaker apparatus;shaking equipment;shaking test apparatus

振动试验室 vibration laboratory
振动试验数据 flutter test data
振动试验台 exciter; vibration test stand; vibrostand
振动试验装置 vibration testing device
振动梳 vibrating comb
振动输料管道 vibrating circuit pipeline
振动输送槽 jigging screen; vibrating trough conveyer[conveyor]; vibratory chute
振动输送带 shaker belting
振动输送机 jigger conveyer[conveyor]; jigging conveyer [conveyor]; shaker conveyer [conveyor]; shaking conveyer [conveyor]; shaking pan conveyer[conveyor]
振动输送料斜槽 vibrating chute
振动数 number of vibrations
振动衰减 damping of vibration; dying away of vibration; flutter fading; vibration damping; vibration attenuation
振动衰减量 vibration attenuation
振动衰减器 vibration absorber
振动衰减曲线 vibration attenuation curve
振动衰减特性 vibration dampening characteristic
振动水冲 vibro-jet
振动水冲(压密)法 vibroflo(a)tation
振动松弛 vibrational relaxation
振动送料器 <浇筑混凝土的> shaking feeder
振动速度级 vibration velocity level
振动速率 rate of oscillation
振动速率调节手柄 handle for regulating vibration speed
振动碎石机 vibration stone crusher
振动损害 vibration damage
振动台 bounce table; bumping table; flow table; jigging platform; joggling table; jolting table; percussion table; plain bumper; plain jolter; platform-type vibrator; platform vibrator; shake table; table vibrator; vibrating stand; vibrating table; vibrating test bench; vibration table; vibratory table; vibro-bench; vibroplatform; trianco <混凝土>
振动台法 shaking table method
振动态回弹性 vibrational resilience
振动弹簧 vibrating spring
振动弹簧继电器 vibrating spring relay
振动探棒 vibrating poker
振动探测导针 vibration lead probe
振动探测计 vibration monitor
振动探测系统 vibration detection system
振动探示器 hunting probe
振动探头 vibration probe
振动探针式磁强计 vibrating sample magnetometer
振动特性 model of vibration; vibrational property; vibration behavio(u)r; vibration characteristic; vibration performance
振动特性温度 vibration characteristic temperature
振动特性系数 vibration characteristic factor
振动特性因数 vibration characteristic factor
振动特征系数 vibration characteristic factor
振动特征因数 vibration characteristic factor
振动体 oscillating body; pendulum; pendulum wire; vibrating body
振动体质量 vibrator mass
振动填料 jolt-packing
振动填实砂桩 vibro-composer
振动条 vibrator bar
振动条筛 vibrating grizzly; vibratory grizzly
振动调节器 vibrating regulator
振动调整 flicker control
振动铁栅筛 vibrating grizzly; vibratory grizzly
振动头 vibrating head; vibrator cylinder; vibrator head; vibratory head; poker shaped case <混凝土振动器的>
振动投料机 vibrating batch charger
振动凸块碾 padfoot vibratory roller
振动图 vibrogram
振动脱水机 dewatering shaker; dewatering vibrator
振动脱水筛 vibrating-dewatering screen
振动脱芯机 vibrating core knockout machine; vibrating decorer
振动陀螺仪 gyrotron; vibratory gyroscope
振动弯曲试验 shock bending test; shock flexure test
振动危害 hazard of vibration; obstacle of vibration
振动微细结构 vibrational fine structure
振动位移 vibration displacement
振动位移级 vibration displacement level
振动喂料机 shaking feeder; vibrating feeder; vibration feeder
振动喂料器 shaking feeder; vibrating feeder; vibration feeder
振动温度 vibration temperature
振动稳定性 stability against vibration; stability of vibration
振动问题 vibration problem
振动污染 vibration pollution
振动污染预测 vibration pollution forecasting
振动误差 vibration error
振动吸力筛 aspirating screen
振动吸收法 vibratory absorption process
振动吸收器 vibration absorber
振动吸收装置 shock arrester[arrestor]; vibro-dampers <在机器下面>
振动洗涤器 jigging washer; vibration washer
振动洗选槽 vibrotrunk
振动系统 vibrating system
振动下沉 sinking by mechanical vibration
振动弦式应变仪 vibrating string extensometer; vibrating string strainometer
振动弦(线) vibrating string; vibrating wire
振动弦型重力仪 vibrating string type gravimeter
振动显示器 vibrograph
振动线 vibrational line
振动线圈 vibrating coil; vibrator coil
振动线圈法 vibrating coil method
振动线扫描器 vibrating wire scanner
振动限度 vibration limit
振动相位超前补偿器 vibrator phase advancer
振动箱 vibrating case; vibrating feed bin
振动响应 vibratory response
振动消除应力法 vibratory stress relief
振动消失 deadening of vibration
振动效应 dither effect; vibration effect
振动楔 vibrating wedge
振动斜槽 shaking chute
振动斜槽输送器 vibrating chute transporter
振动卸车装置 car shakeout
振动卸料系统 vibratory discharge system
振动芯 vibrating core; vibratory core
振动芯模法 vibrating core process
振动形变热处理 thermomechanical vibration treatment
振动性 vibratility
振动性的 vibratory
振动性能 vibrational behavio(u)r; vibration performance
振动序列 shock sequence
振动絮凝试验 jar tests for flocculation
振动选矿筛 vibrating grate
振动靴 vibrating shoe
振动循环 cycle of vibration
振动压滚 vibrating roller; vibratory roller
振动压滚机 vibration roller
振动压机 vibrating press
振动压力 vibratory pressure
振动压裂 <石油开采> fracturing by vibration
振动压路机 vibrating compactor; vibration roller; vibratory compactor; vibratory roller; vibroll(er); vibroroller
振动压密 consolidation by vibration; jigging compaction; vibration compaction; vibratory compaction
振动压密法 compaction by vibration (and compression); vibro-densification
振动压实 compaction by vibration (and compression); vibrating compaction; vibration(al) compaction; vibro-compaction; wobble seal; vibratory compaction
振动压实遍数 dynamic(al) passes
振动压实沉降率 vibrating compaction settlement ratio; vibratory compaction settlement ratio
振动压实的 vibro-compacted
振动压实法 vibratory compacting; vibratory compaction; vibro-densification; wobbler seal
振动压实机 vibratile compacter; vibration compactor; vibratory compactor; vibrating compactor
振动压实技术 vibratory compaction technique
振动压实盘 plate-type vibratory compactor
振动压实平板 vibratory plate compactor
振动压实砂桩 vibratory sand compaction pile
振动压实载重 vibroflot
振动压型机 jolt-ramming machine
振动压制 vibratory pressing
振动压铸 <用超声波或高频声波的> vibrocast
振动压铸法 vibro-casting process
振动轧平机 vibrofinisher
振动岩芯切割器 vibro-core cutter
振动岩芯取样器 <主要用于海底取样> vibro-corer
振动岩芯提取器 vibro-core cutter
振动研究 vibration investigation
振动研磨 vibration grinding
振动研磨机 rock grinder
振动样板 screed board vibrator; screed vibrator
振动液化 liquefaction; thixotropy; vibratory liquefaction
振动液化试验 vibratory liquidation test
振动液压打桩机 silent pile-driver
振动仪 vibration ga(u)ge; vibration indicator; vibration measurer; vibration meter; vibrometer; vibroscope
振动仪器 vibration instrumentation
振动抑制 vibration abatement
振动因子 frequency factor
振动应变测定器 vibration strain pickup
振动应力 dynamic(al) stress; vibration(al) stress; vibratory stress
振动有效控制系统 active vibration control system
振动元件 vibrating element
振动圆盘筛 vibratory disk mill
振动圆盘送料器 vibratory bowl hopper feeder
振动圆频率 circular frequency of vibration
振动源 source of vibration; vibration(al) source
振动跃迁 vibrational transition
振动运动 vibratory motion
振动运输带 shaker conveyer[conveyor]
振动运输机 jigging conveyer[conveyor]
振动运送机 jigger conveyer[conveyor]; jigging conveyer[conveyor]
振动载荷 racking load; vibrating load
振动凿岩 vibrating drilling; vibration drilling
振动造型 vibrocast
振动造型机 jar mo(u)lding machine; vibrating mo(u)lding machine
振动噪声 drumming noise; vibration noise
振动增实带 vibro-compaction zone
振动增实燃烧元件 vibro-compact fuel element
振动找平机 vibrating screed
振动针杆 immersion vibrator; vibrating needle
振动真空法 vibrating vacuum process
振动真空脱水 vibrovacuum dewatering
振动真空脱水工艺 vibrating vacuum dewatering process
振动整面机 vibrofinisher
振动整平机 vibrofinisher; vibratory screed <平整混凝土用>
振动支架 vibrating trestle
振动支座 vibration mounting
振动指示器 vibration detector; vibration indicator
振动指数 ride index
振动制度 vibration regime
振动制模机 jar ram mo(u)lding machine; vibration mo(u)lding machine
振动质量 vibrating mass
振动致实 vibrational compaction
振动中的节点 nodes in vibration
振动中心 center[centre] of vibration; vibrating center[centre]
振动终止 dead beat
振动重叠 vibration overlap
振动周期 complete time of vibration; cycle of vibration; period of vibration; pitch of vibration; vibration period
振动轴 vibrating shaft; vibration axis
振动铸造 vibrational casting; vibromo(u)ld
振动砖壁板 panel wall of vibrated brickwork; vibrated brick panel
振动砖砌护墙板 panel of vibrated brickwork

振动砖墙板 panel wall of vibrated brickwork
振动转动 vibrational rotation
振动转动光谱 vibrational-rotational spectrum
振动转动能量 vibrational-rotational energy
振动转动(谱)带 vibrational-rotational band
振动转动跃迁 vibrational-rotational transition
振动转换 vibratory conversion
振动转盘机 vibratory rotary table machine
振动桩锤 vibrohammer; vibrating hammer
振动装料 vibrated into place
振动装载机 shaking loader
振动装载台 vibrating platform
振动装置 rapping device; rapping gear; shakeout equipment; vibrating device; vibratory device
振动状态 vibrational behavio(u)r; vibrational state
振动子 ticker; trembler
振动子换流器 vibrapack; vibrator converter; vibropack
振动子继电器 pendulum(-type) relay
振动子升压器 vibrating booster
振动子式变换器 vibrator type converter
振动子整流器 vibropack
振动自记计 vibro-record
振动自由度 vibrational degrees of freedom
振动阻抗 vibration impedance
振动阻尼 absorption of vibration; damping of vibration; vibration damping
振动阻尼器 anti-vibration damp; vibration absorber
振动钻 spud vibrator; vibrating drill; vibratory drill
振动钻机 vibration drill; vibrodrill
振动钻进 vibration drilling; vibratory drilling; vibro-boring; vibro-drilling
振动钻井 vibro-drilling
振动钻孔 soil drilling by vibration; vibro-drilling
振动钻孔机 sonic drill
振动钻探 vibration drilling; vibratory drilling; vibro-boring; vibro-drilling
振动钻头 vibration bit
振动钻土 soil drilling by vibration
振动钻土机 vibratory earth borer
振动钻轧 vibrodrill
振动钻眼 vibration drilling
振动钻钻头 vibrating bit
振动最大加速度 maximum acceleration of vibration
振动最佳值 vibrated optimum value
振动最宜值 vibrated optimum value
振动作用 vibration
振动作用力 dynamic(al) force
振动座 shaker plate
振抖式挖掘机 shaker digger
振耳欲聋的 deafening
振浮器 vibroflot
振浮压密 vibrating compaction; vibroflo(a)tation
振浮压实 vibrating compaction
振浮压实法 vibroflo(a)tation; vibroflotating process
振浮压实机 vibroflot machine
振浮压实器 vibroflot
振浮重压式混凝土端承桩 vibroflo(a)tation-type point-bearing concrete pile
振幅 amplitude(of oscillation); amplitude of vibration; range of oscillation; vibration of amplitude
振幅摆动 amplitude fluctuation
振幅包络剖面 amplitude envelope section
振幅比 amplitude ratio
振幅比较器 amplitude comparator
振幅比剖面曲线 profile curve of amplitude ratio
振幅比相位差计 ratio and phase meter
振幅比相位差剖面平面图 profile-plan figure of amplitude ratio and phase difference
振幅比相位差剖面图 profile figure of amplitude ratio and phase difference
振幅比值 quotient of amplitudes
振幅变化 amplitude variation
振幅辨别器 amplitude discriminator
振幅波 oscillatory wave
振幅不稳定度 amplitude instability
振幅测定 amplitude measurement
振幅测井 amplitude log
振幅磁导率 amplitude permeability
振幅电路 amplitude circuit
振幅法 amplitude of vibration method
振幅反差 amplitude contrast
振幅反射率 amplitude reflectance; reflection coefficient
振幅反应曲线 amplitude response curve
振幅范围 amplitude range
振幅范围计数法 range-pair count method
振幅放大图 amplitude magnification factor
振幅放大系数 amplitude magnification factor
振幅放大因数 amplitude magnification factor
振幅放大因子 magnification factor for amplitude
振幅分布 amplitude distribution
振幅分层管 amplitude quantizing tube
振幅分离器 amplitude selector; amplitude separator
振幅分析 amplitude analysis; kicksort(ing)
振幅分析技术 kicksorting technique
振幅分析器 kicksorter
振幅干涉测量 amplitude interferometry
振幅共振 amplitude resonance
振幅光谱 amplitude spectrum
振幅轨迹 amplitude locus
振幅过滤器 amplitude filter
振幅函数 amplitude function
振幅和相移键控 amplitude and phase shift keying
振幅恢复 amplitude recovery
振幅恢复系数值 coefficient value of amplitude recover
振幅畸变 amplitude distortion; attenuation-frequency distortion
振幅计 amplitude meter
振幅计数法 range count method
振幅记录系统 vibration amplitude recording system
振幅加法器 amplitude adder
振幅加强线路 accentuator
振幅鉴别 amplitude discrimination
振幅鉴别电路 amplitude discrimination circuit
振幅鉴别器 amplitude discriminator
振幅鉴频器 amplitude discriminator
振幅键控 amplitude shift keying; on-off keying
振幅降低 lowering of amplitude
振幅阶跃时间 amplitude step time
振幅距离曲线 amplitude versus distance curve
振幅均衡 amplitude equalization
振幅控制 amplitude control
振幅量化 amplitude quantization; amplitude quantizing; quantization of amplitude
振幅量化管 amplitude quantizing tube
振幅灵敏度 amplitude sensitivity
振幅码 amplitude code
振幅脉冲 amplitude pulse
振幅密度谱 amplitude density spectrum; amplitude spectrum density
振幅偏移 amplitude excursion
振幅偏移键控法 amplitude shift keying
振幅频率畸变 amplitude-frequency distortion
振幅频率失真 amplitude-frequency distortion
振幅频率特性 amplitude-frequency characteristic; amplitude response; gain frequency characteristic
振幅频率特性曲线 amplitude-frequency curve; amplitude-frequency response; amplitude versus frequency curve
振幅频率调制 amplitude-frequency modulation
振幅频率响应 amplitude-frequency response
振幅频谱 amplitude-frequency distribution; amplitude spectrum
振幅平衡 amplitude balance
振幅剖分法 amplitude splitting
振幅谱 amplitude spectrum
振幅谱曲线 amplitude spectrum curve
振幅谱因子 amplitude spectrum factor
振幅起伏误差 scintillation error
振幅强度 oscillator intensity
振幅区分 amplitude separation
振幅曲线 amplitude curve
振幅取样器 amplitude sampler
振幅全息摄影栅 amplitude hologram grating
振幅失真 amplitude distortion; attenuation distortion; harmonic distortion; volume distortion
振幅衰减 amplitude attenuation; decay of amplitude
振幅衰减常数 amplitude attenuation constant
振幅衰减特点对比 signature correlation of amplitude attenuation
振幅衰落 amplitude fading
振幅水平剖面 horizontal amplitude section
振幅特性 amplitude characteristic
振幅特性曲线 amplitude response; amplitude response curve
振幅调整 amplitude adjustment; amplitude control
振幅调制 amplitude modulation
振幅透过率 amplitude transmittance
振幅透射率 amplitude transmissivity
振幅稳定激光 amplitude stabilized laser
振幅稳定性 amplitude stability
振幅无关 amplitude independent
振幅系数 amplitude factor; peak factor
振幅限制器 amplitude limiter
振幅相位法 amplitude-phase method
振幅响应曲线 amplitude response curve
振幅选通电路 amplitude-gating circuit
振幅选择器 amplitude selector; amplitude separator
振幅压缩器 amplitude compressor; compressor
振幅异常 amplitude anomaly
振幅异常普查 amplitude anomalous survey
振幅译码器 amplitude discriminator
振幅因数 amplitude factor; crest factor; peak factor
振幅因子值 values of amplitude factor
振幅周期比 amplitude to period ratio
振幅锥度 amplitude taper
振幅自动调整电压 automatic amplitude control voltage
振撼 jolt
振夯式设计 tamping-vibrating design
振痕 chatter mark
振簧 vibrating reed
振簧比较仪 reed-type comparator
振簧法 vibrating-reed method
振簧分析器 vibrating-reed analyser [analyzer]
振簧静电计 dynamic(al) electrometer
振簧频率计 vibrating reed frequency meter
振簧式比较仪 reed comparator
振簧式变压装置 vibrator power pack
振簧式磁强计 vibrating-reed magnetometer
振簧式放大器 chopper amplifier; contact-modulated amplifier; vibrating-reed amplifier
振簧式继电器 vibrating-reed(type) relay
振簧式静电计 vibrating-reed electrometer
振簧式黏[粘]度计 vibrating-reed viscometer
振簧式频率传感器 vibrating-reed frequency sensor
振簧式陀螺 vibrating-reed gyro
振簧式仪表 vibrating-reed instrument
振簧式整流器 vibrating-reed rectifier
振簧式转速计 vibrating-reed tachometer
振簧振荡器 vibrating-reed oscillator
振簧指示器 vibrating-reed indicator
振击打捞筒 jar socket
振击落砂架 jolt knock-out grid
振击能力 jolt capacity
振击器 jar coupling; jar bumper; jar knocker
振击式取土器 <带可提出式的矛式钻头的> retractable plug sampler
振击造型机 jolting machine
振控系统 vibration control system
振量滴定 oscillometric titration
振量法 oscillometry
振铃 ringing; exportation
振铃按钮 sounder push-button
振铃边际 ringing margin
振铃磁石发电机 calling magneto
振铃电键 challenge switch; ringing key
振铃电流 ringing current
振铃电路 call circuit; ringing circuit
振铃电路故障 signal(l)ing fault
振铃电码 ringing code
振铃吊牌 ringing drop
振铃分布 distribution of ringing
振铃分隔继电器 ringing-trip relay
振铃呼叫 bell call
振铃机 ringing set
振铃继电器 ringing relay; ring-up relay
振铃键 assignment key; call-bell key
振铃接通的中继线 ring-down trunk
振铃频率 ringing frequency; signal(l)ing frequency
振铃器 ringer; signal(l)ing set
振铃器盘 ringer panel
振铃切断继电器 ringing-trip relay
振铃设备 calling device
振铃手摇发电机 calling magneto

振铃调整 ringing control
振铃通知 ring-down
振铃位置 ringing position
振铃系统 ringing system
振铃线 ringing wire
振铃效应图像 split image
振铃信号 bell signal;line signal;ring-down signal(l)ing;ringing signal
振铃信号测试器 ringing tester
振铃信号故障报警 ringing fail alarm
振铃信号频率 signal(l)ing frequency
振铃信号器 ringer
振铃信号障碍报警 ringing fail alarm
振铃信号振荡器 ringer oscillator;ringing signal oscillator
振铃信号制 ringing down system
振铃形绝缘器 bell-shaped insulator
振铃音 ringing tone
振铃指示灯 ringing pilot lamp
振铃周期 ringing cycle
振铃转换开关 ringing changeover switch
振铃装置 ring-back apparatus;ringing set
振密混凝土柱法 vibro concrete column
振敏管 vibratron[vibrotron]
振鸣 howling;ringing
振鸣安全系数 singing margin
振鸣边际 singing margin
振鸣边界 singing margin
振鸣点 singing point
振鸣点试验器 sing-point tester
振鸣廊 singing gallery
振鸣容限 singing tolerance
振鸣声 squeal(ing);whistle
振鸣稳定度 singing stability
振鸣信号 singing signal
振鸣压缩器 singing-suppressor
振鸣抑制装置 anti-singing device
振鸣振荡 singing
振鸣作用 singing effect
振凝(现象)rheopexy
振劈哨 resonant wedge whistle
振频特性曲线 amplitude characteristic
振频调节 frequent adjuster
振腔哨 resonant cavity whistle
振筛机 rotap
振筛器 screen vibrator
振筛式给料器 vibrating grizzly feeder
振筛(淘汰机)vanning jig
振实 compaction by vibration (and compression); consolidation; jolt; vibrating compaction; vibration compaction;vibratory compaction
振实拌和机 vibrating compactor mixer
振实的粗骨料混凝土 vibrated coarse concrete
振实的堆石坝 vibrated rock fill dam
振实(的混凝土)柱 vibrated column
振实的接缝 vibrated joint
振实法 <松软地基处理方法之一> vibratory method of compaction; vibro-compaction method; vibro-compressor method
振实混凝土 vibrated concrete;vibrocast concrete;vibration of concrete
振实混凝土构件 concrete compacted element
振实混凝土桩 vibrated concrete pile
振实机 plain bumper;plain jolter
振实挤压 jolt-squeeze
振实力 jolt pressure
振实率 rate of consolidation
振实密度 compacted density;tap density
振实器 consolidation vibrator;jolting vibrator;vibratory compactor
振实设备 compacting equipment
振实式混凝土桩 vibro-concrete pile
振实式砌块制造机 vibrating compactor block making machine
振实台 plain bumper; plain jolter; swing table
振实体积 compact volume
振实系数 compacting factor
振实造型机 bumping mo(u)lding machine;jolt mo(u)lding machine; plain jolting machine
振实制型机 jolter
振实重度 tap density
振水音 splashing sound
振丝伸长计 vibrating wire extensometer
振丝应变计 vibrating wire indicator
振松 decompaction
振速 vibration velocity
振速传声器 open air microphone;velocity microphone
振纹 chatter mark
振弦 vibrating wire;vibrator
振弦加速度计 vibrating string accelerometer
振弦式渗压计 vibrating wire piezometer
振弦式土压力盒 vibrating wire pressure cell
振弦式压力盒 vibrating wire cell
振弦式遥控湿度计 vibrating wire tele-hygrometer
振弦式仪器 vibrating wire instrument
振弦式引伸仪 vibrating string extensometer
振弦式应变计 vibrating string extensometer; vibrating wire strain ga(u)ge
振弦式重力仪 vibrating string gravimeter
振弦应变计 vibratory strain ga(u)ge
振弦应力计 vibrating wire stressmeter
振型 mode of vibration;mode shape; oscillation mode
振型参与系数 mode participation coefficient; mode-participation factor;participation coefficient;participation factor
振型叠加 modal superposition;mode superposition
振型叠加法 modal analysis method; mode superposition method
振型反应 modal response; mode response
振型分解 modal decomposition
振型分解法 modal decomposition technique;mode analysis method
振型分解反应谱法 modal response spectrum analysis
振型分量 modal component
振型分析 modal analysis;mode analysis
振型分析法 modal analysis method
振型幅值 mode amplitude
振型贡献 modal contribution
振型剪切系数 mode shear coefficient
振型解析 modal decomposition
振型矩阵 modal matrix
振型耦合 couple mode; coupling of modes;mode coupling
振型耦联 couple mode; coupling of modes;mode coupling
振型平衡理论 modal balancing theory
振型矢量 modal vector; mode shape vector
振型速度 modal velocity
振型系数 mode factor
振型形式 model shape
振型影响 modal contribution
振型再变换 mode reconversion
振型展开法 modal expansion method;mode expansion(method)
振型综合法 modal synthesis method
振型阻尼矩阵 mode damping matrix
振型组合 modal combination; mode combination
振型作用 modal contribution
振压棒 tamper bar
振压式造型机 jar squeezer; jolt-squeeze mo(u)lding machine; vibrating squeeze mo(u)lding machine;jolt mo(u)lding machine
振摇 pendulate;shake
振摇式振捣器 shock vibrator
振源 oscillation source
振中 center of percussion
振柱器 column vibrator
振筑衬里 jolt-packed liner
振撞击能力 jolt capacity
振子 oscillator;shaker;ticker[tikker]; vibration exciter[excitor];vibration generator;vibrator
振子记录示波器 vibrator recording oscilloscope
振子频率调节器 phonophone
振子强度 oscillator strength
振子球 bob
振子群 oscillator group
振子式超声波黏[粘]度计 rouse ultrasonic viscometer
振子式示波器 loop oscillograph
振子天线 element antenna
振子转换 vibrator conversion
振子坠 bob
振作起来 pull oneself together

赈济 charity

镇尺 paper weight

镇定 suppress
镇定板 smash plate
镇定泵 ballast pump
镇定变压器 damping transformer
镇定电阻 ballast resistor
镇定物 ballast
镇定装置 trigger
镇墩 anchorage block
镇环座 latch holder
镇江大白蚁 <拉> Odontotermes fontanellus
镇脚石 rubble for toe protection;toe-ballasting rubble
镇静 composure
镇静的 deliberate;killed
镇静电路 muting circuit
镇静钢 dead-melted steel; dead mild steel; degasified steel; deoxidized steel;fully killed steel;killed steel; piping steel;solid steel;still steel
镇静钢锭 killed ingot
镇静剂 calmative;killing agent;palliative;quietive;sedative;tranquil(l)izer
镇静期 killing period
镇静室 stilling chamber
镇静碳钢 killed carbon steel
镇块 apron block
镇浪海锚 rough sea anchor
镇浪油 sea calming oil; sea quelling oil; storm oil; wave calming oil; wave quelling oil
镇浪油袋 oil bag
镇浪油弹 oil shell
镇浪油箱 storm oil tank
镇流 ballasting
镇流灯 ballast lamp
镇流电阻 ballast resistance; barretter;steadying resistance
镇流电阻器 amperite
镇流电阻器电桥 barretter bridge
镇流管 amperite; ballastic tube; ballast lamp; ballastron; ballast tube; current stabilizer
镇流管电阻 barretter resistance
镇流漏泄电阻 ballast leakage resistance
镇流器【电】 ballast; barretter; lamp ballast
镇流器噪声的分级 ballast noise rating
镇流系数 ballast factor
镇流线圈 ballast coil
镇气分馏塔盘 ballast tray
镇桥 town bridge
镇区 range line;township <英国旧时的地方行政单位,美国及加拿大的地方行政单位>
镇区测量 town-site survey
镇区道路 township road
镇区公路 township highway
镇区管理 township administration
镇区系统 township system
镇痛剂 balm;balsam
镇痛药 analgesic
镇压 quash;repress;repression;stifle
镇压地 compact the soil
镇压荷重 ballasting
镇压器 compacting machine;compactor;land presser; land roller; press roller
镇压器环 roller segment
镇噪振荡器 noise squelch oscillator
镇长 <美> warden
镇纸 letter balance;letterweight
镇重 ballast weight
镇重物 ballasting
镇重用混凝土 ballast concrete

震波 seismic wave;shock wave

震波测量 seismic survey
震波冲击 seismic shock
震波法 wave propagation
震波反射测量 seismic reflection survey
震波反射剖面法 seismic reflection profiling
震波反射剖面仪 seismic reflection profiler
震波技术 wave propagation technique
震波检测仪 seismic geophone
震波勘测 seismic exploration;seismic survey(ing)
震波勘测船 seismic exploration vessel; seismic explosive vessel; seismic survey vessel
震波勘测法 seismic method of exploration
震波勘测资料 seismic data
震波勘探 seismic prospect(ing)
震波曲线 seismogram
震波水下地形仪 seismic profiler
震波水下回声地质剖面探测仪 hydrosound sub-bottom profiler; hydrosound underwater seismic profiler
震波速度 seismic wave velocity
震波探测 seismic exploration
震波图 earthquake record; seismogram;seismograph
震波图分析 seismic processing
震波图解释 seismogram interpretation
震波微弱带 shadow zone
震波研究 seismic studies
震波折射 seismic refraction
震波折射测量 seismic refraction survey
震波折射剖面法 seismic refraction profiling
震波折射试验 seismic refraction test

震颤 tremor;chattering;fremitus;quaver;quiver;thrill;tremble
震颤波纹 chattering;chatter mark
震颤机 tremulor
震颤描记图 tremorgram
震颤片 trembler blade
震颤声 judder
震颤式电铃 trembling bell
震颤素 tremorine
震颤线圈 trembler coil
震颤性谵妄 potomania
震磁效应 seismomagnetic effect
震旦海侵【地】Sinian transgression
震旦海侵期 Sinian ingression epoch
震旦纪 Sinian period
震旦系 Sinian system
震旦旋回 Sinian cycle
震旦运动 1 Sinian movement 1
震旦走向 Sinian trend
震荡 shock
震荡面 plane of oscillation
震荡培养 shake culture
震荡破坏荷载 shakedown load
震荡音 succusion-sound
震荡运动阶段 oscillatory movement stage
震捣 jolt ramming
震捣过度 overvibration
震捣混凝土 shock concrete
震捣机 jarring machine
震捣梁 vibrating beam
震颠 chatter
震颠声 chatter
震颠行车 chattering drive
震电效应 seismic-electric effect
震动 tremor; jarring motion; shock (motion);tremble
震动爆破 standing shot
震动变形 dynamic(al)deflection
震动变质作用 shock metamorphism
震动冲击 jerk
震动冲击机 jolt ramming squeezer
震动波传输 impact transmission
震动打桩锤 vibratory hammer
震动带 shattering zone
震动捣击 jolt ramming
震动的 quaky
震动法 lash method
震动负载 shock load(ing)
震动钢丝应变仪 vibrating wire strain gauge
震动隔离 shock isolation
震动管 shock tube
震动夯击机 jolt-ramming machine; jolt ramming squeezer
震动荷载 oscillatory load; seismic loading;shock load(ing);vibrating load
震动机 jarring machine
震动激发 shock-excitation
震动加速度 oscillatory acceleration; seismic acceleration
震动间隙 bump clearance
震动灵敏度 sensitiveness to shock; shock sensitivity
震动脉冲 shock pulse
震动面 plane of oscillation
震动能 shock energy
震动频率 frequency of shock
震动破坏效应 failure effect of earthquake
震动器 annihilator;jolter
震动强度 intensity of shaking;shockproofness
震动韧性 jolt toughness
震动筛 reciprocating screen;reciprocating sieve
震动式发火装置 trembler firing device
震动式压路机 wobble wheel roller
震动试验 shock test
震动试验器 shock tester;jar tester <凝集反应试验器械>
震动输送机 jigging conveyer[conveyor]
震动速度 velocity of vibration
震动速率 jolting rate
震动损失 loss due to shock
震动台 bumper;jolt table;shock table
震动填料 jolt-packing
震动危险 shock hazard
震动现象 seismism
震动效应 jarring effect
震动性 concussion blasting
震动引信 concussion fuse
震动运输机 throw transporter
震动造型机 jar-ramming machine; jolt mo(u)lding machine
震动铸锭法 jarring of ingot;jolting of ingots
震动作用 shock effect
震度 degree of seismicity;seismic degree
震耳欲聋的 deafening
震幅 damping
震感半径 radius of perceptibility
震感区 felt area
震害 earthquake catastrophe; earthquake damage; earthquake hazard; seismal hazard
震害等值线图 damage contour map
震害地质调查 geologic(al)survey of seismic calamity
震害调查 earthquake damage survey
震害分布 damage distribution
震害航测 air survey of damage
震害类别 category of earthquake damage
震害类型 type of seismic calamity
震害类型预测 prediction of calamitous type of earthquake
震害率 damage intensity;rate of earthquake damage
震害敏感性 earthquake damage susceptibility
震害强度预测 prediction of intensity of seismic calamity
震害势 damage potential
震害统计 damage statistics
震害图集 earthquake damage atlas
震害现象 damage phenomenon;earthquake damaging phenomenon
震害效应 damage effect; earthquake damage effect
震害预测 earthquake disaster prediction;prediction of seismic calamity
震害预测图 map of prediction of seismic calamity
震害增量 damage increment; earthquake damage increment
震害指数 damage index; earthquake damage index; index of seismic damage
震撼 jolt
震后变形 post-earthquake deformation;post-seismic deformation
震后调查 post-earthquake investigation
震后功能 post-earthquake functionality
震后滑动 post-earthquake slip;post-seismic slip
震后恢复 seismic rehabilitation
震后恢复计划 post-earthquake restoration planning
震后活动 post-earthquake movement
震后火灾 post-earthquake fire
震后检查 post-event inspection
震后救灾 post-earthquake relief
震后数据整理 post-event data reduction
震后土地利用规划 post-earthquake land use planning
震后修复 seismic rehabilitation
震后应变 post-seismic strain
震后应力 post-seismic stress
震后灾害调查 post-earthquake damage survey
震击 jolt(ing)
震击打捞筒 jar crotch socket; jar socket
震击缸 jolting cylinder
震击活塞 jolting piston
震击机 jarring machine
震击平台 anvil jolter
震击器 jars <打捞被卡钻具>;bumper jar <事故处理用>
震击式打捞筒 drive down socket
震击式捞管器 drive down casing spear
震击式韧性试验机 jolt impact tester
震击式韧性试验仪 jolt toughness tester
震击试验仪 jolt impact tester
震击效应 jarring effect
震击行程 jolt stroke
震击作用 jarring action
震激 shock-excitation;shock excite
震激波凿岩 shock wave drilling
震激的 shock-excited
震激振荡器 shock-excited oscillator
震级 earthquake magnitude; magnitude of earthquake;seismicity;seismic magnitude
震级标度 magnitude scale(of earthquake)
震级残差 magnitude residual
震级测定 magnitude determination
震级大小 order of magnitude
震级分布 magnitude distribution
震级分类 classification of magnitude of earthquake
震级公式 magnitude formula
震级估计 magnitude estimate
震级估算 magnitude estimation
震级烈度相关(关系)magnitude intensity correlation
震级烈度关系 magnitude intensity relation
震级频度法则 magnitude frequency law
震级频度关系 magnitude frequency relation
震级频率的平方定律 quadratic magnitude frequency law
震级区间 magnitude interval
震级上限 upper limit earthquake magnitude
震级时间曲线 magnitude time curve
震级统计 magnitude statistics
震级图 magnitude chart
震级稳定性 magnitude stability
震级限值 magnitude threshold
震级异常 magnitude anomaly
震级重现曲线 magnitude intensity correlation; magnitude recurrence curve
震量 earthquake volume
震烈强度 shatter strength
震烈岩石 shattered rock
震裂 shatter crack;shatter(ing)
震裂带 shatter belt;shatter zone
震裂痕 shatter mark
震裂角砾岩 shatter breccia
震裂裂缝 shattercrack
震裂试验 shatter test
震裂系数 shatter index
震裂岩石 shatter rock
震裂指数 shatter index
震裂锥 shatter cone
震裂作用 shattering action
震率 knock rating
震落冲击杆与钻头连接销 knock bit off
震凝度 rheopecticity
震凝流动 rheopectic flow
震凝流体 rheopectic fluid
震凝(现象)rheopexy
震凝性 rheopecticity;rheopexy
震谱 shock spectrum
震前变形 pre-earthquake deformation;preseismic deformation
震前的 preseismic
震前的平静 quiet at approach of earthquake
震前防灾准备 earthquake disaster preparedness
震前滑动 pre-earthquake slip; preseismic slip
震前活动 pre-earthquake movement
震前活动性 preseismic activity
震前畸变 pre-earthquake distortion
震前扩容 pre-earthquake dilatancy
震前位移 preseismic displacement
震前形变 preseismic deformation
震前音响试验 preshock noise test
震情 earthquake situation;seismal regime;seismic regime
震情会商 seismologic(al)consideration
震情预报 earthquake prediction
震区 earthquake province;earthquake region;earthquake zone;seismic area;seismic region
震区图 seismic area map;seismic region map;seismic zone map
震群 earthquake cluster; earthquake series; earthquake swarm; swarm earthquake
震群活动性 swarm activity
震群型 swarm-type
震群型地震 swarms earthquake; swarm type earthquake
震群序列 swarm sequence
震扰 shock
震筛 riddle(r)
震深 depth of focus;focal depth;focal range
震声 barisal guns;earthquake sound
震声反射测量 seismo-acoustic(al)reflection survey
震声区 sound area
震时互易原理 seismic reciprocity
震时活动 earthquake movement
震时效应 coseismic effect
震时形变 coseismic deformation
震时应变 coseismic strain
震时运动 coseismic movement
震实 jolting; jolt-packed; jolt-ram(ming);ram jolt
震实成型机 jarring mo(u)lding machine
震实式造型机 jarring mo(u)lding machine;jolt-type mo(u)lding machine
震实台 jolter
震实膝形阀 jolt knee valve
震实造型 jolt mo(u)lding
震实造型机 jolter;jolting machine; jolt mo(u)lding machine;jolt rammer;jolt-ramming machine
震实制芯机 core jarring machine
震水音 splashing sound
震松地 jarred loose ground; jarred loose land
震淘台 jerking table
震凸变形 chatter bump
震尾 cauda;earthquake coda
震尾波 coda wave
震尾波谱 coda wave spectrum
震尾波振幅 coda wave amplitude

震陷(量) earthquake subsidence
震相 seismic phase
震相突始 impetus
震性 pinking
震压 jolt-squeeze
震压式造型机 jolt-squeeze mo(u)lding machine; jolt-squeezer; jolt-squeezer machine; squeeze rammer
震摇 jolt
震因机理 causal mechanism
震音 percussive sound
震音装置 tremolo
震影 <无震区> earthquake shadow
震源 center[centre] of origin; centring of origin; centrum [复 centrums/centra]; earthquake center[centre]; earthquake foci; earthquake focus; earthquake hypocenter [hypocentre]; earthquake origin; earthquake source; focus; hypocenter[hypocentre](focus); hypocentrum; hypofocus; seismic focus; seismic origin; seismic source
震源波形 source wave form
震源参数 hypocenter[hypocentre] parameter; hypocentral parameter; source parameter
震源测定 hypocentral location
震源船 energy source boat
震源大小 focal dimension; source dimension
震源弹 source bomb
震源地面零点 hypocentre
震源地震学 source seismology
震源点 focal point
震源定位 hypocentral location
震源动力学 earthquake source dynamics; source dynamics
震源范围 focal sphere; source range
震源非球形对称 non-spheric(al) symmetry of the source
震源过程 focal process; source process
震源函数 seismic source function; source function
震源机理 earthquake source mechanism
震源机制 earthquake source mechanism; focal mechanism; mechanism of earthquake foci; source mechanism; focus mechanism
震源机制解 focal mechanism solution
震源机制台阵 source mechanism array
震源计算方法 hypocenter computing process
震源检测 source detect
震源矩 source moment
震源距(离) distance from the epicenter; distance from the focus; focal distance; hypocentral distance; source distance
震源空间分布 spatial distribution of hypocenter
震源类型 source type
震源模型 seismic source model
震源谱 source spectrum
震源迁移 hypocentral shift
震源浅 shallow earthquake focus
震源区 focal area; focal region; focal zone; source region
震源深度 focus depth; depth of focus; depth of origin; depth of seismic focus; depth of shock; earthquake depth; focal depth; source depth
震源深度预测 prediction of depth of focus
震源时间函数 source-time function
震源释放 source release
震源特性 source characteristic
震源体积 earthquake volume; focal volume; source volume
震源体小 small earthquake focus mass
震源同步装置 energy source synchronizer
震源图 hypocentral plot
震源位置 hypocentral location; location of foci; seismic filtering location; seismic focus location
震源位置移动 migration of seismic focus
震源效应 source effect
震源信号 source signal
震源震时 time of shock at the origin
震源至场地距离 focal-to-site distance
震源至台站距离 focal-to-site station distance
震源至震中连线 seismic vertical
震源装置 impulse seismic device
震源资料档案 hypocenter data file; hypocentral data file
震源坐标 focal coordinate
震灾急救计划 earthquake disaster relief planning
震沼 quaking bog
震中 earthquake center [centre]; earthquake epicenter; epicenter [复 epicentre]; epicentrum [复 epicentra]; epifocus; quake center [centre]; seismic center[centre]
震中标绘 plotting of epicentre
震中表 epicenter scale
震中测定 determination of epicenter; epicenter determination
震中常数 constant of earthquake epicentre
震中带 epicentral zone
震中的 epicentral; epifocal
震中定位程序 epicenter location routine
震中对点 anti-center[centre]; anti-center of earthquake; anti-epicenter[anti-epicentre]; anti-epicentrum; epicentral of earthquake
震中方位角 epicenter azimuth
震中分布 epicentre distribution; epicentric distribution
震中集中 epicenter concentration
震中加速度 epicentral acceleration
震中角距 angular epicentral distance
震中经度 longitude of epicenter
震中经纬度 epicenter longitude and latitude
震中距 epicentral distance; epicentre distance
震中烈度 epicentral intensity
震中面积 epicentral area; epicentre area
震中偏差 epicenter[epicentre] bias; epicentral bias
震中迁移 epicenter migration; epicenter shift
震中强度 epicentral intensity
震中区 epicentral area; epicentral region; epicentral zone; epicentre area
震中图 epicenter[epicentre] map
震中纬度 latitude of epicenter
震中位置 epicentral location; epicentre location
震中震时 time of earthquake at epicenter [epicentre]; time of occurrence at the epicenter
震中坐标 epicentre coordinates
震轴 earthquake axis
震筑衬里 jolt-packed liner

争端 dispute

争端裁决委员会 dispute adjudication board
争端当事方(国) party to the controversy
争端的解决 settlement of dispute
争端的仲裁人 arbiter of the dispute
争端和解 alternate dispute resolution; amicable settlement <承包工程不采用仲裁的>
争端审议委员会 dispute review board
争端仲裁 arbitration of disputes
争议的解决 settlement of disputes
争议各方 contesting parties
争用 contention
争用法 contention method
争用时间间隔 contention interval
争用网 colliding station
争用系统 contention system
争用线路 contention
争用站 colliding station
争执方 contestant

征地 expropriation of land; land acquisition; take-over of land

征地补偿 land requisitioning compensation
征地费 fee paid for land; land-use fee
征地红线 borderline of expropriation land; line marking of expropriated land
征地手续 expropriation proceedings
征地水位 requisition level
征地损失申诉 inverse condemnation
征服自然 conquest of nature
征购 acquisition; procurement; requisition by purchase
征购土地 land acquisition; purchase the land; compulsory acquisition
征集 conscription
征聘考试 recruitment examination
征求报标 bid wanted
征求承包 invite bids; invite tenders
征求通讯 solicitation correspondence
征求同意 call for an agreement
征求意见 request for proposal
征求意见表 questionary
征实 levy in kind
征收 acquisition; assess; collect; collection; impose; impose (a duty) on; levy; retaliate; perception <地租等的>
征收比规定费率高的保险费 rate up
征收超额部分收益条款 <美国政府征收超额收益的一半作为基金,由州际商务委员会 ICC 保管> Recapture Clause
征收超额累进所得税 surtax
征收地产税 charges on real estate
征收罚款 levying of fines
征收费用 toll collection
征收附加税 surtax
征收关税 tariff
征收基础 collection basis
征收金额 assessed amount
征收率 assessment percentage
征收年度 year of assessment
征收排污费制度 system of collecting fees for discharging pollutants
征收排污税 imposing discharge fee
征收桥 toll bridge
征收税款 impose
征收所得税 income taxation
征收通行费的高速公路 turnpike
征收土地 compulsory land acquisition
征税 clap on; collect duties; impose tax; imposition; lay on; levy duties on; levy tax; levy toll on; raise tax; tax; taxation; tax collecting; toll collection
征税货物 dutiable goods
征税物品 things liable to duty
征象 signal; signs and symptoms
征象法 charactery
征信机构 credit information service; inquiry agency
征信书 character book; questionnaire
征询方案 request for proposal
征询机构 enquiry agent
征询要求 request for references
征询意见表 questionnaire
征用 acquisition; condemnation; confiscation; dispossession
征用财产 requisitioning of goods
征用财产者 condemner
征用船舶的设备补贴 equipment subsidy
征用估价 condemnation appraisal
征用价值 condemnation value
征用建筑用地 acquisition of building land
征用权 right of eminent domain
征用水位 requisition level
征用条款 condemnation clause
征用土地 acquisition of land; expropriation of land; requisition of land; expropriation; requisition; take-over for use
征用土地补偿费 compensation fees for acquisition of land
征用土地权 eminent domain
征兆 sign; symptom
征兆曲线 symptomatic curve

挣扎 sprawl

蒸出 boil over

蒸镀 coating by vaporization
蒸镀变黑 evaporated black
蒸镀薄膜 evaporated film
蒸镀发黑处理 evaporated black
蒸镀铅膜 evaporated film of alumin(i)um
蒸发 boil(ing) off; ca(u)lking; evaporate; evaporating; perspiration; perspire; raising; steam generation; transpiration; vapo(u)rize; fly-off
蒸发白云岩 evaporative dolomite
蒸发斑状结构 evapoporphyrocrystic texture
蒸发倍率 evaporation ratio
蒸发本领 evaporative capacity; evaporative power
蒸发泵 evaporation pump
蒸发泵作用 evaporative pumping
蒸发比 evaporate ratio
蒸发比率 evaporation rate
蒸发表 atmidometer; evaporation meter; evaporimeter[evaporometer]
蒸发表面 evaporating surface; evaporation surface; evaporator surface; steam disengaging surface; steaming surface; evaporative surface
蒸发表面冷凝器 evaporative surface condenser
蒸发补给水 evaporated make up
蒸发参数 boil-off parameter
蒸发残量 total solids
蒸发残留物 residue on evapo(u)ration
蒸发残留物法 residue-on-evapo(u)ration method
蒸发残余 residue on evapo(u)ration
蒸发残余物 evaporation residue
蒸发残渣 dry residual; evaporated residue; evaporation residue; resi-

Z

due by evapo(u)ration; residue on evapo(u)ration
蒸发残渣总量 total residue of evapo(u)ration
蒸发槽 evaporator tank
蒸发测定法 atmidometry; atmometry; evaporimetry
蒸发测定器 evaporimeter[evaporometer]
蒸发层 evaporation layer
蒸发场 evaporating field
蒸发沉积矿床 evaporate deposit; evaporite deposit
蒸发沉积物 evaporated deposit; evaporite deposit
蒸发沉积作用 evaporate deposition
蒸发成像术 evaporography
蒸发成像仪 evaporograph
蒸发池 evaporation basin; evaporation pond; evaporation tank; tank evapo(u)ration
蒸发带 evaporator strip
蒸发带夹持器 evaporator strip holder
蒸发当量 equivalent evapo(u)ration
蒸发挡板 evaporation shield
蒸发的 evaporative
蒸发的金锑接触 evaporated gold antimony contact
蒸发点 evaporating point; point of evapo(u)ration; steaming point; vapo(u)r point
蒸发掉的水 fly-off water
蒸发度 evaporation degree; evaporativity
蒸发段 evaporation zone
蒸发发散作用 evapotranspiration
蒸发法 evaporation method; method of evapo(u)ration
蒸发负荷 evaporation load
蒸发干燥 drying by evapo(u)ration; evaporation drying
蒸发工段 evaporation section
蒸发工艺 evaporation technology
蒸发观测 evaporation observation; measurement of evapo(u)ration
蒸发管 boiling tube; evaporating pipe; exhale; generating tube
蒸发管束 steam-generating bank
蒸发罐 evaporating pot
蒸发锅 evaporating pan; evaporation boiler; evaporation pan
蒸发过程 evaporation process
蒸发合成 evaporation synthesis
蒸发合金工艺 evaporated alloying technology
蒸发核模型 evaporation nuclear model
蒸发核子 evaporation nucleon
蒸发后剩余残渣 evaporites
蒸发环境 evaporitic environment
蒸发换热系数 evaporative heat transfer coefficient
蒸发回流作用 evaporative reflux
蒸发机 rapid ager; rapid steamer
蒸发机理 evaporation mechanism
蒸发机率 evaporation opportunity
蒸发级 evaporation stage
蒸发计 arm(id)ometer; atmidometer; atmometer; evaporation ga(u)ge; evaporation meter; evaporator ga(u)ge; evaporograph cell; evaporometer; potential evapo(u)rimeter
蒸发记录仪 evaporigraph
蒸发剂 evaporant
蒸发架 evaporation jig
蒸发-降雨比 evaporation-rainfall ratio
蒸发胶质试验 evaporation gum test
蒸发结晶机 evaporated crystallizer; evaporative crystallizer
蒸发结晶器 evaporated crystallizer; evaporative crystallizer
蒸发镜 evaporoscope
蒸发镜面面积 disengagement area
蒸发可能率 evaporation opportunity
蒸发可能性 evaporation opportunity
蒸发坑 evaporation pit
蒸发控制器 evaporation controller
蒸发冷冻 evaporative freezing
蒸发冷凝机理 evaporation condensation mechanism
蒸发冷凝器 evaporative condenser; evaporator condenser; vapotron
蒸发冷却 cold due to evaporation; ebullient cooling; evaporation cooling; evaporative cooled; evaporative cooling; porous cooling; sweat cooling; transpiration cooling; vapo(u)r cooling; vapo(u)rization cooling; vapo(u)r phase cooling
蒸发冷却变压器 vapo(u)r-cooled transformer
蒸发冷却的 transpiration-cooled
蒸发冷却电阻器 vapo(u)r-cooled resistor
蒸发冷却发电机 evaporative cooling generator; vapo(u)r-cooled generator
蒸发冷却法 transpiration cooling
蒸发冷却管 evaporation cooled tube
蒸发冷却机 evaporative cooler
蒸发冷却器 devapo(u)rizer; evaporative cooler; vapotron
蒸发冷却式叶片 transpiration-cooled blade
蒸发冷却塔 wet cooling tower
蒸发冷却系统 evaporative cooling system
蒸发冷却装置 evaporative cooled device
蒸发冷硬 evaporating chilling
蒸发离子泵 evaporation ion pump; evaporator ion pump; evapor-ion pump
蒸发力 evaporative power
蒸发粒子 evaporation particle
蒸发量 amount of evapo(u)ration; capacity; depth of evapo(u)ration; evaporating capacity; evaporation capacity; rate of evapo(u)ration; steam capacity; steam content; steaming capacity; steam output; steam relieving capacity; evaporation; evaporation discharge
蒸发量测定 measurement of evapo(u)ration
蒸发量测定计 evaporating gage; evaporation ga(u)ge
蒸发量测定术 atmometry
蒸发量观测 evaporation capacity observation
蒸发量曲线 evaporogram
蒸发了的 vapo(u)rized
蒸发流 evaporation current
蒸发流量 evaporation discharge
蒸发率 coefficient of evapo(u)ration; evaporation power; evaporative duty; evaporative rate; evaporativity; intensity of evapo(u)ration; percent evaporated; steaming rate
蒸发率测定 evaporation rate determination
蒸发率测定法 atmidometry
蒸发率计 evaporation ratemeter
蒸发率控制 evaporation control
蒸发慢的溶剂 slow solvent
蒸发面 boiling side; disengagement surface; evaporation surface; steam releasing surface; vapo(u)rizing surface
蒸发面积 disengagement area; evaporation area; steam relieving area
蒸发皿 evaporating dish; evaporation disc[dish]; evaporation pan; evaporation tank; evaporator; vapo(u)rating dish
蒸发皿校正系数 pan factor
蒸发皿水面超高 freeboard of evapo(u)ration pan
蒸发皿系数 evaporation-pan coefficient; pan factor
蒸发皿折算系数 evaporation pan reduction coefficient
蒸发模型 evaporation model
蒸发内热 internal heat of evaporation
蒸发能力 evaporating capacity; evaporation capacity; evaporative capacity; evaporativity; potential evapo(u)ration; steaming capacity; evaporation power
蒸发凝结器 boiler-condenser
蒸发凝聚 evaporation condensation
蒸发凝聚传质机理 evaporation condensation material transfer mechanism
蒸发凝聚机理 evaporation condensation material mechanism
蒸发凝气机 evaporator condenser
蒸发凝汽器 evaporative condenser
蒸发浓缩-火焰原子吸收光谱法 evaporative concentration-flame atomic adsorption spectrometry
蒸发浓缩沥青 steam-reduced asphalt
蒸发浓缩作用 evaporation and concentration
蒸发排泄 evaporation discharge
蒸发盘 evaporation pan
蒸发盘管 cryotron; evaporating coil; evaporative coil; expansion coil; vapo(u)rizing coil
蒸发盘组 pan bench
蒸发盆地 <内陆湖流域的> evaporating basin
蒸发盆地模式 evaporate basin model
蒸发平衡 evaporative equilibrium
蒸发气 boil-off gas
蒸发气表 vapo(u)r table
蒸发气冷却 air cooling by evapo(u)ration
蒸发气化器 evaporating carburetor
蒸发器 arm(id)ometer; atmometer; boiler; chiller; evaporating dish; evaporation coil; evaporation ga(u)ge; evaporation hook ga(u)ge; evaporation pan; evaporation unit; flash chamber; steam evaporator; steam raising unit; vapo(u)rator; vapo(u)r riser [vapo(u)rizer]; evapotranspirometer <测土壤的>
蒸发器材料 evaporator material
蒸发器吹风机 evaporator blower
蒸发器的二次蒸气 evaporator vapor
蒸发器的沸腾器 evaporator boiler
蒸发器的上端 upper end of evapo(u)rator
蒸发器灯丝 evaporator filament
蒸发器灯丝支架 evaporator filament support
蒸发器低温泵 evaporator cryopump
蒸发器电流 evaporator current
蒸发器电势 evaporator potential
蒸发器风机 evaporator fan
蒸发器垢层 evaporator scale
蒸发器管 evaporator tube
蒸发器恒温器 evaporator cryostat
蒸发器给水泵 evaporator feed pump
蒸发器加热表面 evaporator heating surface
蒸发器金属 evaporator metal
蒸发器进料 evaporation feed liquor
蒸发器冷凝器壳体 evaporator-condenser shell
蒸发器皿 evaporator boat
蒸发器盘管 evaporator coil
蒸发器身 evaporator body
蒸发器温度调节器 evaporator temperature regulator
蒸发器吸气泵 evaporator getter pump
蒸发器系数 pan factor
蒸发器旋管 evaporator coil
蒸发器压力 evaporator pressure
蒸发器压力控制阀 evaporator pressure control valve
蒸发器压力调节阀 evaporator pressure regulating valve; evaporator pressure regulator
蒸发器液面计 level ga(u)ge for evapo(u)rator
蒸发器与冷凝器分设的空调机组 split system air conditioner with evaporator below condensing unit
蒸发器元件 evaporator element
蒸发器支架 evaporator support
蒸发器中的第二效次 second evapo(u)rator
蒸发潜热 latent heat of evapo(u)rization; latent heat of vapo(u)rization; vapo(u)rization latent heat
蒸发强度 evaporation intensity; intensity of evapo(u)ration
蒸发区 area of evapo(u)ration; evaporation area; steaming zone
蒸发曲线 evaporation curve
蒸发燃烧 evaporate combustion
蒸发燃烧器 evaporation burner
蒸发热 evaporating heat; heat of vapo(u)r(iz)ation; vapo(u)rization heat
蒸发热量 evaporation heat
蒸发热调节 evaporative heat regulation
蒸发容量 evaporation capacity; evaporative capacity
蒸发散发 evapotranspiration
蒸发散热 evaporative heat loss; heat radiating by evapo(u)ration
蒸发蛇管 evaporating coil
蒸发蛇形管 evaporator coil
蒸发设备 calandria; evaporating unit; evaporative apparatus
蒸发深 depth of evapo(u)ration
蒸发深度 depth of evapo(u)ration
蒸发时间 steaming time
蒸发实验小区 evaporation plot
蒸发式的 evaporative
蒸发式干燥 evaporative drying
蒸发式烘燥 evaporative drying
蒸发式化油器 evaporation carburettor; vapo(u)rizing carburet(t)or
蒸发式经济器 steaming economizer
蒸发式冷凝器 evaporative condenser; evaporative type condenser
蒸发式冷却 porous cooling
蒸发式冷却器 evaporative type cooler
蒸发式冷却塔 evaporative type cooling tower
蒸发式离心机 evaporative type centrifuge
蒸发式燃烧器 burner vaporizer
蒸发式燃烧室 evaporative combustion chamber; evaporative combustor
蒸发式制冷剂冷凝器 evaporative refrigerant condenser
蒸发势 evaporation potential
蒸发试验 evaporation test
蒸发试验小区 evaporation pilot
蒸发室 evaporating chamber; evaporation chamber; evaporator room; vapo(u)rarium; vapo(u)rization

room;vapo(u)rizing chamber
蒸发水 evaporable water;evaporating water
蒸发水分 evaporation water;evaporative water;transpiring moisture
蒸发水分吸收的热量 heat absorbed in evapo(u)ration of water
蒸发丝 evaporator wire
蒸发速度 evaporation rate;evaporation velocity;rate of evapo(u)ration
蒸发速度解析法 evaporation rate analysis
蒸发速率 evaporation rate;rate of evapo(u)ration
蒸发速率低的溶剂 slow solvent
蒸发损耗 evaporation loss;invisible loss;loss by evapo(u)ration;loss due to evaporation
蒸发损失 boil-off loss;evaporation loss;invisible loss;loss by evapo(u)ration
蒸发损失控制设施 evaporative loss control device
蒸发损失试验 evaporation loss test
蒸发塔 evaporating column;evaporative tower;evaporator tower
蒸发塘 evaporation pond
蒸发特性 evaporation characteristic
蒸发体积 evaporated volume
蒸发调节 evaporation control
蒸发调节区 evaporation regulation area
蒸发涂敷 evaporation coating;evaporative coating
蒸发尾迹 evaporation trail
蒸发温度 evaporating temperature;evaporation temperature;temperature of evapo(u)ration;temperature of vapo(u)rization;vapo(u)rization temperature;vapo(u)rizing point;vapo(u)rizing temperature
蒸发纹层结构 evapolensic texture
蒸发物 evaporant
蒸发物离子源 evaporant ion source
蒸发雾 evaporation fog
蒸发吸气泵 evaporator getter pump
蒸发吸气离子泵 evaporation getter-ion pump
蒸发吸收热 heat absorbed by evapo(u)ration
蒸发系数 coefficient of evapo(u)ration;evaporation coefficient;evaporation factor;evaporative factor;transpiration coefficient;vapo(u)rization coefficient
蒸发系统 vapo(u)rization system
蒸发箱 evaporation tank
蒸发橡浆 evaporated latex
蒸发消耗量 evaporation discharge
蒸发消散 evaporation dissipation
蒸发效率 evaporation efficiency;evaporative efficiency;vapo(u)rization efficiency
蒸发芯子总成 evaporator core assembly
蒸发型 evaporation type
蒸发性 vapo(u)rability
蒸发性的 exhalent[exhalant]
蒸发性排放 evaporative emission
蒸发性排放控制 evaporative emission control
蒸发性排放控制器 evaporative emission controller
蒸发旋管 evaporation coil
蒸发循环冷却 vapo(u)r cycle cooling
蒸发压力 evapor pressure;evaporating pressure;evaporation pressure
蒸发压力调节阀 back-pressure regulator;evaporating pressure regulating valve
蒸发延迟剂 evaporation retardant
蒸发岩 evaporite;evaporitic rock
蒸发岩沉积 evaporite sediment
蒸发(岩)矿物 evaporite mineral
蒸发岩台地相 evaporite platform facies
蒸发岩相 evaporite facies
蒸发岩序列 evaporite sequence
蒸发盐 evaporate;evaporation salt;evaporite
蒸发盐沉积 evaporite deposition
蒸发盐盆地 evaporite basin
蒸发液 evaporated liquor
蒸发液体式试样放射性测量计 evaporated liquid activity meter
蒸发仪 evaporimeter[evaporometer]
蒸发抑制 evaporation control;evaporation reduction
蒸发抑制剂 anti-evapo(u)rant;evaporation retardant;evaporation suppressant;evaporation suppressor
蒸发阴极 evaporation cathode
蒸发荧光屏 vapo(u)r deposited screen
蒸发用的 evaporating
蒸发余渣法 residue-on-evapo(u)ration method
蒸发雨量比 evaporation-rainfall ratio
蒸发雨量率 evaporation-rainfall ratio
蒸发源 evaporation source;evaporator source
蒸发源活门 evaporation source shutter
蒸发源支持器 evaporator source carrier
蒸发源转盘 evaporation source turret
蒸发晕法 evaporating halo method
蒸发站 evaporate station;evaporation station;evaporator station
蒸发障 evaporate barrier
蒸发罩 evaporation mask
蒸发蒸腾 transpiration-evapo(u)ration
蒸发蒸腾比 evapotranspiration ratio
蒸发蒸腾的积累及降水量 cumulative evapotranspiration and precipitation
蒸发蒸腾量 total evapo(u)ration
蒸发蒸腾能力 potential evapotranspiration
蒸发蒸腾作用 evapotranspiration
蒸发蒸腾作用与水分利用率 evapotranspiration and water use efficiency
蒸发值 evaporation value
蒸发制冷 sweat cooling
蒸发制冷设备 evaporative cooling equipment
蒸发制冷系统 evaporative cooling system
蒸发中子 evaporation neutron
蒸发柱 column evapo(u)ration
蒸发装置 evaporating plant;evaporation installation;evaporation plant
蒸发着的 evaporating
蒸发阻滞 evaporation retardant
蒸发作用 evaporation;steaming;transpiration;vapo(u)rization
蒸釜式养护 mass curing
蒸釜养护 high-pressure steam curing;high-temperature steam curing
蒸干 boil out;evaporation to dryness
蒸干木材 steamed wood
蒸缸熟化 cooking
蒸锅 bain-marie;sauce pan;steamer;steaming pan
蒸烘 steaming
蒸化箱 stream ager
蒸炼燃气化油器 digester gas car
蒸馏 distillate;stilling;draw over
蒸馏残渣 base product;distillation residue;residue by distillation
蒸馏产物 distillate;distillation product
蒸馏厂废水 distillery waste(water)
蒸馏场 distillery
蒸馏出 extract
蒸馏粗甲苯基酸 distilled crude cresylic acid
蒸馏粗碳酸 distilled crude carbolic acid
蒸馏到焦炭 running to coke
蒸馏到汽缸油料 running to cylinder stock
蒸馏的 distillating;distillatory;distilling
蒸馏的木松节油 steam-distilled wood turpentine
蒸馏滴定法 distillation titration method;distillation-titrimetry
蒸馏法 distillate process;distillation method;distillation process;retorting;distillation
蒸馏范围 distillation range
蒸馏分解点 decomposition point of distillation
蒸馏釜 alembic;boiling bulb;distillating still;distillation kettle;distillation pot;distillation still;distilling tank;still;still kettle
蒸馏釜残渣 pressure bottoms
蒸馏釜的蛇形管 timber worm
蒸馏釜顶冷凝器 overhead drum
蒸馏釜馏出物 stillage
蒸馏釜式蒸发器 pot-type evapo(u)rator
蒸馏釜用蒸汽 still steam
蒸馏釜预热器 still preheater
蒸馏盖 still head
蒸馏管 distillation tube;distilling tube;still tube
蒸馏罐 alembic;pot still;retort;retort vessel
蒸馏罐海绵金 retort sponge
蒸馏罐上部 helmet
蒸馏罐炭精 retort carbon
蒸馏罐压制机 retort press
蒸馏锅 boiling bulb;boiling bulk;column boiler;distillating boiler;still;still kettle
蒸馏过程 still process
蒸馏过的 distilled
蒸馏海绵体 retort sponge
蒸馏后的残渣 distillation test residue
蒸馏壶 alembic;distillating still;distillating tank
蒸馏焦 retort coke
蒸馏焦油 distilled tar
蒸馏阶段 phase of distillation
蒸馏精炼 distillation refining
蒸馏净化 distilling purification
蒸馏冷凝器 distilling condenser;distiller condenser
蒸馏冷却器 distillate cooler
蒸馏冷水器 distilling condenser
蒸馏沥青 steam-refined asphalt
蒸馏炼焦 retort coking
蒸馏馏分 distillation fraction
蒸馏炉 distilling furnace;retort oven
蒸馏炉炉室 closet
蒸馏煤沥青 distilled tar
蒸馏膜 distillation membrane
蒸馏母液 distilland
蒸馏瓶 alembic;distillation retort;distilling flask;retort
蒸馏瓶侧管 vapo(u)r tube of flask
蒸馏气(体) distillation gas;retort gas;still gas
蒸馏器 alembic;distillation apparatus;distillator;distillatory vessel;distiller;still;stilling apparatus;vapo(u)rizer;water distillator
蒸馏器旋管 worm
蒸馏区 distillation zone
蒸馏区间 distillation range
蒸馏曲线 distillation curve
蒸馏烧瓶 distillation flask;distilling flask;flask with side arm
蒸馏设备 distillation equipment;distillation head;distilling plant;evaporating installation
蒸馏石油的残油 dead oil;dead-piled
蒸馏石油的残渣 oil residue
蒸馏式检索 search distillation
蒸馏试验 distillation test
蒸馏室 distillation chamber;distillery;stilling chamber;still room
蒸馏水 aqua distillate;distilled water;distilling water;glass distilled water
蒸馏水供给船 distilling ship
蒸馏水器 pavilion distillator;water still
蒸馏水装置车 mobile water distilling plant
蒸馏速率 rate of distillation
蒸馏损失 distillation loss
蒸馏所 distillery;still
蒸馏塔 column still;dephlegmator;distillating column;distillating tower;distillation column;distilling tower;hog still;still column;distilling column
蒸馏塔板 distillating tray;distillation tray;distilling tray
蒸馏塔板的计算 distillation plate calculation
蒸馏塔板框式加热器 tray heater
蒸馏塔的球形盖 bubble car
蒸馏塔中的泡罩 disperser
蒸馏碳 retort carbon
蒸馏头 still head
蒸馏脱硫装置 distillate desulfurization unit
蒸馏温度 vapo(u)rizing temperature
蒸馏温度曲线图 temperature distillation chart
蒸馏业 distilling industry
蒸馏液 distillate;extracting solution
蒸馏仪器 distillation apparatus;distilling apparatus
蒸馏者 distiller
蒸馏脂肪酸 distilled fatty acid
蒸馏柱 column distiller;distilling column
蒸馏装置 distillation apparatus;distilling apparatus;distilling plant;topping unit
蒸馏装置中的预热系统 preheat train in distillation unit
蒸馏装置组 distillation bench
蒸馏作用 distillation;distilling
蒸笼 digester
蒸笼窑 doughnut kiln;ring shelf kiln
蒸木池 hot pond
蒸木料槽 hot pond
蒸木油 creosote
蒸浓 concentration
蒸浓法 inspissation
蒸浓废物 evaporated waste
蒸浓胶乳 revertex
蒸浓器 inspissator
蒸浓塔 evaporating tower
蒸浓柱 evaporating column
蒸气 evaporation;reek;vapo(u)r
蒸气包 vapo(u)r pocket
蒸气饱和的 vapo(u)r-laden;vapo(u)rous;vapo(u)r-saturated
蒸气爆炸 vapo(u)r explosion
蒸气泵 vapo(u)r pump
蒸气薄雾 vapo(u)r mist

蒸气捕集器 vapo(u)r trap
蒸气不渗透性 vapo(u)r impermeability;vapo(u)r imperviousness
蒸气采暖 vapo(u)r heating
蒸气参数 vapo(u)rarium condition; vapo(u)ration parameter; vapo(u)r condition
蒸气测定法 vapo(u)rimetric method
蒸气测试器 vapo(u)r testing apparatus
蒸气层<紧位于水体上面的> vapo(u)r blanket
蒸气沉积镀膜 vapo(u)r deposition
蒸气沉积涂层法 vapo(u)r deposited coating
蒸气(沉积)釉 vapo(u)r glaze
蒸气衬底间界面 vapo(u)r substrate interface
蒸气成分 vapo(u)r composition
蒸气充满的 vapo(u)r-laden
蒸气充满式 vapo(u)r-filled
蒸气抽出气 vapo(u)r extractor
蒸气抽水机 pulsometer; pulsometer pump
蒸气除油 vapo(u)r degreasing
蒸气处理 vapo(u)r treatment
蒸气处治 vapo(u)r cure
蒸气传递 vapo(u)r transfer; vapo(u)r transmission
蒸气囱 vapo(u)r chimney
蒸气的脱脂 degreasing of vapo(u)r
蒸气垫 vapo(u)r cushion
蒸气淀积法 vapo(u)r deposition method
蒸气镀金法 vapo(u)r-plating process
蒸气发生 vapo(u)r generation
蒸气发生器 vapo(u)r generator
蒸气阀 water vapo(u)r valve
蒸气防护栅 vapo(u)r barrier
蒸气放电灯 vapo(u)r discharge lamp;vapo(u)r lamp
蒸气分出 steam bleeding
蒸气分解率 evaporation decomposition rate
蒸气分离 vapo(u)r disengagement; vapo(u)r removal
蒸气分离器 vapo(u)r eliminator
蒸气分馏过程 vapo(u)r rectification process
蒸气分配蒸馏头 vapo(u)r-dividing head
蒸气分压 vapo(u)r pressure; vapo(u)r tension
蒸气封闭 vapo(u)r seal
蒸气腐蚀抑制剂 vapo(u)r corrosion inhibitor
蒸气负荷 rate of evapo(u)ration
蒸气干度 mass quality
蒸气干燥法 vapo(u)r seasoning
蒸气感测 vapo(u)r sensing
蒸气隔板 vapo(u)r barrier
蒸气隔绝 vapo(u)r insulation
蒸气隔膜 vapo(u)r barrier film
蒸气供暖 vacuum vapo(u)r
蒸气供暖系统 vapo(u)r heating system
蒸气供暖装置 vapo(u)r heating equipment
蒸气供热 vapo(u)r heating
蒸气管 vapo(u)r pipe; vapo(u)r tube
蒸气管路 vapo(u)r line
蒸气管线 vapo(u)r line
蒸气过滤器 vapo(u)rarium filter;vapo(u)r filter
蒸气过热循环 vapo(u)r superheated cycle
蒸气含量 vapo(u)r content
蒸气换热的 vapo(u)r-heated
蒸气换热器 vapo(u)r exchanger
蒸气回收 vapo(u)r recovery
蒸气回收系统 vapo(u)r recovery system
蒸气回收装置 vapo(u)r recovery unit
蒸气活门 water vapo(u)r valve
蒸气机式挖掘机 steam digger
蒸气加热器 vapo(u)r heater
蒸气加热双转筒式干燥机 vapo(u)r-heated double-rotary drier[dryer]
蒸气夹套 vapo(u)r jacket
蒸气降温法 vapo(u)r method
蒸气截留量 vapo(u)r holdup
蒸气进样杆 vapo(u)r sampling rod
蒸气精馏器 vapo(u)r rectifier
蒸气净洗 vapo(u)r cleaning
蒸气(静)压力计 isoteniscope
蒸气卷流 vapo(u)r plume
蒸气绝缘 vapo(u)r insulation
蒸气空间 vapo(u)r space
蒸气空间的容积 vapo(u)r-space volume
蒸气空泡 vapo(u)r cavity
蒸气空气混合比 vapo(u)r-air ratio
蒸气空气混合物 vapo(u)r-air mixture
蒸气空气混合物凝结器 devapo(u)rizer
蒸气扩散 vapo(u)r diffusion
蒸气扩散率 vapo(u)r diffusivity
蒸气扩散器 vapo(u)r diffuser
蒸气扩散阻力 vapo(u)r resistivity
蒸气烙器 atmocautery
蒸气冷凝器 vapo(u)r condenser
蒸气炼油厂 wet-rendered
蒸气量 vapo(u)r quantity
蒸气量热计 steam calorimeter
蒸气流 vapo(u)r flow; vapo(u)r stream
蒸气漏入 accidental admission of vapo(u)r
蒸气漏泄 vapo(u)r escape
蒸气密度 vapo(u)r density
蒸气密度法 vapo(u)r density method
蒸气密度器 vapo(u)r density apparatus
蒸气密度球管 vapo(u)r density bulb
蒸气密封薄片 vapo(u)r seal foil
蒸气密封膜 vapo(u)r seal membrane
蒸气灭火 extinction using vapo(u)r
蒸气灭菌器 autoclave
蒸气凝结 devapo(u)ration;vapo(u)r condensation
蒸气凝结的 vapo(u)r-set
蒸气凝聚 devapo(u)ration
蒸气浓度 vapo(u)r concentration
蒸气排出管 vapo(u)r discharge tube
蒸气排放 vapo(u)r release
蒸气抛光 vapo(u)r polishing
蒸气配送管 vapo(u)r delivery tube
蒸气喷净法 steam blast
蒸气喷砂 vapo(u)r blast
蒸气喷砂处理 vapo(u)r blast operation
蒸气喷砂清理法 vapo(u)r blast process
蒸气喷射 vapo(u)r blasting;vapo(u)r injection;vapo(u)r nozzle
蒸气喷射泵 vapo(u)r ejection pump
蒸气喷射器 vapo(u)r steam blaster
蒸气喷射循环 vapo(u)r jet cycle
蒸气喷射制冷循环 vapo(u)r jet refrigeration cycle
蒸气膨胀室 vapo(u)r expansion chamber
蒸气平衡器 vapo(u)r balancer;vapo(u)r balancing mechanism;vapo(u)r balancing unit
蒸气屏层 vapo(u)r barrier
蒸气气塞 vapo(u)r lock(ing)
蒸气-气体混合物 vapo(u)r-gas mixture
蒸气迁移 vapo(u)r migration;vapo(u)r transfer;vapo(u)r transmission
蒸气枪 vapo(u)rchoc
蒸气清洁法 vapo(u)r blasting
蒸气清洗装置 vapo(u)r scrubber
蒸气区域通风 vapo(u)r area ventilation
蒸气去垢 vapo(u)r degreasing
蒸气去油机 vapo(u)r degreaser
蒸气燃气 gas-vapo(u)r
蒸气燃气喷管 steam jet
蒸气容积 vapo(u)r volume
蒸气容量 vapo(u)r capacity
蒸气容器 vapo(u)r vessel
蒸气熵 entropy of evapo(u)ration
蒸气上升管 vapo(u)r uptake
蒸气渗透 vapo(u)r migration;vapo(u)r penetration;vapo(u)r transmission
蒸气渗透系数 coefficient of vapo(u)r permeation;vapo(u)r permeance
蒸气渗透性 permeability to water vapo(u)r;vapo(u)r permeability
蒸气渗透压力计 vapo(u)r pressure osmometer
蒸气渗透阻力 resistance to water vapo(u)r permeability; water vapo(u)r resistance
蒸气湿敷器 wet steam pack apparatus
蒸气式 vapo(u)r system
蒸气室 vapo(u)rizing chamber
蒸气释放速度 vapo(u)r release rate
蒸气收集器 vapo(u)r container
蒸气双缸泵 pulsometer
蒸气速度 vapo(u)r velocity
蒸气损失 vapo(u)r losses
蒸气提取器 vapo(u)r extractor
蒸气提升泵 vapo(u)r lift pump
蒸气体积 vapo(u)r volume
蒸气体积当量 vapo(u)r volume equivalent
蒸气透过性 vapo(u)r transmission property
蒸气图 vapo(u)r chart
蒸气脱脂 vapo(u)r degreasing
蒸气无凝结的 vapo(u)r-free
蒸气雾 sea smoke
蒸气吸附法 vapo(u)r adsorption process
蒸气吸入 vapo(u)r inhalation
蒸气吸收 vapo(u)r absorption
蒸气吸水泵 pulsometer pump
蒸气系统 vapo(u)r system
蒸气相 vapo(u)r phase
蒸气消除器 vapo(u)r eliminator
蒸气效力 vapo(u)r effect
蒸气循环 vapo(u)r cycle;vapo(u)r recirculation
蒸气循环式气液平衡试验器 vapo(u)r recirculating still
蒸气压测定 vapo(u)r pressure test
蒸气压差 vapo(u)r pressure deficit
蒸气压常数 vapo(u)r pressure constant
蒸气压法 vapo(u)r method
蒸气压计 vapo(u)r tension meter
蒸气压力 vapo(u)r pressure;vapo(u)r tension; water vapo(u)r pressure
蒸气压力比 vapo(u)r pressure ratio
蒸气压力差值 vapo(u)r pressure deficit
蒸气压力计 vapo(u)rimeter[vapo(u)rometer]
蒸气压力曲线 vapo(u)r pressure curve
蒸气压力式温度计 vapo(u)r pressure thermometer
蒸气压力势 vapo(u)r pressure potential
蒸气压力梯度 vapo(u)r pressure gradient
蒸气压力图 vapo(u)r pressure chart
蒸气压强 vapo(u)r pressure
蒸气压曲线 vapo(u)r curve
蒸气压渗透法 vapo(u)r pressure osmometry
蒸气压缩法 vapo(u)r compression (method)
蒸气压缩过程 vapo(u)r compression process
蒸气压缩机 vapo(u)r compression machine;vapo(u)r compressor
蒸气压缩冷冻方式 vapo(u)r compression refrigerating system
蒸气压缩热式泵 vapo(u)r compression heat pump
蒸气压缩式制冷机 vapo(u)r compression refrigerator
蒸气压缩式制冷系统 vapo(u)r compression refrigeration system
蒸气压缩系统 vapo(u)r compression system
蒸气压缩循环 vapo(u)r compression cycle
蒸气压缩蒸发器 vapo(u)r compression evapo(u)rator
蒸气压缩制冷方式 vapo(u)r compression refrigerating system
蒸气压缩制冷系统 vapo(u)r compression refrigerating system
蒸气压缩制冷循环 vapo(u)r compression refrigeration cycle
蒸气压缩制冷装置 vapo(u)r compression refrigerating system
蒸气压同位素效应 vapo(u)r pressure isotope effect
蒸气压图表 vapo(u)r pressure chart
蒸气压下降 lowering of vapo(u)r pressure;vapo(u)r pressure lowering
蒸气压抑制剂 vapo(u)r pressure inhibitor
蒸气压真空蒸馏法 vapo(u)r compression-vacuum distillation process
蒸气压指数 vapo(u)r pressure index
蒸气烟囱 water vapo(u)r chimney
蒸气烟雾 vapo(u)r plume
蒸气养护 vapo(u)r cure
蒸气养护试验<测定水泥稳定性> autoclave test
蒸气液 vapo(u)r liquor
蒸气液体平衡 vapo(u)r-liquid equilibrium
蒸气液体平衡常数 vapo(u)r-liquid equilibrium constant
蒸气仪表 vapo(u)r meter
蒸气引射泵 vapo(u)r ejection pump
蒸气与回流液之接触 vapo(u)r-reflux contacting
蒸气浴 vapo(u)r bath
蒸气浴疗室 vapo(u)rarium
蒸气圆顶室 vapo(u)r dome
蒸气源 vapo(u)r source
蒸气再发 vapo(u)r return
蒸气再压缩 vapo(u)r recompression
蒸气增浓 vapo(u)r enrichment
蒸气张力 vapo(u)r tension
蒸气真空泵 vapo(u)r vacuum pump
蒸气制冷循环 vapo(u)r refrigeration cycle
蒸气转移系数 vapo(u)r transfer coefficient
蒸气状的 vapo(u)rific; vapo(u)rish;vapo(u)rous
蒸气阻力 vapo(u)r resistance
蒸气阻率 vapo(u)r resistivity
蒸气阻凝 vapo(u)r barring
蒸气阻凝薄板 vapo(u)r barrier

sheet(ing)
蒸气阻凝薄片 vapo(u)r barrier foil
蒸气组成 vapo(u)r composition
蒸气作用 vapo(u)rization
蒸汽 fume;steam;steam-jacketed mo-(u)ld
蒸汽安全阀 steam security valve
蒸汽熬煮 steam boiling
蒸汽板 steam plate
蒸汽伴热部位 steam trace part
蒸汽伴热管 steam companion;steam trace
蒸汽伴热管路 steam tracing line
蒸汽伴热管线 steam trace line
蒸汽伴随 steam trace
蒸汽伴随加热 steam tracing
蒸汽伴随加热管线 steam tracing line
蒸汽伴随加热小管 steam-heating tracer
蒸汽包 steam bubble;steam dome;steam drum;steam pocket
蒸汽饱和的 steam saturated
蒸汽饱和室 humidor
蒸汽保温加热 steam tracing
蒸汽保温加热的产品 steam traced product
蒸汽爆裂木材纤维 steam exploded wood
蒸汽爆炸 steam explosion
蒸汽备用量 steam reserves
蒸汽泵 donkey;pulsometer pump;pumping engine;steam feed pump;steam-jet air pump;steam pump
蒸汽表 steam ga(u)ge;steam table
蒸汽补胎机 steam vulcanizer
蒸汽采暖 steam heating
蒸汽采暖系统 steam-heating system
蒸汽参数 steam condition;steam parameter;steam quality
蒸汽操舵机 steam steering engine;steam steering gear
蒸汽操纵 steam operation
蒸汽操纵增压器 steam-operated booster
蒸汽操作的工作 workdone by steam
蒸汽槽 steam way
蒸汽侧 steam side
蒸汽产量 steam production
蒸汽产生 generation of steam;raising of steam;steam formation
蒸汽铲 steam shovel
蒸汽铲土机开挖 steam shovel excavation
蒸汽超压开关 excessive steam pressure switch
蒸汽车 steam car;steam coach;steam motor car;truck-mounted steam generator
蒸汽沉积镀膜 vacuum plating
蒸汽成本 steam cost
蒸汽冲击 blast of steam
蒸汽冲压机 steam stamp
蒸汽抽汽器 steam air ejector
蒸汽抽水机 pistonless steam pump;pulsometer
蒸汽抽吸清洗法 steam extraction cleaning method
蒸汽出口 steam discharge;steam outlet
蒸汽出口阀 steam-outlet valve
蒸汽除尘 steam type dust removal
蒸汽除灰 steam soaking
蒸汽除鳞喷嘴 steam-descaling sprayer
蒸汽除油 steam degreasing
蒸汽储存 steam reserves
蒸汽储蓄器 steam accumulator
蒸汽处理 steam(ing) treatment;reconditioning
蒸汽处理的硬木 reconditioned wood
蒸汽处理法 steaming boiling
蒸汽处理沥青 steam asphalt
蒸汽处理木材 reconditioning of timber;steam treated timber
蒸汽处治 steam cure
蒸汽吹出 steam out
蒸汽吹出的气体 steam run gas
蒸汽吹灰器 steam blower
蒸汽吹灰枪 steam lance
蒸汽吹孔 steam-blown poke hole
蒸汽吹炼 steam run
蒸汽吹扫 steam blowing;steam purge
蒸汽吹制 steam blown
蒸汽吹制工艺 steam-blown process
蒸汽吹制机 steam blowing machine
蒸汽吹制沥青 steam-blown asphalt
蒸汽锤 steam forging hammer;steam hammer
蒸汽纯度 steam purity(value)
蒸汽纯度表 steam purity meter
蒸汽淬冷渣 steam-cooled slag
蒸汽打管机 steam pile driver;steam pipe driver
蒸汽打桩锤 steam pile hammer
蒸汽打桩机 steam driver;steam pendulum;steam pile driver;steam pile driving plant
蒸汽打桩绞车 steam piling winch
蒸汽打桩装置 steam pile driving plant
蒸汽带动的 steam-driven
蒸汽带泡沫 foam over
蒸汽带水 carry-over;water entrained by steam
蒸汽单斗挖掘机 steam digger;steam navvy
蒸汽单斗挖土机 steam digger;steam navvy;steam shovel
蒸汽导电率 steam conductivity
蒸汽导电性 steam conductivity
蒸汽导管 jet chimney
蒸汽捣碎机组 steam stamp battery
蒸汽道 steam way
蒸汽的 steamy
蒸汽的干燥度 quality of steam
蒸汽的减温 steam attemperation
蒸汽的脱油器 oil separator for steam
蒸汽的油分离器 oil separator for steam
蒸汽等量线 line of equal steam quantity
蒸汽电力表 steam pressure ga(u)ge
蒸汽电力浮式起重机 steam electric-(al) floating crane
蒸汽电力推进 steam electric(al) propulsion
蒸汽电站 steam electric(al) generating station
蒸汽垫 steam cushion
蒸汽吊车 donkey crane;steam crane;steam hauler;steam hoist
蒸汽吊机 steam hauler;steam hoist
蒸汽动力 steam power
蒸汽动力驳船 steam lighter
蒸汽动力厂 steam plant;steam power plant
蒸汽动力传动 steam drive
蒸汽动力的 steam-driven
蒸汽动力工程 steam power engineering
蒸汽动力汽车 steam-powered(motor-)car
蒸汽动力设备 steam power plant
蒸汽动力拖网渔船 steam trawler
蒸汽动力挖泥船 steam dredge(r)
蒸汽动力循环 steam power cycle
蒸汽动力运输 transport by steam power
蒸汽动力装置 steam power plant
蒸汽动力装置试验 steam trial
蒸汽渡轮 steam-ferry
蒸汽端 steam end
蒸汽断流阀 steam stop valve
蒸汽锻造机 steam header
蒸汽镦锻机 steam header
蒸汽发电 steam electric(al) power generation
蒸汽发电厂 steam electric(al) generating station;steam-generating plant;steam-generating station;steam power plant
蒸汽发电机 steam-driven generator;steam dynamo;steam engine
蒸汽发电站 steam-generating station;steam power station
蒸汽发电装置 steam-generating plant
蒸汽发动机 steam engine
蒸汽发蓝处理 <使钢表面产生四氧化三铁膜> barffing
蒸汽发生 steam generation
蒸汽发生厂 steam generation plant
蒸汽发生堆 steam-generating reactor
蒸汽发生管路 steam-generating circuit
蒸汽发生器 steam boiler;steam can;steam converter[convertor];steamer;steam-generating unit;steam generator;steam generator block;steam raising equipment;steam raising unit;steam transformer
蒸汽发生器隔绝与排放系统 steam generator isolation and dump system
蒸汽发生器排污系统 steam generator blowdown system
蒸汽发生器水罐放水口 steam generator water tank drain
蒸汽发生器水罐进水口 steam generator water tank filler
蒸汽发生器组件 steam generator module
蒸汽发生设备 steaming apparatus
蒸汽发生塔 steam raising tower
蒸汽发生重水反应堆 steam-generating heavy water reactor
蒸汽发生装置 steam-generating equipment;steam-generating plant;steam generator;steam raising plant;primary plant <核反应堆的>
蒸汽伐木绞车 steam logging winch
蒸汽伐木卷扬机 steam logging winch
蒸汽阀(门) steam cock;steam valve
蒸汽法 steaming process
蒸汽法精制地沥青 steam-refined asphalt
蒸汽反向机构 steam reversing gear
蒸汽返回管道 steam return line
蒸汽分布 steam distribution
蒸汽分离器 steam purifier;steam separator
蒸汽分配 steam distribution
蒸汽分配阀 steam-distributing valve
蒸汽分配管 steam-distributing pipe
蒸汽分配管阀 steam-distribution network
蒸汽分配器 steam distributor
蒸汽分配系统 steam-distribution system
蒸汽分子 steam molecule
蒸汽风泵 steam-driven compressor
蒸汽封闭处理 steam sealing treatment
蒸汽浮动吊车 steam floating crane
蒸汽浮动起重机 steam floating crane
蒸汽浮式起重机 steam floating crane
蒸汽辐射采暖 steam radiant heating
蒸汽辐射供暖 steam radiant heating
蒸汽负荷 steam load
蒸汽覆盖 steam blanketing
蒸汽改制 steam-reforming
蒸汽干材法 steam seasoning
蒸汽干度 dryness fraction;dryness fraction of steam;steam mass quality;steam quality
蒸汽干度测量器 steam calorimeter
蒸汽干度计 steam calorimeter
蒸汽干管 main steam pipe;steam main
蒸汽干燥 steaming;steam seasoning
蒸汽干燥度 steam quality
蒸汽干燥法 steam drying;steam seasoning
蒸汽干燥机 steam drier[dryer]
蒸汽干燥器 steam drier[dryer];steam drying apparatus;steam-heated oven
蒸汽缸 steam cylinder;steam pipe
蒸汽缸部件 steam-cylinder assembly
蒸汽缸油 steam-cylinder oil
蒸汽隔层 vapo(u)r control layer
蒸汽跟踪式加热 steam trace heating
蒸汽工程 steam engineering
蒸汽供给管道 steam supply pipeline
蒸汽供暖 steam heating
蒸汽供暖的 steam-heated
蒸汽供暖管 steam-heating pipe
蒸汽供暖锅炉 steam-heating boiler
蒸汽供暖设备 steam-heating apparatus;steam-heating appliance
蒸汽供暖系统 steam-heating system
蒸汽供暖装置 steam-heating apparatus
蒸汽供热 steam heating
蒸汽供热设备 steam-heating apparatus
蒸汽供热系统 steam-heating system
蒸汽供应 steam supply
蒸汽供应线 steam supply pipeline
蒸汽鼓 dry drum;steam drum
蒸汽鼓风 steam blast;steam run
蒸汽鼓风层燃炉 steam-jet stoker
蒸汽鼓风机 steam blower
蒸汽鼓泡搅拌式高压釜 steam-bubbling type autoclave
蒸汽固化 steam cure;steam set
蒸汽管 steam pipe[piping];steam tube[tubing]
蒸汽管道 jet chimney;steam lead;steam line;steam pipeline;steam piping;steam supply;steam supply pipe;steam way;steam conduit
蒸汽管道工人 steam fitter
蒸汽管道配件 steam fittings
蒸汽管丁字头 steam pipe tee head
蒸汽管干燥器 steam-tube drier[dryer]
蒸汽管架式干燥器 steam-rack dryer
蒸汽管件 steam fittings
蒸汽管接头 steam pipe joint
蒸汽管路 steam line
蒸汽管路系统图 steam flow diagram
蒸汽管膨胀圈 steam pipe expansion loop
蒸汽管束 steam-tube bundle
蒸汽管网 steam network
蒸汽管网加湿器 steam grid humidifier
蒸汽管网增湿器 steam grid humidifier
蒸汽管线 steam pipeline;steam piping line
蒸汽管旋转干燥器 steam-tube rotary drier[dryer]
蒸汽罐 steam drum
蒸汽光泽 steam finish
蒸汽轨道车 steam railcar
蒸汽(轨道)动车 <旧式> steam railcar
蒸汽柜 steam chest
蒸汽滚筒 steam roller;sweat roll;water bag
蒸汽锅 steam cooker;steamer;steam-(ing) oven;steam kettle
蒸汽锅炉 boiler;boiler furnace;steam boiler;steam-generating furnace;steam generator;steam raising unit
蒸汽锅炉配件 steam fittings
蒸汽锅炉用煤 navigation-coal;steam coal
蒸汽锅筒 steam drum

Z

蒸汽过滤 steam filtration
蒸汽过滤器 steam filter
蒸汽过热 steam superheating
蒸汽过热可调锅炉 controlled superheat boiler
蒸汽过热器 steam superheater
蒸汽含量 steam content
蒸汽耗用量 steam economy
蒸汽烘缸 steam dryer
蒸汽烘燥机 steam heat dryer
蒸汽壶 steam kettle
蒸汽滑阀泵 steam plunger pump
蒸汽滑门阀 steam gate valve;steam parallel slide stop valve
蒸汽还原法 steam reduction
蒸汽环道 steam loop
蒸汽环管 steam loop
蒸汽换热器 steam converter[convertor];stream heat exchanger
蒸汽回管 steam return line
蒸汽回火 steam tempering
蒸汽回收 steam recovery
蒸汽回收塔 steam recovery tower
蒸汽回水管线 steam return line
蒸汽回转起重机 locomotive crane;steam-driven slewing crane
蒸汽汇集器 steam header
蒸汽混合机 steam-mixer
蒸汽混合物 steam mixture
蒸汽活化 steam activation
蒸汽活塞泵 steam plunger pump
蒸汽活塞环 steam piston ring
蒸汽或热水取暖系统 wet heat
蒸汽或热水入口 steam or hot water inlet
蒸汽货车 steam lorry
蒸汽机 steamer
蒸汽机车 steam locomotive;pig <俚语>;smoker <俚语>
蒸汽机车备件 steam locomotive spare
蒸汽机车倒车手柄 Johnson bar
蒸汽机车段 steam locomotive terminal
蒸汽机车给水站 watering station
蒸汽机车驾驶员 steam-haulage engineer
蒸汽机车牵引的列车 steam train
蒸汽机车司机 throttle jerker
蒸汽机车司机协会 <英> Steam Locomotive Operator's Association
蒸汽机车铁路 steam railroad
蒸汽机车洗修 washout repair of steam locomotives
蒸汽机车整备 hostling
蒸汽机船 steam boat;steam-driven vessel;steamer;steam ship
蒸汽机的坠阀 drop valve
蒸汽机发动的拖拉机 steam tractor
蒸汽机发动的挖掘机 steam excavation
蒸汽机公路车辆 steam-driven road vehicle
蒸汽机滑阀 Allan valve;trick valve
蒸汽机活塞 steam piston
蒸汽机火车 steam engine locomotive
蒸汽机润滑剂 steam engine lubricant
蒸汽机示功器 steam engine indicator
蒸汽机式吊车 steam hoist
蒸汽机式起重机 steam crane
蒸汽机式挖掘机 steam shovel
蒸汽机调节器 steam engine governor
蒸汽机拖轮 steam tug(boat)
蒸汽机务段 steam locomotive terminal
蒸汽机械 steam power machine
蒸汽机械喷油燃烧器 steam-assisted pressure jet burner
蒸汽机引犁 steam plow
蒸汽机引式中耕机 steam cultivator
蒸汽机运货车 steam-driven truck
蒸汽机运转的 steam-operated
蒸汽机桩锤 steam engine pile hammer
蒸汽激波 steam shock
蒸汽及热水散热器 steam-and-water radiator
蒸汽极限曲线 steam-limit curve
蒸汽集材机 yarder
蒸汽集合器 steam collector
蒸汽集气管 steam collector
蒸汽加工 steaming
蒸汽加热 steam heat
蒸汽加热处理 steam seasoned
蒸汽加热的 steam-heated
蒸汽加热的热水器 steam calorifier;steam water heater
蒸汽加热法 steam heating
蒸汽加热管 steam-heating pipe;steam-heating tube
蒸汽加热管道 steam-heated pipe line
蒸汽加热柜 steam-heating tank
蒸汽加热辊 steam-roll
蒸汽加热烘板机 steam-heated plate drier[dryer]
蒸汽加热搅拌 steam bubbling
蒸汽加热炉 steam pipe oven
蒸汽加热模 steam-heated mo(u)ld
蒸汽加热浓缩器 steam-heated concentrator
蒸汽加热耙膛式干燥机 steam-heated rabble-type hearth drier
蒸汽加热盘管 steam coil;steam-heating coil
蒸汽加热器 steam calorifier;steam heater
蒸汽加热圈 steam-heated circle
蒸汽加热设备 steam-heating apparatus
蒸汽加热式干燥机 steam-heated drier[dryer]
蒸汽加热式空气预热器 steam-air heater
蒸汽加热式沥青熔化装置 steam-heating asphalt melting unit
蒸汽加热式沥青贮仓 steam heating storage
蒸汽加热套 steam jacket
蒸汽加热系统 steam-heating system
蒸汽加热蒸发器 steam-heated evapo(u)rator
蒸汽加热蒸馏釜 steam-heated still
蒸汽加湿法 steam humidification
蒸汽加湿器 steam(-jet) humidifier
蒸汽加温盘管 steam coil
蒸汽加温室 steam-heating chamber
蒸汽加温系统 wet heat system
蒸汽加压 steam pressurization
蒸汽加压串列采暖系统 steam pressure type cascade heating system
蒸汽夹层锅 steam-jacked kettle
蒸汽夹带的杂质 steam-borne impurity
蒸汽夹套 steam jacket
蒸汽间 steaming plant
蒸汽减压阀 steam pressure-reducing valve;steam reducing valve
蒸汽减压器 steam pressure reducer
蒸汽交换器 steam transformer
蒸汽胶管 steam hose
蒸汽绞车 steam cargo winch;steam hauler;steam hauling machine;steam winch
蒸汽绞盘 steam capstan
蒸汽搅拌 steam stirring
蒸汽搅拌高压釜 steam-agitated autoclave
蒸汽搅拌机 steam-mixer
蒸汽搅拌器 steam-jet agitator
蒸汽搅动器 steam-jet agitator
蒸汽接管 steam connection
蒸汽节流 steam throttle;steam throttling
蒸汽节流阀 steam stop valve;steam throttle valve
蒸汽节流式热量计 steam throttle calorimeter
蒸汽截煤机 steam channel(l)ing machine
蒸汽截止阀 steam stop
蒸汽解冻 steam thawing
蒸汽解吸段 steam desorption section
蒸汽介质 steam atmosphere
蒸汽界面曲线 steam boundary curve
蒸汽进口 steam inlet
蒸汽进气侧 steam admission side
蒸汽进气管 steam-admission pipe
蒸汽精炼 steam refining
蒸汽精炼过的 steam-refined
蒸汽精制 steam-refined;steam refining
蒸汽精制的沥青 steam-refined asphalt
蒸汽精制法 steam reduction
蒸汽精制过的 steam-refined
蒸汽精制油 steam-refined residuum
蒸汽井 steam well
蒸汽警报器 steam siren
蒸汽警钟 steam bell
蒸汽净化 steam purification
蒸汽净化器 steam purifier
蒸汽卷筒绞车 steam drum winch
蒸汽卷扬机 steam hauler;steam hoist
蒸汽均匀化处理 steam homo-treatment
蒸汽开动 steaming
蒸汽坑 steaming pit
蒸汽空化数 vapo(u)r cavitation number
蒸汽空间 steam space
蒸汽空气鼓风 wet blasting
蒸汽空气混合气 steam and air mixture
蒸汽空气活化 steam-air activation
蒸汽空气模锻锤 steam-air die forging hammer
蒸汽空气气化 steam-air gasification
蒸汽空气烧焦法 steam-air decoking method
蒸汽空气压气机 steam-driven compressor
蒸汽空气自由锻锤 steam-air forging hammer
蒸汽空腔 steam void
蒸汽孔管 steam orifice
蒸汽口 steam orifice
蒸汽烙管 zestocautery
蒸汽冷凝器 steam condenser
蒸汽冷凝区 steam condensation zone
蒸汽冷凝数 steam condensation number
蒸汽冷凝水 steam condensate
蒸汽冷凝液 steam condensate
蒸汽冷却 steam cooling
蒸汽冷却的 steam-cooled
蒸汽冷却器 desuperheater;desuperheating station;steam cooler;vapo(u)r cooler
蒸汽冷却蛇管 steam cooling coil
蒸汽冷却系统 steam cooling system
蒸汽联箱 steam header
蒸汽量热器 steam calorimeter
蒸汽裂化 steam cracking
蒸汽裂化装置 steam cracking unit
蒸汽磷化处理法 steam phosphating;phosteam process
蒸汽流 steam flow
蒸汽流程图 steam flow diagram
蒸汽流道 steam flow channel
蒸汽流动 steam circulation
蒸汽流量 flow of steams;rate of steam flow;steam flow;steam throughput
蒸汽流量表 steam flowmeter;steam meter;steam meter manometer
蒸汽流量计 steam flowmeter;steam meter
蒸汽流量记录器 steam flow recorder
蒸汽流量孔板 steam flow orifice
蒸汽流率 rate of steam flow
蒸汽硫化 steam cure;steam vulcanization
蒸汽硫化器 steam vulcanizer
蒸汽滤净器 steam strainer
蒸汽滤气器 steam strainer
蒸汽滤器 steam strainer
蒸汽路碾 steam roller
蒸汽轮船 steamer
蒸汽轮机 steam turbine
蒸汽落锤 steam drop hammer
蒸汽脉冲 steam impulse
蒸汽铆钉枪 steam riveter
蒸汽煤 navigation-coal;steam coal
蒸汽煤气混合气 steam(-and)-gas mixture
蒸汽密度 steam density
蒸汽密封 steam seal;steam tight
蒸汽密封的 steam sealed
蒸汽灭火 steam fire-extinguishing;steam smothering
蒸汽灭火管道 steam smothering line
蒸汽灭火管系 steam out line;steam out system;steam smothering line
蒸汽灭火系统 steam fire smothering system
蒸汽灭火装置 steam annihilator;steam smothering system
蒸汽灭菌法 steam sterilization;steam sterilizing
蒸汽灭菌器 steam sterilizer
蒸汽膜 steam blanket;steam film
蒸汽摩擦 friction of steam
蒸汽母管 steam main
蒸汽幕 steam curtain
蒸汽内能 internal steam work
蒸汽囊 steam bubble
蒸汽凝固 steam set
蒸汽凝固油墨 steam-set ink
蒸汽凝结 steam condensation
蒸汽暖风机组 steam unit ventilator
蒸汽暖风器 steam-air heater;steam-heated air heater
蒸汽暖气管 steam-heating pipe
蒸汽暖汽器 steam heater
蒸汽排出 overboard steam drain;steam discharge
蒸汽排出时间 steam exhaust period
蒸汽排出周期 steam exhaust period
蒸汽排放 steam discharge
蒸汽排放系统 <汽轮机事故时用的> steam-dump system
蒸汽排空 steam discharge
蒸汽排气口 steam vent
蒸汽排气真空封罐 steam flow sealing
蒸汽盘管 steam coil
蒸汽盘管保暖储油车 steam-coil-heater tank car
蒸汽盘管供暖 steam-coil heated
蒸汽盘管供暖设备 steam-coil heating installation
蒸汽盘管供热 steam-coil heated
蒸汽盘管加热 steam-coil heating
蒸汽盘管加热槽车 steam-coil-heater tank car
蒸汽盘管式储水箱 steam-coil storage tank
蒸汽旁路 steam bypass
蒸汽旁通 steam bypass
蒸汽泡 vapo(u)r bubble
蒸汽配汽 steam distribution
蒸汽喷布拌和工艺 steam dispersion mix
蒸汽喷布拌和机 <一种采用蒸汽洒布沥青或焦油的拌和机> steam dis-

persion mixer
蒸汽喷吹 steam blast
蒸汽喷吹法 steam blowing
蒸汽喷吹纤维 steam-blown fiber[fibre]
蒸汽喷发 phreatic eruption; steam explosion
蒸汽喷管 steam lance
蒸汽喷净(装置) steam blast
蒸汽喷气孔 steam fumarole
蒸汽喷气清洁机 steam-jet cleaner
蒸汽喷气清洁器 steam-jet cleaner
蒸汽喷气清洗 steam-jet cleaning
蒸汽喷枪 steam lance; steam spraying gun
蒸汽喷砂机 steam(-jet) sand blaster
蒸汽喷砂器 steam sand blower; steam sand ejector
蒸汽喷射 steam injection; steam jet; thermojet
蒸汽喷射拔风 steam-jet draft
蒸汽喷射泵 steam-jet pump
蒸汽喷射变形 steam-jet texturing
蒸汽喷射抽汽器 steam-jet air ejector; steam-jet air extractor; steam-jet ash conveyor; steam-jet ejector; steam-jet exhauster
蒸汽喷射除氧器 steam-jet deaerator
蒸汽喷射吹灰器 steam-jet soot blower
蒸汽喷射法 steam-jet method
蒸汽喷射干燥窑 steam-jet blower kiln
蒸汽喷射管 steam-jet siphon
蒸汽喷射加热器 steam-jet heater
蒸汽喷射搅拌器 steam-jet agitator
蒸汽喷射空调系统 steam-jet air-conditioning system
蒸汽喷射拉伸法 steam-jet drawing process
蒸汽喷射冷却器 steam-jet cooler
蒸汽喷射冷却系统 steam-jet cooling system
蒸汽喷射磨细器 steam-jet attriter
蒸汽喷射器 steam blower; steam ejector; steam injector; steam-jet blower; steam-jet injector
蒸汽喷射热水采暖系统 steam-jet hot water heating system
蒸汽喷射热水供暖系统 steam-jet hot water heating system
蒸汽喷射热水系统 injection type steam heating system
蒸汽喷射式混合器 steam-jet mixer
蒸汽喷射式空气泵 steam-jet air pump
蒸汽喷射式空气抽汽器 steam-jet air ejector
蒸汽喷射式冷冻机 steam-jet chiller
蒸汽喷射式燃烧器 burner of steam injector type
蒸汽喷射式热泵 steam-jet heat pump
蒸汽喷射式制冷 steam-jet refrigeration
蒸汽喷射式制冷机 steam-jet refrigerating machine
蒸汽喷射式制冷系统 steam-jet refrigerating system
蒸汽喷射式制冷循环 steam-jet refrigerating cycle
蒸汽喷射吸入器 steam-jet sucker
蒸汽喷射系统 steam ejection system
蒸汽喷射循环 steam-jet cycle
蒸汽喷射压缩 steam-jet compression
蒸汽喷射引风 draft by steam jet
蒸汽喷射真空泵 steam-jet vacuum pump
蒸汽喷射蒸发器 steam ejector evaporator
蒸汽喷射制冷机 steam-jet refrigerating equipment; steam-jet refrigerator
蒸汽喷射制冷系统 ejector cycle refrigeration system; steam-jet refrigerating system
蒸汽喷射制冷装置 steam-jet refrigeration unit
蒸汽喷射装置 steam injection equipment; steam-jet unit
蒸汽喷涂 steam spraying
蒸汽喷涂磷酸盐处理法 steam spraying process method
蒸汽喷雾 steam atomizing; steam spray
蒸汽喷雾法 steam nebulization
蒸汽喷雾机 steam blower; team blower
蒸汽喷雾器 steam atomizer; steam blower; steam-jet sprayer
蒸汽喷洗 steam blast; steam cleaning
蒸汽喷洗装置 steam blast device
蒸汽喷油器 steam atomizer
蒸汽喷嘴 steam cone; steam jet; steam-jet blower; steam nozzle; steam sprayer
蒸汽喷嘴炉 steam-fan furnace
蒸汽膨胀 steam expansion
蒸汽品质 steam quality
蒸汽平板阀 steam gate valve
蒸汽平衡干管 steam balancing main
蒸汽平压 steam platen press
蒸汽屏层 steam barrier
蒸汽破裂事故 steam-break accident
蒸汽起货机 steam winch
蒸汽(起)锚机 steam windlass
蒸汽起重机 steam crane; steam hauler; steam hoist
蒸汽起重机车 steam loco crane; steam locomotive crane
蒸汽气流 steam stream
蒸汽气流喷射装置 steam-jet sprayer
蒸汽气门 steam cock
蒸汽气密性 resistance to water vapo(u)r transmission
蒸汽汽车 steam automobile; steam-powered automobile
蒸汽汽笛 visible air whistle
蒸汽汽轮机 steam turbine
蒸汽汽轮机发电厂 steam-turbine power station
蒸汽汽轮机发电机组 steam-turbine generating set
蒸汽汽轮机发电站 steam-turbine power station
蒸汽汽提 stripping with steam
蒸汽器 steam box
蒸汽牵引 steam traction
蒸汽牵引铁路 steam traction railway
蒸汽轻雾 steam mist
蒸汽清除装置 steam cleaning unit
蒸汽清洁法 steam cleaning
蒸汽清扫 steam soaking
蒸汽清扫装置 steam cleaning unit
蒸汽清洗 steam cleaning; steam flushing; steam scrubbing; steam washing
蒸汽清洗抽油杆 steaming of the rods
蒸汽清洗法 steam scrubbing method
蒸汽清洗机 steam cleaner
蒸汽清洗器 steam cleaner; steam purifier; steam washer
蒸汽清洗设备 steam washer; steam washing unit
蒸汽清洗站 steam cleaning station
蒸汽清洗装置 steam cleaner; steam purifier; steam scrubber
蒸汽清洗作用 steam scrubbing action
蒸汽驱动 steam drive; steam operation
蒸汽驱动泵 steam-driven pump
蒸汽驱动的 steam-driven; steam-operated; steam powered
蒸汽驱动的金刚石钻机 steam-motivated diamond drill
蒸汽驱动挖掘机 steam-driven digger; steam-driven excavator
蒸汽驱动钻机 steam drive drill; steam-operated drill; steam rig
蒸汽驱动钻探设备 steam drive drilling rig
蒸汽取暖管道接头 steam heat connection
蒸汽取暖锅炉 steam heat boiler
蒸汽取样 steam sampling
蒸汽燃气混合物 steam-gas mixture
蒸汽燃气联合循环 steam-gas cycle
蒸汽燃气涡轮泵式 steam-driven pumping system
蒸汽燃气涡轮泵式输送系统 steam-driven pumping system
蒸汽燃油电磁阀 steam oil magnetic valve
蒸汽热 <冷凝时放出的> steam heat
蒸汽热储 steam reservoir
蒸汽热风机 steam-heated calorifier
蒸汽热风器 steam-air heater
蒸汽热交换器 steam heat exchanger
蒸汽热量 steam heat
蒸汽热量计 steam calorimeter
蒸汽热能成本 cost of steam heat
蒸汽热水器 geyser; steam calorifier; steam-heated water heater
蒸汽热网 steam heat-supply network
蒸汽容积 volume of steam
蒸汽容量 steam capacity
蒸汽乳化 steam emulsification
蒸汽乳化试验 steam emulsion test
蒸汽乳化试验仪 steam emulsion testing apparatus
蒸汽乳化值 steam emulsion number
蒸汽入口 steam-in; steam inlet
蒸汽软管 steam hose
蒸汽软管接头 steam hose connection
蒸汽软化的木料 steamed wood
蒸汽润滑 steam lubrication
蒸汽散热片 steam radiator
蒸汽散热器 steam heater; steam radiator
蒸汽扫气 steam purging
蒸汽蛇管 steam pipe coil
蒸汽设备 steam plant
蒸汽设备安装 steaming installation
蒸汽射流 steam jet
蒸汽渗透 steam transmission
蒸汽渗透率 steam permeability
蒸汽升降机 steam-driven lift; steam hoist
蒸汽升压 raise steam; rise steam
蒸汽生产 steam production
蒸汽生产额 steaming power
蒸汽湿度 steam moisture
蒸汽式伴随加热法 steam trace heating
蒸汽式泵 donkey pump
蒸汽式给水预热器 steam feed heater
蒸汽式沥青熔化装置 asphalt steam pipe melter
蒸汽式压路机 steam roller
蒸汽式油加热器 steam oil heater
蒸汽式重油加热器 steam oil heater
蒸汽试验 steam test
蒸汽室 steam cabinet; steam chest; steam header; steam room; boiler steam dome; dry sweating room; sweating room
蒸汽收集器 dry pipe; steam accumulator; steam collector
蒸汽疏浚机 steam dredge(r)
蒸汽疏水器 steam trap
蒸汽输出 steam outlet
蒸汽输入量 steam input
蒸汽熟化 steam cure
蒸汽熟化器 steam ager
蒸汽竖管 steam riser
蒸汽双缸泵 pulsator; pulsometer; pulsometer pump; steam pulsometer pump
蒸汽水系统 steam-water system
蒸汽似的 steamy
蒸汽速度 steam velocity; velocity of steam
蒸汽隧管 steam tunnel
蒸汽损耗 steam loss
蒸汽损失 steam loss
蒸汽塔 fountain; turret
蒸汽塔干汽管 bridge pipe
蒸汽塔总汽门 cab turret valve
蒸汽弹射器 steam catapult; steam-driven catapult
蒸汽弹射系统 steam ejection system
蒸汽烫漂器 steam blancher
蒸汽烫衣机 steam press
蒸汽套的 steam jacketed
蒸汽套管 jacketed pipe; steam pipe sleeve; steam jacket
蒸汽套管子 steam-jacketed pipe
蒸汽套桶 steam-jacketed bucket
蒸汽特性图 steam chart
蒸汽梯度薄层层析法 steaming graded thin-layer chromatography
蒸汽提炼汽缸油 steam-refined cylinder oil
蒸汽提升器 steam lift
蒸汽田 steam field
蒸汽调车机车 steam switcher
蒸汽调节阀 steam control valve; steam regulator
蒸汽调节器 steam regulator
蒸汽调温盘管 steam-heated tempering coil
蒸汽调温蛇管 steam-heated tempering coil
蒸汽调压阀 steam regulating valve
蒸汽调整阀 steam regulating valve
蒸汽铁路的电化 electrification of steam railway
蒸汽铁屑生氢器 steam iron generator
蒸汽停汽阀 steam stop valve
蒸汽停滞 steam stagnation
蒸汽通道 steam channel
蒸汽通入操作 steaming operation
蒸汽筒 steam drum
蒸汽透平 steam turbine
蒸汽透平发电厂 steam-turbine power station
蒸汽透平发电机组 steam-turbine generating set
蒸汽透平发电站 steam-turbine power station
蒸汽透平机车 steam-turbine locomotive
蒸汽透平冷凝器 steam-turbine condenser
蒸汽透水冷凝器 steam-turbine condenser
蒸汽图 steam chart
蒸汽拖拉机 steam tractor
蒸汽拖轮 steam tug(boat)
蒸汽脱除 steam stripping
蒸汽脱漆器 steam stripper
蒸汽挖沟机 steam channel(l)ing machine
蒸汽挖掘机 steam digger; steam navvy; steam shovel
蒸汽挖泥船 steam dredge(r)
蒸汽挖土机 steam dredge(r); steam excavator; steam shovel
蒸汽弯曲法 <木材> steam bending
蒸汽万能挖土机 steam universal excavator
蒸汽往复(给水)泵 donkey pump;

steam-driven reciprocating pump; Worthington pump
蒸汽往复机 steam reciprocating engine
蒸汽温度 steam temperature
蒸汽温度升高 rise in steam temperature;steam temperature rise
蒸汽温度自动控制 automatic steam temperature control
蒸汽温度自动调节 automatic steam temperature control
蒸汽涡轮机 steam turbine
蒸汽涡轮给水箱 feed head of steam turbine
蒸汽污染 steam contamination;steam pollution
蒸汽雾 steam fog
蒸汽雾化 steam atomization; steam atomizing
蒸汽雾化扁平喷嘴 steam atomized flat spray
蒸汽雾化器 steam(-jet) atomizer; steam sprayer
蒸汽雾化燃烧器 steam atomizing burner
蒸汽雾化式燃烧器 steam-jet burner
蒸汽雾化型油喷燃器 steam atomizing oil burner
蒸汽雾化油燃烧器 steam atomizing oil burner
蒸汽雾化圆锥形喷嘴 steam atomizing conic(al) jet
蒸汽吸浆疏浚船 steam suction dredge(r)
蒸汽吸浆疏浚机 steam suction dredge(r)
蒸汽吸入 steam inhalation
蒸汽吸水泵 pulsometer pump
蒸汽吸水机 aquometer;pulsating steam pump
蒸汽洗舱 steam washing
蒸汽洗涤器 steam scrubber; steam washer
蒸汽洗衣房 steam laundry
蒸汽洗衣箱 blowout box
蒸汽系泊绞车 steam mooring winch
蒸汽系统 steam system
蒸汽显色 steam developing
蒸汽线(路) steam line
蒸汽相对出口速度 relative steam exit velocity
蒸汽相对入口速度 relative steam entrance velocity
蒸汽箱 steam box;steam chest
蒸汽消毒 moist-heat sterilization;steam sterilization
蒸汽消毒器 steam sterilizer
蒸汽消毒土壤 steam sterilized soil
蒸汽消耗计 steam consumption meter
蒸汽(消)耗量 steam consumption
蒸汽消耗率 steam rate
蒸汽消声器 steam silencer
蒸汽效率 steam efficiency
蒸汽形成 steam formation
蒸汽性能 steam property
蒸汽需用量 steam requirement
蒸汽蓄热器 steam accumulator
蒸汽悬臂起重机 steam jib crane
蒸汽旋臂起重机 steam slewing crane
蒸汽旋阀 steam cock
蒸汽旋风分离器 steam cyclone; turbo-steam separator
蒸汽旋管 steam coil;steam pipe coil
蒸汽旋管加热的 steam-coil heated
蒸汽旋管加热的油槽车 steam-coil-heater tank car
蒸汽旋塞 steam cock;steam stop cock
蒸汽旋转起重机 steam revolving crane
蒸汽学 atmology
蒸汽熏干 steam seasoning
蒸汽循环 steam cycle
蒸汽循环管 steam circulating pipe
蒸汽循环回路 steam circuit
蒸汽压力 steam pressure;steam tension
蒸汽压力表 steam ga(u)ge;steam pressure ga(u)ge
蒸汽压力罐 steam autoclave
蒸汽压力计 steam ga(u)ge;steam pressure ga(u)ge
蒸汽压力膨胀试验 <用于水泥安定性测定> autoclave expansion test
蒸汽压力曲线 steam pressure curve
蒸汽压力式串列供暖方式 steam pressure type cascade heating system
蒸汽压力图 steam diagram
蒸汽压力仪 steam tensiometer
蒸汽压力指示器 steam pressure detector
蒸汽压力自动调节器 steam pressure automatic regulator
蒸汽压路机 steam road roller;steam roller
蒸汽压迫铸造法 steam casting die
蒸汽压强 steam pressure
蒸汽压缩设备 steam pressure device
蒸汽扬水机 pulsometer plant
蒸汽养护 curing by steam;steam curing;steaming;steam out;steam curing at atmospheric pressure <在大气压下进行的>
蒸汽养护保温罩 insulated steam curing cloche
蒸汽养护的 steam-cured
蒸汽养护的(混凝土)试件 steam-cured unit
蒸汽养护的(混凝土)制件 steam-cured unit
蒸汽养护的加气混凝土 steam-cured gas(-formed) concrete
蒸汽养护的膨胀混凝土 steam-cured expanded concrete
蒸汽养护的(水泥)混凝土 steamed concrete
蒸汽养护覆盖物 steam curing cloche
蒸汽养护混凝土 atmospheric steam cured concrete; steam-cured concrete; steam curing of concrete; steamed concrete
蒸汽养护混凝土砌块 steam harden concrete block
蒸汽养护混凝土桩 autoclaved concrete pile
蒸汽养护间 <混凝土的> steam curing chamber
蒸汽养护前期 <混凝土的> prest-eaming period
蒸汽养护轻质混凝土板 autoclaved light concrete slab
蒸汽养护设备 steam apparatus
蒸汽养护设施 steam curing installation
蒸汽养护升温时间 temperature rise period
蒸汽养护时间 steam-curing cycle
蒸汽养护室 steam box;steam chamber; steam curing room; temperature rise room;steam vapo(u)r curing chamber; steam curing chamber
蒸汽养护水泥净浆 steam-cured paste
蒸汽养护箱 steam-cured box
蒸汽养护循环 steam curing cycle
蒸汽养护窑 curing kiln;steam curing kiln;steam kiln
蒸汽养护罩 curing enclosure for steaming;steam curing hood
蒸汽养护制度 steam curing cycle
蒸汽养护制件 steam-cured unit
蒸汽养护周期 steam curing cycle; steaming cycle
蒸汽养护装置 steam curing installation
蒸汽养生 curing by steam;steam curing
蒸汽氧气气化 steam-oxygen gasification
蒸汽液力传动 steam hydraulic power gear
蒸汽液压机 steam hydraulic press
蒸汽液压剪切机 steam hydraulic shears
蒸汽仪表 steaming apparatus
蒸汽仪器 steaming apparatus
蒸汽引入管 steam inlet
蒸汽印染 steam colo(u)r printing
蒸汽用填密胀圈 steam gland
蒸汽油雾燃烧炉 steam atomizing oil burner
蒸汽余面 steam lap
蒸汽与溶剂冷凝器 steam-and-solvent condenser
蒸汽浴 steam bath
蒸汽浴器 sauna bath;steam bath
蒸汽浴设施 sauna bath installation
蒸汽浴室 sauna;steam room;sudatorium[复 subdatoria/subdatory]
蒸汽浴用的火炉 sauna stove
蒸汽浴装置 sauna installation
蒸汽预热法 <防腐用> Colman method
蒸汽圆筒烘燥机 steam cans
蒸汽圆柱式泵 steam plunger pump
蒸汽云 steam cloud
蒸汽运输机 steam hauler
蒸汽再热器 steam reheater
蒸汽再热装置 reheat plant
蒸汽灶 steam oven
蒸汽增压器 steam intensifier
蒸汽增压式水压机 steam hydraulic press
蒸汽闸阀 steam gate valve; steam parallel slide stop valve;steam stop valve
蒸汽张力 steam tension
蒸汽帐 steam-tent
蒸汽真空泵 steam vacuum pump
蒸汽蒸发 steam raising
蒸汽蒸发器 steam vapo(u)rizer
蒸汽蒸缸 steam digester
蒸汽蒸化 steam-age
蒸汽蒸馏的 steam distilled
蒸汽蒸馏法 steam distillation; steam refining
蒸汽蒸馏器 steam still
蒸汽蒸馏松油 steam-distilled oil
蒸汽支管 branch steam line
蒸汽直接联动式活塞泵 Weir pump
蒸汽直接驱动 steam direct drive
蒸汽止动阀 steam stop valve
蒸汽指示器 steam indicator
蒸汽制动器 steam brake
蒸汽制沥青 steam-refined asphalt
蒸汽制木纸浆 steam mechanical wood pulp
蒸汽质量 quality of steam; steam quality
蒸汽重力捣碎机组 steam gravity stamp
蒸汽重力落锤 airdrop hammer;airlift gravity drop hammer
蒸汽注入 steam infusion
蒸汽注入法 steam injection
蒸汽注入时间 steam-jet time
蒸汽柱塞泵 steam plunger pump
蒸汽转化 steam-reforming
蒸汽转化作用装置 steam-reformer
蒸汽转换开关 steam switcher
蒸汽转筒法 machine-forming process
蒸汽转向装置 steam steering gear
蒸汽转子发动机 steam-powered rotary engine
蒸汽自动车 steam automobile
蒸汽总管 main steam range; steam collector;steam header;steam main
蒸汽阻滞 steam blanketing
蒸汽钻孔装置 steam drilling plant
蒸汽嘴 steam nozzle
蒸汽最大工作压力 maximum steam working pressure
蒸球 rotary spheric(al) digester
蒸去 skimming
蒸去轻油 skimming;topping
蒸热 steaming
蒸热装置 steaming plant
蒸软 boil down
蒸散 evapotranspire
蒸散发 evapotranspiration;fly-off
蒸散计 evapotranspirometer
蒸散潜能 evapotranspiration potential
蒸散式消气剂 flash getter
蒸散损失 evapotranspiration loss
蒸散箱 evapotranspiration tank
蒸散作用 evapotranspiration
蒸渗仪 lysimeter
蒸刷机 steaming and brushing machine
蒸水器 water still
蒸腾 evapotranspire
蒸腾比 transpiration ratio; water-use ratio
蒸腾比率 transpiration rate
蒸腾对淋溶等级的作用 transpiration affection leaching fractions
蒸腾计 phytometer; potometer; transpirometer
蒸腾量观测 transpiration observation
蒸腾流 <由土壤到植物的根茎和叶> transpiration(al) steam; transpiration(al) current
蒸腾率 rate of transpiration; transpiration ratio;water-use ratio
蒸腾能力 potential transpiration
蒸腾强度 intensity of transpiration
蒸腾生产率 productivity of transpiration
蒸腾速度 transpiration rate
蒸腾损失 transpiration loss
蒸腾系数 transpiration coefficient
蒸腾效率 transpiration efficiency
蒸腾作用 transpiration
蒸腾作用的测定 transpiration test
蒸腾作用冷却效应 cooling effect of transpiration
蒸透 pervapo(u)rization
蒸涂 coating by vaporization
蒸涂层 evaporation layer
蒸泄 evaporating excretion
蒸锌炉 zinc distillation furnace
蒸锌炉冷凝器 devanture
蒸压釜 autoclave
蒸压罐模制法 autoclaved mo(u)lding
蒸压混凝土 autoclave-cured concrete
蒸压混凝土桩 autoclaved concrete pile
蒸压加气砌块 autoclaved aerated block
蒸压膨胀 <密封锅中的> autoclave-cured expansion
蒸压膨胀试验 <测定水泥安定性的> autoclave expansion test;autoclave-cured expansion test
蒸压器 autoclave;pressure boiler
蒸压石灰 autoclaved lime
蒸压试验 autoclave-cured test; autoclave test
蒸压水化石灰 autoclave lime
蒸压循环(时间) autoclave cycle
蒸压养护 autoclave curing;high-pressure steam curing
蒸压养护周期 autoclave curing cy-

cle;autoclaving cycle
蒸压砖 autoclave brick
蒸盐锅 fishing pan
蒸养混凝土 steam-cured concrete;steamed concrete
蒸养试验 autoclave test
蒸养制度 autoclaving cycle
蒸煮 cook;digest
蒸煮拌和器 steam-mixer
蒸煮车间 cooking department;digester room
蒸煮处理 <木材的> boiling treatment;cooking
蒸煮的 digestive;steam cooked
蒸煮度 cooking degree
蒸煮法 <木材防腐> boiling process
蒸煮法回收 steam-digestion reclaim
蒸煮罐 cooking kettle;digester
蒸煮锅 boiling-down pan;boiling kier;cooking boiler;digester;digester room;pulp mill digester
蒸煮过的水泥扁饼 boiled pat
蒸煮混合器 steamer-mixer
蒸煮浓缩 boil down
蒸煮器 boiler;boiling vessel;cooker;pulp digester;digester【化】
蒸煮器盘管 digester coil
蒸煮球 spheric(al)digester
蒸煮试验 boiling test;autoclave expansion test <水泥的>
蒸煮室 cure dag
蒸煮桶 cooking vat
蒸煮弯曲法 steam bending
蒸煮消化 steam digestion
蒸煮液 cooking liquor
蒸煮装置 steaming plant
蒸煮作用 digestion

拯

拯救点 rescue point
拯救转储 rescue dump

整

整板车架 slab frame
整板肋骨 solid plate frame
整版 justification;justify
整版的 full-page
整版机 block justifying machine
整备 preparation for running;servicing
整备帮工 hostler
整备车间 servicing shop
整备待班线 locomotive service and temporary rest line;locomotive service and temporary rest track
整备费 <旧机器出售时> make ready cost
整备机 dresser
整备能力【铁】service capacity
整备设备 service facility;servicing facility
整备设施 hostling facility
整备台位 servicing position
整备天数 days of preparation
整备通路 service path
整备线 feed line;feed track;service line;servicing line;serving line;servicing track;service road;service track
整备线配置系数 the allocating factor of service track
整备员 preparer
整备站台 service platform
整备重量 weight in working condition;weight in working order
整备状态 ready condition
整倍数 integral number of times
整笔拨款 blocallocation;block allocation;lump-sum allotment;lump-sum appropriation
整笔付清费用 lump-sum fee
整笔经费分配 lump-sum allotment
整笔运费 lump-sum freight
整笔运费租赁 lump-sum charter
整笔支付 lump-sum payment;single payment
整笔总付办法 lump-sum basis
整闭包 integral closure
整闭环 integrally closed ring
整壁板更换 <集装箱> replacing panel assembly
整边 <木材> full edged
整边钢板 universal mill plate
整边机 <挖沟机的> crumber
整边黏[粘]接杯柄 block handle
整边炮眼 rib hole
整编 compilation;forming
整编流量资料 compiling stream-flow data
整编资料 compilation data;compiling data
整变量 integer variable
整标志 integral denotation
整驳货 lighter loaded cargo
整步【电】synchronization;synchronize;bring into step;synchronizing;timing
整步电抗 synchronous reactance
整步电流 synchronizing current
整步电压 timing voltage
整步电压表 synchronizing voltmeter
整步迭代法 total step iteration
整步功率 rigidity factor
整步脉冲 dating pulse;lockout pulse;synchronizing pulse
整步器 lock unit;synchroniser[synchronizer]
整步试验 pull-in test
整步转矩 synchronizing torque
整材 whole timber
整仓干燥系统 full-bin drying system
整槽绕组 integer-slot winding
整层高的窗 floor-to-ceiling window;full-height window
整层砌的 regular course
整层砌筑 range work
整层砌筑琢石圬工 regular coursed ashlar masonry
整层石工 range work
整常数 integral constant
整厂输出 package plant export
整超越函数 integral transcendental function
整车 carload car;entire car
整车车体检验 body test of whole tank vehicle/trailer
整车等级 carload rating
整车发料 carload shipment
整车分卸 carload freight unloaded at two or more stations;wagon-load goods unloaded at two or more stations
整车荷载 carload
整车货 carload;complete wagonload;full wagon load
整车货场 cargo load team yard;carload lean yard
整车货物 carload freight;carload lot;train load(ing);truckload cargo;carload goods;wagon load
整车货物吨数条件 <最少装货吨数> tonnage condition for complete wagon-loads
整车货物运价率表 scale of rates applied per wagon-load
整车货物运输 carload shipment;transport of wagon-load goods;wagon load traffic
整车货物最低重量 minimum carload weight;weight for carload shipments
整车货物最少装载吨数 carload minimum;wagon load minimum
整车货运价 carload rate
整车交货 carload delivery
整车解体出口 completeness knock-down
整车满载 full-car-load
整车满装剩余货物 freight in excess of full carloads
整车批量 carload lot
整车运价 carload rate;truckload rate
整车运价率 carload rate
整车运行试验 running test of whole tank vehicle/trailer
整车装载量 carload
整齿 commutating tooth;set
整齿工具 jointing tool
整齿料 fit
整齿器 jointer;swage sharper
整齿人字齿轮 continuous double-helical gear
整齿铁钻 set block
整冲压(非金属)轴承 fully mo(u)lded type bearing
整除 contain;exact division;exactly divisible
整除部分 aliquot part
整除的 aliquot
整除数 aliquot
整除性 divisibility
整船包价 lump-sum freight
整船货物 ship's load cargo;shipload lot
整船散货 parcel-bulk shipment
整存 full storage
整存系统 full-storage system
整袋水泥僵块 hard lump of cement
整道【铁】lining
整道机 lining machine;straightening machine
整的 integral
整底吊桶 solid-bottom bucket
整底盘 chassis assembly
整地 ground making;ground preparation;land forming;land grading;land preparation;land preparing;preparation of land;preparation of soil;site preparation;soil preparation;soil tillage
整地工程 grading works
整地机 grading machine;land grader
整地机具 tillage implement
整地机械 grading equipment
整顶板更换 replacing panel assembly
整顶工作 topwork
整定 setting
整定参数 adjustment variable
整定点 set point
整定电流 setting current
整定范围 setting range
整定机构 setting device;setting mechanism
整定精确性 setting accuracy
整定卡规 adjusting ga(u)ge
整定命令 setting command
整定时间 duration of regulation;setting time
整定速度 permanent speed
整定温度 set temperature
整定压力 set pressure
整定仪表 adjusting ga(u)ge
整定元件 setting device
整定值 set point;setting value
整定值跟踪 set point tracking
整定指令 setting command
整定置位 setting range
整定转速 permanent speed;set speed
整定装置 setting device
整锭场 conditioning yard;dressing yard
整斗的 levelled off
整段 single-piece
整段土壤剖面 monolith
整锻 monobloc forging;one-piece-forged
整锻冲动式转轮 integrally forged impulse wheel
整锻鼓形转子 solid drum rotor
整锻辙叉 forged solid crossing
整锻支承辊 solid forged backup roll
整锻轴 forged integrally shaft
整锻转子 gashed rotor;integral disk rotor
整队过闸 fleet lockage;single lockage
整队过闸船闸 barge train lock
整顿 clear up;consolidation;ordering;rearrange;rectification;rectify;rehabilitate;reorganization;straighten(ing)
整顿刷新 face-lift
整二次型 integral quadratic forms
整二元三次型 integral binary cubic forms
整帆待发 stow away a boat
整范数 integral norm
整方 perfect square
整方料 whole timber
整分部分 aliquot
整幅像片【测】full frame
整付经费概算书 estimated for lump-sum appropriations
整改 remedial works
整杆 undivided shaft
整个范围 full range;gamut
整个工程 the whole of the works
整个焊缝熔透 complete joint penetration
整个机器控制系统 total machine control system
整个街坊 whole block
整个截面 full cross-section
整个截面的构件 full cross-sectional element
整个流程 complete flow process;complete line
整个流量 full discharge
整个埋于主夹板内的小型自动锤 miniature rotor
整个剖面 the entire profile
整个戗脊 whole hip
整个区段 whole block
整个生长期间 throughout the growing season
整个使用期间的成本 whole-life cost
整个说来 take all things together
整个蜗壳 full scroll(case)
整个系统 overall system
整个线路的等效电路 overall line equivalent
整个圆盘 solid disc[disk]
整个圆片 solid disc[disk]
整个阅读数字 read-around number
整个运算时间 turnaround time
整功率因数 unity power factor
整管 <非焊接的> homogeneous tube
整管机 reeling machine
整管套 one-piece section
整灌电缆管道 monolithic conduit
整光 fettle;lay the grain
整光的 glossy
整轨锤 gad
整轨机 ga(u)ge setting device
整轨器 gad
整轨钳 gag
整柜载货 full container load
整柜装 full container load
整柜装货物 full container load cargo

整函数 entire function;integral function
整行 full line
整行排铸机 intertype
整行铸排机 line casting machine;linotype
整合【地】 concordance [concordancy];conformity
整合层次 integrative levels
整合层理 conformable bedding;regular bedding
整合层面 concordant bedding
整合地层 conformable stratum [复 strata]
整合地貌 concordant morphology
整合动作 integrative action
整合断层 conformable fault
整合谷 <与地层倾向一顺向谷> concordant valley
整合贯入【地】concordant injection
整合灌入 concordant injection
整合校正 integrative suppression
整合接触【地】conformable contact
整合侵入 concordant injection;conformable injection;concordant intrusion
整合侵入体【地】 concordance intrusive body
整合缺陷型 integration deficient mutant
整合深成岩体【地】 concordant pluton(e)
整合效果 synergy
整合效率 integration efficiency
整合性 conformability
整合岩层 conformable stratum
整合岩体 concordant body;concordant injection
整合障碍 dysintegration
整合褶皱 concordant fold
整合中枢 integration center[centre]
整合作用 integration
整化 integralization;rounding
整环 integral domain
整机 complete appliance;complete set;integral unit
整机的带通特性 overall bandpass response
整机电路 circuitry
整机规模 computer capacity
整机进口 import of complete machines
整机控制部分 system control section
整机宽度 overall width
整机连接杆 assembly rod
整机频率响应 overall frequency response
整机全长【机】overall length
整机全高 overall height
整机全宽 overall width
整机试验 overall test
整机效率 overall efficiency
整机制造厂 system maker
整机重量 total weight
整机总装车间 machine assembly department
整基 integral basis;integrity basis
整基底 integral basis
整剪绿篱 clipped hedge
整件 single-piece
整件缺失 loss of package
整件试验 whole article autoclaving test
整件提货不着 non-delivery of entire package
整件遗失 loss of package
整件铸件 cast unit
整浇 block cast
整浇楼板 monolithic floor
整节距 diameter pitch;diametric(al) pitch;full pitch
整节距绕法 full-pitch winding
整节距绕组 full-pitch winding;diametral winding
整节滤波器 full section filter
整洁的 natty;neat;shipshape;sleek;snug;taut;trig
整洁漂亮的 spruce
整洁小巧商店 snug
整经 beaming;beam-warping;turning on
整经机 ball warping machine;dresser;reeling machine;reeling mill;warper;warping machine
整经机辊筒 warper drum
整距绕组 integral pitch winding
整距线圈 full-pitched coil
整锯 jointing
整锯齿 set
整锯锉 saw file
整锯工 saw fitter
整锯虎钳 saw vice
整锯机 saw setting machine;saw sharpening machine
整锯器 saw set(ting);setting block
整锯钳 saw set(ting) pliers
整锯台 saw sharpener's bench
整卷长度 drum length
整卡车 truck-load
整卡机构 stacking mechanism
整壳式接头 screw-together fitting
整壳体车身 unitary body
整孔器 bit reamer
整孔无碴无枕预应力混凝土梁 monolithic beam without ballast and sleeper
整孔装药 long charge
整孔钻 reamer;rimer
整块 en-block;in block;monoblock;one-piece;single block;single-piece
整块板 monolithic slab;one-piece panel
整块材料 monolithic material
整块材料组装的地板 monolithically assembled floor panel
整块材模子 solid-block mo(u)ld
整块材砌筑墙 solid-block masonry (work)
整块材试验机 solid-block tester
整块材制机 solid-block machine
整块传送 block transfer
整块磁极 solid pole
整块大砖 monoblock
整块的 massive;unsplit
整块地面 seamless floor(ing)
整块端砌 isdmun;isodomon
整块灌注 pour monolithically
整块灌筑 monolithic pouring
整块花岗石块端砌 opus isodomum of granite blocks
整块混凝土 monolithic concrete
整块浇成的地面 screed floor
整块浇灌 pour monolithically
整块浇注 pour monolithically
整块浇筑 pour monolithically;monolithic pour
整块浇铸件 one-piece casting
整块结构 en-block construction
整块旧板吊除法 slab removal
整块锯材 one-piece sawed timber
整块块材 whole block
整块煤 coal in solid
整块模板 solid shuttering
整块切除 en bloc resection
整块染色 <地毯> piece dy(e)ing
整块石端砌 isodomon;isodomum
整块石膏 massive gypsum
整块石料 massive block;monolith
整块石面砌 isodomon
整块石踏步 solid stone step
整块实体屋顶 monolithic solid rood
整块式舱口盖 one-piece hatch cover
整块式码头 monolithic type wharf
整块双窗墙板 two-window one-piece panel
整块双焦点镜 ultex
整块水磨石 monolithic terrazzo
整块松板 <1.25 英寸厚,1 英寸 = 0.0254米> whole deals
整块体 monolith
整块替换 block replacement
整块外壳 monolithic case
整块网眼钢皮板条 expanded metal integral lath(ing)
整块铸造 monoblock cast(ing)
整块砖 four quarters;full clay brick;full brick
整块转移【计】block transfer
整块组件 monoblock unit
整矿柱 solid block
整捆器 bale aligner
整理编辑 reduction
整理表 sorting table
整理操作 housekeeping operation;red-tape operation
整理操作时间 overhead time
整理车间 dressing floor
整理车辆作业 trimming operation
整理程序 arranging routine;collate program(me);collator
整理发货手续费 consolidated rate;management rate
整理费 reconditioning charges;reconditioning fee
整理改装 reconditioning
整理工作 crabbing;housekeeping <房内>
整理过程 pick-up procedure
整理河道 realign(ing)
整理后的重量 dressed weight
整理后统计数字 derivative statistics
整理活动 reorganization
整理机 collator
整理剂 finishing agent
整理勘察资料 consolidation of geotechnical investigation data
整理矿砂机 vanner
整理茅草的工具 legget
整理苗床 preparing seedbed
整理情报 collation of information
整理软件 housekeeping software
整理顺序 collating sequence
整理所搜集的资料 consolidation of collected data
整理所有地质资料 collation of all geologic(al) data
整理通过 collation pass
整理土地 preparation of land
整理位置 justification
整理现场多余石料 clowring
整理账 adjusting entries
整理者 collator;regulator
整理中心 handling center
整理装置 collating unit
整理资料 collation of data;collection of data;process data;sorting data;sort out data
整理组装 housekeeping package
整理作业 readjusting operation
整粒金刚石 whole diamond;whole stone
整粒金刚石钻头 whole diamond bit;whole stone bit
整联箱式锅炉 box header boiler;simple header boiler
整料门窗框 solid frame
整列 fall-in
整列车业务 train load business
整列车装载 complete trainload
整列穿孔 lace
整列穿孔卡片 lace card
整列斗车装碴机 train loader
整列进料机 aligner feeder
整流 commutate;commutation;commute;rectification;rectifying;switching
整流板 eddy plate;rectifying plate
整流倍压器 rectifier doubler
整流比 commutating ratio;rectification ratio;rectifier ratio
整流变压器 rectification switchgear;rectiformer
整流变压器单元 transformer/rectifier unit
整流变压器馈电柜 rectifier transformer feeder cubicle
整流波纹 commutator ripple
整流波形 rectified waveform
整流不良 poor commutation
整流参数 commutating parameter
整流槽 rectifier tank
整流磁场 commutating field
整流磁极极靴 interpole shoe
整流带 commutating zone
整流导叶 straightening vane
整流的 fairing
整流电导 conductance for rectification
整流电动机 motor converter
整流电动势 commutating electromotive force
整流电极 valve electrode
整流电抗 commutating reactance
整流电抗器 commutating reactor
整流电力机车 rectifier locomotive
整流电流 commutating current;rectified current
整流电流波形因素 direct current form factor
整流电路 rectification circuit;rectifier circuit
整流电容器 commutating capacitor;commutation capacitor;commutator capacitor
整流电刷 commutator brush
整流电压 commutating voltage;rectified voltage;unidirectional voltage
整流电压表 rectifier voltmeter
整流电源 eliminator power;eliminator supply;power rectifier
整流电阻 dead resistance
整流度 degree of current rectification
整流堆 rectifier stack;rectistack
整流舵 contra-propeller;contra rudder;counterrudder
整流扼流圈 commutating reactor;commutation reactor
整流二极管 rectifier diode
整流阀 rectifying valve
整流法 rectification method
整流反馈 rectified feedback
整流方程 rectifier equation
整流方式 rectification mode;rectifier system
整流格 honeycomb screen
整流工段 rectification section
整流功率 rectified power
整流故障 commutation failure
整流管 alternating current converter;commutator tube;converter;electric(al) valve;kilotron;rectifier;rectifier cell;rectifier tube;rectifying tube;rectifying valve;rectify tube;tungar
整流管阳极 rectifier anode
整流光电管 rectifier photocell
整流辊 rectifier roll
整流过程 switching process
整流过的脉动电流 pulsating rectified current

整流焊机 metal rectifier welding set; rectifier welding machine; rectifier welding set
整流后的电流 rectified current; redressed current
整流后的电压 rectified voltage
整流弧焊机 commutation arc welding machine; rectifying arc welding machine
整流环 collector ring; ring cowl
整流换流器 rectifier inverter
整流换向 commutation changeover
整流机 motor-generator set
整流机构 rectification mechanism
整流机组 rectifier unit
整流机组效率 efficiency of rectifier unit
整流极 auxiliary pole; commutating pole; compole; interpole; rectifier cascade; rectifier stage
整流极电动机 commutating pole motor
整流极绕组 commutating winding
整流继电器 rectifier-type relay
整流检波器 rectifier detector; rectifying detector
整流接触 rectifying contact
整流扩散 rectified diffusion
整流(螺旋)推进器 contra-propeller
整流脉冲 commutator pulse
整流脉动 commutator ripple
整流脉动系数 rectifier ripple factor
整流能力 detectability
整流片 commutating segment; fairing; fillet; rectiblock; rectifier cell; straightening vane
整流片的竖端连接 riser connection
整流片距 commutator pitch
整流片组 commutator bar assembly; rectifier stack
整流器 current rectifier; current regulator; detector; electric(al) commutator; electropeter; power converter; rectifier
整流器保护 rectifier protection
整流器变电所 rectifier substation
整流器变电站 rectifier substation
整流器变压器 rectifier transformer
整流器参量 rectifier rating
整流器充电 charging rectifier; rectifier charging
整流器存储特性 rectifier storage characteristic
整流器端毡心 commutator end felt wick
整流器堆 rectifier stack
整流器负载 rectifier load
整流器供电 rectifier power supply
整流器关闭 rectifier block
整流器罐 rectifier tank
整流器护板 commutator shield
整流器继电器 rectifier relay
整流器架 rectifier rack
整流器馈电 rectifier feeding
整流器馈电的电动机 rectifier-driven motor
整流器馈电的直流轨道电路 DC rectifier-fed track circuit
整流器漏电流 rectifier leakage current
整流器滤波器 rectifier filter
整流器片 commutator segment; rectifier disk; rectifier plate
整流器桥路 rectifier bridge
整流器热丝 rectifier heater
整流器散热片 heat sink
整流器扇形片 commutator segment
整流器设备 rectifier equipment
整流器式电力传动 rectifier electrical transmission
整流器式轨道电路 valve-type track circuit
整流器式回波抑制器 rectifier-type echo suppressor
整流器式励磁机 rectifier exciter
整流器输出电压调整器 rectifier output voltage regulator
整流器特性 rectifier characteristic
整流器外壳 shell of rectifier; vacuum tank
整流器尾架 commutator end frame
整流器效率 rectifier efficiency
整流器元件 rectifier element
整流器杂音电压 noise voltage of rectifier
整流器真空管 radio valve
整流器轴 commutator shaft
整流器装配 rectifier assembly
整流器子接线叉 commutator riser
整流器组 rectifier set; rectifier stack
整流栅 damping screen; diffuser grid; gate; honeycomb; honeycomb screen; honeycomb straightener; straightener; vortex gate
整流栅叶片 gate leaf
整流设备 commutation equipment; rectifier equipment; rectifying installation
整流射线管 loprotron
整流式 rectifier type
整流式变频 commutator frequency converter
整流式波长计 rectifier wavemeter
整流式差动继电器 rectifier-type differential relay
整流式电动机 brush motor
整流式电机 commutating machine
整流式焊机 welding rectifier
整流式计数器 commutator-type meter
整流式继电器 rectifier-type relay
整流式距离继电器 rectifier-type distance relay
整流式仪表 rectifier instrument; rectifier-type instrument; rectifier-type meter
整流式直流电机 direct current commutating machine
整流式直流弧焊机 rectifier-type arc welder
整流试验 commutation test
整流输出 rectified output
整流探针 rectifying probe
整流特性 rectification characteristic
整流特性函数 rectifying characteristic function
整流条 commutator bar
整流推进器 contra-turning propeller
整流网 gauze screen
整流艉 contraguided stern; contratype stern
整流误差 rectification error
整流吸收 rectisorption
整流系数 rectification coefficient
整流现象 rectifier phenomenon
整流线卷 commutating winding
整流线圈 commutating winding
整流效率 blocking efficiency; efficiency of rectification; rectification efficiency; rectifying efficiency
整流效应 valve effect
整流信号 rectified signal
整流信号接收 rectified recording
整流形辐条 fairing spoke
整流型相位比较器 rectifier phase comparator
整流叶片 straightener stator blade
整流翼 straightening vane
整流因数 commutation factor; rectification factor
整流元件 rectifier cell
整流站 converting plant
整流罩 aerial radome; dome; radom(e)
整流罩舵 bulb rudder
整流罩架 cowling mount
整流罩框架 cowl former
整流罩密封的外导管 external faired pipe
整流罩裙片 cowling shutter
整流罩通风片 cowl(ing) flap
整流罩通风片控制 cowling flap control
整流支柱 fairing spoke
整流值 rectified value
整流质量 quality of commutation
整流周期 commutation cycle; switching cycle
整流轴 axis of commutation
整流装置 commutating device; fairing; rectifier; rectifier device; rectifier unit; rectifying device
整流子 collector; commutator; commuter; ring collector; ring head(er)
整流子绑圈 commutator shrink ring
整流子表面 commutator surface
整流子电刷 dynamo brush
整流子毂 commutator hub; commutator shell
整流子耗散 collector dissipation
整流子(夹)环 commutator ring
整流子接线片 commutator lug; commutator riser; commutator tag
整流子绝缘隔片 commutator insulating segment
整流子绝缘圈 commutator insulating ring
整流子磨光机 commutator grinder
整流子片 commutator bar; commutator segment; segment(ation)
整流子片距 segment pitch
整流子片切槽专用压力机 segment notching press
整流子片试验 back-to-back test
整流子频率变换器 commutator frequency changer
整流子式电动机 commutator motor
整流子式电机 commutating machine; commutator machine
整流子式发电机 commutator generator
整流子竖片 commutator riser
整流子套 commutator shell
整流子套筒 commutator sleeve
整流子铜条 commutator bar
整流子用云母片 segment mica
整流作用 rectified action; rectifying action; valve action
整螺距 even pitch
整买零卖 buy wholesale and sell retail
整面锤 dressing hammer
整面法兰 full faced
整面机 finishing machine; paving finisher; finisher
整面宽度 finishing width
整面涂料 surfacer
整模 mo(u)ld preparation; unsplit pattern
整模间 mo(u)ld yard
整模式 integral pattern
整模型 integral mo(u)ld
整年浇筑量 <混凝土的> year-round pours
整年休闲 black fallow
整盘 full disk[disc]
整盘磁带清洗器 bulk eraser
整抛法 uniform placement
整批 even lot; in bulk; lump sum
整批保险 package insurance; wholesale insurance
整批材料 material in bulk
整批成本 bunched cost
整批成本计算法 batch costing
整批出售 sell by wholesale; sell wholesale
整批的 wholesale
整批定价法 batch pricing
整批购买 basket purchase; lump-sum purchase
整批交货 single consignment
整批交易 packaged; packaged deal
整批买卖杂货 job lot
整批设备交易 package deal
整批运费 lump freight
整坯场 conditioning yard; dressing yard
整坯模 buster
整片 full wafer
整片存储器 full wafer memory
整片大体积混凝土岸壁 mass solid wall of concrete
整片路面 <沥青混凝土等的> sheet pavement
整片面层的 solid-faced
整片铺设木板 close-boarded
整片铺设屋面板 close-boarded roof
整片式 full slice system
整片式楼板 solid floor
整片式路面 sheet pavement
整片式码头 full jetty; precast concrete piles wharf with reinforced concrete whole sole deck
整片式面板 solid deck; solid floor
整片瓦 solid tile
整片叶 one-piece blade
整平 averaging out; blading; dubbing out; equating; flatten(ing); level(l)ing(-up); level off; level off flush; pad; planing; regulating; remove the burr; set level; shaping; trimming; tru(e)ing
整平板 finishing screed; level(l)ing screed(material); strike-off blade; bull nose screed; forming plate; smoother bar; screed; cutting screed <喷射混凝土的>
整平标尺 derby float
整平驳 screeding barge
整平层 drag level(l)ing course; level(l)ing layer; level(l)ing(-up) course; level(regulating) course; regulating course; truing course
整平充填料 level(l)ing mass
整平刀板 strike-off blade
整平捣固机 tamping-level(l)ing finisher
整平道路 pad
整平的 leveled off
整平的基底 level(l)ing base
整平地面 grade
整平地面爆破 bulldoze blasting
整平地面的土方工程 finish grading
整平工具 level(l)ing tool; smoother
整平工人 smoother
整平工作 level(l)ing work
整平刮尺 <粉刷用> derby float
整平规板 finishing board
整平合成物 feathering compound
整平横断面 level(l)ing process
整平后的地面高程 finish grade
整平混凝土层 level(l)ing concrete
整平混凝土路面 bumpcut pavement
整平机 evener; finisher; flattening machine; grade trimmer; planer; tongue scraper; bumpcutter (machine) <整修水泥混凝土路用>; mechanical float <整修水泥混凝土路用>; tenderizer <单板的>
整平机械 planing machine; level(l)-

ing machine
整平梁 <混凝土摊铺机的> strike-off beam; level(l)ing beam
整平料堆 even windrow
整平路肩 graded shoulder
整平路面 bumpcut pavement
整平路型 level(l)ing
整平螺丝 <测量仪的> level(l)ing (foot) screw
整平配接 dressed and matched
整平器 aligner; evener; flattening hammer; flush trimmer; level(l)er; smoother; tongue scraper
整平曲线 faired curve
整平圈 level(l)ing ring
整平设备 smoothing equipment
整平术 planification
整平样板 strike-off screed
整平用刮板 planker
整平匀泥尺 strike-off screed
整平者 level(l)er
整平直线器 screed wire
整平直线绳 screed wire
整平准条 strike-off screed
整坡 boning in; grading; sloping
整坡标桩 guinea
整坡杆 boning rod
整坡机 grade builder; slope grader; sloper; slope trimmer
整坡机具 grading outfit
整坡器 back sloper
整坡曲线 grading curve
整坡作业 grading operation; grading work
整铺混凝土轨道 paved concrete track
整齐 regularity
整齐安放 pattern-place
整齐安装的 flush mounted
整齐层砌的毛石(圬工) regulate-coursed rubble
整齐的 orderly; measured; neat; regular; snug; taut; tidy
整齐方石砌筑 regular coursed ashlar stone work
整齐分层毛石砌体 range masonry
整齐花 regular flower; symmetric(al) flower
整齐间距 regular spacing
整齐石料 regular stone
整墙过梁 through lintel
整区 main plot
整圈打火 ring fire
整圈绕法 full coiled winding
整圈绕组 full coiled winding
整全电视信号 signal complex
整群抽样 chester sampling; cluster sampling; full group sampling
整群代数 integral group algebras
整群取样 cluster sampling
整人循环 complete alternation
整绒 <地毯整理> pile setting
整容 face-lift
整容室 manicure lavatory
整石墙 ashlar masonry
整式 integral expression
整式曲柄箱 unsplit crankcase
整饰 dress; finish; retouch
整饰操作 finishing operation
整饰工作 trimming
整饰过的椽 trimmed rafter; trimming rafter
整饰横梁 trimming joist
整饰化合物 finishing compound
整饰抹子 finishing trowel
整售价格 bundled price
整数 integral; rounded figure; integer; integral number; round figure; round number; round sum; whole number
整数倍(数) integer multiple; integral multiple
整数倍谐波 integer harmonics
整数标志 integral denotation
整数表示 integer representation; integral representation
整数部分 integer part; integral part
整数槽绕组 integral slot winding
整数常数 integer constant
整数的因子 factor of an integer
整数点 integral point
整数定律 law of rationality
整数法 integral process; rounding-off method
整数分割【数】 partition
整数共振 integral resonance
整数规划 integer programming
整数规划问题 integer programming problem
整数环 integral ring; ring of integers
整数基底 basis of integers
整数计算机 integral computer
整数角动量 integer angular momentum
整数解 integer solution; integral solution
整数界限 integral boundary
整数可行解 integral feasibility solution
整数理想 integer ideal
整数理想子环 integral ideal
整数马力电动机 integral horsepower motor
整数马力压缩冷凝机组 integral horsepower condensing unit
整数幂 integral power
整数盘 rounding dial
整数收敛 integral convergence
整数系列与零头系列 series and tail series
整数系数 integer coefficient
整数线性规划 integer linear programming; integral linear programming
整数项 integer item
整数形式 integrated format
整数型标准格式 standard format of type real
整数溢出自陷 integer overflow trap
整数釉式 unity formula
整数预算 lump-sum budget
整数值 integer value; integral quantity; integral value
整数属性 integer attribute
整数转换 integer conversion
整数自旋 integer spin; integral spin
整数组 entire array
整双线性型 integral bilinear form
整速轮 flier[flyer]
整台 complete built-up
整套 complement; complete set
整套部件分析 complete component analysis
整套采购 basket purchase; lump-sum purchase
整套产品 whole set of products
整套衬砌工具 lining service group
整套承包估价 turnkey cost estimate
整套承包合同 <包括规划、设计、施工和管理> package job contract; all-in contract; package deal contract; turnkey contract; turnkey type of contract
整套承包建筑 <包设计施工和包工包料> turnkey building construction
整套承包移交 turnkey handover
整套承包住宅建筑 turnkey housing construction
整套的 complete set of
整套的工厂布置图 completed layout
整套的输送带系统 integrate belt system
整套电器材合电缆 string-a-like assembly
整套房间 suite of rooms
整套工程 packaged deal
整套工程项目 packaged deal project
整套工具 kit
整套购买 basket purchase; lump-sum purchase
整套管 solid sleeve
整套合同 contract package; turnkey contract
整套冷热水箱 combination tank
整套滤器 filter package
整套门 door assembly
整套模板 form-set
整套皮带系统 integrated belt system
整套设备 aggregate; complete equipment; complete set of equipment; equipment complex; equipment package; equipment set; packaged equipment; self-contained unit; set of equipment; unit of equipment
整套设计工具 design aids set
整套输送带系统 integrated belt system
整套太阳能加热器 packaged solar heater
整套提单 set of bills of loading
整套维修零件 parts kit
整套卫生设备 combination fixture
整套卫生线路 complete fixture line
整套系统 turnkey system
整套线路 complete line
整套浴室 <设有浴盆、洗脸器、便器等设备的> complete bath
整套轧辊 set of rolls
整套制成单元 complete built unit
整套装在一起的 self-contained
整套装置 aggregate; complete equipment; equipment package; equipment set; package unit; self-contained unit; unit of equipment
整套钻具 rig
整体 bulk; completion; en bloc; en-block; integer; integrality; integral unit; massif; monoblock; monolithic body; single-piece; total integral; whole body
整体V形凿尖凿 solid type diamond bit
整体安装 integral erection; pre-assemblage erection
整体坝块 monolithic block
整体搬运钻机 skid the rig
整体板 integral plate
整体板牙 solid die
整体半深车体侧壁 integral half depth body side
整体保持架 integral cage; solid retainer
整体爆破 bulk blast
整体逼近公式 overall approximation formula
整体比重 bulk specific gravity
整体表面处理 monolithic surface treatment
整体表面压力计 solid front pressure ga(u)ge
整体补强 integral reinforcement
整体部分 integral part
整体采暖系统 integrated heating system
整体参考系 global reference frame
整体参数 lumped parameter
整体参数模拟模型 lumped parameter simulation model
整体参数系统 lumped parameter system
整体舱壁 intact bulkhead; solid bulkhead; unpieced bulkhead
整体操作 integrated operation
整体测图面 integral plotting surface
整体差动保护装置 overall differential protection
整体拆卸 unit removal
整体超前型钻头 integral pilot
整体车轮 monobloc wheel; solid wheel; wheel rim with tyre
整体车轮轧机 solid wheel rolling mill
整体车身 integral body
整体车身罩 integral body shell
整体沉降 bulk settlement; bulk settling; general settlement
整体沉箱结构 caisson monolith construction; monolith caisson construction
整体衬里 monolithic lining
整体衬砌 monolithic lining
整体衬套 slid bearing
整体成型 bulk forming
整体承口接头 integral bell joint
整体承载车体 monocoque body
整体程序设计 integer programming
整体齿锯 solid-tooth saws
整体齿圆锯 solid-plate circular saw; solid-tooth circular saw
整体翅片管 integrally finned tubes
整体冲头 solid punch
整体传输技术 integrated-transfer technique
整体次 global degree
整体代数函数 integral algebraic(al) function
整体带齿旋转活塞 integral gear rotor
整体单元 monolithic unit
整体单元组分 complete assemblage of element
整体单值解 globally univalent solution
整体挡板 integral bumper; solid shield
整体挡料器 solid stop
整体刀具 completion tool; solid tool
整体刀盘 solid cutter
整体道床 concrete bed; monolithic concrete bed; monolithic roadbed; monolithic track bed; solid bed; solid track bed
整体道床结构钢筋 structural reinforcement in monolithic line bed
整体道牙 integral curb
整体的 integral; in bulk; massive; monobloc; monolithic; packaged; self-contained; unsplit
整体的保护罩 built-in guard
整体的斗柄 one-piece stick
整体的横移托架 <平地机、挖掘机> integral sideshift carriage
整体的水泥仓 integrated cement silo
整体的组合齿轮 cluster gear
整体底板 monolithic floor; unsplit bedpiece; unsplit bedplate
整体底架装载机 rigid frame loader
整体底脚 monolithic footing
整体底座 unsplit bedplate
整体地板 solid bottom
整体地沟 integrated channel
整体电动泵 unipump
整体电荷 bulk charge
整体电加热 integral electrical heating
整体垫板 <交叉中主要一股轨道下的长垫板> integral plate
整体垫片 solid shim
整体吊弦 dropper clip
整体吊装 integral hoisting
整体吊装法 integral hoisting process
整体顶层 monolithic topping
整体顶尖 solid center[centre]
整体定子 one-piece stator
整体动臂 one-piece boom
整体端铣刀 solid-end mill
整体锻件 one-piece forging

Z

整体锻件补强 reinforcement by integrated forging piece
整体锻造 solid forging
整体锻造容器上的缺口标志 chip marks on integrally forged vessel
整体锻造式单层圆筒 integrally forged monolayered cylinder
整体锻造辙叉 integrated forged frog
整体锻制钻头 forged bit
整体墩 monolithic pillar
整体多目标法 integrated multipurpose approach
整体舵架 solid rudder frame
整体舵栓 solid pintle
整体法兰 integral type flange
整体方法 <混凝土中加入粉、液或浆以填充孔隙提高防水性能 > integral process
整体方形板牙手柄 stock for square solid die
整体方形管子板牙 square solid pipe die
整体方形螺栓板牙 square solid bolt die
整体防护料 integral proofer
整体防水 <指由混凝土混合料的组成所产生的不透水性 > integral waterproof(ing)
整体防水法 integral method of waterproofing
整体防水剂 integral water-proofer; integral water repelling agent
整体房间冷却器 self-contained room cooler
整体放射自显影术 whole-body autoradiography
整体飞轮 one-piece flywheel; solid flywheel
整体非金属防火地板 monolithic nonmetallic fireproof floor
整体沸腾式反应堆 integral boiling reactor
整体分析 bulk analysis; complete analysis; global analysis
整体封闭型往复式冷水机组 hermetic reciprocating packaged liquid chiller
整体浮船坞 unit-type drydock
整体浮运的格笼 floated-in crib
整体浮运法 buoyant monolith system
整体腐蚀 uniform corrosion
整体复合材料结构 integral composite structure
整体概念 global concept
整体干燥 volume drying
整体杆系式动力转向系统 integral linkage power steering system
整体感 associative perception
整体刚度 global stiffness; integral rigidity
整体钢筋混凝土墙 monolithic reinforced concrete wall
整体钢轮 monoblock steel wheel
整体钢钎 integral drill steel; integral steel
整体钢铸件 integral steel casting
整体更换法 module replacement
整体更换基底 unit replacement basis
整体供电 bulk supply
整体拱 massive arch; rigid arch
整体构造 integral construction
整体估计 global estimation
整体估计方差 global estimation variance
整体估计值 global estimation value
整体骨架的滚子式单向超越离合器 caged roller clutch
整体故障 complete failure
整体观念 concept of wholism
整体灌注式轨道 paved concrete track
整体规划 general plan; integral programming; integrated plan(ning); overall planning
整体规划方案 unified planning project
整体焊接的铲斗 all-welded bucket
整体焊料试件 all weld metal specimen
整体横梁式中耕机 continuous toolbar cultivator
整体横移式门架 integral side-shifting mast
整体护岸 blanket revetment
整体护(轮)轨 one-piece guard rail
整体护圈 solid retainer
整体滑动 bodily sliding; complete sliding; mass movement; mass sliding
整体滑动泥沙 mass movement deposit
整体滑轮制动器 solid disc brake
整体化 integration
整体化方法 integrated approach
整体化管理计划 integrated management planning
整体化交通控制系统 integral traffic control system
整体化冷却系统 integral cooling system
整体化炼厂 integrated refinery
整体化装置 integrated unit
整体混凝土板墙 monolithic concrete panel wall
整体混凝土衬护 monolithic concrete lining
整体混凝土衬砌 monolithic concrete lining
整体混凝土防波堤 concrete monolith breakwater
整体混凝土轨枕 monoblock concrete sleeper
整体混凝土建筑物 monolithic concrete structure
整体混凝土结构 massive concrete structure
整体混凝土桥台 massive concrete abutment
整体混凝土围护 monolithic concrete encasement
整体混凝土支护 monolithic concrete lining
整体混凝土支座 massive concrete abutment
整体活塞 single-piece piston
整体机身式压力机 single-piece frame press
整体机外壳 unsplit casing
整体机组 packaged unit
整体机组设计 monobloc design
整体机座 unsplit frame
整体积分 integral
整体基础 block foundation; integral basis; mass-type foundation; mat foundation; monolithic footing
整体基础的码头 belled-out pier
整体及分组式槽边侧吸罩及调节阀 integral and unit lateral exhaust hoods and damper
整体集成电路 monolithic integrated circuit
整体计划 comprehensive plan(ning); integrated plan(ning)
整体计算系统 monolithic computing system
整体加热的 integrally heated
整体夹板 full plate
整体夹套 conventional jacket
整体剪切破坏 general(ized) shear failure
整体建筑法 monolithic construction method
整体渐近稳定 global asymptotic stability
整体交通控制 full control of traffic
整体交易污水处理装置 package deal sewage-treatment plant
整体浇的 monolithically concreted; monolithically poured
整体浇灌 integral cast(ing)
整体浇灌楼板 monolithic floor
整体浇灌墙 monolithically cast wall
整体浇注 integral casting; single-piece casting
整体浇注衬砌 monolithic lining
整体浇注的 <混凝土> monolithically poured
整体浇注混凝土 monolithic casting concrete; monolithic concrete
整体浇筑 integral casting; integrally cast
整体浇筑的 cast integral; monolithic; unit-cast
整体浇筑的混凝土结构 monolithic concrete construction
整体浇筑的梁 monolithic beam
整体浇筑混凝土 monolithic concrete
整体浇筑混凝土巨型块体防波堤 monolith concrete block breakwater
整体浇筑混凝土施工法 monolithic concrete construction
整体浇铸 integral cast(ing)
整体浇铸件 one-piece casting
整体接头 unitized joint
整体结构 complete texture; en-block construction; full trailer; integral structure; massive construction; massive structure; massive texture; monolithic structure; one-piece construction; overall structure; structural integrity; unit construction
整体结构船 monolithic ship
整体结构管理 non-figurative management
整体结构模 unit construction mo(u)ld
整体结构模型 complete structure model; unit construction model; whole-structure model
整体结构桥 unit construction bridge
整体结构系统 monolithic structural system
整体结合面层 integral facing
整体结晶器 solid-block mo(u)ld
整体截面 monolithic section
整体截影 global section
整体金刚砂钻头 solid type diamond bit
整体金属型 one-piece die
整体近似 global approximation
整体进气歧管 integral intake manifold
整体浸渍绝缘 mass-impregnated insulation
整体经济潜力 overall economic potential
整体精度 overall accuracy
整体精矿 bulk concentrate
整体聚合物 block polymer
整体绝热层 concentrate insulation cover
整体绝缘 integral insulation
整体开关 global switch
整体空调机 packaged air conditioner
整体空调器 packaged air conditioner
整体控制 integral control; master control
整体控制系统 monolithic control system
整体控制作用 integral control action
整体块 monolithic block
整体块状结构 complete massive texture
整体宽度 overall width
整体矿车 solid car
整体矿柱 rib
整体扩孔钻 solid-core drill
整体拉刀 solid broach
整体拉伸 solid-drawn
整体拉制 solid-drawn
整体拉制的 solid-drawn
整体冷却器 bulk cooler
整体离心式电泵 monobloc centrifugal electro-pump
整体黎曼对称空间 globally symmetric(al) Riemannian space
整体理论 global theory
整体利益 overall benefit
整体连接 monolithic joint
整体连接水管 integral joint tubing
整体连接套管 integral-joint casing
整体连墙基础板 monolithic slab
整体联轴器 integral coupling
整体料 monolith
整体流域规划 basin-wide programme)
整体炉衬 monolithic lining
整体炉床 monolithic furnace hearth
整体路缘(石) <和路面结合在一起的路缘石 > integral curb; integrate(d) curb
整体逻辑视图 over logical view
整体螺丝攻 solid tap
整体螺旋 one-piece auger
整体螺旋桨 solid propeller
整体锚 ordinary anchor; solid anchor
整体帽螺母 integral cap nut
整体镁矿炉底 monolithic magnesite furnace bottom
整体门挡 solid stop
整体门框 integral frame
整体门框碰头 solid stop
整体密封罩 integral enclosure
整体面层 monolithic floor surface; monolithic topping; integral facing
整体面层处理 monolithic surface treatment
整体模板 integral formwork; solid form; solid formwork
整体模具 solid mo(u)ld(ing); block mo(u)ld; one-piece pattern; single pattern; solid die; solid pattern; unit mo(u)ld
整体模塑 solid mo(u)ld(ing)
整体模型 block mo(u)ld; comprehensive model; entire model; lumped model; overall model; solid mo(u)ld(ing); thorough model
整体模型方法 integrated model approach
整体模型试验 comprehensive model test; overall model test; three-dimensional model test
整体模型研究 integrated model approach
整体模压 solid mo(u)ld(ing)
整体模造型 solid pattern mo(u)lding
整体模制 solid mo(u)ld(ing)
整体模铸模具 unit-type die-casting die
整体母合金 integral hardener
整体目标 overall objective; total goal
整体内圈 integral inner ring
整体耐火材料 cast refractory; monolithic refractory
整体耐酸地面 jointless acid resistant flooring
整体黏[粘]结 overall bonding
整体耦合 unity coupling
整体配子 hologamy
整体皮带轮 solid belt pulley
整体皮带轮制动器 solid pulley brake
整体皮带系统 integrated belt system
整体平差【测】 adjustment in one cast; overall adjustment

整体平行分布 parallel distribution in the large
整体平移式滑坡 slab and block slide
整体屏蔽 bulk shield
整体屏蔽堆 bulk shielding reactor
整体屏蔽装置 bulk shielding facility
整体破坏 <桩工> block failure
整体气缸座 monoblock cylinder block
整体气洗 integral air-scour
整体汽缸 block head cylinder;unspilt casing
整体汽缸发动机 monoblock engine
整体汽缸座 monoblock
整体砌块 monolith
整体砌筑码头 masonry quay wall
整体钎 chisel edge steel
整体钎杆 chisel-bit steel;integral drill steel
整体钎钢 <焊有钎头的> stem
整体钎子 integral steel; monoblock drill steel;solid drill steel
整体强度 overall strength
整体墙 intact wall;mass wall;monolithic wall;single-leaf wall
整体桥壳 integral housing
整体侵蚀 mass erosion
整体曲柄 solid crank
整体曲线论 global theory of curves
整体取样法 integral sampling method
整体全深车体侧壁 integral full depth body side
整体燃料箱 integral fuel cell
整体热处理 bulk heat treatment
整体人格 total personality
整体熔模铸造法 block mo(u)ld process
整体散热筋 integral fin
整体砂箱 rigid flask
整体上色的 integrally colo(u)red
整体设计 integrate(d)design
整体渗透试验 <混凝土的> bulk diffusion test
整体施工 monolithic construction
整体实验 integral experiment
整体式 integral side shift;integral type;package type;single-block type
整体式安全栅顶 integral overhead guard
整体式凹模 solid die
整体式坝 monolithic dam
整体式板和基础墙 monolithic slab and foundation wall
整体式表带 integral armband;integral bracelet
整体式侧平石 combined curb-and-gutter
整体式车架 unit frame
整体式车架载重汽车 rigid truck
整体式沉箱 monolithic caisson
整体式沉箱结构 caisson monolith construction
整体式衬砌 cast-in-place lining;construction lining; integral lining;monolithic lining
整体式传动装置 integral drive;integral transmission package
整体式船坞墙 monolith type dock wall
整体式挡土墙 monolithic retaining wall
整体式挡墙 monolithic wall
整体式道床 integrated ballast bed
整体式的 monolithic;one-piece
整体式的循环泵 integral circulating pump
整体式底脚 massive footing
整体式电动泵 unipump;unit pump
整体式电动机 integrated motor
整体式阀 build-in valve
整体式防滑条 <浇筑前嵌入楼梯踏板的防滑条> integral abrasive edging
整体式防火地板 monolithic fireproof floor
整体式风扇 integral fan
整体式缝 <填充后形成的> monolithic joint
整体式浮(船)坞 one-piece floating dock
整体式浮悬地板 integral floating floor
整体式钢筋混凝土 monolithic reinforced concrete
整体式钢筋混凝土结构 monolithic reinforced concrete structure
整体式格床 construction grillage;monolithic grillage
整体式拱 monolithic arch
整体式构件 whole section member
整体式构造 monolithic construction
整体式构筑物 monolithic structure
整体式焊接结构 all-welded unitized construction
整体式混凝土 monolithic concrete
整体式混凝土轨枕 monolithic concrete sleeper;monolithic concrete tie
整体式混凝土建筑 monolithic concrete construction
整体式混凝土结构 monolithic concrete construction
整体式混凝土路面板 full-width slab
整体式机架 unit frame
整体式基础 monolithic foundation
整体式减压阀 integral relief valve
整体式建筑 monolithic construction
整体式接头 monolithic joint
整体式结构 massive footing;monolithic structure; monolithic construction
整体式抗爆结构 monolithic blast protection structure
整体式壳体 integral housing; one-piece housing
整体式客车 coach with integral body
整体式空调调节机组 unitary air conditioner
整体式空气调节机组 self-contained air conditioning unit
整体式空气调节器 package air conditioner;self-contained air conditioner;unitary air conditioner
整体式励磁器 solid-state exciter
整体式炼砖结构 monolithic brick pavement
整体式炼砖路面 monolithic brick pavement
整体式梁 monolithic beam
整体式楼梯 solid stair(case)
整体式路缘(石)monolithic curb
整体式轮毂轴承 integral hub bearing
整体式码头 monolithic type wharf;monolith quay
整体式码头岸壁 monolith quay wall
整体式码头岸墙 monolith quay wall
整体式耐磨镶边 integral abrasive edging
整体式凝气器 integral condenser
整体式喷油器 unit injector
整体式偏心轴 <多缸转子发动机用的> integral one-piece mainshaft
整体式平衡重 integral counterweight
整体式(汽)缸体 monoblock cylinder
整体式钎杆 forged bit
整体式钎头 integral bit;integral drill bit
整体式桥壳 banjo axle
整体式桥台 integral bridge abutment
整体式切割头 one-piece cutter head
整体式曲轴平衡重 integral counterweight of crankshaft
整体式曲轴箱 unsplit crankcase
整体式驱动装置 integral drive
整体式取土器 <取土管不能分开的> solid-barrel sampler
整体式三维框架 integrated three-dimensional frame
整体式射钉枪 integrative nail gun
整体式升降臂 solid lift arm
整体式水泥趸船 monolithic reinforcement cement pontoon
整体式水套 integral water jacket
整体式水箱 integral tank
整体式司机室 cab module; driver module
整体式探头 integral point
整体式天花板 integrated ceiling
整体式挖沟装置 integral type trenching element
整体式圬工墙 monolithic masonry wall
整体式圬工闸室 monolithic masonry lock chamber
整体式屋顶冷风机组 unitary air conditioner for rooftop mounting cooling
整体式线圈 integrated coil
整体式箱形货车 root van
整体式箱形基础 monolithic box foundation
整体式消声器 integral muffler
整体式压力容器 solid wall pressure vessel
整体式饮水冷却器 self-contained drinking water cooler
整体式预应力混凝土轨枕 monoblock prestressed concrete tie
整体式闸底 monolithic chamber floor
整体式找平层 monolithic screed
整体式蒸汽锅炉 package-type steam generator
整体式主机架 unitized main frame
整体式柱 monolithic column; monolithic pillar
整体式转轮 one piece runner
整体式综合吊顶 integrated ceiling
整体式钻杆 monobloc type drill rod
整体式钻头 integral drill bit
整体式坐便器 one-piece water closet
整体试验 bulk testing;integral testing
整体饰面 monolithic finish;monolithic surface treatment
整体收敛 global convergence
整体收敛性定理 global convergence theorem
整体数据处理 integrated data processing
整体数据区 global data area
整体双级滤清器 two-stage combination filter
整体水磨石 monolithic terrazzo
整体顺坡滑动 mass wasting
整体死亡 somatic death
整体速度 bulk velocity
整体塑料踢脚 jointless plastics skirting
整体锁 integral lock
整体碳化钨钻头 integral tungsten carbide bit
整体套管 solid sleeve
整体特性 mass property
整体特征 bulk property
整体踢脚线 jointless skirting
整体调整作用 integral controlled action
整体通信[讯]衔接器 integrated communication adapter
整体凸轮轴 integral cam
整体凸缘 integral type flange
整体退火 integrally annealed
整体托圈 integral ring
整体外壳 integral case; monolithic case;unspilt casing
整体外圈 integral outer ring
整体微分几何 global differential geometry
整体围带 integral cover;integral(tip) shroud
整体围堰 box dam
整体维数 global dimension
整体卫生间 integrated bathroom
整体温度 bulk temperature
整体文件存储衔接器 integrated file adapter
整体稳定 monolithic stability;overall stability
整体稳定性 monolithic stability;overall stability; resistance to overturning
整体稳定性分析 general stability analysis
整体稳定性计算 calculation of stability;computation of stability;general stability computation
整体圬工 monolithic masonry
整体屋顶 monolithic roof
整体屋顶结构 monolithic roof structure
整体误差 global error
整体系统 single-mass system;unitary system
整体先张法 monolithic pretensioning
整体线性规划 integer linear programming
整体相互作用分析 complete interaction analysis
整体橡胶轮胎式压路机 solid rubber-tire roller
整体消磁器 bulk eraser
整体效率 overall efficiency
整体效应 blocking effect; global effect
整体心轴 solid mandrel
整体芯板 solid core
整体芯盒 one-piece core box
整体型空调器 self-contained air conditioner
整体型空调装置 self-contained air conditioning unit
整体型转向加力装置 integral type power steering
整体性 globality; integrality; integrity;massivity;entirety
整体性能 massive character
整体性试验 <使用前对系统的> operational test
整体性质 blocking property; bulk property;global property
整体修饰 monolithic finish;monolithic patch
整体修整 monolithic finish
整体选择 global selection
整体学习 global learning
整体压缩机 integral compressor
整体压型 single cavity die
整体岩块 rock massif
整体岩石 bulk rock;rock bulk;rock massif
整体岩芯 solid drill core
整体岩芯采取率 solid-core recovery
整体颜色 integral colo(u)r
整体堰 solid weir
整体阳极色彩处理 integral colo(u)ring anodizing[anodising]
整体阳极上色程序 integral colo(u)ring anodizing process
整体养护 <密封容器内绝热养护> mass curing
整体样品 bulk sample
整体窑 integral kiln
整体叶片 integral blade
整体仪控制交通 integral(traffic) control system

整体移动 bodily movement; mass movement
整体移动带 whole shift zone
整体移动钻机 skid the rig
整体映射 global mapping
整体硬度 through-hardness
整体硬化剂 integral hardening agent
整体硬质合金刀具 solid carbide
整体油路板 solid manifold
整体预应力混凝土轨枕 monoblock prestressed concrete sleeper
整体预张拉 monolithic pretensioning
整体预制单元 monolithic unit
整体元件电路 integrated component circuit
整体圆盘制动器 solid disc brake
整体运动 bodily movement; mass motion; bulk movement
整体运输网 integrated transportation network
整体运转工作 integrated operation
整体造船法 cast-in-situ shipbuilding
整体轧制 solid roll
整体展望 total view
整体照明开关 integral lighting switch
整体支腿式液压挖掘机 whole-body outrigger hydraulic excavator
整体支柱 monolithic pillar
整体支座 solid bearing
整体止动器 solid stop
整体轴 integral shaft
整体轴承 integral metal; solid bearing; solid pedestal
整体轴承箱体 integral bearing housing
整体轴架 plain pedestal
整体轴颈轴承 solid journal bearing
整体轴心 solid mandrel
整体主义 holism; organicism
整体柱身 monolithic column shaft
整体铸钢件 integral steel casting
整体铸件 monobloc casting; one-piece casting
整体铸铁碟形封头 integral cast-iron dished head
整体铸型 block mo(u)ld; monolithic mo(u)ld; one-piece mo(u)ld; single cavity mo(u)ld; solid mo(u)ld(ing)
整体铸造 en-block cast; monoblock cast(ing); unit-cast
整体铸造的 inblock cast
整体(铸造的)凸轮轴 integral cam shaft
整体铸造构件 inblock-cast member
整体爪 solid jaw
整体转筒 rigid rotor
整体转子 integral rotor; monoblock rotor; mono-rotor
整体桩 solid pile
整体装饰线条 <车制的> solid mo(u)ld(ing)
整体装置 packaged unit; self-contained system; single-unit system
整体锥式 solid cone
整体着色 integral colo(u)ring
整体自撑结构 integral self supporting structure
整体自动化 integrated automation
整体组合式减振器 integral bumper
整体钻杆 solid drill steel
整体钻杆接头 integral tool joint
整体钻探设备 <绞车、转盘、泵及发动机装在一平板上的> consolidated rig
整体钻头 solid drill
整体最佳化 global optimization; total system optimization
整体最佳利益 global optimization
整体最优化 total system optimization; global optimization
整体作用 integral action
整体坐标 global coordinate
整体坐标系 blocking coordinate system; global coordinate system
整天 solid day
整天工数 clear days
整天天数 clear days
整条机 set frame
整条输送机 single-flight conveyer [conveyor]
整帖装订 even working
整土 soil preparation
整位数 round figure
整屋(水管)锅炉 box header boiler
整系数 integral coefficients
整线脚 solid mo(u)ld(ing); struck mo(u)lding
整线圈的 whole coil
整线圈改变极数绕组 whole-coiled pole-changing winding
整线性映射 integral linear mapping
整相器 phase modifier
整相信号 phasing signal
整箱货 container load; full container load; full load
整箱集装箱的荷载 full container load
整箱集装箱货运价 container load rate
整新 make good; renewing
整形 coining; regulating; reshaping; sizing; true-up
整形板 <滑模铺路机的> conforming plate
整形棒 dressing stick
整形标本树 sheared specimen
整形部分 shaping unit
整形草坪 formal lawn
整形的 fairing
整形独立树 sheared specimen
整形锻压 restriking
整形放大器 shaping amplifier
整形分频器 shaper-divider
整形浮板 finishing float
整形钩 plastic hook
整形果树 trained fruit tree
整形花坛 formal flower bed
整形机 shaper
整形技术 shaping technique
整形加工 finish bottoming
整形晶体 idiomorphic crystal
整形块 dressing stick
整形滤波 shaping filtering
整形滤波器 shaping filter
整形路肩 graded shoulder
整形脉冲 shaped pulse
整形模 sizing die
整形镊 plastic forceps
整形刨 truing plane
整形器 dresser; form dresser; reshaper; shaper; truer
整形曲线 fair curve
整形树 shaped tree; trimming tree
整形树木园 topiary garden
整形套管 drifted casing
整形土路 graded earth road; shaped earth road
整形挖泥 configuration dredging
整形挖泥船 profile dredge(r)
整形外科【医】 plastic surgery; orthopedics
整形外科医院 plastic surgery hospital
整形网络 shaping network
整形铣刀 Ingold fraise
整形信号 reshaping signal
整形修剪 pruning
整形修理 form correction repair
整形修整装置 tur(e)ing unit
整形压力机 trimming press
整形艺术 topiary art
整形栽培 form culture
整形种植 architectural planting
整形作用 shaping operation
整型 integer type; integral form; shaping; truing; blading
整型变量 integer variable
整型变量名 integer variable name
整型变量维数 integer variable dimension
整型变量引用 integer variable reference
整型表达式 integer expression
整型量 integer quantity
整型路肩 graded shoulder
整型树 topiary tree
整型数据 integer data
整型土路 graded earth road
整型向量 integer vector
整型形式 integer form
整型压实 blading compaction
整体钎杆 monobloc drill steel
整修 conditioning; dressing; reconditioning; refit; refurbishment; renovation; repairing; touch up
整修岸坡 bank sloping
整修边 <钢板的> finished edge
整修表面 reface; refinishing
整修车间 trimming shop
整修刀 back iron; finishing knife
整修费 <旧机器出售时> make ready cost
整修工 fitter
整修工场 trimming shop; trimming yard
整修工程 modification works
整修工具 dresser; truing tool; refacer
整修工作 reparation
整修过的 dressed
整修花斑 spot finishing
整修货车 rebuilt truck
整修机 mechanical finisher
整修计划 refurbishment program(me)
整修路型 light blading
整修螺孔用丝锥 threading tool chaser
整修码头 refitting quay
整修门面 reface; refacing
整修模 shaving die
整修磨轮 dress grinding wheels
整修平底船 punt
整修平截面 refinishing a flat section
整修平屋顶 redeck
整修铺用碎石 trim ballast
整修墙面 refacing
整修十字形钻头用的对角型铁 diagonal cut dolly for plus bits
整修塑料油漆 plastic dressing paint
整修台 dressing table
整修外表 refacing
整修圬工表面 regrading masonry surfacing
整修铣刀 trimming bit
整修压力机 shaving press
整修余量 shaving allowance
整修运河坡的机器 canal slope trimming machine
整修正 refacer
整修钻头用型铁 dolly for dressing drill bits
整序函数 well-founded function
整序集 well-founded set
整序集法 well-founded-set method
整选择模式 integral choice pattern
整压 seamless
整夜存车 overnight parking
整夜停车 overnight parking
整因子 integral divisor
整有理不变式 integral rational invariant
整有理函数 entire rational function; integral rational function
整有理算子 integral rational operator
整域 integral domain
整元会计 cents-less accounting; whole-dollar accounting
整圆夹板的表 full-plate watch
整圆盘 full disk[disc]
整圆转子 round rotor
整轧车轮 one-piece wheel
整轧轮心 solid rolled center[centre]
整轧坯 solid billet
整闸器【机】 brake adjuster
整张屋面板 individual shingle
整(整几)天 clear days
整正曲线 curve adjusting; curve lining
整正水平 adjusting of cross level
整枝法 <用于园艺> training
整枝剪 secateurs
整直的 orthotic
整直法 straightening; orthosis
整直轨辊 smoothing roll
整直滚板机 plate straightener; plate-straightening machine; plate-straightening roll
整直装置 straightening device
整值多项式 integral valued polynomial
整指数 integral exponent
整治 harnessing; improvement; regulation; training
整治标准 regulation standard
整治措施 corrective measure; regulating measures
整治的河槽 rectified channel
整治的河道 rectified channel
整治的河流 regulated river
整治的水道 improved channel
整治方案 regulation alternative; regulation scheme
整治工程 correction works; improvement works; regulating works; regulation works; regulatory works; rehabilitation works
整治工作 control work
整治航道 dredge waterway
整治河道 adjusted river; realignment
整治河流 regulated stream
整治建筑物 regulating structure; regulating works; regulation structure; regulation works; regulatory works; training structure; training works; rectification structure
整治开挖线 excavation line for regulation
整装空调机 unitary air-conditioner
整治宽度 regulated width
整治了的水道 rectified channel; regulated channel
整治流量 regulation discharge
整治设计 regulation design
整治设施 improvement works; regulation facility; regulatory works
整治水道 training course
整治水位 regulated water stage; regulation stage
整治线 diversion line; regulating line; regulation line
整治线布置 alignment of regulation line; layout of regulating line; layout of regulation line
整治线宽度 regulation width; width of regulating line; width of regulation line
整治性疏浚 corrective dredging; improvement dredging
整周焊缝 all-around weld
整周进水式水轮机 full admission turbine
整周期 complete alternation
整柱 undivided shaft
整柱石 milarite
整铸 block cast; integral cast

整铸电动机 bloc cast engine; block cast motor

整铸缸体 cast cylinder

整铸件 one-piece casting

整铸接头 integral joint

整铸锰钢辙叉 cast manganese steel frog; cast solid manganese steel frog; solid manganese steel crossing; solid manganese steel frog

整铸双面型板 cast-plate

整铸套箱 cast jacket

整铸转轮 cast integrally runner; single-piece cast runner

整砖 four quarters; full clay brick; full-sized brick; whole brick; whole tile

整砖砌拱顶 solid brick vault

整砖砌合 laying to bond

整砖墙 whole-brick wall

整装泵 monoblock pump

整装抽气泵 integral air pump

整装电源机组 power package

整装锅炉 packaged boiler; self-contained boiler

整装货柜 <整装集装箱> full container; load container

整装机组 packaged unit

整装集装箱 load container

整装胶片 bulk film

整装空调机 packaged air conditioner

整装门锁 unit lock

整装式变电所 packaged substation

整装式空气调节 self-contained air conditioning

整装式空气调节器 unit air-conditioner; unitary air conditioner

整装仪表 self-contained instrument

整装运输系统 unit road system

整装(在底板上)的 self-contained

整组照明系统 group lighting system

正

Y 形三通管接头 true Y branch

正埃尔米特算子 positive Hermitian operator

正八面体 octahedron; regular octahedron

正斑花岗岩 invernite

正斑结构 orthophyric texture

正斑状的 orthophyric

正半定矩阵 positive semi-definite matrix

正半定形式 positive semidefinite form

正半轴 positive axis

正伴流 forward wake

正北 due north

正本 original; original copy; principal edition; progenitor; reserved copy; script

正本单据 original documents

正本提单 original bills of lading

正本条款 original or authentic clause

正比传播 normal propagation

正比电离室 proportional ionization chamber

正比计数管 proportional counter tube

正比计数器 direct counter; proportional counter

正比检测器 proportional detector

正比(例) direct proportion(ality); direct ratio

正比例水表 proportional type meter

正比例效应 direct proportional effect

正比量水堰 <流量与水头成正比的> Sutro measuring weir

正比区 proportional band; proportional region; region of proportionality

正比探测器 proportional detector

正比堰 <流量与水头成正比的> proportional weir

正比遥测计 direct relation telemeter

正比于荷重的制动装置 load-proportional braking equipment

正闭包 positive closure

正边 face side

正边界 positive boundary

正边玄武岩 absarokite

正边沿触发时钟脉冲 positive edge clock

正变分 positive variation

正变位 normal shift

正变压中心【气】 isallobaric high

正变质岩 orthometamorphite; orthorock

正变质作用 ortho metamorphism

正标准燃料 primary reference fuel; primary standard fuel

正表 main schedule

正表笔 positive lead

正表面电荷 positive surface charge

正丙苯 n-propylbenzene

正波 positive wave

正波功率使用容量 sine-wave power-handling capacity

正步 positive step

正部分 positive part

正餐 dinner

正残积景观 truly eluvial landscape

正残余结构 ortho-relict texture

正槽 proper channel

正侧面填角焊缝 front fillet weld; front side fillet weld

正测的 canonic(al)

正测度 positive measure

正层型 holostratotype

正铲 crowed shovel

正铲铲刀 shovel blade

正铲铲斗 <挖掘机> shovel bucket

正铲铲斗车 front-end loader

正铲斗卸料角 <挖掘机> shovel bucket dump angle

正铲反铲通用铲斗 reversible bucket

正铲式挖掘 shovel type excavation

正铲式挖掘机 front bucket; shovel-(l)ing machine

正铲推土机 straight dozer

正铲挖掘机 crane shovel; crowd shovel; face shovel; forward shovel; front(-end) shovel; front loading excavation; ; dipper shovel

正铲挖掘机附件 crowd shovel attachment

正铲挖掘机配件 crowd shovel fitting

正铲挖泥船 floating face shovel

正铲挖土机 face excavator; face shovel; forward excavator; forward shovel; front shovel; skid shovel

正铲挖土机铲斗 forward shovel bucket

正铲挖土机斗 face shovel bucket

正铲挖土机附件 face shovel attachment; forward shovel attachment

正铲挖土机配件 forward shovel fitting

正铲挖土机装置 face shovel fitting

正铲挖土位置 face shovel position

正铲型挖掘机 shovel

正铲液压挖掘机 hydraulic front excavator

正铲装置 <挖掘机> shovel attachment; shovel equipment

正长白榴玄武岩 orthoclase leucite basalt

正长斑岩 orthoclase porphyry; orthophyre; syenite porphyry

正长粗面玄武岩 orthoclase trachybasalt

正长花岗岩 orthogranite; syenitic granite

正长辉长岩 syenogabbro

正长脉岩 arizonite

正长闪长岩 orthoclase diorite

正长石 kalifeldspath; orthoclase; orthoclase feldspar; orthose; pegmatolite; sinaite

正长石化 orthoclasization

正长石正长岩 orthoclase syenite

正长伟晶岩 syenitic pegmatite

正长细晶岩 syenitic aplite

正长霞石玄武岩 orthoclase nepheline basalt

正长玄武岩 orthoclase basalt

正长岩 sienite

正长岩类 syenite group

正长岩小方石铺路 syenite sett paving

正常 pH 值 normal pH value

正常安全超高 normal safety freeboard

正常按钮 regular button

正常办理手续 normal procedure

正常包裹体 normal inclusion

正常饱和曲线 normal saturation curve

正常报酬 normal return

正常报废 normal abandonment; normal retirement

正常曝光 constant exposure; normal exposure

正常背景 normal background

正常比降 regime(n) gradient

正常比例混合气 normal mixture

正常比速混流式水轮机 normal specific speed Francis turbine

正常闭合的 normally closed

正常闭合轨道电路 closed track circuit

正常闭合接点 normally closed contact

正常编码 normal encoding

正常变化 regular change

正常变量 normal variable

正常标识符 normal identifier

正常标准成本 normal standard cost

正常表示 normal indication

正常波 ordinary wave

正常波包函数 normal packet function

正常波分量 ordinary-wave component

正常波痕 normal ripple mark

正常波基面 everyday wave base

正常不感潮海滩 normal tideless beach

正常布格梯度 normal Bouguer gradient

正常材料 orthodox material

正常彩色视觉 normal colo(u)r vision

正常参数 normal parameter

正常操作 error free operation; normal operation; regular service; trouble-free operation

正常操作水平 normal operating level

正常操作条件 proper operating condition

正常操作温度 normal running temperature

正常层次 normal gradation

正常层序 normal sequence; normal straight sequence; normal stratification; normal stratigraphic sequence; normal succession; normal superposition

正常差额 normal balance

正常差异 unbias(s)ed variance

正常掺气量 normal aeration

正常产量 normal output

正常产卵 normal brood

正常长度 normal length

正常场 normal field

正常场校正 normal field reduction

正常超球面 proper hypersphere

正常超载 normal overload

正常潮 ordinary tide

正常潮气 normal moisture

正常沉积 normal sedimentation

正常成本 normal cost

正常成本标准 normal cost standard

正常成分 normal component

正常程序 regular program(me)

正常吃水 normal load draught

正常持水量 normal moisture capacity; normal water capacity

正常持续海上运动功率 normal continuous sea-service rating

正常尺寸 just size; normal size

正常尺寸筛分曲线 grading curve representation in normal scale

正常尺度分级曲线 grading curve representation in normal scale

正常尺度级配曲线 grading curve representation in normal scale

正常齿 normal tooth

正常齿距 full pitch

正常充电电流 normal charging current

正常充电时间 normal charging period

正常充气轮胎 proper inflated tire

正常充气压力 normal inflation

正常冲蚀 normal erosion

正常冲刷 normal scour

正常冲刷周期 normal erosion cycle; normal scour cycle

正常抽验 normal sampling inspection

正常抽样检验 normal sampling inspection

正常稠度 normal consistence[consistency]

正常出力 normal output

正常储备 basic stock; normal reserves

正常储量 normal storage capacity

正常川流 normal streamflow

正常传播 normal propagation

正常传播时间曲线 normal travel time curve

正常传输 regular transmission

正常垂线 normal plumb line

正常磁场 normal magnetic field

正常磁导率 cyclic(al) permeability; normal permeability

正常磁感应 normal magnetic induction

正常磁感应强度 normal induction; normal induction strength

正常磁化曲线 magnetization curve; normal magnetization curve

正常磁密度 normal magnetic flux density

正常磁通分布 normal flux array

正常磁照图 normal magnetogram

正常次序 regular turn

正常存量法 normal stock method

正常大潮 normal spring tide; ordinary spring tide

正常大气归算 free air reduction

正常大气压 normal atmosphere

正常大气压力 normal atmospheric pressure; standard atmospheric pressure

正常大气压力下的空气 free air

正常带载操作 normal on-load operation

正常单据 normal documents

正常导磁率 cyclic(al) permeability

正常的 normal; normative; off-peak; regular

Z

正常的城市 health city
正常的基本时间 normal elemental time
正常的结晶顺序 normal order of crystallization
正常的使用条件 regular service condition
正常的通风线路 regular ventilating circuit
正常的因素 regular element; repetitive element
正常等位面 normal spheropotential
正常低潮水位 normal low water; normal low water level
正常低低潮 lowest normal low water
正常地层序(列)【地】normal stratigraphic(al) sequence
正常地层压力 normal formation pressure
正常地磁场 normal magnetic field
正常地磁记录仪 normal magnetograph
正常地电图 normal tellurigram
正常地理坐标 normal geographic(al) coordinate
正常地球 normal earth
正常地球位 normal geopotential; sphero(potential)
正常地球位面上的正常重力 normal gravity on the earth spheropotential
正常地球位数 normal spheropotential number
正常地下潜流 normal base flow
正常地下水亏耗 normal groundwater depletion
正常地下水位 normal groundwater level
正常地震 normal earthquake
正常点亮的信号灯 normally lighted signal lamp
正常电场 normal electric(al) field
正常电导 normal path
正常电流 normal current
正常电压 normal voltage
正常电压下的额定值 normal-voltage rating
正常电子隧道效应 normal electron tunnel(l)ing
正常电阻 normal resistance
正常叠加 normal superposition
正常定时 normal timing
正常定向 normal orientation
正常动高 normal dynamic(al) height
正常动力高 normal dynamic (al) height
正常动作程序 normal sequence of operation
正常动作继电器 normal acting relay
正常渡线 normal crossover
正常断层 normal (-slip) fault; slump fault; gravity fault
正常断开的 normally open
正常断开电路 normally open circuit
正常断开方式 normal disconnected mode
正常断开接点 normally open (ed) contact
正常断裂 normal breaking
正常断面 normal cross section
正常煅烧 normal burning
正常锻造程序 normal forging sequence
正常多变压缩 normal polytropic compression
正常额定功率 normal rated power
正常额定值 normal rating
正常鲕 normal ooid
正常二次曲面 proper quadric
正常发散的 properly divergent
正常发散级数 properly divergent series
正常发散序列 properly divergent sequence
正常帆 plain sail
正常反射率 regular reflectance
正常反跳 ordinary chattering
正常反应 normal reaction
正常反应性 orthocrasia
正常范围 normal range; normal reach
正常方式 normal mode
正常方位图像 normal azimuth picture
正常方向 normal direction
正常放大率 normal magnification
正常放电 normal discharge; regular discharge
正常放牧 normal grazing
正常飞行 normal flight; normal flying
正常飞行角 normal flying angle
正常飞行路线 normal flight-line
正常沸点 normal boiling point
正常费用 natural expenses; normal cost; regular fee
正常分布 normal distribution; proper distribution
正常分界线 normal divide
正常分配率 normal distribution rate
正常分散 normal dispersion
正常分水岭 normal divide
正常丰度 normal abundance
正常风险 natural risk
正常服务 normal service
正常符号 proper symbol
正常辐射 normal radiation
正常负荷 normal duty; normal load; regular burden(ing)
正常负荷机油 normal duty oil
正常负荷率 normal burden rate
正常负荷试验 running test
正常负荷双联扒式分级机 normal duty duplex rake classifier
正常负荷状况 usual loading condition
正常负载 normal load
正常负载的 normally loaded
正常负载期 off-peak period
正常复背斜 normal anticlinorium
正常复正交群 proper complex orthogonal group
正常改正 normal correction
正常甘汞电池 regular calomel cell
正常甘汞电极 regular calomel electrode
正常感觉界限 normal threshold of feeling
正常感觉阈限 normal threshold of audibility; normal threshold of feeling
正常感应 normal induction
正常钢 normal steel
正常高 normal height
正常高潮位 normal high tide level; normal high water (level); normal maximum level; normal top-water level
正常高水位 normal high tide level; normal high water (level); normal maximum level; normal top-water level; retention water level
正常格式通信[讯] normal form message
正常隔热容器 normally insulated container
正常给进 course feed; regular feed
正常耕作 conventional till
正常工况 running at normal level
正常工时 normal hour
正常工资 normal wage
正常工作 normal work(ing); regular work
正常工作的回转窑 unforced rotary kiln
正常工作点 normal working point
正常工作电流 running current
正常工作电压 normal working voltage
正常工作荷载 normal working load; proper working load
正常工作开支 normal operating expenses
正常工作频率 normal working frequency
正常工作情况 normal operation condition; normal working state
正常工作情况下 medium duty
正常工作日 normal working day
正常工作时间 constant failure-rate period; normal operating period; normal working time; ordinary working hours; up-time; good time
正常工作条件 normal running condition; normal working condition; regular service condition; usual service condition
正常工作温度 normal working temperature
正常工作小时数 normal working hours
正常工作压力 normal operating pressure; normal working pressure; service pressure
正常工作状态 normal operating condition; normal upstate
正常公里费用 normal overhead
正常公民义务 normal civil obligation
正常功率 normal horsepower
正常供电 normal power supply
正常供水位 full supply level; normal pool level
正常供应 regular supply
正常共价 normal covalency
正常构造 normal configuration
正常估价人目标精度 <在最低标加10%范围之内> normal estimator's target accuracy
正常谷 normal valley
正常固结 normal consolidation
正常固结饱和黏[粘]性土 saturated cohesive soil for normal consolidation
正常固结沉积土 normally consolidated soil deposit
正常固结的 normally consolidated
正常固结黏[粘]土 normally consolidated clay
正常固结黏[粘]土沉降计算 calculation for normally consolidation clay
正常固结软土 normally consolidated soft soil; soft soil for normal consolidation
正常固结土 normally consolidated soil; normally loaded soil
正常固结线 normal consolidation line
正常关闭的 normally closed
正常关闭的线路 track normally open
正常关闭系统 normal(ly) danger system
正常关闭信号 normal closed signal; normal danger signal; normally closer signal
正常关闭信号机【铁】normally closer signal
正常关闭制 normal(ly) danger system
正常观测 normal observation
正常光电效应 normal photoelectric(al) effect
正常光谱 normal spectrum
正常光线 normal ray; ordinary ray
正常光照 normal illumination
正常轨道 normal orbit
正常过程调用 normal procedure call
正常过渡段海滩 normal crossing shoal
正常海水 normal marine water
正常海滩 normal beach; ordinary beach
正常含水量 normal moisture capacity
正常含水率 normal moisture content
正常函数 normal function; proper function; regular function
正常焊接 normal weld(ing)
正常航标 regular flight
正常航线 rectangular pattern; regular line; regular pattern
正常合金 normal alloy
正常和 normal sum
正常和易性 normal consistency
正常河槽 alveus
正常河床 normal bed
正常河段 normal reach
正常河曲 normal meander
正常河型 regular stream pattern
正常荷载 normal load(ing)
正常荷载设计 working load design
正常荷载条件 normal load condition
正常荷载组合 normal load combination
正常恒星 normal star
正常洪水位 normal top-water level
正常后座距离 normal recoil length
正常厚度 normal thickness; regular thickness
正常呼叫 normal call
正常互易地温梯度 normal reciprocal geothermal gradient
正常化 normalization; normalize; normalizing
正常化降水体积指数 normalized volume of precipitation index
正常化选择 normalizing selection
正常化学岩 orthochemical rock
正常环境 normal environment
正常环境条件 normal environment condition
正常环境直减率 normal environmental lapse rate
正常缓冲区 normal buffer
正常辉光 normal glow
正常辉光放电 normal glow discharge
正常回火 normalised [normalized] tempering
正常绘图精度 normal plotting accuracy
正常混合比 proper mixture ratio
正常混合物 normal mixture
正常混响时间 normal reverberation time
正常活动的 normal action
正常火山碎屑岩类 ordinary pyroclastic rock clan
正常火焰 normal flame
正常机速 normal engine speed
正常积分 normal integral; proper integral
正常基流 normal base flow
正常基线 normal base line
正常激励 normal excitation
正常极大 proper maximum
正常极化 normal polarization
正常极小 proper minimum
正常极性 normal polarity
正常极性下启动电流 direct pick-up
正常集 proper set
正常计费程序 standard procedure
正常加筋混凝土 normal reinforced concrete
正常加速度 normal acceleration
正常加载 normal loading
正常价【化】normal valency
正常价格 normal price; regular price

正常价值 normal value
正常架设姿态 normal sitting position
正常间隔 normal interval
正常间接费率 normal overhead rate
正常间隙 normal clearance
正常检查 health check; normal inspection
正常检查程序 health check program(me)
正常检验 normal inspection
正常奖金 natural premium
正常降水量 normal precipitation
正常降雨量 normal rainfall
正常交错铆接 snake riveting
正常交通增长 normal traffic growth
正常交易 arm's-length transaction
正常交易价格 arm's-length price
正常浇块 normal cast block
正常浇筑混凝土 normal cast concrete
正常角面 normal angle
正常校正 normal correction
正常接触 normal contact
正常结点 proper node
正常结构钢 normal steel; normal structural steel
正常结束 normal termination
正常解 normal solution; proper solution
正常解锁 normal release; regular release; usual releasing arrangement
正常解锁装置 usual releasing arrangement
正常进出 normal entry/exit
正常进度表 normal schedule
正常进给 course feed; regular feed
正常进路 first route; normal route; primary route
正常进路联结处理 normal access connection treatment
正常进行 normal clear; normal proceed
正常进行的 on-stream
正常径节 normal diametral pitch
正常径流量 long-time average annual flow; normal runoff; mean annual runoff; normal flow
正常竞争 normal competition
正常静水压力 normal hydrostatic(al) pressure
正常镜质体 normvitrinite
正常居里点 normal Curie point
正常距 normalized distance
正常绝对最小值 normal absolute minimum
正常开度 normal opening
正常开放 normally at clear
正常开放系统 normally clear system
正常开放信号 normally clear signal
正常开放信号机【铁】 normal-clear signaller
正常开放制 normally clear system
正常开关 sail switch
正常开路制 open circuit system
正常空间 proper space
正常空间异常 normal free air anomaly
正常孔隙 normal opening
正常控制 normal control
正常库存量 normal stock
正常库容 normal storage capacity
正常跨度 normal span
正常跨距 normal span
正常亏水曲线 normal depletion curve; normal recession curve
正常亏损曲线 normal depletion curve; normal recession curve
正常扩散 normal dispersion
正常拉力 correct tension
正常拉伸应力 normal tensile stress
正常浪基面 everyday wave base
正常冷却 normal cooling
正常利率 natural rate of interest
正常利润 normal profit
正常利润率 normal rate of return
正常利用率 normal general utility; normal utility
正常励磁 normal excitation
正常励磁继电器 normally energized relay
正常联锁 normal interlocking
正常亮灯信号机 normally lit signal
正常料 regular burden(ing)
正常临界比降 normal critical slope
正常临界坡度 normal critical slope
正常灵敏度 normal sensibility
正常流 regular flow
正常流过的 on-stream
正常流量 normal discharge; normal (stream) flow
正常流量曲线 normal discharge curve
正常流速 normal flow rate
正常流速分布 normal velocity distribution
正常流态 normal flow pattern
正常流体 normal fluid
正常流向 normal direction flow
正常漏损 ordinary leakage
正常路拱 normal crown
正常路拱路段 normal crown section
正常路由 first route; normal route; primary route
正常路缘石 normal curb
正常轮廓 normal profile
正常轮缘 normal flange
正常螺旋线 normal helix
正常洛仑兹群 proper Lorentz group
正常落差 normal fall
正常落差法 normal fall method
正常落差水位流量关系曲线 limiting-fall rating; normal-fall rating
正常满(负)载 normal full load
正常满库高度 normal full-pool elevation
正常满库水位 normal full-pool elevation
正常贸易途径 ordinary course of trade
正常弥散 normal dispersion
正常密度 normal density
正常瞄准误差 normal aiming error
正常灭灯的 normally dark
正常灭灯系统 normally dark system
正常灭灯信号 normally dark signal; normally extinguished signal
正常灭灯信号机 normally dark signal; normally extinguished signal
正常灭灯制 normally dark system
正常磨耗 fair wear and tear; normal wear; normal wear and tear
正常磨损 fair wear and tear; normal wear; normal wear and tear; wear-out failure period
正常能级 normal energy level; normal level
正常能力 normal capacity
正常能态 normal energy state
正常泥浆钻进 clean drilling
正常逆(矩)阵 normal inverse matrix
正常年(度)<水量、雨量、气温等的> normal year
正常年降水 normal annual precipitation
正常年景 normal yield
正常年径流 normal annual runoff
正常黏[粘]度 normal viscosity
正常捻法 regular lay
正常捻向 regular lay
正常凝固 normal freezing
正常凝固的 normal set
正常凝固方程 normal freezing equation
正常凝固分布 normal freezing distribution
正常凝集素 normal agglutinin
正常凝结 normal-setting
正常凝结的水泥 normal-setting cement
正常浓度家庭污水 normal domestic sewage
正常排水量 normal displacement
正常盘旋 true-banked turn
正常配合料 running batch
正常配偶 orthogamy
正常配置 normal setup
正常偏析 normal segregation
正常频率 normal frequency
正常频率燃烧 normal frequency combustion
正常频散 normal dispersion
正常坡度 regular grade
正常破坏 normal breaking
正常破裂 normal breaking
正常破损 normal breaking; normal spoilage; ordinary spoilage
正常剖面 normal cross section; normal profile
正常谱 normal spectrum
正常期 normal epoch
正常起吊高度 normal lift
正常起动 normal starting
正常起飞总重量 normal gross takeoff weight
正常气候 normal climate
正常气压 normal barometric pressure
正常气压的 normobaric
正常牵引拖曳情况 normal hauling position
正常铅 ordinary lead
正常浅滩 normal shoal
正常强度混凝土 normal-strength concrete
正常桥接 normal bridging
正常切幅犁铧 regular-cut share
正常侵蚀 geologic(al) erosion; normal erosion
正常侵蚀旋回 normal erosion cycle
正常侵蚀周期 normal erosion cycle
正常侵蚀作用 normal erosion action
正常氢量镜质组 orthohydrous vitrinite
正常氢量显微组分 orthohydrous maceral
正常倾斜 normal dip
正常清管工作 normal pigging
正常情况 normal condition
正常区段 normal reach
正常曲线 normal curve; normalized curve
正常渠道 regular channel
正常取代 normal substitution
正常取向 normal orientation
正常去能的 normally deenergized
正常燃烧 non-knocking explosion; normal combustion
正常燃烧速度 normal combustion velocity
正常燃烧条件 non-knocking condition
正常容量 normal capacity
正常入口 normal entry
正常塞曼效应 normal Zeeman effect
正常三色视觉 normal trichromatic vision; normal trichromatism
正常散布 normal dispersion
正常色觉 normal colo(u)r sight; trichromatism; trichromatopsia
正常色散 normal dispersion
正常色视觉 normal colo(u)r sight
正常扇状褶皱 normal fan-shaped fold
正常商品 normal goods
正常商业交易 arm's-length deal; arm's-length transaction
正常商业谈判 arm's-length bargaining
正常上升 normal climb
正常上升高度 normal lift
正常上游水池 normal pool
正常烧成的熟料 correctly burned clinker
正常设计 normal design
正常深度 normal depth
正常深切曲流 normal incised meander
正常生产 normal activity; regular production
正常生产成本会计 normal activity cost account
正常生产程序 normal production program(me)
正常生产量 capacity production; regular output
正常生产率 normal production rate
正常生产能力 normal capacity
正常生产线速度 regular line speed
正常生长 normal growth; regular growing; regular growth
正常生(长环)境 normal habitat
正常生活污水 normal domestic sewage
正常失励的 normally deenergized
正常失励继电器 normally de-energized relay
正常失励吸持继电器 normally de-energized stick relay
正常失效期 normal failure period
正常施工临时费用 normal construction contingency
正常湿度 normal humidity; normal moisture; normal moisture capacity
正常湿气 normal moisture
正常时差 normal moveout
正常时间 base time; leveled time; normal period; normal time
正常时间特性 normal time response
正常时间响应 normal time response
正常时距曲线 normal travel time curve
正常食物链 normal food chain
正常使用 regular service
正常使用功率 normal service rating
正常使用荷载 normal working load
正常使用的极限状态 limit state for normal use; serviceability limit state
正常使用年限 average life
正常使用条件 average service conditions
正常市场需求量 usual market requirement
正常示功图 normal indicator diagram
正常视觉 emmetropia; normal vision
正常试验法的弯沉 normal procedure deflection
正常试验压力 normal test pressure
正常释放 normal release; regular release
正常释放电压 normal dropout voltage
正常释放条件 normal release condition
正常收入 ordinary income
正常收缩 standard taper
正常收益 normal earning
正常收益率 normal income rate
正常收支 normal revenue and expenditures
正常收(支)款 ordinary annuity
正常手动进给 regular hand feed
正常输出 normal output
正常输出功率 firm capacity; firm output; normal output
正常输入 normal entry
正常输入键控 normal input keying

正常输入原因 normal input cause
正常熟料 normal clinker
正常树 proper tree
正常数 positive constant
正常数据行 normal data line
正常数值 regime(n)value
正常衰减屏 normal-speed screen
正常双线 regular doublet
正常水 normal water
正常水化石灰 normally hydrated lime
正常水化水泥 normally hydrated cement
正常水流 <水面坡度与河底坡度平行的> normal flow; normal streamflow
正常水泥 normal cement; regular cement
正常水平 normal level
正常水深 neutral depth; normal depth (of flow)
正常水深线 normal depth line
正常水头 normal head
正常水位 center [centre] position; normal elevation; normal pool level; normal stage; normal water level; ordinary water level
正常水位高程 normal water level elevation
正常水位水面 water standard at its normal level
正常水系 <河网的> regular stream pattern
正常税 regular tax
正常顺序 normal sequence
正常松弛性能 normal relaxation behavio(u)r
正常速度 normal speed; normal velocity; regular speed
正常损耗 normal loss; normal shrinkage; normal wear
正常损坏率 nominal failure rate
正常损失 ordinary loss
正常所见 normal findings
正常索赔 formal claim
正常态 normal state
正常态范围 normal envelope
正常坍落度 true slump
正常坍落型 true slump type
正常弹着效应 normal impact effect
正常梯度 normal gradient
正常提升高度 normal lift
正常体积 normal volume
正常体温 normathermia
正常体重 normal type
正常(田间)持水量 normal field capacity
正常条件 normal condition
正常条件下的使用期限 normal operating time
正常条件下最大能力 normal capacity
正常听觉 normal threshold of audibility
正常听觉范围 acusis range; range of normal audibility; range of normal hearing
正常听力 normal good hearing; normal hearing
正常听力阈值 normal threshold of hearing
正常听阈 normal threshold
正常停车 orderly shutdown; uniform stop
正常停车系统 normal stop system
正常停车制 normal stop system
正常停工 normal shutdown
正常停工检修 turnaround
正常停机 normal shutdown
正常停靠 regular calling
正常通报 normal traffic
正常通报时间 period of normal traffic
正常通报位置 normal traffic position
正常通风的 normally aspirated
正常通气性 normal aeration
正常同步速度 normal synchronous speed
正常同构 proper isomorphism
正常同时水位关系曲线 normal simultaneous stage relationship curve
正常投资回收 normal return of investment
正常图 normogram
正常图像 normal picture
正常土 normal soil
正常土方运输(费) normal haul
正常土壤剖面 normal profile of soil
正常推力 normal rated thrust
正常退带 normal unthreading
正常退水曲线 normal depletion curve; normal recession curve
正常椭球 normal ellipsoid; normal spheroid
正常椭球等位面 normal spheropotential surface
正常挖取效率 free digging rate
正常网点 regular mesh point
正常网络条件 normal network cause
正常危险系统 normally danger system
正常危险信号 normally closer signal
正常危险信号机【铁】 normally closer signal
正常危险制 normally danger system
正常微生物区系 normal flora
正常微震噪声 normal microseismic noise
正常煨火钢 normalized steel
正常维护 normal maintenance
正常维修 ordinary maintenance; ordinary repair
正常尾水位 normal tail water level
正常位 normal potential
正常位函数 normal potential function
正常位移 normal displacement
正常位置 in position; normal position
正常位置表示器 normal position indicator
正常位置的 entopic
正常温度 normal temperature; ordinary temperature
正常温度递减率 normal temperature lapse rate
正常温度分布 normal temperature distribution
正常坞修 normal docking
正常物质 koinomatter
正常误差定律 normal law of errors
正常吸气的 normally aspirated
正常吸收(能力) normalized absorption
正常熄灭的信号灯 normally extinguished signal lamp
正常习惯 normal habit
正常系列车辆 production car
正常系列发动机 production engine
正常细菌 normal bacteria
正常下的 subnormal
正常下降曲线 normal depletion curve
正常下温度 subnormal temperature
正常纤维素 normal cellulose
正常现象 normal phenomenon
正常线路 regular link
正常相 normal phase
正常相序 normal phase sequence
正常响应 normal response
正常销售协定 orderly marketing agreement
正常效应 normal effect
正常信息处理 normal message handling
正常星等 normal magnitude
正常形式 normal form
正常型 normal type
正常修复 normal corrective maintenance
正常需求 normal demand
正常需要 normal requirement
正常蓄水高程 normal water storage elevation
正常蓄水高度 normal water storage elevation
正常蓄水高水位 retention water level
正常蓄水量 normal storage
正常蓄水位 full supply level; normal pond level; normal pool level; normal reservoir level; normal storage water level
正常蓄水位线 normal storage water level line
正常旋转 normal spin
正常循环冲蚀 normal cycle erosion
正常压力 normal pressure
正常压力和温度 normal pressure and temperature
正常压力角 normal pressure angle
正常压力水头 normal pressure head
正常压密土 normal consolidation soil; normally consolidated soil
正常压强 normal pressure
正常压实带 normal compaction zone
正常压实相 normal compaction facies
正常压应力 normal compressive stress
正常眼力【测】 normal eye
正常演替系列 primary succession; prisere
正常氧化土 orthox
正常液体 normal liquid
正常移距 normal shift
正常以上 above normal
正常以下的 subnormal
正常抑制 normal inhibition
正常溢洪道 service spillway
正常阴极电压降 normal cathode drop
正常应答方式 normal response mode
正常溢洪道 normal spillway
正常盈利率 normal income rate
正常营运 normal operation
正常硬化的水泥 normal hardening cement
正常涌水量 normal water yield
正常余辉荧光屏 normal-speed screen
正常预算 normal budget; regular budget
正常原位持水量 normal field capacity
正常运算 normal operation
正常运行 failure-free operation; good running; normal operation; normal running; normal work(ing); regular movement; regular service; trouble-free running
正常运行班次 regular runs
正常运行的 on-stream
正常运行方式 normal operation mode
正常运行负荷 normal running load
正常运行距离 normal reach
正常运行库水位 normal operating level
正常运行期 error free running period
正常运行曲线 normal operation curve
正常运行时间 up-time
正常运行速度 normal running speed
正常运行损失 normal operating loss; normal operation loss
正常运行条件 normal operation condition; normal running condition
正常运行状态 normal travel order
正常运用曲线 normal operation curve
正常运转 failure-free operation; normal operation; normal running; trouble-free running
正常运转安培值 normal operating amperage
正常运转的气温范围 operating temperature range
正常运转电流 normal operating current
正常运转电流量 normal operating amperage
正常运转范围 normal operating limit; normal operating range
正常运转速度 normal running speed
正常运转温度 natural running temperature; normal running temperature
正常运转状态 work(ing) order
正常载荷 normal load
正常再生 regular regeneration
正常增长交通量 normal traffic increment
正常增压 normal supercharging
正常增益 normal gain
正常炸高 normal height of burst
正常张力 normal tension
正常张应力 normal tensile stress
正常照明 normal illumination; normal lighting
正常照明系统 normal lighting system
正常折旧 normal depreciation; ordinary depreciation
正常折射 normal refraction
正常褶皱 normal fold
正常振荡模 normal mode
正常振动 normal vibration
正常振型 normal mode of vibration
正常震源地震 normal-focus earthquake; normal-focus shock
正常蒸发 normal evapo(u)ration
正常整体标准 normality
正常正高 normal-orthometric height
正常正交矩阵 proper orthogonal matrix
正常正交群 proper orthogonal group
正常正交阵 proper orthogonal matrix
正常支承 normal support
正常执行顺序 normal execution sequence
正常执行态 normal execution mode
正常直径的孔 clearing hole
正常直流分量恢复 normal direct current restoration
正常值范围 range of normal value
正常值上限 upper limit of normal
正常植物区系 normal flora
正常指挥系统 normal command channel
正常指示 normal indication
正常制动 normal brake application; normal braking; service brake application
正常制动距离 service braking distance
正常制动停车 normal brake stop
正常终结 normal termination
正常终止 proper termination
正常重力 normal gravity
正常重力场 normal gravity field
正常重力公式 normal gravity formula
正常重力加速度 normal acceleration of gravity
正常重力位差数 normal geopotential number
正常重力线 normal gravity line
正常重力值 normal gravity value
正常重量 normal weight
正常重量骨料 normal weight aggregate
正常注水 normal flood
正常转矩 normal torque
正常转数 normal revolution
正常转速 permanent speed

Z

正常转弯 normal turns
正常转向 normal direction of rotation
正常装卸速度 usual despatch[dispatch]
正常装药 normal charge
正常状况 normal mode;normal state;regime(n)
正常状况荷载 normal load
正常状态 normal condition;normalcy;normality;normal state
正常状态化降水指数 normalized precipitation index
正常准备状态 normal preparedness
正常准(侵蚀)平原 normal peneplain
正常着陆 normal landing
正常总重 normal gross weight
正常走时曲线 normal travel time curve
正常组分 normal component
正常钻进 regular feed
正常最大输出 firm peak capacity
正常最高蓄水位 normal top water level
正常最冷三个月时段 normal coldest three-month period
正常作业 trouble-free operation
正常作业时间 normal working hours
正常坐姿驾驶舱 normal-seated cabin
正车 advancing;ahead running;forward drive
正车惰性滑距 head reach
正车航速 forward speed
正车驱动 positive drive
正车凸轮 ahead cam
正承压水头 positive artesian head
正程间 forward-stroke interval
正程噪声 positive-going noise
正程增辉电路 unblanking circuit
正齿 commutating tooth;spur teeth
正齿背齿轮 spur wheel back gear
正齿齿轮 spur-gear wheel
正齿齿条 spur rack
正齿轮 gear wheel;plain gear;spur;spur gear;spur wheel;straight-cut gear
正齿轮泵 spur-gear pump
正齿轮操纵的阀门 spur-gear-operated valve
正齿轮传动 spur gearbox;spur-gear drive
正齿轮传动的 spur-geared
正齿轮传动机械 spur-geared machine
正齿轮传动装置 spur-gearing;spur-gear set
正齿轮对 spur-gearing
正齿轮分速器 spur differential
正齿轮副轴 spur wheel countershaft
正齿轮和斜齿轮滚齿机 spur and helical hobbing machine
正齿轮滑车 spur-gear pulley
正齿轮回动装置 spur wheel reversing gear
正齿轮机构传动装置 spur gear train
正齿轮减速机 spur-gear speed reducer
正齿轮减速器 cylindric(al) reducer
正齿轮绞车 spur-geared winch
正齿轮坯料 spur-gear blank
正齿轮切削 spur-gear cutting
正齿轮驱动 spur-gear drive
正齿轮润滑 spur-gear lubrication
正齿轮润滑剂 spur-gear lubricant
正齿轮外运传动装置 spur wheel outrigger gear
正齿轮铣刀 spur-gear cutter
正齿轮组 spur-gear set
正齿伞齿轮 spur bevel gear
正齿式行星齿轮 spur-type planetary gear
正齿式行星齿轮装置 spur-type planetary gearing
正齿式行星减速齿轮 spur-type planetary reducer
正赤纬 plus declination
正冲 head-on impact;square impact
正冲波 direct shock wave
正冲件 direct flushing
正冲头退料器 positive punch stripper
正抽头 plus tapping
正除数 positive divisor
正触发脉冲 positive triggering pulse
正穿孔 positive punch
正穿透性 positive penetrability
正传动 positive drive
正传动鼓风机 positively driven blower
正垂面 vertical frontal plane
正垂线 vertical frontal line
正锤线 right plummet
正锤线观测 direct plummet observation
正锤线(观测)法 method of direct plummet observation; method of right plummet observation
正磁极期 normal epoch
正磁矩 positive magnetic moment
正磁异常区 positive magnetic anomaly area
正磁致伸缩 Joule magnetostriction;positive magnetostriction
正磁致伸缩效应 direct magnetostriction effect
正催化剂 positive catalyst
正催化作用 positive catalysis
正淬火法 direct electric(al) process
正搓 hawser laid;right-hand
正搓绳 hawser laid rope;regular lay rope;right-hand rope
正搓绳索 right lay
正大距 positive course pitch
正单光轴结晶 attractive uniaxial crystal
正当 legitimacy
正当补偿 just compensation
正当成本 legitimate cost
正当持票人 holder in due course
正当的 allowable;legal
正当的申请 justifiable complaint
正当法律程序 due process of law
正当费用 justifiable expenditures
正当理由 justification;valid reason
正当纳税人 bona fides taxpayer
正当审判 proper trial
正当收入 legitimate income
正当业务 legitimate business
正当执票人 holder in due course
正导线 positive conductor
正倒车离合器 reversing clutch
正倒镜【测】 change face
正倒镜分中法 <延长直线时> double sighting
正倒镜复测 reiterate;reiteration
正的 positive
正的传号极性 positive mark polarity
正的同步脉冲 positive clock pulse
正的指示 positive indication
正等轴测图 isometric(al) drawing
正低突起 positive low relief
正地槽 orthogeosyncline
正地层学 orthostratigraphy
正地热梯度 positive geothermal gradient
正地热异常 positive geothermal anomaly
正地台 orthoplatform
正地下截水墙 positive underground cutoff
正地形 positive form;positive landform;positive relief
正地形单元 positive topographic(al) unit
正地形单元上的负单元 the negative unit on positive unit
正地形单元上的正单元 the positive unit on positive unit
正递变 normal grading
正递归的 positive recurrent
正碲的 telluric
正点 on schedule;on time;punctuality
正点到达 arrive on time
正点到达时间 due time of arrival
正点的 punctual
正点发车 start on time
正点率 on-schedule rate
正点运行 on-time running
正碘酸 iodic acid
正电 plus electricity;positive electricity;positron scintigraphy;vitreous electricity
正电导线 positive conductor
正电荷 positive charge
正电荷电子射线极 anode ray;canal ray;positive ion rays
正电荷过剩 positive excess
正电荷基 positive charge group
正电荷量 amount of positive charge
正电荷载流子 positive carrier;positive charge carrier
正电极 positive electrode;positive strap
正电接头 positive contact
正电流 copper current;positive current
正电码 normal code
正电射线 positive ray
正电势 positive potential
正电输送带 positive belt
正电刷 positive brush
正电位 positive potential
正电性 electropositivity
正电性的 electropositive
正电性金属 electropositive metal
正电性离子 positively charged ion
正电性凝胶 electropositive gel
正电性元素 electropositive element
正电压 positive voltage
正电子 anti-electron;positive electron;positron
正电子断层照相法 positron tomography
正电子放射 positron emission
正电子辐射 positron radiation
正电子辐射体 positron radiator
正电子还原反应 positron reduction reaction
正电子束 positron beam
正电子衰变 positron decay;positron disintegration
正电子衰变能量 positron-decay energy
正电子素 positronium
正电子湮没辐射 positron-annihilation radiation
正电子湮没能量 positron-annihilation energy
正电子照相机 positron camera
正电阻 positive resistance
正电阻温度系数测辐射热计 positive-α bolometer;positive-α bolometer
正殿 main hall
正迭代法 direct iteration
正叠量阀 overlapped valve
正丁胺 n-butylamine
正丁醇 butyrk alcohol;normal butanol
正丁基 n-butyl;normal-butyl
正丁醚 n-butyl ether
正丁酸 n-butyric acid
正丁烷 normal butane
正顶度规 positive definite meter
正顶风 head to the wind
正顶风航行 tack and tack
正顶尖 male center[centre]
正定 positive definite
正定埃尔米特变换 positive definite Hermitian transformation
正定变换 positive definite transformation
正定第二变分 positive definite second variation
正定对称核 positive definite symmetric(al) kernel
正定对称矩阵 positive definite symmetric(al) matrices
正定二次微分形式 positive definite differential form of degree two
正定二次型 positive definite quadratic form
正定函数 positive definite function
正定核 positive definite kernel
正定积分形式 positive definite integral form
正定矩阵 positive definite matrix
正定内积 positive definite inner product
正定算子 positive definite operator
正定位 pull-off
正定位腕臂 pull-off cantilever
正定系统 positive define system
正定向 positive orientation
正定向曲线 positively oriented curve
正定协方差 positive definiteness covariance
正定型 positive definite form
正定序列 positive definite sequence
正东 due east
正动 positive drive
正动阀 positively actuated valve
正动量误差 positive momentum error
正动态 orthokinesis
正动提升阀 poppet valve
正洞门 orthonormal portal
正读 forward read;reading forward;right reading
正端子 plus end;positive terminal
正断层【地】 normal fault;centripetal fault;downthrow;downthrow fault;tension fault;extensional fault;multiple fault
正断层地堑 ordinary fault rift
正断层谷 rift trough
正断层机制 normal mechanism
正断层裂谷 ordinary fault rift
正断层圈闭 normal fault trap
正断层震源机制 normal-focus mechanism
正断面 normal cross section;normal section
正断面图 orthograph
正断崖 normal fault scarp
正堆积结构 orthocumulate texture
正对 <指相对的物体> enfilade
正对接合 opposite joint
正对面 right opposite
正对准 positive alignment
正多边形【数】 equilateral polygon;isogon;regular polygon
正多边形的垂辐 apothem of a regular polygon
正多边形中心 center[centre] of regular polygon
正多角形 regular polygon
正多面角 regular polyhedral angle
正多面体 regular polyhedron
正多面体群 regular polyhedral group
正多相分类 forward polyphase sort
正舵 helm amidship;right the helm
正鲕绿泥石 orthochamosite
正二面角 positive dihedral angle
正二十面体 regular icosahedron
正二重复形 positive double complex
正二轴测图 dimetric drawing
正二轴的 dimetric
正法线 positive normal
正法向 positive normal
正反变速箱离合器 reversing clutch
正反差 normal contrast;positive con-

trast
正反铲挖掘机 convertible shovel
正反铲挖土机 convertible shovel
正反对 anti-pode;antithetic
正反峰间隔值 peak-to-peak amplitude
正反挤压 backward and forward extrusion
正反控制信号 buck-boost control signal
正反扣的 right-and-left threaded
正反馈 positive feedback;positive reaction; positive regeneration; reaction regeneration;regenerative feedback
正反馈电路 regenerative circuit; regenerative feedback loop
正反馈放大 regeneration
正反馈放大器 positive feedback amplifier; regenerative feedback amplifier
正反馈环路 regenerative loop
正反馈积分器 regenerative integrator
正反力 positive reaction
正反两面可用的 reversible
正反两用铲(斗)convertible shovel
正反螺纹 right-and-left-hand thread
正反面连接 feed-thru connection; front-back connection
正反配板法 book matching
正反射 normal reflection;regular reflection;specular reflection
正反射光泽 objective gloss
正反时代表 normal reversed time scale
正反弯曲试验 back-and-forth bending test; backward and forward bending test
正反向 forward and reverse
正反向比 front-to-back ratio
正反向计数 forward backward counter
正反向计数器 backward forward counter;up-down counter
正反像限角之差<子午线收敛角引起>【测】false bearing
正反应 positive reaction
正反应性 positive reactivity
正反转换 normal-reverse transfer
正反转伺服电动机 reversible electric(al) servomotor
正泛函 positive functional
正方案 direction-determining board
正方的 diametric(al);tetragonal
正方底穹隆 square dome
正方端 milled end
正方断裂线【地】diaclase
正方矾石 aluminite;websterite
正方格网 square grid;square mesh
正方角图幅 quadrangle sheet
正方晶 regular crystal
正方晶格 tetragonal
正方晶系 pyramidal system;quadratic system;tetragon;tetragonal system
正方棱柱体 tetragonal prism
正方棱锥 square pyramid
正方马氏体 tetragonal martensite
正方位艏向 cardinal heading
正方席纹 square basket weave pattern
正方席纹地板 square basket weave parquetry flooring
正方向 forward;positive direction
正方向变化 change positively
正方楔 tetragonal sphenoid
正方形 orthogon; quadrate; quadratic;tetragonal;square
正方形板 square slab
正方形沉砂池 square grit chamber
正方形储罐 square tank
正方形错列 staggered square pitch
正方形单独基础 square individual base;square single base
正方形单独基底 square individual base
正方形单元 tetragonal cell
正方形而带圆角的 subquadrate
正方形分幅 square map-subdivision
正方形隔仓<气压沉箱> square cell
正方形拱顶 cavetto vault
正方形护舷 square fender
正方形开口沉箱 square drop shaft; square open caisson
正方形勘探网 square exploration grid
正方形排列 in-line square pitch; square pitch arrangement
正方形平行构件 squared and parallel element
正方形平行组件 squared and parallel element
正方形企口(接合) square grooving and tonguing
正方形栅板 square grid
正方形舌槽(接合) square grooving and tonguing
正方形伸缩风箱 square bellows
正方形石块墙 squared stone masonry (work)
正方形式搜索 expanding square search;square search
正方形锁 four-square lock
正方形天线 quadrant aerial;quadrant antenna
正方形投影 equirectangular projection
正方形图幅 quadrangle sheet
正方形网格 square grid
正方形网格排列 square-grid pattern
正方形直列 in-line square pitch
正方油石 square oil stone
正方针铁矿 akaganeite
正方柱 tetragonal prism
正方锥 tetragonal pyramid
正房<农村房屋的> main house
正放水波 positive release wave
正非零整数 positive nonzero integer
正沸绿岩 glenmuirite
正分离 effective segregation
正风压控制 positive air pressure control
正封闭层 positive confining bed
正峰信号 positive spike
正峰值 positive peak
正浮力 positive buoyancy
正浮选 positive flo(a)tation
正幅度差 positive separation
正辐射 positive radiation
正俯冲 upright dive
正负 plus minus
正负变换器 sign changer
正负测向 sense finding
正负电子偶的形成 pair creation
正负对向电压 diametral voltage
正负方向 positive negative direction
正负峰幅值 peak-to-peak amplitude
正负峰间 peak-to-peak
正负峰间的倍幅值 double amplitude; peak-to-peak value
正负峰间电压 peak-to-peak voltage
正负峰间幅值 peak-to-peak amplitude
正负峰间振幅值 double amplitude; peak-to-peak value
正负峰(之)间的 peak-to-peak
正负峰值间总幅度 total amplitude
正负符号 sign symbol
正负公差 plus-minus tolerance
正负号 sign
正负号变更 variation of sign
正负号符号变更 sign reversal
正负号规则 rule of sign
正负号函数 signum[复 signa]
正负号交变 sign-alternating
正负号替换 variation of sign
正负校验 check plus minus
正负控制 positive negative control
正负控制器 positive negative controller
正负零偏差 deviation from datum
正负逻辑 mixed logic;positive negative logic
正负三级动作 bang-bang action;positive negative three level action
正负三级作用 bang-bang action;positive negative three level action
正负三位作用 positive negative three level action
正负双极 positive negative bipolar
正负相间磁异常 positive alternating with magnetic anomaly
正负向测定器 sense finder
正负像转换开关 positive negative switch
正负余量 plus and minus tolerance
正复 eutopic reduction
正复形 positive complex
正副二份中之一 counterpart
正干扰 positive interference
正杆 positive bar
正刚度 positive ridigity;positive stiffness
正高 geoidal height; orthometric elevation;orthometric height
正高度角 positive altitude
正高改正 orthometric correction
正高校正 orthometric correction
正高突起 positive high relief
正高误差 orthometric error
正高系统 orthometric system
正割 secant
正割定律 secant law
正割法 secant method
正割公式<分析计算柱体的一种公式> secant formula
正割积分函数 secant integral function
正割检流计 secant galvanometer
正割模量 secant modulus
正割模数 secant modulus
正割曲线 secant curve
正割屈服应力 secant yield stress
正割弹性模量 secant modulus of elasticity
正割圆 secant circle
正割真数【数】natural secant
正镉化合物 cadmic compound
正根 positive root
正庚烷【化】normal heptane
正公差 plus tolerance;positive allowance
正功 positive work
正攻角 positive incidence
正汞的 mercuric
正拱 right arch;sprung arch
正共沸混合物 positive azeotrope
正构醇 n-alkanol
正构化合物 normal compound
正构醛 n-alkanal
正构十六烷 normal cetane
正构烷烃分布 distribution of n-alkane
正构烷烃/异构烷烃 normal alkane/isoalkane
正钴的 cobaltous
正光导性 positive photoconductivity
正光电导性 light positive;photopositive
正光性 positive character
正广义函数 positive distribution
正规 regularity
正规半有限权 normal semifinite weight
正规闭包 normal closure
正规变差 normal variation
正规变分 normal variation
正规变换 normal transformation
正规标架 normal frame
正规表达式 regular expression
正规表示 normal representation
正规表示的代数 algebra of regular represent
正规测地线 normal geodesic line
正规测度 normal measure
正规测量 regular survey
正规测站 authorized station
正规尺寸 regular size
正规穿孔 normal-stage punch(ing)
正规簇 normal variety
正规错误校正 normal error recovery
正规代数 normal algebra
正规单代数 normal simple algebra
正规的 canonic(al);normal;regular; self-conjugate;standard
正规的车用机油 regular motor oil
正规的代数簇 normal algebraic variety
正规的教室 regular classroom
正规的铸铁排水管 normal cast-iron drain(age) pipe
正规灯船 regular lightship
正规电缆 normal cable
正规定单 regular order
正规多角形 normal polygon
正规二进制 regular binary
正规反导数 normalized inverse derivative
正规范畴 normal category
正规方程 normal equation
正规方式 normal mode
正规费用率 regulated fee
正规分程序 normal block
正规峰态 normal kurtosis
正规赋值 normal valuation
正规覆盖 normal covering
正规格 normal lattice
正规格式 normal format
正规工作 regular work
正规工作时间 ordinary working hours;regular work hours;straight time
正规公式 normal formula
正规估计 regular estimate
正规估计量 normal estimator
正规固定 normal bond
正规观测站 authorized station
正规函数 regular function
正规函数族 normal family of functions
正规航空站 regular airport
正规化 formalization;normalization; orthonormality; orthonormalization;regularization
正规化测地线 normalized geodesic
正规化导纳 reduced admittance
正规化的 orthonormal
正规化的方程 normalized equation
正规化的交叉乘积 normalized crossed product
正规化的脚步声传播声级 normalized footstep sound transmission level
正规化定理 normalization theorem
正规化分布 normalized distribution
正规化矩阵 normalised[normalized] matrix
正规化例行程序 normalization routine
正规化链复形 normalized chain complex
正规化(声)级差 normalized level difference
正规化特征函数 normalized eigenfunction
正规化条件 normalization condition

正规化性状 normalized behavior
正规化因子 normalization factor
正规化约束 normalized constraint
正规化转置 regularizing transposition
正规化坐标 normalized coordinates
正规环 normal ring
正规混凝土用砂 regular concrete sand
正规迹 normal trace
正规积分 normal integral
正规基本区域 normal fundamental region
正规集 regular set
正规浇筑 normal cast
正规浇筑产品 normal cast product
正规浇铸 normal cast
正规接头 regular tool joint
正规结构 regular structure
正规紧算子 normal compact operator
正规矩阵 normal matrix;regular matrix
正规可除代数 normal division algebra
正规空间 normal space
正规空心砌块 normal hollow block
正规空心瓦 normal hollow tile
正规扣 regular thread
正规扩域 normal extension field
正规立方晶体 cubic(al) system
正规连分数 normal continued fraction
正规链 normal chain
正规列 normal series
正规列车 normal train
正规邻域 normal neighbo(u)rhood
正规林 normal forest
正规流向 normal flow direction
正规名称 legal name
正规模式 normalized mode
正规模型 normalized mode
正规耐纶 normal nylon; regulation nylon
正规黏[粘]合 normal bond
正规捻 regular lay
正规批发商 regular wholesaler
正规品级 regular grade
正规坡度 regular grade
正规谱测度空间 normal spectral measure space
正规汽油 Q-grade gasoline
正规砌合 normal bond;regular bond
正规砌筑毛石 regular coursed rubble
正规切触黎曼流形 normal contact Riemannian manifold
正规球形穹顶 regular dome
正规曲率半径 normal radius of curvature
正规曲线 normal curve
正规溶液 regular solution
正规溶液理论 regular solution theory
正规溶液模型 regular solution model
正规三线坐标 normal trilinear coordinates
正规扫描 orthodox scanning
正规扇形域 normal sector
正规上链 normalized cochain
正规设备 regular equipment
正规深度 normal depth
正规审判 regular trial
正规生产 set-up production
正规剩余的 normal residual
正规失真 regular distortion
正规实形式 normal real form
正规试验法 ortho-test
正规试验块体 normal block
正规手续 legitimate procedure;regular procedure
正规束 orthodox beam
正规树 normal tree
正规数 normal number
正规数域 normal domain
正规衰减 regularity attenuation
正规水流 normal flow
正规算法 normal algorithm
正规算子 normal operator; normal transformation
正规随机变数 normal random variable
正规随机过程 normal random process; normal stochastic process
正规随机数 normal random number
正规随机数字 normal random digit
正规条件 normalised [normalized] condition
正规桶 regular barrel
正规透射系数 regular transmittance
正规图幅 regular map
正规椭圆积分 normal elliptic integral
正规拓扑空间 normal topological space
正规拓扑群 normal topological group
正规文法 normal grammar; regular grammar
正规误差积分 normal error integral
正规系 normal system
正规系数 normal coefficient
正规相关分析 analysis of canonic-(al) correlation
正规响应 normal response
正规向量 normalized vector
正规斜纹 normal twill
正规形式 normal form; normal format
正规性 normality
正规性质 regularity property
正规序列空间 normal sequence space
正规学校 school of general instruction
正规学校教育 schooling
正规循环群 regular cyclic(al) group
正规样本 normal sample
正规样品 normal sample
正规因子 normalization factor
正规映射 normal mapping
正规有向树 normal directed-tree
正规余树 normal forest
正规余因子 normalized cofactor
正规语言 normal language; regular language
正规预制品 normal precast ware
正规元素 <数学中的> regular element
正规载波系统 normal carrier system
正规整函数 normal entire function
正规正交的【数】normal orthogonal
正规正交化 orthonormalization
正规直因子 normal direct factor
正规中断 regular interrupt
正规属性 normal attribute
正规铸造件 normal cast product
正规砖砌合 regular bond
正规子群 invariant subgroup; normal divisor; normal subgroup; self-conjugate subgroup
正规子系统 normal subsystem
正规自同态 normal endomorphism
正规总体 normal population
正规族 normal family
正规最大车速 nominal speed
正规坐标 normal coordinate
正硅铬酸铅 normal lead silico chromate
正硅酸铁 fayalite
正硅酸盐 orthosilicate
正轨 main track
正轨道闭包 positive orbit closure
正轨的 undeviating
正轨迹 positive rail
正癸酸 n-capric acid
正过载 positive g
正过载容限 positive g tolerance
正过载时间 positive-g period
正过载转弯 positive acceleration turn
正氦 orthohelium
正函数 positive function
正函数化 positivization
正焊 face-up bonding
正好击在钻杆中心 hit the rod square
正好相反的 diametric(al)
正好在临界状态下 just critical
正号【数】positive sign; affirmative sign;plus sign
正号区 plus zero
正合逼近 exact approximation
正合范畴 exact category
正合函子 exact functor
正合加性函子 exact additive functor
正合局部化系统 exact localizing system
正合偶 exact couple
正合平方 exact square
正合上积 exact coproduct
正合上同调序列 exact cohomology sequence
正合时机 high time
正合适的 well-fitting
正合微分 exact differential
正合微分方程 exact differential equation
正合性 exactness
正合性公理 exactness axiom
正合序列 exact sequence
正合序列公理 sequence axiom
正合自同态 exact endomorphism
正和负变量 positive and negative variable
正河口湾 positive estuary
正横 abeam; on the beam; right athwart
正横断面 <垂直于轴线的断面> normal cross section
正横方位 beam bearing
正横方向 abeam direction; off the beam;on the beam;right athwart
正横风 wind abeam
正横后受风 free
正横截面 normal cross section
正横距离 abeam distance
正横前 before the beam
正横前面 forward of the beam
正横向地性 positive diageotropism
正后电位 positive after-potential
正后方 dead astern; rigging aft; right astern
正后倾 plus caster
正后像 positive after-image
正弧 positive arc;regular arc;regular curve
正花岗岩 orthogranite
正滑断层 normal-slip fault
正化 oxidation;positizing
正化多项式 positizing polynomial
正化颗粒 orthochem
正化学照相法 positive chemography
正化组分 orthochem; orthochemical constituent
正环流 direct circulation
正环索线 right strophoid
正黄 positive yellow;process yellow
正回波 positive echo
正回授 positive feedback;positive regeneration;regenerative feedback
正回授电路 regenerative feedback loop
正回授放大 regenerative amplification
正会员 fellow;freeman;senior member
正混合岩 orthomigmatite
正火 normalization; normalize; normalizing
正火处理 normalizing treatment
正火淬火回火 normalizing-quenching-tempering
正火的 normalized
正火回火处理 normalize tempering
正火及回火钢 normalized and tempered steel
正火炉 normalizing furnace
正火区 normalized zone
正火石灰 normally burnt lime
正火温度 normalizing temperature
正火状态 normalized condition
正火组织 normalized structure
正火作业线 normalizing line
正积温 accumulated positive temperature
正基 positive group
正基准面 frontal datum plane
正畸变 pincushion distortion;positive distortion
正畸材料 orthodontic materials
正畸器械 orthodontic appliance
正箕 ulnar loop
正激波 bow wave;normal shock;normal shock wave;normal wave
正激波方程 normal shock relation
正激波关系式 normal shock relation
正激波扩散 normal shock diffusion
正激波压缩 normal shock compression
正极 anode; plus plate; positive electrode;positive plate;positive pole
正极板 positive plate
正极搭铁 positive ground
正极搭铁系统 positive ground system
正极搭铁制 positive ground system
正极大(值) positive maximum
正极的 anodal
正极电流 inflow current
正极端子 positive terminal
正极集电栅 positive collector grid
正极接地 plus earth
正极接地的电流 negative supply
正极接地系统 positive ground system
正极接线柱 positive terminal
正极馈电线 positive feeder
正极连接片 positive strap
正极性 normal polarity; positive polarity; straight polarity; subtractive polarity <变压器的>
正极性触发 positive-going trigger
正极性传输 positive transmission
正极性传送 positive polarity transmission
正极性幅度调制 positive amplitude modulation
正极性调幅 positive amplitude modulation
正极性调制 positive modulation
正极性图像 positive picture phase; positive picture polarity
正极性图像调制 positive picture modulation
正极性图像信号 positive picture signal
正极性衔铁 normal polarity armature;positive armature
正极性消隐脉冲电压 positive pedestal voltage
正极引线 positive wire
正棘爪 positive pawl
正几何规划 positive geometry programming
正己醇 n-hexyl alcohol
正己基 n-hexyl
正己基卡必醇 n-hexyl carbitol
正己基溶纤剂 n-hexyl cellosolve
正己酸酐 n-caproic anhydride
正己烷 n-hexane;normal hexane

正挤压 direct extrusion
正脊 main ridge; upper ridge
正加速度 positive acceleration
正加速相 positive acceleration phase
正钾长石 iso-orthoclase
正价 net price; nominal price; normal valency
正价加附加费合同 cost-plus-fee contact
正价加固定附加费 cost-plus-fixed fee
正价加固定附加费合同 cost-plus-fixed-fee contract
正驾驶员 chief pilot; skipper
正尖波 positive sharp wave
正尖峰信号 positive spike
正尖晶石 normal spinel; positive spinel
正尖晶石型 positive spinel type
正剪力 positive shear(ing)
正检验 positive test
正渐开线 plus involute
正交 cross-cut; normality; orthogonal intersection; orthogonalize; perpendicular; perpendicularity; quadrature【电】
正交安匝 cross ampere-turn
正交比较 orthogonal comparison
正交变换 orthogonal transformation
正交变换群 orthogonal transformation group
正交标架 orthogonal frame
正交标架丛 orthogonal frame bundle
正交表 orthogonal layout; orthogonal table
正交表示 orthogonal representation
正交波痕 pericline ripple mark
正交补 orthocomplement; orthocomplementation; orthogonal complement
正交补的 orthocomplemented
正交补空间 orthogonal complement space
正交部分 quadrature component
正交彩色信号 quadrature chrominance signal
正交参考轴 orthogonal reference axis
正交参数系统 orthogonal parametric system
正交测度 orthogonal measure
正交测斜灵敏度 cross inclination sensitivity
正交叉 right-angle(d) intersection
正交场 cross field; orthogonal field; quadrature field
正交场乘法器 cross field multiplier
正交场电离真空计 crossed field ga(u)ge
正交场放大器 crossed field amplifier
正交场管 crossed field tube
正交场加速管 orthogonal field tube
正交场器件 crossed field device
正交超曲面系 orthogonal system of hypersurfaces
正交冲击波 normal shock wave
正交初始瞬态电抗 quadrature subtransient reactance
正交处理机 orthogonal processor
正交磁场 cross magnetic field; quadrature field
正交磁场发电机 cross connected generator; cross field generator
正交磁化 cross magnetization; cross magnetizing
正交磁化效应 cross magnetizing effect
正交错误控制 orthogonal error control; orthotronic error control
正交代换 orthogonal substitution
正交单位向量 orthonormal vector
正交单位向量组 orthonormal vectors
正交导槽 orthogonal guidance
正交的 normal; orthogonal; orthographic(al); orthometric; perpendicular; right-angle(d)
正交的超曲面系 orthogonal system of hypersurfaces
正交的曲线系 orthogonal system of curves
正交的正交射 orthogonal orthomorphism
正交地图投影 rectangular map projection
正交点 orthogonal point; positive interesting point
正交电磁场 crossed electric(al) and magnetic field
正交电磁场质谱仪 crossed electric(al) magnet fields mass spectrometer
正交电路 quadrature network
正交电刷 cross brush; quadrature brushes
正交电位计 quadrature potentiometer
正交电压 quadrature voltage
正交丁坝 orthogonal spur dike[dyke]
正交定理 orthogonality theorem
正交定律 normality law
正交定相 quadrature phasing
正交对称 rhombic(al) symmetry
正交对称的 ortho-symmetric
正交对合 orthogonal involution
正交对应参数系统 orthogonal corresponding parametric system
正交多项式【数】orthogonal polynomial
正交多项式方法 orthogonal polynomial method
正交多项式展开 orthogonal polynomial expansion
正交多圆锥投影 rectangular polyconic(al) projection
正交反力 normal reaction; normal reactive force
正交反射率 orthogonal reflectivity
正交方格 orthogonal square
正交方式 orthomode
正交方向 normal direction; orthogonal direction
正交放大器 quadrature amplifier
正交分解 orthogonal decomposition
正交分力 normal component; orthogonal component; quadrature component
正交分量 normal component; orthogonal component; perpendicular component; quadrature component
正交分析【法】orthogonal analysis
正交钢索 trajectory cable
正交格(梁) orthogonal grid
正交各向异性 orthogonal anisotropy; orthotropy; perpendicular anisotropy
正交各向异性板 orthogonal anisotropic plate; orthotropic plate; orthotropic slab; perpendicular heterogeneous slab
正交各向异性板法 orthotropic plate method
正交各向异性板面钢箱梁桥 steel box girder bridge with orthotropic deck
正交各向异性材料 cross-anisotropic material; orthotropic material
正交各向异性的 orthogonally anisotropic; orthotropic
正交各向异性非对称板 orthotropic skew(ed) plate
正交各向异性固体 orthotropic solid
正交各向异性加强合成板 orthotropic stiffened composite plate
正交各向异性颗粒 orthotropic particle
正交各向异性楼板 orthotropic plate floor
正交各向异性桥面 orthotropic deck
正交各向异性桥面结构 orthotropic deck structure
正交各向异性壳 orthotropic shell
正交各向异性弹性地面 orthotropic elastic ground
正交各向异性体 anisotropic(al) orthotropic body; orthotropic solid
正交各向异性折叠板 orthotropic folded plate
正交估计量 orthogonal estimate
正交箍缩 orthogonal pinch
正交关系 orthogonality rejection; orthogonality relation(ship); orthogonal relation
正交观察 orthogonal view(ing)
正交广群 orthogonal groupoids
正交归一的 orthonormal
正交规格化的 orthonormal
正交轨(迹)线 orthogonal trajectory
正交过程 orthogonal process
正交函数 orthogonal function
正交函数系 orthogonal function system
正交函数展开 orthogonal expansion
正交函数展开式 orthogonal function expansion
正交函数族 orthogonal family of function
正交涵洞 right culvert
正交合成 orthographic(al) synthesis
正交桁架 orthogonal truss
正交厚度 orthogonal thickness
正交厚度参数 parameter of orthogonal thickness
正交化 orthogonalization
正交化步骤 orthogonalizing process
正交回归 orthogonal regression
正交回归分析 orthogonal regression analysis
正交回归线 orthogonal regression line
正交基 orthogonal basis
正交级数 orthogonal series
正交极 orthopole
正交极化板 cross polarizer plate
正交极化隔离度 directivity separation of orthogonal polarization
正交集 orthogonal set
正交几何(学) orthogonal geometry
正交加边 orthogonal bordering
正交加速度 normal acceleration
正交减震器 orthogonal dashpot
正交检波 orthogonal detection
正交交叉 right-angle intersection; square crossing
正交接管 orthogonal nozzle
正交接合 orthogonal joint; perpendicular joint
正交节平面 orthogonal nodal plane
正交解调 quadrature demodulation
正交解调器 quadrature demodulator
正交晶的 orthorhombic; rhombic(al)
正交晶格 orthorhombic(al) lattice; rhombic(al) lattice
正交晶硫 rhombic(al) sulfur
正交晶系 orthorhombic(al) system; rhombic(al) system
正交晶型铬酸铅 rhombic(al) lead chromate
正交静电磁场 crossed static electric(al) and magnetic field
正交矩阵 orthogonal matrix
正交均等核对方程 orthogonalized parity-check equation
正交均等核对和 orthogonal parity check sum
正交均方拟合 orthogonal mean square fit
正交拉丁长方形 orthogonal Latin rectangles
正交拉丁方 orthogonal Latin squares
正交拉丁立方 orthogonal Latin cubes
正交肋 orthogonal rib
正交棱镜 cross(ed) prisms
正交力 normal force
正交连接 perpendicular joint
正交灵敏度 quadrature sensitivity
正交流 cross current
正交流变仪 orthogonal rheometer
正交流动 cross-flow
正交流动区域精炼炉 cross-flow zone refiner
正交硫 rhombic(al) sulfur
正交滤波 quadrature filtering
正交滤波器 orthogonal filter
正交律 orthogonal law
正交螺旋线 normal helix
正交落差 perpendicular throw
正交码 orthogonal code
正交幂等元 orthogonal idempotent element
正交面 normal surface
正交描述符 orthogonal descriptor
正交模式 orthogonal mode
正交模式转换器 orthomode transducer
正交模态 orthogonal mode; orthonormal mode
正交模态矢量 orthogonal modal vector
正交尼科耳棱镜 cross(ed) Nicols; cross Nicol prism; Nicol crossed
正交拟群 orthogonal quasi-groups
正交黏[粘]弹体 orthogonal viscoelastic body
正交排列 cross spread
正交排列阵 orthogonal array
正交(判别)准则 orthogonality criterion
正交配筋 orthogonal reinforcement
正交配偶 orthogonal mate
正交喷吹灭弧盒 cross-blast explosion pot
正交喷吹油断路器 cross-blast oil circuit-breaker
正交偏光(镜) crossed Nicols
正交偏振 crossed Nicols; cross-polarization; orthogonal polarization
正交偏转 orthogonal deflection
正交频分复用 orthogonal frequency division multiplexing
正交平衡调制 quadrature balanced modulation
正交平面波 orthogonalized plane wave
正交平面框架 orthogonal plane frame
正交平坦形 orthogonal flat
正交剖面 orthogonal section; profile
正交谱 quadrature spectrum
正交谱线密度 cross spectral density
正交奇偶校验 orthogonal parity check
正交奇偶校验和 orthogonal parity check sum
正交桥 orthotropic bridge; right bridge; square bridge
正交切削 orthogonal cutting
正交穹顶 groin vault
正交穹隆 groyne(d) vault
正交球面坐标 orthogonal spherical coordinates
正交曲线 normal curve; orthogonal curve
正交曲线系 orthogonal system of curves
正交曲线系统 orthogonal curve system
正交曲线坐标 curvilinear orthogonal coordinates; orthogonal curvilinear coordinates
正交曲线坐标系 orthogonal curviline-

ar coordinate system
正交取样 quadrature-sampled
正交群 orthogonal group
正交群对 orthogonal group pair
正交绕组 quadrature winding
正交刃 orthogonal kets
正交扫描 orthogonal scanning
正交设计 orthogonal design
正交设计法 orthogonal design method
正交射 orthomorphism
正交射影的 orthographic(al)
正交射影法 method of orthogonal projection; orthogonal projection; orthography
正交失真 quadrature distortion
正交矢量 orthogonal vector
正交视图 orthogonal view
正交试验 orthogonal experiment; orthogonal test
正交试验法 orthogonal design method
正交试验设计 cross experimental design; orthogonal experimental design
正交双向异性钢桥面板 steel orthotropic(al) bridge deck
正交算子 orthogonal operator
正交随机过程 orthogonal stochastic process
正交特征函数 orthogonal characteristic function
正交天球 right sphere
正交天线 orthogonal antennas
正交条件 condition of orthogonality; normality condition; orthogonality condition
正交条件方程 orthogonalized condition equation
正交调幅 quadrature amplitude modulation
正交调节 quadrature-adjust
正交调制 quadrature modulation
正交调制彩色信号 quadrature-modulated chrominance signal
正交调制器 quadrature modulator
正交调制载波 quadrature-modulated carrier
正交跳频 quadrature frequency hopping
正交投影 orthogonal projection; rectangular projection
正交投影点 orthogonal projection point; rectangular projection point
正交投影法 orthography
正交投影图 rectangular chart
正交完全可约性 orthogonal complete reducibility
正交网络 quadrature network
正交维数 orthogonal dimension
正交位移 orthogonal translation; perpendicular displacement; quadrature displacement
正交误差 quadrature error
正交系(统) orthogonal system
正交弦 latus rectum
正交线 cross lines; perpendicular line
正交线分析栅云纹法 grid-analyser moiré method
正交线性变换 orthogonal linear transformation
正交线寻址点矩阵 crossbar addressed dot matrix
正交相伴的 orthogonally associated
正交相干解调器 quadrature coherent demodulator
正交相位 quadrature phase
正交相位分离电路 quadrature phase splitting circuit
正交相位分量 quadrature phase component
正交相位副载波信号 quadrature phase subcarrier signal
正交相移键控 quadrature phase shift key
正交向量 orthogonal vector
正交像片 orthophoto
正交像片地图 orthophotomap
正交信号 orthogonal signal
正交性 orthogonality; orthogonality property
正交压力 pressure at right angles
正交异性板 orthotropic plate; orthotropic slab
正交异性板面桥 orthotropic steel deck bridge
正交异性钢桥面 orthotropic steel bridge deck
正交异性钢桥面板 steel orthotropic(al) decking
正交异性桥面板箱形梁 orthotropic deck box girder
正交异性曲板 orthotropic curved plate
正交异性扇形板 orthotropic sector plate
正交异性上承桥设计 orthotropic deck bridge design
正交因子法 orthogonal factorization method
正交应变 orthogonal strain
正交映射 orthogonal mapping
正交余 orthogonal complement
正交元素 orthogonal elements; orthogonal quantities
正交原理 principle of orthogonality
正交原子轨道 orthogonal atomic orbital
正交圆 orthocycle; orthogonal circle
正交运动 orthogonal motion; quadrature motion
正交运算 orthogonal operations
正交载波 quadrature carrier
正交增量性 orthogonal increment
正交增压器 quadrature booster
正交展开 orthogonal expansion
正交张量 orthogonal tensor
正交阵列 orthogonal array
正交振荡器 quadrature oscillator
正交直和 orthogonal direct sum
正交直线 orthogonal straight-lines
正交直线组 orthogonal lines
正交直线坐标 orthogonal Cartesian coordinates
正交轴 ortho-axis
正交轴齿轮 rigging-angle gear; right-angle gear
正交轴电抗 quadrature axis reactance
正交轴面 orthopinacoid
正交轴线 mutually perpendicular axis; ortho-axis; orthogonal axes; orthogonal axis; quadrature axis
正交主轴组 orthogonal set of principal axis
正交转动对称的 orthotropic
正交装置 orthogonal array
正交锥面 orthogonal cone
正交子空间 orthogonal subspaces
正交(纵横)差错控制 orthogonal error control
正交走向谷【地】 orthogonal valley; diaclinal valley
正交族 orthogonal families
正交阻尼 cross damping
正交坐标 orthogonal coordinates
正交坐标系 orthogonal coordinate system
正胶体 positive colloid
正焦弦 latus rectum
正角 positive angle
正角落 positive corner
正角闪岩 ortho amphibolite
正角式减速器 right-angle speed reducer
正校正 positive correction
正阶 positive exponent
正接 electrode negative; normal polarity; straight polarity
正接触 square bearing
正街 high street
正截口 normal section
正截面 head-on cross-section; normal section; right section
正截水墙 positive cut-off
正截影 right section
正解理【地】 orthoclastic
正经 the twelve channels
正晶粒 subgrain
正晶体 positive crystal
正静液压头 positive static liquid head
正镜【测】 circle left; direct telescope; face left(position); normal position of telescope; telescope direct; telescope in normal position; face left position(of telescope); left circle
正镜测量 direct position of telescope; normal position of telescope
正镜读数 right reading
正镜角 direct angle
正镜位置【测】 direct position of telescope
正矩阵 positive matrices; positive matrix
正距平区【气】 pleion
正距平中心【气】 pleion
正锯齿形电压 positive-going sawtooth voltage
正均衡 positive balance
正抗蚀图 positive resist pattern
正考顿效应 positive Cotton effect
正可透变扭器 positive-permeable torque converter
正空间 positive space
正控制 positive control
正口转炉 concentric(al) converter
正扣 right hand thread
正扣螺纹 right-handed thread
正扣钻杆 right-hand joint
正矿物 plus mineral; positive mineral
正馈电线 positive feeder
正馈线 positive feeder; positive wire
正拉索 positive stay
正类型 positive type
正棱镜 regular prism
正棱柱(体) regular prism
正棱锥(体) orthoprism; orthopyramid; regular pyramid; right pyramid
正离子 cation; cationic; kation; positive ion
正离子发射 positive ion emission
正离子束 positive ion beam
正离子衰变 decay by positron emission
正离子直线加速器 positive ion linear accelerator
正力矩 positive moment
正力矩钢筋 positive moment reinforcement
正力型 orthotonic type
正立方体 normal cube
正立面 façade; front elevation
正立面投影 frontal projection
正立面图 front elevation drawing
正立体 positive stereo
正立体观测 orthostereoscopy
正立体图 normal stereogram
正砾岩 orthoconglomerate
正粒序层理构造 normal graded bedding structure
正粒序构造 normal graded structure
正粒序火山泥球构造 normal graded volcanic mud ball structure
正粒子 positive corpusc(u)le
正粒子束 positive beam
正链 plus strand; positive strand
正链的 normal
正链复形 positive chain complex
正梁 ridge purlin(e)
正亮氨酸 nor-leucine
正量 commercial weight; positive; positive quantity
正裂 normal fracture
正磷酸 orthophosphoric acid
正磷酸钙 tertiary calcium phosphate
正磷酸铝 alumin(i)um orthophosphate
正磷酸镁 tribasic magnesium phosphate
正磷酸盐 normal phosphate; orthophosphate
正零 plus zero; positive zero
正/零/负码速调整 positive/zero/negative justification
正流 drag flow
正流槽型长方形浮选机 oblong-type machine
正流流体裂化装置 orthoflow fluid cracking unit
正流式 orthoflow; orthoforming
正流式过程 orthoflow process
正流式重整过程 orthoforming process
正流重整装置 orthoforming unit
正硫的 sulphuric
正硫化 optimum cure; plateau cure
正硫酸铜 cupric sulfate
正六边形 orthohexagon; regular hexagon
正六方形的 orthohexagonal
正六面体 cube; cubic; hexahedron; regular hexahedron
正路 the correct path; the right way
正绿方石英 lussatite
正逻辑 positive logic
正螺纹 plus thread; positive spiral striation; right-spiral screw
正螺旋面 right helicoid
正落差 normal throw
正码速调整 positive justification
正码子集合 correct subset
正脉冲 positive impulse; positive pulse
正脉冲触发 positive-going trigger
正脉冲计数 positive pulse counting
正脉冲塞入 positive pulse stuffing
正门 entrance porch; front door; front entrance; front gate; gate hono(u)r; grand entrance; main door; main entrance; main portal; portal; gate of hono(u)r <庆典用的>; frontispiece
正门的镂空花纹山墙 portal openwork gablet
正门建筑 portal architecture
正门三角墙端点上的饰物 <希腊建筑> end acroterion
正门柱 portal column
正糜棱岩 orthomylonite
正密度 positive density
正密封衬垫接箍 <井点总管的> positive seal gasket
正面 face side; frontage; frontal area; frontal surface; front face(side); front side; front surface; inner face; leading face; positive side; positive way; felt side <纸的>; scaena fronts <装饰华丽的舞台后房屋>
正面不带电的部件 dead front
正面敞开的商店 open-front store
正面衬纬 frontal weft-insertion
正面冲击 brunt; positive impact
正面冲剪 front shear
正面冲突 front conflict; head-on colli-

Z

sion
正面打印 front print
正面大看台 <运动场等的> grandstand
正面大楼梯 <剧场等公共场所的> grand stair(case)
正面的 face up;frontal
正面的宽度 frontage
正面堤 front embankment
正面吊(运机) reach stacker
正面吊运汽车式起重机 front handling mobile crane
正面吊运移动式起重机 front handling mobile crane
正面吊运自行式起重机 front handling mobile crane
正面钉合 direct nailing
正面端子型 front terminal type
正面对空观察 frontal coverage
正面分度盘 front index plate
正面封闭型钻塔 closed front tower
正面辐射 head-on radiation
正面供给 face-up feed
正面构图法则 law of frontality
正面观测 normal observation
正面观察 normal view;top view
正面光 front(al) light
正面焊 face bonding
正面焊缝 fillet in normal shear
正面焊接 face bonding
正面焊接法 face-up bonding
正面剪切 front shear
正面角焊 frontal fillet welding;transverse fillet weld
正面角焊缝 fillet weld in normal shear;front(al) fillet weld
正面看台 main stand
正面壳 front case
正面宽度 <建筑物> face measure
正面馈送 face-up feed
正面沥青胶块 asphaltic facade slab
正面连接 front connection
正面临街用地 frontal land
正面面积 front face area
正面目标 frontispiece
正面腻子 face putty
正面碰撞 positive impact
正面千斤顶 <盾构工程用> face jack
正面墙 facade wall;faced wall;front wall
正面清洁格栅 front-cleaned rack
正面入口 front entrance
正面入口上方三角墙 <古希腊建筑> eastern pediment
正面塞线 front cord
正面上的 facial
正面式(机坪) frontal system(apron)
正面(视)图 elevation drawing;front view
正面输送 face-up feed
正面双柱式 in antis
正面填角焊 front fillet weld
正面贴角焊缝 flat fillet weld in front
正面贴角连续焊缝 front continuous fillet weld
正面通风 face airing
正面投影 vertical projection
正面投影面积 transverse projected area;frontal area
正面透视 front perspective
正面图 body plan;elevational drawing;elevational front;head-on view;front elevation;front view
正面涂漆 first-surface painting
正面弯曲试验 face-bend test
正面为玻璃的建筑物 glass-facade building;glass-fronted building
正面颜色 face colo(u)r
正面印刷 first run
正面影像 frontal image
正面硬化 hard-faced
正面支撑 <盾构施工中的> face support
正面柱 frontal column
正面砖 front brick
正面装饰 first-surface decorating
正面装卸叉车 front-loading fork
正面阻力 frontal drag;frontal resistance;head resistance;leading-end resistance
正面阻力张线 drag wire
正面钻炮眼 boring shot-holes in the face
正模标本 holotype
正模函数 positive modular function
正模耦合器 orthomode coupler
正模腔 orthomode cavity
正模造船法 upright hull-building
正摩擦 positive friction
正母线 positive bus;positive bus-bar
正目镜 positive eyepiece
正钼的 molybdic
正南 due south
正南偏东 south by east
正南偏西 south by west
正挠钢筋 positive reinforcement
正能波 positive energy wave
正能量 positive energy
正能态 positive energy state
正泥晶灰岩 orthomicrite
正泥晶亮晶灰岩 orthomicrosparite
正拟态长石 orthomic feldspar
正年代化石 orthochronological fossil
正啮合 positive mesh
正镍的 nickelous
正镍化合物 nickelous compound
正镍纹石 orthotaenite
正扭 positive twist
正偶数 positive even numbers
正排量泵 positive-displacement pump
正排量齿轮泵 positive-displacement gear-type pump
正排量电动机 positive-displacement motor
正排量定量液压泵 constant positive displacement hydraulic pump
正排量空气压缩机 positive-displacement air compressor
正排量流量计 <容积式的> positive-displacement meter
正排量螺旋式压气机 positive-displacement screw type compressor
正排量式燃料泵 positive-displacement type fuel pump
正排量式压缩机 positive-displacement compressor
正排量式液压传动装置 positive-displacement hydraulic drive
正排量型 positive-displacement type
正排量液压泵 hydraulic positive displacement pump
正排流量计 positive-displacement meter
正排水泵 positive-displacement pump
正排水齿轮泵 positive-displacement gear-type pump
正排水流量计 <容积式的> positive-displacement meter
正牌子 standard brand
正派的 decent
正碰 central collision;direct impact;head-on collision
正劈锥曲面 right conoid
正片 photographic(al) positive;positive film;positive photograph;positive picture;positive plate;positive print
正片法 positive method
正片反差 density difference of positive
正片记录 positive writing
正片麻岩 orthogneiss
正片面 positive plane
正片拼贴底图 film mosaic
正片显影液 positive developer
正片镶嵌 positive mounting
正片榍石 orthoguarinite
正片岩 orthoschist
正偏差 overga(u)ge;plus deviation;positive deviation
正偏差轧制 overga(u)ge
正偏电阻 pull-up resistor
正偏流 positive bias
正偏态 positive skew(ness)
正偏态对数正常分布 positive skewed lognormal distribution
正偏态分布 positive skewed distribution
正偏析 normal segregation
正偏斜 positive skew(ness)
正偏压 copper bias;positive bias
正偏压电池 positive biasing battery
正偏移 positive skew(ness)
正偏置 positive bias
正偏置的 positively biased
正频率 positive frequency
正品 certified products;custom grade;normal products;product plus;quality goods
正品率 percentage of class A goods
正平方根 positive square root
正平衡 positive balance;positive regime【水文】
正平截头体 regular frustum
正平面 horizontal frontal plane;frontal plane
正平线 frontal line;horizontal frontal line
正平行六面体 right parallelepiped
正平移断层 normal wrench fault
正坡 plus grade
正坡度 positive slope
正坡倾斜隔水底板 positively inclined impervious bottom bed
正坡下降曲线的巴氏函数 Baplorfstz function of depression curve for positive inclination
正剖面 front cross-section;normal cross section;normal section;orthogonal section
正谱 ortho-spectrum
正齐次的 positively homogeneous
正齐次算子 positive-homogeneous operator
正齐性 positive homogeneity
正砌方块立堤 vertical breakwater of normal placed blocks
正砌(水平砌缝)方块 blocks in horizontal bond
正迁移 positive transfer
正铅板 positive plate
正铅板组 positive group
正铅的 plumbean;plumbeous
正前方 dead ahead;rigging ahead;right ahead
正前角 positive rake;positive rake angle;positive top front rake
正强化 positive reinforcement
正墙 head wall
正羟锌石 wiilfingite
正桥 <桥面为长方形> right bridge;main bridge
正切 secant;straight-across-cut;tangent
正切 X 线照相术 tangential roentgenography
正切标尺 tangent scale
正切尺 tangent
正切丛 tangent bundle
正切的 tangential
正切电流计 tangent galvanometer
正切定律 tangential law;law of tangents
正切定线 correct alignment
正切法 tangent method
正切规 Gemowinkel
正切函数 tangent function
正切弧 tangent arc
正切畸变 tangential distortion
正切焦点 tangential focus
正切角 tangent angle
正切卵形线 tangent ogive
正切螺旋 tangent screw
正切瞄准具 tangent sight
正切模量 tangent modulus
正切模数 tangent modulus
正切曲线 tangent curve
正切曲线图 tangent chart
正切损失 loss tangent
正切弹性模数 tangential modulus
正切透平 tangential turbine
正切纬度误差 tangent latitude error
正切误差 tangent error
正切线 tangential line
正切像场 tangential image field
正切真数 natural tangent
正倾型 anacline
正球面投影 normal stereographic(al) projection
正区 plus zero
正曲率 positive camber;positive curvature
正曲率薄壳 shell of positive curvature
正曲率的曲面 surface of positive curvature
正趋光性 photopositive
正趋化性 positive chemotaxis
正趋性 positive taxis
正全色的 ortho-panchromatic
正确曝光部分 correct-exposure portion
正确标出的剂量 weighted portion
正确波形 precision waveform
正确布置 in position
正确操作 proper operation
正确测定故障 pinpoint the trouble
正确尺寸 just size
正确次序 proper order
正确、错误、遗漏计数器 right-wrong-omits counter
正确单位 exact unit
正确到毫厘不差 exact in every particular
正确的 correct;errorless;exact;right;veritable
正确度 order of accuracy
正确发射 clear launch
正确范围 exact category;verification scope
正确隔距长度 true ga(u)ge length
正确混合 proper mixing
正确混合比 correct mixture
正确混合气 correct mixture
正确瞭望到的信号示像 properly observed signal aspect
正确率 validity rate
正确平衡 correct balance
正确评估 good faith estimate
正确铺筑的(道路)表面 good level surface
正确设计的公路 "tailor-made" highway
正确设计的路面板厚度 "tailor-made" slab thickness
正确声调 just intonation
正确时间 orthochronous
正确坍落度 true slump
正确调整 tram
正确图纸 certified drawing

正确位置 tram
正确位置测设 setting-out correct position
正确线向 proper alignment
正确性 truth;correctness;exactitude
正确运行 true-running
正确整型的 <如路基、土路等> well-shaped
正确子集 correct subset
正染色的 orthochromatic
正绕绕组 forward wound winding
正热平衡 direct heat balance
正热平衡法 direct method heat balance
正容差 positive allowance
正溶的 orthotectic
正入射 normal incidence
正入射日射强度表 normal incidence pyrheliometer
正入射吸声系数 normal incident absorption coefficient
正入射吸收系数 normal incidence absorption coefficient
正入射阻抗 normal impedance
正闰秒 positive leap second
正三测投影 trimetric projection
正三角形 equilateral triangle;regular triangle
正三角形标准螺纹 standard full V thread
正三角形错列 staggered triangular pitch
正三角形直立 in-line triangular pitch
正三轴测图 trimetric drawing
正三轴投影 trimetric projection
正伞齿轮减速机 bevel spur gear drive
正色 orthochrome
正色纯保护环 positive guard band
正色的 orthochromatic
正色底片 orthochromatic plate
正色感光膜 orthochromatic coating
正色感光片 orthochromatic film
正色胶卷 orthochromatic film
正色胶片 orthochromatic film
正色滤光器 orthochromatic filter
正色片 orthochromatic photographic(al) material;orthochromatic plate;panchromatic photographic(al) material
正色乳剂 orthochromatic emulsion
正色散 positive dispersion
正色摄影图 orthophoto
正色像纸 orthochromatic film;orthochromatic paper
正色性 orthochromatism
正色性乳剂 orthochromatic emulsion
正砂屑岩 orthoarenite
正筛孔 plus mesh
正栅电流 positive grid current
正栅多谐振荡器 positive grid multivibrator
正栅管 brake-field tube
正栅极 positive grid
正栅极特性 positive grid characteristic
正栅拦图 normal fence diagrams
正栅压闸流管 positive tube
正栅振荡器 positive grid oscillator;retarding field oscillator
正商誉 goodwill positive
正熵 positive entropy
正上链复形 positive cochain complex
正上挠度 positive camber
正烧石灰 normally burning lime;normally burnt lime
正射的 orthographic(al)
正射断路器 orthojector circuit-breaker
正射法 orthography
正射反射率 normal incidence reflectivity
正射负片 orthonegative
正射纠正 orthographic(al) rectification
正射亮度 normal brightness
正射投影 orthogonal projection;orthographic (al) projection;orthophoto-projection
正射投影测图仪 orthocartograph;orthogonal projection cartograph;orthomapper
正射投影地图 orthogonal projection map;orthographic (al) projection map;orthophotomap
正射投影法 orthography;orthophotography
正射投影海图 orthographic(al) chart
正射投影绘图仪 orthophoto plotter
正射投影技术 orthophoto-technique
正射投影纠正仪 orthophotoscope;orthoscopic rectifier
正射投影立体测图仪 <加拿大制造> Stereocompiler
正射投影立体观测 orthostereoscopy
正射投影图 orthograph
正射投影系统 orthophoto-system
正射投影像片 orthophotograph
正射投影像片镶嵌图 orthophotomosaic
正射投影像片组合图 orthophotomap
正射投影仪 orthogonal projector;orthophoto instrument;orthophotomat;orthophoto-projector;orthophotoscope;ortho-projection equipment;orthoprojector;orthoscope
正射投影装置 orthophot
正射透视 orthogonal perspective
正射图 orthograph
正射位置 orthographic(al) position
正射线 anode ray;canal ray;positive ray
正射线抛物径迹 positive-ray parabolas
正射相片 orthophotograph
正射像片 orthographic (al) photograph;orthophoto;orthophotograph
正射像片成像图 orthophoto-restitution
正射像片底图 orthophotobase
正射像片扫描器 orthophoto-scanner
正射像片讨论会 orthophoto-workshop
正射像片图 orthophotoplan
正射像片镶嵌图 orthophotomosaic
正射像片影像 orthophoto-image
正射像片坐标 orthographic(al) coordinates
正射影变换 direct projectivity
正射影像地图 orthophotomap
正射影像图 orthophotoplan;orthophotoquad
正射照片比例尺 orthophoto scale
正射照片成图 orthophoto-restitution
正射坐标 orthographic (al) coordinates
正摄像管 orthiconoscope
正摄像仪 orthodiagraph
正摄影 positive photograph
正砷酸盐 ortho-arsenate
正升力 positive lift force
正失调 positive incoordination
正十边形 regular decagon
正十二面体 regular dodecahedron
正十六烷 n-hexadecane
正十六(烷)酸 palmitic acid
正十四烷醇 n-tetradecanol
正十字形平面 Greek-cross plane
正石英的 orthoquartzitic
正石英砾岩 orthoquartzitic conglomerate
正石英砂岩 quartzitic sandstone
正石英岩 orthoquartzite;quartz arenite
正石英岩质砾岩 orthoquartzitic conglomerate
正石柱 monolith
正时 timing
正时标记 timing mark
正时差 <网络计划中的> positive float
正时齿轮 front gears;timing gear
正时齿轮衬 timing-gear gasket
正时齿轮的摆动 run-out of timing gear
正时齿轮壳 timing-gear housing
正时齿轮室隔板 front timing plate
正时锤 timing chain
正时点火 correct timed ignition;timing ignition
正时间移动 plus time shift
正时扇形 timing sector
正时图 timing diagram
正时系统 timing system
正时钟脉冲 positive clock
正实函数 positive real function
正实矩阵 positive real matrix
正实阵 positive real matrix
正矢【数】versed sine;mid-ordinate;versine
正矢冲击脉冲 versine shock pulse
正矢法 middle ordinates method
正矢计 versine recorder
正矢检查 versine inspection
正矢校平仪 versine equalizing instrument
正式保额 definite order
正式报告 official report;official return
正式标准 official standard
正式餐厅 <古罗马房屋中的> coenatio
正式成员 full member;participating member
正式代理人 official agent;official assignee
正式单位 formal unit
正式道歉 formal apology
正式的 authentic;formal;official
正式地层单位 formal stratigraphic(al) unit;formal unit
正式调查 formal investigation
正式毒性试验 definitive toxicity test
正式发票 definite invoice;official invoice
正式法 formal law
正式访问 formal visit;official visit
正式分类 official classification
正式工人 regular worker;worker officially employed
正式工作时间 straight time
正式工作试验 official service test
正式公报 official bullet
正式公告 formal advertising
正式孤斗 grapple
正式关系 formal relationship
正式广告 formal advertising
正式合法当局 duly constituted authority
正式合同 formal contract;official contract;sealed contract;written contract
正式核准 official authorization
正式环境教育 formal environmental education
正式会谈 colloquy;formal talk
正式会员 full member;member of full standing
正式计划 formal plan;official plan
正式记录 formal record;official record
正式价格 office price
正式检查 formal inspection
正式解除债务 acceptilation
正式进口 formal entry
正式经纪人 inside broker
正式警告 formal warning
正式决定 definite decision
正式决算 official account
正式开通 officially open
正式开张 officially open
正式领导 <拥有组织结构中的正式职位、权力和地位> formal leader
正式命名 definite designation
正式盘存日期 official inventory date
正式批准 formal approval;official approval
正式契约 formal contract;official contract;sealed contract
正式签署 official signature
正式渠道 official channel
正式权力 formal authority
正式确认批准 homologate
正式任命 permanent appointment
正式生产 mature production
正式生效 formal effective
正式声明 official statement
正式使用前试用操作 prelife operation
正式试航 official sea trial;official speed trial;official trial
正式试验 official test
正式试验法 ortho-test
正式收报时间 official time of receipt
正式收据 accountable receipt;formal receipt;official receipt
正式手续 formality
正式书面陈述 affidavit
正式书写 engross
正式讨论 disquisition
正式提单 official bill of lading
正式通车 officially open
正式通过 adoption
正式通知 formal notice
正式投产 mature production
正式投产前的检验 preproduction inspection or test
正式投入运行前的试验 preproduction test
正式途径 official channel
正式委托代理人 authorised agent
正式文本 authorized documents;official copy;official text;transcript
正式文件的交存 deposit of formal instrument
正式协定 formal agreement
正式协议 formal agreement
正式信号示像 proper signal aspect
正式验收 administrative approval;formal acceptance;official acceptance
正式验收试验 formal acceptance test;official acceptance test
正式业务文件 official service documents
正式意见 deliverance
正式印章 <公司的> common seal
正式有资格的独立检查人 duty qualified independent inspector
正式语言 literary language;official language
正式员工 permanent staff
正式原图 office-master drawing;official master drawing
正式运行 commencement of commercial operation
正式运转 production run
正式证明 formal proof;writ
正式证券 definitive bond
正式执照 full license
正式职工 regular staff
正式职员 regular employee
正式仲裁人 official referee
正式组织 formal organization

正势 positive potential
正视 confrontation
正视表示法 orthographic (al) representation
正视的 orthoptic (al)
正视观测 orthogonal observation
正视观察 orthoscopic view
正视画法 orthometric drawing
正视镜 orthoptoscope
正视立体观测 orthoscopic stereo view
正视立体观察 orthoscopic stereo viewing
正视立体效果 orthoscopic effect
正视频信号 positive video signal
正视图 elevation; elevation (al) diagram; elevational drawing; elevation view; front elevation; front elevation view; front outline; front side; front view; head-on view; orthograph; front elevation
正视图投影 orthographic (al) projection
正视颜色 face colo(u)r
正视眼 emmetropia; emmetropic eye
正是此物 the very same thing
正铈的 cereus
正铈异常 positive cerium anomaly
正手刀 right-hand cutter; right-hand tool
正手的 forehand
正手焊 forward welding
正手焊法 forehand welding
正输出 positive output
正输出量 positive output
正输入负输出元件 PINO[positive input negative output] element
正输入值 positive input value
正数 positive number
正数序列 sequence of positive numbers
正双曲线 right hyperbola
正双线性形式 positive bilinear form
正双折射 positive birefringence
正水锤梯度 positive water-hammer gradient
正水头 positive head
正四方块 die square
正四面体 regular tetrahedron
正四面体取向 regular tetrahedral orientation
正速变速箱 positive speed gearbox
正酸 normal acid; ortho acid
正算问题 direct problem
正算子 positive operator
正隧道 main tunnel
正铊的 thallic
正台阶 underhand bench
正台阶采矿法 heading and bench mining method
正台阶回采 heading and stope mining
正台阶回采法 heading and stope system
正台阶开挖 top bench tunnel (l) ing
正台阶开挖法 <隧道> heading and bench system; heading bench method
正台阶开凿法 heading and bench cutting method
正台阶式爆破 underhand blasting
正台阶式的 underhand
正台阶式回采 underhand work
正台阶式掘进 underhand work
正台阶式开挖 underhand work
正台阶式凿岩 underhand work
正态 ortho-state
正态 S 形曲线 normal sigmoid curve
正态逼近 normal approximation
正态比例 natural scale; normal scale
正态变换 normalization transformation; normal transformation
正态不变性 permanence of normality
正态船模 undistorted ship model
正态等差 normal equivalent deviate
正态对数分布 log-normal distribution
正态对数机率级数 normal logarithmic probability series
正态多元分析 normal multivariate analysis
正态法则 normal law
正态方程 normal equation
正态分布 Gaussian distribution; Gaussian normal distribution; normal distribution <一种统计分布,可用一个算术平均数和标准差来表征曲线呈钟形>
正态分布带 zone of normal distribution
正态分布的观测值 normally distributed observations
正态分布法 method with normal distribution
正态分布函数 normal distribution function
正态分布律 normal distribution law; normal law
正态分布区 zone of normal distribution
正态分布曲线 normal distribution curve; normal frequency curve
正态分布图 normal distribution diagram
正态分布圆 circular normal distribution
正态分量 normal component
正态峰 mesokurtosis
正态概率 normal probability
正态概率表 normal probability paper
正态概率分布 normal probability distribution
正态概率密度 normal probability density
正态概率曲线 normal probability curve
正态概率纸 normal probability paper
正态高斯分布函数 normal Gaussian distribution function
正态过程 normal process
正态函数 normal function
正态化 normalize
正态机率曲线 normal probability curve
正态检验 normality test
正态近似 normal approximation
正态距平 normal deviate
正态均方差 unbias(s)ed variance
正态离差 normal deviate
正态量 normal quantities
正态律 normal law
正态密度 normal density
正态密度函数 normal density function
正态面波 normal mode surface wave
正态模式 normal mode; undistorted mode
正态模型 normal model; undistorted model; undisturbed model
正态耐性分布 normal tolerance distribution
正态偏差 normal deviate; normal deviation
正态频率 normal frequency
正态频率分布 normal frequency distribution
正态频率曲线 normal frequency curve
正态频率特性 Gaussian frequency response
正态平稳过程 normal stationary process
正态平稳序列 normal stationary sequence
正态谱 normal spectrum
正态曲面 normal surface
正态曲线 normal curve
正态曲线拟合 normal curve fitting
正态双变量 normal bivariable
正态随机变量 normal random variable
正态随机变数 normal random variable
正态随机过程 normal random process; normal stochastic process
正态随机函数 normal random function
正态随机数 normal random number
正态投影 orthogonal projection; orthographic (al) projection
正态透镜 positive lens
正态误差定律 normal error law; normal law of errors
正态误差法则 normal error law; normal law of errors
正态误差分布曲线 normal error curve
正态误差规律 normal error law
正态误差曲线 normal curve of error
正态线性比(例)尺 normal linear scale; undistorted linear scale
正态相关 normal correlation
正态相关函数 normal correlation function
正态相关曲面 normal correlation surface
正态性 normality
正态性 D 检验法 D-normality test
正态性检验 normality test; test of normality
正态震源地震 normal-focus earthquake
正态直方图 normal histogram
正态中位绝对差 normal deviate medium
正态总体 normal population
正钛 titanic
正坍落 <混凝土> true slump
正弹性模量 normal elastic modulus
正碳离子 carbonium ion
正特性失真 positive characteristic distortion
正梯段掘进 heading and cut
正体的卫生单间 sanitary
正体倾斜体层摄片 tilted anterior tomogram
正体文字 Roman letter
正体学 Romon type
正体字 upright letter; upright type
正调频 positive frequency modulation
正调制 positive modulation
正调制传送 positive transmission
正挑丁坝 orthogonal spur dike [dyke]; perpendicular spur dike [dyke]
正铁的 ferric
正铁淦氧 orthoferrite
正铁辉石 orthoferrosilite
正铁氧体 <计算机等用> orthoferrite
正厅【建】 main hall; aisled hall; atrium[复 atria/triums]; auditorium[复 auditoria]; orchestra; parquet <指剧场正对舞台部分>
正厅后排休息室 pit foyer
正厅后座 parquet; parquet circle; parterre; orchestra circle <剧院或音乐厅前排座位区的>
正厅前排贵宾席 orchestra
正挺杆 positive tappet
正同步脉冲 positive clock
正同系物 normal homologue
正同柱 joggle piece; joggle post
正铜的 cupric
正铜铀云母 ortho-torbernite
正统材料 orthodox material
正统的 orthodox
正统方法 orthodox approach; orthodox method
正统命名 legitimate name
正统派的 classic (al)
正统性 legitimacy
正统原则 principle of legitimacy
正头支柱 jury mast
正投影 orthogonal projection; orthographic (al) projection; third angle projection
正投影法 orthography
正投影面积 frontal projected area
正投影图 orthograph; orthographic (al) drawing
正投影图样说明 orthographic (al) design interpretation
正透穿性 positive permeability
正透镜 convex lens; erecting lens; positive lens
正透射 direct transmission; regular transmission
正透摄影器 orthodiagraph
正透视图 orthogonal perspective
正凸多面体 regular convex solids
正突起 positive relief
正推力 positive thrust
正拖尾 positive streaking
正拖影 positive streaking
正椭圆度 positive ovality
正椭圆柱 cylindroid
正弯管试验 normal bend test
正弯矩 positive bending moment; positive moment; sagging bending moment; sagging moment
正弯矩钢筋 positive moment reinforcement
正弯矩能力 positive moment capacity
正弯矩容量 positive moment capacity
正弯曲 positive bending
正弯曲扇块 positive curvature sector
正弯曲试验 <焊缝的> face-bend test
正弯月透镜 positive meniscus lens
正湾扰 positive magnetic bay
正烷烃 n-alkane; normal alkane
正烷烃的溶解度 solubility of n-alkane hydrocarbon
正微子 positrino
正位 normal; normal state; ortho-position
正位错 positive dislocation
正位的 orthophoric; orthotopic
正位点 normal point
正位反射 righting reflex
正位力臂 righting lever
正位力矩 ortho-position moment; righting couple; righting moment
正位力偶 rightening couple; righting couple
正位移泵 positive-displacement pump
正位移鼓风机 positive-displacement blower
正位移计 positive-displacement meter
正位移压缩机 positive-displacement compressor
正位移液体计量器 positive-displacement meter
正位移增压器 positive-displacement blower
正位置 normal attitude
正温的 thermotropic
正温度变化系数 positive temperature coefficient
正温度系数 positive temperature coefficient
正温度系数半导体元件 kaltleiter
正温度系数热敏电阻 posiode; posistor; sensistor
正温度系数陶瓷 positive temperature coefficient ceramics
正文字 text word

Z

正吻【建】dragon-head main ridge ornament
正稳定性 positive stability
正稳性 positive stability
正屋 main building
正钨的 tungstic
正无级变速传动装置 positive infinitely variable driving gear
正无级变速器 positive infinitely variable unit
正无级变速装置 positive infinitely variable unit
正无穷大 plus infinity;positive infinity
正五边形 regular pentagon
正五面体 regular five-hedron;regular pentahedron
正午 day center[centre];high noon;meridian;meridiem;midday;midnoon;noon;noon tide;noontime
正午标 noon-mark
正午常数 noon constant
正午船位报告 noon report
正午间隔 noon interval
正午瞄准 noon sight
正午线 noon-mark
正戊烷 n-pentane;pentane
正戊烷化 normal pentane
正误表 corrigenda;errata;list of errata
正误差 positive error
正西 due west
正吸附 positive absorption
正吸收 positive absorption
正析摄像管 vericon;orthicon
正析像管 orthicon image tube;orthicon;orthiconoscope
正硒的 selenic
正洗井 direct flushing
正系统 positive system
正霞正长岩 juvite
正纤维蛇纹石 orthochrysotile
正弦 sine;sinus
正弦板 sine plate
正弦半波 half-sinusoid
正弦半波冲击脉冲 half sine shock pulse
正弦包线 sinusoidal envelope
正弦泵 sine pump
正弦边界 sinusoidal boundary
正弦变化 sinusoidal variation
正弦变换 sine transform
正弦变量 sinusoidal variable
正弦表 sine table
正弦波 pure oscillation;sine wave;sine-wave oscillation;sinusoid(al oscillation);sinusoidal trace;sinusoidal wave
正弦波包络线 sine-wave envelope
正弦波发生器 harmonic oscillator;sine-wave generator
正弦波分量 component sine wave
正弦波矩形波转换器 square waver
正弦波频率 sine-wave frequency
正弦波频率特性 sine-wave response
正弦波输入 sine-wave input
正弦波输入功率 sine-wave input power
正弦波数字信号发生器 digital sine wave generator
正弦波特性 sine-wave characteristics
正弦波调制 sine-wave modulation
正弦波通量图 sine-wave flux pattern
正弦波限幅器 sine-wave clipper
正弦波响应 sine-wave response
正弦波信号发生器 sine-wave signal generator
正弦波形 sine-waveform;sinusoidal waveform;wave of sinusoidal form
正弦波形脉冲 sinusoidal pulse
正弦波形曲流 sine-wave shaped meander;sinusoidal meander
正弦波形游荡水流 sinusoidal meander
正弦波音 pure tone
正弦波源 sine-wave sources
正弦波振荡器 sine-wave generator;sine-wave oscillator;sinusoidal oscillator
正弦波转换 sine-wave switching
正弦波状流 sinuous flow
正弦不平顺 sinusoidal irregularity
正弦场 sinusoidal field
正弦尺 sine bar;sine protractor
正弦带 sinusoidal strip
正弦的半周 lobe
正弦的激励 sinusoidal excitation
正弦等效烛光图 sine equivalent candle diagram
正弦电动机 sine motor
正弦电流 harmonic current;simple alternating current;sine-wave current;sinusoidal current
正弦电流计 sine galvanometer
正弦电容器 sine capacitor
正弦电压 sinusoidal voltage
正弦定常分布 sinusoidal stationary distribution
正弦定理 sine rule;sine theorem
正弦定律 simple harmonic law;sine law;sinusoidal law;law of sines
正弦对数 logarithmic sine
正弦对准仪 sine center[centre]
正弦发生器 forcing function(al) generator
正弦发送器 sinusoidal sender
正弦法 sine method
正弦分布 sine distribution;sinusoidal distribution
正弦分度器 sine divider
正弦分频器 sinusoidal frequency divider
正弦副载波输出 sinusoidal subcarrier output
正弦干扰 sinusoidal interference
正弦杆 sine bar
正弦公式 sine formula
正弦拱 sinusoidal arch
正弦关系 sine relation
正弦光波 sinusoidal light wave
正弦规 sine bar;sine plate;sine protractor;sine table
正弦函数 sine function;sinusoidal function;symmetric(al) alternating function
正弦弧 sine curve
正弦缓和曲线 sinusoidal curve
正弦会聚 sine-wave convergence
正弦积 sine product
正弦积分 sine integral
正弦积分函数 sine-integral function
正弦级数 sine series
正弦鉴相器 sinusoidal discriminator
正弦交流 simple alternating current
正弦交流电流 simple sinusoidal alternating current
正弦阶跃 sine step
正弦静电计 sine electrometer
正弦(类)螺线 sinusoidal spirals
正弦量 sinusoidal quantity
正弦量角器 sine protractor
正弦流动 sinusoidal flow
正弦律 sine law
正弦螺线 sine spiral
正弦马达 sine motor
正弦脉冲 sine pulse;sine-shaped impulse;sinusoidal impulse;sinusoidal pulse
正弦平方逼近 sine-squared approximation
正弦平方波测试 sine-squared testing
正弦平方彩色测试信号 sine-squared colo(u)r test signal
正弦平方脉冲 sine-squared pulse
正弦平方脉冲成型网络 sine-squared shaping network
正弦平方脉冲响应 sine-squared pulse response
正弦平方脉冲与方波信号 sine-squared pulse and bar signal
正弦平方网络 sine-squared network
正弦曲线 double continuous curve;sine curve;sinusoid;sinusoidal trace
正弦曲线板 sine bar
正弦曲线吊桥 sinusoidal arch bridge
正弦曲线分布 sinusoidal curve distribution
正弦曲线拱 sinusoid arch
正弦曲线拱桥 sinusoidal arch bridge
正弦曲线开合桥 sinusoidal bascule bridge
正弦曲线量 sinusoidal quantity
正弦曲线扫迹 sinusoidal trace
正弦曲线投影 sinusoidal projection
正弦曲线形的 sine-shaped;sinusoidal
正弦曲线形分布荷载 sinusoidal load(ing)
正弦曲线仰开桥 sinusoidal arch bridge
正弦曲线状细度变化 sinusoidal denier variation
正弦扰动 sinusoidal perturbation
正弦扫描 sine sweep
正弦射频电压 sinusoidal RF potential
正弦失真 sinusoidal distortion
正弦式 sinusoid
正弦式变形 sinusoidal deformation
正弦式波 sinusoidal wave
正弦式的 sinusoidal
正弦式反应性变化 sinusoidal reactivity variation
正弦式反应性扰动 sinusoidal reactivity perturbations
正弦式分布 sinusoidal distribution
正弦式荷载 sinusoidal load(ing)
正弦式荷重 sinusoidal load(ing)
正弦式可变电阻器 sine variohm
正弦式时间变化 sinusoidal time variation
正弦式振荡 sinusoidal oscillation
正弦视差 sine parallax
正弦试验 sinusoidal test
正弦输出 sinusoidal output
正弦输入源 sinusoidal input source
正弦数值 sinusoidal quantity
正弦台 sine table
正弦特性曲线 sinusoidal response
正弦条件 sine condition
正弦调变度 sine flutter
正弦调制 sinusoidal modulation
正弦跳跃函数 sinusoidal jump function
正弦图 sonogram
正弦线轨迹 sinusoidal trace
正弦相位调制 sinusoidal phase-modulation
正弦项 sine term
正弦信号 sine signal;sinusoidal signal
正弦信号发生器 sinusoidal signal generator
正弦信号群 wave-packet portion
正弦信号输入 sinusoidal input
正弦形 sinuidal
正弦形缓和曲线 sine-curve transition
正弦形基波 primary sinusoid
正弦序列 sinusoidal sequence
正弦旋转磁场 sinusoidal rotating flux field
正弦压力 sinusoidal pressure
正弦应变动力试验 sinusoidal-strain dynamic(al) test
正弦余弦编码器 sine-cosine encoder
正弦余弦乘积 sine-cosine product
正弦余弦电势计 sine-cosine potentiometer
正弦余弦电位计 sine-cosine potentiometer
正弦余弦发生器 sine-cosine generator
正弦余弦机构 sine-cosine mechanism
正弦余弦旋转变压器 sine-cosine resolver
正弦运动 harmonic motion;sine motion;sine movement;sinusoidal motion;snake motion
正弦载波 sine-wave carrier;sinusoidal carrier
正弦真数 natural sine
正弦振荡 pure oscillation;sine-wave oscillation;sinusoidal oscillation
正弦振荡输入 sinusoidal input
正弦振动 sinusoidal vibration
正弦振动计 sinusoidal vibrometer
正弦震荡 pure oscillation
正弦状构造 sine-curve shaped structure;sinusoidal structure
正现金流量 positive cash flow
正线【铁】main line;main track;through line;A-wire;plus line;positive line;positive wire
正线承力索 messenger wire for main line
正线道岔 main-line switch
正线的 normolineal
正线电码化 main-line code
正线端 positive terminal
正线钢轨连续式道岔 continuous rail point
正线接触线 contact wire for main line
正线联锁设备 interlocking equipment for main line
正线行车 main-line running movement
正线性泛函 positive linear functional
正线性映射 positive linear mapping
正线延展长度 extended length of main line
正线一侧式布置 uniside layout main tracks
正线中心距离 distance between centers[centres] of lines
正相单元 positive area;positive element
正相电抗 positive phase reactance
正相法 normal phase method
正相反 antithesis;right opposite
正相反的 antithetic;diametric(al)
正相关 direct correlation;positive correlation;positive relativity
正相互作用 positive interaction
正相区 positive area
正相输入 non-inverting input
正相似 direct analogy
正相序电压继电器 positive phase-sequence voltage relay
正相运动 positive movement
正响应区 positive response zone
正向 forward;forward direction;positive direction
正向闭锁期间 forward blocking interval
正向闭锁状态 forward blocking state
正向变换 positive-going transition
正向变换核 forward transformation kernel
正向变换器 forward converter
正向波 forward wave
正向波不稳 positive creep
正向波漂移 positive creep
正向波蠕变 positive creep
正向波形不稳 positive creep
正向波形漂移 positive creep

正向波形蠕变 positive creep
正向插入增益 forward insertion gain
正向产生式系统 forward production system
正向铲 forward shovel；face shovel
正向场 normal field
正向冲击 direct impact；normal impact；normal shock
正向冲击挤压法 Hooker process
正向抽水 pumping test from above to below
正向触发器 positive-going trigger
正向传导 forward conduction
正向传递单元 forward transfer element
正向传递函数 forward transfer function
正向传输功率 forward power
正向传输特性 forward characteristic
正向传输阻容网络 forward transmission resistance capacitor network
正向传信通路 forward signal(l)ing path
正向垂直偏转 normal vertical deflection
正向磁场阻抗 forward-field impedance
正向磁化 normal magnetization
正向错位断层 normal-separation fault
正向错误校正 forward of error correction
正向带 direct band
正向导电区 forward conducting region
正向的 positive-going
正向低偿电压 forward offset voltage
正向地貌 convex relief；positive relief
正向地性 positive getropism
正向电导 forward conductance
正向电流 forward current
正向电路 forward circuit
正向电压 direct voltage；forward voltage
正向电压降 forward drop；forward voltage drop
正向电阻 forward resistance
正向动作温度调节器 positive acting thermostat
正向读出 forward reading；read forward
正向对角线 forward diagonal
正向分带 normal zoning
正向分量 forward component
正向功率计 forward power meter
正向功塞 incident power
正向关断期间 forward blocking interval
正向光性 positive phototropism
正向归并 forward merge
正向规则 forward rule
正向互导纳 forward mutual admittance
正向滑动断层 normal-dip-slip fault
正向话终信号 clear forward signal；disconnecting signal
正向换向 proper crossover
正向恢复电压 forward recovery voltage
正向恢复时间 forward recovery time
正向击穿 forward breakdown
正向机械铲 crowd shovel
正向极限 direct limit
正向挤压 forward extrusion
正向计数器 forward counter
正向监控 forward supervision
正向渐近稳定的 positively asymptotically stable
正向锯齿波 positive-going sawtooth wave
正向开关损耗 forward switching loss
正向开挖侧向装土法 straight excavating and side loading process
正向开挖后向装土法 straight excavating and back loading process
正向空间 forward space
正向控制 forward control
正向跨导纳 forward transadmittance
正向馈电传送 feed forward
正向类 positive sense-class
正向离合器 forward clutch
正向力 normal force；positive force
正向链接 forward chaining；link forward
正向量 positive vector
正向灵敏度 axial sensitivity
正向流(程) normal direction flow
正向流动 forward flow；normal flow
正向漏电流 forward leakage current
正向脉冲 direct impulse；direct pulse；positive-going pulse
正向模 direct die；direct mode
正向模式 forward mode
正向排浆装置 positive displacement
正向排水 positive drainage
正向排液灌浆泵 positive-displacement grout pump
正向偏差 positive deviation
正向偏压 forward bias
正向偏压二次击穿 forward biased second breakdown
正向偏压二极管 forward-biased diode
正向偏压整流器 forward-biased rectifier
正向偏置 forward bias
正向平均电流定额 forward mean current rating
正向平均压降 forward mean voltage drop
正向期 normal epoch
正向气流 positive draft
正向箝位 positive clamping
正向区 forward region
正向散射 forward scatter
正向扫描 forward scan；scan forward
正向施控元件 forward controlling element
正向试验法的弯沉 <与回弹弯沉相对而言> normal procedure deflection
正向输入 positive-going input
正向输送 feed forward
正向衰变 forward decay
正向顺序 forward sequence
正向特性 forward characteristic
正向天体 face-on object
正向通道 forward path
正向通路 forward path
正向同步脉冲 positive-going sync pulses
正向突变 forward mutation
正向外加电压 applied forward voltage；forward applied voltage
正向弯曲模型 cambered model
正向微分电阻 forward differential resistance
正向位移 positive displacement
正向文件 term-on-item file
正向稳定的 positively stable
正向系索 yaw guy
正向削波 positive clipping
正向信道 forward channel
正向信号 forward signal；positive-going signal
正向性 positive tropism
正向修剪 forward pruning
正向旋转泵 positive rotary pump
正向旋转波 forward-rotating wave
正向寻址 forward addressing
正向压挤 forward-extrude；forward extrusion
正向压力波 positive pressure wave
正向阳极峰(值电)压 peak forward anode voltage
正向移动 positive shift
正向阅读 reading forward；right reading
正向运动 positive movement
正向再定位 forward repositioning
正向增益 forward gain
正向照明 orthodromic illumination
正向直流电阻 forward direct current resistance
正向转动力矩 positive rolling moment
正向转接信号 forward transfer signal
正向转矩 forward-field torque；forward torque
正向转移导纳 forward transfer admittance
正向转移特性 forward transfer characteristic
正向转移阻抗 forward transfer impedance
正向转折 forward breakover
正向装料机 front-end loader
正向装载机 front-end loader
正向自动增益控制 positive automatic gain control
正向自动增益控制电路 forward AGC [automatic gain control] circuit
正向阻断峰值电压 positive cut-off peak voltage
正向阻抗 forward impedance
正向阻力 head resistance
正向作业 forward job
正项级数 series of positive terms
正像 erect image；image erection；left-to-right；left-to-right reading；normal image；positive image；positive picture；positive print；right reading；direct positive <直接摄取的>；right way round <左右不反转>
正像本 positive copy
正像比例尺 orthophoto scale
正像复印品 positive print
正像工艺 positive process
正像胶片 positive film
正像接收 positive receiving
正像镜 anascope
正像棱镜【测】erecting prism
正像目镜【测】erecting eyepiece；terrestrial eyepiece；positive eyepiece
正像取景器 erect-image view finder
正像数字 erect number
正像透镜 erecting lens
正像透镜系统 erecting lens system
正像望远镜 erect image telescope；erecting telescope；terrestrial telescope
正像系统 erecting system
正像限 positive orthant
正像形地图 orthopictomap
正像寻像器 erect image view
正像重现 positive image reproduction
正像重影 positive echo；positive ghost
正消隐脉冲 positive blanking pulse
正小齿轮 spur pinion
正效果 plus effect
正效应 direct effect
正胁强 normal stress
正斜率 positive slope
正斜镶嵌头 positive back rake tipper bit
正泄放波 positive release wave
正心拱 axial bracket arm
正心桁 eaves purlin(e)；peripheral column supported purlin(e)
正心软心钢 regular soft-center steel
正辛基硫醇 n-octyl mercaptan
正信号 positive signal
正信号接收 positive receiving；positive reception
正行程 forward stroke
正形 orthomorphy；positive form
正形变换 conformal transformation
正形的 conformal；eumorphic；orthomorphic
正形多圆锥投影 conformal polyconic(al) projection；polyconic(al) conforming projection
正形方位投影 azimuthal orthomorphic(al) projection
正形海图 conformal chart；orthomorphic chart
正形球体 conformal sphere
正形射影 conformal projection
正形双重投影 conformal double projection
正形条件 conformal condition
正形投影 conformal projection；orthomorphic projection；conformable projection
正形投影地图 conformal map；orthomorphic chart
正形投影海图 conformal chart；orthomorphic chart
正形投影图 angle diagram；conformal chart
正形投影纬度 conformal latitude
正形投影制图 orthomorphic mapping
正形图 orthomorphic chart
正形性 conformality；orthography property(of maps)
正形圆柱投影 conformal cylindric(al) projection
正形圆锥投影 conformal conic(al) projection；conic(al) orthomorphic projection
正形坐标系 orthomorphic coordinate system
正型 eurymeric；plus
正型函数 positive type function
正型林分 plus stand
正型落点 positive landing
正性 positivity
正性岸线 positive coastline；positive shoreline
正性感光胶 positive-working photoresist
正性光刻胶 positive photoresist
正性海岸线 positive coastline；positive shoreline
正性胶 positive photoresist
正性乳剂 positive emulsion
正性纵坐标 positive ordinate
正序 plain sequence；positive sequence
正序电流 forward-order current；positive-sequence current
正序多相系统 positive-sequence polyphase system
正序分量 positive-order component；positive-sequence component
正序功率 positive-sequence power
正序集 well-ordered set
正序继电器 positive phase-sequence relay
正序原理 well-ordering principle
正序坐标 positive-sequence coordinate
正旋 direct spinning；forward spinning
正旋转 positive rotation；positive spin
正旋转方向 direct rotational direction
正旋转式泵 positive rotary pump
正循环 direct circulation；normal circulation；positive cycle
正循环冲洗【岩】direct circulation washing；direct flushing
正循环冲洗钻进【岩】direct circula-

tion washing drilling
正循环回转钻孔 rotary drilling by pumping of mud
正循环设计 circulation design
正循环旋转钻进 normal-circulation rotary drilling
正循环钻机 circulation drill
正循环钻进 direct circulation boring; circulation drilling
正循环钻孔 direct circulation boring
正循环钻孔法 circulation boring method
正循环钻探 direct circulation boring
正压【电】positive voltage
正压玻璃纤维袋集尘器 pressure-type glass baghouse
正压操作 operation under positive pressure
正压大气 barotropic atmosphere
正压大气模式 thermotropic(al) model
正压袋集尘器 pressure-type baghouse
正压的【气】barotropic
正压电效应 direct piezoelectric(al) effect; piezoelectric (al) direct effect
正压煅烧 positive pressure burning
正压阀 positive valve
正压风扇 positive pressure fan
正压鼓风机 positive blower
正压锅炉 pressure boiler; pressurized boiler
正压环流 borotropic circulation
正压空气系统 positive pressure air system
正压控制闸流管 positive-control thyratron
正压力 normal pressure; plus pressure; positive pressure; pressure above the atmosphere
正压力法 plus pressure process
正压力控制 positive pressure control
正压料斗 pressurized hopper
正压流量计 positive-displacement meter
正压流体 barotropic fluid
正压煤气燃烧器 pressure gas burner
正压密度 orthobaric density
正压排浆装置 positive displacement
正压排气机 positive pressure exhauster
正压气流 positive draft
正压气体 barotropic gas
正压区 zone of positive pressure
正压曲轴箱通风系统 positive crankcase ventilation system
正压燃烧 pressure combustion; pressurized combustion
正压式呼吸器具 positive pressure breathing apparatus
正压疏散路线 pressurized escape route; pressurized evacuation route
正压送风 forced draft; positive pressure ventilation
正压通风 positive draft; positive pressure ventilation; positive ventilating
正压稳定 positive regulation
正压下 direct draught
正压性 barotropy
正压移动泵 positive-displacement meter
正压应力 normal compressive stress
正压重力波 barotropic gravity wave
正亚铁的 ferriferous
正延长 positive elongation
正延性 length slow; positive elongation
正岩浆的 orthomagmatic
正岩浆矿床 orthomagmatic(mineral) deposit
正岩浆期 orthomagmatic stage; orthotectic stage
正沿 positive-going edge
正掩模图 positive mask pattern
正演 convolution
正演计算 positive-going computation
正演滤波系数 filter coefficient of direct development
正阳电子素 orthopositronium
正阳片 right-reading positive; upright diapositive
正样 basic sample
正叶蛇纹石 ortho-antigorite
正曳力 positive drag
正液面 positive meniscus
正移 shuffle
正乙酸铜 cupric acetate
正异常 normal anomaly; positive anomaly
正因子 positive divisor
正阴片 right-reading negative
正应变 direct strain; normal strain
正应力 direct stress; longitudinal stress; normal stress; positive stress
正迎角 positive incidence
正影描记器 orthodiagraph; orthoroentgenograph; skiagraph
正影描记术 orthodiagraphy; orthoroentgenography; orthoskiagraphy
正影描记图 orthodiagram
正影响面积 positive area of influence
正映射 positive mapping
正涌浪 positive surge
正涌泉水头 positive artesian head
正铀的 uranic
正游标 direct vernier
正铕异常 positive europium anomaly
正诱导 positive induction
正余面 positive lap
正余面阀 overlapped valve
正余像 positive after-image
正玉髓 quartzine
正原始光电流 positive primary photoelectric(al) current
正圆窗 fenestra rotunda
正圆球形的 perispheric
正圆柱体 right cylinder
正圆柱形的 right cylindric(al)
正圆柱形堆 right cylindric(al) reactor
正圆锥体 right circular cone
正月解冻 January thaw
正跃迁 positive transition
正运动圆柱形偏心轮 positive motion cylindric(al) cam
正韵律 positive rhythm
正杂基 orthomatrix
正载波 positive carrier
正载流子 positive carrier
正再生 positive regeneration
正在操作 in operation
正在沉没的船 foundering ship
正在发射电波 on-the-air
正在广播 on-the-air
正在航行 making way; underway
正在滑动滑坡 sliding landslide
正在建造的 abuilding
正在结晶的熔浆 crystallizing magmatic melt
正在进行 in hand
正在进行的 on-going
正在进行的工作 on-going operation
正在进行靠船作业的船只 docking vessel
正在进行中的 underway
正在进行中的项目 on-going project
正在进行中的研究计划 on-going research program(me)
正在开发的地区 developing area
正在靠泊的船 berthing vessel
正在靠码头的船舶 berthing vessel
正在慢性恶化的灾难 slowly developing disasters
正在期 period of loading
正在前面 right ahead
正在生锈 underrusting
正在施工 under construction
正在使用的跑道 active runway
正在退潮 on the ebb
正在下沉的井筒 shaft during sinking
正在修理 under repair
正在运行的地热田 on-going-geothermal field
正则 holomorphy
正则闭集 regular closed set
正则闭子集 regular closed subset
正则边界点 regular boundary point
正则变化 canonic(al) change
正则变换 canonic (al) transformation; contact transformation
正则变量 canonic(al) variable
正则表达式 regular expression
正则表示 regular representation
正则补阵 regular complement matrix
正则参数 regular parameter
正则参数系 regular system of parameters
正则测度 regular measure
正则常数 canonic(al) constant
正则簇 regular variety
正则代换群 regular substitution group
正则代数方程 regular algebraic equation
正则单位矩阵 regular unit matrix
正则单形 regular simplex
正则的 canonic(al); holomorphic
正则点 regular point
正则动量 canonic(al) momentum
正则多面体 regular polyhedron
正则方程 normal equation; normalizing equation
正则方程式 canonic(al) equation
正则仿射变换 regular affine transformation
正则分布 canonic (al) distribution; regular distribution
正则分母 regular divisor
正则分歧 regular ramification
正则分析 canonic(al) analysis
正则分析函数 regular analytical function
正则复函数 regular complex function
正则赋范空间 regular normed space
正则覆盖空间 regular covering space
正则覆盖面 regular covering surface
正则根数 canonic(al) element
正则共轭 canonic(al) conjugate
正则共轭变量 canonic(al) conjugate variable; canonically conjugate variable
正则管 regular tube
正则函数 holomorphic function; regular function
正则函数代数 regular function algebra
正则函数的预层 presheaf of regular functions
正则荷载 regular load
正则弧 regular arc
正则化 regularization
正则化变差函数 regularizing variogram
正则化变数 regularizing variable
正则化模态 normalized mode
正则化转置 regularizing transposition
正则环 regular ring
正则换位子 regular commutator
正则积分 regular integral
正则积分流形 regular integral manifold
正则积分元 regular integral element
正则基数 regular cardinal number
正则极大理想 regular maximal ideal
正则集 canonic(al) ensemble; regular set
正则嘉当空间 regular Cartan space
正则假设 regularity assumption
正则解析曲线 regular analytic curve
正则局部方程 regular local equation
正则局部环 regular local ring
正则矩阵 canonic(al) matrix; regular matrix
正则空间 regular space
正则块 canonic(al) block
正则扩充 canonic(al) extension
正则扩张 regular extension
正则连分数 regular continued fraction
正则链 regular chain
正则列 regular column
正则流形 regular manifold
正则路线 canonic(al) path
正则模 modulus of regularity
正则拟凹性 regular quasi concavity
正则逆阵 regular inverse matrix
正则排序 canonic(al) ordering
正则剖分 regular subdivision
正则谱序列 regular spectral sequence
正则奇点 regular singularity; regular singular point
正则嵌入 regular embedding
正则区间 regular interval
正则曲面 regular surface
正则曲面元素 regular surface element
正则曲线 regular curve
正则容度 regular content
正则射影 regular projection
正则射影变换 regular projective transformation
正则摄动 regular perturbation
正则摄动系统 regular perturbation system
正则时间单位 canonic(al) time unit
正则收敛 regular convergence
正则双曲算子 regularly hyperbolic operator
正则条件 regularity condition
正则图 regular graph
正则外测度 regular outer measure
正则位置 regular position
正则无偏临界域 regular unbia(s)sed critical regions
正则误差 normal law error
正则系统 canonic(al) system; regular system;
正则系综 canonic (al) ensemble; canonic(al) assembly
正则线性映射 regular linear mapping
正则相关 canonic(al) correlation
正则相关系数 canonic(al) correlation coefficient
正则斜纹 regular twill
正则星形线 regular asteroid
正则形式 canonic(al) form
正则型 eigenmode
正则性 regularity
正则性参数 parameter of regularity
正则性条件 regularity condition
正则性质 regularity property
正则序列 regular sequence
正则序数 regular ordinal; regular ordinal number
正则映射 regular mapping
正则有理映射 regular rational mapping
正则有向图 regular directed-graph
正则右理想 regular right ideal
正则展开 canonic(al) extension
正则振动型法 normal mode method
正则振型的正交性 orthogonality property of normal modes

正则置换群 regular permutation group
正则转移概率 regular transition probabilities
正则状态方程式 canonic(al) equation of state
正则子流形 regular submanifold
正则坐标 canonic(al) coordinates; normal coordinate
正增压器 positive supercharger
正照片 positive photograph
正遮盖 plus lap
正遮盖阀 overlapped valve
正折射 positive refraction
正褶皱 nonplunging fold
正锗酸盐 germinate
正整分布 multinomial distribution
正整数 positive integral; signless integer
正整数项 positive integral value
正正横方向 rigging abeam
正正则的 positive regular
正支距【测】normal offset
正直 integrity; probity; rectitude; squareness; straightness
正直的 straight-forward; upright
正直流电极 direct current electrode positive
正直摄影 normal case photography
正直向地性 positive orthogeotropism
正值 positive value
正值限制 positive limiting
正值指示字符 high-positive indicator
正指令 positive order
正指向 positive sense
正指向的三面形的 positively oriented trihedral
正滞后 positive lag
正中 midmost; midst
正中潮(位) mid-extreme tide
正中的 median; mesial
正中缝 median raphe
正中高突起 positive mid high relief
正中空间 <教堂塔楼下的> interstitium
正中隆起 median eminence
正中切开 mid-section
正中切面 median section
正中时刻 midpoint
正中线 lineal median; median line
正中线后的 post meridian
正中心 orthcenter[orthocentre]
正中要害 be to the point
正重叠 positive overlap
正重力加速度容限 positive g tolerance
正周期 positive period
正周期法 positive period method
正轴 direct aspect; normal attitude; ortho-axis; orthodiagonal; positive axis
正轴测投影 normal axonometric projection
正轴等积圆柱投影 normal tangent cylindric(al) equivalent projection
正轴等积圆锥投影 normal tangent conic(al) equivalent projection
正轴等角圆柱投影 normal cylindric(al) conformal projection
正轴等距离圆柱投影 normal cylindric(al) equidistant projection
正轴等距离圆锥投影 normal conic(al) equidistant projection
正轴晶体 orthodiagonal crystal
正轴坡面 orthodome
正轴投影 normal projection
正轴透视圆柱投影 perspective normal cylindric(al) projection
正轴锥 orthopyramid
正柱区 positive column
正铸胶壳 ebonite monoblock
正转 forward rotation; positive rotation
正转点动 normal inching turning
正转间隔挤出机 corotating intermeshing extruder
正转双螺杆混合挤出机 corotating twin-screw mixing extruder
正转自动卸脱双螺杆挤出机 corotating self-wiping twin-screw extruder
正装 right-handed machine
正装卸扣 regular shackle
正锥 positive cone
正锥形尾水管 vertical tapered draft pipe
正字体 corrected form
正自流水头 <地下水> positive artesian head
正自旋取向 positive orientation of spin
正宗古典的 strictly classical
正走一步信号 forward-step signal
正阻抗 positive impedance
正阻力 active draft; active drift; positive drag
正阻尼 positive damping
正作用执行机构 direct acting actuator
正坐标 positive coordinates

证

证件 certificate; certification; support documents; voucher

证件单据 documents
证件副本 attested copy
证件提出人 exhibiter
证据 collateral evidence; evidence; proof; testimony; vestige; witness
证据不全 evidence incomplete
证据不足 evidence insufficient
证据的 evidential; evidentiary
证据审查 administration of evidence
证明 account for; attest; attestation; authentication; aver; beam witness to; bespeak; bring home; demonstration; identifying; justification; proofing; prove; substantiate; substantiation; testify; testimony; verification; verify; vouch; witness
证明程序 justification routine
证明的 evidential; evidentiary; probative
证明对象 object of proof
证明管理费的合理使用 justify control cost
证明合格 qualify
证明合同文件 document evidencing the contract
证明货物合格性文件 document establishing the good's eligibility
证明检验报表 verify a statement
证明牌 identify-disc
证明凭证 evidence voucher
证明人 authenticator; certifier
证明日期 date of certification
证明商品符合合同中规格的证书 certificate of conformity
证明试验 <仲裁的> witness test
证明适合使用 identified for use
证明书 act; attestation; confirmation form; reference; testimonial; certificate; certification
证明投标者的合格性和资格的文件 document establishing bidder's eligibility and qualification
证明文件 documentary evidence; documentation; instrument of certification; supporting documents; testimonial paper
证明无误 certified correct
证明信 letter of reference
证明已纳税印花 denoting stamp
证明有工程价值的储量 proved reserves
证明有石油的地面积 proved acreage
证明有罪 convict; reprove
证明责任 burden of persuasion
证明者 authenticator; certifier; demonstrator; justifier; verifier
证明正确 justification; justify
证讫 <拉> quod erat demonstrandum
证券交易所 bourse; securities exchange; securities house; security and exchange commission; stock exchange
证券金库 security vault
证人 attestor; deponent; subscribing witness; substantiator; voucher; witness; witness' testimony
证认 identification
证认图 finding chart; identification chart
证实 affirmation; attest; confirm; confirmation; justification; substantiate; substantiation; validation; verify; witness
证实程度 degree of demonstration
证实储量 proved reserves
证实发票 certified invoice
证实法 method of identification
证实试验 confirmation test; verification test
证实收到 acknowledge receipt
证实书 letter of confirmation
证实水深 verified depth
证实文件 supporting documents
证实信号 acknowledgement signal
证实性试验 confirmatory test
证实性性能试验 witnessed performance test
证实者 sustainer
证实指示灯 acknowledgement lamp
证示线 witness line
证书 attested documents; charter; credential; deed; instrument; muniment; obligation; voucher
证书本文 body of deed
证书的 documentary
证书检查 inspection of documents
证书检验证 certificate
证言 witness' testimony
证章 insignia

郑

郑重宣布 attest

政

政策参数 policy parameter; political parameter

政府办公房屋 government house
政府办公楼 government house
政府保留用地 government reservation
政府报告通报与索引 <由美 NTIS 出版> Government Reports Announcement and Index
政府标准手册 <美> government standards manual
政府拨给土地 land grant
政府拨款 government appropriation; government grant
政府补助金 government subsidy
政府部门 government departments
政府财政 government finance
政府财政补贴的列车 state-supported train
政府采购法 the government procurement law
政府采购制度 system for government purchase
政府出版物 official publication; government publication <美>
政府船舶 public vessel
政府大楼 government block; government building; government house
政府大厦 government building; government house
政府大宗采购 government bulk-buying
政府单方面终止合同的权利 <美> termination for convenience
政府担保债券 government-guaranteed bond
政府的土地使用计划 government land use plan
政府抵押贷款协会 <美> Government National Mortgage Association
政府电话 government call; government telephone
政府定额 government quota
政府发行有价证券 government paper
政府法令 government ordinance
政府放款 government loan
政府放款净额 net public landing
政府辅助产业 subsidized industries
政府干预 government interference; government intervention
政府各部 ministry
政府给予的经营特权 franchise
政府工程师 government engineer
政府工作人员 public official
政府公报 government appropriation; government bulletin; official gazette
政府公共设施承揽者 government utility undertaker
政府公债 government bond
政府公债分期偿还 amortization of government bond
政府宫殿 government palace
政府估定价值 assessed value
政府股份 government stock
政府鼓励工业投资的地区 development district
政府官样文章 government red tape
政府管理 government administration
政府管理公众事务的权力机构 public authority
政府管制 government restriction
政府广播 government broadcast(ing)
政府规程 government regulation
政府规章 government regulation
政府核准 government concession
政府环境开支 government environmental expenditures
政府会计 government accounting
政府机构 government agency; government apparatus; public agency; government organization
政府机关 government agency; government body; government organ; government service; office
政府基本义务 fundamental duty of states
政府间常设航运委员会 Inter-Governmental Standing Committee on Shipping
政府间的 intergovernmental
政府间海事协商组织 Inter-Governmental Maritime Consultative Organization
政府间海洋委员会 Inter-Governmental Oceanographic(al) Commission
政府间信用证 letter of credit of government to government
政府监督 government regulation; government supervision
政府检查员 government censor
政府检验 government survey
政府建筑 government building
政府建筑物标志 government building

sign
政府经纪人 government broker
政府开发援助 official development assistance
政府控制的港口 government-controlled harbo(u)r
政府律师 crown solicitor
政府贸易代表团 government trade mission
政府贸易协定 government trade agreement
政府免费通信[讯] government message
政府区 municipal ward; municipal zone
政府日常文书工作 government paperwork
政府试验 government test
政府首脑 head of government
政府授给物 <如补助拨款等> government grant
政府双边贷款 government bilateral loans
政府所有的 government-owned
政府所有制 government ownership
政府特派调查员 <美国组织意见听取会的> hearing examiner
政府特派员 commissioner
政府提案 government bill
政府通信[讯]波段 government band
政府投资 government investment; public investment
政府投资计划 public investment program(me)
政府图书馆 government library
政府土地测量法 <美> government survey method
政府委员会 government commission
政府限额 government quota
政府限价 valorize
政府信贷 government credit
政府信用保证 government credit guarantee
政府形式租船 government form chartering
政府许可证 government license
政府因素 government forces
政府(有价)证券 government security
政府预算 government budget
政府援助 government assistance
政府赠款 government grant
政府债券 government bond
政府债务 government obligations
政府债务交易 public debt transaction
政府支持发展的郊区 assisted area
政府支出 government expenditures; government spending
政府执法权 police power
政府指标 government quota
政府主办的研究项目 government-sponsored research
政府主导计划 government-sponsored program(me)
政府驻现场代表 field engineer; project representative
政府专利 government monopoly; government-owned patent
政府资助研究报告 government-funded research report
政纲 politics
政局动荡 policy unstability
政局稳定 policy stability
政令 decree; government decree order
政桥组支点 tandem pivot point
政区地图 administrative map; political map
政区地图集 political atlas
政务参赞 political counselor
政治地理学 political geography
政治罢工 political strike

症

症结 sticking point

之

之字铆接 reeled riveting; staggered riveting; zigzag riveted joint

之字山型 zigzag-mountain type
之字弯头 offset bend
之字线 switchback; switchback track; zigzag; zigzag line
之字线脚 churn mo(u)lding
之字斜角推土法 zigzag and oblique angle earth pushing process
之字形 zigzag wave type
之字形道路 zigzag
之字形的 dog legged; zigzag
之字形定线 switchback alignment
之字形拱 zigzag arch
之字形航法 traverse sailing
之字形航行试验 zigzag test
之字形回转 switchback turn
之字形急弯 chicane
之字形急弯标志 reverse turn sign
之字形连接管 pipe offset
之字形流线 zigzag path
之字形路线 switchback; zigzag line; zigzag route
之字形路线车站 switchback station
之字形路线的转折 zig
之字形爬山路线 <铁路的> switchback
之字形耙 zigzag harrow
之字形砌合工程 herringbone work
之字形桥 zigzag bridge
之字形曲线 zigzag curve
之字形曲线标志 reverse curve sign; zigzag curve sign
之字形弯曲 switchback
之字形位错 zigzag dislocation
之字形线(路) zigzag line; switchback; zigzag route
之字形运动 zigzagging; zigzag motion
之字形轧机 zigzag mill
之字展线 switchback development
之字展线法 zigzag development

支

支背 supporting back

支臂 ally arm
支臂架 bracket; gate arm; support arm
支臂式掘进机 boom type excavating machine
支臂式装车机 shovel loader
支臂提升机 arm conveyer[conveyor]
支臂桅杆起重机 boom derrick crane
支臂轴 support arm shaft
支臂转柱 boom support
支边 set edge
支冰川 tributary glacier
支柄影响 stem effect
支部 affiliated society; subdepartment
支材 ribband; stilt
支槽 branch channel
支汊 anastomosing branch; by-channel; secondary channel; tributary arm
支汊交错 anastomosing branch
支汊锁坝 chute closure dike[dyke]
支撑 abut; abut against; back guy; backup; bolster; bonding; bracing; buckstay; buttressing; crutch; dwang; pinning; poling; pole shore; propping; rafter timbering; range spacer; revetment; shoring column; sprag; strut bracing; strutting; supporting; surround; sustainer; sustain(ing); tie-strut; tom; underpin; underset; waling; block bridging <小梁间加固的>
支撑拔出移位 prop-drawing shift
支撑板 backing plate; backing sheet(ing); batten; bearing plate; bracing panel; decking; facing bar; former; poling board; poling plate; shoe plate; stay plate; strutting board; support(ing) plate; support web; tie strap; pinchers <沟槽开挖中的>
支撑板条 backup strip
支撑板桩 strutted sheet piling
支撑比例 rib ratio
支撑臂 brace; jack boom
支撑表面 stayed surface
支撑并锚固砖石砌体用角钢 shelf angle
支撑玻璃 support-rod glass
支撑不足 under-braced
支撑布置 arrangement of props
支撑布置图 shoring layout
支撑材 fid
支撑材料 supporting material; timbering material
支撑槽钢 bracing channel
支撑叉 stay fork
支撑拆除工作队 prop-drawing gang
支撑超平面 supporting hyperplane
支撑超平面算法 supporting hyperplane algorithm
支撑衬材 support lining
支撑衬里 support lining
支撑从中心向外倾斜的檩支屋顶 purlin(e) roof with struts inclined away from the center[centre]
支撑大立柱 prop of tunnel support
支撑单排板墙 propped single-wall
支撑单位 support a single floret
支撑挡板 braced sheeting
支撑挡土结构 retaining structure with bracing
支撑的 braced; overhand; strutted
支撑地板 sole plate
支撑地层 stand-up ground
支撑点 bearing; brace point; catch point; fixed position; point of support <桩计算用>
支撑电缆 supported cable
支撑垫块 lip block; lipping
支撑垫圈 backup washer
支撑定位器 prop-setters
支撑短截线 supporting stub
支撑短木 crown piece
支撑短柱 strut
支撑耳柄 supporting lug
支撑阀 lift valve; sustaining valve
支撑法兰 backup flange
支撑方法 method of anchoring; method of timbering; shoring procedure
支撑杆 arm tie; bearing bar; brace pole; bracing piece; hurter; oxter piece; stand bar; stay bar; steadying bar; strutting piece; studdle; supporting bar; tension brace; tie-strut; brace strut
支撑刚性梁 buttress brace
支撑钢筋 bearing bar; spacer bar
支撑隔板 bracing diaphragm
支撑工 bolster work; timberman; tunnel jointer
支撑工程 bolster work; shoring work; timbering; shoring
支撑工作 underpinning work
支撑拱体的柱身 <诺曼第式建设中> edge shafts
支撑沟槽挡板的千斤顶 sheeting jack
支撑构架 braced framing; full frame; strut frame
支撑构件 bracing member; strut member
支撑骨架 braced skeleton
支撑管 stay pipe; stay tube; supporting tube
支撑管道杆 mope pole
支撑轨接头 supported rail joint
支撑辊道 depressing table
支撑函数 support function
支撑焊接处 reference junction
支撑痕 <陶器上> pluck
支撑桁架 interconnecting truss; subtruss
支撑横杆 brace(d) rail
支撑环 backup ring; circle brace; circular spacer; jack ring; retaining ring; support ring
支撑机构 supporting mechanism
支撑基坑 braced cut
支撑脊椽梁 angle tie; dragon tie
支撑剂 proppant
支撑加强木 lip-piece
支撑架 backstand; braced frame; bracing frame; carriage; leg assembly; supporting frame(work); tie-in
支撑架的高度 stilted height
支撑架空管道拱形支架 overcast
支撑间距 prop spacing
支撑件 bracing; joggle piece; stretching piece; structuring piece; strutting piece
支撑件及拉杆螺栓 stays and staybolts
支撑角钢 brace angle
支撑角铁 supporting angle iron
支撑脚混凝土垫块 foot concrete block
支撑铰 bolster hinge
支撑结构 backstand; braced structure; bracing structure; outrigger structure; retaining structure; structural support; supporting structure
支撑开挖 braced excavation; excavate with timbering; strutted excavation
支撑块 backing block; backup block; supporting shoe
支撑框 carriage
支撑框架 shoring cage; strut(ted) frame; support frame(work)
支撑拉拔机 prop drawer
支撑缆索 supporting rope
支撑肋 bracing rib
支撑类型 supporting type; type of supporting
支撑篱笆 supporting a fence
支撑力 holding power; support strength
支撑梁 bearing beam; bolster; brace beam; brace summer; bracing beam; propped beam; shoring beam; strut(ting) beam; support beam
支撑檩条 bracing purlin(e)
支撑檩条屋顶 strutted purlin(e) roof
支撑螺杆 studdle
支撑螺栓 expansion bolt; stay bolt
支撑螺栓螺丝 stay bolt screw
支撑螺丝接头 supported screwed joint
支撑密度 prop density
支撑面 bearing surface; binding face; braced face; bracing plane; bracing surface; face; shank base; tread; underset
支撑面积 bearing area
支撑木 stay log; underpinner
支撑木块 soldier
支撑木窝槽 hole for nog
支撑能力 staying power
支撑皮带 backup belt
支撑片 support chip
支撑平面 supporting plane
支撑平面图 shoring plan
支撑起锚机的缆桩 carrick bitts

支撑起重器 outtrigger jack
支撑器 eyelid retractor
支撑千斤顶 supporting jack
支撑强度 support strength
支撑墙 backing wall; bracing wall; carrying wall; knee wall
支撑桥台 bearing abutment
支撑球 fulcrum ball
支撑圈 bracing ring; strutting ring
支撑砂层 buttress sand
支撑砂岩圈闭 buttress sand trap
支撑设备 fastening; Zibell anchoring system
支撑设计 shoring layout
支撑石 staddle stone
支撑式 brace type; strutted
支撑式挡火墙 carrying bridge wall
支撑式起重机 outrigger jack
支撑式起重器 outrigger jack
支撑式千斤顶 outrigger jack
支撑式桥台 supported type abutment
支撑式屋顶 cradle roof
支撑式系船柱 strutted dolphin
支撑式线脚 < 作大括弧形挑出 > brace(d) mo(u)lding
支撑手把 supporting handle
支撑榫 strut tenon
支撑台 brace table
支撑套 strut sleeve
支撑套筒 stop sleeve; support boss
支撑体系 shoring system
支撑条件 support condition
支撑铁 bracing iron
支撑铁架 bracing cage
支撑铁笼 bracing cage
支撑头部 shore head
支撑凸缘 support flange
支撑托座 foot block
支撑挖掘 braced excavation; excavate with timbering; excavation with timbering
支撑弯管 duckfoot bend
支撑文件 supporting paper
支撑稳定千斤顶 stabilizing jack
支撑稳定物 outrigger stabilizer
支撑屋顶 strutted roof
支撑物 bracing; corbel piece; cribber; jack; prop; spur; staddle; stayer; stilt; strutting; supporter; sustenance
支撑物的高度 stilted height
支撑系大梁 braced girder
支撑系节间 braced panel
支撑系结构 braced framing
支撑系统 braced system; bracing system; system of bracing
支撑系主梁 braced girder
支撑下沉 settlement of supports
支撑线 supporting line; supporting wire; thrust line
支撑线函数 supporting line function
支撑向中心倾斜的檩支屋顶 purlin(e) roof with struts inclined towards the center[centre]
支撑斜杆 bracing diagonal
支撑斜立柱 < 深槽开挖中 > back prop
支撑性圬工 backed-up masonry(work)
支撑悬臂梁 propped cantilever beam
支撑压力 abutment pressure; bracing pressure; supporting pressure
支撑叶片 support blade
支撑液膜 supported liquid membrane
支撑移换 replacer of timbering
支撑应力 bearing stress
支撑油缸 support cylinder
支撑元件 support unit
支撑栽培 supporting culture
支撑在墩上 bracing to piers
支撑轧辊 backup roll
支撑轴 prop shaft
支撑轴承 line shaft bearing; pillow block; plummer block; spring bearing; tunnel shaft bearing pillow block
支撑轴(反)力 axial reaction of strut
支撑助 supporting rib
支撑柱 bracing column; dead shore; pillar stiffener; bracing strut; strut
支撑桩 batter pile; bearing pile; brace pile; buttress shaft; jack pile; spur pile(for dredging)【疏】
支撑装定器 prop-setters
支撑装置 backup unit; support device
支撑桌 brace table; supporting table
支撑钻头 bracing the bit
支承 backstand; backup; bearer; bearing; carry(ing); fulcrum bearing; holding; pecker block; pillow block bearing; support(ing); underlie
支承板 backplate; base plate; bearer plate; bearing plate; bolster plate; decking; guide plate; resting plate; rest plate; retaining plate; supporting plate
支承板中上层钢筋的钢筋 support bar
支承板桩 strutted sheet piling
支承板桩的柱 standard pile
支承半径 pivot radius
支承壁炉矮砖墙 fender wall
支承篦条 supporting grid
支承臂 support(ed) boom
支承边 bearing edge
支承边缘 edge supported
支承表面 area supported; support surface
支承表面至喷嘴之间的距离 trunk depth
支承并分布垂直荷载的水平构件 solepiece
支承部分 supporting parts
支承部件 supporting parts
支承舱壁 pillar bulkhead; supporting bulkhead
支承槽 support slot
支承层 bearing layer; supporting course; supporting layer
支承叉架 < 架空索道 > supporting trestle
支承长度 bearing length; length of support
支承车轴 bearing axle
支承沉箱 supporting caisson
支承船舶墩子 ship-bearing block
支承椽子板 rafter joist plate
支承椽子的檩条 rafter-supporting purlin(e)
支承椽子的一部分墙 plate line
支承搭接楼板的砌体 masonry filler unit
支承大梁 stayed girder
支承带 supporting strap
支承导环 slipper path
支承底板 supported bedplate; supporting baseplate; supporting bedplate
支承底垫 bearing shoe plate
支承地板格栅 supporting floor joist
支承地板托梁的镫铁 floor hanger
支承地层 supporting stratum
支承点 bearing bearer; bearing point; carrying point; point of bearing; point of support; supporting point; pivot point < 罗盘磁针的 >
支承点的下陷 sagging of support
支承点间距 bearing distance
支承点运动 support point motion
支承垫 supporting pad; supportive cushioning < 地毯的 >
支承垫板 bearing plate; sleeper plate
支承垫块 bearing block; cradle
支承垫木 < 半圆旋切的 > stay log
支承垫片 backup pad
支承垫圈 supporting bead; supporting washer
支承垫石 bearing pad-stone; bearing plate bed; bearing seat; masonry plate; pad support stone; supporting pad
支承调整 bearing adjustment
支承顶推法 Belgian method; top heading-under head bench
支承端 supported end
支承断块 supporting block
支承断面 supporting section
支承墩 supporting pier
支承耳 support pad
支承耳柄 support lug
支承耳轴 trunnion pivot
支承反力 bearing reaction; reaction of bearing; reaction of supports; supporting force; supporting reaction; supporting resistance;
支承反力竖标距 support reaction ordinate
支承反挠度 camber of bearing
支承范围 support zone
支承方式 support pattern
支承扶壁的墩 pier buttress
支承杆 bearing bar; cramp bar; support(ed) bar; supporting bar; supporting pole; supporting rod
支承杆件 supported member; supporting member
支承杆中心距 bearing bar centers [centres]
支承杆柱 support column
支承钢板 supporting steel plate
支承钢管 steel supporting pipe; steel supporting tube
支承钢筋 raiser bar
支承钢筋的混凝土块 concrete block bar support
支承杠杆 supporting lever
支承高程变化 variation in level of supports
支承高架 supporting scaffold(ing)
支承隔板 supported diaphragm
支承搁板角钢 shelf angle
支承拱 supporting arch < 洞室围岩的 >; pendentive < 方墙四角托圆拱顶的 >
支承拱顶的墙或边 reins of vault
支承拱或其他特殊形状结构施工的木构架 jack logging
支承钩 grappler
支承构架 structural support; supporting frame(work)
支承构件 supported member; supporting element; supporting member
支承构件跨度 span of supporting member
支承构筑物 supporting construction
支承骨架 supporting skeleton
支承固定件 load-bearing fixing
支承刮削 bedding of bearing
支承关系 support relation
支承管 support column; support pipe; support tube
支承管叉头 support tube yoke
支承管道杆 mope pole
支承管耳轴 support tube trunnion
支承轨 supporting rail
支承辊 backing roll; backup roll; carrier roller; carrying roll(er); supporting roll(er); tray roll(er)
支承辊的辊套 backup roll sleeve
支承辊的轴承座 backup chock
支承辊固定夹板 support roll holder
支承辊换辊装置 backup roll extractor
支承辊轴承 backup roll bearing
支承辊轴承座 backup chock
支承滚轮 bogie wheel; fairlead roller
支承滚筒 idler pulley
支承滚轴 idler
支承滚柱 support(ing) roll(er)
支承锅炉的结构 boiler supporting structure
支承和摩擦混合桩 combination of point-bearing and friction pile
支承荷载 bearing load; support load; sustained load(ing)
支承荷载的隔墙 bearing partition (wall)
支承荷载的骨架 bearing skeleton
支承荷载的面 bearing plane
支承荷载的平腹板大梁结构 bearing structure of plain web girders
支承荷载的墙结构 bearing wall structure
支承荷载的实腹板大梁结构 bearing structure of plain web girders
支承荷载能力 bearing property
支承桁架 lattice; subtruss; supporting truss
支承厚墙体的拱心 arriere-voussure
支承环 backup ring; bearing ring; carrier ring; male adapter [adaptor]; supporting ring; chill ring
支承环的制造要点 critical points on the ring fabrication
支承簧 backup spring
支承回采工作面 support stope
支承混凝土 support concrete
支承或固定部件 anchorage
支承机构 bearing mechanism
支承机理 supporting mechanism
支承机械 supporting machinery
支承基面 foot
支承脊椽梁 dragon beam; dragon tie
支承剂 propant
支承剂总用量 amount of proppants
支承加劲肋 bearing stiffener
支承夹具 support clamp
支承夹子 support clamp
支承架 bearing bearer; bearing carrier; bearing frame; mounting set; supporting bracket
支承角 angle of response
支承角钢 shelf angle; supporting angle iron; supporting angle(steel)
支承脚手架 supporting scaffold(ing)
支承铰链 main pivot
支承接头 supportable joint; support(ing) joint
支承结构 bearing structure; bed frame; bed-frame ga(u)ntry; load-bearing structure; standing support; supporting construction; supporting structure; underwork
支承结构构件的垫板 bed plate
支承结构的设计 design of supporting structure
支承介质 supporting medium
支承介质试样 < 道路建筑 > supporting medium sample
支承距(离) bearing distance; length of support
支承绝缘子 support insulator
支承抗力 support resistance
支承壳体 supporting shell
支承空间结构 supporting space structure
支承块 bearing block; plumber block; rest pad; support piece; tab; way block

支承框架 bearer frame;bearing frame;support frame(work);supporting frame(work)
支承拉拔器 prop drawer
支承缆索 supporting cable
支承肋 supporting rib
支承棱体平台 supporting berm
支承力 bearing force;force of support;supporting force
支承力矩 support moment
支承力影响曲线 aline
支承梁 backbar;bearer bar;bearing bar;guide bar;support(ing) beam;rest bar <导卫装置的>
支承梁板 rafter joist plate
支承量 bearing value;supporting bearing;supporting power;support value;support volume
支承疗法 supporting
支承楼梯段的墙 stairwall
支承楼梯踏步三角板 pitch board
支承路面层梁腋 road deck haunch;road haunch
支承轮 bogie[bog(e)y];return roller;support(ing) roll(er);supporting wheel;weight bearing wheel
支承轮毂 support cage
支承轮毂架 support cage carrier
支承轮廓 <履带起重机的> support contour
支承螺钉 bearing screw;supporting screw
支承螺帽 back nut
支承螺母 back nut;holding nut;supporting nut
支承螺栓 bearing bolt;bearing screw;bearing stud;carrying bolt
支承门窗框的框架 subframe
支承面 area of bearing;bearing(bearer);bearing plane;bearing(sur)face;binding face;carrying plane;seating surface;supported plane;supporting area;supporting plane;supporting surface
支承面变位 variation in level of supports
支承面积 area of bearing;bearing face;bearing surface area;supporting area;bearing area
支承摩擦 bearing friction
支承摩擦损失 bearing friction loss
支承模板拉杆的横梁板 waler plate
支承模板立柱 soldier pile
支承木架 brandreth
支承挠矩 moment at support
支承能 supporting energy
支承能力 bearing capacity;carrying capacity;carrying power;supportability;supporting capacity;supporting power;supporting property;supporting quality;sustaining power
支承牛腿 support bracket
支承盘 bearing disc[disk];supporting disk
支承皮带 backup belt
支承平台 support platform
支承起吊杆上的滑轮的重吊杆 boom jack
支承起重机 supported crane
支承器件 <混凝土路面接缝> supporting device
支承强度 bearing strength;supporting strength;ultimate bearing strength
支承墙 retaining wall;supporting wall;sustaining wall
支承墙面板的框架 subbuck;subframe
支承桥面板角钢 <装在腹板的> hitch angle
支承桥式堆垛起重机 top-running stacking crane
支承桥轴 pivot shaft
支承球 fulcrum ball
支承圈 backup ring;support ring
支承圈标高 level of support ring
支承圈高程 level of support ring
支承圈水平度 horizontality of support ring
支承裙筒 support skirt
支承刃 pivot edge
支承三维(空间)结构 supporting three-dimensional structure
支承栅板 support grid
支承设备 bearing apparatus
支承十字架梁 rood beam
支承石板瓦的屋面板 slate board(ing)
支承时间 stand-up time
支承式的 supported
支承式对接 supported butt-joint
支承式封隔器 heavy-duty tail pipe supported packer
支承式钢轨接头 supported rail joint
支承式接头 supportation;supported joint
支承式离心压缩机 pedestal-type centrifugal compressor
支承式内径千分尺 guide inside micrometer
支承式伸缩装置 support type expansion installation
支承式信号灯 support signal
支承式信号装置 support signal;support signal device
支承式支腿 prop leg
支承式支座 prop support
支承丝 support wire
支承索 bearer cable;carrying cable
支承索衬瓦 carrying-cable bearing shoe
支承索拉牢装置 carrying-cable anchorage
支承索闸 carrying-cable brake
支承索支柱瓦 carrying-cable support shoe
支承弹簧 support(ing) spring
支承套 splicing sleeve;supporting sleeve
支承体 bearing body;supporting mass;stamina
支承体系 bearing system
支承条件 condition of support;support condition
支承铁件 supporting iron
支承筒 carrying cylinder
支承凸缘 supporting lug
支承土壤 bedsoil
支承土压力的竖桩 soldier
支承腿架 gate arm
支承托板的有槽砖 pallet brick
支承位置 bearing position;position of bearings
支承圬工 bearing masonry(work)
支承圬工垫板 masonry plate
支承屋架 supporting roof truss
支承物 bearer;supporting body;soffit spacer <决定下模板与钢筋距离的>
支承系杆 <预应力索定位用> supporting tie
支承系数 bearing factor
支承系统 supporting system
支承线 line of support
支承销 fulcrum pin;rest button;rest pin;supporting pin
支承小格栅的地龙墙 sleeper wall
支承小梁 <楼梯平台的> apron piece
支承性能 bearing performance
支承悬臂 holding jib
支承压定【建】bearing set
支承压力 abutment pressure;bearing pressure;supporting pressure
支承压力带 abutment pressure zone
支承压力分布 bearing pressure distribution
支承压力弹性分布 elastic distribution of bearing pressure
支承压实轮 bogie-tire wheel
支承液体 supporting liquid
支承应变 bearing strain
支承应力 bearing stress
支承用大木料 stay log
支承元件 support unit
支承圆顶的墙环梁 wall ring
支承圆环 bearing ring
支承圆柱体 supporting cylinder
支承约束 support constraint
支承在……上 bear on
支承在地上的短柱 ground prop
支承在深海沉箱上的平台 platform supported on deep caisson
支承在枢轴上 pivot on
支承在悬索上的屋面 roof stilted upon suspension cable
支承在桩上的钢筋混凝土垫层基础 reinforced mat foundation supported on piles
支承在桩上的席形基础 mat foundation supported on piles
支承轧辊 backup roller;fixed roll(er)
支承罩 supporting housing
支承值 bearing value
支承质量 supporting mass
支承轴 bearing shaft;carrying axle;carrying shaft;pivot shaft;supporting axis;supporting axle
支承轴承 backup bearing;bearing journal;block bearer;block bearing;idler;pedestal bearing
支承轴颈 bearing journal;supporting journal
支承轴式旋回破碎机 supported-spindle gyratory crusher
支承肘板 support bracket
支承珠 supporting bead
支承柱 bearing post;bearing shaft;supporting column;supporting post
支承转盘 <鞍式牵引车拖挂装置的> fifth wheel
支承桩 bearing pile;column pile;end bearing(of) pile;point-bearing pile;point load pile;stay pile;supporting pile
支承桩的类型 type of bearing pile
支承装置 bearing set;fulcrum arrangement;mounting set;strut attachment;supporting device
支承锥体 supporting cone
支承阻力 bearing resistance;supporting resistance
支承作用 support function
支承坐垫 <预应力索定位用> supporting chair
支承座 bearing bearer;bearing seat
支承座板 sole plate;support plate
支承座立柱 carrick bitts
支持 backing;backstopping;back-up;beam up;holding;hold-up;pedestal;spine;stand for;supporting;sustain(ing);sustenance;susten(ta)tion;uphold;aegis
支持板 mounting panel;supporting lamella
支持臂 supporting arm
支持不住的 untenable
支持层 supporting layer;backing tier <墙面后的>
支持衬垫 bearing pad
支持导架背靠板 supporting heelplate
支持导向背靠板 supporting heelplate
支持的人 backup man
支持点 resting point
支持电介质 supporting dielectric
支持法兰 backup flange
支持反射 supporting reflex
支持反应 supporting reaction
支持盖 inner head cover;inner top cover;internal head cover;runner inner lid <转轮的>
支持杆 hurter;steady arm
支持钢索 bearer cable
支持管 stay pipe
支持管鞋 seating shoe
支持辊辊轴 backup shaft
支持过程 supporting process
支持和支撑面 braced and stayed surface
支持荷载 sustained load(ing)
支持环 spring guide;pivot ring <底环下面的>;support ring <轴箱弹簧>
支持环箍 supporting clip
支持集策略 set-of-support strategy;support set strategy
支持价格 support price
支持架 stand;supporting rack
支持结构 underwork
支持介质 supporting medium
支持金具 support fitting
支持绝缘子 support insulator
支持可持续性城市排放规划 support sustainable urban drainage planning
支持栏杆的楼梯梁顶的线脚 stair shoe
支持力 holding power
支持梁 supported beam
支持炉格的横杆 cross bearer
支持轮 <皮带或链条> jockey pulley;jockey roller
支持螺帽 back nut
支持螺母 back nut
支持螺丝 pilot screw
支持毛细水 sustained capillary water
支持媒质 supporting medium
支持面 seating surface;supporting surface;sustaining plane
支持膜 supporting film
支持能力 sustaining power;tenability;holding power
支持篇 supporting volume
支持器 hander;holder;supporter
支持穹隆的柱顶 vaulting capital
支持软件【计】support software
支持软件包 support package
支持设备 support equipment;support plant
支持水 held retention water;held water
支持速度 sustaining velocity
支持凸轮 supporting block
支持土 supporting soil
支持物 backer;backing;bracer;but buttress;buttress;holder;prop;supporter;susten(ta)tion;underwork
支持线 anchor wire
支持性安排 stand-by arrangement
支持性的 back-up
支持性圬工 backing masonry(work)
支持硬件 support hardware
支持用户软件 user-assistance software
支持圆柱体 supporting cylinder
支持者 prop;proponent;supporter;sustainer
支持轴承 radial journal bearing
支持柱 supporting pillar;supporting pole
支持柱塞 supporting plug
支持座承压垫层 bearing base
支齿点 tooth rest

支出 charge;defrayal;defray(ment);disburse;disbursement;outgo;outlay;payout
支出保留数 appropriation encumbrance;emcumbrance
支出表 account of payments
支出差异 spending variance
支出成本 expenditures;outlay cost
支出单据 payment documents
支出额 disbursement
支出方 less
支出费用 disbursement
支出费用构成 expenditure pattern
支出分配数 appropriation allotment
支出概算 estimate of expenditures
支出概算书 estimated expenditures report
支出节余 underexpenditures
支出结构 expenditure pattern
支出课税 outlay tax
支出控制 control of expenditure;expenditure control
支出款 disbursement
支出款额剖析 breakdown of amount of expenditures
支出扩大 expenditure expansion
支出率法 expenditure rate method
支出毛数 gross charge
支出明细账 appropriation analysis
支出凭单 payment voucher;voucher for disbursement
支出凭证 disbursement voucher;pay order
支出凭证簿 book of original document for payments
支出清单 schedule of disbursement
支出曲线 outlay curve
支出弹簧托架 offset spring bracket
支出限额 expenditure rate
支出项目 head of expenditures;item of expenditures;outgo
支出削减 expenditure cut
支出要素 item of expenditures
支出应计制 accrued-expenditures basis
支出用途 object of expenditures
支出预算 appropriation budget;budget of expenditures
支出预算分类账 appropriation ledger
支出账 account of disbursements
支出账户 account of disbursements;account of payments;outlay account
支出账目 account to give;giving account
支出证明书 certificate of expenditures
支出转移政策 expenditure-switching policies
支出总额 gross expenditures;total outlay
支船架 building cradle
支船木 dog shore
支窗棍 casement stay
支垂杆 anti-sag bar
支锤编带机 carrier round braider
支带轮 top idler
支带轮导轮 jockey wheel
支带轮端盖 roller carrier cover
支挡 retaining works
支挡结构 retaining structure
支挡铁件 iron block
支挡桩 soldier pile
支导线【测】branch traverse;handing traverse;open-end traverse;spur traverse;stub traverse;subsidiary traverse;free traverse;offset line of traverse;unclosed traverse
支道 branch road;by-path
支堤 back levee
支地槽 embayment
支地轮 ground wheel
支点 abut;abutment;anvil;bearing carrier;center[centre] of suspension;fulcrum[复 fulcra/fulcrums];pivot(ing) point;point of bearing;point of support;support;supporting point;sustainer
支点变矩 moment at support
支点沉陷 yielding of supports
支点承座 fulcrum bearing
支点处的力矩 moment about point of support
支点反力 reaction of supports;support pressure;support reaction
支点附近的第三弦 haunch chord
支点滑板 fulcrum slide
支点继电器 pivoted relay
支点架安装座 fulcrum bracket seat
支点间距 length of support
支点离合器圆盘 backplate
支点力 supporting force
支点力矩 support moment
支点千斤顶 fulcrum jack
支点球头 fulcrum ball
支点弯矩 moment about point support;moment at support
支点弯矩的预留量 allowance for moments at support
支点弯矩的裕度 allowance for moments at support
支点线 pivot-point line
支点压力 abutment pressure;dead-end pressure;support pressure
支点移动 support movement
支电流 derived current
支店 subbranch;subdealer
支店账户 branch office account
支垫板条 <焊接操作时的> spacer strip
支垫座 <闸门的> bearing base
支吊架 suspension
支钉 support pin
支顶架 headframe
支顶梁 <临时的> propped beam
支顶木 <屋顶的> hip jack
支顶木撑 filling-in piece
支顶墙 reacting wall
支洞 dogleg tunnel;lateral tunnel
支洞洞口 adit collar
支洞口 adit portal
支端柱 end support column
支断层 branch fault;subsidiary fault
支队 detachment
支队长 division marshal
支墩 attached pier;but buttress;buttress;buttress(ing) pier;counterfort;counterpilaster;cradle;lade;rest pier;support(ing) pier;anterides <古建筑墙体的>
支墩坝 braced framing dam;buttress dam;counterfort dam
支墩坝工程 construction of buttress dam
支墩坝建造 construction of buttress dam
支墩钢筋 buttressed reinforcement
支墩拱 abutment arch;buttressed arch
支墩基石 abutment rock
支墩间撑梁 buttress strut
支墩间距 buttress centers[centres];buttress spacer;buttress spacing
支墩净间距 clear buttress spacer;clear buttress spacing
支墩块 abamurus
支墩宽度 stem width
支墩梁 strut beam
支墩排水 buttress drain
支墩倾斜度 buttress splay
支墩式挡土墙 pier brace retaining wall
支墩式水电站 pier-type hydroelectric(al) station
支墩收缩缝 buttress contraction joint
支墩塔 buttress tower
支墩头 buttress head
支墩系统 abutment system
支墩型坝 counterfort type dam
支墩堰 buttress weir
支墩中心距 buttress centers[centres];distance between buttresses
支墩砖台 abutment masonry
支垛墙 buttress wall
支耳 journal stirrup
支耳和配件 lugs and fitting
支反应 side chemical reaction;side reaction
支风道 branch air duct
支风口 tuyere stock
支付 defrayal;defray(ment);disburse;outlay;pay;paying out;payoff
支付保险 disbursement insurance
支付测量 contract payment survey
支付差额 payment balance
支付传票 charge ticket;pay slip
支付单位 paying unit
支付到期的票据 meet a bill
支付的基础 basis of payment
支付的款项 payment
支付地点 place of payment
支付定额 pay quantity
支付额 disbursement;payment
支付方法 payment line;payoff method
支付方式 form of payment;means of payment;method of payment;modality of payment;mode of payment;payment line;way of payment
支付费用 cover the cost
支付高程 pay level
支付给分包人的金额 amount paid to subcontractors
支付函数 payoff function
支付合同 contract of payment;payment contractor
支付汇票人 drawee
支付货币 currency of payment
支付机构 disbursing officer
支付计划 payment schedule
支付交易应用层 payment transaction application layer
支付矩阵 payoff matrix
支付利息 interest expenses;interest payment
支付量 payment quantities
支付命令 mandate;payment order;pay warrant;warrant
支付能力 ability for pay;ability to pay;capacity to pay;efficiency of making payment;expenditure capability;paying capacity
支付能力学说 ability-to-pay approach
支付能力原则 ability-to-pay principle
支付逆差 adverse balance of payment
支付票据 bills of payment
支付品 construction materials supplied by employer
支付平衡表 balance of payment
支付凭单 disbursement voucher;payment voucher
支付凭证 certificate of payment;payment documents;payment instrument;pay order
支付期票 hono(u)r a bill
支付期票期限 usance
支付期望的 payoff expected
支付期限 terms of payment
支付契约 contract of payment
支付渠道 channel of disbursement
支付权 authority to pay
支付人 drawee
支付日 pay day;term day
支付日期 date of payment;due date
支付申请书 demand note
支付时间 time for payment
支付实物 pay-in kind
支付手段 instrument of payment;means of payment;medium of payment
支付受益人 beneficiary of remittance
支付税款 tax payment
支付所生利息 pay the interest accrued
支付条件 payment term;terms of payment
支付条款 payment term;terms of payment
支付调整 payment adjustment
支付外汇 hand-over foreign exchange
支付委托书 payment order
支付未偿还债务 service outstanding debt
支付误期 be behind in payment;behind with payments
支付现款 paid cash
支付项目 item of payment
支付协定 payment agreement
支付协议 payment agreement
支付意愿 willingness to pay
支付预算 payment estimate
支付约定条款 facility of payment clause
支付账 currency of payment
支付账单期限 term of account bills payable
支付账目 disbursement account
支付者 payer
支付阵 payoff matrix
支付证书 certificate of payment
支付指定 appropriation of payment
支干管 branch main
支干线 service main
支杆 arm tie;bearing rod;braced strut;bracing;distance piece;guy rod;pole support;rack post;spud;strut;support(ed) bar;supporting rod;fulcrum bar【机】
支杆横柱 spreader
支杆螺母 jack nut
支杆天线 strut antenna
支杆销 arm pin
支搁板的木砖 shelf nog
支给费用 defray the expenses
支根 rootlets
支拱板条 jack lagging;lagging pile
支拱板桩 lagging pile
支拱木 <隧道拱顶部> crown lagging
支拱木料 lagging
支沟 field ditch;lateral ditch;quarter ditch;subsidiary sewer;tributary ditch
支钩 offset hitch
支谷 branch valley
支管 branch;branching duct;branching pipe;branching tube;additional pipe;by-pipe;cut-off;foot lug;lateral;lateral pipe;manifold;manifold (branch);off-take;pipe branch;service line;side tube;subsidiary pipe;subsidiary tube;spiral distributor <水斗式水轮机的>;service pipe <由总管通入室内的煤气管、水管等管道>
支管T形管接 branch pipe T
支管泵站 link pumping station
支管道 branch duct;branch pipe;branch tube;lateral duct;small transfer line;subsidiary conduit
支管法分样系统 manifold system

支管裹接 branch wiped joint
支管过滤器 branch pipe strainer; partial flow filter; shunt filter
支管架 saddle
支管间隔 branch interval
支管间距 branch interval
支管接头 branch cell; branch joint; branch piece
支管空气氧化 manifold air oxidation
支管孔 tap
支管量热器 stem calorimeter
支管滤尘器 branch pipe strainer
支管路 by(e)-pass
支管排气孔 by-pass vent
支管千斤顶 pipe jacking
支管渠 lateral
支管烧瓶 side-tube flask; side-neck flask
支管水流 flow-through branch
支管套筒 branch sleeve
支管通 lateral duct
支管系统 link system
支管线 by-pass line
支管压力损失系数 coefficient for branch loss
支管直径 branch diameter
支管装配件 branch fitting
支滚 by roll
支涵 secondary culvert
支航道 branch channel; secondary fairway
支航线 cantilever flying strip
支航线机场 feeder airport
支横挡 waling
支弧 secondary arc
支护 revet; set support; support
支护板 liner plate
支护布置 support pattern
支护材料费 support material cost
支护参数 support parameter
支护承载力 support capacity
支护顶撑 bracing in tunnel support
支护动力费 support power cost
支护反力 support reaction; support resistance
支护方式 support pattern; type of support
支护工人工资 supporter wages
支护工作 support work
支护工作面 supported face
支护构件 support unit
支护规程 support regulation
支护或加固工程 support or reinforcing work
支护结构 supporting structure
支护结构应力监测 stress monitoring for retaining structure
支护力矩 support moment
支护锚杆 retaining anchor
支护设备 support equipment; support facility
支护特征 supported characteristic
支护系统 support system
支护效果 effect of support
支护性能 supporting performance
支护砖垛 buttress
支护桩 soldier pile; tangent pile
支护装置 support device; support unit
支护状况 support state
支化 branching
支化点 branch point
支化度 degree of branching
支化反应 branching reaction
支化高聚物 branched high polymer
支化几率 branching probability
支化聚合物 branched polymer
支化系数 branching coefficient
支化作用 branching action
支环 supporting ring
支回路 branch circuit
支级 sublevel
支架 abut; ally arm; arm; barring; base frame; bay bear; bearer; bear frame; bearing carrier; bolster; bracket; bracket support; cadre; car-carrier; carcase; check piece; console; dolly bar; fixed mount; fixture; gantree; gantry; holder; holdfast; husk; jack leg; lip; mounting; outrigger; over arm support; pecker block; pedestal; pole bracket(cantilever); prop stay; pylon; seating; sole plate; steadier; steady bar; strut frame; support(er); supporting bracket; supporting frame (work); supporting leg; supporting mast; supporting rack; supporting stand; support stand; tappet block; timbering; trestle; cradle <砌拱洞等的>; headstock <旋转部件的>
支架安设 support setting
支架安装 set swinging
支架安装机 timbering machine
支架坝 trestle dam
支架板 cradle plate
支架臂 support arm
支架臂底座 bracket base
支架边隙 lateral overhang
支架布置 arrangement of props; shoring layout
支架参数 support parameter
支架拆除 post-drawing
支架沉陷 support settlement
支架程序 support program(me)
支架单元 support unit
支架导承 fulcrum guide
支架导杆 support guide
支架底梁 footpiece; set sill
支架电路 support circuitry
支架顶部 pylon head
支架顶梁 carrying bar
支架堵塞 <风洞中的> support blockade
支架渡槽 bench flume
支架墩 rest pier
支架法 staging method
支架房屋构造 pole house construction
支架杆件 cradling member; cradling piece
支架工 timberer; timberman
支架工具箱 kist
支架构件 mounting support hardware; support element; support piece
支架毂 spider hub
支架管(壳) support casing
支架滚子 rigid type castor
支架和横撑 sets and lagging
支架横撑 set girt(h)
支架滑木 <船下水滑道的> cradle plate
支架机 timbering machine
支架基础 column foot; column base block <凿岩机的>
支架基座 column
支架集合 support set
支架脚垫 foot pad
支架铰接顶梁 articulated roof bar
支架梁 bearer bar
支架轮缘 spider rim
支架密度 supported density
支架盘 rack
支架筛 bare grizzly
支架上的垫料 lofting
支架上扩大的钻头 Clark's bit
支架式渡槽 bench flume
支架式风钻 post drill
支架式继电器 cradle relay
支架式脚手架 self-supporting scaffold
支架式示波器 rackscope
支架式手推车 jib barrow
支架式水平钻机 column drill
支架式水准尺 rack level(l)ing staff
支架式屋顶 cradle roof
支架式凿岩机 drifter; drifter drill; drifting machine; leg-hammer; post drill
支架式振动运输机 supported shaker
支架式钻机 post drill
支架铁件 rocking iron
支架图(样) formwork drawing
支架弯头 rest bend
支架微丝 cytoskeletal filament
支架窝 brace comb
支架系统 mounting system; supporting system
支架下沉 settlement of supports
支架效应 tare effect
支架形脚手架 horse scaffold
支架形状 pylon shape
支架支撑 cradling
支架种类 kind of support
支架轴 non-leading axle; support shaft
支架柱螺栓 holding stud; support stud
支架砖 <窑碹槽钢支架用> pier
支架桩 trestle pile
支架桌 trestle table
支架钻车 form jumbos
支架钻机 post drill
支角铁 supporting angle iron
支角突 antennifer
支脚 stabiliser[stabilizer](blade); stand bar; stub; supporter; support leg
支脚固定式 leg-secured type
支脚式 leg type
支铰 articulated gate shoe
支铰大梁 trunnion girder
支铰止推座 trunnion thrust block
支铰轴承 main support(pivot) bearing
支街 by-street
支局 minor office; outstation; satellite exchange; subdepartment; subexchange; suboffice; substation
支距 offset; offset distance; offset of lead curve; ordinate
支距标绘法 offset plotting
支距测法 offsetting
支距尺 offset rod; offset scale; offset staff
支距法 offset method
支距法测量 offsetting
支距杆 offset rod; offset staff
支距线 offset line
支距桩 offset stake
支锯片 felling dog
支掘 underpin
支掘基础 underpinning
支掘路堑 underpinning
支抗 anchorage
支壳 supporting shell
支壳层 subshell
支块 retaining block
支款凭证 pay order
支矿脉 branched lode; feeder
支肋 stiffener
支肋拱顶 stellar vaulting
支离破碎 incoherence
支力 bearing pressure; counterpressure
支链 branch; branched chain; forked chain
支链结构 side chain structure
支链缩合 exocondensation
支梁 strut beam
支梁杆 beam holder
支梁钢杆 beam bolster
支梁块 masonry plate
支梁式手推车 jib barrow
支梁双箍 double stirrup for girder
支梁靴 <桥梁> bearing shoe
支量 constituent
支流 affluent(stream); branch(ing); branch river; contributary; contributory; creek; distributary; effluent stream; embranchment; feeder; feeder current; flow by-pass; gut; horn; inflow; inflowing stream; influent river; influent stream; lateral branch; minor river; minor stream; offshoot; on-flow; ramification; river branch; side stream; split current; subsidiary stream; tributary arm; tributary drain; tributary river; tributary stream; tributary flow
支流冰川 distributary glacier
支流储蓄水 lateral storage
支流的 tributary
支流等级 order of tributaries
支流堤 back levee
支流分叉系数 fork factor(of tributary)
支流分水岭 subdivider
支流管配件 branch fitting
支流河槽 branch channel
支流河堤 tributary levee
支流河谷 tributary valley
支流河口 stream inlet; tributary mouth
支流河口浅滩 shoal at a tributary mouth; shoal near a tributary mouth; tributary bar
支流河口沙洲 tributary bar
支流汇合处 confluence; tributary junction
支流汇集 concentration of tributary flow
支流级 <如第一级支流、第二级支流等> stream order; channel order; order of streams
支流集水区 subcatchment
支流开发 river branch development
支流口 tributary inlet
支流来水 tributary inflow
支流流入水量 tributary inflow
支流流域 subbasin; tributary basin; tributary drain; tributary drainage area; subwatershed
支流流域面积 tributary area
支流面积 interdistributary area
支流区间 interdistributary area
支流区域 tributary area
支流三通(管) side outlet tee[T]
支流水道 tributary waterway
支流水库 feeder reservoir
支流水位 affluent level
支流水系 tributary system
支流弯管 side elbow
支流蓄水 lateral storage
支路 access road; branch line; branch path; branch road; by-pass highway; by-pass line; by-pass road; by-pass route; by-path; by-road; collector road; feeder highway; feeder road; local road; minor road; offshoot; secondary road; side road; siding road; skip road; slip road; spur track; subsidiary road; tributary; turn-off; branch circuit【电】; subcircuit【电】; shunpike <避车或疏散车辆>
支路传递系数 branch transmittance
支路单元 tributary unit
支路单元指针 tributary unit pointer
支路单元组 tributary unit group
支路电流 branch current
支路方程 branch equation

支路接收装置 tributary receiver
支路行驶 < 美 > shunpiking
支路异径丁字管 tee reducing on outlet
支路装置 branching unit
支路阻抗 branch impedance
支螺栓 support bolt
支脉 branch range; embranchment; offshoot; ramification; subrange
支脉孔 diverging veined pore
支锚工具 bucking tool
支模 erection of shuttering; form erecting; formwork
支模板 form placing; form setting
支模板梁 service girder
支木 jack timber
支能级 sublevel
支派 outgrowth; ramification; stem
支配 administer; allocate; arrange; budget; dominance; govern (ing); predominate; predomination; presidency; rein; superintend; sway; wield
支配(彩)色 dominant hue
支配彩色的波长 dominant wavelength
支配的 dominant; ruling
支配地位 ascendance[ascendancy]
支配关系 dominance relationship
支配机车【铁】 controllable locomotive
支配流量 dominant discharge
支配木 dominant; dominant tree; prevailing stem; ruling stem
支配误差 governing error
支配性期货 dominant future
支配因素 governing factor
支配自如 at command
支票 bank check; check; cheque; draft
支票背书 indorse
支票簿 bill book; cheque-book; check book < 美 >
支票簿存根 cheque-book stubs
支票持票人 cheque holder
支票持有人 check holder
支平 set level
支气管炎 bronchitis
支器 prop
支墙斜撑 rider; rider shore
支穹柱身 vaulting shaft
支渠 branch canal; by (e) channel; bypass canal; contributary; distributing channel; distributor; field lateral; lateral canal; offset canal; secondary canal; side culvert; spur canal; sublateral; subsidiary canal; subsidiary channel; tributary channel
支渠分水闸 bifurcation headgates of laterals
支渠灌溉 lateral irrigation
支渠过水能力 lateral capacity
支渠进水闸门 lateral headgate
支渠流量 lateral capacity
支渠入流 tributary inflow
支渠系统 lateral system
支取借款 take-down
支圈 mantle ring; rim
支身架 body support
支手杖 maulstick
支枢 fixed pivot; pivot; support trunnion
支枢点 pivot(ing) point
支枢锚碇螺栓 pivot anchor bolt
支枢式闸门 pivoted gate
支枢锁簧 pivoted dog
支枢转动 pivot steering
支枢转向 pivot steering
支竖管 branch riser
支水道 tributary waterway
支水准路线 level (l) ing branch line; spur level(l) ing line; unclosed level(l) ing line
支隧道 diversion tunnel
支所 outstation
支索 guy; guy rope; guy wire; jack stay; reeving; rope guy; stay cord; staying wire; suspending wire; suspension wire; vang
支索端眼【船】 collar
支索帆 staysail
支索固定器 tang
支索绞辘 stay tackle
支索器 rope carrier
支台 post; prop
支套 spider
支条 lattice framing
支铁 iron support
支通道 branch canal; subchannel
支凸轮 offset cam
支突堤 spur jetty
支腿 end frame; landing leg; leg; outrigger; pad width
支腿摆开后运输 < 挖掘机 > stabilizer spread transport
支腿存放腔 outrigger box
支腿底板 outrigger base
支腿底座 earth plate
支腿垫板 outrigger pad
支腿浮动量 outrigger float
支腿架销卸除装置 outrigger box pin removal
支腿间距 outrigger spread
支腿脚板 outrigger pad; pad foot; stabilizer pad
支腿脚板存放仓 float storage compartment
支腿脚板支承宽度 width of pad centers
支腿脚板中心 pad center[centre]
支腿脚板中心距 width of pad centers [centres]
支腿结构 outrigger structure
支腿梁 outrigger beam
支腿螺杆 outrigger jack
支腿千斤顶 outrigger jack
支腿伸降顺序 out-and-down sequencing
支腿伸距 stabilizer spread
支腿式 leg type
支腿式脚手架 bracket scaffold(ing) ; outrigger scaffold
支腿缩回后宽度 < 起重机 > outrigger retracted width
支腿压力 outrigger pressure
支腿液压泵 outrigger pump
支腿油压缸 outrigger hydraulic cylinder
支腿有效支撑距离 effective spread
支腿支撑器 < 汽车式起重机等 > outrigger jack
支腿中心间距 center[centre] distance between outrigger
支托 abut; bracket; chair; console; corbelled out; corbel out; lug
支托板 support bracket; corbel plate
支托钢条 carrier bar
支托拱顶板条的支撑 support of crown lagging
支托痕 pernetti
支托滑轮 back pulley
支托架 pernetti
支托架螺栓 support bracket bolt
支托角钢 carrier angle
支托连接板 supporting joining plate
支托檩条的楔块 purlin(e) cleat
支托轮 jockey wheel
支托托梁的格栅 trimming joist
支托小梁 needle beam
支弯管 branch bend; branch elbow; branch ell
支弯管令 branch ell
支腕杖 mahlstick
支系 collateral series
支下水道 lateral sewer
支线 branch (ed) line; branch (ing); branch road; branch track; by-pass route; derivation wire; feeder line; fraction line; lay by (e); minor line; offset line; offshoot; sidetrack; spur line; spur track; tapped line; tributary line; leg wire
支线长 length of spur line
支线船(舶) feeder ship; feeder vessel
支线电路 branch circuit
支线段 tangent section
支线港 feeder port; feeder service port
支线公共汽车 feeder bus
支线公路 feeder highway; secondary highway; feeder road
支线集装箱船 feeder container ship
支线街道 feeder street
支线开关 branch switch
支线控制器 branch controller
支线连接 branch connection
支线列车运行 branch line service; feeder service
支线滤波器 spur-line filter
支线凭证机 subsidiary token instrument
支线桥梁 feeder bridge
支线损耗 tributary loss
支线铁路 branch line railway; secondary line; secondary railroad; secondary railway; by line
支线弯道 branch bend
支线系统道路 feeder-system road
支线业务 feeder service
支线运输 feeder service; local service
支线运输机 feeder liner
支巷 by-lane; by-street
支销 backing pin
支销槽顶螺母 fulcrum pin castle nut
支销承 rest pin
支销垫圈 fulcrum pin washer
支销盖 fulcrum pin cap
支销悬挂法 pin-on mounting
支销悬挂式液压挖掘机 hydraulic center post excavation
支斜杆 bracing strut
支行 by-branch; subbranch
支悬高度 suspension height
支烟道 branch flue
支腰挡 waling
支腰梁 waling
支叶等的花纹 < 美术 > bocage
支阴沟 branch sewer
支用贷款有效期 availability period
支于凸缘架的 flange-mounted
支援 assist; backstopping; backup; help; support
支援程序 support program(me)
支援船 support vessel
支援方案 backup program(me)
支援火箭登陆艇 rocket support landing craft
支援近海作业的潜水 diving for supporting off shore operation
支援距离 supporting distance
支援热线 support hotline
支援软件 support software
支援系统 support system
支援硬件 support hardware
支援阵地 supporting position
支援组织 supporting organization
支运河 by-pass canal; lateral canal; side canal
支摘窗 prop-up and dismountable window; removable window
支站 substation; tributary station
支帐篷 put up a tent
支重滚筒 weight-supporting roll
支重架 bogey
支重块 bearing piece
支重轮 carrier roller; lower roller; lower tread roller; stabilizing wheel; thrust wheel; traction roller; truck roller; weight-sustaining roller; track roller
支重轮防护板 ground for track roller
支重轮护板 guard for track rollers; track roller guard
支重轮架 roller frame
支重轮架护板 track roller frame guard
支重轮履带下部滚轮 lower track roller
支重轮外端盖 outer track roller cover
支重面积 < 车底板的 > load-bearing area
支重桥 fixed axle
支重台车 undercarriage
支轴 fulcrum[复 fulcra/fulcrums]; fulcrum shaft; support trunnion
支轴承 supporting bearing
支轴式旋回破碎机 supported-spindle (type) gyratory
支轴销 fulcrum pin
支轴型破碎机 supported-spindle (type) crusher
支轴转门 crapaudine door
支住 hold-up
支助拨款 capital grant
支助费用 backup cost
支助环节 support program(me)
支助项目 support program(me)
支柱 supporting block[column/ pillar/ post/ stand/ strut/ leg/ pillar]; anchor jack; backup post; bearing pile; bearing rod; brace (d strut); bracing; buckstay; check piece; cylindric(al) foot hold; jack prop; legged pile; picket; pier (column); pillar(support) ; pole shore; post (column); prop (stay); shoring (column); solid block; stanchion; stilt; strut (bracing); strut leg; strutting piece; stull【矿】; upright stanchion; entablement < 机器部件的 >; leg members < 钻塔的 >; leg piece < 钻塔的 >
支柱拔出移位 prop-drawing shift
支柱比拟法 support analogy
支柱臂 shore arm
支柱布置密实度 post density
支柱布置图 arrangement of props
支柱采矿法 set mining method; timbered stopping method; timber mining
支柱侧面限界 mast ga(u) ge
支柱层 < 机器部件的 > entablature
支柱插座 stanchion socket
支柱拆除工作队 prop-drawing gang
支柱产业 key industry
支柱长度 supporting length
支柱撑杆 prop stay
支柱承载试验压机 testing press for props
支柱尺寸 support dimension; support size
支柱大梁 holding girder
支柱底板 base plate of stanchion; bloom base plate; mat
支柱底座 stanchion base
支柱垫板 foot plate
支柱垫楔 nog
支柱吊桥 shear-pole draw bridge
支柱顶板 crown plate; head plate of stanchion; stanchion cap
支柱定位器 prop-setters
支柱轭架 prop yoke

支柱防波堤 buttress pier
支柱附件 strut attachment
支柱刚度 support rigidity; support stiffness
支柱高 support height
支柱根 brace root; prop root; root support; stilt root
支柱构造体系 studs construction type
支柱冠板 crown plate
支柱管 supporting tube
支柱和框架系统 buttress and frame system
支柱荷载 load on pillar
支柱横撑 dead-shore needle
支柱横木 head tree
支柱机 timbering machine
支柱基础 support footing; support foundation
支柱加固 support reinforcement
支柱架 strut support
支柱间隔 support interval
支柱间间隙 clearance between tie rods
支柱间距 leg of pitch; post-shrinkage spacing; post spacing; prop spacing; span between two posts; support spacing
支柱间跨距 bay span
支柱截面 support section
支柱绝缘子 strut insulator; support insulator
支柱拉拔机 prop drawer
支柱拉拔器 prop drawer
支柱锚碇 column anchor
支柱锚固 support anchorage
支柱帽 support cap
支柱密度 prop density; support density
支柱模板 support forms; support shuttering; power form <电热张拉的>
支柱模壳 support side
支柱拼接 stanchion splice
支柱剖面 strut section
支柱嵌切 seat cut
支柱强度 support strength
支柱桥(台桥) leg bridge
支柱设计 support design
支柱失稳 support instability
支柱石 cantilever block; cantilever vault
支柱式导轴承 pivoted-pad guide bearing
支柱式海底开挖机 seabed drifter
支柱式开山机 drifter; drifter rig
支柱式龙头 pillar tap
支柱式推力轴承 Kingsbury bearing; pivoted-pad thrust bearing; pivoted shoe thrust bearing
支柱式凿岩机 drifter
支柱试验装置 prop test rig
支柱饰面 support facing
支柱梯【船】pillar ladder
支柱头 shore head
支柱稳定式钻井船 column stabilized drilling unit
支柱稳定式钻探船 column stabilized drilling unit
支柱系列 sequence of columns
支柱下沉 settlement of supports
支柱移动 support movement
支柱移动式挖泥船 walking-spud system dredge(r)
支柱意外松动 kickout
支柱应力 support stress
支柱与地梁 prop-and-sill
支柱支撑 propping; strutting
支柱踵 heel of pillar
支柱肘板 pillar knees
支柱装定器 prop-setters
支柱桌 pedestal table
支柱总承载力 aggregate resistance of support
支柱座 bloom base; pedestal carriage
支柱座板 bloom base plate
支爪 holding dog
支桩 bearing piling
支桩梯【船】pillar ladder
支桩肘板 pillar knees
支座 abutment(piece); bearer; bearing(chair); bearing support; bearing support abutment; bolster; bracket support; carrier; chair; cradle feet; duck foot; installer; jammer; maintainer; mounting; pedestal (body); portal frame; prop stay; saddle; seating; shoe; steadier; support; support abutment; support bearing; supporting block; supporting seat; supporting stand; sustainer; landing seat <套管着陆处>; dock shore
支座板 support(ing) plate
支座长度 length of support
支座沉降 depression of support
支座沉陷 depression of support; settlement of supports; support settlement; yielding of supports
支座承垫 bearing pad
支座承压面 abutment cheek
支座尺寸 supporting stand size
支座底板 bearing sole plate
支座垫 bearing pad
支座垫板 bearing plate
支座垫块 bearing block; padstone; pillow block
支座垫石 bearing stone
支座反力 bearing reaction; end reaction; reaction of supports; support(ing) reaction; support pressure
支座风磨机 pedestal pneumatic grinder
支座负载 abutment load
支座高度 height of support
支座夹片 support jaw
支座角钢 seat angle; shelf angle
支座角铁 angle seat
支座铰接的拱(梁) arched girder hinged at the abutment
支座绝缘子 base insulator; stand insulator; stand-off insulator
支座孔距 supporting stand hole distance
支座块 bearing pad
支座框架 seated frame
支座连杆 support link
支座梁 support beam
支座锚固 bearing saddle
支座面变位 variation in level of supports
支座面积 base area; bearing area
支座摩阻力 friction of bearing
支座偏移 bearing offset
支座三通 base tee[T]
支座上的容器活动不受压缩 free movement of vessel on supports
支座设计 support design
支座式电动砂轮机 electric(al) pedestal grinder
支座枢轴 <弧形闸门的> fulcrum pin
支座水箱 pedestal tank
支座调平阀 seat level(l)ing valve
支座图 support graph
支座弯管 base ell
支座弯矩 support moment
支座弯头 base bend; base elbow
支座位移 displacement of abutment
支座系杆 pedestal tie bar
支座系统 support system
支座下沉 depression of support; settlement of supports; settling of supports; yield(ing) of supports
支座压力 abutment pressure; bearing pressure; end pressure; support pressure
支座约束 bearing restraint
支座运动 support motion; support movement
支座轴承 supporting bearing
支座柱 bearing post
支座桩 pedestal pile
支座状态 condition of support
支座座板 base shoe; bed plate

汁

汁 liquid

芝

芝加哥地区运输管理局 <美> Chicago Regional Transportation Authority

芝加哥港 <美> Port Chicago
芝加哥蓝 Chicago blue
芝加哥式大窗 Chicago window
芝加哥式悬臂起重机 Chicago boom derrick
芝加哥学派 Chicago school
芝加哥(闸墩)式沉箱 Chicago caisson
芝麻油 sesame oil

吱

吱嘎嘎地作响 creak

枝

枝孢属 <拉> Cladosporium

枝编棚架 arbo(u)r
枝材 lopwood
枝材形数 branch form factor
枝叉型柱支撑 branch shaped timbering
枝杈 stub
枝茬 snag
枝撑 timbering support
枝管 side tube
枝瑚菌属 <拉> Ramaria
枝迹 branch trace
枝架 trestle works
枝剪 scissor
枝接 grafting proper; scion grafting
枝节 snag
枝节的 irrelative
枝节问题 irrelevance
枝晶 arborescent crystal; dendrite; fern-leaf crystal; fir-tree crystal; pine-tree crystal; tree-like crystal
枝晶长大 dendrite growth
枝晶的 dendritic; tree-like
枝晶间的 interdendritic
枝晶间腐蚀 interdendritic attack; interdendritic corrosion
枝晶间偏析 interdendritic segregation
枝晶间破坏 interdendritic attack
枝晶间侵蚀 interdendritic attack
枝晶间石墨 interdendritic graphite
枝晶间缩松 interdendritic porosity; interdendritic shrinkage; interdendritic shrinkage porosity
枝晶结构 pine-tree structure
枝晶偏析 coring; dendritic segregation
枝晶体生长 dendritic growth
枝晶组织 arborescent structure; pine-tree structure
枝肋 lierne; lierne rib
枝肋拱 lierne vault
枝肋穹顶 lierne vaulting
枝肋穹隆 lierne vault
枝蔓晶 dendrite
枝蔓晶体 dendritic crystal; dendritic crystallization
枝蔓状晶体 dendrite
枝蔓状结构 dendritic morphology
枝梢材 lop and top
枝梢切碎机 brash chopper; slash chopper
枝梢头木 top-end-lop
枝条 shoot; twig; wicker; with; wattle <编篱笆、屋顶等用的>
枝条编的 wicker
枝条编制品 wattle work
枝条材 branchwood; lap wood
枝条低垂 weep
枝条排坝 brush dam
枝条排堰 brush weir
枝条劈裂 branch split
枝脱离 branch abscission
枝稀疏或密集 sparsely or density branches
枝隙 branch gap
枝下高 <木材> timber height
枝形窗格 branch tracery
枝形大吊灯 chandelier; luster[lustre]
枝形灯架 chandelier
枝形吊灯 corona; crystal chandelier
枝形吊灯架 electrolier
枝形(吊式)电灯架 electrolier
枝形花格(窗) branch tracery
枝形煤气吊灯 gaselier
枝形烛架 branched candle holder; girandole
枝形烛台 candelabrum[复 candelabra/candelabrums]
枝桠 slash; twig
枝叶切削机 knife hog
枝叶清理机 stick-and-green leaf machine
枝叶系统 <植物地面以上> shoot system
枝叶状雕饰 branched work
枝叶状装饰 branched work
枝桩接 stubgrafting
枝状 arborescence; arborescent; dendritic; tree-like
枝状虫胶 stick lac
枝状大烛台或灯台 candelabra
枝状的 branched
枝状地衣 fruticose lichen
枝状粉末 arborescent powder
枝状管网 branch(ed)(distribution) system; branched network; branching network
枝状硅 dendritic silicon
枝状脊模 dendritic ridge mo(u)ld
枝状结晶铜粉 dendritic copper powder
枝状金属粉末 dendritic metal powder
枝状晶体 dendritic crystal
枝状晶体半径 dendrite radius
枝状晶形成 dendrite formation
枝状蔓延 dendrite propagation
枝状拟侧丝 dendrophysis
枝状配水管网 branched distribution system
枝状偏析 dendritic segregation
枝状渠道 branching channel
枝状热网 branched heating system
枝状闪电 streak lightning
枝状生长 dendrite propagation
枝状饰 spray
枝状水系 dendritic drainage pattern; dendritic river system
枝状物 spray
枝状形态 dendrite morphology
枝状雪晶 dendritic snow crystal
枝状组合变量 branching composite variable

知

知识产权 intellectual property; intellectual rights

知识产权的国际保护 international

protection of intellectual property
知识工程师 knowledge engineer
知识专家系统 knowledge-based expert system
知识转让 transfer of knowledge

织 plait;plat;weave

织板 textile board
织边 selvage;selvage edge;selvedge
织补 invisible mending
织布机 loom
织成的 textile
织成的网 weave screen
织带 fabric tape; mesh-belt; narrow goods weaving
织带电缆 woven cable
织带式输送器 mesh-belt conveyer [conveyor]
织构 texture;texture anisotropy
织构材料 textured material
织构测角计 textured yarn goniometer
织构化 texturing
织构化热加工 texture developed by hot working; thermomechanical working for texturization
织构技术 texture technique
织构结合料 fabric matrix
织构行为 textural behavio(u)r
织花台布 damask
织机方向 machine direction
织机后梁 back-bearer;whip roll
织机墙板 loom side
织机上梁 loom arch
织机上轴 loom mounting
织机下地轴 bottom shaft
织锦 tapestry
织锦厂 brocade mill
织锦地毯 brocade
织锦缎 brocatel(le);tapestry satin
织料覆盖物 fabric cover
织料接合【机】fabric joint
织料轮胎 fabric tire[tyre]
织料密封轴承 fabric-seal bearing
织女星 Vega
织品 textile;texture;web
织绒地毯 tapestry carpet
织丝方眼筛 woven-wire square mesh screen
织缩 crimping
织网 web
织网钢筋 woven-wire reinforcement
织网钢丝浪弯辊压机 crimping machine
织网筛 woven wire cloth sieve
织网式 woven tape
织网式过滤器 woven filter
织网运输带 mesh-belt
织纹 wavy grain;weavy grain
织纹面砖 textured brick
织纹设计 weave design
织纹状饰面 textured finish
织纹锥螺 textile cone
织纹组合 weave formation
织物 cloth; drapery; fabric; stuff; textile(fabrics); texture; weft; woof; woven fabric
织物板条 cloth lath(ing)
织物背衬 fabric backing
织物变形 fabric distortion
织物材料 fabric material; textile material
织物层 cloth ply;fabric ply
织物层压材料 fabric laminate
织物衬底 fabric underlay
织物衬里 woven lining
织物成型混凝土 fabric-formed concrete
织物冲击强力试验 fabric impact test;fabric shock test
织物除尘器 fabric filter collector
织物处理装置 fabric treating unit
织物传热性试验 fabric heat conductivity test
织物疵点 fabric defects
织物带 webbing
织物的 textile
织物的组成 fabric composition
织物底层 cloth base
织物地毯 textile carpet
织物垫底 fabric base
织物垫片 fabric insert(ing)
织物顶破强力试验 fabric bursting test
织物反面 fabric backing
织物泛水 cloth flashing
织物分析 fabric analysis
织物分析镜 counting glass;fabric analysing glass; pick counter; textile analysing glass
织物酚醛塑胶 textolite
织物酚醛塑料板 textolite
织物缝纫性能试验仪 fabric sewability tester
织物敷层 mat layout;mat layup
织物幅 fabric web
织物幅宽 fabric width
织物覆胶 fabric proofing
织物覆胶装置 fabric coating unit
织物光泽 fabric sheen
织物规格 fabric specification
织物过滤集尘器 fabric-type dust collector
织物过滤器 fabric filter
织物烘燥机 fabric drying machine
织物回弹性 fabric resilience
织物集尘器 fabric dust collector
织物几何学 fabric geometry
织物加固 textile reinforcement
织物加强的 cloth-reinforced
织物加强件 fabric reinforcement
织物加强密封件 fabric-reinforced seal
织物检验 fabric inspection
织物胶轮胎 fabric tire[tyre]
织物结构 construction; fabric construction;fabric structure
织物浸胶装置 fabric dipping unit
织物经纱 web of fabric
织物经纬密度 cloth count; fabric count;thread count
织物经纬纱线滑移性试验仪 fabric shift tester
织物卷取 fabric take-off;fabric take-up
织物卷取和打卷装置 fabric take-up and batching device
织物颗粒过滤器 pleated prefilter
织物可燃性 fabric flammability
织物扣分式检验法 fabric inspection of penalty system
织物料 fabric scrim
织物滤材 textile filtering medium
织物滤层 woven membrane
织物滤尘器 fabric filter collector
织物滤芯 fabric element
织物模板 fabric form
织物磨损试验机 cloth wearing tester;cloth-wear testing machine
织物内渗透性 in-plane permeability
织物耐磨试验仪 textile abrasion tester
织物耐水压试验仪 textile hydrostatic pressure tester
织物挠性万向节 fabric joint
织物黏[粘]附性试验 fabric cling-testing
织物排水板 geodrain
织物片 fabric sheet(ing)
织物剖幅 fabric slitting
织物牵引力 fabric pull
织物嵌填 cloth insert(ing)
织物强力试验 fabric strength test
织物闪光效应 cameleon
织物闪色效应 cameleon
织物设计 fabric design;textile design
织物设计草图 fabric draft
织物伸长 fabric extension
织物丝光整理机 fabric silken machine
织物缩水率 fabric shrinkage
织物缩水试验 fabric shrinkage test
织物胎壳 fabric carcass
织物弹性试验 fabric elasticity test
织物填料 fabric filler
织物条地毯 rag rug
织物透孔性 fabric porosity
织物透气性 fabric breathability
织物透气性测试仪 permeometer
织物透气性试验 fabric air permeability test
织物涂布 textile coating
织物涂胶 fabric coating
织物涂料 cloth-coating; fabric coating;textual coating
织物外观检验 fabric external appearance inspection
织物纹样砌块风格 textile block style
织物吸尘器 cloth filter;fabric filter
织物纤维 fabric fiber[fibre]
织物纤维网 web of fabric
织物悬垂性 fabric fall
织物引张试验机 cloth tester
织物印花 cloth print;textile-printing
织物印刷 cloth print
织物硬挺度试验 fabric stiffness test
织物用漆 textile lacquer
织物与泡沫塑料层压黏[粘]合 fabric-to-foam laminating
织物与织物层压黏[粘]合 fabric-to-fabric lamination
织物与织物黏[粘]合 fabric-to-fabric bonding
织物原料 textile
织物增长 fabric growth
织物增强的 fabric-reinforced
织物照明 fabric light
织物罩盖 baldaquin
织物折断强度试验器 folder
织物真空吸水机 cloth vacuum extractor
织物整理 fabric finish;textile dressing
织物整饰 textile finishing
织物装饰条 valance
织物状饰面【建】textured finish
织物组织 fabric texture
织造 contexture
织造厂 weaving mill
织造品 fabric
织制带 woven belt

肢状物 limb

脂袋<木材的> resin pocket

脂肪 tallow
脂肪醇二异氰酸酯 aliphatic diisocyanate
脂肪的 fatty
脂肪光泽 oily luster
脂肪含量 fat content
脂肪切削油 fatty cutting oil
脂肪酸 fat acid
脂肪酸单甘油酯 glycerin(e) monofatty ester
脂肪酸法 fatty acid process
脂肪酸甘露糖醇酯 fatty acid mannitol ester
脂肪酸酐 fatty acid anhydride
脂肪酸环己六醇酯 fatty acid inositol ester
脂肪酸季戊四醇酯 fatty acid pentaerythritol ester
脂肪酸焦油沥青 fatty acid tar pitch
脂肪酸聚乙烯醇酯 fatty acid polyvinyl alcohol ester
脂肪酸类【化】fatty acids
脂肪酸沥青 wool grease pitch
脂肪酸三甘油酯 triglyceride fatty acid
脂肪酸色淀 fatty acid lake
脂肪酸山梨醇酯 fatty acid sorbitol ester
脂肪酸脱氢酶 fatty acid dehydrogenase
脂肪酸酰胺 fatty acid amide
脂肪酸盐 soap
脂肪酸乙烯酯 fatty acid vinylester
脂肪酸酯 ester of fatty acid
脂肪性的 lipoid
脂肪油 fatty oil
脂肪族 fatty group
脂肪族胺 aliphatic amine;fatty amine
脂肪族醇 fatty alcohol
脂肪族的 fatty
脂肪族化合物 fatty compound; aliphatic compound
脂肪族基【化】fatty group
脂肪族聚酯 aliphatic polyester
脂肪(族)烃 aliphatic hydrocarbon
脂光沥青 berengelite
脂光石【地】elaeolite
脂光正长石 elaeolite-syenite
脂化层压材 resin-treated laminated compressed wood
脂环烃 naphthene
脂环烃的 naphthenic
脂环烃类 alicyclic(al)hydrocarbons
脂环族的 alkyclic
脂环族化合物类 Alicyclic(al) compounds
脂环族环氧树脂 cycloaliphatic epoxy resin
脂类 lipid(e)
脂类分解作用 lipolysis
脂沥青的 bitumastic
脂料 daubing
脂酶 lipase
脂囊<木材年轮间含有树脂的空隙> pitch pocket
脂镍皂石 pimelite
脂溶橙 fat orange
脂溶红 fat red
脂溶黄 fat yellow
脂溶棕 fat brown
脂松节油 gum turpentine
脂松香 gum resin
脂酸冻点(测定法) titer[titre]
脂酸凝点 titer[titre]
脂填缝料 pitch filler
脂烃基缩水甘油醚 aliphatic glycidyl ether
脂样物 butter
脂油环 grease cup
脂油桶 grease bucket
脂质 lipid(e)
脂质组干酪根 liptinite kerogen
脂状光泽 greasy luster[lustre]
脂状琥珀 gedanite
脂状面 fatty surface
脂族的 aliphatic
脂族溶剂 aliphatic solvent

蜘网形缝毛石圬 cobweb rubble masonry

蜘蛛架 spider;spider element;spider frame
蜘蛛架导杆 spider legs
蜘蛛架导杆螺钉 spider leg screw
蜘蛛架外环 spider rim
蜘蛛架中心轴 spider hub

蜘蛛网(交通)分配法【交】spider-web(network traffic) assignment
蜘蛛网状物 cobweb

执柄 tailpiece

执柄门锁 mortice lock with lever handle
执法机构 law enforcement agency
执票人票据 bearer bill
执票人支票 bearer check
执勤车 service walk vehicle
执勤道 staff walkway
执勤廊 service gallery;staff gallery
执勤隧管 service gallery;staff tube
执手 grip;grip handle;handle;knob
执手插锁 mortice lock with handle;mortice lock with knot;mortise latch
执手插销 rim latch
执手弹子锁 knob cylinder lock
执手挡板 push plate
执线人【测】line man
执行 carry into execution;dispensation;enforcement;execute;implement;in pursuance of;prosecution;transact;transaction
执行保护方式 executive guard mode
执行报表 action report
执行报告 executive report
执行编辑 executive editor
执行步骤 execution step
执行部分 operative part
执行部件 execution unit;executive component;power unit
执行部门 executive branch
执行操作系统 executive operating system
执行陈述 executable statement
执行程序 executable program(me);executive;executive program(me);executor;implementation;master routine;monitor routine;steering routine;executive routine
执行程序成分 executive program(me) component
执行程序卡片组 executive deck
执行程序控制 executive program(me) control
执行迟延时间 execution dead time
执行存储器 execute store
执行措施 executive measure
执行错误 execution error
执行错误检测 execution error detection
执行单位 executable unit;executive unit;run unit
执行电动机 actuating motor;actuator motor;force motor;operating motor
执行电路 actuating circuit;execution circuit;executive circuit
执行调度 operation dispatching
执行调度保持 executive schedule maintenance
执行订单 execution of order
执行董事 executive director;managing director
执行董事会 board of executive directors;executive directors
执行端 execution end;executive end
执行段落 operative paragraph
执行法规 code enforcement
执行范围 range of execution;scope of execution
执行方法 method of execution
执行方式 execution mode;executive mode;run mode
执行防护方式 executive guard mode
执行公告 notice of action
执行公务 officiate
执行构件 actuating member
执行官员 executive officer
执行管理程序 executive supervisor
执行管理人 executive
执行规程 agendum[复 agenda]
执行规程卡片 agendum call card
执行过程 implementation
执行合伙人 managing partner
执行合同 carry-out a contract;contract performance;execution of contract
执行合同的计划师 <承包商的> contract planner
执行合同的时间 contract timing
执行和监督 implementation and supervision
执行会计 executing accounting
执行机构 actuating element;actuating mechanism;actuator;enforcement machinery;governing body;implementary machinery;operator;power unit;servo;servo-actuator
执行机构罩 actuator housing
执行机关 executive organ
执行计划 delivery of program;executive plan;operating plan
执行记录 executive logging
执行阶段 execute phase;execution phase;executive phase
执行局 executive board
执行系统应用程序 executive system utility
执行业务的合伙人 active partner
执行业务股东 active partner;managing partner;working partner
执行异常 execute exception
执业会计师 certified public accountant
执业者 practitioner
执业证书 certificate to practice;practising certificate
执照 certificate;charter;licence;license;permit;qualification
执照持有者 licensed applicator
执照费 certification fee;fee of permit;license[licence] fee;license[licence] expenses
执照号码 license[licence] number
执照及许可证 license[licence] and permit bond
执照技术合格证 certificate
执照考试 licensing examination
执照期满 expiration of license[licence]
执照申请 license[licence] application
执照使用者 licensed applicator
执照税 excise;fee of permit;license[licence] tax
执照暂时吊销 license suspended
执照暂时吊销或扣押 license[licence] suspended
执照暂时扣押 license[licence] suspended

直安全阀 straight safety valve

直岸线 rectilinear shoreline
直八式 <八汽缸直排式> straight eight
直八式汽车 straight eight
直把手 straight grip
直板颚式碎石机 straight plate jaw crusher
直板机 plate roll
直板门 batten door;boarded door
直板桥 right slab bridge
直板推土铲 straight-blade dozer
直棒条筛 straight grizzly bar
直杯嘴白蚁 <拉> procapritermes sowerbyi
直背手锯 straight back hand saw
直壁 rupes recta
直壁防波堤 mole with vertical face
直壁空心钻头 straight-wall bit
直壁喷管 straight-wall nozzle
直壁取芯钻头 straight-wall coring bit
直壁式校正扩孔器壳体 straight-wall core shell
直壁式扩孔筒 straight-wall core shell
直壁式岩芯钻头 straight-sided core bit;straight-wall bit
直臂 straight-arm
直臂拌和机 straight-arm mixer
直臂混合机 straight-arm mixer
直臂架 straight cantilever
直臂桨式混合机 arm mixer;arm straight paddle mixer
直臂搅拌机 straight-arm stirrer
直臂调整器 straight-arm compensator
直臂旋桨拌和机 arm straight paddle mixer
直边 constant edge;straight edge;straight flange;straight side
直边玻璃酸瓶 straight-edge glass carboy
直边长度 length of straight flange
直边齿 rack tooth
直边锉 blunt file
直边导规 straight-edged ruler
直边导棱机 straight-edge beveling machine
直边导线 straight-edge guide
直边的唱诗班席位 straight-sided choir
直边的柱 straight-sided column
直边的最小厚度 minimum thickness of skirt
直边段 straight section
直边高度 straight-side length
直边料斗 straight-sided bucket
直边轮胎 straight-edge tire[tyre]
直边轮辋 straight-side rim
直边平轮缘 straight-side flat rim
直边三角形 quadrantal spherical triangle
直边式花键 straight-side flank splines
直边式胎 beadless tire[tyre];straight-side tire[tyre];wired edge tire[tyre]
直边形柱 straight-sided column
直边装岩铲斗 straight-edge rock bucket
直边锥形筒子 straight-edge cone
直柄 straight handle;straight shank
直柄标准齿立铣刀 straight shank standard tooth slotting cutter
直柄冲头 straight shank punch
直柄粗齿立铣刀 straight shank coarse tooth slot-ring cutter
直柄刀杆 straight shank arbo(u)r
直柄刀架 straight shank tool holder
直柄端铣刀 straight shank end mill
直柄锻头刀架 straight shank drop head tool holder
直柄锅炉丝锥 straight boiler tap
直柄键槽铣刀 straight shank keyway cutter
直柄铰刀 straight shank reamer
直柄立铣刀 straight shank cutter
直柄螺母丝锥 straight shank taper tap
直柄螺纹梳刀架 straight shank holder for chaser
直柄螺钻 straight shank drill bit
直柄麻花钻头 straight shank twist drill
直柄梅花扳手 double hexagonal opening box socket wrench with straight handle
直柄三牙钻夹头 straight shank triple grip drill chuck
直柄式的 straight handle
直柄式电钻电机 straight handle drill-motor
直柄式气螺丝刀 straight screwdriver
直柄式气钻 straight drill
直柄式砂轮机 straight grinder
直柄手锯 straight handle hack saw
直柄套筒扳手 straight shank socket wrench
直柄铣刀 straight shank milling cutter
直柄压动式螺丝刀 push-start straight screwdriver
直柄钻夹头 straight shank drill holder
直柄钻(头) straight drill;straight shank drill
直并励电动机 straight shunt-wound motor
直拨电话 direct dial(l)ing
直拨电话交换台 direct telephone exchange
直拨专用自动小交换机 in-dialling private automatic branch exchange
直播 direct seeding;live broadcast
直播节目 live program(me)
直播卫星 direct broadcasting satellite
直播栽培 cultivation by direct seeding
直播种 directed seeding;directed sowing
直布罗陀海峡 Gibraltar Strait;Strait of Gibraltar
直部 straight part;straight portion
直材 straight timber
直槽 straight channel;straight flute
直槽铰刀 straight fluted reamer
直槽刨铁 straight grooving iron
直槽式鱼梯 vertical slot fishway
直槽手用铰刀 straight flute hand reamer;straight slot hand reamer
直槽锥形铰刀 conic(al) straight fluted reamer
直槽锥形销孔铰刀 straight flute taper-pin reamer
直槽钻(头) straight fluted drill;straight-way drill
直叉锚式擒纵机构 straight-line lever
直插 straight cutting
直插入 telescoping
直插式电路 card circuit
直插式封装 in-line package
直插用户 direct plug-in subscriber;subscriber of direct insertion
直茬 indenting;toothing
直铲 straight shovel
直铲式推土机 straight-blade bulldozer;straight bulldozer
直铲推土板 straight blade
直铲推土刀 straight blade
直铲推土机 S-bulldozer;straight-blade dozer;straight dozer
直车 straight turning
直撑 extending bracing
直承式支座 direct bearing
直尺 feather-edge;feather-edge type;rectilinear scale;rectangular scale;ruler;straight edge;straight rule;straight scale;stretch
直尺控制器 ruler inversor
直尺平行运动机构 parallel motion mechanism
直尺水准测量 straight-edge level(l)ing
直尺圆规作图法 construction with ruler and compasses
直尺作图法 linear construction
直齿 spur;straight-tooth
直齿插齿刀 spur shaper cutter

直齿齿轮 straight-cut gear；straight-tooth(ed) gear
直齿齿轮泵 spur-gear pump
直齿齿条 spur rack
直齿的 straight-toothed
直齿端铣刀 end mill with straight teeth
直齿对搭接 short scarf
直齿锯 straight-toothed saw
直齿链锯 scratcher chain saw
直齿轮 face gear；spur gear；straight gear(wheel)；straight spur
直齿面 straight-sided flank
直齿耙 spik(ed)-tooth harrow
直齿伞齿轮 bevel gear；common bevel gear；straight-toothed bevel gear
直齿伞齿轮刨齿机 straight bevel gear generator
直齿伞齿轮刨刀 straight bevel gear cutter
直齿式 spur type
直齿铣刀 straight-toothed (milling) cutter
直齿圆柱齿轮 spur gear；straight-cut gear；straight-toothed spur gear
直齿圆柱齿轮差速器 spur-gear differential
直齿圆柱齿轮行星传动 planetary gearing drive
直齿柱齿轮 straight-toothed conic(al) gear；straight-toothed cylindric(al) gear
直齿锥齿轮 coniflex gear
直齿锥齿轮粗拉法 Revex
直齿锥齿轮副 straight bevel gear pair
直齿锥齿轮刨齿机 straight bevel gear generator
直冲 direct impact
直冲式风速计 direct impact type air speed indicator
直船首柱 vertical stem
直吹 straight blow
直吹管 belly pipe；blast pipe；blow pipe
直吹磨机 direct grinding mill
直吹燃烧方式 direct-fired system
直吹式燃烧 direct firing
直吹式燃烧系统 direct-fired installation
直吹式条形散流器 straight bar
直吹式制粉锅炉 direct-fired boiler
直吹式制粉系统 direct-fired system；unit pulverized-coal system
直吹式制粉系统磨煤机 direct-fired mull
直锤头 straight peen
直粗刨刀 straight roughing shaping tool
直锉法 straight filing
直达 run-through；through；throughout
直达包裹列车 through parcels train
直达波 directive wave；direct wave；forward wave
直达波频谱分析 direct arrival wave spectrum analysis
直达波衰减系数计算 direct arrival wave attenuation factor computation
直达采区巷道 foot rail；foot rill
直达车 through car；through train
直达车(行)道 through roadway
直达成组列车 through freight train；through goods train；through group train
直达船 direct vessel
直达纯蓝 immedial pure blue
直达单据 through document
直达道路 thorough way；through traffic road；throughway；thruway
直达的 non-stop；through-running
直达电路 direct line
直达飞行 non-stop flight
直达干道 major through road
直达港 direct port
直达公路 through highway；through traffic highway
直达国际呼叫 direct international call
直达国际连接 direct international connection
直达过境电报 direct transit telegram
直达航班 through-flight
直达航程 non-stop voyage；through voyage
直达航空线 direct air route
直达话传电报业务 direct telephone-telegram traffic
直达话务 through traffic
直达汇票 straight arrival bill
直达货物 <自起始站至终点站的> through cargo；direct cargo
直达货物清单 through cargo manifest
直达货物运费增额 through cargo arbitrary
直达货物增额 through cargo arbitrary
直达货运 through freight
直达货运路线 through truck route；thrust truck
直达机位式机坪 gate arrival apron
直达基岩的防渗墙 positive barrier to seepage
直达基岩的隔水层 positive barrier to seepage
直达交通 through movement；through traffic
直达交通高速公路 through traffic road
直达进路 direct route
直达距离 throughout distance
直达开关 non-stop switch
直达客票 through ticket
直达快车 through train
直达里程 through kilometrage
直达连接线 tie line
直达联运 combined through transport
直达列车 inner-lock train；non-stop train；rail transit train；through-running train
直达列车和成组装车计划 plan of through trains and wagon-group loadings organized at loading points
直达列车运行 through train run
直达路线 through route
直达路由 direct route
直达旅客 through passenger
直达旅客列车 through passenger train
直达绿 immedial green
直达脉冲 direct pulse
直达耦合线路 tie line
直达汽车运输 through motor transportation
直达青 immedial sky blue
直达声 direct sound
直达声场 direct sound field
直达式 gate arrivals type
直达提单 direct bill of lading；through bill of lading
直达通路 direct channel
直达通信[讯] direct communication；direct transmission
直达通信[讯]线路 direct communication link；tie line
直达卫星传输 direct satellite transmission
直达线 direct line
直达线路 direct line；straight-forward line；tie line
直达线用户 direct line subscriber
直达性 non-stop character
直达运费 throughout carriage charges；through rate
直达运货 drop shipment
直达运价率 throughout rate
直达运输 direct traffic；direct transport；through shipment；through traffic；through transport
直达运输护送队 through convoy
直达整零车 through part-load wagon
直达整装零担车 groupage wagon；through car
直达整装零担货车 groupage wagon
直达总站 through terminal
直达纵波垂直时距曲线 vertical hodograph of direct longitudinal wave
直刀 straight knife
直刀架 straight toolholder
直刀身螺钉钻 through blade screwdriver
直导轨 spur guide；straight closure rail；straight connecting rail；straight guide(way)；straight lead rail
直到符号 up-to symbol
直到目前为止 up to this point
直到现在的 down-to-date；up-to-date
直道 chute
直道的 straight
直道灯标 straight-channel light beacon
直的拱顶 straight vault
直的码尺杆 yard wand
直的帽盖拱顶 straight cap vault
直的黏[粘]土砖过梁 straight clay brick lintel
直的隧道拱顶 straight tunnel vault
直的推料叶片 straight pusher-blade
直的箱形梁 straight box beam
直的圆凸勾缝 string bead
直点式缓行器 tangent point retarder
直动 translation
直动泵 direct acting pump
直动阀 direct acting valve
直动风马达 reciprocating air engine
直动脉 straight artery
直动式煤气恒温器 direct acting gas thermostat
直动调节阀 direct acting regulating valve
直动凸轮 translation cam
直动阻力导数 translatory resistance derivative
直读表 direct reading ga(u)ge
直读传输电平测试器 direct reading transmission level measuring set
直读定标器 direct reading meter
直读法 direct reading method
直读光谱分析仪 direct reading spectrograph
直读计 direct reading meter
直读卷尺 instantaneous-reading tape
直读罗经 direct indicating compass；direct reading compass
直读罗盘 direct indicating compass；direct reading compass
直读扭矩扳钳 direct reading torque spanner；direct reading torque wrench
直读扭矩扳手 direct reading torque spanner；direct reading torque wrench
直读式 direct reading
直读式 pH 计 direct reading pH meter
直读式测厚仪 dial ga(u)ge
直读式测距仪 direct reading tacheometer
直读式测试仪表 direct reading instrument
直读式测试仪器 direct reading instrument
直读式测速仪 direct reading tach(e)ometer
直读式测微仪 dial micrometer；direct reading micrometer
直读式的 readout
直读式电阻表 ohmer
直读式分光计 direct reading spectrograph
直读式分析器 direct reading analyser [analyzer]
直读式分析天平 direct reading analytical balance
直读式光谱仪 direct reading spectrometer
直读式海流计 direct reading current meter；recording current meter
直读式话务记录设备 direct-reading traffic-recording equipment
直读式计算机 direct reading calculator
直读式记录 read and write
直读式刻度 direct-reading scale
直读式频率计 magmeter；readout meter
直读式视距仪 direct reading tach(e)ometer
直读式温度计 direct reading thermometer
直读式相位分析器 direct reading phase analyser[analyzer]
直读式压力计 direct reading manometer
直读式仪表 direct reading meter；magmeter；readout instrument；direct-reading ga(u)ge
直读式仪器 readout meter
直读式油位计 direct reading oil level ga(u)ge；oil level sight ga(u)ge
直读式转速表 direct reading tach(e)ometer
直读式转速计 Jeffcott tachometer
直读试验法 direct reading test method
直读天平 direct reading balance
直读温度计 indicating thermometer
直读文本 clear text
直读压痕硬度计 indentometer
直读液面计 direct reading liquid level ga(u)ge
直读自记海流计 direct recording current meter
直度 straightness
直端唱诗班席位 straight-ended choir
直端导轨 straight end guide
直端洞口 straight endwall
直端锚固 <钢筋> straight anchor
直端墙 straight endwall
直堆 stow fore and aft
直返式布置 direct return scheme
直方带 orthogonal zone
直方的 rectangular
直方图 bar chart；bar diagram；bar graph；histogram
直方图程序 histogram program(me)
直方图规格 histogram specification
直方图海流计 bar chart current meter；bar graph current meter；histogram current meter
直方图计算机 histogram computer
直方图记录 histogram record
直方图均衡化 histogram equalization
直方图平坦化 histogram equalization；histogram flattening
直方图平直化 histogram equalization；histogram flattening
直方图线性化 histogram equalization；histogram linearization
直方图修改 histogram correction；histogram modification
直方图修正 histogram correction；histogram modification
直方图阈值化 histogram thresholding
直方图正态化 histogram normalization
直放式接收机 direct receiver
直锋刀具 straight tool

直缝 straight-line joint;vertical joint
直缝 T 形接头 straight T-joint
直缝对接 straight joint
直缝管 straight pipe; straight seam pipe
直缝焊 straight weld(ing)
直缝或竖缝 perpend or transversal joint
直缝假平顶 false ceiling with straight joints
直缝接头 straight jointing
直缝拼合 <木地板> straight joint
直缝砌法 plumb bond
直缝砌缝 straight joint
直缝砌合 straight joint bond
直缝砌砖法 unit bond
直缝式管 straight seam pipe
直缝砖砌合 unit bond
直氟碳钙钕矿 synchysite-(Nd)
直氟碳钙钇矿 synchysite-(Y)
直辐带轮 straight-armed pulley
直辐射的 orthoradial
直复式分品法 straight multiple grading
直腹板型钢板桩 straight-web steel sheet pile
直腹式截面 straight-web section
直杆 straight lever;straight member
直杆阀 valve with straight stem
直杆结构 <坚硬岩体中的> mullion structure
直杆球形阀 vertical stem globe valve
直杆摇把 crank handle with straight lever
直感 intuition
直感的 audio-visual
直感教具 audio-visual material
直感教学法 audio-visual aids
直感图书 audio-visual book
直感资料 audio-visual material
直缸接力器 linear type servomotor; straight cylinder type servomotor
直钢筋 stor(e)y bar
直钢丝杯形刷轮 straight steel wire cup wheel
直钢丝刷轮 straight steel wire wheel
直沟 chute;straight flute
直沟滚刀 straight fluted hob
直沟铰刀 reamer with straight flutes
直沟渠 straight channel
直沟丝锥 straight fluted tap
直股钢板桩 straight-web piling bar
直挂刀板 straight blade
直挂式推土机 straight bulldozer
直挂推土板 straight blade; straight bulldozer blade
直观 illustrative;visual
直观表面 visible surface
直观表示 visual representation
直观操船训练模拟装置 visual ship handling training simulator
直观测云镜 direct vision nephoscope
直观的 audio-visual;visualized;direct viewing
直观电图像 visual picture
直观定价法 intuitive pricing
直观读数 visual reading
直观法 direct vision method
直观方式 intuitive manner
直观估计 intuitive estimate
直观光度计 subjective photometer
直观红外成像管 directly viewed infrared image tube
直观化 visualize
直观记录 visual record
直观价值 perceived value
直观检查 visual check;visual inspection
直观教材 visual aids
直观教具 audio-visual aids; visible; visual aids
直观教学 audio-visual instruction;intuitional instruction;object teaching
直观控制 visual control
直观模型 physical model
直观判断 intuitive judg(e)ment
直观曲线 visual picture
直观取景器 direct vision viewfinder
直观扫描器 visual scanner
直观识别 visual identification
直观式测距仪 self-contained range finder
直观式存储管 direct viewing storage tube
直观式三色管 tricolo(u)r direct view tube
直观式寻像器 direct-type view
直观试验 macroscopic test;straight-forward test
直观塑性法 viscoplasticity method
直观图 illustrative diagram; intuitive graphic; pictorial diagram; pictorial representation; straight-forward diagram
直观图像 visual picture
直观图形 direct environment;visual picture
直观推断 heuristics
直观文件 visible file
直观误差 apparent error
直观显示 visual display;visual indication
直观显示终端 visual display terminal
直观显示装置 visual display unit
直观显像管 direct viewing tube;direct view kinescope
直观信号 visual signal
直观询问 visual inquiry
直观询问台 visual inquiry station
直观研究法 visual observation
直观淤积管 visual accumulation tube
直观预测 intuitive forecasting
直观运行费 perceived travel cost
直观指示 visual indication
直观终端 visual terminal
直观状态 intuitive manner
直观追踪 direct visual tracing
直管 ascending pipe;ascending tube; chute; piling pipe; piling tube; straight pipe;straight tube
直管垂直照准仪 vertical collimator
直管存水弯 bottle trap
直管段 section of main; section of pipeline;straight run
直管段衬砌 <尾水管的> throat liner
直管箍 straight coupling
直管锅炉 header type boiler; once-through boiler;straight-tube boiler
直管活接头 straight union joint
直管机 pipe straightener machine
直管接 coupling;pipe socket
直管接头 straight connector
直管连接管 straight connector
直管气流式干燥器 tube-type pneumatic drier[dryer]
直管器 pipe straightener
直管式存水弯 bottle trap
直管式发动机 straight motor;straight tubular motor
直管式锅炉 through-flow boiler
直管式挤压灰浆泵 straight tube squeeze pump
直管式喷射器 straight pipe injector
直管式水管锅炉 water-tube boiler with straight tubes
直管式压力表 vertical tube manometer
直管式蒸发器 vertical type evapo(u)rator
直管式蒸汽发生器 straight tube steam generator
直管外螺纹三通接头 street tee
直管线 straight pipe-line
直管轴流泵 tubular axial-flow pump
直管状脉冲器件 straight pulsed device
直规 lute;ruler;straight edge;stretch
直规平面检测器 winding stick
直轨道 through track
直轨工人 gagger
直轨工作 ga(u)ging of rail
直轨机 rail straightener;rail-straightening machine
直轨器 rail straightener;rail-straightening tool
直滚子 straight roller
直焊道 string bead
直航 direct route;direct sailing;stand on
直航程 distance made good
直航船 direct sailing vessel;keep-way vessel;privileged ship;right-of-way vessel;stand on vessel
直航附加费 direct additional charges
直航距离 direct distance
直航向 compound course; direct course;direct course made good;equivalent single course
直航运费 direct freight
直航运输 direct shipment
直合拔轨 straight closure rail;straight lead rail
直合拢轨 straight connecting rail
直和 direct sum
直和分解 direct sum decomposition
直和码 direct sum code
直滑 straight skid
直缓点 tangent to spiral point
直黄铜丝杯刷轮 straight brass wire cup wheel
直黄铜丝刷轮 straight brass wire wheel
直活管节 straight union
直火焚烧炉 direct flame incineration
直火加热 direct fire heating
直击大风 straight-line gale
直击强风 straight blow
直积 direct product
直积集 direct product of sets
直积码 direct product code
直基本轨 straight stock rail
直基线 straight baseline;straight-line basis
直集合管 straight collecting tubule
直脊波痕 straight crested ripple
直夹板 flat splice bar
直夹钳 straight sticking tongs
直夹套 straight collet
直架 straight frame
直尖轨顶部切削 <爬坡直尖轨> top cut
直减率 lapse rate
直剪 direct shear; straight scissors; straight snips
直剪法 direct shear method
直剪强度 direct shear strength
直剪试验 box shear test;direct shear test
直剪试验仪 direct shear apparatus
直剪仪 box shear apparatus; box-shear apparatus with a single surface;direct shear apparatus
直键槽 straight spline
直桨叶 straight-arm
直桨叶片拌和机 straight-blade paddle mixer
直交层积 cross laminate
直交单板 cross band
直交单板层 cross banding
直交的 orthogonal
直交方向 capwise
直交高低穹式交叉拱 Welsh groin
直交隔离 perpendicular separation
直交隔水边界 orthogonal impervious boundary
直交滑距 perpendicular slip
直交化 orthogonalization
直交积分 rectangular integration
直交流电动发电机 DC motor driving AC generator
直交流混合系统 DC and AC combined system
直交落差 perpendicular throw
直交平错 perpendicular heave
直交桥 square bridge
直交收货人提单 <不能转让的> straight bill of lading
直交水准器 cross level
直交提单 straight bill of lading
直交透水边界 orthogonal pervious boundary
直交透水隔水边界 orthogonal pervious-impervious boundary
直交位移 perpendicular displacement
直交性 orthogonality
直交原理 principle of orthogonality
直交直距 perpendicular throw
直交轴 rectangular axis
直交走向 capwise
直浇道 downsprue;sprue
直浇道横浇道内浇道面积之比 sprue-runner-gate area ratio
直浇道模型 sprue pin
直浇道窝 sprue base
直浇口 down gate; down runner; downsprue; pouring gate; running gate;sprue hole;spud;stick gate
直浇口棒 gate stick;runner pin
直浇口拉出器 sprue puller
直浇口模棒 sprue pin
直浇口窝 puddle;sprue base
直浇口下储铁池 cushion
直浇口压痕 sprue base
直角 right angle;vertical angle
直角 T 形接头 right-angle tee[T]
直角板 L-square;right-angle board
直角边 diagonal edge; right-angle side;square edge
直角边盖板 square edge over panel
直角边缘 square edge
直角扁形杆件 rectangular rod
直角扁形金属线 rectangular wire
直角裁割 across
直角测器 survey across
直角尺 bare L-square; carpenter's square <木工用的>;L-square;rectangle ruler;square;square rule;try(ing) square
直角出水口 square-edged outlet
直角初调器 square-on reflector
直角传动 rigging-angle drive; right-angled drive
直角错齿饰 square billet
直角单向阀 angle check;angle check-valve
直角的 normal; orthogonal; orthographic(al); rectangular; right-angle(d)
直角笛卡尔坐标 rectangular Cartesian coordinates
直角点阵 orthogonal lattice
直角度 squareness
直角端面车床 right-angle facing lathe
直角堆料时最小通道宽度 90° stacking aisle width
直角对焊接 jump weld
直角帆 square sail
直角反射镜 corner cube mirror
直角缝式铺砌路面 checker work pavement
直角钢(尺) steel square
直角杠杆 bent lever

直角杠杆调节器 bell crank governor
直角杠杆轴环 bell crank shaft collar
直角固定螺栓 dog bolt
直角刮匙 right-angle curette
直角挂板 right-angle suspending plate
直角挂轮皮带 half-cross(ed) belt; quarter-turn belt
直角拐肘 right-angle crank
直角管 quadrature tube
直角规 try square
直角函数 rectangular function
直角航线进场着陆 rectangular approach
直角弧光灯 Debrun candle
直角换向器 quadrant
直角回转 quarter turn
直角回转传动 quarter-turn drive
直角回转带 quarter-turn belt; quarter-twist belt
直角回转皮带 quarter-turn drive
直角(机头)挤压机 extruder with side delivery head
直角机械传动 hypoid
直角及T角形拐肘用的垫板 base plate for right angle and T-crank
直角脊角屋顶 square roof
直角夹 rigging-angle clamp
直角检查窥镜 right-angle examining telescope
直角检景器 right-angle finder
直角交叉 rigging-angle crossing; right-angle(d) crossing; right-angled intersection; right-angled junction; square crossing
直角交错波痕 rectangular cross ripple mark
直角交错轴双曲面齿轮 hypoid gear
直角交缝式 checkwork
直角交缝式块料路面 checkered surface
直角交缝式(铺砌)路面 checker work pavement
直角交会 right-angle crossing; right-angle intersection
直角浇口 tab gate
直角角域 right-angle-angle domain
直角校正 rectangular alignment
直角接合 head(ing) joint; joint on square
直角接头 angle coupling; elbow; right-angle connector
直角截流阀 angle shut-off valve
直角进水口 square-edged inlet
直角镜 prismatic(al) square; prism square; right-angle mirror
直角开挖 cut square
直角控制器 right-angle inversor
直角宽顶堰 square-edged broad-crested weir
直角棱镜 corner cube prism; double prismatic square; prism square; rectangular prism; right-angle prism
直角棱柱体 corner cube prism; cube corner prism
直角连接 right-angled junction; right-angle joint
直角连接器 rigging-angle connector; right-angle connector
直角联动皮带 quarter belt
直角列线图(解) right-angled nomogram
直角楼梯台 quarter pace stair
直角螺丝钩 square-bend screw hook
直角脉冲 right-angle impulse
直角镘刀 angle float
直角锚固钢缆 <与拉力成90度角> bridle cable
直角模具 crosshead die
直角摩擦轮 rigging-angle friction wheel; right-angle friction wheel
直角诺模图 right-angled nomogram
直角碰撞 angle collision; right-angle collision
直角偏转 normal deflection
直角平行六面体 rectangular parallelepiped(on)
直角器 cross-nailed material; cross staff head
直角牵开器 right-angle retractor
直角切割 square cutting
直角切割的陶瓷空心砖 hollow ga(u)ged brick with right angle cut
直角切削 <木工制作> across
直角切削工具 try square
直角球面三角形 right spheric(al) triangle
直角曲拐 bell crank
直角曲线坐标 rectangular curvilinear coordinate
直角三角形 orthogon; rectangular triangle; right-angle(d) triangle; right triangle
直角三角形两股 legs of a right triangle
直角三通 right-angle(d) tee
直角三通阀 right-angle three way valve
直角山墙顶盖瓦 right gable tile
直角十字杆【测】cross staff
直角式排水 rectangular drainage pattern
直角双曲线 equilateral hyperbola; rectangular hyperbola
直角双曲线坐标 rectangular hyperbolic coordinates
直角丝锥门 tap gate
直角调制串音 quadrature crosstalk
直角停车 right-angle parking
直角停放车 <车辆和路线垂直停放> ranking
直角弯波导 corner waveguide
直角弯管 duck foot; elbow bend; elbow pipe; knee pipe; knee tube; quarter bend; rigging-angle elbow pipe; right-angle(d) bend; right-angled elbow pipe; square elbow
直角弯管接头 elbow union; square elbow; rest bend
直角弯曲 right-angle bending
直角弯头 duck foot; elbow bend; knee bend; normal bend; one-quarter bend; quarter bend(ing); rest bend; right and bent; right-angle(d) bend; square bend; square bent; square elbow
直角弯头阀门 right-angled valve
直角弯头螺钉 square-bend screw hook
直角弯转 right-angled bend
直角碗形砂轮 straight cup(grinding) wheel
直角网络 rectangular cell; right-angle network
直角网状的 orthogonal reticulate
直角温度计 angle thermometer
直角系 rectangular system
直角相位调制 quadrature demodulator; quadrature modulation
直角形 right-angle type
直角形单向阀 angle check-valve
直角形杠杆 bell crank lever; bent lever; knee lever; square lever
直角形平板机 angle moding press
直角形气门嘴 angle valve
直角形弯头 right-angled elbow
直角形弯转接头 bent ferrule
直角形肘管 right-angled elbow
直角旋光器 optic(al) square
直角旋塞 angle cock
直角研磨机 angle grinder
直角檐瓦 right verge tile
直角仪 cross staff; right-angle mirror <用平面镜组成>
直角异径弯管 reducer angle
直角引出线 right-angle lead
直角引线 right-angle lead
直角隅角 square corner
直角照准仪 cross staff
直角支管 right-angle(d) branch
直角支距 right-angle offset
直角止回阀 angle check-valve
直角指示器 right-angle viewer
直角轴 rectangular axis
直角肘管 elbow bend; square elbow
直角转变仪 optic(al) square
直角转光器 optic(al) square
直角转角 right-angled corner; square corner
直角转弯 quarter turn; right-angled bend; right-angled turn
直角转弯的楼梯平台 quarter pace
直角转弯楼梯 dog-leg stair(case); quarter newelled stair(case); quarter-turn stair(case)
直角转弯楼梯过渡平台 quarter landing
直角转弯平台 <楼梯的> quarter landing; quarter-space landing
直角转弯时通道宽度 90° aisle width
直角转弯梯台 quarter pace; quarter pace stair
直角锥 rigging-angle cone
直角组接 corner locking
直角坐标 Cartesian coordinates; orthogonal coordinates; rectangular coordinates; rectilinear coordinates
直角坐标标绘器 X-Y plotter
直角坐标尺 rectangular coordinate card
直角坐标法 rectangular coordinate method
直角坐标法测量(放点) offset measurement; offset method
直角坐标法放点 offset method; setting-out using the perpendicular coordinate
直角坐标方程 rectangular equation
直角坐标格网 grid coordinate system; straight-line graticule
直角坐标矢量图 clock-phase diagram
直角坐标式电位计 rectangular coordinate type potentiometer
直角坐标式机器人 Cartesian coordinates robot
直角坐标式机械手 rectangular coordinate manipulator
直角坐标网 normal grid; rectangular coordinate grid; rectangular grid
直角坐标网格 grid system
直角坐标系 rectangular coordinate system; rectangular setup; system of rectangular coordinates; system of rectangular X and Y; coordinates X-Y
直角坐标系统 Cartesian coordinate notation
直角坐标系轴 rectangular axis
直角坐标仪 coordimeter; rectangular coordinate plotter
直角坐标原点 true origin
直角坐标展点尺 plotting square
直角坐标展点仪 rectangular coordinate plotter; rectangular coordinatograph
直角坐标轴 Cartesian axis; orthogonal axes; rectangular axis; rectangular coordinate axis
直角坐标向量 rectangular vector
直角坐标轴线 rectangular axes
直脚瓷瓶 straight pin porcelain insulator
直脚钉 straight pin
直脚绝缘子 bracket insulator
直脚形绝缘子 spool insulator
直接安排 direct placement
直接安排(私自推销)债券 direct placement of securities
直接安装 direct mounting
直接包装 direct mount; direct packing; immediate packing
直接饱和法 direct saturation method
直接保险 direct insurance
直接报酬 direct compensation; direct consideration
直接比 direct ratio
直接比的 directly proportional
直接比较 direct comparison
直接比较法 direct comparison method
直接比较器 straight comparator
直接比较特点 direct comparison of characteristics
直接比色法 direct colo(u)rimetry
直接编码微指令 direct encoding micro-instruction
直接编址 immediately addressing
直接编制 direct organization
直接编制文件 direct organization file
直接变动成本 direct variable cost
直接变换 collineation; direct mapping; direct transform; direct transformation
直接变态 direct metamorphosis
直接变址 direct indexing
直接标价 direct quotation
直接标价法 direct foreign exchange quotation
直接标志 direct mark
直接表格传送 list-directed transmission
直接表面 immediate surface
直接表演演播室 live talent studio
直接拨号 direct selection; non-register-controlling selection
直接拨号电话系统 direct dial telephone system
直接拨号系统 direct dial(l)ing system
直接波 direct path wave
直接波束 direct path
直接波抑制 main bang suppression
直接播放(录音) play-over
直接播散 direct dissemination
直接补偿 direct compensation
直接补给 direct recharge
直接补贴 direct grants; direct subsidy
直接不分开式扣件 direct unseparated fastening
直接部门 direct department
直接部门成本 direct department cost
直接部门费用 direct departmental expenses
直接材料 direct material
直接材料成本 direct material cost
直接财产 direct goods
直接财产损失 direct property damage
直接财政支出 direct expenditures
直接采购 direct purchase
直接采光 direct daylighting; direct lighting
直接采光系数 direct daylight factor
直接采暖炉 direct space-heating furnace
直接采暖器 direct heating unit
直接采暖设备 direct heating device
直接采暖系统 direct heating system
直接采暖用具 direct warming appliance
直接采样 direct sampling
直接参考地址 direct reference address

直接操纵 direct operated;local control
直接操纵式变速 cane type shift
直接操作 contact servicing;direct operation; direct servicing; direct working
直接操作阀 direct acting valve
直接操作费用 direct operating cost; direct operating expenses
直接操作装置 local control system
直接槽辊涂布机 direct gravure coater
直接侧音 direct side-tone
直接测定 direct assay;direct determination;direct measurement
直接测高程 direct level(l)ing
直接测距 direct distance measurement
直接测距系统 act active ranging system
直接测量 direct measurement
直接测量水速的幕架 travel(l)ing screen
直接测热法 direct calorimetry
直接测深 direct depth sounding
直接测试法 direct probe analysis
直接插板 series connector
直接插入编码 in-line coding
直接插入程序 direct insert routine; in-line process;open routine;open routing
直接插入处理 in-line processing
直接插入的 direct insert;in-line
直接插入函数 intrinsic(al)function
直接插入例行程序 direct insert routine
直接插入式 card edge type
直接插入子程序 direct insert subroutine;in-line procedure;in-line subroutine; open subprogram(me); open subroutine
直接插值法 direct interpolation
直接查表 table look-at
直接差比高程测量 direct differential level(l)ing
直接产量 direct yield
直接产品 direct product
直接产生 directly produce
直接阐明 direct demonstration
直接长途拨号 direct distance dial(l)ing
直接长途拨号电话 direct distance dialing telephone
直接长途拨号网 direct distance dial network
直接长途拨号网络 direct distance dialing network
直接偿付 direct obligation
直接潮(汐)direct tide
直接车轮荷载 direct wheel load
直接成本<包括材料工时在内> direct cost;overhead cost;prime cost
直接成本法 direct costing;direct cost method
直接成本计算法 direct cost calculation method;direct costing
直接成本金额<承包合同中指定分包商或供货商的> prime cost saul
直接成本收益表 direct costing income statement
直接成分 direct composition
直接成粉亚烟煤燃烧装置 direct pulverized sub-bituminous coal firing system
直接成像 direct imagery;direct imaging
直接成像系统 direct imaging optics
直接成像质谱仪 Casting-Slodzian mass analyser [analyzer]; direct imaging mass analyser[analyzer]
直接成型浇口 drop gate
直接承包合同 prime contract
直接承包人 prime contractor
直接承垫 directly seating
直接承载电路 direct bearer circuit
直接程序控制 direct program(me) control
直接迟缓反应 direct delayed reaction
直接持股 direct holding
直接冲量 direct pulse
直接冲洗阀 direct flush valve
直接冲销法 direct charge-off method;direct write-off method
直接抽水 direct pumping
直接筹资 direct financing
直接出口 direct export
直接出口和进口 direct export and import
直接出售 direct sale
直接处理 in-line processing
直接传播 rectilinear propagation
直接传播噪声 air-borne noise
直接传递 direct transmission
直接传动 direct drive gear;direct operation; direct transmission; high gear;positive gearing
直接传动泵 direct acting pump;direct drive pump
直接传动变速装置 with through drive gear
直接传动的 coupling drive;direct connected; direct drive; first-motion drive; integral unit drive; straight drive;straight-through drive;through-drive;direct acting
直接传动的机器 direct drive machine
直接传动的提升机 direct acting hoist
直接传动的转数表 direct drive cyclometer
直接传动度盘 direct drive dial
直接传动固定架 direct drive fixed mount
直接传动辊 direct coupled roll
直接传动机车 direct drive locomotive
直接传动记录仪表 direct acting recording instrument
直接传动加速度 direct drive acceleration
直接传动离合器 direct driving clutch
直接传动螺旋桨 direct drive propeller
直接传动设计 direct acting design; through-drive design
直接传动式剪切机 direct driven shears
直接传动式汽轮机 direct coupled (steam)turbine
直接传动式输送螺旋 direct drive auger
直接传动式拖拉机 direct drive tractor
直接传动速度 direct drive speed
直接传动提升机 first-motion hoist
直接传动条件 straight-through drive condition
直接传动同步机 direct driven synchro
直接传动压力机 plain press
直接传动(液力)变速阀 direct drive shift valve
直接传动(液力)蓄油器 direct drive shift valve accumulator
直接传动振动机 direct drive vibration machine
直接传动装置 direct drive gear;direct drive unit;positive gear
直接传感 direct pick-up
直接传染 direct infection
直接传热炉片 direct radiator
直接传输 direct transmission
直接传输系数 direct transmission factor
直接传送 direct feed;direct transmission
直接传送方式 straight sending system
直接传送工件 direct feed
直接传送探测数据 direct sounding transmission
直接传送系统 direct transmission system;straight sending system
直接串话 direct crosstalk
直接串联再用 direct cascade reuse
直接吹炼法 direct smelting in converters
直接垂直日射 direct normal insolation
直接磁带录像 direct videotape recording
直接次瞬态电抗 direct sub-transient reactance
直接从尺寸起草工程量清单 direct billing
直接从石场运来尚未加工的石料 run-of-pit
直接从数据库获得的资料成果 online information product
直接从属 immediate subordinate
直接淬火 direct quenching
直接存取 primary access;random access
直接存取装置 direct access storage device
直接打捞【救】direct raising;raising directly
直接打捞法 direct raising method
直接大红 direct red
直接带隙半导体 direct-gap semiconductor
直接贷款 direct lending;direct loan
直接单读硬度试验 directional reading hardness test
直接单剪试验 direct simple shear test
直接单位成本 direct drilling cost
直接担保 direct guarantee
直接担保品 direct securities
直接挡 direct high;top gear
直接挡离合器 direct drive clutch
直接导纳 direct mobility
直接导线连接 direct wire link
直接的 direct;first hand;immediate; intermate; straight-forward; substantive
直接的冲击波伤害 primary blast injury
直接的灭火行动 fire duty
直接滴定 direct titration
直接底板 immediate bottom
直接地面径流 direct surface runoff
直接地址 direct address; first address;first-level address;immediate address; one-level address; single-level address
直接地址码 direct address code
直接地址指令 direct instruction;immediate address instruction
直接第二遇险报警系统 direct secondary distress alerting network
直接点火 direct fire
直接电动螺旋锤 percussion screw press
直接电动螺旋压力机 percussion screw press
直接电感耦合 direct inductive coupling
直接电弧炉 arc furnace;direct arc furnace;direct electric(al)arc furnace
直接电弧熔炼法 direct arc melting method
直接电弧熔炼炉 direct arc melting furnace
直接电弧(熔铸)direct arc
直接电路 direct circuit
直接电热 direct electric(al)heating
直接电源馈路 stub feeder
直接电照相 direct electrography
直接电子照相印刷机 direct electrophotography printer
直接电阻加热 direct resistance heating
直接电阻炉 direct resistance furnace
直接调查 direct observation; first hand;first investigation
直接迭代 direct iteration
直接迭代处理(调查)direct iterative procedure
直接钉入 direct nailing
直接顶板 immediate roof
直接订货 direct order
直接订货制度 direct indent system
直接定交点定线 locating by direct locating intersection point
直接定位方式 direct location mode
直接定位仪 direct finder
直接定位装卡 direct mount
直接定线法 method of direct layout of line
直接定向法 direct orientation
直接定向辐射 head-on radiation
直接定址 direct addressing; immediately addressing
直接动力作用推扫耙 direct power acting push sweep rake
直接动作 direct action
直接动作螺管阀 direct acting solenoid valve
直接动作寻线机 direct action finder
直接毒性 direct toxicity
直接毒性评估 direct toxicity assessment
直接读出 direct reading;direct readout
直接读出红外辐射计 direct readout infrared radiometer
直接读数的 direct reader; direct reading
直接读数地震仪 direct digitizing seismograph
直接读数法 direct reading method
直接读数毛细管作用图 direct reading capillary chart
直接读数频率计 direct reading frequency meter
直接读数温度 direct reading temperature
直接度量 direct measurement
直接短路 dead-short circuit
直接短切 direct chopping
直接短切机 direct chopper
直接对内拨号 direct inward dial(l)ing
直接对外拨号 direct outward dial(l)ing
直接对外投资 direct foreign investment
直接多工控制 direct multiplex control
直接二进制 straight binary
直接发电机 direct power generator
直接发货 drop shipment
直接发泡 direct expansion
直接发送 direct readout
直接发送式 straight sending system
直接发送系统 direct transmission system;straight sending system
直接法 direct approach;direct manner
直接法标准化 direct standardization
直接法标准化率 standardized rate by direct method
直接法纠正 direct rectification
直接法氧化锌 American process zinc oxide
直接反馈 direct feedback

直接反馈系统 direct feedback system
直接反射 direct reflection
直接反应 direct reaction
直接反转片 direct reversal film
直接方法 immediate means; straight-forward method
直接方式 direct mode
直接访问地址 direct reference address
直接访问终端 access-rights terminal
直接放大 direct amplification; direct enlargement; straight amplification
直接放大式接收机 straight receiver
直接放大式调谐器 straight tuner
直接放置炸药爆破法 carry on charges blasting
直接费率 direct rate
直接费审计 direct expense audit
直接费（用） direct charges; direct cost; direct expenses; flat cost; out-of-pocket expenses; specific cost; prime cost
直接费用成本 direct expense cost
直接费用的减少 direct cost reduction
直接分度法 direct indexing
直接分度盘 direct index plate
直接分度盘销 direct index plate pin
直接分离法 direct method of isolation
直接分裂 amitosis; direct division; direct segmentation
直接分流汽油 direct gasoline
直接分馏 straight-forward fractional distillation; straight fractional distillation
直接分配 direct allocation; direct distribution
直接分配法 direct allocation method; direct distribution method
直接分色法 direct colo(u)r separation
直接分色加网 direct screening colo(u)r separation
直接分析 direct analysis
直接分支 direct descendant
直接焚化 direct incineration
直接焚烧 direct incineration
直接粉末还原 direct route powder reduction
直接浮动地址 direct address relocation
直接浮选 direct flo(a)tation; straight heat dryer
直接符号识别 direct symbol recognition
直接辐射 direct radiation
直接辐射波 direct ray; ground ray; ground wave; direct wave
直接辐射供热用具 direct radiant heating appliance
直接辐射计 pyrheliometer
直接辐射三角测量 direct radial plot; direct radial triangulation
直接辐射式供暖器 direct radiant heater
直接辐射线 direct radial ray
直接腐蚀 direct corrosion
直接付款 direct payment
直接负荷损失 direct load loss
直接负债 direct debt; direct liability
直接复合 direct recombination
直接复接 straight multiple
直接复示系统 direct repeating system
直接复印 direct reproduction
直接复照法 direct photographic(al) method
直接赋税 direct taxation
直接腹地 backland of straight-forward
直接干扰 direct disturbance
直接干燥 direct dry
直接刚度法 direct stiffness method
直接刚度调和分析法 direct stiffness harmonic analysis
直接钢丝索 direct-rope haulage
直接杠 direct linkage roll feed
直接高程测量 direct level(l)ing
直接个人费用 direct personnel expenses
直接给进调节器 direct supporting type of feed control
直接工程费用 direct construction cost; direct cost; direct engineering cost
直接工时 direct labo(u)r time
直接工值 direct labo(u)r cost
直接工资 direct labo(u)r cost; direct wage
直接工作 direct working; on-line operation
直接公务电路 direct service circuit
直接供电 direct feed
直接供电方式 direct power feeding system
直接供给 direct supply
直接供暖 direct heating
直接供暖面 direct heating surface
直接供暖器 direct heater
直接供暖设备 direct warmer
直接供热设备 direct heating device
直接供热系统 direct heating system
直接供水水库 direct supply reservoir
直接供液蒸发器 direct feed evaporator
直接供应资金 direct financing
直接购买 over-purchase
直接购买旅游卡 over-purchase of travel cards
直接估计 direct estimate
直接估价 direct quotation
直接估算 direct estimate
直接固定 snap-on
直接固定卡 direct-attachment clip
直接固定扣紧件 <铁轨在桥上的> direct fixation fastener
直接固液分离 direct solid-liquid separation
直接雇工 direct employed labo(u)r; direct labo(u)r
直接雇佣劳动力 direct labo(u)r
直接挂接装置 direct hitch
直接挂网片 direct halftone
直接观测 direct measurement; direct observation
直接观测角 <导线中的> direct angle
直接观测平差 adjustment of direct observation
直接观测侦察 reconnaissance by direct observation
直接观察法 direct observation method
直接观察立体镜 direct viewing type stereoscope
直接观看 direct viewing
直接管理费 direct overhead; management direct cost; direct charges; direct expenses
直接管理费账户 direct overhead accounts
直接管头 straight connector
直接贯入法 straight-penetration method
直接灌溉 direct irrigation
直接灌溉面积 direct irrigation area
直接灌区 direct area
直接灌注法 direct dumping
直接光电探测 direct photodetection
直接光反射 direct light reflex
直接光反应 direct reaction to light
直接光探测接收器 direct photodetection receiver
直接光（线） direct light
直接光学测定法 direct optic(al) measurement
直接光照明器 direct light illuminant
直接广播卫星 direct broadcast satellite
直接归约 directly reduce; direct reduction
直接规则 direct rule
直接过驳 overside delivery
直接过滤 direct filtration
直接过账 direct posting
直接焊接 beading
直接焊接头 direct welded connection; direct welded joint
直接焊入线路的电子管 <如超小型笔形管> wired-in tube
直接航线 direct route
直接航运 direct freight
直接合同 direct contract
直接合闸 direct on-line switching; direct switching-in
直接荷载 direct load
直接横向进磨 direct infeed
直接后果 immediate consequence
直接后继块 immediate successor
直接呼叫 direct call
直接呼叫长途线 ring-down-toll line
直接呼叫制 direct challenge system
直接弧焊接 direct arc welding
直接护岸 direct bank protection
直接化学腐蚀 direct chemical attack
直接还原 direct reduction
直接还原法 direct reduction process
直接还原分析 direct reductive analysis
直接还原铁 direct reduced iron
直接环境 direct environment; immediate environment
直接缓冲方式 direct buffering mode
直接缓解式制动机 direct release brake
直接换热 direct heat exchange
直接换装 direct transfer; direct transshipment; overside discharging and loading
直接换装比 percentage of direct transshipment
直接换装作业 direct alongside activities
直接回归方程 linear regression equation
直接回火 direct tempering; prompt tempering
直接回流系统 direct return scheme
直接回输 direct feedback
直接回水系统 direct return scheme; direct return system
直接回水制 direct return system
直接回用 direct reuse
直接汇兑 direct exchange
直接汇兑合约 direct exchange contract
直接汇率 direct rate of exchange
直接汇票 direct draft; direct paper
直接混合法 direct blending method
直接活化 direct activation
直接活菌计数 direct viable count
直接火操作的管式加热炉 fired process tubular heater
直接火烘干燥机 direct-fired drier [dryer]
直接火花点火器 direct spark ignitor
直接火加热的 direct-fired; direct fire heating; direct firing
直接火烤锅 direct-fired kettle
直接火力加热器 direct-fired heater; direct firing furnace
直接火焰加热压力容器 directly fired (pressure) vessel; fired pressure vessel
直接火焰热 direct-flame heat
直接火焰原子吸收法 direct flame atomic absorption method
直接火灾损失 direct firing loss
直接货币损失 direct monetary loss
直接机械传动 straight-forward mechanical drive
直接积分 direct integration
直接基础 direct bearing foundation; direct foundation
直接激冷铸造 direct chill casting
直接激励 direct drive; direct excitation
直接激励天线 directly excited antenna
直接急冷系统 direct quench system
直接集尘器 directional dust ga(u)ge
直接集聚需求模型【交】 direct aggregate demand model
直接集水面积 direct catchment
直接挤压法 direct extrusion
直接计价 direct pricing
直接计件制 direct piecework system
直接计量 direct measurement
直接计数 direct census
直接计数法 direct counting method
直接计算 direction calculation; straight-forward calculation
直接计算法 direction calculation(method)
直接记录 direct record
直接记录的 direct recording
直接记录地震仪 direct recording seismograph
直接记录法 <不经过记录台> direct record working
直接记录方式 direct recording system
直接记录胶片 direct recording film
直接记录器 direct writing recorder; mechanical oscillograph
直接记录示波器 direct writing oscillograph
直接记录式电传真设备 direct recording facsimile equipment
直接记录系统 direct recording system
直接记入方式 direct entry mode
直接技术收益 direct technologic(al) income
直接寄存 direct load
直接加法 direct addition
直接加热 direct heated; direct heating; straight firing
直接加热的热处理炉 direct-fired coil furnace; direct-fired furnace
直接加热法 snead process
直接加热干燥 direct drying
直接加热干燥器 direct heat drier [dryer]
直接加热烘干机 direct heat drier [dryer]
直接加热炉 direct-fired furnace; directly fired kiln; open flame furnace
直接加热器 direct heating apparatus
直接加热热水供应系统 direct heating hot water supply system
直接加热装置 direct heating apparatus
直接加网 direct screening
直接加网分色法 direct screening method
直接加压 free pressing
直接价格 direct price
直接间接抽水 direct-indirect pumping
直接间接式供暖 direct-indirect heating
直接兼容性 direct compatibility
直接监护器 visual monitor
直接监控 direct current supervision
直接剪力 direct shear
直接剪力试验 direct shear test
直接剪力仪 direct shear apparatus
直接剪切法 direct shear method

直接剪切试验 direct shear test
直接剪切仪 direct shear apparatus
直接检波式接收机 direct detection receiver
直接检测强度调制 direct detection intensity modulation
直接检视式储存管 direct vision storage tube
直接检索 direct search
直接检验 direct survey
直接建厂费用 <包括工程设计费和建筑施工费> plant cost
直接降水量 direct precipitation; throughfall
直接交叉连接 direct cross-connection
直接交付 direct delivery; immediate delivery
直接交换 direct exchange
直接交换作用 direct exchange interaction
直接交货 direct consignment; direct delivery
直接交易 direct dealing; direct transaction
直接浇灌 direct dumping
直接浇铸 direct casting; direct pouring
直接角 direct angle
直接教学法 direct teaching method
直接接触 direct contact; immediate contact; physical contact
直接接触曝光 direct contact exposure
直接接触传播 direct contact transmission
直接接触法 direct contact method
直接接触换热器 direct contact heat exchanger
直接接触冷冻方式 direct contact freezing
直接接触膜蒸馏 direct contact membrane distillation
直接接触式 direct contact type
直接接触式换热器 direct contact heat exchanger
直接接触循环系统 direct contact cycle system
直接接触正片 direct positive
直接接到线路 direct to line
直接接地 solid ground
直接接合 direct bonding; direct connect; direct coupling
直接接合的 direct connected; direct coupled
直接接入通信[讯]信道 direct access communication channel
直接接头 straight pipe union; straight-through joint
直接节省额 direct savings
直接结构 direct organization
直接结合 direct bonding
直接结合碱性耐火砖 direct bonded basic brick
直接结合率 direct bonded rate
直接结合耐火材料 direct bonded refractory
直接结合式 machine control unite type
直接结算法 direct closing method
直接解 direct solution; immediate solution
直接解冻 straight thawing
直接解释法 direct interpretation method
直接解译 direct interpretation
直接解译标志 direct interpretation key
直接借出 direct lending
直接借入 direct access; direct debiting
直接金属 direct metal
直接劲度法 direct stiffness method
直接进口 direct import
直接进口路 direct access
直接进口喷射系统 direct import injection system
直接进料焚化炉 direct feed incinerator
直接进路 <与干道直接连接的进口> direct access
直接进水的矿床 mineral deposit of direct inundation
直接浸渍法 direct dipping process
直接经验 first-hand experience
直接经营费用 direct operating cost
直接径流 direct flow; direct runoff; immediate runoff; rainfall excess
直接径流过程线 direct runoff hydrograph
直接径流量 direct runoff
直接浚挖岩石 direct rock dredging
直接开关启动器 direct switching starter
直接抗拉强度 direct tensile strength
直接刻图设备 provision for direct scribing
直接空气冷却 straight air cooling
直接空气循环 direct air cycle; direct cycle
直接控制 direct control; direct operation; in-line control; on-line control; positive governing
直接控制的牵引设备 directly controlled traction equipment
直接控制阀 direct control valve
直接控制机构 direct control facility
直接控制连接 direct control connection
直接控制微程序 direct control microprogram(me); horizontal microprogram(me)
直接控制微程序设计 direct control microprogramming
直接口 straight joint
直接矿物循环理论 direct mineral cycling theory
直接馈电 direct feed
直接馈电电动机 line-fed motor
直接馈电天线 directly excited antenna; directly feed antenna
直接馈电线 direct feeder
直接馈路 direct feeder
直接扩孔 straight reaming
直接扩散 direct diffusion
直接拉后退火 direct patenting
直接拉力 direct tension
直接拉力强度 direct tension strength
直接拉力指示仪 direct tension indicator
直接拉伸法 direct pull method
直接拉伸强度 tensile strength on direct test
直接拉伸试验 direct tension test
直接拉伸试验机 direct tension tester
直接拉应力 direct tensile stress
直接来源 immediate source
直接劳动 direct labo(u)r; state forces
直接劳动与间接劳动 direct labo(u)r and indirect labo(u)r
直接劳工 immediate labo(u)r
直接雷击 direct lightning stroke; direct stroke
直接雷击过电压 direct lightning stroke over-voltage
直接类比 direct analogy
直接类推 direct analogy
直接冷冻淡化 desalination by direct freezing
直接冷冻系统 direct refrigerating system
直接冷却 direct cooling; direct liquid cooling
直接冷却电机 directly cooled machine
直接冷却法 method of direct cooling
直接冷却器 direct cooler
直接冷却水 direct cooling water
直接冷却系统 direct cooling system
直接冷铸法 direct chill process
直接离心浮集法 direct centrifugal flo(a)tation method; Lane method
直接力矩分配法 direct moment distribution
直接立体观察【测】 direct stereoscopic vision; direct stereoscopic view
直接利用 direct use
直接连接 direct link; direct connection
直接连接的 close-connected; direct connected; direct coupled; direct coupling
直接连接器 card edge connecter[connector]
直接连接式涡轮机 directly coupled turbine
直接连接线 link line
直接连系 direct relation
直接连线 tie line
直接联动泵 direct acting pump; direct connected pump; motor direct coupled pump
直接联动往复式泵 direct acting reciprocating pump
直接联动蒸汽泵 direct action steam pump
直接联机处理机 direct on-line processor
直接联结 direct coupling
直接联结的 direct coupled
直接联结设计 <立体交叉的> direct connection design
直接联锁 dead interlocking
直接联系 direct connection
直接联线处理机 direct on-line processor
直接联线启动器 direct on-line starter
直接联想 immediate association
直接联轴节 direct coupling; through coupling
直接炼钢法 direct steel process
直接炼铁法 Dupry process
直接量测 direct measurement
直接列入支出 direct expenditures
直接流量 direct flow
直接流通 direct flow
直接录音 direct pick-up; live pick-up
直接路径 direct path
直接路由 direct route
直接路由选择 direct routing
直接氯化作用 direct chlorination
直接轮载 straight wheel load
直接码 direct code
直接埋管机 direct burial plough
直接埋缆犁 direct burial plough
直接埋设式 direct built-in system
直接买卖 direct business
直接漫射光 direct diffused light
直接漫射(光)照明 direct diffused lighting
直接冒口 runner riser
直接贸易 direct trade
直接煤气供热 direct gas heating
直接瞄准 direct pointing
直接瞄准望远镜 direct sight telescope
直接灭火 direct fire suppression
直接灭火法 direct fire suppression method
直接灭火水压力 direct fire pressure
直接模板学说 direct template theory
直接模拟 direct analogy
直接模型分析 direct model analysis
直接目测法 direct vision method
直接内插法 direct interpolation
直接内存执行 direct memory execution
直接能见区 directly visible area
直接能量转换 direct energy conversion
直接逆转式柴油机 self-reversing diesel engine
直接啮合 direct-geared; positive gearing
直接凝集反应 direct agglutination reaction
直接凝集试验 direct agglutination test
直接耦合 conductive coupling; direct couple; direct coupling; direct pin connection; through coupling
直接耦合的 directly coupled
直接耦合电路 directly coupled circuit
直接耦合方式 direct coupling system
直接耦合放大 directly coupled amplification
直接耦合放大器 directly coupled amplifier
直接耦合功率放大管 directly coupled power amplifier tube
直接耦合机组 directly coupled machine
直接耦合计算机 directly coupled computer
直接耦合晶体管逻辑电路 directly coupled transistor logic
直接耦合驱动三极管 directly coupled driver triode
直接耦合式多谐振荡器 directly coupled flip-flop
直接耦合式天线 plain antenna
直接耦合双三极管 directly coupled twin triode
直接耦联 direct coupling
直接拍摄法 direct photography
直接排代 direct displacement
直接排泥 direct discharge of spoil; direct disposal of spoil
直接排水 direct drainage
直接派生的 direct derivative
直接判读标志 direct key
直接判读样片 direct key
直接判释谱 direct key
直接赔偿 direct compensation
直接配料 direct dosing
直接配制 direct preparation
直接喷洒 directing spray
直接喷射 direct injection
直接喷射法 direct injection method
直接喷射式柴油机 diesel engine with direct injection; direct injection diesel engine
直接喷射式发动机 direct injection engine
直接喷射式喷油器 direct infection nozzle
直接喷射式汽化器 direct injection carburetor[caruret(t)er]
直接喷射式燃料泵 direct injection pump
直接喷油 direct injection
直接膨胀 direct expansion
直接膨胀冷却 direct expansion cooling
直接膨胀盘管 direct expansion coil
直接膨胀式 direct expansion system
直接膨胀网格式盘管 direct expansion grid coil system
直接膨胀蒸发器 direct expansion evapo(u)rator
直接膨胀制冷 direct expansion refrigeration
直接碰撞 central collision; direct im-

Z

pact; head-on collision; knock-on collision
直接偏转比较法 comparison by direct deflection method
直接偏转法 direct deflection method; method of direct deflections
直接拼接 direct splice
直接平板法导热仪 thermal conductivity tester by absolute plate method
直接平版印刷法 autolithography
直接平差 direct adjustment
直接平印 direct lithography
直接破坏效果 physical destructive effect
直接扑火 direct fire suppression
直接铺设楼面 plank-on-edge floor; solid floor
直接铺筑在地基上的混凝土板 floating slab; slab on grade
直接起动 across-the-line starting; direct on-line switching; direct (on) starting
直接起动电动机 across-the-line motor; line-start motor
直接起动法 self-starting method
直接起动器 across-the-line starter; direct on starter; straight-on starter; switching starter
直接起因 direct cause
直接气举 straight gas lift
直接气体还原 direct gaseous reduction
直接气压式热室压铸机 direct air injection die-casting machine; direct pressure hot chamber machine
直接汽化 direct boiling
直接潜水 ambient pressure diving; direct diving
直接切入磨法 direct infeed
直接切碎收集装置 direct cut grass and pickup attachment
直接切碎装置 direct cut attachment
直接倾倒 direct dumping
直接氢冷 hydrogen inner cooling
直接驱动 direct drive
直接驱动的机车 direct drive locomotive
直接驱动电动机 direct drive motor
直接驱动发动机 direct acting engine; direct drive engine
直接驱动螺旋桨 direct driven airscrew
直接取代 direct substitution
直接取景器 direct viewfinder
直接取数 direct access
直接取样 direct sampling
直接全息摄影 direct holography
直接全息图 direct hologram
直接燃烧 direct combustion; direct firing
直接燃烧灯黑 direct combustion black
直接燃烧分解法 method of direct combustion decomposition
直接燃烧式干燥器 direct-fired drier [dryer]
直接燃烧式锅炉 direct-fired boiler
直接燃烧式加热器 direct-fired heater
直接燃烧式暖风机 direct-fired unit heater
直接燃烧式用具 direct-fired appliance
直接燃烧系统的煤磨 direct coal firing mill
直接燃烧循环采暖器 unvented circulator
直接染料 direct dye(stuff); substantive dye(stuff)
直接染色 direct staining; substantive
直接扰动 direct disturbance
直接热耗 direct waste heat
直接热水罐 direct cylinder
直接热水系统 direct hot water system
直接热损失 direct heat loss
直接热源 direct heat source
直接热源取热 direct heating from heat sources
直接热转换反应器 direct heat conversion reactor
直接人工 direct labo(u)r
直接人工成本 direct labo(u)r
直接人工成本法 direct labo(u)r cost method
直接人工成本预算 direct labo(u)r cost budget
直接人工费 direct labo(u)r cost
直接人工费用 direction labo(u)r cost; direct labo(u)r expenses
直接人工费用预算 direct labo(u)r expense budget
直接人工小时 direct labo(u)r hour
直接人工小时法 direct worker-hour method
直接人工小时率 direct labo(u)r hour rate
直接人工预算 direct labo(u)r budget
直接人类经济活动 primary economic activity
直接人员薪金 direct personnel expenses
直接任务 immediate mission
直接韧化处理 direct patenting
直接日射强度表 pyrheliometer
直接日射强度计 pyrheliograph; pyrheliometer
直接日照强度表 pyrheliometer
直接融资 direct-financing
直接融资租赁 direct financing leases
直接(入海)河口 direct mouth
直接软管水流 direct hose stream
直接扫描 directed scan; direct scanning
直接扫描全景摄影机 direct scanning camera
直接扫描摄影机 direct scanning camera
直接色度测量学 direct colo(u)rimetry
直接杀菌剂 eradicant fungicide
直接砂流过滤 direct sand flow filtration
直接晒印 direct copy
直接闪光焊 straight flash welding
直接上级 immediate superior
直接上盘 immediate hanging wall
直接上升 direct ascent
直接上釉 direct on enameling
直接烧 direct fire
直接设计 direct design
直接射入 direct injection
直接射水 direct attack
直接摄像 direct pick-up
直接摄影 direct pick-up
直接摄影术 direct photography
直接审判 direct trial
直接审判原则 principle of direct trial
直接渗透仪 direct permeability apparatus
直接生产 direct production; immediate production
直接生产费用 direct manufacturing expenses; direct operating cost
直接生产工人 direct labor
直接生产过程 direct manufacturing process
直接生产者 immediate producer
直接生化需氧量 immediate biochemical oxygen demand
直接生物毒物效应 direct ecotoxicological effect
直接声测法 direct sounding
直接声场 direct field; direct sound field
直接声呐接触 close sonar contact
直接施工费 direct construction cost
直接施用 directed application
直接湿切机 direct wet chopper
直接石墨炉法 direct graphite furnace method
直接拾波 direct pick-up
直接示值的 direct reading
直接示值器 direct reader
直接式 straight system
直接式爆炸成型 contact type operation(of explosive forming)
直接式程序 straight-line code
直接式程序编制 straight-line coding
直接式传动 direct drive
直接式打料 positive knockout
直接式电路 straight circuit
直接式供暖系统 direct heating system
直接式烤箱 direct-fired oven; directly heated oven; internally heated oven
直接式孔径天线 direct aperture antenna
直接式热交换器 direct contact heat exchanger
直接式热水供暖 direct hot water heating
直接式收发转换器 orthomode transducer
直接式匝道 direct ramp
直接式蒸发器 direct evapo(u)rator
直接式蒸汽泵 direct acting steam pump
直接式蒸汽供暖 direct steam heating
直接式蒸汽机 direct acting steam engine
直接视距传输 horizon transmission
直接视准 direct sight
直接试验 on-line test(ing)
直接适应作用 direct adaptation
直接收益 direct yield
直接受荷轴承 immediate-bearing
直接受拉钢筋 straight tension rod
直接受热面 direct heating surface
直接受压 straight compression
直接受益 direct benefit; direct gain; primary benefit
直接受益系统 direct gain system
直接售货条款 direct sale clause
直接疏浚 direct dredging
直接输出 direct output; on-line output
直接输出程序 direct output writer
直接输入 direct input; substantive input
直接输入电路 direct input circuit
直接输送 direct transmission
直接输送式组合机床自动线 plain transfer line
直接数据 immediate data
直接数据处理 direct data processing
直接数据处理系统 on-line data handling system
直接数据集 direct data set
直接数据检核 direct data checking
直接数据输入 direct data entry
直接数据输入输出 data-directed input-output
直接数据通道 direct data channel
直接数据相关性 immediate data dependency
直接数据组织 direct data organization
直接数控 direct numerical control
直接数值积分 direct numerical integral
直接数值控制 direct numerical control
直接数字编码器 direct digital encoder
直接数字(程序)控制 direct digital control; direct numerical control
直接数字交换 digital switching; direct digital transfer
直接数字接口 direct digital interface
直接数字控制系统 direct digital control system
直接数字控制仪 direct digital controller
直接双相反应 direct biphasic reaction
直接水压式成型法 Kranenburg method
直接水跃 direct jump
直接水蒸气 open steam
直接水准测量 direct level(l)ing
直接税 direct tax; underlying tax
直接顺序文件 direct sequential file
直接死因 cause directly leading to death; direct cause of death
直接搜索 direct search
直接搜索法 direct search method
直接酸度 immediate acidity
直接损害 direct damage
直接损害费用 direct injury cost
直接损失 direct damage
直接锁闭 deadlocking; direct locking
直接太阳受益 direct solar gain
直接摊还法 straight-line method of amortization
直接弹性【交】 direct elasticity
直接探测石油 direct oil detection
直接套汇 direct arbitrage; direct exchange of arbitrage
直接套利 direct arbitrage
直接提单 direct bill of lading; straight bill of lading
直接替代 direct substitution
直接调节 direct control; primary governor control; self-acting control; self-operated control
直接调节器 self-actuated controller
直接调频 direct frequency modulation
直接调频制 direct frequency modulation system
直接调色法 direct toning
直接调整 positive governing
直接调制方式 reactance modulation system
直接跳接线 straight jumper
直接跳闸断路器 direct trip circuit-breaker
直接通报法 ringing junction working
直接通报制 ringing junction working
直接通道 direct path
直接通地 dead ground
直接通电 direct passage of current
直接通风炉 straight-draft furnace
直接通过式加速度 straight-through acceleration
直接通话 direct call; through call
直接通话能力 direct access capability
直接通路 direct path
直接通商 direct transaction in business
直接通信[讯] direct communication; direct transmission
直接通信[讯]联络 direct relation
直接同位素效应 direct isotope effect
直接头 straight joint
直接投入表 direct input table
直接投影 direct projection; front projection
直接投资 direct investment; equity investment
直接投资红利 direct investment dividends
直接投资基金 direct investment fund
直接投资利息 direct investment interest
直接投资流动 direct investment flow

直接透射法 direct transmission method
直接透射密度法 direct transmission density method
直接透射系数 direct transmission factor
直接涂布法 direct roll coating
直接涂片 direct smear
直接涂漆 on-line coating
直接推导 direct derivation
直接推动式 direct acting
直接推动式煤气恒温计 direct acting gas thermostat
直接推动式水泵 direct acting pump
直接推断法 heuristic method
直接推理 immediate inference
直接推力 direct thrust
直接推土板 straight blade
直接推土刀 straight blade
直接托付 direct consignment
直接脱扣 direct trip
直接脱硫 direct desulfurization
直接外运矿石 shipping ore
直接危险地区 immediate danger area
直接为……筹措资金 direct finance for
直接维护 contact servicing
直接维修 contact maintenance;direct maintenance
直接纬纱细砂机 filling frame
直接喂煤回转窑 direct coal fired kiln
直接喂煤系统的磨机 direct feed mill
直接文件 direct file
直接文件存取 direct file access
直接问询调查 direct interview
直接污染物排放系数 direct pollutant discharge coefficient
直接污染源 immediate source
直接物料法 direct materials basis
直接吸泥式挖泥船 moored suction dredge(r)
直接吸引范围 direct attractive range
直接熄弧断路器 plain-break breaker
直接系统 direct system
直接系在工人腰带上的安全绳 drop line
直接下水管道系统 direct sewer system
直接先趋块 immediate predecessor
直接显色分光光度法 direct development of spectrophotometry
直接显示 direct view display
直接显示器 direct display
直接显微镜法 direct microscopic method
直接显像 direct view display
直接显影 chemical development; direct development
直接显影法 direct vision method
直接"线对线"连接型 direct line-to-line connection type
直接线性变换 direct linear transformation
直接相关 direct correlation
直接相加指令 direct add instruction
直接相减指令 direct subtract instruction
直接相邻波道干扰 direct adjacent channel interference
直接相撞 knock-on collision
直接向内拨号 direct inward dial(l)ing
直接向外部拨号 direct outward dial(l)ing
直接像 direct image;true image
直接消防压力 direct fire pressure
直接消费 direct consumption
直接消费税 direct consumption tax; direct tax on consumers
直接消耗 direct consumption
直接消耗系数 direct consumption coefficient
直接消去法 direct elimination method
直接消息 direct message
直接销售 direct sale
直接销售比较法 direct sales comparison approach
直接销售价 direct sale price
直接效果 direct effect
直接效益 direct benefit;on-site benefit;primary benefit
直接效应 direct effect
直接效用 direct utility
直接效用和间接效用 direct utility and indirect utility
直接协商 direct negotiation
直接写入 write direct
直接写通【计】write-through
直接卸船装车 direct ship-to-wagon operation; direct ship-to-wagon working
直接卸货 direct discharging
直接卸料 direct dumping
直接信息 first-hand information
直接信息读出卫星 direct readout satellite
直接信用证 <一次使用的信用证> straight letter of credit; straight credit
直接行动 direct action
直接行销 direct marketing
直接性 directness
直接性模型【交】direct model
直接需求 direct demand
直接需求模型【交】direct demand model
直接需氧量 immediate oxygen demand
直接需要 direct requirement
直接需要品 goods of first order
直接序列扩展频谱 direct sequence spread spectrum
直接续租 direct continuation
直接悬挂 direct suspension
直接悬挂结构 direct suspension construction
直接选择 direct dial(l)ing;direct option; direct selection; non-register-controlling selection
直接选择方式 direct selection system
直接选择器 direct selector
直接选择性浮选 straight selective flo(a)tation
直接选址 direct addressing; immediately addressing; real-time addressing;zero level addressing
直接眩光 direct glare
直接寻址模 direct addressing mode
直接询问 direct question
直接询问系统 direct challenge system
直接循环 direct circulation
直接循环反应堆 direct cycle reactor
直接循环一体化沸水堆 direct cycle integral boiling reactor
直接迅速反应 direct prompt reaction
直接压电现象 direct piezoelectricity
直接压坏试样法 direct crushed sampling
直接压力 direct compression;directe pressure
直接压力波 direct pressure wave
直接压片法 direct compression process
直接压缩 direct compression
直接压应力 direct compression stress; direct compressive stress
直接压制法 straight pressing
直接研究法 visual observation
直接焰 direct flame
直接氧化 direct oxidation
直接氧化法 direct oxidation process
直接氧化反应 direct oxidation reaction
直接氧化浸出 direct oxidation leaching
直接冶炼法 direct smelting process
直接业务 direct activities
直接液化 direct liquefaction
直接液冷 direct liquid cooling
直接液压式板料成型法 hydromatic forming
直接以货币支付的成本 explicit cost
直接以货币支付的利息 explicit interest
直接以贸易货 direct barter
直接抑制 direct inhibition
直接易货 direct barter
直接意义 direct bearing
直接银行保证 direct bank guarantee
直接引导方位 direct bearing
直接印花 application printing; conventional printing;direct printing
直接印染 direct printing
直接印刷 direct printing
直接印刷机 direct printer
直接印刷装饰板 direct printing decorative board
直接应变 direct strain
直接应变仪 direct strain detector
直接应力 direct stress
直接应力试验机 direct stress machine
直接营运设施 direct operational features
直接影响 direct affect; direct bearing;direct influence;primary effect
直接影响因素 direct acting factor
直接影印机 photostat
直接影印(件)photostat
直接影印照片 photosta
直接映象 direct mapping
直接映像高速缓冲存储器 direct mapping cache
直接用户 end-user
直接用日光取暖 direct solar gain
直接用于路面工程的焦油 tar cement
直接优先浮选 straight selective flo(a)tation
直接由柴油机驱动的船 direct drive diesel ship
直接邮电 direct mail
直接邮寄 direct mail
直接有关的款待 directly related entertainment
直接右转匝道 right-turn ramp
直接与混凝土面接触的模板 face contact material
直接与间接过程成本 direct and indirect process cost
直接原材料 direct raw material
直接原材料费用预算 direct material budget
直接原料成本 direct material cost
直接原料购买预算 direct material purchases budget
直接原因 causa; causa causaus; causaus; causa proxima; immediate cause
直接远距离拨号 direct distance dial(l)ing
直接阅读 direct reading
直接阅读式袖珍十字板剪刀仪 direct reading pocket shear vane
直接运动 direct motion;first motion
直接运输 carry-over; direct haulage; direct shipment
直接运输规则 rule of direct consignment
直接再用 direct reuse
直接在下面的 subjacent
直接噪声 direct noise
直接噪声放大器 direct noise amplifier
直接责任 direct obligation
直接增湿器 direct humidifier
直接粘贴 direct glue-down
直接招工 direct solicitation
直接找矿标志 direct prospecting indication
直接照明 accent lighting; direct illumination;direct lighting
直接照明器 direct luminaire
直接照射 direct lighting
直接折旧法 straight-line depreciation method
直接振铃电路 ring-down circuit
直接振铃通话专用线 ring-down trunk
直接振铃制 direct challenge system
直接蒸发的蒸发器 direct expansion evapo(u)rator
直接蒸发冷却 direct expansion cooling
直接蒸发盘管 direct expansion coil
直接蒸发式冷却器 direct expansion cooler
直接蒸发系统 direct expansion system
直接蒸馏 straight distillation
直接蒸气(再生)法 pan process
直接蒸汽 direct steam; live open steam
直接蒸汽发生器 direct steam generator
直接蒸汽循环 straight cycle
直接正片 autopositive
直接正像 direct positive
直接正像材料 autopositive
直接正像胶片 <一次显影的> autopositive film
直接证据 direct evidence; evidence direct
直接证明 direct proof
直接证明的 ostensive
直接证券 direct securities
直接支承 immediate-bearing
直接支承物 direct bearing
直接支出的 out-of-pocket
直接支付 direct payment
直接支援 immediate support
直接指标 direct indicator
直接指令单 manuscript
直接指示电流计 aperiodic(al) galvanometer
直接指示探向器 direct reading direction finder
直接指数化 direct indexation
直接酯化 direct esterification
直接制版法 direct to-plate
直接制冷法 direct method of refrigeration
直接制冷系统 direct expansion refrigeration;direct refrigerating system; direct system of refrigeration
直接致癌物 direct acting carcinogen
直接致癌原 direct acting carcinogen
直接致死地带 zone of immediate death
直接致突变物 direct acting mutagen
直接置换 direct replacement
直接中继器 direct point repeater
直接中继制 direct trunking system
直接重力涵洞 direct gravity culvert
直接轴联节 direct coupling
直接注入法 direct injection
直接注射造型 direct injection mo(u)lding
直接注水系统 direct flooding system
直接转动 direct drive
直接转动变速箱 direct drive transmission
直接转动的起动器 direct cranking starter
直接转换器 direct translator
直接转接国际电路 direct transit in-

ternational circuit
直接转矩 direct torque
直接转移 direct jump
直接转账系统 giro system
直接转账制 giro system
直接装配玻璃 direct glazing
直接装入 direct loading
直接装上的 direct-mounted
直接装桶挤奶室 cow-to-can milking room
直接装卸 direct loading and unloading
直接装运法 direct loading
直接装运通知 direct shipment order
直接装载 direct load
直接装置 direct readout; on-line equipment
直接资本还原率 direct capitalization
直接资料 direct data
直接紫外线 direct sun rays
直接紫外荧光法 direct ultraviolet fluorometric method
直接自动调整器 diactor
直接租借 direct leasing
直接阻抗 direct impedance
直接组分运动 direct componential movement
直接作动发动机 direct acting engine
直接作业 direct operation; direct transshipment
直接作用 direct effect; immediate effect; primary action; direct action
直接作用泵 direct acting pump
直接作用的 direct acting
直接作用的解扣 direct acting trip
直接作用的旋转式选择器 direct acting finder
直接作用荷载 direct acting load
直接作用恒温器 direct-acting thermostat
直接作用控制阀 direct acting control valve
直接作用控制器 direct-acting controller
直接作用式隔膜阀 direct acting diaphragm valve
直接作用式继电器 direct actuating relay
直接作用式减震器 direct acting shock absorber
直接作用式气体温度计 direct acting gas thermometer
直接作用式调节阀 direct acting valve
直接作用式压力调节器 direct action pressure controller
直接作用式蒸汽泵 direct acting steam pump
直接作用式制动机 direct acting brake
直接作用调节器 primary governor control; self-actuated controller
直接作用调温器 self-operated thermostatic controller
直接作用调整器 direct acting regulator
直接作用往复泵 direct acting reciprocating pump
直接作用选择器 direct action selector
直接作用闸 direct acting brake
直接作用蒸汽泵 direct action steam pump
直接作用蒸汽机 <与抽油泵相连的> bull-engine
直接作用蒸汽汽缸 steam ram
直接作用执行器 direct acting actuator
直接作用转向阀 direct acting reversing valve
直结晶器 straight mo(u)ld
直捷 directness
直捷操作 <对被呼台进行自动接通的操作> straight-forward operation
直截了当的 point-blank
直进的 straight-forward
直进给 straight-in feed
直进航线 straight ramp
直进流 direct flow
直进式 dead beat
直进式传送 in-line transfer
直进式擒纵机构 deadbeat escapement
直进式速调管 straight advancing klystron
直进式推土机 straight dozer
直进式振动送料器 vibratory in-line hopper feeder
直进式执行部件 push stem power unit
直进外圆磨床 straight-in grinding machine
直浸没式 vertical submerged type
直茎浆果 standard berry
直井 straight hole
直井炉 shaft furnace
直井式溢洪道 morning-glory spillway; shaft spillway
直颈铆钉 straight-neck rivet; straight shank rivet
直径 diameter
直径比 diameter ratio
直径比率 natural scale
直径变化 change in ga(u)ge
直径长度 diameter length
直径(尺寸)标注 diameter dimensioning
直径的 diametral; diametric(al)
直径的全断面 full diameter
直径方向的 diametric(al)
直径公差 diameter tolerance; ga(u)ge change <钎子组的>
直径和容积的免查 exemptions of diameter and volume
直径级 diameter class
直径级界限 diameter limit
直径级木 diameter class of wood
直径铗 tree compass
直径节距 diameter pitch
直径径向公差 diametral gap
直径卷尺 diameter tape
直径可调节钻头 expanding bit
直径控制 diameter control
直径扩大的烧成带 enlarged sintering zone
直径留隙 diameter clearance
直径率 diameter quotient
直径面 diameter plane
直径磨损 <钻头、钎头> ga(u)ge wear; wear across the ga(u)ge
直径磨小尺寸 ga(u)ge wear size
直径平面 diametral plane
直径曲面 diametral surface
直径曲线 diametral curve
直径上对置的 diametrically opposite
直径损耗 <钻头> ga(u)ge loss
直径缩减率 percentage reduction of diameter
直径缩小 diameter diminution
直径系数 diametral quotient
直径线 diameter track
直径压碎试验 diametral crushing test
直径增量 diameter increment
直径指数安全系统 diameter index safety system
直径中值 median diameter
直径锥度 diameter of taper
直鸠尾榫 straight dovetail
直距线 beeline; beed line
直锯 gang mill; gang saw; pit saw; rip; straight saw
直锯材 flat-sawn timber
直锯法 plain-sawing; through-and-through sawing
直锯平纹木材 flat-grained lumber
直锯四开木材 <对生长年轮约成直角> quarter-sawn timber
直锯四开木料 Comanchic grain timber; quarter-sawn timber <对生长年轮约成直角>
直锯条式锯轨机 straight-blade type rail sawing; straight-blade type rail sawing machine
直卷边接缝管 lock seam pipe; lock seam tube
直觉校正过程 intuitive recovery procedure
直觉判断 intuitive judg(e)ment
直觉数据采集 visual data acquisition
直觉显示器 visual display
直觉显示终端 visual display terminal
直觉信息显示 visual information display
直开浇口管 punch pipe
直靠把 straight fender
直靠背椅子 straight chair
直孔 straight hole
直孔闭孔型 tongue and groove pass
直孔空心钻头 straight-sided core bit
直孔型 straight pass
直控搪瓷 direct on enamel
直快列车 through train
直框架结构船 deadrise boat; double-wedge boat; hard chine boat; straight frame(d) ship
直框式闸门 straight frame(d) gate; vertical frame(d) gate
直馈式天线 directly fed antenna
直馈式线圈 direct fed coil
直拉窗 <上下拉的> sash window
直拉条 straight brace
直拉条架 straight-braced frame
直缆束 straight cable
直廊 straight corridor
直肋骨船 straight frame(d) ship
直棱 straight edge
直棱纹织物 soleil
直楞【建】bar grating; dog bar <门下部的>
直楞炉箅 bar grading
直冷 bar grating
直冷炉箅 bar grating
直冷式冰箱 direct cooling refrigerator
直冷式蒸发器 direct evapo(u)rator
直犁刀 hanging cutter; knife colter; knife cutter; sliding colter
直立 bristle; endway; endwise; erection; stand on end; vertical occurrence
直立安放 to be kept upright
直立岸壁 quay with vertical face
直立岸壁码头 quay with vertical face
直立背斜 erect anticline
直立臂式挖沟机 vertical boom type trencher
直立边墙 vertical sidewalk
直立储筒 pillar reservoir
直立船首柱 square stem; upright stem
直立床身式铣床 vertical bed type milling machine
直立单干形 upright cordon; vertical cordon; vertical palmette
直立的 erect; standing; stand-up; upright; erective; orthostatic; perpendicular; upstanding; vertical
直立的厕所污水管 soil stack
直立的叉式起重车 stand-up fork truck
直立的剪力撑系统 vertical bracing system
直立的土壁 standing bank
直立的消防主管 rising fire main
直立堤堤身 body of vertical breakwater; stem of upright sea wall
直立堤上部结构 superstructure of vertical breakwater
直立地层 stand-up formation
直立电缆 riser cable
直立法 uprighting; upright method
直立反应 orthostatic reaction
直立防波堤 upright breakwater
直立覆盖植物 erect cover
直立杆 down rod; up-and-down rod; upright
直立杆架 upright rod guide
直立钢筋 vertical bar
直立共面装置 vertical common-plane array
直立共轴装置 vertical common-axis array
直立拐肘 vertical crank
直立管 uprise
直立焊缝板 standing seam plate
直立接缝 standing seam
直立节理 upright joint
直立截槽法 shearing
直立浸没式 vertical submerged type
直立茎 erect stem
直立矩形棱柱法 vertical rectangular prismy forward calculation method
直立矩阵 column matrix
直立卷扬机 vertical hoist
直立棱形 upright diamond
直立棱柱 right prism
直立棱锥 right pyramid
直立梁 upstand(ing) beam
直立菱形 upright diamond
直立炉 vertical furnace
直立煤层 upright coal seam
直立耐性 orthostatic tolerance
直立胚 erect embryo; straight embryo
直立起拱点 vertical springing
直立墙 upright wall; vertical wall
直立墙防波堤 vertical wall breakwater
直立墙面大理石 standing wall-facing marble
直立墙式防波堤 vertical face breakwater
直立墙头旗杆 vertical wall-mounted flagpole; vertical wall set flagpole
直立桥墩 standing pier
直立倾伏褶皱 upright plunging fold
直立设备 vertical vessel
直立式岸壁 vertical wall
直立式沉箱防波堤 upright caisson breakwater
直立式传动装置 stand-up-drive
直立式导缆滚轮 vertical roller of fairleader
直立式发动机 vertical engine
直立式防波堤 upright breakwater; vertical(face) breakwater; vertical sea wall
直立式防波堤堤身 body of upright breakwater; body of vertical breakwater
直立式防波堤上部结构 superstructure of vertical breakwater
直立式风洞 vertical wind tunnel
直立式海堤 vertical(-type) sea wall
直立式海塘 vertical(-type) sea wall
直立式海洋考察平台 perpendicular oceano-graphic(al) platform
直立式加速器 vertical accelerator
直立式金刚石刀架 diamond cutter held upright
直立式临水结构物 vertical waterfront structure
直立式路缘石 vertical curb
直立式码头 quay wall; vertical faced wharf
直立式气缸 upright cylinder
直立式墙 vertical wall
直立式水尺 vertical ga(u)ge
直立式炭化炉 vertical retort
直立式样条 stand transect

直立式蒸发器 vertical type evapo-(u)rator
直立式钻 upright drill(er)
直立水平褶皱 upright horizontal fold
直立穗 erect head
直立梯 standing ladder;vertical ladder
直立铁芯 upright core
直立图像 erecting image
直立外堤 upright wall
直立往复锯 gig saw
直立位置 orthostatic position;upright position;vertical position
直立物 tedge
直立吸水式 vertical suction type
直立消波块堤 upright breakwater with dissipating concrete blocks
直立消波式码头 upright type quay-wall with wave dissipating structure
直立旋转式停车楼 solid rotary parking tower
直立压缩机 straight-line compressor
直立岩层 vertical bed
直立堰 vertical weir
直立咬口 standing seam
直立阴影角 vertical shadow angle
直立圆筒混合机 end-over-end mixer
直立圆柱 right circular cylinder
直立圆锥 right circular cone;right cone
直立褶皱 erect fold;upright fold
直立支撑 post shore
直立支承 direct bearing
直立枝 erect branch;upright growth;vertical branch
直立中央内龙骨 vertical center keelson
直立柱 right cylinder
直连安装 in-line installation
直连泵 direct connected pump
直连插头座 feed-through connector
直连电动机 direct connected motor
直连杆 straight connecting rod
直连励磁机 direct connected exciter;direct coupled generator;direct coupling exciter
直连式动力转向装置 in-line linkage
直连式发电机 direct coupled generator
直连式立(体)交(叉) direct connection interchange
直连式励磁机 direct coupled exciter
直联 direct connection
直联泵 direct connected pump;monoblock pump
直联检测器系统 on-detector system
直联式振动器 rigid type vibrator
直联型发电机 coupled-type generator
直链 linear chain;normal chain;straight chain
直链化合物 linear compound;straight chain compound
直链聚合物 straight chain polymer
直链聚乙烯 straight-linear polyethylene
直链石蜡 straight chain paraffin
直链烃 linear hydrocarbon;normal hydrocarbon;straight chain hydrocarbon
直链烷化磺酸盐 linear alkylate sulfonate
直链烷基苯 linear alkylbenzene
直链烷基苯磺酸盐 linear alkylbenzene sulfonate
直链烷烃 linear paraffin;straight chain paraffin
直链形悬挂 polygonal equipment
直链藻 melosira
直链脂肪烃 straight chain aliphatic hydrocarbon
直梁 straight beam
直梁半挂车 straight frame semitrailer
直梁单元 straight beam element
直梁法 straight beam method
直梁接合 splice web
直梁支护 straight girder support
直列八缸发动机 straight-eight engine
直列的 in-line;orthostichous
直列二进制 Chinese binary;column binary
直列汽缸发动机 straight-line engine
直列汽缸排列 in-line cylinder arrangement
直列山脉 montes recti
直列式 in-line arrangement
直列式柴油机 straight diesel
直列式传动(装置) in-line drive
直列式多缸发动机 straight cylinder
直列式发动机 in-line engine;row engine;straight engine
直列式发动机曲柄 in-line engine crankshaft
直列式建筑物 line building
直列式冷却器 <烧结矿的> straight cooler
直列式连杆系 in-line(loader) linkage
直列式料箱 linear bin;linear silo
直列式填装炸药 extended charge
直列式柱塞泵 in-line plunger pump;in-line pump
直列式柱塞电动机 in-line plunger motor
直列式柱塞马达 in-line plunger motor
直列式装置 linear plant
直列四冲程发动机 vertical in-line 4-stroke engine
直列线 orthostichy
直列向量 column vector
直列型联合收割机 in-line combine
直列指示管 pandicon
直列制表 vertical tabulation
直列制表符号 vertical tabulation character
直列装(炸)药 extended charge
直列组合 in-line combination
直裂纹 vertical crack
直棂 <门窗上的> monial;minion
直棂窗 munnion window;muntin window;vertical bar window
直棂式幕墙 mullion-type curtain wall
直流 cocurrent;constant current;continuous current;one-through;single-flow;straight current;straight flow;unidirectional flow;zero frequency current
直流安培计 direct current ammeter
直流泵 straight-way pump
直流闭塞法 direct current blocking
直流边 <全波整流器的> direct current side
直流变流器 continuous current transformer
直流变压器 direct current transformer
直流拨号 direct current selection
直流拨号脉冲 direct current impulse;direct current impulsing
直流布置 <燃气轮机的> straight-through arrangement
直流部分 direct component;steady component
直流部分的恢复 direct current restoration
直流操作 once-through operation
直流槽型浮选机 level type flo(a)tation machine
直流测速发电机 direct current tachogenerator
直流成套设备 complete set of direct current equipment
直流的 direct flow;once-through;straight-through;straight-way;uniflow;uniflow
直流等离子体板 varisymbol
直流等离子体喷枪 direct current plasma torch
直流电 direct current[DC];galvanic current;unidirectional current;galvanism <电池等利用化学反应产生的>
直流电报 direct current telegraphy
直流电表 direct acting current meter
直流电磁泵 conduction pump
直流电电力牵引 direct current traction system
直流电动供暖 direct electric(al) heating
直流电动机 direct current motor
直流电动机驱动泵 direct current motor driven pump
直流电动起动系统 direct electric(al) starting system
直流电动势 direct electromotive force
直流电动转辙机 direct current point machine;direct current switch machine
直流电法 direct current survey
直流电法场源类型 type of field source in DC electrical prospecting
直流电焊 direct current welding
直流电焊机 direct current welder
直流电弧 direct current arc
直流电弧焊 arc welding with DC;DC arc welding
直流电弧焊机 direct current arc welder
直流电弧焊接 direct current arc welding
直流电换流器 direct current inverter
直流电机 direct current machine
直流电绞刀 DC electric(al) reamer
直流电缆 direct current cables
直流电力回路器具 direct current power circuit apparatus
直流电力机车 direct current electric(al) locomotive
直流电力起动系统 direct electric(al) starting system
直流电力牵引 direct current electric(al) traction
直流电力牵引制 direct current electric(al) traction system
直流电铃 direct current ringer;trembler bell;trembling bell
直流电流放大器 direct current amplifier
直流电流互感器 DC current transformer
直流电路 direct current circuit
直流电码 direct current code
直流蓄电池充电引出端 direct current battery charging outlet
直流电气牵引 direct current electric(al) traction
直流电桥 direct current bridge
直流电驱动 direct current drive
直流电设备 direct current equipment
直流电锁器 direct current electric(al) lock
直流电压 direct current voltage;direct voltage
直流电压探测单元 direct current voltage detection unit
直流电压探测装置 direct current voltage detection unit
直流电源 direct current power supply;direct current source;direct current supply;direct supply
直流电源屏 direct current power supply panel
直流电子管收音机 direct current tube electric(al) set
直流电阻 direct current resistance;ohmic resistance
直流发电机 constant current dynamo;direct current dynamo;continuous current dynamo;continuous current generator;direct current generator;dynamo;steam dynamo;metadyne(generator) <由电枢反应励磁的>
直流阀 straight tubing valve;straight-way(line) valve
直流法 <勘探> direct current method;rectiflow system
直流放大器 direct current amplifier
直流放电 direct current discharge
直流分量 direct component;direct current component;zero frequency component
直流分量插入 direct current inserter
直流(分量)恢复电路 direct current restoration circuit;direct current restoring circuit
直流(分量)恢复二极管 direct current restorer diode
直流分量失真 direct current component distortion
直流负载 direct current load
直流复励发电机 direct current compound generator
直流复位器 direct current restorer;reinserter
直流工作电压 direct current working voltage
直流功率 direct current power
直流功率放大器 amplidyne
直流供电 direct current supply
直流供电制 direct current power supply system
直流管 battery tube;dry-cell tube
直流轨道电路 direct current track circuit
直流锅炉 Benson boiler;concurrent boiler;monotube boiler
直流锅炉设备 monotube installation
直流过流释放装置 direct over-current release device
直流过滤 in-line filtration
直流互感器 continuous current transformer
直流换气 uniflow scavenging
直流汇流排 direct current bus
直流积复激电动机 direct current cumulative compound motor
直流激发极化法 direct current induced polarization
直流计算机 direct current computer
直流继电器 direct current relay
直流键控 direct current keying
直流交流变换器 direct current-alternating-current converter
直流接收机 direct current receiver
直流截面 straight-through section
直流介电分离法 direct current dielectric(al) separation method
直流进馈线快速开关柜 fast switch cabinet with direct current in feeder
直流均衡器 direct current balancer
直流开关柜 direct current switchgear
直流可控电抗器 direct current controllable reactor
直流控制电路 direct current control scheme
直流控制方案 direct current control scheme
直流快速断路器 direct current high-speed circuit-breaker
直流馈电 direct current feeding
直流馈电线 direct-current tie
直流冷却 once-through cooling;one-through cooling

直流冷却水 once-through cooling water;one-through cooling water
直流冷却水系统 once-through cooling water system;one-through cooling water system
直流冷却系统 one-through cooling system
直流马达轮 direct current motorized wheel
直流脉冲 direct current pulse
直流抹音器 direct-current eraser
直流耦合 direct-current coupling
直流配电 direct current distribution
直流偏磁 direct current bias
直流偏置 direct current bias
直流平衡机 direct current balancer
直流平衡器 direct current balancer
直流屏 direct current panel
直流汽轮发电机 turbodynamo
直流牵引 direct current traction
直流箝位 direct current clamp
直流驱动的交流发电机 DC motor driving AC generator
直流人工闭塞 direct current manual block
直流扫气 uniflow scavenging
直流式泵 straight pump;straight-flow pump
直流式沉淀池 vertical flow settling basin
直流式阀 straight-flow valve
直流式风洞 direct action wind tunnel;straight-through-type wind tunnel
直流式锅炉 once-through boiler;uniflow boiler
直流式加热器 flow-type calorifier
直流式空气调节系统 direct air conditioning system
直流式空心阀 direct flow hollow spool valve
直流式冷却 once-through cooling
直流式冷却器 uniflow cooler
直流式煤气热水器 flow-type gas water heater
直流式燃气轮机 series flow gas turbine
直流式燃烧室 straight-flow combustor; straight-through (flow) combustion chamber
直流式热水器 flow-type water heater
直流式脱盐 direct flow demineralization
直流式旋风除尘器 straight-through cyclone
直流受电器 direct current pantograph
直流输出功率 direct current output
直流水轮发电机 turbodynamo
直流水系统 direct flow water system;once-through water system
直流伺服发动机 direct current servomotor
直流伺服放大器 direct current servo amplifier
直流调节器 direct current regulator
直流调谐电容器 direct current tuned capacitor
直流调整器 direct current regulator
直流通报 closed circuit working
直流通道 direct current channel
直流通路 direct current channel
直流图像传输 direct-current picture transmission
直流稳定器 direct current stabilizer
直流稳压器 direct current stabilizer
直流系统 direct current system;once-thru system; one-through system; straight-flow system;straight-through arrangement; once-through system <一次性排放的>
直流系统冷却水处理 cooling water treatment of once-through system
直流线性充电 direct current linear charger
直流消磁器 direct-current eraser
直流消磁头 DC erasing head
直流谐振充电 direct current resonance charges
直流旋流分流器 straight-through cyclone
直流选择 direct current selection
直流循环 once-through circulation; one-through circulation;open cycle
直流斩波器 direct current chopper
直流振动换流器 direct current chopper
直流振铃 signal(l)ing
直流直流变换器 DC-DC converter
直流制 direct current system
直流主电源 direct current main
直流转极电码 DC polarity-changing code;polarity-changing code
直馏 straight distillation;straight run; straight-run distillation
直馏柴油 straight-run diesel oil
直馏柴油馏分 straight-run diesel distillate
直馏产品 straight run; straight-run product
直馏粗汽油 virgin naphtha
直馏灯油 once-run kerosene
直馏地沥青 <直接自蒸馏石油取得> straight-run asphalt; straight-run bitumen;straight-run pitch
直馏发动机燃料 straight-run motor fuel
直馏法 straight-run distillation process
直馏矿物油 straight mineral oil
直馏沥青 straight asphalt;straight reduced asphalt;straight-run bitumen
直馏沥青胶泥 straight-run asphalt cement
直馏沥青填(塞)料 straight bituminous filler
直馏馏分 straight run; straight-run distillate;straight-run virgin
直馏煤焦硬沥青溶液 straight-run coal tar solution
直馏煤焦硬沥青乳液 straight-run coal tar emulsion
直馏煤焦硬沥青悬浮液 straight-run coal tar dispersion
直馏煤焦油 straight-run coal tar
直馏煤焦油脂 straight-run coal tar pitch
直馏煤沥青 straight-run coal tar; straight-run pitch
直馏煤油 virgin kerosene
直馏汽油 straight gasoline; straight-run gasoline;straight-run spirit
直馏轻馏分 straight-run light fraction
直馏轻汽油 straight-run tops
直馏燃油 straight-run fuel
直馏石脑油 straight-run naphtha
直馏无铅汽油 straight-run clear gasoline
直馏硬煤焦油脂 straight-run pitch
直馏硬煤沥青 straight-run coal tar pitch
直馏油 straight-run oil;virgin oil
直馏油料 straight-run stock; virgin stock
直馏油品 straight-run oil
直馏渣 straight-run residue
直馏重馏分 straight-run heavy fraction
直录磁带 direct recording magnetic tape
直路 ah line; enfilade; straightaway; tangent
直路距离 airline distance
直路取样机 straight-path sampler
直路异径丁字管 tee reducing on run
直路缘石 straight curb;straight kerb
直率 directness;flatness
直率的 bluff;outright
直轮 straight wheel
直轮式平地机 straight wheel grader
直轮式平路机 straight wheel grader
直螺脚绝缘子 bracket insulator
直螺栓 through bolt
直螺纹 straight thread
直落差 vertical throw
直落式溢洪道 straight drop spillway; vertical drop spillway
直埋电缆 direct burial cable; direct buried cable; direct embedded cable;optic(al) fibre for direct burial
直埋法 direct burial method
直埋敷设 directly buried installation; underground piping
直埋管 direct buried pipe
直埋式电缆 direct bury cable
直埋式光缆 plow-in optic(al) cable
直脉型 straight vein
直毛 broad wool
直幂 direct power
直面粗车刀 straight faced roughing tool
直面搭接(木材) box scarf
直面轮齿 rectilineal face tooth; rectilinear face tooth
直面缘石 normal curb
直模标本 orthotype
直母线 straight-edge line;straight-line generator
直木材 stick;straight timber
直木纹 edge grain;straight grain
直捻钢丝绳 lang-lay line
直啮合 conductive coupling
直镊 straight forceps
直扭构造体系【地】 normal shear structural system
直耦式放大器 direct coupled amplifier
直爬梯 cat ladder; rung ladder; step ladder;vertical ladder
直爬梯横挡 rung
直排八汽缸 eight-in-line;straight eight
直排的 collinear
直排发动机 in-line engine
直排汽缸 cylinders in line
直排式 straight line
直排式燃气轮机 straight compound gas turbine
直排天线阵 collinear array
直排五缸发动机 in-line five-cylinder engine
直判法 direct interpretation method
直跑楼梯 straight flight stair(case); straight-run stair;straight stairway
直跑式楼梯 straight stair(case)
直跑休息平台 straight-run landing
直配(钢)筋 straight reinforcement bar
直配孔型轧制法 straight flange method;straight method
直膨式冷却器 unit cooler
直碰 direct impact
直碰垫 straight fender
直劈的 straight-split
直偏法 direct deflection method
直拼Z形撑门 ledged and braced door
直拼撑门 ledged door
直拼地板 straight joint floor
直拼缝地板 straight joint floor
直拼斜撑框构门 braced door; ledged door
直拼斜撑框架门 framed door
直拼斜撑门 ledged and braced door
直坡 free face;straight grade
直坡道 straight(-run) ramp
直坡度 straight grade
直企口接缝 straight-tongued jointing
直砌砖 plain-wire-cut brick
直牵引力 directional pull
直前法 straight-ahead method
直钳 friction(al) clamp;straight tongs
直嵌接头 straight scarf joint
直墙 <钢筋混凝土挡土墙的> stalk
直墙衬砌内轮廓线 inner periphery of lining for vertical wall
直墙堤堤身 body of upright wall; stem of upright wall
直墙式防波堤 vertical wall breakwater;breakwater with vertical faces; upright wall breakwater; vertical face breakwater; vertical type breakwater; wall-breakwater; wall-type breakwater
直墙水上部分 superstructure of upright wall
直墙水下部分 submerged part of upright wall
直桥 straight bridge
直桥桥面板 right bridge deck(ing)
直切 scarp
直切环 stepped cut ring
直切砖 plain-wire-cut brick
直倾型 orthocline
直曲柄板 straight crank web
直取比重 percentage of direct transshipment
直燃机燃烧吸收式制冷机 direct-fired absorption type refrigerating unit
直燃式溴化锂吸收式制冷机 direct-fired lithium-bromide absorption type refrigerating machine
直热灯丝 directly heated filament
直热式的 directly heated
直热式电子管 heaterless tube
直热式覆氧化物阴极 coated filament
直热式空气加热器 direct-fired air heater
直热式热变电阻器 direct heated thermistor
直热式热敏电阻 direct heated thermistor
直热式热敏电阻器 directly heated thermistor
直热式阴极 direct heated cathode;directly heated cathode; filament cathode; filament type cathode; heater cathode
直热阴极接线 heater-cathode connection
直刃刮刨 straight spoke shave
直刃虹膜刀 straight iris knife
直刃剪 normal straight blades snips
直刃剪床 squaring shears
直刃口 straight cutting edge
直刃式雪犁 straight-blade snow plough
直刃推土机 straight-blade bulldozer
直刃弯旋槽刀 straight curved grooving cutter
直刃铣槽刀 straight grooving cutter
直刃雪犁 straight-blade snow plough
直认 avow
直熔锭 dingot;direct ingot
直熔锭块 dingot regulus
直熔锭清理 dingot cleaning
直熔锭修整 dingot scalping
直熔法拉丝 direct-melt process
直熔金属锭 dingot metal
直三通 straight tee
直伞齿轮 straight bevel gear
直沙嘴 epi
直筛条 straight grizzly bar
直晒纸 printing-out paper

直闪浅粒岩 anthophyllite leuco granoblastite
直闪石 anthophyllite;bidalotite
直闪石石棉 anthophyllite asbestos
直闪云母片岩 anthophyllite-mica schist
直扇状流 straight fan
直上的 rising
直上焊接能力 capability of welding vertically upwards
直上楼梯 single-flight stair(case)
直上总管 rising main
直烧干燥器 direct-fired drier[dryer]
直烧旋转干燥器 direct-fired rotary drier[dryer]
直烧窑 directly fired kiln
直烧蒸发器 direct-fired evapo(u)rator
直勺 straight scoop
直射 perpendicular incidence;projectivity;shoot at right angles;collineation
直射变换【数】collineatory transformation
直射充实水柱 direct jet solid stream
直射光 direct light;sun light
直射光束 straight beam
直射光照明 direct-light luminaire
直射距离 battle-sight range
直射脉冲 direct impulse
直射式喇叭 straight horn
直射式汽缸盖 direct injection cylinder head
直射式速调管 straight advancing klystron
直射束 straight beam
直射线 direct ray
直射炫目光线 direct glare
直射阳光 direct sunlight
直射阳光的进入 entry of direct sunshine
直射阳光照度 direct sunlight illumination
直射影像 direct image
直射照明 direct lighting
直伸导线 straight-line traverse
直伸三角网 straight triangulation network
直伸线 stretch-out line
直升的 vertical lift
直升电梯 elevator
直升活动桥 vertical lifting bridge
直升机 autogiro;copter;egg beater flying windmill;gyrodyne aircraft;helicopter;heliogyro;hoverplane;rotorcraft;verticraft;vertiplane;whirly-bird;autogyro;flat riser;vertical takeoff landing
直升机登陆运输舰 amphibious assault ship
直升机地平面图 pinwheel ground plan
直升机吊放式声呐 helicopter-dunked sonar
直升机吊箱 helitank
直升机飞机场 helidrome
直升机和空降灭火 helitack
直升机滑道 helicopter skid
直升机机场 helicopter ground;helipad;heliport;sky-port
直升机基地 helibase
直升机基地总管 helibase manager
直升机降落场 helispot
直升机降落垫板 helicopter landing-pad
直升机交通巡逻 traffic patrolling(by helicopter)
直升机灭火领班 helitack foreman
直升机喷雾飞机 helicopter sprayer
直升机普查 helicopter-borne survey
直升机起落平台 helicopter platform
直升机起落坪 helicopter pad
直升机起重机 sky crane
直升机撒播飞机 helicopter sower
直升机升降场 helipad
直升机升降甲板 helicopter deck
直升机声呐 helicopter sonar
直升机停机坪 heliport deck;helispot
直升机停机台 helideck
直升机消声器 helicopter muffler
直升机援救 helicopter rescue
直升机运输 helicopter traffic
直升机站 airstop;heliport
直升机自动驾驶仪 helicopter autopilot
直升机临时降落点 helispot
直升浸没式 vertical submerged type
直升跨度 vertical-lift span
直升拉门 vertically sliding door
直升门 vertical-lift door
直升灭火飞机 helitanker
直升平板闸门 flat vertical gate
直升平台 vertical lift platform
直升汽车(房)屋 helihome
直升栅门 vertical-lift gate
直升式储气罐 column-guided holder;guide-framed gas holder
直升式船闸闸门 vertical-lift lock gate;vertical-rising type of lock gate
直升式定轮闸门 vertical-lift fixed wheel gate;fixed-wheel vertical lift gate
直升式(矩形)泄水闸门 penning gate
直升式桥 vertical-lift bridge
直升式闸门 vertical-lift gate
直升吸水式 vertical suction type
直升闸门堰 draw door weir
直升装置 vertical riser
直生论 orthogenesis
直生现象 orthogenesis
直示高温计 demonstration pyrometer
直式管用丝锥 straight pipe tap
直式滑道 end haul slipway
直式桥台 stub abutment
直式推(土)铲 straight bulldozer
直式推土机 straight-blade bulldozer
直视 direct view;direct vision;staring blankly forward
直视的 direct viewing;visualized
直视地平线 optic(al) horizon
直视分离术 division under direct vision;exclusion under direct vision
直视缝 direct vision port
直视光度学 visual photometry
直视检景器 direct vision finder
直视距离 horizon range;optic(al) range
直视距离传输 horizon transmission
直视可达范围 line-of-sight coverage
直视控制 sight control
直视棱镜 direct vision prism
直视棱镜分光镜 constant deviation spectroscope;direct vision spectroscope
直视望远镜 direct telescope
直视物端棱镜 direct vision objective prism
直视物方棱镜 direct vision objective prism
直视显示 visible display;visual display
直视正像物镜 direct vision erecting prism
直手柄 straight hand lever
直手指手套 straight finger gloves
直首(柱)【船】straight bow;straight stem
直首桩【船】straight bow;straight stem
直属办公室 immediate office
直属上级委员会 parent committee
直竖的 bolt-upright
直竖管 rising pipe;rising piping;rising tube;rising tubing
直水道 reach;straight channel
直水口 peg gate
直丝 raw silk rings
直丝扣 straight thread
直送方式【机】headless system
直榫 straight tenon;straight tongue
直榫接合 joining with peg-shoulder
直榫接头 joining with peg-shoulder
直塔 vertical derrick
直弹性轴 axis of direct elasticity
直探头 normal probe;straight beam probe
直碳链 normal carbon chain
直套管 straight connector;straight sleeve
直套筒 straight sleeve
直梯 straight ladder
直体步行的 orthograde
直体字【计】vertical letter(ing)
直条地板 strip flooring
直条耕作 straight-row farming
直条夹 straight strap clamp
直条图 bar chart;bar diagram
直跳板 straight-type rampway
直挺 ramrod;straight elevator
直通 direct connection;feed-through;straight bore;straight in;through
直通按钮 through button
直通拨号 through dial(l)ing
直通侧线 through siding
直通测量(线路) end-to-end measurement
直通插销 passage set
直通岔线 through siding
直通场【铁】through yard
直通车场【铁】transit yard
直通车道 through lane
直通车辆限界 transit vehicle ga(u)ge
直通程序 straight-line coding
直通程序设计 straight-line coding
直通大街 through street
直通单向阀 in-line check-valve;straight check-valve;straight-way check-valve
直通道路 through highway;through road
直通的 once-through;straight-through;through and through;through-running;throughway;thruway
直通等级运价 through class rates
直通地 dead earth;dead ground
直通地段 through lot
直通电话线路 hot-line
直通电话增音器 interphone amplifier
直通电键 through switching key
直通电空制动机 direct acting electro-pneumatic brake
直通电流 through current
直通电流偏置 through current bias
直通电路 direct circuit
直通垫外底 through midsole
直通对讲扬声器通信【讯】two-way direct calling loudspeaker communication
直通阀 full-way valve;pass valve;straight-through valve;straight-way valve;through way valve
直通扶梯 through ladder
直通干馏炉 through retort
直通干线 through main line
直通公路 pass highway;throughway
直通轨 through rail
直通轨道 through track
直通过程 straight-through process
直通换料 once-through refuelling
直通回路 straight circuit
直通汇接呼叫制 straight-forward junction working
直通活门 straight-way stopcock
直通货物运输 through goods transportation
直通(机头)挤压机 extruder with straight delivery head
直通监控 through supervision
直通接头 straight coupling;straight fitting
直通接线装置 through joint
直通街道 through expressway;through street;throughway;thruway
直通进路 straight-through route
直通进位 standing carry
直通开关 globe cock;passage cock;straight-through cock
直通客车 through coach;through passenger car;through carriage <英>
直通客车行驶计划 <西欧> European through coach working plan
直通客车组 through carriage group
直通客流 through passenger flow
直通客流图 through passenger flow diagram
直通客票 through ticket
直通空气制动 straight air brake application
直通空气制动阀 straight air brake valve
直通快动阀 straight-through fast acting valve
直通快速车道 through traffic road
直通连接 straight-through joint;through connection
直通列车 through train;transit train
直通裂化 straight-through cracking
直通流动 straight-through flow
直通楼梯 straight stair(case)
直通炉 belt furnace;through furnace
直通旅客列车 through passenger train
直通旅客列车速度 <公里/日,包括沿途各站停留时间在内> through passenger train speed
直通旅客特别快车 through express train
直通配线 direct distribution
直通棚车 <西欧标准轨距铁路与西班牙、葡萄牙两国宽轨铁路间可换轮轴的> Transfesa closed wagon
直通偏流 through bias
直通票价 through fare
直通气阀 straight air brake
直通气力制动器 straight air brake
直通清除 through scavenging
直通区段 main section
直通人孔 straight-through manhole
直通式车站 through station
直通式锅炉 once-through boiler
直通式货站 through-type freight station
直通式空气制动机 straight air brake
直通式空气制动器 straight air brake
直通式炉 continuous passing furnace
直通式迷宫密封 straight-through labyrinth
直通式喷头 straight advancing spray-head
直通式疏干钻孔 directly lead borehole for dewatering
直通式洗澡用锅炉 instantaneous geyser
直通枢纽 through terminal
直通丝锥 non-reversing tap
直通榫眼 through-mortice[mortise]
直通通道 through channel
直通通话位置 through position
直通途径 through path
直通线 direct line;through line;through track;transit track
直通线路 direct circuit;direct line;straight circuit;straight-forward circuit;through track;transit track

直通线群 direct group
直通信号 through signal
直通行 direct line
直通型 through-type
直通型检测器 flow-through detector
直通旋塞 straight-through cock; straight-way cock; through way cock
直通压出 straight-through extrusion
直通雨水管 straight down gutter
直通运价 through rate
直通运输 through transport
直通运输收入 through traffic revenue
直通运行作业收入 through transportation operation income
直通真空制动机的额定衰减 rated depression of a through vacuum brake
直通制 straight-forward system
直通中继 direct trunking; straight-forward trunking
直通中继线 direct trunk
直通中继站 repeater without drop
直通作业清算 settlement of income from through transportation
直筒教会式屋面瓦 straight-barrel mission roofing tile
直筒楼梯 single-flight stair(case)
直筒型回转窑 straight rotary kiln
直筒阴阳屋面瓦 straight-barrel mission roofing tile
直头 straight peen
直头半圆 straight half round
直头半圆刮刀 straight half round scraper
直头车刀 straight tool
直头机 opener
直头尖口锤 straight-peen hammer
直头尖嘴锤 straight-peen hammer
直头手锤 straight peen hand hammer
直图 straight graph
直推式执行机构 push stem power unit
直腿型钢斜轧法 diagonal straight flange method
直挖 straight cut
直弯机 opener
直尾翅梢 fin-tip
直尾床子鸡心夹头 straight tail lathe dog
直尾鸡心夹头 straight tail dog
直尾夹头 straight tail dog
直尾箭形铲 straight-stem sweep
直尾轧头 straight tail dog
直纹 straight burr; vertical stripe; straight grain < 木材的 >
直纹布 parallel fabric
直纹带 cord belt
直纹的 straight grained
直纹滚花 straight-line knurling
直纹刻石刀 straight-cut burr
直纹理 straight grain
直纹理的 straight grained
直纹(理)木板 straight grained wood; straight timber
直纹(理)木材 straight grained wood; straight timber; vertical grained timber
直纹面的母线 ruling of a ruled surface
直纹面的奇异母线 singular generator of a ruled surface
直纹面的线列 line series ruled surface
直纹面的准锥面 director cone of a ruled surface
直纹母线 rectilinear generator
直纹(木)甲板 straight deck
直纹曲面 ruled surface
直纹曲面图 ruled-surface map
直纹凸榫 straight tongue
直纹网 laid wire
直纹线汇 rectilinear congruence
直纹线条 laid line
直纹形母线 generator of a ruled surface
直纹纸 laid paper
直蜗杆 straight worm
直吸式挖泥船 moored suction dredge(r)
直系家庭 directly related family
直系亲属 lineal consanguinity
直隙微波加速器 dromotron
直辖的 ordinary
直辖市 centrally administered municipality; municipality; municipality directly under the central authority; municipality of centrally administered
直下 down right
直纤维 straight fiber
直弦桁梁 parallel-chord truss
直舷船 fat-sided ship; vertical side ship
直线 beeline; right line; straight-forward line; straight line; tangent
直线 V 形底 straight-line vee[V] bottom
直线 Z 箍缩 linear Z-pinch
直线岸壁 straight quay
直线岸壁式码头 straight quay
直线把 bundle of lines
直线坝头 straight abutment
直线坝座 straight abutment
直线倍增加速器 linear multiple accelerator
直线本身 straight-line body
直线比 linear ratio
直线比例尺 bar scale; divided scale; graphic(al) scale; linear scale; make line; scale line
直线比例尺分划 scale mark
直线比网络法 corresponding grid method
直线笔 crow-quill pen; drawing pen; line pen; lining pen; ruling pen; border pen < 绘图廓用 >
直线笔划线 pen ruling
直线编码 straight-line code; straight-line coding
直线变速传动 < 机动车 > straight shift transmission
直线标度 linear scale; line scale
直线标志 line marking
直线拨正 lining of straight line
直线波 linear wave
直线波长式 straight-line wavelength
直线不重合度 misalignment
直线布水器 rectilinear distributor
直线布置 straight-through arrangement
直线步进电动机 linear pulse motor; linear step(ping) motor
直线部分 straight-line portion; straight-way
直线槽缝 straight-line joint
直线侧缝 straight-side slot
直线测量 linear measurement
直线产生器 line generator
直线潮流 rectilinear tidal current
直线车削 straight turning
直线衬里 straight lining
直线程 crow-flight path
直线程序 in-line procedure
直线尺 straight-edge(d) ruler
直线齿形鼠牙盘离合器 hirth coupling
直线冲程 head reach
直线抽样法 line sampling
直线传播 linear propagation; rectilinear propagation
直线传动 straight-line drive
直线船程 straight course
直线导体天线 linear conductor antenna
直线导线 rectilinear wire
直线导线导轮 straight pulley
直线的 lineal; linear; orthographic(al); rectilineal
直线的法线式 normal form of a straight-line
直线地震 linear earthquake
直线点减速器 tangent point retarder
直线点组 series of points
直线电动机 linear electric(al) motor; linear motor
直线电机 linear motion actuator
直线电机绘图机 linear-motor plotter; linear-motor table
直线电流 straight-line current
直线(电子)加速器 linac
直线吊线 straight ligament
直线定测法 alineation
直线定律 linear law; straight-line law
直线定子 linear stator
直线定子电机 linear stator machine
直线度 straightness
直线度量 linear measure
直线段 straightaway; straight reach; straight section; straight stretch; straight-way; tangential path < 曲线间的 >; through section < 转辙器的 >
直线断裂 straight-line breaking
直线堆场 in-line stockpile
直线对准 adjusting to a line; align; alignment
直线对准盘 alignment disc[disk]
直线对准器 aligner
直线多边形 rectilinear polygon
直线发展 unilinear development
直线发展的 unilinear
直线法 linear method; straight-line method of stress-strain
直线反映 line reflection
直线方程式 linear equation; straight-line equation
直线方向 rectilinear direction
直线放大率 linear magnification
直线飞行 rectilinear flight; straight-line flight
直线飞行距离 crow-flight distance; crow-fly distance
直线分布 straight-line distribution
直线分布理论 theory of straight line distribution
直线缝隙天线阵 linear-slot array
直线浮筒导轨 linear float guide
直线俯冲 straight dive
直线俯冲攻击 straight-line diving attack
直线盖度 linear cover-degree
直线杆 straight rod
直线杆塔 straight-line pole
直线感应电动机 linear induction motor
直线感应加速器 linear induction accelerator
直线隔水边界 linear impervious boundary
直线公理 linear axiom
直线公式 < 设计柱的 > straight-line formula
直线功(职)能式组织 line-functional organization
直线攻击 straight-line attack
直线箍缩 linear pinch
直线箍缩机 linear pinch-machine
直线挂线器 straight-line hanger
直线拐肘 straight-arm crank; straight crank
直线关联 straight-line correlation
直线关系 linear relation(ship); rectilinear relation; straight-line relation
直线关系(公)式 straight-line formula
直线规 liner
直线轨道间菱形交叉 diamond crossing on straight track
直线轨迹 straight path
直线轨迹振动筛 linear-path screen
直线滚焊机 straight-line seam welder
直线滚花 straight-line knurling; straight-line knurls
直线滚花头螺钉 screw with straight-line knurls
直线海岸 straight coast
直线函数 linear function
直线焊道 string bead
直线焊缝 straight bead
直线焊缝焊接 line weld(ing)
直线焊接 straight bead welding
直线航线 straight-line course
直线和谐运动 straight-line harmonic motion
直线河段 straight reach
直线横切锯 straight-line cross-cut saw
直线护坦水墙 straight-line apron boom
直线护坦水栅 straight-line apron boom
直线花型 straight-line pattern
直线滑槽连杆 straight link
直线滑车连杆 straight link
直线化 linearization
直线化电阻 linearizing resistance
直线画笔 straight-line pen
直线画线针 straight point scriber
直线环 straight ring
直线回归 linear regression
直线回归方程 linear regression equation
直线火花计数管 linear spark counter
直线机构 straight-line mechanism
直线基线 straight-line basis
直线基阵 linear base
直线基准 linear base
直线加工工作 < 天然石材 > straight-line work
直线加速度 linear acceleration
直线加速器 rectilinear accelerator
直线加速器长度 linac length
直线加速器负载因子 linac duty factor
直线加速器回旋加速器组合 linac-cyclotron combination
直线加速器级 linac stage
直线加速器加速腔 linac tank
直线加速器接收度 linac acceptance
直线加速器结构 linear-accelerator structure
直线加速器介子工厂 linear-accelerator meson factory
直线加速器聚焦 linac focusing
直线加速器枪 linac gun
直线加速器输出 linac output
直线加速器调试 linac adjustment
直线加速器运行组 linac operation group
直线加速器轴线 linac axis
直线加速器注入器 linac injector; linear-accelerator injector
直线尖轨 straight switch rail
直线尖轨式转辙器 straight split switch
直线剪切装置 linear shear apparatus
直线建筑 straight-line block
直线渐变 rectigradation
直线交错层理构造 rectilinear cross-stratification structure
直线交会法 linear intersection(method)
直线角 rectilinear angle
直线叫入 direct in-line
直线校准用夹具 aligning jig

直线接合 straight joint
直线接头 straight joint
直线节距 linear pitch
直线结构 linear structure
直线解 rectilinear solution
直线近似 straight-line approximation
直线近似法 straight-line approximation method
直线进场 straight-in approach
直线进/出闸 entering/leaving lock in a rectilinear way
直线进入目标 straight-in run
直线进入着陆 straight-in landing
直线晶格 line lattice
直线晶界 straight-edge boundary
直线精度 linear precision
直线井排 linear well rows
直线迴归 linear regression
直线距离 crow-fly distance; linear distance; slant distance; straight-line distance; straight stretch
直线聚焦 linear focus(ing)
直线可变电容器 straight-line condenser
直线刻度长度 lineal scale length
直线刻划 linear graduation
直线刻绘器 straight-line graver
直线控制 linearity control; straight-line control
直线控制器 straight-line controller
直线宽口接杆 straight wide jaw
直线理论 straight-line theory
直线力 linear load
直线励磁同步电动机 direct current excited synchronous motor
直线连杆运动 straight-line link motion
直线连接杆 straight link
直线量测 linear measure; linear measurement
直线列 alignment array
直线裂缝 straight slit
直线流 linear flow
直线流动 linear flow; rectilineal flow; straight-line flow; straight-line motion
直线流水作业 straight conveyer[conveyor]
直线流水作业法 straight-line-flow method
直线流速 linear rate of flow
直线流域 reach
直线炉 straight-line furnace
直线路近似法 straight-path approximation method
直线路面拱 straight-legged roadway arch
直线轮廓的结构 straight-line design
直线轮胎 straight-side tire[tyre]
直线螺纹磁阻电动机 rectilinear screw-thread reluctance motor
直线脉冲电动机 linear pulse motor
直线米 running meter
直线面 plan area
直线描绘 straight-line plotting
直线描绘测试仪 drawing line tester
直线磨边机 straight-line edger
直线磨具 straight grinder
直线幕僚式组织 line-staff organization
直线内插 linear interpolation
直线内插法 straight-line interpolation
直线拟合【数】 fitting of a straight-line
直线爬行 rectilinear creeping
直线排列 linear arrangement; tandem position
直线派 secession
直线派建筑家 secessionist
直线跑道 straightaway; straight-way
直线喷流 rectilinear jet
直线膨胀 linear expansion
直线膨胀系数 coefficient of linear expansion
直线劈理 linear cleavage
直线偏振波 linearly polarized wave
直线偏振光 rectilinearly polarized light
直线频率式 straight-line frequency
直线平地海岸 flat straight coast
直线平均计算法 straight-line method
直线平坦海岸 flat straight coast
直线平移 rectilinear translation
直线坡 straight slope
直线起飞滑跑 straight takeoff run
直线迁移 linear migration
直线桥 straight bridge
直线切割机 straight-edge cutting machine
直线切割线交点误差曲线 cross-over point error curve of line and cross line
直线切削 straightaway cut
直线球轴承 linear ball bearing
直线区 linearity sector
直线区间 < 铁路、道路的 > straight-line section; tangent
直线区域 linearity region
直线驱动 linear actuation
直线驱动原理 linear motor principle
直线趋势 rectilinear trend
直线三角形 rectilinear triangle
直线扫描 linear scan(ning); linear sweep; rectilinear scan(ning)
直线扫描发生器 linear sweep generator
直线上升 straight climb
直线射程 straight-line flight
直线射频质谱仪 linear radio-frequency mass spectrometer
直线生成元 rectilinear generator
直线时基 linear time base
直线时基管 line-time base valve
直线式 linear type; orthoscopic; secession; secessionism
直线式布钻孔 line boring
直线式程序 linear programming
直线式窗花格 rectilinear tracery
直线式磁阻电动机 linear reluctance motor
直线式叠接的倾斜航空照片 oblique line overlap
直线式感应传感器 linear inductosyn
直线式感应同步器 linear inductosyn
直线式钢筋拔丝机 line type wire-drawing machine
直线式机械手 rectilinear manipulator
直线式建筑风格 rectilinear style
直线式控制器 rectilinear manipulator
直线式拉丝机 straight-line machine
直线式起重机 < 船岸间用缆索或上行式轨道 > straight-line crane
直线式权责图表 linear responsibility chart
直线式时代 rectilinear period
直线式天线阵 linear array
直线式销售组织 linear sales organization
直线式压缩机 straight-line compressor
直线式自动线 linear transfer
直线受压 straight compression
直线输送机 straight conveyer[conveyor]
直线水流 rectilinear flow
直线顺坡 straight ramp
直线速度 linear velocity; space rate < 沸腾焙烧炉的 >
直线速度测算法 linear speed method
直线隧道式窑 straight-line tunnel kiln
直线塔 tangent tower
直线摊销法 average method of amortization; straight-line method of amortization
直线弹性的 linearly elastic
直线镗床 line borer
直线镗孔 line boring
直线镗削的 line-bored
直线掏槽 straight cut
直线掏槽爆破 burned cut blasting
直线特性 linear characteristic; straight-line characteristic
直线天线阵 collinear array
直线条式建筑 rectilinear style
直线跳板式拦河埂 straight-line apron boom
直线通过 straight-line pass
直线透射比 rectilinear transmissivity; rectilinear transmittance
直线透视图 linear perspective
直线透水边界 linear pervious boundary
直线图 linear chart; line graph
直线图解法 linear-graphic(al) method
直线图形 alignment chart; rectilinear figure; line pattern
直线外插法 linear extrapolation
直线外推长度 linear extrapolation length
直线弯曲 straight-line bending
直线往复潮汐流 < 每 6 小时流向改变一次 > rectilinear tidal current
直线往复运动 straight reciprocating motion
直线位移 straight-line displacement
直线位移差动变压器 linear variable differential transformer
直线涡 rectilinear vortex
直线涡流 rectilinear vortex
直线污泥吸集器 straight-line sludge collector
直线铣削 line milling
直线系 rectilinear system; system of lines
直线下滑 straight glide
直线下降 linear decrease; straight-in penetration
直线下限 straight-line lower bound
直线线路 tangent track; tanget
直线相关 linear correlation; rectilinear correlation; straight-line correlation
直线相关系数 linear correlation coefficient
直线斜率法 straight slope method
直线泄漏通路 straight-line leakage path
直线行车 straight movement
直线行程 rectilinear path
直线行进 straightaway
直线行驶性能试验 straight-line running test
直线形超高顺坡 straight superelevation ramp
直线形磁轴 straight magnetic axis
直线形灯丝 line filament
直线形轨道 linear path
直线形航线 rectilineal(outline) course
直线形建筑 linear building
直线形结构 linear structure
直线形列柱式(建筑) orthostyle
直线形排水系统 rectify drainage; rectilinear drainage
直线形墙段 straight panel
直线形式装饰 linear style of ornamentation
直线形水系 rectilinear drainage
直线型 linear pattern; linear type
直线型坝 straight dam
直线型爆破 linear pattern shooting; straight pattern blasting; straight pattern shooting
直线型边坡 linear slope
直线型车身 straight-line body
直线型船 straight-lined vessel
直线型船首 straight bow; straight stem
直线型的 linear; unidimensional
直线型颚板 straight jaw plate
直线型分子 linear molecule
直线型钢丝束 < 预应力钢筋结构中的 > straight cable
直线型河流 straight stream
直线型滑动面 linear slip surface
直线型结构 linear structure
直线型结构体系 linear structure system
直线型空气分级机 rectilinear air classifier
直线型料斗 linear hopper
直线型流线 straight streamline
直线型推拉(百叶)门 straight sliding (shutter) door
直线型预应力钢索 straight-line prestressed steel tendon
直线型装船机 linear shiploader
直线型自动取样器 straight-line type automatic sampler
直线性 linearity; rectilinearity
直线性测量波形 linearity measuring wave form
直线性测试振荡器 linearity test generator
直线性粗调 coarse linearity control
直线性的 linear; rectilinear
直线性电流 rectilinear current
直线性电阻器 peaking resistor
直线性度 straightness
直线性放大器 linear amplifier
直线性光 linear light
直线性过程 linear process
直线性换向 linear commutation
直线性加速器 linac; linear(ity) accelerator
直线性检查仪 linearity checker
直线性校正电路 linearity correction circuit
直线性内插法 linear interpolation
直线性能跑道 < 供汽车性能测试用 > straight check road
直线性强风 line blow
直线性透镜 rectilinear lens
直线性系数 curvature
直线性线圈 linearity coil
直线序列 linear order
直线旋涡 rectilinear vortex
直线压力 straight compression
直线压制 linear pressing
直线演化 rectilinear evolution
直线叶片式 straight blade
直线叶栅 rectilinear cascade
直线叶式搅拌机 straight-arm mixer
直线移侧显微镜 linear traverse microscope
直线移测技术 < 测定混凝土空隙特征等的 > linear traverse technique
直线移动 translatory motion
直线移动式起重机 straight-line crane
直线移动式系统 < 喷灌的 > linear move system
直线移动式座椅 linear motion type seat
直线应力图 rectilinear stress diagram
直线硬芯箍缩 linear hard core pinch
直线预应力(法) rectilinear prestressing
直线运动 linear motion; linear movement; rectilinear motion; rectilinear movement; rectilinear translation; straight-line motion; straight movement; vectilinear motion
直线运动传感器 rectilinear transducer
直线运动促动器 linear actuator

直线运动导向器 straight guiding
直线运动的击锤 straight-line hammer
直线运动的伺服控制 servo-control for linear motion
直线运动感应式传感器 Nultrax; straight-line inductive transducer
直线运动机构 straight-line motion mechanism
直线运动能量 translational energy
直线运动起重机 straight-line crane
直线运动球轴承 ball bushing
直线运动式取出装置 straight-line type unloader
直线运动速度 point-to-point speed
直线运动液压缸 linear actuator
直线运行型有轨巷道堆垛机 straight-way travel(l)ing S/R machine
直线再生引出 linear regenerative extraction
直线增长 orthometric growth
直线张拉 linear tensioning
直线丈量 linear measurement
直线折旧法 average method of depreciation; depreciation-straight-line method; straight-line depreciation method; straight-line method of depreciation
直线辙叉轨 straight crossing rail
直线阵 line array
直线振动 rectilinear vibration; straight-line oscillation
直线整流 straight-line commutation
直线正齿 straight spur gear tooth
直线支承体系 linear bearing system
直线轴承 linear bearing
直线主运动 straight main motion; straight main movement
直线注管 linear beam tube
直线装船机 linear ship loader
直线装置 linear device
直线状 straight-linear
直线状光源 straight-line light source
直线状注流 well-collimated beam
直线总收缩 linear total shrinkage
直线走刀曲面仿形铣 stroke milling
直线组成 rectilinear composition
直线组列 linear array
直线钻孔 linear drilling; line boring; line drilling
直线坐标 linear coordinates; rectilinear coordinates
直线坐标系统 rectilinear system of coordinates
直镶式钻头 vertical tipped bit
直向船台 longitudinal slipway
直向道岔 straight track
直向地心性 orthotropism
直向地性 orthogeotropism; parallelogeotropism
直向迭珠焊接 straight bead welding
直向发生的 orthogenic
直向股道 straight track; tangent track
直向过岔速度 speed through turnout main
直向进化 orthoevolution
直向性 rectipetaly
直向选择 orthoselection
直向演化【地】orthogenesis
直向阳性 orthoheliotropism
直像管 orthiconoscope; vericon
直消光 straight extinction
直销 straight pin
直销锚栓 flat-stock anchor
直销头滚针 straight trunnion end needle roller
直楔机构 straight-wedge mechanism
直写电流计 direct writing galvanometer
直卸料式拌和机 non-tilting-drum mixer
直卸式拌和机 non-tilting mixer
直芯墙土石坝 central core earth-rockfill dam
直芯墙式堆石坝 vertical core rockfill dam
直行 straight run
直行布置 in-line arrangement
直行车车道 through lane
直行车当量 through car equivalent
直行车道 <交叉口的> direct through lane
直行车辆 straight-through traffic; straight-through vehicle; through vehicle
直行交通 straight-going traffic; straight-through traffic
直行楼梯段 straight flight stair (case); straight-run stair
直行绿箭头灯 green straight-through arrow
直行纹理 <木材的> edge grain; quarter-sawn grain; straight grain
直行小车单位 through car unit; through ear unit
直形槽与舌 straight tonguing and grooving
直形杆 straight profiled bat
直形拱顶 straight wagon vault
直形企口 straight tonguing and grooving
直形卸扣 straight shackle
直形砖 straight brick
直雄榫 straight tongue
直旋塞 straight cock; through cock
直压 vertical compression
直压板 plate clamps
直压板夹具 plate clamps
直压成型机 corrugator
直压法 direct pressure closing
直压夹具 plate clamps
直压式暖汽装置 direct pressure heating apparatus; direct steam heating system
直压与大气压取暖混合装置 pressure and vapor heating system
直眼掏槽 burn cut; burnt-cut; cylinder cut; parallel cut; straight hole cut
直眼中空掏槽 burn-cut pattern
直焰管式加热炉 updraft type heater
直焰炉 direct-fired heater; direct firing furnace; updraft furnace
直焰式锅炉 direct flame boiler
直焰窑 updraft kiln; up-draught kiln
直叶桨 straight blade; straight oar
直叶离心式风扇 straight-blade centrifugal fan
直叶片 prismatic (al) blade; straight blade
直叶片拌和机 straight-arm mixer; straight-blade paddle mixer
直叶水力飞轮 straight-vaned fluid flywheel
直移 translation
直移断层 translational fault; translatory fault
直移滑动 translational gliding
直移滑坡 translational landslide
直移幂函数 translated power function
直移运动 translational motion; translational movement
直译 literal translation; transliterate; transliteration; verbal translation
直翼推进器 cycloidal propeller; rotating blade propeller; Voith-Schneider propeller
直涌旋涡 boil-vertical eddy
直圆点 point of curve; tangent to curve point
直圆盘犁 wheatland harrow plow
直圆柱孔道 straight cylindric (al) duct
直圆柱体 right circular cylinder
直缘 straight edge
直缘拉制件 straight drawn shell
直缘型 rectimarginate
直越航线 straight crossing course
直越式速调管 monotron
直运或装船 director with transshipment
直运提单 direct bill of lading; straight bill of lading; through bill of lading
直展云 cloud with vertical development; heap cloud; high vertical development cloud
直照【物】direct lighting
直照法 normal illumination method
直照光源 direct luminaire
直照晕渲【测】hill shading; photographic(al) hill shading
直照晕渲法 shading with zenithal lighting; vertical system of shading
直褶皱 standing fold
直针 straight needle
直蒸馏到沥青 reduction to flux
直证的 deictic
直支撑 straight brace
直支管 straight branch
直支柱 straight strut
直指的 deictic; orthodactylous
直指伏特计 aperiodic(al) voltmeter
直至取消 till countermanded
直轴 vertical shaft
直轴磁化效应 direct magnetizing effect
直轴磁化作用 direct magnetizing effect
直轴的 d-axis; direct axis; straight axle; straight shaft
直轴电路 direct axis circuit
直轴阀 valve with straight stem
直轴分量 direct axis component
直轴起始瞬态电抗 d-axis subtransient reactance
直轴式 caryophyllaceous type; diacytic type
直轴式起落架 straight-axle landing gear
直轴同步电抗 direct axis synchronous reactance
直轴同步阻抗 direct axis synchronous impedance
直轴旋辊 vertical roller
直属办公室 immediate office
直属上级委员会 parent committee
直砖 straight brick
直砖过梁 straight brick lintel
直桩 vertical pile; plumb pile
直桩栈桥式码头 trestle type wharf with vertical piles
直装 stow fore and aft
直锥 right cone
直锥形尾水管 straight cone draft tube; straight conic (al) draft pipe [tube]; straight taper draft pipe; straight taper draft tube; vertical flaring draft pipe[tube]
直钻头 straight drill

值【数】valuation

值班 attendance; on watch; radio watch; shift; watch
值班报务员 radio watcher; wire watcher
值班表 watch bill
值班簿 log; logbook
值班采暖 stand-by heating
值班待修时间 standby time
值班的 on duty
值班登记表 roster
值班登记簿 duty record book
值班房面积 dog house area
值班工长 driver; shift-boss; shift foreman
值班工程师 engineer in charge; shift engineer
值班供暖 stand-by heating
值班官员 duty officer; on-duty officer
值班管制员 controller in charge
值班驾驶员 commanding officer; duty officer; officer in charge; officer of the watch; officer on watch; watch officer
值班科长 chief of division on duty
值班轮机员 engineer on duty; watch engineer
值班轮机员室 watch engineer room
值班人 roundsman; station agent
值班人员 attendant; watchkeeper; operator; operator in charge; operator on duty
值班人员表 rota
值班时间 attended time; length of shift; on-duty time; watch
值班时间表 duty work schedule
值班室 attendant cabin; duty room; room for staff on duty; staff on duty room; watch house; watch room
值班搜索 roster scan
值班台 attended station; logging desk
值班台长 chief operators desk
值班艇 duty boat
值班小时 hours of duty
值班休息室 rest room
值班员 attendant; operator on duty; watcher; watch officer
值班员操纵控制 attendant-operated control
值班员控制的电梯 attendant-controlled lift
值班员室 attendant's cabinet; caretakers' room
值班站 attended station
值班站长 station master on duty
值班指挥员 duty-shift commander
值班制度 watch system
值班主任 on-duty director; duty head
值半夜班的人 middle night watcher
值部分 value part
值参数 value parameter
值得的破坏 compensatory damage
值得开采的(矿层)厚度 paying thickness
值得考虑 merit consideration
值得考虑的污染风险 significant pollution risk
值得探索的新技术分析 innovations deserving exploratory
值得探索分析的新技术 innovations deserving exploratory analysis
值得重视的 appreciable
值得注意 claim attention; merit attention
值得注意的 remarkable
值的范围 range of values
值调用 value call
值分布 value distribution
值分布理论 value distribution theory
值岗交通警 point-duty policeman
值机室 operator's cab
值机员 attendant; radio operator
值机员再次呼叫信号 operator recall signal(l)ing
值锚更 anchor watch
值勤 on duty; point-duty
值勤泵 service pump
值勤表 duty chart; duty roster
值勤记录本 run book
值勤交通 journal to work

值勤救生艇 ready lifeboat
值勤名单 roster;watch and duty list
值勤跑道 duty runway
值勤旗 guard pennant
值勤人员工作台 desk
值勤艇 duty row(ing) boat
值群 value group
值日员 officer of the day
值替换 value substitution
值守人员 watch man
值夜 vigil
值夜人员 jack-o-lantern
值域 codomain;range
值域分析 range analysis
值域空间 range space
值指数 value index number

职

职别 job title;level of position
职称 job title;positional title;professional rank and title;professional title
职称等级 professional qualification
职称人员 competent personnel
职工 employee;journal-man;journey-man;personnel;staff and labo(u)r;staff and worker;workers and staff members;workman
职工保险基金 employee insurance fund
职工保证保险 fidelity guarantee insurance
职工补偿保险 worker compensation insurance
职工参与管理法 participative management
职工餐厅 mess hall;mess house;staff dining room
职工厕所 personnel toilet;staff toilet
职工酬劳金 employee bonus
职工储蓄金 employee saving fund
职工存款 deposit(e) from employees
职工大楼 staff building
职工大学 staff block
职工带眷比 percentage of employees living with dependents
职工单身宿舍 employee dormitory
职工导向 employee orientation
职工的筹款费用 employee's flo(a)tation expenses
职工调查法 employee questionnaire method
职工调岗 transfer of staff
职工队伍 staff ranks
职工饭堂 mess hall;mess house;mess room
职工饭厅 mess hall
职工费用 personal cost;personnel cost
职工分配指标 allocate income ratio for workers and staff members
职工福利 employee benefit
职工福利费 employee services and benefit
职工福利基金 employee's benefit fund;staff welfare funds;workers and staffs welfare fund
职工福利基金准备 benefit fund reserve
职工抚恤金 employee pension fund
职工更衣室 personnel changing room;staff changing room
职工工资率 staff rates of pay
职工股 staff shares
职工股票购买权 employee stock option
职工关系 employee relations
职工盥洗室 personnel toilet
职工红利 bonus to employees
职工积极性 employee incentive
职工集体承包制 workers-staff contracting system
职工技术培训费 technical training cost for employees
职工简历 employee profile
职工奖金制度 premium bonus system
职工奖励基金 employee incentive fund source;staff and worker's bonus;staff award fund;workers and staffs bonus fund
职工交通量 personnel traffic
职工交通运输 staff traffic
职工教育费 employee's educational expenses
职工救济基金 employee relief fund
职工卡片索引 staff card index
职工开支 staff payment
职工抗议行动 job action
职工考核 evaluation of employee
职工考核表 employee rating form
职工劳动生产额 staff output
职工临时食堂 temporarily staff canteen
职工流动率 employee turnover
职工轮班 rotation of staff
职工培训 staff training;training and education of staff and workers
职工培训费 employee-training expenses
职工培训计划 employee-training plan
职工配备 staffing
职工平均收入 average income of nonagricultural workers
职工欠款 due from officers and employees
职工区段 personnel block
职工人数计划 staff amount plan
职工入口 personnel entrance
职工入口处 staff entrance
职工膳食供应区 personnel accommodation portion
职工设备 staff facility
职工身份证 employee identification
职工升调 promotion and transferring of employee
职工生活费指数 index number of cost of living of staff and workers
职工生活福利 employee welfare
职工生活区 staff block
职工食宿供应区 staff accommodation portion
职工食堂 mess hall;mess house;mess room;personnel canteen;personnel dining room;staff's cafeteria
职工室 staff room
职工宿车 camp car
职工宿舍 living quarters for staff and worker;staff quarter;personnel accommodation
职工宿舍楼 staff building
职工通道 staff aisle
职工投资款 employee's investment fund
职工退休基金 employee's pension fund;employee's retirement fund
职工脱产培训 vocational training for the employees
职工小卖部 personnel canteen;staff canteen
职工休息室 staff lounge
职工寻找系统 staff location system
职工询问(调查)法 employee questionnaire(method)
职工业务支出 employee's business expenses
职工衣帽箱 <带锁的> personnel lock
职工衣物锁柜和午餐室 locker and lunch room
职工医院 worker's hospital
职工意见箱 staff suggestion box
职工意外事故保险法 insurance law of labo(u)r accident
职工用房 personnel building
职工优先认股权 stock option
职工预支 advance to officers and employees;advance to staffs and workers
职工职务和值班时间登记表 staff roster
职工制服补贴 uniform subsidy
职工忠诚担保书 fidelity bond
职工忠诚一揽子担保书 blanket fidelity bond
职工周转率 employee turnover ratio
职工住房 staff house
职工住宿部分 personnel accommodation portion
职工住宅 apartment house for employees;dwelling house for staff and workers;housing of staff;personnel house;staff house
职工专业训练费用 familiarization cost of workers and staff
职工租赁公寓 staff rental apartment;staff rental flat
职工租赁起居单元 staff rental living unit
职工租赁生活单元 staff rental living unit
职工租赁套房 staff rental flat
职工租赁住房单元 staff rental dwelling unit
职涵 exposed culvert
职能 competence;function
职能部门 functional department
职能的 functional
职能范围 functional area;terms of reference
职能符号 functional character
职能管理 functional management
职能合并 functional consolidation
职能化 functionalization
职能会计 functional accounting
职能机构 functional organization;staff function
职能机器人 heuristics robot
职能区 functional zone
职能区域 functional area;terms of reference
职能权力 functional authority
职能权限 functional authority
职能式分权制 functional decentralization
职能式组织管理机构 functional structure
职能图 symbolic circuit
职能完整性 functional viability
职能训练 functional training
职能要求 functional requirement
职能预算 performance budgeting
职能资本 functioning capital
职能资本家 entrepreneurial capitalist;industrialist and businessman
职能组织 functional organization
职能组织形式 functional type of organization
职前训练 vestibule training
职前指导 job orientation
职权 authority;competence;incumbency
职权范围 functional area;limits of functions and powers;province;terms of reference
职权工作范围 term of reference
职权或任期 presidency
职位 appointment;duty;job;position;post;rank;procuratorate <代理人或代诉人、检察官等的>;presidency <总统、校长等的>
职位出缺 voidance
职位分类 position classification
职位空缺 job vacancy
职位轮调 job rotation
职务 duty;employment;job;post;task
职务称呼 position title
职务乘车证 duty pass
职务的 functionary
职务等级工资制 post-rank salary system
职务调动 transference
职务分离控制室 work duty separation control room
职务分析 job analysis
职务工资 pay according to one's post
职务工资率 <指某具体职务的最低工资率> job rate
职务和值班时间表 roster
职务级别 service grade;service rank
职务津贴 duty allowance
职务轮流 job rotation
职务培训 job training
职务上的 functional;official
职务说明 job description
职务一览表 duty list
职务一览图 duty chart
职务以外的 extraofficial
职务执行令 mandamus
职务终身制 lifetime employment system
职业 employment;job;occupation;profession;pursuit;vocation
职业癌 cancer due to occupation;professional cancer
职业安全 occupational safety
职业安全卫生管理局 <美> Occupational Safety and Health Administration
职业安全与卫生法 Occupational Safety and Health Act
职业安全与卫生条例 Occupational Safety and Health Act
职业白内障 occupational cataract
职业保健 occupational health
职业保障 job security
职业保障保险 professional indemnity insurance
职业变化 occupational shift
职业病 employment disease;industrial disease;industrial illness;occupational disease;professional sickness
职业病防治所 occupation disease prevention and treatment center [centre]
职业病门诊 occupation disease clinics
职业才能测试仪 vocational aptitude tester
职业查账员 professional auditor
职业道德 ethics of profession;professional ethics
职业道德规范 standards of professional practice
职业道德准则 code of professional ethics
职业的 occupational;professional
职业等级 professional grade
职业地区分布 geographic(al) employment distribution
职业调查 employment survey
职业毒理学 occupational toxicology
职业发展训练 career development training
职业房地产经理 professional real estate executive
职业分布 employment distribution
职业分类 occupational classification
职业分析 job analysis
职业辐射剂量 occupational radiation

dose
职业妇女 career-oriented women;career woman
职业概念 professional image
职业工程师 professional engineer
职业工作规范 on-the-job specification
职业规划 career planning
职业过失 malpractice
职业和社会结构 occupational and social structure
职业化 professionalize
职业环境 occupational environment
职业会计师 professional accountant;public accountant
职业驾驶员协会 Professional Drivers' Council
职业建筑师 professional architect
职业健康 occupational health
职业教育 professional education;vocational education
职业结构 occupational structure
职业介绍所 employment agency;employment office;hiring hall;house of call;intelligence office;labo(u)r exchange;employment exchange <英>
职业禁忌症 occupational contraindication
职业精神病学 occupational psychiatry
职业开发规划 career development plan
职业考试 trade test
职业疗法 occupational therapy
职业领事 professional consul
职业领域 occupational area
职业流动性 occupational mobility
职业培训 vocational training
职业评估操纵统一标准 uniform standards of professional appraisal practice
职业前程发展管理 career management
职业前程辅导 career counseling
职业前途 career prospect
职业前训练 job orientation and training
职业潜水 professional diving
职业人口 active population;classified population
职业上的 professional;vocational
职业神经症 occupational neurosis
职业实习 professional intership
职业史 occupational history
职业守则 on-the-job specification
职业税 occupational tax
职业特征 occupational stigmata
职业统计调查 occupation census
职业卫生 occupational health
职业卫生标准 standard of occupational hygiene
职业卫星 occupational satellite
职业消防队 professional fire brigade
职业消防局 paid fire department
职业消防战斗员 paid firefighter;paid man
职业心理学 occupational psychology;professional psychology;vocational psychology
职业心志 job psychograph
职业性癌 occupational cancer
职业性变态反应 occupational allergic reaction
职业性尘末 occupational dust
职业性传染病 occupational infectious disease
职业性耳聋 occupational deafness
职业性辐射防护 occupational radiation protection
职业性辐射工作人员 occupational radiation worker
职业性辐射危险 occupational radiation hazard
职业性过失 occupational negligence
职业性化学侵蚀 occupational chemical erosion
职业性接触 occupational exposure
职业性接触皮炎 occupational contact dermatitis
职业性紧张 occupational stress
职业性近视 occupational myopia
职业性雷诺现象 occupational Reynaud's phenomenon;Reynaud's phenomenon of occupational origin
职业性判断 professional judg(e)ment
职业性皮肤病 dermatergosis;industrial dermatosis;occupational dermatosis;professional dermatosis
职业性皮炎 occupational dermatitis
职业性曝露 occupational exposure
职业性铅中毒 occupational lead poisoning
职业性神经官能症 occupational neurosis
职业性湿疹 occupational eczema
职业性听力损失 occupational hearing loss
职业性听力障碍 occupational dysaudia
职业性危害 occupational hazards;occupational risk
职业性矽肺病 industrial pneumoconiosis
职业性哮喘 occupational asthma
职业性眼病 occupational ophthalmopathy
职业性意外事故 occupational accident
职业性噪声接触 occupational noise exposure
职业性噪声源 occupational noise source
职业性照射 occupational exposure;occupational radiation
职业性致癌危险性 occupational cancer risk
职业性致癌物 occupational carcinogen
职业性中毒 occupational intoxication;occupational poisoning
职业许可证 occupancy permit
职业选择 choice of occupation
职业学校 college;school of general instruction;training school;vocational school;vocational-technical school
职业训练 career training;vocational training
职业训练学院 vocational training college
职业引起的 occupational
职业映象 professional image
职业噪声 occupational noise
职业责任保险 professional liability insurance
职业执照 professional licensing
职业治疗 occupational therapy
职业中毒 occupational poisoning
职业中学 secondary modern school
职业专门化 occupational specialization
职业转换 job conversion;occupational shift;occupational transformation
职业组 occupational class
职员 clerk;officer;office worker;official;staff(member);blackcoat <英>
职员的更换 changes in personnel
职员的雇佣 engagement of staff
职员的提供 supply of personnel
职员更替调整 adjustment for turnover of staffs
职员工作服 career apparel
职员公共休息室 staff room
职员纪律 staff morale
职员奖金 bonus to officer
职员劳动生产率 staff productivity
职员设置 staffing
职员宿舍 staff quarter
职员薪金 office salaries
职员专用车 staff vehicle
职责 business;duty;responsibility
职责范围 limitation of liability;responsibility range
职责条例 job description
职责学 deontology
职掌规定 job description

植

植棒 planting-peg
植保 plant protection
植保服务 plant protection service
植保机具 pest control equipment
植保站 plant protection unit
植被 cover plant;green covering;ground vegetation;mantle of vegetation;plant cover;vegetal cover;vegetation;vegetational cover;vegetational growth;vegetation feature;vegetative cover <利于植物生长的表土层>;ground cover
植被百分数 percentage of vegetation
植被保护 plant protection;vegetative protection;vegetative treatment
植被边坡 <有植物覆盖的边坡> planted slope
植被测量 vegetation survey
植被层 vegetable layer;vegetation stratum
植被抽象单位 nodum
植被垂直地带性 vertical vegetational zonation
植被带 vegetation belt
植被单位 nodum
植被单元 vegetation unit
植被的 vegetated
植被的连续 vegetation continuum
植被地带 florizone;vegetation zone;zone of vegetation
植被地带性 vegetational zonation
植被地理学 vegetational plant geography
植被调查 vegetational survey
植被度 degree of vegetation
植被防护层 protective cover of vegetation
植被分类 vegetative breakdown;vegetative classification
植被符号 vegetable pattern;vegetation pattern
植被覆盖百分率 percentage of vegetation
植被覆盖度 vegetation coverage
植被覆盖密度 density of cover
植被固坡 planting protect slope
植被固沙 fixed sand by vegetation
植被管理 vegetation management
植被灌溉 turf irrigation
植被海岸 vegetation coast;vegetation shore
植被和生态系统制图 vegetation and ecosystem mapping
植被护坡 soil cover(ing);ground cover
植被恢复 revegetation
植被监测 vegetation monitoring
植被截留 interception by vegetation
植被经度地带性 longitudinal vegetational zonation
植被景观 vegetational landscape
植被类别 vegetation classification
植被类型 cover type;type of vegetation;vegetation form
植被类型图 vegetation chart;vegetation map
植被连续群概念 vegetation continuum concept
植被连续群指数 vegetation continuum index
植被连续体 vegetation continuum
植被密度 density of vegetation cover
植被面积减少 decrease of vegetation cover
植被排水量 vegetal discharge
植被屏蔽影响 vegetation shielding influence
植被剖面图 vegetation profile
植被强度 vegetation cover strength
植被切碎机 vegetation disintegrator
植被区域 vegetation region
植被圈 circle of vegetation
植被群落 vegetational type
植被生长季节 vegetation season
植被生长期 vegetation period
植被水平地带性 latitudinal vegetational zonation
植被损失 vegetation loss
植被图 vegetation map
植被弯曲强度 vegetation bending strength
植被吸收 absorption of vegetation
植被稀疏的 sparsely vegetated
植被系统 vegetation system
植被箱 <测量蒸发散发用的> vegetation tank
植被型 cover type
植被性质 vegetation nature
植被亚系统 vegetation subsystem
植被研究 vegetation study
植被要素透明片 vegetation overlay
植被再生 regeneration of vegetation
植被再造 revegetation
植被蒸发 evaporation from vegetation
植被种类 vegetation type
植被重建 revegetation
植被阻滞 vegetal retardance
植草 grass planting;seeding;sprig-grass;sward;turf(ing)
植草保护 grass protection
植草边道 grass margin
植草边坡 seeded slope;sodded slope
植草边缘带 grassed margin
植草衬护 <渠道的> vegetative lining
植草带 <道路的> seeded strip
植草的中央分隔带 grassed central reserve
植草地带 grass strip;grassed area
植草法 sod culture
植草反渗透 seeded reverse osmosis
植草沟 grassed swale
植草沟道 vegetated channel
植草加固路肩 vegetated stabilized shoulder
植草路肩 grass shoulder;turf shoulder;vegetated shoulder
植草路缘 grass margin;sod curb
植草面 seed(ed) surface
植草明渠 grassed channel
植草皮 lining turf;planting turf
植草皮边坡 sodded slope
植草皮地面 grass surface
植草坡 sodded slope;turfed slope
植草渠道 grass-lined channel
植草水道 grass(ed) waterway;vegetated channel;vegetated waterway
植成土 phytogen(et)ic soil
植虫 zoophyte
植床 plant bed
植丛 socies;thicket;virgulata

植丛群落 lochmium
植冠密度 canopy density
植灌木 planting bush
植花草 planting flowers and plants
植甲藻类 Phytodiniales
植距 espacement;planting distance
植林 forestation;forest culture;foresting
植林地 woodlot
植林区 forest plantation
植苗林 forest planting
植木胶 phytoxylin
植爬蔓植物 planting climber
植畦 planting-line
植绒 flocking;pile coating
植绒壁纸 flock paper
植绒薄绸 flock printed sheer
植绒地毯 flocked carpet
植绒点子 flock dot
植绒法 flocking process
植绒帆布 foxes;thrum
植绒机 flocking machine
植绒黏[粘]合剂 flock binder
植绒黏[粘]合强力 flock bond strength
植绒喷枪 flock gun
植绒墙纸 flock
植绒纱 flocked yarn
植绒涂层 flock coating
植绒涂料 flock coating
植绒涂装 flock finishing
植绒涂装法 flock finish method;flocking
植绒印花 flock printing
植绒整理 flock finishing
植绒织物 flocked fabric;flocked goods
植绒纸 flock paper
植鞣沉淀 vegetable tanning waste
植鞣液 tan liquor;vegetable tan liquor
植入 implant;implantation
植渗流带 grassed percolation area
植生单位 plant association
植生图 vegetation map
植生系列 sere
植树 afforest(ation);planting;tree planting
植树保护 forestation protection;tree planting protection
植树草地 tree lawn
植树成行 avenue planting
植树带 planting belt;planting strip
植树的最好季节 best season for planting trees
植树点 planting point
植树方案 tree plan;tree program(me)
植树镐 planting hoe
植树护岸 tree planting protection
植树护堤 copping
植树机 tree planter;tree planting machine;wood-falling machine;hoedag
植树计划 planting plan
植树季节 season for planting
植树节 Arbor-Day;tree planting day
植树拦砂丁坝 tree groin
植树螺旋钻 tree planting auger
植树面积 forested area
植树木 planting tree
植树套 planting sleeves;planting tube
植树箱 planting box
植树运动 movement of planting trees
植树造林 afforest(ation);forest planting;plantation
植树造林法 forest plantation
植树站 afforestation station
植树滞流 tree retards
植树滞流挂柳 tree retards
植酸 phytic acid
植烷 phytane
植烷酸 phytanate;phytanic acid
植烷/正十八烷 phytane/n-18 alkane ratio
植物 plant;vegetable wax;vegetal;vegetation
植物保护 crop protection;plant protection;vegetation protection
植物保护措施 vegetative practice
植物保护带 vegetative screens
植物保护的喷射技术 spray application techniques for crop protection
植物保护服务 plant protection service
植物保护器 plant protector
植物被覆 vegetation cover
植物变种 botanical variety
植物标本采集者 plant hunter
植物标本干燥器 plant dryer
植物标本集 herbarium
植物标本室 herbarium
植物表面的微植物群落 plant surface microglora
植物病 phytopathy
植物病虫害 plant diseases and insect pests
植物病毒 plant virus
植物病害 disease of plants;insect pests and plant disease;plant disease
植物病害防治原则 plant disease control principle
植物病害检验 phytopathological inspection
植物病害流行预测 forecast of epiphytotic disease
植物病理学 phytopathology
植物病原微生物 plant pathogenic microorganisms
植物病原真菌 plant pathogenic fungi
植物玻璃温室 cold frame
植物博物馆 botanical museum
植物采样部位 vegetation sampling position
植物采样记录 vegetation sampling record
植物采样器官 vegetation sampling organ
植物采样器官年龄 age of vegetation sampling organ
植物采样位置 vegetation sampling locality
植物残体 dead plant part;plant residue;plant trash
植物残屑 vegetal debris
植物残余物 plant remain
植物测法 phytometry
植物测量 vegetation survey
植物产地 habitat
植物产品 plant product
植物尘肺 phytopneumonoconiosis
植物沉淀素 phytoprecipitin
植物成因 vegetable origin
植物成因的 phytogenic
植物成因矿物质 vegetolene mineral matter
植物成因论 vegetable theory
植物醇 phytol
植物丛 vegetation bed
植物萃取法 phytoextraction
植物带 botanical zone;floral zone;floristic zone;florizone;vegetational zone
植物单宁 natural tannin;vegetable tannin
植物单元 plant unit
植物导管 vessel
植物的 floristic;vegetable wax
植物的根 the root of the plant
植物的视觉效果 visual effect of plants
植物的水分吸收 uptake of water by plant
植物地理区 phytogeographical region
植物地理学 floristic geography;floristics;geographic(al) botany;phytogeography;plant geography
植物地理学家 phytogeographer
植物地球化学 botanogeochemistry
植物地球化学样品 botanogeochemical sample
植物地毯 plant carpet
植物地下部分 foot end of plant
植物调查 botanizing
植物顶部滴下的雨水 crown drip
植物冻 vegetable jelly
植物冻害 freezing injury of plant
植物毒 plant poison;vegetable poison
植物毒性 phytotoxic
植物毒性空气污染物 phytotoxic air pollutant
植物繁茂 thrive
植物防寒温室 cold frame
植物分布 plant distribution
植物分类学 phytotaxonomy;plant taxonomy;systematic botany
植物分类园 systematic grade
植物副产品 plant by-product
植物覆层 vegetal cover
植物覆盖 foliage cover;plant cover;plant mulching;vegetal cover;vegetation cover
植物覆盖层 vegetable cover
植物覆盖率 rate of plant coverage;rate of vegetation
植物覆盖物 cover degree
植物干燥样品 dried plant samples
植物隔声 vegetable insulation
植物根系 root system of plant
植物工厂 plant factory
植物公园 labeled plants park
植物固定法 phytostabilization
植物固砂<又称植物固沙> sand fixation by plantation;vegetative sand control
植物光呼吸 photorespiration
植物耗水量 plant consumption;plant consumption of water
植物合成鞣剂鞣法 vegetable-synthetic tannage
植物黑 vegetable black
植物黑颜料 vegetable black pigment
植物花纹图案 plant motiff
植物化 vegetalization
植物化石 dendrolite;phytolite[phytolith]
植物化石学 phytopal(a)eontology
植物化学 phytochemistry;vegetable chemistry
植物化学的 phytochemical
植物环境 plant environment
植物灰 plant ash
植物挥发法 phytovolatilization
植物基料黏[粘]合剂 vegetable adhesive
植物检疫 phyto-sanitary;plant health;plant quarantine;plant sanitation
植物碱 alkaloid;plant alkaloid;vegetable base;vegeto-alkali
植物降解 phytodegradation;phytotransformation
植物胶 vegetable gelatin;vegetable glue;vegetable gum
植物胶粘剂 vegetable adhesive
植物焦油 vegetable tar
植物截留 fly-off<降水的>;interception<水量的>
植物截留的水 fly-off water
植物截留降水量 interception of precipitation by vegetation
植物截留雨水 canopy interception
植物解剖学 phytotomy
植物界 plant kingdom;regnum vegetable;vegetable kingdom
植物浸出液 plant extract
植物景观 floral landscape
植物净化 plant purification
植物净化活性污水处理系统 phytodepurational activated sludge system
植物绝热 vegetable insulation
植物科学 plant science
植物壳 plant bowl
植物拉丁名 botanical name
植物蜡 vegetable wax
植物来源 plant origin
植物(乱草)侵路 encroaching growth
植物毛粪石 phytotrichobezoar
植物毛根 fibril
植物茂密地区 densely vegetated area
植物煤素质 phyteral
植物密度 plant density
植物名标牌 plant label
植物名称 botanical name
植物名牌 plant label
植物名签 garden label
植物泥质结构 phytogenic argillaceous texture
植物黏[粘]胶 mucilaginous gum
植物配植 plant arrangement
植物皮 plant bowl
植物铺盖 vegetative cover
植物气候 phytoclimate;plant climate;vegetation climate
植物气候带 phytoclimatic zone
植物气候地区 phytoclimatic district
植物气候学 phytoclimatology;plant climatology
植物区系 flora
植物区系学 florology
植物区域气候界线 biochore
植物群 flora;ground flora
植物群丛 association;plant association
植物群丛复合体 association complex
植物群带 floral zone
植物群阶【地】floral stage
植物群类 phyto-group
植物群落 formation of plant;phytocoenosis;phytocoenosium;phytocommunity;plant association;plant community;vegetation community
植物群落地理学 geographic(al) synecology
植物群落型 phytocoenosium type
植物群落学 phytocoenology;phytocoenostics;phytosociology
植物群区 floral province
植物群体 plant population
植物群系 plant formation
植物群系成分 floral element
植物染料 vegetable dye;wood dye
植物人工气候室 phytotron(e)
植物鞣剂 natural tannin;vegetable tannin
植物鞣料 vegetable tanning material
植物润滑油 vegetable lubricant
植物散发耗水量 vegetal discharge
植物色素 plant colo(u)ring matters;plant pigment
植物杀虫剂 botanical insecticide;vegetable insecticide
植物沙爆危害 plant sandblast damage
植物上陆 plant disembarkation
植物生长 vegetation
植物生长激素 auxin
植物生长季节 vegetation season
植物生长茂盛 luxuriant growth
植物生长期 period of vegetation;vegetal period;vegetation period
植物生长在墙上的花园 wall garden
植物生化分类学 plant chemotaxonomy

植物生活力 vegetation vitality
植物生理生态学 plant physiologic(al) ecology
植物生态地理学 ecologic(al) plant geography
植物生态学 ecologic(al) botany; phytoecology; plant ecology
植物生物化学 phytobiochemistry; plant biochemistry
植物时间分布史 synchrology
植物受灾情况 vegetation damaging condition
植物树脂 propolis
植物水分应力 <植物体水分盈亏指标> plant water stress
植物水消耗 plant consumption
植物酸 plant acid
植物碎屑 <灌浆用填料> canebreak
植物损害 plant damage; vegetation damage
植物损害图 plant damage map
植物损失 vegetation loss
植物炭黑 Frankfort black
植物提取物 plant extract
植物体 plant material
植物填充材料 vegetable filling material
植物图解指数 phytograph index
植物土 vegetable soil
植物土壤学 phytopedology
植物微生物 phytomicroorganism
植物萎枯点 wilting point
植物萎蔫 plant wilt
植物卫生管理 phyto sanitary control
植物温度 plant temperature
植物无性繁殖系 clone
植物物候学 phytophenology
植物物质 plant material
植物细胞渗透率 permeability of cell
植物纤维 plant fiber[fibre]; vegetable fiber[fibre]; vegetable hair
植物纤维板 vegetable fiber[fibre] board
植物纤维材料 vegetable fiber[fibre] material
植物纤维滤材 plant fiber[fibre] filtering medium
植物纤维疏松织物 <覆盖墙面用的> grass cloth
植物纤维填缝材料 cane filler
植物相克(现象) allelopathy
植物象牙 vegetable ivory
植物橡胶 vegetable rubber
植物消波 dissipation of wave energy by plantation
植物小气候学 phytoclimatology
植物形态变化及突变 plant morphologic(al) and mutational changes
植物形态学 plant morphology
植物形态学参数 <遥测> plant morphologic parameter
植物型原生动物 holophytic protozoa
植物性的 vegetal
植物性动物 zoophyte
植物性毒素的 phytotoxic
植物性粉尘 dust of vegetable origins; plant dust; vegetable dust
植物性浮游生物 phytoplankton
植物性垃圾 vegetable garbage
植物性黏[粘]合剂 vegetable adhesive
植物性屏障 vegetative barrier
植物性松软沃土 vegetable mo(u)ld
植物性填料 vegetable dust; vegetable filler
植物性物质 vegetable matter
植物修复 phtoremediation
植物需水量 plant requirement
植物需水系数 water-use ratio
植物需要的氧分 plant food
植物学 botany; phytology
植物学的 botanical
植物学家 botanist
植物学名 botanical name
植物岩 phytogeneous rock; phytogenic rock; phytolite; phytolith; phytophoric rock
植物岩溶 phytokarst
植物沿茎(降雨)水流 stem flow
植物颜料 vegetable pigment
植物演替 plant succession; succession of plant
植物羊皮纸 vegetable parchment
植物阳面采样 sampling at the sunny side of plant
植物养分污染 plant nutrient pollution
植物样品 vegetation sample
植物样品类型 vegetation sample type
植物蚁巢 myrmecodomatia
植物异常 vegetation anomaly
植物阴面采样 sampling at the shadowy side of plant
植物营养 plant food; plant nutrient; plant nutrition
植物营养物 plant nutritious substance
植物营养物污染 pollution by plant nutritious substance
植物永久枯萎系数 <用土壤含水量表示> permanent wilting coefficient
植物油 seed fat; vegetable oil
植物油脂 vegetable fat
植物有机体 plant organism
植物与气候的相互作用 plant-climate interaction
植物育种与驯化研究所 Plant Breeding and Acclimatization Institute
植物育种站 plant-breeding station
植物园 arboretum; botanical garden; botanical park
植物甾醇类 phytosterol
植物栽培 culture of plants; plant culture; plant(ing) growing
植物栽植 plant setting
植物蒸发 evaporation from plant; evaporation from vegetation
植物蒸发作用 plant transpiration
植物蒸腾 plane transpiration
植物蒸腾表 phytometer
植物蒸腾耗水量 vegetal discharge
植物蒸腾计 phytometer
植物蒸腾仪 phytometer
植物整形 topiary
植物脂肪酸 vegetable fatty acid
植物志 flora
植物质 vegetable matter
植物质松节油 vegetable turpentine
植物种类的 floristic
植物种子 plant material
植物状动物 zoophyte
植楔 planting-wedge
植屑泥炭 chaff peat
植于……之下 underplanting
植枝培土 hilling
植种机 mechanical planter
植株篱笆 green fence
植字 stick-up lettering

殖

殖积土 cumulose soil

殖民 colonization
殖民城市 colonial city
殖民地 colony
殖民地的 colonial
殖民地化 colonization
殖民地建筑 colonial architecture
殖民地建筑风格 colonial architecture
殖民地建筑风格重现 colonial revival
殖民地时期木墙板 <美> colonial siding
殖民地时期外墙板 <美> colonial siding
殖民地式 colonial style
殖民地住房建筑 colonial house
殖民地总督 <英> governor
殖民团 colony
殖民者 settler

止

止摆杆簧 whip operating lever spring

止摆杆夹板 whip bridge
止摆杆套 whip tube
止摆杆压片 whip cover
止摆杆柱 whip stud
止摆杆座 whip support
止摆心杆 whip hammer
止摆装置 anti-oscillation device
止闭点 catch point
止步 off-limit
止步接点 off-limit contact
止潮层 humidity stop
止车挡 stop block
止车器 train stop
止车楔 scotch block
止车楔子 scotch
止尘 dust suppression
止尘材料 dust laying material
止尘的 dust laying
止尘伞 deflector
止冲器 buffer stop
止单元 stop element
止挡 arresting device; back stop; boss; holding catch; stop
止荡锚 yaw-checking anchor
止荡锚泊法 hammerlock moor
止荡绳 frapping line
止滴物 stopback
止点 dead center[centre]; dead point; death point; refusal <桩的>
止钉 <继电器的> residual shim; non-magnetic shim
止顶 knock pin
止动 chock; locking; stopping
止动按钮 locking key; stop button
止动板 check plate; stop board; stop plank; stop plate
止动柄 clamping handle; locking handle; locking lever
止动薄片 stopping plane; stopping plate
止动操作杆 whip operating lever
止动操作杆销 whip operating lever pin
止动槽 stop notch
止动掣子 stop pawl
止动传感器 stop pick-up
止动磁铁 stop magnet
止动带 locking strip
止动挡 stop catch
止动挡块 stop-adjusting screw; stop tab
止动垫圈 lock washer; stop washer
止动阀 stopping brake; stop valve
止动阀球 stop valve ball
止动杆 arresting lever; blocking lever; gag lever; gripping lever; kick-out lever; locking plunger; locking-up lever; retaining bar; stop arm; stop lever; stop rod; stop spindle; throw-out lever
止动杆臂 stop rod arm
止动杠杆 locking lever; whip lever
止动钩 trip dog
止动鼓 stop drum
止动管 stop tube
止动环 check ring; closing ring; ring spring; snap ring; stop collar; stop ring
止动环槽 snap-ring groove
止动环提取器 snap-ring extractor
止动环装卸器 snap-ring driver
止动黄灯信号 stopping amber
止动回动杆 stop-and-reverse rod
止动机构 lock(ing) mechanism; stop gear; stop motion mechanism
止动及锁紧螺钉 set and locking screw
止动继电器 drive stopping relay; keep relay
止动夹环 clamping ring stop
止动监听按钮 check key
止动键 locking key; stop key
止动开关 stop cock; stop switch
止动块 backing block; stop bracket; stop dog
止动块挡铁 arresting stop
止动流速 sediment stopping velocity; velocity of breaking movement
止动螺钉 anchoring screw; attachment screw; backing-up screw; banking screw; check screw; fastening screw; fixing screw; limit screw; lock screw; retainer bolt; retention screw; screw-in compression; set screw; stopper(ed) screw; stop screw; thrust screw
止动螺钉扳手 set-screw spanner; set-screw wrench; set wrench
止动螺帽 cam nut; jam nut; keeper; locking cap; stop nut
止动螺母 block nut; cam nut; check nut; jam nut; keeper; lock nut; stop nut
止动螺栓 catch bolt; stop screw
止动螺丝 attachment screw; set screw; stop bead screw; stop screw
止动螺旋 backing-up screw; banking screw; fastening screw; fixing screw; stop screw
止动密封套 retainer gland
止动能力 stopping capacity
止动盘 arresting disc[disk]
止动片 locking plate; stop plate
止动片板 stop plate
止动器 chock; clog; detent; dog; dog driver; limit stop; retainer; retaining device; stop dog; turret lock
止动器架 pawl rack
止动圈 stopper ring
止动扇形板 locking quadrant
止动扇形齿轮 locking quadrant
止动设备 shut-down device; stop contrivance
止动手柄 locking of a lever
止动栓 stop bolt
止动索 preventer; stop rope
止动锁杆 whip lock
止动台肩 retaining ledge
止动弹簧 check spring; retaining spring; stop spring
止动套管 stop bush
止动套筒 stop sleeve
止动凸爪 clutch stop
止动线 stop line
止动销(钉) locking stud; stop pin; arresting pin; detent pin; fixing pin; latch; lock bolt; normal pin; preventer pin; retaining pin; retention pin; shotpin; stop; stopper pin; thrust plunger
止动销弹簧 stop pin spring
止动楔 chock block; grip wedge
止动楔块 chock
止动心杆簧 whip-hammer spring
止动心杆桩 whip-hammer stud
止动元件 stop element
止动闸 fixing brake; holding brake; stopping brake
止动指针 stop finer

止动轴衬 stop bush
止动柱塞 stop plunger
止动柱塞弹簧 stop plunger spring
止动爪 click stop; holding detent; lock pawl; retaining pawl
止动装置 arresting device; locking device; motion cut equipment; retainer; retaining device; shut-down device; stop motion; stopping device
止动总泵缸 kickout master cylinder
止端量规 no-go-ga(u)ge
止端轴承 head bearing
止舵器 rudder stop(per)
止舵楔 rudder stop(per)
止阀 stop valve
止阀球 stop valve ball
止沸 defervescence
止付 cessation of payment; countermand payment; stop payment
止付命令 stop payment order
止付通知 stop order; stop payment note
止杆 stopping bar; stopping lever
止杆螺钉 step screw
止工木条 <混凝土浇制中的暂设的> stop shutter
止滑 limited slip
止滑结 back hitch; slide knot
止滑木 dagger
止滑器 dog shore; ship stopper; trigger
止滑斜撑 dogger; dog shore
止环 guard ring
止簧 stop spring
止回瓣 non-return flap
止回的 non-return
止回底阀 non-return flap valve; non-return foot valve
止回垫圈 check washer
止回阀 check(out) valve; anti-flooding interceptor; automatic back valve; backflow pressure valve; backflow(stop) valve; back(-pressure) valve; backwater valve; clack valve; clapper valve; flap trap; flap valve; foot valve; holding valve; inverted valve; non-return damper; non-return flap; non-return valve; oil outlet valve; one-way valve; paddle valve; reaction trap; rebound valve; reflex valve; retaining valve; retention valve; return (flow prevention) valve; reverse flow check valve; vacuum breaker
止回阀瓣 clapper
止回阀导座 check-valve guidance
止回阀盖 check-valve cap
止回阀杆 check valve lever
止回阀杆弹簧 check-valve lever spring
止回阀夹 check-valve clamp
止回阀壳 check-valve case; clack box
止回阀孔衬圈 check-valve hole liner
止回阀配流泵 check-valve pump
止回阀喷嘴 check-valve nozzle
止回阀球 check valve ball; pump check-valve ball
止回阀球弹簧 check-valve ball spring
止回阀球座 check-valve ball seat
止回阀弹簧 check-valve spring
止回阀体 check-valve body
止回阀体盖 check-valve body cap
止回阀调节的泵 seated valve pump
止回阀箱 check-valve case
止回阀箱密封垫 check-valve case gasket
止回阀罩 clack box
止回阀至歧管管 check-valve to manifold pipe
止回阀珠 check valve ball; pump check-valve ball
止回阀柱塞 check-valve plunger
止回阀柱塞导座 check-valve plunger guide
止回阀座 check-valve seat
止回杆 check lever
止回活门 reaction trap
止回棘轮机构 hold-back ratchet
止回棘爪【机】 check pawl; ratchet pawl
止回棘爪支枢 check pawl pivot
止回减压阀 non-return and pressure reducing valve
止回接头 check joint
止回节流阀 restrictor check-valve
止回排水阀 storm valve
止回气阀 air back valve
止回汽水阀 non-return trap
止回器 back stop; trap; back stop
止回球阀 ball check valve
止回蜗杆 irreversible worm
止回旋塞 one-way cock
止回爪 check detent
止回爪枢 check pawl pivot
止回装置 device to eliminate running back
止火板 draft stop; fire stop
止浆片 grout stop
止浆岩盘 mass for stopping grout
止卡 pawl
止卡皮垫 pawl leather
止卡托(卧铺) latch bracket seat
止块 dog; stop
止块传动盘 dog drive plate
止块夹头 dog clamp
止棱 locking cone
止链醇酸 chain stopped alkyd
止裂 crack arrest; fracture arrest
止裂铆 crack arrester[arrestor]
止裂能力 crack arrest capability
止裂器 crack arrester[arrestor]
止裂试验 crack arrest test
止裂温度 crack arrest temperature
止裂小孔 stopper hole
止铃开关 bell silencing switch
止流阀 stop valve
止流旋塞 stop cock; stop tap
止流止回两用阀 combined stop and check value
止漏管箍 leak clamp
止漏环间隙 wearing ring clearance
止漏或止水侵 stop water loss or entrance
止轮垫 sprag
止轮块 wheel chock
止轮器 clog; hand-scotch; scotch; scotch block; stop block; wheel-stop; wheel stopper; brake retainer <高坡地区列车上携带的>
止轮铁鞋 wheel skid
止轮楔 stop block
止煤深度 end depth of drilling coal
止门器 door stop; terminated stop <开启四十五度或九十度>; hospital stop <医院>
止门石 stop stone
止门弹簧 door check and spring
止逆的 non-return
止逆阀 back valve; check valve; pressure retaining valve; retaining valve
止逆球阀 check ball
止逆闸门 lock check gate
止喷按钮 abort station
止喷开关 abort switch
止气化作用 devapo(u)ration
止气塞 locking out cock
止水 plugging; seal; sealing up; shut-off of water; standing water; water exclusion; water seal; water shut-off; water-tight seal
止水板 seal plate; stop log; water-stop plate; water bar
止水边条 ledge waterstop
止水薄膜法 membrane method of water-proofing
止水材料 leakproofing material; sealant; sealing compound; sealing material; water-bar material; water stop; water-stop material
止水槽 sealing groove
止水层 seal coat
止水层位 stratum of water sealing
止水衬垫 sealing gasket
止水虫胶 seal lac
止水带 cut-off; sealant tape; water bar; water-stop strip; weather bar; water stop
止水垫圈 staunching bead
止水阀 stop cock; stop valve
止水法兰 packing flange
止水方法 method of water sealing
止水粉料 stopper powder
止水缝 sealed joint; sealing joint; staunching piece
止水工程 sealing works
止水管柱 water string
止水横挡 clapping sill
止水化学(处理)剂 water control agent; water shutoff chemicals
止水环 seal(ing) ring
止水剂 plugging agent; water stop(per)
止水键 water-stop key
止水胶垫 rubber gasket; water-tight gasket
止水角铁 seal angle(iron)
止水接缝 water-sealed joint
止水接头 water-sealed joint
止水龙头 curb stop
止水面 <坞口> meeting face
止水面层 sealing coat
止水木条 log seal; timber meeting piece; timber seal
止水目的 purpose of water sealing
止水盘根 circular-shaped seal
止水片 staunching piece; water-stop strip
止水片伸缩缝 plate-type contraction joint
止水器 stopwater
止水墙 barrier wall; seepproof screen; water seal wall
止水圈 sealing collar
止水塞 filler block; water plug; water stop
止水栅条 dam-needle
止水设备 sealing device
止水设施 sealing installation
止水深层 sealing coat
止水深度 depth of water sealing
止水试验 water test
止水栓 curb stop; stop cock
止水添加剂 wall-sealing compound
止水条 draught bead; sealing rod; sealing strip; staunching rod; stop-water; water bar; water stop; water-stop strip; weather bar
止水铜板 copper plate
止水铜片 copper seal; copper water seal; copper waterstop; copper plate
止水涂层 stopping coat
止水帷幕 water-tight screen
止水压条 clapping sill
止水闸板 stop plank; stop plate
止水质量 quality of water sealing
止水装置 anti-priming; sealing device
止松垫片 nut lock washer
止松垫圈 nut lock washer
止索 check stopper
止碳 blocking; stopping
止碳硅铬铁 blocking chromes
止跳 debouncing
止痛的 analgesic; balmy
止痛剂 obtundent
止推 thrust
止推板 thrust strip
止推宝石轴承 end stone
止推承座 thrust block
止推挡板 thrust sheet; thrust plate
止推挡边后端面 flange back face
止推的 anti-thrust
止推垫圈 end washer; pressure disc[disk]; thrust collar; thrust washer
止推杆 throw-out lever; thrust bar
止推辊 tappet roller
止推滚球轴承 ball collar thrust bearing
止推滚珠轴承 ball thrust bearer; end thrust ball bearing; thrust ball bearing
止推滚锥轴承 conic(al) roller thrust bearing
止推护环 guard ring
止推环 collar thrust; retaining ring; stop ring; thrust collar; thrust ring; thrust washer
止推环定炮眼位置 collar thrust
止推环键 thrust-washer key
止推混凝土管座 thrust block concrete
止推肩 stop shoulder
止推块 stop block; gudgeon <一种加强衬套的>
止推面 thrust face
止推盘 thrust plate
止推片 thrust plate
止推器 pusher cleat; thrust terminator
止推球(滚珠)轴承 ball-thrust bearing; thrust ball bearing
止推枢轴承 kingpin bearing
止推瓦 thrust shoe
止推圬工管座 thrust block masonry
止推销 thrust plunger
止推销轴承 kingpin bearing; slewing journal
止推支座 thrust bearing; thrust block
止推轴 thrust shaft
止推轴承 angular bearing; anti-thrust bearer; anti-thrust bearing; axial bearing; axial load bearing; axial thrust bearing; block bearing; dead abutment; end thrust bearing; foot-step bearing; journal bearing; locating bearing; thrust bearing; toe bearing
止推轴承板 bearing disc[disk]
止推轴承衬 thrust pad
止推轴承定位环 thrust collar
止推轴承环 collar of thrust bearing
止推轴承调节垫 thrust pad
止推轴承瓦 thrust bearing shoe
止推轴承外圈 shaft washer
止推轴承箱 thrust box
止推轴颈 blocking journal; collar journal; journal; thrust button; thrust journal; thrust pin
止推装置 thrust brake; thrust device; thrust gear; thrust unit
止推座 thrust block; thrust block seat(ing)
止推座板 thrust block plate
止退板 anti-thrust plate
止退垫圈 check washer; floor clip
止响板 <讲台上的> abat-voix
止响物 deafening
止响装置 deafening
止销 detent plug; set pin; stop pin
止血 stanch
止血剂 hemostat; styptic
止血器 hemostat
止血药 astringent; styptic

止移板 shifting board;shifting plank
止制钩 trip dog
止住 hold;keep back;snub
止爪 claw stop;retaining click;retaining pawl;stop claw
止爪片 click banking stop
止转 standstill locking
止转棒 scotch
止转棒轭 dog link;Scotch yoke
止转电流 standstill current
止转杆 dog link
止转楔 spline
止转楔键槽 spline
止转转矩 static torque
止转转子 blocked rotor
止转转子电流 blocked rotor current;locked rotor current

只

只包人工的承包 labo(u)r only subcontracting
只包扎端部 ends protected only
只保船舶全损 total loss of vessel only
只表示结构各构件中心线的图 linear diagram
只承担火险 fire risk only
只读出 read only
只读光盘存储器 compact disk-read-only memory[CD-ROM]
只读文件 read only file
只分期付息而到期后一次偿还本金的抵押方式 interest only mortgage
只负责财产损失 property damage only
只负责自然损失 physical damage only
只供使用而不住人的结构 purely functional structure
只供行人使用 pedestrianize
只雇佣某工会会员的工厂 closed shop
只接收 receive only;receiving only
只能采用指定的一种工程材料的严格技术标准 <美国建设技术标准协会规定的> narrow scope specifications
只能转账的支票 check only for account
只赔全损 free of all average
只配置收缩钢筋的砌体 plain masonry
只收 receive only
只收地面站 receive-only earth station
只收器 receive-only device
只收设备 receive-only equipment
只收终端 receive-only terminal
只数 number of elements
只退还保险金 cancelling returns only
只为赚钱的(商品) catchpenny
只限行人活动的商业区 shopping mall
只限于地面(陆地)上的 earthbound
只限于水运 water-borne only
只写存储器【计】write-only memory
只压坡度的转弯 turn by banking
只应答 answer only
只用于申报 for declaration purpose only
只有百分比的收益表 common-size income statement
只有百分比的资产负债表 common-size balance sheet
只有帆船避免航行 navigation of sailing vessels prevented only
只有合价 lump sum
只有空气制动管的车辆 piped vehicle
只有行车钢轨的线路交叉 <与两种或三种钢轨组成的铁路交叉有区别> single rail crossing
只有一边的 one-sided
只有一次的 one-shot
只在车场使用的货车 wagon for internal yard use only
只在高峰小时使用的公共汽车 tripper bus
只在砖的周边用灰浆 <指在圬工接缝处> lipped
只造成损坏的事故 damage-only accident
只准右转【交】right-turn only

纸 paper

纸柏板 masonite board
纸斑 paper speck
纸板 board paper;heavy paper;kraftboard;paper;paperboard;paper mo(u)ld; paper sheeting; pappe;pasteboard;pressboard;presspahn;slab paper;strawboard
纸板安装机 slab paper installer
纸板安装者 slab paper installer
纸板裁切机 board cutter
纸板厂 paperboard mill
纸板厂废水 paperboard mill wastewater;waste liquor for pulping mill
纸板衬垫 fiber[fibre] packing
纸板串连带 card lacing; jacquard cord
纸板串联机 card lacing machine
纸板打孔器 punch card machine
纸板垫 pressboard padding
纸板垫圈 fiber[fibre] washer
纸板法兰垫 cardboard flange
纸板工业 paperboard industry
纸板盒 carton
纸板厚度 cal(l)iper
纸板护墙 cardboard wall
纸板环 ring of compressed cardboard
纸板机 board machine; board mill;wet press
纸板螺栓套 bolt carton; cardboard bolt sleeve
纸板模型 cardboard mo(u)ld
纸板排水 cardboard drain; paper drain
纸板排水井 wick drain
纸板排水预压法 paper drain preloading method
纸板石蜡 cardboard wax
纸板竖向排水 wick drain
纸板套管 presspahn sleeve
纸板筒 fiber[fibre] core
纸板屋顶 cardboard roof
纸板箱 cardboard box;karton;paperboard container;carton
纸板性能 paperboard property
纸板性能测定 paperboard test
纸板养护 paper cure;paper curing
纸板用 paperboard grade(stock)
纸板制品 paperboard products
纸板重铸铅板 lead cast
纸版印花 paper stencil printing
纸包漆皮绝缘电缆 paper-core enamelled type cable
纸包装 paper wrapping
纸杯 paper drinking cup;paper glass
纸背 paper backing
纸背金属条板 paper-backed metal lath
纸背铝箔 alumin(i)um foil with paper backing;paper-backed alumin(i)um foil
纸背面蹭脏 set-off of ink
纸背饰面壁柱 mersida
纸背贴面板 veneer with paper backing
纸币制度 paper money system
纸边 deckle edge
纸边铃机构 end-of-line bell mechanism
纸边收缩 tight-edged
纸病 paper defect;paper flaw
纸材 paper wood
纸草 papyrus
纸草花式柱 papyrus column
纸层 paper layer
纸层析法 paper chromatography
纸常数 paper constant
纸衬 gasket paper;paper sheathing
纸衬垫 paper gasket
纸衬片 paper liner
纸充填物 paper filler
纸处理器 paper disposer
纸带 paper tape;tape
纸带自动穿孔机 automatic tape puncher(er)
纸袋 paper bag;paper sack
纸袋包装 paper bag package;sack paper package
纸袋封口机 bag sealing machine
纸袋机 paper bag machine
纸袋卷盘 reel of sacks
纸袋库 bag storage; magazine for empty sacks
纸袋托架 bag chair;bag saddle
纸的 paper
纸的饱和性 saturating capacity of paper
纸的生产 paper-making
纸的吸墨程度 absorbence[absorbency]
纸的循环 paper recycling
纸灯罩 paper lamp-shade
纸底油毛毡 roofing paper
纸电泳 paper electrophoresis
纸电泳法 paper electrophoresis; paper electrophoresis method
纸垫 gasket paper;paper backing
纸垫固定 fixation with paper pad
纸垫片 paper gasket
纸垫圈 paper washer
纸蝶 butterfly
纸镀铝 paper-backed alumin(i)um
纸堆 paper-pile
纸堆式收纸 pile delivery
纸房状构造 card-house structure
纸分配色谱法 paper partition chromatography;partography
纸粉 paper meal
纸粉堆积 collecting of lint
纸粉聚积 collecting of lint
纸幅 paper web
纸幅成型 sheet forming
纸幅切边水针 knock-off shower
纸覆盖 paper mulching
纸盖模型 paper cover mo(u)lding
纸盖线脚 paper cover mo(u)lding
纸革 leatheret(te)
纸隔扇 paper sliding-screen
纸挂毯 paper tapestry
纸管 paper tube
纸辊 paper roll
纸滚筒搬运附件 <叉车> paper roll handling attachment
纸过滤器 paper air filter;paper filter
纸和纸板制造 paper and board manufacture
纸盒式小建筑 bandbox
纸厚 body paper
纸糊窗 paper window
纸糊顶棚 paper ceiling
纸糊墙 paper wall
纸黄金 <特别提款权> paper gold
纸黄(色) paper yellow
纸回收 paper recovery; reclamation of paper
纸基 paper base
纸基金箔 paper foil
纸夹 folder;paper clip;paper finger;portfolio;sheet-holder
纸浆 cellulose; paper pulp; paper stuff;pulp;stock;wood pulp
纸浆泵 pulp pump;stock pump
纸浆布 pulped cloth
纸浆残液 lignin(e) liquor
纸浆厂 paper pulp mill;pulp mill
纸浆厂废水 pulp liquor from pulping mill; pulp mill effluent; pulp mill waste;pulp mill wastewater
纸浆厂废物 pulp mill waste
纸浆厂污染 pulp mill pollution
纸浆池 pulp chest
纸浆的试验 pulp assay
纸浆法纸 pulp process paper
纸浆废水 pulp wastewater
纸浆废液 black liquor;paper-making waste; paper mill waste; pulp waste;sulfite lye;sulphite liquor
纸浆工业 pulp industry
纸浆过滤器 paper pulp filter
纸浆黑液 pulp black liquor
纸浆回收机 pulp saver
纸浆浸渍模塑 pulp mo(u)lding
纸浆精磨机 dynofiner
纸浆密度 pulp density
纸浆模 papier-mâche
纸浆木材抓取器 pulpwood handler
纸浆捏合机 kneading pulper
纸浆浓缩机 pulp thickener
纸浆排气装置 deculator
纸浆揉混机 pulp kneader
纸浆筛 pulp screen
纸浆筛滤器 pulp strainer
纸浆筛选 pulp sifting
纸浆筛余物 pulp mill screenings
纸浆水泥板 pulp cement board
纸浆洗涤机 pulp-dresser
纸浆堰 pulp sluice
纸浆液 pulping liquor
纸浆原材 pulpwood
纸浆原料储存场 concentration yard
纸浆造纸厂 pulp and paper mill
纸浆造纸厂黑液 pulp and paper mill black liquor
纸浆造纸工业 pulp and paper industry
纸浆蒸煮锅 pulp mill digester
纸浆(纸)板 pulp board
纸浆制造机 macerater[macerator]
纸介质电容器 film and paper capacitor;paper condenser;paper capacitor
纸巾分配器 paper towel dispenser
纸巾售卖箱 towel vendor
纸筋灰 lime plaster with straw pulp;paper pulp fibered plaster; paper strip mixed lime mortar;staff
纸筋灰粉刷 paper strip mixed mortar plaster
纸筋灰面罩 paper pulp lime plaster finishing
纸筋石灰 paper strip plaster
纸卷 paper-holder;paper in reel;paper roll;scroll
纸卷模拟记录 paper analog(ue) recording
纸卷筒夹 paper roll clamp
纸绝缘 paper insulation
纸绝缘的 paper-insulated
纸绝缘地下电缆 paper-insulated underground cable
纸绝缘电缆 paper cable;paper-core insulated cable;paper-insulated cable
纸绝缘屏蔽电缆 paper-core screened cable
纸绝缘漆包线 paper-insulated enamel wire
纸绝缘铅包电缆 paper-insulated lead-sheathed cable
纸绝缘线 paper-covered wire
纸空气过滤器 paper air filter
纸孔带加工装置 linasec

纸库 paper warehouse
纸垃圾 paper refuse
纸料 paper stock
纸篓 waste basket
纸滤气器 paper air filter
纸滤器 paper filter
纸滤清器 paper filter
纸滤芯 paper filter element
纸煤 dysodile;leaf coal
纸棉绝缘电缆 paper-and-cotton insulated cable
纸面本 paper back
纸面的 paper-faced
纸面加光机 paper-glosser
纸面绝缘材料 quilt insulation
纸面石膏板 gypsum lath(ing);gypsum plasterboard; gypsum wallboard;paper-faced gypsum board;plasterboard; sheetrock; Thistle board
纸面石膏板钉 plasterboard nail
纸面涂布 paper coating
纸面效率 paper surface efficiency
纸模 paper form;paper mo(u)ld
纸模片 paper template
纸模型 paper model
纸模纸板 papier-mâche
纸盘 paper dish
纸盘记录器 recorder with paper disc
纸胚 base paper
纸盆膜片 cone diaphragm
纸盆式扬声器 cone-type loudspeaker
纸坯 ground paper
纸皮 leatheroid
纸皮桦 canoe birch;paper birch;silver birch
纸片 paper sheet
纸片法 paper strip method
纸片抛光 polishing paper disc[disk]
纸粕 pulp
纸签 tally
纸桥 paper bridge
纸绕法(包装)paper winding
纸色层电泳法 paper chromatoelectrophoresis
纸色谱法 paper chromatography
纸色谱法-荧光分光光度法 paper chromatography spectrophotofluorometry
纸色谱分析法 paper chromatographic-(al)method
纸色谱扫描器 paper chromatographic-(al)scanner
纸色谱图 paper chromatogram
纸色谱浊度分析法 paper chromatographic(al)-nephelometric method
纸莎草 papyrus
纸莎草饰柱 papyrus column
纸莎草饰柱头 papyrus capital
纸莎草形半柱 papyrus half-column
纸莎草形柱 papyrus column
纸莎草形柱头 papyrus capital
纸莎草芽状柱帽 papyrus-bud capital
纸莎草芽状柱头 papyrus-bud capital
纸上的 paper
纸上电泳法 paper electrophoresis
纸上电泳分离法 paper electrophoretic separation
纸上电泳仪 paper electrophoresis apparatus
纸上定线【测】route location on paper
纸上定线法 paper location of line
纸上分析 paper analysis
纸上过滤 filtration on paper
纸上计算 paper calculation
纸上热色谱法 paper thermochromatography
纸上色层分离法 paper chromatography
纸上色层分析法 paper chromatography
纸上色谱法 paper chromatography
纸上谈兵 desk study
纸上作业法 paper method
纸速 paper speed
纸塑复合袋 paper cloth bag
纸胎油毡 asphalt-saturated felt;malthoid;saturated bitumen felt
纸毯 paper blanket
纸套 paper tube
纸套粉尘取样器 paper-thimble dust sampler
纸套管 paper bushing;paper sleeve
纸填 paper wad
纸填料 paper wadding
纸条 label;label paper tape;slip
纸条打字机控制设备 paper tape control
纸条电谱法 stripping electrography
纸条法 paper strip method
纸条复凿孔机 paper tape reperforator
纸条盒 < 邮政车 > slip case
纸条盘 cantilever
纸条中继 tape relay
纸条转报 tape relay
纸条转接中心 tape relay center[centre]
纸条转移法 strip transfer method
纸贴缝 tape joint
纸桶 fiber[fibre] drum
纸筒 paper tube
纸筒过滤器 cartridge filter
纸托盘 paper pallet
纸纤维污泥 paper fiber sludge
纸线 paper yarn
纸线编织 paper knitting
纸线织物 paper-yarn fabric
纸箱 paper box
纸箱包装 paper box package
纸箱货 carton cargo
纸屑 paper meal;paper scrap
纸芯电缆 paper-core(d)cable
纸芯管 paper core
纸芯胶合板 paper-core plywood
纸芯滤清器 paper-element filter
纸型 matrix[复 matrixes/matrices];paper matrix;paper mo(u)ld;papier-mâche
纸型干燥机 scorcher
纸型加湿机 humidifier installation
纸型空气滤清器 paper air filter
纸型用纸 flong
纸压铁 paper weight
纸页成型器 sheet former
纸页传真记录器 page facsimile recorder
纸页干燥器 sheet dryer[drier]
纸页式打字机 receive-only page printer
纸页式电传打字机 page teleprinter
纸页式副本 page copy
纸页式印字电报机 page(-at-a-time) printer
纸页式印字机 receive-only page printer
纸页送出 page feed-out
纸印 paper stain
纸毡 paper felt;parchment
纸站 cardboard tube
纸张 stock
纸张变形 paper distortion
纸张表面强度 surface strength of paper
纸张尺寸 paper size
纸张处理 paper conditioning
纸张处理井筒 paper disposal shaft
纸张穿孔器 paper-punch
纸张的正面 recto
纸张对油墨的吸收性 strike-in
纸张泛黄 after of paper
纸张焚化炉 paper incinerator
纸张光边尺寸 trimmed size
纸张规格 paper size
纸张横向 cross direction of paper
纸张厚度 paper thickness
纸张加工 paper conversion
纸张洁白度 whiteness of paper
纸张静电消除 destaticizing of paper
纸张开数 paper size
纸张控制器 sheet control
纸张扣钉 paper fastener
纸张黏[粘]结剂 paper adhesive
纸张平滑度 smoothness of paper
纸张起毛 fuzzing
纸张强度 paper strength
纸张染色 paper colo(u)ring
纸张上光面 coated side
纸张伸长率 paper stretch
纸张生产 paper manufacture
纸张施胶度 ink diffusivity of paper
纸张湿度计 paper hygroscope
纸张收缩 paper shrinkage
纸张透油性测量仪 vancometer
纸张涂层 paper coating
纸张涂料 paper coating
纸张外观质量 apparent quality of paper
纸张吸墨试验 pen-and-ink test
纸张吸墨性 ink absorptivity of paper
纸张纤维方向 grain of paper fiber[fibre]
纸张消耗量 paper consumption
纸张性能测定 paper test
纸张性能测定仪 paper tester
纸张颜料 paper colo(u)r
纸张颜色 paper colo(u)r
纸张与油墨适(应)性试验 paper-and-ink-compatibility test
纸张罩光漆 gloss varnish for paper
纸张罩光清漆 overprinting varnish
纸张整饰机 paper curing machine
纸张重量 basis weight;basis weight of paper;paper weight
纸张装饰 paper finishing
纸正面 felt side
纸织品 paper textile
纸制的雪栅栏 paper snow fence
纸制品加工机 paper converter
纸制织品 paper textile;sheet paper textile
纸制装饰 paper sculpture
纸质 paper characteristic
纸质表面层 paper blanket
纸质层压板 paper-based laminate
纸质层压贴面板 all-paper laminate
纸质磁带 paper tape
纸质底片 paper negative
纸质垫片 paperboard gasket;paper gasket
纸质分隔器 paper separator
纸质绝缘电缆 paper-insulated cable
纸质绝缘漆包线 paper-insulated enamelled wire
纸质绝缘铅包电缆 paper-insulated lead-covered cable
纸质空气滤清器 paper air cleaner
纸质蓝图 paper print
纸质滤盘 filter paper disk
纸质贴面板 paper overlay board
纸质斜槽 paper chute
纸质照片 paper print
纸重 basis weight
纸状页岩 bibliolite;paper shale

咫

咫尺莫辩的 zero-zero

指

指 index board

指按(门窗)闩 thumb bolt
指按门栓 thumb latch
指按锁栓 thumb latch
指板 finger-board;finger plate
指北 northing;red
指北参考脉冲 north reference pulse
指北花纹 lily
指北极 north-seeking pole
指北箭头 north arrow
指北力系数 north-seeking force factor
指北偏差 northing
指北陀螺【航海】north-seeking gyro
指北陀螺仪 fettered gyroscope
指北仪 north-finding instrument;north-seeking instrument
指北针 compass;north arrow;north-stabilized indicator
指臂 index arm;index bar
指臂头 < 六分仪的 > index arm head
指标 index(ing);index mark;indicator
指标差改正 correction for index error
指标窗 index window
指标的短缩 contraction of indicates
指标等高线 index contour
指标订正量 index correction
指标改正量 index correction
指标杆 calibration bar;index bar;index rod
指标过程线 characteristic hydrograph
指标函数 indicator function
指标集 index(ing)set
指标价格 norm price;target price
指标镜 index mirror;movable reflector
指标镜色片 index shade
指标刻度 index mark
指标空间 index space
指标林 index forest
指标临界系数 mission criticality factor
指标零 index zero
指标面积法 < 径流预报的 > index-area method
指标全重 index weight
指标设定 target setting
指标生物 index organism;index species;indicator organism
指标速率 target rate
指标特性 index property
指标特性试验 index property test;index test;index-type test
指标体系 index series;index system;indicator system;system of indices;target system
指标图 index chart;indicatrix
指标温度计 index thermometer
指标误差 scale error;index error
指标系统 index system
指标显微镜 index microscope
指标线 index line;indicatrix
指标相关 correlation of indices
指标泄漏率 target allowable leak rate
指标性能 index property
指标序列 index series
指标因子 index factor
指标增长率 target rate of growth
指标站 index station
指标值 desired value;index value
指标制度 index system;system of indices
指标字 index word
指拨控制 finger-tip control
指泊 berth designation
指泊计划 berth schedule
指泊命令 berth order
指泊通知 berth order
指泊员 berthing master;berthing officer
指槽 < 玻璃扯门的 > finger pull
指称概念 denotational concept
指出 point;point out
指出的违约行为 designed offence
指触干 dry tack-free;dry to handle;dry-to-non tacky;setting to touch;touch dry
指触干时间 dry-to-touch time;set-to-

Z

touch time
指触干燥 tack-free dry
指触干燥时间 < 涂层 > tack-free time
指挡 finger stop
指导 engineer; guide; pilot
指导标志 regulating sign
指导测验 directions test
指导错的 misguided
指导错误 misguide
指导的 directive
指导灯 tutorial light
指导法则 guiding rule
指导价格 guide price; guiding price
指导教师 instructor; tutor < 英国大学 >
指导卡片 instruction card
指导路线 guideline
指导片 instructional film
指导人员 supervising staff
指导书 guidebook
指导通则 guiding rule
指导委员会 steering committee
指导文件 policy paper
指导系统 guidance system
指导性 guiding
指导性残留量 guideline level
指导性的 tutorial
指导性规范 guide specification
指导性化石 guide fossil
指导性计划 guidance plan; guiding plan
指导性技术文件 technical guidance documents
指导性施工组织设计 directive construction organization design
指导性文摘 indication abstract
指导学说 instructive theory
指导员 conductor
指导原则 governing principle; guideline; guide principle
指导者 adviser; director; directorate; instructor; manager; rudder
指点 pointing
指点标 marker
指点信标天线 marker antenna
指垫 finger pad; pulp of finger
指订货物 cargo destined
指定 allocate; appoint; appointment; assign; assignation; destine; mandate; nominate; specify; state
指定搬运装卸设备位置者 spotter
指定保险 named peril(s)
指定边境地点交货价 delivery at frontier
指定编码基数 destination code base
指定泊位 allocated berth assign a berth; appointed berth; designated berth
指定采取行动 action to take
指定侧向负荷 specified side load
指定差距跨期单 basis order
指定场强 legal required field intensity
指定车站 destined station
指定承包工程 special appointment work
指定承包人 nominated contractor; specified contractor
指定承包商 nominated contractor; specified contractor
指定承包者 nominated contractor; specified contractor
指定船名保险单 named policy
指定存款 designated deposit
指定大修站 designated overhaul point
指定代理 authorized agency; nominated agency
指定代理人 authorized agent; designated person(al)
指定代理人条款 consignment clause
指定的 authorized; specified
指定的半径 specified radius
指定的车站 designated station
指定的订单转换系统 designated order turn-around system
指定的公司不动产代理人 designated real estate broker
指定的路由 prescribed route
指定的权限 specified power
指定的试验载荷 specified test load
指定的运输 assigned service
指定的直达电路 nominated direct circuit
指定的主管人 designated officer
指定的总量单位 assigned amount unit
指定地付款 payment domiciled
指定地区 designated area
指定发货地点 named departure point
指定法域的法律 legal order
指定范围 stated limit
指定分包合同 nominated sub-contract; special appointment contract
指定分包人 nominated sub-contractor
指定分包商 nominated sub-contractor
指定分洪区 designated diversion area
指定服务 assigned service
指定符号差 allocation signatures
指定港(口) designated port; named port; picked port
指定高度 specified altitude assignment
指定格式 specified format
指定工期 designated term of works
指定供货商 nominated supplier
指定供应商 nominated supplier
指定贯入度 specific penetration; specified penetration
指定海岸待命 coast for orders
指定荷载 specific load
指定护舷 specified fender
指定货币 designated currency
指定货棚 assigned shed
指定级配 given grading; instructed grading; specified grading
指定价值 specified value
指定检验人 appointed surveyor
指定界限 delimit
指定金额 specified amount
指定经营实体 designated operational entity
指定竞争投标 closed tender
指定来回运行的货车 wagon on shuttle service
指定路 specified circuit
指定锚泊区 designated mooring
指定锚地 designated anchorage
指定目标 intended target
指定目的地 named destination
指定目的地完税后交货价 delivery duty paid
指定配方 specified mix
指定配合比 instructed mix; specified mix
指定配合法 arbitrary proportion method
指定频率 assigned frequency; designated frequency
指定器 designator
指定区(域) defined area; designated area
指定人提单 order bill of lading
指定人员 designated person(al)
指定人支票 <俗称抬头人支票> order check; order cheque
指定日期 appointed day; characteristic date; scheduled date
指定色 bulletin colo(u)r
指定设计 point design
指定时间 appointed time; set time
指定时期 designated period of time
指定实验室试验 prescribed laboratory test
指定式背书 endorsement to order
指定式汇票 bill drawn to order
指定收货人 named consignee
指定收信人住址代码 call directing code
指定受益人 designated recipient
指定受益人信用证 straight letter of credit
指定数 designation number(al)
指定水体 designated water body; designated waters
指定水域 designated water body; designated waters
指定水质标准 designated water quality standard
指定态势 specified attitude
指定替代通路 alternate routing
指定条件 specified condition
指定停泊区 designated mooring
指定通信[讯]线路 routing
指定投标 private tender; tender of specified contractors
指定外汇银行 authorized foreign exchange bank
指定完工日期 project-due date; projected-due date
指定位置 assigned position; implacement
指定系数 directivity factor
指定项目 designated project
指定项目分析水样 analysis sample of appointing item
指定验船师 surveyor appointed
指定一次行程免票 trip pass for one specific journey
指定一定位置 emplacement
指定一个缓冲区 assign a buffer
指定一个通道 assign a channel
指定一条径路有效的客票 ticket available over one fixed route
指定遗赠 specific legacy
指定遗嘱执行人 executor
指定银行 appointed bank; designated bank; authorized bank
指定用途 earmark(ing)
指定用途留存收益 appropriated retained earnings
指定用途盈余 appropriated surplus; earmarked surplus
指定优先 assigned priority; assignment priority
指定运行进路 routing
指定载荷条件 specified loading condition
指定站点 designated point
指定者 designator
指定值 designated value
指定终点站 selection terminal
指定终端站 selection terminal
指定仲裁人 nomination of arbitrator
指定仲裁员的机构 appointing authority
指定周期 period demand
指定装船港 named port of shipment
指度针 index pin
指阀 finger valve
指法符号 fingering
指方规 alidade
指缝中挤出泥炭 squeezing peat out of soil
指缝中挤出清水 squeezing clear water out of soil
指幅 fingerbreadth
指杆架 finger rack
指杆筛 finger grate; finger grid
指管 vial
指画 finger-paint
指画法 finger painting
指环剂量计 finger-ring badge
指挥 conduct; conductor; direct; superintend; superintendence; wield; command
指挥按钮 direction button
指挥部 head office; headquarters
指挥舱 command module; control room
指挥操舵 conn
指挥车 command car; staff car
指挥车辆的交通信号 vehicle signal
指挥船 control vessel
指挥的 commanding; directive
指挥官 commandant; commander; commanding officer
指挥机 director
指挥驾驶 conn
指挥舰 commander ship
指挥控制 command and control
指挥控制装置 command and control device
指挥轮 command module
指挥码 directing code
指挥器 director
指挥器记数器 director meter
指挥权 command
指挥人 spotter < 交通 > ; superintendent
指挥所 commanding house; command post; control center[centre]
指挥塔 command tower; control tower
指挥塔台 operations tower
指挥台 command set; conning tower; director's console; mouse station
指挥台水平升降舵 sail diving plane
指挥台围壳 fair water
指挥系统 command echelon; director system
指挥线 command line; command link
指挥信标 mouse beacon
指挥信号 command signal; demand sign
指挥仪 directive; director
指挥仪瞄准系统 director-sight system
指挥仪型计算机 director-type computer
指挥用通信[讯] commanding communication
指挥(与)控制系统 command and control system
指挥语言 command language
指挥员 commander; commanding officer; leader
指挥站设备 command post equipment
指挥者 cammander; captain; director
指挥职能 directorship
指挥制 director system
指挥中心 command center[centre]; control center[centre]
指挥组 command echelon
指极星 Pointers; Pole Star
指甲白色横纹 transverse white bands of nails
指甲槽 < 折刀上便于拉开的 > nail hole
指甲刻痕试验 finger-nail indentation test
指甲硬度 nail hardness
指甲油 nail enamel
指尖操纵装置 < 驾驶汽车时利用电的作用等可以手不离转向盘而操纵变速杆的装置 > finger-tip control
指接 finger joint
指居住街坊 island site
指距 finger distance
指孔 dactylopore; finger hole; vent hole
指孔盘 finger plate
指控声呐浮标系统 command activated sonobuoy system
指粒 fingers
指梁 finger

指梁台 finger-board
指零 nulling
指零表 null meter;zero meter
指零电桥电路 null-type bridge circuit
指零读数 null reading
指零法 null method;zero method
指零器 null indicator
指零锐度 null sharpness
指零调置钮 zero adjustment knob
指零旋钮 zero control
指零仪表用放大器 null amplifier
指令 command; control word; discriminating order; injunction; instruction; order; prescribe; decode【计】;directive <美>
指令安排 order structure
指令安排形式 order format
指令按钮 command button; order button
指令包 instruction packet
指令编号 order number
指令编码方案 command coding scheme
指令编码器 command encoder
指令变化 modification of orders
指令变换 instruction map
指令变换点 entry point
指令变量 command quantity
指令表 code repertoire;code repertory; code table; instruction catalogue; instruction list; instruction repertoire; instruction repertory; instruction table; machine code; repertoire;repertory;set of instructions
指令步 instruction step
指令部分 operation part
指令部件 instruction unit
指令舱 command module
指令操作码 instruction operation code; operation part
指令长度 instruction length
指令长度码 instruction length code
指令常规 instruction routine
指令常数 instruction constant
指令程序 instruction program(me); instruction repertoire
指令处理 command processing; instruction processing
指令处理程序 command processor
指令处理机 instruction processor
指令处理器 instruction processing unit
指令作用距离 instruction interval
指路标 marker
指路标志 informational sign;informatory sign; route-indicating signal; route indicator; direction(al) sign; guide post; guide sign; information sign
指路标志上的彩色反光底板 colo(u)-red reflective background on guide signs
指路牌 destination board; direction arm; finger-nail post; finger post; guard board; guide board; road delineator; route definition sign; route indicator
指路信号 route-indicating signal
指轮 <图形显示终端上定位用的指轮> thumbwheel
指轮开关 thumbwheel switch
指轮十进制开关 thumbwheel decade switch
指轮式搂草机 finger-wheel rake; pullrake; rotary wheel-type rake; spider-wheel rake;wheel-type rake
指瞄准 pointing
指名保险证券 named policy
指名收货人 named consignee
指名投标 tender contractor
指名委托人 named principal
指明 indexing
指明储量 indicated reserve
指明原因 assignable cause
指目 feeling the pulse with the finger-tip
指南 companion; directory; finger post; guidebook; guideline; handbook;instruction book;southing
指南针 bearing bar;compass;magnetic needle
指黏[粘]试验 finger tab-out test
指拧螺旋 thumb screw
指派 appointment; assignment; designation
指派机构 appointing authority
指派式背书 endorsement to order
指派为代表 co-opt
指派问题 assignment problem
指配频带 assigned frequency band
指配频道 allocated channel;assigned channel
指配频率 assigned frequency
指配信道 allocated channel
指揿开关 push-button; push-button switch
指栅条 <空隙为一指宽> finger trap
指时针 stylus;gnomon <日晷>
指示 denote; denotement; direction; indication;point out;prescribe
指示 24 小时的电子钟 day clock
指示板 indicated board; indicator case;indicator plate
指示臂 indicating arm
指示标 cursor symbol
指示标记 cue mark
指示标拉绳 indicator cord
指示标志 directory sign; indicative mark; indicative sign; informative sign;indication sign;indicator sign; informational sign
指示标柱 indicator post
指示标桩 identification post; mandatory sign; supplemental post for survey monuments
指示表 dial ga(u)ge;index ga(u)ge; indicating ga(u)ge; indicator ga-(u)ge
指示表传动机构 indicator gear
指示表刻度盘 indicator dial
指示表拉绳 indicator cord
指示波浪周期 significant wave period
指示玻璃柱 indicated glass column
指示薄煤带 guiding bed
指示不准确 misregister
指示簿 order-book
指示测微计 indicated micrometer;indicating micrometer
指示层 index bed;indicator horizon; marker; marker bed; marker horizon; marker lamination; marker zone;reliable marker
指示差改正 correction for index error
指示产量 indicated output
指示长度 indicating length;indicator length
指示车速 indicated speed
指示秤 self-indicating scale
指示尺 cuing scale
指示出量 indicated output
指示储存 indication memory
指示储量 indicated reserve
指示传动轮摩擦簧 indicator driving wheel friction spring
指示磁铁 indication magnet
指示错误 error of indication
指示带 index strip
指示的 indicated;indicative
指示的方向 indicated direction
指示灯 approach light;display lamp; guiding light; indicated lamp; indicating lamp;indicating light;indication lamp;indication light;indicator lamp; indicator light; indicatory lamp; light indicator; miniature lamp;monitor lamp;pilot lamp;pilot light;pilot tube;position lamp; scale lamp;sign lamp;tellite
指示灯插座 pilot bracket
指示灯检查电路 lamp check circuit
指示灯检查试验 lamp check test
指示灯接线柱 indicator lamp terminal
指示灯驱动器 lamp driver
指示灯罩 indicator lamp cover
指示等高线 accented contour; index contour
指示地植物学 indicative geobotany
指示地轴架 indicating floor stand
指示电极 indicator electrode
指示电流 indicator current
指示电路 display circuit; indicating circuit; indication circuit; indicator circuit
指示电位计 indicating potentiometer
指示吊牌 indicator drop
指示定义性出现 indication-defining occurrence
指示度盘 display disc[disk]
指示阀 indicator cock;indicator valve
指示法 finger-length measurement
指示范围 indicated range; indicating range;indication range;range of indication
指示服务 finger
指示浮标 indicator buoy;marker buoy;position buoy
指示浮游生物 indicator plankton
指示符 designator;indicant;indication symbol;pointer;indicator
指示符号 designated symbol
指示符图 indicator chart
指示幅度 indicator range
指示干燥剂 indicating desiccant
指示杆 index arm;indicating arm
指示功 indicated work; indicator work;working input;input work
指示功率 dynamic(al) horsepower; indicated efficiency; indicated horsepower;indicated output;indicated power;true horsepower
指示管 indicator tube;inditron
指示光 identification light
指示规 alidade;sight rule
指示函数 indicator function
指示航线 indicated course line
指示航线方向误差 indicated course directional error
指示航线曲率 indicated course curvature
指示航线弯曲 indicated course bend
指示航向误差 indicated course error
指示航向象限 indicated course sector
指示荷载 indicated load
指示洪水 index flood
指示滑翔道 indicated glide path
指示滑翔道曲率 indicated glide path curvature
指示滑翔道弯曲 indicated glide path bend
指示滑翔道象限 indicated glide path sector
指示汇票 bill drawn to order
指示火花 pilot spark
指示机构 indication mechanism
指示计 index ga(u)ge;indicated ga-(u)ge;indicating ga(u)ge;indicating meter; indicator; indicator ga-(u)ge
指示记录器 indicated recorder;indicating recorder
指示剂 indicating paper; indicator; visual indicator
指示剂变色范围 colo(u)r range of indicator
指示剂常数 indicator constant
指示剂抽水试验法 method of tracer pumping test
指示剂碱性色 indicator basic colo-(u)r
指示剂试验 indicator test
指示剂颜色转变点 colo(u)r transition point
指示继电器 indicated relay;indicating relay;indication relay
指示寄存器 indicator register
指示价 indicative price
指示价格 indication price
指示简介 indicative abstract
指示箭头 guide finger
指示角 indicated angle
指示精度 indicating accuracy
指示镜 <测角器的> index glass
指示菌 indicator bacteria; indicator strain
指示卡规 dial snap ga(u)ge;indicating calipers;registering calipers
指示卡(片) indicator card
指示卡钳 register calipers
指示空气速度 indicated air speed
指示空速 indicated air speed
指示控制 indicating control
指示控制器 indicating controller
指示矿量 indicated ore
指示矿物 guide mineral;index mineral;indicator mineral
指示棱镜 index prism
指示量 indicatrix;trace quantity
指示灵敏度 display sensitivity
指示铃 indicating bell
指示零位交叉指针 null reading cross pointer
指示流量计 indicated meter;indicating flowmeter
指示流域 index basin
指示路面标志 mandatory road marking
指示轮 indicator wheel
指示轮定位杆 indicator wheel jumper
指示轮夹板 indicator wheel bridge
指示马力 dynamic(al) horsepower;indicated horsepower; indicated output; real horsepower; true horsepower
指示马力小时 indicated horse-hour; indicated horsepower hour
指示脉 indicator vein
指示脉冲 marker pulse
指示面 indicatrix
指示牌 destination board; direction board; direction plate; directory board;index(rag) plate;indicator; indicator board; indicator sign post;sign post
指示盘 index disc[disk];indicating dial;indicator disc;scope face
指示培养 indicator culture
指示票据 order instrument;order paper
指示平均有效压力 indicated mean effective pressure
指示平均有效应力 indicated mean effective stress
指示屏 indicator board;indicator panel
指示坡度的标志牌 gradient board
指示坡度的路标牌 gradient board
指示起浮的标尺 ascending indicated rod
指示气温 indicated air temperature; outside air temperature
指示器 automatic drop;cursor;designator; director; enchancer; glass runner;indexer;index mark;indicated device;indicated meter;indica-

ting ga(u)ge; indicating meter; indicating unit; indicator (scale); marker; monitor; pointer; pointer arm; reader; rear finder; sighting marker; telltale; viewer
指示器按钮 indicator button
指示器板 indicator board
指示器保险装置 indicator safety
指示器标度盘壳 indicator dial cover
指示器齿轴 indicator pinion
指示器传动机构 indicator gear
指示器传动轮 indicator driving wheel
指示器传动轮衬套 indicator driving wheel ring
指示器传动轮座 indicator driving wheel core
指示器传动装置 indicator gear
指示器的记录笔 indicator pencil
指示器垫 indicator seating
指示器定位板隔片 indicator maintaining plate distance piece
指示器动作 indicator motion
指示器方程 indicator equation
指示器隔片 indicator distance piece
指示器固定簧片 indicator clip
指示器关闭 indicator-off
指示器管 indicator pipe
指示器过轮传动轮 indicator setting wheel driving wheel
指示器活塞 indicator piston
指示器记录头 indicator pencil
指示器件 indicating device
指示器接头塞 indicator connection plug
指示器卷筒 indicator drum
指示器开关 indicator switch
指示器开启 indicator-on
指示器孔 telltale hole
指示器离合操作杆 indicator clutch operating lever
指示器离合杆 indicator clutch
指示器离合簧 indicator clutch spring
指示器门电路 indicator gate
指示器盘 indicator board
指示器图表 indicator diagram
指示器显示 indicator display
指示器线圈 director coil; search coil
指示器信号灯 indicator light
指示器压片 indicator maintaining small plate
指示器荧光屏 indicator screen
指示器闸门 indicator gate
指示器支杆 <闸门的> indictor post
指示器指针 indicator pointer
指示器组 display bank
指示器组件 indicator module
指示牵引力 indicated tractive effort; indicated tractive power
指示倾斜航线 indicated slant course line
指示倾斜航线(方向)误差 indicated slant course directional error
指示曲线 indicative curve
指示热电偶 indicating thermocouple
指示热消耗率 indicated heat specific consumption
指示热效率 indicated thermal efficiency
指示熔断器 indicating fuse
指示色标 sea dye
指示设备 indicating equipment; indicator plant
指示射线位置 index signal amplifier
指示生物 bio-indicator; biologic(al) indicator; index organism; indicating organism; indicator organism
指示生物法 indicator organism method
指示时间 instruction time
指示时钟 telltable clock
指示式 indicating type
指示式计划 indicative planning
指示式开关 indicating switch
指示式量测仪器 indicating measuring instrument
指示式示波器 indicating oscillograph
指示式债权 claim to order
指示式支票 cheque to order
指示式自同步机 indicating selsyn
指示式自整角机 indicating selsyn
指示试验法 indicating test method
指示书 instruction book
指示数 indicated number
指示数字 designation number
指示水位 indicated water level
指示水准器 index level
指示速率 indicated speed
指示锁 indicating lock; indicator lock
指示特征 index property; indicative character
指示天线 marker antenna
指示调节器 indicating controller
指示停车 mandatory stop
指示停止 mandatory stop
指示通向地方支路的标志 local approach sign
指示图 arrow diagram; index chart; index map; indicated diagram; indicating diagram; indicator card
指示推力 indicated thrust
指示瓦特计 indicated wattmeter
指示瓦特计法 indicating wattmeter method
指示望远镜 director telescope
指示微生物 indicator microorganism
指示位 indicating bit
指示位移误差 indicated displacement error
指示位置的继电器 relay for position indication
指示温度化合物 temperature indicating compound
指示温度计 indicating thermometer
指示物 biologic(al) indicator; indicator; indicator species
指示物质 index substance
指示物种 indicative species
指示误差 error of indication; index error; indicating error; indication error; pointing error
指示系统 indicating system; indicator system; information system
指示细菌 bacterial indicator
指示先行信号 priority at traffic signal
指示显微镜 index microscope
指示线 index line; indicatrix
指示(线)图 indicator diagram
指示线性象限 indicated linearity sector
指示项 indicator
指示销 index pin
指示效率 adiabatic efficiency; indicated efficiency
指示信 letter of instruction
指示信号 indication sign; indicator signal; pilot signal; visual signal
指示信号灯 clearing lamp
指示信托 directory trust
指示性标志 indicative mark
指示性财务计划 indicative financial plan
指示性的 indicative
指示性化石 guide fossil
指示性计划 indicative planning
指示性价格 indicative price
指示性简述 indicative abstract
指示性培养基 indicator medium
指示性数字 indicative figure
指示性微生物 index microorganism
指示性物种 indicator species
指示性仪表 indicating instrument
指示性引用标准 indicative reference
指示性增长率 indicative growth
指示性着色 indicator colo(u)ring
指示学 pointer
指示压力 indicated pressure; indicator pressure
指示压力表 indicated pressure ga(u)ge; indicating pressure ga(u)ge
指示压力计 indicating pressure ga(u)ge; indicator ga(u)ge
指示延迟 indication lag
指示液面计 indicating liquid level ga(u)ge
指示仪 indicator ga(u)ge
指示仪表 indicated instrument; indicating meter; reading instrument
指示仪表板 indicating panel
指示(仪表刻度)盘 indicator case; indicator dial
指示仪器 displaying instrument; indicating instrument; indication unit; reading instrument
指示应变 indicated strain
指示应力 index stress
指示应用性出现 indication applied occurrence
指示用电池 pilot cell
指示有效平均压力 indicated mean effective pressure
指示与指导性文摘 indicative-informative abstract
指示语句 directive statement
指示元素 indicator element
指示摘要 indicative abstract
指示者 shower
指示针 finger; index hand; indicating end; indicating finger; indicator pointer
指示正确性 indicating accuracy
指示证 letter of instruction
指示值 index value; indicated value
指示植物 indicating plant; indicator plant; phyto-indicator; plant indicator
指示植物地理学 indicative phytogeography
指示植物群丛 indicator assemblage
指示(植物)群落 indicator community
指示指令 directive command
指示指令断开 indicator-off
指示指令接通 indicator-on
指示制 pilot block system
指示装置 index unit; indicating device; indicating mechanism; indicating unit; telltale device
指示装置传动轮保护隔片 indicator device driver maintaining washer
指示装置夹板 indicating device bridge
指示装置离合轮 indicator device clutch wheel
指示装置驱动轮 indicator device driver
指示字 pointer
指示字变量 pointer variable
指示字地址 pointer address
指示(字)图 indicator chart
指示组 key group
指示钻孔 indicator hole
指式耙肥器 finger feed
指数 exponent(ial); factor; index indicative; index(ing); index mark; index number; index value; modulus [复 moduli]; superscript
指数倍增时间 exponential doubling time
指数逼近 exponential approximation
指数比例尺 exponential scale
指数变化 index movement
指数变化电压 exponential voltage change
指数变换 exponential transformation
指数标准成本 index standard cost
指数表架 dial ga(u)ge stand
指数表示 exponential representation; exponentiation
指数波 exponential wave
指数部分 exponential part; exponent part
指数部分格式 exponential part format; exponent part format
指数参考线 index reference line
指数插值 exponential interpolation
指数稠度试验 index-consistency test
指数传递函数 indicial transfer function
指数传输线 exponential line; exponential transmission line
指数存活曲线 exponential survival curve
指数大气 exponential atmosphere
指数的 exponential; indicial
指数递减 exponential decline; exponential decrease; exponential taper
指数递减方程 exponential decline equation
指数定理 exponential theorem; index theorem
指数定律 exponential law; law of exponents; law of indices
指数定义 index definition
指数法 exponential method; indexation; method of number index; index method <水文预测>
指数反应堆 exponential reaction pile; exponential reactor
指数方程 exponent equation; exponential equation; indicate equation; indicial equation
指数分布 exponent distribution; exponential distribution
指数分布服务 exponential service
指数分布区 exponential distribution region
指数分布族 exponential family of distributions
指数分析法 analytic(al) method by index
指数服务时间分布 <服务完成时刻与服务持续时间无关,如顾客到达的模型服从泊松分布,则顾客到达之间的时间分布是指数分布> exponential service-time distribution
指数符号 exponent sign
指数赋值 exponential valuation
指数公式 exponential formula; index formula
指数估算法 estimating method of index
指数关系 exponential relationship; power relation
指数函数 quantity in the exponent; exponential function
指数函数表 table of exponential function
指数函数的底【数】 base of exponential function
指数函数近似 exponential function approximation
指数函数模型 exponential function model
指数函子 exponential functor
指数和 exponential sum
指数化 exponentiate; indexation
指数化贷款 indexed loan
指数化债券 indexed bond; stabilized bond
指数积分 exponential integral
指数积分函数 exponential integral function
指数基 base index
指数级 exponential order
指数级数 exponential series
指数计数法 exponential notation
指数计数器 exponential counter; in-

dex counter
指数计数制 exponential notation
指数记号 exponential notation
指数记数法 exponential notation
指数寄存器 indexing register; index register
指数加权 exponential weighting
指数价值函数 exponential cost function
指数减幅量 exponentially damped quantity
指数减弱 exponential attenuation
指数阶 exponential order
指数解 exponential solution
指数近似法 exponential approximation
指数棱镜 index prism
指数量 exponential quantity
指数滤波 exponential filtering
指数滤波器 exponential filter
指数律 index law
指数律放电 exponential discharge
指数律上升 exponential rising
指数律失真 power law distortion
指数律阻尼 exponential damping
指数率 index percent
指数密度函数 exponential density function
指数模型 index model
指数内插法 exponential interpolation
指数匹配 exponential matching
指数平滑 exponential smoothing
指数平滑法 exponential smoothing (method); exponent smoothing
指数平滑预测法 index-smoothing forecasting
指数平均(数) exponential average
指数期 exponential phase
指数器 telltale
指数倾斜 exponential ramp
指数区分符 exponent specifier
指数曲线 exponent curve; exponential curve; index curve; indicator curve
指数曲线波形 exponential waveform
指数曲线拟合 exponential curve fitting
指数曲线尾 exponential tail
指数曲线形工具架 exponential tool table
指数曲线形喇叭 exponential horn
指数曲线形振幅扩大棒 exponential horn
指数曲线预测法 exponential curve forecasting
指数趋势法 exponential trend
指数扫描振荡器 exponential sweep generator
指数生长 exponential growth; logarithmic growth
指数生长期 exponential phase of growth
指数生长曲线 exponential growth curve
指数时基 exponential time
指数时间轴 exponential time
指数实验 exponential experiment
指数实验装置 <反应堆的> exponential experiment assembly
指数式保持器 exponential holder
指数式成长 exponential growth
指数式渐近稳定 exponential asymptotical stability
指数式流动 exponential flow
指数式脉冲 exponential pulse
指数式扫描 exponential sweep
指数式时基 exponential time base
指数式衰减 exponential decay
指数式稳定的 exponentially stable
指数式扬声器 exponential loudspeaker
指数式增加 exponential increase
指数式阻尼 exponential damping
指数势阱 exponential potential well; exponention potential well
指数收敛 exponential convergence
指数衰变 exponential disintegration
指数衰变律 exponential decay law
指数衰减 exponential damping; exponential discharge
指数衰减法 exponential attenuation method
指数衰减律 exponential decay law; negative exponential law
指数衰减率 exponentially decaying rate
指数衰减曲线 exponential decay curve
指数衰减时间常数 exponential decay time constant
指数衰减线路 exponentially tapered line
指数衰减因素 exponential damping factor
指数瞬变过程 exponential transient
指数算子 exponential operator
指数随机变量 exponential random variable
指数随机数表 table of exponential random numbers
指数特性 indicial response
指数特性曲线管 exponential response curve tube
指数特性时间延迟 exponential time delay
指数梯度 exponential gradient; index gradient
指数梯度装置 exponential gradient device
指数调平 exponential smoothing
指数调整 exponential smoothing; index setting
指数调整制度 indexing system
指数图像字符 exponent picture character
指数外推法 exponential extrapolation
指数污染物损失 exponential loss of pollutant mass
指数误差 index error; indication error
指数吸收 <按指数规律吸收> exponential absorption
指数吸收定律 exponential absorption law
指数稀释法 exponential dilution method
指数相关 correlation of indices
指数项 exponential term
指数形脉冲 exponential pulse
指数形式 exponential form
指数形状 exponential form
指数型 exponential type; type of index numbers
指数型分布 exponential type distribution
指数型过渡历程 exponential transient
指数型加权移动平均(值) exponentially weighted moving average
指数型母函数 exponential generating function
指数型曲线 exponential type curve
指数型趋势曲线 <预估远景第 30 高峰小时系数用> exponential trend curve
指数型衰减 exponential decay
指数性衰减 exponential decay
指数修匀法 exponential smoothing
指数序模 exponential order
指数循环 index cycle
指数因子 exponent factor
指数增长 exponential growth; exponential increase
指数增长级 exponential growth level
指数增益校正 exponential gain correction
指数振荡 exponential oscillation
指数值 exponential quantity; exponential value
指数滞后 exponential lag
指数重现周期 repetition rate of the exponential
指数转移函数 indicial transfer function
指数锥削线 exponential tapered line
指数锥削形传输线 exponentially tapered line
指数子程序 exponential subroutine
指数自回归模型【数】 exponential autoregressive model
指数阻尼量 exponentially damping quantity
指数阻尼正弦曲线 exponentially damped sinusoid
指套 finger-bag; finger-cot; finger guard; finger-stall
指体界面 finger interface
指头 finger
指涂 finger-paint
指涂法 finger painting
指脱落 dactylolysis
指外符号 outer marker
指望 bank on; bank upon; bargain for; build on
指纹 dactylogram; finger mark
指纹打印 finger printing
指纹法 dactylography
指纹技术 fingerprint technique
指纹色谱图 fingerprint chromatogram
指纹图 fingerprint map
指纹图案 finger print
指纹型路面裂纹 finger printing
指纹学 dactylography
指纹状烃 fingerprinted hydrocarbon
指物照明 directional lighting
指相化石 facies fossil
指相矿物 diagnostic mineral
指向 indexing; point; sense of orientation
指向标 beacon; course beacon; directing sign; localizator; localizer; marker beacon
指向标触发发射机 beacon trigger transmitter
指向标记 arrow mark
指向标失答 beacon skipping
指向标失迹 beacon stealing
指向标识板 finger-board
指向标天线 beacon antenna
指向标天线设备 beacon antenna equipment
指向标显示 beacon presentation
指向标志 arrow mark; directing sign; directional sign; guide sign
指向测定器 sense finder
指向出口标记 directional exit sign
指向传声器 directional microphone
指向导洞 pilot heading
指向的 directional; directive
指向浮标 direct float; direction float
指向改正 pointing correction
指向构造 directional structure
指向河流下游的丁坝 groyne pointing downstream
指向活动仿真 activity-directed simulation
指向箭头 directional arrow
指向角 angle of direction
指向校准 pointing calibration
指向控制 pointing control
指向类 sense-class
指向力 directive force
指向力矩 directive moment; meridian seeking torque
指向脉冲 marker pip; marker pulse
指向脉冲选择器 marker selector
指向牌 direction arm; direction post
指向器 direction indicating device; direction indicator
指向设备 sensing equipment
指向射线 directed ray
指向事件的仿真 event-directed simulation; event-oriented simulation
指向事件模拟 event-directed simulation
指向水听器 directional hydrophone
指向塔 lead tower
指向特性 directional property
指向天线 directional antenna
指向图 directive diagram
指向误差 error in pointing; pointing error
指向系数 directivity factor
指向效应 beam effect
指向信标 directional beacon; direction-giving beacon
指向性 directionality; directive property; directivity
指向性反射 retro-reflection
指向性函数 directional function; directivity function
指向性可调传声器 polydirectional microphone
指向性声呐 directional sonar
指向性图(案) beam pattern; directional response pattern; directivity pattern; directional pattern; directive pattern
指向性无线电标志 radio range
指向性因数 directivity factor; directivity gain
指向性增益 directional gain; directive gain
指向性指数 directional gain; directional index; directivity index
指向选择器 marker selector
指向运输终点的运输 terminal routing
指向障板 directional baffle
指向柱 direction(al) post; guide post; sign post; finger post <道路交叉口的>
指销 stylus pin
指形板 finger stoped plate
指形棒 cap bar finger
指形齿轮铣刀 finger gear cutter; finger-type gear-milling cutter
指形触点 finger(-type) contact
指形船坞 finger dock
指形灯 finger lamp
指形电离室 thimble ionization chamber
指形钢板 <用于伸缩缝> jaw plate
指形港池 <突堤码头区的> finger basin; finger dock
指形格条天井 finger raise
指形禾草 finger grass
指形湖 finger lake
指形滑槽 finger chute
指型机坪 finger system apron
指形接合 finger joint
指形晶 dactylite
指形晶状的 dactylitic
指形抗裂试验 finger cracking test; fining test
指形控制棒 finger-type control rod
指型廊道 pier finger
指形冷冻器 cold finger
指形冷凝管 finger-shape condenser
指形螺母 finger nut
指形码头 <与岸边垂直或构成某种角度的码头> finger pier; jetty wharf
指形排水沟 finger gull(e)y(ing)
指形刨 thumb plane
指形膨胀缝 finger-type expansion joint
指型平房体系 one-stor(e)y finger system

指型平房系统 one-stor(e)y finger system
指形平面 finger plan
指形手柄 finger lever
指形刷 finger brush
指形突(堤)码头 finger jetty;finger pier
指形铣刀 finger cutter
指形压板 finger clamp
指形叶 palm-like lobe
指形叶片搅拌机 finger blade agitator
指形折布抛光轮 finger buff
指形制动销 finger stop
指旋螺钉 thumb screw
指旋螺帽 thumb nut
指旋螺母 thumb nut
指旋螺丝 thumb screw
指旋销 thumb turn
指压干 tack free
指压干时间 tack-free time
指样订货 order by sample
指引 guide;index;manuduction;pilot
指引电路 indexing circuit
指引罐 index tube
指引卡 guide card
指引卡片 guide card
指引力 directive force
指引脉冲 index pulse
指引射束 index beam
指引信号 index signal
指引元素 pathfinder element
指印 finger mark;finger print
指印现象 <烧结多孔材料> fingering
指印消除型防锈油 fingerprint remover
指责 censure;reproach
指责者 accusant
指掌形(平面设计)系统 finger system
指针 arm;arrow;cursor;dial;finger;guide finger;index;index arm;index finger;index hand;index pin;indicating needle;indicator;indicator needle;needle;needle pointer;pointer;rudder;trammel
指针摆动 <测量仪器> beat of pointer
指针板 dial plate
指针表 pointer ga(u)ge
指针测微计 microcator;minimeter
指针测微器 microindicator
指针挡销 pointer stop
指针电流计 pointer galvanometer
指针丢失 loss of pointer
指针读数 dial counter;pointer reading;reading of a pointer;total indicator reading
指针读数器 reading of pointer
指针回转齿轮机构 counter train
指针机构 point gear
指针尖 pointer tip
指针类型 pointer type
指针罗盘 pointed compasses
指针面板 dial needle
指针盘 indicating dial
指针偏斜 needle deflection
指针偏转 throw of pointer
指针偏转式仪表 deflectional instrument
指针平衡锤 pointer counterbalance
指针平衡口 equalizing port for needle
指针示号器 needle annunciator
指针式 pointer type
指针式摆幅仪 amplimeter
指针式计量器 pointer register
指针式计量装置 dial register
指针式计数盘 pointer dial
指针式计数器 dial counter;pointer counter
指针式检流计 pointer galvanometer
指针式精密校表仪 precicheck
指针式刻度盘 pointer dial
指针式频率计 pointer frequency meter
指针式同步器 rotary synchronizer
指针式温度计 dial thermometer
指针式仪表 dial instrument;pointer instrument
指针式指示器 needle indicator;pointer-type indicator
指针枢纽 needle armature
指针天平 indicator balance
指针调整计数 pointer justification count
指针停摆 dead beat
指针温度计 pointer thermometer
指针振摆现象 power swing phenomenon
指针指示器 needle indicator
指针重合式刻度盘 follow-the-pointer dial
指针重合式指示器 follow-the-pointer indicator
指针轴 indicator needle shaft;needle pivot
指针最大偏转 swing
指重表 indicated weight meter;weight ga(u)ge;weight indicator
指重表传感片 diaphragm of the weight indicator
指重表传压膜 diaphragm of the weight indicator
指轴 spindle
指状 fingered
指状凹口 finger notch
指状冲刷痕 scour finger
指状大透镜砂体 bar-finger sand
指状的 digital;digitate
指状电极 finger electrode
指状分叉背斜褶皱 digitation
指状分散作用 digitation
指状格条天井 finger raise
指状沟蚀 finger gull(e)y
指状轨条溜口 finger chute
指状湖 finger lake
指状缓速器 finger retarder
指状簧齿除草机 finger weeder
指状交错 interfinger
指状交通 interfingering
指状浇口 finger gate
指状联结 interdigitation
指状溜口闸门 finger chute gate
指状排水 drainage fingers
指状排水沟 finger drain
指状排水系统 finger drain
指状三角洲趾 digitate margin of delta
指状沙坝 bar-finger sand
指状沙坝圈闭 finger-bar trap
指状手柄 finger lever
指状弹簧油封 finger spring seal
指状物 finger;finger piece
指状铣刀 finger cutter
指状限位器 finger stop
指状岩芯提取器 finger lifter;finger-type core lifter
指状元件 finger
指状闸门 finger door
指状闸门间距 finger spacing
指状抓取器 finger grip
指状抓手 finger-action tool
指状组合型 interdigital
指状钻头 finger bit

枳 trifoliate orange

趾 toe

趾板 toeplate;toe board;toe slab
趾部焊缝 toe weld
趾部铺盖 toe blanket
趾长度 toe length
趾端垫块 toe block
趾连接螺栓 toe joint bolt
趾墙 toe wall
趾压力 toe pressure
趾延长【铁】 toe extension
趾应力 toe stress

酯 ester

酯二醇 esterdiol
酯化 esterify(ing)
酯化催化剂 esterification catalyst
酯化当量 esterification equivalent weight
酯化的 esterified
酯化度 degree of esterification
酯化树脂 esterified resin;resinter
酯化松香 esterified resin
酯化速度 esterification rate
酯化天然树脂 esterified natural resin
酯化天然树脂清漆 esterified natural resin varnish
酯化油 esterified oil
酯化脂肪酸总量 total esterified fatty acid
酯化值 ester number;esterification number;ester value
酯化作用 esterification
酯基转移作用 transesterification
酯键 ester linkage
酯交换作用 ester interchange;interesterification
酯胶 glycerin(e)ester;rosin ester
酯蜡 ester wax
酯膨润土润滑脂 ester-bentonite grease
酯强混凝土 estercrete
酯式合成干性油 ester type synthetic-(al)drying oil
酯树胶 ester gum
酯树胶清漆 ester gum varnish
酯酸 ester acid
酯缩合作用 ester condensation
酯烷基 ester alkyl
酯系溶剂 ester solvent
酯增塑剂 ester plasticizer
酯值 ester value
酯转移 transesterification

至

至撤消为止 till countermanded

至此 thus much;up to this point
至点 solstice;solstitial point
至点潮 solstitial tide
至点潮流 solstitial tidal current
至点圈 solstitial colure
至点时 solstitial point
至多 not more than
至目标距离 distance to object
至上主义 suprematism
至上主义派建筑 suprematist architecture
至圣处 inner sanctum
至圣堂 <古典教堂中的> delubrum

志

志贺氏杆菌 Shigella

志贺氏杆菌痢疾 Shigellosis
志贺氏菌病 Shigellosis
志留纪【地】 Silurian;Silurian period
志留纪大海退【地】 Silurian great regression
志留纪岩石 Silurian rock
志留系【地】 Silurian;Silurian system;Siluric
志趣水平 level of aspiration
志田数 Shida's number

制

制白云石砖的机器 dolomite brick press

制板 retaining plate;slabbing
制板厂 board mill;board sawmill;plate mill;plate shop(mill)
制板工厂 plate shop(mill)
制板机 plate mill;slab forming machine
制板照相机 press camera
制版 block-making;plate-making
制版工艺 mask-making technology
制版墨 tusche
制版摄影机 reticle camera
制版用光电自动装置 scan-agraver
制版照相机 process camera;reproduction camera;reticle camera
制棒和制管机 bar and tube machine
制备 fabricate;fabricating;magister;preparation;prepare
制备超速离心机 preparative ultracentrifuge
制备方法 preparation method
制备分配色谱法 preparative partition chromatography
制备辐射化学 preparative radiation chemistry
制备干草成套机具 hay-making range
制备好的型砂 conditioned mo(u)lding sand;prepared mo(u)lding sand
制备级薄层色谱法 preparative thin-layer chromatography
制备级气相色谱法 preparative gas chromatography
制备级试样 preparative scale sample
制备级塔板数 preparative scale plate number
制备级纸色谱法 preparative paper chromatography
制备磷光体的材料 phosphor material
制备品 preparation
制备气氛 prepared atmosphere
制备区带电泳法 preparative zone electrophoresis
制备色谱法 preparative chromatography
制备石膏线脚 <在原来位置> running mo(u)lding
制备型色谱法 preparative scale chromatography
制备性光化合成 preparative photosynthesis
制备性能 processability
制备液相色谱 preparation scale liquid chromatogram
制备液相色谱仪 preparative liquid chromatograph
制备柱 preparative column
制备柱电泳 preparative column electrophoresis
制币厂 mint
制标签机 label(l)ing machine
制标志机 marking equipment
制表 charting;print totals only;tabling;tabulate;tabulation
制表挡板 tabulating stop
制表法 tabulating method
制表符号 tabulation character
制表机 tabulating machine;tabulator
制表卡刀具 tab-card cutter
制表人 lister;tabulator
制表设备 tabulating equipment
制表序列格式 tabulation sequential format
制表仪 tabulator
制表员 tabulator
制表装置 tabulation set
制冰 ice manufacture

制冰厂 ice factory; ice house; ice-making plant; ice plant; refrigerating plant
制冰场所 ice house
制冰池 ice-making tank
制冰机 ice engine; ice machine; ice maker; ice-making equipment; refrigerator
制冰机房 ice house
制冰机械 ice-making machinery
制冰块盘子 ice tray
制冰能力 ice make capacity; ice-making capacity
制冰器 ice maker
制冰设备 ice plant
制冰室 ice-making compartment
制冰水设备 ice water installation
制波纹机 corrugation machine
制玻璃 founding
制玻璃的槽形耐火设备 debiteuse
制玻璃原料 frit
制钵用黏[粘]土 saggar clay
制箔 foliation
制箔机 foil maker
制箔碾压工厂 foil rolling mill
制材 cleaving timber; conversion of timber; lumbering; lumber sawing; sawing lumber; sawn wood; wood making
制材厂 lumber mill; sawmill; timber mill; timber sawmill
制材厂废料 saw mill residue; saw mill waste
制材厂废屑 saw mill refuse
制材厂商 lumber manufacturer
制材车间 sawmill; timber mill
制材削片联合机 chipping heading
制材斜纹 diagonal grain
制裁 impose sanctions against
制层 preparative layer
制层电泳分离法 preparative zone electrophoretic separation
制层色谱法 preparative layer chromatography
制成 manufacturing; turn-off
制成把手形 knobbing
制成板状的建筑石料 ragstone
制成标准组件的 modular
制成玻璃珠 beading of glass
制成薄板 laminate
制成薄层木板 veneering
制成成本 cost of manufacture
制成的 made-up; manufactured; ready-made
制成焦炭 coke
制成拒付证书 drawing up a protest
制成拒绝证书的汇票 bill duly protested
制成粒 corning
制成零件分户账 finished parts ledger
制成率 pull rate
制成泥浆 <陶瓷料> blunging
制成品 end products; finished goods; finished products; finished stock; manufactured article; manufactured goods; manufactured products; ready-made; workmanship
制成品存货账 finished goods inventory account
制成品分户账 finished goods ledger
制成品盘存 finished products inventory
制成品预计成本 predicted cost goods manufactured
制成品周转率 turnover of finished goods
制成平板的 platten
制成平箔的 platten
制成斜面 bevel
制成一体的 in-built
制齿轮机 gear cutting machine
制出 turnout
制船驳船 mooring barge
制船柱 mooring dolphin
制瓷方法 china clay method
制瓷工艺 porcelain process
制瓷过程 porcelain process
制刺钢丝机 barbed wire machine
制刺线机 barbed wire machine
制带机 pocket builder
制淡水设备 freshwater distilling plant
制氮站 nitrogen generation station
制挡机件 stop work
制荡板 wash board
制导 guidance; guidance control; guiding; vector
制导波束 guidance beam
制导波束宽度 guidance beamwidth
制导场 guidance site
制导传感器 guidance sensor
制导滚轮 guide roller
制导火箭 control rocket
制导精确度 guiding accuracy
制导雷达 guidance radar; rider
制导脉冲 controlling pulse
制导设备 guidance equipment; guidance unit; missile guidance set
制导设备失灵 guidance malfunction
制导网 guidance network
制导误差 guidance error; guidance miss distance
制导系统 chain home beamed; control system; guidance system; guiding system
制导系统设备检查 guidance control
制导信息 guidance information
制导与导航计算机 guidance and navigation computer
制导与导航控制系统 guidance and navigation control system
制导站 guidance site; guidance station
制导中继装置 guidance link
制导中心 direction center[centre]
制导装置 guidance apparatus; guide apparatus
制垫片的模型 bedder
制钉材料 nail material
制钉厂 nailery
制钉刀 pointing knife
制钉钢丝 nail wire
制钉工人 nailer
制钉机 nail(ing)(making) machine
制钉模 nailing pattern
制钉铁丝 nail wire
制钉用凿 nail smiths chisel
制订 draw up; work(ing) out
制订财务计划 financial plan(ning)
制订成本降低计划 institute cost reduction program(me)
制订法律 legislation
制订防范措施 institute precautions
制订规范 preparing specification
制订模块 evoke module
制订潜水计划 diving planning
制订人 implementer
制订项目 project formulation
制订项目计划 project programming
制订者 framer
制定 enact; enactment; establishing; formulate; laydown; work out
制定……配方 formulate
制定单位 organization for working out
制定法 artificial law
制定法律的 enactive
制定费率 rate-making
制定规章的机构 regulatory authority
制定航线 made course
制定机车、车辆和职工工作班次及时间的方法 rostering
制定及控制发展规划的机构 planning authority
制定计划 frame a plan
制定价格 pricing; pricing practice; rate marking
制定键 locking key
制定可行性方案 formulate alternative
制定两年期计划 biennial programming
制定目标 target making
制定时间 time of working out
制定预算 budgeteering
制定远景规划 advanced planning
制定运价 rate-making
制定运价表 tariffing
制定者 constituent; maker
制定综合平衡方案【给】 compromise programming
制锭机 performer; preforming press
制动 application of brakes; braking path; damp; deboost; drag; escapement; locking; manufacturing; put on brake; retardation; stopping
制动安全性 service stability
制动按钮 locking key; stop button
制动把 clamp pin
制动百分率 brake percentage
制动扳手 brake spanner
制动板 braking vane; catch plate; check plate; dive flap; keep plate; retarding disc[disk]; slipper
制动棒 scotch
制动保护继电器 brake-protecting relay
制动保压 service lap
制动报表 braking sheet
制动报警继电器 brake warning relay
制动倍率 amplification ratio; brake leverage; braking leverage; leverage ratio; multiplication of the brake gear ratio
制动泵主筒 brake master cylinder
制动比 brake percentage; brake ratio
制动臂 brake arm; retaining arm
制动臂轴 brake arm shaft
制动柄 clamping handle
制动波 brake propagation; propagation of brake action
制动波速 brake propagation rate; brake propagation speed; propagation rate; rate of propagation; speed of (brake) application; speed of propagation of brake action; transmission speed of brake action
制动不良 poor stop
制动操纵 braking maneuver
制动操纵装置 brake operation device
制动操作 brake application
制动操作杆 stop lever
制动操作规则 brake operating conditions
制动操作系统 brake operating system
制动操作楔 brake operating wedge
制动测功机 brake dynamometer
制动测功计 brake dynamometer
制动测功器 absorption dynamometer
制动测功仪试验 dynamometer test
制动测力计 brake dynamometer
制动测量法 brake method
制动差动机构 retaining differential mechanism
制动差速转向装置 brake differential steering
制动车 brake van
制动车辆 abrupt deceleration vehicle; braked wagon
制动掣子 brake latch
制动掣子柄 brake latch spoon
制动掣子弹簧 brake latch spring
制动衬带 brake bush; brake lining
制动衬带更换机 brake reliner
制动衬里 brake covering
制动衬面 brake lining
制动衬片 brake facing
制动齿轮 spragging gear
制动齿轮罩 spragging gear housing
制动传动装置 brake rodding
制动传送速度 brake transmission speed
制动唇 locking lip
制动磁阀 brake application magnet valve
制动磁铁 brake magnet; braking magnet; damping magnet; retarding magnet
制动带 back strap; bracketed string; brake band; brake cord; check band; friction(al) band
制动带衬里 brake band lining
制动带衬面 brake band lining
制动带架 brake and bracket
制动带摩擦片 brake lining
制动带调整器 brake band adjuster
制动带系杆 brake band clevis
制动带张紧工具 brake band tightener
制动带支座 brake band anchor clip
制动单元 brake unit
制动到规定的速度 retardation to target speed
制动道岔 catch points
制动的 banking; retaining
制动的管线 brake piping
制动的货车 braked wagon
制动的机械构造 brake mechanism
制动的手拉杆 brake hand lever
制动的支承 brake support
制动的重力 braked weight
制动灯 brake lamp
制动等级 braking grade
制动点火时间 retrofire time
制动电磁铁 brake electromagnet
制动电动机 brake motor
制动电极 decelerating electrode; retarding electrode
制动电流 braking current; stalling current
制动电流变换器 brake current transducer
制动电路 brake circuit; braking circuit
制动电路控制器 brake circuit governor
制动电阻 brake resistance; braking resistance; stopping resistor
制动电阻器 braking resistor
制动电阻器风扇 braking resistor arrangement
制动电阻栅 braking grid
制动垫片 brake pad
制动垫片夹 brake pad securing clip
制动叠层板 brake plate
制动动作的传递 propagation of brake action
制动墩 brake pier; braking pier; abutment pier <多孔拱桥承受不平衡推力的>
制动发动机 brake engine; retro-engine; retromotor; retropack; retro-rocket
制动发动机点火 retrofire; retroignition
制动发动机轮 retro module
制动发动机起动时的机动 retrofire maneuver
制动阀 brake control valve; brake valve; stopping brake; stop valve
制动阀传动装置 brake valve actuator
制动法 deceleration method
制动反应时间 brake reaction time; brake response time
制动方式 braking method; braking mode; method of braking

Z

制动放空 emptying
制动分泵 brake cylinder; wheel cylinder
制动风缸 brake supply reservoir
制动风缸压力 brake cylinder pressure
制动风速表 bridled anemometer
制动风压板 bridled pressure plate
制动风叶 braking vane
制动幅度 braking range
制动辅助转向 secondary brake steering
制动负荷 brake load; drag load
制动杆 brake bar; brake lever; brake rod; braking bar; gripping lever; locking-up lever
制动杆端叉形铁 brake-rod crevice
制动杆扇形板 brake lever segment
制动杆弹簧 brake-rod spring
制动杆调整器 brake-rod adjuster
制动杆爪 brake lever latch; brake lever pawl
制动感应线圈 brake induction coil
制动缸 braking cylinder; checking cylinder; hydrocheck
制动缸安装座 brake cylinder plate
制动缸充气时间 brake cylinder filling rate
制动缸充气速度 brake cylinder filling speed
制动缸垫 cylinder gasket
制动缸垫木 brake cylinder block; brake cylinder wood filler
制动缸吊 brake cylinder hanger
制动缸防护器 brake cylinder protector
制动缸盖 brake cylinder head
制动缸盖垫 cylinder cap gasket
制动缸杠杆 brake cylinder lever; cylinder lever
制动缸杠杆和均衡杠杆的拉杆 cylinder lever and equalizing lever connecting rod
制动缸杠杆和联结杠杆的拉杆 cylinder lever and fixed lever connecting rod
制动缸杠杆和游动杠杆的连杆 cylinder lever and floating lever connecting rod
制动缸杠杆拉杆 cylinder lever connecting rod
制动缸杠杆链 cylinder lever chain
制动缸杠杆托架 cylinder lever bracket
制动缸杠杆支架 cylinder lever support
制动缸管 brake cylinder pipe
制动缸后盖 back cylinder head; pressure head
制动缸活塞推力 brake cylinder force
制动缸漏气沟 brake cylinder leakage groove
制动缸排量 brake cylinder displacement
制动缸皮碗 piston cup leather; piston packing leather
制动缸润滑器 brake cylinder lubricator
制动缸润滑油 brake cylinder lubricant
制动缸体 cylinder body
制动缸推杆 push rod
制动缸托 brake cylinder support
制动缸压力 brake cylinder pressure
制动缸增压 build-up in brake cylinder
制动缸支点 brake coupling guide
制动钢薄层板 brake steel plate
制动钢丝绳用钢丝 brake cable wire
制动杠(杆) brake arm; brake bar; brake lever; brake rod; stop lever; working beam
制动杠杆倍率 multiplication of rigging
制动杠杆臂长比 brake leverage
制动杠杆标识牌 brake lever badge plate
制动杠杆导板 brake lever guide
制动杠杆的偏移 angularity of brake lever
制动杠杆拉杆 brake lever connecting rod; brake lever tie rod
制动杠杆推杆 brake lever thrust bar
制动杠杆托架斜撑 brake lever bracket brace
制动杠杆下拉杆 brake lever strut
制动杠杆销 brake pin
制动杠杆支点 brake lever fulcrum
制动杠杆支点联结板 brake lever fulcrum tie plate
制动杠杆支点托 brake lever bracket
制动杠杆轴 brake lever axle
制动杠杆爪 brake lever jaw
制动隔离 isolation of the brake
制动工况 brake operation
制动工人 brake(s)man
制动工人小室 brakeman's box
制动公式 retardation formula
制动功率 brake horse power; brake horsepower; brake power; braking power; stopping power
制动钩 arrester hook; wheel hook
制动钩操纵机构 arrester hook actuation gear
制动箍 stop ring
制动鼓衬垫 brake drum liner
制动鼓车床 brake drum lathe
制动鼓防尘罩 brake drum dust cover
制动鼓(筒) brake drum
制动鼓砧 brake drum anvil
制动关闭的车辆 dead car
制动管 brake pipe; main brake pipe
制动管充气 filling of brake pipe
制动管吊架 brake pipe hanger
制动管吊卡 brake pipe anchor
制动管端接套 brake pipe nipple
制动管端塞门 brake pipe end cock
制动管风压 brake pipe pressure
制动管减压 brake pipe reduction
制动管滤尘器 brake pipe strainer
制动管排空加速器 brake pipe emptying accelerator
制动管泄漏 brake pipe leakage
制动管压力梯度 brake pipe gradient
制动惯性 braking inertia
制动轨道 braking orbit
制动柜 brake cabinet
制动辊 brake roll
制动过热剥落 brake burn combustor
制动过载系数 damped overload ratio
制动夯具 locking ram
制动和空压机检修车间 brake and air compressor repairing workshop
制动荷载 brake load; braking load
制动盒 brake box
制动桁构 brake truss
制动桁架 brake truss; retarding truss
制动横轴 brake cross shaft
制动滑块 brake slipper; shoe block
制动环 brake hoop; brake ring; stop ring
制动缓解 brake release; release of brakes
制动缓解汽笛 release brake whistle
制动缓解曲线 brake release curve
制动缓解时间 brake release time
制动缓解速度 < 驼峰编组场 > brake release speed
制动缓解踏板 brake release pedal
制动换能器 braking chopper
制动簧片 pawl spring
制动毁坏 brake failure
制动活动印迹 braking skid mark
制动机 < 机车车辆两级速度控制的 > R brake
制动机车 brake locomotive
制动机充气 filling the brake
制动机的风缸 brake reservoir
制动机的制动位 position for gradual application of brake
制动机定检期内的检查 indate test
制动机动 retrograde maneuver; retro maneuver
制动机构 brake actuator; brake mechanism; braking mechanism; detent mechanism; lock mechanism; retaining mechanism; stop gear
制动机过充气 overcharging of the brake
制动机检修工 air monkey
制动机试验 brake test
制动机试验台 brake rigging
制动机适度制动位 position for moderate application of brake
制动机性能 brake performance
制动机压力 brake pressure
制动机压强 brake pressure
制动机运行试验 running brake test
制动机再充气 refilling the brake
制动机自动发生快动紧急制动的毛病 dynamiter
制动激磁接触器 braking excitation contactor
制动级别 step of retardation
制动级(缓行器) brake step; stage of retardation
制动棘轮 brake ratchet wheel
制动棘轮机构 brake ratchet
制动集中器 brake centralizer
制动继电器 brake relay; braking relay
制动加力器 brake assistor; brake booster
制动加速器 brake accelerator
制动夹 brake clip
制动夹钳 braking clamp
制动夹钳试验装置 braking clamp test bench
制动架 detent plate stop
制动间 < 车辆 > brake cabin
制动减速 brake deceleration; braking deceleration
制动减速度 brake deceleration; braking deceleration
制动减速计 brake decelerometer
制动减速率 brake deceleration; braking deceleration
制动减振器 skid shock absorber
制动减振作用 dampening effect
制动检查 brake attention
制动检查所 brake inspection depot
制动检修所 brake inspection depot
制动键 brake key
制动角 drag angle
制动绞盘 brake reel; pall
制动阶段 deboost phase
制动接触器 braking contactor
制动截面 stopping cross-section
制动界面装置 brake interface unit
制动警示继电器 brake warning relay
制动静止试验 standing brake test; stationary brake test
制动距离 brake distance; brake-stopping distance; braking distance; braking length; braking path; braking way; stopping distance; absolute braking distance【铁】
制动距离测量仪 stopmeter
制动距离记录仪 stopmeter
制动均衡梁 brake balancing arm
制动均压 brake mean effective pressure
制动卡销 detent plunger
制动卡销弹簧 detent plunger spring
制动卡子 brake staple
制动开关 brake switch; tappet switch
制动开关组 braking switch group
制动壳发热灯 hot brake housing light
制动空气阀 brake air valve; brake shuttle valve
制动空气供给箱 brake air supply tank
制动空转轮 braked idler
制动空走距离 equivalent virtual braking distance
制动控杆 brake chain connecting rod
制动控制 brake control
制动控制变阻器 brake control rheostat
制动控制单元 brake control unit
制动控制阀 brake control valve; brake selector(valve)
制动控制轴 brake control shaft
制动块 arresting stop; brake block; brake clamp; brake pad; brake-shoe; brake stop; braking clamp; drag shoe; hem shoe; pad; scotch block; shoe brake; skate brake; skid shoe; slipper; stop block; stop-motion block; stop plate; wheel block
制动块间隙 brake-shoe clearance
制动块压力 brake-shoe pressure
制动块支撑 brake anchor pin
制动块支撑板 brake anchor plate
制动块支座 brake anchorage
制动拉杆 brake rod; connecting rod
制动拉杆叉 brake clevis; brake-rod yoke
制动拉杆叉形头 brake lever clevis
制动拉杆导承 brake-rod guide
制动拉杆调整器 brake-rod adjuster
制动拉杆头 brake jaw
制动拉杆销 brake connection pin
制动拉簧 brake retracting spring
制动拉索 brake cable; brake wire
制动缆索 brake cable
制动类型(缓行器) type of retardation
制动冷却器 brake cooler
制动离合器 brake clutch; brake coupling
制动力 brakeage; brake effort; brake force; brake power; brake resistance; braking effort; braking force; braking power; catching load; drag force; retarded force; retarding effort; retarding force
制动力对车辆重量的比 ratio of retardation to weight of car
制动力矩 brake moment; moment of braking force; retarding torque; brake torque; braking moment
制动力曲线 braking force curve
制动力衰减 brake-fade
制动力调节 brake force regulation
制动力系数 brake force coefficient; braking coefficient; braking force coefficient
制动力最佳分布 optimization of braking force distribution
制动励磁接触器 braking excitation contactor
制动连接器 brake coupling
制动连接器软管 brake coupling hose
制动联杆 brake linkage
制动联结系 braking bracing
制动链(条) brake chain; chain cable compressor; chain cable controller; curb chain; ground chain
制动梁 brake beam; brake cross bar; braking beam; retarding girder; skid girder
制动梁安全链 brake beam safety chain; brake safety link
制动梁安全链吊 brake beam safety chain hanger

制动梁安全托 brake beam safety guard;brake beam safety hanger;brake safety strap
制动梁吊 brake beam hanger
制动梁吊挂 brake beam suspension
制动梁吊环 brake beam adjusting hanger;parallel brake hanger
制动梁端轴 stub-end of brake beam
制动梁端轴套 brake beam stub end bush
制动梁杠杆的拉杆 brake beam and lever connection
制动梁弓形杆 brake beam tension rod;brake beam trussy rod
制动梁滑槽 brake beam bracket
制动梁滑铁 sliding chair
制动梁缓解弹簧 brake beam release spring
制动梁四点支托 brake beam four points support
制动梁托兼安全托 combination support and safety device
制动梁压缩构件 brake beam compression member
制动梁有眼螺栓 brake eye bolt
制动梁支点 brake beam fulcrum
制动梁支柱 brake beam king post;brake beam strut;brake lever clevis;brake lever fulcrum
制动梁安全托 brake beam safety
制动灵敏度 braking sensibility
制动路 brake effectiveness road
制动率 brake ratio;braking lever;braking percentage;braking rate;braking ratio;percentage of brake power;percentage of braking
制动率不足 braking ratio deficiency
制动轮 brake drum;brake pulley;brake ratchet wheel;brake wheel;braking wheel;head block
制动轮泵 wheel cylinder
制动轮导油器 brake drum oil deflector
制动轮毂 brake hub
制动轮棘爪架 pawl rack
制动螺钉 clamp screw;lock(ing) screw;stop screw
制动螺钉六分仪 clamp screw sextant
制动螺杆 brake screw
制动螺帽 jam nut;retaining nut;set nut
制动螺母 jam nut;lock nut;retaining nut;set nut;stop nut
制动螺栓 bolt with stop;chock bolt
制动螺丝 clamp screw;lock(ing) screw;set screw;stop screw
制动螺旋 backing(-up) screw;clamp(ing screw);fastening screw;fixing screw;lock(ing) screw;stop screw
制动螺旋摇把 brake screw handle
制动马力 brake horse power;brake horsepower
制动马力小时 brake horsepower-hour
制动面 brake surface
制动面材料 brake lining material
制动面摩擦片 brake lining disc
制动摩擦 drag friction
制动摩擦衬面 brake lining
制动摩擦力 braking friction
制动摩擦面积 brake friction area
制动摩擦片拉伸器 brake lining stretcher
制动摩擦片修磨机 brake doctor
制动摩阻力 braking friction
制动能 braking energy
制动能高【铁】energy head of retarder location;velocity head of a retarder location;velocity hump crest of retarder
制动能高的损耗 <驼峰溜放车辆> loss due to braking;braking loss
制动能力 braking ability;braking capacity;braking power;stopping capacity;stopping power
制动扭矩 brake torque
制动扭(矩)环 brake torque collar
制动盘 brake disc[disk];brake plate;braking disc[disk];disc brake;retarding disc[disk]
制动盘测径仪 disc brake calliper
制动喷流 braking jet
制动喷嘴 brake nozzle;reverse flow nozzle
制动片 brake block;brake disc[disk];brake-holder block;brake lining;brake-shoe;catch;detent plate stop;retaining piece;stop plate
制动片复位弹簧 shoe retracing spring
制动偏心轮 brake eccentric;brake eccentric wheel
制动平衡器 brake balancer;brake equalizer
制动平衡重 brake counterweight
制动平衡装置 brake equalizer
制动平均有效压力 brake mean effective pressure
制动坡度 braking gradient;braking slope
制动气泵 brake compressor
制动气缸 brake cylinder
制动气管路 brake line
制动气室 brake chamber
制动汽缸 brake cylinder
制动器 actuator;arrester[arrestor];arrester gear;arresting gear;brake;brake actuating system;braking device;damper;decelerator;detent;drag;hand-scotch;negative booster;stop for rotary driving pawl;stopper;thumber;trigger
制动器板 detent plate
制动器泵筒 brake cylinder
制动器操作 brake service
制动器操作轴 brake operating spindle
制动器掣子 brake pawl
制动器衬垫 brake lining pad
制动器吹风电动机 brake blower motor
制动器从衰减中恢复的试验 fade-and-recovery test
制动器带 brake band;brakeband
制动器带调整器机构 brake band adjuster mechanism
制动器的动作 brakeage
制动器的作用缸 actuating brake cylinder
制动器垫 brake disc[disk]
制动器动作机构 brake actuator
制动器短路开关 brake short switch
制动器防护装置 damper guard
制动器分离弹簧 brake release spring
制动器缸 brake cylinder
制动器杠杆系统 brake beams
制动器工作表面 braking surface
制动器挂架 brake carrier
制动器罐 brake reservoir
制动器横杆受拉条 brake beam tension member
制动器接合盘 brake ring
制动器接合器 brake ring
制动器控制 brake control
制动器控制阀 brake control valve
制动器控制管道 brake control line
制动器扩张器 brake expander
制动器拉杆弹簧 brake pull rod spring
制动器拉杆组件 brake sub-assembly
制动器拉索 brake cable
制动器连接 brake coupling
制动器联杆 brake linkage
制动器摩擦衬片 brake bush
制动器内衬 brake lining
制动器平衡锤 brake counterweight
制动器汽缸活塞 brake cylinder piston;brake piston
制动器汽缸活塞皮碗 brake cylinder piston cup
制动器汽缸夹紧器 brake cylinder clamps
制动器热衰退 brake-fade
制动器润滑脂 brake dressing
制动器渗漏 bleeding of the brake
制动器试验 brake test
制动器试验器 brake tester
制动器手柄 brake handle
制动器手杆 brake hand lever
制动器伺服活塞 brake servo piston
制动器松放弹簧 brake release spring
制动器松开位置 release brake position
制动器锁销弹簧 brake lock pin spring
制动器踏板杠杆 brake pedal arm
制动器踏板轴 brake pedal shaft
制动器弹簧 brake spring
制动器调节器 brake adjuster;brake regulator
制动器停机 brake stop
制动器托架 brake truss
制动器箱 brake housing
制动器性能 brake performance
制动器压缩空气罐 brake air reservoir
制动器液压缸 brake cylinder
制动器用软管 brake hose
制动器闸 brake
制动器闸瓦 brake-holder block;brake-shoe
制动器闸瓦调整装置 brake block adjusting gear
制动器支座 brake support
制动器直角杠杆 brake bell crank
制动器止推环 brake ring;brake thrust ring
制动器止推调节螺钉 brake thrust screw
制动器致动方法 actuating system
制动器中间轴 brake intermediate shaft
制动器轴 brake shaft
制动器轴小齿轮 fly pinion
制动牵引主控制器 braking traction master controller
制动钳工 brake mechanician
制动腔 brake chamber
制动强度 severity of braking
制动强烈程度 rate of braking
制动切断塞门 brake cutout cock
制动区 service application zone
制动曲线 braking curve
制动圈 trig loop
制动全部作业时间 full application time of braking
制动缺口 braking notch
制动热裂纹 brake burn cracks
制动热效率 brake thermal efficiency
制动软管 brake hose
制动软管标志 air brake hose label
制动软管垫圈 air brake hose coupling gasket
制动软管夹 brake hose clamp;brake hose clip
制动软管接头 air brake hose nipple;brake hose coupling head;hose nipple
制动软管卡子 air brake hose clamp
制动软管连接 brake hose joint
制动软管连接器 air brake hose coupling;gland-hand
制动软管螺纹接套 brake hose connecting nipple
制动三角木 brake scotch;trig
制动栅 barrier grid
制动设备 arrestment
制动设施 braking device
制动射流 braking jet;bremsstrahlung
制动升压器 brake booster
制动生产的力 force due to braking
制动生效时间 <从运用制动器时起至制动生效时止> brake lag
制动绳 brake rope;check band
制动失灵 brake failure
制动失效 brake failure
制动时的啸声 brake squeak
制动时段 braking period
制动时间 brake time;braking time;deceleration time;duration of braking;retardation time;stop(ping) time;time of braking
制动时铰接车架的弯折作用 jackknifing
制动实验 braking test
制动式测功器 prong brake dynamotor
制动式摩阻测定仪 skiddometer
制动事故 brake incident
制动试验 brake test;brake trial;braking test;immobilization test;retardation test;wheel locked test(ing)
制动试验台 brake tester;braking bench
制动试验用地面 braking surface
制动试验用路面 braking surface
制动室 brake chamber
制动手把 binding handle;brake lever;braking lever;hand brake handle
制动手柄 brake handle;brake lever;ground-hog <火车上的>
制动术 immobilization
制动衰减【机】fading of brake
制动栓 locking pin;retaining pin
制动水轮机 braking turbine
制动水域【船】ship stopping area;water area for braking ship;water area for stopping ship
制动顺序 retroaction sequence;retrosequence
制动司机 brake(s)man
制动速度 braking speed
制动损耗 retarding loss
制动索 lanyard stopper;rope stopper;brake rope <架空缆道>
制动锁栓 trigger bolt
制动踏 brake pedal
制动踏板 brake paddle;brake pedal;brake step;brake treadle
制动踏板传动比 pedal ratio
制动踏板阀 brake pedal valve
制动踏板力计 foot brake pedal pressure ga(u)ge
制动踏板轴衬 brake pedal bushing
制动弹簧 brake spring;tripping spring
制动弹簧垫圈 retaining snap ring
制动特性 braking characteristic
制动特性表 braking chart
制动提升的磁铁 brake lifting magnet
制动蹄 brake-shoe
制动蹄复位弹簧 shoe spring
制动蹄回位弹簧 shoe spring
制动蹄块直径 shoe diameter
制动蹄(摩擦)衬面 brake-shoe lining
制动蹄摩擦片快速黏[粘]结器 minute bonder
制动蹄磨床 brake-shoe grinder
制动蹄片接触压力 shoe contact pressure
制动蹄片磨削机 brake-shoe grinder
制动铁带 brake iron
制动铁鞋 brake block;brake-shoe;brake slipper;drag shoe;hem shoe;shoe brake;skate;skate brake;skid;skid shoe;slide shoe;slipper;wheel rim wedge
制动铁鞋安置机 skate placing machine
制动铁鞋夹 skate clip
制动调节 brake set(ting)
制动调节阀 adjust valve;brake metering valve

制动调节机构 brake adjusting gear
制动调节模块 braking chopper module
制动调节器 brake governor; brake regulator
制动调整的螺母 brake adjusting nut
制动调整的压力油缸 brake adjusting jack
制动调整楔 brake adjusting wedge
制动停车装置 arresting gear
制动透平 brake turbine
制动透平空气循环 brake turbine air cycle
制动凸轮 brake cam; brake toggle
制动凸轮轴 brake camshaft
制动凸轮轴环 brake camshaft collar
制动推杆 brake pushrod
制动推力 retrothrust
制动推力器 thrust reverser
制动瓦 brake scotch; brake-shoe; clamping shoe; shoe brake
制动微机控制单元 electronic brake control unit
制动位 brake position; retarder location
制动位置 application position; braking position; retarder location; retarding position <缓行器>
制动温度试验 thermal brake test
制动涡轮 brake turbine
制动误差 clamping error
制动系杆 braked tie rod
制动系数 coefficient of braking; retardation factor
制动系统 brake assembly; brake staff; brake system; braking system; retrosystem
制动系统管路 brake line; brake pipe
制动系统试验 parking brake verification test
制动系统压力 brake system pressure
制动系统最低性能准则 minimum performance criterion for brake system
制动现象 brake phenomenon
制动线圈 amort winding; brake coil; restraining coil
制动箱 banking pin; brake housing; detent plug; retaining pin; shotpin; stop
制动销（钉）retaining pin; shotpin; stop pin; locking pin
制动销套 brake pin bushing
制动效果 brake effect; braking effect; damping effect
制动效力 brake effect; braking effect; damping effect
制动效率 brake efficiency; braking efficiency; retardation efficiency
制动效能 brake effect; braking effect; damping effect
制动效应 brake effect; braking effect; damping effect; trigger action
制动楔 brake wedge; lock; scotch; wedge brake
制动斜坡 jinny
制动信号 brake signal; speed slackening signal
制动信号开关 braking signal switch
制动性能 braking performance; stopping performance
制动蓄压器 brake accumulator
制动靴片回位弹簧 brake-shoe release spring
制动压力 brake pressure
制动压力阀 brake pressure valve
制动压力机 brake pressure ga(u)ge
制动压力线 brake pressure line
制动延时 <踏下制动器至实际开始制动的时间> brake lag
制动延时距离 brake lag distance
制动演习 stopping maneuver
制动摇臂 brake rocker arm
制动叶片 brake vane
制动曳力 brake pull
制动液 brake fluid; brake oil
制动液管 brake pipe
制动液箱 brake fluid header
制动液压缸 brake master cylinder
制动液压管 brake pipe
制动仪 <测定路面抗滑力> skiddometer
制动引擎 retroengine
制动应力 braking stress
制动用的风缸充气 filling of brake reservoirs
制动用喷气发动机 reverse jato
制动用砂 track sand
制动油 brake oil; hydraulic brake oil
制动油缸 brake cylinder
制动油管路 brake line
制动油路排气装置 brake bleeder
制动有效均压 brake mean effective pressure
制动有效平均压力 brake mean effective pressure
制动元件 braking element
制动员 car catcher; brake(s)man <俚语>; pinhead <俚语>
制动员室【铁】brakeman's cabin
制动圆板 brake disc[disk]
制动圆鼓 brake drum
制动圆盘 brake disc [disk]; brake puck; braking disc[disk]
制动圆筒 brake cylinder; brake drum
制动运行 running under braking; running while braking
制动增力装置 brake assistor
制动闸 brake band; brake gate; brake lock; damper brake; disc brake; holding brake
制动闸边皮 stitch brake lining
制动闸衬 brake lining
制动闸的控制杆杠 brake lock control lever
制动闸集尘器 brake dust collector
制动闸块压力 shoe pressure force
制动闸调整 brake adjustment
制动闸压力积储器 brake accumulator
制动胀圈 brake sealing cup
制动遮盖物 brake housing
制动真空加力泵 brake vacuum booster cylinder
制动证实指示灯 brake proving indicator
制动支管 brake branch pipe; brake pipe branch
制动支架 back brake support
制动支索 jumper stay
制动值 brake value
制动止点 brake stop
制动指令 braking instruction
制动指数 braking index
制动质量 braked mass
制动重力的百分率 braked weight percentage
制动重量 brake weight
制动重量转换 brake changeover weight
制动周期 braking period
制动轴 brake axle; brake shaft; brake spindle; braking axle
制动轴架 brake shaft bracket
制动轴数 number of axles braked
制动轴轴承 brake shaft bearing
制动肘节 brake toggle
制动主缸盖 brake master cylinder cover
制动主缸加油器塞 brake master cylinder filler plug
制动主缸总成 brake master cylinder assembly
制动主管 braking main; main brake pipe
制动主管滤尘器 brake pipe air strainer
制动助力器 brake booster
制动爪 arrestment; brake jaw; brake latch; brake ratchet; holding detent; lathe dog; locking dog; locking pawl; paul; pawl; retaining pawl; stop pawl
制动爪安装板 dog plate
制动爪板 pawl plate
制动爪装配图 dog chart
制动转换 brake changeover
制动转换器箱体 case for changeover braking device
制动转矩 brake torque; braking moment; braking torque; damping torque; retarding torque; stalling torque
制动转子力矩 locked rotor torque
制动转子试验 blocked rotor rest
制动装置 arrester[arrestor]; arresting device; arresting gear; arrestment; backset; brake apparatus; brake device; brake equipment; brake-gear; brake rigging; braking device; cataract; catcher; clamping device; damper gear; damping device; drag brake; dragging equipment; dynamic(al) braking; hold-back; locking device; motion cut equipment; skidding device; stopping device
制动装置复位弹簧 brake-gear return spring
制动装置效率 brake-rigging efficiency
制动装置支承 brake-rigging support
制动装置支架 brake-gear support
制动状态 braking mode; braking position; on position
制动锥(体) brake cone; drag cone
制动总泵 master cylinder
制动总泵缸 main braking cylinder
制动阻力 brake drag; brake resistance; braking resistance
制动作用 brake action; brakeage; brake application; braking action; braking effect; cushioning action; damping action; check action <门窗的>
制动作用分割装置 <真空制动> dividing attachment
制动作用面 braking surface
制动作用时间 brake application time
制洞 cavity preparation
制度 institution; rules and regulations; system
制度化 institutionalization
制度经济学 institutional economics
制度台日数 machine days in accordance with working days
制度限制 institutional constraint
制度影响 institutional influence
制度与程序 systems and procedures
制度约束 institutional control
制断面图 profiling
制锻模工 die maker
制锻模铣床 profiler
制堆肥 composting
制法 recipe
制帆缝法 sailmaker's seaming
制帆工具 sailmaker's tool
制帆术 sailmaking
制枋 slabbing
制粉厂 miller
制粉机 cornmill
制粉设备 powder manufacturing apparatus
制服 garb; habiliments; uniform; vanquish
制服垫款 payment incurred for uniform
制服官员 enforcing authority
制服帽 uniform cap
制服上衣 blouse
制干草块 wafering hay
制钢筋混凝土桩 prefabricated reinforced concrete pile
制钢模法 die making
制高点 commanding elevation; commanding height; commanding point; critical elevation; dispatching point; dominating point; highest elevation; key point; key position; maximum elevation; summit of height
制高高程 critical elevation
制革厂 curriery; tannery; tanning factory
制革厂废料 tannery waste
制革厂废水 tannery waste(water); tanyard waste(water)
制革厂废水处理 tanyard wastewater treatment
制革厂污泥 tanyard sludge
制革场 tan yard
制革法 tanning
制革废水 leather industry wastes; leather waste; leather wastewater; tanning wastewater
制革废物 leather waste
制革工人 currier
制革工业废水 leather industry sewage; tannery waste; tannery wastewater
制革工业废水处理 treatment of tannery wastes
制革设备 tanning equipment
制革业 curriery
制革综合废水 synthetically tan wastewater
制管 tubing
制管白黏[粘]土 pipe clay
制管薄板 duct sheet
制管厂 pipe mill; pipe plant; tube(-rolling) mill
制管钢板 pipe plate; skelp; steel pipe plate
制管机 pipe-making machine; pipe section machine; tube-drawing machine; tube extruder; tube machine; tube mill; tuber; tubing machine
制管土 pipe clay; terra alba
制罐头工业废水 cannery waste
制光装置 dimmer
制海权 command of the sea; maritime power; sea power; sea supremacy; thalassocracy
制合金 alloying
制衡原则 check and balance
制花样线脚模 wadded
制滑卡环 mousing shackle
制滑器 trigger
制环机 ring forming machine
制灰砂砖机 lime sand brick machine; sand-lime brick machine
制混凝土板的细木工 joiner for concrete moulds
制混凝土模板用的木材 concrete-timber
制火线 hold line
制际串音 intersystem crosstalk
制剂 drug grade; formulation; preparation; prepn
制剂分析 analysis of pharmaceutical dosage forms
制尖 metal pointing; pointing
制碱工业 basic industry
制碱(碳酸钠)废液 Solvay liquor
制碱用灰岩 limestone for alkali industry
制浆 pulp(ing); slurrying

制浆材 pulpwood
制浆材抓钩 pulp hook
制浆槽 pulping tank
制浆厂 pulp mill
制浆车间 slip house
制浆方法 pulping process
制浆过程 pulping process
制浆机 pulper
制浆木材 pulpwood
制浆漂白剂 pulp bleaching agent
制浆设备 slurry equipment
制浆系统 <造纸的> pulping system
制浆造纸废水 pulping and paper-making effluent
制浆中段废水 pulp making intermediate wastewater
制胶版 offset
制胶废水 gum waste
制胶合板用模板 Plyform
制胶囊剂 encapsulant
制金属模工人 scratcher
制景人员 decorator
制镜 mirror making
制镜合金 speculum[复 specula]
制镜用板玻璃 silvering plate glass
制酒厂 wine producing factory
制酒厂废水 winery wastewater;winery waste
制菌剂 anti-bacterial agent;bacteriostat
制拷贝 exemplify
制空权 air supremacy
制块 clamp dog
制块材机 block-making machine
制块机 block-making machine;cuber
制蜡过程 wax manufacturing
制蜡器 wax extractor
制酪厂废物 dairy plant waste
制酪场 dairy;dairy farm
制酪场肥料 dairy manure
制酪场排污 dairy-farm slurry
制酪废水 dairy wastewater
制酪工业 dairy industry
制酪工业废水 dairy industry wastewater
制冷 cold production
制冷安装工程师 refrigeration erection engineer
制冷部件 refrigeration component
制冷产量 refrigerating output
制冷厂 cooling plant; refrigerating plant;refrigeration plant
制冷单元 refrigeration unit
制冷的 cryogenic;freezing;frigorific;refrigerating;refrigeratory
制冷电路 refrigerating circuit
制冷吨位 refrigeration tonnage
制冷负荷 refrigeration duty; refrigeration load; refrigeration requirement
制冷工程 refrigerating engineering;refrigerating work;refrigeration engineering
制冷工程承包者 refrigeration contractor
制冷工程师 refrigeration engineer
制冷工管道 refrigeration piping
制冷工业 cooling industry;refrigerating industry
制冷工作者 refrigerationist
制冷供暖热泵 cooling and heating heat pump
制冷构件 refrigeration component
制冷管道 refrigeration piping
制冷管路 refrigeration pipe line
制冷管网 pipe cooling grid
制冷管组 cooling battery;refrigerating battery
制冷过程 cooling process
制冷回路 refrigerating circuit
制冷机 cold generator;cryocooler;freezer;refrigerating engine;refrigerating machine;refrigeration apparatus;refrigerator;cooling plant
制冷机房 refrigerating plant room;refrigerating station;refrigeration machinery building
制冷机均压箱 accumulator
制冷机器 refrigeration machine
制冷机械 refrigeration machinery
制冷机油 refrigerating machine oil
制冷机组 factory assembled system of refrigeration machine;refrigerating unit
制冷技工 refrigeration mechanic
制冷技师 refrigeration mechanic
制冷技术员 refrigeration technician
制冷剂 cold producing medium;cooling material; cryogen; freezing agent;kinetic chemicals;primary refrigerant; refrigerant; refrigerating medium
制冷剂侧 refrigerant side
制冷剂充罐 refrigerant charge
制冷剂储液器 liquid refrigerant receiver
制冷剂导管 refrigerant vessel
制冷剂分配器 refrigerant distributor
制冷剂观察玻璃 refrigerant sight glass
制冷剂管路 refrigerant line;refrigerant pipe
制冷剂过滤器 refrigerant filter
制冷剂换向 refrigerant change-over
制冷剂回流速度 return refrigerant velocity
制冷剂计量装置 refrigerant metering device
制冷剂控制器 refrigerant controller
制冷剂流量控制 refrigerant flow control
制冷剂流速 refrigerant velocity
制冷剂排出速度 discharge refrigerant velocity
制冷剂盘管 refrigerant coil
制冷剂瓶 refrigerant cylinder
制冷剂热回收盘管 refrigerant heat recovery coil
制冷剂容槽 refrigerant vessel
制冷剂容量 refrigerant charge
制冷剂软管 refrigerant hose
制冷剂送出速度 delivery line refrigerant velocity
制冷剂温度 refrigerant temperature
制冷剂吸入速度 suction line refrigerant velocity
制冷剂泄漏 refrigerant leakage
制冷剂循环 refrigerant cycle
制冷剂循环率 refrigerant circulating rate
制冷剂压力容器 refrigerant pressure vessel
制冷剂预热器 refrigerant heater
制冷剂蒸发器 refrigerant evaporator
制冷剂蒸气 refrigerant vapo(u)r
制冷剂贮液器 liquid refrigerant receiver
制冷季节 cooling season
制冷介质 refrigerant medium;refrigerating medium
制冷冷凝器 refrigerant condenser;refrigeration condenser
制冷冷却器 refrigeration cooler
制冷量 refrigerant capacity;refrigerating capacity; refrigerating effect;refrigeration capacity
制冷能力 duty;refrigerant capacity;refrigerating capacity;refrigerating effect;refrigeration capacity
制冷凝汽器 refrigeration condenser
制冷盘管 refrigeration coil
制冷气体 refrigerant gas
制冷器 cryostat;freezer;refrigerator
制冷器鼓筒 freezer drum
制冷设备 cooling facility; refrigerating installation;refrigerating plant;refrigerating unit; refrigeration apparatus;refrigeration plant
制冷式杜瓦瓶 refrigerating Dewar vessel
制冷室通风管堵头 breather plug
制冷水 chilled water
制冷维修工程师 refrigeration service engineer
制冷维修技工 refrigeration service man
制冷温度 refrigerating temperature
制冷物质 ice mass
制冷吸入速度 return refrigerant velocity
制冷系数 coefficient of performance;performance energy ratio;performance factor
制冷系统 cryogenic system;refrigerating system;refrigeration system
制冷系统等级 class of refrigerating system
制冷系统高压边 high-pressure side;high side
制冷系统配管 refrigerant piping
制冷系统性能系数 refrigerating system performance factor
制冷系统制冷量 refrigerating system capacity
制冷效果 refrigerating effect
制冷效率 refrigerating efficiency
制冷效应 refrigerating effect;refrigeration effect
制冷性能系数 refrigerating coefficient of performance
制冷学 refrigeration
制冷循环 refrigerating circuit;refrigerating cycle;refrigeration cycle
制冷压气机 refrigeration compressor
制冷压缩机 cold compressor;refrigerant compressor; refrigerating compressor;refrigeration compressor
制冷压缩机润滑剂 refrigerating compressor lubricant;refrigerator compressor lubricant
制冷压缩机组 refrigerant compressor unit;refrigeration compressor unit
制冷压缩装置 refrigerant compressor unit
制冷液 coolant; refrigerant fluid; refrigerating fluid
制冷原理图 circuit diagram of refrigeration
制冷装置 cooler;refrigerating installation;refrigerating plant;refrigerating unit; refrigeration apparatus;refrigeration plant
制冷装置泵 refrigerating circulating pump
制冷装置冷凝器泵 refrigerating condenser pump
制冷总厂 central refrigerating plant
制立面图和剖面图 drawing of elevations and sections
制粒 nodulizing;palletizing;pelletization[pelletisation]; pelletize; shotting
制粒法 granulation
制粒机 briquet(te); cuber; cubing machine; grainer; granulating machine; granulator; nodulizer; pelleter;pelletizer
制粒盘 balling disc[disk]
制粒器 granulating screen
制炼厂 process plant
制链机 chain making machine
制链器 chain stopper; anchor chain stopper【船】
制漏嘴压机 tip press
制铝工业 alumin(i)um industry
制轮 ratch
制轮工厂 wheelwright
制轮木 sprag
制轮器 skid-pan;slipper;trigger
制轮器铁鞋 skid
制轮凸轮 trigger cam
制轮楔 linchpin[lynchpin]
制轮业 wheelwright
制轮业者 wheelwright
制轮爪 handle pawl; jumper; pawl <防齿轮倒转的爪>
制螺钉机 screw machine;screw making machine
制螺母及螺栓头机 screw shaving machine
制螺栓机 bolt-making machine
制螺栓螺帽机 bolt and nut making machine
制螺纹 threading
制螺旋机 screw chasing machine
制螺旋钻机的模 die of auger machine
制麻袋 hemp sacking
制麦芽糖废水 malting wastewater
制毛面 mottling
制毛坯 roughcast
制铆钉机 rivet machine;rivet making machine
制煤粉装置 powdered fuel plant
制煤气 coal gas
制门窗用板材 factory planks
制门器 door buffer;door stop
制门器式应变计 <用于测定岩体双向应力的> doorstopper cell
制皿机 plate-machine
制模 all-core mo(u)lding; mo(u)ld construction; mo(u)lding; replication
制模班组 forming crew;forming gang;forming party;forming team
制模板 pallet;pallet
制模板材 pattern lumber
制模薄板 formwork sheet(ing)
制模车间 die shop; model(l)ing shop;pattern-maker's shop;pattern room;pattern shop
制模尺 pattern-maker's rule
制模叠层木料 laminated pattern lumber
制模工 die sinker; mo(u)lder; mo(u)ld mark
制模工厂 model plant
制模工具 form aid;formwork aid
制模工人 pattern maker;patterner
制模工作 formwork;pattern-making
制模合金 pattern metal
制模河段 modeled reach
制模机 moling machine
制模剂 form aid; forms agent; formwork agent; formwork aid; formwork paste
制模锯 pattern-maker's saw
制模壳的板 shuttering panel;shuttering plate
制模壳的薄板 shuttering sheet
制模蜡 forms wax;formwork wax
制模密封剂 forms sealer; formwork sealer
制模木工具 model(l)ing tool
制模黏[粘]土 model(l)er's clay;model(l)ing clay
制模石膏 pattern plaster
制模收缩 pattern-maker's contraction
制模树脂 pattern resin
制模型 mock-up
制模压力 forming pressure
制模用泥子 mo(u)ld compound
制模者 pattern maker
制模周期 mo(u)lding cycle

制膜器 film applicator(blade)
制(木)材机械 lumbering machine
制木线(脚)机 wood mo(u)lding machine
制木屑机 wood wedge cutter
制木屑器 wood wedge cutter
制内胎机 tuber
制奶废水 milk waste
制逆 antireverse
制坯 biscuit;burnish
制坯板 stillage
制坯工 blanker
制坯工序 preforming
制坯模膛 blocker
制片 flaking
制片机 flaking mill; pelleter; pellet mill; tablet compressing machine; tablet(t)ing press
制片人 producer
制片术 kinematograph
制品 article; artifact; facture; goods; products;ware
制品尺寸 product size
制品固定板 cramping table
制品合格率 job efficiency
制品性能 product property
制品用途 product-use
制瓶玻璃 bottle glass
制瓶机 bottle-making machine
制漆过程中的健康危害 health hazards in paint making
制气炉 producer
制气甑 gas making retort
制气周期 make-run
制铅(工)厂 lead works
制氢站 hydrogen generation station
制清漆豆油 varnish soybean oil
制球车床 ball lathe
制球机 briquetting pressing; granulater;marble machine
制球机床 ball forming machine
制取封口蜡的木材 sealing-wax wood
制取火漆的木材 sealing-wax wood
制圈【电】reactor
制热系数 performance factor
制乳厂 dairy
制绳厂 rope works
制绳场地 rope yard
制绳机 rope machine;strander;stranding machine
制绳狭长走道 rope walk
制绳纤维 cordage fiber[fibre]
制时间表 scheduling
制式运输工具 organic transport
制试块机 briquet(te)
制熟铁坯 nobbling
制刷纤维 brush fiber
制栓杆 lever tumbler
制栓杆锁 lever-tumbler lock
制霜 making drugs into frostlike powder
制水舱壁 wash bulkhead
制水泥用的石灰岩 cement rock
制水泥用矿渣 cement slag
制丝 throwing
制丝厂 filature;reeling mill
制丝工程 raw silk production process
制速器 check piece
制酸 relieving hyperacidity
制酸厂 acid plant
制酸工业 acid industry
制酸剂 antacid;anti-acid
制榫机 dovetailer; dovetailing machine; dovetail machine; matcher; mortising machine; tenon-cutting machine; tenoner; tenoning machine; tenon-making machine; tonguing and grooving machine
制榫开槽两用机 tenoning and trenching machine
制梭 shuttle checking
制锁 lockmaking
制锁材料 locking material
制锁工 locksmith
制膛线 rifling
制糖厂 sugar refinery
制糖厂废水 sugar factory waste(water);sugar mill effluent
制糖厂污水 sugar mill effluent
制糖废水 sugar effluent
制糖废液稳定土工艺(或技术) molasses stabilization
制糖工业 sugar industry
制糖工业废水 sugar industry wastewater
制糖工业污水 iron manufacture industry sewage
制糖作物 sugar-producing crop
制陶 potting
制陶瓷型材料 ceramic mo(u)lding material
制陶工艺 pottery
制陶管土 pipe clay
制陶设备 pottery making equipment
制陶术 ceramics
制条机 rod machine;set frame;slivering machine
制铁坯 knobbling
制铁丸 shotting
制铜技工 brazier
制桶板材 heading;stave wood
制桶板用的短木材 stave bolt
制桶工人 cooper
制桶业 cooperage
制筒铁皮 tank sheet-iron
制凸凹榫接机 rabbeting machine
制图 charting; drafting; draughting; drawing;drawing practice;drawing up;draw to scale;graphic(al) plot; map making; mapping; mapping drawing; plotting; protraction; protract <用量角器或比例尺>
制图 CAD 环境变量 ACAD environment variable
制图板 drafting board; draughting board;drawing board;trestle board
制图比例尺 drawing scale; plotting scale
制图笔 drawing pen
制图编号 drawing number
制图编辑系统 cartographic(al) editing system
制图标准 drafting standard; mapping standard
制图标准规范 drawing standard specifications
制图部门 chart-making institution
制图材料 graphic(al) medium
制图参数 mapping parameter
制图操作 cartographic(al) manipulation
制图测量 <地图> cartographic(al) survey(ing)
制图常数 chart constant
制图衬影 hatching
制图成果 cartographic(al) product; map production
制图程序库 cartographic(al) program(me) bank
制图程序组件 graphics software package
制图尺 drafting scale;drawing scale
制图处理 cartographic(al) manipulation
制图单位 cartographic(al) unit; organization of mapping
制图队 cartographic(al) unit; chart-making unit;map-making unit
制图对象 map subject
制图法 cartography; chartography; graphics
制图方法 cartographic(al) technique; mapping method
制图分级 cartographic(al) hierarchy
制图分类 cartographic(al) classification
制图概率 cartographic(al) probability
制图概论 map study
制图钢笔 drafting pen;drawing pen
制图格网 cartographic(al) grid; graticule line;map graticule;map grid
制图格网数字注记 ladder grid numbers
制图工程师 cartographic(al) engineer
制图工程学 cartographic(al) engineering
制图工序 fragmented parts of the mapping profession; sequence of cartography
制图工作 cartographic(al) operation; chart work
制图功能键盘 cartographic(al) function key
制图规程 drafting standard
制图规范 specification of cartography
制图规划 cartographic(al) planning
制图规则 cartographic(al) convention
制图过程 map-making process
制图机 draft(ing) machine; draught(ing) machine; drawing machine; graphic(al) plotter;graph plotter
制图机构 chart-making institution; map-making organization
制图机械 drafter;draughter
制图计划 mapping schedule
制图技巧 cartographer's craft;cartographic(al) craft
制图技术 cartographic(al) technique; drawing technique; draftsmanship; draughtsmanship
制图监控软件 cartographic(al) monitor software
制图简化 cartographic(al) simplification
制图教学 cartographic(al) training
制图精度 accuracy factor of mapping; cartographic(al) accuracy; mapping accuracy
制图局 map-making office
制图均匀性 homogeneity of drafting
制图科 map section
制图夸大 cartographic(al) exaggeration
制图雷达设备 cartographic(al) radar equipment
制图类别 type of drawings
制图美术式记录系统 cartographic(al) and graphic(al) arts typesetting system
制图蒙片 cartographic(al) mask
制图明线 object line
制图模板 drafting template
制图培训 cartographic(al) training
制图剖面 cutting plane
制图器 chartometer
制图区(域) covered surface; ground cover(age)
制图区域示意图 coverage diagram
制图人 draftman;drawer;mapper
制图日期 mapping date
制图软件 graphics software
制图设备 cartographic(al) installation;drawing device;graphics face; graphics facility
制图生产 map production
制图室 cartographic(al) house; cartographic(al) room; cartographic(al) unit; chart-making unit; drafting department; drafting office; drafting room; draughting scale; drawing office; map-making room; map-making unit
制图手册 draft room manual
制图算法 mapping algorithm
制图台 drafting table; drawing table; layout table
制图网格 cartographic(al) grid; map graticule;map grid
制图网格板 graticule plate
制图网格展绘模片 graticule template
制图文件 cartographic(al) file
制图误差 cartographic(al) error
制图系统 mapping system
制图信息库 cartographic(al) information storage
制图虚折线 break line
制图选取 cartographic(al) selection
制图学 cartography;graphics
制图学领域 cartographic(al) field
制图学术语 cartographic(al) terminology
制图学投影 cartographic(al) projection
制图仪 diagraph;drawing equipment; graphic(al) plotter
制图仪器 drafting instrument;graphic(al) instrument;graph plotter
制图影线 section line
制图硬件 graphics hardware
制图用刮刀 drawing knife
制图用光泽纸板 drawing bristol
制图语言 graphics language
制图员 cartographer; cartographic(al) draftsman;cartographic(al) draughtsman; chartographer; chartographer drawer;designer;detailer;draftsman; draughtsman; mapmaker; mapper; mappist
制图原则 mapping standard
制图圆规 drawing compasses
制图照相机 mapping camera
制图者 delineator; describer; draughter;drawer;drawer/compiler;mapper;mappist
制图纸 drawing paper;kent
制图指引线 leader
制图质量 draftsmanship; draughtsmanship
制图质量技术规程 instruction in draughtmanship
制图中心 mapping center[centre]
制图专家系统 cartographic(al) expert system
制图桌 drawing bench;drawing desk
制图资料 cartographic(al) documents; cartographic(al) record; source material of mapping
制图字体 atlas type;cartographic font; map type
制图自动化 automated cartography
制图自动化研究小组 automated cartography development team
制图综合 cartographic(al) generalization; cartographic(al) generation; map cartographics
制图综合比例尺 generalization scale
制图综合标准 comprehensive cartographic(al) standard
制图组长 draftsman chief
制土分民之律 population carrying capacity of land according to quality
制土坯砖用土 adobe soil
制团 briquet(te); briquetting; kneading
制团机 briquet(te); briquetting machine
制退机 recoil absorber;recoil apparatus
制退筒 recoil cylinder
制瓦 forming of tiles

制瓦厂 building tile factory;building tile plant;tilery
制瓦机 clay plate press
制瓦模 clay plate mo(u)ld
制瓦木段 shingle bolt
制瓦碎片 spalt
制外单位 off-system unit;unit outside system
制丸 pelletization[pelletisation]
制丸设备 pelletizer
制网铁丝 barbed wire;barbwire
制屋面卷材的沥青 roofing saturant
制锡箔 tin foiling
制纤维作物 fiber-producing crop
制线厂 wire work
制线机 twine machine
制线脚针叶材 mo(u)lding stock
制箱木料 boxing
制销 cotter
制销成本 cost of make and sell
制销联轴节 cottered joint
制芯 core;coremaking
制芯车床 core turning lathe
制芯工段 cored department;core shop
制芯工人 cotemaker
制芯工作台 coremaker's bench
制芯机床 core turning lathe
制芯热合 hot box
制芯台 core bench
制芯铁砂床 grid bed
制芯油 core oil
制芯作业 core-making operation
制锌版 zincograph
制锌版术 zincography
制信封机 envelope machine
制型芯机 core-making machine;core mo(u)lding machine
制型纸 carton-pierre;papier-mâche
制压 neutralize
制压箱 break pressure tank
制烟废水 tobacco waste(water)
制烟业 tobacco manufacture
制盐 salt manufacture
制盐工场 saline
制氧车厂 oxygen-producing plant
制氧车间 oxygen generating plant;oxygen-making plant; oxygen plant; oxygen-producing plant
制氧机 oxygenerator
制氧设备 oxygen manufacture equipment
制样板 mock-up
制药 pharmacy
制药厂 pharmaceutical factory; pharmaceutical plant
制药废料 pharmaceutical waste
制药废水 pharmaceutical wastewater
制药工业 pharmaceutical industry
制药工业废水 wastewater from pharmaceutical industry
制药工艺学 pharmaceutical technology
制药化学 pharmaceutical chemistry
制药学 pharmaceutics
制衣业 garment industry
制阴模 die sinking
制饮料作物 beverage crop
制印 map process and mapping
制印原图 final original
制油厂 oil mill
制油工 oilman
制雨器 rainmaker
制约 confinement
制约长度 constraint length
制约超车视距 restricted passing sight distance
制约沉降 hindered settling
制约程度 degree of restraint
制约式继电器 restraint relay
制约式声辐射器 constrained acoustic-(al)radiator
制约停车视距 restricted stopping sight distance
制约性 conditionality
制约与平衡 check and balance
制约振荡 constrained oscillation
制约振动 constrained vibration
制约装置 restraint device
制皂废水 soap-making wastewater
制皂片辊轧机 soap flaking rolls
制造 fabricate; machining; making; production
制造泵轴的钢材 steel for sucker rods
制造标准 manufacturer's standard
制造标准手册 manufacturing standards manual
制造部件清单 manufacturing parts list
制造部门 manufacturing department
制造操作 manufacturing operation
制造长度<电缆的> completed length
制造厂 builder;factory;manufactory; manufacturing plant; manufacturing works;producer plant
制造厂保证 manufacturer's guarantee
制造厂编号 manufacturer's serial number
制造厂标记 stamp of the maker
制造厂标志 maker's mark
制造厂标准 manufacturer's standard
制造厂的部分资料报告 manufacturer's partial data report
制造厂的测定 maker test
制造厂的数据报告 manufacturer's data report
制造厂的数据报告格式 form of manufacturer's data report
制造厂的印记 manufacturer's stamp
制造厂家 maker;manufacturer
制造厂家标记 manufacturer's mark
制造厂家的资格声明 manufacturer's qualification statement
制造厂家试验 manufacturer's trial
制造厂家证明书 certificate of manufacturer
制造厂检测 maker test
制造厂检查员 manufacturer's examiner
制造厂名 manufacturer's name;name of manufacture
制造厂名代号 sign of manufacturer
制造厂名牌 name of manufacturing plant
制造厂铭牌 builder's name plate; name-plate of manufacturing plant
制造厂棚 manufacturing hangar
制造厂商 builder
制造厂商标 maker's mark
制造厂说明书 maker instruction
制造厂提供的工作曲线 manufacturer's working curve
制造厂型号 manufacturer's model number
制造厂样本 manufacturer's brochures
制造厂印记 stamp of the maker
制造厂证明书 certificate of manufacturer
制造厂证书 maker's certificate
制造厂装配图 manufacturer's fabrication drawing
制造车床 produce lathe
制造车间 manufacturing shop
制造成本 cost of manufacture;factory cost; manufacturing cost; production cost
制造成本报告书 factory cost report
制造成本表 manufacturing statement
制造成本单 factory cost sheet
制造成本核算 manufacturing costing
制造成本预算 manufacturing cost budget
制造承重胶合板用的单板 structural veneer
制造程序 fabrication schedule; fabrication sequence; maker; manufacturing procedure; manufacturing process; manufacturing program(me)
制造尺寸 fabricating dimension;manufacturing dimension;manufactured size
制造大理石花纹 marbleize
制造淡水船 desalination ship
制造岛<指一组独立完整的制造体系> production platform
制造的 manufactural; manufactured; manufacturing
制造的结构产品 manufactured structural product
制造的可能性 manufacturing feasibility
制造电缆的机械 cable making machinery
制造锻铁用灰口铁 grey forge pig
制造法 making
制造方法 manufacturing method;method of manufacture;methods of fabrication;manufacturing process
制造费用 burden; fabrication cost; production cost; cost of production; fabricating cost; manufacturing burden; manufacturing cost; manufacturing expenses
制造费用表 burden statement
制造费用差异 factory expense variance; manufacturing expense variance
制造费用分户总账 manufacturing expense ledger
制造费用分类账 factory expense ledger;manufacturing expense ledger
制造费用分配 burden apportionment; manufacturing expense distribution
制造费用分配的人工小时法 labo(u)r hour basis for overhead application
制造费用分配基础 basis for application of overhead
制造费用分析 factory expense analysis
制造费用耗费差异 manufacturing overhead spending variance
制造费用率的机器工时法 machine-hour method based a manufacturing expense rate
制造费用明细表 schedule of manufacturing overhead
制造费用数量差异 manufacturing overhead volume variance
制造费用通知单 manufacturing expense order
制造费用预算 manufacturing expense budget
制造费用预算报告 manufacturing expense budget report
制造更改说明 manufacturing change note
制造钢结构构件工程师 constructional engineer
制造工厂 manufacture plant; manufacturing works
制造工程 process engineering
制造工程师 production engineer
制造工程学 industrial engineering; manufacturing engineering
制造工序 manufacturing procedure; manufacturing process
制造工业 manufacture;process industry;manufacturing industry
制造工艺 fabricating technology;fabrication craft; fabrication process; fabrication technology; manufacture engineering; manufacture process; manufacturing engineering;manufacturing process; manufacturing technique; processing technique; workmanship
制造工艺保证 manufacturing practice guarantee
制造工艺规程 mill sheet
制造工艺卡 fabricating dispatch
制造工艺培训 manufacturing discipline
制造公差 fabrication tolerance; machining allowance;manufacture tolerance;manufacturing tolerance
制造公司 fabricator; manufacturing company; manufacturing establishment
制造管子的喇叭口 belling
制造过程 course of manufacture;fabricating process; fabrication procedure;fabrication process; manufacturing process;process of manufacture
制造过程表 manufacturing flow chart
制造好的 fabricated
制造或使用机械的技术 enginery
制造机 maker;producer
制造机构 maker
制造极限 manufacturing limit
制造计划改变 manufacturing planning change
制造技术 fabricating technology;manufacturing engineering; manufacturing technology;process engineering;techniques of manufacture
制造技术规范 fabrication specification
制造技术指令 manufacturing technical directive
制造加工 fabrication
制造加工废水 wastewater from manufacturing process(ing)
制造加工废物 waste from manufacturing processing
制造间接费 factory burden
制造间接费分摊余缺 over-and-under absorbed overhead
制造间接费用 factory overhead;manufacturing overhead expense
制造胶合板原胶 primary gluing
制造进度表 time schedule(chart)
制造精度 accuracy of manufacture; accuracy of manufacturing; manufacture accuracy;manufacturing accuracy
制造精密设备房间 clean room
制造可锻铸铁的生铁 malleable(pig) iron
制造块体场地<防波堤用> casting yard
制造缆索通道的机器 cable duct making machine
制造肋材的弯木 futtock
制造连接板用异形钢材 splice bar
制造量规 manufacturing ga(u)ge
制造路线表 routing
制造命令 manufacturing directive
制造模型 mock;mock-up;model(l)-ing
制造木商品的 wood working
制造木制品的 wood working
制造目的 manufacturing purpose
制造年份 year built
制造泡沫隔热层 site-foamed insulation
制造品 manufactured article
制造企业 manufacturing concern
制造铅珠的塔 shot tower
制造区间 manufacturing bay
制造拳石的石料 blockage stone
制造拳石或小方石的石料 block and

cross bond
制造缺陷 fabricating defect; manufacturing failure; manufacturing imperfection
制造人 fabricant
制造日期 date of manufacture
制造日期标记 date mark
制造商 manufacturer
制造商厂牌 manufacturer's brand
制造商出具的质量证明 quality certificate issued by manufacturers
制造商担保 guarantee by manufacturer
制造商的分销机构 manufacturer's sales branches
制造商的新订货 manufacturer's new orders
制造商行 manufacturing firm
制造商品质证明书 manufacturer's certificate of quality
制造商数量证明书 manufacturer's certificate of quantity
制造商验收试验压力 <在安装前对管子及其配件进行的> works test pressure
制造商证书 certificate of manufacture
制造上的缺陷 manufacturing deficiency
制造设备 manufacturing equipment
制造时间 production time; time of production
制造试验程序 manufacturing test procedures
制造手册 manufacturing manual
制造说明书 fabrication specification
制造所需时间 manufacturing lead time
制造条例 manufacturing regulation
制造图(纸) manufacturing drawing; shop drawing
制造完整性试验 manufacturing integrity test
制造网络监视器 manufacturing network monitor
制造物 workmanship
制造误差 foozle; manufacturing error
制造小方石的石料 blockage stone
制造型心用草带编织机 hay band spinning machine for cores
制造许可证 certificate of manufacture
制造学 technology
制造亚硫酸盐产生的酒精副产品 sulfite spirit; sulphite spirit
制造样机 mock-up; production model
制造样品 preparation of specimen
制造业 manufacturing business
制造业方面的优势 dominance in manufacturing
制造(业)废水 manufacturing wastewater
制造(业)废物 manufacturing waste
制造(业)废液 manufacturing waste
制造业会计 manufacturing accounting
制造业开工率 rates of capacity utilization
制造业区 manufacturing district
制造业生产指数 manufacturing output index
制造用料 materials being fabricated
制造与检查记录 manufacturing and inspection record
制造预算 manufacturing budget
制造裕度 manufacturing tolerance
制造账户 manufacturing account
制造者 builder; constructor; fabricator; framer; maker; manufacturer; producer; wright
制造者名牌 patent plate
制造者市场 mill market
制造者职责 responsibility of manufacturer
制造直径 working diameter
制造纸浆原木 pulp log
制造质量 quality of conformance
制造质量控制 manufacturing quality control
制造中心 manufacturing center[centre]
制造周期 manufacturing cycle; manufacturing period
制造装配 fabrication
制造装配零件表 manufacturing assembling parts list
制造装配图 manufacturing assembly drawing
制造装置 manufacturing installation
制渣设备 detritus equipment
制毡材料 felting; felting material
制毡法 felling; felting
制毡机 felt forming machine; felting machine; mat machine
制毡生产线 mat line
制针钢丝 wire for needle products
制针用钢丝 needle wire
制止 checking; keep in check; lid; refrain; repress; restrain; restrict; snub; suppression; withhold
制止背信原则 principle of anti-fraud
制止(车轮)滚动 scotch
制止(传动)装置 arrester[arrestor]
制止的 corrective; deterrent
制止电流 curbing current
制止阀 stop valve
制止反应 stopped reaction
制止令 inhibition
制止螺杆 check bolt
制止螺栓 bolt with stop
制止模型 interdiction model
制止器 arrester[arrestor](catch); inhibitor; stopper
制止侵蚀的 erosion deterrent
制止损坏 arrested failure
制止物 deterrence; deterrent
制止线圈 restraining coil
制止信号 inhibiting signal
制止因素 deterrence
制止爪 stop claw
制止装置 holdout device
制纸浆木材 pulpwood
制轴材料 shafting
制轴车床 axle turning lathe; shafting lathe
制住 trig
制砖 adobe soil; brick manufacture; tile making
制砖厂 brick making plant; brick plant; building tile factory; building tile plant; tile factory
制砖传送带 apron feeder
制砖法 brick making method
制砖工 brickmaker
制砖工场 brickfield; brick yard
制砖机 brick making machine; brick mo(u)lding machine; brick press
制砖机械 brick making machinery
制砖框架 <专门制壁龛需要的砖> reducing box
制砖模 tile pattern
制砖黏[粘]土矿床 brick clay deposit
制砖取土场表面层土 encallow
制砖设备 brick making equipment
制砖业 brick industry
制砖(用)黏[粘]土 brick clay; adobe clay
制砖(用)土 adobe soil; brickearth
制转楔 scotch block
制桩 pile manufacture
制桩材料 pile material
制桩场地 pile fabricating yard; pile yard
制桩混凝土 pile concrete
制作 construction; execution; fabricate; fabrication; make; make-up; making; manufacture; production; tailor
制作备份 back-up
制作布线图(案) patterning
制作场 fabricating yard; fabrication yard
制作尺寸 fabrication dimension
制作法 fabricating method; facture
制作方法 fabrication method; method of fabrication
制作分色版 break-up for colo(u)rs
制作工地 fabricating yard
制作过程 manufacturing process
制作混凝土 confection concrete
制作建筑模型 architectural modelling
制作块体场地 casting yard
制作蜡型器械 instrument for wax pattern preparation
制作类别 type of fabrication
制作立体地图 line relief mapping
制作模板的木板 shuttering board
制作模型 analog(ue) formation; model building; simulation
制作模型方法 model(l)ing
制作模型者 model builder; model(l)er
制作软土砖的泥浆处理法 slurry process for making soft-mud brick
制作图 constructional drawing; shop drawing; working drawing
制作弯头 construction bend
制作物 fabrication
制作线划图 line mapping
制作详图 production drawing
制作者 fabricant; fabricator
制作指导 production director
制作质量 workmanship

治安 public order

治安保护 police protection
治安信息系统 policing information system
治安员 peace officer
治标剂 palliative
治导工程 training works
治导线 diversion line
治河 regulation of river; river control; river improvement; river regulation; river training; stream training
治河措施 river regulation measure
治河工程 regulation works; river construction; river engineering; river improvement works; river project; river training works; fluvial civil engineering
治河工程师 river engineer
治河规划 river project
治河建筑物 training structure
治洪 flood control
治理地表水和地下水 treating surface water and ground water
治理工程 implementation project; river project
治理规划 planning of water logging control; river project
治理河道 canalization
治理蝗灾 locust control
治理三废拨款 appropriation for pollution control
治理项目 conservation project
治理效果 effectiveness of treatment; processing result
治理原则 principle of regulation
治疗房屋 <医院> treatment building
治疗闸室 medical lock
治疗中心 therapeutic center[centre]
治沙 desertification control; sand control; training for sediment
治疝花楸 checker tree
治水 regulate watercourses; water control
治水措施 method of prevent floods by water control
治水工程 riparian works; water control work
治水管理区 water district
治水和输水 water control and delivery
治水系统 water disposal system
治滩 rapids regulation
治泽化冰沼土 swampy tudra soil

质保金 performance guarantee

质变 change in quality; qualitative change; transmutation
质变阶段 stage of qualitative change
质变温度 critical temperature
质变温度范围 critical temperature range
质变温度限程 critical temperature range
质次价高 inferior quality-high price
质的观察 qualitative observation
质的规定性 qualitative prescription of
质的量 amount of substance
质的调节 qualitative regulation
质的统计 statistics of attributes
质的限制 qualitative limitation
质的因素 qualitative factor
质的属性 qualitative attribution
质底法 qualitative background method; quality base method
质地 body; fibre; substrate; texture
质地不均 unevenness in texture
质地的严重土壤 heavy-textured soil
质地等级 texture grade
质地分级 texture classes; texture separate
质地分类 texture grading
质地感 texture effect
质地厚实的 thick-set
质地名称 texture name
质地剖面 texture profile
质地涂层 body coat
质地优良 high quality
质地指标 texture index
质点 mass point; material point; particle; point mass; point particle; point unit
质点沉积 deposition of particles
质点传热系数 point unit heat transfer coefficient
质点大小分布 particle-size distribution
质点动力学 particle dynamics
质点轨迹 path of particle
质点厚度测量技术 particle-thickness technique
质点集 set of particles
质点加速度 particle acceleration
质点加速器 particle accelerator
质点力学 mechanics of particles; particle mechanics
质点能量 particle energy
质点速度 particle velocity
质点位移 particle displacement
质点系 system of particles
质点运动 particle motion; mass transport <波浪的>
质点运动轨迹 particle path
质点运动滤波器 particle motion filter
质反应 qualitative response
质分类 qualitative classification

质光半径关系 mass-luminosity-radius relation
质光比 mass-luminosity ratio; mass-to-light ratio
质光定律 mass-luminosity law
质光关系 mass-luminosity relation
质光曲线 mass luminosity curve
质荷比 mass/charge ratio; mass-to(-electric)-charge ratio
质化 materialization
质级 quality class
质检员 quality inspector
质径关系 radius-mass relation
质控(回)填土 engineered fill
质理证明 quality certificate
质粒 plasmid
质粒工程 plasmid engineering
质粒嵌合体 plasmid chimera
质量 mass; quality
质量百分比 mass fraction; mass percent
质量半径关系 mass-radius relation
质量保持率 quality retention
质量保险系统 quality assurance system
质量保证 quality assurance
质量保证操作程序 quality assurance operating procedure
质量保证规定 quality assurance directive
质量保证计划 quality assurance plan; quality control plan(ning); quality control system; quality program(me)
质量保证技术出版物 quality assurance technical publications
质量保证检查 quality assurance inspection
质量保证控制程序 quality assurance operating procedure
质量保证模式 model of quality assurance
质量保证期 period of quality guarantee
质量保证实验室 quality assurance laboratory
质量保证试验 quality assurance test
质量保证手册 quality control(assurance) manual
质量保证书 certificated quality level; guarantee against defects; guarantee of quality
质量保证数据系统 quality assurance data system
质量保证体系 quality assurance(control) system; quality certification system
质量保证条例 quality assurance provision
质量保证验收标准 quality assurance acceptance standard
质量保证验收人员 quality assurance personnel
质量保证制度 quality certification system
质量比 <锤重与桩重之比> mass ratio <ratio of hammer weight to pile weight>
质量比例 part by weight
质量比流率 mass flow rate
质量变化 quality change; quality variation
质量变坏 debase
质量标度 mass scale
质量标号 quality mark
质量标志 quality mark
质量标准 measurement standard; qualitative criterion; quality criterion; quality specification; quality standard; standard of quality; performance criterion
质量表示法 qualitative representation
质量补偿 mass balancing
质量不符规定记录 quality discrepancy record
质量不高 inferior quality; low quality; undergrade
质量不好 not good quality
质量不灭 conservation of mass
质量不灭定律 law of conservation of mass
质量不平衡 mass unbalance
质量不平衡补偿器 mass unbalance compensator
质量不齐 dispersion in quality
质量不匀的 spotty
质量不足 mass deficiency; mass deficit
质量参考指数 figure of merit; quality reference index
质量参数 figure of merit; mass parameter; quality parameters
质量测定 mass estimation; mass measurement; quality determination
质量测定插入法 bracketing method of mass measurement
质量测定系统 quality measurement system
质量测量 mass measurement
质量策划 quality planning
质量层化 mass loss segregation
质量层析谱 mass chromatogram
质量差的 poor quality; subquality
质量差的工程 shoddy works
质量差价 quality price differential
质量差异 mass discrepancy
质量产额 mass abundance; mass yield
质量产额分布 mass-yield distribution
质量产额曲线 mass-yield curve
质量成本 quality cost; quality-related cost
质量成本比率 quality cost ratio
质量重整化 mass renormalization
质量传递 mass transfer; mass transmission; mass transport
质量传递机理 mass transport mechanism
质量传递系数 coefficient of mass transfer
质量传递 mass transfer
质量单位 mass unit; unit of mass
质量担保 quality assurance; warranty
质量担保函 quality guarantee
质量当量 mass equivalent
质量倒数 reciprocal mass
质量的米制技术单位 metric-technical unit of mass
质量的相对论性变化 relativistic change in mass
质量的相对论性增加 relativistic increase of mass
质量等级 quality grade; quality level; quality class
质量等级说明 graded description
质量等级因素 grade factor
质量等效系数 equivalent coefficient of mass
质量低劣 inferiority of quality; inferior quality
质量低劣的 bum
质量第一 quality first
质量电感模拟 mass-inductance analogy
质量电阻率 mass resistivity
质量定律 mass law
质量度量 quality measure; quality meter
质量发光度半径关系 mass-luminosity-radius relation
质量法规 qualitative regulation
质量反馈 quality feedback
质量反应性系数 mass coefficient of reactivity
质量范围 mass range; mass region; quality range
质量方程 mass equation
质量方针 quality policy
质量费用 quality cost
质量分辨本领 mass resolution
质量分辨率 mass resolution; mass resolving power
质量分布 mass distribution
质量分布函数 mass distribution function
质量分布图 quality distribution chart
质量分等 quality rating
质量分光计 mass spectrometer
质量分级 quality rating
质量分类试验 attribute test(ing)
质量分离 mass separation
质量分离器 mass separator
质量分散 mass dispersion
质量分数 mass fraction
质量分析 mass analysis; qualitative analysis; quality analysis
质量分析法 mass analysis method
质量分析计 mass analyser
质量分析离子动能度谱术 mass analyzed ion kinetic energy spectrometry
质量分析器 mass analyser
质量分析图 quality map
质量分析仪 mass synchrometer
质量丰度 mass abundance
质量峰 mass peak
质量幅度 quality latitude
质量辐合 mass convergence
质量辐散 mass divergence
质量辐射阻止本领 mass radiative stopping power
质量负荷 mass load(ing)
质量负荷变化 mass loading variation
质量负荷曲线 mass loading curve
质量负载 mass load(ing)
质量改进 quality improvement
质量改善剂 modifying agent; quality booster
质量干燥速率 mass drying rate
质量感 qualitative perception
质量工程学 quality engineering
质量工序 quality process
质量公式 mass formula
质量功率比 mass-to-power ratio
质量管理 mass control; quality control; quality control engineering; quality engineering; quality management
质量管理标准 quality control standard
质量管理部门 quality management department
质量管理方针 quality management program(me)
质量管理工程 quality control engineering
质量管理工程师 quality control engineer
质量管理计划 quality management plan
质量管理监督 quality control audit
质量管理检查 quality control check(ing)
质量管理检验 quality control check(ing)
质量管理体系 quality management system
质量管理系统 quality control system
质量管理小组 quality control circle
质量管理小组活动 quality control circle activity
质量管理员 quality supervisor
质量管理政策 quality management policies
质量管理制度 quality control system
质量管理专家 quality expert; quality management specialist
质量惯性矩 mass moment of inertia
质量惯性力 mass-inertia force
质量规定 quality clause
质量规范 quality specification; specifications of quality
质量规格 quality requirement; quality standard; specification of quality
质量规格说明书 quality specification
质量过滤器 mass filter
质量过剩 mass excess
质量函数 mass function; quality function
质量好 good quality
质量好的材料 quality material
质量好的原状土(样) good-quality undisturbed sample
质量好坏程度 degree of quality
质量合格标准 acceptable quality level
质量合格的 accredited
质量合格的实验室 accredited laboratory
质量合格证 certificate of fitness; certification of fitness
质量合同 quality agreement
质量和可靠性 quality and reliability
质量和可靠性目标 quality and reliability objective
质量和运距的乘积 <车辆运输的> mass-distance
质量很差 bad quality
质量很好 very good quality
质量厚度 mass thickness
质量环 quality loop
质量换算因子 mass conversion factor
质量混合比 mass mixing ratio
质量级 quality class
质量极惯性矩 mass polar moment of inertia
质量集中分析 qualitative investigation
质量计划 quality plan; quality planning
质量计量单位 mass units
质量计量室 quality and measurement office
质量计量学 qualimetry
质量技术说明 quality specification
质量技术条件 quality specification
质量加速度 mass acceleration
质量监测台 monitoring desk
质量监察员 quality supervisor
质量监督 quality monitoring; quality supervision; quality surveillance
质量监督和保证体系 quality control and quality assurance system
质量监控 quality control check(ing); quality monitoring
质量监控台 quality monitor console
质量监视 quality check
质量监视控制台 quality monitor console
质量减弱系数 mass attenuation coefficient
质量减损 mass decrement
质量检测器 mass detector
质量检查 quality audit; quality check(ing); quality control; quality inspection; quality test(ing)
质量检查标准 quality inspection criterion
质量检查计划 quality examination program(me)
质量检查科 quality control division
质量检查系统 quality control system
质量检查员 defect marker
质量检查证明书 quality inspection certificate
质量检定 quality arbitration; quality

Z

survey
质量检验 quality check(ing);quality control surveillance;quality determination;quality examination;quality inspection;quality test(ing)
质量检验标准 quality control standard
质量检验表 quality control table
质量检验阶段 quality inspection phase
质量检验卡 quality control chart
质量检验器 quality checker
质量检验员 quality inspector
质量检验证书 certificate of quality test;inspection certificate of quality
质量鉴定 appraisal of quality;merit rating; quality appraisal; quality certification; quality detection; quality determination
质量鉴定试验 qualification test
质量鉴定试验程序 qualification test procedure
质量鉴定试验规范 qualification test specification
质量奖 quality bonus
质量奖金 quality bonnet
质量降解 mass degradation
质量降解速度因素 mass degradation rate factor
质量交换 mass transfer
质量交换率 mass transfer rate
质量交换速率 mass transfer rate
质量交换系数 mass transfer coefficient
质量交流 mass exchange
质量较低的工业用金刚钻 boart-bortz
质量经理 quality manger
质量矩 moment of mass
质量矩阵 mass material;mass matrix
质量均匀分布梁 distributed mass beam
质量卡片 quality card
质量抗 mass reactance
质量壳 mass shell
质量可靠性保证 quality reliability assurance
质量控制 mass control;quality control;quality monitoring
质量控制标准 quality control specification;quality control standard
质量控制表格 quality control form
质量控制成本 quality cost control
质量控制传感器 quality control pick-up
质量控制的混凝土 quality controlled concrete
质量控制等级 quality control level
质量控制方法 method of quality control
质量控制分析 quality control analysis
质量控制工程师 quality control engineer
质量控制和验收依据 quality control and basis of acceptance
质量控制活动 quality control activity
质量控制机构 quality control organization
质量控制计划 quality control plan(ning)
质量控制技术报告 quality control technical report
质量控制检查 quality control check(ing)
质量控制可靠性 quality control reliability
质量控制流变仪 quality control rheometer
质量控制取样器 quality control selector
质量控制人员 quality control officer; quality control personnel
质量控制手册 quality control manual
质量控制数据 quality control data
质量控制体系 quality control system
质量控制条 quality control strip
质量控制条例 quality regulations
质量控制投资 quality controlled investing
质量控制图表 quality control chart
质量控制委员会 quality control committee
质量控制系统 quality control system
质量控制样品 quality control representative
质量控制员 quality control staff
质量控制振荡器 mass-controlled oscillator
质量控制中心 quality control center [centre]
质量控制专家 quality control expert
质量控制装置 quality control package
质量控制资料 quality control data
质量亏损 mass defect;mass deficiency;mass loss;packing effect
质量扩散 mass diffusion
质量扩散率 mass diffusivity
质量扩散系数 mass diffusivity
质量累积检测器 mass integral detector
质量力 body force;mass force
质量利用系数 coefficient of mass utilization
质量灵敏值 mass-sensitive quantity
质量流动 mass flow
质量流动连续条件 condition of mass flow continuity
质量流动速度 mass velocity
质量流动学说 mass flow theory
质量流过速率 mass rate of flow
质量流控制 mass flow control
质量流量 flow mass;mass flow;mass flow rate;flow of mass
质量流量比 mass flow ratio
质量流量常数 mass flow constant
质量流量方程 mass flow equation
质量流量计 mass flowmeter
质量流量检测 mass flow detection
质量流量控制器 mass flow controller
质量流量率 mass rate of flow
质量(流量敏感)型检测器 mass flow detection
质量流量敏感性检测器 mass flow rate sensitive detector
质量流量扰动 mass flow rate perturbation
质量流量系数 mass flow coefficient
质量流量-压差关系 mass flow-pressure difference relation
质量瘤 mascon
质量螺旋 quality spiral
质量密度 mass concentration; mass density;specific mass
质量密集 mascon
质量面 mass surface
质量面积比 mass area ratio
质量敏感分离器 mass-sensitive deflector
质量摩尔浓度 molality;molar concentration;mol concentration
质量目标 quality goal;quality objective
质量能量传递系数 mass-energy transfer coefficient
质量能量换算系数 mass-energy conversion coefficient
质量浓度 mass concentration
质量浓度系数 mass concentration factor
质量耦合 mass coupling
质量耦联 mass coupling
质量排放流量 mass rate of discharge
质量排放速率 mass emission rate
质量判据 quality criterion
质量抛射 mass ejection
质量配料 proportioning by weight
质量碰撞阻止本领 mass collision stopping power
质量偏差 mass deviation
质量品级 quality grade
质量品级说明书 quality grade description
质量平衡 balancing of masses;mass balance
质量平衡方程 mass balance equation;mass balancing equation
质量平衡计算 mass balance calculation
质量平均线法 quality curve method
质量平均值 mass average value
质量评定 grade estimation;grade evaluation; merit rating; qualification;qualitative assessment;quality evaluation
质量评定规则 rules for quality evaluation
质量评估 evaluation of quality
质量评估图 control chart; quality control chart
质量评价 grade evaluation;merit rating; qualitative assessment; quality check(ing); quality evaluation; quality rating
质量评价系统 quality rating system
质量歧视 mass discrimination
质量契约 quality agreement
质量迁移 mass transport
质量迁移率 mass transfer rate
质量迁移模式 model of quantity migration
质量曲线 mass curve;mass diagram
质量趋势 quality trend
质量全息术 mass holograghy
质量缺陷 mass defect
质量燃烧速度 mass burning rate
质量燃烧速率 mass burning rate
质量热容比 ratio of the massive heat capacity
质量人口学 qualitative demography
质量容积 mass volume
质量散度 mass divergence
质量散射系数 mass scattering coefficient
质量扫描 mass scan
质量色谱法 mass chromatography
质量色谱分析 mass chromatographic analysis
质量色谱图 mass chromatogram
质量烧蚀速率 mass ablation rate
质量审查 quality audit; quality review
质量审查制度 quality auditing system
质量审核 quality audit
质量审核观察(结果) quality audit observation
质量审核员 quality auditor
质量实质 quality entity
质量试验 qualitative test;quality test(ing)
质量试验规范 quality test schedule
质量试验器 quality tester
质量收支 mass budget
质量手册 quality manual
质量守恒 conservation of mass;mass conservation
质量守恒定律 law of conservation of mass;law of conservation
质量守恒定律方程 equation of mass conservation
质量守恒方程 mass conservation equation
质量守恒原理 principle of mass conservation
质量输送 mass transport
质量输送速度 mass transport velocity
质量输运 mass transport
质量属性 qualitative attribute
质量数 mass number;nucleon number;unclear number
质量数标度 mass scale
质量数测定 mass assignment
质量数据 qualitative data
质量(数)指示器 mass marker
质量衰减 mass attenuation
质量衰减系数 mass attenuation coefficient
质量双线 mass doublet
质量水平 quality level
质量水平鉴定合同 quality level certification agreement
质量水平证明书 quality level certificate
质量水准 merit grade
质量顺坡移动 mass erosion
质量说明 quality description
质量说明书 quality specification;quality standard
质量速度 mass velocity
质量速度比 mass-velocity ratio
质量算符 mass operator
质量碎片(谱)法 mass fragmentography
质量损失 loss in mass;mass loss;quality loss
质量损失率 mass loss rate
质量弹簧缓冲器组合 mass-spring-dashpot combination
质量-弹簧体系 mass-spring system
质量-弹簧系统 mass-spring system
质量-弹簧-阻尼器体系 mass-spring-dashpot system
质量弹性特性 mass-elastic characteristics
质量特性 mass characteristic;quality characteristic;quality coefficient
质量特征 quality characteristic
质量特征和参数 quality characteristic and parameter
质量体系 quality system
质量体系评审 quality system review
质量条例 qualitative regulation
质量调节 ratio governing
质量通量 mass flux
质量统计 qualitative statistics
质量统计检验 statistic(al) quality control
质量温度关系 mass-temperature relation
质量稳定性评价标准 quality retention rating
质量物理学 macrophysics
质量吸收 mass absorption
质量吸收系数 coefficient of mass absorption; mass absorption coefficient
质量吸收效应 mass absorption effect
质量系数 coefficient of mass; figure of merit;mass coefficient;mass factor;quality coefficient; quality factor
质量系数值 quality factor value;Q-value
质量系统 quality system
质量下降 quality reduction
质量限额 mass limit
质量相等 equality of mass
质量响应 mass-basis response
质量项 mass term
质量消融速率 mass ablation rate
质量销蚀 mass wasting
质量效果 mass effect
质量效应 mass effect
质量协议 quality agreement
质量信息分析 quality information analysis

Z

质量性状 qualitative character;quality character
质量性状相关表 contingency table
质量循环 quality circle
质量循环速度 mass circulation velocity
质量压力 mass pressure
质量研究 quality research
质量验证试验 quality verification test
质量要求 prescription; quality requirement
质量一般 middle quality
质量一般良好 fair average quality
质量一致性检验 inspection of quality conformity
质量移动姿态控制系统 mass-motion attitude control system;mass-shifting attitude control system
质量以买方样品为准 quality as per buyer's sample
质量以卖方样品为准 quality as per seller's sample
质量意识 quality consciousness;quality mind
质量因数 factor of quality; quality factor
质量因数计 quality meter
质量因数频率 figure of merit frequency
质量因素 factor of quality;goodness;quality factor
质量因素比 quality factor ratio
质量因子 factor of merit
质量引力 mass attraction
质量引力垂线 mass attraction vertical
质量影响系数 mass influence coefficient
质量优良程度 merit factor
质量优良的小客车 quality car
质量与检验一致 quality conformance inspection or test
质量预报 qualitative forecast(ing)
质量元素 mass element
质量运动 mass movement
质量再循环率 mass recirculation
质量增加 gain in mass;mass addition
质量甄别 mass discrimination
质量诊断 quality diagnosis
质量证(明)书 certificate of quality;quality certificate
质量职能 quality function
质量指标 index of quality; performance figure; performance index; qualitative index;qualitative indicator;quality index(number);quality indicator
质量指标曲线 figure of merit curve
质量指南 quality guideline
质量指示器 mass indicator
质量指数 quality index(number);quality factor <等于抗拉极限与拉伸百分比的乘积>
质量中等 fair average quality
质量中和 quality neutralization
质量中位直径 mass median aerodynamic(al) diameter; mass median diameter
质量中心 barycenter[barycentre];center[centre] of mass; centroid; mass center[centre];mass centroid;quality center[centre]
质量中心系统 center[centre] of mass coordinate
质量中心轴线 centroidal axis
质量中值 median mass
质量转移 mass transfer
质量转移反应 mass transfer reaction
质量转移理论 mass transfer theory
质量转移率 mass transfer rate
质量转移速率 rate of mass transfer
质量转移系数 mass transfer coefficient
质量资料系统 quality data system
质量阻止本领 mass stopping power
质量组分 mass component
质量作用 mass action
质量作用定律 law of mass action; mass action law
质量作用效应 mass action effect
质量作用原理 mass action principle
质量坐标系统 mass-energy coordinate system
质流量 mass flow
质膜 plasma membrane
质能传递系数 mass-energy transfer coefficient
质能当量 mass-energy equivalence; mass-energy equivalent
质能当量原理 mass-energy equivalence principle
质能关系 mass-energy relation
质能换算公式 mass-energy conversion formula
质能抛物线 mass-energy parabola
质能曲线 mass energy curve
质能守恒 mass-energy conservation
质能守恒定律 law of conservation of mass and energy
质能吸收系数 mass-energy absorption coefficient
质能相当性 mass-energy equivalence
质能相当原理 mass-energy equivalence principle
质能循环 mass-energy cycle
质能转换 mass-energy conversion
质凝法 questionnaire method
质朴的 rustic
质谱 mass spectrogram; mass spectrum
质谱测定法 mass spectrometry
质谱测定计 mass spectrometer
质谱测量 mass-spectrometer measurement
质谱测量法 mass spectrometry
质谱差热分析 mass-spectrometric(al) differential thermal analysis
质谱的似平衡原理 quasi-equilibrium theory of mass spectrum
质谱的自动记录 automatic recording of mass spectra
质谱法 mass spectrography; mass-spectrometric (al) method; mass spectrometry
质谱分析法 mass spectrographic analysis; mass spectrography; mass-spectrometric (al) analysis; mass spectrometry
质谱分析计算 mass-spectrometric(al) computation
质谱分析数据中心 mass spectrometry data center[centre]
质谱化 mass spectrometer
质谱技术 mass-spectrometric (al) technique
质谱检测 mass-spectrometric(al) detection
质谱检漏仪 mass-spectrometer leak detector
质谱检索系统 mass spectral search system
质谱热分析 mass-spectrometric(al) thermal analysis
质谱数据 mass-spectrometric(al) data
质谱数字转换器 mass spectrum digitizer
质谱同位素稀释分析 mass-spectrometric(al) isotope-dilution analysis
质谱图 mass spectrogram
质谱稳定同位素稀释分析 mass-spectrometric (al) stable isotope-dilution analysis
质谱线 mass spectrum line
质谱线劈裂 mass splitting
质谱学 mass spectrometry;mass spectroscopy
质谱仪 isatron; mass spectrograph; mass spectrometer; mass spectroscope;velocitron
质谱仪管 mass-spectrometer tube
质谱仪检定 mass-spectrometric(al) detection
质谱仪探测器 mass spectrometer detector
质权人 pledgee
质色调 mass tone
质数 prime(number);simple quantity
质数定理 prime number theorem
质数偶 prime couple
质体 plastid
质体基粒 grana
质体流动学说 mass flow theory
质调节 constant flow control
质相关 qualitative correlation
质心 barycenter[barycentre]; center[centre] of mass; centroid; mass center[centre]
质心标记 centroid label
质心的 barycentric;centroidal
质心矩 centroidal moment
质心力学时 barycentric dynamical time
质心速度 systemic velocity
质心系 center[centre] of mass frame
质心要素 barycentric element
质心主轴 centroidal principal axis; principal axis of centroid
质心坐标 barycentric coordinates; center[centre] of mass coordinate; mass centric coordinates
质心坐标系 center[centre] of mass coordinate system;center[centre] of momentum coordinate system;mass-centered[centre] coordinate system; mass-centered[centre] system
质询 interpellation
质押 pledge
质押公司债 collateral bond
质押票据 bill as security
质押品 collateral security
质押书 letter of hypothecation
质押资产 pledged assets
质因数 prime factor
质域 prime field
质元素 prime element
质蕴涵 prime implicant
质蕴涵覆盖表 prime implicit covering table
质子 hydrion;proton
质子泵 proton pump
质子传递 proton transfer
质子磁共振光谱法 proton magnetic resonance spectroscopy
质子磁强计 proton magnetometer
质子地磁仪 proton magnetometer
质子电子分光仪 proton-electron spectrometer
质子电子谱仪 proton-electron spectrometer
质子惰性的 aprotic
质子俘获 proton capture
质子给体溶剂 proton donor solvent
质子共振地磁仪 proton resonance magnetometer
质子活度 proton activity
质子活化法 proton activation
质子活化分析 proton activation analysis
质子激发 X 射线光谱法 proton excited X-ray spectrometry
质子激发 X 射线荧光分析 proton induced-X-ray emission fluorescent analysis
质子接受体 proton acceptor
质子平衡 proton balance
质子受体溶剂 proton acceptor solvent
质子数 proton number
质子探针 scanning proton microprobe
质子探针显微分析 proton probe microanalyzer[microanalyser]
质子望远镜 proton telescope
质子稳定常数 proton-stability constant
质子吸收能力 proton absorption capacity;proton uptake capacity
质子显微镜 proton microscope
质子线性加速器 proton linear acceleration
质子性介质 protonic medium
质子性溶剂 protonic solvent
质子性酸 protonic acid
质子序数 proton number
质子旋进磁力仪 proton precession magnetometer
质子衍射图 protonogram
质子质量 protonatomic mass;proton mass
质子中子分光仪 proton-electron spectrometer
质子自由旋进磁力仪 proton free precession magnetometer

栉

栉壳状胶结 crustified cement

栉梳亚麻粗纤维 heckling hards
栉梳亚麻短纤维 heckling tow
栉状水系模式 pectinate mode

秩

秩次统计量 rank order statistics

秩法 rank technique
秩分解 rank factorization
秩和检验 rank-sum test;rank test
秩和检验法 rank sum test method
秩检验 rank test
秩空间 rank space
秩亏 rank defect
秩亏网 free network with rank deficiency
秩亏自由网 free network with rank deficiency
秩力元素 rank force element
秩评定 ranking
秩统计量 rank statistics
秩相关 rank correlation
秩相关系数 rank correlation coefficient
秩序 cosmos;order
秩序化 systematization
秩序紊乱 disorderly

致

致癌工业化学品 carcinogenic industrial chemicals

致癌化合物 carcinogenic compound
致癌环境 carcinogenic environment
致癌健康风险 carcinogen health risk
致癌污染物 carcinogenic contaminant;carcinogenic pollutant
致癌物 carcinogen(e)
致癌物质 carcinogenic compound; carcinogenic substance
致癌性 carcinogenicity
致癌作用 cancer-causing power;carcinogenesis
致变物 mutagen
致病的 disease producing;pathogenic
致病环境因素 pathogenic environmental factor
致病菌 pathogenic bacteria;pathogen-

ic organism;pathogens
致病生物体 pathogenic organism
致病微生物 pathogenic microorganism
致病物 etiologic agent
致病性 pathogenicity
致病性固体废物 etiologic solid waste
致病原体固体废物 etilogic solid waste
致病原因 pathogenesis;pathogenicity
致病职业因素 pathogenic occupational factor
致承包商的通知 notice to contractor
致动的 actuating
致动器 actuator
致动系统 actuating system;brake actuating system
致动液压缸 actuating cylinder
致动装置 actuating unit
致断力试验机 breaking machine
致断容量 breaking capacity
致断伸长度 elongation at break
致断延伸率 breaking elongation;breaking extension
致断应力 breaking stress
致腐的 putrefective
致腐菌 decay-causing fungi
致雇主和工程师的通知 notice to employer and engineer
致光敏物质 contactant photo sensitizer6
致害环境 damage on environment
致幻的 psychedelic
致幻剂 hallucinogen
致畸作用 teratogenesis
致极函数 extremal
致冷 chilling;refrigerate;refrigeration
致冷参量放大器 cooled parametric amplifier;cryogenically cooled parametric amplifier
致冷层 cooling zone
致冷的 frigorific;refrigerant;refrigeratory
致冷和冷却塔 chillers and cooling towers
致冷恒温槽 coolant thermostat
致冷环 ice ring
致冷机 kinetic chemicals;refrigerator
致冷级联过程 cascade process of refrigeration
致冷剂 cooling agency;cooling medium;cryogen;frigorific mixture;primary refrigerant;refrigerant;refrigeration mixture
致冷剂通道 coolant channel
致冷面 refrigeration surface
致冷能力 refrigeration output
致冷器 cryostat;freezer;refrigerator
致冷设备 chilling unit
致冷物体 ice mass
致冷效应 chilling effect
致冷循环 refrigeration cycle
致冷液 refrigerant fluid
致冷照相机 cryogenic camera
致冷装置 chiller;chilling unit;condenser; cooler; refrigerating system; refrigeration equipment; refrigeration installation;refrigerator unit
致力于 address oneself to;apply one's mind to;apply one's self
致裂应力 rupture stress
致密 condensation; densification; thickening
致密白云岩 dense dolomite
致密包膜层 compact coating
致密层 compact layer;dense layer; stratum compactum;tectum
致密长石 felsite;felstone
致密充填 pack compression
致密的 compact;dense;douke;sound; tight
致密地层 dense formation
致密度 degree of compaction;density;tightness
致密镀层 close coating
致密粉末 dense powder
致密封烁质岩石 eurite
致密锆英石耐火材料 dense zircon refractory
致密锆英石砖 dense zircon block; dense zircon brick
致密构造 compact structure; dense structure
致密硅岩 gan(n)ister
致密硅岩系 gan(n)ister measures
致密硅页岩 phthanite
致密焊缝 seal weld;tight weld
致密化【生】pyknosis
致密化动力学 densification kinetics
致密灰岩 compact limestone
致密接头 tight joint
致密结构 close texture; compact structure; compact texture; fine structure;fine texture
致密结晶状石墨 compact crystalline graphite
致密结块 dense cake
致密晶粒 compact grain
致密(晶粒)组织 compact-grain structure
致密砾石 compact gravel
致密路面 tight surface
致密钠云母 cosa mica
致密耐火材料 dense refractory
致密黏[粘]土 heavy clay;leck
致密凝胶体 eugelinite
致密砌块 dense block
致密区 dense area
致密熔渣 compact slag;solid slag
致密砂层 close sand;tight sand
致密烧结 tight burning
致密射电源 compact radio source
致密石膏 compact gypsum
致密石灰岩 compact limestone
致密石墨棒 low-porosity graphite rod
致密石英砾岩 banket(te)
致密石英岩 novaculite
致密速率 densification rate
致密天体 compact object
致密突起 dense projection
致密涂层 close coating;compact coating
致密物质 dense matter
致密系数 compacting factor
致密型 dense form
致密性 compactability;compact(ed)ness;consistence;soundness
致密性试验 seal tightness test
致密性试验压力 tightness test pressure
致密岩 dense rock;gan(n)ister;tight rock
致密岩层 compact formation; tight formation
致密岩石 compact rock;intact rock; tight rock
致密氧化铬 dense chrome oxide
致密氧化铬耐火材料 dense chrome oxide refractory
致密耀斑 compact flare
致密褶皱 compaction fold
致密质地 close texture
致密中心 dense core
致密铸件 sound casting
致密装填 solid loading
致密组织 compact tissue
致密作用 compaction
致敏作用 sensitization
致命错误处理 disaster handling
致命打击 deathblow
致命的 lethal
致命地 vitally
致命故障 critical failure
致命量50%试验【给】LD50 test
致命缺陷 critical defect
致命伤 death-wound
致命失效 catastrophic failure
致命时间 time-to-death
致命损伤 critical damage
致命物 deathblow
致命效果 lethal effect
致命性 mortality;vulnerability
致命性错误 fatal error
致偏板 abat-vent;deflection plate;deflector
致偏场 bending field;deflecting field
致偏磁铁 deflecting magnet
致偏挡板 deflecting baffle
致偏电极 deflecting electrode
致偏电流放大器 deflection current amplifier
致偏电路 deflecting circuit;deflection circuit
致偏电压 deflecting voltage
致偏角 deflection angle
致偏棱镜 deflecting prism
致偏器 deflector
致偏失真 deflection distortion
致偏系统 deflecting system
致偏衔铁 deflecting yoke; deflection yoke
致偏线圈 deflecting coil; deflection winding;sweeping coil
致偏谐振腔 deflecting cavity
致偏转磁体 deflecting magnet
致偏装置 deflecting device;deviator; deviatoric
致热的 pyretogenic;pyretogenous; pyrogenic
致热物 pyretogen;pyrogen
致热原 pyrogen
致热作用 pyrogenic action; thermogenic action
致伤的 injurious
致死当量 lethal equivalent
致死的 fatal; killing; lethal; mortiferous;virulent
致死低温 fatal low temperature
致死低温带 zone of low fatal temperature
致死毒气 lethal gas
致死毒素 lethal toxin
致死范围 lethal range
致死辐射 lethal irradiation
致死辐射剂量 lethal radiation dose
致死干(燥)度 fatal dryness; lethal dryness
致死高温 fatal high temperature
致死高温带 zone of high fatal temperature
致死光线 lethal ray
致死化合物 deadly compound
致死积 lethal index;mortality product
致死剂 lethal agent
致死剂量 fatal dose;lethal dosage
致死角 trapping corner
致死界量 limes death
致死量 lethal dose
致死临界温度 zero point
致死率 fatal rate; lethality; lethality rate
致死浓度 fatal concentration; lethal concentration
致死区 lethal zone
致死湿度 fatal humidity
致死时间 killing time;lethal time
致死速率 fatality rate
致死突变 lethal mutation
致死温度 deadly temperature; fatal temperature; killing point of temperature; killing temperature; lethal temperature;thermal death point
致死物质 killing substance;lethal substance
致死限 lethal range
致死限值 lethal threshold
致死效应 lethal effect
致死性合成 lethal synthesis
致死烟雾 killing smog
致死因素 lethal factor
致死阈(值) lethal threshold
致死照射 lethal exposure
致死指数 lethal index;mortality product
致死中浓度 lethal concentration required to kill 50% of the test animals;median lethal concentration
致酸物质 acid-causing substance
致损原因 cause of damage
致温室效应气体 greenhouse gas
致温室效应气体的吸热能力 heat-trapping ability
致吸尘物质 absorbing dust mass
致谐线圈 syntonizing coil
致眩强光 disability glare;discomfort glare
致雨云 rain-bearing cloud;rain producing cloud
致噪声材料 noise-inducing material

掷 throw

掷锤人 leadsman
掷线 line of departure

窒息 asphyxiation;choke;smother; stifle;stuffiness;suffocating

窒息的 asphyctic;asphyctous;asphyxial
窒息毒气 blackdamp;suffocating gas; choke damp
窒息而死 be choked to death
窒息剂 asphyxiant
窒息灭火 extinguishing by smothering;fire smothering
窒息灭火法 extinguishments by smothering;smothering method
窒息灭火管系 smothering line
窒息灭火气体 fire-smothering gas
窒息灭火系统 fire-smothering system
窒息灭火用蒸汽 fire-smothering steam
窒息灭火装置 smothering arrangement
窒息死亡 death by suffocation
窒息性毒剂中毒 asphyxiant poisoning
窒息性毒气 asphyxiating gas; blood gas;choking gas
窒息性空气 blackdamp
窒息性气体 asphyxiant gas;asphyxiating gas; blackdamp; choke damp; suffocating gas
窒息性气体灭火装置 fire-smothering gear
窒息性物质 suffocating substance
窒息装置 asphyxiator
窒息状态 asphyxia;asphyxy;smother
窒息作用 asphyxiant action; suffocation

智慧车辆 smart vehicle

智慧道路 smart highway
智慧女神雅典娜的神像 Palladium
智慧型车辆 smart car
智慧型道路 smart road
智慧型小汽(轿)车 smart car
智力工业 brain industry
智力资产 intellectual property
智利 <拉丁美洲> Chile

智利硝 caliche; Chilean nitrate; sodium nitrate
智利硝石 Chile niter; Chile slatpeter [slatpetre]; nitratine; soda nitrite
智利中部棋盘格式构造 central Chile chess-board structure
智囊 brain power
智囊班子 thinktank
智囊团 brain trust; think factory; think group; thinktank
智能 brain power; intellect; intellectual ability; intellectual faculty; intelligence
智能办公楼 smart building
智能彩色技术【计】colo(u)r smart
智能常速行驶控制【交】intelligent cruise control
智能超声波探伤仪 intelligent ultrasonic detector; intelligent ultrasonic prober
智能车辆 intelligent vehicle
智能车辆道路系统 intelligent vehicle highway system
智能车辆逻辑单元器 intelligent vehicle logic(al) unit
智能车路系统 intelligent vehicle highway system
智能迟钝 intellectual retardation
智能传感器 intelligent sensor
智能打印机 intelligent printer
智能大楼 <包括电脑控制、局部网络的微波光纤通信[讯]和文字处理、大容量多功能交换机等> intelligent building
智能大厦 intelligent building
智能代理 intelligence agency; intelligent agent
智能电缆 intelligent cable
智能放大器 intelligent amplifier
智能分布系统 distributed intelligence system
智能复合材料 intelligent composite material
智能复印机 intelligent copier
智能钙铁分析仪 intelligent calcium-ferrite analyser[analyzer]
智能工艺学 intellectual technology
智能工作区 smart work zone
智能公路 automated highway
智能光电探测器 true alarm photo sensor
智能光电烟雾报警器 true alarm photoelectric(al) smoke sensor
智能化城市 intellectualized city
智能化计算机 humanized computer
智能化建筑 intellectualized building
智能混凝土 intelligent concrete
智能机 intelligence machine
智能机器人 intelligence robot; intelligent robot
智能积分 intelligent integration
智能计算机 intelligent computer; intelligent machine
智能计算机辅助设计 intelligent computer-aided design
智能技术 intellectual technology
智能加工 intelligent processing
智能检查 intelligence test
智能键盘系统 intelligent keyboard system
智能交叉口 intelligent intersection
智能交通运输基础设施 intelligent transportation infrastructure
智能交通运输系统 intelligent transportation system
智能结构材料 smart structure material
智能卡 <收费、付费卡> smart card
智能卡片阅读器 intelligence card reader
智能科学 intelligence science; intelligent science
智能控制 intelligent control
智能控制机 intelligent controller
智能控制器 intelligent controller
智能流量计 intelligent flowmeter
智能密集型产业 intelligence-intensive industry
智能模拟 artificial intelligence; human simulation
智能模拟法 artificial intelligence approach
智能模式 intelligent schema
智能气候预测 intelligent weather prediction
智能汽车 <车内电脑可以接受各种信息,能节省时间,减少车祸> intelligent car
智能汽车控制系统 advanced vehicle control system
智能热探测器 true alarm heat sensor
智能人机接口 intelligent man-machine interface
智能商数 intelligence quotient
智能数据采集 intelligent data acquisition
智能数据输入端 intelligent data entry terminal
智能水平 level of intelligence
智能探测器底座 true alarm base
智能陶瓷 intelligent ceramics
智能铁路系统 intelligent railway system
智能图 intelligent graphic
智能图形系统 intelligent graphic system
智能网络 intelligent network
智能微机【计】intelligent micro
智能系数 coefficient of intellectual ability
智能系统 intelligence system
智能显示器 intelligent display
智能消费品 intelligent consumer product
智能巡行控制 intelligent cruise control
智能延伸器 mindstretcher
智能仪器 intelligent instrument
智能终端 intelligence terminal; intelligent terminal
智能终端设备 smart terminal equipment
智能终端图形显示 intelligent graphic display
智能住宅小区 intelligent residence district
智能资料收集 intelligent data acquisition

滞变 hysteresis

滞变包线 hysteresis envelope
滞变回线 hysteresis loop
滞变能量耗散 hysteresis energy dissipation
滞变能量消散曲线 hysteresis energy dissipation curve
滞变曲线 hysteresis curve
滞变(曲线)图 hysteresis diagram
滞变损失 hysteresis loss
滞变特性 hysteresis characteristic
滞变效应 hysteresis effect
滞变型共振曲线 hysteresis-type resonance curve
滞变性能 hysteretic behavio(u)r
滞变应变能 hysteresis strain energy
滞变阻尼 hysteresis damping
滞变阻尼器 hysteresis damper
滞变阻尼系数 hysteresis damping coefficient
滞变阻尼因数 hysteresis damping factor
滞差运动 lost motion
滞产 prolonged labo(u)r; protracted labo(u)r
滞超补偿器 lag lead compensator
滞潮 stack water
滞车 <等待信号> stop for signal
滞尘植物 dust holding plant
滞点密度 stagnation point density
滞点温度 stagnation point temperature
滞点温度传感器 stagnation-temperature probe
滞点线 stagnation line
滞点压力 impact pressure; stagnation pressure
滞点压力传感器 stagnation pressure probe
滞钝 ganosis
滞付利息 interest for delay; interest for delinquency; interest in arrears
滞港费 demurrage charges
滞光瓷砖 unglazed tile
滞滚作用 in roll damping
滞海沉积 euxinic deposit(ion)
滞海盆地 euxinic basin
滞海相 euxinic facies
滞洪 detain flood; flood detention; flood retention
滞洪坝 detention dam; flood detention dam
滞洪池 retarding basin
滞洪工程 flood detention project; flood detention work; flood retarding project
滞洪建筑物 floodwater retarding structure; water flow retarding structure
滞洪库容 detention storage
滞洪流量 <水库> holdout
滞洪率 rate of retention; retention rate
滞洪能力 retention ability; retention capacity
滞洪期 period of retardation
滞洪区 detention area; detention basin; flood detention area; flood detention basin; flood detention district; flood retarding basin; flood retention basin; rainwater retention basin; retarding basin; retarding pool; retention basin; flood retention area
滞洪容量 detention volume; flood detention capacity
滞洪水池 floodwater-detention pool; flood control pool
滞洪水库 detention basin; detention reservoir; flood detention reservoir; retardation basin; retarding basin; retarding pool; retarding reservoir
滞洪水深 detention depth
滞洪洼地 detention basin; flood retention basin
滞洪小水库 floodwater-detention pool
滞洪效果 retarding effect
滞洪效应 flood retarding effect; retarding effect; retention effect
滞洪蓄水 detention storage
滞洪蓄水库 floodwater retarding reservoir
滞洪作用 detention effect; retarding effect; retention effect; retention effect of lake
滞后 lag(ging); lag in phase; retardation; time delay; delay; drag; file over; carry-over
滞后崩落 retarded caving
滞后变形 hysteresis set
滞后补偿 lag compensation; quadrature-drop compensation
滞后补偿器 lag compensator
滞后部件 lag unit
滞后材料响应 hysteretic material response
滞后操纵 control lag
滞后常数 hysteresis constant
滞后超前补偿 lag lead compensation
滞后沉积 lag deposit
滞后充填 delayed fill
滞后窗 lag window
滞后存储元件 hysteretic memory cell
滞后单元 lag unit
滞后的供给调整 lagged supply adjustment
滞后地面距离 delay ground distance
滞后点火 delayed firing
滞后电流 lagging current
滞后电路 lagging circuit
滞后电位法 retarding potential method
滞后电压 hysteresis voltage; lagging voltage
滞后定形 hysteresis set
滞后断层 lag fault
滞后法 lag method; retardation method
滞后反应 delayed reaction; hysteresis response
滞后负荷 <电流> lagging load
滞后功率 afterpower
滞后功率因数 lagging power factor
滞后环 hysteresis loop
滞后环节 delay component
滞后缓冲作用 hysteresis damping
滞后换向 lagging commutation
滞后回弹后效 delayed recovery after effect
滞后回线 hysteresis loop
滞后计 hysteresimeter
滞后加演算法 lag-and-route method
滞后间隔后区间 lag interval
滞后角 angle of delay; angle of lag; delay angle; lag(ging) angle; retardation angle
滞后校正 correction of lag
滞后阶段 lag phase
滞后经济指标 lagging indicator
滞后经济指标综合指数 composite index of lagging indicators
滞后径流 delayed flow; delayed run-off
滞后九十度 quadrature-lagging
滞后绝对值 absolute lag value
滞后模量 hysteresis module; hysteresis modulus
滞后排水 <土壤的> delayed drainage
滞后平潮 lag slack
滞后破坏 delayed failure
滞后曲线 hysteresis curve; lagging curve
滞后屈服时间 delayed yield
滞后燃烧 delayed combustion
滞后燃烧爆震 lag knock
滞后容量 lagging capacity
滞后生热性 hysteresis heat build-up
滞后时基 ratchet time-base
滞后时间 delay time; detention time; lag time; retardation time
滞后试验机 hysteresis tester
滞后算子 backward operator
滞后损耗 hysteresis loss
滞后损失 hysteresis loss
滞后弹性 hysteresis elasticity
滞后弹性恢复 delayed elastic recovery
滞后弹性模量 <应力与弹性后效应变之比值> modulus of elasticity after effect
滞后弹性应变 delayed elastic strain
滞后特性 hysteresis characteristic; lag behavio(u)r; retarding characteristic
滞后特性曲线 lag curve

滞后调整 lag
滞后突水 delayed bursting water
滞后网络 lag network
滞后无功负荷 lagging wattless load
滞后误差 lag error; hysteresis error <自动安平水准仪的>
滞后系数 hysteresis coefficient; hysteresis constant; hysteresis factor; hysteresis loss coefficient; lag coefficient
滞后现象 hysteresis; hysteresis effect; hysteresis phenomenon
滞后现象的 hysteretic
滞后线圈 lagging coil
滞后相供电臂 lagging phase feeding section
滞后相关 lag correlation
滞后相位 lagging phase; retarding phase
滞后相位角 lagging phase angle
滞后相(移)角 hysteretic phase angle; lagging phase angle
滞后响应区 hysteresis response zone
滞后效应 after effect; carry-over effect; hysteresis effect; hysteretic effect; lagging effect
滞后信号 delay signal
滞后型 lagged type
滞后型方程 lagging-type equations
滞后性 hysteresis quality; hysteretic nature
滞后性能 hysteresis property; hysteretic behavio(u)r; hysteretic property
滞后压屈强度 postbuckling strength
滞后压屈现象 postbuckling behavio(u)r
滞后淤积 lag deposit
滞后元件 lag element
滞后运动 lost motion
滞后周线 hysteresis loop
滞后转弯 lagging turn
滞后装置 lagging device
滞后状态 hysteretic state
滞后阻力的 late-bearing
滞后阻尼 hysteresis damping; hysteretic damping
滞后阻尼器 hysteresis damper; lag damper
滞后作用 delay(ed) action; hysteresis (effect)
滞后作用的 hysteretic
滞环误差 hysteresis error
滞缓 creep(ing); shunt running
滞缓流动 creeping flow
滞缓指数 delay index
滞回环 hysteresis loop
滞回模量 hysteresis modulus
滞回曲线 hysteresis curve; hysteresis loop; hysteretic curve
滞回特性 hysteretic characteristic
滞回压实 hysteresis compaction
滞火 hangfire
滞火处理 fire-retardant treated; fire-retarding treatment
滞火剂 retardant
滞积比 lag ratio
滞积层 lag deposit
滞积距离 lag distance
滞积砾石 lag(ging) gravel
滞积砂 lag sand
滞积质 lag deposit
滞链 snub the chain
滞料 feed slug
滞流 current retard; ineffective flow; laminar flow; laminar motion; misrun; slugging; stagnation; viscous flow
滞流坝 detention dam
滞流玻璃 dead glass
滞流沉淀池 detention tank
滞流带 stagnant zone
滞流点 stagnation point
滞流阀 snubber valve
滞流河 alluvial river; perched stream
滞流河川 sluggish stream
滞流空气 stagnant air; trapped air
滞流盆地 ponded basin
滞流屏 detention screen
滞流气体流量计 viscous flow air meter
滞流区 dead-air pocket; detaining zone; stagnant area; stagnant pocket; stagnant wake
滞流设施 current retard
滞流温度 stagnation temperature
滞流系数 retardance coefficient
滞流岩 bafflestone
滞留 demurrage; detention; residence
滞留板 retention board
滞留槽 detention tank
滞留沉积 lag deposit
滞留衬垫法 retained gasket system
滞留成本 sunk cost
滞留池 detention tank
滞留船舶 detention of ship
滞留的 perched
滞留费 demurrage; demurrage charges; detention charges; detention fee
滞留份额 retention fraction
滞留锋 stationary front
滞留降水 retention
滞留金 retention money
滞留砾石 lag gravel
滞留量 retention volume; hold-up (weight); retainage
滞留率 retention rate
滞留喷雾 residual spray
滞留期 demurrage; period of retardation; period of retention
滞留期间 detention period
滞留气 entrapped gas
滞留气旋 stationary cyclone
滞留区 slack-water area
滞留曲线 retention curve
滞留容积 retention volume
滞留杀虫剂 residual insecticide
滞留时间 demurrage time; detention time; hold-up time; residence time; resistance time; retention time
滞留时期 detention period
滞留水 logging water; retained water
滞留体积 hold-up volume
滞留天数 days of demurrage
滞留条款 detention clause
滞留瓦斯 standing gas
滞留涡 stagnant vortex
滞留物 lag
滞留系数 coefficient of retardation; retardation coefficient; retention factor
滞留系统 delay system
滞留效应 retention effect
滞留性 anelasticity
滞留性喷洒 residual spray
滞留演化 arrested evolution
滞留演进 arrested evolution
滞留在港内的 port bound
滞留在漆膜内的气泡 air entrapment
滞留指数 retention index
滞留周期 detention period
滞留作用 detention effect; retention effect
滞膜 stagnant film
滞纳加征税款 tax for default
滞纳金 fine for delaying payment; fine for paying late; overdue fine; surcharge for overdue payment
滞纳利息 arrear of interest; defaulted interest
滞纳税 delinquent tax
滞纳税款 delinquent tax; overdue tax payment; tax in arrears
滞纳特赋 delinquent special assessment
滞期费 demurrage(charges); demurrage money
滞期费和速遣费 demurrage and despatch
滞期费和速遣费同等 demurrage same despatch
滞期交货 backlog
滞期留置权 demurrage lien
滞期日数 days on demurrage; demurrage days
滞期天数 demurrage days; days of demurrage
滞期条款 detention clause
滞区 dead zone
滞燃玻璃 fire-retarding glazing
滞燃处理 fire-retardant treatment
滞燃化学剂 fire-retardant chemicals
滞燃结构体系 fire-retardant structural system
滞燃木材 fire-retardant wood
滞燃黏[粘]合剂 fire-retardant adhesive
滞燃饰面 fire-retardant finish
滞燃树脂 fire-retardant resin
滞燃系统 fire-retarding system
滞燃罩面 fire-retardant coating
滞塞 bind-seize; freeze; seizure
滞塞点 choke-point
滞塞给料 choke feeding
滞塞给料的 choke-fed
滞湿 persistent retention of moisture
滞时 lag time; time of lag
滞时法 lag method
滞水 perch(ed) groundwater; slack water; stagnant water; standing water; stickwater
滞水标志 backwater mark
滞水层 aquiclude; aquitard
滞水池 aquifer; retardation basin; stagnant basin; stagnant pool
滞水带曝气系统 hypolimnion aeration system
滞水含水层 perched aquifer
滞水河 perched stream
滞水湖 perched lake
滞水库 retention storage
滞水面 perched water table
滞水盆地 ponded basin
滞水期 period of stagnation
滞水区 detention basin
滞水曲线 water-retention curve
滞水泉 perched spring
滞水容量 retention capacity
滞水时间 retention time
滞水时期 retention period
滞水水库 retention reservoir
滞水水位 perched water table
滞水塘 retention pond
滞水体 stagnum
滞水土 poorly drained soil
滞水洼地 slacking basin
滞水位 perched water table
滞水性 water-retentivity
滞水岩层 aquiclude; aquifuge
滞弹性 <弹性变形在卸荷后一个时间才恢复> delayed elasticity; delaying elasticity; retarding elasticity; anelasticity
滞弹性变形 delayed elastic deformation
滞弹性层 viscoelastic layer
滞弹性弛豫 anelastic relaxation
滞弹性的 anelastic
滞弹性衰减 anelastic attenuation
滞弹性松弛 anelastic relaxation
滞弹性效应 delayed elastic effect; elastic after effect
滞弹性形变 anelastic deformation
滞弹性性能 anelastic behavio(u)r
滞弹性性质 anelastic property
滞相调整 lag adjustment
滞相运行 lagging phase operation
滞相装置 lagging device
滞销 be dull of sale; draggy sales; dull sale
滞销的 dull; slow moving
滞销货 dead storage; slow seller; unsalable goods
滞销品 carry-over; sticker
滞销商品 unsalable goods
滞效 residual effect
滞效喷洒 residual effect spray
滞效温度计 lag thermometer
滞卸费 congestion surcharge
滞压测流管 stagnation tube
滞延成本 balance-delay cost
滞延指数 lagging index
滞颐 wet cheek
滞育 diapause
滞站时间 delay time at stop
滞胀(变形) dilatation
滞胀现象 dilatancy
滞止 diffuse; stagnancy; stagnation
滞止参数 reservoir value; stagnation parameter
滞止层 retarded layer
滞止焓 stagnation enthalpy
滞止激波 shock stall
滞止假潜育土 stagnopseudogley
滞止流 stagnation flow
滞止密度 stagnation density
滞止能 stagnation energy
滞止气流 stagnant air
滞止条件 stagnant condition
滞止温度 impact temperature; reservoir temperature; stagnation point temperature; stagnation temperature; total temperature
滞止温度调节 stagnation-temperature control
滞止线 stagnation line
滞止压力 stagnation pressure
滞止压力减小 stagnation pressure reduction
滞止值 stagnation value
滞止状态 stagnation condition; stagnation state
滞阻 entrapment

蛭

蛭沸石 scolerite
蛭石 <一种绝热材料> pelhamite; roseite; vermiculite
蛭石板 vermiculite slab
蛭石的片状剥落 vermiculite of exfoliation
蛭石骨料 vermiculite aggregate
蛭石灰浆 vermiculite plaster
蛭石灰浆抹面 vermiculite plastering
蛭石灰泥抹面 vermiculite plaster finish
蛭石混凝土 vermiculite concrete
蛭石混凝土刮板 vermiculite concrete screed
蛭石混凝土块 vermiculite concrete block
蛭石混凝土砖 vermiculite concrete brick
蛭石矿床 vermiculite deposit
蛭石抹面灰泥 vermiculite plaster
蛭石砂浆 vermiculite mortar
蛭石砂浆抹面 vermiculite plaster finish
蛭石石膏抹灰层 vermiculite gypsum plaster
蛭石松散填充绝热材料 vermiculite loose fill insulation

蛭石吸声粉刷 vermiculite absorbent plaster
蛭石砖 vermiculite brick

稚鱼期 alevin stage

置1脉冲 set 1 pulse

置备 furnish
置产费用 capital expenditures
置产人以抵押款付给出售人作为部分产价 purchase money mortgage
置产支出 capital expenditures;capital outlay
置车板 <停车场> pallet
置……城堡中 castle
置尺点 sighting point
置存价值 carrying value
置定航向罗经 course-setting compass
置堆栈 set heap
置放式测斜仪 reset clinometer
置放式挠度仪 setting-type deflectometer
置放式水平测斜仪 setting-type horizontal declinometer
置放装置 containment
置换 permutation; permute; replace; replacement; substitute; substitution;transpose;transposition
置换比 replacement ratio
置换变化方式 replacement changes way
置换表 permutation table
置换剥离 transposition foliation
置换剥理 transposition foliation
置换不足当量法 replacement substoichiometry
置换部件 substitution box
置换材料 replacement material
置换测量法 substitution method of measurement
置换产物 substitution product
置换长度 replacement length
置换沉淀 cementation; displacement precipitation
置换出来 cement out
置换次序 displacement series; replacement series
置换的 substitutive
置换滴定 displacement titration; replacement titration
置换地址 relocated address
置换电镀 displacement plating
置换定律 displacement law
置换法 displacement method; method of substitution; replacement method;substitution method
置换反应 reaction of replacement; replacement reaction; substitutional reaction
置换反映 displacement reaction
置换方式 substitute mode
置换分类技术 replacement sorting technique
置换分离 separation by development; separation by displacement
置换分析 replacement analysis; substitutability analysis
置换分置 separation displacement
置换符 substitute character
置换灌浆 displacement grouting
置换规则 replacement rule
置换机理 replacement mechanism
置换基础 displaced foundation
置换剂 displacer
置换继电器 substitution relay
置换价 displacement value
置换检测装置 substitution detector
置换检查 checking by resubstitution
置换交联 displacement crosslinking
置换浇注 mechanical displacement pouring
置换接替 supersede
置换结核 replace concretion
置换解码 permutation decoding
置换净化 cementation purification
置换矩阵 permutation matrix
置换空位 displaced vacancy
置换控制 replacement control
置换块 substitution box
置换扩散 substitutional diffusion
置换理论 replacement theory
置换力 replacing power
置换量 replacement quantity
置换率 replacement ratio
置换螺(旋)钻 displacement auger
置换码 permutation code; permuted code
置换面理 transposition foliation
置换模型 replacement model
置换能 displacement energy
置换能力 diadochy;replaceability;replacing power
置换泥浆 displace mud
置换器 displacer;replacer
置换群 permutation group
置换容积 swept volume
置换色层法 displacement chromatography
置换施工法 <软土的> replacement method
置换时间 replacement time
置换速度 displacement velocity
置换算法 replacement algorithm
置换索引 permutation index
置换体积 displacement volume
置换调制 permutation modulation
置换脱附 desorption by displacing
置换网络 permutation network
置换温度 replacement temperature
置换问题 replacement problem
置换物 substitute
置换吸附 substitution adsorption
置换系数 coefficient of displacement
置换系统气体发生器 recession gas generator
置换显影 displacement development
置换效率 displacement efficiency
置换型固溶体 substitutional (type) solid solution
置换型杂质 substitution impurity
置换性 permutability
置换性灌浆 displacement grouting
置换性盐基 replaceable base
置换修复体 prosthesis
置换序列 constant series; displacement series; electromotive series; Volta series
置换液 displacing fluid
置换异构体 pertmuational isomer
置换应力 replace stress
置换元素 substitutional element
置换圆锥 precipitation cone
置换阵列 replacement array
置换柱 column precipitator
置换桩 replacement pile
置换阻抗测量装置 substitution impedance measuring set
置换作用 displacement;metathesis[复 metatheses]
置景工 griphand
置镜点 instrument point
置零 adjust to zero;reset;set to zero; unset;zero setting
置零开关 zero gate
置零控制装置 zero set control
置零脉冲 reset pulse
置零门 zero gate
置零速度 reset rate
置零旋扭 zero setting knob
置零置一触发器 reset-set flip-flop
置乱 scrambling
置乱器 scrambler
置埋电缆 buried cable
置忙 make-busy
置锚板 billboard
置模器 die set
置平【测】 horizontalization; setting true to perpendicular; level(l)ing adjustment
置切值 mute value
置圈器 band adapter
置入位置 implantation site
置闰 intercalation
置石 stone arrangement;stone layout
置位 set(ting)
置位保持电路 set-hold circuit
置位点 set point
置位定位方式 set location mode
置位复位 set-reset
置位复位触发器 reset-set flip-flop; set-reset flip-flop
置位复位控制 set-reset control
置位复位脉冲 set-reset pulse
置位复位双稳 set-reset bistable
置位复位字 set-reset word
置位开关 setting switch
置位控制 set control
置位脉冲 set(ting) pulse
置位门 set gate
置位名 set name
置位时间 setting time;time of setting
置位时间计数器 set time counter
置位时钟 set clock
置位线圈 set coil
置位阈 setting threshold
置位指令 set command
置信带 confidence band; confidence belt;fiducial belt
置信的 fiducial
置信点 <统计学> fiducial point
置信度【数】 degree of confidence; confidence;confidence level
置信度分布函数 confidence distribution function
置信范围 confidence band;confidence interval;fiducial limit;fiducial range
置信分布 confidence distribution; fiducial distribution
置信分布法 confidence distribution method
置信概率 confidence probability; fiducial probability
置信概率函数 confidence probability function
置信级 confidence level;level of confidence
置信极限 confidence limit; fiducial limit
置信间隔 confidence interval
置信界限 confidence bound; confidence interval; confidence limit; fiducial limit
置信密度 confidence density
置信区间 confidence interval;fiducial interval
置信区间范围 fiducial interval range
置信区域 confidence region
置信曲线 confidence curve
置信时距 confidence interval
置信水平 confidence level; level of confidence
置信水准 confidence level; level of confidence
置信推断 confidence inference
置信椭圆 fiducial confidence ellipse
置信系数 confidence coefficient
置信因数 confidence factor
置信圆 circle of confidence
置信指数 confidence index; fiducial index
置有抗风支撑的吊杆 wind-braced boom
置于凹处的方形浴盆 square recessed bath tub
置于侧面的 flanked
置于罐中 pot
置于零位 initialize
置于起始值 initialize
置于墙内的烟道 built-in chimney
置中 centering[centring]; center[centre] adjustment
置中器 centering device
置中销 centering pin
置桩方法 pile-placing method

雉堞 battlement; crenel(l)ated; crenel(l)ation; embattlement; merlon

雉堞角塔 battlement turret
雉堞墙 battlement;castellation;defensive wall
雉堞式射击口【建】 machicolation
雉堞塔楼 battlement tower
雉堞状桥栏 crenel(l)ated bridge parapet
雉堞状桥墙 crenel(l)ated bridge parapet
雉墙 ballast wall; bridge seat back wall;retaining backwall

中 L 跨线桥 <道路枢纽的> central flyover

中阿尔卑斯期地槽 middle Alpine geosyncline
中鞍 medial saddle
中凹的 concave
中凹度磨削 camber grinding;cambering
中凹度磨削装置 crowning set
中奥陶世【地】 Middle Ordovician
中奥陶统 Middle Ordovician series
中白垩世 Middle Cretaceous
中班 <下午四点到半夜> middle shift;swing shift
中班工人 swing shift;swing shiftman
中斑晶的 mediophyric
中板 core; cross band; crossband veneer; medial plate; median plate; median septum;medium plate;middle plate
中板材 medium sheet
中板厂 medium plate mill
中板块【地】 mesoplate;mid-plate
中板离缝 center gap
中板轧机 jobbing sheet-rolling mill; light plate mill;medium plate mill
中半圆锉 middle half-round file
中保 mediator
中北天山海槽 Mid-north Tianshan marine trough
中比例 mean proportion(al)
中比例尺地图 <一般指 1:20 万至 1:50 万地图> intermediate scale map; medium scale map
中比例尺地质测量 medium scale geologic(al) survey(ing)
中比例尺地质调查 medium scale geologic(al) survey(ing)
中扁锉 middle flat file
中扁木锉 second cut flat wood rasp
中(变质)带 mesozone
中变质的 medium grade metamorphosed

中变质烟煤 mid-grade metamorphic bituminous
中变质作用 mesometamorphism
中表层 mesexine
中宾夕法尼亚世【地】Middle Pennsylvania
中冰块 medium floe;medium ice floe
中冰盘 medium ice floe
中冰隙 medium fracture
中冰源 medium ice-field
中波 medium frequency wave(row) <波长为 200～1000 米的电磁波>;medium wave;mid-wave;moderate sea <海浪三级>
中波波段 medium wave band
中波长的红外区 middle infrared
中波发射机 intermediate wave transmitter;medium wave transmitter
中波海面 <波高 3～5 英尺> moderate sea
中波接收机 medium wave receiver
中波浪 <深水波与浅水波之间的> intermediate wave
中波能 intermediate wave energy
中波频率 mid-band frequency
中波天线 medium wave antenna
中波通信[讯] medium wave communication
中薄板 board;jobbing sheet
中部 center[centre] portion; center[centre] section; intermediate section;middle;middle part
中部标准时间 central standard time
中部衬 middle gasket
中部冲积扇【地】mid-fan
中部冲水式 <水轮的> breasting
中部船体 mid-body
中部地方 midland
中部地区 middle part;midland
中部吊杆 middle hanger
中部多孔性 central porosity
中部风暴沉积 intermediate storm deposit
中部拱起 arching
中部荷载 <混凝土路面的> interior load
中部厚度 interior thickness
中部加强的运输带 straight-ply belt
中部接坡 center[centre] ramp
中部开门客车 <市郊客车> centre-door coach
中部拉门 center[centre] sliding door
中部肋骨 middle frame; midship frame
中部面层 middle surface
中部炮眼 breast hole
中部牵引单位 <列车合并时的> mid-train slave unit
中部水口 centre nozzle
中部台阶 middle bench
中部台隆【地】central platform uplift
中部凸出的钎头 high-center[centre] bit
中部凸出的十字钻头 pilot-and-reamer bit
中部凸出的钻头 high-center[centre] bit
中部凸出式扩孔钻头 pilot-reaming bit
中部凸出的钻头 high-centre bit
中部(为)平顶的反水槽式拱顶 cavetto vault
中部(为)平顶的反水槽式穹顶 cavetto vault
中部舷侧 quarter
中部卸载式挂车 center[centre] dump trailer
中部悬挂机具架 center toolbar;mid-mounted toolbar
中部叶 middle leaf
中部闸门 intermediate gate
中部栅栏 center[centre] barrier
中餐馆 Chinese food;Chinese restaurant
中舱 center[centre] hopper; central hopper
中槽 medium sinus;sinus;top cut
中槽空气 <容器中层的蒸汽空气空间> middle tank air
中草本层 middle herbaceous layer
中草层 middle field layer
中草酸 mesoxalic acid
中层 intermediate; mesosphere; middle level
中层大气 middle atmosphere
中层带 mesopelagic(al)zone
中层的 medial
中层顶 mesopause
中层定置网 midwater trap net
中层浮游生物 mesoplankton
中层固定网 midwater trap net
中层管理 middle management
中层湖水 metalimnion
中层滑坡 moderate layer landslide
中层环流 intermediate circulation
中层甲板 between decks;mid deck;middle deck
中层间托座 intermediate sill
中层建筑 medium rise building;mid-rise building
中层街坊 medium rise block
中层矿体 medium bedded orebody
中层流 intermediate current
中层流网 meso-layer drift net
中层楼板格栅 intermediate floor joist
中层帽木 intermediate sill
中层面 middle surface
中层平台有护栅的钻塔 California derrick
中层漆 flat coat
中层桥面 intermediate deck
中层乳剂 middle emulsion
中层摄食者 midwater feeder
中层水 intermediate water; middle water;midwater
中层水潜水器 midwater vehicle
中层水区 midwater zone
中层水团 intermediary water mass;midwater mass
中层涂料 intermediate coating;middle coating
中层纤维 mean fibre[fiber]
中层鱼类 mesopelagic(al)fishes;midwater fishes
中层云 medium cloud;middle cloud
中层状构造 medium bedded structure
中插板【船】center board;dagger board
中插等高线 half-interval contour
中差插值法 interpolation by central difference
中差式针形调压阀 interior differential needle valve
中产阶级化 gentrification
中长安山岩 andesine andesite
中长反铲斗柄 medium backhoe stick
中长拉长斜长岩 andesine labradorite anorthosite
中长坡 middle-long slope
中长期贷款 term loan
中长期计划 middle and long term plans
中长期水文预报 medium and long range hydrologic forecast;mid and long range hydrologic forecast;mid and long term hydrologic forecasting;mid and long term hydrology forecast
中长期投资 middle and long term investment
中长期信贷 medium term and long-term credit
中长期信用 middle and long term credit
中长石 andesine;pseudo-albite
中长输电线路 medium transmission line
中长透辉角闪岩 andesine diopside amphibolite
中长纤维 medium length fibre;medium staple fibre
中长斜长岩 andesine anorthosite
中长玄武岩 andesinite basalt;hawiite
中长岩 andesinite
中常波 moderate sea
中常年 median year;medium year
中常水流 moderate current
中常雾 moderate fog
中场地震运动 medium field earthquake motion
中超耐磨炉黑 intermediate superabrasion furnace black
中超耐磨炉炭黑 intermediate superabrasion furnace carbon black
中超音速 moderate supersonic speed
中朝地块成矿区 Sino-Korean massif metallogenetic province
中朝地台北侧深断裂系 northern marginal deep fracture zone of the Sino-Korean platform
中朝古陆 Sino-Korean old land
中朝准地台 Sino-Korean axis paraplatform
中潮 mean tide
中潮差 mesotidal range
中潮带 zone of intermediate tide;zone of mean tide
中潮面 half tide level;mean tide level
中潮曲线 medium tide curve
中潮汐环境 medium tidal environment
中沉校准 <测量带尺的> sag correction
中成相 intermediate facies
中承矮梁桥 half-through bridge
中承式拱桥 half-through arch bridge
中承式桁架 half-through truss
中承式桥 half-through bridge;intermediate bridge; midheight-deck bridge
中承式桥面 intermediate deck
中承压水 mesopiestic water
中程 intermediate range;medium distance
中程导航系统 medium range system
中程的 medium range;middle range
中程调度 medium term schedule
中程定位系统 medium range positioning system; middle-range positioning system
中程飞机 medium distance aeroplane;medium haul aeroplane; medium range plane
中程红外光谱学 middle infrared spectroscopy
中程声呐 intermediate range sonar
中匙 dessert spoon
中尺度低压 meso-low;mesoscale low
中尺度高压 mesohigh;mesoscale high
中尺度扩散 mesoscale diffusion
中尺度滤膜分离 pilot-scale nanofiltration membrane separation
中尺度气候 mesoclimate
中尺度气候学 mesoclimatology
中尺度气象学 mesometeorology
中尺度气旋 medium scale cyclone
中尺度(天气)系统 mesosystem
中尺度涡动 mesoscale eddy
中尺度系统 mesoscale system
中尺度旋涡 mesoscale eddy
中尺手 intermediate tapeman
中齿半圆木锉 second cut half round wood rasp
中稠度 mean consistency
中稠黏[粘]土 mean clay
中储棉箱 mid reserve box
中穿孔 middle punch
中垂 sagging
中垂力矩 sagging moment
中垂线 midperpendicular;perpendicular bisector
中垂应力 sagging stress
中垂状态 sagging condition
中粗 <指锉刀> middle
中粗锉 coarse file
中粗钢丝 <1.5～3.0 毫米> medium thick steel wire
中粗级的 coarse middling
中-粗粒结构 medium coarse texture
中粗磨石 medium-grained grinding stone
中粗砂 medium coarse sand
中粗树木的铲除 intermediate clearing
中脆沥青 meso-impsonite
中锉 middle file;second cut file
中大陆-安第斯有孔虫地理区系 mid-continent-Andean foraminiferal realm
中大陆的 mid-continental
中大气圈 mesosphere
中大西洋带 mid-Atlantic belt
中大西洋海岭 mid-Atlantic ridge
中大西洋裂谷 mid-Atlantic rift valley
中带 median band;mesosphere
中带标准矿物 mesonorm
中带测光 intermediate band photometry;medium band photometry
中带宽度 medium band width
中带片麻岩 mesogneiss
中带岩 meso rock
中氮茚系(颜料) pyrrocoline
中挡 <门的> intermediate rail
中挡速度 median speed; medium speed;middle gear
中挡速率 middle gear rate;mid-gear rate
中刀锉 middle knife file
中刀架 intermediate slide
中导板 parting slip
中导的 sprocket
中导孔 center[centre] hole; center[centre] feed hole; guiding hole;sprocket hole
中导脉冲 sprocket pulse
中导位 sprocket bit
中到中 between centers;centers[centres]; centre[centre]-to-centre[centre]
中到中距离 center[centre]-to-center[centre] distance; distance form center[centre] to center[centre]
中得利亚斯冰阶【地】older Dryas stade
中等 intermediate type;middle level;moderation
中等倍(数)物镜 medium power objective
中等焙烧的耐火瓷器 medium burned refractory ware
中等比例尺 medium scale
中等比例尺海图 medium scale chart
中等比重 medium gravity
中等变色 medium stain
中等变色边材 medium stained sapwood
中等冰块 glacon
中等剥落 medium scaling
中等操作条件 moderate operating condition
中等产量指标 middling yield

中等长度扫描 medium scan
中等潮差海岸带 <潮差 2~4 米> mesotidal belt
中等城市 medium sized city; medium urban; middle city
中等程度的潮解 <石灰> medium slaking
中等程度的放射性废弃物 medium level radioactive waste
中等程度的消化 <石灰> medium slaking
中等程度开裂 intermediate cracking
中等程度污水 average sewage
中等程度轧碎 intermediate breaking; intermediate crushing
中等尺寸 medium dimension; medium size; middle size
中等尺寸的 medium sized; middle-sized; moderate-size(d)
中等尺寸的版本 medium sized format
中等尺寸级 <小客车等的> middle-of-the-road class
中等尺度 mesoscale
中等稠度 medium consistency
中等臭味 moderate odor
中等粗糙 middle rough
中等粗糙的结构面 medial rough discontinuity
中等粗齿的切刀 medium rough toothed cutter
中等粗度凿石面 six-cut finish
中等粗砂 medium coarse sand
中等粗质地 moderately coarse texture
中等粗质土 moderately coarse texture
中等存取时间存贮器 medium access memory
中等存取时间记忆 medium access memory
中等锉 middle rasp
中等锉刀 middle-cut rasp
中等大小 median size; moderate-size; medium-size
中等大小的彩色镶嵌瓷砖 medium sized mosaic tile
中等大小的颗粒 middle-sized grains
中等大小的砾砂 middle-sized gravelly sand
中等大小的砾石 medium-grained gravel
中等大小的粒度 middle-sized grains
中等大小的粒子 middle-sized particles
中等大小的卵石 middle-sized gravel
中等大小的煤 chews
中等大小的砂岩 middle-sized sandstone
中等大小的石子 middle-sized gravel
中等大小的微粒 middle-sized particles
中等大小铺路块石 medium paving sett
中等大形变 moderately large deformation
中等代数 intermediate algebra
中等道路 intermediate road
中等的 dominated; intermediate; mediocre; medium; middle; middling; moderate
中等的雾 moderate fog
中等的岩体 middle-season rock mass
中等地面 moderate-duty floor(ing)
中等地势 medium relief; moderate relief
中等地震 intermediate magnitude earthquake; moderate shock; moderate size earthquake; moderate strong earthquake
中等电压 medium voltage
中等电压输电 subtransmission
中等发热量煤 medium calorific value coal; mid class calorific value coal
中等发热量燃气 medium calorific value gas
中等反差 intermediate contrast; medium contrast; middle contrast
中等方木材 medium square
中等防火工业建筑物 moderate-hazard industrial buildings
中等放射性操作 medium level work
中等放射性工作室 intermediate level cave; intermediate level cell
中等分辨率光谱仪 medium resolution spectrometer
中等分选的 moderate-sorted
中等分子量 medium molecular weight
中等风化冰碛 mesotill
中等风化的 moderately weathered; moderate weathering
中等风力 moderate wind
中等风速 moderate wind
中等服务条件 moderate-duty service
中等浮雕 middle relief
中等幅度 moderate range
中等腐败性水 medium rotten water
中等腐水性地区 mesosaprobic(al) zone
中等负荷 moderate duty
中等负荷区 <电线路> medium loading district
中等负荷运行 moderate-duty service
中等负载 medium weight load
中等负载运行 medium duty service
中等复杂场地 medium complexity site
中等富水溢出带 middle watery overflow zone
中等感光速度硬片【测】medium fast plate
中等高层建筑 medium rise building
中等高层街坊 medium rise block
中等高度隔板 medium high partition
中等高度隔墙 medium high partition
中等高度间壁 medium high partition
中等高度建筑 medium rise building
中等格栅 intermediate weave
中等工况 medium heavy-duty
中等工作条件 moderate operating condition
中等功率堆 medium power reactor
中等功率放大器 intermediate power amplifier
中等功率汽车 medium powered car
中等功率拖拉机 medium tractor
中等功能密封膏 medium performance sealant
中等构造活动 moderate tectonic activity
中等骨料 intermediate stone
中等固体量 intermediate solid
中等光洁度 medium finish(ing)
中等规模 mesoscale
中等规模的 medium scale
中等规模格式 medium sized format
中等锅炉 medium boiler
中等含水层 middle aquifer
中等含水量 intermediate water content
中等含水量乳化液 medium irrigated emulsion
中等含盐的 moderately brackish
中等含盐量 medium salinity
中等含盐量的咸水 mesohaline water
中等航高航空摄影 mean-altitude photography
中等航高航摄 mean-altitude photography
中等荷载 medium duty; medium weight load
中等横向加速转向运动 mid-range lateral acceleration handling maneuver
中等洪水 moderate flood
中等喉道宽度 medium throat width
中等厚度金属套管 intermediate metal conductor
中等护理疗养设施 intermediate care facility
中等环 middle ring
中等缓解 middle relief
中等灰尘 <能见度为 1 千~2 千米> moderate dust
中等灰色 medium gray[grey]
中等挥发性燃料 medium volatile fuel; moderately volatile fuel
中等挥发性燃油 moderately volatile fuel oil
中等回波 medium return
中等回火钢 medium temper steel
中等活动 moderate activity
中等活动的 moderately active
中等火灾危险 moderate fire hazard
中等货 medium quality; middle class goods; middling
中等基础 medium base
中等技术水平的司机 average operator
中等技术学校 <英> secondary modern school
中等剂量 median dose; mild moderate dose
中等加工槽黑 medium processing channel black
中等加载 medium heavy loading
中等加载电路 medium heavy loaded circuit
中等价格 medium price
中等间距平行跑道 intermediate parallel runways
中等碱度 intermediate alkalinity
中等交通量 average traffic; moderate traffic
中等交通量道路 intermediate traffic volume road
中等胶结 moderate cementation
中等角度冲断层 medium angle thrust
中等角度晶界 medium angle boundary
中等阶层住宅区 middle class residential zone
中等节 <直径小于 1/2 英寸, 1 英寸 = 0.0254 米> medium knot
中等解理 medium cleavage
中等进近灯光系统 medium-approach lighting system
中等晶粒 medium grain
中等晶体 medium crystal
中等精度惯性导航系统 medium accuracy inertial navigation system
中等距离 moderate distance
中等距离停车站 moderate-length stop
中等飓风 mesocyclone
中等卷曲 intermediate crimp
中等抗硫酸盐水泥 cement with moderate sulphate resistance; moderate sulfate resisting cement; moderate sulphate resisting cement
中等科技 intermediate technology
中等颗粒 medium grain
中等颗粒尺寸 median size
中等颗粒的 medium granular; medium sized
中等颗粒的泥沙 middle-sized silt
中等颗粒的砂质砾石 middle-sized sandy gravel
中等颗粒的淤沙 middle-sized silt
中等颗粒骨料 medium-sized aggregate
中等可选 mid washability
中等跨度 intermediate grade span
中等块状结构 fragizone
中等宽度带钢 medium width steel strip
中等矿化 middle mineralized water
中等矿体 medium orebody
中等栏杆 intermediate rail
中等粒度的 medium-grained; moderately coarse
中等粒度结构 medium structure
中等粒度研磨 medium grinding
中等粒径 median particle diameter
中等粒径的 medium-grained
中等粒状的材料 medium-grained material
中等粒状的砂砾 medium-grained sandy gravel
中等粒状的砂石 medium-grained sandstone
中等粒状的砂子 medium-grained sand
中等粒状的微粒 medium-grained particle
中等粒状结构 medium-grained structure
中等亮度煤 intermediate coal
中等量促进剂 medium accelerator dosage
中等量交通 medium traffic
中等裂缝边 mid crack edge
中等裂隙化 medium fissured
中等流速 mid velocity
中等楼面 moderate-duty floor(ing)
中等螺旋扇块 moderate spirality sector
中等毛石填充 medium rubble fill
中等冒落顶板 less easily falling roof
中等密度干草压捆机 medium density hay press
中等密度交通 medium heavy traffic
中等密度聚乙烯 medium density polyethylene
中等密度砂 medium compact sand
中等密实纤维板 medium density fiberboard
中等明度 medium brightness
中等模量比岩石 medium ratio of modulus
中等磨粒 medium grain
中等木 dominated; intermediate
中等耐火(黏[粘]土)砖 intermediate duty fireclay brick
中等耐久的 medium lived
中等耐用的 medium lived
中等能级 intermediate level
中等能见度 <能见度范围为 4~7 千米> moderate visibility
中等能力的公共交通系统 intermediate capacity transit system
中等能量 medium energy; moderate energy
中等能量的 medium duty
中等能量海岸 moderate energy coast
中等黏[粘]度燃油 medium viscosity fuel oil
中等黏[粘]土 medium clay
中等黏[粘]滞性 medium viscosity
中等捻度 medio-twist
中等浓度 medium concentration
中等排出 intermediate discharge pressure
中等跑道 intermediate course
中等配筋率的 medium reinforced
中等品位 medium grade
中等品质 fair average quality
中等平滑圆锉刀 medium smooth circular-cut file
中等屏幕 medium screen
中等坡度 intermediate grade; medium gradient; moderate grade
中等破坏 moderate damage; moderate failure
中等企业 medium size enterprise

中等起伏 moderate relief

中等起伏地形 indeterminate form; intermediate form

中等气泡 medium air bubble

中等强度 intermediate strength; medium intensity; medium tenacity; moderate intensity; moderate strength

中等强度的放射性废料 intermediate level radioactive waste

中等强度的岩石 middle strength rock

中等强度电流 medium current

中等强度放射性废物 intermediate level waste

中等亲水的 medium hydrophilicity

中等侵蚀 moderate attack

中等倾伏褶皱 moderately plunging fold

中等倾斜 medium bank; medium pitch

中等清晰度 intermediate resolution

中等热 moderate fever

中等热裂黑 medium thermal black

中等容量 medium capacity; midcapacity

中等容量电站 medium capacity plant

中等溶度的 moderately soluble

中等乳浊液 medium emulsion

中等软度的 medium soft

中等软弱岩石 moderately weak rock

中等色 medium shade

中等伤害碰撞强度 moderate injury level

中等烧过的瓦 medium-baked tile

中等深度的泉水 spring of intermediate depth

中等深度基槽 <英国标准为深度 1.5 ~ 6.0 米> medium trench

中等深度坑 <英国标准为深度 1.5 ~ 6.0 米> medium pit

中等深水 <英国，相对水深为 0.05 ~ 0.5 波长的> intermediate water

中等渗透度 medium permeability; moderate permeability

中等渗透率 medium permeability

中等生产 medium duty

中等生产量 average production

中等湿度潜水区 medium moist underground water area

中等湿陷 medium collapsible

中等石块 intermediate stone

中等时间存车 medium term parking

中等时间停车 medium term parking

中等使用寿命的 medium lived

中等使用条件 medium operating condition; moderate-duty service; moderate operating condition

中等试验水平 moderate testing level

中等收入 median income

中等收入水平住宅 moderate-income housing

中等收入者住宅 moderate-income housing

中等收入组别 middle-income group; moderate-income group

中等寿命的 medium lived

中等输出功率 moderate output

中等输出压力 intermediate discharge pressure

中等树节 medium knot

中等水 general water

中等水化热抗硫酸盐水泥 moderate heat of hydration and sulphate-resistant cement

中等水化热水泥 moderate heat of hydration cement

中等水流 moderate current

中等水深密度 medium water depth density

中等水头 moderate head

中等水硬石灰 moderately hydraulic lime

中等松弛 middle relief

中等速度 intermediate speed; median speed; medium speed; moderate speed; moderate velocity

中等速度润滑 moderate-speed lubrication

中等塑性土 medium plastic soil

中等碎片 medium chip(ping)s

中等体力劳动 moderate physical labor

中等条件 moderate condition

中等透水层 middle permeable stratum

中等托运长度 medium length haul

中等弯曲半径弯头 medium sweep elbow

中等威力 intermediate yield

中等威力炸药 middle-strength explosive

中等维修 medium maintenance

中等温度 medium temperature; moderate temperature

中等温度凝结胶 intermediate-temperature-setting adhesive

中等温度凝结胶黏[粘]剂 medium temperature setting adhesive

中等温度热水供热 medium-temperature water heating

中等纹理 medium grain

中等稳定矿物 moderate resistance mineral

中等污染带 mesosaprobic(al) belt; mesosaprobic(al) zone

中等污染区 mesosaprobic(al) zone

中等污染水 moderately polluted water

中等污水 moderate effluent

中等污水稳定池系统 pilot waste stabilization pond system

中等吸附 moderate adsorption

中等稀释沥青 medium curing asphalt

中等细粒 moderate fines

中等细质地 moderately fine texture

中等细质土 moderately fine texture

中等纤维板 medium board

中等咸度的 <指含盐量 5‰ ~ 18‰范围内> mesohaline

中等线体字 clarendon

中等小块 medium chip(ping)s

中等斜纹 middle twill

中等卸载 middle relief

中等行车条件 moderate-duty service

中等性土壤 moderately acid loamy soil

中等学校 secondary school

中等学校建筑 secondary school buildings

中等压力 intermediate pressure; medium pressure; moderate pressure

中等压入配合 medium force fit

中等压头范围 medium head range

中等研磨性岩石 medium abrasive rocks

中等盐渍化的 moderately brackish

中等颜色 medium colo(u)r

中等以上的 average-to-good

中等应变 medium strain

中等营养化水 mesotrophic water

中等硬度 medium hardness

中等硬度的 medium hard

中等硬度钢 medium hard steel

中等硬度土 moderately firm ground

中等硬水 moderate hard water

中等涌浪 moderate swell

中等有机污染河流 mesosaprobic(al) river; mesosaprobic(al) stream

中等余辉的磷光体 medium persistance phosphor

中等原木 average log

中等原子量同位素 intermediate isotope

中等运量 medium heavy road traffic

中等运量轨道交通系统 intermediate capacity rail transit system

中等运输密度线路 medium traffic density line; moderate density line

中等运转条件 moderate operating condition

中等运转性能 <发动机的> mid-range performance

中等载货汽车 medium truck

中等载重等级 medium heavy-duty class

中等载重货车 medium lorry

中等载重量货车 medium duty truck

中等张开的 medial opening

中等胀缩地基 moderately swelling-shrinkage foundation

中等胀缩性土 medium swelling-shrinkage soil

中等折裥 intermediate crimp

中等震级地震 intermediate magnitude earthquake

中等治理 medium control

中等质地 <土壤的> medium texture

中等质地土壤 medium texture soil

中等质量 average quality; fair average quality; mid specification quality

中等致死剂量 mid lethal dose

中等重件货物 medium heavy piece of cargo

中等重量 intermediate weight; medium weight

中等重量锚 stream anchor

中等重量铁锤 <英国标准为 0.5 ~ 2 千克的锤> lump hammer

中等周转量 moderate turnover

中等皱缩 intermediate crimp

中等住宅区 bedroom town

中等柱叶形泥刀 middle post leaf

中等专科学校 polytechnic school; secondary technical school; technical secondary school

中等专业学校 secondary specialized school; specialized high school

中等阻力 moderate resistance

中等阻尼 moderate damping

中低浓度有机废水 organic wastewater with medium-low concentration

中低山 <绝对高度 500 ~ 1000 米> medium low mountain

中低收入住房 low-and-moderate-income housing

中低温热卤水 medium to low temperature brine

中低温釉 soft glaze

中低压【气】 meso-low

中低压容器 middle and low-pressure vessel

中地板 medium floor

中地形 mesorelief

中地震 moderate earthquake

中点 center [centre] middle; median point; medium point; mid-circle; middle point; midpoint; midpoint circle

中点保护系统 midpoint protective system

中点比例尺 scale at mid-point

中点多边形 central point figure

中点多边形系 central polygon system

中点法 midpoint method

中点法则 midpoint rule

中点荷载 center [centre]-point load; central point load

中点回波抑制器 intermediate echo suppressor

中点渐屈线 mean evolute

中点键控 centertap keying

中点交叉 midpoint crossing

中点接地 centre-point earthing; neutral earthing

中点馈电天线 center-fed aerial; centre-driven antenna

中点连接 midpoint connection

中点水流 <水源与出口的中点> midstream

中点凸泛函【数】 midpoint convex functional

中点线 midpoint line

中点引线 centre tap

中点圆 midpoint circle

中点值 midpoint value; mid-range

中点指标 middle marker

中电流 intermediate current

中电流等离子弧焊 intermediate current plasma arc welding

中电气岩 mesotourmalite

中垫 arolium

中殿【建】 nave

中顶盖 intermediate head cover

中定伸强力 normal modulus

中东 Middle East; Mideast

中洞 moderate cave

中陡坡 middle-steep slope

中度 moderate

中度地震 moderate shock

中度风 moderate wind

中度挥发烟煤 medium volatile bituminous coal

中度极压润滑剂 mild extreme pressure lubricant

中度开发 intermediate in the scale of development

中度开发国家 intermediate developed country

中度开裂 moderate crack

中度冷冻 moderate refrigerating

中度裂缝 medium check

中度裂化 medium break(ing); moderate cracking

中度敏感 medium sensitivity

中度钠质水 medium sodium water

中度侵蚀 moderate erosion

中度倾斜坡 mesocline

中度软焦油脂 medium soft pitch

中度深水波 intermediate depth wave

中度衰落 moderate fading

中度污染 median pollution; meso-pollution; moderately contaminated

中度污染带 mesosaprobic(al) zone

中度削凿加工的 medium-pointed

中度盐水 medium salinity water

中度盐渍化 middle salinization

中度氧化 moderate oxidation

中度有效性 moderate availability

中度预应力 <混凝土拉应力超过容许值> moderate prestressing

中度紫色 parma

中短波 medium short wave; medium high frequency wave <频率从 1500 千赫兹到 6 兆赫兹、波长 50 ~ 200 米>

中短波收音机 double range receiver

中短期贷款 short and intermediate term credit

中短期(地震)预报 medium-to-short range prediction

中短期专项贷款 middle and short term special loan

中段 centre section; middle reach

中段标高 level elevation

中段储量计算平面图 reserve-calculating plan on mining level

中段地质平面图 geologic(al) plan of mining level

中段废水 intermediate wastewater; middle stage wastewater

中段高程 level elevation

中段拱起 hogging

中段号 level number

中段间距 level interval

中段起拱 hogging
中段取样平面图 sampling plan of mining level
中段挺紧的挡土板 middling board
中段下垂 sagging
中断 break(ing) up;breakout;break short;blackout;blocking;chopping;discontinuance;discontinuation;discontinue;disruption;intermission;intermittence;interrupt;interruption;out of action
中断按钮开关 interrupt push switch
中断百分率(电话) percentage of trunking
中断包 interrupt packet
中断保存 interrupt stacking
中断保留 interruption pending
中断比较器 break comparator
中断编码 gap coding
中断标记 interrupt flag
中断标志 interrupt identification
中断捕获 interrupt trap
中断操作 interrupt operation
中断程序 interrupt routine
中断程序工业 interruptable process industry
中断程序信号 interrupt program-(me) signal
中断程序作业 interruptable process industry
中断初始化序列 interrupt initialization sequence
中断处 chasm
中断处理 interrupt handling;interruption handling;interrupt processing
中断处理程序 interrupt handling routine;interrupt processing routine
中断处理机 bank interrupt processor
中断处理例行程序 interruption handling routine
中断处理逻辑 interrupt handling logic
中断触发器 interrupt flip-flop;interrupt trigger
中断触发信号 interrupt trigger signal
中断存储区 interrupt storage area
中断错误 interrupt error
中断的 batchwise;discontinuous;interrupted
中断的地下水位 interrupted water table
中断的侵蚀旋回 interrupted cycle of erosion
中断的三角挑檐 interrupted arch
中断的三角檐饰 interrupted arch
中断等待(时间) interrupt latency
中断地址 interrupt address
中断地址向量 interrupt address vector
中断点 breakaway point;breaking down point;breakout point;point of interruption
中断点地址 break address
中断电话通知 blocked precedence call announcement
中断电流器 rheotome
中断电钮 interrupt button
中断动作 stop motion
中断冻结状态 interrupt freeze mode
中断堆叠 interrupt stacking
中断队列 interruption queue
中断发射机 slave transmitter
中断反馈信号 interrupt feed-back signal
中断返回 return from interrupt
中断返回控制字 interrupted return control word
中断返回指令 interrupted return instruction
中断方式 interrupt(ed) mode
中断分析 interrupt analysis
中断服务 interrupt servicing
中断服务程序 interrupt service routine
中断干线 trunk main
中断功能 break-in facility;look-at-me function
中断拱的装饰线脚 interrupted arch mo(u)lding
中断管理程序 interruption supervisor
中断过程 interrupt procedure
中断河 interrupted river;interrupted stream
中断河流纵剖面 interrupted river profile
中断级 interrupt class;interrupt level
中断级配 jump grading
中断记录 interrupt logging
中断接口过程 interrupt interface procedure
中断禁止 interrupt inhibit
中断控制程序 interrupt control routine
中断扩展器 interrupt expander
中断路 interrupt road
中断码 interrupt code;interruption code
中断面 medium cross-section
中断面法 method of middle area
中断面积法 mid-section method
中断模件 interrupt module
中断能力 interrupting capacity
中断排队 interruption queue
中断屏蔽 interruption mask;interrupt mask;unmask
中断屏蔽标志 interrupt mask flag
中断屏蔽触发器 interrupt mask flip-flop
中断屏蔽输出 interrupt mask out
中断屏蔽位 interrupt mask bit
中断屏蔽状态【计】 interruption masked status
中断剖面 interrupted profile
中断期 interruption
中断期间 interval;layover <旅行中的>
中断器 interrupter
中断潜水面 interrupted water table
中断嵌套 interrupt nesting
中断侵蚀旋回 interrupted cycle of erosion
中断请求 break request;interrupt request
中断请求信号 break interrupt signal
中断趋势 broken trend
中断三角楣饰 broken pediment
中断设施 <操作系统中> break-in facility
中断时间 downtime;off-time;outage time;time of intermittence
中断识别 interrupted identification
中断式拱 broken arch;interrupted arch
中断式山墙 broken pediment
中断诉讼 discontinuance of legal proceedings;discontinue an action
中断洗提 interrupted elution
中断系统 interrupt(ion) system
中断/陷阱应答处理 interrupt/trap acknowledge transaction
中断响应 interrupt response
中断向量 interrupt vector
中断信号 interrupt(ion) signal
中断信号反馈 interrupt signal feedback
中断行车 suspend the traffic
中断用系统 interrupt-oriented system
中断优先表 interrupt priority table
中断优先级 interrupt priority;interrupt prior level
中断优先级信号 interrupt priority signal
中断优先片 interrupt priority chip
中断优先权【计】 interrupt priority
中断优先系统 interrupt priority system
中断优先阈值 interrupt priority threshold
中断原因 interruption source
中断源 interrupt source
中断源触发器 interrupt source flip-flop
中断允许与禁止(指令) interrupt enable and disable
中断运行 interrupt run
中断运转时间 idle time
中断振荡 blocking
中断指定 interrupt assignment
中断指定策略 interrupt assignment strategy
中断指令 break-point instruction;interrupt(ed) instruction
中断指示器 interrupt indicator
中断重指定 interrupt reassignment
中断周期 interrupt cycle
中断装置 cut-out;interrupting device
中断状态 interruptable state;interruption status;interrupt mode;interrupt phase
中断状态字 interruption status word;interrupt status word
中断着陆 refused landing
中断子(例行)程序 interruption subroutine
中断字寄存器 interrupt word register
中断自动同步机 relay selsyn
中断总线 interrupt bus
中断租让合同 annul a concession agreement
中堆积结构 mesocumulate texture
中队 squadron
中墩 intermediate pier;center[centre] pier;central pillar;splitter wall <水闸>
中垛 parma
中舵 center[centre] line rudder;middle rudder;mid rudder
中耳【医】 ear drum;tympanum drum
中耳气压伤 ear squeeze;middle ear barotrauma;otitic barotrauma
中二迭世【地】 Middle Permian
中二楼 mezzanine
中反差 medium contrast
中方 <约6~12英寸见方,1英寸=0.0254米> medium square
中方材 medium hewn squares
中方锉 middle square file
中方格 medium square
中枋 die square;medium square
中放部分 intermediate frequency strip
中放废物 intermediate level waste;medium level waste
中放晶体滤波式超外差接收机 radiostat
中放射性废水 medium lever radioactive wastewater
中放射性废液 medium activity liquor;medium lever radioactive waste liquor
中非关税经济同盟 Central-African Customs and Economic Union
中非铜矿床 Central-African copper deposit
中沸点溶剂 medium boiling solvent
中沸石 mesolite;mesotype
中分辨率分光计 intermediate resolution spectrometer
中分辨率红外辐射计 medium resolution infrared radiometer
中分辨率红外辐射仪 medium resolution infrared radiometer
中分辨率图像传输 medium resolution picture transmission
中分脉 median sector
中分面 mid-plane;split
中分面飞轮 split-arm flywheel;spoke-divided flywheel
中分面密封 mid-separate surface seal
中分面密封的螺栓力计算 calculation for bolt force of mid-separate closure
中分面式汽缸 split casing
中分面支持 centring support
中分碛 interlobate moraine;intermediate moraine
中分式系统 middle feed system;mid-feed system
中分式中心岛 split central island
中分双扇门 center-opening door panel
中分纬度 mid latitude;middle latitude
中分纬度改正表 tables for correcting mean mid latitude
中分纬度改正量 correction middle latitude
中分析 mesoanalysis
中分选的砂 moderate-sorted sand
中粉砂(土) medium silt
中风 paralysis
中风化 medium weathering;moderate weathering
中缝 centre line joint;mid seaming;raphe
中缝板 centre joint plate
中浮冰块 medium ice floe
中浮雕 mezzo relievo
中浮雷 neutral buoyant mine
中腐生带 mesosaprobic(al) zone
中腐熟腐殖质 medium mull
中腐水性地区 mesosaprobic(al) zone
中腐殖酸煤 mid humic acids coal
中负荷脚手架 medium duty scaffold
中负荷生物滤池 intermediate rate biological filter
中负突起 mid-negative relief
中富矿石 medium grade ore
中腹 <桶等的> bilge
中干 middle trunk;mesome【地】
中杆 king rod
中感应测井 medium investigation induction log
中感应测井曲线 medium investigation induction log curve
中钢板 light plate
中港浮标 mid-channel buoy
中高草 medium height grass;midgrass
中高度气象卫星 intermediate altitude meteorological satellite
中高度通信[讯]卫星 medium altitude communication satellite
中高度卫星 intermediate altitude satellite;medium altitude satellite
中高发热量煤 mid high calorific value coal
中高费用的清洁生产方案 middle-high cost clean production option
中高固体分 medium high solid
中高河水位 mid and high river level
中高空通信[讯]卫星 intermediate altitude communication satellite
中高频 medium high frequency
中高频波 medium high frequency wave
中高频测向仪 medium high frequency direction finder
中高频端 front end
中高山 <绝对高度3500~5000米> medium height mountain
中高卸料通道 middle-height discharge tunnel
中割型切割器 cutterbar for middle cut;middle-cut cutter
中隔 septum[复 septa]

中隔壁 median lamella; medium septum
中隔沥青毡层 black sheeting felt
中隔门 partition door
中隔墙 intermediate wall; mid-board
中铬黄 medium chrome yellow
中根次序 inorder
中更新世【地】Medio-Pleistocene; mid Pleistocene epoch; middle Pleistocene
中更新统【地】mid Pleistocene series
中耕 cultivate; cultivation; inter-cultivation; intertill; intertillage
中耕除草机 cultivating implement; extirpator; trash cultivator
中耕机 cultivator; hoeing machine; rearer
中耕作物 intertilled crop; row crop
中拱 camber; hogging【船】
中拱力矩 hogging moment
中拱弯矩 hogging moment
中拱应力 hogging stress
中拱状态 hogging condition
中沟 gutter; medial groove; median furrow; median groove; median gutter
中构造 mesotectonics
中古 mesoid
中古的 mediaeval
中古构造体系 mesoid tectonic system
中古建筑 mediaeval architecture
中古期拱和柱层层内叠 arch order
中古期要塞墙上的瞭望孔 arrow loop
中古期要塞墙上的竖射箭孔 arrow loop
中古期阴沟管或垃圾箱 archivoltum
中古时代的城墙 mediaeval city wall
中古时代的建筑 mediaeval architecture
中谷 median valley
中关 second trial
中观环境政策 meso-environmental policy
中观结构 meso texture
中观经济 medium economy
中管提升 central lift
中管提升搅拌器 central lift agitator
中贯挡 transom bar; transom(e)
中光度藻类 mesophotic algae
中规 equator
中规模的 medium sized
中规模集成电路 medium scale integration; medium scale integrated circuit
中硅酸 mesosilicic acid
中硅质的 mediosilicic
中辊 central roll; intermediate roll; middle roll
中国白 Chinese white; zinc white
中国白蜡 Chinese tree wax
中国保险条款 China insurance clause
中国标准化年鉴 China Standardization yearbook
中国侧柏 oriental arbo(u)r-vitae
中国测绘学会 China Surveying and Mapping Society; Chinese Society of Surveying and Mapping
中国长城 great wall of China
中国城 Chinatown
中国(虫)蜡 Chinese insect wax
中国传统风格 traditional Chinese style
中国传统山水画 traditional Chinese painting of mountains and water
中国传统园林 traditional Chinese garden
中国船级社 China Classification Society
中国大漆 Chinese lacquer
中国灯笼 Chinese lantern
中国地台 Chinese platform
中国地毯 China carpet; China rug
中国地震烈度 Chinese seismic degree
中国地震学会 China Society of Seism
中国地质学会 China Society of Geology
中国东部滨太平洋活动区 marginal-Pacific active area in Eastern China
中国东部间歇性隆升区【地】intermittent uplifting area in East China
中国东海 East China Sea
中国东海沉降海盆 East China Sea subsiding basin
中国东海东部-新竹西部拗陷地带 Eastern East China Sea-Western Xinzhu depression region
中国东海陆架坳陷带 Downwarping zone of East China sea shelf
中国对外承包工程商会 China International Contractors Association
中国对外贸易运输(集团)公司 China National Foreign Trade Transportation Corp.; Sinotrans
中国多花紫树 Chinese tupelo
中国风俗 Sinicism
中国风味食物 Sinicism
中国港口<杂志名> China Ports
中国港湾建设总公司 China Harbo(u)r Engineering Company
中国革命博物馆 Museum of the Chinese Revolution
中国工程建设标准化协会 China Association for Engineering Construction Standardization
中国工业标准 Chinese Industrial Standards
中国工艺品用铅锡黄铜 Chinese art metal
中国公路桥梁工程公司 China Road and Bridge Engineering Company
中国公路运输协会 China Highway and Transportation Society
中国公路运输学会 China Highway and Transportation Society
中国公用电子信箱系统 Chinamail
中国公用(计算机)互联网 CHINANET
中国古代建筑 ancient Chinese architecture
中国古代园林 ancient Chinese garden
中国古典园林 Chinese classical garden; classic(al) Chinese garden
中国灌溉排水委员会 Chinese National Committee on Irrigation and Drainage
中国硅酸盐学会 Silicate Society of China
中国国际贸易促进会经贸仲裁委员会 Economic and Trade Arbitration Commission of CCPTT
中国国家标准 National Standard of PR China
中国国务院所属环境保护领导小组 Leading Group Environment (al) Protection under the State Council of China
中国海上搜索救助中心 China Maritime Search and Rescue Center
中国海事仲裁委员会 Maritime Arbitration Commission of China
中国海洋湖沼学会 Chinese Society of Oceanology and Limnology
中国海洋学会 Chinese Society of Oceanology
中国红 Chinese red
中国红茶 congou
中国红十字会 Red Cross Society of China
中国华南-东南亚板块【地】South China-Southeast Asia Plate
中国画 Chinese painting
中国环保基本政策 the basic policies of China's environmental protection
中国环境教育影视资料中心<北京> Environmental Education Television Project for China
中国环境与可持续发展资料研究中心<北京> China Environment and Sustainable Development Reference and Research Center
中国黄 Chinese yellow
中国机械工程学会 China Society of Mechanical Engineering
中国计量单位制 China Official System of Units
中国技术进出口总公司 China National Technical Import & Export Corporation
中国建设工程造价管理协会 China Association for the Management of Construction Cost
中国建筑<杂志名> Chinese Architecture
中国建筑工程公司 China Construction Engineering Corporation
中国建筑工程机械公司 China National Construction Machinery Company
中国建筑工业公司 China Building Industrial Corporation
中国建筑设备配件出口总公司 China Building Equipment and Parts Export Corporation
中国建筑史 history of Chinese Architecture
中国建筑学会 Architectural Society of China
中国建筑装饰 Chinese architecture decoration
中国交互网络协会 China Interactive Network Association
中国交通运输协会 China Communications, Transportation Association
中国教育和科研(计算机)网【计】Chinese Education & Research Network
中国教育科研网 China Education and Research Network
中国近代公园 modern gardens in China
中国近代花园 contemporary gardens in China
中国镜铜 Chinese speculum metal
中国科学技术协会 China Association for Science and Technology
中国科学院 Academia Sinica; Academy of Sciences of China; Chinese Academy of Sciences
中国科学院兰州冰川冻土沙漠研究所 Institute of Glaciology
中国科学院学部委员 division member
中国蜡 Chinese wax
中国蓝 China blue; Chinese blue
中国历史博物馆 Museum of Chinese History
中国柳 cathay willow
中国绿 China green; Chinese green; lakao
中国煤气学会 China Gas Society
中国庙 joss house
中国民航管理局 Civil Aviation Administration of China
中国墨(水) China ink; Chinese ink
中国南海北部陆缘坳陷带 Downwarping zone of northern continent margin of South China Sea
中国南海南部陆缘坳陷带 Downwarping zone of southern continental margin of South China Sea
中国农学会 Chinese Society of Agriculture
中国判小蠹<拉> Scolytoplatypus sinensis
中国屏风 Chinese screen
中国七号信令【铁】Chinese No.7 Signal(l)ing
中国七叶树 Chinese horsechestnut
中国漆 Chinese varnish
中国青铜 Chinese bronze
中国人民保险公司 The People's Insurance Company of China
中国人民建设银行 People's Construction Bank of China
中国柔性路面设计法 China flexible pavement design method
中国山水园 Chinese mountain and water garden
中国商品检验局 China Commodity Inspection Bureau
中国生漆 Chinese lacquer
中国剩余定理 Chinese remainder theorem
中国式 Chinese style
中国式二进制 Chinese binary
中国式建筑 Chinese Architecture
中国式水车 Chinese dragon pump
中国式算盘 swanpan
中国式装饰艺术 Chinese decorative art; Chinoiserie
中国数字数据网 China Digital Data Network
中国水车 Chinese noria
中国水利部 The Ministry of Water Resources of China
中国水利学会 China Society of Hydraulic engineering; Chinese Hydraulic Engineering Society; Chinese Society of Hydraulic Engineering
中国水墨画 Chinese ink-and-wash painting
中国特产 special product of China
中国天文学会 Chinese Astronomical Society
中国铁路标准活载 standard railway live load
中国铁道科学研究院 China Academy of Railway Sciences
中国铁路工业 China railway industry
中国桐油 wood oil
中国土木工程公司 China Civil Engineering Corporation
中国土木工程学会 China Society of Civil Engineering
中国外轮代理公司 China Ocean Shipping Agency
中国外轮理货公司 China Ocean Shipping Tally Company
中国网 China network
中国线规 Chinese Wire Ga(u)ge
中国谢库多铜合金 Chinese Shaku-do bronze
中国型 sinotype
中国医疗队 Chinese Medical Team
中国医学科学院 Chinese Academy of Medical Sciences
中国已加入的国际公约 international conventions into which China has accessed
中国艺术风格 Chinoiserie
中国银合金 China silver alloy
中国银行 Bank of China
中国远洋运输(集团)总公司 China Ocean Shipping(Group) Company
中国招商局轮船股份有限公司 China Merchants Steam Navigation Co.LTD
中国折扇形磨瓦 Chinese folding-fan shaped abrasive tile
中国制冷学会 Chinese Association of Refrigeration
中国中部半干旱干旱亚热带 central China semiarid to arid subtropical zone
中国钟蚼【动】Campamuria chinesis
中国朱砂 Chinese vermil(l)ion
中国珠茶 gunpowder

中国竹斗戽水车 Chinese bamboo water bucket
中海 intermediate sea
中函数关系分析 analysis of functional relationship
中寒 cold stroke; cold syndrome of middle-Jiao
中寒武世【地】Middle Cambrian
中寒武统 Middle Cambrian series
中航程控制 midcourse control
中航道 mid-channel
中号锉 second cut file
中号的 medium sized
中号金刚石钻头 medium stone bit
中合金 medium alloy
中合金钢 medium alloy steel
中合金钢钢丝 medium alloy steel wire
中和 balance out; counteract; counteraction; neutralize; stand-off
中和比率 neutralization ratio
中和变压器 neutralizing transformer
中和操作规程 neutralization procedure
中和槽 neutralization chamber; neutralization tank; neutralizing bath; neutralizing tank; neutralizing well
中和层 neutral layer
中和产品 neutralized products
中和沉淀 neutralizing precipitation
中和沉淀池 neutralizing precipitation tank
中和沉降 neutralization settling; neutralizing settling
中和池 balancing tank; neutralization pond; neutralization tank; neutralizer; neutralizing tank; stand-off basin
中和处理 neutralization treatment; neutralizing; neutralizing treatment
中和当量 neutralization equivalent
中和的 corrective; neutral; neutralizing
中和滴定曲线 titration curve of neutralization
中和地带 neutral zone
中和点 balance point; neutralization point; neutral point
中和点法 neutral point method
中和点轨迹 null circle
中和电荷 neutralizing charge
中和电路 neutralizing circuit; neutrodyne circuit
中和电容器 neutralizing capacitor; neutrodon
中和电压 neutralizing voltage
中和段 neutralizing zone; neutral zone
中和法 neutralization; neutralization method; neutralization process; neutrodyne; neutrodyne system
中和法脱硫 neutral process desulfurization
中和反应 neutralization reaction; neutral reaction
中和放大器 neutralized amplifier
中和废水处理 neutralization wastewater treatment
中和浮力 neutral buoyancy; neutralizing buoyant
中和工具 neutralizing tool
中和罐 neutralizing tank
中和轨温 rail neutral temperature
中和过滤 through neutralite filtration
中和过酸度 neutralizing overacidity
中和化学药品 neutralizing chemicals
中和缓冲器 neutralized buffer
中和黄颜料 neutralizing yellow tone
中和混凝 neutralization coagulation
中和纪元 epoch of neutralization
中和剂 absorbent; absorber; averager; neutralization agent; neutralization reagent; neutralizer; neutralizing agent
中和继电器 neutral relay; non-polarized relay
中和角进气口 neutral-angle air intake
中和接收法 neutrodyne
中和接收机 neutralized receiver; neutrodyne receiver
中和接受法 neutrodyne reception
中和接受机 neutrodyne receiver
中和截面 neutral cross-section
中和界 neutral level; neutral pressure level; neutral zone
中和孔径 neutralizing aperture
中和力 counteragent
中和硫酸盐废液 neutralized sulfite liquor
中和滤波器 neutralizing filter
中和滤池 neutralizing filter
中和面 neutral flow plane; neutral plane; neutral surface
中和面法 neutralized surface method
中和膜滤法 neutralizing-membrane filtration process
中和凝聚法 neutralizing coagulation
中和平面 neutral plane
中和器 averager; neutralizer
中和区 neutralizing zone; neutral zone
中和曲线 neutralization curve
中和热 neutralization heat
中和色 neutral colo(u)r; neutralized colo(u)r
中和色调 neutral-tone
中和设备 neutralizing device; neutralizing equipment
中和射频放大器 neutralized radio-frequency amplifier
中和射频级 neutralized radio-frequency stage
中和升力线 neutral lift line
中和式超外差接收机 neutrosonic receiver
中和式高频调谐放大器 neutrodyne
中和试剂 neutralization reagent
中和试验 neutralization test
中和数 neutralization number
中和水池 neutralizing water tank
中和水深 neutral depth
中和塔 neutralizing tower
中和条件 neutrality condition
中和调整 neutral adjustment; neutralization adjustment
中和通道 neutralizing canal
中和土壤 sweetening of soil
中和脱臭装置 malodo(u)r counteraction equipment
中和脱硫酸法 neutral desulfating process
中和位置 neutral position
中和温度 neutral temperature
中和污泥泵 neutralized sludge pump; neutral sludge pump
中和系统 neutralized system
中和线 dividing line; neutralizing wire; neutral line
中和线圈 neutralization coil; neutralizing coil
中和效应 neutralizing effect
中和压力 neutral pressure
中和亚硫酸盐液 neutralized sulfite liquor
中和应力 neutral stress
中和值 neutralization number; neutralization value; neutralizing value
中和轴 zero line; neutral axis
中和轴高度 neutral axis depth
中和轴深度 <梁的> depth of neutral axis
中和轴线 neutral axis
中和装置 neutralizing equipment; neutrodyne system
中和作用 neutralization; neutralizing effect
中横挡 parting rail; middle rail <门的>; check rail <双悬窗的>
中横隔梁 mid-span diaphragm
中横框 transom
中横缆 waist breast
中横梁 middle cross beam; middle transom <三轴转向架>
中红外 medium infrared; middle infrared
中红外干涉光谱仪 mid-infrared interferometric(al) spectrometer
中红外光谱 middle infrared spectrum
中红外线 medium infrared
中泓 <河道的> middle thread; midstream; stream centerline
中泓浮标 central buoy; central float; middle thread float
中泓水面流速 central surface velocity
中泓水深 midstream depth
中泓线 channel line; stream centre line[center line]; thread of stream; line of fastest flow; line of maximum velocity; thread of channel
中后桥驱动 center[centre] and rear axle drive
中后桥驱动式铲运机 center[centre] and rear axle drive scraper
中后桥驱动式后卸卡车 center[centre] and rear axle drive rear-dump
中后轴平衡悬架 <汽车> bogie unit
中后轴驱动 center[centre] and rear axle drive
中厚 <3 毫米厚的窗玻璃> demi-double thickness
中厚板 cut deal; medium plate
中厚板的厚度 ga(u)ge of plate
中厚板剪切机 plate cutter
中厚板矫直机 plate level(l)er; plate level(l)ing machine; plate-straightening machine
中厚板拉伸矫直机 plate stretcher
中厚板双边剪边剪切机 double-sided plate trimming shears
中厚板轧机 heavy and medium plate mill
中厚玻璃 <2.8 ~ 3.0 毫米> semi-double strength
中厚层 medium bed
中厚层的 medium bedded
中厚层土 medium bedded soil
中厚层状 medium bedded
中厚层状结构 medium thick bedded texture
中厚窗玻璃 <3 毫米厚> demi-double strength window glass; demi-double thickness sheet glass
中厚钢板 medium plate; plate iron; plate steel; steel plate
中厚钢板轧制 plate rolling
中厚黄铜板 brass plate
中厚矿层 medium thickness ore seam
中厚煤层 medium thickness coal seam
中弧线 mean camber line; medial camber line; skeleton line
中弧线弯度 camber curvature
中湖 mesolimnion
中华白海豚 Sousa chinensis
中华斗鱼 Macropodus opercularis chinensis
中华古陆 Zhonghua old land
中华海鲇 Arius sinensis
中华河蚓 Rhyacodrilus sinicus
中华柳 Salix cathayana
中华门 Zhonghua Gate
中华全国总工会 All-China Federation of Trade Unions
中华人民共和国 People's Republic of China、
中华人民共和国测绘法 Law of Surveying and Mapping of the People's Republic of China
中华人民共和国船舶检验局 Register of Shipping of the People's Republic of China
中华人民共和国防止船舶污染海域管理条例 Regulations of the People's Republic of China on Administration for Prevention of Pollution in Sea Areas by Vessels
中华人民共和国海上交通安全法 Maritime Traffic Safety Law of the People's Republic of China
中华人民共和国海洋环境保护法 Marine Environmental Protection Law of the People's Republic of China
中华人民共和国海洋倾废管理条例 Regulations of the People's Republic of China on Administration of Wastes Dumping to Ocean
中华人民共和国海洋石油勘探开发环境保护管理条例 Regulations of the People's Republic of China on Administration for Environmental Protection of Offshore Oil Exploration and Exploitation
中华人民共和国航道管理条例 River Navigation Administration Regulations of the People's Republic of China
中华人民共和国河道管理条例 River Course Administration Regulations of the People's Republic of China
中华人民共和国核材料管理条例 Nuclear Material Management Regulations of the People's Republic of China
中华人民共和国环境保护标准管理条例 Regulations on Administration of Environmental Protection Standards of the People's Republic of China
中华人民共和国环境保护法 Environmental Protection Law of the People's Republic of China
中华人民共和国环境影响评价 Law of the People's Republic of China on Environmental Impact Assessment
中华人民共和国环境噪声污染防治条例 Regulations of the People's Republic of China on Prevention and Control of Environmental Noise Pollution
中华人民共和国进出口动植物检疫条例 Regulations of the People's Republic of China on Quarantine of Animals and Plants' Import and Export
中华人民共和国经济合同法 Economic Contract Law of the People's Republic of China
中华人民共和国矿产资源法 Mineral Resources Law of the People's Republic of China
中华人民共和国民法通则 General Principles of the Civil Law of the People's Republic of China
中华人民共和国民用核设施安全监督管理条例 Safety Supervision Administration Regulations of Civil Nuclear Device of the People's Republic of China
中华人民共和国人民法院 people's Courts of the People's Republic of China
中华人民共和国森林法 Forest Law of the People's Republic of China

中华人民共和国森林实施细则 Detailed Rules and Regulations for the Implementation of the Forest Law of the People's Republic of China
中华人民共和国食品卫生法 Food Hygiene Law of the People's Republic of China
中华人民共和国水法 Water Law of the People's Republic of China
中华人民共和国水土保持法 Law of the People's Republic of China on Water and Soil Conservation
中华人民共和国水污染防治法 Law of the People's Republic of China on Prevention and Control of Water Pollution
中华人民共和国水污染防治实施细则 Detailed Regulations for Implementation of the Law of the People's Republic of China on Prevention and Control of Water Pollution;Implementation Rules for Law of the People's Republic of China on Prevention and Control of Water Pollution
中华人民共和国水污染控制法 Law of the People's Republic of China on Prevention and Control of Water Pollution
中华人民共和国水污染控制实施细则 Detailed Regulations for Implementation of the Law of the People's Republic of China on Prevention and Control of Water Pollution;Implementation Rules for Law of the People's Republic of China on Prevention and Control of Water Pollution
中华人民共和国铁路标准活载 standard railway live load specified by P.R. China
中华人民共和国土地管理法 Land Administration Law of the People's Republic of China
中华人民共和国文物保护法 Law of the People's Republic of China on Protection of Cultural Relies
中华人民共和国宪法 Construction of the People's Republic of China
中华人民共和国消防条例实施细则 Rules for Implementation of the Fire Regulations of the People's Republic of China
中华人民共和国行政诉讼法 Administration Procedure Law of the People's Republic of China
中华人民共和国野生动物保护法 Law of the People's Republic of China on Wild Animal Protection
中华人民共和国渔业法 Fisheries Law of the People's Republic of China
中华人民共和国治安管理处罚条例 Regulations of the People's Republic of China on Administration Penalty against Public Security
中华绒螯蟹 Eriocheir sinensis
中华铈矿 zhonghuacerite
中华双尖藻 Hammatoidea sinensis
中华水韭 Isoetes sinensis Palmer
中华水蚤 Mesocyclops
中华夏式 Meso-Cathaysian
中华夏系【地】Meso-Cathaysian
中华小长臂虾 Palaoinonestes sinensis
中华新米虾 Neocaridina denticulate sinensis
中华鲟 Acipenser sinensis; Chinese sturgeon
中华织锦村 China fabrics village
中化石灰 medium slaking lime
中环 middle ring; discharge ring < 转桨式水轮机的 > ; intermediate kit
中环带 intermediate belt
中环化合物 medium ring compound
中环路 intermediate belt; intermediate ring road
中环螺栓 toggle bolt
中簧继电器 neutral-tongue relay
中灰 medium gray[grey]
中灰滤光片 neutral density disc;neutral density filter
中灰滤光屏 black screen; dark-tint screen
中灰煤 mid-ash coal
中挥发分煤 medium volatile coal
中挥发性物 medium volatile
中活载 medium live loading;standard railway live load specified by P.R. China【铁】
中火釉 intermediate fire(d) glaze
中击式水轮机 middle shot water wheel
中基性喷出岩 medium basic eruptive
中基质 mesostroma
中级 intermediate grade;intermediate type;medium rank
中级变质作用 medium metamorphism
中级玻璃纤维 medium grade glass fibre
中级材 middlings
中级处理 intermediate treatment
中级处理的 medium curing
中级的 intermediary; intermediate; middling;mesic < 土温夏冬温差大于5℃、年平均5~15℃ >
中级法院 intermediate court
中级刚度 intermediate stiffener
中级钢 intermediate grade steel
中级钢杆 intermediate grade billet
中级钢坯段 intermediate grade billet
中级过滤汽缸油 medium filtered cylinder oil
中级会计师 semi-senior accountant
中级金刚砂粉 medium emery powder
中级晶族 intermediate category
中级精度配合 medium fit
中级精加工 medium finish(ing)
中级净化 medium cleaning
中级绝缘 medium insulation
中级颗粒 intermediate grain
中级课程 intermediate course
中级矿砂 middling
中级粒度的 medium granular
中级裂纹 medium torn grain
中级路面 intermediate class pavement;intermediate type pavement; intermediate type surface
中级磨碎 medium grinding
中级(耐水)胶黏[粘]剂 intermediate glue
中级黏[粘]度润滑脂 medium hair grease
中级(黏[粘]土质)耐火砖 medium duty fireclay brick; intermediate-duty fireclay brick
中级跑道 intermediate course
中级配合 medium fit
中级品 middling
中级品位 medium grade
中(级破)碎机 intermediate crusher
中级汽油 intermediate gasoline
中级燃油 intermediate fuel oil
中级筛分 medium screening;medium sizing
中级筛选 medium screening;medium sizing
中级伤害剂 moderate casualty agent
中级熟油 middle stand oil
中级无烟煤 medium rank anthracite
中级线路 middle class line
中级性 semi-polarity
中级压(紧)配合 medium force fit
中级压碎 intermediate crushing; medium crushing
中级烟煤 medium rank bitumite
中级研磨 intermediate grinding
中级研磨膏 mean paste
中级涌浪 moderate swell
中级油 middle oil
中级重量 medium weight
中级重燃油 medium heavy fuel
中级砖 medium brick
中极潮(位) mid-extreme tide
中极性 semi-polarity
中脊 median ridge
中脊重力异常 gravity anomaly over mid-ridge
中计 intermediary total;intermediate total;inter summary【计】
中继 junction;relay(ing);repeating; retransmission;retransmit;trunking
中继泵 relay pump
中继泵站 booster pump station;relay pump(ing) station
中继泵转输 relay pumping
中继变速器 relay box
中继变压器 relay transformer
中继布置 trunking scheme
中继储存 relay storage
中继处理机 trunk processor
中继传输 relay transmission
中继地面站 relay earth station
中继点 relay point
中继电缆 coupling cable; feeder cable;interconnecting cable;intermediate cable; junction cable; trunk cable
中继电路 junction circuit; link circuit;trunk circuit
中继电视 relay television
中继电台 linking-up station
中继段 repeater section
中继发射机 relay transmitter;repeater transmitter;retransmitter
中继阀 relay valve
中继方式 trunking;trunking scheme
中继放大器 relay amplifier
中继分割区段 relayed cut section;repeating cut section
中继港 transit port
中继港口 junction port
中继工作场 staging area
中继机测试台 repeater test board
中继机台 repeater board
中继计划 switching scheme
中继继电器 repeater relay;repeating relay
中继间法 relay chamber method
中继交换机 tandem exchange
中继接收机 link receiver
中继接收系统 ball reception
中继接续 trunk connection
中继局 junction exchange; switching office;tandem office
中继空气阀组 relay air valve unit
中继连接 trunk connection
中继连接继电器 trunk connecting relay
中继链路网 trunk link network
中继平台 relaying platform
中继器 repeater;trunk circuit
中继器测试架 repeater test raceway
中继器盘 repeater panel
中继器振鸣 singing of repeater
中继塞绳 connecting cord;disconnecting link;junction cord;patch cord
中继扫描器 trunk scanner
中继设备 repeating instrument;trunk equipment
中继台 junction position;linking station; repeater station; trunk line board
中继台话务员 junction operator
中继通信[讯] relay communication
中继透镜 relay lens
中继图 junction diagram
中继卫星 relay satellite;relay station satellite
中继线 connecting circuit; junction; junction line; trunk; trunk line; trunk main
中继线测试台 trunk test desk
中继线方式图 trunking diagram
中继线分品法 grading
中继线呼叫状态信息 trunk call status information
中继线继电器 trunk relay
中继线架 bay trunk
中继线交换 trunk exchange
中继线路 relay line
中继线齐次分品 homogeneous grading
中继线圈 repeating coil
中继线全部占线信号 overflow signal(l)ing
中继线全忙 all trunks busy
中继线全忙计次器 all-trunks-busy register
中继线群 junction group;trunk group
中继线群号码 trunk group number
中继线塞孔 trunk jack
中继线塞子 trunk plug
中继线设备 junction equipment; relayed line unit
中继线示闲灯 idle trunk lamp
中继线网 <市内电话> junction line network
中继线效率 trunk efficiency
中继线寻线机 trunk finder; trunk hunting connector
中继线占线 trunk busy
中继线占用指示灯 trunk congestion lamp
中继线组 junction group
中继信号 repeater signal; repeating signal
中继信号机【铁】 transition signal
中继业务 junction traffic
中继音响器 relaying sound;repeating sound
中继油库 relay bulk plant;relay bulk station; relay depot; relay tank farm;relay dep < 英 >
中继站 linking station; minor relay station; rebroadcast station; relaying station;relay point;relay station;repeater station;repeating station;through station
中继站发送 retransmit
中继制 relay system
中继中心 relay center[centre]
中继状态 relay state
中继子系统 trunk subsystem
中继作用 relaying action
中加成分 infix
中加里东期地槽 middle Caledonian geosyncline
中间 bosom;interim;interspace;mean; middle;midst;midway
中间安全岛 <高速公路的> median
中间(安全)扶手 midrail
中间按钮 intermediate button
中间搬运 intermediate handling
中间板 intermediate bay;intermediate lamella;intermediate plate
中间板块 intermediate plate; mesoplate
中间板梁 middle girder
中间半径 intermediate radius
中间帮电机 through repeater
中间包 bakie;pouring basket;pouring box;trough;tundish

中间包浇注 basket pouring
中间包浇铸 trough casting
中间保险链 intermediate safety chain
中间报告 interim report;intermediate report
中间倍数 intermediate multiple
中间泵 process pump
中间泵站 intermediate pumping station
中间比例模型 mesoscale model
中间闭塞机 intermediate block instrument
中间壁 intermediate partition; midfeather;midfellow;partition
中间避雷器 intermediate lightning arrester
中间臂节 insert jib section
中间臂靠 division arm
中间边帮角 intermediate angle
中间边坡角 intermediate slope angle
中间变量 intermediate variable
中间变流器 current transformer; intermediate converter
中间变压器 intermediate transformer
中间波 mid-wave;transitional wave
中间玻璃 intermediate glass
中间玻色子 intermediate boson
中间剥削 interposition
中间补偿器 intermediate compensator
中间补燃加力燃烧室 interburner
中间部分 center[centre] section;central section; intermediate section; intermediate part
中间部件 centre[centre] piece
中间材 intermediate wood
中间采光 center[centre] light
中间采区 mid-workings
中间仓 basket; buffer hopper; midbin;surge bin
中间仓和喂煤机燃烧系统 bin and feeder firing
中间仓库 intermediate depot
中间舱 center[centre] tank
中间舱壁 intermediate bulkhead
中间层 interbedding;interface layer; interleaf; intermediate belt; intermediate course;intermediate layer; intermediate zone; interstratification;medium bed;meso-layer;mesosphere;middle layer;mid-shaft
中间层壁板 <冷藏车用> blind lining
中间层臭氧 mesospheric ozone
中间层地下水 intermediate groundwater
中间层顶 mesopause
中间层顶板 <冷藏车> blind ceiling
中间层改正量 intermediate correction
中间层厚度图 thickness figure of mid-layer
中间层校正 intermediate correction; mid layer correction; stone slab correction
中间层校正值 intermediate layer correction value
中间层墙板 intermediate lining
中间层水体 intermediate water
中间层影响 the effect of intermediate layer
中间层最高温度点 mesopeak
中间产品 between product;in-process product;intermediate goods;intermediate product; intermediates; semifinished product;semi-processed unit
中间产物 between product;in-process product; intermediate product; intermediates;middling
中间产物理论 intermediate product theory
中间产物品位 cascade grade
中间厂规模 pilot plant scale
中间场 intermediate field
中间潮 intermediate tide
中间车道 center[centre] lane;middle lane
中间车间 semi-work
中间车辆 intermediate car
中间车桥驱动式卡车 center-axle drive type
中间车速 median speed
中间车站 intermediate station; wayside station
中间沉淀 intermediate sedimentation
中间沉淀槽 intermediate sedimentation tank
中间沉淀池 intermediate sedimentation tank; intermediate settling tank;dilly hole <中型的>
中间沉降 intermediate sedimentation
中间衬料 intermediate lining material
中间成像能谱仪 intermediate-image spectrometer
中间承木 intermediate bearer
中间承托 intermediate bearer
中间程序 interlocutory proceeding; interlude
中间程序块 intermediate block
中间程序块检查 intermediate block checking
中间澄清池 intermediate clarifier
中间池 fill and draw basin
中间尺寸 intermediate size
中间尺寸的 medium sized
中间尺度 intermediate scale; medium scale
中间齿轮 idler gear; idler sprocket (wheel); intermediate gear; intermediate pinion; internal gear; midgear;neutral gear;translating gear; transposing gear
中间齿轮套筒 intermediate gear sleeve
中间齿轮托架 intermediate gear bracket
中间齿轮箱 intermediate gear box
中间充气汽轮机 induction turbine
中间充填带 intermediate strip pack
中间冲洗 intermediate rinse
中间抽汽室 intermediate bleeding chamber
中间抽水站 intermediate pumping station
中间抽头 center tap;intermediate tap
中间抽头的 tapped
中间抽头扼流圈 centre-tapped choking coil
中间出车台 mid-door
中间出料管 intermediate delivery pipe
中间初轧机 intermediate bloomer
中间储存 intermediate storage
中间储存仓 buffer bin
中间储罐 intermediate storage tank; relay reservoir
中间储棉箱 intermediate hopper
中间处理 intermediate handling; intermediate treatment
中间处理部件 intermediate processing element
中间处理机 interprocessor
中间触点 intermediate contact
中间传动 intermediate drive;intermediate gearing; intermediate transmission
中间传动齿轮 intermediate drive gear
中间传动杆 intermediate shaft
中间传动轴 intermediate propeller shaft
中间传送 intermediate transport
中间传送机 intermediate conveyer [conveyor]
中间船舶 middle vessel
中间椽 common rafter; intermediate rafter
中间窗 center[centre] light; center [centre]-to-center[centre] window
中间窗梃 intermediate mullion
中间锤座 intermediate block
中间磁极 commutating pole; consequent pole;interpole
中间磁极绕组 commutating winding
中间催化剂薄膜 intermediate catalyst film
中间存储保持器 intermediate memory storage
中间存储器 buffer storage;intermediate memory; intermediate storage; intermediate store; scratch-pad storage;temporary storage
中间存取存储器 intermediate access storage
中间打点轮 intermediate strike wheel
中间大梁 intermediate girder
中间(代)码 intermediate code
中间代谢物 intermediate metabolite
中间带 central strip; median; vadose water zone <渗水带>【地】
中间带地下水 intermediate water
中间带护栏 median guard bar
中间带宽 intermediate frequency bandwidth
中间带排水 median drainage
中间带喷嘴排气管 exhaust pipe with intermediate nozzles
中间带上层滞水 intermediate vadose water
中间单位 intermediate unit
中间单元 temporary location
中间挡板 intermediate diaphragm
中间挡速率 intermediate speed
中间刀杆支架 intermediate type arbo(u)r support
中间导管 middle vessel
中间导线 central conductor;intermediate conductor
中间导叶 intermediate guide blade
中间倒位 intercalary inversion
中间道次 pony-roughing pass
中间的 intermediary;intermediate;interstage;medial;mesial;middle
中间的二分之一 middle half
中间的三分之一 middle third
中间等高线 intermediate contour; mediate contour;medium contour
中间等级的 intermediate grade;middle-bracket
中间底片 internegative
中间底图 intermediate guide key
中间地板 intermediate floor
中间地槽 mediterranean; mesogeosyncline
中间地带 intermediate area;intermediate belt; intermediate zone; middle strip
中间地带渗流水 intermediate vadose water
中间地块 betwixt mountain; median mass;median massif
中间地面标志 middle ground buoy
中间地平圈 intermediate horizon
中间地下水 intermediate ground water
中间地下水湿生植物 mesophreatophyte
中间地质报告 intermediate geologic(al) report
中间点 intermediate point
中间电池供给装置 intermediate battery supply board
中间电话局 intermediate exchange; intermediate office
中间电极 intermediate electrode
中间电缆 intermediate cable
中间电缆盒 intermediate cable terminal box
中间电路 intermediate circuit
中间电路控制分站 intermediate circuit sub-control station
中间电容器 intercondenser
中间电位 intermediate potential
中间电阻 interlaminated resistance; intermediate resistance
中间垫片 intermediate washer
中间吊灯 interpendent
中间吊饰 interpendent
中间调车 intermediate switching
中间丁坝 intermediate groin
中间钉(瓦)法 <屋面石板瓦> center[centre] nailing
中间定位 interfix
中间定位器 middle locator
中间定向 <在一导线点上向一远方控制点观测的方向> intermediate orientation
中间读出 intermediate readout
中间读数【测】 intermediate siding
中间独立舱 independent tank center [centre]
中间段 through zone;middle section
中间墩 intermediate pier
中间舵栓 intermediate pintle
中间二分之一 center[centre] half
中间发券 intermediate arch
中间发送器 half-time emitter
中间阀 intercepting valve;intermediate valve
中间阀盖接头 intercepting valve head connection
中间阀室 intermediate valve chamber
中间阀体 <压缩机的> intermediate valve cage
中间法兰 spacer flange
中间帆 interpositum;velum interpositum
中间反射率 medium reflectivity
中间反向进位 intermediate negative carry
中间反应塔 intermediate reaction tower
中间范围 intermediate range
中间防擦块 intermediate chafing block
中间放大级 interstage amplifier; interstage amplifier section
中间放大器 intermediate amplifier
中间分车带 median strip;medium strip
中间分隔带 median(separator);median strip;medium strip
中间分类处理中心 intermediate processing center[centre]
中间粉磨仓 intermediate grinding compartment; medium grinding compartment
中间粉碎 intermediate comminution; intermediate reduction
中间封锁 interblock
中间浮雕 middle relief
中间浮子 intermediate float
中间负荷 intermediate load
中间负荷火力发电站 middle load thermal power plant
中间负荷机组 cyclic(al) unit; intermediate load unit
中间负荷装置 intermediate load unit
中间负进位 intermediate negative carry
中间负重轮 intermediate road wheel
中间附着 intercalary attachment
中间赋值 intermediate assignment
中间盖 mid-cover
中间干线配线架 intermediate trunk distributing frame
中间干燥 intermediate drying

中间杆 center[centre] pole; middle pole
中间缸 intermediate part
中间钢丝<待热处理和继续拉拔的> process wire
中间港 intermediate port
中间隔板 common bulkhead
中间隔墙 mid-feather
中间工厂 mini-plant; pilot plant; replica plant; semi-plant
中间工厂规模 pilot scale
中间工厂规模生产 semi-commercial production; semi-works production
中间工厂规模试验 semi-works production
中间工厂试验 pilot plant test
中间工厂试验炉 pilot furnace
中间工厂研究 pilot plant work
中间工业生产规模 semi-industry scale
中间公差限度 median tolerance limit
中间功率放大器 intermediate power amplifier
中间功能 intermediate function
中间拱 center arch
中间拱廊 middle stoa
中间共巷道掘进 intermediate drivage
中间共振 intermediate resonance
中间沟谷 median valley; medium valley
中间构架 intermediate frame
中间构件 intermediate member
中间构造层次 intermediate tectonic level
中间骨料 intermediate aggregate
中间鼓起的镶板 raised and fielded panel
中间固定 interim fixing
中间固溶体 intermediate solid solution
中间挂车<旅客动车列车组> center [centre]-trailer
中间挂架 mid-mounted frame
中间挂饰 interpendent
中间观测 intermediary observation; intermediate observation; intermediate sight
中间管 intervalve
中间管板 sagging plate
中间管口加粗异径三通管接 reducing on outlet tee[T]
中间管理图 median chart
中间罐 basket; intermediate container; intermediate tank; pouring basket; pouring box; receiving tank; surge tank; tundish
中间罐底 tundish bottom
中间罐浇注 basket pouring; tundish casting
中间罐浇铸 trough casting
中间罐区储液能力 intermediate tankage
中间归约 intermediate reduction
中间规模堆 pilot plant reactor
中间轨道 interim orbit; intermediate orbit
中间轨距板 center gage plate
中间辊 transfer roll
中间辊道 delay table
中间滚轮<铲土机> intermediate idler
中间过程导航 midcourse navigation
中间过道 central corridor
中间过渡地带 limbo
中间过渡地段 limbo
中间过滤器 intermediate
中间过轮 intermediate setting wheel
中间过热器 interheater; intermediate superheater; reheater
中间航道<船闸> intermediate navigation channel; intervening channel
中间壕 intermediate trench
中间合金 interalloy; intermediate alloy; rich alloy; tempering metal
中间核(心) central core
中间荷载 intermediate cycling load
中间桁架 intermediate truss
中间桁条 middle purlin(e)
中间横撑 intermediate strut
中间横隔板 intermediate diaphragm
中间横联<桁架桥> sway bracing
中间横联杆 intermediate cross frame
中间横梁 cross beam; intermediate cross girder; intermediate transverse girder
中间横向框架 intermediate transverse frame
中间烘干 intermediate drying
中间湖沼 mesolimnion
中间滑动球节 sliding center[centre] ball
中间滑轮 guide pulley; transmitting pulley; intermediate pulley; intermediate idler<铲土机>
中间化合物 intermediate compound
中间化学品 intermediate chemicals
中间环 adapter ring
中间缓冲器 intermediate buffer; intersnubber; interstage surge tank
中间缓行器 intermediate retarder
中间灰度扩张<图像处理中> mid region stretch
中间灰质 medicinerea
中间回流 intermediate reflux
中间回路 intermediate loop
中间汇率 medium rate of exchange
中间混合法 medial allegation
中间活塞 intermediate piston
中间货品 intermediate goods
中间机场 staging field
中间机构 interagency
中间机座 intermediate stand; interstand; run-down mill
中间积分 intermediate integral
中间基 intermediate base
中间基本方位 intercardinal point; quadrantal point
中间基石油 intermediate base crude (oil); intermediate crude
中间基原油 intermediate base crude (oil)
中间激发 split shooting
中间级 intermediate stage; intermediate step; transtage
中间级配 intergrade
中间极 dynode; interpole
中间极绕组 consequent poles winding
中间集尘器 intermediate dust collector
中间集结区 transit area
中间集乳器 interceptor jar
中间给水站 intermediate water supply station
中间计数 intermediate total
中间计算 subtotaling
中间计算机 intermediate computer
中间计算器 intermediate counter
中间技术 intermediate technology
中间技术检查 interperiod inspection
中间技术检验 intermediate technical examination
中间剂 intermediate agent
中间继电器 auxiliary relay; intermediate relay
中间寄主 intermediate host
中间加劲杆 intermediate stiffener
中间加劲肋 intermediate stiffener
中间加劲柱角钢 intermediate stiffener angle
中间加强杆 intermediate stiffener
中间加强肋 intermediate stiffener
中间加热 intermediate heating; interstage heat
中间加热器 booster heater; interheater; intermediate heater; reheater
中间夹层 intermediate course; intercupola<圆屋顶的>
中间甲板 between decks; intermediate deck
中间价 middle rate
中间价格 middle price
中间架杆 intermediate support spar
中间假肋 intermediate false rib
中间间隔 mid-feather
中间减速过齿轮 intermediate reduction pinion
中间减速轮 intermediate reduction wheel
中间减速器 intermediate gear box; intermediate retarder
中间检查 interim inspection; intermediate inspection; interperiod inspection
中间检查点 intermediate inspection point
中间检订 intermediate revision
中间检验 half-time survey; intermediate inspection; intermediate survey
中间件 intermediate part
中间渐屈线 intermediate evolute
中间槛 intermediate sill
中间交工验收 acceptance in intermediate construction stage; intermediate acceptance
中间交换计算机 intermediate switching computer
中间胶片 intermediate film
中间焦点 intermediate focus
中间较厚的镶板 raised panel
中间阶段 interim stage; intermediate stage; mesophase
中间阶段报告 interim stage report
中间阶段复制 intermediate reproduction
中间阶段规划 intermediate plan
中间阶段设计 middle phase of planning
中间接触轨 center[centre] contact rail; center[centre] conductor rail
中间接合盘 intermediate flange
中间接环 connect collar
中间接力器 pilot servomotor
中间接片 intermediate jack
中间接受器 intermediate receiver
中间接头 compromise joint; transition joint; union
中间接线盒 intermediate jointing box
中间接运工具 intermediate carrier
中间街区 mid-block
中间节 intermediate nodule
中间结果 intermediate result
中间结果程序 intermediate object program(me)
中间结果存储区 working area
中间结果寄存器 scratch-pad register
中间截面 intermediate cross-section; mid-section
中间截止阀 intercept valve; intermediate stop valve
中间解冻时期 intermediate thaw(ing) period
中间介体 vehicle base
中间介质气化器 intermediate vaporizer
中间金属 intermetallic metal
中间金属相微粒 intermetallic phase particle
中间浸渍 intermediate impregnation
中间净化 medium cleaning
中间径 pitch diameter
中间局 cut-in station; intermediate administration; middle station<电话>
中间局振铃器 intermediate ringer
中间距离 intermediate range
中间开动架锯机 center driven log frame sawing machine
中间开关 intermediate switch
中间开关站<输电线的> intermediate switching station
中间开坯机座 intermediate bloomer
中间拷贝 intermediate copy
中间颗粒 intermediate grain
中间壳 middle case
中间壳体 middle casing
中间空气过滤器 intermediate air filter
中间控制 intermediate control
中间控制变更 intermediate control change
中间控制变换 intermediate control change
中间控制尺寸 intermediate controlling dimension
中间控制分站 intermediate sub-control station
中间控制级 centre gate
中间控制设备 neutral-controlled plant
中间控制数据 intermediate control data
中间扣件 intermediate fastening
中间库存 interim stock; interval stock
中间库防锈 rust prevention in interstore
中间跨 central span; interior span; intermediate bay
中间跨度 central spalling; intermediate span
中间跨弯矩 mid-span moment
中间块 intermediate mass; massa intermedia
中间矿层 middle seam
中间矿柱 interstall pillar
中间扩大部分<隧道开挖> upraise
中间拉杆 intermediate draw bar
中间拉杆销 intermediate draw bar pin
中间蜡 intermediate wax; motor oil wax
中间栏杆 middle rail
中间浪 buckle
中间肋 intermediate rib; tertiary rib<哥特式拱上的>; tierceron<穹的>
中间肋骨 intermediate frame
中间类型 intermediate form; passage form
中间类型桩 friction(al) and end-bearing pile
中间冷凝器 intercondenser; intermediate condenser
中间冷却 cooling back; intercooling; intermediate cooling; interstage cooling; intervening cooling
中间冷却管 intercooler
中间冷却级 intercooling stage
中间冷却剂 intercoolant; intercooler; intermediate coolant
中间冷却器 intercooler; intermediate cooler; interstage cooler; intervening cooler
中间冷却设备 intercooling device
中间冷却循环 intercooled cycle
中间冷却压气机 intercooled compressor
中间冷却装置 cooling back installation; intercooler unit
中间理论州<对于抵押贷款，美国各州立法不同，一些州立法机构认为抵押时所有权归属受押方(称所有权理论州)，另一些州认为受押方只有留置权(称留置权州)> interme-

diate theory states

中间立柱 intermediate post; middle prop; middle standing pillar; door mullion <双门口的>

中间粒级 intergrade; intermediate gradation

中间粒径 median diameter

中间连杆 intermediate connecting rod

中间连接 intermediate junction

中间连接板 intermediate link

中间连接螺母 interconnecting nut

中间连接器 intermediate connector

中间联结零件 intermediate fastening

中间链传动轮 intermediate chain wheel

中间梁 intermediate beam; intermediate girder; middle girder; center [centre] sill <车的>

中间量 intermediate quantity

中间料仓 intermediate bin; intermediate bunker

中间料斗 middle hopper

中间裂缝 <混凝土路面的> intermediate cracking

中间裂谷 median rift

中间裂纹 median crack

中间檩 intermediate purlin(e)

中间零位继电器 center zero relay

中间领料单 intermediate requisition

中间流 interflow

中间流槽 tundish

中间留空地的建筑 building with central space

中间馏分 intermediate cut; intermediate distillate; midbarrel; middle fraction; middle(oil) distillate; middle runnings

中间馏分油 middle oil

中间馏份 intermediate fraction

中间隆起 median rise

中间楼板 intermediate floor(slab)

中间楼板梁 intermediate floor beam

中间楼面 intermediate floor(slab)

中间楼面间梁 intermediate floor beam

中间楼梯 intermediate stair(case)

中间楼梯梁 centre stringer

中间露天采场设计 intermediate pit design

中间律 medial law

中间率 <相邻筛号之间的成分百分率> percentage between screens

中间轮 breast-shot water wheel; cock wheel; dead pulley; loose roller; middle wheel; third wheel

中间轮换向夹板 intermediate wheel reverser bridge

中间轮锁杆 intermediate wheel lock

中间螺丝攻 intermediate tap

中间冒头 intermediate rail

中间贸易 intermediate trade

中间门柱 intermediate gate post

中间面 intermediate surface; median surface

中间磨粉仓 medium grinding compartment

中间磨细 intermediate reduction

中间目标程序 intermediate object program(me)

中间耐火材料 neutral refractory(material)

中间挠度 midway deflection

中间拟群 medial quasi-group

中间黏[粘]结剂 binder bridge

中间凝汽器 intercondenser

中间耦合 intermediate coupling

中间排架 intermediate bent

中间盘 intermediate pan

中间判决 interim judgement

中间旁承垫 center plate extension pad

中间抛泥区 interim disposal site

中间炮孔 intermediate hole

中间泡沫层 intermediate foam layer

中间配电盘 medium distributor

中间配线架 intermediate board frame; intermediate distributing [distribution] frame

中间配线盘 intermediate distributing board

中间喷嘴 intermediate nozzle

中间皮带传动装置 intermediate belt gearing

中间皮带轮 counter pulley; intermediate pulley

中间片 intermediate

中间频率 intermediate frequency; mean frequency

中间平面 intermediate plane

中间平台 intermediate landing; intermediate platform; intermediate stage

中间平巷 auxiliary level; interdrive; intermediate entry; intermediate heading; intermediate level; midworkings; subdrift; sublevel

中间平行填充带 intermediate parallel pack

中间屏蔽 intershield

中间坡度 intermediate grade; intermediate gradient; intermediate slope

中间破碎 intermediate crushing; intermediate reduction

中间破碎机 intermediate breaker; intermediate crusher

中间破碎振动棒磨 intermediate crushing vibrating rod mill

中间铺砌部分 paved median

中间期 intermediate stage

中间气压 mesobar

中间汽柜 intercepting steam chest

中间器材储备 intermediate storage

中间钎子 intermediate drill

中间墙 mid-board

中间墙壁 mid-feather

中间墙衬纤维板 intermediate fiberboard sheathing

中间桥 center-axle drive

中间桥墩 intermediate pier

中间桥桩 intermediate pile

中间球 semi-pellet

中间区 intermediate region; mesozone; neutral zone

中间区段 centre section

中间曲柄销 intermediate crank pin

中间曲柄装置 center-crank arrangement

中间渠道 intermediate channel

中间缺失 intercalary deletion

中间群 intermediate group

中间燃料油 intermediate fuel oil

中间绕组 interwinding

中间热变换器 intermediate heat exchanger

中间热储 intermediate reservoir

中间人 broker; finder; go-between; intermediary; intermediator; link man; middleman

中间人的赢利 jobber's turn

中间人佣金 finder's fee

中间忍受水平 median tolerance level

中间容积 fill and draw basin

中间容器 middle vessel

中间溶液 intermediary solution; intermediate solution

中间熔剂 intermediate flux

中间融化时期 intermediate thaw(ing) period

中间入口 central entrance

中间三等分 middle third

中间三通接头 union tee

中间散热器 intercooler

中间散射函数 intermediate scattering function

中间扫描 interscan

中间扫描分类法 intermediate pass sorting

中间色 neutral tint; pastel colo(u)r; demitint

中间色调 half tint; half-tone; middle colo(u)r; semi-tone

中间色散 mean dispersion

中间砂箱 cheek; intermediate flask; middle box

中间筛(分) intermediate screening; medium sizing

中间商 commission agent; commission merchant; intermediary business; jobber; middleman

中间商品 intermediate goods; intermediate product

中间商人代理商 factor

中间上条轮 intermediate winding wheel

中间烧结 intermediate sintering

中间设备 intermediate equipment; intermediate means

中间设计阶段 middle phase of planning

中间射向 intermediate line of fire

中间渗流流水 intermediate vadose water

中间生产 semi-production

中间生产设备 semi-production equipment

中间生物膜生物 intermediate biofilm organism

中间盛钢桶衬砖 tundish lining brick

中间盛钢桶耐火砖 tundish brick

中间十字接头 union cross

中间时间发送器 half-time emitter

中间时刻解 middle-time solution

中间实验厂设备 pilot plant installation

中间实验厂装置 pilot-scale device; pilot-scale equipment; pilot-scale facility

中间矢量玻色子 intermediate vector boson

中间市场 intermediate market; jobbing market

中间视 intermediate sight

中间视距 intermediate horizon

中间视觉 mesopic vision

中间视觉光谱光视效率 mesopic spectral luminous efficiency

中间试剂 semi-commercial production; semi-works production

中间试验 intermediate scale experiment; pilot experiment; pilot plant test; pilot test(ing); semi-plant; semi-plant scale test

中间试验(工)厂 pilot plant; semi-plant; semi-work; pilot unit; semi-commercial plant

中间试验设备 semi-plant scale equipment

中间试验研究 pilot study

中间试验窑 pilot plant kiln

中间试验柱 pilot column

中间试验桩 pilot pile

中间试验装置 pilot plant; pilot-scale facility

中间试样 middle sample

中间试制 semi-commercial production; semi-works production

中间释放轮 intermediate unlocking wheel

中间收获 intermediate yield

中间收集器 intercepter[interceptor]

中间收益 mesne profit

中间枢轴 intermediate pivot

中间熟料破碎机 intermediate clinker breaker; intermediate clinker crusher

中间竖井 intermediate shaft; medium shaft; medium shaft

中间数据 intermediate data

中间数据处理机 interdata processor

中间数据集 intermediate data set

中间数位 sandwich digit

中间数(字) median; sandwich digit

中间双臂曲柄 intermediate bellcrank

中间双滚筒驱动 <胶带机> intermediate wrap drive

中间水 intermediate water

中间水轮 breast water wheel

中间水平 intermediate communication level; mid-shaft

中间水平井底车场 intermediate landing-station

中间水区 transitional waters

中间水深波浪 intermediate depth wave

中间水质量 intermediate water mass

中间水准点 intermediate benchmark

中间税 intermediate tax

中间丝 intermediate filament

中间丝锥 intermediate tap

中间速度 mid-range speed

中间速率 median speed

中间宿主 bridge host

中间锁闭 intermediate locking

中间踏板 center foot pedal

中间台 intermediate station

中间态度 neutral attitude

中间滩地浮标 middle ground buoy

中间掏槽 middle cut

中间套管 intermediate casing

中间套管柱 intermediate string

中间梯度法 mid-gradient method

中间梯度曲线 mid-gradient profiling curve

中间提升杆 <拖拉机半悬挂系统的> mid-lift link

中间体 center[centre] piece; intermediary; intermediate body; intermediate compound; intermedium [复 intermedia]; interstitial material; mid-body

中间体玻璃形成物 intermediate glass former

中间体氧化物 intermediate oxide

中间填充 intermediate pack

中间填充带 intermediate pack strip

中间调整器 intermediate compensator

中间铁路 <联运货物的> intermediate carrier

中间庭院 central court(yard)

中间停车场 way station

中间停车点 intermediate stop-off point

中间停车站 intermediate stopping station; way station

中间筒 intermediate cylinder

中间投入 intermediate input

中间透镜 intermediate lens

中间图像 intermediate image

中间涂层 anchor coat; body coat; flat coat; intercoating; intermediate coat; middle coating; undercoat

中间涂覆金属封接 intermediate metallic sealing

中间土壤 intermediate soil

中间退火 annealed in process; annealing in process; commercial annealing; interanneal(ing); intermediate annealing; interprocess annealing; interstage annealing; process annealing

中间退火钢丝 annealed-in process wire; interannealed wire

中间托梁 intermediate bearer; inter-

mediate joist
中间托钎器 intermediate drill steel support
中间弯段 intermediate bend
中间弯曲 intermediate bend
中间弯头 intermediate bend;union elbow
中间网络 go-between
中间网络节点 intermediate network node
中间微分 intermediate differential
中间位 sandwich digit
中间位置 center[centre] position; cross-over position;in-between position; intermediate position; interposition; medium position; middle position; mid-gear; mid-position; changing station <测量站的>; mid-travel <指开关位置>
中间位置串通阀 tandem center valve
中间位置封闭阀 closed center valve
中间位置控制系统 proportioning control system
中间位置旁通换向阀 centre bypass valve
中间喂给装置 intermediate feed
中间坞门 intermediate gate; inner dock gate
中间物 compromise;intermediate;intermedium[复 intermedia]; medium;middle
中间物质 intermediate species
中间误差 median error
中间误差范数 median error norm
中间系杆 intermediate tie
中间系统 intermediate system
中间隙 intermediate gap
中间纤维 median fibre[fiber];medium fibre[fiber];neutral fiber[fibre]
中间线 intermediate line;median line
中间线路所 <在两个车站之间> intermediate block post
中间相 intermediate phase; interphase;interstitial phase;mosophase <离地面 15~50 英里,1 英里 = 1609.34 米>; intermediate facies【地】
中间相沥青 mesophase pitch
中间箱 intermediate box
中间镶板 intermediate panel
中间巷道掘进 subdrifting
中间项 mean term;middle entry
中间消费者 intermediate consumer
中间消耗 intermediate consumption
中间小齿轮 intermediate speed pinion
中间小椽 intermediate jack rafter
中间小两头大 pattee
中间歇负载 medium intermittent duty
中间歇荷载 medium intermittent duty
中间斜支撑 middle shore
中间卸料 intermediate throw off
中间卸料的链斗式提升机 bucket elevator with central discharge
中间卸料式提升机 internal discharge bucket elevator
中间心轴 idler axle
中间辛烷值 intermediate octane rating
中间信号 intermediate signal
中间信号点 intermediate signal location
中间信号机【铁】intermediate signal
中间信号楼【铁】intermediate signal box
中间信号配时方案 intermediate setting options
中间信号位置 intermediate signal location
中间信号箱【铁】intermediate signal box
中间行程 middle of stroke; mid-stroke;mid-travel
中间形成 intermediate formation
中间形式 intergrade; intermediate form
中间形态 intermediary
中间形状 intermediate shape
中间型 intermediate type; medial type;medium type;middle type
中间型灯头 intermediate base
中间型路面 intermediate type pavement
中间型盆地 median basin
中间型营养 mesotrophy
中间型植物 intermediate plant
中间性 intermediateness
中间性筹款 bridge financing;interim financing
中间性更新 interim replacement
中间性状 intermediary character; intermediate character
中间休息板 half landing; half-space landing
中间悬挂 mid-mounting
中间悬挂抖动输送筛式挖掘机 mid-mounted delivery shaken sieve digger
中间悬挂割草机的手动起落机构 hand lift for-mounted mower
中间悬挂式 centrally mounted
中间悬挂式割草机 mid-mounted mower
中间悬挂式双向犁 mid-mounted reversible plough
中间选组器 intermediate group selector
中间循环 intercycle
中间循环冷却器 intercycle cooler
中间压力 intermediate pressure; interstage pressure
中间压力室 intermediate pressure chamber
中间压碎 intermediate crushing
中间烟道 intermediate flue
中间烟道闸板 middle flue damper
中间研磨 intermediate grinding
中间掩膜 retic(u)le
中间验收 interim certificate
中间洋流 intermediate ocean current
中间业务 intermediation
中间业主 mesne lord
中间液体 intermediate liquid
中间引水闸门 intermediate head
中间应变轴 intermediate strain axis
中间应变轴方位 orientation of intermediate strain axis
中间应力 intermediate stress
中间营养的 mesotrophic
中间营养状态 mesotrophic state
中间影调 midtones
中间油罐 intermediate(oil) tank; relay tank
中间油品 intermediate oil
中间油水分离器 intermediate oil/water segregator
中间铀氟化物 intermediate uranium fluoride
中间游动盘 floating center[centre]
中间游离基 intermediate radical
中间有焊剂的空芯焊丝 flux-cored solder wire
中间有效范围 intermediate coverage range
中间语言 intermediate language
中间预测 median forecast
中间预热 intermediate heating
中间预热器 intermediate preheater
中间元素 neutral element
中间运价 intermediate rate
中间运输 auxiliary haulage
中间运输工具 intermediate carrier
中间运输工艺系统 intermediate transport technology system
中间运输机 intermediate haulage conveyer[conveyor]
中间载波 mean carrier
中间载热剂 intermediate coolant
中间载热体降温器 indirect cooled desuperheater
中间再热多级膨胀 reheat staged expansion
中间再热级 stage of reheat
中间再热式发动机 reheater engine
中间再热式汽轮机 double reheat steam turbine
中间再热式涡轮机 reheat(ing) turbine
中间择一律 medial alternative law
中间增音机 intermediate repeater; through line repeater
中间增音站 intermediate repeater station
中间轧辊 intermediate roll
中间轧机 intermediate mill; intermediate rolling mill; pony-roughing mill
中间轧机组 intermediate rolling train
中间轧碎机 intermediate breaker
中间轧制 intermediate rolling
中间轧制孔型 intermediate pass
中间闸门 intermediate gate
中间闸墙 middle lock wall
中间闸首 intermediate lock head
中间站 cut-in station; interface location;intermediate depot;intermediate tank farm; repeater station; repeating station; roadside station; through station;wayside station
中间站列车会让计划 train crossing plan on intermediate station
中间站台 intermediate platform
中间站停车 intermediate stop
中间站线路和房屋布置标准图 standard track layout at intermediate stations
中间站再生中继机 regenerative repeater for intermediate station
中间照准【测】intermediate siding; intermediate sight
中间照准觇标 intermediate sight target
中间折返站 interim reversion station
中间砧木 interstock
中间蒸汽 intermediate steam
中间证明 intermediate proof
中间支撑 center[centre] mount support; intermediate strut; intermediate support
中间支承 center[centre] mount support; central support; intermediate support; middle support; midship mounting
中间支承式鞍座 center[centre] pivot cradle
中间支点 intermediate pivot;intermediate support
中间支点的开合桥 pivot bridge
中间支杆 supporting mast
中间支架 intermediate support
中间支柱 intermediate post;intermediate stanchion;intermediate strut; intermediate studdle <井框间>
中间支座 intermediate support
中间织纹 intermediate weave
中间直接形杠杆 intermediate bellcrank
中间直径 fractional diameter; medium diameter
中间直视距离 intermediate horizon
中间值 intermediate value;mean value; mid-score; mid-value; median【数】
中间值电流 intermediate current
中间植草带 grassed central strip
中间止动环 middle stop
中间止回阀 line check valve
中间纸板 mid-board
中间指令 intermediate command;metainstruction
中间制动片 middle brake lining
中间制造商 intermediate manufacturer
中间致死量 mean lethal dose;median lethal dosage
中间中继器 intermediate repeater; through line repeater
中间中继线配线架 intermediate trunk distributing frame
中间周边卸料 center peripheral discharge
中间周期 intercycle;intermediate cycle
中间周相 interphase
中间轴 connection shaft; counter; countershaft; countershaft unit; dummy shaft; interaxis; intermediate axle; intermediate shaft; intermediate spindle;jackshaft;layshaft; line shaft;main shaft;middle shaft; tunnel shaft
中间轴承 countershaft bearing;intermediate bearing; layshaft bearing; middle bearing;neck bearing
中间轴承盖 plumber block keep
中间轴承台 tunnel stool
中间轴承罩 intermediate bearing casing
中间轴齿轮 countershaft gear
中间轴齿轮组 layshaft gear cluster
中间轴的万向轴 jackshaft propeller shaft
中间轴惰轮 intermediate underdrive gear
中间轴盖 intermediate shaft cap
中间轴颈 intermediate journal
中间轴驱动式 center-axle drive
中间轴套 middle shaft bushing
中间轴五挡齿轮 countershaft fifth speed gear
中间轴支承总成 countershaft mounting assembly
中间轴支架 intermediate shaft supporting bracket
中间轴轴承 intermediate shaft bearing; midship shaft bearing; pillow block; spring bearing; tunnel shaft bearing
中间轴轴承盖 midship shaft bearing cover
中间轴轴承套筒 intermediate shaft bearing sleeve
中间轴主动齿轮 jackshaft drive pinion;layshaft drive gear
中间主构架 interprimary frame
中间主体 intermediate host
中间主应变 intermediate principal strain
中间主应力 intermediate principal stress;mean principal stress
中间主应力方位角 direction of the intermediate principal stress
中间主应力面 intermediate principal plane
中间主应力倾角 dip of the intermediate principal stress
中间主应力影响系数 parameter of the influence of intermediate principal stress
中间主应力值 intermediate principal stress value
中间主应力轴 axes of mean stress
中间主轴承 intermediate main bearing
中间贮存 intermediate storage
中间贮水池 interim storage(tank)
中间柱 intermediate column;stud

中间柱廊 middle stoa
中间砖墙 intermediate brickwork
中间转报站 intermediate repeater station
中间转播站 intermediate repeater station
中间转动齿轮轴 intermediate shaft; secondary gear shaft
中间转动轴 second-motion shaft
中间转换 buffering; intermediate conversion
中间转换器 buffer module
中间转换用计算机 buffered computer
中间转让 mesne assignment
中间转向臂 intermediate steering arm
中间转运点 intermediate transfer point
中间桩 center[centre] pile; intermediate pile
中间装配地点 intermediate assembly point
中间装填 intermediate pack
中间状态 intermediary; intermediate state; interphase; mediacy; neutral condition; transition state
中间锥形圆库 central cone silo
中间自动闭塞信号机【铁】 intermediate automatic block signal
中间自位阀 self-neutralizing valve
中间纵梁 intermediate longitudinal girder
中间阻抗 medium impedance
中间组 centre section
中间组成物 intermediate constituent; intermetallic compound
中间组合箱桩 intermediate box pile
中间钻臂 cut-boom
中间钻杆 intermediate drill
中间钻孔 intervening boring
中间作物机具 row-crop equipment
中间坐标 middle ordinate
中间座 interstand
中检台位【铁】 intermediate technical examination position
中碱玻璃 medium alkali glass
中碱玻璃纤维 medium alkali glass fibre
中渐屈线 middle evolute
中槛 transom bar
中交水运规划设计院 China Communications Planning and Design Institute for Water Transportation
中胶层 mesogl(o)ea
中焦距 mid-focal length
中焦距光学跟踪器 intermediate focal length tracking telescope
中焦油 tar medium oil
中阶 scala media
中阶梯光栅 echelle grating
中阶梯光栅摄谱仪 echelle grating spectrograph
中接线头 centre tap
中截面 intermediate section; middle section
中介次系 intermediate subsystem
中介的 mesomeric
中介地区 betwixt land
中介电解质 mesolyte
中介机构 intermediary institution
中介结束 escrow closing
中介联想 mediate association
中介脉冲 interpulse
中介贸易 intermediate trade
中介耦合 intermediate coupling
中介耦合理论 intermediate-coupling theory
中介漂移 intermediate polar
中介漂移暴 intermediate drift burst
中介人 escrow officer; mediator; middleman
中介日冕 intermediate corona
中介态 mesomeric state
中介体 mesogen
中介委托书 escrow instruction
中介物 intermediary
中介现象 mesomerism; resonance
中介效应 mesomeric effect
中介轴 stub
中介作用 intermediation
中介坐标 intermediate coordinates
中界轮 idler
中界面 median surface
中界山脉 betwixt mountain
中金刚钻压花滚轮 medium diamond-point knurling rolls
中金檩 intermediate principal purlin(e)
中近海带 mesoneritic
中近景 two-shot
中近景镜头 medium close shot
中近距摄影 close medium shot; intermediate short range photography
中晶白云岩 medium crystalline dolomite
中晶灰岩 medium crystalline limestone
中晶结构 medium crystalline texture
中晶颗粒【地】 mesocrystalline particle
中晶质的 mesocrystalline
中景 short shot
中景镜头 close medium shot
中净化 medium cleaning
中径 effective diameter; intermediate diameter; median diameter; pitch diameter
中镜片 <取景器的> mean element
中巨砾 medium boulder
中距 center[centre]-to-center[centre] distance; center[centre] spacing; middle ordinate; mid-range; space centers[centres]; spaced centers[centres]
中距导航台 middle locator
中距电子定位系统 medium range electronic positioning system
中距法 method of middle ordinates; mid-ordinate method
中距护刃器切割器 medium finger spacing cutter
中距离 medium range
中距离电子定位系统名 Raydist
中距离跑 middle distance race
中距离喷药 middle distance spray
中距离频率继电器 intermediate frequency relay
中距离无线电导航仪 medium distance aids
中距为…… spaced…centers[centres]
中开口提花机 center[centre] shed jacquard
中康酸 mesaconic(al) acid
中抗硫油井水泥 moderate sulfate-resistant oil well cement
中颗粒 medium sized grain
中颗粒的 medium-grained; medium sized
中颗粒骨料 medium-sized aggregate
中颗粒集料 medium-sized aggregate
中颗粒砂 medium granular sand
中空坝 hollow dam
中空波导管 hollow waveguide
中空玻璃 double glazing glass; insulation glass unit; sealed unit
中空玻璃微珠 hollow glass sphere
中空玻璃纤维 hollow glass fiber[fibre]
中空冲裁 hollow cutting
中空导线 hollow tubing conductor
中空的 hollow
中空电子束 hollow beam
中空锻造 hollow forging
中空多面体 <一种防波堤块体> shed
中空耳轴 hollow trunnion bearing
中空方柱体 hollow pyramid
中空飞机 medium altitude aircraft
中空干法回转窑 hollow dry process rotary kiln
中空钢 cored steel; hollow steel
中空钢制螺旋桨 hollow steel propeller
中空隔热玻璃 sealed glass unit
中空机油道连杆 connecting rod with oil passage
中空挤压锭 reamed extrusion ingot
中空夹壁墙 double wall
中空胶合板 void plywood
中空壳 cavity shell
中空螺旋管 solenoid
中空内管 hollow bowl
中空砌块墙 masonry cavity wall
中空砌筑墙 masonry cavity wall
中空器皿 hollow ware
中空(钎)钢 hollow drill steel
中空碳微珠 hollow carbon microphere
中空掏槽眼 burn-cut hole
中空体 ducted body
中空凸模冲孔 hot trepanning
中空纤维超滤膜 hollowed fiber[fibre] ultra-filtration membrane
中空纤维反渗透膜 hollow fiber reverse osmosis membrane
中空纤维膜 hollowed fiber[fibre] membrane
中空纤维膜接触器 hollowed fiber[fibre] membrane contactor
中空纤维膜生物反应器 hollowed fiber[fibre] membrane bioreactor
中空纤维微滤 hollowed fiber[fibre] microfiltration
中空纤维微滤膜 hollowed microfiltration fibre membrane
中空效应 hole-in-the-center effect
中空形芯 hollow core
中空型荷载计 centre-hole type load cell
中空圆柱 hollow circular cylinder
中空圆柱体护舷 hollow cylindric(al) fender
中空圆柱体轴向受压护舷 hollow cylindrical axially loaded fender
中空圆柱形护舷 hollow cylindric(al) fender
中空制品 hollow article
中空轴 female spindle; hollow shaft(ing); quill shaft
中空轴承 neck bearing
中空轴颈 hollow trunnion
中空轴轴承 hollow trunnion bearing; main trunnion bearing
中空铸型法 hollow casting
中空铸造 slush casting
中空砖 inside-emptied brick
中空钻探钢 hollow drill steel
中孔 center span; mid-span; central space; central span
中孔凸模冲机 hot trephining
中孔隙 mesopore
中孔型 mesothyrid
中孔型穿刺用探头 puncture transducer with center channel
中孔凿岩机 burn hole drilling machine
中控室 center[centre] control room
中控主管 control room supervisor
中口径 medium caliber
中枯水急滩 minor and low level rapids
中枯水位河床 minor bed; minor river bed
中跨 center[centre] arch span; center bay; center[centre] opening; center[centre] span; central spalling; intermediate space; intermediate span; middle span; mid-span; central space; central span
中跨板带 middle strip
中跨度屋顶 medium-span roof
中跨度屋盖 medium-span roof
中跨断面 mid-span section
中跨钢筋 mid-span reinforcement
中跨挠度 mid-span deflection
中跨挠曲 mid-span deflection
中跨桥梁 intermediate span bridge; medium-span bridge
中块煤 medium sized coal
中块石 one-man stone
中块石混凝土 cobble mix
中狂涌 average heavy swell; average high swell
中矿 middles; middling
中矿采样 middling sampling
中矿带 middle band; middlings band; middling zone
中矿粒 middling particle
中矿品位 middles grade
中矿提升机 middle elevator; middlings elevator
中矿再磨回路 middling regrinding circuit
中框网络 backbone network
中眶型 mesoconch
中廊式客车 center aisle coach
中浪 <三级风浪> moderate sea
中浪海况 medium wave condition
中肋骨条藻 Skeletonmacostatum
中冷器 after-cooler; charge air cooler; intercooler; interstage cooler
中冷器心子 intercooler core
中冷增压柴油机 intercooled diesel engine
中立保证 warranty of neutrality
中立操纵杆 neutral stick
中立船舶证 sea letter
中立船舶证(明)书 sea brief; seapass
中立的 neutral
中立地带 neutral zone
中立港 neutral harbo(u)r
中立关闭 <阀的> all-ports block; closed center[centre]
中立国 neutral country
中立国财产征用权 right of angary
中立国船舶 neutral ship; neutral vessel
中立国港口 neutral port
中立国(国)旗 neutral flag
中立国领海 neutral waters
中立国水域 neutral waters
中立旁通 tandem center[centre]; center by-pass <阀的>
中立区 neutral area; neutralized area; neutral zone
中立体 neutral body
中立位 lap position
中立位置 isolation position; neutral position
中立位置方向舵 neutral rudder
中立温度 neutral temperature
中立泄流 exhaust center[centre]
中立油缸口关闭 cylinder port block
中立者 neutral
中立柱 B-pillar; center pillar
中立状态 neutral condition
中砾 cobble; pebble(stone); medium gravel
中砾和细砾 medium gravel and fine gravel
中砾角砾岩 pebble breccia
中砾结构 pebble texture
中砾石 boulderet; cobble gravel; cobble boulder; cobblestone; medium gravel
中砾岩 pebble stone

中粒 medium grain
中粒变晶结构 medium granular crystalloblastic texture
中粒度的 medium-grained; medium granular; mesograined
中粒度破碎室 <破碎机的> intermediate chamber
中粒粉砂 medium silt
中粒粉砂和细粒粉砂 medium silt and fine silt
中粒花岗岩 medium-grained granite
中粒灰岩 medium-grained limestone
中粒级标度 phi grade scale
中粒结构 intermediate granular texture; medium-grained texture; medium structure
中粒金刚石钻头 medium stone bit
中粒浸染 medium dissemination
中粒径的 medium sized
中粒径粉土 medium-grained silt
中粒沥青混凝土路面 medium graded bituminous concrete pavement
中粒砾石合细粒砾石 <美国标准 9.52 ~ 25.4 毫米为中粒,2.00 ~ 9.52 毫米为细粒> medium gravel and fine gravel
中粒曲流带沉积 medium-grained meander belt deposit
中粒热裂法炭黑 medium thermal carbon black
中粒砂 medium-grained sand
中粒砂石 medium-grained sandstone
中粒砂状结构 medium granular psamitic texture
中粒碎屑 medium clastics
中粒玄武岩 anamesite; graystone
中粒雪 mean grain firn; medium-grained snow
中粒岩石 medium granular rock
中粒状的 medium granular
中粒子热裂炉黑 medium thermal black
中连合 mediocommissure
中链(员) intermediate chainman; intermediate tapeman
中梁 center[centre] sill
中梁补强铁 center sill stiffener
中梁槽钢 center sill channel
中梁盖鞍座 <底开门车> center sill saddle
中梁隔板 center sill separator
中梁截换 center sill splicing
中梁拼接 center sill splicing
中梁上角钢 center sill top(flange) angle
中梁式车架 central frame
中梁下角钢 center sill bottom(flange) angle
中量交通 medium traffic; moderate traffic
中量人口密度 moderately populated density
中量元素肥料 secondary nutrient
中列数 mid-range
中裂的 <乳化沥青> medium breaking; medium set(ting)
中裂沥青乳液 medium-setting asphaltic emulsion; semi-stable emulsion <英>
中裂面 median cleavage plane
中裂乳化地沥青 medium-setting emulsified asphalt <美>; medium breaking emulsified asphalt <英>
中裂乳化沥青 medium breaking emulsified asphalt; medium breaking emulsified bitumen; medium-setting emulsion; semi-stable bituminous dispersion
中裂乳液 medium breaking emulsion
中裂阳离子路用乳液 <英> medium acting cationic road emulsion
中林 composite forest; middle forest; reserve sprout forest
中林作业 reserve sprout system
中林作业法 coppice with standard system; coppice with standard
中磷煤 mid-phosphorus coal
中磷闸瓦 medium phosphor brake-shoe
中磷铸铁 medium phosphorus cast iron
中磷铸铁闸瓦 medium phosphorus content cast iron brake shoe
中檩条 central purlin(e)
中零刻度 center zero scale
中流 mid-channel; middle course; midstream
中流泊位 <河流的> anchor berth
中流量 medium discharge
中流量喷嘴 medium volume nozzle
中流沙洲 mid-channel bar
中流速纸 medium flow rate paper
中流作业转运港 mid-stream transfer terminal
中硫煤 mid-sulfur coal
中龙骨立板 centre girder
中隆线 median carina
中隆褶皱 median fold
中楼 mezzanine
中漏斗板 <漏斗车> tenter ridge plate
中绿 medium green
中卵石 cobble; medium gravel
中螺口茄形指示灯泡 medium eggplant screw indicating lamp bulb
中马格达莱纳盆地 middle Magdalena basin
中埋 medium depth
中埋深隧道 tunnel in medium depth
中冒头 center rail; central rail; lock rail; middle rail; intermediate door rail <门的>; frieze rail <最上一块门芯板下的>
中楣 frieze
中煤 middlings
中煤含量 middling content
中煤含量百分比分级 middling content per cent graduation
中煤含量法 method of middling content per cent
中煤样 middling sample
中美海路 middle America seaway
中美洲 Central America
中美洲共同市场 Central America Common Market
中美洲海沟 mid-America trench
中美洲红木 Central America mahogany
中美洲建筑 Central America architecture; Meso-American architecture
中美洲经济合作委员会 Central American Economic Cooperation Committee
中美洲自由贸易区 Central American Trade Area
中门 central door; intermediate gate
中门横挡 intermediate door rail
中门梃 meeting stiles
中蒙古深断裂系 Central Mongolian deep fracture zone
中锰钢 medium manganese steel
中密的 intermediate dense; medium dense; moderate dense
中密度级硬质纤维板 medium hardboard
中密度捡拾压捆机 medium density baler
中密度建筑纤维板 medium density building fiberboard
中密度聚乙烯 medium density polyethylene
中密度聚乙烯塑料 medium density polyethylene plastics
中密度刨花板 medium density particle board
中密度碎料板 <比重 0.6 ~ 0.8> medium density particle board; medium density particle wood
中密度碎木板 medium density wood chipboard
中密度贴面胶合板 medium density overlay; medium density plywood
中密度纹理 medium grain
中密度纤维板 intermediate fiberboard[fibreboard]; medium density fiberboard; medium density hardboard
中密度纤维板废水 medium density fiberboard wastewater
中密度亚麻板 medium density flax-based board
中密度硬质纤维板 medium density hardboard
中密西西比世【地】Middle Mississippian
中密状态 intermediate condition
中绵 medium draft
中冕 middle corona
中面积法 method of middle area
中面角 nasal prognathism
中面(壳体) middle plane; middle surface; neutral surface
中面型 mesoprosopy
中面型的 mesoprosopic
中模 center[centre] form
中磨 intermediate grinding; medium grinding
中末比【数】extreme and mean ratio
中木工锉 middle carpenter file
中木桶 cask
中幕中央撑住 middle line awning stanchion
中内龙骨 centre line keelson; centre through plate; main keelson; middle line keelson; vertical center keelson
中耐磨炉黑 medium abrasion furnace black
中耐受水平 median tolerance level
中耐受限度 median tolerance limit; medium tolerance limit
中南部 south central
中南亚珊瑚地理大区 south-central Asian coral region
中能堆 intermediate reactor
中能海岸 moderate energy coast
中能回旋加速器 medium energy cyclotron
中能级辐射 intermediate level radiation
中能加速器 medium energy accelerator
中能力驼峰 middle capacity of hump
中能(量)区 intermediate energy region
中能型染料 medium energy type dyes
中能中子反应堆 intermediate neutron reactor; intermediate reactor; moderate energy neutron reactor
中能中子谱堆 intermediate spectrum reactor
中泥盆世【地】Middle Devonian
中泥盆世绝灭【地】mid-Devonian extinction
中泥盆统【地】Middle Devonian series
中黏[粘]度 medium viscosity
中黏[粘]度醇酸树脂 medium oil alkyd resin
中黏[粘]度油 medium viscous oil
中黏[粘]结性煤 mid caking coal
中黏[粘]壤土 medium clay(ey) loam
中黏[粘]土 medium clay
中凝 <指液体沥青或轻制沥青> medium curing; medium set(ting)
中凝沥青 medium curing asphalt
中凝轻制(地沥青) <由煤油类掺膏体地沥青制成> medium curing cutback asphalt
中凝轻制液体沥青 medium curing asphalt
中凝时间 medium-setting time
中凝水泥 semi-slow cement
中凝液体地沥青 medium curing liquid asphalt
中凝液体石油沥青 medium curing liquid asphalt
中欧砂岩相 mid-European sandstone facies
中欧山松 Swiss mountain pine
中欧型岩溶 central European karst
中耙 <自航耙吸式挖泥船的> centre drag
中盘内套 middle disc lever
中刨 foreplane
中批量生产 medium duty
中皮型重力构造 mesodermal type of gravitative tectonic
中片材 medium sheet
中片堆垛机 medium size sheet stacker
中频 <300 千赫 ~ 3 兆赫> intermediate frequency; medium frequency; mid-frequency; mid-range frequency
中频变换器 medium frequency changer
中频变压器 intermediate frequency transformer
中频波 medium frequency wave(row)
中频部分 intermediate frequency section
中频淬火 intermediate frequency induction hardening
中频带 mid-band
中频带宽 intermediate frequency bandwidth
中频电路 intermediate frequency channel
中频放大 <外差接收机中> supersonic amplification
中频放大管 intermediate frequency tube
中频放大级 intermediate frequency stage
中频放大器 intermediate frequency amplifier; medium frequency amplifier; mid-range frequency amplifier
中频分支站 intermediate frequency branching station
中频干扰 intermediate frequency interference
中频干扰比 intermediate frequency interference ratio
中频感应电炉 medium frequency induction furnace
中频感应炉 intermediate frequency furnace
中频功率放大器 intermediate frequency power amplifier
中频合并 intermediate frequency combining
中频互连 intermediate frequency interconnection
中频加热弯管 hot elbowing by inductive heating
中频接收机 intermediate frequency receiver
中频介电分离仪 medium frequency dielectric(al) separator
中频率 medium frequency; mid-range frequency
中频率的隔声 medium frequency sound insulation
中频率响应 medium frequency response

中频偏移 intermediate frequency deviation
中频频带 medium frequency band
中频前置放大器 IF preamp; intermediate frequency preamplifier
中频扫频仪 intermediate frequency sweep signal generator
中频声脉冲 ping
中频失真 intermediate frequency distortion
中频输出 intermediate frequency output
中频损耗 medium frequency loss
中频陶瓷滤波器 intermediate frequency ceramic filter
中频特性测试仪 intermediate frequency characteristics measuring set
中频调制 intermediate frequency modulation
中频通带 intermediate frequency passband
中频通路 intermediate frequency channel
中频拖尾 mid-band streaking
中频系统 intermediate frequency system
中频响应 intermediate frequency response
中频谐波干扰 intermediate frequency harmonic interference
中频信号 intermediate frequency signal; medium frequency signal; mid-range frequency signal
中频抑制 intermediate frequency rejection
中频抑制比 intermediate frequency rejection ratio
中频抑制系数 intermediate frequency rejection factor
中频载波 intermediate carrier
中频凿岩机 medium frequency rock drill
中频噪声 intermediate frequency noise; mid-frequency noise
中频增益 intermediate frequency gain; mid-frequency gain
中频振荡 medium frequency oscillator
中频振荡器 intermediate frequency oscillator
中频直通 intermediate frequency feedthrough
中频转换 intermediate frequency switching
中频转接 intermediate frequency interconnection
中频自动增益调整 intermediate frequency automatic gain control setting
中平【测】center line stake level(l)ing
中平方法 middle square method; mid-square method
中平方生成程序 mid-square generator
中平面 mid-plane
中平面场 median-plane field
中平面轨道 median-plane orbit
中平面校正 median-plane correction
中平面内聚集 median-plane focusing
中平面位置 median-plane location
中坡(度) mesoslope; moderate grade; mid-slope
中剖 bisection
中剖定理【计】bisection theorem
中铺 intermediate berth
中期 intermediate term
中期报告 interim report
中期测定 medium term measurement
中期测量 interim survey
中期产品 intermediates
中期成岩作用 middle diagenesis
中期贷款 extended facility; intermediate loan; medium term loan
中期的 medium range; medium term
中期地震预报 medium range earthquake prediction; medium term earthquake prediction
中期放款 intermediate loan; medium term loan
中期分裂 metakinesis
中期腐朽 intermediate decay
中期腐朽阶段 intermediate stage of decay
中期付款 interim payment
中期付款证书 interim payment certificate
中期改造 mid-life update
中期更新 interim replacement
中期河曲 full meander
中期计划 interim plan; medium term plan; middle-range plan; middle-term plan
中期计量 interim measurement
中期检验 intermediate inspection
中期客运计划 medium term passenger-transport program(me)
中期目标 medium objective
中期欧洲贷款市场 medium term Eurocredit market
中期气候 mesoclimate
中期强度 intermediate strength
中期审查 mid-term review
中期审计 interim audit
中期生产规划预测 mid-term production-programming forecast
中期(天气)预报 medium range forecast
中期污染 middle term pollution
中期物资计划 middle-term material plan
中期信贷 intermediate(term) credit; medium term credit
中期延期决定 interim determination of extension
中期验收证书 interim acceptance certificate
中期(英格兰)哥特式 middle pointed style
中期预报 extended forecast(ing); medium range forecasting; mid-range forecasting
中期预报方法 intermediate term approach
中期预测 medium term forecast; mid-range estimate
中期预算 interim budget
中期债券 medium term bonds
中期证书 interim certificate
中期植物 intermediate plant
中期资本 medium term capital
中期资金筹措 intermediate term financing
中祁连岛链【地】mid-Qilian island chain
中起伏 mesorelief
中气候 mesoclimate
中气候学 mesoclimatology
中气泡曝气 medium air bubble aeration
中气泡曝气系统 medium bubble aeration system
中气田 middle gas field
中气象学 mesometeorology
中碛【地】medial moraine; middle moraine
中碛堤 medial moraine bar
中铅玻璃 medium lead crystal glass
中铅毒的 saturnic
中铅铜合金 medium leaded brass
中签 bonds drawn
中浅变质的多期变形带 multiply deformed belts of low and medium metamorphic grade
中强地震 intermediate magnitude earthquake; moderate(size) earthquake; moderate strong earthquake
中强地震动 moderate ground shaking
中强地震运动 moderate-to-strong earthquake motion
中强度电焊条 medium tensile strength steel welding rod
中强度煤 mid-strength coal
中强钢丝 medium-strength steel wire
中强褶皱背斜聚集带 accumulation zone of middle-strong folding anticline
中强震仪 intermediate strong motion instrument
中墙 < 双线船闸 > middle wall
中墙板 dead light; inside window panel; window panel
中墙板木梁 window panel furring
中墙板镶条 window panel moulding
中橇架 central skid chassis
中桥 medium bridge
中桥差速器 jackshaft differential
中桥差速器壳 middle axle differential case
中桥差速箱 jackshaft differential carrier
中桥驱动式侧卸卡车 center-axle drive side-dump
中桥驱动式铲运机 center-axle drive scraper
中桥驱动式底卸卡车 center-axle drive bottom-dump
中桥驱动式后卸卡车 center-axle drive rear-dump
中桥箱 jackshaft housing
中桥轴中心到铲刀的距离 < 平地机 > blade base
中桥主动及被动伞齿轮 intermediate axle bevel gear and pinion
中桥左轮制动管路 intermediate axle left wheel brake tube
中切机 middle filling machine
中轻交通 medium light traffic
中倾斜 medium dip
中倾斜矿层 medium steep seam
中倾斜岩层 medium dipping bed
中擎凸出式十字钻头 pilot-and-reamer bit
中区 median space
中区地形改正值 terrain correction value of medium distance
中区地形改正精度 accuracy of terrain correction on medium distance
中区牵伸 intermediate draft; mid zone draft
中曲率 mean curvature
中曲面 mean camber
中躯宽深 roomy middle
中圈 centre washer; mesosphere
中圈顶 mesopause
中壤土 medium loam
中热 < 水泥 > moderate heat of hardening
中热波特兰水泥 moderate heat Portland cement
中热带 middle tropical zone
中热硅酸盐水泥 moderate heat Portland cement
中热抗硫酸盐水泥 moderate heat of hydration and sulphate-resistant cement
中热水泥 moderate heat cement; moderate heat of hydration cement; moderate heat Portland cement; modified Portland cement; modified cement < 抗酸盐性能较好的标准水泥,水化热低 >
中热油井水泥 moderate heat oil well cement
中热值煤气 middle-heat calorie gas
中日照植物 middle-day plant
中绒 medium staple
中容量存储器 medium capacity storage
中容量电站 medium capacity plant
中容量喷雾 medium volume spraying
中溶盐 medium soluble salt; moderately soluble salt
中溶盐含量 content of medium soluble salt
中溶盐试验 medium soluble salt test; moderately soluble salt test
中溶盐含量试验 moderately soluble salt content test
中软的 medium soft
中软沥青 medium soft pitch
中三迭世【地】Middle Triassic epoch
中三迭统【地】Middle Triassic series
中三分 middle third
中三分定律 middle third rule
中三分法 < 指压力线不超出截面核心范围的法则 > middle third rule
中三角 intermediate cam
中三角锉 middle triangular file
中色 medium colo(u)r
中色的 mesocratic; mesotype
中色调 medium shade
中色伟晶斑岩 mesopegmatophyre
中色岩 mesocrate; mesocratic rock
中砂 < 又称中沙 > medium(granular) sand
中砂轮 medium plain emery wheel
中砂内端标 junction mark
中砂内端浮标 junction buoy; upper end buoy
中砂黏[粘]土 medium sandy clay
中砂外端 bifurcation point
中砂外端浮标 bifurcation buoy
中砂下端浮标 lower end buoy
中砂箱 cheek flask; raising middle flask
中砂屑 medium sandy clast
中砂岩 medium sandstone
中砂岩储集层 medium reservoir
中筛 medium mesh; medium screen
中山 < 绝对高度 1000 ~ 3500 米 > medium mountain
中山地森林 mid-montane forest
中山陵 Sun Yat-sen's Mausoleum
中山区 middle mountain
中山植物园 Zhongshan botanical garden
中扇 middle fan; mid-fan
中扇沉积 middle fan deposit
中上 above average; better than average
中上层区 pelagic(al) region; pelagic(al) zone
中上层鱼类 pelagic(al) fishes
中上层鱼类群体 pelagic(al) fish stocks
中上层鱼群 pelagic(al) fish concentration
中上层资源 pelagic(al) resources
中上等的 average-to-good
中上流社会 belgravia
中上能见度 fair visibility
中射式水轮 breast water wheel
中射式水轮机 middle shot water wheel
中深(变质)带【地】mesozone
中深层次 middle-deep level
中深成的 mesogene
中深地震 earthquake of intermediate depth; intermediate earthquake; intermediate shock
中深海 < 2000 ~ 4000 米水深的 > mesopelagic(al)
中深海沉积物 moderate sea deposit

中深海底的 mesobenthos
中深含沙量 mid-depth concentration
中深井 medium deep well
中深井油井水泥 moderate deep well oil well cement
中深孔 medium depth (bore) hole; medium length(bore) hole
中深浓度 mid-depth concentration
中深炮孔 medium length holing
中深热液的 mesothermal
中深热液矿床 mesothermal deposit
中深热液矿脉 mesothermal vein
中深热液作用 mesothermal process
中深水流系统 mid-deep floe system
中深型钻机 medium depth rig
中深源地震 intermediate focus earthquake
中深钻孔 medium deep drill hole
中生代【地】Mesozoic era; Secondary era
中生代的 Mesozoic
中生代构造带 mesoide
中生代海洋 Mesozoic ocean
中生代盆地 Mesozoic basin
中生代山脉 mesoiden
中生代原油 Mesozoic crude
中生代褶皱带 mesoide
中生代植物 mesophyte
中生的 mesic
中生浮游生物 mesoplankton
中生界 Mesozoic erathem; Mesozoic group
中生茎叶植物 mesocormophyta
中生林 mesophytic forest
中生生境 mesophytic habitat
中生性 mesophytism
中生演替系列 mesarch sere
中生植物环境 mesophytic environment
中生植物群系 mesophytic formation
中湿 syndrome due to attack of pathogenic dampness
中湿类型 medium dampness type
中湿沼泽 moderate moist swamp
中石器时代 Mesolithic Age; Middle Stone Age; Transitional Age
中石器时代的 Mesolithic
中石器时期 Mesolithic period
中石色油漆 mid stone paint
中石炭统 middle carboniferous series
中石铁陨石 mesosiderite
中世纪 Middle Ages; the middle ages
中世纪堡垒中的空地 ballium
中世纪城堡 tower house
中世纪城堡的主塔 donjon
中世纪城市 mediaeval town
中世纪大学的学生宿舍 bursa
中世纪道路模式 medi(a)eval road pattern
中世纪的 mediaeval
中世纪极大期 medieval maximum
中世纪建筑 mediaeval architecture
中世纪教堂东端半圆室上的拱券 arcus presbyteri
中世纪教堂圣坛节日悬挂圣物的梁 pertica
中世纪教堂中殿与圣台的拱券 arcus ecclesiae
中世纪奶酪房 arcella
中世纪式的玻璃 < 有色玻璃窗 > antique glass
中世纪晚期建筑艺术 late-mediaeval architecture
中式布置 < 机电 > medium type layout
中视 intermediate foresight
中视镜 mesoscope
中试 medium scale test; pilot test
中试泵 pilot pump
中试厂 pilot plant
中试厂研究 pilot plant investigation
中试车间 semi-work

中试规模 pilot scale
中试规模系统 pilot-scale system
中试过程 pilot process
中试试验 pilot-scale test
中试研究 pilot study
中试窑 pilot kiln
中试装置 pilot plant
中室 medial cell; median cell; medium cell
中枢 hub; center pin; central pivot; main center [centre]; pivot; spine; axis[复 axes]
中枢查询 hub polling
中枢导承 centre pin guide
中枢的 central; centric; pivotal
中枢端 central end; proximal end
中枢管理费用 central administrative expenses
中枢灌溉系统 centre-pivot irrigation system
中枢神经系统 · central nervous system
中枢销 king bolt; king pin
中枢兴奋机制 central excitatory mechanism
中枢摇枕 pivot bolster
中枢抑制 central inhibition
中枢运转港 pivot port
中枢支承旋转窗 center pivoted window
中枢轴式转向 center pin steering; center pivot steer(ing)
中枢轴支承 center pin bearing
中竖框 < 门窗 > mullion; pillar
中竖龙骨 center[centre] keel
中竖樘 middle stile
中数 mean; median
中数粒径 median diameter; median particle diameter; median particle size
中数所得 median income
中数所在组 median class
中数直径 median diameter
中数致死浓度 median lethal concentration
中双纹锉 middle double cut file
中水 non-potable reclaimed water
中水导治 mean water training
中水道 intermediate water course
中水供水 intermediate water supply
中水供水系统 intermediate water supply system
中水河槽以上段 overbank area
中水回用 intermediate water reuse; middle water reuse
中水急滩 mid-level rapids
中水年 average year; median flow year; moderate-flow year; normal year
中水期 median water period
中水头 medium head; moderate head
中水头船闸 medium lift lock; moderate lift lock
中水头电站 medium head plant; medium head scheme; moderate head plant
中水头水电站 hydroelectric(al) plant of moderate head
中水头水轮机 medium head (hydraulic) turbine
中水位 mean water; median water level; median water stage; medium (water) level; medium water stage; midwater level; ordinary water level
中水位河床 mean water river bed; minor bed; ordinary water level
中水位流量 median discharge
中水位桥 medium water level bridge
中水位整治 mean water regulation

中水位导治 mean water training
中水治理 mean water regulation
中水装置 intermediate water facility
中丝 central hair; central thread; central wire
中丝锥 plug tap; second-hand tap; second tap
中苏门答腊盆地 central Sumatra basin
中速 median speed; medium speed; middle velocity; moderate speed; moderate velocity
中速 V 形发动机 medium speed V-engine
中速柴油机 medium speed diesel; medium speed engine; middle speed diesel engine
中速齿轮 intermediate gear; medium gear
中速齿轮付 mediate counter gear
中速船(舶) medium speed ship; medium speed vessel
中速促进剂 medium accelerator; moderate accelerator
中速存储器 medium speed memory; medium speed storage
中速存取存储器 medium access memory
中速电梯 medium speed elevator; medium speed lift
中速度 medium speed
中速发动机 medium speed engine
中速辐向轴流式水轮机 medium speed Francis turbine
中速干燥的 < 指油脂 > medium drying
中速钢球座圈式磨煤机 ball-and-race type pulverizer mill
中速混流式水轮机 medium speed Francis turbine
中速计算机 medium speed computer
中速接近示像 medium-approach aspect
中速接近信号 medium-approach signal
中速进行 medium-clear
中速进行低速接近示像 medium-approach slow aspect
中速进行示像 medium-clear aspect
中速进行信号 medium-clear signal
中速离心机 mid-speed centrifuge; swing-type centrifuge
中速率 intermediate speed
中速凝固的 medium curing
中速凝结浆 intergrout
中速球磨机 race pulverizer
中速乳胶 medium speed emulsion
中速扫描 medium fast scan; medium fast sweep
中速行驶运输工具 semi-fast transport
中速硬化的 medium hardening
中速预告接近示像 medium-advance approach aspect
中速远距继电器 medium speed distance relay
中速云 intermediate velocity cloud
中速运转 half-throttle
中塑性 medium plasticity
中塑性粉土 fair-plastic silt
中塑性黏[粘]质土 fair-plastic clay soil
中酸凝灰岩 ignimbrite
中酸性火山喷发组合 medium acidic volcanic assemblage
中酸性熔岩 neutral-acidic lava
中酸性岩 intermediate acidity rock
中碎 intermediate grinding; intermediate size reduction; medium crushing; middle crushing
中碎机 middle crusher; secondary crusher

中碎石 < AASHO 规定粒径 3/8 ~ 1 英寸,1 英寸 = 0.0254 米 > medium stone
中碎室 < 碎石机的 > intermediate crushing chamber
中碎用圆锥式破碎机 secondary gyratory crusher
中索 < 双杆作业用的 > span
中索滑车 span block
中塔克拉玛干构造段 Mid Taklimakan tectonic segment
中塔里木隆起区 median Tarimu upwarping region
中太平洋 mid-Pacific Ocean
中太平洋调查计划 mid-Pacific project
中太平洋海盆 mid-Pacific basin
中太平洋海盆巨地块 mid-Pacific Ocean basin block
中滩 central bar
中碳钢 medium carbon steel
中碳钢钢丝 medium carbon steel wire
中碳钢丝 bullet wire
中堂窗 < 教堂 > middle vessel window
中堂的拱顶 < 教堂的 > middle vessel vault
中堂的拱顶跨度 < 教堂 > middle vessel vault bay
中堂的列柱排列 < 教堂的 > middle vessel row of columns
中堂的支柱 < 教堂的 > middle vessel pier
中堂的柱列 middle vessel row of columns
中堂的柱列间的间距 < 教堂的 > middle vessel range of columns
中堂连拱廊 nave arcade
中梯曲线半极值点弦长 bowstring length between half-maximum point of curve mid-gradient
中梯曲线零值点间距 distance between zero point of mid-gradient curve
中梯曲线弦切距 bowstring tangent distance of mid-gradient curve
中体 mesosome
中体积溶胀 medium volume swelling
中天 meridian passage; meridian transit; transit; culmination
中天迟滞 lag of culmination
中天顶距 meridian zenith distance
中天法 culmination method; transit method
中天高度 culmination altitude; meridian altitude
中天高度求纬度 latitude of meridian altitude
中天高度修正 reduction to the meridian
中天观测 meridian observation
中天光度计 meridian photometer
中天记录器 transiter
中天面时 clock time of transit
中天山地峡 mid-Tianshan isthmus
中天射电望远镜 transit radio telescope
中天时 transit time
中天时间逐次逼近算法 first estimate-second estimate method
中天时刻 transit time
中天望远镜 transit telescope
中天星 culminant star
中条 median bar
中条板 middle strip
中条带状结构 middle-banded structure
中条带状煤 medium banded coal

中条灯芯绒 mid-wale corduroy
中厅 central nave;nave
中庭 cavaedium;patio;atrium[复 atria/triums] <古罗马建筑物的>;displuviatum <屋顶从房顶采光井向外倾斜的>
中庭式建筑 atrium architecture
中庭式门厅 atrium lobby
中停站 stop-over station
中梃 ledge;minion;mullion[munnion]
中同轴电缆 standard coaxial cable
中统【地】Middle series
中统靴 ankle boot
中凸 convexity
中凸的四瓣花装饰的线脚 tooth ornament
中凸龙骨 hump keel
中凸起轧辊 roll crown
中凸弯矩 hogging moment
中凸线 crown line
中凸形 camber
中凸形镶板 fielded panel
中涂 middle coating
中涂层 floating coat
中途 midcourse;midway;halfway
中途报关手续 immediate transportation entry
中途泵站 relay pump(ing) station
中途不停的 non-stop
中途测试 middle way test
中途测试时间 measure-while-drilling time
中途测站 reference transit station
中途岛环礁 midway atoll
中途调度站 intermediate control office
中途短暂停留 layover
中途返回 midway return operation
中途分道 stop-off
中途分卸整车 break-bulk car
中途港 intermediate harbo(u)r;port of call;intermediate terminal;way point
中途关停 <房地产开发项目> belly-up
中途换班制【铁】changing crew at midway system
中途换车出行 transit trip
中途货物加工的优惠 <适用直通运价> in-transit privilege
中途机场 staging post
中途加压站 relay pump(ing) station
中途加油 refueling
中途加载 additional shipment
中途接线盒 intermediate junction box
中途勘测站 reference transit station
中途康复站 halfway house
中途口岸 intermediate port
中途旅馆 halfway house
中途枢纽 intermediate terminal
中途停泊港 intermediate port
中途停车 intermediate stop
中途停车喂饲(牲畜) feeding in transit
中途停留 stop-off;stop over
中途停泊港 port of route
中途停运 stoppage in transit
中途下车 break of journey;to break the journey
中途下车站 stop-over station
中途下车站戳 break-of-journey station stamp
中途运输变更 diversion in transit
中途站 intermediate point;intermediate station;intermediate terminal;staging point;staging station;station in transit;way station;staging post <英>
中途折返 midway return operation
中途重装 intermediate reloading
中途装卸 intermediate handling
中外合营船舶 Chinese and Foreign joint-venture vessel
中外合资企业 Chinese-foreign joint ventures;Sino-foreign joint ventures
中外合资项目 joint venture project
中外联营企业 Sino-foreign joint ventures
中外贸易 Sino-foreign trade
中晚石炭世海浸 mid-late carboniferous transgression
中晚侏罗世气候分带 mid-late Jurassic climatic zonation
中微子 neutrino
中微子探测器 neutrino detector
中微子天文学 neutrino astronomy
中微子透镜 neutrino horns
中微子望远镜 neutrino telescope
中桅 <五桅船的第三桅> middle mast;topmast
中桅帆 topsail
中桅桅肩 topmast hounding
中维斯康辛【地】middle Wisconsin
中纬度 mid(dle) latitude
中纬度大气 middle-latitude atmosphere
中纬度地带 extra-tropic(al) belt;middle-latitude belt;middle-latitude zone
中纬度地区 middle-latitude region
中纬度法 middle-latitude method
中纬度风暴带 roaring forties
中纬度干旱气候 dry middle-latitude climate
中纬度公式 formula of mean latitude
中纬度航迹计算法 middle latitude sailing
中纬度航行法 middle latitude sailing
中纬度荒漠 middle-latitude desert
中纬度气候 middle-latitude climate
中纬度气旋 extra-tropic(al) cyclone;middle-latitude cyclone
中纬切面 equatorial section
中纬线 equator(of lens);mid-parallel
中位 meso-position
中位变速锁销控制器 centershift lock pin control
中位变速直角曲柄 centershift bell crank
中位差 median deviation
中位稠度拌和 medium consistence mixing
中位触点 mid-position contact;neutral position contact
中位地点车速 median speed
中位断开式极化继电器 center-off polarized relay
中位方向 median direction
中位换挡 center shift
中位换挡齿条 centershift rack
中位换挡导轨 centershift rail
中位换挡拉杆 centershift link
中位换挡连杆 centershift link
中位换挡液压缸 centershift cylinder
中位换挡指示器 centershift indicator
中位径 median diameter
中位空挡 neutral
中位离差 median deviation
中位流量 median discharge
中位泥炭沼泽 soligenous
中位浓度 median concentration
中位钳 mid forceps
中位潜育土 mid-gley soil
中位舌簧继电器 neutral-tongue relay
中位数 median
中位(数)粒径 median diameter
中位数无偏估计量 median unbias(s)ed estimator
中位数无偏性 median unbias(s)edness
中位(数)组 median class
中位水道 <河流高水位和低水位中间的> intermediate water channel
中位心 mesocardia
中位星等 median magnitude
中位溢流门 middle level overflow gate
中位照相星等 median photographic(al) magnitude
中位指数 median index number
中位周期 median period
中魏克塞尔冰期 middle Weichselian glacial epoch
中温沉积物 intermediate temperature deposits
中温带 mesophilic zone;middle temperate zone
中温的 mesophilic
中温发酵 mesophilic digestion
中温焚化炉 medium temperature incinerator
中温固化胶粘剂 intermediate temperature setting adhesive
中温过程 mesophilic process
中温和高温同相厌氧氧化 mesophilic and thermophilic co-phase anaerobic digestion
中温回火 average tempering;medium temperature tempering
中温加工 hot-cold work(ing);warm working
中温甲烷发酵 mesophilic methane fermentation
中温菌 mesophilic bacteria
中温炉 moderate oven
中温煤焦油 coal-tar middle oil
中温黏[粘]土砖 moderate heat duty fireclay brick
中温凝固胶合剂 intermediate temperature setting glue
中温凝固胶粘剂 intermediate temperature setting adhesive
中温暖系统 medium temperature water-heating system
中温气候 mesothermal climate;mesothermic climate
中温清漆 intermediate temperature varnish
中温燃料电池 medium temperature fuel cell
中温热水储 moderate temperature water reservoir
中温热液矿床 mesothermal deposit
中温润滑脂 medium temperature grease
中温烧成砖 medium burned brick
中温石英 medina quartz
中温水供暖系统 medium temperature water-heating system
中温退火 medium anneal
中温微生物 mesophile;mesophilic microorganism
中温温室 intermediate plant house;temperate house
中温污泥消化 <温度控制在33~35℃> mesophilic sludge digestion
中温细菌 mesophilic bacteria;mesophilous bacteria
中温消毒法 median temperature digestive treatment
中温消化 <污泥处理,一般在40℃以下> mesophilic digestion
中温消化处理 medium temperature digestive treatment;middle temperature digestive treatment
中温消化范围 mesophilic range
中温烟囱 medium temperature chimney
中温硬化胶 warm-setting glue
中温硬化胶着剂 intermediate temperature setting adhesive
中温硬化黏[粘]合剂 warm-setting adhesive
中温硬化黏[粘]结剂 intermediate temperature setting agent
中温油箱 medium temperature fuel cell
中温釉 intermediate fire(d) glaze;medium glaze
中温淤渣 intermediate temperature sludge
中温浴室 <古罗马> tepidarium
中温植物 mesotherm
中温植物型 mesothermal type
中文电传机 Chinese character telephotography
中纹锉 middle-cut file
中稳索 midship guy;schooner guy
中稳态曲线 neutral stability curve
中污带指示种 mesosaprobic(al) zone indicator
中污腐生物 mesosaprobium
中污染的 mesosaprobic(al)
中污(生物)带 mesosaprobic(al) zone
中午 meridian;midday;noon;noonday;noon tide;noontime
中误差 error of mean square;mean square error;medium error;quadratic mean deviation;quadratic mean error
中误差标准 mean square-error criterion
中误差椭圆 mean error ellipse;standard deviation ellipse;standard error ellipse
中雾 moderate fog
中西部地区 the central and western regions
中稀土 mesial rare earth
中喜马拉雅构造结 mid-Himalaya tectonic knot
中喜马拉雅古陆 mid-Himalaya old land
中细锉 medium smooth file;second cut file
中细粉磨 medium fine grinding
中细粉砂 medium silt and fine silt
中细粒花岗岩 interior fine granite
中细粒结构 medium fine texture
中细纹(锉) second cut
中下 below average;less than average
中下水深带 <水深在300~600英尺,即90~180米间的海底> inframedian zone
中弦线支距数据 mid-chord offset data
中线 mean line;median line;medium line;middle line;mid-line;neutral(main);neutral conductor【电】
中线安装 center line erection;center line mount
中线板 center[centre] plate
中线标杆 center[centre] post
中线标高 elevation of center-line
中线标划机 center[centre] line marker;center[centre] line marking machine
中线标志 mid-cut mark
中线标桩【测】center[centre] post;center stake
中线补偿器 neutralator;neutral compensator
中线参考标志 centre line witness mark
中线测量 center[centre] line survey(ing);central line surveying
中线的 mesal
中线地面标高 <黑色标高> center[centre] line ground elevation
中线点 center[centre] line point
中线电流 current in middle wire
中线电流继电器 neutral current relay
中线对策论 median game theory
中线法 <尾矿坝> center[centre] line

method
中线法洞内测量 hole survey by centerline method
中线缝 center[centre] joint; central joint; centre line joint
中线符号 center mark
中线高程 elevation of center-line
中线沟 center[centre] trench
中线惯性矩 equatorial moment of inertia
中线黑色标高 center line ground elevation
中线红色标高 center line grade elevation; centre line grade elevation
中线间距(离) distance between center[centre] lines
中线接地 neutral earthing; neutral ground(ing)
中线接地制 grounded neutral system
中线接缝 centre line joint
中线空气剂量 midline air dose
中线面 central plane; center line plane
中线面浸水部分中心 center of submerged lateral area
中线木钉 center line peg
中线木桩 center line peg
中线偏移 deviation of alignment
中线平均高度 center[centre] line average height
中线平均值 center[centre] line average
中线平台 center[centre]-line platform
中线坡道标高 <红色标高> center[centre] line grade elevation
中线坡度 center[centre] line grade; center[centre]line slope
中线外移桩导线测绘 survey of center line offsetting traverse
中线弯度 centre line camber
中线系统 mean line system
中线相配 center matched
中线相位 neutral phase
中线心桩 centre stake
中线照射量 midline exposure
中线支柱 middle line pillar
中线指示信标 middle marker beacon
中线桩【测】 alignment pole; alignment stake; center[centre] line stake; intermediate line point
中线桩高程测量 <简称中平>【测】 center[centre] line stake level(l)-ing
中陷船首尾 hollow ended
中陷龙骨 hollow keel
中箱 middle flask; mid-part
中向性格 ambiversion
中项 mead term; mean term; medium term; middle term
中消石灰 medium slaking lime
中小比例尺 median and small scales
中小厂商 small and medium sized business
中小城市 small-medium city; small-medium town
中小房 middle housing
中小工业区 zone of middle and small industry
中小工业用地 industrial estate for small and medium size industry
中小口径顶管机 <一种遥控泥水土压平衡的施工设备> tele-mole
中小企业 minor enterprise; small and medium sized enterprise
中小桥梁 small and medium bridge
中小型企业 medium and small scale enterprises
中小型沙发 settee
中小型型钢轧机 jobbing mill
中小型装置 medium and small sized unit
中小植物 indeterminate plant
中效过滤器 intermediate efficiency filter
中效空气过滤器 medium efficiency air filter
中斜线 diagonal
中卸粉磨 double rotator mill
中心 centrum[复 centrums/centra]; epicenter[复 epicentra]; epicentre; epicentrum[复 epicentra]; heart center[centre]; hub; intermediate line; navel; nucleus[复 nuclei/nucleuses]; spine
中心 *N* 边形 centred *N*-sided polygon
中心安全岛 central island; safety island; safety isle
中心按钮 center button
中心凹 central fovea
中心板 center[centre] plate; center[centre] plank
中心板材 center[centre] board
中心板橡木垫块 centre plate oak block
中心拌和厂 central mixing plant
中心拌和机 central mixer
中心拌和站 centre mixer
中心瓣 central lobe
中心泵站 central power
中心闭锁阀 closed center valve
中心边缘变化 center limb variation
中心变电所 center transformer
中心标度 centre scale
中心标高 central elevation; central level
中心标记 center[centre] mark
中心标志 center mark
中心标桩【测】 center stake; centre peg; hub stake
中心表现 central representation
中心播送室 main studio
中心不正 off-center[centre]
中心布线方式 wire center system
中心部分 central part; core
中心仓库 central warehouse
中心槽制 centre-slot system
中心侧向平面 central lateral plane
中心层 central layer; core sheet
中心层硬化 case tension; reverse case hardening
中心插口 central spigot; centre spigot
中心插口接合 hub and spigot joint
中心插销吊卡(堆开式) center-latch elevator
中心差 equation of the center[centre]; great inequality
中心差分 centered difference; central difference
中心差分插值法 central difference interpolation
中心差分求积公式 quadrature formula with central differences
中心厂搅拌法 central plant mixing
中心场所 central place
中心车站 key station
中心承枢 pivot pin
中心城市 central city; corduroy city; core city; hub city; key city
中心城镇 central town
中心齿轮 central gear; heart gear; sun gear
中心齿轮驱动 output sun gear
中心齿轮输出 output sun gear
中心齿轴 centre pinion; intermediate wheel and pinion
中心冲击 centric(al) impact
中心冲孔 center[centre] punching; central punching
中心冲孔器 center punch
中心冲头 center[centre] punch; central punch; dotting punch; prick punch
中心冲洗式钎头 centre-hole bit
中心冲凿 center punching
中心冲装的轮爪 center punched grouser shoe
中心冲装履带抓地齿 center punched grouser shoe
中心冲子 center punch
中心抽头 center tap; centre tap; mid-point tap; mid-tap; tapped center[centre]
中心抽头变压器 center-tapped transformer
中心抽头次级线圈 center[centre]-tapped secondary
中心抽头的 centre-tapped
中心抽头电抗器 centre-tapped reactor
中心抽头电阻器 centre-tapped resistor
中心抽头平滑扼流圈 interphase reactor
中心抽样检验 core test
中心出料式卷线机 laying reel
中心出水 centre take-off water
中心厨房 central kitchen
中心厨房加热 central kitchen heating
中心处理机 central processing unit
中心处理装置 central processing unit
中心处圆磨削 centre-type cylindric(al) grinding
中心传动 center drive; central shaft drive
中心传动车床 centre drive lathe
中心传动和边缘卸料的多仓磨 central drive peripheral discharge type mill
中心传动球磨机 central shaft drive ball mill
中心传输设备 central transmission equipment
中心床身式连续自动工作机床 centre column transfer machine
中心床身式钻床 centre column drilling machine
中心垂线定线 vertical centering alignment
中心锤 center[centre] weight; central weight
中心错位 off-center[centre]
中心带 central zone; core
中心带宽 midbandwidth
中心带杂质 mid-band impurity
中心刀片框架锯 centre-blade frame saw
中心导坑 centre-drift
中心导坑隧道掘进法 center drift tunnel(l)ing
中心导坑隧洞掘进法 center drift tunnel(l)ing
中心导孔隧道掘进法 center drift tunnel(l)ing
中心导孔隧洞掘进法 center drift tunnel(l)ing
中心导线 center[centre] conductor
中心导向 centre pilot
中心导向式金刚石扩孔钻头 pilot-reaming bit
中心岛 center[centre]-island; central island
中心岛地带 <道路的> neutral zone
中心到表面 center-to-face
中心到端面 center-to-end
中心到中心 centre-to-centre; center-to-center
中心的 central; centric
中心低标号混凝土 low-grade hearting concrete
中心低压锅炉房 central low-pressure boiler station
中心地 metropolis
中心地带 core area; core zone; midland
中心地(带)理论 central place theory
中心地区 corduroy area; core area; heart land
中心地位 centrality; central space
中心地学说 central place theory
中心点 central point; center[centre] point; focus; navel; omphalos[复 omphali]; pivot; pivoting point
中心点电抗接地系统 reactor grounded neutral system
中心点电阻接地方式 resistance grounded neutral system
中心点火 center fire
中心点火法 center firing
中心点火器 central igniter
中心点间距 pivot pitch
中心点接地 center-point earth
中心点转向 <车辆> center-point steering
中心电极 radial deflection terminal
中心电站 central power; central station
中心电钟 clock synchronizer
中心店 center[centre] shop
中心调度室 central dispatching room
中心对称的 central-symmetric(al); centrosymmetric(al)
中心对称晶体 centrosymmetric(al) crystal
中心对称性 central symmetry; centrosymmetry
中心墩 center[centre] pier; central pier
中心二次曲线 central conic
中心发电厂 central electric(al) power plant; central power plant
中心发电站 central electric(al) power station; central power station
中心发火 center fire
中心阀 center[centre] valve
中心法则 central dogma
中心方向比 centroid aspect ratio
中心放气孔 center[centre] vent
中心放射型(城市)规划 centrally planed
中心分辨率 center resolution
中心分配板 central distribution switch board
中心封合 central involution
中心缝 <路面的纵向缝> center joint
中心服务站 service center[centre]
中心辐射式全景建筑 panopticon
中心负荷 central concentrated load
中心负载 central loading; centric load
中心附属礼拜堂 <哥普特教堂三个圣所的> haikal
中心干管 central manifold
中心杆 centibar; tie rod
中心杆衬套 centre bushing
中心缸 suction
中心高度 center[centre] height
中心高度规 centre-height ga(u)ge
中心高炉 central furnace
中心隔板 central partition plate
中心给冲洗水 internal flushing
中心工厂 central factory
中心工作平台 central operation platform
中心工作站 central working station
中心供电 centre feed
中心供给 central feed
中心供气式喷头 converging diverging sprayhead
中心供热厂 central heating plant
中心供热系统 central heating; central heating system
中心供水凿岩机 internal water-feed machine

中心拱 center[centre] arch
中心构形 central configuration;central figure
中心构筑物 central structure
中心固定螺旋 central clamping screw
中心刮板 center scraper
中心馆 central repository
中心管 central conduit;central tube;header pipe
中心管道 central tube
中心管道现象 chimneying
中心管的根部 root of a central tube
中心管方法 central tube method
中心管进气喷燃器 center-diffusion tube gas burner
中心惯性矩 central moment of inertia;central second moment
中心光线 central ray
中心广场 central square;concourse
中心规 center[centre] ga(u)ge;centering ga(u)ge
中心硅藻 centric diatom
中心轨迹线 <拱圈的> line of centers[centres];locus of centres[centers]
中心辊 breast roll
中心锅炉房 central boiler house
中心锅炉装置 central boiler installation
中心合榫材 center matched
中心合榫构件 center matched
中心核 amphinucleus;central core;centronucleus
中心荷载 central(line) load(ing);center[centre](line) load(ing);center[centre]-point load(ing);centric load;concentric(al) load
中心喉管 piping
中心后处理厂 central reprocessing plant
中心厚度 center[centre] thickness
中心厚度路面 center thickness
中心花岗岩 central granite
中心花饰 center flower;center piece
中心滑环 center[centre] collector ring
中心滑块式连接 center-block type joint
中心化 centralization
中心化系统 centralized system
中心化子 centraliser[centralizer]
中心环岛 central island;roundabout island
中心环节 central link;key link
中心缓冲器 central buffer
中心换热式均热炉 vertically fired pit
中心回转接头 central rotating joint
中心汇率 central(exchange) rate;pivotal rate
中心会聚 center convergence
中心火警探测器 central fire detecting equipment
中心机房 central machine room
中心激励天线 center-driven antenna
中心级 central level
中心极限定理【数】central limit theorem
中心集电环 center[centre] collector ring
中心集电器 center[centre] collector ring
中心集流环 center[centre] collector ring
中心集水井 central caisson
中心给水的尾钎 shank rod for central flushing
中心计时系统 common timing system
中心记录加速度仪 central recording accelerograph
中心记录器 central recorder
中心记录系统 central recording system
中心记录站 centre recording station
中心加热 centrally heated
中心价格 central price
中心架 center[centre] frame;center[centre] rest;steady rest;work rest
中心尖端 centre point
中心间 <古建筑> central bay
中心间隔效应 central spacing effect;core spacing effect
中心间距(离) distance between centers[centres];distance between center to center;on centers[centres]
中心监护器 central monitor
中心监视 centralized monitoring
中心剪丝钳 end cutting nippers
中心检查站 central examination station
中心检验室 central laboratory
中心件 center piece
中心键 center key
中心交换局 central office exchange
中心交换转移 central exchange jump
中心交易所 central exchange
中心浇口 center[centre] gate;trumpet;trumpet assembly
中心焦距 center[centre] focus
中心角 angle at center[centre];centering[centring] angle;central angle;center[centre] angle
中心角尺 center[centre] square
中心铰 centering hinge;centring hinge
中心搅拌厂 center mixing plant;central mixing plant
中心搅拌机 central mixer
中心搅拌站 center mixing plant;central mixer;central mixing plant
中心校正 centering[centring]
中心阶乘矩 central factorial moment
中心接触的灯座 centre-contact holder
中心接线盒 central junction box
中心街区 central block
中心结 centre junction
中心结构 division center[centre]
中心结合 heart bond
中心截面 central cross-section
中心进料孔 central well
中心进料塔 central dumping tower
中心进气通道 central gas inlet duct
中心进水沉淀池 central sedimentation basin
中心进水圆形沉淀池 centre-feed circular sedimentation basin
中心井 center[centre] well
中心局 central office;central station;group center[centre];primary outlet;center operator <经营可视信息检索业务的机构>
中心局交换系统 central office switching system
中心局之间的中继线 primary-to-primary circuit
中心矩 central moment;moment about the mean
中心距(离) center distance;central distance;centre distance;centre-to-centre distance;centre-to-centre spacing;distance between centers[centres];width between centers;center-to-center distance;center-to-center spacing
中心距曲面 surface of centres[centers]
中心聚焦 centre focus
中心橛 centre peg
中心开口 open center[centre]
中心开口控制 open center[centre] control
中心开裂 central burst
中心开挖 center cut(ting)
中心开挖法 center[centre]-cut method;central excavating process
中心空调设备 central conditioning plant
中心空调站 air-conditioning station
中心空调装置 central conditioning installation
中心孔 center[centre] bore;center[centre] hole;center holing;center opening;central hole;central opening;centre opening;countersinking
中心孔机床 centre-hole machine
中心孔加工机床 centering machine
中心孔铰刀 center[centre]-hole reamer
中心孔径 central aperture
中心孔磨床 centre-hole grinder
中心孔抛光轮 center buff
中心孔式喷嘴 central-hole nozzle
中心孔铣刀 center fraise
中心孔隙 central porosity
中心孔研磨机 center hole lapping machine
中心孔直径 diameter of center hole
中心孔钻床 centre-drilling lathe
中心孔钻(头) centre drill;starting drill;center hole drill
中心控制 central control;master control;positioning control;centralized control
中心控制单元 centralized control unit
中心控制器 master controller
中心控制设备 central control equipment
中心控制室 central control room;master control room
中心控制台 central control board;console of centralized control;master control board
中心控制站 key station
中心库 central storage
中心跨度 center[centre] span;central span
中心跨距 center[centre] span;central span
中心块 center[centre] piece;central block;central piece
中心快门 central shutter;lens shutter;between(-the)-lens shutter <俗称镜间快门>
中心馈电 apex drive;center[centre] feed
中心馈电的 centre fed
中心拉紧螺栓 central tension bolt
中心拉力 axial tension
中心离子 central ion
中心里程 center mileage;central mileage
中心力 central force;centre force
中心力场 central force field
中心力矩 central moment;moment about the mean
中心立轴 center journal;king bolt;king pin;master pin
中心立轴盖 kingpin cover
中心立轴盖衬 kingpin cover gasket
中心立轴活动键 kingpin draw key
中心立轴毡 kingpin felt
中心立轴止推轴承 kingpin thrust bearing
中心立柱 pillar
中心立柱墙 stud wall(partition)
中心励磁式 central-excitation system
中心粒 central granule;centriole
中心粒团 microcentrum
中心粒小轮 centriole pinwheel
中心连接法 center-to-center method;centre-to-centre method
中心连接器 centre coupler
中心连线 line of centers[centers]
中心联杆 center[centre]-bound tie
中心联线 line of centers[centers];line through center[centre]
中心量柱 button ga(u)ge
中心裂缝 centre cleavage
中心裂纹 centre burst;chev(e)ron crack;heart check;heart shake
中心零位电流计 center-point galvanometer
中心零位式仪表 centre-zero instrument;zero center instrument
中心零位压力计 center-zero ga(u)ge
中心零位仪表 center zero instrument
中心溜分 heart cut
中心流道模 center-gated mo(u)ld
中心流钢砖 king brick;spider brick
中心轮 center wheel;central gear
中心轮下降 depressed center wheel
中心螺钉 central screw
中心螺杆式挤出机 central screw extruder
中心螺栓 center bolt
中心螺旋 central screw
中心螺旋角 mean spiral angle
中心盲点的 cecocentral;centrocecal
中心锚结 midpoint anchor
中心锚结线夹 mid-anchor clamp
中心锚柱 midpoint pole
中心门挡 center gate stop
中心密集度 central concentration
中心面 central plane
中心秒针 center second;direct sweep second
中心磨床 center[centre] grinder
中心磨床附件 centre grinder attachment
中心木质 duramen
中心木质部 juvenile wood
中心内龙骨 center(line) keelson
中心凝聚物 central condensation
中心排料式棒磨机 center discharge rod mill
中心排水沟 center[centre] drain;central drain
中心排水管 center[centre] drain;central drain
中心排水隧洞 central drainage tunnel
中心盘 spider
中心配料厂 central proportioning plant
中心配流 centrally ported
中心喷发 central eruption
中心喷管 center burner;central burner;puffer-pipe
中心喷射式钻头 center[centre] jet drill bit
中心喷雾旋风除尘器 cyclonic spray scrubber
中心批发市场 central wholesale market
中心偏移 center drift;off centering;shift
中心频段 mid-band
中心频率 center[centre] frequency;central frequency;idle frequency;mid-frequency
中心频率脉冲 center frequency pulse
中心频率偏移传感器 osciducer
中心频率剩余曲线 residual curve of middle frequency
中心平面图 central plan
中心平旋桥 centre bearing swing bridge
中心破碎强度 radial crushing strength
中心剖面 center section;centre section
中心气压【气】central pressure
中心器 centralizer
中心器绕组 centre quad

中心千分尺 hub micrometer
中心桥墩 centre pier
中心切割 heart cut
中心清晰度 central definition; sharp central curve
中心球 centrosphere
中心球桩齿 center button
中心区 central area; central city; central district; central zone; centre region; mid-continental region
中心区分布 centre zone profile
中心区划区 planning of central district
中心区线网密度 central district line net density
中心曲柄 center crank
中心曲杆 centre crank
中心曲轴 centre crank shaft
中心驱动集泥器 centre drive sludge collector
中心驱动排锯机 central driven log frame
中心趋向(势)度量【数】central tendency measurement
中心取样 plot sampling core; sampling core
中心缺口试样 centrally notched specimen
中心燃烧设备 central burning appliance
中心燃烧式均热炉 bottom center-fired pit
中心燃烧式热风炉 centre combustion stove
中心染色过浅 central pallor
中心热水站 central water heating
中心人物 keyman
中心润滑 central oiling
中心润滑法 centralized lubrication
中心塞钻头 center plug
中心伞齿轮 centre bevel gear
中心商店 central store
中心商务区 center business district; central business district
中心商业区 central business area; central business district; central commercial district
中心设施 central facility
中心射线 central ray; centre ray
中心射影 central projection
中心石 choke stone
中心实验室 center laboratory; centralab; central lab; central laboratory
中心蚀 central eclipse
中心蚀曲线 central eclipse curve
中心食 central eclipse
中心市 nodal city
中心市场 central market; central plaza
中心式 center type
中心式空气调节 central air conditioning
中心式喷发 central eruption
中心式水热喷发 central hydrothermal eruption
中心式舞台 arena stage; central stage
中心视觉 central vision
中心视野 central field of vision
中心试验器 center[centre] tester
中心试验室 centralab; central laboratory
中心试验站 center[centre] test station
中心饰 epergne
中心室 central chamber
中心室搅拌库 central chamber blending silo
中心收敛 center convergence
中心受拉强度 uniaxial tensile strength
中心受压构件 centrally compressed member
中心受压柱 central compressed column; centrally compressed column
中心枢纽建筑物 centralized building
中心枢轴 center gudgeon; center pivot; central pivot; centre joint; center pin
中心枢轴式喷灌机 center pivoted sprinkler
中心枢轴销 center[centre] pivoted pin
中心疏松 center porosity; central porosity
中心输送机 central conveyer
中心数据处理机 central data processing unit
中心数据处理器 central data processing unit
中心数据库 central data bank
中心水口式钻头 center[centre]-hole bit
中心丝 < 光学仪器的 > central hair
中心四绕组 center quad
中心榫接 center matched
中心缩管 central pipe
中心缩孔 central porosity
中心索 axial cable
中心锁定 center-lock
中心台 zone center[centre]
中心弹簧杆 centring spring rod
中心弹簧座 centering spring seat
中心掏槽 center[centre] cut(ting)
中心掏槽爆破 center peg; center[centre] shot
中心掏槽炮眼 center[centre] shot
中心特别低的车架 kick-up frame
中心体 center[centre] body; central body; centrum[复 centrums/centra]
中心条件 neutral condition
中心条纹 seam line
中心调节 centering control; centring control
中心调节电流 centring control current
中心调节放大器 centering amplifier
中心调整 centering (adjustment); centering control; centre adjustment; centr(e)ing; positioning control
中心调整管 centering tube; centre tube
中心调整系统 centring system
中心调整线圈 centering winding; frame coil
中心跳跃 center jump
中心铁 center iron
中心停车区 central parking district
中心通道 central passage
中心通道客车 center[centre] gangway coach
中心通风孔 centre vent
中心通行洞 central access hole
中心筒墙 central core wall; centre core wall
中心投影【测】central projection
中心投影法 central projection method; method of central projection
中心透视 central perspective
中心透视不变式 invariant of central perspective
中心透视成像 central-perspective image formation
中心透视相关 central-perspective correlation
中心凸出的多级钻头 crowned bit
中心凸出钻头 convex bit
中心凸轮 centering jaw
中心凸起 < 板材的 > center buckle
中心突出的三翼铸钢钻头 cast-steel 3-wing pilot bit
中心图解法 centrography
中心图书馆 central library
中心图像 central image
中心土墩 dumpling
中心椭圆半轴 semi-axis of the central ellipse
中心外圆磨床 centre-type cylindric(al) grinder
中心外圆磨削 center-type cylindric(al) grinding
中心网络管理系统单元 central net system unit
中心为零的安培计 central zero ammeter
中心未对准 disalignment; disalinement
中心位移齿轮 centershift pinion
中心位移连杆 centershift link
中心位移驱动轴 centershift drive shaft
中心位置 center[centre] position; central place; central position; midposition
中心位置冲小孔 centre punching
中心位置调节电路 centring circuit
中心位置调整磁铁 centering magnet
中心位置调整电路 positioning circuit
中心温度 central temperature
中心无风带 central calm
中心无风区 central calm
中心舞台剧场 arena theatre[theater]; theatre-in-the-round
中心洗衣房 central laundry
中心系统 central system
中心弦 centre chord
中心线 intermediate line; line of centers[centers]; mean line; middle line; pitch line; center[centre] line; central line; axis[复 axes]; bisector; bisectrix; camber line; central wire; inner conductor
中心线板条 centre line strake
中心线半径 center line radius; radius of central line
中心线测量 center-line survey
中心线端子 zero terminal
中心线校正 centre line adjustment
中心线节 center line knot
中心线流速 center[centre] line velocity
中心线浓度 center[centre] line concentration
中心线偏离 disalignment
中心线偏斜 lack of alignment
中心线偏移 disalignment
中心线平均值 centre line average
中心线收缩 center line shrinkage cavity
中心线速度 center line velocity
中心线弯曲角 camber angle
中心线稀释度 center[centre] line dilution
中心相 central facies
中心相带 center facies zone
中心镶嵌塞 < 金刚石钻头 > center plug
中心向联轴节 hub and spigot joint
中心销 center[centre] pin; core pin; king bolt; king pin; main pin; fifth wheel king pin < 车的 >
中心销承窝 centre pin socket
中心销盖板 center pin floor plate
中心销支承 center pin support
中心小齿轮 sun pinion
中心小木桩 center[centre] peg; central peg
中心小水口全面金刚石钻头 pencil core bit
中心小轴杆 centre staff
中心楔块 < 造桥用的 > slack block
中心卸料 center[centre] discharge; central discharge
中心卸料磨 trunnion discharge mill
中心心轴 centre arbor
中心信号室 central dispatching room
中心星 central star
中心性 centrality
中心修理厂 central repair shop
中心悬臂桁架 central cantilever truss
中心旋流器 centriclone
中心旋转接头 center revolving joint
中心选矿厂 customs plant
中心压力调节器 central pressure regulator
中心压缩 central compression
中心焰 neutral flame
中心阳极光电管 central anode photocell
中心遥控 centralized telecontrol
中心移动球头螺栓 ball stud for centershift
中心移动支架 centershift rack
中心引爆装置 central igniter
中心引线的 centre-tapped
中心浴场 central bath
中心圆 limiting circle
中心圆管灌浆 circuit grouting
中心圆锥线 central conic; centre conic
中心匝道 center ramp
中心杂岩 central complex
中心载荷 center line load
中心站 central site; centre site; junction center[centre]; control station
中心胀大 centre bulb
中心辙叉 < 菱形交叉或交分道岔的 > center[centre] frogs
中心震 central earthquake; centre earthquake
中心支撑 center support
中心支撑系统 center support system
中心支承 center bearer; center[centre] bearing; center support; pivot(ing) bearing
中心支承平旋桥 center bearing swing bridge; central bearing swing bridge; pivot-bearing swing bridge
中心支承式 center[centre] bearing type; central bearing type
中心支承旋转窗 center[centre]-pivoted window; central pivoted window
中心支承旋转门 center pivoted door
中心支持系统 centre support system
中心支枢喷灌系统 central pivot irrigation system
中心支销座架 center pivot mount
中心支轴式喷灌机 center pivoted sprinkler
中心支轴旋开门 crapaudine door
中心支柱 center pole; central post; centre strut
中心支座式喷灌 center pivot sprinkler irrigation
中心值【数】central value; midpoint
中心值移值 removal value of central value
中心止门装置 center gate stop
中心指零式测量仪表 center zero meter
中心指零式仪表 zero centre meter
中心至边缘 center-to-edge; centre to edge
中心制 central system
中心制冷站 central refrigeration plant
中心致密颗粒 central dense granule
中心置信区间【数】central confidence interval
中心重叠法 central overlap technique
中心洲 central bar
中心轴 center axle; center shaft; central shaft; central spindle; layshaft; neutral axis; spigot shaft; center pin < 旋转电铲的垂直固定轴 >; cen-

tre pintle < 指旋转电铲的垂直固定轴 >
中心轴承 center bearer; central bearing; pivot bearing; centre bearing
中心轴承除泥罩 centre bearing mud slinger
中心轴承壳 centre bearing housing
中心轴承壳盖 centre bearing housing cap
中心轴承壳软垫 centre bearing housing insulator
中心轴承连轭螺母 centre bearing companion yoke nut
中心轴承托架 centre bearing housing bracket; centre bearing housing carrier
中心轴枢 central pivot
中心轴线 central axis
中心轴销 central pivot of king bolt
中心轴柱 < 旋转型楼梯 > spindle
中心主题 central motif
中心主轴 central principal axis; king bolt; principal central axis
中心主轴线 central principal axis
中心注管 git; trumpet assembly
中心注视点 central fixation point
中心柱 mid-span mast; newel; stele; central post
中心柱形阀 centre spool valve
中心桩【测】center line stake; centre line peg; peg of crossing centerline; center peg; centre pile; centre stake; king pile
中心桩号 center[centre] station
中心装 center mount
中心装料 center[centre] filling; centre charging
中心装料法 method of center[centre] charging
中心锥 center cone; center key
中心锥体 conic(al) inner body
中心资料馆 central repository
中心字 key word
中心纵缝 longitudinal center[centre] joint
中心钻 center[centre](hole) drill; centering drill; nicker; spotter
中心钻臂 center boom
中心钻孔 center[centre] drilling (bore) hole
中心钻头 center[centre] bit; central bit
中心钻削 center[centre] drilling
中心最大磨削直径 maximum grinding diameter on centers
中新生界盆地地下水 groundwater in Mesozoic-Cenozoic basin
中新世【地】Miocene epoch
中新世末期事件【地】terminal Miocene event
中新世黏[粘]土 Miocene clay
中新统【地】Miocene series
中星仪 astronomic(al) transit; meridian instrument; transit; transit instrument
中星仪盘东 level east
中星仪式干涉仪 meridian interferometer
中行车 middle rolling car
中型 bergy bit; medium pattern; medium size; mesotype
中型板 jobbing sheet
中型板材 medium size panel
中型材 medium section
中型槽 medium cell
中型车床 medium heavy lathe
中型出租汽车 midi-bus
中型船 intermediate ship
中型的 medium duty; medium scale; medium sized; medium weight; middle scale; middle-sized; moderate duty; moderate-size(d)
中型灯头 medium base
中型登陆艇 < 海洋钻探用 > landing ship medium
中型地貌 middle-scale landform
中型电动机 medium sized motor
中型电解槽 medium cell
中型电站 medium capacity plant
中型动力触探 mesotype dynamic(al) sounding
中型斗柄 < 挖掘机 > medium stick
中型发动机 medium duty engine
中型浮游生物 mesoplankton
中型公共汽车 light bus; midi-bus
中型构件 intermediate section
中型构造 mesoscopic structure
中型构造尺度 mesoscopic scale
中型管纱 bastard cop
中型管柱 intermediate tubular column
中型管座 intermediate base
中型规模 medium scale
中型和轻型气焊工具配套单位 medium and light gas welding unit
中型和重型气焊工具配套单位 medium and heavy gas welding unit
中型滑坡 medium landslide
中型货物集装箱 cargo sized container
中型计算机 medium scale computer; medium size computer
中型加班公共汽车 para-transit vehicle
中型建设项目 medium construction item; medium construction project
中型脚手架 medium duty scaffold
中型轿车 < 美 > intermediate
中型结构 medium texture
中型均匀反应堆 intermediate size homogeneous reactor
中型卡车 medium duty truck; medium type truck
中型客车 light bus; van pool
中型客车共乘 van pooling
中型宽带材轧机 medium wide strip mill
中型矿床 medium size ore deposit
中型矿山 medium tonnage mine
中型露天矿 medium sized quarry
中型煤矿 medium size coal mine
中型品种 medium weight breed
中型平碾 middle weight smooth wheel roller
中型企业 medium lot producer; medium sized enterprise
中型汽轮发电机 medium size turbine generator
中型砌块 medium block
中型轻便汽车 medium car
中型区系 mesofauna
中型舢板 galley
中型设备 medium plant
中型生物群 mesobiota
中型双开弹簧门 medium duty flexible(swing) door
中型双柱灯头 medium bipost base
中型水利工程 medium hydraulic project
中型水源地 middle-sized water source
中型塑料细管 medium straw
中型突水 medium scale of bursting water
中型拖拉机 medium sized tractor
中型温室 medium sized green house
中型问题 intermediate scale problem
中型污水稳定塘系统 pilot waste stabilization pond system
中型细菌 mesobacteria
中型箱【机】cheek
中型箱铸模 cheek pattern
中型小客车 intermediate(size) car
中型小区试验 medium sized plot test
中型型材 intermediate section
中型型钢轧机 medium section mill
中型压缩机 moderate-duty compressor
中型油轮 < 常指 6 万 ~ 12 万吨级的 > medium sized tanker
中型预聚焦灯座 medium prefocus base
中型运输带 normal duty conveyer [conveyor]
中型运输汽车公司 medium transportation truck company
中型载货汽车 medium duty truck; medium truck; medium type truck
中型载重货车 medium lorry
中型载重汽车 medium duty truck; medium pickup car; medium type truck; medium weight truck
中型凿岩机 medium drill
中型中砾 medium pebble
中型轴承 medium size bearing
中型柱 intermediate column
中型抓斗 medium grab bucket
中型装载机 medium loader
中型钻机 medium drill
中性 neutrality
中性 pH 值 neutral pH
中性岸 neutral coast
中性白土 neutral clay
中性柏油 tar medium oil
中性包装 neutral packing
中性表面物种 neutral surface species
中性滨线 neutral shoreline
中性波 neutral wave
中性波长 neutral wavelength
中性玻璃 neutral density glass; neutral glass
中性玻璃滤光片 neutral filter glass
中性材料 neutral material
中性层 neutral area; neutral axis; neutral layer; neutrosphere
中性层顶 neutropause
中性场线 neutral field line
中性超流继电器 neutral over-current relay
中性成粒作用 neutral granulation
中性冲击波 neutral shock wave
中性传输 neutral transmission
中性刺激 neutral stimulus
中性大气 neutral atmosphere
中性大气密度 neutral atmosphere density
中性大气模式 neutral atmosphere model
中性大气温度 neutral atmosphere temperature
中性导体 neutral conductor
中性导线 neutral conductor
中性的 eutrophic; neutral; normal
中性地带 neutral zone
中性点 midpoint; neutral point; non-slip point; no-slip point; star point; neutrality point < pH 值为 7.0 的 >
中性点变压器 neutral point transformer
中性点补偿器 neutral compensator; neutrator
中性点不接地系统 isolated neutral system; neutral point no-earthing system
中性点接地 neutral ground(ing); neutral point earthing
中性点接地地阻 neutral point earthing resistance
中性点接地网络 network with earth-connected neutral
中性点接地系统 earth neutral system
中性点接地自耦变压器 neutral auto-transformer
中性点接线端 neutral terminal
中性点绝缘网络 network with insulated neutral
中性点绝缘制 isolated neutral system
中性点理论 neutral point theory
中性点调整 neutral adjustment
中性点位移 < 星形接线 > neutral point displacement
中性点位移变压器 neutral displacement transformer
中性点引出线 neutral lead
中性点直接接地 neutral point solid ground
中性碘化钾 neutral potassium iodide
中性电极 neutral electrode
中性电离流 neutral ionized stream
中性电流 neutral current
中性电流片 neutral current sheet
中性电位 neutral potential
中性电阻 neutral resistance
中性端 neutral end
中性反应 neutral reaction
中性方程 neutrality equation
中性分光镜 neutral beam splitter
中性分子 neutral molecule
中性风切变 neutral wind shear
中性浮子 neutrally buoyant float
中性辐射三角测量 centre triangulation
中性干线 neutral main
中性感受器 neutroceptor
中性钢 neutral steel
中性工作 neutral operation
中性共轭 neutral conjugation
中性共振 neutral resonance
中性光劈式滤光片 neutral wedge filter
中性光楔 neutral wedge; sensitometric wedge; step wedge
中性轨 neutral rail
中性过电流继电器 neutral over-current relay
中性海岸 < 无升降变化的 > neutral coast; neutral shoreline
中性海滨线 neutral shoreline
中性焊剂 neutral flux
中性合成洗涤剂 neutral synthetic washing agent
中性河口(湾) neutral estuary
中性红 neutral red; toluylene red
中性化 neutralization
中性化合物 neutral compound
中性化器 neutraliser
中性化深度 depth of neutralization
中性化室 neutralizing cell
中性缓冲碘化钾溶液法 neutral buffered potassium iodide method
中性灰 < 指图像 > neutral-tone
中性灰滤光片 neutral density filter
中性灰色 neutral gray[grey]
中性灰色滤光玻璃 neutral grey glass
中性汇流排 neutral bus-bar
中性活化法 neutral activating method
中性火成岩 intermediate igneous rock
中性火焰 neutral flame
中性基 neutral radical
中性级数 neutral series
中性极 neutral pole
中性继电器 neutral relay
中性键控 neutral keying
中性胶料 neutral size
中性胶束 neutral micelle
中性胶体 neutral colloid
中性角 angle of nonslip point; neutral angle < 轧制用语 >
中性阶梯减光板 neutral step filter; neutral step weakener
中性阶跃劈 neutral step wedge
中性接触 neutral contact
中性接触体 hermaphroditical contact

Z

中性接地电阻器 earthing resistor of neutral point; neutral earthing resistor; neutral ground resistor
中性接点 neutral contact
中性结构 mesomorphism
中性结构的特性 mesomorphy
中性介质 neutral medium
中性介质中退火的钢丝 lime bright annealed wire
中性介子【物】 neutral meson; neutretto
中性金属 neutral metal
中性浸出 neutral leach(ing)
中性空位 neutral vacancy
中性矿石 neutral ore
中性矿物 neutral mineral
中性矿渣 neutral slag
中性馈线 neutral feeder
中性扩散体 neutral diffuser
中性蓝 neutral blue
中性类模型 neutral model
中性类型 neutral type
中性离子注入器 neutral ion injector
中性粒子谱仪 neutral spectrometer
中性粒子质谱仪 neutral mass spectrometer
中性流相互作用 neutral current interaction
中性炉衬 neutral lining
中性炉气 neutral atmosphere
中性炉渣 neutral slag
中性滤波 neutral filtration
中性滤光镜 black glass; neutral density filter; neutral-tint filter
中性滤光片 black glass; light-balancing filter; neutral colo(u)r filter; neutral density disck[disk]; neutral (density) filter; non-selective absorbent; non-selective absorber
中性滤光片工作标准 neutral filter working standard
中性滤光器 neutral density filter; neutral filter
中性滤色镜 neutral colo(u)r filter; neutral density filter
中性绿 neutral green
中性络合物 neutral complex
中性密度光楔 neutral density wedge
中性密度滤光片 gray filter; neutral density filter; neutral filter
中性密度楔 neutral density wedge; wedge of neutral glass
中性面 neutral area; neutral flow plane; neutral surface
中性母线 neutral bus
中性耐火材料 neutral refractory(material)
中性耐火砖 neutral fire brick; neutral refractory brick
中性偶 neutral pair
中性劈 neutral wedge
中性片 neutral sheet
中性片重连接 neutral sheet reconnection
中性平衡 indifferent equilibrium; neutral equilibrium
中性平面 neutral plane
中性气氛 neutral atmosphere
中性气体 indifferent gas; neutral gas
中性气团 indifferent air-mass; neutral air mass
中性气旋 neutral cyclone
中性轻子流 neutral lepton current
中性请求信号 interrupt request signal
中性区 differential gap; neutral region; neutral zone
中性曲线 indifference curve; neutral curve
中性去污剂 neutral detergent
中性去污(剂)溶解物 neutral detergent soluble
中性去污剂纤维 neutral detergent fiber
中性泉 neutral spring
中性燃烧 neutral burning
中性燃烧曲线 neutral-burning curve
中性染剂正染性的 orthochromophil; orthoneutrophil
中性染料 neutral dye; neutral stain
中性溶剂 neutral solvent
中性溶液 neutral solution
中性溶质 neutral solute
中性熔剂 neutral flux
中性熔岩 intermediate lava
中性熔岩流 intermediate lava flow
中性润滑油 neutral lubricating oil
中性润滑油配料 neutral lubricating stock
中性润湿 neutral wetting
中性散射 neutral scattering
中性色 neutral colo(u)r
中性色调像片 paper with a neutral image tone
中性色区域 achromatic region
中性石 mesodialyte
中性石蕊试纸 litmus neutral test paper
中性试剂 neutral reagent
中性试验 neutralisation test
中性束 neutral beam
中性束源 neutral beam source
中性束注入 neutral beam injection
中性束注入加热 neutral beam injection heating
中性树脂 neutral resin; resinene
中性衰变产物 neutral decay product
中性水 neutral water
中性水玻璃 neutral water glass
中性水处理 neutral water treatment
中性水流 neutral current
中性苏打 neutral soda
中性碎片 neutral fragment
中性体 neutral body
中性填料 inert filler; neutral filler
中性条件 neutral condition; neutrality condition
中性调色剂 neutral toner
中性突变 neutral mutation
中性土(壤) neutral soil
中性土植物 neutral soil plant
中性拖曳不稳定性 neutral drag instability
中性位置 neutral position
中性稳定度 indifferent equilibrium; indifferent stability; neutral stability
中性稳定性 indifferent equilibrium; indifferent stability; neutral stability
中性物 neutral matter
中性吸收 neutral absorption
中性吸收剂 neutral absorbent
中性吸收器 neutral absorber
中性洗涤剂 neutral detergent; neutral washing agent
中性洗液 neutral wash solution
中性纤维 neutral fiber[fibre]
中性衔铁 neutral armature
中性现象 neutralism
中性线 axis of commutation; common wire; fourth wire; neutral (conductor); neutral line; neutral main; neutral wire; neutral fiber
中性线电流 natural current
中性线电流继电器 neutral current relay
中性线端 neutral terminal
中性线放电理论 neutral line discharge theory
中性线回路 neutral return path
中性线继电器 neutral relay
中性线加载 neutral loading
中性线接地电抗器 neutral reactor
中性线接地电阻器 neutral grounding resistor
中性线开关 neutral switch
中性线路 neutral circuit
中性线圈 neutral coil
中性像纸 neural paper; normal paper
中性型 neutral type
中性性状 neutral character
中性溴化钾 neutral potassium bromide
中性悬浮颗粒 neutrally buoyant particle
中性悬浮物质 neutral suspended substance
中性压力 neutral pressure; neutral stress
中性亚硫酸盐半化学的 neutral sulphite semi-chemical
中性亚硫酸盐纸浆 neutral sulphite pulp
中性岩 intermediate rock; medium rock; mesite; neutral rock
中性岩浆元素 elements of neutral magma
中性岩类 intermediate rocks; medium rocks; neutral rocks
中性盐 neutral salt
中性颜料 neutral pigment
中性颜色 neutral colo(u)r
中性氧化作用 neutral oxidation
中性液体 neutral liquid
中性异染的 metaneutrophil
中性应力 neutral stress
中性油 neutral oil
中性油清漆 medium oil varnish
中性油脂 neutral fat
中性有机化合物 neutral organic compound
中性有色玻璃 neutral-tinted glass
中性元素 neutral element
中性杂质 neutral impurity
中性载体膜电极 neutral carrier membrane electrode
中性渣 neutral slag
中性褶皱 neutral fold
中性振荡模式 neutral oscillation mode
中性蒸化 neutral ag(e)ing
中性蒸馏 neutral distillation
中性脂肪 neutral fat
中性指示器 neutralizing indicator
中性质子物种 proton-neutral species
中性轴(线) neutral axis
中性状况 neutrality condition
中性状态 neutral state
中修 intermediate overhaul; intermediate repair; medium maintenance; medium overhaul; medium repair
中修工程 intermediate maintenance works
中修组 medium maintenance group
中序遍历 inorder traversal
中序后继块 inorder successor
中悬 centre-hung; horizontally pivot hung
中悬窗 center[centre]-hung sash window; center [centre]-hung swivel window; horizontally pivot (ing) (hung) window
中悬窗扇 center[centre]-hung window
中悬翻窗 center[centre]-hung window
中悬挂式 centre mounted type
中悬门 center-hung door
中悬式窗 horizontally pivoted hung window
中悬索 centre suspension cord
中悬折叠门 centafold door
中悬转轴 center-hung pivot
中旋 centre-hung
中旋窗 center [centre]-pivoted window; pivot-hung window
中旋窗窗压条 sash bead
中旋窗框 center-hung sash
中旋回 mesocycle
中旋门 center[centre]-hung door; center[centre] pivoted door
中旋涡 mesoscale(eddy); mideddy
中旋式气窗框 center-hung sash; centre-hung sash; swivel frame
中选 middling
中选标 accepted bid
中选的投标者 selected bidder
中学 high school; middle school; secondary school
中学建筑 secondary school buildings
中学一年级学生 <美> freshman
中压 intermediate pressure; medium pressure; medium voltage
中压氨吸收塔 medium pressure ammonia absorber
中压备用气门 reheat emergency valve
中压泵 medium lift pump; medium pressure pump
中压变质作用 medium pressure metamorphism
中压部分 intermediate pressure section
中压槽 medium pressure tank
中压电力网 medium voltage network
中压电力系统 medium voltage network
中压段 intermediate pressure section
中压发动机 medium pressure engine
中压反应器 middle-pressure reactor
中压供电网 medium voltage network
中压汞灯 medium pressure mercury lamp
中压锅炉 medium pressure boiler; middle-pressure boiler
中压胶板 medium pressure rubber sheet
中压开关柜 medium voltage switchgear
中压开关装置 medium voltage switchgear
中压离心鼓风机 intermediate pressure centrifugal blower
中压联合气门 combined reheat stop and intercept valve; combined reheat valve; reheat stop interceptor valve
中压轮胎 medium pressure tire[tyre]
中压喷焊器 medium pressure blowpipe
中压喷淋器 medium pressure sprinkler
中压喷洒器 medium pressure sprinkler
中压气流式输送器 medium pressure pneumatic conveyer[conveyor]
中压汽缸 intermediate pressure cylinder
中压汽轮机 intermediate pressure turbine; medium pressure steam turbine; middle-pressure steam turbine
中压汽门 reheat emergency valve
中压区域变质相系 medium pressure regional metamorphic facies series
中压区域变质作用 medium pressure regional metamorphism
中压燃烧器 medium pressure burner
中压绕组 medium voltage winding
中压热水 medium pressure hot water
中压容器 medium pressure vessel
中压洒水车 medium pressure sprinkler
中压输电线路 middle voltage transmission line
中压水蒸气 medium pressure steam
中压隧道 medium pressure tunnel
中压隧洞 medium pressure tunnel

中压缩性土 middle compressible soil
中压套管 <变压器的> medium tension bushing;medium voltage bushing
中压调节阀 reheat control valve;reheat(ing) interceptor valve
中压透平 intermediate pressure turbine
中压涡轮机 intermediate pressure turbine;middle-pressure turbine
中压吸收塔 medium pressure absorber
中压系统 medium voltage system
中压线路 medium voltage line
中压橡胶石棉板 middle pressure rubber asbestos
中压型 medium pressure type
中压蓄力器 medium pressure accumulator
中压压气机 intermediate pressure compressor
中压压缩机 intermediate pressure compressor
中压液面计 liquid level ga(u)ge of medium pressure
中压乙炔发生器 medium pressure acetylene generator
中压油吸收过程 medium pressure oil absorption process
中压元件 medium voltage unit
中压蒸汽系统 medium pressure steam system
中亚-蒙古地槽系【地】central Asia Mongolian geosyncline system
中亚蒙古迁移区 middle Asia Mongolian migration region
中亚黏[粘]土 medium clay(ey)loam;medium textured loam;middle clay(ey)loam;middle loam
中亚热带 middle subtropical zone
中亚(细亚)Central Asia
中延烧性面 surface of medium flame spread
中研磨性地层 middle abrasive formation
中盐度的 mesohaline
中盐度热储 moderate salinity reservoir
中盐度水 medium salinity water
中盐性 mesohaline
中盐鱼 medium salted fish
中盐渍 medium salting
中盐渍土 moderately salified soil
中燕山亚旋回【地】Middle Yanshanian subcycle
中央 centre;medium;navel
中央凹的 central concave
中央摆铁 center bearing plate
中央背斜带 central anticlinal belt
中央泵送功率 push and pull pumping power
中央壁 median wall
中央标 mid-channel mark
中央标准时 central standard time
中央并列式开拓系统 center-parallel development program(me)
中央薄板 central sheet
中央补助 Federal aid
中央部 central portion
中央部分 middle body
中央部位之装饰品 center piece
中央材干围 mid-timber girth
中央操纵计算机 central control computer
中央操纵室 cab(in)
中央操纵台 center pedestal;central control console;pedestal controller
中央操作系统 central operation system
中央操作员 center operator
中央层 central stratum
中央差动机构 center differential
中央长途电话局 toll center[centre] lane;center[centre]-lane traffic stripe;middle lane;traffic stripe
中央车道 center[centre] lane;center[centre]-lane traffic stripe;middle lane;traffic stripe
中央车道交通 center[centre]-lane traffic
中央程序 center[centre] routine
中央抽真空系统 central vacuum system
中央出口式贮仓 center[centre]-outlet bunker
中央储槽 central tank;pivot tank
中央处理 central processing
中央处理机 central processing unit;central processor(unit);main processor
中央处理机结构 central processor organization
中央处理机主循环 central processing unit loop
中央处理器 central process unit
中央处理器群 central complex
中央处理系统 central processing system
中央处理装置 central processor(unit);central process unit
中央触发器 center trigger
中央穿孔 central perforation
中央存储器 central memory;central storage
中央大道 central avenue
中央大殿 <社寺建筑的> presbyterium
中央大断层 main center thrust
中央大厅 central hall;centre half;rotunda;megaron <古希腊建筑的>;concourse <美>
中央带 central zone;middle strip
中央挡 <侧倾车> center stop
中央导洞 central heading
中央导洞法 center[centre]drift method;central pilot tunnel(l)ing method
中央导坑 center[centre] drift;center[centre] heading;central heading
中央导坑法 center[centre] drift method;central drift method
中央导坑式隧道施工法 core method of tunnel construction
中央道路研究院 Central Road Research Institute
中央的 central;medial;middle
中央灯具 center light
中央地块 median mass
中央电池组 central battery;common battery
中央电话局 central exchange;central telephone bureau;telephone central office;central office
中央电信局 Central Telecommunication Office
中央调度所 central dispatching station
中央断块断层 central basin fault
中央断面 central section;section at key
中央对角式开拓系统 center-opposition development program(me)
中央阀 center valve
中央方位点 <即4个象限点的中间点> intercardinal point
中央方形三面半圆形建筑中的歌唱班席位 triconch choir
中央方形三面半圆形教堂 triconch church
中央分布润滑法 servo-lubrication
中央分车带 central reservation;median divider
中央分车岛 medial strip;divisional island;medial island;median island
中央分车栅栏 <公路的> median barrier
中央分隔带 central median;central separator;central strip;medial divider;medial strip;median divider;median separator;median strip;medium strip;central reservation;central reserve
中央分隔带车道 median lane
中央分隔带护栏 median barrier
中央分隔带禁入标志 keep off median sign
中央分隔带开口 median opening
中央分隔带栽植 median strip planting
中央分隔岛 medial island;medial strip
中央分隔线 central separate line
中央峰 central mountain;central peak
中央缝缀 centre-stitching
中央浮筒 central float
中央复合体 medial complex
中央干围 mid-girth
中央隔离线 central separate line
中央工业 central industry
中央公园 central park
中央供电 central current supply;central power supply
中央拱廊 central archway
中央骨料室 central collecting compartment
中央故障存储单元 center failure storage unit;central failure storage unit
中央管 central canal
中央管板 central sheet
中央灌溉电力局 <印度> Central Board of Irrigation and Power
中央罐托 center anchor
中央光带 central band
中央广播电台 central broadcasting
中央广播控制站 central broadcasting controller microphone call
中央广场 central square;concourse;hall-nave <车站的>
中央轨 centre rail
中央滚动 center disk cutters
中央滚柱轴承 center pin bearing
中央海脊 mid-oceanic ridge
中央海岭 mid-oceanic ridge
中央航道 central course;central track;mid-channel
中央航道浮标 mid-channel buoy
中央合同委员会 Central Committee for Contacts
中央和中部通过台 <市郊客车> central and mid-way vestibules
中央荷载 central load
中央横梁 center[centre] cross member
中央横剖面 middle cross section;middle transverse section
中央花饰 central flower;centre flower
中央花饰块 <顶棚的> center piece
中央缓冲杆 center buffer stem
中央缓冲弹簧 center buffer spring
中央缓冲自动钩 automatic central buffer coupling
中央火车站 central railway station
中央基座 center[centre] pedestal
中央集合 centralization
中央集权式 functionalization
中央集权式采购 functionalization in purchasing
中央集权下的经济统制 statism
中央集权制 centralism
中央集中载荷 central concentrated load
中央集中制 centralization system
中央给水规划单位 central water planning unit
中央计划经济 centrally planned economy
中央计划经济国家 countries having centrally planned economics
中央计时设备 central timing equipment;central timing unit
中央计时装置 central timing unit
中央计算机 central computer;centralized computer
中央计算机输入输出 central computer input-output
中央记录 central log
中央记录站 central recording station
中央寄存器 central register
中央加料 center[centre] feed;central feed
中央夹紧机构 central clamping mechanism
中央监督控制室 central supervisory control room
中央减速车道 median deceleration lane
中央减速器 central reducer(unit)
中央降液管蒸发器 central downcomer evapo(u)rator;central well down take evapo(u)rator
中央交通控制计算机 central traffic control computer
中央搅拌 centre mix(ing)
中央接坡 center ramp;central ramp
中央接收器 central receiver
中央接收器技术 central receiver technology
中央接收器系统 central receiver system
中央截口 medial section
中央金库 central treasury
中央经济计划 central economic planning
中央经线 central meridian
中央经线比例尺 scale along central median
中央经线投影 meridian central projection
中央镜箱 <多镜头航摄仪> centre chamber;central chamber
中央局用户线 central office line
中央开口 center opening
中央空间 central space
中央空气调节器 central air conditioner
中央空气调节系统 central air conditioning system
中央空气装置 central air conditioning installation
中央控制 central control;centralized control;centring control;master control;federal control <美>
中央控制点 central control point;centralized control point
中央控制机构 central controlling organization;centralized control mechanism
中央控制计算机 central control computer
中央控制键 central master key
中央控制空调系统 central air conditioner
中央控制楼 central control tower;centralized control box
中央控制面板 central control desk;centralised [centralized] control panel
中央控制盘 centralized control panel;centre panel
中央控制器 central controller
中央控制设备 central control equipment
中央控制室 central control room;centralized control room;main control room;master control room
中央控制数字化系统 digital computer system of central control
中央控制所 central control post(aboard)

中央控制锁 central master-keyed lock
中央控制台 central control board; central control console; central control deck; central control desk; central controller; central control unit; centralized control desk; master control console; pedestal controller
中央控制系统 central control system
中央控制钥匙 central master key
中央控制站 central control station
中央控制装置 central control unit; centralized control device
中央跨(孔) main opening
中央馈电 center feed; center feeding; centre feed
中央馈电的 centre fed
中央馈电式轨道电路 center-fed track circuit
中央扩充工作码程序 center owncoding routine
中央棱镜 centre prism
中央冷却 central cooling
中央立龙骨 through plate center keelson
中央立柱 gin pole; pedestal
中央量油口 centre ga(u)ge
中央列车控制 central train control
中央林荫道 central avenue
中央林荫分隔带 central mall
中央流卸 center flow
中央流卸式干散货物车 center flow dry-bulk commodity car
中央隆起 central uplift
中央楼梯侧板 central string(er)
中央履带车道 central crawlway
中央落料管 central drop duct
中央面 median plane
中央面板 center panel
中央木 mean sample tree
中央排水 central drainage
中央排水管 central drain
中央排水渠 central drain
中央排水系统 central drainage system
中央配料车间 central proportioning plant
中央披水板 center board; center keel
中央劈开 median chorisis
中央平行圈 central parallel circle
中央起源区 central initiation zone
中央气象局 Central Bureau of Meteorology; Central Meteorological Bureau
中央气象台 Central Meteorological Observatory
中央气旋 central cyclone
中央牵引杆 center draft drawbar
中央倾向 central tendency
中央穹隆式教堂 domed central-plan church
中央驱动动力割草机 centre drive motor mower
中央驱动式动力割草机 motor mower with center drive
中央群 central group
中央燃烧设备 central fire
中央日期 central date
中央润滑 servo-lubrication
中央设备室 central apparatus room
中央射线 central ray
中央深浅调节器 center[centre] depth regulator
中央升管 central riser
中央时刻 clock time of the middle time signal
中央实验室 central laboratory
中央市场 central market
中央式通风 central ventilation; middle ventilation
中央室 central compartment
中央枢轴 central pivot
中央输导束 central bundle
中央竖井 central shaft
中央数据变换设备 central data conversion equipment
中央数据处理机 central data processor
中央数据处理计算机 central data processing computer
中央数据处理系统 central data processing system
中央数据处理装置 central processing unit
中央数据收集系统 central data collection system
中央数据性能监控器 central data performance monitor
中央数据站 central data station
中央数据终端 central data terminal
中央水 central water
中央水文测验局 Hydrographic(al) Central Bureau
中央丝 central wire
中央送风 central air supply
中央索面 central cable plane
中央台 central station
中央条纹 central fringe
中央调整 centralized regulation
中央铁路工厂 central railroad shop
中央通道客车 center[centre] aisle coach; coach with center[centre] gangway
中央通风系统 centralized ventilation system
中央通信[讯]线路 central communication line
中央统计部 Central Statistics Department
中央头灯 central headlamp
中央脱位 central dislocation
中央挖槽施工法 central trenching method
中央湾 sinus median
中央微管 central microtubule
中央纬线比例尺 scale along central median
中央位置 mid-position
中央文件系统 central file system
中央无线电实验室 central radio laboratory
中央系统 center[centre] system; central system
中央细部 <提琴等的> waist
中央显示器 central display unit
中央显示装置 central display unit
中央现货市场 central spot market
中央卸货货车 center dump car
中央卸货漏斗门 center discharge gate
中央卸货有盖漏斗车 center discharge cover hopper
中央信号楼 central control tower; central signal box
中央(信息)处理机 central processor
中央星 central star; nuclear star
中央行车程序处理设备 central processing unit
中央型 central type
中央蓄电池站 common battery station
中央悬臂断面 <拱坝> section at key
中央悬挂 central mounting
中央循环管 central circulating tube; central down-comer
中央循环管式蒸发器 calandria
中央循环管蒸发器 central downcomer evapo(u)rator; central well down take evapo(u)rator; standard vertical tube evapo(u)rator
中央岩株 central stock
中央液泡 central vacuole
中央伊利诺湾铁路实验所 <美> Illinois Central Gulf Experiment
中央银行 banker's bank; bank of banks; central bank
中央圆屋顶 central dome
中央运输咨询委员会 <英> Central Transport Consultative Committee
中央匝道 <立体交叉的> center[centre] ramp
中央渣口 breast hole
中央站 central station
中央真空清洁系统 central vacuum cleaning system
中央蒸汽供热 central steam heating
中央直径 mid-diameter
中央直辖市 special municipality under direct central control
中央值 median value
中央指零式仪表 zero center meter
中央制动器 center[centre] brake; central brake
中央制动闸 center[centre] brake; central brake
中央制冷设备 central refrigerating plant
中央制冷装置 central refrigerating plant; central refrigeration plant
中央中断寄存器 central interrupt register
中央中断制 central interrupt system
中央中间灰质 central intermediate grey matter
中央终点站 central terminal station
中央终端 central terminal
中央终端设备 central terminal unit
中央终端装置 central terminal unit
中央轴 axis of centres; central axis
中央轴丝 central axial filament
中央主计算机 central main computer
中央贮槽 pivot tank
中央柱 center pole; central rod; trumeau <门窗口的>
中央状态 centrality
中央资料库 central data bank
中央子午面 central meridianal plane
中央子午线 central meridian
中央子午线中天 central meridian transit
中央自动信息记录 centralized automatic message accounting
中央纵缝 center[centre](line) joint
中央纵距 middle ordinate; mid-ordinate
中央纵距法 method of middle ordinates; mid-ordinate method
中央走廊 central corridor; centre corridor
中央作垄型体 centre ridger body
中扬程水泵 medium lift pump
中腰窗 window center line
中摇枕 <三轴转向架> center bolster; center bearing bridge; side bearing arch; side bearing bridge; truck bolster; center bearing beam
中摇枕拱板 <三轴转向架> center bearing anchor bar
中药 herb-medicine
中药店 shop of tradition Chinese medicines; store of tradition Chinese medicines
中药废水 traditional Chinese medicine wastewater
中医科 traditional Chinese medicine department
中医学院 college of traditional Chinese medicine
中译 Chinese translation
中翼 middle limb
中音乐扬声器 squawker
中音频 sound intermediate frequency
中音提琴 baryton
中印度洋海盆 mid-Indian basin
中英混合式庭园 <十世纪英国流行的> English-Chinese garden
中英混合式园林 Anglo-Chinese style garden
中营养河 mesotrophic stream
中营养河系 mesotrophic river system
中营养湖 mesotrophic lake
中营养泥沼 mesotrophic mire
中营养水体 mesotrophic water body
中硬的 half-hard
中硬地层 medium hard formation
中硬度白灰罩面 medium hard white coat
中硬钢 half-hard steel; medium hard steel
中硬金属 half-hard metal
中硬黏[粘]土 medium stiff clay
中硬水 moderate water
中硬岩 semi-head rock
中硬岩层 medium ground
中硬岩层钻进 moderate drilling
中硬岩石 medium hard rock
中庸干燥度 moderate drying
中油度 medium oil; middle oil
中油度醇酸树脂 medium oil alkyd
中油度清漆 copal varnish; medium oil varnish
中油度树脂 medium oil resin
中油度油漆 medium oil paint; middle oil paint
中油田 middle oil field
中油性油漆 medium oil paint; middle oil paint
中游 middle course; middle reach; midstream
中游地段 valley tract
中游河段 middle course; middle reach
中宇宙 mesocosm
中雨 moderate rain
中雨量 moderate rainfall
中域 medial area; ulnar area
中域组合 mesoassociation
中元古代【地】 Mesoproterozoic era
中元古亚代【地】 middle proterozoic subera
中元古界【地】 Mesoproterozoic erathem
中元古亚界【地】 middle proterozoic suberathem
中原木 medium size log
中圆锉 middle round file
中源 mesogene
中源地震 intermediate focus shock
中远景镜 medium long shot
中远洋的 mesopelagic(al)
中云 medium(level) cloud; middle cloud
中云族 middle family of cloud
中陨铁 mesosiderite
中运量 middle transport volume
中运量轨道公交系统 people mover
中运量客运系统 intermediate capacity transit system
中运量快速客运系统 medium capacity transit system
中载公路 <硬质路面,有 2～4 条车道的公路> medium road
中载荷 moderate duty
中载应力 interior load stress
中轧室 <轧石机的> intermediate crushing chamber
中站电气集中联锁 all-relay interlocking for medium station
中针距 mid-cut
中真空 medium vacuum
中真空泵 intermediate pump
中震 middle earthquake

中震源地震 intermediate focus earthquake
中支枢 pivot
中支枢轴 center hinge pivot
中支柱 center strut;king tower
中直线 cathetus
中值 median(value)medium(value);mid-value
中值等高线 mid-contour
中值电平 median level;medium level
中值定理 intermediate value theorem;mean value theorem;theorem of mean;law of mean
中值定律 law of the mean
中值反射率 medium value reflectivity
中值方法 median method
中值粒度 median(particle)diameter;median(particle)size;medium diameter
中值粒径 median(particle)diameter;median(particle)size;medium diameter
中值流量 median discharge;median stream flow
中值滤波 median filtering
中值年径流 medium yearly runoff
中值浓度 median concentration
中值容许浓度 median tolerance limit
中值随机变量 median random variable
中值有效剂量 medium effective dose
中值有效浓度 medium effective concentration
中值月径流 median monthly runoff
中值直径 <分成重量相同的两部分的直径> median diameter;medium diameter
中止 abeyance;breakdown;break off;cease;cutback;cut-out;discontinuance;discontinuation;discontinue;dwell;early termination;intermittence;interruption;shortstop;stop;stoppage;suspend;suspension;terminate;unexpected halt;cycle stealing;hang-up
中止的 suspended;suspensive;terminative
中止的呼叫 suspended call
中止定时器 abort timer
中止反应 stopped reaction
中止付款 suspension of payment
中止工作 knock off
中止功能 abort function
中止滚转 roll recovery
中止合同 suspension of contract
中止履行(合同)义务 suspend performance of obligations
中止命令 stop order
中止模式 abort pattern
中止时间 suspension time;intermission
中止时限行为 act of interruption
中止诉讼手续的申请 caveat
中止谈判 suspend talks
中止项目 suspended item
中止效应 off-effect
中止信号 stop signal
中止序列 pause sequence
中止循环 off-cycle;stop loop
中止循环法除霜 <冷藏库> off-cycle defrosting
中止氧化 blocked heat;blocking
中止语句【计】abort statement
中止运输 <行使停运权> stoppage in transit
中止支付 stop payment
中止执行 suspension of execution
中止执行令状 supersedeas
中指 digitus medius;medius;middle finger
中指标点 <机场> middle marker
中至中 centre-to-centre
中至中距离 center-to-center distance;distance between centers[centres];distance between center to center
中志留世【地】Middle Silurian
中志留统【地】Middle Silurian series
中质柏油 medium tar
中质差 isotopic number
中质机械油 middle machine oil
中质焦油(沥青)medium tar
中质路油 medium road oil
中质煤馏油 tar medium oil
中质铺路油 medium road oil
中置(式)发动机 central engine;midship engine
中置翼 midwing
中中涌 average moderate swell;moderate average-length swell
中重的 medium weight
中重型的 medium heavy
中重质石油 medium heavy oil
中周期地震仪 intermediate period seismograph
中洲浮标 middle ground buoy
中轴 axis[复 axes];center shaft;centre shaft;intermediate axle;mean axis;middle axle;cardan shaft <汽车的>
中轴承 middle bearing
中轴的 axial;central axial
中轴距离 <船的> half breadth
中轴壳 <汽车的> intermediate axle housing
中轴式整枝 central leader training
中轴索 axial cable
中轴胎座式【医】axile placentation
中轴线 axle wire;backbone curve;central axis
中轴压力盒 <三轴压力仪用的> axial cell
中轴压力筒 <三轴电力仪> axial cell
中轴转门 crapaudine door
中侏罗世【地】Middle Jurassic
中侏罗统【地】Middle Jurassic series
中主梁 middle girder
中主平面 intermediate principal plane
中主应力 diate principal stress
中注管 fountain;funnel
中柱 center column;central pillar;king's piece;king's post;king's rod;king bolt;kingpost;kingpost truss;king rod;newel post;stela[复 stelal/steles]【植】
中柱的大头 king head
中柱接头 newel joint
中柱节点 kingpost joint
中柱螺旋式楼梯 <中柱支承悬臂踏步> solid newel stair(case);spiral stair(case);staircase of helical type with newel
中柱螺旋梯 newel stair(case)
中柱石 chelmsfordite;fuscite;mizzonite;ontariolite;porcelain spar;sodaite;wernerite
中柱式桁架 kingpost truss
中柱式螺旋楼梯 stair of helical type with newel
中柱式旋梯 newel type spiral staircase
中柱、双柱与风撑 king-and-queen post and wind filling
中柱挑出长形屋顶 umbrella roof
中柱原 plerome
中铸件 medium casting
中铸铁 medium cast-iron
中爪 median claw
中专生 secondary specialized
中砖 medium brick
中转 transferring;tranship(ment);transit <客货运输>
中转保管费 storage in transit
中转泵 relay pump
中转仓库 central warehouse;transit warehouse
中转场所 transshipment platform
中转车 transfer car;transit car
中转车平均停留时间 average detention time of car in transit;average time of detention per goods operation
中转处 transfer correspondence
中转岛 transshipment island
中转地点 transfer point;transition point;transit point;transship point
中转点 intermediate transit point;transfer point
中转调运设备 transit distribution facility
中转费 transfer charges;transit charges;transshipment charges
中转费率 transit rate
中转腹地 trans-backland
中转港(口)entrepot port;junction port;port of transit;port of transshipment;transfer port;transit port;transshipment port;turnround port
中转公司 transit company
中转(换乘)层 transfer floor
中转货仓 transfer house
中转货场 transfer yard;transshipment yard
中转货车 transship wagon
中转货进口报单 transship entry
中转货棚 transit shed
中转货物 goods to be transshipped;transit cargo;transship cargo
中转货物摘录 transit abstract
中转集装箱 transshipment container
中转库 entrepot storage
中转劳务包干费 transit labour package charge
中转利息 transit interest
中转列车 relay train;transit train
中转列车轮渡 transfer steamer
中转旅馆 transient hotel
中转旅客 transfer passenger
中转旅客候车室 transfer passengers' waiting room
中转码头 marine transfer terminal;transfer terminal;transshipment terminal;transshipping wharf
中转贸易 entrepot trade
中转票据 transfer document
中转时间 transfer time;put-through time <列车或车辆>
中转收据 forwarding receipt
中转枢纽 central terminal station;transfer hub;transshipment terminal
中转税 transit duty
中转条款 transit terms
中转停留时间 time of detention in transit
中转线 relay track;transfer line;transfer track;transshipment line
中转线路 transit route
中转斜坡台 transshipping ramp
中转性冷库 transit cold store
中转油库 terminal
中转运费率 cutback rate
中转运价表 transfer tariff
中转运输 transshipment traffic
中转站 intermediate repeater;relay station;transfer depot;transfer station;transit depot;transshipment station;transshipping station;transit station <两路间>
中转站台 transfer platform;transshipment platform
中转整零车 transshipment part-load wagon
中转中心 intermediate center[centre];relay center[centre];transshipment center[centre]
中转装备 transshipment installation
中转作业费 transshipment charges
中桩 center[centre] stake【测】;ring-pile
中桩测设 setting-out of centre peg;staking out center line
中桩填挖 cut-and-fill at center stake
中桩填挖高度 height of cut(-and)-fill at center[centre] stake
中缀 infix
中缀表达式 infix expression
中缀表示法 infix notation
中缀式 infix form
中缀运算符 infix operator
中缀摘录 infix extract
中浊 moderately turbid
中滋育泥炭 mesotrophic peat
中滋育沼泽 mesotrophic swamp
中子 neutron
中子波 neutron wave
中子波长 neutron wavelength
中子参数 neutron parameter
中子测井 neutron well logging
中子测井记录 neutron log(ging)
中子测井刻度器 neutron log calibrator
中子测井刻度值 neutron log calibration value
中子测井曲线 neutron log curve
中子测井系统 neutron logging system
中子测谱术 neutron spectrometry;neutron spectroscopy
中子测湿仪 neutron moisture ga(u)ge;neutron moisture meter
中子测水法 neutron moderation method
中子测水分仪 neutron moisture meter
中子产额 neutron productivity;neutron yield
中子产生 production of neutron
中子产生率 neutron production rate
中子超流性 neutron superfluidity
中子-超热中子测井 neutron-epithermal log
中子-超热中子测井曲线 neutron-epithermal neutron log curve
中子冲击 neutron bombardment
中子传感器 neutron sensor
中子单粒子能量 neutron single-particle energy
中子弹 N-bomb;neutron bomb
中子发射 emission of neutrons
中子发射体 neutron emitter
中子反射 neutron reflection
中子反射层 neutron reflector
中子反射剂 tamper
中子反射镜 neutron-reflecting mirror
中子反射器 tamper
中子反向散射装置 neutron backscatter
中子反应堆 neutron(ic)reactor
中子防护屏 neutron shield(ing)
中子防护墙 neutron wall
中子防护箱 neutron shield tank
中子放射线照相术 neutron radiography
中子非弹性散射反应 neutron inelastic scattering reaction
中子俘获 neutron capture;neutron death
中子俘获截面 neutron capture cross section
中子辐射分析 neutron radiation analysis
中子辐射俘获 non-fission neutron absorption
中子辐射监测器 neutron irradiation

Z

monitor
中子辐射损伤 neutron-induced damage;neutron radiation damage
中子辐照 neutron irradiation
中子伽马测井 neutron-gamma log(ging)
中子伽马测井曲线 neutron-gamma log curve
中子伽马灵敏度比 neutron-to-gamma sensitivity ratio
中子固化 neutron curing
中子含水量(测定)仪 neutron moisture meter
中子互相作用 neutron event
中子缓速法 neutron moderation method
中子活化 neutron activation
中子活化测井 neutron activation log
中子活化测井曲线 neutron activation log curve
中子活化法 neutron activation method
中子活化法装置 apparatus of neutron activation analysis
中子活化分析 neutron activation analysis
中子活化技术 neutron activation technique
中子激活能 neutron activation energy
中子计算器 neutron counter
中子记录器 neutron monitor
中子剂量玻璃 neutron dose glass
中子监测 neutron monitoring
中子监测器 neutron monitor
中子减速法 neutron moderation
中子检测仪表 neutron instrumentation
中子胶片剂量计 neutron film badge
中子角流量 neutron angular current
中子角密度 neutron angular density
中子角通量 neutron angular flux
中子结合 neutron binding
中子结合能 neutron binding energy
中子结晶学 neutron crystallography
中子截面 neutron cross section
中子静止质量 neutron rest mass
中子壳层 neutron shell
中子孔隙度 neutron porosity
中子跨导 neutron transconductance
中子宽度 neutron width
中子扩散 neutron diffusion
中子扩散法 <测土壤湿度的> neutron scattering method
中子扩散方程 neutron diffusion equation
中子扩散冷却现象 neutron diffusion cooling phenomenon
中子扩散流 neutron diffusion current
中子扩散系数 neutron diffusion coefficient
中子冷却系数 neutron cooling coefficient
中子链式反应 neutron chain reaction
中子链式(反应)堆 neutron chain reactor
中子量测 neutron log(ging);neutron measurement
中子量测记录法 neutron logging method
中子裂变闪烁探测器 neutron fission-scintillation detector
中子裂变碎片角关联 neutron-fragment angular correlation
中子裂变探测器 neutron fission detector
中子灵敏层 neutron sensitive coating
中子灵敏室 neutron sensitive chamber
中子流 neutron current; neutron streaming
中子流密度 neutron current density; neutron flux density
中子流密度矢量 neutron current density vector
中子流切断屏 neutron curtain
中子流矢量 neutron current vector
中子漏泄 escape of neutron;neutron escape;neutron leakage
中子路径 paths of neutrons
中子脉冲 neutron pulse
中子慢化 moderation of neutrons
中子慢化剂 neutron moderator
中子慢化湿度计 neutron-moderation moisture meter
中子密度 density of neutrons; neutron density
中子密度不利因子 neutron density disadvantage factors
中子密度分布 neutron density distribution
中子密度起伏 neutron density fluctuation
中子敏感 neutron sensitive
中子敏感元件 neutron-sensing element
中子能级 neutron level
中子能级间距 neutron level spacing
中子能量 neutron energy
中子能量分散 neutron energy spread
中子能谱 neutron energy spectrum
中子能谱强度 neutron spectrum intensity
中子能群 neutron energy groups
中子碰撞 neutron collision
中子碰撞半径 neutron collision radius
中子平衡 neutron balance
中子平衡表 neutron balance sheet
中子平衡方程 neutron balance equation
中子平均寿命 mean neutron lifetime
中子屏蔽 neutron shield(ing)
中子屏蔽用涂料 neutron shield paint
中子谱 neutron spectrum
中子谱测量 neutron spectrum measurement
中子谱学 neutron spectroscopy
中子谱仪 neutron spectrograph;neutron spectrometer
中子谱硬化 neutron spectrum hardening
中子热电堆 neutron thermopile
中子热化 neutron thermalization
中子热化理论 neutron thermalization theory
中子-热中子测井 neutron-thermal neutron log
中子-热中子测井曲线 neutron-thermal neutron log curve
中子散射 neutron scattering
中子散射法 <探测土壤湿度的> neutron scattering method
中子散射技术 neutron scattering technique
中子散射湿度表 neutron scattering moisture meter
中子散射湿度计 neutron scattering moisture meter
中子色散曲线 neutron dispersion curve
中子闪光管 neutron flashtube
中子闪烁计数器 neutron scintillation counter
中子闪烁器 neutron scintillator
中子闪烁体 neutron-detecting phosphor
中子射出 neutron emerging
中子射线处理表面涂饰法 dynacote process
中子射线照相 neutron radiograph
中子射线照相法 neutron radiography
中子射线照相检查 neutron radiographic inspection
中子生成率 neutron formation rate
中子湿度计 neutron moisture ga(u)ge;neutron moisture meter
中子事故剂量计 neutron emergency dosimeter
中子事件 neutron event
中子势阱测井 neutron(-gamma)(well)logging
中子视 neutrovision
中子守恒原理 neutron conservation principle
中子寿命 neutron life time
中子寿命测井(记录) neutron lifetime log
中子寿命测井曲线 neutron lifetime log curve
中子寿命循环 neutron life cycle
中子输运 neutron transport
中子束 neutron beam
中子束断续器 neutron shutter
中子束反应堆 beam reactor
中子束孔 neutron port
中子束流 neutron streaming
中子束流率 neutron streaming rate
中子束衰减 neutron beam attenuation
中子束准直 neutron beam collimation
中子数 neutron number
中子数密度 neutron number density
中子衰变 neutron decay
中子衰减 neutron attenuation
中子水分测定仪 neutron moisture ga(u)ge
中子水分计 neutron moisture ga(u)ge
中子水分探头 neutron moisture probe
中子水分仪 neutron moisture meter
中子速度 velocity of neutrons
中子速度谱仪 neutron velocity spectrometer
中子速度选择器 neutron velocity selector
中子速率 neutron speed
中子损伤 neutron damage; neutron injury
中子探测 neutron detection;neutron log(ging)
中子探测法 neutron log(ging);neutron logging method
中子探测反应 neutron detection reaction
中子探测器 neatron probe;neutron(ic)detector;neutron sensor
中子探测系统 neutron detection system
中子探测仪 neutron detector;neutron probe
中子探伤法 neutron radiography
中子天然伽马射线测井记录 neutron-natural gamma log
中子厅 neutron hall
中子通量 neutron flux
中子通量过高停堆 excess neutron flux shutdown
中子通量监测器 neutron flux monitor
中子通量密度计 neutron flux density meter
中子通量密度监测器 neutron flux density monitor
中子通量密度扫描装置 neutron flux density scanning assembly
中子通量密度指示器 neutron flux density indicator
中子通量敏感元件 neutron flux sensor
中子通量强度 neutron flux intensity
中子通量时间常数 neutron period
中子通量水平 neutron flux level
中子通量探测器 neutron flux detector
中子通量图 neutron flux pattern
中子通量修正设备 neutron flux modifying device
中子通量展平 neutron flux flattening
中子通量转换器 neutron flux converter
中子土壤湿度计 neutron soil moisture ga(u)ge;neutron soil moisture meter;neutron soil moisture probe
中子土壤湿度计量器 neutron soil moisture ga(u)ge; neutron soil moisture probe
中子土壤水分计 neutron soil moisture meter
中子土壤水分仪 neutron soil moisture meter
中子退极化效应 neutron depolarization effect
中子危害性 neutron hazard
中子温差电堆 neutron thermopile
中子温度 neutron temperature
中子温度计 neutron thermometer
中子物理学 neutronics
中子物质 neutron matter
中子吸收 neutron capture
中子吸收玻璃 neutron-absorbing glass
中子吸收材料 neutron-absorbing material
中子吸收混凝土 neutron absorbent concrete
中子吸收剂 neutron absorber; neutron-absorbing material
中子吸收截面 neutron absorption cross section
中子吸收作用 neutron absorption
中子相互作用 neutron interaction
中子响应 neutron response
中子像增强管 neutron image intensifier tube
中子削裂 neutron stripping
中子泄漏 neutron leakage
中子泄漏谱 neutron leakage spectrum
中子信号 neutron signal
中子星吸积 accretion by neutron star
中子形貌术 neutron topography
中子形状因子 neutron form factor
中子选择器 chopper;neutron chopper
中子循环 neutron cycle
中子衍射 neutron diffraction
中子衍射分析 neutron diffraction analysis
中子衍射计 neutron diffract meter
中子衍射器 neutron diffraction apparatus
中子衍射图 neutron diffraction pattern
中子衍射研究 neutron diffraction study
中子衍射仪 neutron diffract meter; neutron diffractometer
中子衍射照相机 neutron diffraction camera
中子仪器 neutron instrument(ation)
中子移出截面 neutron-removal cross-section
中子引起的热应力 neutron-induced thermal stress
中子引起损伤 neutron-induced damage
中子硬化 neutron hardening
中子有效寿命时间 effective neutron cycle time
中子有效寿期 neutron effective lifetime
中子诱发的 neutron-induced
中子诱发反应 neutron-induced reaction
中子诱发放射性同位素 neutron-induced radioisotope
中子诱发过程 neutron-induced process
中子元件 neutron element
中子原子质量 neutron atomic mass
中子源 neutron emitter;neutron producer;neutron source
中子源反应堆 source reactor
中子源柜 neutron source box

中子源夹持器 neutron source holder
中子源刻度 neutron source calibration
中子源密度 neutron source density
中子再反射 neutron reemission
中子再生 neutron reproduction
中子噪声 neutronic noise
中子涨落 neutron fluctuation
中子照射 neutron exposure; neutron irradiation
中子照射室 neutron-exposure chamber
中子照射装置 neutron-exposure facility
中子照相 neutron photography
中子照相法 neutronography
中子照相机 neutron camera
中子照相术 neutron graphy; neutron radiography
中子质量 neutron mass
中子质子比 neutron-proton ratio
中子质子关联效应 neutron-proton correlation effect
中子质子交换力 neutron-proton exchange forces
中子质子扩散 neutron-proton diffusion
中子质子碰撞 neutron proton collision
中子质子散射 neutron-proton scattering
中子质子系统 neutron-proton system
中子质子相互作用 neutron-proton interaction
中子质子质量差 neutron-proton mass difference
中子-中子测井 neutron-neutron log
中子-中子测井曲线 neutron-neutron log curve
中子转换器 neutron converter
中子转移反应 neutron transfer reaction
中子准直器 neutron howitzer
中子准直仪 neutron collimator
中子总数 neutron inventory; neutron population
中子作用下可裂变的 neutron-fissionable
中紫外区 middle ultraviolet
中纵舱壁 center line bulkhead; central longitudinal bulkhead
中纵浸水平面 central immersed longitudinal
中纵隆线 median longitudinal carina
中纵剖面 <船体的> central fore-and-aft vertical plane; central lateral plane
中纵通板 center line through plate
中纵线面 central longitudinal plane
中纵坐标 mid-ordinate
中组值 mid-class mark
中作业 middle work

忠

忠烈祠 martyrium

忠实协定运价表 fidelity agreement tariff
忠于专业 professional loyalty
忠于组织 organizational loyalty

终

终拔前尺寸 base size; common draw size

终板 end-flake; end plate; terminal lamina; terminal plate
终板栅 end plate grid
终棒 terminal ledge
终变期 diakinesis; diakinesis stage
终变制定 final set
终冰 complete ice clearance; final clearing of ice; last ice
终冰期 last ice date
终冰日期 last ice date
终部 terminal part
终测 final survey
终产品 end products
终产物 end products
终程港 port of destination
终池 terminal cistern
终传动 final drive
终传动护板 final drive guard
终传动减速装置 final reduction gear
终传动链轮 final drive sprocket
终传动内侧 inner section of final drive
终传动箱 case of final drive; final drive box
终传动轴 final drive shaft
终传动装置 final drive gear; final drive unit
终到旅客列车 passenger train arriving at destination station
终地形 ultimate form; ultimate landform
终点 destination (point); end point; final destination; finishing point; home; stopping point; terminal (point); termination (dead end); terminus [复 termini]; threshold; winning-post; blaenau [复 blaen]; dead end
终点泵站 terminal pump station
终点比率法 termination rate method
终点辨认 recognition of end-point
终点场 terminal field
终点车站 terminal station
终点迟滞 terminal delay
终点冲量 terminal impulse
终点挡板 end stop
终点地址 end address
终点电线杆 end wire pole
终点电压 end point voltage
终点调查 destination-based survey
终点调查表 destination-based questionnaire
终点端 terminus[复 termini]
终点法 end point method
终点反向进位 final negative carry
终点反应 terminal reaction
终点飞机场 terminal aerodrome
终点杆 terminal rod
终点港 final port; last port; shipping terminal; storage port; terminal port; port of destination
终点广场 <车站的> terminal area
终点过滤器 ultimate filter
终点航(空)站 destination airport; home airport; terminal airport
终点和距离标志 destination and distance sign
终点化 terminalization
终点环线 terminal loop
终点回答包 end-reply packet
终点基调查 destination-based survey
终点及其距离标志 destination and distance sign
终点交通 terminal traffic
终点角【测】terminal angle
终点开关 limit stop; terminal switch
终点控制 end point control
终点馏分 end point of fraction
终点能量 end point energy
终点起爆 final mass
终点气象预报 terminal area forecast
终点汽油 end point gasoline
终点区 destination zone
终点容量 terminal capacity
终点设施 terminal facility
终点时间指示器 terminal-time indicator
终点市场 terminal market
终点事件 <统筹方法中,标志一个或多个活动完成的事项> end(ing) event
终点速度 terminal velocity
终点条件 end condition
终点停车场 terminus[复 termini]
终点停车处 <公共交通的> layover
终点停机坪 <机场> terminal apron
终点停止滴定法 dead stop process
终点突变 end point mutation
终点位置 end position; final position
终点温度 end point temperature; final temperature; terminal temperature
终点线 finishing line
终点消毒站 terminal sterilization facility
终点信号 terminal signal
终点行【计】end line
终点压力 terminal pressure
终点油库 final tank; terminal tank
终点油量 terminal impulse
终点运输业者 <指在到达站交付货物的运输业者> terminal carrier
终点站 dead(-end) terminal; end bull plant; end bull station; end dep; end depot; end stop; end tank farm; reversing station; stub-end station; terminal (depot); termination; terminus [复 termini]; terminus station; terminal area <机场>
终点站百货商店 terminal department store
终点站的 terminal
终点站房 passenger terminal building; terminal budding
终点站费用 <如装卸、搬运、驳运、调车等费用> terminal charges
终点站候车室 passenger terminal building; terminal building
终点站建筑 terminal building; principal terminal building【铁】
终点站情况报告 terminal situation report
终点站设备 terminal facility
终点站枢纽区域 terminal hub area
终点站停车时间 layover time
终点站通过能力 terminal throughout capacity
终点站作业 terminal service
终点止动装置 end stop
终点指挥部 terminal command
终点指示标志 destination sign
终点指示符 end point indicator
终点制动 end brake
终点转换开关 travel-reversing switch
终点装配 end assemblage point
终点作业 <如调车、装卸、搬运等> terminal operation
终电开关 terminal switch
终读数 end readout; final reading
终端 dead end; ending; end point; stop end; terminal end; terminating; trailing end
终端安全开关 emergency terminal switch
终端安全释放机构 end release; final safety trip
终端按钮 destination button; exit button
终端按钮继电器 exit button relay
终端板 tag block; threshold plate <电动走道的>; terminal sheet; terminal plate
终端棒 end bond
终端包 terminating packet
终端宝石 end stone
终端绷紧(固定)度 terminal degree
终端闭塞的一头 <管子等> dead end
终端闭塞机 end block
终端闭塞机构 end block mechanism
终端闭塞区间 block terminus
终端闭塞所 end block station
终端闭塞站 end block station
终端标记 terminal mark(ing)
终端标志 end mark(ing)
终端表 terminal list
终端冰碛【地】end moraine
终端不通的 dead-ended
终端布局 terminal layout
终端部件 end fitting
终端采光 end lighting
终端操作目标 terminal performance objectives
终端操作人员 terminal operator
终端操作系统 terminal operating system
终端测试 terminal test
终端产品 final product
终端长途电话中心局 terminating toll center
终端沉淀池 final tank
终端沉降速度 terminal fall velocity
终端齿轮 extreme gear
终端出入口 terminal port
终端处理 terminal handling
终端处理机 terminal handler; terminal processor
终端触点 terminal contact
终端触发器 terminating trigger
终端传动 final drive
终端传输接口 terminal transmission interface
终端传输设备 terminal transmission equipment
终端串 terminal string
终端存储器 terminal memory
终端打印机 terminal printer
终端打字机 terminal writer
终端单元 terminal unit; terminating unit
终端导航 terminal(phase) guidance
终端导航设备 terminal aids
终端导接线 terminal bond
终端导引 terminal guidance
终端的 dead-ended
终端灯 termination light
终端等时线 terminal synchrone
终端地区弹着点 terminal area impact point
终端地址 terminal address
终端地址选择口 terminal address selector
终端电传打字机 teletypewriter terminal
终端电缆 terminal cable
终端电缆盒 cable termination box
终端电流曲线 arrival current curve
终端电路 terminating circuit; termination circuit
终端电容性负载 terminal capacitive load
终端电压 receiving-end voltage; terminal voltage
终端电压降 terminal voltage drop
终端电阻 terminal resistance; terminate resistance; terminating resistance
终端电阻材料 end-resistance material
终端电阻器 terminating resistor
终端垫板 terminal shoe
终端吊钩 end hook
终端吊线夹板 <电缆> terminal spindle
终端读出 end sensing
终端短接 no-go nipple
终端段 end section
终端队列 terminal queue
终端队列溢出 termination queue overflow
终端对角斜撑 end diagonal
终端对接 end-to-end
终端多路转换器 terminal multiplexer
终端发射机 terminal transmitter

Z

终端发射设施 terminal launch facility
终端房杆的 end frame
终端仿真 terminal emulation
终端仿真程序 terminal emulator
终端访问 terminal access
终端访问网络 terminal access network
终端放大器 final amplifier; last amplifier; terminal amplifier
终端分配器 terminal distributor
终端服务器 terminal server
终端符号 terminal symbol
终端辐射计划 terminal radiation program(me)
终端负载 terminal load; terminate load; terminating load; termination; terminator
终端负载式功率计 termination load type power meter
终端干线 terminal trunk
终端杆 dead-end pole; end pier; end pole; stayed pole; stay rod; terminal pole
终端格式 terminating format
终端功率负载 power termination
终端功能键 terminal function key
终端拱 extreme arch
终端固定接头 end-fixture splice
终端固定性 end fixity
终端故障 terminal fault
终端管理程序 terminal supervisory program(me)
终端国 terminal country
终端过滤器 after-filter; final filter; ultimate filter
终端盒 end box; end connector; final connector; terminal box
终端盒密封 end box sealing
终端横撑杆 extreme cross girder
终端横大梁 extreme cross girder
终端横隔墙 end diaphragm
终端话(报)务 terminal traffic; terminating traffic
终端环 end loop; end ring
终端绘图仪 terminal graphic(al) plotter
终端活动 end activity
终端机 terminal equipment; terminal machine; terminal set; terminating machine
终端机的功能 terminal function
终端畸变 end distortion
终端极限开关 terminal limit switch
终端集液器 dead-end trap
终端计算机 terminal computer
终端计算机通信[讯] terminal computer communication
终端记发器 terminal register; terminating register
终端继电器 exit relay; final impulse operating relay
终端架 final frame; terminal bay; terminating bay; end frame
终端假负载 terminal dummy load
终端监督程序 terminal monitor program(me)
终端减速开关 terminal slow-down switch
终端检测器 end point detector
终端检查 terminal check
终端检验 end-of-pipe test
终端件 termination piece
终端键盘 terminal keyboard
终端交叉钢板 terminal iron
终端交通管制 terminal traffic control
终端校正 end correction
终端接口 terminal interface
终端接口处理机 terminal interface processor
终端接口控制器 terminal interface controller
终端接口模件 terminal interface module
终端接口设备 terminal interface equipment
终端接收站 receiving terminal station
终端接头 plug end fitting; terminal bond; terminal fitting
终端接线 terminal connection
终端接线片 terminal lug
终端接续 final connection
终端进位 final carry digit
终端警告标志 destination warning marker
终端局 cusp station; end office; end station; terminal exchange; terminal office; terminal station; terminating exchange; terminating office; termination office; terminus station
终端局间 end-to-end
终端局间测试 end-to-end test
终端局间通信[讯] end-to-end communication
终端局间通信[讯]业务 end to end service
终端局通话 terminal call
终端绝缘子 end insulator; terminal insulator
终端开断线路 open-end line
终端开关 dead-end switch; limit switch; over-travel-limit switch; terminal switch
终端开路线(路) open-ended line
终端可靠性 terminal reliability
终端空气过滤器 final air filter
终端控制 terminal control
终端控制单元 terminal control element; terminal controller; terminal control unit
终端控制分站 terminal sub-control station
终端控制器 terminal controller; terminal control unit
终端控制设备 terminal control device; terminal control equipment
终端控制台 terminal console
终端控制问题 end point control problem
终端控制系统 terminal control system
终端块件 end pot
终端框架 end framing
终端馈电 end feeding
终端馈电式轨道电路 end-fed rail circuit; end-fed track circuit
终端馈线 dead-end feeder
终端拉线 dead-end guy
终端类程 terminal class
终端力矩 terminal moment
终端立面 end facade
终端立柱 terminal post
终端立柱帽 terminal post cap
终端连杆 end link
终端连接板 offset; strap
终端连接点 termination connection point
终端连接器 end connector; terminal connector; terminator
终端连接系统 end connecting system; terminal linking system
终端联机键 attention key
终端联机中断 attention interrupt(ion)
终端联结 made circuit
终端链(路) final link
终端流 terminal stream
终端楼梯 end stair(case)
终端轮询 terminal polling
终端脉冲 terminal pulse
终端锚定 dead-end anchor
终端锚墩 end anchor block
终端锚固毁坏 end anchorage failure
终端锚块 end concrete block
终端门 end door
终端面板 terminal panel
终端模件 terminal module
终端目的地 terminal destination
终端能级 terminal level
终端能力 terminal capability
终端耦合 end-on coupling
终端盘 terminal panel
终端配电板 end panel
终端配电盘 terminal distributor
终端配线部件 distributing terminal assembly
终端匹配 reflexless terminal
终端片 end piece; pill; terminal plate
终端平台 terminal platform
终端屏 end panel
终端前的 preterminal
终端墙洞 end loop
终端区 termination environment
终端区导航系统 terminal area navigation system
终端区空中交通管制系统 terminal air traffic control system
终端区域 terminal area
终端区域管理 terminal area control
终端区域雷达控制 terminal area radar control
终端曲面 terminal surface
终端人机对话 terminal interaction
终端容量 terminal capacity
终端容量矩阵 terminal capacity matrix
终端入口 terminal port
终端塞子 terminating plug
终端散热器 ultimate sink
终端扇出端 terminal fanout
终端设备 end device; ending; end installation; end instrument; terminal; terminal device; terminal equipment; terminal installation; terminal unit; terminating unit; termination
终端设备测试架 terminal test bay
终端设备测试器 tester for terminal equipment
终端设备成分 terminal component
终端设备定时询问 polling
终端设备接口 terminal device interface
终端甚高频全向信标区 terminal very high-frequency omnirange area
终端甚高频全向指向标 terminal very high-frequency omnirange
终端升降曲线 terminal contour
终端声阻 terminal acoustic(al) resistance
终端绳轮 terminal wheel
终端失真 end distortion
终端时间 terminal time
终端使用办公室 end office
终端式 dead-end type
终端室 terminal room
终端适配器 terminal adapter; terminator
终端梳形板 <电动走道的> threshold comb
终端输出 final output
终端输入电流 arrival current
终端衰耗线 terminal dissipation line
终端速度 terminal velocity
终端隧道 dead-end tunnel
终端隧洞 dead-end tunnel
终端损耗 terminal loss
终端损失 terminal loss
终端塔 terminal tower
终端塔架 dead-end tower
终端摊分额 terminal share
终端弹着预测 terminal impact prediction
终端塘 terminal pond
终端套管 dividing box; pot head; solid end; terminator
终端体系结构 terminal architecture
终端条件 terminal condition
终端调节剂 terminal telomer
终端调形 terminal contour
终端调整器 end compensator
终端调制解调器接口 terminal modem interface
终端停车平台 terminal landing
终端停车装置 machine final-terminal stopping device
终端停机设施 terminal stopping device
终端通信[讯] terminal communication
终端通信[讯]方式 terminal signal(l)ing
终端通用性 terminal versatility
终端筒形联轴器 end sleeve
终端透明度 terminal transparency
终端图案 terminal pattern
终端弯矩 end bending moment; terminal bending moment
终端网络 terminating network; termination network; termination rack
终端温差 terminal (temperature) difference
终端误差 terminal error; end error
终端吸收 end absorption
终端系统 terminal system; rear-end system
终端显示方式 terminal display mode
终端显示装置 display terminal
终端线 tag wire; terminal line
终端线夹 dead-end clamp; terminal clamp; terminal clip; terminal connector
终端线路 terminated line
终端线圈 end coil
终端线束 terminal cluster[clustre]
终端限位开关 terminal limit switch
终端相互作用 terminal interaction
终端箱 extreme case
终端消散带 <冰川> terminal dissipation zone
终端小窗 end loop
终端效应 end effect; terminal effect
终端协议 terminal protocol
终端谐振腔 end cavity
终端信号 terminal signal
终端信号装置 terminating signal(l)ing unit
终端信息 final word; termination message
终端信息包 terminating packet
终端旋转 end rotation
终端压力池 terminal reservoir; terminal tank <管道的>
终端业务 terminal traffic; terminating traffic
终端仪表 end instrument
终端仪表进场程序 terminal instrument procedure
终端引线 terminal lead
终端印刷 end print
终端硬件 terminal hardware
终端硬锚导线 fixed equipment; fixed termination conductor
终端用户 end-user; terminal user
终端用户费率表 end-user tariffs
终端用户线 final exchange line
终端用户语言 end user language
终端油罐储量 terminal tankage
终端油水分离器 ending oil-water segregator
终端游标 terminal cursor
终端与主机连接 terminal to host connection
终端源编辑程序 terminal source editor

终端约束 end constraint;end restraint;terminal constraint
终端再热 terminal reheat
终端再生 terminal regeneration
终端增音机 terminal repeater
终端扎结<电缆的> end sealing
终端扎结法 terminal tie
终端占用信号 terminal seizing signal
终端站 cusp station;end station;terminal;terminal station;terminus station
终端站间通信[讯] end-to-end communication
终端站控制计算机 terminal control computer
终端站设备 equipment for terminal station
终端站甚高频全向(导航)信标 terminal very high-frequency omnirange
终端站再生中继机 regenerative repeater for terminal station
终端站总体控制 total terminal control
终端照明 end lighting
终端支持 terminal support
终端支电路 final sub-circuit
终端支架 terminal bracket
终端支柱上的弯折座圈 folded base ring on end supports
终端指示器 terminal indicator
终端制导系统 terminal guidance system
终端滞后 terminal delay
终端中继器 end repeater
终端中继线容量 last trunk capacity
终端中继站 terminal repeater station
终端中心 terminal center[centre]
终端轴 terminal shaft
终端主管部门 terminal administration
终端柱 end column;terminal pillar
终端转播机 terminal repeater
终端转接器 termination adapter
终端装置 end device;terminal assembly;terminal device;terminal installation;termination
终端着陆系统 terminal landing system
终端子系统 terminal subsystem
终端字母表 terminal alphabet
终端自动开关装置 terminal stopping device
终端自由振荡幅度 residual betatron amplitude
终端阻抗 load impedance;terminal impedance;terminating impedance;termination impedance
终端作业 terminal job;terminal service
终端作业处理 termination job processing
终段 terminal section
终段澄清器 final clarifier
终锻 finish-forging
终锻模 finisher
终锻模膛 finish(ing) impression
终锻温度 final forging temperature;terminal forging temperature
终堆石 end moraine;terminal moraine
终伐 final cutting;final felling
终沸点 end boiling point;final boiling point
终粉磨 finish grinding
终粉磨机 finish mill
终固化 final curing
终轨 final orbit
终辊道 finishing roll
终过滤器 final filter
终焊 final welding
终航向 final course
终合温度 sol-air temperature
终话信号灯 clearing lamp
终话振铃装置 ring-off system
终话指示器 ring-off indicator
终回转直径 final turning diameter
终击贯入度<桩的> final penetration
终极沉速 terminal velocity
终极基准面 ultimate base level
终极降解 ultimate biodegradation
终极颗粒尺寸 ultimate particle size
终极侵冲性 ultimate buffer
终极侵蚀基准面 ultimate erosion base level
终集(合) final set
终寄主 definitive host;final host
终加工 end-process
终价 final value
终检 finish ga(u)ge;final inspection
终检值 censored observation
终降 final decline
终降雪 latest snowfall
终接 terminate;terminating;termination
终接电缆 terminate cable
终接电路 final circuit;terminating circuit;termination circuit
终接电平 terminated level
终接机 final selector
终接器 end connector;final connector;final selector;line selector
终接器继电器 connector relay
终接器架 connector board;connector shelf
终接器设备 connector system
终接器线弧 connector bank
终接失配 mistermination
终接选择器 final selector
终结 closure;conclusion;end-all;end conclusion;grand final;terminating;wind-up
终结标识符 terminal identifier
终结裁决 award in final
终结的 ultimate
终结队列 terminating queue
终结反应 end reaction
终结符 terminal symbol;terminating symbol;termination symbol;terminator
终结符记录 terminator record
终结股利 final dividend
终结记录 final entry
终结期 end period;wear-out period
终结气体 end gas
终结情况 terminal situation
终结式 resultant
终结相位 terminating phase
终结消毒 terminal disinfection
终结账簿 book of final entry
终结账户 terminal account
终结字符 terminal character
终孔 distant hole;finally formed hole
终孔层位 terminal layer of drilling
终孔尺寸 finished hole size
终孔钎头 finishing bit
终孔钎子 finishing steel
终孔深度 depth of drilling terminal
终孔直径【岩】 diameter of distant hole;diameter of finally formed hole;final diameter of hole;final hole diameter
终孔质量验收制 drill hole quality inspecting rule
终冷却器 after-cooler
终了 expiration
终了符号 end mark
终了年度 year ending
终了期 terminal stage
终了时间 end time
终了位置 ultimate position
终了温度<焊接的> finishing temperature
终了信号 ending sign
终了信息 termination message
终了增益时间 end fain time
终滤器 final filter
终锚线夹 dead-end clamp
终磨 final grinding;finish grinding;secondary grinding
终末波 ending wave;terminal wave
终末部 terminal portion
终末沉降速度 terminal settling velocity
终末奋进 end spurt
终末感受器 peripheric receptor
终末激发 terminal motivation
终末向量 terminal vector
终年 all the year round;the year round;through the year
终年的 perennial
终年放牧 year-round grazing
终年积雪(地区)的 nival
终年积雪的高山 mountains perennially covered with snow
终碾 final rolling
终凝<水泥及混凝土> permanent set;final set(ting);full set
终凝曲线<水泥剂混凝土> final setting curve;curve of final set(ting)
终凝时间<水泥及混凝土的> final setting time;time of final setting
终钮 end-feet;terminal button
终判决函数 terminal decision function
终平原 ultimate plain
终期 concluding stage
终期的 terminal
终期准平原 end peneplain
终碛 marginal moraine;terminal moraine
终碛堤 end-moraine bar
终碛湖 end moraine lake
终碛盆地 terminal basin
终碛前缘 terminal front
终器 end organ
终切 final shear stress;egress【天】
终球 end bulb
终日 all day(long);last date
终筛 finishing screen
终栅 wire grid
终身 life time
终身保险 insurance for life;insurance under a life policy;perpetual insurance;whole life insurance
终身保证期 lifetime warranty
终身财产所有权<指非世袭的财产> life interest
终身残疾 permanent disability
终身的 lifelong
终身雇佣制 career-long employment;lifetime employment system
终身剂量 lifetime dose
终身教育 life-long education
终身就业 lifetime employment
终身年金 life annuity;perpetual annuity;perpetuity
终身权益 life interest
终身丧失工作能力的损伤 permanent impairment
终身受益人 life tenant
终身一次照射量 once-in-a-lifetime exposure
终身优恤 life pension
终身占有 tenancy for life
终身职务 job for life
终身属有的地产 estate for life
终身租户 life tenant
终身租赁 life leasehold
终审法院 court of final jurisdiction;court of last jurisdiction
终审结果 final view
终审判决 final decree;final judgment
终渗能力 final infiltration capacity
终升 rise
终生毒性试验 life-span toxicity test
终生浮游生物 holoplankton
终生鸡舍 brood-grow-lay house
终生寄生物 permanent parasite
终生事业 lifework
终生畜舍 birth to finishing housing
终生制 day old to death system
终时 terminal hour
终始温度 terminal temperature
终饰 decorative finish;final finishing;finishing;finish work
终饰材 finish
终饰层【建】 finishing layer;finishing coat
终饰插刀 finishing tool
终饰等级 finished grade
终饰地板 overlay flooring
终饰工具 finishing tool
终饰工作 finishing operation;finishing process
终饰坡度 finished grade
终饰条件 finishing condition
终饰行业 finishing trade
终饰性 finishability
终霜【气】 latest frost
终丝 filum terminale
终速度 final speed;final velocity;terminal speed
终宿主 definitive host;final host
终塑化 final curing
终碎产品 final crushed product
终态 final state;terminal state
终态流形 terminal-state manifold
终态相互作用 final state interaction
终体 end-body
终位 final bit;home
终位移 total movement
终温差 final temperature difference
终纹 stria terminalis;terminal stria[复 striae]
终线 terminal line
终相 last phase
终向馈给 endwise feed
终雪 last snow
终压 final rolling;finish rolling;terminal pressure
终压力 end pressure;final pressure
终夜 overnight
终夜灯 whole night lamp;all night lamp<客车>
终油酸 terminolic acid
终于得到 come up with
终渣 final slag
终轧 finish to ga(u)ge
终轧尺寸 delivery ga(u)ge
终轧道次 last pass
终轧机 finishing mill;finishing rolling mill
终轧孔型 final pass;last finishing pass
终轧前孔型 leading pass
终轧温度 finishing temperature
终站 dead terminal
终值 final cost;final quantity;final value;finite value;future worth;posterior value;terminal value
终值参数 terminal parameter
终值定理 final value theorem
终值控制 terminal value control
终值强度 ultimate strength
终止 cessation;end(ing);end off;stop;termination
终止按钮 mute key
终止标记 end mark
终止参数 terminal parameter
终止程序 terminator;terminator program(me)
终止代理协议 termination of agency agreement
终止的 ended
终止地点 ending place

终止电压 cut-off voltage; end-voltage; final voltage
终止反应 cessation reaction; stopping reaction; termination reaction
终止符号 terminal symbol; terminating symbol
终止合同 terminate a contract; termination of contract
终止合同费用 termination charges; termination expenses
终止后的付款 payment after termination
终止活动时间 terminative time of fault growth
终止剂 stopping agent; termination agent; terminator
终止键 mute key
终止角 end angle
终止节点 terminal node
终止卡片 terminal card
终止款额 termination amount
终止码 stop code
终止密码子 stop codon; termination codon; terminator codon
终止任职 cease to hold office
终止日(期) date of expiry; expiry date; date of termination
终止时的支付 payment on termination
终止时间 end time; finish time
终止时效索赔 barred claim
终止通知 termination notice
终止位 stop bit
终止温度 final temperature
终止纹槽 finishing groove
终止协议事件 events of termination
终止信号 stopping signal; termination signal
终止性证明 termination proof
终止因子 termination factor
终止责任 cessation of liability
终止者 terminator
终止周期 last cycle
终止状态 final state
终致癌物 ultimate carcinogen
终柱 terminal cylinder

钟摆 clock pendulum; pendulum (wire); ticker

钟摆力 pendulum force
钟摆律 pendulum rhythm; tic-tac rhythm
钟摆式 pendulum model
钟摆式的 pendular
钟摆式流速仪 pendulum current meter
钟摆调节螺母 rating nut
钟摆运动 pendular movement
钟摆状节律 pendulum rhythm
钟摆钻具 pendulum assembly
钟表 horologe; horologium; timekeeper; time piece
钟表摆轮 verge ring
钟表车床 turn bench; watch-maker's lathe
钟表的摆轮心轴 verge
钟表的长针 minute-hand
钟表店 watch-maker's shop
钟表机构 clockwork; timing watch < 测斜仪 >; watch section < 测斜仪 >
钟表机构计数器 clock meter
钟表机构控制摄影机 clockwork-motor camera
钟表机械装置 clock mechanism
钟表机心 movement
钟表检查员证书 watch inspector's certificate
钟表检查员执照 watch inspector's certificate
钟表起子 jeweler's screw-driver
钟表千分尺 comparator micrometer
钟表润滑剂 timepiece lubricant
钟表上发条 clockwork feed
钟表上弦手把 < 测斜仪 > watch winding stem
钟表式计时器 clock meter; clock recorder
钟表式计数器 clock meter
钟表式转速计 chronometric tach(e)-ometer
钟表台式车床 micro-lathe
钟表用螺丝刀 jeweler's screw-driver
钟表油 watch lubricant; watch-maker's oil; watch oil
钟表装置 clockwork
钟表装置指示器 clockwork indicator
钟差 clock correction; clock error; correction for clock
钟差测定 determination of clock correction
钟差精度因子 time dilution of precision
钟锤 clock weight
钟锤杠杆 bell crank
钟的安装 bell installation; bell system
钟点工 part-time job
钟顶天篷 bell canopy
钟斗装置 < 高炉的 > bell and hopper arrangement
钟房 clock room
钟浮标 bell buoy
钟盖 bell
钟盖虹吸水箱 bell cistern
钟杆 bell rod
钟杆推力轴承 bell bearing
钟花树属 trumpet tree
钟黄铜 bell brass
钟机限时解锁器 clockwork time release
钟架 bell cage; bell cot
钟角楼 bell-turret
钟开阀 bell valve
钟控 clock control
钟控系统 clock control system
钟控制系统 clock system
钟口 P 形存水弯 Buchan trap
钟口接头 bell joint; female joint
钟口漏斗 bell and hopper
钟口式滴水槽 bell cast
钟口式接榫 bell and spigot joint
钟口式接头 bell and spigot joint
钟口式接头瓦管 bell and spigot tile
钟口屋檐 bellcast eaves
钟口溢流 bell-mouthed overflow
钟口溢流坝 bell-mouthed spillway
钟口溢流堰 bell-mouthed spillway
钟口总深 total depth of bell
钟楼 belfry; bell chamber; bell house; bell tower; bell-turret; carillon; clocker; clock tower; clock turret; campanile < 车站 >; Tower of Winds < 古希腊 >
钟楼墙垣 wall belfry
钟帽 bell cap
钟面玻璃 < 美 > crystal
钟面读数 clock indication
钟面时 clock time
钟面式风向指示器 wind clock
钟内按刻报时装置 quarter-jack
钟盘测微计 clock dial micrometer
钟盘时间等分 temporal hour
钟片 time card
钟频 clock frequency; clock rate
钟潜水技术 bell diving technology
钟擒纵叉 clock fork
钟青铜 bell bronze; bell metal
钟人 bellman
钟乳拱 stalactite vault(ing)
钟乳石 dripstone; stalactite
钟乳石长度 length of stalactite
钟乳石洞 stalactite grotto
钟乳石饰 stalactite ornament
钟乳石直径 diameter of stalactite
钟乳石质的 stalactitic(al)
钟乳石状的 stalactitic(al); stalactiform
钟乳石状石膏 stalactitic(al) gypsum
钟乳饰穹顶 stalactite vault(ing)
钟乳体 cystolith
钟乳形穹顶 stalactite vault(ing)
钟乳形饰 stalactite ornament; stalactite work; maqarna < 伊斯兰教建筑中天花板装饰 >
钟乳形细工 muqarnas; stalactite work
钟乳状沉积物 stalactite
钟乳状雕饰 stalactite work
钟乳状方解石 dripstone; drop stone
钟乳状霰石 stalactitic(al) aragonite
钟声装置 chime unit
钟绳 bell cord
钟时序数 clock-hour figure
钟式冲洗水箱 bell-type flushing cistern
钟式存水湾 bell-type trap
钟式电解池 bell cell; glocken cell
钟式滤水器 hopper filter
钟式速度计 clock-type speedometer
钟式旋转布料器 bell-type distributing gear
钟式穴坑挖掘铲斗 bellhole bucket
钟式压力计 bell-type manometer
钟式闸板 bell damper
钟室 bell room; clock chamber
钟速 clock rate; rate of clock
钟速校正 rate correction
钟塔 campanile; clock tower
钟头式装料装置 cup-and-cone arrangement
钟响报时 chime
钟响导航信号船 bell boat beacon
钟响浮标 bell buoy
钟形 bell-shaped; campaniform
钟形柄头防水垫圈 bell-shaped crown gasket
钟形玻璃制品 bell glass
钟形铲斗 belling bucket
钟形沉箱 bell caisson
钟形冲孔器 bell punch
钟形冲洗水箱 bell-type flushing cistern
钟形出水口 outlet bellmouth
钟形储蓄器 bell-type accumulator
钟形存水弯 bell trap
钟形打捞筒 wide mouth socket
钟形的 bell-mouth; flared; mitrate; tulip-shaped
钟形的玻璃制品 bell jar
钟形底脚 bottom dome
钟形底料罐 drop bottom bucket
钟形地坑桩机 bell pit
钟形电解池 bell cell
钟形端板 end bell
钟形墩 bell(ed) pier
钟形阀 bell-shaped valve; bell valve; cup valve
钟形浮标 bell buoy
钟形感器 dome organ; sensillum campaniformium
钟形缸 bell jar
钟形拱 bell arch
钟形构架 bell cage; bell frame
钟形冠的 bell crowned
钟形管口 bell-mouth
钟形焊接 bellhole welding
钟形护墩桩 bell dolphin
钟形护舷木 bell fender
钟形基础 socket-type footing; bell-type foundation
钟形加料装置 charging ball gear
钟形接头 bell-mouth
钟形结构设计 bell design
钟形进水口 bell-mouth inlet; bell-mouth intake
钟形均衡器 bump equalizer
钟形卡盘 bell chuck; cup chuck
钟形开挖 < 开挖断面上大下小 > belled excavation
钟形靠船墩 bell dolphin
钟形壳 bell housing; bell-shaped shell
钟形坑 bellhole
钟形孔 bell-mouth; bell-mouthed opening
钟形孔口 bell-mouth orifice
钟形口 bell-mouthed opening; bell-mouth orifice
钟形口的 belled; bell-mouthed
钟形口溢流 bell-mouth overflow
钟形扩大地基 belled base
钟形扩大基础 underreamed bell footing
钟形联结器 bell coupling
钟形螺钉 bell screw
钟形凝汽阀 bell trap
钟形排气机 bell exhauster
钟形喷管 bell nozzle
钟形喷嘴 bell-shaped nozzle
钟形气流调节器 bell damper
钟形汽笛 bell whistle; dome whistle
钟形汽室 dome
钟形汽室垫圈 dome collar
钟形潜水器 diving bell
钟形桥墩 bell pier
钟形穹顶 bell-shaped dome
钟形曲柄 bell crank; bell-shaped crank
钟形曲线【数】bell-shaped curve
钟形曲线模型 bell-shaped curve model
钟形入口 bell-mouth inlet; flaring inlet
钟形山 domed mountain
钟形烧结炉 bell jar; sintering jar
钟形失真 cobs
钟形示潮器 tidal clock; tide clock
钟形套管 bell-mouth casing
钟形套筒 < 打捞工具 > bell socket
钟形天篷 bell canopy
钟形铜炉罩 < 烧结炉 > copper furnace bell
钟形涡轮 bell-shaped turbine
钟形屋顶 bell roof
钟形物 bell
钟形系船墩 bell dolphin; Baker bell dolphin
钟形系船柱 bell dolphin
钟形箱 bell housing
钟形胸件 bell-type chest piece
钟形穴坑 bellhole
钟形窑 bell kiln
钟形乙炔发生器 bell-type generator
钟形溢流口 overflow bellmouth
钟形溢水口 bell-mouth overflow
钟形印度塔墓 bell stupa(mound)
钟形圆饰 < 古希腊陶立克柱头顶板下 1/4 圆饰 >【建】echinus
钟形帐篷 bell tent
钟形罩 bell jar; bellhole housing; bell housing; dome cap; petticoat
钟形中心冲孔器 bell center punch
钟形中心冲头 bell center punch
钟形柱头【建】bell(-shaped) capital; calathus; bell of capital
钟形组合锚头 bell-shaped anchorage; bell-shaped assembly anchorage
钟形钻孔锥 < 钻扩底桩用 > belling bucket
钟穴形(挖掘)铲斗 bellhole bucket
钟罩 immersion bell; phosphoriser; plunger
钟罩排气 bell jar exhaust
钟罩式储气罐 bell-type gasholder

钟罩式虹吸水箱 bell cistern
钟罩式脉冲器 bell-type pulsator
钟罩式调压器 bell-type governor
钟罩式窑 top hat kiln;truck chamber kiln
钟罩式自记流量计 bell-type recording flowmeter
钟罩试验 bell jar testing
钟罩形计数管 end window counter
钟罩形扩散器 dome diffuser
钟罩形气封 gas seal bell
钟罩窑 bell top kiln
钟制风速表 clockwork anemometer
钟状 mitriform
钟状挡板 bell damper
钟状的 campanulate;mitrate
钟状拱【建】bell arch
钟状花冠 campanulate corolla
钟状火山 cupola;tumulus[复 tumuli]
钟状小屋 bell cot
钟状压力空间 dome-shaped pressure space
钟状印度塔墓 bell-shaped stupa (mound)
钟状罩 bell housing
钟状柱帽 bell capital
钟组 carillon
钟座 potence

舯 the middle of ship

舯部舱壁 midship bulkhead
舯吃水 draft amidship
舯后机型船 aft-engine ship
舯机型船 amidships-engined ship
舯宽 midship beam
舯耙 center drag
舯剖面积 area of midship section
舯剖面图 midship section plan
舯前机型船 fore-engined ship the middle of ship
舯切面系数 midship section coefficient

肿大 intumescence; tumefy; tumo(u)r

肿瘤医院 oncologic hospital; tumour hospital
肿缩式窗棱 pinch-swell mullion
肿缩型石香肠 pinch-swell boudin
肿胀 bloating;swelling;turgescence
肿胀土 swelling soil
肿胀物 intumescence

种 genus

种类 breed; category; class; genus; kind;sort;species;stripe;variety
种类复杂的废物 bulky waste
种类组成 species composition
种料 seeding material
种绿肥的休耕地 green fallow
种苗场 nursery
种泡 <一种磁泡> seed bubble
种圈 artenkreis
种群 population
种群爆发 outburst of a population
种群波动 oscillation of a population
种群调查法 census method
种群动态 population dynamics
种群分析 population analysis
种群估算 population estimate
种群过密 overpopulation
种群减退 population depression
种群均衡 population equilibrium
种群控制 population control
种群扩散 population dispersion
种群力学 population dynamics
种群密度 population density
种群疏散 population dispersal
种群数量 population quantity
种群效应模型 population effect model
种群演替 population succession
种群蕴藏量 standing population
种群增长 population growth
种群指数 population index
种型群 hypodigm
种型群概念 hypodigm concept
种源 epicenter[复 epicentra]
种属差异 species difference; species variation
种子 seed
种子繁殖 seedling
种子基金 seed money
种子基金贷款 seed money loans
种子库 seed storage
种子乳液 seed emulsion
种子田 seedbed
种子筒底 hopper bottom
种子箱底 hopper bottom
种子移植工艺 sprigging operation
种子、幼苗或嫩枝移植工艺 sprigging operation
种子纸育草法 covering by seed paper
种族分布图 map of mankind

冢 sepulchral mound

踵板 <挡土墙的> heel-board;heel slab

踵角 angle of heel
踵形接触 heel contact
踵压力 heel pressure

中标 acceptance of the bid;acceptance of the tender;award of contract;win a bid;win a tender; winning bid;contract awarding; successful bid

中标厂家 winning bidder
中标厂商 winning bidder
中标度 phi scale
中标方 contractor
中标公司 winning bidder
中标函 letter of acceptance
中标合同 contract awarded;contract gained
中标合同价 awarded contract price
中标价格 tender price
中标人 successful bidder; winning bidder
中标通知 notice of award
中标通知书 letter of acceptance;notification of award; acceptance of tender
中标者 successful bidder; successful tenderer
中标正式通知书 official notice of award
中毒 intoxication; poisoning; toxicosis;venenation
中毒的 toxic
中毒谷粒 intoxicant grain
中毒机理 <树脂> mechanism of poisoning
中毒剂量 poisonous dose;toxic dose; ultimate total dosage
中毒浓度极限 limit of toxic concentration
中毒树脂 poisoned resin
中毒危险 toxic hazard
中毒物质 toxic substance
中毒性皮病 toxicoderma[toxicodermia]
中毒性皮炎 toxicodermitis[toxicodermatitis]
中毒性气体 nerve gas
中毒灾害 toxic hazard
中毒症状 toxicity symptom
中肯 be to the point
中肯厚度 critical thickness
中肯耦合 critical coupling

仲氦 parhelium

仲胺 secondary amine
仲胺碱 secondary amine base
仲胺值 secondary amine value
仲班酸 parabanic acid
仲裁 arbitrage;arbitration;umpirage; umpire
仲裁办公室 adjudicating office
仲裁裁决 arbitral award;arbitral decision; arbitration award; arbitration decision
仲裁裁决的执行 enforcement of arbitral awards
仲裁裁决书 arbitral award
仲裁裁决注明理由 give reasons for an award
仲裁长 umpire
仲裁程序 arbitral procedure;arbitral proceedings;arbitration procedure; arbitration proceedings
仲裁程序规则 arbitral rules of procedure
仲裁的 arbitrate
仲裁的地点 place of arbitration
仲裁的时间限制 time limit for arbitration
仲裁的语言 language of arbitration
仲裁地 place of arbitration
仲裁法 act of arbitration; arbitration act;arbitration law;referee method
仲裁法试验 referee method test
仲裁法庭 arbitral court; arbitral tribunal;arbitration court;arbitration tribunal
仲裁费用 arbitration expenses; arbitration fee;fee for arbitration
仲裁分析 umpire analysis
仲裁规则 arbitration rule; rules for arbitration
仲裁合同 contract of arbitration
仲裁机构 arbitral authority; arbitral body; arbitral institution; arbitration agency;arbitration body;arbitration organization
仲裁检验 umpire examination
仲裁解决 arbitral settlement
仲裁决定 arbitrament
仲裁立法 arbitration legislation
仲裁判定书 arbitral award
仲裁人 arbiter;arbitrator;moderator; referee
仲裁上诉裁决 arbitration award on appeal
仲裁身份 moderatorship
仲裁实验室 laboratory for umpire analysis
仲裁使用的程序法 procedural law
仲裁试验 arbitration test; referee test;refery test
仲裁条款 arbitral clause; arbitration clause;reference clause
仲裁条例 act of arbitration
仲裁庭 arbitration court
仲裁通知书 arbitration notice
仲裁团 arbitral college
仲裁委员会 appeal committee; arbitration board; arbitration commission; arbitration committee; board of arbitration
仲裁小组 panel of arbitrators
仲裁协定 arbitral agreement;arbitration agreement
仲裁协议 arbitral agreement;arbitration agreement
仲裁样品数 number of sample for umpire analysis
仲裁员 arbiter; arbitrator; arbitrator de jure
仲裁争端 arbitrate a dispute
仲春 second month of spring; the middle of spring
仲醇 secondary alcohol
仲醇和叔醇混合物 pontol
仲丁醇 secondary butanol;secondary butyl alcohol
仲冬 midwinder; second month of winter
仲法线【数】binormal
仲分子 paramolecule
仲氦 parahelium
仲己醇 secondary hexyl alcohol
仲甲醛 paraformal-dehyde
仲聚焦 parafocus
仲谱 para-spectrum
仲氢 parahydrogen
仲秋 midautumn;second month of autumn
仲醛醇 paraldol
仲叔醇 secondary tertiary alcohol
仲酸 para acid
仲缩臂液压缸 telescoping cylinder
仲缩钻臂 extension boom
仲态 para-state
仲碳原子 secondary carbon atom
仲烷基磺酸盐 secondary alkyl sulfonate
仲烷基硫酸酯 secondary alkyl sulphate
仲夏 midsummer; second month of summer
仲硝基化合物 secondary nitro-compounds
仲盐 secondary salt
仲乙醛 paraldehyde
仲载者 overman
仲胀囊 <加气混凝土的小气泡,消除混凝土毛细孔中在受冻时所发生的液压> expansion chamber

众多 plurality;throng

众数 modal value;mode【给】
众数的 modal
众数粒径 modal diameter
众数频率 modal frequency
众数组 modal class
众议院各种委员会 <美> House Committees
众值 mode
众值平均 mode average

种板 blank;mother blank

种板材料 blank material
种板槽 starting sheet cell; stripper cell
种柄 seed pedicel;seed stalk
种草护堤 grass dike
种草护坡 grass slope
种草滩肩 grass berm
种的多样性 species diversity
种的特征 species specificity
种花 floriculture
种花工 floriculturist;gardener
种胶 <紫胶> brood lac
种晶 seed crystal
种树 planting

种树合同【植】 planting contract
种下 underplanting
种下多样性 infra-specific diversity
种植 cropping; cultivation; implant; plant; plantation
种植保护 plant protection
种植玻璃房 cold frame
种植槽 plant trough
种植草皮 planting sods; planting turf
种植场 plantation farm
种植池 planting bed
种植大样图 detail planting design
种植地 planted area
种植地段 crop tract; vegetation block
种植地界 plant canopy community
种植方法 planting
种植工程 planting work
种植海岸 vegetation coast
种植计划 planting scheme
种植犁 deep-plough
种植蔓生植物 plant scroll
种植密度 cultivation density
种植面积 acreage under cultivation; area under crops; growing area
种植盆 plant trough
种植期 planting season
种植区 ground cover areas
种植顺序 cropping sequence
种植土 agricultural soil; planting soil; vegetable soil
种植屋面 planted roof
种植箱 planting box
种植穴挖掘机 plant hole digger
种植与降低噪声 planting and noise suppression
种植与小气候 planting and micro-climate
种植园 estate; hacienda; plantation
种植园道路 estate road
种植园房屋 plantation house
种植园事务所 cutcherry
种植园橡胶 plantation rubber
种植园中草皮 plantation sods
种植者 cropper; planter
种植指数 cropping index

重

重白铅 London white

重板龙骨 center keel; centre board
重磅皮 heavy weight hide
重磅平布 heavy shirting
重磅铁锤 sledgehammer
重钡火石玻璃 dense barium flint glass
重钡冕玻璃 dense barium crown glass
重苯 heavy benzene; heavy benzol
重苯胺 semidine
重比 anharmonic ratio; cross ratio; double ratio
重冰 <固态重水> heavy ice
重冰靶 heavy ice target
重兵器 hardware
重病 serious disease
重病病房 intense care ward
重病护理单元 intensive care unit
重病监护室 intensive care unit
重材 dense timber
重残油 residuum[复 residua]
重侧 <壁厚不均匀管子的> heavy side
重差测高计 gravimetric(al) altimeter
重差高度计 gravimetric(al) altimeter
重差计 <测比重用> gravity meter; gravimeter[gravimetre]; gravi(to) meter
重柴油 diesel fuel oil; heavy diesel oil; heavy fuel oil; heavy oil
重掺杂 heavily dope
重掺杂层 heavily doped layer
重掺杂基区层 heavy base layer
重掺杂籽晶 heavily doped seed
重车 laden car; loaded car; loaded wagon; recondition; redress
重车车流表 loaded wagon flow table
重车动载量 average dynamic(al) load of loaded wagon
重车动载重 dynamic(al) load for a loaded car
重车方向 loaded wagon direction
重车公里 loaded wagon kilometers
重车公里占总车公里百分率 percentage of loaded wagon kilometres to total wagon kilometers
重车列车 loaded train
重车流 loaded flow
重车磨 reconditioning
重车平均净载重 net ton-kilometers per loaded car kilometers
重车平均净重量 net ton-kilometers per loaded car kilometers
重车调整 adjustment of loaded car
重车位 loaded position
重车制动缸 load brake cylinder
重车制动力 loaded braking force
重车制动作用 braking of the load
重车重心 center of gravity for car loaded
重车重心高 center height of gravity for car loaded
重车走行公里 loaded wagon kilometers
重沉积物 dense sediment
重沉箱 heavy caisson
重齿的 duplicident
重锤 ball breaker; counter-balance; counterweight; heavy bob; heavy hammer; heavy punch; heavy rammer; help weight; movement weight; weight jaw
重锤摆动装置的机构 oscillating device mechanism
重锤测深 hydrographic(al) cast
重锤拆除法 skull cracker demolition; wrecking ball demolition
重锤导向筒 weight cylinder
重锤吊架 weight hanger
重锤阀 gravity closed damper; gravity gate; weighted valve
重锤杆 balance weight arm; counterweight arm; movement weight rod; weight bar; weight lever; weight rod
重锤杠杆调节器 weight-lever regulator
重锤钩 movement weight hook
重锤夯 heavy ram
重锤夯实 heavy tamping; deep dynamic compaction; dynamic (al) compaction; ground bashing
重锤夯实法 compaction by tamping; heavy tamping method
重锤滑轮 movement weight pulley
重锤幌动式汞管继电器 <电压调整用> astatic relay
重锤击实试验 modified Proctor's compaction test
重锤架 weight bracket; weight holder
重锤井 weight pit
重锤扣 weight buckle
重锤块 movement weight mass
重锤链 movement weight chain
重锤落距 heavy hammer fall distance
重锤平衡吊架 counterweight hanger
重锤平衡式接轴托架 counterweighted balanced type spindle carrier
重锤平衡式联结轴 weight-balanced spindle
重锤平卫板 guard balanced by a counter weight
重锤少击法 few strikes with a heavy hammer
重锤绳 movement weight cable; movement weight cord
重锤式安全阀 dead-loaded safety valve; deadweight safety valve; weighted(lever) safety valve
重锤式扳道器 weighted switch
重锤式波纹收报机 weight driven undulator
重锤式道岔 counterweight type points; weighted points
重锤式关闭风门 gravity closed damper
重锤式料位计 pilot detecting level meter
重锤式平衡 counterweight balance
重锤式调节阀 weight-loaded regulator
重锤式万能试验仪 deadweight testing machine
重锤式蓄能器 weighted accumulator; weight-loaded accumulator
重锤式岩芯取样器 gravity core sampler
重锤式张紧装置 ballast tightening device; gravity take-up
重锤式转辙机 weighted switch
重锤式转辙握柄 switch lever with counter-weight
重锤弹簧 weight spring
重锤筒 movement weight cartridge
重锤箱 weight box
重锤压力计 deadweight piston page
重锤圆盘 weight disc[disk]
重锤扎钩 weight hook
重锤闸 load brake
重锤张力垫圈 oscillating weight tension washer
重锤装置 weight-driven
重淬火 reharden
重存取 reaccess
重锉 refile
重锉法 heavy filing
重大 magnitude; materially
重大的 grave; momentous
重大的差别 material difference
重大地震 significant earthquake
重大发现 breakthrough
重大改变 <态度意见等的> about-face
重大故障 major failure; primary failure
重大过失 gross negligence
重大河床 major river bed
重大火灾 fire disaster
重大技术装备国产化 domestic development of major technical equipment
重大紧急警报 critical emergency
重大科学技术成就 breakthrough in science and technology
重大利益 vital interest
重大偏离 material deviation
重大缺陷 significant deficiency; major defect
重大事故 catastrophe; grave accident; heavy accident; major accident; major breakdown
重大事故损失 major accident loss
重大事件 milestone
重大损害 material damage
重大损失 material damage
重大突破 quantum jump
重大违反 flagrant violation
重大违约(行为) fundamental breach; material breach
重大修改 substantial change
重大支出项目 major items of expenditures
重大阻断 major breakdown
重带电粒子 heavy charged particle
重氮氨苯脒乙酰甘氨酸盐 berenil; diminazene aceturate
重氮氨基化合物 diazoamino compound
重氮胺基苯 diazoaminobenzene
重氮胺亮黑 diazamine brilliant black
重氮苯磺酸 diazobenzene sulfonic acid
重氮靛蓝 diazo indigo blue
重氮二硝基苯酚 diazodinitrophenol
重氮分解物 diazosoplit
重氮复印 diazo-print
重氮复印法 diazo copying process; diazo process; dyeline
重氮复印机 diazo copier; diazo printer; dyeline printer
重氮复印件 diazo-print
重氮复印纸 diazo paper
重氮感光片 diazo film
重氮合成仪 diazo compositor
重氮化本领 diazotability
重氮化处理 diazotization process
重氮化滴定法 diazotization titration method
重氮化反应【化】 diazo-reaction
重氮化合物 diazo compound; diazonium compounds
重氮化作用 diazotization
重氮基 diazo group; diazo; diazonium
重氮基苯 diazobenzene
重氮(基)处理 diazo process
重氮基醋酸盐 diazoacetate
重氮基醋酸乙酯 ethyl diazoacetate
重氮基醋酸酯 diazoacetate
重氮基的 diazo
重氮基乙酸盐 diazoacetate
重氮甲基 diazomethyl
重氮甲基苯磺酰胺 diazomethylbenzenesulfonamide
重氮甲烷 azimethane; diazomethane
重氮坚牢枣红 diazo fast bordeaux
重氮胶卷 diazo film
重氮精蓝 diazogen blue
重氮刻图片 diazo scribing coating film
重氮亮绿 diazo brilliant green
重氮氰化物 diazocyanide
重氮染料 diazo colo(u)r; diazo dye; diazotable dye; diazotizing colo(u)rs; diazotizing dyes
重氮热敏成相法 diazo-thermography
重氮晒图 diazo-print
重氮晒图原纸 diazo paper raw base; original paper for diazo reproduction
重氮晒图纸 diazo paper
重氮晒印装置 diazo printing apparatus
重氮树脂 diazo resin
重氮水杨酸 diazosalicylic acid
重氮撕膜片 diazo stripping film
重氮酸 diazoic acid
重氮酸盐 diazotate
重氮缩微片 diazo microfilm
重氮酮 diazo-ketones
重氮烷 diazoparaffins
重氮稳定剂 diazo stabilizer
重氮烯 diazene
重氮硝基酚 <引爆炸药> diazonitrophenol
重氮盐 diazol; diazonium salt; diazosalt
重氮盐成相法 diazotype
重氮盐坚牢橙 diazol fast orange
重氮盐类 diazols
重氮盐浅橙 diazol light orange
重氮盐浅黄 diazol light yellow
重氮盐晒图 diazo copying; diazotype
重氮盐晒图工艺 diazo copying process
重氮盐印相法 diazotype process; ozalid process
重氮盐枣红 diazol bordeaux
重氮氧代己氨酸 diazooxonor leucine
重氮照相法 diazo process
重氮正片 diazo positive
重氮纸 diazo paper

重氮资料复印机 diazo document copying machine
重的 loaded;weighty
重的泥浆 high weight mud
重地沥青 heavy asphalt
重点 focal point;highlight;key point;priority;double point【数】
重点摆平 major shaker
重点病害 major fault
重点舱舱口 main hatch(way)
重点抽样 importance sampling
重点储备项目 major stocking item
重点词 heavy duty word
重点大学 key university
重点调查 focal point investigation;key point investigation
重点方案 major project
重点分析 selective analysis
重点改造 high-spot reform
重点工程 key construction;key project;major project;priority construction;priority project
重点工程指挥部 key project headquarters
重点工业 major industry
重点公园 basic park
重点货舱 controlling hold;key hold;main hold
重点集体面谈话法 focus group interview
重点计划 intensified program(me);major project
重点建设项目 major construction project;priority project
重点建筑 priority construction
重点建筑工程计划 major construction project
重点考查 high-spot review
重点考察 high-spot review
重点列车 emphasis train
重点色彩 accent colo(u)r
重点审查 high-spot review
重点试验 major test
重点文化保护单位 important protected cultural sites
重点污染物 priority pollutant
重点物资 key point products
重点项目 accented term;key item;key project;major item;major project;priority project
重点修(养路)spot maintenance
重点研究课题 major project
重点研究领域 priority research areas
重点养护 inface
重点找平 spot surfacing
重点照明 accent lighting
重点整顿 major shaker
重点支出 funding for some key projects
重电子 < 即介子 > heavy electron;meson;mesotron;penetron
重吊船 heavy lift vessel
重吊杆 boom set;heavy lift boom
重吊杆基座 heavy derrick step
重度 absolute density;gravity of volume;unit weight
重度标度 gravity scale
重度过滤 X 射线束 heavily filtered X-ray beam
重度疏伐 acceretion
重度污染 heavy pollution
重吨 long ton
重吨碾压 heavy rolling
重吨位车辆的交通线 heavy-tonnage lines of communications
重二型动力触探 heavy(2)dynamic(al)sounding
重罚税款 heavy penalty tax
重犯 recidivism
重防腐蚀涂料 heavy-duty anti-corrosive coating
重飞行器 aerodyne;aeronef
重废材 heavy slash
重废钢 heavy melting scrap
重粉壤土 heavy silty loam
重浮冰 heavy pack ice
重辐射系数 reradiation factor
重腐蚀面 heavily etched surface
重负 tax
重负荷 heavy load(ing)
重负荷标准定额 heavy load nominal rating
重负荷补偿装置 heavy-load compensating device
重负荷侧传动 heavy-duty final drive
重负荷的 heavy-duty;high duty
重负荷的使用条件 heavy-duty operating condition
重负荷多片式转向离合器 heavy-duty multiple disk steering clutch
重负荷分动箱 heavy-duty transfer case
重负荷工作的发动机 heavy-duty engine
重负荷工作状态 hard service
重负荷机油 heavy-duty oil
重负荷料 heavy burden
重负荷区 < 电线路 > heavy loading district
重负荷润滑脂 heavy-duty lubricating grease
重负荷橡胶弹簧 heavy-duty rubber spring
重负荷型制动液 heavy-duty type brake fluid
重负荷运行 heavy service
重负路面 < 承担繁重运输的 > heavy-duty pavement
重负载 heavy-duty;heavy load(ing);severe duty
重负载补偿装置 heavy-load compensating device
重负载触头 heavy contact
重负载电接触器材 heavy-duty contacts
重负载断续试验 heavy intermittent test
重负载继电器 heavy load relay
重负载接点 heavier-duty contact;heavy(-duty)contact
重负载调整 heavy load adjustment
重负载循环 heavy duty cycle
重负载运行 heavy-duty service;heavy service
重附着 reattachment
重高温曲线 WAT[weight,altitude and temperature] curve
重铬钾石 lopezite
重铬明胶 dichromated gelatin
重铬明胶膜 dichromated gelatin film
重铬酸 dichromic acid
重铬酸铵 ammonium bichromate
重铬酸钡 barium dichromate
重铬酸处理 dichromate treatment
重铬酸的 dichromic
重铬酸电池 bichromate cell
重铬酸法 < 测土中有机物含量 > dichromic acid method
重铬酸根离子 bichromate ion;dichromate ion
重铬酸汞 mercuric bichromate
重铬酸钾 potassium bichromate[dichromate]
重铬酸钾法 potassium dichromate method
重铬酸钾耗氧量 potassium dichromate oxygen demand
重铬酸钾滤光器 potassium dichromate filter
重铬酸钾消解液 potassium dichromate digestive solution
重铬酸钾中毒 potassium bichromate poisoning
重铬酸钾紫外分光度法 potassium dichromate ultraviolet photometry
重铬酸锂 lithium bichromate[dichromate]
重铬酸铝 alumin(i)um bichromate
重铬酸镁 magnesium bichromate
重铬酸钠 sodium bichromate[dichromate]
重铬酸铅 lead dichromate;plumbous bichromate
重铬酸铯 cesium bichromate
重铬酸铜 cupric bichromate
重铬酸锌 zinc dichromate
重铬酸盐 bichromate[dichromate]
重铬酸盐表面处理 dichromate treatment
重铬酸盐处理层 bichromated coating
重铬酸盐处理法 bichromate process;dichromate treatment
重铬酸盐电池 bichromate cell
重铬酸盐法 bichromate method
重铬酸盐光敏树胶 gum bichromate
重铬酸盐耗氧量 dichromate oxygen consumed
重铬酸盐化 bichromated
重铬酸盐化值 bichromate value
重铬酸盐还原 bichromate reduction
重铬酸盐浸渍处理 bichromate dipped finish
重铬酸盐明胶 dichromated gelatin
重铬酸盐明胶板 dichromated gelatin film
重铬酸盐物种 bichromate species
重铬酸盐需氧量 dichromate oxygen demand
重铬酸盐氧化法 dichromate oxidation method
重铬酸盐氧化性 dichromate oxidizability
重铬酸盐值 dichromate value
重铬酸银 silver bichromate[dichromate]
重工程车间 heavy machine shop
重工业 basic industry;heavy industry;large-scale industry
重工业厂矿 heavy plant
重工业景观 heavy industrial landscape
重工业区 heavy industrial district
重工作 heavy-duty
重工作循环 heavy duty cycle
重骨料 heavy(weight)aggregate;high-density aggregate
重骨料混凝土 heavy weight concrete;high-dense concrete;high-density concrete;heavy-aggregate concrete
重挂白合金 < 轴承 > relining
重灌 repour
重硅酸钙 dicalcium silicate
重轨 heavy rail
重轨交通 heavy rail transit
重轨快速交通 heavy rail rapid transit
重轨螺栓扳手 heavy rail track bolting wrench
重辊 rider
重过磷酸钙 acid phosphate of lime;concentrated superphosphate
重夯 heavy tamping
重荷间歇试验 heavy intermittent test
重荷模 load cast
重核 heavy nucleus
重核反应 heavy nuclear reaction
重荷条件下使用 heavy-duty service
重荷载 heavy load(ing);heavy weight load
重化学工业 heavy chemical industry
重化学品 heavy chemical
重混合 reblending;reblunge
重混凝土 < 用于防辐射 > high-density concrete;dense concrete;heavy weight concrete;loaded concrete
重混凝土防护层 heavy-aggregate concrete shield
重混凝土砌块 heavy concrete block;heavy weight concrete block
重活 dead works;task work
重活动 high activity
重活薪金 dirty money
重火石玻璃 dense flint glass;heavy flint glass
重货 deadweight cargo;high-density cargo;weight cargo
重货船 heavy cargo carrier;heavy cargo ship;heavy lift vessel
重货吊绳 burton pendant
重货及集装箱泊位 heavy lift cum container berth
重货列车信号 tonnage signal
重货物 heavy cargo;heavy freight;heavy goods
重货物列车 heavy freight train
重货运 heavy freight
重击 dunting;rapping;thump;thwack
重击声 thud
重级飞机跑道 heavy-duty runway
重级工作制 heavy-duty
重级工作制(起重)吊车 heavy-duty crane
重级轨道车 heavy-duty car
重级路面 < 用于繁重交通要道 > heavy-duty pavement
重级耐火砖 high-duty(fire)clay brick
重极化 repolarization
重集料 heavy(weight)aggregate;high-density aggregate
重脊屋顶 ridge and furrow roof;ridge and valley roof
重寄生的 hyperparasitic;superparasitic
重寄生物 hyperparasite;superparasite
重寄生(现象)hyperparasitism;superparasitism
重加感 < 对电缆 > heavy load(ing);heavy coil-loading
重加感区 heavy loading district
重甲板船 heavy decker;heavy deck vessel
重钾矾 mercallite
重间歇荷载 heavy intermittent duty
重碱 dense soda
重件 heavy lift;heavy load(ing);heavy piece;heavy weight
重件货(物) heavy cargo;heavy goods;heavyweight
重件货物运输系统 heavyweight transport system
重件货物装卸 heavy lift handling
重件运输船 heavy cargo carrier
重件专用船 heavy lift carrier;heavy lift vessel;heavy load ship
重件装卸设备 heavy lift equipment
重浆 heavy sizing
重浆整理 heavy finishing
重交通 heavy traffic
重交通结合料 heavy duty binder
重交通沥青碎石路面 heavy duty macadam
重焦油 coal-tar heavy oil
重解石 baricalcite
重介法 suspensoid process
重介选 dense medium separation
重介选矿法 heavy-medium separation
重介质 dense medium;heavy-medium
重介质分离器 dense-media separator
重介质分选 sink-float separation

Z

重介质分选槽 dense medium bath; separating bath
重介质分选机 heavy-medium separator
重介质分选箱 dense medium bath; separating bath
重介质固体 medium solid
重介质回路 heavy medium circuit
重介质跳汰 heavy-medium jigging
重介质跳汰机 dense medium jig
重介质洗选 dense medium washing
重介质悬浮液 heavy-medium suspension
重介质旋流器 heavy-medium cyclone
重介质选矿法 dense-media process; float-and-sink process separation; heavy-medium process
重介质选矿分离器 sink-float separator
重介质振动溜槽 dense medium vibrating sluice
重介子 heavy meson; K meson
重金属 heavy metal
重金属表征 heavy metal specification
重金属捕集剂 heavy metal chelating agent
重金属沉淀 heavy metal precipitation
重金属防爆盒 heavy metal cassette
重金属废水 heavy metal-containing wastewater; wastewater containing heavy metal
重金属废水处理 treatment of wastewater containing heavy metal
重金属废水净化剂 heavy metal wastewater purifying agent
重金属铬 heavy metal chromium
重金属含量 heavy metal content
重进刀 heavy feed
重金属进入 heavy metal input
重金属进入污水处理厂 heavy metals entering sewage treatment works
重金属矿产 heavy metal commodities
重金属离子 heavy metal ion
重金属离子吸附剂 heavy metal ion adsorbent
重金属量 amount of heavy metals
重金属浓度 heavy metal concentration
重金属生物毒性 heavy metal biotoxicity
重金属生物可利用性 heavy metal bioavailability
重金属生物吸附剂 biosorption of heavy metals
重金属示踪 heavy mineral tracer
重金属污染 heavy metal pollution; pollution of heavy metal
重金属污染物 heavy metal pollutant
重金属吸附 heavy metal adsorption; heavy metal sorption
重金属形态 heavy metallic form; species of heavy metal
重金属盐 heavy metallic salt
重金属(元素)污染 heavy metal contamination
重金属皂 heavy metal soap
重金属指标 heavy metal index
重金属指标法 heavy metal index method
重金属中毒 heavy metal poisoning
重金属总量 total heavy metal
重经组织 double fabric
重晶石 baria; barite; baritite; baroselenite; baryte; bologna spar; bononian stone; boulonite; cawk; heavy spar; liverstone; mineral white; schohartite; heavy earth
重晶石板 barite slab; barytes slab
重晶石方解石萤石矿石 barite-calcite-fluorite ore
重晶石粉 barite powder; blanc fix(e)
重晶石骨料 barite aggregate; barytes aggregate
重晶石骨料混凝土 barite aggregate concrete; barytes aggregate concrete
重晶石化 baritization
重晶石黄色 barite yellow; barytes yellow
重晶石混合颗粒 barytes grain mix(ture)
重晶石混凝土 barite concrete; barytes concrete; barytite
重晶石集料 barite aggregate; barytes ore aggregate
重晶石集料混凝土 barite aggregate concrete; barytes aggregate concrete
重晶石加重泥浆 barite-weighted mud
重晶石胶结物 barite cement
重晶石角砾岩 barite breccia; barytes breccia
重晶石颗粒混合料 barite grain mix(ture)
重晶石矿 barite ore
重晶石矿床 barite deposit
重晶石矿石 barytes ore
重晶石矿石骨料 barite ore aggregate
重晶石矿石混凝土骨料 barite ore concrete aggregate
重晶石矿石混凝土集料 barite ore concrete aggregate; barytes ore concrete aggregate
重晶石矿石集料 barite ore aggregate
重晶石类白色体质颜料 strontium white
重晶石玫瑰花结 barite rosettes
重晶石玫瑰花状 petrified rose concretion
重晶石墙板 barite wall slab; barytes wall slab
重晶石砂浆 barite mortar; barytes mortar
重晶石砂浆抹面 barytes mortar finish
重晶石砂浆饰面 barytes mortar finish
重晶石水泥 barite cement; baritic cement
重晶石异常 anomaly of heavy spar
重晶石萤石复合矿化剂 barite fluorite composite mineralizer
重晶石萤石矿石 barite fluorite ore
重晶土 heavy earth
重径积 weight-diameter product
重酒石酸铵 cream of tartar
重聚合物 heavy polymer
重颗粒 heavy particle
重空气 heavy air
重空穴 heavy hole
重空穴带 heavy hole band
重空穴态 heavy hole state
重空子 heavy hole
重块 pouring weight; weight
重块锚 <混凝土重块或某些无杆锚> clump anchor
重块式锚 deadweight anchor
重块体锚 deadweight anchor
重矿车 full tub
重矿石 heavy ore
重矿物 heavy mineral; dense mineral
重矿物分布图 distribution map of heavy mineral
重矿物颗粒 heavy suite
重矿物区 heavy mineral province
重矿物淘选盘 pan
重矿物油 black mineral oil
重矿物质土(壤) heavy mineral soil
重矿物质异常 heavy mineral anomaly
重矿物组分 heavy suite
重矿渣 <用干法冷却> dry slag
重矿渣混凝土 heavy slag concrete
重镧火石玻璃 dense lanthanum flint glass
重劳动 sweat
重离子【物】 heavy ion
重力 body force; force of gravitation; force of gravity; gravitation; gravitational force; gravitative attraction; gravity; gravity force; power of gravity; weight force
重力坝 dam of gravity type; gravitational dam; gravity dam; gravity type dam; solid dam; solid gravity dam
重力摆 gravitational pendulum; gravity bob; gravity pendulum
重力摆测定 pendulum determination
重力摆动缓冲板 swinging gravity fender
重力摆观测 pendulum observation
重力摆偏差 pendulum deviation
重力拌和 gravity mixing
重力崩塌构造 gravity collapse structure
重力泵 gravity pump; sight pump
重力闭合差 closing error of gravimetry
重力边缘效应 gravity edge effect
重力编组场 gravitation(al) yard; gravity classification yard; gravity marshalling yard; gravity yard; hump yard; summit yard
重力编组作业 gravity marshalling operation
重力扁率 gravitational flattening; gravity flattening
重力变电感器 gravity variometer
重力变化 variations of gravity
重力变化测定仪 gravimetric(al) variometer
重力变化改正 correction for variation of gravity
重力变化记录计 gravimetric(al) variometer
重力波 gravitational wave; gravity wave
重力波过渡区 transitional of gravitational wave
重力波面 gravitational wave surface
重力波阻 gravity wave drag
重力补偿 gravity compensation
重力补偿托架 balance weight chair
重力不稳定性 gravitational instability
重力不足 gravity deficiency
重力采样 gravity coring
重力操纵控制系统 gravity steering control
重力槽 gravity feed tank
重力槽形不稳定性 gravitational flute instability
重力测定 determination of gravity; gravimetric(al) observation; gravitational measurement; gravity determination; gravity measurement
重力测定法 gravimetry
重力测高计 gravimetric(al) altimeter
重力测井 gravity log
重力测力计 gravity dynamometer
重力测量 gravimetric(al) determination; gravimetric(al) measurement; gravitational determination; gravitational survey; gravity determination; gravity measurement; gravity survey
重力测量导线 gravimetric(al) traverse
重力测量的 gravimetric(al)
重力测量地区 gravity coverage
重力测量点 gravimetric(al) station
重力测量法 gravimetric(al) method
重力测量方法 gravity surveying methods; method of gravity measurement
重力测量分类 classification of gravity survey
重力测量工作方法 method of field work of gravity survey
重力测量检核基线 control gravimetric base
重力测量精度 accuracy of gravity anomaly
重力测量任务 survey task of gravity
重力测量数据 gravimetric(al) data
重力测量图 gravity map
重力测量卫星 gravisat
重力测量学 gravimetry
重力测量仪器 instrument of gravity measurement
重力测量因子 gravimetric(al) factor
重力测量用表 gravimetric(al) table; gravity table
重力测量员 gravimetrist
重力测量资料 gravimetry data
重力测量总精度 total accuracy of gravity survey
重力测速仪 gravitach(e) ometer
重力测探法 gravitational sounding
重力测网 gravimetric(al) network
重力测线 gravity traverse
重力层化 gravity bedding
重力差 gravity difference
重力差分作用 gravitational differentiation
重力掺和机 gravimetric(al) blender
重力常数 constant of gravity; gravitational constant; gravity constant; Newtonian constant of gravitation; Newtonian universal of gravity
重力场 field of gravity; gravitational field
重力场变化图 gravimetric(al) map
重力场常数 constant of gravitation
重力场方程 gravitational field equations
重力场观测 gravity field observation
重力场速度 intensity of gravity field
重力场位能 gravitational energy
重力场学 gravics
重力场异常 anomaly of gravitational field
重力潮汐改正 correction of gravity measurement of tide
重力沉淀 gravitational sediment(ation)
重力沉淀池 gravity settler; gravity settling tank
重力沉淀器 gravity settler
重力沉淀室 gravity deposit chamber; gravity settling chamber
重力沉降 gravitate; gravitational settling; gravity settlement
重力沉降槽 gravitation settler; gravity settler; gravity settling tank
重力沉降澄清 gravity settling clarification
重力沉降澄清池 gravity settling tank
重力沉降池 gravity sedimentation tank; gravity settling basin
重力沉降分离器 gravity separator
重力沉降浓缩 gravity thickener; operational throughput
重力沉降器 gravity settling chamber
重力沉降室 gravity settling chamber
重力沉陷 gravitational settling
重力秤 gravity balance
重力充填 gravity stowing
重力冲灰系统 gravity sluicing system
重力冲击球 drop ball; head ball; wrecking ball <自由下落击碎混凝土或岩石>
重力冲蚀 gravity erosion
重力抽吸式钻头 gravity aspirator bit
重力出料球磨机 gravity discharge ball mill

重力除尘 dust removal by gravity; gravity dedusting
重力除尘器 gravitational dust collector; gravity dust separator; gravity separator
重力除尘室 gravity setting chamber
重力除水阀 gravity relief trap
重力储藏器 gravity bag
重力储蓄器 gravity accumulator
重力触探试验 weight sounding test
重力触探仪 weight sounding apparatus
重力传动 gravity drive
重力传感开关 gravity sensing switch
重力传感器 gravity sensor
重力床 gravity bed
重力垂线偏差 gravimetric(al) deflection(of the vertical)
重力垂向二阶导数图 second vertical gravity derivative map
重力垂直二阶导数 second vertical derivative of gravity
重力垂直取样管 gravity corer
重力垂直梯度 vertical gradient of gravity
重力锤 gravitational hammer; gravity hammer
重力锤蓄力器 gravity loaded accumulator
重力纯度 gravity purity
重力打桩机 drop hammer pile driver
重力大地测量基本方程 principal equation of gravimetric geodesy
重力大地测量学 gravimetric(al) geodesy
重力大地水准面 gravimetric(al) geoid
重力大地水准面起伏 gravimetric(al) undulation
重力袋滤器 gravity bag filter
重力单位 gravitational unit; gravity unit
重力单位制 gravitational system of units; gravitational unit system
重力挡土墙 mass retaining wall
重力导出量 derived quantity of gravity
重力导管 gravity conduit; gravity line
重力导数 derivative of gravity
重力捣碎机组 gravity stamp battery
重力的 gravimetric(al); gravitational
重力等差线 isogam
重力等位面 equipotential surface of gravity
重力低 gravity low
重力低的 gravitational low
重力低压蒸汽供暖 gravity low-pressure steam heating
重力地面沉降 gravitate settling
重力地堑 gravity graben
重力地图 gravity map
重力地下水 gravity groundwater; gravity water
重力点 gravimetric(al) point; gravimetric (al) station; gravimetry point; gravitational point; gravity station
重力点标志 gravity station mark
重力点网 gravitational system
重力电池 gravity battery; gravity cell
重力调查 gravitational survey
重力调车 gravity shunting; gravity switching; switching by gravitation
重力调车场 gravity yard
重力调车场驼峰 hump in gravity yard
重力调车法 <在坡道牵出线上> shunting by gravitation
重力调车线 gravity shunting siding
重力定向 gravity direction
重力动力学 gravitation dynamics
重力动作闸门 gravity-operated shutter
重力斗式输送机 gravity bucket conveyer[conveyor]
重力陡度计 gravity gradiometer
重力断层 gravity fault
重力断层作用 gravitative faulting
重力堆积物 colluvial deposit
重力堆积型 gravity accumulation type
重力对流 gravitational convection; gravity convection; free convection
重力对流器高度 draft head
重力二次导数异常 second derivative gravity anomaly
重力阀 weight valve
重力法测定 gravimetric(al) survey
重力法测量 gravimetric(al) survey
重力法的 gravimetric(al)
重力法检查 gravimetric(al) survey
重力翻板阀 gravity tipping valve
重力翻车机 gravity dumper
重力翻斗 gravity acting skip
重力反常 gravity anomaly
重力反向 gravity reversal
重力反演法 gravity inversion method
重力范围 sphere of gravity
重力方法 <加气混凝土配料法的一种> gravimetric (al) method; gravitational method
重力方向 gravity direction
重力放矿法 gravity system
重力放矿溜井 gravity ore pass
重力分布 gravity distribution
重力分层 gravitational separation; gravity segregation
重力分级 gravitational classifying; gravitational separation; gravity classification; gravity classifying
重力分解(列车) gravity sorting
重力分界 gravipause; neutral point
重力分界线 null circle
重力分离 gravitational segregation; gravity separation
重力分离处理法 gravitational separation processing
重力分离法 gravimetric(al) separation method
重力分离作用 gravitational separation; gravity separation; gravity fractionation
重力分量 component of gravity; gravity component; weight component
重力分配 gravity distribution
重力分析 gravimetric(al) analysis
重力分析法 gravimetric (al) procedure
重力分选 gravity separation
重力分选器 gravity separator
重力分异作用【地】 gravitational differentiation; gravitative differentiation
重力风 canyon wind; drainage wind; gravity wind; katabatic wind; mountain breeze; mountain wind
重力风门 flop damper
重力封闭 gravity closing
重力浮选装置 gravity-flo(a)tation plant
重力辐射 gravitational radiation
重力负载和地应力 gravitational load and geostress
重力复测 gravity reiteration
重力复原旁承 gravity side bearing
重力富集法 gravity concentration
重力改化 gravimetric(al) reduction
重力改正 gravimetric(al) correction; gravity correction
重力改正分类 classification of gravity corrections
重力干管 gravity main
重力感 graviperception; gravity feed tank
重力感受器 gravirecepter[graviceptor]; gravitational receptor
重力高 gravity head; gravity high
重力高差给料器 gravity head feeder
重力高的 gravitational high
重力高度计 gravimetric(al) altimeter
重力高度校正因数 gravity elevation correction factor
重力给料 gravity feed
重力给料的 gravity-fed
重力给料装置 gravity feed apparatus
重力给油 gravity fuel feed
重力给油化油器 gravity feed carburetor
重力给油汽化器 gravity fed carburet(t)or; gravity feed carburetor
重力给油润滑器 gravity feed lubricator
重力给油系统 gravity fuel system
重力公式 gravity formula
重力供料 gravity feed
重力供料方法 gravity process
重力供料机 gravity feed machine; gravity flow machine
重力供料箱 gravity tank
重力供水 gravity feed
重力供水系统 gravity water supply system
重力供水箱 gravitational tank; gravity tank
重力供油 gravity feed; gravity feed oil
重力供油箱 gravitational tank; gravity tank
重力拱 arch-gravity
重力拱坝 massive arch dam; gravity arch dam
重力构造 gravity tectonics
重力构造类型 type of gravity tectonics
重力构造说 gravitational tectonics theory
重力观测 gravimetric (al) observation; gravity observation
重力观测方法 gravity observation method
重力管 gravity duct
重力管道 gravity conduit
重力管路 gravity pipeline
重力惯性分级机 gravitational inertia classifier
重力灌溉 gravity irrigation
重力灌浆 gravitational grouting
重力灌浆的 gravity grouted
重力灌注砂浆 gravity placed sanded grout
重力灌装 <灌装货物> filling by gravity
重力罐 gravitational tank
重力归算 gravity correction; gravity reduction
重力归算用表 gravity reduction table
重力柜 gravitational tank
重力辊道 gravity roller
重力辊式输送机 gravity roller conveyer [conveyor]; gravity wheel conveyer[conveyor]
重力滚道运输机 gravity roller conveyer[conveyor]
重力滚轮输送机 gravity wheel conveyer[conveyor]
重力滚轴运输机 gravity roller carrier
重力过滤 gravity filtration
重力过滤法 gravity filtration process
重力过滤器 percolation filter; weighted filter
重力海波 gravity ocean wave
重力焊接 deck welding; gravity type arc welding; gravity welding
重力焊条 gravity electrode
重力夯 drop weight tamper
重力耗散不稳定性 gravitational dissipative instability
重力荷载 gravity load(ing); gravitational load
重力厚浆池 gravity thickener
重力滑槽 gravity discharge chute
重力滑动 gravitational gliding; gravitational sliding; gravitational slip(ping); gravity slide; gravity sliding; gravity slipping
重力滑动板(块) gravitational sliding plate; gravitational slide plate
重力滑动断层 gravity-glide fault
重力滑动堆积 olistostrome
重力滑动构造 gravity gliding tectonics
重力滑动构造特征 feature of gravitative gliding structure
重力滑动盆地 gravitational sliding basin; gravitational slipping basin
重力滑动作用 gravity gliding
重力滑坡 ecoulement; gravitational slip(ping); gravity slide
重力滑曲褶皱 cascade fold
重力滑移 <发生于造山运动时大范围岩体>【地】 gravitational gliding; gravitational sliding; gravitational slip (ping); gravity slide; gravity sliding; gravity slipping
重力滑移作用 gravity gliding
重力滑油系统 gravity oil system
重力环流 gravity circulation
重力换算 gravity reduction
重力恢复力矩 gravity restoring moment
重力回程柱塞 gravity return ram
重力回流 gravity reflux; gravity return
重力回燃喷管 gravity reinjection nozzle
重力回水 gravity (water) return; unassisted gravity return
重力回水锅炉 gravity return boiler
重力回水进给 gravity return feed
重力回水系统 gravity return system
重力昏暗 gravity darkening
重力混合 gravity blending; gravity mixing
重力混合机 gravity mixer
重力活荷载 live gravity load
重力活塞采样器 gravity piston sampler
重力机理 gravitational mechanism
重力积聚作用 gravity accumulation
重力基础 gravity foundation
重力基点网 gravity base station network
重力基点网平差 adjustment of gravity base station network
重力基线 gravity base(line)
重力基准 gravimetric(al) datum
重力基(准)点 gravity base station; gravity reference point; gravity reference station
重力集尘 gravity dust collection
重力集尘器 gravitational dust collector
重力集尘装置 gravity dust separator
重力给水管路 gravity feed line
重力给水系统 gravity water system
重力计 gravi(to)meter[gravi(to)metre]
重力计算机 gravity computer
重力继电器 gravitation type relay
重力加料 gravity filling
重力加料器 gravity feeder
重力加料设备 gravity carrier
重力加煤机 gravity feed stoker
重力加汽油 gravity gasoline feed; gravity petrol feed
重力加速度 acceleration due to gravity; acceleration of gravity; gravitational acceleration; gravity acceleration
重力加油 gravity oiling

重力间接效应 indirect effect on gravity
重力间歇式拌和机 gravity batch mixer
重力间歇式混合机 gravity batch mixer
重力检测器 weight detector
重力检定基线 calibration line
重力降水 gravity dewatering
重力浇注 gravity-assist pouring
重力浇筑混凝土 concrete placing by gravity
重力浇铸法 gravity casting
重力铰链 gravity hinge
重力校正 correction for gravity; gravimetric(al) correction; gravity correction
重力截流管 gravity interceptor
重力截流器 gravity interceptor
重力进给 gravity feed
重力进给杯 gravity feed cup
重力进给滑油系统 gravity circulating oil system
重力进给循环系统 gravity feed circulating system
重力进料 gravity feeding
重力进料器 gravity feeder
重力进料线 gravity feed line
重力阱 gravity well
重力镜 gravity mirror; weighted mirror
重力矩 gravitational moment
重力均衡 gravitational equilibrium
重力均衡改正值 gravity isostatic correction value
重力均衡异常 isostasy gravity anomaly
重力勘探 gravitational exploration; gravitational prospecting; gravity exploration; gravity prospecting
重力勘探法 gravimeter method (of exploration); gravitational exploration method
重力空气分级机 gravity type air classifier
重力控制 gravity control
重力控制点 control gravimeter point; gravity control station
重力控制基点 gravity control base station
重力控制陀螺 gravity controlled gyro
重力控制网 gravity control network
重力控制仪器 gravity controlled instruments
重力扩展 gravitational spreading; gravity spreading
重力扩张 gravitational spreading; gravity spreading
重力拉紧 gravity tensioning
重力缆(车车)道 <下坡重车带动上坡空车> gravity railroad
重力冷却器 gravitational cooler; gravity cooler
重力离析 gravity segregation
重力理论 <泥沙运动的> gravitational theory
重力连接杆 gravity bar
重力联测 gravity conjunction
重力量值 gravitational magnitude
重力裂隙 gravitational fissure
重力溜槽 gravity chute
重力溜放 gravitating
重力流 free surface flow; gravity current; weight flow; gravity flow
重力流变 gravitational flow
重力流层序 gravity flow sequence
重力流动 gravity motion; flow by gravity; gravity flow
重力流动床 gravity flowing bed
重力流动式冻结机 gravity froster
重力流动褶皱 gravity flow folding
重力流供热系统 gravity flow heating system
重力流管道 free surface flow line
重力流灌溉 gravitational flow irrigation
重力流设备 gravity flow installation
重力流失 gravitational erosion
重力流水系统 gravity water system
重力流隧道 free surface flow tunnel
重力流装置 gravity flow installation
重力滤层 gravity filter
重力滤池 gravitational filter; gravity type filter
重力滤器 gravitational filter; gravity type filter
重力落棒快速停堆 gravity scram
重力落差进料器 gravity head feeder
重力毛管孔隙 gravitation-capillary pore space
重力弥散 gravitational differentiation
重力密度 <单位体积材料所受的重力> unit weight; weight density
重力面 plane of gravity
重力模拟 gravity simulation
重力模式 gravity mode
重力模型 gravity model
重力模型法 <预测交通量的一种方法,与吸引力成正比,与距离成反比> gravity model approach; gravity model method
重力磨 gravity mill
重力内插 interpolation of gravity
重力能 gravitational energy; gravity energy; power of gravity
重力凝聚 gravitational coagulation
重力浓集 gravity concentration
重力浓缩 gravity thickening
重力浓缩池 gravity thickener
重力浓缩法 gravity concentration method
重力浓缩机 gravity thickener
重力暖气锅炉 gravity furnace
重力排储层 gravity drainage reservoir
重力排放 gravity discharge; weight discharge
重力排矿球磨机 gravity discharge ball mill
重力排气式蒸汽供暖 steam-heating with gravity air removal
重力排气通风 gravity exhaust ventilation
重力排砂 grit discharge by gravity
重力排水 drainage by gravity; gravity discharge; gravity drain; gravity drainage; non-pressure(d) drainage
重力排水法 gravity water system
重力排水涵洞 gravity culvert
重力排水线 gravity drain line
重力排泄堵塞 gravity drain plug
重力排油曲线 gravity drainage curve
重力排注 gravity transfer
重力喷布器 gravity distributor
重力喷灌机 gravity sprinkler
重力喷泉 gravity spring
重力喷洒 gravity distribution
重力喷洒车 gravity distributor
重力喷雾洗涤器 gravity spray tower
重力偏差 gravity deflection
重力偏析 gravitational segregation; gravity segregation
重力漂移 gravity drift
重力平衡 gravitational equilibrium; gravity balance; gravity equilibrium
重力平面 geoid
重力平台 gravity platform
重力屏障 <油水分离机的> gravity dam
重力坡 gravity slope; wash slope; gravity incline <使货车自由运行的下坡>; haldenhang
重力坡滑 ecoulement
重力坡脚 haldenhang; wash slope
重力破碎机 gravity crusher
重力剖面 gravity profile
重力铺装刨花板 gravity speed fine particle board
重力普查 gravity reconnaissance
重力普查勘探 gravity reconnaissance survey
重力气压表 weight barometer
重力气压计 weight-barograph
重力汽油箱 gravity gasoline tank; gravity petrol tank
重力潜水钟 gravity driving clock
重力潜移 gravitational creep
重力强度 intensity of gravity
重力墙式 gravity wall type
重力墙式船闸 gravity wall type of lock
重力侵蚀 gravitational erosion; mass erosion; gravity erosion
重力轻便索道 gravity ropeway; gravity wirerope way
重力倾翻式手推车 gravity tipple
重力倾尽式油槽车 gravity tank truck
重力倾卸车身 gravity dump body
重力倾卸斗 gravity tipping skip
重力倾卸式运输机 gravity tipping conveyer[conveyor]
重力倾卸小车 gravity tipping skip
重力驱动 gravity drive; gravity impulse
重力取样 gravity sampling
重力取样器 gravity sampler
重力取样器取样 sampling by gravity-sampler
重力圈 gravisphere
重力泉 gravity spring
重力扰动 gravity disturbance
重力扰动矢量 gravity disturbance vector
重力热风采暖装置 gravity warm-air heating plant
重力热水供暖系统 gravity hot water heating system
重力蠕变 gravitational creep
重力蠕动 gravitational creep
重力润滑 gravity feed lubrication; gravity lubricating
重力润滑法 gravity lubrication
重力撒布器 gravity spreader
重力撒布机 gravity spreader
重力扫选 gravity scavenging
重力砂滤池 gravity sand filter
重力砂滤器 gravity sand filter
重力上行下给式蒸汽供暖系统 steam-heating down-feed gravity system
重力升降机 gravity elevator
重力失稳 gravitational instability
重力-时间曲线 drift curve
重力矢量 gravitational vector; gravity vector
重力式 gravity type
重力式安培计 gravity ammeter
重力式岸壁 gravity bulkhead; gravity (type) quaywall; mass(ive) quaywall
重力式岸墩 gravity ability; gravity abutment
重力式岸墙 gravity type quaywall
重力式拌和厂 gravity plant
重力式拌和机 gravity mixer
重力式驳岸岸壁 gravity quay wall
重力式补充回路 gravity replenished circuit
重力式采样管 gravity core sampler
重力式场强度 gravity field strength
重力式沉淀池 gravity settling basin
重力式沉箱结构 gravity caisson structure
重力式沉箱码头 gravity caisson wharf
重力式船闸 gravity type lock
重力式(船闸)闸墙 gravity lock wall; gravity type lock wall
重力式锤 gravity hammer
重力式粗选风力分级机 gravitational roughing air classifier
重力式打桩锤 drop pile hammer; gravity pile hammer
重力式打桩机 gravity hammer
重力式大体积混凝土船闸 gravity mass concrete lock
重力式单管采暖 gravity type one-pipe heating
重力式单管供暖 one-pipe gravity heating
重力式单管供热 gravity type one-pipe heating
重力式单向阀 gravity-held check-valve
重力式挡水墙 gravity bulkhead
重力式挡土结构 gravity-type earth retaining structure
重力式挡土墙 gravity bulkhead; gravity(type) retaining wall; solid gravity wall
重力式底基结构 gravity-based structure
重力式地下连续墙 gravity diaphragm wall
重力式电弧焊 gravity arc welding; gravity welding
重力式吊艇杆 gravity davit
重力式吊艇杆上滑轨 track way of gravity davit
重力式吊艇柱 gravity davit
重力式斗式提升机 gravity bucket elevator
重力式墩 gravity-type dolphin
重力式墩台 gravity abutment; gravity pier; gravity pier and abutment
重力式防波堤 gravity breakwater
重力式防冲物 suspended fender
重力式防冲装置 suspended fender
重力式防舷材 gravity fender
重力式防撞装置 gravity(type) fender; suspended fender
重力式分离器 gravitational separator
重力式干船坞 gravity drydock
重力式干坞底板 gravity drydock floor
重力式干坞坞墙 gravity drydock wall
重力式隔墙 gravity bulkhead
重力式供给系统 gravity feed system
重力式供料 gravity supply
重力式供水 gravity water supply
重力式供油柜 gravity feed tank
重力式拱坝 arch-gravity dam; gravity arch dam
重力式拱座 gravity arch abutment
重力式构筑物 gravity type construction; gravity type structure
重力式关闭风门 gravity closed damper
重力式滚轴输送机 gravity roller conveyer [conveyor]; gravity wheel conveyer[conveyor]
重力式过滤器 gravity filter
重力式(海上)平台 gravity platform
重力式护舷 gravity fender
重力式滑油系统 gravity lubricating oil system
重力式缓行器 gravity type retarder; load pressure retarder
重力式回水 gravitational backwater; gravitational return
重力式混凝土坝 mass concrete dam
重力式混凝土墩 gravity type concrete pier
重力式混凝土溢流坝 concrete gravity overfall dam; concrete gravity overflow dam; mass concrete overflow dam

重力式集尘器 gravitational dust precipitator
重力式给水 gravity water supply
重力式加料装置 gravity carrier
重力式减速器 gravity type retarder; load pressure retarder
重力式搅拌机 gravity mixer
重力式搅拌站 gravity plant
重力式结构 gravity structure; massive construction; massive structure
重力式结构设计 design of gravity structure
重力式进料设备 gravity carrier
重力式进排气系统 gravitational supply and exhaust system; gravity supply and exhaust system
重力式快砂滤池 rapid gravity sand filter
重力式缆索道 gravity wirerope way
重力式(沥青等)喷洒机 gravity distributor
重力式料斗 gravity bin
重力式滤池 gravity filter
重力式滤器 gravitation filter; gravity filter
重力式落锤 board hammer; gravity drop hammer
重力式码头 gravity(type) quay; gravity(type) wharf
重力式码头岸壁 gravity quay wall
重力式锚碇 gravity anchor
重力式磨煤机 gravity mill
重力式(自动)撤布 gravity distribution
重力式凝汽阀 gravity trap
重力式平台 gravity platform
重力式墙 gravity(type) wall
重力式桥墩 gravity type abutment; gravity(type) pier; massive abutment
重力式桥墩土墩 gravity pier
重力式桥台 gravity(type) abutment
重力式擒纵机构 gravity escapement
重力式取土样器 gravity core sampler
重力式燃烧设备 gravity burning appliance
重力式燃烧装置 gravity burning appliance
重力式热风采暖 gravity warm water heating
重力式热风采暖系统 gravity warm-air heating system
重力式热风供暖 gravity warm water heating
重力式热风供暖系统 gravity warm-air heating system
重力式热风炉 gravity furnace
重力式热风炉供暖 gravity furnace heating
重力式热水中央供暖 gravity hot water central heating
重力式洒布器 gravity spreader
重力式洒水车 gravity tank truck
重力式设备 gravity installation
重力式渗滤 gravity percolation
重力式输送机 gravity conveyer[conveyor]
重力式输送器 gravity conveyer[conveyor]
重力式水车 gravity water wheel
重力式水力旋流沉淀池 gravity hydrocyclone settling tank
重力式水力旋流器 gravity hydrocyclone
重力式水轮 gravity water wheel
重力式锁定器 gravity lock
重力式通风系统 gravity type venting system
重力式外海钻井平台 gravity(offshore) platform
重力式屋顶通风器 gravity roof ventilator
重力式坞底板 gravity drydock floor
重力式坞墙 gravity(dry) dock wall
重力式纤维束滤池 gravity type fibre-bind filter
重力式衔铁 gravity drop armature
重力式卸载升运器 gravity discharge elevator
重力式悬挂碰垫 gravity type suspension fender
重力式循环供暖 gravity circulating heating
重力式溢流坝 gravity spillway dam
重力式翼墙 gravity wing wall
重力式引水堰 gravity type diversion weir
重力式蒸汽供暖 steam-heating by gravity
重力式正压供热 gravity plenum heating
重力式直墙防波堤 gravity upright wall
重力式注油器 gravity feed oiler
重力式抓钩 gravity type grab
重力式爪 gravity type pawl
重力式转轮 gravity water wheel
重力式装置 gravity apparatus
重力式自动铺料装置 gravity automatic paving apparatus
重力式自动洒水车 gravity tank truck
重力势 gravity potential; weight potential; gravitational potential
重力势能 gravitational potential energy; natural potential
重力试验 gravity test
重力试验卫星 satellite for gravimetric experiment
重力收集系统 gravity flow gathering system
重力收缩 gravitational contraction
重力疏干 gravity dewatering
重力输水管 gravity aqueduct; gravity main
重力输水设施 down service
重力输水总管 gravity main
重力输送器 gravitation transporter
重力输送索道 gravity transporting cableway
重力输送系统 gravity transporting system
重力数据 gravity data
重力栓 gravity bar
重力水 bulk water; free water; gravitational; gravitational water; gravitative water; mobile water; vadose water; wandering water
重力水层 zone of suspended water
重力水车 gravity wheel
重力水储罐 gravity tank
重力水分 gravitational moisture
重力水柜 gravity tank
重力水冷 gravity-system water cooling
重力水力除灰系统 gravity sluicing system
重力水流 gravitational flow; gravity flow
重力水轮 gravity wheel
重力水平坡度 horizontal gradient of gravity
重力水平梯度 horizontal gradient of gravity
重力水平梯度剖面图 horizontal gradient of gravity profile
重力水头 gravitational head; gravity head
重力水箱 gravity water tank
重力送料 natural feed(ing)
重力送料道 gravity track
重力送料管 gravity feed line
重力送料器 gravimetric(al) feeder
重力送料式喷枪 gravity feed type gun
重力送料箱 gravitational feed tank; gravity feed tank
重力速测仪 gravitach(e)ometer
重力隧道 gravity tunnel
重力索道 gravity cable(way)
重力索道集材 gravity cable logging
重力塌陷构造 gravity collapse structure
重力弹簧刀 gravity knife
重力探测法 gravitational method of exploration
重力探矿法 gravimeter method (of exploration)
重力梯度 gradient of gravity; gravitational gradient; gravity gradient
重力梯度测量 gradiometry
重力梯度计测量 gradiometer survey
重力梯度离心法 gravity gradient centrifugation
重力梯度试验 gravity gradient test
重力梯度试验卫星 gravity gradient test satellite
重力梯度卫星 gravity gradient satellite
重力梯度稳定 gravity gradient stability; gravity gradient stabilization
重力梯度仪 gradiometer
重力梯度制导 gravitational gradient guidance
重力梯度姿态控制仪 gravity gradient attitude control
重力体系 gravity system
重力填充法 gravitational packing method; gravity packing
重力调节 gravitative adjustment
重力调节器 gravity regulator
重力铁路 gravity railway
重力通风 gravitational ventilation
重力通量 gravitational flux
重力投料设备 gravity carrier
重力图 gravitational map
重力脱蜡 gravity dewaxing
重力脱水机 gravity thickener
重力驼峰 gravity hump; hump in gravity yard
重力外推 extrapolation of gravity
重力网 gravitational network; gravity network
重力为零的 agravic
重力位 gravity potential
重力位二阶导数 second-order derivative of gravity potential
重力位高阶导数 higher-order derivative of gravity potential
重力位各阶导数 many order derivative of gravity potential
重力位能 gravitational potential energy
重力位能积聚 gravity storage
重力位势 gravitational potential
重力位势高度 geopotential height
重力位水准面 level of gravity potential
重力位移 gravitational displacement
重力喂料 gravimetric(al) feed; gravity feed
重力喂料机 gravimetric(al) feeder; gravity feeder
重力稳定容器 gravity settling chamber
重力稳定陀螺仪 gravity-stabilized gyroscope
重力稳定卫星 gravity-stabilized satellite
重力稳定系统 gravity stabilization system
重力污泥浓缩池 gravity sludge thickener
重力污泥浓缩器 gravity sludge thickener
重力吸气钻头 gravity aspirator bit
重力吸下 gravitate downwards
重力吸引 gravitate; gravitating
重力析离 gravitative adjustment
重力系统 gravimetric(al) system
重力下沉 gravity settling
重力下降 gravity drop; gravity lowering
重力下水道 gravity sewer
重力线 gravity line; gravity vertical; line of gravity
重力相似准则 gravitational similarity rule
重力详查 detailed gravity survey
重力向量 gravity vector
重力效应 effect of gravity; gravitational effect; gravity effect
重力效应逐层去除法 gravity stripping
重力斜面 go-devil plane
重力斜坡道 gravity incline
重力泄水管 gravity main
重力卸车 empty by gravity; gravity discharge; gravity unloading
重力卸料 gravity discharge; gravity dumping
重力卸料车身 gravity dump body
重力卸料式散装水泥车 bulk cement car discharged by gravity
重力卸空 empty by gravity
重力型电量计 gravity voltameter; weight voltameter
重力型风力分级机 gravity type air
重力型继电器 heavy armature relay; safety circuit relay
重力型胶结物结构 gravity cement texture
重力型结构 gravity type structure
重力型桥台 gravity type abutment
重力型倾斜指示器 gravity type bank indicator
重力型调节器 weight type regulator
重力型堰 gravity type weir
重力型增稠剂 gravity type thickener
重力型增稠器 gravity type thickener
重力型增厚器 gravity type thickener
重力蓄力器 gravity loaded accumulator; weight-loaded accumulator
重力悬(着)水 gravity suspended water
重力选 gravity preparation
重力选分 gravitational classifying; gravitational separation
重力选粉机 gravity type separator
重力选精矿 gravity concentrate
重力选矿法 gravimetric(al) concentration; gravity concentration; gravity dressing; gravity separation (method); gravity concentration method
重力选矿机 gravity flow concentrator
重力选矿流程 gravity flowsheet
重力选种机 gravity seed cleaner
重力循环 gravitational circulation; gravity circulation; natural circulation
重力循环采暖 gravity central heating
重力循环供热系统 gravity circulation heating system
重力循环供热装置 gravity circulation heating installation
重力循环加热器 gravity circulation heater; gravity heater
重力循环热水供暖系统 gravitational hot water heating system; gravity circulation hot water heating system
重力循环润滑 gravity circulation lubrication
重力循环式集中炉 gravity type central furnace

重力循环式空气盘管 valance
重力循环式空气盘管系统 valance system
重力循环式空气盘管装置 valance unit
重力循环系统 gravity circulation system;gravity system
重力压力 gravitational pressure
重力压强 gravitative pressure
重力压实 gravitational compaction
重力压实排驱机理 expulsion mechanism of gravitational compaction
重力压缩型垃圾车 gravity packer vehicle
重力压头 gravity head
重力延拓 continuation of gravity
重力岩芯提取器 gravity corer
重力堰 gravity weir
重力仪 gravi(to)meter[gravi(to)metre]; gravity instrument; gravity meter
重力仪标值 scale value of gravimeter
重力仪常数标定 calibration of gravimeter constant
重力仪掉格 gravimeter drift
重力仪辅助设备 additional device of gravimeter
重力仪杆 gravimeter beam
重力仪观测精度 observation accuracy of gravimeter
重力仪观测总精度 total observation accuracy of gravimeter
重力仪光线灵敏度 beam sensitivity of gravimeter
重力仪混合零点改正值 compound drift correction of gravimeter
重力仪角灵敏度 angular sensitivity of gravimeter
重力仪精度 accuracy of scale value determination
重力仪灵敏度 sensitivity of gravimeter
重力仪零点改正值 drift correction of gravimeter
重力仪零点位移 drift of gravimeter
重力仪器 gravity apparatus
重力仪一致性精度 congruity accuracy of gravimeter
重力夷平作用 graviplanation
重力移动消摆装置 moving weight stabiliser
重力移位沉积 gravity-displaced deposit
重力异常 gravitational anomaly;gravity anomaly;gravity disturbance
重力异常场 gravity anomaly field
重力异常垂向二阶导数换算 second vertical derivative calculation of gravity anomaly
重力异常代表误差 representation error of gravity-anomaly
重力异常地质类型 geologic(al) type of gravity anomaly
重力异常反演问题 inverse problem of gravity anomaly
重力异常分类 classification of gravity anomalies
重力异常峰值 gravity anomaly peak
重力异常划分 gravity anomaly dividing
重力异常幻视 oculogravic illusion
重力异常极大值 gravity anomaly maximum
重力异常极小值 gravity anomaly minimum
重力异常阶方差 degree variance of gravity-anomaly
重力异常解释 interpretation of gravity anomaly
重力异常类型 type of gravity anomaly
重力异常面 gravity anomaly surface
重力异常模拟器 gravity anomaly simulator
重力异常内插误差 interpolation error of gravity-anomaly
重力异常剖面图 gravity anomaly profile
重力异常数据处理 data processing of gravity anomaly
重力异常水平梯度 horizontal gradient of gravity anomaly
重力异常梯度带 gradient zone of gravity anomaly
重力异常图 gravity anomaly map
重力异常向量 gravity anomaly vector
重力异常向上延拓 upward continuation of gravity anomalies
重力异常向下延拓 downward continuation of gravity anomalies
重力异常形态 gravity anomaly configuration
重力异常延拓 gravity anomaly continuation
重力异常圆滑 gravity anomaly smoothing
重力异常正演问题 forward problem of gravity anomaly
重力异常走向 gravity anomaly strike
重力翼墙 <分水闸的> gravity wing
重力引力 gravitational attraction
重力应力分配 gravitational stress distribution
重力影响 gravity effect;influence of gravity
重力硬模铸造 gravity die foundry
重力油柜 gravity oil tank;oil gravity tank;oil head tank
重力油水分离技术 gravity oil-water separation technology
重力预测 gravity prediction
重力预搅拌 gravity preblending
重力圆盘 gravity disc[disk]
重力运动 gravitational motion;gravity flow
重力运料箱 gravity tank
重力运输 gravity haulage;self-acting incline
重力运输机 gravitation transporter
重力运输器 gravity carrier
重力运行 gravity run
重力再注入喷嘴 gravity reinjection nozzle
重力(造成的)偏转 gravity deflection
重力造山作用 gravity orogenesis
重力增稠 gravity thickening
重力闸 weight brake
重力张紧(装置) take-up by gravity
重力找矿 gravity prospecting
重力折刀 gravity knife
重力褶皱 gravity fold
重力真空地下管道交通系统 gravity-vacuum tube system
重力真空地下管道列车 gravity vacuum capsule train using all underground tube
重力真空公共交通系统 <在研究中的一种新式地下铁道系统> gravity-vacuum transit
重力真空运输 gravity-vacuum transit
重力正常位 normal potential of gravity
重力正演法 gravity direct method
重力支墩坝 cellular dam
重力值 gravity value;value of gravity
重力值归算 reduction of gravity value
重力制 gravity system
重力中心 centre of gravity
重力钟 gravity clock
重力轴 gravity axis
重力助推 gravity assist
重力注入压铸 gravity die-casting
重力注油示油器 sight gravity feed oiler
重力注油式轴承 gravity oil filling bearing
重力驻波 standing gravity wave
重力铸造 gravitational casting
重力抓钩 gravity grab
重力爪 gravity pawl
重力砖石墙 gravity masonry wall
重力转动浓缩器 rotating gravity concentrator
重力转矩 gravitational torque
重力转向 gravity turn
重力转仪钟 gravity driving clock
重力(桩)锤 gravity hammer
重力装车法 <粒状和粉状货物> loading by gravity
重力装车坡(或斜坡台) gravity loading incline
重力装袋 bagging-off by gravity
重力装罐 gravity caging
重力装料 gravity load(ing)
重力装载 gravity load(ing)
重力装置 gravitational apparatus
重力坠落褶皱 gravity drop fold
重力资料 gravimetric(al) data;gravity data
重力子 graviton
重力(自动)加料 gravity feed
重力自动倾斜装置 tipple
重力自流 run by gravity
重力自流管 gravity pipe
重力自流加油 gravity oil feed
重力自流进料管 gravity feed line; gravity feed pipe
重力自流泉 gravity spring
重力自流润滑系统 gravity circulating oil system;gravity oil system;gravity lubricating system
重力(自流)通风 gravity ventilation
重力自流转油管线 gravity rundown lines
重力自由空间异常 free air anomaly
重力总基点 gravity main base station
重力最大值 gravity maximum
重力最小值 gravity minimum
重力作业区 gravity chance
重力作用 action of gravity; gravity action
重力作用滚轴运输机 gravity roller conveyer[conveyor]
重力作用输送器 gravity conveyer [conveyor]
重力作用下沉 gravitate
重力作用爪 gravity type pawl
重力坐标 gravimetric(al) coordinates
重沥青(基)油 heavy asphaltic base oil
重连 reconnection
重联 double heading
重联阀 double heading valve
重联机车【铁】 double heading locomotive; double-headed locomotives;double header
重联机车走行公里【铁】 double-locomotive moving kilometers
重联继电器 multiple unit connection relay
重联开关 multigang switch;multipleswitch;multiple unit switch;multiplex switch
重联控制阀 multiunit control
重联塞门 double heading cock
重联位 double heading position;multiunit position
重联运行 coupled running
重炼铁 double refined iron
重梁 bearing piling
重量 heaver;weight;weight transfer adjuster
重量安全阀 weighted safety valve
重量百分比 percent by weight
重量百分比率 percentage by weight; weight percentage
重量百分含量 weight percentage content
重量百分浓度 concentration expressed in percentage by weight
重量百分数 percentage by weight; percentage in weight; weight percent(age)
重量保安阀 deadweight valve
重量报表 weight account
重量本位 weight basis
重量比 proportion by weight;ratio by weight;ratio of weight;weight percent(age);weight ratio
重量比例 part by weight
重量比配合法 proportioning by weight
重量比配料 weight-mix
重量比配料法 weight method
重量比热 specific heat by weight; weight specific heat
重量变化 change in weight
重量变化测定 weight change determination
重量变化指示器 weight change indicator
重量变异耐受性 weight variation tolerances
重量标志 weight mark
重量剥采比 weight stripping ratio
重量不明 weight unknown
重量不匀率 weight unevenness
重量不足 deficiency in weight;loss in weight; shortage in weight; short weight
重量不足的 underweight
重量参数 weight parameter
重量测定 gravimetry;weight determination
重量测定的 gravimetric(al)
重量差异 weight differential; weight variation
重量差异试验 weight variation test
重量超过 outweight
重量成分 weight item
重量充装系数 coefficient of charge weight
重量传感器 weight sensor
重量代换罐 weight replacing tank
重量大的物件 heavyweight
重量单 weight list;weight memo
重量单位 unit of weight;weight unit
重量当量 weight equivalent
重量当量浓度 weight normality
重量的增加 gain in weight; weight gain
重量电导率 weight conductivity
重量电量计 weight coulometer
重量定额 weight rating
重量定量分析 gravimetric(al) analysis
重量定量器 weighing controller
重量定量式加料器 weight-actuated filler
重量动力比 weight-power ratio
重量短少 weight shortage
重量对容量的比 weight-to-capacity ratio
重量吨 deadweight ton;ton weight; weight ton
重量吨或容积吨 weight (ton) or measurement(ton)
重量吨位 weight tonnage
重量多项式 weight polynomial
重量阀 weighted valve
重量法 <加气混凝土配料> gravimetric(al) method;weighing method
重量分布 weight distribution
重量分布方式 weight distribution pattern
重量分布函数 weight distribution function

Z

重量分布曲线 weight curve; weight distribution curve

重量分级机 weight grader; weight sizer; weight sorter

重量分级计 weight grader

重量分配 proportioning by weight; weight distance; weight distribution

重量分配比 weight distribution ratio

重量分配计算 calculation of weight distribution

重量分配系数 weight distribution coefficient; weight distribution factor

重量分数 weight fraction

重量分析 gravimetric(al) analysis; weight analysis

重量分析步骤 gravimetric(al) procedure

重量分析测定法 gravimetric(al) determination

重量分析的 gravimetric(al)

重量分析法 gravimetric(al) method; method of weight analysis; gravimetry

重量分析粉尘采样器 gravimetric(al) dust sampler

重量分析因数 gravimetric(al) factor

重量份 amounts by weight; part by weight

重量份数 parts by weights

重量丰度 abundance by weight

重量功率比 weight-to-power ratio

重量估计 weight estimation

重量关系 weight-relation

重量惯性矩 weight moment of inertia

重量规度 weight normality

重量含水量 moisture weight percentage

重量含水率 <土壤的> moisture weight percentage; moisture content ratio; water content by weight

重量函数 weighting function

重量核对 checking of weight

重量核实 verification of weight

重量环 weight ring

重量回收率 weight recovery

重量货物 deadweight cargo; high-density cargo; weight cargo

重量基位 weight basis

重量及尺码标志 weight and measurement mark

重量及计量条例 weights and measures act

重量级(别) weight class

重量集中在背部 weight concentrated in the back

重量计 poidometer; weightimeter [weightometer]; weight machine

重量计算法 <加气混凝土配料> weight calculation method

重量记录计 weighting recorder

重量记录器 weighting recorder

重量加权平均灰分产率 weighted mean ash production rate by weight

重量架设 weight hypothesis

重量假设 weight assumption

重量减轻 weight loss; weight reduction; weight saving

重量检定证书 certificate of weight

重量检验书 inspection certificate of weight

重量节省 weight saving

重量均匀度 weight uniformity

重量可称性 ponderability

重量克分子浓度 modal concentration

重量空间速度 weight-space velocity

重量控制 weight control

重量控制器 weight controller

重量库仑计 weight voltameter

重量累积计算仪器 weight totalising instrument

重量力学 barodynamics

重量利用系数 coefficient of wear efficiency; weight utilization factor

重量量测 weight measurement

重量量配 batching by weight

重量流量 weight discharge; weight flow; weight rate

重量流率 weight flow rate; weight rate of flow

重量流体 heavy fluid

重量马力比 weight-to-horsepower ratio

重量密度 gravimetric(al) density; weight density

重量摩尔的 molal

重量摩尔浓度 molal concentration; molality; weight molality; weight molar concentration

重量摩尔浓度的 molal

重量摩尔溶液 molal solution

重量浓度 weight concentration

重量浓度法 weight concentration method

重量排水量 weight displacement

重量配比 proportioning by weight

重量配比补偿器 weight ratio compensator

重量配合 weigh mixing

重量配合比 mixing ratio by weight; mix proportion by weight; parts by weights

重量配合法 weigh mixing; proportion(ing) by weight

重量配料斗 weigh-batching hopper

重量配料法 batching by weight; gravimetric(al) batching; weigh batching; weight mix(ture) method

重量配料机 weigh batcher

重量配料器 weigh batcher; weighing batch box

重量配料装置 batching weigh gear

重量平衡 weight balance

重量平衡表 weight and balance sheet

重量平衡测定 weight balance measurement

重量平均分子量 weight-average molecular weight

重量平均聚合度 weight-average degree of polymerization

重量平均粒径 weight mean diameter

重量平均直径 weight mean diameter

重量强度 weight strength

重量强度比 weight strength ratio

重量轻 in light weight; light in weight

重量轻的 lightweight

重量曲线 weight curve

重量容量比 weight-to-volume ratio

重量容器 batcher box; batcher bucket; batcher hopper

重量时空速度 weight hourly space velocity

重量式保压阀 weight type pressure retaining valve

重量式称量设备 weight batcher

重量试验 balance measurement

重量试样 gravimetric(al) sample

重量数量比 weight-number ratio

重量水溶度 gravimetric(al) moisture content

重量送料器 weight feeder

重量速度 weight velocity

重量损耗 loss in weight

重量损耗率 weight loss ratio

重量损失 loss in weight; loss of weight; weight loss

重量损失率 weight loss ratio

重量体积 weight volume; bulking value

重量体积百分比 percent weight in volume

重量体积比 weight-to-volume ratio

重量体积法 weight volume method

重量体积关系 weight-volume relationship

重量调节器 weight regulator

重量图 weight chart

重量推力比 thrust loading

重量威力比 weight-to-yield ratio

重量位面计 weighing level ga(u)ge

重量喂料仓 weight bin

重量温度计 weight thermometer

重量吸收法 gravimetric(al) absorption method

重量系数 weight coefficient; weight factor

重量细度 weight fineness

重量限度 weight limit

重量限制 weight limitation; weight restriction

重量限制标志 weight limit sign

重量相当 weight equivalent

重量箱 weight case

重量箱折算系数 weight case conversion factor

重量消耗量 weight flow

重量效率 gravimetric(al) efficiency; weight efficiency

重量一览表 weight schedule

重量异常的 dysponderal

重量因数 weight factor

重量增长因数 weight-growth factor

重量增加 weight gain; weight increase

重量增加法 added weight method

重量证(明)书 certificate of weight; weight certificate

重量支承机构 weight bearing mechanism

重量转动惯量 weight moment of inertia

重量转化 weight conversion

重量转移 <即前轴减载> weight shift; weight transfer(ence)

重量转移附属装置 weight transfer attachment

重量转移装置 weight transfer device; weight transfer unit

重量自动分选机 automatic weight classifier

重量自动记录器 <起重机> automatic weight recording device

重料 heavy burden; sink material <浮选中的>

重列 rearrangement

重列柱【建】 super-columniation

重列柱式的 supercolumnar

重裂变碎片 heavy fission fragments

重磷冕玻璃 dense phosphate crown glass

重流 density current; density flow

重硫铋铅铜矿 hammarite

重馏 double distilled

重馏分 heavy distillate; heavy cut

重垆坶 heavy loam

重炉渣 heavy slag

重滤料 heavy filtering medium; heavy filter material

重路由 <通信[讯]> heavy route

重路由长途电话 heavy route toll telephone

重路由地面站 heavy route earth station

重路由电话通信 [讯]heavy route telephone message

重路由电路 heavy route circuit

重路由站 heavy route station

重轮载(汽车)交通 heavy wheel-load traffic

重脉 dicrotism

重脉波 dicrotic wave

重镁氧 heavy magnesia

重密度介质 heavy-density medium

重冕玻璃 dense crown glass

重敏车辆减速器 weight proportional retarder; weight response type retarder; weight responsive (car) retarder

重敏缓行器 weight proportional retarder; weight responsive (car) retarder

重摹(衍射)光栅 replica grating

重模的 molal

重模溶液 molal solution

重膜状矿脉 sheeted-zone veins

重摩(浓度) molality

重木 heavy timber

重木材 heavy wood

重钠矾 matteuccite

重泥浆 heavy mud

重铌铁矿 mossite

重黏[粘]结土 heavy-textured soil

重黏[粘]壤土 heavy clayey loam; heavy clay loam

重黏[粘]土 fat clay; ga(u)ging clay; ga(u)ging gault; gumbo (clay); heavy clay soil; plastic clay; rich clay; strong clay; unctuous clay; gault(clay) <烧砖用>

重黏[粘]土产品 heavy clay product

重黏[粘]土的 clay-rich

重黏[粘]土地砖 heavy clay flooring tile

重黏[粘]土工业 heavy clay industry

重黏[粘]土制品 heavy clay article

重黏[粘]土(砖瓦)窑 heavy clay kiln

重黏[粘]质土 heavy-textured soil

重扭力荷载 heavy torque load

重浓度区 area of concentrated emphasis

重平板 flat heavy plate

重平楼盖板 flat heavy plate

重起动负荷 hard starting load

重气体模型 heavy gas model

重汽油 heavy petrol

重牵引铁路 heavy haul railway

重潜水 heavy weight diving

重潜水服 hard hat dress

重潜水盔 heavy weight helmet

重潜水作业 deep-sea dive work

重切伤 diacope

重切削 heavy cut

重侵蚀 appreciable attack

重氢 deuterium; heavy hydrogen

重氢核 deuteron; deuton

重丘区 heavy hilly area; hilling terrain; hilly terrain

重轻油切换阀 heavy oil-light oil transfer valve

重球 ball weight

重区地改值 terrain correction value of farther distance

重圈 barysphere; bathysphere; centrosphere

重缺陷 major defect

重燃料油 heavy fuel oil

重壤土 heavy loam

重软流圈【地】 asthenosphere

重软泥 heavy sludge

重润滑油 black oil; heavy lubricating oil

重砂 heavy sand

重砂采样 heavy mineral sampling

重砂测量 heavy placer survey

重砂测量布局方式 distribution pattern of heavy placer prospecting

重砂测量成果表示法 expression method of heavy placer prospecting result

重砂鉴定 identification of heavy mineral sand

重砂鉴定报告 identification report of heavy placer; report on identifica-

Z

tion heavy mineral sand
重砂鉴定样品 sample for heavy mineral sands determination
重砂浆 heavy mortar
重砂矿物 heavy placer mineral; placer mineral
重砂矿物定量测定法 heavy sand mineral quantitative determination method
重砂矿物分离及鉴定仪器 heavy sand mineral separation and identification instrument
重砂矿物分析 heavy placer mineral analysis
重砂矿物含量 content of heavy placer mineral
重砂矿物鉴定 heavy sand mineral identification
重砂矿物组合 heavy placer mineral association
重砂矿物组合法 method of heavy mineral association
重砂壤土 heavy sandy loam
重砂样品 heavy mineral concentrate
重砂异常 heavy placer anomaly
重砂找矿法 heavy mineral prospecting method; method of heavy minerals
重砂质黏[粘]土 heavy loam
重商主义 merchantilism
重赏 handsome reward
重十字沸石 harmotome
重石 scheelite
重石脑油 heavy naphtha
重石油 heavy petroleum oil
重石油醇 heavy petroleum spirit
重石油精 heavy petroleum spirit
重石油蜡 heavy petroleum wax
重石油沥青 heavy asphalt
重石油醚 heavy mineral spirit; heavy petroleum spirit
重蚀地 badland
重矢量介子 heavy vector meson
重视 care about; make account of; make a point of
重视风险 risk premium
重视效率的 effect-oriented
重水 deuterated water; deuterium oxide; deut(er)oxide; heavy water
重水靶探头 heavy water probe
重水冰 heavy ice
重水反射层 heavy water reflector
重水反应堆 heavy water reactor
重水沸腾堆 heavy water boiling reactor
重水工厂 heavy water plant
重水合物 deuterate
重水减速反应堆 heavy water moderated reactor
重水减速剂 heavy water moderator
重水冷却反应堆 heavy water cooled reactor
重水慢化剂 heavy water moderator
重水平摆 heavy horizontal pendulum
重水系统 heavy water system
重水铀反应堆 deuterium-uranium reactor
重水中化堆 deuterium-moderated reactor
重苏打 heavy soda
重苏打灰 heavy ash; heavy soda ash
重燧石玻璃 dense flint glass; heavy flint glass
重缩毡 felt buff
重钽锰矿 manganotapiolite
重碳地蜡 koenlite
重碳钾石 kalicinite
重碳酸钙 calcium bicarbonate
重碳酸钙镁型热水 calcium magnesium bicarbonate thermal water
重碳酸钙泉 calcium bicarbonate spring
重碳酸钙水 calcium bicarbonate water
重碳酸钙型热水 calcium bicarbonate thermal water
重碳酸钙盐离子 bicarbonate ion concentration
重碳酸根 bicarbonate radical
重碳酸钾 kalicinite
重碳酸镁 heavy magnesium carbonate
重碳酸镁水 magnesium bicarbonate water
重碳酸钠 sodium bicarbonate
重碳酸钠泉 sodium bicarbonate spring
重碳酸钠型 bicarbonate sodium type
重碳酸钠型水 sodium bicarbonate water
重碳酸水带 hydrocarbonate water zone
重碳酸盐 bicarbonate[dicarbonate]; nahcolite
重碳酸盐带 dicarbonating zone
重碳酸盐缓冲系 bicarbonate buffer system
重碳酸盐碱度 bicarbonate alkalinity
重碳酸盐矿泉水 bicarbonate mineral spring water
重碳酸盐类水 bicarbonate groundwater
重碳酸盐离子 bicarbonate ion
重碳酸盐硫酸盐体系 bicarbonate sulphate system
重碳酸盐浓度 bicarbonate concentration
重碳酸盐树脂 bicarbonate resin
重碳酸盐型离子交换柱 bicarbonate anion exchange column
重碳酸盐硬度 bicarbonate hardness
重体力劳动 hard physical labo(u)r; heavy physical labo(u)r
重体模型 massive object model
重体油 heavy-bodied oil
重铁钽矿 tapiolite
重铁铤 door breaker
重烃 heavy hydrocarbons; higher hydrocarbon
重梃宽门 double-margined door
重同位素 heavy isotope
重同位素积累 heavy isotope build-up
重土 heavy earth; terra ponderosa
重土壤 heavy soil
重退火 reannealing
重砣 <吊在钢丝绳端部用以加速其下降的> cow sucker
重瓦斯 serious gas
重弯隐形眼镜 dual curvature contact lens
重尾馏分 heavy tails
重纹 twinned grooves
重纹孔 bordered pit
重稳距 metacentric height
重污染 severe contamination; substantial pollution
重污染感潮河道 heavily polluted tidal river
重污染河流 heavily polluted river
重污染水域 heavily polluted water body; heavily polluted waters
重污水的 polysaprobic
重屋顶 heavy roof
重屋面 balk roofing
重物 heavy; load; weight
重物安全结构 load-guard structure
重物保持架 load maintainer
重物保持闸 load holding brake
重物保持制动器 load holding brake
重物保护装置 loadguard
重物锤式储能器 weight-loaded accumulator
重物的中心线 center of mass
重物的装卸 handling of load(s)
重物后挡板 load backrest
重物起吊具 parbuckle
重物升降卷筒 load hoist drum
重物悬距 load overhang
重雾 heavy fog
重稀土 heavy rare earth
重稀土矿 heavy rare earth ores
重稀土亏损 heavy rare earth deficiency
重稀土类 heavy rare earths
重衔铁继电器 weighted armature relay
重陷试验 slump test
重相 heavy phase
重相入口 heavy liquid inlet
重相位 rephase
重相堰 heavy phase weir
重箱 <集装箱> loaded van; loaded container
重箱吊具 loaded container spreader
重心 barycenter[barycentre]; center[centre] of gravity; centroid; gravity center[centre]; mass center[centre]; median point; medium point; orthcenter[orthocentre]
重心波长法 centroid method
重心布置 gravity allocation
重心垂直位置 vertical center of gravity
重心的 centrobaric
重心法 gravity model approach
重心范围 centre-of-gravity range
重心高差 <由比重不同引起的压力差> gravity head
重心高度 height of center of gravity; height of gravitational centre
重心箍 balance band; gravity band
重心惯性主轴 central principal axis
重心横向位置 transverse center of gravity
重心积分 gravity center integral
重心基准点 centre of gravity datum
重心极限 centre of gravity limits
重心距离 centroidal distance
重心面 plane of gravity
重心能量 barycentric energy
重心偏移 centre-of-gravity shift
重心平衡(确定)法 balancing method
重心扰动 centre-of-gravity disturbance
重心铊 gravity nut
重心位置 centre-of-gravity position
重心下移 squat
重心线 center of gravity line
重心悬置法 centre-of-gravity suspension
重心压应力 centroidal compressive stress
重心运动 centre-of-gravity motion
重心运动的守恒 conservation of movement of the centre of gravity
重心运动轨迹 center of gravity path
重心支承系统 centre-of-gravity mounting system
重心轴(线) axis of gravity; gravity axis; mass axis
重心轴油 heavy spindle oil
重心纵向位置 longitudinal center of gravity
重心坐标 areal coordinates; barycenter coordinate; barycentric coordinates
重心坐标系 barycentric coordinate system
重锌锑矿 ordonezite
重型 gravity type; heavier-duty; heavy(-duty) type
重型变矩器 heavy-duty converter
重型叉车 heavy-duty truck; heavy lift fork
重型柴排 woven lumber
重型柴油 heavy-duty diesel oil
重型柴油车辆 heavy-duty diesel vehicle
重型柴油机用柴油 heavy-duty diesel oil
重型铲斗 heavy-duty bucket; mass excavation bucket; severe-duty bucket
重型长导程道岔 <便利双向高速运行> heavy-duty long-lead switch
重型车床 heavy-duty engine lathe; heavy-duty lathe; heavy-duty turning lathe
重型车(辆) heavy(-duty) vehicle
重型车辆底盘 heavy-duty chassis
重型车辆技工 heavy-duty mechanic
重型车辆用的柴油 derv fuel
重型冲凿 slogging chisel
重型船体修理舰 heavy hull repair ship
重型锤 striking hammer
重型粗加工刀具 heavy-duty roughing tool
重型粗砂摇床 record table
重型带齿推板 <推土机的> rock rake
重型的 heavy-duty; high duty
重型电缆 heavy cable
重型电气设备 heavy electrical plant
重型电钻 heavy-duty electric drill
重型吊车 heavy-duty crane; sling cart
重型吊杆 heavy derrick; jumbo boom
重型吊杆座 heavy derrick stool
重型吊钩 snag hook
重型钉齿耙 heavy-duty spike tooth harrow; heavy-tined harrow
重型发动机 heavy-duty engine
重型发动机油 heavy-duty(engine) oil
重型阀 block valve
重型防泡沫轴承 heavy-duty anti-foam bearing
重型防水帆布 oil belt duck
重型飞机 heavy-duty plane; ship
重型飞机跑道 heavy-duty runway
重型分节式掩护支架 heavy type sectionalized shield
重型分组光面镇压器 heavy segmented flat roller
重型风动式钻机 heavy pneumatic drifter
重型风钻 heavy drifter; heavy pneumatic hammer drifter
重型负载 heavy-duty load
重型刚架 <建筑用的> structurals
重型钢管支架 heavy steel tube frame
重型钢轨 heavy rail
重型钢架结构 heavy fabricated steel construction
重型钢筋弯曲 heavy bending
重型割捆机 heavy drawn binder
重型格栅 heavy joist
重型工业厂房基础 heavy foundation
重型公路 heavy-duty highway; heavy-duty road
重型钩(子) heavy-duty hook; bull hook
重型构架 heavy-duty frame
重型构件 heavy unit
重型构件断面 heavy profile; heavy shape
重型构件预制 heavy precasting
重型固定手弓锯框 heavy fixed hand hack saw frame
重型挂车 heavy-duty trailer
重型管子虎钳 heavy-duty pipe vice
重型轨(道) heavy(-duty) track; heavy rail
重型轨道车 heavy-duty motor troll(e)y; heavy rail motor troll(e)y; heavy section car
重型辊碎机 heavy rolls

重型海底作业潜水器 heavy-duty seabed vehicle
重型夯具 heavy compactor
重型桁架 heavy truss
重型护刃器 heavy-duty guard
重型滑车 <有三~四个滑轮的> heavy block
重型环 heavy ring segments
重型混凝土 massive concrete
重型混凝土结构 massive concrete structure
重型活动修理车间 heavy mobile machine shop
重型货车 heavy goods vehicle
重型货车交通 heavy-duty traffic; heavy truck traffic
重型货物专门运输公司 heavy specialized carrier
重型货运机车 heavy freight locomotive
重型击实试验 modified Proctor test
重型机车 heavy-duty engine
重型机床 heavy-duty machinery; heavy-duty machine tool; heavy-duty tool; power tool
重型机床厂 heavy machine tool plant
重型机器 heavy-duty machine; Hercules
重型机器厂 heavy machinery works
重型机器润滑油 heavy-duty engine lubricating oil
重型机器制造工业 heavy machine building industry
重型机械 heavy-duty machinery; donkey doctor
重型机械操纵器 heavy-duty manipulator
重型机械厂 heavy-duty machinery plant; heavy-duty machinery shop; heavy machinery shop
重型基础 heavy foundation
重型夹盘 heavy-duty chuck
重型甲板 heavy deck
重型架式凿岩机 heavy drifter
重型剪切机 heavy-duty shears
重型剪切机作业线 heavy shear line
重型建筑 heavy construction
重型建筑动力学 barodynamics
重型绞车 heavy-duty winch
重型脚手架 builder's staging; heavy-duty scaffold
重型结构 heavy construction; heavy structure; massive structure
重型结构力学 barodynamics
重型结构系统 heavy structural system
重型金属丝筛网 <用于筒筛> heavy-wire screen
重型浸轧机 heavy padder
重型锯木机 timber mill
重型卡车 heavy-duty truck; heavy load truck; half track <前后轮推动的卡车>
重型卡车货运 heavy trucking
重型卡车交通 heavy truck traffic
重型铠装运输机 heavy-duty armored conveyer[conveyor]
重型空投 heavy drop
重型块石护坡 heavy grade stone protection
重型拉门滑动装置 coburn fitting
重型犁 heavy plough; heavy plow
重型链 heavy chain; heavy-duty chain
重型链板喂料机 heavy-duty apron feeder
重型零件 heavy parts
重型龙门式起重机 goliath(crane)
重型楼地面 heavy-duty floor(ing)
重型履带式推土机 heavy crawler dozer
重型履带式拖拉机 traxcavator
重型轮胎 heavy-duty tire[tyre]
重型螺纹 heavy-duty thread
重型锚臂 heavy anchor arm
重型铆钉机 bull riveter
重型门闩 tower bolt
重型摩托车 heavy motorcycle
重型木建筑 heavy timber construction
重型木结构 heavy timber construction
重型泥泵 heavy-duty slurry pump
重型捻线机 heavy doubling frame
重型碾压 heavy rolling
重型耙 field drag; heavy-duty harrow
重型平板挂车 heavy haul trailer
重型平板式震动夯实机 vibrating plate compactor
重型平地机刮刀 heavy-duty cutting blade
重型平开门 heavy-duty flexible(swing) door
重型平面铣刀 heavy-duty slab milling cutter
重型平碾 heavy smooth wheel roller
重型平展匀湿两用机 heavy-duty samming and setting machine
重型破碎机 heavy-duty crusher
重型铺板 solid apron
重型铺地砖 heavy-duty tile
重型普通车床 heavy-duty geared head lathe
重型普通铣刀 heavy-duty plain cutter
重型起吊设备 heavy-lift equipment
重型起动工况 heavy starting duty
重型起落架 heavy undercarriage
重型起重货船 heavy lift ship
重型起重机 giant crane; hammerhead; heave lift (crane); heavy crane; heavy-duty crane; titan crane
重型起重机钢轨 heavy-duty crane rail
重型起重机轨道 heavy-duty crane rail
重型起重机械 heavy-duty lifting machine
重型起重设备 heavy-duty hoisting equipment; heavy lift equipment
重型气锤 stoper
重型气动风挡吊 heavy type air-operated screen wiper
重型气动风屏帚 heavy type air-operated screen wiper
重型气动振动器 heavy-duty air vibrator
重型气割工具配套单位 heavy gas cutting unit
重型汽车 heavy-duty car
重型器材修理厂 heavy unit shop
重型牵索 <两船间传送用的> heavy jackstay
重型牵引车 heavy hauler; heavy traction vehicle
重型潜水服 deep-sea diving suit; weight vehicle
重型墙体 massive wall
重型桥 heavy bridge
重型桥梁桥面板 heavy bridge deck panel
重型桥台 heavy abutment
重型切割器 heavy-duty cutter
重型切片器 heavy-duty cross-section device
重型倾卸车 heavy-duty tipper
重型球轴承 heavy-duty ball bearing
重型缺口圆盘耙 heavy-duty cutaway disc harrow
重型燃气轮机 heavy-duty gas turbine
重型容器 heavy vessel
重型润滑脂 heavy-duty lubricating grease
重型三脚架 heavy-duty tripod
重型设备 essential facility; heavy (-duty) equipment
重型设备制造厂家 heavy equipment manufacturer
重型设计(方案) heavy-duty design; heavy version
重型深梁 deep heavy beam
重型施工机械 Hercules
重型施工设备 heavy construction equipment; heavy construction plant
重型石凿 pitching chisel
重型手持式风钻 heavy hand-held rock drill
重型手持式凿岩机 hand-held drifter; heavy hand-held rock drill
重型水压机 heavy hydraulic press
重型松土机 heavy ripper
重型碎石机 primary crusher
重型碎石镇压器 rolling chopper
重型塔式起重机 heavy tower crane; Hercules crane
重型弹簧 heavy-duty spring
重型弹性铲柄 heavy-spring standard
重型坦克 heavy metal
重型通信[讯]卫星 heavy communication satellite
重型土方机械 heavy earthmoving equipment
重型托梁 heavy joist
重型拖车 full trailer; heavy trailer
重型拖车车组 full trailer combination
重型拖拉机 heavy(-duty) tractor
重型拖网渔船 heavy-duty trawler
重型挖掘机 mass excavator
重型挖掘抓斗 heavy digging grab
重型弯沉仪 heavy weight deflectometer
重型卫星 heavy satellite
重型卫星平台 heavy satellite platform
重型屋面板 heavy ga(u)ge roofing sheet
重型物件 heavy ware
重型铣床 forge milling machine
重型橡套电缆 heavy type cab-tire cable
重型消防炮车 heavy-duty apparatus
重型小梁 <美国木材规格,厚4英寸、宽8英寸以上的制材,1英寸=0.0254米> heavy joist
重型斜撑 heavily inclined strut
重型卸料车 ingoldsby car
重型型材 heavy section
重型型钢 heavy section
重型雪犁 heavy-duty snow plough
重型巡洋舰 heavy cruiser
重型压力机 heavy-duty press; supercompactor
重型压实机 super-compactor
重型压实试验 heavy compaction test
重型压碎机 primary crusher
重型压缩机 heavy-duty compressor
重型摇臂钻床 heavy-duty radial drilling machine
重型液压支臂 heavy hydraulic boom
重型有轨交通 heavy rail transit
重型预选摇臂钻床 heavy-duty preselect radial drilling machine
重型圆盘犁 heavy-duty disk plow
重型圆盘耙 heavy-duty disc harrow
重型圆柱 heavy-duty roller bearing
重型圆锥动力触探 heavy-duty dynamic(al) penetration test
重型运货车 heavy-duty truck
重型运货汽车 heavy goods vehicle
重型运输机 heavy-duty transport machine
重型运输卡车 heavy haulage lorry
重型载货车 heavy-duty freight car
重型载货车辆 heavy goods vehicle
重型载货汽车 heavy-duty truck; heavy lorry; heavy motor truck; heavy(type) truck
重型载重卡车 heavy-duty truck
重型载重汽车 heavy-duty truck
重型凿井凿岩机 heavy sinking drill
重型凿岩机 driller; heavy (rock) drill; large drill; Burleigh <需要两人抱住的>
重型轧车 high-pressure mangle
重型闸刀 guillotine
重型窄锄铲 bull tongue shovel
重型沼泽地缺口圆盘耙 bog harrow
重型振捣器 heavy-duty vibrator
重型振动格筛 heavy-duty vibrating grizzly
重型振动压路机 heavy vibratory roller
重型整治(工程) heavy-duty regulation work
重型支臂 heavy boom
重型支承筛管 heavy-duty supporting screen pipe
重型支柱 heavy section prop
重型纸模板 paper form
重型制动器 heavy-duty brake
重型制品 heavy weight rubber product
重型制榫机 heavy tenoning machine
重型中继海底电缆 heavy intermediate submarine cable
重型柱 heavy-load-carrying column; pilotis <使建筑物在地面上架空的>
重型铸铁井盖 heavy-duty cast-iron well cover
重型铸铁井盖座 heavy-duty cast-iron well cover seating
重型抓斗 heavy clamshell; heavy type grab; heavy-duty bucket; mass excavation bucket; severe-duty bucket; heavy-dug bucket
重型砖石墙 massive masonry wall
重型装备 weight equipment
重型装备抢救排 heavy recovery section
重型装车机 heavy loader
重型装卸设备 <起重能力大于50吨> heavy lift equipment
重型装载机 heavy-duty loader; supercharger
重型自(动倾)卸车 heavy tipper
重型自卸车身 severe-duty dump body
重型自卸汽车 heavy-duty dump truck; heavy lorry; heavy type truck
重型自卸载货汽车 heavy dumper
重型钻床 heavy-duty drill; heavy-duty drilling machine
重型钻杆 heavy steel drilling rod; stem
重型钻机 heavy-duty drill; heavy rock drill
重型钻塔 high-tensile steel derrick
重型作业工作台 heavy-duty bench
重悬浮液 dense medium; heavy-medium
重悬浮液再生 heavy-density recovery
重选 gravity treatment; reelect
重选浮选技术 gravity-flo(a)tation technique
重选呼叫路由 call rerouting
重选机 gravity flow concentrator
重选矿法 gravitational treatment
重选路由 rerouting
重选尾矿 gravity tailings
重选摇床 gravity table
重选装置 gravity concentration apparatus
重选作业线 reclassifier line
重镟轮箍 returning the tyre
重循环油 heavy cycle oil
重压 dead milling; oppress; repress(ing); weigh
重压板 loading slab; loading coat <地下室抗水压地板>

重压盖 heavy cap
重压紧配合 heavy force fit
重压力 heavy pressure
重压入条料坯料 push-back blank
重压橡皮 dead milled rubber; dead-rolled rubber
重压再结晶 load recrystallization
重压轧制 dead rolling
重芽 double bud
重亚黏[粘]土 heavy(clayey)loam
重亚砂土 heavy mild sand; heavy sandy loam
重岩浆化作用 anamigmatism
重研磨 heavy cutting
重盐碱水 heavy saline-alkaline water
重盐土 strongly salined soil
重盐渍的 heavily salted
重氧 heavy oxygen
重要备件计划 important spare-parts program(me)
重要部分 part-and-parcel
重要部件 vital part
重要成就 breakthrough
重要城市防洪体系 flood-control systems for major cities
重要程度 degree of importance; degree of significance; significance level
重要抽样 selective sampling
重要的 critical; momentous; significant; weighty
重要的错报 material misrepresentation
重要地势 key terrain
重要地位 importance
重要地形 critical terrain
重要地震 significant earthquake
重要承包商 principal contractor
重要电路 vital circuit
重要电路继电器 vital circuit relay
重要发明 breakthrough
重要工业税 key industry duty
重要共振 important resonance
重要关键 important key
重要会议文献选辑 highlights
重要机具 work horse
重要结构 important structure
重要连接点 important juncture
重要零件 strength member
重要目标体系 <军事判读> component target system
重要农业基地 important agricultural base
重要排放源 major emitting facility
重要商品 staple
重要设备 essential facility
重要事件 milestone
重要试件 research class
重要特征 essential feature
重要天气 significant weather
重要文章 key paper
重要性 fundamentality; importance; materiality
重要性程度 level of significance
重要性抽样 importance sampling
重要性函数 importance function
重要性态 significant behavio(u)r
重要性系数 factor of importance; important factor
重要样品 research class
重要仪器 key instrument
重要因素 important factor; key factor
重要优先权(项目) high priority
重要优先项目 high priority project
重要原料 critical material
重要陨石 important meteorite
重要战略物资 critical material
重要值 importance value
重要字 key word
重液 dense liquid; gravity solution; heavy fluid; heavy liquid; heavy liquor; heavy solution
重液法 heavy liquid method
重液分离 heavy fluid separation
重液分离法 heavy-medium separation method; separation with heavy liquid
重液分选 heavy liquid concentration
重液浮选厂 heavy-media plant
重液浮选法 heavy-media operation; heavy-media separation; heavy-medium operation; heavy-medium separation
重液入口 heavy liquid inlet
重液体气泡室 heavy liquid bubble chamber
重液洗选机 heavy fluid washer
重液相 heavy liquid phase
重液悬浮法 heavy-medium method
重液选矿 heavy liquid separation
重液选矿法 heavy liquid process
重一型动力触探 heavy(1) dynamic-(al) sounding
重硬黏[粘]土 gault(clay)
重油 axle oil; black mineral oil; dead oil; fuel oil; heavy fuel; heavy naphtha; heavy oil; mas(o)ut[maz(o)ut]; naphtha residue; recued fuel oil; reduced oil; rock tar
重油泵 heavy oil pump; petroleum pump
重油齿轮泵 gear pump for heavy oil service
重油储罐 heavy oil service tank
重油船 dirty ship; foul vessel
重油点火嘴 oil warm-up torch
重油发动机 crude oil engine; heavy fuel burning engine; heavy fuel engine; heavy fuel power plant; heavy oil engine
重油分离机 heavy fuel oil clarifier
重油分离器 heavy fuel separator
重油分水机 heavy fuel oil purifier
重油供给泵 heavy fuel feeding pump
重油锅炉 heavy oil fired boiler
重油化油器 heavy oil carburet(t)or
重油机 heavy oil engine; oil engine
重油加热炉 heavy oil heater
重油裂化 heavy oil cracking
重油裂化原料 black stock
重油馏分 heavy oil fraction
重油轮 dirty ship
重油黏[粘]度 Furol viscosity
重油喷嘴 heavy oil burner
重油气化 heavy oil gasification
重油燃烧器 heavy oil burner
重油砂 heavy oil sand
重油设备 heavy fuel equipment
重油升压泵 heavy fuel booster pump
重油脱硫 heavy oil desulfurization
重油稳压装置 pressure accumulator
重油油船 dirty tanker
重油浴式空气滤清器 heavy-duty oil bath air cleaner
重油原油发动机 heavy and crude oil engine
重油运输 dirty oil trade
重油中沥青膏渣浮油 asphaltic residual oil
重油轴助油箱 heavy oil service tank
重油转子发动机 heavy fuel rotating combustion engine
重铀酸铵 ammonium diuranate
重铀酸钠 sodium diuranate; uranium yellow
重铀酸盐 diuranate; uranate
重有色金属 heavy non-ferrous metal
重釉色彩 overglaze colo(u)r
重淤泥 heavy sludge
重于 outbalance; out poise; outweigh; overbalance
重元素 heavy element
重原子法 heavy-atom method
重载 deadweight; over-freight
重载驳船停泊区 loaded barge storage
重载齿轮 heavy-duty gear
重载传动 heavy-duty drive; severe duty drive
重载的 heavily-loaded; heavy-duty; heavy-laden
重载的市郊客运业务 heavily-loaded commuter service
重载的市郊旅客列车运行 heavily-loaded commuter service
重载定点循环列生 heavy duty unitrain
重载钢丝绳 load rope
重载公路 <硬质路面,有3~4条以上车道,承担繁重交通的公路> heavy-duty road
重载轨道 heavily loaded track
重载荷 heavy load
重载货物列车 tonnage freight train
重载货物线路 heavy haul freight line
重载机械 heavy-duty machinery
重载夹盘 heavy-duty chuck
重载链 heavy chain
重载列车 heavy haul train; heavy weight train
重载轮胎 heavy-duty tire[tyre]
重载排水量吨位 loaded displacement tonnage
重载平板车 heavy-duty bogie; heavy-duty wagon
重载启动负载 heavy starting duty
重载气 heavy carrier gas
重载汽车运输 heavy wheel-load traffic
重载牵引 heavy drag
重载桥梁 <承担繁重交通的桥梁> heavy-duty bridge
重载绳 head rope
重载试验 heavy-duty test; severe duty test; severe test
重载铁路 heavily loaded line; heavy-duty traffic railroad; heavy haul railway
重载消化池 heavy digester; heavy-duty digester
重载型楼板 heavy-duty floor(ing)
重载型碎石路面 heavy duty macadam
重载圆锥式破碎机 heavy-duty cone crusher
重载运输 heavy haul
重载运输机 heavy-duty transport machine
重载运行速度 loaded running speed
重载轴承 heavy-duty bearing
重载柱 heavy-load-carrying column
重载状态 load condition
重摘心 heavy pinching
重质柏油 heavy tar; tar heavy oil
重质苯 heavy benzene
重质材料制成的混凝土 <防辐射用> heavy concrete
重质残燃渣料油 heavy residual fuel oil
重质残油 heavy residual stocks; heavy still bottoms
重质残油铺地 oil pavement
重质残渣 heavy residue
重质的 heavy-duty; high duty
重质灯油 mineral seal oil
重质粉土 heavy silt
重质混凝土 heavy weight concrete
重质货物 heavy freight
重质货物列车 tonnage freight train
重质货物列车编组 heavy train composition
重质集料制成的混凝土 <防辐射用> heavy concrete
重质碱式碳酸镁 heavy magnesium subcarbonate
重质焦油 heavy tar oil
重质焦油沥青 heavy tar
重质焦油馏出物 heavy tar distillate
重质蜡 heavy wax; phroparaffine
重质冷凝产物 heavy condensation products
重质沥青 heavy asphalt; heavy bitumen
重质沥青基原油 heavy asphaltic crude
重质沥青油 heavy asphaltic oil
重质沥青质残渣油 heavy asphaltic residues
重质沥青质原油 heavy crude asphalt petroleum
重质量的市场 quality market
重质馏分 heavy ends
重质煤焦油 heavy tar
重质煤馏油 creosote oil
重质煤油 mineral seal oil
重质黏[粘]土 heavy clay
重质黏[粘]土铺地瓷砖 heavy clay flooring tile
重质黏[粘]土瓦 heavy clay tile
重质黏[粘]土砖 heavy clay tile
重质(铺)路油 heavy road oil
重质汽油 heavy gasoline
重质燃料 heavy fuel; low-volatility fuel
重质燃烧油 heavy fuel oil
重质溶剂油 heavy mineral oil; heavy mineral spirit; heavy naphtha; heavy petroleum spirit; heavy solvent naphtha
重质乳液 <法国用于道路基层的乳化沥青> grave emulsion
重质润滑油原料 heavy lube stock
重质石脑油 heavy naphtha
重质石油 heavy crude; heavy oil
重质树脂油 heavy resin oil
重质碳氢化合物 heavy hydrocarbon
重质碳酸钙 calcium carbonate; ground-and-washed chalk; ground calcium carbonate; ground chalk; ground limestone; ground whiting; heavy calcium carbonate
重质碳酸镁 heavy magnesium carbonate
重质陶瓷 heavy ceramics
重质天然汽油 heavy natural gasoline
重质填料 high gravity filler
重质烃 heavy hydrocarbon
重质烃类气体 heavy hydrocarbon gas
重质烃油 heavy hydrocarbon oil
重质土 heavy soil
重质拖拉机煤油 heavy-end power distillate
重质瓦斯油 heavy gas oil
重质烷烃 heavy paraffinic hydrocarbon
重质尾部馏分 heavy tails
重质圬工 heavy masonry
重质污染物 heavy pollutant
重质循环油 heavy recycle stock
重质氧化镁 heavy-burned magnesia
重质液体 heavy liquid
重质液体残油 heavy liquid residuum
重质液体分离法 heavy liquid separation
重质液体蜡膏 heavy liquid petrolatum
重质有机垃圾 heavy organic muck
重质浴 heavy tar
重质原料 heavy charge; heavy feed stock
重质原油 heavy crude oil
重质中性油 heavy neutral oil
重质中性油料 heavy neutral stock
重质重油 reduced fuel oil
重质子 heavy particle
重质钻井泥浆 heavy weight drilling fluid
重质钻孔泥浆 heavy weight drilling

fluid
重质钻液 heavy weight drilling fluid
重/重百分比 percent by weight in weight
重重加强的 heavily reinforced
重周转距离 average loaded wagon kilometers in one complete turnaround of wagon
重轴 solid axle
重轴荷载 heavy axle load
重柱 <基础用> bearing pile
重柱式 super-columniation
重筑路基 resubgrade
重铸 double teem(ing);recast;rehandle;remint;remo(u)ld
重铸巴氏合金 rebabbit
重铸红铜 heavy cast red brass
重铸铅板 lead cast
重桩 king pile
重装料 heavy burden
重浊 highly turbid
重子 baryon
重钻泥 heavy mud

州

州财政 state finance

州产权登记法 Torrens title
州长 warden;governor <美>
州道 provincial highway; provincial trunk highway;state highway <美>; state road <美>;state route <美>
州道路线标志 <美> state-line sign
州道网 <美> state highway system
州道系统 state highway system;state road system
州的规划 provincial planning
州地图集 state atlas
州二级道路系统 <美> state secondary highway system
州二级系统 <美> state secondary system
州分等(货物) <美> State classification
州干道系统 state primary system
州工程师 <美> state engineer
州公路局 <美> State Highway Department;State Highway Division
州公路署 State Highway Agency
州公路系统 <美> state highway system
州计划 state planning
州际大港 world port
州际道路系统 <美> interstate highway system
州际的 interstate
州际公路 <美> interstate highway; interstate road;interway
州际公路运输 interway
州际公园 interstate park
州际管线 interstate line
州际货运 interstate trucking
州际基线 intercontinental base
州际交通 <美> interstate traffic
州际贸易 <美> interstate commerce
州际汽车货运 <美> interstate trucking
州际商务条例 <美> Interstate Commerce Act
州际商务委员会 <设在美国纽约,主管运输监督工作> Interstate Commerce Commission
州际通路交换网络设备 exchange network facility of interstate access
州际土地出售 interstate land sales
州际土地出售注册办公室 <美> Office of Interstate Land Sales Registration
州际网 intercontinental network
州际协定 <美> interstate compact
州际运价 <美> interstate rate
州际运输 interstate traffic
州间航空通信[讯]电台 interstate airways communications station
州界 <美> state line
州路 <美> state road
州路网 <美> state road system
州内公司 domestic corporation
州内汽车交通 <美> intrastate traffic
州内运价 <美> intrastate rate
州内运输 <美> intrastate traffic
州平面坐标网 state plane coordinate grid
州三级道路系统 <美> state tertiary highway system
州三级系统 <美> state tertiary system
州辖公路 <美> state highway
州县区划示意图 county diagram
州行政(管理)办公大楼 provincial administration building
州养公路 state-maintained highway
州一级道路系统 <美> state primary highway system
州一级系统 <美> state primary system
州营码头 state dock
州营铁路 state-owned railway
州有铁路 state-owned railway
州展览会 provincial exhibition
州政府 state government
州政府拥有的港口 <美> state-owned port
州直角坐标网 state rectangular coordinate grid
州字型 thirty-type
州最高法院 <美> Supreme Court

舟

舟皿 boat

舟桥 bateau bridge;boat bridge;pontoon bridge
舟桥筏 raft of pontoons
舟形反应器 boat reactor
舟形漏板 boat-shaped bushing
舟形褶皱 canoe fold
舟状槽 mulde
舟状向斜 canoe fold
舟状褶皱 canoe fold

周

周摆线 pericycloid

周包膜 peripheral envelope
周报 weekly report
周报表 weekly returns
周壁导坑 multiple drift; peripheric drift
周边 circumference; perimeter; periphery;skirt
周边保温 edge insulation
周边爆破 blasting of profiles;perimeter blasting;periphery blasting
周边泵 peripheral pump
周边标志机 peripheral-marking machine
周边采暖 <建筑物> perimeter heating
周边长度 perimeter;peripheral length
周边尺寸 peripheral dimension
周边出料 peripheral discharge
周边出水堰 peripheral weir
周边传动 peripheral drive
周边传动式浓缩机 traction thickener
周边瓷砖 rounded edge tile
周边挡板 toe bead;toe board
周边挡土墙 skirt retaining wall
周边道路 perimeter road
周边的 circumferential;peripheral
周边地带 perimeter zone
周边电路 peripheral circuit
周边断层 circumferential fault
周边对接圈 circumferential butt joint ring
周边泛水 perimeter flashing
周边封口 perimeter seal
周边缝 peripheral joint;round joint
周边缝排料 peripheral-slot discharge
周边负荷 perimeter load
周边供暖 board heating;border heating;perimeter heating
周边供暖系统 perimeter heating system
周边灌浆 edge grouting;perimeter grouting;periphery grouting
周边光线 marginal ray
周边滚花的纹饰 knulling
周边焊 boxing
周边荷载 peripheral loading
周边加强 ring stiffening
周边加强的 ring-stiffened
周边夹紧的平板 clamped edges plate
周边间隙 peripheral clearance
周边减光 limb darkening
周边接缝 perimetric joint
周边结构系统 perimeter structural system
周边界影响 perimeter effect
周边进水沉淀 peripheral feed settling tank;rim-feed sedimentation basin
周边均布压力 uniform peripheral pressure
周边空气管 peripheral air ducting
周边空隙内灌浆 grouting of annular space
周边孔 rim holes
周边框架 perimeter frame
周边拉杆 peripheral bar
周边拉力 ring tension
周边拉应力 circumferential tensile stress
周边力 peripheral force
周边连接 perimeter connection
周边联系梁 peripheral tie beam
周边链 peripheral chain system
周边梁 perimeter beam
周边流 peripheral current;peripheral flow <圆池周边流>
周边密焊元件 element welded tight all around
周边排矿式磨碎机 screen-type mill
周边排水 margin drainage
周边排水系统 <建筑物> perimeter drainage system
周边炮孔 circuit hole;contour hole; peripheral hole
周边炮眼 contour hole; peripheral hole;rim holes;trimmer
周边喷射 peripheral jet
周边喷射气垫艇 peripheral jet craft
周边平面 peripheral plane
周边砌块 peripheral block
周边墙 perimeter wall; peripheral wall
周边墙衬砌 perimeter wall lining
周边墙构件 perimeter wall(building) component
周边墙施工 perimeter wall construction
周边墙饰面 perimeter wall facing
周边球齿 ga(u)ge button
周边区 perimeter zone
周边驱动型的浓缩池 circumferential driven thickener
周边取暖系统 perimeter heating system
周边绕流 peripheral flow
周边热风供暖系统 perimeter warm air heating system
周边射流气垫船 peripheral jet hovercraft
周边式 perimetric pattern
周边式布置 enclosure plan
周边式系统 perimeter system
周边式压缩机 peripheral compressor
周边视野检查法 perimetry
周边受拉伸的混凝土柱 perimeter tensioned concrete column
周边水深 side water depth
周边速度 circumferential velocity;peripheral speed;peripheral velocity; peripheric velocity
周边挑出 perimeter bracket
周边停车场 perimeter car park
周边投料 peripheral feed
周边突伸 perimeter bracket
周边围堤 peripheral dike[dyke]
周边位置 circumferential position
周边圬工墙 perimeter masonry wall
周边圬工墙衬砌 perimeter masonry wall lining
周边圬工墙贴面 perimeter masonry wall facing
周边圬工墙柱 perimeter masonry wall column
周边铣齿 peripheral teeth
周边系统 perimeter system;peripheral system
周边下水系统 circumferential drainage system
周边效率 circumferential efficiency
周边效应 border effect
周边卸料 peripheral discharge
周边卸料带筛球磨机 screen discharge ball mill
周边卸料孔 peripheral discharge opening
周边卸料碾磨机 rim discharge mill
周边卸料筛 peripheral discharge screen
周边性 peripherality
周边压力 circumferential pressure; peripheral pressure;periphery pressure
周边眼 perimeter hole;peripheral hole; periphery hole;side hole
周边堰 peripheral weir
周边抑制 peripheral inhibition
周边应力 peripheral stress;circumferential stress
周边影响 perimeter effect
周边诱导系统 perimeter induction system
周边预加应力的混凝土柱 perimeter prestressed concrete column
周边张力 ring tension
周边照明 boundary light
周边支撑 drum support; peripheral mount support
周边支承的 edge supported
周边柱 perimeter column
周边砖石砌合技术 perimeter bonding technique
周边转动 peripheral drive
周边钻探 perimeter drilling
周变应力比 cyclic(al) stress ratio
周波 cycle;periodicity
周波表 cycle counter
周波跳跃 cycle skip(ping)
周波重合 cycle matching
周层 pericline;perisphere
周查比例尺 scale of investigation
周查区面积 area of investigation region
周长 circumference;girt(h);perimeter
周程 trip
周丛生物 periphyton
周丛藻类 periphyton
周得来 <一种木材防腐剂,对病虫害

有效 > Jodelite
周/分 cycles per minute
周幅关系 cycle-amplitude relation
周工资 weekly wage
周供量水罐 week-supply tank
周供量调节水库 week-supply reservoir
周供量蓄水池 week-supply reservoir
周供量油罐 week-supply tank
周光关系 period-luminosity relation
周光滤光器 ambient light filter
周光曲线 period-luminosity curve
周光色关系 period-luminosity-colo(u)r relation
周光照明 ambient light illumination
周护沟 <防止基础受冲刷> perimeter trench
周剪力 perimeter shear
周检制 weekly check
周交通形式 <显示连续七天内每天的交通量> weekly traffic pattern
周角【数】round angle;perigon <360°>
周节 circumferential pitch; circular pitch
周节齿距【机】circular pitch
周节累积误差 accumulated pitch error; accumulative pitch error; cumulative pitch error; pitch accumulative error
周节误差 circular pitch error
周界 circumference; perimeter
周界保安 alarming
周界灌浆 containment grouting
周界剪 perimeter shear
周界剪力 perimeter shear
周界连接 perimetric joint
周界面积比 perimeter(over)area ratio
周界线 periphery
周界限制灌浆 <即先在外周用低压灌浆,使周界具有一定的强度,以便在内部进行高压灌浆> containment grouting
周界效应 perimeter effect
周界影响 perimeter effect
周径关系 period-radius relation
周颗粒 peripheral granule
周口店沉积【地】Zhoukoudian deposit
周口店灰岩 Zhoukoudian limestone
周口店阶【地】Zhoukoudian(stage)
周口店期【地】Zhoukoudian period
周历拨爪 day finger
周历计数器 day counter
周历快速调整 day quick setting
周历盘 day disc; day indicator; day roller
周历针 day hand
周裂的 circumscissile
周领料单 weekly order
周流 circumfluence
周率 frequency
周率表 frequency meter
周每秒 cycles per second
周密调查 through investigation
周密分析 close analysis
周密计划 well-conceived plan
周密计划的 well-planned
周密设计 deliberate design
周密踏勘 deliberate reconnaissance
周密选线 deliberate reconnaissance
周/秒 cycles per second; period per second
周末车次 weekend schedule
周末交通 weekend traffic
周末客票 weekend ticket
周末列车时刻表 weekend schedule
周末旅游车 recreational vehicle; recvehicle
周末停止运转 weekend shutdown
周末休假住宅 weekend house
周末休假住宅区 weekend house zone
周内日变化 day-of-week variation
周年变化 annual change; annual variation
周年变化率 annual rate of change
周年变异 annuation
周年波动 annual wave
周年不等性 <由气象条件造成的潮汐和水位的季节性变化> annual inequality
周年差 annual equation
周年磁变化 annual magnetic change
周年的 annual
周年等磁变线 isopor; isoporic line
周年光行差 annual aberration
周年纪念展览会 jubilee exhibition
周年减量 annual decrease
周年偏差 annual variation
周年热波动 annual thermal wave
周年视差 annual parallax; heliocentric parallax; stellar parallax
周年视运动 annual apparent motion
周年岁差 annual precession
周年增加 annual increase
周年值 annual value
周年钟 anniversary clock; four-hundred-day clock
周平均交通量 weekly average traffic volume
周平均日交通量 week average daily traffic
周期 complete alternation; cycle period; loop cycle; period; periodic(al) time; time period
周期摆动 periodic(al) oscillation; period oscillation
周期保护停堆 period shut-down
周期比 period ratio
周期边界条件 periodic(al) boundary condition
周期变幅关系 period-amplitude relation
周期变化 aspection; period change; periodic(al) variation
周期变换 periodic(al) transformation
周期变距操纵 cyclic(al) pitch control
周期变距螺纹 drunken thread
周期变量 periodic(al) variable
周期变形 cyclomorphosis
周期表 periodic(al) chart; periodic(al) table <化学元素>
周期表主族 main group
周期表属 group in periodic(al) table
周期波 periodic(al) wave
周期波谱 period spectrum
周期不规则性 cyclic(al) irregularity
周期步长逻辑 cycle step logic
周期步长同步脉冲 cycle step synchronizing pulses
周期操作 cycle operation
周期测量计 period meter
周期测量试验 period run
周期测量仪 period meter
周期差 periodic(al) inequality
周期场 cyclic(al) field
周期场聚焦 periodic(al) field focusing
周期抽样 periodic(al) sampling
周期磁导率 cyclic(al) permeability
周期存货 cycle inventory
周期单剪试验 cyclic(al) simple shear test
周期的 cyclic(al); periodic
周期的模 modulus of periodicity
周期的期刊 periodic
周期的原始平行四边形 primitive parallelogram of period
周期地 periodically
周期点 periodic(al) point
周期电磁阀 cycling solenoid valve
周期电势 periodic(al) potential
周期顶板压力 periodic(al) roof pressure
周期顶极 cycle climax
周期定律 law of period; law of periodicity
周期定时继电器 relay cycle timer
周期定值 period demand
周期动量 momentum of cyclic(al) motion
周期动作 periodic(al) motion
周期动作开关 periodically operated switch
周期断流器 rheotome
周期断面 die rolled section
周期断面钢材 deformed reinforcing bar; deformed steel bar; period section steel
周期断面型材轧机 die rolling mill
周期断面型钢 periodic(al) section
周期断面型钢轧机 reinforcing bar mill
周期断面轧槽 deforming groove
周期断面轧制 deformed rolling; die rolling; periodic(al) rolling
周期反向电流 <电镀的> periodic(al) reverse current
周期反应式活性污泥法 sequencing batch reactor
周期方程 periodic(al) equation
周期方向图 periodic(al) pattern
周期分辨率 period resolution
周期分布薄膜波导 periodically distributed thin-film waveguide
周期分布曲线 periodic(al) distribution curve
周期分类法 periodic(al) classification
周期分配 cycle split
周期分配作 cycle division(split)
周期风 periodic(al) wind
周期负载 periodic(al) duty
周期复原 cyclic(al) recovery
周期干燥器 intermittent dryer[drier]
周期干燥窑 compartment kiln
周期格子 period parallelogram
周期工作的 cycling
周期工作的电磁阀 cycling solenoid valve
周期工作方式 periodic(al) duty
周期共用存储器 cycle-shared memory
周期关系 period-luminosity relation
周期轨道 periodic(al) orbit
周期过程 cycle event; cyclic(al) process
周期函数 periodic(al) function
周期和绿信比型 cycle and split pattern
周期荷载试验 cyclic(al) load test
周期荷载作用下应力强度范围 stress intensity range under cyclical loading
周期化 periodization
周期环 weekly cycle
周期换向 periodic(al) reversal
周期回升 cyclic(al) recovery
周期基准起动控制 period-and-level start-up control
周期激振 periodic(al) excitation
周期激振反应 response to periodic(al) excitation
周期计 cycler; cyclometer; period meter
周期计量器 cycle(rate) counter
周期计时器 cycle timer
周期计数器 cycle(rate) counter
周期计算器 period counter
周期加荷单剪试验 cyclic(al) simple shear test
周期加荷三轴试验 cyclic(al) triaxial test
周期加荷至断裂 cycles to failure
周期加热试验 cyclic(al) heat test
周期加热窑 periodic(al) kiln
周期加载三轴试验 cyclic(al) triaxial test
周期加载试验 cyclic(al) test
周期间线 interperiodic(al) line
周期检查 cyclic(al) check; periodic(al) inspection
周期检查工作卡 period inspection work card
周期检验 cycle check
周期交通能量概率设计 cycle capacity probability design
周期解 periodic(al) solution
周期控制 periodic(al) control
周期库容 cyclic(al) storage
周期亏耗 cyclic(al) depletion
周期来压 periodic(al) coming pressure
周期类 periodic(al) group
周期离子交换 batch ion exchange
周期力 cyclic(al) force
周期力的力幅 amplitude of periodic force
周期历时 duration of cycle; periodic(al) duration
周期连锁 cycle load(ing); cycle locking
周期链 periodic(al) chain
周期量 harmonic quantity; periodic(al) quantity
周期流 periodic(al) flow
周期流量图 cyclic(al) flow profile
周期律 law of periodicity; periodic(al) law
周期螺旋 periodic(al) spiral
周期脉冲列 periodic(al) pulse train
周期脉冲重复时间标记 cycle repeat timer
周期密度关系 period-density relation
周期模型 periodic(al) model
周期挪用 cycle stealing
周期挪用方式 cycle steal mode
周期排水 periodic(al) drainage
周期排污 periodic(al) drainage
周期疲劳应力 cycle fatigue stress; cyclic(al) fatigue stress
周期偏心率关系 period-eccentricity relation
周期漂移 periodic(al) drift
周期频率特性 period-frequency characteristic
周期频率调制 periodic(al) frequency modulation
周期平均密度关系 period-mean density relation
周期迁移 periodic(al) migration
周期窃用 cycle stealing
周期区段 period range
周期曲面 periodic(al) surface
周期曲线 cyclic(al) curve
周期曲线图 periodogram
周期泉 periodic(al) spring
周期热 periodic(al) fever
周期上升波 periodic(al) up-wave
周期上限 up-time
周期摄动 periodic(al) perturbation
周期生产率 cycle output; cycle productivity
周期生长量 periodic(al) increment
周期时长 cycle length time
周期时长变换 cycle length change
周期时间 cycle length; cycling time; memory cycle; periodic(al) time; cycle time
周期式 periodic
周期式拌和机 batch mixer
周期式带钢轧机 Kessler mill
周期式烘干炉 batch-type drying oven
周期式灰浆搅拌机 batch mortar mixer
周期式混砂机 batch sand mixer

周期式搅拌机 batch mixer
周期式滤清器 discontinuous filter
周期式四摇臂酸洗机 four-arm pickler
周期式退火炉 periodic(al) annealing furnace
周期式轧管法 pilgrim rolling process
周期式轧管机 rotary forging mill
周期式轧机 intermittent-acting mill
周期式制冰机 cyclic(al) maker
周期式作业炉 periodic(al) furnace
周期事件 cycle event
周期势 periodic(al) potential
周期试验 periodic(al) test
周期输送 cyclic(al) feeding
周期数 interval hours;period frequency of cycle
周期衰化 cyclic(al) degradation
周期衰减 periodic(al) attenuation
周期水流 periodic(al) current
周期随机过程 periodic(al) random process
周期索引 periodic(al) key
周期替续器 cycle timer
周期条件 periodicity condition
周期调节 period control;periodic(al) regulation;period regulation
周期调节器 period regulator
周期同步 cycle load(ing);cycle locking
周期同步法 periodic(al) resynchronization
周期突变 period discontinuity
周期图 cycle diagram;periodogram
周期图表 cyclogram
周期图分析 periodogram analysis
周期图示法 cyclegraph technique
周期图仪 periodograph
周期退火 cyclic(al) annealing
周期网 period parallelogram
周期维修 cycle maintenance
周期涡流 periodic(al) vortex
周期误差 circular error;cyclic(al) error;periodic(al) error
周期系(统) cyclic(al) system;periodic(al) system
周期下降波 period down-wave
周期现象 periodic(al) phenomenon
周期线 periodic(al) line
周期相量 phase of periodic(al) quantity
周期项 cyclic(al) term;periodic(al) term
周期小计 total cycle
周期谐振 periodic(al) resonance
周期信号 interval signal;time-cycle signal
周期信用证 periodic(al) credit
周期型 batch-type;periodic(al) type;periodism
周期性 cyclicity;intermittence;periodicity;periodism
周期性暴发 periodic(al) outbreak
周期性变动 cyclic(al) variation;periodic(al) variation;rhythm
周期性变负荷 cycling load
周期性变化 cyclic(al) variation;periodic(al) change;periodic(al) variation
周期性变形 cyclic(al) deformation
周期性拨款 recurrent appropriation
周期性波动 cyclic(al) change;cyclic(al) fluctuation;cyclic(al) oscillation;cyclic(al) swing;periodic(al) fluctuation;periodic(al) oscillation;periodic(al) undulation
周期性不稳定水流 periodic(al) unsteady flow
周期性不稳定性 periodic(al) instability
周期性部分 periodic(al) component
周期性潮流常数 harmonic current constant
周期性沉积 cyclic(al) deposit
周期性沉积作用【地】 cyclic(al) sedimentation
周期性沉降 periodic(al) settling
周期性成分 periodicity component
周期性冲淤 periodic(al) silting and scouring
周期性传染 cyclic(al) infection
周期性传热 periodic(al) heat transfer
周期性搓板 rhythmic(al) corrugation
周期性导磁率 cyclic(al) permeability
周期性的 cyclic(al);cycling;intermittent;periodic
周期性的波状起伏 <一种路面缺陷> periodic(al) undulation
周期性灯光 rhythmic(al) light
周期性电磁波 periodic(al) electromagnetic waveguide
周期性电动势 periodic(al) electromotive force
周期性电流 periodic(al) current
周期性电流回路 periodic(al) circuit
周期性动作 cycling
周期性读出 cyclic(al) readout
周期性断续器 periodic(al) interrupter
周期性发病 cyclic(al) disease development
周期性发生的故障 recurrent failure
周期性发作 periodic(al) attack;periodic(al) seizure
周期性泛滥 periodic(al) flood
周期性放电 alternating discharge;periodic(al) discharge
周期性放牧 periodic(al) grazing
周期性分量 periodic(al) component
周期性峰荷 periodic(al) peak load
周期性浮游生物 periodic(al) plankton
周期性负荷 load cycling;periodic(al) load
周期性负载 cyclic(al) load(ing);repeated load(ing);periodic(al) duty;periodic(al) load
周期性负载额定强度 periodic(al) rating
周期性负载工作能力 periodic(al) rating
周期性复查 periodic(al) resurvey
周期性干旱 periodic(al) drought
周期性干扰力 periodic(al) interfering force
周期性感染 cyclic(al) infection
周期性工作(方法) periodic(al) duty;cyclic(al) duty;periodic(al) operating;periodic(al) work
周期性过程 batch-(like) process;periodic(al) process
周期性海流 periodic(al) current
周期性耗竭 <地下水的> cyclic(al) depletion
周期性河床 oshana
周期性河床变化 cyclic(al) river-bed change;cyclic(al) river-bed variation
周期性荷载 cyclic(al) load(ing);periodic(al) load;load cycling;oscillator load
周期性荷载频率 frequency of cyclic(al) loading;frequency of periodic(al) load
周期性洪泛区 periodically flooded area
周期性洪水 periodic(al) flood
周期性换向 periodic(al) reversal;periodic(al) reverse
周期性活动 periodic(al) movement
周期性活性污泥系统 periodic(al) activated sludge system
周期性火山 periodic(al) volcano
周期性积涝的 periodically waterlogged
周期性积水的 periodically waterlogged
周期性继续进给 jump feed
周期性加荷 cycle load(ing);cyclic(al) load(ing)
周期性间歇(喷)泉 standing geyser
周期性检测 periodic(al) testing
周期性检查 periodic(al) inspection
周期性检定 periodic(al) verification
周期性检修 cycled recondition
周期性检验 periodic(al) survey
周期性交变荷载 periodic(al) alternate load
周期性结构 periodic(al) structure
周期性结果 periodicity fruiting
周期性进坞 cyclic(al) docking;periodic(al) docking
周期性经济复苏 cyclic(al) economic recovery
周期性经济危机 cyclic(al) economics crises;periodic(al) economic crises
周期性空蚀 periodic(al) cavity
周期性空穴 periodic(al) cavity
周期性库存系统 periodic(al) inventory system
周期性馈送 cyclic(al) feeding
周期性流 periodic(al) current
周期性流动 periodic(al) flow
周期性流水 periodic(al) current
周期性流行 periodic(al) recurrence
周期性脉冲 recurrent pulse
周期性脉冲群 periodic(al) pulse train
周期性偶极起伏 periodic(al) dipole fluctuation
周期性排水 periodic(al) drainage
周期性喷出的油井 belching well
周期性起伏 periodic(al) undulation
周期性气体升液器 periodic(al) gas lift
周期性气象因素 periodic(al) meteorologic(al) cause
周期性强度 periodic(al) intensity
周期性切向加荷 cyclically tangential loading;cyclic(al) tangential loading
周期性窃用 cycle stealing
周期性曲线 cyclic(al) curve
周期性趋向 cyclic(al) trend;periodic(al) tendency
周期性取样 periodic(al) sampling;intermittent sampling
周期性泉 periodic(al) spring
周期性燃烧过程 periodic(al) combustion process
周期性入射波 periodic(al) incident wave
周期性闪光 cyclic(al) flashing
周期性上升和下降 cyclic(al) upturn and downturn
周期性失业 cyclic(al) unemployment
周期性受灼试验 intermittent-flame-exposure test
周期性输入 periodic(al) input
周期性输入输出 periodic(al) input/output
周期性衰减 periodic(al) damping
周期性水流 periodic(al) water flow
周期性随机过程 periodic(al) stochastic process
周期性天线 periodic(al) antenna
周期性调节 period regulation
周期性通货膨胀 cyclic(al) inflation
周期性退水 cyclic(al) depletion
周期性外力 external periodic(al) force
周期性维护 periodic(al) maintenance
周期性温度变化 cyclic(al) temperature change;periodic(al) temperature variation
周期性污染 periodic(al) pollution
周期性误差 periodic(al) error
周期性下沉 periodic(al) settling
周期性下降 cyclic(al) downturn
周期性消耗 cyclic(al) depletion
周期性消融 <冰川> cyclic(al) depletion
周期性型式 repeating pattern
周期性选择 cyclic(al) selection
周期性循环 periodic(al) return
周期性循环指示数字 cyclic(al) indication
周期性压缩荷载 periodic(al) compressive load
周期性窑 periodic(al) kiln
周期性应力 cyclic(al) stress;periodic(al) stress
周期性拥挤 <指高速公路,在预计的地区,一定时段内,有规律地发生的拥挤> recurrent congestion
周期性拥塞【交】 recurring congestion
周期性涌波 cyclic(al) surge
周期性涌浪 cyclic(al) surge
周期性运动 periodic(al) motion
周期性运输机械 intermittent movement machinery
周期性运行 cycling service;periodic(al) operation
周期性运转 periodic(al) operation
周期性载荷装载法 cycle load(ing);cyclic(al) load(ing)
周期性折流板厌氧反应器 periodic(al) anaerobic baffled reactor
周期性振荡 cyclic(al) oscillation;periodic(al) oscillation;intermittent oscillation
周期性振动 periodic(al) vibration
周期性植物流行病 cycle epiphytotic
周期性指标 cyclic(al) index
周期性质 periodic(al) nature;periodic(al) property
周期性中断 cycled interrupt;cyclic(al) interrupt;periodic(al) interrupt
周期性重复 rhythmic(al) succession
周期性转储 periodic(al) dumping
周期性转储分时 periodic(al) dumping time-sharing
周期性自动重合闸装置 periodic(al) automatic-reclosing equipment
周期性阻尼 periodic(al) damping
周期性作业 periodic(al) operating
周期序列 periodic(al) sequence
周期蓄水 cyclic(al) storage
周期旋涡 periodic(al) vortex;vortex periodic
周期选择 cycle selection
周期选择作用 periodic(al) selection
周期循环事件 periodic(al) recurrent event
周期压力强度 cyclic(al) pressure strength
周期延长 periodic(al) elongation
周期应变 cycle strain
周期应变软化 cyclic(al) strain softening
周期应力 intermittent stress
周期元素 period element
周期运动 cycle motion;cyclic(al) motion;cyclic(al) movement;periodic(al) movement
周期运算式模拟计算机 repetitive analog computer;repetitive type analog(ue) computer
周期运行 cycle operation;cyclic(al) operation
周期运行方式 periodic(al) duty
周期载荷试验 cyclic(al) load test
周期振荡 rectilinear oscillation

周期振荡器 period oscillator
周期振动 periodic(al) variation; periodic(al) vibration; period of vibration
周期振铃 periodic(al) ringing
周期征询 polling
周期值 periodic(al) quantity
周期指示器 period indicator
周期重复脉冲 repetition pulse; repetitive pulse
周期转储备量 current reserve
周期状态 periodic(al) state
周期自然冒落 periodic(al) spontaneous pressure
周期族 periodic(al) family
周期作业 discontinuous running; periodic(al) job; periodically running
周期作业窑 periodic(al) kiln
周碛 peripheral moraine
周切力 perimeter shear
周圈的 circumferential
周刃隙角 circumferential clearance; land clearance
周日变化 daily variation; day-of-week variation; diurnal change; diurnal variation
周日波动 daily wave; diurnal wave
周日不均衡系数 day-of-week unbalance factor
周日潮 diurnal tide
周日潮流 diurnal tidal current
周日潮汐订正 diurnal tidal correction
周日潮汐校正 diurnal tidal correction
周日潮汐循环 daily tidal cycle
周日秤动 diurnal libration
周日磁变 daily magnetic change; magnetic daily variation; magnetic diurnal variation
周日的 diurnal
周日地平视差 diurnal horizontal parallax
周日分潮 diurnal constituent
周日光行差 daily aberration; diurnal aberration
周日航程 day's run
周日航行定位工作 day's work
周日弧【天】 diurnal arc
周日加速度 diurnal acceleration
周日漂移 diurnal drift
周日平行圈 diurnal parallel circle
周日起潮力 diurnal tide-producing force
周日圈 diurnal circle
周日热波动 daily thermal wave; diurnal thermal wave
周日生潮力 diurnal force
周日视差 diurnal parallax
周日视动 apparent diurnal motion
周日视运动 diurnal apparent motion
周日天平动 diurnal libration
周日运动 diurnal motion
周日章动 diurnal nutation
周日钟速 diurnal clock rate
周日转动 diurnal rotation
周容许摄入量 acceptable weekly intake
周视描准镜 panoramic sight
周视图 panorama
周丝单孢锈菌属 <拉> Olivea
周斯坦固土法 Joosten process
周斯坦灌浆法 <一种双液灌浆法> Joosten process
周斯坦硅化加固法 Joosten process
周斯坦化学加固土壤法 Joosten process
周斯坦化学溶液凝固掘进法 Joosten chemical solidification process
周斯坦加固法 Joosten consolidation process
周斯坦式(溶液固土)掘进法 Joosten process
周斯坦双液化学加固法 <以水玻璃和氯化钙两种液体注入土壤> Joosten two fluid chemical consolidation process
周速(度) circular velocity; peripheral speed; circumferential velocity
周调节 weekly regulation
周调节池容量 weekly pondage
周调节电站 weekly storage plant
周调节水库 weekly storage
周围 ambience; around; atobit; periphery; precinct; round
周围材料 surrounding material
周围层 peripheral layer
周围大气 ambient atmosphere; surrounding atmosphere
周围带 peripheral zone
周围单元 surround unit
周围的 ambient; circumjacent; environmental; peripheral; surrounding
周围的基督教小礼拜堂 surrounding chapel
周围的流行性传染病 environmental epidemic
周围的气流区 <运动物体> periptery
周围的情况 surrounding
周围的人 entourage
周围的事物 surrounding
周围的土 <隧洞> surrounding material
周围的土地 entourage
周围的影响 effect of surroundings
周围地区 environs; peripheral area; surrounding area; surrounding region
周围地形 surrounding topography
周围断层 peripheral fault
周围房地产 adjacent property; adjoining property
周围封闭 periphery seal
周围杆 peripheral rod
周围光(线) ambient light
周围焊接 boxing
周围恒压 constant ambient pressure
周围环境 ambient environment; community air; entourage; environment; surrounding
周围环境条件 ambient condition
周围环境噪声 neighbo(u)rhood noise
周围环境状态 ambient condition
周围加冰 body icing
周围间隙 peripheral clearance
周围减压断裂试验 decreasing axial pressure fracture
周围脚手架 circumferential scaffolding
周围介质 ambient medium; environment(al) factor; surroundings
周围介质接触腐蚀 environmental media contact corrosion
周围介质密度 ambient density
周围介质条件下的试验 ambient test under surrounding media conditions
周围介质温度 ambient temperature; external temperature
周围介质状态 ambient condition
周围空气 ambient air
周围空气水平 ambient air level
周围空气质量 ambient air quality
周围空气质量标准 ambient air quality standard
周围列柱的朝廷 open peristyle court
周围列柱的法院 open peristyle court
周围列柱的庭院 open peristyle court
周围列柱的议会 open peristyle court
周围轮廓 surround profile
周围能见度 all-round visibility
周围排水系统 <建筑物> perimeter drainage system
周围剖面 surround profile
周围起爆 peripheral initiation
周围气流的 peripteral
周围(倾)斜层理 periclinal bedding
周围情况 ambient condition; condition
周围区 peripheral region
周围神经系统 peripheral nervous system
周围湿度 ambient moisture
周围视觉 indirect vision; peripheral vision
周围双层侧廊 two-stor(e)yed surrounding aisle
周围双层耳房 two-stor(e)yed surrounding aisle
周围水质 ambient water quality
周围条件 ambient condition
周围土层 surrounding soil stratum
周围土壤 surrounding soil
周围温度 ambient temperature; environmental temperature
周围温度传感器 external temperature sensing device
周围温度影响 <仪表的> external temperature influence
周围线 circumferential wire
周围镶边 surround trim
周围镶有玻璃的 glass-enclosed
周围向斜 rim syncline
周围削磨 circumdenudation
周围小道 path surround
周围性的 peripheral
周围修饰 surround trim
周围压力 all-round pressure; confined pressure; confining pressure; ambient pressure <三轴试验的>
周围压力灌浆 perimeter grouting
周围岩石 surrounding rock
周围液体压力 surrounding fluid pressure
周围因素 environmental factor
周围有裂缝的石材 stone accompanied by fracture
周围有气流的 peripteral
周围噪声 ambient noise
周围状况 ambient condition
周线 circumference; contour; periphery
周线递归编码 recursive contour coding
周线图 alignment chart
周线应力 hoop stress
周限增长率 finite rate of increase
周相 phase
周相常数 phase constant
周相的无规分布 random phase distribution
周相速度 phase velocity
周相延迟 phase-delay
周相移动 phase shift
周向缠绕 circumferential winding; hoop winding
周向钢筋 circumferential reinforcement
周向钢筋束 circumferential tendon
周向焊缝 girth seam
周向畸变 circumferential distortion
周向拉应力 circumferential tensile stress; tensile circumferential stress
周向连接 circumferential joint
周向拧紧力 circumferential make up
周向应力 circumferential stress; hoop stress; peripheral stress
周薪制 stab
周应力 hoop stress
周游 circular journey; circular tour; cruise
周游(客)票本 booklet of circular tour tickets
周游票 round trip
周余角 explement of angle
周缘 circumference; pericline; rim
周缘凹口盘 notched disk[disc]; notched plate
周缘出料磨 peripheral discharge mill
周缘的 circumferential
周缘地槽 peripheral geosyncline
周缘断层 peripheral fault
周缘分界线 peripheral divide
周缘盆地 peripheral basin
周缘速度 circumferential speed; circumferential velocity; tip velocity
周缘铣 peripheral milling
周缘系数 peripheral coefficient
周缘向斜 rim syncline
周缘向斜特征 feature of rim syncline
周缘压力 circumferential pressure
周缘支承 rim bearing
周月章动 monthly nutation
周张力 hoop tension
周值 revolution value; value of the turn
周柱廊式的 perstylar
周柱廊式建筑 perstylos
周柱廊中庭 <古罗马> peristylium [复 peristylia]
周柱式 peristyle
周柱式房屋 peristyle house
周柱式建筑 peristyle building
周柱式建筑物 peripteros
周柱式平顶 peristyle ceiling
周柱式天花板 peristyle ceiling
周柱式天棚 peristyle ceiling
周柱式庭院 peristyle court
周柱式中的柱 peristyle column
周转 complete revolution; complete rotation; recycling; revolution; revolve; turnover; turn round; turnaround
周转臂 epicyclic(al) arm
周转仓库 transfer shed; transit shed
周转差额 working balance
周转齿轮【机】 epicyclic(al) gear
周转齿轮传动 epicyclic(al) drive
周转齿轮传动线 epicyclic(al) driveline
周转齿轮系 epicyclic(al) train of gears
周转储备 turnover stock
周转传动 epicyclic(al) drive; epicyclic(al) gear drive
周转次数 times of turnover
周转贷款 revolving loan
周转的周期 cycle turnover
周转房 housing transition; relocation housing
周转股票 turnover stock
周转轨 inventory stock rails
周转过程 circular flow
周转回动齿轮组 epicyclic(al) reversing gear set
周转货币 vehicle currency
周转基金 circulation fund; revolving fund; working capital fund
周转减速装置 epicyclic(al) reduction gear
周转金 deposit(e) fund; revolving fund
周转晶体法 rotating crystal method
周转库存量 cycle inventory; live storage
周转离合器闸 epicyclic(al) clutch brake
周转量 turnover; volume of the circular flow; traffic mil(e)age <英里程>
周转量换算 conversion of freight mileage
周转率 rate of turnover; turnover; turnover rate; turnover ratio
周转率常数 turnover rate constant
周转率调查 turnover study
周转轮传动【机】 epicyclic(al) gear gearing
周转轮系 cyclic(al) gear train; cyclic(al) train; epicyclic(al) gearing; ep-

icyclic (al) train; epicycloidal or planetary gear train; epicyclic (al) gear train
周转能力 turnover capacity
周转期 turnaround
周转圈 epicycle
周转容量 live storage
周转时间 turning-round time; turnover time; turnaround time
周转使用的瓶子 returnable bottle
周转式起重机 full circle crane; full-rotating derrick
周转税 receipt tax; tax on turnover; turnover tax
周转速度 circular velocity; handling speed; rate of turnover; turnaround speed; turnaround velocity; velocity of turnover
周转速率 turnaround speed
周转天数 daylight of turnover
周转文件 turnaround documents
周转系统 turnaround system
周转现金 working cash
周转现款 petty cash
周转箱 pass box
周转信贷额度 revolving line of credit
周转信贷计划 revolving credit plan
周转信用(卡) revolving credit
周转性包销便利 revolving underwriting facility
周转循环 cycle of turnover
周转圆【数】epicycle
周转圆的 epicyclic(al)
周转圆运动 epicyclic(al) motion
周转账户 account turnover
周转专家 <能使未完工项目或无利可图的项目变为有利可图的人> turnaround specialist
周转准备金 working capital and reserve fund
周转资本 working capital; working fund
周转资金 circulating capital; circulating fund; operating fund; revolving fund; rotary fund; turned-over capital; working capital; revolving fund; working fund
周转资金账 revolving funds account
周转资料 turnaround documents
周转总成 replacement unit; reversible unit

洲 continent

洲地图 continental map
洲际传输 oversea(s) transmission
洲际传送 intercontinental transmission
洲际的 intercontinental
洲际电路 intercontinental circuit
洲际航空港 intercontinental airport
洲际航空站 intercontinental airport
洲际基线 intercontinental baseline
洲际检疫 interstate quarantine
洲际经转交换台 intercontinental transit exchange
洲际经转接电路 intercontinental transit circuit
洲际距离 global range
洲际连接 intercontinental connection
洲际通信[讯] transcontinental communication
洲际同步 intercontinental synchronization
洲际维护 intercontinental maintenance
洲际业务 intercontinental service
洲际直达货物 intercontinental through-shipment
洲际中继 intercontinental trunking
洲头分汊浮标 bifurcation buoy
洲头顺坝 longitudinal dike at head; training dike at head of central bar
洲尾 tail of central bar
洲尾分汊浮标 junction buoy
洲尾顺坝 longitudinal dike at downstream end of island; longitudinal dike at end; training dike at tail of central bar

粥 样斑块【医】atheromatous plaque

粥状 mushly state
粥状冰 ice gruel

轴 pin; axis; shaft

轴安装误差 axle misalignment
轴鞍 axle saddle
轴摆度 shaft displacement; shaft runout
轴摆度过大 excessive shaft runout
轴棒 mandrel
轴棒弯曲试验 mandrel bend test
轴包板 coffin plate; shaft plate
轴包架 propeller shaft bossing; shaft bossing; shell bossing
轴包架整流罩 contra bossing
轴包套 bossing; shaft bossing
轴比 axial ratio; ellipticity
轴比测量法 method of axial ratio measurement
轴壁 axial wall
轴臂销 shaft arm pin
轴柄安全型吊钩 shank safety hook
轴柄轴承 hub bearing
轴材 axial wood
轴材料 shaft material
轴槽 axial trough
轴测(量)法 axonometry
轴测投影 axonometric(al) projection
轴测投影面 axonometric(al) projection plane
轴测投影三向图 axonometric(al) perspective
轴测投影透视图 axonometric(al) perspective
轴测投影图 drawing by axonometric projection; axonometric projection
轴测投影轴线 axonometric(al) projection axis
轴测透视图 axonometric(al) perspective; radial projection
轴测图 axonometric(al) drawing; isometric(al) view
轴测斜投影 oblique axonometry
轴叉 axle yoke
轴差应力 <三轴压缩试验最大主应力与最小主应力之差> deviator(ic) stress
轴长 axial length
轴车床 shaft lathe
轴衬 axle bush(ing); bearing backing; bearing bush(ing); bearing liner; bearing segment; boss; bouchon; bush; follower; journal bearing; shaft bushing; shaft liner; bush(ing)
轴衬拆卸器 bushing puller
轴衬滚轮链 bush roller chain
轴衬套 screw-shaft liner
轴衬销 bearing bushing pin
轴衬装置 gland packing
轴承 axis bearing; axle bearing; bearing; bearing axle; bearing bearer; box bearing; gudgeon block; mechanical bearing; pillow; shaft bearing
轴承安装工具 bearing installing tool
轴承暗销 bearing journal
轴承巴氏合金 bearing babbit; bearing brass
轴承白合金 bearing white metal
轴承板 shaft bearing plate
轴承包装油 petrolatum
轴承保持架 retainer
轴承报警器 bearing monitor
轴承杯 bearing cup
轴承备件 bearing replacement
轴承编号 bearing number
轴承标志 bearing designation
轴承表面防护油膜 tempered oil
轴承布置 bearing arrangement
轴承材料 bearing material
轴承槽磨床 groove grinder
轴承拆卸工具 bearing extractor
轴承拆卸器 bearing extractor; bearing puller; shaft bearing replacer
轴承拆装工具 bearing replacer
轴承长销 long dowel
轴承常数 bearing constant
轴承厂 bearing plant
轴承衬 bearing backing; bearing lining; shaft bearing liner
轴承衬层 lining of bearing
轴承衬合金 lining alloy
轴承衬里金属 lined bearing
轴承衬料 bearing bush(ing); bearing liner
轴承衬套 bearing insert
轴承衬支座 liner backing
轴承承窝【机】bearing socket
轴承冲出器 bearing driver
轴承垂球 <机械的> shaft plumb bob
轴承代号 bearing symbol
轴承挡盖 bearing housing
轴承挡圈 end ring; retaining ring for bearing
轴承的出口端 off-side of bearing
轴承的拉线找正 alignment of bearings by taut wire
轴承的内隔圈 bearing inner spacer
轴承的配合 bedding-in of bearings
轴承的下瓦 bottom brass
轴承的压入和取出 pressing in and pulling out of bearing
轴承低油压脱扣装置 bearing oil pressure trip device
轴承低油压遮断装置 low bearing oil pressure governor
轴承底座 shaft base
轴承电流 shaft current
轴承垫 bearing gasket; bearing pad
轴承垫片【机】bearing shim
轴承垫圈 bearing washer
轴承吊架 bearing hanger; bearing suspension
轴承端部间隙 shaft-end clearance
轴承端部漏油 bearing end leakage
轴承端部油压 bearing end pressure
轴承端盖 bearing ball cover
轴承短销 short dowel
轴承法兰(盘) bearing flange
轴承反力 bearing reaction; reaction of bearing
轴承反压力 bearing counter pressure
轴承反作用力 bearing reaction; reaction of bearing
轴承方位角 azimuth bearing angle
轴承防尘圈 bearing shield
轴承放松杠杆 bearing release lever
轴承粉料 bearing-type mix
轴承腐蚀试验 bearing corrosion test
轴承负荷 bearing load
轴承负载 bearing load
轴承盖 bearing cap; bearing cover; bearing keep; step cover
轴承盖垫圈 bearing cap gasket
轴承盖弹簧圈拆装工具 bearing cap snap ring replacer set
轴承盖油封拆卸器 bearing cap oil seal remover
轴承钢 <钢结构> bearing steel
轴承钢滚珠 bearing ball
轴承钢珠 bearing ball
轴承隔环 bearing spacer ring
轴承隔离圈 bearing spacer
轴承隔圈 bearing cage; bearing spacer
轴承工作面 bearing face
轴承工作系数 bearing coefficient
轴承公差 bearing slackness
轴承沟道磨床 groove grinder
轴承构造 bearing formation
轴承毂 bearing hub
轴承固定板 bearing setting plate
轴承固定环 bearing race snap
轴承刮刀 bearing scraper
轴承管 bushed bearing
轴承滚道 ball track; bearing race; coulisse; rolling race
轴承滚道磨床 race grinder; raceway grinder; raceway grinding machine
轴承滚针 bearing needle; needle roller
轴承滚珠 bearing ball
轴承滚柱 bearing roller
轴承滚子 bearing roller
轴承滚子夹圈 roller cage
轴承过热 bearing running hot
轴承过载 bearing overloading
轴承合金 bearing alloy; bearing metal; box metal; bush metal; Corronium; star alloy; white metal
轴承荷载 bearing load
轴承护圈 bearing retainer; bearing retaining ring; retainer band
轴承护油圈 ring-oiled bearing
轴承滑轮 bearing pulley
轴承环 bearing collar; bearing ring; guide track; neck collar; race; racer
轴承环形密封 bearing annular seal
轴承换装器 bearing replacer
轴承黄铜 bearing brass; journal brass
轴承黄铜合金 bearing brass; journal brass alloy
轴承回油 scavenge oil
轴承架 bearing bracket(stand); bearing carrier; bearing frame; bearing pedestal; foot step; headstock; pedestal bearing
轴承架绝缘 pedestal bearing insulation
轴承间隔圈 bearing spacer
轴承间距离 distance between bearings
轴承间隙 bearing clearance; bearing slackness
轴承间隙校验 bearing clearance check
轴承结构 bearing structure
轴承金属 bearing metal
轴承金属磨损 bearing metal wear
轴承颈 bearing neck
轴承净空 bearing slackness
轴承绝缘 bearing insulation
轴承抗磨伤特性 bearing non-scoring characteristics
轴承壳 bearing case; cartridge housing; shaft casing; shell bearing
轴承壳车孔机 shell bearing boring machine
轴承壳套 bearing shell
轴承空隙 bearing clearance
轴承拉拔器 bearing puller
轴承拉出器 bearing puller; bearing remover
轴承拉紧 bearing take up; tightening of bearing
轴承类型 bearing type

轴承冷却水 bearing cooling water
轴承冷却水门(煤水车)flood cock
轴承零件探伤机 detector for bearing parts
轴承笼 bearing cage
轴承螺帽 bearing nut
轴承螺母 bearing nut
轴承螺栓 bearing bolt
轴承迷宫密封 bearing labyrinth
轴承密封 bearing seal;seal of bearing
轴承密封填料 bearing packing
轴承面 bearing surface
轴承面剥落 bearing flaking
轴承面碎裂 bearing disintegration
轴承模数 bearing module
轴承摩擦 bearing friction
轴承摩擦损失 bearing friction loss
轴承摩擦阻力 bearing drag
轴承摩阻力 bearing resistance
轴承磨损 bearing wear
轴承磨损级别 grade of bearing wearing
轴承磨损模式 bearing wearing mode
轴承磨损速度 bearing wearing rate
轴承内孔 bearing bore
轴承内圈 bearing cone;bearing inner race;inner race
轴承内压痕 Brinell
轴承黏[粘]合剂 bearing bond
轴承配合 bearing fit
轴承疲劳强度极限 bearing fatigue point
轴承偏转限制器 bearing guard
轴承铅 babbit(t)
轴承铅青铜 plastic bronze
轴承腔 bearing bore
轴承强度 bearing strength
轴承青铜 bearing bronze
轴承圈 ball race
轴承容许荷载 allowable bearing load;safe bearing load;safe bearing power
轴承润滑 bearing lubrication
轴承润滑剂 bearing lubricant
轴承润滑油 bearing luboil;bearing oil
轴承润滑油管 bearing feed pipe;bearing oil pipe
轴承润滑脂 anti-friction grease
轴承上推挡圈 bearing disc[disk]
轴承上轴瓦 upper bearing
轴承烧熔 bearing burning out
轴承生锈 bearing rust
轴承失灵 bearing failure
轴承试验 bearing test
轴承寿命 bearing life
轴承损坏 bearing failure
轴承锁紧螺帽垫圈 bearing nut lock washer
轴承锁紧螺母 bearing lock nut
轴承锁圈 bearing locking collar
轴承台 bearing stand;pedestal
轴承弹簧 bearing spring
轴承套 bearing case;bearing cover;bearing holder;bearing housing;bearing sleeve
轴承套圈 bearing race;bearing ring
轴承套筒 bearing sleeve
轴承体 bearing body;bearing box;bearing housing
轴承调整 bearing adjustment
轴承铜衬 bearing brass
轴承投影面积 projected bearing area
轴承凸缘 bearing flange
轴承推力 bearing thrust
轴承托 bearing retainer
轴承瓦 bearing liner;bearing shoe
轴承外圈 bearing outer ring;cup;outer race
轴承外圈拔出工具 bearing cup puller
轴承外套 bearing jacket
轴承外座圈 bearing outer race;outer cage
轴承位置 position of bearings
轴承温度报警器 bearing temperature alarm
轴承稳管夹 bearing saddle
轴承窝 spindle socket
轴承系列 bearing series
轴承系数 bearing module
轴承下瓦 lower half bearing
轴承箱 bearing box;bearing housing;journal box
轴承销 bearing pin;pin bearing
轴承泄油面 bearing off side
轴承型号 bearing designation
轴承性能 bearing performance
轴承压紧螺母 bearing nut
轴承压力 bearing pressure
轴承压力的抛物线形分布 parabolic(al) bearing pressure distribution
轴承眼圈 dead eye
轴承咬死 bearing seizure
轴承应用 bearing application
轴承用滚珠 bearing ball
轴承用减磨金属 soft metal
轴承用韧性锡青铜 Tourun Leonard's metal
轴承油 bearing oil
轴承油泵 bearing oil pump
轴承油槽 bearing groove;bearing oil sump
轴承油封 bearing enclosure;bearing oil seal
轴承油孔 bearing oil hole
轴承油冷却器 bearing oil cooler
轴承油膜 bearing film
轴承油压降低继电器 low bearing oil pressure trip device
轴承油压降低跳闸试验 low bearing oil pressure trip test
轴承油压下降报警试验 low bearing oil pressure alarm test
轴承油脂 bearing grease
轴承与润滑油的接触面 bearing oil interface
轴承与轴面的配合性 bearing conformability
轴承预热 burning of bearing
轴承预先受载 bearing preloading
轴承预载指示器 bearing tension indicator
轴承允许荷载 bearing capacity
轴承毡圈 bearing felt
轴承罩 bearing cage;bearing guard;bearing housing;bearing lock
轴承罩套 bearing lock sleeve
轴承枕垫 pillow block
轴承振颤 bearing chatter
轴承支撑 bearing support
轴承支承面 bearing face;bearing surface
轴承支架 bearing bridge;bearing support;support of bearing;support stand with bearing
轴承支圈 bearing rim
轴承支柱 shaft column
轴承脂 bearing grease
轴承止环 bearing stop ring
轴承止推作用 bearing thrust
轴承中心 bearing center[centre]
轴承重挂白合金 relining of bearing
轴承重加润滑脂 bearing repacking
轴承轴线 bearing axis
轴承铸铁 bearing cast-iron
轴承装配 bearing assembly
轴承装配工具 Babbitting jig
轴承锥形内圈 bearing cone
轴承锥形外圈 bearing cone cup
轴承阻力 bearing resistance
轴承最终磨损量 final bearing wear
轴承座 adapter;bearing block;bearing box;bearing bracket;bearing cage;bearing carrier;bearing chock;bearing housing;bearing pedestal;bearing saddle;bearing seat;bearing support;beating pedestal;cage;chuck;end bracket;pillow
轴承座垫片 ball-bearing washer
轴承座盖 pedestal cap
轴承座架 bearing stool
轴承座孔 bearing saddle bore;bearing spigot
轴承座圈 bearing race
轴承座圈滚道 raceway track
轴承座圈拉出器 bearing race puller
轴承座套 bearing sleeve
轴承座套垫片 bearing sleeve gasket
轴承座套盖 bearing sleeve cover
轴承座套式密封 bearing sleeve seal
轴承座套销衬套 bearing sleeve pin bushing
轴承座位移指示器 bearing pedestal movement indicator
轴齿轮 shaft gear
轴传动 shaft drive
轴传动操舵系统 shaft steering system
轴传动的 driven off by shaft
轴传动发电机 shaft generator
轴传动回转台 shaft-driven rotary table
轴传动伞齿轮 axle drive bevel gear
轴传动移动起重机 shaft-driven travel(l)ing crane
轴带发电机 shaft generator
轴带发电机系统 shaft generator system
轴单位 axis unit distance
轴单位燃料消耗量 brake specific fuel consumption
轴挡 axle bumper
轴刀式茎秆切碎器 spindle-knife shredder
轴导 spindle guide
轴导承 shaft guide
轴道 axle track
轴的 axial;axle
轴的摆动 shaft throw
轴的摆度 shaft displacement;shaft run-out
轴的变换 transformation of axis
轴的传动装置 shaft drive
轴的单位长度扭转角 angle of twist per unit length of shaft
轴的对准 shaft alignment
轴的分隔距离 bay of a shaft
轴的棘轮 dog end
轴的径向跳动 shaft run-out
轴的离心旋动 centrifugal whirling of shaft
轴的临界转速 critical speed of shaft
轴的密封装置 shaft seal
轴的末端 the tip of the axis
轴的平衡 shaft balancing
轴的曲拐部分 crank portion of shaft
轴的驱动端 drive end
轴的润滑 O 形密封 lubricated O-ring shaft seal
轴的弹性挡圈 circlips for shaft
轴的延长端 extension of the shaft
轴的支架 shaft support
轴的中线 axis of spindle
轴点 axial point
轴电流 shaft current
轴电流保护(装置)shaft current(circulating) protection
轴电位 shaft potential
轴垫 axle pad
轴垫圈 shaft filler washer
轴吊架 axle suspension
轴定位 shaft locating
轴定心卡规 shaft alignment ga(u)ge
轴动机器脚踏车 shaft-drive motorcycle
轴动移动起重机 shaft-driven travel(l)ing crane
轴动支点 rolling fulcrum
轴端 axial-tag terminal;axle end;axle head;bar head;end axle;shaft end;shaft tip;stub
轴端编码盘 shaft encoder
轴端出力 shaft-end output
轴端传动<发电机>axle end drive
轴端浮动 end float
轴端功率 shaft-end output
轴端护板 gland retainer plate
轴端间隙 end play
轴端间隙装置 end play device
轴端漏泄 shaft leakage
轴端曲柄 side crank
轴端曲柄轴 side crank shaft
轴端曲柄装置 side crank arrangement
轴端套圈 end collar
轴端推力 end thrust
轴端压垫 shaft end shim
轴端移动 end travel
轴端游隙 end play
轴端余隙 end play;side play
轴端余隙器件 end play device
轴断裂面 axial fracture
轴断面 axial section
轴对称 axial symmetry;axisymmetry;rotational symmetry
轴对称变形 axially symmetric(al) deformation
轴对称变形条件 axisymmetric(al) deformation condition
轴对称的 axisymmetric(al);en axe;rotationally symmetrical
轴对称地震反应 axisymmetric(al) seismic response
轴对称地震响应 axisymmetric(al) seismic response
轴对称锻造 axisymmetric(al) forging
轴对称分析 axisymmetric(al) analysis
轴对称固结 axially symmetric(al) consolidation
轴对称荷载【物】axisymmetric(al) load
轴对称加速器 axially symmetric(al) accelerator
轴对称壳体 axisymmetric(al) shell
轴对称课题 axisymmetric(al) problem
轴对称扩大 axisymmetric(al) expansion
轴对称扩散 axially symmetric(al) diffusion
轴对称流动 axial flow;axially symmetric(al) flow;axisymmetric(al) flow
轴对称穹隆 dome of rotational symmetry
轴对称射流 axially symmetric(al) jet;axisymmetric(al) jet
轴对称式重力仪 axially symmetric(al) gravimeter
轴对称水流 axially symmetric(al) flow;axisymmetric(al) flow
轴对称体 axially symmetric(al) body;axisymmetric(al) solid
轴对称弯曲 axially symmetric(al) bending
轴对称紊动射流 axisymmetric(al) turbulent jet
轴对称物体 axisymmetric(al) body
轴对称行列式 axisymmetric(al) determinant
轴对称应力 axisymmetric(al) stress
轴对称应力分布 axially symmetric-

(al)stress distribution
轴对称应力条件 axisymmetric(al) stress condition
轴对称应力与变形 axisymmetric(al) stress and deformation
轴对称有限元法 axisymmetric(al) finite element method
轴对称圆顶 cupola
轴对称振动 axisymmetric(al) vibration
轴对正 shaft alignment
轴对轴模 concentric(al)die
轴轭 axle yoke
轴法兰 shaft flange
轴封 bearing seal; crankcase seal; gland seal; radial packing; radial seal(ing); shaft seal(ing)
轴封泵 shaft seal pump
轴封抽气系统 gland leak-off arrangement
轴封处数据 data at seal
轴封供汽调节器 gland steam regulator
轴封环 packing ring
轴封漏气 shaft-packing leakage
轴封排汽 leakage steam; leak-off steam
轴封润滑系统 shaft seal grease system
轴封水泵 seal water pump
轴封套 shaft gland
轴封体 shaft seal housing
轴封填料 shaft packing
轴封型式 type of shaft sealing
轴封蒸汽 gland steam
轴封装置 mechanical seal
轴腐病 spike rot
轴负重 weight per axle
轴盖 axle cap; journal lid; shaft cap; shaft-cup
轴盖簧 shaft-cup spring
轴干 axletree
轴杆 axle tree; axostyle; axostylus; coulisse; shaft lever
轴杆架 three-arm base; tripod for base measurement
轴杆头水准 tripod-head level(l)ing in base measurement
轴钢 axial steel; axle steel; shaft steel
轴梗 spindle
轴梗制动器 spindle brake
轴公里 axle-kilometer[kilometre]
轴公里运价 tariff per axes kilometer
轴功率 axial power; axle power; brake horsepower; brake power; input power; power-on shaft; shaft horsepower; shaft power
轴毂端支架 boss end bracket
轴箍 shaft clip
轴挂电动机 axle-hung motor
轴管 axle casing; axle tube; newel tube; shaft tube
轴光速 axial bundle
轴规 axle ga(u)ge
轴号 axle(serial) number; number of axles
轴荷 axle load(ing)
轴荷载 axle load(ing); axle loading force; load per axle; shaft loading
轴荷载分布 <列车运行中> traffic axle load distribution
轴荷载极限 axle weight limit
轴荷载天平 axle load scale
轴荷载调节装置 axle load adjusting device
轴荷载限度 axle weight limit
轴荷重秤 axle load scale
轴花键 external spline; male spline
轴滑轮 <止下推拉窗开关的> axle pulley
轴环 arbor collar; attachment clip; axle collar; axle ring; banded collar; collar; cylinder jaw; retainer collar; ruff; shaft collar
轴环端 collar end
轴环端轴承 collar end bearing
轴环机 shaft ring machine
轴环接合 collar joint
轴环接头 collar joint
轴环润滑轴承 collar oiled bearing
轴环套 collar bush
轴环支承 collar receiver
轴环注油 collar oiling
轴汇 axis congruence
轴级度 axial gradient
轴挤压 axial compression
轴计数 axle count
轴加工自动机 shaft machine
轴夹 axle clamp
轴架 distance bar; pedestal(body); pedestal frame; shaft hanger; steady bar
轴架导板 pedestal horn
轴架底座 pedestal base
轴架盖 pedestal cap
轴架横拉条 pedestal cross brace
轴架夹板 pedestal binder
轴架拉条 pedestal brace
轴架螺栓 pedestal bolt
轴架体 pedestal body
轴架系杆螺栓 pedestal tie bar bolt
轴架系紧螺栓 pedestal tie bolt
轴尖 pivot
轴尖式量规 point ga(u)ge
轴尖修理 repivoting
轴尖悬置式 strip-suspension type
轴尖轴承 pivot bearing
轴间 between centers; interaxis
轴间差速器 centre differential; interaxle differential
轴间的 interaxle
轴间角 angle between axes; axis angle; axial angle
轴间距(离) axle base; between centers; center distance; center-to-center; axle spacing
轴肩 fillet; shaft shoulder; shoulder of spindle
轴肩车床 shoulder turning lathe
轴肩车削 shoulder turning
轴肩挡圈 protecting collar; retaining ring for shaft shoulder; shaft collar; shoulder ring
轴肩端面 shoulder face
轴肩间隙 pivot clearance
轴肩有眼螺栓 shoulder eyebolt
轴检【铁】 axle-box lubrication check-up; repacking
轴键 axle key; shaft key; shaft spline
轴浆 axoplasm
轴浆流 axoplasmic flow
轴浆运输 axoplasmic transport
轴交叉 axial cross
轴交角 crossed axes angle
轴角 axial angle; shaft angle
轴角编码器 shaft position encoder
轴角模(拟)数(字)转换器 shaft position digitiser[digitizer]
轴矫直机 shaft straightener
轴铰链 pivot hinge
轴校正 shaft alignment
轴校直机 axle straightener
轴接点 shaft contact
轴节 boss; cardo
轴节间片 lorum
轴节理 axial joint
轴结构 axle construction
轴截面 axle section
轴晶体 axial crystal
轴颈 axle end; axle journal; axle stub end; bearing location; cross-rod; fulcrum pin; gudgeon journal; journal neck; journal of shaft; neck journal; neck of axle; neck of shaft; pivot journal; shaft collar; shaft neck; stub-end of axle
轴颈不等 non-conformity of the pivots
轴颈不一致 non-conformity of the pivots
轴颈车削及抛光车床 axle journal turning and burnishing lathe
轴颈衬套 journal bush
轴颈挡 journal stop
轴颈挡环 axle journal collar
轴颈负荷 journal load
轴颈工作面 working surface of the pivot
轴颈辊 journal roll
轴颈合金 journal metal
轴颈环 journal collar; neck collar
轴颈检测器 journal detector
轴颈警报热继电器 journal alarm thermal relay
轴颈警报信号继电器 journal alarm signal relay
轴颈密封垫 journal packing
轴颈摩擦 journal friction
轴颈磨床 axle journal grinder
轴颈抛光机 pivots polishing machine
轴颈圈 shaft collar
轴颈润滑油试块 neck grease briquette
轴颈润滑脂 journal compound
轴颈速度 journal velocity
轴颈套筒 journalling sleeve
轴颈填料 journal packing
轴颈填密 journal packing
轴颈铜衬 journal brass
轴颈头 axle journal head
轴颈误差 error of pivot
轴颈箱 journal box; outside axle-box
轴颈箱盖 journal box cover; journal box lid
轴颈箱盖铰链 journal box lid hinge
轴颈箱盖弹簧 journal box lid spring
轴颈箱簧座 journal box spring seat
轴颈箱密封装置 journal box seal
轴颈箱楔 axle-box wedge; journal box wedge
轴颈销 journal pin
轴颈信号继电器 journal signal relay
轴颈旋床 axle journal turning and burnishing lathe
轴颈压力 journal pressure
轴颈油 journal oil
轴颈支承 journal rest
轴颈支承的大轴 journal shaft
轴颈支承的磨机 trunnion-mounted mill
轴颈中心距 distance between acting center[centre]
轴颈轴承 bearing of journals; journal bearing; neck-journal bearing; bearing journal
轴颈轴承黄铜 journal bronze
轴颈轴承面 journal bearing surface
轴颈轴承套 journal bearing sleeve
轴颈阻力 journal resistance
轴径 diameter of axle; shaft diameter
轴距 axle base; axle distance; axle spacing; distance between shafts; spread of axles; track base; wheel-(axle) base <指车>; wheel-base of motor vehicle <指车>
轴距离长度 length of wheel base
轴距与轮距比 wheel-base to track ratio
轴锯箱 miter box; mitre-box
轴开关滑轮 axle pulley
轴开滑轮 axle pulley
轴开裂的 septifragal
轴壳 axle housing; shaft casing
轴壳十字头 axle housing trunnion
轴壳万向节 axle housing trunnion
轴孔 axle hole; hollow of shaft
轴孔观察设备 borescope
轴孔填密 axle packing
轴孔指示器 borescope
轴孔座 boss
轴扣【机】 thrust collar
轴类加工自动机 shaft machine
轴棱镜 axicon
轴力波 axial force wave
轴力矩 braking moment
轴力图 axial force diagram
轴连接 shaft coupling
轴联杆端 axle stub end
轴联励磁机 axle-driven exciter
轴梁 axial girder
轴料料斗 shaft hopper
轴列式 axle arrangement
轴领 axle collar; banded collar; bearing collar; shaft collar
轴流 axial blower; axial flow
轴流泵 axial-flow pump; propeller pump; radial-flow pump; shaft pump; straight-flow pump
轴流冲击式水轮机 axial-flow impulse turbine; axial-flow type wheel
轴流冲击式涡轮机 axial-flow impulse turbine
轴流单向扫气系 uniflow scavenging system
轴流定桨式水轮机 fixed blade propeller(type) turbine; fixed-pitch propeller turbine; propeller turbine
轴流定桨式转轮 fixed propeller type wheel; propeller
轴流反击式透平 axial-flow reaction turbine
轴流风 axial air
轴流风机 axial fan; axial-flow blower; axial-flow fan
轴流风机叶轮 axial-flow impeller
轴流风扇 axial-flow fan; propeller fan; tube-axial fan
轴流辐流泵 combined axial and radial flow pump
轴流鼓风机 axial-flow blower; axial fan
轴流火焰 axial flame
轴流螺旋桨 axial-flow propeller
轴流排气通风机 axial-flow exhaust fan
轴流排气装置 propeller release
轴流式 axial-flow type
轴流式背压汽轮机 axial back pressure turbine
轴流式泵 axial-flow pump; propeller pump; screw-impelled pump
轴流式冲动汽轮机 axial-flow impulse turbine
轴流式反转燃气轮机 axial-flow reversing gas turbine
轴流式风机 axial blower; screw ventilator; tube-axial fan
轴流式风机叶片 axial-flow vane
轴流式风扇 axial(-flow) fan; tube-axial fan
轴流式风扇转子 axial-flow fan rotor
轴流式高压风扇 axial-flow high-pressure fan
轴流式空气压缩机 axial-flow(type) air compressor
轴流式马达 axial-flow motor
轴流式暖风机 unit heater with axial fan
轴流式排气扇 axial-flow exhaust fan
轴流式喷气发动机 axial-flow jet engine

轴流式气体分离器 axial-flow gas separator
轴流式汽轮机 axial-flow steam turbine;axial-flow turbine
轴流式强制循环泵 axial forced circulating pump
轴流式燃气轮机 parallel flow engine
轴流式燃气透平 axial-flow gas turbine
轴流式扇风机 radial-flow fan;screw fan
轴流式水力发电机组 axial-flow hydroelectric(al) unit
轴流式水轮机 axial-flow hydraulic turbine; axial-flow turbine; axial turbine;axial water turbine;Kaplan turbine;parallel flow turbine;axial-flow wheel;axial wheel
轴流式通风机 axial-flow aeration machine;axial fan;axial-flow blower;axial-flow fan; axial-flow ventilating fan;axial-flow ventilator;propeller fan
轴流式透平 axial-flow turbine;axial turbine;parallel flow turbine
轴流式透平机 axial-flow turbomachine
轴流式透平机械 axial turbomachinery
轴流式涡轮机 axial-flow turbine;parallel flow turbine
轴流式涡轮压缩机 axial-flow turbo compressor
轴流式涡轮液力变矩器 hydraulic torque converter with axial-flow runner
轴流式涡轮增压器 axial turbo-blower
轴流式压气机 axial-flow compressor;axi-compressor;axiradial compressor
轴流式压缩机 axiradial compressor;axial(-flow) compressor
轴流式叶轮 axial-flow impeller
轴流式叶片泵 propeller pump
轴流式油泵 axial-flow oil pump
轴流式增压器 axial-flow blower
轴流式转子 axial-flow rotor
轴流送风机 axial-flow fan
轴流通风机 aerofoil fan;axial-flow blow;propeller fan
轴流叶轮 axial wheel;propeller-type impeller
轴流转桨式水轮机 movable propeller turbine
轴隆区 culmination
轴律 axis law
轴率 axial rate;axial ratio
轴轮 arbor wheel
轴螺母 spindle nut
轴螺栓 axle bolt
轴螺旋弹簧 shaft helical spring
轴马力 brake horse power;power input rating;shaft horsepower
轴帽 axle cap
轴迷宫密封 shaft labyrinth
轴密封 shaft seal
轴密封垫圈 Simmer gasket
轴密封盖 gland retainer plate;shaft gland
轴密封环 Simmer ring
轴面 axial plane;axial region;axial surface;meridian plane <转轮的>
轴面波 axial wave
轴面产状 attitude of axial-plane
轴面的 pinacoidal
轴面固 meridianal section
轴面间距 axial plane separation
轴面解理 pinacoidal cleavage;pinacoidal joint
轴面劈理 axial plane cleavage
轴面刃倾角 axial tilt
轴面速度分布 meridianal velocity distribution
轴面体 axial plane;axial surface
轴面叶理 axial plane foliation
轴面褶皱 axial-surface fold
轴磨床 travel(l)ing wheel-heed roll grinder
轴逆变换 inverse transformation of axis
轴偶极矢量 axial dipole vector
轴刨 spokeshave
轴配合 shaft fit
轴配合间隙 clearance
轴坯 shaft blank
轴偏心度 shaft eccentricity
轴偏移 axle offset
轴平面 axial plane
轴器 axial organ;organum axiale
轴前的 preaxial
轴丘 axon hillock
轴曲线 axial curve
轴驱动 axle drive
轴驱动的 axle-driven;shaft-driven
轴驱动发电机 axle-driven dynamo;axle-driven generator;axle electric(al) generator
轴驱动发电机调节器 axle generator regulator
轴驱动频率发生器 axle-driven frequency generator
轴圈 blowout patch
轴热扫描装置 wheel thermo-scanner unit
轴润滑油 spindle oil
轴润滑油管 axle lubrication pipe
轴山 betwixt mountain
轴上的 epaxial
轴上荷载 load on axle
轴上许用荷载 allowable axle loading
轴上旋转调制盘 rotating on-axis reticle
轴梢 shaft tip
轴伸出 shaft extension
轴伸出壳(体)部分 tail shaft
轴伸贯流式水轮机 shaft extension type tubular turbine; S turbine; standard tube turbine;tubular water turbine with extended axis
轴身 axle body;axle middle;body of axle
轴矢量 axial vector
轴矢量耦合 axial vector coupling
轴式 axle arrangement; shaft type; wheel arrangement
轴式撑螺丝锥 spindle stay bolt tap
轴式过滤器 shaft filter
轴式混合器 shaft mixer
轴式混料机 shaft mixer
轴式搅拌机 shaft-type mixer
轴式旋流器 axial cyclone
轴视图 axonometric(al) drawing
轴枢螺栓 pivot bolt
轴枢瓦 pivoted shoe
轴输出 shaft output
轴输出功率 shaft output;shaft power
轴输入功率 shaft input
轴束 axial strand
轴数 number of axles
轴双面 sphenoid
轴隧 shaft trunk;shaft(tunnel) alley
轴隧出入口 tunnel opening
轴隧道 screw alley;tunnel alley
轴隧端室 tunnel recess
轴隧坑 tunnel well
轴隧肋骨 tunnel frame
轴隧平台 tunnel flat;tunnel platform
轴隧式船尾 <浅水船> tunnel form escape;tunnel form stern
轴隧太平洞 tunnel escape
轴隧通道 tunnel trunk
轴隧推进器 tunnel screw propeller
轴隧尾室 stuffing box recess;tunnel recess
轴隧泄水阱 tunnel well
轴隧应急出口 fireman's escape;tunnel escape
轴榫 pivot
轴索 axon
轴锁闭 shaft locking
轴锁紧 shaft locking
轴锁紧螺帽 spindle jam nut
轴锁紧螺母 axle lock nut
轴台 pillow block;plumber(block)
轴台盖 pillow block cover;plummer block cover
轴台盖衬 pillow block cover gasket
轴台螺栓 plumber block bolt
轴台通气管 pillow block breather
轴台凸轮 pillow block flange
轴台轴承杯 pillow block bearing cup
轴套 axle-box;axle casing;axle housing;axle sleeve;bearing bush;bearing liner;box coupling;flange bearing;muff;shaft casing;shaft sleeve;spindle sleeve;shaft installing
轴套管 pipe spreader
轴套螺母 sleeve nut
轴套帽 axle housing cap
轴套式连接 male connection
轴套轴承 hub bearing
轴调节器 shaft-type governor
轴调速器 shaft governor
轴调整 axial adjustment
轴调整器 axle adjuster
轴同心度 shaft concentricity
轴铜衬 step brass
轴筒 beam barrel
轴头 axle journal(neck);pivot bracket;spindle nose
轴头衬套 swivel bushing
轴头式铰链 pivot hinge
轴头销 gudgeon(pin)
轴头装置 boom attachment
轴投影 axial projection
轴凸肩 shaft shoulder
轴突 axis cylinder;axon
轴推进器 shaft propeller
轴推力 axial thrust
轴推力平衡装置 axial thrust balancing apparatus
轴瓦 bearing backing;bearing bush(ing);bearing insert;bearing liner;bearing lining;bearing metal;bearing shell;bush;bush(ing) bearing;follower;journal bearing;liner;pivoted shoe;step brass;bushing
轴瓦半圆周长 half peripheral length
轴瓦背面 back of bearing
轴瓦衬套 bush
轴瓦垫 journal bearing key;journal box wedge
轴瓦垫板 bearing wedge
轴瓦垫片 bearing strip;packing strip
轴瓦钩 uncoupling lever
轴瓦固定螺钉 bearing anchor
轴瓦刮刀 bearing scraper
轴瓦厚度 half bearing thickness
轴瓦块轴承 segment(al) bearing
轴瓦盆 spool box
轴瓦调整垫片 fitting strip
轴瓦凸起 bearing lip
轴瓦销 bearing shell pin
轴瓦圆周压缩量 crush
轴外的 abaxial;off-axis
轴外点 off-axis point
轴外光线 abaxial ray
轴外交叉极化 off-axis cross-polarization
轴外模 non-axial mode;off-axis mode
轴外色差 off-axis chromatic aberration
轴外伸部 shaft extension
轴外像差 off-axis aberration
轴外像散 abaxial astigmatism
轴外旋转调制盘 rotating off-axis reticle
轴位 axle order
轴位数字变换器 shaft position-to-digital converter
轴位移 shaft displacement
轴温 axle temperature
轴温探测器 wheel detector
轴系 axis system;set of axes;shafting;system of axes
轴系的横振动 lateral vibration of shafting
轴系临界转速 combined critical speed
轴线 axial cord;axial line;axial trace;axis[复 axes];axis(shaft) line;axistyle;center line;ground line;shaft line
轴线不对准 axis malalignment
轴线测定 <晶体的> axonometry
轴线点 centre line point
轴线对称弯曲 axisymmetric(al) bending
轴线方向 axis direction
轴线和倾斜度校正 transfer of line and grade
轴线间的 interaxle
轴线角 axis angle
轴线校准 boresighting
轴线校正 boresighting
轴线尽端雕像 <街道布置> terminal statue
轴线尽端雕像台座 terminal pedestal
轴线控制桩【测】 pegs of wall centre line
轴线力 axial force
轴线偏位 deviation of alignment
轴线偏斜密封 seal with axis deviation
轴线偏置螺旋伞齿轮 spiral bevel gears with offset axes
轴线曲度 axial curvature
轴线曲率 axial curvature
轴线取向 axis orientation
轴线试用形状 trial shape of axis
轴线水道 axial stream
轴线缩孔 center[centre] line shrinkage cavity
轴线投测 transfer of building line
轴线误差 axis based error
轴线重合 axis dead in line;dead in line
轴陷 axial depression
轴箱 axle-box;axle casing;axle housing; housing box; jewel; journal box;journal housing;pedestal box;spindle box
轴箱 U 形均衡梁 journal box yoke
轴箱补充润滑 additional lubrication of axle-boxes
轴箱拆卸 dismantling of the axle-boxes
轴箱衬 hub plate
轴箱衬垫 liner
轴箱传送带 shaft box conveyer[conveyor]
轴箱导架 axle-box guide
轴箱导框 axle-box case;axle-box guide;box guide;journal box guide;pedestal
轴箱导框撑 horn stay
轴箱导框间隙 axle-box to horn guide clearance
轴箱导框螺栓 pedestal bolt
轴箱底座 axle-box seating
轴箱垫 journal bearing wedge
轴箱定位 box guide;guidance of axle

Z

box;location of axle box
轴箱定位装置 box guidance
轴箱耳 box hinge lug
轴箱防尘罩 axle-box dust-guard
轴箱盖 axial box cover;axle-box cover;axle-box lid;box cover;box lid;journal box cover;journal box lid
轴箱盖连接 axle-box cover joint
轴箱盖螺栓 journal box cover bolt
轴箱盖弹簧 journal box lid spring;lid spring
轴箱盖销 journal box lid pin
轴箱毂衬 journal box hub liner
轴箱箍筋 stirrup of axle-box
轴箱滑动轴承 plain bearing axle-box
轴箱机械润滑 mechanically lubricated axle-box
轴箱加速度 axle-box acceleration
轴箱架 journal box yoke
轴箱螺栓 <拱板转向架> box bolt
轴箱螺栓系杆 journal-box bolt tie bar
轴箱切口 pedestal jaw
轴箱润滑器 journal box lubricator
轴箱上下拉杆 Althom link
轴箱式英里程计 axle-box mileage counter
轴箱弹簧 axle-box spring; journal spring
轴箱弹簧支柱 axle-box spring strut
轴箱弹簧装置 journal spring device
轴箱弹簧座 axle-box spring seat
轴箱体 axle-box body
轴箱凸缘 box flange
轴箱托板 axle-box binder; axle-box bracket; box support; journal box tie bar;box tie bar
轴箱楔螺栓 frame wedge bolt
轴箱斜铁 axle-box wedge
轴箱悬挂 single-axle truck
轴箱油池 keep of axle box
轴箱油垫 axle packing;box packing;journal box packing
轴箱油垫挡 journal packing guard
轴箱油垫工具 packing tool
轴箱油棉纱 journal box cotton waste;journal packing waste
轴箱游隙 play in axle-box
轴箱罩 axle-box cage
轴箱轴承 axle-box bearing
轴箱轴承的机械润滑 mechanically lubricated axle-box bearing
轴箱轴承分解清洗流水线 dismounting and cleaning flowline for bearings of shaft box
轴箱轴承组装流水线 mounting flowline for bearings of shaft box
轴箱轴瓦 axle-box liner
轴箱注油员 journal oiler
轴箱装配 assembling of axle-boxes
轴箱座 axial box seating; axle-box bracket
轴响应 axial response
轴向 axial direction
轴向安匝 axial ampere turn
轴向安装 axial mount
轴向凹槽 axial notch
轴向摆差 axial run-out
轴向摆动 axial oscillation; axial run-out;axial wobble
轴向变螺距 axially varying pitch
轴向变形 axial deformation
轴向变形体理论 axial strained-body theory
轴向变形系数 coefficient of axial deformation
轴向波型 axial mode
轴向补偿 nose balance
轴向布置 axial plan;axle arrangement
轴向布置凝汽器 axial condenser
轴向部分 axial component
轴向缠绕 axial winding
轴向场 axial field
轴向沉积 axial deposition
轴向承载(能)力 axial carrying capacity
轴向承载受压杆件 axially loaded compression element
轴向齿厚 axial tooth thickness
轴向齿廓 axial profile
轴向齿形角 axial profile angle
轴向冲击式水轮机 axial-flow action turbine
轴向抽出 axial withdrawal
轴向传热 axial heat conduction
轴向磁通电动机 axial flux motor
轴向次级 axial secondary
轴向窜动 axial float;axial movement;end play;uphill and downhill travel
轴向单元 axial element
轴向刀面角 <铣刀片的> helical rake angle
轴向导板 axle-box guard
轴向导流器 axial exducer
轴向道路 axial road
轴向的 direct axis;longitudinal
轴向的壁 axial wall
轴向地 axially
轴向地心偶极子场 axial geocentric dipole field
轴向递增螺距 axially increasing pitch
轴向电枢 axial armature
轴向动作 axial action
轴向端游隙 end float
轴向断裂 axial rift
轴向断面 axial section
轴向堆料 axial depositing;axial stockpiling
轴向对称 axial symmetry
轴向对称薄壳 axially symmetric(al) shell
轴向对称辐射器 axially symmetric(al) radiator
轴向对称荷载 axially symmetric(al) load
轴向对称流动 axially symmetric(al) flow
轴向对称射流 axially symmetric(al) jet;axisymmetric(al) jet
轴向对称形 axially symmetric (al) shape
轴向对称性 axial symmetry
轴向对称应变 axial symmetric (al) strain
轴向对称应力 axisymmetric (al) stress
轴向对称运动 axially symmetric(al) motion
轴向发电机 shaft generator
轴向发动机 shaft generator
轴向发展理论 axial theory of growth
轴向放大率 axial magnification;longitudinal magnification
轴向分布 axial distribution
轴向分带 axial zoning
轴向分力 axial component
轴向分量 axial component
轴向风区 axial air section
轴向缝 axial joint
轴向浮筒式举船机 lift with axial floater
轴向浮筒式升船机 lift with axial floater
轴向辐流可反转的水泵水轮机 Francis reversible pump turbine
轴向辐流式水泵水轮机 reversed Francis turbine
轴向辐流式水轮机 Francis (water) turbine
轴向辐流式透平 Francis(water)-turbine
轴向辐流式涡轮机 Francis turbine
轴向辐射天线 end-on fire antenna
轴向辐射天线阵 alignment array; end-fire (aerial) array; end-on directional array
轴向负荷 axial load(ing);thrust load
轴向负荷承受能力 thrust capacity
轴向负荷型芯 axial load carrying core
轴向负荷支承能力 thrust capacity
轴向负荷柱 axial loaded column
轴向负载 axial load(ing);axis bearing
轴向负载轴颈 journal for axial load
轴向刚度 axial rigidity
轴向工作压力 axial working thrust
轴向供给 axial admission
轴向构造带 axial tectonic belt
轴向鼓风机 helical blower
轴向固定 axial fixity;axial restraint
轴向固结 axial consolidation
轴向惯性矩 axial moment of inertia
轴向光束 axial pencil
轴向光行差 axial aberration
轴向滚铣 through milling
轴向荷载 axial load (ing); centric load
轴向荷载 A 形井架 axial loading A-frame;direct thrust A-frame
轴向荷载疲劳试验 fatigue testing under axial loading
轴向荷载试验 axial load test
轴向荷载轴承 journal for axial load
轴向后角 axial relief angle
轴向厚度 axial thickness
轴向滑动 endwise slip
轴向滑动阀 spool valve
轴向滑脱 axial slip
轴向滑移面 axial glide plane
轴向回转柱塞泵 Janney pump
轴向回转柱塞式液压电动机 Denison motor;Janney motor
轴向回转柱塞液压马达 axial rotary plunger motor;Janney motor
轴向混合 axial mixing
轴向活塞泵 axial piston pump
轴向活塞机 axial piston machine
轴向活塞机器 axial piston engine
轴向活塞马达 axial piston motor
轴向活塞式气动马达 axial piston air motor
轴向活塞式液压传动 axial piston transmission
轴向活塞式(液压)马达 axial piston type motor
轴向活塞引擎 axial piston engine
轴向活塞装置 axial piston unit
轴向极化比 axial polarization ratio
轴向极限沉淀速度 axial limit deposit velocity
轴向计数 axle count
轴向加速度 axial acceleration
轴向加速器 in-line booster
轴向加载 axial load(ing)
轴向间隙 axial clearance; axial play; end clearance;end play;side play
轴向减压断裂试验 decreasing axial pressure fracture test
轴向键 a bond;axial bond
轴向角 axial angle
轴向接头 longitudinal joint
轴向节距 axial pitch;linear pitch
轴向截面 axial section
轴向进刀 axial feed
轴向进刀法 axial feed method
轴向进给 axial admission
轴向进口气旋分离器 axial-inlet cyclone
轴向进口叶轮 axial entry impeller
轴向进汽 axial admission
轴向距离 axial distance
轴向均匀输出 hometaxial output
轴向抗拉强度 axial tensile strength
轴向空隙 axial clearance
轴向孔隙 axial pore
轴向孔隙率 axial porosity
轴向扩散 axial diffusion;axial dispersion
轴向扩散式外壳 axial diffusion casing
轴向拉力 axial tensile force;axial tension
轴向拉伸 axial tension
轴向拉伸强度 tensile strength on direct test
轴向拉伸试验 axial extension test;axial tension test
轴向拉应力 axial tensile stress
轴向拉张应力 axial prestressing
轴向肋 axial rib
轴向离合器 axial clutch
轴向离焦 axial defocusing
轴向力 axial force;thrust force
轴向力图 axial force diagram
轴向力系数 axial force factor
轴向量 axial vector;axiator
轴向裂谷 axial rift
轴向临界压力 axial critical thrust
轴向灵敏度 axial sensitivity
轴向流动 axial flow
轴向螺距 axial pitch
轴向螺旋 coaxial spiral
轴向马力 shaft horse
轴向密封 axial seal
轴向密封垫 axial gasket
轴向模数 axial module;axial modulus
轴向磨削力 axial grinding force
轴向内力 axial internal force
轴向扭矩 axial torque
轴向偶极测深曲线 curve of axial dipole-dipole sounding
轴向偶极频率测深曲线 frequency sounding curve of axis dipole array
轴向偶极装置 axial dipole-dipole array
轴向排气 axial exhaust
轴向喷流速度 axial jet velocity
轴向膨胀 axial expansion
轴向膨胀密封 axially expansive seal
轴向偏移 axial deviation
轴向平衡 axial balance
轴向平面 axial plane
轴向平行光速 axial parallel beam
轴向平移 axial translation
轴向剖面 axial section
轴向剖面图 axial cross-section
轴向气流 axial flow
轴向前角 axial rake angle
轴向嵌固 axial fixity;axial restraint
轴向强度 axial strength
轴向曲度 axial curvature
轴向曲路密封 axial labyrinth seal
轴向全息摄影 in-line holography
轴向燃气轮机 axial gas turbine
轴向绕组 axial winding
轴向色差 axial chromatic aberration
轴向射流 axial jet
轴向伸长 axial elongation;axial stretching
轴向伸展 axial stretching
轴向渗透 axial dispersion
轴向声源级 axial source level
轴向矢量 axial vector
轴向式 axial type
轴向式的 axial;in-line
轴向式电动油泵 axial electric(al) oil pump
轴向式发动机 axial engine
轴向式活塞液压设备 axial piston hydraulic equipment
轴向式活塞液压系统 axial piston hy-

draulic system
轴向式压缩机 axial-flow blower;axial-flow compressor
轴向式压缩器 axial-flow blower;axial-flow compressor
轴向受拉 direct tension
轴向受力空心圆柱体 axially loaded hollow cylinder
轴向受力中空圆柱形护舷 axially loaded hollow cylindrical fender
轴向受力柱 axially loaded column
轴向受压 axial compression;direct compression
轴向受载柱 axially loaded column
轴向水推力 axial hydraulic thrust force
轴向速度 axially directed velocity;axial velocity
轴向速度比 axial velocity ratio
轴向速率 axial rate
轴向调整螺母 take nut
轴向通风 axial ventilation;longitudinal ventilation
轴向通量 axial flux
轴向通量分米波超高四极管 axial-flow resnatron
轴向凸轮 axial cam
轴向推力 axial pressure;axial thrust (force);end thrust
轴向推力平衡 axial thrust balancing
轴向推力轴承 axial thrust bearing
轴向弯曲 longitudinal bow
轴向位移 axial displacement;axial translation;center distance modification <指光>
轴向位移系数 coefficient of axial displacement;center distance modification coefficient <指光>
轴向位移指示器 axial position indicator;shaft position indicator
轴向涡流 axial eddy;axial vortex;axial whirl
轴向蜗杆轴承外环 steering worm bearing outer ring
轴向物点 axial object point
轴向误差 axial error;axis error
轴向系统 axial system
轴向响应度 axial response
轴向向量 axial vector
轴向像差 axial aberration
轴向像点 axial image point
轴向形变 axial deformation
轴向旋涡 axial eddy;axial vortex
轴向压力 axial compression;axial compressive force;axial pressure;end pressure;thrust brake;thruput <拱的>
轴向压力轨道 thrust rail
轴向压力线 <拱的> thrust line
轴向压缩 axial compression
轴向压缩应力 axial compressive stress
轴向延长 axial elongation
轴向延伸 axial elongation;axial stretching
轴向叶片旋流分离器 vane-axial cyclone
轴向移动 axial displacement;axial movement;longitudinal movement
轴向引线 axial lead
轴向应变 axial strain
轴向应力 axial stress
轴向游动 axial play
轴向游隙 axial clearance;axial play;end play
轴向余隙 axial clearance
轴向预加负荷 axial preload
轴向预加拉力 axial prestressing;axial pretensioning
轴向预张力 axial pretensioning
轴向圆盘隔板泵 axial cylinder swashplate pump
轴向约束 axial constraint;axial fixity;axial restraint
轴向运动 axial motion;axial movement
轴向运动子空间 axial subspace
轴向增益 on-axis gain
轴向张拉 axial tensioning
轴向张拉强度 axial tensile strength
轴向张力 axial tensile force;axial tension
轴向找正 axial spotting
轴向振动 axial vibration;in-line vibration
轴向振动减振器 axial vibration damper
轴向正应力 axial normal stress
轴向止推轴承 axial thrust bearing
轴向中心线 longitudinal center[centre] line
轴向中心销 axle king pin
轴向重复荷载试验 repeated load axial test
轴向轴承 axial bearing;thrust bearing
轴向轴流式气轮机 radial axial flow turbine
轴向轴流式水轮机 radial axial flow turbine
轴向主销 axle king pin
轴向主应力 axial principal stress
轴向住房建筑 axle housing
轴向住房建筑伸臂柱头 axle housing bracket cap
轴向住房建筑枢轴 axle housing trunnion
轴向助推器 tandem booster
轴向柱 axial column
轴向柱塞泵 axial piston pump;axial plunger pump
轴向柱塞马达 axial piston motor;axial plunger motor
轴向柱塞式 axial piston
轴向柱塞式传动装置 axial piston transmission
轴向柱塞式液压马达 axial plunger type hydraulic motor
轴向柱塞液压马达 air-piston hydraulic motor
轴向转动 axial rotation
轴向转力矩 axial torque
轴向自紧密封 axial seal energized by internal pressure;axial self-seal
轴向自卸漏斗车 axial discharge hopper wagon
轴向自由度 axial freedom
轴向阻力 axial resistance
轴像 axial figure
轴销 axle pin;dowel;pin;pivot;shaft pin
轴心 axis[复 axes];axle center[centre];bobbin
轴心高度 pitch of centers
轴心荷载 axial load(ing);centric load;concentric(al) load
轴心加荷 load on axle
轴心进线的卷线机 deadhead coiler
轴心距 distance between axial centers
轴心抗压强度 axial compressive strength
轴心拉力 axial tensile force
轴心轮询 hum polling
轴心偏斜 axis deviation
轴心偏移 desaxe
轴心受拉 axial tension
轴心受压 axial compression
轴心受压柱 axial compression column
轴心线 axial lead;line shaft
轴心油 axle oil
轴心预加拉力 axial tensioning
轴性 axial character
轴悬牵引电动机 axle-hung traction motor
轴悬置 axle suspension
轴旋转 rotation of axes
轴压 axle load(ing)
轴压比 axial compression ratio
轴压力 pulldown
轴应变 axial strain
轴英马力 shaft horsepower
轴用材 hub block
轴用钢 axle steel
轴用钢棒 axle steel bar
轴用钢配筋 axle steel reinforcing
轴用滑脂 axial grease
轴用卡簧钳 external pliers
轴用润滑脂 axle grease
轴用油 axle oil
轴油 axle oil;journal oil;shafting oil
轴油道 slinger
轴油枪总成 oil pump assembly
轴羽 axial feather
轴圆端 nose circle
轴圆锥段 screw-shaft cone
轴载 axle mass
轴载当量因子 axle load equivalence factor
轴载荷 axle capacity
轴载荷限度 axle weight limit
轴载限制 axle load limitation
轴载重 axle load(ing);axle weight
轴载重极限 axle weight limit
轴载重刻度(尺) axle load scale
轴载重组 axle load weight group
轴载转移 load transfer
轴闸储气筒调准小齿轮轴 wheel cylinder adjusting pinion shaft
轴罩 spindle-cap
轴折断 axle fracture
轴褶升区 axial culmination
轴针 pin
轴针式喷嘴 pintaux nozzle
轴枕 pillow;pillow block;rail bearing
轴振动 shaft vibration
轴整直机 shaft straightening machine
轴整直器 shaft straightener
轴支承 axle suspension
轴支持装置 spindle carrier
轴支架 shaft support
轴支柱 axle strut
轴制动率 braking ratio per axle
轴制动器 hub brake
轴踵 toe
轴踵输送机 toe conveyer[conveyor]
轴重 axle load(ing);axle load on rail;axle mass;axle weight;wheel set load;gross rail load on axle <轴颈上>
轴重比率 ratio for weight per axle
轴重分配图 weight diagram
轴重检测器 axle weight detector
轴重调整装置 axle load adjusting device
轴重限度 axle weight limit
轴重限制 axle weight limit;rail load limit
轴重仪 axle load meter
轴重转移 load transfer
轴柱 axostyle;gudgeon;jack post;peg
轴转动 shaft drive
轴转关门扣 fall bar
轴转关门闩 fall bar
轴转矩 shaft torque
轴转式输送机 fan conveyer[conveyor]
轴转数计 shaft counter;shaft revolution indicator
轴转速 shaft revolution;speed of spindle
轴转速表 shaft counter
轴转向销 axle swivel
轴装单级减速齿轮箱 axle-mounted single reduction gearbox
轴装电动机 axle-mounted motor
轴装交流发电机 axle-mounted alternator
轴装励磁机 shaft-mounted exciter
轴装小齿轮 shaft-mounted pinion
轴锥 axial cone
轴组构 axial fabrics
轴座 axle bed;axle fit;axle seat;bracket;pillow block;wheel bore

肘

肘板 bracket plate;knee plate;toggle plate;wrist plate
肘板承座 toggle seat
肘板框架肋骨 bracket frame
肘板式 bracket system
肘板压机 knuckle press
肘板压力机 toggle plate press
肘板趾 bracket toe;toe of bracket
肘臂 <漏斗车的> toggle arm
肘部投影高 elbow-seat height
肘的 anconeal
肘钉 staple
肘动把手 elbow action handle
肘动轮 knee action wheel
肘端 point of elbow
肘反射 elbow jerk
肘杆 toggle link;toggle rod
肘杆传动 knuckle-lever drive
肘杆曲柄动作 toggle action
肘杆式拉深压力机 toggle drawing press
肘杆式离合器 toggle clutch
肘杆式压力机 embossing press;knuckle joint press
肘杆式压下压力机 knuckle-lever coining press
肘杆式压印机 toggle press
肘杆式压榨机 toggle press
肘杆锁定的 toggle locking
肘杆压机 knuckle press
肘杆压力机 knuckle-lever press
肘杆压片机 toggle-type tablet machine
肘管 angle branch;bent pipe;elbow bend;elbow pipe;elbow tube;elbow tubing;side leg
肘环套接 knee joint;knuckle jointing;toggle
肘环套接板 toggle plate
肘环套接卡盘 toggle chuck
肘间距 akimbo span
肘接式压力机 knuckle joint press
肘接(头) knuckle joint;elbow joint;knee joint;toggle joint
肘接销钉 toggle pin
肘接制动器 toggle-joint brake
肘节 elbow joint;knuckle;toggle joint;wrist【机】
肘节臂 toggle arm
肘节柄 toggle grip
肘节动作 wrist action
肘节颚式破碎机 toggle jaw crusher
肘节(杠)杆 angle lever;elbow jointed lever;toggle(-joint) lever;toggle link
肘节钩 toggle hook
肘节机构 toggle mechanism
肘节角 toggle angle
肘节接合 toggle joint
肘节开关 toggle switch
肘节扩眼器 <弧线钻进用钻具> ball-and-socket reamer
肘节联结 toggle linkage
肘节链 toggle chain
肘节螺母 toggle nut
肘节螺栓 toggle bolt

肘节铆机 toggle-joint riveting machine
肘节铆接机 toggle-joint riveter
肘节扇形锁 <转辙器> toggle-joint sector lock
肘节式颚式破碎机 toggle-type jaw crusher
肘节式夹钳 toggle clamp
肘节式压床 knuckle joint press
肘节销 wrist pin
肘节销螺母垫圈 wrist-pin nut washer
肘节销铜衬楔止动螺钉 wrist-pin brass wedge set screw
肘节销铜楔垫 wrist-pin brass wedge washer
肘节销油杯 wrist-pin oil cup
肘节形接头 knuckle joint; rule joint【建】
肘节型闪光器 toggle-type flasher
肘节压碎机 toggle crusher
肘节压砖机 toggle press
肘节运转 wrist action
肘节闸 toggle-joint brake
肘节支座 toggle seat
肘节制动器 toggle brake
肘节钻具 wiggle tail
肘节作用 wrist action
肘开关水嘴 elbow operated tap
肘开式龙头 elbow operated faucet
肘连接的 knee-jointed
肘式开关 toggle switch
肘托 ancon(e); console
肘销 knuckle pin; wrist pin
肘销铜衬楔垫块 wrist-pin brass wedge block
肘销铜衬楔螺栓 wrist-pin brass wedge bolt
肘形板 elbow board
肘形插口 elbow socket
肘形承口 elbow socket
肘形搭钩 elbow catch
肘形灯托架 elbow lamp bracket
肘形阀门 toggle valve
肘形杆 articulated lever
肘形关节 knuckle joint
肘形管 pipe bend
肘形轨 <钝角辙叉> knuckle rail
肘形焊接 knuckle soldered joint
肘形铰接 knuckle joint; rule joint
肘形接缝 <双折屋面中两个斜面的交线> knuckle joint
肘形接合 knuckle joint
肘形接头 knuckle joint
肘形扣 elbow catch
肘形连接 knuckle joint; knuckling
肘形梁 knee girder
肘形龙头 elbow cock
肘形瞄准具 elbow sight
肘形配件 elbow piece
肘形燃烧室 elbow combustion chamber
肘形锁闭器 elbow-type lock (ing mechanism)
肘形锁扣 elbow catch
肘形淘弯 sweep elbow
肘形突出部分 <门窗框角上的> crosette
肘形弯管 elbow; elbow bend (pipe); elbow pipe; el (l); knee bend; sweep elbow; service ell <卫生管道中的>
肘形弯管的外半径 heel radius
肘形弯流 sweep elbow
肘形望远镜 elbow telescope
肘形尾水管 bent draft tube; elbow draft tube
肘形锥 knuckle cone
肘型道岔锁闭器 toggle-joint sector lock
肘状关节 knuckle
肘状河曲 ancon(e)
肘状梁 elbow-type girder; knee girder
肘状梁式竖旋桥 knee-girder bascule
肘状突出部【建】 ancon(e)
肘状物 elbow
肘状支柱 ancon(e)
肘座位置 toggle seat

帚柄 broomstick

帚处理 broom finish
帚面 broom finish
帚形喇叭辐射器 hoghorn
帚型构造【地】 brush-type structure
帚状 broom-pattern
帚状断层 brush faults
帚状构造 brush-type structure
帚状节理【地】 brush joint
帚状结构 broom texture
帚状矿脉 broom-like veins; brush veins
帚状褶皱 brush-type folds

宙斯保护区 sanctuary of Zeus

宙斯祭坛 altar of Zeus
宙斯禁猎区 sanctuary of Zeus
宙斯圣地 sanctuary of Zeus

绉布 creped cotton

绉缎 crepe-backed satin
绉毛线 crape wool
绉棉布 crape cotton
绉棉线 crape cotton
绉呢 crape wool; creped wool
绉缩处理 creping treatment
绉缩的 creped
绉条纹薄织物 seersucker
绉纹板 channeled plate
绉纹边缘 flash ridge
绉线 creped cotton
绉橡胶 crape rubber
绉纸 crape paper; creped paper

昼 daylight

昼标 day beacon (mark); day beacon signal; daytime signal; unlighted mark; unlighted beacon <无灯光>
昼灯 daytime light
昼风 wind by day
昼光 daylight
昼光玻璃 daylight glass
昼光光度计 hemeraphotometer
昼光觉 daylight vision; photopic vision
昼光滤色镜 daylight filter
昼光系数 daylight factor
昼光效应 daylight effect
昼光因数 daylight factor
昼光照明 natural illumination
昼弧 day arc
昼花夜香树 day jessamine
昼间变动 diurnal fluctuation
昼间标 day mark
昼间的 diurnal
昼间电路 daytime circuit
昼间活动范围区 daytime sphere
昼间瞭望距离 day sighting range; day visibility range
昼间流量 day flow
昼间人口 daytime population
昼间示像 day aspect
昼间显示 day indication
昼间显示距离 day visibility range
昼间信号 day signal; daytime signal
昼间障碍标识 guide mark
昼间作用距离 daylight range
昼盲 day blindness
昼盲症 hemeralopia
昼射程 daylight range; day range
昼视度 daylight visibility
昼视觉 daylight vision
昼现周期性 diurnal periodicity
昼行性的 diurnal
昼夜 civil day; day and night
昼夜比 ratio of day and night
昼夜变动 diurnal fluctuation
昼夜变化 diurnal change; diurnal variation; diurnal behavio(u)r <电离层>
昼夜变化的强光带 day-and-night changed bright zone
昼夜变化幅度 diurnal amplitude
昼夜变异 diurnal variation
昼夜不停的 all-round the clock; around-the-clock; round-the-clock
昼夜不停地作业 operate around-the-clock
昼夜差异性 diurnal variability
昼夜长短表 horoscope
昼夜潮(汐) diurnal tide
昼夜垂直移动 diurnal vertical migration
昼夜灯光开关 day-and-night light switch
昼夜等效声级 day-night equivalent sound level
昼夜点灯按钮 day-and-night light button
昼夜放牧 whole day grazing
昼夜服务 night-and-day service; round-the-clock work; twenty-four hours service; uninterrupted service
昼夜负荷图 daily load diagram; diurnal load diagram
昼夜负载 daily load
昼夜工作 night-and-day service; twenty-four hours service; round-the-clock work
昼夜工作的船舶电台 ship station carrying on continuous day and night service
昼夜和季节变化 diurnal and seasonal variation
昼夜活动 diel movement
昼夜加热 diurnal heating
昼夜交换量 diurnal capacity
昼夜节律 around rhythm; circadian rhythm; day-night rhythm; diurnal rhythm
昼夜节奏 day-night rhythm
昼夜轮班工作制 fully shifted system
昼夜平分点 equinox
昼夜平分点小潮 equinoctial neap tide
昼夜平分日 equinoctial day
昼夜平分线 equator; equinoctial circle; equinoctial line
昼夜平分(线)的 equinoctial
昼夜迁移 diurnal migration
昼夜曲线法 diurnal curve method
昼夜入口率 ratio of daytime to night-time population
昼夜施工 around-the-clock job; round-the-clock job; round-the-clock work
昼夜事故率之比 N/D rate ratio
昼夜通用信号 day-and-night signal; signal for day and night
昼夜温差 diurnal amplitude
昼夜温度 day-night temperature; diurnal temperature
昼夜温度变化 change of diurnal temperature
昼夜温度范围 diurnal temperature change
昼夜温度自动调节器 day-night thermostat
昼夜小时数 day-and-night hours
昼夜循环 diurnal cycle
昼夜涌水量 water-make in day and night
昼夜值勤 day-and-night duty
昼夜周期 daily cycle; day-night cycle
昼夜转换设备 day-night switching equipment
昼夜转换装置 day-night switching equipment
昼用叠标 day beacon range

皱 crimple; crinkle; ruckle; wrinkle crease

皱边 frilling
皱波 ripple
皱波导 corrugated waveguide
皱波状的 crispate
皱层 rennet
皱的 crumpled; rugose; rugous
皱耳 earing
皱缝 slack seams
皱革菌属 folding leader dry rot; Cladoderris <拉>
皱谷 ravine
皱合法兰 shrunk-and-peened flange
皱痕 crease; wrinkle mark
皱痕层理 crinkled bedding
皱脊 wrinkle ridge
皱晶结构 crystallitic texture
皱卷法兰 shrunk-and-rolled flange
皱孔菌 Merulius lacrymans
皱裂 alligatoring; chap; cockle cuts; cockle <薄板边缘的>
皱裂缝 crow foot crack
皱裂纹 <薄板边缘的> cockle
皱铝箔 crumpled alumin(i)um foil
皱面 drawn grain
皱皮 cockle; elephant's peel; elephant skin; flow mark
皱片机 mascerating machine; mascerator
皱起 pucker; rivel
皱起来 wrinkle
皱曲 buckling; rugosity
皱曲钢板 buckle plate
皱缩 crease-shrinkage; crimp; shrinkage; shrinking; shrivel; oil-canning <金属屋面材料>
皱缩光栅 shrunken raster
皱缩纹 shriveling
皱缩压制 crimping
皱桐油 Abrasin oil
皱铜 crimped copper
皱纹 buckling; cockle; corrugation; crease; crispature; crumple; furrow; plication【地】; pucker (ing); ruffle; rugosity; wrinkle; crinkle; riffle <板带材侧缘的>; cold shut <铸件的>
皱纹板 buckle plate; channel (l) ed plate
皱纹板岩 puckered slate
皱纹表面 corrugated surface
皱纹表皮 <铸件的> creasy surface
皱纹玻璃 rough plate glass
皱纹薄橡皮板 crepe rubber
皱纹不稳定性 rippling instability
皱纹层理 crinkled bedding
皱纹层流火焰 wrinkle laminar flame
皱纹瓷漆 wrinkled enamel
皱纹的 corrugated; ribby; wrinkled
皱纹的表面 creasy surface
皱纹的形成 wrinkle formation
皱纹灯泡 corrugated bulb
皱纹钢板 gof (f) ered plate; rifled sheet
皱纹钢管圆顶 <俚语> slump bottom
皱纹钢丝 crimped wire
皱纹革 shrink leather; wrinkled leather

皱纹构造 plicated structure; plication structure
皱纹管 bellows; sylphon
皱纹管式压力计 bellows-type manometer
皱纹火管 corrugated flue
皱纹金属丝网格 crimped wire screen
皱纹理 curly grain
皱纹沥青 tabbyite
皱纹滤纸 gof(f)ered filter
皱纹面漆 cockle finish; shrivel finish; crinkle finish
皱纹面饰 ripple finish
皱纹膜片 corrugated diaphragm
皱纹木材 wavy-grained wood
皱纹漆 alligatoring lacquer; crepon finish; plastic paint; wrinkle finish
皱纹清喷漆 wrinkle lacquer
皱纹清漆 crepe varnish; rivel varnish; shrivel varnish; wrinkle varnish
皱纹饰面 riveling; wrinkle finish; wrinkling
皱纹套管 spirally wound sheath
皱纹铁 gauffered iron; gof(f)ered iron
皱纹铁板 gof(f)ered sheet iron; rifled sheet iron
皱纹铁管涵 Armco culvert; corrugated steel culvert
皱纹桶 corrugated drum
皱纹图案 crumple pattern
皱纹瓦 ripple finish tile
皱纹橡胶踏板料 corrugated rubber matting
皱纹形成 crowding
皱纹叶 rugose leaf
皱纹釉 curtain glaze
皱纹罩面 elephant skin
皱纹罩面漆 ripple finish; wrinkle finish
皱纹织造 crimp weave
皱纹纸 crape paper; crepe paper; gauffered paper; gof(f)ered paper
皱纹纸板 gauffered board
皱纹状 ripple figure
皱纹状波痕 corrugated ripple mark
皱纹状构造 crease structure
皱纹状修饰 crimped finish
皱纹组织 tuck weave
皱形裂纹 Hertz(ian) fracture
皱折 corrugation; crease; crinkling; fold; lap; wrinkle
皱折波长 buckle wave length
皱折的 corrugated; crinkled
皱折度 wrinkle rating
皱摺钢板 buckle plate
皱褶 crease; fold; gauffer; gof(f)er; pucker; ruck; rugosity; wrinkling; webbing <一种漆病>
皱褶板 corrugated plate
皱褶机 gauffer machine; gof(f)er machine
皱褶饰面 wrinkle finish
皱褶缘 ruffed border
皱褶状态 regosity
皱状纹 curly figure

骤

骤变 <地壳> cataclysm

骤变点 discontinuity
骤变水流 rapid varied flow
骤发洪水 flash flood
骤风 gust
骤风荷载 gust load
骤风荷载系数 gust load factor
骤风因素 gustiness factor
骤加负荷 sudden load(ing)
骤加荷载 sudden load(ing); suddenly applied load
骤加应力 sudden stress
骤加载荷 sudden load(ing)
骤降 <水位的> sudden drawdown
骤降试验 drawdown test
骤扩 sudden enlargement
骤扩系数 coefficient of sudden expansion
骤冷 emergency cooling; quench(ing); shock chilling; shock cooling; sudden chilling
骤冷泵 quench pump
骤冷篦床 quenching grate
骤冷槽 quenching tank
骤冷法 quenching process
骤冷辊 chill roll
骤冷剂 quenching agent; quenching compound; quenching medium
骤冷裂纹 hardening crack(ing); quenching crack
骤冷凝 quenching condensation
骤冷气 quench gas
骤冷器 expander; quencher
骤冷室 shock chamber
骤冷塔 quench(ing) tower
骤冷效应 quenching effect
骤冷应力 quenching stress
骤冷原料 quench stock
骤冷作用 shock chilling function
骤凝 flash set
骤然扩大损失 loss due to sudden enlargement
骤然收缩损失 loss due to sudden contraction
骤然下降(气温、价格等) plummet
骤然泄降 rapid drawdown
骤缩 sudden contraction
骤缩系数 coefficient of sudden contraction
骤退 crisis[crises]; sudden fall
骤脱毛发 trichorrhea
骤雨 gust of rain; rain gush; shower; torrential rain; torrent of rain
骤雨水头 drencher head
骤蒸 flash evapo(u)ration
骤止 quick stoppage

朱

朱庇特 <罗马神话中的主神> Jupiter

朱庇特神庙 Temple of Jupiter
朱庇特性 column of Jupiter
朱顶红 barbedos lily
朱格拉周期 <中期经济周期> Juglar cycle
朱红 ponceau; bright red; vermil(l)ion red
朱红的 cinnabaric; cinnabarine
朱红木雕 deep red wood work
朱红色 Chinese red; cinnabar(ite); horse chestnut; imperial red; minium; vermeil
朱红色的 cadmium vermil(l)ion; vermil(l)ion
朱红色淀 vermil(l)ion red
朱红图像瓷瓶 red-figure vase
朱红图像式样 red-figure style
朱红涂料 vermil(l)ion paint
朱红云芝 Polystictus cinnabarinus
朱利特纳地体【地】 Chulitna terrane
朱漆 red(japan) lacquer; red paint
朱漆柱 vermil(l)ion column
朱色的 Chippendale
朱砂 cinnabar(ite); mercuric sulfide; Spanish red; vermil(l)ion cinnabar; zinnober
朱砂代用品 vermil(l)ion ate; vermil(l)ionette
朱砂的 cinnabaric; cinnabarine
朱砂硫化汞 hepatic cinnabar
朱伊夫阶【地】 Djuifian
朱庄矿体 ore chimney

侏

侏罗白垩 Jura chalk

侏罗纪【地】 Jurassic period
侏罗纪礁石 Jurassic reef
侏罗纪砂岩 Jurassic sandstone
侏罗纪石灰 Jurassic lime
侏罗纪石灰岩 Jurassic limestone
侏罗礁 Jurassic reef
侏罗-三叠纪【地】 Jura-Trias
侏罗式褶皱【地】 Jurassic type of folding; Jura-type folds; Jura type of folding
侏罗系【地】 Jurassic system

株

株距 between tree distance; distance between hills; distance between holes; seed spacing; spacing in the row

株木属 cornel; dog wood; Cornus <拉>
株式会社 <日本的股份有限公司> kabuskiki kaisha

珠

珠宝 bijou

珠宝保险 jewelry insurance
珠宝店 jeweler's shop
珠宝工的弓锯 jeweler's piercing saw
珠宝商 jeweller
珠宝室 bullion room
珠滴 globule
珠断 bead break
珠光 pearl glossy
珠光壁 nacreous wall
珠光玻璃 mother-of-pearl glass; nacreous glass
珠光粉 fish scale powder; pearl essence
珠光光泽 mother-of-pearl lustre[luster]; pearl luster[lustre]
珠光灰 pearl gray[grey]
珠光剂 pearlescent agent; pearling agent
珠光结构 shelly structure
珠光漆 opalescent lacquer; pearlescent lacquer; pearl finish; pearl lacquer
珠光石 argentine
珠光石英 cotterite
珠光体 pe(a)rlite[pe(a)rlyte]; semi-steel
珠光体的 pearlitic
珠光体钢 pearlite steel; pearlitic steel
珠光体结构 pearlitic texture
珠光体晶粒 pearlitic grain
珠光体可锻铸铁 pearlite iron; pearlite malleable cast-iron; pearlitic iron
珠光体临界冷却速度 pearlite critical cooling rate
珠光体内的 intrapearlitic
珠光体耐热钢 pearlitic structure heat resisting steel
珠光体耐热钢焊条 pearlitic heat resistant steel electrode
珠光体渗碳体 pearlitic cementite
珠光体生铁 pearlitic iron
珠光体团 pearlite colony
珠光体铸铁 Lanz cast iron; pearlitic cast-iron; perlit
珠光体组织 pearlitic structure
珠光涂料 pearlescent coating
珠光效应 pearlescent effect
珠光颜料 nacreous pigment; pearlescent pigment
珠光颜色 nacreous pigment
珠光油墨 pearl ink
珠光云母 pearl mica
珠光组织 perlitic texture
珠焊 bead(ing) weld(ing); light closing weld
珠灰色 pearl gray[grey]
珠江 Pearl River
珠江基面 Pearl River datum
珠卷饰 bead and reel enrichment
珠蓝色 pearl gray[grey]
珠粒体的 pearlitic
珠链 bead chain
珠链饰 bead and reel
珠明料 asbolite
珠磨机 bead mill; pearl mill
珠母贝 nacre
珠母层 mother-of-pearl
珠母云 mother-of- nacreous cloud; mother-of-pearl clouds; nacreous clouds
珠穆朗玛峰 Mont Qomolangma
珠盘饰凸圆线脚 bead and reel
珠皮呢织物 bouclé
珠球反应 bead reaction
珠球萤光分析 bead fluorescence analysis
珠式线脚 pearl mo(u)lding
珠饰 bead and reel; reel and bead
珠饰环中心 beaded centre; center beaded
珠饰平顶 beaded ceiling
珠饰天顶 beaded ceiling
珠饰天花板 beaded ceiling
珠水云母 margarodite
珠算 abacus computation
珠网形缝毛石圬工 cobweb rubble masonry
珠形苏格兰胶 bead glue
珠缘 beading
珠缘平面嵌板 beadflush panel
珠脂 margarine; oleomargarine
珠脂针 margrain-needles
珠状壁 beading wall
珠状的 beadlike; pearly
珠状封接 bead seal
珠状构造 pearlitic structure
珠状花边 beadwork
珠状结构 perlitic texture
珠状聚合法 bead polymerization
珠状聚合物 bead polymer
珠状冷凝 drop condensation
珠状流挂 drip bead
珠状流纹玻璃 marekanite
珠状凝结 dropwise condensation
珠状热敏电阻 bead resistance
珠状热敏电阻器 bead; bead thermister[thermistor]
珠状闪电 bead(ed) lightning
珠状饰 beaded
珠状饰方砖 spattered tile
珠状炭黑 drop black; pearl black
珠状物 pearl
珠子 pearl

猪

猪背海岸 <浪蚀软硬间层而成> hogback coast

猪背脊【地】 galera; kettle back
猪背岭【地】 hogback
猪背山 hogback; swine back
猪场 pig farm; piggery
猪场废水 piggery waste(water)
猪场废水消化液 digested piggery wastewater
猪厩肥 hog manure
猪圈 hogpen; pigpen; pigsty(e)
猪圈废水 piggery waste(water)
猪李【植】 hog plum

猪牛粪水 pig and cattle slurry
猪舍 hog house;pig house
猪饲料 hog feeding
猪饲养场 hog ranch
猪尾形线 <连接电刷与刷握的软电缆> pigtail
猪油 lard oil
猪脂泥煤 lard peat
猪脂泥炭 lard peat
猪鬃 bristle

蛛

蛛丝 <仪器的> spider lines

蛛丝视距仪 wire tach(e)ometer[tachymeter]
蛛丝网 spider web
蛛丝状 arachnoid
蛛网 spider web;web
蛛网定理 cobweb theorem
蛛网分析法 cobweb analysis
蛛网缝毛石砌体 cobweb rubble masonry
蛛网焊 spider bonding
蛛网模型 cobweb model
蛛网绕组 spider-web winding
蛛网式(道路)系统 web-like system
蛛网式天线 spider-web antenna
蛛网式天线反射器 spider-web reflector
蛛网线迹 mocca
蛛网线圈 spider-web winding
蛛网形道路系统 spider-web type of street system
蛛网形缝乱毛石砌体 polygonal random rubble masonry
蛛网形缝乱石圬工 cobweb random rubble masonry
蛛网形缝毛石圬工 cobweb rubble masonry
蛛网形接合 spider bonding
蛛网形裂隙 spider's web of cracks
蛛网形模片 spider template; spider templet
蛛网形散兵坑 spider fox-hole
蛛网形铁丝网 spider wire entanglement
蛛网形网眼 spider net
蛛网形线 slab line
蛛网形线圈 slab coil; spider(-web) coil
蛛网状 cobwebbing
蛛网状城市 spider-web city
蛛网状的 arachnoid
蛛网状裂纹 <玻璃及搪瓷缺陷> spider's web
蛛网状图案 cobweb pattern
蛛网状涂层 cobwebbing
蛛网状物 spider web
蛛网组织 spider weave
蛛心裂 <木材缺陷> spider heart
蛛形的 arachnoid

槠

槠corkoak

潴

潴水屋面 submerged roof

竹

竹暗渠 bamboo drain

竹板条 bamboo lath;furring of bamboo
竹编条板 bamboo lathing
竹材 bamboo;bambusaceae
竹材顶棚 bamboo ceiling
竹材房子 bamboo house
竹材工作 bamboo work
竹材构架 bamboo frame; bamboo framing
竹材回廊 bamboo corridor
竹材脚手架 bamboo scaffold(ing)
竹材天花板 bamboo ceiling
竹材亭 bamboo pavilion
竹材细工 bamboo work
竹尺 bamboo scale;bamboo tape
竹筹理货 bamboo tally
竹椽 bamboo rafter
竹丛 canebrake
竹雕 bamboo carving
竹钉 bamboo nail
竹筏 bamboo raft
竹房子 bamboo house
竹杆 bamboo-pole
竹杆脚手架 bamboo-pole scaffold(ing)
竹杆子 bamboo carrying pole
竹工 bamboo worker
竹构架 bamboo framing
竹刮刀 bamboo spatula
竹棍 bamboo reeds
竹基 bamboo cane
竹家具 bamboo furniture
竹浆 bamboo pulp
竹浆纸 Indian paper
竹胶合板 bamboo plywood
竹脚手架 bamboo(-pole) scaffold(ing)
竹节钢 indented(ribbed)bar;knotted bar iron; bamboo steel; corrugated bar
竹节钢筋 corrugate(d) steel(bar); deformed reinforcement; deformed (reinforcing) bar; deformed steel bar;high-bond bar; ribbed(indented) bar; ribbed rebar; ribbed reinforcing bar;ribbed rod
竹节石绝灭 tentaculitid extinction
竹节式墙系杆和鸠尾形锚固件 corrugated anchor
竹节瓦 bamboo-notch tile
竹节形炻器 bamboo ware
竹结构 bamboo framing
竹筋 bamboo reinforcement
竹筋混凝土 bamboo(-reinforced) concrete
竹筋混凝土桥 bamboo (reinforcement) concrete bridge
竹茎沉排 bamboo mattress;cane mattress
竹刻 bamboo carving; bamboo engraving
竹篚 gabion
竹框架工程 <用于河道束水的> bandal
竹捆浮碰垫 floating bamboo fender
竹篱 bamboo(rail)fence
竹帘 bamboo blinds;bamboo curtain; bamboo screen;chick <印度东南亚的>
竹帘门 bamboo blind door
竹凉席 bamboo mat
竹林 bamboo forest;bamboo groves
竹檩条 bamboo purlin(e)
竹笼 bamboo basket; bamboo cylinder;bamboo gabion
竹笼围堰 bamboo gabion cofferdam
竹楼 multistoried bamboo pavilion
竹篓 bamboo basket
竹绿色 green blue
竹麦杆 culm
竹篾 bamboo splits
竹木工 bamboo worker
竹/木排流放 downstream floating of bamboo or log raft
竹木镶嵌 bamboo-wood inlay
竹排 bamboo raft;cane mattress
竹片固定 fixation with bamboo splints
竹片马赛克 bamboo mosaic
竹器 bamboo ware
竹签 prod;tally
竹墙 bamboo walling
竹桥 bamboo bridge
竹扫帚 bamboo broom
竹栅 bamboo lath screen
竹饰 bamboo ornament
竹栓 cane bolt
竹丝 bamboo filament
竹笋 bambooshoot
竹索桥 bamboo cable bridge
竹梯 bamboo ladder
竹条气窗 bamboo lath transom
竹亭 bamboo pavilion
竹土箕 bamboo earth pan
竹瓦 bamboo tile
竹围墙 bamboo fence wall
竹屋架 bamboo roof truss
竹席 bamboo mat
竹纤维 bamboo fiber[fibre]
竹销 bamboo bolt
竹轩(餐厅)bamboo lofty restaurant
竹压片 bamboo spatula
竹叶状灰岩 wormkalk
竹叶状砾屑结构 wormy gravel clastic texture
竹叶状砾岩 edgewise conglomerate
竹园 bamboo garden;bamboo plantation
竹制 bamboo rapter
竹制轨槽 <推拉门用> bamboo rail
竹制回廊 bamboo corridor
竹制家具 bamboo furniture
竹制胶合板 bamboo plywood
竹制螺栓 bamboo bolt
竹制碰垫 bamboo fender
竹制亭 bamboo pavilion

烛

烛尺 candle-foot

烛光 candela[ed];candle power;candle <旧发光强度单位>
烛光·米 candle-meter
烛光·小时 candle-hour;candle power hour
烛光灯 candle lamp
烛光度 lighting strength
烛光功率分布曲线 candle power distribution curve
烛光力 candle light(ing)
烛光亮度 candle power brilliance
烛光英尺 <1 英尺 = 0.3048 米> candle-foot
烛光照明 candle light(ing)
烛果油 candlenut oil;carapa oil;katio oil;Kukui oil
烛架 candelabrum[复 candelabra/candelabrums]
烛沥青煤 pelionite
烛煤 black jack; candle coal; cannel (coal); cannelite; jet long flame coal;kennel
烛煤页岩 cannel shale
烛米 <照度单位> candle-meter
烛木 candle wood
烛式过滤 candle filtration
烛式过滤机 canned-candle filter
烛式过滤器 candle filter
烛台 candle holder;candlestick;chandler;lustre;taper-stick;jesse <教堂用,呈树枝形>
烛台架 candelabrum[复 candelabra/candelabrums];candlestand
烛台托盘 bobeche
烛芯纱盘花簇绒织物 candlewick
烛形灯 candle lamp
烛焰形装饰灯 candle lamp
烛藻煤 cannel boghead
烛枝 candle
烛脂柏油 candle tar
烛脂焦油沥青 candle tar
烛脂煤焦油 candle tar
烛脂硬(焦油)沥青 candle pitch
烛状物 candle
烛状柱 tapering column

逐

逐板 step-by-step tray

逐板计算 plate-to-plate calculation
逐步板 step-by-step tray
逐步饱和法 step-by-step saturation method
逐步逼近法 method of approach; method of successive approximation;method of trials and errors; step-by-step approach; step method;successive approximation
逐步逼近装置 successive approximation unit
逐步比较 stepwise comparison
逐步比较法 method of successive comparison;successive comparison
逐步变化的 stepping
逐步采用 phase in
逐步测量法 step-by-step method
逐步测试程序 step-test procedure
逐步测试法 step-test procedure
逐步撤出 phase-out
逐步成型拉刀 generating broach
逐步澄清 progressive defecation
逐步打捞 raising in steps
逐步打捞法 method of raising in steps
逐步代替法 successive substitution
逐步的 step-by-step;stepwise
逐步迭代法【数】step-iterative method
逐步多项式回归 stepwise polynomial regression
逐步法 step-by-step method;step-by-step procedure;stepwise method
逐步反复加工机 step-and-repeat machine
逐步反应 step-by-step reaction; step reaction
逐步方法 step-by-step method; step-by-step procedure
逐步分解 stepwise decomposition
逐步分流式水带铺设 progressive hose lay
逐步分期施工法 progressive stage construction
逐步改善 incremental improvement; progressive improvement
逐步干燥器 progressive drier[dryer]
逐步过程 step-by-step procedure
逐步回归(法)stepwise regression
逐步回归分析 stepwise regression analysis
逐步积分 step-by-step integration
逐步积分法 step-by-step integration procedure
逐步计算 stepwise computation
逐步加成聚合物 step addition polymer
逐步加热 stage heating
逐步加载试验 progressive loading trial;progressive trial
逐步间隔法 successive intervals method
逐步减少 step down
逐步减缩 phase down
逐步简化 successive reduction
逐步渐近 successive approximation
逐步渐近法 cut-and-trial method; method of successive approximation; step-by-step approximation; successive approximation method
逐步渐进法 cut-and-try method; cut-and-try process
逐步降低的成本 diminishing cost
逐步降级 de-escalate
逐步降级的 de-escalatory
逐步校正法 method of successive

corrections
逐步接近 cut-and-trial;successive approximation
逐步解码 step-by-step decoding
逐步近似法 method of successive approximation; step-by-step approximation; successive approximation method
逐步进给 step-by-step feed
逐步聚合 stepwise polymerization; transition polymerization
逐步扩大 gradual enlargement
逐步扩大管压力损失系数 coefficient for gradual enlargement
逐步冷冻法 progressive freezing
逐步冷却退火 step annealing
逐步励磁 step-by-step excitation
逐步模拟 step-by-step simulation
逐步判别 stepwise discrimination
逐步破坏 successive failure
逐步曝气 step aeration
逐步求解法 step-by-step method
逐步求解过程 step-by-step process
逐步求精 successive refinement
逐步求精法 stepwise refinement
逐步取代法 successive substitution
逐步筛选 step sizing
逐步上升 escalation
逐步上升价格 escalating price
逐步设计法 step-by-step design
逐步升级 escalation;pyramid
逐步失效 degradation failure;gradual failure
逐步实施 gradual execution
逐步实施计划 phased program(me)
逐步使用 phase in
逐步试凑法 trial-and-error procedure
逐步试验 point-by-point test;step-by-step test
逐步试验法 step-test procedure
逐步算法 step-by-step algorithm
逐步缩减 phase down
逐步缩小的 de-escalatory
逐步淘汰法 sieve method;successive sweep method
逐步替换法 successive displacement method
逐步停止生产 phase-out
逐步投产 phase in
逐步投入 phase into
逐步推进施工法 hand over hand construction
逐步退焊法 backstep welding
逐步限速示像 graduated restrictive aspect
逐步向后法 stepwise backwards method
逐步向前法 stepwise forwards method
逐步寻址 stepped addressing
逐步应用 stepwise application
逐步杂凝聚 stepwise heterocoagulation
逐步征用 creeping expropriation
逐步执行 step execution
逐步置换法 method of successive displacement
逐步钻法 step-by-step drilling
逐层 consecutive layer;layer to layer
逐层后退 step back
逐层密实 compaction in layers
逐层排出的砖墙 creasing course
逐层切削法 <测残余应力> Mach's method
逐层绕法 layer-by-layer winding
逐层挑出的砖墙 creasing course
逐层挑出屋面瓦 tile creasing
逐层挑瓦 creasing
逐层向上运土法 cast after cast
逐车道拌和(施工法) lane-by-lane mixing
逐齿分度法 intermittent indexing
逐出 dislodge;drive out;eviction
逐次 by series;gradualness;successive
逐次逼近法 approximate approach; approximate method; approximate solution;cut-and-try method;method of approach; progressive approximation; sequential method; step-by-step approximation; try-and-error method; successive approximation
逐次逼近解法 trial-and-error solution
逐次逼近求解过程 step-by-step process
逐次逼近型模拟数字转换 successive approximation type analog(ue)-digital
逐次比较 sequential comparison
逐次比较法 method of successive comparison
逐次测定 sequential test
逐次测量法 step-by-step method
逐次差分 successive difference
逐次差拍法 beat-down method
逐次超松弛 successive overrelaxation
逐次超松弛法 successive overrelaxation method
逐次撤退 successive withdrawal
逐次沉淀 successive sedimentation
逐次成群超松弛 successive group overrelaxation
逐次抽样法 successive sampling method
逐次抽样检查 sequential sampling inspection
逐次代换法 method of successive substitution;successive substitution
逐次代入法 method of successive iteration; successive iteration method;successive substitution process
逐次导数 successive derivative
逐次点超松弛法 successive point overrelaxation method
逐次迭代法 step-iteration [iterative] method
逐次对比 successive contrast
逐次范围 successive range
逐次分析 sequential analysis
逐次改良过程 process of successive improvement
逐次改正法 method of successive corrections
逐次共轭法 method of successive conjugates
逐次归纳法 successive induction
逐次焊接 step soldering
逐次行松弛法 successive row relaxation method
逐次积分法 successive integration
逐次极小点 successive minimum points
逐次简化 successive reduction
逐次渐近法 cut-and-trial method; cut-and-trial process; cut-and-try method; method of approach; method of successive approximation; successive approximation
逐次渐近法分析连续梁 analysis of continuous girders by successive approximation
逐次校正法 step-by-step calibrated procedure; step-by-step calibration method; step-by-step calibration procedure; step-by-step correction procedure
逐次接近法 approximation method; cut-and-try procedure; cut-and-try process;trial-and-error procedure
逐次接近解法 solution by successive approximation
逐次结构 successive structure
逐次近似 progressive approximation
逐次近似法 convergence method; iterative procedure; method of successive approximation; relaxation method; step-by-step method; successive approximation
逐次近似过程 relaxation process
逐次近似值 successive approximate values
逐次进位 sequential carry;successive carry
逐次进行 succession
逐次距离 successive range
逐次聚焦 gradual convergence
逐次扣款卡 decrementing card
逐次偏导数 successive partial derivative
逐次平差 adjustment in successive steps
逐次启闭快门 progressive shuttering
逐次倾斜法 gradual incline method
逐次求导 successive derivative
逐次求积分 successive quadrature
逐次求偏微分法 successive partial differentiation
逐次区间 successive intervals
逐次趋近法 successive approximation
逐次试探法 successive trial
逐次衰变 successive decay
逐次松弛 stepwise relaxation
逐次松弛法 stepwise relaxation method
逐次弹性分析法 method of stepwise elastic analysis
逐次弹性杆法 method of stepwise elastic bar
逐次替换 successive displacement
逐次替换法 method of successive displacement;successive displacement method
逐次网格加密 successive mesh reduction
逐次网格加密法 successive mesh reduction method
逐次微分 successive differentiation
逐次微分系数 successive differential coefficient
逐次微商 successive derivative
逐次位移法 successive displacement method
逐次线超松弛 successive line overrelaxation
逐次线性滤波 successive linear filtering
逐次相加 over-and-over addition
逐次消除法 successive elimination
逐次消去法 method of successive elimination
逐次消元法 successive elimination; successive elimination method
逐次循环 successive cycles
逐次验收抽样法 sequential sampling plan
逐次移位法 method of successive displacement;successive displacement method
逐次约化 successive reduction
逐次跃进 successive bound
逐次跃迁 gradual transition
逐次增量(交通)分配法 incremental assignment method
逐次增区(交通)分配法 quantal assignment method
逐次值 successive value
逐次置换法 method of successive displacement; successive displacement method
逐次综合法 successive synthesis
逐次最小二乘法 iterative least square method
逐滴供给 drip feed
逐点 point-to-point;pointwise
逐点爆炸法 shot popper
逐点测定法 point-by-point method; point-to-point method
逐点测量 point-by-point measurement;point-by-point survey
逐点测图 point-by-point mapping
逐点传送法 dot-sequential method
逐点的 point-to-point;pointwise
逐点发送图像 dot-sequential image
逐点法 point-to-point method
逐点航行 point-by-point sailing
逐点积分 point-by-point integration
逐点计算 point-by-point calculation; point-by-point computation
逐点计算法 point-by-point calculation method
逐点加速试航 progressive speed trials;standardized trial
逐点纠正 point-by-point rectification
逐点可应用度 pointwise availability
逐点控制 point-to-point control
逐点内插 point-by-point interpolation
逐点扫描 point-by-point scanning
逐点示功器 point-by-point indicator
逐点松弛法 successive point relaxation method
逐点析像 point-by-point scanning
逐点自动匀光 point-to-point dodging
逐段逼近 piecewise approximation
逐段逼近法 piecewise approximation method
逐段迭代 piecewise iteration
逐段回归方法 stagewise correlation procedure; stagewise regression procedure
逐段检验法 cut researching method
逐段浇注 progressive placement
逐段精选法 stage concentration
逐段就位 progressive placement
逐段连续的 continuous bit by bit
逐段联路信令【铁】 link-by-link signal
逐段确定法 piecemeal determination
逐段线性 piecewise linear
逐段线性迭代 piecewise linear iteration
逐段线性回归 piecewise linear regression
逐段寻找法 cut searching method
逐对分开 pair off
逐稿轮 check drum;concave beater; cylinder beater;grain carrier beater
逐稿器 separator riddle;strawwalker
逐稿器挡帘 check board
逐稿器曲轴 shaker crank;strawwalker;strawwalker crank
逐稿器鱼鳞筛 straw-rack louver
逐稿器逐稿轮 strawwalker beater
逐个点火起爆 firing one by one
逐个审查 case by case screening
逐行倒相彩色电视制 phase alternation line colour system
逐行读出方式 row-by-row system
逐行交错 progressive interlace
逐行箝位 line-by-line clamping
逐行扫描 line-by-line scan(ning); progressive scan(ning);sequential scan(ning);successive scan
逐行扫描成像 line-by-line mapping
逐行扫描光栅 non-interlaced raster
逐行水平读出 line-by-line horizontal read out
逐行同步 line-by-line synchronization
逐行印刷装置 line-at-a-time printer
逐航线纠正 strip-by-strip rectification;stripwise rectification
逐户兜售推销员 knocker
逐户分送 door-to-door delivery
逐户收集垃圾 collection of refuse at every door

逐级 step-by-step
逐级曝光试验 step exposure test
逐级测量法 cascade method; cascading method
逐级的 hierarchic(al)
逐级滴定 stepwise titration
逐级点支承的 support pointwise
逐级分解 stepwise decomposition
逐级过程 stagewise process
逐级合理利用热能 realizable utilization of heat energy stage by stage
逐级计算法 stage-by-stage method; stepwise computation
逐级加荷试验 incremental loading test
逐级加载(应力)试验 step stress test
逐级进位 cascaded carry
逐级控制器 step-by-step controller
逐级平差 adjustment in successive steps
逐级破碎 stage breaking; stage crushing; stage reduction
逐级筛选 step sizing
逐级审批程序 one-step authorization
逐级生成常数 stepwise formation constant
逐级试验 step-by-step test
逐级试验法 step-test procedure
逐级衰变 successive decay
逐级水解 stepwise hydrolysis
逐级松弛 stepwise relaxation
逐级缩径钢壳与钢管组合桩 pipestep-taper pile
逐级调节 step-by-step regulation
逐级跳闸 cascade tripping
逐级稳定常数 stepwise stability constant
逐级消化 <用两个或多个池子进行污泥的> stage digestion
逐级选择标记 step strobe marker
逐级增厚翼板 garboard planking
逐级轧碎 stage crushing; stage breaking; stage reduction
逐级展开 gradual development
逐减计数器 down counter
逐件检查 screening inspection
逐件装卸法 break-bulk
逐渐 in-process of time
逐渐崩坏 successive collapse
逐渐逼近 successive approximation
逐渐逼近平差法 step-by-step approach to final
逐渐变薄 gradually thinning
逐渐变化 gradual variation
逐渐变宽 progressive widening
逐渐变弱 thin down
逐渐剥落 progressive scaling
逐渐不用规定 phase-out regime
逐渐沉降 gradual settlement; progressive settlement
逐渐沉陷 gradual settlement; progressive depression; progressive settlement; progressive subsidence
逐渐导致 lead up
逐渐地 by degrees; gradually; little by little
逐渐跌落 gradual roll off
逐渐放宽 gradual widening; progressive widening
逐渐分离 gradual-separation
逐渐过程 piecemeal process
逐渐滑动 progressive glide; progressive slide
逐渐恢复健康 convalescence
逐渐毁坏 sap
逐渐集中 gradual massing
逐渐加高式尖轨 <跟部落平> switch point with graduated risers
逐渐加荷 entry into power
逐渐加强 gaining strength
逐渐加载(法) gradual load
逐渐减薄层 graduated courses
逐渐减弱 taper
逐渐减少 pare; wear through
逐渐减小的 convergent
逐渐减压室 <压缩空气工作环境的> decant lock
逐渐接合 gradual engagement
逐渐接近法 approximate method; method of approach
逐渐接近平差法 step-by-step approach to the final adjustment
逐渐近似法 step-by-step construction
逐渐近似模/数转换器 successive approximation analog/digital converter
逐渐靠拢到看得见 hove in sight
逐渐枯竭 peter out
逐渐扩张 gradual expanding; gradual widening
逐渐冷凝法 gradual freezing
逐渐冷却 gradual cooling
逐渐连续不规则 gradual continual non-regular variation
逐渐连续有规则 gradual continual regular variation
逐渐(零零碎碎)改进 piecemeal improvement
逐渐磨细的 taper ground
逐渐黏[粘]合 progressive bonding
逐渐破坏 progressive failure; successive failure
逐渐趋近法 method of approach
逐渐趋近解法 solution by successive approximation
逐渐深宽的水槽 <在钻头底唇上的> expanding waterway
逐渐渗出 exfiltration
逐渐失效 gradual failure
逐渐施加荷载 gradually applied load
逐渐适应温度带 zone of increasingly favorable temperature
逐渐释放 gradual release
逐渐收缩 gradual contraction
逐渐收缩管 convergent tube; diminishing pipe
逐渐收狭的车行道断面(或路段) tapered carriageway section
逐渐损坏 progressive failure
逐渐缩小 gradual contraction; successive reduction; taper
逐渐缩窄 gradual narrowing; progressive contracting; progressive narrowing
逐渐提成 initial lump sum
逐渐停止 phase-out
逐渐停止消耗臭氧物质 phase-out of ozone-depleting substances
逐渐退化 gradual degradation
逐渐稀薄 thin down
逐渐下跌的物价 receding prices
逐渐消耗 wear through
逐渐消失 fade-away; fading
逐渐消隐 fade down
逐渐要废弃的 obsolescent
逐渐引到 lead up
逐渐增加 step up
逐渐增压式破碎机 gradual-pressure crusher
逐渐展平 <水波在扩散过程中> flattening out
逐渐制动 graduate braking
逐渐转变的轮廓 gently modulated profile
逐渐转变的剖面 gently modulated profile
逐渐转化(到相邻的层次) gradate
逐节变直径管 stepped taper tube
逐句编译程序 incremental compiler
逐孔施工法 progressive erection method; span-by-span construction
逐孔现浇法 span-by-span cast-in-place method
逐跨施工法 progressive erection; span-by-span construction; span-by-span method <移动支架的>
逐例进行 case-by-case basis
逐列相乘 multiply column by column
逐面分类法 faceted classification
逐面分析法 facet analysis
逐年比较 year to year comparison
逐年变化 year-to-year change
逐年计算折旧 compute depreciation on an annual basis
逐年平摊偿还计划 level annuity plan
逐年延展计划 rolling plan
逐年游高的潮漫滩 aggradated flood plain
逐年租赁 year-to-year tenancy
逐鸟把戏 bird dogging
逐批结算 load-to-load
逐批容差 lot tolerance
逐期还本减息抵押贷款 direction reduction mortgage
逐期一致性 historic(al) consistency
逐期折旧明细表 lapsing schedule
逐日编列流水号数 numbering in daily series
逐日变化 daily variation; day-to-day variation
逐日差程 daily range
逐日覆盖 daily cover
逐日掘土方量 daily yardage
逐日平均 daily mean
逐日取样 daily sample; day sample
逐日试验 day-to-day test
逐日温度 degree-day
逐日预报 daily forecasting
逐日最高温度平均值 mean daily maximum temperature
逐升的弯曲斜坡 helicline
逐时变化系数 hourly variation coefficient
逐时潮高 hourly height
逐时负荷 hourly load
逐时观测 hourly observation
逐时降水量 hourly precipitation
逐时冷负荷 hourly cooling load
逐时冷负荷综合最大值 maximum sum of hourly cooling load
逐时量 hourly amount; hourly yield
逐时综合温度 hourly sol-air temperature
逐条拌和【道】 lane-by-lane mixing
逐条车道搅拌法 lane-by-lane mixing
逐条列举 itemize
逐位 successive
逐位操作 bitwise operation; step-by-step operation
逐位存储 bit-by-bit memory
逐位法 digit by digit method
逐位跟踪 audit trial
逐位计算 step-by-step computation
逐位进位 cascaded carry; step-by-step carry
逐位进位加法器 ripple adder
逐位运算 bitwise operation
逐相起动 phase-after-phase start-up
逐项 member by member; part by part; successive term
逐项比较法 item-by-item method
逐项承包(工程等) job-by-job
逐项的 project-by-project; term-by-term
逐项登记表 roster
逐项地 termwise
逐项给予 case-by-case
逐项积分 term-by-term integration; termwise integration
逐项计算 calculated throughout
逐项可积级数 termwise integrable series
逐项可微的 termwise differentiable
逐项连续取样 item-by-item sequential sampling plan
逐项微分 termwise differentiation
逐项微分法 term-by-term differentiation
逐一处理 detail
逐月 month to month
逐月变化 monthly variation
逐月方向频率 monthly wind direction and frequency
逐月付款的累积购买法 layaway plan
逐月降水量分配 monthly distribution of precipitation
逐帧照射 frame-by-frame exposure
逐帧直接再现 frame-to-frame playback
逐字记录 verbatim record
逐组操作 block-by-block operation

主安全阀定位 main relief setting

主安全机构 primary safety mechanism
主安全控制器 primary safety control
主坝 key dam; major dam; main dam
主摆 master pendulum
主板 mainboard; mother board
主板加长段 base plate extension
主板条 king plank
主板桩结构 pile and sheet-pile
主版 key plate
主办 auspices; sponsoring; to sponsor
主办单位 host unit
主办公室 principal office
主办公司 sponsoring firm
主办人 entrepreneur; sponsor
主办事处 principal office
主办银行 main funding bank
主办者 entrepreneur; sponsor
主半空间 principal half-space
主半页 prime half page
主瓣 main lobe; major lobe
主瓣立体角 main-lobe solid angle
主保护装置 main protection
主保险丝 main fuse; principal fuse
主报警 <信号器> main alarm
主报警电路 primary alarm circuit
主报警器 main alarm
主报警系统 primary alarm system
主曝光 main exposure
主备降机场 master diversion airfield; primary diversion field; primary divert field
主、被叫用户双方话终拆线 called and calling subscribers release
主本征值 dominant eigenvalue
主泵 main pump; primary pump
主泵机组 main pump unit
主比较器 main comparator
主比例尺 basic scale; general scale; main scale; nominal scale; principal scale
主闭环 main closed loop
主篦式雨水口 inlets with vertical gratings
主臂 main arm; main boom; main jib; boom column <起重机>
主边带 main sideband
主边界(逆冲)断层 main boundary thrust
主编 chief editor; general editor; managing editor
主编单位 compiled unit in chief
主编人 editor in chief
主变电所 main transformer station
主变电站 main power substation; main transformer station
主变电站环控 environment control of

main substation
主变电站设备 main substation equipment
主变量 main variable;master variable
主变数 principal variable
主变速机 main gear box
主变速箱 main gear box
主变形 principal deformation;principal distortion
主变形方向 principal direction of distortion
主变压器 high-voltage transformer;main transformer
主变压器零序阻保护 power transformer zero phase sequence impedance protection
主变压器气体保护 power transformer gas protection
主变阻器 master rheostat
主标 principal mark
主标度线 main scale mark
主标记 main pip
主标题 main title
主标志 main mark;principal mark
主表面 first type surface; primary surface
主表现 principal representation
主冰期 major glacial epoch
主波 dominant wave; fundamental wave;main wave;major wave;predominant wave;principal wave;superior wave
主波瓣 main lobe;major lobe
主波长 principal wave length
主波道 main channel
主波门 main gate
主波束 main beam
主波束效率 main beam efficiency
主播送区 primary service area
主播送室 main studio
主部分离 separation of principal part
主部件 mass unit;master unit
主部件总成更换 major unit assembly replacement
主材 principal
主菜单 main menu;root menu
主参考机 master reference machine
主参数 principal parameter;rating parameter
主舱 main(living)chamber
主舱壁 king bulkhead;main bulkhead
主舱口 main hatch(way)
主操舵装置 main steering equipment;main steering gear
主操纵杆 master handle;master lever
主操纵杆棘齿板 master lever ratchet
主操纵轮 main wheel;master wheel
主操纵室 main control room
主操纵手柄 main control lever
主操纵手柄联锁释放装置 main control lever interlock release
主操纵台 main control board; main control console; master control console
主操纵系统 primary control circuit
主操作 main operation;master operation
主操作控制程序 main operation control program(me)
主操作控制台 master operational console
主操作控制装置 master operations controller
主操作内务程序 main operation house-keeping routine
主操作盘 main controlling board
主操作系统 master operating system
主槽 main channel
主槽流量 main channel flow
主槽线 channel line
主侧 main side;primary side
主测点 master station
主测距电路 main range circuit
主测量点 main measuring point
主测深线间距 main sounding line interval
主测试段 principal test section
主测试台 primary test board
主测线 dominant line
主测线走向 rend of main survey line
主测油孔 main metering jet
主测站 major station;primary station
主测站标桩 main station peg;primary station peg
主层 main stor(e)y
主层集料 <贯入式道路等的> main-course aggregate
主层楼面 principal floor
主层石料 main-course ballast
主插件板 master board
主汊道 main branch(line)
主产量 principal yield
主产物 principal product
主厂房 machine hall; main building; main generator hall;main generator room; main house; main machine hall;power house
主场聚焦 main field focusing
主唱片 master
主唱诗班 main choir
主潮 main flood;primary tide;principal tide
主潮港 reference station; standard port;standard station
主车道 main lane
主车道内移 lane shift
主车架 main frame
主车间 main shop
主车梁 main frame
主车辆甲板 main vehicle deck
主车门 main body entrance door
主车站 major station
主衬套 master bushing
主衬铁 main filler
主撑 main brace
主成分 principal component;principal constituent
主成分变换 principal component transformation
主成分分析 principal component analysis
主成分回归 principal component regression
主成圈区 main knitting station
主成圈系统 leading knitting system
主承包商 principal contractor
主承力索 main carrier cable; main catenary wire【电】
主承载框架 main bearing frame
主承载系统 main load-bearing system
主承重结构 principal bearing structure
主承重系统 main load-bearing system
主城堡 upper citadel
主程序 main program(me); main routine; mass program(me); master routine
主程序带 master instruction tape; master program(me) tape
主程序段 mother block
主程序控制器 master program(me) controller
主程序库 master library
主程序日程表 master programming schedule
主程序数据区 main program(me) data area
主程序文件 master program(me) file
主程序序列 main program(me) sequence
主程序循环 main program(me) cycle
主程序钟 master program(me) clock
主程序装置 master sequencer
主程序组 chief program(me) team
主澄清 main defecation
主持 aegis;officiate;preside
主持单位 appraising unit in chief
主持建筑师 architect in charge
主持开幕仪式者 inaugurator
主持落成仪式者 inaugurator
主持人 emcee <电台节目的>;moderator <学术会议的>
主尺 main scale;primary division <比例尺的>
主尺度 principal dimension
主齿轮 bull gear;master gear
主齿轮箱 main gear box
主齿条(机构) master rack gear
主冲程 main stroke
主冲击 main stroke
主抽气泵 main scavenging pump
主抽水泵 permanent pump
主抽头 main tapping
主出水槽 main outlet channel
主出水口 main outlet
主储藏场 main storage yard
主储存器 main reservoir; primary memory;primary store
主储气器 main gas container; main reservoir
主储气筒 main air reservoir
主储油器 main oil reservoir
主处理 main treatment
主处理部件 main processing block
主处理程序 master processor program(me)
主处理机 host processor;main processer
主处理污水 main process wastewater
主处理系统 host processing system
主触点 main contact
主触点组 main contact group
主触发器 driving flip-flop; master flip-flop
主触簧片 main contact spring
主触头 main contact
主传动 final transmission;main drive;main transmission;master drive
主传动操纵手柄 main drive control
主传动齿轮 main drive gear
主传动齿轮轴承 main drive gear bearing
主传动电动机 main drive motor
主传动机构 main drive; main drive gear
主传动机构齿轮 main drive gear wheels
主传动件 master driver
主传动轮 main driving wheel
主传动轮毂 final drive hub
主传动设备 main drive
主传动系 main transmission
主传动箱 main transmission box
主传动小齿轮 main drive pinion
主传动轴 final drive shaft;line shaft;main driving axle
主传动轴轴承的固定螺栓 stud-horse bolt
主传动装置 main drive gear
主传动锥齿轮 axle drive bevel gear
主传感器 master sensor
主传送带 main belt;primary conveyor
主传装置 master driver
主船闸 main lock
主椽 principal rafter;main rafter
主椽椽脚系梁 footing beam
主椽木 principal rafter;main rafter
主椽体 main rafter
主窗 prime window
主垂面方位角 azimuth of principal plane
主垂线 principal vertical(line)
主垂直圈 principal vertical circle
主锤 main ram
主磁场 main magnetic field
主磁场绕组 main field winding
主磁点 primary magnetic standard
主磁极 main pole
主磁极间隙 main pole gap
主磁极铁芯 main pole core
主磁力阀 master magnet valve
主磁铁环 main magnet ring
主磁通 main magnetic flux
主磁滞回线 major hysteresis loop
主次道路交叉口 arterial-local type intersection
主次环结构 major/minor loop organization
主次梁构架 beam and girder framing
主次梁接合 girder and beam connection;girder-to-beam connection
主次梁连接 beam-to-girder connection; girder and beam connection; girder-to-beam connection
主次梁式结构 beam and girder construction
主次梁式楼板 beam and girder floor
主次数 principal degree
主次双柱式 prick post
主刺激 cardinal stimuli
主从操纵器 master-slave manipulator
主从触发器 jack-king flip-flop; master-slave flip-flop
主从传感器距离 distance between master and slave sensor
主从带式传动 master-to-slave tape drive
主从的 master-slave
主从方式 master-slave mode;master-slave system
主从关系 master slave relation
主从计算机 master-slave computer
主从计算机系统 master-slave computer system
主从累加器 master slave accumulator
主从配置 master-slave configuration
主从式 client-server
主从式吊具 master-slave spreader
主从式多道程序设计 master-slave mode multiprogramming
主从同步 master-slave synchronization
主从系统 master-slave system
主从运转 master-slave operation
主丛 principal bundle
主存储 main file
主存储键 storage key
主存储内的数据传输 transferring data in storage
主存储器 main memory; main storage;primary memory;primary storage;primary store;dedicated memory;main store
主存空间 primary memory space
主打印机 master printer
主大齿轮 main gear wheel
主大梁 cephalophorous;main girder
主大圆 fundamental circle; primary great circle
主带 master tape
主单纯形法 primal simplex method
主单位 primary unit
主单元 main unit
主导层 guide seam
主导的 prevailing
主导电动机 capstan motor
主导电索 main messenger
主导洞 main tunnel heading
主导阀 director valve; pilot valve; master valve

主导风 prevailing wind; predominant wind
主导风跑道 prevailing wind runway
主导风向 cardinal wind; leading wind direction; main wind; most frequent wind direction; predominant wind; prevailing direction of wind; prevailing wind direction; principal wind direction
主导工业 key industry
主导构架 leading frame
主导构造体系 leading tectonic system
主导观念 guiding idea
主导管 main conduit; main manifold
主导轨 main guide
主导河岸 dominant bank; predominant bank
主导化石 guide fossil; leading fossil
主导活门 director valve
主导机械 key machine
主导勘探线 leading exploration line
主导矿物 guide mineral
主导浪 predominant wave
主导轮 capstan
主导脉冲 master pulse
主导能见度 prevailing visibility
主导漂沙 predominant drift
主导数 principal derivative
主导水流 prevailing current
主导思想 ruling idea
主导条件 prevailing condition
主导误差 governing error
主导线【测】main traverse(line); primary traverse; main guide
主导线网 main traverse net; principal polygonometric network
主导因子 key factor; leading factor
主导语言 ruling language
主导振荡器 capstan oscillator
主导指令 key instruction
主导轴 capstan axle; capstan shaft; drive capstan
主导轴承 main guide bearing
主导轴空转轮 capstan idler
主导轴输送机构 capstan feed
主导装置 master
主岛弧【地】primary arc
主灯(光) key light; king light; main light
主灯丝 main filament
主灯丝断丝报警 alarm for burnout of a main filament
主等高线 primary contour; principal contour
主低温换热器 main cryogenic heat exchanger
主低压 primary depression; primary low
主堤 main dam; main dike[dyke]; main levee; mainline levee; primary dike
主底板 main base
主底板总成 main floor unit
主底梁 main sill
主底片 master negative
主底色 basic shade
主地产 dominant tenement
主地球站 main earth station
主地下水 main ground water
主地下水面 main water-table
主地下水位 main ground water table
主点 cardinal point; center point; central plant; main point; major controlled point; photo-centre; principal point <空摄影测量相片的>
主点导线 principal point traverse
主点法 principal point method
主点辐射线 principal point radial ray
主点辐射线绘图【测】principal point radial line plotting
主点位移 principal displacement
主点误差 principal point error
主点像片三角测量法 principal point method
主点阵 host lattice
主电波 main current
主电池 main battery; main cell
主电池电路 main battery circuit
主电动机 main motor; primary motor
主电动机联轴器 lead spindle
主电动机置台 main motor storage rack
主电动接触器 main motor contactor
主电极 main electrode
主电抗 principal reactance
主电缆 main cable; principal cable
主电缆沟 main cable trench
主电力电缆 primary power cable
主电流 main current; principal current
主电流继电器 main current relay
主电流终端开关 main current terminal switch
主电路 main circuit; power circuit; trunk feeder
主电路电流 main current
主电路断路器 main circuit breaker
主电路开关 main circuit switch
主电路连接 main circuit connection
主电路组合开关 power switch group
主电门 main switch
主电平控制 master level control
主电视天线 master television antenna
主电刷 energy brushes; main brush
主电线束 main wire bundle
主电压 principal voltage
主电压选择钮 main voltage selector
主电源 main power source; main power supply; main source of electric(al) power; main supply; major power supply; primary source
主电源同步 mains hold
主电源线 principal wire feeder
主电闸 main gate
主电站 major station
主电子计算机 master computer
主垫圈 main washer
主殿 main shrine; nanos <古希腊、古罗马庙宇的>
主吊杆 basic boom; basic jib
主吊钩 main hoist
主吊架 main gantry
主吊索 main suspension cable
主吊运车 main trolley
主调 chief dispatcher
主调度 master scheduling
主调度程序 main scheduler; main scheduling routine; master scheduler
主调度程序任务 master scheduler task
主调度中心 main dispatching center [centre]
主叠合素 principal coincidence
主顶板 main roof
主定时脉冲 master clock
主定时频率 master timing frequency
主定时器 master timer
主定时台 master timing station
主定时系统 master timing system
主定位线圈 major positioning coil
主动 initiative
主动安定面 active stabilizer
主动安全性 active safety
主动保护 active protection
主动报价 unsolicited offer
主动臂 master arm
主动边 active edge; tension side
主动变换器 active transducer
主动部件 active component; driving part
主动侧 drive side; driving side; master end
主动测距系统 active ranging system
主动场屏蔽 active field shielding
主动齿轮 bull gear; drive gear; driving(toothed) gear
主动齿轮架 pinion cage
主动齿轮键 driving gear key
主动齿轮套 pinion carrier
主动齿轮调整 <用垫片> pinion shimming
主动齿轮箱 drive gear box; drive gear carrier; drive gear casing; driving gear box
主动齿轮轴 driving gear shaft
主动齿轮轴承盖 main drive gear bearing cap
主动齿轮轴承隔圈 pinion bearing spacer
主动齿轮轴承护圈 main drive gear bearing retainer
主动齿轮轴承抛油环 main drive gear bearing oil slinger
主动传递 active transfer
主动船舶交通服务 active vessel traffic service
主动大陆边缘 active continental margin
主动导航 active guidance; active homing
主动导向 active guidance
主动的 active; driving
主动等待 active wait
主动地位 whip hand
主动地震 active earthquake
主动地震压力 active seismic pressure
主动端 drive end; drive side; driving end; master side
主动端盖板 drive end head
主动端抛油环 drive end plate slinger
主动端毡垫圈护罩 drive end felt washer guard
主动端轴承 drive end bearing
主动段 powered phase; propulsion branch
主动段速度 speed under power
主动段以后 postboost
主动段制导 powered phase guidance
主动多普勒速度传感器 active Doppler velocity sensor
主动舵 active rudder
主动法 active method
主动放大系统 active augmentation system
主动(杠)杆 driving lever
主动隔振 active vibration isolation
主动跟踪 active following; active tracking
主动跟踪系统 <将信号发射到卫星上,并接收由卫星返回的回波> active tracking system
主动拱 active arch
主动构件 active member; driving member
主动管 drive pipe
主动光学 active optics
主动辊 drive roll; guide roll
主动横减摇装置 active roll damping facility
主动红外探测系统 active infrared detection system
主动滑车 driving pulley
主动滑动面 active surface of sliding
主动滑动区 active slide area
主动滑轮 driving pulley; feed pulley
主动换能器 active transducer
主动回忆 active recall
主动活动 active movement
主动或被动土压 active or passive thrust
主动机构 drive mechanism; driving mechanism; main drive
主动机构位移 carry shift
主动机械运动方向 driving direction
主动激光跟踪系统 active laser tracking system
主动激励 driving
主动级轮 stepped driving cone
主动棘轮 driving ratchet
主动记录设备 active recording device
主动加热式潜水服 actively heated suit
主动假目标 active decoy
主动减摇装置 active stabilizer
主动剪应力 active shear stress
主动件 drive part
主动抗摇系统 active antiroll system
主动控制系统 active control system
主动朗肯破坏区 Rankine active failure zone
主动朗肯区 active Rankine zone
主动雷达航标 radar marker; ramark
主动雷达探测器 active radar probe; active radar sensor
主动雷达物标 active range target
主动力 active force; drive force; driving force
主动力系统 main power system
主动力系统辅助设备 main power auxiliaries
主动力源 primary power source
主动力装置 main power plant; primary power plant
主动链 drive chain; driving chain; propelling chain
主动链轮 driving chain sprocket; head sprocket; drive sprocket; driving sprocket(wheel)
主动链轮轴 drive sprocket axle
主动轮 action wheel; driver; drive wheel; driving pulley; driving wheel; leading wheel; motion work wheel; traction wheel; primary wheel <离心泵的>
主动轮传动 capstan feed
主动轮电动机 capstan motor
主动轮刹车 driver brake
主动轮误差 capstan error
主动轮叶 runner blade
主动轮闸 driver brake; driving wheel brake
主动轮闸油环 driver brake oil cup
主动轮制动器 driver brake
主动轮重 driver weight
主动轮重量 weight on drivers
主动螺母 driving nut
主动门扇 active leaf
主动摩擦圈 driving friction ring
主动扭矩 driving torque
主动耙(吸)头【疏】active draghead
主动盘 driving disc[disk]
主动皮带轮 drive pulley; driver; driving pulley; fast pulley; head pulley; main pulley
主动破坏 active failure
主动破坏楔体 active failure wedge
主动桥的主动轴颈 output axle nose
主动曲柄 driving crank
主动圈 driving ring
主动人造卫星 active man-made satellite
主动冗余 active redundance[redundancy]
主动伞齿轮 driving bevel gear
主动声呐检测 active sonar detection
主动声呐系统 active sonar system
主动施工法 <施工前融化或挖去基础内冻层的> active method
主动式 active model; active system
主动式传感器 active sensor
主动式存储器 active memory; active storage

主动式红外跟踪系统 active infrared tracking system
主动式救生(衣)服 active lifejacket; active survival suit
主动式目标 active target
主动式平交道口防护 active highway grade crossing protection
主动式声呐 active sonar
主动式水雷 torpedo mine
主动式太阳房 active solar house
主动式太阳能采暖 active solar heating
主动式太阳能采暖系统 active solar heating system
主动式太阳能系统 active solar system
主动式太阳能装置 active solar energy system
主动式显示 active display
主动式遥感 active remote sensing
主动式夜视仪 active infrared equipment
主动式仪器 active instrument
主动式指令声呐浮标系统 command active sonobuoy system
主动式转换器 active transducer
主动式自动引导头 fully active homing head
主动输送 active transport
主动水硬性结合料 active hydraulic binder
主动塑性平衡状态 active state of plastic equilibrium
主动损坏 active failure
主动塔轮 driving condition
主动太阳能系统 active solar energy system
主动弹簧 drive spring; driving spring
主动停止 positive stop
主动头 driving head
主动凸轮 actuating cam
主动凸缘 drive flange
主动土力 active soil force
主动土推力 active earth thrust; active thrust of earth
主动土压力 active earth pressure; active soil pressure; active thrust of earth
主动土压力系数 active earth pressure factor; coefficient of active earth pressure
主动土压力重分布 active earth pressure redistribution; active soil pressure redistribution
主动土压力状态破坏面 active earth pressure rupture; active rupture
主动托轮轴 driving friction wheel
主动微波 active microwave
主动微波系统 active microwave system
主动违拗 active negativism
主动吸收 active absorption
主动系统 active system
主动相关器 active correlator
主动消防系统 active fire protection
主动小齿轮 drive pinion; driving pinion
主动小齿轮及轴 drive pinion and shaft
主动小齿轮油封 drive pinion oil seal
主动小齿轮锥形轴承 drive pinion cone bearing
主动小伞齿轮 driving bevel pinion
主动小正齿轮 spur driving pinion
主动小锥齿轮 bevel drive pinion
主动小锥齿轮滚珠轴承 bevel drive pinion ball bearing
主动型耙头【疏】 active draghead
主动寻址 active position address
主动压力 active pressure
主动压油齿轮 driver pressure gear
主动抑压法 active containment alternatives
主动抑制 active inhibition
主动引力质量 active gravitational mass
主动油缸 actuating cylinder
主动圆盘【机】 driving disc[disk]
主动源法 active source method
主动运动 active exercise; active movement
主动运输 active transport
主动再吸收 active reabsorption; active resorption
主动照明 active illumination
主动者角色 agent case
主动制导 active guidance
主动轴 axle shaft; drive axis; drive shaft; driving axis; driving axle; driving shaft; head shaft; leading axle; main drive shaft; main driving shaft; motive axle; drive axle
主动轴套 driving axle housing; driving axle sleeve
主动轴箱 drive shaft housing; driving axle box
主动轴重量 weight on driving axles
主动轴轴承 drive shaft bearing
主动轴轴承套 main drive shaft bearing sleeve
主动爪 driving claw
主动转向架 priving bogie
主动转移 active transport
主动桩 active pile
主动装置 aggressive device
主动状态 active state
主动状态破坏面 active rupture
主动追踪 active following
主动锥形(皮带)轮 driving cone pulley
主动准双曲线伞齿轮 hypoid drive pinion
主动自整角机 driving synchro
主动阻尼器 active damper
主动钻杆 drive pipe; driving stem; grief kelly; kelly
主动作用 active role
主斗桥 main bucket ladder
主度盘 master dial
主端子 main terminal; major terminal; principal terminal
主段 root segment
主断层 main fault; major fault; master fault; master fracture; principal fault
主断层带 main fault belt; main fault zone; major fault zone; principal fault belt; principal fault zone
主断层线 dominant fault line
主断裂组 principal sets of fault
主断路器试验台 main breaker test bench
主断面 principal section
主堆场 main storage yard
主堆垛机 master stacker
主堆积面 principal surface of accumulation
主对比度 master contrast
主对比度调整 master contrast control
主对称面 principal plane of symmetry; principal symmetry plane
主对称轴 principal axis of symmetry
主对话端 primary half-session
主对角多项式 principal diagonal polynomial
主对角平巷 main angle
主对角线 leading diagonal; main diagonal; principal diagonal
主对角(线)加劲肋 principal diagonal rib
主对角线矩阵 diagonal matrix
主对角线元素 element of principal diagonal
主对偶型 primal dual type
主对偶性定理 main duality theorem
主墩 main pier
主遁点 main vanishing point
主多谐振荡器 driving flip-flop
主舵 main rudder
主扼流圈 master retarder
主发电厂 main power station
主发电机 main generator
主发电机室 main generator room
主发电机通风机 main generator blower
主发电机组 main generating set
主发电站 main generating station
主发动机 cruising engine; main engine; prime engine; sustainer engine; sustainer motor
主发动机燃烧室 sustainer chamber
主发动机推力 cruising thrust
主发动机自动停车 automatic sustainer cut-off
主发动机组 sustainer unit
主发光峰 main glow peak
主发射机 master transmitter
主发射台 main transmitter; main transmitting station; master station
主发条 main spring
主发条钩 mainspring hook
主发条卷线器 mainspring winder
主伐 final clearing; harvest cutting; principal felling; regeneration-cutting
主伐林 final crop
主伐林木 main crop
主伐龄 final clearing age
主伐木 crop tree
主伐收获 final yield
主伐收益 return from final clearing
主阀 king valve; main valve; primary valve
主阀衬套 main valve bush(ing)
主阀大活塞 large main valve piston
主阀大活塞环 large main valve piston ring
主阀盖有头螺钉 main valve head cap screw
主阀杆 main valve stem
主阀杆螺母 main valve stem nut
主阀环 main valve ring
主阀簧 main valve spring
主阀盘 main valve disc
主阀室 main valve chamber
主阀小活塞 small main valve piston
主阀柱塞 main valve plunger
主法线 principal normal
主法线方向 principal normal direction
主法向应力 principal normal stress
主帆 main course
主反馈比 primary feedback ratio
主反馈系数 primary feedback coefficient
主反射率 main reflectivity
主犯 main culprit; principal offender
主范畴 fundamental category
主范式 principal normal form
主范数 principal norm
主方差 principal variance
主方程 master equation
主方式 master mode
主方位角 principal azimuth
主方向 predominant direction; principal direction
主方向风 prevailing wind
主防波堤 main breakwater; weather breakwater
主防护墙 primary protective barrier
主房 main house
主放大器 main amplifier; prime amplifier
主放大器电路 main amplifier circuit
主放电电流 main discharge current
主分布函数 principal distribution function
主分接头 principal tapping
主分界线 main divide; major divide
主分量 principal component
主分量变换 principal component transformation
主分脉 principal sector
主分配阀 main distributing valve
主分配运输机 main distributing conveyer[conveyor]
主分区 main partition
主分散 primary dispersion
主分式理想 principal fractional ideal
主分水界 main divide; major divide
主分水岭 major divide
主分水线 main divide; major divide
主分支 main branch(line); main split
主分支点 subcenter[subcentre]
主分支排水管 primary branch drain
主份酸 body acid
主风 main air; prevailing wind
主风道 air main; main air duct; main duct
主风洞 main tunnel
主风阀 main air valve
主风缸 main air reservoir; main reservoir
主风缸压力 main reservoir pressure
主风缸压力双重调节 duplex main reservoir regulation
主风管 air main
主风机 main air blower; primary fan
主风口 main airport
主风流隔开风门 main separating door
主风巷 main airway
主峰 basic peak; dominant peak; main peak; prominent peak
主峰最大值 main peak maximum
主锋【气】 principal front
主扶手 main runner
主服务区 primary service area
主辐射 primary radiation
主辐射器 primary feed
主辅 king pin
主辅助开关 master subswitch
主复位 master reset
主副瓣抑制比 main and auxiliary lobes restraining rate
主副分类 major minor sorting
主干 backbone; leader; spine; trunk(stem)
主干冰川 trunk glacier
主干材 bodywood
主干车站【铁】 principal railroad station
主干导爆线 main blasting lead
主干导线 primary traverse
主干道 trunk road; arterial road; backbone artery; backbone road; main artery; main road; main street; major artery
主干道绿灯(信号)【交】 main street green
主干的 arterial
主干堤岸 stem bank
主干电极电缆 trunk cable
主干电缆 main cable
主干电缆接入 primary cable access
主干分支 offshoot
主干钢筋 main bar
主干公路 arteria[复 arteriae]
主干管 main collector; principal

main;trunk duct
主干管道 principal line
主干管服务面积 area with main services
主干管线 principal line
主干控制 primary control
主干路 arterial road;main trunk highway;main trunk road;trunk line;turnpike road;trunk line road < 英国最高级公路 >
主干渠 parent channel;arterial drainage
主干隧道 major tunnel
主干天线 mains antenna
主干通信[讯]电话电缆 trunk call
主干网(络)backbone network
主干无枝的 clear boled
主干线 arterial primary main;main collector;mains;main trunk circuit;main trunk line;principal distribution line;trunk main;main line
主干线部分 main-line section
主干线路 main circuit;main track line
主干线配线柜 backbone closet
主干修剪 stem pruning
主干专用线 < 连接铁路线与专用线区的 > trunk siding
主干子系统 backbone subsystem
主杆 kingpost(truss);main pole;main post;master rod;main rod
主杆护架 main rod guard
主杆机 main rod
主杆三角测量 main triangulation;major triangulation
主杆套 main rod strap
主杆套螺栓 main rod strap bolt
主杆铜楔螺栓 main rod brass wedge bolt
主杆轴承 main rod bearing
主缸出口塞 master cylinder outlet plug
主缸出油接头垫片 master cylinder outlet fitting gasket
主缸底阀 master cylinder bottom valve
主缸垫密片 main cylinder gasket
主缸阀 master cylinder valve
主缸阀簧 master cylinder valve spring
主缸阀座 master cylinder valve seat
主缸盖 master cylinder cover
主缸盖垫片 master cylinder head gasket
主缸杆 master cylinder lever
主缸杆衬套 master cylinder lever bushing
主缸杆轴 master cylinder lever shaft
主缸供给箱 master cylinder supply tank
主缸活塞 master cylinder piston
主缸活塞副皮碗 master cylinder(piston)secondary cup
主缸活塞回动簧 master cylinder piston return spring
主缸活塞皮碗 master cylinder piston cup
主缸活塞推杆 master cylinder connecting link;master cylinder piston push rod
主缸加油口盖垫片 master cylinder filler cap gasket
主缸塞 master cylinder plug
主缸托架 master cylinder bracket
主缸罩 master cylinder boot
主钢筋 main reinforcement(bar);main reinforcing steel;principal reinforcement;principal reinforcing bar;principal(round)bar;main steel < 混凝土中的 >;main bar
主钢梁 principal steel beam;principal steel girder
主钢索 main cable
主钢桩【疏】main spud
主港 standard port
主港池 main basin
主港口房屋 principal terminal building
主杠杆 main lever
主高压透平 primary high-pressure turbine
主高压涡轮机 primary high-pressure turbine
主割理(方向)headway
主割面 < 用于测量学和应用力学 > principal plane
主格仑 main cell
主格栅 main joist;main runner;principal grid
主格网 major grid
主隔(舱)板 king bulkhead
主根 main root;principal root;taproot
主根系 tap root system
主功率 primary power
主功率开关 master power switch
主攻方向 main thrust
主供风管 main feed pipe
主供给基地 air parent base
主供水管 main supply
主供水立管 rising main
主供水区 main district of supply
主供油管 main supply line;oil supply bar
主拱 main arch;major arch;principal arch
主拱道 main archway
主拱顶 main vault;principal vault
主拱圈 main arch ring
主钩 main hook;principal sulcus
主构舱壁 structural bulkhead
主构架 basic structure;king truss;main frame;primary frame;principal frame
主构件 leading frame;primary component;primary member
主构造线 trend line
主构造应力场 main tectonic stress field
主谷 main valley
主鼓风机 main blower
主固定架 principal supporting structure
主固定站 main fixed station
主固定支架 main anchor
主固结 main consolidation;primary consolidation
主固结沉降 primary consolidation settlement;principal consolidation settlement
主顾 customer;patron
主关键字 major key;primary key
主观 judgmental probability
主观斑纹 subjective speckle
主观斑纹问题 subjective speckle problem
主观标准 subjective criterion;subjective standard
主观参考格网 subject frame of reference
主观测孔 main observation hole
主观测听法 subjective audiometry
主观测线 primary line of sight
主观的 subjective
主观的领域原则 subjective territorial principle
主观反映 subjective response
主观方法 subjective method
主观赋权变量 subjective weighting variable
主观概率 personal probability;subjective probability
主观概率法 subjective probability method
主观概率模型 subjective probability model
主观感觉 subject feeling;subjective sensation
主观感受 subjective response
主观光度计 subjective photometer
主观光泽 subjective gloss
主观混响时间 subjective reverberation time
主观计光术 visual photometry
主观价值 subjective value
主观亮度 luminosity
主观量度 subjective measurement
主观能动性 subjective activity
主观判断 absolute judgement
主观评分 subjective(panel)rating
主观评级 subjective rating
主观评价 subjective evaluation
主观色彩 subjective colo(u)r
主观(上的)不确定性 subjective uncertainty
主观视角 subjective angle of view
主观疏伐 subjective thinning
主观数据 subject data
主观温度 subjective temperature
主观误差 human error;subjective error
主观相等点 point of subjective equality
主观相等平均点 average point of subjective equality
主观性 subjectivity
主观样本 subjective sample
主观因素 subjective element
主观噪声计 subjective noise meter
主观证据 subjective evidence
主观资料 subjective information
主观自我 subject ego
主观坐标系统 subject coordinate system
主管 authoritative organ;person in charge;superintend;superintendence;the boss;the chief;top management;walking boss;governor < 美 >;house riser < 指供气供水管 >;main duct【给】;main pipe【给】;stack pipe【给】
主管部门 competent authority;competent department;department in charge;qualified institution;responsible institution
主管处 appropriate department
主管当局 administrative authority;competent authority
主管当局的证明书 certification by the competent authority
主管道 header;main pipe(line);principal pipe(line)
主管的 competent;master;in charge
主管断流阀 sectionalizing valve
主管法庭 competent court
主管工程师 engineer in charge;project engineer;staff engineer;supervising engineer
主管官员 officer in charge
主管机构 qualified institution;responsible organ
主管机关 agency in charge;appropriate body;authority having jurisdiction;competent authority;qualified institution;responsible institution;the authorities concerned
主管技师 technologist-in-charge
主管经理 department manager
主管路 main circuit;main pipeline;operating(pipe)line
主管路流量调节阀 restrictive flow-regulator
主管人 controller;principal-in-charge;superintendent;supervisor
主管人的职位 superintendency
主管人员 executive staff;officer in charge
主管水流 flow-through run
主管通气管 stack vent
主管线 common line;main circuit;main pipeline
主管业务的股东 managing partner
主惯量 principal moment
主惯性半径 principal radius of inertia
主惯性矩 main moment of inertia;principal moment of inertia
主惯性平面 principal plane of inertia
主惯性轴 main axis of inertia
主灌溉渠 main feeding ditch;main irrigation canal
主光 hot light;key light
主光谱 principal spectrum
主光线 chief ray;principal ray
主光源 primary light source
主光照明 hot lighting
主光轴偏角 averted angle of photographic(al)axis
主光轴倾角 tilt angle of photographic(al)axis
主光轴(线)principal optic(al)axis;primary optic(al)axis
主规 master ga(u)ge
主轨 main rail;main track
主轨道 main orbit;main trajectory;principal orbit
主辊 home roll
主滚筒 master rotor
主锅炉 main boiler
主锅炉检测 main boiler survey
主过程 main procedure
主过滤器 main filter
主过去文件 master history file
主海水系统 main sea water system
主涵洞 main conduit
主焊缝 main weld
主航程指示器 master distance indicator
主航道 fairway;main course;main fairway;main(navigable)channel;main navigation channel;trunk line route
主航道线 base course;fairway line;truck line
主航道中心线 center line of main channel;neutral line of main channel;talweg
主航路信标 principal airway beacon
主航天飞船 primary space vehicle
主航线 main(shipping)track;trunk line;trunk route
主航向 principal course
主合点 main vanishing point;principal vanishing point
主合点控制器 vanishing point controller
主合金 master alloy(ing)
主合同 main contract;major contract;master contract;prime contract
主河 main river;main stream;master river
主河槽 channel of main stream;main river bed;main river channel;major river bed
主河床 body of river;main bed;major bed
主河道 channel of main river;channel of main stream;main bed;main river channel;main stem;major river bed;principal channel;drainage line < 水系中的 >
主河流 mass river;main stream
主核线 principal epipolar ray
主荷载 main load;primary load;principal load

主荷载转移 major load transfer
主黑子 main spot
主恒温器 master thermostat
主桁架 basic structure; main couple; main truss; primary truss
主横舱壁 main transverse bulkhead
主横断面 principal transversal
主横截面 principal cross-section
主横梁 cross beam; main crossbar; main crossbeam; main cross girder; main beam
主横线 principal horizontal line
主横像差 principal lateral aberration
主后角 primary clearance; end relief <刀具的>
主候机室 principal terminal building
主弧 major arc; minor arc; positive arc <六分仪的>
主护面层 primary cover layer
主护面块 main armo(u)r
主滑动体 main slide mass
主滑阀 main slide valve; primary spool
主滑脚 main shoe
主滑脚铸件 main shoe casting
主滑轮 head pulley; head sheave; main sheave
主滑面 main gliding surface
主滑面产状 attitude of main glide plane
主滑条 main runner
主滑脱面深度 depth of main detachment
主化合价 chief valence
主化油器 main carburetor
主划分 main partition
主环 main ring
主环路 main circuit
主缓行器 main retarder; master retarder; primary retarder
主缓行器控制 main retarder control; primary retarder control
主缓行器位置 main retarder position; primary retarder position
主簧 main spring
主簧凹缺 recess for main spring
主回答表 master answer sheet
主回答单 master answer sheet
主回风道 main return-airway
主回风流 main return air current
主回路 main circuit; main loop; major loop; outer loop; primary loop
主回路失压 voltage failure of main circuit
主回路系统 primary heat transport system
主回授 primary feedback
主回授系数 primary feedback ratio
主回凸轮 main and return cam
主回转运动 circular main motion; circular main movement
主汇流排 primary bus
主会议厅 main assembly hall
主混凝剂 primary coagulant; primary coagulating agent
主活塞 main piston; main plunger
主活塞衬套 main piston bushing
主活塞套 cylinder bushing
主活塞涨圈 main piston ring
主火成岩 host-pyrolite
主火山口 main crater
主货舱 main hold
主机 bare machine; base machine; basic machine; host computer; host machine; key machine; main (frame) engine; main unit; master unit; principal machine; main frame computer
主机板 central processing unit
主机操纵程序 main engine control program(me)
主机操纵台 main engine maneuvering stand; main engine operation console; main machinery console
主机测试 engine test
主机程序 main frame program(me)
主机传输 host transmission
主机存储器 main frame memory
主机到主机通信[讯]量 host-to-host traffic
主机定速 ring-off engine
主机缸数 number of diesel cylinder
主机工作装置 machine attachment
主机功率 main engine power
主机航行试验 main engine portion of trial test
主机号 host number
主机机柜 main frame
主机(机)架 main frame
主机接口 host interface; interface of host computer
主机接口程序 host interface routine
主机接口块 host interface block
主机节点 host node
主机解体检修用吊车 main engine overhauling crane
主机壳 main case
主机控制机构 main machinery controlling mechanism
主机控制台 main engine control desk
主机馈电键 main set feeding key
主机连接设备 host attachment facility
主机每分转数航程 distance by engine revolutions per minute
主机名称 host name
主机命令 host command
主机设计者 engine designer
主机失常 main engine abnormal
主机室 machine hall
主机室地面高程 machine floor elevation
主机适应性 host adaptation
主机特征数 machinery number
主机筒 main barrel
主机系泊试验 main engine portion of bollard test
主机遥控系统 main engine remote control system
主机用毕【船】finish with engine
主机用户 host subscriber; host user
主机油道 main oil conduit
主机与辅机 main and auxiliary
主机与主机协议 host-to-host protocol
主机执照 engine licence[license]
主机制造商 engine builder
主机制造者 main frame maker
主机中央操纵台 main engine control center[centre]
主机重量 weight of main machine
主机轴 boss rod
主机转数表 main engine revolution indicator; receiver of main engine revolution
主机转数航程 distance by engines revolution
主机转数记录器 engine revolution counter
主机转速表 main engine revolution indicator
主机准备【船】ring standby; stand-by engine
主机自动控制装置 automatic control appliance of main engine
主机组 main set; major unit; principal mechanical components
主机组总成 major unit assemblies
主迹 principal trace
主基点 main base
主基脉冲群 main pulse reference group
主基准 master reference; primary standard
主基(准)面 grand base level; primary datum
主激光器 main laser
主激活剂 dominant activator
主激励器 master driver
主级 main stage
主级发动机 sustainer; sustainer chamber
主级数 principal series
主级站 primary station
主极 active pole; main pole; primary pole
主极大 primary maximum; principal maximum
主极化率 principal polarizability
主极化面 principal polarization plane
主极绕组 main pole winding
主极小 primary minimum; principal minimum
主极靴 main pole piece
主集材架杆 head spar(tree)
主集尘器 principal collector
主集装箱 master container
主给水泵 main feed pipe; main feed pump; main feed water pump
主给水管 principal water supply pipe
主给水管路 main feed line
主给水止回阀 main feed check(valve)
主脊 backbone; principal crest
主脊梁 main rafter
主计部 controller's department
主计长 chief controller; paymaster <英>
主计官 controller
主计划预算 master planning budget
主计权责 controllership
主计时装置 master timekeeper
主计数器 basic counter
主计算 main calculation
主计算机 center computer; central computer; host computer; main frame computer; master computer; principal computer
主记录 master record; owner record
主剂 host crystal
主继电器 main relay
主祭坛 central altar
主祭坛后空间 back choir
主加法标准展开定理 principle addition canonic(al) expansion theorem
主加热 basic heating
主加热器 main heater; primary heater
主加速器 main accelerator
主夹板 bottom plate; lower plate; lower wall; main plate
主夹板衬框 main plate enlargement ring
主夹板螺母 main plate nut
主夹盘 key chuck
主甲板 main deck
主甲板下方的甲板 between decks
主甲板舷墙栏杆 main bulwark rail
主甲板缘 main deck stringer
主甲板中横梁 main deck beam
主价【化】principal valence
主价键 primary bond; primary valence bond; primary valence[valency]
主架 base frame; body frame; main frame; main support
主架大梁 crab girder
主尖锋 major peak
主间的 intercardinal
主间平面 intercardinal plane
主间隙 main gap
主监测台 master supervisory and test console
主监察器 on-the-air monitor; transmission monitor
主监控器 master monitor
主监视器 master monitor
主监视台 main monitor rack
主监视仪表板 main supervisory panel
主减速机 main fan
主减速器 main reducing gear; main retarder; master retarder; primary retarder
主减速器位置 primary retarder position
主减速箱 main speed reduction box
主减震支柱 main shock strut
主剪力面 primary shear plane
主剪力线 main shear wire; primary shear wire
主剪应力 principal shear(ing) stress
主剪应力平面 plane of principal shearing stress
主检修厂房 main inspection and repair shed
主检验井 main inspection chamber; principal inspection chamber
主检验室 principal inspection chamber
主件 main unit; major section; master part master device
主件程序修改 program(me) master-file update
主件副本 subject copy
主建筑 main building
主建筑工地 main site
主建筑拱顶 principal building vault
主键 main chain; major key; master key; primary key; principal bond; principal linkage
主键控制系统 master-keying system
主交叉 main crossing
主交换机 host exchanger; main switch
主交换台 principal switch board
主焦点 meridianal focus; primary focus; prime focus; principal focus; tangential focus
主焦点光电管 prime focus cell
主焦距 principal focal distance
主焦笼 prime focus cage
主焦煤 main coking coal
主角 major angle; principal angle
主角度基准 master angular reference
主叫电话局 calling exchange; originating exchange
主叫方 calling party
主叫方类别信号 calling party's category signal
主叫局 office of origin; originating office
主叫控制复原方式 calling party release; calling subscriber release
主叫显号器 calling subscriber's number indicator
主叫用户 caller; calling party; calling subscriber; call party
主叫用户单方话终拆线 calling party release; calling subscriber release
主叫(用户)号码 calling number
主教<天主教的> ostiary
主教邸宅 bishop's palace
主教官邸 bishop's palace
主教十字架 archiepiscopal cross; patriarchal cross
主教座 cathedra
主接触器 main contactor
主接地点 main earthing terminal
主接点 main contact
主接点组 main contact group
主接力器 main servomotor
主接头 host tap; main joint
主接线 electric(al) main
主接线图 main wiring diagram
主街(坊) main block
主街区 main block
主节点 host node

主节理 main joint;major joint;principal joint;master joint
主节流阀 main throttle valve;primary throttle valve
主节流口 primary choke
主节气门 main throttle
主结点 junction;major node
主结构 primary construction;primary structure;principal frame;principal structure;main frame
主结构件 primary structural component
主结构凸出物的支柱 outrigger shore
主结线 main circuit
主结线布置 busbar connection arrangement
主结线单线图 main single line diagram
主结线图 one-line diagram
主截面 main cross-section;principal cross-section;principle cross-section
主截止阀 main stop valve
主截止手柄 main shut-off handle
主解 principal solution
主解理面 face cleat
主解理直角工作面 end-on
主解析集 principal analytic set
主介电常数 principal dielectric(al) constant
主筋 main bar;main reinforcement
主筋保护层 cover of main bars
主筋悬伸搭接 <焊接网> overhang lap
主进厂通道 main entrance
主进程 host process;master process
主进风道 main gate;main intake
主进风管 <双管制动> main feed pipe
主进风平巷 blowing road
主进口 principal entrance
主进气道 main air intake
主进气阀 master air suction valve
主进气节流截止阀 main inlet throttle-stop valve
主进气控制阀 main inlet control valve
主进气口 primary air inlet;primary inlet port;primary stage port
主浸 main impregnation
主经 main warp
主茎 stalk
主晶 host crystal;oikocryst
主晶格 host lattice
主井 riser shaft
主井提升机 main hoist;production hoist
主井眼 original hole
主景 main feature
主景石 trump stone
主景树 trump tree
主景植物 accent plant
主警告信号 master warning signal
主净化系统 main air cleaning system
主镜 main telescope;primary mirror
主镜室 primary mirror cell
主局 control exchange;controlling office
主矩阵 principal matrix
主距 principal distance
主距改正 reduction for principal-distance
主距调焦 principal distance focusing
主距误差 principal distance error
主锯 head saw
主锯机 head rig
主聚焦透镜 main focusing lens
主卷 master volume
主卷筒 main drum
主卷扬 main hoist
主卷扬机 main winch
主绝缘 earth insulation;grounding insulation;main insulation;major insulation
主卡片 master card
主开关 main breaker;main cock;main switch;principal cock
主开关盘室 main switchboard room
主开关设备合闸 main switchgear closing
主看台 grandstand;main stand
主考人 examinant;examiner
主考人的 examinatorial
主壳 main casing
主壳层 major shell
主壳体 main housing
主刻度线 main scale mark
主客户 customer
主坑道 main tunnel
主空间 principal space
主空气 primary air
主空气阀 master air valve
主空气管道 blast main
主空气过滤器 main air filter
主空气进口 primary air enter
主空气室 major air cell
主空气通路 primary air duct
主空心轴 main hollow spindle
主孔 main span;original hole;principal boring
主控 main control
主控按钮板 master push button panel
主控按钮盘 master push button panel
主控板 main control panel
主控薄片 master slice
主控操作技术员 master control operator
主控程序 master control;master control program(me);primary control program(me)
主控程序工序 master control program(me) procedure
主控船 primary control ship
主控磁铁 main control magnet
主控电钟 control clock;controlling electric(al) clock
主控定时脉冲 master timing pulse
主控多谐振荡器 master multivibrator
主控方式 master mode
主控分开关 master subswitch
主控火花隙 master spark gap
主控机 master controller
主控计算机 host computer
主控继电器 master relay
主控寄存器 main control register
主控驾驶仪 master pilot
主控减速器 master retarder
主控开关 master control switch;master switch
主控脉冲 master pulse
主控盘 main control board
主控平面位置显示器 master plan position indicator
主控屏 master control panel
主控起闭按钮 master start-stop button
主控起动按钮 master start button
主控器 principle controller
主控切换 master switching
主控室 master control room
主控手柄 key lever;king lever;master controller handle;master lever
主控数据 major control data
主控双稳态 master bistable
主控台 key station;master control;master station;master station service;rate station
主控调节阀 master regulating control valve
主控调速器 topping governor
主控握柄 key lever;king lever;master lever
主控线路 master control circuit
主控信号 master signal
主控寻线机 master switch
主控仪表 master instrument
主控增益 master gain
主控闸门 main control gate;master control gate
主控站 master control station
主控振荡槽路 exciter[excitor]
主控振荡器 driving oscillator;exciter;pilot oscillator;master oscillator
主控振荡器功率放大器 master oscillator dower amplifier
主控振荡器频率 master oscillator frequency
主控制板 master control board;master control panel
主控制变化 major control change
主控制部件 main control unit
主控制程序 master control program(me);primary control program(me);master control routine
主控制阀 main control valve;master control valve
主控制杆 master control lever
主控制滚筒 main control drum
主控制回路 main control loop
主控制建筑物 main controlling structure
主控制链 main control chain
主控制盘 main control board;master control panel;main control unit
主控制屏 main panel
主控制器 main controller;main control unit;master control unit;principal controller;master controller
主控制器环 master controller ring
主控制室 main control room
主控制台 main console;master control board;master(control) console;net control station;primary console
主控制(仪表)板 <信号系统的> master board
主控制元件 main controlling element;master controller
主控制站 main control station;master station;M-station
主控制中断 master control interrupt
主控制中心 main control center[centre];master control center[centre]
主控制装置 main control unit
主控装置 master control
主控自动化 master control automation
主口门【港】 main entrance
主库数据库 subject data base
主跨 main span
主跨孔 main opening
主矿层 main reef;main seam
主矿井 main pit;main shaft
主矿脉 main branch(line);main gallery
主矿体 main orebody
主矿物 host mineral;palasome
主框架 advance setting;main frame;primary framing;principal frame;chief frame
主亏格 principal genus
主馈 main feed
主馈电路 trunk feeder
主馈电器 primary feed
主馈电线 main feeder;trunk feeder
主馈线 main feed
主扩散 drive-in diffusion
主拉杆 main pull rod;main tie
主拉力 diagonal tension
主拉伸比 principal extension ratio
主拉伸应力的极限 principal tensile stress limit
主拉索 main stay
主拉应变 principal tensile strain
主拉应力 diagonal tensile stress;principal tensile stress;diagonal tension stress
主缆 main hauling line
主缆检修道 cable parapet
主缆扩散铸件 splay casting
主缆索 principal cable
主廊道 main gallery
主雷达站 master radar station
主肋 main rib;prime rib
主肋板【船】 main floor
主肋骨 chief frame;main frame
主类 main classes
主累加器 main store;primary accumulator
主冷阱 primary cold trap
主冷库 roll-in freezer;roll-in refrigerator
主冷凝器 main condenser
主冷却剂循环泵 primary coolant pump
主冷却剂总量 primary coolant inventory
主离合磁铁 main clutch magnet
主离合器 main clutch;master clutch
主离合器拨叉支架 main clutch shifting lever bracket
主离合器操纵机构 main clutch control
主离合器分离叉 master clutch release yoke
主离合器零件 main clutch part
主离合器闸 master clutch brake
主离合器总成 main clutch assembly
主理想 principal ideal
主理想定理 principal ideal theorem
主理想环 principal ideal ring
主理想整环 principal ideal integral domain
主理想子环 principal ideal minor ring
主力 main load;principal load
主力冲击 main thrust
主力舰 battle ship;capital ship
主力矩 main moment;primary moment;principal moment
主力纵队 main column
主立方根 principal cubic root
主立管 main riser;main standpipe;principal riser;standpipe main
主力舰 capital ship
主立面 main facade;principal facade
主立柱 principal stud
主利率 ruling rate of interest
主励磁机 main exciter
主连杆 main connecting rod;master connecting-rod;master link;mother rod
主连杆十字头 main crosshead
主连接 main connection
主连接杆 <吊具的> master link
主联动曲拐销 main coupled crank pin
主联杆 master link
主联管箱 main header
主联结器 principal coupling
主联轴器 main coupling
主链 backbone chain;fundamental chain;principal chain;trunk chain
主链节 main chain
主链路站 primary link station
主链索 main chain
主链条 main chain
主链液晶聚合物 main chain liquid crystal polymer
主梁 deep beam;girder(beam);jack beam;king girder;king piece;kingpost;main frame;main supporting bar;main supporting beam;primary beam;primary girder;principal beam;principal girder;spar boom;

Z

spine beam;stub pole
主梁间距 girder spacing;beam spacing
主梁桥 bridge beam
主梁柱 king piece
主量规 principal ga(u)ge
主量孔 high-speed jet;main(metering)jet;power jet
主量孔喷管 main discharge tube
主量孔喷油管 main jet tube
主量孔喷油孔 main jet spray hole
主量子数 main quantum number;primary quantum number;principal quantum number
主列 chief series;main series
主列行程序 master routine
主林 major forest
主林班线 main compartment line;main frame;major ride
主林层 main stor(e)y
主林产物 major forest produces
主林道 main forest-road;main ride
主林分 residual stand
主林木 chief species;crop tree;major forest tree;principal crop;superior stand
主林木疏伐 crop-tree thinning
主林荫大道 principal mall
主令开关 control switch
主令线路 command link
主溜线【铁】thread of stream
主流 main current;main flow;main stream flow;master stream;mother current;predominant current;primary current;primary flow;principal river;stem stream;trunk stream
主流槽 sow channel
主流程线 main process stream
主流管 primary flow pipe
主流河槽 main bed of stream
主流浆箱 primary headbox
主流交通 main traffic
主流交通量 main stream flow
主流量喷嘴 main flow orifice plate
主流流量 main stream flow
主流流向 main current direction
主流抛泥 flow-lane disposal
主流倾向线 main current trend line;major directional desire line
主流线 drift line;main current line;thread of stream;thread of the current
主流向 prevailing direction of current
主流与回流 primary flow-and-return
主流中泓线 main thread
主流中心线 main thread
主馏分 main distillate fraction
主龙骨 carrying channel;channel runner;main joist;main keel;main runner
主龙头 main cock;main tap
主楼 central block;main building;principal building
主楼板 principal floor
主楼层 principal floor;principal level;principal stor(ey)
主楼梯 main stair(case);principal stair(case)
主炉身 main shaft
主滤器 main filter
主滤油器 main fuel filter
主路径 main path;primary path
主路由 main route;primary route
主轮 main wheel
主轮舱盖作动筒 main wheel door actuating cylinder
主轮距 wheel track
主轮收放作动筒 main gear actuating cylinder
主轮缘 main rim
主轮支柱 main wheel strut
主罗经 main compass;master compass
主罗盘 main compass;master compass
主螺杆 driving screw;main driving screw
主螺母 mother nut
主螺栓 king bolt;king rod;principal stud
主螺线管 main solenoid
主马达联轴器 lead spindle
主码 primary key
主埋面积 principal surface of accumulation
主脉 backbone range
主脉冲 main pulse;master clock
主脉冲发生器 basic pulse generator;master synchronizer
主脉冲/气泡比 main pulse/bubble ratio
主脉冲信号 main bang
主脉冲信号抑制 main bang suppression
主锚 bower anchor;main anchor
主锚碇墙 main anchor wall
主锚链 main chain
主门 principal portal
主门发生器 main gate generator
主门面 principal facade
主门厅 principal entrance hall
主密封 primary seal
主面 primary plane
主面包房 host bakery
主灭火点 principal vanishing point
主命令 main command
主模 dominant mode;main mode;master pattern;principal mode
主模激励 principal-mode excitation
主模块 main module;major module
主模型 master cast;pattern master
主母线 main bus-bar
主目标 major heading
主目录 master catalogue;master directory;root directory
主内存储器 main internal memory
主内导航原型器件 vehicle navigation prototype unit
主逆掩断层面 major thrust plane
主逆止阀 main check-valve
主黏[粘]度计 master viscometer
主黏[粘]合剂 primary binder
主凝结泵 main condensate pump
主凝水泵 main condensate pump
主排出管 main discharge
主排风机 main exhaust fan
主排气 main exhaust
主排气泵 main scavenging pump
主排气管 main stack
主排水泵 main dewatering pump
主排水沟 leader drain;main ditch;main drain
主排水管 leader drain
主排水管道 trunk sewer
主排水口 main outlet
主排水渠道 main outlet channel
主盘 master disk
主盘模型 master matrix
主盘正片 master positive
主刨 spindle
主炮塔 main turret
主跑道 main runway;major runway;primary runway
主跑道的基本长度 basic length of basic runway
主配电板 main distribution board;main distribution frame
主配电点 main distributing point
主配电间 main switchroom
主配电盘 main distributing board;main(service)panel;main switchboard;principal distribution panel;principal switch board
主配电盘室 main switchboard room
主配电室 principal switchroom
主配电线路 distribution trunk line;principal distribution line
主配电箱 main junction box
主配电站 main distributing center [centre]
主配电装置 main switching compound
主配筋 main reinforcement bar
主配水器 main distribution;main distributor
主配线架 main distributing frame;main distribution frame;main frame
主配压阀 main control valve;main gate valve
主喷发口 principal vent
主喷口 main jet;primary jet
主喷口塞 main discharge jet plug
主喷射阀 main injection valve
主喷射管 main ejector
主喷油嘴 main fuel spray nozzle
主喷嘴 main discharge jet;major jet;principal nozzle
主喷嘴扣紧簧 main nozzle retainer spring
主喷嘴螺帽 main discharge jet nut
主片 main leaf
主偏移 master shift
主偏转 main deflection
主频 basic frequency;fundamental frequency;master frequency
主频道 main channel
主频率 basic frequency;master frequency;predominant frequency;primary frequency;dominant frequency
主频微震 primary frequency microseism
主平方根 principal square root
主平面 cardinal plane;main plane;principal level;principal plane
主平面方向图 principal plane pattern
主平面上的应力 stress on principal plane
主平巷 level road;main drift;main drive;main gallery;main gateway;main heading
主平巷矿柱 main entry pillar
主平巷平面图 main level plan
主平行线 principal parallel line
主屏蔽 main shield;primary shield(ing)
主屏蔽层 primary shield(ing)
主坡 principal slope
主破裂 main fracture;major fracture
主剖面 principal section
主歧管 main manifold
主起动阀 <又称主启动阀> main starting valve
主起动空气阀 main starting air valve
主起落架 main undercarriage
主起升(机构)main hoist
主起重绞车 main crab
主气垫管道 main cushion duct
主气阀 emergency stop valve;main gas valve;main throttle valve
主气流 main current
主气门 throttle
主气门压力 stop valve pressure
主气体动力平衡 main aerodynamic(al) balance
主气旋 primary cyclone;primary low
主汽阀芯子 stop valve spindle
主汽缸放出接头 master cylinder outlet connection
主汽缸加油口盖 master cylinder filler cap
主汽缸输入接头 master cylinder inlet connection
主汽缸体 master cylinder body
主汽管 live steam pipe
主汽管道 live steam piping
主汽门 main inlet throttle-stop valve;main throttle valve;main steam valve;over-speed valve;steam-outlet valve;stop valve
主千斤顶 main jack <盾构工程用>;beaver-tail <俚语>
主牵伸 main draft
主牵引梁千斤顶 drawbar jack
主牵引系统 main traction system
主牵引运输机 main drag conveyer [conveyor]
主前角 front top rake angle
主潜水面 main water-table
主腔 main cavity
主墙 chief wall;main wall
主墙筋 main runner
主桥 main bridge
主桥墩 main pier
主桥箱 main housing
主切曲线 asymptotic(al)curve;primary tangent;principal tangent
主切刃法截面 cutting edge normal plane
主切线 primary tangent;principal tangent
主切削分力 main cutting force
主切削角 primary cutting angle
主切削金刚石 <在冲洗槽主动边的> track diamonds
主切削刃 edge;lead cutting edge;main cutting edge;major cutting
主切削运动 principal cutting movement
主切应力 principal shear(ing)stress
主倾斜轴 main tilting axis
主清除开关 master clear(switch)
主穹隆 main dome
主区 primary area
主曲柄 main crank
主曲柄箱 main crank case
主曲柄销 main crank pin
主曲率 principal curvature
主曲率半径 principal curvature radius;principal radius of curvature;radius of principal curvature
主曲率方向 direction of principal curvature;principal direction of curvature
主曲率线 line of main curvature;line of principal curvature;principal curvature line
主曲率中心 principal centers of curvature
主曲线 master curve;principal curve
主曲轴箱后部 crankcase rear main section
主曲轴箱前部 crankcase front main section
主曲轴箱中部 crankcase center main section
主驱动电动机 driver motor;master driver;master motor
主驱动器 master driver
主驱动装置 main driver
主渠 main canal;main cannel;principal canal
主渠道 principal conduit
主取向方向 principal orientation direction
主圈 primary circle
主权 dominion;equity;jurisdiction;mastership;ownership;sovereign(ty)
主权比率 equity ratio
主权国家 sovereign state
主权资本 equity capital

主泵 main spring
主群 master group;principal group
主群排 master group bank
主群调制 master group modulation
主群线路 master group link
主燃料 prime fuel
主燃料喷嘴 main burner; main fuel spray nozzle
主燃料调节器 main fuel control
主燃料系统 normal fuel system
主燃气轮机 main propulsion gas turbine
主燃区 primary combustion zone
主燃烧器 main burner
主燃烧室 main(combustion) chamber;primary(combustion) chamber
主燃油泵 main fuel pump
主燃油阀 main fuel valve
主绕组 main winding;principal winding
主人卧室 master bedroom
主任 chief;director;manager;officer in charge;superintendent;supervisor
主任调车员 foreman shunter; head shunter
主任工程师 chief engineer; department chief engineer; division engineer; engineer in charge; engineer in chief; principal engineer; project engineer
主任会计 chief accountant
主任会计师 accountant in charge; incharge accountant
主任绘图员 chief drafter;chief draftsman
主任货物调度员 chief of goods dispatchers
主任机车调度员 chief of locomotive dispatchers
主任机械师 chief mechanic
主任建筑师 architect in charge;chief architect; chief resident architect; department chief architect
主任客运调度员 chief of passenger train controller
主任列车长 chief conductor
主任列车调度员 chief dispatcher
主任事务员 chief clerk
主任收货员 chief inwards clerk
主任务 main task;major task
主任信号员 chief signalman; chief towerman
主任助理秘书 principal assistant secretary
主韧带 cardinal ligament
主容器 primary tank; primary container
主容许条件 primal feasible condition
主熔断器 main fuse
主蠕变 primary creep
主入口 main entry door;principal entrance
主入口地址 main entry address
主入口点 main entry point
主入口线 main lead
主入射角 principal angle of incidence
主软件 host software
主润滑油泵 main lubricating oil pump
主润滑油管 main delivery pipe;main oil pipe
主三角形 principal triangle
主伞齿轮 master bevel gear
主扫描 main scan(ning);main sweep
主色 main colo(u)r; essential colo(u)r;mass colo(u)r;mass tone; dominant colo(u)r;hue
主色调 dominant hue
主色浓度 depth of mass tone
主色散区 primary dispersion
主色微差比色计 colo(u)r master differential colo(u)rimeter
主色相 dominant hue;predominant hue
主山岛弧 primary arc
主山谷 principal valley
主栅 main grid
主扇 primary fan
主扇风机 main fan
主梢 leading shoot
主设计截面 principal design section
主射口 power jet
主射流 main power jet; power stream;principal jet
主射束 main beam;major beam
主射线 main ray;principal ray
主射影性质 principal projection property
主伸长 principal elongation
主神龛 central shrine
主神龛塔 <印度> principal stupa (mound);major stupa(mound)
主神座 principal altar
主审人 appraiser in chief
主渗透系数 primary coefficient of permeability
主升降机构 main hoist
主升降控制 master lift control
主升降口 trunk hatch
主升降装置 main hoist
主升运器 main lifting elevator
主生成规则 principal generating rule
主生活区 principal living area
主声拾音器 main vocal microphone
主绳 main rope
主时间效应 primary time effect
主时钟 main clock; master clock; master synchronizer;master timer; primary clock
主时钟脉冲发生器 master clockpulse generator
主时钟频率 master clock frequency
主蚀带 principal etched zone
主食 staple diet;staple food
主食加工间 staple food preparation room
主食品 main meal products; staple food
主矢量 master vector
主始群丛 primary association
主式原则 donor principle
主视差表 master parallax table
主视频混合器 central vision mixer
主视区 primary vision area
主视图 front view;principal view
主视线 main-line of sight; primary line of sight;principal visual ray
主试验架 main test frame
主试验台 master test station
主试验站 master test station
主室 main chamber
主收报台 main receiving station
主收尘器 principal collector
主收发站 main transmit and receive station
主收获 major produces
主收敛的 principal convergent
主输出 primary output
主输电线 electric(al) main; power transmission line; trunk transmission line
主输入 main input;primary input
主输送管道 main delivery pipe
主输送机 main conveyer[conveyor]; mother conveyer[conveyor]; trunk conveyer[conveyor]
主树干 main shaft
主竖井 main shaft
主竖区 <消防分隔制> main vertical zone
主数 pivot number
主数据 main data;master record
主数据表 master data sheet
主数据表检查器 master mark feature
主数据单 master data sheet
主数据区 main data area
主衰减器 master fader
主水舱 main sump
主水道 main canal;main channel
主水解 main hydrolysis;principal hydrolysis
主水流 main current
主水平线 principal horizontal line
主水平应力 principal horizontal stress
主水源 main suit
主水准器 main level; master level; east-west level <中星仪>
主水准线 primary level line
主水准仪 main level;master level
主四面体 principal tetrahedron
主寺庙 main temple
主伺服电动机 main servomotor
主伺服马达 main servomotor
主送风机 primary air fan
主送线器 main wire feeder
主速度 principal velocity
主速率接口 primary rate interface
主隧道 haulage way
主榫头 principal stud
主索 carrier cable; main cable; main rope
主索滑车 bull block;main-line block
主索矢跨比 sag ratio of cable
主索引 master index
主索引图 master index map
主锁闭力 main locking force
主塔 central tower;main column;king tower <塔式起重机的>;main tower
主台 key station;master station(service); parent station; primary set; main unit;master unit;master set
主台地波重复副台天波的改正量 reading of master ground-wave and slave sky-wave match
主台阶 principal altar
主台脉冲 master pulse
主台座 master pedestal
主抬升浮筒 main lifting pontoon
主态 master mode
主坛 high altar
主弹簧 main spring
主弹簧板 main leaf spring
主弹簧片 main leaf of spring; mainspring leaf;master spring leaf
主探井 main pit
主套筒 master collet
主特性曲线 master characteristic (curve)
主特征标 principal character
主梯架 <挖土机的> main ladder
主提升机构 main hoisting mechanism
主提升器 principal riser
主题 lemma[复 lemmata/lemmas]; motif;subject;subject matter
主题报告 keynote address; keynote presentation;keynote speech
主题报告人 keynoter;keynote speaker
主题标目 subject heading
主题标准档 subject authority file
主题表 subject heading list
主题词 feature word; key word; subject terms;topical words
主题导卡 subject guide card
主题地图 special subject map; thematic map
主题发言人 keynote speaker
主题法 subject indexing method
主题分类 subject classification
主题分类表 subject classification table
主题公园 theme park
主题花纹 principal motif
主题检索 subject retrieval
主题理解测试 thematic apperception test
主题排架法 subject arrangement
主题判读样片 <遥感摄影> subject key
主题判图 thematic mapping
主题数据库 subject data base
主题索引 subject index
主题统觉测验 thematic apperception test
主题图案墙纸 subject wallpaper
主体 bulk; case; corpus; key body; main block;main body;main bulk; main part; primary body; principal part;subject
主体爆破 primary blasting
主体波 subject wave
主体测图 machine plotting
主体长度 principal length
主体分段建造法 block system
主体工程 main structure; carcassing work;main job;main project;main works
主体工程费 subject construction cost
主体工程开挖 required excavation
主体工程项目 principal work item
主体骨料 carcass
主体航空摄影绘图仪 cartographic(al) plotter
主体化合物 host compound
主体计算机 main frame computer
主体加料 body feed
主体建筑 main body building;principal building
主体建筑工程时间 principal construction time
主体建筑开间 principal building bay
主体建筑跨度 principal building bay
主体建筑区 principal block
主体结构 main structure;major structure
主体结构体系 main structure system; major structure system
主体金属 base metal
主体进料 body feed
主体距离状态的轴荷分配 axial loading distribution of principal distance conditions
主体框架 main body frame
主体排气孔 main body bleeder
主体平行股道 <编组场> body track
主体群落 major community
主体色 body colo(u)r
主体图 block diagram
主体土方工程 major earthwork
主体托换 main underpinning
主体外壳 main body cover
主体完工 topping out
主体相 bulk phase
主体信号 main(running) signal;principal signal
主体信号灯 main signal light
主体信号灯光 main signal light
主体信号机【铁】 main(running) signal;principal signal
主体信号示像 main signal aspect
主体行车信号机【铁】 main running signal;principal signal
主体性质 bulk property
主体岩石 host rock
主体样品 bulk sample
主体意识 subject-consciousness
主体铸造试棒 cast integral test bar; cast-on test piece
主体转动铰 body hinge
主体坐标系 principal body axes
主天线 master antenna
主天线电视系统 master antenna television system
主天线分配系统 master antenna dis-

tribution system
主调节 master control
主调节阀 main governor valve; main inlet control valve
主调节器 main controller; master selector
主调节器-副调节器 master-submaster controller
主调谐 main tuning
主调压器 main governor
主调整 master control
主调整台 main control console
主调制解调器 master modem
主跳板【船】 main ramp
主铁水沟 main runner
主厅 principal hall
主厅屏蔽 main-vault shielding
主停车装置 master trip
主停机 main shut-down
主停机系统 main shut-down system
主停气阀 main stop valve
主通道 main channel; major path
主通道卫星 primary path satellite
主通风管通道塞 main vent tube passage plug
主通风井 main shaft
主通风孔 main vent
主通路 main channel; main path; primary path
主通气管 main stack; main vent
主通气主管 main vent stack
主通信[讯]站 main traffic station
主通信[讯]中心 main communication center[centre]
主同步脉冲 main synchronization pulse; master synchronization pulse
主同步器 master synchronizer
主投影 principal project
主透镜 main lens
主透气管 main vent
主凸轮 drive cam; main cam; master cam
主凸轮轴 main camshaft
主突波 main bang
主图 main map; master map
主图像监视器 main picture monitor
主推进 main propulsion
主推进器 main propeller
主推进装置 main propelling machinery; main propulsion unit
主推进装置负载 main propulsive load
主推力 active thrust
主拖缆 main tow rope
主陀螺罗经 master gyrocompass
主陀螺仪 master gyroscope
主椭圆项 principal elliptic term
主挖土机 basic excavator
主外存储器 master file
主外结点 primary external nodal point
主弯矩 primary moment
主弯曲 principal curvature
主弯曲平面 principal plane of bending
主网路 main network
主网络 main network; major network; principal network
主望远镜 primary telescope; principal telescope
主微商 principal derivative
主桅 main course; main mast
主桅杆 main mast
主桅冠 main trunk
主桅前支索 main mast fore stay; main stay
主桅纵帆 brigantine
主位移 primary displacement
主温度控制器 main temperature controller
主文件 main file; master file
主文件带 master file tape
主文件联机 central file on-line
主文件目录 master file directory
主文件清单 master file inventory
主文件索引 master file index
主文件修改程序 master file updating program(me)
主问题 primal problem
主涡轮机 main turbine
主卧室 primary bedroom
主握柄 key lever; king lever; master lever
主污水管 main sewer
主屋谷 principal valley
主屋架 main couple
主屋面天沟 principal valley
主屋前庭园 main garden
主无线电天线 master radio antenna
主无线电指标信号 main entrance signal
主物面 principal object plane
主吸口根部 main suction foot
主吸入阀 Kingston valve
主吸收波长 principal absorption wavelength
主溪谷 principal valley
主席 chairman; chairperson; president; presiding officer
主席台 rostrum[复 rostra]
主席团 presidium[复 presidia/presidiums]
主席团执行主席 executive chairman
主席职位 chairmanship
主洗加热期 main wash heating phase
主洗跳汰机 main washbox
主系木 main stay
主系索 main hauling line; main mooring line
主系统 host system; main system; major system; primary system
主系统带 master system tape
主系统连接装置 host attachment
主系统准备功能 host preparation facility
主系线 principal series line
主下部结构 principal supporting structure
主下水道 main sewer; principal collector; principal sewer; trunk main; trunk sewer
主下水管道 sewer main
主纤维 principal fiber[fibre]
主纤维长度 length of dominant fiber
主纤维丛 principal fiber bundle
主显示台 main display console
主线 cardinal line; main track; principal line; lead wire
主线轨道 center line track
主线控制 main-line control
主线路测试台 master line test console
主线圈 main coil; main winding; principal winding
主线圈电流 main coil current
主线束 principal wire bundle
主线隧道 center line tunnel
主线系 principal series
主线占线 master busy
主限制开关 main limit switch
主相 magnafacies; main phase; principal phase; principal portion
主相环电流 main phase ring current
主箱 main tank
主响应 main response
主向交通流 main traffic flow
主向量 principal vector
主巷道 main tunnel; main gallery <地下工程的>
主项 basic term; principal term; subject
主像 main image; primary image
主像面 principal image plane
主像限角 principal bearing
主消声器 main muffler
主销 king bolt; king journal; king pin; main pin; main pivot; master pin; pivot bolt; pivot pin
主销衬套 king bolt bush; kingpin bush; master bushing
主销定位螺钉 kingpin stop screw
主销负后倾角 minus caster; negative caster
主销盖 kingpin cap
主销后倾 castor; castorite; master castor; reverse caster
主销后倾角 caster angle; master angle
主销后倾作用 master action
主销内倾角 kingpin inclination; swivel pin angle
主销倾角 caster; kingpin angle; kingpin inclination
主销倾角止推销轴承 kingpin bearing
主销球窝 king bolt ball and socket
主销式转向机构 kingpin steering assembly
主销铜衬 main pin brass
主销铜衬扁销止动螺钉 main pin brass cotter set screw
主销外壳 vertical pivot pin housing
主销轴承 kingpin bearing
主销座 kingpin bearing
主小车 main crab
主小齿轮 main pinion
主小区 main plot
主效波 significant wave
主效波高 significant wave height
主效波周 significant wave period
主效果 main effect
主斜撑 main brace; main diagonal; principal diagonal
主斜杆 main diagonal; principal diagonal
主斜沟坡谷 principal valley
主斜井 main slant; main slope
主斜井卷扬机司机 main-slope engineer
主斜肋 main diagonal rib
主谐波 main harmonic
主谐振 main resonance
主泄水管 main drain
主泄水系统 main drain system
主芯 body core
主芯骨架 core crab
主信标 leading beacon
主信道 main channel
主信号 master signal; primary signal face
主信号度盘 master dial
主信号机【铁】 main signal
主信号楼 control signal box
主星 primary body; primary component; primary star
主星形 principal star
主星序 dwarf sequence; main sequence
主行人区 main pedestrian zone
主行星 primary planet; principal planet <九大行星的>
主形 dominant shape
主形变 principal deformation
主型 dominant mode; principal mode; prototype
主修车间 main repair shop
主徐变 primary creep
主序 main sequence; principal order
主序后阶段 post-main sequence stage
主序模 principal order module
主序拟合 main-sequence fitting
主序前 premain sequence
主序折向点 turn-off point from main sequence
主蓄热室 primary regenerator
主旋回 major cycle
主旋塞 main cock; principal cock
主旋叶(传动)轴 main-rotor shaft
主旋翼 main rotor
主旋翼传动 main-rotor drive
主旋转运动 principal rotating motion
主选 primary cleaning
主选波器 master selector
主选水力旋流器 primary hydrocyclone
主选跳汰机 primary washbox
主选通脉冲 main gate; main gating pulse
主选择脉冲 main gate
主选择器 master selector
主循环 main circulation; main cycle; main loop; major cycle; major loop; primary cycle
主循环泵 main circulating [circulation] pump
主循环发送 main dispatch loop
主压力 diagonal compression; main pressure; principal pressure
主压力管 principal force main
主压力管道 main delivery pipe
主压力控制器 master pressure controller
主压气管道 airline main
主压缩 primary compression
主压缩机 main compressor
主压缩线圈 main compression coil
主压应变 principal compressive strain
主压应力 principal compression[compressive] stress
主压载水舱 main ballast tank
主压载水通海阀 main ballast Kingston(valve)
主压榨 main press
主烟囱 flue collector; main flue; main shaft
主烟道 flue collector; main flue
主延迟线 main delay line
主岩 country rock; host rock; palasome
主檐 <古典建筑的> principal cornice
主验潮站 primary tide station
主阳极 main anode
主样品 master sample
主窑 <生产水泥用> vertical kiln
主摇动式运输机 mother shaker
主遥控器 master remote controller
主要保险单 master policy
主要报表 principal statement
主要备件 major spare parts
主要比例(尺) primary scale
主要比率 key ratio
主要避难通道 primary means of escape
主要编组场【铁】 main classification yard; main marshalling yard
主要编组站【铁】 main marshalling station
主要便道 main access road
主要标本 key sample
主要标记 main mark
主要波长 dominant wavelength
主要波浪 dominant wave
主要补给点 primary supply point
主要补给品分类 major materiel category
主要补给线路 main supply road
主要部分 integral part; main block; main body; main part; main portion; major component; major part; part and parcel(of); trunk
主要部件 critical parts; critical piece; essential component; main parts; main unit; major component; major unit assemblies; master unit
主要部件参考表 master parts reference list

主要材料 main material;primary material;priming material
主要材种 prevailing wood species
主要采矿 basal level
主要参比燃料 primary standard fuel
主要参数 key parameter; main parameter;major parameter
主要餐馆 principal restaurant
主要测量仪器 main meter
主要测试检验 major examination
主要产品 basic products;main products;major products;staple
主要产区 major production areas
主要产物 primary product;staple
主要产业 major industry
主要长途货物列车 < 在各编组站间运行 > major line-haul train
主要常数 primary constant
主要厂商 leading manufacturer
主要超挖地段 major overbreak zone
主要潮流观测站 control tide station; primary tide station
主要潮位站 control tide station;primary tide station
主要潮汐观测站 primary tide station
主要车(行)道 main carriageway
主要成本 prime cost;first cost
主要成本法 < 指负荷分配的方法 > prime cost method
主要成本项目 prime cost item
主要成槽流量 dominant formative discharge
主要成分 backbone;essential component; essential ingredient; fundamental component; main ingredient;major constituent;predominant constituent;predominating constituent;principal constituent;principal ingredient;staple
主要成分分析 principal component analysis
主要成膜物 film former
主要成品 major item
主要承包人 prime contractor;principal contractor
主要承包商 prime contractor;principal contractor
主要承包者 prime contractor;principal contractor
主要承载结构 main load-carrying structure
主要承重部件 principal load-carrying parts
主要承重结构 main bearing structure;main load-carrying structure;principal load-carrying structure
主要承租人 major tenant
主要程序 master program(me)
主要程序步骤 main program(me) sequence
主要尺寸 general dimension;key dimension; leading dimension; main dimension;major dimension;principal dimension; significant dimension
主要尺寸检查结果 inspection result of main dimension
主要出口国 leading exporter
主要出力 primary output
主要船厂 principal shipyard
主要船坞 principal dock
主要大街 main street;main thoroughfare
主要代理商 main agent
主要单元 formant;basic unit
主要担保人 principal underwriter
主要导航设施 prime navaids facility
主要导线【测】 main traverse;bus line【电】
主要道路 king's highway;main road;main route; major road; primary road;principal road
主要道路干线 principal road line
主要道路网 primary road network
主要的 cardinal; chief; main; major; master; predominant; prevailing; primary;principal;staple;stellar
主要的承载结构 principal load-bearing structure
主要的或第一次侵染 main or first infection
主要的加载情况 principal loading case
主要的受载结构 principal loaded structure
主要的外界条件 main environments
主要的专业服务公司 prime professional
主要的专业服务人 prime professional
主要等高线 principal contour
主要抵押 master mortgage
主要地 chiefly;mostly
主要地层 master stratum; predominant formation
主要点计数 key count
主要电力 primary power
主要叠标 main transit marks
主要订约人 prime contractor
主要订约商 prime contractor
主要定时开关 master timer
主要定时装置 master timer
主要动机 domain motive
主要洞室 leading room
主要段 main section;major section
主要断裂线 major lineaments
主要反应 key reaction; main reaction;primary reaction;principal reaction
主要反应模型 major reaction model
主要饭店 principal restaurant
主要方案 basic version
主要方式 fundamental mode
主要方向 direction of orientation
主要飞机跑道 basic runway
主要飞行高度 cardinal altitude
主要费用 capital cost;capital expenditures; major charges; major expenditures
主要分包商 primary subcontractor
主要分接头 main tapping
主要分类账 key ledger
主要分量 essential component
主要分流道 primary distribution road;primary distributor road
主要分流(道)路 primary distribution road; primary distributor road; main distributor road
主要分水岭 great divide;main divide
主要分支线 main branch connections
主要风 prevailing wind
主要风险 principal risk
主要风向 cardinal wind;main direction of wind;main wind direction; predominant wind
主要服务中心 principal services center[centre]
主要辐射线(路) major radial
主要负荷 principal load
主要负载 basic load(ing);principal load
主要负责人 principal-in-charge; primary debtor
主要附表 leading schedule
主要改善 major betterment
主要概念 major concept
主要干道 main trunk road; principal trunk
主要干路 backbone road
主要干线 main artery
主要干线道路 main trunk road
主要干线公路 main trunk highway; principal arterial highway; major highway
主要干线管路 main trunk line
主要杆件 main member;primary member;principal member
主要岗位 key post
主要港池 principal dock
主要港口 main harbo(u)r; major port;primary port;principal harbo(u)r;principal port;main port
主要港口建筑 main terminal building
主要港湾 main harbo(u)r
主要港务管理机构 leading port authorities
主要根系吸水深度 effective rooting depth
主要更改记录 master change record
主要工厂 capital plant
主要工程 main job;main works;project works
主要工程计划 major project
主要工程试验项目 major engineering test item
主要工程数量表 quantities of major works
主要工程项目 key engineering project
主要工具 basic tool
主要工序 master operation
主要工业 key industry;main industry;major industry
主要工业产品 principal industrial products
主要工艺设备 main process equipment
主要工作 capital works;main job; major job
主要工作参数 basic operating condition
主要工作附件 major attachment
主要工作人员 key personnel
主要工作装置 major attachment;primary equipment
主要公路 arterial road; main highway; main road; major highway; primary highway; primary road; principal road;trunk road
主要公路干线 major highway artery
主要公司 dominant company
主要功率 primary power
主要功能 key function; major function
主要供货商 principal supplier
主要供应来源 main supply source
主要供应线路 principal supply road
主要构件 main component part;primary member; principal member; main member
主要构件的结构材料 construction materials of main parts
主要构造体系与震中分布图 map of distribution of dominant tectonic systems and epicenters
主要构造线方位 azimuth of principle structural line
主要构造要素 principle structural elements
主要购物中心 main shopping center[centre]
主要股东 principal shareholder
主要固定(污染)源 major stationary source
主要故障 major failure
主要故障部位 key trouble spot
主要关键部分 major critical component
主要关键部件 major critical component
主要观测站 main station;primary station
主要管线 < 指水管 > header line
主要灌木林(土墩) major tope(mound)
主要光源 primary light source;principal source of light
主要规范 master specification;principal particulars
主要规格 condensed specification; main specification
主要轨道 main track
主要国家 major country
主要海港 principal harbo(u)r
主要害虫 primary pest
主要含水层 basal water
主要焊缝 principal weld
主要焊接工班 < 焊接管道接缝的 > firing-line
主要行列式 principal minor
主要行业 major industry
主要航线 main shipping route
主要航站楼 main terminal building
主要合伙人 senior partner
主要合金成分 main alloying constituent;principal alloying constituent
主要合同 main contract; master agreement; master contract; prime contract
主要河流 base stream; main river; master river;master stream;principal river;trunk river;main stream
主要荷载 basic load(ing);main load; primary load;principal load
主要荷载承重结构 main load-bearing structure
主要横街 major cross street
主要环节 key link
主要环境化学品数据网 data on environmentally significant chemicals network
主要环境因素 essential environment(al) factors
主要回程 primary return
主要绘图员 leading draughtsman
主要货币 main currency; principal currency
主要货件 principal commodity
主要货流 predominant cargo flow
主要货物 leading commodities;major commodities;prime cargo
主要货运方向 predominant traffic direction
主要货运站 major freight station
主要货种 principal commodity
主要机场 main airport;major airport
主要机件 critical part
主要机具 main machinery
主要机理 dominant mechanism
主要机能 primary function
主要机器 basic machine
主要机械 base machine
主要机组 major combination
主要基本站 main base station
主要基本指标 main basic indicator
主要基面 principal base
主要激发地震活动 principal stimulated seismic activity
主要给水管 principal supply pipe
主要计划目标 main target of a plan
主要计划指标 main target of a plan
主要计时器 master timer
主要记录 principal data
主要技术 major technique
主要技术标准 master specification
主要技术参数 main technical parameter
主要技术规格 main technical specifications
主要技术经济指标 main techno-economic targets
主要技术数据 basic technical data; main technical data;main technical details
主要技术性能 basic mechanical de-

Z

sign feature; main technical behavio(u)r
主要技术指标 main technical index; key technical index
主要技术资料 basic technical data
主要祭坛 main altar
主要价值 chief value
主要驾驶仪表【船】primary steering instrument
主要间壁 main bulkhead
主要间隔铁 main filler
主要检验 main test
主要建筑 principal building
主要建筑单元 major construction element
主要建筑拱顶 main building vault
主要建筑跨度 main building bay
主要建筑立面 main facade
主要建筑设备 major plant
主要建筑师 principal architect
主要建筑物 building principal; key construction; key structure; principal building; main building
主要建筑物面积 area of principal building
主要建筑翼部 main transept
主要降水中心 main precipitation center[centre]; main precipitation core
主要交叉口 main crossing; major junction
主要交通干道 main traffic artery
主要交通计数 master traffic count
主要交通路段 major weaving section
主要交通路线 principal traffic route
主要交通设计 master traffic plan
主要交通站点数 key count
主要阶段 primary period; primary stage
主要阶段图 master phasing chart
主要接线图 primary circuit diagram
主要街道 arterial street; main stem; main street; major street
主要节理【地】master joint
主要结点地面终端机 major nodal earth terminal
主要结构 main structure; primary structure; principal structure
主要结构架 main frame
主要结构系数 main textural coefficient
主要结晶轴 principal crystallographic axis
主要金属 major metal
主要进口 main body entrance
主要进料器 main feeder
主要进路 main route
主要经费 major expenditures
主要经济指标 main economic indicators; selected economic indicators
主要经济指标综合指数 composite index of lagging indicators
主要经销商 principal business
主要井框 bearing ring
主要掘进工作面 main end
主要卡盘 prime cartridge
主要开采水平 principal mining level
主要勘探工作 main investigation
主要可采煤层 principal workable coal seam
主要坑道 gangway
主要空间目标 primary space target
主要控制 major control; primary control
主要控制尺寸 principal controlling dimension
主要控制点 major control(led) point; principal control(led) point
主要控制断路 major control break
主要控制改变 major control change
主要控制网 major control net; major framework
主要矿产 main commodities
主要矿产地区 principal mining region
主要矿产国 principal mining nation
主要矿山 principal mines
主要矿体 principal orebody
主要矿物 principal mineral; essential mineral
主要矿物储量 major mineral reserves
主要矿物或元素 major mineral or elements
主要拉条 principal brace
主要类型 predominant type
主要离子 leading ion; major ion
主要里程碑 key milestone
主要利率 prime rate
主要利益 principal benefit
主要利用指标种 key utilization species
主要联结点 major junction
主要粮食 staple food
主要裂缝 primary crack
主要裂片 principal lobe
主要林荫道 main mall
主要零(部)件 main parts; major parts
主要流量 dominant discharge; dominant flow
主要楼层 main floor; piano nobile
主要露头岩石 classed of mainly exposed rock
主要路 primary road; principal road
主要路线 main avenue; main route; principal line; principal path; principal route; trunk road
主要路由 primary circuit routing
主要路与次要路环形交叉 major/minor rotary
主要矛盾 principal contradiction
主要矛盾线方法 critical path method
主要矛盾线分析 critical path analysis
主要贸易区 principal trade partness
主要煤巷 main entry
主要庙宇 principal temple
主要木材 primary timber
主要木框架 <房屋的> timber stud
主要目标 fundamental purpose; major objective; primary target; prime objective; prime target
主要目标区 primary target area
主要目标线 primary target line
主要目的 basic objective; fundamental purpose; major objective; primary target; prime objective; prime target
主要目的层厚度 thickness of main desired layer
主要内容 main content; primary coverage
主要能流 main power path
主要能源 basic power source; main energy
主要年龄 preponderant age
主要排水 arterial drainage
主要排水管 conducting drain
主要排污点 key discharge locations
主要排泄口 main outfall
主要跑道长度 basic runway length
主要配水渠 major distributary
主要配置 major disposition
主要劈理 face cleat
主要频率 main frequency
主要平地作业 major grading
主要平面图 main working plan
主要平巷 main gate
主要剖面 major profile
主要谱线 principal line
主要起重设备 main hoist
主要气候条件 prevailing weather condition
主要气象站 principal synoptic (al) station
主要契约 master deed; single contract
主要倾向 main current
主要情况 main condition
主要缺陷 major defect
主要燃料 main fuel
主要燃料供给 principal fuel supply
主要燃烧期 controlled combustion period
主要人员 main staff
主要任务 essential role; primary mission
主要溶剂 primary solvent
主要入口 main entrance; principal entrance
主要三角测量 main triangulation
主要三面形 principal trihedral
主要散步之处 principal walk
主要山脉 backbone
主要商品批发商 stapler
主要商业中心 main commercial center[centre]
主要设备 capital equipment; capital facility; main equipment; main facility; main installation; major equipment; primary equipment; principal facility; principal installation; vital plant
主要设备材料表 list of materials and main equipment
主要设备材料清单 list of materials and main equipment
主要设备的逐渐替换 piecemeal replacement of major units
主要设备(明细)表 list of main equipment
主要设备清单 master equipment list
主要设计功能 primary design function
主要设计指标 the main design figures
主要设施 capital facility
主要生产国 principal productive nation
主要生产率 primary productivity
主要生产能力 primary productivity
主要生产商 principal productive commerce
主要生产装置 main production plant
主要圣殿 principal temple
主要圣坛 principal temple
主要失效模式 major failure mode
主要施工机械 basic work machine
主要施工设备 major plant
主要十字路口 main crossing
主要时间延迟调节器 master timer
主要时期 primary period
主要食物 major foodstuff; staple food
主要驶出口 <互通式立体交叉的> major exit(of interchange)
主要世界货币 major world currencies
主要市场 leading market; primary market; staple market
主要事故 major accident
主要收益 primary income
主要手段 main means of transport
主要受抚养人 primary dependent
主要受拉钢筋 principal tensile reinforcement; main tensile reinforcement
主要受力部件 principal stress-carrying part
主要受力构件 primary load-supporting member
主要受压破坏 primary compression failure
主要受载结构 main loaded structure
主要受载情况 main loading case
主要枢纽 main terminal; major terminal
主要梳理机 proper card
主要输出 primary output
主要竖杆 main vertical
主要数据 key data; main data; major data; principal data; salient date; master data
主要数字 main number
主要水坝 major dam
主要水处理厂 major water treatment plant
主要水道 main navigable waterway
主要水流 primary current
主要水媒疾病 principal water-associated disease
主要水平 basal level; main level
主要水准测量 principal level(l)ing
主要死因 underlying cause of death
主要搜索方向 principal search direction
主要素 principal constituent
主要损失 main loss
主要台站 key station; main station
主要台柱 main stay
主要特点 central feature; main feature
主要特色 principal motif
主要特性 key property; leading feature
主要特征 key feature; leading feature; main characteristic; major character; principal character
主要梯阶 main altar
主要体系特性 key architectural characteristics
主要条件 main condition
主要条款 main clause
主要铁道 main railroad[railway]
主要铁路 leading railroad; main railroad[railway]
主要通道 main thoroughfare; main fairway <油、气等>
主要通路 main route; major avenue of approach
主要通信[讯]路由 principal traffic route
主要统计值 major total
主要图纸 principal drawing
主要途径 main path; major route; principal pathway
主要土层 key horizon
主要土类 main soil group
主要挖掘机 basic shovel
主要挖土机 basic shovel
主要危险 primary hazard
主要维修可行性 major maintenance availability
主要问题 primal problem; principal problem
主要污染 key pollution
主要污染物 key pollutant; principal pollutant
主要污染源 key pollution source; primary pollution source; main pollution sources
主要污染指标 focused pollution index
主要污染指数 focused pollution index
主要无机物 major inorganism
主要物 dominant
主要误差 governing error
主要系统 primary system
主要系统道路 main system road
主要线脚 principal mo(u)lding
主要线条 principal mo(u)lding
主要巷道 main opening; main shaft
主要巷道运输机 trunk conveyer[conveyor]
主要项 dominant term
主要项目 main item; major project; principal particulars
主要项目检查结果 inspection result of main item
主要项目修理零件单 major item re-

pair parts list
主要效果 main effect
主要效应 main effect
主要心理能力 primary mental ability
主要行车道 main carriageway
主要行车地面 main travel area
主要行车方向 major movement
主要行人区域 principal pedestrian zone
主要行政管理费用 main office expense
主要形状 dominant shape
主要性能 key feature;main performance;primary function
主要性能指标 main performance index
主要需求国 principal demandable nations
主要序列 master suite
主要学派 main schools
主要循环管道 primary now-and-return pipe
主要压力 brunt
主要延长枝 extension leader
主要岩层 predominant formation; prevailing rock formation
主要岩浆 basic magma
主要岩石 fundamental rock
主要岩性单位 principal lithologic unit
主要研究者 principal investigator
主要颜色 fundamental colo(u)r
主要药包 main charge
主要要求 major requirement
主要要求线 <根据城市交通流量而标定的> major desire line
主要要素 principal feature
主要业务 primary service
主要依据 main basis
主要仪器 key instrument;primary instrument
主要因素 critical factor; dominant factor; key factor; major factor; predominating factor
主要应急离机口 primary emergency escape hatch
主要应力 main;major stress;primary stress;stress
主要应力差 <土工试验> deviator(ic) stress
主要应力迹线 principal stress trajectory
主要应力极限 main tensile stress limit
主要影响 dominating influence; primary effect
主要用户 bulk client
主要用量 principal use tonnage
主要用量占百分比 principal use of mineral commodities at percentage
主要用途 main application; principal use
主要用于起飞的跑道 primary departure runway
主要优点 major advantage; principal advantage
主要优势 major advantage
主要优先项目 high priority project
主要有机危险成分 principal organic hazardous constituent
主要有用矿物 principal useful mineral
主要有用组分 essential useful component;principal useful component
主要渔区 primary division
主要预算 main budget;master budget
主要元件 major component
主要元素 essential element;major element
主要元素分析 major-element analysis
主要原材料 main raw material; principal raw material
主要原材料成本 main material cost
主要原生构造 major primary structure
主要原因 key causes
主要原因之一 one of the prime reasons
主要圆顶塔 <印度> principal tope (mound)
主要愿望线 <根据城市交通流量而标定的> major desire line
主要运量 predominant traffic
主要运输 key traffic;main haulage
主要运输道 trunk roadway
主要运输机 main haulage conveyer [conveyor]
主要运输路线 main shipping route
主要运输平硐 main haulage tunnel
主要运输平巷 main haulage road; main lateral
主要运输石门 main haulage cross-cut
主要运输手段 main means of transport
主要运输水平 main haulage horizon; main haulage level
主要运输巷道 main haulage drift
主要运行条件 basic operating condition
主要杂质 major impurity
主要载流子 majority carrier
主要造床流量 capital formative discharge; dominant formative discharge
主要责任 primary responsibility;ultimate liability
主要炸药 primary blasting explosive
主要债务 primary liability
主要站 key station
主要站交通量计数 key-station volume count
主要站卸车安排 unloading arrangements at main station
主要账簿 head book;principal book
主要账户 main account;primary account;principal account
主要账目 primary account
主要照明 main lighting
主要折光率 principal refraction [refractive] index
主要阵地 key position
主要震动 principal shock
主要证据 primary evidence
主要症状 cardinal symptom
主要支持 main stay
主要支出 major expenditures
主要支出用途 main object of expenditures
主要支墩 main column;main support
主要支路 main branch road
主要职员 key personnel
主要指标 main index
主要指标的指数 index of leading indicator
主要指示 key instruction
主要制造设备 main manufacturing plant
主要终点站 main terminal;major terminal
主要种(类) dominant species;essential species
主要周期 predominant period
主要铸件 main casting
主要专用线 main access road
主要装药 principal charge
主要装载水平 main loading level
主要装置 principal installation
主要着陆架 basic travel(l)ing gear
主要资产 primary assets
主要资料 primary data;principal data
主要子午线 prime meridian
主要自动关门器 principal stud
主要自动驾驶仪 primary autopilot
主要总成 principal assembly
主要总成装配 principal unit assembling
主要走道 main walk;principal walk
主要租户 prime tenant
主要组成(部分) chief component; key component
主要组分 key component; primary constituent
主要作物 chief crop;dominant crop;dominant plant;leading crop;main crop;staple crop
主要作业 key operation
主要作业表 master schedule
主要作业进度表 master schedule
主要作用 essential role
主叶片 main wing
主页 home page
主液压泵 main hydraulic pump
主液压缸 master cylinder; primary cylinder
主液压系统 main hydraulic system
主液压系统泵装置 main hydraulic power plant
主溢洪道 main spillway
主引出线 main lead;main outlet line
主引导站 primary control station
主引水沟渠 main feeding ditch
主引线端子 line terminal
主应变 main strain;principal strain;principle strain
主应变方向 direction of principal strain;principal strain direction
主应变空间 principal strain space
主应变面 principal plane of strain
主应变轴 axis of principal strain; principal axis of strain
主应变轴与单剪方向夹角 angle between principal strain axis and direction of shear
主应答表 master answer sheet
主应急馈电线 main emergency feeder
主应力 basic stress;main stress;primary stress;principal stress;principle stress
主应力比 principal stress ratio;ratio of principal stresses
主应力差 difference of principal stress;deviator stress <三轴压缩试验中的>
主应力承重部件 principal stress-carrying part
主应力法 <形变力计算用> slab method
主应力方向 direction of main stress; direction of principal stress
主应力轨迹 principal stress trajectory; trajectory of principal stress; trajectory of stresses
主应力轨迹图 map of principal stress trajectories
主应力轨迹线 isostatics;trajectories of principal stresses
主应力迹线 diagonal stress trajectory; line of principal stress; main stress line;principal stress trajectory
主应力空间 principal stress space
主应力络网 principal stress trajectory
主应力面 principal plane of stress; principal stress plane
主应力系数 coefficient of principal stress
主应力线 line of principal stress; principal stress line
主应力圆 circle of principal stress; cite of principal stress; principal stress circle
主应力轴 axis of principal stress; main axis of stress;principal axis of stress
主应力轴图解 diagram of principal stress axes
主营业务收入 prime business income
主映射 principal mapping
主用泵 duty pump
主用波道 main channel
主用户同线 calling party's line relay
主用接收机 main receiver
主用卫星 primary satellite
主优阳极 preponderating anode
主油泵 main oil pump
主油道 main oil distributing passage; main oil gallery
主油缸 master cylinder
主油管 main loading or discharging line;main oil pipe;oil main line
主油管堵塞阀 master and block valve
主油路 main circuit;working connection
主油系 high-speed circuit;power feed system
主油箱 main fuel tank;main oil tank
主余震 major aftershock
主余震型地震 main-aftershock earthquake
主语言数据库 host language data base
主语言系统【计】 host language system
主遇险报警网络 primary distress alerting network
主元 pivot element
主元件 primary element
主元件分析 principal component analysis
主元素 host element;pivot element
主元选择 pivot selection
主元运算 pivot operation
主原子价键 primary valence bond
主圆 principal circle
主圆棒 principal round bar
主缘 cardinal margin
主源 main source
主源程序 master source program(me)
主源泉 master well
主约 master contract
主钥匙 master key
主运动 main motion; main movement; primary motion; principal motion
主运输大巷 main haulage
主运输道 main haulage track; main haulageway
主运输机 mother conveyer [conveyor];trunk conveyer[conveyor]
主运输平巷 main gangway
主运输平巷运输机 gangway conveyer[conveyor]
主运算处理机 main arithmetic processor
主宰空间 master space
主载波 main carrier
主载流子流 primary-carrier flow
主造山运动 primary orogeny
主增塑剂 primary plasticizer
主增益控制 master gain control
主轧辊 king roller;tread roll
主轧机 main mill
主闸缸 brake main cylinder
主闸轮 centre brake drum
主闸门 main gate;work gate
主债务人 principal debtor
主站 master station;primary station
主张 affirmation;allegation;assert;aver;claim;make a point that;proposition;submission
主张量 principal tensor
主张应力 principal tensile stress
主张中央集权下经济统制的人 statist
主章动 principal tensor
主涨流 main flood
主照明 key lighting

主照明光 key light
主折射率 principal refraction[refractive] index
主褶曲 major fold
主褶皱 main fold; major fold; principal fold; prominent fold
主真空管道 main vacuum manifold
主振部分 driver unit
主振荡 main oscillation; primary oscillation; principal oscillation
主振荡电路 driver circuit
主振荡器 king oscillator; main oscillator
主振荡器系统 master oscillator system
主振动 principal vibration
主振动形式 principal mode of vibration
主振动运输机 mother shaker
主振功率放大器 master oscillator power amplifier
主振管 exciter tube
主振频率 driving frequency
主振器 driver; driving unit
主振器控制雷达系统 master oscillator radar system
主振式振荡器 independent drive oscillator
主振型 fundamental mode; principal mode; principal type of vibration
主振子 main element
主震 main earthquake; main shock; major earthquake; major shock; principal earthquake; principal shock; principle earthquake
主震时间 time of principal earthquake
主震相 principal part
主震余震类型 main shock-aftershock type
主震余震型 main shock-aftershock type
主震震级 dimension of principal earthquake
主震最大波 maximum wave of principal shock
主蒸汽 main steam
主蒸汽阀 main steam valve
主蒸汽管 main steam pipe; steam main
主蒸汽管道 main steam line; steam generator lead; steam main
主蒸汽管道系统 main steam-piping system
主蒸汽集管 main steam header
主蒸汽系统 main steam system
主蒸汽与给水系统 main steam and feed water system
主蒸汽闸阀 steam-outlet valve
主蒸汽止回阀 main steam isolation valve
主整流器部件 main rectifier block
主正交钢板桥 major orthogonal steel-plate bridge
主正交各向异性板 major-orthotropic plate
主正交各向异性板法 major-orthotropic plate method
主正交各向异性钢板桥 major-orthotropic steel-plate bridge; principal orthogonal steel-plate bridge
主正交各向异性桥 major-orthotropic bridge
主正交各向异性桥面结构 major-orthotropic deck structure
主正交系(统) principal orthogonal system
主正截面 principal normal section
主正面 principal facade
主正向通路 main forward path
主帧 prime frame
主支承结构 principal supporting structure
主支承面 main supporting surface
主支承轴承 main step bearing
主支管 primary branch
主支架 master bracket
主支流 <河道的> main branch (line)
主支气管 main bronchus
主支索 main stay
主支线 main branch(line)
主支柱 main pier; main pole
主直径 full diameter
主直线运动 principal straight motion
主值 principal value
主止回阀 main check-valve
主旨 general tenor; keynote; motif; motive
主指令带 master instruction tape
主指令缓冲器 main instruction buffer
主指令流水线 primary instruction pipeline
主指令码 basic order code
主指数 pivot index; principal exponent
主制动储气缸 main braking reservoir
主制动触点 main brake contact
主制动缸 main brake cylinder; main braking cylinder; master brake cylinder
主制动轮 centre brake drum
主制动螺线管 main brake solenoid
主制动器 maxi-brake; primary brake
主制动凸缘 centre brake flange
主制动位置 primary retarder position
主质 parenchyma
主致密线 major dense line
主中继段 main repeater section
主中继站 main repeater station; major relay station
主中频放大器 main intermediate frequency amplifier
主中频感应加热设备 mains medium frequency induction heating equipment
主中桅 main-topmast
主中心 principal center[centre]
主中心教堂 principal church
主中心局 main center[centre]
主中心轴线 main centerline[centreline]
主中央计时系统 master central timing system
主钟 primary clock
主钟控制网 despotic network
主(重)要的 overriding
主周期 primary period; principal period
主轴 arbor; axle rod; backbone; basic shaft; collet chuck; drive spindle; driving spindle; line shaft; main arbor; main axis; main shaft; major axis; perch; primary axis; primary shaft; principal shaft; principle axis; quill; rim shaft; vertical pin; vertical shaft; female spindle <气螺刀>
主轴臂 spindle arm
主轴变换 principal axis transformation
主轴变速齿轮套 mainshaft speed gear sleeve
主轴变速齿轮止动环 mainshaft speed gear retainer ring
主轴部件 spindle unit
主轴差速齿轮 mainshaft differential gear
主轴差速器十字架 mainshaft differential spider
主轴差速十字架小齿轮 mainshaft differential spider pinion
主轴超速齿轮衬套 mainshaft overdrive gear bushing
主轴超速(传动)齿轮 mainshaft overdrive gear
主轴衬 main bush
主轴衬套 mainshaft bushing
主轴承 base bearer; base bearing; crankcase bearing; journal bearing; main bearing
主轴承扳手 main bearing wrench
主轴承拆卸 main bearing removal
主轴承盖 main bearing cap
主轴承盖销钉 main bearing cap dowel
主轴承后油衬 rear main bearing oil seal
主轴承浇巴氏合金夹具 main bearing babbitting jig
主轴承铰刀 main bearing reamer
主轴承精镗削工具 main bearing reboring device
主轴承壳 main bearing shell
主轴承螺柱 main bearing stud
主轴承敲击声 main bearing knock
主轴承润滑 main bearing lubrication
主轴承润滑油管 main gallery pipe
主轴承镗刀 main bearing boring tool
主轴承镗杆 main bearing boring bar
主轴承油封 main bearing oil seal
主轴承油管 main bearing oil pipe
主轴承油压计 main bearing oil pressure ga(u)ge
主轴承轴颈 main bearing journal
主轴承轴瓦 main box; main steps; main brasses
主轴承柱螺栓 main bearing stud
主轴齿轮 mainshaft gear
主轴齿轮衬套 mainshaft gear bushing
主轴传动机构 spindle drive
主轴传动小齿轮 mainshaft drive pinion
主轴从动齿轮 mainshaft driven gear
主轴从动齿轮轴承 mainshaft driven gear bearing
主轴挡油圈 mainshaft oil baffle
主轴导向轴承 mainshaft pilot bearing
主轴低速滑动齿轮 mainshaft low speed sliding gear
主轴低速及倒车 mainshaft low and reverse sliding gear
主轴第三速齿轮 mainshaft 3rd speed gear
主轴第三速齿轮套 mainshaft 3rd speed gear bushing
主轴第三速齿轮轴承滚针 mainshaft 3rd speed gear roller
主轴第五挡齿轮 mainshaft 5th speed gear
主轴电动机 spindle drive motor; spindle motor
主轴电缆 mainshaft cable
主轴吊架 line shaft hanger
主轴端部 spindle nose
主轴(端)盖 mainshaft cap
主轴对准 spindle alignment
主轴分析 principal axis analysis
主轴俯仰现象 topple
主轴俯仰轴 topple axis
主轴高速滑动齿轮 mainshaft high speed sliding gear
主轴鼓轮 spindle drum
主轴滚珠轴承 mainshaft ball bearing
主轴滚柱轴承 mainshaft roller bearing
主轴横动 spindle traverse
主轴后端 rear-end of spindle
主轴后轴承 mainshaft rear bearing
主轴后轴承油封 mainshaft rear bearing oil seal
主轴滑动齿轮 mainshaft sliding gear
主轴滑动座架 spindle saddle; spindle slide
主轴架 main pedestal
主轴肩 spindle shoulder
主轴键 mainshaft key
主轴颈 main journal
主轴开口环 mainshaft snap ring
主轴壳体 kingpin housing
主轴孔 spindle hole
主轴孔径 spindle bore
主轴控制进水管 spindle operated penstock
主轴离合杆 spindle control lever
主轴螺帽 shaft nut
主轴螺母 mainshaft nut
主轴螺栓 upper shaft bolt
主轴门 main throttle
主轴面 plane of principal axis; spindle face
主轴莫氏锥度 spindle Morse taper
主轴气泵 main scavenging pump
主轴前端 front-end of spindle
主轴倾斜度 spindle inclination
主轴球窝 king bolt and socket
主轴润滑油 spindle oil
主轴上的节数 nodes in main axis
主轴式 kingpost system
主轴式天线 kingpost antenna
主轴式天线座架 kingpost antenna mount
主轴手动进给 main spindle hand feed
主轴套 collar bush; mainshaft sleeve; spindle sleeve
主轴套管 main spindle quill
主轴套筒夹紧手柄 mainshaft sleeve clamp lever
主轴同步齿轮 mainshaft synchronizer gear
主轴同步齿轮套 mainshaft synchronizer gear sleeve
主轴头 spindle head
主轴头部 spindle nose
主轴突出部分 spindle projection
主轴推压盖 mainshaft thrust cap
主轴托架 spindle carrier; spindle holder
主轴瓦耐磨合金层 main bearing lining
主轴外表面 outer face of spindle
主轴下套 lower bushing
主轴线 main axis; spindle axis; principal axis
主轴线测设 setting-out of main axis
主轴箱 headstock; machine head; main spindle box; main spindle head; spindle box; spindle head(stock)
主轴箱滑动导轨 rail plate
主轴向力 principal longitudinal force
主轴销 king pin
主轴旋涡 vertical eddy
主轴压盖 mainshaft gland
主轴芽 leader bud
主轴运动传动机构 spindle driving gear
主轴找中心 alignment of shafts
主轴直接传动 direct spindle drive
主轴直径 major diameter
主轴中速齿轮 mainshaft intermediate gear
主轴中心 spindle center[centre]
主轴轴承 mainshaft bearing
主轴轴承盖 mainshaft bearing cap; mainshaft bearing cover
主轴轴承盖垫密片 mainshaft bearing cap gasket
主轴轴承滚柱 mainshaft bearing roller
主轴轴承护圈 mainshaft bearing retainer
主轴轴承抛油环 mainshaft bearing oil slinger

主轴轴承油封 mainshaft bearing oil seal
主轴转速 shaft speed;speed of rotation of spindle
主属性 prime attribute
主柱 king piece;kingpost;main post;pillar;principal column;principal post
主柱基础 main column foundation;principal-column foundation
主柱正面 front of column
主转动惯量 principal moment of inertia
主转换开关 master switch
主转换中心 master switching center [centre]
主转向臂 main steering arm
主转子 main rotor
主转子叶片 main-rotor blade
主桩 key pile;king pile;main pile;principal post;ringpile
主桩挡土墙 king post wall
主桩架 pile frame
主桩套板结构 pile with horizontal timber
主装备 head rig
主装料运输带 main charging belt
主装配线 main assembly line
主状态 major state
主状态发生器 major state generator
主状态逻辑发生器 major state logic generator
主锥 main cone
主锥形驱动轮 main cone driving wheel
主着陆跑道 main landing runway
主资料 master file
主子方阵 principal square submatrix
主子空间 principal subspace
主子式 leading minor;principal minor
主子图 major subgraph
主子午线 principal horizontal line;principal meridian
主字码 descriptor
主自同构 principal automorphism
主纵梁 main longitudinal girder
主纵线 principal line
主纵线比例尺 principal line scale;Y-scale
主纵向力 principal longitudinal force
主纵像差 principal longitudinal aberration
主走廊 main aisle;main gallery
主租借契约 master lease
主族元素 main group element
主组成列 principal composition series
主组成子群列 principal composition series of subgroups
主组分 major constituent;principal component
主组分分析 principal component analysis
主组分图像 principal component image
主最大(值) main maximum;principal maximum
主坐标 master coordinate;principal coordinates
主坐标分析 pivotal coordinate analysis
主坐标平面 principal coordinate plane
主座板 principal saddle
主座舱盖 main canopy
主座舱罩 main canopy
主座封 primary seal

煮 bucking kier

煮白 blanch
煮布锅 kier
煮布锅堆布器 kier piler
煮布锅精炼 kier-boiling;kier scouring
煮茶工人 <英国工地上> drummer
煮沸 boiling
煮沸安定性试验 boiling water soundness test
煮沸不能消除的硬度 non-alkaline hardness
煮沸沉淀 coctoprecipitation
煮沸沉淀原 coctoprecipitinogen
煮沸的 boiled
煮沸干燥 boiling seasoning
煮沸鼓 boiler drum
煮沸灭菌 boiling sterilization
煮沸灭菌法 boiling sterilization
煮沸灭菌器 boiling sterilizer
煮沸器 boiling vessel
煮沸试验 <测定水泥安定性的> boiling test
煮沸消毒 boiling water sterilization
煮沸消毒器 boiling sterilizer
煮干 boil away;boil dry
煮锅 boiler
煮解 digestion
煮解罐 digestion tank
煮解器 digester[digestor]
煮开 boil up
煮练 boiling-off
煮炼斑渍 kier stain
煮炉 boiling out
煮呢 crabbing;roll boiling
煮呢机 crabbing machine
煮浓 boil(ing) down
煮器 boiler
煮散 decoction made from powder
煮生石灰坑 boiling hole
煮熟过的 boiled
煮丝 silk boiling
煮盐锅 salt furnace;salt oven
煮皂釜 caldron
煮皂锅 ca(u)ldron
煮制 boil
煮制亚硫酸盐纸浆 sulfite pulping

住舱 berthing compartment;berthing room accommodation;accommodation quarter,accommodation space

住舱舱壁 accommodation bulkhead
住舱甲板【船】 berth deck;mess deck
住处 dwelling(house);dwelling place;domicile;living space;residence;residency
住处不受侵犯 inviolability of domicile
住处面积 accommodation area
住处自动喷洒灭火系统 occupancy sprinkler system
住地 dwelling place;habitat
住房 dwelling;habitable house;habitable room;housing;lodging;residence building;residence housing;rooming occupancy;whare <新西兰毛利人的>
住房标准 housing standard
住房标准化 housing standardization
住房补贴 housing subsidy
住房补贴凭证 housing voucher
住房补助付款 housing assistance payment
住房补助计划 housing assistance plan
住房财务 housing financing
住房产出量 housing output
住房产权 housing tenure
住房(稠)密度 housing density
住房筹资 financing of housing
住房筹资机构 housing finance agency
住房传统 housing tradition
住房单位 housing unit
住房单元 housing unit
住房的声学质量 liveness
住房等级 grades for building;residential building rate
住房抵押 housing mortgage
住房抵押贷款 home loan
住房抵押贷款手册 settlement book
住房地板面积 floor area of a dwelling
住房调查 housing survey
住房调查表 housing questionnaire
住房定额 residential building rate
住房短缺 housing shortage;shortage of housing
住房法规 housing act;housing code;housing law;housing ordinance
住房方案 housing scheme
住房房屋 domestic building;dwelling building
住房费用 housing expenses;housing finance
住房分配 housing allocation;obligation of organizations to provide housing
住房分配货币化 money oriented housing allocation
住房复兴拨款 housing rehabilitation grant
住房复兴贷款 housing rehabilitation loan
住房改建 house alteration;house remodel(l)ing
住房改善 home improvement;house improvement;housing betterment;housing improvement
住房改善贷款 home improvement loan
住房改善计划 housing improvement program(me)
住房供给 housing
住房供暖 house heating
住房供应量 housing supply
住房顾问 housing consultant
住房管理 housing administration;housing management
住房管理部门 housing authority
住房管理局 housing authority
住房管理试验设施 residential custodial care facility
住房管理所 housing management office
住房规划 housing planning;housing program(me)
住房过多的 overhoused
住房合作公司 housing cooperative
住房合作社 housing cooperative
住房基本需求 basic need for shelter
住房基地 housing site
住房基地净面积 net site area
住房及城市开发部 <美> Department of Housing and Urban Development
住房集资 housing finance
住房计划 housing program(me)
住房价格 residential building rate
住房兼厨房 living kitchen
住房检查记录 dwelling inspection record
住房建设 house building;housing output <指建设过程>
住房建设部 Department of Building and Housing;Ministry of Housing
住房建设工程 housing project
住房建设规划 housing program(me)
住房建设实施 housing undertaking
住房建设者 residential developer
住房建造计划 housing plan
住房建造量 housing output;housing production
住房建筑 housing
住房建筑标准 housing standard
住房建筑工业 housebuilding industry
住房建筑规范 housing code
住房建筑计划 housing project
住房建筑协会 housing society
住房金融 housing finance
住房津贴 accommodation allowance;house allowance;housing allowance
住房竣工量 housing completions
住房开发 housing development
住房开发公司 housing development corporation
住房开发商委员会 Council of Housing Producers
住房开工量 housing starts
住房开支 housing expenses
住房空间 housing space
住房类型 housing type
住房立法 housing legislation
住房利率 residential building rate
住房联合开发 mixed housing development
住房量 stock of housing
住房零售价格 house purchase price
住房面积 floor area for residential dwellings
住房面积过大的 overhoused
住房普查 census of housing;housing census
住房起居室 dwelling living room
住房前门 fore door
住房区段 block of houses
住房区域 housing area
住房商品化 housing commercialization
住房设备 dwelling equipment
住房申请权 entitlement to housing
住房失修项目一览表 schedule of dilapidations
住房使用面积 net dwelling area;usable area of dwelling
住房市场 housing market
住房市场分析 housing market analysis
住房体系 housing system
住房厅堂 residence hall
住房统计 housing statistics
住房投资 investment in housing
住房拖车 mobile home
住房卫生设备(安装) house plumbing
住房问题 housing problem
住房闲置率 vacancy ratio
住房协会 housing association
住房需求 housing demand;housing need
住房需要 housing need
住房选择 housing preference
住房要求 housing requirement
住房业主中禁止歧视法 open housing law
住房议案 housing bill
住房营造业者 house builder
住房拥挤程度 extent of housing overcrowding
住房浴室 dwelling bathroom
住房援助计划 housing assistance plan
住房债券 housing bonds
住房占用率 occupancy rate;residential occupancy
住房占用权 housing tenure
住房占有率 occupancy rate
住房遮篷 residential awning
住房折旧 housing depreciation
住房政策 housing policy
住房质量 housing quality
住房质量标准 housing quality standard

住房主管部门 housing authority
住房状况 housing condition; housing situation
住房资金 housing funds
住房资助的例外收入极限 exception income limits
住房资助计划 housing assistance plan
住房综合体 complex of houses
住房总量 housing stock
住房租赁房地产 residential rental property
住房租售的开放(政策) open housing
住户 apartment dweller; dweller; household; householder; inhabitant; occupier; resident
住户抵押人 occupant mortgagor
住户电表 tenant's meter
住户访问法 dwelling unit interview method
住户供热 house heating
住户交叉分类法【交】 household cross classification method
住户连接线 house connection
住户消费 household consumption
住户自有公寓 condo; condominium
住户自有公寓抵押金 condominium mortgage insurance
住户自有公寓建设项目 condominium project
住户自有公寓式住房 condominium housing
住户自有公寓式住宅 condominium dwelling
住家式旅店 residential home
住居区隔离【交】 dwelling insulation
住留谱线 persistent line
住人房屋 accessory building
住人岩洞 cliff dwelling(settlement)
住宿 lodge; lodg(e)ment; lodging
住宿船 accommodating barge; accommodating vessel; accommodation ship; dormitory ship; living boat; living quarter; quarter boat; quarter module <海上钻井平台的>; floatel; floating quarter vessel; houseboat
住宿短缺 housing shortage
住宿费 quarterage
住宿汽车 motor home
住宿区 dormitory block
住宿人员密度 accommodation density
住宿拖车 mobile home
住宿者 live in
住所 abode; boziga; dwelling place; habitable house; habitable room; habitation; lodg(e)ment; lodging; nest; quarter; residence; digs <俚语>
住所不定人口 floating population
住所不定者 floating population
住所的 domiciliary; residential
住所地址 home address
住所记录 home record
住屋 tenement
住屋边圈地 in-by land
住院部 admission department; in-patient department; ward
住院隔离 isolation in hospital
住在工作地点的 live in
住在河边的人 riverain
住宅 abiding place; abode; domicile; dwelling house; family dwelling unit; family mansion; habitable house; habitation; homestead; house; housing stock; inhabitation; messuage; residence building; residential home; domus <古罗马或中世纪时期的>
住宅标准 residential standard
住宅布局 housing layout
住宅布线 residential electric(al) wiring
住宅部分 residential portion
住宅采光 residential lighting
住宅采暖 residential heating
住宅采暖系统 residential heating system
住宅采暖装置 domestic heating installation
住宅车房 residential garage
住宅城镇 dormitory town
住宅厨房 apartment kitchen; domestic kitchen; flat kitchen
住宅窗扉 residence casement
住宅存水湾 house trap
住宅大楼型 domestic building type
住宅大门 flat entrance door
住宅大厦 domestic block
住宅贷款信托债券 trust of housing loan bonds
住宅贷款银行 home loan banks
住宅单位面积 neighbo(u)rhood unit area
住宅单元 dwelling unit; housing unit; residential dwelling unit
住宅单元面积 neighbo(u)rhood unit area
住宅担保 home warranty
住宅担保计划 home warranty program(me)
住宅的 residential
住宅的翻新 rehabilitation of housing
住宅抵押贷款 residential mortgage
住宅地块 residential plot
住宅地下排水管 house subdrain
住宅电话 residence telephone
住宅电梯 elevator residential
住宅发展区 residential development
住宅翻建 remodel(l)ing a house
住宅房地产 housing estate
住宅房屋 domestic building
住宅废水 house wastewater
住宅废物 residential waste
住宅分配 residential allotment
住宅焚烧炉 residential incinerator
住宅改善 housing improvement
住宅更新 housing renewal
住宅工程 housing project
住宅公债 housing funds
住宅规划 housing project
住宅规模 dwelling house scale; dwelling size
住宅和公共上下水道连接管 house connection
住宅和环境设计 housing and environment(al) design
住宅后院 backside
住宅花园 home ground garden
住宅基址 home site
住宅及基地 messuage
住宅及屋基 messuage
住宅集中供暖系统 residential central cooling system
住宅检测 home inspection service
住宅检测员 home inspector
住宅建设 construction of housing; domestic construction; dwelling construction; home building; home construction activity; housing construction; housing development; residential construction
住宅建设拨地 residential allotment
住宅建设公司债券 housing corporation bond
住宅建设管理人 administrator of a home
住宅建设管理员 administrator of a home
住宅建设规划 housebuilding program(me)
住宅建设基金 housing funds
住宅建设计划 housebuilding program(me); housing development plan
住宅建设经济 economy of housing construction
住宅建设用地 housing land
住宅建造 domestic construction; home building; residential construction
住宅建造督察 home inspector
住宅建造视察员 home inspector
住宅建造商 homebuilder; house builder
住宅建筑 house building; residential construction; residential structure
住宅建筑标准 housing standard
住宅建筑方案 housing scheme
住宅建筑合作社 building society
住宅建筑商 homebuilder; house builder
住宅建筑学 domestic architecture
住宅街 residential frontage
住宅经济学 housing economics
住宅居住单元 bed dwelling unit; bed-sitting room dwelling unit
住宅开发商 residential developer
住宅空气调节 residential air-conditioning
住宅老化 housing obsolescence
住宅类型 housing type
住宅连接管 house connection pipe
住宅楼层 apartment floor
住宅旅馆 residential hotel
住宅煤气管道 house gas piping
住宅门 landing door
住宅密度 dwelling density; housing density
住宅免税 homeowner's tax exemption
住宅内房间 <供热、供应热水等公用事业设备的> utility room
住宅内室 ben
住宅内室门 domestic room door
住宅排水沟 house drain
住宅排水管 house drain; house drain pipe; residential drainage piping; residential sewage pipe
住宅汽车间 domestic garage
住宅前室存积雨水的方形贮水池 <古罗马> impluvium
住宅区 residential area; residential district; residential quarter; residential zone; bedroom block; bedroom building; bedroom community; bedroom house; bedroom unit; block of houses; block of housing; domestic quarter; dwelling district; housing estate; neighbo(u)rhood unit; populated area; residence area; residence district; residence quarter; residence range; residence region; residence section; residence zone; uptown; zone of residential; living quarter
住宅区道路 housing estate road
住宅区的发展 residential development
住宅区的消音装置 residential silencer
住宅区发展计划 residential development project
住宅区功能 function of settlement
住宅区供热 residential heating
住宅区规划 planning of residential area
住宅区街道 residence street; residential(local) street
住宅区垃圾 residential waste
住宅区停车场计划 residential parking program(me)
住宅区(域) housing area
住宅区噪声 dwelling area noise
住宅群 block of flats; clump of houses; cluster; cluster houses; cluster housing; group dwelling; house complex; housing complex; housing estate; housing group
住宅群公用地综合 estate planning
住宅群规划 estate planning
住宅群体分区规划 cluster zoning
住宅群选址定点 siting of houses
住宅热水采暖 domestic water heating
住宅热水供热系统 domestic hot-water system
住宅热水需要量 residential hot-water demand
住宅入口 flat entrance; residential entrance
住宅商店组合建筑 combined dwelling house
住宅设计卫生 residential planning hygiene
住宅设施 residential accommodation
住宅施工 domestic construction
住宅实用内院 domestic utility patio
住宅实用走廊 domestic utility corridor
住宅使用面积 residential usable floor area
住宅市场 housing market
住宅式司机室插销 <起重机> house lock
住宅数量短缺 quantitative housing shortage
住宅水管尺寸 residential water pipe size
住宅水管系统 domestic water piping system
住宅所有权 home ownership
住宅所在位置 home site
住宅塔楼 housing tower
住宅厅堂 residence hall
住宅庭园 courtyard of dwelling house; curtilage
住宅通风 ventilation of residences
住宅投资 housing investment
住宅土地临街界线 residence lot frontage
住宅团地 housing complex
住宅为主的地区 predominantly residential area
住宅维护设备 residential maintenance equipment
住宅卫生 residential hygiene; residential sanitation
住宅卫生设备 house plumbing
住宅位置 residential location
住宅问题 housing problem
住宅污水管 house connection; house sewer
住宅污水流量 residential sewage discharge
住宅下水出口存水弯 house trap
住宅下水道 house sewer
住宅下水道的卫生密封设施 house trap
住宅小区 housing colony; residential square
住宅小区布置 minor residential layout
住宅兴建方案 housing scheme
住宅型窗 residential window
住宅蓄(雨)水池 house cistern
住宅烟囱 domestic chimney
住宅以外的固定投资 non-residential fixed investment
住宅用玻璃棉绝热材料 fiberglass home insulation
住宅用地 housing estate; residential plot; residential portion
住宅用焚化炉 residential incinerator
住宅用铰链 residential hinge
住宅用设备 residential-type equipment
住宅用水泵 house pump
住宅杂用室 domestic utility room; domestic workroom
住宅噪声 domestic noise

住宅占用率 residential occupancy
住宅照明 residential illumination
住宅正大门 main entrance
住宅中生活社交活动区域 big house
住宅柱厅 < 印度 > tibari
住宅专用区 exclusive residential district;restricted residential district
住宅资产抵押贷款 home equity loan
住宅资金供应 housing financing supply
住宅自动化 house automation
住宅综合 dwelling house comprehensive
住宅综合体 residential complex
住宅组群 group house
住宅组群用地 grouped site
住宅组团 housing cluster; housing group;housing of a grouped site
住帐篷 tent
住址 dwelling place
住址变动 change of residence;housing mobility

助爆 boost

助爆破药 booster charge;booster explosive
助爆器 booster
助产室 obstetrical room
助产学校 midwifery school
助铲-铲运机组 push-scraper combination
助铲法 helping earth-scraping process
助铲机 push car; pusher (tractor); push-loading tractor
助铲架 < 铲土机的 > push frame
助铲前推板 < 铲土机的 > front push-plate
助铲式铲运机 push loaded scraper; push-type scraper
助铲推板 < 铲土机的 > push plate
助铲拖拉机 pusher tractor
助铲装土 < 铲土机的 > push-loading
助沉淀剂 settling aids agent
助沉剂 settling agent
助沉剂高位槽 settling accelerator head tank
助促进剂 secondary accelerator
助催干剂 drier activator
助催化剂 catalyst promoter;promoter[promotor]
助导航 eyesight navigation
助定理 lemma[复 lemmata/lemmas]
助动车 autobike; boosting bike; moped[motor pedal]
助动车自动弯沉仪 autocycle
助动的 motion aiding
助动重力仪 astatic gravimeter;astatic meter
助飞器 jato
助浮剂 flo(a)tation agent
助干剂 drying aid
助拱 discharging arch
助钩角 canting piece;tripping palm
助焊板 runoff plate
助焊剂 scaling powder;slagging medium;soft solder flux;soldering flux; welding flux
助航标志 aids-to-navigation; navigating mark;navigation(al)mark;sea mark < 设置在浅水区的 >
助航灯 navigation light
助航设备 aids-to-navigation; navigation aid facility; navigation (al) aids;navigation instrument
助航设施 aids-to-navigation; navigation aid facility; navigation aids; navigation instrument
助航拖轮 helper tug
助航仪表 navigation instrument
助航装置 navaid
助记操作码 mnemonic operation code
助记地址 mnemonic address
助记地址码 mnemonic address code
助记符号 mnemonic symbol
助记可变名 mnemonic variable name
助记码 mnemonic code
助记名 mnemonic name
助记术 mnemonics; mnemotechnics; runemotechny
助记忆操作码 mnemonic operation code
助记忆码 mnemonic code
助记指令码 mnemonic instruction code
助剂 additive(agent);assistant;auxiliary agent
助驾系统 driver-aid system
助驾信息及路线导行系统【交】driver aid information and routing system
助检查 sight ga(u)ge
助解程序 heuristic routine; heuristic program(me)
助金 subsidy
助聚剂 promoter
助卷机 wrapper
助卷机辊 wrapper rolls
助理 coadjutant;coadjutor
助理编辑 assistant editor
助理城市设计师 assistant town planner
助理翻译 assistant translator; assistant interpreter
助理工长 straw boss
助理工程师 assistant engineer; engineer-in-training
助理管理员 assistant supervisor
助理管事 assistant purser
助理化学师 assistant chemist
助理记者 assistant reporter
助理驾驶员 assistant mate; assistant officer
助理监督 assistant superintendent
助理监督员 assistant supervisor
助理讲师 assistant lecturer
助理经纪 broker associate
助理会计师 assistant accountant
助理轮机员 assistant engineer
助理农业师 assistant agronomist
助理人员 assistant;helpmate
助理实验师 assistant experimentalist
助理事务长 assistant purser
助理收货员 extension clerk
助理统计员 assistant statistician
助理委员 assistant commissioner
助理研究员 assistant research fellow; research associate
助理引航员 assistant pilot
助理凿岩工 chuck tender
助理值班员 assistant master
助理专家 associate expert
助力泵 booster pump
助力操纵机构 power control
助力操纵离合器 servoclutch
助力传动 servo-drive
助力传动装置 servo-link
助力磁电机 booster magneto
助力杆 assister bar
助力缸 booster cylinder
助力机构 servo-mechanism
助力机构伺服机件 servo-unit
助力汽缸罩 booster cylinder boot
助力器 assistor; augmenter; booster; power-control servo; servo-actuator; servo-unit; strengthener; thrust augmenter
助力器牵引力 booster tractive force at speeds
助力器牵引坡度 booster-traction gradient
助力器软管夹 booster hose clamp
助力器软管箍 booster hose clip
助力刹车 servobrake
助力式的 power assisted
助力式转向 power assisted
助力压力 boost pressure
助力液压缸 booster cylinder
助力油缸 power cylinder;servocylinder;slave cylinder
助力运动 assist exercise
助力制动 servobrake
助力制动器 servobrake
助力转向 boosted steering;power-assisted steering;power steering
助力转向系统 proportional demand steering system
助力转向轴 power-steering shaft
助力装置 work saving device
助沥滤剂 leaching agent
助流剂 flow aid;glidant
助流区 entrance region
助留剂 retention aid
助滤剂 filter(ing)aid;filtration aid; leaching agent;agglomeration aid
助滤器 filter aid
助磨剂 grinding aid
助磨料 grinding aid
助凝剂 agglomeration aids;aid-coagulant; coagulant aids; hardening admix(ture)
助凝剂投配管 coagulant dosing tube
助配位体溶解过程 ligand-promoted dissolution process
助漆溶剂 lacquer solvent
助强波纹 stiffening bead
助曲线 extra contour
助燃的 comburant; combustion-supporting
助燃风机 combustion air blower
助燃剂 burning-rate accelerator;combustion adjuvant; combustion improver;supporter combustion
助燃空气 combustion(-supporting) air;oxidizing air;secondary air
助燃空气入口 combustion air inlet
助燃气体 combustion-supporting gas
助燃设备 auxiliary combustion equipment
助燃添加剂 perfect combustion catalyzer
助燃物 comburant
助扰器 turbulence promoter
助热器 heat booster
助溶基团 solubilizing group
助溶剂 latent solvent;solutizer
助熔 fluxing
助熔剂 fluxing agent; fluxing medium; flux material; flux oil; furnace addition;fusing agent;slagging medium
助熔剂秤量器 stone batcher
助熔剂供料定量器 stone batcher
助熔剂计量箱 stone batcher
助熔金属 flux metal
助熔矿石 fluxing ore
助熔能力 fluxing power
助熔石料 < 石灰石、白云石等 > fluxstone
助熔性 fluxibility
助熔作用 fluxing action
助柔剂 flexibility agent;flexible agent
助色基团 auxochrome;auxochromous group
助色团 auxochrome; auxochromic group;auxochromous group
助声器 baffle
助式离合器 power-actuated clutch
助视力的 specular
助手 adjunct; adjuvant; ancillary; assistant; coadjutant; coadjutor; helper;helpmate;mate
助水溶物 hydrotrope
助塑剂 co-plasticizer
助听放大器 hearing aid amplifier
助听器 acousticon;aid-hearing;audiophone; audiphone; deaf-aid; dentophone; ear-trumpet; eartrumpet; hearing aid; listening device; osophone[otophone];aerophone < 探测飞机的 >
助听器耳机 hearphone aid earphone
助推 boosting
助推动力装置 booster power plant
助推发动机 booster;booster engine
助推滑翔机 boostglider vehicle;boostgliding vehicle
助推机 assistor
助推机车 helper locomotive
助推级 booster stage
助推器 assistor; auxiliary booster; booster; boost motor; propelling booster;thruster
助推器冲量 boosting impulse
助推拖拉机 push(er)tractor
助洗剂 builder;detergency promoter; detergent builder
助响瓮 echea
助消化的 digestive
助絮凝剂 flocculant aid
助悬浮剂 suspending agent
助学金 education allowance; grant-(s)-in-aid; living allowance; study grant
助氧化剂 pro-oxidant
助增塑剂 secondary plasticizer
助长 promote
助注剂 intrusion aid
助抓突角 spur

苎麻 china grass;hemp

苎麻布垫 hemp burlap mat
苎麻废水 ramee wastewater
苎麻纤维 hemp fiber[fibre]; ramie hemp;ramee;ramie
苎麻纤维浸解废水 ramee retting wastewater

注不混溶气 immiscible gas injection

注册 book;enrol(l);enrollment;inscribe;ledger;log-in;log on;register;registration
注册编号 registered number
注册标记 monomark
注册簿 register
注册测量师 chartered surveyor
注册产权 record title
注册车数 car registration
注册承包商 licensed[licenced] contractor
注册处 registrar's office;registration office;registry
注册船 registered ship
注册代号 < 英 > monomark
注册的土地 registered land
注册地界图 recorded plat
注册地块 registered plot
注册地址 registered office
注册吨 < 商船注册(或登记)的容积单位,一个注册吨相当于 100 立方英尺,1 立方英尺 = 0.02832 立方米 > registered ton
注册吨位 net tonnage;registered tonnage
注册法 registration law
注册房地产 licensed[licenced] premises

Z

注册费 registration fee
注册港 port of registry;port of register
注册工程师 chartered engineer; licensed[licenced] engineer; professional engineer;registered engineer
注册公司 company incorporated(Inc.); incorporation
注册股本 registered capital
注册股票 registered certificate of shares
注册官 registrar
注册国际财产专家 certified international property specialist
注册过程 logon procedure
注册号 registration mark
注册号码 herd number; registered number;registration number
注册记录 statement of record
注册建筑师 licensed[licenced] architect;registered architect
注册交易人 registered trader
注册净吨位 <船舶> net registered tonnage;registered net tonnage
注册会计师 certified (public) accountant; chartered accountant; licensed(public)accountant
注册马力 registered horsepower
注册码 poll code
注册器 logger
注册人口 registered population
注册容量 marked capacity
注册入学 matriculation
注册商标 registered brand;registered trademark
注册商标示意图 registered certification trade mark scheme
注册水深 chartered depth
注册图案 registered design
注册土地 book land
注册委员会 registration board
注册问题 registration problem
注册信息提供当局 registered information provider
注册业主 owner of record
注册用户 registered user
注册员 register;registrant
注册证书 certificate of registration; registering certificate; registration certificate
注册执照 certificate of registration; certificate of registry; registering certificate
注册主管 registrar
注册资本 registered capital
注尺寸 dimension; insert the dimensions
注尺寸的线 dimensional line
注肥器 <灌溉管道的> fertilizer injector
注富气 enhanced gas injection
注钢 teem
注光 spotlight
注氦 helium injection
注记 annotation
注记板 lettering plate
注记编排 lettering
注记等高线 figured contour
注记分类 name classification
注记负载 scope of the name contents
注记盖章 stamping tripod
注记剪贴 stick-up lettering
注记空白位置 title block
注记空位 <供记注用空白位置> background removal for lettering
注记透明片 names overlap
注记透写图 names trace
注记文件 registration files
注记轴配置 arrangement of axes
注记字列 placement of lettering
注记坐标网 lettered grid
注件模缝 casting seam
注件内部浆痕 wreathing
注浆 grouting
注浆斑点 casting spot
注浆泵 grout pump;injection pump; mud jack
注浆材料 grout agent;injecting paste material;injection material
注浆参数 injection parameters for grouting
注浆产生的应变 casting strain
注浆产生的应力 casting stress
注浆成型 casting; slip casting; slurry casting
注浆成型法 casting process;slip casting process
注浆稠度 grout consistency [consistence]
注浆堵水 grout to seal water
注浆阀 cement valve
注浆法 injection process; injector method; slurry-casting method; gouting
注浆法工艺 technical of injector method
注浆方法 grouting method; grouting system
注浆分层厚度 thickness of grouting layer
注浆管千斤顶 injection pipe jack
注浆过程 <陶瓷> casting process
注浆混凝土 intrusion concrete
注浆机 casting machine
注浆孔 grout hole; injected hole; injection hole
注浆孔工艺参数 technical parameter of injector
注浆孔间距 distance between grouting hole
注浆孔径 diameter of grouting hole
注浆孔圈数 circle number of grouting hole
注浆孔深 depth of grouting hole
注浆孔数目 number of grouting (bore) holes
注浆孔吸浆量 absorption volume of grouting hole
注浆孔允许偏斜率 allowable deviation of grouting hole
注浆孔直径 initial diameter of injection hole; orifice diameter of grouting hole
注浆量 grout volume
注浆料斗 casting hopper
注浆料入口 casting head
注浆锚杆 anchor bolt with grout
注浆冒口 riser
注浆模 casting mo(u)ld
注浆设备 grout injection apparatus; injection apparatus
注浆升压封堵 seal and block with grouting and pressure heightening
注浆生产线 casting production line
注浆施工 chemical grouting;injection
注浆顺序 sequence of injection
注浆速率 casting rate
注浆陶瓷 cast ware
注浆涂搪 enamel(l)ing by pouring
注浆纤维混凝土 slurry-infiltrate fiber concrete
注浆型锚杆 grouted(rock) bolt
注浆压力 grouting pressure;injection pressure
注浆硬化 cast hardening
注浆用泥浆 casting slip
注浆终压 final injection pressure
注角变量 subscripted variable
注脚 bottom note;footnote;subscript
注解 annotation; comment; illustration;postil
注解卡片 comment card
注解栏 comment field
注解码 comment code
注解项 comment entry
注解行【计】comment line
注解约定 comment convention
注井法 well-injection method
注口 drain;stopper nozzle
注口颈圈 filler neck strap
注口塞 spout plug
注口砖 nozzle brick
注口座 nozzle pocket
注拉法 injection drawing process
注量 fluence
注料口 sprue
注料量 shot
注料嘴接头弯管 adaptor bend
注流边界 beam boundary
注流孔 orifice
注满 brim;fill;inject;topping
注皿培养法 pour plate culture
注明 by notation
注明尺寸 marking out dimension
注明尺寸的图 dimensional drawing
注明尺寸等级 step the sizes
注明额定值 marked rating
注明日期 dating
注明日期的 dated
注明日期公积金 dated earned surplus
注明无追索权票据 drawing without resource clause
注明运费预付 marked freight prepaid
注模 injection mo(u)ld
注模法 injection mo(u)lding process
注模机 injector
注目显示 attention display
注目效果 catch eye effect
注泥浆下套管法 mud in
注排淘洗 fill and draw elutriation
注坯吹塑 injection blow mo(u)lding
注频 injection
注频调制器 injection modulator
注气 gas injection
注气保持压力 gas pressure maintenance
注气法 injection method;insufflation
注气恢复地层压力 gas repressuring
注气井 gas injection well
注铅检测 lead proof
注入 breathe; disgorge; implant; implantation; impregnate; impregnation; influx; infusion; inject; injection;instillation;instil(l)ment; disembogue <湖或海>
注入泵 filling pump;injection pump
注入变熔作用方式 ectexis way
注入冰川的河流 englacial stream
注入参数 injection parameter
注入场 injection field
注入程度 degree of impregnation
注入磁铁 injection magnet
注入到油层 infusion in seam
注入的载流子 <半导体> added carrier
注入点 decanting point
注入电极结构 injecting electrode structure
注入电流 injection current
注入电流值 injected value of current
注入电路(信号) injection circuit
注入电平 injection level
注入电压 injecting voltage
注入二极管 injection diode
注入发光 injection luminescence
注入阀 fill(ing) valve
注入法 injection method; method of impregnation
注入方式 injection mode;injection regime
注入工艺 injection technology
注入管 ascending pipe;charging tube; filler pipe; filling pipe; injection tube
注入管线 injection line
注入光纤 injection fiber[fibre]
注入光学 injection optics
注入光学系统图 injection optics diagram
注入轨道 injection orbit; injection trajectory
注入轨道半径 injection orbit radius
注入化学药剂 grouting chemicals
注入环 injected annulus
注入灰浆 injection mortar
注入混合岩 injection migmatite
注入混合作用 entexis
注入混凝土 concreted in
注入机 injector
注入技术 injection technique
注入剂 injection agent; intrusion agent
注入剂前缘 injection-fluid front
注入加热 injection heating
注入浆液 injection grout
注入井 injection well; intake well; key well
注入井堵塞 injection well plugging
注入井间隔 injection interval
注入井井口压力 intake well head pressure
注入井试井 injectivity test
注入空气 blow in
注入孔 filler hole;filler opening;filling aperture;filling hole;filling opening;fill-in well;fill opening;injecting hole
注入孔道 injector tunnel
注入孔用悬挂式孔壁封隔器 hook wall flooding packer
注入口 filler box;filler opening;filling opening; filling port; handhole; throat;sprue <铸型的>
注入扩散法 Cobra process;gun-injection process
注入冷却 injection cooling
注入粒子 injecting particle
注入量测量 injectivity survey
注入料 filler
注入流 injected stream;injection flow
注入流体 injection fluid
注入滤网 filling sieve
注入率 injection rate;rate of injection
注入脉冲 injected pulse
注入锚 injection anchor
注入能力 injection capacity
注入片麻岩 injection gneiss
注入频率 injected frequency
注入剖面 entry profile
注入气体 injected gas;input gas
注入气油比 injection gas-fluid ratio
注入器 filler;injector;inspirator
注入器部分 injector section
注入器发射 injector emittance
注入器高压电极 injector terminal
注入器管道 injector tube
注入器级 injector stage
注入器加速器 injector accelerator
注入器枪 injector gun
注入器输出 injector output
注入器束流 injector current
注入器调制器 injector modulator
注入器效率 injector efficiency
注入器械 injection apparatus
注入器阳极 injector anode
注入器阴极 injector cathode
注入器柱 injector column
注入区 injection domain
注入绕转频率 injection revolution frequency

注入热量 heat influx
注入热液 hot fluid injection
注入软管 filing hose
注入塞 filler plug;filling plug
注入栅极 injector grid
注入时间 injection length;injection time
注入式电致发光 injection electroluminescence;Lossev effect;recombination electroluminescence
注入式放大器 injection amplifier
注入式栅极 injection grid
注入水 injected water;injection water
注入瞬间 injection instant
注入速度 injection rate;input rate;input speed;rate of injection;speed of impregnation
注入锁定 injection locking
注入锁相 injection locking;injection phase locking
注入条件 injection condition
注入通道 injection channel
注入头 <使钻液注入钻杆的水龙头接箍> injection head
注入位置 injection phase;injection point
注入物 implant;infusion
注入系数 injection ratio
注入系统 injecting system
注入相位角 injection phase angle
注入象限 injection quadrant
注入效率 injection efficiency
注入型半导体激光器 injection semiconductor laser
注入型激光器 injection laser
注入性 injectivity
注入压力 grouting pressure;injection pressure
注入延续时间 infusion time
注入液 impregnating liquid;injection fluid
注入油气比 gas input factor
注入与泄放阀 fill and drain valve
注入站 <色液测流法的> injection station
注入蒸汽 steaming
注入直线段 injection straight section
注入中性束 injection neutral beam
注入桩 injection pile
注入装置 injection device;refiller
注塞 pour plug
注砂井 mine fills storing and mixing shaft
注上星号 asterisk
注射 inject;introduction;jetting;syringe
注射泵 injection pump;injector pump;squirt pump
注射成型机 injection mo(u)lding machine;injection press
注射成型喷嘴 jet mo(u)lding nozzle
注射吹塑(成型) injection blow mo(u)lding
注射导管 injection catheter
注射导引器 syringe guide
注射滴定管 syringe buret(te)
注射阀 introduction valve
注射法 injection
注射法隧道施工 <开凿土质隧道时,土质松软,易于坍落,须先用稳定加固剂如水泥浆,注入土层中,然后再钻凿> injection method in tunnel construction
注射管 injection pipe;injection syringe;injection tube;jet pipe;nozzle pipe;syringe
注射管嘴 injection nozzle
注射灌浆 injection grouting;jet grouting
注射硅酸钠及氯化钙加固地基法 silicate injection
注射混合器 injector mixer
注射活塞 injection ram
注射基础 injected foundation
注射技术 injection technique
注射剂 injection
注射加热部位 injection hotspot
注射加热器 injection heater
注射孔 injection hole
注射孔口 injection orifice
注射口 injector port
注射块 injection block
注射冷凝器 injector condenser
注射力 injection force
注射量 shoot capacity
注射流动方法 injection flow method
注射模 injection mo(u)ld
注射模法 injection mo(u)lding
注射模塑 injection mo(u)lding
注射模塑压力 injection mo(u)lding pressure
注射模型成型机 injection mo(u)lder
注射能力 injectability;injection capacity
注射喷嘴 injection nozzle
注射谱带扩展 injection band spreading
注射起动机 injection starter
注射器 gun;hypodermic syringe;injection gun;injection packaging;injection unit;injector;inspirator;scooter;squirt;syringe
注射器反应 syringe reaction
注射器活塞 syringe piston
注射器进样 syringe sampling
注射器刻度 syringe graduation
注射器喷嘴 syringe nozzle
注射器吸液管 syringe pipette
注射器煮沸消毒器 syringe boiling sterilizer
注射枪 filling gun
注射燃料管 injector fuel pipe
注射燃料雾体 injected fuel spray
注射燃烧器 inspirator burner
注射润滑 shot lubrication
注射设备 injection installation
注射深度 injection depth
注射式绝缘 extrudable insulation
注射式通风道 injection air vent
注射式煮布锅 injector kier
注射寿命 injection life
注射水 injection water
注射水压力 injection pressure
注射速度 injection speed
注射塑制 injection mo(u)lding
注射塔 injection tower
注射通风 injector ventilation
注射温度 injection temperature
注射误差 injection error
注射旋塞 injection cock
注射液调节器 injection adjusting apparatus
注射用灭菌粉剂 sterilized powder for injection
注射用水 water for injection
注射用油 oil for injection
注视 contemplate;contemplation;fixate;observation;remark;watching
注视点 fixation point;point of attention
注视时间 time of fixation
注释 annotation;explanatory note;notation
注释编号 numbering of note
注释栏 comment field
注释目录 annotated catalogue
注释人 commentator
注释说明 explanatory notes
注释索引 annotated index
注释条款 interpretation section
注释项 comment entry
注释行 comment line
注释语句 comment statement
注水 flooding;flood water;syringe;topping up;water;water-flooding;water infusion;water jetting
注水爆破 water infusion blasting
注水泵 priming pump
注水舱 flooded tank
注水处理 injection water treatment
注水地震关系 injection earthquake relationship
注水电池 inert cell
注水断面 injection profile
注水阀 charging valve;injection valve;water injection valve
注水阀连接管头 water valve spud
注水阀塞 water valve piston
注水阀压盖 water valve gland
注水法 water injecting method
注水方式 water injection pattern
注水盖 filler cover
注水高度 height of water injection
注水供航 flashing
注水管 filler tube;filling pipe;water infusion tube <岩层注水用>
注水管 U 形接头 <温水取暖装置> filling pipe U connection
注水管接头 water stem nut
注水管连接套管 water hose stem
注水管连接弯头 water connection L tube
注水管路 fill-up line;water flood line
注水管线 fill-up line;water flood line
注水脊 injecting ridge
注水计划 injection project
注水加密 compaction by watering
注水井 diffusing well;flooding well;injection well;input well;water injection well;recharge well
注水井受水剖面 profile-log of water injection
注水井网 flood pattern
注水井用悬挂式堵塞器 hook well floating packer
注水孔 filler hole
注水孔凿岩机 injection drill
注水孔钻机 injection drill
注水口 filling pipe end
注水口旋塞 feed-water plug;filling cock
注水冷凝塔 elevated jet condenser
注水连接弯管 water connection sleeve
注水连接弯头 water connection sleeve
注水量 quantity of tracer injection
注水漏斗 water funnel
注水密实法 compaction by watering
注水泥 cementing job
注水泥方法 means of injecting slurry
注水泥浮箍 cementing float collar
注水泥浮靴 cementing float shoe
注水泥固井 cementing of well
注水泥管 grout pipe
注水泥管鞋 set shoe
注水泥技术 cementing practice
注水泥浆 cementing;grout injection
注水泥浆钻孔钻进 grout-hole drilling
注水泥井段 cementing point
注水泥设备 cementation pumping equipment;cementing equipment;cementing outfit;shoe squeeze tool <加压下>
注水泥设计 cementing design
注水泥时间 cementing time
注水泥套管 cement casing
注水泥套管头 cementing casing head
注水泥套管靴 cementing shoe
注水泥用塞 cementing plug
注水泥止水 grout off
注水泥装置 cement equipment;cementer
注水器 can filler;injector;refiller
注水器汽阀 injector steam valve
注水器托架 injector bracket
注水塞 filler plug;fill(ing) plug;priming plug;water filling plug
注水渗透试验 pumping-in permeability test
注水时间 duration of charging
注水使土密实 compaction by watering
注水式潜水器 flooded submersible
注水式乙炔发生器 water to carbide (type) generator
注水试验 constant head test;flood pot experiment;injection test;pumping-in test;water inflow test;water injecting test;water injection test
注水试验孔 water injection test hole;water rejection test hole
注水速度 water injection rate
注水桶 <混凝土搅拌用> inundator
注水系数 water injection rate
注水系数测定 injectivity index test
注水系数测试 injectivity index test
注水系数试验 injectivity index test
注水系统 flood pattern;water filling system
注水效果的经验预测法 empiric(al) water flood prediction method
注水性能 flood performance
注水压实法 compaction by watering
注水油水比 water-oil ratio in flooding
注水诱发地震 injection induced earthquake;water injection-triggering earthquake
注水周期 watering period
注水柱 <钢柱防火> water-filled column
注水钻机 injection drill
注水钻井 water injection
注水作业 water flooding operation
注塑成型 cast mo(u)lding;injection mo(u)lding
注塑成型机 injection mo(u)lding machine
注塑法 injection mo(u)lding
注塑酚醛树脂 cast phenolic resin
注塑缸 injection cylinder
注塑后续压力 injection follow-up pressure
注塑机 injection press
注塑静止时间 dead time in injection mo(u)lding
注塑冷料 cold slug
注塑量 shot;shot capacity
注塑模具 injection mo(u)ld
注塑树脂 cast resin
注塑铸造法 injection mo(u)lding
注酸泵量 pumping rate of acidizing
注酸泵压 pump pressure of acidizing
注填充因数 beam filling factor
注吸器体 injector body
注下 downpour
注销 annulment;cancellation;charge off;cross-off;log-down;logoff;logout;retirement;write off
注销保险单退费 cancelling returns
注销船级 expunge the class
注销戳 cancellation stamp
注销的 crossed
注销的支票 cancelled check
注销电报 cancellation of telegram
注销订货单 cancel order
注销费 cancellation charges
注销分保合同 cancellation of reinsurance contract

注销合同 cancellation of treaty
注销金额 cancellation amount
注销客票 cancel a ticket; deface a ticket
注销领料单 cancellation of material requisition
注销契约 cancellation of treaty
注销日 cancellation date
注销条款 cancelling clause
注销通知 cancellation notice; notice of cancellation
注销退款 cancelling returns
注销文件 deleted file
注销遗失的票据 cancellation of a lost instrument
注销债务 cancellation of indebtedness
注销支票 cancelled check; cancelled cheque
注星号 asterisk
注型管 beam-type tube
注压接头机 injection splicer
注压制品 injection mo(u)lded item
注液泵 topping-up pump
注(液泵装)油泵 filling pump
注液电池 intercell
注液法 injection process
注液加固基础 injected foundation
注液胶机 rubberizing machine
注液孔塞 filler plug; filling plug
注液漏斗 filling funnel
注液站 <测流用> injection station
注液蒸煮法 injection cooking
注液装置 priming device
注意标记 care mark
注意标志 care mark; cautionary mark
注意不足 hypoprosexia
注意操舵 mind your helm
注意措施 cautionary measures
注意点 lime light
注意反应 attentive response
注意浮标 watch buoy
注意符号 attention sign
注意观察天气变化情况 keep one's weather eye open
注意降低成本的 cost-conscious
注意牌 attention plate
注意区 caution zone
注意实效的 pragmatic
注意示像 caution aspect
注意事项 matters need attention
注意危险 risk of danger
注意位置 caution position
注意显示 caution position
注意信号 attention signal; caution signal
注意增强 accentuation of attention
注意中断 attention interrupt(ion)
注意装置 attention device
注音字母 phonetic letter
注油 charging; fuel injection; greasing; lubrication; oil filling; oiling; oil injection
注油泵 injection pump
注油不足 underpriming
注油槽 oiling groove
注油点 lubrication point
注油法 creosoting process
注油盖 filling oil cover
注油工作空间 oil-filled working space
注油管 filler pipe; handhole door; nipple; oil filler pipe; oil filling pipe; oiling tube; pouring tube
注油壶 dope can; lubricating can; sight feed lubricator
注油壶继电器 oil dash pot relay
注油环 oiling ring
注油机 oiling machine
注油孔 grease hole; lubricating hole; oil(filler) hole; oil filler point; oilway
注油孔盖 oil-hole cover
注油口 filler; oil filling opening; pouring orifice
注油口塞 priming plug
注油龙头 oil faucet
注油漏斗 priming funnel
注油螺杆压缩机 oil flooded screw compressor
注油器 grease chamber; grease cup; grease squirt; lubricator; oil ejector; oiler; oil feeder; oil filler; oil gun; oil injector; oil liner; oil lubricator; priming can
注油器接头 oil-filling adapter
注油器驱动(装置) lubricator drive
注油枪 filling gun; grease(pressure) gun; oiling gun; oil syringe
注油塞 filler plug; fill(ing) plug; oil filler plug; oil filling plug
注油栓 lubricating plug
注油台 oil filling platform
注油装置 oiling device; oiling system
注油总管 oil injection header
注油嘴 grease nipple; lubricating nipple; lubricator fitting; nipple
注有标高的纵断面图 profile in elevation
注有数字等高线 index contour
注重标志 care mark
注重喷撒时间 emphasize application season
注资 capital injection
注子 pourer

贮备 stockpiling

贮备给水 reserve feed
贮备胶乳 preserved latex
贮备清水池 reserve feed water tank
贮备物 garner
贮冰仓容量 ice storage bin rating
贮冰库 ice house
贮冰室 bunker
贮材蓝变 yard blue
贮菜仓 vegetable storage; vegetable store
贮菜库 vegetable storage; vegetable store
贮仓压力 silo pressure
贮仓组 group of bins; group of silos
贮藏 bestow; coffer; hoarding; repertory; reposition; stocking; storage; store; stow
贮藏槽 storage tank
贮藏车 accumulator car
贮藏处 storage space; stowage
贮藏地位 storage space
贮藏斗 storage hopper
贮藏费 stowage charges
贮藏柜 <医药、梳洗、化妆品的> bathroom cabinet
贮藏货币 hoarding of money
贮藏库 storage
贮藏量 store content
贮藏模量 storage modulus
贮藏棚 storage shed; store shed
贮藏期 keeping period
贮藏期限 shelf life
贮藏容积 stowage
贮藏设备 bunkerage
贮藏室 storage room; storage compartment; storage plant; store closet; store house; storeroom; a(u)mbry; repository; stock room
贮藏所 depository; receptacle; deposit <美>
贮藏稳固性 storage stability
贮藏物 stowage
贮藏系数 storage factor
贮藏箱 storage bin
贮槽 bunker; collecting tank; hut(ch); receiver; receiving tank; receptacle; storage tank
贮槽容量 tankage
贮槽贮藏费 tankage
贮场面积 stoking area
贮尘室 dust pocket
贮尘箱 trash can
贮池 holding pond
贮存 carry-over; depot; reposit; stockpiling
贮存场装载输送机 reclaimer
贮存费 storage cost
贮存窖 bunker in the ground
贮存老化 bin ag(e)ing
贮存期 pot life; storage period
贮存期限 shelf life
贮存器 container
贮存器中不能卸出的水泥 sticky cement
贮存时间 period of storage
贮存食品的场所 larder
贮存宿主 reservoir host
贮存所 depot
贮存条件 condition of storage
贮存稳定性 bin stability; can stability
贮存稳定性试验 storage stability test
贮存系数 storage coefficient
贮存箱 container; storage tank
贮存信托公司 depository trust company
贮放干燥 seasoning
贮放时间 pot life
贮粉室 pollen chamber
贮罐防风梁 wind girder
贮灰仓 ash silo
贮灰场 ash disposal area
贮积水 impounded water
贮浆池 drainer
贮浆搅拌机 agitator for stored pulp
贮浆筒 paint storage tank
贮料仓 surge silo
贮料场 bunker yard; storage yard
贮料称量装置 receiving weigher
贮料橱架 storage rack
贮料斗 filler bin; hopper
贮料斗存料指示器 bin level indicator
贮料堆 stock dump; stockpile; storage pile
贮料堆(中材料)含水量 stockpile moisture
贮料坑螺旋 pit auger
贮料棚 storage shed; store shed
贮料器 bank
贮料塔 accumulator
贮料(筒)仓 storage silo
贮留池 retention basin; retention lagoon
贮埋槽 <放射性废物> burial tank
贮煤场 coal storage yard
贮煤船 coal hulk
贮煤库 coal storage
贮煤量 coal storage
贮煤室 coal room
贮煤塔 coal storage tower
贮木场 depot; forest depot; land depot; landing; log dump; lumber yard; mill pond; storage area; timber basin; timber depot; timber yard; wood depot; wood storage; wood yard
贮木场内变色 yard stain
贮木池 pond; rafting reservoir; timber basin; timber pond; mill pond <制材厂的>
贮木池(原木)截锯 pond saw
贮木港池 timber pond
贮能焊 percussion welding
贮能摩擦焊 flywheel type friction welding; inertia welding
贮能元件 energy storage unit
贮漆槽 bottom pan
贮气罐 air vessel; gasometer
贮气器 gas collector; gas holder; gasometer; holder; pressure chamber
贮气容积 reservoir volume
贮气箱 air tank
贮器 receptacle
贮氢材料 hydrogen storage material
贮燃油罐 fueling receiver
贮热阀门 thermal storage valve
贮热能力 heat storage capacity
贮热器 heat reservoir
贮热容器 thermal storage vessel
贮热因数 thermal storage factor
贮散装(石)油建筑物 petroleum bulk storage buildings
贮砂池 debris basin
贮砂库 debris basin; debris storage basin
贮水 impounded water
贮水槽 hopper
贮水槽门 handhole door
贮水层时代 age of water-storing bed
贮水池 body of water; catch basin; catchment basin; cistern; impounded body; lochan; ponded lake; reception basin; retention basin; stock pond; storage pool; storage tank
贮水池底垫块 tank bottom block
贮水池式水力发电厂 hydropower plant with reservoir
贮水船闸 lock with storage chamber
贮水量 pondage
贮水面积 reservoir area; water surface area
贮水泡 water vesicle
贮水器 cistern; water back; water cistern
贮水塘 storage lagoon
贮水桶 water butt
贮水箱 feed tank
贮酸槽 acid(storage) tank
贮酸罐 acid(storage) tank
贮酸塔 acid tower
贮铁包 receiving ladle
贮铁水炉 forehearth
贮纬器 pick accumulator
贮物柜 wanigan
贮物箱 wanigan
贮箱 conduit head
贮液杯 cistern
贮液槽 sump
贮液罐 receiver
贮液器 accumulator; liquid receiver; surge drum
贮液桶 cistern tank
贮油槽 oil storage tank; oil sump
贮油场 bulk-storage plant
贮油池 oil pond; oil tank
贮油罐 oil reservoir; oil storage tank; oil tank
贮油柜 oil-expansion vessel
贮油库 oil store
贮油器 oil receiver
贮油设备 oil storage facility
贮油箱 oil storage tank
贮油箱密封 grease retainer seal
贮油站 tank farm
贮雨水槽 rainwater storage tank
贮雨水箱 rainwater storage tank

驻班制【铁】 crew changing at turn-around depot system

驻波 clapotis; conjunction(al) wave; permanent wave; seiche; standing swell; standing wave; stationary

wave
驻波保护电路 standing-wave protecting circuit
驻波比<吸声系数用的> ratio of standing wave;standing-wave ratio
驻波比测量 standing-wave ratio measurement
驻波比测量器 standing wave ratio meter
驻波比指示器 standing-wave ratio indicator
驻波波长 standing-wave length
驻波波导 standing-wave guide
驻波波压 standing-wave pressure
驻波槽<测流量用的> standing-wave flume
驻波测定器 standing-wave meter
驻波测量 standing-wave measurement
驻波测量仪 standing-wave measuring instrument
驻波产生器 standing-wave producer
驻波场 standing-wave field
驻波潮 standing water tide;standing-wave tide
驻波潮汐理论 stationary wave theory of tide
驻波存储器 standing-wave memory
驻波电压比 standing-wave voltage ratio
驻波法 standing-wave method; stationary wave method
驻波放大器 standing-wave amplifier
驻波分析 standing-wave analysis
驻波幅 amplitude of seiche
驻波功率比 standing-wave power ratio
驻波管 standing-wave tube;stationary wave tube
驻波管法 standing-wave tube method; stationary wave tube method; impedance tube method <测定材料吸声系数的一种方法>
驻波激发器 standing-wave producer
驻波加速器 standing-wave accelerator
驻波检测器 standing-wave detector; standing-wave indicator; standing-wave meter; standing-wave ratio meter
驻波节点 standing-wave node
驻波节数 nodality of seiche
驻波结构 standing-wave structure
驻波理论 stationary wave theory
驻波能量 energy of the standing wave
驻波强度 intensity of standing wave
驻波情况 standing-wave condition
驻波式鉴别器 standing-wave type discriminator
驻波损耗因数 standing-wave loss factor
驻波天线 standing-wave antenna
驻波条件 standing-wave condition
驻波图(案) standing-wave pattern; stationary wave pattern
驻波系数 standing-wave ratio
驻波系统 standing-wave system
驻波线 standing-wave line
驻波选矿机 standing-wave separator
驻波压力 standing-wave pressure;stationary wave pressure
驻波仪器 standing-wave apparatus
驻波运行 standing-wave operation
驻波振荡 standing oscillation
驻波直线加速器 standing-wave linac; standing-wave linear accelerator
驻波指示器 standing-wave indicator; standing-wave meter
驻泊条款 liberty to touch and stay
驻泊油轮 station tanker
驻测 stationary ga(u)ging
驻场代表 field representative;project representative
驻场工程师 field engineer; resident engineer
驻场监察员 resident inspector
驻场建筑师 resident architect
驻场员 site agent
驻场主任工程师 chief resident engineer
驻场主任建筑师 chief resident architect
驻潮波 stationary tidal wave
驻车 parking
驻车道 parking lane
驻车累计调查 accumulation study
驻地补给品储存量 station stock level
驻地代表团 resident mission
驻地监工员 resident inspector
驻地(盘)工程师 resident engineer
驻地(盘)总工程师 chief resident engineer
驻地实验室 campsite laboratory
驻地主任工程师 chief resident engineer
驻点 arrest point;critical point;stagnation point;stationary point
驻点传热 stagnation point heat transfer
驻点温度 stagnation point temperature
驻点压力 stagnation pressure
驻定点 stationary point
驻段工程师 chief resident engineer
驻段线务员 mains man
驻段主任工程师 chief resident architect
驻港船长 harbo(u)r master; port captain
驻港轮机长<船方或货方> port engineer
驻港日记 harbo(u)r log
驻港值班 harbo(u)r watch
驻工地代表 field representative
驻工地工程师 resident engineer
驻工地建筑师 resident architect
驻工地主任工程师 chief resident architect
驻工地主任建筑师 chief resident architect
驻工地总工程师 chief resident engineer
驻工段工程师 resident engineer
驻激波 standing shock wave;stationary shock wave
驻极电介体 electret
驻极体传声器 electret microphone
驻极体电容传声器 electret capacitor microphone
驻极体耳机 electret earphone
驻极体换能器 electret transducer
驻节部长 minister resident
驻军营房 canton
驻立长波 standing swell
驻流 standing current
驻留单元 resident element
驻留的 resident
驻留轨道 parking orbit
驻留过程【数】 stationary process
驻留控制程序 resident control program(me)
驻留谱线 ultimate lines
驻留时间 dwell time
驻留资源 resident resource
驻面 stagnation surface
驻人观测站 attended station
驻栓 bolt bar
驻退杆固定突座 recoil piston rod lug
驻退簧 recoil spring
驻退活塞 recoil piston
驻退活塞杆 recoil piston rod
驻退机 buffer; recoil breaker; recoil mechanism
驻退机阀 recoil valve
驻退机调节杆 recoil throttling rod
驻退筒 recoil cylinder
驻退筒密封塞 recoil cylinder closing plug
驻退筒调节器 recoil cylinder replenisher
驻退系统 recoil system
驻退液 recoil fluid
驻外办事处代表价 central FSO [foreign service office] price
驻外办事处特定代表价 central special FSO price
驻外办事处帝国定价 imperial FSO price
驻外办事处特定价 imperial special FSO price
驻外官员 outposted officer
驻外国办事处 Foreign Service Office
驻外职员 expatriate employee
驻涡 standing eddy
驻相法 method of stationary phase
驻云【气】 crest cloud;standing cloud; stationary cloud;cap cloud
驻在的 resident
驻扎 cantonment
驻站维修 station-based maintenance
驻值 stationary value

柱 pillar;pole;post;stake;stud

柱板连接 support-to-floor connection;support-to-slab connection
柱板式挡土墙 column-plate retaining wall; pile (-and)-plank retaining wall
柱板式墙 column-and-panel wall
柱板围墙 post-and-block fence; post and post fence
柱半径 column radius
柱半径倍数比 modular proportion
柱保护钢板 column guard
柱比法<结构应力计算方法> method of column analogy;column analogue
柱比法设计 column analogy method of design
柱比拟法 support analogy;column analogy method
柱比设计法 column analogy method
柱壁 post jamb
柱变位 column(ar) deflection
柱表贴面 column box
柱波面 cylindric(al) wave
柱材料 column material
柱参数 column parameter
柱槽 cannelure; flute; fluting (of column);striga[复 strigae]
柱槽间 interglyphe
柱槽间脊 ridge fillet
柱槽筋 facet(te);reglet
柱槽突面 facet(te);listel
柱槽琢面石<平行的> tooled ashlar
柱侧脚 batter of column
柱侧倾机理 column sideway mechanism
柱层析法 column chromatographic method;column chromatography
柱插 stake pocket
柱长 column length; length of column;support length
柱沉陷 column settlement
柱撑式三角桁架 kingpost truss
柱撑榫 strut tenon
柱承梁上水平木<一排支柱或短柱支承梁上的水平木> reason
柱承楼座 column-supported gallery
柱承式阳台 altana
柱齿轮 stud gear;stud wheel
柱齿钎头 button bit
柱齿式扩孔钻头 button reaming bit
柱础 column base;column plinth;column socle; plinth (stone); support base
柱础基座 subplinth
柱床 poster bed
柱丛拱墩 shafted impost
柱丛构造方式<中世纪建筑的> shafting
柱丛区 column zone
柱锉 pillar file
柱带 girdle;strip
柱的 columnar
柱的不稳定 support instability
柱的不稳定性 column instability
柱的布置 column layout;support layout
柱的侧倾(移) column sidesway
柱的衬砌 column lining
柱的尺寸 column dimension; column size;support dimension
柱的断面 support section
柱的腹部 column web
柱的覆面 column facing
柱的刚度 column rigidity; column stiffness; support rigidity; support stiffness
柱的刚度系数 column stiffness coefficient
柱的钢筋 column reinforcement
柱的格式布置 support grid pattern
柱的核心部分<混凝土> core of column
柱的横向钢筋 lateral reinforcement
柱的基座 dado
柱的极限强度 ultimate column resistance;ultimate column strength
柱的极限稳定性 ultimate column stability
柱的尖头 beak
柱的间距 column interval
柱的截面 column cross-section; support section
柱的劲度 column stiffness; support stiffness
柱的宽度 column width
柱的连接 column connection;support connection
柱的锚碇 column anchorage
柱的面饰 column facing
柱的模板<混凝土> column form; post mo(u)ld
柱的模壳 column shuttering
柱的母线 elements of a cylinder
柱的强度 column strength
柱的蠕变 column creep
柱的蠕动 support creep
柱的设计 column design
柱的设置 post setting
柱的式样 columnar style
柱的台座 column pedestal
柱的凸缘 column flange
柱的外壳 column jacket
柱的弯曲线 column curve
柱的无支(撑段)长度 unbraced length of column
柱的徐变 column creep
柱的压曲 column buckling
柱的压曲应力 column buck ling stress
柱的一段 column section
柱的一节 column section
柱的一览表 column schedule
柱的移动 column movement
柱的应力 column stress
柱的有效长度 column effective length;effective length of column
柱的有效高度 effective column height
柱的缘饰 support guard

柱的支承 column bearing
柱的自由长度 unbraced length of column
柱的纵向挠曲 column deflection
柱的纵向弯曲 columnar deflection
柱灯 post light
柱底 base of column;column base;toe
柱底板 column base plate
柱底部直径 inferior diameter
柱底放大部分 scape
柱底机构 <指电动臂板信号机的> base-of-mast mechanism
柱底脚 column footing
柱底梁 patand
柱底特大部分 scape
柱底支承板 column base plate
柱底座 foot stall;patin;patten
柱底座部分 zoccola
柱电泳法 column electrophoresis
柱电阻 columnar resistance
柱垫木 dolly
柱钉电弧焊 arc stud welding
柱钉焊接 stud weld(ing)
柱顶 capital;column cap;column capital;column head;column top;head mast;head of column;masterhead;masthead; mast top; bell capital <早期英国建筑的>
柱顶板 abacus[复 abaci/abacuses]
柱顶灯具 post-top luminaire
柱顶电动臂板信号机【铁】top mast motor signal
柱顶垫板 crown plate
柱顶垫木 raising piece
柱顶垫石 pulvin
柱顶方块材 <支撑顶梁的> pila
柱顶飞檐顶棚 plafond
柱顶过梁 architrave;epistyle;epistylium
柱顶过梁凸线脚 crossette
柱顶过梁最下一条 lower fascia
柱顶横梁的顶条 upper fascia
柱顶横梁柱顶部 <包括飞檐、雕带及横梁三部分> trabeation
柱顶横檐梁 trabeation
柱顶横檐梁式 trabeated style
柱顶横檐梁式构造【建】trabeated construction
柱顶横檐梁式结构 trabeated construction
柱顶横檐梁式结构系统 trabeated system
柱顶机构 top of mast mechanism
柱顶加厚托板用模板 drop panel form
柱顶交织拱 interlacing arches
柱顶(脚)饰 newel cap
柱顶砾 pedestal boulder
柱顶木托块 head tree
柱顶盘 entablature;trabeation
柱顶盘座面 fascia[复 fa(s)ciae/fa(s)cias]
柱顶上部 trabeation
柱顶石 capital stone; pedestal boulder; pedestal rock; sommer; summer(beam)
柱顶挑架 bracket scaffold(ing)
柱顶条板 column strip
柱顶托板 cap plate of stanchion;capping piece;drop panel
柱顶托板花 abacus flower
柱顶托块 impost block
柱顶线盘 entablement;top rail
柱顶悬挑 geison
柱顶圆角装饰 orle;orlo
柱洞 posthole
柱端凹线脚【建】spophyge
柱端力矩 column end moment
柱端弯矩 support end moment
柱段 shell of column
柱段形屋顶 cylinder segment roof
柱断面 column section;section of column
柱墩 column bent pier; dado; foot stall;pilaster;stub;stump
柱墩坝 columnar buttress dam
柱墩的斜压顶 amortizement
柱墩划板组【建】pier and panel system
柱墩交替系统 system with alternating columns and piers
柱墩幕墙系统 pier and panel wall system
柱墩砌块 pilaster block
柱墩式码头 cylinder-pier jetty
柱墩式栈桥 column-supported trestle
柱墩心 pier core
柱墩最下层 zoccola
柱阀 column valve
柱反压力 column back-pressure
柱沸石 epistilbite;reissite
柱分类 column classification
柱丰度 column abundance
柱负荷 column load
柱杆 post pole;post shore;stake;tige
柱杆头 tie-rod knuckle
柱钢 column steel
柱钢筋箍 column tie
柱高 column height; length of column;support height;support length
柱供暖系统 main heating system
柱沟【建】stria[复 striae]
柱箍 ornamental tie;tie
柱箍条 cimbia
柱固定底座 fixed base
柱固式吸扬挖泥船 spud-type suction dredge(r)
柱冠 abacus[复 abaci/abacuses]
柱冠球【建】balloon
柱管 column jacket;column tubing
柱硅钙石 <一种水泥矿物质> afwillite
柱过载 column overloading
柱函数 cylindric(al) function
柱号 column number
柱荷载 column load(ing)
柱荷载量 loading capacity of column
柱桁架 post truss
柱红石 priderite
柱后反应器 post-column reactor
柱环 anchor ear;band of column;collar;pile collar
柱(环)带 cincture
柱环节链 stud chain;stud link chain
柱环饰 annulet;band of column;corbel ring;shaft ring
柱环支撑 collar brace
柱辉铋铅矿 bursaite
柱辉锑铅矿 fuloppite
柱基【建】column footing;pillar-box;patten; pillar support; stanchion base; stereobate; stybolate; subplinth <柱底座下的>
柱基凹弧边饰 scotia
柱基础 pier footing;column base;column footing; column pad; footing foundation; foot stall; foot stall of column;pillar support;plinth;single foundation;support base
柱基础板 column base plate
柱基础块 column foundation block
柱基础锚固件 column anchor
柱基础石 the stone of a column
柱基底 column socle
柱基底座 orlo
柱基连接 support-to-footing connection
柱基石 base block
柱基饰 surbase
柱基线脚的最下座 hypobasis
柱基支承板 column baseplate
柱基座 base of a column;column socle;pillar support
柱基座线脚饰 surbase
柱及柱型构件的箍筋 lateral reinforcement
柱极焊 stud weld(ing)
柱夹 column clamp
柱夹块式围墙 post-and-block fence; post and post fence
柱夹镶板式墙 column-and-panel wall
柱钾铁矾 goldichite
柱架 pillar support; trestle; column mounting
柱架牛头刨床 pillar shaping machine
柱架刨床 pillar shaper
柱架式风钻 bar-rigged drifter
柱架式光学比测仪 pedestal optic(al) comparator
柱架式结构 column bracket structure
柱架式起重机 column hoist
柱架式千斤顶 column jack
柱架式潜孔钻机 down-the-hole drill rig
柱架式凿岩机 column-mounted rock drill;cradle drifter;drifter hammer; drifting machine;post-mounted drill; stand-mounted drill
柱架式钻机 post-mounted drill
柱架钻床 pillar drill(ing machine)
柱尖 column cap(ital)
柱间带 middle strip
柱间的 intercolumnar
柱间隔墙 bail
柱间拱 arcading
柱间横挡板 interpile sheeting
柱间间断 intercolumnar screen
柱间较狭的 systyle
柱间距(定比)【建】intercolumniation
柱间距离 column spacing;interval of columns
柱间空隙 intercolumn
柱间墙 filler wall;choir wall <拱廊下分隔唱诗座与走道的>
柱间桥门联结系 portal bracing
柱间水平支撑 interpile sheeting
柱间填充墙 filler wall
柱间为一个三槽板的 <陶立克式> monotriglyphic
柱间镶板墙 column-and-panel wall
柱间支撑 column bracing; interpile sheeting
柱建筑群 forest of columns
柱角护 column guard
柱脚 alligator jaw;column base;column foot(ing);column pedestal; heelpost; patand; patten; pedestal; pedestal foot(ing); post bracket; seating shoe; socle; support foot; toe;zoccolo;zocle
柱脚板 heel slab
柱脚保护罩 <防止车辆冲击> pilot piece
柱脚撑木 needle timber
柱脚底座的墩台 solidium
柱脚垫木 footer;footlid;ground plate
柱脚雕带 pedestal frieze
柱脚花线 surbase
柱脚角钢 shoe angle
柱脚块 base block;foot block
柱脚石 base block;foot block
柱脚体 solidium
柱脚凸圆线脚 spira
柱脚下的第二阶方石座 scamillus
柱脚圆盘线脚 ba(s)ton;torus[复 tori]
柱铰 column hinge
柱截面 column section
柱筋 column bar;column reinforcement;pillar reinforcement
柱晶 column crystal
柱晶磷矿 jezekite
柱晶石 kornerupine;prismatine;prismatite
柱晶松脂石 flagstaffite
柱颈 collarine;gorgerin;neck;cellarino <托斯卡纳或陶立克式柱头的馒形饰下的>;trachelion <希腊陶立克式柱的>;hypotrachelium <柱端与拇指圆饰环纹间的>
柱颈花边饰 <古典圆柱的> colarim; collaring
柱颈螺栓 necked bolt
柱颈饰 trachelium
柱颈线脚 necking; neck mo(u)ld(ing)
柱颈线饰 neck mo(u)ld(ing);necking
柱径 column diameter;support radius
柱径计 cylindrometer
柱径渐减的柱身 diminished shaft
柱径渐减的柱样板 diminished rule; diminishing rule
柱距 column space;column spacing; intercolumniation; interval of columns;post spacing;spacing of columns;support spacing
柱距式集中荷载 bay-type collection
柱壳 column casing
柱坑 posthole
柱坑挖掘机 posthole digger
柱空气 column of air
柱空体积 column void volume
柱孔(螺旋)钻 posthole auger
柱块云母 algerite
柱宽 support width
柱廊 antium; colonnade; double colonnade; portico; portico of columns;porticus;stoa
柱廊或拱廊 <希腊建筑中的> stoa
柱廊空间 pteroma
柱廊式 prostyle
柱廊式建筑 colonnaded building;prostyle building
柱廊围成的广场 colonnade-enriched piazza
柱廊巷道采矿 stoop-and-room
柱廊巷道采石 stoop-and-room
柱老化 ag(e)ing of column;column conditioning
柱肋 column rib;rib of column
柱类构件 column type member
柱利用度 degree of column utilization
柱连接 support connection
柱梁构架 post-and-beam framing
柱梁结构 post-and-beam construction;post-and-beam structure;post-and-lintel construction
柱梁框架结构 post and beam framing
柱梁式房屋 post-and-beam building
柱梁式棚子 prop-and-sill
柱梁系统 post-and-beam system
柱梁之间的焊接 welded column-girder connection
柱列 colonnade;line of columns;peristyle;range of columns
柱列带 <无梁楼板的> column strip
柱列灌注桩 grouted column pile wall
柱列基础 colonnade foundation
柱列基座 stylobate
柱列间隔为三米的 diastyle
柱列脚座【建】stylobate
柱列式地下连续墙 column type diaphragm wall;drilled wall;pillar-row type under-ground diaphragm wall
柱列轴线 axis of column row
柱林 forest of columns
柱磷铝石 fischerite
柱磷铝铀矿 phuralumite
柱磷锶锂矿 palermoite
柱流量 column flow rate

柱流失 column bleed(ing)
柱硫铋铅矿 ustarasite
柱硫铋铜铅矿 gladite
柱硫锑铅银矿 freieslebenite
柱氯铅矿 daviesite
柱螺栓 bolt stud;double-end bolt;screw stud;stud bolt;stud pin
柱螺栓车床 stud lathe
柱螺栓传动轮 stud driver
柱螺栓隔片 stud spacer
柱螺栓键 set key
柱螺栓孔 stud bolt hole
柱螺栓六角螺母 stud hexagonal nut
柱螺栓螺钉 stud screw
柱螺栓螺母 stud nut
柱螺栓切削机 stud removal machine
柱螺栓线接头 terminal stud connector
柱螺栓销 stud pin
柱螺栓轴 stud shaft
柱螺栓装置 stud mount
柱螺纹 male screw
柱埋深度 depth of a column
柱帽 capital;column cap(ital);column head;haunch;post cap;stanchion cap;support cap;support head
柱帽石 summer stone
柱帽锡 cap tin
柱帽子 cap piece
柱镁石 pinnoite
柱密度 column density
柱面 column face;cylinder surface
柱面编号 cylinder number
柱面波 cylindric(al)wave
柱面的 cylindric(al)
柱面格子型花饰 column grid pattern
柱面光反应器 cylindric(al)photoreactor
柱面光栅 lenticular screen;lenticulated screen
柱面光栅板 lenticular plate
柱面函数 cylindric(al)function
柱面花饰 column-figure
柱面极坐标 cylindric(al)polar coordinates
柱面极坐标系 polar cylindric(al)coordinate system
柱面加工夹具 cylindric(al)attachment
柱面开挖 cylinder cut
柱面量爪 cylindric(al)jaw
柱面螺旋线 cylindric(al)helix
柱面磨光器 cylindric(al)polisher
柱面母线 segment of a cylinder
柱面内顺序布局 sequential-within-cylinder layout
柱面屏 cylindric(al)surface(d)screen
柱面数 cylinder number
柱面透镜 cylindric(al)lens
柱面透镜光栅板 cylindric(al)lens raster
柱面性 cylindricality;cylindricity
柱面压缩透镜 hypergonar lens
柱面叶片 cylindric(al)blade
柱面照度 cylindric(al)illuminance;cylindric(al)luminance
柱面装饰 support facing;support surfacing
柱面坐标 cylindric(al)coordinates
柱面坐标系(统)cylindric(al)coordinate system
柱模(板)column clamp;column form(work);pillar form(work);support forming;support form(work)
柱模板箍铁 column clamp
柱模侧板 column side
柱木 spar
柱内壁 column wall
柱内箍筋 binder;column binder
柱内力矩 column moment
柱钠铜矾 kroehnkite
柱排架 support bent
柱盘孢属 <拉> Cylindrosporium
柱硼镁石 pinnoite
柱平环链 bushing chain
柱前 pre-column
柱墙间距 pteroma
柱群 bundle of columns;cluster of columns;column cluster;forest of columns
柱群基础 colonnade foundation
柱容量 column capacity
柱入口压力【化】inlet pressure of column
柱塞 piston plug;plunger(piston);pommel;trunk piston
柱塞泵 piston plunger pump;plug pump;plunger(type)pump;ram(-type)pump
柱塞泵式喷雾器 sprayer-slide
柱塞泵撞头 plunger pump ram
柱塞超冲程 plunger overtravel
柱塞衬筒 plunger bushing
柱塞尺寸 plunger size
柱塞冲程变化 change in length of stroke
柱塞传动曲柄 plunger crank
柱塞传动松脱器 ram drive release
柱塞传动油缸 ram drive cylinder
柱塞传动装置 ram drive
柱塞传动装置分离机构 ram drive release mechanism
柱塞导承 plunger guide
柱塞导筒弹簧 plunger follower spring
柱塞阀 plunger valve
柱塞杆 plug rod;plunger lever;plunger pole;plunger rod;plunger tappet;stopper rod
柱塞杆销 plunger lever pin
柱塞杆轴 plunger lever shaft
柱塞缸 plunger case;plunger(type)cylinder
柱塞隔膜泵 plunger diaphragm pump
柱塞给料机 plunger feeder
柱塞回动弹簧 plunger return spring
柱塞回位弹簧 plunger return spring
柱塞剪切刀 plunger shear knife;ram knife
柱塞卡住 seizure of plunger
柱塞开关 plug cock
柱塞控制簧 plunger control spring
柱塞扣眼 plunger eye
柱塞力 plunger force
柱塞连杆 plunger connecting rod;plunger link;plunger pitman
柱塞流 piston flow
柱塞螺旋槽 plunger helix
柱塞面积 ram area
柱塞模 plunger mo(u)ld
柱塞磨损 plunger wear
柱塞配合 fit of the plunger
柱塞皮封 plunger leather
柱塞上螺边 plunger upper helix
柱塞深度 depth of plunger
柱塞升程 ram lift
柱塞式动力油缸 ram-type cylinder
柱塞式高架<海上补给用> ram tensioned highline system
柱塞式滑阀 spool valve
柱塞式灰浆泵 piston(-type)mortar pump
柱塞式浸注压铸机 submerged plunger die-casting machine
柱塞式控制阀 plunger type control valve
柱塞式炉用推料机 stripper plunger
柱塞式马达 plunger motor
柱塞式泥炮 plunger type clay gun
柱塞式黏[粘]度计 plunger type visco-(si)meter
柱塞式喷粉机 plunger duster
柱塞式汽缸 ram cylinder
柱塞式热压铸机 submerged plunger die-casting machine
柱塞式手提灭火器 plunger type portable extinguisher
柱塞式双缸灰浆泵 double piston mortar pump
柱塞式酸洗机 plunger type pickling machine
柱塞式铁芯调整 plunger setting
柱塞式团矿机 plunger type briquetting machine
柱塞式推钢机 plunger pusher
柱塞式蓄力器 ram accumulator
柱塞式压出机 ram extruder
柱塞式压捆机 plunger press;ram-baler
柱塞式液压电动机 plunger motor
柱塞式溢流阀 plunger relief valve
柱塞式油缸 ram cylinder
柱塞式凿岩机 drifter
柱塞式周期酸洗装置 plunger mast-type batch pickler
柱塞式注压机 ram injection machine
柱塞室 plunger compartment
柱塞弹簧 plunger spring
柱塞弹簧皮碗 plunger spring cup
柱塞套 plunger barrel;plunger bushing;plunger sleeve
柱塞套圈 plunger eye
柱塞提升机 plunger elevator
柱塞挺杆 plunger tappet
柱塞筒 plunger barrel
柱塞头 stopper head
柱塞推进时间 plunger forward time
柱塞下螺边 plunger lower helix
柱塞销 plunger pin
柱塞行程 plunger lift;plunger stroke;plunger travel;ram travel
柱塞行程长度 travel of plunger
柱塞旋钮 plunger knob
柱塞叶片 plunger vane
柱塞液动机 plunger motor
柱塞液压马达 plunger type hydraulic motor
柱塞液压提升机 plunger hydraulic elevator
柱塞油泵 plunger type fuel pump
柱塞运动暂停机构 plunger stop
柱塞直接驱动的升降机 direct-plunger elevator
柱塞直径 diameter of plunger;plunger diameter
柱塞注模机 plunger machine
柱塞注压机 projected area mo(u)lding
柱塞转换开关 plunger switch
柱塞作用的 plunger-actuated
柱色层 column chromatograph
柱色层法 column chromatography
柱色谱法 column chromatographic method;column chromatography
柱上凹槽间的隆起线 strix
柱上板带 column strip
柱上变压器 pole type transformer
柱上的开孔 column opening
柱上的微凸线 entasis[复 entases]
柱上分级 column fractionation
柱上工作台 gallery
柱上铰链 support hinge
柱上靠栏 safety ring
柱上楣构【建】entablature;entablement
柱上水准点 level mark on column
柱上踏台 gallery;landing
柱上微凸线【建】entasis[复 entases]
柱上微凸线样板 eutasis reverse;eutasis rule
柱设计 support design
柱身【建】trunk;column shaft;frustum;fust;scape;scapus;shaft of column;tige;vivo;boatel<集柱的>
柱身凹槽 striga[复 strigae];strix
柱身凹槽饰 striges
柱身的收分 entasis of a column
柱身扣环 retaining ring for shaft
柱身棱纹 rib of column
柱身模板 column side board
柱身收分测量器 stylometer
柱身条形圆箍线脚 shaft ring
柱身凸筋 striga[复 strigae];strix
柱身突筋 stria[复 striae];strix
柱身斜度量测器 stylometer
柱渗透性 column permeability
柱升起 rise of columns
柱施工一览表 column schedule
柱石 base block;corner stone;pedestal stone;pillar stone;the stone of a column;zacco
柱石面真高 true elevation of stone monuments
柱石洗气器 column type scrubber
柱石已损坏盘石完好 upper stone monument damaged lower stone monument perfect
柱式 column order;post-type;eustylos<古希腊、古罗马神庙的>
柱式扳道座 column-throw stand
柱式边墙 column type sidewall
柱式边墙衬砌 lining with column-type sidewalls
柱式采矿 pillar-and-panel work
柱式采矿法 pillar method of mining;pillar method of stoping;pillar method of working;pillar system
柱式成型机 pillar shaper
柱式萃取器 column extractor
柱式打印机 rotating cylinder printer
柱式灯架 post light support
柱式吊架轴承 post hanger bearing
柱式墩 columnar pier;column(-like)pier
柱式高压釜 column autoclave
柱式管子虎钳 post pipe vise
柱式过滤器 column filter
柱式桁架 post truss
柱式护栏 post fence
柱式回转支承(环)central post slewing ring
柱式基础 spot footing;spot foundation
柱式给水栓 pillar hydrant
柱式间歇泉 columnar geyser
柱式建筑 columnar architecture
柱式接头 column splice
柱式结构 column structure;pole construction
柱式举升机 post lift
柱式绝缘子 post insulator
柱式开采法 pillar mining
柱式开关 pillar switch
柱式龙头 pillar tap
柱式轮廓标 post delineator
柱式码头 pillar quay
柱式门架 columnar portal
柱式磨床 housing grinder
柱式牛头刨床 pillar shaper
柱式暖气片 column type radiator
柱式耙 pillar harrow
柱式起重机 column crane;pillar crane
柱式汽炉片 column of radiator
柱式千斤顶 column jack
柱式桥墩 column pier;shaft pier
柱式散热器 column(type)radiator
柱式伸臂 strut boom
柱式提升机 column elevator
柱式体形 column body type

Z

柱式图解 histogram
柱式消防栓 pillar hydrant;post(fire) hydrant;standpost hydrant
柱式消火栓 pillar hydrant;post(fire) hydrant;standpost hydrant
柱式信号灯 pedestal signal
柱式胸塑像 terminal figure
柱式悬轴承 post hanger bearing
柱式旋转支承装置 central post slewing supporting device
柱式压力机 column press;pillar press
柱式衣架 clothes tree
柱式饮水池 pillar fountain
柱式油压机 columnar type oil hydraulic press;open rod press
柱式照明 pole lighting
柱式支架 strut support
柱式制动器 back brake;post brake
柱式轴承架 pillar bracket bearing
柱式装饰 column charge
柱式钻床 pillar drill(ing machine)
柱试验 column test
柱饰 band of column
柱饰的细圆部分 newel collar
柱收分 entasis[复 entases];entasis of column
柱寿命 column life
柱竖筋 column vertical
柱栓 bitt pin
柱栓齿轮 stud wheel
柱栓齿轮板 stud wheel plate
柱水钒钙矿 sherwoodite
柱水耙 pillar harrow
柱损失 column bleed(ing)
柱台 pylon
柱套 column casing;column sleeve
柱体 barrel;cylinder;podetium;stylidium;support shaft
柱体凹槽 architectural flute
柱体的综合参数 synthetic(al) parameter of cylinder
柱体贯入试验 cylinder penetration test
柱体立模 column forming;support shuttering
柱体模板 column shuttering;support forms
柱体模壳 support forms
柱体平衡 cylindric(al) equilibrium
柱体条棱 rib of column
柱填充 packing of column
柱填充材料 column-packing material
柱填充程序 column-packing procedure
柱填充法 column-packing method
柱填充剂 column-packing agent
柱填充结构 column-packing structure
柱填充漏斗 column-packing funnel
柱填充物 column filling;column-packing
柱条 lesene
柱调节 column conditioning
柱厅 hall of columns;hypostyle
柱统计表 columnar statistics table
柱筒 column casing;support casing
柱头 chapiter;column cap(ital);column head;dado cap(ping);head (piece) of column;pilaster strip
柱头凹线脚 acapus;apophyge
柱头板 cap plate;head slab;support head
柱头变尖 diminution
柱头虫 acorn worm
柱头的叶形雕饰 cil(l)ery
柱头垫木 cap;cap block
柱头垫石 pulvinus
柱头雕塑 capital carving
柱头拱墩 impost block
柱头构造 head attachment;head construction
柱头横木 bolster;head tree
柱头卷杀 column-head entasis treatment
柱头卷叶饰 cil(l)ery
柱头科 bracket set on column
柱头科斗拱 column-top corbel-bracket set
柱头梁座 column-head cater
柱头螺栓 stud bolt
柱头模板 column head formwork
柱头盘座面 facia
柱头铺作 bracket set on column
柱头上盘座面 fascia[复 fa(s)ciae/fa(s)cias]
柱头饰<陶立克柱> trachelium
柱头凸圆线脚 echinus and astragal
柱头托板 dropped panel
柱头桅顶装饰 acorn
柱头蜗卷式柱 acclivous column
柱头形式 column capital form
柱头幼虫 tornaria
柱头圆球饰 acorn
柱头装饰 capital ornament
柱涂层 column coating
柱腿内倾的A形塔架<桥塔> A-shaped pylon with inward leaning legs
柱外壳 column box
柱网 column grid
柱网布置 column arrangement;column layout;layout of column grid
柱网排列 column arrangement
柱网平面 plan of column grid
柱网线 column grid line
柱围房屋 peripteге[periptery]
柱围篱 post fence
柱围墙 post-and-block fence
柱位线 line of posts
柱稳定性 column stability
柱窝 needling
柱吸附色谱法 column adsorption chromatography
柱洗提程序 column elution program(me)
柱系杆 column tie
柱系列 pillaring
柱系统 column system
柱下部圆柱体 collarino
柱下单独基础 single foundation in pillar
柱纤状结构 prismatic(al) fibrous texture
柱销 dowel pin;plug
柱销拆卸器 pin drift
柱效率 column efficiency
柱效能 column efficiency
柱斜撑杆 let-in brace
柱谐函数 cylindric(al) harmonics
柱心力矩 corduroy moment;core moment
柱芯 core of column;support core
柱芯混凝土 kern concrete
柱芯(混凝土)断面 kern cross-section
柱芯混凝土强度 Kern strength
柱星叶石 neptunite
柱形 columnar order;column shape
柱形摆动阀 oscillating cylindric(al) valve
柱形测孔规 internal cylindric(al) ga(u)ge
柱形沉井 columnar pit sinking
柱形带储油罐单点系泊系统 SPAR buoy system
柱形灯桩【航海】 pile dolphin
柱形阀 cylindric (al) valve;pillar valve;spool valve
柱形浮标 beacon buoy;column;pillar buoy;spar buoy
柱形坩埚 skittle pot
柱形管 column tube
柱形基础 column foundation
柱形捐献箱<教堂里的> offertory box
柱形绝缘子 pillar insulator;post-(-type) insulator
柱形开关装置 switchgear pillar
柱形空间 cylindric(al) space
柱形控制台打字机 cylinder console typewriter
柱形拉伸弹簧 cylindric(al) tension spring
柱形立标【航海】 pile beacon;pole beacon
柱形连接口 push nipple
柱形梁 column beam
柱形螺栓 pillar-bolt
柱形螺旋弹簧 cylindric(al) helic(al) spring
柱形面 cylindroid
柱形模板 column box
柱形平衡阀 cylindrically balanced valve
柱形平台 pillar platform
柱形凭证闭塞机 spheric(al) token instrument
柱形凭证机 spheric(al) token instrument
柱形桥墩 column-like pier
柱形擒纵机构板状部分 chariot
柱形砌体工程 column blockwork
柱形曲面 cylindric(al) surface
柱形散热器 column radiator
柱形塔式起重机 mast crane
柱形弹簧 cylindric(al) spring
柱形凸轮 drum cam
柱形图 column diagram;histogram
柱形涡流 cylindric(al) vortex
柱形蜗杆 parallel-type worm
柱形物 scapus
柱形吸风筒 aspiration leg
柱形系船柱 pillar type bollard
柱形橡胶防舷材 cell type rubber fender
柱形斜脊屋顶端部 cylindric(al) hip (ped end)
柱形压缩弹簧 cylindric(al) compression spring
柱形炸药包 powder column
柱形支撑外钩 pivoted brace
柱形直齿轮 spur pinion
柱形自由涡流 cylindric(al) free vortex
柱形钻床 pillar drill(ing machine)
柱型 columnar order;column type;support formwork
柱型材 column section
柱型色层分离法 column chromatography
柱性能 column performance
柱锈菌属<拉> Cronartium
柱蓄量 prism storage
柱穴挖掘机 posthole digger
柱液色谱【化】 column liquid chromatography
柱应力 support stress
柱用钢筋笼 column cage
柱用螺旋钢筋 column spiral
柱用预制钢筋骨架 column cage
柱铀矿 schoepite
柱与板的连接 column-to-slab connection
柱与地板的连接 column-to-floor connection
柱与基脚的连接 column-to-footing connection
柱与柱的接合 column-to-column joint
柱预老化程序 column preconditioning program(me)
柱圆筒段 drum of column
柱载体 column support
柱载重 column load
柱展开色谱法 column development chromatography
柱展宽率 spreading rate of column
柱支撑 post shore
柱支撑梯<便于开窗采光> columnated window stair(case)
柱支承的 column supported
柱支承梁桥 leg bridge
柱支楼梯 columnated window stair (case)
柱支座 column support;pillar bearing;pillar support
柱指运动 columnar finger motion
柱滞留 column hold-up
柱中钢筋 column bar
柱中竖直钢筋微弯 offset bend
柱中线 column center line
柱中心 column center[centre]
柱中心混凝土 kern concrete
柱中心线上格梁加强的双向板 slab stiffened with beams on column centerlines
柱周包围物 support casing
柱轴 column axis;support shaft
柱轴线间距离 buttress spacing
柱柱连接 support-to-support connection
柱桩 colum(nar) pile;pile acting as a column;stump
柱桩间板桩 interpile sheeting
柱桩运输架 post carrier
柱装绞车 waughoist
柱状 stylolitic
柱状变晶结构 prismatic(al) blastic texture
柱状表 bar graph
柱状采泥器 core sampler
柱状采样 columnar sampling
柱状窗 columnar window
柱状的 columnar;cylindric(al)
柱状断口 columnar fracture
柱状粉末 sprills
柱状工作台压力机 horn press
柱状构造【地】 columnar structure;prismatic(al) structure;pillar structure;prismoidal structure;stylolitic structure
柱状活塞闸缸 trunk piston brake cylinder
柱状火药包 column load
柱状碱土 columnar alkali soil
柱状浇筑<混凝土的> pouring with longitudinal joint
柱状节理 columnar joint(ing);prismatic(al) joint(ing)
柱状节理煤 columnar coal
柱状节理下段 colonnade
柱状结构 columnar structure;columnar texture;column structure;prismatic(al) structure;prismatic(al) texture
柱状结构面 surface of columnar structure
柱状结晶过程 column crystallization
柱状结晶区 columnar zone
柱状晶钢 scorched steel
柱状晶粒 columnar grain
柱状晶体 columnar crystal;directional crystal;fringe crystal;styloid
柱状绝缘子 post insulator
柱状壳体 cylindric(al) shell
柱状壳体的过渡 transition in cylindric(al) shells
柱状块体 column blockwork
柱状矿体 chimney orebody;columnar ore body
柱状梁<即等惯矩梁> prismatic(al) beam
柱状滤池 column filter
柱状滤器 column filter

柱状煤样 columnar coal sample
柱状模 cylindric(al) mould
柱状排水 <土石坝内的> pillar drain
柱状排水管 column pipe
柱状坯块 cylindric(al) compact
柱状劈理 columnar cleavage
柱状劈裂 columnar cleavage
柱状频率图 histogram
柱状平面图【地】graphic(al) log
柱状剖面 geologic(al) column
柱状剖面图 column cross-section; column profile; graphic (al) log; graphic(al) strip log; columnar section
柱状剖面图程序 columnar section program(me)
柱状取心钻 column drill
柱状施工法 column method of execution
柱状石 chimney rock
柱状石纪念碑 pillar stone
柱状试样 test cylinder
柱状铁素体 columnar ferrite
柱状图 bar chart; columnar figure; column chart; stick plot; bore log <土层>; columnar section <土层>
柱状图表 bar graph
柱状图程序 column program(me)
柱状团聚体 column aggregate; columnar aggregate
柱状物 column; drum; pillar
柱状习性 prismatic(al) habit
柱状形 pillar type
柱状玄武岩 columnar basalt
柱状旋涡 cylindric(al) vortex
柱状压屈 column buckling
柱状岩取芯钻 column drill
柱状岩(芯)样 core shaped sample
柱状药包 extended charge; pellet charge
柱状炸药包 column cartridge; column charge; column load; stick of dynamite; stick of explosive
柱状装药爆破 column blasting
柱状装(炸)药 column charge; column load; long charge
柱状坐标 columnar coordinate
柱锥混合式卷筒 cylindroconic (al) drum
柱子 column; post; stanchion
柱子尺寸 support size
柱子的安全限度 boundary pillar
柱子的方石座 zocco
柱子底盘饰 spira
柱子地带 column zone
柱子吊装测量 plumbing column survey
柱子护面 casing of column
柱子混凝土 column concrete
柱子排架 column bent
柱子排列成向外突出的弧形的回廊 cyrtostyle
柱子上的长条形凹槽 striga[复 strigae]
柱子坍陷法 block caving method
柱子系列 sequence of columns
柱子细部结构 column detailing
柱子与支架 post and timber
柱子支撑 buttress stiffener
柱子中心线 column line
柱子装饰 columnar ornament
柱(纵向)挠曲 columnar deflection
柱纵向弯曲 column deflection
柱阻力 column resistance
柱组 clustered column
柱组合 column combination
柱钻 cylindric(al) drill
柱坐标 cylindric(al) coordinates
柱座 acapus; column base; pillar bearing; pillar-box; pillar stand; stylobate; support base
柱座凹线 apophyge; scape
柱座板 column base plate
柱座标高 plinth level
柱座标志 pedestal sign; stanchion sign <可移动的>
柱座或柱头凹线脚 apophyge
柱座盘 ba(s)ton
柱座(式)信号(灯) <装在路面上的固定座脚上,也可装在安全岛上> pedestal signal
柱座叶形角饰 <欧洲中古建筑> angle leaf

著录的原始资料 original materials for description

著名的古式风格 distinguished antique style
著名建筑物 prestige building
著名论著 disquisition

蛀虫 bark borer; borer; boring insect; bristletail; copper-worm; moth; worm

蛀虫孔 shothole; pin-hole <木材的>
蛀船虫 bankia; limnoria; ship borer; ship-worm; teredo; marine borer
蛀洞 shothole
蛀坏 eat away
蛀混凝土海虫 concrete borer
蛀孔 borer hole; worm channel; worm hole
蛀木虫 beetle; pileworm; sea worm; ship borer; ship-worm; wood borer; wood-boring insect; wood fretter; xylophagan; marine borer
蛀木的 xylophagous
蛀木动物 wood fretter
蛀木海虫 chelura; martesia; xylotya
蛀木海虱属 limnoria
蛀木海蚤 chelura; gribble
蛀木水虱 limnoria; limnoria lignorum
蛀木水蚤 gribble
蛀木简虫 timber beetle
蛀木小甲虫 death watch beetle
蛀石(海)虫 boring clam; concrete borer; pholad (idae); rock borer; stone borer
蛀蚀的 gnawed
蛀食的 eaten
蛀屑 Frass
蛀心虫 borer
蛀眼 pin(worm) hole; worm channel; worm eye; worm hole

筑坝材料 embankment material

筑坝堵水 dam in
筑坝法 damming
筑坝方案 impounding scheme
筑坝技术 dam construction technique
筑坝拦水 dam out; pound; damming
筑坝排水 dam out
筑坝人员 dam builder
筑坝形成的流域 basin due to damming
筑坝蓄水河道水流 impounding streamflow
筑坝壅水 dam(ming) up
筑坝壅水结构 damming structure
筑巢区 nesting site
筑成 building-up
筑成从内侧至外侧向上倾斜的弯道 banked bend
筑成堤状的 banked
筑成台地的圣坛 pura
筑成梯田的地 terraced land
筑城 fortification
筑城堡 castle
筑城学 fortification
筑岛沉井 caisson built on artificial island on shallow water; sinking open caisson on built island
筑岛沉箱 caisson built on artificial island on shallow water
筑岛施工法 artificial island construction method
筑堤 banking; bordering; build a dike [dyke]; bunding; damming; embank elongation; embanking; embankment fill; levee construction; mound
筑堤材料 banking materials; embanking material; embankment material; material composing the embankment
筑堤挡水 embanking
筑堤堵水 dam
筑堤堆放机 stacker for building up fills
筑堤防 embank
筑堤防堵 embank
筑堤防护 embank
筑堤防护的港口 artificial sheltered harbo(u)r
筑堤工 diker
筑堤工程 dam construction works; embankment
筑堤工人 dike master; diker[dyker]
筑堤护岸 leveed bank
筑堤机 diker[dyker]
筑堤区域 diked area; embanked area
筑堤疏干围垦法 polder method; polder reclamation
筑堤输送机 stacker for building up fills
筑堤挖渠机 diker
筑堤围池 leveed pond
筑堤围湖 leveed pond
筑堤围垦 land reclamation by enclosure
筑堤围垦区 polder region
筑堤围入 bank in
筑堤蓄水段 overland section
筑堤用石笼 gabion
筑堤造地 land reclamation by enclosure
筑丁坝 groining[groyning]
筑陡坡 escarp(ing)
筑防波堤 groining[groyning]
筑防波堤防护的港口 artificially sheltered harbo(u)r
筑港工程 harbo(u)r works
筑港机具 harbo (u) r construction plant
筑港设备 harbo (u) r construction plant
筑埂 ridging
筑埂机 banker; land shaper; ridger
筑埂机器 earth banking apparatus
筑埂犁 checker plow
筑埂平地机 terracing grader
筑埂器 banker; border making implement; earth banking apparatus; ridge-former; ridger
筑基础 underpin
筑井工具 well construction tool
筑井工人 well builder
筑井工作 well construction work
筑井机具 well construction rig
筑井模架 well shuttering
筑井设备 well construction plant; well construction rig
筑垒 rampart
筑垒术 vallate papilla; vallation
筑篱工人 hedger
筑垄 ridging
筑炉工程 furnace engineering
筑路 road construction; road laying; road making
筑路柏油 road oil; road tar
筑路拌和机 road mixer
筑路拌料机 road mixer; road-mix machine
筑路材料 highway material; road (building) material
筑路材料拌和 road-mix(ture)
筑路材料测量 road materials survey
筑路材料加工厂 road plant
筑路材料能量含量 energy content of road-building materials
筑路铲运机 road scraper
筑路成本 road cost
筑路承包人 road contractor
筑路承包商 road contractor
筑路承包者 road contractor
筑路堤 embanking
筑路斗式运料车 highway hopper
筑路方针 road policy
筑路工 road builder; road paver
筑路工程 road construction works; road making; road works; road building
筑路工地 road construction site
筑路工人 road-man; roadsman; bootman <用特别装备的卡车给公路涂油的>
筑路工作 road work
筑路工作面 roadhead
筑路工作者 road builder
筑路公司 paving company; road building firm
筑路合同章节 road building contract section
筑路混合料 road-making mix(ture); road-mix(ture)
筑路混合料拌和设备 road-mix(ing) plant
筑路机 road builder; road machine
筑路机械 highway machinery; mechanical highway equipment; road building equipment; road building machinery; road (building) machinery; road construction machine
筑路机械队列 train of equipment for road construction
筑路机械全过程管理 all-life period management of road machine
筑路技术 road building technique
筑路加热炉 road cooker
筑路监工 road overseer
筑路焦油(沥青) road tar
筑路搅拌机 road mixer
筑路结合料 road binder
筑路捐 pavage
筑路康拜因 all-in-one pav(i)er
筑路块料 road block
筑路沥青 road oil
筑路沥青乳胶 road asphalt emulsion
筑路砾石 road gravel
筑路路碾 path roller
筑路模板 road form; road formwork; road shuttering
筑路黏[粘]合料 road binder
筑路器材 road material
筑路器材测量 road materials survey
筑路权 right of way
筑路权的取得 right-of-way acquisition
筑路乳液 highway emulsion
筑路撒料机 road spreader; spreader device
筑路设备 road building equipment; roadway construction equipment
筑路石料 road stone
筑路石屑 highway chip(ping)s; highway stone chip(ping)s; road chip(ping)s; road stone

Z

筑路石油沥青 road asphalt
筑路碎石 metal(l)ing;road metal
筑路碎石料 road macadam;road metal;road metal
筑路土壤 road construction soil
筑路现场 road building site
筑路业 road industry
筑路用底卸式自卸汽车 highway hopper
筑路用地 right of way
筑路用焦油沥青 road tar
筑路用搅拌机 mixer for road construction
筑路用沥青 asphaltic-bitumen for road purposes; bitumen for road purposes
筑路用砾石 highway gravel
筑路用平地机 road grader
筑路用平路机 road grader
筑路用青石 <美> blue stone
筑路用拖拉机 highway tractor
筑路用挖掘机 road excavator
筑路用油 road oil
筑路用凿岩机 paving breaker drill
筑路振动整面机 road vibrating and finished machine
筑路振实机 road vibrating machine; road vibration machine
筑墙材料 fencing
筑墙锤 walling hammer
筑墙工 waller
筑墙过水法 walling for procedure
筑墙围地 walled enclosure
筑墙圬工 walling masonry
筑穹隆或拱 inbow
筑人孔用工程砖 engineering brick for manholes
筑山庭园 hill garden
筑石墙 stone wall(ing)
筑坛 terracing
筑塘 ponding
筑土埂 earth banking
筑围堤 diking
筑围墙 fence;fencing
筑围堰 cofferdamming
筑围堰施工法 construction with cofferdam
筑小拱的弧形板 turning piece
筑以城垛 embattle
筑有堤的 banked
筑有护岸的河弯 banked bend
筑有良好道路的 well-roaded
筑有围堤的地区 embanked area
筑于山上的阿拉伯要塞或据点 qala'a
筑运河的机械 canal building machinery
筑运河用移动混凝土模板 canal concrete paver
筑在墙内的部分墩顶 blind pier capital
筑在斜地面上的运河 canal cut on sloping ground
筑造(上的)滑动面 plane of tectonic slip
筑主坝 key dam

铸

铸疤 blister
铸币 coin;coinage;mintage
铸币(硬币)平价 specie par
铸币准备 specie reserve
铸杓 ladle
铸成槽 casting groove
铸成成对的汽缸 cylinders casting-pair
铸成凸缘 cast flange
铸成整体的发动机 monobloc engine
铸尺 contraction rule
铸床 casting bed;pig bed
铸疵 casting defect; foundry defect; rough surface
铸带机 band extruder
铸道 runner pipe
铸道形成针 casting sprue former
铸道针 sprue wire
铸电版工 batteryman
铸锭 cake;casting;cast ingot;foundry pig; ingot casting; ingotting; keel block
铸锭车 casting bogey;casting bogie;ladle barrow
铸锭车床 ingot lathe
铸锭车过道 ingot run
铸锭工 ladleman;pourer
铸锭工段 casting aisle
铸锭机 pig-casting machine; pig mo(u)lding machine
铸锭间 pouring bay
铸锭均热坑 ingot pit
铸锭坑 casting pit
铸锭跨 casting aisle; ingot casting bay;pouring aisle;teeming aisle
铸锭炉 ingot furnace
铸锭模 ingot mo(u)ld
铸锭耐火砖 pouring pit brick
铸锭破碎机 pig breaker
铸锭起重机 ingot charging crane
铸锭设备 casting unit
铸锭台 teeming platform
铸锭型 ingot mo(u)ld
铸锭冶金法 ingot metallurgy
铸锭用耐火材料 pouring pit refractory
铸锭砖 casting-pit brick
铸锭组织 ingot structure
铸封材料 potting compound
铸封用树脂 potting resin
铸缝 casting seam
铸钢 cast(carbon)steel;casting steel;form steel;ingot steel;steel casting
铸钢板 cast-steel plate
铸钢厂 steel-casting foundry; steel foundry
铸钢车间 steel-casting foundry; steel foundry
铸钢(衬砌)管片 cast-steel segment
铸钢齿状钻头 castellated bit
铸钢船尾材 stern casting
铸钢的 steel-cast
铸钢轭 cast-steel yoke
铸钢杆件 cast-steel member
铸钢隔板 cast-steel separator
铸钢构架 cast-steel frame
铸钢构件 cast-steel work
铸钢管 cast-steel pipe
铸钢管法兰 cast-steel pipe flange
铸钢管凸缘 cast-steel pipe flange
铸钢滚轴支座 cast-steel rocker
铸钢横梁 cast-steel separator
铸钢缓冲索引梁 cast sill extension pocket
铸钢机壳 cast-steel case
铸钢夹板 cast gripping block
铸钢件 cast steel;steel cast(ing)
铸钢两轴转向架 cast-steel four wheel truck
铸钢轮 cast-steel wheel
铸钢锚杯(套筒)cast-steel socket
铸钢锚碇块 cast gripping block;cast-steel anchor block
铸钢锚链 cast-steel chain cable
铸钢能摇动底座 cast-steel rocker
铸钢泥斗 cast-steel bucket
铸钢品种 cast-steel grade
铸钢桶 pouring pot
铸钢桶衬砖 ladle brick
铸钢桶挂钩 ladle bail
铸钢桶手推车 ladle barrow
铸钢涂料 steel mo(u)lder's paint
铸钢蜗壳 cast spiral case;cast-steel case
铸钢系船柱 cast steel bollard
铸钢楔块 cast-steel wedge
铸钢楔扬料板 cast-steel flight; cast-steel lifter
铸钢楔扬料器 cast-steel flight; cast-steel lifter
铸钢摇动底座 cast-steel rocker
铸钢叶片 cast-steel leaf
铸钢用耐火材料 casting-pit refractory
铸钢闸门 cast-steel gate
铸钢辙叉 cast crossing;cast-steel crossing
铸钢枕座 reaction casting
铸钢支座 cast-steel support
铸钢转向架 cast-steel truck
铸钢桩靴 cast steel pile shoe
铸钢座圈 cast-steel support
铸工 caster; foundry hand; foundryman;mo(u)lder
铸工铲 casting shovel
铸工铲凿 mo(u)lder's peel
铸工场 foundry(shop)
铸工车间 captive foundry;casting department; casting shop; foundry shop;iron-foundry(shop)
铸工尘肺 foundry worker's pneumoconiosis
铸工捣实工具 brasq(ue)
铸工工作 foundry work
铸工鼓风机 foundry fan
铸工热 foundryman's fever
铸工实习 foundry practice
铸工用黑色碳粉 founder's black
铸刮膜 film casting
铸管 cast pipe;cast tube
铸管模 segment block
铸罐 ladle pot
铸罐车 foundry car
铸罐顶头 stopper head
铸锅 skillet
铸焊 aluminothermic welding; burn(ing)-in; cast weld (ing); flow welding;liquid metal welding
铸焊钢轨接头 cast welded rail joint
铸焊焊缝 poured weld
铸焊接 cast joint
铸焊结构 cast-weld construction
铸后锻造 precast-forging
铸黄铜 cast brass
铸件 castings; cast parts; founding; foundry goods;foundry pig iron
铸件壁厚 casting section thickness
铸件壁厚的敏感度 sectional sensitivity
铸件表面 casting surface
铸件表面积 casting surface area
铸件表面黏[粘]砂 scab
铸件表面缺陷 casting surface defect
铸件表皮 peeling;skin of casting
铸件补缩 feeding a casting
铸件测绘缩放仪 casting layout machine
铸件场 casting bay
铸件成品检查仪 casting analyser[analyzer]
铸件翅边 fins
铸件的鉴定 identification of castings
铸件的清理 dressing of a casting
铸件的氧化物色斑 flower
铸件顶出器 mechanical casting extractor
铸件飞边 casting fin
铸件分析 cast analysis
铸件工艺图 finished machine drawing
铸件黑皮 crust
铸件检验 castings examination
铸件浸补 sealing of a casting
铸件精整 casting finish
铸件孔穴 draws
铸件冷激深度 depth of chill
铸件冷却装置 casting cooling system
铸件裂痕 flaw in castings
铸件鳞片 foundry scale
铸件毛刺 hard edges
铸件冒口 riser
铸件冒头 deadhead
铸件模数 casting modulus
铸件模样 casting form
铸件内表面检查 endoscopy
铸件内表面检查仪 endoscope
铸件内腔 core cavity
铸件抛光 chasing
铸件披缝 casting fin
铸件起皮 sand buckle
铸件气孔 air bubble
铸件气泡 blow-hole;cavity
铸件清理 casting cleaning; casting clean-up;fettling
铸件清理机 casting cleaning machine; fettling machine
铸件清砂机 casting cleaning machine
铸件缺陷 casting defect
铸件砂眼 pocket
铸件设计机 casting layout machine
铸件渗补剂 casting sealer
铸件渗漏 casting leakage
铸件生产 cast production
铸件修补 salvage of casting
铸件修补的批准 approval of casting repairs
铸件油 casting oil
铸件振动脱砂 shakeout
铸件周缘翅片 casting fin;feather
铸件皱纹 casting lap
铸件自然时效 auswittering
铸件最后抛光 chase;chasing
铸壳煤气表 cast case meter; hard case meter;iron case meter
铸坑 center[centre] casting; foundry pit;jacket
铸坑操作 pit practice
铸坑耐火材料 pouring pit refractory
铸坑起重机 centre casting crane
铸坑样品分析 pit analysis
铸坑样品试验 pit test
铸孔 blowhole; cast hole; core (d) hole
铸口 cast gate; casting nozzle; font cavity;geat;nozzle;sprue
铸块 butt;cast block;ingot(bar);pig
铸粒 cast grain
铸铝 alumin(i)um casting; cast alumin(i)um
铸铝车间 alumin(i)um foundry
铸铝合金 alumin(i)um casting alloy; cast alumin(i)um alloy
铸铝件砂型涂料 alumin(i)um mo(u)ld paint
铸铝散热器 cast alumin(i)um radiator
铸铝字母 cast alumin(i)um letter
铸轮 cast wheel
铸模 cast die; casting form; casting mo(u)ld; charging box; die; foundry mo(u)ld;ingot
铸模补助注口 dozzle
铸模车 pan car
铸模的上型箱 cope
铸模粉末 mo(u)lding powder
铸模工 mo(u)ldman
铸模工具 gagger
铸模合缝 burr
铸模盒 die case
铸模机 casting machine; mo(u)lding machine
铸模面料 foundry facing
铸模润滑剂 die lubricant
铸模砂 foundry sand
铸模石膏 casting plaster;gypsum mo(u)lding plaster

铸模树脂 casting resin;potting resin
铸模头 mo(u)lded head
铸模涂料 adhering mo(u)lding material;mo(u)ld-proofing wash
铸模修理工具 spatula
铸模用润滑油 light slushing oil
铸模用砂 mo(u)lding sand
铸排机 composing machine;hot-metal composing machine
铸跑火 run-out
铸坯 casting blank
铸坯导架 strand guide
铸坯轧压 strand reduction
铸皮 black skin;casting skin
铸品 casting
铸铅 cast lead;pig lead
铸铅存水弯 cast lead trap
铸铅片 cast sheet lead
铸嵌钻头 cast-insert bit
铸青铜 cast bronze
铸熔术 founding
铸入 cast-in;injection
铸入钢管 cast-in steel pipe
铸入钢套 cast-in steel bush
铸入加热器 cast-in heater
铸入式的 cast-in
铸入式叶片 cast-in blade
铸石 artificial stone; molten-rock casting;mo(u)ld stone;reconstituted stone; stone casting; synthetic(al)stone
铸石表面 cast stone skin
铸石车间 cast stone shop
铸石地面饰面 patent stone floor covering
铸石地坪 cast stone floor(ing)
铸石地坪覆面 cast stone floor cover(ing)
铸石防水剂 cast stone waterproofer
铸石工 cast stone work
铸石和天然块石清洗剂 cleaner for cast and natural stone
铸石块 cast stone
铸石(块)机 cast stone machine
铸石楼梯 cast stone stair(case)
铸石铺路板 cast stone paving flag
铸石贴面砖 patent stone tile
铸石用白云岩 dolomite for cast stone
铸石用辉绿岩 diabase for cast stone
铸石原料矿产 raw material commodities for cast stone
铸石砖 cast stone tile
铸石砖地坪覆面 cast stone tile floor cover(ing)
铸速 teeming speed
铸塑 casting
铸塑法 casting method
铸塑酚醛树脂 cast phenolic resin
铸塑酚醛塑料 catalin
铸塑机 casting machine
铸塑浆 casting syrups
铸塑模 drawn mo(u)ld
铸塑品 casting
铸塑软模具 flexible mo(u)ld
铸塑石膏 casting plaster
铸塑树脂 casting resin
铸塑塑料 cast plastics
铸塑用混合料 casting compound
铸缩管 casting crinkling pipe
铸缩应变 casting strain
铸态 as cast condition;as-cast state
铸态结构 as-cast structure
铸态金属 as-cast metal
铸态硬质合金 cast carbide
铸态铸件 greensand casting
铸态组织 as-cast structure;cast structure
铸体 body of casting
铸体冷缩 shrinkage of a casting
铸条机 line casting machine
铸铁 cast-iron;foundry(pig)iron;ingot(iron);pig iron;pot metal
铸铁T字管节 cast-iron tee
铸铁安全臼 iron pad
铸铁板 cast-iron plate
铸铁保护板 cast-iron shield
铸铁箅子 cast-iron grating;cast-iron strainer
铸铁变阻器栅 cast-iron resistor grid
铸铁标准部件 cast-iron standard parts
铸铁标准水箱 cast-iron sectional tank
铸铁表面渗铝法 metacolizing
铸铁(厕所用)冲洗水箱 cast-iron flushing cistern
铸铁长大 cast-iron swelling
铸铁厂 iron-foundry(shop)
铸铁厂废水 iron-foundry wastewater
铸铁厂废物 iron-foundry waste
铸铁场地栅栏 cast-iron area grating
铸铁车间 iron-foundry(shop)
铸铁车轮 cast-iron wheel
铸铁衬段 cast-iron lining segment
铸铁衬里 cast-iron lining
铸铁衬砌 cast-iron lining
铸铁衬砌管件 cast-iron lining segment
铸铁衬圈 cast-iron washer
铸铁承插管 cast-iron bell and spigot pipe
铸铁承插管弯头 angle collar; bevel collar
铸铁承口三通管 cast-iron socket tee
铸铁承口三通管接 cast-iron socket T
铸铁冲洗水柜 cast-iron flushing tank
铸铁窗 cast-iron window
铸铁窗花格 cast-iron tracery
铸铁床板 cast-iron bedplate
铸铁存水弯 cast-iron trap
铸铁大梁 cast-iron girder
铸铁导缆钩 opencast-iron chock
铸铁电阻器 cast-iron resistor
铸铁垫圈 cast-iron washer
铸铁锭 foundry ingot
铸铁短配件 short-body cast-iron fittings
铸铁(盾构)弓形支撑 cast-iron segment
铸铁发动机部件 cast-iron engine part
铸铁阀(门) cast-iron valve
铸铁法兰 cast-iron flange
铸铁法兰管 cast-iron flange(d)pipe
铸铁法兰管件 cast-iron flange(pipe) fittings
铸铁方垫箱 cast-iron box
铸铁防辐射屏蔽墙 cast-iron radiation shielding wall
铸铁废料 cast-iron scrap
铸铁废品 off-iron
铸铁废水管 cast-iron waste pipe
铸铁粪便污水管 cast-iron soil pipe
铸铁腐蚀 cast-iron corrosion
铸铁盖 cast-iron cover; ingot iron cover;manhole head
铸铁盖板 cast-iron cover(ing)plate; stop iron <下水道进入孔口用>
铸铁坩埚 cast-iron pot;iron crucible
铸铁构造 cast-iron construction
铸铁管 cast(ing)-iron pipe;cast-iron tube[tubing];ingot iron pipe[piping]
铸铁管段 cast-iron pipe segment
铸铁管件 cast-iron pipe fittings
铸铁管接头 cast-iron pipe joint
铸铁管连接端头 connected end of cast - iron;hub end
铸铁管配件 cast pipe fittings
铸铁管片 cast-steel lining
铸铁管片衬砌 cast-iron segmental lining
铸铁管片段 cast-iron segment
铸铁管柱 cast-iron tubular column
铸铁罐 cast-iron pot
铸铁锅 casting iron pan
铸铁锅炉 cast-iron boiler
铸铁锅炉额定蒸发器 cast-iron boiler rating
铸铁焊料合金 castolin
铸铁焊条 cast-iron electrode
铸铁合金 alloy cast-iron
铸铁护栏 cast-iron rail
铸铁花管 cast-iron perforated drain pipe
铸铁化工设备 cast-iron chemical equipment
铸铁环 cast ring;cast rink
铸铁机 casting machine;pig machine
铸铁机器件 machinery casting
铸铁(给)水管 cast-iron waste pipe
铸铁加肋垫圈 cast-iron ribbed washer
铸铁件 iron cast(ing)
铸铁(接头)配件 cast-iron fittings
铸铁结构管材 cast-iron structural pipe;cast-iron structural tube
铸铁结构管子 cast-iron structural pipe;cast-iron structural tube
铸铁井圈 cast-iron manhole ring;ring of cast-iron manhole
铸铁开沟器 cast-iron boot
铸铁块 casting pig;mo(u)ld pig iron
铸铁块衬砌环 cast-iron ring
铸铁块铺面 iron block pavement; iron paving
铸铁块铺砌 cast-iron paving
铸铁块(铺砌)路面 iron block pavement
铸铁拉手 cast-iron pull(handle)
铸铁犁 cast plough
铸铁立面 cast-iron front
铸铁立式冷壁 cast-iron stave
铸铁楼面板 cast-iron floor plate
铸铁螺口接件 cast-iron screwed fittings
铸铁螺纹管 cast-iron screw pipe
铸铁螺纹接头 Durham fitting
铸铁螺旋桩头管桩 cast-iron screw cylinder
铸铁马桶弯头 cast-iron closet bend
铸铁模型 iron pattern
铸铁木材结合环 cast-iron timber connector
铸铁木材连接器 cast-iron timber connector
铸铁排水管 cast-iron discharge pipe; cast-iron drain; cast-iron sewer pipe
铸铁排水管零件 drainage fitting
铸铁喷铝法 Metcolising
铸铁盆 cast-iron tub
铸铁皮带轮 cast-iron belt pulley
铸铁片式锅炉 cast-iron sectional boiler
铸铁平板闸门 cast-iron flat gate
铸铁平板闸门式泄水闸 cast-iron flat-gate-type sluice
铸铁屏蔽墙 cast-iron shielding wall
铸铁铺板 cast-iron apron
铸铁铺地砖 cast-iron block
铸铁铺路块 cast-iron road paving
铸铁桥梁 cast-iron bridge
铸铁容器 cast-iron vessel
铸铁熔液 metal
铸铁散热器 cast-iron radiator;radiator
铸铁砂 iron sand
铸铁栅 cast-iron grating
铸铁升井和盖 cast-iron riser and cover
铸铁生长 growth of cast-iron
铸铁生物屏蔽墙 cast-iron biologic(al)shielding wall
铸铁石英砂 cast-iron grit
铸铁鼠尾 mapping;rat tail
铸铁水管和配件 cast-iron water pipe and fittings
铸铁水落管 cast-iron rainwater pipe
铸铁水中污物预防器 cast-iron water waste preventer
铸铁踏步 cast-iron step
铸铁碳当量测定仪 eutectometer
铸铁搪瓷 cast-iron enamel
铸铁套管 cast-iron socket
铸铁铁芯 cast-iron core
铸铁筒柱 cast-iron cylinder
铸铁凸缘 cast-iron flange
铸铁凸缘管 cast-iron flange(d)pipe
铸铁凸缘管件 cast-iron flange(pipe) fittings
铸铁(托)架 cast-iron bracket
铸铁弯管 cast-iron bend
铸铁弯头 cast-iron bend
铸铁窝接式废水管接头 cast-iron spigot and socket waste pipe
铸铁窝接式排水管接头 cast-iron spigot and socket discharge pipe; cast-iron spigot and socket draining-pipe
铸铁窝接式污水管接头 cast-iron spigot and socket waste pipe
铸铁蜗壳 cast-iron spiral casing;cast spiral case
铸铁蜗形机壳 cast-iron volute casing
铸铁污水管 cast-iron waste pipe
铸铁污水管和配件 cast-iron soil pipe and fittings
铸铁污水管支架 cast-iron soil pipe support
铸铁系船柱 cast iron bollard
铸铁箱形排水沟 cast-iron box gutter
铸铁箱形屋顶天沟 cast-iron roof gutter
铸铁箱形雨水排水沟 cast-iron box rainwater gutter
铸铁屑 cast-iron scrap
铸铁压力管 cast-iron pressure pipe
铸铁压重 <浮标的> cast-iron ballast weight
铸铁叶片 cast blade
铸铁引靴 cast-iron guide shoe
铸铁鱼腹式大梁 cast-iron fish-bellied girder
铸铁雨水斗 cast-iron rainwater outlet
铸铁雨水管件制品 cast-iron rainwater product
铸铁雨水排水制品 cast-iron rainwater goods
铸铁浴盆 cast-iron bath tub
铸铁圆形碟形封头 cast-iron circular dished heads
铸铁轧辊 <砂型的> grain roll
铸铁闸阀 cast-iron gate valve
铸铁闸瓦 cast-iron brake shoe
铸铁辙叉 cast crossing
铸铁正面 cast-iron facade
铸铁支柱 cast-iron mast
铸铁制的建筑物正面 iron front
铸铁柱 cast-iron column
铸铁砖铺砌 iron brick paving
铸铜 cast(ing)brass;cast(ing)copper
铸铜场 brass foundry
铸铜车间 brass foundry
铸铜电焊条 electric(al)welding cast copper solder
铸铜合金 casting copper alloys
铸桶 ladle;skeo;skip
铸桶叉形夹 ladle carrier
铸桶车 casting carriage;ladle carriage
铸头硬度试验器 mo(u)ld hardness tester
铸镶材料 cast-setting material
铸镶金刚石钻头 cast bit;cast-set diamond bit
铸镶取芯钻头 cast-set coring bit
铸镶叶片 cast blade
铸像 statue

铸像龛 statue niche
铸芯焊条 electrode with cast core
铸芯油 casting oil
铸型 cast form; casting mo(u)ld; foundry form; foundry mo(u)ld; mo(u)ld a form; proplasm
铸型壁 mo(u)ld wall
铸型表面 mo(u)ld face
铸型补强剂 cushioning material
铸型尺寸 mo(u)ld dimension
铸型尺寸增量 rapping allowance
铸型充填性 mo(u)ld filling ability
铸型冲蚀 mo(u)ld erosion
铸型出气 venting
铸型错箱 mismatch in mo(u)ld
铸型棍 strickle
铸型合箱 mo(u)ld assembling and closing
铸型合箱工 mo(u)ld fitter
铸型烘干 mo(u)ld drying
铸型烘干炉 mo(u)ld drier[dryer]
铸型滑泥 casting slip
铸型化石 cast
铸型浇注 casting-up
铸型浇注口 mo(u)ld opening
铸型紧固夹 mo(u)ld clamps
铸型裂口 metal break out
铸型裂纹 mo(u)ld veining
铸型螺旋孔冷却 spiral mo(u)ld cooling
铸型落砂冲锤机 press for mo(u)ld extrusion
铸型密封 sealing of a mo(u)ld
铸型黏[粘]土 foundry clay
铸型黏[粘]土矿石 foundry clay ore
铸型品 cast
铸型破裂 veining
铸型起模 mo(u)ld lifting; mo(u)ld stripping
铸型腔 mo(u)ld cavity
铸型倾斜浇注法 pouring tilt
铸型圈 casting ring
铸型砂 foundry sand
铸型试合箱 tryoff
铸型寿命 mo(u)ld life
铸型输送器 mo(u)ld conveyer[conveyor]; mo(u)ld transfer
铸型树脂 cast resin
铸型透气性 mo(u)ld permeability
铸型涂料 adhering mo(u)lding material; casting slip; mo(u)ld facing
铸型稳定性 mo(u)ld stability
铸型箱 mo(u)ld(ing) box
铸型修理工具 spatula
铸型硬度 mo(u)ld hardness
铸型用黏[粘]土 clay for foundry
铸型用砂 sand for foundry
铸型用销钉 mo(u)lding pin; sand pin; stake
铸型预热浇注法 Lanz-pearlite process
铸型原料矿产 foundry raw material commodities
铸型注入口 sprue
铸型装配 mo(u)ld assembly; mo(u)ld closing
铸型装配机 mo(u)ld closer
铸玄武岩 cast basalt
铸压成型 die-casting mo(u)lding
铸造 cast; coin; found(ing); foundry; mo(u)lding
铸造巴比合金轴承 cast babbit metal bearing
铸造板坯 block
铸造匾牌 cast tablet
铸造表面 casting surface
铸造玻璃 cast glass
铸造不锈钢 cast stainless steel
铸造部 casting station
铸造材料 cast material
铸造厂 casting bay; casting factory; commercial foundry; foundry shop; jobbing foundry; sand foundry
铸造厂废水 foundry wastewater; wastewater from foundry
铸造场 casting bed
铸造车间 caption foundry; captive foundry; casting room; casting shop; foundry shop; casting bay
铸造车间的三废 foundry effluent
铸造车间设备 foundry equipment
铸造车间设计 foundry layout
铸造车轮 cast wheel
铸造成的 fluid origin
铸造成型 cast form
铸造齿轮 cast gear
铸造的 as cast; molten
铸造的芯撑 cast chaplet
铸造电解铁 cast electrolytic iron
铸造电炉 teeming furnace
铸造锻铁工厂 casting and forging factory
铸造法 casting
铸造法兰 cast flange
铸造法兰接头 cast flange adapter
铸造法镶嵌 cast setting
铸造方法 casting method; casting process
铸造废钢 casting scrap
铸造废料 foundry scrap
铸造废品 foundry loss
铸造废水 founding process wastewater
铸造废铁 foundry scrap
铸造分析 cast analysis
铸造改良黄铜 Oker
铸造高温合金 cast superalloy
铸造隔板 cast-in diaphragm
铸造工 foundryman
铸造工人 foundry worker
铸造工业 foundry industry
铸造工艺 casting technique; foundry technique
铸造工作 foundry work
铸造钩头链 cast detachable chain
铸造构件 cast member
铸造冠 casting crown
铸造冠桩 casting crown post
铸造管 cast tube
铸造合金 foundry alloy; cast alloy
铸造合金工具 cast alloy tool
铸造合金冠 casting alloy crown
铸造烘炉 foundry stove
铸造黄铜 cast(ing) brass
铸造货币 coin
铸造机 caster; casting machine; mo(u)lding machine
铸造机械 foundry machinery
铸造基托 cast metal base
铸造间 casting department
铸造浇注场 cast house; pouring house
铸造焦炭 cupola coke; foundry coke; melting coke
铸造结构 cast structure
铸造金 casting gold
铸造金属 cast(ing) metal
铸造卡环 casting clasp
铸造坑 casting pit
铸造蜡 casting wax
铸造裂纹 casting crack
铸造零件 mo(u)lded piece
铸造炉 founding furnace; foundry furnace
铸造铝合金 alumin(i)um casting alloy; Birmasil
铸造毛刺 feather
铸造冒口 feeder
铸造锰黄铜 manganese casting brass
铸造模 casting die
铸造模型 casting pattern
铸造泥芯 mo(u)ld core
铸造批量 casting volume
铸造披缝 casting scull; feather
铸造铅丸 mo(u)ld shot
铸造青铜 cast(ing) bronze
铸造缺陷 casting defect; casting flaw; foundry defect
铸造热 foundryman's fever; fume fever
铸造熔化炉 foundry furnace
铸造砂箱 foundry flask
铸造设备 casting equipment; casting machine; foundry equipment
铸造设计 casting design; casting plan
铸造生产 foundry practice
铸造生产的废水 founding process wastewater
铸造生产线 foundry production line
铸造生铁 cast-iron; foundry pig
铸造生铁炉渣 casting pig slag
铸造试棒 cast-on test bar
铸造试块 cast test block
铸造收缩 casting shrinkage
铸造术语 foundry term
铸造速度 casting speed
铸造缩尺 shrinkage ga(u)ge
铸造胎体 cast metal matrix
铸造特性 casting characteristic
铸造温度 casting temperature
铸造蜗壳 cast spiral case
铸造物 all-core mo(u)lding; mo(u)lding
铸造型心 cast core
铸造性 castability; coulability
铸造性能 castability; casting character; casting characteristic; casting property; founding property
铸造性试验 castability test
铸造学 foundry practice
铸造阳极 cast anode
铸造业 foundry industry
铸造叶片 cast blade
铸造应变 casting strain
铸造应力 casting stress
铸造用不锈钢块 stainless steel disc for casting
铸造用鼓风机 foundry fan
铸造用合金 casting alloy
铸造用黑粉料 blacking
铸造用化铁炉 foundry cupola
铸造用黄铜 brass for casting
铸造用焦(炭) casting coke; foundry coke
铸造用炉 casting furnace
铸造用耐火材料 foundry refractory
铸造用黏[粘]土 foundry clay
铸造用器具 ingot manipulator
铸造用砂 casting sand; foundry sand; heap sand
铸造用生铁 foundry pig iron; raw pig iron
铸造用树脂 cast resin; foundry resin
铸造用铁 foundry cast-iron
铸造用铜 casting copper
铸造用锡青铜 Perking brass
铸造支承板 cast bearing plate
铸造轴承 cast bearing
铸造钻头 cast metal bit; cast-set bit
铸造作业废水 foundry process wastewater
铸制钉 cast nail
铸制极板 formed plate
铸制耐火材料 castable refractory; casting(-pit) refractory
铸制树脂 cast resin; foundry resin
铸制型材 cast profile
铸制座 cast stand
铸钟铜 bell metal
铸字 typecasting; typefounding
铸字车间 typefoundry
铸字工 founder
铸字工人 typefounder
铸字机 <印刷> caster
铸字铅 type metal
铸字业 typefounding
铸嘴 lip
铸座 casting base

抓 扒用具 scratcher

抓扒者 scratcher
抓板器 plate grab
抓包机 plucker
抓柄 grab handle; grip end; grip handle
抓叉 grapple fork
抓铲 grab jaw
抓铲的布置 clamshell arrangement
抓铲的设计 clamshell arrangement
抓铲挖掘机 chamshell(excavator)
抓齿转动油缸 tine ram
抓出 seizure
抓挡件 catch member
抓底能力 holding ability; holding power
抓地板 ground lug
抓地齿 spur; strake
抓地齿长度 lug length
抓地花纹轮胎 traction-type tire
抓地爪 ground lug
抓钉 clasp nail
抓顶皮带 top grasping belt
抓斗 bucket(grab); clam bucket; clamshell(jaw); clamshell scoop; claw bucket; crab bucket; flapper; glove bucket; grab; grab(bing) bucket; grapple bucket; grappler; grappling bucket; halfscoop; pair of half scoops; pinchers <其中的一种>; grab clamshell <挖土机的>
抓斗闭斗索 <挖掘机> closing line
抓斗闭合电路 grab closing circuit
抓斗闭合绳 closing line
抓斗闭合索盘 grab closing drum
抓斗闭合系统 grab closing circuit
抓斗闭锁钢丝绳 grab closing rope
抓斗臂距 grab outreach
抓斗采样器 bottom grab
抓斗铲土机 clamshell shovel
抓斗齿 grab teeth; rockover rake; bucket teeth
抓斗的抓取量 mouthful of grab
抓斗吊车 clamshell-equipped crane; grabbing crane
抓斗吊钩 grab hook
抓斗吊索 holding line
抓斗附件 grab attachment
抓斗钩 grappling bucket hook
抓斗荷载 grabbing load
抓斗横移机构 grab traverse mechanism
抓斗回转机构 grab slewing mechanism
抓斗回转液压马达 grab rotation hydraulic motor
抓斗机 grab claws; grab machine
抓斗机工作半径 working radius of grab machine
抓斗机具 grab rig
抓斗机旋转半径 slewing speed of grab machine
抓斗加料器 clampshell hopper
抓斗夹紧力 clamping force
抓斗绞车 grab hoist
抓斗进料式混凝土搅拌机 scraper-fed concrete mixer
抓斗开斗速度 opening speed
抓斗开度 grab spread
抓斗启闭机构 grab bucket opening and closing mechanism
抓斗起重工作 grabbing work
抓斗起重机 bucket crane; clamshell (grabbing) crane; grab bucket

crane;grapple equipped crane
抓斗起重设备 grabbing rig
抓斗起重小车 bucket trolley
抓斗取样 sampling by grabber;grab sample
抓斗取样器 grab sampler
抓斗容量 grab capacity
抓斗设备 grabbing equipment
抓斗升降启闭钢缆 hoist and hold wires
抓斗升降索 holding line
抓斗式 grab type
抓斗式采样器 grab sampler
抓斗式地下连续墙挖掘机 bucket diaphragm wall excavator
抓斗式吊车 clamshell crane
抓斗式集材机 grappler skidder
抓斗式料桶 clamshell-type dump bucket
抓斗式起重机 clamshell(-equipped) crane;clamshell grab crane;crab derrick;grabbing crane;grab crane
抓斗式疏浚船 clamshell dredge(r)
抓斗式输送机 grab bucket conveyer [conveyor]
抓斗式挖掘机 clamp excavator;clamshell bucket dredge(r);clamshell excavator;grab excavator
抓斗式挖泥船 chuck with holdfast;grab bucket dredge(r);grapple dredge(r);shell clam dredge(r);grab boat;grab-dredge(r)
抓斗式挖泥机 clamshell dredge(r);grab bucket dredge(r);grab excavator;grapple dredge(r);grab-dredge(r)
抓斗式挖土机 clamshell excavator;clamshell shovel;grabbing excavator
抓斗式运送机 grab bucket conveyer [conveyor]
抓斗式蒸汽挖泥船 clamshell bucket steam dredge(r)
抓斗式装卸起重机 transporter grab crane
抓斗式装卸桥 transporter grab crane
抓斗式装岩机 cable-operated mucker;grab-type loader
抓斗式装载机 clam type loader;grab-type loader
抓斗式自航挖泥船 grab hopper dredge(r)
抓斗挖掘船采矿 grab-dredging
抓斗挖掘机 clamshell
抓斗挖泥 clampshell digging;grab-dredging
抓斗挖泥船 clamshell dredger;grab (hopper) dredger <带泥舱的>;grab pontoon dredger <无泥舱的>
抓斗挖土机 clamshell shovel
抓斗稳定器 grab stabilizer
抓斗稳定索 grab stabilized line;tag line
抓斗小车 grab troll(e)y
抓斗卸船机 grab ship unloader
抓斗卸货 grab discharge
抓斗行走吊车 grapple travel(l)ing crane
抓斗旋钩 grab rotation
抓斗液压缸 grab hydraulic cylinder
抓斗移行吊车 grapnel travel(l)ing crane
抓斗运砂船 sand carrier with grab bucket
抓斗载荷 grab load
抓斗重 grab weight
抓斗抓取量 mouthful
抓斗爪 rockover rake;rock rake
抓斗装载起重机 loading grabbing crane
抓斗装置 grab(bing) equipment;clamshell equipment
抓斗最大吊高 maximum hoisting height of grab
抓钩 bowl slip-sockets;catch hook;dog hook;dog iron;grab hook;grab iron;grapple
抓钩叉子 grapple fork
抓钩动力集材 grapple yarding
抓钩器 grappler
抓钩式集材机 grappler skidder;skidder-grapple
抓钩提升器 anchor lift
抓钩制动器 prong brake
抓管工具 pipe grab
抓痕 cat scratch;deep sleek;scratch mark
抓戽 <挖土机的> grab bucket
抓货附具 grab attachment
抓货钩 load binder
抓货夹具 grab attachment
抓机 grappler;grappling iron
抓紧器 grasper
抓紧千斤顶 <盾构工程用> gripper jack
抓举强度 grab strength
抓举延伸率 grab elongation
抓具 grab(handle);grapple;gripping apparatus
抓掘机 grabbing excavator
抓卡盘 chuck with holdfast
抓孔桩 piles formed by grabbing
抓拉试验 grab tensile test
抓牢 anchor hold;clutch;grip
抓力 holding capacity
抓力不良锚地 poor holding ground
抓梁钩 beam grab
抓煤机 coal grab
抓木工具 wood grabber
抓木器 log grapple;wood grabber
抓泥斗 pair of half scoops
抓泥机 grapple;ream grab
抓泥器 snapper
抓爬式挖泥船 grapple dredge(r)
抓片机构 in-and-out movement
抓破试验 scratch test
抓起荷载 grabbing load
抓器 catcher;gripper
抓钳 snap-on tongs
抓取 grab
抓取法取样 grab sampling
抓取方法 grasping means
抓取工具 gripping tool
抓取机 grabs
抓取面 gripping surface
抓取器 grabber;grabble;gripping device
抓取钳 grasping tongs
抓取式取土器 <用于采取海底土样> grab sampler
抓取试样 grab sample;snap sample
抓取水样 grab sample
抓取样品 grab sample;snap sample
抓取重量 weight capacity
抓取装置 gripping device
抓圈 quoit
抓升钩 grapple hook;grappling hook;grappling iron
抓石斗 stone grab
抓石机 clamshell;rock grapple;rock rake
抓石耙 rock rake
抓石器 bucket grab
抓石钳 rock grapple
抓石叉 stone grapple
抓式采泥器 snapper sampler
抓式采样器 snapper sampler
抓式起重机 grabbing crane
抓手 <机械手的> grip
抓手凹槽 gripping device;hollow grip;tongs
抓条 grab bar
抓筒 junk basket
抓头 grappling fixture
抓土斗 earth grab(bing) bucket
抓土机 ream grab
抓挖机 hammer grab
抓握反射 grasp(ing) reflex
抓握器 clasper
抓岩 mucking
抓岩铲斗 crab bucket
抓岩机 bucket grab;clamshell mucker;grab clamshell;grab loader;mucking unit;rock rake
抓岩机吊车 grab crane
抓岩机绞车 grab hoist
抓岩机抓斗 crab bucket
抓岩机抓岩 grabbing
抓岩机装岩 clamshell loading
抓扬机 grab;grapple
抓扬机装载 grabbing
抓样机 grab sampling machine
抓样器 grab sampler
抓样强度 grab strength
抓样强力试验 grab test
抓者 nipper
抓制纸浆圆木的抓斗 pulpwood grapple
抓重比 holding power to weight ratio
抓住 catch hold of;grapple;grasp;gripe;hold on;prehension;seize;seizure;snatch;bite <指锚抛稳>
抓住管子 <提引器> latch on
抓住机会 take occasion to
抓住落鱼 engagement with the fish
抓住要点 keep to the point
抓爪 catch;catcher;grab hook;grip
抓爪机构 gripper mechanism
抓爪进给机构 gripper-feed mechanism
抓爪器 gripper
抓爪钳 gripping pliers
抓爪式草捆堆提升机 grab-type bale stock lifter
抓爪式草捆提升机 pincer-type bale lifter
抓爪式装载机 tined loader
抓爪系统 gripper system
抓砖器 brick grab
抓子钩 catch pawl

爪 claw;foot lug;pawl;trip latch

爪板 claw plate
爪锤 claw hammer
爪垫 palmula
爪斧 claw hatchet
爪杆 claw bar;pinch bar
爪钩 grabber;prongs hook
爪棍 pinch bar
爪簧 pawl spring
爪簧式岩芯提断器 basket core lifter
爪极发电机 claw pole generator
爪极式电动机 Lundell motor
爪件 bluff piece
爪接链 attachment chain
爪具 dog
爪卡盘 chuck with holdfast;dog chuck
爪块 rest shoe
爪轮 ratchet wheel
爪内片 subunguis
爪盘 claw disc
爪盘联轴节 jaw coupler;jaw coupling
爪片的 claval suture
爪式除草器 claw weeder
爪式打捞筒 dog type overshot
爪式垫圈 pawl washer;tab washer
爪式胶带输送机 claw-rubber belt conveyer
爪式卡盘 claw chuck;jaw chuck
爪式冷床 pawl type cooling bed
爪式离合器 claw chuck;claw clutch;claw coupling;dog clutch;dog coupling;jaw clutch;jaw coupling;ratchet coupling
爪式连接器 claw coupling;dog coupling
爪式联结器 jaw coupling;ratchet coupling
爪式联轴节 claw coupling;dog coupling;jaw coupling;ratchet coupling
爪式联轴器 claw coupling;dog coupling;jaw coupling;ratchet coupling
爪式链节 lug link
爪式起重机 claw crane;crab crane
爪式心轴 jaw mandrel
爪式原始岩芯钻头 poor boy
爪式制动器 jaw brake
爪式抓斗 jaw grab
爪饰 talons
爪栓 claw bolt
爪哇白拉胶 java para
爪哇比率 Java ratio
爪哇达理木 Java stonewood;lumbayao
爪哇海 Java Sea
爪哇海沟 Java trench
爪哇黑腐病 Java black-rot
爪哇棉 capoc
爪哇熔融石 javaite
爪哇重阳木 katang
爪维尔水 Javel water
爪纹 crow's foot mark
爪销 pawl pin
爪心锤 hammer-claw
爪形扳手 jaw spanner
爪形冲击锤 cross chipper hammer
爪形磁铁 claw magnet
爪形钩 claw-like hook;dog hook
爪形夹盘 dog chuck
爪形离合器 claw chuck;claw clutch;claw coupling;dog clutch;dog coupling;jaw clutch;jaw coupling;ratchet coupling
爪形离合器套筒 dog clutch sleeve
爪形连接器 dog coupling
爪形联结器 claw coupling;jaw coupling;ratchet coupling
爪形联轴节 claw coupling;dog coupling;jaw coupling;pawl clutch;ratchet coupling;shifting sleeve
爪形联轴器 claw chuck;claw coupling;dog coupling;jaw coupling;ratchet coupling
爪形裂缝 crow foot crack
爪形炉箅 dog;dog grate
爪形撬棍座 bench of crowbar
爪形手 claw hand
爪形条纹 talons
爪形镇压器 crowfoot packer;toothed roller
爪形止退垫圈 lug washer
爪形皱裂 claw-foot crack;crow foot crack
爪形皱纹 <油漆> crow's foot(ing)
爪悬架 claw suspension gear
爪凿 <石工用的> claw chisel
爪闸 jaw brake
爪罩 pawl casing
爪状(钩)钳 claw
爪状晶纹 crow's foot(ing)
爪状裂纹 crow foot crack
爪状饰 griffe
爪状物 claw;talons
爪状细裂纹 crowfoot checking
爪状趾 claw toe
爪子 paw

专案法官 ad hoc judge
专案管理 project management
专案进口签证 special import licensing
专长 expertise; know-how; special(i)ty; special skill
专车 private car; special; special car
专断行为 arbitrary action
专功能水库 simple purpose reservoir
专供销售的农作物 cash crop
专供……之用 appropriated to
专管经济区 <200 海里内> exclusive economic zone
专化特性 specialized feature
专化性 specialization
专机 special plane
专家 adept; consummator; expert; professional; proficient; specialist; technician; trained worker
专家报告 expert's survey; expert report
专家程序系统 expert program(me)s
专家酬金 expert fee
专家代表团 delegation of authority
专家的意见 expert opinion
专家调查 expert enquiry
专家估值费 expert's valuation fee
专家顾问团 brain trust
专家官员 technocrat
专家管理 technocracy
专家会议 expert meeting
专家建议 specialist advice
专家鉴定 expert's appraisement; expert's examination; expert appraisal; expertise[expertize]; written expert testimony
专家鉴定结论 expert's conclusion
专家鉴定书 expert's statement
专家经验 expertise
专家经营市场制 specialist market-making system
专家决策 expert decision-making
专家评价 expert appraisal; expertise [expertize]; expert opinion
专家评判 professional judg(e)ment
专家讨论会 seminar
专家统治 meritocracy
专家团 expert panel
专家委员会 board of experts; expert committee; specialist commission
专家问题求解软件 expert problem solver
专家系统 expert system
专家系统模块 expert system module
专家小组 expert panel; expert sub-group; panel of experts; specialist sub-group; technical panel
专家小组成员 panelist
专家小组讨论 panel discussion
专家小组预测法 panel of experts
专家证据 expert testimony
专家证明 expert testimony
专家证人 expert witness
专家政治 technocracy
专家政治论者 technocrat
专家治国论者 technocrat
专家主张 expert opinion
专家咨询费 fee for expert opinion
专家咨询系统 expert consultancy services system; expert consulting system
专家组 specialist committee
专见种 exclusive species
专开列车 special train; train run as a special; special running
专开列车的警戒 watch for special train
专刊 special issue
专科病房 specialty ward
专科图书馆 departmentalized library
专科性百科全书 special encyclopaedia
专科学校 college; college for professional training; institute; school of general instruction; training school
专科学院 <综合性大学内的> cluster college
专科医院 special hospital; specialty hospital
专科院校 academy
专款 special fund
专款授权 obligation authority
专款账户 rubricated account
专款专用 earnmarking; special appropriation for special use
专款准备 funded reserve
专栏 specialized column; special column
专栏分类账 split ledger account
专栏日记账 split column journal
专类花园 special flower garden
专利 monopolization; patent
专利板材 patent board
专利板料 patent plate
专利保护 patent protection
专利报告 patent report
专利背衬地毯 patent-back carpet
专利标记 patent marking
专利玻璃窗木嵌条 timber patent glazing bar
专利玻璃装配的屋顶 patent glazing roof
专利材料 patent material; proprietary material
专利产品 proprietary product
专利厂 proprietary plant
专利持有人 patentee; patent holder
专利持有者 patentee; patent holder
专利代理人 patent agent; patent attorney
专利的 monopolistic; patent; proprietary
专利的玻璃装配隔条 patent glazing bar
专利的地板 patented floor
专利的石材防水剂 patent stone waterproofer
专利的无油灰窗玻璃隔条 patent glazing bar
专利的无油灰窗玻璃屋顶 patent glazing roof
专利登记簿 patent rolls
专利地板 patent floor
专利独占 patent monopoly
专利对照 patent concordance
专利对照索引 patent concordance index
专利发明 patented invention
专利法 patent law; patents act
专利法案 patent act
专利费 patent expenses; patent fee
专利斧 patent axe
专利公报 patent bulletin
专利公司 monopoly
专利功能 proprietary function
专利共同使用制度 patent pooling
专利共享 patent pool
专利共享企业 patent pool
专利共享协定 patent-pool agreement
专利号对照索引 patent number concordance
专利号码分类索引 patents allocation index
专利号索引 numeric(al) patent index
专利和商标公告 patent and trademark notices
专利鹤嘴锄 patent pick
专利护板 patent protector
专利化学品 proprietary chemical
专利黄 patent yellow
专利混合料 proprietary mix
专利混合物 proprietary mix
专利技术 patent technology; proprietary technology
专利技术许可合同 know-how license contract
专利检索 patent retrieval
专利建筑板材 patent board
专利建筑材料 proprietary construction material
专利局 patent and trademark office; patent office
专利蓝 patent blue
专利垄断 patent monopoly
专利律师 patent attorney
专利绿 patent green
专利门 proprietary door
专利配方 proprietary formulation
专利品 monopoly; patent article; proprietary article; proprietary material
专利平板玻璃 patent plate
专利期限已满 patent runs out
专利墙板 patent board
专利侵权 patent infringement
专利情报 patent information
专利情报服务 patent information service
专利情报活动 patent information activity
专利权 chatter; exclusive right; patent (right); rights of patent; rights of monopoly; subject of numerous patents; monopoly
专利权持有人 holder of the patent right; patentee; patent holder
专利权持有者 holder of the patent right; patentee; patent holder
专利权的标示 patent marking
专利权的放弃 surrender
专利权的国际保护 international protection of patent
专利权的国际公约 International Convention on Patents
专利权的交换实施权 cross license[licence]
专利权的侵犯 infringement of patent rights
专利权的维持 renewal of licensed patents
专利权登记 registration of patent
专利权法 patent law
专利权合作条约 patent cooperation treaty
专利权名牌 patent plate
专利权期满 expiration of patent
专利权使用费 patent royalty
专利权授与人 patentor
专利权授与者 patentor
专利权税 royalty
专利权所有人 patentee
专利权所有者索引 index of patentees
专利权限 patent claim
专利权许可证 patent licence[license]
专利燃料 patent fuel
专利燃料厂 patent-fuel plant
专利软件 proprietary software
专利设备 proprietary device; proprietary equipment
专利申请 patented claim
专利申请表 application for a patent
专利申请范围 patent claim
专利申请服务 patent application service
专利申请权限 patented claim
专利申请人 patent applicant
专利申请人服务 patent applicant service
专利申请书 patent application
专利申请说明书 patent application specification
专利式 patent type
专利室 patent room
专利收录范围 patents coverage
专利手镐 patent pick
专利水权 appropriative water rights
专利税 franchise tax
专利说明书 description of patent specifications; patent specifications; proprietary specification
专利说明书节录本 patent abridgement
专利索引 patent index
专利特许使用权 patent licence[license]
专利特许证 letter patent; patent
专利通告事项 patent notices
专利图书馆 patent library
专利图样 patent drawing
专利文件 patent documents
专利文献 patent documentation; patent documents; patent literature
专利文摘 patent abstract
专利文摘丛刊 patent abstract series
专利文摘目录 patent abstracts bibliography
专利协议 patent agreement
专利性 patentability
专利许可 patent grant; patent licensing
专利许可证 patent licence[license]
专利许可证协定 patent license agreement
专利有效期 patent pending
专利杂志 patent journal
专利折旧 depreciation on franchises
专利证 patent
专利证书 certificate of patent; letter of patent; patent certificate
专利执照 certificate of patent; patent certificate
专利制 patent system
专利专藏 patent collection
专利装玻璃配件 patent glazing
专利资料档案 patent file; patent specification file
专利自动失效 automatic lapse of a patent
专利自动终止 automatic lapse of a patent
专利作用 proprietary function
专列 special train
专列火车 unit train
专列货车 unit train
专论 monograph; tract
专买便宜货的人 close buyer
专卖 exclusive dealing; monopoly sale
专卖材料 proprietary material
专卖的 proprietary
专卖公司 departmentation specialty stores
专卖混合料 proprietary mix
专卖混合物 proprietary mix
专卖货物 monopolized commodities
专卖沥青 <一种木块地板用的> Ebanoid
专卖品 exclusive lines
专卖权 exclusive right to sale; monopoly
专卖政策 exclusive agency policy
专门操作者 specialist operator
专门齿轮箱 special gear box
专门出口 special export
专门词汇 lexicon; onomasticon
专门的 professional; special; specialized; technical
专门地貌图 special geomorphic map
专门地图 special map; special-purpose map
专门调查官员舞弊情况的政府官员 ombudsman
专门定制 job made

专门分析 ad hoc analysis
专门附件 special attachment
专门工种承包人 specialty contractor
专门工组 special gang of workers
专门观测 special observation
专门观测孔 special observation borehole
专门合同 special agreement
专门化 specialisation[specialization]; specialize; technicalization
专门化品系 specialized strain
专门机构 functional body; special agency; specialized agency; special organ
专门计划 ad hoc program(me)
专门技能 expertise[expertize]; expert skill; know-how; professional skill; technical know-how; technical skill
专门技术 expertise[expertize]; expert skill; know-how; professional skill; proprietary technology; technical know-how; technical prowess; technical skill; technics
专门技术合同 know-how contract
专门技术名词 nomenclature; technical term
专门技术文献 special technic(al) publication
专门建造的 purpose-built
专门鉴定 expertise
专门进口 special import
专门论文 professional paper
专门逻辑 ad hoc logic
专门贸易 special trade
专门名词 technical terminology; term; terminology
专门名词词源学 onomastics
专门全套部件 speciality packet
专门人员 technician; technicist
专门设备 specific installation
专门设计的 purpose-designed
专门设计的码头 purpose-designed terminal; specially designed terminal
专门设施 special provision
专门事项 technicality
专门试验 specific test
专门室内系统 special indoor system
专门术语 nomenclature; technical term
专门水文地质勘察 special hydrogeological investigation
专门图 thematic map
专门委员会 ad hoc committee; departmental committee
专门系统 dedicated system
专门项目 specialty
专门小组 functional group; panel; special task force; workshop
专门小组委员会 expert panel
专门小组、学部 workshop
专门性 technicality
专门性工程地质测绘 special engineering geology mapping
专门性劳动 specialized labo(u)r
专门性水文地质调查 special hydrogeologic(al) survey
专门性水文地质图件 specific hydrogeology maps
专门学科 monoscience
专门学术性质 technicality
专门学校 college; special school
专门岩芯挤出器 purpose-made core extruder
专门研究 speciality; specialization; special study
专门意义 technical meaning
专门用品商店 specialty shop; specialty store
专门用语 buzz word; nomenclature; special notes
专门语 technicality
专门预报 special forecast
专门运输工具 technical transport
专门知识 expertise[expertize]; expert knowledge; know-how; specialized knowledge; technical know-how
专门职业 professionalization
专门指导 special instruction
专门装运爆破石料的尾卸车 rocker
专门组织 special task force
专名部分 specific element
专区 administrative region; prefecture; subprovincial administrative region
专人递送 delivered in person
专任会计人员 private accountant
专任验船员 exclusive surveyor
专设户外娱乐区 recreation development
专收门诊病人的医院 day hospital
专送电报 express telegram
专送费 express charges
专题 special subject; special topic
专题报告 position paper; professional paper; report on a special topic; seminar; specialist report; special report; theme lecture
专题成像仪 thematic mapper
专题处理 theme processing
专题档案 case history
专题地图 special subject map; topical map; thematic map
专题地图集 thematic atlas
专题地图制图学 thematic cartography
专题地图资料 thematic data; thematic information
专题地图自动制图系统 automated cartographic(al) system of thematic map; thermatic map auto-mapping system
专题调查 case survey
专题服务 special subject service
专题工作组 task force; task group
专题规划 functional plan
专题海图 thematic chart; thematic marine map
专题检索 specific search
专题节目 feature program(me)
专题论述 dissertation
专题论文 disquisition; monograph; treatise
专题论文的作者 monographist
专题论证 monographic(al) demonstration
专题索引 special index
专题讨论 seminar
专题讨论会 specialty session; symposium; teach-in; workshop
专题图 special topic figure
专题文件 theme files
专题文献 document collection
专题文章 monograph
专题性世界会议 ad hoc world conference
专题研究 case study; monographic(al) study; special study
专题研究班 joint topical workshop; seminar
专题研究报告 special research report
专题研究会 seminar; symposium; task session; workshop
专题研究组 workshop
专题要素抽出 theme extraction
专题制图 thematic mapping
专题制图仪 thematic mapper
专题著作 monograph
专题资料 case history; case record
专题资料集 casebook
专为某公司运输的货物 company material
专文 feature article
专喜高温微生物 obligate thermophilic organism
专线 dedicated circuit; individual line; private line; private wire; special line
专线报警器 leased line annunciator
专线触排 private bank
专线触排接点 private bank contact
专线电报机 telex
专线电传机 private line teletype
专线电传系统 private wire teletypewriter system
专线电话 private telephone
专线互联业务 private line interconnection service
专线连接 radial selector
专线连线 private line arrangement
专线入口 local access
专线双向中继电路 private line twoway trunk circuit
专线通信[讯]网 leased line network
专线业务 private line service
专线用户 individual line subscriber
专项保险 insured peril
专项拨款 itemized appropriation
专项存款 special deposit
专项贷款 special-purpose loan; tied loan
专项费用 specific cost
专项工程承包招标 inviting bids of special engineering contracting
专项工程支出 special construction expenditures; special fund; special work expenditures
专项规划 functional plan; special-purpose planning
专项监督检查 special supervision and examination
专项经验 expertise[expertize]
专项留置权 special lien
专项试验路 special experimental road
专项体检 special physical examination
专项通报 special call
专项投资 specific investment
专项投资额 specific investment cost
专项外汇 special sum of foreign exchange
专项治理 special control
专项资金 special fund
专销 special pin
专效性农药 narrow spectrum pesticide
专效性杀虫剂 narrow spectrum insecticide
专性 specificity
专性的 obligate; obligative; specific
专性腐生物 obligate saprophyte
专性共生 obligate symbiosis
专性共生物 obligate symbiont
专性光能自养生物 obligate photoautotroph
专性好气细菌 obligate aerobic bacteria
专性好氧菌 obligate aerobes
专性好氧微生物 obligate aerobes
专性化能自养生物 obligate chemoautotroph
专性活动 specific activity
专性寄生物 obligate parasite; obligatory parasite
专性离子电极 specific ion electrode
专性离子效应 specific ion effect
专性力 specific force
专性嗜热菌 obligate thermophilic bacteria
专性吸附 specific adsorption
专性嫌气细菌 obligate anaerobic; obligate anaerobic bacteria
专性需氧的 obligate aerobic
专性需氧菌 obligate aerobes
专性需氧微生物 obligate aerobes
专性盐生植物 obligate halophyte
专性厌气细菌 obligate anaerobic bacteria
专性厌氧 obligate anaerobic
专性厌氧微生物 obligate anaerobe
专性厌氧细菌 obligate anaerobe
专性异养生物 obligate heterotroph
专性异养细菌 obligative heterotrophic bacteria
专性自养 obligate autotrophy
专性自养生物 obligate autotroph
专业 career; craft; discipline; profession; special(i)ty; specialized post
专业版本说明 special interest edition statement
专业标准 professional standard; specialized standard
专业标准部件 speciality packet
专业操作 professional practice
专业厂 special manufacturer
专业承包人 specialist contractor
专业承包商 specialist contractor; specialized contractor
专业承包者 specialist contractor
专业程序设计员 professional programmer
专业词汇 specialized vocabulary; technical term
专业词书 lexicon
专业的 professional; technical
专业等级 professional grade
专业队 gang; professional team
专业队伍 professional contingent
专业范围 world view
专业费用 professional fee
专业服务 professional service
专业服务费 professional fee
专业负责人 principal in professional practice; speciality sponsor
专业港 specialized port
专业工长 trade foreman
专业工程师 expert engineer; professional engineer
专业工具 special-purpose tool
专业工种承包商 trade contractor
专业工作组 squad cohesive
专业公司 firm of specialists; specialist firm
专业顾问 professional adviser[advisor]
专业管理 professional management
专业户 specialized household producer
专业化 professionalization; specialization; specialize
专业化处理 specialized processing
专业化管理 management specialization
专业化码头 specialized(marine) terminal; special-purpose wharf; specialized quay
专业化情报 specialized information
专业化生产车间 manufacture shop; manufacturing shop
专业化生产系统 manufacture system; manufacturing system
专业化数据处理 specialized data processing
专业化协作 co-ordination among specialized departments; specialization and cooperation
专业化协作价格 prices for specialization and cooperation
专业化修制【铁】 specialized repair system
专业会计师 professional accountant
专业会议 congress
专业技术人员 professionals & techni-

Z

cal
专业计算机 special-purpose computer
专业技能 profession skill
专业鉴定 expertize
专业教育 professional education
专业经济核算 specialized economic accounting
专业经理人 professional manager
专业经验 professional practice
专业经营 professional management
专业刊物 trade journal
专业冷藏库 specialized cold store
专业论文 professional paper
专业码头 specialized marine terminal
专业门类 specialized fields
专业名词 technical term
专业农场 ranch
专业农业 specialized farming
专业判读 special-purpose interpretation
专业培训 professional training; special training; vocational training
专业评判 professional judg(e)ment
专业情报 specialized information
专业情报检索 special information retrieval
专业情报源 specialized information sources
专业情报中心 special information center[centre]; specialized information center[centre]
专业人员 expertise; personnel; practitioner; professional personnel; specialized person; specialized staff personnel
专业人员在职培训 on-the-job professional training
专业人员助理 paraprofessional
专业人员助手 paraprofessional; subprofessional
专业商店 single-line store
专业设备 special plant
专业设计工程师 discipline design engineers
专业施工队(伍) specialized team
专业实践 professional practice
专业实践经验 professional practice
专业实习 specialized practice
专业熟练程度 professional qualification
专业术语 professional terminology
专业调绘【测】 special vocation annotation
专业图书馆 special library
专业图书室 departmentalized library
专业团体 professional society; special technology group
专业文献 professional literature
专业文献综述 review on special information
专业橡胶 special rubber
专业小组 functional group
专业协会 professional Association
专业性 specialty
专业性的 specialized
专业性港口 specialized harbo(u)r; specialized port; special-purpose port
专业性货运站 specialized freight station; specialized goods station
专业性码头 specialized wharf; special-purpose wharf
专业性质 technicality
专业学校 professional school; specialized school
专业训练 career training; specialized training; special training
专业养猪场 specialized pig-farm
专业业务 professional practice
专业业务保险 professional liability insurance
专业用语 technical word
专业阅览室 faculty reading room
专业运输 professional transport; transport on account of third parties
专业杂志 trade magazine
专业责任保险 professional liability insurance
专业支路 private driveway
专业知识 expertise; expert knowledge; professional knowledge; specialized knowledge
专业制造厂家 specialist manufacturer
专业著作 professional literature
专业准则 professional standard
专业咨询公司 professional corporation
专业资格 professional competence
专业组件 speciality packet
专业组织 trade association
专一乘坐 <出租汽车从起点到终点的单一行程> exclusive ride
专一的 exclusive
专一反应 exclusive reaction
专一试剂 specific reagent
专一性 specificity
专一指示剂 specific indicator
专营国库券政策 bills-only policy
专营合同 exclusive contract
专营权 exclusive right; franchise
专营性条款 exclusive clause
专营许可证 exclusive licence[license]
专用 appropriation; dedication; exclusive use; special application; special-purpose; special use
专用安全装置 exclusive safety device
专用白炽灯 incandescent special-service lamp
专用半圆键 special woodruff key
专用包 special-purpose packet
专用包网 private packet network
专用报表 single-purpose statement
专用报警系统 private alarm system
专用报汛站 special flood-reporting station
专用报汛站网 special flood-reporting network
专用编码 own coding; specific coding
专用变电所 house substation
专用标志 special(-purpose) mark
专用表 special table
专用表壳 special watch case
专用拨款 appropriation for special uses; earnmark; grant; special allocation; special appropriation(fund)
专用泊位 appropriate berth; dedicated berth; exclusive use berth; single-user berth; berth specialized
专用泊位计划 appropriated berth scheme
专用布置 hookup
专用操作系统 proprietary operating system
专用操作异常 special operation exception
专用测试设备 special test equipment
专用插件插座 personality card socket
专用插座出线口 special-purpose receptacle outlet
专用产品生产系统 closed system of product
专用铲斗 special bucket
专用长途中继电路 special toll trunk circuit
专用车 private car; specia(lized) car; special-purpose car
专用车床 special-purpose lathe
专用车道 accommodation lane; accommodation line; exclusive lane; separate lane; dedicated lane
专用车道分流标志 arrow direction exclusive lane sign
专用车库 private(parking) garage
专用车辆 special-purpose vehicle; specific vehicle
专用程序 special program(me); special routine; specific program(me); specific routine
专用程序包 special package
专用程序带输入机 dedicated tape reader
专用程序库 private library
专用出线口 special-purpose outlet
专用储能 dedicated storage
专用处理机 dedicated processor
专用触排 private bank
专用传真台 private phototelegraph station
专用船 pure vessel; specialized vessel; special-purpose ship; special service ship; special service vessel; specified cargo vessel
专用船舶 specialized vessel
专用船队 specialized fleet
专用窗 special-purpose window
专用存储媒体 private volume
专用存储器 private memory; private storage
专用存储区 dedicated memory; dedicated storage
专用存款 special deposit; tied deposit
专用代码 private code; unique code
专用贷款 earmarked loan; loan for exclusive use; special-purpose loan
专用刀杆 special holder
专用道路 <指定为一种交通专用的道路> accommodation road; exclusive road; occupation road; private road; reserved road; single-purpose road
专用道路与铁路平交道口 private crossing
专用的 dedicated; individual; private; single-purpose; special(-purpose); tailor-made
专用地 reservation
专用地段 accommodation area
专用地区 exclusive district; restricted district
专用地图 applied map; special-purpose map; special use map; specific purpose map
专用地址 specific address
专用地质数据库 special geologic(al) data base
专用电报 exclusive telegram
专用电报和电话业务 private telegraph and telephone service
专用电动机 special-duty electric(al) motor
专用电话 private telephone; special-purpose telephone
专用电话交换机 private branch exchange
专用电话交换网 private switched telephone network; private switching network
专用电话网 private telephone network
专用电话线 private telephone line
专用(电话)线电路 private line circuit
专用电缆 private cable; restricted use cable; special cable
专用电路 dedicated circuit; exclusive circuit; special circuit
专用电台 private station
专用吊架 special hoisting bar
专用吊柱 drop tube
专用调车机车全周转时间 turnover of special marshalling locomotive
专用调压井 specific surge tank
专用调用程序库 private call library
专用调制器 special modulator
专用叠标 special-purpose range; special-purpose transit
专用定义符 delimiter
专用短程通信[讯] dedicated short-range communication
专用断路器 definite-purpose circuit breaker
专用发电装置 isolated plant
专用阀 special service valve
专用法规 appropriation law
专用防护(金属)薄片 special protecting foil
专用房间 function room; shack; special room
专用房屋 function(al) building; single-purpose building
专用放大机 special enlarging printer
专用飞机 general aviation aircraft
专用焚化炉 private incinerator
专用浮标 special-purpose buoy
专用浮动式存储库 private relocatable library
专用附件 special accessories; special components
专用覆盖金属箔片 special covering foil
专用钢筋 special reinforcement(steel)
专用港(口) single-user harbo(u)r; single-user port; special(-purpose) port; industrial port <工厂企业所属的>
专用隔板式门 special stopping
专用工具 built-for purpose tool; specialist tool; special tool; specialty tool; specific tool; quirk cutter <开挖圆形地沟的>
专用工具机 special-purpose machine
专用工艺设备 special technological equipment
专用工作服 special clothing
专用公路 accommodation highway; accommodation road; single-purpose highway; special highway; special road
专用功能 dedicated function
专用供水设备 private water supply equipment
专用供水系统 private water supply system
专用构件 special unit
专用构件生产系统 closed system
专用固定件 special fastener
专用管路 individual line
专用归航收讯机 specialized homing receiver
专用过程控制计算机 dedicated process control computer
专用海图 special-purpose chart
专用函数产生器 special function generator
专用航道 private channel; special-purpose waterway
专用航海图 nautical chart for special-purpose
专用合同条款 special conditions of contract
专用横向电话系统 special telephone system for transverse linkage
专用虎钳 special vise[vice]
专用化程序过高 overspecialization
专用化的车辆 specialized wagon utilization
专用化码 unique code
专用化学品 specialty chemicals
专用混凝土制品 special concrete

product
专用火警箱 private box
专用货车 special-duty wagon;specialized freight car;special type wagon
专用货棚 appropriate shed
专用机车 private locomotive;special locomotive
专用机床 production machine;special locomotive
专用机器 special(-purpose) machine
专用机械 production machine;special machinery
专用机油 mobile oil;mobiloil
专用基金 funds for special use;special assessment fund;special fund;special-purpose fund;specific fund
专用基金拨款 appropriation for special fund
专用基金存款 special fund deposit
专用基金明细表 detailed schedule of special fund
专用基金项下的实物资产 physical assets under specific fund
专用基金银行存款 bank deposit on specified fund
专用基金增长率 growth rate of earmarked fund
专用基金资产 assets of special fund
专用集成电路 application specific integrated circuit
专用集装箱 specialized container;specific purpose container
专用给水 private waterworks
专用计算尺 special slide rule
专用计算机 dedicated computer;single-purpose computer;special-purpose computer
专用计算机程序 computer dependent program(me)
专用计算技术 personal computing technology
专用记号 special token
专用继电器 definite-purpose relay
专用加热炉 special heating furnace
专用夹具 special fixture;unit clamp
专用夹盘 special-purpose chuck
专用夹盘爪 special chuck jaw
专用夹头 special carrier
专用减压室 special decompression chamber
专用检索 specific search
专用件 special-purpose item
专用建筑 functional architecture
专用建筑材料 special construction(al) material
专用建筑构件 special construction unit
专用建筑机械 single-purpose construction machine
专用建筑体系 closed system
专用建筑型材 special construction profile;special construction shape
专用交换 private branch exchange
专用交换分机 private branch exchange
专用交换分局 private branch exchange
专用交换机 private branch exchange;private exchange
专用交换机中继线 private branch exchange line
专用交换机终接器 private branch exchange final selector
专用交换台 private branch switchboard
专用交换线路 exclusive exchange line
专用浇灌混合料 special pouring compound
专用胶片 professional film
专用教育设施 special educational facility
专用节 private section
专用结构材料 special structural material
专用结构构件 special structural unit
专用结构型材 special structural profile;special structural section;special structural shape
专用结构装饰件 special structural trim
专用借款 special loan
专用紧固件 special fastener
专用进程栈 private process stack
专用局部存储器 private local memory
专用卷帘 private volume
专用卷宗 private volume
专用决算表 special-purpose statement
专用蝌蚪图 single-purpose tadpole plot
专用可视信息检索终端 dedicated videotex terminal
专用客车运输 private transportation
专用控制台 uniset console
专用款项 specific cost
专用旷地 private open area
专用立筋卡 individual stud clip
专用连接 dedicated connection
专用连线 dedicated connection
专用联动机床 special power pack set machine
专用量 private volume
专用列车 conditional train
专用零件 parts peculiar
专用流动式起重机 specific purpose mobile crane
专用龙头 private tap
专用楼梯 private stairway
专用路 reserved road
专用路权 exclusive right-of-way
专用路线 individual line;leased circuit;leased line;private wire;reserved route;tie line
专用绿地 special garden plot
专用罗盘测角元件 special compass angle units
专用逻辑记录 special logic record
专用螺栓 special bolt
专用马桶 private stable
专用码 own code;private code
专用码头 appropriated berth;appropriated wharf;captive terminal;dedicated terminal;private terminal;private wharf;special wharf;private dock
专用锚地 special-purpose anchorage
专用锚具 special fixture
专用密码 special-purpose system
专用密码系统 specific crypto-system
专用名词 nomenclature;specific term
专用名词表 onomasticon
专用模 die for special purpose
专用模板 special shuttering
专用模件板 special-purpose module board
专用模块测试 special module testing
专用排气口 coughing
专用培训设备 special training equipment
专用配方 proprietary formula
专用配件 special fittings
专用起重机 captive crane
专用汽车 special-purpose vehicle
专用牵引发电站 special traction power station
专用桥(梁) accommodation bridge;service bridge;occupation bridge
专用桥式起重机 special overhead crane
专用清洗设备 purpose-built washing equipment
专用情报检索系统 special-purpose information retrieval system
专用区 exclusive use district
专用区域规划 exclusive use zoning
专用权 private right
专用人工交换机 private manual exchange
专用人工支线交换机 private manual branch exchange
专用溶剂 proprietary solvent
专用塞孔 special jack
专用砂箱 special-shaped mo(u)lding box
专用商标名 proprietary name
专用设备 dedicated set of equipment;isolated plant;optional equipment;proprietary appliance;special equipment;special hardware;specialized equipment;specialized facility;special-purpose equipment;task equipment
专用设备多头铣床 special milling machine
专用设备检验规范 specified equipment inspection code
专用设备清单 special list of equipment
专用设施 private facility;special-purpose facility;specialized facility
专用摄影机 special-purpose camera
专用审计日程表 special-purpose audit program(me)
专用升降机 special lift
专用声呐(装置) special sonar system
专用施工机械 single-purpose construction machine
专用时刻表的列车 dedicated timetabled train
专用实用程序 special-purpose utility program(me)
专用式 server-based
专用试验设备 special-purpose test equipment
专用室 single use room
专用术语 designatory term
专用数据 dedicated data;exclusive data;private data
专用数据集 dedicated data set
专用数据通信[讯]服务网 leased data communication service
专用数据网 private data network
专用数据终端 personal data terminal
专用数据组 private data set
专用数字电话交换机 private branch exchange
专用双闸瓦制动器 claw shell brake
专用水库 single-purpose reservoir
专用水权 appropriative water rights;water appropriation right
专用水栓 private tap
专用水文站 hydrometric(al) station for special purposes
专用速冻装置 individually quick freezer
专用酸 <地板上或油桶内用以抗沥青的> acetas
专用索引分析器 special index analyser[analyzer]
专用台阶 special step
专用台(站) special station
专用太平梯 special emergency stair(case)
专用填泥料 special spackling compound
专用条件 special condition;conditions of particular application
专用条款 private terms
专用铁道 access railroad;access railway;dedicated line;exclusive railway;special-purpose railway
专用铁路 access railway;access railroad;dedicated line;exclusive railway;special-purpose railway
专用铁路平交道 occupating crossing
专用铁路线 industrial line;industrial railway;industrial siding;private line;private railway;railway siding
专用铁路支线 private siding
专用停车场 special parking place
专用通风管 individual vent
专用通气立管 specific vent stack
专用通气竖管 specific vent stack
专用通信[讯]处 private address
专用通信[讯]网 private communication network;private wire network
专用通信[讯]系统 special communication system
专用通信[讯]信道 private communication channel
专用图 appropriate chart;special(-purpose) chart;special(-purpose) map
专用涂料 special coating
专用瓦 individual tile
专用网(络) dedicated network;private network;special network
专用微处理机系统 dedicated microprocessor system
专用维修设备 special equipment for maintenance
专用文件 private file
专用污水处理厂 private sewage disposal
专用污水处理系统 individual sewage disposal system
专用污水管 private sewer
专用屋顶排水沟槽 special valley gutter
专用物资 special store
专用稀砂浆 special slurry
专用稀释硬煤沥青混合物屋顶 special fluxed pitch composition roofing
专用稀释硬煤沥青混合物屋面 special fluxed pitch composition roofing
专用稀释硬煤沥青预制屋面 special fluxed pitch ready roofing
专用洗涤剂 special-purpose detergent
专用铣床 manufacturing miller;production miller
专用系统 dedicated system;proprietary system
专用细目概念 special isolate idea
专用下水道 private sewer
专用线 accommodation line;assigned siding;business lines;business track;industrial line;industry track;private line;private siding;private wire;siding;special line;dedicated line
专用线费用 private siding charges
专用线服务 private line service
专用线共用 share of industrial siding
专用线路 dedicated line;feeder road;individual line;leased line;personal circuit;private circuit;private line;tie line
专用线路连接 private wire connection
专用线契约 private siding contract
专用线使用主 owner of private siding
专用线业务 private line service;private wire service
专用线至专用线间的整列车运输 siding-to-siding train load traffic
专用线至专用线间为某单位专编的货物循环直达列车 siding-to-siding unit train
专用线终接器 individual line connector
专用箱形雨水槽 special-purpose box rainwater gutter

Z

专用箱形雨水沟 special-purpose box rainwater gutter
专用项目 speciality
专用橡胶 special-purpose rubber
专用消防栓 private hydrant
专用消防系统 private fire protection system;private fire service system
专用小交换机 private branch exchange; private branch exchange switchboard
专用小交换机中继线 private branch exchange trunk
专用楔子 special wedge
专用信道 clear channel; dedicated channel; private channel; special channel
专用信号 special signal
专用信号量 private semaphore
专用信箱 private letter-box
专用行走架 special application undercarriage
专用型 tailored version
专用型材 special profile;special shape
专用修理设备 special repair and maintenance equipment
专用许可证 special use permit
专用叙词表 specialized thesaurus
专用循环直达列车 company train
专用压载舱 segregated ballast tank
专用压载舱保护位置 protection location of segregated ballast tank
专用压载水舱系统 segregated ballast water tank system
专用遥感卫星 specialized remote sensing satellite
专用业务中继电路 special service trunk circuit
专用银行借款 special bank loans
专用引线 dedicated pin
专用硬件 special hardware
专用硬质合金头 special in carbide tipped
专用用地区划 exclusive use zoning
专用于传递的座席 position specializing in transmission
专用于收受的座席 position specializing in reception
专用语言 special-purpose language
专用域 specific field
专用元件 professional component; professional element
专用运河 private canal; special-purpose canal
专用运输 special transit
专用运输工具 private carriage; private carrier
专用运输航空 general transport aviation
专用轧辊 single-purpose roller
专用轧制构件 special rolled unit
专用轧制型材 special rolled profile; special rolled section;special rolled shape
专用站 specialized station; special-purpose station; specific purpose station
专用照明 task illumination
专用罩面熟石灰 special finishing hydrated lime
专用者 appropriator
专用整体化列车 integrated train
专用证 identification card
专用支架 special stand
专用支路 private driveway
专用支票 special-purpose check
专用支线 bay-line;private siding
专用支线交换机 private branch exchange
专用支座 special carrier
专用执行程序 dedicated executive
专用纸张 specific paper
专用指令 special instruction
专用制动器 special stopper
专用智能终端 special-purpose intelligent terminal
专用中继线 individual trunk
专用筑路油 special road oil
专用铸件 special casting
专用砖 individual tile
专用转弯车道交叉口 crossing with separate turning lane
专用装备数量表 special-purpose table of allowance
专用装玻璃法 patent glazing
专用装饰构件 special rolled trim
专用装饰件 special trim
专用装饰砌块 special decorative block; special ornamental block
专用装饰砖 special decorative tile; special ornamental tile
专用准则 specific criterion
专用资本 special capital
专用资本货物 specialized capital goods
专用资费标志 special tariff indicator
专用资金 special capital
专用资金额 earmarked fund quota
专用子程序 special type subroutine
专用字段 specific field
专用字符 special character
专用自动电话交换总机 private automatic exchange
专用自动电话系统 private automatic telephone system
专用自动(电话)小交换机 private automatic branch exchange
专用自动交换分机 automatic private branch exchange; dial private branch exchange;private automatic branch exchange
专用自动交换机 private automatic exchange
专用自动交换机局 private automatic exchange
专用自动数据处理 specialized automatic data processing
专用自卸卡车 special dump truck
专用自主部件 dedicated autonomous unit
专用自主模件 dedicated autonomous module
专用阻抗 special resistance
专用组合机 special-purpose combination machine
专用组合系统 special assembly system
专用组件 personal module
专用钻 special jewel
专用钻削头 special drilling unit
专用坐车 seating accommodation
专有 appropriation
专有程序 proprietary program(me)
专有供水 private water supply
专有环境方差 special environmental variance
专有技术 exclusively owned technology; know-how; proprietary technology
专有技术情报 know-how information
专有技术协议 know-how agreement
专有技术许可 know-how license[licence]
专有名称 proper name
专有权 appropriated right; exclusive right;patent right
专有设计 proprietary design
专有使用 exclusive use
专有水权 appropriative water rights
专有特权 exclusive privilege
专有图案 patent drawing
专有图样 patent drawing
专有消耗品 proprietary consumables
专有准则 doctrine-of-appropriation
专员 commissioner
专运木材小船 tosher
专责 specific responsibility
专账 separate accounts
专职的 all-time
专职工作组 squad cohesive
专职管理的委员会 full-time administrative commission
专职管理人员 full-time administrator
专职教师 full-time teacher
专职人员 full-time staff
专职消防队 professional fire brigade
专职训练 career training
专制人格 authoritarian personality
专制制度 authoritarianism
专属捕鱼管辖权 exclusive fishing jurisdiction
专属捕鱼区 exclusive fishery zone
专属工业区 exclusive industrial district
专属管辖权 exclusive jurisdiction
专属经济区 exclusive economic zone
专属经济区概念 exclusive economic zone concept
专属渔区 exclusive fisheries zone;exclusive fishery limits; exclusive fishing zone
专属渔业区 exclusive fishery zone
专属职权 exclusive competence
专属住宅区 exclusive residential district
专著 monograph; specialized publication
专座 special seat
专做定货的 bespoke

砖 brick;pup

砖凹槽 frog
砖坝 brick dam
砖板内衬 tile lining
砖半厚墙 brick-and-a-half wall
砖保护层 brick protective skin;brick skin
砖背衬 brick backing
砖背贴纸瓷砖 tile papered on the back
砖壁木架 nogging piece
砖壁座 bricking ring
砖边 slot lips
砖标号 brick ga(u)ge;ga(u)ge of brick;grade of brick;strength grading of brick
砖表面 brick skin
砖冰 briquet(te)ice
砖薄壳 brick shell
砖侧面 stretcher of brick
砖层 brick course; bricklayer; brick mason;course of bricks;course of brickwork;masonry tier
砖层凹齿 bonding pocket
砖层高度标准尺 brick ga(u)ge
砖层接缝 brick joint
砖层凸出 bonding pocket
砖茶 brick tea
砖碴 brick ballast
砖厂 brickfield; brick making plant; brickworks;brick yard
砖场 brick yard
砖衬 bricking; brick set; brickwork casing;lining;steening;tile lining
砖衬背 brick backing
砖衬的 brick(ed)-in
砖衬的砍砖砧 brick set
砖衬砌 brick lining
砖衬砌渠道 brick-lined canal
砖衬圬工 brick-lined masonry
砖承重结构 load-bearing brickwork construction
砖承重墙 brick bearing wall
砖窗槛 brick sill;ceramic window sill [cill]
砖窗台 brick mo(u)lded sill; brick sill
砖疵 brick defect
砖大方脚 brick footing
砖挡土墙 brick retaining wall
砖道牙 brick curb
砖的 bricky
砖的标号 strength grading of brick
砖的表面 brick skin
砖的薄涂层 salt dip
砖的长边 stretcher
砖的尺寸 brick format;brick size
砖的固定工作 tile fixing work
砖的横向砌合 running brickwork bond
砖的级别 brick grade
砖的加湿 wetting brick
砖的鉴定 identification of brick
砖的浸湿 brick wetting;wetting brick
砖的精加工 tile finish
砖的宽面朝外的砌砖法 bull stretcher
砖的码砌序列 course of blocks
砖的黏[粘]结 bonding of brickwork
砖的砌合 bonding of brickwork
砖的砌结 bonding of brickwork
砖的强度 brick strength
砖的清洗机 brick cleaning machine
砖的清洗剂 tile cleaning agent
砖的砂裂 <砖面出现的裂缝> sand crack
砖的上釉 saltern
砖的烧透程度 degree of clay brick burning
砖的烧制 brick burning
砖的湿膨胀 moisture expansion of bricks
砖的试验 brick test(ing)
砖的顺边 stretcher of brick
砖的提升器 brick lift
砖的吸收率 suction rate of bricks
砖的洗净剂 brick cleaner
砖的一端 splay end
砖的制造 manufacture of brick
砖的装饰线列 brick ornamental string (course)
砖底座 brick plinth;tile base
砖地 tile floor
砖地面 brick floor(ing);brick paving
砖雕 brick carving
砖叠结构 corbelled brick construction
砖叠涩 brick corbel; corbelled brickwork;corbelling of bricks
砖钉 masonry nail
砖顶压檐墙 <砖压顶女儿墙> brick-cap brick face;brick-cap parapet
砖定型剂 tile fixing agent
砖斗 hod
砖堵塞 bricking-up
砖端 end of brick
砖堆 brick clamp
砖墩 brick pier;brickwork pier
砖垛 brick buttress; brick pier; stack of bricks
砖防潮层 brick damp course
砖房 brick house
砖放大脚基础 brick spread foundation
砖粉 brick dust; brick flour; ground (clay) brick
砖风道 wall duct
砖缝 brickwork joint;slot
砖缝盖条 brick mo(u)lding
砖缝接合混合物 tile jointing compound
砖缝结合剂 tile jointing composition
砖缝木嵌条 brick mo(u)lding

砖缝线 tile line
砖扶梯 tiled stair(case)
砖浮雕 ajarcara;carved brickwork
砖复面的 brick-clad
砖哥特式(建筑)backsteingotik
砖格 checker;chequer
砖格孔道 checker passage
砖格气道 checker flue
砖格蓄热室 checkerboard regenerator
砖格烟道 checker flue
砖格子 checker work grillage
砖隔断 tile partition(wall)
砖隔屏 brick baffle
砖隔墙 baffle brick;brick baffle;brick block;brick partition;masonry partition;tile partition(wall)
砖工 brick-laying;brick masonry;brick setting;brickwork
砖工锤 bricklayer's hammer;brick(layers)hammer;scutch hammer
砖工工程 bricklayer's work
砖工工作 bricklayer's work
砖工灰浆 brickwork mortar
砖工加筋 brickwork reinforcement
砖工脚手架 bricklayer's scaffold(ing)
砖工脚手台 siege
砖工结构 brick masonry structure
砖工镘(板)bricklayer's trowel;brick layers trowel
砖工砌体增强材料 masonry reinforcement
砖工砌造 laying brick
砖工刷灰 brickwork casing
砖工镶嵌 brickwork insert
砖工用的脚手架 bricklayer's scaffold(ing)
砖工用的以方木构成的脚手架 bricklayer's square scaffold
砖工圆头凿 mallet-headed chisel
砖工罩面 brickwork casing
砖工中的撞击增压 bumping
砖拱 arch of brickwork;brick arch;brick vault;brickwork arch
砖拱顶楼面 brick vault roof
砖拱管 brick arch sewer
砖拱过梁 soldier arch
砖拱楼板 brick arch floor
砖拱楼盖 brick arch floor
砖拱楼面 brick arch floor
砖拱桥 brick arch bridge
砖拱屋盖 arched brick roof covering
砖拱柱螺栓 brick arch stud
砖骨料混凝土 brick aggregate concrete
砖管道式污水管 brick conduit-type sewer
砖过梁 beam brick;brick beam;brick lintel;tile lintel
砖涵洞 brick culvert(pipe)
砖号 brick grade
砖和砂浆棱柱体 prism
砖红黏[粘]土 lateritic clay
砖红壤 kabouk
砖红壤成土作用 laterite type of soil formation
砖红壤化 lateri(ti)zation
砖红壤土 cabook;laterite;laterite soil;lateritic soil;latosol
砖红壤性土 laterite soil;lateritic soil;latosol;reddish brown latosol
砖红壤性土壤及砖红壤区 brown laterite soil and brown laterite area
砖红色 brick-red
砖红土 latosol
砖红土结壳 lateritic crust
砖红土卵石 lateritic gravel
砖红土型风化壳 residuum of laterite type

砖护面 brick facing
砖滑道 brick slip
砖灰 brick dust
砖灰缝涂白 pencil(l)ing
砖混结构 brick-and-concrete composite construction;brick-concrete structure
砖基 brick base
砖基础 brick footing;brick foundation
砖集料 brick aggregate
砖集料混凝土 brick aggregate concrete
砖加灰缝的尺寸 brick format
砖夹 brick clamp;brick grip
砖甲房屋 brick clad building
砖建住宅 brick dwelling
砖建筑阀 tile construction method
砖建筑物 brick building;brick structure
砖接缝 brickwork joint
砖节点 <空斗墙> collar joint
砖结构 brick construction;brick fabric;brick structure;brickwork
砖锯 brick saw;tin saw
砖龛 brick niche
砖靠砖砌砖法 brick and brick
砖块 brick bat;fragment of brick
砖块侧砌路面 brick on edge pavement
砖块构造 block construction
砖块路面 brick pavement
砖块面层 brick facing
砖块平铺 flat brick paving
砖块铺砌 brick paving
砖块砌面 brick facing
砖块输送机 travel(l)ing block machine
砖块吸收率 absorption rate
砖块镶面 brickwork casing
砖块状 bricky
砖捆 brick pack
砖勒脚顶层 plinth course
砖肋 brick rib
砖立面 brick facade
砖连接块件 tile jointing mass
砖梁 tile beam;tile girder
砖笼 bulk container
砖楼面 brick floor(ing)
砖炉围 brick fender
砖路 brick road;tile path
砖路面 brick pavement;brick paving;tile paving
砖墁地 brick paving
砖镘 bricklayer's trowel
砖锚件 brick anchor
砖门面 brick facade
砖面 face of brick
砖面凹槽 frog
砖面凹坑 frog
砖面板 tile panel
砖面剥落 flaking of brickwork;spalling of brick surface
砖面泛白 brick efflorescence
砖面沟纹 scoring of brick;scoring of tiles
砖面黑斑 <硅酸铁形成的> fayalite
砖面夹墙 brick-veneered wall
砖面贴纸瓷砖 tile papered on the front
砖面圬工墙 opus mixtum
砖面修刮 combing of brick face
砖面修整 brick trimmer
砖面盐迹 scum
砖面缘石 brick curb
砖模 brick die;brick mo(u)ld;stock mo(u)ld
砖模底板 stock board
砖磨损试验 brick rattle
砖磨损试验机 brick rattler

砖木房屋 post-and-panel;post-and-pan house;post and petrail
砖木混合结构 half-timber(ed) construction
砖木混合结构的 half-timbered
砖木建筑 post and petrail
砖木结构 brick-and-timber construction;brick wood construction;ordinary construction;post-and-panel structure
砖木结构住宅 wood and brick-clad home
砖木墙壁 brick-and-stud;brick nog(ging)
砖木墙壁工作 brick-and-stud work
砖内衬 brick lining
砖黏[粘]土 brick clay
砖配筋结构 reinforced brick construction
砖坯 adobe(brick);adobe clay;air-dried brick;clay brick;dobie;green brick;raw brick;unfired brick
砖坯车 hack barrow
砖坯呆空干燥 scintling
砖坯干燥槽架 hake
砖坯机 brick mo(u)lding machine
砖坯留孔风干 scintling
砖坯留孔排放 <排放方式以利通风> scintling
砖坯留孔通风干燥 scintling
砖坯帽盖 lew
砖坯模 horse mo(u)ld
砖坯模型放尺率 green brick size enlargement
砖坯压制机 brick-pressing machine
砖坯再压机 brick repress
砖皮 masonry tier
砖片 brick bat
砖平拱 brick flat arch;brick lintel
砖屏 brick barricade
砖铺底层 bed of brick
砖铺底浇灌砂浆的预应力槽形板 <瑞士创制的> Stahlton slab
砖铺地面 brick floor(ing);brick pavement;tile floor
砖铺垫层 bedding of brick;brick bedding
砖铺基床 bedding of brick
砖铺路面 brick road;brick surfacing
砖铺面 brick pavement;tile paving
砖砌 bricking-up;brick setting;laying brick;steining
砖砌坝 brick dam
砖砌壁炉 brickwork fire place
砖砌层 brick course
砖砌层水平接缝 brick bed joint;brick course joint
砖砌承重墙 brick masonry bearing wall
砖砌城堡 brickwork castle
砖砌厨房 tiled kitchen
砖砌窗花格 brick tracery
砖砌窗台 brick mo(u)lded sill
砖砌大方脚 brick spread foundation
砖砌挡墙 brick backing
砖砌道路 brick road
砖砌的菱形图案 diaper work of bricklaying
砖砌地下排水沟 brick underdrain
砖砌独立基础 brickwork individual base
砖砌二内心挑尖拱 brickwork pointed arch
砖砌二心内心挑尖拱 pointed arch of brickwork
砖砌二心同似挑尖拱 brick pointed arch
砖砌防水水泥 brick cement
砖砌房屋 all-brick building
砖砌坟墓 <古埃及> brick mastaba
砖砌隔墙 brick stopping

砖砌工程 brick masonry;mason construction
砖砌工作 tiling work
砖砌拱 arch of brickwork;brickwork arch;masonry arch
砖砌拱模 brick core
砖砌沟管 brick sewer
砖砌过梁 brick lintel
砖(砌)涵(洞)brick culvert(pipe)
砖砌合 header joint
砖砌烘箱 tiled stove
砖砌厚缝 clip joint
砖砌护堤 brickwork mound
砖砌花格 brick grill(e)
砖砌机 brick cutter
砖砌基础 brick base;brickwork base
砖砌基座 brickwork base
砖砌检查井 brick manhole;brickwork manhole
砖砌建筑物 brick architecture
砖砌结构 brick construction;brick masonry structure;brickwork
砖砌井壁 brick walling
砖砌井坑 brick pit
砖砌井筒 brick pit
砖砌军事建筑 military brick architecture
砖砌坑 brick pit
砖砌空斗墙 clay brick cavity wall
砖砌空心墙 brick cavity wall;brick hollow wall
砖砌勒脚 brick-paved plinth
砖砌料仓 masonry bin
砖砌楼梯间 brick stair(case)
砖砌炉顶 masonry arch
砖砌炉围 brick fender
砖砌炉子 masonry fireplace
砖砌轮形窗扇 brick wheel window
砖砌镘刀 brick trowel
砖砌密闭墙 masonry bulkhead
砖砌面层 brick facing;brick surfacing;brickwork casing
砖砌面式样 surface pattern
砖砌面样 surface pattern
砖砌内衬 lining
砖砌平拱 Dutch arch
砖砌墙板 wallboard masonry
砖砌墙体 brick walling
砖砌穹顶 brick vault
砖砌渠道 brick conduit
砖砌人行道 brick sidewalk
砖砌山墙 brick gable;brickwork gable
砖砌盛钢桶 bricked ladle
砖砌实心拱 brick wall arch
砖砌水库 brick reservoir
砖砌体 brick built;brick masonry;brick setting;brickwork bond;brickwork casing;holdfast;pack of bricks;arches <在窑口或拱脚间的>
砖砌体齿形待接插口 toothing of brick wall
砖砌体齿形接口 toothing of brick wall
砖砌体的一种砌合法 blind bond
砖砌体钢筋垫网 brick reinforcement fabric;brick reinforcement mat
砖砌体裂缝 cracks in brickwork
砖砌体留齿插口 toothing of brick wall
砖砌体箍制接头 clip joint
砖砌体嵌凸缝 tuck pointing
砖砌体伸缩缝 brickwork movement joint
砖砌体因硫酸盐而引起的膨胀 sulfate expansion of brickwork
砖砌体中的硫酸盐膨胀 sulphate expansion of brickwork
砖砌挑檐 brick cornice
砖砌筒形穹顶 brick tunnel vault

Z

砖砌图样 surface pattern
砖砌圬工 brick masonry
砖砌污水沟 brick sewer
砖砌污水管 brick sewer
砖砌下水道 brick conduit-type sewer; brick sewer
砖砌线脚 brick mo(u)ld
砖砌镶面层 brick veneer
砖砌旋梯的中柱 closed newel
砖砌烟囱 masonry stack; tile chimney; tile stack
砖砌烟道 brick flue; brickwork chimney; masonry flue
砖砌檐墙 parapet masonry(wall)
砖砌窑洞屋盖 arched brick roof covering
砖砌窑门 brick wicket
砖砌阴沟 brick sewer; brick underdrain
砖砌用砂浆 mortar for(clay) brickwork
砖砌浴室 tile bathroom
砖砌圆顶 brick vault
砖砌圆拱 rowlock arch
砖砌缘石 brick curb
砖砌住宅楼 domestic clay brick block; domestic clay brick building
砖砌柱 brickwork column
砖砌柱身 brick shaft
砖砌筑砂浆 brickwork mortar
砖强度 brick strength
砖墙 brick wall; tile wall
砖墙变形缝 brickwork movement joint
砖墙承重 brick wall load bearing
砖墙承重结构 brick wall bearing construction
砖墙到顶 brickwork to top out
砖墙丁头砌合 inbond
砖墙发券 brick wall arch
砖墙风化 bloom
砖墙刮缝 struck joint
砖墙横缝 longitudinal joint
砖墙划粗纹 stabbing
砖墙留茬 toothing of brick wall
砖墙面划粗纹 stabbing
砖墙内的辅助拱 arriere-voussure
砖墙裙 brick dado wall
砖墙伸缩缝 brick wall expansion joint
砖墙渗漏 penetration through brickwork
砖墙收分 scarcement
砖墙外角砌合 angle bond
砖墙外皮 <外层墙> external skin
砖墙外挑压顶 tile creasing
砖桥 brick bridge
砖穹隆 brick dome
砖渠 brick canal
砖塞孔 bricking-up
砖散水 brick apron
砖石边墙门槽 masonry wall recess
砖石薄层 masonry work leaf
砖石材料 masonry material
砖石层 masonry work course; masonry work layer
砖石衬砌 masonry(work) lining
砖石衬砌隧道 masonry-lined tunnel
砖石承重墙 masonry bearing wall
砖石承重墙的最小厚度 minimum thickness of masonry bearing walls
砖石城堡主塔 masonry donjon; masonry work donjon
砖石地牢 masonry work dungeon
砖石雕刻 carved work
砖石钉 masonry nail
砖石墩 masonry pier
砖石房屋 masonry building
砖石干砌体 dry masonry
砖石隔板 masonry work diaphragm
砖石隔墙 masonry bulkhead
砖石工 brick mason; mason; mason-work; bricklayer
砖石工承包商 masonry contractor
砖石工程 brick-and-stone work; brick construction; brick engineering; mason construction; masonry construction; masonry work
砖石工程表层 masonry work skin; masonry work surfacing
砖石工程处理 masonry work treatment
砖石工程的清洁处理 masonry work cleaning
砖石工程防潮密封剂 masonry work moisture sealing agent
砖石工程风格 masonry work style
砖石工程管理 masonry work keep
砖石工程加腋 masonry work haunching
砖石工程结构 masonry work structure
砖石工程开洞 masonry work opening
砖石工程码头 masonry work pier
砖石工程墓穴 masonry work tomb
砖石工程配筋 masonry work reinforcement
砖石工程屏幕 masonry work screen
砖石工程破坏性试验 masonry work failure test
砖石工程强度 masonry work strength
砖石工程桥墩 masonry work pier
砖石工程筒仓 masonry work silo
砖石工程维护 masonry work keep
砖石工程用花岗岩 masonry granite
砖石工程中的嵌入件 build-in items in masonry
砖石工程住宅塔楼 masonry work residence tower
砖石工锤 mason's hammer
砖石工砌墙用U形木块 preacher
砖石工用砂浆桶 bricklayer's hod; mason's hod
砖石(工)用熟石灰 mason's hydrated lime
砖石工用水准尺 mason's level
砖石工准线 mason's line
砖石公寓塔楼 masonry apartment tower
砖石拱 masonry arch
砖石拱桥 masonry archy bridge
砖石和钢筋混凝土组合砌体 <在垂直墙中的> quetta bond
砖石和混凝土组合砌体 quetta bond
砖石或混凝土衬砌 <井或污水池周壁的> steening
砖石基础 masonry foundation
砖石加筋砂浆平卧砌合 quetta bond
砖石建筑 architectural masonry (work); masonry; masonry(work) building
砖石建筑的防水防潮处理 waterproofing and damp-proofing of masonry(work)
砖石建筑面漆 masonry finish
砖石建筑饰面 masonry veneer
砖石建筑用漆 masonry paint
砖石接缝 masonry work joint
砖石接头 masonry joint
砖石结构 brick construction; brick masonry structure; brick-stone construction; brick-stone masonry; brick structure; mason construction; masonry construction; masonry structure; stone structure
砖石结构房屋 masonry building
砖石结构类型 masonry work construction type; masonry work type of construction
砖石居住塔楼 masonry work dwelling tower
砖石锯 masonry saw
砖石抗水处理 hydrophobic treatment of masonry(work)
砖石类型 masonry work type
砖石里壁 <加固墙壁的> backup; masonry back-up
砖石梁 masonry beam
砖石啮合小过梁 joggled lintel
砖石平接灰缝 flush masonry jointing
砖石铺路 mason flagger
砖石砌拱 voussoir arch
砖石砌合空心墙 masonry bonded hollow wall
砖石砌墙 opus listatum
砖石砌体干砌 ground joint
砖石砌体基础 brick mass foundation
砖石砌体角接缝 mason's miter[mitre]
砖石砌体配筋 masonry reinforcing
砖石砌体砌合 bonding
砖石砌体水泥 masonry cement
砖石砌体增强材料 masonry reinforcing
砖石砌筑的实墙 dead masonry wall
砖石砌筑防水处理 water-repellent treatment of masonry
砖石砌筑灰浆 masonry mortar
砖石墙 chippings from masonry ruins; masonry wall
砖石墙丁头砌合 bonding
砖石墙顶拱 masonry wall crown
砖石墙勾缝刀 wall channeler
砖石墙金属泛水 flexible metal masonry wall-flashing(piece)
砖石墙金属片材泛水 sheet-metal masonry wall flashing
砖石墙裂开 masonry wall dissolution
砖石墙面清洗剂 cleanser for masonry
砖石墙入口 entrance masonry wall
砖石墙上金属凸圆线脚 masonry work metal bead
砖石墙体临时夹具 masonry wall temporary clamp
砖石墙挑出块 tailing
砖(石)墙凸出块【建】tailing
砖石墙压顶板 masonry wall coping slab
砖石桥 masonry bridge
砖石桥梁 masonry work bridge
砖石穹隆 masonry dome
砖石山墙 gable masonry wall
砖石饰面 masonry work facing
砖石术 masonry
砖石水塔 masonry work water tower
砖石碎片 Irish confetti
砖石榫接 keying-in
砖石填充框架 masonry infilled frame
砖石填充墙 masonry filler wall
砖石填充墙板结构 masonry in-fill panel construction
砖石挑出层 overtailing course
砖石挑出块 tailings
砖石通道 masonry gallery
砖石凸圆线脚 masonry work bead
砖石圬工 brick masonry; plain masonry
砖石污水沟渠 masonry sewer
砖石行会 mason's guild
砖石蓄水池 masonry reservoir
砖石烟囱 masonry work chimney
砖石用涂料 masonry paint
砖石圆顶 masonry work cupola
砖石支护 line with bricks
砖石支护井筒 masonry shaft
砖石重力坝 masonry gravity dam
砖石住宅楼 masonry dwelling tower
砖石柱 masonry column
砖石筑墙 opus listatum
砖石装饰横线脚 ledg(e)ment
砖石装饰性砌法 ornamentation
砖石装饰腰线 ledg(e)ment
砖石走廊 masonry gallery
砖式卷材 <沥青浸涂毡材凸出似砖状> brick roll
砖试验机 brick testing machine
砖饰面 brick facing; brick veneer
砖饰面的 brick faced; brick set
砖饰面结构 brick facing construction
砖饰墙隅 brick quoin
砖碎片 brick chip(ping)s
砖塔 brick pagoda
砖踏步 brick step
砖台阶 brick step
砖提升机 brick lift
砖体 tile body
砖体结构 brick texture
砖体系建筑 brick system building
砖填木构架墙 brick-nogged timber wall
砖填木架房屋 brick nogging building
砖填木架隔墙 brick-nogged; brick nog(ging)
砖挑檐 brick cornice
砖贴面 brick facing; brick veneer
砖贴面墙 brick-veneered wall
砖筒拱结构 brick barrel-vault construction
砖头 bat; brick bat
砖头组装 brick package
砖土 brickclay; brick earth
砖托 brick holder; brick supporter
砖托梁 brick trimmer
砖托座 <砖层挑出圬工> corbelling of bricks
砖瓦 tile arch
砖瓦厂 brick yard; tile plant; tilework
砖瓦尺寸 tile format
砖瓦顶棚 tile ceiling
砖瓦堆栈 tile store
砖瓦垛 tile stack
砖瓦工 masonry; mud scraper <俚语>
砖瓦工程 bricking
砖瓦工的 masonic
砖瓦工业 brick and tile industry; clay industry; heavy clay industry
砖瓦固定材料 tile fixer
砖瓦货仓 tile store
砖瓦密封混合物 tile sealing compound setting
砖瓦密封料 tile sealant
砖瓦密封组合物 tile sealing composition
砖瓦黏[粘]合剂 tile cementing agent
砖瓦黏[粘]土矿石 brick clay ore
砖瓦坯 green tile
砖瓦坯干燥槽架 hake
砖瓦坯压制机 tile and slab press
砖瓦铺设 tile fixing
砖瓦砌筑 tile setting
砖瓦砌筑黏[粘]结剂 tile setting adhesive
砖瓦强度 tile strength
砖瓦切割刀具 tile cutter
砖瓦切割机 clipping machine for brick and tile
砖瓦清边 back edging for brick and tile
砖瓦清边法 back edging
砖瓦上釉 tile glazing
砖瓦石工用的金属刮板 <俚语> cockscomb
砖瓦圬工 tile masonry
砖瓦窑 heavy clay kiln
砖瓦用黏[粘]土 clay for brick-tile
砖瓦用砂岩 sandstone for brick-tile
砖瓦用页岩 shale for brick-tile
砖瓦用釉料 majolica glaze
砖瓦制品 heavy clay product
砖瓦装配 tile fitting
砖瓦装饰术 tilery
砖外壳 <空心墙> leaf of blocks
砖围墙 brick enclosing wall; brick enclosure; brick fencing wall; brick

wall fence
砖圬工 brick masonry
砖圬工墙 tile masonry wall
砖线脚 brick mo(u)ld(ing)
砖镶房屋 brick clad building
砖镶面 brick veneer; veneered with brick
砖镶面的 brick-veneered
砖镶面构造 brick-veneered construction
砖镶面墙 brick-veneered wall
砖楔块砌合 block-in-course bond
砖屑 brick dust; brick flour; ground clay brick
砖心 <不承重的> brick core
砖形 brick shape
砖形的 brick shaped
砖形视口 tiled viewpoint
砖形物 brick
砖型 brick type
砖压顶 brick coping
砖压顶女儿墙 brick-cap parapet
砖烟囱 brick chimney; masonry chimney
砖烟道 brick flue
砖窑 brick furnace; brick kiln
砖窑隔火墙 bag wall
砖窑炉壁上层的砖 arches
砖油 brick oil
砖釉 brick glaze
砖圆层顶 brick cathedral
砖缘石 brick curb
砖凿 brick set
砖正面 brick facade
砖支座 brick seat
砖柱 brick pier; brick(work) column; masonry column; pilla
砖筑水泥 brick cement
砖铸 loam casting
砖座 brick base; brickwork base

转氨酶【生】 transaminase

转氨作用 transamination
转白试验 blanching reaction
转(摆)动轴线 pivot center[centre]
转板 flap
转板式震压造型机 turnover jolt squeeze mo(u)lding machine
转包 assign a contract; assignment of contract; subcontracting; sublet(ting)
转包办法 subcontracting arrangement
转包给第三者 subcontract
转包工作 subcontract
转包合同 subcontract
转包合同更改通知 subcontract change notice
转包合同项目 subcontract item
转包人 subcontractor
转报 transit telegram
转报底 transit form
转报台 relay station; repeater station; repeating station
转杯风速表 cup(-cross) anemometer
转杯流速仪 cup current meter
转杯喷嘴 rotary cup burner
转杯式风速仪 Robinson's anemometer
转杯式燃烧器 rotary cup burner
转杯式雾化器 rotary cup atomizer
转笔刀 pencil sharpener
转壁炉 rotary wall furnace
转壁式双向犁桦 reversible share
转壁双向犁 kentish plow; turnwrest plough
转臂 derrick jib; jib arm; pivoted arm; rotating arm; swivel(l)ing jib; tumbler
转臂操纵油缸 jib cylinder
转臂传动装置 luffing unit
转臂吊车 derrick car
转臂吊机 derrick crane
转臂分配器 rotating-arm distributor
转臂高架起重机 derrick tower gantry
转臂回转角度 boom slew
转臂起重车 derrick car
转臂起重机 derrick; derrick(ing) crane; swivel(l)ing crane; whip line
转臂起重机柱 derrick post
转臂式吊车 derrick crane
转臂式垛草机 swing-around(hay) stack; swinging boom stack
转臂式干草堆垛机 derrick hay stack
转臂式混砂机 paddle-type mixer
转臂式螺旋卸载机 sweep arm auger unloader
转臂式起重机 derrick(ing)(jib) crane; turnstile crane; jib crane
转臂式起重桅杆 crane post
转臂式拖拉机装载器 jib-type tractor loader
转臂式挖沟机 jib-type ditcher
转臂式液力装载机 hydraulic jib-type loader
转臂式液压装载机 hydraulic jib-type loader
转臂式装载机 boom loader; crane-type loader; jib-type loader; rotating loader; swing loader
转臂式钻机 jib-mounted drill
转臂收割机 dropper
转臂塔式起重机 luffing tower crane
转臂塔式起重机架 derrick tower gantry
转臂弹簧 tumbler spring
转臂型超重机柱 derrick post
转臂液压缸 jib cylinder
转臂油气缸 boom cylinder
转臂支柱 slew post
转臂转动油缸 swing cylinder
转变 change(-over); inversion; transform; transition; transmutation
转变比 transition ratio
转变层 transition(al) layer
转变产物 transmutation product
转变长度 transition length
转变次序 transition order
转变带 transition zone
转变点 arrest point; inversion point; point of contraflexure; point of inflection; point of inflexion; point of transition; transformation point; transition point; transit point; transship point
转变点电池 transition cell
转变点电位 transition potential
转变段 transition; conversion
转变反应 transformation reaction
转变范围 transformation range; transition range
转变方向 face-around
转变过程 transition process
转变加速素 convertin
转变间隔 transition interval
转变交通 shifted traffic
转变阶段 transition phase
转变链 transformation chain
转变孪晶 transformation twin
转变能 transition energy
转变期 period of transformation; transformation period; turn
转变区 transition(al) zone; zone of transition
转变曲线 transition curve
转变热 heat of transformation; transition heat
转变热焓 enthalpy of transition
转变熵 entropy of transition
转变时期 period of transition
转变双晶 transformation twin
转变宿主 transfaunation
转变损失 transition loss
转变态度 face-around
转变态理论 transition state theory
转变位置 transition position
转变温度 inversion temperature; transformation temperature; transition temperature
转变温度范围 transformation temperature range
转变温度区 transformation temperature range
转变性 convertibility
转变因素 transforming principle
转变值 switching value
转变中的湖泊 transition lake
转变中的农业 agriculture in transition
转变状态 transition stage
转变阻力 steering resistance
转柄 stem
转柄螺栓 pivoted bolt
转柄装置 rotating crank gear
转柄钻(头) center[centre] bit
转拨付款 transfer payment
转拨款项 transfer of appropriation; transfer payment
转拨清单 schedule of transfer
转播 rebroadcast; rediffusion; relay broadcast(ing); relaying; repeating; retransmission; retransmit
转播车 mobile control room
转播点 interception point
转播电台 retransmitting station
转播发射机 rebroadcasting transmitter; retransmitter
转播接收机 rebroadcasting receiver; retransmission receiver
转播台 rebroadcasting station; relay(ing) station; retransmitting station; translator station
转播信号通道 relay channel
转播用接收机 ball receiver; direct pick-up receiver; relay receiver
转播用收音机 radio set for rebroadcasting
转播噪声 translation noise
转播站 interception point; rediffusion station; relay base; relay point; relay station; repeating station; translator station
转播中心 repeating center[centre]
转播装置 repeating mechanism; retransmission unit
转驳运输 overside delivery
转材车 butt-carriage
转槽 turn trough
转叉式装卸车 articulated fork lift truck
转插板 jack panel; patchboard; plugboard
转插站 repeater station
转差 slippage
转差测定器 slip meter
转差计 slip counter; slip meter
转差继电器 slip relay
转差率 revolutional slip; slip; slip ratio
转差率调节器 slip regulator
转差能量 slip energy
转差频率 slip frequency
转差曲线 slip curve
转差损耗 slip loss
转产 change the line of production
转产契 deed release
转厂试验 transplant test
转场 transfer to another track or yard
转场备用机场 redeployment airfield
转场感应电动机 revoluting field induction motor
转场进路 yard-to-yard route
转场式磁电机 rotating pole type magneto
转场式电动机 revolving field type motor
转场式电机 revolving field type machine; rotating-field machine
转场式发电机 rotating-field generator
转场式励磁机 rotating-field exciter
转潮 change of tide; turning of tide
转潮点 amphidromos
转潮流 turn of tidal current; turn tidal stream
转车板 turn sheet
转车车票 transfer
转车地点 change point
转车机 jacking engine; turning engine; turning gear
转车盘 railway turntable; turning circle; turning jack; turntable
转车盘梁 turntable girder
转车盘入库线 turntable track
转车盘锁闭器握柄 turntable lock lever
转车盘组 series of turntables
转车台 carriage turn table; car transfer table; curb ring; engine turntable; rolling circle; slewing gear ring; sliding bridge; swinging circle; transfer table; travel(l)ing platform; traverser; traverse table; turning plate; turnplate; turntable; articulated turntable【铁】
转车台底盘 turntable undercarriage
转车台发动机 turntable engine
转车台坑 traverse pit
转撤机外壳 casing of switch machine
转成面 surface of revolution
转成体 revolving solid
转成责任 vicarious responsibility
转承包(合同)成本 subcontracting cost
转承的 vicarious
转承责任 vicarious responsibility
转出 exit turn; roll-out; turn-off
转出工具 turning out device
转出列车(让另一列车越行) turn off a train
转出率 switching-out rate
转出投资的审计 transfer investment audit
转出主存储器 roll-out
转储 memory transfer; unload
转储程序 dump program(me); dump routine
转储点 dump point
转储方法 dump method
转储方式 dump mode
转储检验 dump check
转储校验 dump check
转船 transship; trans(s)hipment
转船池 turning basin; turning circle
转船费 transshipment charges
转船附加费 transshipment surcharge
转船货 transship cargo; transshipment cargo
转船货报关单 transshipment entry
转船货装船单 transshipment shipping bill
转船交货单 transshipment delivery order
转船提单 transshipment bill of lading
转船通知 transshipment note
转船许可证 transshipment permit
转船运输 trans(s)hipment
转船栈单 transshipment delivery order
转窗 pivoted window; revolution window; revolving window; swivel sash

转床刀具 touple
转锤破碎机 swing hammer crusher
转磁 magnetize
转次页 balance carried down;carried forward
转刺(点)【测】 point transfer; prick transfer;transfer point
转脆温度 transition temperature
转存冷库 roll-in freezer; roll-in refrigerator
转存区 transient area
转达 convey(ing);transmit
转带式试验台 track treadmill
转袋装置 bag turner
转刀 revolving cutter
转刀式盾构 cutter shield
转导 transduction
转导因子 transduced element
转导子 transductant
转倒舷内 swing in
转倒舷外 swing out
转到侧线 side-track(ing)
转到它线 <将机车、车辆或列车> switch over to another track
转道沟 transfer pit
转底光面机 bottom facing machine
转底(环状隧道)窑 rotary hearth kiln
转底炉 rotary hearth furnace
转底式连续(加热)炉 rotary hearth continuous furnace
转抵押 remortgage;re-pledge
转点【测】 transfer point; turning point;change point;point transfer; turning station
转点(测流)定位 pivot-point layout
转点尺 <画路线转点用> point rule
转点仪 point transfer device;transcriber;transfer device
转点桩 <水准测量用> turning point pin
转点装置 point transfer device
转电线圈 conversion transformer; line repeating coil;repeating coil
转电线圈架 repeating coil rack
转电线圈桥接软线 repeating coil bridge cord
转吊机 slewing crane
转订合同 subcontract
转动 angular motion; rotary movement; rotate; rotation; rotational movement;turn
转动安装角 roll setting angle
转动把手 cam handle
转动板 rotor plate
转动半径 radius of gyration
转动棒条筛 moving bar grizzly
转动臂 cursor; rotor arm; steering arm;swing jib
转动变换 rotational transform
转动变形 angular deformation; rotational deformation
转动表 stem
转动柄 turning handle
转动铂电极 rotating platinum electrode
转动薄膜式萃取器 rotary film contactor
转动不稳定性 rotational instability
转动不正常 run untrue
转动布朗运动 rotary Brownian motion;rotatory Brownian motion
转动部件 rotatable parts;rotating parts
转动餐盘 lazy Susan
转动差速 non-slip differential
转动常数 rotational constant
转动车钩 swivel draw bar
转动车架 pivoted bogie
转动承架 rolling segment
转动承梁 rolling segment
转动承座 rolling segment
转动齿轮 arm revolving gear
转动传感器 rotation sensor
转动锤击式振筛器 rotap shaker
转动带 rotating band
转动挡板 flap shutter
转动的 rotary;rotating;rotational;rotatory
转动的角振幅 amplitude of angular rotation
转动的楣窗 pivoting transom(e)
转动的起重工作 hoist works for rotary
转动灯标 revolving light
转动等离子体装置 rotating plasma device
转动底扳 pivot floor
转动底板臂 arm of pivoting floor
转动底板倾卸油缸 dump cylinder of floor
转动地震仪 rotation seismograph
转动点 run-on point
转动电极 rotating electrode
转动吊桥 rotating draw
转动定位器 positioner;swing stop
转动动能 rotational kinetic energy
转动动作 rotary motion
转动度盘的手柄 rotating dial knob
转动断层 pivotal fault;rotatory fault
转动惰性 rotational inertia
转动颚式破碎机 roll-jaw crusher
转动发射装置 traverser
转动阀(门) rotating valve; rotary valve; spheric (al) valve; straight-flow rotary valve
转动法 method of rotations; rotation method
转动范围 slewing area
转动方向 direction of rotation; rotation direction;sense of rotation
转动分量 component of rotation; rotational component; rotative component
转动分速度 rotational velocity component
转动浮子式直读流量计 rotor meter
转动辐 moving spider
转动负荷 rotating load
转动干燥器 rotary dryer[drier]
转动杆 dwang
转动刚度 rotational stiffness; true stiffener
转动刚度系数 rotational stiffness factor
转动刚性系数 rotational stiffness factor
转动隔仓 rotating cell
转动给料刮板 rotatable feeding shoe
转动工具 turning tool
转动功率 rotative power
转动刮板 rotating blade
转动关节 cradle head
转动管子 chain dog
转动惯矩 rotary moment of inertia
转动惯量 moment of inertia; moment of gyration; processional moment; rotary inertia;rotational inertia;rotatory inertia
转动惯量系数 coefficient of moment of inertia; rotational inertia coefficient
转动惯性力矩 rotary moment of inertia
转动光阑 rotating stop
转动光谱 rotation spectrum
转动光谱学 rotational spectroscopy
转动光闸 rotating shutter
转动辊 live roller
转动核 rotational nucleus
转动横进刀装置 rotating crossfeed unit
转动滑阀 rotary spool;turning slide valve
转动滑块 turning block
转动滑移 rotational slip
转动滑座 rotational slide
转动环节 rocking link
转动换位 rolling transposition
转动换位器 positioner
转动机构 rotation gear
转动机观察仪 rotascope
转动机械阻抗 rotation mechanical impedance
转动激发 rotational excitation
转动架 turret
转动间隙 <轴和轴承间的> running clearance
转动拣选台 revolving sorting table
转动件 revolving member
转动件旋转轴 pivot center[centre]
转动件轴承 round bearing
转动桨(叶) <水轮机的> movable blade
转动角 angle of rotation; striking angle;tilt angle
转动角位移 rotational displacement
转动搅拌 rotary mixing
转动搅拌机 rotary mixer; rotating mixer
转动搅拌器 rotating mixer
转动搅拌筒 churn
转动接触器 rotating contactor
转动接头 rotary joint
转动节流阀 rotary throttle valve
转动结构 rotational structure
转动劲度比 rotational stiffness ratio
转动劲率 rotational stiffness factor
转动锯 rotary saw
转动卡块 <扭卸钻头用> breakout block
转动开关 rotating switch; rotative switch
转动块 turning block
转动框形天线 rotating frame antenna
转动扩散 rotary diffusion; rotatory diffusion
转动扩散常数 rotation diffusion constant
转动犁 removing plow
转动犁式给矿机 rotary plow feeder
转动力 rotary force; rotatory force; torsional force;turning effort
转动力矩 driving moment; driving torque;moment of gyration;moment of rotation; pivoting moment; rotary moment; rotative moment; turning moment;rotation moment
转动连接 rotary joint
转动链 swivel chain
转动梁 cant beam
转动量子数 rotational quantum number
转动流变仪 rotary rheometer
转动流槽 rotary launder
转动硫化 tumble curing
转动炉算 travel(l)ing grate
转动路帚 rotary broom
转动率 rotation rate
转动轮 running wheel
转动螺旋分级机 revolving spiral classifier
转动落球黏[粘]度计 rolling sphere instrument
转动门 rotary door;pivot gate
转动门钮 thumb turn
转动门闩 lift latch
转动密封 rotary seal
转动面 plane of rotation
转动面积 slewing area
转动模量 moment of inertia
转动磨石 grinding runner
转动能级 rotational energy level
转动能力 rotation capacity; turning power
转动能量 energy of rotation;rotational energy
转动能谱 rotational spectrum
转动能障 rotational energy barrier
转动黏[粘]度计记录器 rotational visco(si)meter recorder
转动黏[粘](滞)度计 rotational visco(si)meter
转动扭矩 turning torque
转动配合 clearance fit; loose fit; normal running fit; working fit; running fit
转动配水器 rotary distributor
转动喷水式洗涤器 rotor-spray washer
转动喷嘴 rotating nozzle
转动皮带 continuous belt
转动皮带轮 rotating pulley
转动偏振光 rotatory polarized light
转动漂动 plain drifting
转动频率 rotation(al) frequency
转动平衡 rotary balance; rotational balance;rotational equilibrium
转动平稳 run true
转动平移装置 rotational and translator device
转动谱 rotation spectrum
转动谱带 rotational band
转动起重臂 swivel(l)ing jib
转动起重机 slewing crane
转动器 cursor;rotator
转动桥式刮泥器 rotating-bridge scraper
转动轻敲式机械摇动器 rotap mechanical shaker
转动求和定则 rotational sum rule
转动曲柄 rotating crank
转动圈 rotor coil
转动筛 rotary screen
转动筛分机 rotex-screen
转动设备 rotating equipment
转动设备数据 rotating equipment data
转动式储藏容器 rotating storage container
转动式吹灰器 rotary soot blower
转动式磁铁 rotating magnet
转动式吊艇杆 radial davit; swivel davit
转动式吊艇柱 radial davit
转动式多喷嘴涂装装置 rotary multiple-spray nozzle coating head
转动式风挡 swivel damper
转动式风门 revolving damper;swivel damper
转动式工作平台 swivel platform
转动式谷壳撒布器 rotating chaff distributor
转动式硅整流器 rotating silicon rectifier
转动式滑坡 rotational landslide
转动式挤奶厅 rotary stall
转动式拦鱼栅网 rotating fish screen
转动式联轴器 swivel coupling
转动式滤网 revolving screen
转动式门 turnable gate
转动式喷管 jetevator
转动式气泵 rotary air pump
转动式桥堰 swing bridge weir
转动式燃料溜槽 fuel swivelling chute
转动式输送机 fan conveyer[conveyor]
转动式淘洗盘 rotary washing pan
转动式停车标志 rotary stop sign;rotating stop sign
转动式筒子架 turn creel
转动式烟道调节板 swivel damper
转动式闸门 pivoted gate; pivot-leaf gate;revoluting gate;turnable gate

转动式张力装置 rotating tensioning device
转动式遮光窗 pivoted shutter
转动式装土斗 tipping bucket
转动式座椅 <汽车中能放倒的> tilting seat
转动手柄 turning handle
转动双曲柄 rotating double crank
转动送进盘管 feed-through spool
转动送进筒 feed-through spool
转动速度 rate of turn;rotational speed;rotational velocity;speed of turn
转动速率 rate of angular motion;rotation speed
转动速率控制 turning rate control
转动算符 rotation operator
转动损失 rotational loss
转动索 slew line
转动台 rotary table;turntable
转动台式压砖机 rotating table press
转动态 rotational state
转动套 rotary sleeve;rotating sleeve;rotation chuck;rotation sleeve;spline drive socket <凿岩机的>
转动套筒 chuck sleeve
转动特性 rotational characteristic
转动体 rotator;rotor;tumbling mass
转动体轴 rotor shaft
转动天线的电动机 aerial turning motor
转动同轴圆筒测黏[粘]法 rotational coaxial-cylinder viscometry
转动筒牵引 cord-and-drum drive
转动头 digging head
转动凸缘 rotating flange
转动图样 rotation pattern
转动推力环 rotating thrust collar
转动挖斗 rotating excavating bucket
转动挖掘铲斗 rotating excavating bucket
转动弯管 swing bend pipe
转动位置 turned position
转动稳定 roll stabilization
转动稳定性 roll stability;rotational stability
转动稳定作用 slow-roll stabilization
转动无序 rotational disorder
转动系船柱 fairlead;fair leader
转动系数 rotation coefficient
转动系统 driving system
转动衔铁继电器 rotating-armature relay
转动线 rotational line
转动线圈 rotary coil;rotating coil
转动限位螺钉 rotation set screw
转动效率 rotational efficiency
转动效应 rotational effect;turning effect
转动型锻机 rotary swaging machine
转动型护舷 rotating fender
转动型橡胶护舷 roller-type rubber fender
转动修整 swirl finish
转动延性 rotational ductility
转动样品磁强计 rotating sample magnetometer
转动叶轮 rotor impeller
转动叶轮定量给料机 rotary feeder
转动叶片 movable blade;moving blade;rotating blade;rotor blade;turning vane
转动叶片泵 rotary-vane pump
转动叶片级 moving blade stage
转动应变 rotational strain
转动元件 rotating element
转动钥匙 turnkey
转动跃迁 rotational transition
转动运动 rotary motion;rotational motion
转动运动密封件 rotating seal
转动闸门 rotating shutter
转动张弛 rotational relaxation
转动张量 rotation tensor
转动振动谱 rotation-vibration spectrum
转动振动谱带 rotation-vibration band
转动振动态 rotation-vibration state
转动质量 rotary mass
转动中心 center[centre] of rotation;rotary center[centre];rotational center[centre]
转动重力不稳定性 rotational gravitational instability
转动周期 period of rotation
转动轴 axis of rotation;live shaft;movable shaft;rotating shaft;rotation(al) axis;train axis;swing axle
转动轴承密封 rotating bearing seal
转动轴的环形密封止水 clearance seal
转动轴颈轴承 rotating journal bearing
转动轴线 rotation axis
转动轴箱 rotation chuck
转动铸条法 rotary strip casting process
转动装置 rotary actuator;rotator;slewing gear;wheelwork <机器中的>
转动自由度 rotational degree of freedom
转动阻抗 rotational impedance
转动阻力 rotary resistance;rotational resistance
转动阻尼 rotary damping
转动坐标 rotational coordinates
转动坐标系 rotating coordinate system
转斗 revolving bucket;rotating bucket
转斗定位器 bucket positioner
转斗杠杆 tilt lever
转斗力 bucket curling force
转斗连杆【机】tilt link
转斗器 rotator
转斗式输送机 pivoted bucket conveyer[conveyor]
转斗式提升机 rotary bucket elevator
转斗式运输机 pivoted bucket carrier
转斗液压缸【机】tilt cylinder
转度虎钳 <牛头刨用> index center [centre]
转垛 relocation of storage pile
转舵吃力 taut helm
转舵矩 rudder torque
转舵力矩 rudder moment
转舵链 rudder chain
转舵索 steering rope;steering wire;tiller line;tiller rope
转舵索导环 tiller rope guide
转舵装置 helm gear;helm rudder gear;steering device
转发 repeating;retransmission;retransmit;transpond;transponding
转发电流 repeat current
转发电路 translation connection
转发发射机 retransmitter
转发机 repeater transmitter
转发继电器 repeater relay;repeating relay;translating relay
转发连接 translation connection
转发器 follower;interpolator;repeater;sink;translator;transponder
转发台 booster
转发音响器 repeating sound
转发站 repeater station;retransmitting station
转发指示器 repeater-indicator
转阀 plug cock;rotary plug valve;turning valve
转阀式风力跳汰机 plumb pneumatic jig
转阀芯 rotary plate
转帆索 counter brace
转方向 veer
转/分 revolutions per minute;rotations per minute
转分保 retrocede
转分保分出公司 retroceding company
转分保合同 retrocession treaty
转分保接受人 retrocessionaire
转分配 suballotment
转分手续费 overrider
转风点 amphidromos
转干式闸机 turnstile gate
转杆 bull stick
转杆器 cant hook;cantihook
转杆式冲击疲劳试验 rotating bar impact fatigue test
转杆式黏[粘]度计 rotating rod viscometer
转杆式疲劳试验机 rotating beam type machine
转杆疏通 rod turning clean
转感器 transducer
转缸发动机 rotary combustion engine
转缸式发动机 revolving cylinder engine;rotary engine
转割机 rotary cutter
转弓 swivel bow
转钩 swivel hook
转购人 subpurchaser
转鼓 basket;bowl;revolving drum;rotary drum;rotating drum;rotor drum;wobbling drum
转鼓齿轮 drum gear
转鼓传热 rotating drum heat transfer
转鼓唇缘 basket lip
转鼓打磨机 cylinder sander;drum sander
转鼓底 basket bottom
转鼓吊车 drum hoist
转鼓毂 basket hub
转鼓滚涂法 barreling
转鼓过滤机 rotary drum filter
转鼓加料器 rotary drum feeder
转鼓加油 oil wheeling
转鼓搅拌机 rotary drum mixer
转鼓浸没度 drum submergence
转鼓抛光 drum polishing;tumble polishing
转鼓抛光机 drum polisher
转鼓撇油器 drum skimmer
转鼓清洗螺塞 plug for rotor cleaning
转鼓裙 basket skirt
转鼓鞣革 drum tannage
转鼓式拌和机 drum mixer;revolving drum mixer;rotary (drum) mixer;rotary(speed) mixer
转鼓式测功机 chassis dynamometer
转鼓式掺和机 tumble blender
转鼓式肥料斗 revolving drum hopper
转鼓式分离器 drum separator
转鼓式焚化炉 rotating drum incinerator
转鼓式粉碎机 tumbling crusher
转鼓式干燥器 drum drier[dryer];rotary drum drier[dryer]
转鼓式过滤机 rotary drum filter
转鼓式过滤器 rotary drum filter
转鼓式混合器 tumbling mixer
转鼓式继电器 rotating drum relay
转鼓式搅拌机 rotary speed mixer
转鼓式连续加压过滤机 rotary drum type continuous pressure filter
转鼓式连续加压过滤器 rotary drum type continuous pressure filter
转鼓式筛网 rotary drum screen
转鼓式洗涤 drum washing
转鼓式洗涤机 tumbler-type washing machine
转鼓式压机 revolving drum press
转鼓式研磨机 tumbling mill
转鼓式照相 rotating drum camera
转鼓式真空过滤机 drum-type vacuum filter
转鼓试验 tumbler test
转鼓试验台 tread mill
转鼓涂布 drum coating
转鼓涂漆 barrel enamelling;tumble enamelling;tumbling barrel process
转鼓涂漆法 barrel finishing
转鼓涂装 tumbling barrel finishing
转鼓外壳 rotor cover
转鼓制动器 drum brake
转轨【铁】switch
转轨器 rail shunter
转轨枕木 switch sleeper
转辊组 <输送带的> idler set
转滚 tumbler
转滚轴链轮 chain gear for drum shaft
转锅式拌和机 rotary pan mixer;rotating pan mixer
转航【船】fetch
转桁索 brace pendant
转户 transfer of domicile registration
转护喷溅 sloppy
转化 conversion;convert;inversion;invert;transmutation
转化变量 reduced variate
转化产品 conversion product
转化常数 transformation constant
转化处理 conversion process
转化催化剂 reforming catalyst
转化点 inflecture point;inversion point
转化定量法 trans-quantitative method
转化段 conversion zone
转化法 conversion method
转化反应 conversion reaction
转化过程 conversion process
转化阶段 transformation stage
转化介质 transforming agent
转化开裂温度 dunting point
转化炉 converter;reburner;receiver;reformer;reforming furnace
转化炉的结构 structure of reborner
转化炉管 reformer tube
转化炉预热器 reformer desuperheater
转化率 conversion rate;percent conversion;rate of conversion;transformation rate
转化气 reformed gas;reforming gas
转化气废热锅炉 reforming gas boiler
转化器 converter[convertor](controller);reformer
转化曲线 inversion curve
转化热 critical heat;exchanged heat;heat of conversion;heat of transition;transition heat
转化乳剂 invert emulsion
转化设备 conversion apparatus;conversion equipment
转化试验 conversion test
转化数 turnover number
转化速度 conversion rate
转化体 transformant
转化调制器 transmodulator
转化土 vertisol
转化为石质 petrify
转化温度 conversion temperature;inversion temperature;temperature of inversion;transit temperature
转化污水处理后出水 converting sewage effluent
转化物含量 invert content
转化物质 transformation substance
转化系数 transfer coefficient;transformation factor
转化型 convertible type
转化型涂料 conversion coating;con-

vertible coating; convertible paint
转化性 convertibility
转化压力 inversion pressure
转化因素 transformation principle; transforming agent
转化因子 transforming factor
转化周期 transformation period
转化装置 reforming plant; reforming unit
转环 pivot bracket; revolving ring; rotary swivel; slack puller; swivel; thrust bearing runner; runner collar; runner plate <推力轴承的>
转环地面 looping floor
转环钩 swivel hook
转环滑车 swivel block
转环接合 swivel joint
转环节 swivel link
转环卡环 swivel shackle
转环连接端 union swivel end
转环连接杆 swivel couple
转环联结器 swivel coupling
转环联轴节 swivel coupling
转环链段 swivel piece
转环锚 swivel anchor
转环式车钩 swivel coupling
转环式连接器 swivel coupling
转环提引钩 heave hook
转环卸扣 swivel shackle
转换 change (-over); changing (-over); commutation; conversion; convert; cut-over; diversion; exchanging; switch(ing); switchover; throwback; throw over; transfer; transform; transition
转换板块边界【地】 transform plate boundary
转换保险 convertible insurance
转换比率 conversion ratio; transition ratio
转换比值 conversion ratio
转换边界 transform boundary
转换边缘 transform boundary
转换变压器 conversion transformer
转换表 conversion chart; conversion table; transfer table; transition table; translation table
转换波 converted wave
转换波处理 converted wave processing
转换波叠加剖面 stacked section of converted wave
转换部分 conversion fraction
转换材料 transition material
转换舱 changing cabin
转换舱口 <装卸时用> hatch shifting
转换舱口时间 hatch shifting time
转换操纵 control transfer
转换插头 change-over plug
转换常数 conversion constant; transfer constant
转换车道 cross-over lane; crow over lane
转换成本 conversion cost
转换成套件 conversion kit
转换程序 conversion program(me); conversion routine; converter; switching procedure; switching sequence
转换齿轮 change gear
转换抽提 transfer extraction
转换处 transfer station
转换触点 transfer contact
转换触排 switch bank
转换贷款 conversion loan
转换导管 transition pipe
转换到代用材料 conversion to alternative fuels
转换的 non-reversible; transitional; transitionary
转换的拉巴雷公式 <用于计算群桩效率> converse-Labarre equation
转换低速 downshifting
转换抵押 pass-through security
转换地址 reference address
转换点 breaking point; change (-over) point; inversion point; switchover point; transformation point; transition point
转换电话局 transit exchange
转换电路 change-over circuit; switching circuit; translation circuit
转换电压比 transfer voltage ratio
转换电子 conversion electron
转换电子光谱 conversion spectrum
转换电阻器接触器 transition-resistor contactor
转换定位机 transfer posting machine
转换端 end of conversion
转换段 transition section
转换断层【地】 transform fault
转换队形 turn formation
转换阀 change(-over) valve; conversion valve; cross(-over) valve; diverter valve; on-off valve; reversing valve; selector valve; switch valve; transfer valve
转换反射波垂直时距曲线 vertical hodograph of convert reflection wave
转换反应堆 advanced thermal reactor; converter reactor
转换方程 transfer equation; transformation equation
转换方向装置 miter gear
转换分辨力 conversion resolution
转换分销商 distributor switching
转换杆 change-over lever; selector bar
转换杠杆 shifter lever
转换高度 transition altitude; transition level
转换故障 translation exception
转换管 cross-over connection
转换管线 cross-over line
转换光谱 inversion spectrum
转换光束 commutating optic(al) beam
转换轨迹 contrail
转换过程 change-over process; conversion process; switching process
转换过程时间 change-over time
转换函数 transfer function; transformation function
转换函数T拟合 comparison in the resistivity transform function T
转换盒 box car
转换后的方程 modified equation
转换机 interpreter
转换机构 shifter; switching mechanism
转换基面 transformation base
转换级 switching stage
转换挤压 transpression
转换计数器 event counter
转换技术 converter technique
转换继电器 cut-over relay; switching relay; transfer relay
转换键 change-over key; conversion key
转换交通量 conversion traffic volume; converted traffic volume
转换角 arc-to-chord correction; conversion angle; half convergency
转换角度 indexing
转换接触器 reversing contactor
转换接点 change-over contact; transfer contact
转换接头 adapter; sub
转换截面 <将钢筋混合梁截面转换为一种材料的当量截面> transformed section
转换介质 transfer medium
转换进料 switch the feed
转换矩阵【数】 transition matrix; transfer matrix
转换矩阵法 method of transition matrices; transfer matrix method
转换卡片 transfer card; transition card
转换开关 alternation switch; change-over cock; change-over switch; changer; change-switch; changing-over; commutator; diverted switch; diverter switch; gang switch; permutator; reset switch; selector knob; selector switch; switcher; switch key; switch unit; transfer lever; transfer switch; transwitch; tumbler; uniselector
转换开关电键 change-over switch key
转换开关盒 switch box
转换开关活动臂 switch arm
转换开关控制器 change-over switch controller
转换开关凸轮 cathead
转换开关匣 switch box
转换控制 switching control
转换拉伸 transtension
转换拉张盆地 transtensional basin
转换例行程序 conversion routine
转换连接 transition link
转换联轴器 converter coupling
转换流网 transformed flow net
转换滤色镜 conversion filter
转换率 transfer rate; transformation rate; translation rate; turnout rate; turnover rate
转换码 conversion code
转换码(字)符 escape character
转换脉冲 switch impulse; switching pulse
转换门 change-over gate
转换模 modulus of conversion
转换模块 modular converter
转换母线 transfer busbar
转换能 transfer energy
转换盘 change-over panel
转换配电盒 cross board hut
转换配电箱 cross board
转换频率 inversion frequency
转换谱 conversion spectrum
转换器 board; changer; commutator; converter[convertor]; electropeter; interconnect device; switch; switchboard; switch box; switch stand; transducer; translating unit; translator; transverter
转换区 transitory section
转换曲线 transfer curve; transformation curve
转换权利 conversion privilege
转换热 transition heat
转换三通 cross-over tee
转换设备 changement; conversion apparatus; conversion device; conversion equipment; translating equipment
转换失效 translation exception
转换失真 slewing distortion
转换时间 change-over period; conversion time; switching interval; switching time; transfer time
转换时期 change-over period; period of transition
转换实验 transition experiment
转换式温度调节器 change-over thermostat
转换式指示仪表 transfer instrument
转换事件 transition event
转换室 switch room
转换手把 change-over handle
转换数 turnover number
转换速度 translatory speed
转换速率 conversion rate; slewing rate; switching rate
转换损耗 conversion loss
转换损失 conversion loss
转换锁 permutation lock
转换锁闭动作 switch and lock movement
转换锁闭机构 switch and lock mechanism
转换锁闭器 switch and lock mechanism
转换台 transfer table
转换特性 swivel(l)ing characteristic
转换特性曲线 conversion diagram
转换通道 switching channel
转换通路 translated channel
转换透镜 convertible lens
转换位置 dislocation; switching position
转换温度 inversion temperature; turnover temperature
转换误差信号 commutator error signal
转换系数 conversion coefficient; conversion effectiveness; conversion factor; conversion fraction; conversion ratio; reduction coefficient; shift factor; switching coefficient; transfer coefficient
转换系统 change-over system; conversion system; switching system; transition system
转换显示 display change
转换线 switching line; transitional track
转换线圈 reset coil
转换线性 transfer linearity
转换相关系数 alienation coefficient
转换箱 cross box
转换像片 transformed print
转换效率 conversion efficiency; transfer efficiency
转换信号 change-over signal; switching signal; switch signal
转换型喷洒头 change-over spray head
转换性 convertibility; transitivity
转换旋塞 change-over cock
转换压缩 transpression
转换延时 transfer delay
转换仪 transformer
转换仪表分路器 transition meter shunt
转换仪表指示灯 transition indicating meter light
转换仪器 transformation instrument
转换(移)模型 transmodel
转换因数 conversion factor
转换因子 conversion factor; reproduction factor
转换元件 interface element; transition element
转换增益 conversion gain; translation gain
转换站 switching station
转换直线性 transfer linearity
转换指令 conversion instruction
转换滞后 switching hysteresis
转换周期 change-over period; switching cycle
转换柱 transition mast
转换转轮 <电路的> rack wheel
转换装置 change-over arrangement; change-over device; change-over gear; changer; conversion device; conversion equipment; conversion unit; cross-over assembly; function

table; interface element; reversing arrangement; shift transmission; switching device; switch unit; transfer device; transfer equipment
转换装置轴销 steering pitman
转换准则 switching criterion
转换子程序 conversion subroutine
转换作用 switching action
转回 roll back
转回分录 reversing entry
转汇 cross exchange
转绘 rendition
转绘仪 apparatus for transformation by drawing; camera lucida; transferscope
转绘仪器 transcope; transformation apparatus
转击式筛析试验机 rotap
转机 turning point
转机旅客 transferring passenger
转迹线 roulette
转积土 travel(l)ed soil
转极 polarity reversal
转极电码 reversing direction code
转极继电器 pole changer relay; pole-changing relay
转极接点 pole-changing contact
转极开关 pole-changing switch
转极片 pole-changing piece
转极器 commutator; polarity change over mechanism; polarity changer; pole changer; pole reverser
转极时间 pole-changing time
转极式磁电机 rotating magnet magneto; rotating pole type magneto
转极值 pole-changing value; polar working value
转极座 pole-changing base
转记账簿 book of secondary entry
转记装置 carry-over facility
转寄 forward
转架 revolving frame; rotating stand; turner
转架车 bogie car
转嫁 shifting
转嫁于 impute
转嫁责任 vicarious responsibility
转筒式清洗机 drum-type washer
转件枢轴 swivel sheaves pivot shaft
转键离合器 rotating key clutch
转键前面板 turnkey front panel
转键系统 turnkey system
转桨 washing gates
转桨机构 blade adjusting mechanism
转桨式水轮机 adjustable blade propeller turbine; adjustable blade wheel; adjustable vane turbine; feathering vane wheel; Kaplan turbine; variable-pitch turbine
转桨式转轮 controllable pitch propeller; feathering vane runner; feathering vane wheel; Kaplan runner
转桨叶片 feathering vane
转交 care of; farm out; onward transmission
转交条款 carry-over clause
转角 angle of rotation; corner; degree of turn(ing); intersection angle; return; rotating angle; slope; striking angle; turn angle; turning corner; deflection angle
转角 L 形块体 quoin block
转角爱奥尼亚式柱头 angular Ionic capital
转角板桩 corner and connection pile; corner and junction pile; corner sheet pile
转角半径 corner radius; knuckle radius; radius of corner
转角半露方柱 corner pilaster
转角壁炉 corner fireplate
转角壁柱 corner pilaster
转角变位法 method of slope deflection
转角侧房 <建筑物> return wing
转角处 nook
转角处导向叶片 corner vane
转角处的加固钢筋 corner reinforcement
转角处的加固角钢 corner reinforcement
转角处的联珠线脚 return bead
转角处的弯矩 bending moment at a corner
转角处墩柱 squint pier
转角处削角砖 angle brick
转角传动法 angle drive
转角窗梃 corner post
转角窗柱 corner post
转角搭接 lapped corner joint
转角搭接盖面燕尾榫 lapped dovetail
转角导线导轮 side wheel
转角地段 corner lot
转角点 intersecting point; point of intersection; intersection point
转角电杆 angle pole
转角电线塔 angle tower
转角斗拱 corner corbel-bracket
转角法 method of rotations
转角房屋 corner house
转角分辨率 corner resolution
转角缝 corner joint
转角扶壁 diagonal buttress
转角扶垛 angle buttress; diagonal buttress
转角盖缝(木)条 angle stile
转角杆 corner pole; corner post
转角钢 lug angle
转角钢筋 corner reinforcement
转角钢丝网板条 corner lath
转角公寓 corner apartment
转角公寓单元 corner apartment unit
转角公园边饰 corner garden edging
转角管塞 angle cock
转角横杆 buck arm
转角滑轮 quarter block
转角加宽式交叉口 intersection with widened corners
转角架 <架电线的> corner pole
转角建筑 corner building
转角接缝 angle seam
转角接头 knee joint
转角绝缘子 angle insulator; corner insulator
转角拉条 corner brace
转角连接用波纹铁片 corrugated fastener
转角联系条 corner brace
转角梁 angle beam
转角楼梯 quarter-turn stair(case)
转角路缘石 curb return
转角螺丝刀 round-the-corner screwdriver
转角锚墙 return anchor wall
转角帽头 angle capital
转角门 round-the-corner; round-the-corner door
转角摩阻系数 friction(al) curvature coefficient
转角蘑菇石剁边 angle-drafted margin
转角牛腿 corner bracket
转角皮带运输机构 angular belting
转角皮带装置 angular belting
转角频率 corner frequency
转角平衡 equilibrium of angles of rotation
转角平衡法 angle-balancing method
转角铺作 bracket set on corner
转角砌块 return corner block
转角嵌接 corner halving
转角墙 return wall
转角墙间填石 rubble between return walls
转角切除 corner cut
转角切除长度 length of corner cut
转角区土地 corner zone lot
转角式温度计 angle thermometer
转角视角 corner sight angle
转角视距 corner clearance
转角饰条【建】 return
转角输送螺旋 round-the-corner auger
转角塔 corner pole; corner post; angle tower
转角塔楼 angle tower
转角踏步 angle step
转角套间 corner flat
转角梯级 angle step
转角跳板 angle web
转角庭院 rounded property corneryards
转角停止 <建筑或装饰线脚的> return
转角通行面积 <人行道> circulation area
转角凸圆线 return bead
转角推拉遮光窗 around-corner sliding shutter
转角位移方程 slope-deflection equation
转角位置 <架空索道的> angle station
转角误差 error of swing
转角(系)条 corner bracing
转角线脚【建】 return; corner mo(u)lding
转角限制器 train stop
转角小教堂 corner chapel
转角协调方程 equation of compatibility of rotation
转角烟囱 angle chimney
转角延伸侧墙 return wall
转角阳台 corner balcony
转角椅 corner chair
转角鱼尾板 angular fish plate
转角浴缸 corner bath tub
转角圆线 rabbet bead
转角缘石半径 corner curb radius
转角约束 angular restraint
转角折叠 dog ear; dog-ear(ed) fold
转角支杆 angle support
转角止水 angular water seal
转角指示器 angle indicator
转角中梃 corner mullion
转角柱 angle post; angular column; corner post; corner stub
转角柱顶 angle capital
转角柱墩 corner pillar
转角柱帽 angle capital; corner capital
转角柱头 angle capital; corner capital
转角柱头装饰 angle capital decoration
转角转动研磨机 angle drive grinder
转角桩墩 corner pillar
转角琢边 angle-drafted margin
转角走向 run at angle
转接 adapter coupling; built-up connection; interchange; interconnect(ion); reset; switching; switchover; transfer; transit; turning joint
转接板 key set; patching board; patchpanel; pinboard
转接插件 adapter card; patchboard; patching board
转接插头 patch-plug
转接插头座 rack and panel connector
转接插座 multitap
转接处理机 switching processor
转接点 switching point; transit point
转接电键 transfer key
转接电路 built-up circuit; repeater circuit; through circuit; via circuit
转接电涌 switching surge
转接段 group section; switching section
转接方式 substitute mode
转接分线箱 transfer box
转接工作 transit working
转接呼叫 relayed call; transit call
转接话务量 volume of transit traffic
转接汇流条 transfer bus
转接机电路 repeater circuit
转接机构 change-over mechanism
转接记发器 transit register
转接交换台 through switchboard; transfer board; transfer switchboard
转接节点 switching node
转接禁止信号 transfer-prohibited signal
转接局 switching office; tandem office; through switching exchange; tie-station; transit exchange; transit office; transit station
转接开关 change-over switch; changing switch; setting switch; switcher; transfer switch
转接孔 switch port
转接控制计算机 switch control computer
转接路由 transit route; transit routing
转接母线 transfer bus
转接能力 switchover capability
转接盘 patching panel
转接器 adapter[adaptor]; adapter coupling; ball adaptor; circuit changer; commutator
转接器插头 adapter plug
转接器电缆 adapter cable
转接塞孔 patch jack; through jack; transfer jack
转接设备 change-over; interconnecting device; transit equipment; transit facility
转接时间 transfer time
转接时钟 <仪表的> change-over clock
转接室 transfer chamber
转接释放 transfer trip
转接衰减器 switching pad
转接顺序图 <开关的> sequence chart
转接损耗 transit loss
转接台 attendant board; attendant desk; lending position; tandem position; through position; tie-station
转接台话务员 tandem operator
转接套 reducing sleeve
转接通报 transfer message
转接通话 tandem working
转接通信[讯] transfer message
转接网 transit network
转接线 tie line
转接箱 cross board
转接信号 change-over signal
转接业务 transfer service
转接拥塞 switching congestion
转接占用信号 transit seizing signal
转接站 relay point; switching center [centre]; transit exchange
转接站载波 group through-connection station
转接指令 reference order
转接中继线 transit trunk
转接中心 relay center[centre]; switching center [centre]; transit center [centre]
转接装置 change-over arrangement; switching system
转接着手发码信号 transit proceed-

to-send signal
转节 swivel joint;trochanter
转节点 trochanterion
转节距 trochanter spur
转借 onward lending;sublease;underlease; underlet (ting); subtenancy <不动产的>
转借援助 onward-lending aid
转晶法 rotating crystal method
转井式煤气发生炉 rotary body producer
转镜 tilting mirror
转镜经纬仪【测】 transit compass; transit theodolite
转镜式仪表 mirror instrument
转镜系统 rotating mirror system
转镜仪 transit compass
转镜照准仪 tube-in-sleeve alidade
转矩 angular force;deflecting couple; deflecting torque;moment of rotation; rotation moment; running torque;torque(moment);torsional moment; turning moment; twisting moment;rotation torque <钻机的>
转矩扳手 tension wrench;torque spanner;torque(tension)wrench
转矩臂 torque arm
转矩变换 torque conversion
转矩变换器 torque converter
转矩变换器的换挡继电器 torque converter shifting relay
转矩变换器反应器 torque converter reactor
转矩变换器液体 torque converter fluid
转矩标定值 torque rating
转矩波动系数 ripple torque coefficient
转矩操作机构 torque motion
转矩测量 torque measurement
转矩测量装置 torque measuring device
转矩产生器 torquer
转矩常数 torque constant
转矩传感器 torque master; torque pickup;torque transducer
转矩磁力计 torque magnetometer
转矩电动机 torque motor
转矩电流 torque current
转矩电流常数 torque current constant
转矩发送器 torquer
转矩法 rotor-torque method
转矩反作用 torque reaction
转矩放大器 torque amplifier
转矩分布曲线 torque grading curve
转矩分配 torque division
转矩分配器 torque divider
转矩分配器传动控制阀 torque divider transmission control valve
转矩分配器传动(装置)torque divider transmission
转矩分配器星形装置 torque divider planetary set
转矩峰值 torque peak
转矩负载 torque load(ing)
转矩杆 torque rod
转矩隔绝器 torque insulator
转矩管 torque stay;torque tube
转矩管传动 torque-tube drive
转矩管螺旋桨轴 torque-tube propeller shaft
转矩管球接 torque-tube ball
转矩管式悬置 torque-tube type suspension
转矩惯性比 torque inertia ratio
转矩环路 torque loop
转矩换向器 torque converter
转矩极限 torque limit
转矩计 torque dynamometer;torquemeter;torsi(o)meter
转矩检测装置 torque master
转矩角 angle of torsion;torque angle
转矩抗力 torque resistance
转矩控制法 torque control method
转矩拉紧螺母 torquing of nuts
转矩理论 theory of torsion
转矩量测 torque measurement
转矩流变仪 torque rheometer
转矩马力 torque horse power
转矩脉动 torque pulsation; torque ripple
转矩曲线 moment curve
转矩曲线图 torque curve
转矩容许误差 torque allowance
转矩蠕动 torque creep
转矩式磁力仪 torque-type magnetometer
转矩式功率计 torque-type power meter
转矩输出 torque output
转矩双压电晶片 torque bimorph
转矩速度变换器 torque-speed converter
转矩速度特性曲线 torque-speed characteristic;torque-speed curve
转矩损耗 torque loss
转矩特性 torque characteristic
转矩梯度 torque gradient
转矩调节扳手 torque control wrench
转矩调整器 torque regulator
转矩同步机 torque synchro
转矩图 turning moment diagram
转矩污水泵 torque sewer pump
转矩系数 moment coefficient
转矩限制 torque limitation
转矩限制器 torque limiter
转矩相加器 torque summing member
转矩效率 torque efficiency
转矩允许误差 torque allowance
转矩噪声 torque noise
转矩振幅 torque amplitude
转矩值容许误差 torque allowance
转矩指示扳手 torque indicating(hand) wrench;torque measuring wrench
转矩指示器 torque indicator
转矩滞缓 torque creep
转矩柱 torque pillar
转矩转变 torque conversion
转矩转换 torque conversion
转矩转速测量仪 tach(e)ometer torquemeter
转锯 rotary saw
转开 turn-off
转开信用证 back-to-back credit
转科记录 transfer note
转口 entrepot;transit
转口港 entroport;junction port;port of reshipment; port of transshipment; transfer port; transit harbo(u)r;transit port
转口虎钳 swivel jaw vice
转口货仓库 entrepot warehouse
转口货港 entrepot port
转口货物 floating goods;transit goods
转口贸易 carrying trade; entrepot trade <法语>;switch trade;transit trade
转口商品 merchandise in transit
转口税 entrepot duty;transit duty
转口信用证 transit letter of credit
转库 transfer of cargo to another storage
转亏为盈 come out of the red
转拉电钮 turn-pull button
转来的余额 balance brought forward
转力矩 torque(moment);torsion(al) moment
转帘烘干机 brattice drying machine
转链轮装置 <自行车多速挡的> derailleur
转梁铰接式松土机 pivot-beam ripper
转梁式疲劳试验机 rotating cantilever beam type machine
转料隔板 transfer diaphragm
转捩点 crunch
转流 turn of tidal current;turn of tidal stream
转流室 reversion chamber
转流术 shunt
转笼 cage rotor
转笼烘燥机 tumble drier[dryer]
转笼磨碎机 disintegrator
转笼式除尘机 cone duster
转笼式翻车机 rotary car dumper
转笼式洗衣机 tumbler washer
转炉 air-blown converter; converter [convertor]; converting furnace; reel oven; retort; revolver; rotary furnace
转炉倍吹 turn down
转炉残渣 converter residue
转炉厂 converter shed
转炉车间 converter mill; converter plant
转炉吹炼 converter blowing
转炉吹冶炼法 bessemerizing
转炉风口箱 converter tuyere box
转炉风箱 converter air box
转炉风嘴清孔机 tuyere puncher
转炉钢 Bessemer pig;Bessemer steel; converted steel; converter steel; pneumatic steel
转炉钢锭 converter steel ingot
转炉钢渣 Bessemer furnace slag
转炉工 blower
转炉供风系统 converter air delivery system
转炉过吹 overblow
转炉后吹 after blow
转炉加料台 converter charging platform
转炉炼钢 pneumatic steel-making
转炉炼钢法 converter process; converting process
转炉炼钢用生铁 converter pig
转炉锍 Bessemer matte
转炉炉鼻 converter nose
转炉炉衬 converter lining
转炉炉衬砖 lining brick for converter
转炉炉底 converter bottom
转炉炉壳 converter shell
转炉炉口 converter mouth
转炉炉料 converter charge
转炉炉气 converter gas
转炉炉身 converter body
转炉炉台 converter platform
转炉炉体 converter body
转炉炉渣 Bessemer furnace slag;converter slag
转炉托圈 converter trunnion ring
转炉制铁法 <同时可得炉渣水泥> Basset method
转录 re-recording; sync; synchronization
转录程序 transducer
转录器 transcriber
转录设备 transcription device
转录图 transcription map
转录系统 rerecording system
转录装置 rerecording device
转路插头 change-over plug
转率 rate of rotation
转率计 turnmeter
转轮 impeller; revolver; rotating wheel; rotator; runner; swivel wheel;turning wheel
转轮安装高程 runner level
转轮泵 rotary pump
转轮比值 wheel ratio
转轮草图 runner outline
转轮测距仪 perambulator
转轮拆卸通道 runner removal access
转轮除湿机 rotary dehumidifier
转轮顶盖 runner cap
转轮割刀 roller cutter
转轮割管器 roller cutter
转轮过流能力 wheel discharge capacity
转轮喉(口直)径 throat diameter of runner
转轮(花)窗 rose window
转轮环 runner belt
转轮检修通道 runner removal access
转轮(件)支座 swivel sheaves support bracket
转轮节圆 <转轮上与射流轴线相切的圆> runner pitch circle
转轮进口 impeller eye
转轮进口直径 runner eye diameter
转轮进水口 runner inlet
转轮壳 runner housing
转轮轮毂 impeller hub;runner boss; runner hub
转轮轮冠 runner crown
转轮轮环 impeller belt
转轮轮廓线 runner outline
转轮轮盘 runner disc[disk]
转轮轮锥 runner cone
转轮密封环 runner seal ring
转轮内顶盖 runner inner lid
转轮喷砂机 wheel abrator machine
转轮入口直径 runner eye diameter
转轮上的铲斗 bucket on revolving wheel
转轮上冠 runner crown;upper rim; runner hub <混流式转轮的>
转轮上冠密封 runner crown seal
转轮上止漏环 runner crown seal
转轮式换热器 heat wheel;rotary heat exchanger;thermal wheel
转轮式挤压灰浆泵 rotary squeeze pump
转轮式计程仪 propeller log; rotary balance log
转轮式耐磨性试验机 wheel abrader
转轮式喷砂机 wheel abrader
转轮式漆膜测厚仪 rotary paint film ga(u)ge
转轮式热回收器 heat-recovery wheel; rotary heat recovering unit
转轮式扫路机 rotary road brush
转轮式水泥浆供料机 ferris wheel slurry feeder
转轮式推土机 turnadozer
转轮式挖沟机 wheel ditcher
转轮式印刷机 rotary-printing press
转轮室 runner chamber;runner housing;runner wall ring;throating ring <转桨式水轮机的>
转轮室上段 runner chamber upper ring
转轮室中环 discharge ring
转轮提斗机 raff wheel
转轮提升机 rotary elevator
转轮体 runner boss
转轮体装配 runner hub assembly
转轮外壳 runner casing;wheel case
转轮外廓线 runner outline
转轮系列 runner series;series of runner
转轮下环 impeller band;runner band; runner shirt; runner shroud; runner skirt;throating ring
转轮下密封环 runner band seal
转轮下止漏环 runner band seal
转轮效率 wheel efficiency
转轮泄水锥 runner cone
转轮泄水锥上段 upper rubber boss cover

转轮泄水锥下段 lower runner boss cover
转轮型谱 runner series;series of runner
转轮叶片 impeller vane;rotor blade;runner blade;runner bucket;runner vane
转轮叶片操作接力器 blade operating servomotor
转轮叶片接力器 runner blade servomotor
转轮叶片锯条 runner blade
转轮叶片调节接力器 blade adjusting servomotor
转轮叶片转臂 blade lever
转轮罩<水斗式转轮的> runner casing;runner housing
转轮折向器 runner deflector
转轮支持盖 runner inner lid
转轮直径比<冲击式水轮机的节圆直径与射流直径之比> wheel ratio
转轮止水环 runner seal ring
转轮中心高程 runner level
转轮中心体 impeller hub;runner boss
转落螺旋 dippy twist
转码器 function table
转买 resale;subpurchase
转买人 subpurchaser
转卖 resell;transference
转卖合同 resale contract
转门 revolution door;revolving door;revolving gate;rotary door;swing(ing)door
转门立轴 center[entre] shaft
转门门扇 revolving door wing;revolving leaf;revolving wing
转门门枢 swing bar
转门门套 revolving door casing;revolving door enclosure
转门中轴 center[centre] shaft
转面式风速计 rotating surface type air speed indicator
转/秒 revolutions per second
转名 transfer of names
转名日 transfer day
转模 revolving die
转模折弯机 rotary bender
转膜蒸发器 wiped film evapo(u)rator
转膜蒸馏器 wiped wall still
转磨清理滚筒 rumbling cleaning barrel
转耙式浓密机 rotary-rake thickener
转盘 turning plate;swinging circle;merry-go-round;movable disc;revolving bedplate;revolving leaf;revolving platform;rolling circle;rotary table;rotating disk[disc];rotating table;surface traverser;swing circle;swivel(l)ing table;traverse table;turnaround;turning table;turnplate;turntable;sliding bridge<铁路>;guide shell<圆盘式导轨桨>
转盘补芯 drive bushing;rotary bushing
转盘补芯阻力 rotary bushing drag
转盘衬套 rotary drive bushing;rotary table bushing
转盘齿圈 circle ring gear
转盘传递带 carousel
转盘传动 table drive
转盘传动箱体 circle drive housing
转盘萃取器 rotary disk extractor;rotating disk extractor
转盘萃取塔 rotating disk extraction tower
转盘单独驱动 separate table drive
转盘单元 rotating disk unit
转盘挡数 seeds of rotary table
转盘的通孔 table opening
转盘地带 turnaround area
转盘地坑 traverser pit
转盘电极 rotated disk electrode
转盘反循环钻机 rotary table reverse circulation drill
转盘风扇 fan wafter;wafter
转盘负荷 load of rotary table
转盘给矿机 rotary disk feeder
转盘给矿器 rotary feed table
转盘给料机 revolving plate feeder;rotary disk feeder;rotary table feeder
转盘给料器 rotary disk feeder;rotary feed table
转盘给煤机 rotating turn table
转盘功率 rotary table power
转盘灌装机 circular filling machine
转盘过滤机 revolving-leaf type filter;travel(l)ing pan filter
转盘横梁 skid beam
转盘(横移)装置<平地机> circle sideshift
转盘后托板 rear guide shoe
转盘虎钳 swivel bottom vice
转盘混合器 rotating pan mixer
转盘基础 rotary base
转盘挤奶台 rotalactor
转盘加料器 revolving feed table;revolving table feeder
转盘架 turntable mounting
转盘角板 swivel angle plate
转盘接触器 rotary disk contactor;rotating disk contactor
转盘精选机 rotary fettling table
转盘坑 turntable-pit
转盘脉冲抽提器 rotary disk pulsed extractor
转盘摩擦起电机 rotating disk friction generator
转盘平衡摇臂 rotary balanced jack
转盘铺砂器 spinner disk spreader
转盘曝气 rolling circle aeration;swash plate aeration
转盘起重机 bull wheel crane;bull wheel derrick;curb ring crane
转盘牵引架<平地机> circle drawbar
转盘前托板 front guide shoe
转盘驱动齿轮 circle drive pinion
转盘驱动马达 circle drive motor
转盘驱动小齿轮箱体 circle drive pinion housing
转盘驱动装置 rotating blade drive
转盘筛式给矿机 rotary disk grizzly feeder
转盘式 gyradisc
转盘式变传动副轴 rotary table transmission counter-shaft
转盘式表 disk meter
转盘式场存储器 rotating disk field store
转盘式存储器 rotating disk memory
转盘式粉碎机 gyradisc crusher
转盘式给料器 rotary plate feeder
转盘式过滤机 rotary disk filter;rotating disk filter
转盘式过滤器 rotary disk filter;rotating disk filter
转盘式互通电话机 rotary line selector type interphone
转盘式换筒装置 revolving turret
转盘式回转支承 turntable slewing ring
转盘式加料器 dial type feed mechanism
转盘式加湿器 spinning disk humidifier
转盘式交叉道路 gyratory intersection
转盘式交叉(口) gyratory junction;rotary intersection;roundabout (crossing)
转盘式交叉中心岛 roundabout island
转盘式交流换热器 disc type rotary regenerator
转盘式搅拌机 roller pan mixer;rotating mixer;rotating pan mixer
转盘式静电喷涂机 disk-type electrostatic sprayer
转盘式快门 rotating disk shutter
转盘式立体交叉 rotary intersection;roundabout crossing
转盘式流量计 rotary disk meter
转盘式破碎机 gyradisc crusher
转盘式清沟机 rotary disk ditch cleaner
转盘式求积仪 rolling disc planimeter
转盘式撒布机 spinning-disk distributor
转盘式扫描器 rotating disk analyser[analyzer];rotating disk scanner
转盘式生物膜反应器 rotating disk biofilm reactor
转盘式水表 disc[disk] type watermeter;disc[disk] water meter
转盘式松砂机 spike desintegrator
转盘式松土机 rototiller
转盘式碎石机 gyradisc crusher
转盘式隧道窑 rotary hearth kiln
转盘式淘汰盘 rotary-tray table
转盘式细轧机 gyradisc crusher
转盘式斜面升船机 ship incline with turntable bogie truck;turnplate ship incline
转盘式旋转支承装置 turntable type slewing supporting device
转盘式压力机 rotary table press;rotating table press
转盘式压(团)机 revolving table press;rotary table press
转盘式压砖机 rotary disk press
转盘式液固萃取塔 rotating blade column for liquid-solid extraction
转盘式造型生产布置 carousel type mo(u)lding lay-out
转盘式轧石机 gyradisc crusher
转盘式中心岛【道】 roundabout island
转盘式装煤车 rotary table charging car
转盘式自动搪瓷 automatic rotating pan enamelling
转盘式钻机 rotary table rig;rotary(bucket)drill<用于钻软岩层或土层>
转盘送料装置 rotary disk feeder
转盘塔 rotary disk column;rotating disk column
转盘台虎钳 swivel base bench vice
转盘铁格筛 rotary disk grizzly
转盘托板 guide shoe
转盘喂料机 disk feeder;rotary table feeder
转盘雾化法 rotating disk method
转盘雾化器 rotating disk atomizer
转盘线 turntable track
转盘型道岔 turntable switch
转盘压(砖)机 turnable press
转盘研磨机 rotating disk mill
转盘噪声 rumble;turntable mumble
转盘噪声电平 rumble level
转盘噪声滤波器 rumble filter
转盘真空计 rotating disk vacuum ga(u)ge
转盘支架 slide rest
转盘中心移动装置<平地机> centre shift
转盘中心指示器 centershift indicator
转盘转矩 table torque
转盘转数 rotary table rotations per minute
转盘转速指示器 table speed indicator
转盘装置 rating-disc unit;rotating disk device;rotation-disc unit
转盘钻机 rotary drilling machine;rotary machine;rotary table drill
转盘钻机绞车 rotary draw works
转盘钻进扩孔钻头 rotary underreamer
转盘钻进设备 rotary drill rig;rotary outfit
转盘钻进用钻杆 rotary drill pipe
转盘钻具 rotary tool
转盘钻司钻 rigger;rotary driller;rotary runner
转盘钻探 rotary drilling
转盘最大转速 maximum rate of rotary table
转盘座架 rotary support frame
转片 rotor;rotor of condenser
转片型衰减器 rotary-vane attenuator
转坡点 turning point;turning station
转期利率 carry-over rate
转期票据 note renewals
转期调节 carry-over storage
转期信贷技术 revolving credit technique
转求法黏[粘]度计 rotating ball viscometer
转球 swivel ball
转球求积仪 rolling ball planimeter;rolling sphere planimeter
转球式(光标)指示器 track ball pointer
转屈 rotexion
转圈 circumrotation
转圈挖土法 revolving excavating processing
转圈运动 circus movement
转让 alienate;assign;assign a contract;assignation;assignment;attorn;convey(ing);devolution;dislocation;make over;negotiate;negotiation;transference;transfer【计】;alienation<财产或产权的>;cession<权利、财产等的>
转让部分投资 divestment
转让产权 assignment of property;conveyance of property;livery
转让成本 assigned cost
转让船租合同 demise charter party
转让代理人 transfer agent
转让方 transferer
转让房地产 demised premises
转让费用 assignment charges;cost of transfer;transfer fee
转让合同 assignment of contract
转让价格 transfer price
转让价值 resale value
转让建筑 alien structure
转让交易 transfer transaction
转让经济 alienation economy
转让利益 assignment of interests
转让配额 transfer of quota
转让契约 deed of assignment;deed of transfer;quit claim;quit claim deed
转让取得 transferred acquisition
转让权 assignment(of) right
转让权利 assignment privilege
转让权协议 covenant of right to convey
转让人 assigner[assignor];endorser;transferor[transferrer]
转让收入 transfer income
转让手续费 negotiation commission;transfer commission
转让书 letter of assignment;transfer form;letter of appointment<财产权利的>
转让税 transfer tax
转让所有权 transfer of title
转让条件 transfer condition

转让条款 assignment clause
转让土地 demised premises
转让危险 assigned risk
转让限制 restraint of alienation
转让项目 transfer item
转让性付款 transfer payment
转让一揽子 transfer package
转让移交 make over
转让佣金 transfer commission
转让债权 assignable claim
转让折扣 assignment allowance
转让者 assigner[assignor];transferor
转让证书 deed of transfer;grant deed;transfer deed
转让支付 transfer payment
转让支票 negotiable check
转让资产 livery
转让租船合同 demise charter party
转容量 swept volume
转熔 peritectics
转熔的 peritectic
转熔点 peritectic point
转熔反应 peritectic reaction
转熔过程 peritectic process
转熔体 peritectoid
转熔温度 peritectic temperature
转熔作用 peritectic reaction
转入 entrance turn;roll-in
转入成本 transfer-in cost
转入俯冲 turn into dive
转入管道 turn into the line
转入率 switching-in rate
转入排水状态 touch down
转入配合 wringing fit
转入下期 carry-over
转入下页 brought forward; brought over;c/f[carried forward]
转入箱 transfer-in box
转入转出 roll-in and roll-out;roll-in/roll-out
转入总账 post in the ledger
转塞 rotoplug
转塞式 rotoplug type
转塞式龙头 plug tap
转塞污泥脱水机 rotor-plug sludge concentrator
转色试金 annealing
转筛 band screen;belt screen;revolving screen;travel(l)ing screen
转筛式给料机 revolving grizzlies feeder
转栅 turnstile
转舌形道岔 tongue switch
转身楼梯 halfpace stair(case);half-space stair(case)
转升门 turning-lifting gate
转湿系数 coefficient of moisture transition
转石 erratic block(of rock);float;rubble
转式铲土机 rotating shovel
转式打气泵 rotary air pump
转式电铲 rotary bucket excavator
转式浇灌 rotary sprinkler system
转式挖土铲 rotating shovel
转式挖土机 rotating shovel
转式栅门 turnstile
转手货物 switch cargo
转手交易 switch operation
转手买卖地产 double escrow
转手贸易 switch trade;transfer trade
转手掮客 switch dealer
转手销售 switch selling
转手资料 feedback
转售 resale;resell
转售价格 resale price
转售收入 resale proceeds
转售协议 agreement to resell
转枢式电动机 rotating-armature motor
转枢式电机 rotating-armature machine
转枢式发动机 rotating-armature generator
转枢式励磁机 rotating-armature-type exciter
转输 custody transfer
转输管线 transit pipeline
转输站 custody transfer point;relaying station;transfer station
转数 convolution;number of revolution;revolution;rotational frequency;rotative speed
转数比 ratio of revolutions
转数表 revolution counter;revolution indicator; revolution meter; rotor meter;speed counter;turn indicator
转数表传动 tach(e)ometer drive
转数表传动轴 revolution counter drive quill
转数表传感器 tach(e)o-generator;tach(e)ometer generator
转数测定法 tach(e)ometry
转数范围 range of revolution
转数/分 revolutions per minute
转数航速表 revolution speed table
转数机 motometer
转数计 cycle counter;cyclometer;cycloscope; motometer; rate indicator; rev counter; revmeter; revolution counter; revolution indicator; revolution meter; speedometer; tach(e)ometer(ga(u)ge)
转数计传动齿轮 tach(e)ometer drive gear
转数计数器 revolving counter
转数记录器 gyrograph;revolution recorder
转数降低 revolution drop
转数/秒 revolutions per second
转数器 revolution counter
转数图示器 gyrograph
转数稳定器 revolution stabilizer
转数指示器 gyrograph;revolution indicator
转刷曝气 brush aeration;rotary brush aeration;rotating brush aeration
转刷曝气池 brush aeration tank
转刷曝气机 rotor-brush aerator
转刷曝气系统 brush aeration system
转刷曝气装置 rotary brush aerator
转刷式分离器 rotating brush separator
转刷式转子 brush-type rotor
转送层 transport level
转送函 letter of transmittal
转送器 transfer network
转送热室 transfer cell
转送文件信 letter of transmittal
转送装置 grass hopper
转速 rate of revolution;rate of rotation;rating of evolution;revoluting speed; revolving speed; rotary speed; rotary velocity; rotating speed;rotating velocity; rotational speed;rotational velocity;rotation rate;rotation speed;rotative speed;rotative velocity;speed of revolution;speed of rotation;velocity of rotation
转速比 ratio of transmission;reduction rate;speed ratio;transmission (-qear) ratio
转速变化率 relative speed variation
转速表 cycle counter;motometer;revolution counter;revolution indicator;speed counter;speed ga(u)ge;speed indicator;speed meter;speedometer;tach(e)ometer;tach-(e)oscope
转速表表盘 tach(e)ometer dial
转速表传动 tach(e)ometer drive
转速表传动软轴 tach(e)ometer drive cable
转速表传动装置 tach(e)ometer sending unit
转速表传感器 tach(e)ometer adapter;tachogenerator
转速表电气指示器 tach(e)ometer electric(al) indicator
转速表发电机 tachometer generator
转速表接头 tach(e)ometer connection
转速表空心传动轴 tach(e)ometer drive quill
转速表式流量传感器 tach(e)ometric-(al) flowmeter sensor
转速表试验台 tach(e)ometer testing table
转速表用发电机 tach(e)ometer dynamo
转速表止动装置 tach(e)ometer stop gear
转速表指示器 tach(e)ometer indicator
转速表组 tach(e)ometer unit
转速波动 fluctuation of speed;speed fluctuation;speed oscillation
转速不灵敏度 speed insensibility
转速测定法 tach(e)ometry
转速测量 tach(e)ometric(al) survey
转速测量法 tach(e)ometry
转速测量仪 rotameter
转速测量仪表 speedometer for measuring rotation
转速传感 speed sensing
转速传感可变的定时装置 speed sensing variable timing unit
转速传感器 revolution speed transducer; rotary transducer; rotative velocity transducer; speed sensing device
转速导线 tach(e)ometric(al) traverse
转速电流特性曲线 speed-current characteristic curve
转速定点 speed setpoint
转速定值器 speed setter
转速读出 speed sensing
转速读出可变的定时装置 speed sensing variable timing unit
转速读出装置 speed sensing device
转速发电机 tachogenerator
转速范围 range of speeds of rotation;rotary speed range
转速负荷特性 speed-load characteristic
转速给定装置 speed setting gear
转速级数 number of speeds
转速极限 rev limit
转速计 cycloscope;kinemometer;operameter;revolution counter;revolution indicator; revolution meter; speed ga(u)ge;speed meter;speedometer;tach(e)ometer
转速计标尺 tachometer sight
转速计的传动 tachometer drive
转速计发送器 tach(e)ometer transmitter
转速计反馈 tach(e)ometer feedback
转速计瞄准器 tach(e)ometer sight
转速计时器 tach(e)ometer chronograph
转速计数器 spin counter
转速计算表 tach(e)ometric(al) table
转速计算尺 tach(e)ometric(al) ruler
转速计稳定系统 tach(e)ometer stabilized system
转速计用发电机 tach(e)ometer dynamo
转速记录 rate-of-turn record
转速记录器 rate-of-turn recorder
转速记录图 tach(e)ogram
转速继电器 tach(e)ometric(al) relay
转速加快 speed pick-up
转速监视器 speed monitor
转速检测器 revolution detector
转速检测装置 speed detector
转速降低 <主动元件的> drive down
转速降落率 relative speed drop
转速界限 speed threshold
转速禁区 red band
转速警报器 revolution alarm
转速静电计 tach(e)ometric(al) electrometer
转速开关 speed switch
转速可控选粉器 speed controlled separator
转速控制 rotary speed control;speed control
转速控制范围 range of speed control
转速控制机构 speed control mechanism
转速控制器 rotary speed controller;rotational speed governor; speed controller
转速扭矩特性曲线 speed-torque characteristic curve
转速频度 speed-frequency
转速上升 speed rise
转速设定点 speed setpoint
转速升高 speed rise
转速升高度 relative speed rise
转速试验 speed test
转速死区 speed dead band
转速特征曲线 speed characteristic curve
转速调节杆 speed lever;speed setting lever
转速调节螺钉 speed setting screw
转速调节器 speeder;speed governor;speed regulator
转速调节器传动 governor drive
转速调节伺服马达 speed setting servomotor
转速调节凸轮 speed setting cam
转速调节装置 speed adjusting device;speed setting unit
转速调整缸 speed setting cylinder
转速调整横梁 speed setting walking beam
转速调整机构 speed adjusting device
转速调正活塞 speed setting piston
转速通量 transit flow
转速图表 kinemograph
转速稳定 stabilization of speed
转速稳定指数 speed stability index
转速系数 speed coefficient
转速下降 revolution drop
转速显示 speed sensing
转速显示可变的定时装置 speed sensing variable timing unit
转速显示器 speed sensing device
转速限制器 revolution stop
转速相对升高 relative increase of speed
转速信号装置 tach(e)ometer singalling device
转速压力传感器 speed-to-pressure transducer
转速仪 cycloscope
转速增量 incremental speed
转速照准仪 tach(e)ometric(al) alidade
转速振荡 speed oscillation
转速整定 speed setting
转速指示计 rate indicator
转速指示器 revolution indicator;shaft revolution indicator; speed indica-

tor;velocity indicator
转速指数 rotary speed exponent
转速转矩特性 speed-torque characteristic
转锁 twist lock
转塔 turret;turret lock
转塔车床 turret lathe
转塔车床六角头 rotary turret lathe head
转塔刀架 capstan head slide;revolving tool box saddle
转塔顶门 turret hatch
转塔环形齿轮 turret ring gear
转塔回转机构 gyrotraverse mechanism
转塔门 turret door
转塔式冲床 turret multiple punch
转塔式风力发动机 merry-go-round windmill
转塔式六角冲床 turret punch press
转塔式六角式钻床 turret vertical drilling machine
转塔式压力机 turret multiple punch; turret press
转塔丝杠 capstan screw
转塔装置 turret mounting
转台 revolving bedplate;roll-over;rotating platform; rotating table; swivel(l)ing table;swivel mount; tarn table;turning table;turnplate; turntable; turret; revolving superstructure【机】;rotary table
转台翻车机 turntable tip
转台挂接式拖车 fifth wheel trailer
转台机构 table-turning mechanism
转台磨床 rotary grinding machine
转台喷砂机 rotary table sand blasting machine
转台式混凝土砌块压制机 revolving table concrete block press
转台式落条筒机构 turntable doffing mechanism
转台式屏蔽 rotatable shield
转台式天线座架 turntable antenna mount
转台式系泊 turret moor
转台稳定器 rotary table stabiliser
转台型喂料器 turret type feeder
转台压制机 revolving table press
转台支架 turntable support
转台自动进给装置 transfer turn table ribbon feeder
转态过程 polling
转躺椅 rota-cline seat
转套 teleflex
转套感应式磁电机 rotary sleeve inductor type magneto
转套式磁电机 rotating sleeve type magneto
转梯 helical stair(case);screw stair (case);spiral stair(case)
转体 swivel
转体架桥法 bridge erection by swinging method;construction by swing
转体施工 swinging erection
转体施工法 construction by swing
转桶法 <加气混凝土配料法的一种> rolling method
转桶式筛面 rolling-tub screen deck
转筒 revolving barrel; revolving drum;rotary drum;rotary tumbler; rotating bowl; rotating drum; rotor;rumble;trommel;turn barrel
转筒拌和机 tumbling mixer
转筒波纹记录器 kymograph
转筒长度 <卷扬机的> winding face
转筒除锈法 tumbling barrel method
转筒煅烧炉 rotary calciner
转筒多级结晶器 rotating drum multistage crystallizer
转筒阀 rotary sleeve valve
转筒法 rotating cylinder method;rotating drum method
转筒粉磨机 drum pulverizer;rotary drum pulverizer
转筒干燥器 cylinder drier[dryer];revolving drier [dryer]; rotary drum drier[dryer]
转筒给料机 drum-type feeder;pulley type feeder;roll feeder
转筒烘干机 cylinder drier[dryer];revolving drier[dryer]
转筒混合 rumbling
转筒混合机 rotating drum mixer
转筒混合器 tumbling mixer
转筒记录器 cymograph;kymograph
转筒记录纸 drum chart
转筒加工法 drumming
转筒卷绕的 drum winding
转筒卷绕机 drum winder
转筒快门 rotating barrel shutter
转筒磨 drum mill
转筒泥浆筛 rotary mud screen
转筒黏[粘]度计 rotating cylinder visco(si)meter
转筒浓度测定法 rotary cylinder viscometry
转筒浓缩带式压榨过滤机 rotary cylinder thickening filter belt press
转筒喷砂 rotoblast;tumblast
转筒曝气系统 cage-rotor aeration; cage-rotor aeration system
转筒切坯机 reel cutter
转筒筛 bolting reel; drum sieve; revolving screen; riddle drum; rolled screen; rotary cage; rotary drum screen;rotary sieve;rotating cage; shaft screen; trommel; trommel screen(ing);rotary screen;revolving drum screen
转筒筛分 trommelling
转筒筛式洗矿机 revolving screen washer
转筒筛选 trammel(l)ing
转筒筛选法 drum screening
转筒式安瓿机 horizontal ampoule forming machine
转筒式拌和机 drum mixer;revolving drum mixer;rotary drum mixer;rotary mixer; rotary speed mixer; rotating mixer
转筒式标志 rotating drum sign;rotation-drum sign
转筒式成球机 roll mill granulator
转筒式磁铁分离器 magnetic drum
转筒式粉碎机 tumbling crusher
转筒式干燥法 rotary drying
转筒式干燥机 revolver drier[dryer]; revolving drum drier[dryer];rotary drum drier [dryer]; tumble drier [dryer]
转筒式干燥器 revolver drier[dryer]; revolving drum drier [dryer]; tumble drier[dryer]
转筒式干燥与加热方法 rotary drying and heating process
转筒式干燥与加热过程 rotary drying and heating process
转筒式过滤机 rotating filtering drum
转筒式过滤器 drum filter; rotating filtering drum
转筒式烘干机 revolver drier[dryer]; revolving drum drier [dryer]; tumble drier[dryer];rotary drum drier [dryer]
转筒式混合机 barrel mixer
转筒式混凝土拌和机 drum-type concrete mixer; revolving drum concrete mixer; rotary drum concrete mixer
转筒式混凝土搅拌机 drum-type concrete mixer; revolving drum concrete mixer; rotary drum concrete mixer;rotating concrete mixer
转筒式交流换热器 rotating drum regenerator
转筒式搅拌车 revolving drum truck
转筒式搅拌机 rotary drum mixer;revolving drum(type) mixer;rotating drum mixer
转筒式可变指示标志 rotating drum type variable message sign
转筒式快速拌和机 rotary speed mixer
转筒式冷却机 rotary cooling drum
转筒式冷却器 rotary cooler
转筒式流化干燥器 rotary trammel fluidized drier[dryer]
转筒式滤器 rotary drum filter
转筒式滤清器 rotary drum filter
转筒式滤网 rotary drum screen
转筒式磨损试验机 rattler
转筒式铺路拌和机 rotating drum paver mixer
转筒式摄影机 revolver camera
转筒式输送车 revolving drum truck
转筒式喂料机 rotary drum feeder
转筒式洗涤器 revolving drum washer
转筒式卸料装置 rotary drum discharge unit
转筒式运输机 rotary tubular conveyer[conveyor]
转筒式真空过滤机 rotary vacuum drum filter
转筒式真空过滤器 rotary vacuum drum filter
转筒速度 drum speed
转筒洗筛 trommel washer
转筒淹没度 drum submergence
转筒研磨机 tube mill
转筒堰 drum weir
转筒窑 rotary calciner; rotary kiln; rotating drum furnace
转筒印花 drum printing
转筒印刷 drum printing
转筒真空计 rotating cylinder vacuum ga(u)ge
转筒直径 drum diameter
转筒纸 cartridge paper
转头 rotary swivel
转头滑槽 swivel head chute
转头水域 turning basin(circle)
转途经理人 forwarding agent
转腿铰接松地机 swivel ripper;swivel shanks binge-type ripper
转托支架 pivoting cradle
转弯 cornering;obversion;pitchover; round a corner;turn(ing)
转弯半径 radius of bending;radius of turn; turn (circle) radius; turning radius;turn radius
转弯避让其他机车 turnout of traffic
转弯不足 <汽车> understeer
转弯部位 turning site
转弯操纵 turn control
转弯侧力 lateral force
转弯车带 turn slots
转弯车道 turning lane;turning roadway;carriageway for turning traffic
转弯车道岔口 turning roadway terminal
转弯车流 turning flow
转弯处 turning area
转弯处路缘石的模板 radius kerb mo(u)ld
转弯导流叶栅 corner-vane cascade
转弯的 turning
转弯的线脚 returned mo(u)lding
转弯点 turning point
转弯动作 turning moment; turning movement
转弯度 bendiness
转弯段 corner section
转弯扶手 handrail wreath;wreathed handrail(ing);wreath(ed string)
转弯扶手柱 <楼梯的> angle newel
转弯浮标 turning buoy
转弯规则 turning rule
转弯轨迹 turning path
转弯轨迹中心 turning center[centre]
转弯横向力 lateral force
转弯滑轮 quarter block
转弯计算器 revolution counter
转弯驾驶杆 turn controller
转弯交通 turning traffic
转弯角度 angle of bend; angle of turn;turning angle
转弯宽度 turning width
转弯离开车道 turn-off the runway
转弯流槽 curved-trough flow device
转弯楼梯 angled stair(case)
转弯路线 turning path
转弯面积 sweep area; driving area <汽车>
转弯能力 cornering ability;turnability
转弯平台式楼梯 half-space stair(case)
转弯坡道 L-shaped ramp;switchback ramp
转弯墙 canted wall
转弯倾斜指示器 turn and bank indicator;turn-and-slip indicator
转弯曲度 turning curvature
转弯时的倾翻负荷 turned tipping load
转弯时间 turning time
转弯驶出 <从过镜线转出坡道> exit turn
转弯驶入 <从匝道转入过境线> entrance turn
转弯示向灯 turn signal lamp
转弯数 number of turns
转弯速度指示器 rate-of-turn indicator
转弯速率 rate of turn;turning speed
转弯损失 corner loss;turn loss
转弯踏板 turn tread
转弯踏步 balanced winders; danced stair(case); dancing steps; dancing winders
转弯特性 turning performance
转弯外倾 toe-out on turns
转弯限制 turn restriction
转弯小径 turning path
转弯协调 turn coordination
转弯信号灯 turn signal light
转弯行车 turning movement
转弯行车的车道 lane for turning traffic
转弯性能 turnability; turning ability; turn performance
转弯休息平台 half-space landing
转弯圆周间隙 clearance circle
转弯运动 turning movement
转弯运行 turning movement
转弯直径 turning diameter
转弯指示器 bank-and-turn indicator; turn indicator
转弯专用车道 exclusive turning lane; reserved turning lane; special turning lane
转位 dislocation; indexing; inversion; positioning; station index; toe index;transposition
转位冲模 index die
转位刀具 index tool
转位工作台 indexing table
转位工作台机床 index table machine
转位机 indexing machine
转位机构 indexing mechanism
转位加工机床 station index machine

转位时间 index time
转位因子 transposable elements
转位中心架和工件自动装卸器 swing-position steady rest and ejector
转位子 transposon
转下页 carried forward
转现货交易 exchange for physical
转线 transfer to another track; transfer to another track or yard
转线调车 transfer track service
转线调车钩 transfer trip
转线轨道 cross-over; cross-over track
转线换乘 converse routing
转线路 cross-over road
转线行程 transfer trip
转相 phase inversion
转相谱 quadrature spectrum
转相温度 phase inversion temperature
转向 alter course; change course; cornering; recurvature; slew; steering; sway; swing; turn about; turning; turn-off; veer(ing); recurve <风、水等>
转向板 deflecting plate; deflector plate
转向半径 turning radius
转向保险 steering lock
转向保险角 steering lock angle
转向泵 steering pump
转向臂 drop arm; steering (drop) arm; steering lever; tie-rod arm; pitman arm
转向臂杠杆式拉杆机构 spindle lever linkage
转向臂固定螺帽 steering arm nut
转向臂规 spindle arm ga(u)ge
转向臂螺栓 steering arm bolt
转向臂球头柱螺栓 steering arm ball stud
转向臂调节螺钉扳手 steering arm adjusting screw wrench
转向臂销 steering arm pin
转向臂轴 pitman arm shaft; steering arm shaft
转向臂轴油封毡垫 steering pitman arm shaft felt washer
转向变速比 steering reduction ratio
转向变速器盖 steering box cover
转向标志 sign of rotation; turn marking
转向不够 understeering
转向不足 understeer
转向部件 steering component
转向操纵阀壳 steering valve housing
转向操纵杆 steering arm; steering rod
转向操纵机构振动 shaking of steering
转向操纵性能 steering quality
转向操纵装置 steering control
转向操作力 steering force
转向测试 slew test
转向差 difference of readings at the opposite orientation
转向差速器 steering differential
转向车【铁】 bogie [bogey/bogy]; bogie bolster; bogie car; bogie truck; bogie wagon
转向车道 steering lane; turning lane
转向车道岔口 turning roadway terminal
转向车架 bogey; bogie truck; bogy
转向车流 turning flow
转向车站 reversing station
转向齿弧 steering sector
转向齿轮 steering gear; tooth sector
转向齿轮机构降速比 steering gear reduction ratio
转向齿轮夹板 steering gear clamp plate
转向齿轮速比 steering gear ratio
转向齿轮调整螺母 steering gear adjusting nut
转向齿轮调整螺母保险 steering gear adjusting nut lock
转向齿轮箱 steering gear box
转向齿轮轴 steering gear shaft
转向齿轮柱衬套 steering gear column bushing
转向齿轮柱夹紧夹子 steering gear column clamp
转向齿条 steering rack
转向处 turning place
转向处罚【交】 turning penalty
转向传动比 steering gear ratio
转向传动杆系 steering linkage
转向传动间隙 steering gear back lash
转向传动轴 steering drive axle
转向传感器 rotation direction sensor
转向垂臂 pitman arm; drop arm
转向垂臂轴 pitman arm shaft
转向垂臂轴毡垫盖 steering arm shaft felt washer cup
转向带销 pivot shaft
转向导板 diverter
转向导航台 turn inbound
转向导流管 steering nozzle
转向导网辊 return wire leading roll
转向导向片 reversing blade
转向的 diverted
转向的增压泵 steering booster pump
转向灯 direction signal light; turn light
转向点 inflection point; point of deviation; point of diversion; recurvature point; turn(ing) point; change point
转向点半径 turnover radius
转向点检验 turning point test
转向叠标 turning range marks
转向度 degree of turn(ing)
转向端悬臂外伸长 steer end overhang
转向墩 <帮助船舶回转的> turning dolphin
转向舵机 steering engine
转向舵盘 steering pitman arm
转向阀 conversion valve; cross valve; diversion valve; diverter valve; steering valve
转向帆 steering sail
转向反面 turnover
转向范围 steering range
转向方法 steering system
转向方式 steering method
转向浮标 turning buoy
转向杆 deflecting bar; deflecting piece; deflecting rod; steering lever; steering rod; turning bar
转向杆臂 steering lever arm
转向杆柄 steering lever handle
转向杆件 steering link
转向杆球端 steering arm ball
转向杆扇形齿轮 steering lever quadrant
转向杆手柄 steering lever handle
转向杆套 steering steer sleeve
转向杆系 steering gear connection; steering linkage
转向杆系降速比 steering linkage reduction ratio
转向杆销 steering lever shaft pin
转向杆轴 steering lever shaft
转向杆柱 steering lever column
转向构架 bogie frame
转向关节 steering connection; steering knuckle
转向关节臂 steering knuckle arm
转向关节衬套 steering knuckle bushing
转向关节杆 steering knuckle lever
转向关节横拉杆接头油嘴 steering knuckle tie rod end pressure lubricator
转向关节横拉杆接头总成 steering knuckle tie rod end assembly
转向关节横拉杆球节 steering knuckle tie rod ball joint
转向关节枢 steering knuckle pivot; steering knuckle spindle; steering stub; steering yoke bolt
转向关节栓 spindle bolt
转向关节系杆 steering knuckle tie rod
转向关节系杆端 steering knuckle tie rod end
转向关节系杆端塞 steering knuckle tie rod plug
转向关节系杆球端 steering knuckle tie rod ball
转向关节系杆球座 steering knuckle tie rod ball seat
转向关节系杆弹簧 steering knuckle tie rod spring
转向关节销 steering knuckle pin
转向关节油封座圈 steering knuckle case oil seal retainer
转向关节止动螺钉 steering knuckle stop screw
转向关节止推垫圈 steering knuckle thrust washer
转向关节主动轴承 steering knuckle thrust bearing
转向关节主销 steering knuckle king pin
转向关节主销衬套 steering knuckle king pin bushing
转向关节主销盖 steering knuckle king pin dust cover
转向关节主销加油器 steering knuckle king pin lubricator
转向关节主销键钉 steering knuckle king pin locking pin
转向关节主销上轴承 steering knuckle king pin upper bearing
转向关节主销下轴承 steering knuckle king pin lower bearing
转向管 deflecting pipe
转向管接头 swivel union
转向管凸缘 torque-tube flange
转向管柱 steering column
转向轨迹宽度 turning track width
转向轨距 turning track
转向辊 bending rolls; guide roll; steering roll; turning roll
转向滚轮 guide roll; steering drum; steering roll
转向滚筒搅拌机 reversing drum mixer
转向过渡 oversteer(ing)
转向和制动的联动系统 steering and brake linkage
转向河 diverted river; diverted stream
转向河绠 swing boom
转向横拉杆 spindle connecting rod; steering cross rod; steering tie rod; tie rod
转向横拉杆臂 guide lever
转向横拉杆轭 steering knuckle tie rod yoke
转向横拉杆接头 steering cross rod jaw
转向横拉杆球铰接头 tie-rod end
转向滑车 angle block; corner block
转向滑轮 angle pulley; deflecting pulley; deviation pulley <架空缆道>
转向环(形)道 turnaround loop
转向回正 steering reversal
转向回转杆 steering pivot lever
转向机 steering engine
转向机侧盖 steering box side cover
转向机场 turn inward
转向机车【铁】 bogie engine; bogie locomotive; locomotive for negotiating curve; radial locomotive
转向机构 steering device; steering gear; steering hardware; steering mechanism
转向机构传动比 steering ratio
转向机构机油管道 steering oil line
转向机构减震器 steering gear check; steering shock eliminator
转向机构调整 steering gear adjustment
转向机构蜗轮轴承 steering gear cam bearing
转向机构箱 steering box
转向机构性能 steering performance
转向机构摇动 shaky steering
转向机构中的减振器 steering damper unit
转向机构轴 steering gear shaft
转向机壳底盖 steering gear housing lower cover
转向机壳支架 steering gear housing bracket
转向机械 steering machinery
转向机支架 steering gear bracket
转向基准位置 steering reference position
转向极限 steering lock
转向架 bogie [bogey/bogy]; caster frame; fork head; pivoted bogie; steering frame
转向架安放台 bogie storage rack
转向架安全链 check chain; truck safety chain
转向架边界裙板 bogie skirt
转向架侧架 bogie frame; truck side; truck side frame
转向架侧架立柱 truck bolster guide bar
转向架拆卸工具 tools for assembly and disassembly of bogie
转向架车辆 truck car
转向架承梁 bogie bolster; track bolster; truck bolster
转向架导框 truck pedestal
转向架的平稳特性 riding characteristic of bogie
转向架定心装置 truck centering device
转向架对角线 bogie diagonal
转向架耳轴 bogie gudgeon
转向架防侧滚装置 anti-roll device
转向架复原装置 bogie pitch
转向架杠杆间隙调整器 truck lever slack adjuster
转向架更换整备 bogie-changing installation
转向架构架 bogie frame; truck frame
转向架构件 bogie-frame member; bogie sole bar <底梁>
转向架固定杠杆 dead truck lever
转向架挂车 bogie truck
转向架横向弹簧控制 bogie lateral spring control
转向架滑座 bogie slide
转向架基础制动装置 bogie brake rigging
转向架间距 truck spacing
转向架间连结 interbogie coupling
转向架减振器 bolster damping
转向架检修间 inspection and repair shed for bogie
转向架检修台 inspection bench for bogie
转向架交换方式 bogie exchange system

转向架角位移 angular displacement of bogie
转向架脚板 bogie skirt
转向架结构 truck structure
转向架截断塞门 bogie cut-out cock
转向架均衡梁 truck equalizer
转向架库存 bogie stock
转向架联结装置 center draft drawbar;radial drawgear
转向架轮耳轴销 bogie wheel gudgeon pin
转向架轮托座 bogie wheel bracket
转向架扭曲刚度 bogie against deflection rigidity
转向架旁承 bogie bolster
转向架偏转 truck swing
转向架清洗机 bogie cleaning machine
转向架全轴距 total wheel base of bogie assembly
转向架上铁路货车 bogie wagon
转向架式车辆 bogie vehicle
转向架式公路铁路两用挂车 bogie roadrailer
转向架式机车 bogie locomotive
转向架弹簧 bogey spring;bogie spring;bogy spring;truck spring
转向架调节装置 bogie adjusting gear
转向架推拉小车 truck push/pull for bogies
转向架小车 bogie stock
转向架卸取梁 beam for dismounting bogie
转向架心盘 bogie center plate
转向架心盘中心间距离 truck centers
转向架心盘中心距 bogie pivot pitch
转向架心轴 bogie pin
转向架旋转枢轴 bogie pivot
转向架摇枕 center pin bolster
转向架移动杠杆 live truck lever
转向架闸缸 truck brake cylinder
转向架支承板 bogie sole bar
转向架支承梁 bogie bolster
转向架制动杠杆 truck lever
转向架制动杠杆的拉杆 truck lever connection
转向架制动梁头 bogie beam headstock
转向架制动装置 bogie brake rigging;truck brake rigging
转向架中心距 distance between bogie centers[centres]; distance between center to center of bogies
转向架中心销 bogie pivot
转向架轴 truck axle
转向架轴距 bogie pitch;bogie wheelbase; truck wheelbase; wheel-base of a bogie
转向架轴箱 truck journal box
转向架转盘 slewing ring for bogie
转向架转向机理 bogie steering mechanism
转向架转轴 bogie pin
转向架锥形弹簧 bogie volute spring
转向架自由转动 free rotation of bogie
转向架纵向悬梁 longitudinal suspension beam
转向减速比 steering reduction ratio
转向减震杆 steering damper rod
转向减震器 steering anti-kickback snubber;steering damper
转向减震器摩擦片组合 steering damper friction assembly
转向减震器组合 steering damper assembly
转向角 deflection angle; steering angle;turning angle
转向角范围 steering range
转向角限止器 steering gear lock
转向角指示器 steering angle indicator
转向铰接头 steering link
转向接点 change-over contact
转向节 knuckle pivot; slewing journal;steering knuckle;stub axle
转向节臂 knuckle arm;steering knuckle gear rod arm;steering knuckle lever
转向节臂接头 knuckle end
转向节叉 steering yoke
转向节长度 knuckle length
转向节衬套 knuckle bushing
转向节的限位螺钉 steering knuckle stop screw
转向节短轴 steering stub axle
转向节横拉杆接头油嘴 steering knuckle tie rod end pressure lubricator
转向节横拉杆接头总成 steering knuckle tie rod end assembly
转向节横拉杆球节 steering knuckle tie rod ball joint
转向节上的轮毂轴 steering knuckle spindle
转向节枢 steering stub
转向节枢销 knuckle trunnion
转向节套筒 steering knuckle sleeve
转向节系杆 steering knuckle tie rod
转向节系杆端 steering knuckle tie rod end
转向节系杆端塞 steering knuckle tie rod plug
转向节系杆球端 steering knuckle tie rod ball
转向节系杆弹簧 steering knuckle tie rod spring
转向节销 knuckle pin;knuckle pivot;knuckle spindle; steering knuckle pivot;swivel pin
转向节型式 steering knuckle type
转向节旋转中心线 knuckle pivot center[centre]
转向节支架 knuckle support;steering knuckle bracket; steering knuckle support
转向节止推垫片 steering knuckle thrust washer
转向节止推轴承 knuckle thrust bearing
转向节中心轴 fork center[centre] shaft
转向节轴 knuckle spindle
转向节主销 spindle bolt
转向节主销盖 steering knuckle king pin dust washer
转向节主销轴线 steering axis
转向节柱 knuckle post
转向节总成 knuckle assembly
转向进路 diverted route; indirect route
转向锯 radial arm saw;radial saw
转向开关 pole changer
转向坑道 diversion gallery
转向控制 steering control
转向控制杆 steering control level
转向控制指令 steering order
转向控制轴 steering controlling shaft
转向控制装置 steering control
转向拉杆 steering track rod;tie bar;tie rod;steering link
转向拉杆机构 drag link mechanism
转向拉杆机构总成 steering linkage assembly
转向棱镜【测】deviating prism;deviation prism
转向离合器 steering clutch;steering clutch lever
转向离合器拨叉 steering clutch release yoke;yoke
转向离合器部件 steering clutch assembly
转向离合器从动盘 steering clutch driven disk
转向离合器分离拨叉 steering clutch release fork
转向离合器分离叉 steering clutch release
转向离合器分离环 steering clutch release ring
转向离合器分离摇臂 steering clutch release rocking lever
转向离合器钢片 steering clutch plate
转向离合器杠杆挡块 steering clutch control butt
转向离合器开关曲柄轴 steering clutch release crank axle
转向离合器壳 steering clutch housing
转向离合器控制 steering clutch control
转向离合器控制钮 steering clutch control button
转向离合器摩擦片 disc; steering clutch lining
转向离合器弹簧 steering clutch spring
转向离合器调整螺钉 steering clutch adjusting screw
转向离合器推杆 steering clutch push rod
转向离合器推杆臂 steering clutch push rod arm
转向离合器压板 steering clutch pressure plate
转向离合器闸轮 steering drum
转向离合器制动器 clutch brake; steering clutch brake
转向离合器轴 steering clutch shaft
转向离合器主动鼓 steering clutch driving drum
转向力 steering effort
转向力矩 turning moment
转向力矩测定仪 steering torque indicator
转向立轴 king pin
转向连杆 steering link
转向连杆系统 steering linkage
转向连接杆 steering tie rod
转向联动装置 steering linkage
转向联杆 steering gear connecting rod;steering linkage
转向联杆端 steering gear connecting rod end
转向联锁装置 running direction interlock
转向联轴节 steering coupling
转向梁 <纵向摇枕> steering beam
转向灵敏度 steering sensitivity
转向路感 steering response
转向轮 bend pulley;deflecting roller;fifth wheel;fifth wheel attachment;gear tumbler; hand wheel; steering hand wheel; steering roll; tumbler;guide roll <三轮滚轧机>
转向轮安装几何位置 steering wheel geometry
转向轮安装角测定仪 aligner
转向轮摆头 wobble of wheel
转向轮摆振 steering wheel flutter
转向轮定位 front-wheel alignment
转向轮辐 steering wheel arm
转向轮后倾效果 caster effect
转向轮减振装置 shimmy damper
转向轮键 steering wheel key
转向轮抗摆头装置 shimmy stop
转向轮轮胎 steering tire
转向轮十字架 steering wheel spider
转向轮手柄 tumbler holder
转向轮胎 steering tire
转向轮辋 steering wheel rim
转向轮隙 steering wheel tilting
转向轮中心距 steering track; steer wheel track
转向轮轴 steering wheel shaft
转向轮专用轮胎 steering wheel tire
转向螺杆 steering screw; steering spindle
转向螺母 steering nut
转向螺旋桨 screw-rudder
转向门 diverter gate
转向面 steering surface
转向模块 steering module
转向能力 steering capability; turning ability
转向盘 hand wheel;steering disc[disk]; steering wheel;wheel spinner
转向盘操纵力 steering wheel rim effort
转向盘操纵器 steering wheel actuator
转向盘定位 steering wheel placing
转向盘辐条 steering wheel arm
转向盘钢架芯 steering wheel steel core
转向盘拉具 steering wheel puller
转向盘轮辐 steering wheel spider
转向盘轮毂 steering wheel hub
转向盘轮缘 steering wheel rim
转向盘螺母 steering wheel nut
转向盘扭矩 steering wheel torque
转向盘锁 steering wheel lock
转向盘套 steering wheel cover
转向盘下变速杆 steering wheel gear change
转向盘游隙 steering wheel play
转向盘运动 steering wheel movement
转向盘轴 steering column; steering wheel shaft;steering wheel spindle
转向盘柱变速 handle change
转向盘柱变速杆 steering column gear control
转向盘柱换挡 handle change
转向盘柱上支架 steering column upper bracket
转向盘柱锁 steering column lock
转向盘自由间隙测量器 steering wheel clearance checking scale
转向皮带 pivot belt
转向平台 transfer platform
转向平台轨道 rails for transfer platform
转向鳍 <用于盾构纠偏的> steering fin
转向器 deflector;diverter;steering device;steering gear;switch;tumbler
转向器臂 pitman
转向器侧盖 steering gear housing side cover
转向器夹箍 steering gear clamp
转向器壳 steering gear housing
转向器壳垫密片 steering gear housing cover gasket
转向器壳端盖 steering gear housing end plate
转向器壳盖 steering gear housing cover
转向器连接架 bogie link
转向器扇形蜗轮 steering worm sector
转向器扇形蜗轮轴承 steering worm sector bearing
转向器上盖 steering gear housing top cover
转向器轴 steering column shaft
转向桥 steering axle
转向桥支座 steering axle support
转向桥轴 steer axle
转向桥轴座架 steer axle mount
转向球 steering ball
转向球节 steering ball joint
转向球节盖 steering ball cap
转向曲柄 steering crank
转向曲线 steering curve

转向驱动桥 driving and steering axle
转向驱动轴 steering driving axle
转向三角线 reversing triangle; triangular track; turnaround wye; wye track; Y-track
转向三角线测量 wye-track survey
转向三通阀 deflecting damper
转向扇形齿板 slewing arc
转向扇形齿轮 steering quadrant
转向设备 steering device; turning equipment【铁】
转向时间 time to turn
转向时前轮越轨 compensating drive
转向试验 steering test
转向试验场 steering pad
转向枢轴 steering pivot
转向水域 swing area; turning basin
转向伺服主油缸 servo master cylinder
转向隧道 dogleg tunnel
转向损失 bend loss
转向踏步 diminishing step; wheeling steps
转向台车 shunt carriage; transfer table
转向台风 recurving typhoon
转向梯形 steering trapezium
转向梯形机构 Arkerman steering gear
转向条纹 truck stripe
转向停止器 steering stop
转向筒子架 truck creel
转向头 steering head; steering yoke
转向头保险 steering head lock
转向头衬套 steering head bushing
转向头托架 steering head bracket
转向头钥匙 steering head lock key
转向凸轮 steering cam
转向托架 bogie bracket
转向弯头 reversible deflector
转向万向接头销 steering knuckle pin
转向万向联轴节 steering universal joint
转向稳定器 steering stabilizer
转向稳定性 steering stability
转向蜗杆 steering screw; steering worm
转向蜗杆齿形齿轮轴 steering worm sector shaft
转向蜗杆滚柱轴承 steering worm roller bearing
转向蜗杆偏心调整套 steering worm eccentric adjusting sleeve
转向蜗杆凸轮 steering cam
转向蜗杆止推轴承 steering gear worm thrust bearing
转向蜗杆轴 steering worm shaft
转向蜗杆轴承 steering worm bearing
转向蜗杆轴承外环 steering worm bearing outer ring
转向蜗轮 steering gear worm; steering worm gear; steering worm wheel
转向蜗轮轴 steering gear worm shaft
转向误差 turning error
转向西面 wester
转向系杆 steering tie rod
转向系检验工间 steering analysis department
转向系统 steering system
转向系统的振动 steering system vibration
转向系统压力 hydraulic steering system pressure
转向峡谷 gore of diverted river
转向下风 falling off
转向线 switchback; turn-back; passenger coaches turn-around siding【铁】
转向限角 steering lock
转向限位角 < 车轮的 > angle of lock
转向限制角 angle of lock
转向箱 steering box
转向销 king pin
转向销负倾角 negative caster
转向销支撑凸块 swivel pin boss
转向销轴线 knuckle pin center[centre] line
转向斜踏步 kite winder
转向信贷 swing credit
转向信号 steering signal
转向信号开关 direction signal switch
转向信号闪光灯 turn signal flasher
转向信号指示灯 turn signal control lamp
转向行星轮系 steering planetary
转向性能 cornering ability; steering behavio(u)r; steering quality
转向性能试验 turning test
转向旋盘 steering handle
转向旋转面 steering swivel
转向选择方法 steering selection
转向液压缸 steering cylinder
转向液压缸枢销 steering cylinder pivot pin
转向液压马达 steering hydraulic motor
转向液压系统 steering hydraulic system
转向液压油泵 steering hydraulic oil pump
转向一边 turn away
转向用液压缸 steering hydraulic cylinder
转向油阀 steering oil valve
转向油缸 steering hydraulic cylinder
转向余隙 steering play
转向语句【计】 go to statement
转向圆柱体枢销 steering cylinder pivot pin
转向匝道 reversing loop
转向闸 steering brake
转向闸带 steering band
转向闸瓦 steering brake-shoe
转向褶皱 deflection fold; deflexion fold
转向直拉杆 drag rod
转向值 turning value
转向指令 steering order
转向指示灯 blinker; direction indicator lamp
转向指示器 arm indicator; running direction indicator; trafficator
转向制动杆 steering brake lever
转向制动鼓 steering brake drum
转向制动鼓轴轴承 steering brake drum shaft bearing
转向制动器 steering brake
转向制动器放油开关 steering brake bleed nipple
转向制动器横轴 steering brake cross-shaft
转向制动器控制阀 brake control valve; break control valve
转向中心 turning center[centre]
转向轴 radial axle; steering axle; steering camshaft; steering shaft; stub axle
转向轴管 column tube
转向轴护管 jacket tube; steering column
转向轴护环 steering shaft grommet
转向轴颈 steering journal
转向轴枢销 steering axle pivot pin
转向轴套 steering column jacket; steering spindle bushing
转向轴套管衬套 steering tube bush
转向轴套油孔盖 steering column jacket oil hole cover
转向轴外管 jacket tube
转向轴蜗杆 steering shaft worm
转向轴止推轴承 steering shaft thrust bearing
转向轴轴承 steering shaft bearing
转向轴轴承填密 steering shaft bearing packing
转向主销 king pin; pivot pin; pivot stud; steering king pin
转向主销内倾角 kingpin inclination
转向主销倾角 pivot stud angle
转向主销推力轴承 steering knuckle thrust bearing
转向主销轴线与路面的交叉点 knuckle-pin center[centre]
转向助力缸 steering(actuation) cylinder
转向助力器 steering booster
转向助力油泵 steering pump
转向助力油缸 steering cylinder
转向助力装置 steering booster
转向柱 steering column; steering mast; steering pillar; steering post
转向柱变速 steering column gear change; steering post gear shift
转向柱隔片 steering post spacer
转向柱固定架 steering gear frame bracket
转向柱管 steering column tube; steering gear tube
转向柱夹 steering column clamp; steering post clamp
转向柱夹绝缘体 steering post clamp insulator
转向柱壳 steering pillar shell
转向柱控制 steering column control
转向柱套 steering post jacket; steering(wheel) tube
转向柱套管 mast jacket; steering column case
转向柱套管固定架 steering column bracket
转向柱套管托架 steering column mounting
转向柱套管轴环 steering post collar
转向柱调整手柄 adjustable steering column lever
转向柱万向节 steering column universal joint
转向柱斜度 steering rake
转向柱支架 steering column support
转向柱轴齿条 steering column shaft spline
转向柱轴花键 steering column shaft spline
转向柱座托架 steering post bracket
转向装置 abat-vent; deflector; steerage gear; steering apparatus; steering arrangement; steering component; steering device; steering gear; steering mechanism; transfer; turning gear
转向装置保险 steering gear lock
转向装置臂 steering gear arm
转向装置检验秤 steering gear checking scale
转向装置壳塞 steering gear housing plug
转向装置壳注油塞 steering gear housing(oil) filler plug
转向装置空间凸轮 steering gear tube cam
转向装置箱 steering gear case
转向装置销定杆 steering lock lever
转向装置轴销 steering pitman
转向装置总成 steering gear assembly
转向纵拉杆 drag link; steering drag rod; steering drag link
转向纵拉杆后端油嘴 steering drag link rear end grease fitting
转向纵拉杆前端油嘴 drag link front end grease fitting; steering drag rod front end pressure lubricator
转向纵拉杆球接头 steering drag rod ball
转向纵拉杆弹簧 steering drag rod spring
转向纵拉杆橡胶套 steering drag link rubber boot
转向阻力 steering drag; steering resistance
转向阻力矩 steering resisting moment
转像 reversed image
转像差 phase quadrature; quadrature
转像差成分 quadrature component
转像棱镜 Amici prism; image rotation prism; inverting prism; reversing prism; rotating prism
转像谱 quadrature spectrum
转像相差 phase quadrature
转销 write off
转/小时 revolutions per hour
转效点 break point
转效时间 break time
转楔测距仪 rotating wedge range-finder
转写法 method of transliteration
转写油墨 transfer ink
转芯架 trestles core
转形变异 transformation
转续宿主 paratenic host
转续线圈 repeating coil
转延侧面 < 指墙壁、嵌线 > return
转延侧面的压条 return bead
转延装饰嵌线 returned mo(u)lding
转眼钩 reverse eye hook
转焰炉 roller hearth continuous rotaflame furnace
转窑 ring kiln; rotary kiln
转窑法 rotary kiln method
转窑干燥器 rotary kiln dryer
转窑水泥 rotary kiln cement
转窑水泥飞灰 flue cement; flue dust
转窑运料车 tunnel kiln car
转窑(支)座 kiln pier
转叶 revolving leaf; rotating vane
转叶风速仪 deflecting vane anemometer
转叶给料器 rotary-vane feeder
转叶加料器 rotary-vane feeder
转叶料斗 pocket star
转叶轮 rotary impeller
转叶式拌和机 revolving blade mixer
转叶式供料器 pocket feeder
转叶式加料器 rotary-vane feeder
转叶式搅拌机 revolving blade mixer
转叶式水泵 swinging-vane pump
转叶式油泵 swinging-vane oil pump
转叶式增压器 vane-type supercharger
转一圈 take a turn
转一整圈 turn right round
转仪钟 driving clock
转移 shifting; avert; divert(ing); jump transfer; metastasis[复 metastases]; relegate; transceiving; transfer(ence); transferral; transferring; translocation
转移泵 shifting pump
转移表 branch table; jump table; transfer table
转移表向量 transfer table vector
转移舱口 transfer hatch
转移操作 branch operation; jump operation; shift operation; transfer operation; branching operation【计】
转移产权 transferred title
转移常数 transfer constant
转移成本 transfer cost
转移船位线 line of position run; position line transferred; replaced position line

转移存储 transfer memory
转移导纳 transfer admittance
转移的 transitional
转移的跑道入口 displaced threshold
转移的位置线 transferred position line
转移地址 branch address; jump address; transfer address
转移点 branching-off point; branch point; entry point; point of transition; transfer point; transition point
转移电极 transfer electrode
转移电流 transfer current
转移电流比 transfer current ratio
转移电路 carry circuit; transfer scheme
转移电压 transfer voltage
转移电阻 transfer resistance
转移调用 branch calling
转移发射装置 translauncher
转移法 transfer method
转移翻译机 transfer interpreter
转移反应 shift reaction; transfer reaction
转移方案 transfer scheme
转移方程 equation of transfer; transfer equation
转移分级层次 jump hierarchy
转移风 transferred wind
转移风管 transferred air duct
转移风口 transferred air grill(e)
转移封锁 jump lock
转移符号 transition symbol
转移概率 transition probability
转移概率矩阵 transfer probability matrix; transition probability matrix
转移概率律 transition probability law
转移概率系统 transition probability system
转移跟踪程序 branch tracer
转移工艺 transfer of technology
转移拱顶线<在交叉口或超高处的> cross-over crown line
转移规则 transition rule
转移轨道 transfer orbit; transfer path
转移轨道太阳传感器 transfer orbit sun sensor
转移辊 transferring roll(er)
转移过程 transfer process
转移函数 transfer function
转移函数测量仪 transfer function measuring instrument
转移横列 transferring course
转移弧 transferred arc
转移剂 transfer agent
转移继电器 transfer relay
转移交通量 transfer traffic volume; diverted traffic volume
转移胶合机 transfer laminator
转移校验 transfer check
转移接点 break-make contact
转移矩阵【数】transfer matrix; transition matrix
转移卡 transfer card; transition card
转移卡片 transfer card
转移控制 jump control; transfer control
转移控制指令 transfer-control instruction
转移律 transition law
转移率 transport rate
转移螺栓闸刀 transfer bolt cam
转移命令 transfer command
转移盘 transfer table
转移频率矩阵 transfer frequency matrix
转移器 diverter; translator
转移请求处理 transfer request handling
转移曲线 transfer curve; diversion curve
转移曲线分配法【交】diversion curve assignment
转移群 transferred group
转移人工 transferred labor
转移三角 selecting transfer cam
转移设备 transfer equipment
转移时的临时连接 transfer joint
转移时间 transfer time
转移室 transfer chamber
转移收益 transfer earning
转移速度 transfer velocity; translational velocity
转移(速)率 transfer rate
转移算法 branching algorithm
转移损耗 transfer loss
转移所有权 passage of title
转移特性 transfer characteristic
转移特性校正 transfer characteristic correction
转移特性曲线 transfer characteristic curve
转移条件 jump condition
转移同态 transfer homomorphism
转移投资 shift in investment
转移图 transform diagram
转移涂布 transfer coating
转移涂层机 transfer laminator
转移途径 route of metastasis
转移网络 transfer network
转移网络文法 transition network grammar
转移文件 movement documents
转移系数 removal factor; transfer coefficient; transfer factor; transfer ratio; transmission coefficient
转移线 line of transference; transition(al) line
转移线圈 transfer stitch
转移向量 jump vector; transfer vector
转移效率 transfer efficiency
转移信号 transfer signal
转移信息 transferred information; transinformation
转移信息量 mutual information
转移信息组 jump field
转移信用 transferred credit
转移性【交】transferability
转移性支出 transfer expenditures
转移延迟 transfer lag
转移印花 transfer printing
转移应力地区 region of transitional stress
转移语句 branch statement
转移预测 jump prediction
转移运算 branching operation
转移造型 transfer mo(u)lding
转移责任 hold harmless
转移责任之约定 hold harmless agreement
转移轧液 transfer padding
转移张量 transport tensor
转移支付 transfer payment
转移支付办法 procedures for transfer payments
转移值 branch value
转移指令 blank instruction; branch order; control transfer instruction; derailer; jump instruction; reflexive order; transfer command; transfer instruction; branch instruction【计】; transfer order
转移指令优化 optimization of branch instruction
转移指示器 branch on indicator; carry-over indicator
转移资金 transfer of financial resources
转移子程序 jump subroutine
转移阻抗 interaction impedance; transfer impedance; transition impedance
转移作用 transference
转椅 revolving chair; swivel chair
转椅座席 revolving seat
转义 trope
转义(字)符 escape character
转翼式操舵装置 rotary-vane steering gear
转翼式液压操舵装置 turning-vane steering gear
转翼式液压马达 rotary-vane motor
转印 rendition
转印油墨 transfer ink
转印装饰法 transfer decoration
转用 divert
转用公制 metrication
转用十进制 metrication
转油线换热器 transfer line exchanger
转运 carting; reforwarding of luggage; reshipment; transfer; transport; transportation; transship; trans(s)hipment
转运报单 transhipment entry
转运泵 transfer pump
转运仓 transfer hopper
转运仓库 transit depot; transit storage
转运舱单 transhipment manifest
转运车 transfer buggy; transfer car; transport car(t)
转运车站 transfer depot
转运传送机 transfer conveyer[conveyor]
转运道 transfer track
转运地点 transshipment platform
转运点 relay point; reloading point; transfer point; transition point
转运吊机 transport crane; transshipment crane
转运堆栈 transit shed
转运费 transfer cost; transfer fee
转运港 port of transshipment
转运工具 carrier
转运公共汽车 feeder bus
转运公司 forwarder; forwarding agent; forwarding company; freight agent; freight forwarder; transfer company; transit company
转运公司运输 forward traffic
转运轨道 transfer track
转运货 transit cargo
转运货车 transfer lorry
转运货棚 transit shed; transshipment shed; transship shed
转运货物 goods to be transshipped; transshipment cargo
转运货物的箱子 packing case
转运机构 transshipment activity
转运机械 transloader
转运基地 terminal
转运浇包 transfer ladle
转运口岸 entrepot
转运廊道 transfer corridor
转运量 trans(s)hipment
转运溜槽 transfer slide; transfer chute <皮带运输机用的>
转运溜井 transfer chute; transfer raise
转运码头<尤指煤码头> staith(e)
转运门 transfer door
转运能力 turnover capacity
转运平巷 transfer drift
转运起重机 transport(er) crane; transposed crane; transshipment crane; transfer crane <从货车至卡车或相反>
转运桥 transporter bridge
转运区(域) transit area; transshipment area
转运容器 transfer container
转运筛 trommel screen
转运平台 transfer platform
转运设备 rehandling plant
转运设施 transfer facility
转运式半拖挂车 transfer-type semitrailer
转运水仓 transfer sump
转运台 conveyer table; transfer table
转运台车 transfer troll(e)y
转运提单 transshipment bill of lading
转运条款 transit clause
转运铁路货车的多轮式公路牵引车<两条铁路的车站间在城市内未能连接时使用> wagon transporter
转运途中 in transit
转运系统 movement system; transfer system
转运箱 transfer box
转运小车 transfer car
转运形式 transport form
转运许可证 transhipment permit; transshipment license
转运业 forwarding business
转运医院 evacuation hospital
转运运输 secondary haulage
转运站 rehandling plant; relay station; transfer point; transfer station; transshipment point
转运证明书 diversion certificate
转运装置 drag-over skid; transfer gear
转载 reproduction; reshipment; transship
转载仓 transfer bin
转载点 transfer point; transshipment point
转载方法 transfer method
转载机 reloader; reversed loader
转载设备 rehandle facility; rehandling facility
转载运输机 reclaiming conveyer[conveyor]; transfer conveyer[conveyor]
转载装置 rehandling device; rehandling facility
转账 transfer of accounts; transferred account; account transfer
转账拨款账户 transfer appropriation account
转账传票 transfer slip
转账单 bill of transfer account; voucher
转账付款 transfer payment
转账付款系统 giro system
转账机构 transfer mechanism
转账价格 transfer price
转账卡 transferred account card
转账凭单 cost memo; transfer document; transfer voucher
转账凭证 transfer documents; transfer voucher
转账通知单 account transfer memo
转账性支付 giro-cheque; transfer payment
转账账簿 book of secondary entry
转账账户 account on transfer transactions
转账账目 account transfer memo
转账支票 check for transfer; giro-cheque; transfer account cheque only for account; transfer check; transfer payment slip
转账支票支付 pay in cross check
转折 breakover; break <特性曲线的>
转折点 break-even point; break point; breakthrough; buckling point; critical point; crunch; inflection point; inflexion point; kick point; knick point; pivotal point; point of cont-

raflexure; point of inflection; point of transition; transition location; turning point; turning station
转折点横坐标 abscissa of turning point
转折电压 breakover voltage
转折端 hinge end
转折角 angle of bend
转折棱镜 deflecting prism
转折时间 time transition
转折梯段的踏步 French flier
转辙【铁】switching; turnout track
转辙柄 point lever
转辙船闸 switch lock
转辙点 switching point
转辙电动机 electric(al) point motor; electric(al) switch motor; switch motor
转辙电路控制器 electric(al) point circuit controller; electric(al) switch circuit controller
转辙杆 switch bar; switch lever; point bar; point slide; switch slide; switch column <道岔就地操纵>
转辙工人 pointsman
转辙轨 connection rail; switch rail
转辙回路管制器 electric(al) point circuit controller; electric(al) switch circuit controller; point control switch; point instrument; switch box
转辙机 goat; switch machine
转辙机安装装置 switch machine installation
转辙机构 point mechanism; switch mechanism
转辙机控制器 switch machine controller
转辙机箱 casing of switch machine
转辙检测器 point detector
转辙检查器 plunger detector
转辙角 <铁路道岔> angle of switch; switch(crossing) angle
转辙联动装置 switch gear
转辙轮 <双导线制> point motion wheel
转辙器 cross-over switch; derailer; rail shunter; runaway switch; shunter; switch(er); goat; shunt
转辙器侧向区段 <与直通区段相对应> switching section
转辙器长垫板 switch soleplate
转辙器的止动器 switch-stop
转辙器垫板 switch plate
转辙器防爬铁撑 switch anchor
转辙器附件 switch fixtures
转辙器跟 <尖轨跟> heel of points; heel of switch
转辙器跟撑 brace for heel of switches
转辙器轨枕 switch tie
转辙器后连接杆 back connecting rod
转辙器拉杆 throw rod
转辙器拉杆座 throw switch stand
转辙器理论尖端 theoretic(al) point of switch
转辙器轮缘槽 clearance point; point trough
转辙器密贴调整杆 threw rod with switch adjustment
转辙器平衡铁 points counterbalance
转辙器前连接杆 front connecting rod
转辙器融雪器 switch heater
转辙器实际尖端 actual point of switch
转辙器锁闭器 plunger lock
转辙器信号 switch signal
转辙器闸座 throw switch stand
转辙器趾 toe of switch
转辙器中心 center[centre] of switch
转辙器座 switch stand
转辙器座锁闭闩 switch stand latch
转辙锁闭器 face point lock; facing point lock; mechanical plunger; plunger lock; point bolt; point lock; switch and lock movement
转辙握柄重锤 switch weight
转辙握柄座 switch stand
转辙信号 switch signal
转辙员 switchman
转针 rotating the needle
转针延展机 rotary gill box
转振 rotational oscillation
转振光谱 rotational vibration spectrum
转振谱带 band of rotation-vibration; rotation-vibration band
转正棱镜 erecting prism
转职日期 date of transfer
转质模型 transferring mass model
转致轮廓 rotation(al) contour; rotation(al) profile
转置 transpose; transposition
转置伴随矩阵 adjugate matrix; transposed matrix
转置伴随行列式 adjugate determinant
转置表示 transposed representation
转置并向量 transposed dyadic
转置方程 transposed equation
转置积分方程 transposed integral equation
转置矩阵【数】transposed matrix; associated matrix; transformation matrix; transpose of a matrix
转置平面 plane of transposition
转置算子 transposed operator
转置微分算子 transposed differential operator
转置文件 transposed file
转置线性映射 transposed linear mapping
转置性 permutability
转置映射 transposed mapping
转轴 center journal; pivot shaft; revolution axis; rotating shaft; rotation axis; rotor(shaft); spindle
转轴安装的 trunnion-mounted
转轴编码器 shaft encoder
转轴吊车 rolling truck
转轴浮筒 swivel pontoon
转轴固定螺钉 rotary shaft set bolt
转轴滑轮 swivel mount pulley
转轴颈 trundle
转轴密封 rotary shaft seal
转轴镊 Knapp's forceps; roller forceps
转轴切砖器 reel cutter
转轴上的键槽 spindle keyway
转轴式拌和机 rotary(type) mixer
转轴式窗 pivoted window
转轴式窗遮帘 roller blind
转轴式吊杆 gooseneck boom
转轴式吊艇柱 gooseneck davit; swannecked davit
转轴式搅拌机 rotary(type) mixer; rotating mixer
转轴式门 pivoted door
转轴式黏[粘]度计 rotating spindle visco(si)meter
转轴式松土拌和机 rotovator
转轴式松土机 rototiller
转轴头 <提运货的> nigger head
转轴叶片 rotor blade
转轴用磁粉密封 magnetic powder seal for rotating shafts
转轴支墩 pivot pier
转轴转速 rotor speed
转轴装置 pivot mechanism; turning gear
转主寄生 tropoparasite
转贮点 dump point
转贮校验 <内存信息> dump check
转柱 rotary column
转柱门 post gate
转柱起重机 pillar crane
转柱式回转支承 slewing ring with rotary column
转柱式起重机 pillar crane; slewing pillar crane; post crane; column crane
转柱式塔式起重机 tower crane with inner mast
转柱式挺杆起重机 pillar jib crane
转柱悬臂起重机 pillar jib crane
转爪锚 stockless anchor; swinging fluke anchor
转装费 transloading cost
转桌 turntable
转桌轴套 table bushing
转子 armature; face roller; rotator; rotor; runner; wheel runner; impeller <水泵的>; cup vane assembly <旋杯流速仪>
转子安匝 rotor ampere turns
转子安装场地 rotor assembly space
转子半径 rotor radius
转子半径偏心距比 radius-to-eccentricity ratio
转子绑线 rotor bandage
转子本体槽 rotor-body slot
转子泵 drum pump; impeller pump; orbit pump; rotary pump; rotatory pump; rotor pump
转子变阻器 rotor rheostat
转子波阻 rotor wave resistance
转子槽楔 rotor slot wedge
转子侧板 impeller side plate
转子柴油机 rotary diesel engine
转子长度 rotor length
转子成型术 trochanter plasty
转子齿 rotor tooth
转子齿距 rotor slot-pitch
转子齿轮 rotor gear
转子齿轴 rotor pinion
转子冲片 rotor punching; rotor sheet
转子出线盒 rotor terminal box
转子传动筛 rotor driven screen
转子串电阻起动 rotor-resistance starting
转子磁场 rotor field
转子磁轭 rotor rim
转子磁铁 rotor magnet
转子刀形开关 roller leaf actuator
转子导动装置 rotor guidance
转子导流锥体 rotor guide cone
转子导前角顶 rotor's leading apex
转子导体 rotor conductor
转子导线 rotor wire
转子的 trochanterian; trochanteric
转子电流 rotor current
转子电路 rotor circuit
转子电路调节 regulation in rotor circuit
转子电压 rotor voltage
转子电阻 rotor resistance
转子电阻起动器 rotor-resistance starter
转子叠片组 rotor plate assembly
转子动力泵 roto-dynamic(al) pump
转子动力学 rotor dynamics
转子端箍 rotor end-cap
转子端环 rotor end ring
转子端帽 rotor end-cap
转子端面油封装置 rotor side oil seals
转子短路 rotor short-circuit
转子对偏心轴的转速比 rotor-to-crankshaft speed ratio
转子发动机 epitrochoidal engine; rotary expansion engine; rotary piston machine; rotary piston motor; rotating combustion engine; rotomotor
转子发动机的速比 rotor gearing ratio
转子发动机缸体 rotor housing
转子发动机工作循环 rotary engine combustion cycle
转子发动机货车 rotary truck
转子发动机小客车 rotary engined car
转子发动机主轴轴承 rotary engine bearing
转子发热 rotor heating
转子阀门 discharge rotary valve
转子反应 rotor reaction
转子风孔 rotor vent
转子风阻损耗 rotor windage loss
转子辐臂 rotor spider
转子负载 rotor loading
转子工作面 rotor peripheral face
转子工作面凹坑 recess in rotor; rotor (face) recess; rotor pocket; rotor depression <转子发动机的>
转子工作面空腔 rotor face cavity
转子工作面浴盆状凹坑 rotor face tub
转子故障保护 rotor earth fault protection
转子惯性冷却法 rotor inertia cooling
转子过载保护 rotor overload protection
转子护环 carrying rod
转子护环装配 rotor banding
转子环 rotor ring
转子活塞泵 rotopiston pump; runner piston pump
转子极靴 rotating pole-piece
转子挤压作用 rotor squish action
转子夹板座 rotor bridge support
转子间嵴 intertrochanteric crest
转子间径 bitrochanteric diameter
转子间线 intertrochanteric line; linea intertrochanterica
转子角 rotor angle
转子角顶凸岸 raised apex lands
转子角动量 rotor angular momentum
转子角度检测仪 rotor angle detector
转子角位指示器 rotor angle indicator
转子接地 rotor ground
转子节距 rotor pitch
转子进气角 rotor inlet gas angle
转子开关 roller actuator
转子壳 rotor case
转子宽度对创成半径的比值 rotor width-engine generating radius ratio
转子馈电式电动机 rotor feed type motor
转子馈电式多相并激电动机 rotor feed type polyphase shunt motor
转子类型 rotor type
转子冷却 rotor cooling
转子量的折算 referring of rotor quantity
转子流量计 flowrator; purgemeter; rotameter; rotary flowmeter; rotor meter; variable area flowmeter
转子流速计 rotameter; rotary meter
转子流速仪 rotor flowmeter
转子漏磁场 rotor leakage field
转子轮辐 rotor spoke
转子轮毂 rotor hub
转子轮缘 rotor rim
转子密封 rotor seal
转子密封环 rotary piston sealing ring
转子密封系统 rotor sealing system
转子挠度 rotor bow; rotor sagging
转子膨胀 rotor expansion
转子偏移 rotor displacement
转子频率 rotor frequency
转子平衡 rotor balancing
转子曝气 rotor aeration
转子曝气渠 rotor aeration ditch
转子起动器 rotor starter
转子绕组 armature loop; rotor winding
转子上的条形风力分级孔 airway slits in rotor

转子式测流(速)仪 purgemeter;rotor type current meter
转子式除雪机 rotary snow remover
转子式除雪犁 rotary snow plow
转子式定量给料秤 dosing rotor weighfeeder
转子式发动机 rotor motor
转子式翻车机 rotary car dumper;rotary tipper;rotating dumper
转子式翻斗车 rotary car dumper
转子式杠杆开关 roller lever actuator
转子式继电器 rotor-type relay
转子式节气门 rotary throttle
转子式空气压缩机 rotary air compressor
转子式沥青乳化机【机】rotary bitumen emulsifying machine
转子式流量计 rotary flowmeter
转子式流速仪 rotameter;velocity-type flowmeter
转子式滤清器 rotary filter
转子式煤气表 Connersville meter;Root's meter;rotary displacement meter
转子式喷燃器 rotary burner
转子式喷射机 rotor spraying machine
转子式曝气法 rotor aeration
转子式燃烧器 rotary burner
转子式输油泵 rotary fuel-feed pump
转子式水表 roto-dynamic(al) water meter
转子式碎石机 rotator crusher
转子式稳定土搅拌机 rotary type soil stabilizer
转子式细粒造粒机 rotor fine granulator
转子式压气机 rotary compressor
转子式液压马达 rotary hydraulic motor
转子收缩 rotor contraction
转子水箱 rotor water box
转子速度加减键 <陀螺罗经用> motor speed switch
转子损耗 rotor loss
转子体 rotor block
转子体积 rotor volume
转子铁芯装配 rotor-core assembly
转子停转电压 locked rotor voltage
转子铜耗 rotor copper loss
转子铜条 rotor bar
转子铜线 rotor copper
转子头 rotor head
转子凸棱 rotor nog
转子推力平衡活塞 rotor thrust balance piston
转子外环 rotor exterior ring
转子外接电阻起动 secondary resistance starting
转子外径 rotor diameter;rotor outer diameter
转子弯曲 rotor bow
转子尾后角顶 rotor's trailing apex
转子位移角 rotor displacement angle
转子位置传感器 rotor-position sensor
转子温度计 <自动发出温度临界点信号的仪器> rototherm
转子温度梯度 rotor temperature gradient
转子窝 fossa trochanterica
转子线棒 rotor bar
转子线槽 rotor slot
转子线圈 rotor coil;rotor spool;rotor winding
转子线圈主绝缘 rotor slot armor
转子线匝 rotor turn
转子相对移动 relative rotor displacement
转子相位 rotor phase
转子相位齿轮机构 rotor timing gears
转子箱 rotor case;rotor casing
转子芯 rotor core
转子型高落式浮选机 k-and-k machine
转子压缩机 rotor compressor;straight-lobe compressor
转子叶轮组合件 rotor impeller
转子叶片 impeller;rotor blade;rotor vane;runner vane
转子叶片泵 rotating vane pump
转子叶片组 rotor blader
转子叶栅组 set of rotor vanes
转子应力 rotor stress
转子应力程序 rotor stress program(me)
转子圆周速度 rotor peripheral speed
转子运动轨道 rotor orbit
转子杂散磁场 rotor stray field
转子噪声 rotator noise;rotor noise
转子罩 rotor cap
转子振动 rotor oscillation
转子振动筛 rotor-vibrated screen
转子支架 field spider <水轮发电机的>;spider;rotor field spider
转子止动杆 rotor stop lever
转子止退垫圈 rotor check washer
转子止转试验 rotor-blocked test
转子制动器 rotor brake
转子制动着陆 rotor-type landing
转子中心 rotor center[centre]
转子中心体 rotor center part;rotor hub
转子重量 rotor weight
转子周边曲线 rotor flank contour
转子周面 rotor flank
转子周期凹坑 rotor pocket recess
转子轴 rotor axis;rotor shaft;rotor spindle
转子轴承 rotor bearing
转子轴承油槽 rotor bearing reservoir
转子轴孔 rotor shaft hatch
转子转差率 rotor slip
转子转速 rotor speed
转子装配 rotor assembling;rotor assembly
转子装配台 rotor erection pedestal
转子组 rotor set
转子组合锁 puzzle lock
转子组合体 rotor assembly
转字锁 dial lock
转字锁转轮 combination tumbler
转租 assignment of leases;relet;sandwich lease;sublease;sublet(ting);underlease;subtenancy <不动产的>
转租承受人 sublessee
转租承租人 undertenant
转租船舶 subchartering
转租合同承租人 subcharterer
转租人 sublessor;subtenant
转租人所付租金 subrent
转租通知书 letter of attornment
转租土地 land subleased
转足羊足碾 turn foot roller
转足(羊足)压路机 turn foot roller
转钻 rotary boring;rotary drilling
转作资本 capitalization
转座 swing-around seat;swivel base;swivel mount;swivel seat
转做承包的工作 subcontract

啭

啭声 warble tone

赚

赚得纯利 net profit earned

撰

撰稿人 contributor
撰稿员 <尤指广告> copywriter

庄

庄稼保护 crop protection
庄司顿土壤灌浆加固法 Johnsten process
庄严的风格 solemnity of style
庄严的集会 consistory
庄园 chateau;estate;grange;manor;plantation;villa garden
庄园建筑 schloss
庄园(内)道路 estate road
庄园中农场 barton
庄园主的大住宅 manor-castle
庄园主住宅 mansion(house);manor house;manor place;hacienda <美国南部的>
庄重的窗 monumental window

桩

桩 pile
桩坝 stake dam
桩板 pile sheathing;sheeting plank
桩板式挡土墙 sheet-pile retaining wall
桩背后钉木板的护岸 back-boarding
桩被打成弯曲形 dogleg(ged) pile
桩壁 pile lining
桩标点 pegged station
桩表面摩擦 pile friction
桩表皮 pile skin
桩材 piling;post
桩侧壁阻力 lateral pile resistance
桩侧表面负摩擦力 negative skin friction along the pile
桩侧摩擦(阻)力 skin friction along the pile;pile friction resistance
桩长 pile height
桩撑浮筏 pile-supported
桩承驳岸 piled quay
桩承底脚 pile(d) footing;pile foundation
桩承筏基(础) pile-supported raft;piled raft
桩承浮阀基础 pile-supported raft (foundation)
桩承浮筏 pile-supported raft
桩承基础 pile-supported footing
桩承基脚 pile-supported footing
桩承基础底板 piled raft
桩承连续基础 pile-supported continuous footing;pile-supported continuous foundation
桩承码头 piled quay
桩承桥 pile bridge
桩承式岸壁 quay wall on piles;quay wall on pile type
桩承受的上浮力 uplift on pile
桩承受荷载的弹性缩短 elastic shortening of loaded pile
桩承台 pile cap(ping);piling platform;platform on piles;relieving-type platform supported on piles;rider cap
桩承台梁 pile capping beam
桩承台式岸壁 pile-supported quay-wall;platform quay wall;platform wall;quay wall on piles;quay wall with platform on piles;relieving platform type of quaywall
桩承台式码头岸壁 pile-supported wharf wall;platform wharf wall
桩承载力 pile load capacity
桩承载力的歇后增长 pile freeze
桩承载力公式 pile(capacity) formula
桩承载量 pile bearing capacity;pile capacity
桩承载能力 pile capacity
桩承载试验 load test on pile;pile test
桩承栈桥 pile trestle
桩承重力 pile bearing capacity;pile load capacity
桩承作用 pile action
桩锤 beetle head;driving hammer;monkey of pile driver;pile driver hammer;pile driving hammer;pile hammer;pile monkey;ram;rammer;ramming weight;Vulcan
桩锤侧面导向角钢 angle iron guide
桩锤尺寸公式 hammer-size formula
桩锤冲程 hammer stroke;ramming stroke
桩锤冲扩桩法 impact displacement column method
桩锤锤击点 ram point
桩锤导杆 ram guide
桩锤导架 pile hammer guide;piling guide;piling lead;ram guide;ram guiding
桩锤导柱 ram guide
桩锤底 ram point
桩锤垫 hammer cushion;hammer grab
桩锤吊钩 dog
桩锤吊索 hammer line;hammer rope
桩锤钩 monkey hook
桩锤夯 hammer tamper
桩锤活索 trip rope
桩锤击 pile hammer blow
桩锤架 hammer stand
桩锤卷筒 pile hammer drum
桩锤缆索 pile hammer rope
桩锤落高 hammer drop;stroke
桩锤帽 pants
桩锤台 hammer stand
桩锤头 pile hammer head
桩锤围栏 pile hammer rail
桩锤砧 anvil of piling hammer
桩锤罩 hammer case
桩锤重 pile hammer weight
桩锤重量 pile hammer weight;ram weight;weight of monkey
桩锤总质量 total mass of pile hammer
桩簇坝 pile-clump dike[dyke]
桩簇丁坝 pile-clump spur dike[dyke]
桩簇捆绑 pile-clump lashing
桩打的防波堤 stockade
桩打入地下时的冲击应力 driving stress
桩挡土墙 piled dike[dyke]
桩的拔出试验 pull-out test for pile
桩的保护层 pile coating
桩的保护套 pile encasement
桩的表面负摩擦力 negative skin friction of piles
桩的表面摩擦力 pile adhesion;skin friction of pile
桩的表面阻力 pull-out resistance of pile
桩的布置 arrangement of piles;pile layout
桩的残根 stump
桩的侧向荷载试验 lateral pile load test
桩的侧向土压力 lateral earth pressure against piles
桩的颤动 pile flutter
桩的沉降 pile settlement
桩的承载机(理) bearing mechanism of pile
桩的承载力 carrying capacity of pile;carrying power of pile;pile capacity
桩的承载力公式 pile capacity formula
桩的承载力快速试验 quick test
桩的承载力试验 pile load(ing) test;pile test(ing)
桩的承载量 pile bearing capacity;pile capacity

Z

桩的承载能力 pile bearing capacity; pile capacity; supporting capacity of a pile; supporting power of a pile
桩的承载试验 pile load(ing) test
桩的尺寸 pile size
桩的垂直反力系数 coefficient of vertical pile reaction
桩的垂直锚碇 vertical anchor for pile
桩的粗端 blunt end of pile; butt of pile; pile butt <一般指原木桩>
桩的打入深度 hold of pile
桩的打入阻力 driving resistance
桩的大小 pile size
桩的等速贯入试验 constant rate penetration test of pile
桩的底座 footing on piles
桩的定位 pile positioning
桩的动测法 dynamic(al) measurements of pile
桩的动测试验 TNO 方法 TNO method of pile testing
桩的动荷载试验 dynamic pile load test
桩的动力荷载试验 dynamic(al) load test of pile
桩的动力(小应变)测试 shock or transient response method of pile testing
桩的动力学 pile dynamics
桩的堆放(场) pile staging (storage yard)
桩的钝头 blunt end of pile; butt of pile
桩的反力 pile reaction
桩的非破损性试验 pile integrity test
桩的动力试验 dynamic pile testing
桩的端承载力 end-bearing capacity of pile
桩的分部安全系数 partial safety factors of pile
桩的腐蚀 pile corrosion
桩的负摩擦力 negative skin friction along the pile; negative skin friction of pile
桩的复打 redriving of pile
桩的刚度 stiffener of pile
桩的钢筋骨架 pile cage
桩的钢靴 steel shoe
桩的根端 butt of pile
桩的工作荷载 pile working load
桩的鼓状部位 pile drum
桩的贯入 penetrating pile; pile penetration
桩的贯入度 penetration of pile; pile penetration
桩的贯入量 penetration of pile; pile penetration
桩的贯入试验 driving test of pile
桩的灌浇法 pile-placing method
桩的荷载试验 pile load(ing) test
桩的横向窜动 lateral springing of pile
桩的横向构架 pile bent
桩的横向抗力 lateral resistance of pile
桩的横向弹动 lateral springing of pile
桩的回升 pile heave
桩的回弹 rebound of pile; spring-back of pile
桩的回跳 rebound of pile
桩的回跃 rebound of pile; spring-back of pile
桩的基脚 footing on piles
桩的极限侧阻力 ultimate skin friction of pile
桩的极限承载力 pile ultimate (bearing) capacity; ultimate capacity of pile; ultimate bearing value of a pile
桩的极限端阻力 ultimate tip resistance of pile; pile ultimate tip resistance
桩的极限荷载 limit load of pile; ultimate load of pile; ultimate pile load
桩的加荷 pile loading
桩的尖端 tip of pile
桩的检验荷载 pile proof load
桩的间隔 pile spacing
桩的间距 pitch of piles; spacing of piles
桩的截断处 cut-off
桩的截面面积 pile cross-section
桩的接长 pile extension
桩的劲度 <桩的弹性模量和桩长之比> stiffness of piles
桩的静力公式 static formula of pile
桩的静载荷试验 static-load pile testing
桩的静载试验 pile load(ing) test
桩的就位 setting pile
桩的抗拔力 pulling resistance of a pile; pull-out resistance of pile; upright resistance
桩的抗沉 refusal of pile; refuse of pile
桩的抗拉试验 pulling test on piles
桩的拉力试验 pulling test on piles
桩的类型 type of piles
桩的连接 pile joint
桩的联系杆 bracing to piling
桩的临界承载力 critical bearing capacity of pile
桩的临界荷载 critical load of pile
桩的隆起 pile heave
桩的黏[粘]着系数 adhesion factor of pile
桩的凝固 pile freeze
桩的扭剪试验 pile torque shear
桩的蓬裂顶部 broom head of pile
桩的偏位 pile eccentricity
桩的偏心距 pile eccentricity
桩的拼接 pile splicing
桩的拼接物 pile splice
桩的平均贯入度 mean penetration of pile
桩的起吊长度 handling length of piles
桩的起吊点 pick-up point; lifting point of pile
桩的起吊方式 pick-up arrangement of piles
桩的倾度 pile rake
桩的倾斜度 pile eccentricity; pile rake
桩的屈服 yielding of pile
桩的容许承载力 allowable bearing capacity of pile
桩的容许荷载 allowable pile (bearing) load
桩的容许偏位 pile tolerance
桩的入土长度 depth of penetration; embedded length
桩的入土深度 penetration of pile; pile penetration
桩的入土总深度 total penetration of pile
桩的上拔力 uplift of pile
桩的设计荷载 pile design load
桩的设置 installation of pile
桩的深度-阻力曲线 depth resistance curve of pile
桩的声波测试 acoustic(al) pile test
桩的实际安全荷载 actual safe load of pile
桩的试验荷载 pile proof load
桩的竖向反力系数 coefficient of vertical pile reaction
桩的水平承载力试验 lateral pile load test
桩的水平反力 horizontal piling reaction
桩的水平反力系数 coefficient of horizontal piling reaction; coefficient of horizontal pile reaction
桩的套管 pile casing(tube)
桩的套护 pile jacketing
桩的特征图像 signature of pile
桩的弯曲力矩 pile bending moment
桩的完整性试验 pile integrity test
桩的维持荷载试验 maintained load pile test
桩的相互干扰 mutual interference of piles
桩的斜度 pitching of pile
桩的卸载 pile unload
桩的压应力 pile compressive stress
桩的验证荷载 pile proof load
桩的有效埋深 effective embedment of sheet pile
桩的预制场 pile-casting yard
桩的预制段 pile segment
桩的遮帘面 <前板桩高桩码头> shielding surface
桩的遮帘作用 screen effect of pile
桩的支承能力 pile-bearing capacity
桩的支承作用 pile action
桩的直径 pile diameter
桩的止点 <桩的最后打入深度> refusal of pile; pile stoppage; refusal (point); refuse of pile
桩的制作 pile manufacture
桩的中性点 neutral point of pile
桩的重量 weight of pile
桩的总长度 overall length of pile
桩的总贯入量 total penetration of pile
桩的总下沉量 gross penetration
桩的最终下沉 final set of pile
桩堤 piled dike[dyke]
桩底 pile toe
桩底沉渣检测 sludge measurement of bored pile
桩底脚 pile pedestal
桩底圆趾 pile bulb
桩点 peg(ged) point
桩点法 picket-point method
桩点法测图 pegged point mapping
桩垫 anvil block; cap block; cushion; pile block; pile cushion; pile pad
桩垫木 dolly
桩垫头 cushion head
桩吊点 pick-up point
桩钉 dowel pin
桩顶 head of pile; pile block; pile top
桩顶标高 pile tip elevation
桩顶大板 capping slab
桩顶大梁 pile cap girder
桩顶垫层 cushion block
桩顶垫块 doll(e)y; dollie
桩顶发裂 brooming
桩顶箍 pile(-crown) ring
桩顶横梁 capping beam; capping girder
桩顶开花 brooming
桩顶开花抗力 resistance to broom
桩顶抗裂能力 resistance of pile brooming during driving; resistance to broom
桩顶联系梁 pile cap beam
桩顶帽 pile helmet
桩顶面积 area of pile head
桩顶散裂 brooming
桩顶填芯混凝土 filling concrete for "tube-pile" head
桩顶位移 pile top displacement
桩顶自由桩 free end pile
桩定 pegging out; peg out
桩定坡面(点)【测】 slope staking
桩定位筋 pile fork; pile spacer bar
桩动力公式 dynamic(al) pile formula
桩端 pile tip; point of pile; tip of pile
桩端标高 pile butt elevation; pile head elevation; tip grade
桩端(部)支承力 end bearing resistance of pile
桩端承载量 end bearing capacity of a pile; point-bearing capacity
桩端承载能力 end bearing capacity of a pile; point-bearing capacity of a pile
桩端承重能力 end bearing capacity of a pile; point-bearing capacity
桩端高度 pile butt elevation; pile head elevation
桩端阻力 pile toe resistance; point resistance of a pile; tip resistance; toe resistance of a pile; end resistance
桩段 pile cutoff; pile section
桩堆放场 pile yard
桩墩 pile; pile pier; stub; stump
桩墩盖梁 cap wale
桩筏基 pile(d) raft
桩筏基础 pile-raft foundation
桩负摩擦力 downdrag; negative friction of pile
桩复打试验 pile redriving test; pile retapping test
桩盖顶 pile top
桩杆 stake
桩钢筋骨架保护层垫环 pile cage spacer
桩钢筋笼 pile cage
桩钢罩 bonnet
桩高 pile height
桩格排 pile grillage
桩格排基础 pile grillage foundation
桩隔板 pile diaphragm
桩隔墙 pile diaphragm
桩根 stump
桩工 palification; pile work; piling
桩工波动方程分析 wave equation analysis of piling behavio(u)r
桩构堵壁 pile bulkhead
桩构排架 pile trestle
桩构栈道 pile trestle
桩菇属 <拉> Paxillus
桩箍 drive band; driving band; ferrule of pile; follow block; lagging of pile; pile band; pile collar; pile driving helmet; pile ferrule; pile hoop; drive collar <桩头铁圈>; pile ring <木桩的>
桩固地基法 <用砂桩、木桩、灰土桩等加固地基> palification
桩固横撑丁坝 pile and waling groynes
桩冠 dowel crown; post crown
桩管 cutting shoe; driving tube; pile casing(tube); piling pipe
桩管刃脚 cutting shoe
桩贯入度 penetration of pile
桩贯入试验 pile penetration test
桩号 chainage【测】; station number【测】; pile No.
桩和板桩打桩工作 pile and sheet-pile driving work
桩和桩基础 pile and pile foundation
桩荷载 load on pile; pile load
桩荷载试验 load test on pile; pile load(ing) test; pile test
桩护壁 pile sheathing
桩环 anchor ear; pile ring
桩回升 pile heave
桩回跃 spring-back of pile
桩基 piled foundation; pile bottom; pilotis <水下或陆地群桩基础>
桩基岸壁 quay wall on a pile foundation; pile bulkhead
桩基板型基础 piled raft
桩基层 piling layer
桩基沉降 settlement of pile foundation
桩基承台 footing slab; pile grillage; pile-supported platform
桩基承台码头 quay wall on piles
桩基承台岸壁 quay wall on piles

桩基承载能力 bearing capacity of pile foundation
桩基础 piled footing; piled foundation; pile footing; pile foundation; piling footing; piling foundation
桩基础破坏 failure of a pile foundation
桩基础钻探机 boring machine for pile foundation
桩基的 piled; piling-supported
桩基等效沉降系数 equivalent settlement coefficient for calculating settlement of pile foundations
桩基低应变动测法 low strain integrity testing of pile foundation; pile integrity test
桩基底灌浆 pile base grouting
桩基动力学 pile dynamics
桩基墩 piled dolphin
桩基防波堤 piled jetty; piled breakwater
桩基刚度 pile foundation stiffness
桩基钢筒 cylinder
桩基高应变动测法 high strain integrity testing of pile foundation; pile driving analyser[analyzer] method
桩基格排 pile foundation grillage
桩基工程 pile foundation work
桩基荷载试验 pile load(ing) test
桩基后压浆 post grouting for pile
桩基计算 K 值法 design of pile foundation by K-method
桩基计算 M 值法 design of pile foundation by M-method
桩基检测 pile test
桩基减压平台 relieving platform supported on bearing piles
桩基减载平台 open relieving platform
桩基建筑物 pile structure
桩基脚 post footing
桩基结构 pile(foundation) structure; pile-supported structure
桩基靠船墩 pile(d) dolphin
桩基块体码头 pile-supported block wall quay
桩基连续基础 pile-supported continuous footing; pile-supported continuous foundation
桩基码头 open pier; piled jetty; piled pier; piled wharf; pile pier; pile quay; pile wharf; quay on piles
桩基平台 piles driven platform; piles supported platform
桩基桥 pile(d) bridge
桩基桥台 piled bridge abutment
桩基设计计算的方法 alpha method of pile design
桩基式 open-piled type
桩基式近海平台 pile structure offshore platform
桩基受最大荷载的计算沉降量 calculated settlement of pile foundation under maximum loading
桩基突堤(码头) piled jetty; piled pier; piled trestle
桩基与混凝土基脚的锚固 incorporation of piles into a concrete footing
桩基载荷试验 pile load(ing) test
桩基栈桥 pile(d) pier; pile(d) trestle
桩基钻机 drilling machine for pile foundation
桩及钢丝网石笼护岸建筑 stake-and-stone sausage construction
桩及横撑丁坝 pile and waling groin; pile and waling groynes
桩及石笼护岸建筑 stake-and-stone sausage construction
桩极限承载力 ultimate bearing capacity of pile
桩极限荷载 limit pile load
桩加固地基 palification
桩夹 pile holder; pile keeper
桩架 pile driver; pile frame; pile helmet; piling rig; ram guide
桩架基座 bed-frame of pile driver
桩架桥 pile bridge
桩架座基 bed(ding) of a pile driver
桩尖 foot of pile; pile point; pile tip; pile toe; point of pile; tip of pile; toe of pile; point of spud【疏】
桩尖变形 point deformation
桩尖标高 tip grade; toe level of pile; pile tip elevation; tip elevation
桩尖标高线 toe line
桩尖承载力 end bearing of pile; pile point bearing capacity
桩尖端 point of pile
桩尖封闭式桩 <指空心桩> closed-end pile
桩尖加强 pile tip reinforcement
桩尖抗力 pile end-resistance
桩尖扩大 under-reaming of pile
桩尖扩大基础 enlarged pile base
桩尖扩大桩 expanding pile; belled pile; enlarged-base pile
桩尖扩孔机 belling bucket; under-ream bucket
桩尖头 point of pile
桩尖斜度 tip grade; toe line of pile
桩尖支承 pile point bearing
桩尖阻力 end resistance; pile point resistance; pile tip resistance; point resistance of pile; tip resistance of pile
桩尖阻力系数 end-resistance coefficient; point damping factor
桩间板壁 intermediate bearer
桩间板墙 interpile sheeting
桩间标准距离 <在英美通常用 100 英尺,过去也有用一链即 66 英尺的,1 英尺 = 0.3048 米>【测】station
桩间插板 interpile sheeting
桩间挡土板 interpile sheeting
桩间的水平挡土板 interpile sheathing
桩间距 pile spacing
桩间水平支撑 interpile sheeting
桩间水平支撑板 interpile sheeting
桩角 foot of pile
桩脚 foot of pile; pile foot; pile tip; pile toe; socle
桩脚附加钢筋 toe steel
桩脚线 toe line
桩接 staking
桩接的 pin-connected
桩接头 pile splice
桩截段 pile cutoff
桩(截流的)砂笼板 sheet piling gabion filled with sand
桩截水墙 pile cutoff
桩径 pile diameter
桩距 pile spacer; pile spacing; spacing of pile; spacing between pile
桩距两用汽车立方码 <计算土方运输的单位,英制桩间距离 100 英尺,1 英尺 = 0.3048 米> station yard
桩橛 deadman
桩抗拔力 pulling resistance
桩抗力 resistivity
桩壳 shell of pile
桩孔 pile hole
桩盔 driving helmet; pile helmet
桩拉伸应力 pile tensile stress
桩篱 pile and fascine
桩篱式围篱 pale fence; paling fence
桩力计 pile force ga(u)ge
桩螺栓 pile bolt
桩锚 pile(d) anchor(age)
桩锚碇 pile anchoring
桩帽 anvil block; cap block; cap of pile; dolly; drive head; driving cap; driving head; driving helmet; driving hood; false pile; follow block; pile cap(ping); pile cover; pile crown; pile driving helmet; pile head; pile helmet; piling cap; piling crown; rider cap; pile-driving cap <美>
桩帽导向板 guiding for pile helmet
桩帽垫 head packing
桩帽梁 pile capping beam
桩帽模型 <制造桩帽用的> cap form
桩模 pile mo(u)ld
桩摩擦力 pile friction
桩木 pile; pile wood
桩木挑水坝 pile dike
桩排 curtain of piles; pale fencing; pile curtain; row of piles; pile row
桩排架 pile bent
桩排架盖木 captimber for pile bent
桩排架刚度系数 stiffness coefficient of pile bent; pile bent stiffness coefficient
桩排架夹桩木 pile clamp
桩排架间距 spacing between pile bents
桩排架桥 pile bent bridge
桩排架桥墩 pile bent pier
桩排头 pile bent
桩排围篱 pale fencing
桩偏位距离 pile eccentricity
桩偏心距 pile eccentricity
桩平台 cushion cap
桩墙 pile(d) wall; piling wall
桩桥 pile bridge
桩桥台 pile bent
桩桥钻井 pier drilling
桩群 bundle of piers; clump; cluster of piles; group of piles; pile cluster; pile grillage; pile group; pile network; piling
桩群抗力 resistance of piles
桩群整体破坏 block failure of pile group
桩群作用 group action of piles
桩入土长度 embedded length of pile
桩上铺板 pile-planking
桩设计的 P-Y 法 P-Y method of pile design
桩身 body pile; pile body; pile shaft; pile stem
桩身承载力 bearing capacity of pile shaft
桩身附着力 shaft adhesion
桩身结构承载力 bearing capacity of pile shah
桩身截面完整性类别指数 the exponent of pile cross section's integrity grade
桩身摩擦力 shaft friction
桩身摩阻力 friction(al) force of shaft; pile shaft frictional forces
桩身缺陷 pile defect
桩身上举 <由于邻近打桩而引起的> heaved pile
桩身缩颈 necking of piles
桩身完整性 pile integrity
桩深 pile depth
桩式岸壁 quay wall of piles
桩式驳岸 pile bulkhead
桩式堤 pile(d) dike[dyke]
桩式防波堤 elevated barrier on pile foundation; pile breakwater; piled jetty
桩式防舷材 pile fender
桩式横堤 pile cross-dick
桩式护舷 piled fendering
桩式接合 staking
桩式靠船墩 pile type dolphin
桩式靠船建筑物 dolphin type breasting structure
桩式码头 piled wharf; pile pier; pile quay; pile wharf
桩式锚 piled anchor
桩式木墩 timber pile bent
桩式木桥台 pile bent abutment
桩式耙 pile harrow; pillar harrow
桩式桥 pile bridge
桩式桥墩 pile(d) pier; pile-type pier
桩式桥台 pile(bent) abutment; pile-type abutment
桩式水尺 pile ga(u)ge; stake ga(u)ge; stake pole; stake staff
桩式突堤 piled jetty
桩式突码头 piled jetty
桩式围堰 pile cofferdam
桩式消防栓 pile hydrant; pillar hydrant
桩式栈桥码头 piled jetty
桩式直立堤 piled vertical breakwater
桩束 clump of piles; cluster of piles; group of piles
桩束群 clustered piles
桩数量 number of piles
桩栓 fastening bolt
桩松弛 pile relaxation
桩台 cap of pile; pile cap(ping); pile platform; pile cap girder
桩台横撑 brace of a pile grating
桩台式闸 pile platform type lock
桩台式闸墙 pile platform type lock wall
桩套 pile casing(tube); pile protection; pile encasement
桩套管 casing; shell of pile
桩套管的颚口 alligator point
桩体 pile body
桩调整器 pile adjuster
桩筒式码头 pile and cylinder jetty
桩筒式突堤 pile-cylinder pier
桩筒系船柱 pile-cylinder dolphin
桩头 butt(of pile); head of pile; pile crown; pile head; pile butt
桩头保护 pile head protection
桩头被锤击开裂 mushing
桩头变形 point deformation
桩头处理 finish pile head
桩头垫 scarp
桩头附加钢筋 head steel
桩头钩 monkey hook
桩头箍 drive cap; driving cap
桩头开花 broom head of pile; brooming
桩头落锤 beetle head
桩头帽 cushion head
桩头面积 area of pile head
桩头切割高程 cut off level
桩头铁箍 pile band; pile ring
桩头铁圈 ferrule of pile
桩土荷载比 pile-soil loading ratio
桩土体系 pile-soil system
桩土相互作用 pile-soil interaction
桩土应力比 pile-soil stress ratio; stress ratio of pile to soil
桩腿 leg
桩围栏 pile stockade
桩围堰 curtain of piles; piled dike [dyke]
桩围渔网 stake net
桩位 peg point; pile location; pile position
桩位布置 pile layout
桩位布置图 piling plan
桩位公差 pile tolerance
桩位平面图 piling plan
桩位图 arrangement of piles
桩位线 line of piles
桩席屏 <河道束水用> bandalls
桩席屏系统 bandalling system
桩系统 Balkan piling system Balkan

桩线 anchoring wire;anchor wire
桩芯 pile core;pile mandrel[mandril]
桩形 pile shape
桩形立标 pile beacon
桩靴 alligator jaw;drive shoe;driving shoe;jaw of pile;pile shoe;shoe of pile;pile-driving shoe;pile point;driving shoe of pile;rock shoe
桩靴附件 shoe attachment
桩靴钢板 shoe plate
桩压 pile jacking
桩堰 pile weir
桩应力 pile stress
桩用振动钢丝测力计 pile force ga(u)ge
桩与支撑桁架相联结 pile to jacket link
桩垣 piling wall
桩载公式 pile formula
桩载荷 pile load
桩载试验 load test on pile
桩在岩石中锚固 anchorage of pile in rock
桩栅(栏)pile stockade;pile screen
桩栅栏工程 <防护坡面的> pile-hurdle work
桩张应力 pile tensile stress
桩砧 anvil stake
桩正法【测】peg adjustment method
桩支承 pile bearing
桩支承的 pile-supported
桩支承结构 pile-supported structure
桩支承平台 piles supported platform
桩支承桥 pile bridge
桩支房屋 stilt house
桩支码头 pile;pile pier
桩直径 pile diameter
桩止点 <桩的最后打入深度> pile stoppage point;refusal of pile;refusal point
桩趾水平线 toe line
桩质声学检验法 sound pile test
桩中心线 pile centre line
桩中应力 stresses in pile
桩重新打入试验 pile redriving test
桩周围变形区 zones of deformation around piles
桩轴向荷载试验 axial loading test of pile
桩柱 pier stud;stilt <水上房屋的>
桩柱灯塔 pile lighthouse
桩柱底脚 pole footing
桩柱顶板 pier template
桩柱公式 pile formula
桩柱碰垫 pile fender
桩柱栈架 pile trestle
桩子 spile;stake
桩阻力 refusal
桩组 bundle of piers;cluster of piles;group of piles;pile cluster;pile group;piling cluster
桩钻孔机 pile boring rig;pile drill;pile drilling machine

装

装暗盒 magazine
装板 boarding
装板衬料 justifier
装板及细木工 carpenter and joiner
装版 justify;lockup
装版补偿器 compensating plate lock-up
装版工人 justifier;stoneman
装版顺序 plating sequence
装版台 imposing stone
装棒机 rod charger
装包 bagging
装包机 bag-filling machine;bagging machine
装包机插嘴 bagging spout
装包平台 bagging platform
装包入箱 encase
装备 accouterment;equipage;equipment;fit(ting) out;fit-up;furnish;habiliments;installation;outfit;rig;set out;setting up
装备表 table of equipment
装备不良 ill-equipped
装备长梯的救火车 ladder truck
装备船舶 rig a ship
装备船舶索具 ship's tackling
装备的 equipped;equipping
装备电子计算机 computerisation [computerization]
装备费用 equipment outlay
装备工业 equipment manufacturing industry
装备工作 preliminary operation
装备供给表 basis of issue
装备过度 over-equipment
装备好的 well-appointed
装备滑车系统 string up
装备火箭的舰船 rocket ship
装备加工机械 tool up
装备驾驶仪的测试车 drivometer-equipped test car
装备减速器的编组场 retarder-equipped yard
装备雷达反射器的浮标 buoy with radar reflector
装备良好的船 taut ship
装备两台发动机的 tandem powered
装备履带式拖拉机的推土机 dozer-equipped track-type tractor
装备齐全的 full rigged;well-appointed
装备齐全的发动机 fully equipped engine
装备生产厂 armament factory
装备图 installation diagram;installation drawing
装备完善的 snug;well-equipped
装备现代化 facility modernization
装备性能试验报告 equipment performance report
装备用具 <房屋> plenish
装备有 equipped with
装备与设施控制台 equipment and facility console
装备增压泵的消防车 fire pumper
装备质量 mass in working order
装备状况报告 status-of-equipment report
装备状况登记板 status-of-equipment board
装备总体 outfit of equipment
装壁板 wainscot(t)ing
装边缘 rim
装表玻璃压力器 press for fitting watch glass
装冰孔 icing hatch
装冰口 icing door
装冰口门 <冷藏车> ventilator plug
装柄 haft
装病 flight into disease;malingering
装玻璃窗的瞭望台 glassed-in observation deck
装玻璃的 glazed
装玻璃的洞口 glazed opening
装玻璃的孔穴 glazed opening
装玻璃的研磨台 laying end;laying yard
装玻璃垫条 heel bead
装玻璃工(人)glazier
装玻璃用钉 glazing brad
装驳船 in lighter
装驳站 loading point
装材车 kiln truck
装仓 binning;bunkering
装舱不当 badly stowed
装舱操作 topping off
装舱阀 hopper-loading valve;lander (discharge) valve;valve in landers
装舱方量 volume on board
装舱管系 discharge distribution pipes;hopper-loading piping;lander
装舱货物 hold cargo
装舱货物系数 <40 立方英尺每吨,1 立方英尺 = 0.02832 立方米> stowage factor
装舱及平舱 stowed and trimmed
装舱记录曲线 load diagram
装舱检查 loading survey
装舱开始日 stem(ming) date
装舱链斗挖泥船 hopper bucket dredge(r)
装舱率 filling rate;hopper-loading rate
装舱施工 loading dredging
装舱速度 hopper-loading rate
装舱吸扬挖泥船 hopper suction dredge(r)
装舱效率 hopper-loading efficiency
装舱溢流 loading and over-flowing dredging
装舱抓斗 grab hopper
装舱抓斗挖泥船 grab hopper dredge(r)
装舱自卸吸扬挖泥船 self-dumping hopper suction dredge(r)
装舱纵面图 stowage chart
装舱作业 hopper-loading operation;loading dredging
装槽罐 tankage
装槽罐存储费 tankage
装槽箱 tankage
装槽箱存储费 tankage
装草机 hay loader
装草耙架 hayrack
装碴 mucking;muck loading
装碴车 loader
装碴工具 debris loading means
装碴和钻眼液压支臂联合机组 loader and drill boom combination
装碴机 ballast loader;loader;mucking loader;mucking shovel
装碴机械 mucking machine(ry)
装拆架 mounting rack
装拆式飞机起降跑道 landing mat
装拆式钢桥 fabricated and detachable steel bridge;fabricated steel bridge
装车 car filler;car load(ing);loading;entrucking <载重汽车的>
装车板 loading board
装车铲斗 loading bucket
装车场 loading yard
装车车位 loading bay;loading berth
装车地点 point of loading
装车地直达运输计划完成百分率 percent of actual wagon loadings carried by through trains organized at loading points to those planned
装车点 draw point;embussing point;loading point
装车发运状态 tucked transport position
装车费 loading charges
装车费率 loading rate
装车工 docker;loader
装车工艺系统 car loading technology system
装车钩 becket
装车规则 loading rule
装车滑车 loading tackle
装车活门 loading trap
装车(或发送)吨数 tons loaded
装车机 car loader;loader;mine car loader;track loader;train loader;truck loader
装车机械 loading machine
装车计划 scheme of loading
装车计划完成百分率 percentage of actual wagon loadings to total planned
装车架 car deck;truck-loading rack
装车口 wagon spout
装车跨 loading bay
装车量 truck-load
装车料仓 truck-load bin
装车料斗 truck-loading hopper
装车领工员 loading foreman
装车漏斗 <上料斗> loading hopper
装车漏口 haulage box
装车配矿 ingredient ore of load car
装车起重机 loading crane;lorry loading crane;truck loading crane
装车起重器 loading jack
装车前轮荷载比 laden weight ratio on front wheel
装车桥 loading bridge
装车清单 list of loaded wagons;loading list
装车区 loading area
装车设备 car loading facility;truck-loading facility
装车设施 wagon loading equipment
装车升运器 truck elevator;wagon elevator
装车时间 loading time
装车时轮胎荷载比 laden weight ratio on tyre
装车输送机 car elevator
装车输送机组 loading conveyor system
装车数 car loadings;number of wagons loaded;rolling stock amount of loading;wagon loadings
装车台 deck;loader slide;loading jack;loading platform;loading table;log deck;truck fill stand;truck-loading platform
装车台图库 outgoing(loading) space
装车调整 adjustment of car loading
装车通路 loading gangway
装车托盘 loading pallet
装车线(路)loading siding;loading track
装车斜坡台 loading ramp
装车用环形线 circular line for loading
装车站台 loading deck;loading platform
装车装置 loading installation
装车作业 loading operation
装成袋的 sacked
装承套的墩柱 socketed stanchion
装齿 toothing
装齿轮 cog
装齿形物 tusking
装储食物 <船只> victuals
装船 embarkment;ship;shiploading;shipment;stevedore;loading of ship
装船标志 shipping mark
装船泊位 loading berth;shipping berth
装船舱单 shipping manifest
装船秤 shipping weigher
装船尺寸 intake measure
装船单据 document of shipping;shipping documents
装船的货物数量 intake
装船地点 point of shipment;shipping point
装船费用 loading charges;shipping charges
装船付款 cash on shipment
装船付现 cash on shipment
装船港(口)shipping port;port of shipping;port of shipment;port of loading

Z

装船工艺系统 shiploading system
装船后不负责任 no risk after shipment
装船货物 hold cargo
装船货样 shipping sample
装船或卸船 loading/unloading
装船机 shiploader
装船机臂架 shiploader boom
装船机机头 loading head
装船机驾驶室 shiploader operator's cab
装船机可伸缩臂架 telescoping shiploader boom
装船机落料管 shiploader spout
装船机(伸)臂 shiploader bridge; shiploading bridge
装船机械 shiploading machinery
装船记载 on-board notation
装船净重量 net shipping weight
装船廊道 shipping gallery
装船累积数 shipping accumulation number
装船量条款 shipped quality terms
装船码头 embarkation quay; loading dock; loading quay; loading wharf; place of embarkation
装船能力 shiploading capacity
装船品质 quality shipped; shipping quality
装船期 period for shipment
装船期限 date of shipment; date of shipping
装船气力输送泵 ship bulk-load pump
装船前 before shipment
装船前不保险 non-risk till waterborne
装船前不担保 non-risk till on board
装船前付款 cash before shipment; pay before shipment
装船前付现 cash before shipment; pay before shipment
装船前货损 country damage; land damage
装船前检查 preshipment inspection
装船前已检验 surveyed before shipment
装船日期 shipping date
装船设备 loading out facility; shiploading plant
装船设施 shiploading facility
装船申请书 application for shipment
装船时间 time of delivery; loading time
装船输送机 shipping conveyer [conveyor]
装船速率 loading rate
装船塔架 shiploading tower
装船提单 on-board bill of lading; shipped bill of lading
装船条件 terms of shipment
装船条款 shipping clause
装船通知 notice of shipment; shipment advice; shipping advice
装船通知单 shipping note; shipping order; shipping permit
装船通知书 advice of shipment
装船外运机械设备 outloading mechanical equipment
装船误期 fall down
装船须知 shipping instruction
装船许可证 loading permit
装船序号 shipping number
装船延期 delay in shipping
装船用燃油 oil fuel bunkering
装船用燃油港口 oil fuel bunkering port
装船证书 certificate of shipment
装船指示 shipping instruction
装船重量 loading weight; shipping weight; weight shipped
装船准单 permit to shipping
装船作业 shiploading operation
装窗壁缘 windowed frieze
装窗玻璃 face glazing
装窗的 windowed
装窗间距 windowed bay
装窗节间 windowed bay
装带 head harness; tape loading
装带过程 loading process
装带号 installation tape number
装带环 loading ring
装袋 bagging(-off); sack
装袋部门 bag-packing department
装袋车间 bag filling plant
装袋秤 sacker weigher
装袋秤重机 bagging and weighing machine
装袋的 sacking
装袋工作 bagwork; sacking operation
装袋滑槽 bagging-off chute
装袋机 bagger; bagging machine; bag packer; sack filling machine; sack packer
装袋机械 bagging machinery
装袋计量秤 bagger weigher
装袋口 bagging-off spout; sacking spout
装袋螺旋推运器 bagging auger; sacker auger
装袋能力 bagging capacity
装袋器 bagger; sacker
装袋升运器 bagger elevator; sacker elevator
装袋水泥 sacked cement; sacking cement
装袋装置 bagging apparatus; bagging attachment; bagging point
装袋作业 bagwork; sacking operation; bagging operation
装挡风条 draught stripping
装刀杆 tool holder
装倒刺 barb
装倒钩 barb
装到满载吃水 down to her marks
装灯的 lighted
装灯笛声浮标 lighted whistle buoy
装灯配件 lighting fitting
装灯钟铃浮标 lighted bell buoy
装电力牵引机组 electric(al) tractive unit
装电线 wire
装电压电源逆变器的电力机车 electric(al) locomotive with voltage source inverter
装垫 bolster
装吊带 rigging band
装吊起重机 loading crane
装钉 stitch
装钉钢丝 bookbinder wire
装顶盖者 topper
装顶棚 ceil
装订 binding; book bind; bookbinding; draw unit; severity sew; sew
装订厂 bookbindery
装订车间 binding department; bookbindery
装订成册 a bound volume; bound a volume
装订工 binder
装订机 binder; binding press; bookbinding machine; machine stitcher
装订卷数 physical volume
装订式分类账 bound ledger
装订书本式 bound book form
装订所 bindery; bookbindery
装订用线 binding thread
装订整理 collate
装定的航程 set run
装定方位角 set the azimuth
装定光圈 iris setting
装定光学瞄装器 scope setting
装定机构 set-up unit
装定距离 range setting
装定器 setter
装定前灯 setting of headlights
装定误差 setting error
装定叶片角 blade setting
装定装置 setting device
装锭吊车 ingot charging crane
装锭设备 ingot charging gear
装动车轴货车 wagon for carriage of axles
装斗角 filling angle
装斗系数 bucket fill factor; bucket load factor
装肥料铲斗 fertilizer bucket
装粪吊车 dung crane
装封隔器的井 well set on packer
装封口油灰的斗 <管子接头配件> dope bucket
装钢窗用混凝土砌块 metal sash block
装钢轨配件 rail fastenings
装搁板于 shelve
装格架 trellis
装格子 trellis
装格子的窗 latticed window
装工件时间 mounting time
装工具时间 set-up time
装箍的人 tagger
装谷斜槽安装钩 grain spout hook
装骨架 bone
装管 box; tabulation; tubing; tubulate
装管工 swabber
装管工程 building plumbering; pipe laying
装管工具 pipe tool; tube setting tool
装管工人 internal plumber; pipe fitment; pipe fitter
装管机械 tubing machine
装管用钩 hook for pipe fixing
装灌泵 filling pump
装灌管 filling pipe
装灌密度 filling density
装灌贮存 tankage
装罐 caging; can filling
装罐储存 tankage
装罐的 canned
装罐发动机 decking engine
装罐机 caging machine; can filler; decking gear
装罐设备 caging unit
装罐水平 banking level; decking level
装罐台 cage platform
装罐用推车机的推杆 cager rocker shaft
装滚轮的配料器 troll(e)y batcher
装锅炉外壳用钻机 boiler shell drilling machine
装焊工 bandsman
装合车 assembled car
装合螺钉 attachment screw
装合铆钉 dummy rivet
装荷变化 load fluctuation
装盒机 cartoning machine
装后备的工作重量 equipped operating weight
装滑车 pulley
装潢 decorate; decoration; upholster; furnishing
装潢材料 decorating media; decorative material; upholstery material
装潢的大理石 decorative marble
装潢灯闪烁器 electrolier switch
装潢地板 ornamental floor(ing)
装潢工件 decorative work
装潢过的 upholstered
装潢华丽的营业室 parlo(u)r
装潢砌合 decorative bond
装潢砌筑 ornamental bond
装潢式吊灯架 electrolier
装潢式开关 electrolier switch
装潢式样 ornamental form
装潢贴面 decorative coating
装潢艺术 ornamental art
装潢用铝件 decorative alumin(i)um
装潢元件 decorative element
装簧空转轮 spring-loaded idler
装回原位 reinstall
装活塞器 piston inserter
装火药者 primer
装货 loading cargo; freight; freightage; lading; loading; shipment; shipping; on-load(ing)
装货班轮 cargo liner
装货报表 loading report
装货比 loading ratio
装货臂架 loading boom
装货标志 shipping mark
装货并加衬垫防止移动工作的完成 rounding-off
装货驳船 cargo barge
装货驳船用的运河 canal for goods-carrying barges
装货泊位 loading berth
装货舱口 filling hatch; loading hatch; porthole
装货场 loading bay; loading yard; place of embarkation
装货车的拖车 wagon carrying trailer
装货成本 stowing cost
装货储料器 goods loading banks
装货处 place of loading; stowage
装货次序 loading turn
装货大绳网 sling
装货单 shipment order; shipping list; shipping order; packing list
装货单据 document of shipping
装货地点 loading place; loading spot; place of loading
装货电梯 trunk elevator
装货垫板 cargo board
装货吊车 loading crane
装货顶推机 load ejection
装货发票 shipping invoice
装货法 stowage
装货方式 loading pattern
装货费(用) loading charges; stowage
装货港 loading port; port of loading; port of shipment
装货港检查 loading port survey; survey in loading port
装货工 stower
装货国 country of embarkation
装货过多的 over-laden
装货滑槽 loading chute; loading pipe; loading spout
装货环 loading ring
装货机 loading machine
装货记录簿 cargo book
装货间 loading bay
装货检查卡 loading inspection card
装货检查牌 loading inspection card
装货经纪人 loading broker
装货口 hatch; porthole; charging hatch <集装箱>
装货口盖 hatch cover; roof door
装货口盖托臂 hatchcover lifter
装货口加强肋骨 port frame
装货口塞 hatch plug
装货口下盖 hatch plug
装货量 shipment
装货溜管 loading pipe; loading spout
装货率 loading rate
装货落空损失费 compensation for failure of loading
装货码头 loading quay; loading terminal; loading wharf; shipping terminal
装货门 cargo door; port
装货密度 <车、船> shipping density

装货能力 cargo(carrying) capacity
装货票号 shipping number
装货平台 loading dock; loading platform
装货起点 place of loading
装货清单 freight account; freight bill; freight note; invoice; list of cargo; loading list; manifest; shipping bill; shipping order
装货人 loader
装货容量 cargo capacity; shipping capacity
装货设备 cargo accommodation; charging equipment; loading facility; loading plant
装货申请书 shipment request
装货时间 loading time; time allowed for loading
装货收据 shipping receipt
装货顺序单 loading list
装货速度 loading speed
装货速率 loading rate; rate of loading
装货塔架 loading tower
装货台 loading bank; loading stage
装货天数 loading days
装货跳板 dock level(l)er; landing skid; loading gang memo; loading ramp
装货通知单 shipping order
装货通知书 shipping advice; shipping memo; shipping note
装货突出部分 <集装箱> protrusion
装货网 loading net
装货桅杆 <码头> cargo mast
装货箱 loading case; packing box
装货效率 efficiency of loading; loading rate
装货许可证 shipping permit
装货延误 delay in loading
装货站 loading station
装货站台 loading depot; loading platform
装货证明书 loading certificate
装货执照 stamp-note
装货重量 shipped weight; shipping weight
装货重量条款 shipping weight terms
装货贮料器 goods loading banks
装机发动机 installed engine
装机费用 installation charges
装机功率 installed horse power
装机净功率 installed net horsepower
装机净扭矩 installed net torque
装机轮容量 installed wheel capacity
装机马力数 installed horse power
装机容量 equipped capacity; installation capacity; installed engine power; installed(generating) capacity
装机容量利用系数 capacity factor
装机容量收费率 demand charges rate
装机容量选择 selection of capacity
装机重量 installed weight
装(集装)箱 vanning; filling container
装夹模具用气缸 die clamping cylinder
装夹式车刀 clamp bit
装甲 armo(u)r(ing); armo(u)r protection; metal clad; armature <电缆的>
装甲暗盒 <带飞行记录的> heavy metal cassette
装甲板 armo(u)r(ed) plate; green plate; plate armo(u)r; sheet of armor
装甲板倾斜度 obliquity of armo(u)r plate
装甲板轧机 armo(u)r plate rolling mill
装甲板轧制厂 armo(u)r plate rolling mill
装甲背板 plate behind armo(u)r
装甲波导管 vertebra
装甲玻璃 armo(u)red(plate) glass; toughened glass
装甲玻璃板 flat wired glass
装甲舱壁 armo(u)red bulkhead
装甲车 armo(u)red car; panzer
装甲车交通道路 road for tank traffic
装甲车身 armo(u)red body; armo(u)red hull
装甲车停车处 laager
装甲船 armo(u)red ship
装甲带 armo(u)r belt
装甲的 armo(u)r-clad; armo(u)r-(ed) plated; armo(u)red; iron cased; metal-clad; panzer; steel-clad; iron clad
装甲电缆 armo(u)red cable; iron-clad cable; sheathed wire; sheathing wire; shielded cable
装甲防火门 armo(u)red fire proof door
装甲钢板 armo(u)r plate
装甲钢板轧机 armo(u)r plate mill
装甲甲板 armo(u)red deck; ballistic deck
装甲舰 armo(u)red vessel
装甲胶合板 armo(u)r ply
装甲轿车 armo(u)red sedan
装甲列板 armo(u)r strake
装甲列车 armo(u)red train
装甲门 armo(u)r(ed) door
装甲磨损试验机 tank wear machine
装甲披挂 applique armor
装甲铺面板 <工业厂房地坪> armo(u)red paving slab
装甲铺面砖 <工业厂房地坪> armo(u)red paving tile
装甲汽车 armo(u)red automobile; armo(u)red(motor) car
装甲箱 armo(u)red cassette
装甲巡洋舰 armo(u)red cruiser
装甲曳引车 dragon
装甲运输车 armo(u)red carrier; cavalry carrier
装甲运输汽车 combat carrier
装甲侦察车 armo(u)red scout car
装甲铸件 armo(u)r casting
装甲装备 armo(u)red equipment
装甲装载机 armo(u)red loader
装架冲挖 trestle hydraulicking
装架焊接 match assemble welding
装架环 mounting circle
装架螺栓 mounting bolt
装架起重机 erection crane
装架损失 rack loss
装架阻力 rack resistance
装尖塔 spire
装尖头 cusp; nib; tipping
装脚隔电子 post-type insulator
装脚绝缘子 pin type insulator; post insulator
装铰链板条 hinge strap
装铰链的门梃 shutting stile
装铰链门柱 harr
装铰链条 hanging strip
装铰门梃 shutting stile
装搅卸料系统 loading mixing and discharging system
装接容量 connected load; customer connected load
装接橡皮管 rubber tubing
装金属百叶窗 metal shuttering
装金属板盖 metal sheet cover(ing)
装金属板壳 metal sheet cover(ing)
装金属板套 metal sheet cover(ing)
装金属板罩 metal sheet cover(ing)
装金属箍 ferrule; tag
装金属箍头的 shod
装金属栏杆 metal railing
装紧 fastening
装井口 well-head installing
装井口时间 time for mounting well head
装具 harness
装卷帘式铁门 metal shuttering
装竣日期 installation date
装卡 chucking
装开尾销 cotterpinning
装块根机 root loader
装矿港 ore-loading port
装矿机 muck loader
装矿机放矿点 mucking machine drawpoint
装矿石滑槽 grain shoot
装矿(作业) muck loading
装框 enframe
装拉线 anchoring
装栏杆 rail
装牢的 hard set
装雷管 prime; priming
装梁 mounting frame
装粮袋机 sack loader
装粮机 grain loader; grain-o-vator; grain piler
装粮箱 grain loading bin
装两层的 double-stacked
装两批料卡车 two-batch truck
装了玻璃的 glazed
装了尖头的 nibbed
装了套筒的 bushed
装了炸药的爆破孔 shot hole
装料 burden; charging; feed charge; feeding; filling; loading; stocking
装料板 charging plate; firing plate; loading sheet
装料磅秤 charging scale
装料泵 charging pump; loading pump
装料臂 filling arm; loading arm
装料不均匀 non-uniform charging
装料不足 underloading
装料仓 charging hopper; loading hopper
装料槽 charge chute; charging spout; feeding chute; flume; loading chute
装料槽起重绞车 trough crab
装料叉 loading fork
装料铲 charging scoop; loading shovel
装料场 charging area; loading bay
装料车 charge car; charging car; charging lorry
装料称量装置 loading weigher
装料程度 filling degree
装料程序 charging schedule; feeding sequence
装料传送机 feeder conveyer[conveyor]; stage loader
装料次数 charging frequency
装料存仓 loading bin
装料单 loading note; loading voucher
装料点 filling point; loading point
装料吊车 charging crane; loading crane
装料斗 batch(er) hopper; charging bucket; charging funnel; charging hopper; charging scoop; feed bin; feed box; feed hopper; hopper; loader shovel; loading skip; loading trapper; loading hopper
装料斗车 charging lorry
装料斗衬里 hopper lining
装料斗冷却套 hopper cooling jacket
装料端 charging end; head end
装料端护罩 feed end housing
装料阀门 fill valve
装料翻斗 charging skip; loading skip <混凝土>
装料方式 charging method; loading pattern
装料杆 charge peel; charging peel; peel
装料高度 filling height; loading height
装料工 charger; charging hand; filling operator; loading operator; stocker
装料工平台 stoker's platform
装料管 charging pipe; filling pipe; filling spout; filling tube
装料管线 filling pipeline; loading pipeline
装料罐 charge can; charging bucket; loading cask
装料横梁 loading boom
装料环 loading ring
装料活门 fill(ing) valve
装料机 charge carriage; charger; charging carriage; charging machine; excavator-loader; filling engine; filling machine; fuel charger; material loader; stocking machine
装料机顶杆 charging bar
装料机构 charge mechanism; charging gear; charging mechanism; charging system; fill mechanism
装料机起重臂 loader arm
装料机推杆 charging peel
装料机械 charging mechanism; loader; loading machine
装料计划 loading plan; load(ing) program(me); loading schedule
装料计数器 furnace filling counter
装料记录 charge book
装料夹钳 loading clamp
装料架 skid bed
装料间 loading bay
装料卷扬机 loading winch
装料孔 charged hole; charging hole; charging opening; charging port; fill opening; loading aperture; receiving opening
装料口 charged hole; charge opening; charging hole; charging opening; door; engorgement; feed opening; fill port; loading groove; loading inlet; receiving opening
装料跨度 charging bay
装料量 charge; charge amount; innage; inventory forced filling rate
装料溜槽 receiving chute; charging chute
装料漏斗 batch hopper; charging cone; load hopper; receiver cone; charging hopper; loading hopper
装料螺旋 loading auger
装料码头 loading quay; loading terminal; loading wharf
装料门 charging door; charging opening; lading door; loading door; charging gate
装料门坎 door bank
装料密度 loading density
装料面 charging face
装料盘 charging tray; loading tray
装料皮带带式装载机 loading belt
装料皮带机 loading belt
装料皮带托架 loading belt jib
装料平台 charging deck; filling place; charging floor; loading platform
装料破碎机 feeder-breaker
装料期 period of loading
装料起重机 charging crane; feeding crane; loading bridge
装料器 bunk; charging machine; loader; loading machine
装料器秤 loader scale
装料器斗 loader skip
装料器提升机构 loader raising gear
装料桥 charge bridge
装料人 charger
装料容量 connected load; installed capacity

装料容器 charge cask
装料入料斗 bunker
装料软管 filling flexible conduit; filling hose; loading flexible conduit; loading hose
装料筛 loading screen
装料设备 change direction gear; charger; charging apparatus; charging appliance; charging arrangement; charging device; charging facility; charging installation; loading facility; loading installation; tipping cradle
装料设施 filling facility; filling installation; filling plant; loading facility; loading installation
装料时 when filling with substance
装料时间 charge time; charging time; duration of charging; loading duration
装料室 charging chamber; loading space
装料输送带 loading belt
装料输送机 charging conveyer[conveyor]; conveyor loader; loading conveyer[conveyor]
装料输送器 filling conveyer[conveyor]
装料速率 rate of charging
装料台 charge level; charging area; charging magazine; charging platform; depiler; de-piling magazine; loading rack
装料台的格栅 unscrambling grid
装料台架 charging skid; loading platform
装料桶 charging ladle
装料筒仓 loading silo
装料图 loading schedule
装料图表 loading chart; loading table
装料位置 charging position; filling position; loading position
装料系数 coefficient of charge
装料箱 feeder hopper; packing box
装料小车 charging carriage
装料(斜)台 loading ramp
装料悬臂 loading arm
装料悬臂管 loading arm tube
装料因子 fill factor
装料用空气压缩机 bulkload compressor
装料运输机 charging conveyer[conveyor]
装料站 filling station; loading station; loading terminal
装料指示器 stock indicator
装料中的金属粒料 metallics of charge
装料中的金属物质 metallics of charge
装料装置 charging apparatus; charging device; charging gear; charging mechanism; feeder; loading attachment; loading device
装料嘴 filling spout
装料作业 filling operation; loading operation
装料作业人员 loading operator
装零星货物的吊盘 loading tray
装楼梯踏步 stave in
装炉 charge; charging; filling; shove
装炉辊道 furnace charging table
装炉量 batch
装炉料 furnace burdening
装炉时的水分 moisture as charged
装炉推料器 stock pusher
装炉温度 charging temperature
装辘轳 tackle
装轮的起重机 wheeled crane
装轮机 wheel mounting press
装轮圈 rim
装轮式喷雾机 wheel-mounted sprayer
装轮胎 tire[tyre]
装轮胎叉 tire[tyre] fork
装(轮)胎杆 tire[tyre] lever
装轮通用机架 wheel-mounted carrier
装轮子的栅栏 fence of wheel
装锣浮标 gong buoy
装马蹄铁 horse-shoeing
装满 brim; fill in; filling to capacity; loading to capacity; pile-up; top fill up; topping off; top up
装满的 filling
装满的管线 live line; loaded line
装满的集装箱 container load
装满系数 coefficient of admission; coefficient of fullness
装满液体 filled with fluid
装铆钉 drive the rivets
装煤 coaling
装煤槽 coal chute; coal shoot
装煤车 charging car
装煤车煤斗 larry bin
装煤船 coaling vessel
装煤单绞辘 coaling whip
装煤导筒 coaling trunk
装煤吊桶 coal loading bucket
装煤港(口) coaling port; coaling station; port of coaling
装煤工 coal passer
装煤工具 coaling gear
装煤灰船 ash boat
装煤灰船驳 ash barge; ash lighter
装煤机 coal loader
装煤孔座 charging hole seat
装煤量 coal capacity
装煤偏导板 plough deflector
装煤起重机 coaling crane; coal loading crane
装煤器具 coal loading apparatus
装煤桥 coaling bridge
装煤日 coaling day
装煤设备 coaling installation
装煤台 trestle
装煤舷门 coal door; coal port
装煤栈桥 coal pier
装煤站 coaling station
装煤装置 coal gear; coal unit
装门架 door jack
装门面措施 window dressing
装门锁横挡 lock rail
装门心板 panel(l)ing
装模 die-filling; packing compact
装模料 load; mo(u)ld charge
装模料温度 loading temperature
装模台 filling bench
装木横架补强铁 <运木材车> bunk truss
装泥机 dirt loader; mucking unit; muck loader
装盘 loading plate
装配 assemblage; assemble; assembling; assembly; binding; boxing-in; build(ing)-up; construction; driving fit; erecting; erection; fabricate; fabrication; fit(out); fitted assembly; fitting(assembling); install; installation (work); instalment; match; mating-up; mixed rags; mount(ing); outfitting; packaging; placing; put together; rig(ging)
装配板 assembly plate; buck-plate; mounting plate
装配备件 mounted spares
装配编号 match marking
装配标记 assembly mark; fitting mark; location mark; match marking
装配标志 match marking
装配玻璃 glassing; glazing
装配玻璃尺寸 glazing size
装配玻璃尺度 glazing dimension
装配玻璃的部件 glazing shape
装配玻璃的窗孔 glassed-in opening
装配玻璃的开孔 glassed-in opening
装配玻璃的瞭望台 glassed-in viewing platform
装配玻璃的木门 glassed-in timber door; glassed-in wood(en) door
装配玻璃的型材 glazing shape
装配玻璃的走廊 glazed corridor
装配玻璃定位块 glazing block
装配玻璃工作 glazier's work; glazing work
装配玻璃面积 area of glazing
装配玻璃修饰物 glazing trim
装配玻璃用金属杆 metal glazing bar
装配玻璃用金属条 metal glazing bar
装配泊位 fitting-out berth
装配薄膜 built-up membrane
装配步骤 set-up procedure
装配部件 assembled part; assembling unit; assembly unit; build-up member
装配部件清单 assembled parts list
装配部门 assembling department; assembly department
装配材料 fixing material
装配层 assembly floor
装配/拆卸 assembly/disassembly
装配拆卸间 erection and dismantling bay
装配产品 assembled product
装配常规 assembly routine
装配厂 assembling factory; assembling plant; assembly shop; erecting shop; fitting shop
装配场 assembly area; assembly bay; assembly floor; assembly yard; fabrication yard; formation yard; make-up yard; place of assembly; assembly depot
装配场地 erecting floor
装配场运输轨 assembly trolley rails
装配车 assembled car
装配车间 adjusting shop; assembling plant; assembly department; assembly shop; erecting bay; erecting shop; erection shop; fabricating shop; fabrication shop; fitting shop; assembling department; assembling workshop; assembly floor; assembly plant; fitter's shop; setting-up shop; shop fitting
装配车辆 prefabricated car
装配成本 assembly cost
装配成部件大修 assembly or unit overhaul
装配成的防火窗 fire window assembly
装配程序 assembling process; assembly procedure; assembly process; balance of assembly; fitting procedure; set-up procedure
装配程序表 assembly schedule
装配尺寸 assembly dimension; fitted position; fixing dimension
装配冲模压床 die setting press
装配船厂 assembly shipyard
装配船只 fitting-out a ship
装配次序 assembly sequence
装配错误 loading error
装配带 rigging band
装配带号 installation tape number
装配单位 assembly unit
装配单元 assembly unit
装配单元预制厂 unit maker
装配导轨 mounting channel
装配的 assembled; built-up; fabricated; installed; prefabricated
装配地坑 assembly pit
装配电路 wiring harness
装配电路图 interconnecting wiring diagram
装配吊车 erecting crane; jib-boom crane
装配吊杆 erecting jib
装配定额 assembling rate
装配段 load segment; service bay
装配法 pattern match
装配法兰【电】 mounting flange
装配范围 fitting limit
装配方法 assembling method; assembly method; fabrication method; method of erection; method of fabrication
装配费 assembly cost; processing cost
装配分度线 match mark
装配分图 partial assembly drawing; partial general view
装配符号 match marking
装配负载 erection load
装配附件 fitting attachment; mounting fittings
装配杆 assembly rod
装配港池 fitting-out basin
装配根据线 rigging datum line
装配工 adjuster; assembler; assemblyman; erecting machinist; erector; fabricator; fitter; mounter; operator; outfitter; repairman; rigger
装配工班 rig-up crew; rigging crew
装配工厂 assembly plant
装配工场 assembly yard
装配工程师 engineer for erection (work)
装配工锉 fitter file
装配工的脚手架 fitter's scaffold(ing)
装配工间 fitting workshop
装配工脚手架 rigger's scaffold(ing)
装配工具 assembly tool; fitter's tools; installation kit; replacement tool
装配工人 fitter fitment; millwright; rigger; matcher
装配工业 assembly industry; fabricating industry
装配工艺 assembling process; assembling technology
装配工用的脚手架 mechanic's scaffold(ing)
装配工长 chief fitter
装配工种 assembling work; assembly work; erection work; fitting work
装配工作 assembly work; erecting work; erection work; fitter's work; fitting work; installation work; setting work
装配工作队 fitting-up gang
装配工作台 fitter's bench
装配公差 build-up tolerance; fitting allowance; fit tolerance; tolerance on fit
装配构件 building block; build-up member
装配构件记号 piece mark
装配过程 assembling process; fitting process; process of setting
装配过紧的轴承 preloaded bearing
装配焊接 erecting welding; erection welding; match assembly welding
装配和测试 assembly and checkout
装配和检验 assembly and checkout
装配和校正 assembly and checkout
装配和修理 assembly and repair
装配荷载 erection load
装配后的 post-installation
装配划线法 pattern scribing
装配环 mounting ring
装配货物 component goods
装配机 assembly machine
装配机架 assembling jig
装配机具 erector
装配机器 assembling machine

Z

装配机械 make-up machinery
装配及拆卸车间 erection and dismantling bay
装配记号 assembling mark; assembly mark; auxiliary aiming mark
装配技术规程 instruction for assembly; instruction for erection work
装配夹 mounting clip
装配夹架 assembly jig
装配夹具 assembling jig; assembly jig; setting block
装配架 assemble frame; assembly fixture; assembly frame; assembly rack; assembly rig; jig; mounting block; mounting bracket; mounting jig
装配间 assembly bay; assembly building; assembly room; building room; erecting shop; erection area; erection bay; loading bay; service bay
装配间隙 fit-up gap
装配检查 fit-up inspection
装配简图 scheme of erection
装配件 built-up member; fabricated product; fitting parts
装配建筑 prefabricated architecture
装配键 assembly key
装配胶 assembly glue; mounting cement
装配胶合 assembly gluing
装配胶粘剂 assembly adhesive
装配角度 setting angle
装配阶段 assembly stage
装配接头 erection joint; fitting joint
装配紧密的部件 compact unit
装配距离 mounting distance
装配可拆式梁 collapsible beam
装配孔 matching hole; matching holing
装配立即执行 load-and-go
装配连接 fitting joint
装配链接 assembly chaining
装配梁 built-up beam
装配(零)件 assembly parts
装配零件预选分组 preassembly selection
装配流水线 assembling line; assembly line
装配流水作业 assembly line flow process
装配流水作业法 assembly line method; assembly line process
装配轮式拖拉机的推土机 dozer fitted to wheel tractor
装配螺钉 attachment screw; fitting screw; rigging screw; tacking screw
装配螺帽 mounting nut
装配螺栓 assembling bolt; erection bolt; fitting-out bolt; fitting-up bolt; mounting bolt; site bolt; tack bolt; tack(ing) screw
装配螺旋 rigging screw
装配码头 fitting-out quay
装配铆钉 tack rivet
装配铆钉线 match mark
装配密封 fitting tight
装配面 fitting surface
装配面积 erection space
装配模块 load module
装配模块库 load module library
装配泥芯 core setting
装配黏[粘]合剂 assembly adhesive
装配黏[粘]胶剂 assembly adhesive
装配平面图 collection plan; setting plan
装配平台 erecting deck; erection platform; fabrication platform; flat block; mounting plate; erecting stage
装配剖视图 sectional arrangement drawing
装配铺轨式拖拉机的推土机 dozer fitted to track-laying type tractor
装配期 erecting stage
装配器 assembler
装配器具 mounting device
装配前试验 preinstallation test
装配区 assembly section
装配任务 fittage
装配容量 fitted capacity; plant capacity
装配容许偏差 erection allowance
装配砂浆 screed mortar
装配上 fit on
装配上的误差 inaccuracies of fabrication
装配设备 assembling equipment; erection plant; nipple up; processing facility
装配设施 fabrication facility
装配生产线 assembling line
装配时间 assembly time; installation time; set-up time; time of setting up
装配时间表 assembly schedule
装配式 assembly type
装配式坝 fabricated dam
装配式板柱结构 assembled plate column structure
装配式办公楼 prefabricated office
装配式衬砌 precast lining; prefabricated lining
装配式成品 assembled product; fabricated product
装配式厨房 kitchen building block module
装配式带式运输机 preengineered belt conveyer[conveyor]
装配式单杆支柱 fabricated single-post shore
装配式的 fabricated
装配式房屋 assembly building; fabricated building; open system; prefab; prefabricated building; prefabricated house
装配式分节浮船坞 bolted sectional dock
装配式刚架桥 portable frame bridge
装配式钢管脚手架 fabricated tubular flame scaffold
装配式钢筋混凝土房屋 precast reinforced concrete building; precast reinforced concrete house; prefabricated reinforced concrete building
装配式钢筋混凝土结构 assembled reinforced concrete structure; precast reinforced concrete structure
装配式钢筋混凝土桥梁 precast reinforced concrete bridge
装配式钢模 collapsible steel shuttering
装配式钢桥 fabricated steel bridge
装配式钢体 fabricated steel body
装配式钢坞墩 fabricated steel block
装配式隔墙系统 partition system
装配式构架桥 portable frame bridge
装配式构造 articulated construction
装配式混凝土房屋 precast concrete house
装配式混凝土围墙 precast concrete fence
装配式活性污泥单元 activated sludge package unit
装配式机架 fabricated frame
装配式家具 sectional furniture
装配式建筑 assembly building; fabricated construction; prefabricated building; prefabricated construction; prefabrication engineering; systematic building
装配式建筑方法 system building method
装配式建筑施工法 camus
装配式建筑物 assembly type structure; fabricated building; fabricated structure
装配式脚手架 built-up type scaffolding
装配式洁净间 prefab clean room
装配式结构 all-dry construction; assembled structure; assembly type structure; construction of assembly units; fabricated building; fabricated construction; fabricated structure; fabrication structure; precast construction; prefabricated building; prefabricated construction
装配式金工车间 fabricated machine shop
装配式空调系统 modular air conditioning system
装配式框架梁板码头 precast reinforced concrete frame and slab platform type wharf
装配式框架型脚手架 fabricated frame scaffold
装配式冷间 sectional cold room
装配式旅馆 prefabricated hotel
装配式模板 <浇筑混凝土梁板用的> pan
装配式木建筑 prefabricated timber house
装配式木屋 drybilt; prefabricated timber building; ready cut house
装配式墙 prefabricated wall
装配式桥(梁) built-up bridge; fabricated bridge; precast bridge; unit construction bridge
装配式施工 all-dry construction; fabricated construction; package-type construction; prefabricated construction
装配式施工法 all-dry construction method; systematic building
装配式施工现场 fabricating yard
装配式隧道窑 package tunnel kiln
装配式体系 open system
装配式亭子 prefabricated pavilion
装配式污水处理厂 package sewage-treatment plant
装配式屋架 prefabricated truss
装配式舞台 demountable stage
装配式铣刀 assembled milling cutter
装配式窑 modular type kiln
装配式预应力砖过梁 prefabricated prestressed brick lintel
装配式预应力组合构件 prefabricated prestressed compound unit
装配式预制的 precast-segmental
装配式预制沟渠管 precast-segmental sewer
装配式预制混凝土环 precast concrete segmental ring
装配式辙叉 fabricated crossing
装配式支撑 assembling support
装配式支架 segmental timbering
装配式主机组 major unit assemblies
装配式住房 sectionalized house
装配式住房建筑 system building construction
装配试验 rig test(ing)
装配室 assembly room; make-up room
装配顺序 assembly sequence; setting-up sequence
装配说明书 fitting instruction
装配速度 assembling speed; assembly rate; assembly speed
装配速率 rate of assembly
装配锁紧垫圈 mounting lock washer
装配索具 rig up
装配台 assembling stand; assembly floor; assembly platform; assembly stand; erecting bay; erecting bed
装配弹簧的 mounted elastically
装配搪瓷 enamel furniture
装配调整 set-up
装配透镜 mounted lens
装配图 arrangement diagram; erecting drawing; installation drawing; load map; shop drawing; wire layout; wiring layout; assembly drawing; erection drawing
装配图表 set-up sheet
装配图样 assembly diagram; assembly drawing; collective drawings; erecting diagram; erection diagram; erection drawing; fitting drawing; mounting diagram; mounting drawing; setting drawing
装配外墙玻璃 external glazing
装配桅杆 erecting mast
装配位置 rigging position
装配温度 fitting temperature
装配圬工墙板 masonry panel wall
装配舞台 erecting platform
装配误差 inaccuracy erection; rigging error
装配系列 assembly chaining
装配系统 assembly system
装配系统程序 load system program(me)
装配现场 assembling site; erecting yard
装配线 assembling line; mountain channel; production line
装配线平衡 assembly line balancing
装配线生产 assembly line production
装配线修理 on-line repair
装配箱 set-up box
装配镶板 fielded panel
装配详图 <面砖、板材等> distribution map
装配销 rigging pin
装配楔块 assembly key
装配信号灯 signalization
装配压床 assembling press
装配要求 matching requirement; requirements of assembling
装配应力 assembly stress; erection stress; misalignment stress
装配用吊 erection crane
装配用辅助装置 assembly aids
装配用构架 erecting frame
装配用滑车 building block
装配用火 building fire
装配用胶粘剂 assembly adhesive
装配用起重机 fitting-out crane
装配用悬架 mounting lug
装配余量 fitting allowance
装配与测试 assembly and test
装配与拆卸 assembly and disassembly
装配与检查 assembly and checkout
装配与检验 assembly and checkout
装配与维修 assembly and repair
装配预制的分段船体 prefabricate
装配站 assembly station; station
装配者 assembler; erector
装配整流电源的整套零件 power supply kit
装配证号 erection mark
装配支架 mounting holder; mount support
装配支柱 form brace; soldier
装配指导 instruction for erection work
装配指导书 installation specification
装配置 connecting up
装配周期 assembly time; placing period
装配主件 major assembly

装配铸型 assembling mo(u)ld
装配组 assembly crew;make-up unit
装配组件 load module;subassembly
装配作业线 assembly operation line
装配作业线生产 assembly line operation
装配座 assembling stand
装坯用三角支架 stilt
装片 film-loading;inserting film;mounting
装瓶 bottling;potting
装瓶机 bottle filling machine;bottling machine
装瓶间 bottling hall
装瓶区 bottling area
装瓶设备 bottling installation
装瓶在桶内 bottle off
装七斗 dirt bucket
装漆机 filling machine
装气门机 valve reseating tool
装汽车轮胎的 rubber-mounted
装嵌木门窗樘的混凝土砌块 wood sash jamb block
装球缺口 filling slot
装球室 ball collector
装燃料 bunkering;refueling
装燃料机 refueling machine
装燃料区 charging area
装燃料入舱 bunker
装燃料设备 bunkering facility
装燃油 oil bunkering
装热套 shrink
装人行道采光玻璃 king's glazing
装容量 installation capacity
装入 built-in; carry-in; embed; fit into; loading; load up; seal in; stow down
装入凹槽的门拉手 case handle
装入并执行 load-and-go
装入舱内 bunker
装入场址 load address
装入程序 loader(routine);load(ing) program(me);loading routine
装入程序块 loader block
装入传动器 built-in clutch
装入次序 loading order
装入存储指令 load store instruction
装入大桶 vat
装入的 embedded;set in
装入的液压千斤顶 built-in hydraulic jacks
装入的照明 built-in lighting
装入(电缆内)式增音机 built-in repeater
装入堵头 insertion of plug
装入方式 load mode
装入和互换 load and exchange
装入机头 loading unit
装入寄存器 load register
装入架子 shelving
装入控 load control word instruction
装入雷管的 capsulate(d)
装入立即执行(程序)load-and-go
装入例行程序 loading routine
装入连接编辑程序 load and linkage editor
装入马达搅拌杆 built-in motor poker
装入模块 load module
装入区 loading area
装入时间 load time
装入式电动机 built-in motor
装入式铺砂器 built-in bottom type gritter
装入填料管 loading pipe
装入桶内 running into tank
装入图像 load image
装入文件 load file
装入系数 loading factor
装入线槽 boxing-in
装入向量 load vector
装入新程序 load new program(me)
装入因子 loading factor
装入载重汽车交货 delivery in truck
装入指令 load
装入中断屏蔽 load interrupt mask
装入装配图 load map
装色谱柱 packing
装纱网压条 <门窗> screen spline
装砂机 sand loader
装上 fit over
装上侧面 side
装上活塞环 ring plugging
装上卡车 on truck
装上框格 sash
装上索具 rig up
装上铁路敞车的集装箱 container on flat car
装上楔子 cleat
装设点 placement point
装设电路 circuiting
装设阀门 valving
装设轨道电路 track circuiting
装设加感线圈的进入孔 <地下线路> loading manhole
装设铅管 plumb
装设仪器 instrumentation
装设在带定位桩驳船上的钻机 barge-mounted drill rig
装设在下锚驳船上的钻机 barge-anchored drill rig
装设在下锚木筏上的钻机 float-mounted drill rig
装设在自升式施工驳船(平台)上的钻机 barge-mounted drill rig
装牲畜用的集装箱 livestock container
装石工 lasher
装石筛石厂 loading and screening plant
装实心轮胎的 solid-tired[tyred]
装饰 decorate; dress; adorn; beautification; beautify; decking; embellish; embossment; figuration; finery; garnish; garniture; habiliments; investiture; mounting; ornament; ornamentalize; prank; purfle; setoff;set out;trappings;trim(ming); tufting;upholster(ing)
装饰板 clad plate; decoration sheet; decorative board; filling-in panel; micarta board; ornamental board; ornamental plate; ornamental sheet; panel decoration;panel(l)ing;wainscot
装饰板材 decorative panel;dal(l)e
装饰板内墙的装饰接缝 crowned joint
装饰板条 crown mo(u)lding;mo(u)lding
装饰包层 decorative cladding
装饰保护漆 decorative paint
装饰碑 decorative tablet
装饰(壁)板 decorative laminate
装饰壁龛 ornamental niche
装饰边 <用于室内装饰和家具套> welted nosing; welting; obscuration band
装饰边缘 ornamental margin
装饰匾 decorative tablet
装饰表面 decorated surface; ornamental surface
装饰玻璃 architectural glass; decorative glass;ornament glass
装饰玻璃窗 ambetti
装饰玻璃制品 architectural glass article
装饰玻璃砖 decorative glass block; decorative glass tile; ornamental glass brick
装饰箔 ornamental foil
装饰薄板 fancy veneer
装饰材料 decorating medium;decorative material;finishing material
装饰材料成分 finish composition
装饰彩绘 decorative painting
装饰侧砌铺面 ornamental rowlock paving
装饰层 decorative layer;ribbon <屋面瓦的>
装饰层压板 decorative laminate; ornamental laminate(d) board
装饰成的三心拱 false ellipse arch
装饰成有立体感的拉毛粉刷 decorative model(l)ed stuccowork
装饰成有立体感的面层 decorative model(l)ed coat
装饰承包商 decorating contractor
装饰城门 ornamental town gateway
装饰池 ornamental lake
装饰椽头 ornament rafter end
装饰椽子 trimmed rafter
装饰窗 decorative window;ornamental window
装饰窗玻璃 ambetti;carreau
装饰窗框 decorative window frame
装饰窗帘 decorative window curtain
装饰窗纱 decorative window screening
装饰瓷砖 decorated tile; decorative tile;ornamental tile
装饰粗陶浮雕用的料浆 barbotine
装饰大理石 ornamental marble
装饰大型陶瓷 ornamental heavy ceramics
装饰带条 decorative band
装饰单板 decorative veneer; fancy veneer;veneering
装饰单板样片 swatch
装饰单元 decorative unit
装饰的 adorned;decorative;ornamental
装饰的壁灯座 torchere
装饰的椽木 trimmed rafter
装饰的门框边框 ornamental jamb
装饰的山墙封檐板 fly rafter
装饰的外观 decorative appearance
装饰的形象 decorative appearance
装饰的阳极化铝 decorative anodized alumin(i)um
装饰的主题 subject for decoration
装饰灯 architectural lamp;decorative lamp;ornamental lantern
装饰灯具 ornamental light fitting;ornamental luminaire(fixture)
装饰雕刻法 <一种将背景平雕或将花纹平雕的方法> opus interrasile
装饰雕塑 ornamental sculpture
装饰吊顶棚 ornamental hung ceiling; ornamental suspended ceiling
装饰钉 ornamental nail;stud
装饰顶带 head band
装饰顶棚砖 decorative ceiling tile
装饰顶棚花砖 decorative ceiling tile
装饰顶棚瓦 decorative ceiling tile
装饰堆砌 riotous welter of ornament
装饰法 decoration method
装饰范围 decorated area; decorative area
装饰费用 decoration cost
装饰粉饰 stucco
装饰风格 decorative style;ornamental style;ornamental touch;style of decoration;style of ornamentation
装饰缝 ornamental joint
装饰扶手 ornamental railing
装饰浮雕 ornamental relief
装饰盖板 deck board
装饰钢材 ornamental steel
装饰高压叠层胶合板 decorative high pressure laminate
装饰革 fancy leather
装饰格调 decorative touch;ornamental touch
装饰格栅 ornamental grill(e)
装饰格式 ornamental form
装饰隔断 ornamental partition
装饰隔墙玻璃 partitioning glazing
装饰隔音石膏镶板 decorative acoustic(al)gypsum
装饰工程 embellishment work;ornamental work
装饰工具 finishing tool
装饰工人 decorator
装饰工作 decorative work;ornamental work
装饰功能 decorative capacity
装饰拱 blind arch;broken arch;ornamental arch
装饰拱顶 decorative vault;ornamental tunnel vault;ornamental wagon vault
装饰拱廊 <不开洞的> arcature; blind arcade; ornamental arcade; wall arcade
装饰构件 architectural element;decorative element; ornamental element;ornamental unit
装饰骨料 ornamental aggregate
装饰固定装置 ornamental fixture
装饰轨 garnish rail
装饰过的椽子 trimmed rafter
装饰焊缝 sealing run
装饰合金 fancy alloy
装饰横挡 Balconet(te)
装饰湖 ornamental lake
装饰(护)壁板 ornamental panel
装饰花纹 ornamental motif;ornamental relief
装饰花纹玻璃 pattern glass
装饰华表 ornamental pillar
装饰华丽的 highly decorated;ornate
装饰画 ornamental painting
装饰混凝土 architectural concrete; facing concrete;faircrete;fair-faced concrete;ornamental concrete
装饰混凝土砌块 trimmed block
装饰混凝土外加剂 admixtures to architectural concrete
装饰或保护性覆盖 coverture
装饰机械 decorating machinery
装饰夹芯板 face(d) plywood
装饰家 decorator;ornamentalist
装饰建筑 decorative architecture;ornamented architecture
装饰建筑材料 architectural building material
装饰建筑单元 ornamental unit
装饰胶合板 decorative glued laminated slab; ornamental laminate(d) board;decorative plywood
装饰铰链 decorative link;ornamental link
装饰节点 decorative joint
装饰结构 ornamental structure
装饰结构陶瓷 ornamental heavy ceramics; ornamental structural ceramics
装饰金属 decorative metal;ornamental metal
装饰金属板 fancy sheet metal
装饰精修 ornamental finish
装饰井 ornamental shaft;ornamental well
装饰镜板 cartouch(e)
装饰刻痕 score
装饰刻花 engraved decoration
装饰孔眼 ornamental perforation
装饰栏板 ornamental balustrade panel
装饰肋 ornamental rib
装饰篱笆 trimmed hedge
装饰连接 ornamental link
装饰链环 decorative link
装饰梁 fascia beam;jesting beam

Z

装饰裂纹 processed shake
装饰零件 ornamental fittings
装饰铝(材) ornamental alumin(i)um
装饰螺栓 garnish bolt
装饰门 architectural door; blank door; blind door; decorative door; ornamental door
装饰门头 ornamental portal
装饰面 decorative surface
装饰面层 decorative overlay; ga(u)ged skim coat; ornamental coat(ing)
装饰面积 decorated area; decorative area; ornamental area
装饰面下陷 finish sag
装饰面砖 ornamental tile
装饰抹灰 pargetry; stucco; placard <有凸起图案的>
装饰抹灰用的拉展金属网 stucco mesh
装饰木工 decorative woodwork
装饰木条 ribband
装饰木条锯割机 sticker machine
装饰木纹镶板 figured veneer
装饰牌 ornamental tablet
装饰派 ornamentalism
装饰配件 decorative fittings
装饰片材 decorative sheet
装饰拼合地板 decorative composition floor(ing)
装饰品 adornment; bibelot; cosmetic; decoration; dofunny; fineries; finery; garnish; garnishment; garniture; ornament; ornamentation; set-off; trimming; embellishment
装饰品搁板 plaque rail
装饰品格架 plate rail
装饰品搪瓷 ornamental enamel
装饰平顶 <圣坛上面的> celure
装饰屏 decorative screen; ornamental screen
装饰铺面 ornamental paving
装饰漆 decorative coating; fancy paint; trim paint
装饰砌合法 decorative bond
装饰砌块 decorative(faced) block
装饰器皿 ornamental ware
装饰嵌丝玻璃 decorative wire(d) glass; ornamental wire(d) glass
装饰嵌条 garnished mo(u)lding
装饰嵌线 mo(u)lding
装饰墙 ornamental wall
装饰墙板 decorative panel
装饰墙壁的材料 <瓷砖等> wainscot
装饰墙壁的天然石料 fieldstone
装饰墙壁用材料 wainscot
装饰墙壁用瓷砖 wainscot
装饰墙基 ornamental bracket
装饰墙牛腿 ornamental wall bracket
装饰墙托架 ornamental wall bracket
装饰青铜 jewelry bronze; trim bronze
装饰区 ornamental area
装饰券 ornamental arch
装饰染料 ornamental lake
装饰人造石 ornamental reconstructed stone
装饰容量 ornamental capacity
装饰乳化涂料 ornamental water(-carried) paint
装饰三聚氰胺层压板 decorative melamine laminate
装饰色彩的面积效应 areal effect of colo(u)r
装饰砂浆 decorative mortar
装饰山墙 ornamental gable
装饰栅 grill(e); louver[louvre]
装饰上檐口 cheneau
装饰烧结玻璃 decorative sintered glass; ornamental sintered glass
装饰设计 decorated design
装饰师 decorator
装饰施工 decorative application
装饰石 <装修门口用的> antepagmenta
装饰石板 decoration stone
装饰石膏 decorated gypsum; ga(u)ge(d) plaster; ga(u)ge(d) staff; ga(u)ge(d) stuff; hydraulic plaster; putty and plaster
装饰石膏板 decorative plaster board
装饰式拱 decorated arch
装饰术 ornamentation
装饰水池 ornamental pool
装饰水泥 decoration cement
装饰塑料 decorative plastics
装饰塑料板 decorative plastic board; ornamental plastic board
装饰塑料薄板 ornamental plastic sheet
装饰隧道拱 decorative tunnel vault
装饰锁 ornamental lock
装饰塔 ornamental tower
装饰塔楼 ornamental turret
装饰台总成 platform assembly
装饰台座 gaine
装饰特色 ornamental feature
装饰特性 decorative property; ornamental property
装饰特征 decorative feature; ornamental power
装饰体 ornamental masonry
装饰天花板 decorated ceiling; decorative ceiling (board); decorative ceiling sheet
装饰贴脸板 ornamental trim
装饰贴面板 decorative veneer
装饰铁件 ornamental ironwork
装饰铁件工艺 ornamental ironwork technique
装饰铁器 ornamental iron
装饰亭 decorative pavilion; ornamental pavilion
装饰铜合金 ornamental brass alloy
装饰桶拱 decorative barrel vault
装饰筒拱 decorative barrel vault
装饰筒形拱顶 ornamental tunnel vault
装饰图案 ornamental pattern; vignette
装饰图案设计 ornamental design
装饰涂层 decorative coating; finishing coat; ornamental coat(ing)
装饰涂料 finish coating; ornamental paint
装饰瓦 decorated tile; ornamented tile
装饰五金 ornamental hardware
装饰物 adornment; ornament; trim; decoration
装饰吸音板 absorptive coffer
装饰细工 buhl and counter; ornamented work
装饰细条线 band(e)let
装饰细线条 bandlet
装饰匣 <奉献给罗马教堂主持人的> pix
装饰线脚 ca(u)lked edge; mo(u)lding
装饰线脚用灰泥 mo(u)lding plaster
装饰线条 mo(u)lding; solid mo(u)lding <木材上雕刻的>
装饰线条板 band mo(u)lding
装饰线条成型 sticking
装饰线条型板 mo(u)ld board
装饰线条用熟石膏 mo(u)lding gypsum
装饰镶板 decorative panel; pattern(ed) veneer
装饰镶边 ornamental trim
装饰镶面板 fancy veneer
装饰镶嵌 garnished mo(u)lding
装饰镶条 finishing strip; garnish mo(u)lding; trim strip
装饰小兽 zoophorism in ornament
装饰小五金 decorative hardware; finish hardware; trim hardware
装饰效果 decorative effect; ornamental effect
装饰效力 decorative power
装饰斜楼梯梁 finish string
装饰斜梯梁 finish string
装饰形式 decorated style; decorative form; ornamental form
装饰型材 decorative profile
装饰性壁龛 decorative niche
装饰性玻璃块 decorative glass block
装饰性玻璃砖 decorative glass brick
装饰性薄板 decorative sheet
装饰性层压板 decorative laminated board
装饰性窗帘 drapery
装饰性窗帘钩 tieback
装饰性瓷漆 trim enamel paint
装饰性大型陶瓷 decorative heavy ceramics
装饰性带条 ornamental band
装饰性的 cosmetic
装饰性地板面层 decorative floor cover(ing); decorative floor finish
装饰性地板抛光 ornamental floor finish
装饰性地板终饰 ornamental floor finish
装饰性吊顶 decorative suspended ceiling
装饰性顶棚板 decorative ceiling (board)
装饰性顶棚薄板 decorative ceiling sheet
装饰性镀金 ornamental gilding
装饰性房屋用制品 decorative building unit
装饰性粉刷 decorative coating
装饰性浮雕 decorative emboss(ment)
装饰性钢件 decorative steel
装饰性格栅 decorative grill(e)
装饰性拱 decorated arch; decorative arch
装饰性拱门 decorative archivolt
装饰性拱缘 ornamental archivolt
装饰性骨料 decorative aggregate
装饰性挂帘 lambrequin
装饰性绗缝 trapunto
装饰性花边加工 matting process
装饰性花格 decorative lattice
装饰性花色砖工 ornamental pattern brickwork
装饰性绘画 ornamental painting
装饰性混凝土 architectural concrete; decorative concrete
装饰性混凝土薄板 decorative concrete tile
装饰性混凝土块 decorative cast concrete block
装饰性混凝土制品 decorative cast concrete product; decorative concrete product
装饰性混凝土砖 decorative concrete tile
装饰性集料 decorative aggregate
装饰性脊瓦 crest tile; decorative ridge tile
装饰性建筑 florid architecture
装饰性建筑制品 decorative building unit
装饰性浇制混凝土砖 decorative cast concrete tile
装饰性浇制块 ornamental cast block
装饰性浇制品 decorative cast product
装饰性浇制石 ornamental cast stone
装饰性胶合板 decorative glued plywood; decorative laminated board; fancy plywood; kraftwood <内墙面用>
装饰性角楼 diminutive tower
装饰性接合 decorative joint
装饰性结构 decorative structure
装饰性结构陶瓷 decorative structural ceramics
装饰性孔眼 decorative perforation
装饰性拉杆 decorative tie; ornamental tie
装饰性肋 decorative rib
装饰性连接 ornamental bond
装饰性楼板 decorative floor(ing)
装饰性楼板面层 decorative floor cover(ing)
装饰性漏窗 decorative perforation
装饰性美术涂装 decorative coat
装饰性门窗侧板 decorative jamb linings
装饰性面层 decorative coating
装饰性抹灰 decorative plaster
装饰性牛腿 ornamental bracket
装饰性喷泉 architectural fountain
装饰性拼花地板 decorative composition floor(ing)
装饰性拼花砖工 decorative pattern brickwork
装饰性铺地 <古希腊和罗马住宅> lithostrotum opus
装饰性铺面 decorative paving
装饰性铺砌 decorative paving; opus lithostratum <古希腊和罗马的>
装饰性砌合 ornamental bond
装饰性砌块 decorative block
装饰性墙 decorative wall
装饰性墙上牛腿 decorative wall bracket
装饰性墙上托架 decorative wall bracket
装饰性人工石 ornamental artificial stone
装饰性人造石 decorative cast stone
装饰性入口 decorative portal
装饰性三角墙 ornamental gable
装饰性散热器 ornamental radiator
装饰性色漆 decorative paint
装饰性山墙 decorative gable; ornamental gable
装饰性设备 decorative fixture
装饰性石膏板 decorative gypsum board
装饰性石块图案铺砌 decorative sett paving
装饰性石块镶花铺砌 decorative sett paving
装饰性石束带层 stone ornamental string course
装饰性饰件 decorative hung ceiling
装饰性书写柜 vargueno
装饰性竖铺面 decorative rowlock paving
装饰性水池 decorative pool
装饰性塑料贴面墙板 decorative plastic-faced wall board
装饰性锁 decorative lock
装饰性天花板材 ornamental ceiling board
装饰性天花板块 ornamental ceiling board
装饰性条纹 ornamental band
装饰性贴面板 decorative laminate; decorative sliced veneer; decorative timber veneer
装饰性铁件 decorative iron
装饰性铁制品 decorative ironwork
装饰性铁制品工艺 decorative ironwork technique
装饰性筒形拱顶 decorative wagon

vault;ornamental barrel vault
装饰性图案 decorative pattern
装饰性涂料 decorative coating
装饰性涂漆 decorative painting
装饰性托架 ornamental bracket
装饰性托座 decorative bracket
装饰性外观 cosmetic appearance
装饰性围栏 decorative railing
装饰性屋脊 ornamental ridge covering
装饰性无缝楼地板 decorative jointless floor(ing)
装饰性无缝楼面 decorative seamless floor(ing)
装饰性吸声华夫干状石膏板 ornamental acoustic(al) gypsum waffle slab
装饰性现场浇制地板 decorative in-situ floor(ing)
装饰性现浇地板 decorative in-situ floor(ing)
装饰性乡镇入口 decorative town gateway
装饰性小块地毯 area rug
装饰性小五金 decorative fittings
装饰性小圆柱 colonnette
装饰性修整 decorative finish;decorative trim
装饰性音响石膏镶板 ornamental acoustic(al) gypsum waffle slab
装饰性印刷 decorative painting
装饰性油画 decorative painting
装饰性油漆 decorative paint
装饰性预制混凝土薄板 decorative precast concrete tile
装饰性预制混凝土块 decorative precast concrete block
装饰性预制混凝土制品 decorative precast concrete product
装饰性预制混凝土砖 decorative precast concrete tile
装饰性照明设备 decorative light fitting;decorative light fixture
装饰性照明装置 decorative luminaire (fixture)
装饰性罩面 decorative coating
装饰性支柱 vaulting shaft
装饰性柱箍 decorative tie
装饰性柱式 suborder
装饰性柱子 decorative column
装饰性砖工 decorative brickwork; ornamental brickwork
装饰性砖石砌合 decorative masonry bond
装饰性砖瓦铺贴术 tilery
装饰性转角小柱 angle shaft
装饰性装置 decorative fixture
装饰压缝条 ornamental trim
装饰压花粉刷 ornamental model(l)ed coat
装饰压花拉毛粉刷 ornamental model(l)ed stuccowork
装饰压条 crown mo(u)lding
装饰液压泵 load cylinder
装饰艺术 art of design; decorating art;decoration art;decorative art; ornamental art
装饰艺术风格 art deco
装饰音管 montre
装饰硬质纤维板 decorative hardboard
装饰用壁灯 wall scone
装饰用玻璃 decoration glass; ornamental glass
装饰用箔 decorative foil
装饰用锤 decorator's hammer
装饰用大烛台 flambeau [复 flambeaus/flambeaux]
装饰用的黄色合金箔 Florence leaf
装饰用钉 decorative nail
装饰用锻铁件 ornamental wrought iron work
装饰用高锌黄铜 rich low brass
装饰用拱 blank arch
装饰用合金 jewel(l)ery alloy
装饰用黄铜 mosaic gold;nu-gold < 锌 12.2% >;nu-gild < 锌 13% >
装饰用混凝土 architectural concrete; decorative concrete
装饰用混凝土板 faircrete [fair-air concrete]
装饰用集料 architectural concrete aggregate;decorative aggregate
装饰用挤制叶型 architectural extruded section
装饰用建筑材料 architectural structure material
装饰用胶合板 decorative plywood; malenite < 呈木纹面的 >
装饰用胶合层积板 decorative glued laminated wood
装饰用胶合层积材 decorative glued laminated wood
装饰用金属箔 ornamental foil
装饰用金属薄件 decorative foil
装饰用金属或陶瓷制板 plaque
装饰用冷水涂料 decorative cold water paint
装饰用铝件 decorative alumin(i)um
装饰用毛织物 decorative felt(ed fabric)
装饰用木材 coramandel; Amboyna < 产于东印度 >
装饰用木线脚 applied mo(u)lding; ornamental wood mo(u)lding
装饰用品 ornamental fittings
装饰用青铜制品 ornamental bronze
装饰用人造石 decorative artificial stone;decorative patent stone;decorative reconstructed stone
装饰用水基涂料 decorative water(-carried) paint
装饰用水溶性涂料 ornamental cold water paint
装饰用水性漆 ornamental cold water paint
装饰用塑料板 ornamental plastic slab
装饰用塑料薄板 decorative plastic sheet(ing)
装饰用塑料薄膜 decorative plastic film
装饰用塑料贴面板 decorative plastic-faced board
装饰用天花板面砖 ornamental ceiling tile
装饰用铁件 ornamental iron
装饰用铜 ornamental bronze
装饰用铜合金 gilding alloy
装饰用铜铝合金 < 含铝 3% ~ 5% > imitation gold
装饰用五金 trim hardware
装饰用小塔 decorative turret
装饰(用油)漆 decorative paint
装饰用预制混凝土 architectural casting(ing);architectural concrete casting
装饰用预制混凝土构件 architectural concrete element
装饰用预制混凝土制品 architectural concrete product
装饰用预制块 ornamental cast block
装饰用毡 decorative felt(ed fabric)
装饰用毡制织物 ornamental felted fabric
装饰用铸块 decorative cast block
装饰用砖 decorative brick
装饰优点 ornamental feature
装饰优质石材 ornamental patent stone
装饰油漆 ornamental paint
装饰与制备工程 dressing and preparation engineering
装饰雨篷 < 门窗顶上的 > hood mo(u)ld
装饰预制混凝土构件 architectural casting concrete element
装饰预制混凝土面砖 ornamental precast concrete tile
装饰预制混凝土制品 architectural cast concrete product; ornamental precast concrete product
装饰元素 decorative element
装饰造型 ornamental mo(u)ld(ing)
装饰照明 decorative illumination; decorative lighting; ornamental lighting
装饰照明壁灯 decorative light wall lamp
装饰照明设施 decorative light fitting
装饰者 garinsher
装饰者用的抹灰板条 decorator's lath
装饰正面 false front
装饰织物 decorative fabric
装饰植物 decorative plant
装饰制品 decorative article
装饰质量 decorative quality; ornamental quality
装饰雉堞墙 decorated with battlements
装饰中的小块彩色金属箔取得闪烁效果 paillette
装饰终饰 ornamental finish
装饰烛台 flambeau [复 flambeaus/flambeaux];ornamental candlestick
装饰主题 decorative motif;ornamental motif
装饰柱 decorative column;ornamental column
装饰砖 ornamental brick
装饰砖雕 ajarcara
装饰砖砌体 decorative brick masonry (work)
装饰砖石砌法 ornamental masonry bond
装饰作业 decorative work
装枢轴 pivot
装双端螺栓工具 stud setter
装水 take in water
装水表用户 metered service
装碎石小车 spoil wagon
装榫 cross cogging
装锁边梃 closing stile; strike stile; striking stile
装锁的 lock fitted
装锁的门窗扇侧边 strike edge
装锁的门梃 slamming stile
装锁横挡 lock rail
装锁横木 lock rail
装锁夹具 lockset
装锁家具 lock furniture
装锁锯 locksaw
装锁冒头 lock rail
装锁门窗扇边框 strike edge
装锁门冒头 lock door rail
装锁门梃 latch jamb;lock(ing) stile; shutting stile
装锁木块 lock block
装锁舌板横挡 lock rail
装锁舌盒的门框边梃 lock jamb
装锁竖框 lock stile
装弹簧座椅的四轮马车 buckboard
装套座椅 upholstered seat
装梯级 stave
装添燃料 fuel feeding; fuel investment
装填 charging;filling;loading;pack;stowing
装填比 charge ratio;loading ratio
装填槽 < 轴承滚珠 > filling notch; filling slot
装填持续时间 loading time
装填袋 charging bag
装填方式 type of feed
装填过度 overstuff
装填过多 surcharge
装填机 bagging machine; impactor; loader;stemming machine
装填键 embedded key
装填角 loading angle
装填空隙度 packing void
装填料管 loading pipe
装填率 packing ratio
装填门 filling door
装填密度 loading density; packing density
装填器 loader;packing pawl
装填时间 filling time;loading interval
装填手门座圈 loader's door race
装填数据 padding data
装填体积 admission space; tamped volume
装填位置 loading position
装填物 priming
装填系数 charge ratio;fill factor
装填因数 fill factor
装填者 filler
装填滞后 loading lag
装填重量 charge weight
装填装置 filling device
装贴边 welt
装铁甲的 steel-clad
装听 can filling
装通风设备 ventilate
装桶 barreling
装桶机 barrel filler; barrel packing machine;cask filling machine;cask packing machine
装筒黑色炸药 black stix
装凸边 flange
装土草袋围堰 earth-filled straw bag cofferdam
装土斗 dirt bucket
装土机 clay loader;earth loader
装土箱 soil tank
装拖车用平板车 flat car for trailer
装网格的加热器 strip heater
装桅 masted;masting
装尾 tail
装屋角石 quoining
装匣钵 saggar placing;saggar setting
装狭板 stave
装线 wiring
装线工 wire man
装线规则 electric(al) wiring regulation
装线圈车 coil car
装箱 vanning < 指集装箱 >;box in; box(ing)(up);casing;encase;encasement;packaging;stuffing;package < 美 >;filling container
装箱报告单 vanning report
装箱标志 box marking
装箱/拆箱费 stuffing/unstuffing cost
装箱拆箱库 < 集装箱码头的 > consolidation building; consolidation shed; freight consolidation building;stuffing and stripping shed
装箱衬垫 casing liner
装箱尺寸 packing dimensions; packing measurement
装箱待运 be cased up for transport
装箱单 container packing list; container loading list < 集装箱 >; packing list; packing slip; packing specification; shipping list; unit packing list
装箱的 cased
装箱费 crating charges;packing charges
装箱分舱箱位图 loading bay plan

装箱固体渗碳 box carburizing
装箱后毛重 crated weight
装箱机 caser
装箱率 percent pack
装箱门件总成 <一种建筑材料> Evos door
装箱明细表 packing list
装箱破损率 failure percentage
装箱清单 packing list
装箱烧结 pack-sintering
装箱渗碳 pack-hardening
装箱渗碳炉 pack-hardening furnace
装箱渗碳硬化 pack-hardening
装箱退火 <板材的> box annealing; coffin annealing; pack annealing; pot annealing
装箱退火薄钢板 box annealed sheet
装箱小结 loading summary
装箱指示 vanning order
装箱重 crated weight
装箱重量 crated weight
装箱总尺寸 boxed dimension
装镶木板工作 planking
装橡胶轮胎的 rubber-tyred
装斜玻璃 inclined glazing
装卸 handling; loading and discharging; loading(-and)-unloading
装卸安排 material handling arrangement
装卸安全灯 cargo cluster
装卸搬运工 piggy packer
装卸搬运设备 stevedoring facilities
装卸搬运作业 handling operation
装卸板 dock board
装卸爆炸性货物的锚地 explosive anchorage
装卸操作区 freight handling area
装卸铲 loading shovel
装卸铲斗 loading bucket
装卸长 chief stevedore; loadmaster
装卸场 handling yard; loading ramp; loading yard
装卸车 ladder truck; loader-unloader; lorry loader
装卸车轮水压机 hydraulic wheel press
装卸车设施 loading-unloading facility
装卸车位 cargo-handling bay; cargo-handling berth; unloading bay; loading bay
装卸车站费率 terminal rate
装卸成本 handling cost
装卸程序 handling procedure; loading sequence
装卸出勤率 rate of attendance in loading/discharging
装卸除外 exclusive of loading and unloading
装卸单位 cargo bay
装卸单位成本 unit cost of loading/discharging
装卸岛 loader island; loading island
装卸地点 loading and unloading point; point of loading and unloading
装卸吊车 handling crane; charging and drawing crane; charging and driving crane
装卸定额 cargo-handling norms
装卸定额完成率 coefficient of realization of cargo-handling norms
装卸斗 dump skip
装卸队长 chief foreman
装卸吨位 stowage tonnage
装卸阀门 charge and discharge valve; filling and drainage valve
装卸放射线货物容许作业时间 allowable time for ensuring safety in loading and unloading of radioactive goods
装卸废钢铁的吊货工具 scrap box sling
装卸费率 handling rate; terminal rate
装卸费(用) amount of handling; cargo expenses; handling charges; loading and unloading expenses; loading charges; stevedorage; stevedore charges; terminal (handling) charges; coolie hire; stowage; handling expenses; handling fee
装卸服务 handling service
装卸干混合料的分批运料车 batch truck
装卸港 port of loading
装卸港池 cargo-handling bay; handling bay; loading and unloading bay
装卸港口费率 terminal rate
装卸港通告油船的信息 terminal advice to the tanker
装卸工 charge hand
装卸工班定额 man-shift quota for cargo handling
装卸工班数 man shifts for cargo handling
装卸工班效率 efficiency of cargo-handling per man-shift; man-shift rate
装卸工具 handling tool; stevedore's gear
装卸工具储藏室 stevedore's warehouse
装卸工具房 stevedore's warehouse
装卸工具证书 cargo gear certificate
装卸工具租金 cargo gear hire
装卸工人 wharfman; porter; dock walloper; longshoreman; stevedore <码头>; lumper; freight handle; loader; loading member
装卸工人生产率 labo(u)r productivity of stevedore
装卸工人休息棚 longshoremen's shelter
装卸工时产量 cargo-handling output per man-hour
装卸工时定额 man-hour quota of cargo handling
装卸工时利用率 utilization factor of man-hours in loading-discharging
装卸工手钩 docker's hook
装卸工序 cargo-handling unit operation; procedure of cargo handling
装卸工序吨 procedure tons of cargo-handling
装卸工艺 cargo-handling process; cargo-handling technology; handling technology; material handling arrangement
装卸工艺管理 technologic(al) management of cargo handling
装卸工艺流程 technologic(al) process of cargo handling
装卸工艺设计 design of cargo-handling technology
装卸工艺系统 cargo-handling system; cargo-handling technology system
装卸工作 stevedoring
装卸工作半机械化 semi-mechanization of cargo-handling
装卸工作半自动化 semi-automation of cargo-handling
装卸工作机械化 mechanization of cargo-handling
装卸工作全盘机械化 full mechanization of cargo-handling
装卸工作时数 ship hours
装卸工作系数 ratio of operational tons to ports throughput
装卸工作自动化 automation of cargo-handling
装卸公司 freight handler; stevedoring company; stevedoring firm; stevedoring organization
装卸共用时间 all-purposes
装卸钩 stevedore hook
装卸过程 cargo-handling process; process of cargo-handling
装卸河湾 loading bay
装卸滑槽 cargo chute
装卸滑车 cargo block
装卸滑轮 loading tackle
装卸货驳 stevedoring barge
装卸货固定伸臂 burtoning
装卸货滑车 cargo block
装卸货绞辘 garnet purchase
装卸货日期 lay days
装卸货时间 laytime; running laydays
装卸货跳板 cargo-handling ramp; ramp
装卸货物 loading and unloading cargo; stevedore
装卸货物的驳船 stevedoring barge
装卸货物的侧舷门 cargo port
装卸货物的船舶时数 ship hours of cargo handling time
装卸货物港口的优化 option for loading/discharging port
装卸货物用的起重绞车 cargo-handling winch
装卸货物站台 goods platform
装卸货油操纵台 <油轮> cargo loading control console
装卸机 charging crane; conweigh belt; forcing jack; forcing machine; handler; loader; loader-unloader; off-loader; reloader; reloading machine
装卸机的卸料运输带 loader front conveyer
装卸机加料起重机 charging crane
装卸机具 handling equipment
装卸机械 cargo-handling equipment; cargo-handling gear; cargo-handling machine; cargo-handling machinery; cargo-handling plant; handling machinery; loading and unloading machine; stevedoring machine; handling plant
装卸机械非完好台时 unworkable machine-hours
装卸机械工作台时 machine hours at work
装卸机械化 loading and unloading mechanization
装卸机械化程度 extent of mechanization in loading and unloading
装卸机械化系统 mechanization system of cargo handling
装卸机械技术生产率 technical productivity of handling machinery
装卸机械利用率 utilization factor of cargo handling machinery
装卸机械平均台时产量 mean hourly output of handling machinery
装卸机械起运量 handling tonnage of handling machinery
装卸机械日历台时 calendar machine-hours of cargo handling machinery
装卸机械生产定额 output norm of handling machinery; output quota of handling machinery
装卸机械台时产量 output per machine-hours
装卸机械停工台时 machine hours in idleness
装卸机械完好率 coefficient of readiness of cargo handling machinery; percentage of perfectness of handling machinery
装卸机械完好台时 workable machine-hours
装卸机械维修基地 repair base for loading and unloading machineries
装卸机械作业量 output of cargo-handling machinery
装卸机械作业台时 machine hours at cargo work
装卸计划 cargo-handling plan; cargo planning
装卸技术 handling technique
装卸监督员 landwaiter; port reeve; port warden
装卸检修所【铁】 in-service freight car maintenance depot
装卸交货总管理员 cargo sheet clerk; ship's clerk
装卸绞车 loading winch
装卸节省时间的合计 all working time saved both ends
装卸进度 progress of cargo work; progress of loading/discharging
装卸进度表 progress chart of cargo work; progress chart of loading/discharging
装卸矿车 decking
装卸两港 both ends
装卸量 handling volume
装卸料场 loading dock
装卸料点 load unload point
装卸料吊车 charge hoist
装卸领班 foreman
装卸码头 discharging quay; shipping dock; dock pier; loading quay
装卸煤设施 coal handling facility
装卸门 cargo door
装卸能力 cargo-handling capacity; handling ability; handling capacity; loading and unloading capability; loading capacity
装卸排作业收入 load and unloading and discharging task revenue
装卸棚 loading shed
装卸皮带 <输送机的> loading belt
装卸平均品质条件 average or mean shipped and landed quality terms
装卸平台 loading and unloading platform; loading dock; loading platform; loading stage
装卸平台棚 loading dock shelter
装卸期间 lay days; laytime
装卸期间损失 loss sustained during loading and unloading
装卸期限 handling term
装卸起重机 charging and driving crane; stamp work's crane; stevedoring crane; stevedoring unloader
装卸钎杆 steel handling
装卸桥 handling bridge; bridge transporter; gantry crane; loading bridge; transfer crane; transport(er) crane; unloading tower
装卸桥的小车牵引索 troll(e)y rope
装卸桥顶部十字架 top cross beam
装卸桥分配表 crane splitting advice
装卸桥使用密度 crane intensity
装卸桥式卸货机 gantry unloader
装卸桥悬搁时间 crane hang
装卸桥作业顺序 crane sequence
装卸区 loading zone
装卸区操作管理费 terminal handling charges
装卸区到装卸区业务 terminal to terminal service
装卸区堆场操作费 transfer charges
装卸区间运费 terminal to terminal rate
装卸区内结关 <集装箱> clearance terminal
装卸区内装箱 <集装箱> terminal vanning
装卸人 lumper
装卸日数可调剂使用 time reversible

装卸容器 handling cask
装卸软管 cargo hose; charge and discharge rubber(lined) hose
装卸散装货料斗 loading hopper
装卸设备 handling appliance; cargo gear; cargo-handling equipment; cargo-handling gear; cargo-handling plant; freight handling facility; handling equipment; handling facility; handling machinery; handling plant; loading and unloading facility; load rig; material handling equipment; transshipment facility
装卸设备布置图 cargo gear arrangement
装卸设备合格证 cargo gear register
装卸设施 loading facility; loading/unloading facility
装卸施工设备用的半拖车 semi-trailer for construction equipment
装卸石油产品的栈桥 oil-loading rack
装卸时货叉俯仰液压缸 laydown
装卸时间 handling time; loading and unloading time; terminal time; loading time
装卸时间表 time sheet
装卸实际工作工日产量 efficiency of cargo-handling per man-shift; output per actual man-day for loading/discharging
装卸实际工作日数 actual man-days for loading/discharging
装卸式大看台 <预制构件的> portable grandstand
装卸式隔断 demountable partition; relocatable partition
装卸式货盘 stevedore type pallet
装卸事故 loading and unloading accident
装卸手 boom man
装卸手钩 cargo hook
装卸疏运设备 facility for dispatch
装卸顺序单 handling sequence list
装卸速率 rate of loading and discharge; rate of loading and discharging; loading rate
装卸损坏 stevedore damage
装卸损失 handling loss
装卸塔 dump tower; loading tower
装卸塔司机 loading tower operator
装卸塔系泊 tower mooring
装卸台 cargo stage; distributing ramp; freight platform; landing; loading stage; platform; shipping dock
装卸台车挡 dock bumper
装卸台雨棚 loading dock shelter
装卸台组件 platform assembly
装卸天数 ship's days
装卸天数扣除星期日 Sunday excepted in lay days
装卸条件 handling condition
装卸跳板 bridge plate; gang board; gang plank
装卸铁板起重机 sheet-iron crane
装卸停车时间 contact time
装卸通路 loading driveway
装卸突堤码头 loading pier
装卸推土机 C-frame
装卸托板 stevedore's pallet
装卸托盘 stevedore's pallet; stevedoring pallet
装卸湾 loading bay
装卸网兜 cargo net; loading net
装卸污染货物所增加的装卸费 dirty money
装卸误期费 demurrage
装卸线 loading and unloading track; loading-unloading siding; loading siding; loading track; team track; transfer track【铁】; loading-unloading line
装卸线脚 dressing
装卸箱 lift van
装卸箱货物 lift van
装卸效率 efficiency in loading and unloading; efficiency of cargo handling; handling efficiency; handling rate
装卸斜坡道 loading and unloading ramp; loading ramp
装卸许可天数 days purpose
装卸液压缸 laydown cylinder
装卸用大功率照明灯 cargo flood light
装卸用斗柄 <挖掘机> material handling stick
装卸用浮式起重机 floating crane for cargo handling
装卸用起重机 loading and unloading crane
装卸油泵 cargo pump
装卸油管线 boat line; boat pipeline
装卸油架 <码头上的> dock riser
装卸逾期日 day on demurrage
装卸与运输 handling and transportation
装卸月台 loading dock
装卸运输费用 handling and shipment expenses
装卸运输路线 loading and unloading road
装卸运输系统 handling transportation system
装卸运输与储藏设备 handling and storage equipment
装卸责任 stevedore's liability
装卸栅门 landing door
装卸栈桥 loading stage
装卸站 loading and unloading station
装卸站台 loading dock; loading platform; loading-unloading platform; loading/discharging platform
装卸振动 handling shock
装卸指标标志 pictorial marking for handling of non dangerous goods
装卸指导员 foreman
装卸指示 handling instruction
装卸制纸浆木材的叉车 pulpwood fork
装卸中心 handling center[centre]
装卸抓斗 loading bucket
装卸抓钩 handling grab
装卸转运码头 <英国方言,尤指煤炭码头> staith(e)
装卸装置 handler; handling attachment; handling device
装卸自然吨 physical ton of handling
装卸总成本 total cost of loading/discharging
装卸组 loading brigade; longshoreman's gang; stevedore gang
装卸组长 chief stevedore; head stevedore; ship foreman; shipworker
装卸最重限额 heavy handy dead weight
装卸作业 cargo-handling operation; cargo work; handling operation; loading and unloading operation; stevedoring; stevedoring operation; terminal operation
装卸作业半自动化 semi-automation of cargo-handling
装卸作业灯 cargo light
装卸作业方案 program(me) of loading and unloading operations
装卸作业费 loading and unloading expenses
装卸作业分析 analysis of transfer operation
装卸作业分析路线 analysis of transfer path
装卸作业工时 an-hours for loading/discharging
装卸作业机械化 handling mechanization; mechanization of cargo-handling; mechanization of loading and unloading operation
装卸作业机械化程度 degree of mechanization of cargo handling
装卸作业区主任 wharf manager
装卸作业人时 man-hour for loading-discharging
装卸作业线 cargo-handling operation line; operating line of cargo-handling
装卸作业中央控制屏 central loading/unloading control panel
装卸作业自动化 automation of cargo-handling
装卸作业组织 organization of cargo handling
装芯 cored
装新阀子 revalve
装信管 fusing
装型盒 flasking
装修 dressing; finish; fixture
装修……的外表 reprofile
装修标准 standard of finishes
装修材料 finishing material; surfacing material
装修层 finished face; finished layer
装修的窗框 window casing
装修吊篮 basket flying scaffold; basket hanging scaffold; lifting basket for decoration
装修分站 conditioning sub-station
装修工 decorator; finisher
装修工厂 conditioning plant
装修工程 decoration works; finishing works; fitting works
装修过的 enriched
装修过的面积 decorated area
装修过的天然石料 dressed with natural stone
装修后尺寸 finished size
装修加固混凝土 fixing concrete
装修煤气管的工人 gas fitter
装修面 finishing face
装修面积 decorative area
装修磨光作业 finishing off
装修木地板作业 knifing
装修木工 joiner; joinery work
装修木工场 joining yard
装修木工车间 joining workshop
装修木工作业 casework
装修木料 joinery wood
装修抛光 finishing polish
装修平台 lifting platform for decoration
装修时间 lay days; modification time
装修用板 wrought board
装修用钉 finishing nail
装修用石灰 finishing lime
装修站 conditioning station
装修作业 decorative application
装畜通道 loading chute
装旋梯柱的井孔 hollow newel
装靴的 shod
装雪车 snow loader
装雪机 snow loader
装岩 loading; mucking; muck loading; mullocking
装岩铲斗 rock bucket
装岩动力消耗 consumed power of loading
装岩斗 loading bucket; loading edge
装岩方法 loading method
装岩方式 mucking arrangement
装岩工 groundman; mucker
装岩工长 muck boss
装岩工时消耗 consumed man-hour of loading
装岩工作量 amount of work of loading
装岩机 dirt loader; loading grab; mechanical mucker; mucker; mucking equipment; mucking machine; mucking unit; muck loader; rock loader
装岩机传送带 mucker belt
装岩机司机 mucker operator
装岩机特性 muckerism
装岩机与转载机联合装岩 combined loading by loader and conveyer [conveyor]
装岩技术经济指标 technical-economic index of loading
装岩溜槽 mucking pan
装岩设备 lashing equipment; loading equipment; mucking equipment
装岩设备折旧摊销及大修费 loader equipment depreciation apportion and overhaul charges
装岩生产定额 loading production quota
装岩生产率 loading productivity
装岩时间 loading time; mucking time
装岩檐口 mucking slot
装岩用垫板 mucking plate
装岩用挖掘机 rock loading shovel
装岩运输机 rock loading conveyer [conveyor]
装研磨体 media charging
装窑 carry-in kiln <退火窑的>; charge kiln; kiln placing
装窑垫板 setter slab
装窑高度 setting height
装窑工 setter; carry-in boy <玻璃退火时装送制品的人员>
装窑机 kiln feeder
装窑密度 placing density; setting density
装窑棚架 kiln shelf
装窑容量 setting capacity
装窑窑具 kiln setter
装窑用叉子 carry-in fork; take-in fork
装药 charge; powder charging; projectile filling
装药棒 tamping stick
装药不耦合系数 not coupling coefficient of loading
装药不足 undercharge
装药长度 charge length; explosives-loading length; loaded length
装药车 loading truck
装药程度 degree of packing
装药点火管 cartridge igniter
装药洞室 powder mine
装药方式 pattern of charges
装药放炮工人 shotfirer
装药高度 shot elevation
装药工 charge hand
装药工人 cartridge loader; powder monkey
装药管 <岩石爆破的> cartridge tube
装药棍 loading stick
装药号 zone charge
装药机 charger
装药结构 charging construction; pattern of charges
装药巷道封墙 walling up
装药孔 breech opening
装药孔塞柱 stemming
装药量 charge weight; explosive load; powder charge
装药量计算公式 blasting formula
装药量确定 charge establishment
装药量-岩石体积比 charge rock volume ratio
装药率 <爆破岩石单位体积岩石的炸药量> loading ratio
装药密度 charge concentration; char-

ging density; density of charge; packing degree; powder-loading density
装药炮孔 charged hole; loaded hole
装药平硐 coyote tunnel(l)ing method
装药平巷 coyote hole
装药平巷法 coyote tunnel(l)ing method
装药器 loading vessel
装药软管 loading hose
装药深度 charge depth; shot depth
装药筒 cartridge tube
装药系数 charge coefficient; charge ratio; coefficient of charge
装药限度 charge limit
装药巷道 powder drift; powder mine
装药引爆工人 shotfirer
装药用具 charging accessories
装药重量 weight of charge
装药铸造 grain casting
装药组合 charge assembly
装药作业 charging
装一半 half-full
装以柴油机 dieselization [dieselisation]
装翼模板 winged mo(u)ld board
装油 charge of oil
装油泵 cargo oil pump; filling pump
装油臂 oil-loading arm; loading arm
装油车 oil wagon
装油房 filling house
装油干线 filling main
装油管 marine loading arm
装油管线 filling line
装油平台 loading platform
装油软管吊柱 hose davit
装油设备 oil-loading facility
装油图表 filling table
装油效率 oil-loading rate
装油/卸油软管 loading hose
装油遥控系统 cargo oil remote control system
装油栈桥 oil-loading rack; overhead oil loading rack
装油站 filling point; oil-loading terminal
装有凹槽的门框 buck frame
装有凹形铲刀的推土机 bowldozer
装有宝石的 jewelled
装有保险丝的板 cut-out board
装有拨号盘和送受话器的试验器 buttinski
装有布帘的 valanced
装有车轮的 wheel-mounted
装有衬垫的病房 padded cell
装有衬垫的门 padded door
装有冲压式喷气发动机的飞行器 ramjet
装有处理机的终端 intelligent terminal
装有窗帘框架的 valanced
装有磁铁分离器的磨煤给煤机 pulverizer feeder with magnetic separator
装有催化净化废气系统的小客车 catalyst-equipped passenger car
装有大型发动机的船舶 large-engined vessel
装有单向信号机的线路 track signalled for one-way operation
装有导流管的螺旋桨 ducted propeller
装有电动机的车轮 power wheel
装有电动机的振动棒 <混凝土捣实用> poker with motor
装有电键发送器的座席 <电话> semi-mechanical position
装有电线的 wired
装有吊舱 pod
装有调度集中和列车运行描述器的行车调度所 traffic regulation office equipped with CTC and train describer system
装有多孔滤板的(气力)输送机 porous block conveyer[conveyor]
装有发动机的 motorised[motorized]
装有法兰的 flanged
装有翻车保护架的司机室 roll-over protective structure cab; roll-over protection structure cab
装有反射器的灯 reflectorized lamp
装有防滑钉的轮胎 studded tyre[tire]
装有辐射管的炉 radiant tube furnace
装有干燥设备的拌和机 drying mixer; drying mixer combination
装有干燥设备的搅拌机 dryer-mixer combination
装有钢箍的 steel-shoed
装有钢筋的钢筋混凝土桩 bound pile
装有钢靴的 steel-shoed
装有割刀的 knife-equipped
装有轨道电路的股道 circuited track
装有滚珠轴承的对接铰链 ball-bearing butt hinge
装有滚柱的螺旋重力溜槽 roller-fitted spiral gravity shoot
装有滚柱轴承的牙轮钻头 cone-type roller bearing rock
装有护套的 sheathed; shielded
装有滑道垫木的 skid-mounted
装有滑脚的开沟器 runner-attached boot
装有滑块的支重块 <支承跨装货物> bearing piece with sliding piece
装有机械螺旋钻的卡车 lorry-mounted(mechanical) auger
装有几个模型的型板 card of patterns
装有记数器的匝道 metered ramp
装有驾驶室的车辆 vehicle with driving cab
装有尖顶的 steepled
装有减速器的 retarder-equipped
装有减速器的驼峰 retarder-equipped hump
装有交叉支架的电杆 bracket pole
装有角撑板的多层纸袋 gussetted multiwall paper bag
装有绞车的机车 crab locomotive
装有绞车的小船 bum boat
装有绞关索具的平台 headworks
装有脚轮的小台 troll(e)y-table
装有铰盘的卡车 lorry winch
装有金属箍头的 steel-shod
装有晶体滤波器的差频分析器 crystal filter heterodyne analyser[analyzer]
装有空气调节器的 air-conditioned
装有空气制动机的列车 train fitted with air brake
装有空调的建筑 air-conditioned building
装有空调的平顶 conditioned ceiling
装有空调的天花板 conditioned ceiling
装有空调设备的 air-conditioned
装有雷达的惯性导航系统 radar-equipped inertial navigation system
装有雷达定向仪的头盔 radar helm
装有雷达反射器的目标 radar reflecting target
装有雷管的导火线 coiled capped fuse
装有立柱的卡车 lorry with stanchions
装有沥青熔化炉的四轮车 fire wagon
装有链式输送机的上倾式装载机 overhead loader with armo(u)red chain conveyer[conveyor]
装有两种牵引电流的车站 station provided with two types of traction current
装有量测仪器的桩 instrumented pile
装有量测元件的桩 instrumented pile
装有履带的 crawler-mounted
装有履带的起重机 crawler-mounted crane
装有轮胎的 wheel-mounted
装有明轮的船 paddle wheeler
装有木构件的 timbered
装有尼龙纤维的 nylon filled
装有泥砂泵的挖泥船 sand pump dredge(r)
装有配重块的 weights-installed
装有喷气发动机的 jet-powered
装有起重杆升降设备的桥式起重机 bridge crane with level luffing crane
装有起重滑车设备的房屋 tackle-house
装有起重机的 crane-equipped
装有气门的 ported
装有切草刀的柱塞 knife-equipped plunger
装有切割轮的玻璃切割刀 glass cutter with cutting wheels
装有区别机的卫星局 <自动电话> discriminating satellite exchange
装有区别机的支局 <自动电话> discriminating satellite exchange
装有曲柄的 cranked
装有(散热)叶片的电热器 strip electric(al) heater
装有纱窗的洞口 openings screened
装有闪光灯或电铃的道口自动门栏 automatic crossing gate with flashing light of bells
装有甚高频无线电控制的机车 locomotive equipped with VHF [very high frequency] radio control
装有渗漏水配件的阻止反虹吸作用设备 backsiphonage preventer with leakage water fitting
装有升降玻璃的车门 sash door
装有枢轴的棘爪 pivoted detent
装有双向信号机的双线插入段 <为组织不停车会车> short lengths of double track, bidirectionally signal(l)ed
装有双向信号机的线路 track signalled in both directions
装有水泵的消防车 pumper
装有水表的公寓 apartment metered
装有水电等线路的顶棚 serviced ceiling
装有撕碎机的垃圾收集汽车 shredder truck
装有套管的钻孔 cased bore hole
装有提升机的卡车 lift truck
装有填料的加劲杆 filled stiffener
装有铁淦氧的空腔谐振器 ferrite-loaded(resonant) cavity
装有停车计时仪的路缘石 metered curb
装有推土板的拖拉机 tractor equipped with bulldozer
装有推卸器的铲斗 bucket with ejector
装有挽缆插栓的桅箍 futtock band; spider band; spider hoop; spider iron
装有碗形铲刀的推土机 bowldozer
装有望远镜的矿山罗盘 improved dial
装有帷幔的 valanced
装有无线电的惯性导航系统 radio-equipped inertial navigation system
装有无线电通信[讯]设备的列车 radio-equipped train
装有吸砂泵的挖泥船 sand pump dredge(r)
装有橡胶轮胎的 rubber-tired mounted
装有橡胶轮胎的装卸机 rubber-tired loader
装有小发动机的自行车 moped
装有卸料漏斗的楼板 hoppered floor
装有压气机的履带式自行钻机 drill-cat
装有叶轮的回转选粉机 bladed rotor separator
装有叶片的 bladed
装有叶片的叶轮 bladed wheel
装有叶片的转子 bladed rotor
装有液压斜铲的履带式拖拉机 crawler tractor fitted with hydraulic angle dozer
装有仪器的 instrumented
装有仪器的试桩机 instrumented pile driver
装有再点火器的火焰安全灯 relighter flame safety lamp
装有闸门的反弧形溢流堰 gated ogee structure
装有闸门的堰 gated weir
装有闸门的溢洪道 gate-controlled spillway
装有闸门的溢洪口 gate-controlled spillway
装有制动台的货车 <即守车> wagon with brake cabin
装有重型齿条的推土机刮刀 rock rake
装有抓斗吊车的储库 grabbing pit
装有自动信号区段线路占线路总里程百分率 percentage of sections having automatic signals to total route kilometers
装有自动制动机的车辆 fitted vehicle
装有自动制动机的列车 fitted train
装有嘴子 in-tap
装于舱内 keep in hold
装于顶棚的设备 ceiling mounted fixture
装于后部的 rear-mounted
装于后部的液压挖掘装置 rear-mounted hydraulic digger attachment
装于后部的抓斗 rear grab
装于后部的装置 rear rig
装于后部可拆换装置 rear interchangeable equipment item
装于滑动底板的箱形喷洒器 skid-mounted box-like spreader
装于滑动底板的箱形撒料机 skid-mounted box-like spreader
装鱼码头 fish landing quay
装原木的车辆 bunk; log bunk
装原木的横架 bunk; log bunk
装原木的卡车 bunk; log bunk
装缘植物 edging plant
装源 loading of source
装运 blockade; carry; handling and loading; load and transport; shipment; shipping
装运保险单 shipment policy
装运报告 report of shipment
装运备忘录 shipping memorandum
装运变形 handling strain
装运标志 shipping mark
装运冰块用的车箱 ice body
装运擦痕 transit rub
装运擦伤 handling scratch
装运车 loader transporter
装运尺寸 shipping dimension
装运单 shipping list; shipping order; shipping ticket
装运单据 shipping bill; shipping documents
装运的数量 quantity of shipment
装运地点 point of shipment; point of shipping
装运点 shipping point
装运堵塞 shipping plug
装运方法 handling technique; manner of shipping; shipment method
装运方式 means of transport(ation)

装运废碴 mucking
装运费(用) shipping cost; shipping expenses; cargo expenses; shipping charges
装运符号 shipping mark
装运付款 cash on shipment
装运干燥度 <为防止木材运输途中发生病虫害,在装运前进行干燥> shipping-dry
装运钢轨设备 rail-loading plant
装运港 loading port; port of shipping; shipping port
装运港离岸价 FOB [free on board] port of shipment
装运罐 coffin
装运货量 shipment
装运货物输送地点 point of shipment delivery
装运机 load-and-carry machine; load haul-dump unit; transloader
装运集装箱的挂车装车数 trailer loadings
装运集装箱能力 container-carrying capacity
装运集装箱平车 container wagon; flat car for container
装运口岸 port of loading; port of shipment
装运口岸及目的地 loading port and destination
装运连挂的公路挂车的背负式平车 <每辆车装有自动侧向装卸机,不需用起重机> minipiggy
装运两层汽车的车辆 two tier car carriage
装运两层汽车的货车 bi-level(rack) car
装运量 shipping amount
装运码头 <用于装运重型设备> load-out dock
装运碰撞 shipping shock
装运期限 deadline for shipment
装运汽车的列车 car-carrier train
装运汽车的平车 autoveyor
装运汽车的特种长大敞车 car transporter
装运汽车的专用船 autocarrier
装运前品质条件 preshipment quality terms
装运前资金通融 preshipment financing
装运热钢锭货车 wagon for carriage of hot ingots
装运日价格 day of shipment price
装运日期 date of shipment; shipping date
装运容器 shipping cask; transfer cask
装运三层汽车的货车 tri-level car
装运三层汽车的货车卸车区 tri-level car unloading area
装运三脚架 loading tripod
装运设备 carrier loader
装运石块的后翻斗卡车 rear-dump rocker
装运石料的车身 quarry body; rock body
装运手续 shipping formality; shipping process
装运双层汽车的货车 double deck car carrier; rach car
装运索赔 shipping claim
装运条件 shipping term; terms of shipment
装运通知 shipment advice; shipping advice; advice of shipment
装运通知单 shipping notice; shipping permit
装运图 shipment drawing
装运托架 shipping bracket
装运托盘化 palletization [palletisation]
装运挖土机用的半拖车 semi-trailer for transport of excavating machines
装运物 tote
装运误期费 demurrage
装运限期 due of shipment
装运箱 shipping box; shipping case; shipping container; tote pan
装运小汽车的大型载重汽车 autocarrier
装运卸碴联合机 load haul dump
装运卸多用途车辆 loading-handling-displacement vehicle
装运卸功能 load haul dump
装运卸设备装岩 loading by load-haul-dump equipment
装运卸用途 load-and-carry application
装运须知 shipment instruction
装运许可证 certificate of shipment; shipping permit
装运岩石的自卸车身 rock body
装运应力 handling stress
装运员 dispatcher
装运载有集装箱挂车的背驮式平车 piggyback car of trailers
装运站 point of shipment; shipping point
装运支座 shipping bracket
装运只有司机室而无车身的卡车时采用后车爬上前车的方法 trucks with cabs only, loaded in "saddle-back" fashion
装运指令 shipping instruction
装运指示 shipping instruction
装运滞期费 demurrage
装运中的破裂 handling breakage
装运重量 shipped weight; ship(ping) weight
装运重质散装货物的货车 wagon for carriage of heavy bulk traffic
装运砖的笼子 brick cage
装运装置 shipping device; shipping unit
装运状态 shipping condition
装运总量 gross shipping weight
装杂物的容器 catch-all
装载 load(ing); handling; cargo stowing; charging; lade; lading; on-load(ing); stow(age); take load up
装载安排 loading arrangement
装载班 loading shift
装载板 loading plate
装载半径 <起重机> load radius
装载比 loading ratio
装载臂杆 loading boom
装载表 load(ing) table
装载不良 bad stowage; improper stowage
装载不平衡检测器 unbalanced loading detector
装载不足 short shipment; underloading
装载仓 loading bin; loading pocket
装载槽 loading chute; loading trough
装载(侧)线【铁】 loading siding
装载铲 loader shovel; loading shovel
装载铲斗 loading shovel
装载铲斗车 loading shovel bucket
装载场 loading area
装载超重 over load
装载车 wheeled pusher
装载车轮胎 loader tire[tyre]
装载程度 loading degree
装载程序 load module
装载吃水线 loading line
装载尺寸 loading ga(u)ge
装载处理 handling of load(s)
装载存储 load store instruction
装载大体积货物的驳船 hoy
装载的 laden
装载的材料 material to be loaded
装载的货物 lading
装载地点 loading point; loading site
装载地区 loading area
装载调查 stowage survey
装载斗 excavating bucket; loader-mounted shovel; loader shovel; loading bucket; loading hopper
装载斗仓 loading hopper; loading skip
装载堆垛机 stacker-loader
装载队 loading team
装载吨 <1 装载吨 = 40 立方英尺, 1 立方英尺 = 0.02832 立方米> freight ton; shipping ton; capacity ton
装载法 stowage
装载方法 loading method
装载方式 loading method; loading type; load mode
装载分布 loading distribution
装载分布图 loading diagram
装载负荷 live load; movable load; moving load
装载港 port of embarkation
装载高度 height of loading; loading height; loading line; ride height
装载高度低 low loading height
装载工平巷 <采矿> loader gate road
装载构件 loading unit
装载规程 loading instruction; load regulation
装载轨道 loading track
装载过程 process of loading
装载过多 overcharge; overlade; overloaded
装载合格证 stowage certificate
装载戽斗 dump bucket
装载混凝土 concrete loading
装载货物 stowage
装载货物长度 length of load
装载货物处 loading place; stowage
装载货物的复查 cross-check of the load
装载货物的移动 shifting of load
装载货物(在车上)的倒塌 tipping over of load
装载机 car loader; excavator-loader; loader; loading machine; loading shovel; material loader; mechanical loader; mechanism loader; mucking unit; shovel loader; tractor shovel; traxcavator
装载机铲斗 basket; loading shovel
装载机导轨 loader runway
装载机的定义 loader definition
装载机的破碎杆 cutter-loader bar
装载机的前部运输机 loader front conveyer[conveyor]
装载机的尾部运输机 loader rear(ward) conveyer[conveyor]
装载机的卸载运输机 loader discharge conveyer[conveyor]
装载机的运输机 loader conveyer [conveyor]
装载机工型铲斗连杆 Z-bar loader linkage
装载机工作范围 loader-reach
装载机工作循环时间 loader cycle time
装载机构 loader mechanism
装载机骨料机头 loader gathering head
装载机后端运输带 loader rear(ward) conveyer[conveyor]
装载机后座管理员 loader end man
装载机滑道 loader runway
装载机集料机头 loader gathering head
装载机驾驶员 loader engineer; loading machine runner
装载机前端运输带 loader discharge conveyer
装载机倾覆荷载 static tipping load
装载机提升到最高点的最大装载能力 lift capacity to maximum height
装载机稳定度 stability gradient of loader
装载机卸料悬臂 loader discharge boom
装载机行走操纵杆 loader tramming lever
装载机用轮胎 loader tire[tyre]
装载机主架 loader frame
装载机转臂 loader boom
装载机组 loading unit
装载集装箱的挂车及集装箱的平车运输枢纽 trailer on flat car/container on flare car terminal
装载集装箱和装载挂车式集装箱的平车组成的半固定车底的列车 containers on flat car/trailers on flat car semi-permanently coupled train
装载计划 loading plan
装载记录仪 <挖泥船的> load recorder
装载检查 loading survey; stowage survey
装载键 load key
装载角 angle of loading
装载结束时刻 station time
装载界限 loading ga(u)ge
装载净重能力 delivered payload capability
装载距离 loading reach
装载空间 stowage space
装载控制 stowage control
装载宽度 loading width
装载力 charge capacity; holding power
装载良好 shipped in good condition
装载粮食 victual(l)ing
装载两层或三层汽车的货车 auto-rack car
装载两用推土机 loader-dozer
装载量 charging capacity; holding capacity; laden load; loadage; loading capacity
装载量定额 burden rate
装载料车 loading skip
装载溜槽 loading chute; mucking pan
装载漏斗 batcher hopper; loading hopper
装载螺旋 filling auger
装载码头 loader wharf; loading wharf
装载门 loading door
装载密度 charging density; load density
装载面 load(ing) surface
装载面积 loading area; loading space
装载明细单 bill of lading clause
装载模式 load module
装载能力 loading capability; loading capacity
装载平巷 loader drift; loading drift
装载坡道 loading ramp
装载坡台 loading ramp
装载期限 load(ing) term
装载起重机 loading crane; toter
装载起重设施 loading installation
装载汽车的货车 car-carrier
装载汽缸 loading cylinder
装载器 loading bin
装载器械 loading appliances
装载强度 loading intensity
装载桥 loading bridge
装载桥式吊车 loading bridge(hoist)
装载情况 loading condition
装载区 load area
装载容积 stowage
装载容量 charge capacity; heap(ed) capacity; stowage space; struck ca-

pacity <挖掘机的>
装载容量利用系数 capacity factor
装载散货 bulk loading
装载散料 loading in bulk
装载设备 change direction gear; charging appliance; loading device; loading equipment; loading facility; loading plant; loading system
装载生产率 output on loading
装载石块铲斗 bucket for rock
装载时间 loading time; time of loading
装载式 load mode
装载式货物集装箱 stowable cargo container
装载数据 loading data
装载顺序 loading sequence
装载说明书 loading instruction
装载松散物体的船只 bulk vessel
装载速度 loading rate; rate of loading; speed of loading
装载隧道 load-out tunnel
装载索 carrying cable; track rope
装载索具 loading jack
装载塔 loading tower
装载塔驾驶员 loading tower operator
装载台 loading stage
装载提升机 loading elevator; loading stage
装载图【船】stowage plan
装载推土机 loader-dozer
装载推土两用机 loader-dozer
装载挖掘机 loader-excavator
装载位置 loading point
装载系数 charge coefficient; charge ratio; coefficient of charge; coefficient of load(ing); load(ing) coefficient; load(ing) factor
装载限度 loading limit
装载限界 loading clearance; loading ga(u)ge
装载(小)车 charging carriage
装载小车的斗 buggy-loading hopper
装载效率 efficiency of loading
装载卸料悬臂 loader discharge boom
装载循环 filling cycle; loading run
装载循环次数 number of load cycles
装载液压系统 loading hydraulic system
装载一次的时间 time per load(ing)
装载已搅拌好混凝土的卡车 non-agitating unit
装载因数 load factor; stowage factor
装载因子 load factor; stowage factor
装载用皮带输送机 loading belt conveyer[conveyor]
装载用气闸 loading airlock
装载用气闸室 loading airlock chamber
装载用设备 loading system
装载运输车 truck loader
装载运输带 loader conveyer[conveyor]
装载运输机 loading conveyer[conveyor]
装载杂货选择权 dreadage; dreading
装载站 loading depot; loading location; loading station
装载站台 loading platform
装载支架 loading dock
装载指示 shipping instruction
装载指示器 stocking indicator
装载重 burden
装载重量 laden weight; loaded weight; loading weight; weight of load
装载周期至破损阶段 loading cycles to failure
装载抓斗 loading grab
装载爪 gathering arm
装载装置 charging gear; loading attachment; loading mechanism; loading station
装载最高点 load peak
装载作业 loading operation; loading work
装在变速箱上的油泵 transmission-mounted pump
装在侧架上的配重块 sprocket mounted counter-weight
装在侧面的 side-mounted
装在车顶上的空气调(节)吊箱 root-mounted air conditioning pod
装在车架上的电动机 frame-mounted motor
装在车上的 carriage mounting; mounted in a vehicle; vehicle-mounted
装在车上的超高测量仪 supermeter
装在车上用地磅过秤 carload
装在储粉箱上的磨粉机 mill over meal bin
装在船舱内 shipped in the hold
装在带式输送机旁的引带 slobber-belt
装在电杆转向架上的灯 truck light
装在电缆内的柔性放大器 built-in flexible amplifier
装在发动机上的 engine-mounted
装在发动机上的油泵 engine mounted pump
装在浮驳上的混凝土搅拌设备 pontoon-mounted concreting plant
装在钢轨上的扫描器 rail-mounted scanner
装在挂车上的绞盘 trailer-mounted winch
装在轨道上的 rail-mounted
装在轨道上的起重机 rail-mounted crane
装在后部的松土装置 rear scarifier attachment
装在后部的振动器 rear vibrator
装在回转窑上冷却器 integral cooler
装在货车上的公路车辆 <背驮式运输> road vehicles on wagon
装在机架上的护板 frame-mounted guard
装在机架上的护罩 frame-mounted guard
装在机内的 built-in
装在机器上的电子计算机 on-board computer
装在吉普车前面的推土刀 bulldozer attachment for jeeps
装在甲板 shipped on deck
装在绝缘子上的电缆 line installed on insulators
装在卡车上的路面加热器 lorry-borne road heater
装在卡车上的撒布机 truck spreader
装在卡车上的液压挖泥机 hydraulic lorry-mounted excavator
装在卡车上的液压挖土机 hydraulic lorry-mounted excavator
装在控制台上的仪表 dash-mounted ga(u)ge
装在框架上的筛网布 framed screen cloth
装在框内 enframe
装在冷却板上的 cold plate-mounted
装在链轮上的配重块 sprocket mounted counter-weight
装在列车上的 train-borne
装在列车上的双杆液压升降机 <为装运长焊接钢轨用> train-mounted twin-jib hydraulic hoist
装在履带底盘上的机械 crawler-mounted machine
装在履带上独立工作的马达 independent track motor
装在轮毂上的马达 hub motor
装在轮胎上的 wheel-mounted
装在门底与地板之间的铰链 floor hinge
装在面板上的 panel-mounted
装在平车上的集装箱的半挂车 semi-trailers on flat car
装在(平底)船上的 barge-mounted
装在汽车上的翻斗装置 lorry-mounted tipping unit
装在汽车上的起重机 motor-truck mounted crane; truck-mounted crane
装在汽车上的挖土铲 truck shovel
装在汽车上的液压装载机 lorry-mounted hydraulic loader
装在墙裙上的木(轨)条 dado rail
装在墙上的壁橱 wall closet
装在墙上的便桶 wall-mounted closet
装在墙上的旗杆 wall flagpole
装在三点悬挂装置上 three-point linkage mounting
装在三脚架上的 tripod-borne; tripod-mounted
装在上面 top loading
装在适当位置的 well-placed
装在竖柜内的加热器 mullion heater
装在铁路平车上运输的载重汽车 piggyback truck
装在通用机架上的机具 toolbar machine
装在同轴管中的偶极天线 sleeve-dipole antenna
装在托盘上 palletization[palletisation]
装在拖车上的 trailer-mounted
装在拖车上的绞盘 trailer-mounted capstan
装在拖车上的消防泵 trailer pump
装在拖拉机上的泵 tractor-mounted pump
装在拖拉机上的升运器 tractor-mounted elevator
装在拖拉机上的振动式铲板 tractor-mounted vibrating blade
装在拖拉机上的振动式刮刀 tractor-mounted vibrating blade
装在外壳内的叶轮 cased impeller
装在箱内 incase
装在橡胶软垫上的 rubber-mounted
装在橡皮垫上 bedding on rubber cushions
装在斜槽上的闸门 grizzly
装在压气管路上的润滑器 airline lubricator
装在载重汽车上的撒布机 lorry-mounted spreader
装在支架上的钻管 racked drill pipe
装在转向架上的集装箱 container with bogie
装在转向柱上的变速杆 steering column selector
装在装载机上的钻车 loader-mounted jumbo
装渣 muck loading
装渣工具 debris loading means
装渣机 mucker
装渣平台 muck loading platform
装渣箱 muck box
装炸药 charging; loading; powder charge
装炸药车 charging car
装炸药的爆破工人 powder monkey
装炸药方法 explosive charge method
装炸药工人 powder monkey
装炸药密度 loading density
装炸药前清除炮眼的工具 sludger
装炸药塞柱 stemming
装炸药体积 charge volume
装炸药系数 coefficient of charge
装罩灯 shaded lamp
装蒸馏水的盆 pan of distilled water
装帧 physical make-up
装帧精美的 rich-bound
装帧设计 graphic(al) design
装帧样本 thickness dummy
装纸箱 cartooning
装制动摩擦夹片 screw clamp for brake-shoe lining
装制动器张紧弹簧用的夹子 brake spring pliers
装制公司 stevedoring organization
装置 apparatus; appliance; array; assemblage; assembly; connecting up; contrivance; device; equipage; equipment; facility; feature; gadget; installing; maker; mount; placing; rigging; seating; set(ting)-up
装置玻璃 setting glass
装置布置 configuration of installation
装置尺度 plant bulk
装置抽气量 volume of unit to be evacuated
装置处理量 unit capacity
装置传动带 belting
装置的操作 plant operation
装置的物理构型 physical plant configuration
装置地址 unit address
装置电线的金属套管 electrical metallic tubing conductor
装置定员 plant personnel required
装置独立性 device independence; device-independent
装置法 installation method
装置费 installation cost
装置浮标 buoyage
装置改进因数 unit improvement factor
装置高度 mounting height
装置工程 unit engineering
装置工具 installation kit
装置工作周期 turnaround of unit
装置构型 unit configuration
装置故障 plant failure
装置焊接 erection welding
装置好 rig up
装置护舷的靠船栈桥 fendered berthing beam
装置活动百叶窗的浮石混凝土匣 pumice concrete roller box
装置夹 alignment clip
装置检修 turnaround of unit
装置角 setting angle
装置绞车及拉杆的卡车 <可拖带其他车辆> wrecker
装置铰链 erection hinge
装置控制 device control; plant unit control
装置控制符号 device control character
装置控制开关 equipment control switch
装置利用系数 plant utilization factor
装置帘格 grid-ironing
装置马力 installed horse power
装置模板 form setting
装置模壳 <浇混凝土用> form setting
装置排风机 system fan
装置起重机的卡车 truck-mounted crane
装置汽蚀系数 plant cavitation factor
装置器 fixture
装置轻型吊机的卡车 lorry loader
装置容量 installation capacity; installed capacity; plant capacity
装置设备 installation
装置设计 plant design; unit design
装置寿命 device lifetime
装置台 charging floor
装置陶瓷 mounting ceramics
装置图(样) erection drawing; instal-

lation drawing; setting drawing; equipment drawing
装置妥(当)set-up
装置外 off-site
装置外控制 off-site control
装置外设施 off-site facility
装置位 unit bit
装置温度 equipment temperature
装置系数 coefficient of configuration
装置系统图 circuit of installation
装置下水道 plant sewer
装置效率 plant efficiency; system efficiency; unit efficiency
装置叶片 blading
装置叶片工作 blading work
装置因数 unit factor
装置因素 unit factor
装置运转率 unit service factor
装置在车辆上的转筒式绞车 drum winch mounted on a vehicle
装置在结构上的设备 built-in device
装置在履带式拖拉机尾部的重型卷扬机 towing winch
装置在拖车上的钻机 wagon drill
装置占地 plant area
装置整体 plant bulk
装置直接数字控制系统 unit direct digital control system
装置重量 installation weight
装置柱螺栓防松螺母 mounting stud lock nut
装置组成 constitution of the plant
装置最优化 unit optimization
装置座 erection seat
装轴承用的全套工具 bearing driver set
装轴台 poppet
装柱技术 column-packing technique
装砖工人 brick burner
装砖手推车 brick buggy
装转铰的 swivel mounted
装转座的 swivel mounted
装桩帽桩 capped pile
装装门面的 cosmetic
装着 laden
装着货的 loaded
装自动连接器的客车 coach with automatic coupling
装钻机的机架和滑橇 frame and skid mounted drill

壮大 swell

壮工 general labo(u)r; hod carrier; hodman; unskilled labo(u)r
壮健的 ruddy
壮丽景色 noble sight
壮丽冷杉 noble fir
壮年 full; maturity
壮年滨线 mature shore line
壮年的 mature
壮年地面 mature land
壮年地区 mature land
壮年地形 mature form; mature topography; topographic(al) maturity
壮年谷 mature valley
壮年海岸 mature coast; secondary coast
壮年海滨线 mature shore line
壮年河 mature river; mature stream
壮年河谷 mature river valley
壮年河曲 full meander
壮年喀斯特 mature karst
壮年期 full mature stage; full maturity; mature stage; maturity phase
壮年期地貌 mature-stage landform
壮年切割平原 maturely dissected plain
壮年曲流 full meander
壮年熔岩【地】mature karst
壮年山 mature mountain
壮年中期地形 full mature relief
壮数 cone number

状观的 spectacular

状活塞 plunger piston
状况 condition; estate; position; status
状况变化报告 status change report
状况表 <资产负债的> position statement
状况分析 regime(n) analysis
状况监测系统 condition monitoring system
状况良好 fair in place
状况良好平均时间 mean up time
状况文法 case grammar
状况显示 situation display
状况约束 behavio(u)r constraint
状态 condition; position; predicament; standing
状态边界面 state boundary face; state boundary surface
状态变化 change of phase; change of states
状态变量 state variable
状态变量法 state variable method; state variable technique
状态变量方程式 state variable equation
状态表 state table
状态表分析 state table analysis
状态表面 state surface
状态不良 bad order
状态参数 state parameter; variable of state
状态测试 state verification
状态称模型 state-transition model
状态程序 status routine
状态触发器 state flip-flop
状态传递矩阵 state-transition matrix
状态带符号对 state tape symbol pair
状态灯 status lamp
状态等价 state equivalent
状态点 state point; conditional point <曲线图上的>
状态电路 status circuit
状态读出 status read
状态对应误差 state-correspondence error
状态反馈 state feedback
状态反射 attitudinal reflexes
状态方程(式)【数】 equation of state; state equation
状态分类 state classification
状态分配 state assignment
状态符合条件 state matching condition
状态改变 status change
状态概率 state probability
状态估计 state estimation
状态估值 state estimation
状态观测器 state observer
状态过程 state procedure
状态过渡法 state-transition method
状态函数 function of state; state function
状态行 status line
状态合并图 merger diagram
状态缓冲器 status buffer
状态寄存器 state register; status register
状态监测 condition monitoring
状态监控 status supervision
状态检查 condition check
状态检验 condition survey
状态简化 state reduction
状态开关 status switch
状态开关指令 status switching instruction
状态可变滤波器 state variable filter
状态可观测性 state observability
状态可控性 state controllability
状态空间 state space
状态空间理论 state space theory
状态空间模型 state-space model
状态空间系统矩阵 state space system matrix
状态控制 mode control; state control
状态控制寄存器 mode control register; state control register
状态控制字 mode control word
状态历程 state path
状态量 quantity of state
状态滤光片 status filters
状态逻辑 phase logic; state logic
状态码 condition(al) code; status code
状态描述 state description
状态模拟程序 state simulator
状态模型 state model
状态判定器 state-estimator
状态平面 state plane
状态器 stater
状态迁移 state transition
状态迁移函数 state-transition function
状态迁移图 state-transition diagram
状态求反触发器 complementing flip-flop
状态曲线 condition(al) curve
状态矢量 state vector
状态矢量方程 state vector equation
状态受限 state constraint
状态输入指令 status input instruction
状态算符方程 operator equation of state
状态随机变化的 state-dependent
状态特征位 status flag
状态填充 shape fill
状态条 status bar
状态图 bubble diagram; constitutional diagram; equilibrium diagram; stable diagram; state diagram; state graph; status diagram; status map; structural diagram
状态途径 state path
状态位 mode bit; status bit
状态文法 state grammar
状态系数 coefficient of regime(n)
状态显示 situation display; status display
状态显示方式 status display mode
状态显示器 status indicator
状态显示区域 status display area
状态线 <空气调节> condition line
状态向量 state(of) vector
状态向量空间 state vector space
状态信号 status signal
状态信息 status information
状态形式 status format
状态修 condition-based maintenance; repair according to status
状态修改符 status modifier
状态选择 state selection
状态选择器 mode selector
状态选择性 state selectivity
状态询问 status enquiry
状态约束 state constraint
状态跃迁矩阵 state-transition matrix
状态值 status value
状态指定 state assignment
状态指示灯 position light
状态指示符 status indicator
状态指示器 positioning indicator; status indicator
状态指示信号 state indicating signal
状态重构 state reconstruction
状态转换 status switching
状态转换表 state-transition table
状态转移方程 state-transition equation
状态转移概率 state-transition probability
状态转移矩阵 state-transition matrix
状态转移图 state-transition diagram
状态字 status word
状态坐标 state coordinates

撞冰船头 ice ram

撞擦伤 impact abrasion
撞车 demolition derby
撞车保险 collision insurance
撞车报警系统 collision warning system
撞车比赛 demolition derby
撞车分析图 collision diagram
撞车事故 collision accident
撞车损失 collision damage
撞尺【机】striker
撞捶法 rammer method
撞槌 claying bar; ram
撞锤 ball breaker; battering ram; hammer tup; rammer; tup; wreck ball
撞唇 <锁舌孔板的> lip strike
撞倒 knock down
撞冻 accretion
撞翻 canting
撞杆式舵机 ram steering gear
撞轨器 hump
撞痕【地】percussion mark
撞坏 crash
撞回距离 <撞锁> strike backset
撞击 attack collision; batting; brunt; bump(ing); clash; concussion; impact(ion); impingement; knock(ing); percussion; strike
撞击安全装置 crash safety
撞击臂 trip arm
撞击变质岩 impact metamorphic rock
撞击玻璃 impact glass
撞击采样 sampling by impaction
撞击传声 impact sound transmission
撞击传声器 impact microphone
撞击槌 bumping mallet
撞击锤 bumping mallet; jumper hammer; wreck ball <拆房屋用>
撞击脆性 impact brittleness
撞击刀 breaking knife
撞击点 impact point; point of impact
撞击法 collision method; hitting method
撞击范围 impingement area
撞击粉碎 comminution by impact reduction
撞击杆 trip lever; trip rod
撞击杆臂 trip arm
撞击杆式(列车)制动器 trip train stop
撞击杆式列车自动停车装置 trip train stop device
撞击隔声 impact insulation; impact isolation
撞击隔声级 impact noise rating
撞击工具 striking tool
撞击构造 impact structure
撞击构造编号 number of impact structure
撞击构造类型 type of impact structures
撞击构造名称 name of impact structure
撞击构造特征 feature of impact crater
撞击构造位置 location of impact structure
撞击轨迹 hammer path
撞击焊接 jump weld; percussion welding; percussive welding
撞击荷载 bump stroke; collision load
撞击痕 impact striae
撞击后测试 after-trip test

Z

撞击活塞 percussion piston
撞击机构 knocking gear
撞击积算仪 < 测定路况的,英国 > bump integrator
撞击集尘器 impinger
撞击计 impactometer;impactor
撞击加速度 impact acceleration
撞击假说 impact hypothesis
撞击角 impingement angle
撞击角砾岩 impact breccia
撞击坑 impact crater
撞击坑的构造 structure of impact-crater
撞击坑深度 depth of impact crater
撞击坑凸边 convex fringe of crater
撞击坑形成年代 age of impact-crater
撞击坑直径 diameter of impact crater
撞击块 bumping block
撞击力 force of percussion; impact force; percussive force; striking force; berthing impact; collision load; docking impact load (ing) < 船靠泊时 >
撞击流反应器 impinging stream reactor
撞击(滤尘)器 impinger
撞击路线 hammer path
撞击铆钉机 percussion riveting machine
撞击铆接法 percussive method of riveting
撞击帽 percussive cap
撞击面 impact face;impact surface
撞击敏感度 impact sensitivity
撞击模 impact cast
撞击能 percussive energy
撞击排料 shock discharge
撞击破坏 pounding damage
撞击起爆药包 percussion primer
撞击器调节杆 butter adjuster lever
撞击切割机 ramming cutter
撞击取样 sampling by impaction
撞击取样计 impactometer
撞击取样器 impactor
撞击韧度试验机 impact toughness machine
撞击韧性 impact ductility
撞击熔融物 impact melt
撞击熔岩 impact lava
撞击熔渣 impact slag
撞击筛 bumper type screen; impact screen; thump
撞击伤 collision injury
撞击声 impact noise;impact sound
撞击声的发生 creation of impact sound;structure-borne sound
撞击声改善量 improvement of impact sound
撞击声隔绝 impact sound insulation
撞击声隔声 impact sound insulation
撞击声隔声材料 impact sound insulation material
撞击声隔声参考曲线 curve of reference values for impact sound; impact sound insulation reference contour
撞击声级 impact sound level
撞击声强度 impact sound intensity
撞击声透射级 impact sound transmission level
撞击声选择 insulation of impact sounds
撞击声指数 impact sound index
撞击式除尘器 impingement collector
撞击式分类器 impingement separator
撞击式改锥 impact driver; impact screwdriver
撞击式攻击 ram(ming) attack
撞击式检尘器 impinger
撞击式溜槽 bumping trough
撞击式破碎机 impact crusher
撞击式曝气器 impingement aerator
撞击式气液分离器 impingement separator
撞击式球磨机 jar mill
撞击式驱动旋转喷洒头 impact-driven rotary head
撞击式燃烧器 target burner
撞击式凿岩机 percussion type of drill
撞击式振动 jolt vibration
撞击试验 blow-test; bump test; collision test; impact test(ing); ram test
撞击试验机 ram tester
撞击试验台 bump rig
撞击试验仪 ram tester
撞击水锤 knocking
撞击说 impact theory
撞击速度 impact velocity
撞击碎屑散布面积 divergence area of impact-clastics
撞击弹回式分选器 bumper mill
撞击图像 impact figure; percussion figure
撞击系数 impact factor
撞击性护舷 impact fender
撞击岩类型 type of impactite
撞击仪 bumpometer
撞击引起的火灾 postcrash
撞击应急开关 impact crash switch
撞击应力 bump stress; collision stress; jerking stress
撞击噪声 impact noise
撞击噪声级 impact noise rating
撞击震颤声 axial knock; chatter
撞击中心 center[centre] of impact; center[centre] of percussion
撞击重锤 ball breaker
撞击铸型 impact cast
撞击装置 percussion device
撞击自停装置 knock-off device
撞击钻孔 percussion drill
撞击钻探 percussive boring
撞击钻头 percussion bit
撞击作用 effect of impact; percussion action; percussive action
撞角 ram
撞角型船首 ram bow; ram stem; snout bow; tumble-home bow
撞紧的 butted tight
撞拉装置 trip device
撞破 stave in
撞球 < 拆除建筑物用 > breaker ball; headache ball
撞燃的引火物 pyrophore
撞伤 bruise; injury accident from collision
撞上 dash against; run into
撞上浮标 bull the buoy
撞碎机 percussion grinder
撞损 damaged owing to collision
撞锁 spring lock; warded lock
撞锁钥匙 passkey
撞头 bunter; ram
撞压机 percussion(power) press
撞针 bouncing pin; bullet; firing pin; plunger; priming wire; striker; anvil block < 凿岩机活塞的 >
撞针杆 plunger
撞针簧 action spring
撞针枪 needle gun
撞针头 needle pellet
撞针托板 striker plate
撞纸机 jogger
撞着 blunder against
撞钻 hammer drill; jack hammer drill; jack-percussion drill; jump drilling; jump(er) bar; jumper boring bar; percussion boring; percussion drill
撞钻法 percussion boring method
撞钻孔 jump drilling
撞钻装置 percussion boring apparatus

追补试验 penalty test

追补资金 retroactive financing
追捕者 tracker
追测式测波仪 studio wave meter; surface tracing wave meter
追查 follow-up
追查故障用的数据 trouble shooting data
追查通信[讯] trace correspondence
追偿 recovery
追偿合约 recourse arrangement
追偿诉讼 recovery action
追偿损失 recover losses
追偿条款 recourse arrangement
追偿债务 recover debt
追肥 additional fertilizer; after manure; top application manure
追肥铲 applicator blade
追肥铲刀 application knife
追肥器 fertilizer placement drill
追肥作业 additional fertilization
追赶 chase; chasing; overrun; pursue
追赶的货车 < 溜放时 > wagon catching up
追赶法 pursuance method; shooting method
追赶前钩车 overrun a previous cut
追光灯 follow spotlight
追过 outrun; outstrip
追过他船 drop a vessel
追回 recovery
追回款 recoveries
追回欠债 recovery debts
追回全部款项 recover funds in full
追回损失 recovery of loss
追迹流 ship's wake; stern wake
追计加入 retroactive admission
追记 retrospect
追记入账的项目 post entry
追加 addendum[复 addenda]; subjoin
追加保险 additional insurance; additional premium
追加保险额 additional cover
追加保险费 additional call; supplementary calls of premium
追加保证金 additional cover; additional margin; additional security; call margin; marginal calls; margin requirement
追加报酬 additional return
追加备用金 additional remittance
追加拨款 deficiency appropriation; supplemental appropriations; supplementary appropriation
追加成本 extra cost; supplemental cost; supplement cost
追加贷款 complementary financial facility
追加担保品 additional collateral
追加的 supplementary
追加的服务 additional service
追加的工程项目 additional work
追加的购买力 added purchasing power
追加的劳动力 supplementary labo(u)r power
追加垫款 additional advance
追加订单 additional order
追加订货 additional order
追加定货 additional order
追加费用 additional charges; additional cost; additional expenditures; additional expenses
追加付款 additional payment
追加概算 supplementary estimate
追加工程 additional work
追加工作 additional work
追加股款 supplementary call; supplementary calls of remittance
追加股息 supplementary dividend
追加关税 additional duty; additional tax
追加汇款 additional remittance
追加借款限制 additional debt restriction
追加流动资产投资 investment in additional current assets
追加清车时间 < 按照交通状态自动调节的时间 > extra clearance period
追加区 additional area
追加生产资料投资 additional investment in the means of production
追加生活费 additional living expenses
追加事业费用款计划 additional payment plan of undertaking expense
追加试验 penalty test
追加税 additional tax
追加岁入预算 supplementary budget for annual receipts
追加所得税 additional income tax; supplementary income tax
追加条款 rider
追加投资的限值 present value of the additional investment
追加涂层 additional coating
追加信贷 supplemental credit
追加需求 additional demand
追加押金 additional margin; call margin
追加佣金 overriding commission
追加预算 additional budget; supplementary budget; supplementary estimate; additional appropriation
追加预算拨款 additional budget allocation
追加运费费率 incidental fees tariff
追加折旧 additional depreciation
追加支出 added items of expenditures; additional expenditures
追加支出预算 supplementary budget for expenditures
追加支付 additional payment
追加准备金 additional reserve
追加资本 additional capital
追加租金 < 按营业额追加的租金 > overriding royalty
追究 research
追究询问 following up inquiries
追浪 overtaking wave
追猎 chase
追求 pursue; pursuit; run after
追求最大利润 maximize profit
追认代理人所订合同 ratification of agent's contract
追认的代理 ratification
追日镜 heliostat
追收 dun
追税 back duty
追诉 prosecution
追溯 backdate; retrieve; retrospect; trace back to; tracing
追溯标记 retroactive notation
追溯法 retroactive method
追溯费率 retrospective rating
追溯检索法 retrospective search
追溯检索情报资料 retrospective information
追溯交款 retroactive contributions
追溯调整 retroactive adjustment
追溯效力 relation back; retroaction; retroactive effect; retrospective application
追算 hindcasting
追算技术 hindcasting technique
追随目标时间 time on target
追随者 follower
追索法 tracing method

追索合约 recourse arrangement
追索路线 tracing route
追索请求 recourse claim
追索权 recourse; right of recourse; with recourse
追索权基础 recourse basis
追索权协定 recourse agreement
追索诉讼 recourse action; recovery action
追索通知 recourse note
追索协议 recourse agreement
追逃对策 pursuit evasion game
追尾 catch-up
追想 call to mind
追寻 run after
追越 outdistance; over-take
追越船 overtaking vessel
追越局面 overtaking situation
追越他船 overhaul; overtaking the other
追债 dun for debt; recover debt
追债代理人 collecting agent
追债公司 collecting company
追逐齿 hunting tooth
追逐对策 pursuit game
追逐利润 profit-seeking; quest for profit
追逐时距曲线系统 chasing control hodograph
追撞 overtaking collision
追踪 pursuit; routing; trace; tracing; tracking; trail
追踪标志 tracking mark
追踪测量 follow-up survey
追踪沉积间断 tracing sediment discontinuity
追踪程序 flow tracer routine; trace program(me); tracer; tracing routine; tracking program(me)
追踪的 fleeting
追踪地震反射面 tracing seismic reflected interface
追踪调研 follow-up study
追踪断层的归并 incorporation for trailing of faults
追踪法 backtracking method
追踪航线 dog course
追踪积压工作 tracking backlog work
追踪节理 tracing joints
追踪解释程序 trace interpretive program(me)
追踪列车 following movements; train spaced by automatic block signals
追踪列车间隔时间 time interval between trains spaced by automatic block signals
追踪迁移 tracing migration
追踪摄影 following shot
追踪式构造 track structure
追踪式显示 pursuit display
追踪数据 tracking data
追踪行为 tracking behavio(u)r
追踪运行 fleeting move; fleeting operation
追踪运行控制 fleeting control
追踪运行控制按钮 fleeting control button
追踪运行控制手柄 fleeting control lever
追踪运行控制系统 fleeting control system
追踪运行控制制 fleeting control system
追踪运行图 train diagram for automatic block signals
追踪站 tracking station
追踪者 tracer; tracker; trailer
追踪指标 tracer
追踪装置 hunting gear; tracer

椎

椎模 prod cast

锥

锥 cone; broach taper; prod; stabber

锥摆 conic(al) pendulum
锥摆调节器 conic(al) pendulum governor
锥摆调速机 conic(al) pendulum governor
锥板式流变性测定仪 cone plate rheogoniometer; cone plate rheometer
锥板式黏[粘]度计 cone plate visco(si)meter
锥杯形砂轮 taper cup(grinding) wheel
锥壁喷管 conic(al) wall nozzle
锥冰晶石 arksutite; chiolite
锥柄 awl haft; taper(ed) shank
锥柄铰刀 taper-shank reamer
锥柄扩孔钻 taper-shank core-drill
锥柄立铣刀 taper-shank end-milling cutter
锥柄麻花钻 taper drill
锥柄麻花钻头 taper-shank twist drill
锥柄桥工铰刀 taper-shank bridge reamer
锥柄三牙钻夹头 taper-shank triple-grip drill-chuck
锥柄铣刀 taper-shank cutter
锥柄直槽机用铰刀 taper shank straight flute jobber's reamer
锥柄轴 taper-shank arbor
锥柄自调钻夹头 taper-shank self-adjustable drill-chuck
锥柄钻夹头 taper-shank drill chuck
锥柄钻头 taper-shank drill
锥柄钻托 taper-shank drill-holder
锥部 petrous pyramid
锥槽阀杆 tapered-groove valve stem
锥层岩席【地】 cone sheet
锥沉量 cone penetration index; cone penetration number
锥承试验 cone bearing test
锥齿 bevel gear differential
锥齿轮 angle gear; angular wheel; bevel gear; bevel wheel
锥齿轮变速装置 bevel wheel change gear
锥齿轮差动装置 bevel gear differential
锥齿轮齿根锥 outside cone; root cone
锥齿轮传动 bevel gear drive
锥齿轮传动灯泡式水轮机 bevel geared bulb turbine
锥齿轮传动箱 bevel gear housing
锥齿轮副 bevel gear pair
锥齿轮换向法 bevel gear reverse
锥齿轮减速器 conic(al) reducer
锥齿轮检查仪 bevel gear tester
锥齿轮角齿顶高 <外径至节圆直径之差> angular addendum
锥齿轮校准套 bevel pinion adjusting sleeve
锥齿轮啮合调整衬垫 bevel gear mesh adjusting gasket
锥齿轮坯料 bevel gear blank
锥齿轮千斤顶 bevel gear jack
锥齿轮切齿机床 bevel gear cutting machine
锥齿轮推力铜挡板 bevel gear thrust bronze plate
锥齿轮铣刀 bevel gear cutter
锥齿轮系 bevel gear system
锥齿轮主传动机构 bevel gear main drive
锥齿轮装置 bevel gearing
锥齿磨床 bevel gear grinder
锥齿形摘锭 tapered-tooth spindle
锥的母线 elements of a cone
锥底 tapered bottom
锥底槽 conic(al) bottom tank
锥底混凝土锚栓 conic(al) bottom concrete anchor bolt
锥底库 hopper bottomed bin
锥底料斗 conic(al) bottom hopper
锥底面积 area of penetrometer
锥底气体弥散搅拌器 cone-bottom gas-dispersion agitator
锥底式料桶 cone-bottom bucket; cone discharging bucket; cone door type charging bucket
锥底陶器 pointed bottom pottery
锥底直径 diameter of penetrometer
锥顶 conic(al) node; vertex of a cone
锥顶(储)罐 cone roof tank
锥顶点 conic(al) node; conic(al) point
锥顶尖 conic node
锥顶角 cone-apex angle; flaring angle <喇叭>
锥顶射影 conic(al) projection
锥顶涨杆 socket mandrel
锥堵 choke; choke plug
锥度 angle of taper; cone distance; conicity; coning; flare; gradient; taper
锥度比 concentration ratio; contraction ratio; taper ratio
锥度表面规 flush surface ga(u)ge
锥度测量器 taper tester
锥度尺 tapering scale
锥度大的选粉机 high gradient separator
锥度的 tapered
锥度垫圈 limpet washer
锥度附件 taper attachment
锥度管子配件 flare fitting
锥度光纤 conic(al) optic(al) fibre
锥度规 conic(al) ga(u)ge; taper ga(u)ge
锥度珩磨机 taper honing machine
锥度环规 taper ring ga(u)ge
锥度级数 number of taper
锥度计算 taper calculating
锥度量规 cone ga(u)ge; taper ga(u)ge
锥度螺纹 tapered screw
锥度配合 conic(al) fit; taper fit(ting)
锥度配合油压拆卸工具 oil hydraulic dismounting tool for conic(al) fit
锥度塞规 taper plug ga(u)ge
锥度砂箱 taper flask
锥端接管螺母 nipple nut
锥端紧定螺钉 plint set screw
锥堆取样 sampling of cones
锥堆四分法 coning; quartering
锥阀 cone valve; miter valve
锥盖式油箱 cone roof tank
锥镉硒矿 cadmoselite
锥构刻压仪 cone indenter
锥鼓离心机 conic(al) drum centrifugal
锥管 taper pipe; thimble
锥管螺纹 pipe tap
锥管螺纹环规 taper-pipe-thread ring-ga(u)ge
锥管螺纹柱形 taper-pipe-thread plug-ga(u)ge
锥贯入试验 cone penetration test
锥贯入稳定性试验 <测定沥青混合料用> cone penetration stability test
锥光 conic(al) ray
锥光干涉图 optic(al) axial interference figure
锥光观察 conoscopic observation; konoscopic observation
锥光镜 conoscope
锥光偏振仪 conoscope; hodoscope
锥光仪 konoscope
锥滚轴承内圈 bearing cone
锥滚轴承内外圈付 cone & cup set of taper roller bearing
锥果木属 buttontree
锥黑铜矿 paramelaconite
锥辉石 aegirine; acmitee[akmite]
锥击试验 cone impact test; cone penetration impact test
锥尖 cone point; conic(al) tip; tapered point
锥尖角 point angle
锥尖阻力 cone resistance
锥角 cone angle; wedge slope; taper angle
锥角拱 truncated angular arch
锥铰刀 bit reamer; broach reamer; taper reamer
锥晶石 lacroixite
锥颈螺母 conic(al) necked nut
锥镜 axicon lens
锥距 pitch cone radius
锥壳 conic(al) shell
锥壳堆料 cone shell stacking
锥壳堆料法 cone shell stacking method
锥壳钢 Coniconchia
锥壳形堆料法 cone shell stockpiling
锥孔 counterbore; countersink; female cone; taper holing
锥孔的管塞 filler plugs for trepanned holes
锥孔滚珠轴承 ball-bearing with taper bore
锥孔盘针入度计 tapered-hole disk penetrometer
锥孔器 bodkin; piercer
锥孔镗孔 taper boring
锥孔镗削 conic(al) boring
锥孔圆柱滚子轴承 taper-bore cylindrical-roller bearing
锥孔轴 taper-bored spindle
锥孔装药 cavity charge; hollow charge; shaped charge
锥孔钻 tap borer
锥孔钻套 sleeve for taper shank drill
锥口孔 counterbore; countersink
锥口钻 countersink drill; countersinking
锥栗木 chestnut oak
锥氯铜铅矿 cumengeite; cumengite
锥轮 cone pulley; cone wheel; conic(al) pulley; pulley cone
锥轮传动 cone pulley drive; conic(al) pulley drive
锥轮传动装置 cone gear
锥螺纹 conic(al) thread; taper thread
锥螺纹板牙 die for taper thread
锥锚式千斤顶 tapered pin type jack
锥面 cone; cone area; pyramidal face
锥面薄片 conic(al) diaphragm
锥面的形成 coning
锥面阀 conic(al) valve
锥面方程 equation of cone
锥面环 taper-face ring
锥面活塞环 piston ring with taper face; taper-face piston ring
锥面活塞压环 taper-face compression-ring
锥面配合 conic(al) fit
锥面形穹顶 conoidal vaulting
锥面展开 development of cones
锥磨头 tapered grinding head
锥盘黏[粘]度计 Zahn viscosimeter
锥坡 conic(al) slope; truncated cone banking
锥桥形铰刀 taper bridge reamer
锥球体 cone-spheroid
锥曲线 conic(al) curve
锥刃【岩】 bit of drill
锥入度测定 penetration test
锥入度仪 penetration test apparatus
锥塞 drift-ping; wedge plug
锥塞管头 taper plug header
锥塞喷嘴 spigot jet nozzle

锥石类 conularida
锥式 tapering shape
锥式差动齿轮 bevel-type differential gear
锥式触探仪 <能测孔隙水压力的> piezocone
锥式高温计 prod-type pyrometer
锥式离合器 cone clutch
锥式链接钎头 taper socket bit
锥式喷嘴 converging-diverging cones injector
锥式曲面 conic(al) surface
锥式行星齿轮 bevel-type planetary gear
锥式行星传动装置 bevel-type planetary gearing
锥式液限(贯入)仪 balance cone;cone penetrometer for liquid limit test
锥式仪液限试验法 Vasiliev(cone) liquid limit method
锥枢(销) drift pin
锥台式屋顶 mansard roof
锥探试验 cone penetration test
锥探仪 cone penetrometer
锥套 bevel shell; taper (reducer) sleeve
锥套管 taper sleeve
锥套形喷嘴 drogue nozzle
锥体 cone;conoid;conus;taper
锥体变径段 conic(al) shell-reducer section
锥体部分 conic(al) section
锥体侧束 lateral pyramidal tract
锥体淬硬性试验 cone test for hardenability
锥体法 cone method
锥体分布理论 cone theory
锥体高温计 cone pyrometer
锥体贯入度 cone penetration
锥体贯入阻力 cone resistance
锥体护坡 conic(al) embankment protection;conic(al) filling;conic(al) pitching
锥体滑动 cone slip(ping)
锥体交通标志 traffic cone
锥体截面 conic(al) section
锥体静力触探 static cone penetration test
锥体静力触探仪 static cone penetrometer
锥体拉出试验插入物 pull-insert
锥体棱镜 cone prism;conic(al) prism
锥体前的 prepyramidal
锥体切割 pyramid cut
锥体塞规 taper plug ga(u)ge
锥体锁止 cone-lock
锥体锁止冲头护理 cone-lock punch retainer
锥体外束 extrapyramidal tract
锥体外系统 extrapyramidal system
锥体系 pyramidal system
锥体形成 pyramid formation
锥体压入阻力 cone resistance
锥体叶 pyramidal lobe
锥体与筒体连接 cone-to-cylinder junction
锥条形导线 tapered bar conductor
锥筒筛 perforated cone
锥头传感器 cone sensor
锥头电极 truncated tip electrode
锥头角度 angle of penetrometer
锥头孔 countersink
锥头螺钉 cone point screw
锥头螺栓 cone-head(ed) bolt
锥头铆钉 bat rivet;bevel head rivet;cone-head rivet;conic(al) head rivet; conic(al) rivet; rivet steeple head
锥头索 tapered wire rope
锥头系数 cone factor
锥头形榫接 tapered-end joint
锥头(圆)柱体 cone-cylinder(body)
锥头阻力 cone point resistance
锥碗断口 cup-and-cone fracture
锥纹石 kamacite
锥窝接合 cone-and-socket joint
锥窝接合装置 cone-and-socket arrangement
锥窝接头 cone-and-socket joint
锥稀土矿 tritomite
锥线法 conics
锥线厚度组合规 taper wire and thickness ga(u)ge
锥线求长 rectification of a conic
锥削 taper
锥削传输线 taper line
锥削度 conicity
锥销 tapered pin;wedge pin
锥销孔铰刀 taper pin-hole reamer
锥心 tap web
锥锌矿 matraite
锥形 conic(al) contour; coning; conoid;flaring;pyramid;taper
锥形暗销 coned dowel;tapered dowel
锥形凹槽 conic(al) socket
锥形凹槽桩 tapered fluted pile
锥形凹槽桩组合 tapered fluted pile union
锥形板 tapered plate
锥形拌和机 conic(al) mixer
锥形棒 tapered rod
锥形棒式针 taper baton hand
锥形棒条筛 taper-bar grizzly
锥形包络面 conic(al) envelope surface
锥形爆破炸药 shaped demolition charge
锥形爆破装药 shaped demolition charge
锥形杯 cone cup;taper cup
锥形杯突深冲极限值 conic(al) cup value
锥形杯形砂轮 conic(al) cup wheel
锥形边 cone edge
锥形边沿衬条 tapered-edge strip
锥形变薄旋压 cone spinning
锥形变换器 tapered transformer
锥形变截面柱 battered post
锥形标 tower beacon
锥形标准磨口接头 tapered interchangeable ground joint
锥形柄 conic(al) grip; tapered grip; taper handle;taper shank
锥形柄轮 stepped cone pulley
锥形波 pyramidal wave
锥形波导 tapered transmission line
锥形波导(管)递变截面波导管 tapered waveguide
锥形波束 conic(al) beam
锥形波纹铁皮 tapered corrugated sheet
锥形薄板构成的柱 column of tapered sheet construction
锥形薄壳基础 cone shell foundation; conic(al) shell foundation
锥形不取芯钻头 taper bit
锥形裁剪 tapered cut
锥形残丘 tepee butte
锥形槽卷筒 conic(al) barrel
锥形侧铣刀 conic(al) side milling cutter
锥形车体 tapered hull
锥形车削 taper turning
锥形沉淀管 settling cone
锥形沉淀箱 cone settler
锥形沉降 conic(al) settlement
锥形沉降槽 cone settling tank;conic(al) settling tank
锥形沉陷 cone of depression;cone of drawdown;cone of influence;conic(al) depression
锥形衬垫 cone washer
锥形衬套 tapered bush(ing)
锥形衬条 tapered liner;tapered slip
锥形成粒机 cone granulator
锥形承口套管 bell collar
锥形承窝 taper socket
锥形澄清池 conic(al) type clarifier
锥形澄清器 conic(al) settler
锥形池 conic(al) tank
锥形齿轮 angular wheel;bevel gear; transmission pinion
锥形齿轮差动器 bevel differential
锥形齿轮滚刀 bing-cutter;taper hob
锥形齿轮减速器 conic(al) reducer
锥形冲头 angle drift;drift punch
锥形冲头联结 angle drill attachment
锥形除渣器 centric cleaner; centriclone;hydrocyclone
锥形触探仪 cone-shaped sounding apparatus
锥形传动齿轮 bevel wheel
锥形串接腔 tapered cavities
锥形床 conic(al) bed;tapered bed
锥形打捞器 cone fisher;tapered tap
锥形大小头 taper pipe
锥形挡板分级机 cone-baffle classifier
锥形刀具 bevel tool
锥形导杆 taper guide-bar
锥形导管 diverging duct
锥形导水机构 inclined guide apparatus
锥形导叶板 conic(al) liquid diversion baffle
锥形的 cone-type;conic(al);coniferous;coniform;fastigiated;subulate; taper(ed);V-shaped;V-type
锥形灯罩 conic(al) lamp-shade
锥形低凹硬度 conic(al) indentation hardness
锥形底 cone bottom;conic(al) bottom;hoppered bottom;tapered bottom
锥形底仓 conic(al) bottom bin
锥形底储罐 cone-bottom tank
锥形底冷凝器 cone-bottom condenser
锥形底面 conic(al) floor
锥形地形 cone
锥形电缆网 cone cable net;conic(al) cable net
锥形垫 <暖汽调整阀的> rider
锥形垫板 thimble bat
锥形垫环连接 <钻头和钻杆之间的> taper-and-shim drive fit
锥形垫密封 seal with metal cone ring
锥形垫圈 taper(ed) washer
锥形垫圈连接 taper-and-shim drive fit
锥形顶 cone roof
锥形顶储罐 conic(al) tank
锥形顶尖 conic(al) center[centre];tapered center[centre]
锥形顶油罐 cone roof reservoir
锥形定位销 conic(al) dowel pin
锥形镀铬表玻璃圈 tapered chrome plated bezel
锥形端 tapering point
锥形短管接头 tapered nipple
锥形断路器 horned circuit breaker
锥形断面 taper section
锥形堆积体 conic(al) pile
锥形堆取样法 cone sampling
锥形堆四分法 cone quartering
锥形墩 tapered pier
锥形阀(门) cone valve; conic(al) valve; miter [mitre] valve; mushroom valve; conic(al) plug valve; Howell-Bunger valve; needle type valve;wing valve
锥形反射器 conic(al) reflector
锥形废气管 exhaust cone
锥形分布 tapered distribution
锥形分级机 cone classifier;conic(al) classifier
锥形分级器 cone classifier;conic(al) classifier
锥形分散器 conic(al) dispersion
锥形风暴信号 storm cone
锥形风标 <机场的> drogue
锥形风帽 conic(al) cowl; conic(al) ventilator;tapered cowl
锥形封隔器 tapered packer
锥形封头 conic(al) head
锥形浮标 cone buoy;conic(al) buoy; conic(al) float;nun buoy
锥形浮子 conic(al) float
锥形盖 cone-shaped cover
锥形干混机 conic(al) dry blender
锥形钢体 conic(al) steel body
锥形钢销 taper pin
锥形钢桩 steel tapered pile
锥形割槽 conic(al) cut
锥形格筛 conic(al) screen
锥形格条 cone grate
锥形给料器 cone-type feeder
锥形工件 tapered workpiece; taper work
锥形工件磨削 taper-work grinding
锥形工具 cone tool;taper tool;taper-type dropper <向井内下开口楔的>
锥形骨料斗 collector cone;cone collector
锥形管 conic(al) pipe; conic(al) tube;continuous taper tube;diminishing piece;diminishing pipe;diurnal tube; diverging pipe; diverging tube; diversing tube; increaser; tapered conduit; taper(ed) pipe; taper(ed) tube
锥形管接头 tapered coupling
锥形管螺纹 tapered pipe thread
锥形管套 horn socket
锥形管柱式旋转流量计 tapered-tube rotameter
锥形管子丝锥 taper-pipe tap
锥形管嘴 conic(al) nozzle
锥形贯入仪 cone penetrometer;conic(al) penetrometer
锥形贯入硬度计 ram penetrometer
锥形罐 conic(al) tank; tapered shaped can;tapered tin
锥形光路聚光器 cone-channel condenser
锥形光束 cone-shaped beam; pencil beam;pencil light beam
锥形光纤 conic(al) fibre[fiber]
锥形光纤波导 tapered fiber waveguide
锥形滚刀 conic(al) hob;roller cutter <隧道掘进机用>
锥形滚碎机 conic(al) roll
锥形滚筒 conic(al) drum;conic(al) roll;taper drum;tapered roller
锥形滚筒筛 conic(al) trommel
锥形滚轴 conic(al) roller
锥形滚轴止推轴承 taper roller thrust bearing
锥形滚珠轴承座套 taper sleeve of ball bearing
锥形滚柱 conic(al) roller;taper-roller
锥形滚柱和保持架 tapered roller and cage
锥形滚柱止推轴承 tapered roller thrust-bearing; thrust taper roller bearing
锥形滚柱轴承 taper(ed) roller bearing
锥形滚柱轴承的减摩 anti-friction of tapered bearing
锥形滚柱轴承毂 taper-roller-bearing hub
锥形滚柱轴承内座圈 cone
锥形滚柱轴承外座圈 cup
锥形滚子轴承套管打钩 cone bearing

casing hook
锥形锅炉 conic(al)boiler
锥形锅炉管 taper shell course
锥形锅炉筒 conic(al) shell course; conic(al)shell ring
锥形过滤器 hopper filter
锥形桁架钢柱 lattice braced tapered steel mast; lattice braced tapered steel pole
锥形厚度规 tapered thickness ga(u)-ge
锥形护坡 truncated cone banking
锥形护弦 cone fender
锥形花键 tapered spline
锥形滑阀 tapered slide valve
锥形滑块 conic(al)sliding block
锥形环 conic(al) ring; impeller rim; impeller shroud;wheel cone
锥形簧压缩柱螺栓 volute spring compressor stud
锥形混合机 cone-type mixer; cone-type blender;conic(al)blender
锥形混合器 cone blender; conic(al) mixer
锥形混合式干燥器 blender-type drier;conic(al)blender type dryer
锥形混合筒 conic(al)blender
锥形锪钻 bevel wheel drill
锥形活门 miter[mitre] valve
锥形活塞 tapered piston
锥形火山 conic(al)volcano
锥形火山岩体 conic(al)pile
锥形机身 taper fuselage
锥形机用丝锥 taper machine screw tap
锥形基础 cone(-type) foundation; conic(al)foundation
锥形激波 cone shock; conic(al) shock wave
锥形极 tapered pole
锥形极端 tapered tip
锥形挤压型材 tapered extruded shape
锥形脊坡 <四坡屋顶的> cone hip (ped end)
锥形夹紧装置 taper grip
锥形夹具 tape attachment; tapered clamp
锥形夹盘 cone chuck
锥形架旋臂起重机 pyramidal slewing crane
锥形尖顶储料棚 longitudinal ridge roofed shed
锥形尖顶装饰 conic(al)spire
锥形尖端 tapered tip
锥形尖头木材工具 timber broach
锥形减小器 tapered reducer
锥形件 diminishing piece
锥形渐缩体 tapered shank
锥形键 cone key;conic(al)key;taper serration
锥形桨毂盖 conic(al)spinner
锥形交叉拱 conic(al)groin
锥形交通路标(组织交通用) traffic cone
锥形角阀 right-angle cone tap
锥形角形水嘴 right-angle cone tap
锥形绞刀 broche
锥形脚羊蹄压路机 taper-foot sheep-foot roller
锥形铰刀 bit reamer;taper bit
锥形搅拌机 cone agitator;conic(al) mixer
锥形搅拌器 cone agitator
锥形阶梯钻头 tapered step-core bit
锥形接触 conic(al)contact
锥形接缝 tapered joint
锥形接缝条 <铅皮屋面> batten roll;conic(al)roll
锥形接合 tapered fitting
锥形接头 taper(ed)joint;taper(ed) tab;flush coupling【岩】
锥形接头钎头 tapered socket bit
锥形接头套筒 tapered adapter-sleeve
锥形截面 conic(al)section
锥形金刚砂扩孔器 tapered ledge reamer
锥形金刚石压头 brale
锥形金属壳 metal cone
锥形进口 tapered inlet
锥形进气管 tapered snorkel
锥形进位 pyramid carry
锥形精磨机 jordan
锥形矩阵 pyramid matrix
锥形卷刨的木刨片 cone cut veneer
锥形卷绕 taper winding
锥形卷扬机筒 conic(al) hoisting drum
锥形卡环 tapered clamping ring
锥形卡盘 cone chuck
锥形喀斯特 cone karst
锥形开关 horn-break switch
锥形开口 tapered opening
锥形开口销 tapered split-dowel
锥形开尾销 split taper cotter;split taper pin
锥形抗裂试验 tapered type cracking test
锥形颗粒分级器 classifying cone
锥形壳(体)conic(al)shell
锥形空心窑具 thimble
锥形孔 bell-mouth; flare; taper(ed) hole
锥形孔安装 taper-bore mounted
锥形孔道 tapered channel
锥形孔筛布置 conic(al) screen arrangement
锥形块石 Telford;Telford stone
锥形块石基层 Telford base; Telford foundation
锥形快门 conic(al)shutter
锥形扩大<管道> cone out;tapered enlargement
锥形扩大器 tapered increaser
锥形扩孔 countersink(ing)
锥形扩孔器 reaming edge taper; tapered(edge)reamer;taper reaming shell
锥形扩散管 conic(al)diffuser
锥形扩压器 conic(al) diffuser;tapered diffuser
锥形拉杆<混凝土模板的> taper tie
锥形拉模 conic(al)drawing die
锥形拉伸 tapered extrusion
锥形喇叭口 conic(al)horn
锥形喇叭透镜 conic(al)horn lens
锥形离合器 bevel clutch;cone clutch
锥形离心式除渣器 centri-cleaner
锥形连接 slip-on attachment;tapered connection; tapered fitting; tapered junction
锥形连接的可卸式钻头 slip-on bit
锥形连接口 push nipple
锥形连接钎杆 tapered rod
锥形连接钻头 tapered socket bit;taper fitting bit
锥形联轴 cone coupling
锥形联轴器 cone coupler
锥形链轮 taper sprocket
锥形量杯 conic(al)graduate
锥形量规 cone ga(u)ge
锥形料堆 conic(al)stockpile
锥形零配件 cone assay
锥形流 conic(al)flow
锥形漏斗 cone
锥形炉用丝锥<锥度为1/16> taper boiler-tap
锥形滤(网)conic(al)screen
锥形路标<路上施工时临时导向用的> cone
锥形轮 cone pulley; taper cone pulley;cone wheel;tapered wheel
锥形轮箍 conic(al)tire[tyre]
锥形轮胎 conic(al)tire[tyre]
锥形螺钉 conic(al) screw; tapered screw
锥形螺杆 conic(al) screw; tapered screw
锥形螺帽 cone nut;conic(al)nut
锥形螺帽拉杆 cone nut tie
锥形螺母 conic(al) nut;tapered cup nut<装皮碗用>
锥形螺盘管 spiral taper pipe
锥形螺塞 taper(ed)plug
锥形螺栓 cone bolt; conic(al) bolt; tapered bolt
锥形螺丝攻 taper tap
锥形螺纹 conic(al)thread;taper(ed) thread
锥形螺纹堵头 tapered screw plug
锥形螺纹夹头 taper screw chuck
锥形螺纹接驳器<捞取钻孔中失落钻杆用> die overshot
锥形螺纹量规 screw ga(u)ge for taper thread
锥形螺纹丝锥 tap for taper thread
锥形螺旋 conic(al)screw
锥形螺旋槽铰刀 conic(al)spiral fluted reamer
锥形螺旋磁体 conic(al)helimagnet
锥形螺旋阀簧 conic(al) spiral valve spring
锥形螺旋搅拌机 mixer with conic(al)screw
锥形螺旋塞 tapered screw plug
锥形螺旋式分离器 spiral cone
锥形螺旋线 conic(al)helix
锥形螺旋压榨机 tapered screw press
锥形埋头键 taper sunk-key
锥形锚 cone(-shaped)anchorage
锥形锚碇装置 cone anchorage
锥形锚具 cone anchorage(device); conic(al) wedge anchorage; Freyssinet cone anchorage
锥形锚锚塞 male cone
锥形锚圈 conic(al)anchor ring
锥形锚塞<预应力钢筋混凝土用> conic(al)plug
锥形锚楔 conic(al)anchor wedge
锥形铆钉 taper(ed)rivet
锥形铆钉头 conic(al)rivet head
锥形密封面 conic(al) seal(contact) face
锥形密封式高压釜 cone closure autoclave
锥形模<冷拔钢丝用> tapered die
锥形膜片 cone diaphragm
锥形摩擦传动装置 cone friction gear
锥形摩擦减速器 cone friction gear
锥形摩擦离合器 bevel cone friction clutch;cone friction clutch
锥形摩擦联轴节 cone friction coupling
锥形摩擦轮 friction(al) bevel gear; friction(al)cone
锥形摩擦闸 cone friction brake
锥形磨 cone mill
锥形磨机 conic(al)mill;gyratory mill
锥形磨浆机 conic(al)refiner
锥形磨削 taper grinding
锥形磨削头 tapered grinding head
锥形抹子 conic(al)trowel
锥形木塞 turning pin
锥形挠曲试验机 conic(al) mandrel tester
锥形黏[粘]度计 cone visco(si)meter
锥形啮合 conic(al)grip
锥形排风罩 taper of cone hood
锥形盘簧 conic(al)spring
锥形盘式离心机 nozzle-disc[disk] centrifuge
锥形旁路 tapered branch
锥形炮眼 conic(al)mind
锥形配件 tapered fitting
锥形喷管 conic(al)jet nozzle
锥形喷流 spray-cone
锥形喷雾嘴 conic(al)spray nozzle
锥形喷嘴 conic(al)nozzle
锥形匹配 taper matching
锥形片 cone sheet
锥形偏光 conic(al)polarized light
锥形坡度 diminution
锥形破碎机 cone(-type) crusher; conic(al) breaker; conic(al) crusher;gyratory cone crusher
锥形曝气池 cone aerator
锥形曝气器 cone aerator
锥形钎头 cone bit
锥形钎尾 tapered shank
锥形嵌齿轮 bevel mortise wheel
锥形戗脊 conic(al)hip
锥形戗脊瓦 conic(al)hip tile
锥形切削单板 cone cut veneer
锥形侵蚀丘 tepee butte
锥形倾翻出料式搅拌机 tilting mixer
锥形穹顶 cone dome; conic(al) vault;conoid(al)vault
锥形穹棱 conic(al)groin
锥形穹隆 cone-type vault; conic(al) dome
锥形丘 conic(al)hill
锥形球磨机 conic(al)ball mill;conic(al)mill;Hardinge ball mill
锥形取芯钻 tapered core bit
锥形取样器 cone sampler
锥形去毛刺手绞刀 burring reamer
锥形圈 taper ring
锥形燃烧炉 conical burner
锥形燃烧室 conic(al)combustor
锥形绕组 tapered winding
锥形容器 conic(al)vessel
锥形熔岩山顶间凹地 intercollins
锥形入口 tapered inlet
锥形塞 conic(al)plug
锥形塞垫 taper line
锥形塞阀 conic(al)plug valve
锥形塞规 taper plug ga(u)ge
锥形塞头 bull plug nose
锥形三角洲 cone delta
锥形三片式夹 two-pieced tapered clamp
锥形三片式夹具 three-pieced tapered clamp
锥形扫描 conic(al)scanning
锥形扫描仪 conic(al)scan
锥形扫描轴 axis of conic(al)scan
锥形砂轮 cone wheel;tapered wheel
锥形砂碛 sand cone
锥形筛 cone screen
锥形筛磨 conic(al)screen mill
锥形筛条 tapered grizzly bar
锥形烧杯 conic(al)beaker
锥形烧瓶 conic(al) flask;Erlenmeyer flask
锥形射束 conic(al)beam
锥形省油器阀 economizer conic(al) valve
锥形实壳体转筒 helical-shell conic(al)shaped bowl
锥形视觉 cone vision
锥形视域 cone vision
锥形视轴 axis of visual cone
锥形室搅拌库 cone compartment silo
锥形收集器 conic(al)collector
锥形收敛的 conic(al)convergent
锥形手铰刀 hand taper reamer
锥形手用丝锥 taper hand tap
锥形枢 cone pivot
锥形输水管 tapered conduit
锥形竖井 tapered shaft
锥形数据 cone data

锥形栓 tapered plug
锥形栓式千斤顶 tapered pin type jack
锥形双螺杆挤出机 conic(al) twin-screw extruder
锥形双螺杆挤塑机 conic(al) twin-screw extruder
锥形双头螺柱 tapered stud
锥形水塔 conic(al) tank
锥形水箱 conic(al) tank
锥形丝锥 taper tap
锥形四分法 coning quartering
锥形搜索 tapered search
锥形碎矿机 conic(al) breaker
锥形碎石机 cone(-type) crusher
锥形榫 tapered dowel;tapered tenon
锥形锁接头的梯形丝扣 tapered acme
锥形锁销 locking cone
锥形塔 tapered tower;ziggurat
锥形塔尖 conic(al) spire
锥形塔轴 taper-shank arbor
锥形踏面 conic(al) tread
锥形弹簧 cone spring;conic(al) spiral spring;conic(al) spring;volute spring
锥形探头 conic(al) probe
锥形镗孔 taper boring
锥形掏槽 center cut(ting);cone cut;conic(al) cut;pyramid(al) cut
锥形套管 conic(al) sleeve;conic(al) socket;tapered thimble
锥形套尖 conic(al) socket point
锥形套节 conic(al) socket
锥形套筒 conic(al) sleeve;tapered sleeve
锥形体 bullet;conic(al) body;spire
锥形体接头 swedged nipple
锥形天窗 conic(al) light
锥形天沟 tapered parapet gutter
锥形天线 cone antenna;conic(al) antenna;tapered antenna
锥形条 tapered strip
锥形调速器 cone governor
锥形通风筒 mushroom ventilator
锥形筒 cone
锥形筒体 conic(al) shell
锥形筒子 tapered bobbin
锥形筒子络纱机 cone winder;cone winding machine
锥形头 pan head;mushroom head <支撑无梁板的>
锥形头部 conic(al) nose
锥形透度计 cone penetrometer;conic(al) penetrometer
锥形凸轮 conic(al) cam
锥形突 conoid process
锥形脱水机 cone dewaterer
锥形洼地 conic(al) depression
锥形瓦楞铁皮 tapered corrugated sheet
锥形弯管 tapered bend;tapered draft tube;tapered elbow
锥形弯曲 conic(al) camber
锥形弯头 tapered bend;tapered draft tube
锥形尾水管 conic(al) draft tube;hydraucone;tapered draft tube
锥形尾水管喇叭式尾水管 hydraucone type draft tube
锥形屋顶 conic(al) (broach) roof;tent roof
锥形屋脊 conic(al) hip;conic(al) hipped end
锥形无岩芯钻头 <回转钻进用的> cone bit
锥形物 cone
锥形吸出管 conic(al) draft tube
锥形铣刀 cone milling cutter;conic(al) cutter;taper cutter;taper mill
锥形线密度镜 taper line gratings
锥形镶条 taper gib
锥形销 cone pin;taper(ed) cotter;taper(ed) pin
锥形销齿式镇压器 taper foot roller
锥形销钉 conic(al) dowel;tapered dowel
锥形销钉紧固 taper pin fixing
锥形销孔 taper pin-hole
锥形小屋 <印第安人用树皮盖的> wigwam
锥形效应 cone effect
锥形楔 conic(al) wedge
锥形楔钥匙 taper wedge key
锥形斜面墩 tapered pier
锥形卸料口 conic(al) outlet
锥形心轴 conic(al) mandrel;socket mandrel;taper(plug) mandrel
锥形心轴弯曲柔韧性 flexibility conic(al) mandrel
锥形芯柱式旋转流量计 tapered center-column rotameter
锥形行浆管 tapered flow header
锥形行轮 taper-tread wheel
锥形悬挂屋顶 conic(al) suspension roof
锥形旋塞 cone cock
锥形旋筛 conic(al) rotating screen
锥形压力降低区 <即承压水层受水井抽水影响区> pressure-relief cone
锥形压碎机 cone crusher
锥形压制缝 batten seam
锥形牙轮 conic(al) cutter;taper cutter;taper mill
锥形牙嵌离合器 bevel(l)ed claw-clutch
锥形烟囱 tapered shaft
锥形烟羽 coning plume
锥形延伸段 tapered extension
锥形岩脉 cone sheet
锥形岩席 cone dike[dyke];cone sheet
锥形岩芯钻头 <换径时用> tapered core bit
锥形羊蹄碾 taper-sheepfoot roller
锥形羊蹄压路机 taper-sheepfoot roller
锥形羊足碾 tapered-foot type sheepsfoot roller;taper-sheepfoot roller
锥形阳螺纹 taper male thread
锥形叶轮 disk of conic(al) profile
锥形叶片 taper(ed) blade
锥形液膜密封 cone-type liquid-film seal
锥形异径管 taper reducer
锥形翼缘梁 tapered flange beam
锥形阴螺纹 taper female thread
锥形硬度试头 cone penetrator
锥形有缝夹套 conic(al) collet
锥形圆储库 cone-shaped circular storage hall
锥形圆盘 tapered disk
锥形圆盘导翼阀 conic(al) wing valve
锥形圆屋顶 cone dome
锥形云母环 mica taper ring
锥形凿岩钻头 cone rock bit
锥形轧辊 cone-shaped roll;crushing cone
锥形闸 cone brake
锥形帐篷式料堆 coned tent pile
锥形照明 conic(al) illumination
锥形罩 cone-shaped cover;tapered hood
锥形褶皱 conic(al) fold(ing)
锥形支墩坝 conoidal dam
锥形支路 tapered branch
锥形支柱 tapered strut
锥形支座 cone assay
锥形织物 conic(al) weave fabric
锥形直浇道 wedge sprue
锥形止推轴承 taper thrust bearing
锥形制动器 cone brake;conic(al) brake
锥形重力平台 conic(al) gravity platform
锥形洲 curpate bar
锥形轴 tapered shaft;tapering spindle
锥形轴衬 conic(al) bushing
锥形轴承 cone bearer;cone bearing;conic(al) bearing;tapered bearing
锥形轴承内环 bearing cone
锥形轴承内圈拔出工具 bearing cone puller
锥形轴承内圈更换工具 bearing cone replacing tool
锥形轴端 coned shaft end
锥形轴环 conic(al) collar;tapered collar;tapered shoulder
锥形轴颈 conic(al) journal;conic(al) pivot;countersunk spigot;pointed journal
锥形轴套 tapered hub
锥形主轴 taper spindle
锥形柱 conic(al) column;diminished shaft;tapered column;tapering column;tapering shaft
锥形柱基 cone foundation
锥形柱平台 conic(al) tower platform
锥形柱身 tapered shaft
锥形爪式离合器 bevel(l)ed claw-clutch
锥形砖 compass brick
锥形转鼓 rotary drum (cone-element)
锥形转筒筛 conic(al) rotating screen
锥形转向盘壳 tapered steering wheel hub
锥形桩 conic(al) pile;taper(ed) pile
锥形桩的大头 butt of pile;pile butt
锥形桩头 conic(al) shoe
锥形桩靴 conical point;drive shoe;driving point
锥形装配 cone assay
锥形装药 cone-shaped charge
锥形准直器 cone-type collimator
锥形钻刃 cone-shaped drill point
锥形钻头 miser;point bit;splayed boring tool;taper(ed) bit <回转钻进中金刚石不取芯钻头,用于坚硬的岩石>;corncob bit <换径用>
锥形座 conic(al) seat;taper seat
锥形座阀 taper-seat valve
锥形座面阀 miter valve
锥形座喷孔 nozzle with conic(al) seat
锥旋角 coning angle
锥元素 elementary cone
锥罩式窑 hovel kiln
锥轴 axis of a cone
锥轴链 conic(al) chain
锥柱 tapering column
锥柱堵头 blind taper joint
锥柱体模型 cone cylinder model
锥状变形研究 cone deformation study
锥状冰山 pyramidal iceberg
锥状测触探仪 cone-shaped sounding apparatus
锥状沉陷 conic(al) depression
锥状冲积层 alluvial cone
锥状的 pyramidal
锥状地形 cone
锥状断裂 cone fracture
锥状火山 pyramidal volcano
锥状基础 tapered foundation
锥状解理 pyramidal cleavage
锥状落水洞 cone-shaped sink hole
锥状塞子 tapered plug
锥状三角洲 cone delta
锥状洼地 conic(al) depression
锥状陷穴【地】conic(al) depression
锥状岩墙 cone dike[dyke]
锥状岩屑 <有裂缝> shatter cone
锥子 awl;bodkin;gimlet;piercer;pricker;wimble
锥足路碾 taper foot roller
锥钻 bradawl;sprig bit
锥钻头 miser[mizer]
锥钻轴套 drill chuck barrel
锥座阀 conic(al) seat valve

坠

坠尘掩蔽处 <放射性的> fallout shelter

坠锤 drop ball
坠锤破碎 drop crushing
坠阀 drop valve
坠阀装置 drop valve gear;poppet-valve gear
坠灌隔流阀 down-draught diverter
坠毁 crash;prank
坠积的 gravistatic
坠积土 drop accumulation soil
坠积物 cliff debris;colluvium
坠落 purler;rock fall
坠落高度 height of fall
坠落伤 injury by falling
坠落试验 drop shutter test;drop test
坠石 drop pebble;falls of stone
坠损 trauma due to a fall
坠台 catch platform
坠砣 balance weight;rig weight
坠砣抱箍 guide strap of balance weight
坠砣补偿器 balance weight tensioner;weight tension balance
坠砣杆 balance weight rod
坠砣夹板 weight strap
坠砣限制架 limit frame for balance weight
坠屑冰川 debris glacier
坠重 lower weight
坠重冲击试验 falling weight impact test
坠撞器 bullet;go-devil
坠子 buoy stone;drogue;sinker;mouse <上下推拉窗使用>

缀

缀板 batten plate;connecting plate;gusset(plate);lacing board;lacing plate;lattice plate;stay plate;tie plate

缀板构件 tie plate member
缀板与缀条系统 lacing system
缀板柱 batten(ed) column;tie plate column
缀材 lattice strut
缀钉 rivet
缀缝铆钉 stitch rivet
缀焊 stitch weld
缀合 conjugate;lace-up;lacing;latticing
缀合板 batten plate;brace plate
缀合板壁 batten wall
缀合的 latticed
缀合杆 latticed bar;latticed member
缀合钢带 batten
缀合隔墙 latticed partition wall
缀合力 stitching force
缀合梁 laced beam
缀合铆钉 stitch riveting
缀合支撑 latticed strut
缀合支柱 battened plate column;battened strut;latticed stanchion;stanchion
缀合柱【建】batten(ed)(plate) column;laced column;lattice column
缀结 rivet
缀锦砖墙 tapestry-brick wall
缀模 louver mo(u)lding
缀墙 countermure
缀饰位错 decorating dislocation
缀条【建】lace bar;lacing(bar);lattice bar;stay rod;tension piece;batten

缀条支撑 straining beam
缀砣补偿器【电】balance weight tensioner
缀字电报 cipher telegram;telegram in cipher

赘杆 redundant member

赘生物 excrescence
赘述 tautology
赘余 excess baggage
赘余的 redundant
赘余度 degree of redundancy
赘余反力 redundant reaction
赘余反力法 method of redundant reaction
赘余杆件 redundant member
赘余构架 redundant frame
赘余荷载力 redundant load
赘余结构 redundant structure
赘余力 redundant force
赘余内力 redundant stress
赘余约束 redundant constraint
赘余支承 redundant support

准埃洛石 halloysite;metahalloysite

准安全区 quasi-safe area
准稗花 Dallis grass
准斑脱岩 metabentonite
准饱和土 quasi-saturated soil
准爆电流 safe firing current
准爆发日珥 quasi-eruptive prominence
准备安放 ready-to-place
准备安装 ready-to-place
准备并车的发电机 incoming generator
准备不够的 half cooked
准备部署 ready disposition
准备车间 beamhouse;preparing shop
准备成本 setting-up cost
准备乘自行车者 potential cyclist
准备程序 preparation routine;set-up procedure
准备出发状态 about to depart
准备出海 ready for sea;ready to put-off;ready to sail
准备出航时期 <船大修后> readiness for sea period
准备出售 for order
准备储量 prepared reserves
准备处理 stand-by process
准备的 preparative;preparatory
准备电路 preference circuit
准备队列 ready queue
准备发送 ready for sending
准备方式 stand-by mode
准备费 mobilization fee;provisional sum
准备费用 preparation cost;setting-up expenses
准备符号 preparation symbol
准备工程 headworks;preliminary works;preparatory works
准备工时 preparation time
准备工序 preliminary operation
准备工作 preliminaries;preliminary work;preparatory work;head work;home work;make ready activity;ready for work;roughing-in
准备工作的规划 preparedness planning
准备工作费 preparation cost
准备工作(框)图 set-up diagram
准备勾缝 racking out
准备过程 set-up procedure
准备过冬 winterization
准备航行 get under way;ready for sail
准备好安装 ready for installation;ready for mounting
准备好拌和 ready-to-mix
准备好的 ripe
准备好的灰浆 prepared plaster
准备好的建筑设计 drawing issued for construction
准备好的砾石 prepared gravel
准备好的状态 preparedness
准备好架设 ready-to-erect
准备和辅助工作时间 time spent in preparatory and ancillary operations
准备回采区段 stoping ground
准备货币 reserve currency
准备机能 preparatory function
准备基金 provident fund
准备计划 preparatory plan
准备绞锚 put windlass into gear
准备阶段 preparation stage;preparatory stage
准备阶段账户 preparatory stage account
准备接收信号 ready-to-receive signal
准备金 appropriation reserves;emergency fund;reserve funds
准备金恢复使用 desterilization of reserves
准备金水平 reserve level
准备金投资 investment of reserves
准备金要求 reserve requirement
准备进路 preparation of the route
准备就绪 in readiness;readiness
准备就绪的 ready
准备就绪通知书 notice of readiness
准备就绪指示灯 ready indicator
准备靠岸 stand-by to alongside
准备脉冲 enabling pulse
准备模制的叠层材料 blanketing
准备牛栏 prep stall
准备抛锚 stand-by anchor
准备喷射 ready-to-spray
准备撇缆 have heaving lines ready
准备期 preparation period;stand-by period
准备起动 ready to start
准备起动按扣 starting preparation push-button
准备起动按钮 starting preparation push-button
准备起锚 stand by weighing;stand by heaving up
准备情况检查操作 readiness monitoring operation
准备区 preparation bay
准备上漆 ready-to-paint
准备上需要的 preparatory
准备设计 preliminary design
准备深潜 rig for deep submergence
准备时间 get-away time;lead time;make ready time;make-up time;preparatory time;readiness time;set-up time;stand-by time;time of setting up
准备使用 ready for operation;ready for use
准备室 anteroom
准备数据报告 preliminary data report
准备台 prep stand
准备停泊条款 ready berth clause
准备投标的过程 tendering
准备投入业务 ready for service
准备物 backlog
准备吸附剂 model adsorbent
准备下潜 rig ship for dive
准备线 ready line
准备巷道 preparatory workings;subsidiary development drivage
准备小锚 stand by kedge anchor
准备信号 ready signal
准备信号时间 warning period
准备信息 ready message
准备性负债 reserve liability
准备银行 reserve bank
准备与结束时间 time of preparation and wind-up
准备运动 warm up
准备运行 ready for operation
准备支付到期票据 meet a bill
准备制动 brake ready for operation
准备装货通知 preparation notice
准备装配 ready-to-assemble
准备装卸货通知书 notice of readiness
准备装运 ready to be picked up
准备状态 readiness;ready condition;ready state
准备状态字 ready status word
准备资本 reserve capital
准备资产 reserve assets
准备钻进 hitch up
准备左/右锚/双锚 stand by port/starboard/both anchors
准备作业 preparation operation;preparatory operation
准钡砷铀云母 metasandbergerite
准边界资源 paramarginal resources
准变数 quasi-variable
准变量 pseudo-variable
准遍历假设 quasi-ergodic hypothesis
准遍历性的 quasi-ergodic
准标 collimating mark;fiducial mark
准冰碛岩 para-tillite
准不变量 quasi-invariant
准不动产<指租借地权等> chattel real
准不法行为 quasi-delict
准不连续体 quasi-discontinuity
准不燃材料 semi-non-combustible material
准差系数 coefficient of standard deviation
准常定的 quasi-steady
准常定低压 quasi-permanent low
准超固结 quasi-overconsolidation
准超重元素 quasi-superheavy element
准尺【测】object staff
准储备量 prepared reserves
准触变性 quasi-thixotropy
准翠砷铜铀矿 metazeunerite
准大地水平面 quasi-geoid
准单 clearance paper
准单边带发送制 quasi-single sideband transmission system
准单分子的 quasi-monomolecular
准单频脉冲 quasi-monochromatic pulse
准单色辐射 quasi-monochromatic radiation
准单色光 quasi-monochromatic light
准单色束 quasi-monochromatic beam
准单稳态电路 quasi-monostable circuit
准导体 quasi-conductor
准等热层 quasi-isothermal layer
准等压过程 quasi-isobaric process
准地槽 parageosyncline
准地台【地】meta-platform;paraplatform
准地震区 peneseismic country
准地转孤立波 quasi-geostrophic solitary wave
准地转近似 quasi-geostrophic approximation
准点 fiducial point;just in time
准点可靠性 on-time reliability
准点阵 quasi-lattice
准电场 quasi-electric(al)field
准电荷平衡方程 modified charge balance equation
准电介质 quasi-dielectric
准电中性 quasi-electroneutrality
准定常流 quasi-steady flow
准定常态 quasi-steady state
准定量时 quasi-sidereal time
准动力的 pseudo-dynamic
准动力试验 quasi-dynamic(al)test
准动态机器人 quasi-dynamic(al)robot
准对称脉冲 quasi-symmetric(al)pulse
准对角矩阵 quasi-diagonal matrix
准二度体 planar body
准二级动力学模型 pseudo-second order kinetic model
准二级反应动力学 pseudo-second order reaction kinetics
准二级反应动力学模型 pseudo-second order reaction kinetic model
准二阶平稳假设 hypothesis of quasi-stationarity of order 2
准二元流 quasi-two-dimensional flow
准反射 quasi-reflection
准方案 quasi-project
准非线性特性 quasi-nonlinear characteristic
准非线性特征 quasi-nonlinear characteristic
准非周期干扰 quasi-aperiodic(al)interference
准费密能级 quasi-Fermi level
准分期付款 semi-installment
准分子 quasi-molecule
准分子激光器 excimer laser
准风暴 substorm
准峰值 quasi-peak(value)
准峰值测量仪 quasi-peak meter
准峰值检测器 quasi-peak detector
准峰值输出 quasi-peak output
准峰值型伏特计 quasi-peak type voltmeter
准峰至峰电压 quasi-peak-to-peak voltage
准峰至峰幅度 quasi-peak-to-peak amplitude
准改组<指对公司的> quasi-reorganization
准钙钒铀矿 metatyuyamunite
准杆沸石 metathomsonite
准高岭土 metakaolin
准高速列车 quasi-high-speed train
准高性<水准仪的> anallatism
准各态经历假说 quasi-ergodic hypothesis
准各向同性层合板 quasi-isotropic laminate
准各向同性的 quasi-isotropic
准各向异性分层媒质 quasi-anisotropic layered medium
准工业区 quasi-industrial zone
准公共物品 quasi-public goods
准共振 quasi-resonance
准购证 purchase permit
准固定路径 quasi-fixed route
准固结压力【岩】pseudo-consolidated pressure;pseudo-consolidation pressure
准固体的 quasi-solid
准光波 quasi-optic(al)wave
准光频 quasi-optic(al)frequencies
准光学波段 pseudo-optic(al)band
准光学的 quasi-optic(al)
准光学技术 quasi-optic(al)technique
准光学馈源系统 quasi-optic(al)feed system
准硅 eka-silicon
准轨 standard track
准轨尺 rail square
准国家 pseudo-state
准国家实体 para-state entity

Z

准合伙人 quasi-partner
准合同 quasi-contract
准褐铁矿 metahohmannite
准恒定的 quasi-steady
准恒定流 quasi-steady flow
准横波 quasi-transverse wave
准横向传播 propagation; quasi-transverse
准衡方程 regime(n) equation
准衡概念 regime(n) concept
准衡理论 regime(n) theory
准红磷铁矿 metastrengite; phosphosiderite
准红土 lateritoid
准化学方法 quasi-chemical method
准化学近似 quasi-chemical approximation
准化学平衡 quasi-chemical equilibrium
准化学溶液模型 quasi-chemical solution model
准环 almost ring; guide ring
准环礁岛 almost-atoll island
准辉锑矿 metastibnite
准会员 associate member
准活动中心 quasi-center of action
准火山的 phreatic
准火山瓦斯 phreatic gas
准货币 quasi-money
准货物 quasi-goods
准基波型 quasi-fundamental mode
准基性岩 metabasite
准畸变 quasi-distortion
准极地太阳同步轨道 quasi-polar sun-synchronous orbit
准计算机 semi-computer
准假 leave of absence
准简并模式 quasi-degenerated mode
准简波 quasi-simple wave
准简谐系统 quasi-harmonic system
准简谐振荡 quasi-harmonic oscillation
准交联 quasi-crosslink
准结合态 quasi-bound state
准解理断口 quasi-cleavage fracture
准解理断裂 quasi-cleavage crack
准解析函数【数】 quasi-analytic function
准金属 metalloid
准紧 precompact
准进口证 import permit
准进速度 feed speed
准经典的 quasi-classical
准晶格 quasi-crystalline lattice
准晶格理论 quasi-lattice theory
准晶格模型 quasi-lattice model
准晶模型 quasi-crystal model
准晶体 quasi-crystal
准晶质 crystalloid
准竞争解 quasi-competitive solution
准静荷载 quasi-static load
准静横向力 quasi-static lateral force
准静力触探贯入度 quasi-static cone penetration
准静力的 quasi-static
准静力法 pseudo-static approach
准静力分析 quasi-hydrostatic(al) analysis; quasi-static analysis
准静力过程 pseudo-static process; quasi-static process
准静力近似 quasi-hydrostatic(al) approximation; quasi-hydrostatic(al) assumption
准静力模型 quasi-static model
准静力试验 quasi-static test
准静力位移 quasi-static displacement
准静水压力近似 quasi-hydrostatic-(al) approximation
准静态的 quasi-static
准静态地震力学 quasi-static earthquake mechanics
准静态电场 quasi-static electric(al) field
准静态动力试验 quasi-static dynamic test
准静态功率系数 quasi-static power coefficient
准静态过程 pseudo-static process; quasi-static process
准静态计算 quasi-static calculation
准静态解 quasi-static solution
准静态近似 quasi-static approximation
准静态强度试验 quasi-static strength test
准静态试验 qualification test; quasi-static test
准静态探测 quasi-static sounding
准静态形变 quasi-static deformation
准静态应力场 quasi-static stress field
准静态越障 obstacle quasi-static crossing
准静载 quasi-static load
准静止锋 quasi-stationary front
准矩形波 quasi-rectangular wave
准据法 applicable law
准距点 anallatic center[centre]; anallatic point
准距校正 anallatic correction
准距线 stadia lines
准距性 anallatism
准距中心【测】 anallatic center[centre]; center[centre] of anallatism
准聚变【物】 quasi-fusion
准绝对方法 quasi-absolute method
准绝热 quasi-adiabatic
准绝热对流 quasi-adiabatic convection
准绝缘体 quasi-insulator
准绝缘子 quasi-insulator
准军用产品 off-the-shelf item
准军用设备 off-the-shelf equipment; off-the-shelf gear
准均衡 quasi-equilibrium
准均衡位移 quasi-isostatic displacement
准均匀波 quasi-homogeneous waves
准均匀材料 quasi-homogeneous material
准均匀的 quasi-homogeneous
准均匀反应堆 quasi-homogeneous reactor
准均匀辐射 quasi-homogeneous radiation
准均匀混合物 quasi-homogeneous mixture
准均匀流 equilibrium flow; quasi-uniform flow
准均匀流量 quasi-uniform flow
准均匀时 quasi-uniform time
准均匀振动 quasi-homogeneous vibration
准可重入 quasi-reentrant
准克拉通【地】 paracraton
准块状构造 para-massive structure
准矿物 mineraloid
准拉丁方 quasi-Latin square
准蓝磷铝铁矿 metavauxite
准粒子 quasi-particle
准连续的 quasi-continuous
准连续能级 quasi-continuum of level
准连续速度控制 quasi-continuous speed control
准两年振荡 quasi-biennial oscillation
准两税制 partial double tax system
准裂变 quasi-fission
准临界的 pseudo-critical
准临界阻尼 quasi-critical damping
准磷铝石 metavariscite
准磷铁锰矿 metatriplite
准流动 quasi-flow
准流纹岩 metarhyolite
准陆壳 quasi-continental crust
准螺旋形轨道 quasi-spiral orbit
准毛矾石 metaalunogen
准免运费的 franco
准模拟模型 quasi-analogue model
准模拟系统 quasi-analogue system
准能隙 quasi-energy gap
准黏[粘]聚力 pseudo-cohesion
准黏[粘]性流 quasi-viscous flow
准黏[粘]性流体 quasi-viscous fluid
准黏[粘]性蠕变 quasi-viscous creep
准黏[粘]滞流动 quasi-viscous flow
准黏[粘]滞流体 quasi-viscous fluid
准黏[粘]滞蠕变 quasi-viscous creep
准黏[粘]滞性流变 quasi-viscous flow
准牛顿型流动 quasi-Newtonian flow
准抛物形反射器 quasi-parabolic reflector
准平衡 pseudo-equilibrium; quasi-equilibrium
准平衡过程 quasi-equilibrium state
准平衡理论 quasi-equilibrium theory
准平衡状态 quasi-equilibrium state; state quasi-equilibrium
准平滑水流 quasi-smooth flow
准平面 directrix plane
准平面波 quasi-plane wave
准平稳分析 quasi-stationary analysis
准平原【地】 almost plain; peneplane; rolling country; rolling terrain; peneplain
准平原残丘【地】 monadnock
准平原沉积 peneplain deposit
准平原面交切 morvan; skiou
准平原作用 peneplanation
准齐次【数】 quasi-homogeneous
准弃船 quasi-derelict
准契约 <契约以外的债务关系> quasi-contract
准前期固结压力 <土的> quasi-preconsolidation pressure
准球面【数】 director sphere
准球形的 toripherical
准曲线【数】 directrix curve
准全息的 quasi-holographic
准确 nicety
准确长度 exact length
准确成型 accurate forming
准确的 accurate; exact; punctual
准确的时间 correct time
准确的外形 accurate shape
准确的样本 unbias(s)ed sample
准确地方时 exact local time
准确调出图像 sharp focusing
准确定位 accurate fix; pinpoint
准确定向 accurate pointing
准确定性模型 quasi-deterministic model
准确读数 accurate reading
准确度 degree of accuracy; accuracy; degree of preciseness; degree of precision; measure of accuracy; percent of accuracy; precision
准确度百分数 percentage of accuracy
准确度和能源强度 accuracy and source strength
准确度级(别) accuracy grade
准确度控制系统 accuracy control system
准确度控制字符 accuracy control character
准确度试验 accuracy test(ing)
准确度系数 accuracy factor
准确度限值 accuracy tolerance
准确度用户 accuracy user
准确翻译 close translation
准确高度 exact height
准确焦点 exact focus
准确距离标志 accurate range marker
准确聚焦的 fine-focus(s)ed
准确控制 close control
准确宽度 exact width
准确率 accuracy rating
准确脉冲延迟电路 phantastron circuit
准确切割戗脊瓦 close-cut hip
准确切割斜沟瓦 close-cut valley
准确数据 accurate data
准确调节 close regulation
准确图像 accurate picture
准确性 accuracy; correctness; punctuality
准确仪表 sharp instrument
准确值 accurate value
准确自差系数 exact coefficient of deviation
准群 pregroup
准燃料 reference standard fuel
准热等静压 pseudo-hot isostatic pressure
准柔性流动 quasi-plastic flow
准三角矩阵 almost triangular matrix
准三角洲 paradelta
准三维计算 pseudo-three-dimensional computation; quasi-three dimensional computation
准三向固结理论 pseudo-three-dimensional consolidation theory
准三元解 pseudo-three-dimensional solution; quasi-three dimensional solution
准社会环境 quasi-social environment
准砷铁矿 parasymplesite
准砷铀钡矿 metaheinrichite
准砷铀铁矿 metakahlerite
准绳 knotted wire; measuring wire; plumb line
准失业 quasi-unemployment
准石墨 subgraphite
准时 just in time; keep time; on schedule
准时按值 on-time and within budget
准时的 punctual
准时(点)送货 just in time delivery
准时看到陆地 make a good landfall
准实时 quasi-real time
准实验估计 quasi-experimental estimate
准势 quasi-potential
准势方程 quasi-potential equation
准数位 quasi-digit
准双工 quasi-duplex
准双曲轨道 quasi-hyperbolic orbit
准双曲面齿轮 hypoid gear
准双曲面齿轮加工机床 hypoid generator
准双曲面齿轮润滑剂 hypoid lubricant
准双曲面的 hypoid
准双曲面伞齿轮 hypoid bevel wheel
准双曲面伞齿轮传动 hypoid bevel
准双曲面伞齿轮副 hypoid gear and pinion
准双曲面锥齿轮【机】 hypoid bevel gear
准双曲线齿轮 hypoid gear
准双曲线伞齿轮 hypoid bevel gear
准双曲线小锥齿轮 hypoid pinion
准双曲形反射器 quasi-hyperbolic reflector
准双稳态电路 quasi-bistable circuit
准水钒钙石 metarossite
准水硅钙铀石 metaranquilite
准水磷钡铝石 metaschoderite
准水平微指令 quasi-horizontal microinstruction
准水平运动 quasi-horizontal motion
准瞬时变化 quasi-instantaneous change

Z

准素分支空间 primary component space
准塑性流体 pseudo-plastic fluid
准随机存取 quasi-random access
准随机连续波 quasi-random continuous wave
准随机数 pseudo-random number; quasi-random number
准随机数序列 pseudo-random number sequence; pseudo-random sequence of numbers; quasi-random sequence of numbers
准随机噪声 pseudo-random noise
准所有权 quasi-proprietary
准弹塑性流动 quasi-elastoplastic flow
准弹性的 hypoelastic; quasi-elastic
准弹性定律 hypoelastic law; quasi-elastic law
准弹性公式化 hypoelastic formulation
准弹性光散射 quasi-elastic light scattering
准弹性力 quasi-elastic force
准弹性偶极子 quasi-elastic dipole
准弹性散射 quasi-elastic scattering
准弹性岩石 quasi-elastic rock
准弹性振动 quasi-elastic vibration
准弹性振子 quasi-elastic oscillator
准碳酸铅矿 metascarbroite
准条<定墙上灰泥厚度的> float pan; plaster screed; running screed; screed
准条粉光机 smoothing beam finisher; smoothing screed finisher
准调和的 preharmonic
准通货膨胀 semi-inflation
准同步 quasi-synchronization
准同步测量法 para-synchronous surveying
准同步的 quasi-synchronous
准同期变形构造 penecontemporaneous deformation structure
准同期的 penecontemporaneous
准同期断裂 penecontemporaneous faulting
准同期褶皱 penecontemporaneous fold
准同生变形构造 penecontemporaneous deformation structure
准同生的 penecontemporaneous
准同时的 quasi-instantaneous
准同形性 morphotropism
准铜铀云母 metatorbernite
准透长石 metasanidine
准推进系数 quasi-propulsive coefficient
准椭圆形的 quasi-elliptic
准弯曲褶皱 quasi-flexural fold
准完备码 quasi-perfect code
准完全 quasi-full
准危机 near-crisis
准卫星 satelloid
准卫星城 quasi-satellite city
准稳岛 quasi-stable island
准稳定场方程 quasi-stationary field equations
准稳定的 metastable; quasi-steady
准稳定法 quasi-steady method
准稳定过程 quasi-stationary processing; quasi-steady process
准稳定空气动力 quasi-steady aerodynamic forces
准稳定矿物 metastable mineral
准稳定流 quasi-stable flow; quasi-stationary current; quasi-stationary flow
准稳定谱 quasi-stationary spectrum
准稳定随机过程 quasi-stationary random process
准稳定态 quasi-stable state
准稳定同位素 quasi-stable isotope
准稳定性 quasi-stability
准稳定运动 quasi-stationary motion
准稳定状态 metastable state; quasi-stationary state
准稳级 quasi-stationary level
准稳流 quasi-stationary flow; quasi-steady flow
准稳能级 quasi-stationary energy level
准稳态 quasi-stationary state; quasi-steady state; steady state
准稳态的 quasi-stationary
准稳态地震力学 quasi-static earthquake mechanics
准稳态电路 metastable circuit
准稳态动力波 quasi-steady dynamic wave
准稳态工作 quasi-steady operation
准稳态氦气式磁力仪 metastable helium magnetometer
准稳态流量分析 quasi-steady flow analysis
准稳态脉冲噪声 quasi-steady impulsive noise
准稳态声压 quasi-stationary sound pressure
准稳态振动 quasi-steady-state vibration
准稳元素 quasi-stable element
准稳振荡 quasi-stationary oscillation
准无向性 quasi-isotropy
准物理环境 quasi-physical environment
准系统误差 quasi-systematic error
准细目 quasi-isolates
准细目概念 quasi-isolate idea
准先期固结压力<土的> quasi-preconsolidation pressure
准纤纳铁矾 metasideronatrite
准纤维的 quasi-fibrous
准现金储备 secondary cash reserves
准现金资产 near cash assets
准线【测】base line; alignment; ga(u)ge line; grade line; guideline; guide wire; neat line; directrix [复 directrixes/directrices]【数】; plumb line; net line
准线偏斜 lack of alignment
准线图 alignment chart
准线性程序算法 quasi-linear programming algorithm
准线性的 almost linear; quasi-linear
准线性反馈控制系统 quasi-linear feedback control system
准线性放大器 quasi-linear amplifier
准线性化 quasi-linearization
准线性化方法 quasi-linearization method
准线性理论 quasi-linear theory
准线性模型 quasi-linear model
准线性弹 quasi-linear elasticity
准线性系统 quasi-linear system
准线修正 alignment correction
准相干束 quasi-coherent beam
准相位空间 quasi-phase space
准削拉刀 size broach; sizing broach
准销 pilot pin
准谐波场 quasi-harmonic field
准谐波系统 quasi-harmonic system
准谐波振荡 quasi-harmonic oscillation
准谐的 quasi-harmonic
准星 front sight
准星片 sight blade
准许 empower; permission; permittance
准许偿付期间 permissible payment period
准许代用品 approved equal (substitution)
准许的方法 approved method
准许机构 permission mechanism
准许价值 permissible value
准许交货的 free delivered
准许进口通知 entrance notice
准许开业公司 admitted company
准许列车进入占用的线路 call on
准许落地<飞机> cleared to land
准许免费通过 frank
准许免税输入商品 admission of goods free of duty
准许入港旗 entrance permit flag
准许上船 taking on board
准许示像 permission aspect
准许停车的通知 notice giving authority to stop
准许握柄 permission lever
准许下降(高度) clear down
准许显示 permission aspect; permission indication
准许应力 permissible stress
准许用地 permit occupancy
准许运输凭证<禁装、禁运时期由有关部门发给的> shipping permit
准许运送期间 period allowed for conveyance[conveyancy]
准许证 permit
准许证书 certificate of authorization
准许指示 permission indication
准许专利 patent
准许装车或卸车期间 period allowed for loading or unloading
准许装货单 shipping permit
准选择性 pseudo-selectivity
准循环码 quasi-cyclic(al) code
准样本 quasi-sample
准移动闭塞 distance-to-go block
准音器 tonometer
准饮用水 near-potable water
准永久荷载 quasi-permanent load
准用频率 allowed frequency
准优解 quasi-optimal solution
准予成交 admitted to dealings
准予赔偿 indemnify entertained
准预浸料 pseudo-prepreg material
准元素 eka-element
准原地花岗岩 parautochthonous granite
准原子 quasi-atom
准圆【数】director circle
准圆环坐标 quasi-toroidal coordinates
准圆形 quasi-circular
准圆形轨道 quasi-circular orbit
准圆柱投影 modified cylindric(al) projection
准圆柱形文件 quasi-cylindric(al) documents
准圆锥投影 modified conic(al) projection
准允 admission
准运单 shipping permit
准运证 navicert
准则 criterion; figure of merit; formula [复 formulae/formulas]; guideline; guide rule; guiding rule; maxim; norm
准则草案 rough rule
准则代数 primary algebra
准则稳定性 criteria stability; criterion stability
准则系列 series of standards
准张量 quasi-tensor
准针钒钙矿 metahewettite
准震区 peneseismic country; peneseismic region
准整合【地】paraconformity; non-depositional unconformity; pseudo-conformity
准整数 preintegral
准整体的 quasi-monolithic
准正常水流 quasi-normal flow
准正方形 dead square
准正交的 quasi-orthogonal
准正交各向异性板法 quasi-orthotropic plate method; quasi-orthotropic slab method
准正交各向异性的 quasi-orthotropic
准正交线 quasi-orthogonal
准正射投影形式 quasi-orthographic(al) projection form
准正弦波 quasi-sine-wave
准正弦曲线 quasi-sinusoid
准正则函数 quasi-regular function
准政府性组织 quasi-public organization
准直 collimate; collimating; collimation
准直标志 collimation mark
准直差 error of collimation
准直尺 level(l)ing rule
准直锤体 collimating cone
准直带 collimation band
准直单色光 collimated monochromatic light; combined monochromatic light
准直灯 collimated lamp
准直点 collimating point
准直缝 collimating slit
准直伽马射线闪烁分光计 collimated γ-ray scintillation spectrometer
准直管 collimator
准直光 collimated light
准直光束 collimated light beam
准直角 angle of collimation
准直精度检查 alignment test
准直镜 collimating mirror; collimator mirror
准直孔径 collimating aperture
准直快速接头 in-line quick coupling
准直联备用信号线路 quasi-associated reserve signal(l)ing link
准直面 collimation plane
准直目镜 collimating eyepiece
准直器 collimating device; collimator
准直容差 alignment tolerance
准直栅格 collimating grid
准直声束 collimated beam
准直十字线 collimator
准直束 collimated beam
准直水平 collimation level
准直水平仪 aligning level; collimating level; plumb level
准直台 collimation bench
准直透镜 collimation lens
准直望远镜 autocollimator; collimating telescope; collimator field glass
准直(误)差 collimation error
准直系统 collimation system
准直线 collimating ray; collimation line; line of collimation
准直仪 aligner; collimation converter; collimator(set)
准直仪棱镜 collimator prism
准直轴 collimating line; collimation axis
准值 guidance value
准指令【计】quasi-instruction
准指令形式 quasi-instruction form
准指数衰减 quasi-exponential attenuation
准质环 primary ring
准中性 quasi-neutrality
准中性凝聚 quasi-neutral cluster
准仲裁员 quasi-arbitrator
准重力 quasi-gravity
准周期 quasi-period
准周期的 quasi-periodic(al)
准周期性 quasi-periodicity
准周期性变化 quasi-periodic(al) variation
准周期振动 quasi-periodic(al) vibration

准周日变化 quasi-diurnal variation
准轴电极结构 quasi-axial electrode configuration
准柱面形正交坐标 quasi-cylindric(al) orthogonal coordinates
准柱面坐标 quasi-cylindric(al) coordinates
准柱铀矿 metaschoepite
准资本货物 quasi-capital goods
准纵波 quasi-longitudinal wave
准纵向传播 quasi-longitudinal propagation
准租金 quasi-rent
准租金函数 quasi-rent function
准最大值 quasi-maximum value
准坐标 quasi-coordinates

拙劣画者 dauber

捉住 grappling;seizure

桌布 table cloth

桌机 desk stand; desk (-type) telephone set;table telephone set
桌框 table-rack
桌面 tabletop
桌面管理接口 desk-top management interface
桌面检查 desk checking
桌面排版系统 < 利用电脑 > desk-top publishing system
桌面倾斜的讲台 < 寺院 > lectern
桌面输送机 table-top conveyer[conveyor]
桌面(台式)出版 desk-top publishing
桌墙两用式电话机 desk-wall type telephone set
桌球房 billiard room
桌上安装 desk-top mounting
桌上按钮 table push
桌上按钮操纵机 table button machine
桌上按钮控制机 table button machine
桌上搬闸开关 desk key switch
桌上操纵机 table-top control machine
桌上电路控制器 table circuit controller
桌上电子计算机 desk top
桌上话筒 desk stand
桌上回路管制器 table circuit controller
桌上计算机 desk computer
桌上检验 desk check
桌上控制机 table-top control machine
桌上联锁机 table interlocking machine
桌上联锁装置 table interlocker
桌上设备 desk-top equipment
桌上握柄 table interlocker;table lever
桌上握柄联锁机 table lever machine
桌上研究 < 没有实际调查或试验的研究 > desk study
桌上用的 tabletop
桌式操纵台 desk control unit;desk-type console
桌式底模 table formwork
桌式电话机 desk set telephone;desk-type telephone set;table telephone set
桌式电扇 desk fan
桌式调度台 desk-type control panel
桌式继电器 desk-type relay
桌式控制盘 desk-type control panel
桌式控制台 desk control unit;desk keyboard;desk-like console panel;desk-like control panel;desk-type console;desk-type machine
桌式梯 step table
桌式子钟 desk type secondary clock
桌毯 tapis
桌帷 valance
桌形礁 table reef
桌状冰山 table iceberg;tabular iceberg
桌状构造 desk structure
桌状海丘 guyot;table knoll
桌状礁 table reef
桌状山 table mountain
桌状台地 table platform
桌子 < 下可贮物的 > doughboy table
桌子山 mesa

卓越 excellence;prominence[prominency]

卓越频率 dominant frequency;predominant frequency
卓越振型 predominant mode
卓越周期 eminent period;predominant period

灼减量 loss on ignition

灼烙 cauterization
灼烙剂 moxa
灼烙试验 < 屋面材料 > burning-brand test
灼皮 sun scald
灼燃试验 ignition test
灼热 glow;scorching hot;red heat < 500~900℃ >
灼热崩落 glowing avalanche
灼热变质 paroptesis
灼热的 burned;flaming
灼热点温度指定器 hot spot indicator
灼热电流 glow current
灼热度 intensity of incandescence
灼热感 burning heat sensation
灼热力 intensity of incandescence
灼热试验 ignition test
灼热丝 glowing filament
灼热天 dayglow
灼热余渣 ignition residue
灼人 scorching
灼伤 burn(a);heat injury;scorch(ing)
灼烧 burn(a);calcination;firing
灼烧残渣 combustion residue;ignition residue;residue on ignition
灼烧残渣总量 total residue on ignition
灼烧沉淀 ignition precipitate
灼烧滴定法 < 测定水泥中硫酸根含量 > combustion-titrimetric apparatus
灼烧减量 calcination loss;ignition loss
灼烧帽盖 firing hood
灼烧失量 loss on ignition
灼烧试验 ignition test
灼烧损失 calcination loss;ignition loss;loss on ignition
灼烧样品 ignited sample
灼失量 ignition loss
灼失总量 total loss on ignition

茁长素 auxin

斫开 chop;hack

浊斑 opacity

浊层搬运 turbid layer transport
浊层流 turbid layer flow
浊点【化】cloud(ing) point
浊点试验 < 润滑油类 > cloud test
浊度 cloudiness;turbidity;turbidness
浊度变化 turbidity fluctuation
浊度变送器 turbidity transmitter
浊度标 turbidity scale
浊度标准液 turbidity standard
浊度表 turbidimeter;turbidity meter
浊度测定 turbidimetric analysis;turbidimetry;turbidity measurement
浊度测定法 nephelometric method;nephelometry;turbidimetric assay;turbidimetric method;turbidimetry
浊度测量仪 turbidimetric apparatus
浊度测流 cloud velocity gauging
浊度穿透 breakthrough of turbidity
浊度单位 turbidity unit
浊度的 nephelometric
浊度滴定法 turbidity titration;turbidometric titration
浊度点 turbidity point
浊度法终点检测 nephelometric end-point detection
浊度分析 nephelometric analysis;photoextinction method;turbidimetric analysis
浊度分析法 turbidometry
浊度机理 turbidity mechanism
浊度计 neoprene ring;nephelometer;scopometer[skopometer];turbidimeter[turbidometer];turbidity meter
浊度计(测得的)细度 turbidimeter fineness
浊度计的 turbidimetric
浊度计试验 < 测定水泥比表面积的 > turbidimeter test
浊度计浊度 nephelometer turbidity
浊度监测计 opacity monitor;turbidity monitor
浊度监测器 opacity monitor;turbidity monitor
浊度流 turbidity current
浊度去除 turbidity removal
浊度去除速率 rate of turbidity removal
浊度试验 turbidity test
浊度水平 turbidity level
浊度系数 coefficient of turbidity;turbidity coefficient;turbidity factor
浊度仪 nephelometer;opacimeter;turbidity meter
浊度影响 turbidity effect
浊度值 turbidity value
浊沸石 laumonite;lomontite
浊沸石-葡萄石绿纤石-硬柱石相组 laumontite-prehnite pumpellyite lawsonite facies group
浊沸石相 laumontite facies
浊沸石型变火山岩建造 metavolcanite formation of laumonite type
浊锋 nose
浊管头 tube nipple
浊积灰岩 turbidity limestone
浊积泥 turbidite mud
浊积盆地 turbidite basin
浊积扇 deep-sea fan;turbidite fan
浊积岩 fly(s)ch;turbidite
浊积岩韵律 turbidite rhythm
浊积作用 turbidite sedimentation
浊流 suspension current;turbidity current;turbidity flow
浊流搬运作用 transportation turbidity current
浊流剥蚀作用 turbidity current denudation
浊流层 turbidite
浊流层序 turbidite sequence
浊流沉淀 turbidity current deposit;turbidity sediment
浊流沉积 turbidity current deposit;turbidity sediment
浊流沉积模式 sedimentation model of turbidity current
浊流沉积岩 resedimented rock
浊流沉积作用 turbidity current deposition
浊流的 CM 图像 CM pattern of turbidity current
浊流动力学 turbidity dynamics
浊流粒度分析 turbidity size analysis
浊流粒径分析 turbidity size analysis
浊流扇 turbidity fan
浊流相 turbidite facies
浊流岩 turbidite
浊气 air exhaled;flatus discharged;foul air
浊气道 foul air flue
浊气归心 conveyance of turbid essence to the heart
浊气井 gas trap
浊色 dirty colo(u)r;dull colo(u)r
浊水 cloudy water;turbid water
浊水河流 turbid river
浊水泉 turbid-water spring
浊雾 preformed casse
浊响 dullness;dull resonance
浊循环水 turbid circulating water
浊液测流(速)法 cloud velocity gauging
浊音 dull
浊值 turbidity value

酌量 make allowance for

酌情发给许可证 discretionary licensing
酌情管理 contingency management

啄 peek

啄掘 peek
啄孔 < 木材瑕疵 > peck
啄木鸟 pecker
啄形接头 bird-head bond
啄斩石面 pecking

着长靴工人 < 新浇混凝土工地的 > bootman

着床期 implantation period
着淡色的 light-colo(u)red
着地 touch down
着地拉杆 land tie
着地蔓生的植物(或草) draped trailing plants
着地面 < 拖曳原木时的 > ride
着地压力 ground pressure
着发引信 contact fuse[fuze];impact fuse
着发作用 effect of impact
着陆 grounding;land fall;landing;planet-fall;planet-landing;touch-down(of aircraft)
着陆舱 lander
着陆场 landing ground;landing site
着陆程序 landing procedure
着陆冲击 ground impact
着陆带 air strip;landing strip
着陆导航波束 landing beam
着陆导航设备 approach aid(s);landing aids
着陆导向灯 landing direction light
着陆道标灯 threshold light
着陆灯 approach light;landing light;strip light
着陆地 landing field;landing ground;landing place
着陆点散布 touch-down dispersion
着陆点上的高度 height above touch-down
着陆方向指示器 landing direction in-

dicator
着陆费 landing charges
着陆辅助设备 landing aids; landing system
着陆航线 landing strip
着陆航向信标 localizer beacon
着陆航向信标台 runway localizer
着陆航向信标台天线 localizer antenna
着陆荷载 landing load
着陆滑跑 landing run
着陆滑跑操纵 roll-out control; roll-out guidance
着陆回收场 recovery field
着陆机场 landing port
着陆减速滑跑距离 run-out distance
着陆减速设备 landing deceleration aids
着陆减震 landing shock absorption
着陆阶段 landing stage
着陆襟翼 landing flap
着陆距离 landing distance
着陆拦阻装置 arrester gear
着陆轮架的荷载 landing gear load
着陆跑道 landing cross; landing runway; landing strip
着陆跑道照明 landing runway lighting
着陆器 lander
着陆区 touch-down zone
着陆区最高点 touch-down zone elevation
着陆十字路 landing cross
着陆时间 alighting time
着陆速度 landing speed
着陆信标 landing beacon
着陆信标射束 landing beam
着陆仪器系统 <无线电护航安全措施> instrument landing system
着陆仪器装置 <无线电保护航空安全措施> instrument landing system
着陆用照明弹 landing flare
着陆噪声 landing noise of aircraft
着陆照明弹 landing flare
着陆震动 <飞机着陆时对道面的冲击震动> landing impact
着陆指示灯 approach light
着陆装置 landing gear
着落 alight
着墨 blackening; black overlay; ink drafting; load with ink
着青镀色 bronzing
着色 colo(u)r rendition; colo(u)r tintage; dye; painting; stained glass; stain(ing); tincture; tintage; tint(ing)
着色斑 lentigo stain
着色斑的 lentiginose[lentiginous]
着色玻璃 colo(u)r coating glass; colo(u)red glass; pigmented glass; stained glass
着色玻璃细珠 colo(u)red glass beads
着色玻璃纤维 colo(u)red glass fibre
着色不当 miscolo(u)r
着色不均 offset colo(u)r
着色不足 hypochromatism; hypopigmentation
着色材料 colo(u)ring material; colo(u)ring substance
着色层 coat of colo(u)r
着色程度 degree of dyeing
着色外加剂 colo(u)r admixture
着色瓷釉 painted enamel
着色挡风玻璃 shaded windscreen; shaded windshield
着色的 colo(u)red; paint-coated; pigmented; stained; tinctorial; tinge
着色的铝 colo(u)red alumin(i)um
着色底漆 undercoat
着色地板清漆 pigmented floor(ing) varnish
着色度 degree of pigmentation; degree of staining
着色法 colo(u)ring; scheme of colo(u)r; staining method
着色工 stainer
着色过度 hyperchromatism; hyperpigmentation
着色过浓 over colo(u)r
着色过深 hyperchromatosis
着色花园砖瓦 pigmented garden tile
着色混合料 colo(u)r compound
着色混凝土 colo(u)red concrete; pigmented concrete
着色混凝土人行道 tinted concrete walk
着色剂 colo(u)rant; colo(u)ring admixture; colo(u)ring agent; colo(u)ring material; colo(u)ring matter; colo(u)ring pigment; dyed agent; dyestuff; dyeware; stainer; staining agent; tinctorial substance; tinter; tinting material
着色剂的分散 colo(u)rant dispersion
着色剂分配器 colo(u)rant dispenser
着色剂分散体 colo(u)rant dispersion
着色剂配色 colo(u)rant match
着色检查 penetrant inspection
着色检验 dye penetrant inspection; dye penetrator inspection
着色浆 tinting paste
着色力 colo(u)rability; colo(u)ring power; colo(u)r intensity; colo(u)r strength; tinting power; tinting strength
着色力测定 determination of tinting strength
着色沥青 colo(u)r(ed) asphalt
着色沥青玛蹄脂 pigmented mastic asphalt
着色量 colo(u)r pickup
着色料 staining compound
着色铝网 alumin(i)um colo(u)red fabric
着色木材 stained wood
着色能力 colo(u)rability; tinctorial power; tinting strength; staining power
着色浓度极限 combustible limit
着色浓度时间曲线 dye time concentration curve
着色喷砂 shaded sandblast
着色匹配单位 colo(u)r matching unit
着色拼花小块 abasciscus
着色汽油 dyed gasoline
着色强度 colo(u)ring power; colo(u)r intensity; staining power; tinting strength
着色清漆 colo(u)red varnish; stain varnish
着色溶剂 stained flux
着色砂浆 pigmented mortar
着色师 colo(u)rist
着色示踪物 colo(u)red tracer
着色试验法 <观测水流形状> paint test
着色饰面 colo(u)r finish; pigmented finish
着色水泥 colo(u)red cement
着色水体 dyed volume
着色特深的 hyperchromatic
着色涂料 pigmented finish
着色外加剂 colo(u)ring admixture
着色稳定性 retention of colo(u)r
着色无光漆 pigmented flatting varnish
着色物 stainer
着色物体 colo(u)ring substance
着色纤维素材料 dyed cellulosic materials
着色限度 colo(u)r limitation
着色效果 tinting effect
着色效力 tinting effect
着色效应 tinting effect
着色性 colo(u)r acceptance
着色性能 pigmenting property; tinctorial property
着色颜料 colo(u)red pigment; colo(u)ring pigment; tinting pigment
着色氧化物 colo(u)r oxide
着色样张 colo(u)red form
着色液 stainer
着色液体 dyed fluid
着色印刷 direct colo(u)r print
着色用染料 tinting colo(u)r
着色用炭黑 colo(u)r black
着色原图 colo(u)red original
着色者 painter
着色值 colo(u)ring value
着色装饰 pigmented finish
着色作用 colo(u)ration; pigmentation
着水板 hydrofoil
着水点 catch
着重 accentuate; lay emphasis; place emphasis
着重成矿地质研究 ore deposition research strengthen
着重点 stress
着重工程的 engineering-oriented
着重力流的设计概念 <混凝土结构的> flow of forces-oriented conception
着重市场的经济 market oriented economy
着重说明 underscore
着重于 attach importance to

琢 <用尖利小锤砍平石料> nidge; nig

琢边 drafted margin
琢边锤 drafting chisel
琢边块石 drafted stone; draughted stone
琢边石工 pitched dressing
琢边凿 draft chisel
琢成面 tooled finish
琢方 square yard
琢方石 ashlar; pitched dressing; plain ashlar; squared stone
琢痕 peck
琢痕石板 punched work
琢混凝土机 scabbler
琢孔方石 prison ashlar
琢毛 chipping; roughen; roughening (by picking); roughening by picking chipping; sparrow peck
琢毛的 chipped
琢毛(块)石路面 roughened sett road surface
琢面 dressing
琢面锤 dressing hammer; face[facing] hammer; scaling hammer
琢面的 faced
琢面方石 dressed ashlar
琢面毛石墙 bastard ashlar; bastard masonry
琢面石 stone surfacer
琢面石板 plain ashlar
琢面(石)锤 faced hammer
琢面饰面 ashlar facing
琢面圬工 dressed masonry; tool-facing masonry
琢面砖 ashlar brick; rock-faced stone
琢磨 polishing
琢磨花样 cut
琢木斧 chip axe
琢平 knobbling
琢去砂浆外露混凝土表面骨料 exposed aggregate finish
琢石 ashlar; ashlering; axed work; broad stone; chipped stone; crandalled stone; cut stone; dimension stone; dressed stone; natural stone; pitchstone; scotching; square (d) stone; astler <古英语>
琢石板 ashlar veneer; stone ashlar slab
琢石板工 slater
琢石边的凿子 drafting chisel
琢石层 ashlar pavement; chippings stone course
琢石厂 stone-dressing plant
琢石成材 stone
琢石窗 ashlar window; stone ashlar window
琢石锤 chipping hammer; crandall; face hammer; flat pane hammer; flat peen hammer; flat pein hammer; bush hammer; patent hammer; scaling hammer
琢石墩 ashlar pier
琢石法 stone dressing
琢石扶壁 ashlar buttress
琢石斧 chip ax(e); hack hammer; stone axe
琢石工 banker mason; block chopper; stone dresser
琢石工场 stone-dressing plant
琢石工场的标志 stone dresser's sign
琢石工场的牌号 stone dresser's sign
琢石工程 ashlar work; cut-stone work
琢石工具 broad tool
琢石拱 ashler arch; stone ashlar arch
琢石拱顶 ashlar vault; stone ashlar vault
琢石机 stone-dressing machine; stone mill
琢石建筑物 ashlar structure
琢石接缝 ashlar joint
琢石结构 ashlar structure; stone ashlar structure
琢石块 stone ashlar
琢石露头 fair ends
琢石路面 ashlar pavement; ashlar paving
琢石面 axed brick; chipped surface; rock face; stone dressing; stone facing; tooled surface; tooled finish of stone <平行的>
琢石面的 rock-faced
琢石面毛石墙 ashlar-faced rubble wall; rubble ashlar
琢石铺装 ashlar pavement
琢石砌合 ashlar bond; stone ashlar bond
琢石砌面 ashlar(stone) facing
琢石砌体 ashlar masonry; cut-stone masonry; tool-faced masonry; ashler
琢石砌体建筑 ashlar masonry work
琢石砌体露头顶石 ashlar bonder; ashlar bonding header
琢石砌体露头锚石 ashlar bonder; ashlar bonding header
琢石砌体露头石 ashlar bonder; ashlar bonding header
琢石砌体砌合丁砖 ashlar bonder
琢石砌筑 ashlar masonry work
琢石砌筑工程 stone ashlar masonry
琢石墙 ashlar wall
琢石墙砌块 stone ashlar bonder
琢石桥墩 ashlar pier
琢石穹隆 ashlar vault
琢石(石)板 ashlar slab
琢石饰面 ashlar facing; ashlar finish; ashlar veneer
琢石束石 ashlar bond stone

琢石贴面 ashlar facing
琢石圬工 ashlar masonry; cut-stone masonry
琢石线 ashlar line
琢石镶面 ashlar facing; ashlaring
琢石修边垫铁 dressing iron
琢石凿子 drafting chisel
琢石柱 ashlar pier
琢石砖 ashlar brick; rock-faced brick
琢石装饰面 stone dressing
琢石锥 crandall
琢石作业 stonework
琢纹石 figurate stone
琢纹饰面 tooled finish
琢细面砖 ashlar brick
琢有平行槽的石面 tooled finish
琢有斜边板块的结构 prismatic(al) pitched-slab structure
琢有斜边的板块 prismatic(al) slab
琢凿加工 pointed work
琢凿加工好的建筑石材 stone dressed ready for building
琢凿石面 dabbing; daubing

兹良冰期【地】 Zyrianka glacial stage

兹司康铝锌合金 ziskon
兹西高强度铝合金 zisium

咨询 briefing; consultation; consulting

咨询办公室 consultation office; information office
咨询部 advisory department; consultative department
咨询部门 consultancy service; consultant service
咨询程序 consulting process
咨询处 advisory service; consulting room
咨询代理 consulting agent
咨询法律顾问 consulting lawyer
咨询范围 terms of reference
咨询费 advisory fee; consultant fee; consultation charges; consulting fee
咨询服务 adviser service; advisory service; consultancy service; consultant service; consulting service
咨询服务小组 advisory group; consultative panel
咨询工程公司 consulting engineering firm
咨询工程师 advisory engineer; consultant engineer; consulting engineer
咨询工程师的责任 liability of the consultant
咨询公司 consultant (engineering) firm; consultants corporation; consulting company; consulting firm
咨询机构 advisory body; machinery for consulsations
咨询机关 advisory body
咨询记录 search record
咨询建筑师 consulting architect
咨询人的责任 liability of the consultant
咨询人(员) consultant
咨询式计算机控制系统 advisory computer control system
咨询室 consulting room
咨询委员会 advisory board; advisory commission; advisory committee; advisory council; board of reference; consultation committee; consul(ta)tive committee; prudential committee
咨询文件 consultant paper
咨询系统 request system
咨询协议 consultant agreement
咨询业务 advisory service; consultancy
咨询意见 advisory opinion
咨询应用 inquiry application
咨询者 consultant
咨询中心 consulting center[centre]
咨询专家 consultancy expert; expert consultant
咨询组 advisory group; consulting team

姿开关 posture switch

姿态传感器 <车身的> attitude sensor
姿态回转仪 attitude gyro
姿态可控卫星 controlled-attitude satellite
姿态控制 attitude control
姿态控制系统 <卫星的> attitude control(ling) system
姿态控制子系统 attitude control subsystem
姿态喷射器 attitude jet
姿态陀螺仪 attitude gyro; attitude gyroscope
姿态陀螺组 attitude gyro package
姿态显示 attitude indication
姿态指示器 attitude director indicator

资本 capital

资本保险 capital insurance
资本报酬率定价法 target return method
资本比率 capital ratio
资本边际收益递减 declining marginal efficiency of capital
资本变动表 capital statement
资本变动分析 analysis of capital changes
资本标准 capital standard
资本拨款 capital appropriation
资本补偿 capital repair
资本不流动情况 capital immobility case
资本财力 capital resources
资本财物的定义 definition of capital goods
资本财物的生产力 productivity of capital goods
资本财物的限定 definition of capital goods
资本参与 equity participation
资本差额 capital balance; capital gap
资本产出 capital-output
资本产出比率 capital-output ratio
资本产值比率 capital-output ratio
资本偿还期 capital pay-off
资本成本 capital cost; cost of capital
资本成本因素 cost-of-capital facing
资本承诺 capital commitment
资本充足 abundance of capital
资本抽回 revulsion of capital
资本筹措 fund raising
资本储备(金) capital reserve
资本的边际效益 marginal efficiency of capital
资本的附加费用 carrying cost of capital
资本的杠杆作用 capital leverage
资本的国际化 internationalized capital
资本的价值构成 value-composition of capital
资本的价值增值 expansion of capital value
资本的利得与损失 capital gains and losses
资本的社会机会费用率 social opportunity cost of Capital
资本的运用 employment of capital
资本抵补 replacement of capital
资本调回 capital repatriation
资本定额 capital rating
资本独占 capital monopoly
资本对固定负债比率 ratio of capital to fixed liabilities
资本对流动负债比率 ratio of capital to current liabilities
资本发展基金 capital development fund
资本费用 capital charges; cost of capital
资本费用分析 capital cost analysis
资本费用和经营费用 capital and running cost
资本分担 equity participation
资本分配制度 capital sharing system
资本负债 capital liabilities
资本负债比率 capital liability ratio; ratio of capital to liabilities; worth debt ratio
资本公积 capital reserve
资本公积金 capital reserve; capital surplus; contributed surplus
资本构成 capital composition; capital gearing
资本构成决策 capital structure decision
资本股利 capital dividend
资本股票 capital stock
资本过多 overcapitalization
资本过高估价 overcapitalization
资本过户税 capital transfer tax
资本过少 under-capitalization
资本过剩 overcapitalization; surplus of capital
资本耗损 capital destruction
资本核算率 capitalizing rate
资本红利 capital bonus; dividend
资本花费 capital spending
资本化 capitalization; plowing-back
资本化成本 capitalized cost
资本化费用 capitalized cost; capitalized expenses
资本化还原比率 capitalized ratio
资本化价值 capitalized value
资本化利率 capitalization rate; capitalized interest rate
资本化利润 capitalized profit
资本化利息 capitalized interest
资本化率 capitalization rate
资本化年费用法 capitalized annual cost method
资本化证券 capitalization issue
资本还本期 payback period
资本还原成本 capital recovery cost
资本还原过程 capitalization
资本还原率 capitalization rate
资本恢复系数 capital recovery factor
资本恢复因子 capital recovery factor
资本回收 capitalization recapture
资本回收加利息 capital-recovery-plus-interest
资本回收系数 capital recovery; equal-payment series
资本会计 accounting for ownership equities
资本货物 capital goods
资本货物需求 demand for capital goods
资本积累 accumulation of capital; capital accumulation; concentration of capital
资本积累率 rate of capital accumulation
资本基础 capital base
资本基金 capital fund
资本集聚 aggregation of capital
资本集中 centralization of capital
资本加大 accretion of the capital
资本价 capital value
资本价值 capital value
资本减让 capital allowance
资本交易 capital transaction
资本交易账户 capital transaction account
资本接受国 capital recipient country
资本节省 capital saving
资本结构 capital structure
资本结构比率 capital structure ratio
资本结构调整 <因子公司合并等> recapitalization
资本结合率 capital gearing
资本借贷 capital loan
资本金 capital in cash
资本金额 capital amount
资本金利用率 usage factor of capital
资本净额 net capital
资本净利润率 net profit ratio of registered capital
资本净收益 net capital gains
资本净值与负债比率 net worth to debts ratio
资本净值与固定资本比率 net worth to fixed capital ratio
资本聚集 aggregation of capital
资本(课)税 capital levy
资本控制 capital control
资本亏损 capital deficit; capital loss
资本扩大 capital widening
资本来源 capital supply
资本劳力比 capital-labo(u)r ratio
资本理财 capital financing
资本利得税 tax on capital profit
资本利润 capital profit
资本利润率 rate of capital return; return rate on capital employed
资本利润平均化 equalization of profit rates on capital
资本利息 capital interest; interest on capital
资本利息税 capital interest tax
资本联合 capital combination
资本流出 capital exodus
资本流动 capital movement; capital transfer; flow of capital
资本流动差额 balance of capital movement
资本流动情况 capital mobility case
资本流量 capital flow
资本流入 capital inflow; capital influx; inflow of capital
资本流通 capital flow
资本垄断 capital monopoly
资本密集 deepening of capital
资本密集程度 capital intensity
资本密集的 capital intensive
资本密集的投资 capital intensive investment
资本密集度 capital intensity
资本密集工业 capital intensive industry
资本密集货物 capital intensive goods
资本密集企业 capital intensive enterprise
资本密集投资计划 capital intensive investment scheme
资本密集型工业 capital intensive industry
资本密集型项目 capital intensive project
资本密集性 capital intensity
资本配额 capital rationing
资本设备 capital equipment
资本深化 deepening of capital
资本升值收入 capital gains
资本生产率 capital productivity
资本市场 capital market

Z

资本市场报告系统 capital market reporting system
资本市场承诺费 capital markets department commitment fee
资本市场的不完整性 capital market imperfection
资本市场管理制度 capital market control system
资本收回系数 capital recovery factor
资本收入 capital income;capital receipts;capital revenue
资本收益 capital gains;capital income;capital receipts;capital revenue;yield of capital
资本收益分配 capital gains tax
资本收益率 capital return;return on capital
资本收益税 capital gains tax
资本收支账 receipt and expenditures on capital account
资本输出 capital export;export of capital
资本输出国 capital export country
资本赎回 capital redemption
资本刷新计划 capital improvement program(me)
资本税 capital tax
资本损耗成本曲线 capital wastage cost curve
资本损耗设备抵 wear and tear allowance of capital
资本损失 capital loss
资本损益 capital gains or loss
资本所有权 capital ownership
资本摊缴 capital contribution
资本逃避 flight of capital
资本调节表 capital reconciliation statement
资本调整账户 capital adjustment account
资本投入 capital input
资本投入与扩展 capital investment and expansion
资本投资 capital investment
资本投资回收年数法 payoff method for capital investment
资本投资决策 capital investment decision
资本投资评价 capital investment appraisal
资本投资项目 capital project
资本投资效果 efficiency of capital investment
资本投资中项目的选择 project selection in capital investment
资本外流 capital outflow;flight of capital
资本未平均化的利润率 unequalized profit rate of capital
资本系数 capital coefficient
资本现值 present capital value
资本限额 capital optimum;investment limitation
资本项目 capital item
资本消耗 capital consumption
资本消耗补偿 capital consumption allowance
资本消耗扣除 capital consumption allowance
资本效率 capital efficiency
资本信用 capital credit
资本信用证明书 capital credit certificate
资本形成 capital formation
资本形成总值 gross capital formation
资本形式 formation of capital
资本性的投资 equity-type investment
资本性账户 capital account
资本性资产 capital assets
资本雄厚的 solid
资本需要量 capital demanded
资本循环 circuit of capital
资本业务 capital operation
资本溢价 capital premium
资本引入 capital infusion
资本盈利 capital surplus
资本盈利水平 level(ling) of capital gains
资本盈余 capital surplus
资本盈余表 capital surplus statement
资本盈余准备 capital surplus reserve
资本有机构成 organic composition of capital
资本与产量比率 capital-output ratio
资本与固定负债比率 worth to fixed debt ratio
资本与固定资产比率 worth to fixed assets ratio
资本与流动负债比率 worth to current debt ratio
资本预算 capital budget
资本预算编制 capital budgeting
资本再生产 reproduction of capital
资本再转移 capital-re-switching
资本增密 capital deepening
资本增益 equity build-up
资本增值 appreciation of capital;increase in capital;self-expansion of capital;capital appreciation
资本增值税 capital gains tax
资本债券 capital debenture
资本账户 capital account;proprietorship account
资本账户结余 balance of capital account
资本账户科目 capital account items
资本账户平衡表 capital account balance sheet
资本账户外流 capital account outflow
资本账目 capital account
资本账目资料 capital account data
资本折算 capitalization
资本支出 capital charges;capital expenditures;capital outlay;capital spending;charge against capital
资本支出对销售的比率 capital expenditure to sales ratio
资本支出分析 capital expenditure analysis
资本支出决策 capital expenditure decision
资本支出控制 capital expenditure control
资本支出与收入 capital expenditure and receipt
资本支出与收益支出 capital expenditure and revenue expenditure
资本支出预算 capital budget
资本支付 capital payment
资本值 capital cost;capitalized cost
资本重定 recapitalization
资本周转 capital turnover;turnover of capital
资本周转率 capital turnover rate;capital turnover ratio;rate of capital turnover
资本周转期 capital turnover period
资本主义市场经济 capitalist market economy
资本转化为产品 convert the capital into a product
资本转让 capital transfer
资本转让税 capital transfer tax
资本转移 capital transfer
资本准备 capital reserve
资本资产 capital assets
资本资产基金 capital assets fund
资本资产计价模式 capital asset pricing model
资本资产评价模型 capital asset pricing model
资本总额 capitalization;capital pool;capital sum;total capital
资本总额收益率 all capital earnings rate
资本总额与利润比率 total capital profit ratio
资本总额周转率 total capital turnover rate
资本总周转率 total capital turnover
资本租赁 capital lease;financial lease
资本租赁与财务租赁的比较 capital lease compared with financing lease
资本租赁债务 obligations under capital leases
资财 exchequer;purse
资产 assets;property
资产保险 property insurance
资产报废 assets retirement
资产变价损失 loss on realization of assets
资产变现能力分析 analysis of financial liquidity
资产陈旧 obsolescence of assets
资产持有费用 carrying charges
资产储备金 assets reserves
资产处置权 disposal right of assets
资产担保 assets cover
资产担保率 assets coverage
资产的购置成本 acquisition cost of assets
资产抵偿 assets cover
资产抵减账户 assets reduction account
资产抵押贷款 equity loan;equity mortgage
资产抵债顺序 marshal
资产冻结 freezing of assets
资产对销账户 contra-asset account
资产废置损失 loss on property retired
资产负债 assets liabilities;capital liabilities
资产负债报表 balance sheet statement
资产负债比例管理 rant loans in accordance with applicant's assets-liabilities ratios
资产负债比率 assets-liabilities ratio;capital and liability ratio;equity-debt ratio;worth debt ratio
资产负债表 balance sheet;financial position statement;statement of assets and liabilities
资产负债表分析 analysis of balance sheet;balance sheet analysis
资产负债表结构 construction of balance sheet
资产负债表日期 balance sheet date;date of balance sheet
资产负债表审计 balance sheet audit
资产负债表外 off-balance-sheet
资产负债表外的资金融通 off-balance sheet financing
资产负债表外投资 off-balance sheet financing
资产负债表项目对冲 balance sheet hedge
资产负债表项目外筹资 off-balance sheet financing
资产负债表账户 balance sheet account
资产负债表证明书 certificate of balance sheet
资产负债对冲 offsetting assets and liabilities
资产负债估算表 estimated balance sheet
资产负债汇总表 consolidated statement of resources and obligations
资产负债率 ratio between assets and liabilities;ratio between liabilities and assets
资产负债数据 balance sheet data
资产负债总表 summary statement of resources and obligations
资产改良 betterment of assets
资产更迭 assets alteration
资产构成 composition of assets
资产购置成本 book cost
资产购置和退废预算 assets acquisition and retirement budget
资产购置预算 assets acquisition budget
资产估价 appraisal of assets
资产股份 assets stock
资产股息率 percentage cash flow
资产管理者 assets manager
资产计价 assets valuation
资产价值的调整 adjustment of assets value
资产价值对销货额比率 ratio of asset value to sales
资产价值法 property value method
资产减项 assets reduction account
资产结构 assets structure
资产经营责任制 assets management responsibility system
资产经营责任制审计 audit of assets management responsibility system
资产净额 net assets
资产净值 equity;net assets value
资产净值的盈利 return on equity
资产决算 assets settlement
资产控制 assets control
资产扩充 addition of assets
资产利润率 return on assets
资产利用情况类比率 asset-utilization ratios
资产目录 statement of assets
资产盘亏 assets inventory shortage
资产盘盈 assets inventory surplus
资产平衡表 balance sheet
资产评估 assets valuation
资产评估过程 appraisal process
资产评估基金会 Appraisal Foundation
资产评估学会 appraisal institute
资产评估原则 appraisal principles
资产清理分配 remaining property liquidation and distribution
资产审计 assets audit
资产收益 assets income
资产收益比率 assets-income ratio
资产收益率 return on assets
资产损失 property damage
资产所得 assets income
资产所有权 equity ownership
资产台账 accounts of assets
资产摊提基金 amortization fund
资产项目 description of property
资产消耗会计处理 accounting for assets expirations
资产信贷额度 equity line of credit
资产形态 assets form
资产与负责 assets and liabilities
资产再生成本 reproduction cost
资产增值 appreciation in asset value
资产账户 active account;assets account
资产账面价值 depreciated value
资产折旧 assets depreciation
资产折旧基准期 assets guideline period
资产折旧(年限)幅度 assets depreciation range
资产置换基金 retirement fund
资产置留权 charge on assets
资产重估(价) assets revaluation;revaluation of assets;assets appraisal
资产重估税 assets revaluation tax;tax on reappraisal of tax
资产重估增值税 tax on the write-up

资产重组 reorganization of assets
资产周转 assets turnover
资产周转率 assets turnover ratio
资产转让分类账 assets transfer ledger
资产总额 gross assets;total assets
资产总额利润报酬率 rate of return on total assets
资产总额周转率 total assets turnover; turnover of total assets
资产组合 portfolio
资费表 tariff list
资费等值 equivalent(s) of charges
资格 competence[competency]; qualification;status
资格股<就任重要职务时的必要股份> qualification stock
资格后审 post-qualification
资格鉴定考试 qualifying licensing examination
资格考核 qualification test
资格确认 qualification affirmative
资格审查 approbation; prequalification
资格审查文件 qualification statement
资格声明 qualification statement
资格文件 qualification documents
资格预审 prequalification
资格预审合格公司 prequalified firm
资格预审合格商号 prequalified firm
资格预审申报表 prequalification application
资格预审申请 prequalification application
资格预审文件 prequalification documents;qualification documents
资格证件 qualification affirmative
资格证(明)书 certificate of competence [competency]; qualification certificate
资格证明文件 qualification documents
资格执照考试 qualifying licensing examination
资化成本 capitalized cost
资金 bankroll; capital stock; exchequer; finance; financial resources; fund; munition; principal sum; purse
资金安排 capital arrangement
资金报酬率定价法 rate-of-return pricing
资金表 fund statement; statement of fund
资金不足 capital scarcity; insufficient funds
资金偿还 capital pay-off;refundment
资金成本 capital cost; cost of fund; cost of money
资金承诺 commitment of fund
资金充足 abundance of capital
资金筹措 financing;fund raising;money mobilization
资金筹措方案 financing alternative
资金筹措费用 financing cost
资金筹措活动 fund raising activities
资金筹集 capital financing; fund raising
资金筹集报表 financing statement
资金筹集方法 financing process
资金筹集方式 mode of financing
资金筹集人 raiser
资金担保 fund guarantee
资金的分配 distribution of capital
资金的积累 accumulation of funds
资金的流入 inflow of fund
资金的时间价值 time value of money
资金的统筹 pooling of capital
资金的现时成本 current cost of fund
资金的游离 disintermediation
资金的运用 operation of capital; utilization of fund
资金调拨 transfer of financial resources
资金调度 fund procurement
资金动用 tapping of resources
资金短缺 shortage of money
资金短少 shortage of capital
资金返还价值 reversion value
资金返还系数 reversion factor
资金分配 allocation of funds allocation of profits
资金分析 capital analysis;fund analysis
资金负债表 statement of resources and liabilities
资金搁死 lock;lockup
资金供应 financing
资金供应办法 financing arrangement
资金供应时间表 capital supply schedule
资金管理 capital control
资金管理信息系统 capital management information system
资金耗损 capital destruction
资金回收率 percent return on investment;recapture rate
资金回收率评估法 rate of return method
资金回收年成本 annual cost capital recovery
资金汇报 financial return
资金积压 tying-down of capital
资金稽查 capital check
资金集中 concentration of capital
资金解冻 deblocking of funds
资金净差额 net shortfall in resources
资金亏损 loss capital
资金来源 capital sources;financial resources;sources of fund
资金来源与运用 source and dispositions of funds; source and use of funds
资金来源与运用表 source of funds and usage statement; statement of funds source and utilization; statement of sources and application of funds; sources and application of funds statement
资金累积 accumulation of funds
资金利润率 profit rate on funds;rate of capital net profit
资金利用效果 effective utilization of funds
资金流(动) cash flow;financial flow; resource flow
资金流(动)分析 cash flow analysis
资金流动折现评估法 discounted cash flow method
资金流动周期 cycle of fund movement
资金流量 resource flow; stream of money
资金流量分析 flow of funds analysis
资金流量净增额 incremental cash flow
资金流通 flow of funds
资金流转 flow of funds;funds flow
资金流转表 funds flow statement
资金密集的 capital intensive
资金平衡表 balance sheet
资金遣返回国 repatriation
资金缺口 gap financing
资金融通费 commitment fee
资金使用 use of funds
资金使用者 user of finance; user of funds
资金收入率 rate of income from the funds
资金外流 capital flight; capital outflow; drainage of capital; drain of capital;fund outflow;reflux
资金外逃 capital flight
资金限额 capital rationing
资金需求 demand for funds
资金需要量 capital requirement
资金移动 movement of funds
资金预算 capital budget
资金运动的径路 route of fund movement
资金运动规律 law of fund movement
资金运用 fund application;use of funds
资金运用表 statement of fund (and its) application
资金运用的估计 valuation of fund application
资金运用的评价 valuation of fund application
资金运用效率 efficiency of fund operations
资金运用账户 account of application of fund
资金运作 operation of fund
资金占用 occupation of funds; occupied funds
资金占用费 payment for the use of state funds
资金重估 reevaluation of capital
资金周转 turnover of capital; turnover of funds;cash flow
资金周转函数 velocity function of fund
资金周转率 turnover rate of funds; velocity of money
资金周转期 fund turnover period
资金周转时期 turnover time
资金专用 earnmarking of fund
资金转拨 funds appropriation;transfer of financial resources; transfer of funds
资金转形效应 asset-transmutation effect
资金转移 transfer of financial resources;transfer of money
资金状况 status of funds
资金最低标准 minimum funding standard
资力雄厚的银行 strong bank
资历 longevity;seniority
资历长的 longevity
资历较浅的 junior
资历介绍 qualification
资历预审 prequalification
资历预审调查表 prequalification questionnaires
资料 data; datum; information; literature;material;stuff
资料保存 file maintenance
资料报告 data report
资料变换 transformation of data
资料表 information sheet
资料不充分 inadequate information
资料不全 inadequate information
资料采集 data acquisition
资料查找方法 data retrieval
资料产品 information product
资料成果 information product
资料程序 documentor
资料程序包 information package
资料出版物 information product
资料出处 source
资料储存 storage of data
资料处理 data processing; data treatment;processing of data
资料处理日期 date of data processing
资料传送 data transmission
资料存储自动记录【计】 data storage register
资料档案 data file
资料的报告 reporting of data
资料的不可靠性 unreliability of data
资料的查阅 access to data
资料地图 derived map
资料订正 adjustment of data
资料定位面 document plane
资料短缺的地区 sparse data area
资料短缺的区域 sparse data area
资料发放范围 distribution limitation
资料发送 data transmission
资料分发 information dissemination
资料分析 analysis of data;data analysis
资料分析图像初译 data analysis/preliminary image interpretation
资料分析中心 data analysis center [centre]
资料分享仪器 interpreting device
资料服务中心 document service center [centre]; information service center[centre]
资料工具书 data book
资料共享 data sharing
资料馆 data repository
资料管理员 librarian
资料管理站 document control station
资料归纳 data reduction;reduction of data
资料归算 data reduction;reduction of data
资料函索单 gift request
资料积累 accumulation of data
资料记录表 data form
资料剪辑 clipping
资料检索 data retrieval;document retrieval;information retrieval
资料检索方法 data retrieval method
资料鉴定 data identification
资料交换 exchange of information
资料交流 information flow
资料校正 adjustment of data
资料井 data well
资料卷宗 data file
资料卡 data card
资料可靠性 data reliability
资料库 bank of information; bank of material;data bank;database;information bank
资料来源 data origination;data source; information source; source of data; source of information
资料量 quantity of information
资料流通站 delivery station
资料名称 name of materials
资料年表 chronological tabulation
资料评价说明 credit note;description of data evaluation
资料库设计 database design
资料室 morgue; recording room; reference room
资料收集 collection of data;data acquisition;data collection;data gathering
资料收集程序 data gathering procedure
资料收集系统 data capture system; data collection system
资料收集与核对系统 data acquisition and check system
资料收集装置 data acquisition equipment
资料数据 feedback
资料数据的准确性 accuracy of data
资料搜集 collect data;gather material
资料索引 source index
资料提供者 informant
资料通报 information circular
资料图 data map;source map
资料图纸清单 data drawing list
资料稀缺的地区 sparse data area
资料稀缺的区域 sparse data area
资料稀少的地区 sparse data area
资料稀少的区域 sparse data area
资料性附录 information annex
资料学 informatics; information science

资料压板 document platen
资料有限的钻孔 semi-tight borehole; semi-tight well
资料折算 data reduction;reduction of data
资料整编 data handling;data processing;processing of data;treatment of data
资料整理 data handling;data processing; data reduction; processing of data; reduction of data; treatment of data
资料整理仪器 interpreting device
资料质量控制 data quality control
资料中心 data center[centre]
资料主要来源 central source
资料准备 data preparation
资料自动编辑中继装置 automatic data editing and switching system
资料自动传输 automatic transmission of data
资料自动检索系统 automatic document retrieval system
资深船长 extramaster
资信报告 status report
资信调查 credit investigation
资信可靠度 creditworthy
资信评价 credit rating
资信情况 credit information; credit position
资用长度 effective length; working length
资用储量 available storage(capacity)
资用断裂强度 <缆索或钢链的> working limit
资用功 available work
资用荷载 service load; useful load; work(ing) load
资用流量 utilizable discharge
资用绿地 utility green
资用强度 working strength
资用水深 available depth
资用水头 useful head
资用压力 available pressure;working pressure
资用应变 working strain
资用应力 working stress
资用应力设计 working stress design
资用应力设计法 working stress design method
资用重量 working weight
资用坐标 provisional coordinate
资源 resource
资源保持 conservation
资源保护 conservation of resources; resource conservation
资源保护法 conservation law
资源保护和回收法案 resource protection and recovery act
资源保护区 conservancy; conservation zone;resource reserves
资源保护与恢复法 Resource Conservation and Recovery Act
资源边际替代率 marginal rate of substitution for resources
资源波动 fluctuation of stock
资源不足 resource scarcity
资源成本 resource cost
资源冲突 resource contention
资源储备前景 anticipation of mineral reserve
资源的保存价值 preserving value for resources
资源的地理分布 geographic(al) distribution of resources
资源的定价政策 pricing policies of resources
资源的辅助性使用 complementary use of resources
资源的竞争使用 competitive use of resources
资源的开发和利用 exploiting and using of natural resources
资源的浪费 wasting of resources
资源调查 inventory survey; resource investigation;resource survey
资源调查卫星 resource survey satellite
资源调度 scheduling of resources
资源防护工程 conservancy engineering
资源分布 distribution of resource;resource allocation
资源分布图 resource map
资源分配 distribution of resource;resource allocation;resource partitioning
资源分配及多计划调度 resources allocation and multiple project scheduling
资源分配微处理机 resource allocation microprocessor
资源分配问题 allocation problem
资源分析 resource analysis
资源分享 resource sharing
资源丰度系数低 lower coefficient of resource abundance
资源丰度系数高 height coefficient of resource abundance
资源丰富 abound resources
资源丰富地区 repository
资源共享 resource sharing
资源共享多处理机 resource-sharing multiprocessing
资源共享网络 resource sharing network
资源共享执行程序 resource sharing executive
资源估价 resource appraisal
资源管理 resource management
资源管理程序 resource management program(me);resource manager
资源管理情报系统 resource management information system
资源管理系统 resource management system
资源管制 resources regulation
资源规划 resource-based planning
资源函数 resource function
资源耗竭补偿税 resource depletion allowance
资源耗竭补偿税率 rate of depletion allowance
资源合理利用 ratio use of resources
资源化 reclamation
资源化技术 technology for effective utilization of waste
资源还原 source reduction
资源环境水文地质学 resource environmental hydrogeology
资源环境政策 resources environmental policy
资源回收 resource recovery
资源回收厂 resource recovery plant
资源回收利用 resource reclamation; resource recovery utilization;waste reclamation
资源回收系统 resource recovery system
资源会吸入肺部的空中尘埃 respirable dust
资源基础 resource base
资源急缺 most short
资源集中管理 resource centralized management
资源技术卫星 earth resources technology satellite
资源结合应用 conjunctive use of resources
资源经济学 resource economics
资源净损耗 net resource depletion
资源均匀 scramble competition
资源开采 resource exploitation
资源开发 development of resources; exploitage; exploitation of resources; resource development; resource exploitation
资源开发规划 resources exploitation planning
资源开发微生物学 microbiology of resource development
资源开发正常枯竭 resource depletion by mining
资源勘探 resource exploration; resource reconnaissance
资源考察卫星 resource survey satellite
资源科学 resource science
资源控制块 resources control block
资源枯竭 depletion of stock;resource depletion
资源库 resource pool
资源浪费 resource waste; wasting mineral resources
资源利用 resource utilization;utilization of resources
资源利用竞争型 competition resource use type
资源量 quantity of mineral resources; stock number;stock size
资源量分级 grade of resource
资源量预测方法 assessment method of resource
资源量预测和计算方法 assessment and computational method of resource
资源流动 flow of resources
资源贸易 resources trade
资源密集 resource-intensive
资源密集型工业 resource-intensive industry
资源排队 resource queue
资源配置 resources allocation
资源配置及优化模型 resources allocation and optimization model
资源配置员 resource allocation clerk
资源配置主管 resource allocation supervisor
资源平衡 equilibrium of stock
资源平衡税 resource equalization tax
资源评价 resource appraisal; stock assessment
资源清单 resource inventory
资源趋势的分析 analysis of resource trends
资源生物量 stock biomass
资源使用费用 resource accounting
资源税 resources tax
资源危机 resource crisis
资源物质 resource material
资源闲置情况 idle status of resources
资源效率 resource efficiency
资源信息 resource information
资源形势分析范围 analytic(al) region of resources situation
资源形势因素分析 resolution into factors of the resources situation
资源遗弃物质分类 source separation for derelicts
资源应用控制程序 resource application controller
资源预测 prediction of mineral resources
资源再分配 resource deal location
资源再生 resource recovery
资源再生利用率低 low-rate of reusability
资源再生利用率高 high rate of reusability
资源增长量 increment to the stock
资源占有税 taxes on the possession of resources
资源战 resources war
资源质量提高 resource quality enhancement
资源转让 transfer of resources
资源状况修改 resource status modification
资源综合利用 multipurpose use of resources
资源总量 total amount of resources
资源最佳分配 optimum allocation of resources
资源最适分派 optimum allocation of resources
资质 human quality
资质测验 aptitude test
资质证书 qualified certificate
资助 bankroll;endow(ment)
资助单位 granting agency
资助国 contributor;donor country
资助机构 financing body
资助人 patron

孳生地 breeding place

滋化率仪 susceptibility meter

滋墨 spreading
滋墨测定仪 spreadometer
滋润的土地 moist soil
滋养共生 trophic symbiosis
滋养湖 eutrophy
滋养品 nourishment
滋养物 nutrition
滋育湖 eutrophic lake
滋育木本沼泽 eutrophic swamp
滋育泥炭 eutrophic peat
滋育沼泽 eutrophic swamp

辎重 baggage;impedimenta

锱酸钡 manganese green

仔细测试 close control

仔细检查 close examination; scrutinize;scrutiny;double-check
仔细平整土地 fine grading
仔细刷过的混凝土表面 fine scrubbed concrete surface
仔鱼期 alevin stage

籽晶 crystallon;inoculating crystal; seed crystal

籽晶取向 seed orientation
籽苗 seedling
籽实产量 yield of kernel

子半群 subsemigroup

子表 sublist;subtable;subtabulation
子波 wavelet
子波变换 wavelet transform
子波采样间隔 sampling interval of wavelet
子波处理 wavelet processing
子波处理剖面 wavelet processing section
子波处理时间剖面 wavelet process time section
子波反演滤波 inverse wavelet filtering
子波反褶积 wavelet deconvolution
子波提取 extraction of wavelet

子驳 daughter ship; shipborne barge
子步 substep
子部分 subdivision
子菜单 submenu
子参数 subparameter
子操作数 suboperand
子槽 pilot channel
子层 sublayer
子插件 daughter board; subboard
子查询 subquery
子产物 daughter product
子场 auxiliary yard
子潮 component tide
子程序 functional element program(me); subprogram(me); subroutine(cell)
子畴 subdomain
子处理机 subprocessor
子串 substring
子丛 subbundle
子簇 subvariety
子存区组合 subpool
子存区组合序号 subpool number
子代产物 daughter product
子代数 subalgebra
子单元 subcell; subelement
子弹 bullet
子弹车 bullet train
子弹库 ammunition depot
子弹式穿孔机 bullet perforator
子弹式穿孔器 bullet perforator
子弹头式贷款 bullet loan
子弹头似的 bullet-headed; bullet-nosed
子弹型列车 bullet train
子堤 summer dike[dyke]
子地址 subaddress
子电钟 controlled electric clock
子定义 subdefinite
子动调节 hand regulation
子队列 subqueue
子范畴 subcategory
子分类 subclassification
子分配 suballocation
子分式 subfraction
子分支 subbranch
子港 feeder port
子格 sublattice
子格局 subconfiguration
子工程 sub-project
子工作 subtask
子公司 affiliated company; allied company; ancillary firm; constituent company; constituent corporation; daughter company; related company; subcompany; subsidiary; subsidiary business; subsidiary company; underlying company
子公司试算表换算 conversion of subsidiary trial balance
子拱 subarch
子沟 cunette; pilot channel
子构架 subexchange; subframe
子函数 subfunction
子核 daughter
子画面 sprite
子环 subring
子环路 subloop
子回路 subloop
子火山 volcannello
子级数 subseries
子集 subaggregate; subclass; subensemble; subset
子集合 subclass; subset
子集语言 subset language
子接收机 slave receiver
子结构 minor structure; substruction; substructure
子结构法 <有限元法用> substructural method; substructure method
子进程 subprocess
子晶格 sublattice
子矩阵 submatrix
子句集 clause set
子壳层 subshell
子空间 subspace
子空间迭代 subspace iteration
子空间迭代法 subspace iteration method
子口 coty ledon
子跨度 subspan
子矿物 daughter mineral
子矿物比率 the percentage of daughter minerals
子类 subclass
子例程 subroutine
子例行程序 subroutine
子链 subchain
子码 subcode
子模 submodule
子模块 submodule
子模式 subschema
子模型 submodel
子模型复制 submodel copying
子母车 finger-car
子母车运输 piggyback operation
子母船 lighter aboard ship
子母罐 composite tank
子母锚 successive anchors
子母式吊具 master beam spreader
子母钟 primary-secondary clock
子母钟系统 clock system
子目 subitem
子目标 subgoal
子目录 subdirectory
子囊 ascus[复 asci]
子囊孢子 ascospore
子囊层被 epithecium
子囊果 ascocarp
子囊菌纲 <拉> Ascomycetes
子囊壳 perithecium[复 perithecia]
子囊盘 apothecium
子埝 sub-cofferdam
子女补助金 children's benefit
子女和孪生指示符 child-and-twin pointer
子女津贴 children's allowance
子配件 subassembly
子墙 ballast wall; retaining backwall
子区间 subinterval; subregion
子区间之和 union of subintervals
子区域 <即单元体>【数】subregion; subblock
子群 blockette; subunit; subgroup【计】
子任务 subtask
子任务优先级 priority of subtask
子三角洲 subdelta
子扫描程序 subscanner
子设备 subset
子生物系统 subbiosystem
子实层 hymenial layer; hymenium
子实层体 hymenophore
子实体 carpophore; conk; fructification; sporocarp; sporophore
子式【计】minor
子树 subtree
子数据库 subdata base
子孙 seed
子台阵 subarray
子体 daughter; descendant
子体放射性 daughter radioactivity
子体物质 descendant
子体系 subsystem
子条款 subclause
子条目 subentry
子通道 subchannel
子图 subgraph
子图差 difference of subgraph
子图形 sprite; subfigure
子网(络) subnet(work)
子围堰 subcofferdam
子卫星 subsatellite
子文件 subfile
子问题 subproblem
子午标 meridian mark
子午潮 spring tide; syzygial tide
子午潮幅度 range of spring tides
子午带 longitude zone; meridianal zone; meridianal belt
子午订正 reduction to the meridian
子午方向力矩 <陀螺经纬仪> directional moment to the meridian
子午分量 meridianal component
子午高潮 high water spring
子午高度角 <天体中天的高度角> meridianal height
子午观测 meridian observation
子午光度计 meridian photometer
子午合股结构 radial ply construction
子午环 meridian circle; transit circle
子午环地球仪 meridian ring globe
子午焦线 meridianal focal line
子午角 polar angle
子午角差 hour angle difference
子午角距 meridian angle
子午截面 meridianal section
子午距 meridianal distance
子午刻度环 <地球仪> meridian ring
子午肋 <圆屋顶的> meridian rib
子午流 meridianal flow
子午面 longitude plane; meridian(al) plane
子午面光线 meridianal ray
子午面曲率半径 radius of meridianal section
子午面投影 meridianal projection of the sphere
子午圈 circle of longitude; longitude circle; meridian circle; observer's meridian; principal vertical circle; transit meridian
子午圈的 meridianal
子午圈的法线 normal of meridian
子午圈订正 reduction to meridian
子午圈高度 meridian altitude
子午圈环流 meridian circulation
子午圈角 hour angle; meridian angle
子午圈角差 hour angle difference; meridian angle difference
子午圈曲率半径 radius of curvature in meridian
子午圈天顶距 meridian zenith distance
子午丝 meridian wire
子午天体测量学 meridian astrometry
子午天文学 meridian astronomy
子午望远镜 meridian telescope; transit telescope
子午卫星 transit satellite
子午线 hour circle; hour curve; meridian(line); meridianus; meridian of longitude
子午线倍值 meridianal parts
子午线标 <子午仪上方位角读数标志> meridian mark
子午线测定 meridian determination
子午线测定仪 meridian finder
子午线尺度 scale in meridian
子午线赤经 right ascension of meridian
子午线的 meridianal
子午线的 X 射线衍射图 meridianal X-ray diffractogram
子午线断面 meridianal plane
子午线反射 meridian reflection
子午线方向 meridianal direction
子午线方向强度 meridianal intensity
子午线高度 meridian altitude
子午线航行 meridian sailing
子午线弧 arc of meridian; meridian(al) arc
子午线弧长表 table of meridianal parts
子午线弧度测量 grade meridian measurement; meridianal arc measurement
子午线弧法 meridian(al) arc method
子午线会聚 meridian convergence; meridianal convergence
子午线间距 meridianal interval
子午线截面 axial net cross section
子午线经圈 meridian
子午线精密指示器 precision indicator of meridian
子午线距离 meridianal distance
子午线轮胎 radial(ply) tire[tyre]
子午线平面 plane of the meridian
子午线平面图 diagram on the plane of the celestial meridian; projection on meridian
子午线剖面轮廓 meridianal contour
子午线曲率 curvature of meridian
子午线曲率半径 meridian radius of curvature
子午线曲率改正 correction for meridian curvature
子午线曲率校正 correction for meridian curvature
子午线上的 culminant
子午线矢距 meridianal offsets
子午线收敛 gisement
子午线收敛角 convergence of meridian; declination of grid north; grid convergence; grid declination; meridian(al) convergence; theta angle
子午线四分仪 meridian quadrant
子午线探寻仪 meridian finder
子午线投影 meridian projection
子午线椭圆 meridianal ellipse
子午线弯曲度 meridianal bending
子午线象限弧长 arc length of meridian quadrant
子午线坐标系 meridian coordinate
子午星表 meridian catalogue
子午仪 astronomic(al) transit; meridian circle; meridian instrument; meridian telescope; meridian transit; transit telescope; transit circle
子午仪测微器 transit micrometer
子午仪导航系统 transit navigation system
子午仪改进卫星 transit improvement program(me) satellite
子午仪卫星 transit satellite
子午仪卫星多普勒定位 transit Doppler positioner; transit Doppler positioning
子午正切光线 meridianal tangential ray
子系 daughter
子系产物 daughter
子系统 daughter system; subsystem; subset【数】
子系统库 subsystem library
子系统信息检索设施 subsystem information retrieval facility
子系物质 daughter substance
子细胞 subcell
子弦 subchord
子相 product phase
子项目 subitem; sub-project
子信息块 blockette
子星 component star
子行列式 minor(of a) determinant; subdeterminant; underdeterminant
子序列 subsequence
子序列数据 subsequent data
子旬形式 clause form
子循环 subloop
子盐 <水泥中最主要的固溶物> alite

子样本 subsample
子样方差 sample variance
子样品 subsample
子叶【植】coty ledon
子夜 midnight
子夜太阳 <极地的> midnight sun
子夜星组 <用于纬度测量> intermediate group
子一代 first filial generation
子因子 subfactor
子语言 sublanguage
子域【数】subdomain;subfield
子域子码 subfield subcode
子元件 subelement
子增量 subincrement
子阵 subarray
子整环 subdomain
子帧 subframe
子指令 subcommand
子指数 subindex
子中心 subcenter[subcentre]
子中心的 subcentral
子钟 secondary clock;slave clock
子子程序 subsubsystem
子总体 subpopulation
子作业 subjob

姊妹公司 sister company

姊妹国家 sister country

梓树 Chinese catalpa

梓(树属) <拉> catalpa
梓油 catalpa oil;stillingia oil
梓油酸 catalpic acid
梓属 bean tree

紫变 purple stain

紫赤色 puniceous
紫脆石 ussingite
紫灯 purple light
紫丁香 lilac
紫椴 Amur linden
紫蒽酮 violanthrone
紫方钠石 <方钠石的变种> hackmanite
紫刚玉 Oriental Amethyst
紫梗胶 stick lac
紫光 purple light
紫光电池 violet photocell
紫硅碱钙石 charoite
紫硅镁铝石 yoderioite
紫果冷杉 red fir
紫褐色 purple brown;purplish brown;violet-brown
紫褐色的 puce
紫褐色硬木 <中美洲产> moruro
紫黑色 purple black;purplish-black;violet-black
紫红 crimson
紫红货舱调和漆 red oxide hold paint
紫红木 purple heart
紫红漆 red oxide paint
紫红色 amaranth;aubergine;claret(brown);fuchsia;heliotrope;hyacinth;maroon prunosus;mauve;murrey;purplish red;reddish purple;wine;violet red
紫红色淀 Florentine lake;maroon lake
紫红色原 maroon toner
紫红调色剂 maroon toner
紫红铁粉 crocus
紫花景天 orpine
紫花苜蓿【植】alfalfa;lucern(e)
紫花泡桐 karri-tree;kiri;Princess tree
紫花树 chinaberry
紫灰色 cloud gray;purplish grey;violet-grey
紫酱色 maroon
紫胶 lac;shellac(k)
紫胶棒 stick shellac
紫胶虫 lacca;lac insect
紫胶的别名 gum lac
紫胶酚 laccal
紫胶块 lump lac
紫胶蜡 lac wax;shellac wax
紫胶蜡酸 lacceroic acid
紫胶粒 graining lac
紫胶盘菌属 <拉> Coryne
紫胶片 shellac(k)
紫胶漆片橙(液) orange shellac
紫胶清漆 French polish;lac varnish;shellac varnish
紫胶染料 lac dye
紫胶色淀 lac lake
紫胶树脂 sclerolac
紫胶塑料 shellac plastics
紫胶酸 lac acid;laccaic acid;shellolic acid
紫胶桐酸 aleuritic acid
紫金城 the Forbidden City
紫金山天文台 Zijinshan observatory
紫堇色 corydalis green
紫茎 Chinese stewartia
紫荆篱 thorn hedge
紫晶 amethystine quartz
紫晶色 amethyst
紫晶色的 amethystine
紫矿石 purple ore
紫蓝色 hyacinth;Oxford-blue;violet blue
紫锂辉石 kunzite
紫磷铁锰矿 purpurite
紫硫镍矿 violarite
紫柳 bitter willow;purple osier;purple willow;red osier;red willow
紫罗兰 violet
紫罗兰花 gillyflower;stock
紫罗兰色 violet
紫罗兰色的 violaceous
紫茉莉树脂 jalap resin
紫木 kingwood
紫苜蓿阀 <位于地下管道灌溉供水系统的竖管顶端> alfalfa valve
紫苜蓿闸门 <灌溉工程中用> alfalfa gate
紫钼铀矿 mourite
紫纳闪石 bababudanite
紫茜素色靛 violet alizarin(e) lake
紫群青 violet ultramarine
紫色 purple;purpure
紫色板岩 purple slate
紫色的 purple
紫色淀 purple lake;violet lake
紫色分界线 purple boundary
紫色蓝 violetish blue
紫色木 kingwood
紫色染料 purple dye
紫色四方形信号机 <车站岔线上停车信号机>【铁】violet square signal
紫色凸透镜 purple convex lens
紫色涂料 <划线用> purple ink
紫色土 purple soil
紫色颜料 purple pigment;violet pigment
紫色氧化物颜料 purple oxide
紫色硬木 violet wood
紫砂 red porcelain
紫杉 Japanese yew;yew
紫杉属 yew;Taxus <拉>
紫射线 violet ray
紫石英 amethyst;violet quartz
紫石英色的 amethystine
紫树 cotton gum;tupelo gum
紫树属 black gum;tupelo;Nyssa <拉>
紫水晶 amethyst;amethystum
紫水晶色颜料 amethyst colo(u)r
紫四环镍矿 abelsonite;nickel porphyrin
紫苏安玄岩 alboranite
紫苏花岗闪长岩 enderbite
紫苏花岗岩 birkremite;charnockite
紫苏花岗岩相 Aldan facies
紫苏辉长岩 hypersthene gabbro
紫苏辉石 fieinite;germarite;hypersthene;szaboite
紫苏辉石安山玢岩 hypersthene granular porphyrite
紫苏辉石安山岩 hypersthene andesite
紫苏辉石二长岩 hypersthene porphyry
紫苏辉石橄榄辉长岩 hypersthene olivine gabbro
紫苏辉石橄榄岩 hypersthene peridotite
紫苏辉石花岗岩 hypersthene granite
紫苏辉石苦橄岩 hypersthene picrite
紫苏辉石麻粒岩 hypersthene granulite
紫苏辉石浅色麻粒岩 hypersthene light-colo(u)red granulite
紫苏辉石球粒陨石 hypersthene chondrite
紫苏辉石闪长岩 hypersthene diorite
紫苏辉石无球粒陨石 diogenite
紫苏辉石玄武岩 hypersthenes basalt
紫苏辉石岩 hypersthenite
紫苏石 hypersthenite
紫苏石榴黑云斜长片麻岩 hypersthene garnet biotite plagioclase gneiss
紫苏透辉角闪斜长麻粒岩 hypersthene diopside amphibole plagioclase granulite
紫苏透辉斜长麻粒岩 hypersthene diopside plagioclase granulite
紫苏岩 hypersthenite
紫苏耀石粒玄岩 hypersthene dolerite
紫苏英闪岩 bugite
紫苏子油 perilla oil
紫穗槐 bastard indigo
紫穗槐属(固沙植物) <拉> Amorpha
紫檀 red sandal wood;red sanders
紫檀木 narra
紫檀香木 redsandal
紫檀硬木 <产于非洲、缅甸、印度和安达曼群岛> padauk[padouk]
紫檀属 padauk[padouk];Pterocarpus <拉>
紫藤 Chinese wistaria
紫藤花色 wistaria violet
紫藤颜料粉 powder blue
紫铁矾 paracoquimbite;quenstedtite
紫铁铝矾 millosevichite
紫铜 copper;red brass;red copper
紫铜板 copperplate;sheet copper
紫铜带 copper belt
紫铜的 coppery
紫铜钉 copper nail;copper tack;rivet nail;roove clinker nail
紫铜管 copper tube
紫铜烙铁 copper bit;copper bolt;soldering copper
紫铜铝锑矿 cyanophyllite
紫铜毛滤清器 copper wool filter
紫铜排水管 copper drainage tube
紫铜盘条 copper wire rod
紫铜皮 copper sheet;sheet copper
紫铜片 copper sheet
紫铜水管 copper water tube
紫铜印染辊筒 copper printing roller
紫土 purple soil
紫外成像 ultraviolet imagery
紫外二次电子导电管 uvicon
紫外反射 ultraviolet reflectance
紫外反射光谱法 ultraviolet reflectance spectrometry
紫外分光法 ultraviolet spectrometry
紫外分光光度测定法 ultraviolet spectrophotometry
紫外分光光度测量 ultraviolet spectrophotometry
紫外分光光度法 ultraviolet spectrophotometric method
紫外分光光度学 ultraviolet spectrophotometry
紫外分光计 ultraviolet spectrometer
紫外分光术 ultraviolet spectrometry
紫外分析 ultraviolet analysis
紫外辐射 ultraviolet irradiation;ultraviolet light
紫外辐射强度 ultraviolet radiation intensity
紫外辐射损伤 ultraviolet radiation damage
紫外辐射探测器 ultraviolet radiation detector
紫外光 ultraviolet light
紫外光比吸光度 specific ultraviolet light absorbance
紫外(光)灯 ultraviolet lamp
紫外光电元件 ultraviolet cell
紫外光电子分光计 ultraviolet photoelectron spectrometer
紫外光电子光谱法 ultraviolet photoelectron spectrometry
紫外光电子能谱学 ultraviolet photoelectron spectroscopy
紫外光度测量 ultraviolet photometry
紫外光度滴定 ultraviolet photometric titration
紫外光度法 ultraviolet photometry
紫外光度计 ultraviolet photometer
紫外光发光法 ultraviolet luminescent method
紫外光分光镜 ultraviolet spectroscope
紫外光感受器 ultraviolet receptor
紫外光及可见光检测器 ultraviolet and visible detector
紫外光镜头 ultraviolet lens
紫外光可见光检测器 ultraviolet-visible detector
紫外(光)谱 ultraviolet spectrum
紫外光谱法 ultraviolet spectrography
紫外光谱仪 ultraviolet spectrograph
紫外光区 ultraviolet region
紫外光显微镜 ultraviolet microscope
紫外红外线漆膜固化装置 ultraviolet infrared film curing apparatus
紫外监测器 ultraviolet detector;ultraviolet monitor
紫外可见光比色监测器 ultraviolet-visible colo(u)rimetric monitor
紫外可见光分光光度法 ultraviolet-visible spectrophotometry
紫外可见光分光光度计 ultraviolet-visible spectrophotometer
紫外可见光流出物监测器 ultraviolet-visible effluent monitor
紫外可见光吸收光谱法 ultraviolet-visible absorption spectrometry
紫外敏感成像管 ultraviolet-sensitive image tube
紫外敏化剂 ultraviolet sensitizer
紫外目标辐射计 ultraviolet target radiometer
紫外强度 ultraviolet intensity
紫外强化臭氧氧化工艺 ultraviolet enhanced ozonation process
紫外区 ultraviolet band;ultraviolet zone
紫外扫描器 ultraviolet scanner
紫外射线吸光度 ultraviolet ray adsorbance
紫外视像管 ultraviolet vidicon
紫外太阳辐射 ultraviolet sun radia-

tion
紫外天文望远镜 celestial ultraviolet telescope
紫外天文学 ultraviolet astronomy
紫外图像 ultraviolet image
紫外望远镜 ultraviolet telescope
紫外无极放电灯 ultraviolet electrodeless discharge lamp
紫外吸光度 ultraviolet adsorbance
紫外吸收分光法 ultraviolet absorption spectroscopy
紫外吸收分光光度学 ultraviolet absorption spectrophotometry
紫外吸收光谱法 ultraviolet absorption spectrophotometry
紫外吸收光谱分析 ultraviolet absorption spectroscopy
紫外吸收光谱图 ultraviolet absorption spectrogram
紫外吸收检测器 ultraviolet absorption detector
紫外吸收滤光镜 ultraviolet absorbing filter
紫外吸收曲线 ultraviolet absorption curve
紫外吸收装置 ultraviolet absorption device
紫外显微法 ultraviolet microscopy
紫外显微照相术 ultraviolet photomicrography
紫外显像测密术 ultraviolet densitometry
紫外线 ultraviolet(ray);violet ray;alpine light;vitalight
紫外线 A 波动 ultraviolet A radiation
紫外线背景 ultraviolet background
紫外线波长 ultraviolet wavelength
紫外线波段 ultraviolet band
紫外线玻璃 ultraviolet glass
紫外线成像 ultraviolet image
紫外线处理 ultraviolet light treatment
紫外线处理法 ultraviolet treatment
紫外线穿透性 ultraviolet transmission
紫外线的 ultraviolet
紫外线灯 finsen light; ultraviolet lamp; ultraviolet radiator; ultraviolet ray lamp; vitalight lamp; ultraviolet light
紫外线灯装置 germitron
紫外线电视显微镜 ultraviolet television microscope
紫外线发光 ultraviolet luminescence
紫外线发生器 ultraviolet generator
紫外线法水质有机污染监测仪 ultraviolet organic pollution monitor
紫外线范围 ultraviolet ray range
紫外线防护纸 ultraviolet ray resisting paper
紫外线分光光度计 ultraviolet spectrophotometer
紫外线辐射 ultraviolet radiation; ultraviolet
紫外线辐射(光)源 ultraviolet radiation sources
紫外线辐照法 ultraviolet irradiation
紫外线感光记录纸 ultraviolet recording paper
紫外线固化 ultraviolet curing
紫外线光电子光谱法 ultraviolet photoelectron spectroscopy
紫外线光度计 ultraviolet photometer
紫外线光记录器 ultraviolet light recorder
紫外线光敏树脂固化器 ultraviolet light activator for resin polymerization
紫外线光谱法 ultraviolet spectroscopy
紫外线光谱光度计 ultraviolet spectrophotometry
紫外线光谱图 ultraviolet spectrogram
紫外线光谱学 ultraviolet spectroscopy
紫外线光谱仪 ultraviolet spectrometer
紫外线光源 ultraviolet light source
紫外线红外线灯 ultraviolet and infrared ray lamp
紫外线激光器 ultraviolet laser
紫外线计 ultraviolet meter
紫外线记录器 <记录应变用> ultraviolet recorder
紫外线记录仪 ultraviolet recorder
紫外线技术 ultraviolet technique
紫外线剂量计 ultraviolet dosimeter
紫外线剂量仪 ultraviolet dosimeter
紫外线监视 ultraviolet surveillance
紫外线检测器 ultraviolet detector
紫外线检查 ultraviolet inspection
紫外线检偏镜 ultraviolet analyser[analyzer]
紫外线降解 ultraviolet degradation
紫外线老化 ultraviolet ray ageing
紫外线滤光镜 ultraviolet filter
紫外线灭菌法 ultraviolet sterilization
紫外线灭菌作用 rentschlerization
紫外线能量 ultraviolet energy
紫外线屏蔽剂 ultraviolet screener; ultraviolet screening agent
紫外线谱带 ultraviolet band
紫外线气体分析器 ultraviolet gas analyser[analyzer]
紫外线区(域) ultraviolet range; ultraviolet region
紫外线杀菌 ultraviolet sterilization
紫外线杀菌灯 ultraviolet germicidal lamp
紫外线摄像管 Ubicon
紫外线摄影(术) ultraviolet photography
紫外线示波器 ultraviolet oscillograph
紫外线示波图 ultraviolet oscillograph
紫外线试池 ultraviolet cell
紫外线损伤 ultraviolet injury
紫外线太阳灯 ultraviolet alpine sun lamp
紫外线探测 ultraviolet detection
紫外线探漏器 ultraviolet leak detector
紫外线探伤器 ultraviolet ray flaw detector
紫外线特征 ultraviolet signature
紫外线通信[讯] ultraviolet communications
紫外线透过率 ultraviolet transmittance
紫外线透射 ultraviolet ray transmission
紫外线望远镜 ultraviolet telescope
紫外线稳定剂 ultraviolet stabilizer
紫外线稳定性 ultra-violet stability
紫外线吸收法真氧监测仪 ultraviolet absorption ozone monitor
紫外线吸收光谱 ultraviolet absorption spectrum
紫外线吸收光谱法 ultraviolet absorption spectrometry
紫外线吸收剂 ultraviolet absorbent; ultraviolet absorber
紫外线吸收作用 ultraviolet absorption
紫外线显微镜 ultraviolet microscope
紫外线消毒 disinfection by ultraviolet ray; disinfection with ultraviolet rays; ultraviolet disinfection; ultraviolet sterilizing
紫外线消毒法 disinfection method by ultraviolet ray
紫外线抑制剂 ultraviolet inhibitor
紫外线荧光仪 ultraviolet fluorescence meter
紫外线影响 influence of ultraviolet ray
紫外线诱导 ultraviolet induction
紫外线灾难 ultraviolet catastrosphe
紫外线照明 ultraviolet lighting
紫外线照片 ultraviolet photograph
紫外线照射处理 ultraviolet ray treatment
紫外线照射灯 ultraviolet light gun
紫外线照射法 ultraviolet exposure; ultraviolet illumination; ultraviolet irradiation; ultraviolet radiation
紫外线照射检查法 examination under ultraviolet light
紫外线照相机 ultraviolet camera
紫外线照相(术) ultraviolet photography
紫外线真空计 ultraviolet light vacuum ga(u)ge
紫外-氧化氮分光计 ultraviolet nitric oxide spectrometer
紫外氧化法 ultraviolet-oxidation system
紫外仪器 ultraviolet instrumentation
紫外仪器法 ultraviolet instrumentation method
紫外荧光 ultraviolet fluorescence
紫外诱变 ultraviolet mutagism
紫外源 ultraviolet source
紫外照度计 ultraviolet power meter
紫外直接影印照片 ultraviolet photostat
紫外自遮摄影法 ultraviolet self-mattering process
紫菀属植物 aser
紫葳花 eriogonum inflatum
紫薇 crape myrtle; grape-myrtle
紫霞 purple light
紫心木 <产于圭亚那> purple heart
紫馨【植】jessamin(e)
紫(叶)山毛榉 copper beech; purple beech
紫贻贝 cypraea macula; Mytilus edulis
紫移 hypsochromic effect
紫玉兰 purple yulan
紫竹 black bamboo
紫棕色 purple brown

字

字板纸 flong paper
字表 word table
字长 <每个存储字的彼特数>【计】 machine length; word capacity; word length
字盒式打印机 moving type box printer
字画 kakemono
字图图解方式 alphageometric

自

自岸伸出最远的礁 reef farthest from a bank
自摆系统 hunting system
自办工程账 <政府自办工程的估价计算> force account
自办设计 in-house design
自伴边值问题 self-adjoint boundary value problem
自伴变换 self-adjoint transformation
自伴的 self-adjoint
自伴方程式 self-adjoint equation
自伴扩张 self-adjoint extension
自伴偏微分算子 self-adjoint partial differential operators
自伴谱测度 self-adjoint spectral measure
自伴算子 self-adjoint operator
自伴随系统 self-adjoint system
自伴椭圆型算子 self-adjoint elliptic operator
自伴椭圆型微分算子 self-adjoint elliptic differential operator
自伴微分方程 self-adjoint differential equation
自伴微分算子 self-adjoint differential operator
自伴线性微分方程 self-adjoint linear differential equation
自拌式混凝土厂 self-propelled concreting plant
自饱和 self-saturation
自饱和磁放 self-saturated magnetic amplifier
自饱和电抗器 autotransductor; self-saturated reactor
自保 block; own risk; self-hold(ing); self-insurance; self-preservation; sticking up
自保保险公司 captive insurance company
自保持 self-perpetuating
自保持触点 self-holding contact
自保持的 self-sustaining
自保持电路 lock-in circuit; lock-on circuit
自保持电源 self-contained supply
自保持二极管 locking diode
自保持寄存器 latching register
自保持接点 seal-in contact; self-holding contact
自保持式信号 stick signal
自保单元 stick unit
自保电流 holding current
自保电路 holding circuit; self-holding circuit; stick(ing) circuit
自保护 self-shield
自保护变压器 self-protected transformer
自保护的 self-protecting; self-protective
自保护电弧焊 self-shielded arc welding
自保护焊丝 self-shielded welding wire
自保护式变压器 self-protective transformer
自保继电器 holding-on relay; latched relay; relay with latching
自保式按钮 stick type push button
自保式的 self-lapping
自保弹簧 stick spring
自保线 stick wire
自保线圈 stick coil
自保线选择 stick wire selection
自保作用 stick effect
自报 self-reporting
自报所得税额 income tax self-assessed
自爆 spontaneous detonation
自爆的 self-bursting
自爆熔断器 self-bursting fuse
自爆系统 self-destruction system
自爆引信 autodestructive fuse
自爆装置 destructive mechanism
自卑感 inferiority feeling
自卑情结 inferiority complex
自备的 self-contained
自备电池 self-contained battery
自备电源式机车 self-propelled electric(al) locomotive
自备动力厂 self-supply power plant
自备动力工作平台 self-contained platform
自备动力设备 self-contained plant
自备动力钻机 self-contained drill
自备发电厂 isolated generating plant; self-supply power plant
自备发电机 house generator
自备发电站 isolated generating sta-

tion
自备工业给水工程 exclusive industrial water supply
自备供电设备 auxiliary power supply equipment; self-contained power supply equipment
自备火灾报警 private fire alarm
自备货车 shipper-owned wagon
自备能源的 self powered
自备泥舱 <挖泥船> self-contained hopper
自备汽车家庭 car owning household
自备食品 <矿工> bait
自备式导航 self-aid navigation
自备式导航仪 self-contained navigator
自备式抢救车 self-contained crash tender
自备式自动导航系统 self-contained automatic navigation system
自备水井 private well
自备水源 privately owned water-supplies
自备外汇 self-provided foreign exchange
自备涡轮机 house turbine
自备污水管 private sewer
自备消火栓 private fire hydrant
自焙 self-baking; self-roasting
自比较法 <短寿命核测量方法> self-comparison method
自比知觉 automorphic perception
自闭 sticking; sticking up
自闭电路 holding circuit; stick circuit
自闭多功能扁嘴钳 self-gripping general purpose pliers
自闭阀 self-locking valve; automatic isolating valve
自闭防火门窗配件 self-closing fire assembly
自闭防火门窗五金 self-closing fire assembly
自闭盖 snappy spring cover
自闭合 self-closing
自闭合导线 closed-on-itself traverse
自闭合空气门 self-closing air-door
自闭继电器 stick relay
自闭继电器电路图 stick relay scheme
自闭铰链 self-shutting hinge
自闭接点 stick contact
自闭门 automatically closing door; self-closing door
自闭塞 self-blocking
自闭塞门 self-closing cock
自闭式电动闸门 self-closing electric(al) gate
自闭式阀 self-closing valve
自闭式防火门 automatically closing fire door; self-closing fire door
自闭水阀 self-closing faucet
自闭水龙头 self-closing faucet
自闭水密门 automatic closing door; self-closing door
自闭条件 stick requisition
自闭线 stick wire
自闭线选择 stick wire selection
自闭性土料 <坝工> self healing soil
自闭旋塞 self-closing faucet
自闭闸门 self-closing gate
自闭装置 automatic closing device; self-closing device
自蔽性能 self-screening property
自编程序 self-programming
自编程序软件 self-programming software
自编出发列车 outbound train made up
自编机 self-organizing machine
自编辑程序 self-editor
自编密码 private code
自编数据 self-editing data
自编系统 self-organizing system
自变换量程 automatic ranging; autoranging
自变量 argument; independent variable
自变量表 argument list
自变量字节 argument byte
自变数 argument; independent variable
自变息债券 drop-lock bond
自变形丝 self-texturing yarn
自变质作用【地】 autometasomatism; autometamorphism; automorphosis; synantexis
自并励 self-shunt excitation
自播植物 self-sown flora
自补偿 self-compensation
自补偿触点 self-replenishing contact
自补偿磁平衡 self-compensating magnetic balance
自补偿电机 self-compensated machine
自补偿式电动机 self-compensated motor
自补偿系统 self-compensating system
自补代码 self-compensating code
自补的 self-complementary; self-complementing
自补计数器 self-complementary counter; self-complementing counter
自补码 self-complementary code; self-complementing code
自补强 self-strengthening
自擦拭双螺杆挤出机 self-wiping twin-screw extruder
自测试 self-test(ing)
自测试继电器 self-test relay
自测试因数 self-testing factor
自测试自修复计算机 self-testing-and-repairing computer
自插导管 autocatheterism
自差 autodyne; autoheterodyne; endodyne; magnetic deviation; self-heterodyne
自差表 deviation table; magnetic compass table
自差补偿装置 deviation compensation device
自差电路 autodyne circuit
自差分析 analysis of deviation
自差改正量 correction of deviation
自差记录簿 deviation log book
自差角 angle of deviation
自差校正场 swinging ground
自差校正磁铁 adjusting magnet; correcting magnet
自差校正试验仪 deviascope
自差接收法 autodyne reception
自差接收机 autodyne receiver
自差接收器 autodyne
自差近似系数 approximate coefficient of deviation
自差力 deviating force
自差路 autodyne circuit
自差曲线 deviation curve; deviation diagram
自差曲线图 dygoram; namogoniogram
自差式接收 autodyne reception
自差式接收机 autodyne receiver
自差收音机 autodyne
自差图 deviation card
自差系数 coefficient of deviation; deviation coefficient; magnetic coefficient
自差消除 compensation of deviation
自差振荡器 autodyne oscillator
自产共管公寓 freehold flat
自产原料 originating material
自产自用项目 captive item
自铲 self-loading
自铲式 self-loaded
自铲式铲运机 self-loading scraper
自场 self-field
自沉淀作用 autoprecipitation
自沉积漆 autodeposition paint
自陈法 self-report method
自撑能力 self-supporting tendency
自撑式 self-supporting style
自撑式脚手架 self-supporting scaffold
自成单一体货物 <如钢锭、钢棒、纸板装货物、纸滚筒等> self-unitized load
自成的 autogenetic
自成地形 autogenetic topography
自成景观 independent landscape
自成模板 self-forming
自成模型 self-formed model
自成水系 autogenetic drainage
自成碎屑岩相 autoclastic fragmental facies
自成土 mesomorphic soil
自成系统的 self-contained
自成悬浮体 self-made quicksand
自成一体的单元 self-contained unit
自成正交 self-orthogonal
自承的 self-supporting; self-sustaining
自承拱 self-supporting arch; self sustaining arch <洞室围岩>
自承脚手架 self-supporting scaffold
自承结构物 self-sustaining structure
自承能力 ability in self-support; self-supporting property
自承式 self-supporting
自承式分段浮船坞 self-docking dock
自承式分段坞修 self-docking repair
自承式钢管 self-bearing steel pipe
自承式钢框架 self-supporting steel frame
自承式锅炉 bottom-supported boiler
自承式楼梯 self-supporting stairway
自承水泥罐 self-supporting cement container
自承重的 self-supporting
自承重隔断 self-supporting partition
自承重隔热 self-supporting insulation
自承重结构 self-supporting structure
自承重空心砖隔墙 structural clay tile partition(wall)
自承重墙 self-supporting wall; self-sustaining wall
自承重墙板 self-bearing wall panel
自乘 involution; involve; raise to a power
自澄清 self purificating
自持的 self-supporting; self-sustained; self-sustaining
自持堆 self-maintaining reactor; self-sustaining reactor
自持发射冷阴极 self-sustained emission cold cathode
自持反应 self-sustaining reaction
自持放电 self-maintained discharge; self-maintaining discharge
自持放电检测器 self-maintained discharge detector
自持光栅 self-supporting grating
自持过程 bootstrap process; self-sustaining process
自持介质 self-sustaining medium
自持链式反应 self-sustaining chain reaction
自持链式核反应 self-maintaining nuclear chain reaction
自持起动 self-contained start
自持时间 stand-up time
自持式气动凿岩机 autostoper
自持水下呼吸器 self-sustained underwater breathing apparatus
自持天线杆 self-supporting antenna mast
自持系统 bootstrap
自持线圈 self-supporting coil
自持振荡 self-sustained oscillation
自持转速 self-sustaining speed
自充电 self-charging
自充水面目标 self-inflation surface target
自充水式风窗玻璃洗涤装置 self-filling windscreen washer
自冲击射流喷嘴 self-impinging injector
自冲击阻抗 self-surge impedance
自冲滤池 self-backwashing filter
自冲坡度 <水管能自动冲沙的坡度> self-cleaning gradient
自冲洗的墙壁 self-washing wall
自重置循环 self-resetting loop
自稠密的 dense in itself
自筹的 self-financed
自筹投资 self-investment; self-provided investment
自筹资金 organization-raised funds; self-finance; self-financing; self-raising fund
自筹资金比率 self-financing ratio
自筹资金企业 enterprise self-raised fund
自初始化 self-initialize
自除铝粉钻头 self-cleaning bit
自触发程序 self-triggered program(me)
自穿透 self-gating
自传 autobiography
自吹鼓式过滤机 self-blowing drum filter
自炊处 self-cooking place
自垂椭圆裂隙 self-similar elliptical crack
自磁场 self-magnetic field
自磁(感)的 self-magnetic
自磁能 self-magnetic energy
自猝火 squegging
自催化反应 self-catalyzed reaction
自淬灭 self-quenching
自淬灭计数器 self-quenching counter
自(淬)硬钢 self-hardening steel
自淬硬钢钢轨 rail of self-hardening steel
自带呼吸系统 built-in breathing system; self-contained breathing system; self-contained underwater breathing apparatus
自带力 possessiveness
自带潜水呼吸器 self-contained breathing apparatus; self-contained underwater breathing apparatus
自带熔剂的硬钎料 self-fluxing brazing alloy
自带芯 own core
自担风险 at one's own risk
自导程序系统 bootstrap system
自导纳 self-admittance
自导式的 self-guided
自导向的重锤式蓄能器 self-guided weight-loaded accumulator
自导向转向架 self-steering (arm) truck
自导引 seek; self-aiming; self-guidance
自导引系统雷达 self-contained radar
自导引系统作用距离 seeking range
自底向上 bottom up
自底向上程序设计 bottom-up programming
自底向上的 bottom up
自底向上法 bottom-up approach
自底向上设计 bottom-up design
自底向上选择 selective bottom-up
自底向上研制法 bottom-up development
自地槽【地】 autogeosyncline
自点火焰安全灯 relighter flame safety lamp

自电导 self-conductance
自电离作用 autoionization;self-ionization
自电势 self-potential
自电势勘探 self-potential prospecting;self-potential survey
自电势曲线 self-potential curve
自电梯度测量剖面平面图 profiling-plan figure of self-potential gradient measurement
自电梯度测量剖面图 profiling figure of self-potential gradient measurement
自电位测井 self-potential log
自电位差 self-potential difference
自电位勘探 self-potential prospecting;self-potential survey
自电位曲线 self-potential curve
自电源车辆检测器 self-powered vehicle detector
自电源线来的干扰 conducted interference
自电阻 self-resistance
自垫高板钉 self-furring nail
自垫高板条 self-furring
自垫高金属拉网 self-furring metal lath
自垫高金属条板 self-furring metal lath
自垫高条钉 self-furring nail
自垫国外关税销售法 factoring
自叠加性 self-superposability
自顶部注入油罐内 overshot tank filling
自顶升趸船 self-elevating barge
自顶升平台 self-elevating platform
自顶式铆枪 self-bucking hammer
自顶向下的 top-down
自顶向下方式 top-down system
自顶向下分析程序 top-down parser
自顶向下分析法 top-down approach;top-down method;top-down parsing
自顶向下分析过程 top-down process
自顶向下分析技术 top-down technique
自顶向下分析算法 top-down analyser[analyzer];top-down parsing algorithm
自顶向下规划 top-down planning
自顶向下快速分析程序 top-down fast-back parser
自顶向下设计 top-down design
自顶向下识别算法 top-down recognizer
自顶向下选择 selective top-down
自顶向下研制法 top-down development
自定界码 self-demarcating code
自定时记录法 self-clocking recording method
自(定)位 self-alignment
自定位效应 self-aligning effect
自定心冲孔器 self-centering punch
自定心内表面测量仪 self-centering internal measuring instrument
自定心三爪卡盘 three-jaw self-centering chuck
自定信用项目 discretionary credit limit
自定义的 self-defining
自定义函数 user defined function
自定义数据 self-defining data
自定义项 self-defining term
自定义值 self-defining value
自定中心耳轴 self-centering trunnion
自定中心振动筛 self-centering vibrating screen;self-centering vibratory screen
自动 auto;automatic nature;self-act;self-action;self-motion;self-movement
自动安平水准仪 autolevel;automatic level;autoset level;compensator level;compensator level(l)ing instrument;pendulum level;self-adjusting level;self-aligning level;self-level(l)ing instrument;self-level(l)ing level
自动安平装置 self-setting device;self-setting unit
自动安全阀 automatic safety valve
自动安全负载指示器 automatic safe load indicator
自动安全装置 automatic safety device
自动安装 self-erecting
自动安装的 self-mounting
自动岸基接收检测设备 automatic shorebased acceptance checkout equipment
自动按量分配装置 automatic batching system
自动按钮控制 automatic push-button control
自动按钮控制的 automatic push-button controlled
自动按钮控制的电梯 lift with automatic push-button control
自动按钮控制的升降机 lift with automatic push-button control
自动拔出器 automatic puller
自动百叶窗 automatic louvers
自动摆动 self-sustained oscillation
自动扳机 automatic cocking lever
自动扳手 autowrench
自动搬运方式 automatic transfer car system
自动搬运台 moving platform
自动板牙头 automatic die head
自动磅秤 automatic weigher
自动包装 automatic packaging
自动包装机 automatic packaging unit;automatic packer;automatic packing machine;automatic wrapping machine
自动保持装置 automatic holding device
自动保护 automatic protection
自动保护X射线管 autoprotective tube
自动保护和联合跟踪 automatic detection and integrated tracking
自动保护金属极电弧焊 shielded metal arc welding
自动保护开关 automatic protection switching
自动保护切换 automatic protection switching
自动保护系统 automatic protective system
自动保护装置 automatic protector;automatic safety device
自动保密电话通信[讯] automatic secure voice communication
自动保险 voluntary insurance
自动保险螺钉 captive screw
自动保压阀 self-lapping valve
自动保真度控制 automatic fidelity control
自动保证安全设计 fail-safe design
自动报火警设备 automatic fire warning device
自动报价机 ticker
自动报警 autoalarm
自动报警错误率 automatic false alarm rate
自动报警接收机 autoalarm;automatic alarm receiver
自动报警器 automatic alarm;automatic call device;automatic call point
自动报警设备 automatic alarm;automatic alarm equipment
自动报警系统 autoalarm system;automatic warning system
自动报警信号键控设备 automatic-alarm-signal keying device
自动报警信号接收机 autodistress signal apparatus
自动报警装置 autoalarm device;automatic alarm device;automatic vigilance device;automatic warning device
自动报警状态 automatic telling status
自动报时器 chronopher
自动报时装置 program(me)clock
自动报文交换 automatic message switching
自动报文交换中心 automatic message switching center[centre]
自动抱闸 self-braking
自动曝光补偿 automatic exposure compensation
自动曝光控制 automatic exposure control
自动曝光设备 automatic exposure equipment
自动曝光装置 automatic exposure device
自动爆口机 automatic crack-off machine
自动备份 automatic backup
自动备水消防系统 automatic wet-pipe sprinkler system
自动背景(噪声)控制 automatic background control
自动背景调整 automatic background control
自动泵 automatic pump;self-acting pump
自动泵站 automatic pump station;self-acting pump station
自动逼真度控制 automatic fidelity control
自动比尺控制 automatic scale control
自动比例水样采集器 automatic proportional water sampler
自动比率控制 automatic ratio control
自动比色计 autocolo(u)rimeter;automatic colo(u)rimeter
自动闭合 <裂缝的> self-healing
自动闭环过程控制 automatic closed-loop process control
自动闭环控制系统 automatic monitored control system
自动闭塞 automatic block(ing);self-blocking
自动闭塞电线路 automatic block line
自动闭塞供电线路 automatic block power line
自动闭塞号志 automatic block signal
自动闭塞计划 automatic block project
自动闭塞联系电路 connecting circuit with automatic blocks
自动闭塞区段 automatic block district;automatic signal(l)ing section
自动闭塞区域 automatic block territory;automatic signal territory
自动闭塞设备 automatic block equipment
自动闭塞系统 automatic block system
自动闭塞信号 automatic block signal
自动闭塞信号法 automatic block signal(l)ing
自动闭塞信号防护法 automatic block signal(l)ing protection
自动闭塞信号机【铁】 automatic block signal;automatic block signal(l)er
自动闭塞信号机的间隔【铁】 spacing of automatic block signals
自动闭塞信号区域 automatic block signal territory
自动闭塞信号系统 automatic block signal system
自动闭塞信号线 automatic block signal line
自动闭塞信号制 automatic block signal system
自动闭塞行车规则 automatic block signal(l)ing rule
自动闭塞运用 automatic block operation
自动闭塞制 automatic block system
自动闭塞装置 automatic block equipment;automatic block installation;automatic block system;self-locking gear
自动闭锁 automatic lock;coupler lock;track block
自动闭锁电路 automatic shut-off circuit
自动闭锁器 self-locking catch
自动闭锁系统 automatic block system;automatic lock system
自动闭锁装置 self-locking gear
自动避碰设备 automatic anti-collision aids
自动避碰系统 automatic anti-collision system;automatic collision avoidance system
自动编程工具 automatic programmed tool
自动编程加工语言 automatically programmed tools language
自动编程(序) automatic programming;autoprogram(ming)
自动编程序系统 self-programming system
自动编程序装置 automatic programming unit
自动编号发射机 automatic numbering transmitter
自动编号机 automatic numbering machine
自动编码 autocode;autocoding;automatic code;automatic coding
自动编码程序 autocoder
自动编码机 autocoder;automatic coder;automatic coding machine
自动编码技术程序 program(me) for automatic coding technique
自动编码器 autocoder;automatic coder;automatic coding machine;compiler
自动编码系统 automatic coding system
自动编码语言 autocoder;automatic coding language
自动编排系统 automatic patching system
自动编索引 autoindex(ing);automatic index(ing)
自动编图 computer-aided map compilation
自动编图系统 <美国制造> Calmagraphic system
自动编译程序 automatic compiler;autopiler
自动编译程序装置 autopiler
自动编译语言 compiler language
自动编运行图 automatic scheduling
自动编制 automatic abstract
自动编制程序 automatic programming;self-programming
自动编制文摘 automatic abstracting
自动编组调度员【铁】 automatic marshalling controller
自动编组控制设备【铁】 automatic marshalling controller
自动编组线 <箭翎形线路布置>【铁】 automatic making-up siding
自动变电所 automatic substation
自动变电站 automatic substation

自动变光开关 automatic beam changer
自动变荷调压器 automatic loading regulator
自动变化传动 automatic transmission
自动变换器 autochanger; autoconverter
自动变量 automatic variable
自动变量输送泵 automatic variable delivery pump
自动变速 automatic gear shift(ing); automatic transmission; self-shifting transmission
自动变速齿轮 self-changing gear
自动变速器 automatic speed selector; automatic transmission
自动变速器控制系统 automatic transmission control system
自动变速器遥控装置 remote control for automatic transmission
自动变速箱 self-changing gearbox
自动变速箱工作液 automatic transmission fluid
自动变速装置 automatic transmission
自动变向闸板 automatic change-over damper
自动变向阻尼器 automatic change-over damper
自动变压器 automatic transfer
自动变址 autoindex(ing); automatic index(ing)
自动变址寄存器 autoindex register
自动变阻器 automatic rheostat
自动标灯 automatic signal lamp
自动标定图板 automatic plotting board
自动标度盘 autodial
自动标刻度盘 autodial
自动标引 automatic index(ing)
自动标志识别和分类 automatic object recognition and classification
自动表征 automatic attribute
自动并列装置 automatic paralleling device
自动并行装置 automatic paralleling device
自动拨号报警系统 automatic dial alarm system
自动拨号单元 automatic dialing unit
自动拨号电话机 automatic dial telephone set; call-a-matic telephone set
自动拨号机 automatic dialer
自动拨号盘 autodial
自动拨号双向电话 automatic dial-up two-way telephone
自动拨号装置 automatic dialing unit
自动波特率检测 automatic baud rate detection
自动驳船 self-propelled lighter
自动薄层涂布器 automatic thin layer spreader
自动补偿 autocompensation; automatic compensation
自动补偿的 self-compensating; self-reacting; self-balancing
自动补偿地下水位电测仪 self-compensating underground water detector
自动补偿拉紧装置 self-compensating take-up
自动补偿离合器 self-compensating clutch
自动补偿器 automatic compensator; automatic tensioner
自动补偿游丝 self-compensating hairspring
自动补偿装置 self-compensating device; self-compensating unit
自动补充存货 automated replenishment
自动补充调测量 automatic additional adjust
自动补风 automatic recharge
自动补给 normal issue
自动补足保额条款 automatic reinstatement clause
自动布局 automatic layout; autoplacement
自动步长调整 automatic step adjustment
自动步道 <输送式> moving passenger conveyer [conveyor]; moving walk; travel(l)ing platform; travel(l)ing sidewalk; trawolater
自动步道坡度 slope of moving walk
自动步进 automatic stepping
自动步进再启动 automatic step restart
自动部件选择 automated component selection
自动擦拭式 self-wiping
自动采集(数据) autoacquisition
自动采样 automatic sampling
自动采样器 automatic sampler
自动采样器分析仪 automatic sampler analyser[analyzer]
自动采样系统 automatic sampling system
自动采样装置 automatic sampling device
自动彩色校正 automask
自动彩色控制 automatic chrominance control; automatic colo(u)r control
自动餐车 cafeteria car
自动仓库 automatic warehouse
自动舱单系统 automated manifest system
自动舱盖 autohatch; automatic hatch cover
自动舱口盖 automatic hatch cover
自动操舵 automatic steering
自动操舵船 automatic steering ship; autopilot steered ship
自动操舵机 automatic helmsman
自动操舵开关 pilot selector switch
自动操舵装置 automatic course-keeping gear; automatic pilot; automatic steering device; automatic steering gear; autopilot; autosteerer
自动操纵 automatic control; automatic operation; steering control
自动操纵的 automatic
自动操纵的风门 automatically operated damper
自动操纵阀 automatically operated valve
自动操纵飞机 robot plane; autoplane
自动操纵器 mechanical pilot
自动操纵装置 automatic run device
自动操作 automatic operation; automatic working; automation; instrumental operation; non-stick working; operator less; self-control; unattended operation
自动操作板 automatic operations panel
自动操作的 governor-operated
自动操作和调度程序 automatic operating and scheduling program(me)
自动操作进给阀 automatically operated inlet valve
自动操作气体分析器 automatically operating gas analyser[analyzer]
自动操作器 automatic operator
自动操作台 automatic operations panel
自动操作系统 automatic operation system
自动操作与调度程序 automatic operating and schedule program(me)
自动侧倾出渣车 automatic side-dump muck car
自动侧倾卸载装置 automatic side dumping arrangement
自动侧卸铁路货车 automatic side tipping wagon
自动测波浮标 pitch and roll buoy
自动测定记录计 automatic measuring recorder
自动测定路面弯沉值车 <其中的一种> Lacroix deflectograph
自动测厚规 automatic thickness ga(u)ge
自动测距导航设备 distance indicating automatic navigation equipment
自动测距-方位角法定位 automated range-azimuth positioning
自动测距跟踪 automatic range tracking
自动测距镜筒 range finder scope
自动测距仪 automatic range finder; automatic range only; odograph; Autotape <一种用于海上距离测量的仪器>
自动测距装置 automatic range unit
自动测量 automated measurement; automated survey; automatic measurement; automatic sizing; automatic survey
自动测量计 autometer
自动测量器 automatic measuring unit
自动测量系统 automated survey system
自动测量仪 <一种惯性测量装置> auto-surveyor
自动测量装置 automatic sizing device; autosizing device; self-operated measuring unit
自动测试 automatic test(ing); autotest(ing)
自动测试设备 automatic test(ing) equipment; built-in test equipment
自动测试系统 automated test system
自动测试仪 automatic test(ing) equipment
自动测试与检查设备 automatic test and checkout equipment
自动测试与控制系统 automatic checkout and control system
自动测试装置 automatic checkout test equipment; automatic test(ing) equipment
自动测水准仪 self-level(l)ing level
自动测速仪 automatic speed measuring device
自动测图仪 autocartograph; automat; automatic plotting instrument; automatic plotting machine; Aviograph
自动测微计 automatic micrometer
自动测微密度计 automicro densitometer
自动测向 automatic direction finding
自动测向器 automatic direction(al) finder; radio compass
自动测向系统 automatic direction finding system
自动测向仪 automatic direction(al) finder
自动测向仪方位指示器 automatic direction finder bearing indicator
自动测向仪指针反转 automatic direction finder reversal
自动测压计 mechanical manometer
自动测验机 automatic ga(u)ging and detecting machine
自动叉车 automatic fork lift
自动插补 automatic interpolation
自动插袋机 automatic sack applicator
自动插入 Autoinserter
自动插销 automatic bolt; self-latching bolt
自动查错 automatic debugging
自动查询 auto-query
自动查找系统 automatic detecting system; automatic detection finding system
自动拆线 automatic release
自动拆卸装置 automatic disconnecting device
自动掺气水流试验 experiment on self aerated flow
自动掺杂 autodope
自动铲运机 motorized scraper; motor scraper; self-loading scraper
自动长途通信[讯] automatic long-distance service
自动抄车号 automatic car identification
自动抄录 autoabstract
自动抄针器 automatic stripper
自动超高速传动装置 cruising gear
自动超速挡 automatic selective overdrive
自动超越控制转向人工操纵的系统 automatic back-up and over-ride system
自动超载控制 automatic overload control
自动潮位计 automatic tidal ga(u)ge; automatic tide ga(u)ge; automatic tide recorder; recording tide ga(u)ge
自动潮位记录计 automatic recording tide ga(u)ge
自动潮位记录器 automatic recording tide ga(u)ge; automatic tide recorder
自动潮位仪 automatic (recording) tide ga(u)ge; automatic tidal ga(u)ge; automatic tide recorder; recording tide ga(u)ge
自动车 automobile; automotive; motorcycle
自动车床 automatic lathe; screw machine; self-acting lathe
自动车道控制系统 automatic lane control system
自动车发动机 automotive engine
自动车钩 autocoupler; automatic car coupler; automatic coupling device; coupler guard arm
自动车钩抵角 coupler horn
自动车钩钩头 drawhead
自动车钩头 automatic coupler head
自动车号识别 automatic car identification
自动车机械学 automobile mechanics
自动车空气调节 automotive air conditioning
自动车辆 self-propelled car
自动车辆定位系统 automatic vehicle location system
自动车辆分类 automatic vehicle classification
自动车辆跟踪 automatic vehicle tracking
自动车辆跟踪系统 automatic vehicle tracking system
自动车辆缓行器系统 automatic retarder system
自动车辆缓行器制 automatic retarder system
自动车辆记录器 automatic traffic recorder
自动车辆监控 automatic vehicle monitoring
自动车辆减速器系统 automatic retarder system
自动车辆减速器制 automatic retarder system
自动车辆识别 automatic car identification
自动车内燃机 automobile engine

自动车无线电设备 autocar radio installation
自动车闸 motorcar brake
自动撤销 voluntary withdrawal
自动沉淀析出 autoprecipitation
自动沉积 autodeposition
自动沉降描绘器 automatic settlement plotter
自动沉降仪 automatic settlement plotter
自动沉箱式闸门 automatic pontoon lock gate
自动称量记录装置 weightometer
自动称量搅拌机 combined weighing and mixing machine
自动称量配料 automatic weighing batching
自动称量设备 automatic weighing equipment;automatic weighing plant
自动称量系统 automatic weighing system
自动称量装置 automatic weighing apparatus; automatic weighing device; automatic weighing equipment;automatic weighing unit
自动称料 automatic weighing
自动称重拌和机 combined weighing and mixing machine
自动称重分散器 autoweigh disperser
自动称重累加器 autoweigh totalizer
自动称重设备 automatic weighing plant
自动称重式(土壤)测渗仪 automatic weighing lysimeter
自动称重式(土壤)渗漏仪 automatic weighing lysimeter
自动称重式土壤蒸发器 automatic weighing lysimeter
自动称重仪 weightometer
自动成本估计 automated cost estimates
自动成槽托滚 self-troughing idler
自动成卷装置 automatic lap former
自动成图 automatic mapping
自动成图程序文件 AutoCAD program(me)file
自动成图软件包 AutoCAD software package
自动成像处理 automated imagery processing
自动成型 automatic forming
自动成型机 automatic mo(u)lding machine
自动承保 automatic cover
自动乘客检票机 automatic passenger gate
自动程序 automated program(me); automatic process; automatic sequence
自动程序监控器 automonitor
自动程序检查装置 automatic programmed checkout equipment
自动程序控制 automated program(me) control; automatic program(me) control; scheduling control automation
自动程序控制单元 automated program(me)control unit
自动程序控制工具 automatically programmed tool
自动程序控制计算机 automatic sequence-controlled computer
自动程序控制计算器 automatic sequence-controlled calculator
自动程序控制通信[讯]设备 programmed automatic communication equipment
自动程序控制系统 automatic programmed control system
自动程序设计 automatic programming;autoprogram(ming)
自动程序设计机 automatic programming machine
自动程序设计器 automatic programmer
自动程序设计试验系统 automatic programmer-and-test system
自动程序调整 automatic preset
自动程序中断 automatic program(me) interruption
自动程序主控器 automatic master sequence
自动秤 automatic scale; automatic weigher; automatic weighing machine; automatic weighting plant; self-acting scale; weigher; weightograph;weightometer
自动秤量 autoweighing
自动秤量器 automatic batcher; self-adjusting weigher
自动秤量填装机 automatic weighing filler
自动尺寸监控 autosizing
自动尺寸控制 automatic size control
自动充满的 self-priming
自动充气救生衣 gas inflation automatically operated type life-jacket
自动充气器 snifter
自动充水 automatic priming
自动充填器 autoplugger
自动冲程调整器 automatic stroke adjuster
自动冲床 automatic center punch;automatic punching machine
自动冲锤打桩机 automatic ram pile driver
自动冲沙进水口 intake with automatic flushing
自动冲水槽 <公共厕所> flushing trough
自动冲水系统 automatic flushing system
自动冲水箱 automatic cistern; automatic flushing cistern
自动冲水贮水器 automatic flushing cistern
自动冲洗 automatic flushing(out)
自动冲洗槽 automatic flushing trough
自动冲洗储水器 automatic flushing cistern
自动冲洗大便器 automatic water closet
自动冲洗机 automatic developing machine
自动冲洗水箱 automatic flushing cistern; automatic flushing tank; self-flush tank;automatic washing tank
自动冲洗系统 automatic flushing system
自动冲洗贮水器 automatic flushing cistern
自动冲压硬度试验机 autopunch
自动重调 automatic reset
自动重发器 autorepeater;synchro repeater
自动重发请求 automatic request for retransmission
自动重发通信[讯]方式 automatic repeat request system
自动重复发送方式 automatic repetition system
自动重(复)发(送)系统 automatic repetition system
自动重复要求 automatic repeat request
自动重合 autoregistration; automatic reclosing
自动重合断路器 autoreclose circuit breaker
自动重合闸 automatic reclosure; autoreclose circuit breaker; recloser
自动重合闸断路器 autoreclose breaker
自动重合(闸)开关 automatic reclosing switch
自动重合闸起动继电器 autorecloser starting relay
自动重接 automatic reclosing
自动重接继电器 automatic reclosing relay
自动重接器 recloser
自动重振控制 automatic recruitment control
自动抽出 automatic extraction
自动抽取 autoabstract
自动抽水系统 automatic pumping system
自动抽水站 self-acting pump station
自动抽头切换设备 automatic tap changing equipment
自动抽样 autoabstract;auto accident
自动抽油井 unattended pumper
自动抽逐器 self-acting ejector
自动出力式制动器 self-energizing; self-energizing brake
自动出料 self-emptying
自动出纳银行 drive-in bank
自动出烟口 automatic fire vent
自动除粉钻进 self-cleaning drilling
自动除粉钻眼 self-cleaning drilling
自动除极化 spontaneous depolarization
自动除霜 automatic defrosting
自动除霜器 automatic decrystalization device;automatic defrosting machine
自动除渣 self-skimming
自动除渣器 automatic de-sludger
自动储存放料装置 automatic magazine loader
自动储存送料 magazine feed
自动储存送料附件 magazine feed attachment
自动储存送料装置 magazine loader
自动处理机 automatic processor
自动处理设备 automatic processing equipment
自动处理数据 datamation
自动触发 automatic triggering; self-triggering
自动触发器 automatic trigger; self-trigger
自动穿带的 self-threading
自动穿孔 automatic punch
自动穿孔机 automatic punch(er)
自动穿连纹板机 automatic lacing machine
自动传播 automatic dissemination
自动传播的 self-propagating
自动传布的 self-propagating
自动传导 autoconduction
自动传动 automatic drive
自动传动机构 automatic driver
自动传动箱 automatic transmission
自动传动装置 automated gear; automatic driver; automatic gear; automatic transmission;drive master
自动传感 autoconduction
自动传感器 automated sensor
自动传感设备 automatic transfer equipment
自动传感装置 automatic sensing device
自动传输 automatic conveying; automatic transmission
自动传送机 autotransmitter
自动传送设备 automatic transfer equipment
自动传真 autofax
自动传真系统 automatic picture transmission system
自动船舶定位和姿态测量系统 automated ship location and attitude measuring system
自动船首基准系统 automatic heading reference system
自动窗孔 autowindow
自动窗锁 automatic window catch
自动垂直指标 automatic vertical index
自动垂准仪 autoplumb
自动锤衬套 oscillating weight ring
自动锤衬套簧 oscillating weight ring spring
自动锤齿轴 oscillating weight pinion
自动锤带钻隔片 oscillating weight jewels distance piece
自动锤导套 oscillating weight guiding tube
自动锤击打桩机 automatic ram pile driver
自动锤夹板 oscillating weight bridge
自动锤栓 oscillating weight bolt
自动锤栓簧 oscillating weight bolt spring
自动锤凸轮 oscillating weight cam
自动锤下夹板 oscillating weight lower bridge
自动锤轴承螺钉 screw for bearing wheel of oscillating weight
自动磁带卡盘 autohub
自动催化反应 autocatalytic reaction
自动催化剂 autocatalysis
自动催化作用 autocatalysis
自动存车场 autosilo
自动存储传送装置 magazine
自动存储器 automatic storage; inherent store
自动存储器配置 automatic storage allocation
自动存储区 automatic storage area
自动存储区类 automatic storage class
自动存入 automatic log(ging)
自动错误校正法 automatic error correction
自动打包机 automatic packing machine
自动打夯机 automatic compactor
自动打结器 automatic knotter
自动打开 automatic opening
自动打壳机 automatic crust breaker
自动打捆机 self-binder
自动打捆机构 self-tying mechanism
自动打磨机 air sander
自动打印 automatic printing
自动打印机 automatic marker; automatic stamper
自动打桩锤 automatic pile hammer
自动打桩机 automatic pile driver
自动打字机 automatic typewriter
自动大门 automatic gate
自动代码 autocode;automatic code
自动带宽控制 automatic bandwidth control
自动带式砂光机 automatic belt sander
自动单层压机 automatic single daylight press
自动单斗装料机 bucket loader
自动单速调整 automatic one speed control
自动单行收割机 self-propelled single-row harvestor
自动挡 automatic catch
自动挡板 automatic damper
自动挡火闸 automatic fire damper
自动挡铁 automatic dog
自动导轨公共交通【交】 automated guideway transit
自动导航 automatic navigation; automatic pilot;self-navigation
自动导航和数据收集系统 automatic

navigation and data acquisition system
自动导航计算机和航迹描绘仪 automatic navigational computer and plotter; automatic pilot computer and plotter
自动导航控制装置 automatic navigation control equipment
自动导航设备 homer
自动导航卫星通信[讯] satellite telecommunication with automatic routine
自动导航系统 automatic navigation system
自动导航仪 automatic navigation device; automatic navigator; autonavigator; autopilot
自动导航装置 automatic pilot device
自动导轮 reverser mounted
自动导频电路 automatic pilot channel
自动导向的 self-guided
自动导向交通(系统) automated guided transit; automated guideway transit
自动导向控制 autonomous control
自动导向履带式工业拖拉机 self-laying track-type industrial tractor
自动导向系统 automatic guidance system; self-sufficient navigation system
自动导向运输系统 automatic guided system; automatic transport system
自动导向装置 automatic pilot device
自动导星 automatic guiding
自动导星装置 autoguider; automatic guider
自动导引 homing guidance
自动倒空装置 automatic drainage device
自动倒塌 self-collapse
自动捣固机 automatic rammer
自动到站停车 automatic station stop
自动道钉机 automatic spiker
自动道路分析仪 automatic road analyzer
自动的 autogenous; automatic; automobile; automotive; non-stick; offhand; powered; power-operated; self-acting; self-actuating; self-mobile; self-moving; self-operated; self-powered; self-travel(l)ing; spontaneous; unmanned; self-driven <机器、车辆等>
自动灯 non-attended light
自动灯光信号 automatic light signal
自动登录 autolog-on
自动登录器 automatic register
自动等高线仪 automatic contour liner
自动等径控制 automatic diameter control
自动等静压制 mechanized isostatic pressing
自动等深线数字转换器 automatic contour digitizer
自动等时染色机 automatic isochronous staining machine
自动低噪 automatic noise reduction
自动滴定计 automatic titrimeter
自动滴定器 automtitrator
自动底层水取样器 free vehicle bottom-water sampler
自动底盘 implement carrier; implement porter; power frame tractor; self-propelled (tool) carrier; self-propelled tool chassis; toolbar carrier; tool-carrier chassis; tool frame tractor
自动底卸车 automatic drop bottom car
自动抵抗力 active resistance
自动地 automatically
自动地面航行 automatic terrain recognition and navigation
自动地图制图与公用设施管理 automated mapping/facilities management
自动地形测图仪<德国蔡司制造> Topomat
自动地震检测 automatic earthquake detector
自动地震检测器 automatic event detector
自动地震数据处理 automatic earthquake processing
自动地址修改 automatic address modification
自动递减 autodecrement
自动递增 autoincrement
自动点灯 automatic lighting
自动点焊 automatic spot welding; stitch bonding; stitch welding
自动点焊法 automatic spot weld
自动点火 autoignite; autoignition; automatic ignition; self-ignition
自动点火的<火箭发动机> self-priming
自动点火定时器 automatic spark timer
自动点火器 autoigniter
自动点火提前设备 automatic advanced element
自动点火温度 autoignition temperature; self-ignition temperature
自动点火装置 automatic pilot device
自动点燃锅炉 automatically lighted boiler
自动电报传输 automatic telegraph transmission
自动电报机 automatic telegraph; dial telegraph
自动电传测试 automatic telex test
自动电磁导向 automatic magnetic guidance
自动电磁制导 automatic magnetic guidance
自动电点火 automatic electric(al) ignition
自动电动机测试器 automatic motor tester
自动电镀(工艺) automatic plating
自动电焊 automatic electric(al) welding; automatic weld(ing)
自动电焊机 automatic arc welding machine; automatic welder; automatic welding machine
自动电弧焊 automatic arc welding; cyc-arc welding
自动电弧焊缝管 automatic arc welded tube
自动电弧焊接机 automatic arc welding machine
自动电弧焊(接机)头 automatic arc welding head
自动电话 automatic telephone
自动电话总局 dial central office
自动电键 automatic key
自动电空制动机 automatic electro-pneumatic brake
自动电力气动制动机 automatic electro-pneumatic brake
自动电流试验 automatic current testing
自动电流调节器 automatic current regulator
自动电路分析器 automatic circuit analyser
自动电路开关【计】 automatic circuit breaker
自动电路重合开关 automatic circuit recloser
自动电平补偿 automatic level compensation
自动电平记录器 automatic level recorder
自动电平控制 automatic level control
自动电平控制器 automatic lever controller
自动电平调节 automatic level regulation
自动电平调节系统 automatic level regulating system
自动电桥 automatic bridge
自动电热烘箱 automatic electric(al) oven
自动电热水加热器 automatic electric(al) water heater
自动电热水器 automatic electric(al) water heater
自动电势计 automatic potentiometer
自动电梯 automatic elevator; automatic lift; escalator; self-service elevator; moving staircase
自动电位测定 automatic potentiometry
自动电位滴定法 automatic potentiometry
自动电信中心 automatic telecommunication centre
自动电压控制 automatic voltage control
自动电压调节器 automatic voltage regulator
自动电压调整器 automatic tension regulator; automatic voltage regulator
自动电压稳定器 constac
自动电渣焊 automatic slag pool welding
自动电子发射 autoelectronic emission
自动电子发射的 autoelectronic
自动电子计算机 automatic electronic computer
自动电子数据交换中心 automatic electronic data-switching center [centre]
自动电子仪 brain
自动垫纸机 automatic paper interleaving machine
自动吊车 power hoist
自动吊罐机 bailing machine
自动吊卡 automatic slip elevator
自动吊桶 automatic bucket
自动调车 automatic shunting; automatic switching
自动调度 automatic scheduling
自动调度呼叫设备 automatic calling equipment
自动调度集中 automatic control concentrating
自动调度设备 automatic calling equipment
自动调度系统 automatic dispatch system
自动调度装置 autodispatcher
自动蝶形阀 self-closing butterfly valve
自动定边 automatic sense
自动定标 automatic scale
自动定标器 autoscaler
自动定尺寸内圆磨床 sizematic internal grinder
自动定点停车 automatic spotting
自动定距钻孔台 automatic spacing table
自动定量冲洗阀 flushometer valve
自动定量关闭阀门 automatic metering valve
自动定量关闭龙头 automatic metering tap
自动定量喂给机 automatic weighing and feeding machine
自动定平水准仪 self-level(l)ing instrument
自动定期操作器 automatic routineer
自动定期测试装置 automatic routineer
自动定时齿轮 automatic timing gear
自动定时磁电机 automatic timed magneto
自动定时的 self-timing
自动定时开关 automatic time switch; automatic timing switch; automatic clock switch
自动定时控制 automatic timing control
自动定时器 automatic timer; autotimer
自动定时锁 time lock
自动定时装置 automatic timing device
自动定位 automatic positioning; automatic station finding; automatic station keeping; autopositioning; self-alignment; self-positioning
自动定位的柱座 self-centering column seating
自动定位方法 automatic position setting
自动定位圆犁刀 pivoting colter; swivel disk colter
自动定位装置 automatic positioning device; automatic positioning equipment
自动定温器 self-contained thermostat
自动定向 automatic direction finding; self-orientation
自动定向的 self-directing; self-orientating
自动定心 self-centering[centring]
自动定心的进口导板 self-centering entry guide
自动定心机构 self-centering unit
自动定心夹盘<立轴的> self-centering chuck
自动定心夹头 self-centering collet; self-centring chuck
自动定心卡盘 concentric(al) jaw chuck; self-centering chuck
自动定心轮 self-centering wheel
自动定心密封 self-centering seal
自动定心模板 self-centering formwork
自动定心能力 self-centering capability
自动定心式筛分机 self-centering screen
自动定心装置 self-centering apparatus
自动定心作用 self-centering action
自动定序 automatic sequencing
自动定义文件 automatic defining file; self-defining file
自动定值 autopact
自动定中心轴 self-centering mandrel
自动动力厂 automatic power plant
自动动态平衡 homeostasis
自动动态评价 automatic dynamic(al) evaluation
自动斗车 gunboat
自动斗门 drop bottom seal
自动毒品识别 automated drug identification
自动读出的 self-reading
自动读出器 automatic reading machine
自动读带机 automatic tape transmitter
自动读卡机 automatic card reader; automatic card transmitter
自动读数 automatic reading
自动读数水准尺 serf-reading leveling rod
自动读数仪器 self-reading instrument
自动堵缝 automatic seal
自动镀锡设备 automatic tinning equipment
自动短路保护 automatic short-circuit

Z

protection
自动短路器 automatic short-circuiter
自动断电器 automatic circuit breaker;automatic cut-out;automatic interrupter
自动断火闸 automatic fire shutter
自动断开 automatic opening; automatic shutdown;knock off
自动断流 automatic shut-off
自动断流阀 automatic cut-out valve; automatic shut-off valve;automatic stop valve;self-closing stop valve
自动断流器 automatic cut-out
自动断路 autocut-off; autocut-out; automatic break; automatic cut-out;automatic cutting off;auto off
自动断路器 auto-circuit breaker; autocut-out; automatic circuit breaker;automatic interrupter;automatic switch;free handle breaker
自动断路器线圈 breaker trip coil
自动断路压路机 mobile breaker-roller
自动断路装置 automatic shut-off device
自动断箱 battery cutout
自动断续开关 flasher
自动断续器 automatic interrupter; autotrembler;self-interrupter
自动断续疏水器 flash trap
自动断续装置 flasher
自动堆包机 automatic bale stacker
自动堆垛机 automatic piler;stacker
自动堆垛台 unscrambler
自动堆焊 automatic surfacing
自动堆料机 automatic stacker
自动堆码设施 automatic stacking device
自动堆纸机 automatic sheet stacker
自动堆装机 < 库场用 > automatic guided vehicle
自动对比度控制 automatic brightness contrast control
自动对答 automatic answering
自动对光反射投影仪 autofocus reflecting projector
自动对光纠正仪 autofocus rectifier; automatic focus rectifier
自动对光设备 autofocus apparatus; self-focusing apparatus
自动对光投影仪 autofocus reflecting projector;automatic focus projector
自动对光系统 automatic focusing system
自动对光装置 automatic focusing apparatus;self-focusing apparatus
自动对话 automatic crosstell
自动对焦 autofocusing
自动对零 automatic zero setting
自动对流 autoconvection; automatic convection
自动对流直减率 autoconvective gradient;autoconvective lapse rate
自动对位 automatic positioning;automatic station finding;automatic station keeping; autopositioning; self-positioning
自动对位的液压缸 self-aligning hydraulic ram
自动对心的液压缸 self-aligning hydraulic ram
自动对心辊 self-aligning centring roll
自动对正平接铰链 Hurlhinge
自动对中 automatic centering; self-centering[centring]
自动对中导轨 self-centering guide rail
自动对中三脚架 self-centering tripod;automatic centering tripod
自动对准 autocollimate;autocollimation; automatic alignment; self-alignment
自动对准的 automatic aligning;self-aligning
自动对准经纬仪 autocollimating theodolite
自动对准中心 automatic centering
自动多梭梭箱 checking motion
自动多元素摄谱仪 autometer; autrometer
自动舵 autohelm;automatic pilot;autopilot
自动舵操作方式 operation modes in autopilot
自动舵分压器 rudder potentiometer
自动舵管制器 rudder setter
自动舵航迹积算仪 pilot automatic dead reckoning equipment
自动舵接触器触轮 rudder contactor trolley
自动舵接通与调整指示器 autopilot engage and trim indicator
自动舵控制系统 autopilot control system
自动舵驱动装置 autopilot driver
自动舵伺服装置 autopilot servo unit
自动舵速率控制 autopilot rate control
自动舵调整面板 autohelm adjustment panel
自动发报机 automatic transmitter
自动发报机头 autohead
自动发电厂 automatic generating plant;automatic power plant
自动发电控制 automatic(al) generation control
自动发电站 automatic power station
自动发电装置 automatic generating plant
自动发码 automatic coding
自动发射 autoemission
自动发生紧急制动的故障 dynamiter; kicker
自动发声浮标 automatic sound buoy
自动发送接收机 automatic send-receive set
自动发送器 automatic sender;transmitter distributor
自动发现 automatic diagnosis
自动发芽室 automatic generation chamber
自动阀 action valve; automatically operated valve;autovalve;prepayment valve;self-acting valve;self-action valve
自动阀传动杆 motorized valve actuator
自动阀门 automatic valve;balanced gate
自动阀型避雷器 autovalve lightning arrester[arrestor]
自动阀执行机构 motorized valve actuator
自动翻板门 self-turn-over plate gate
自动翻板闸门 balanced wicket;flashboard check gate
自动翻倒式闸门 self-collapsing gate
自动翻斗秤 automatic skip scale
自动翻落式 self-tipping
自动反冲洗 automatic flushback
自动反偏压 automatic back bias
自动反应行为 automatic response behaviour
自动反转 autoreverse
自动反转轮 reverser
自动反转片 automatic reversal film; autoreversal film
自动返回 self-recovery
自动返回地面的铲斗 self-level(l)ing bucket
自动返回机构 automatic return mechanism
自动返回中间位置的肘节开关 spring-return-to-center toggle switch
自动返油作用原理 automatic fuel oil returning principle
自动方法 automated method
自动方位指示仪 automatic bearing indicator
自动防爆装置 automatic explosion-proof device
自动防火帘 automatic shutter
自动防火门 automatic fire door;heat-actuated fire door
自动防火门组合装置 fire assembly automatic
自动防水消防栓 self-draining hydrant
自动防止故障的 fail safe
自动防止故障危害的 fail safe
自动防撞系统 automatic collision avoidance system
自动仿形控制 contouring control
自动仿形轧辊车床 automatic contouring roll lathe
自动放出救生筏 autoreleasing life-craft
自动放大 autoenlarge
自动放大机 autoenlarging apparatus
自动放大器 autoenlarging apparatus; automatic amplifier
自动放电 automatic discharge;self-discharging;spontaneous discharge
自动放电器 automatic discharge device
自动放开的夹具 self-releasing grappling device
自动放空 self-emptying
自动放平装置 autolevel(l)er
自动放气阀 air relief valve;automatic blow off valve; automatic vent; valve air relief
自动放气装置 automatic air eliminator
自动放弃 waiver
自动放射性测量 car-borne radioactivity survey
自动放水阀拍板 automatic clapper
自动放水开关 automatic drain cock
自动放水塞 automatic drain cock
自动放水闸 automatic dam;automatic sluice
自动放水闸门 flop gate
自动放索装置 automatic paying gear
自动放泄阀 automatic drain valve
自动放油阀 automatic fuel drain valve
自动飞行 automotive vehicle
自动飞行器 automotive vehicle
自动废气净化厂 automatic waste-gas purifying plant
自动分度法 automatic index(ing); self-indexing
自动分度头 automatic dividing head
自动分段减压 automatic split reduction
自动分段信号系统 automatic block signal(l)ing system
自动分光光度计 automatic spectrophotometer
自动分光计 automatic spectrometer
自动分级 self-grading
自动分级机 automatic grader
自动分解 autodecomposition
自动分解脱水凝聚 autocoacervation
自动分类 automatic categorization; automatic classification
自动分类机 automatic classifier;automatic sorter;automatic sorting machine;autosorter
自动分类计数机 automatic differential count machine
自动分离器 automatic segregator
自动分离装置 automatic separation device
自动分量机 automatic dose dividing apparatus
自动分配呼叫制 straight-forward junction working
自动分配器 automatic distributor
自动分批拌和 automatic batch mixing
自动分批拌和设备 automatic batch-(er)plant;automatic batcher
自动分色机 automatic colo(u)r separation device
自动分析 automatic analysis
自动分析程序 autoanalyzer procedure
自动分析流程图 autoanalyzer flow chart;autoanalyzer flow diagram
自动分析器 autoanalyzer[autoanalyser];automated analyser[analyzer]
自动分析仪 autoanalyzer[autoanalyser];automated analyser[analyzer]
自动分箱机 automatic box splitter
自动分选机 automatic fraction collector
自动分选装置 automatic sorting equipment
自动粉末检波器 autocoherer
自动粉末喷枪 automatic powder spray gun
自动粉碎机 autogrinding machine
自动风暴观测站 automatic storm observation service
自动风挡调节器 automatic damper regulator
自动风速表 anemometrograph
自动风闸 automatic air brake
自动封闭 self-sealing
自动封闭塑胶接合剂 self-sealing plastics
自动封闭桶 self-sealing tank
自动封缝 automatic seal
自动封缄机 automatic closure machine
自动封口水泥包装机 automatic valve bag packer
自动封口装置 automatic closing device
自动封漏(口)轮胎 puncture-proof tire[tyre]
自动封锁 holding
自动峰值限制器 automatic peak limiter
自动缝合器 automatic suturing machine
自动敷设 autolay
自动扶梯 escalator;motorstair;moving stair(case);moving stairway;travel(l)ing stair(case);travel(l)ing stairway
自动扶梯垂直高度 vertical rise of escalator
自动扶梯护板 motorstair shutter
自动扶梯坑 escalator pit;pit for escalator
自动扶梯行程 travel of escalator
自动扶梯鱼尾板 motorstair shutter
自动扶正救生艇 self-righting lifeboat
自动扶正艇 self-righting boat
自动服务商店工作人员 aisle-man
自动服务站 automated service station
自动浮标 automatic buoy;automatic float
自动浮船 automatic pontoon
自动浮船式闸门 automatic pontoon
自动浮动 free float
自动浮桥 automatic pontoon
自动浮球阀 automatic air valve
自动浮式闸门 < 船闸船坞用的 > automatic pontoon
自动幅度控制 automatic amplitude control

自动幅度控制电路 automatic amplitude control circuit
自动幅度调整电路 automatic amplitude control circuit
自动俯冲趋势 <在临界马赫数附近时的> tuck-under
自动付费计算 automatic message accounting
自动付款机 automatic paying machine
自动负载调节器 automatic load regulator
自动复归指示器 self-restoring indicator
自动复位 automatic reset; self-reset
自动复位的 self-resetting; self-righting
自动复原 automatic recovery; self-recovery
自动复原继电器 automatic reset relay
自动复原手动释放控制 self-reset manual release control
自动复原性 self-restorability
自动复凿机转换开关 automatic reperforator switching
自动改变地址 automatic address modification
自动改正 automatic compensation
自动干草叉 automatic grapple fork
自动干手器 automatic hand drier [dryer]
自动干油集中润滑系统 centralized grease automatic system
自动干燥 self-desiccation
自动干燥机 automatic drier[dryer]
自动干燥器 automatic drier[dryer]
自动杆式车床 automatic bar type machine
自动感应 autoinduction
自动感应烘手器 autoinductive hand drier
自动感应联动(信号)系统 automatic induction coordinated system
自动感应门 autoinductive door
自动感应式联动信号(系统) automatic induction coordinated signal
自动高逼真度 automatic height fidelity
自动高速车床 automatic high speed lathe
自动高速分散机 automatic dissolver
自动高速螺母制造机 nut planking machine
自动高温计 self-recording pyrometer
自动高压控制 automatic high voltage control
自动高压调节器 automatic high-voltage regulator
自动高压稳压电路 automatic high-voltage regulator
自动告警装置 automatic alarm
自动割草机 motor mower
自动割捆机 self-binder
自动搁架 automatic stack
自动搁浅 voluntary stranding
自动格式化 automatic formatting
自动隔离阀门 automatic isolating valve
自动隔离控制电路 automatic amplitude control circuit
自动隔膜式圆锥分级机 automatic diaphragm cone
自动给电系统 automatic feed system
自动给进换挡把手 autofeed selector
自动给进控制 automatic feed control
自动给进凿岩机 autofeed drill; automatic feed drill
自动给进钻进 autofeed drilling; automatic feed drilling; push-feed drilling
自动给料 automatic feed; self-feed(ing)
自动给料的 automatic feeding
自动给料机 self-feeder
自动给料控制 automatic feed control
自动给料盘 automatic feed tray
自动给料器 automatic feeder; self-feeder
自动给料系统 automatic feed system; gravity system
自动给料装置 automatic feeding device
自动给煤炉箅 automatic feed grate
自动给气阀 automatic feed valve
自动给气控制阀 automatic input control valve
自动给油器 self-oil feeder
自动跟踪 autofollowing; automatic follower; automatic following; automatic tracking; autotrack(ing); lock-on; target-seeking guidance
自动跟踪避碰报警装置 automatic tracking and anti-collision warning
自动跟踪分析数学计算机 automatic digital tracking analyzer computer
自动跟踪激光雷达 autofollowing laser radar; automatic tracking laser radar; automatic tracking lidar
自动跟踪控制 automatic following control; automatic tracking control
自动跟踪控制装置 automatic tracking controller automatic following controller
自动跟踪雷达 automatic tracking radar; autoradar
自动跟踪器 automatic tracker; autotracking unit
自动跟踪水位计 automatic following water level meter
自动跟踪天线 automatic tracking antenna
自动跟踪调谐 automatic follow-up tuning
自动跟踪头 automatic tracing head
自动跟踪系统 automatic tracking loop; automatic tracking system; self-tracking system
自动跟踪显微镜 automatic following microscope
自动跟踪遥测接收天线 automatic tracking telemetry receiving antenna
自动跟踪装置 automatic tracking device; autotracker; homing device; homing system of guidance
自动跟踪作用 self-tracking action
自动更换滤材机构 automatic media renewing mechanism
自动更换式空气过滤器 automatic removable air filter
自动更换型空气过滤器 self-renewable air filter
自动更换折叠式空气过滤器 automatic folding type air filter
自动工厂 automatic plant
自动工程 automatic engineering
自动工程设计 automated engineering design
自动工具磨床 automatic cutter grinder
自动工艺规程设计系统 automated process planning system
自动工作 automatic working
自动工作循环 automatic work cycle
自动工作仪表 self-service instrument
自动公共交通系统 self-transit system
自动公路系统 automatic highway system
自动功率补偿器 automatic power compensator
自动功率控制 automatic power control
自动功率调节器 automatic power regulator
自动功率调整器 device for automatic power
自动攻螺母机 automatic nut tapping machine
自动攻丝 automatic tapping; self-tapping
自动攻丝机 automatic tapping machine
自动攻丝螺杆 self-tapping screw
自动攻螺丝座 socket for self-tapping screw
自动供电调节器 automatic flow controller
自动供给燃料 automatic fuel(l)ing; automatic fuel supply
自动供料 automatic feed(ing); gravity feed; self-feed
自动供料机 automatic feeder
自动供料器 self-feeder
自动供暖 automatic heating
自动供水泵 automatic feed water pump
自动供水水源 automatic water supply source
自动供氧系统 demand system
自动钩舌车钩 automatic knuckle coupler
自动估价及价值指示器 automatic rate(-and)-price indicator
自动固定电信网络 automatic fixed telecommunication network
自动固定式夹头 self-clamping chuck
自动固结 automatic consolidation; self-consolidation
自动固结仪 autoodometer
自动固有存储器 automatic inherent storage
自动故障定位 automatic fault location
自动故障隔离 automatic fault isolation
自动故障检测器 automated flaw detector
自动故障探测 automatic fault detection
自动故障位置类型测定 automatic fault location
自动故障信号 automatic fault signal
自动故障寻查器 automatic trouble locator
自动故障寻找 automatic fault finding
自动故障(原因)测定 automatic fault isolation test
自动故障(原因)检测 automatic fault isolation test
自动故障(原因)检测器 automatic fault isolation tester
自动刮面器 automatic scraper
自动刮土机 self-propelled scraper
自动挂车联结器 self-acting trailer coupling
自动挂钩 automatic coupler; automatic coupling; automatic hitch; automatic hook
自动挂钩器 autocoupler
自动挂钩装置 auto-coupling
自动挂接装置 automatic coupling device; automatic pickup hitch
自动挂脱梁 automatic hooking beam
自动挂闸 automatic application
自动关闭 automatic closing; automatic shut-off; self-blocking; self-closing
自动关闭阀 automatic self-closing valve; automatic shuttle valve; self-closing valve
自动关闭防火门 self-closing fire door
自动关闭盖 self-closing lid
自动关闭铰链 self-closing hinge
自动关闭节 self-closing link
自动关闭龙头 self-closing faucet
自动关闭门 self-closing door
自动关闭喷嘴 automatic check nozzle; check nozzle
自动关闭设备 latching
自动关闭式蝴蝶阀 self-closing butterfly valve
自动关闭式拉门 self-closing door; self- sliding door
自动关闭式量油口盖 self-closing ga(u)ge seal
自动关闭水龙头 self-closing tap
自动关闭泄水阀 automatic self-closing sluice valve
自动关闭装置 automatic closing gear; automatic shut-off device; door closer
自动关门 automatic closing
自动关门机 door engine
自动关门器 closer; door check; door closer
自动关门装置 automatic closing device; automatic door closer; closing device; self-closing device
自动观测台 automatic observatory
自动观测仪 automatic observatory; automatic observer
自动观测站 automatic observatory; automatic station
自动管理 automatic control
自动管理电台 automatic station keeping
自动管理电站 automatic station keeping
自动管理台站 automatic station keeping
自动惯性导航仪 inertial autonavigator
自动灌封机 automatic filling and sealing machine
自动灌溉 automated irrigation; automatic irrigation
自动灌溉系统 automated irrigation system
自动灌浆浮鞋 automatic fill-up float shoe
自动灌浆机 automatic grouter; automatic grouting machine
自动灌浆器 automatic grouter; automatic grouting machine
自动灌瓶机 automatic cylinder loader
自动灌水器 automatic primer
自动灌水系统 automated irrigation system
自动光电滴定器 automatic photoelectric(al) titrater
自动光电定时器 photoelectric(al) timer
自动光度计 automatic photometer
自动光阑透镜 lens with automatic diaphragm
自动光量调节 automatic light control
自动光圈 automatic diaphragm
自动光色补偿 automatic colo(u)r compensation
自动光弹性仪 auto polariscope
自动光学曲线追踪气割机 monopole automatic gas cutter
自动归算 automatic reduction; autoreduction
自动归算测速仪 self-reducing tach(e)ometer
自动归算的 self-reducing
自动归算电子测距照准仪 autoreduction electronic distance measuring alidade
自动归算视距仪 automatic reduction tach(e)ometer; autoreduction tachymeter; self-reducing tach(e)ometer; self-reducing tachymeter
自动归算速测仪 automatic reduction tach(e)ometer; self-reducing tach(e)ometer; self-reducing tachymeter
自动归心 automatic centering; self-

Z

centering
自动归约测量仪 reduction tacheometer
自动规 spring bow compasses
自动轨道车 trackmobile
自动轨道衡 automatic scale
自动轨条洗净器 automatic rail washer
自动柜员机 automatic teller machine
自动滚动式滚柱输送机 live-roll conveyer[conveyor]
自动滚轮 automatic carriage; tape-controlled carriage
自动滚筒纤维过滤器 automatic roller fabric filter
自动锅炉 automatic boiler
自动锅炉控制 automatic boiler control
自动过程控制 automatic process control
自动过负荷控制 automatic overload control
自动过滤 automatic filtration
自动过滤器 automatic filter; automatic mechanical filter
自动过载电路 automatic overload circuit
自动海洋旋光测定仪 automatic marine polarimeter
自动焊缝 automatic weld deposit
自动焊(割)炬 mechanical blowpipe
自动焊工 welding operator
自动焊焊把 mechanically manipulated electrode holder
自动焊机 automatic welder; welding machine
自动焊机走架 carriage of automatic welding machine
自动焊接 autobond; automated weld; automatic soldering; automatic weld(ing)
自动焊接程序 automatic welding procedure
自动焊接方法 automatic welding procedure
自动焊接工艺规程 automatic welding process
自动焊接机 automatic welding machine
自动焊接系统 robot welding system
自动焊抗裂试验 cracking test for automatic welding
自动焊料焊接 automatic soldering
自动焊头 automatic welding head
自动夯土机 automatic rammer
自动航标灯交换器 automatic lamp changer
自动航程记录仪 odograph
自动航海通告 automatic notice to mariners system
自动航线稳定器 automatic course stabilizer
自动航线追随装置 automatic chart line follower
自动号码识别 automatic number identification
自动号志 automatically operated signal
自动合拢 self-healing
自动合箱机 automatic closer
自动和人工双重控制 dual operation
自动核对 automatic self-verification
自动荷载控制 automatic load control
自动荷载稳定器 automatic load maintainer; constant load maintainer
自动恒温器 automatic thermostat; autothermostat; self-acting thermos; self-acting thermostat; thermotostat
自动恒温箱 thermautostat
自动横流充电机 automatic constant current charger
自动横切割链锯 automatic cross-cut chain saw
自动衡器 automatic weigher
自动虹吸(管) automatic siphon[syphon]
自动虹吸作用 self-siphonage
自动厚度控制 automatic ga(u)ge control
自动呼叫 autocall; automatic call(ing)
自动呼叫和应答装置 automatic calling and answering unit
自动呼叫畸变 automatic call distortion
自动呼叫器 automatic calling equipment; automatic calling unit; automatic call sign unit
自动呼叫设备 automatic calling unit; automatic call sign unit
自动呼叫失真 automatic call distortion
自动呼叫应答器 automatic calling and answering equipment
自动呼叫装置 automatic calling equipment; automatic calling unit; automatic call sign unit
自动呼吸器 automatic respirator
自动弧焊 automatic arc welding
自动弧焊机 automatic arc welding machine
自动弧形节制阀 automatic radial check gate
自动弧形节制闸门 automatic radial check gate
自动互助船舶营救系统中心 automated mutual vessel rescue center [centre]
自动戽斗 automatic bucket
自动戽斗定量秤 bucket scale; hopper scale; skip scale
自动滑动锁闭机构 automatic slide lock gear
自动滑架 automatic carriage
自动滑门 automated sliding door; automatic sliding door
自动化 robotization
自动化办公室设备 automated office equipment
自动化泵站 automatic pumping station; automatic pump station; unmanned pumping station
自动化编译程序装置 autopiler
自动化编组场【铁】 automatic marshalling yard; automatic classification yard
自动化编组场计算机 yard automatic computer
自动化编组站【铁】 automatic marshalling station; automatic classification station
自动化布置设计程序 automated layout design program(me)
自动化材料处理系统 automated material processing system
自动化测站 automatic station
自动化超级市场 automatic supermarket
自动化车间 automated workshop; automatic systems workshop
自动化程度 automaticity; degree of automation; level of automation
自动化冲压生产线 automatic press line
自动化船 automated ship; automatic ship
自动化船载高空探测计划 automated shipboard aerologic(al) program(me)
自动化存货管理系统 automated inventory control system
自动化大量生产 automatic large scale production
自动化导向线路 <用于自动控制运输系统> automated guideway
自动化道路系统 automated highway system
自动化的 automated; automatic; robot
自动化地图 automatic map
自动化地下铁道列车 automated subway train
自动化电站 unmanned power station
自动化调车场【铁】 automatic classification yard
自动化定座系统 automated seat reservation system
自动化锻造装置 automated forged system
自动化发电厂 automatic plant
自动化发电站 automatic generating station; automatic station
自动化分类 mechanized classification
自动化分析 automated analysis
自动化高速公路 high-speed automated highway
自动化高速运输 automated rapid transit
自动化工厂 automatic plant; automatic production line; push-button plant; robotized plant
自动化工程学会 <美> Society of Automotive Engineers
自动化工序 automation procedure
自动化公路 automated highway
自动化管理 automated management; automatic management
自动化管理控制中心 automated management control centre; automation management control center
自动化管理系统 automated management system; automatic management system
自动化管子加工机 pipe managing and working device
自动化灌溉 automated irrigation
自动化轨道检查 automated track inspection
自动化轨道交通车 automated guideway transit vehicle
自动化轨道交通运输 automated guide transportation
自动化锅炉 unattended boiler
自动化过程 automation process
自动化海图索引夹 automated nautical chart index file
自动化环境模拟室 automated environment(al) chamber
自动化货柜终站系统 automatic container terminal system
自动化货物快速处理系统 automated cargo expediting system
自动化机械 automatable machine
自动化机械加工 automatic machining
自动化机械加工设备 automatic machining equipment
自动化机械装置 automechanism
自动化集装箱起重机 automatic container crane
自动化计算机辅助设计 automatic computer-aided design
自动化技术 automation technique
自动化监测系统 automated monitoring system; automonitoring system
自动化监测仪 automatic monitor
自动化检索词典 automatic dictionary
自动化交通控制 automated traffic control
自动化开采 automated mining
自动化控制 automatic control
自动化控制楼 automated control tower
自动化库存分配系统 automated inventory distribution system
自动化快速交通系统 automatic rapid transit
自动化缆索操作 automated line handling
自动化雷达跟踪系统 automated radar tracking system
自动化理论 automata theory
自动化立体存车库 automatic stereopark
自动化立体停车库 automatic stereopark
自动化立窑 automatic shaft kiln; automatic vertical kiln
自动化量测设备 automated measuring equipment
自动化列车防护 automatic train protection
自动化列车控制 automatic train control
自动化列车运行系统 automatic train operation system
自动化流程图程序 autochart
自动化逻辑图 automated logic diagram
自动化模式 automatic mode
自动化配料器 automatic batcher
自动化气象站 robot weather station
自动化桥梁厂 <一种桥梁施工方法在桥墩上安装一种能滑动的封闭装置,全部桥梁上部结构的施工均在该装置内进行,当该装置向前滑动时,后面已是建成的桥梁> auto-bridge factory
自动化轻型快速交通线路 automated light rapid transit line
自动化软化水厂 automatic softening plant
自动化砂处理装置 automatic sand plant
自动化设备 automatic appliance; automation equipment; push-button plant
自动化设计 automation design
自动化生产 automatic production; automation production
自动化生产记录系统 automatic production recording system; automatic production record system
自动化生产线 automatic production line
自动化施工 robot-controlled construction
自动化时代 automatic era
自动化书刊目录 automated bibliography
自动化数据处理 automatic data processing
自动化数据系统 automated data system
自动化数字 automatic digit
自动化水处理厂 automatic water treatment plant
自动化水电站 automatic hydropower station
自动化水力发电厂 automatic hydroelectric(al) plant; automatic hydroelectric(al) station
自动化水力发电站 automatic hydroelectric(al) station
自动化水平 automation level
自动化水文测量系统 automated hydrographic survey system
自动化台站管理系统 automatic station keeping system
自动化提升机 automatic winder
自动化天文定位装置 automated astronomic(al) positioning device
自动化铁路 automatic railway
自动化铁路电话系统 automatic railway telephone system

自动化铁路管理 automated railroading;automatic railroading
自动化图书馆 automatic library;electronic library
自动化托盘移动机 automated pallet mover
自动化驼峰【铁】automatic hump
自动化驼峰编组场【铁】automatic gravity classification yard;automatic gravity marshalling yard;automatic hump yard
自动化无人驾驶地铁 automated driverless metro
自动化洗舱装置【船】automatic cleaning equipment
自动化系统 automated system;automatic system
自动化信息系统 automated information system;automatic information system
自动化悬链式链斗卸驳/装驳机 automated catenary bucket barge unloader/loader
自动化遥测系统 remote observation system automation
自动化仪表 automation instrument
自动化仪器 self-reacting device
自动化银行业务 automated banking
自动化用具 automatic appliance
自动化预拌混凝土工厂 automatic ready mix(ed) concrete fabricator
自动化运输 automation transportation
自动化指令系统 automated command system
自动化制图 automatic cartography
自动化中心 automation center[centre]
自动化属性 automatic attribute
自动化铸造车间 automatic foundry
自动化铸造机 automatic casting machine
自动化装配 automated assembly
自动化作业 automation operation;automation procedure
自动化作用 automation;automatization
自动划分装置 automatic demarcating device
自动划线装置【道】autoset road marking
自动划行器 automatic marker
自动画图 autochart
自动还原 automatic reduction;self-reset
自动还原按钮 non-locking press button;press-button key;push-button key
自动还原电键 non-locking key
自动还原继电器 non-locking relay or key
自动缓闭器 door check
自动缓冲牵引车钩 automatic buffering and draw coupler
自动缓解 automatic release;false release by itself;spontaneous remission
自动缓行控制器 automatic retardation control
自动缓行器 automatically operated retarder
自动缓行器驼峰编组场 automatic retarder classification yard
自动换(唱)片装置 automatic record charger
自动换挡 automatic gear change;automatic shift;automatic switching;automatic transmission;automatic turning;self-shifting;automatic gearshift;autoranging
自动换挡的超速传动 automatic selective overdrive
自动换灯泡机 automatic lamp changer
自动换排 self-shifting transmission
自动换排挡 automatic gear shift(ing)
自动换盘放送器 record changer
自动换片器 autochanger;automatic record charger;record changer
自动换钱器 coin changing device
自动换筒抽丝机 automatically changing winder
自动换向 automatic reverse;automatic reversing;autoreverse;self-shifting
自动换向的 self-reversing
自动换向逆变器 self-commutated inverter
自动换向握柄 self-reversing lever
自动换行 automatic line feed;automatic new line
自动换样器 automatic sample changer
自动换用 automatic switch over;autoswitch over
自动黄油枪 power-operated grease gun
自动恢复 automatic recovery;automatic restoring;self-recovery
自动恢复程序 automatic recovery program(me)
自动恢复的 self-resetting;self-restoring
自动恢复运转 automatically restored to operation
自动恢复指示器 self-restoring indicator
自动回答 automatic answering
自动回答机构 answer-back mechanism;automatic answer back unit
自动回答问题 automated answering question
自动回答装置 answer-back unit
自动回归分析 automatic regression analysis;autoregression analysis
自动回归过程 automatic regression process
自动回叫 automatic callback
自动回流闸门 automatic return trap
自动回收设施 automatic retrieving device
自动回收系统 automatic recovery system
自动回水管路 automatic return of water
自动回位 self-return
自动回正转向机构 self-centering steering
自动回正作用 self-right(ing) effect
自动回执 automatic acknowledg(e)ment
自动回中装置 self-centering unit
自动回转工作台 automatic revolving table
自动回转缓冲装置 automatic swing cushion
自动回转伸缩式凿岩机 self-rotated stoper
自动绘流程图程序 autoflow
自动绘图 autodraft(ing);automatic data plotting;automatic drafting;automatic draughting;automatic mapping;automatic plot(ting);autoplaceme;machine plotting
自动绘图机 autodrafting system;automatic drafting machine;automatic drafting system;automatic drawing system;automatic mapper;automatic plotter;automatic plotting machine;automatic plotting system;autoplotter;plot(o)mat;Coradomat <商品名>
自动绘图记录仪 chart recorder
自动绘图阶段 automatic drawing stage
自动绘图器 autodrafting system;automatic drafting machine;automatic mapper;automatic plotter;autoplotter
自动绘图系统 automated drafting system;automated tracing system;automatic drafting system;automatic drawing system;automatic plotting system
自动绘图显示器 automatically plotted display
自动绘图仪 autograph;automatic graph plotter;automatic plotter;autoplotter;computerized plotter;computer plotter;plot(o)mat
自动混合 self-mixing
自动混合阀门 automatic mixing valve
自动混合气控制 automixture control;automatic mixture control
自动混凝土泵 self-propelled concrete pump
自动混凝土运送工厂 automatic depot
自动混色机 automatic colo(u)r dispersing machine;mechanical colo(u)rant dispenser
自动活瓣闸门 automatic flap gate
自动活化 autoactivation
自动活门 clapper;self-acting valve
自动火花提前 automatic spark advance
自动火警报警器 fire detector
自动火警警报器 automatic fire detector
自动火警警报系统 automatic fire detecting system
自动火警警报装置 automatic fire detecting system;fire detecting system
自动火警系统 automatic fire detection system
自动火焰光度计 automatic flame photometer
自动火焰清理 automatic scarfing
自动火灾报警系统 automatic fire alarm
自动火灾检测系统 automatic fire detecting system
自动火灾警报系统 automatic fire alarm system
自动火灾警报装置 automatic fire alarm system
自动货单系统 automated manifest system
自动机 automat;automation;automaton[复 automata/automatons];locomobile;robot;self-acting mule;self-actor
自动机车 locomobile;automatic machine
自动机床拖拉环设计 two ring layout
自动机床用钢 automatic steel
自动机的 automotive
自动机的范畴 category of automaton
自动机的极小化 minimization of automaton
自动机搅拌 mixed-in-transit
自动机润滑油 automatic oil
自动机室 autoroom
自动机械 automatic;automation;automaton[复 automata/automatons]
自动机械雕刻机 Keller machine
自动机械加工法 automatic machine process
自动机械卡瓦 automatic power slips
自动机械送料器 automatic mechanical feed
自动机械装置 automatic machine device
自动机转速 speed of motor
自动积分仪 automatic integrator
自动激光定位仪 automatic laser ranging and direction instrument
自动激光控制坡度系统 automatic laser grading system
自动极性变换 autopolarity
自动急泄阀 automatic blow off valve
自动集成电路测试器 automatic integrated circuit tester
自动集中联锁 automatic interlocking
自动集装箱装卸机构 autopallet swinger
自动给水泵 automatic feed water pump
自动给水控制 automatic feed water control
自动给水调节器 automatic feed water regulator
自动给水系统 automatic feed system
自动挤紧 self-compacting
自动计程仪 odograph;odometer
自动记录膨胀计 autographic(al) dilatometer
自动计力器 dynamo-graph
自动计量 autometering;autoweighing
自动计量秤 automatic weighing scale
自动计量器 automatic ga(u)ge register;automatic measuring device;ga(u)ge register
自动计量设备 automatic measuring plant
自动计量式箱 self-measuring tank
自动计量仪 automatic ga(u)ge
自动计量装置 automatic measuring device;self-measuring device
自动计时 automatic time marking;self-clocking
自动计时的 self-timed;self-timing
自动计时风速计 self-timing anemometer
自动计时开关 automatic time switch
自动计时器 automatic timer;autotimer;recording chronometer;timing register
自动计数 auto-counting;automatic counting
自动计数系统 automatic counter system
自动计算 automatic calculation;automatic computation;automatic computing
自动计算部件 automatic counterpart
自动计算车辆到峰底连挂点距离的系统 automatic car count distance-to-coupling system
自动计算机 automatic computer;brain machine;robot brain
自动计算机器 automatic computing engine
自动计算器 automatic calculator;comptograph;robot calculator
自动计算装置 automatic computing equipment;robot brain;robot scaler
自动记力器 dynamo-graph
自动记录 autographic(al) record;automatic log(ging);automatic registration;recording
自动记录安培计 curve drawing ammeter
自动记录表 autographic(al) recorder
自动记录测高计 recording altimeter
自动记录测听计 automatic recording audiometer;self-recording audiometer
自动记录潮位计 registering tide ga(u)ge
自动记录潮位仪 automatic recording tide-ga(u)ge
自动记录秤 recording balance
自动记录磁天平 automatically recording magnetic balance
自动记录磁通计 recording fluxmeter

自动记录带 record chart
自动记录的 automatic recording;self-recording;self-registering
自动记录滴定器 automatic recording titrator;automatic recording titrimeter
自动记录电流计 automatic current recording meter
自动记录伏特表 graphic(al) recording voltmeter
自动记录伏特计 graphic(al) recording voltmeter
自动记录高程计 automatic recording altimeter;recording altimeter
自动记录高温计 autographic(al) pyrometer;automatic graphic(al) pyrometer
自动记录光度计 autorecording photometer
自动记录衡量器 weightograph
自动记录红外线光谱仪 automatic recording infrared spectrograph
自动记录机构 self-recording mechanism
自动记录计 automatic recording ga(u)ge;self-acting recording meter;self-registering ga(u)ge; writing pointer
自动记录加速器 self-recording accelerator
自动记录卡片 recording card
自动记录立体坐标量测仪 recording stereocomparator
自动记录立体坐标仪 automatic record stereo-coordinator; Stecomter <德国蔡司厂制造>
自动记录流速计 recording tachometer
自动记录秒表 self-timer
自动记录器 automatic logger; automatic recorder; automatic register; automatograph; autorecorder; autoscope; kymograph; pen-and-ink recorder; register; registering apparatus; self-recording device; self-recording unit; self-register; setting device; train grapher; writing pointer; self-recorder
自动记录器笔 recorder pen
自动记录器记录纸 recorder chart
自动记录热量计 recording calorimeter
自动记录设备 self-recording apparatus
自动记录伸长试验仪 recording extensometer
自动记录湿度计 self-recording hygrometer
自动记录式称重仪 weightograph
自动记录式磁秤 automatically recording magnetic balance
自动记录式的 graphic(al) recording
自动记录式分光光度计 automatic recording spectrophotometer
自动记录式高温计 autographic(al) pyrometer
自动记录式计算机 automatic recording calculating machine
自动记录式瓦特乏尔计 recording watt and varmeter
自动记录式仪器 self-registering instrument
自动记录式转速计 recording tachometer
自动记录水尺 automatic recording ga(u)ge
自动记录天平 recording balance
自动记录调节器 recording controller
自动记录瓦特计 graphic(al) recording wattmeter; self-recording wattmeter
自动记录位移地震仪 self-recording displacement seismograph
自动记录仪 autographic(al) recording apparatus; automatic observer; automatic recording apparatus; automatic recording instrument; automatic register; automatic registering ga(u)ge; curve drawing instrument; recorder; self-recording device; self recording ga(u)ge; self-registering apparatus; self-registering ga(u)ge
自动记录仪表 automatic graphic(al) instrument; graphic(al) instrument; self-recording meter; self-registering meter; automatic recording ga(u)ge
自动记录仪表车 recording bus; recording car
自动记录仪表用的记录带 record chart
自动记录仪校准 recorder adjustment
自动记录仪器 grapher; graphic(al) instrument; self-recording instrument; graphic(al) meter; graphic(al) recording instrument; recording instrument; self-recording apparatus
自动记录照相机 photorecorder
自动记录针 automatic recording pointer
自动记录指示器 automatic recording indicator
自动记录转速计 recording tachograph; recording tachometer
自动记录装置 automatic recording apparatus; automatic recording device; automatic recording instrument; self-recording unit
自动记录坐标 automated coordinate recording; automatic coordinate recording
自动记时装置 automatic timing device
自动继电器 automatic relay; autorelay
自动继电器式计算机 automatic relay calculator
自动加固 <集装箱> automatic fastening
自动加连字符 autohyphenation
自动加料 autofeed; automatic charging; feed control; self-feeding
自动加料机 automatic charging equipment; automatic filling machine
自动加料炉 autoladle furnace
自动加料器 automatic feeder; self-feeder
自动加煤 automatic stoking
自动加煤锅炉 automatic stoking boiler
自动加煤机 automatic stoker; mechanical stoker; stoker
自动加煤机罩 stoker casing
自动加煤炉 magazine stove
自动加煤器 automatic stoker
自动加煤(小)锅炉 magazine boiler
自动加气油站 <俚语> gasateria
自动加球机 automatic marble feeder
自动加热 self-heat(ing)
自动加湿机 automatic humidifier
自动加速表 recording accelerometer
自动加速(度) autoacceleration; automatic acceleration
自动加速度记录仪 accelerograph
自动加索引 automatic index(ing)
自动加压 follow-up pressure; self-pressurize
自动加页码 autopagination
自动加油 automatic fuel supply; automatic oiling
自动加油杯 automatic oil cup
自动加油的 self-oiling
自动加油计量泵 series gasoline meter pump
自动加油器 automatic oiler; self-oiler
自动加油旋塞 automatic oil cock
自动加油站 self-serve station
自动加纡机构 cop changer; cop loader
自动加载 autoloading
自动夹持器 <钻杆的> automatic spider
自动夹紧 automatic hold; chucking automatic
自动夹紧拔销 self-holding taper
自动夹紧的 automatic holding; self-hold(ing)
自动夹紧模具装置 automatic die clamps
自动夹紧十字叉 automatic spider
自动夹紧装置 automatic clamping device
自动夹具 automatic grab; automatic holding device
自动夹钳 automatic tongs
自动夹爪 self-gripping jaws
自动驾驶 automatic driving; autopilot; self-steering; automotive steering
自动驾驶道路 automatic driving highway
自动驾驶控制 automatic driving control
自动驾驶领航仪 autopilot navigator
自动驾驶命令 automatic driver command
自动驾驶设备 automatic driving equipment
自动驾驶随动装置 autopilot servo unit
自动驾驶同步电动机 autopiloted synchronous motor
自动驾驶系统 automatic driving system; automatic pilot system; automatic steering system; robot guiding system
自动驾驶仪 attitude control system; automatic gyroscope; automatic pilot; autopilot; co-pilot; gyropilot; pilot monitor; robot pilot
自动驾驶仪操纵器 autopilot controller
自动驾驶仪计算机 autopilot computer
自动驾驶仪控制器 autopilot controller
自动驾驶仪耦合器 autopilot coupler
自动驾驶员 autopilot
自动驾驶员辨识 automatic driver identification
自动驾驶制 automatic driving system
自动驾驶装置 automatic pilot; autopilot
自动驾驶装置控制器 autopilot controller
自动架 self-propelled mount(ing)
自动架辊器 automatic roller float device
自动架辊装置 automatic roller float device
自动架设的 self-erecting
自动间隔计时器 self-interval timer
自动间隙调整器 automatic slack adjuster
自动间歇制动器 self-acting intermittent brake
自动监测 automatic monitoring; automatic surveillance; automonitoring
自动监测器 automatic monitor; automonitor
自动监测系统 automated monitoring system; automatic monitoring system
自动监测仪 automated monitor; automonitor; unmanned surveillance equipment
自动监测装置 automatic monitoring device
自动监督 automatic supervision
自动监督程序 automonitor; automonitor routine
自动监督器 automonitor
自动监控 automatic monitoring
自动监控程序 automonitor
自动监控器 automata; automatic monitor; automaton [复 automata/automatons]; automonitor
自动监控系统 automatic monitored control system; automatic supervisory control system
自动监视器 automatic monitor; automonitor
自动监视系统 automonitoring system
自动监听器 automonitor
自动拣选机 automatic selecting machine
自动捡拾压捆机 automatic pickup baler
自动减磁 automatic field weakening
自动减(负)荷 automatic load curtailment
自动减量 autodecrement
自动减量寻址 autodecrement addressing
自动减速控制 automatic retardation control
自动减速器编组场 automatic retarder classification yard
自动减速设备 automatic retardation equipment
自动减速系统 automated retarder system
自动减压阀(门) automatic pressure-reducing valve; automatic pressure relief valve
自动减载装置 automatic load limit; automatic load limitation
自动减震悬挂装置 automatically damped suspension
自动剪板机 automatic clipper
自动检波 automatic detection
自动检波器 autodetector
自动检测 automatic check(ing); automatic detection; automatic ga(u)ging
自动检测程序 autotest(ing)
自动检测和估算系统 automatic checkout and evaluation system
自动检测和调整装置 automatic checkout and adjustment equipment
自动检测和综合跟踪 automatic detection and integrated tracking
自动检测记录仪 automatic checkout and recording equipment
自动检测技术 automatic measurement technology
自动检测设备 automatic checkout equipment
自动检测系统 automatic checkout system; automatic detecting system
自动检测仪 automatic checkout equipment
自动检测与鉴定系统 automatic checkout evaluation system
自动检测装置 automatic checkout device
自动检查 automatic inspection
自动检查测量机 automatic inspection and ga(u)ging machine
自动检查点再启动 automatic checkpoint restart
自动检查分析装置 analmatic

自动检查和准备装置 automatic checkout and readiness equipment
自动检查器 automatic monitor;robot inspector
自动检查试验装置 automatic checkout test equipment
自动检查数据存储器 <工艺过程的> automatic inspection data accumulator
自动检查与鉴定系统 automatic checkout and evaluation system
自动检出系统 automatic checkout system
自动检核 automatic check(ing);self-checking
自动检票 automatic inspection of tickets
自动检票系统 checking tickets system
自动检热器 automatic heat detector
自动检索 automatic index(ing)
自动检索法 automatic request method
自动检索系统 automated retrieval system
自动检验 automatic check(ing);automatic ga(u)ging;autoverify;machine check;self-check(ing)
自动检验的 self-verifying
自动检验电路 self-checking circuit
自动检验分选机 automatic checking and sorting machine
自动检验技术 automatic checkout technique
自动检验器 automatic checker
自动检验设备 automatic test(ing) equipment
自动检验诊断与预报系统 automatic inspection diagnostic and prognostic system
自动检验装置 automatic checkout set
自动简化视距仪 autoreducing tachymeter
自动建筑师 autoarchitect
自动键控器 autokeyer
自动键控装置 automatic keying device
自动降板机 automatic lowerator
自动降杆器 pole retriever
自动交互作用 autocorrelation
自动交换 automatic switching
自动交换分机 automatic branch exchange
自动交换机 automatic exchange;automatic switchboard; automatic switching system; machine switching system
自动交换机设备 dial system equipment
自动交换器 autochanger
自动交换情报 automatic crosstell
自动交换区 dial exchange
自动交换设备 automatic switching equipment
自动交换装置 automatic clearing apparatus
自动交通管制信号灯 automatic traffic control light signal
自动交通监测 automatic traffic surveillance
自动交通监视 automatic traffic surveillance
自动交通信号 automatic traffic signal;robot
自动浇版机 automatic plate casting machine
自动浇注装置 automatic pouring device
自动绞车 autohoist;automatic hoist;automatic winch; motor cable winch
自动绞缆机 autotension mooring winch
自动搅拌 automatic agitation
自动搅拌机 transit mixer
自动搅拌器 automatic mixer; automatic stirrer
自动搅拌装置 automatic operating batch plant; power-operated stirring gear;self-poking arrangement
自动搅抖混合饲料车 self-mixing trailer
自动校核 machine check
自动校频管 transitrol
自动校平方式 autolevel mode
自动校平装置 autolevel(l)ing assembly
自动校平状态 autolevel mode
自动校验 automatic check(ing);automatic machine;built-in check
自动校验系统 automatic checkout system
自动校验中断 automatic check interrupt
自动校验装置 automatic checking equipment
自动校正 autocalibration;autocorrection;automatic compensation;automatic correction;self-correcting
自动校正快慢仪 Ratematic
自动校正偏振仪 Repere matic
自动校正器 automatic correction device
自动校正系数 autocorrection coefficient; automatic correction coefficient
自动校正系统 automatic correction system
自动校准 automatic calibration;self-calibration
自动校准系统 self-calibrating system
自动校准装置 self-checking device
自动接地保护复原继电器 automatic ground protective reset relay
自动接合 autobond
自动接合焊机 autobond welder
自动接入 reclosure
自动接入多信道无绳电话 automatic multiple channels cordless telephone
自动接入继电器 recloser
自动接收机 automatic receptor
自动接替 <代替人工操作> automatic backup
自动接通 autochangeover
自动接线 automatic connection
自动接线器 automatic connector
自动接纸装置 automated paster;automatic paster
自动节流阀 automatic throttle(valve)
自动节流活门 autothrottle
自动节律性 autorhythmicity
自动节气门 automatic choke
自动节省燃料设备 automatic fuel economizing device; automatic fuel saving device
自动结捆式拣拾压捆机 self-tying pick-up baler
自动结算的指令 automatic tally order
自动结算机 automatic balance
自动结算转账系统 clearing house automated transfer system
自动结账系统 automatic checkout system
自动截止止回阀 automatic stop and check valve
自动解扣 trip-free
自动解缆钩 automatic releasing hook;autoreleasing hook; self-releasing hook
自动解列装置 automatic disconnecting device; automatic separation device
自动解释 automated interpretation
自动解锁 automatic release
自动解锁进路 automatically released route
自动解锁器 automatic release
自动解译 automated interpretation
自动进场 automatic approach
自动进场耦合器 automatic approach coupler
自动进出料仓系统 <材料> bin drainage system
自动进出料辊道 automatic feeding and catching table
自动进刀 autofeed; automatic feed; self-acting feed
自动进刀装置 automatic feeder
自动进给 autofeed; automatic feed; self-acting feed;self-feeding
自动进给机构 automatic feeding mechanism
自动进给控制机构 automatic feed control mechanism
自动进口限额 automatic import quota
自动进料 autofeed; automatic feed; power feed;self-acting feed
自动进料器 autofeeder;self-feeder
自动进料式粉碎机 automatic feed grinder
自动进料系统 automatic feed system
自动进料装置 automatic feeder
自动进片机构 automatic film advance mechanism
自动进气燃烧器 atmospheric induction burner;self-induced air burner
自动进气装置 automatic intake device
自动进水控制 automatic feed water control
自动进水装置 automatic intake device
自动进退刀离合器 feed clutch
自动进位 self-instructed carry
自动进样系统 automatic sample handling system
自动进钻器 autofeeder
自动浸油滚筒过滤器 automatic viscous roller filter
自动晶体切片机 automatic crystal slicing machine
自动精炼 autofining
自动精密坐标展点仪 <蔡司商品名> Cartimat
自动精磨机 autogrinding machine
自动精压 automatic coining
自动警报 automatic warning
自动警报器 autoalarm; automatic alarm
自动警报信号 autoalarm signal;automatic alarm
自动警报装置 autoalarm device
自动警戒信号继电器 automatic vigilance signalling relay
自动净化 self-purging
自动净化槽 self-purging trap
自动净化池 self-purging cell
自动净化的 self-clearing
自动净化能力 self-cleaning capacity;self-purification capacity
自动净水机 automatic purifier
自动净水器 automatic purifier
自动静噪器 automatic noise canceller
自动纠错 automatic error correction
自动纠正仪 automatic rectifier
自动救生信标 automatic rescue beacon
自动举船机 autolift
自动巨型起重机 titan crane
自动距离跟踪 automatic remote control
自动距离跟踪器 automatic range-tracking unit
自动距离控制 automatic range control;automatic remote control
自动锯台 self-acting saw bench
自动聚合 automatic polymerization;autopolymerization
自动聚焦 autofocus;automatic focus
自动聚焦纠正仪 autofocus rectifier
自动聚焦雷达投影仪 autofocus radar projector
自动聚焦投影仪 autofocus reflecting projector
自动聚焦作用 automatic focusing action
自动卷带 automatic coiling of tape
自动卷回遮篷 automatic roll-up awning
自动卷帘式门控制器 automatic shutter door operator
自动卷片 automatic film advance
自动卷片照相机 magazine camera
自动卷染机 autojig;automatic dye jig
自动卷绕式过滤器 automatic roll filter
自动卷绕式空气过滤器 moving curtain air filter;roll-type air filter
自动卷条装置 roll former
自动卷旋函数 autoconvolution function
自动卷扬机 automatic hoist;automatic winch;motor cable winch
自动掘进控制机构 automatic feed control mechanism
自动均衡 automatic equalization
自动均衡器 automatic equalizer
自动均衡前置放大器 self-equalizing preamplifier
自动卡盘 automatic chuck
自动卡盘车床 automatic chucking lathe;automatic chucking machine
自动卡瓦 power slips
自动卡芯器 automatic core-breaker
自动开闭窗 automatic window
自动开闭式夹纱器 self-releasing gripper
自动开动的 automatic starting
自动开关 autolay;automat;automatic circuit breaker;automatic cut-out;automatic interrupter; automatic switch;automaton[复 automata/automatons];autoswitch;battery cut-out;keying
自动开关板 automatic switchboard;automatic switching panel
自动开关窗 automatic window
自动开关地弹簧 floor springs and checks
自动开关刮水器 self-parking wiper
自动开关门 <光电控制的> autodoor
自动开关器 autotimer
自动开关式喷水灭火系统 fire cycle system
自动开关式燃烧器 automatic on-off type burner
自动开关室内消防洒水器 indoor on-off sprinkler
自动开关释放器 trip coil
自动开关(双扇)风门 self-acting door
自动开关装置 automatic closing device;automatic operator;recloser
自动开关组合 automatic switch unit
自动开合板牙 opening die
自动开合螺丝板牙切头 self-opening die head
自动开环控制 automatic open loop control
自动开门器 automatic opener
自动开门装置 automatic door operator;automatic door unit
自动开启的 self-opening

自动开启的门 automated (opening) door

自动开启闸门 automatic (opening) gate

自动开塞机 autopour

自动抗干扰电路 automatic antijam circuit

自动可转动打磨机 automatic turning sander

自动刻螺纹 tapping screw

自动刻图 automatic scribing

自动客票检查机 automatic ticket inspection machine

自动空气断路器 automatic air circuit breaker

自动(空)气阀 automatic air valve

自动空气加热器 autoair heater; automatic air heater

自动空气开关 automatic air-break switch

自动空气调湿装置 humidostat

自动空气循环供暖 gravity air heating

自动空气压缩机 automatic air compressor

自动空气制动机 automatic air brake

自动空气制动器 automatic air brake

自动空调器 self-contained air conditioner

自动空心纤维超滤法 automated hollow fibre ultrafiltration method

自动空中交通管理 automatic air traffic control

自动空重车制动机 automatic empty-and-load brake

自动空重车制动器 automatic empty-and-load brake

自动控制 autocontrol; automatic control; automatic regulating; automatic regulation; cybernation; self-acting control; self-action control; self-governing; self (-operated) control; self-regulation; automatic weight control <钻头压力的>

自动控制搬运车 automatically controlled transfer car

自动控制测距仪 self-contained range finder

自动控制测量系统 robotic surveying system

自动控制秤 automatic control balance; check weigher

自动控制的平底绘图仪 auto-troll flat bed plotter

自动控制的轧机 automatically controlled rolling mill

自动控制的制动铁鞋 skate with automatic control

自动控制电路 automatic control circuit

自动控制发电机 automatic generator

自动控制发生器 automatic generator

自动控制阀 autocontrol valve; automatic(ally) control(led) valve

自动控制阀体 automatic control valve body

自动控制浮子放水阀(塞) automatic floater-controlled bleeder tap

自动控制杆 automatic control rod

自动控制高程 automatic grade control

自动控制给料器 automatic feeder control

自动控制工程学 automatic control engineering

自动控制航向 automatic steering

自动控制烘干机 automatic drier[dryer]

自动控制弧长的钨极电弧焊 controlled tungsten-arc welding

自动控制环境畜舍 controlled environment house

自动控制机床 automatic control machine tool

自动控制集电器 automatically controlled current collector

自动控制技术 automatic control engineering

自动控制交通系统 automatic controlled transportation system

自动控制框图 automatic control block diagram

自动控制理论 automata theory; automatic control theory; theory of automatic control

自动控制螺旋给进 automatically controlled screw feed

自动控制盘 automatic control board

自动控制频率响应 automatic control frequency response

自动控制平底绘图仪 autotruck control flat bed plotter

自动控制评价系统 automatic control evaluation system

自动控制气压灌肠器 self-control enemator

自动控制气闸 automatic control damper

自动控制器 automat; automatic controller; automatic regulator; controller regulator; open and shut controller; over-and-under controller; pilot-operated controller; self-actuated controller; self-controller; self-operated controller; single contact controller

自动控制器的差示 differential of a controller

自动控制牵引装置 automatic traction equipment

自动控制设备 automatically controlled plant

自动控制设定值 set point desired value

自动控制式受电弓 automatically controlled pantograph

自动控制受电弓 automatically controlled current collector

自动控制受电器 automatically controlled current collector

自动控制瞬时分析 automatic control transient analysis

自动控制伺服阀 automatic control servo valve

自动控制台 automatic control panel

自动控制稳定性 automatic control stability

自动控制误差系数 automatic control error coefficient

自动控制系统 auto-control system; automatic control system; auto-troll system; non-stick system of control; regulating system; robot control system

自动控制消防泵 automatically controlled fire pump; automatic fire pump

自动控制畜舍 controlled house

自动控制学 automatics; robotics

自动控制学说 theory on autocontrol

自动控制仪表 automatic control instrumentation

自动控制用热离子变阻器 evatron

自动控制运输系统 automatic controlled transportation system

自动控制暂态分析 automatic control transient analysis

自动控制轧机 electronically operated mill

自动控制指示板 autopanel

自动控制制 non-stick system of control

自动控制专用组件 professional group on automatic control

自动控制装置 automatic control assembly; automatic control device; automatic control equipment; automatic control gear

自动口料 automatic charging

自动扣 ball catch; bullet catch

自动库存调度系统 automatic inventory dispatching system

自动库存控制 automated stock control

自动快门 automatic gate; automatic shutter; self-opening gate

自动快速返回 automatic fast return

自动快速曝光 instantaneous exposure

自动框架螺钉 screw for framework of automatic device

自动馈给装置 autofeeder

自动捆结式捡拾压捆机 self-tying baler

自动捆扎 self-tying

自动捆扎机 automatic flexible strapping machine

自动捆扎装置 automatic bonding unit

自动扩口接头 self-flaring fitting

自动扩展的 self-propagating

自动扩展式点火器 expanding pilot

自动拉出器 automatic puller

自动拉出用绞车 self-recovery winch

自动拉紧设备 automatic tensioning equipment

自动拉平 autoflare

自动栏木 automatically controlled gate; automatic barrier; automatic crossing gate

自动栏木道口 level-crossing with automatic barrier

自动雷达 automatic radar; robot radar

自动雷达标绘仪 automatic radar plotting aids

自动雷达跟踪系统 automatic radar tracking system

自动雷达航站系统 automated radar terminal system

自动雷达绘图仪 automatic radar plotter apparatus

自动雷达控制与数据设备 automatic radar control and data equipment

自动雷达数据测量设备 automatic radar measuring equipment

自动雷达信标 automatic radar beacon

自动雷达信标定序器 automatic radar beacon sequencer

自动雷达终端系统 automated radar terminal system; automatic radar terminal system

自动雷管 automatic detonator; automatic primer

自动冷锻机 automatic cold former

自动冷凝作用 self-refrigeration

自动冷却 self-cooling

自动冷热试验计 autocalorimeter

自动离合器 automatic clutch; automatic governor; self-acting clutch

自动离职 self-dimission; dragging-up <俚语>

自动力 ultromotivity

自动立体测图 automatic stereocompilation; automatic stereomapping

自动立体测图系统 automatic stereomapping system

自动立体测图仪 automatic stereo compiler; automatic stereomapping instrument; automatic stereo-plotter; automatic stereoplotter; stereoautograph; Aeromat <商品名>

自动立体断面记录仪 automatic stereo-profiler; autostereo-profiler

自动立窑 automatic shaft kiln

自动励(磁)作用 automatic excitation

自动沥青喷洒机 <装在汽车上的> truck-mounted pressure distributor

自动例行测试器 automatic routine apparatus

自动例行程序 automatic routine

自动粒度分析 automatic particle size analysis

自动粒度分析器 automatic particle size analyser[analyzer]

自动粒径分析仪 automated particle size analyser[analyzer]

自动粒子计数器 automatic particle counter

自动连杆 automatic pitman

自动连接 autolink; automatic connection; self-coupling; snap coupling

自动连接轮挡 automatically engaging catch

自动连接器 automatic connector

自动连接器助接 emergency head backup connection

自动连续汇报制 automatic serial reporting system

自动联结器 autocoupler; self-couplin

自动连续监测系统 automatic and continuous monitoring system

自动连续卷布机 continuous batching machine

自动连续空气监测系统 automatic and continuous air monitoring system

自动连续输送设备 automated continuous conveying equipment

自动联合器环 pick-up ring

自动联合收割机 self-propelled combine harvester

自动联机 automatic log-on

自动联结 automatic connection; automatic coupling

自动联结器 automatic coupling; self-coupling; self-coupling hitch; snap coupling

自动联结装置 automatic coupling device; self-coupling device

自动联锁 automatic interlock; self-blocking

自动联锁法 automatic interlocking

自动联锁系统 emergency trip wire system

自动联锁装置 automatic interlock; automatic interlocking device

自动联轴节 automatic coupler

自动链接 autobond

自动链条式制榫机 automatic chain mortising machine

自动亮度反差调整 automatic brightness contrast control

自动亮度控制 automatic brightness control

自动亮度调整 automatic brightness control

自动亮度限制电路 automatic brightness limiter circuit

自动量 automatic measurer

自动量测 automated measurement; automatic measurement

自动量程选择器 automatic range selector

自动量程选择数字显示 autoranging digital display

自动量度仪 automatic measuring machine

自动量计 automatic ga(u)ge

自动量水表 <混凝土搅拌机的> automatic water ga(u)ge

自动量水装置 automatic water measuring device

自动量算 automated measurement

自动列车 driverless train
自动列车进路 automatic routing system
自动列车运行 automatic train operation
自动裂化 spontaneous fission
自动灵敏度校正 automatic sensitivity correction
自动灵敏度控制 automatic sensibility control
自动零点调整电路 automatic balancing circuit
自动零回路 autozero loop
自动零位器 nullifier
自动领航仪 autonavigator;aviograph;robot navigator;automatic pilot
自动溜放 automatic shunting
自动(溜放车辆的)保持距离间隔 automatic distance keeping
自动流程图 autoflow chart
自动流程图程序 autochart
自动流出的 artesian
自动流量观测站 automatic discharge-observation station
自动流量计 automatic discharge ga(u)ge
自动流量控制 automatic flow control
自动硫化 auto-vulcanization;self-curing
自动硫化胶 self-curing cement
自动硫化胶料 auto-vulcanizing stock
自动馏分收集器 automatic fraction collector
自动六分仪 automatic sextant
自动六角车床 automatic turret lathe
自动楼梯 motorstair;mover stair(case);moving stair(case);moving stairway;travel(l)ing stair(case)
自动楼梯百叶门 escalator shutter
自动楼梯布置 escalator arrange
自动楼梯交叉布置 criss-cross arrangement of moving stair(case)
自动楼梯平行布置 parallel arrangement of moving stair(case)
自动楼梯隧道 escalator tunnel
自动楼梯通道 inclined motor stair(case) shaft
自动楼梯斜道 inclined escalator shaft;inclined moving stair(case) shaft
自动楼梯照明计时开关 time switch for automatic staircase lighting
自动漏量控制 automatic leak control
自动炉门零件 automatic fire door parts
自动炉排 Robot fireman
自动炉子 automatic furnace
自动录波器 automatic recording oscillograph
自动滤气器 automatic air filter
自动路径选择 automatic routing
自动路面弯沉仪 automatic Benkelman beam apparatus
自动路碾 self-propelled roller
自动路由设置 automatic route setting
自动路由选择 automatic route selection
自动铝盖卷边机 automatic alumin(i)um cap fitting machine
自动率定 autocalibration
自动轮 independent axle
自动轮毂 autohub
自动轮胎打气泵 automatic tire[tyre] pump
自动轮胎式间隙式铺路机 self-propelled rubber-mounted twin batch paver
自动轮询 automatic poll
自动螺帽扳手 nut-runner
自动螺丝扳手 nut-runner
自动螺丝车床 automatic screw machine
自动螺纹车床 automatic screw lathe
自动落锤 automatic trip-hammer(monkey)
自动落辊器 automatic roller dropping device
自动落辊装置 automatic roller dropping device
自动落锥探测 <动力圆锥触探> automatic ram sound
自动摞包机 automatic palletizer
自动埋弧电碴焊 automatic submerged-slag welding
自动埋弧焊 automatic submerged-arc welding
自动脉冲 autopulse
自动脉冲水表 autoimpulse water meter
自动脉冲调制发射机 self-pulsed transmitter
自动脉宽控制单稳电路 automatic width control one-shot
自动铆(钉)机 automatic riveter
自动铆(钉)枪 automatic riveting gun;pom-pom;rivet gun
自动煤秤 coal meter
自动门 automated door;automatic door
自动门底防风隔声附件 automatic door bottom
自动门底防风隔声设施 automatic door bottom
自动门感应设备 automatic door sensing equipment
自动门钩 automatic door hook;automatic door stay
自动门槛 automatic door bottom;automatic threshold door reopening closer
自动门开关 automatic door switch
自动门开启装置 automatic shutter door operator
自动门控制器 automatic door operator
自动门扣 catch bolt
自动门闩 snap-on bolt
自动门栓 snap-on bolt
自动门锁栓 automatic door lock;automatic gate lock
自动门再启装置 automatic door reopening device
自动密封垫 automatic door seal;automatic gland packing
自动密封接头 self-sealing coupling
自动密封生产线 automatic sealing line
自动密封式快速接头 self-seal type quick-joint
自动免疫 autoimmunity
自动免疫性 active immunity
自动面积计 automatic area meter
自动描绘等高线 automatic contouring
自动描绘器 automatic plotter
自动描迹仪 automatic track plotter
自动描图系统 autotracing system
自动瞄准 automatic aiming;homing control
自动瞄准仪 autocollimator
自动灭磁 automatic field suppressing
自动灭磁开关 suicide contactor
自动灭磁装置 automatic excitation cut off device
自动灭火 self-extinguish(ing)
自动灭火花机 automatic spark extinguisher
自动灭火机 automatic fire sprinkler
自动灭火喷淋器 automatic fire sprinkler
自动灭火喷头 automatic fire sprinkler;automatic sprinkler head
自动灭火区 sprinklered
自动灭火系统 automatic fire protection system;automatic fire suppression system
自动灭火装置 automatic fire sprinkler
自动模 automatic mo(u)ld;transfer die
自动模锻压力机 automatic stamping press
自动模拟呼叫器 automatic call simulator
自动膜片阀 diaphragm automatic valve
自动摩擦铁鞋式缓行器 automatic friction skate retarder
自动摩擦铁鞋式减速器 automatic friction skate retarder
自动磨板及磨瓷砖机器 automatic slab and tile grinder
自动磨床 autogrinding machine;self-acting grinding machine;self-grinding machine
自动磨刀片装置 self-sharpening device
自动磨盖板机 automatic flat grinding machine
自动磨锐 self-sharpening
自动(木材)检尺机 autoscaler
自动木栅 boomer
自动目标选择 automatic target selection
自动内部诊断 automatic internal diagnosis
自动内插 automatic interpolation
自动内燃夯土机 automatic compactor
自动能量控制 automatic energy control
自动逆流分配器 automatic counter flow distributer
自动逆向阀 automatic non-return valve
自动逆止阀 automatic non-return valve
自动逆转器 automatic reverser
自动黏[粘]度计 automatic viscometer
自动黏[粘]度控制器 automatic viscous controller
自动黏[粘]结的 self-binding
自动黏[粘]液过滤器 automatic viscous filter
自动凝固 self-solidifying
自动凝结 autocoagulation
自动凝聚 autocoacervation;autocoagulation
自动扭绞 autolay
自动耦合 auto-coupling
自动耦合器 autocoupler
自动耦合设备 automatic coupling device
自动耦合装置 automatic coupling device
自动拍发器 autokeyer
自动排版 automatic type setting
自动排尘装置 automatic dust ejector
自动排除 automatic exclusion
自动排放 automatic discharge;automatic drainage
自动排风活门 automatic exhausting window
自动排浇机 Monotype
自动排缆绞车 self-stowing mooring winch
自动排料 self-discharging
自动排路 automatic route setting
自动排齐 self-aligning
自动排气 automatic air release
自动排气的 self-bleeding
自动排气阀 automatic air bleed valve;automatic ball valve;automatic exhaust air valve;automatic exhaust steam valve
自动排气分析器 automatic exhaust gas analyser[analyzer]
自动排气管 automatic escape;automatic exhaust air pipe
自动排气过滤器 automatic exhaust gas filter
自动排气活门 automatic air exhausting window
自动排气口 automatic air vent
自动排气门 automatic pressure vent
自动排水 self-bailing;self-draining
自动排水泵 float-controlled drainage pump
自动排水阀 automatic discharge valve
自动排水管 automatic discharge pipe;self-draining pipe
自动排水器 automatic water discharger
自动排水设施 automatic flushing system
自动排水装置 automatic bailer;self-draining arrangement
自动排污阀 automatic blowdown valve
自动排污管 automatic sludge discharge pipe
自动排污系统 automatic blowdown system
自动排烟口 automatic fire vent;automatic smoke vent
自动排烟天窗 automatic fire vent;automatic smoke(or fire) vent
自动排液分离器 automatic drain separator
自动排渣净油机 self-discharging purifier;self-opening purifier
自动排整插棒式预选机 <旧式史端桥自动电话机> self-aligning plunger line switch
自动盘车装置 automatic barring gear
自动盘存控制 automatic inventory control
自动判别 automatic discrimination
自动判读 automated interpretation
自动判读技术 computer assisted interpretation technique
自动旁录 automatic intercept
自动旁路 autoby pass
自动旁通阀 automatic bypass valve
自动旁通滤油器 self-bypassing filter
自动抛锚装置 automatic anchoring device
自动刨版机 autoshaver
自动刨床 <木工用> automatic surfacer
自动跑风门 automatic air vent
自动赔偿 <行李包裹、货物破损时> voluntary indemnity
自动配电站 automatic substation
自动配合 self-consistent
自动配衡的 automatically tared
自动配料拌和 automatic batch mixing
自动配料称量 automatic batch weighing
自动配料秤 automatic weigh batcher
自动配料的 self-proportioning
自动配料混合 automatic batch mixing
自动配料计量器 automatic batcher
自动配料控制 automatic proportioning control
自动配料器 automatic batcher
自动配料系统 automatic batching system
自动配料装置 automatic batching system
自动配平离合器 autotrim clutch
自动配气阀 automatic valve
自动配水槽 dosing tank
自动喷粉器 automobile duster
自动喷灌机 automatic sprinkler
自动喷净回转台 automatic shot blasting rotating table

自动喷淋器 motor-sprinkler
自动喷漆机 automatic spraying machine
自动喷泉 automatic geyser; self-operating fountain
自动喷洒灭火装置 automatic fire sprinkler
自动喷洒器 automatic sprinkler
自动喷洒系统 automatic sprinkler system
自动喷射机 self injector
自动喷射器 self-acting injector
自动喷射速度调节阀 automatic speed change valve
自动喷射调整阀 automatic jet control valve
自动喷水灭火机 automatic water spraying fire extinguisher
自动喷水灭火器 sprinkler; sprinkler (fire) extinguisher
自动喷水灭火区 sprinklered
自动喷水灭火系统 automatic sprinkler system; autosprinkling system; motor-sprinkler system
自动喷水灭火装置 automatic fire-sprinkler system
自动喷水器 automatic sprinkler
自动喷水器立管 automatic sprinkler riser
自动喷水系统 automatic water sprinkler system
自动喷水消防系统 preaction sprinkler system; sprinkler system
自动喷涂 automatic spraying
自动喷涂机 automatic coating machine
自动喷涂装置 automatic spray apparatus
自动喷雾器 automatic sprayer; automobile sprayer
自动喷油车 tank truck
自动喷油定时调整机构 automatic injection timing adjuster
自动喷油阀 automatic fuel spray valve
自动喷油调节器 fuel oil injection self-regulator
自动喷嘴 automatic nozzle
自动膨胀阀 constant pressure expansion valve; automatic expansion valve
自动碰撞检测与避撞系统 automatic collision detection and avoidance system
自动批量混合法 automatic batch mixing
自动匹配 automatic matching
自动偏心夹紧卡盘 work driver
自动偏压 autobias; automatic bias
自动偏压补偿 automatic bias compensation
自动偏压截止 self-biased off
自动偏移计算机 automatic traverse computer
自动偏振光镜 auto polariscope
自动偏置 autobias
自动漂浮测站 automatic floating station
自动贫化燃烧混合物 autolean mixture
自动频率分析器 automatic frequency analyser[analyzer]
自动频率校正 automatic frequency correction
自动频率控制 automatic frequency control
自动频率调谐器 automatic frequency tuner
自动频率调整振荡器 automatic frequency control generator
自动频率微调 automatic frequency fine control; automatic trimming
自动频率微调管 transitrol
自动频率稳定 automatic frequency stabilization
自动频谱分析仪 automatic spectrum analyser[analyzer]
自动平舱舱口 self-trimming hatchway
自动平舱船(舶) self-trimming ship; self-trimming vessel
自动平舱运煤机 self-trimming collier
自动平地机 motor-driven (blade) grader; motor grader; motorized road grader; self-propelled blade grader; self-propelled blader; self-propelled grader
自动平衡 automatic balance; self-balanced; self-balancing; self-equilibrating; self-equalizing; self-poise
自动平衡保护 self-balance protection
自动平衡倒相器 self-balancing phase inverter
自动平衡电路 automatic balancing circuit
自动平衡电位计 self-balancing potentiometer
自动平衡机构 autobalance
自动平衡记录器 self-balancing recorder
自动平衡起重机 autobalancer crane
自动平衡器 autobalancer
自动平衡式辐射热计 self-balancing bolometer
自动平衡式记录器 self-balancing type recorder
自动平衡水柜 self-trimming tank
自动平衡脱水机 self-balancing extractor
自动平衡压力计 autobalance manometer; automatic balance manometer
自动平衡仪表 automatic balance instrument; self-balancing instrument
自动平衡应力 self-balancing stress
自动平路机 motor grader; self-propelled blader
自动平磨机 automatic muller
自动平行机构 automatic parallel holding
自动凭证闭塞机 automatic token instrument
自动凭证机 automatic token instrument
自动坡道 moving ramp
自动破碎设备 self-propelled crushing plant
自动铺料机 motopaver
自动铺路机 motopaver
自动铺砌机 motopaver
自动铺砂机 self-propelled gritter
自动铺纸机 automatic paper interleaving machine
自动铺筑机 motopaver
自动普通车床 automatic engine lathe
自动曝气水流 self-aerated flow
自动启闭电路装置 automatic circuit recloser
自动启闭闸板 automatic flash board
自动启闭栅 automatic crossing gate
自动起爆 self-cocking action
自动起爆器 automatic primer; self-primer
自动起爆装置 automatic primer; self-primer
自动起钓机 autoline hauler
自动起动 automatic priming; automatic start (-up); autostart; self-start; self-prime
自动起动泵站 automatic starting pump station
自动起动的 automatic starting; self-starting
自动起动法 self-starting method
自动起动机 automatic starter; self-starter
自动起动机环形齿轮 self-starter ring gear
自动起动控制 automatic starting control
自动起动器 automatic starter; autostarter; self-starter
自动起动器齿环 self-starter ring gear
自动起动设备 automatic starting equipment; automatic trigger
自动起动水泵 automatic starting pump
自动起动同步电动机 self-start synchronous motor
自动起动线路 automatic trigger
自动起动小齿轮 self-starter pinion
自动起动延迟 autostart delay
自动起动注油 self-priming
自动起动注油器 automatic primer
自动起动装置 automatic starter; automatic starting device; self-starter
自动起落机构 power lift
自动起落犁 self-lift plow
自动起落器 self-lift
自动起落器离合操纵杆 power trip lever
自动起重机 automatic crane
自动起重小吊车 automatic crab
自动气动控制 automatic pneumatic control
自动气阀 automatic air brake; relief valve
自动气割 automatic gas cutting
自动气割机 automatic gas cutting machine
自动气化器 automatic carburetor
自动气刨附加装置 autogenous gouging attachment
自动气体发生器 automatic gas producer
自动气体分析器 automatic gas analyser[analyzer]
自动气体进样阀 automatic gas sampling valve
自动气相色谱仪 automatic gas chromatography
自动气象观测 automatic meteorological observation
自动气象观测系统 automatic meteorological observation system
自动气象观测站 automatic meteorological observation station; automatic meteorological observing station
自动气象台 automatic meteorological station; automatic weather station
自动气象系统 automatic meteorological system; automatic weather system
自动气象站 automatic meteorological station; automatic weather station
自动气压虎钳 automatic air vice [vise]
自动气压计 recording barometer
自动气闸 automatic ah brake; automatic air brake
自动气胀式救生筏 self-inflating lifeboat
自动汽泵 automatic tire[tyre] pump
自动砌墙机 automason
自动器 automation
自动千分尺 automicrometer
自动牵引控制设备 automatic traction control equipment
自动前端重力翻车机 car tipper
自动枪 pom-pom
自动强度控制 automatic strength control
自动强化 autogenous healing; self-healing
自动强行励磁 automatic field forcing
自动切齿机 automatic gear cutting machine
自动切除浇口 automatic despurcing
自动切断 automatic break; automatic cut-out; automatic disconnection; automatic release; automatic shut-off
自动切断阀门 automatic shut-off valve
自动切断工具 automatic shut-off tool
自动切断加热连接 automatic release heating coupling
自动切断装置 automatic disconnecting device
自动切割机 automatic cutter; automatic cutting machine
自动切割器 automatic cutter
自动切换 autochangeover; automatic switch over; autoswitch over
自动切换开关 automatic transfer switch
自动切换设备 automatic transfer equipment
自动倾侧槽 automatic tipper
自动倾侧闸门 automatic tilting gate
自动倾倒槽 automatic tipper
自动倾倒式 automatic tipping; self-tipping
自动倾翻 power-operated skip
自动倾翻车 air-dump wagon
自动倾翻卸货机 tiltster
自动倾式堰 balanced weir
自动倾卸槽 automatic tipper
自动倾卸车 automatic tripping car; car dumper; dumper; dump (ing) car; dump (ing) wagon; hopper wagon; self-tipping lorry; self-tripping car; tipper; tipping lorry; tipping truck; tripping car; self-dumping car
自动倾卸车身 tripper body
自动倾卸的 self-dumping
自动倾卸机构 automatic tipper
自动倾卸卡车 automatic dump truck; dump lorry; dump truck; tip truck
自动倾卸汽车 dumper car; motortilter
自动倾卸器 automatic tipper
自动倾卸升降机 tippler lift
自动倾卸石渣车 automatic ballast tipping wagon
自动倾卸式半拖车 dump semi-trailer
自动倾卸式货车 damp-body truck
自动倾卸式输送机 tripper conveyer [conveyor]
自动倾卸式拖车 dump trailer; dump carrier
自动倾卸输送机 tripper
自动倾卸输送机廊道 tripper-conveyer gallery
自动倾卸装置 tipple
自动清偿贷款 self-liquidating loan
自动清除 automatic clear; self-cleaning; self-clearing
自动清除输入 autoclear input
自动清除装置 automatic clearing apparatus
自动清灰装置 automatic ash remover
自动清洁的火花塞 self-cleaning spark plug
自动清洁的履带 self-cleaning track
自动清洁器 automatic cleaner
自动清洁式油箱 self-cleaning tank
自动清洗毛机 automatic self-cleaning wash bowl
自动清丝装置 autoclipping apparatus
自动清污栅(网) self-cleaning screen
自动清晰度控制 automatic resolution

control
自动清洗 automatic cleaning; self-cleaning; self-clearing
自动清洗过滤器 automatic self-cleaning strainer
自动清洗离心机 self-cleaning centrifuge
自动清洗软管过滤器 automatic cleaning flexible filter
自动清洗式双螺旋反应器 self-cleaning twin-screw reactor
自动清洗转筛过滤器 self-cleaning rotation screen filter
自动清洗装置 automatic cleaning machine; automatic flushing device
自动情报数据系统 automatic intelligence data system
自动请求检索 automatic request for repetition
自动请求重发 automatic request for repetition
自动请求重复发送 automatic request for repetition
自动球阀 automatic ball valve
自动区截制 automatic block system
自动曲线跟踪装置 automatic curve follower
自动曲线绘算器 automatic curve plotter
自动曲线绘制器 variplotter
自动取款机 automatic teller machine
自动取平的 automatic level(l)ing; self-level(l)ing
自动取消 automatic cancellation
自动取样 automatic sampling
自动取样分析器 automatic sampling analyser[analyzer]
自动取样计数器 automatic sampler-counter; automatic sampling counter
自动取样器 automatic sampler; autosampler
自动取样设备 automatic sampling equipment
自动取样系统 automatic sampling system
自动取样装置 automatic sampling device
自动去皮重 autotare
自动去载 automatic unload
自动确定 automatic diagnosis
自动确认 automatic acknowledg(e)ment
自动燃料供给系统 automatic fuel(l)ing system
自动燃烧风门板 automatic fire shutter
自动燃烧风门片 automatic fire shutter
自动燃烧控制 automatic combustion control
自动燃烧控制炉子 automatic furnace
自动燃烧控制装置 automatic combustion control system
自动燃烧器 automatic burner
自动燃烧设备 automatic burning appliance; automatic burning equipment
自动燃烧系统 autocombustion system
自动燃烧装置 automatic burning appliance; automatic burning equipment; automatic fire device; automatic furnace
自动绕栅机 grid lathe
自动热水器 automatic water heater
自动热水箱 automatic heating tank
自动人 self-actualizing man
自动人行带的槛板 moving walk threshold plate
自动人行道 moving sidewalk; pallet-moving walk; passenger conveyer [conveyor]; moving ramp; moving walkway
自动人行道系统 electric(al) walk system; moving walk system
自动认址【机】 autoidentify
自动绒集作用 autoflocculation
自动溶解器 automatic dissolver
自动熔封器 automatic melt-sealer
自动软化水厂 automatic softening plant
自动软件工程 automated software engineering
自动软水器 automatic water softener
自动软水装置 automatic softening installation; automatic water softener
自动润滑 self-lubricate; self-lubrication; self-oil
自动润滑的 self-lubricated; self-lubricating
自动润滑法 automatic lubrication
自动润滑机理 mechanism of self-lubrication
自动润滑器 automatic lubricator; self-acting lubricator
自动润滑式链轨支重轮 self-lubricating lower track roller
自动润滑循环 automatic greasing cycle
自动润滑脂枪 power-operated grease gun
自动润滑轴承 oilless bearing; self-lubricating bearing; self-oiling bearing
自动润滑轴瓦 oilless bushing
自动撒布机 automatic distributor
自动撒砂 automatic sanding
自动洒水车 tank truck
自动洒水机 automatic sprinkler
自动洒水灭火装置 automatic fire sprinkler; automatic sprinkler system
自动洒水喷头 sprinkler head
自动洒水器 automatic sprinkler
自动洒水系统 automatic water system
自动洒水消防系统 automatic sprinkler system
自动洒水装置 automatic watering system
自动塞 automatic plug
自动塞棒机 autopour
自动三元电子扫描阵列 automatic three-dimensional electronic scanned array
自动扫描 automatic scan(ning)
自动扫描电路 automatic search circuit
自动扫描器 automatic scanner
自动扫描望远辐射计 autoscan teleradiometer
自动扫描远距离光谱辐射计 automatic scanning telespectroradiometer
自动色度控制 automatic chroma control
自动色度控制电路 automatic chrominance control circuit
自动色度控制误差 automatic chroma control error
自动色度信号电路 autochroma circuit
自动色谱扫描器 automatic chromatogram scanner
自动刹车 autobrake; self-acting brake
自动砂料撒布机 automatic sand distributor
自动砂轮整形器 automatic wheel truer
自动砂轮整形装置 automatic wheel truing device
自动闪光灯 automatic flash light
自动闪烁装置 flasher
自动上带装置 hypertape knit
自动上发条 self-wind
自动上发条的 <钟、表等> self-winding
自动上浮采泥器 free fall rocket core sampler
自动上料移动式运输机 self-feeding portable conveyer[conveyor]
自动上升趋势 tuck-up
自动上锁装置 automatic locking mechanism
自动上舷快门 single action shutter
自动上油的 self-oiling
自动上釉 self-glazing
自动舌瓣闸门 automatic flap gate
自动设备 autoequipment; automatic equipment; automatic plant; autoplant; robot equipment; unattended equipment
自动设备识别 automated equipment identification
自动设计 autodesign; automadesign; automatic design; design automation
自动设计程序 autolayout; automatic coding; codified procedure
自动设计工程学 automated design engineering; automatic design engineering
自动设计工具 automated design tool
自动设计技术 automated design engineering; automatic design technique
自动设计中心 automatic design center[centre]
自动设置 automatic setup
自动射线照相 autoradiograph; radio autogram
自动射线照相术 autoradiography; radionautography
自动摄冲机 photomaton
自动摄影机 automatic camera; electric(al) eye camera; photomaton; serial camera; strip-film camera
自动伸缩活塞环 self-expanding piston ring
自动伸缩推进机 <凿岩机的> step feeder
自动升板机 automatic elevator
自动升降机 autohoist; autolift; automatic elevator; automatic hoist; automatic winch; motor cable winch
自动升降犁 self-lift plough
自动升降梯 moving stair(case)
自动升降椅 <沿楼梯的> inclinator
自动升降装载车 elevating wagon
自动升降装置 auto-elevator device
自动升速装置 automatic speed run-up equipment
自动升温 automatic rise temperature
自动升压调整 automatic booster control
自动升运装载车 elevating wagon
自动生产 automatic production
自动生产管理 automated production management
自动生产计划系统 automatic production planning system
自动生产线 automatic production line
自动生产线监督 automatic line supervision
自动生成 automatic generation
自动生成剖面 automatic profiling
自动生氧器 autogenor
自动声道转换装置 automatic track shift
自动声响浮标 automatic sounding buoy
自动剩余静校正 autoresidual static correction
自动失效 automatic avoidance
自动失效闸板 flashboard
自动湿度滴定器 automatic moisture titrator
自动湿度计 automatic moisture meter; automatic recording hygrometer
自动湿度控制器 automatic humidity controller
自动湿度调节器 automatic humidity controller
自动石板瓷砖研磨机 automatic slab and tile grinder
自动石板研磨机 tile grinder
自动时标 automatic time marking
自动时差电路 automatic time difference circuit
自动时间校正器 automatic timing corrector
自动时间校准器 automatic timing corrector
自动时控存储特征 automatic timed save feature
自动时序操作 automatic sequence operation; automatic sequential operation
自动时序控制 automatic sequence control
自动识别 automatic diagnosis; automatic identification; automatic recognization
自动识别目标技术 automatic target recognition
自动识别设备 automatic identification equipment
自动识别系统 automatic identification system
自动识别箱号装置 systems for the identification of numbers
自动识别装置 automatic identification equipment
自动实时(数据)处理 automatic real time processing
自动拾取剖面 autopicking section
自动蚀版机 automatic stencil etching machine
自动食品售货机 automatic food vending machine
自动示波器 automatic oscillograph
自动示数装置 automatic indexing device
自动式 self-acting
自动式播音记录装置 audiometer[audiometre]
自动式车钩 automatic coupler
自动式底层水取样器 free vehicle bottom water
自动式混料计量斗 batcher mixture scale
自动式计量给料拌和机 gravity batch mixer
自动式计量器 batcher scale
自动式磨粉机 automatic grinding mill
自动式喷洒沥青机 motorized tar spreader
自动式牵引缓冲装置 automatic buffering and draw coupler
自动式人行道 moving pedestrian
自动式铁丝打结装置 automatic wire tying device
自动式装卸设备 mobile handling equipment
自动事故信号装置 automatic fault signal(l)ing
自动事件检测器 automatic event detector
自动视距仪 automatic tach(e)ometer; automatic tachymeter
自动视准法 autocollimation method
自动视准仪 autocollimator
自动试验设备 automatic test(ing) equipment

Z

自动室 autoroom
自动拭浆机 automatic packing machine
自动适应地形的链耙 self-relieving chain harrow
自动释放 automatic release; trip-free release
自动收报机 pen-and-ink recorder; tape-recording machine; ticker[tikker]
自动收发 automatic send(-and)-receive
自动收发器 automatic send-receive unit
自动收发设备 automatic send receiver
自动收放拖缆绞车 self-rendering towing winch
自动收费 automatic fare collection
自动收费的 automatic pay
自动收费公用电话 pay telephone
自动收费管理 automatic tolling enforcement
自动收费机 automatic toll collecting equipment
自动收费系统 automatic fare collection system
自动收集 automatic collection
自动收集器 automatic collector
自动收紧绞车 automatic rendering winch; self-rendering winch
自动收票款装置 automatic fare collection
自动手 autohand
自动手表 automatic watch; self-winding watch
自动手动按钮 automanual switch
自动手动的 automanual; automatic manual
自动手动开关 automan; automan switch
自动手动式 automanual system
自动手动转换柄 non-follow-up controller
自动手动转换开关 automatic and hand-operated changeover switch; automan; automanual switch
自动首向基准系统 automatic heading reference system
自动售饭机 <旅客列车> automatic food vending machine
自动售货 automatic selling
自动售货车 <美> sales van
自动售货的 self-service
自动售货机 automat; automatic slot machine; coin-in-the slot machine; dispenser; penny-in-the-slot; slot-machine; vending machine; vendor
自动售货商店 self-service shop; supermarket
自动售检票 automatic fare collection
自动售票 automatic ticketing
自动售票机 automatic fare collection machine; automatic ticket dispenser; automatic ticket issuing machine; automatic ticket vending machine; automatic ticket vendor; ticket-issuing automat; ticket vending machine; passimeter <车站的>
自动售票器 passimeter[passometer]
自动售票系统 automatic fare collection system
自动输出控制 automatic output control
自动输电 self-discharging
自动输料控制阀 automatic input control valve
自动输入 automatic input
自动输入数据 automatic input data
自动输送机 automatic conveyer[conveyor]
自动输送站 robot station
自动竖立 self-erecting
自动数据编辑转接系统 automatic data editing and switching system
自动数据采集与控制系统 automatic data acquisition and control system
自动数据测取装置 automatic data acquisition
自动数据处理 datamation
自动数据处理程序 automatic data processing program(me)
自动数据处理程序报告系统 automatic data processing program(me) reporting system
自动数据处理方法 automatic data processing resources
自动数据处理机 automatic data processing machine; automatic data processor
自动数据处理设备 automatic data processing equipment
自动数据处理系统 automatic data handling system; automatic data processing system
自动数据处理中心 automatic data processing center[centre]
自动数据处理装置 automatic data processing equipment
自动数据传输线路 automatic data link
自动数据分析系统 automatic data analysis system
自动数据服务中心 automatic data service centre
自动数据管理情报系统 automated data management information system
自动数据互换系统 automated data interchange system
自动数据获取装置 automatic data acquisition
自动数据集编辑程序 automatic data set editing program(me)
自动数据记录 automatic data-recording
自动数据记录器 automatic data logger
自动数据简化(处理) automatic data reduction
自动数据简化器 automatic data reducer
自动数据简化设备 automatic data reduction equipment
自动数据交换系统 automatic data exchange system; automatic data interchange system
自动数据交换中心 automatic data switching center[centre]
自动数据介体 automated data medium; machine-readable medium
自动数据介质 machine-readable medium
自动数据录取 automatic data acquisition
自动数据录取系统 automatic data acquisition system
自动数据迁移 automatic data migration
自动数据设备 automatic data set
自动数据收集 automatic data acquisition
自动数据收集器 autodata gather
自动数据数字化系统 automatic data digitizing system
自动数据搜集浮标 automatic data collection buoy
自动数据译码器 automatic data translator
自动数据整理器 automatic data reducer
自动数据贮存与传输装置 automatic data accumulator and transfer
自动数据转换器 automatic data translator
自动数据转换系统 automatic data switching system
自动数据转换中心 automatic data switching center[centre]
自动数据转接中心 automatic data switching center[centre]
自动数据装置 machine-readable medium
自动数控程序 automatically programmed tool; automatically programmed tools language
自动数控绘图 automatic digital drafting
自动数片机 automatic counter for tablets
自动数学计算机 automatic digital computer
自动数字编码系统 automated digital encoding system; automatic digital encoding system
自动数字测试装置 automated digital test unit
自动数字跟踪 automated digital tracking
自动数字跟踪分析计算机 automated digital tracking analyzer computer
自动数字化 automatic digitization
自动数字记录与控制 automated digital recording and control
自动数字键 automatic number key
自动数字命令 automatic digital command
自动数字识别 automated digital identification
自动数字输入输出系统 automated digital input/output system
自动数字数据汇编系统 automated digital data assembly system
自动数字网(络) autodin[automatic digital network]
自动数字信息交换中心 automated digital message switching centre
自动数字制图系统 automated digital cartographic(al) system
自动栓门 snap bolt
自动双叠门堰 automatic double trap weir
自动双弧焊机 two-head automatic arc welding machine
自动双作用闸瓦间隙调整器 automatic double acting slack adjuster
自动水泵 automatic pump
自动水车 automatic water-wheel
自动水处理装置 automatic water softener
自动水环流 automatic return of water
自动水力发电站 automatic hydraulic station
自动水幕系统 automatic drencher system
自动水平 automatic horizon
自动水平调节吊具 self-level(l)ing spreader
自动水平调节器具 self-level(l)ing spreader
自动水平仪 automatic level
自动水位表 <混凝土搅拌机的> automatic water ga(u)ge
自动水位计 automatic ga(u)ge; automatic level recorder; mareograph
自动水位记录仪 automatic water stage recorder
自动水下考察船 self-propelled underwater research vehicle
自动水样采集器 automatic water sampler
自动水栅 boomer
自动水质监测 automated water-quality monitoring; automatic water-quality monitoring
自动水质监测控制系统 automated water-quality control system; automatic water-quality control system
自动水质监测系统 automated water-quality monitoring system; automatic water-quality monitoring system
自动水质监测仪 automated water-quality monitor; automatic water-quality monitor
自动水质监测站 automated water-quality monitoring station; automatic water-quality monitoring station
自动水准仪 automatic level; autoset level
自动顺序 automatic sequence
自动顺序操作 automatic sequential operation
自动顺序计算机 automatic sequence computer
自动顺序控制 automatic sequence control
自动顺序控制计算器 automatic sequence-controlled calculator
自动顺序控制器 programming controller
自动顺序控制仪表 instrument for automatic sequence control
自动顺序设备 automatic sequencing equipment
自动顺序投入的吹灰装置 automatic sequential soot blower
自动司锤装置 hammer automatic operating device
自动司机 automatic driver
自动松机构的捞矛 spear with releasing mechanism
自动松紧绞车 autotension winch
自动松紧调整器 automatic slack adjuster
自动松缆装置 automatic paying out gear
自动送带机构 tape-controlled carriage
自动送锭车 self-driven ingot buggy; self-propelled ingot buggy
自动送卡穿孔 automatic feed punch
自动送卡穿孔机 automatic feed punch
自动送料 automatic feed; feed control; self-acting feed; self-feeding
自动送料槽 gravity tank
自动送料带锯机 band-saw machine with autofeed carriage; band-saw mill with autofeed carriage
自动送料分类机 automatic feeding and sorting machine
自动送料杆 automatic rod magazine
自动送料器 autofeeder; automatic feeder; self(-acting) feeder
自动送料饲槽 feed conveyor trough
自动送料装置 automatic feeding
自动送纸机构 automatic carriage
自动搜索 automatic scan(ning); automatic search(ing)
自动搜索按钮 automatic search button
自动搜索电路 automatic search circuit
自动搜索干扰器 automatic search jammer
自动搜索干扰台 beagle
自动搜索干扰振荡器 broom
自动速度控制 automatic speed control
自动速度控制器 automatic speed con-

troller
自动塑模 automatic mo(u)ld
自动榫合 < 混凝土路面的 > self-dowelling
自动缩微胶片信息系统 automatic microfilm information system
自动索引 autoindex(ing);automatic index(ing)
自动索引木干 self-draught beam;self-drawbar
自动锁 automatic locking
自动锁紧装置 forced locking device
自动锁线机 automatic booksewing machine
自动锁相 automatic phase lock
自动台式计算机 automatic desk computer
自动抬刀机构 tool-relief mechanism
自动抬刀架 tool pick-up
自动抬刀装置 automatic tool lifter;automatic tool pick-up;tool lifter
自动弹簧阀 automatic spring loaded valve
自动弹簧锁 clasp lock
自动弹塑性分析 automatic elastic plastic analysis
自动弹性分析 automatic elastic plastic analysis
自动探测 automatic detection
自动探测和跟踪 automatic acquisition
自动探测系统 automatic detection system
自动探测仪 autodetector
自动探伤器 automated flaw detector
自动探向器 automatic direction(al) finder
自动探询 automatic poll;autopoll
自动探针测试 automatic probe test
自动探针检测机 autoprobing machine
自动镗床 self-acting boring machine
自动套 automatic cathead
自动特色 automatic feature
自动特征 automatic feature
自动梯 autostair;motorized ramp
自动梯传动装置 escalator drive
自动提动钻具设备 < 钻孔收缩时 > automatic retraction device
自动提缴保费贷款 automatic premium loan
自动提款机 < 美 > cashomat
自动提前(点火或喷油) automatic advance
自动提前点火机 automatic spark control
自动提前点火装置 automatic advanced element
自动提前断电器 automatic advanced breaker
自动提前断路器 automatic advanced breaker
自动提前喷射装置 automatic injection advanced device
自动提升机 self-lift;automatic elevator
自动提升器 self-lift
自动提升施工法 automatic lifting construction
自动提升系统 self-lifting system
自动提升装载车 elevating wagon
自动提升装置 power-operated hoist
自动提示器 autocue
自动体积计量设备 automatic volume batching plant
自动体位 active posture
自动体重秤 automatic weighing scale
自动替续增音器 autorepeater
自动天平 automatic scale;automatic weigher
自动天气预报 automatic weather forecasting
自动天气站 robot weather station
自动天体导航 automatic celestial navigation
自动天体导航仪 automatic celestial navigator
自动天文导航系统 automatic astronavigation system
自动天文定位系统 automatic astronomic(al) positioning system
自动添水式饮水器 self-filling drinking bowl
自动填空铅机 automatic quadder
自动填料机 automatic filling machine
自动填塞 autopack
自动填实 self-packing
自动调幅控制 automatic amplitude control
自动调弧氩弧焊 aircomatic welding
自动调浆箱 automatic stuff box
自动调焦 automatic focusing
自动调焦机构 autofocus mechanism
自动调焦纠正仪 self-focusing rectifying apparatus
自动调焦系统 automatic focusing system
自动调焦转绘仪 zoom transferscope
自动调焦装置 automatic focusing apparatus;self-focusing apparatus
自动调节 automatic adjust;automatic adjustment;automatic control;automatic governing;automatic regulating;automatic regulation;autoregulation;self-acting control;self-adjustment;self-governing;self-operated control;self-regulation
自动调节坝
automatic regulation dam;self regulating dam;self-regulation dam
自动调节导向轮 automatic training idler
自动调节的 autostable;self-adjusting;self-regulating
自动调节电弧焊 Argonaut welding;self-adjusting arc welding
自动调节动力装置 self-conditioned power plant
自动调节发电机 self-regulating generator
自动调节阀 automatic expansion valve;automatic regulating valve;automatic regulation valve
自动调节反馈机理 autoregulatory feedback mechanism
自动调节反馈控制 self-regulating feedback control
自动调节风门 automatic damper
自动调节给进速度 automatically regulated feed speed
自动调节灌浆机 automatic grouter
自动调节灌浆器 automatic grouter
自动调节过程 self-regulating process
自动调节绞缆机 automatic tension mooring winch;tension winch
自动调节拦河坝 self-regulating barrage
自动调节黏[粘]稠水的水罐车混合装置 demand viscous water tanker-mixer
自动调节膨胀室 automatic adjusting expansion chamber; self-adjusting expansion chamber
自动调节器 automatic controller;automatic governor;automatic regulating apparatus; automatic regulator;autoregulator;self-regulator
自动调节燃料泵 self-regulating fuel pump
自动调节式交通管制系统 adaptive system
自动调节式座位 power seat
自动调节水平仪 compensator instrument
自动调节水准仪 autolevel
自动调节温度装置 thermotank system
自动调节系统 automatic adjusting system;automatic regulating system;automatic regulation system;varitrol
自动调节系统质量准则 control criterion
自动调节相位线路 quadricorrelator
自动调节堰 self-adjusting weir
自动调节堰坝 self-regulating barrage
自动调节液位箱 self-priming tank
自动调节站台梯板 automatic dock leveller;dock level(l)er
自动调节直流车轴发电机 direct current axle generator with self regulating mechanism
自动调节轴承 self-aligning bearing
自动调节轴承座 self-adjustable bearing
自动调节装置 automatic controller;automatic regulating device;automatic regulation device
自动调节自动机 self-adjusting automaton
自动调距螺旋桨 automatic propeller
自动调缆绞车 automatic tension winch
自动调力拉杆 draught sensing link
自动调零 automatic zero-point correction;automatic zero setting;auto zero;autozeroset
自动调频 automatic frequency control
自动调频装置 automatic frequency regulating device
自动调平阀 < 空气悬架的 > level(l)ing valve
自动调平法 self-level(l)ing
自动调平激光器 self-level(l)ing laser
自动调平器 autolevel(l)er
自动调平式减震器 self-level(l)ing shock absorber
自动调平式筛架 self-level(l)ing shoe
自动调平水准仪 automatic level
自动调平悬架 self-level(l)ing suspension
自动调平装置 autoset level(l)ing device;self-level(l)ing device
自动调气阀 relief damper
自动调气闸 relief damper
自动调色机 mechanical colo(u)rant dispenser
自动调时齿轮 injection timer;self-timing gear
自动调时的 self-timing
自动调时器 automatic timing device
自动调速 automatic speed control;automatic speed regulation
自动调速器 automatic governor;automatic speed controller;automatic speed governor
自动调位的液压机构 depth-o-matic
自动调位回转段托滚 self-aligning return idler
自动调温 automatic heat regulation
自动调温开关 on-off thermostat
自动调温器 automatic heat regulator;self-operated thermostatic controller;thermostat
自动调隙挺杆 self-adjusting tappet
自动调相器 quadricorrelator
自动调相线路 quadricorrelator
自动调谐 automatic tuning;autotune
自动调谐短波发射机 self-tuning short-wave transmitter
自动调谐系统 automatic tuning system
自动调谐制 automatic tuning system
自动调心 self-centering[centring]
自动调心滚珠轴承 self-aligning ball-bearing
自动调心滚柱 automatic self-aligning roller
自动调心球面滚子轴承 self-aligning spheric(al) rolling bearing
自动调心式 self-centering type
自动调心轴承 self-aligning bearing
自动调压 automatic voltage regulation
自动调压阀 pilot valve
自动调压器 automatic pressure controller;automatic voltage regulator;pressure stat;pressuretrol
自动调用 autocall
自动调用程序库 automatic call library
自动调整 adjustment of itself;autocontrol;automatic adjustment;automatic regulation;automatic setting;automatic tuning;self-adjusting;self-adjustment;self-align;self-control;self-correcting;self-regulation;self-setting;voluntary adjustment
自动调整车钩 self-aligning coupler
自动调整抽汽式汽轮机 automatic extraction turbine
自动调整的 automatic adjusting;self-adjustable;self-regulating;self-adjusting
自动调整的输送带导向托辊 self-aligning belt idler
自动调整的水平仪 self-adjusting level
自动调整的制动器 self-adjusting brake
自动调整定时 automatic variable timing
自动调整机构 < 转盘钻进,钢丝绳下放时 > automatic feed off
自动调整机理 automatic adjusting mechanism;automatic regulation mechanism;self-adjusting mechanism
自动调整机制 servo-mechanism
自动调整减震器 self-adjustable shock absorber
自动调整结算 swing account
自动调整连续过程 automatically controlled continuous process
自动调整联轴节 self-aligning coupling
自动调整平衡锤 automatically adjusting counter weight
自动调整器 automatic regulating apparatus;automatic regulator
自动调整式推力轴承 pivoted shoe thrust bearing
自动调整式制动器 self-adjusting brake
自动调整式轴承 self-adjusting bearing
自动调整限动器 self-adjusting stop
自动调整楔形夹 self-aligning wedge grip
自动调整型喷管 self-adjusting type nozzle
自动调整堰 self-adjusting weir
自动调整液面水箱 self-priming tank
自动调整仪 control(ling) gear
自动调整闸瓦托 self-adjusting brake head
自动调整张力绞缆机 load limiting winch
自动调整轴承 pivoted bearing;self-adjustable bearing
自动调制 self-modulation
自动调制解调器选择 automatic modem selection
自动调制控制 automatic modulation

control
自动跳合开关 kickdown switch
自动跳越 automatic skip
自动跳闸 automatic tripping;trip-free
自动跳闸断路器 automatic trip circuit breaker
自动贴花机 automatic decal machine
自动贴商标机 automatic labeling machine
自动停闭 automatic shutdown
自动停车 automatic stop(ping);automatic train stop【铁】;autostop;automatic shutdown <发动机>
自动停车场 autosilo
自动停车滑阀 shut-down slide valve
自动停车机构 stop motion mechanism
自动停车器 check piece;trip dog
自动停车系统 automatic stop system
自动停车装置 automatic stopping device; automatic train stop equipment
自动停车撞击装置 tripping device
自动停带开关 automatic shut-off
自动停供燃料 automatic fuel cut-off
自动停机 automatic stop(ping);autostop;knock off
自动停机的临界载荷 deadman load
自动停止 automatic shutdown;automatic stop;self-braking
自动停止阀 automatic stop valve;trip valve
自动停止进刀挡块 trip block
自动停止进入燃料 automatic fuel shut off
自动停止控制 automatic stop control
自动停止器 automatic stop(ping);autostop
自动停止水位 stop level
自动停止套筒 automatic stop sleeve
自动停止装置 automatic stop arrangement; automatic stopping device;autostop;autostoper
自动通断 automatic switching
自动通风 automatic ventilation
自动通风机 automatic ventilator
自动通风孔 automatic air vent
自动通风煤气炉 universal combustion burner
自动通风煤气燃烧器系统 atmospheric gas-burner system
自动通风气体燃烧器系统 atmospheric gas-burner system
自动通风器 automatic ventilator
自动通风室 self-ventilated chamber
自动通风调节器 automatic draft regulator
自动通风舷窗 automatic ventilating side scuttle
自动通过按钮电路 automatic passing button circuit
自动通信[讯] automated communication
自动通信[讯]和信息处理系统 automated communications and message processing system
自动通信[讯]网 automatic communication network
自动通信[讯]装置 automatic communication device
自动通知 automatic advice
自动同步 automatic synchronization;self-lock(ing);self-synchronism;self-synchronizing
自动同步传动装置 selsyn driver;selsyn train
自动同步传感器 self-synchronizing transmitter;transmitting selsyn
自动同步传输 self-synchronous transmission
自动同步传输系统 self-synchronous transmission system
自动同步的 automatically synchronous; self-synchronizing; self-synchronous
自动同步电动机 autosynchronous motor;selsyn motor;synchromotor
自动同步发电机 selsyn generator
自动同步发射机 self-synchronous transmitter;selsyn generator;selsyn transmitter
自动同步发送机 synchro
自动同步感受器 autosyn receiver
自动同步机 autosyn;self-syn system;selsyn;synchro;synchromagslip
自动同步机传动 selsyn drive
自动同步机传动机构 selsyn drive gear
自动同步机控制 selsyn control
自动同步机盘 selsyn disc
自动同步机数据系统 selsyn-data system
自动同步机系统 selsyn system
自动同步机主动齿轮 selsyn drive gear
自动同步接收机 receiving selsyn;selsyn receiver
自动同步接收器 mechanical repeater;receiving selsyn
自动同步控制 self-synchronized control
自动同步控制的差动式发送器 synchrocontrol differential transmitter
自动同步离合器 self-synchronizing clutch
自动同步模型 selsyn model
自动同步器 automatic synchronizer;autosyn
自动同步设备 selsyn device
自动同步系统 automatic synchronous system; autosyn system; self-synchronizing system; self-synchronous system; self-syn system; selsyn train
自动同步仪表 selsyn instrument
自动同步噪声抑制器 automatic synch-noise suppressor
自动同步振动器 self-synchronizing vibrator
自动同期系统 automatic synchronous system; autosyn system; self-synchronizing system
自动同期装置 automatic paralleling device
自动投量 automatic dosing
自动投料机 automatic feeder; self-feeder
自动投料器 automatic feeder; self-feeder
自动投配池 dosing chamber
自动投配的 self-proportioning
自动投配虹吸管 dosing siphon
自动投配式拌和机 self-erecting batch type mixer
自动投配式搅拌机 self-erecting batch type mixer
自动图标记录器 autographic(al) apparatus
自动图表程序 autochart
自动图示记录器 autographic(al) apparatus
自动图示记录仪 autographic(al) apparatus
自动图示仪 autographometer
自动图线跟踪器 automatic chart line follower
自动图像传输 automatic picture transmission
自动图像传送 automatic picture transmission
自动图像发送系统 automatic picture transmission system
自动图像检索系统 automatic image retrieval system
自动图像控制 automatic picture control
自动涂复机 automatic coating machine
自动涂漆机 automatic coating machine
自动涂油器 oiler
自动推动器 self-mover
自动推杆器 nudger; pusher installation;stick pusher
自动推进 automatic feed;self-drive
自动推进冲击采场 stoper
自动推进的 locomobile;self-driven;self-moving;self powered;self-propelling;self-propelled
自动推进的除草机 self-propelled grass mower
自动推进的轮胎式起重机 self-propelled wheeled crane
自动推进滑座 power feed shell
自动推进机 self-propelling machine
自动推进架式凿岩机 autofeed drifter;automatic feed drifter
自动推进架式钻机 autofeed drifter;automatic feed drifter
自动推进爬行起重机 self-propelled crawler mounted crane
自动推进器导向滑板 <凿岩机> shell power feed
自动推进倾料器 self-propelling tripper
自动推进式柴油机有轨车 self-propelled diesel railcar
自动推进式车辆 self-propelled vehicle
自动推进式打壳机 self-propelled crust breaker
自动推进式底盘 self-propelled chassis
自动推进式渡轮 self-propelling ferry
自动推进式街道洒水车 self-propelled street sprinkler
自动推进式煤沥青喷布机 self-propelled tar distributor
自动推进式黏[粘]土拌和机 travel mixer
自动推进式气腿 pusher leg
自动推进式(汽车)底盘 self-propelled chassis
自动推进式塔架 self-propelled tower
自动推进式挖掘机 self-propelled digger
自动推进式钻车 self-propelled drill mounting
自动推进式钻机 automatic feed drill;push-feed drill
自动推进式钻井机 self-propelled rig
自动推进挖泥机 self-propelling dredge(r)
自动推进凿岩机 self-feed drill
自动推进装置 <钻机> automatic feed unit;rotofeed
自动推理 automated reasoning
自动退刀装置 automatic tool retracting unit
自动退轮机 automatic dismount press
自动退水闸 automatic escape(gate)
自动托盘装载机 automatic pallet loader
自动拖车联结器 self-acting trailer coupling
自动拖挂装置 self-acting trailer coupling
自动脱钩 self-releasing hook
自动脱钩吊具 self-releasing hanger
自动脱钩装置 tripper
自动脱开 self-disengaging
自动脱开钩 trip hook
自动脱扣(器) automatic trip; free trip
自动脱扣装置 tripper
自动脱粒机 automatic thresher
自动脱落渣 self-releasing slag
自动脱模聚合物 autodemo(u)lding polymer
自动脱模装置 pick-off device
自动脱绳钩 self-detaching hook
自动脱水 self-desiccation
自动陀螺驾驶仪 automatic gyropilot
自动陀螺仪 autogyro
自动挖土斗 automatic excavating bucket
自动挖土机 self-propelled excavator
自动瓦斯报警器 automatic firedamp alarm
自动弯沉仪 autodeflectometer
自动弯机 automatic bending machine
自动网路分析器 automatic network analyser[analyzer]
自动往复阀 automatic shut-off valve
自动微调 automatic fine control;automatic fine turning
自动微调控制 automatic fine-tuning control
自动维护 self-maintenance
自动尾水管充水管路 automatic draft-tube fill line
自动尾水管充水管线 automatic draft-tube fill line
自动尾卸车辆 end-tipping vehicle
自动尾卸载重卡车 end-tipping motor lorry
自动尾卸载重汽车 end-tipping motor lorry
自动卫星监视系统 satellite automatic monitoring system
自动卫星通信[讯]控制系统 automatic satellite communication control system
自动位置对准 autotrack(ing)
自动位置调节器 automatic position controller
自动位置调节系统 positioning system
自动位置指示器 automatic position indicator
自动喂料篦子 automatic feed grate
自动喂料机 self-feeder
自动喂料系统 automatic feeding system
自动温度记录调节器 automatic temperature recorder controller; automatic temperature recording controller
自动温度记录控制器 automatic temperature recorder controller
自动温度控制 automatic temperature control;comfortron
自动温度控制器 automatic temperature controller
自动温度器 thermograph
自动温度调节器 automatic temperature regulator;autothermoregulator
自动温控电炉 automatic temperature-controlled electric(al) furnace
自动温湿度记录计 automatic thermohygrograph
自动文件管理 automatic document control
自动文献分类 automatic document classification
自动文献检索系统 automatic document retrieval system
自动文献请求服务处 automatic document request service
自动文字切换 autoword wrap
自动稳定岸墙 autostable wall
自动稳定的 autostabilized;autostable
自动稳定航向装置 automatic course stabilizer
自动稳定和控制系统 automatic stabilization and control system
自动稳定器 automatic stabilizer;au-

Z

tostabilizer
自动稳定设备 automatic stabilization equipment
自动稳定系统 automatic stabilization system; autostabilization system
自动稳定性 automatic stability; autostability; self-stability
自动稳定转向系 self-stabilizing steering
自动稳定装备 autostabilizer unit
自动稳定装置 automatic stabilization equipment; autostabilizer
自动稳态机制 homeostatic mechanism
自动稳相原理 phase stability
自动稳压器 automatic voltage regulator; autostabilizer
自动稳压装置 automatic voltage regulator
自动无阀重力式滤池 automatic valveless gravity filter
自动无栏平交道口 automatic ungated level crossing
自动无线电测距器 automatic-range-only radar
自动无线电测向仪 automatic radio compass; automatic radio direction finder
自动无线电定向系统 automatic radio direction finding system
自动无线电气象漂浮系统 drifting automatic radio meteorologic(al) system
自动误差校正 automatic error correction
自动误差校正系统 automatic error correction system
自动雾笛 automatic fog whistle
自动雾号 automatic fog signal
自动雾号控制 automatic fog control
自动雾化喷嘴 inspirited burner; self-induced air burner
自动雾炮 automatic fog gun
自动雾钟 automatic fog bell
自动吸入喷射器 self-starting injector
自动吸入器 automatic choke
自动吸(移)管 automatic pipet(te)
自动袭夺 auto-capture
自动洗车机 automatic car washer
自动洗车设备 automatic car washer
自动洗涤机 automatic washing machine
自动洗矿槽 automatic strake
自动洗衣店 laundromat
自动洗衣店废水 laundromat waste
自动洗衣机 <投币机器自行开动> coin-up
自动铣床 automatic milling machine; self-acting milling machine
自动铣削 milling automatic
自动系统 automated system
自动系统故障分析 automatic system trouble analysis
自动系统控制 automated system control; automatic system control
自动系统起始功能 automatic system initiation
自动系统生成 automatic system generation
自动下沉采泥器 free fall rocket core sampler
自动下落指示器 automatic drop indicator
自动下落制动器 drop brake
自动纤维光导生物传感器 automated fibre optic biosensor
自动显示 automatic display
自动显示标绘系统 automatic display and plotting system
自动显示呼叫指示器 automatic display call indicator
自动显示与绘图系统 automatic display and plotting system
自动显微照相机 automatic microphotographic(al) camera
自动显影机 automatic developing machine
自动显影胶片 automatic developing film; self developing film
自动显影设备 automatic film developing equipment
自动现金出纳机 automatic cashier
自动现象 autokinetic phenomenon
自动线 autoline; automatic line; transfer machine; transfer-matic
自动线绘图机 automatic line plotter
自动线路分段开关装置 automatic line sectionalizer
自动线性定位系统 automatic linear positioning system
自动限额 voluntary ceiling
自动限时解锁器 automatic time release
自动限载装置 automatic load limit(ation); automatic load limitation device
自动限制出口 voluntary restriction of export
自动相对运动作图器 automatic relative plotter
自动相关器 automatic correlator
自动相位比较电路 automatic phase comparison circuit
自动相位控制 automatic phase control
自动相位控制回路 automatic phase control loop
自动相位同步 automatic phase synchronization
自动镶嵌 <钻头的> automatic setting
自动详图制作 automated detailing
自动响墩 automatic detonator
自动(向加热炉)送坯装置 bundle buster
自动向下进给 automatic down feed
自动消除信号方向指示器 self-cancelling direction indicator
自动消除讯号方向指示器 self-cancelling trafficator
自动消除隐藏线 automatic hidden line removal
自动消磁 autodegauss; self-demagnetization
自动消磁电路 automatic degaussing circuit
自动消毒 self-sterilizer
自动消毒器 automatic sterilizer
自动消防 <包括发出火警信号和传感系统的装置> active fire defense
自动消防泵 automatic fire pump
自动消防喷水系统 standpipe system
自动消防水泵 automatic fire pump
自动消防系统 automatic fire-fighting system; sprinkler system
自动消声器 auto-muffler
自动消失振荡器 squegger
自动销保 <当战争爆发时> automatic termination cover
自动销紧弹簧螺母 quick-adjusting nut
自动小交换机 automatic private branch exchange; dial private branch exchange; private automatic exchange
自动斜坡 gravity haulage; self-acting slope
自动泄洪活动坝 self-acting moveable flood dam
自动泄洪闸门 automatic spillway gate
自动泄水 automatic discharge; self-discharging
自动泄水道 automatic spillway
自动泄水阀 self-acting discharge valve
自动泄水救生艇 self-bailing boat
自动泄水尾阱 self-bailing cockpit; self-draining cockpit; self-emptying cockpit
自动泄水闸 falling sluice
自动卸道砟机 ballast self-unloader
自动卸荷器 automatic unloader
自动卸货 self-discharging; self-unloading
自动卸货驳船 self-unloading barge
自动卸货车 automatic dump truck; self-discharging wagon; tip car; tip lorry
自动卸货船 self-unloading ship; self-unloading vessel
自动卸货的 self-clearing
自动卸货机 automatic unloading machine
自动卸货漏斗敞车 auto-discharge open top hopper car
自动卸货拖车 self-emptying trailer
自动卸货与搬运系统 automated cargo release and operations system
自动卸开的 self-releasing
自动卸料 automatic discharge; automatic dump; automatic unloading; self-dumping
自动卸料泵 self-unloading pump
自动卸料车 automatic discharge wagon; dump car; self-cleaning car; self-clearing car; self-discharging wagon; self-unloading vehicle; tripper; tripping car
自动卸料车体 hopper car body
自动卸料的 automatic discharging
自动卸料机 automatic unloading machine; self-unloader
自动卸料离心分离机 self-cleaning type centrifugal separator
自动卸料配料秤 automatic dumping batch scale
自动卸料压滤机 automatic cake discharge type filter press
自动卸料装置 automatic discharge unit
自动卸载 self-discharge; self-discharging; self-dumping; self-emptying
自动卸载泵 self-unloading pump
自动卸载车 automatical self-discharger
自动卸载船 gunboat; self-unloading vessel
自动卸载挂车 automatic unloading trailer
自动卸载器 automatic unloader
自动卸载曲轨 self-dumping scroll
自动卸载式干涉沉降分级机 Consenco classifier
自动卸载(小)车 gunboat
自动芯棒轧管机 automatic plug rolling mill
自动信道选择 automatic channel selection
自动信号 automatic signal; autosignal(l)ing
自动信号安全设备 automatic signal protection
自动信号保护(设备) automatic signal protection
自动信号灯 automatic signal lamp; automatic signal light
自动信号电压 automatic signal voltage
自动信号发送 automatic signal(l)ing
自动信号法 automatic signal(l)ing
自动信号继电器 automatic signal relay
自动信号器 automatic signal(l)ing apparatus
自动信号区段 automatic signal(l)ing section
自动信号区域 automatic block territory; automatic signal(ling) territory
自动信号识别装置 automatic signal recognition unit
自动信号锁闭 automatic signal locking
自动信号装置 automatic signal(l)ing device
自动信件盖销装置 automatic letter facing
自动信托 active trust
自动信息处理 automatic information processing; automatic message handling
自动信息处理机 automatic processor
自动信息传播系统 automated information dissemination system
自动信息分配系统 automatic message distribution system
自动信息管理系统 automated information management system
自动信息计算 automatic message counting
自动信息记录 automatic message recording
自动信息检索系统 automatic information retrieval system
自动信息交换中心 automatic message switching center[centre]
自动信息交换(装置) automatic message exchange
自动信息显示 automatic information display
自动信息显示设备 automatic information display equipment
自动行车道 autoroad
自动行程 self-acting travel; self-act travel
自动行进喷灌系统 self-travel(l)er irrigation system
自动行人传送带 escalator
自动行人道 moving sidewalk; moving walk
自动行驶式混凝土车 self-propelled concrete car
自动性 automaticity; automatic nature; automatism
自动性能分析系统 automatic performance analysis system
自动修补系统 automatic patching system
自动修理机 automatic trimmer
自动修坯 autofettle
自动虚警率 automatic false alarm rate
自动序列 automatic sequence
自动序列控制系统 automatic train control system
自动续纸装置 automatic feeder; machine feeder
自动絮凝作用 automatic flocculation; self-flocculation
自动宣告破产 voluntary bankruptcy
自动悬浮 autosuspension
自动旋压成型车床 automatic spinning lathe(machine)
自动旋凿 spiral ratchet screwdriver
自动旋转 autorotation
自动旋转菜碟架 dumb waiter
自动旋转冲缩式凿岩机 automatically rotated stopper drill
自动旋转打磨机 brush backed sander
自动旋转机构 self-turning mechanism
自动旋转伸缩式凿岩机 automatic rotation stopper; self-rotated stoper
自动旋转式小型柱架式钻机 automat-

Z

ically rotated stopper drill
自动旋转向上式凿岩机 self-rotated stoper
自动选废 automatic rejection
自动选路 automatic routing;selective automatic route control
自动选路控制 automatic routing control
自动选配装置 matching control
自动选通脉冲 automatic strobe pulse
自动选行 automatic line selection
自动选样器 automatic sampler
自动选择 automatic selection
自动选择的 autoselective
自动选择节的 self-adjustable
自动选择控制 automatic selection control
自动选择脉冲 automatic strobe pulse
自动选择系统 automatic selective system
自动选择性控制 automatic selectivity control
自动选址 automatic addressing
自动选中继线 trunk hunting
自动学 automatization
自动寻查故障装置 automatic trouble-locating arrangement
自动寻的 active homing
自动寻的传感器 homing sensor
自动寻的控制 automatic homing control
自动寻光装置 light-seeking machine
自动寻呼系统 automation paging system
自动寻求最优控制 optimalizing control
自动寻线 automatic hunting;self-hunting
自动寻优控制 optimalizing control
自动寻找 group hunting
自动寻址系统 automatic addressing system
自动巡查故障装置 automatic trouble-locating arrangement
自动巡路平地机 autopatrol grader
自动巡路平路机 autopatrol grader
自动巡逻的 autopatrol
自动巡视的 autopatrol
自动巡行平路机 <养路机> autopatrol grader
自动询问 automatic interrogation
自动循环 autocycle;automatic cycle
自动循环控制 autocycle control;automatic cycle control;automatic cycling control
自动压道车 autonomous car
自动压盖机 automatic capping machine
自动压合热焊 autogenous pressure welding
自动压机 mo(u)lden press
自动压力机 automatic press
自动压力控制 automatic pressure control
自动压力喷灯 automatic pressure jet burner
自动压力调节器 automatic pressure controller;automatic pressure regulator
自动压滤机 automatic filter press
自动压滤器 automatic pressure filter
自动压路机 self-propelled roller
自动压模 automatic mo(u)ld
自动压片机 automatic tableting press
自动压缩 autocondense
自动压缩空气测探仪 automatic air-compressed depth ga(u)ge
自动压缩空气传动开关 autopneumatic circuit breaker
自动压形 automatic pressing
自动压型 automatic tamping
自动压榨机 automatic squeezer machine
自动压制 automatic pressing
自动氩弧焊 Argonaut welding
自动烟尘采样器 automatic smoke sampler
自动烟尘取样器 automatic smoke sampler
自动烟感器 automatic smoke detector
自动烟雾报警器 automatic call point
自动烟雾报警装置 automatic call point
自动延期贷款 revolving loan
自动延期租赁 tenancy from year to year
自动延时补偿器 automatic time relay compensator
自动研光 self-glazing
自动研磨 autogenous grinding
自动研磨机 autogenous grinding mill;autogrinding machine; automatic grinder
自动演奏装置 player
自动验潮仪 automatic tide ga(u)ge
自动验潮站 automatic tide ga(u)ge station
自动验算 automatic check
自动验证系统 automated verification system
自动堰 automatic weir
自动扬谷机 <码头谷物卷扬> automatic distributor
自动阳离子交换设备 automatic cation exchange unit
自动养护 <混凝土的> self-curing
自动养护机 <混凝土> automatic curing machine
自动养路机 motor maintainer
自动氧 autoxidator
自动氧化期 autooxidation phase
自动氧化相 autooxidation phase
自动氧化作用 automatic oxidation;auto(o)xidation;spontaneous oxidation
自动氧气安全阀 automatic oxygen safety valve
自动氧气切割 automatic oxygen cutting
自动窑 step-kiln
自动摇摆链式清除器 automatic swinging chain cleaner
自动遥测跟踪系统 automatic telemetry tracking system
自动遥测气象仪 automatic telemetering weather system
自动遥测设备 remote automatic telemetry equipment
自动遥测仪 remote automatic telemetry equipment
自动遥控 automatic remote control
自动遥控开关 automatic teleswitch
自动遥控装置 teleautomatics
自动业务 automatic service
自动夜间服务 automatic night service
自动液传动 automatic fluid transmission
自动液力传动 automatic hydraulic transmission
自动液面计 automatic level ga(u)ge
自动液位记录器 liquid level recorder
自动液压大型捣固车 autolevel(l)ing-lifting-lining-tamping machine
自动液压喷雾器 automatic hydraulic sprayer
自动液压制动装置 automatic hydraulic brake device
自动仪 robot
自动移位 automatic shift
自动移位数据编码机 automatic shift position data encoder
自动移位同步离合器 self-shifting synchronizing clutch
自动移相器 automatic phase shifter
自动抑阻 spontaneous resistance
自动译码器 automatic code translator
自动溢洪道 automatic spillway
自动溢流 automatic safety relief
自动溢流阀 automatic bypass valve
自动溢流堰 automatic weir
自动翼缝隙 autoslot
自动音量控制 automatic volume control
自动音量扩展 automatic volume expansion
自动音量压缩 automatic volume compression
自动音响浮标 automatic sounding buoy
自动银行出纳 drive-in teller
自动引出通路 automatic pilot channel
自动引导 self-direction;homing
自动引导导弹 homer
自动引导的头部 seeker
自动引导信息 homing intelligence
自动引导装入程序 automatic bootstrap loader
自动引导装置 homing mechanism
自动引水泵 self-priming pump
自动饮料出售机 beverage dispenser
自动饮水杯 automatic drinking cup
自动饮水器 autodrinker;automatic drinking bowl;automatic drinking cup;automatic water bowl;automatic waterer;drinking fountain;self-waterer;self-watering device
自动饮水装置 drinking fountain;self-waterer;self-watering device
自动印刷机 automatic press;automatic printing machine
自动应答 auto-answer;automatic answer
自动应答器 automatic answer back unit;automatic answering back device;automatic answering unit
自动应答调制解调器 automatic answering modem
自动应力扫描仪 automatic stress scanner
自动硬度测定仪 automatic hardness tester
自动硬化 autogenous hardening
自动硬模铸造机 automatic chill casting-machine;automatic gravity die casting machine
自动用户线测试器 automatic subscriber-line tester
自动油罐站 automatic tank battery
自动油浆浮箍 automatic fill-up float collar
自动油开关 automatic oil switch
自动油门 autothrottle
自动油润轴承 self-oiling bearing
自动油脂集中润滑系统 automatic centralized grease system
自动油脂中心站 automatic central grease station
自动有轨巷道堆垛机 automatic S/R [storage/retrieval] machine
自动与直通空气联合制动机 combined automatic and straight air brake
自动雨量计 automatic rainfall recorder
自动雨水采样器 automatic rain sampler
自动语音识别【计】automatic speech recognition
自动预报系统 automatic forecasting system
自动预告 automatic warning
自动预选采样器 automated pre-concentration sampler
自动预占 automatic camp-on
自动阈限控制 autothreshold control
自动阈值变更 automatic threshold variation
自动愈合趋势 self-healing tendency
自动远程声呐 automatic long-range sonar
自动匀光印像机 autododge;automatic dodge
自动允许闭塞沿途联锁 automatic permissive block wayside interlock
自动运货单填写与信息系统 automated tariff filing and information system
自动运输 active transport
自动运输秤 automatic conveyor scale
自动(运输)斜坡道 self-acting incline
自动运算 automatic operation
自动运算系统 automatic operation system
自动运行 automatic operation;automatic run;instrumental operation
自动运行设施 automatic operator
自动运转 self-prime
自动运转起重机 self-erecting crane
自动晕渲 automatic hill shading
自动载波控制 automatic carrier control
自动载货辨识 automatic cargo identification
自动载运器 automatic carriage
自动再闭合 automatic reclosing
自动再充电 automatic recharge
自动再关闭 automatic reclosing
自动再接通断路器 automatic reclosing circuit breaker
自动再启动 automatic restart;autorestart;autorestarter
自动再生式空气过滤器 automatic renewable air filter
自动凿岩 auto drilling; automatic drilling
自动造形设备 automatic mo(u)lding plant
自动造型机 automatic mo(u)lding machine
自动造型机组 automatic mo(u)lding unit
自动造型线 automatic mo(u)lding line
自动噪声降低 automatic noise reduction
自动噪声控制器 automatic noise controller;automatic noise suppressor
自动噪声数字指示器 automatic noise figure indicator
自动噪声图形指示器 automatic noise figure indicator
自动噪声限制器 automatic noise limiter
自动噪声消除器 automatic noise canceller
自动噪声抑制器 automatic noise suppressor
自动增大螺距装置 autocoarse pitch
自动增力制动器 self-energizing brake
自动增量 autoincrement
自动增量标志 autoincrement flag
自动增量减量 autoincrement decrement
自动增量式 autoincrement mode
自动增量寻址 autoincrement addressing
自动增压调节器 automatic boost control
自动增益 autogain
自动增益控制 autogain control;automatic gain control

自动增益控制电路 automatic gain control circuit
自动增益控制管 automatic gain control tube
自动增益调节器 automatic gain regulator
自动增益调整 autogain control; automatic gain control
自动增益稳定器 automatic gain stabilization
自动扎管法 automatic plug mill process
自动轧钢机 automatic rolling mill
自动轧管 plug rolling
自动轧管法 automatic plug mill process; plug mill process; Stiefel process
自动轧管机 piercer; plug tube-rolling mill; Swedish mill
自动轧石厂 self-propelled crushing plant
自动轧石机 self-propelled crushing plant
自动闸 automatic brake
自动闸杆 automatic brake rod
自动闸门 balanced gate; self-shooter
自动栅门 automatic gate; automatically controlled gate
自动摘钩 automatic coupler detached; automatic releasing hook
自动摘钩设备 automatic uncoupling and air brake hose
自动摘录 autoabstract
自动摘要 autoabstract; auto accident
自动展期贷款 revoluting loan
自动张紧承力索 automatically tensioned catenary
自动张紧绞车 autotension winch
自动张紧缆索 automatically tensioned carrying cable
自动张紧装置 automatically tensioned equipment; automatic take-up device
自动张力补偿器 automatic tension regulator
自动张力计 autotensiometer
自动张力记录仪 automatic tension recorder
自动张力控制 automatic tension control
自动张力平衡器 automatic tension balancer
自动张力调节器 automatic tension controller
自动着火 autogenous ignition
自动着火点 autogenous ignition temperature
自动着火温度 autogenous ignition temperature; autoignition temperature
自动找平 autolevel; self-level(l)ing
自动找平补偿器 automatic level(l)ing compensator
自动找平工程水准仪 automatic engineer's level
自动找平密封胶 self-level(l)ing sealant
自动找平器 automatic load compensator
自动找平式滑模摊铺机 autograde (type) slipform machine
自动找平饰面层 self-level(l)ing floor finish
自动找平饰面剂 self-level(l)ing floor finish
自动找平水准仪 automatic level; self-level(l)ing instrument
自动找平装置 auto-elevator device; automatic level(l)ing device
自动找寻故障装置 automatic trouble-locating arrangement
自动照明控制器 automatic lighting regulator
自动照相机 automat
自动照相排字机 automatic photocomposer; automatic photosetter
自动照准的 self-alignment; self-aligning
自动照准仪 autocollimator
自动遮断阀 automatic intercepting valve
自动遮光 automatic dodging
自动折叠机 automatic layboy
自动折返 driverless reversing of train
自动折返功能 automatic reverse function
自动折线指示器 automatic clearing indicator
自动折叶门 automatic folding leaf door
自动折叶塑料线烫订机 automatic folding and thread-sealing machine
自动折纸机 layboy
自动真空阀 automatic vacuum valve
自动真空封罐机 automatic vacuum can sealing machine
自动真空组织处理机 vacuum automatic tissue processor
自动诊断 automatic diagnosis
自动诊断程序 automatic diagnostic program(me)
自动诊断监控器 auto-diagnostic monitor
自动诊断系统 automatic diagnosis system
自动振打装置 automatic mechanical rapping system
自动振荡 autonomous oscillation; auto-oscillation; self-sustained oscillation
自动振荡法 self-oscillation method
自动振动器 autotrembler; trembler
自动振动式沟槽压路机 self-propelled vibratory trench roller
自动振动式双轮压路机 self-propelled vibratory tandem roller
自动振动探测计 automatic vibration monitor
自动振幅控制 automatic amplitude control
自动振铃 automatic call; automatic ringing; machine ringing
自动振筛机 automatic sieving machine
自动镇流灯 blended lamp
自动震动器 trembler
自动蒸馏器 automatic distillator
自动蒸汽锤 steam automatic hammer
自动蒸汽发生器 automatic steam generator
自动蒸汽温度控制 automatic steam temperature control
自动整步系统 automatic synchronous system; autosyn system; self-synchronizing system
自动整流器 automatic rectifier
自动整排式 self-aligning type
自动整平的 self-adjusting
自动整平控制装置 automatic screed control unit
自动整平起道拔道捣固车 autolevel(l)ing-lifting-lining-tamping machine
自动整平装置 automatic screed
自动整修机 automatic trimmer
自动正压状态【气】 autobarotropy
自动支付 voluntary payment
自动执行 automatic performance
自动直接长途拨号系统 automatic direct distance dialling system
自动直接作用 automatic direct action
自动直通空气制动机 automatic straight air brake
自动直通两用制动机 double brake
自动值守 auto-watch
自动止回阀 automatic check-valve; automatic non-return valve
自动止水阀 automatic stop valve
自动纸带穿孔机 automatic tape punch(er); output tape punch; reperforator
自动纸带穿孔机转换中心 reperforator switching center[centre]
自动纸带穿孔器 automatic tape punch(er)
自动指令信号 autocommand signal
自动指示 self-indicating
自动指示高温计 automatic indication pyrometer
自动指示器 automatic indicator
自动制 dial system working
自动制动 automatic brake application; self-actuating brake; self-lock(ing)
自动制动传动机构 self-stopping gear
自动制动带 self-energizing brake band
自动制动的 automatic braking; self-braking
自动制动阀 automatic brake valve; automatic stop valve
自动制动阀手把 automatic brake valve handle
自动制动滑行装置 automatic skidding device
自动制动机 automatic skidding device
自动制动机构 self-sustaining gear
自动制动控制 automatic stop control
自动制动器 autobrake; autostoper; self-acting brake; automatic brake
自动制动调节器 automatic brake adjuster
自动制动闸 self-energizing brake
自动制动装置 self-braking device
自动制动装置调整器 automatic brake-gear adjuster
自动制动作用 automatic braking
自动制管机 automatic pipe(making) machine
自动制瓶机 automatic flask machine
自动制榫机 automatic dove tailing machine
自动制图 autodraft(ing); automated drafting; automatic drafting; automatic mapping; computer-aided cartography
自动制图程序 automatic cartographic(al) program(me)
自动制图法 automatic cartography
自动制图工艺 technology of automated cartography
自动制图机 drafting machine
自动制图软件 automated cartography software; software of automated cartography
自动制图系统 automated cartographic(al) system; automated mapping system
自动制图学 automatic cartography
自动制图仪 autocartograph; automatic autocartograph
自动制图硬件 hardware of automated cartography
自动制图综合 automated generalization in cartography
自动制氧机 automatic oxygen generator
自动质量控制 automatic quality control
自动置零 automatic zero setting
自动置中 autocentering; automatic centering
自动中断 automatic interrupt(ion); self-blocking
自动中继 automatic relay
自动中继计算机 automatic relay computer
自动中继站 automatic relay station
自动中心拌和厂 automatic central mixing plant
自动中心冲床 automatic center punch
自动中心计算机系统 autocentral computer system
自动中央缓冲车钩 automatic central buffer coupling
自动终端信息服务 automatic terminal information service
自动终端信息业务 automatic terminal information service
自动终接器 automatic connector
自动终止条款 automatic termination clause
自动钟表开关 automatic clock switch
自动重量称量器 automatic weigh batcher
自动轴瓦垫调整 automatic wedge adjustment
自动逐年延期租赁权 estate from year to year
自动注册 automatic log-on
自动注射板 autoinjection panel
自动注油启动的 self-priming
自动注油起动系统 automatic priming system
自动注油器 line oiler; impermeator
自动贮藏 automated storage
自动筑路机 automatic road-builder
自动抓斗 automatic grab
自动抓紧 autoclawing
自动抓梁 automatic grab beam; automatic hooking beam
自动抓岩机 automatic grab; automatic rock grab
自动爪 trip dog
自动专用交换机 electronic private automatic branch exchange
自动转报系统 telegram retransmission system; telegraph automatic system
自动转报中心 telegram retransmission centre
自动转换 autochangeover; automatic switching(over); automatic tandem working; automatic transition; self-shifting
自动转换板 automatic switching panel
自动转换道岔 automatically positioned switch
自动转换活塞 relay piston
自动转换开关 attent-unattent switch; automatic change-over switch; automatic transfer switch; hospital switch
自动转换开关变速箱 self-shifting transmission
自动转换器 autochangeover unit; automatic switch; automatic translator; switching over unit
自动转换设备 automatic switching equipment
自动转换调节器 automatically variable governor
自动转换系统 automatic switching system
自动转接 automatic switch over; autoswitch over; through dial(l)ing
自动转接的操作 dial system tandem operation
自动转接的动作 <自动或半自动通信[讯]中> dial system tandem opera-

tion
自动转接设备 automatic transfer equipment
自动转接衰减器 switching pad
自动转接制工作 mechanical tandem operation
自动转接作用 automatic tandem working
自动转让 voluntary transfer
自动转让财产 gift
自动转让财产证书 deed of gift
自动转速控制 automatic speed control
自动转塔式六角车床 automatic turret lathe
自动转台 automatic turntable
自动转台式换卷机 autoturret winder
自动转向 automatic steering; self-steering
自动转向电动机 auto-reversing electric(al) motor; autoreversing electric(al) motor
自动转向机构 automatic change-over switch
自动转向架 motor bogie
自动转向装置 autosteerer
自动转移 automatic branching
自动转移呼叫 automatic transfer of an answering call
自动转运装置 automatic transfer device
自动转辙器 automatic switch
自动桩锤 automatic monkey
自动装袋封口机 automatic bag filling and closing machine
自动装袋式联合收获机 bagger combine
自动装弹器 automatic gun charger
自动装膏机 automatic paste filler
自动装管机 automatic tube filling machine
自动装货车 autoloader; automatic freight handling car; self-leading vehicle; self-loading vehicle
自动装货式汽车 self-loading vehicle
自动装料 gravity load(ing)
自动装料槽 automatic chute
自动装料的 self-loaded
自动装料机构 automatic loading mechanism
自动装料器 self-loading device
自动装料预成型机 autoload preformer
自动装轮机 automatic wheel press
自动装配 automated assembly; automatic assembly; automatic packaging; self-assembly; self-erecting
自动装配机 automatic assembly machine
自动装配技术 automatic assembly technique
自动装片 automatic threading
自动装片暗盒 autoload cassette; automatic magazine
自动装片摄影机 automatic magazine camera
自动装入程序 autoloader; automatic loader
自动装入器 autohandler
自动装填 self-loading
自动装填机 autoloader; automatic filling machine; automatic loader
自动装填机构 self-loading mechanism
自动装填器 autoloader
自动装桶机 barrel packing machine
自动装卸材料的设备 automated material handling equipment
自动装卸车 autoloader; lift truck
自动装卸船 self-loading and unloading ship
自动装卸机 autoloader; automatic loader and unloader; lorry loader; truck lift; truck loader
自动装卸设备 automatic handling equipment
自动装载 gravity load(ing)
自动装载铲运机 self-loading scraper
自动装载船 automatic loading ship; self-loading ship
自动装载机 autoloader; automatic loader; truck loader
自动装载机动手推车 self-loading motorized buggy
自动装载轮式铲运机 self-loading wheeled scraper
自动装载牵引斗式提升机 self-loading tractor bucket-elevator
自动装载拖车 self-loading trailer
自动装置 autodevice; automat; automatic device; automatic equipment; automatic gear; automatic nature; automatic plant; automatic system; automation; automaton[复 automata/automatons]; autoplant; complete automatic device; recording device; robot; self-acting device; self-contained plant
自动装置升降机 self-loading elevator
自动装置下夹板螺钉 screw lower bridge of automatic device
自动装置与自动控制 automat and automatic control
自动追踪全站仪 automated self-tracking total station
自动准合<混凝土路面的> self-dowelling
自动准进式凿岩 push-feed drilling
自动准进式钻眼 push-feed drilling
自动准直 autoalignment; autocollimate; autocollimation; automatic collimation
自动准直附件 autocollimation accessory; automatic collimation accessory
自动准直角 autocollimation angle
自动准直摄影机 autocollimation camera
自动准直望远镜 autocollimating telescope; automatic collimating telescope
自动准直仪 autocollimator
自动准直装置 autocollimation device
自动着陆系统 autoland system; automatic landing system
自动着陆装置 automatic landing system
自动资料处理 automatic data processing
自动资料检索系统 automatic data retrieval system; automatic document retrieval system
自动资料整理机 automatic data processor
自动子午环 automatic transit circle
自动自行式缓行器 automatic mobile retarder
自动自行式减速器 automatic mobile retarder
自动走廊 moving walkway
自动租借汽车 public automobile service
自动阻车器 automatic stop block
自动阻断器 autocut-out
自动阻尼 automatic damping
自动阻尼作用 autoinhibition
自动阻气门 automatic choke
自动阻气门至歧管连接管 automatic choke to manifold pipe
自动阻汽具 automatic steam trap
自动阻塞 self-blocking
自动钻 automatic drill
自动钻床 automatic drilling machine; pantodrill; self-acting drilling machine
自动钻进 auto drilling; automatic drilling
自动钻进系统 automatic drilling system
自动钻孔 automatic drilling
自动钻孔机器人 automatic drilling robot
自动钻头夹盘 automatic chuck; hydraulic chuck
自动最佳控制 self-optimizing control
自动最优化系统 self-optimizing system
自动作图仪 variplotter
自动作业 automatic operation; automatic working
自动作用 automatism
自动作用的 automatic
自动作用继电器 automatism relay
自动坐标绘图机 automated coordinate plotter; automatic coordinate plotter
自动坐标绘制器 automatic coordinatograph
自动坐标记录仪 coordimeter
自动坐标记录装置 automatic coordinate recording unit
自动坐标数字化器 automated coordinate digitizer
自动坐标展点仪 automatic coordinate plotter; coordimat
自抖 own jitter
自读标尺 self-reading rod; self-reading staff; speaking rod
自读测距仪 automatic readout distance meter
自读的 self-reading
自读剂量计 self-reading dosimeter
自读器 automatic readout
自读式石英丝剂量计 self-reading quartz fiber dosimeter
自读式水平标尺 Sopwith staff
自读式水准标尺 self-reading level(l)ing rod; self-reading staff; speaking rod
自读式水准尺 self-reading level rod
自读式袖珍剂量计 self-reading pocket dosimeter
自读数水准尺 self-reading level(l)ing rod
自读水准尺 self-reading (level) rod; self-reading staff; speaking rod
自读水准杆 self-reading (level) rod
自读型水准仪 self-reading level(l)ing-staff
自读仪 self-reading device
自镀液加热时间 self-plating solution heating time
自镀液浓度 self-plating solution concentration
自镀液温度 self-plating solution temperature
自断安全销 breaking pin
自断式安全阀 shear pin type relief valve
自对偶的 self-dual
自对偶函数 self-dual function
自对偶恒等式 self-dual identities
自对偶拟阵 self-dual matroid
自对偶群 self-dual groups
自对偶系 self-dual system
自对偶线性空间 self-dual linear space
自对应的 self-corresponding
自对中模板 self-centering formwork
自对中心弹簧盒 centring spring case
自对准的 self-aligned; self-aligning
自对准工艺 self-registered technology
自对准系统 self-aligning system
自轭的 self-conjugate
自发崩裂 spontaneous spalling
自发变化 spontaneous change; spontaneous variation
自发参数 autonomous parameter
自发成核 spontaneous nucleation
自发成核作用 spontaneous nucleation
自发磁化 spontaneous magnetization
自发磁化强度 spontaneous magnetization
自发磁致伸缩 spontaneous magnetostriction
自发催化剂 spontaneous catalyst
自发的 autonomic; idiomorphic; spontaneous; voluntary
自发电池 galvanic cell
自发电位测井 spontaneous potential well logging
自发电重量测定法 spontaneous electrogravimetry
自发对称性破坏 spontaneous symmetry breaking
自发对流 spontaneous convection
自发发射 spontaneous emission
自发发射体 spontaneous emitter
自发发射系数 spontaneous emission coefficient; A-coefficient
自发反射 idioreflex
自发反应 spontaneous reaction
自发放电 spontaneous discharge
自发放射 spontaneous emission
自发分解 spontaneous decomposition
自发分散 spontaneous dispersion
自发粉 self-raising flour
自发辐射 spontaneous activity emission; spontaneous radiation
自发复合 spontaneous recombination
自发复合辐射 spontaneous recombination radiation
自发复原 spontaneous restoration
自发观念 autochthonous idea
自发光 autoluminescence; spontaneous light
自发光材料 self-luminescent material
自发光的 self-luminous
自发光灯 self-luminescent lamp
自发光度 self-luminosity
自发光度盘 self-luminous dial
自发光发射器 spontaneous light emitter
自发光化合物 self-luminous compound
自发光色 self-luminous colo(u)r
自发光体 self-luminescent material; self-luminous body
自发光涂料 autoluminescent paint; self-luminous paint
自发光物质 self-luminescent material
自发光颜料 self-luminous pigment
自发光油漆 self-luminous paint
自发光源 self-emitting light source
自发光(知觉)色 self-luminous (perceived) colo(u)r
自发过程 autogentic process; spontaneous process
自发核反应 spontaneous nuclear reaction
自发核化 spontaneous nucleation
自发核转化 spontaneous nuclear transformation
自发活动 spontaneous activity
自发活性 spontaneous activity
自发火 endogenous fire; self-ignition
自发火的 self-triggering
自发火发动机 self-ignition engine
自发火温度 self-ignition temperature
自发畸变 spontaneous aberration
自发激光脉冲 spontaneous laser pulse
自发极化 spontaneous polarization
自发极化法 spontaneous polarization

method
自发极化强度 intensity of spontaneous polarization
自发加倍 spontaneous doubling
自发建设的贫民窟 happening shanty town
自发交易 autonomous transaction
自发酵作用 autofermentation
自发结晶 spontaneous crystallization
自发卷曲 spontaneous crimp
自发孔 idiomorphic pore
自发块体运动 spontaneous mass movement
自发量 independent variable
自发裂变 spontaneous decay;spontaneous fission
自发裂变过程 spontaneous fission process
自发裂变计数 spontaneous fission counting
自发裂变径迹密度 spontaneous fission track density
自发裂变率 spontaneous fission rate
自发裂变源 spontaneous fission source
自发凝集(反应) spontaneous agglutination
自发凝集素 idioagglutinin
自发凝聚 spontaneous condensation
自发票日后 after date
自发破坏<由腐蚀等造成的> spontaneous failure
自发破裂 cracking;spalling;spontaneous breaking
自发侵染 spontaneous infection
自发燃烧 spontaneous combustion
自发乳化作用 spontaneous emulsification
自发乳状液 spontaneous emulsion
自发散射 spontaneous scattering
自发射电极 self-emission electrode
自发射噪声 spontaneous emission noise
自发伸长 spontaneous elongation
自发剩余磁化 spontaneous remanent magnetization
自发衰变 spontaneous decay;spontaneous disintegration
自发弹性恢复 spontaneous elastic recovery
自发体荧光 autofluorescence
自发投资 autonomous investment
自发突变 spontaneous mutation
自发突变性 automutagenicity
自发蜕变 spontaneous decay;spontaneous disintegration
自发消退 spontaneous regression
自发信号电子选矿 self-signal(l)ing electronic separation
自发性 autonomy;spontaneity
自发演替 autogenic succession
自发氧化 spontaneous oxidation
自发液化 spontaneous liquefaction
自发跃迁 spontaneous transition
自发运动 autogenic movement;autonomic movement;spontaneous motion
自发运动错觉 autokinetic illusion
自发再结晶 spontaneous recrystallization
自发振荡 spontaneous oscillation
自发转化 spontaneous transformation
自发着火 spontaneous ignition
自发着火点 spontaneous ignition temperature
自发自收测试 back to back
自罚机制 self-punishment mechanism
自翻车 self-discharging wagon
自翻吊桶 alligator
自繁殖系统 self-reproductive system
自反赋范空间 reflexive normed space
自反关系 reflexive relation
自反函数 self-reciprocal function
自反局部凸空间 reflexive locally convex space
自反空间 reflexive space
自反馈 self feedback
自反馈放大器 self-feedback amplifier
自反馈式磁放大器 amplistat
自反律 reflexive law
自反模 reflexive module
自反应的 self-reacting
自返式沉积取芯器 boomerang sediment corer
自方差 autovariance
自放电 self-discharge;self-discharging;local action
自费电话 paid call
自费顾问 self-employed consultant
自分 autotilly
自分层涂料 self-stratifying coating
自分解 self-decomposition
自分界码 self-demarcating code
自分离 self-detaching
自分馏油泵 self-fractionating oil pump
自分配性 self-distributivity
自分散 self-dispersing
自粉磨 autogenous grinding
自封包装 self-sealing wrapping
自封闭 self-sealing
自封闭地热田 self-sealing geothermal field
自封闭地热系统 self-sealed geothermal system
自封(闭)漆 self-sealing paint
自封闭容器 self-sealing container
自封闭涂料 self-sealing paint
自封袋包装机 valve bag packer
自封袋插口 valve bag opening
自封袋装填机 valve bag filling machine
自封顶 self-topping
自封口袋 valve bag;valve sack
自封口纸袋 paper valve bag;self-closing paper valve sack
自封口纸袋贮库 magazine for valve sacks
自封连接器 self-sealing coupling
自封联轴节 self-sealing coupling
自封轮胎 puncture-proof tire[tyre];self-sealing tire[tyre]
自封燃料箱 self-sealing fuel tank
自封式 self-sealing
自封式内胎 self-sealing inner tube
自封式轴承 self-sealed bearing
自封箱 self-sealing tank
自封油系统 self-sealing oil system
自峰顶至计算停车点的距离 crest-to-clearance distance
自浮 self-floating
自浮动 self-relocation
自浮动程序 self-relocating program(me)
自浮管线 self-floating pipeline
自浮海底取样器 self-floating seabed sampler;free instrument
自浮式钢构架桩基平台 self-floating steel jacket and pile platform
自浮物 self-floater
自浮装置 self-buoyant unit
自浮浊流 auto-suspending turbidity flow
自辐射分解 self-radiolysis
自辐射阻抗 self-radiation impedance
自辐照 self-irradiation
自辐照损伤 self-irradiation damage
自辐照效应 self-irradiation effect
自负风险 owner's risk;own risk
自负盈亏 self-financing;sound financial footing
自复 spring return
自复操作 non-stick working
自复的 non-locking;self-healing
自复吊牌 self-restoring drop
自复断路器 self-reclosing breaker
自复归继电器 self-restoring relay
自复激励 self-compound excitation
自复继电器 reclosing relay;self-restoring relay
自复绝缘 self-restoring insulation
自复绕式电流互感器 autocompounded current transformer
自复式按钮 non-locking push button;non-stick push button;pushing button;spring return push button
自复式按钮控制 non-stick button control
自复式的 non-stick
自复式进路 non-stick route
自复式排列进路 non-stick working of a route
自复式拖挂设备检测器 self-restoring dragging equipment detector
自复位按钮 non-stick button
自复位回线 self-resetting loop
自复位循环 self-resetting loop
自复原 self-resetting;self-restoration
自复原回路 self-resetting loop
自复原效应 self-healing effect
自复制 self-reproduction
自复制的 self-replicating
自复作用 self-healing action
自覆盖土壤 self-mulching soil
自干 air dry;air setting;cold curing
自干晶纹(喷)漆 crystallizing lacquer
自干清漆 air drying varnish
自干扰 self-infection;self-interference
自干(色)漆 air drying paint
自干砂 no-bake sand
自干时间 air drying time
自感 auto-induction;self-feeling
自感磁通 self-induced magnetic flux
自感磁通量 magnetic flux of self-induction
自感电动势 self-induction electromotive force
自感电压 inductance voltage
自感量 self-inductance
自感耦合 autoinductive coupling
自感漂移 self-induced drift
自感系数 electrodynamic(al) capacity;self-inductance coefficient;self-induction coefficient
自感线圈 choke coil;choking coil;self-induction coil;self-inductor coil
自感线圈避雷器 self-induction coil lightning arrestor
自感应 electric(al) inertia;self-inductance;self-induction
自感应电流 self-induction current
自感应电压 self-induction voltage
自感应透明 self-induced transparency
自感应系数 coefficient of self-induction;self-inductance
自感振动 self-induced vibration
自给的 autotrophic;self-feeding;self-supplying;self-supporting;self-sustained
自给滴定管 automatic burette
自给电池组 local battery
自给计划 project plan
自给冷却机组 self-contained cooling unit
自给率 degree of self-sufficiency
自给能中子探测器 self-power neutron detector
自给农业 subsistence agriculture;subsistence farming
自给暖炉 base burner
自给偏压 self-bias
自给偏压电路 auto bias circuit
自给偏压放大器 self-biased amplifier
自给偏压截止 self-biased off
自给企业 self-contained plant
自给器 self-feeder
自给润滑泵 self-lubricating pump
自给生态系统 self-contained ecosystem
自给式平板机 unitized press
自给式平台 self-contained platform
自给式潜水器 autonomous underwater vehicle
自给式水下运载工具 autonomous underwater vehicle
自给式携带电灯 self-contained portable electric(al) lamp
自给式遥控潜水器 autonomous remotely controlled submersible
自给系统 autonomous system;self-contained unit
自给籽晶 self-seeding
自给自足 autarky;self-sufficiency
自给自足的 self-centered[centred];self-sufficient;self-sustaining
自给自足经济 self-sufficient economy
自耕农 owner-peasant
自耕农场<美国、加拿大分给移民的> homestead
自功率谱 autopower spectrum
自攻螺钉 self-drilling screw;sheet-metal screw;steel drill screw;tapping screw
自攻螺栓 tap bolt
自攻螺丝 self-tapping screw;tapping screw
自攻螺纹 automatic tapping;self-tapping thread
自攻螺纹的 self-threading
自攻螺针 self-threading screw;tapping screw
自攻丝螺钉 self-tapping screw;self-threading screw;sheet-metal screw;tapping screw
自供电探测器 self-powered detector
自供电自炸装置 self-powered destructor
自供能量的 self-energized
自供气式潜水装具 self-contained diving apparatus
自供热的 autothermic
自供水系统 personal water system
自共轭的 self-adjoint
自共轭二次曲面 self-conjugate quadric
自共轭二次曲线 self-conjugate conic
自共轭分类 self-conjugate partition
自共轭核 self-conjugate nuclei
自共轭粒子 self-conjugate particle
自共轭算子 self-conjugate operator
自共轭图 self-conjugate graph
自共轭向量空间 self-conjugate vector space
自共轭性 self-adjointness
自共轭元素 self-conjugate element
自箍式压力水管<水力发电> self-hooped penstock
自固定系统 self-standing system
自固化 self-vulcanizing
自固化树脂砂 self-setting resin sand
自固黏[粘]合剂 self-curing adhesive
自固黏[粘]结剂 self-curing adhesive
自关闭阀 self-closing valve
自关闭防爆阀门 autoclosing explosion flap
自关闭防火门 self-closing fire door
自关风门 automatic closing door;self-closing door
自关铰链 self-shutting hinge
自关龙头 self-closing faucet

自关门 swing door
自关门弹簧 spring gate
自灌 self-prime;self-priming
自灌充 self-priming recharge
自灌式泵房 self-priming pumping house
自光电效应器件 self-electrooptic(al) effect unit
自过滤 inherent filtration
自焊的 autogenic; autogenous; self-brazing
自航 self-propulsion
自航半潜钻探船 self-propelled semi-submer sible drilling rig
自航边耙挖泥船 self-propelled side drag hopper dredge(r)
自航驳(船) self-powered cargo vessel; self-propelled barge; self-propelling barge;self-propulsion barge
自航(驳)式的 self-propelled; self-propelling;self-propulsed
自航船 self-propelled ship; self-propelled vessel
自航船模 self-propelled ship model; self-propulsion model ship
自航的 automotive;self-propelled
自航浮式起重机 self-propelled floating crane
自航回旋浮吊 self-propelled floating revolving crane
自航集矿机 self-propelled mine-collecting machine
自航绞吸式挖泥船 self-propelled cutter suction dredge(r)
自航救生艇 self-propelled lifeboat
自航开底泥驳 self-propelled bottom dump barge(r)
自航链斗挖泥船 self-propelled bucket dredge(r)
自航泥驳 self-propelled mud barge; self-propelled spoil barge
自航耙吸式挖泥船 free floating suction dredge(r);hopper dredge(r); self-propelled hopper suction dredge(r); trailer; trailing suction hopper dredge(r)
自航器 auto-pilot;auto-steerer
自航式兵营船 self-propelled barrack ship
自航式带泥舱链斗挖泥船 serf-propelled hopper bucket dredger
自航式开底泥驳 self-propelled hopper barge
自航式链斗挖泥船 serf-propelled bucket dredge(r)
自航式料仓采砂船 self-propelled hopper dredge(r)
自航式平台 self-propelled platform; self-propelled unit
自航式潜水器 self-propelled submersible
自航式深海探测艇 self-propelled bathyscaphe
自航式水下研究潜水器 powered underwater research vehicle
自航式挖泥船 hopper dredge(r);seagoing dredge (r); self-propelled dredge(r);self-propelling dredge(r)
自航式抓斗挖泥船 serf-propelled grab dredge(r)
自航式装料采砂船 self-propelled hopper dredge(r)
自航式钻井平台 self-propelled drilling platform
自航试验 self-propelled test
自航水驳 self-propelled water barge
自航挖砂船 self-propelled sand dredge(r)
自航旋转浮吊 self-propelled floating revolving crane
自航装舱抓斗挖泥船 self-propelled grab hopper dredge(r)
自耗电弧炉 consumable electroarc furnace
自耗电极 consumable electrode;consutrode
自耗电极成型 consumable electrode-forming
自耗电极电弧熔炼法 consumable arc-melting process
自耗电极炉 consumable electrode furnace
自耗电极熔化 consumable melt;consutrode melting
自耗电极熔炼 consumable electrode melting
自耗电极熔炼的 consumably melted
自耗电极水冷坩埚电弧熔炼 consumable electrode cold-mold arc melting
自耗电极真空电弧炉 cold-mo(u)ld furnace
自耗定向电渣焊 consumable guide electroslag welding
自耗弧熔锭 consumable ingot; consumably arc-melted ingot
自耗嵌块 consumable insert
自耗系数 decay coefficient
自耗阳极 consumable anode
自合式抓斗 self-closing grab
自互补天线 self-complementary antenna
自互感 self-mutual inductance
自护式的 self-guarded
自护式辙叉 self-guarded frog
自护型泵 self-protecting type pump
自滑道 go-devil plane
自还原 autoreduction
自环绕系数 self-linking coefficient
自恢复的 serially reusable
自恢复电路 self-repairing circuit
自恢复能力 self-recovery capability
自恢复式粉末检波器 self-restoring coherer
自恢复循环 self-resetting loop
自回复定时器 self-resetting timer
自回归 autoregression
自回归表达式 autoregressive representative
自回归公式 autoregressive formula
自回归过程 autoregression process
自回归级数 autoregressive series
自回归累积移动平均(模型) autoregressive integrated moving average
自回归流动平均数 autoregressive moving average
自回归模型 autoregressive model
自回归平均滑动模型 automatic regression mean slide model
自回归系数 autoregressive coefficient
自回火 self-tempering
自回路 self-loop
自回授 self feedback
自回授放大器 self-feedback amplifier
自回位手柄 deadman's handle
自回正的 self-righting
自回正力矩 self-aligning torque
自回作用 self-healing action
自毁 destruct; self-destroying; self-mutilation
自毁器 self-destructor
自毁系统 destruct system
自毁线 destruct line
自毁引信 self-destroying fuse; self-destruction-type fuze
自毁装药 self-destruct charge
自毁装置 destructor;self-breakup unit; self-destroying device; self-destruction device;self-destruction equipment; self-destruction feature; sterilizer
自毁作用 self-destruction
自激摆振 self-excited oscillation
自激本机同步振荡器 free running local synchronizer oscillator
自激差速器 free running differential
自激磁放大器 self-excitation transductor;self-excited transductor
自激的 free running; self-energized; self-energizing; self-excited; self-exciting;self-sustained
自激低频发电机 self-excitation low-frequency generator
自激电动机 motor with self-excitation;self-exciting motor
自激电焊发电机 arc welding generator with self-excitation
自激电弧焊发电机 self-excited arc welding generator
自激电路 autonomous circuit; free running circuit;self-excited circuit
自激多谐振荡器 free running multivibrator; self-excited multivibrator; slave flip-flop
自激发 autoexcitation;build itself-up; self-excitation
自激发电机 free running generator; self-excited dynamo; self-exited generator
自激发射机 self-excited transmitter
自激放大器 self-excited amplifier
自激放电器 self-activated switch
自激共振器 independently excited cavity
自激行扫描电路 self-driven line-scanning circuit
自激弧焊发电机 arc welding generator with independent excitation
自激活 self-activation
自激活动 self-activating
自激激光器 free running laser
自激间歇振荡器 free running blocking generator; free running blocking oscillator
自激交流发电机 self-excited alternator
自激锯齿波发生器 free running sawtooth generator
自激空气动力 self-excited aerodynamic force
自激励 autoexcitation;self-excitation
自激励的 self-energizing
自激励系统 self-excited system
自激喷雾洗涤器 impinging entraining scrubber
自激频率 free running frequency
自激绕组 self-excitation winding;self-exciting winding
自激扫描 free running scan;free running sweep;self-running sweep
自激式除尘器 self-induced scrubber
自激式磁流体发电机 self-excited magnetohydrodynamic generator
自激式进相机 self-excited phase advancer
自激式升压机 self-excited booster
自激式水下船 autonomous underwater vehicle
自激式调制放大器 self-exciting modulating amplifier
自激式温度线数据采集系统 autonomous temperature line acquisition system
自激式再生发电机 self-excitation method regeneration generator
自激式振动密度计 self-exciting vibration densimeter
自激条件 self-exciting condition
自激现象 self-excitation phenomenon
自激型 self-excitation type
自激移位寄存器电路 autonomous shift register circuit
自激再生制动方式 self-excited regenerative braking
自激振荡 autoexcitation; autonomous oscillation; self-excitation oscillation;self-excited oscillation;self-induced oscillation; self-oscillating; self(-sustained) oscillation
自激振荡的 self-oscillatory
自激振荡电路 astable circuit; self-maintained circuit
自激振荡器 free running generator; self-excited oscillator; self-exited generator;self-oscillator
自激振荡器式发报机 self-oscillator radio telegraphy transmitter
自激振荡式换能器 self-oscillatory transducer
自激振动 self-excited vibration;self-induced vibration;self-sustained oscillation
自极化继电器 self-polarizing relay
自己的行为 factum[复 facta/factums]
自己动手建造者 self-builder
自己动手施工者 self-builder
自己动手做 do-it-yourself
自己供电的 self powered
自己驾驶自备汽车的人 own driver
自己仪表测井 recorder well
自己支持的 self-supporting
自挤压 self-compaction
自挤压螺旋铆件 thin-wall(ed) fixing; cavity fixing <薄墙壁用>
自计风速仪 anemometrograph
自计器 automatic recording ga(u)ge; automatic registering ga(u)ge; recording ga(u)ge;self recording ga(u)ge;self-registering ga(u)ge
自计视距仪 self-reducing stadia instrument;self-reducing tach(e)ometer
自记 self-registering
自记安培计 curve drawing ammeter
自记曝光计 actinograph
自记比色计 recording colo(u)rimeter
自记笔 recording pen;recording pointer;tracing pen
自记波长计 cymograph
自记波高计 automatic wave(height) recorder
自记波高仪 automatic wave(height) recorder
自记波浪仪 automatic wave(height) recorder
自记波频计 cymograph
自记补偿器 self-recording compensator
自记测波器 undograph
自记测波仪 automatic wave(height) recorder; self-recording wave meter;undograph;wave recorder
自记测风器 wind recorder
自记测高仪 altigraph;altitude recorder
自记测深仪 bathygraph; depth recorder;recording sounder
自记测渗计 recording lysimeter
自记测渗仪 recording lysimeter
自记测试车 recording truck
自记测微光度计 recording microphotometer
自记测微计 recording micrometer; self-recording micrometer
自记测微器 recording micrometer; registering micrometer;self-recording micrometer;self-registering micrometer
自记超声测波仪 automatic ultrasonic wave recorder
自记潮位计 automatic tide ga(u)ge;

automatic tide recorder; marigraph; recording tide ga(u)ge; self-recording tide ga(u)ge; self-registering tide ga(u)ge
自记潮位仪 mareograph; tide-recording gauge
自记潮汐计 automatic tide ga(u)ge; automatic tide-meter
自记潮汐仪 mareograph
自记称重仪 weightograph
自记传输测试器 recording transmission measuring set
自记磁天平 self-recording magnetic balance
自记的 autographic(al); self-recording
自记滴定 self-recording titration
自记地中渗透仪 recording lysimeter
自记电力表 recording wattmeter
自记电流表 recording ammeter
自记电流计 writing galvanometer
自记电压表 recording voltimeter
自记分光光度计 self-recording spectrophotometer
自记分光计 automatic recording spectrometer; self-recording spectrometer
自记分光仪 automatic recording spectrometer; self-recording spectrometer
自记风力表 anemograph; recording airspeed indicator; recording anemometer; self-recording anemometer; self-registering anemometer
自记风力计 anemogram; anemograph; recording airspeed indicator; recording anemometer; self-recording anemometer; self-registering anemometer
自记风速表 anemograph; recording airspeed indicator; recording anemometer; self-recording anemometer; self-registering anemometer
自记风速计 anemograph; recording airspeed indicator; recording airspeed meter; recording anemometer; self-recording anemometer; self-registering anemometer
自记风速仪 anemograph; self-recording anemometer
自记风向标 recording wind vane
自记风向计 recording wind vane
自记风向仪 wind direction(al) recorder
自记风压计 anemobiagraph
自记浮标液位计 tape float liquid-level ga(u)ge
自记浮子 recorder float
自记浮子水位计 tape ga(u)ge
自记干涉式气体分析器 self-recording interferometric gas analyser[analyzer]
自记干湿球湿度计 psychrograph
自记高差仪 recording statoscope; registering statoscope
自记高温计 recording pyrometer
自记光度计 recording photometer
自记海流计 recording current meter; self-recording current meter
自记海洋水文站 oceanographic(al) recording station
自记红外分光度计 recording infrared spectrometer
自记湖泊水位计 limnigraph [limnograph]
自记湖泊水位图 limnigram
自记回声测深仪 recording echo sounder; echograph
自记积深仪 integrating depth recorder
自记计程仪 odograph
自记计时器 registering chronograph
自记计数器 recording meter
自记记录 autographic(al) record
自记记录图 recording tachograph
自记记录仪 autographic(al) recording apparatus
自记加速表 recording accelerometer
自记加速度计 accelerograph
自记加速度图 accelerogram
自记接受器 record receiver
自记经纬仪 recording theodolite; stadi(o)meter
自记空盒气压表 aneroidograph
自记空盒气压计 recording aneroid barometer
自记空速计 recording airspeed meter
自记里程计 trochometer
自记量雪计 automatic snow ga(u)ge; self-recording snow ga(u)ge
自记量雪器 automatic snow ga(u)ge; self-recording snow ga(u)ge
自记流量表 self-recording flow meter
自记流量附属装置 flow recording attachment
自记流量计 flow recorder; recording flow-meter; self-record discharge meter; self-recording flow meter; water flow recorder
自记流量仪 recording stream ga(u)ge
自记流速表 recording current meter; self-recording current meter; self-recording flow meter
自记流速计 recording hydrometer
自记流速仪 recording current meter; self-recording current meter; self-recording flow meter
自记录高速仪表 self-recording high-speed instrument
自记录固结仪 autographic(al) oedometer
自记录式称重计 recording scale; recording weighing machine
自记录式秤 recording scale; recording weighing machine
自记录式仪表 grapher
自记录水位计 limnigraph
自记录压力计 registering manometer
自记录雨量计 autographic(al) rain ga(u)ge; recording rainfall ga(u)ge; pluviograph
自记录雨量器 ombrograph; pluviograph; recording rainfall ga(u)ge
自记脉冲计数器 scaler-printer
自记挠度计 self-recording deflection ga(u)ge
自记黏[粘]度计 recording visco(si)meter
自记扭力计 torsiograph
自记气体分析器 recording gas analyser[analyzer]
自记气体分析仪 recording gas analyser[analyzer]
自记气象记录仪 self-recording meteorograph
自记气压表 barograph; recording barometer; self-recording barometer; self-registering barometer
自记气压计 aneroid barograph; barograph; recording barometer; self-recording barometer; self-registering barometer
自记气压曲线 barogram; barograph trace
自记气压温度计 barothermograph
自记气压温度湿度计 barothermohydrograph
自记器 automatic recorder; registering apparatus; registering instrument; self-recorder
自记器笔杆 pen arm
自记器记录 autographic(al) record
自记热膨胀计 self-recording thermal expansion meter
自记溶度计 recording lysimeter
自记设备 recording equipment
自记渗漏仪 recording lysimeter
自记湿度表 hygroautometer; hygrograph; self-recording hygrometer
自记湿度计 hygroautometer; hygrograph; hygrometer recorder; self-recording hygrometer
自记湿度仪 hygrometer recorder
自记时的 self-timed
自记示波器 automatic recording oscillograph; recording oscillograph
自记示波仪 automatic recording oscillograph; recording oscillograph
自记示差热膨胀计 self-recording differential thermal expansion meter
自记示振计 vibrograph
自记示振仪 vibrograph
自记式测流仪 direct recording current meter
自记式电位计 recording potentiometer
自记式浮子水位计 recording float ga(u)ge
自记式海流计 recording current meter
自记式加速计 self-recording accelerometer
自记式金箔验电器 self-recording gold-leaf electroscope
自记式流量控制器 recording flow controller
自记式湿度表 recording hygrometer; recording psychrometer
自记式湿度计 recording hygrometer; recording psychrometer
自记式温度计 recording thermometer; thermograph; thermometrograph
自记式遥测仪 telerecorder
自记式应变记录器 autographic(al) strain recorder
自记式真空计 vacuum recorder
自记水表 water flow recorder
自记水尺 recording ga(u)ge
自记水位测井 self-recording water level log
自记水位观察井 ga(u)ge well
自记水位计 automatic fluviograph; automatic ga(u)ge; automatic water(level) ga(u)ge; automatic water stage recorder; fluviograph; hydrograph; level recorder; level recorder ga(u)ge; limnograph; recording(water) ga(u)ge; self-recording water level ga(u)ge; self-registering ga(u)ge; stage recorder; water-level meter; water-level recorder; water-stage recorder; water-stage register
自记水位计台 stage recorder installation
自记水位井 stilling well
自记水位井滞后 stilling well lag
自记水位图 limnograph
自记水位仪 automatic fluviograph; automatic ga(u)ge; automatic water(level) ga(u)ge; automatic water stage recorder; fluviograph; hydrograph; level recorder; level recorder ga(u)ge; recording ga(u)ge; self-recording water level ga(u)ge; self-registering ga(u)ge; water-level meter; water-stage recorder; water-stage register
自记水位仪井 ga(u)ge box; ga(u)ge well; ga(u)ging well
自记水位仪室 ga(u)ge chamber; ga(u)ge house
自记水位站 automatic water stage recording station; stage recorder station; water-stage recorder
自记水文资料仪 hydrological recording ga(u)ge
自记速度表 recording speed indicator
自记速度计 recording tach(e)ograph; tach(e)ograph
自记速率表 recording speed indicator
自记速率计 recording speed(o)meter
自记调节器 recording controller
自记图 record chart
自记温度表 automatic temperature recorder; recording thermometer; registering thermometer; self-recording thermometer; self-registering thermometer; thermograph
自记温度计 autographic(al) recording thermometer; automatic temperature recorder; registering thermometer; self-recording thermometer; self-registering thermometer; temperature recorder; thermograph
自记温度气压计 thermobarograph
自记温度气压记录器 self-recording thermo-barograph
自记温度图 thermogram
自记温湿度表 thermohygrograph
自记温湿度计 self-recording thermo-hygrometer; thermohygrograph
自记温湿计 hygrothermograph
自记无线电气象仪 automatic radiometeorograph
自记无液气压计 recording aneroid barometer; self-recording aneroid barometer
自记无液气压图 self-recording aneriodograph
自记型水压测波仪 direct recording pressure type wave meter
自记压力表 pressure-type type of recording ga(u)ge; recording pressure ga(u)ge
自记压力计 recording air ga(u)ge; recording manometer; recording pressure ga(u)ge
自记验潮仪 automatic(recording) tide ga(u)ge; marigraph; recording ga(u)ge; recording tide ga(u)ge; self-recording tide ga(u)ge; tide recorder
自记遥测 automatic recording and telemetering
自记遥测浮标 automatic recording and telemetering buoy
自记液度计 recording lysimeter
自记液面计 liquid level recorder; recording liquid level ga(u)ge
自记液体比重计 recording hydrometer
自记仪 autographic(al) recorder; self-recorder; self-recording unit
自记仪表 recording meter; self-recording meter
自记仪器 graphic(al) instrument; recorder; recording apparatus
自记仪器测井 recorder well; self-recorder log
自记仪器房 recorder shelter
自记仪器浮子 recorder float
自记仪器记录纸 recording instrument paper
自记仪器图纸 drum chart
自记仪器箱 recorder shelter
自记仪仪器房 recorder shelter
自记仪仪器箱 recorder shelter
自记应变仪 recording strain ga(u)ge
自记应力仪 stress recorder
自记雨量计 hyetograph; ombrograph; rainfall recorder; rainfall self-re-

corder; rain recorder; recording rain-ga(u)ge; registering pluviometer; self-recording rain-ga(u)ge; udomograph; pluviograph
自记雨量记录曲线 rainfall recorder chart
自记雨量器 rainfall recorder; recording rain-ga(u)ge
自记雨量站 recording precipitation station
自记雨雪量计 snow-rain recorder
自记雨雪量器 snow-rain recorder
自记站网 recording network
自记蒸发计 recording evapo(u)rimeter; self-recording evapo(u)rimeter
自记蒸发量记录(曲线) evaporogram
自记蒸发器 recording evapo(u)rimeter; self-recording evapo(u)rimeter
自记蒸散发仪 recording lysimeter
自记转速计 tach(e)ograph
自记转速图 self-recording tachogram
自记转速仪 self-recording tach(e)ograph
自记装置 recorder unit; recording device; recording instrument
自记纵断面测绘器 <测平整度用> profilograph
自加力制动带 self-activating brake band; self-energizing brake band
自加力制动器 self-activating brake
自加料自动称重掺和机 self-loading autoweighing blender
自加速度 autoacceleration; self-acceleration
自加速分解 self-accelerating decomposition
自加速分解温度 self-accelerating decomposition temperature
自加压断路器单元 self-compression interruption unit
自加压系统 self-pressurized system
自甲站通过之时起至乙站通过之时止的运转时分 pass to pass time
自驾汽车出租业务 self-drive car-hire service
自驾汽车旅行者 motorist
自驾驶的 self-drive
自监控式自动驾驶仪 self-monitored autopilot
自监控系统 self-monitoring system
自监控装置 self-monitoring device
自减型编址 autodecrement addressing
自检 self-checking; self-examination; self-inspection
自检编码 self-checking code
自检测故障与告警 self-test and failure warning
自检测试 self-test(ing)
自检查 self-check
自检程序 self-check procedures; self-check program(me)
自检法区域网平差 self-calibrating block adjustment
自检校法 self-calibration method
自检校区域网平差 self-calibrating block adjustment
自检控制电压 self-detecting control voltage
自检码 self-demarking code
自检能力 self-test capability
自检器 autoscope
自检系数 self-testing factor
自检信号频率 self-detecting signal frequency
自检验代码 self-checking code
自检验码 error checking code; error detecting code
自检(验)数 self-checking number
自检氧系统 self-calibrating oxygen system
自检装置 automatic trouble-locating arrangement; self-checking unit
自建 direct undertaking
自建电场 build-in field
自建房屋协会 self-build housing society
自建工程 direct undertaking works
自建公助 owner-occupation with public subsidy
自建公助住房 publicly aid private housing
自建业主 owner-builder
自建住房 self-help housing
自建住宅 self-help housing
自建自营(模式) build-own-operate
自降分级器 free settling tank classifier
自交代作用【地】 autometasomatic process; autometasomatism
自绞 self-heaving
自校 self-correcting
自校编码 self-correcting code; self-correcting coding; self-correction code
自校稳态 self-correcting homeostasis
自校验 self-checking
自校验电路 self-checking circuit
自校验方式 self-checking system
自校验号码 self-checking number
自校验码 self-checking code
自校验数 self-checking numeral
自校验数控装置 self-checking number feature
自校验数位 self-checking digit
自校正 self-correction; self-repairing
自校正记忆 self-correcting memory
自校正控制系统 self-aligning control system
自校正能力 self-correcting capability
自校正器 self-tuner
自校正调节器 self-turning regulator
自校正系统 self-correcting system
自校正自动领航仪 self-correcting automatic navigator
自校正自适应控制 self-tuning adaptive control
自校正自适应调节器 self-tuning adaptive controller; self-tuning adaptive regulator
自校准开关 self-check key
自校准水平仪 self-level(l)ing level
自校自动导航 self-correcting automatic navigation
自校钻夹 self-adjustable drill chuck
自接通 self-closing
自节油 <用于油漆的一种石油溶剂> white spirit
自洁 self-cleaning
自洁螺杆 self-cleaning screw
自洁式过滤器 self-cleaning strainer
自洁式烤箱 self-clean oven
自洁式钻头 self-cleaning bit
自洁搪瓷 self-cleaning enamel coating
自洁陶瓷 self-cleaning enamel
自结合材料 self-bonding material
自结皮泡沫 self-skinned foam
自解码 self-demarking code
自金内燃凿岩机 breaker contact point
自紧的 self-tightening
自紧法 autofrettage
自紧防松螺母 self-tightening locknut
自紧夹头 self-clinching fastener
自紧螺母 self-tightening nut
自紧密封 automatic seal; seal energized by medium pressure; self-acting seal; self-adjusting seal; self-sealing packing; self-tightening seal
自紧式夹杆 self-tightening lever clip
自紧式金属空心 O 形环 metallic hollow O-ring for self seal; metallic hollow self-energizing O-rings
自紧式密封 self-energized seal
自紧式密封环 self-sealing ring
自紧式弹性金属垫片 self-energizing resilient metal gaskets
自紧油挡 self-sealing oil seal
自紧油封 self-tightening oil seal
自紧油封圈簧 garter spring
自紧制动蹄片 primary brake-shoe
自进给 self-feed
自进式平台升降机 mobile hoist
自进式楔垫块 self-advancing chock
自进式支架 self-advancing support
自进式支柱系统 self-advancing prop system
自精制 autofining
自净参数 self-purification parameter
自净(的) self-cleaning; self-purifying; self purificating
自净电极 self-cleaning electrode; self-purification electrode
自净过程 self-cleaning process; self-purification process
自净化 automatic depuration; autopurification; self-purification
自净化过滤器 self-cleaning filter
自净化机理 mechanism of self-purification
自净化空气过滤器 self-cleaning air filter
自净化流速 self-cleaning velocity
自净化滤池 self-cleaning filter
自净化能力 autopurification power; self-cleaning ability; self-purifying ability; self-purification ability; self-purifying capacity
自净化水体 self-purification of water body
自净化特性 natural purification characteristic
自净化系统 self clarification system; self-cleaning system
自净化作用 autopurification; self-purification
自净活性 self-purification activity
自净机理 mechanism of self-purification
自净拦污栅 self-cleaning screen
自净量 amount of self-purification; self-purification amount
自净流速 <污水管内保持悬浮物不沉淤的最小流速> self-cleaning (flow) velocity
自净能力 self-cleaning ability; self-cleaning capacity; self-purification activity; self-purification capability
自净黏[粘]液过滤器 self-cleaning viscous filter
自净坡度 self-cleaning grade; self-purification grade
自净区 zone of self-purification
自净容量 self-cleaning capacity
自净软管过滤器 self-cleaning flexible pipe filter
自净式火花塞 self-cleaning spark plug
自净水体 self-purification water body
自净速度 self-purification velocity
自净速率 rate of self-purification
自净特性 self-purification characteristic
自净特征 self-purification characteristic
自净梯度 self-cleansing gradient
自净系统 self clarification system
自净效力 effect of self-purification
自净效应 self-purifying effect
自净性能 self-purification characteristic
自净转筛过滤器 self-cleaning rotating screen filter; self-cleaning rotation screen filter
自净作用 autopurification; effect of self-purification; natural purification
自净作用常数 self-purification constant
自净作用数据 data of self-purification
自救 self-help
自救呼吸器 self-rescue apparatus
自救绞车 self-recovery winch
自救绞盘 recovery winch
自救灭火 first-aid extinguishing
自救器 escapes apparatus; self-rescuer
自举程序 bootstrap routine
自举电路 bootstrap circuit
自举放大器 <阴极输出放大器> bootstrap amplifier
自举激励器 bootstrap driver
自举驱动器 bootstrap driver
自举作用 bootstrap function
自具电荷 self-charge
自具能量 self energy
自聚合作用 self-polymerization
自聚焦 self-focusing
自聚焦光束 self-trapped optic(al) beam
自聚焦光纤 self-focusing optical fiber
自聚焦光学纤维 self-focusing optic(al) fiber
自聚焦机制 self-focusing mechanism
自聚焦激光器 self-focusing laser
自聚焦激光束 self-focused laser beam
自聚焦束 self-focusing beam
自聚焦束流 self-focusing stream
自聚焦纤维 self-focusing fiber; selfoc fibre[fiber]
自聚焦整形放大镜 autofocus rectifier
自聚物 autopolymer
自卷的 self-winding
自卷积 self-convolution
自觉执行 self-enforcing
自觉遵守 voluntary observance
自均衡 self-recovery
自卡式溜口 self-choking chute
自开电梯 self-service elevator
自开孔鞍形三通 self-tapping saddle tee
自开翼梢缝 automatic wing tip slot
自控 internally piloting
自控的 self-controlled
自控阀 internally piloted valve
自控工程 automatic engineering
自控管理 management by self-control
自控恒温器 self-thermostat
自控能力缺失【心】 acrasia
自控器 automatic controller
自控栅 self-aligning gate
自控生态系统 self-contained system
自控式的 self-controlled
自控式园圃拖拉机 automatic garden tractor
自控式振动 self-controlled vibration
自控水下监听站 robot submarine listening post
自控微波天线阵 self-steering microwave array
自控系统 automatic control system; autonomous system
自控系统年检 annual automated control system survey
自控仪表 automatic instrument
自控油压表 automatic oil pressure ga(u)ge
自控与遥控 auto and remote
自控振荡器 self-controlled oscillator
自溃坝 emergency dam; fuse plug dam
自溃堤 breaching dike
自溃式安全溢洪道 exploding plug
自溃式非常溢洪道 fuse plug emergency spillway
自溃式闸板 collapsible flash board
自溃闸门 <非常洪水时应急之用> self-collapsing gate
自扩散系数 self-diffusion coefficient
自扩散作用 self-diffusion
自来夹头 <工人口语>【机】 three-

jaw chuck
自来水 city water(supply);main water;running water;tap water;water supply
自来水笔 fountain pen
自来水表 cold water meter
自来水厂 public water works;water plant;water purification station;water supply plant;water(treatment)works
自来水厂建设 waterworks construction
自来水厂进水口 intake of water works
自来水厂设计 waterworks design
自来水厂运作 waterworks operation
自来水的臭氧处理 ozonation;ozonization
自来水费 charges for water;water rate
自来水工程 public water works;waterworks
自来水工程师 water engineer
自来水工程师学会 Institution of Water Engineers
自来水工程系统 public water works system
自来水公司 water(supply)company;water undertaking
自来水供应服务人 water undertaker
自来水管 water main
自来水救火龙头 water hydrant
自来水配水管网 distribution system of water supply
自来水普及率 water supply pervasion
自来水塔 water tower
自来水箱 city water tank
自来水消毒 disinfection of water supply
自来水消防栓 water hydrant
自来水消费者 water consumer
自来水用户 water consumer;water user
自来水闸阀 water gate valve
自冷 self-ventilation
自冷变压器 self-cooled transformer
自冷的 non-ventilated;self-cooled
自冷电动机 non-ventilated motor
自冷凝作用 self-condensation
自冷却的 self-cooled
自冷式 self-cooling
自冷式变压器 self-cooled transformer
自冷式电机 self-cooled machine
自冷式喷管 self-cooled nozzle
自冷式汽缸 still-cooled cylinder
自冷式叶片 self-cooling vane
自冷速度 self-cooling speed
自理充电蓄电池 self-servicing battery
自理的 self-service
自理停车场 self parking
自理洗衣店 self-service laundry
自理行包寄存锁柜 self-service checking locker
自理装卸 cargo-handled by owner
自力动作阀 self-acting valve
自力更生 regeneration through one's own efforts;self-reliance
自力更生的 self-made
自力供料装置 gravity feed
自力接触 self-contact
自力控制 self-operated control
自力式恒温器 self-acting thermostat
自立的 self-supported;self-supporting;self-sustaining
自立横板式密仓 horizontal plank free standing silo
自立脚手架 self-supporting scaffold
自立式板桩码头 cantilever sheet pile type wharf
自立式瓷砖隔墙 free-standing partition(wall)
自立式挡土结构 non-supporting retaining structure
自立式地下连续墙 non-supporting diaphragm wall
自立式电杆 self-supporting mast
自立式钢板网 self-centering lath
自立式钢结构 self-supporting structure
自立式柜 free-standing cubicle
自立式梯子 standing step ladder
自立式天线杆 self-supporting antenna mast
自立式铁塔 self-supporting tower
自立式线圈 self-supporting coil
自立式烟囱 free-standing chimney
自立型燃料元件 free-standing fuel element
自立支架梯 trestle ladder
自立柱 isolated column
自励并激磁场 self-excited shunt field
自励串激绕组 self-excited series winding
自励磁发电机 self-exciting dynamo
自励(磁)绕组 self-excitation winding
自励的 self-driven;self-excited;self-exciting
自励电动机 self-excited motor
自励电机 self-excited machine
自励电路 self-excited circuit
自励发电机 self-excited dynamo;self-excited generator;self-exciter
自励恒压装置 self-exciting constant voltage device
自励交流发电机 self-excited alternative current generator
自励式(发)电机 self-excited dynamo
自励式振动 self-excited vibration
自励式直流(发)电机 self-excited dynamo
自粒化污泥 self-granulated sludge
自连续法 endochronic approach
自连续理论 endochronic theory
自亮标志 self-illuminated sign
自裂 spontaneous cracking
自裂变 self-fission
自裂缝 endokinetic fissure
自裂化 autothermic cracking
自裂开排驱机理 autofracturing expulsion mechanism
自溜供料 gravity feed
自溜角 angle of slide
自溜式过滤器 gravity filter
自溜式喷洒机 gravity distributor
自溜式撒布 gravity distribution
自溜式索道 gravity ropeway;gravity wireway
自溜式装置 gravity apparatus
自溜输送 self-slipping transportation
自流 downgrade flow;flow by gravity;gravity current;gravity flow
自流产水量 artesian capacity
自流产水率 artesian capacity
自流承压水 flowing artesian
自流充灌 fill by gravity
自流充填 controlled-gravity stowing
自流充填法 flow stowing;gravity stowing;mobile filling
自流储罐 gravity tank
自流的 artesian
自流地带 artesian land
自流地下水 artesian groundwater;confined groundwater;piestic water
自流地下水层 artesian aquifer
自流发电站 natural flow station
自流法 gravity method
自流干管(路) gravity main
自流工程方案 gravity scheme
自流供料 gravity feed
自流供料储槽 gravitational tank
自流供料罐 gravitational tank
自流供水 gravity feed;gravity(water)supply
自流(供水)系统 gravity system
自流管 free flow conduit
自流管道 gravity line
自流管井 artesian bored well
自流管路 gravity main
自流管线 gravity line;gravity pipeline
自流灌溉 gravity irrigation;irrigation by gravity
自流灌溉田 commanded land
自流灌浆 gravitational grouting;gravity grouting;self-flowing grout
自流含水层 artesian aquifer;artesian formation;confined aquifer
自流(含水层)建造【地】 artesian formation
自流回灌 artesian recharge
自流加(汽)油 gravity gasoline feed;gravity petrol feed
自流减压井 artesian relief well
自流浇注料 free flowing castable
自流井 artesian(flowing)well;blow(ing)well;bore well;flowing artesian well;flowing bore;flowing water well;flow(ing)well;free flowing bore;free flowing well;gravity well;gusher;self-flowing well
自流井泵 artesian well pump
自流井产水率 artesian well capacity
自流井出水量 artesian capacity of well;artesian well capacity;capacity of artesian well
自流井出水率 artesian well capacity
自流井盆地 artesian basin
自流井竖管 drive pipe;drive tube
自流井水 artesian water
自流井水量损耗 artesian waste
自流井水水面 artesian surface
自流井水水头 artesian surface;artesian head
自流井水头 artesian pressure
自流井压力 artesian pressure
自流井压力水位 artesian surface
自流井用抽水机 artesian well pump
自流流程图 gravity scheme
自流流出量 artesian discharge
自流流量 artesian discharge;artesian flow
自流排水 artesian discharge;artesian waste;drainage by gravity;free drainage;free draining;gravity discharge;gravity drain;gravity drainage
自流喷泉 artesian fountain
自流盆地 artesian basin
自流平地面石膏 self-level(l)ing floor plaster
自流平地面涂料 self-level(l)ing floor coating
自流平密封膏 self-level(l)ing sealant
自流坡度及流域 artesian slope and basin
自流畦灌 gravity check irrigation
自流区 artesian province
自流区域 artesian-flow area
自流泉 artesian fountain;artesian spring;blow-well;flowing spring;gravity spring;pressure spring
自流泉盆地 artesian basin
自流润滑法 gravity lubrication
自流筛 gravity flow screen
自流渗漏 artesian leakage
自流式干燥机 gravity flow drier
自流式供给 natural feed(ing)
自流式计量给料拌和机 gravity batch mixer
自流式凝汽器 scoop condenser
自流式排水法 artesian drainage method
自流式(排种)节流活门 self-flow reducer
自流式输送器 gravity conveyer[conveyor]
自流式输油箱 gravity feed tank
自流式水口 free running nozzle
自流式水冷却 gravity-system water cooling
自流式(水力)发电站 run-of-river type power station
自流式隧道 free surface tunnel
自流式卸料槽 gravity discharge chute
自流式溢洪道 open spillway
自流式溢流堰 free(fall)weir;uncontrolled spillway weir;weir with free fall;uncontrolled weir
自流输水管 gravity conduit
自流输送系统 gravity system
自流水 artesian flow;artesian water;confined groundwater;confined water;gravitational water;piestic water;gravity water
自流水层 artesian aquifer
自流水出流量 artesian discharge
自流水地区 artesian area
自流水动力 artesian water power
自流水管网的干管 gravity main
自流水井 artesian flowing well
自流水静压力 artesian hydrostatic(al)pressure
自流水力发电站 natural flow station
自流水量 <自流井的> artesian discharge
自流水流 artesian flow;artesian water flow
自流水面积 artesian-flow area
自流水排泄量 artesian discharge
自流水盆地 artesian basin
自流水情况 artesian condition
自流水区域 artesian-flow area
自流水水头 artesian head
自流水水压头 artesian pressure head
自流水损失 artesian loss
自流水头 geodetic pressure head
自流水头高度 altitude of artesian water head
自流水系统 gravity system
自流水斜地 artesian monoclinal stratum
自流水循环 artesian water circulation
自流水压力 artesian pressure
自流水压面 artesian pressure surface
自流水压头 artesian pressure
自流水源 artesian sources
自流筒仓 live silo
自流系统 artesian system
自流斜地 artesian arrisway
自流斜坡 artesian slope
自流卸料门孔 gravity gate
自流卸载式谷物车厢 gravity flow grain box
自流循环 gravity circulation
自流压力 artesian pressure
自流压力面 <地下水> artesian pressure surface
自流压力坡度 artesian pressure gradient
自流压头 natural head
自流液计 free running juice
自流油罐 gravitation tank
自流注射器 fountain syringe
自流装/卸油 oil-loading/discharging by gravity flow
自流钻井 gusher type well
自流钻孔 flowing borehole
自留地 family land plot;private plot;small holding
自留资金 retained fund
自硫化 self-vulcanizing
自漏 natural leakage
自录气压计 barometrograph;self-re-

cording barometer

自录式固结仪 autographic(al) oedometer

自录式连续比色计 self-recording serial colo(u)rimeter

自律 autonomy;self-discipline

自律分散同步系统 autonomous decentralized synchronization system

自律控制 internal control;non-interacting control

自律式控制 non-interacting control

自律式调节 independent control

自律系统 self-contained system;self-containing unit

自律性 automaticity;automatic rhythmicity

自律性条件 non-interacting condition

自轮滚动装卸 roll on/roll off

自轮运转(的货物) running on its own wheels

自落闭锁装置 gravity lock

自落连续式混凝土搅拌机 continuous gravity concrete mixer

自落式拌和机 free fall mixer

自落式混凝土浇筑管 pipe for concrete placing by gravity

自落式混凝土搅拌机 free drop concrete mixing plant;rotating concrete mixer

自落式混凝土搅拌设备 gravity type concrete mixing plant

自落式搅拌机 free drop mixer;free fall mixer;gravity type mixer

自落式气功涂料喷射机 gravity paint sprayer

自落式取芯器取样 sampling by free fall coring

自落式混凝土搅拌机 free drop concrete mixing plant

自落式稳定土厂拌设备 gravity type stabilized soil mixing plant

自落式稳定土搅拌设备 gravity-stabilized soil mixing plant

自落式抓斗取样 sampling by free fall grabber

自落填缝反滤层 collapsible filter

自码头货棚至水边间的空间 wharf apron

自埋锚 self-burying anchor

自埋装卸 free in and out

自脉冲系统 self-pulsed system

自蔓延高温合成法 self-propagating high-temperature synthesis

自锚 < 即黏[粘]结锚固 > self-anchor(ing)

自锚碇 self-anchorage

自锚碇的 self-anchored;self-anchoring

自锚固 self-anchoring

自锚拉索 self-anchored stay cable

自锚式 self-anchored;self-anchoring

自锚式吊桥 self-anchored suspension bridge

自锚式斜拉桥 self-anchored cable-stayed bridge

自锚式悬索桥 self-anchored suspension bridge

自锚式悬索体系 self-anchored suspension system

自锚体系 self-anchored system

自密封 self-sealing

自密封气压管路系统 self-sealing pneumatic tube system

自密封条 self-sealing strip

自密封涂层 self-sealing coating

自密封橡胶隔片 self-sealing rubber septum

自密集 dense-in-itself-set

自密切 self-osculation

自密切点 osnode

自密实 self-compaction

自密实混凝土 self-compacting concrete

自描述 self-description

自灭河 suicidal stream

自灭弧断路 self-extinguishing circuit-break

自灭式计数管 self-quenching counter tube

自灭系统 self-quenching system

自灭信号 self cancelling signal

自灭作用 autosterization

自明的 self-evident;self-explanatory

自鸣噪声 ringing noise

自模化 self-modeling

自模拟方程 self-simulating equation

自磨 autogenous grinding

自磨机 autogenous grinding mill;autogenous tumbling mill

自磨刃犁桦 self-sharpening share

自磨刃式锄铲 self-sharpening blade

自磨刃式刀片 self-sharpening blade

自磨式钻头 self-sharpening bit

自磨效果 self-sharpening effect

自磨钻头 free cutting bit

自磨作用 self-grinding action;attrition

自某地点至某地点的特价 point-to-point special rate

自母岩向上运移 upward migration from source rock

自母岩向下运移 downward migration from source rock

自南来 auster

自逆电容 self-stiffness

自黏[粘]层压板 self-adhesive laminate

自黏[粘]带 self-adhesive tape

自黏[粘]的 self-adhesive

自黏[粘]的薄膜 self-adhesive film

自黏[粘]的面砖 self-adhesive tile

自黏[粘]法 self-adhesion method;self-bonding

自黏[粘]合 self-adhesion

自黏[粘](合)材料 self-adhesive material

自黏[粘]合碳化物 self-bonded carbide

自黏[粘]胶带 self-adhesive tape

自黏[粘]聚合物改性沥青油毡 self-adhering polymer modified bituminous sheet

自黏[粘]力 autohesion

自黏[粘]铝箔屋面卷材 self-stick alumin(i)um roll roofing

自黏[粘]性 autohesion;self-adhesive

自黏[粘]性绝缘胶带 self-adhesive insulating tape

自黏[粘]作用 autohesion;self-adhesion

自凝(固)的 self-setting;air setting;self-solidifying

自凝灰浆 self hardening slurry;self-setting slurry

自凝灰浆防渗墙 self-hardening slurry cut-off wall

自凝灰浆截水墙 self-hardening slurry cut-off wall

自凝泥浆 air-settling mortar

自凝黏[粘]结剂 air-setting binder

自凝膨润土泥浆 self-setting bentonite slurry

自凝树脂液 liquid for self-curing resin

自凝液体 self-solidifying liquid

自凝液体树脂 solidifying liquid resin

自耦变压器 auto connected transformer;autoconverter;autoformer;auto-jigger;compensator starter;divisor;transtat;varitran;volt box;autotransformer

自耦变压器反馈振荡器 autotransformer coupled oscillator

自耦变压器供电方式 autotransformer feeding system

自耦变压器供电线 autotransformer feeder

自耦变压器框架 compensator frame

自耦变压器耦合 autoinductive coupling

自耦变压器启动 compensator starting

自耦变压器式起动器 autostarter;compensator starter;autotransformer starter

自耦变压器所 autotransformer post

自耦磁放大器 autotransductor

自耦起动器 autostarter

自耦调压变压器 joystick transformer;variac

自拍 autodyne;autoheterodyne;self-heterodyne

自拍电路 autodyne circuit

自拍接收机 autodyne receiver

自排气的 self-purging

自排水 self-drain

自排水工程 self-drainage

自排水救生艇 self-baling(life) boat;self-draining lifeboat

自配 autogamy

自配级的 self-conjugate;self-polar

自配极曲面 self-polar surface

自配极曲线 self-polar curve

自配极三角形 self-conjugate triangle;self-polar triangle

自配极四面形 self-conjugate tetrahedron;self-polar tetrahedron

自配极拓扑 self-polar topology

自配准 self-registration

自喷 blow in;flow by heads

自喷层测试 flowing zone test

自喷产量 flowing production rate

自喷管柱 flow string

自喷井 gusher;flowing artesian well

自喷井管头压力 flowing tubing head pressure

自喷井口装置 flow head

自喷射 self-injection

自喷生产 flowing production

自喷压力 flowing pressure

自喷油井 flowing well;gusher(well);unloading well

自喷油井管 well under control

自喷油气比 flowing gas factor;flowing gas-oil ratio

自喷油嘴 flow bean choke;flow plug

自喷闸门 flow gate

自喷装置 well-control equipment

自膨胀水泥 self-expanding cement;stressing cement

自偏电阻器 self-biasing resistor

自偏放大器 self-biased amplifier

自偏管 self-biased tube

自偏压 self bias(sing)

自偏压电子管 self-biased tube

自偏压电阻器 self-bias resistor

自偏压截止效应 self-bias cut-off effect

自偏压三极管 self-biased triode

自偏压效应 self-biasing effect

自偏阻抗 self-biasing impedance

自频 self-frequency

自平衡 self-balancing;self-equilibrating

自平衡补偿测量 self-balancing compensation measurement

自平衡测量仪器 self-balancing measuring equipment

自平衡磁伺服放大器 self-balancing magnetic servo amplifier

自平衡放大器 self-balancing amplifier

自平衡分压器 self-balancing potentiometer

自平衡接收记录控制器 self-balancing receive-recorder controller

自平衡力 self-balancing force;self-poise

自平衡热线风速计 self-balancing hot wire anemometer

自平衡式电位计 self-balancing potentiometer

自平衡调节 inherent regulation

自平衡系统 self-balancing system

自平衡性 self-regulation

自平衡应变仪 self-balancing strain ga(u)ge

自平水准仪 self-level(l)ing level

自平行曲线 self-parallel curves

自平预算 balanced budget

自屏蔽 self-screening;self-shielding

自屏蔽范围 self-screening range

自屏蔽干扰 self-screening jamming

自屏蔽系数 self-shielding factor

自屏蔽线圈 self-shielding coil

自屏蔽效应 self-shielding effect

自屏蔽因子 self-shielding factor

自屏毒物 self-shielded poison

自屏函数 self-shielding function

自起电 autogenous electrification

自起动 < 又称自启动 > self-start(ing)

自起动泵 self-priming pump;self-starting pump

自起动的 self-triggering

自起动电动机 self-drive motor;self-start(ing) motor

自起动负荷 self-start-up load

自起动虹吸管 self-priming siphon;self-starting siphon

自起动离心泵 self-starting centrifugal pump

自起动能力 self-start-up capacity

自起动器 self-starter

自起动容量 self-start-up capacity

自起动水泵 self-prime pump;self-priming pump

自起动同步电动机 self-starting synchronous motor

自起动旋转变流机 self-starting rotary converter

自气化 autopneumatolysis

自气化作用 autopneumalolysis

自汽化 self-evaporation

自洽力场 self-consistent field

自洽性 self-consistency

自嵌入 self-embedding

自嵌入特性 self-embedding property

自切点 tac-point

自切点轨迹 tac locus

自侵入作用 autoinjection;autointrusion

自倾漏斗车 self-dumping hopper

自倾汽车 dump truck

自倾式泥驳 self-dumping mud barge

自倾式洗矿槽 self-tilting concentrator

自倾双轮(小)车 dump buggy

自倾卸驳(船) dump barge;self-tipping barge

自倾卸货卡车 tipping lorry

自倾卸拖(挂)车 dump body trailer

自清铲斗 self-cleaning bucket

自清(除) self-purging

自清除冗余 self-purging redundance

自清流速 self-clean(s)ing velocity

自清能力 self-purification ability

自清扫 self-cleaning

自清扫接点 self-cleaning contact;self-wiping contact

自清扫式烟箱 self-cleaning smokebox

自清式格筛 self-cleaning grizzl(e)y

自清式钻头 self-cleaning bit
自清速度 self-clean(s)ing velocity
自清洗 self-cleaning
自清洗式分离器 self-cleaning separator
自清洗式滤器 self-cleaning strainer
自清系统 self clarification system
自清作用 self-purification
自驱点 self-propulsion point
自驱动 self-driven
自驱动仪表 self-driven instrument
自屈曲 self-inflection
自取式陈列柜 self-service display case
自去磁 self-demagnetization
自去磁力 self-demagnetization force
自去磁损耗 self-demagnetization loss
自去磁因数 self-demagnetization factor
自去油过滤器 self-degreasing oil filter
自然 naturalness;nature
自然白度 natural whiteness
自然半宽 natural half-width
自然保持 nature conservation
自然保护 conservation of nature;natural conservation;nature shelter
自然保护法 law of natural protection;natural conservation law
自然保护教育 natural conservation education
自然保护区 area of natural reserves;area of nature reserves;conservancy area;conservation area;natural conservation area;natural conservation zone;natural protection area;natural reserve area;natural reserves;nature protection area;nature reserves;nature sanctuary;protected natural area;protected natural feature
自然保护区的类型 categories of natural conservations;categories of reservations
自然保护区管理 nature reserve management
自然保护区管理机构 office of nature reserves
自然保护区或军事禁区 unmined of natural area or military area
自然保护区评价标准 criterion for reservation assessment
自然保护统计 statistics of natural conservation;statistics of natural protection
自然保护委员会 Nature Conservation Council
自然保护宪章 charter of natural conservation
自然保护运动 natural conservation campaign
自然保护政策 natural conservation policy
自然保留地 nature reserves
自然保障 natural conservation;natural shelter
自然背景 natural background;physical setting
自然背景辐射 natural background radiation
自然背景值 natural background level
自然倍数 natural multiple
自然本底 natural background
自然本底分析 natural background analysis
自然本底辐射 natural background radiation
自然本底值 natural background level
自然崩落 free breakage;natural quarrying;spontaneous caving
自然比例 natural proportion
自然比例尺 natural scale;representative fraction
自然比例极限 natural proportional limit
自然比特 natural bit
自然庇护区 nature reserves
自然边界 natural boundary
自然边界测量 natural boundary survey
自然边界条件 natural boundary condition
自然边界线 physical boundary
自然贬值 natural depression
自然变化 spontaneous change
自然变换 natural transformation
自然变质 inherent vice
自然辩证法 natural dialectic
自然标 natural rod
自然标高 elevation of natural ground;natural grade
自然标架 natural frame
自然标志 natural monument
自然表面 curved physical surface;physical surface
自然波 natural wave
自然波动 natural oscillation
自然播种植物 self-sown flora
自然博物馆 Museum of Natural history
自然步长 natural step length
自然材料 natural material
自然财富 natural wealth
自然裁弯 autopiracy;natural cutoff
自然裁弯段 natural cutoff
自然裁弯河流 avulsion river
自然采光 daylighting;natural lighting
自然采光时间 service period
自然采光系数 coefficient of natural lighting
自然采光照度系数 natural illumination factor
自然采暖 natural heating
自然参考架 natural reference frame
自然参考面 physical reference surface
自然测度 natural measure
自然层 natural layer
自然层理状况 in-place conditions
自然层土的含水量 field capacity
自然产生的离散电磁波 atmospherics
自然产水量极限 <地下水的> physical yield limit
自然产物 fructus naturales;natural product
自然产油井 natural well
自然潮 natural tide
自然尘土 natural dust
自然沉淀 natural sedimentation;physical sedimentation;plain coagulation;plain sedimentation
自然沉淀池 natural sedimentation tank;natural settling tank;physical sedimentation tank;plain settling tank
自然沉淀法 natural sedimentation method
自然沉积 plain sedimentation
自然沉降 free settling
自然陈化 natural ag(e)ing
自然成熟 natural maturity
自然乘法 natural multiplication
自然澄清 natural clarification
自然橙色烟 self-activating orange smoke signal
自然池净化 lagooning
自然尺寸 natural size
自然尺度 natural scale
自然充气 natural aeration
自然冲刷 geologic(al) erosion;natural erosion;normal erosion;self-scouring
自然冲刷周期 normal erosion cycle
自然冲洗 natural flushing
自然冲淤 self-maintaining;self-scouring
自然抽风 chimney ventilation;natural draft
自然抽风锅炉 natural draft boiler
自然抽风效应 chimney effect
自然抽风烟囱 natural draft chimney
自然抽气机 naturally aspirated engine
自然穿孔取直 avulsive cutoff
自然穿透取直 avulsive cutoff
自然传粉 natural pollination;open pollination
自然传输 natural transmission
自然床层 natural bed
自然垂度 natural sag
自然磁畴 spontaneous magnetic domain
自然磁阻 natural reluctance
自然(粗)糙率 natural roughness
自然村 dispersed settlement;dispersed village
自然单位 natural unit
自然到期制 running-out system
自然的 native;natural;naturalistic;spontaneous;unaffected;unearned;unforced
自然的非构造运动 natural non-tectonic movement
自然的人化 humanization of nature
自然的自我恢复能力 self-restoring capacity of nature
自然等价 natural equivalence
自然堤 natural levee;raised bank
自然堤后滩地 backland
自然敌害 natural enemy
自然底数测定值 measurement value of nature base number
自然抵抗力 natural resistance
自然地槽 self-geosyncline
自然地层 natural ground
自然地带 natural belt;natural zone
自然地电势 natural earth potential
自然地基 natural base
自然地理参数 physiographic(al) parameter
自然地理的 physio(geo)graphic(al)
自然地理规划 physical geographic(al) planning
自然地理环境 physiographic(al) environment
自然地理气候 physical geographic(al) climate
自然地理区划 physical geographic(al) regionalization;physical geographic(al) zoning
自然地理群系 physiographic(al) formation
自然地理生态学 physiographic(al) ecology;physiography ecology
自然地理条件 physiographic(al) condition
自然地理图(集) physical map;physiogeographic(al) map;physiographic diagram
自然地理学 physical geography;physiography
自然地理循环 physiographic(al) cycle
自然地理演替 physiographic(al) succession
自然地理要素 ground feature;land feature;natural feature
自然地理因素 physiographic(al) factor
自然地理志 scenery
自然地貌 natural feature;natural geomorphy
自然地面 natural ground
自然地面沉降 natural land subsidence
自然地平 natural grade
自然地区 <未受人类活动影响的> natural area
自然地图 physical map
自然地下水 native groundwater
自然地形 natural feature;physical relief;physiographic form
自然地质过程 natural geologic(al) process
自然地质环境 physical geologic(al) environment
自然地质现象 physical geologic(al) phenomenon
自然地质学 physical geology
自然地质因素 natural geologic(al) factor
自然地质灾害 natural geologic(al) hazard
自然地质作用 natural geologic(al) process
自然地租 natural rent of land
自然递增 natural increasing
自然递增过程 natural increasing process
自然碲 tellurium
自然点火 autogenous ignition
自然电场 natural electric(al) field
自然电场测量方法 way for measurement self-potential method
自然电场法 method of electric(al) prospecting section;natural electric(al) field method;natural electric(al) exploration
自然电场法电位测量 potential measurement in self-potential method
自然电场法梯度测量 gradient measurement in self-potential method
自然电场法推断成果图 speculated result figure of self-potential method
自然电场法异常曲线 anomaly curve of self-potential method
自然电场法综合平面图 comprehensive plan figure of self-potential method;comprehensive profile of self-potential method
自然电场作用 action of natural-electric(al) field
自然电磁场起伏 natural electromagnetic field fluctuation
自然电磁能量 natural electromagnetic energy
自然电磁现象 natural electromagnetic phenomenon
自然电动势 self-electromotive force
自然电离 spontaneous ionization
自然电流 natural current;natural electric(al) current;spontaneous electrical current
自然电流勘探法 natural-current method
自然电位 natural potential;spontaneous potential;self-potential
自然电位测井 spontaneous potential log
自然电位测井干扰因素 spontaneous potential log noise factor
自然电位测井记录 spontaneous potential log;self-potential log
自然电位测井曲线 curve of spontaneous potential logging;spontaneous potential log curve;self-potential log
自然电位测量 stir-potential survey
自然电位测量剖面平面图 profiling-plan figure of self-potential measurement
自然电位测量剖面图 profiling figure of self-potential measurement

自然电位等值线平面图 contour plan of self-potential method; contour profile of self-potential method
自然电位法 self-potential method; spontaneous potential method
自然电位录井曲线 self-potential log
自然电位探测法 natural potential method
自然电位梯度测井曲线 spontaneous potential gradient log curve
自然电位仪 natural electric(al) potential instrument
自然冻结 spontaneous freezing
自然度量 natural metric
自然断连 spontaneous amputation
自然断裂系统 natural fracture system
自然堆积角 angle of natural heaping; angle of natural stacking
自然堆积土 natural soil deposit
自然对合 natural involution
自然对流 free convection; gravitational convection; natural convection
自然对流传热 free convection heat transfer
自然对流放热器 natural convector
自然对流沸腾 free convection boiling
自然对流风冷器 natural convection air cooler
自然对流换热 natural convection heat transfer
自然对流空气冷却器 natural convection air cooler
自然对流空气冷却式冷凝器 natural convection air-cooled condenser
自然对流冷却 cooling by natural convection; free convection cooling; natural convection cooling
自然对流器 natural convection unit
自然对流式热风供暖 natural convection furnace heating
自然对流循环 natural convection circulation
自然对流再循环 natural convection recirculation
自然对数 hyperbolic logarithm; Napierian logarithm; natural logarithm
自然对数表 table of natural logarithm
自然对数的底 base of natural logarithm; Napierian base; natural base
自然对应 natural correspondence
自然恶化 natural deterioration
自然二-十进制 <自然二进制编码的十进制> natural binary-coded decimal
自然发电站 natural flow station
自然发火温度 spontaneous ignition point
自然发酵 natural curing; spontaneous fermentation
自然发热 autogenous heating; natural heating
自然发射 spontaneous emission
自然发生 accrue; archigony; autogenesis; self-generation; spontaneous generation; spontaneous origin
自然发生的 autogenic; emergent
自然发生的污染物 corollary pollutant; natural pollutant
自然发生论 abiogenesis
自然发生说 abiogenesis; spontaneous generation theory
自然发展 outgrowth
自然发展的地区 natural area
自然发展地区 natural development area
自然法学家 naturalist
自然法则 law of nature; natural law
自然法则限制 physical restraint
自然方 dank measure
自然防护 natural defence
自然防水性 physical waterproofing
自然防治 natural control
自然放射现象 natural radioactivity
自然放射性 natural radioactivity
自然肥料 natural fertilizer
自然分布范围 natural range
自然分层 natural stratification
自然分次 natural gradation
自然分风 natural splitting
自然分解 natural degradation; physiolysis
自然分类 natural classification
自然分类法 natural classification; natural system
自然分类特征 natural key
自然分类系统 natural classification system
自然分离 natural separation
自然分区 natural division; natural regionalization
自然丰度 natural abundance
自然风 natural wind
自然风干 natural drying out; natural air drying; natural seasoning
自然风干法 natural air drying; natural seasoning
自然风化 natural weathering
自然风化破碎作用 natural detrition
自然风景 natural landscape
自然风景保护区 natural protected landscapes area
自然风景区 area of outstanding natural beauty
自然风冷变压器 natural draft transformer
自然风流分支 natural splitting
自然风险 natural risk
自然风险保护计划 natural hazard plan
自然缝干法 natural seasoning
自然辐射 natural radiation
自然腐化 natural decay
自然腐蚀 spontaneous corrosion
自然负荷 nature load
自然复原 recovery of nature
自然富营养化 natural eutrophication
自然覆盖(植)物 natural cover
自然伽马测井 gamma-ray log
自然伽马测井曲线 gamma-ray log curve
自然伽马刻度器 gamma-ray log calibrator
自然伽马刻度值 gamma-ray log calibration value
自然伽马能谱测井 gamma-ray spectrum log
自然伽马能谱测井曲线 gamma-ray spectrum log curve
自然伽马射线测井 natural gamma ray logging
自然改变 physical alteration
自然干化 sun drying
自然干扰 natural interference
自然干燥 air drying; air seasoning; natural curing; natural desiccate; natural draught; natural dry(ing)
自然干燥的 naturally seasoned
自然干燥法 natural seasoning
自然干燥木材 air-dried lumber; natural seasoned lumber
自然干燥强度 air strength
自然刚性 natural rigidity
自然港(口) natural harbo(u)r; natural port
自然高程 natural grade
自然镉 cadmium
自然铬 chromium
自然更新 natural regeneration; natural reproduction; self-restoration
自然更新资源 self-renewal resources
自然工资率 natural rate of wages
自然公差 natural allowance
自然公园 natural park
自然公园观赏路 nature trail
自然汞 mercury
自然拱 natural arch
自然共鸣 natural resonance
自然共振 natural resonance
自然构造和组织 physique
自然骨料 natural aggregate
自然固化 spontaneous curing
自然光 natural daylight; skylight
自然光的 naturally lighted
自然光圈 natural stop
自然光线 natural light
自然光源 daylight source
自然光照 natural lighting
自然光照地貌立体表示法 natural colo(u)r-relief presentation; plastic colo(u)r-relief presentation
自然归化植物 natural naturalized plant
自然归还 natural restitution
自然规划 natural planning
自然规律 law of nature; natural law
自然规律学 physiognomy
自然过程 natural process
自然过滤法 natural filtration
自然过滤速率 natural filtration rate
自然海平面 natural sea-level; physical sea-level
自然海洋调查 physical oceanographic(al) survey
自然函数发生器 natural function generator
自然焊接 natural welding
自然合龙 natural closure
自然河槽 natural channel
自然河工模型 natural river engineering model
自然荷载 environmental load
自然横摇 natural rolling
自然滑坡 natural landslide; natural landslip
自然化学浓度 natural chemical concentration
自然划分 natural division
自然环境 natural cycle; natural environment; physical environment; physical setting; habitat <动植物生长的>
自然环境保护 landscape conservation; natural conservation
自然环境保护调查 green survey; vegetation survey
自然环境保护法(令) Natural Environmental Conservation Act
自然环境保护区 natural environment preservation area
自然环境背景图 map of natural environmental background
自然环境恶化 natural environment deterioration; physical environment deterioration
自然环境复原 natural environment rehabilitation
自然环境规划立法 physical planning legging; physical planning legislation
自然环境结构 natural environmental structure
自然环境压力 natural environment stress
自然环境要素 natural environmental element
自然环境异常 natural environmental anomaly
自然环境影响 natural environment influence
自然环境致癌物 natural environmental carcinogen
自然环境中的特性 behavio(u)r in natural environment
自然环境资料 physical data
自然环流法 natural circulation
自然缓和曲线 natural transition curve
自然缓解 self-releasing; spontaneous release
自然换向 natural commutation
自然恢复 spontaneous recovery
自然回水 positive prime
自然火 spontaneous fire
自然或半自然景色 wildscape
自然或半自然景象 wildscape
自然机械性 natural mechanical property
自然基础 natural foundation
自然级 size fractions of raw coal
自然级数 natural scale
自然极化 spontaneous polarization
自然极化法 spontaneous polarization method
自然极限 natural limit
自然计数 natural count
自然纪念碑 natural monument
自然季节现象 natural seasonal phenomenon
自然加热 natural heating; spontaneous heating
自然价格 natural price
自然价值 natural value
自然减损 natural diminution
自然建造 natural formation
自然疆界 natural boundary
自然降低 natural sink
自然降解 natural degradation
自然降解处理 natural attenuation
自然接地体 natural earthed body
自然结构 natural structure
自然结构的丧失 loss of natural structure
自然介质 natural medium
自然界 natural world; nature; realm of nature
自然界保护 protection of nature
自然界存在的同位素 naturally occurring isotope
自然界的 natural
自然界的合理性 natural fitness
自然界(定)律 law of nature
自然界奇迹 miracle
自然界水循环 natural water cycle
自然界噪声 noise in natural
自然界中最丰富的物质之一 one of the most abundant substances
自然金 gold
自然金属 natural metal; primary metal
自然进气量 natural air inflow
自然浸出 natural leaching
自然浸入 natural immersion
自然浸渍 natural impregnation
自然禁猎地 native sanctuary; nature sanctuary
自然禁猎区 native sanctuary; nature sanctuary
自然经济 natural economy; subsistence economy
自然景观 landschaft; natural landing; natural landscape; physical landscape
自然景色 landschaft; natural landscape; physical landscape
自然景物区 <未开发的> natural scenic area
自然景致园 naturalistic garden
自然净化 natural purification
自然净化动力学 kinetics of natural purification
自然净化过程 natural purification process
自然净化能力 natural assimilating power; natural purification capacity

自然净化特性 natural purification characteristic
自然净化作用 natural purification; self-cleaning
自然静止角 angle of rest
自然菌群 indigenous flora
自然开裂 ag(e)ing crack
自然科学 hard science; natural science; physical science
自然科学博物馆 Museum of Natural Science
自然科学工作者 naturalist
自然科学家 natural scientist
自然科学生物中心 center for biology of natural systems
自然科学心理学 natural science psychology
自然空冷式发动机 naturally air-cooled engine
自然空气 natural air; free air
自然空气干燥 open air drying
自然空气冷却 natural air cooling
自然空气冷却方式 natural air cooling system
自然空气循环 natural air circulation
自然控烟 natural smoke control
自然控制 natural control
自然宽度 natural breadth; natural width
自然矿层形的石墙面 seam face
自然扩张 natural extension
自然拉伸比 natural draw ratio
自然铼 rhenium
自然老化 natural ag(e)ing; natural weathering; self-aging; weathering age
自然老化试验 atmospheric exposure test
自然铑 rhodium
自然类群 natural group
自然类型边界线 boundary between natural types of ores
自然冷却 cooling by radiation; natural cooling; self-cooling; free cooling
自然冷却池 natural cooling pond
自然冷却电子管 radiating-cooled tube
自然冷却式变压器 transformer with natural cooling
自然冷却式电机 machine with natural cooling
自然冷却水 natural cooling water
自然力 natural agent; natural forces; natural power
自然历史博物馆 natural history museum
自然历史过程 historic(al) process of nature
自然利率 natural interest rate; natural rate of interest; ordinary interest rate
自然励磁 free current operation; natural excitation
自然沥青色的 natural asphalt colo(u)r
自然联锁 natural interlock
自然量 natural scale
自然钌 ruthenium
自然列线图 natural alignment diagram
自然劣化 natural deterioration
自然裂变年代测定法 spontaneous fission dating
自然裂缝 natural cleft
自然裂开 ag(e)ing crack
自然裂纹 ag(e)ing crack; natural crack
自然裂隙 self-open
自然林 wild wood
自然流 spontaneous current
自然流道 natural escape
自然流动 flow by gravity; gravity flow; natural flow
自然流动式热水 gravitational hot water; natural flow hot water
自然流量 natural flow; nature flow
自然流速 natural velocity
自然流域 natural basin
自然硫 native sulphur
自然垄断 natural monopoly
自然露天风化试验 natural outdoor weathering test
自然露头 day stone
自然铝 alumin(i)um
自然轮廓 natural skyline
自然美 natural beauty
自然美化工作 landscaping
自然模 natural norm
自然模式 example derived from nature
自然模态 natural mode
自然模型 natural model; self-formed model
自然磨损 natural wear and tear
自然目标 natural target
自然内积 natural inner product
自然内摩擦角 angle of true internal friction
自然内射 natural injection
自然耐久性试验 natural durability test
自然能量循环 natural energy cycle
自然能源 natural energy source
自然泥浆 natural mud
自然年 calendar year; civil year; natural year; solar year
自然年度 natural business year
自然黏[粘]结力 natural binding force
自然镍 nickel
自然凝固 natural coagulation; spontaneous coagulation
自然凝集价 natural agglutination valence
自然凝集素 natural agglutinin
自然农业 natural farming; subsistence agriculture
自然浓度 natural concentration
自然弄干 natural drying out
自然排风 natural exhaust
自然排风系统 natural exhaust system
自然排水 natural drainage
自然排水材料 free draining material
自然排水沟 natural drainage-channel
自然排水系统 autogenetic drainage
自然排污率 natural discharge ratio
自然排烟 natural smoke discharge
自然喷出 flow naturally
自然频率 free running frequency; natural frequency
自然平舱的 self-trimming
自然平衡 balance of nature; natural balance; natural equilibrium
自然平衡拱 natural self-supporting arch
自然平衡拱理论 natural balancing dome theory
自然平衡状态 natural stability
自然坡度 depositional gradient; equilibrium slope; ground-line gradient; natural fall; natural grade; natural gradient; natural slope
自然坡角 natural angle of slope
自然剖面 natural profile
自然剖面烃产率曲线 hydrocarbon productivity curve on natural section
自然曝气 natural aeration
自然曝气池 natural aeration pond
自然起伏 physical relief
自然气流 natural current
自然气体 natural gas
自然铅 lead
自然嵌入 natural embedding
自然强度 natural strength
自然侵蚀 geologic(al) erosion; natural erosion
自然清淤 self-dredging; self-cleaning
自然清淤坡度 self-cleaning gradient
自然清淤速度 self-cleaning velocity
自然倾角 angle of crater
自然倾向 aptitude
自然倾斜角 angle of rest; natural angle of slope; natural inclination angle; nature lean angle
自然倾斜坡度 angle of friction
自然区 natural area
自然区划 natural division; natural regionalization; physiographic(al) demarcation; physiographic(al) regionalization
自然区域 natural region
自然区域环境 natural regional environment
自然曲面 curved physical surface
自然趋势 course of nature
自然缺陷 natural defect
自然群落 natural community; natural population
自然群体 natural group; natural population
自然燃料发动机 hypergol
自然热释光曲线 natural heat release light curve
自然人 natural person; physical person
自然人口 natural population
自然人行为能力 disposing capacity of natural person
自然容许度 natural allowance
自然熔剂 natural flux
自然三角函数 natural trigonometrical function
自然散热 natural heat dissipation
自然散热器 natural convection unit
自然扫描 self-scan
自然色 natural colo(u)r; self-colo(u)r
自然色的 natural colo(u)red
自然色散 natural dispersion
自然森林 natural forest; natural woods
自然沙漠化 desertization
自然砂 natural sand
自然上升效应 chimney effect
自然设计 naturalistic design
自然社会系统 nature-society system
自然砷 arsenolamprite
自然渗气水流 self-aerated flow
自然渗透性 natural permeability
自然渗析 natural dialysis
自然升华 spontaneous sublimation
自然生产力 natural productiveness
自然生境 native habitat; natural habitat
自然生境的保护 conservation of natural habitats
自然生态系保全 conservation of natural ecosystem
自然生态系(统) natural ecosystem
自然生态学 ecology of nature; natural ecology
自然生图像 self-generated graphics
自然生物处理法 natural biologic(al) treatment method
自然生物地球化学循环 natural biogeochemical cycling
自然生物法 <污水处理的> natural biologic(al) method
自然生物区 natural biotic area
自然生物群落 natural biotic community
自然生物污水处理法 natural biologic(al) waste(water) process
自然湿度 natural humidity; natural moisture; natural water content
自然石面的 self-faced
自然时效 natural ag(e)ing; natural seasoning; seasoning; weathering
自然实物单位 unit of natural kinds
自然实验 full-scale experiment
自然实验室 natural laboratory; nature laboratory
自然食品 natural food
自然史 natural history
自然式 natural style
自然式道路系统 informal road system
自然式花篱 natural flowering hedge
自然式设计 naturalistic design
自然式庭园 naturalistic garden
自然式园林 natural garden style
自然势能 natural potential
自然适应 natural adaptation
自然收缩 autogenous shrinkage; natural shrinkage
自然寿命 natural life-span
自然疏干 natural drying out
自然数【数】 natural number
自然数据级数 series of natural numbers
自然刷深 natural degradation; self-dredging
自然衰变 natural decay
自然衰减 natural attenuation; organic decay
自然衰减单位 <如奈培或分贝等> natural attenuation unit
自然双折射 natural birefringence
自然水道 natural water course
自然水力驱动 natural water drive
自然水(利)资源 natural water resources
自然水流 natural drainage flow
自然水泥 natural cement
自然水体 natural water body; natural waters
自然水位 natural water level
自然水温 natural water temperature
自然水温区 natural water temperature region
自然水文曲线 natural hydrograph
自然水文灾害 natural hydrologic hazards
自然水污染 natural water pollution
自然水系 natural water system
自然水样 natural water sample
自然顺序 natural order; natural sequence
自然顺序关系 natural ordering
自然死亡 natural mortality
自然死亡率 natural mortality rate
自然死亡系数 coefficient of natural mortality
自然隧道风 natural tunnel wind
自然损耗 natural loss; natural wastage; normal loss; ordinary wear-and-tear
自然态 natural mode
自然弹性极限 natural elastic limit
自然弹性限度 natural elastic limit
自然淘汰 natural elimination; natural selection
自然特征 <包括目标的方位、大小结构等> physical characteristic; physical feature
自然特征水温 natural characteristic water temperature
自然天气季节 natural synoptic(al) season
自然天气区 natural synoptic(al) region
自然天气现象 natural weather phe-

Z

nomenon
自然天气周期 natural synoptic(al) period
自然添附 natural accretion
自然条件 natural condition; physical condition
自然条件荷载 environmental load
自然条件试验 natural condition test
自然条件下的强度 in-situ strength
自然条件限制 physical constraint
自然条件障碍 adverse physical congestion
自然调节 natural regulation
自然调节河(流) naturally regulated river; naturally regulated stream
自然调谐 just tuning
自然铁 ferrite; sideroferrite
自然通风 chimney ventilation; draft ventilation; draught ventilation; gravitational ventilation; natural draft; natural draught; natural ventilation
自然通风单元 natural draught ventilation unit
自然通风地坑炉 natural draft pit furnace
自然通风电机 natural draft machine
自然通风干燥机 natural draft drier
自然通风干燥器 free air dryer
自然通风干燥窑 natural ventilating dry kiln
自然通风锅炉 natural draft boiler
自然通风混流冷却塔 natural draught mixed cooling tower
自然通风间接触式冷却塔 natural draught indirect contact cooling tower
自然通风冷却 native air cooling; natural air cooling
自然通风冷却的 air natural cooled
自然通风冷却器 cooler with natural draught
自然通风冷却塔 atmospheric cooling tower; cooling stack; natural draft cooling tower; natural draught cooling tower
自然通风冷水塔 natural draft cooling tower
自然通风炉 air furnace
自然通风滤池 airflow through filter
自然通风逆流冷却塔 natural draught counterflow cooling tower
自然通风气体燃烧器 natural draft gas burner
自然通风器 natural ventilator
自然通风燃烧器 natural draft burner
自然通风湿塔 natural draft wet tower
自然通风式电动机 self-ventilated motor
自然通风式炉 wind furnace
自然通风水冷器 atmospheric water cooler
自然通风吸入口 air scoop
自然通风系统 natural ventilating system
自然通风压力 natural ventilating pressure; natural ventilation pressure
自然通风烟囱 natural draught chimney
自然通风组件 natural draught ventilation unit
自然通量 natural flux
自然通气 natural draft ventilation
自然同构 natural isomorphism
自然同态 natural homomorphism
自然铜 native copper
自然投影 natural projection
自然突变 sporadic mutation
自然图案大理石 forest marble
自然土层的含水量 field capacity
自然土壤 natural soil
自然土壤图 soil map
自然团聚体 natural aggregate
自然推进剂 hypergolic propellant
自然推理演算 calculus of natural inference
自然退化 natural degradation
自然挖取效率 free digging rate
自然外观 natural look
自然弯线 easement
自然维护 self-maintaining
自然位能 natural potential
自然位置 natural position
自然温湿度调节 natural conditioning
自然稳定极限 natural stability limit
自然稳定性 natural rigidity; natural stability
自然污染 natural pollution
自然污染物 natural pollutant
自然污染源 natural pollution source
自然污染源水域 natural pollution source waters
自然误差 natural error
自然吸气 natural aspiration
自然吸气式柴油机 natural aspirated diesel engine
自然吸气式发动机 natural aspiration engine; naturally aspirated engine
自然吸湿率 natural regain
自然稀疏 natural thinning
自然系统 natural system
自然现象 natural phenomenon
自然现象的特征 <河流等> regime(n)
自然现象异常 natural anomaly
自然线宽 natural line width
自然线性映射 natural linear mapping
自然限制 natural restriction
自然响应 natural response
自然橡胶 natural rubber
自然消除 natural elimination
自然消解 natural slaking
自然效应 native effect
自然谐振 natural resonance
自然形成的 naturally formed; air-formed
自然形成的边坡 naturally developed slope
自然形态 natural feature; natural pattern
自然形状 physical form
自然休止角 angle of natural repose; natural angle of repose
自然序数编号法 natural number designate
自然絮凝 natural coagulation
自然选择 natural selection
自然选择说 natural selection theory
自然循环 natural circuit; natural cycle
自然循环的水冷却 natural circulation water cooling
自然循环法 natural circulation
自然循环反应堆 natural circulation reactor
自然循环干燥窑 natural circulation kiln; natural draft kiln
自然循环供暖 natural circulation heating
自然循环锅炉 natural circulation boiler
自然循环冷却 natural circulation cooling
自然循环冷却系统 natural circulation cooling system
自然循环汽化冷却 pressureless type evapo(u)rative cooling
自然循环系统 natural cycle system
自然循环蒸发器 natural circulation evapo(u)rator
自然压力过程 natural pressure cycle
自然延伸 natural prolongation
自然岩沥青 natural rock asphalt
自然研究 nature study
自然演变冲刷 natural scour
自然演替 natural succession
自然演替系列 ferralarch
自然演绎 natural deduction
自然厌氧消化 natural anaerobic digestion
自然养护 air season; autogenous curing; dry curing; natural curing; natural maintenance
自然氧化法 natural oxidation method
自然氧化还原滴定法 natural redox titration
自然氧资源 natural oxygen resources
自然要素 natural element
自然要素名称 feature name; natural feature name
自然叶饰 natural foliage
自然铱 iridium
自然遗产海岸 heritage coast
自然遗产维护 heritage preservation
自然疫源地 natural infection focus
自然疫源地带 zone of natural foci
自然因素 natural cause; natural factor; physical factor
自然音阶 just scale; scales of just temperament
自然铟 indium
自然银 sliver
自然引风 blow torch action; buoyancy
自然引风装置 chimney operated system; hot draft apparatus
自然应变 logarithmic strain; natural strain
自然应力 native stress; natural stress
自然应力消除 natural stress relief
自然荧光 natural fluorescence
自然营业年度 natural business year
自然影响 original influence
自然硬度 natural hardness; natural rigidity
自然硬化 spontaneous hardening
自然硬化的 air set
自然油冷却 natural oil cooling
自然油循环 natural oil circulation
自然淤积 natural accretion; natural siltation
自然雨 natural rain
自然元素 natural element
自然圆弧频率 natural circular frequency
自然源遥感 passive remote sensing
自然运动 proper motion
自然灾变 convulsion of nature
自然灾害 act of God; act of nature; geophysical hazard; natural calamity; natural catastrophe; natural disaster; natural hazard
自然灾害损失 loss from natural calamity
自然灾害异常 natural hazard anomaly
自然灾害预告系统 natural disaster warning system
自然再充气 natural reaeration
自然再用 natural reuse
自然暂态稳定极限 natural transient stability limit
自然噪声 natural noise
自然增长 natural increase
自然增长率 natural rate of growth; rate of natural increase
自然增长系数 coefficient of natural increase
自然增长资产 natural growth assets
自然增值 natural increase of value; unearned increment
自然涨落 spontaneous fluctuation
自然照度系数 coefficient of natural illumination
自然照明 natural lighting
自然罩面 natural finish
自然折旧 natural depreciation; natural depression
自然哲学 natural philosophy
自然振荡 natural oscillation
自然振动 free vibration; natural vibration
自然振型 natural mode
自然振源 natural resources
自然蒸发 physical evapo(u)ration; spontaneous evapo(u)ration
自然蒸发作用 natural evapo(u)ration
自然蒸汽 natural steam
自然正锥 natural positive cone
自然支撑采矿法 natural propping
自然植被 natural vegetation
自然制动 self-braking; spontaneous braking
自然致宽 natural line broadening
自然终止 natural termination
自然重量 natural weight
自然重砂样品 sample for natural heavy mineral
自然周频率 natural circular frequency
自然周期 free period; natural period
自然主义 <建筑、艺术品> naturalism
自然主义处理 naturalistic treatment
自然主义形式 naturalistic form
自然主义形态 naturalistic form
自然主义者 naturalist
自然状况 natural condition
自然状态 natural state; nature; state of nature
自然状态的 natural
自然状态露头【地】 day stone
自然资源 natural resources; natural stock; natural wealth; physical resources
自然资源保护 conservation of natural resources; natural resource protection; natural resources conservation
自然资源保护测量 conservation survey
自然资源保护促进会 Society for the Promotion of Natural Reserves
自然资源保护技术 conservation technology
自然资源保护论者 conservationist
自然资源保护区 conservancy
自然资源保护委员会 natural resources defense council
自然资源保护者 preservationist
自然资源保护主义者 convervationist
自然资源的管理 conservancy
自然资源的浪费使用 wasteful use of natural resources
自然资源的利用 utilization of natural resources
自然资源法 law of natural resources
自然资源分布 natural resources distribution
自然资源管理 natural resources management
自然资源管理人员 conservator
自然资源规划 natural resources planning
自然资源经济学 natural resource economics
自然资源开发 development of natural resources
自然资源生态系统 natural resource ecosystem
自然资源委员会 Commission on Natural Resources
自然资源消耗 natural resources consumption
自然自净能力 natural self-purification capacity

自然自净效力 natural self-cleaning effect
自然自净作用 natural self-purification
自然综合体 natural complex
自然阻抗 natural resistance
自然阻力 natural resistance
自然阻尼 natural damping
自然阻塞 normal congestion; normal jam
自然组＜其范围小于目＞ cohort
自然坐标 natural coordinates
自然坐标系 natural system of coordinates
自燃 autoignite; autoignition; breeding fire; fire-fanging; self-igniting; burnt spontaneously
自燃报警器 combustible gas alarm
自燃的 autogenic; autogenous; self-combustible; spontaneous combustible; hypergolic＜燃料＞
自燃的极限 flammability limit
自燃的原因 self-combustion reason
自燃点 autoignition point; self-ignition point; spontaneous ignition point
自燃点火器 hypergolic igniter
自燃发火周期 self-combustion firing period
自燃混合气 free-burning mixture; hypergolic mixture
自燃混合物 self-inflammable mixture
自燃火箭燃料 hypergol; hypergolic fuel
自燃火灾 autoignition conflagration; freely burning fire
自燃金属 pyrophoric metal
自燃煤 spontaneous combustion coal
自燃倾向等级 self-combustion tendency class
自燃倾向指标 self-combustion tendency index
自燃热 spontaneous heating; spontaneous combustion
自燃双组分推进剂 self-igniting bipropellant
自燃通风煤气燃烧器 atmospheric gas-burner
自燃推进剂 pyrophoric propellant
自燃温度 autoignition temperature; self-ignition point; spontaneous ignition temperature
自燃物 pyrophoric material; pyrophorus
自燃物品 spontaneous combustion articles
自燃物质 self-inflammable material; spontaneous combustible substance
自燃险 risk of spontaneous combustion
自燃性 pyrophoricity
自燃预防 spontaneous fire prevention
自燃造成的火灾 fire ignition due to spontaneous combustion
自燃着火 self-ignition; spontaneous ignition
自燃着火点 self-ignition point temperature; spontaneous ignition temperature
自燃着火发动机 natural ignition engine
自燃着火温度 self-ignition temperature; spontaneous ignition temperature
自燃指示器 combustible gas indicator
自燃指数 self-ignition index
自燃作用 self-combustion
自热 self-heat(ing)
自热的 autothermic
自热高温好氧废水处理系统 autothermal thermophilic aerobic waste (water) treatment system
自热高温好氧消化 autoheated thermophilic aerobic digestion; autothermal thermophilic aerobic digestion
自热回火 self-tempering
自热缺陷 natural defect
自热熔炼 autogenous smelting; pyritic smelting
自热式 self-heat(ing)
自热式热电偶 self-heated thermocouple
自热式热离子管 self-heating thermionic tube
自热填料床反应器 autothermal packed bed reactor
自热退火 self-annealing
自热脱硫作用 pyritic action
自容量 self capacitance; self-capacity
自容式数字记录资料收集装置 self-contained digital recording data acquisition system
自容式图表记录剖面仪 self-contained graphic(al) recording profiler
自溶 autolyze; self-digestion
自溶产物 autolysate
自溶酶 autoenzyme
自溶作用 autodigestion; autolysis
自熔的 self-fusible
自熔矿 self-fluxing ore
自熔性 self-fluxing nature
自熔性矿石 self-fluxing ore
自熔性烧结矿 self-fluxing sintered ore
自乳化 self-emulsifying
自锐式钻头 self-sharpening bit
自锐性 self-sharpening
自润滑材料 self-lubricating material
自润滑车 self-lubricating block; self-oiling block
自润滑传动滑槽 self-lubricating gear notch
自润滑导套 self-oiling guide bushing
自润滑的 self-oiling
自润滑电刷 self-lubricating brush
自润滑合成材料 self-lubricating synthetic material
自润滑合金 self-lubrication alloy
自润滑滑块式轴承箱组 self-lube take-up unit
自润滑环形轴承箱组 self-lube cartridge unit
自润滑轮 self-lubricating sheave
自润滑双螺栓菱形法兰式轴承箱组 self-lube two-bolt flange unit
自润滑四螺栓方形法式轴承箱组 self-lube four-bolt flange unit
自润滑性能 self-lubricating property
自润滑悬挂式轴承箱组 self-lube hanger unit
自润滑圆形法兰式轴承箱组 self-lube flange cartridge unit
自润滑枕式轴承箱组 self-lube pillow block unit
自润滑轴衬 self-lubricating bush
自润滑轴承 maintenance-free bearing; oilless bearing; self-lubricated bearing
自润滑轴承箱组 self-lube bearing block
自润滑轴瓦 oilless bushing
自润轴颈 self-lubricating journal
自撒料料斗 self-spreading bucket
自散射 self-scattering
自散射效应 self-scattering effect
自扫描成像器件 self-scanned image sensor
自扫描固态成像传感器 self-scanned solid-state image sensor
自扫描光敏阵列 self-scanned photosensing array
自扫描条图 self-scan barograph
自扫描图像传感器 self-scanned image sensor
自扫描图像显示板 self-scan imaging display panel
自扫描线性阵列 self-scanned linear array
自扫描阵列 self-scanned array
自色的 idiochromatic
自杀位＜汽车中驾驶员侧的座位，在交通事故中，坐在此位上的人死亡率甚高＞ suicide seat
自杀性竞争 cut-throat contest
自闪光 self-flashing
自上而下的设计 top-down design
自上而下方式 top-down approach
自上而下分析 top-down analysis
自上而下近似 top-down approach
自上而下面向目标分析步骤 top-down goal-oriented analysis procedure
自上而下剖析 top-down parsing
自上而下强制预算 imposed budget
自上而下取样 up the hole sampling
自上而下试验 up the hole testing
自上而下无塞灌浆法 downstage without packer method
自上而下有塞灌浆法 downstage packer method
自上发条表 self-winding watch
自上发条钟 self-winding clock
自射式井点 self-jetting (-type) well point
自身安全的 inherently safe
自身安装塔式起重机 self-erecting tower crane
自身变态反应 autoallergy
自身标记假说 self-marker hypothesis
自身不稳定的钢框架结构 nonself-supporting steel frame
自身缠绕试验＜钢丝的＞ button test
自身冲刷河 self scouring river
自身除湿 self-desiccation
自身的 autologous; in-house
自身电感 self-inductance
自身电抗 self-reactance
自身电容 self capacitance; self-capacity
自身电阻 self-resistance
自身断离 autoamputation
自身对称破坏【力】 spontaneous symmetry breakdown
自身反应 idreaction
自身放电 self-discharge; self-discharging
自身分裂 self-division
自身辐射 self-radiation
自身改良土壤 self-amelioration
自身干扰 autointerference
自身供给能量的 self-energizing
自身故障 primary failure
自身回火 self-tempering
自身活化 autoactivation
自身加速度 self-acceleration
自身检查 self-verifying
自身检眼镜 autoophthalmoscope
自身抗弯强度 self-bending strength
自身控制 non-interacting control
自身密封式内腔 self-sealing inner chamber
自身耐受性 self-tolerance
自身能量 self energy
自身凝集反应 autoagglutination; autohemagglutination
自身凝集作用 autoagglutination
自身喷雾器 autospray
自身溶解 autolysis
自身润滑 self-lubrication
自身润滑性 self-lubricity
自身生利还本的 self-liquidating
自身识别 self-recognition
自身式调整 non-interacting control
自身饰面的屋面毡 self-finished roofing felt
自身说明的 self-explanatory
自身调节 autoregulation
自身退火 self-annealing
自身稳定性 inherent stability; stability under own weight
自身相关 semi-correlation
自身相互反应 self-interaction
自身消化 autodigestion
自身氧化 autooxidation
自身荧光镜 autofluoroscope
自身荧光图 autofluorograph
自身映射 onto mapping
自身硬化 hardening by itself
自身约束 self restraint
自身噪声 self-noise
自身照明觇标 self-illuminating sight
自身振荡 self-oscillation
自身振荡超外差接收器 tropadyne
自身支撑能力 self-supporting property
自身装卸 self-loading
自身装卸卡车 self loader
自身装置有装卸机械的 self-sustaining
自升的 self-elevating; self-hoisting
自升浮筒 self-elevating pontoon
自升高模板 self-raising forms
自升工作平台 self-elevating working platform
自升降平台 self-elevating platform
自升落式工作平台 self-powered platform
自升式 self-elevation
自升式安装用塔式起重机 self-erecting tower crane
自升式铲运机 elevating scraper
自升式趸船 self-elevating platform; self-elevating pontoon; jack-up pontoon
自升式钢模 self-climbing steel-form
自升式海上钻井平台 self-elevation offshore platform; self-jacking offshore drilling vessel
自升式海上钻探平台 self-elevation offshore platform; self-jacking offshore drilling vessel
自升式井架 jack-up rig
自升式模板 self-elevating shutting; self-raising forms
自升式平底船 self-elevating barge
自升式平台 jack-up platform; self-elevated platform; self-elevating platform
自升式平台液压腿 jack-up leg
自升式起重机 climbing crane; jack-up crane
自升式起重小车 man trolley
自升式施工驳船 jack-up construction barge
自升式施工趸船 jack-up pontoon
自升式施工平台 jack-up construction barge; self-elevating barge【港】; jack-up construction rig
自升式水上(作业)平台 self-elevating platform
自升式塔吊 climbing crane; kangaroo crane
自升式塔吊辅助构架 climbing frame
自升式塔式起重机 self-climbing tower crane; self-hoisting tower crane; self-raising tower crane
自升式钻井船 offshore jack-up drilling vessel; self-elevating drilling ship
自升式钻井平台 jack-up drilling platform; self-elevating drilling platform; jack-up(drilling) rig
自升式钻探船 offshore jack-up drilling vessel; self-elevating drilling ship
自升式钻探平台 jack-up drilling plat-

form; jack-up rig; self-elevating drilling platform
自生 autogeny
自生沉积物 authigenic sediment
自生成矿变化阶段 phyllomorphic stage
自生的 authigenous; autogenic; autogenous; self-grown; spontaneous; authigenic【地】
自生动态膜 self-forming dynamic membrane
自生动态膜生物反应器 self-forming dynamic membrane bioreactor
自生固氮 non-symbiotic nitrogen fixation
自生河 autogenic river; autogenic stream; self-grown river; self-grown stream
自生河流 autogenous stream
自生换能器 self-generating transducer
自生胶结作用 autocementation
自生角砾岩化 autobrecciation
自生介质 autogenous medium
自生矿床 authigenic deposit
自生矿物 authigene; authigenic mineral; authigenous mineral; autogenic mineral; autogenous mineral
自生矿物变化图 variation map of syngenetic mineral
自生力 idiosthenia
自生砾岩 autoconglomerate
自生裂纹 delayed failure
自生脉冲 self-pulsing
自生泥晶灰岩 automicriteic rock
自生偏压 self-bias
自生熔剂 self-fluxing
自生熔渣杂质 self-slagging impurity
自生生物动态膜 self-forming bio-dynamic membrane
自生石化作用 autolithification
自生世代 free living generation
自生收缩 autogenous shrinkage
自生碎屑 autoclast
自生碎屑岩 autoclastic rock
自生体积变化 autogenous volume change
自生同心剪切面 spontaneous concentric(al) shear plan
自生物质 authigenic material
自生啸声 self whistle
自生兴奋 autogenic excitation
自生压力 self-generated pressure
自生演变 autogenic change
自生抑制 autogenic inhibition
自生愈合 autogenous healing
自生再结晶 spontaneous recrystallization
自生振荡 self-oscillation
自生作用【地】authigenesis
自识别 self-identification
自蚀 self-corrosion
自蚀底漆 self-etch primer
自蚀光谱 reversal spectrum
自蚀(光谱)线 reversed line
自始至终 throughout
自式行沟挖机 self-propelled ditch digger
自势 self-potential
自视读数 visible reading; visual reading
自饰面屋面油毡 self-finished roofing felt
自适应 self-adap(ta)tion
自适应变换码 adaptive transform coding
自适应操舵模型 adaptive steering module
自适应测试 adaptive test
自适应测试器 adaptive tester
自适应差分脉冲编码调制 adaptive differential pulse code modulation
自适应常速巡航控制系统 adaptive cruise control system
自适应程序 self-adapting program(me)
自适应的 self-adapting; self-adaptive
自适应叠加 adaptive stack
自适应定阈值 adaptive thresholding
自适应反馈控制 adaptive feedback control
自适应(负荷)预报 adaptive(load) forecasting
自适应复用器 adaptive multiplexer
自适应工业机器人 self-adapting industrial robot
自适应管理 adaptive management
自适应过程 adaptive process
自适应计 adaptometer
自适应技术 adaptive technique
自适应交通控制系统 adaptive traffic control system
自适应交通控制优化技术 adaptive signal control optimization technique; adaptive traffic control optimization technique
自适应交通信号控制系统 adaptive traffic signal control system
自适应阶段 adaptive phase
自适应接收 adaptive reception
自适应结构 <数据库的> adaptive optimization
自适应解码 adaptive decoding
自适应均衡 adaptive equalization
自适应均衡器 adaptive equalizator; automatic adaptive equalizer
自适应控制 adaptive control; self-adaptation control; self-adaptive control; threshold control
自适应控制程序 adaptive control procedure
自适应控制的 self-adaptive
自适应控制器 self-adaptive controller
自适应控制系统 adaptive control system; self-adaptive control system
自适应控制序列 adaptive control sequence
自适应控制优化 adaptive control optimization
自适应控制装置 adaptive controller; adaptive control unit
自适应控制作用 adaptive control action
自适应雷达控制自动舵 adaptive radar-controlled autopilot
自适应力 autoforce
自适应链耙 self-relieving chain harrow
自适应滤波器 self-adapting filter; adaptive filter
自适应滤料 adaptive filtering material
自适应路径选择 adaptive routing
自适应路径选择算法 adaptive routing algorithm
自适应路由选择 adaptive routing
自适应逻辑线路 logic(al) circuit for self-adaptation
自适应模型 adaptive model
自适应评价 adaptive assessment
自适应式自动驾驶仪 self-adaptive(-adjusting) autopilot; self-adaptive-correcting autopilot
自适应衰减 adaptive attenuation
自适应伺服机构 adaptive servo
自适应伺服系统 adaptive servo-system
自适应调节 self-adaption
自适应调节的 self-adaptive
自适应通信[讯] adaptive communication
自适应系统 adaptive system; self-adaptation system; self-adapting system; self-adaptive system
自适应系统理论 adaptive system theory
自适应线性滤波 adaptive linear filtering
自适应线性元件【计】adaptive linear element
自适应消卷积 adaptive deconvolution
自适应信号处理 adaptive signal processing
自适应信号控制优化技术 adaptive signal control optimization technique
自适应信息处理 adaptive information processing
自适应性 self-adaptivity
自适应性响应 adaptive response
自适应性阻尼 self-adaptive damping
自适应译码 adaptive decoding
自适应预测 adaptive forecasting; adaptive prediction
自适应预测编码 adaptive forecasting coding; adaptive prediction coding
自适应阈 adaptive thresholding
自适应阈值控制 adaptive threshold control
自适应增量调制 adaptive delta modulation
自适应阵列 adaptive array
自适应指令接收技术 self-adaptation command receiving technique
自适应自动舵 adaptive autopilot
自适应自动驾驶仪 adaptive autopilot; self-adaptation autopilot
自适应(自动)均衡器 adaptive automatic(al) equalizer
自适应最佳控制 adaptive optimal control
自适应最佳系统 adaptive optimal system
自适应最优化 adaptive optimization
自适应最优控制 adaptive optimal control
自适转换电路 adaptive switching circuit
自释救生筏 autoreleasing lifeboat
自释联轴节 self-release coupling
自收缩 autogenous shrinkage
自收缩束 self-constricted beam
自守的 automorphic
自受激发射 self-stimulated emission
自售商店服务员 aisle-man
自疏水 self-draining
自输入 self-load
自熟化嵌缝膏 self-curing sealant
自署 autograph
自署的 autographic(al)
自述语言 unstratified language
自数 self-counting
自顺向河 autoconsequent river; autoconsequent stream
自顺向流 autoconsequent stream
自顺向瀑布 autoconsequent fall
自碎斑晶 autoclastic phenocryst
自碎的【地】autoclastic
自碎的片岩 autoclastic schist
自碎集块熔岩 autoclastic agglomerate lava
自碎角砾熔岩 autoclastic breccia lava
自碎结构 autoclastic texture
自碎凝灰熔岩 autoclastic tuff lava
自碎岩 autoclast; autoclastic rock
自缩合作用 self-condensation
自锁 self-hold(ing); self-jamming; self-lock(ing); self-retention
自锁安全吊货钩 safety latch(cargo) hook
自锁差速器 no-spin lockup
自锁传动 self-sustaining gear
自锁电路 latching circuit; self-locking circuit
自锁阀 latching valve; lock-up valve
自锁继电器 latching relay
自锁角旋塞 self-locking angle cock
自锁紧固件 self-locking fastener
自锁紧螺栓 self-locking bolt
自锁螺帽 lock nut; self-loading nut; self-locking nut
自锁螺母 ca(u)lking nut; self-locking nut
自锁螺旋 self-locking screw
自锁模 self-mode-locking
自锁式差速器 self-locking differential (gear)
自锁式凸轮 latching cam
自锁网络 latching network
自锁蜗杆 irreversible worm
自锁销 self-lock pin
自锁旋塞 self-locking cock
自锁循环器 latching circulator
自锁原理 self-lock mechanism
自锁转向装置 irreversible steering assembly; irreversible steering gear; self-locking steering gear
自锁作用 self-locking action
自提升模板 self-lifting forms
自体导电法 autoconduction
自体发生 autogenesis
自体放射照片 autoradiograph; radioautograph
自体放射照相术 autoradiography; radioautography
自体分解 autodecomposition
自体免疫性 autoimmunity
自体凝集素 autoagglutinin
自体凝集作用 autoagglutination
自体燃烧 autogenous combustion
自体溶解 autocytolysis; autodigestion; autolysis; autolyze; self-digestion
自体调节 autoregulation
自填塞环形填密件 autopack ring packing
自调 self-regulation
自调变压器 autoformer
自调的 self-acting; self-adjustable
自调定形机 autosetter
自调多孔轴承 self-aligning porous bearing
自调发动机 self-tuning engine
自调恒温器 self-acting thermostat
自调间隙制动器 self-adjusting brake
自调胶卷盒 self-adjusting magazine
自调节 inherent regulation
自调(节)的 self-regulating
自调节电机 self-regulated machine
自调节堆 self-regulating reactor
自调节反应堆 self-limiting reactor
自调节弧焊变压器 self-regulating arc welding transformer
自调节交流发电机 self-regulating alternator
自调节速度 rate of inherent regulation; rate of self-regulation
自调节装置 self-regulation device
自调口虎钳 self-adjusting jaw vice
自调拉力绞车 self-tensioning winch
自调滤波法 autoregulation filtering method
自调脉冲振荡器 self-pulsed oscillator
自调耙吸头 self-adjusting draghead
自调平 self-level(l)ing
自调平衡 self-adjusting balance
自调起绒机 autozero cloth raising machine
自调式程序 self-debugger
自调式电弧焊 self-adjusting arc welding
自调式减振器 self-adjustable shock absorber
自调式挺杆 self-adjusting tappet
自调式钻卡 self-adjustable drill chuck

自调水平仪 self-adjusting level
自调水准仪 self-adjusting level
自调速率 speed of adaptation
自调特性 self-regulating characteristics
自调通风机 colt
自调推土机 tilting bulldozer
自调托滚 self-acting idler
自调位承载托滚 self-aligning carrying idler
自调位滚柱轴承 self-aligning roller bearing
自调位托辊 guider; self-aligning idler; self-aligning roller
自调位轴承支座 self-aligning support
自调系统 self-regulating system
自调系统程序控制 autonomics
自调系统程序控制研究 autonomics
自调谐线圈检测器 self-tuning loop
自调性 self-stabilization
自调液面水箱 self-priming tank
自调匀整练条机 autolevel(l)er draw box
自调匀整双圈条练条机 autolevel(l)er bicoil draw box
自调匀整针梳机 autolevel(l)er gill box
自调蒸发冷却系统 self-regulated vaporization cooling system
自调整 self-adapting; self-aligning; self-adjustment
自调整的 self-adjusting
自调整功率变换器 self-regulating power transformer
自调整滑动轴承 self-adjustable sliding bearing
自调整控制 self-adjusting control
自调整控制系统 self-aligning control system
自调整率 self-regulation rate
自调整模型 self-adjusting model
自调整式 self-align type
自调整系统 self-adjusting system
自调整因子 self-tuning factor
自调整照准线 self-adjusting line of sight
自调整自适应调节器 self-tuning adaptive regulator
自调整自适应控制 self-tuning adaptive control
自调支座 self-aligning support
自调制 automodulation; self-modulation
自调制振荡器 self-modulated oscillator; self-modulating oscillator
自调质量转移系统 self-regulating mass transfer system
自调轴承 tilting bearing
自调转向架 self-aligning truck
自调装置 self-adjusting device
自调准 self-aligning
自铁工 sheet-metal working
自停 automatic stop(ping)
自停机构 self-stopping gear; stopping motion
自停装置 automatic stop arrangement; knock off <机器故障>
自通电报交换台 gentex exchange
自通电报局 gentex office
自通电报系统 gentex system
自通风 self-ventilation
自通风电机 self-ventilated machine
自通风断路器 self-blast circuit-breaker
自通排水 free drainage
自同步 motor synchronizing; self-clocking; self-synchronizing
自同步的 self-synchronous; self-timing
自同步电动机 selsyn motor; self-synchronous motor
自同步发射机 selsyn generator; synchrogenerator; synchrotransmitter; transmitter synchro
自同步机 automatic synchronizer; self-synchronous device; selsyn; synchro; synchromagslip
自同步接收机 receiver synchro; selsyn transformer; synchroreceiver
自同步码 self-synchronizing code
自同步仪表 autosyn instrument
自同步装置 self-synchronous device
自同构 automorphism
自同构的【数】 automorphic
自同期 random synchronizing
自同态 endomorphism
自同态环 ring of endomorphisms
自投 autoswitch on; autoswitch over
自涂标 whitening mark
自推进式翻斗车 self-propelled skip
自推进式洒水车 self-propelled sprinkler lorry
自推式铲运机 self-propelled scraper
自退磁系数 self-demagnetizing factor
自退火 <锻后的> self-annealing
自脱水 self-desiccation
自卫 self-defensiveness
自位 self-potential; self-align
自位板 self-angling discs
自位导向支承滚轴 self-aligning guide idler
自位的 self-aligning
自位对排滚珠轴承 self-aligning double row ball bearing
自位滚珠 position finder ball
自位滚珠轴承 self-aligning ball-bearing
自位轮 caster [castor]; swivel-type castor; swivel wheel
自位式圆盘小前犁 swivel disk jointer
自位轴承 pivoted bearing; self-aligning bearing; tilted pad bearing
自位转矩 self-aligning torque
自稳定 self-adjusting
自稳定性 self-stabilization
自稳定性能 autostability; inherent stability
自稳机理 homeostasis; homeostatic mechanism
自稳机制 homeostasis
自稳时间 stand-up time
自稳岩石 competent rock
自稳作用 autostabilization
自我重组 reconfigure itself
自我风险 own risk
自我服务设备 self-service facility
自我服务设施 self-service facility
自我服务装置 self-service installation
自我呼吸装置 self-contained breathing apparatus
自我检查 self-check
自我检验 self-certification
自我净化系统 self-purifying system
自我目标管理形 self-fulfilment management style
自我援救装置 self-rescuer
自我再生过程 self-regenerative process
自我支撑 self support
自污染源 source of self-pollution
自坞修船坞 self-docking dock
自吸 self-prime
自吸泵 self-priming pump; self-sucking pump
自吸的 self-priming
自吸电路 self-holding circuit
自吸和自蚀 self-absorption and self-reversal
自吸离心泵 self-priming centrifugal pump
自吸能力 suction capacity
自吸起动离心泵 self-priming centrifugal pump
自吸取 self-reduction
自吸入的 self-priming
自吸设备 self-priming device
自吸式离心泵 self-priming centrifugal pump; self-suction centrifugal pump
自吸式汽化器 self-feeding carburettor
自吸式水泵 self-prime pump; self-priming pump
自吸收 self-absorption; self-reduction
自吸涡轮 self-aspirating turbine
自吸效率 efficiency of imbibitions
自吸引 self-gravitation
自吸值 rate of self-absorption
自熄的 self-extinguish(ing)
自熄灭性 self-extinguish(ing)
自熄式转向指示灯 self-cancelling trafficator
自熄树脂 self-extinguishing resin
自熄性材料 self-extinguish(ing)
自熄性树脂 self-extinguishing resin
自洗式过滤器 self-washing filter
自洗轴承 washer bearing
自下承接的 underhung
自下而上方式 bottom-up approach
自下而上分段灌浆法 ascending stage grouting
自下给(燃)料 under-stoke
自下上煤机 under-stoker
自下向上凿井法 raising raiser
自下支撑的 under-braced
自下支承的 underhung
自显指示剂 owner indicator
自现干涉圈 idiophanism
自限协定 self-restraint agreement
自限性 property of self-confinement
自陷级 trap level
自陷效应 self-trapping effect
自相抵销的 self-cancelling
自相对的 self-relative
自相对地址 self-relative address
自相对寻址 self-relative addressing
自相对子程序 self-relative subroutine
自相感应 auto-induction
自相关 autocorrelation; natural correlation; self-correlation
自相关比较波形法 deltic method
自相关定理 autocorrelation theorem
自相关分析 autocorrelation analysis
自相关分析系数【数】 autocorrelation analysis coefficient
自相关函数 autocorrelation function; self-correlation function
自相关函数平面图 plane curve of autocorrelation function
自相关计算机 self-relative computer
自相关接收机 autocorrelation receiver
自相关接收器 autocorrelation receiver
自相关频谱计 autocorrelation spectrometer
自相关器 autocorrelator; self-correlator
自相关式计算机 autocorrelation computer; autocorrelogram computer
自相关图 autocorrelogram
自相关系数 autocorrelation coefficient; coefficient of autocorrelation
自相关型辐射计 autocorrelation type radiometer
自相关作用 autocorrelation
自相矛盾 antilogy; antinomy; self-contradiction
自相矛盾的说法 paradox
自相似 self-similarity
自相似解 self-similar solution
自相似天线 self-similar antenna
自相位共轭 self-phase conjugation
自相位调制 self-phase modulation
自相一致的 self-congruent
自相一致的测量单位 self-consistent units
自相一致的曲线族 self-consistent family of curves
自相制控制(调节) non-interacting control
自响浮标 automatic sound buoy
自项向下法 top-down approach
自消弧 self-extinction of arc
自消灭 self-quenching
自消洗 self-cleaning
自销 self-sale
自销扣 latch hook
自协变 autocovariance
自协方差【数】 autocovariance
自协方差函数 autocovariance function
自携式呼吸器 self-contained breathing apparatus
自携式水下呼吸器 self-contained underwater breathing apparatus
自卸驳(船) hopper barge; self-discharging barge; self-dumping barge; tumbler barge; dump barge
自卸驳船队 self-unloading barge unit
自卸仓 gravity bin
自卸车 automatic tipper; dump cart; self-discharging car; tripping car; dump truck; tipping lorry
自卸车车厢 dump body
自卸车辆 dump car; tipping wagon
自卸车身 <载重汽车的> gravity dump body; dump(ing) body
自卸车身水平延伸板 running board
自卸车身外壳 body shell
自卸车尾开门 chute gate
自卸车箱 dump car; dump cart
自卸车重型车身 heavy-duty dump body
自卸船 self-discharging ship; self-unloader; self-unloading ship; self-unloading vessel
自卸大卡车 mechanized lorry
自卸的 self-discharging; self-dumping; self-emptying; self-tipping
自卸吊车 clamshell car
自卸吊斗 self-dumping skip
自卸翻斗车 dump car
自卸式干散货船 bulk carrier sel unloader
自卸挂车 dump trailer
自卸罐笼 self-dumping cage
自卸荷载的 self-clearing
自卸货车 automatic dump truck; dump (-bed) truck; self-clearing car; self-discharging wagon; self-tipping wagon; self-unloading wagon; self-dumping truck
自卸角 drawdown angle
自卸卡车 dumper(truck); dumping car; dump(ing) truck; end-dump truck; self-discharging truck; tip lorry; tippler
自卸矿车 self-dumping car
自卸垃圾车 tip truck
自卸料仓 self-unloading hopper
自卸料车 dumper
自卸漏斗 dump hopper
自卸泥驳 barge hopper; dump scow
自卸泥驳船 hopper barge
自卸汽车 dump truck; self-dumping truck; tipper; dumping truck; self-discharging wagon; self-unloading wagon; tipping lorry; tipping truck; tipping vehicle; tipping wagon
自卸汽车倾卸机构 dump mechanism
自卸散装货轮 self-discharging bulk carrier
自卸砂驳 hopper barge; self-unloading sand barge
自卸式 dump-type; self-tipping type;

self-unloading
自卸式车厢 self-discharging body
自卸式车厢的拖车 dump carrier
自卸式吊桶 self-discharging bucket
自卸式货船 self-unloading ship
自卸式卡车 dump cart
自卸式料仓 self-emptying hopper
自卸式料车 dump wag(g)on
自卸式饲料分送车 self-unloading forage box
自卸式饲料分送器 forage unloader box
自卸式拖车 dump trailer
自卸式拖运器 autosledge
自卸式运煤船 self-discharging collier
自卸式运木船 self-dumping log carrier
自卸式载重卡车 automatic dump truck
自卸拖驳船队 self-unloading tug-barge unit
自卸拖车 dump wagon; self-discharging trailer; self-unloading wagon; dump trailer
自卸系统 self-discharging system
自卸卸净角 <车辆> self-cleaning angle
自卸运煤船 self-discharging collier
自卸运输机 tripper belt conveyer
自卸抓斗 self-dumping grab
自卸装置 automatic tipper; self-discharging system; self-emptying facility
自信 assurance; positivism
自信息 self-information
自行 self-propelled travel(l)ing; proper motion <指天体自行>
自行安装 do-it-yourself
自行安装塔吊 self-erecting tower crane
自行包装 self-packing
自行保护 self-protecting; self-protection
自行测量部件 self-operated measuring unit
自行测量单元 self-operated measuring unit
自行测试 self-test(ing)
自行铲装式 self-loading
自行车 bicycle; cycle; pedal bicycle; pedal cycle; push-bicycle; push-bike; push cycle; roadster <英旧时的>
自行车比赛场 bicycle-racing arena; cycle-race course; velodrome
自行车比赛跑道 bicycle racing track; cycle race track
自行车测力计 bicycle ergometer
自行车厂 bicycle plant
自行车场 bicycle parking; cycle stand
自行车车道 bicycle lane; bike lane
自行车车道加边 cycle path edging
自行车出行 cycle trip
自行车存车场 cycle park
自行车存车架 bicycle rack
自行车存放处 bike park
自行车存放率 total bicycle parking ratio
自行车存放设施 bicycle parking facility
自行车存(停)车架 cycle rack
自行车道 bicycle path; bicycle route; bicycle way; bike-linking; bike route; cycle path; cycle route; cycle track; cycleway
自行车道隔离线 separate line for cyclist
自行车道网 cycleway network
自行车的 bicyclic(al)
自行车发电机 bicycle dynamo
自行车房 bicycle room
自行车飞轮 free wheel
自行车功量计 bicycle ergometer
自行车和人行(专用)道 bicycle and pedestrian path
自行车活动范围 bicylce activities sphere
自行车计程仪 cyclometer
自行车驾驶熟练证书 cycle proficiency certificate
自行车架 bicycle rack; bicycle stand; cycle stand
自行车交通 bicycle traffic; cycle traffic; cyclist traffic
自行车教育配套卡 bicycle education kit
自行车库 indoor bicycle parking area
自行车流 cycle flow
自行车路段通行能力 bicycle traffic capacity on uninterrupted flow
自行车路(径) bicycle track; cycle path; cycle track
自行车路线 bicycle route; bike route; cycle route; pedal cycle track
自行车轮式屋顶 bicycle wheel roof
自行车内胎 bicycle tube; cycle tube
自行车跑道 bicycle track; cycle race track
自行车跑道交叉 cycle track crossing
自行车棚 bicycle shed; bicycle shelter
自行车前叉子的搁脚处 coaster
自行车桥 bicycle bridge; cycle-bridge
自行车赛场 cycling stadium
自行车赛道 bicycle racing track
自行车赛跑道 cycle racing track
自行车式起落架 bicycle gear
自行车事故 bicycle accident
自行车隧道 cycling tube
自行车所有权 bicycle ownership
自行车胎 cycle tyre
自行车停车场 cycle parking area
自行车停车换乘 bike and ride
自行车停放处 bicycle park; cycle park
自行车停放架 bicycle stand
自行车外胎 cycle casing
自行车违章事件 bicycle violation
自行车小路 bicycle track; cycle track
自行车行驶线 cycle lane
自行车型轴承 bicycle type bearing
自行车与人行道合用 bicycle pedestrian path
自行车运行条件 cycling condition
自行车专用车道 bike lane
自行车专用道 bicycle lane; bicycle track; bicycle way; bike lane; bike way; cycle path
自行车专用路 cycle road; cycle track
自行成员 proper motion membership
自行充气 self-inflating
自行充填开采巷道 self-stowing gate
自行处理 discretion
自行处理的开支 discretionary spending
自行处置 self-help
自行传送带 mobile belt conveyer[conveyor]
自行的 self-actuated; self-actuating; self powered; self-propelled; self-propelling; self-travel(l)ing
自行点火 self-ignition
自行点燃法 self-ignition method
自行垫高料 self-furring
自行吊车 self-travel(l)ing hoist; mobile crane <常指无轨起重机>
自行断续电路 self-interrupting circuit
自行断续器 self-interrupter
自行翻斗卡车 self-tipping lorry
自行分裂 dichastasis
自行风冷电动机 self-ventilated motor
自行浮起的 self-buoyant
自行干燥 self-desiccation
自行更新 self-renewal
自行攻(丝)的 self-tapping
自行夯击式压路机 self-propelled tamping roller
自行还本生利的投资 self-liquidating investment
自行缓解 spontaneous remission
自行换气 self-ventilation
自行恢复的 self-healing
自行激磁 self-excitation
自行挤碎 own-accord crushing
自行加宽(路面)修整机 self-widening finisher
自行加力制动器 self-energized brake
自行架 self-propelled mount(ing)
自行架设塔式起重机 self-erecting tower crane
自行监控的 self-monitoring
自行净化 self-cleaning
自行开体泥驳 self-propelled hopper barge; self-propelled split barge
自行控制 self-actuated control; self-operated control
自行矿车 mobile wagon
自行扩散 self-diffusion
自行缆车集装箱起重机 container crane with self-propelled trolley
自行冷却的 self-cooling
自行履带式运载装填车 self-propelled tracked loader
自行轮式铲运机 wheel tractor-scraper
自行轮式缆索起重机 cable rubber-tired mobile base crane
自行爬高卷扬机 self-climbing hoist
自行平衡的 self-balancing
自行破坏 active failure
自行启动 self-starting
自行启动泵 self-acting pump; self-starting pump
自行起动的 self-running
自行起重机构台 mobile gantry
自行起重门架 self-launching gantry
自行牵引绞车 self-propelled puller winch
自行桥式起重机 moving bridge crane
自行清灰袋式集尘器 self-cleaning cloth screen dust collector
自行清洁的履带 self-cleaning track
自行清洁作用 self-cleaning effect
自行驱动的 self-driven; self powered
自行群 proper motion group
自行认证 self-certification
自行入坞 self-docking
自行入坞式 self-docking type
自行散料抛掷码垛机 lorry dump piler
自行上水水箱 self-priming tank
自行设备 mobile equipment
自行设计的程序 roll-your-own program(me)
自行设计的系统 roll-your-own system
自行申请破产 voluntary bankruptcy
自行渗气水流 self-aerated flow
自行生效 automatic approval
自行施釉 self-glazing
自行式 self-traction
自行式搬运设备 mobile handling equipment
自行式拌制混凝土装置 low-profile plant
自行式铲运机 motor scraper; motorized scraper; self-powered scraper; self-propelled scraper; tractor scraper
自行式车辆 automotive vehicle; driving carriage; locomobile; motor vehicle
自行式打壳机 self-propelled crust breaker
自行式单马达吊车 one motor travel(l)ing crane
自行式的 self-propelled; self-traveling
自行式底盘 self-propelled (power) frame; self-propelled mount (ing) <挖掘机的>
自行式地面作业机械 mobile land machine
自行式堆 self-propelled mobile reactor
自行式堆木机 mobile yarder
自行式发电机组 mobile generator set
自行式割草机 motor mower
自行式工程机械 mobile equipment; self-propelled work machine
自行式工地搬运设备 mobile site handling unit
自行式刮土机 self-propelled scraper; truck scraper
自行式焊接机 mobile welding plant
自行式互路机 motor roller
自行式混凝土浇灌机 self-propelled concreting plant
自行式混凝土浇灌设备 self-propelled concreting plant
自行式混凝土浇灌装置 self-propelled concreting plant
自行式混凝土浇筑机 self-propelled concreting plant
自行式混凝土浇筑设备 self-propelled concreting plant
自行式混凝土浇筑装置 self-propelled concreting plant
自行式机械 mobile unit
自行式机械起重机 mobile mechanical crane
自行式架桥车 self-propelled assault bridge
自行式井下无轨矿车 koalmobile
自行式沥青混合料摊铺机 self-propelled asphalt-paver
自行式联合收割机 self-propelled combine harvester
自行式林业机械 mobile forestry machine; self-propelled forestry machine
自行式龙门架起重机 travel(l)ing gantry tower
自行式龙门起重机 serf-propelled gantry crane
自行式路面破碎压实机 mobile breaker-roller
自行式路面破碎压实器 mobile breaker roller
自行式路缘石制造机 curb machine
自行式农业机械 self-propelled agricultural machine
自行式配套凿岩钻车 self-propelled and self-contained mounting
自行式平地机 motor grader; motorized grader; power grader; self-propelled grader; wheel grader; autopatrol grader <养路用平路机>
自行式平路机 motorized road grader; self-contained blade grader; self-propelled blader
自行式起重机 mobile crane; motorcrane; motor gravity; self-propelled crane
自行式气胎碾 self-propelled pneumatic roller
自行式气胎压路机 self-propelled pneumatic roller
自行式桥梁检测架 self-propelled bridge inspection cradle
自行式撒砂机 self-propelled sand spreader
自行式伸臂回转起重机 mobile revolving boom crane
自行式伸缩塔身 mobile telescoping tower
自行式升降平台挖泥船 walking serf-elevating dredge(r)
自行式石屑撒布机 self-propelled chip

spreader
自行式手喷雾器 self-propelled sprayer for hand
自行式水泥卸载机 mobile cement unloader
自行式挖掘机 self-propelling dredge-(r)
自行式挖掘装载机 tractor-loader-backhoe
自行式稳定土搅拌机 self-propelled pulvi-mixer
自行式吸扬挖泥船 suction hopper dredge(r)
自行式悬臂起重机 mobile cantilever boom type crane
自行式压路机 ride on roller;self-propelled(road)roller
自行式羊足碾 self-propelled sheep-foot roller
自行式液压起重机 hydraulic mobile crane;mobile hydraulic crane
自行式液压挖掘机 mobile hydraulic excavation
自行式原木装载机 self-propelled log loader;travel to load log loader
自行式越野工程机械 off-highway self-propelled work machine
自行式运输车 self-propelled transfer car
自行式运输机 mobile conveyer[conveyor]
自行式凿岩机 self-propelled drill-(er);self-propelled rock drill(er)
自行式凿岩台车 self-propelled jumbo
自行式照明设备 self-propelled lighting equipment
自行式振捣机 power-propelled vibrating machine
自行式振动板 self-propelled vibroplate
自行式振动夯实板 self-propelled vibratory-plate compactor
自行式振动压路机 self-propelled vibratory compactor; self-propelled vibrating roller
自行式支索器 mobile rope carrier
自行式装车机 mobile loader
自行式装料斗采砂船 self-propelling hopper dredge(r)
自行式装料机 mobile-charging machine
自行式装砂机 mobile sand loader
自行式装载机 mobile loader; self-propelled loader;tractor loader
自行式钻车 mobile jumbo; mounted self-propelled drill; self-propelled jumbo;self-propelled wagon drill
自行式钻机 mobile drill;self-hauling drill;self-propelled drill(er);self-propelled rig; self-towing equipment
自行式作业升降机 self-propelled aerial work platform
自行疏干 automatic dewatering
自行速度 self-propelled travel(l)ing speed
自行塔式起重机 self-propelled tower crane
自行台车 mobile jumbo
自行填实 self-packing
自行调心制动器 self-centering brake
自行调整式 self-aligning type
自行调直式 self-aligning type
自行通风 self-ventilation
自行推进的振动压路机 self-propelled vibrating roller
自行推进式起落架 self-propelling landing gear
自行退火 self-annealing;spontaneous annealing
自行维持和调节 self-maintenance and regulation
自行小车集装箱起重机 container crane with self-propelled trolley
自行小车门式起重机 gantry crane with self-propelled trolley
自行卸载的货运汽车 dump body truck
自行压碎 own-accord crushing
自行硬化 self-hardening
自行支撑的地下建筑物 self-supporting underground structure
自行制动 self-brake;self-retention
自行主机 self-propelled main machine
自行抓斗挖泥船 self-propelled grab dredge(r)
自行装配 self-erecting
自行装载机 mobile loader; mobile loading machine
自行装置 motorized cart;self-towing device
自形变晶【地】idioblast
自形变晶的 idioblastic
自形变晶结构 idioblastic texture
自形变晶系列 idioblastic series
自形成的 self-formed
自形的【地】automorphic; automorphous;euhedral;idiomorphic
自形结晶 euhedral crystal
自形晶 idiomorph
自形晶粒状结构【地】euhedral-granular texture; idiomorphic granular texture
自形晶体 idiomorphic crystal
自形粒状 automorphic granular
自形粒状的 euhedral-granular; idiomorphic granular
自形粒状结构 automorphic-granular texture
自形组构 idiotopic
自形作用 idiomorphism
自型 autotype
自修复 self-repair(ing)
自修复计算机 self-repairing computer
自修复控制 self-repairing control
自修复能力 self-repairing capability
自修复系统 self-repairing system
自修改 self-modification
自修式浮船坞 self-docking floating drydock
自修正自动导航 self-correcting automatic navigation
自絮凝作用 autoflocculation;self-flocculation
自悬浮作用 autosuspension
自旋 spin;spinning
自旋变相 spin flipping
自旋标记 spin label(l)ing
自旋标记物 spin label
自旋波 spin wave
自旋波共振 spin-wave resonance
自旋波函数 spin wave function
自旋波谱 spin wave spectrum
自旋玻璃 spin glass
自旋补偿 spin compensation
自旋捕获 spin trapping
自旋弛豫时间 spin relaxation time
自旋磁感应波 spin density wave
自旋磁共振 spin magnetic resonance
自旋磁化率 spin susceptibility
自旋磁矩 spin magnetic moment
自旋磁性 spin magnetism
自旋代数 spin algebra
自旋的极化 spin polarization
自旋点阵 spin lattice
自旋点阵弛豫 spin-lattice relaxation
自旋点阵相互作用 spin-lattice interaction
自旋动量矩 spin momentum
自旋动量矢量 spin-momentum vector
自旋多重性 spin multiplicity
自旋翻转碰撞 spin-flip collision
自旋翻转散射 spin-flip scattering
自旋反向碰撞 spin-flip collision
自旋反向散射 spin-flip scattering
自旋反转激光器 spin-flip laser
自旋方向 spin direction
自旋分裂 spin splitting
自旋共振 spin resonance
自旋关联系数 spin-correlation parameter
自旋回波 spin echo
自旋回波存储 spin-echo storage
自旋回波法 spin echo method
自旋回波谱法 spin-echo spectroscopy
自旋回声 spin echo
自旋回声技术 spin echo technique
自旋极化 spin polarization
自旋加快 spin-up
自旋减慢 spin down
自旋减速力矩 spin-decelerating moment
自旋简并度 spin degeneracy
自旋角动量 spin angular momentum
自旋进动法 spin-precession method
自旋进挖泥船 self-propelled dredge(r)
自旋晶格弛豫时间 spin-lattice relaxation time
自旋晶格耦合 spin-lattice coupling
自旋晶格耦合常数 spin-lattice coupling constant
自旋晶格相互作用 spin-lattice interaction
自旋矩 moment of spin;spin moment
自旋矩矢量 spin moment vector
自旋矩阵 spin matrix
自旋空间 spin space
自旋离域 spin delocalization
自旋量子数 spin quantum number
自旋裂分 spin-spin splitting
自旋密度 spin density
自旋能 spin energy
自旋耦合 spin coupling
自旋耦合图解法 spin-orbit coupling diagram method
自旋排列 spin alignment
自旋取向 spin orientation
自旋取样技术 spin-probe technique
自旋去偶 spin decoupling
自旋去偶法 spin decoupling method
自旋去偶图 spin decoupling pattern
自旋扫描摄云照相机 spin scan cloud camera
自旋顺磁性 spin paramagnetism
自旋顺序 spin sequence
自旋速率 spin rate
自旋态 spin state
自旋投影 spin projection
自旋投影算符 spin projection operator
自旋微扰法 spin microharass method
自旋温度 spin temperature
自旋稳定 spinning stability;spin stabilization
自旋稳定地球同步卫星 spin-stabilized geostationary satellite
自旋稳定性 spin stability
自旋相关 spin dependence
自旋相关力 spin-dependent force
自旋效应 spin effect
自旋有关 spin dependence
自旋与轨道相互作用 spin-orbit interaction
自旋运动 spin motion
自旋张量 spin tensor
自旋轴进动 spin precession
自旋轴(线)spin axis
自旋转 autogyration
自旋转稳定卫星 spin-stabilized satellite
自旋转向转变 spin-flop transition
自旋自旋弛豫时间 spin-spin relaxation time
自旋自旋耦合常数 spin-spin coupling constant
自旋坐标 spin coordinates
自选 self-selection
自选副食品 groceteria;self-service grocery
自选商店 convenience store;self-service shop;self-service store
自选商店 <美> groceteria
自选市场 marketeria;self-service market;supermarket
自选最佳通信[讯] self-optimizing communication
自炫 self-display
自学系统 self-learning system
自寻系统接收机 homing receiver
自寻制导 homing guidance
自寻最佳点 self-optimalizing
自寻最佳化控制 self-optimizing control
自寻最佳化控制系统 self-optimizing control system
自寻最佳化滤波器 self-optimizing filter
自寻最佳决策系统 self-optimizing decision system
自寻最佳系统 self-optimizing system
自寻最优点控制 optimalizing control
自循环 self-loop
自压安全阀 deadweight safety valve
自压喷灌机 gravity sprinkler
自压实混凝土 <浇筑时无需振捣的混凝土> self-compacting concrete
自压实性 <自压实混凝土的一个特性> self-compactability
自验程序 self-test routine
自扬链斗挖泥船 pump bucket dredge-(r)
自养 autotrophic nutrition; autotrophism
自养代谢 autotrophic metabolism
自养氮 autotrophic nitrogen
自养的【生】autotrophic
自养反硝化 autotrophic denitrification
自养护 autogenous curing;autogenous cutting
自养生物 autotroph; autotrophic organism
自养微生物 autotrophic microbe;autotrophic microorganism
自养细菌 autotrophic bacteria
自养营养 autotrophic nutrition
自养指数 autotrophic index
自养组分 autotrophic component
自氧化 autooxidation
自氧化期 autooxidation phase
自移式支架 self-advancing support; walking prop
自抑制 self-inhibition
自益放大电路 boostrap circuit
自益放大器 bootstrap amplifier
自因推果 a priori
自引泵 self-priming pump
自引导测试序列 self-initializing test sequence
自引导天线 homing antenna
自引风 blow torch action
自应力 autostressing;self-stress(ing)
自应力大小 magnitude of self-stress
自应力的预应力混凝土 self-stressed prestressed concrete
自应力恢复 self-stress recovery
自应力混凝土 self-stressed concrete; chemically prestressed concrete; contained concrete; self-stressing concrete
自应力混凝土管 self-stressing concrete pipe

自应力浆液 self-stressing grout
自应力砂浆 self-stressing mortar
自应力水泥 <即膨胀水泥> self-stressing cement; chemically prestressing cement; expansive cement; self-expanding cement; self-prestressed cement; stressing cement
自应力水泥浆 self-stressing grout
自应力损失 loss of self-stress
自营工程 force-account work
自营公司 operating company
自营商 dealer
自营商业 do business on one's account
自营设备 <顾客自有并维护的设备> customer owned and maintained equipment
自营推销业务 business of dealer
自营项目开支 force account
自硬 self-hardening
自硬的 no-bake
自硬法 cold-setting process
自硬法造型 air set mo(u)lding
自硬钢 air hardening steel; Mushet('s) steel
自硬化 cold-setting
自硬化黏[粘]结剂 self-set binder
自硬黏[粘]结剂 air-setting binder; cold-setting binder; self-curing binder
自硬砂 cold-setting sand; no-bake sand; self-hardening sand(mixture)
自硬砂芯 no-bake core
自硬砂型 self-curing sand mo(u)ld
自硬芯 air set core
自硬型 air set mo(u)ld
自硬性 self-hardening property; self-setting
自硬油 air-setting oil
自泳 autophoresis
自泳涂装 autodeposition paint; autophoretic coating
自用电话机 private telephone set
自用电量 own demand
自用辅助电源切换 switching over of auxiliary power system
自用负载 works load
自用货物运输 private haulier
自用冷藏库 private cold warehouse
自用生产 captive production
自用水量 water consumption in water-works
自用污水管 private sewer
自用下水道 private sewerage
自用运输 private transport; private transportation; transport on own account
自用蒸汽 self-used steam
自用蒸汽管 auxiliary steam pipe
自用住房 owner-occupied house
自用自配涂料 do-it-yourself paint
自优化 self-optimization
自优化的 self-optimizing
自优化控制 self-optimizing control
自优化适应特性 self-optimizing adaptive feature
自优控制 self-optimizing control
自由安置 free-standing
自由摆 free(swinging) pendulum
自由摆动 free swing; natural oscillation
自由摆动多谐振动器 free running multiple vibrator
自由摆动式甩刀 free swinging flail
自由摆动周期 free swing period
自由板长度 free slab length
自由半群 free semi-group
自由保有地产 freehold property
自由比例尺 indefinite scale; undefined scale
自由比例尺像平面图 free scale photoplan
自由壁 free wall
自由边界 <不承受荷载的> free boundary
自由边界流动理论 free streamline theory
自由边界问题 free boundary problem
自由边(缘) free edge
自由边缘膜片 free edge diaphragm
自由变分 free variation
自由变换车道 free lane change
自由变换群 free transformation group
自由变量 free variable; real variable
自由变数 free variable; real variable
自由变项 free variable
自由变异 free variation
自由变异性 free variability
自由表 free list
自由表面 free surface; available surface <海水>
自由表面波 free surface wave
自由表面放大作用 free surface amplification
自由表面能 free surface energy
自由表面水分 free surface moisture
自由并集 free union
自由波 free wave
自由波吸收系数 free wave absorption coefficient
自由波型 free wave pattern
自由博弈 free game
自由布尔代数 free Boolean algebra
自由裁量权 discretion
自由掺气 ample aeration
自由缠绕 <卷筒> free spooling
自由长度 compute length; free length; overhanging length; unbraced length; unsupported distance; unsupported length
自由偿还制度 free redemption
自由场地面运动 free field ground motion
自由场电流灵敏度 free field current sensitivity; receiver current sensitivity
自由场电流响应 free field current response
自由场电压灵敏度 free field voltage sensitivity; receiving voltage sensitivity
自由场电压响应 free field voltage response
自由场功率谱 free field power spectrum
自由场功率响应 free field power response
自由场互易校准 free field reciprocity calibration
自由场校准 free field calibration; free field correction
自由场灵敏度 free field sensitivity
自由场频率响应 free field frequency response
自由场设计谱 free field design spectrum
自由场室 free field chamber; free field room
自由场特性曲线 free field characteristic
自由场条件 free field condition
自由场响应 free field response
自由场修正曲线 free field correction curves
自由场应力 free field stress
自由场运动 free field motion
自由场质点速度 free field particle velocity
自由超车 free passing of cars
自由超滤子 free ultrafilter
自由潮汐波 free tidal wave
自由车流 free flow of motor
自由车速 free flow operating speed
自由沉降 free sedimentation; free settling
自由沉降分级 gravity classifying
自由沉降分级机 free falling classifier; free settling classifier
自由沉降管 free settling tube
自由沉降颗粒 free settling particle
自由沉降水力分级机 free settling hydraulic classifier
自由沉降速度 free falling velocity
自由陈列逻辑 uncommitted array logic
自由成型 free forming; off-hand process
自由城市 free city
自由程 free path
自由尺寸 free size
自由冲孔 free piercing
自由出流 free discharge; free efflux; free outflow
自由出流孔 free orifice
自由出流口 free outlet
自由出入 in-and-out
自由出入的河口 free outfall
自由出现 free occurrence
自由处理成本 discretionary cost
自由处理(权限) discretion
自由穿流分子 free draining molecule
自由穿流模型 free draining model
自由传递式空调机组 free delivery-type air conditioning unit
自由传输范围 free transmission range
自由吹送式空气冷却器 free delivery-type air cooler
自由吹送式空调机组 free delivery-type air conditioning unit
自由吹送式空调器 free-blow air conditioner
自由吹送式通风空调器 free delivery-type air conditioner
自由吹制 free blown
自由吹制玻璃 free-blown glass
自由吹制成型 chair work; free blowing
自由磁荷 free magnetic charge
自由磁化状态 free magnetization condition
自由磁极 free magnetic pole; free pole
自由磁性 free magnetism
自由存储表 free storage list
自由存储块 free storage block
自由存储区 free core pool; free storage area
自由存储栈 free store stack
自由大气 free air
自由大气层 free atmosphere
自由大气温度 free air temperature
自由带 free strip
自由导纳 free admittance
自由导数 free derivative
自由导线【测】 free traverse; opening station <开导线>; arbitrary marching of line; arbitrary routing of line <不用足限坡定线>
自由导线端点【测】 opening station
自由导叶 free guide vane
自由的 free; autonomous; unattached; unbound; unconfined; unobstructed; unrestrained; unrestricted; unsupported; untrammelled
自由的多边贸易 free multilateral foreign trade
自由的料 unbound
自由地下水 available groundwater; free ground water; gravitative water; gravity groundwater; unconfined groundwater
自由地下水位 free water table
自由电磁场 free electromagnetic field
自由电荷 free charge; free electric(al) charge; free electricity
自由电荷传递 free charge transfer
自由电解质 free electrolyte
自由电位栅极 free potential grid
自由电泳法 free electrophoresis
自由电子 free electron(ic); roaming electron; unbound electron
自由电子活度 activity of free electrons
自由电子浓度 free electronic concentration
自由电子跃迁 free-electron(ic) transition
自由掉调 <船舶> free turning
自由跌价 free depreciation
自由跌落曝气器 free fall aerator
自由跌水 free drop
自由动态阻抗 free motional impedance
自由独异点 free monoid
自由度 degree of freedom; freedom degree; freedom of motion; freeness; variance
自由度的重分 subdivision of degrees of freedom
自由度数 number of degrees of freedom
自由度陀螺仪 free gyroscope
自由度轴(线) axis of freedom
自由端 free end; freely supported end
自由端三轴试验 <端部滑动的> free end triaxial test
自由端位移 free end travel
自由端游丝 free end spring
自由端支承 free end bearing
自由端柱 free end column
自由段格式 free field format
自由断面掘进机 partial boring machine
自由断面隧道掘进机 line header; partial face machine
自由锻 hammer forging
自由锻锤 flat-die hammer
自由锻造 flat-die forging; free forging; hammer forging; open forging; smith forging
自由锻造水压机 hydraulic press for free forging
自由堆积密度 free pouring density
自由对流 free convection; natural convection
自由对流边界层 free convection boundary layer
自由对流高度 level of free convection
自由兑换 convertibility; free convertibility
自由兑换货币 convertible currency; freely convertible currency
自由放电 free discharge
自由放牧场 free range
自由放牧地 open range
自由放任政策 laissez-faire policy
自由放索 on power reel out
自由分布法 distribution-free method
自由分解 free decomposition; free resolution
自由分量 free component
自由分轮 free cannon pinion
自由分配格 free distributive lattice
自由分子扩散 free molecule diffusion
自由分子流 free molecular flow; free molecule flow; Knudsen flow
自由分子流作用在物体上的力 free molecular flow force
自由分子弥散 free molecule diffusion
自由焚风 free foehn
自由粉碎机 free mill
自由风格 freely adapted style

自由缝 free joint
自由浮动 free float <工程进度上的>；clean float <汇率的>
自由浮动沉箱 freely floating pontoon
自由浮动沉箱式闸门 freely floating pontoon
自由浮动古塞 free floating piston
自由浮动汇率 free exchange rate；freely floating exchange rate
自由浮动汇率制度 freely floating system；variance-exchange-rate system
自由浮动平台 free floating platform
自由浮动生长 free floating growth
自由浮动时间 <工程网络计划中的> free float
自由浮式驳船 freely floating pontoon
自由符号 free symbol
自由符号序列 free symbol sequence
自由辐射 free radiation
自由辐射器 free radiation；free radiator
自由复形 free complex
自由杆 free bar
自由港 free harbo(u)r；free port；open harbo(u)r；open port；optional port
自由港区 free port zone
自由高度 free height；unsupported height
自由高空气球 free aerostat
自由格 free lattice
自由格式 free format
自由格式输入 free format input
自由格式源代码 free format source code
自由给料 free feeding
自由工作面 free working
自由雇佣企业 <指雇佣职工不必限于工会会员> non-union shop；merit shop；open shop
自由关税区 tariff-free zone
自由观察 unobstructed view
自由管钳 universal pipe wrench
自由惯性流 free inertial flow
自由灌水舱 free flooding ballast
自由光谱区 free spectral range
自由广群 free groupoid
自由滚动闸门 free rolling gate
自由滚动作用 free rolling action
自由过境 free transit
自由含水层 free aquifer；unconfined aquifer
自由含水量 free moisture；free aquifer；free water content；unconfined aquifer
自由函数 free function
自由焓 free enthalpy
自由航速 sailing speed
自由航行 free navigation；free running
自由航行船模 free sailing ship model
自由航行试验 free sailing test
自由和谐振动 free harmonic vibration
自由横滑板 free cross slide
自由横向加速度 free lateral acceleration
自由横摇 free rolling
自由洪峰溢洪道 uncontrolled crest spillway
自由后座 free recoil
自由后座炮架 free recoil mount
自由弧 free arc
自由滑动 free gliding；slide free
自由滑动距离 free run
自由滑动支承 freely movable bearing
自由环 free ring
自由环境 free environment
自由簧 free spring
自由簧片 free reeds
自由回声 free echoes
自由回转法兰 lap-joint flange
自由回转滚柱式离合器 freewheeling roller clutch
自由回转机构 freewheel mechanism
自由回转离合器 freewheeling clutch
自由活动 run free
自由活动托架 freely moving carriage
自由活塞 floating piston；free piston
自由活塞泵 free piston pump
自由活塞打入式取土器 free piston drive sampler
自由活塞发动机 free piston engine
自由活塞机械 free piston machinery
自由活塞煤气发生炉发动机 free piston gas generator engine
自由活塞煤气发生炉透平 free piston gas generator turbine
自由活塞内燃压气机 free piston diesel compressor
自由活塞气体发生器 free piston gas generator
自由活塞燃气发生器 free piston gas generator；free piston gasifier
自由活塞式 free piston
自由活塞式发电机 free piston generator
自由活塞式发动机 free piston engine
自由活塞式空气压缩机 free piston air compressor
自由活塞式煤气发生器 free piston generator
自由活塞式气体发生器 free piston gasifier
自由活塞式取土器 free piston sampler
自由活塞式燃气轮机 free piston gas turbine
自由活塞式燃气轮机装置 free piston gas turbine
自由活塞式燃气涡轮机车 free piston gas turbine locomotive
自由活塞式压缩机 free piston compressor
自由活塞压缩机 free piston compressor
自由活塞装置 free piston installation；free piston plant
自由火焰 free flame
自由货币 free currency
自由货物 open cargo
自由积 free product
自由基 free radical
自由基反应 free radical reaction
自由基交联 radical cross linking
自由基聚合 radical polymerization
自由纪念碑 freedom memorial
自由加速试验 free acceleration test
自由价【化】free valence
自由价格 free price
自由剪切紊流 free turbulent shear flow
自由降落 free fall；free overfall
自由降落法 free fall method
自由降落时间 time of free fall
自由降落水舌 free overflow jet
自由降落速度 free fall(ing) velocity；settling velocity
自由降落系数 free settling ratio
自由交变电流 free alternating current
自由交换群 free Abelian group；free commutative group
自由交谈 open-end interview
自由交通流 free traffic flow
自由教育 liberal education
自由节 universal coupling
自由结点 free knot
自由结合代数 free associative algebra
自由结合环 free associative ring
自由截面 free cross-section
自由截面积 free cross-section(al) area
自由界面电泳 free boundary electrophoresis
自由进近 clear approach
自由进料 free fed；free feed
自由进流 free inflow；free influx；unobstructed influx
自由进入 free access
自由进入市场 liberal market access
自由浸沉式换能器 free flooded transducer
自由经济区 free economic zone
自由竞争 free competition
自由就位 free positioning
自由距离 free distance
自由开架式 free-open-stack system
自由开挖面 free working
自由可动性 free mobility
自由空间 free air；free space；headroom
自由空间爆破 free space blasting
自由空间波 free space wave
自由空间波长 free space wavelength
自由空间场强 free space field intensity
自由空间传播 free space propagation
自由空间传播条件 free space propagation condition
自由空间导磁率 permeability of free space
自由空间的辐射图 free space pattern
自由空间电场 free space field
自由空间电场强度 free space field intensity
自由空间电容率 permittivity of free space
自由空间辐射方向图 free space radiation pattern
自由空间改正 free space correction
自由空间改正值 free air correction value
自由空间功率 free space power
自由空间光通信[讯] free space optic-(al) communication
自由空间校正 free air correction
自由空间雷达方程 free space radar equation
自由空间路径损耗 free space path loss
自由空间衰减 free space attenuation
自由空间搜索距离 free space range
自由空间损耗 free space loss
自由空间中的作用距离 free space range
自由空间阻抗 free space impedance
自由空气 free air；free atmosphere
自由空气超压力 free air overpressure
自由空气电离室 free air ionization chamber
自由空气改正 free air correction
自由空气剂量 free air dose
自由空气校正 free air correction
自由空气条件 free air condition
自由空气压力 free air pressure
自由空气异常 free air anomaly
自由空气重力异常 free air gravity anomaly
自由空投物 free drop
自由空隙 free space
自由空穴 free hole
自由空穴吸收 free hole absorption
自由孔径 free aperture
自由孔隙 free pore space
自由控制的筒仓容量 free running silo capacity
自由跨距圆顶 free span dome
自由块 free block
自由块队列元素 free block queue element
自由宽度 unsupported width
自由扩散 free diffusion
自由类群 free monoid
自由冷却 free cooling
自由冷缩 free cooling shrinkage
自由力矩 free moment
自由粒子 uncoupled particle
自由联结空间 free coupling space
自由联想 free association
自由链复形 free chain complex
自由梁 free beam
自由领域 free zone
自由溜行 freewheeling
自由流 free motion of a liquid；free motion of a stream；free (stream) flow
自由流变定理 free rheological law
自由流出 free discharge
自由流的动压力 free stream dynamic-(al) pressure
自由流动 free flow(ing)；free running；free outflow；unconfined flow；uncontrolled flow；unrestricted flow
自由(流动)车速 free flow speed
自由流动的 freely flowing；free pouring；free running
自由流动阀 free flow valve
自由流动粉末 free flowing powder
自由流动颗粒 free flowing granules
自由流动黏[粘]度计 free flow visco-(si) meter
自由流动水 free water；mobile water
自由流动速度 free stream velocity
自由流动状的 free flowing
自由流动状态的 free flowing
自由流静压力 free stream static pressure
自由流量 free discharge
自由流马赫数 free stream Mach number
自由流态 free flow condition
自由流体指数 free fluid index
自由流通面积 free flow area
自由流湍流 free stream turbulence
自由流紊动 free stream turbulence
自由流线 free streamline
自由流线(流动)理论 free streamline theory
自由流压力 free stream pressure
自由滤子 free filter
自由路径选择 free routing
自由路线 free path
自由轮 liberty ship
自由轮部件 freewheel unit
自由轮传动 <指汽车后轮速率较引擎高时，后轮与引擎联系自动分离，惯性滑行> freewheeling
自由轮毂 loose boss
自由轮机构 freewheel mechanism
自由轮廓 free outline
自由轮离合器 freewheel device
自由轮轮胎 free rolling tyre
自由轮式离合器 freewheeling clutch
自由轮体系 freewheeling system
自由轮转筒 freewheeling drum
自由轮装配 freewheel assembly
自由螺旋形旋涡 free spiral vortex
自由落差 height of free fall
自由落程 height of free fall
自由落锤 free fall hammer；free fall of ram；drop hammer；free-hanging hammer；drop pile hammer <桩工>
自由落料 free fall
自由落水救生艇 free fall lifeboat
自由落体 free faller；free falling body；freely falling body
自由落体定律 free fall law
自由落体加速度 free fall acceleration
自由落体模型 free fall model
自由落体实验装置 free falling body experiment instruments
自由落体速度 free fall velocity

Z

自由落下 free fall;gravitate
自由漫灌 free flooding irrigation
自由蔓延火 free-burning fire
自由贸易 free trade
自由贸易港 free port
自由贸易港口码头 free trade wharf
自由贸易区 free trade area;free(trade) zone;zone of free trade
自由贸易协定 free trade agreement
自由贸易政策 free trade policy
自由贸易主义 manchesterism
自由棉 free fiber
自由面 free face;open face
自由面爆破 free burden
自由面崩落岩(石)层 free burden
自由面不稳定性 free surface instability
自由面积比 free area ratio
自由面能量 free surface energy
自由面水流 free surface flow
自由面水流渠 free surface flow channel
自由面水流隧道 free surface flow tunnel
自由面水流线 free surface flow line
自由面速度 free surface velocity
自由面紊流 free surface turbulent flow
自由面涡漩 free surface vortex
自由面运动状态 free surface kinematic condition
自由民 freeman
自由模 free module
自由模式 free schema
自由磨粒 loose grain
自由内接 free inscribing
自由能变化 free energy change
自由能标度 free energy scale
自由能泛函 free energy functional
自由能获得效率 efficiency of free energy capture
自由能量 free energy
自由能判据 free energy criterion
自由能曲线 free energy curve
自由能最小原理 principle of free energy minimum
自由扭振频率 natural torsional frequency
自由扭转 free torsion
自由女神像<美> Statue of Liberty
自由排气能力 free air capacity;free air displacement
自由排水 free draining
自由排水堆石 flee-draining rock-fill
自由排水规范 free draining specification
自由配额 unilateral import quota
自由配合 free fit;free running fit
自由配合的 fitted freely;fitting freely
自由喷流 free jet
自由喷射流 free discharging jet
自由膨胀 free expansion;free-swell(ing)
自由膨胀率 free swelling rate;free swelling ratio
自由膨胀率试验 free swelling ratio test;coefficient of heat expansion test
自由膨胀试验 free swell test
自由膨胀水箱 open expansion tank
自由膨胀序数 free swelling index
自由膨胀指数 free swelling index
自由偏动涡轮机 free deviation turbine
自由平面边界 plane-free boundary
自由屏蔽 free screening
自由破碎 free crushing
自由企业经济 economic of free enterprise
自由企业制度 free enterprise system
自由起升高度 free lift
自由气饱和率 free gas saturation
自由气流 free air flow;free stream
自由气流方向 free stream direction
自由气流湿空气的湿含量 humidity ratio of moist air in the free stream
自由气流速度 free stream velocity
自由气流总压头 free stream total head
自由气球 free balloon
自由气生产能力 free air delivery capacity
自由气(体) free gas
自由气体离子 free gas ion
自由气隙 free air space
自由前进波 free progressive wave
自由钳 come-along
自由潜水 free diving;free ground water
自由潜水面 free water table;unconfined water table
自由嵌固端条件 free-clamped end condition
自由切削 free cutting
自由切削钢 free cutting steel
自由切削黄铜 free cutting brass
自由区 free zone
自由区表 free area list
自由区程序 free area routine
自由区域 free field
自由曲流 free meander
自由曲面 free form surface
自由曲线 free curve
自由取舍 take-it or leave-it
自由群 free group
自由燃烧率 free-burning rate
自由绕线盘 freewheeling drum
自由热 free heat
自由任务控制块 free task control block
自由溶液电泳法 free solution electrophoresis
自由软件 free software
自由筛分 free screening
自由筛选 free screening
自由栅极 free grid
自由上浮 free ascent
自由上浮脱险通道 access trunk for free escape and free ascent
自由上升对流 freely-rising convective stream
自由上升杆铲车 free-lift mast forklift truck
自由上升力 free lifting force
自由上升桅杆 free-lift mast
自由设计图纸 free design drawing
自由射流 free efflux;free jet;free jet stream;free nappe;open jet;apparent free nape<水流出口的>
自由射流波 jet wave
自由射流喷嘴 free jet nozzle
自由射流式水轮机 constant pressure turbine
自由射流式涡轮机 free jet turbine
自由射流斜槽 free jet chute
自由伸缩的 free-going
自由伸缩接头 free expansion joint
自由伸展 free expansion;unrestricted spread
自由渗碳体 free cementite
自由升力 free lift
自由生活类型 free living form
自由生物 nekton organism
自由声场 acoustic(al) free field;free field of sound;free sound field
自由湿气 free moisture
自由石灰 free lime
自由时差 free float;positive float
自由矢量 free vector
自由使用 at command
自由市 free city
自由市场 free market;open market
自由市场价格 free market price
自由市场经济 free market economy
自由市场制拍卖 freely market system auction
自由式 free style
自由式布置 open layout
自由式挡土墙 free retaining wall
自由式道路网 free style road system
自由式道路系统 free road system;free style road system
自由式底层水取样器 free vehicle bottom-water sampler
自由式浮箱门 free floating gate
自由式哥德建筑风格 freely adapted Gothic manner
自由式路网 free line net
自由式平面布置 open planning
自由式擒纵机构 free escapement
自由式三点悬挂装置 three-point free linkage
自由式室内布置平面 open plan
自由式溢洪道 free spillway
自由式溢流堰 free fall weir;free spill weir
自由式造型设计 free form design
自由释放 release-free
自由释放继电器 release-free relay;trip free relay
自由释放试验 release-free test
自由收缩 free contraction;free shrinkage
自由收缩缝 free contraction joint
自由手柄 free lever
自由手柄进出式进路联锁信号楼【铁】entrance-exit free lever signal box
自由手柄进出式进路联锁信号箱【铁】entrance-exit free lever signal box
自由手柄式集中联锁机 free lever (inter)locking machine
自由手柄式继电集中联锁 free lever type relay interlocking
自由手柄式继电联锁 free lever type relay interlocking
自由手柄式继电联锁信号楼 all-relay free lever cabin
自由手柄式联锁机 free lever(inter) locking machine
自由手柄式信号楼【铁】free lever signal box
自由手柄式信号箱【铁】free lever signal box
自由售酒商店 free house
自由输入输出方式 free input-output mode
自由束缚因子 free bound factor
自由竖向应变 free vertical strain
自由衰减 free damping
自由衰减的 free damped
自由衰减振荡示振仪 free damping oscillation vibrometer
自由衰减振动 free attenuation vibration
自由水 flee water;mobile water;phreatic water
自由水分 free moisture
自由水高程 free water elevation
自由水灰比 free water/cement ratio
自由水流 free flow
自由水面 free water level;free water surface
自由水面比降 free surface slope;free water surface slope
自由水面高度 free level
自由水面冷却 cooling through water surface
自由水面(坡)线 piezometric line
自由水面坡线图 free water surface graph
自由水面汽化热 heat of vapo(u)rization for free water
自由水面系统 free water surface system
自由水面线 pressure grade line
自由水面蒸发 free water surface evapo(u)ration
自由水清除率 free water clearance
自由水舌 free nappe
自由水舌溢流堰 free nappe weir
自由水体 free water body
自由水头 free head
自由水位 free level;free water elevation;free water level;free water surface
自由水溢流堰 free waste weir
自由水跃 free hydraulic jump;free jump
自由速率<指不受其他交通影响时的车速> free speed
自由缩尺的模型 single contraction pattern
自由态 free state
自由碳 free carbon;uncombined carbon
自由碳酸酐 free carbon dioxide
自由提升高度 free-lift height
自由体 free body
自由体分析图 free body diagram
自由体积 free volume
自由体积理论 free volume theory
自由体受力图 flee-body diagram
自由体图解 free body diagram
自由调节的储仓容量 live silo capacity
自由跳闸 free handle;free trip
自由停车 driver parking
自由停车房建筑 driver parking building
自由通风 free ventilation;free venting;uncontrolled ventilation
自由通风式空调机组 free-blow air conditioner
自由通过 free ride;free transit
自由通航权<国际河流或公海上的> freedom of navigation
自由投放 free fall
自由投稿 free-lance writing
自由投票 free voting
自由投资 free investment
自由图案 free pattern
自由图式 free scheme
自由土地保有权 freehold
自由土地保有人 freeholder
自由退解 free unwinding
自由脱扣型 trip-free type
自由拓扑代数 free topological algebra
自由拓扑群 free topological group
自由外汇 convertible currency;convertible foreign exchange;free foreign exchange
自由外汇贷款 convertible credit
自由弯矩图 free bending moment diagram
自由弯曲波 free flexural wave
自由弯曲波速 speed of free flexural wave
自由弯曲试验<不用模型> free bend test
自由弯曲延展性 free bend ductility
自由蜿蜒河道 free meandering channel
自由王国 realm of freedom
自由网 free net(work)
自由涡流 free vortex(flow);point vortex;potential vortex
自由涡流面 free vortex sheet
自由涡流式压气机 free vortex compressor
自由涡流型 free vortex-type
自由涡流叶片 free vortex blade
自由涡轮 free power turbine;free turbine
自由涡轮燃气轮机 free turbine gas

turbine engine
自由涡轮设计 free turbine design
自由(无压)流体射频记录<物探> free-fluid log
自由(无压)流体射频指数<物探> free fluid index
自由系泊 free-mooring
自由下沉 free subsidence
自由下沉的 free falling
自由下沉式分级机 free falling classifier
自由下沉式水力分级机 free settling hydraulic classifier
自由下降挡 free drop catch
自由下降加速度 free fall acceleration
自由下落 free fall
自由下落挡 free falling catch
自由下落的 free falling
自由下落搅拌机 free fall mixer
自由下落取心管 free fall corer
自由下落取样管 free fall corer
自由下落时间 free fall time
自由下落式拌和鼓 free fall type mixing drum
自由下落式拌和机 free fall mixer
自由下落试验 run-down test
自由相互碰撞模拟装置 liberty mutual automotive crash simulator
自由响应 free response
自由向量 free vector
自由向量空间 free vector space
自由项 free term
自由谐振 free harmonic vibration
自由泄流 free discharge;free flow
自由泄流阀 free discharge valve
自由心证 free evaluation of evidence through inner conviction
自由信托 voluntary trust
自由信息区 free field
自由行波 free progressive wave;free travel(l)ing wave
自由行车 free traffic
自由行车速率 free flow operating speed
自由行程 free path;free running;free stroke;tree travel
自由行动 free hand
自由行动的 uninfluenced
自由行动者 free-lance
自由行驶 free running
自由行驶车速 free flow operating speed
自由行走集装箱装卸设备 free running container handling equipment
自由形变 free deformation
自由形晶 eleutheromorph
自由形式 free form
自由形式的屋顶结构 free form roof structure
自由形式数据 free form data
自由型编码 free form coding
自由性 freedom
自由蓄水层 free aquifer
自由悬臂 free cantilever
自由悬臂与临时拉索法 free cantilever and provisional stay method
自由悬浮法 free levitation method
自由悬挂的重物 freely suspended load
自由悬挂活套 free hanging loop
自由悬挂卷筒 unsupported barrel
自由悬挂装药 freely suspended charge
自由旋进磁力仪 free precession magnetometer
自由旋涡 free eddy;free vortex
自由旋涡型 free vortex-type
自由旋涡叶片 free vortex blading
自由旋涡运动的环量常数 circulation constant
自由旋转 free rotation
自由旋转导流罩 free swiveling hydrodynamic(al) fairing
自由旋转作用 free revolving action
自由选取的试样 grab sample
自由选择 take it-or-leave-it
自由询问 open question
自由循环 free circulation
自由循环群 free cyclic(al) group
自由压碎 free crushing
自由淹灌 uncontrolled flooding;wild-flooding irrigation
自由堰 free weir
自由摇摆 free rocking
自由液面 free surface of liquid
自由液面泵 free surface pump
自由液面水 free water
自由液面水损害 free water damage
自由液面水修正系数 free water surface correction coefficient
自由液面水影响 free water effect
自由液面效应 free surface effect
自由液面修正 free surface correction
自由液面阻力 free surface resistance
自由液体 free liquid
自由移动 free motion;free translation
自由移动支承 freely movable bearing
自由溢流 free overfall;free overflow
自由溢流坝 self-spillway dam
自由溢流口 flee outfall;free outfall
自由溢流式静水池 free overfall jet stilling basin
自由溢流式消力池 free overfall jet stilling basin
自由溢流水舌 aerated nappe;aerated sheet of water;free overfall jet
自由溢流堰 free fall weir;free flow weir; free overfall weir; uncontrolled weir;weir with a free fall;free(discharge)weir
自由溢流堰顶 free crest of spillway;free overfall crest; unobstructed crest of spillway
自由溢水门 free overfall gate
自由溢水射流 free overflow jet
自由音场 free field
自由引水 optional pilotage
自由营业 freehold
自由涌浪 free surge
自由运动 free motion; free movement;freedom of motion
自由运费率 open rate
自由运行 free travel
自由运转的活塞 free running piston
自由载量 disposable load
自由载流子 free carrier
自由载流子浓度 free carrier concentration
自由载流子吸收 free carrier absorption
自由胀缩 free to contract and expand
自由振荡 free alternation;free oscillation;free running operation;free vibration; natural mode; natural motion;natural oscillation
自由振荡的 free running
自由振荡电流 free alternating current
自由振荡共振激发 resonance excitation of betatron oscillations
自由振荡激光模式 free running laser mode
自由振荡频率 oscillation frequency
自由振荡系统 free oscillating system
自由振荡周期 free oscillation period
自由振动 free vibration; natural vibration
自由振动测试 free vibration test
自由振动法 free vibration method
自由振动频率 free frequency
自由振动试验 free vibration test
自由振动形式 free vibration mode
自(由)振柱试验 free vibration column test
自由之碑 monument of liberty
自由支撑端 merely supported end
自由支承 free end bearing
自由支承的 freely supported
自由支承式钢蜗壳 free-standing steel scroll
自由支承式梁 freely supported beam
自由支承式压力钢管 freely supported penstock
自由支承调压塔 free-standing surge tank
自由支配收入 disposable income
自由支座 free bearing;free support
自由执行法 free steering method
自由执行序列 free steering sequence
自由职业 liberal profession
自由职业者 free-lance
自由指标 free index
自由指定径路 free routing
自由终沉速度 unhindered terminal settling velocity
自由重组 free recombination
自由轴 free axis;free axle
自由轴向窜动 free end-play
自由柱容 free column volume
自由柱状漩涡 free cylindric(al)vortex
自由转动构件 unrestrained member
自由转换法 free drift method
自由(转)轮 free wheel
自由转子陀螺稳定的惯性参考平台 free rotor gyro-stabilized inertial reference platform
自由转子陀螺仪 free rotor gyroscope
自由撰稿人 free-lance
自由桩 free pile;unstrained pile
自由桩头 unstrained pile head
自由状态 free condition;free state
自由坠落 free falling
自由准备金 free reserve
自由准粒子近似 free quasi-particle approximation
自由资产 free assets
自由字段 free field
自由字段存储 free field storage
自由阻抗 free impedance
自由组合件 freewheel assembly
自由作业 open shop
自由作用 free action
自游捕食者 nektonic redator
自游动物网 nekton net
自游生物 nekton
自有的 self-owned
自有货车队 owned fleet
自有流动资金 self-owned circulating fund
自有施工机械 owned equipment
自有装卸机械的船舶 self-sustained vessel;self-sustaining vessel
自有资本 capital owned;equity capital;own capital;ownership capital
自有资金 internally generated fund;owned capital;own fund;self-hold capital
自右乘 post multiplication
自预应力水泥 self-prestressed cement
自愈<裂缝的> self-healing
自愈合 autogenous healing;self-healing
自愈合裂缝<混凝土> self-sealing crack
自愈合土层 self-healing coating
自愈混合环 self-healing hybrid ring
自愈能力 self-healing ability
自愈网 self-healing network
自愈性<裂缝的> autogenous healing;self-healing
自愈性土料<坝工> self healing soil
自愿保险 voluntary insurance
自愿储蓄 voluntary savings
自愿搁浅 voluntary stranding
自愿贡献 voluntary contribution
自愿观测船 voluntary observing ship
自愿合伙 partnership at will
自愿合作 voluntary cooperation
自愿救助 salvage by voluntary action;voluntary salvage
自愿救助船 voluntary salvaging ship
自愿捐款 voluntary contribution;voluntary fund
自愿连锁系统 voluntary chain
自愿联合公司 voluntary chain
自愿留置权 voluntary lien
自愿清理 voluntary liquidation
自愿认证制 voluntary system of certification
自愿申请破产 voluntary bankruptcy
自愿税收 voluntary tax
自愿限制协定 self-restraint agreement
自愿性协议 voluntary compliance agreement
自愿援助方案 voluntary assistance program(me)
自愿援助计划 voluntary assistance program(me)
自愿约束 voluntary restraint
自愿仲裁 voluntary arbitration
自愿转让 voluntary alienation;voluntary conveyance
自愿组织 voluntary societies
自云母石英片岩 muscovite-quartz schist
自匀货 pulverulent
自运转的 self-operated
自运转电路 free running circuit
自载铲运机 self-loading scraper
自载导航设备 self-contained navigation aid
自再生过程 self-regenerative process
自再生自动机 self-reproducing automaton
自噪声 self-noise
自责观念 idea of self-accusation
自增的 self-energizing
自增力式制动器 self-energizing brake
自增力作用 self-servo action
自增强 autofrettage;self-reinforcing
自增强复合材料 self-reinforced composite material
自增强高压圆筒 autofrettaged high-pressure cylinder; self-shrinkage high-pressure cylinder
自增强圆筒 autofrettaged cylinder;self-shrinkage cylinder
自增型编址 autoincrement addressing
自增压 self-pressurization
自增殖 self-reproduction
自炸装置传感器 self-destruct sensor
自展的 bootstrap
自展式航天站 self-deploying space station
自张式扩孔器 self-opening reamer
自照装置 self-timer
自照准 self-alignment
自照准法 autocollimation method
自折叠条形记录纸带 self-folding strip chart
自诊断 self-diagnostics
自诊断的 self-diagnosing; self-diagnostic
自振 autonomous oscillation;inherent vibration; natural oscillation; natural vibration;self-oscillation;fundamental vibration
自振潮 self-established tide
自振荡 auto-oscillation
自振荡系统 self-oscillating system
自振荡消除 self-oscillation elimination
自振荡直线感应电机 self-oscillating linear induction motor

Z

自振点 point of self-oscillation
自振动 self-excited oscillation
自振动系数 coefficient of self-oscillation
自振模态 mode of free vibration; mode shape
自振频率 free vibration frequency; frequency of free vibration; frequency of self-excited vibration; frequency to free vibration; natural frequency of vibration; natural vibration frequency
自振铷汽磁强计 self-oscillating rubidium vapo(u)r magnetometer
自振特性 dynamic(al) characteristic; dynamic(al) property; free vibration characteristic
自振特性参数 free vibration characteristic parameter
自振振型 mode of free vibration; mode shape
自振周期 natural period of vibration; natural vibration period; period of natural vibration; fundamental period of vibration
自振状态 self-oscillating regime
自镇静反应 auto-killing
自镇流灯 self-ballasted lamp
自震<指隧洞开挖后因应力解除而引起的地震> bump
自蒸发 flash; self-evaporation; spontaneous evapo(u)ration
自蒸发能力 self-evapo(u)rative quality
自整的 self-consistent
自整定系统 self-adjusting system
自整角差绕变压器 synchrodifferential transformer
自整角传动系统 selsyn train
自整角电动机 selsyn motor; selsyn-type electrical motor; synchromotor
自整角电机 self-synchronous motor; selsyn-type electric(al) machine
自整角发电机 selsyn generator
自整角机 autosyn; self-synchronous device; selsyn; synchro
自整角机差动装置 selsyn differential
自整角机驱动电动机 selsyn drive electric(al) motor
自整角机式同步系统 selsyn-type synchronous system
自整角机式轧辊开度指示器 selsyn-type roll-opening indicator
自整角机组 synchrosystem
自整角接收机 selsyn receiver; synchroreceiver
自整流 self-rectifying
自整流 X 射线管 self-rectifying X-ray tube
自整流的 self-righting
自整流发射机 self-rectified transmitter
自整流式交流发电机 self-rectifying alternator
自整流型 X 射线装置 self-rectifier type x-ray apparatus
自正交分组码 self-orthogonal block code
自正交卷积码 self-orthogonal convolutional code
自正交块码 self-orthogonal block code
自支撑的 self-sustaining
自支撑隔墙 self-supporting partition
自支撑脚手架 self-supporting scaffold
自支承的波形瓦 self-supporting pantile
自支承的分隔墙 self-supporting partition wall
自支承隔墙 self-supporting partition
自支承踏步 self-supporting step
自支持的 self-supported; self-sustaining
自支持架空电缆 self-supporting aerial cable
自支护岩石 self-supporting rock
自支式充填带 self-supporting pack
自支援系统 self-support system
自制 abstention
自制材料 worked material
自制的 home-made; self-controlled; self-made
自制动 self-retention
自制零件费 one's own making parts cost
自制凭证 self-made evidence
自制设备 home equipment
自制系统 automatic restraint system
自治 autonomy
自治城市 autonomous city; corporate town; self-governing city; borough <英>
自治村镇<美> borough
自治的 autonomic; autonomous; self-governing
自治的阈值元件网络 autonomous network of threshold elements
自治港 autonomous port
自治机关 organ of self-government
自治控制 autonomous control
自治联邦 autonomous federal state
自治领 self-governing dominion
自治区 autonomous region; municipality
自治区域 autonomous area
自治权 autonomy
自治市 burgh; municipality
自治系统 autonomous system
自治县 autonomous county
自治殖民地 self-governing colony
自治州 autonomous prefecture; autonomous state
自治州界 autonomous state boundary
自置初值的 self-initializing
自置生产能力 ownership capacity
自中和频率 self-neutralizing frequency
自重 dead load; deadweight; gravity load(ing); lightweight; own load; own-weight; self-load; self-weight; sole weight; tare weight; tare mass <皮重>
自重冲洗 gravity flushing
自重对载重能力的比<自重系数> tare-to-load capacity
自重吨公里 tare ton-kilometers
自重翻斗小车 gravity tipple
自重非湿陷性黄土 self-weight non-collapse loess
自重固结 self-consolidation
自重烘弯 sagging
自重滑行坡 jinny
自重减压阀 dead weight relief valve
自重进料 gravity feed
自重力矩 dead-load moment; deadweight moment; own-weight moment
自重流动 free flowing
自重流下装车 on-the-run
自重挠度 dead-load deflection
自重碾压机 deadweight roller
自重破裂 weight break
自重轻、快速、舒适列车 lightweight, rapid and comfortable train
自重倾卸车身 gravity dump body
自重湿陷 self-weight collapse
自重湿陷量 self-weight collapse settlement
自重湿陷系数 coefficient of self-weight collapse; coefficient of self-weight collapsibility; coefficient of collapsibility due to overburden pressure; coefficient of overburden collapsibility
自重湿陷系数 self-weight collapsibility test; overburden collapsibility test
自重湿陷性 self-weight collapsibility; overburden collapsibility
自重湿陷性黄土 self-weight collapse loess; self-weight collapsible loess
自重送料锅炉 gravity feed type boiler
自重塌陷黄土 self-weight collapse loess; loess collapsible under overburden pressure
自重塌陷系数 coefficient of self-weight collapsibility
自重天平 dormant scale
自重弯矩 self-weight moment
自重系数 coefficient of dead weight
自重下降制动器 drop brake
自重向量 self-weight vector
自重泄油<管路> gravity drain
自重型非关节式流线型机车 non-articulated streamline train of light weight
自重压力 geostatic pressure; overburden pressure
自重压力安全阀 deadweight safety valve
自重应力 body stress; dead-load stress; deadweight stress; geostatic stress; inherent stress; self-weight stress; internal stress
自重运输 gravity-operated haulage; self-acting rope haulage
自重运输井筒 drop shaft
自重运输斜坡道 gravity runway; self-acting gravity incline
自重运行线路 gravity line
自重载重比 weight-payload ratio
自重注水 gravity flood
自重装料 gravity load(ing)
自重组别 unladen weight group
自主 autonomy; self-help
自主部件 autonomous unit
自主操作 autonomous working
自主导航 independent navigation
自主导航技术 autonomous navigation technology
自主的 autonomic; autonomous; independent
自主地产 allodium
自主地产制 allodial system
自主分化 self-differentiation
自主管理权 self-management
自主经营 full power of management
自主决定权限 discretionary limits
自主控制 autonomous control
自主权 decision-making power; power of decision
自主式潜水器 autonomous submersible
自主式水下船 autonomous underwater vehicle
自主式遥控潜水器 autonomous remotely controlled submersible
自主税则 autonomous tariff
自主通道 autonomous channel
自主系统 autonomous system
自主性 autonomy; independence
自主性投资 autonomous investment
自住的房屋 owner-occupied housing
自住业主 owner-occupant
自助 self-help; self-service
自助冰箱 self-service refrigerator
自助餐馆 automat; cafeteria; luncheteria; restauranteria <美>; self-service restaurant; vending cafeteria; buffet
自助的 do-it-yourself
自助的房子 self-help house
自助(公建)住宅 self-help housing
自助计划 self-help program(me)
自助快餐店 self-service snack bar
自助快餐馆 self-service snack bar
自助冷藏箱 self-service refrigerator
自助力作用 self-energizing action
自助零售商店 self-retailing shop
自助熔合金 self-fluxing alloy
自助商店 self-service store
自助食堂 automat; cafeteria; self-service restaurant
自助食堂两用大厅<学校或大楼内> cafetorium
自助洗汽车店 washeteria
自助洗衣场 laundromat
自助洗衣店 launderette; laundromat; self-service laundry; washeteria
自助洗衣间 self-help laundry
自助制度 help-yourself-system
自助住宅方案 self-help housing program(me)
自注的 self-priming
自转 autogiration; autorotate; axial rotation; rotation; slew
自转变星 rotating variable; rotational variable
自转不稳定性 rotational instability
自转的 self-rotated
自转动 autorotation
自转翻转 spin flip
自转方向 sense of rotation
自转基准轴 spin reference axis
自转极 pole of rotation
自转胶片和相纸<一种直接从正片得出正片,或从负片得出负片的照相材料> autopositive film and paper
自转角动量 rotation momentum
自转角速度 spin velocity
自转角速率 rotational angular velocity
自转矩 self-torque
自转力矩 spin moment
自转率调节 spin-rate control
自转喷灌 center[centre] pivot(ed)(sprinkler) irrigation
自转曲线 rotation curve
自转式喷灌机 center pivoted sprinkler
自转式洒布机 spin-spreader
自转式转向架 self-steering(arm) truck
自转速度 velocity of rotation
自转速度光谱型关系 rotational velocity-spectral type relation
自转速率 autorotation speed; speed of autorotation
自转特快 glitch
自转天体 rotator
自转停止器 self-rotated stoper
自转突变 glitch
自转温度 rotation temperature
自转向机构 self-steering mechanism
自转效应 rotation effect
自转旋翼飞机 gyro
自转旋翼机 gyroplane
自转移反应 spontaneous transfer reaction
自转折 self-breakover
自转致宽 rotational broadening
自转周期 period of rotation; period of transformation; rotation period
自转轴 axis of rotation; rotating axis; spinning axis
自转轴北极 mean pole of rotational axis
自转轴进动 rotational axis precession
自转综合孔径 rotation synthesis bore
自装 self-chambering
自装铲运机 self-loading scraper
自装的 self-loading
自装低轮集材车 self-loading skidder
自装公路车 self loading road vehicle
自装刮土机 self-loading scraper
自装集材机 forwarder
自装交出车 wagons loaded locally

and delivered at junction station
自装搅拌车 self-loading mixer
自装料输送机 self-feed conveyor
自装入 self-loading
自装入程序 self-loading program(me)
自装式箕斗 self-loading skip
自装式龙门起重机 self-loading gantry crane
自装式轮胎防滑链 self-mounting tyre chain
自装式运输机 loader conveyer[conveyor]
自装卸沿海船 self load/discharge coaster
自装仪器 home-made
自装运输机组 self-loading hauling unit
自装载铲运机 self-loading scraper
自装载挖土机 self-loading excavator
自装载运输机 self-loading hauler
自装自卸车 wagons loaded and unloaded locally
自坠(落) by gravity
自准光谱仪 autocollimatic spectroscope
自准直 autocollimation
自准直点 principal point of autocollimation
自准直惰轮 guider
自准直法 autocollimation method
自准直分光计 autocollimating spectrometer
自准直分光仪 autocollimating spectrometer
自准直管 autocollimator
自准直目镜 autocollimating eyepiece; autocollimation eyepiece
自准直摄谱仪 autocollimating spectrograph; autocollimation spectrograph
自准直望远镜 autocollimator
自着色材料 self-colo(u)red material
自走犁 motor plough
自走链式侧向搂草机 self-propelled chain-type side-rake
自走式铲运机 truck scraper
自走式单行收割脱粒机 self-propelled single-row harvester-thrasher
自走式动力割草机 self-propelled power mower
自走式动力喷雾机 self-propelled motor atomizer
自走式分批配料设备 low-profile batching plant
自走式干草捡拾压捆机 haycruiser; hay liner
自走式割草机 self-propelled mower
自走式横向搂草机 self-propelled dump rake
自走式集中配料拌和机 low-profile central plant
自走式捡拾压捆机 self-propelled pickup baler
自走式搅拌设备 low-profile mixing plant
自走式厩肥高架起重机 self-propelled gantry manure crane
自走式门架起重机 travel(l)ing gantry tower
自走式喷灌 travel(l)ing irrigation
自走式喷灌机 self-propelled sprinkler
自走式喷灌系统 self-travel(l)er irrigation system
自走式喷雾器 self-propelled sprayer
自走式平地机 self-propelled grader
自走式起重机 self-propelled crane
自走式输送机 shuttle conveyer[conveyor]
自走式挖掘机 self-propelled ditch digger
自足式助航设备 self-contained navigation aid
自阻抗函数 self-impedance function
自阻尼 self-damping
自组合纳米复合材料 self-assembling polar nano-composite
自组织 self-organisation; self-organize; self-organizing
自组织程序 self-organizing program(me)
自组织计算机 self-organizing computer
自组织控制 self-organization control; self-organizing control
自组织控制器 self-organizing controller
自组织控制系统 self-organizing control system
自组织设备 self-organizing equipment
自组织系统 self-organizing system
自钻孔攻丝螺钉 self-drilling screw; self-tapping screw
自钻孔自攻丝螺钉 self-drilling tapping screw
自钻式横压仪 self-boring pressuremeter
自钻式扭剪仪 self-boring type torsion shear apparatus
自钻式旁压试验 self-boring pressuremeter test
自钻式旁压仪 autoforeur probe; self-boring pressuremeter
自左至右线扫描 linear left-to-right scan
自作文摘 autoabstract
自作用 self-action
自作用的 self-acting
自作用力 self-force
自作主张 self-assertion

恣

恣意破坏 wanton destruction

渍

渍害 water-logging disaster

渍涝水 water-logging
渍水 water-logging
渍水草甸 water meadow
渍水的 water-logged
渍水地层 water-logged stratum
渍水地区 water-logged ground
渍水土(壤) flooded soil; water-logged soil
渍水沼泽排水 swamp drainage
渍水沼泽土 swamp soil

枞

枞胺 abietylamine

枞醇 abietinol
枞胶 balsam
枞木制的 firry
枞醛 abietinal
枞树 fir tree
枞树形天线 Christmas tree antenna
枞树形叶根固定 fir-tree connection
枞树脂 abietin
枞松 fir pine; spruce pine
枞松木 fir wood
枞酸 abietic acid; abietinic acid; sylvic acid
枞酸基 abietyl
枞酸甲酯 methyl abietate
枞酸型酸 abietic-type acid
枞酸酯 abietate
枞萜 sylvestren(e)
枞烯 abietene
枞香脂 balsam of fir; Canada balsam; Canada turpentine; Canadian balsam
枞油烯 sylvestren(e)
枞脂 abietic resin
枞脂石 fichtelite

宗

宗教房屋 ecclesiastical building

宗教哥特式 religious Gothic(style)
宗教绘画 religious painting
宗教机构 religion institute
宗教纪念建筑 religious monument
宗教建筑 ecclesiastical architecture; religious architecture; religious building
宗教建筑物 sacred building
宗教节日 religious holidays
宗教生态学 religious ecology
宗教习惯 religious customs
宗教洗礼用泉水 <清真寺> fountain for ceremonial ablutions
宗教象征(手法) religious symbolism
宗教信仰 religion
宗教仪式用水盆 liturgical water vessels
宗教影响 religious influence
宗谱 genealogy
宗谱纹章图案 eastern crown
宗主国本土 metropolitan homeland

综

综钩 heald frame hook

综合 aggregate; colligate; comprehension; generalization; integration; synthesize
综合安全系数 global safety factor
综合办法 comprehensive approach
综合办公室 general affairs department
综合保险 blanket insurance; comprehensive insurance; insurance against all-risks
综合保险单 comprehensive policy
综合报表 consolidated statement
综合报告 comprehensive report; consolidated report; consolidated return
综合比较分析 trade-off analysis
综合比率 composite ratio
综合编录 final documentation
综合变量 composite variable
综合标志图 complex attributes map
综合标准化 integrated standardization
综合表 aggregative table; synthetic(al) table
综合波 complex wave; composite wave
综合波束 synthesized beam
综合铂族矿石 Pt-group element ore
综合布线网络系统 integrated distribution network system
综合布置 combined scheme
综合部门 mixed department
综合材料表 list of the comprehensive materials
综合材料险 comprehensive material damage insurance
综合财务报表 conglomerate financial statement
综合财务计划 overall financial plan
综合财政计划 composite financial plan
综合采购 basket purchase; lump-sum purchase
综合采水器 integrating water sampler
综合彩色图 comprehensive colo(u)red map
综合测井长度 total length of synthetic(al) logging well
综合测量仪 general measuring instrument
综合测试 integral test; integration testing
综合测试设备 integral test system; integrated test system
综合测试系统 integral test system; integrated test system
综合测试要求大纲 integrated test requirement
综合查勘 comprehensive investigation
综合差错控制 hybrid error control
综合偿债保障率 overall coverage ratio
综合车站 integrated station
综合成本 combined cost; overall cost
综合成本项目 composite item of costs
综合成果图 composite result figure
综合成图法 synthetic(al) drawing method
综合承包人 integrated contractor
综合承包商 integrated contractor
综合承包者 integrated contractor
综合城市 combined city; synthetic(al) city
综合程度 degree of integration
综合程序 assembly program(me); general program(me)
综合尺寸 united inches
综合抽气系统 comprehensive pumping system
综合出口 blanket export
综合除法 synthetic(al) division
综合储备 commingled fund
综合储蓄 package saving
综合处理 comprehensive approach; comprehensive treatment; integrated treatment; joint disposal; joint treatment
综合处理系统 total treatment system
综合处置 <工业废水与生活污水的> joint disposal
综合传热系数 overall coefficient of heat transfer
综合传输协议 integrated transport protocol
综合船舶管理系统 integrated vessel management system
综合船舶仪表系统 integrated ship instrumentation system
综合存款方案 deposit(e) combined account
综合措施 comprehensive measure
综合大词典 complete dictionary; comprehensive dictionary
综合大修折旧率 compositive rate of major repair depreciation
综合大学 multiversity; polyversity; university
综合单价 comprehensive unit price
综合单位过程线 composite unit graph; composite unit hydrograph
综合单位过程线法 synthetic(al) unit graph method
综合单位水文过程线 synthetic(al) unit hydrograph
综合单位水文图 synthetic(al) unit hydrograph
综合导航系统 integrated navigation system
综合导航显示器 integrated navigation display
综合道路交通运输环境 integrate road transport environment
综合的 combined; complex; compositive; comprehensive; integrated; integrative; overall; synthetic(al)
综合的材料处理 integrated material handling
综合的交通需要 synthesized travel desires
综合的任务指标 integrated task index
综合的运输系统 comprehensive transportation system
综合等高线 generalized contour

Z

综合地层柱状剖面图 general-stratigraphic(al) column
综合地层柱状图 synthetic(al) columnar section of strata
综合地球物理勘探 comprehensive geophysical survey
综合地图 aggregate map;comprehensive map
综合地图集 comprehensive atlas
综合地下管道 main conduit
综合地下水水质模型 integrated groundwater quality model
综合地址 general address
综合地质编录 comprehensive geologic(al) logging
综合地质剖面 generalized geologic(al) section; zonal geologic(al) profile
综合地质数据库 integrated geologic(al) data base
综合点污染源和非点污染源最近评价技术 better assessment science in integrating point and non-point sources
综合电波 synthesis wave
综合电耗 combined power consumption;overall power consumption
综合电缆 composite cable;compound cable
综合电力网 integrated power grid
综合电子设备 integronics
综合电子信息处理机 integrated electronic processor
综合调查 comprehensive investigation;comprehensive study;comprehensive survey; overall investigation
综合调查船 comprehensive investigation ship; comprehensive survey vessel; integrated research vessel; synthetic(al) research ship
综合调水 comprehensive water transfer
综合调研船 multipurpose research vessel; multipurpose research vessel ship
综合定位 composite fixing
综合定位法 position by mixed observations
综合动作电位 resultant action potential
综合都市 synthetic(al) city
综合毒性 comprehensive toxicity;integrated toxicity
综合毒性指数 integrated toxicity index
综合多目标法 integrated multipurpose approach
综合多目标水管理系统 integrated multipurpose water control system
综合二级污水处理后出水 synthetic(al) secondary wastewater effluent
综合发展 integrated development; comprehensive development
综合发展计划 comprehensive plan(ning)
综合法 method of synthesis;synthesis method;synthetic(al) method
综合法测绘地物 photo planimeter
综合法测图【测】 planimetric(al) photo;complex mapping;photo-planimetric(al) method; photogrammetric mapping(method)
综合法单张像片测图 photo-planimetric(al) method of monophotogrammetry;photo-planimetric(al) method of single photograph mapping
综合法绘图 planimetric photo
综合法全野外布点 photo-planimetric(al) method for field point layout
综合法像片测图 photo-planimetric(al) method of photogrammetry; photo-planimetric(al) method of photomap mapping
综合方案 comprehensive approach;overall project
综合方法 combined method;comprehensive approach;holistic approach
综合方向图 synthesized pattern
综合防护法 comprehensive method of protection
综合防治 complex control; comprehensive prevention and cure; integrate(d) control
综合防治措施 comprehensive prevention and control measures;comprehensive preventive health measure
综合防治技术 integrated control technique
综合仿真板 integrated simulation panel
综合废水 synthetic(al) wastewater
综合费率 consolidated rate
综合费用 general expenses
综合费用指标 overall cost rate
综合分布的楼面系统 integrated distribution floor system
综合分类学 biosystematy
综合分析 aggregate analysis;comprehensive analysis; general equilibrium analysis; generalized analysis; integrated analysis;multipurpose analysis
综合分析法 analysis by synthesis method;complex analysis procedure;comprehensive analysis method
综合风险 overall risk
综合风险评价指标体系 integrated risk assessment indexes system
综合封隔 combination packer
综合服务费 miscellaneous service charges
综合服务项目 comprehensive services
综合负荷 diversified load
综合概率 compound probability
综合概率法 hybrid probabilistic method
综合干燥磨碎机 kiln mill
综合钢铁厂 integrated iron and steel plant
综合个例研究 integrated case studies
综合跟踪 integrated tracking
综合跟踪系统 integrated tracking system
综合耕作 combine tillage operation
综合工程 integrated project
综合工程地质图 comprehensive engineering geological map;general engineering geologic(al) map
综合公司 comprehensive company
综合公园 comprehensive park; synthetic(al) park
综合功能 complex function
综合固体垃圾管理 integrated solid waste management
综合观测结果 observational synthesis
综合观测系统 integrated observation system
综合管道 collector duct; composite duct; composite pipe line; tunnel for utility mains;main conduit <各种地下管线的共用管道>
综合管道线图 synthesis plan of pipelines
综合管廊 collector duct; composite pipe line
综合管理 comprehensive management; general management;integrated management;total management
综合管理方法 integrated management approach; integrated management method
综合管理任务 integrated management function
综合管理系统 integrated management system;total management system
综合管理信息系统 integrated management information system
综合管网 utility engineering
综合管线图 combined service drawing
综合光变曲线 synthetic(al)light curve
综合光缆 synthetic(al) optic(al) cable
综合光纤系统 integrated optic(al) fiber[fibre] system
综合广播 general broadcast
综合规划 comprehensive plan(ning); comprehensive programming; integrated network; integrated plan(ning);overall plan;unified plan
综合规划援助 comprehensive planning assistance
综合过程线 complex hydrograph
综合焊条 stranded electrode; stranded welding wire
综合航行制 integrated navigation system
综合核算 comprehensive accounting
综合荷载 combined load(ing)
综合厚度 combined thickness
综合湖泊生态系统分析器 comprehensive lake ecosystem analyser [analyzer]
综合化 totalization
综合化学指数 general chemical index
综合环节 summation loop
综合环境控制 integrated environmental control
综合环境(条件) combined environment
综合环境学 comprehensive environmental science
综合环境影响 standort
综合环境影响、补偿和责任法令 Comprehensive Environmental Response, Compensation and Liability Act
综合环境影响评价模型 comprehensive model for environmental impact assessment
综合环境原理 principle of holocoenotic environment
综合环境指数 composite environment index
综合回路 summation loop
综合货船 composite vessel
综合机械化 comprehensive mechanization;system mechanization
综合机械化采煤法 comprehensive mechanized coal mining method
综合机械化设备 fully mechanized equipment
综合机制 comprehensive mechanism; synthetic(al) mechanism
综合基本折旧率 compositive rate of basic depreciation
综合基地 multipurpose base
综合基金 integrated fund
综合基金表 compositive fund statement
综合几何 synthetic(al) geometry
综合计划 comprehensive plan(ning); comprehensive program(me); comprehensive scheme; integrated plan(ning); integrated project; overall planning; overall project; package plan
综合计划方案 overall project
综合计算机中心 all round computer center[centre]
综合计算系统 integrated computing system
综合技术室 general technical office
综合加权指数 aggregative weighted index
综合加速测量仪 integrating accelerometer
综合加速度 integrating acceleration
综合加速度仪 integration accelerometer
综合价格 blend price;composite price
综合价格换算系数 implicit price deflator
综合架 composite rack
综合监测 integrated monitoring
综合建设规划 comprehensive development plan
综合建筑群 building complex; complex
综合建筑系统 integrate(d) building system;total building system
综合交通管理 comprehensive traffic management
综合交通规则 comprehensive communication planning
综合交通控制 multiple traffic control
综合交通量 synthesised[synthesized] flow
综合交通预测模型【交】 comprehensive forecasting model
综合结构 compages; complex; complex structure
综合结构设计 <委托人、建筑师和结构工程师密切合作下进行的设计> integrate(d) structural design
综合介绍 review paper
综合进口价格指数 composite import price index
综合经济基础论 theory of a synthesized economic basis
综合经济临界品位 break-even point of composite economic grade
综合经济平衡 comprehensive economic equilibrium
综合经济实体 comprehensive economic entity
综合经济效果 overall economic effect
综合经济效益 composite economic result;comprehensive economic results
综合经济学 synthetic(al) economics
综合经验 generalized experience
综合经营 comprehensive management; mixed farming
综合决策 decision-making package
综合开采工艺 combined mining technology
综合开发 all-round development;complex development;comprehensive development; comprehensive exploitation; comprehensive harnessing; integrated development; multiple objective development; multiple-purpose development; overall development
综合开发工程 multiple-purpose project;multipurpose project
综合开发计划 all-out development plan; comprehensive exploitation plan; multiple-purpose project; multipurpose development plan
综合开拓法 open up by combined methods
综合勘察 comprehensive investigation; comprehensive survey; integrated survey
综合勘探 comprehensive exploration of mine district
综合抗剪钢筋 combined shear reinforcement

Z

综合考察 comprehensive investigation; comprehensive survey; integrated survey; multipurpose investigation
综合考虑 compromise
综合可焊性 overall weldability
综合可靠性数据系统 integrated reliability data system
综合课程 integrated course
综合空间 <城市的> total space
综合空间计算设备 space computing complex
综合空气状态表 synthetic(al) air chart
综合孔径 aperture synthesis; synthetic(al) aperture
综合孔径射电望远镜 aperture synthesis radio-telescope; synthesis radio-telescope
综合孔径望远镜 synthesis telescope
综合控制 combination control; general control; integrated control
综合控制供油式化油器 integrated fuel control carburetor
综合控制技术 integrated control technique
综合控制台 integrated console
综合控制系统 complex control system; integrated control system
综合库存 integrated storage
综合快滤器 composite fast filter
综合宽带通信[讯]系统 integrated wideband communication system
综合矿石 complex ore
综合棱镜 resultant prism
综合历时曲线法 synthetic(al) duration curve method
综合利用 comprehensive application; comprehensive harnessing; comprehensive utilization; integrated use; integrated utilization; joint use; many-sided utilization; multiple use; multiple utilization; multipurpose use; multipurpose utilization; synthetic(al) utilization
综合利用坝 multiple-purpose dam; multipurpose dam
综合利用的 multipurpose
综合利用工程 multiple-purpose project; multipurpose development; multipurpose project
综合利用规划 multiobjective planning
综合利用规划方案 comprehensive planning program(me)
综合利用、化害为利的原则 principle of comprehensive utilization of turning harm into good
综合利用利润提成 multipurpose use profit-sharing
综合利用水库 multiple-purpose reservoir; multipurpose reservoir
综合利用水利工程 hydraulic engineering complex; hydroengineering complex; multipurpose water control project
综合利用项目 multipurpose project
综合利用效益 comprehensive utilization benefit
综合利用循环处理过程 integrated reuse-recycle treatment process
综合利用研究 compositive use research
综合利用资源 multiple use resources
综合联调 comprehensive integrated test
综合联动试验 unit interlock test
综合量规 receiver ga(u)ge
综合量指数 aggregative quantity index
综合流程 general flow; integrated flowsheet
综合流程图 general flow chart
综合流量 synthesised[synthesized] flow
综合流量曲线 integrated flow curve
综合流域规划 comprehensive river basin planning
综合楼 complex building; multiuse building
综合旅游区 recreation complex; resort complex
综合履约、劳工及材料的付款担保 penal bond
综合率 composite rate
综合脉码调制交换网络 integrated PCM-switching network
综合模量 <即把抗压的弹性模量和抗拉的弹性模量化合在一起的模量> reduced modulus
综合模式 aggregated model; aggregative model
综合模型 collective model
综合目录 comprehensive list
综合内部通信[讯]设备 integrated intercom set
综合能耗 comprehensive energy consumption
综合能力 aggregate capacity; combined capacity
综合年限 composite life
综合农业 integrated agriculture
综合农业污染管理系统 integrated system for the management of agricultural pollution
综合排水设施 combined drain facility
综合平差法 combined adjustment method
综合平衡 aggregate balancing; comprehensive balance; overall balance; overall balance and unified arrangement; overall balancing
综合平衡法 population balance method
综合平均年(限)折旧法 depreciation-composite life method
综合平均数 compound average
综合平均值 compound average
综合平面图 composite plane
综合评定试验 composite evaluation test; integrated evaluation test
综合评价 comprehensive appraisal; comprehensive assessment; comprehensive evaluation; integrated appraisal; integrated evaluation; overall evaluation; overall merit
综合评价方法 comprehensive assessment method; comprehensive evaluation; multipurpose evaluating method
综合坡度 resultant gradient
综合剖面(图) composite profile; generalized section; zonal profile
综合谱带 combination band
综合企业 complex
综合启动器 combination starter
综合气候学 complex climatology; synthetic(al) climatology
综合气象仪器 meteorologic(al) multiinstrument
综合汽车控制 comprehension automobile control; comprehensive automobile control
综合契约 general contract
综合器 synthesizer
综合潜水系统 composite diving system
综合情报 summarized information
综合情报系统 integrated information system
综合区 complex zone; comprehensive area; mixed-use area
综合区划 composite zoning
综合曲线 compound curve; resultant curve; resulting curve; synthetic(al) curve
综合取样器 combination sampler; composite sampler
综合全球导航监视系统 integrated global navigation and surveillance system
综合全球海洋站系统 integrated global ocean station system
综合全险责任 comprehensive general liability
综合群体生产模式 generalized stock production
综合热分析法 combined thermal analysis
综合热耗 overall heat consumption
综合热效率 overall thermal efficiency
综合人员配备表 consolidated manning table
综合任务指标 integrated task index
综合容量 aggregate capacity
综合蠕变 combined creep
综合软件包 comprehensive software package
综合撒布机 combination spreader
综合森林利用 integrated forest utilization
综合沙埋 compound sand cover
综合商品指数 composite merchandise index number
综合设备 integrated equipment
综合设计 composite design; comprehensive design; general design; integral design; unitized design
综合设计洪水 synthetic(al) project flood
综合社区规划 comprehensive community plan
综合生产函数 aggregate production function
综合生产流程 integrated line
综合生产率 aggregate productivity
综合生产系统 integrated manufacturing system
综合生产线 integrated production line
综合生态风险 comprehensive ecological risk
综合生态风险指数 comprehensive ecological risk index
综合生态环境评价 synthetic(al) assessment of eco-environment
综合生态学 synthetic(al) ecology
综合生态影响 holocoenosis
综合生物系统 integrated biosystem
综合生物效应 combined biologic(al) effect
综合声呐系统 integrated sonar system
综合声呐站 triple treat sonar
综合湿度 combined moisture
综合实验室 complex laboratory; laboratory complex
综合市场 aggregate market
综合式堆垛机 combination stack; combination sweep rake and stack
综合式梁 integrated beam
综合式路网 integrated line network
综合式液力变矩器 torque convertor coupling
综合式遮阳 comprehensive sunshade; synthetic(al) sunshade
综合视频信号 composite video signal
综合视频终端 integrated video terminal
综合试验 call through test; comprehensive experiment; comprehensive test
综合试验设备 test complex
综合试运行 unit trial operation
综合试运转 unit trial operation
综合室 general office
综合收益 composite income
综合收益率 overall rate of return
综合数据 generated data; summarized information
综合数据处理法 integrated data processing
综合数据处理文件 integrated data handling file
综合数据处理中心 complex data handling center[centre]; integrated data processing centre
综合数据管理系统 integrated data management system
综合数据库 integrated data bank; integrated data base; integrated data storage
综合数据网 integrated data network
综合数据系统 integrated data system
综合数据显示及控制系统 integrated data presentation and control system
综合数字传输和交换 integrated digital transmission and switching
综合数字服务网用户模块 integrated service digital network subscriber module
综合数字服务网中继模块 integrated service digital network trunk module
综合数字网 integrated digital network
综合水管理 holistic water management
综合水文地质图 composite hydrogeologic(al) map; synthetic(al) hydrogeologic map
综合水文过程线 synthetic(al) hydrograph
综合水文曲线 synthetic(al) hydrograph
综合水文学 statistic(al) hydrology; synthetic(al) hydrology
综合水质数据 integrated water quality data
综合损益表 composite income sheet; consolidated income sheet
综合所得税 tax on aggregate income; unitary income tax
综合索赔 compound claim
综合塔式吊车 combined tower crane
综合弹性模量 overall elasticity modulus
综合特性 combined characteristic; comprehensive characteristics
综合特性曲线 combined characteristic curve; efficiency hill diagram; hill curve; hill diagram; mussel diagram
综合特性图表 composite rating chart
综合体复数【数】 complex
综合体育馆 games hall complex
综合体制 comprehensive institutional framework
综合天气预报 general forecast
综合条款 comprehensive provisions; omnibus clause
综合调和分析 generalized harmonic analysis
综合调节 combined governing; combined regulation; composite conditioning; integrated control
综合调解系统 integrated control system
综合调整 overall system adjustment
综合调整方案 aggregate adjustment option
综合通信[讯] integrated communication
综合通信[讯]适配器 integrated communication adapter
综合通信[讯]网 integrated communication network
综合通信[讯]系统 integrated communication system
综合统计 benchmark statistics; com-

Z

prehensive statistics
综合统计数据 benchmark statistic(al) data; benchmark statistics
综合统计资料 combinatorial statistic(al) data
综合图 collective diagram; complex chart; complex map; composite map; comprehensive chart; general view
综合图表 cross-plot
综合图解 cumulative diagram
综合土地开发 composite land development
综合土木工程 integrated civil engineering
综合土木工程系统 integrated civil engineering system
综合网络 complex network; integrated network
综合危害系数 synthetic(al) harmfulness coefficient
综合卫星导航设备 integrated satellite navigation equipment
综合卫星系统 synthetic(al) satellite system
综合温度 resulting temperature
综合温度计 resultant thermometer
综合稳定 comprehensive stabilization
综合稳定基层 comprehensive stabilized base
综合稳定系统 synthetic(al) stability system
综合污染数据 synthetic(al) pollution data
综合污染指标 synthetic(al) pollution quota
综合污染指数 synthetic(al) pollution index
综合污染资料 synthetic(al) pollution data
综合污水分流 comprehensive sewer separation
综合污水渠 combined sewer
综合物 combination; complex; synthesis
综合物价指数 overall price index
综合物料平衡 overall material balance
综合物探 comprehensive geophysical survey
综合物探(方)法 comprehensive geophysical exploration method; comprehensive geophysical method; integrated geophysical method
综合误差 combined error; composite error; composition error; general error; overall error; resultant error; total error
综合系数 overall coefficient; overall factor
综合系统 comprehensive system; integrated system; synthetic(al) system; system assembly; system assemble <统计力学>
综合先张法和后张法预应力 prepost-tensioning prestress
综合纤维素 holocellulose
综合显示系统 integrated display system
综合显微镜 universal research microscope
综合险 all risks(insurance); comprehensive risks; overall risks
综合险保单 all risks policy
综合险条款 all risks clauses
综合向量 resultant vector
综合项目 composite items; integrated project
综合消除污染 comprehensive pollution abatement
综合销售系统 integrated marketing system
综合效率 combined efficiency
综合效应 comprehensive effect
综合谐波分析 generalized harmonic analysis
综合信号 integrated signal
综合信息处理 integrated information processing
综合信息系统 integrated information system; total information system
综合行动方案 comprehensive program(me) of action
综合形势预报图 composite prognostic chart
综合型澄清池 comprehensive clarifier
综合型船闸 combined-type lock
综合型终端设备 integrated terminal equipment
综合型袋装及散货两用装船机 combined-type shiploader
综合型资产保险 property-comprehensive form insurance
综合性 complexity; globality
综合性保健 comprehensive health care
综合性泊位 multipurpose berth
综合性采伐 integrated logging
综合性产品 composites
综合性城市 compound function city
综合性传输系统 integrated transmission system
综合性辞典 general dictionary
综合性大学 comprehensive university
综合性的 synthetic(al)
综合性动物地理区 comprehensive faunal province
综合性多用途港 complex multipurpose port
综合性服务数据网 integrated services digital network
综合性服务系统 complex utility routine
综合性服务站 superservice station
综合性改建 comprehensive redevelopment
综合性钢铁厂 integrated iron and steel works
综合性港口 multipurpose harbo(u)r; port complex; multipurpose port; multiterminal port; all-round port; general port
综合性港区 port complex
综合性工厂 multiple-producing factory
综合性工程地质测绘 comprehensive engineering geology mapping
综合性工程公司 integrated engineering-construction firm
综合性工科大学 polytechnic college
综合性工业城市 diversified industrial city
综合性工艺学校 polytechnic
综合性和多方面的 synthetic(al) and many-sided
综合性货场【铁】 composite(team) yard; comprehensive(team) yard
综合性货运站 comprehensive goods station; general freight station; general goods station
综合性机场 airfield complex
综合性冷库 multipurpose cold storage; multipurpose refrigerator
综合性码头 multipurpose quay; multipurpose terminal
综合性能 all-round property; combination property; overall performance; overall property
综合性企业 integrated enterprise
综合性情报 combinational information
综合性全球海洋服务系统 integrated global ocean services system
综合性群集 global cluster
综合性群梁 global clustered beams
综合性商业建筑 commercial complex
综合性施工组织设计 comprehensive construction organization design
综合性石油公司 integrated oil company
综合性数据 synthetic(al) data
综合性水管理 integrated water management
综合性水文地质图件 comprehensive hydrogeology maps
综合性文件 general document
综合性医院 synthetic(al) hospital
综合性指标 composite target
综合性中学 <英> comprehensive school
综合修饰 composite trim
综合修正系数 combined correction factor; combined correctness factor
综合需求 composite demand
综合选择 complex selection
综合选择样片 integrated selective key
综合学校 comprehensive school
综合压水 comprehensive injecting water
综合研究 comprehensive research; comprehensive study; synthetic(al) study
综合研究报告 aggregative study; synthetic(al) study; integrated study
综合衍射图 hologram
综合衍射学 holography
综合养护车 combined maintenance truck
综合样本 composite sample
综合样品 integrated sample
综合遥测系统 integrated telemetering system
综合遥控监测装置 synthetic(al) remote control and supervisory device
综合业务数字网的接入能力 integrated service digital network access capability
综合业务数字网的连接 integrated service digital network connection
综合业务数字网(络) integrated service digital network
综合业务(通信[讯])网 integrated service network
综合液力变矩器 combined hydraulic torque converter
综合医院 general hospital; pandocheum; polyclinic
综合仪表盘 integrated instrument panel
综合仪器箱 apparatus casing
综合因子 multistress; summarized index
综合应力 combined stress; complex stress; compound stress
综合盈利率 comprehensive profit rate
综合盈余 consolidated surplus
综合盈余表 consolidated surplus statement
综合影响 combined influence
综合影响系数 synthetic(al) effect coefficient; synthetically effect coefficient
综合用途楼宇 composite building
综合游乐场所 resort complex
综合有效税率 overall effective tax rate
综合预报 composite forecast
综合预算 comprehensive budget; overall budget; unified budget
综合预算定额 composite budget norm
综合预算方案 integrated budget formula
综合愿望线 major directional desire line
综合运费率 freight all kind rate
综合运输 <水运、铁路、公路等> composite transport; mixed transport; combined transport; coordinated transportation; integrated transportation; overall transportation; intermodal transport
综合运输计划 overall transport plan
综合运输体系 overall transportation complex
综合运输网 integrated transport network
综合运输系统 integrated transport system; overall transportation complex; unified transportation system
综合运输研究 comprehensive transportation study
综合运输制度 integrated transport system
综合运行测试 integrated running test
综合运营培训系统 integrated operation training system
综合再利用 integrate reuse
综合责任(总保)险 comprehensive general liability insurance
综合增长指数法 synthetic(al) growth index method
综合摘要 comprehensive summary
综合账户 combined account
综合障碍 resultant fault
综合找矿 comprehensive prospecting
综合照明 combined lighting
综合折旧 composite depreciation
综合折旧率 composite depreciation rate
综合征 syndrome
综合整理 composite trim
综合整治 comprehensive treatment; synthesis dredge
综合整治措施 comprehensive harnessing measure; comprehensive improvement measure
综合支付 aggregate payment
综合指标 aggregative indicator; aggregative indicatrix; comprehensive indicator; sum parameters; overall indicator <多指路面质量>
综合指标值 comprehensive index value
综合指示器 aggregative indicator; overall indicator; overall target
综合指数 aggregative index(number); composite index; general index; overall index <多指路面质量>
综合制图 combine system; complex mapping
综合制图法 cartographic(al) generalization
综合治理 comprehensive harnessing; comprehensive treatment; integrated control
综合治理措施 comprehensive improvement measure
综合质量 quality of comprehensive boring
综合周转额 composite turnover
综合周转率 composite turnover
综合注浆 combined casting
综合注模控制系统 comprehensive injection mo(u)lding control system
综合柱状剖面图 composite columnar section
综合柱状图 synthesis cylindric(al) diagram; composite columnar section
综合专利索引 general patent index
综合资本成本 overall cost of capital

综合资产负债表 aggregate balance sheet;complete balance sheet;composite balance sheet
综合资产商誉 composite assets goodwill
综合资料 selected information;synthetic(al)data
综合资源管理 integrated resource management
综合自动化 integrated automation;integrated automatization
综合自动化系统 integrated automatic control system
综合自动控制系统 complex automatic control system;integrated automatic control system
综合组织 integrative organization
综合钻机 combination drill;combination outfit;combination rig
综合钻进设备 combination drilling equipment;combination drilling outfit
综合钻头 combination bit
综合作用 combined action;integral action;resultant action;synthetic(al)action;synthesis
综合作用泡沫 synactic foam
综摄法 <一种集思广益的决策、解决问题、发展新思想等方法> synectics

棕板 palm plate

棕垫 coir mat
棕儿茶丹宁 Gambier tannin
棕腐化作用 ulmification
棕腐酸 ulmic acid;ulmin
棕腐殖酸 hymatomalenic acid
棕腐质 ulmin;vegetable jelly
棕腐质的 ulmic
棕腐质无结构镜煤 ulmain
棕钙土 brown desert steppe soil;brown semi-desert soil;brown soil;calcic brown soil
棕褐煤 brown lignite;lignite B
棕褐色 dark brown;sepia
棕褐色变 <砌体灰缝> brown stain(ing)
棕褐色的 chocolate-brown
棕褐色土 brown earth
棕黑的 brownish black
棕黑色 brownish black;brownish-black colo(u)r;sepia
棕黑色调 brown black tone
棕黑颜料 sepia
棕黑蛭石 basstonite
棕红 brownish red
棕红色 henna;red-brown
棕环试验 brown-ring test
棕黄色 brown-yellow;nankeen;nankin;tan
棕黄色硬木 chengal <产于马来西亚>;degame wood <一种产于西印度群岛的木材>
棕黄试验 brown-ring test
棕灰色粗纹石灰石 Ancaster stone
棕缆(绳)coir rope
棕榈 Chinese coir palm;hemp palm;palm;windmill palm
棕榈核油 palm kernel oil
棕榈科 palm
棕榈蜡 palm wax
棕榈仁油 palm kernel oil
棕榈饰 palmette
棕榈树杆般的柱 palm shaft column
棕榈树叶状 palmiform
棕榈酸 cetylic acid;palmic acid;palmitic acid;palmitinic acid
棕榈酸铝 alumin(i)um palmitate
棕榈酸酰胺 palmitamide
棕榈酸酯 palmitate
棕榈形柱头饰 <古埃及的> palm capital
棕榈叶式柱 palm column
棕榈叶饰 <古罗马和古希腊建筑> anthemion(mo(u)lding)
棕榈叶屋面之简易房 nipa house
棕榈叶纤维 piassava
棕榈叶形拱顶 palm vault(ing)
棕榈叶装饰 palm-leaf ornament
棕榈油 palm butter;palm grease;palm(pulp)oil
棕榈油厂废水 palm oil-mill effluent
棕榈油酸 palmitoleic acid
棕榈油烯酸 zoomaric acid
棕榈园 palmetum
棕漠土 brown desert soil
棕皮 palm bark
棕壤(土)brown earth;brown loam;brown soil;brunisolic soil;braunerde;brown forest soil
棕仁 palm kernel
棕色 brownness
棕色半荒漠土 brown semi-desert soil
棕色玻璃 brown glass
棕色草原土 brown soil
棕色尘雾 brown fume
棕色冲积土 alluvial brown soil
棕色的 brown;palm colo(u)r
棕色调 brown cast;brown tone
棕色腐朽 brown rot
棕色钙质土 brown calcareous soil
棕色合剂 brown mixture
棕色褐煤 brown coal
棕色环试验 brown-ring test
棕色荒漠草原土 brown desert steppe soil
棕色荒漠土 brown desert soil
棕色灰化土 brown podzolic soil
棕色碱土 brown alkali soil
棕色轻雾 brown haze
棕色染料 brown dye
棕色森林土 braunerde;brown forest soil
棕色湿草原 prairie-steppe brown soil
棕色树脂 rosthornite
棕色酸 brown acid
棕色陶器 brown ware
棕色图 brownprint
棕色土 brown soil
棕色线晒图 dialine
棕色线条图 brownline
棕色烟气 brown smoke
棕色烟雾 brown fume
棕色氧化铝 brown alumin(i)um oxide
棕色氧化物 brown oxide
棕色硬木 bombway <产于安达曼群岛>;coolibah <一种产于澳大利亚的木材>
棕色硬质虫胶清漆 brown hard varnish
棕色釉 brown glaze
棕色炸药 brown powder
棕闪石 barkevikite
棕闪斜煌岩 camptonite
棕闪正煌岩 sannaite
棕绳 coir rope;palm(fiber[fibre])rope
棕石灰 brown lime
棕树皮 palmae bark
棕丝板条 coir board lath
棕丝材料 coir material
棕丝隔热板 coir insulating sheet(ing)
棕丝隔热板条 coir insulation strip
棕丝门毯 coir door mat
棕丝墙板 coir wallboard
棕丝型石膏板 coir-type gypsum plank
棕索 coir rope;palm rope
棕桐树 palm
棕图 brownprint
棕土 umber
棕土棕 raw umber
棕席 coir mat
棕巷指示器 brown box
棕雪 brown snow
棕叶花饰【建】palmette;anthemion(mo(u)lding)
棕叶饰 <古希腊建筑> anthemion ornament
棕叶装饰 palm-leaf ornament;anthemion ornament <古希腊建筑>
棕叶状花饰 palmette
棕纸板 brown wood board

踪迹 scent;whereabouts

踪迹化石 ichnofossil;trace fossil

鬃bristle;hard hair

鬃毛 mane
鬃漆刷 bristle brush
鬃丘 hogback
鬃刷 bristle brush
鬃帚 hair brush

总α 和总 β 放射性 gross alpha and gross beta radioactivity

总阿尔发(放射性)测量 total alpha-radiation survey;total alpha-radioactivity survey
总安排 general arrangement
总安全控制器 primary safety control
总安全系数 <用于极限状态设计> global safety factor
总安装 general assembly
总安装时间 overall erection time
总安装图 general assembly drawing;general installation drawing
总氨氮 total ammonia nitrogen
总氨基甲酸酯残基 total carbamate residue
总罢工 general strike
总扳手 master wrench
总板式计数 total plate count
总办公室 general office
总办事处 general office;head office;main office
总半衰期 total half-life
总包单价付款合同 combined lump-sum and unit price contract
总包单位 main contractor
总包干(工作)lump work
总包干合同 lump-sum contract
总包工 main contractor
总包工程 lump-sum work
总包合同 lump-sum contract;single contract
总包合约 general contract;single contract
总包价契约 lump-sum agreement
总包价协议 lump-sum agreement
总包项目 lump-sum item
总包与分包 general contractor and subcontractor
总包整套新建公共住房 turnkey-new construction
总包租 gross charter
总保单 blanket policy
总保额 gross line
总保费 office premium
总保留时间 total retention time
总保险单 floating policy;general insurance policy;running policy
总保险丝 main fuse
总保险值 total insured value
总报酬 overall consideration
总报告人 reporter general
总报价 all-in-bid;general offer
总报务量 total traffic
总贝塔(放射性)测量 total beta-radiation survey;total beta-radioactivity survey
总崩溃 <指市场> crash
总泵站 base pump station
总比降 general slope;overall gradient;overall slope
总比率 overall ratio
总壁 general wall
总臂距 total span
总编号 master serial number
总编辑 chief editor;editor-in-chief;general editor
总变差 total variation
总变电站 main substation
总变动成本 total variable cost
总变量 global variable
总变位 net slip;resultant displacement
总变形 general yielding;global deformation;total deflection;total deformation
总变形长度 total deformation length
总变形量 total deformation
总标准需水量 ultimate standard water demand
总表 general schedule;general table;summarized table;summary table;synoptic(al)table;master ga(u)ge <指仪表>
总表观吸附密度 total apparent adsorption density
总表面 total surface
总表面积 total surface area
总表面剂量 total surface dose
总波浪力 total wave force
总波力 total wave load
总剥采比 general stripping ratio
总补给量 overall logistic(al)demand
总不饱和物 total unsaturates
总布线法 global wiring
总布置图 general layout;general outline;general plan;layout(sheet);master plan;plan of site;principle layout;site drawing
总部 head office;headquarters
总部费 head office cost
总财产 gross property
总裁 director-general;managing director;president
总采 general cutting-plan
总残留量 total residue
总残留物 total residue
总残溴 total residual bromine
总残氧化剂量 total residual oxidants value
总残余氯 total residual chlorine
总残渣 total residue
总仓库主任 general storekeeper
总操作程序图 general flow chart
总操作费用 total operation expense
总操作速度 overall operating speed
总操作台 general service desk
总槽蓄量 total channel storage
总草酸盐吸附 total adsorption of oxalate
总侧面阻力 total shaft resistance
总侧向压力 omnilateral pressure
总测量 overall measurement
总测量员 general surveyor
总差额 overall balance
总产出 aggregate output;total output;total yield
总产冷量 condensing unit capacity;overall refrigerating effect

总产量 aggregate output; cumulative production; full production capacity; gross output; gross product; gross production; overall output; overall production; overall yield; total generation; total make; total output; total production; total quantity of output; total yield; ultimate production
总产率 gross production rate; overall yield
总产品 gross product
总产品率 percentage of pass
总产物 gross product
总产值 gross output; gross value; total output value; total values of output
总长 entire length; overall length
总长度 aggregate length; footage; length overall; overall length; overlength; summed-up length; total length; out-to-out; total run <椽梁跨度水平距离>; total terminal centres <运输机组的>
总常数 cumulative constant; gross constant
总厂 central plant; central station; complex; general factory; general work; main workshop; principal station
总场 total field
总超额赢利 total excess profit
总超高 <坝顶> gross freeboard
总超越度 total overshoot
总车站 union station
总车重 gross vehicle weight
总车轴距 total wheel base
总尘量 total dust
总尘雾密度 gross fog density
总沉降 settlement total
总沉降量 total settlement; gross settlement
总称 generic term
总成 assemblage; assembly; unit assemblage; unit assembly
总成本 aggregate cost; all-in cost; assembling cost; complete cost; final cost; overall cost; prime cost item; prime cost sum; total cost; ultimate cost; total drilling cost【岩】
总成本费用 sales cost
总成本核算 total costing
总成本现值法 present worth of total cost
总成本与总市价孰低 lower of total cost or total market
总成的底盘架 complete chassis underframe
总成的磨合 running-in of assembly unit
总成分 bulk composition; overall composition; total components
总成更换 unit replacement
总成小修 unit current repair
总成修理法 unit repairing method
总成装配 set assembling
总承包 master contract; overall contract; packaged deal; turnkey; turnkey contract
总承包公共住房 turnkey public housing
总承包公共住房的修复 turnkey rehabilitation
总承包合同 all-in contract; prime contract; general contract; main contract
总承包-互助-新住房建设 turnkey-mutual help-new construction
总承包人 general contractor; main contractor; original contractor; package dealer; prime contractor
总承包商 general contractor; main contractor; original contractor; package dealer; prime contractor
总承包商订货单 general contractor order
总承包商固定价格开价单 general contractor's fixed price offer
总承包施工 original contract construction
总承包型建筑 turnkey type building
总承包者 general contractor; main contractor; original contractor; package dealer; prime contractor
总承付租赁 gross charter; gross terms
总承载能力 aggregate resistance
总弛度 full dip
总迟延时间 total dead time
总尺寸 aggregate dimension; boxed dimension; joint measurement; out-to-out dimension; out-to-out distance; overall dimension; overall size; total dimension
总齿轮传动比 total gear ratio
总冲击强度 aggregate impact value
总冲力 total reactive force
总冲刷量 gross erosion
总抽水头 total pump(ing) head
总抽水站 main pumping station
总臭氧单位 total ozone unit
总出口 general export
总出量 total make; total output
总出纳 general casher
总出水高度 gross freeboard
总出水下水道 outfall sewer
总(出水)扬程 <水泵的> total dynamic(al)(discharge) head
总出站信号机【铁】 advance starting signal
总初级生产力 gross primary productivity
总初级生产量 gross primary production
总除尘效率 overall collection efficiency
总除去率【给】 overall reduction
总储藏量 aggregate storage; gross reserves; gross storage; total reserves; total storage(capacity)
总储(存)量 aggregate storage; gross reserves; gross storage; total reserves
总储气管 main reservoir pipe
总处理能力 throughput capacity
总处理时间 total processing time; total time of treatment
总传导率 total conductivity
总传递函数 overall transfer function
总传动比 overall ratio; resultant gear ratio; total gear reduction; total reduction
总传动减速比 total reduction gear ratio
总传热 overall heat transmission
总传热率 total heat transfer rate
总传热面 total-heating surface
总传热系数 coefficient of overall heat transmission; overall coefficient of heat transfer; overall heat transfer coefficient; overall heat transmission coefficient; overall transfer coefficient; total heat transfer coefficient; U value
总传热阻 resistance of overall heat transmission
总传输特性 overall transfer characteristic
总传质系数 overall mass coefficient
总床沙 total bed sediment
总床沙质 total bed sediment load
总垂度 total sag; full dip <电线的>
总磁场 total magnetic field
总磁场强度 total magnetic intensity
总磁场相位 phase of total magnetic field
总磁场振幅 amplitude of total magnetic field
总磁导率 total permeability
总磁化率 total magnetic susceptibility
总磁力 total magnetic force
总磁力强度 total magnetic intensity
总磁强图 total magnetic intensity chart
总存款 deposited sum
总存取时间 total access time
总存水弯 main(drain) trap
总大肠杆菌 total coliforms
总大肠杆菌检测 total coliform detection
总大肠杆菌浓度 total coliform concentration
总大肠杆菌群 total coliform bacteria population
总大肠杆菌群数 total coliform population number
总大肠菌类 total coliforms
总代表 chief representative
总代理 general agency
总代理人 general agent; sole agent; universal agent
总代理商 general agent; sole agent; universal agency; universal agent
总代理行 general agent
总代谢 total metabolism
总带宽 total bandwidth
总贷款 overall loan
总单体 total monomer
总担保 general warranty
总氮 total nitrogen
总氮氧化物 total nitrogen oxide
总氮氧化物量 total oxides of nitrogen
总氮与总磷量比 ratio of total nitrogen to total phosphorus
总当量 total equivalence; total equivalent amount
总当量比(值) overall equivalence ratio
总导电率 total conductivity
总导管 common excretory duct
总导纳 resultant admittance; total admittance
总道窗宽 total channel window width
总道异常 total channel anomaly
总的 general; gross(main); overall; total
总的产生速度 total generation rate
总的合压力 total resultant pressure
总的计数 cover count
总的降落 <指水头、水压> gross fall
总的接近 global approach
总的净流入量 total net inflow volume
总的净提升能力 total net lifting capacity
总的牵引吨数 gross hauled tonnage
总的色位移 resultant colo(u)r
总的色泽 integral colo(u)r
总的水量均衡方程 balance equation of total groundwater quantity
总的弹模量 overall elasticity modulus
总的体系结构 overall architecture
总的外汇头寸 overall position of exchange
总的物资需求量 overall logistic(al) demand
总的应变状态 general state of strain
总的应力状态 general state of stress
总的住房 overall housing
总登记吨(位) gross register(ed) ton(nage)
总等效比(值) overall equivalence ratio
总底泥需硝酸盐量 total sediment nitrate demand
总底数 total base number
总抵抗力 aggregate resistance
总抵押 general mortgage
总地磁场强度 total geomagnetic intensity
总地区 general locality
总地热梯度 general geothermal gradient
总地震荷载 base shear force
总第一级生产量 gross primary production
总电表板 main switchboard
总电导率 total conductivity
总电荷 total charge; total electric(al) charge
总电化学势能 total electrochemical potential energy
总电键 master key
总电键控制 master key control
总电解反应 overall cell reaction
总电解质 total electrolyte
总电缆沟 main cable channel
总电离截面 total ionization cross section
总电力需要量 total load demand
总电流 summed current; total current; joint current <全部串联电源的电流>
总电流调节器 total current regulator
总电路断流器 primary breaker
总电平 overall level
总电容 integral capacitance; total capacitance
总电势 combined potential
总电压 total voltage
总电源 general supply
总电源保险丝 main electric(al) power supply fuse
总电源熔断丝 main electric(al) power supply fuse
总电站 central station
总电站装机容量 total plant capacity
总电阻 total resistance
总店 head office; headquarters
总调度所 central traffic control office
总调度员 central traffic controller
总叠加次数 total stacking fold
总订货单 blanket order; general order
总定额 gross rating
总定时设备 master timing device
总动力 total output
总动力厂 central power plant
总动力出流水头 total dynamic(al) discharge head
总动力密度 total power density
总动力水头 total dynamic(al) head
总动力压头 total dynamic(al) head
总动量 aggregated momentum
总动式汽锤 single-acting steam hammer
总动态误差 total dynamic(al) error
总动员 total mobilization
总督宫 <意大利威尼斯> Doge's Palace; palazzo
总端阻力 point resistance force
总段长 superintendent
总段工程师 district engineer
总断距【地】 net slip
总断面 gross section
总断面(面)积 gross sectional area; total basal area
总锻工车间 main forge shop
总吨 gross ton
总吨公里 gross ton-kilometer[kilometre]
总吨数 aggregate tonnage; gross ton; gross tonnage
总吨位 gross tonnage measurement; total tonnage; gross tonnage <船舶总容积吨>

总额 all the amount;entirety;footing;gross amount;lump sum;omnium;sum;total amount;total sum
总额承包合同 lump-sum contract
总额承包投标 lump-sum bid
总额承包协定 lump-sum agreement
总额抵押 floating mortgage
总额定出力 gross rated capacity
总额定容量 gross rated capacity
总额费用 lump-sum freight
总额和超额积算表 excess and total meter
总额收费 lump-sum fee
总额指拨 lump-sum appropriation
总发电量 gross generation;total generation
总发电站 central power station
总发光量 total light output;total light yield
总发货量 total shipment
总发热量 gross calorific power;gross heating value;total calorific value
总发射度面积 total area of emittance
总发射率 total emissivity
总发射能力 total emissive power
总发射强度 total emission strength
总发行额 gross circulation
总阀 king valve;main valve;master valve
总法律顾问 general counsel
总法向力 total normal force
总翻修 general overhaul
总反差 overall contrast
总反馈 total feedback
总反应 overall reaction;total reaction
总反应速率 overall reaction rate
总反作用力 total reactive force
总范围 total size
总方差 population variance
总方针 general policy
总放大倍数 <显微镜> total magnification
总放大率 overall amplification;resultant magnification
总放大系数 overall gain
总放射测量 gross radioactivity measurement
总放射性 gross activity;gross radioactivity
总放射性测量 gross activity measurement
总放水阀 main discharge valve
总非残渣 total non-residue
总非过滤性残渣 total non-filterable residue;total unfilterable residue
总非碳酸盐碳 total non-carbonate carbon
总费用 all-in cost;all-in use;general expenses;gross business expenses;overall cost;total cost;total expenses
总分辨率 overall resolution
总分辨效率 total resolution efficiency
总分类账 general ledger accounts;key ledger
总分类账户 general ledger accounts
总分离效能 overall resolution efficiency
总分离效能指标 overall resolution efficiency
总分配系数 bulk partition coefficient
总分析 bulk analysis
总风道 main air duct;main duct;main trunk(duct)
总风缸管 main air reservoir pipe
总风管 blast main;main air supply pipe;main blast line
总风力 total wind force
总风量 total air quantity
总峰值失真 total peak distortion
总服务台 general service counter;reception desk
总氟化物调节缓冲器 total-fluoride adjustment buffer
总浮力 gross buoyancy
总幅度 net amplitude
总辐射 global radiation;total radiation
总辐射度 integrated radiant emittance
总辐射功率 total radiant power;total radiation power
总辐射计 pyranometer
总辐射剂量 total radiation dosage;total radiation dose
总辐射量 total radiation
总辐射能 integrated radiant emittance
总腐殖酸 total humic acid
总腐殖酸含量% 分级 total humic acids content per cent graduation
总付合同 lump-sum contract
总负荷 diversified load;gross load;total loading
总负荷需求 total load demand
总负荷需要量 total load demand
总负载 gross load;total loading
总负载能力 total capability for load
总负载强度 gross loading intensity
总负责 overall charge
总负责人 general superintendent
总负债 gross liability;total liability
总复位 general reset
总富裕(时间)<网络进度计划中的> total float
总改正量 total correction
总盖度 total cover-degree
总概率 general probability;total probability
总概算 general estimate;overall approximate estimate;total estimated cost
总概算费用 overall approximate estimate cost
总概算造价 overall approximate estimate cost;total estimated cost
总概要说明 integral schematic description
总干固物 <干燥法测得> total solids by drying
总干管 main collector
总干管隧道 tunnel for utility mains
总干渠 arterial canal
总干生物质 total dry biomass
总干事 director-general
总干物质 total dry matter;total solids
总干物质量 total solid yield
总干舷高 gross freeboard
总干线 main collector
总干燥时间 total drying time
总干燥余渣 total dry residue
总刚度 global rigidity;overall rigidity;overall stiffness
总刚度矩阵 total stiffness matrix
总纲 general arrangement;general outline;superclass
总高差 total head
总高度 aggregate height;overall height;overall depth;total height
总镉 total cadmium
总铬 total chromium
总工程承包人 general engineering contractor
总工程承包商 general engineering contractor
总工程承包者 general engineering contractor
总工程计划 master program(me)
总工程师 chief engineer;engineer in charge;engineer in chief;general engineer;superintendent;underground hog <俚语>
总工程造价 total work cost
总工料造价承包契约 lump-sum contract
总工料造价投标 lump-sum bid
总工期 time for completion of contract;time for completion of whole works;total project duration
总工头 general foreman
总工务员 general foreman
总工业储量 total ultimate reserves
总工艺卡 master route chart
总工长 general foreman
总工资 total wages
总工作费用 total work cost
总工作时间 net cycle time;total operation time
总工作重量 total working weight
总公差 gross tolerance
总公司 controlling company;general office;head office;home office;main office;master office;parent company;parent firm
总公司工作人员 head office staff
总公司经费 head office overheads
总公司统驭账 head office control account
总公约 umbrella convention
总功 total work
总功耗 total power consumption
总功率 aggregate capacity;gross power;total output;total power;total volt-ampere;ultimate capacity;gross horsepower <不带附件的>
总功率表 totalizing wattmeter
总功率记测板 total output panel
总功率接收机 total power receiver
总功率损耗 total power loss
总功率望远镜 total power telescope
总供电能力 total capability for load
总供给 aggregate supply
总供给管 service main
总供给函数 aggregate supply function
总供给价格 aggregate supply price
总供气管 supply main
总供热系统 total energy system
总供水管 supply main
总供水量 total supply;total water supply
总供应 aggregate supply
总汞分析 total mercury analysis
总汞量 total mercury
总共 all together;all told;crop and root;figure out at;in all;in the aggregate;sum;total
总估计 total estimate
总股本 general capital;shareholding equity
总骨料 total aggregate
总固定成本 total fixed cost
总固定固体 total fixed solid
总固态残渣 total solid residue
总固体 total solids
总固体测定 total solids test
总固体含量 total solid capacity;total solid content
总固体量 total solid
总固体浓度 total solid concentration
总固体污泥 total sludge solid
总固体物 total solid matters;total solids
总固形物含量 total solids content
总关闭阀 main shut-off valve
总管 collecting main;common main;header pipe;main conduit;main line;main pipe;trunk;central tube;line pipe;main collector;manifold;principal pipe <气、水等>;general superintendent;overman;manager;superintendent <指管理人员>
总管道 main line;main pipeline;main piping;man pipe
总管的 master
总管管架 header support
总含水量 total moisture
总管理部门 general management
总管理处 general administration;general office;head office;headquarters
总管理费 general caretaking expenses;general charges;general cost of administration;general expenses
总管理费用 general overhead cost
总管理机构开支 overhead charges
总管逆止阀 manifold check valve
总管数 a total of tubes
总管损失 main loss
总管系 main pipeline
总管线 transit pipeline
总管装置 manifolding
总贯入量 total penetration
总惯性模量 gross moment of inertia
总灌溉率 gross duty(of water)
总光合作用 gross photosynthesis
总光圈大小 overall f-number
总光通量 total light flux;total luminous flux
总规划图 layout sheet
总滚压宽度 overall rolling width
总过冲偏移量 total over shoot offset
总过硫酸盐氮 total persulfate nitrogen
总过调量 total overshoot
总含分量 total capacity for water
总含量 total content
总含硫量 total sulfur content;total sulphur content
总含气量 total air;total air content
总含沙量 total load;total sediment load
总含水量 total moisture content
总含水量测试仪 total water content instrument
总含烃量 total hydrocarbon content
总含盐量 total salinity;total salt content
总函数 generic function
总行 <银行的> home office;head office;main office
总号 total number
总耗汽量曲线 Willans line
总耗热量 total heat consumption
总耗水量 total water consumption
总耗氧量 total oxygen-consuming capacity
总耗氧量速率 total oxygen utilization rate
总耗油量 cumulative fuel consumption;total burn-off
总合 integrate;integration
总合草苔虫 Bugula neritina
总合成力 generalized force
总合成误差 total composite error
总合电位 summating potential
总合价值 aggregate value
总合碱量 total contain alkali volume
总合平衡 overall balancing
总合容量 joint capacity
总合式定量预测 aggregate estimation of mineral resources
总合同 blanket contract;macrocontract
总合有效需求 aggregate effective demand
总合折扣 aggregate discount
总合轴向应力 combined axial stresses
总和 aggregate;grand sum;grand total;major total;sum;summation;totalling
总和比率 ratio of aggregates
总和的 summational
总和的百分之五 five percent of the

sum
总和的二十分之一 one-twentieth of the sum
总和电路 summing circuit
总和法 method of summation
总和峰校正值 total peak corrected value
总和符号 summation sign
总和号 sign of summation
总和积分器 summing integrator
总和精度 overall accuracy
总和流量过程线 summation hydrograph
总和年数折旧法 sum-of-the-year digits depreciation
总和配电盘 summation panel
总和器 summation instrument
总和曲线 <单位过程线的> summation graph;summation curve
总和数字法 sum-of-digits method
总和水文过程线 summation hydrograph
总和图 summation graph
总和误差 composite error
总和仪表 summation instrument
总和指示器 total indicator
总和指数 summary index
总荷载 all-up weight;compound loading;gross load;resultant load;total load(ing)
总荷载强度 gross loading intensity;gross loading strength
总荷载区 total loading area
总荷重 gross load;resultant load;total loading
总横截面积 gross cross-sectional area
总横距 total departure
总烘干时间 total drying time
总厚 aggregate depth
总厚度 aggregate thickness;overall thickness;total sum;total thickness
总厚度变化 overall thickness variation
总候车室 general waiting room
总互调 overall intermodulation
总滑距 net slip;total slip
总化学势能计算法 total chemical potential energy approaches
总化学需氧量 chemical oxygen demand total;total chemical oxygen demand
总还原硫 total reduced sulfur
总还原能力 total reduction capacity
总环境 general environment;total environment
总环境科学 science of total environment
总环境控制 total environmental control
总环境排放量 total environmental emission
总环境评价 total environmental value
总环境质量指数 total environmental quality index
总缓冲强度 total buffer intensity
总灰分 total ash;total percent of ash
总灰分百分含量 total percent of ash
总灰雾密度 gross fog density
总挥发酸类 total volatile acids
总挥发性固体 total volatile solids;volatile total solids
总挥发性悬浮固体 total volatile suspended solid
总回归 total regression
总回归曲线 total regression curve
总回收价值 gross recoverable value of ore
总回收率 overall recovery;overall yield
总汇 concourse
总汇流条 common bus
总会 headquarters
总混合物 total mix(ture)
总混合液挥发性悬浮固体 total mixed liquor volatile suspended solids
总混炼周期 total-mixing cycle
总活度 gross activity;total activity
总活度测量 gross activity measurement
总活菌计数 total viable count
总活菌数 total viable numbers
总活性磷 total reactive phosphorus
总火线保险器 main service fuse
总货价收入 gross proceeds
总机 master station;office equipment;switchboard
总机存储匣 office storage group
总机动时间 <网络计划的> total float
总机构 head office
总机记录保持继电器 office stick register relay
总机线路电码匣 office line coding group
总机械师 chief mechanic;master mechanic
总机械阻抗 total mechanical impedance
总基数 total base number
总基耶达氮 total Kieldah nitrogen
总基准面 general base level
总绩效 total performance
总畸变 overall distortion
总稽核 auditor-general;comptroller
总吉布斯能量 total Gibbs energy
总级间流量 total interstage flow
总级效率 gross stage efficiency
总极化误差 total polarization error
总集尘效率 overall(dust) collection efficiency
总集成 integration
总集料 total aggregate
总集平均 ensemble average
总集气管 main header
总集水深井 collector well
总计 amount(to);aggregate;all told;figure up;final total;foot up;global sum;grand total;gross;gross amount;gross total;lump sum;major summary;major total;sum;summary calculation;summation;summing-up;sum total;total amount;totalize;totalizing
总计操作时间 total operating time
总计穿孔 summary punch(ing)
总计穿孔机 gang summary punch;summary puncher;totalizing punch
总计传递 total transfer
总计打印 summary printing
总计的 lump sum;all-in;resultant
总计吨位 aggregate tonnage
总计复穿孔机 summary gang punch
总计合成位移 total cumulative resultant displacement
总计核对 summation check
总计划 central plan;general layout;master plan;overall planning;overall program(me);overall project;site plan
总计及过账的验证 verification of footings and postings
总计及过账的检验 verification of footings and postings
总计记录器 summary recorder
总计检验 sum check
总计键盘穿孔机 totalizing key puncher
总计卡片 summary card
总计卡片穿孔机 summary card punch
总计曲线 cast line
总计时间 total time
总计试验时间 total testing time
总计数 gross count
总计数测量 total-count measurement;total-count surveys
总计数法 total-count method
总计数率 gross-count rate
总计数率标准偏差图 total-count rate standard deviation map
总计数器 total counter
总计算速度 overall computing speed;overall computing velocity
总计验证 proof total
总计运费 lump-sum freight
总计转移 total transfer
总剂量 accumulated dose;cumulative dose;integral dose
总剂量攻击 total dosage attack
总剂量停留时间 total dose-stay time
总加速度 total acceleration
总加速时间 total acceleration time
总价 aggregate cost;lump sum;sum total;total price;total sum
总价不变合同 fixed-price contract
总价采购 lump-sum acquisition
总价付款 lump-sum payment
总价格 gross price
总价格生长率 total value increment percent
总价合同 all-in contract;lump-sum contract
总价交易 packaged deal
总价(入账)法 gross price method
总价投标书 lump-sum tender
总价项目 lump-sum item
总间隔 total spacing
总间隙 total backlash;total clearance
总监 director-general
总监督员 general superintendent
总监工 central foreman;director general;general foreman
总监理工程师 general supervision engineer
总监视器 master monitor
总减少量 overall attenuation
总减速 total reduction
总减速比 overall reduction ratio;total reduction ratio
总减缩 total reduction
总剪力 total shear
总剪切 total shear
总剪切力 total shearing force
总剪切破坏 general shear failure
总剪切应力 total shearing stress
总检 overall check
总检查长 inspector-general;inspector-in chief
总检查师 chief quality engineer
总检查员 chief inspector
总检修 major overhaul
总碱 total alkali
总碱度 total alkalinity
总碱耗 total alkali consumption
总碱量测定 determination of total base
总碱值 total base number
总建造体积 overall volume of construction
总建筑承包人 general building contractor
总建筑承包商 general building contractor
总建筑承包者 general building contractor
总建筑基地面积 gross site area;total site(construction) area
总建筑毛面积 total gross floor area
总建筑密度 gross building density;total building density
总建筑面积 gross building density;overall floorage;overall floor area;total floor area
总建筑师 architect in charge;chief architect
总建筑使用面积 overall floorage
总建筑占地面积 gross building area
总建筑(占地)面积系数 <总建筑占地面积和建筑用地总面积之比> total coverage
总降低量 overall attenuation
总降低作用 overall reduction
总降深度 total drawdown
总降水 total precipitation
总降水量 general rain;gross precipitation;total precipitation
总降雨量 general rain;general rainfall;quantum of rainfall
总交换能力 total exchange capacity
总交换台(电话) main switchboard
总交通量 total traffic volume
总交通流量 main traffic flow
总胶量 total rubber
总角 accumulated angle
总角动量 total angular momentum
总角动量矩数 total angular momentum quantum number
总角动量值 J-value
总校正(值) cumulative correction
总接受器 catch-all
总接线图 general connection diagram
总节距 resultant pitch
总结 conclusion;final report;generalize[generalise];summarize;summary;summation;sum(ming) up
总结报告 concluding report;final report;general summary;summary report;termination report
总结表 summarized table
总结构 general construction;gross structure
总结构强度 general structural strength
总结构图 scantling plan
总结合能 <核的> total binding energy
总结经验 generalized experience
总结算 lump sum
总结文件 final act
总结账单 final settled account statement
总结指令 tally order
总截留损失量 <植物对降水的> gross interception loss;total interception loss
总截面 bulk cross-section;gross section;overall cross-section;total cross-section
总截面变形 gross section yielding
总截面面积 gross cross-sectional area
总金额 aggregate amount;amount;sum total;up to an aggregate amount of
总金属含量 total metal contents in soil
总金属浓度 total metal concentration
总金属输入量 total metal input
总劲度 overall stiffness
总劲度矩阵 overall stiffness matrix
总进尺 total footage
总进度 overall program(me);overall schedule
总进度表 overall plan
总进度计划 aggregate scheduling;master schedule
总进度图表 overall schedule
总进给 total feed
总进口 general import
总进入碳酸盐 total dissolved carbonate
总进水量 total inflow;total intake
总经纪人 principal broker
总经理 chief executive officer;chief manager;executive head;general manager;managing director;top

manager
总经理室 general manager's room
总经销 exclusive distribution; general agency
总经销人 general sales agent; wholesale distributor
总经营成本 total operating cost
总经营计划 general plan of working; general working plan
总精度 overall accuracy; resultant accuracy
总精确度 over accuracy; overall accuracy
总景勘探线 complete exploration line
总警报开关 general alarm switch
总警告信号电路 general alarm circuit
总警监 commissioner general
总净空 gross freeboard
总净累积率 total net accumulation rate
总净收支 total net budget
总净消融量 total net ablation rate
总净消融率 total net ablation rate
总净重 gross dry weight; total net weight
总径流量 total runoff; yielding flow; mass runoff
总静力高差 total static head
总静压联合探头 combined pitotstatic probe
总静压探测管 combined pitotstatic probe
总静压头 total static head
总就业人数 total employment
总居住面积 gross residential area
总局 central station; controlling operator; general office; head office; headquarters; main office; master office
总局局长 director general
总矩阵法 total matrix
总距离指示器 total distance indicator
总聚焦 gross focusing
总决算 general final accounts; general summary of account
总绝热效率 overall adiabatic efficiency
总开关 control switch; main cock; main(supply) switch; master cock; master switch; passing key
总开关板 main switchboard
总开关盘 main switchboard
总开空隙率 total void ratio
总开销 overhead; overhead charges
总开支 cost record summary; general overhead cost; overall charge; overhead cost
总抗剪力 total shearing resistance
总抗力 aggregate resistance
总科 super-family
总颗粒磷 total particulate phosphorus
总颗粒面(积) total grain surface
总颗粒浓度 total particle concentration
总颗粒物 total particulate matter
总可灌溉面积 gross irrigable area
总可回收苯酚化合物 total recoverable phenolic compound
总可鉴别的氯化烃 total identifiable chlorinated hydrocarbon
总可靠度 cumulative reliability; overall reliability
总可靠性 cumulative reliability; overall reliability
总可溶解锰 total dissolvable manganese
总可溶物 total soluble matter; total solubles; total soluble solid
总可色谱有机物 total chromatographable organics
总可氧化氮 total oxidizable nitrogen
总空气管 main air pipe
总空气量 total air; total amount of air
总空气污染 total air pollution
总空隙 total pore space
总空隙率 total void ratio
总孔隙 total pore space
总孔隙度 total porosity
总孔隙率 total porosity
总孔隙体积 total pore volume; total volume of voids
总控键 turnkey
总控钥匙 turnkey
总控制 master control
总控制板 main switchboard
总控制计数器 total control counter
总控制盘 general control panel
总控制室 main control room; master control room
总控制中心 master control center [centre]
总筘抛光机 reed and harness polishing machine
总库 pool
总库容 aggregate storage(capacity); gross capacity of reservoir; gross reservoir capacity; gross storage; reservoir volume; storage volume; total capacity of reservoir; total reservoir capacity; total storage (capacity); level-full capacity
总跨度 total space; total spalling; total span
总会计 chief accountant; controller
总会计师 general accountant; general accounting officer; treasurer
总会计师室 chief accountant room; general accounting office
总会计室 general accounting office
总会计职能 general accounting function
总宽 <指船> beam overall; overall breadth; overall width
总宽度 aggregate width; full width; overall breadth; overall width; total width; out-to-out
总宽度比 total-width ratio
总矿化度 total mineralization
总矿化度深度关系曲线 relative curve of total solids with depth
总框图 general flow chart
总馈电线 main feeder
总括 colligate; colligation; en bloc; sum up
总括保险 blanket insurance
总括保险单 blanket policy
总括保险费率 blank(et) rate
总括保证保险 blank(et) bond
总括贷款 omnibus credit
总括的 omnibus; overhead
总括定(货)单 blanket order
总括法 lump-sum method
总括价 lump price
总括式损益表 all-inclusive income statement
总括索赔 blanket claim
总括条款 omnibus clause; umbrella article
总括条文 umbrella article
总拉力 gross tractive effort; total tractive effort
总拉伸力 total stretching force
总来流量 total inflow
总揽承包商 package dealer
总累积量 total accumulation
总累积率 total accumulation rate
总冷凝器 main condenser
总冷却功率 total cooling power
总冷却功效 total cooling effect
总冷却水泵 main cooling water pump
总冷却效应 overall refrigerating effect; total cooling effect
总离散 population variance
总离子记录器 total ion monitor
总离子交换容量 total ion exchange capacity
总离子流 total ion current
总离子流检测器 total ions detector
总离子流剖面图 total ion current profile
总离子流色谱图 total ions chromatogram
总离子浓度 total ion concentration
总离子强度调节缓冲器 total ionic strength adjustment buffer
总理 chancellor; prime minister
总理事 director general
总力矩 resultant couple
总力偶 resultant couple
总立管 main riser; main standpipe; rising main
总利润 gross profit; total profit
总利润法 gross profit method
总利润效益 total profit contribution
总利息 gross interest
总利用率 overall percent utilization
总连接环 <吊具的> gathering ring
总连通电荷 total connected electric(al) load
总链产额 total chain yield
总亮度 total brightness
总亮度传输特性 overall brightness-transfer characteristic
总量 gross; gross amount; main bulk; population parameter; quantum [复 quanta]; total amount; total load
总量差额 undercoverage
总量分析 macroanalysis
总量互补商品 gross complements
总量经济学 aggregate economics
总量控制 total quantity control of pollutant discharge
总量模型 aggregative model
总量取样法 total bag sampling system
总量生产函数 aggregate production function
总量收费 total quality charges
总量替代商品 gross substitutes
总量指标 total amount index; total amount indicatrix
总量子数 total quantum number
总裂变产物 gross fission product
总裂缝强度 total fracture density
总临界荷载 total critical load; total critical pressure
总磷 total phosphorus
总磷负荷 total phosphorus load
总磷浓度 total phosphorus concentration
总磷水平 total phosphorus level
总磷酸盐 total phosphate
总灵敏度 total sensitivity
总领班 chief foreman; general foreman
总领工员 general foreman
总领事 consul general
总领事馆 consulate general
总流程 main process stream
总流程图 general flow chart
总流量 aggregate discharge; cumulative discharge; full flow; gross discharge; total discharge; total flow; total flux; total runoff
总流量表 quantity meter; total flow meter
总流量指示器 summation flowmeter indicator
总流明效率 total lumen efficiency
总流域污染物负荷 total watershed contaminant load
总留量 total allowance
总留置权 general lien
总硫 total sulfur
总硫量 total sulfur content; total sulphur content
总馏出率 percent recovery
总隆起量 total heave
总楼层空间 total stor(e)y space
总漏磁通 total leakage flux
总漏气量 overall air leakage rate
总漏泄试验 gross leak test; total leak test
总卤代乙酸 total haloacetic acid
总卤素浓度 total halogen concentration
总滤过性残渣 total filterable residue
总滤系统 omnifiltration system
总路拱 total crown
总路径损耗 total path loss
总路线 general line
总氯 total chlorine
总略图 general sketch
总轮机长 engineer superintendent; general chief engineer; port engineer; superintendent engineer
总论 general information; general instruction; general remark
总落差 gross head; total head; total drop <地质断层的>; total throw <地质断层的>
总马力 gross horsepower
总码穿孔机 totalizing puncher
总码数 gross yards
总脉冲 overall pulse
总毛重 total gross weight
总贸易 general trade
总贸易差额 general balance
总煤耗 total consumption of coal
总煤气管 main gas pipe
总蒙蔽本领 total obscuring power
总锰 total manganese
总秘书处 general secretarial; general secretariat(e)
总密度 gross density
总面积 gross area; total area; total surface
总面积宽度 width of coverage
总名称 generic name
总摩擦(水头)损失 total friction head
总母线 bus line
总木 hammer butt
总目 control file; general view; superorder
总目标 general objective
总目录 composite catalog(ue); general catalog(ue); master table of contents
总钼酸活性磷 total molybdate-reactive phosphorus
总纳污水体污染物平衡 total receiving water pollutant budget
总挠度 combined deflection; total deflection
总能级宽度 total level width
总能力 gross capacity; total capacity
总能量 gross energy; overall energy; total energy
总能量法 total energy method
总能量方程式 total energy equation
总能量衡算 total energy balance
总能流 total energy flow
总能通量 total energy flux
总能头线 total head(grade) line
总泥沙量 total sediment
总黏[粘]着对速度的限制 speed limit providing total adhesion
总扭矩 gross torque
总浓度 total concentration
总浓度分布图 total concentration distribution graph
总排放标准 total emission standard

总排放控制 total emission control
总排(放)量 total displacement; total release
总排放量管理规则 regulation of total emission
总排气量 gross exhaust gas
总排汽管 exhaust steam main
总排水沟 main drainage ditch
总排水管 main discharge pipe; main drain; main sewer; sewer main
总排水量 total displacement
总排水体积 total discharge volume
总排污量控制 total emission control
总配车所 centralized car distribution office
总配电板 main distribution board; main switchboard
总配电电缆 main distribution cable
总配电盘 main control board; main distribution board
总配电站 main substation
总配水管 main distribution pipe; main service
总配线架 main distributing frame; main frame
总配线架终端块 terminating block of main distributing frame
总配置图 general layout
总膨胀比 overall expansion ratio
总批发 wholesale
总批发商 general distributor
总批量 total lot amount
总皮重 gross empty weight
总偏差 total departure; total deviation
总偏角 total deflection angle
总偏移 total drift
总漂沙量 total drift
总漂移 total drift
总贫化率 overall dilution ratio
总频率 sum frequency
总频率计 master frequency meter
总频率特性 resultant frequency characteristic
总频率响应 overall frequency response
总频率响应曲线 total frequency response curve
总频谱发射率 total spectral emissivity
总平错 total heave
总平方和 total sum of squares
总平衡 overall balance
总平衡浓度 equilibrium total concentration
总平均 grand average; population mean
总平均功率 total mean power
总平均生长量 total mean increasing; total mean increment
总平均竖标距 grand average ordinate
总平均值 general average; total average; overall average
总平均纵距 grand average ordinate
总平面 general plan
总平面布置 general arrangement; plot planning
总平面布置图 general arrangement plan
总平面规划 site planning
总平面设计 main plan design; site planning
总平面设计条件 <如地形、地质、水文等> conditions for site planning; site planning condition
总平面图 general arrangement drawing; general outline; key plan; layout plan; location plan; plan of site; site plan of site; situation plan of site; general layout; master plan; site plan; general plan
总评价 general evaluation; overall merit
总坡度 overall slope
总破碎比 gross reduction ratio
总起动电门 master starting switch
总起升高度 total lift
总起止畸变 gross start-stop distortion
总起重量 suspended load capacity; total suspended load
总气管 gas main
总气孔率 total porosity
总汽门 steam regulator; throttle (valve); throttling of steam
总汽门 U 形夹 throttle clevis
总汽门臂 throttle arm
总汽门阀杆 throttle valve stem
总汽门拉条 throttle rod
总汽门曲拐 throttle bell crank
总汽门手把 throttle lever
总汽门填料盒 throttle stuffing box
总汽门箱 throttle box; throttle chamber; throttle valve case
总汽门直立管 throttle pipe
总汽门座 throttle base
总契约 master deed
总牵引(能)力 gross tractive effort; gross tractive force; total draft; gross tractive power
总牵引运输量 gross traffic hauled
总铅量 total lead
总潜热 total latent heat
总强度 omnidirectional intensity; overall strength; resultant intensity; total strength
总强度测量 omnidirectional measurement
总切削面积 total area cut
总侵蚀量 gross erosion; total (amount of) erosion
总清单 general list; master list
总情况码 condition summary code
总氰化物 total cyanide
总曲率 main curvature; total curvature
总曲率线 main curvature line
总曲线 combined curve; compound curve; master curve
总屈服强度 general yield strength
总趋向 common trend
总渠 arterial canal; collecting channel; main canal
总取样 bulk sampling
总去除率 overall reduction
总去污系数 gross decontamination factor
总圈数与抽头圈数比 tapping ratio
总群落耗氧量 total community oxygen consumption
总燃料消耗量 total fuel consumption
总燃烧法 total combustion method
总燃烧热 gross heat of combustion
总热 total heat
总热传导 gross-heat-conductivity
总热传递 total heat transfer
总热分析 total thermal analysis
总热负荷 total heat flux; total heat load
总热功率 total thermal power
总热函 total enthalpy; total heat content
总热焓 total enthalpy
总热焓量 total enthalpy
总热耗 gross heat consumption; overall heat consumption
总热耗率 gross heat rate
总热历程 total heat history
总热量 grand total heat; gross heat budget; gross load; total amount of heat; total heat
总热流量 total heat flux
总热平衡 overall heat balance; total heat balance
总热熵图 total heat-entropy diagram
总热剩余磁化强度 total thermal remanent magnetization
总热通量 total heat flux
总热效率 gross thermal efficiency; overall thermal efficiency
总热值 gross calorie value; gross calorific value; gross heating value; gross heat of combustion; gross thermal value; overall heating; total heating value
总热阻 air-to-air resistance; entire thermal resistance; overall thermal resistance
总人工回灌量 total artificial recharge
总人口密度 gross population density
总任务 general assignment
总日产量 total daily production
总日程表 master schedule; master scheduling
总日射表 pyranometer
总日照 global insolation
总容积 bulk volume; gross volume; level-full capacity; total measurement (volume); total volume
总容积的 full-sized
总容量 aggregate capacity; full load capacity; gross capacity; gross demonstrated capacity; gross fuel capacity; gross volume; total capacity; total tankage; total unit weight
总容量的 full-sized
总容许承载力 allowable gross bearing pressure
总容许水头 overall head allowance
总容许限度 total tolerance
总容重 total weight
总溶度 total soluble matter
总溶解沉积物 total dissolved sediment
总溶解氮 total dissolved nitrogen
总溶解铬 total dissolved chromium
总溶解固体 total dissolved solid
总溶解金属 total dissolved metal
总溶解金属浓度 total dissolved metal concentration
总溶解离子 total dissolved ions
总溶解磷 total dissolved phosphorus
总溶解气体 total dissolved gas
总溶解气体浓度 total dissolved gas concentration
总溶解热 total heat of solution
总溶解碳酸盐浓度 total dissolved carbonate concentration
总溶解碳酸盐物种 total dissolved carbonate species
总溶解铜 total dissolved copper
总溶解无机固体 total inorganic dissolved solid
总溶解无机碳 total dissolved inorganic carbon
总溶解物 total soluble matter
总溶解物质 total dissolved materials; total dissolved substance
总溶解盐 total dissolved salt
总溶解氧 total dissolved oxygen
总溶解游离碳水化合物分数 total free dissolved carbohydrate fraction
总溶解有机固体 total organic dissolved solid
总溶解有机碳 total dissolved organic carbon
总溶解有机物 total dissolved organic matter
总溶性镉浓度 total soluble cadmium concentration
总溶性固体 total soluble solid
总溶性化学需氧量 total soluble chemical oxygen demand
总溶性磷 total soluble phosphorus
总溶性物质 total soluble matter
总溶性盐 total soluble salt
总溶性阳离子 total soluble cation
总熔断器 main cut-off
总柔度 general flexibility
总柔度矩阵 total flexibility matrix
总蠕变 total creep
总蠕动 total creep
总入风(道) main intake
总入口 fore gate; main entrance
总入口名 generic entry name
总入流量 total inflow
总三卤代甲烷生成势 total trihalomethane formation potential
总散发量 total evapo-transpiration
总散射 total scattering
总散射概率 total scattering probability
总散射系数 total scattering coefficient
总扫描时间 total scanning time
总色差 total colo(u)r difference
总筛分面积 overall screen surface area
总筛余 total residue
总商业面积 gross commercial area
总熵 total entropy
总上覆压力 total overburden pressure
总设备利用率 plant factor
总设备马力 total installed horsepower
总设计 chief design
总设计师 chief designer; designer in chief; master designer
总设计图 master design drawing; site plan
总设计者 chief designer
总社 head office
总摄入量 total intake
总伸长 general extension; total extension
总伸长度 overall elongation
总伸展 total extension
总砷 total arsenic
总深度 total depth
总审计师 chief auditor
总渗漏量 total leakage
总渗透率 overall permeability
总升力 gross lift; overall lift; total lift
总升力面积 total lifting area
总生产 aggregate production
总生产成本 total work cost
总生产额 aggregate output
总生产规划 aggregate production planning
总生产计划 aggregate production planning
总生产进度表 master production schedule
总生产量 gross production; total growth
总生产量预测 total-products forecast
总生产率 total output
总生产能力 total productive capacity
总生产赢利 total operating profit
总生长量 total increment
总生化需氧量 ultimate biochemical oxygen demand
总生活耗水量 total domestic water consumption
总生物需氧量 biologic(al) oxygen demand total; total biological oxygen demand
总声功率 total acoustic(al) power
总失真 total (telegraph) distortion
总施工费用 final construction cost
总施工管理员 spread superintendent
总施工进度表 master schedule
总施工进度程序表 master programme
总施工图 general engineering set
总湿度 total humidity; total moisture
总湿密度 total wet density
总湿陷量 total collapse
总时差 total float
总时计 time totaliser[totalizer]

总时间 cumulative time;total time <168 小时/周>
总时限 all hours
总使用水头 gross operating head
总势能 total potential energy
总势头 gross head;total head
总视差 <把横视差和纵视差合并起来考虑时的视差>【测】total parallax
总试样 bulk sample;gross sample
总收尘效率 overall(dust)collection efficiency
总收额 gross revenue
总收获表 general yield table
总收获量 throughput;total yield
总收集处 catch-all
总收集器 catch-all
总收集效率 overall collection efficiency
总收率 overall yield;total yield
总收入 aggregate income;gross income
总收入量 general income;gross credit; gross earnings; gross income; gross receipt;gross revenue
总收入提成折旧法 gross earning depreciation method
总收入系数 gross income multiplier
总收入中扣除数 deduction from gross income
总收缩 total shrinkage
总收缩量 gross shrinkage
总收益 gross benefit; gross income; gross receipt;gross revenue;overall yield;total income;total revenue
总收益率 gross rate
总收益提成折旧 depreciation-gross earning
总收益提成折旧法 depreciation-gross earning method
总收支率 <指冰川> total budget rate
总寿命 entire life;gross life;total life
总受热面 total-heating surface
总受压应力 main compression stress
总受益收入 total benefits
总售价 gross selling price
总枢纽 toll communication center [centre]
总枢纽站 central terminal station
总疏水器 main trap
总输出 total output
总输出电屏 total output panel
总输出功率记测板 summation panel; total output panel
总输出量 aggregate output
总输入信号 total input
总输沙量 total discharge of sediment; total load discharge; total sediment discharge;total sediment load;total solids;total stream load
总输沙率 total sediment discharge; gross transport rate
总输沙能力 total sediment transport capability
总输水水头 dynamic(al)delivery head;total delivery head
总输水效率 total diversion efficiency
总输送电缆 main cable
总束流 total beam
总述 general review
总数 aggregate(number); amount; lump sum;major total;overall number;population parameter;sum(total);tot;total amount;totality;total number;tote
总数包干标价 lump-sum bid
总数大小 size of population
总数计数器 total counter
总数量 total quantity
总数密度 total number density
总衰减 complete attenuation;total attenuation
总衰减测量 overall attenuation measurement
总衰减电平 overall attenuation level
总衰减反射 total attenuated reflection
总衰减量 overall attenuation
总水 total ashed water soluble
总水泵站 main pumping station
总水分 total moisture
总水封 main trap
总水管 common main;hydraulic main; main;mains water;water main
总水管增压泵 water main booster pump
总水灰比 total water/cement ratio
总水况 surface flow and underground storage water
总水力比降 total head gradient
总水力坡降线 total hydraulic gradient
总水力梯度 total head gradient
总水量 total water content
总水量地下水 surface yield
总水龙头 main cock;master cock
总水平水压力 total horizontal water pressure
总水溶物 total water solubles
总水溶性碳水化合物 total water-soluble carbohydrate
总水深 total depth
总水头 geodetic head;gross fall;gross head; overall head; total (hydraulic)head
总水头比降 total head gradient
总水头差 total head differential
总水头富裕量 overall head allowance
总水头勘测 total head exploration
总水头容许值 overall head allowance
总水头损失 total head loss
总水头梯度 total head gradient
总水头线 energy grade line; line of total head;total head line
总水头线坡度 total hydraulic gradient
总水位差 gross available head
总水位削减值 total influence value of water table
总水量与水泥量之比 <包括骨料吸附水> total water/cement ratio
总水闸门 head gate
总水柱高度 total water ga(u)ge
总税 lump-tariff
总税额 total taxes
总说明 general description; general remark
总说明书 general specification
总死亡率 total death rate;total mortality rate
总松弛 total relaxation
总送风道 main supply duct
总送水扬程 total delivery head
总速比 total gear ratio
总速度 overall velocity
总速率 <路段内车辆总路程除以总时间所得之商,总时间包括行车及停车时间> overall rate; overall speed overall travel speed
总酸 total acid
总酸度 total acidity;total titratable acidity
总酸度测定 titration of total acidity
总酸性基 total acidic group
总酸值 total acid number; total acid value
总岁差 general precession
总损耗 overall loss;total loss
总损耗测量 overall loss measurement
总损失 overall loss;total loss
总损失率 overall losses ratio
总损益 consolidated profit and loss
总所得税 gross income tax
总索引 general index
总塌陷量 total collapse
总塔板效率 overall plate efficiency
总台 main station
总态 total state
总坍陷量 total collapse
总弹簧 main spring
总弹回性 proof resilience
总弹性 global elasticity
总碳 total carbon
总碳量 overall carbon content
总碳量测定器 total carbon apparatus
总碳氢化合物 total hydrocarbon
总碳水化合物 total carbohydrate
总碳酸 total carbonic acid
总碳酸盐浓度 total carbonate concentration
总特性 integrated characteristic
总特性曲线 combined characteristic (curve); overall characteristic (curve);total characteristic(curve)
总梯度流 total gradient current
总提单 omnibus bill of lading
总提前角 total lead angle
总提前时间 total lead time
总提升高度 total lift
总体 entirety;general population;integer;integral;population;totality; universe
总体安装 bulk installation
总体百分位数 population percentile
总体搬运 total passing
总体本征值 population characteristic value
总体比较 architectural comparison
总体比例 overall proportion;population proportion
总体标准化率 population standardize rate
总体标准(离)差 population standard deviation
总体布局 entire allocation;entire distribution;general planning
总体布置 block layout planning;general arrangement (plan); general layout;general plan;layout;overall layout;master planning
总体布置图 general(arrangement)diagram; general drawing; general plan;master plan;general arrangement drawing; detailed drawing; key plan
总体部署方案 overall arrange system
总体采样 bulk sampling
总体参数 parameter of a population; population parameter
总体参数估计 estimation of population parameters
总体参数空间 population parameter space
总体参数模 parametric(al)mode
总体测量程序 general survey procedure
总体测试 overall test
总体查找与替代 global search and replace
总体尺寸 general dimension;general specification;overall dimension
总体传输测试 overall transmission test
总体磁场 overall magnetic field
总体的 overall
总体抵押 package mortgage
总体冻胀 mass heaving
总体发展规划 master plan
总体反射 mass reflex
总体方案 general planning; overall plan;overall project
总体方差 population variance; variance of population
总体方差估计 estimation for population variance
总体防空 total air defense
总体防水 integral waterproof(ing)
总体防水法 integral method of waterproofing;method of integral waterproofing
总体防水剂 integral water-proofer
总体分布 population distribution
总体分析 aggregate analysis; macroanalysis;bulk analysis
总体服务 bulk service
总体负荷量 total body burden
总体工程 system engineering;systems enrichment
总体规定 general specification
总体规划 master plan(ning); aggregate plan (ning); comprehensive plan(ning);general plan(ning);overall plan(ning);total plan(ning)
总体规划图 master plan
总体规划限制 general plan restriction
总体规划研究 master planning study
总体合同 <即一揽子合同或交钥匙合同> all-in contract;main contract
总体环境可持续性 general environmental sustainability
总体环境条件 general environmental condition
总体环境通报 general environmental message
总体环境影响 general environmental impact
总体积 bulk volume;cumulative volume;total volume;volume of total
总体积距 total haul
总体极大 global maximum
总体极小 global minimum
总体计划 aggregate plan(ning);master plan;overall design;overall plan(ning);general plan
总体计划与规划 aggregate plans and programs
总体剪切滑动 general shear slide
总体剪切模式 general shear pattern
总体结构 architectural structure; global structure;gross structure;overall structure; system architecture
总体结构设计原理 integrated structure design philosophy
总体解法 overall solution
总体经济 macroeconomy
总体经济不平衡 macrodisequilibrium
总体经济学 macroeconomics
总体经济政策 macroeconomic policy
总体经营处理系统 <原为美国南太平洋铁路行车、调度和处理信息系统,英国铁路已正式采用> Total Operations Processing System
总体矩 population moment
总体矩阵 global matrix
总体决定 aggregate decision
总体均衡 general equilibrium
总体均数 population mean
总体均数估计 estimation for population mean
总体均值 population average;population mean
总体均值的置信界限 confidence limits for population mean
总体勘测 general survey
总体考虑 general considerations;aggregate considerations
总体控制 integral control
总体流动 bulk flow
总体流动过程 bulk fluid process

总体路线 general route
总体率 population rate;rate of population
总体率估计 estimation for population rate
总体模拟 ensemble modeling
总体模拟试验 general simulation test
总体模式 aggregated model
总体模型 overall model
总体排污标准 total emission standard
总体培养 gross culture
总体配套系统化 total systematization
总体平衡 aggregative equilibrium
总体平衡法 population balance method
总体平均 ensemble mean;ensemble average
总体平均数 mean of population;population mean
总体平均值 ensemble average;mean of population;population mean
总体谱密度 ensemble spectral density;general density;integral density
总体企业 macrobusiness
总体强度 bulk strength;overall strength
总体情报控制 total information control
总体人口分布 overall population distribution
总体人口均值 overall population mean
总体人口情况 aggregate population measure
总体杀灭 total eradication
总体设计 alignment design;general layout;integral design;integrated design;master design;overall design;overall plan;system design;unit design;general design
总体设计方案 overall planning scheme
总体设计概念 total design concept
总体设计工程师 total design engineer
总体设计员 system designer
总体设想 general considerations
总体市场 aggregate market
总体式增压器 integral intensifier system
总体试验 bulk test
总体数据规划 total design data planning
总体数量 overall quantity
总体水质 general water quality
总体算术均数 arithmetic(al)mean of population
总体特征值 population characteristic value
总体调试 general test
总体图式 overall pattern
总体网络 integrated network
总体误差 overall error
总体系统 total system
总体相关函数 ensemble correlation function
总体相关系数【数】population correlation coefficient
总体相依变化 dependent variation in general dense
总体销售 macromarketing
总体协方差 population covariance
总体形态学 gross morphology
总体循环法 integrated cyclic(al)process
总体岩性估计 gross lithology estimates
总体应力反应 gross stress reaction
总体预测 macroforecast
总体运输研究 comprehensive transport study
总体中位数 population median
总体中心矩【数】population central moment
总体装卸设备 <关于集装箱和载有集装箱的拖车的> fully integrated lift-on/lift-off facility
总体组成 main assembly
总体最优值 global optimum
总体坐标 global coordinate
总调节器 main governor
总调试 overall commissioning
总调整 total trim
总铁 total iron
总铁结合力 total iron binding capacity;total iron combining power
总烃 total hydrocarbon
总烃含量 content of total hydrocarbon
总停留时间 total residence time;total retention time
总通风功率 total ventilating power
总通风效率 overall ventilation efficiency
总通风压力 total ventilation pressure
总通量 total flux
总同化 gross assimilation
总同位素成分 total isotopic composition
总铜 total copper
总统府 office and residence of the president;office of the president;presidential palace;residence of the president
总统宫 president's palace
总统型压砖机 <商品名> President press
总投入 total input
总投资 total estimated cost;total investment
总投资额 aggregate investment;gross investment;total investment
总投资率 rate of gross investment
总透光度 total transmittance
总透过损失 overall transmission loss
总透过系数 total transmittance
总透气量 total porosity
总透射率 total transmittance
总图 assembly drawing;complete schematic diagram;general arrangement layout;general chart;general diagram;general drawing;general map;general surface plan;general view;key diagram;key map;master plan;overall drawing;plot plan;skeleton diagram
总图布置 general arrangement
总图布置设计 layout plan
总图设计 site planning
总图图册 booklet of general drawing
总图运输 layout for road and transportation in plant area
总土压力 gross earth pressure
总推动 total push
总推进效率 overall propulsive efficiency
总推力 gross thrust;total impellent force;total thrust
总推销费 total salable expenditures
总拖运量 gross traffic hauled
总挖方量 gross cut
总外形 overall form
总(万能)钥匙 grandmaster key
总网络 master network
总危险度 total risk factor
总微生物生物质 total microbial biomass
总尾矿取样机 general tailing sampler
总位能 total potential
总位数 total bit
总位移 net displacement;net shift;resultant displacement;total displacement
总位置图 general location sheet
总温 stagnation temperature;total temperature
总稳定常数 overall stability constant
总稳定性 overall stability
总涡量 total vorticity
总污染 total contamination
总污染负荷 total pollution loading
总污染物 total contaminant;total pollutant
总污染物负荷 total contaminant loading;total pollutant loading
总污染物量控制 total contaminant quantity control;total pollutant quantity control
总污水含氮浓度 total effluent nitrogen concentration
总污水磷酸盐浓度 total effluent phosphate concentration
总污水需氧量 total effluent oxygen demand
总无机氮 total inorganic nitrogen
总无机碳 total inorganic carbon
总务 general affairs;general service
总务部 general affairs department
总务处 general department
总务费 general expenses
总务及管理费用 general and administrative expenses
总务科 general affairs section;miscellaneous division
总务长 dean of general affairs
总物价水平的变动 general price level change
总物价指数 general price index;index number of all commodities
总物料平衡 overall material balance
总误差 aggregate(d)error;combined error;general error;global error;gross error;overall error;resultant error;total error
总吸附 total adsorption
总吸附量 gross absorption;total sorbed amount
总吸附密度 total adsorption density
总吸附能力 total adsorption capacity
总吸附容量 total adsorptive capacity
总吸热量 amount of heat absorption
总吸升水头 <水泵的> dynamic(al)suction head;total suction head
总吸声率 total sound absorption
总吸收表面 total absorbing surface
总吸收分光计 total absorption spectrometer
总吸收剂量 total absorbed dose
总吸收量 gross absorption;total absorption;overall absorption
总吸收率 total absorptivity
总吸收能力 total absorption capacity
总吸收系数 total absorption coefficient
总吸水量 total suction
总吸音能力 total absorption
总烯含量 total olefins
总稀土 total rare earth
总系数 overall coefficient;overall factor
总系统容量 total system capacity
总系统增益 overall system gain
总下水道 common sewer
总显热系数 grand sensible heat factor
总现金流量 gross cash flow
总线 bus(wire);common bus;concentration line;trunk
总线标准 bus standard
总线裁决器 bus arbitrator
总线插接器 bus connector
总线插座 bus hub
总线存取模块 bus access module
总线导电条 trunk
总线底板 bus mother board
总线电缆 bus cable
总线电路 bus circuit
总线多路传送 bus multiplexing
总线访问冲突 bus access conflict
总线访问争用 bus access conflict
总线分离器 bus separator
总线分配器 bus allocator
总线互锁通信[讯] bus interlocked communication
总线计时器 bus clock
总线寄存器 bus register
总线兼容的 bus compatible
总线监控器 bus monitor
总线接插件 bus connector
总线接口 bus interface
总线接口电平 bus interface level
总线接口仿真 bus interface emulation
总线接口连接(技术)bus interfacing
总线接口时序 bus interface timing
总线结构 bus organized structure;bus structure
总线就绪 bus ready
总线可用 bus available
总线可用脉冲 bus available pulse
总线可用信号 bus available signal
总线空闲机器工作周期 bus idle machine cycle
总线控制部件 bus control unit
总线控制逻辑 bus control logic
总线控制器 bus controller;bus master
总线控制者 bus master
总线宽 highway width
总线类别 bus category
总线路 main line
总线路图 complete schematic diagram
总线母板 bus mother board
总线请求 bus request
总线请求周期 bus request cycle
总线驱动器 bus driver
总线式结构 bus organized structure;bus organization
总线收发器 bus transceiver
总线受控 bus slave
总线受控者 bus slave
总线输出 bus out
总线输出校验 bus-out check
总线输入 bus in
总线衔接器 bus adapter
总线型拓扑 bus topology
总线性胀缩率 total linear swelling shrinkage
总线应答 bus acknowledge
总线优先级结构 bus priority structure
总线优先结构 bus priority structure
总线优先输出 bus priority out
总线优先输入 bus priority in
总线周期 bus cycles
总线主控 bus master
总限额 <配额的> aggregate limit
总相关 total correlation
总相关系数 coefficient of total correlation;total correlation coefficient
总箱位数 <集装箱等> gross room number
总响应 overall response;total response
总消耗 aggregate consumption;overall consumption;wastage in bulk or weight
总消耗量 cumulative consumption;total consumption;total flow
总消耗率 total consumption rate
总消化养分 total digestible nutrients
总消化营养 total digestible nutrients
总消融 total ablation rate
总消融量 <冰川> gross ablation;total ablation
总消蚀作用 summed ablation
总效概率密度 ensemble probability density
总效果 cumulative effect;ensemble
总效率 aggregate efficiency;com-

bined efficiency; effective efficiency; gross efficiency; overall efficiency; total efficiency
总效益 general benefit
总效应 gross effect
总效用 general utility
总协议 umbrella agreement
总协议书 general act
总谐波比 total harmonic ratio
总谐波失真 total harmonic distortion
总谐波失真电平 total harmonic distortion level
总泄漏耗损 total leakage
总泄水量 total release
总锌 total zinc
总信贷 gross credit; overall credit
总信号 resultant signal
总信号灯 master pilot
总信号楼 central signal box
总信号设备 main signal(l)ing equipment
总信息 total information
总信息量 gross information content
总星系 metagalaxy
总行程 total run; total travel
总行程时间 total travel time
总行驶里程 total mileage
总行驶阻力 <车辆> total motion resistance
总行政管理费 general administrative expenses
总形成常数 overall formation constant
总修理厂 central repair shop
总修正角 prediction angle
总修正角计算机 prediction computer
总需求函数 aggregate demand function
总需求量 aggregate demand; integrated demand; total demand
总需求清单 ground requirement list
总需求热量 total heat requirement
总需水量 gross water requirement; total water requirement; total water demand
总需氧量 total oxygen demand; ultimate oxygen demand
总需氧量自动记录仪 total oxygen demand automatic recorder
总需要功率 total power requirement
总需要量 total requirement
总许可证 blanket license[licence]
总蓄水量 gross reservoir capacity; total storage(capacity)
总悬浮固体 total suspended solid
总悬浮固体量 total suspended solids
总悬浮固体浓度 total suspended solid concentration
总悬浮颗粒 total suspended particles; total suspended particulates
总悬浮颗粒物 total suspended particle matter; total suspended particulate matter
总悬浮微粒 total suspended particles; total suspended particulates
总悬浮微粒量指数 total suspended particulate index
总悬浮物 total suspended matter
总旋塞 main cock
总选择性 overall selection
总雪荷载 total snow load
总压 stagnation pressure
总压比 overall pressure ratio
总压测量器 Pitot pressure ga(u)ge
总压传感器 total pressure probe
总压法 total pressure method
总压管 ram-air pipe
总压和静压差 Pitot-static difference
总压孔 Pitot hole
总压力 general pressure; gross pressure; overall pressure; resultant pressure; total pressure
总压力测管 total pressure probe
总压力测量管 Pitot probe
总压力测头 total pressure probe
总压力差 total pressure difference
总压力传感器 total pressure probe
总压力盒 total pressure cell
总压力恢复系数 total pressure recovery coefficient
总压力减小 stagnation pressure reduction
总压力降 general pressure drop; overall pressure drop; total pressure drop
总压力曲线 Pitot curve
总压力损失 loss of total pressure; total pressure loss
总压力探头 total pressure probe
总压密量 total sum compressibility
总压强 total pressure
总压强张量 total pressure tensor
总压曲线 Pitot curve
总压实量 total compaction
总压梳状管 comb pitot
总压损失 Pitot loss
总压缩系数 total compressibility
总压头 gross head; total head
总压头测定仪 facing ga(u)ge
总压头管 total head tube
总压头线 line of total head
总压头效率 total head efficiency
总氩量 total argon
总烟囱 chimney stack
总烟道 collecting flue; main (chimney) flue
总烟道闸板 main chimney damper; main flue damper
总延迟 total delay
总延迟时间 overall time delay
总延伸量 overall elongation; total elongation
总延伸率 breaking elongation
总延伸系数 total coefficient of elongation
总延时 total delay
总岩芯采取率 total core recovery
总岩芯回收率 overall core recovery
总盐度 total salinity
总扬程 <水泵的> gross lift; gross pumping head; total delivery head; total pump(ing) head; total head
总扬程差 total head differential
总扬水水头 total pump(ing) head
总氧 total oxygen
总氧化氮 total nitrogen oxide
总氧化剂 total oxidant
总氧化剂浓度 total oxidant concentration
总样(品) bulk sample; gross sample
总液压缸 master cylinder
总移距 net shift; total shift
总移位 total displacement
总议定书 general protocol
总异养菌 total heterotrophic bacteria
总抑制点 total inhibition point
总因数 overall factor
总银浓度 total silver concentration
总引水管 main pipe
总应变 bulk strain; gross strain; overall strain; total deformation; total strain
总应变场 total strain field
总应变理论 total strain theory
总应力 gross stress; overall stress; total stress; total stressing force
总应力法 method of resultant stress; total stress approach; total stress method
总应力分析 total stress analysis
总应力盒 total stress cell
总应力盒法 total stress cell method
总应力集中系数 gross stress concentration factor
总应力路径 total stress path
总应力强度参数 total stress strength parameter
总应力强度指标 total stress strength index
总应力损失 loss of total stress
总(英)尺码 footage
总盈余 consolidated surplus
总营业收入 gross operating income
总营运资金 gross working capital
总硬度 <水的> total hardness
总涌水量 total flux
总用泵 general service pump
总用水量 gross water use
总用水率 gross duty(of water)
总油气比 total gas-oil factor; total gas-oil ratio
总游离石灰量 total free lime
总游离氧化钙量 total free calcium oxide
总有机氮 total organic nitrogen
总有机负荷 total organic loading
总有机结合氯 total organically bound chlorine
总有机卤化物 total organic halide; total organic halogen
总有机卤化物生成势 total organic halide formation potential
总有机卤素 total organohalogen
总有机卤素量 total organic halogen
总有机卤素浓度 total organic halogen concentration
总有机卤素生成势 total organic halogen formation potential
总有机氯 total organic chlorine
总有机氯化合物 total organic chlorine compound; total organo-chlorine compound
总有机碳含量 total content of organic carbon; total organic carbon content
总有机碳量 total organic carbon
总有机碳量分析仪 total organic carbon analyser[analyzer]
总有机物 total organic matter
总有机物浓度 total organics concentration
总有机溴化合物 total organic bromine compound
总有机组分 total organic component
总有效荷载 gross payload
总有效碱容量 total available alkali
总有效利用系数 overall efficiency
总有效氯 total available chlorine
总有效落差 gross available head
总有效马力 effective horsepower
总有效能量 total available energy
总有效水头 gross available head
总有效载荷 gross payload
总有效载重量 gross payload
总有一个 one or the other
总余量 total allowance
总余氯量 total residual chlorine
总雨量 gross precipitation; gross rainfall; total rainfall; total volume of rain
总雨量计 <测定相当长时间的总降水量用> totalizer
总预报时间 gross forecasting time
总预估沉降量 total estimated settlement
总预见期 gross forecasting time
总预算 brutto-budget; general budget; gross budget; main budget; master budget
总预算草案 drafted general budget
总预应力 total pre-stressing force
总月数 total months; total drilling months【岩】
总钥匙 master key; passkey
总钥匙能开的锁 master-keyed lock
总钥匙锁 grandmaster-keyed lock
总钥匙系列 grandmaster-keyed series
总钥匙系列锁 master-keyed lock
总钥锁 master-keyed lock
总云量 total amount of cloud; total cloud cover; total cloudiness
总运费 gross freight; total transportation expenses; total transport cost; total transport expense
总运价册 rate book
总运距 total haul; total haulage
总运输吨哩 total traffic ton-miles
总运输费和保险费 cost-freight-insurance
总运输量 total gross traffic; total transportation volume
总运行成本 total operating cost
总运行费 overall operational cost
总运行时间 total operating time
总运行水头 gross operating head
总蕴藏量 aggregate storage
总杂质 total impurities
总杂质级【冶】 total impurity level
总载荷 gross load; load in bulk; total load
总载货车重 gross laden weight
总载货量 gross shipping weight
总载量 gross load
总载重 gross load; gross vehicle load
总载重对净载重的比率 proportion of gross to net load
总载重吨位 deadweight ton(nage); deadweight capacity <货物加船用燃物料>
总载重量 dead load; deadweight tonnage; gross deadweight; total deadweight; deadweight capacity <船舶>
总藻类密度 total algae density
总造价 final cost; gross cost of manufacture; total cost; total cost of building
总噪声 overall noise
总噪声暴露指数 composite noise exposure index
总噪声级 overall noise level
总噪声谱 total noise spectrum
总噪声温度 overall noise temperature
总则 general articles; general condition; general considerations; general guideline; general principle; general provisions; general rules; summarized principle
总增焓 total enthalpy gain
总增量 total increment
总增益 full gain; overall gain; total gain
总闸门 main valve
总占地面积 gross coverage
总占地指标 gross floor space index
总站 chief station; main office; main station; master office; master station; port; union station; principal railroad station【铁】
总站电子计算机 master computer
总站数目 total number of stations
总账 control account; general account; general ledger; journal ledger; ledger account
总账簿 ledger
总照度 total illumination
总照明 general ambient light
总蒸发 evapotranspiration; fly-off
总蒸发量 gross evapo(u)ration; total evapo(u)ration
总蒸发热量 total heat of evapo(u)ration

总蒸发-蒸腾量 total transpiration-evapo(u)ration
总蒸散发量 total transpiration-evapo(u)ration
总政策 general policy
总支出 aggregate expenditures; gross charge; gross expenditures; gross outlay
总支出协定 overall payment agreement
总支出协议 overall payment agreement
总直径 overall diameter
总值 aggregate value; amount; grand total; gross value; main value; overall cost; total cost; total value
总值加权法 aggregate value method of weighting
总指标 overall performance
总指挥室 general director's room
总指示灯 master pilot lamp
总指数 combined index; general index (number); summation index
总制动功率试验 gross brake power test
总制动马力 gross brake horsepower
总制冷量 gross refrigerating capacity; overall refrigerating effect
总质量 gross mass; lumpy mass; total mass
总质量控制 total quality control
总质量流量 total mass flux
总质量摩尔浓度 total molar concentration
总质量运输 total transport of mass
总质量阻止本领 total mass stopping power
总质心能量 total centre-of-mass energy
总滞留量 total holdup
总置零键 master zero key
总中断站 main repeater station
总中剖面【船】midship profile
总重 gross weight; gross vehicle load <车的>; packed weight
总重吨公里 gross ton-kilometer[kilometre]
总重范围 gross weight range
总重分布 gross weight distribution
总重货运密度 gross ton-kilometers per kilometer of line
总重金属 total heavy metal
总重量 gross weight; all-up weight; full weight; gross load; overall weight; total weight; nominal weight
总重密度 density of total passing tonnage
总重皮重比 ratio of gross to tare weight
总重组别 weight group
总周围压力 total confining pressure
总周转资金 gross working capital
总轴 line shaft; main shaft
总轴架 main pedestal
总轴距 total wheel base
总轴马力 total shaft horse-power
总轴系 line shafting
总(主)风缸切断塞门 main reservoir cut-out cock
总主应力 total active soil force
总注册吨(位) gross register(ed) ton(nage)
总注入层段 entire injection interval
总贮存量 total stocks
总柱面积 gross column area
总转储 total dump
总转换时间 total conversion time
总转换误差 total conversion error
总转矩 total torque
总转向杆 main steering arm
总装 final assembly; general assembling; general assembly
总装备表 master equipment list
总装车间 assembling workshop; assembly shop
总装机容量 installed gross capacity; overall installing capacity; total installed capacity
总装配 general assembly; total assembly; major assembly
总装配带技术 assembly belt technique
总装配架 main frame; master rig
总装配输送带 gathering conveyer [conveyor]
总装配图 general assembly drawing
总装配线 general assembly line; main assembly line
总装式汽轮机装置 integrated steam turbine plant
总装式蒸汽动力装置 integrated steam power plant
总装填时间 total loading time
总装卸率 gross crane rate
总装窑密度 overall kiln setting density
总装药量 total charge; total net charge
总装载量 gross load
总装载液体混合样品 gross liquid mixed cargo samples
总装装配架 master jig
总状分枝式 recemose branching
总准确度 overall accuracy; resultant accuracy
总灼烧损失 total loss on ignition
总资本 aggregate capital; general capital; total capital
总资本化率 overall capitalization rate
总资本利润率 total capital profit ratio
总资本利益率 rate of earnings on total capital employed
总资产 total assets
总资产报酬率 rate of return on total assets
总资产负债表 general balance sheet
总资产周转率 turnover of total assets
总资源回收 total resource recovery
总资源量 total resources
总自行 total proper motion
总自由面 <浮选> total free surface
总综合进料比 overall combined feed ratio
总综合喂料比 overall combined feed ratio
总纵距 total latitude
总纵弯曲 longitudinal bend(ing)
总纵向力 main longitudinal force
总走向【地】common strike; common trend
总租船契约 gross charter
总租约 master lease
总阻抗 total impedance
总阻力 drag overall; overall drag; resultant drag; total draft; total drag; total resistance
总组合熵 total configurational entropy
总钻进时间 gross drilling time
总钻进速度 overall drilling rate
总钻孔时间 gross drilling time
总钻探进尺 total drilling footage
总钻探时间 total drilling time
总钻眼时间 overall drilling time
总作用 resultant action
总坐标(系) global coordinate
总坐标系(统) global coordinate system

纵

纵板道 fore-and-aft road
纵半砖【建】queen closer
纵比降 longitudinal gradient; longitudinal slope
纵比降 <河流> longitudinal river slope
纵壁 longitudinal wall
纵边接缝 landing edge; longitudinal seam
纵标 ordinate
纵标度 vertical scale
纵标目 vertical subheading
纵波 compression-dila(ta)tion wave; longitudinal wave; primary seismic wave; P wave; waves of compression
纵波波速 p-wave velocity; velocity of longitudinal wave; velocity of pressure wave
纵波法 <探伤> longitudinal beam technique; normal beam technique
纵波速度 longitudinal velocity
纵波运动 longitudinal wave motion
纵材 fore-and-aft runner; longitudinal stringer
纵材木 <依木纹锯开的木材> grain wood
纵舱壁【船】fore-and-aft bulkhead; longitudinal bulkhead
纵槽 cannelure; longitudinal slot; pod <钻头的>
纵槽钻头 pod auger
纵侧面线 grade line
纵测量器 ordinatometer
纵测线 extended line
纵层 longitudinal layer; stratum longitudinale
纵差保护 line differential protection
纵产式 longitudinal lie
纵长 fore and aft
纵长的 lengthways; lengthwise
纵长构造 elongated structure
纵长海岸 longitudinal coast
纵长量度 running measure
纵长米 running meter
纵长三角洲 elongate delta
纵长沙丘 longitudinal dune
纵长十字架 Latin cross
纵长褶皱 elongated fold
纵长座 lengthwise seat
纵撑 brace; longitudinal stay
纵尺 lineal foot; linear ruler
纵次序 longitudinal order
纵大梁 under girder
纵挡板 longitudinal wall
纵荡【船】surge; surging
纵刀架 main carriage; plain turning slide
纵的一致性 vertical consistency
纵底板条(舢板) footling
纵电动势 <由强电感应引起的> longitudinal electromotive force
纵吊门 vertically suspension door
纵叠砂 multistor(e)y sands
纵叠砂体 multistor(e)y sand body
纵顶撑 longitudinal crown stay
纵顶杆 longitudinal crown bar; longitudinal roof bar
纵动比较仪 longitudinal comparator
纵度 vertical range
纵断层 longitudinal fault
纵断距 vertical separation
纵断面 lengthwise section; longitudinal surface; profile; sagittal surface; surface curve; vertical profile; vertical section; edge grain <木材的>; longitudinal section
纵断面包线设计 rolling grade design
纵断面测绘器 <测路面平整度用> profilograph; profilometer
纵断面测量 longitudinal section measurement; measurement of profile; profile survey
纵断面分析仪 longitudinal profile analyzer
纵断面改进 profile revision
纵断面改善 profile improvement
纵断面高程 elevation in profile
纵断面观测镜 profiloscope
纵断面计量 profile measurement
纵断面记录本 profile book
纵断面略图 profile scheme
纵断面坡度 profile grade
纵断面曲线 vertical curve
纵断面设计 design of longitudinal section; design of vertical alignment; profile design
纵断面设计线【道】vertical alignment
纵断面水平测量 profile level(l)ing
纵断面水准测量 longitudinal level(l)ing; profile level(l)ing
纵断面缩图 condensed profile
纵断面图 longitudinal section (profile); profile (diagram); profile map; skiagraph; vertical profile map; longitudinal profile <尤指道路>; profile in elevation <注有标高的>
纵断面图幅 profile sheet(ing)
纵断面图纸 profile paper; profile sheet(ing)
纵断面位移 profile displacement
纵断面线 profile line
纵断面仪 profile meter
纵断器 slitter
纵堆装法 longer
纵队 line astern
纵队队形 trail formation
纵二分裂 longitudinal binary fission
纵帆 fore-and-aft sail
纵帆船 fore-and-after; fore-and-aft rigged vessel; schooner
纵帆前缘上的环 hank
纵帆下风缘 leech
纵帆下桁【船】boom
纵分水界 longitudinal divide
纵风 range wind
纵风测定误差 range wind error
纵缝 lane joint; longitudinal joint; longitudinal seam; side seal
纵缝焊接 longitudinal seam welding
纵缝开槽机 longitudinal joint groove former
纵缝缺口弯曲试验 longitudinal bead notched bend test
纵缝顺直度 <混凝土路面> straightness of longitudinal joint
纵缝弯曲试验 longitudinal bead bend test
纵高 lifting
纵割 slit cutting
纵割锯 rip saw
纵割用拉锯 vertical cross cutting saw
纵隔 mediastinum
纵隔 X 线照相术 mediastinography
纵隔凹陷【医】depression of mediastinum
纵隔摆动【医】pendular movement of mediastinum
纵隔板 longitudinal partition
纵隔板支承的隔墙 mid-feather
纵隔堵 longitudinal bulkhead
纵隔墙 quartered
纵拱窑 longitudinal arch kiln
纵沟 cannelure; longitudinal ditch; vertical drainage
纵构架 longitudinal bent
纵构架式 longitudinal system
纵构架式船体 longitudinally framed hull
纵构件 longitudinal member
纵谷 longitudinal valley

纵骨 fore-and-aft runner
纵骨架 longitudinally frame
纵骨架结构系统 longitudinal framing system
纵骨架式 longitudinal system of framing
纵骨架式船 longitudinally framed vessel
纵管 longitudinal tube
纵轨枕【铁】stringer
纵过道 longitudinal aisle
纵焊缝 longitudinal jointing; longitudinal welding seam; longitudinal weld
纵痕 tramline
纵桁 fore-and-aft beam; longitudinal girder; web longitudinal stringer
纵桁板【船】stringer plate
纵桁布置图 girder plan
纵桁架 stringer girder
纵桁上翼板 girder top plate
纵桁下翼板 girder bottom plate
纵行 longitudinal row
纵行的 longitudinal; vertical
纵行密度 wale count
纵行添纱 wale thread plaiting
纵行歪斜 wale deflection
纵行转移 wale slippage
纵横 crossbar
纵横比 aspect ratio
纵横变换器 crossbar transformer
纵横波并行处理 parallel processing efforts of P-and S-wave
纵横尺寸比 aspect ratio
纵横尺寸标注 ordinate dimensioning
纵横出中继电路 outgoing trunk circuit to crossbar office
纵横刀架滑板 cross and top slides
纵横定位设备 coordinate location device
纵横混合 criss-cross mixing
纵横机 crossbar switch
纵横交叉的交通 criss-cross traffic
纵横交错 criss-cross
纵横接线器 crossbar switch
纵横接线器(自动)测试器 tester for crossbar switch
纵横开关 crossbar switch
纵横梁联结 longitudinal girder and cross beam connection
纵横两用缝焊机 universal seam welder
纵横两用锯 mitre saw
纵横流动区域精炼炉 cross-flow zone refiner
纵横排列 vertical and horizontal arrangement
纵横奇偶监督码 horizontal and vertical parity check code
纵横奇偶校验 cross parity check
纵横倾调节系统【船】trim-heel regulating system
纵横人工出中继电路 outgoing trunk circuit in toll office to crossbar exchange
纵横入中继电路 incoming trunk circuit from crossbar office
纵横式交换器 crossbar switch
纵横水准仪 cross test level
纵横图 magic square
纵横系统 crossbar system
纵横向进刀量种数 number of both longitudinal and cross feeds
纵横向调平 four-way level(l)ing; two-way level(l)ing
纵横行植 check row
纵横摇摆 labo(u)r
纵横摇舵自动记录仪 gyro pitch and roll recorder
纵横移动工作台 cross slide table
纵横制 crossbar system; longitudinal-transverse system < 自动电话交换机的一种制式 >
纵横制电话交换系统 crossbar telephone switching system
纵横制交换机 crossbar switch; cross connecting board
纵横制交换台 cross connecting board
纵横制接线机 crossbar connector
纵横制连接器 crossbar connector
纵横制市内自动电话局 crossbar local office
纵横制选择器 crossbar selector
纵横制自动电话交换机 crossbar type automatic telephone switchboard; crossbar automatic telephone exchange
纵横制自动电话局 crossbar telephone center[centre]
纵横制自动交换方式 crossbar automatic exchange system
纵横自动电话交换机 crossbar automatic telephone system
纵横走刀 compound motion
纵护顶背板 roof strainer
纵簧 longitudinal spring
纵火 incendiarism; incendiary fire
纵火材料 incendiary material
纵火犯 firebug
纵火管 incendiary pencil
纵火狂 pyromaniac
纵火器 incendiary file destroyer
纵火焰 longitudinal flame
纵火焰池窑 longitudinal flame tank furnace; unifloe tank furnace
纵火焰窑 longitudinal flame furnace
纵火油料 incendiary oil
纵火者 firer; incendiary
纵加强筋 longitudinal reinforcement
纵间隔宽度 longitudinal interval
纵间隙 axial clearance
纵剪和横剪作业线 slitting and shearing line
纵剪和卷取作业线 slitting and coiling line
纵剪和切边作业线 slitting and trimming line
纵剪机刀片 slitter knife
纵剪切机 slitting shears
纵槛 longitudinal sill
纵接 longitudinal joint
纵接地板 edge joint of flooring board
纵接头 lane joint
纵节理【地】longitudinal jointing
纵结构式 Isherwood framing; Isherwood system; longitudinal framing
纵结合 longitudinal joint
纵截面 lengthwise section; longitudinal section; profile section
纵截面积 longitudinal section area
纵截木 long cut wood
纵解 ripping
纵筋 longitudinal bar
纵进刀 longitudinal feed
纵进给 longitudinal feed
纵距 latitude difference; longitudinal distance; longitudinal spacing; ordinate
纵距闭合差 closing error in latitude
纵距量测器 ordinatometer
纵距误差 error in latitude
纵距(总)和【测】total latitude
纵锯 resaw; rip sawing
纵开取样器 sample splitter
纵拉杆 longitudinal tie
纵拉条 longitudinal bracing
纵肋 longitudinal rib; vertical rib
纵肋骨 Isherwood frame; longitudinal frame
纵肋架式 Isherwood framing; longitudinal framing
纵肋架式船 longitudinal framed vessel
纵肋结构 longitudinal system
纵肋结构系统 Isherwood framing; longitudinal framing
纵肋片 longitudinal fin
纵棱锤 straight-peen sledge
纵连合索 longitudinal commissure
纵联差动保护 longitudinal differential protection
纵联复式汽轮机 tandem compound turbine
纵联杆 longitudinal strut
纵联涡轮机 tandem turbine
纵梁 fore-and-aft beam; interior stringer; longeron; longitudinal beam; longitudinal girder; stringer; string piece; way beam; side member < 机车的 >
纵梁拉杆 longitudinal bracket; stringer bracket
纵梁联结系 stringer bracing
纵梁碰撞 rail impact
纵梁桥 stringer bridge
纵梁托架 stringer bracket
纵梁斜撑 stringer bracing
纵列 tandem
纵列传动箱体 tandem drive housing
纵列的 in tandem
纵列多丝埋弧焊 tandem sequence submerged arc welding
纵列二进码 column binary code
纵列钢筋 axial reinforcement; longitudinal reinforcement
纵列进行 defile
纵列平板拖车 fiat-bed tandem trailer
纵列砌合法 < 美 > longitudinal bond
纵列驱动轮 tandem drive wheel
纵列式车间 longitudinal type of shop
纵列式到发线 longitudinal arrangement of station tracks
纵列式二进位表示法【计】column binary
纵列式会车站 lap siding
纵列式会让站 longitudinal type passing station
纵列式区段站 longitudinal type district station
纵列式停车 parallel parking
纵列式(站线)布置 longitudinal arrangement
纵列旋转枢体 tandem pivot housing
纵裂 longitudinal division; longitudinal fission; longitudinal fissure; longitudinal segmentation; longitudinal split; split crack; vertical split head
纵裂缝 longitudinal cracking; longitudinal seam; rod crack; slitting
纵裂纹 rod crack
纵裂隙 longitudinal crevasse
纵流 lengthwise ventilation
纵脉 longitudinal vein
纵镘板 longitudinal float
纵面车钩 vertical plane coupler
纵面坡度 profile grade
纵面线形 vertical alignment
纵面站台 dead-end platform; longitudinal platform
纵磨 longitudinal grinding
纵木滑道 pole chute
纵排 < 货物装舱 > lengthwise
纵排列 extended spread; vertical disposition
纵劈的薄片 splinter
纵劈理 longitudinal crevasse
纵偏差 longitudinal deviation
纵偏振波 longitudinal polarization wave
纵平衡 longitudinal balance
纵坡 profile grade; running slope < 车辆行驶方向的坡度 >
纵坡标桩 grade stake
纵坡层 grade course
纵坡度 longitudinal grade; longitudinal gradient; longitudinal slope; profile grade
纵坡度转折点 point of change of gradient; point of intersection
纵坡角 top rake
纵坡坡降点 knick point
纵坡设计线 vertical alignment
纵坡稳定 < 河床的 > grade stabilization; grade stability
纵坡线 grade line; longitudinal slope line
纵坡折变 breadthwise in grade; break in grade
纵坡折减 < 曲线上的 > grade compensation; compensation of grades
纵坡转折点 point of intersection
纵坡桩 grade stake
纵剖法 section topography
纵剖锯 rip saw
纵剖锯材 ripped lumber
纵剖面 buttock; longitudinal cross-section; longitudinal section; profile; surface curve; vertical section; longitudinal profile
纵剖面饱和度分布图 saturation profile
纵剖面布置图 outline inboard profile
纵剖面测绘器 profilograph; profilometer
纵剖面测绘仪 profiloscope
纵剖面回声测深仪 profile echograph
纵剖面图 longitudinal plan; longitudinal profile; long-profile; profile; profile in elevation; sectional arrangement drawing; longitudinal section; inboard profile【船】
纵剖小圆锯机 short-log bolter
纵剖砖【建】queen closure
纵汽包锅炉 longitudinal drum boiler
纵汽包水管锅炉 longitudinal drum water tube boiler
纵墙 longitudinal wall
纵墙承重 longitudinal wall load bearing
纵墙承重结构 longitudinal bearing wall construction
纵切 length cutting; rip; slab cut; slit cutting; strip cutting; vertical cut; width cutting
纵切单轴自动车床 longitudinal automatic single spindle lathe
纵切刀 knife bar; longitudinal knife
纵切的 straight cut
纵切复卷机 slitter rewinder; slitting and re-reeling machine; slitting and re-winding machine
纵切工 slitter operator
纵切辊 slitting roll
纵切机 longitudinal scoring machine; slitter; slitting mechanism
纵切机组 slitting machine
纵切剪(切)机 dividing shears; slitting shears
纵切锯 rip saw
纵切锯条 rip saw blade
纵切开 longitudinal incision
纵切裂痕 slitter crack
纵切流程 slitting line
纵切面 longitudinal section
纵切线 slitter line
纵切圆锯 circular rip saw
纵切圆锯机 bolter
纵切圆盘刀心杆 slitting roll
纵侵蚀作用 longitudinal erosion
纵倾【船】longitudinal trim; trim
纵倾和稳定计算书 trim and stability

booklet
纵倾计算表 trim sheet
纵倾计算图表 trim diagram;trimming table
纵倾角 angle of trim(ming) <船的>; pitch;trim angle <船的>;angle of pitch(ing) <飞机>; pitch angle <飞机>
纵倾力矩 trimming moment
纵倾排水量 trimmed displacement
纵倾平衡系统泵 trimming pump
纵倾平稳水舱 peak tank; trimming tank
纵倾曲线 trim curve
纵倾水线 trimmed waterline
纵倾调整 trim correction
纵倾斜角前倾角 top rake
纵倾修正 alternation of trim
纵倾指示器 trim indicator
纵倾状态 trimming condition
纵沙丘 <又称纵砂丘> sand levee
纵沙滩 longitudinal bar
纵沙洲 longitudinal bar
纵深堆货法 deep stowage
纵深配置 distribution in depth
纵声速 longitudinal sound velocity
纵矢量 column vector
纵式滨线 concordant shoreline
纵式海岸 concordant coast;longitudinal coast
纵式海岸线 concordant coastline;longitudinal coastline
纵式联合 vertical combination merger
纵式锁闭 longitudinal locking
纵视差 longitudinal parallax; vertical parallax;Y-parallax
纵视差螺旋 longitudinal parallax screw
纵视图 longitudinal view
纵顺管 longitudinal pipe
纵顺混合 longitudinal mixing
纵顺梯度 longitudinal gradient
纵顺向河 longitudinal consequent river;longitudinal consequent stream
纵丝 vertical thread
纵锁条 longitudinal locking bar
纵弹簧 longitudinal spring
纵弹性模数 modulus of longitudinal elasticity
纵通道波 longitudinal channel wave
纵挖【疏】longitudinal dredging
纵挖法 longitudinal dredging method
纵弯 crooking
纵弯曲 buckle;buckling
纵弯褶皱 buckle folding
纵纹 longitudinal grain; longitudinal striation;side grain
纵稳定心曲线 curve of longitudinal metacenter[metacentre]
纵稳定性 longitudinal stability
纵稳定性过大 pitch stiffness
纵稳定(中)心 longitudinal metacenter [metacentre]; metacenter for longitudinal inclination
纵稳心高度 longitudinal metacentric height
纵铣 lengthways milling
纵铣削 longitudinal milling
纵弦 longitudinal chord
纵线向 longitudinal alignment
纵相 longitudinal phase
纵相关系数 longitudinal correlation coefficient
纵相位石棉脉 asbestos vein of slip fibre
纵向 fore-and-aft direction; lengthwise direction; longitudinal direction; longitudinal judder; machine direction
纵向安定性 pitch stiffness
纵向安全性准则 longitudinal stability criterion
纵向按序排列 column order
纵向暗沟 <路面下的> longitudinal underdrain
纵向凹槽 longitudinal fluting
纵向百叶窗板条 vertical slat
纵向摆动 end shake; fore-and-after motion;longitudinal oscillation
纵向摆动迁移 longitudinally swing migration
纵向搬运 longitudinal haul
纵向爆破 longitudinal blast
纵向比例尺 vertical scale
纵向比率分析 vertical ratio analysis
纵向闭合差 closing error in latitude
纵向边 longitudinal edge
纵向边墙支梁 wall plate
纵向边(缘)梁 longitudinal edge beam
纵向变形 axial deformation; linear deformation; longitudinal deformation; longitudinal strain; longitudinal warping
纵向标高 longitudinal elevation
纵向标线【道】longitudinal road marking
纵向表面流速 longitudinal surface velocity
纵向冰缝 longitudinal crevasse
纵向冰隙 longitudinal crevasse
纵向波 compressive wave; longitudinal wave
纵向波荡器 longitudinal undulator
纵向波状的 longitudinally corrugate
纵向补偿 longitudinal compensation
纵向不平顺 longitudinal irregularity
纵向不平整性 longitudinal level irregularity
纵向不稳定性 instability in pitch;longitudinal instability
纵向布局 longitudinal layout
纵向布置的发动机 longitudinally mounted engine
纵向布置发动机 engine mounted longitudinally
纵向步桥 fore-and-aft gangway
纵向部分 longitudinal component
纵向舱壁 longitudinal bulkhead
纵向操纵 fore-and-aft control
纵向操纵机构 longitudinal controller
纵向操纵装置 fore-and-aft motion
纵向槽 cannelure;longitudinal groove; longitudinal slot
纵向差动 longitudinal differential
纵向缠绕 longitudinal winding;planar winding
纵向缠绕环扣强度 longitudinal winding loop tenacity
纵向长凳座席 longitudinal bench seat
纵向长度 lineal foot
纵向长度伸缩振动模式 longitudinal length extension vibration mode
纵向长度振动模式 longitudinal length modes
纵向长座椅 longitudinal seat
纵向场 longitudinal field
纵向场式 longitudinal field type
纵向场透镜 longitudinal field lens
纵向潮流力 longitudinal current force
纵向(车距)控制 longitudinal control
纵向车辙 wheel track rutting
纵向撑杆 spreader bar; longitudinal bracing
纵向承力索 longitudinal carrier cable
纵向承重 load-bearing in longitudinal direction; weight-carrying in longitudinal direction
纵向弛豫 longitudinal relaxation
纵向冲动 longitudinal impulse
纵向冲断层 longitudinal thrust fault
纵向冲击 impact of collision
纵向重叠 longitudinal overlap
纵向重复 longitudinal repeat
纵向触簧梳 vertical contact comb
纵向传播 longitudinal propagation
纵向传导率 longitudinal conductivity
纵向传输校验 longitudinal transmission check
纵向传送装置 longitudinal conveyer [conveyor]
纵向船排滑道 longitudinal railway slip; longitudinal slipway with rail tracks
纵向垂直支撑 longitudinal vertical bracing
纵向磁场 longitudinal magnetic field
纵向磁场发电机 longitudinal magnetic field generator
纵向磁化 longitudinal magnetization; solenoidal magnetization
纵向磁迹 longitudinal track
纵向磁记录 longitudinal magnetic record
纵向磁偶极 longitudinal magnetic dipole
纵向磁通 longitudinal magnetic flux
纵向磁通磁体 longitudinal flux magnet
纵向磁通势 longitudinal magnetomotive force
纵向磁通线圈 longitudinal flux coil
纵向磁致电阻 longitudinal magnetoresistance
纵向磁致伸缩 longitudinal magnetostriction
纵向次序 longitudinal sequence
纵向错移 endwise mismatch
纵向大梁 longitudinal girder
纵向单向拉伸模量 modulus of simple longitudinal extension
纵向挡板 longitudinal baffle
纵向刀架 longitudinal tool rest; main saddle
纵向导管 longitudinal duct
纵向的 direct axis; lengthwise; longitudinal;fore and aft
纵向堤 <导流设施> longitudinal dyke
纵向堤岸 longitudinal embankment
纵向底部廊道 longitudinal floor culvert
纵向地震 longitudinal earthquake
纵向电场 longitudinal electric (al) field
纵向电磁力 longitudinal electromagnetic force
纵向电动势 longitudinal electromotive force
纵向电流 longitudinal current
纵向电路 longitudinal circuit
纵向电阻缝焊 longitudinal resistance seam welding
纵向调查(研究) longitudinal study
纵向叠加 longitudinal stack
纵向叠加次数 number of vertical stack
纵向顶梁 running balk;running bar
纵向定程进刀 end-feed
纵向定程进刀磨削 end-feed grinding
纵向定线 longitudinal alignment
纵向动量 longitudinal momentum; longitudinal play
纵向动作 longitudinal action
纵向端墙墙面 end of longwall face
纵向断层 longitudinal fault
纵向断层带 longitudinal fault belt
纵向断裂 longitudinal fracture
纵向堆垛法 end-piled loading
纵向堆料 axial stockpiling
纵向多层砌合 longitudinal bond
纵向多层砌体 longitudinal bond
纵向多层顺砖砌墙法 longitudinal bond
纵向多道隔墙 multiple withe[wythe]
纵向多角经营 vertical diversification
纵向多样化 vertical diversification
纵向多余数位检验 longitudinal redundancy check
纵向多种行业合并 vertical and conglomerate merger
纵向扼流线圈 longitudinal choke coil
纵向发裂 longitudinal hair crack
纵向发散 longitudinal divergence
纵向发展 vertical development
纵向翻斗车 longitudinal overturned car
纵向范围 longitudinal extent
纵向方式延迟线 longitudinal mode delay line
纵向仿形 longitudinal copying
纵向仿形切削 longitudinal copying
纵向放大率 axial magnification;longitudinal magnification
纵向分辨力 longitudinal resolving power
纵向分辨率 along-track resolution;axial resolution;longitudinal resolution
纵向分布 longitudinal distribution
纵向分带 longitudinal zoning
纵向分隔板 longitudinal divider
纵向分级器 vertical classifier
纵向分类器 vertical classifier
纵向分量 longitudinal component
纵向分裂电压 longitudinal voltage
纵向分散 longitudinal dispersion
纵向分散系数 longitudinal dispersion coefficient
纵向分水界 longitudinal divide
纵向分水岭 longitudinal divide
纵向分析 vertical analysis
纵向风(荷)载 longitudinal wind load
纵向风力 longitudinal wind force
纵向风力系数 longitudinal wind force coefficient
纵向封垫 longitudinal seal
纵向缝 longitudinal seam
纵向缝焊 longitudinal seam welding
纵向浮心 longitudinal center of buoyancy
纵向幅度 flight
纵向负荷 longitudinal load(ing);thrust load
纵向负载 longitudinal load(ing)
纵向复原力矩 longitudinal restoring moment
纵向腹板加劲肋 longitudinal web stiffener
纵向干扰 common-mode interference; longitudinal interference
纵向杆 longitudinal rod
纵向杆件 longitudinal member
纵向刚度 longitudinal rigidity; longitudinal stiffness
纵向刚性试验 rigidity-longitudinal test
纵向钢筋 axial bar; axial reinforcement; longitudinal bar; longitudinal reinforcement; longitudinal rod; longitudinal steel; vertical reinforcement
纵向钢筋配筋率 longitudinal steel content
纵向钢丝束 <预应力配筋结构> longitudinal cable
纵向钢丝索 <预应力配筋结构> longitudinal cable
纵向割边机 longitudinal trimmer
纵向隔板 fore-and-aft diaphragm; longitudinal baffle; longitudinal bulkhead; longitudinal pass partition plate;swash plate
纵向隔风墙 line brattice

Z

纵向隔墙 mid-feather
纵向工作缝 longitudinal construction joint
纵向工作后角 working back clearance angle
纵向工作台移动 longitudinal table travel
纵向工作楔角 working back wedge angle
纵向供给 longitudinal feed
纵向拱(券) longitudinal arch
纵向共振 longitudinal resonance
纵向沟脊 longitudinal furrows mark
纵向沟纹 longitudinal groove
纵向构架 longitudinal bent;longitudinal framing;longitudinal frame
纵向构架船 longitudinal framed ship
纵向构件 longitudinal member
纵向构造迁移方向 direction of longitudinally tectonic migration
纵向构造体系 longitudinal tectonic system
纵向管距 depth tube spacing;longitudinal tube spacing
纵向惯矩 longitudinal moment of inertia
纵向光磁电效应 longitudinal photo-magnetoelectric(al) effect
纵向光学基频率 longitudinal fundamental optic(al) frequency
纵向轨道梁 waybeam
纵向轨条压曲 vertical track buckling
纵向轨枕 longitudinal sleeper; longitudinal tie; longitudinal timber; stringer
纵向轨枕线路 longitudinal sleeper track
纵向滚动舱口盖 end-rolling hatch cover
纵向滚(线)焊接 longitudinal seam welding
纵向海岸 longitudinal coast
纵向海岸线 longitudinal coastline
纵向焊缝 longitudinal seal;longitudinal weld(seam);longitudinal seam
纵向焊缝的允许偏移 allowable offset in longitudinal joints
纵向焊缝管 longitudinally welded pipe
纵向焊接 longitudinal welding
纵向焊接钢管桩 longitudinally welded steel tube pile
纵向焊接机 longitudinal welding machine
纵向焊接顺序 longitudinal sequence
纵向合并 vertical combination;vertical consolidating;vertical merger
纵向河 longitudinal river;longitudinal stream
纵向河谷 longitudinal valley
纵向荷载 longitudinal load(ing)
纵向桁架 longitudinal truss
纵向横截面 longitudinal cross-section
纵向厚度模式 longitudinal thickness modes
纵向湖震 longitudinal seiche
纵向滑板 longitudinal slide
纵向滑道 end-on slipway;longitudinal slipway; longitudinal marine railway【港】
纵向滑动 longitudinal sliding motion
纵向恢复力矩 longitudinal righting moment
纵向回路 vertical loop
纵向汇管 longitudinal manifold
纵向混合 longitudinal mixing
纵向货舱口围板 cargo hatch side coaming
纵向机电耦合系数 longitudinal electromechanical coupling factor
纵向畸变 longitudinal distortion
纵向激电 induced polarization longitudinal mid-gradient array
纵向激励 longitudinal excitation;vertical excitation
纵向极化 longitudinal polarization
纵向极化束 longitudinal polarized beam
纵向集合管(道) longitudinal manifold
纵向给水主管 <建筑物内> system riser
纵向挤压力 longitudinal compression
纵向脊 longitudinal ridge
纵向脊肋 longitudinal ridge-rib
纵向记录 longitudinal recording;longitudinal register;longitudinal registration
纵向技术转移 vertical technology transfer
纵向加工 longitudinal process
纵向加工成波纹或槽型的屋面板 profiled sheet
纵向加劲 vertical reinforcing
纵向加劲板梁 longitudinal stiffened plate girder
纵向加劲杆 longitudinal stiffener
纵向加劲系梁 longitudinal girt
纵向加强肋 stringer
纵向加热 longitudinal heating
纵向加速传感器 longitudinal acceleration sensor
纵向加速度 fluctuating acceleration; fore-and-aft acceleration; forward acceleration; longitudinal acceleration
纵向加载 loaded in longitudinal direction
纵向夹紧联轴器 clamp coupling
纵向假潮 longitudinal seiche
纵向间隔 longitudinal interval;longitudinal separation
纵向间距 longitudinal pitch;longitudinal spacing
纵向间隙 axial clearance
纵向减速 longitudinal deceleration
纵向剪力 longitudinal shear
纵向剪切 longitudinal shear
纵向剪切裂隙 longitudinal shear crack
纵向剪切流 longitudinal shearing flow
纵向剪切试验 longitudinal shearing test
纵向剪(切)应力 longitudinal shearing stress
纵向检验 longitudinal check
纵向建筑缝 longitudinal construction joint
纵向角 <刀面的> back rake angle
纵向角振动 pitching
纵向校正磁棒 fore-and-aft correctors
纵向校准 longitudinal alignment
纵向接缝 longitudinal joint(ing)
纵向接合板 longitudinal fish plate
纵向接头 landing edge; longitudinal joint(ing)
纵向接头效率 longitudinal joint efficiency
纵向节距 longitudinal pitch
纵向节理【地】 longitudinal joint
纵向结构 longitudinal construction
纵向结合 vertical integration
纵向截面 longitudinal cross-section
纵向劲度 longitudinal stiffness
纵向进刀 length feed
纵向进刀量范围 range of longitudinal feeds
纵向进刀装置 longitudinal feed gear
纵向进给 length feed; length travel; long feed; longitudinal feed; traversing feed
纵向进给的 end-feed
纵向进给量种数 number of longitudinal feeds
纵向进给无心磨 end-feed centerless grinding
纵向井排数目 number of longitudinal well rows
纵向静力稳定性 longitudinal static stability
纵向静态稳定性 longitudinal static stability
纵向距离 fore-and-aft distance;longitudinal distance; longitudinal separation
纵向锯成四块 quarter saw
纵向锯木 ripping
纵向锯切 ripping
纵向聚焦 longitudinal focusing
纵向开裂 longitudinal crack
纵向抗压强度 longitudinal compression strength
纵向可焊性试验 longitudinal bead test
纵向坑道 longitudinal gallery
纵向空间电荷效应 longitudinal space charge effect
纵向空气运动 longitudinal movement of air
纵向空腔 longitudinal cavity
纵向空穴 longitudinal void
纵向快度 longitudinal rapidity
纵向扩散 longitudinal diffusion;longitudinal dispersion
纵向扩散系数 longitudinal diffusion coefficient;longitudinal diffusivity
纵向拉杆 longitudinal tie rod;longitudinal bracing
纵向拉结砌法 longitudinal bond
纵向拉力 longitudinal stretching force;longitudinal tensile force
纵向拉木机 longitudinal log haul-up
纵向拉伸 longitudinal expansion
纵向拉线 longitudinal stay
纵向拉应力 fiber stress; longitudinal tensile stress; longitudinal tension stress
纵向肋片 longitudinal rib
纵向肋条 longitudinal rib
纵向力 fore-and-aft(er) force; longitudinal force
纵向力操纵装置 poaptor
纵向力矩 longitudinal moment
纵向力系数 coefficient of longitudinal force;longitudinal force coefficient
纵向立面图 longitudinal elevation
纵向连接系统 longitudinal connection system; longitudinal brace system <闸门的>
纵向连接装置 one-above-the-other layout
纵向连续带光源 longitudinal continuous band illuminant
纵向连续反应器 longitudinal flow reactor
纵向联杆 flange bracing
纵向联合 vertical combination;vertical integration
纵向联合成整体 vertical integration
纵向联结杆 longitudinal tie
纵向链流动化 longitudinal chain mobilization
纵向梁 longitudinal beam
纵向梁腹加劲杆 longitudinal web stiffener
纵向梁支柱 <隧道支柱式支撑的> underpinning post
纵向裂缝 longitudinal crack(ing); longitudinal crevasse; longitudinal fracture
纵向裂痕 longitudinal cracking
纵向裂纹 longitudinal crack;longitudinal crevasse;longitudinal fracture
纵向裂(隙) longitudinal cracking
纵向临界载荷 buckling load
纵向流 longitudinal stream
纵向流变 longitudinal direction
纵向流动 longitudinal flow
纵向流速 longitudinal current velocity
纵向漏斗车 longitudinal(door) hopper
纵向滤波器 longitudinal filter
纵向路面摩擦系数 longitudinal friction factor of pavement
纵向路面线 longitudinal pavement line
纵向轮廓 longitudinal profile
纵向螺栓连接系统 system of longitudinal bolt
纵向镘板整平机 longitudinal (bull-float) finishing machine
纵向弥散 longitudinal dispersion
纵向弥散度 longitudinal dispersivity
纵向弥散系数 longitudinal dispersion coefficient
纵向密封 longitudinal seal
纵向面调查 longitudinal survey
纵向模量 longitudinal modulus
纵向模数 longitudinal modulus
纵向摩擦力 longitudinal frictional (force)
纵向木薄壁组织 longitudinal strand parenchyma;longitudinal wood parenchyma
纵向木纹 longitudinal grain
纵向木纹板 longitudinal grain board
纵向挠曲 longitudinal bend(ing); flexure produced by axial compression <轴向压力引起的>
纵向挠曲变形 cripple
纵向挠曲稳定性 buckling stability
纵向黏[粘]滞性 longitudinal viscosity
纵向排斥 longitudinal repulsion
纵向排号 longitudinal number;longitudinal numeral
纵向排列 end arrangement; end-to-end arrangement;end-to-end setup
纵向排列电解槽 end-to-end placed cells
纵向排水(沟) longitudinal drainage
纵向排水管 longitudinal drainage pipe
纵向配筋 longitudinal reinforcement
纵向配筋柱 column with lateral reinforcement
纵向膨胀 longitudinal dilatation;longitudinal expansion
纵向膨胀缝 longitudinal expansion joint
纵向偏差 longitudinal deviation
纵向偏斜 longitudinal deflection
纵向平衡 fore-and-aft balancing
纵向平衡补偿 longitudinal counterbalancing
纵向平面 fore-and-aft plane;longitudinal plane
纵向平面波 longitudinal plane wave
纵向坡度 grade in longitudinal direction; head fall; longitudinal grade; longitudinal slope; longitudinal fall <混凝土溜槽的>
纵向坡降 longitudinal fall
纵向破裂 <平板玻璃轧制时的> snake
纵向剖面 longitudinal profile;longitudinal section
纵向奇偶检验 longitudinal parity check
纵向奇偶位 longitudinal parity
纵向奇偶性 longitudinal parity
纵向奇偶性检验 horizontal parity check;longitudinal parity check
纵向畦长 irrigation run
纵向起重机梁 longitudinal crane girder
纵向气流 longitudinal air flow;longitudinal current
纵向气楼 clear stor(e)y
纵向砌合 longitudinal bond
纵向迁移 longitudinal removal;longitudinal transport
纵向前角 back rake angle

纵向强度 longitudinal strength
纵向强力构件 longitudinal strength member
纵向翘曲 longitudinal warp(ing)
纵向切分机组 slitting unit
纵向切割 longitudinal cutting
纵向切割机 longitudinal cutter
纵向切割面 compressional dissected plane
纵向切口 cannelure
纵向切片机 slitter
纵向切线 longitudinal tangent; profile tangent
纵向切削 longitudinal cutting; straight cut
纵向切削操作 straight-cut operation
纵向侵蚀谷 longitudinal erosion valley
纵向倾角分量-深度图 longitudinal dip component versus depth diagram
纵向倾角调节 pitch adjustment
纵向倾角调整 <电动刨路机的> pitch adjustment
纵向倾斜 fore-and-aft tilt; head fall; longitudinal tilt; Y-tilt
纵向倾斜船台 longitudinal inclined building berth
纵向倾斜计 longitudinal clinometer; longitudinal inclinometer
纵向倾斜指示器 pitch indicator
纵向球(面像)差 longitudinal spheric(al) aberration
纵向驱动 longitudinal motion
纵向扰动电压 longitudinal disturbed voltage
纵向热传导 longitudinal thermal conduction
纵向熔珠试验 longitudinal bead test
纵向冗余 longitudinal redundancy[redundance]
纵向冗余检验 longitudinal redundancy check
纵向冗余检验符号 longitudinal redundancy check character
纵向润滑油孔 gallery hole
纵向散布 range scattering
纵向散聚 longitudinal debunching
纵向扫描 longitudinal scan
纵向色差 longitudinal chromatic aberration
纵向沙坝沉积 longitudinal-bar deposit
纵向沙垄 longitudinal sand-ridge
纵向沙丘 longitudinal dune; seif (dune); sword dune
纵向沙洲 longitudinal bar
纵向筛网张力 longitudinal screen tension
纵向上船台的升船机 end transfer shiplift
纵向上平联 top lateral bracing; top laterals
纵向设计 longitudinal layout
纵向伸缩 longitudinal extension
纵向伸缩缝 longitudinal expansion joint; longitudinal warping joint
纵向升降机 longitudinal elevator
纵向生长 longitudinal growth
纵向施工缝 longitudinal construction joint
纵向式通风 longitudinal ventilation
纵向式系统 longitudinal system
纵向视差螺旋 longitudinal parallax screw
纵向视电阻率曲线 longitudinal apparent resistivity curve
纵向视距 vertical sight distance
纵向试验 longitudinal testing
纵向收缩 longitudinal contraction; longitudinal shrinkage; longitudinal shrinking
纵向收缩裂纹 longitudinal shrinkage crack
纵向收缩率 linear shrinkage
纵向受力结构 longitudinal structure
纵向受力绳 longitudinal carrier cable
纵向受弯应力 longitudinal bending stress
纵向输水涵洞 longitudinal culvert
纵向输水廊道 longitudinal culvert
纵向输送 longitudinal feed
纵向输送机 longitudinal conveyer[conveyor]
纵向束流截面 longitudinal beam section
纵向竖联(结系) longitudinal bracing
纵向数 longitudinal number(al)
纵向水库水面波动 longitudinal seiche
纵向水力助推 longitudinal hydraulic assistance
纵向水流 longitudinal current
纵向水流力 longitudinal current force
纵向水流力系数 longitudinal current force coefficient
纵向水平摆动 fore-and-aft shift
纵向水平移动 fore-and-aft shift
纵向水平支撑 longitudinal horizontal bracing
纵向水系 longitudinal drainage
纵向水准器 longitudinal level
纵向水准仪 fore-and-aft level
纵向撕破 lengthwise tear
纵向四极 axial quadrupole; longitudinal quadrupole
纵向送进 longitudinal feed
纵向速度 longitudinal velocity
纵向速度分量 range velocity component
纵向碎木机 long grinder
纵向缩尺放样 longitudinal scale lofting
纵向缩缝 lengthwise shrinkage crack
纵向锁闭 longitudinal locking
纵向锁条 longitudinal locking bar
纵向弹性 longitudinal elasticity
纵向弹性模量 modulus of longitudinal elasticity
纵向梯度 longitudinal gradient
纵向剃齿 convectional shaving; crown shaving
纵向天窗或气楼 clear stor(e)y; clerestor(e)y
纵向天窗或气楼檩条 clerestor(e)y purlin(e)
纵向天桥 ca(u)lk walk bridge; fore-and-aft bridge
纵向天线引信 longitudinal antenna fuze
纵向填平板 pitch damper
纵向挑水坝 longitudinal dike[dyke]
纵向条痕 lengthwise streaking
纵向调平 fore-and-aft level(l)ing; longitudinal level(l)ing
纵向调整 longitudinal adjustment
纵向(贴)角焊 longitudinal fillet weld
纵向通风 axial ventilation
纵向通风系统 longitudinal ventilation system
纵向通过角 ramp breakover angle
纵向同位 longitudinal parity
纵向透水性 longitudinal permeability
纵向推力 longitudinal thrust
纵向拖拉法 erection by longitudinal pulling method; longitudinal pulling method
纵向挖掘 end digging; straightaway digging
纵向挖掘多斗挖掘机 longitudinal multibucket excavator
纵向外加荷载 longitudinal imposed load
纵向弯矩 longitudinal bending moment
纵向弯力 longitudinal bending strength
纵向弯翘 longitudinal warping
纵向弯曲 buckling; longitudinal bend(ing); longitudinal bow; longitudinal curvature
纵向弯曲强度 buckling strength; longitudinal bending strength
纵向弯曲强度计算 buckling analysis
纵向微观弥散 longitudinal microscopic dispersion
纵向围梁 longitudinal girt
纵向围堰 longitudinal cofferdam; longitudinal dam; longitudinal dike [dyke]
纵向位移 length travel; longitudinal displacement; longitudinal travel
纵向位移阻力 resistance to longitudinal displacement
纵向位置 lengthwise position
纵向温度缝 longitudinal warping joint
纵向纹理结构 <大理石中的> styolite
纵向稳定的 longitudinally stable
纵向稳定喷口 longitudinal stabilizing nozzle
纵向稳定器 longitudinal stabilizer
纵向稳定性 longitudinal stability
纵向稳定性试验 longitudinal stability test
纵向稳定中心 longitudinal metacenter[metacentre]
纵向稳心半径 longitudinal metacentric radius
纵向涡流 longitudinal turbulence
纵向误差 longitudinal error
纵向系杆 longitudinal tie
纵向系缆力 <船闸> longitudinal cable force
纵向细微裂纹 longitudinal hair crack
纵向下平联 lower lateral bracing
纵向下水 longitudinal launching; end launching
纵向纤维 longitudinal fiber[fibre]
纵向线 vertical line
纵向线缩率 longitudinal shrinkage rate
纵向相空间 longitudinal phase space
纵向相位曲线 longitudinal phase curve
纵向像差 axial aberration; longitudinal aberration
纵向小椽 longitudinal jack rafter
纵向效应 longitudinal effect
纵向斜架滑道 longitudinal slipway and wedged chassis
纵向斜面升船机 longitudinal inclined shiplift; longitudinal ship incline
纵向泄水沟 <碴床> longitudinal trench
纵向形变 longitudinal deformation
纵向形状 longitudinal shape
纵向修船滑道 longitudinal slipway
纵向修整机 longitudinal finisher
纵向漩涡 longitudinal turbulence
纵向压电效应 longitudinal piezoelectric(al) effect
纵向压力 longitudinal compressive force; vertical pressure; longitudinal pressure
纵向压缩 longitudinal compression
纵向压弯后强度 postbuckling strength
纵向延迟线 longitudinal-made delay line
纵向延伸力 longitudinal stretching force
纵向延性 longitudinal ductility
纵向研究 longitudinal study
纵向摇摆 <车辆> oscillation in the pitch mode
纵向一体化 <指跨国公司的垂直型经营> vertical integration
纵向移船车 longitudinal transfer car
纵向移船的滑道 marine railway with longitudinal transfer
纵向移船台 longitudinal transfer table
纵向移动 longitudinal motion; longitudinal movement; longitudinal shift(ng); longitudinal traverse
纵向移动的 longitudinally traversable
纵向应变 longitudinal strain
纵向应力 longitudinal stress
纵向应力波 longitudinal wave
纵向游隙 longitudinal play
纵向有肋扭合联轴器 ribbed clamp coupling
纵向余隙 longitudinal clearance
纵向鱼尾板 longitudinal fish plate
纵向预(加)应力 longitudinal prestress(ing)
纵向预应力钢筋束 longitudinal tendon
纵向预应力腱 longitudinal tendon
纵向运动 lengthwise movement; longitudinal (mode of) motion; longitudinal movement; pitching motion
纵向运动频率特性 longitudinal response characteristic
纵向运动特性 longitudinal characteristic
纵向载荷 longitudinal load
纵向载重 weight-carrying in longitudinal direction
纵向增强 longitudinal reinforcing; vertical reinforcing
纵向栅栏 <古罗马运动场的> spina
纵向张力 longitudinal stretching force
纵向照度 longitudinal illumination
纵向照明 longitudinal illumination
纵向照明设备 vertical lighting device
纵向折叠 lengthwise fold; longitudinal fold
纵向折流板 longitudinal baffle
纵向折射率 longitudinal refractive index
纵向振荡 fore-and-aft oscillation; lengthwise oscillation; longitudinal oscillation
纵向振荡模延迟线 longitudinal mode delay line
纵向振动 axial vibration; longitudinal vibration; extensional vibration; in-line vibration
纵向振动减振 longitudinal damping
纵向振动频率 frequency in pitch
纵向振动器 longitudinal vibrator
纵向振动特性 longitudinal oscillatory behavio(u)r; longitudinal vibration characteristics
纵向振型 longitudinal mode
纵向振子 longitudinal vibrator
纵向振子换能器 longitudinal resonator transducer
纵向整平板 longitudinal strike-off (blade)
纵向支撑 longitudinal bracing
纵向支撑杆 longitudinal strut
纵向支承 supporting in longitudinal direction
纵向支承荷载 bearing in longitudinal direction
纵向支重块 <支承跨装货物> lengthwise bearing piece
纵向止动器 length stop; longitudinal stop
纵向制动器 length stop
纵向质量 longitudinal mass
纵向中缝 longitudinal center[centre] joint; longitudinal center [centre] line
纵向中梯装置 array for longitudinal

mid-gradient
纵向中线 longitudinal center[centre] joint
纵向中线焊接缝 longitudinal center [centre] joint
纵向中心距 longitudinal pitch
纵向中心铅垂面 longitudinal plane of symmetry
纵向中心线 longitudinal center[centre] line
纵向重心 longitudinal center of gravity
纵向周期性 longitudinal periodicity
纵向轴 longitudinal axis
纵向轴线 direct axis; longitudinal centerline
纵向珠焊试验 longitudinal bead test; slow bead test
纵向转移 vertical transfer
纵向姿态 longitudinal attitude
纵向自差校正磁铁 fore-and-aft magnet
纵向自由度 longitudinal degrees of freedom
纵向自走机构 longitudinal traverse gear
纵向阻抗 longitudinal impedance
纵向阻力 longitudinal resistance
纵向阻力曲线 nose-to-tail drag curve
纵向阻尼 longitudinal damping
纵向最大倾斜度 maximum longitudinal inclination
纵小管 longitudinal tubule
纵效应 longitudinal effect
纵形沙丘 seif(dune)
纵型 longitudinal type
纵型最大最小探索次序 depth minimax search sequence
纵性放大率 longitudinal magnification
纵压加固(耐弯)材 precompressed wood
纵压轴承 thrust bearing
纵摇 pitch;pitching motion;tangage
纵摇幅度 amplitude of pitch
纵摇加速度 acceleration in pitch
纵摇角 pitch(ing) angle
纵摇强烈 ride hard
纵摇轻微 ride easy
纵摇误差 pitching error
纵摇周期 period of pitching; pitching period
纵摇轴 axis of pitch
纵摇纵向惯性矩 longitudinal moment of inertia
纵移 longitudinal shift
纵应力 longitudinal stress
纵英尺 linear foot
纵载构件 axial loaded member
纵轧 longitudinal rolling
纵褶皱 longitudinal fold
纵枕木 longitudinal tie
纵振动 compressional vibration
纵振换能器 longitudinal vibration transducer
纵支撑 longitudinal bracing
纵支杆 longitudinal girt; longitudinal strut
纵直纵平面 longitudinal vertical plane
纵重心 longitudinal center of gravity
纵轴 axis of bank; axis of ordinate; fore-and-after axis; line of departure; longitudinal axis; longitudinal shaft; ordinate axis; vertical axis; Y-axis; fore-and-aft axis【测】; X-line <高斯投影>
纵轴承 thrust bearing
纵轴次瞬变电抗 direct axis subtransient reactance
纵轴的 direct axis
纵轴环流 longitudinal axis circular current
纵轴卷筒式起锚机 capstan
纵轴排列 longitudinal arrangement
纵轴式干燥窑 longitudinal shaft
纵轴瞬变电动势 direct axis transient electrodynamic(al) potential
纵轴瞬变电抗 direct axis transient reactance
纵轴同步电抗 direct axis synchronous reactance
纵轴线 longitudinal axis
纵轴旋耕机 rotary cultivator with longitudinal shafts
纵轴压顶试验 anvil test
纵转【测】 transit
纵转望远镜 replace in the supports; reverse a telescope; telescoping
纵总强度 total longitudinal strength
纵坐标 axis of ordinate; longitudinal coordinate; ordinate; vertical axis; vertical ordinate; vertical scale; Y-axis; Y-coordinate; northing <高斯-克吕格坐标系>
纵坐标法分析 vertical coordinate assessment; vertical grid ordinate assessment
纵坐标法评价 vertical coordinate assessment; vertical grid ordinate assessment
纵坐标增量 latitude
纵坐标中间 mid-ordinate
纵坐标轴 axis of ordinate; line of departure; ordinate axis; vertical axis; Y-axis; vertical coordinate

走

走板 <机车的> running platform; running board
走板安装托 running board saddle bracket
走板端架 running board end bracket
走板托 running board saddle
走板托梁 running board support
走板托梁系铁 running board support tie piece
走板延伸端 running board extension
走不通的 passless
走查 walk-through
走车牵伸输出运动 jacking delivery motion
走车牵伸运动 jacking motion
走车运动凸轮 carriage cam
走出来的小路 beaten(-coben) path
走带机构 tape deck; tape transport; tape transport mechanism; transport【计】
走刀 feed
走刀杆托架 feed rod bracket
走刀痕迹 tooth marks
走刀箱 feed box; quick change gear
走刀箱变速杆 feed selection lever
走道 ai(s)le; aisleway; gangway; path of travel; walkway; inner side aisle <座位间的>
走道板 walk board
走道拱顶 ambulatory vault
走道平台 <桥式吊车的> footpath platform
走道最小宽度 <室内车辆直角拐弯时> minimum intersecting aisle
走调 out of tune
走调的 off-key
走动 locomote
走动的 ambulant
走读半径 school attendance sphere
走航采样(戽)斗 scoopfish
走航底质取样器 underway bottom sampler
走合 running-in
走合期 bedding-in period; running-in period
走合期驾驶 driving in running-in period
走合时间 bedding-in period; running-in time; running-in period
走合维护 running-in maintenance
走滑断裂 strike slip fault
走回头路 backtrack
走火 accidental discharge; misfiring
走极端 carry too far
走廊 ai(s)le; alley way; ambulatory aisle; breezeway; cloister; dogtrot; galleria forest; gallery; hall; lanai; lobby(area); passage; passageway; tresuante; veranda(h); dalan <波斯或印度建筑中的>; loggia [复 loggias/loggie] <建筑物一侧俯临庭院的>; porch <美>
走廊采光 corridor lighting
走廊层 half stor(e)y
走廊窗 flanking window
走廊灯 corridor lamp
走廊地板 corridor floor
走廊地带 corridor
走廊(地带)交通 corridor traffic
走廊地洼系【地】 corridor Diwa system
走廊地毯 matting runner
走廊风道 corridor duct
走廊过道 aisle passage
走廊回风 air return through corridor
走廊交通管理计划 corridor traffic management program
走廊进深 depth of passageway
走廊林 gallery forest
走廊楼层 corridor stor(e)y
走廊墙壁 corridor wall
走廊式发展模式 ribbon development pattern
走廊式木桥 covered timber bridge
走廊式人行道 cloister walk
走廊屋顶 aisle roof
走廊衣帽柜 corridor locker
走廊照明 corridor illumination
走廊折叠椅 folding aisle seat
走廊支柱 aisle pier
走廊终端面积 dead-end areas in corridors
走了气的 stale
走路人 wayfarer
走路石子 path gravel
走轮 travel(l)ing wheel
走马灯式列车 block train; merry-go-round train
走马灯式旋转木马 merry-go-round
走马灯式预应力钢筋连续张拉设备 merry-go-round equipment
走马灯式展示行李包裹 merry-go-round for baggage
走马灯似的打转 merry-go-round
走锚 dragging; drag(ging) of anchors
走锚自动报警器 dragging alarm apparatus
走入旁路 turn-off
走沙水流 heavy sediment-laden flow; sand scouring flow
走时 travel time
走时变化 rate change
走时表 travel timetable
走时残数 travel-time residual
走时测定器 rate measuring instrument
走时差 travel-time difference
走时反演 inversion of travel time; travel-time inversion
走时检验 rate checking
走时曲线 travel-time chart; travel-time curve
走时异常 travel-time anomaly
走时指示器 up-and-down indicator
走兽 ridge-mounted beast ornament
走私 contraband; illicit and clandestine trafficking; owling; smuggle; smuggling
走私船 smuggler; smuggling ship; owler <尤指夜间>
走私贩私活动 smuggling and sale of smuggled goods
走私货 contraband; smuggled goods
走私货物 run goods; smuggled goods
走私集团 smuggling ring
走私漏税 bootlegging; loss revenue through smuggling
走私者 owler; smuggler
走弯路 come round
走线材 miscut lumber
走线架 chamfer; chute
走线纹 snake
走线问题 routing problem
走向 assignment of direction; level course; bearing of trend【地】; line of strike【地】; strike【地】; trend【地】; direction of strike【地】
走向变位 strike shift
走向不连续 stratigraphic(al) discontinuity along strike
走向超覆 strike overlap
走向地质剖面图 strike geologic(al) profile
走向断层 direction of strata; strike(-slip) fault
走向断层作用 strike faulting; wrench faulting
走向分析 trend analysis
走向割理 strike(d) cleat
走向谷 strike valley
走向河 strike river; strike stream
走向滑动【地】 strike slip
走向滑动地震 strike slip earthquake
走向滑动断层 strike slip fault; strike-shift fault; wrench fault
走向滑动盆地 fault-strike slip basin
走向滑断层作用 strike slip faulting
走向滑距 horizontal displacement; horizontal separation; strike slip
走向滑移冰碛层 shove moraine
走向滑移断层 shove fault; strike slip fault
走向滑移盆地 fault-strike slip basin
走向滑运动 strike slip motion
走向滑震源机制 strike slip focal mechanism; strike slip source mechanism
走向机构 running mechanism
走向间距 strike separation
走向角【地】 angle of strike; strike angle
走向节理 strike joint
走向勘探线 strike exploration line
走向离距 strike separation
走向玫瑰图 strike rose diagram
走向平行 strike paralleling
走向平移断层 strike slip fault
走向迁移 strike migration
走向上升断层 strike lift fault
走向水平掘进 drifting
走向投影法 strike projection method
走向线 line of bearing; line of strike; strike line; trend line
走向线方位角【地】 angle of strike
走向移距 strike shift
走向移位【地】 strike shift
走向褶皱 strike fold
走向追索法 investigating method of following the trend
走行部灯 bogie lamp
走行部分 running gear
走行传动 running transmission
走行方向 direction of travel
走行公里 running kilometrage

走行轨 running rail
走行轨道 <起重机的> running track
走行机构 running mechanism
走行距离 walking distance
走行控制稳定性 ride control stability
走行摩阻力 running friction
走行式搅拌机 transit mixer
走行台车 running truck
走行装置 running gear
走形的 outmoded
走样的 out of shape
走纸 form feed(ing)
走纸速度 paper going velocity
走砖 stretcher
走砖面 long face

奏出和谐乐声 chime

奏乐堂 <古希腊、罗马的> odeum

租车 hired car;hired carriage

租车处 <旅客自己驾车> car hire
租车机构 car rental service
租车契约 charter
租车人 charterer
租出固定资产 fixed assets for rent
租船 chartered ship;chartering;chartering of tonnage;freightage;ship chartering;affreight
租船按航次计费 charter by voyage;voyage charter
租船按时计费 charter by time;time charter
租船成交备忘录 fixture note
租船承运商 charter's operator
租船代理人 chartering agent
租船代理(商)chartering agent
租船单 chartering order
租船费 charter money;charterage
租船费率 charter rate
租船附则 side clause of charter party
租船公司 chartering company
租船航次 chartered voyage;voyage charter
租船合同 affreightment;charter;charter party;contract of affreightment;charter contract
租船合同标准格式 charter party forms
租船合同提单 charter party bill of lading
租船合同租金计算标准 charter base
租船合约 affreightment
租船货运 affreightment
租船基价 hire base
租船经纪人 chartering broker
租船经纪人公会 institute of chartered ship brokers
租船老板 boatman
租船期限 term of charter
租船契约 affreightment;charter(party);contract of affreightment
租船契约中允许的装卸期的每一日 lay days
租船人 charterer;freighter
租船人支付费用 charterer pays dues
租船市场 chartering market
租船委托书 charter order
租船佣金 charter commission
租船运费 chartered freight
租船运货 affreightment
租船运价 chartered freight;tramp freight and charter hire
租船运输 affreightment;shipping by chartering
租船证 chartering order
租船证书 charterer certificate
租船支付条款 freight clause;hire and payment clause
租船中止条款 off-hire clause
租船主 chartered owner
租船驻船代表 supercargo
租得物 <尤指土地> leasehold
租的 chartered
租地 lease;tenant
租地建造权 building lease
租地金 acreage rent
租地契约 land lease agreement
租地权 tenant right
租地人 leaseholder;tenant
租地造房权 building lease
租地造屋合同 building lease
租地造屋者 <每年付租金的> superficiary
租地造物权 build lease
租额 rental rate
租房人 tenant
租房暂住 shack up
租费 charter hire;hire charges;rental (charges);rental payment;rent charges
租费清单 list of hiring charges
租费需要 rental requirement
租费账 hire charge account
租购 hire purchase
租购法 hire purchase system
租购合同 agreement for hire purchase
租购信贷公司 hire purchase company
租购信用保险 hire purchase credit insurance
租户 leaseholder;lessee;tenant
租户保险 renter's insurance
租户加装的设施 tenant's fixtures
租户迁出通知 notice to quit
租户之间的补差款 owelty of partition
租户自行维修 tenant maintenance
租户组合 tenant mix
租还地 redelivery area
租还期 redelivery date
租回 sale-leaseback
租回出售财产 lease back
租回已售出产业 lease back
租界 concession
租借 lease;lend lease
租借的 leasehold
租借的车队 hire fleet
租借的船队 hire fleet
租借的转让 assignment of leases
租借地 concession;holding;leased territory;settlement;leasehold
租借法案 lend-lease act
租借购买 lease purchase
租借合约 tenancy agreement
租借货车 leased fleet
租借期 leasehold
租借期限 lease term;period of lease;time limit of the lease
租借契约 contract of tenancy;leasehold;tack
租借权 lease;leasehold;tenancy
租借权继续 continuance of tenancy
租借权终止 termination of tenancy
租借人 leaseholder
租借屋宇 tenement house
租借物 lease
租借物抵押贷款 leasehold mortgage
租借物业 leasehold property
租借者 hirer;leaseholder
租金 hiring;leasing;rent;rentage;rental(value);rental charges;rent expenses;renting;reprise;rack rent <大于年产值的三分之二的>
租金表 tariff
租金补助 rent subsidies
租金乘数 rent multiplier
租金调整 rent escalation
租金法定支付期 term time
租金费用 rent expenses
租金负担 rent charges
租金过户 assignment of rents
租金和维修费 rental market value
租金计算率 lease rate quotes
租金市价 rental market value
租金收入 rent(al)income;rent(al)receipts
租金退还 rent rebate
租金拖欠 rent arrear
租金限价 ceiling rent
租金账 rent account
租金账单 rent roll
租金转让 assignment of rents
租金总收入 gross possible rent
租金总收入系数 gross rent multiplier
租矿金 acreage rent
租来的 leasehold
租来的船 leasehold ship
租赁 charter;hire(purchase);lease;leasehold;leasing;rent;tenancy;tenantry
租赁财产 leasehold property
租赁财务公司 lease finance company
租赁车 rental car
租赁成本 cost of lease;hiring cost
租赁筹资 lease financing
租赁代理人 rental agent
租赁担保 lease bond
租赁的地产 estate taxation
租赁房屋 rental housing
租赁费(用) hiring cost;lease rent(al);rental fee;rent expenses
租赁港 landlord port
租赁购置协议 lease-purchase agreement
租赁固定资产 lease(d)fixed assets
租赁合并 consolidation by lease
租赁合同 agreement for hire;contract of lease;contract of tenancy;lease agreement;lease contract
租赁价值 leasehold value
租赁建筑设备 renting of construction equipment
租赁贸易 lease trade;leasing trade
租赁期 rental period
租赁期间 lease term
租赁期限 charter period
租赁汽车 rent-car
租赁契约 agreement for hire;agreement of lease;contract of lease;contract of tenancy;lease agreement;lease contract
租赁契约背书 lease back
租赁权 leasehold
租赁权益保险 leasehold interest insurance
租赁权转让 lease back
租赁人 hirer;leaseholder;renter
租赁商 hirer
租赁条件 tenancy condition;term of tenancy
租赁协定 agreement for hire;agreement of lease;rental agreement;lease agreement
租赁协议 agreement for hire;agreement of lease;rental agreement;lease agreement
租赁协议书 agreement to hire
租赁信贷 lease
租赁续约 renewal of lease
租赁制 rental system
租赁制度 rentalism
租赁转让 assignment of leases
租赁装置 lease installation
租期 lease term
租期到期 fall-in
租让地 concession
租让合同 concession agreement;concession contract;contract of concession
租让合同承租人 concessionaire
租让协议 concession agreement
租售无障碍的住房(政策)open housing
租书处 book rental;lending library;rental library
租税 gavel
租税负担法则 law of incidence
租屋 tenant
租线电报 leased telegraph
租用 charter;hire;tenement;tenancy <土地或房屋的>
租用场地 rented space
租用场所 rented space
租用车 hired car;leased car;rental car
租用船 chartered ship
租用的汽车 <美> rent-a-car
租用的设备 hired plant
租用的专用通路 leased private channel
租用的专用信道 leased private channel
租用地 tenement
租用地方 rented space
租用电路 leased circuit
租用电路链接 leased circuit connection
租用电路数传业务 leased circuit data transmission service
租用定额 hire rate
租用房屋 tenement
租用费 lease charges
租用货车 hired wagon;leased freight car
租用机车 leased locomotive
租用机械 hired plant
租用基群线路 leased group link
租用价格 hire rate
租用宽频带通路 <计算机和数据传输用> Telpak
租用列车 chartered train
租用面积 rental space;rented area;rented space
租用起居室 bed sit(ter)
租用契约 contract of hire
租用人 lessee
租用设备 leased facility
租用设备费率 equipment rental rate
租用设施 leased facility
租用施工机械 leased construction equipment
租用时间 reimbursed time
租用水费 lease water rate
租用条件 conditions of hire
租用外轮 charter foreign vessels
租用网路业务 leased network service
租用卫星 leased satellite
租用物 hireling
租用线路 leased channel;leased line
租用线路网 leased line network
租用线路信道 leased line channel
租用线网络 leased line network
租用信道 leased(dedicated)channel
租用邮电局电线 hired post-office line
租用住房 rental housing
租用专线通信[讯]网 leased line network
租有 lease
租约 lease;lease agreement;leasehold;tenancy agreement
租约补订书 addendum to charter party
租约到期后仍不归还房产的承租户 hold-over tenant
租约附录 addendum to charter party
租约附则 rider to charter party
租约格式 pro forma charter party
租约期满 expiry of tenancy
租运 freightage

足标 subscript

足尺 full size;whole size
足尺比例 full-scale
足尺寸研制 full-scale development
足尺寸样机试验 full-scale experiment;full-scale test
足尺大样 full-scale detail;full-size detail
足尺大样图 special detail drawing
足尺的 full size(d);full-scale
足尺放样 full-scale layout;full-scale template;prototypic(al)layout
足尺复印 true-to-scale print
足尺结构 prototype structure
足尺结构动力试验 dynamic(al)test of full-scale structure
足尺模拟 full-scale modeling
足尺模型 full-scale model;full-size model;prototype;template
足尺模型试验 full-scale model test;mock-up test
足尺挠度 full-scale deflection
足尺设计图 full-size design
足尺实验 full-scale experiment;full-scale test
足尺试验 full-scale experiment;full-scale test
足尺图 full-size drawing
足尺图案 cartoon;epure(asphalt);full-scale pattern
足尺样板 full-scale template;full-scale templet
足赤 red gold
足够的安全系数 ample margin
足够的(池)容量出 adequate pondage
足够的面积 ample area
足够样本 adequate sample
足迹 combination pliers;footfall;foot mark;footprint;foot step;track;trail
足迹化石 ichnofossil;trace fossil;vestigiofossil
足迹化石学 ichnology
足价 full price
足量 full measure;good measure
足量灌溉 heavy irrigation
足龄混凝土 matured concrete
足耦合 sufficient coupling
足球场 football field;soccer field
足岁 age at last birthday
足趾 digit
足重 lumping weight
足状冰川 foot glacier
足状三角洲 foot delta

卒塔婆 <印度佛塔> stupa;tope

族名 generation number

族系 family

镞 gad

阻 interception

阻碍 barricade;ba(u)lk;blockade;counteract;deter;fouling;handicap;hinder;hindrance;impede;rebuff;retard;road block;stand in the way of;stoppage;thwart;setback
阻碍的 deterrent
阻碍航道 deaden the way;impede the passage
阻碍航行 impede navigation
阻碍河流的碎岩 debris jam
阻碍交通 obstruct the traffic
阻碍角 angle of obstruction
阻碍生长作用 growth-retarding effect
阻碍视距的弯路 blind curved road
阻碍视线 obstruct the view
阻碍物 deterrent;hold-back;encumbrance;hamper;obstruction;trammel
阻碍物防护 barrier shielding
阻碍物质 obstructive matter
阻碍消声器 obstruction muffler
阻碍型裂缝 restriction crack
阻碍岩粉排出 impeded discharge of cuttings
阻碍因子 obstruction factor
阻碍运动 impede motion
阻碍运行 obstacle to running
阻板 dasher
阻爆剂 inhibitor of ignition
阻爆器 detonation sealer
阻波板 wave blade
阻波凸缘 choke flange
阻颤器 damper
阻车 stop for obstacles
阻车堤 <厂矿道路> stopping truck heap
阻车(进入)式中央分隔带 deterring type median
阻车器 car arrester;kick-up block;safety stop;wagon arrester
阻车时间 <交叉口红灯时间> obstruction period of traffic
阻尘剂 dust retardant
阻臭管 stink trap
阻穿堂风 draught-proofing;draft-proofing
阻带 attenuation band;rejection band;stop band
阻带中的损耗 loss in suppressed range
阻带阻抗 rejector impedance
阻弹板 bunter plate
阻挡 bar off;countercheck;entrap;obstruct;prevent;trapping
阻挡层 barrier film;blocking layer;blocking level;blocking lever;trapping layer
阻挡层电容器 barrier-layer capacitor
阻挡层电压 barrier voltage
阻挡层光电池 barrage cell;barrier film cell;barrier-layer cell;blocking-layer cell;electronic photovoltaic cell;photoelement;photovoltaic cell
阻挡层光电管 barrage photocell;barrier film photocell;barrier layer;barrier-layer photocell;barrier-plane;blocking layer photocell
阻挡层区域 barrier region
阻挡层整流器 barrier film rectifier;barrier-layer rectifier
阻挡道岔 trap points
阻挡电阻 blocked resistance
阻挡方向 inverse direction;reverse direction
阻挡环 catch ring
阻挡交通的山谷 obstructed valley
阻挡膜 barrier film
阻挡气流的穿堂 draft excluder;draught excluder
阻挡气流的门廊 draft excluder;draught excluder
阻挡区 key-out region
阻挡栅极 barrier grid
阻挡栅极马赛克 barrier grid mosaic
阻挡式护栏 block out type safety fence
阻挡压条 check stop
阻挡者 stopper
阻挡阻抗 blocked impedance
阻钉 gang nail
阻冻剂 anti-freeze;anti-freezing agent
阻冻溶液 anti-freeze solution
阻冻涂料 anti-freezing painting
阻断 blockade;blocking;cut-out;intercept;interdict;kill
阻断波 shut-off wave
阻断河 blocked stream
阻断剂 blocking agent;repressor
阻断开关 disconnecting switch
阻断滤片 block filter glass
阻断通知 notice of interruption
阻断闸板 shut-off gate
阻断状态 blocking state
阻墩式通气器 baffle aerator
阻遏 baffling;repression
阻遏物 repressor
阻遏系统 repressible system
阻风 choke;wind bound <不能航行>
阻风穿堂 draft lobby;draught lobby
阻风带 air lock strip
阻风门 air choke;choker;choke valve;draftproof door
阻风门蝶阀 choker fly
阻风门廊 draft lobby;draft preventer;draftproof lobby;draught lobby;draught preventer
阻风门通风 choke
阻风器轴 air shutter shaft
阻风设施 draft preventer;draught preventer
阻风障体 draftproof barrier;draughtproof barrier
阻干剂 drying inhibitor
阻槁杆 cut-off bar
阻隔 orifice;stopping;severance <交通公害之一,阻隔路线两侧穿越>
阻隔板 baffle
阻隔层 barrier layer
阻隔潮湿 moisture stop
阻隔栏杆 barrier railing
阻隔墙 barrier wall
阻隔区 baffle area
阻垢 anti-precipitation;scale inhibition
阻垢分散 scale inhibition and dispersion
阻垢分散剂 scale inhibition and dispersion agent
阻垢缓蚀剂 scale and corrosion inhibitor
阻垢剂 anti-precipitant;anti-scalant;scale inhibition substance;scale inhibitor;scaling inhibitor
阻垢率 scaling inhibition ratio
阻垢试验 scaling inhibition test
阻垢性能 scale inhibition performance
阻光的 light-resistant
阻光度 opacity
阻光化玻璃 anti-actinic glass
阻光或滤热玻璃 calorex
阻光率 opacity
阻光染料 sharpening dye
阻光通道 light trap
阻焊加压 welding force
阻焊膜对位偏差 restrict misalignment
阻航 obstruction to navigation
阻化剂 inhibitor;negative catalyst;negative catalyzer;retarding agent;stop-off agent;stopping agent
阻化油 inhibited oil
阻环 balk ring
阻活剂 anti-activator
阻火 fire stop
阻火舱壁 fire-resisting bulkhead
阻火产品 fire stop product
阻火的 fire-retardant
阻火法 fire stopping
阻火隔板 fire block
阻火剂 fire-retardant chemicals
阻火器 fire arrestor;fire check;fire trap;flame arrester[arrestor];flame damper;flame trap
阻火墙 flame resisting wall
阻火通道 fire aisle
阻火位置 anchor point
阻火屋盖 fire-retardant roof coverings
阻极检波系数 transrectification factor
阻胶剂 gum inhibitor
阻截耗损 interception loss
阻截泉水的结构 spring intercepting structure
阻截物 intercepted matter
阻举比 drag lift ratio
阻举系数 drag lift coefficient
阻距 torsional resistance
阻聚剂 inhibiting agent;inhibitor;polymerization inhibitor;polymerization retarder;polymerization stopper;retarder
阻抗 impedance;reactance;resistance
阻抗保护系统 distance protection;impedance protective system
阻抗保护装置 impedance protection
阻抗比 impedance ratio
阻抗变化法 impedance variation method
阻抗变换 impedance conversion
阻抗变换电路 impedance inverter circuit
阻抗变换滤波器 impedance transforming filter
阻抗变换网络 impedance transformer network
阻抗变压器 impedance transformer
阻抗标度 impedance scale
阻抗标准 impedance standard
阻抗标准化 impedance normalization
阻抗拨号 impedance dial(l)ing
阻抗补偿器 impedance compensator
阻抗不规则性 impedance irregularity
阻抗不均匀性 impedance irregularity
阻抗部件 impedance component;impedance part
阻抗参数 impedance parameter
阻抗测量 impedance measurement
阻抗测量设备 impedance measuring equipment
阻抗测试器 impedance checker
阻抗程度 resistance degree
阻抗传感器 resistive transducer
阻抗导纳图 <导抗图> immittance chart
阻抗倒置网络 impedance inverting network
阻抗的 resistant
阻抗滴定 impedance titration;impedimetric titration
阻抗滴定池 impedimetric cell;impedimetry cell
阻抗滴定法 impedimetry
阻抗递变线路 impedance tapered line
阻抗电桥 impedance bridge
阻抗电容耦合 impedance-capacitance coupling
阻抗电压 impedance voltage
阻抗(电压)降 impedance drop
阻抗法 impedance method;mesh method
阻抗法校准 impedance method calibration
阻抗法勘探 impedance exploration
阻抗放大器 impedance amplifier
阻抗分量 impedance component
阻抗复合消声器 combination muffler

Z

阻抗管法 impedance tube method
阻抗归一化 impedance normalization
阻抗轨隙连接器 impedance bond
阻抗函数 impedance function
阻抗恒订单元 constant impedance unit
阻抗换置法 impedance transposition method
阻抗计 electric(al) impedance meter; impedance meter; impedimeter[impedometer]
阻抗继电器 distance relay; impedance relay
阻抗渐变线 impedance tapered line
阻抗角 angle of resistance; impedance angle
阻抗校正器 impedance corrector
阻抗校正网络 impedance correction network
阻抗接合 inductive bond
阻抗结合【电】impedance bond
阻抗静电计 impedance electrometer
阻抗矩阵 impedance matrix
阻抗类型比 impedance type analogy
阻抗力 resistivity
阻抗力矩 moment of resistance; resisting moment
阻抗联结变压器 auto-impedance bond; transformer impedance bond
阻抗联结变压器 impedance bond
阻抗联结器 impedance bond; auto-bond; inductive(track) bond; reactance bond
阻抗脉波计 impedance plethysmograph
阻抗模量 modulus of resistance
阻抗欧姆 impedance ohm
阻抗耦合 impedance coupling
阻抗耦合的 impedance coupled
阻抗耦合放大器 impedance coupled amplifier
阻抗匹配 impedance match(ing)
阻抗匹配变换器 impedance converter
阻抗匹配变压器 impedance matching transformer
阻抗匹配电抗线圈 impedance matching reactor
阻抗匹配短线 impedance matching stub
阻抗匹配负载箱 impedance matching load box
阻抗匹配线圈 impedance matching coil
阻抗频率 impedance frequency
阻抗平衡 impedance balance
阻抗起动器 impedance starter
阻抗器 impedor
阻抗千伏安培 impedance kilovolt-amperes
阻抗曲线 impedance curve
阻抗三角形 impedance triangle
阻抗失败衡测定装置 impedance unbalance measuring set
阻抗失衡 impedance unbalance
阻抗失衡测定器 impedance unbalance measuring set
阻抗失衡的 impedance unbalanced
阻抗失配 impedance mismatch(ing)
阻抗式消声器 impedance sound absorber
阻抗试验器 impedance tester
阻抗适配器 impedance adapter
阻抗损耗 impedance loss
阻抗探索 impedance exploration
阻抗特性 impedance characteristic
阻抗体积描记图 impedance plethysmogram
阻抗调配器 impedance tuner
阻抗调谐器 impedance tuner
阻抗图 impedance diagram; rheogram
阻抗图示仪 impedance plotter
阻抗稳定法 impedance stabilization method
阻抗稳定性 impedance stability
阻抗系数 coefficient of impedance; coefficient of resistance [resistancy]; impedance factor; resistance factor
阻抗显示装置 visual impedance meter
阻抗线圈 impedance coil
阻抗陷波器 impedance wavetrap
阻抗消声器 impedance muffler; impedance silencer
阻抗谐波器 rejector
阻抗性能 resistance property
阻抗压降 impedance voltage drop
阻抗压降试验 impedance-drop test
阻抗仪 impedance meter
阻抗因数 impedance factor
阻抗元件 impedor
阻抗圆 impedance circle
阻抗圆图 impedance chart; impedance circle diagram
阻抗值<一种测定基层强度的> resistance value
阻抗值试验法 resistance value method
阻抗转换器 impedance transducer
阻抗转接器 impedance adapter
阻抗作用 hindrance function
阻口拉索 choke cable
阻块 stop block
阻扩散层 diffusion barrier
阻扩散剂 diffusion barrier
阻拦式路缘石 barrier curb
阻力 counter force; drag (force); holding power; pullback; resistance (drag); resistance force; resistance pressure; resisting force; resisting power
阻力必要条件 rental requirement
阻力边 resistive side
阻力变弱 hollow of resistance
阻力参数 drag parameter
阻力测定 measure of resistance
阻力撑杆 drag strut
阻力传动器 friction(al) clutch
阻力单位 resistance unit; unit of resistance
阻力地块 resistant block
阻力动力 drag power
阻力法则 resistance law
阻力方向 drag direction
阻力费用 rental expenses
阻力分布 drag distribution
阻力分离 drag separation
阻力负载 resistance load; resistive load
阻力钢筋 resistance reinforcement
阻力功 work of resistance
阻力函数 resistance function; resisting function
阻力桁架 drag truss
阻力计算 drag calculation
阻力加速度 drag acceleration
阻力减摆器 drag dampening device
阻力铰 drag hinge
阻力矩 moment of resistance
阻力距 resistance moment
阻力孔 throttle opening; throttling orifice
阻力孔口式调压塔 restricted orifice surge tank
阻力孔式调压塔 restricted orifice surge tank
阻力轮式冷拉机 resisting pulleys type cold-drawing machine
阻力面积 drag area; rental area
阻力模型 drag model
阻力能 drag energy
阻力能高 resistance energy head
阻力能高的损耗<驼峰编组场> resistance loss
阻力平方区流动 fully rough flow
阻力区 restriction section
阻力曲线 curve of resistance; drag axis; drag curve; drag polar; resistance curve; resistivity curve
阻力扰动 drag perturbation
阻力矢量 drag vector
阻力试验 drag experiment; resistance test
阻力数据 drag data
阻力水头 resistance(water) head
阻力顺序 resistance sequence
阻力损失 resistance head; resistance loss
阻力特许权 rental concession
阻力体积描记法 impedance plethysmography
阻力体流量计 drag-body flowmeter
阻力天平 torsion force balance
阻力停车 resistance stopping
阻力图 drag diagram; resistance diagram
阻力系数 coefficient of drag; coefficient of resistance [resistancy]; damping coefficient; drag coefficient; drag factor; drag force coefficient; resistance coefficient; resistance factor
阻力线 line of resistance
阻力线原理 principle of resistance line
阻力因数 resistance factor
阻力增长 drag rise
阻力增长系数<打桩> refusal increasing factor
阻力增大 drag increment
阻力增加 argument of resistance
阻力增量 drag increment
阻力闸 drag brake
阻力张线 drag wire
阻力支柱 drag strut
阻力指示器 resistance indicator
阻力制动系统 drag brake system
阻力中心 center[centre] of drag; center[centre] of resistance; drag center[centre]; draught center[centre]
阻力轴 drag axis
阻力转矩 drag torque
阻力装置<桩试验> reaction system
阻力作用 retardation
阻裂 resistance to separation
阻裂材 crack arrestor
阻裂材防裂肘板 crack arrester[arrestor]
阻流 choked flow; choking up
阻流板 baffle board; spoiler
阻流电容器 blocking capacitor
阻流阀 choke valve; resistance valve; throttle valve
阻流环 block ring
阻流环密封 restrictive ring seal
阻流活门 throttle valve
阻流剂<焊接> stop-off agent
阻流脉冲 inhibit current pulse
阻流膜片 diaphragm baffle
阻流器 flow plug
阻流圈 choke coil; choking coil
阻流圈避雷器 choke coil lightning arrestor
阻流式透平 choked-flow turbine
阻流式涡轮机 choked-flow turbine
阻流箱 baffle box
阻流蓄水池<用以阻滞和储蓄地面径流> detention reservoir
阻留 detention
阻留雨量计 interceptometer
阻沫的 anti-foam
阻沫剂 anti-foam
阻纳 immittance
阻挠 baffle
阻挠式消声器 impedance sound absorber
阻尼 amortization; amortize; anti-vibration; attenuation; buffering; damp(en); damping; deboost
阻尼摆 damped pendulum
阻尼摆动 damped oscillation
阻尼摆动部分 damped periodic(al) element
阻尼比 damping ratio; ratio of damping; relative damping ratio; subsidence ratio
阻尼变压器 anti-hunt transformer; damping transformer
阻尼波 damped wave; damping wave; decadent wave; shrinking wave
阻尼不足 underdamping
阻尼材料 dampening material; damping material
阻尼参变数 damping parameter
阻尼测量 damping measure(ment)
阻尼层 damping course; damping layer
阻尼差 ballistic damping error; damping error
阻尼常量 damping constant
阻尼常数 attenuation constant; damping constant
阻尼场 recall field
阻尼(程)度 degree of damping
阻尼冲击检流计 damped ballistic galvanometer
阻尼磁体 damping magnet
阻尼磁铁 brake magnet; damped magnet; damping magnet
阻尼大气 sensible atmosphere
阻尼的 anti-hunt; damped
阻尼电感 damped inductance
阻尼电键 damping key
阻尼电流 damping current
阻尼电路 anti-hunt circuit; buffer circuit; damper circuit; damping circuit
阻尼电位 counter potential
阻尼电阻 buffer resistance; damped resistor; damping resistance; despiking resistance
阻尼电阻器 damping resistor
阻尼垫圈 damping bush
阻尼度 extent of damping
阻尼对数衰减率 damping log(arithmic) decrement
阻尼二极管 booster diode; efficiency diode
阻尼阀 damper valve; damping value
阻尼法 retardation method
阻尼分布 damping distribution
阻尼风分流器 damping air diverter
阻尼风机 damping air blower
阻尼辐射 braking radiation; damping radiation
阻尼固有(振动)频率 damped natural frequency
阻尼管 damper tube; damping tube; damping value
阻尼横摇 resisted rolling
阻尼环 balk ring; damping ring
阻尼缓冲 amortization
阻尼机 damping machine
阻尼机构 damping mechanism
阻尼继电器 brake relay; braking relay; dashpot relay
阻尼架 damping frame
阻尼假潮 damped seiche; damping seiche
阻尼减幅 damping decrement
阻尼减振 damping vibration attenuation
阻尼节流孔 damping orifice
阻尼结构 damped structure; damping construction; damping structure

阻尼介质 damping medium; resisting medium
阻尼矩阵 damping matrix
阻尼空气 damping air
阻尼孔 damping hole
阻尼控制 damping control
阻尼控制面 damping control surface
阻尼宽度 damping width
阻尼框 damper frame
阻尼力 damped force; damping force
阻尼力矩 damping moment
阻尼力矩系数 damping moment coefficient
阻尼力矢量 damping force vector
阻尼量 damped capacity
阻尼率 damping capacity; damping ratio; rate of damping; ratio of damping
阻尼脉冲 damping impulse; damping pulse
阻尼毛细管 damped capillary
阻尼媒质 resisting medium
阻尼模量 damping modulus
阻尼模型 damping model
阻尼能力 damping capacity; damping power
阻尼耦合 damping coupling
阻尼泡沫材料 damping foam material
阻尼片 damping fin
阻尼期 damping period
阻尼器 damp(en)er; damping apparatus; damping arrangement; damping device; absorber; amortisseur; annihilator; anti-hunt means; anti-vibration damp; anti-vibrator; attenuator; baffler; buffer; bumper; counterbuffer; dashpot; deadener; deoscillator; depressor; eliminator; shock absorber; shock damper; restrainer < 用于桥梁抗震等 >
阻尼器环 buffer ring
阻尼强度 damping strength
阻尼曲线 damping curve
阻尼绕组 amortisseur; amortisseur winding; damper winding; damping winding
阻尼设备 damping equipment; damping plant
阻尼时间 damping period; damping time; relaxation time
阻尼式测力计 absorption dynamometer
阻尼式调压室 throttled surge chamber
阻尼势指数 damped potential index
阻尼室 damping chamber
阻尼水面波动 < 水库湖内海的 > damped seiche
阻尼(速)率 damping rate
阻尼踏板 damper pedal
阻尼弹簧 buffer spring; damping spring
阻尼特性 damping characteristic
阻尼天平 damped balance; damping balance
阻尼调整 damping adjustment; damping control
阻尼调整器 damping control
阻尼涂料 damping coating
阻尼陀螺仪 rate gyro(scope); secondary gyroscope
阻尼网络 damping network
阻尼稳定式仪表 damped periodic(al) instrument
阻尼吸收 braking absorption
阻尼系数 coefficient of damping; coefficient of resistance [resistancy]; damp(er) coefficient; damping coefficient; damping factor; damping parameter; dashpot constant
阻尼系数值 damping coefficient value
阻尼系统 damped system
阻尼线 damping wire
阻尼线圈 choking winding; damper coil; damping coil; restraining coil
阻尼线匝 damping turn
阻尼效应 attenuative effect; damp(ing) effect
阻尼谐和运动 damped harmonic motion
阻尼性 damping quality
阻尼性质 damping property
阻尼翼 damping vane
阻尼因数 damped factor; damping coefficient; damping factor
阻尼影响系数 damping influence coefficient
阻尼油 damping oil
阻尼油罐 damping oil vessel
阻尼云母片 damping mica
阻尼运动 damp(ed) motion; damping motion
阻尼振荡 damped oscillation; damping oscillation; dying oscillation
阻尼振荡器 damped oscillator
阻尼振荡运动 damped oscillatory mode of motion
阻尼振荡周期 damping period
阻尼振动 damped vibration; damping vibration
阻尼振动荷载试验 damped shock load test
阻尼正交性 damping orthogonality
阻尼正弦量 damped sinusoidal quantity
阻尼正弦曲线 damped sinusoid
阻尼支点 damping fulcrum
阻尼值 damping value
阻尼致宽 broadening by damping; damping broadening
阻尼重量 damping weight
阻尼驻波 damped seiche
阻尼装置 damper gear; damping apparatus; damping arrangement; damping device
阻尼锥体 damping cone
阻尼阻抗 damped impedance
阻尼阻力 damping resistance
阻尼最小二乘法 damped least square method; damping least square method
阻尼作用 anti-hunt action; buffer action; buffer effect; damping action; damping effect
阻黏[粘]剂 anti-plastering agent
阻凝 anti-coagulation; anti-freeze
阻凝层 vapo(u)r barrier
阻凝剂 anti-coagulant; dispersion stabilizer
阻凝作用 anti-coagulant action
阻泡剂 foam inhibiting agent
阻(泡)沫 anti-foam
阻(泡)沫的 anti-foaming
阻频带 suppressed frequency band
阻气 throttling
阻气板 anti-aeration plate
阻气单向阀 choker check-valve
阻气阀 choke(r) valve; pneumatic valve
阻气阀操作按钮 choke button
阻气阀弹簧 choke valve spring
阻气阀托架 choke valve bracket
阻气阀轴 choke valve shaft
阻气管 choke tube
阻气活塞 balance piston; relief piston
阻气具 air trap
阻气门 choke(r)【机】; strangler
阻气门电缆 < 汽化器的 > strangler cable
阻气门控制 choke control
阻气排水器 steam trap
阻气喷嘴 choke nozzle
阻气器 air shutter
阻气塞门 choke; strangler
阻气室 choke chamber
阻汽排水器 steam trap
阻汽器 condensate trap; steam trap
阻屈器 buffer
阻燃 ABS 聚合物 flame-retarded ABS polymer
阻燃丙烯腈-丁二烯-苯乙烯聚合物 flame-retarded ABS polymer
阻燃材料 fire-proofing material; fire-retardant material; flame-resistant material
阻燃舱壁 fire-retarding bulkhead
阻燃层 fire-retardant coating
阻燃处理 fire-retarding treatment; flame-retardant treatment; fire-retardant treated lumber
阻燃处理木材 fire-retardant-treated wood; flame-retardant-treated wood
阻燃的 fire-resistant; fire-resisting; flame-retardant; slow-burning
阻燃地毯垫 flame-retardant carpet cushion
阻燃电缆 flame-retarding cable
阻燃顶板 fire-retardant ceiling
阻燃顶棚 fire-retardant ceiling
阻燃飞行服 fire-retardant suit flight
阻燃分隔 fire-retarding division
阻燃盖布 flame-resistant tarpaulin
阻燃隔离层 fire-retardant barrier
阻燃隔墙 flame-retardant partition
阻燃管道的绝热 fire-retardant duct insulation
阻燃化学 fire-retardant chemistry
阻燃化学品 fire-retardant chemicals
阻燃化学物质 fire-retardant chemicals
阻燃机理 fire-retardant mechanism
阻燃剂 fire-proofing agent; fire-retardant; fire-retardant agent; fire-retardant chemicals; fire-retarding chemical; flameproofing agent; flame resistant; flame-retardant chemical
阻燃剂处理木材 wood fire retardant treatment
阻燃剂压力浸渍 fire-retardant pressure impregnation
阻燃建筑材料 incombustible building material
阻燃结构 fire-retardant construction
阻燃浸渍纸 fire-retardant paper
阻燃聚合物 flame-retarding polymer
阻燃聚酯树脂 fire-retardant polyester resin
阻燃门 fire check door
阻燃木材 fire-retardant lumber; treated wood
阻燃木料 fire-retardant lumber; fire-retardant wood
阻燃黏[粘]合剂 fire-retardant adhesive
阻燃泡沫材料 fire-retardant foam
阻燃漆 fire-retardant paint; fire-retarding paint; flame-retardant paint
阻燃砂 fire-retardant sand
阻燃烧剂 flame retarder; flame-retardant
阻燃设施 flame retardant
阻燃试验 fire-retardancy test; fire-retardant test
阻燃树脂 flame-retardant resin; flame-retarded resin
阻燃天花板 fire-retardant ceiling
阻燃添加剂 fire-retardant additive
阻燃涂层 flame-retardant coating
阻燃涂层抛光 fire-retardant finish
阻燃涂料 fire-retardant coating; flame-retardant coating; flame-retardant paint
阻燃屋面覆盖层 fire-retardant roof cover
阻燃物 flame-retardant
阻燃纤维 fire-retardant fibre; flame-retardant fiber[fibre]
阻燃纤维板 flame-retardant
阻燃芯(体) fire-retardant core
阻燃型条板 fire-retardant lath
阻燃性 burning resistance; fire retardance [retardancy]; flame resistance; flame resistant; flame spread retardancy
阻燃性等级 flame-retardant grade
阻燃性树脂 fire-retardant resin
阻燃性塑料 flame-retardant plastics
阻燃性贴墙布 fire-retardant wall covering
阻燃性油毡 flame-retardant asphalt felt
阻燃性增塑剂 fire-retardant plasticizer
阻燃整理 flame checking
阻燃织物 flame-retardant fabric; fire foe < 商品名 >
阻燃装修 fire-retardant finish
阻燃作用 fire-retardation
阻抗价值 nuisance value
阻热传动性能 resistance to heat transmission
阻热墙板 thermal wallboard
阻容常数 resistance-capacitance constant
阻容电路 resistance-capacitance circuit; resistance-capacitance network; resistance-capacitive circuit
阻容分压器 resistance-capacitance divider
阻容滤波器 resistance-capacitance filter
阻容耦合 resistance-capacitance coupling
阻容耦合放大器 resistance-capacitance coupled amplifier
阻容时间常数 capacity resistance time constant; resistance capacitance time constant
阻容网络 resistance-capacitance network; resistance-capacitance set
阻容元件 resistor-capacitor unit
阻容振荡器 resistance-capacitance oscillator; resistor-capacitor oscillator
阻塞 blockage; blocking; block off; bottleneck; back-up; balk; ball up; barrage; barricade; choke up with; choking; clog(ging); close off; congestion; dam; damp; emcumbrance; encumber; hanging-up; jamming; logjam; obstruction; stem; stoppage; stop(ping); throttle
阻塞不前 logjam
阻塞程度型 congestion mode
阻塞的 blinding; choked; clogged; obstructed
阻塞的滤清器 loaded filter
阻塞的筛子 choked screen
阻塞的山谷 blocked up valley
阻塞的水沟 choked gull(e)y
阻塞点 sticking point; choke-point < 交通枢纽点 >
阻塞电容器 block-condenser; blocking capacitor
阻塞电压 blocking voltage
阻塞阀 choke
阻塞阀体 choke valve body
阻塞反气旋 blocking anticyclone
阻塞分派问题 bottleneck assignment problem
阻塞锋 clogging front
阻塞干扰 barrage jamming; blocking interference
阻塞高压 blocking anticyclone; bloc-

king high
阻塞管理系统 congestion management system
阻塞航道 block the fairway
阻塞河 ponded river;ponded stream
阻塞河槽 obstructed channel
阻塞河道 obstructed channel
阻塞河谷 blocked up valley
阻塞呼叫清除 blocked calls cleared
阻塞环 choke ring
阻塞极限 choking limit
阻塞继电器架 block relay frame
阻塞交通 congestion;jam
阻塞交通示威 lie-down
阻塞警报 <使接近警报器装置地点的列车自动停车> fouling alarm
阻塞控制的旋钮 choke control knob
阻塞控制模式 jam mode
阻塞力 stopping power
阻塞了的管线 clogged line
阻塞了的滤池 clogged filter
阻塞了的筛网 clogged screen
阻塞流 blocking flow
阻塞滤池 clogged filter
阻塞脉冲 disabling pulse
阻塞门 choke
阻塞密度 jammed density
阻塞喷注 choked jet
阻塞气流 choked stream
阻塞气蚀 chocking cavitation
阻塞强光 disability glare
阻塞时间 blocking time
阻塞衰减 blocking attenuation
阻塞损失 loss due to obstruction
阻塞探测装置 jam detection device
阻塞套 feed cutoff
阻塞网络 blocking network
阻塞物 choker;constriction;stopper
阻塞现象 <风洞的> choking phenomenon
阻塞线 choke line
阻塞信号 block(ing) signal
阻塞性空洞 obstructed cavity
阻塞性空化 choking cavitation
阻塞性空蚀 choking cavitation
阻塞性汽蚀 choking cavitation
阻塞预防【交】congestion prevention
阻塞振荡 blocking-oscillator driver
阻塞振荡器 blocking oscillator
阻塞蒸汽设备 steam traps
阻塞指数 congestion index
阻塞装置 blocking device;choking device
阻塞状态 blocking state
阻塞阻力 choking resistance
阻塞作用 blocking action;choking action
阻沙措施 measure of obstructing sand
阻沙设施 sand prevention device
阻渗板桩 water-tight screen
阻声屏 sound baffle
阻蚀剂 corrosion inhibiter[inhibitor]
阻蚀添加剂 corrosion inhibiting admixture
阻视距离 blind distance
阻视曲线 blind curve
阻水板桩 cut-off piling
阻水材料 moisture barrier material
阻水槽 water check groove
阻水层 aquiclude;confining bed;confining layer;confining stratum;impermeable diaphragm; water-tight diaphragm
阻水窗框 water checked casement
阻水地层 confining stratum
阻水断层 water-resisting fault
阻水构造 water-resisting structure
阻水环 flashing ring
阻水活栓 water check
阻水接缝 water-tight joint
阻水面层 water-tight facing
阻水面积 current-obstruction area
阻水片 water stop
阻水墙 diaphragm wall
阻水帷幕 water-tight screen
阻水小坝 astyllen
阻水岩脉 water-resisting rock vein
阻水堰 stop-log weir
阻水栅栏 water bar
阻息 damp out
阻性负载 resistive load
阻性消声器 resistance silencer;resistive muffler
阻锈剂 anti-corrosion admixture;anti-corrosive agent;corroding inhibitor;corrosion inhibiting admixture;corrosion inhibitor;rust inhibitor
阻盐岩石(柱) salt barrier pillar
阻堰装置 flame trap
阻焰器 flame arrester[arrestor]
阻焰性 flame resistivity
阻摇联结系 sway bracing
阻摇支撑 sway bracing
阻摇支撑系统 sway bracing
阻曳理论 <根据板底摩擦力设计混凝土路面钢筋的理论> drag theory
阻抑 damp;dampen;depression
阻抑常数 inhibition constant
阻抑动力学分光光度法 inhibitory kinetic spectrophotometry
阻抑机理 inhibition mechanism
阻抑剂 co-inhibitor;inhibiting agent;inhibitor
阻抑卤化 obstructive halogenation
阻抑水跃 check jump
阻抑素 impedin
阻抑现象 inhibition phenomenon
阻音屏 noise barrier
阻油圈 chip washer;oil interceptor
阻越流系数 anti-leakage factor
阻障产生的水头损失 loss of head due to obstructions
阻障隐蔽 barrier shielding
阻值允许误差 resistance tolerance
阻止 arrestment;blockade;choke;deprive; deter; estop; estoppage; forbid; foreclose; hold-back; hold in check; impede; inhibition; resist; stop;stopping off
阻止本领 linear energy transfer;stopping power
阻止层 trapping layer
阻止当量 stopping equivalent
阻止的 disincentive
阻止对方呼叫 bar for originating call
阻止阀 interceptor valve
阻止杆 arresting lever
阻止开裂 stop a crack
阻止能力 linear stopping power
阻止物 preventer;stayer
阻止楔 lock wedge
阻止钥匙 stop key
阻止运动 retardation motion
阻滞 blocking; drag; hinder; inhibition;retardation;stoppage
阻滞表面 detention surface
阻滞沉降 hindered settling
阻滞沉降率 hindered-settling ratio
阻滞沉落 hindered settling
阻滞电极 retarded electrode
阻滞反应 retarded reaction
阻滞干扰 friction(al) interference
阻滞剂 blocking agent;paralyzer[paralyser]; retardant; retarder; retarding agent
阻滞交通 delays to traffic
阻滞阶段 retention stage
阻滞期 retention period
阻滞溶剂 retarder solvent
阻滞筛 retaining screen
阻滞时间 residence time; retention period;retention time
阻滞式链板运输机 retarding chain conveyer[conveyor]
阻滞式圆盘运输机 retarding disk conveyer[conveyor]
阻滞水 retained water
阻滞水流 retarded flow
阻滞素 retardin
阻滞酸液 retarded acid
阻滞弹簧 retarder spring
阻滞弹性 retarded elasticity
阻滞稀释剂 retarder thinner
阻滞系数 coefficient of retardation; impedance factor;retardance coefficient;retardation of discharge;retarding coefficient;retention coefficient
阻滞系统 retarding system
阻滞相 retardance phase
阻滞相位 retention phase
阻滞效应 retarding effect
阻滞学说 block theory
阻滞因数 retardation factor
阻滞作用 encroachment;retarding action

组

组 bank; brigade; cluster; gang; group; multitude; population; squad;squadron;formation【地】

组氨酸 histidine
组壁柱 grouped pilasters
组标记 group mark
组标志 field mark;group mark
组补充 group compensation
组长度 field length
组沉淀 group precipitation
组沉淀剂 group precipitant
组成 building-up;compo;constituent;formation;forming;make-up
组成百分比 percentage composition
组成比 ratio of components
组成变化曲线 composition history curve
组成波 composition wave
组成部分 component(parts);constituent;ingredient;integral part;integrant;moiety;structural parts
组成部件 constituent element
组成成分分析 component analysis
组成吹管 divided blast pipe
组成代谢 anabolic metabolism;anabolism;constructive metabolism
组成代谢物质 anabolic substance
组成的 built-up;compositive;constitutional;unitized
组成的连接 sectioning link
组成的塞绳 sectioning link
组成的统一性 unity of composition
组成分布 composition profile
组成分类 textolite textural classification;textural classification
组成格构式的 cancellated
组成工形柱 built H-column
组成过程 anabolic process
组成滑车 made block
组成俱乐部 club
组成肋 built rib
组成肋材 built rib
组成列 composition series
组成三角形 composition triangle;triangulate
组成砂 synthetic(al) sand
组成式 constitutional formula
组成水分 essential water; water of composition
组成投标文件 documents coupling the bid
组成图 composition diagram;constitution(al) diagram
组成网格状 composition grid-form
组成未明的产物 unidentified product
组成屋面料 ready roofing
组成物 composition;constituent;construct
组成要素 component element
组成一队 team
组成一体的 all-in-one piece
组成音 component tones
组成元素 elementary composition
组成整体式的 all-in-one piece
组成作用 composite action
组处理机 set processor
组传送 group propagate
组丛法 <分析沥青用> group method
组锉 key file;key files set
组灯 <航标> group light
组地址 group address
组段 class range
组对 assembly
组对公差 alignment tolerance
组对检查 fitting-up inspection
组反应 group reaction
组分离 group separation
组分 component;constituent;ingredient
组分采样 constituent sampling
组分成熟度 compositional maturity
组分法 <分析沥青用> group method
组分反应 component reaction; constituent reaction
组分分析 analysis component;proximate analysis
组分隔符 group character;group separator
组分隔符号 group separator character
组分过冷 constitutional supercooling
组分恒定 composition constant;constant of composition
组分活度 activity of component
组分可溶性参数 component solubility parameter
组分浓度 component concentration
组分评价 component evaluation
组分梯度 composition gradient
组分物质定律 component substances law
组分物种 component species
组分性质 constitutive property
组分逸度 fugacity of component
组分元素 constituent element
组分运动 componential movement
组分转变 constitutional change
组符 group mark
组符号 group code
组阁 cabinet making
组格式 cellular system
组格式板系船墩 cellular sheet pile dolphin
组格式板桩码头 cellular sheet pile wharf
组格式沉箱 cellular caisson
组格式货柜 <集装箱货船> cellular container vessel
组格式双层底 cellular double bottom
组构【地】 fabric; petrofabric; rock fabric;structural fabric
组构的对称性 symmetry of fabric
组构分析 fabric analysis
组构各向异性 fabric anisotropy
组构数据 fabric data
组构图 fabric diagram;petrofabric diagram
组构线理 fabric lineation
组构要素 fabric element
组构要素统计图 statistic(al) diagram of fabric elements
组构域 fabric domain

组构轴 reference axis
组号 group indication;group number
组合 assembling; assembly; association; bank; build-up; combination; configuration; consociation; corporation; grouping; interlace; merging;mixed rags;unit block
组合安装 block mount
组合扳手 combination wrench
组合板 component panel; composite board; composite plate; composite slab; composition board; composition panel;composite panel
组合板材 built-up board
组合板大梁 built-up plate girder; flitched(plate)girder;flitch girder
组合板梁 built-up plate girder; flitched beam;sandwich beam
组合板梁加强钢板 flitch plate
组合板牙平板梳刀 Namco chaser
组合包装 combination packaging
组合保持继电器 retainer neutral polarized combination relay
组合保持架 composite cage
组合保险丝 company fuse
组合爆破 pattern shooting
组合爆破孔 multiple shothole
组合爆炸 pattern shooting
组合泵 combination pump; Sundstrand pump;unipump;unit pump
组合比率 component ratio
组合臂架 double link jib
组合编码 combination code
组合表 combinative table; table of combination
组合波衰减 attenuation of combination
组合薄膜 built-up membrane
组合部分 built-up section
组合部件 <建筑框架> subassemblage;unit construction
组合材 composite wood
组合材料 combination of materials; combined material; composite material;composition material
组合材料瓦 composition shingle
组合采暖系统 integrated heating system
组合采样 composite sampling
组合舱壁 composite bulkhead
组合舱盖板 built-up hatch board
组合操作 combination operation
组合槽钢 <用钢板及角钢拼接而成> built-up channel
组合侧铣刀 mesh side cutter
组合测角器 combination angle ga(u)ge
组合插座 plate connector
组合潮 compound tide
组合车 combination vehicle; composite car
组合车床 combined lathe
组合车辆 combination car
组合车轮 built-up wheel
组合车身 unitized body
组合衬砌 composite lining
组合程序 assembly program(me); built-up sequence; packaged program(me)
组合齿轮 built-up gear; combination gear;compound gear;fraction gear
组合齿轮铣刀 duplex gear cutter
组合冲击式涡轮机 combined-impulse turbine
组合传动 group drive;series drive
组合传送带系统 composite belt system
组合船 built-up boat; fabricated vessel
组合船队 composite ship; dismountable ship
组合船桥系统 integrated bridge system
组合船首材 plate stem
组合船尾肋骨 fabricated stern frame
组合窗 combination window; composite window; composition window;fenestrato
组合错齿槽铣刀 interlocking milling cutter
组合大梁 built-up girder; composite girder; sandwiched girder; compound girder
组合带 assemblage zone
组合带卷 built-up coil;composite coil
组合单元 composite unit; compound unit;unitized unit
组合刀架 combination tool block
组合刀具 box tool; combined tool; gang head;rake tool
组合刀锯 dado head saw
组合刀头 dado head
组合导管 associated conductor
组合导轨 composite guideway
组合导航 integrated navigation
组合导航及避碰系统 integrated navigation and collision avoidance system
组合导航计算机 integrated navigation computer
组合导航设备 integrated navigation equipment
组合导航系统 integrated navigation system
组合导航与自动驾驶系统 integrated navigation and autopilot system
组合导火线 company fuse
组合导体 associated conductor
组合导线 associated conductor
组合的 built-up; combinative; combinatorial; combined; composite; resultant;unitized
组合的厨房设备 combined kitchen equipment
组合的电磁阀及座 magnet bracket
组合的拉比兹钢丝网布和芦苇板条 combined Rabitz type wire cloth and reed lath(ing)
组合等温处理 combined isothermal treatment
组合等温淬火 combined austempering
组合地板 composite floor(board); composition floor; composition floor cover(ing)
组合地板铺装工 composition block layer;composition floor layer
组合地板砌块 composition block floor
组合地面砖 composite floor cover(ing)tile
组合地震反应谱 combined earthquake response spectrum
组合地震计 multiple seismometer
组合地震检波器 multiple geophone
组合地震系数 combined earthquake coefficient
组合地砖 cleaving tile
组合电极 combination electrode
组合电集尘器 combined electric(al) dust collector; combined electrofilter
组合电缆 <通信[讯]用电缆,一般由100对线组成> unit cable; combination cable
组合电流 combination current
组合电路 combination(al) circuit; combinatorial circuit;combiner circuit;combining circuit
组合电容器 multiple anode unit capacitor
组合电子导航系统 composite electronic navigational system
组合电阻连接法 combined resistance connection
组合电阻器 composition resistor
组合垫板 flitch plate
组合垫片 compound gasket
组合吊货钩 union hook
组合吊综装置 pressure harness
组合顶梁 <金属支架的> linked roof bar
组合顶推驳船 integrated tug-barge
组合定位 integrated positioning
组合端子 terminals of a unit block
组合断层 multiple faults
组合锻模 split-die
组合堆石坝 composite rockfill dam
组合对偶 combinational dual
组合墩 compound pillar
组合舵柄 sectional tiller
组合舵架 built-up rudder frame
组合发动机 combination engine
组合阀 combination valve; compound valve;group valve;make-up valve; multitandem valve
组合阀体 valve module
组合法 built-up method; combination method;method of combination
组合反射镜 facetted mirrors
组合方法 combined method
组合方式 compound mode
组合房间 combined rooms
组合放大器 group amplifier; unit amplifier
组合飞轮 sectional flywheel
组合分解 combination decomposition
组合分配灭火系统 combined distribution fire-extinguishing system
组合分析 combinational analysis; combinatory analysis; composite analysis
组合风阀 combined damper
组合蜂窝板 composite honey-comb board
组合扶手 combination railing
组合浮标 combination buoy
组合概率 combinatorial probability; combined probability; compound probability
组合杆 built mast; compound mast; compound rod
组合杆件 built-up member; composite member
组合杆子 hollow spar
组合刚度矩阵 combined stiffness matrix
组合刚性构架 built-up rigid frame
组合钢架 composite rigid frame
组合钢梁 compound girder
组合钢梁结构 compound steel beam structure
组合钢梁桥 composite steel beam bridge
组合钢柱 compound stanchion
组合杠杆式可调比较仪 Mikrotast
组合格构 built-up lattice
组合工具 tool pad
组合工况 combination operating condition
组合工形柱 built H-column
组合工艺学 group technology
组合工字柱 built-up H-column
组合弓形拱 built-up segmental arch
组合攻丝机 tapping unit
组合拱 built-up arch;compound arch
组合构架 built-up frame; composite member;segmental member
组合构件 build-up member; combined member;segmental member
组合购买 group purchase
组合骨料 combined aggregate
组合关系 cam relationship; constitutive relation;on cam relationship
组合管(道)<变断面的> compound tube; composite tubing; compound pipe[piping]
组合管接头 junction block
组合管口 bicomponent nozzle
组合广义刚度 combined generalized stiffness
组合规 combination ga(u)ge
组合规则 rule of combination
组合滚珠轴承 compound ball bearing
组合涵洞砌块 built-up culvert block
组合焊缝 composite weld
组合焊接型材 fabricated welded shape
组合焊柱 built spar
组合和 combinatorial sum
组合荷载 combination load;combined load(ing);resulting load
组合桁架 built-up truss; combination truss; composite frame; composite truss;compound truss
组合桁架梁 flitch-trussed beam
组合横断面 built-up cross-section
组合厚板 flitch chunk
组合弧线 composite curve
组合弧形木楼梯梁 built-up string
组合滑车 built block; compound pulley;made block
组合滑动面 composite sliding surface
组合化学 combinatorial chemistry; comichem
组合灰胶纸柏板 composite plasterboard
组合回声测距系统 combined echo ranging echo sounding
组合混凝土骨料 combined concrete aggregate
组合混凝土挠曲结构 composite concrete flexural construction
组合混凝土受弯构造 composite concrete flexural construction
组合混凝土弯曲构件 composite concrete flexural member
组合活塞 assembled piston; built-up piston
组合机床 aggregate machine-tool; standard unit type machine
组合机床的动力头 screw unit
组合机床的组件 modular machine components
组合机结构 unit machine construction
组合基距 array length
组合激发炮数 number of shorts in array
组合集料 combined aggregate
组合计算机辅助环境评价法 combination computer aiding method of environmental impact assessment
组合技术 build-up technique
组合继电器 combination relay; neutral polarized relay; neutral-polar relay;polarised relay
组合继电器接点 polarized relay contact
组合加工 unitized tooling
组合加工步骤 assembly processing step
组合加工法 group method
组合加工中心机床 machining centers
组合夹板 combined bridge
组合夹具 built-up jig; combined clamp
组合家具 combination furniture; modular furniture
组合价格 package price
组合驾驶台系统 integrated bridge system
组合架 combination beam; compound beam;unit block assembly rack
组合剪力 combined shear

组合检波器数 array geophone number
组合件 composite member; molectron; package(unit); subassembly
组合建造船 composite-built vessel
组合建筑 component building; composite building
组合建筑板 composite building sheet
组合建筑材料 combined building materials
组合建筑单元 composite building unit
组合交叉 combination crossing
组合胶带运输系统 integrated belt system
组合焦距 combined focal length
组合角尺 combination set; combination square
组合角度深度尺 combination angle depth ga(u)ge
组合角条件 condition for sum of angles
组合铰刀 gang reamer
组合接头 built-up joint
组合接线装置 combination tap assembly
组合节点 grouped nodes
组合结构 built-up construction; built-up section; combination construction; composite construction; composite structure; hybrid structure <钢与混凝土,混凝土与木等>
组合结构桥 combination bridge
组合结构系统 unit construction system
组合结构原理 unit construction principle
组合截面 built-up section; composite section; compound section; fabricated section <铆接式焊接>
组合进站信号示像 combination home signal aspect
组合晶体分辨率 resolution of composite crystal
组合晶体体积 composite crystal volume
组合井 <各高程直径不同> compound well; combination well; gang of wells
组合矩尺 combination square
组合剧增 combinatorial explosion
组合锯 combination saw
组合锯条 nest of saws
组合均化库 combination silo
组合卡片 composite card; package card
组合开关 combined switch; pocket type switch; small control switch; switch group; unit switch
组合开关板 cubicle switchboard
组合开关电路 combinational switch circuit
组合开关箱 combination switch board
组合开关装置 cubicle switchgear
组合壳 sectional casing
组合刻度 combination scale
组合空气调节器 air handler
组合空心模架 built-up pattern
组合控钮式气扳机 combination torque and control wrench
组合控制器 cluster controller
组合控制装置 grouped controls
组合矿物 associated mineral
组合框架 combination frame
组合框架结构 composite framed structure
组合框缘 built-up architrave
组合拉刀 built-up broach; combination broach; combined broach
组合栏杆 combination railing
组合缆 combine rope

组合雷管 composition caps
组合肋 built-up rib
组合肋板 bracket floor; open floor; skeleton floor
组合肋骨 built-up frame; compound frame
组合棱镜 component prism
组合理论【数】combination theory; combinatorial theory
组合力 combining ability
组合力矩 combined moment
组合立管 <供消防和生活用水> combination standpipe
组合沥青屋面覆盖层 built-up asphalt roof covering
组合连杆 jointed connecting rod
组合连接 built-up connection
组合连接器 combination sleeve; unit connector
组合连轴器 combination sleeve
组合联结 composite joint
组合链系 compound chain
组合梁 built-up beam; combination beam; composite beam; composite girder; compound beam; hybrid beam; integrate beam; integrated beam; keyed girder; lenticular beam; lentiform beam; split beam
组合梁夹铁 <夹在组合板梁中起加强作用的钢板> flitch plate
组合梁桥 composite beam bridge; composite girder bridge
组合梁式斜拉桥 composite deck cable stayed bridge
组合梁效应 beam building effect
组合量板 fitch
组合量规 built-up ga(u)ge; combination ga(u)ge
组合量角规 combination bevel
组合列车 combination train; combined train
组合临时支撑 <隧道开挖钢拱肋和钢或木制背板的> rib and lagging
组合檩条 built-up purlin(e)
组合灵敏度 array sensitivity
组合零件 assembly parts; component parts; multiple component units
组合楼板 compofloor; composite floor (board); composite floor slab; compound floor
组合楼板装饰 composition flooring finish
组合楼梯斜梁 built-up stair horse; built-up string
组合芦苇板 composite reed board
组合炉灶 combination range
组合滤波器 composite filter
组合滤气单向阀 combined strainer and check-valve
组合滤气止回阀 combined strainer and check-valve
组合滤色片 filter pack
组合轮 combination wheel
组合轮毂 sectional hub
组合轮胎 combination tire[tyre]
组合轮辋 built-up rim
组合轮辋式盘轮 divided type disk wheel
组合轮缘 built-up rim; divided rim
组合轮转子 built-up disk rotor
组合论 combinatorics
组合逻辑 combinatory logic
组合逻辑元件 combinational logic(al) element
组合螺栓 assembling bolt; assembly bolt
组合螺旋桨 built-up propeller
组合螺旋桨轴 divided propeller shaft
组合落料模 gang blanking die
组合门 composite door

组合门窗框缘 built-up architrave
组合门窗头线条板 built-up architrave
组合门框 combined frame
组合面 plane of composition
组合面积 built-up area
组合命令 merge command
组合模 built-up mo(u)ld; built-up pattern; segmental die; segment mold; split-die
组合模板 built-up form; built-up plate; gang forming; gang form(work); insert pattern; panel form; segmental pattern plate
组合模板施工 ganged form construction
组合模衬 split-die liner
组合模具 assembling die; collapsible die; combination die; gang die; multicomponent die; segmental mo(u)ld; segment die; single split mo(u)ld; split(-ring) mo(u)ld; split-segment die; splitting die
组合模模块 die segment
组合模式 integrated mode
组合模型 composite model; constituent model; split pattern
组合模样 sectional pattern
组合磨料 compound abrasive
组合母线 group bus
组合木板桩 Wakefield pile
组合木材 built-up timber; pieced timber
组合木构件 built-up timber
组合木料 built-up timber
组合内底的结构 unishank construction
组合内底涂胶机 unishank cementing machine
组合内底制作工艺 unishank process
组合耙 jointed harrow
组合排气管 group vent
组合排水系统 composite drainage system
组合盘形离合器 combined disk[disc] clutch
组合刨 combination plane
组合配线架 combination distributing frame; combined distribution frame
组合皮带轮 split pulley
组合片簧 compounded plate spring
组合频率 combination frequency
组合频率制 grouped-frequency operation
组合品 assemblage
组合谱带 combination band
组合谱线 combination line
组合企口板桩 Wakefield sheet(ing) pile
组合企口木板桩 Wakefield piling
组合企业 associated enterprise
组合砌块 bond beam block
组合器 combiner; combiner tree
组合器轴【机】driving shaft
组合钎子 jointed rod
组合嵌岩桩 pile-socketed and anchored in rock
组合墙 composite wall(ing); compound partition(wall); compound wall(ing)
组合墙体的加固 <配平行钢筋> composite wall reinforcing[parallel type]
组合桥(梁) <钢梁与混凝土板的> composite bridge; hybrid bridge
组合桥面板 composite bridge deck(ing)
组合桥台 compound abutment
组合曲柄 built-up crank
组合曲柄心棒 built-up crank axle

组合曲柄轴 built-up crank axle; built-up crank shaft
组合曲尺 combination square
组合曲面侧面刨床 combined curve-and-side planing machine
组合曲轴 unit-type crank shaft
组合取样器 composite sampler
组合圈闭 combination trap
组合全球导航与监视系统 integrated global navigation and surveillance system
组合燃料 composite fuel; composite propellant
组合沙发 modular sofa; sectional sofa
组合砂箱 built-up flask; built-up mo(u)lding box
组合筛 nest of screens
组合设备 array device; built-up apparatus; combination unit; composite set; unit equipment
组合设备的特殊标志 special stamping of combination units
组合设备项目 combination equipment item
组合设计 modular design; unit design; unitized design
组合深度角度规 combination depth-and-angle ga(u)ge
组合时间 assembly time
组合式 combined style; combined type; segmental type
组合式凹模 combination die; combined die
组合式坝 integrated structure dam
组合式板桩 sheet pile
组合式驳船 integrated barge
组合式驳船队 integrated fleet; integrated tows
组合式测量仪表 combination ga(u)ge
组合式车轮 multipiece wheel
组合式车体 composite car-body
组合式衬套 split bushing
组合式船队 composite craft
组合式电气集中联锁 unit block type all-relay interlocking
组合式吊具 composite spreader
组合式钝角辙叉 built-up obtuse crossing
组合式耳环 <三部分组成的> three-piece bail
组合式发动机 composite engine
组合式防波堤 composite breakwater
组合式房屋 modular house
组合式浮船坞 integrated floating dock; pontoon dock
组合式腹板 compound web plate
组合式钢梁桥 composite steel beam bridge
组合式隔断 combination space divide
组合式锅炉 combined boiler; sectional boiler
组合式过滤器 combined filter
组合式毫伏安计 onemeter
组合式桁架桥 composite truss bridge
组合式换乘 compound transfer
组合式换热器 unitized exchanger
组合式混合计算机 combined hybrid computer
组合式混凝沉淀池 combination coagulation
组合式混凝土挠曲结构 composite concrete flexural construction
组合式混凝土受弯构件 composite concrete flexural member
组合式活塞 built-up piston
组合式活塞环 compound piston ring; multipiece piston ring
组合式集装箱船 composite container ship
组合式继电集中联锁 panel type relay

interlocking
组合式继电联锁 panel type relay interlocking
组合式加热器 heater battery
组合式家具 sectional furniture
组合式绞吸挖泥船 built-up cutter suction dredge(r); dismountable cutter suction dredge(r)
组合式搅拌器 multimixer
组合式接插件 modular connector
组合式洁净室 partition clean room
组合式结构 built-up section; modular construction; section construction
组合式金属模板 sectionalized steel form
组合式绝热材料 sectional insulation
组合式开关 packet type switch
组合式开孔钻头 combination spudding bit
组合式壳体 split housing
组合式空调机 air-handling unit
组合式空调器 packaged air conditioner
组合式空气过滤器 unit-type air filter
组合式空气调节机组 modular air handling unit
组合式空气调(节)器 combined air handling units; self-contained air conditioner; self-contained air conditioning unit
组合式缆索体系 combined cable system
组合式冷凝器 direct contact condenser; multishell condenser
组合式冷却器 hybrid cooler
组合式立管 <供消防和生活用水> combination standpipe
组合式立交 combination interchange
组合式连接件 modular connector
组合式链轮 split sprocket
组合式流量调节阀 combination flow regulator
组合式楼梯 combination stair(case)
组合式轮辋 multipiece wheel
组合式螺母扳手 multiple nutrunner
组合式落料模 composite section blanking die
组合式门 combination door
组合式模 combination die; combined die
组合式模具 sectional die
组合式模具的构成元件 composite die sections
组合式木梁 Clarke beam
组合式拧螺母 multiple nutrunning
组合式浓度流量指示器 integrated concentration and flow indicator
组合式喷管 segment nozzle
组合式喷烧工具 combination burner tool
组合式喷射器 unit injector
组合式气箱 built-up casing
组合式汽水分离器 combination trap
组合式汽水疏水器 combination trap
组合式桥梁 combination bridge
组合式桥台 composite abutment
组合式曲柄 built-up crank; divided crank
组合式曲轴 build-up crankshaft; combination type crank shaft
组合式驱动链轮 segmented drive sprocket
组合式设计 modular design
组合式十进制 packed decimal system
组合式十进制数格式 packed decimal number format
组合式手纸盒 combination toilet tissue cabinet
组合式水柜与炉灶单元 combination sink cabinet and cooker
组合式水冷壁 block water wall
组合式司机室 modular cab
组合式梯 sectional ladder
组合式调速器 compounded governor
组合式土坝 composite earth(en) dam
组合式土工织物 composite geotextile
组合式土石坝 composite rockfill dam
组合式推力车 push section car
组合式托盘 built-up pallet
组合式拖拉机 sectionalized tractor
组合式挖沟机 integral type trencher
组合式挖掘铲 divided shovel; split-digger blade
组合式挖掘机 divided shovel
组合式挖泥船 built-up dredge(r); dismountable dredge(r)
组合式无缝地板 petropine
组合式铣刀 block type cutter
组合式楔密封 closure with multiple wedge gasket; seal by wedge-like gasket and O-ring
组合式斜拉桥 composite cable-stayed bridge
组合式信号机构 modular design; unit-type signal mechanism
组合式悬索桥 composite suspension bridge
组合式旋风分离器 multiple cyclone
组合式压路机 combination roller; roller combination roller
组合式压实机 combination compactor
组合式烟囱 concentric(al) stack
组合式预应力钢筋混凝土大梁 composite prestressed concrete girder
组合式预应力钢筋混凝土梁 composite prestressed concrete beam
组合式圆盘 comprising disc
组合式再热蒸汽阀 combined reheat valve
组合式轧辊 composite backup roll; sleeved backup roll
组合式轧辊的辊轴 roll mandrel
组合式轧机 combination mill
组合式辙叉 assembled frog; built-up crossing
组合式制止器 combination stopper
组合式主轴 fabricated shaft
组合式住宅 compound type apartment house
组合式助铲机 push section car
组合式柱子 combination column
组合式铸铁锅炉 sectional cast-iron boiler
组合式转轮 fabricated runner
组合式转台六角车床 combination turret lathe
组合式转辙器 composite switch
组合式自动操纵 group automatic operation
组合式纵向通风 composite longitudinal ventilation
组合式钻车 combination jumbo
组合试样 composite sample
组合(室内)开关 cubicle switch
组合室容器 combination chamber vessel
组合手柄 combination handle
组合受弯构件 composite flexural member
组合受压构件 built-up compression member; composite compressive member
组合数据库 combined data base
组合数学 combinatorial mathematics; combinatorics
组合双工 combination duplex
组合水枪 array water gun
组合顺序 built-up sequence; sequence of combination
组合丝锥 sectional tap
组合算法 combinational algorithm; combinatorial algorithm
组合榫 gang grooved
组合索引 combined index
组合锁 combination padlock
组合塔式六角车床 combination turret lathe
组合台 compound table
组合台阶 bench group
组合太阳望远镜 spar
组合弹簧 cluster spring
组合探针 combination probe
组合套管柱 combined string of casing
组合套筒 combination socket
组合特性 combined characteristic
组合特性曲线 combination characteristic curve
组合体 aggregate
组合体系 combined system
组合体系拱桥 combined system arch bridge
组合体系桥 built-up bridge
组合天花板电线匣 rosette
组合条作动弹簧 combination-bar operating spring
组合贴板 flitch
组合通放带宽度 array pass-band width
组合头 combination head; gang head
组合头钻床 unit head machine
组合投标 package bid
组合投影 combined projection; composite projection; condensed projection
组合透镜 compound lens
组合透气管 group vent
组合凸轮 compound cam
组合突缘 built-up flange
组合图 constitutional diagram
组合图像信号 composite picture signal
组合图形 composite figure
组合推进器 detachable propeller
组合推力径向轴承 combined thrust and radial bearing
组合拓扑学【数】combinatorial topology; piecewise linear topology
组合万能机床 combined universal machine tool
组合网 combinational network
组合网架 composite space truss
组合围堰 composite cofferdam
组合桅杆 built-up mast; made mast
组合卫生设备 combination fixture
组合文件 combined file
组合问题 combinatorial problem
组合圬工墙 composite masonry wall
组合屋顶桁架 composite roof truss
组合屋桁架 composite roof truss
组合屋面 composite roof(ing); composition roof(ing); gravel roof(ing)
组合屋面板 composition shingle
组合屋面材料 built-up roof(ing); composition roofing material
组合屋面材料制造设备 installation for manufacturing prepared roofing
组合屋面层 built-up roof covering; built-up roof(ing)
组合屋面的生产 composition roofing manufacture
组合屋面的制造 composition roofing manufacture
组合屋面盖片 asphalt shingle
组合屋面木片瓦 composition roofing shingle
组合物 composition
组合吸收器 unit absorber
组合洗涤设备 combination fixture
组合铣床 combination-type miller
组合铣刀 built-up cutter; cutter block; gang cutter; gang mill(ing) (cutter); group milling cutter
组合系列 combination range
组合系数 combination coefficient; composite coefficient
组合系统 combinatorial system; group system
组合弦杆 composite chord; packed chord
组合线路 combinational circuit; combination circuit; combinatorial circuit
组合相关系数 coefficient of interclass correlation
组合箱梁 composite box beam
组合箱形梁 composite box beam
组合项 group item
组合效应 array effect
组合斜角规 combination bevel
组合芯盒 combined core box; multiple-core box
组合芯件 composition core
组合信号示像 combination signal aspect
组合行列 combination range
组合形变热处理 combined thermomechanical treatment
组合形式 array pattern; composite style
组合形状 composite shape
组合型半潜式平台 composite semisubmersible platform
组合型钢(截面) built-up steel section
组合型填料 combined packing
组合型土工织物 composite geotextile
组合型微控制器 component type microcontroller
组合性质 combinatorial property
组合选择 combinational selection
组合选择器开关 unit-selector switch
组合循环 composed cycle
组合压杆 built-up compression member
组合压力杆 built-up compression member
组合压力计 multiple manometer
组合压铸模 built-up mo(u)ld; unit die
组合压铸型 unit die
组合烟囱 compound chimney
组合扬声器 columnar speaker
组合样品 combinated sample; compound sample
组合仪表 cluster ga(u)ge
组合音 combination tone
组合音拍 resulting beat
组合应力 combined stress
组合油封 combination oil seal
组合油气田 oil-gas field complex
组合有效余氯 combined available residual chlorine
组合遇险报警网络 combined distress alerting network
组合圆弧拱 built-up segmental arch
组合云母 built-up mica
组合运动 aggregate motion
组合运动链 compound kinematic chain
组合运具【交】assembled unit
组合运输系统 intermodal transportation system
组合载重 combined load(ing)
组合灶具 incorporated appliance
组合增强 combination reinforcement
组合闸刀 group knife
组合站 combined station
组合折旧 composition depreciation
组合辙叉 bolted rigid frog; built-up crossing
组合振动 combination vibration

组合支承辊的心轴 backup roll arbor
组合支柱 <钢的或木的> compound stanchion; compound pillar
组合指数 group index
组合制 group system; panel system
组合质量 combined mass
组合中梃 built-up mullion
组合中心钻 combination center drill; combined drill and countersink
组合轴 sectional shaft; subdivided shaft
组合轴承 combination bearing; composition bearing; compound bearing; multipart bearing
组合柱 built-up column; built-up pile; built-up pillar; clustered columns; composite column; compound column; reinforced column; combination column
组合柱颈 composite capital
组合柱式 composite order
组合柱头 composite capital
组合砖 composite brick
组合砖砌体 composite brick masonry
组合转子 combined rotor
组合转子的振动波型 mode of combined rotor
组合桩 built-up pile; composite pile; compound pile
组合装配 block mount
组合装置 aggregate unit; combination fixture; rig; unit equipment
组合自动控制器 combination automatic controller
组合阻力 composite resistance
组合钻 combination drill
组合钻床 combination drilling machine
组合钻床钻头 unit-type drill head
组合钻杆 jointed rod; sectional rod; sectional steel
组合钻具 <由稳定器、钻铤、扩孔器组成> packed hole assembly; way-type drilling unit
组合钻探剖面图 compound graphic(al) log
组合钻头 combination drill; combined drill
组合(最)优化 combinatorial optimization
组合作用 combined action; composite action; group action
组合座脚 built shoe
组呼 group call(ing)
组呼键 group calling key
组换 <隧道灯泡> group replacement
组记录 group record
组继电器 group relay
组间比较 comparison among groups
组间比较设计 grouped comparison design
组间的 interblock
组间方差 inter-class variance
组间间隔 interblock gap
组间间距 interlock gap; interlock space
组间相关 inter-class correlation
组间相关系数 inter-class correlation coefficient
组件 cluster; component (parts); component ware; constituent element; module; pack unit; subassembly
组件安装 unit mount(ing)
组件测试 module testing
组件测试程序 module test program(me)
组件测试机 module tester
组件测试设备 component testing equipment
组件承受短路能力 component short-circuit withstand ratings
组件底座 unitized substructure
组件封装 functional package
组件功耗 module dissipation
组件化 modular
组件化设计 modular design
组件夹具 module fixture
组件结构 modular design; unit construction
组件矩阵 module matrix
组(件)开关 cluster switch
组件密度 component density
组件耐振性试验 vibration testing of components
组件清洁 unit clean
组件式供暖器 unit heater
组件式构造 modular construction
组件式计算机 unit construction computer
组件式喷水灭火系统 package-type sprinkler system
组件式歧管系统 modular manifolding systems
组件式通风器 unit ventilator
组件特性 component characteristic
组件效率 component efficiency
组件信息 module information
组件选通 chip enable; chip select
组件引线 package lead
组件直径 module diameter
组件装配 subassembly; subgroup assembling
组建费 organizational expenses
组界 class boundary; class limit
组警告信号电路 group alarm circuit
组镜 arrangement of mirrors
组据数列 class interval series
组距 class interval; grouping interval
组锯 gang saw
组开关 group switch
组块地震仪 block-type seismograph
组块石平台 block platform
组梁砌块 bond beam block
组梁贴板 flitch
组码 group code
组忙 group busy
组忙灯 group-busy lamp
组忙继电器 group-busy relay
组忙音信号 group-busy signal
组名 group name; set name
组模 gang form(work)
组内方差 intraclass variance
组内改正 reduction to group mean
组内距 elements interval
组内相关 intraclass correlation
组内相关系数 coefficient of intraclass correlation
组坯 assembly
组(拼)装原理 building block principle
组频谱带 combination band
组平均纬度 group mean latitude
组平均值 group average; group mean
组群 cluster
组群方式 grouping; grouping plan
组群分析法 panel analysis
组群回叫 group automatic callback
组闪灯 group-flashing light
组式布置 block-type arrangement
组式犁 independent beam plow
组首项指示 group indication
组首项指示表 first item list
组数 class number
组锁 hotel lock
组态 configuration
组态相互作用 configuration interaction
组态自由能 configurational free energy
组套 stack
组特征 group property
组调 gang adjustment; gang control
组调电路 ganged circuit
组桶式开关 bucket brigade switch
组桶式器件 bucket brigade device
组团式建设 cluster development
组团式区划 cluster zoning
组团式住宅 cluster housing
组系 series
组匣 modular block; unit block
组匣包装 modular package
组匣互连 module interconnection
组匣化 modularize
组匣架 modular rack
组匣结构 block construction
组匣联锁系统 modular interlocking system
组匣联锁制 modular interlocking system
组匣式电气集中 unit block interlocking
组匣式集中联锁 unit block interlocking
组匣式结构 modular construction
组匣式联锁 modular interlocking; unit block interlocking
组匣系统 modular system; unit block system
组匣信号系统 modular signaling system
组匣信号制 modular signaling system
组匣原则 modular principle
组匣制 unit block system
组下限 lower class limit
组限 class limit
组相联缓冲存储器 set associative buffer storage
组芯芯座 tie bar print
组芯造型 all-core mo(u)lding; cored-up mo(u)ld; core mo(u)lding; mo(u)lding in cores; core assembly
组芯造型法 core assembly process
组芯铸型 core assembly mo(u)ld; core(sand) mo(u)ld; core-up mo(u)ld
组芯铸造 multiple-core casting
组信号设备 group signal(l)ing equipment
组选择 group selection
组选择器 group selector
组页器 page composer
组元 component; constituent element; group component
组闸 modular block
组闸架 modular rack
组闸式电气集中联锁 modular all-relay interlocking
组长 charge hand; chief operator; gang-boss; gang pusher; teamleader
组织 agency; contexture; fabric; frame work; framing; organism; organize; organizing; regimentation; set-up; texture
组织程序流程图 organizational process flow chart
组织措施 organizational measures
组织大纲 memorandum of an association
组织的 constitutional
组织的等级制度 organizational hierarchy
组织钝感性 structure insensitive property
组织发展 organization development
组织法 basic instrument; constituent instrument
组织方案 organization scheme
组织放射自显影术 histoautoradiography
组织费 cost of organization; establishment charges
组织改建 tissue reconstruction
组织革新 organizational change
组织(管理)心理学 organizational psychology
组织光度测定法 histophotometry
组织号数 structure number
组织和方法 organization and methods
组织化 systematization; systematize
组织化学 histochemistry
组织环境 organizational setting
组织机构 institutional framework; organizational framework; organizational structure; organization
组织机构层次示意图 organization chart
组织机构上的障碍 institutional hang-up
组织机构图 organigram; organization chart
组织机构图表 organization chart
组织计划 organization scheme
组织间隙 interstitial space; tissue space
组织简介 general information
组织阶层 organizational hierarchy
组织结构 histology; level of organization; organizational framework; organizational structure; unitized construction
组织结构致密 fine texture
组织结算 organize clearing
组织紧密的 compact grained
组织均匀性 structural homogeneity
组织开发 organizational development
组织类型 pattern of organization
组织敏感性 structure sensitive property
组织模式 enterprise schema
组织培养 tissue culture
组织情况变化报告 organization status change report
组织缺陷 <木材的> defect of tissue
组织人 organization man
组织上的 structural
组织水 constitution water
组织投标 request for tenders
组织图 histogram
组织图表 organization chart
组织图谱 histography
组织委员会 organizing committee
组织效率 organizational effectiveness
组织协调 organizational coordination
组织学 histology
组织学论文 histology
组织研究 fabric study
组织应力 structural stress; structure stress
组织语句 organizational statement
组织原则 principle of organization
组织者 framer; organizer
组织(指挥)系统中的等级 echelon
组植 group planting
组指标 <路基土壤分类> group index
组指数 <路基土壤分类> group index
组中值【数】 mid value of class; class median; class mid value; mid-class mark
组重复间隔 group repetition interval
组重复频率 group repetition frequency
组重复周期 group repetition interval; group repetition period
组柱支墩坝 columnar buttress dam
组桩折减系数 reduction factor for piles in group
组装 assemblage; assembling; assembly; building; build-up; erection; fitting; pack; package; packaging; set-up
组装部件 assembling parts; assembly unit; assembly parts

组装场 assembly yard;fabricating yard;assembly area;assembly depot
组装车间 assembly shop;composing room
组装程序 assembly program(me)
组装尺寸 assembly dimension
组装船 fabricated ship
组装单元 assembly unit;module unit;packaged unit
组装的 built-up;gang mounted
组装的车辆 assembled car
组装的房屋构件 knocked down
组装法 pattern match
组装轨道(工作)fabricated track work
组装锅炉 packaged boiler
组装后 post assembling
组装划线法 pattern scribing
组装基础 built-up foundation
组装记号 assembly mark
组装技术 package technique
组装加工 assembly processing
组装加工步骤 assembly processing step
组装夹架 assembly jig
组装结构 package assembly;packaged design
组装坑 assembly pit
组装空调机 packaged air conditioner
组装空调器 packaged air conditioner
组装连接件 assembled fitting
组装零件 group assembly parts;knock down parts;parts
组装螺栓 assembly rod;erection bolt
组装密度 packaging density;packing density
组装模板 rigging
组装模具 assembly jig
组装模具法 piece mo(u)ld process
组装平台 modular platform
组装墙面连接件 structural grid elements
组装设备 packaged equipment;packaged plant
组装式泵 packaged pump
组装式布料杆 erecting placing boom
组装式厨房 package kitchen
组装式给水处理厂 modular water treatment plant
组装式机组 packaged unit
组装式汽轮机 block-type turbine
组装式燃气轮机装置 modular gas turbine power plant
组装式双法兰空心轴 assembled double-flanged hollow shaft
组装式水轮机 block-type turbine
组装式提升机 built-up pallet
组装式透平 block-type turbine
组装式挖泥船 built-up dredge(r);dismountable dredge(r)
组装式涡轮机 block-type turbine
组装式窑 modular type kiln
组装式自动专用带锯 special automatic modular type band saws
组装数组 packed array
组装水箱 sectional tank
组装胎型 assembling jig;assembly jig
组装图 arrangement diagram;assembly diagram;assembly drawing
组装外壳 package shelf
组装微型模件 assembled micromodule
组装系统 modular system
组装线脚木模 horsing-up mo(u)ld
组装油箱 sectional tank
组装之前检查 inspection before assembling

祖

祖代特征 palingenetic character
祖母绿 emerald;smaragd
祖先 ancestor;forerunner

钻

钻 bore;broach taper

钻版机 routing machine
钻棒 auger stem
钻爆参数 parameter of drilling and blasting
钻爆法 drill(ing) and blast(ing) method;drilling and firing method
钻爆(炸礁)船 underwater drilling and blasting ship
钻贝生物 shellboring organism
钻臂 boom arm;drill boom;drill jib;main boom;working beam
钻臂摆动缸 <凿岩机> boom swing cylinder
钻臂举升缸 boom lift cylinder
钻臂倾斜缸筒 tilt cylinder
钻臂倾斜液压缸 tilt cylinder
钻臂扇形(定位)板 segment
钻臂伸缩液压缸 boom extension cylinder
钻臂体 boom arm
钻臂头 <凿岩机> bulkhead
钻臂推进器 <凿岩机> feed extension
钻臂旋转体 boom swivel head;slewing head
钻臂旋转头 slewing head
钻臂转柱 boom support
钻臂座 boom yoke;holder
钻冰孔 ice bore
钻柄 bit stock;drill shank;drill stock
钻柄和钻头间的连接 <宾夕法尼亚索钻> sub for bit and auger-stem
钻采机 auger miner
钻采平台 drilling platform
钻槽 drill flute
钻侧孔眼钻头 side-tracking bit
钻测 borehole survey(ing);drill-hole survey
钻场 drill site
钻车 banjo;blast-hole rig;boring rig;drill(ing) carriage;drilling machine;drill mobile;drill rig;drill truck;drill wagon;jumbo(drill);jumbo truck;truck-mounted drill(ing rig);wagonette drill;wagon-mounted drill;jumbolter <安装锚栓和打眼用的>;wagon <凿岩机>;drill carriage
钻车的钻机托臂 jumbo arm
钻车底盘 wagon drill carriage
钻车动臂 rig boom
钻车上部结构 drilling superstructure
钻车上的钻机托杆 jumbo column;rig column
钻车上钻机悬臂 column
钻尘 drilling dust
钻成带套的管子 bored casing pipe;bored casing tube
钻成带套的桩 bored cased pile
钻成的灌浆孔 drilled grout hole
钻成的缆索隧道 bored cable tunnel
钻(成)井 drilled well
钻成空心的 hollow-bored
钻成排桩 bored diaphragm
钻程长度 drilling length
钻冲击钻进法 pole-tool method
钻出 drill(ing)out
钻出的孔 drilled holing;drill way
钻出自流井 artesian bored well
钻穿 <将某一地层> make through
钻穿水泥填塞物 drilling out of the cement plug
钻船 rock drill barge;rock drill vessel
钻床 baby drill;driller;drill(ing) machine;drill lathe;drill press;drill stock;machine drill
钻床床轴头箱 drill head
钻床立轴箱 drill head
钻床立柱 drilling machine column
钻床零件 drill press parts
钻床磨削 drill press grinding
钻床主轴 drilling press spindle
钻床柱 drilling machine upright;drilling stand
钻锤 drill hammer
钻唇 cutting lip
钻刀角 drill lip angle
钻刀余隙角 drill lip clearance
钻导孔 preboring
钻到 *n* 米深 bottomed at *n* meters
钻到规定深度 bottom out
钻到设计深度 drilling to completion
钻到预定深度 drill to predetermined point
钻的方套管 grief stem
钻的筒形头 drill barrel head
钻的筒形靴 drill barrel shoe
钻地沉箱 <一种开口圆筒形沉箱> drilled-in caisson
钻地沉箱挡土墙 drilled-in caisson retaining wall
钻地沉箱的混凝土填方 concrete fill of drilled caisson
钻点 lance point
钻垫 drill pad
钻掉 drill out
钻锭 cobalt ingot
钻洞机 borehole auger;hole borer
钻斗【岩】drilling bucket
钻粉 bore-dust;bore meal;borings;core borings;cuttings;drill dust from boring;drillings;drilling dust;drilling meal;sludge
钻粉排出 drilling dust extraction
钻粉排出空隙 chipway space
钻锋圆边 margin of drill
钻干孔 carry a dry hole
钻杆 auger extension;auger stem;bit-rod;bit shank;bore bar;bore rod;boring bar;boring rod;bulling bar;bull rod;drill bar;drill bit-rod;drilled rod;drilling beam;drill(ing)rod;drill pipe;drill steel;drill(ing) stem <钢丝绳冲击>;drill string;jack rod;shaft bar;shank of bit
钻杆摆放架 laydown rack
钻杆泵压损失 pressure loss in drilling rod
钻杆壁厚 drill pipe wall thickness
钻杆长度 length of drill rod;run of steel
钻杆长度耐蚀年限 foot well life
钻杆衬套 drill bushing;Kelly(bar) bushing
钻杆尺寸 drill rod size
钻杆冲击钻机 pole drill;rod-tool rig
钻杆冲击钻进 jigging down
钻杆冲击钻进法 pole-tool method
钻杆冲击钻进工具 pole tool
钻杆冲击钻进装置法 rod and drop-pull system
钻杆串 rod string
钻杆打捞母锥 female drill rod tap
钻杆打捞器 screw bell
钻杆打捞爪 grapnel tube
钻杆打头 drive head
钻杆代用品 drill pipe sub
钻杆导槽 rod slide;slide
钻杆导向滑槽 rod slide
钻杆的内平连接 flush-joint of rods
钻杆的平头接合 flash joint of rods
钻杆(地层)测试 drill stem test(ing)
钻杆垫板 <钻塔内的> tube support
钻杆吊索的升降卷筒 spudding reel
钻杆定心器 rod guide
钻杆断裂 breaking of the drill stem;wringer neck
钻杆浮阀 drill pipe float valve
钻杆浮鞋 drill pipe float
钻杆刚度 stiffness of drill pipes
钻杆钢 jumper steel
钻杆钢级 steel degree of drill pipe
钻杆割刀 drill pipe cutter
钻杆公母接头 drill pipe box and pin
钻杆公母扣 drill pipe box and pin
钻杆挂钩 rod hanger
钻杆环空返速 drill pipe annular return velocity
钻杆环空压降 annular pressure drop of drill pipe
钻杆环圈 <位于钻杆与钻头连接处> drill collar
钻杆回旋速度 drill rod rotation speed
钻杆或钻套的安全夹子 bulldog
钻杆记录 drill stem logging
钻杆夹 boring rod clamp;drive clamp;finger grip <塔上的>
钻杆夹板 rod clamp
钻杆夹持器 drill pipe clamp;floor clamp;foot clamp;rod clamp;rod lifter
钻杆夹头 drill adapter;rod clamp
钻杆架 rack for rods;rod rack;steel rack
钻杆间隙 rod clearance
钻杆胶塞 drill pipe plug
钻杆接杆 lengthening rod
钻杆接箍 drill pipe coupling;drill rod coupling;rod coupling
钻杆接头 drill pipe sub;drill rod coupling;drill rod joint;rod adapter;rod coupling;spudding shoe
钻杆接头打捞母锥 female rod coupling tap
钻杆接头端面 shoulder of the joints
钻杆截面积 section area of drill pipe
钻杆具组 drill string
钻杆卡盘 drill rod drive quill;drive quill
钻杆卡盘扳手 rod chuck wrench
钻杆卡头 drill adapter
钻杆卡瓦 drill pipe slips
钻杆控制井底测试阀 drill stem controlled down hole test valve
钻杆捞管 grapnel tube
钻杆捞矛 rod spear
钻杆捞取器 fishing tool
钻杆立根 drill pipe stand;rod stand;stand of drill pipe;stand of rods
钻杆立根拧卸作业 laydown job
钻杆立根数 rod pulls
钻杆立根指梁 pipe finger
钻杆连接插口 drill socket
钻杆连接套筒 drill sleeve
钻杆(螺钉)扳手 <凿岩机> hand dog
钻杆螺丝缩接 drill stem bushing
钻杆螺丝缩接夹具 drill stem bushing holder
钻杆螺纹攻 rod tap
钻杆螺纹规 drill pipe ga(u)ge
钻杆螺旋津 scroll
钻杆锚定 drill string anchor
钻杆名义重量 nominal weight of drill pipe
钻杆摩擦力 rod drag
钻杆母接头 box of tool joint
钻杆内经 inside diameter of drill pipe;pipe inside diameter
钻杆内流速 flow-velocity of drill rod
钻杆内压降 pressure drop in drill pipe
钻杆逆止阀 rod check-valve

钻杆扭矩 drill pipe torque
钻杆排放 stacking of drill pipe
钻杆盘根盒 tight head
钻杆碰撞 rod slap
钻杆钳 drill pipe tongs;extension tongs;rotary tongs
钻杆敲击孔壁 slapping
钻杆驱动 Kelly bar drive
钻杆润滑剂 rod grease
钻杆润滑脂 rod dope
钻杆渗透试验 drill stem permeability test
钻杆试井 drill stem test(ing)
钻杆试验 drill stem test(ing)
钻杆寿命 drill rod life
钻杆水泥头 drill pipe cement head
钻杆丝扣护箍 drill pipe protector
钻杆丝扣油 drill pipe lubricant
钻杆缩径接头 rod reducing coupling
钻杆探伤器 tuboscope
钻杆套管 Kelly bar bushing
钻杆提取器 rod puller;steel puller
钻杆提引器 rod extractor
钻杆提引塞 rod plug
钻杆填料函 rod stuffing box
钻杆涂料 rod dope
钻杆外径 outside diameter of drill pipe;pipe outside diameter
钻杆弯曲 stem deflection
钻杆稳定器 drill pipe stabilizer;rod stabilizer
钻杆下部装的测斜仪 plain clinometer
钻杆橡胶护箍 casing protector
钻杆橡胶护圈 casing protector
钻杆异径接头 rod reducing coupling <两头都是公扣>;pipe reducer of drill rod <两头一公一母>;rod reducing bushing <两头一公一母>
钻杆折断 rod snap;wringer neck
钻杆振动 rod vibration
钻杆直径 diameter of drill pipe
钻杆注浆 conventional injection;rod injection
钻杆柱 drilling shaft;drilling string;drill pipe string;drill rod string;hollow rodding;string of drill rods
钻杆柱定向 orienting drill pipe
钻杆柱摩擦 rod friction
钻杆柱扭矩 drill string torque
钻杆柱受高转矩负荷 high-torque on the drill string
钻杆柱弯曲 drill swaying;rod sag
钻杆柱稳定器 rod string stabilizer
钻杆柱振动 rod string vibration
钻杆转动手把 rod-tiller
钻杆装运车架 pipe basket
钻杆锥形提升器 drill pipe cone elevator
钻杆自动夹持器 automatic rod holder
钻杆自动提升器 automatic rod lifter
钻杆组 drilling column;drill(pipe) string;drill steel set;string of drill pipes
钻杆组顶端转动装置 brace head
钻杆组转动手把 <冲击钻> tiller
钻杆钻 rod borer
钻杆钻进 rod boring;rod system boring
钻杆钻头 <不带岩芯管的> drill rod bit;rod bit
钻钢 drilling steel
钻钢接长 drill steel extension
钻钢接长直径 drill steel extension diameter
钻钢组 drill steel set
钻工 borer;drilling machine operator;drilling worker;drill man;drill runner;rough hand;well digger;wrencher;pipe racker <操作管子的>;floorman <井口>;pipe pusher <拧管的>
钻工工头 foreman driller
钻工工作平台 foothold of driller
钻工夹具 drilling attachment
钻工领班 foreman driller;tool pusher
钻箍 drill chuck
钻冠的尖度 bit cutting angle
钻冠的角度 bit cutting angle
钻管 bit tube;bore pipe;boring pipe;casing tube;corduroy barrel;drill pipe;drill tube;drill tubing;lance pipe;sinking tube
钻管浮球阀 drill pipe float valve
钻管荷载 drill pipe load
钻管荷重 drill pipe load
钻管夹(板) drive clamp
钻管接头 drill pipe joint
钻管接头松紧索 <起扳手作用的> spinning line
钻管连接器 drill pipe coupling
钻管螺纹 drill pipe thread
钻管螺纹润滑 drill pipe thread dope
钻管拧紧 drill pipe twist-off
钻管体积校正系数 pipe constant;pipe factor
钻管托夹 tubing clamps
钻管下入深度 casing point
钻管销 drill pipe pin
钻管校正系数 pipe factor
钻管运转 running-in of the drill pipe
钻管支架 stand of drill pipe
钻管钻入 running-in of the drill pipe
钻盒 box;boxing
钻横隔墙 bored diaphragm
钻环 drill collar
钻环稳定器 drill collar stabilizer
钻机 auger rig;bore hammer;borehole drilling machine;borer;boring installation;boring machine;boring rig;drilling rig;drill unit;machine drill;rig;rig base;trepan
钻机安装工的工具 rig builder outfit
钻机安装柱 mounting column
钻机搬迁 moves of rig
钻机臂杆 rig boom
钻机(布置)密度 drill density
钻机部件 components of drill rig;drill member
钻机操作费 rig cost
钻机车 gadder;gadding car;gadding machine;wagon drill
钻机尺寸 dimension of drilling rig
钻机冲锤 drilling hammer
钻机穿进速度 penetration speed
钻机吹粉器 drill blower
钻机的铁附件 rig irons
钻机的座架 drill casing
钻机底盘 drill carriage;drill carrier
钻机底座 drill base
钻机电动机的变阻控制器 drilling control
钻机吊钩 drilling hook
钻机顶柱 drill column jack
钻机定位时间 boom positioning time
钻机定位装置 drill positioner
钻机房 shanty
钻机附件 drill attachment;drill fittings
钻机钢绳滑轮 crowned pulley
钻机钢丝绳 drill cable;rig line
钻机钢索 drill cable
钻机给进装置 drill feed
钻机工人 rough neck
钻机回转器主动齿轮 drill head drive gear
钻机机架 drill frame
钻机机组 assembly drill;drilling unit
钻机机组人员 drilling crew
钻机记录仪 stool pigeon
钻机夹具 drill chuck
钻机架 boring frame;boring tower;gadder;carriage mounting <凿岩机>;drilling derrick
钻机绞车 boring winch;drilling winch
钻机壳 drill shell
钻机类型 type of drilling machine
钻机立轴 drive rod
钻机立轴卡盘 drive chuck;drive quill
钻机立轴套筒 drive-rod bushing
钻机立柱基座 brace socket
钻机利用率 rate of operating rig
钻机马达 drill motor
钻机马达转子 drill-motor rotor
钻机能力 drill capacity
钻机配件 drill fittings
钻机皮带管 drilling house
钻机平台 drilling island
钻机启动器 starter drill
钻机起重机架 drill derrick
钻机气腿 air leg
钻机取杆器舌门 clapper
钻机日钻孔报表 driller's tour report
钻机软管接头帽 drill hose connection
钻机设备 drill fixture
钻机设计人员 drilling rig developer
钻机失速 drill stalling
钻机实际运作时间 drilling time
钻机台班 rig-shift
钻机台月 rig-month
钻机台月利用率 time availability of rig-month
钻机台座 boring pillar
钻机套管装卸绞盘 calf reel
钻机头部空间 drill headroom
钻机托臂 drill jib;rig column
钻机桅杆 drilling mast
钻机维修人员 spanner man
钻机稳定器 drill stabilizer
钻机系列 drill series
钻机下套管 casing off
钻机芯轴 drill spindle
钻机型号 rig type
钻机修理车间 boring repair shop;drill repair shop;drill sharpening shop
钻机修理工 drilling doctor
钻机选配 rig selection
钻机岩芯 drill core
钻机研制者 drilling developer
钻机用软管 drill hose
钻机用双支柱 <坑道钻进用> double drill column
钻机闸瓦 brake lining for drill
钻机支架 drill stand;drill frame
钻机支腿 jack leg
钻机支柱 drill column;drill(ing) post;jack bar
钻机重量 weight of drilling rig
钻机柱架 drill column;rig column
钻机转速 rotational speed of drilling machine
钻机装载 drill mounting
钻机自动推进装置电动机 power feed motor of drilling machine
钻机总功率 total rig power
钻机总台时 total rig time
钻机钻进深度 drill capacity
钻机钻速监控器 penetration rate monitor
钻夹尺寸 collet capacity
钻夹头 drill holder;drill chuck
钻夹子 collet
钻架 auger board;bore frame;bore jackleg;boring rig;boring tripod;drill carrier;drill column;drill derrick;drilling cramp;drill mounting;drill stand;drill tower;jumbo;mounting column;rig;rock drill mounting
钻架车 drill carriage;drill rig;drill(ing) jumbo
钻架大梁 drill tower girder
钻架地下室 drill tower cellar
钻架吊车 drill tower crane
钻架杆臂 column arm
钻架工作人员 drill tower man
钻架基础 drill tower foundation
钻架基础格床 drill tower grillage
钻架基柱 drill tower foundation post
钻进角度及长度检验器 drill grinding ga(u)ge
钻架平台 tower platform
钻架上飞檐 drill tower cornice
钻架塔顶 drill tower crown
钻架台车 drill carriage;drilling jumbo;drilling platform;jumbo
钻架腿夹 derrick leg clamp
钻架下部结构 drill tower substructure
钻架支撑 drill tower brace
钻架支撑座 rig base
钻架支架 mounting bar
钻架支柱 drill tower leg
钻尖 apex point;drill point;drill pointer;drill tip;tip of drill
钻尖后角 primary relief
钻尖厚度 thickness of drill tip
钻尖角 drill point angle
钻尖偏移距离 apex distance
钻件夹具的导套 jig bush(ing)
钻浆 bore sludge
钻角规 drill ga(u)ge
钻铰复合刀具 drill reamer
钻铰两用刀 combined drill and reamer
钻进 boring;drilling;headway;kennel;plunge
钻进班 boring party
钻进参数 drilling parameter
钻进层位 drilling level
钻进除粉管 drill blower
钻进导向管 drill director
钻进导向器 drill director
钻进动力头 drilling head
钻进队 boring party;drill crew
钻进法 cable tool drilling
钻进反循环 reverse circulation drilling
钻进方法 drilling method
钻进方向 direction of drilling
钻进费用按英尺计算 footage cost
钻进给进力 weight capacity
钻进工艺 drilling technology
钻进管道进行清理的工人 swab man
钻进规程 drilling practices;drilling regulation;drill regime
钻进规程参数 balanced factor
钻进规范 drilling recommendation
钻进过程 drill progress
钻进过度 over-drilling
钻进计划 drill program(me)
钻进记号 shaft identification
钻进记录 <记录每尺钻进的时间> drilling time log;boring journal;boring log
钻进技术 drilling technique
钻进角 cutting angle;drilling angle
钻进介质 drilling medium
钻进进尺 drilling footage
钻进进尺数 drill footage
钻进控制 drilling control
钻进率图表 boring performance diagram
钻进米数 drilling footage
钻进磨料 drilling material
钻进难易程度 drillability
钻进能力 drillability;drill(ing) capacity;drilling duty
钻进黏[粘]土层用钻头 gumbo bit
钻进器材 drilling material
钻进切削具 boring cutter
钻进情况 drilling condition
钻进取样 drive sampling

钻进日志 boring journal
钻进设备 drilling apparatus; drill rig <成套的>
钻进深度 drilling depth; feed travel
钻进时间 boring time; drilling time; penetration time
钻进时间记录仪 drilling time recorder
钻进时间利用率 availability of penetrating time
钻进时间曲线 drilling time log
钻进时震动 drilling chatter; drilling shock
钻进实验 drilling experimentation
钻进水平 <地下钻进时> drilling level
钻进水平巷道 drilling floor
钻进速度 advance rate; drilling rate; drill(ing) speed; penetration rate; penetration speed; rate of advance; rate of drilling; rate of penetration; rate of sinking
钻进速度记录仪 tattle-tale
钻进速度值 drill rate value
钻进速度自动记录仪 geolograph
钻进突变 drill break
钻进系数 drilling factor
钻进效率 drilling efficiency
钻进性 drillability
钻进循环 drilling cycle
钻进岩芯直径 diameter of the core
钻进液 drilling fluid; drilling mud
钻进一个立轴长度 run a screw
钻进仪表 <负荷和钻速指示器> drillometer
钻进用冲洗水接头 water coupling
钻进用水 drill water
钻进原理 boring principle
钻进振击器 drilling jars
钻进中的异常现象 anomaly while drilling
钻进转率 drilling duty
钻进阻力 driving resistance
钻进组 boring party
钻井 auger shaft; bored tube well; bored tubular; bored well; borehole well; bore well; boring shaft; boring well; draw well; drill hole; drilling well; shaft boring; shaft drilling; well boring; well drilling; well lowering
钻井泵 borehole pump; driven well pump
钻井编号 drilling number
钻井驳船 drilling barge
钻井布置 arrangement of wells
钻井采矿法 borehole mining
钻井采油 mud oil
钻井参数 drilling parameter
钻井操作 drilling operation; drilling works
钻井成本 cost of drilling
钻井承包商 well drilling contractor
钻井程序 drilling program(me)
钻井船 drill boat; drilling vessel; drill ship
钻井导孔 pioneer hole
钻井的底点 bottom well location
钻井电测 electric(al)(well) logging
钻井电测记录 electric(al) log
钻井电缆 borehole cable
钻井定名 well naming
钻井队 boring crew; boring gang; boring party; boring team; drilling crew; drilling gang; drilling party; drilling team
钻井队长 boring master
钻井法 sinking method
钻井法凿井 sinking by boring
钻井方案 well program(me)
钻井方法 putting-down method
钻井浮船 floating drilling vessel
钻井浮船升降机 floating barge elevator
钻井附属船 drilling tender
钻井工 sinker; spudder; well borer; well sinker
钻井工程 drilling works; well works
钻井工程全面控制 total drilling control
钻井工程质量 drilling quality
钻井工作 sinking work
钻井公式 well equation; well formula
钻井钩 drilling hook
钻井估计产量 rating of well
钻井过程 drilling process
钻井过滤器 well strainer
钻井合同 boring contract
钻井机 drill rig; shaft boring machine; shaft drill jumbo; well borer; rig; well drill
钻井机架 trepan; well-rig
钻井机位置 driller's station
钻井机械 casing machine
钻井机站 driller's station
钻井机专用车厢 rig body
钻井记录 borehole log; bore log; boring log; driller's log; drilling log; drill time log; profile of a well; well log(ging); well record
钻井技术 drilling technology; well drilling technique
钻井架 boring tower; derrick; rig; drilling derrick
钻井架格排 derrick grillage
钻井间距 wells spacing
钻井简要记录 driller's log
钻井进度 drilling progress
钻井勘探 borehole inspection
钻井类别 well type
钻井磨料 drill material
钻井泥浆 drilling fluid; drilling mud
钻井泥浆失水量 water-loss in drilling mud
钻井泥浆在环状空间的回流速度 upward annular velocity of drilling mud
钻井排水 bored drain
钻井平台 boring platform; casing platform; derrick platform; drill(ing) platform; sinking platform; tower platform; well platform
钻井剖面 bore profile; drill log; drill record; profile of a well; well log
钻井起重机 drill tower crane
钻井起重机顶部 tower crown
钻井起重机架 drilling derrick
钻井起重机架支撑 drilling derrick brace
钻井倾角 pitch of holes
钻井取(土)样 well sampling
钻井取样记录 sample log
钻井软管 drilling hose
钻井设备 boring equipment; boring outfit; boring rig; casing machine; drilling equipment; drilling outfit; drilling rig; well-rig
钻井设计 drilling design
钻井深度 drilling well depth
钻井施工测量 construction survey for shaft-drilling
钻井事故 drilling failure
钻井数控记录 digital log
钻井水力压裂 well fracturing
钻井水面 phreatic water surface
钻井水平位移 well horizontal departure
钻井水位降低 drawdown of well
钻井塔 frame of boring rig; drilling tower
钻井台 boring block outfit
钻井套管鞋 well casing starter
钻井通路 sinker pathway
钻井完成 completion of well
钻井位置 drilling position; drilling site
钻井污水 drilling wastewater; well drilling sewage
钻井现场 sinking site
钻井斜度 pitch of holes
钻井许可证 well permit
钻井压力灌浆堵漏 squeeze a well
钻井岩芯 well core
钻井液的沉砂系统 setting system for drilling fluid
钻井液漏失控制 drilling fluid loss control
钻井液面声学测深仪 acoustic(al) well sounder
钻井液添加剂 drilling fluid additive
钻井液污染 drilling fluid contamination
钻井用阀门 drilling valve
钻井用钢丝绳 drilling cable; drilling rope
钻井用工程模拟器 engineering simulator for drilling
钻井用捞钩 boot jack
钻井用水 drilling water
钻井支柱 well column
钻井直接成本 pure drilling cost
钻井柱状剖面图 well log columnar section
钻井柱状图 drill log
钻井装置 boring block; boring device; boring unit; drilling block; drilling device; drilling rig; drilling unit; rig
钻井装置事故 rig accident
钻井准备 prepare for drilling
钻井总费用 total drilling cost
钻井作业 drilling operation
钻径规 drill ga(u)ge
钻径及钢丝规 drill and steel wire ga(u)ge
钻具 auger; borer; drilling accessory; drilling device; drilling rig; drilling tool; drill (steel) set; drill tool; string of tools
钻具部件 parts of drilling tool
钻具参数 parameter of drilling tool
钻具打捞器 overshot assembly; tool extractor
钻具定方位角 azimuth tool orientation
钻具浮重 <钻孔中充满冲洗液时的> buoyant weight
钻具给进 advance of tool
钻具更换 change of tools
钻具滚筒 bull wheel
钻具滚筒端的轴颈套 bull wheel box
钻具滚筒刹车 bull wheel brake
钻具滚筒轴套 bull wheel shaft clamps
钻具护丝 thread protector of drill tool
钻具检查时间 checking drill string time
钻具接头 boring tool joint; drilling tool substitute; tool joint
钻具卡槽钻孔 keyseated hole
钻具卡住部位指示器 freeze point indicator
钻具捞取器 grab iron
钻具连接母扣 box of tool joint
钻具临界钻压 critical weight on drill string
钻具扭曲度 tensional deflection of drill tool
钻具切点 tangent point of drill string
钻具稍提 short run
钻具弯曲度 bending deflection of collars
钻具为泥包住 ball up
钻具下部稳定器 lower drill stabilizer
钻具下落 dropping of the tool
钻具陷落 drop of drill tools
钻具陷落深度 depth of drill tool drop
钻具修理工 tool dresser; toolie; tool shiner
钻具修整 tool dressing
钻具牙轮部分 cutter-mounting ring
钻具遇阻 hit an obstruction
钻具运动 <钢绳冲击钻进时> drilling motion
钻具支点 supporting point of drill string
钻具中途遇阻 hit a bridge
钻具重量 drilling weight; weight of drill tool
钻具转速 drill speed
钻具自动提升装置 <缩孔时> retraction device
钻具总重 total weight of drill tool
钻具组 drilling set; drill string; string
钻具组合 combination of drill tool
钻具组升降程序 drill string tripping
钻距 drill change
钻距钎子组的长度公差 drill change
钻开 drill out; drill over
钻开油层 top the oil sand
钻坑及竖柱杆的工程车 earth borer and pole setter truck
钻坑探结合法 exploration method by combined drilling-opening engineering
钻孔 borehole; bore well; boring; drill hole; drilling borehole; sinking of borehole; auger hole; blast hole; bore; bored-hole; bored tube well; bored well; bouche; countersink; down-the-hole; cut hole; drill(ing); auger boring; holing; sink a hole; sink of bore hole; solid boring; spudding
钻孔包裹式应力计 borehole inclusion stressmeter; borehole inclusion stressometer
钻孔包体式应力计 borehole inclusion stressometer
钻孔爆扩桩 cored pedestal pile
钻孔爆破 borehole blasting; drilling and blasting
钻孔爆破数据 boring and blasting data
钻孔爆破资料 boring and blasting data
钻孔爆破作业 drilling and blasting operation
钻孔泵 bore hole pump; borehole pump
钻孔壁 borehole wall; drill hole wall; wall of borehole
钻孔壁糊泥 bulling
钻孔壁上螺纹槽 rifling
钻孔壁与套管之间的间隙 annular space
钻孔编号 borehole number[No.]; drill log
钻孔编录 borehole logging
钻孔编录表 record table of drilling hole
钻孔变位仪 borehole extensometer
钻孔变形 borehole deformation
钻孔变形计 borehole deformation meter
钻孔变形仪 borehole deformation ga(u)ge; borehole deformation meter
钻孔变形应变计 borehole deformation strain cell
钻孔标高 elevation of hole
钻孔剥采比 overburden ratio of drill hole
钻孔驳船 drill barge
钻孔布置 arrangement of holes; hole placement; hole placing; layout of boreholes; pattern of wells; drilling pattern

钻孔布置方案 pattern layout
钻孔布置模式 hole pattern
钻孔布置(平面)图 drilling map; drilling pattern
钻孔布置效率 pattern efficiency of boreholes
钻孔布置型式 borehole pattern; boring pattern; drilling pattern
钻孔采样 drill hole sampling
钻孔测氡法 determining radon method in drill
钻孔测厚 ga(u)ge by drilling
钻孔测井仪 borehole caliper
钻孔测量 borehole survey(ing); drilling survey; hole survey
钻孔测试装置 borehole measurement device
钻孔测温曲线 curve of temperature measuring in drill hole
钻孔测斜 boring deflection measurement; drill hole inclination survey; drilling inclination survey
钻孔测斜方法 inclination measurement method
钻孔测斜仪 borehole clinometer; borehole inclinometer; clinograph
钻孔插入桩 bored pile
钻孔长度 boring length
钻孔沉井 drilled(-in) caisson
钻孔沉箱 drilled(-in) caisson
钻孔衬管 drilled lining tube
钻孔成本 hole cost
钻孔承包人 boring contractor
钻孔程序 boring program(me)
钻孔尺寸 size of the(bore) hole
钻孔充电法 method of charge in well
钻孔冲洗 irrigating of borehole; irrigating of drill hole
钻孔冲洗术 trephination and irrigation
钻孔冲洗水 drilling water
钻孔冲洗液压力 drilling fluid pressure
钻孔冲洗周期 purge period
钻孔出水量 borehole yield
钻孔船 blue water vessel; boring ship; boring vessel
钻孔垂(直)度检测 bored hole verticality measurement
钻孔磁力仪 drilling magnetometer
钻孔打入桩 drilled and driven pile
钻孔大设备 boring jumbo
钻孔刀具 drilling tool
钻孔的 perforated
钻孔的封堵 sealing of boreholes
钻孔的进程 course of bore
钻孔的菱形布置 diamond-shaped pattern of well hole spacing
钻孔的水平位移 horizontal drift of a (bore) hole
钻孔的吸水能力 water intake capacity of a well
钻孔底 bottom of hole; end of drillhole; hole back; hole bottom; point
钻孔底板 perforated baseplate
钻孔底部 borehole bottom
钻孔地下水测定仪 dipmeter[dipmetre]
钻孔地下水记录仪 dipmeter[dipmetre]
钻孔地质编录 drill hole geologic(al) record; geologic(al) documentation of drill hole
钻孔地质分层图 boring log
钻孔地质技术指导书 geologic(al) technical instruction manual of drill hole
钻孔地质剖面 geologic(al) logs of drill hole
钻孔地质柱状图 borehole log

钻孔点 drilling point
钻孔电视 borehole televiewer; borehole television; borehole TV
钻孔电视摄像机 borehole television camera
钻孔电视系统 borehole television system
钻孔电阻记录 laterology
钻孔垫木 bored tie
钻孔调查 boring test
钻孔顶角偏斜率 changing rate of hole zenith angle deviation
钻孔顶角与方位角 inclination and direction of boreholes
钻孔定位 attack cut; hole position; touching point of bit
钻孔定位的精定 fine positioning
钻孔定位器 hole locator
钻孔定位支架 hole director
钻孔定向 drill hole orientation
钻孔定向偏差 directional deviation of borings
钻孔定向偏离 directional deviation of borings
钻孔定向器 borehole director; hole director
钻孔定向仪 oriented core barrel
钻孔定向装置 drill orientation device
钻孔定中心冲头 centering drill
钻孔毒杀 borehole poisoning
钻孔堵塞 ball; blockage of borehole; blockage of boring; build-up in the (bore) hole
钻孔断面 bore cross-section
钻孔断面测定仪 cal(l)iper log
钻孔断面图 borehole profile; boring profile
钻孔队 boring team
钻孔队员 boring crew
钻孔墩 drilled caisson; drilled pier
钻孔多点伸长仪 multiple point borehole extensometer; multiple position borehole extensometer
钻孔多点位移计 multiple point borehole extensometer; multiple position borehole extensometer
钻孔法 borehole method; method of drilling
钻孔法凿井 shaft-sinking by borehole
钻孔反流 drill hole returns
钻孔范围 boring field
钻孔方法 boring method; boring system; putting-down method
钻孔方位角 azimuth of hole; hole azimuth angle
钻孔方位角偏斜率 changing rate of hole azimuth angle deviation
钻孔方向 boring direction; direction of the borehole; drill hole direction
钻孔防腐法 borehole anti-corrosion method
钻孔费用 drill cost
钻孔封孔 bore plug
钻孔封口 bore hole seal
钻孔复杂情况 hole problems
钻孔盖板 hole cover
钻孔盖塞 borehole plugging
钻孔感应装置 borehole impression device
钻孔钢板桩 sheet-pile placed by boring
钻孔钢管桩 <钢管桩内钻孔下沉> drilled-in steel tube pile
钻孔钢螺钉 steel drill screw
钻孔钢绳 boring rope
钻孔高压旋喷桩 boring and high pressure jet grouting pile
钻孔给水 borehole water supply
钻孔工程 boring engineering
钻孔工地 boring field

钻孔工具 boring equipment; boring outfit; boring tool; drilling implement; drilling tool
钻孔工人 borer
钻孔工作 boring(work); drilling; hole drilling
钻孔工作船 drilling pontoon
钻孔工作队 boring gang; boring party
钻孔攻丝复合刀具 drill tap
钻孔观测站 borehole observation station
钻孔观察镜 borescope
钻孔管理 boring control
钻孔贯入器 borehole penetrometer
钻孔灌浆 hole cementation; boring grout <沉井外围>
钻孔灌注 cast-in-drilled hole
钻孔灌注的混凝土桩 cast-in-place concrete pile
钻孔灌注混凝土 cast-in-place concrete
钻孔灌注桩 bored(and) cast-in-place pile; borehole cast-in-place (concrete) pile; cast-in bored pile; cast-in-drilled pile; cast-in-situ concrete pile; drill hole grouting pile; hole grout pile; non-displacement pile; packed-in-situ pile; pile built in place; pile in pre-bored holes; uncased concrete pile; cast-in-place pile; bored pile; drilled pile
钻孔灌注桩打桩机 cast-in-situ piling machine; in-situ piling machine
钻孔灌注桩机 in-situ piling machine
钻孔(光学)探测器 introscope
钻孔规 borehole caliper
钻孔合同 boring contract
钻孔和攻丝两用机床 drilling and tapping machine
钻孔荷载试验 drill hole loading test
钻孔横向荷载试验 borehole lateral load test
钻孔横向岩体位移观测 drillhole lateral displacement measurement [measuring]
钻孔横向应变计 borehole diametral strain indicator
钻孔划线 layout for drilling
钻孔回填 backfilling of boring; plugging
钻孔锪孔两用钻头 combined drilling and counter sinking drill
钻孔或试坑勘探 prospecting by boring or trial pits
钻孔机 aiguille; boring machine; boring rig; gadding machine; perforator; piercing machine; tapping machine; boring and mortising machine; driller; drill(ing) machine; pipe drill; boring mill <尤指大型钻孔的>
钻孔机构 drilling mechanism
钻孔机轴 drilled shaft
钻孔几何因素 geometric(al) factors of hole
钻孔挤扩支盘桩 broach squeezed pile
钻孔计量器 drilled metre[meter]
钻孔记录 borehole log(ging); borehole record; bore log; bore record; boring log; boring record; drill log; drill record; log(of borehole); operational log
钻孔记录表 boring record sheet
钻孔技术 boring engineering
钻孔技术档案 drill hole technical file
钻孔加固机 casing machine
钻孔加固设计 casing program(me)
钻孔加深 subdrilling
钻孔加深法 borehole deepening method for stress measurement
钻孔加深用捞砂筒 bailer conductor

钻孔夹具 dowel(l)ing jig; drilling attachment; tapping clamp
钻孔间距 borehole spacing; boring spacing; drill hole spacing; drilling spacing; hole pitch; hole spacing; pitch of boreholes; pitch of holes; pitch spacing of(bore) holes; spacing of wells
钻孔间歇时间 boring time break
钻孔监工员 boring superintendent
钻孔剪切试验 borehole shear test
钻孔剪切仪 borehole shear apparatus
钻孔检查显示器 borescope
钻孔角度 angle of hole
钻孔结构 borehole structure; drill hole structure; hole structure
钻孔结构剖面图 borehole construction profile
钻孔结构图 follower chart
钻孔结果 boring result
钻孔进尺 drilling meterage
钻孔进度 boring progress; penetration advance
钻孔进度图表 boring progress chart
钻孔进入矿层 hit the pay
钻孔进展图表 boring performance diagram
钻孔井 bored well
钻孔径 bore size
钻孔径迹测量 track survey in borehole
钻孔径向松胀仪 <一种旁压仪> borehole dilatometer
钻孔径向应变计 borehole diametral strain indicator
钻孔径向应变显示器 borehole diametral strain indicator
钻孔径向应变指示计 borehole diametral strain indicator
钻孔距离 hole spacing
钻孔卡规 borehole caliper
钻孔开采 broach
钻孔开采法 broaching
钻孔开孔 starting the borehole
钻孔勘察 borehole logging
钻孔勘探 borehole survey(ing); drill hole exploration; prospecting by boring
钻孔孔壁糊泥 bulling
钻孔孔壁摄影机 borehole camera
钻孔孔隙 boring porosity
钻孔控制 boring control
钻孔口 borehole collar; drill collar; hole collar
钻孔口涌水 water flush at well mouth
钻孔扣环 retaining ring for bore
钻孔扩大 reaming of a borehole
钻孔扩大不足 under-ream(ing)
钻孔扩端法 boring-and-underreaming method
钻孔类型 hole type
钻孔累积偏距 hole overall deviation distance
钻孔历时 boring time
钻孔量测仪 boring ga(u)ge
钻孔量规 bit ga(u)ge
钻孔流量测定工具 equipment of boring flow measurement
钻孔流量仪 borehole velocity instrument
钻孔流速测量 velocity measure in borehole; flow velocity measurement of borehole
钻孔流速仪 borehole flow instrument; borehole flow meter; well current meter
钻孔漏失 loss of circulation; lost circulation of drill hole
钻孔轮班 boring shift
钻孔螺帽 bored nut

钻孔锚碇装置 drilled anchorage
钻孔锚固桩 drilled-in-caisson
钻孔密度 density of bore holes
钻孔目的 drilling target
钻孔内径 bore size
钻孔内径量规 ga(u)ge for boreholes
钻孔内填实炸药 solid loading
钻孔挠度计 borehole deflectometer
钻孔能力 drilling capacity
钻孔泥浆 borehole mud;boring mud;drilling mud
钻孔泥浆试验 bored slurry test
钻孔排放 drill hole discharge
钻孔排列 borehole array;hole array;hole pattern
钻孔排列法 boring pattern;hole pattern
钻孔排列型式 boring pattern
钻孔(旁压)千斤顶试验 <两块金属反向对孔壁加压> borehole jack test
钻孔配件 downhole equipment;drill fittings
钻孔膨胀仪 borehole dilatometer;borehole expansion probe
钻孔偏差 borehole deviation;borehole throw;hole deviation
钻孔偏差角 drift angle of hole
钻孔偏差角度 drift angle
钻孔偏距 hole deviation distance
钻孔偏斜 borehole throw;deflection of borehole;deflection of hole;hole deflection;hole deflexion;hole deviation;wedge off
钻孔偏斜测量 hole deviation survey
钻孔偏斜测量资料的整理 process of surveying data on hole deviation
钻孔偏斜度 hole curvature
钻孔偏斜计算方法 computation process of deflection
钻孔偏斜角 hole deviation angle
钻孔偏斜与测量 hole deviation and surveying
钻孔偏斜指标 deflecting index of hole
钻孔偏斜指示器 deflection indicator
钻孔偏斜钻具 deflecting drill tool
钻孔偏移 drift of a borehole
钻孔平台 boring island
钻孔平行布置 parallel arrangement of (bore) holes
钻孔剖面 drill record
钻孔剖面图 section of drill hole
钻孔器 aiguille;attack drill;bit brace;borer;brace bit;broach;crank brace;gadder;hand drill(ing machine);hole digger;motor starter;piercer;posthole auger;punch;sinker;sinker drill;starter
钻孔千斤顶 <钻孔内荷载试验用的> borehole jack
钻孔潜望镜 borehole periscope
钻孔潜望镜检查 periscopic inspection of drill holes
钻孔潜在产量测定试验 potential test
钻孔侵蚀 <池窑耐火材料> pitting
钻孔侵水原因 source of water troubles
钻孔倾角 dip angle of drilling hole;dip angle of borehole;inclination of the borehole
钻孔倾斜方位角 azimuth of borehole dip
钻孔清除器 wimble
钻孔清洗液 drilling fluid
钻孔清渣器 wimble
钻孔取土器 borehole sampler
钻孔取样 borehole sampling;boring sampling;drill hole sampling;fishing;subsurface sampling;tube sample boring
钻孔取样器 messenger;posthole auger
钻孔取样试验 core drill test
钻孔全测 complete survey of the well
钻孔雀石 kolwezite
钻孔燃烧器 drilled burner
钻孔人员 driller
钻孔容量 drilling capacity
钻孔塞孔法 borehole jack method
钻孔三脚架 boring tripod
钻孔扫描器 borehole scanner
钻孔上的闭塞机构 cellar control
钻孔设备 borehole equipment;borehole facility;borehole installation;borehole rig;drilling equipment;drilling facility;drilling implement;sinking equipment
钻孔设备三脚架 borehole equipment tripod;Michigan tripod
钻孔设计图 drill hole layout map
钻孔摄影 borehole photography
钻孔摄影测绘 photographic(al) borehole survey(ing)
钻孔摄影测量 photographic(al) borehole survey(ing)
钻孔摄影检查 borehole camera inspection
钻孔摄影仪 borehole(pick-up) camera
钻孔伸长仪 borehole extensometer
钻孔伸缩仪 borehole extensometer
钻孔深部排污 waste disposal in the depths of borehole
钻孔深度 boring depth;depth of bore (hole);depth of drill(ed) hole;depth of hole;drill hole depth;drilling depth;drilling length
钻孔深入率 drilling performance
钻孔渗透试验 borehole permeability test
钻孔生物 boring organism
钻孔绳索 boring rope
钻孔时间 boring time
钻孔实际时间 actual boring time
钻孔式泵 borehole-type pump
钻孔试验 auger hole test;boring test;drilling experiment;drill(ing) test;puncture test <非岩土工程勘察方面的钻孔>
钻孔试验器 hole tester
钻孔试验段特征 characteristics of borehole test section
钻孔试样 borehole sample;bore specimen;subsurface sample
钻孔竖井 drilled shaft
钻孔数 number of boreholes;number of borings
钻孔水 drill hole water
钻孔水力循环系统 borehole hydraulic circulating system
钻孔水平间距 hole spacing
钻孔水平偏距 horizontal departure
钻孔水平位移 hole horizontal departure
钻孔水位降低 drawdown of hole
钻孔水压测试 water pressure tests in borings
钻孔顺序 drilling sequence
钻孔速度 drilling speed
钻孔速率 boring rate;penetration rate
钻孔缩径 reduction of drill hole
钻孔所在位置的经度 longitude of the well position
钻孔所在位置的纬度 latitude of the well position
钻孔台车 drill(ing) carriage;drilling jumbo;jumbo
钻孔台车轨道 jumbo track
钻孔台车桩 jumbo column
钻孔探查术 exploratory trephination
钻孔套管 borehole casing;borehole lining;borehole tube;boring casing;drilled casing;drilled lining tube;drive pipe;perforated casing
钻孔套管防坠器 casing catcher
钻孔条件 conditions of borehole
钻孔停歇时间 boring break
钻孔通道 passage of borehole
钻孔通风板 hit-and-miss ventilator
钻孔图式 boring pattern;drill(ing) pattern
钻孔图型 drill hole pattern
钻孔土样 borehole plug;borehole sample;borehole specimen;boring sample
钻孔外壳 borehole casing
钻孔弯度计 borehole deflectometer
钻孔弯曲 curve of borehole;hole deviation;well deflection
钻孔弯曲平面 hole deviation plane
钻孔弯曲强度 hole deviation intensity
钻孔完成 finish a borehole;finish a well
钻孔位观测点 observation point of drill hole
钻孔位置 borehole location;boring location;boring position;drilling position;drilling site;hole site
钻孔位置测量 borehole position survey;boring location survey
钻孔稳定性 stabilization of borehole
钻孔吸收(泥浆)能力 intake of the hole
钻孔系列 series of holes
钻孔下潜望检视 periscopic inspection of drill holes
钻孔下塞 bridge the hole
钻孔现浇混凝土桩 bored and cast-in-situ concrete pile
钻孔销 drawbore pin
钻孔斜度测量 surveying of borehole
钻孔循环水漏失 circulation loss
钻孔压力 borehole pressure
钻孔压力恢复试验 borehole pressure recovery test
钻孔压力仪 borehole pressuremeter
钻孔压水试验 packer test
钻孔岩粉 bore meal
钻孔岩石取样器 verifier
钻孔岩屑 drill cuttings;well cuttings
钻孔岩芯 borehole core;boring core;drill core;well core
钻孔岩芯采取 core recovery
钻孔岩芯测井 core logging
钻孔岩芯分析 well-core analysis
钻孔岩芯回收总长度 corduroy recovery
钻孔岩芯记录 drill(hole) log
钻孔岩芯抗碎强度 drilled core crushing strength
钻孔岩芯损耗 loss of(drill) core
钻孔岩性柱状图 borehole log;boring log
钻孔样板 boring pattern
钻孔样品 borehole sample
钻孔样品的夹具 boring clamshell
钻孔要求 boring requirement
钻孔要素 essential elements of drill hole
钻孔液压应力计 hydraulic borehole stressmeter
钻孔引流法 trepanation and drainage
钻孔引伸计 borehole extensometer
钻孔引伸仪 borehole extensometer
钻孔引水 preboring for drainage
钻孔应变测量法 method of borehole strain measurement
钻孔应变计 borehole extensometer;borehole strain ga(u)ge;borehole strainometer
钻孔涌砂 sand flush of bore hole
钻孔涌水 gone to water;hole over flow
钻孔用浆 <俚语> mud flush
钻孔用缆索 boring cable;boring line
钻孔用压缩空气 compressed-air for drilling
钻孔鱼尾板 drilled fishplate
钻孔与爆破数据 drilling and blasting data
钻孔与车削两用刀夹 drill and turning tool holder
钻孔预算价格 field rate
钻孔预想柱状图 prediction column of drillhole
钻孔预制桩 bored precast pile
钻孔原始地质编录 borehole initial geologic(al) logging
钻孔原位测试技术 drill hole testing technique in situ
钻孔鏨 boring chisel
钻孔凿(子) boring chisel;hammer eye
钻孔造斜器打捞器 whipstock grab
钻孔炸药 auger hole charge;borehole charge
钻孔障碍 boring obstacle
钻孔照相机 borehole camera;boring camera
钻孔照像测绘 photographic(al) borehole survey(ing)
钻孔者 piercer
钻孔枕木 bored tie
钻孔支架 boring mast;drill carriage
钻孔直剪仪 borehole direct shear device
钻孔直径 bore diameter;borehole diameter;drilling diameter;hole diameter;hole size;well diameter
钻孔直径的选择 choice of(bore) hole diameter
钻孔直径记录图 cal(l)iper log
钻孔直径检测 bored hole diameter measurement
钻孔直线度检查 <人工偏斜前> straight hole test
钻孔职责 boring obligation
钻孔指数 drilling index
钻孔质量指标 index of hole quality
钻孔中钋 210 法 210Po method in borehole
钻孔中沉积的泥饼 mud cake
钻孔中的样品 subsurface sample
钻孔中二节套管之间的密封 bradenhead
钻孔重力仪 borehole gravimeter
钻孔轴线 center[centre] of borehole;hole axis
钻孔轴向岩体位移观测 axial displacement measurement[measuring] of drillhole
钻孔轴向应变计 borehole axial extensometer;borehole axial strain indicator
钻孔轴向应变显示器 borehole axial strain indicator
钻孔轴向应变指示计 borehole axial strain indicator
钻孔注射法 auger hole injection
钻孔注水泥下木塞 bottom plug
钻(孔)柱 drilling column
钻孔柱状剖面 cal(l)iper log
钻孔柱状剖面图 borehole columnar section;drill hole columnar section;graphic log
钻孔柱状图 bore(hole) log;boring log;columnar section of drilling;drill(hole) column;drill hole log;log(of borehole);log of drill hole;log sheet;drill column

钻孔柱状图录 drill log
钻孔转子叶片 drill-motor rotor vane
钻孔桩 bored cast-in place pile; bored piling; digging pile; drilled caisson; drilled pier; drilled pile; drilled shaft; drilling pile; non-displacement pile; piles formed by drilling
钻孔桩基础 drilled in foundation
钻孔桩墙 bored pile wall
钻孔装药 borehole charge
钻孔装药封泥 stem a hole
钻孔装置 borehole device; boring installation; drill feed; sinking installation
钻孔综合成果图表 graph of synthetic(al) result of drill hole
钻孔阻力 boring resistance; hole resistance
钻孔组 round of holes
钻孔钻进规程的调整 handling a well
钻孔钻塔 borehole rig
钻孔最大出水量 maximum yield of drillhole
钻孔最大进尺 maximum meterage of hole
钻孔最大直径 maximum drilled diameter
钻孔最深处的炸药 bottom charge
钻孔最小进尺 minimum meterage of hole
钻孔作业活动 boring campaign
钻孔作业周期 boring campaign
钻孔坐标 borehole coordinates; hole coordinates
钻口 cutting lip; drilling point
钻矿机 trepan
钻扩机 hole expanding boring machine
钻缆 drilling cable
钻粒 adamantine shot; small shot
钻粒给进装置 shot feed
钻粒供给 shot feeding
钻粒供给漏斗 shot hopper
钻粒供给器 shot feeder
钻粒消耗量 consumption of shots
钻粒岩芯钻机 shot core drill
钻粒岩芯钻进 shot core drilling
钻粒钻机 calyx drill; chilled-shot drill; shot-boring drill; shot-drilling machine
钻粒钻机钻井 shaft drilling
钻粒钻进 calyx drilling; chilled-shot drilling; shot boring; shot drilling
钻粒钻进的井筒 shot-drilled shaft
钻粒钻进的炮孔 shot drill hole
钻粒钻进法 chilled-shot system; shot drill method
钻粒钻进用钻杆 calyx rod
钻粒钻进钻机 adamantine drill
钻粒钻进钻孔 shot borehole; shot drill hole
钻粒钻孔钻机 shothole drill
钻粒钻头 chilled-shot bit; crown for chilled shot; shot bit; shot drilling bit
钻粒钻头斜水口 diagonal slot
钻粒钻眼法 shot drilling
钻埋头孔 countersink; countersinking
钻埋头孔夹具 countersinking fixture
钻铆钉孔锥口 countersinking of the rivet holes
钻煤用螺旋钻 break auger
钻煤钻头 coal bit
钻面 drilling area
钻模 drill(ing) jig; jib; jig; toolmaker's button
钻模板 bushing plate; drill plate; jig plate; plate jig
钻模键 jig key
钻模镗床 jig borer
钻模钻床 jig drill
钻模钻机 jig drill
钻磨两用机(床) drilling(and) grinding machine; drilling grinding machine
钻末 drill cuttings
钻木虫 wood borer
钻木虫木屑 bore dust
钻木虫钻的孔洞 bore hole; worm hole
钻木机 wood boring machine; wood drilling machine
钻木昆虫 chelura terebrans
钻木生物 wood boring organism
钻泥 bore mud; bore slime; bore sludge; drilling mud; sludge
钻泥抽取器 spoon bit
钻泥分析 sludge assay
钻泥管 sediment pipe; sediment tube
钻泥浆冲洗试验 mud flush test
钻泥块 drill sludge cake
钻泥皮 cake; drilling mud cake
钻泥砂泵 dirt bailer
钻泥提取管 bailer
钻泥提取器 shoe shell
钻泥提桶 dirt bailer
钻盘补心 drill stem bushing
钻旁孔 <在完成井中的> fork the hole
钻炮眼段 drilling bench
钻炮眼设备 blast rig; shothole rig
钻坯 bit blank; bit shank; shank
钻平孔底 shoulder to square up
钻扦吊索的升降卷筒 bull reel
钻钎车间 drill steel shop
钻钎修尖机 drill sharpening machine
钻前工程 drilling preengineering
钻前工程费 preengineering cost
钻前准备 prepare for drilling
钻前验收 precheck for drilling
钻桥塞 milling out a bridge plug
钻取法 boring method
钻取混凝土芯 core drilling
钻取混凝土芯样 coring operation
钻取土样 coring
钻取岩芯 corduroy-drilling; core drilling
钻取岩芯设备 core drill rig
钻取岩(芯)样 core drill sampling; drill sample
钻取岩芯样试验 core drill test
钻取样孔 spring borehole
钻刃 bit edge; cutting lip; drill point
钻入 bite into; drilling in; plunge; spudding in
钻入沉箱 drilled-in caisson
钻入速度 penetration rate
钻深计 boring ga(u)ge
钻深井 boring well
钻深孔装置 long hole boring apparatus
钻深仪 bit depth ga(u)ge
钻绳 drilling line
钻湿孔 carry a wet hole
钻湿土用大型钻头 miser
钻石 brilliant; diamond
钻石安装工 diamond setter
钻石玻璃刀 diamond cutter
钻石(齿圆)锯 diamond saw(splitter)
钻石的 lithodomous
钻石工具 diamond tool
钻石机 churn drill; rock borer; rock boring machine; rock drill; rock drilling machine; stone drill
钻石精修钻 diamond finishing bur
钻石粒 bort(z)
钻石粒度 diamond size
钻石磨蚀剂 diamond abrasive
钻石器 aiguille
钻石器械 diamond instrument
钻石砂轮 diamond wheel
钻石砂石针 diamond point
钻石式支墩坝 diamond-head buttress(ed) dam
钻石砣盘 diamond plate
钻石形塔架 diamond-shaped pylon
钻石型激波 shock diamond
钻石凿井 diamond drilling
钻石凿井机 diamond drill(ing) rig
钻石针头 diamond tool
钻石轴承润滑油 pivot bearing lubricant
钻石柱 jewel post
钻时记录 drilling time log
钻时录井间距 drilling hours log interval
钻式挤出器 auger-type extrusion(unit)
钻视探头 hole televiewer
钻水泥塞 drilling out a cementing plug
钻速 drill rate
钻速表 penetration rate meter
钻速方程 equation of drilling rate
钻速记录 penetration log; rate-of-penetration log; drilling time log
钻速记录仪 drilling time recorder; geolograph
钻速控制 drilling control
钻速曲线 drilling curve; drilling time log
钻速系数 bit speed coefficient
钻速指数 bit speed exponent
钻碎 chop up; drilling out; drilling up
钻碎塞堵 drill the plug
钻碎水泥塞 drilling up cement retainer
钻榫眼机 mortising slot machine
钻塔 blast-hole rig; boring tower; derrick; drill derrick; drilling derrick; drilling rig; drill tower; headgear; bugle derrick <立根架在一侧的>
钻塔安装工作 work derricks
钻塔绑绳 derrick guy
钻塔大梁 drill tower girder
钻塔大门 window opening
钻塔的扶管平台 <高 20～40 英尺, 1 英尺 = 0.3048 米> stabbing board
钻塔的金属部件 <天车、游动滑车、钩环、螺栓、锻件等> derrick irons
钻塔的静荷载 dead-load of derrick
钻塔底梁 derrick sill
钻塔底座 derrick substructure; derrick support
钻塔地板高度 elevation of derrick floor
钻塔地下室 drilling derrick cellar; drill tower cellar
钻塔吊车 drilling derrick crane; drill tower crane
钻塔顶 drilling derrick crown
钻塔顶层平台 attic
钻塔顶工作台 crown platform
钻塔顶框 derrick crown
钻塔顶挑檐 derrick cornice
钻塔高度 derrick height
钻塔工 derrick man
钻塔工作人员 drill tower man
钻塔滑车 derrick pulley
钻塔基础 derrick base; derrick footing; derrick foundation; drilling derrick foundation; drill tower foundation; rig base
钻塔基础底梁 derrick foundation post
钻塔基础格床 drilling derrick grillage; drill tower grillage
钻塔基柱 drilling derrick foundation post; drill tower foundation post
钻塔楼板 drilling derrick floor
钻塔内斜撑 interior leg braces of drilling tower
钻塔棚 drill shack
钻塔平台 derrick platform; drilling derrick platform
钻塔起重机 drilling derrick crane
钻塔起重量 lifting capacity of derrick
钻塔上飞檐 drill tower cornice
钻塔上檐口 drilling derrick cornice
钻塔上支架水平横梁 header of the gin pole
钻塔设备 derrick equipment
钻塔损坏 derrick failure
钻塔塔顶 drill tower crown
钻塔台板 kelly board; kelly platform
钻塔天车台 derrick crown
钻塔腿 leg of derrick
钻塔围梁 derrick girt; drilling derrick girder
钻塔下部结构 drilling derrick substructure; drill tower substructure
钻塔正面 rig front; V-of a derrick
钻塔支撑 derrick brace; drilling derrick brace; drill tower brace
钻塔支柱 drilling derrick leg; drill tower leg
钻塔主腿 running legs
钻塔贮放钻杆容量 racking capacity of derrick
钻塔自重 dead-load of derrick
钻台 boring island; boring platform; derrick floor; drill floor; drilling platform; drilling rig; drill stand; jumbo <凿岩机>; platform
钻台设备 drill floor equipment
钻探 borehole survey(ing); boring; drilling; drilling survey by boring; exploration (boring); exploration drilling; exploratory boring; misering; probing; prospecting; scout drilling; test boring
钻探班报表 driller's tour report
钻探班长 drill foreman
钻探班次 drilling shift
钻探班组 drilling crew
钻探报表 boring journal; drilling record
钻探报表外进尺 lay-by footage
钻探报告 drilling(exploration) report
钻探爆破工 drill-blaster
钻探驳 drilling pontoon
钻探驳船 boring barge; drill(ing) barge; drill(ing) scow
钻探部门 drilling department
钻探采样 boring sample
钻探参数 drilling parameter
钻探操作 boring operation; drilling operation
钻探操作规程 drilling operating instruction
钻探场地 boring site
钻探车 drilling jumbo; drill truck; pneumatic-tyred wagon drill; wagon drill
钻探车架 wagon drill frame
钻探成本 drilling cost
钻探承包人 drilling contractor
钻探承包商 boring contractor; drilling contractor
钻探程序 boring program(me); drilling program(me)
钻探冲洗液 drilling fluid
钻探冲洗液返流 drilling return
钻探船 boring ship; drill(ing) barge; drill(ing) boat; drill(ing) ship; drill(ing) vessel; floating drill barge; floating rig
钻探船钻杆滑动接头 bumper sub
钻探导管 guide tube
钻探地点 boring point
钻探电磁测井 electromagnetic log-

ging
钻探电动机 boring motor;drilling motor
钻探定向偏差 directional deviation of borings
钻探队 drilling crew;drilling gang;drilling master;drilling party;drilling team
钻探队长 foreman driller
钻探趸船 drilling pontoon
钻探发动机 boring engine
钻探范围 boring range;drilling range
钻探方案 boring plan;boring program(me);drilling program(me)
钻探方法 drilling method;drilling system
钻探方向 drilling direction
钻探费用 cost of drilling;drilling cost
钻探浮船 floating drilling vessel
钻探辅助船 drilling tender
钻探附件 boring accessories;drilling accessories;drilling attachment
钻探杆 boring rod
钻探工 borer;driller
钻探工班 drilling crew
钻探工长 drilling master
钻探工程 drilling engineering;drilling works; exploratory drilling engineering
钻探工程法 exploration method by drilling engineering
钻探工程师 drilling engineer
钻探工程系统 drilling system
钻探工程质量 quality rank of drilling engineering
钻探工地 boring point; boring site; drilling field;drilling site
钻探工地设备 drilling field equipment
钻探工具 boring tool;drilling instrument; drilling tool; earth boring tools;rig irons
钻探工具接头 drilling tool joint
钻探工人 drilling worker
钻探工头 drilling foreman;drillmaster
钻探工业 drilling industry
钻探工艺 drilling technology
钻探工作量 amount of drill working
钻探公司 boring firm;drilling firm
钻探供水管路 drill water line
钻探管材 drill tubing
钻探规程 drilling specification
钻探规范 drilling specification;drilling standard
钻探合同 drilling contract
钻探护壁液 drilling fluid
钻探回流泥浆 gas-cut mud
钻探机 boring machine;drilling installation; miser; reaming machine; sounding borer;drilling engine
钻探机船 floating rig
钻探机具 boring rig;drilling rig
钻探机械 drilling machinery; boring machinery
钻探机组 drill unit
钻探及爆破 drill and blast
钻探计划 boring plan;boring program(me);drilling program(me)
钻探记录 boring log;daily boring report; daily drilling report; driller's log;drilling record
钻探记录(报)表 drilling record
钻探技工 boring master
钻探技师 drilling foreman;drillmaster
钻探技术管理 drilling technical management
钻探技术要求 drilling technical specification
钻探架 boring frame;boring tower
钻探结果 borehole evidence; boring result;drilling result
钻探解释剖面 interpretative log
钻探解译剖面 interpretative log
钻探金刚石 drilling diamond
钻探进尺 drilled footage; length of drilling
钻探进尺分布定额 distributed quota of drilling depth
钻探进度 boring progress; boring rate;drilling progress
钻探进度(图)表 boring progress chart;drilling progress chart
钻探井架工 drilling derrick man
钻探井筒 drilling well
钻探坑 bore pit
钻探孔 borehole;test hole
钻探控制 drilling control
钻探框状图 boring log
钻探零件 boring accessories
钻探流体 drilling fluid
钻探轮班 drilling shift
钻探某一厚度所需时间 drilling time at given length
钻探能力 drilling capacity
钻探泥浆 bore mud; driller's mud; drilling mud
钻探泥浆表面活化剂 drilling mud surfactant
钻探泥浆材料 drilling mud materials
钻探泥浆计量泵 drilling mud metering pump
钻探泥浆喷射 drilling mud jet
钻探泥浆使用技术 drilling mud practices
钻探泥浆重量 weight of drilling mud
钻探配件 boring accessories
钻探平底船 boring barge
钻探平台 boring tripod; drilling island; drilling platform; production platform
钻探剖面 drill dog
钻探剖面图 boring profile
钻探契约 drilling obligation
钻探器具 drill set
钻探取样 boring sample;drilling sample
钻探取样重量指示器 drilling indicator
钻探刃磨 drill sharpener
钻探日报表 daily boring report;daily drilling report
钻探日志 drilling log
钻探设备 boring apparatus;boring installation;boring rig;drilling equipment;drilling outfit;drilling rig
钻探设备安装 rigging up
钻探设备安装技师 rig builder;rig fixer
钻探设备参数 specification of drilling equipment
钻探设备类型 type of drilling equipment
钻探设备每日成本 rig day rate
钻探设备设计者 rig developer
钻探设备使用期限 drilling life
钻探设备事故率 accident rate of drilling equipment
钻探深度 depth of borings
钻探深度计 boring ga(u)ge
钻探生产管理 drilling management; management of drilling production
钻探绳索 drilling rope
钻探绳索采样系统 drilling and wireline coring system
钻探绳索取芯系统 drilling and wireline coring system
钻探施工 drilling construction
钻探施工计划 drilling operation planning
钻探施工期 drilling time
钻探时间 drilling time;rig time
钻探时间中断 drilling time break
钻探事故 boring accident;drilling accident;drilling trouble
钻探试验 boring test;drilling test
钻探水文地质观测数据 hydrogeologic(al) observation data in drilling
钻探水文地质观测项目 hydrogeologic(al) observation items in drilling
钻探速度 drilling rate;drilling speed
钻探速率 drilling rate
钻探隧道 bored tunnel
钻探塔 boring tower
钻探塔架 boring tower;drilling tower
钻探台车 drill carriage;drill jumbo
钻探套管 boring casing;corduroy shell;guide tube
钻探条件 drilling condition
钻探图 boring scheme
钻探徒工 apprentice-driller
钻探土样 bore plug;bore specimen; boring sample;drilling sample;soil core
钻探网格 exploratory grid
钻探位置 boring position
钻探系统 boring system
钻探现场 drilling yard;drill site
钻探巷道 prospecting entry
钻探效率 penetration per rig-month; penetration rate
钻探芯样 drilling core
钻探型式 drilling pattern
钻探性测试 drillability test
钻探延伸油井 field extension well
钻探岩芯 bore plug;bore specimen
钻探岩样 bore specimen;boring core
钻探样品 boring sample
钻探仪表 drilling instrument
钻探用泵 borehole pump; boring pump;drill pump
钻探用扁钢 flat drill steel
钻探用钢材 drilling steel
钻探用钢丝绳 drill cable; drilling rope;drilling line
钻探用浆叶式钻头 boring blade bit
钻探用金刚石 drill bo(a)rt; drill bortz; drill diamond; drill quality diamond
钻探用轮转机 vane borer
钻探用泥浆 drilling fluid;drilling mud
钻探用泥浆坑 mud pit
钻探用黏[粘]土 drilling clay
钻探用品 drill goods
钻探用绳(索) drilling cable; drilling rope
钻探用双缸发动机 twin-cylinder drilling engine
钻探用无缝钢管 seamless steel tube for drilling
钻探用钻头 boring chisel;cold chisel;drill bit
钻探与物探结合法 exploration method by combined drilling-geophysical engineering
钻探炸礁船 drilling and blasting ship
钻探障碍物 drilling obstacle
钻探者 drilling people
钻探指挥人 drilling superintendent
钻探质量 quality of boring
钻探柱状图 boring log
钻探装置 boring block;drilling installation;drill setup
钻探总成本 total cost of drill
钻探总进尺 total drilling footage of drilling
钻探总台时 total rig time
钻探纵剖面图 boring profile
钻探阻力 boring resistance
钻探钻具规格 specification of drill tool
钻探钻具种类 kind of drill tool
钻探钻头 boring bit;boring chisel; drilling bit
钻探钻头参数 drill bit parameter
钻探钻头类型 drill bit type
钻探作业 drilling operation
钻探作业图 drilling performance diagram
钻探作业周期 drilling campaign
钻掏泥砂泵 <美国制造的> combination bit and mud socket
钻套 bit arbor;collet;drill bush;drill chuck;drill holder;drill sleeve;drill socket;jig bush(ing)
钻梯 drill ladder
钻体余隙 drill body clearance
钻天然石机 natural stone drilling machine
钻天杨 lombardy polar
钻挺扳手 tool wrench
钻铤 drill collar
钻铤壁厚 wall thickness of collars
钻铤单位重量 weight per unit of collars
钻铤钢度 stiffness of collars
钻铤环空返速 collar annular return velocity
钻铤环空压降 annular pressure drop in collar
钻铤矫直器 stem straightener
钻铤截面积 section area of collars
钻铤内径 inside diameter of collars
钻铤内流速 collar flow velocity
钻铤内压降 pressure drop in collar
钻铤丝扣油 drill collar lubricants
钻铤外径 outside diameter of collars
钻铤稳定器 drill collar stabilizer
钻通的 drilled in
钻头 aiguille;bit(head);bit of drill; bore bit;borer;boring bit;boring crown;boring head;chisel point; cope cutter;cutting bit;drill;drill(ing)bit;drilling crown;drill(ing) head;drill point;first bit;router bit;solid borer;stop drill;boring block;boring cock
钻头巴掌 bit leg
钻头保持环 retaining collar
钻头泵压损失 pressure loss in bit
钻头比水马力 unit bit hydraulic horse power
钻头壁 <丝扣与胎体钻头的部分> bit wall
钻头壁厚 bit wall thickness
钻头编号 bit number
钻头变径套 drill socket
钻头柄 bit stock;bit stub;drill shank; shank of drill
钻头插口 drill socket
钻头长度 bit footage;bit length
钻头超速 bit overfeed
钻头成本 bit cost
钻头尺寸 bit size
钻头尺寸参数 parameter of bit rate
钻头齿刃 bit teeth;digging teeth
钻头冲击取得的样品 <钢丝绳冲击钻进时> bit sample
钻头冲洗孔 flushing hole
钻头出厂家 bit manufacture
钻头出厂日期 production date of bit
钻头出刃 bit clearance
钻头穿透作用 bit penetration
钻头唇部外形 bit contour
钻头打捞公锥 <金刚石的> bit recovering tap
钻头打捞钩 bit holder;bit hook
钻头刀 bit wing
钻头刀口 drill bit edge
钻头导向器 <不取芯金刚石钻头中央凸出端部> bit pilot
钻头的齿 teeth of the bit
钻头的导向翼 lead blade of bit

钻头的横刃 chisel edge
钻头的结构要素 structural elements of bit
钻头的凿尖 chisel edge
钻头的纵槽 pod
钻头定程停止器 bit stop;drill stop
钻头定位孔 collared hole
钻头动力作用试验仪 chisel impact pressure chamber
钻头端部外缘金刚石 shoulder stone
钻头端部外缘磨损 shoulder wear
钻头对孔底压力 down pressure
钻头额定进尺数 drill bit cutoff
钻头费 bit cost
钻头负荷 bit weight; drilling load; pressure on the bit
钻头负荷调节 adjustment of bit load
钻头负载 bit load;drill pressure
钻头附件 drill equipment;bit attachment
钻头钢 drill steel
钻头钢坯 steel blank
钻头高度 bit height
钻头割刀 bit of drill head
钻头给进 bit feed
钻头给进机构 bit feeding mechanism
钻头根面 drill heel
钻头、工具检修工 <俚语> drill doctor
钻头工作性能 <总进尺,单位进尺成本> bit performance;performance of a bit
钻头构槽 drill groove
钻头箍 drill collar
钻头规 bit ga(u)ge
钻头和岩石的接触面 bit-rock interface
钻头荷载 bit loading
钻头夹持器 bit holder
钻头夹具 drill chuck;drilling attachment
钻头夹盘 drill chuck
钻头夹盘轴 arbor for drill chuck
钻头价格 bit cost
钻头尖 bit prong
钻头尖度 bit cutting angle
钻头角 point angle
钻头角度规 drill-bit ga(u)ge;drill point ga(u)ge
钻头脚 bit feet
钻头接柄 bit extension
钻头接杆 bit extension
钻头接套 drill socket
钻头接头 bit adapter
钻头接头表面硬化 hard-facing of tool joint
钻头进程 feed of drill
钻头进尺 bit footage; bit meterage; feet per bit;footage(drilled)per bit
钻头进尺速度 bit feed
钻头进给装置 drill feed
钻头径规 drill template
钻头卡具 drill holder
钻头卡盘 drill chuck
钻头卡头 chuck bushing;drill sleeve
钻头卡住 bit freezing; steel seizure; steel sticking
钻头空转 drill free
钻头扩大部分 skirt of bit
钻头捞取器 drill extractor
钻头雷诺数 bit Reynolds number
钻头类型 bit type
钻头类型系数 factor of tooth wear
钻头连接端 bit end; striking end of shank
钻头连接法 bit-attachment method
钻头连接(装置)bit attachment
钻头量规 bit ga(u)ge
钻头轮廓 bit contour
钻头落井 bit falling down hole
钻头密封轴承 sealed bearing of a bit
钻头面冲洗式金刚石钻头 face discharge bit;face injection bit
钻头面冲洗式钻头 bottom-discharge bit
钻头模具 bit die;bit mo(u)ld
钻头磨床 bit grinder; bit sharpener; drill(bit)grinder;drill grinding machine;drill point grinder
钻头磨光 bit polishing
钻头磨耗 bit wear and tear
钻头磨机 drill grinding machine
钻头磨尖和磨细机 drill sharpening and shanking machine
钻头磨尖机 drill pointing machine
钻头磨尖器 drill sharpener
钻头磨角 bit cutting angle
钻头磨刃装置 grinding attachment
钻头磨锐机 bit sharpener; drill steel sharpener
钻头磨损 bit wear
钻头磨损级别 grade of bit wearing
钻头磨损检测器 drill bit wear detector
钻头磨损试验 <测定岩石抗磨性> bit wear test
钻头磨削量规 drill grinding ga(u)ge
钻头内径 bit inside diameter; center [centre] bore
钻头内锥度 <放置岩芯提断器的> bit taper
钻头内锥面磨损 cone worn out
钻头泥包 bit bailing
钻头拧卸器 bit breaker; bit puller; breakout plate
钻头拧卸器提手 breaker lugs
钻头排屑槽 flute of drill;router
钻头喷嘴 bit nozzle
钻头偏离钻孔位置 deflection of the bit
钻头偏斜 bit deflection
钻土器 earth borer
钻头前刃面 face of tool
钻头钳 bit wrench
钻头切割边 bit cutting angle
钻头切口 cutting edge of bit;drill bit cutting edge
钻头切削部分损坏 <因压力过大或转速过高的> twist-off on the bottom
钻头切削刀 drill edge
钻头切削角 bit cutting angle
钻头轻压慢转钻井地层 easing the bit in
钻头倾角 bit inclination
钻头情况 boring condition
钻头刃 bit blade;bit edge;bit face;bit wing
钻头刃带 margin of drill
钻头刃尖角 bit face angle
钻头刃磨 drill sharpening
钻头刃磨机 drill point grinder; drill sharpening machine
钻头上金刚石数量 bit count
钻头上下晃动 raising and lowering of the bit
钻头设计 bit design
钻头使用期限 bit life;crown life
钻头式采样器 drill sampler
钻头寿命 bit life;crown life;drill bit life
钻头寿命模式 bit life mode
钻头枢轴 bit pin
钻头水力特性 bit hydraulics
钻头水马力 bit hydraulic horsepower
钻头水眼 bit nozzle; circulating hole of the bit; circulating openings; slush nozzles
钻头丝扣部分 bit shank
钻头送修工 <美> bit hustler
钻头损坏 bit breakage
钻头索 boring cable;boring line;boring rope
钻头胎体 bit crown;bit matrix
钻头胎体金属 bit crown metal
钻头套 bit holder
钻头套筒 drill sleeve;drill socket
钻头提取器 drill extractor
钻头体 bit body; bit frame; bit leg; body of the bit;frame of the bit
钻头条件 boring condition
钻头跳动 bit bounce
钻头头部 drilling head
钻头外侧磨损 junk damage
钻头外侧刃金刚石 outer stone
钻头外径 bit outside diameter
钻头铣刀 drill cutter
钻头镶嵌 bit setting;crown setting
钻头镶嵌块 bit insert;bit slug
钻头镶嵌物 drill studs
钻头消耗 bit consumption
钻头销 finger of bit
钻头型号 bit symbol;bit type
钻头形式 bit style
钻头修理工 bit dresser;drillsmith
钻头修理间 drill building
钻头修理器具 bit dresser
钻头修理钳 dressing bit tongs
钻头修理铁砧 dressing block
钻头修配间 drill sharpening shop
钻头修配作业 drill sharpening practice
钻头修整 bit dressing;bit recondition
钻头修整机 bit dresser; bit sharpener;dressing machine
钻头修整器 <凿岩机> swage
钻头旋转速度 bit rotation speed
钻头压降 pressure drop on bit
钻头压降系数 coefficient of bit pressure drop
钻头压力 <钻进时的> bit load(ing);bit pressure;downhole pressure; weight of the bit; drilling pressure; drilling thrust; pressure on the bit
钻头压模 crown die;crown mo(u)ld
钻头牙轮 roller cone
钻头牙轮齿 bit prong
钻头牙轮架 bridge of the bit
钻头研磨机 bit grinder; drill bit grinder
钻头研磨量规 drill bit grinding ga(u)ge
钻头样板 drill-bit ga(u)ge;drill point ga(u)ge
钻头翼 leg of the bit
钻头翼片 bit wing
钻头翼片厚度 bit wing thickness
钻头翼片角度 bit wing angle
钻头与孔壁间隙 bit clearance
钻头与孔底接触面积 bottom-hole coverage
钻头造型 crown mo(u)ld
钻头整修 drill sharpening
钻头整修工 drillsmith
钻头整修机 drill sharpener;drill sharpening equipment
钻头整修器 drill sharpener;drill sharpening equipment
钻头支承部 bit leg
钻头直径 bit diameter;bit ga(u)ge; drill bit diameter;ga(u)ge of bit
钻头直径量规 drill ga(u)ge;bit ring
钻头直径磨损量 drill bit ga(u)ge loss;ga(u)ge loss of bits
钻头轴距 pitch of drills
钻头(轴向)压力 drill thrust;weight applied to the bit
钻头铸型 crown die
钻头爪 bit leg
钻头转速 bit speed;drill speed
钻头装卸器 bit breaker; breakout plate
钻头装置 bit attachment
钻头自转速度 bit speed
钻头总荷载 total bit load; total bit pressure
钻头纵槽 pod
钻头钻管 bit nozzle
钻头钻进限制器 bit stop
钻头钻进压力 feeding pressure
钻头座 crown holder
钻透 sinking through
钻土 earth boring
钻土法 earth drill method
钻土机 earth drill; earth screw; ground drill;ground-hog
钻土螺旋钻 continuous flight auger
钻土螺钻 earth auger;earth screw
钻土器 earth borer;earth drill
钻土器具 earth boring outfit
钻土器卡车 earth auger truck
钻土取样法 <套管螺旋钻的> shell-and-auger boring
钻土设备 earth boring outfit
钻土钻机 earth auger
钻完 drill up
钻完的井 sunken well
钻下表面套管的钻孔 surface hole drilling
钻限偏斜 <凿岩机> drilling scattering
钻削动力头 drill unit
钻削力 drill thrust
钻削效应 drilling effect
钻销 lead plug
钻销孔 drawbore
钻小直径超前孔 mouse ahead
钻斜孔 rat holing
钻斜孔法 off-angle drilling
钻屑 bit cuttings; bore meal; boring breakers; boring dust; borings; boring sludges; cuttings of boring; drill cuttings; drilling breakers; drilling dust;drillings;drilling sludges
钻屑出口 drill exhaust
钻屑分离器 drilling cuttings separator
钻屑试样 test drilling
钻屑收集器 swarf collector for drills
钻屑收集筒 bucket; calyx; sediment pipe; sediment tube; sludge barrel; sludge bucket
钻心虫 borer
钻芯 core bit;auger core;web
钻芯测定法 core testing
钻芯法 core drilling method
钻芯法检测 core drilling inspection
钻芯混凝土检测仪 concrete core-drilling testing apparatus
钻芯强度试验 core strength test
钻芯试件 core specimen;test core
钻芯试件强度 core specimen strength
钻芯试验 core test
钻芯试样 core sample
钻芯型机 core cutting machine
钻芯折算混凝土立方体强度 in-situ cube strength
钻形螺栓 bit bolt
钻穴 boring
钻压 bit load; bit pressure; bit weight; drilling load; drilling pressure; drilling weight; weight of the bit; weight on bit; pressure per stone <每颗金刚石所受的>
钻压表 drilling pressure meter
钻压传递 transfer of drilling weight
钻压指数 bit weight exponent
钻压转移 transfer of drilling weight
钻岩船 rock drill barge; rock drill vessel; drill boat <船侧作业,进行海底爆破>
钻岩锤滑轨 feed rails

钻岩的 lithotomous
钻岩机 machine rock drill;rock borer;rock boring machine;rock drill;rockdriller; rock drilling machine; stone drill
钻岩机合金衬片 rock drill insert
钻岩器 gadder
钻岩设备 rock drill
钻岩梯段 drilling bench
钻岩指数 drilling index
钻岩钻头 rock bit
钻研 delve; devote oneself to; dig; dive;exploration
钻眼 blast hole
钻眼爆破 drilling and blasting; hole drilling and blasting
钻眼爆破作业 drilling and blasting operation
钻眼底 hole back
钻眼工 doctor drill;drill man
钻眼进度 penetration advance
钻眼偏斜 hole scattering
钻眼设备 < 采石场 > quarry blast-hole rig
钻眼试验 drilling experiment
钻眼顺序 drilling sequence
钻眼型 shothole pattern
钻眼延米数 drillmeter
钻眼用供水筒 water-feed tank for drill
钻眼直径 drilling hole diameter
钻眼转率 drilling duty
钻眼装岩平行作业 drilling synchronized to mucking
钻眼装药 drilling and charging
钻眼装置 drilling device
钻眼作业 drilling
钻野猫探井 wildcat drilling;wildcatting
钻叶片 bit blade
钻液 drilling fluid
钻硬岩钻头 hard formation bit
钻用冲洗液 drilling fluid
钻用金刚石屑 drill bortz
钻用振击器 drilling jars
钻缘后角 lip relief angle
钻缘隙角 lip clearance
钻月 drill working-month
钻月进尺 meterage per drill working-month
钻月数 amount of drill working-month
钻月效率 meterage per drill working-month
钻越过 drill around
钻凿 boring;drilling out
钻凿表土层钻头 overburden bit
钻凿管井 bored tubular well
钻凿(井)管 drilling pipe
钻凿手锤 hand hammer rock drill
钻凿竖井 shaft boring
钻凿速度 drilling rate
钻渣 bore mud;drilling mud
钻轧头 chuck
钻至基岩的钻孔 spudded-in hole
钻至设计深度 drill to design depth
钻制喷管 drilled nozzle
钻中心孔 drill centers
钻周速 drill peripheral speed
钻轴 auger spindle; drilling machine spindle;drill spindle
钻轴滑座 drill carriage
钻轴机 line boring machine
钻轴支撑 drill carriage
钻轴支架 drill carriage;drill spindle support
钻柱 drill stem;drill string
钻柱防喷回压阀 internal preventer
钻柱升沉补偿器 drill string heave compensator
钻蛀虫 borer
钻抓斗式挖掘机 boring and clamming hole trenching machine
钻装机 jumbo loader
钻状的 subulate
钻锥 burr-drill
钻子 awl;brog;first bit;scratch awl
钻子柄把 < 木工、皮革工用 > awl haft
钻钻 < 破碎岩石或砖砌石砌体的 > bull point
钻座 boring anchor

嘴 beak;spout

嘴板 nozzle plate;tip plate
嘴唇间步 lip-sync
嘴管垫圈 nozzle tube washer
嘴角 corners of mouth
嘴脸 mug
嘴状物 beak
嘴子 mouthpiece
嘴子排成直列的固定式装包机 in-line packing machine

最

最保守时间 < 即最长时间 > most pessimistic time
最便利径路 most favo(u)rable route
最表层鱼 top skimmer
最薄的 final thin
最薄氧化层厚度 minimum oxide thickness
最不发达国家 least developed country
最不利的含水量 pessimum moisture content
最不利荷载 the most disadvantageous load
最不利荷载组合 the most disadvantageous combination of loads
最不利环路 index circuit
最不利情况 worst-case condition
最不利情况试验 most-unfavorable-condition test;worst-case test
最不利值 worst-case value
最差收益率 yield to worst
最长传输时间 maximum transmission time
最长的 longest
最长工作寿命 maximum service life
最长局内电缆 maximum in-station cabling
最长绿灯换相 < 感应信号用 > maximum green change
最长绿灯时间 maximum green period
最长区间 < 按运转时分而不是按距离计算 > longest section
最长时间 longest time;maximum duration;maximum interval
最长时间估计 < 统筹方法中,在最不利条件下(但不能考虑意外的不幸)对活动所需要的最长时间的一个估计 > pessimistic time estimate
最长使用期限 maximum service life
最长使用寿命 maximum service life
最长限制 maximum interval
最长序列 maximum length sequence
最长作业时间探索法 longest-operation-time heuristic method
最长作业顺序 longest operation sequence
最常见的 modal
最常见风向 most frequency wind direction; most frequent wind direction
最常见水位 most frequent water-level
最常见速率 < 观测车速中出现频率最高的数值 > modal speed; model speed
最常见值 modal value
最常水位 most frequency water level
最常用比例尺 most common scale
最常用记录 most-frequently-used record
最常用污染物控制技术 best conventional pollutant control technology
最常钻到的地层 predominant formation
最常钻到的岩层 predominant formation
最常作用的扭矩 prevailing torque
最迟节点时间 latest node time
最迟结束时间 latest finish time
最迟竣工时间 latest finish time
最迟开工时间 latest starting time
最迟开始时间 latest starting time
最迟时间 latest time
最迟完成日期 latest finish date
最迟完成时间 latest finishing time
最迟完工时间 latest finishing time
最迟许可日期 < 统筹方法中,指事项不致影响按时完工的最迟日期 > latest allowance date
最迟允许日期 < 统筹方法中,指事项不致影响按时完工的最迟日期 > latest allowable date
最初参数 initial parameter
最初产量 original yield
最初成本 first cost;initial cost;prime cost
最初成本法 first cost method
最初成分 initial component
最初承运人 initial carrier
最初迟钝 initial torpor
最初稠度 original consistency
最初的 initial;primal;prime;primordial
最初的负载 early loading
最初费用 first cost;initial cost
最初付费 initial lump sum
最初付款 initial payment
最初复杂模型 initially complex model
最初干缩 initial dry(ing)shrinkage
最初固结 first consolidation
最初轨道 preliminary orbit
最初滑溜 incipient skidding
最初滑行 incipient skidding
最初回收率 primary recovery
最初简单模型 initially simple model
最初静止期 initial stationary phase
最初馏分 tops
最初瞄准点 initial aiming point
最初浓度 initial concentration
最初起动期 initial start-up period
最初切线模量 initial tangent modulus
最初侵染 earliest infection
最初日产量 initial daily production
最初申请 initial application
最初投入 initial input
最初投资 initial capitalization; initial investment
最初温度 initial temperature
最初稳定期 initial stationary phase
最初效应 ancestor
最初效用 initial utility
最初颜色 initial colo(u)r
最初养恤金 initial pension
最初预算 initial budget
最初重量 initial weight
最初装备费 initial installation expenses
最初撞击点 initial impact point
最纯的石墨 finest grade of graphite
最纯的石英 spectrosil
最大 maximum[复 maximums/maxima]
最大阿贝耳扩张 maximal Abelian extension
最大安全承载能力 maximum safe bearing capacity
最大安全地震烈度 maximum safe earthquake intensity
最大安全电流 maximum safe firing current
最大安全定值 maximum safety setting
最大安全度 maximum safe capacity
最大安全荷载 maximum safe load
最大安全空速指示器 maximum safe air-speed indicator
最大安全密度 maximum safe density
最大安全浓度 maximum safe concentration
最大安全容量 maximum safe capacity
最大安全水深 maximum safe water-depth
最大安全速度 maximum safe velocity;maximum safe speed
最大安全速率 maximum safe rating; maximum safe speed
最大安全系数 maximum safety factor;ultimate factor of safety
最大安全泄量 maximum safe capacity
最大安全装药量 maximum safety charge
最大百分比超调量 maximum percent overshoot
最大摆动 full swing
最大板极输入 maximum plate input
最大半径 maximum radius
最大半径时的卸料高度 dumping height at maximum radius
最大棒料直径 maximum bar diameter
最大包迹线 maximum envelope curve
最大包络线 maximum envelope curve
最大包装箱尺寸 maximum packing case size
最大饱和度 maximum saturation
最大饱和潜水条件 maximum saturation diving condition
最大保留时间 maximum retention time
最大保水量 maximum water-holding capacity
最大保水能力 maximum water-holding capacity
最大保证出力 maximum guaranteed capability
最大保证功率 maximum guaranteed capability
最大暴露极限 maximum exposure limits
最大暴雨量 maximum storm rainfall
最大北纬 greatest north latitude
最大背景等值照度 maximum background equivalent illumination
最大泵量 maximum flow rate
最大比例(尺)maximum scale
最大比率 maximum ratio
最大比率并合器 maximal ratio combiner
最大比推力 high specific thrust
最大比值合并 maximum ratio combining
最大比重梯度层 < 分层水体中的 > picnocline
最大臂距 boom reach
最大变差值 maximum value variogram
最大变幅 maximum amplitude of variation;maximum range
最大变化 maximum change
最大变形理论 maximum strain theory
最大变形能理论 theory of maximum strain energy
最大标定声功率级 maximum rated sound-power level
最大标度 maximum scale

Z

最大标准差 maximum standard deviation
最大标准需水量 ultimate standard water demand
最大表观误差 maximum apparent error
最大冰川作用 glacial maximum
最大波动 maximum fluctuation
最大波高 maximum wave height; peak wave height
最大波高均值 average height of the highest wave
最大波浪 heaviest seas; highest wave; maximum wave
最大波振幅 maximum trace amplitude
最大驳船队 maximum barge train
最大补强 maximum reinforcement
最大不变统计量 maximal invariant statistic
最大不纯度 maximum impurity
最大不失真功率输出 maximum undistorted power output
最大部件重量 maximum part weight
最大彩色分辨率 maximum colo(u)r acuity
最大残留(限)量 maximum residue limit
最大测距 farthest range; maximum range
最大差异沉降 maximum differential settlement
最大差值 maximum difference
最大产量 maximum output; maximum yield
最大产卵区 maximum spawning area
最大铲齿长度 maximum relieving length
最大铲齿深度 maximum relieving depth
最大铲掘深度 maximum digging under ground
最大铲土角 maximum grading angle
最大铲土深度 maximum depth of cut
最大长度 extreme length; maximum overall length
最大长度反馈移位寄存器码 maximal-length FSR[feedback shift register] code
最大长度码 maximum length code
最大长度序列 maximum length sequence
最大偿付能力 ultimate solvency
最大偿债能力 ultimate solvency
最大超高度 maximum superelevation
最大超高率 maximum superelevation rate
最大超挖地段 maximum overbreak
最大超压 peak overpressure
最大潮差 extreme tidal range; extreme tide range; maximum tide range
最大潮流 maximum tidal current; strength of current
最大潮流间隙 current hour
最大潮流流速 maximum tidal current velocity; peak tidal current velocity
最大潮流速度 maximum tidal current velocity; peak tidal current velocity
最大车削长度 maximum turning length
最大车轴荷载 maximum axle load
最大车轴重 maximum axle load
最大沉淀(速)率 maximum settling rate
最大沉积中心 depocenter [depocentre]
最大沉降 maximum subsidence
最大沉降量 maximum subsidence rate
最大沉降率 maximum settling rate
最大沉降时间 maximum subsidence time
最大沉降速率 maximum subsidence rate
最大沉水大型植物生物量 maximum submerged macrophyte biomass
最大沉陷 maximum subsidence
最大称量能力 total capacity of scale
最大成孔深度 maximum drilling depth
最大承载力的极限状态 ultimate limit state
最大承载量 ultimate bearing capacity
最大承载能力 maximum carrying capacity
最大诚信原则 principle of utmost good faith
最大程度 at utmost; extreme; maximum extent
最大程度工厂组装 maximum shop assembly
最大吃水(深度) extreme draft; extreme draught; maximum draft; maximum draught
最大吃水船舶 deepest-draughted vessel
最大持水量 <土壤> maximum moisture capacity; maximum water-holding capacity
最大持水能力 maximum water-holding capacity
最大持续产水量 maximum sustained yield
最大持续出力 maximum continuous output
最大持续功率 maximum continuous rating
最大持续运行功率 maximum continuous service rating
最大持续转速 maximum continuous revolution
最大尺寸 boxed dimension; full size; maximum dimension; maximum size; out-to-out; overall dimension; peak size; top size; upper limit of size boxed dimension
最大尺寸的集料 top-size aggregate
最大尺度 maximum dimension; maximum size
最大齿轮切削模数 maximum module of gear cut
最大齿轮切削外径 maximum outside diameter of gear cut
最大充电容量 maximum charging capacity
最大充装重量 maximum charge weight
最大冲击电流 making current
最大冲击力 maximum jet force
最大冲击压力 maximum shock pressure; peak shock pressure
最大冲刷深度 limiting scour depth; maximum depth of scour; maximum erosion depth; maximum scour depth
最大冲刷线 maximum scoured line
最大抽吸速度 peak pumping speed
最大出力 maximum capacity; maximum output; peak power
最大出力运行 maximum capability operation
最大出水量 maximum yield
最大除雪距离 maximum snow clearing distance
最大除雪量 maximum snow clearing capacity
最大储藏 maximizing storage
最大储量 maximum reserves
最大处理效率 maximum treatment efficiency
最大传动比 maximum transmission ratio
最大传动功率 maximum driving power
最大传热系数 maximum coefficient of heat transfer
最大传送(速)率 maximum transfer rate
最大船舶长度 length of the largest vessel
最大船长 extreme vessel length
最大船宽 beam of ship; extreme vessel breadth
最大椽子距离 maximum rafter distance
最大吹程 maximum blow-distance
最大垂度 maximum sag
最大垂直速度 maximum vertical speed
最大磁导率 maximum permeability
最大磁感应强度 magnetic limit
最大磁化力 maximum magnetizing force; peak magnetizing force
最大磁化率 maximum magnetic susceptibility
最大次级排放量 maximum secondary discharge
最大次降水量 maximum precipitation at a time
最大粗糙宽度 roughness-width cutoff
最大存储电路 maximum-remembering circuit
最大存储时间 maximum storage time
最大存储系数 maximum storage coefficient; maximum storage factor
最大存活时间 maximum survival time
最大打料行程 <压力机的> maximum knock-out stroke
最大大肠菌污染量 maximum amount of fecal coliform pollution
最大大气浓度 maximum atmospheric concentration
最大带宽 maximum bandwidth
最大单位流量 maximum unit flow; peak unit flow
最大单位应力 maximum unit stress
最大单向功率增益 maximum unilateralized power gain
最大单值区域 maximum unambiguous range
最大导纳频率 maximum admittance frequency
最大倒棱 maximum chamfer
最大得热量 maximum heat gain
最大得热率 maximum heat gain
最大的 maxi; maximal; utmost
最大的赔偿责任 maximum liability
最大的前期压力 maximum past pressure
最大的允许气孔显示 maximum permissible porosity indication
最大的噪声 worst-case noise
最大等待时间 maximum latency
最大等效日操作量 maximum equivalent daily handling quantity
最大低压区 peak suction
最大底宽 maximum base width
最大底质利用率 maximum specific substrate utilization rate
最大抵抗力 maximal resistance
最大地表径流 maximum surface runoff
最大地面面积 maximum floor area
最大地面浓度 maximum ground concentration
最大地球弧线 great-circle line
最大地震变位 maximum earthquake deflection
最大地震力 maximum earthquake force
最大地震烈度 maximum seismic intensity
最大电力需求量 maximum electric demand
最大电流 peak current
最大电流负荷 maximum current capacity
最大电流密度 maximum current density
最大电流限制 maximum current limitation
最大电流值 maximum current value; peak value
最大电流自动断路器 maximum cutout
最大电压 crest voltage; peak voltage; spike voltage
最大电阻 maximum resistance
最大吊幅 boom-out
最大吊杆 largest derricks
最大吊杆长度 maximum crane boom
最大叠加次数 maximum stacking fold
最大动力的混合气 best power mixture
最大动力高度 maximum power altitude
最大动能 maximum kinetic energy
最大动水压 minimum hydrodynamic pressure
最大动压头 maximum dynamic(al) head
最大冻结深度 maximum depth of frozen ground; maximum depth of frozen soil
最大冻土深度 maximum depth of frozen ground; maximum depth of frozen soil
最大陡度 maximum gradient; maximum steepness
最大读数 full-scale reading; maximum reading
最大断开容量 maximum breaking capacity
最大断裂负荷 maximum breaking load
最大断裂负载 maximum breaking load
最大断面客流量 maximum cross-section passenger volume
最大堆码高度 maximum stacking height
最大对比度 maximum-contrast
最大舵角限制钮 stopper snug
最大额定工作压力 maximum normal working pressure
最大额定荷载 maximum rated load
最大额定值 maximum rating
最大额定转速 maximum rated revolutions per minute
最大二次排放量 maximum secondary discharge
最大二次污染容许量 secondary maximum contaminant level
最大发电量 maximum generating watt
最大发热量 maximum heat output
最大反复速度 maximum toggle speed
最大反射率 maximum reflectivity
最大反向电压 maximum reverse voltage; peak reverse voltage
最大反向峰压 maximum peak inverse voltage
最大反向峰值电流 maximum inverse peak current
最大反向峰值电压 maximum peak reverse voltage
最大反向阳极电压 peak inverse anode voltage
最大反应 peak response
最大返回信号 maximum return signal
最大范围 maximum extent; maximum range
最大方差法 varimax method
最大方向转动界 maximum total traverse

最大放热速率 maximum heat-release rate
最大飞行速度 maximum flying speed
最大非负边界空间 maximal non-negative boundary space
最大分布荷载 maximum distributed load
最大分级排水速率 the greatest step discharge rate
最大分泌量 maximal secretory capacity
最大分子持水度 maximum molecular specific retention
最大分子持水量 maximum molecular moisture(holding)capacity
最大分子容水量 maximum molecular moisture(holding)capacity
最大分子水溶量 minimum absorbed water volume
最大分子吸水量 maximum molecular moisture(holding) capacity;maximum molecular water content
最大风力 maximum wind force
最大风量 maximum quantity of wind
最大风蚀量 maximum deflation quantity
最大风速 extreme wind velocity;maximum wind speed;maximum wind velocity
最大风速半径 radius of maximum wind
最大风速和风切变图 maximum wind and shear chart
最大风速英里 maximum wind velocity mile;fastest mile
最大峰值 maximum peak
最大峰值电压 maximum peak voltage
最大服务车流量 maximum service volume
最大服务流量【交】maximum service flow
最大幅度 amplitude peak;maximum crest;maximum format
最大幅度信号 maximum amplitude signal
最大俯角 maximum depression
最大俯仰分角线 <推土机> mid-pitch
最大俯仰角中线 <推土机> mid-pitch
最大负担 peak load
最大负荷 breaking load;load on top;load peak;maximum load(ing);maximum utilization;peak(ing) load;ultimate load
最大负荷测试器 demand meter
最大负荷估计周期 demand-assessment period
最大负荷利用小时数 maximum loading hours
最大负荷(容)量 maximum carry-(ing)capacity
最大负荷小时 busy hour(traffic);heavy hour
最大负荷指示器 maximum demand meter
最大负压区 peak suction
最大负载 breaking load;load on top;load peak;maximum load(ing);maximum utilization;peak load;ultimate load
最大负载点 point of maximum load
最大负载控制(装置)demand control
最大负载能力 maximum load capacity
最大负载状态 full load condition
最大负值 negative peak
最大附着扭矩 maximum adhesion torque
最大复位时间 maximum reset time
最大复原 ultimate recovery
最大复原力臂对应角 angle for maximum righting lever
最大覆盖面 maximum coverage
最大概差 most probable error
最大概率 maximum probability
最大概率位置 most probability position
最大干密度 maximum dry density
最大干容重 maximum dry unit weight
最大干涉 maximum interference
最大干重 maximum dry weight
最大刚度 maximum rigidity
最大刚性 maximum rigidity
最大高程 highest elevation;maximum elevation
最大高度 maximum altitude;maximum height
最大高峰交通位置 highest traffic peak
最大高水位 maximum high water(level)
最大跟踪误差 maximum tracking error
最大工作场半径 maximum good field radius
最大工作电压 maximum operating voltage;maximum working voltage
最大工作高度 maximum working height
最大工作功率 maximum service rating
最大工作荷载 maximum working load
最大工作空间 maximum working space
最大工作面积 maximum working area
最大工作频率 maximum operation frequency
最大工作压力 maximum working pressure;peak working pressure
最大工作张力 maximum working tension
最大工作阻力 maximum operation resistance
最大公测度 greatest common measure
最大公称破碎比 nominal maximum reduction ratio
最大公共子群 greatest common subgroup
最大公因数 greatest common factor
最大公因子 greatest common factor;highest common divisor;greatest common divisor
最大公约数 greatest common divisor;highest common divisor;greatest common factor;greatest common measure
最大功 maximum work
最大功定额 maximum power rating
最大功耗 maximum power dissipation
最大功率 maximum power;maximum rating;peak power;ultimate capacity;ultimate output
最大功率传输定理 maximum power-transfer theorem
最大功率定额 maximum power rating
最大功率输出 maximizing power output
最大功率输出频率特性 power response
最大功率限制 peak power limitation
最大功率与平均功率的比 maximum-to-average-power ratio
最大功率运行方式 maximum power operation mode
最大功效的 most powerful
最大功效检定 most powerful test
最大供风量 maximum air supply
最大供水量 full supply duty;maximum amount of water supply
最大供油量限止杆 maximum feed stop lever
最大估计 maximum estimation
最大估计负荷 maximum estimated load
最大估计流量 maximum computed flow
最大估算负荷 estimated maximum load
最大估算流量 maximum computed flood
最大估算溢流 maximum computed flood
最大骨料尺寸 maximum aggregate size;top aggregate size;top-size aggregate
最大骨料粒径 maximum diameter of aggregate;maximum size of aggregate
最大固结压力 maximum consolidation pressure
最大观测降水量 maximum observed precipitation
最大惯性轴 axis of maximum inertia
最大灌溉率 maximum duty of water
最大灌水定额 maximum duty of water
最大光谱灵敏度 maximum spectral sensitivity;spectral sensitivity peak
最大光圈 full aperture;max aperture
最大光输 maximum light transmission
最大光通量 highlight flux
最大滚丝距 maximum pitch rolled
最大过冷度 maximum subcooling
最大过热温度 maximum overtemperature
最大过水能力 maximum flow capacity
最大过调量 maximum overshoot
最大过盈 maximum interference
最大海拔高度 maximum height above sea level
最大海浸阶段 thalassocratic period
最大海浸面 thalassocratic sea level
最大含尘量 maximum dust content
最大含筋量 maximum reinforcement content
最大含沙量 maximum sediment concentration
最大含沙浓度 maximum sediment concentration
最大含水量 maximum moisture content;maximum water capacity;maximum(water-)holding capacity
最大含盐量 maximum salinity
最大航程 maximum range;ultimate run
最大航程巡航 range cruise
最大航高【航测】maximum flight height
最大航行附加水深 maximum overdraught
最大耗水量 maximum(water)consumption
最大耗水率 peak use rate
最大耗氧量 maximum oxygen consumption
最大合成地面位移 maximum resultant ground displacement
最大和最小密度 <散体材料> maximum and minimum density
最大河道流量 maximum stream flow
最大河流流量 maximum stream flow
最大荷载 all-up weight;fully factored load;maximum load(ing);peak load;ultimate load(ing)
最大荷载工作量 work to maximum load
最大荷载量 maximum load(ing)
最大荷载设计法 maximum load design
最大荷载压强 peak load pressure
最大荷载重 maximum load(ing)
最大(横)截面 maximum cross-section
最大横剖面系数 maximum cross-section coefficient;maximum transverse section coefficient
最大横向调整表 maximum cross adjustment
最大横摇角 max roll(angle)
最大洪峰 maximum flood peak
最大洪峰流量 largest peak discharge;maximum peak discharge
最大洪流流量 maximum flood flow
最大洪水 maximum flood
最大洪水流量 highest flood discharge;maximum flood discharge;maximum flood flow
最大后备反应性状态 most reactive condition
最大厚度 <木材或线材的> maximum ga(u)ge
最大花型范围 maximum pattern area
最大滑动 maximum slip
最大滑行重量 maximum ramp weight;maximum taxing weight <飞机>
最大化判定 decision of maximization
最大化问题 maximization problems
最大还款额 payment cap
最大回填力 maximum backfill force
最大回填力矢量 maximum backfill vector
最大混合层厚度 maximum depth of mixed layer
最大混合深度 maximum mixing depth
最大活剪力 maximum live shear
最大或然率洪水(位)maximum possible flood
最大或然误差 most probable error
最大或然性 maximum probability
最大或然值 most probable value
最大积雪量 maximum accumulation of snow
最大基底剪力 maximum base shear
最大畸变 maximum distortion
最大畸度差 max distortion
最大极化强度 maximum polarization
最大极限 greatest limit;maximum limit
最大极限沉积流速 maximum limit deposit velocity
最大极限尺寸 maximum limit of size
最大极限应力 limiting maximum stress
最大集料尺寸 top aggregate size
最大集中函数 maximal concentration function
最大挤压力 maximum extrusion pressure
最大计算洪水(流量)maximum computed flood
最大加工长度 maximum machining length
最大加工高度 maximum machining height
最大加工直径 maximum machining diameter
最大加速层次 maximum speed-up hierarchy
最大加速度 maximal acceleration;maximum acceleration;peak acceleration
最大加速度反应 maximum acceleration response
最大加温转数 maximum warm-up speed
最大假设事故 most assumed accident
最大间隙 maximal clearance;maximum clearance
最大剪力 maximum shear
最大剪力矩 ultimate shear(ing) moment
最大剪力理论 maximum shear theory
最大剪力区 region of maximum shear
最大剪(切)强度 maximum shearing strength;peak shear strength
最大剪(切)应变 maximum shear

Z

strain
最大剪(切)应变能量理论 critical shear strain energy theory; maximum shear strain energy theory
最大剪(切)应力 maximum shear(ing) stress; ultimate shear(ing) stress
最大剪(切)应力理论 maximum shear stress theory
最大剪(切)应力面 plane of maximum shearing stress
最大剪(切)应力强度理论 maximum shear stress criterion
最大剪(切)应力屈服条件 maximum shear stress yield criterion
最大剪(切)应力值 magnitude of maximum shearing stress
最大剪(切)应力准则 maximum shear stress criterion
最大剪切阻力 peak shearing resistance
最大检测限度 ultimate detection limit
最大降落重量 <飞机> maximum landing weight
最大降深 maximum drawdown
最大降水带 zone of maximum precipitation
最大降水量 maximum(quantity of) precipitation; maximum rainfall
最大降水强度 maximum rainfall intensity
最大降水日 day of maximum rainfall
最大降雪量线 maximum snow fall line
最大降雨量 maximum(quantity of) precipitation; maximum rainfall; extreme rainfall
最大降雨面积 area of maximum rainfall
最大(降)雨强度 maximum rainfall intensity
最大降雨区 area of maximum rainfall
最大降雨日 day of maximum rainfall
最大交会角 maximum intersection angle
最大角度变形 maximum angular distortion
最大角度下降 steepest descent
最大角频率偏移 maximum angular frequency deviation
最大矫顽磁力 maximum coercivity
最大搅拌深度 maximum mixing depth
最大接地压力 maximum earth contact pressure; maximum ground contact pressure
最大接受角 maximum acceptance angle
最大接受浓度 maximum acceptable concentration; maximum admissible concentration
最大结构有效载重 maximum structural payload
最大结合 maximum combination
最大结晶区 zone of maximum crystal formation
最大结温度 <晶体管的> maximum junction temperature
最大截击距离 ultimate interception range
最大截面 frontal area; peak cross-section
最大截面的等效面积 equivalent frontal area
最大截止频率 maximum cut-off
最大解 maximal solution; maximum solution
最大解冻深度 maximum thaw depth
最大介电强度 maximum dielectric(al) strength
最大金额 maximum amount
最大紧密度 maximum tight
最大近回潮龄 age of parallax inequality
最大近震 maximum near earthquake
最大进尺 controlled footage
最大进口速度 maximum velocity of entrance
最大经济(上)升坡 maximum economic(al) ascending grade; maximum economic(al) plus grade
最大精度 limit of precision; maximum accuracy
最大井径 maximum hole size
最大井径变化率 maximum changing rate of hole size
最大井斜角 maximum bole inclined angle
最大井斜角变化率 maximum changing rate of hole inclination
最大净空 maximum clearance
最大净水头 maximum net head
最大径尺寸 absolute bore size
最大径流 maximum runoff
最大径流量 highest runoff; maximum runoff; peak(rate of) runoff; peak runoff
最大径流速度 maximum runoff rate
最大静挠度 maximum static deflection
最大静水压力 maximum hydrostatic(al) pressure
最大静态额定功率 static maximum rating
最大静止期 maximum stationary phase
最大举升高度 maximum lifting height above ground
最大距角 greatest elongation
最大距离 maximum distance; maximum range; ultimate range
最大距离可分码 maximum distance separable code
最大聚值集 greatest lower cluster set
最大卷板厚度 maximum thickness of bending plate
最大掘土角 maximum grading angle
最大掘土深度 maximum depth of cut
最大开采(产)量 maximum mining yield
最大开采速度 maximum recovery rate
最大开度 maximum opening
最大开挖宽度 maximum width of cut
最大勘探深度 maximum depth of exploration
最大抗滑势能 maximum skid-resistance potential
最大抗剪强度 peak shearing resistance; ultimate shear(ing) stress
最大抗剪阻力 maximum shear resistance
最大抗拉应力 ultimate tensile stress
最大抗磨力 maximum abrasion resistance
最大靠泊角 maximum berthing angle
最大颗粒 the largest particles
最大颗粒聚集体 <不超过1/4英寸,1英寸=0.0254米> coarse aggregate
最大颗粒尺寸 maximum particle size
最大可采深度 admissible maximum of minable depth
最大可测距离 maximal detectable range
最大可承受压力 maximum acceptable pressure
最大可持续收获量 maximum sustained yield
最大可大增益 maximum available gain
最大可得增益 maximum available gain
最大可调速度 maximum governed speed
最大可分区域 maximal separable field
最大可浮粒度 maximum floatable size
最大可见度 maximum visibility
最大可接受的有毒物质浓度 maximum acceptable toxicant concentration
最大可接受交货量 maximum quantity acceptable
最大可接受浓度 maximum acceptable concentration
最大可靠出力 maximum dependable capacity
最大可能饱和浓度 maximum possible saturation concentration
最大可能暴风雨 maximum probable storm
最大可能波浪 maximum possible wave
最大可能出水量 potential yield
最大可能的压力 maximum feasible pressure
最大可能地震 maximum possible earthquake
最大可能分布 most probable distribution
最大可能洪水(量) maximum admissible flood; maximum probable flood; probable maximum flood; maximum possible flood
最大可能降水(量) maximum possible precipitation; maximum probable precipitation; maximum probable rainfall; probable maximum precipitation
最大可能降雨(量) maximum possible precipitation; maximum probable precipitation; maximum probable rainfall; possible maximum precipitation
最大可能利用率 maximum possible operating factor
最大可能量法 maximum likelihood method
最大可能烈度 maximum possible intensity
最大可能数 most possible number
最大可能数量 most probable quantity
最大可能速率 maximum possible rate
最大可能损失 maximum probable loss
最大可能误差 maximum probable error
最大可能吸附强度 maximum possible adsorption density
最大可能性 maximum likelihood
最大可能阈 maximal hearing threshold
最大可能允许量 maximum possible magnitude
最大可能震级 maximum possible magnitude
最大可平面图 maximal planar graph
最大可取风量 maximum allowable draft
最大可容水平 maximum permissible level
最大可视误差 maximum apparent error
最大可信地面运动 maximum credible ground motion
最大可信地震 maximum credible earthquake
最大可信地震动 maximum credible ground motion
最大可信烈度 maximum credible intensity
最大可信事故 maximum credible accident
最大可用存储量 maximum available storage
最大可用风量 maximum available draft
最大可用功率 maximum available power
最大可用流量 maximum stream flow
最大可用频率 maximum usable frequency
最大可用增益 maximum available gain
最大可预测费用 maximum foreseeable cost
最大可预测价格 maximum foreseeable cost
最大可预期地震 maximum expectable earthquake
最大刻度 maximum scale
最大刻度值 maximum scale value
最大空气浓度 maximum air concentration
最大空载转速 maximum non-load speed
最大孔径 maximum diameter of hole; maximum size
最大孔隙比 maximum void ratio; void ratio in loosest state
最大孔隙度 maximum porosity
最大孔隙率 maximum porosity
最大孔隙体积 maximum pore volume
最大控顶距离 maximum distance of face control
最大库存量 maximum inventory
最大跨径 maximum span
最大跨距 maximum outreach
最大跨深比 maximum span-depth ratio
最大块尺寸 maximum lump size
最大宽度 extreme breadth; maximum breadth; maximum width; overall width
最大宽度处横梁 main beam
最大亏量 maximal deficiency
最大扩张 maximum extension
最大拉力 maximum draft; maximum draught; maximum pull; maximum tension
最大拉伸比 maximal draw ratio
最大拉伸强度 maximum tensile strength
最大拉伸强力 true tensile strength
最大拉应力 maximum tension stress
最大拉应力理论 maximum tensile stress theory
最大拉应力区范围 range of maximum tension area
最大来源 the largest source
最大浪高 maximum wave height
最大浪涌 maximum surge
最大浪涌电流 maximum surge current
最大冷(却)负荷 maximum cooling load
最大离陆重量 <航测> maximum take-off weight
最大理论流量 maximum theoretic(al) yield
最大理论压缩比 max possible compression ratio
最大力矩 maximum moment
最大力矩制动器 maxi-brake
最大利率 interest rate cap
最大利润 maximum profit
最大利用的线路容量 maximum utilized line capacity
最大利用率 peak use rate
最大利用系数 maximum utilization factor
最大粒度 limiting grain; maximum grain size; maximum particle diameter; maximum particle size
最大粒径 maximum granular size;

maximum particle diameter; maximum particle size; maximum size; size of greatest particle
最大粒径骨料 maximum size of aggregate
最大连通集 maximal connected set
最大连续 maximum continuous
最大连续产量 maximum continuous output
最大连续出力 maximum continuous rating
最大连续出水量 maximum continuous rating
最大连续电压 maximum continuous voltage
最大连续负载 maximum continuous load
最大连续工作负载 maximum continuous duty
最大连续功率 continuous maximum rating
最大连续荷载 maximum continuous load
最大连续输出功率 maximum continuous output
最大连续速率 maximum continuous rate;maximum continuous rating
最大连续转速 maximum continuous revolution
最大联锁范围 maximum interlocking range
最大亮度 greatest brilliancy;highlight bright;highlight brightness
最大量 maximum[复 maximums/maxima];maximum amount;peak;peak amount
最大量程 maximum range;meter full-scale
最大烈度 maximum intensity
最大烈度图 maximum intensity map
最大裂缝宽度 maximal crack width; maximum crack width
最大临界孔隙比 peak critical void ratio
最大临界状态的 peak critical
最大灵敏度 maximum sensitivity; peak response;peak sensitivity
最大流 high flux;max-flow;maximum flow
最大流出量 maximum output; peak output
最大流动度 maximum fluidity
最大流动温度 maximum flow temperature
最大流量 highest discharge;maximum (high) discharge; maximum rate of flow; maximum stream flow; peak discharge; peak flow; peak rate of flow;ultimate flow;maximum flow
最大流量算法 maximal flow algorithm
最大流量问题 maximum flow problem
最大流率 maximum flow rate
最大流速 maximum current velocity; maximum strength of current;maximum velocity;maximum velocity of flow;peak rate
最大流速轨迹 maximum velocity locus
最大流速线 line of fastest velocity of flow; line of maximum velocity; line of maximum stream
最大流问题 maximum flow problem
最大流最小截定理 max-flow min-cut theorem
最大轮班 maximum shift
最大轮廓面积 admissible maximum of outline area
最大轮胎负载 maximum tyre load
最大轮载 maximum wheel load
最大螺纹直径 maximum thread diameter
最大落潮流 ebb strength
最大落潮流间隙 strength of ebb interval
最大落潮流流速 maximum ebb strength;strength of ebb
最大落潮流速 maximum ebb current
最大马力 maximum horsepower
最大马力转矩 torque at peak horsepower
最大脉冲电流 maximum surge current
最大脉冲功率 peak pulse power
最大脉冲率 maximum pulse rate
最大脉冲振幅 peak pulse amplitude
最大毛管持水量 maximum capillary (water) capacity
最大毛管水量 maximum capillary capacity
最大毛水头 maximum gross head
最大毛细管力 maximum capillarity
最大毛细水量 maximum capillary capacity
最大毛细吸湿量 maximum capillary capacity
最大毛细作用 maximum capillarity
最大锚固力 maximum anchorage
最大每秒流量 maximum discharge per second;maximum secondly discharge
最大每日用水量 maximum daily (water) consumption
最大每小时交通量 maximum hourly volume
最大每小时消耗量 maximum consumption per hour;maximum hourly consumption
最大每小时需水量 maximum demand per hour
最大门电压 maximum gate voltage
最大门功率 maximum gate power
最大密度 maximum density
最大密度级配规范 maximum density gradation criterion
最大密度流 high flux
最大密实度 maximal density; maximum density
最大密实度和最佳含水量 maximum density and optimum moisture
最大密实度级配 maximum density grading
最大免费运距 limit of free haul
最大秒流量 maximum discharge per second; maximum secondly discharge
最大模 maximum norm
最大模估计量 maximum-norm estimates
最大模数 maximum modulus
最大模原理 maximum modulus principle
最大摩擦角 maximum angle of friction
最大摩擦力 maximal friction; maximum friction(al force)
最大磨矿效率 maximum grinding efficiency
最大磨削长度 maximum grinding length
最大磨削行程 maximum grinding travel
最大磨削直径 maximum grinding diameter
最大耐受(剂)量 maximum tolerated dose;maximal tolerance dose;maximum tolerance limit
最大耐受浓度 maximal tolerable concentration
最大南黄纬 greatest south latitude
最大挠曲长度 maximum flexural strength
最大能见度 maximum visibility
最大能力 maximum capacity;ultimate capacity
最大能量 ceiling capacity; maximum capacity; maximum energy; peak energy;ultimate capacity
最大能量的共振 highest resonance of energy
最大年洪水量 maximum annual flood
最大年降水量 maximum annual precipitation
最大年径流量 maximum annual runoff
最大年限 limit of age
最大黏[粘]着力 maximum adhesion
最大扭矩 acrotorque;maximum torque
最大扭矩储备系数 maximum torque rise
最大扭矩的最小点火提前角 minimum advance for best torque
最大扭矩点火正时 maximum torque spark timing
最大扭矩转速 maximum torque speed
最大扭力 acrotorque
最大浓度 maximum concentration; peak concentration
最大浓度点 maximum concentration site
最大浓度时间 peak concentration time
最大浓度水平 maximum concentration level
最大努力承担义务 best efforts commitment
最大爬坡度 maximum gradeability; maximum negotiable gradient
最大爬坡率 gradability
最大爬坡能力 gradability limit;maximum gradeability
最大爬升能力 <起重机> maximum climbing capacity
最大耙土深度 maximum scarifying depth
最大排出量 maximum throughput
最大排队长度【交】 maximum queue length
最大排放量 maximum discharge
最大排放率 maximum emission rate
最大排放浓度 maximum emission concentration
最大排高排距时的泥浆排量 mixture output at maximum height and distance
最大排量 maximum circulation rate of pumps; maximum pump discharge
最大排泥高度 maximum spoil-discharge height
最大排泥距离 maximum spoil-discharge distance
最大排水沟 <古罗马> cloaca maximum
最大排污量 maximum discharge
最大排污浓度 maximum emission concentration
最大排泄率 maximal excretion rate
最大排泄面积 ultimate spacing pattern
最大刨削长度 maximum planing length
最大刨削高度 maximum planing height
最大刨削宽度 maximum planing width
最大喷射高度 maximum jet height
最大喷射宽度 maximum jet width
最大喷速 maximum nozzle velocity
最大膨胀 maximum swelling
最大膨胀度 maximum dilatation
最大偏差 maximum deviation; maximum offset
最大偏度方向舵 limiting rudder
最大偏析 maximum segregation
最大偏心距 maximum eccentricity
最大偏移距 maximum offset
最大偏转 maximum deflection
最大偏转角 maximum deflection angle
最大频率 maximum frequency
最大频率工作 maximum frequency operation
最大频率偏(移) maximum frequency deviation
最大平均不确定性原理 maximum average uncertainty principle
最大平均功率输出 maximum average power output
最大平均日潮差 greatest mean diurnal range; maximum mean diurnal range
最大平均时间 maximum averaging time
最大平均温度 maximum mean temperature
最大平坦 maximally flat
最大平坦滤波器 maximumly flat filter
最大平坦特性 maximally flat characteristic
最大平坦响应 maximum flat response
最大屏亮度比率 maximum screen luminance ratio
最大坡道 extreme slope
最大坡道之长度 length of maximum grade
最大坡度 limiting grade;limiting gradient; maximum grade; maximum gradient;maximum slope
最大坡度线 line of greatest slope
最大坡降 limiting grade;limiting gradient;maximum gradient;maximum slope
最大期望地震 maximum expectant earthquake
最大期望利润 maximizing expected profit
最大启动电流 maximum starting current
最大起动电流 maximum starting current
最大起飞重量 maximum take-off weight
最大起重高度 maximum lifting height
最大起重荷载 maximum starting load
最大起重机 largest crane; maximum lifting capacity
最大起重量 maximum hoisting capacity;maximum lifting load;maximum load of lifting
最大气泡压力法 maximum bubble pressure method
最大汽缸容积 maximum cylinder volume
最大牵引车架 maximum traction truck
最大牵引负荷 maximum trailed load
最大牵引荷载 maximum trailed load
最大牵引力 maximum drawbar pull; maximum pull; maximum traction; maximum tractive effort; maximum tractive force;peak traction
最大前期表观压力 maximum past apparent pressure
最大潜在强度 maximum potential strength
最大嵌接斜度 maximum slope of scarf
最大腔内压力 maximum cavity pressure
最大强度 full strength; maximum intensity; maximum rate; maximum strength;peak strength
最大强度设计法 limit design(ing)

最大切削度 maximum depth of cutting edge
最大切削量 maximum cut
最大切削模数 maximum module to be cut
最大切削容量 maximum cutting capacity
最大切胁变 maximum shear strain
最大切胁强 maximum shearing stress
最大倾角 inclination maximum；maximum angle of inclination
最大倾斜角 maximum angle of inclination
最大倾卸角 maximum angle of dumping
最大清除率 maximal clearance
最大清理宽度 maximum cleanup width
最大区最小区 maxima minima
最大曲度 maximum(degree of) curvature
最大曲率 maximum(degree of) curvature
最大曲线率 point of maximum curvature
最大取水量 maximum intake water
最大圈距点 maximum turn separation point
最大全长 maximum overall length
最大燃火率 maximum firing rate
最大燃烧率 maximum combustion rate
最大燃油限制螺钉 maximum fuel limit screw
最大热负荷 maximum heating load
最大热量排放量 maximum thermal discharge
最大热通量 maximum heat flux
最大日潮差 great diurnal tidal range
最大日供气量 maximum day sendout；peak day sendout
最大日降雨量 maximum one-day rainfall
最大日较差 greatest variation of rates
最大日平均流量 maximum daily average discharge
最大日用水量 maximum daily consumption of water
最大容积 maximum volume
最大容量 heaped capacity；maximum capacity；ultimate capacity
最大容水量 maximum moisture capacity
最大容许暴露量 maximum permissible exposure
最大容许标准 maximum permissible standard
最大容许采收率 maximum permissible rate
最大容许尺寸 maximum admissible dimension；maximum permissible dimension
最大容许大气浓度 maximum permissible atmosphere concentration
最大容许电流 maximum permissible current
最大容许毒物浓度 maximum acceptable toxicant concentration；maximum allowable toxicant concentration
最大容许非效应水平 maximum allowable no-effect level
最大容许辐射量 maximum permissible exposure to radiation
最大容许负载 maximum permissible load(ing)
最大容许个体浓度 maximum permissible individual concentration
最大容许工作压力 maximum allowable working pressure
最大容许功率 maximum permissible power
最大容许轨距 maximum permissible ga(u)ge
最大容许荷载 maximum admissible load(ing)；maximum permissible load(ing)
最大容许机体浓度 maximum permissible body burden
最大容许极限 maximum permissible limit
最大容许剂量 maximum allowable dose；maximum permissible dosage；maximum permissible dose
最大容许剂量当量 maximum permissible dose equivalent
最大容许剂量率 maximum permissible dose rate
最大容许累积剂量 maximum permissible accumulated dose；maximum permissible cumulated dose；maximum permissible integral dose
最大容许量 maximum permissible amount
最大容许流速 maximum allowable velocity；maximum allowance flow velocity
最大容许满载总重 maximum permissible gross laden weight
最大容许摩擦系数 maximum allowable coefficient of friction
最大容许能级 maximum permissible level
最大容许浓度 maximum acceptable concentration；maximum admissible concentration；maximum allowable concentration；maximum permissible concentration；maximum permitted concentration；maximum tolerable concentration
最大容许排放 maximum permissible release
最大容许排放量 maximum allowable discharge
最大容许排放质量浓度 maximum allowable discharge concentration
最大容许偏离 maximum permissible deviation
最大容许日摄入量 maximum permissible daily intake
最大容许溶解氧缺量 maximum allowable dissolved oxygen deficit
最大容许摄取量 maximum permissible intake
最大容许摄入量 maximum permissible intake
最大容许生态排放量 maximum allowable ecological discharge
最大容许水面比降 maximum allowance surface gradient
最大容许水平 maximum allowable level；maximum permissible level
最大容许速度 maximum admissible velocity；maximum allowable velocity；maximum permissible speed；maximum permissible velocity；maximum permitted speed
最大容许通量 maximum permissible flux
最大容许弯沉 maximum tolerable deflection
最大容许弯曲半径 maximum admissible radius of curvature
最大容许位移 maximum allowable misalignment；maximum permissible displacement
最大容许污染物浓度 maximum allowable contamination concentration
最大容许物质浓度 maximum allowable concentration of substance
最大容许误差 limit of error；margin of error；maximum allowable error；maximum allowance error；maximum permissible error
最大容许线量<放射线> maximum permissible exposer
最大容许限度 maximum permissible level；maximum permissible limit；maximum tolerance limit
最大容许限期 maximum permitted period
最大容许效应水平 maximum allowable effect level
最大容许压力 maximum allowable pressure；maximum permissible pressure
最大容许压力降 maximum allowable pressure drop
最大容许样本尺寸 maximum allowable sample size
最大容许杂波 just tolerable noise
最大容许载重量 maximum loading capacity；maximum payload
最大容许噪声 just tolerable noise；maximum tolerable noise
最大容许噪声电平 maximum permissible noise level
最大容许增压 maximum permissible boost pressure
最大容许照射 maximal allowance exposure；maximal allowance irradiation；maximum allowance exposure；maximum allowance irradiation
最大容许照射水平 maximal allowance irradiation level
最大容许值 maximum permissible value
最大容许重量 maximum permissible weight
最大容许周剂量 maximum permissible weekly dose
最大容许转数 maximum permissible revolution
最大容许装药量 maximum permitted charge
最大容许总装载重 maximum permissible gross laden weight
最大容许纵坡度 maximum permissible gradient
最大容许阻力 maximum allowable resistance
最大溶解度 maximum solubility
最大熔化能力 maximum melting capacity
最大入渗量 ultimate infiltration capacity
最大锐度 maximum sharpness
最大撒布宽度 maximum spreading width
最大塞入率 maximum stuffing rate
最大散布 extreme spread
最大散发物浓度<空气中容许的> maximum emission concentration
最大扫描 maximum scan
最大扇出数 maximum fan-out
最大伤害度 most severe injury degree
最大熵 maximum entropy
最大熵法 maximum entropy method
最大熵反褶积 maximum entropy deconvolution
最大熵原理 principle of maximum entropy
最大上坡速度 maximum uphill speed
最大上升率 maximum climbing
最大上涌浪 maximum upsurge
最大设计波高 maximum design wave height
最大设计风速 maximum designed wind speed
最大设计距离 maximum design spacing
最大设计靠泊角度 maximum design berthing angle
最大设计流速 maximum designing velocity
最大设计坡降 maximum designing gradient
最大设计压力 maximum design pressure
最大设计轴马力 maximum designed shaft horse-power
最大设想事故 maximum credible accident
最大射程 extreme range；maximum range；range maximum
最大射角 maximum elevation
最大射速 maximum rate of fire
最大涉水深度 maximum fording depth
最大摄氮量 maximum nitrogen uptake
最大伸臂距 boom-out
最大伸长 maximum elongation；ultimate elongation
最大伸距 boom-out；boom reach；maximum outreach
最大伸缩率 maximal dilatation
最大深度 depth capacity；extreme depth；maximum depth
最大深度定律 law of greatest depth
最大深度-面积-历时资料 maximum depth-area-duration data
最大深度潜水 maximum dive
最大渗透率 maximum permeability
最大升力冲角 angle of incidence of maximum lift
最大升力角 angle of maximum lift
最大升压马力 maximum boost horse power
最大生产量 maximum output
最大生产率 full output；peak performance
最大生产能力 maximum productive capacity；peak capacity
最大生长率 maximum growth rate
最大生化需氧量 ultimate biological oxygen demand
最大生境容量 maximum habitat volume
最大生境稳定度 maximum habitat stability
最大生物容许浓度 maximum allowable biological concentration
最大生物吸附容量 maximum biosorption capacity
最大生物制氢率 maximum biohydrogen production rate
最大声级 maximum sound level
最大声级的平均能量级 mean energy level dB(A) of maximum sound level
最大声强度 maximum sound intensity
最大声压 maximum sound pressure；peak sound pressure
最大声压级 maximum sound pressure level
最大剩余消毒剂浓度指数值 maximum excess disinfectant concentration index value
最大剩余消毒剂水平 maximum excess disinfectant level；maximum residual disinfectant level
最大剩余消毒剂水平目标 maximum excess disinfectant level goal；maximum residual disinfectant level goal
最大剩余压力 peak overpressure
最大失真 maximum distortion；peak distortion
最大湿度 maximal humidity；maxi-

mum humidity

最大时用水量 maximum hourly consumption

最大实测每小时交通量 maximum observed hourly volume

最大实际流量 maximum working flow

最大实体条件 maximum material condition

最大实体原则 maximum material principle

最大实体状态 maximum material condition

最大使用荷载 limit load

最大使用流量 maximum useful discharge

最大使用频率 maximum useful frequency

最大使用压力 maximum service pressure;maximum working pressure

最大试样量 maximum allowable sample size

最大收入 marginal gain

最大收缩度 maximum contraction

最大收益 maximum return

最大舒适度的横向摩擦力 maximum comfortable side friction

最大输出 maximum output

最大输出电流 maximum output current

最大输出功率 maximum power output;peak power output

最大输出信号 spiking output

最大输入电平 maximum input level

最大输入信号电平 maximum input signal level

最大输入重复频率 maximum input repetition rate

最大束流 maximum beam

最大树【数】maximal tree

最大竖曲线半径 maximum vertical curve radius

最大双折射率 largest birefringence

最大水流 maximum stream; ultimate stream

最大水马力 maximum hydraulic horsepower

最大水面流速线 thread of channel

最大水平射程 maximum horizontal range

最大水汽压力 maximum vapo(u)r pressure

最大水汽张力 maximum vapo(u)r tension

最大水深 maximum depth of water

最大水深线 line of maximum depth

最大水头 maximum(hydraulic)head; maximum water head

最大水位降深 maximum drawdown

最大水压力 maximum water pressure

最大瞬时风速 maximum instantaneous wind speed

最大瞬时功率 maximum instantaneous power

最大瞬时流量 maximum instantaneous discharge

最大瞬时速度变化率 maximum momentary speed variation

最大似然抽取 maximum likelihood extraction

最大似然从属度函数 maximum likelihood membership function

最大似然方法 maximum likelihood method

最大似然分类 maximum likelihood classification

最大似然估计法 maximum likelihood estimate

最大似然估计量 maximum likelihood estimator

最大似然检测 maximum likelihood detection;maximum likelihood test

最大似然检验 maximum likelihood test

最大似然解码 maximum likelihood decoding

最大似然量法 maximum likelihood method

最大似然率 maximum likelihood

最大似然率准则 maximum likelihood criterion

最大似然判别法则 maximum likelihood decision rule

最大似然误差校正剖析程序 maximum likelihood error correcting parser

最大似然性 maximum likelihood

最大似然译码 maximum likelihood decoding

最大似然原理 maximum-likelihood principle

最大松散度 maximum loose

最大松土深度 maximum scarifying depth

最大送料长度 maximum feeding length

最大速度 critical velocity; maximal rate;maximum speed;maximum velocity;top speed

最大速度比 maximum velocity ratio

最大速度调节器 maximum speed governor

最大速度反应 maximum velocity response

最大速度航程 range at maximum speed;range of maximum speed

最大速度升力 top-speed lift

最大速率侧辊 maximum rate roll

最大塑性功原理 principle of maximum plastic work

最大塑性阻力 maximum plastic resistance

最大摊铺厚度 maximum depth of spread;maximum paving thickness; maximum spreading thickness

最大摊铺宽度 maximum paving width; maximum spreading width

最大弹性力矩 maximum elastic moment

最大探测距离 maximum detectable range

最大梯度 maximum gradient

最大提供的线路容量 maximum offered line capacity

最大提升高度 maximum lift height

最大提升力 <指钻机立轴> feed pull maximum

最大提升能力 maximum lifting capacity

最大体积 maximum volume

最大体积变化 peak volume change

最大天线电流 maximum antenna current

最大田间持水量 maximum field moisture capacity

最大田间容量 maximum field capacity

最大条件 maximal condition

最大调整率 maximum justification rate

最大调制度 maximum modulation; maximum percentage modulation

最大调制范围 modulation capability

最大通风量 maximal ventilatory capacity;maximal ventilatory volume

最大通过能力 maximum throughput capacity;maximum tonnage capacity;maximum traffic capacity

最大通航流量 maximum navigation discharge

最大通量 flux peak

最大通气量 maximal breathing capacity; maximal ventilatory capacity; maximal ventilatory volume

最大透过容量 maximum penetration volume

最大凸率 maximum convexity

最大突水量 maximum water bursting yield

最大图像高度 maximum picture height

最大图像宽度 maximum picture width

最大图形信号频率 maximum picture frequency

最大涂层厚度 maximum depth of spread

最大土压力边 <基础倾覆时> pressed edge

最大土压力的库仑式楔 Coulomb's wedge of maximum earth pressure

最大推力 full power;full thrust;maximum thrust;top thrust

最大退潮流(速) ebb strength;maximum ebb strength;strength of ebb

最大退磁场 maximum demagnetizing field

最大挖掘半径 boom-out; maximum cutting radius;maximum digging radius

最大挖掘高度 maximum cutting height

最大挖掘深度 maximum digging depth

最大挖宽【疏】maximum cut(ting) width;maximum dredging width

最大挖深 maximum dredging depth

最大外径 maximum outside diameter

最大外可平面图 maximal outerplanar graph

最大外伸工作半径 maximum reach

最大外伸距离 maximum reach

最大外显误差 maximum apparent error

最大弯沉 peak deflection

最大弯道半径 maximum negotiable radius

最大弯度 maximum camber

最大弯矩 maximum bending moment

最大弯矩包络线 maximum bending moment envelope

最大弯矩图 maximum bending moment diagram

最大弯矩应力 maximum bending stress

最大弯曲点 point of maximum curvature

最大维护重量 maximum maintenance weight

最大纬向西风带 maximum zonal westerlies

最大位差 greatest difference of rates

最大位移反应 maximum displacement response

最大位移量 maximum displacement

最大温度变化量 maximum temperature variation

最大温度梯度 maximum thermal gradient

最大温升 maximum temperature-rise

最大稳定倾角 maximum stable tilting angle

最大稳定性参数 peak stability number

最大稳定增益 maximum stable gain

最大稳性力臂 angle of maximum righting arm

最大握钉力 maximum nail holding power

最大污染程度 maximum polluted level

最大污染浓度带 zone of maximum pollution concentration

最大污染水平 maximum contaminant level;maximum contamination level

最大污染水平目标 maximum concentration goal

最大污染物浓度 maximum concentration of pollutant

最大污染限度 maximum contaminant limit

最大无畸变输出功率 maximum undistorted output

最大无觉察到的效应浓度 maximum non-observed effect concentration

最大无觉察到的效应水平 maximum non-observed effect level

最大无失真输出功率 maximum undistorted power output

最大无影响剂量 maximum non-effect dose

最大无影响量 maximum non-effect level

最大无影响浓度 maximum non-observed effect concentration

最大无作用浓度 maximum non-effect level

最大误差 maximal error; maximum error

最大吸程 maximum sucking height

最大吸附 pH 值 pH of maximum adsorption

最大吸附量 maximum adsorption quantity

最大吸附密度 highest adsorption density

最大吸附强度 maximum adsorption density;maximum sorption density

最大吸附容量 maximum adsorption capacity;maximum sorption capacity

最大吸气量 maximal inspiratory capacity

最大吸入高度 maximum suction lift

最大吸入压力 maximum suction pressure

最大吸湿度 maximum hygroscopicity

最大吸湿水 maximum hygroscopicity

最大吸湿性 maximum hygroscopicity

最大吸收 absorption maximum;maximum absorption

最大吸收波长 maximum absorption wavelength

最大吸收剂量 maximum absorbed dose

最大吸着水溶量 maximum absorbed water volume

最大系数 greatest coefficient

最大系柱拖力 maximum bollard pull

最大下界 greatest lower bound;great lower boundary;infimum【数】

最大下潜深度 maximum diving depth

最大下限 greatest lower bound;great lower boundary

最大纤维应力 maximum fibre stress

最大显示面大小 maximum display surface size

最大线流速 maximum filamental flow

最大线路通过能力 maximum line capacity;maximum track capacity

最大线性过滤 optimum linear filtering

最大限度 full range; highest limit; high limit;maximal limitation;maximally;maximum limit;superior limit;upper limit

最大限度利润 profit maximization

最大限度载重线 deepest loadline; deep load-line

最大限价 ceiling price

最大限界 maximum clearance

最大限载开关 maximum load limit switch

最大限制示像 most restricted aspect

最大限制信号 most restrictive signal

最大限制应力 limiting maximum stress

最大相 maximal phase

最大相对变位 net shift

Z

最大相对孔径物镜 lens of extreme aperture
最大相对时间间隔出错 maximum relative time interval error
最大相似性 maximum likelihood
最大相位 maximum phase; uppermost phase
最大向量 maximum vector
最大项 maximal term; maxterm
最大项展开 maxterm expansion
最大消光角 largest extinction angle
最大消耗量 consumption peak; maximum consumption
最大小时供气量 maximum hourly sendout
最大小时耗水量 maximum hourly consumption
最大小时降雨量 maximum hourly rainfall
最大小时交通量 maximum hourly volume
最大小时需量 maximum hourly demand
最大小时需水量 maximum hourly requirement; maximum hourly water demand
最大小时用水量 maximum hourly water consumption
最大小型系统 maxi-mini system
最大效果 maximum efficiency
最大效率 maximum efficiency
最大效率流量 normal discharge
最大效率时的负荷 best load
最大效率时的负载 most efficient load
最大效用 maximum utility
最大挟砂力 limiting tractive power
最大卸载半径 maximum dumping radius
最大卸载高度 maximum dump clearance; maximum dumping height; maximum unloading height
最大卸载角 maximum dumping angle
最大卸载角时的卸载高度 height at full dump
最大信号电平 maximum signal level; peak signal level
最大信号法 maximum signal method
最大信号识别时间 maximum signal recognition time
最大信息组长度 maximum block length
最大行程 maximum range; maximum stroke
最大行驶速度 maximum travel(l)ing speed
最大需量功率计 maximum demand power meter
最大需量记录器 maximum demand recorder
最大需量收费制 maximum-demand tariff
最大需量瓦时计 maximum demand watt-hour meter; watt-hour demand meter
最大需量指示器 maximum demand indicator
最大需求量 maximum demand; peak demand
最大需热量 maximum heat demand
最大需水量 maximum water consumption; maximum water need
最大需水日 day of maximum demand
最大需要量 maximum demand; peak demand
最大需用量记录器 demand register
最大许可速度 maximum allowable velocity
最大许可压力 maximum allowable pressure
最大许可直径 maximum admitted diameter
最大许可重量 maximum admitted weight
最大许可转数 maximum permissible revolution
最大许用风力 maximum wind condition
最大许用工作压力 maximum allowable working pressure
最大许用应力 maximum permissible stress
最大序模 maximal order(module)
最大续航距离 maximum range
最大续航力 range maximum
最大续航时间 maximum endurance
最大蓄量水位 height of the maximum storage
最大蓄水高度 capacity level
最大蓄水水位 maximum storage level
最大旋回纵距 maximum advance
最大旋转速度 maximum rotative speed
最大旋转压实次数 maximum number of gyrations
最大雪崩 climax avalanche
最大雪量 maximum snow
最大循环剪应力 maximum cyclic(al) shearing stress
最大压力 maximum pressure; peak pressure
最大压力调节器 maximum pressure governor
最大压力阀 maximum pressure valve
最大压力拱 maximum pressure arch
最大压力线 line of maximum pressure; maximum pressure line
最大压实干密度 maximum compacted dry density
最大压实干容重 maximum compacted dry unit weight
最大压碎强度 crushing strength at maximum load; maximum crushing strength
最大压缩变形 maximum compression deflection
最大压缩轴 axis of maximum compression; maximum compression axis
最大压头 maximum head
最大压位防松螺母 nut for maximum pressure head check
最大压下量 maximum reduction
最大压应力 maximum crushing stress
最大压应力轴方位 orientation of maximum compressive stress
最大烟囱高度 maximum stack height
最大延长 maximum extension
最大延长绿(灯)设置 <感应信号用> maximum extension setting; maximum vehicle-extension green period
最大岩石压碎强度 maximum rock crushing strength
最大盐度层 maximum salinity layer
最大盐碱度 maximum salinity
最大阳极反向电压 crest inverse anode voltage
最大阳极正向电压 crest forward anode voltage
最大仰角 maximum elevation
最大野外持水量 maximum field moisture capacity
最大叶宽比 maximum blade width ratio
最大已知洪水 maximum known flood
最大异常值 maximum abnormal value
最大溢流量 maximum overflow
最大音响 acoustic(al) maximum
最大应变 maximum strain
最大应变理论 maximum strain theory
最大应变能 maximum strain energy
最大应变能理论 theory of maximum strain energy
最大应变破坏理论 maximum strain theory of failure
最大应变轴 maximum strain axis
最大应变轴方位 orientation of maximum strain axis
最大应力 fundamental stress; maximum stress; ultimate stress
最大应力的荷载 loading for maximum stress
最大应力极限 maximum stress limit
最大应力理论 maximum stress theory
最大盈利运距 limit of profitable haul
最大硬度 peak hardness
最大涌水量 maximum(water) yield
最大用电日 day of maximum demand
最大用水量 maximum(water) consumption
最大优越性 prime advance
最大有感距离 maximum distance of perceptibility
最大有利运距 limit profitable haul
最大有效采收率 maximum efficient rate
最大有效含水量 maximum available moisture content; maximum available water; readily available water
最大有效荷载 payload capacity
最大有效记录速度 maximum usable writing speed
最大有效库容 maximum available storage
最大有效力 maximum effect force
最大有效率 maximum effective rate
最大有效日输送量 maximum load available daily
最大有效射程 maximum effective range
最大有效生产率 maximum efficient rate of production
最大有效水分 maximum available moisture; maximum available water
最大有效网宽 maximum deckle
最大有效位 most significant bit; most significant digit
最大有效蓄水 maximum available storage
最大有效载荷比 maximum payload ratio
最大有效增益 maximum available gain
最大诱导流量 maximum flow induction
最大预估(计)沉降量 maximum anticipated settlement
最大预应力筋拉应力 jacking stress
最大远震 maximum far earthquake
最大越壕宽 maximum width of ditch
最大越野行驶速度 maximum speed on cross-country
最大允许曝光量 maximum permissible exposure
最大允许表速 maximum permissible indicated speed
最大允许操作温度 maximum allowable operating temperature
最大允许电流 current-carrying conductor
最大允许电压波动 maximum permitted voltage pulsation
最大允许定位 maximum permissible setting
最大允许辐照量 maximum permissible exposure
最大允许高度 maximum allowable level
最大允许工作压力 maximum allowable working pressure
最大允许含尘量 maximum permissible dustiness
最大允许机体负荷 maximum permissible body burden
最大允许极限 maximum permissible limit
最大允许剂量 maximum permissible dose
最大允许加速度 maximum permissible acceleration
最大允许量 maximum magnitude; maximum permissible amount
最大允许流速 maximum permissible velocity
最大允许摩擦系数 maximum allowable; maximum allowable coefficient of friction
最大允许浓度 exempt concentration; maximum admissible concentration; maximum allowable concentration; maximum permissible concentration
最大允许偏差 maximum allowable offset; maximum permissible deviation
最大允许牵引荷载 maximum permissible drawbar load
最大允许倾斜度 permissible maximum inclination
最大允许人体负荷 maximum permissible body burden
最大允许使用温度 maximum permissible service temperature
最大允许试样量 maximum allowable sample size
最大允许水平 maximum allowable level
最大允许水位下降值 maximum allowable water table decrease
最大允许速度 maximum permissible speed; maximum permissible velocity
最大允许通风损失 maximum allowable draft loss
最大允许污染物 maximum allowable contaminant
最大允许误差 margin of error
最大允许信号 most favo(u)rable signal
最大允许压力 maximum permissible pressure
最大允许应力 maximum permissible stress
最大允许有效负荷 maximum allowable payload
最大允许载重 maximum allowable load
最大允许值 maximum permissible value
最大允许轴荷重 maximum admissible axle load
最大允许抓取重量 maximum allowable weight capacity
最大允许转速 maximum permissible speed
最大运量 peak load
最大运输重量 maximum transport weight
最大运算电压 maximum machine voltage
最大运行尺寸 maximum moving dimension
最大运行限界 maximum moving dimension
最大运转(效)率 maximum operating efficiency
最大载波电平 maximum carrier level
最大载荷 maximum gross weight; maximum load
最大载荷力矩 maximum load moment

Z

最大载客段(公交) maximum load point
最大载油量航程 range with maximum tankage
最大载重 load limit; maximum load(ing)
最大载重吨位 maximum dead weight
最大载重量 ceiling load; maximum gross; maximum payload
最大载重水线 deep-water line
最大噪声等级 maximum noise level
最大增压 full boost; maximum boost; maximum pressure boost
最大增压功率 maximum boost horse power
最大增益 maximum gain; ultimate gain
最大增重的 top-gaining
最大站间距 maximum station interval
最大张力 maximum tension
最大张应力 maximum tensional stress
最大涨潮流 strength of flood
最大涨潮流间隙 flood interval; strength of flood interval
最大涨潮流速 maximum tidal flood strength; tidal flood strength; flood strength; maximum flood
最大照度 maximal illumination
最大真空度 maximum suction; peak suction
最大阵风 maximum gust; peak gust
最大振荡 full swing
最大振幅 maximum amplitude; maximum crest
最大振幅滤波器 maximum amplitude filter
最大振幅脉冲 maximum amplitude pulse
最大震相 maximal phase
最大蒸发量 evaporation capacity
最大蒸发率 maximum evapo(u)ration rate
最大蒸气压力 maximum vapo(u)r pressure; top steam
最大蒸汽工作压力 maximum steam working pressure
最大正弯矩 maximum positive moment
最大正向峰值电流 maximum peak forward current
最大正向峰值电压 maximum peak forward voltage
最大正应变理论 maximum normal strain criterion
最大正应力理论 theory of maximum normal stress
最大支承间隙 maximum support clearance
最大支配时间 maximum dominating time
最大直径 maximum diameter; peak diameter; maximum ga(u)ge <木材或线材的>
最大值 crest value; maximum [复 maximums/maxima]; maximum rating; maximum value; peak(ing) value; ultimate value
最大值比 maximum ratio
最大值的 maximal
最大值法 maximization[maximisation]
最大值和平均值之比 peak-to-mean ratio; peak-to-average ratio
最大制动功率 maximum brake power; peak brake power
最大制动力矩 maximum brake torque
最大制动坡道 maximum braking gradient
最大制动坡度 maximum braking gradient
最大致死剂量 maximum lethal dose
最大致死浓度 maximum lethal concentration
最大致死氧浓度 maximum lethal oxygen concentration
最大滞后距 maximum lag
最大置信事故 maximum credible accident
最大中心距 maximum distance between centers
最大中止时间 maximum suspension time
最大重量 all-up weight; maximum weight
最大重吸收率 maximal reabsorption rate
最大轴负载 maximum axle load
最大轴功率 maximum shaft power
最大轴荷载 maximum axle load; maximum load per axle
最大轴荷重 maximum axle load; maximum load per axle
最大主能量 maximum principal energy
最大主应变 major principal strain; major principle strain; maximum principal strain; maximum principle strain
最大主应变理论 maximum principal strain criterion
最大主应力 major principal stress; major principle stress; maximum principal stress
最大主应力方位角 direction of the maximum principal stress
最大主应力理论 maximum principal stress criterion
最大主应力倾角 dip of the maximum principal stress
最大主应力值 maximum principal stress value
最大主应力轴 axes of major stress
最大住房密度 maximum net residential density
最大贮存量 ultimate storage
最大转舵位置 hard over position
最大转角 hard over
最大转矩 maximum rotary torque; maximum (running) torque; peak torque
最大转数 maximum revolution
最大转数旋转的发动机 engine turning at peak revolution
最大转速 maximum rotary speed; top speed
最大转弯能力 turning capacity
最大转向角 steering locking angle
最大转移能量 maximum energy transfer
最大装夹直径 <卡盘的> chucking capacity
最大装模高度 maximum die set height
最大装卸率 <油轮> maximum loading rate
最大装药量 maximum charge
最大装载量 maximum load(ing); maximum payload
最大装载限界 load-limit ga(u)ge; load limiting ga(u)ge; maximum loading ga(u)ge
最大装载状态 fully laden state
最大准则 largest value guideline
最大自共轭子群 maximum self-conjugate sub-group
最大总长度 maximum overall length
最大总垂度 maximum total sag
最大总高度 maximum total height
最大总量 maximum gross weight
最大总效率 maximum overall efficiency
最大总盈利 maximum overall profitability
最大总质量 maximum gross mass
最大总重量 maximum gross weight
最大总资源量 maximum total resource
最大纵距 maximum longitudinal distance
最大纵坡(度) maximum longitudinal grade; maximum longitudinal gradient; ruling grade; ruling gradient; rolling gradient
最大纵坐标 maximum ordinate
最大阻抗频率 maximum impedance frequency
最大阻力 hump drag; maximum drag; maximum resistance
最大钻进深度 drilling capacity
最大钻孔深度 maximum drilling depth
最大钻孔直径 drilling capacity; maximum drilling diameter
最大钻速 maximum penetration rate
最大钻头水马力 maximum bit hydraulic horse power
最大钻压 maximum bit weight
最大最大准则 guideline of largest of largest value
最大最小平均值 extreme mean
最大最小系统 minimax system
最大最小限度 high low limit
最大最小原理 maximin principle; maximum principle
最大最小值计数法 mean-crossing peak count method
最大最小值原理 maximum-minimum principle
最大最小准则 guideline of largest of smallest value
最大作业宽度 maximum operating width
最大作业深度 maximum operating depth
最大作业压力 maximum operating pressure
最低 minimum[复 minima/minimauls]
最低安全高度 minimum safe altitude
最低安全水位 lowest safe waterline
最低报价 best offer; lowest bid; lowest quotation; lowest tender
最低标 lowest bid; lowest tender sum
最低标高 lowest elevation; lowest level
最低标价 lowest bid price
最低标准 minimal standard; minimum standard
最低波动价位 minimum price fluctuation; minimum price movement
最低波谷 lowest wave trough
最低部分 lowest part; rock bottom【地】
最低残留饱和度 irreducible saturation
最低草温 grass minimum
最低草温表 grass minimum thermometer
最低层 rock bottom
最低层大气 lowest atmospheric layer
最低层节点 lowest level node
最低偿债力 margin of solvency
最低潮 dead tide; neap; neap tide
最低潮位 lowest water level
最低潮位标志 lowest tide water mark
最低车速标志【交】 minimum speed sign
最低成本 cost minimization; least cost
最低成本法 least cost method; minimum-cost rule
最低成本估算 least cost estimating
最低成本估算与调度法 least-coat estimating and scheduling system
最低成本估算与调度系统 least-coat estimating and scheduling system
最低成本组合 least cost combination
最低成膜温度 minimum film-forming temperature
最低持续速度 minimum continued speed
最低抽样频率 minimum sampling frequency
最低出口价格 floor export price
最低出力 minimum output
最低储备定额 minimum reserve quota
最低储备金 minimum reserves
最低触发频率 minimum toggle frequency
最低船运通知 final shipping instruction
最低存活时间 minimum survival time
最低存量法 minimum stock method
最低大气层 lowest atmospheric layer
最低单位成本 minimum unit cost
最低单位费用 minimum unit cost
最低的 lowest; minimal; rock bottom; undermost; minimum [复 minima/minimauls]
最低的频率 underfrequency
最低的投标 lowest tender
最低低潮 dead low water; lowest low tide; lowest low water; extreme low water
最低低潮位 lowest low water level; extreme low water
最低低水位 lowest low water level
最低地下水 basal water; lowest ground water
最低地下水层 basal water
最低地下水位 basal water level; lowest stage of groundwater table; phreatic low
最低点 lower-most point; lowest point; minimum [复 minima/minimauls]; nadir(point)【天】
最低电压 minimum voltage
最低吊窗 bottom hung window
最低订货量 minimum quantity per order
最低定价法 penetration pricing
最低端 least significant end
最低额应缴股本 minimum subscription
最低发火电流 minimum firing current
最低反复频率 minimum toggle frequency
最低反应剂量 minimum reaction dose
最低反应浓度 minimum response concentration
最低飞行高度 minimum flight altitude
最低费用 minimum charges
最低费用运转 least cost operation
最低风险 prime risk
最低负荷 base-load <发电厂>; minimum load
最低负荷运行 minimum load operation
最低负载 minimum load
最低感应浓度 minimum response concentration
最低高潮位 lowest high water
最低高程 lowest level
最低高度 minimum altitude
最低高水位 lowest high water
最低工资 minimum wage
最低工资水平 minimum earnings level
最低工作电流 minimum running current
最低工作频率 lowest operating frequency
最低工作水位 bottom water level
最低公倍数 lowest common multiple
最低公分母 lowest common denominator
最低供应量 minimum supply
最低共沸混合物 minimum boiling

point azeotrope
最低鼓风量 critical air blast
最低观测频率 lowest observed frequency
最低规格 minimum specification
最低国家标准 minimal national standard
最低含量 minimum content
最低含氧层 layer of oxygen minimum
最低函数 minimum function
最低航速 dead slow(speed)
最低合格标单 lowest responsive bid
最低合格投标人 lowest qualified bidder;lowest responsible bidder
最低合格值 minimum acceptance value
最低环境标准 minimum environment standard
最低击穿电压 minimum breakdown voltage
最低激发电位 minimum excitation potential
最低级的 first degree
最低级的板材 wrack
最低极限值 liminal value
最低几个振型 first several modes
最低计费字数 minimum number of chargeable words
最低记录 all-time low
最低记录值 lowest value on record
最低记载水位 lowest recorded stage
最低剂量 lowest dose level
最低价的 knock down
最低价的报盘 lowest offer
最低价(格) bedrock price; bottom price; floor price; minimum price; price floor; reserve price; rock bottom price; the lowest price; reservation price
最低价格定价法 penetration pricing
最低价投标人 lowest bidder
最低价投标者 lowest bidder
最低检出放射性强度 minimal detectable activity
最低检出量 limit of identification; lower limit of detectability; minimum detectable limit
最低检出浓度 concentration limit
最低降雨量 minimum precipitation
最低阶段 minimum stage
最低结算费率 minimum accounting rate
最低借款利率 minimum lending rate
最低径流量 minimum runoff
最低居住标准 minimum standards of occupancy
最低均匀 lowest evenness
最低可采厚度 admissible minimum of minable thickness; minimum minable thickness
最低可采极限 pay limit
最低可采宽度 admissible minimum of minable width
最低可采平均品位 admissible minimum of average grade
最低可达排放(速)率 lowest achievable emission rate
最低可见色度差 least perceptible chromaticity
最低可见有害作用水平 lowest observed adverse effect level
最低可见作用水平 lowest observed effect level
最低可能低潮面 lowest possible low water
最低可能价格 lowest possible price
最低可听频率 lower limit of hearing
最低可听强度 minimal audible intensity
最低可听阈 minimum audible threshold
最低可嗅度 minimal identifiable odo(u)r
最低可用高频 lowest useful high frequency
最低可用频率 lowest possible frequency; lowest usable frequency; minimum usable frequency
最低客舱甲板 lowest passenger deck
最低课税限度 lowest taxable limit
最低空位凝聚速率 minimum rate of vacancy condensation
最低枯水位 extreme low water; lowest ever known water; lowest everknown water level
最低库存 base stock bin
最低库存量 minimum quantity of stores
最低库存量投资 minimum inventory investment
最低库容 inactive storage
最低库水位 lowest reservoir level; lowest storage level; minimum pool level
最低冷态起动温度 minimum temperature of cold starting
最低离地高度 minimum terrain clearance altitude
最低利率 prime rate
最低利润 minimum profit
最低量 minimum[复 minima/minimauls]
最低量两部式预付电度表 minimum two-part prepayment meter
最低临界速度 lowest critical velocity
最低流量 minimum discharge; minimum flow
最低流速 minimum(current) velocity
最低率 minimum rate
最低脉冲调制频率 minimum sampling frequency
最低米百分数 admissible minimum of average meter percent
最低耐火花电压 minimal sparking voltage resistance
最低能见度 minimum visibility
最低能量 lowest energy; minimum energy
最低年工资保证制 guaranteed annual wage plan
最低拍卖价 price lining
最低拍卖限价 upset price
最低配筋率 minimal percentage of reinforcement
最低票价 cheapest fare; lowest fare
最低品位 marginal grade; minimum grade
最低平均数 lowest average
最低评标价投标 lowest evaluated bid
最低起爆电流 minimum firing current
最低起动试验 pull-up test
最低起动转矩 pull-up torque
最低气象条件 weather minimum
最低气压 baric minimum
最低清偿能力 minimum liquidity
最低燃料消耗率 best economy rating
最低人口 minimum population
最低容许可靠性 minimum acceptable reliability; minimum allowable reliability
最低容许溶解氧量 minimum allowable dissolved oxygen
最低溶解氧浓度 minimum dissolved oxygen concentration
最低上游库水位 lowest upper pool elevation
最低射角 minimum quadrant elevation
最低生产成本 lowest production cost
最低生长温度 minimum growth temperature
最低生活保障制度 minimum living standard security system
最低生活水平 minimum standard of living
最低生活水准 absolute standard of living
最低生境要求 minimum habitat
最低时延 minimum-time lag
最低实测水位 lowest recorded stage
最低使用频率 lowest useful frequency
最低收益率 minimum rate of return
最低售价 minimum selling price; upset price
最低数位 least significant digit
最低水价 minimum water rate
最低水泥含量 minimum cement content
最低水泥用量 minimum cement content
最低水平 rock bottom
最低水平容许水质标准 lowest level allowing water quality standard
最低水头 minimum head
最低水位 idler level; lowest water level; minimum water level; minimum (water) stage; gulder <枯水期>
最低水位出现时间 time of minimum water level
最低水位基准面 lowest low water datum
最低水位水尺 minimum stage ga(u)ge
最低税率 minimum tariff
最低死亡率 minimum mortality
最低送风温度 minimum delivered air temperature
最低速度 creep speed; dead slow (speed); lowest speed; minimum speed
最低速率 minimum rate
最低速率限制 minimum speed limit
最低提单费 minimum bill of lading charges
最低天文潮 lowest astronomical tide
最低天文潮位 lowest astronomical tide level
最低天线高度 minimum antenna height
最低条件 lowest term; minimum condition
最低通航流量 lowest navigable discharge
最低通航水深 minimum navigable depth; minimum navigation depth
最低通航水位 lowest navigable stage
最低通航水位保证率 guaranteed rate of the lowest navigable stage; guaranteed rate of the lowest navigable water level
最低尾水位 minimum tailwater level
最低位 least significant bit; lowest order
最低位数字 least significant digit; low-order digit
最低位有效数字 least significant digit
最低位置 extreme lower position
最低位字符 least significant character
最低温差 lowest temperature; minimal temperature; minimum temperature difference; temperature approach
最低温度 lowest temperature; minimum temperature; nadir
最低温度表 minimum thermometer
最低污染 minimum pollution
最低下的 bottom-most
最低下涌浪 extreme downsurge; lowest downsurge
最低现金余额 minimum cash balance
最低限刺激 liminal stimulus
最低限度 inferior limit; minimum[复 minima/minimauls]; rock bottom; threshold
最低限度的 minimal
最低限度的保养 minimum maintenance
最低限度耕作法 minimum till
最低限度律 law of minimum
最低限度培养基 minimal medium
最低限额 zero norm
最低限额价值 floor value
最低限额缴款数 least payment
最低限价 floor price; lowest price limit; minimum price; price floor
最低限速标志 minimum stated-speed sign
最低限值 threshold limit value
最低响应标价 lowest responsive bid
最低响应浓度 minimum response concentration
最低项 lowest term
最低消费量 minimum consumption
最低消落水位 minimum drawdown level
最低销售额 minimum sale
最低小潮 dead neap
最低泄降水位 minimum drawdown level
最低辛烷值 minimum octane rating
最低信号电平 minimum signal level
最低信号识别时间 minimum signal recognition time
最低性能标准 minimum performance criterion
最低需水量 minimum water requirement
最低需要量 minimum requirement
最低许用金属温度 minimum permissible metal temperature
最低压力 minimal pressure; minimum pressure
最低岩粉用量 least quantity of rock powder
最低要求 minimum requirement
最低要求辐射功率 lowest required radiating power; minimum required radiation power
最低要求接收电场强度 minimum required receiving field intensity
最低要求效率 minimum required efficiency
最低一级楼梯踏步 <两侧鼓状的> drum head
最低溢流高程 <溢洪道顶部的> top water level
最低溢流量 overflow crest
最低银行利率 lowest interest rate
最低应存款 compensating balance; compensatory balance
最低应税所得额 minimum taxable income
最低应税限度 minimum taxable ceiling
最低盈利 minimum profit
最低用水量 minimum water consumption
最低油位 idler level
最低有效功率 lowest effective power
最低有效鼓风 <相邻单元炉缸及相对单元炉缸刚好接触时的鼓风> working minimum blast
最低有效剂量 minimum effective dose
最低有效量 minimal effective dose
最低有效数位 least significant digit
最低有效位 least significant bit; least significant position
最低有效位数 least significant digit
最低有效位组 least significant character

Z

最低有效温度 minimum effective temperature
最低有效信号 minimum useful signal
最低有影响浓度 lowest observed effect concentration
最低预定收益率 minimum reserved rate of return
最低预付的保险费 minimum deposit premium
最低预付押金 minimum deposit
最低预期报酬率 minimum desired rate of return
最低允许温度 minimum permissible temperature
最低运费 minimum freight
最低运费率 minimum freight rate
最低运费提单 minimum bill of lading
最低运价率 cheapest rate; minimum rate
最低运行水位 minimum operating level; bottom water level <挡水结构后面水体的>
最低载波电平 minimum carrier level
最低照度 minimal illumination; minimum illumination
最低振型 lowest mode of vibration
最低正常潮位 lowest normal tide
最低正常低潮 lowest normal low water
最低正常水位 lowest normal tide
最低正常运行水位 minimum normal operational level
最低值 minimum[复 minima/minimals]
最低值温度计 minimum thermometer
最低制动速度 minimum braking speed
最低致死剂量 minimum lethal dose
最低致死浓度 minimum lethal concentration
最低致死温度 fatal low temperature
最低致死氧浓度 minimum lethal oxygen concentration
最低住房标准 minimum housing standard
最低住房(标准)法规 minimum housing code
最低着火点 lowest ignition point
最低着火能 minimum ignition energy
最低自留额 minimum retention
最低自燃温度 lowest self-ignition point
最低总标价 lowest tender sum
最低总成本法 least total cost method
最低租金 minimum rental
最低租赁支出 minimum lease payment
最低钻探工作量 <美国法律规定每年的> assessment work
最低最佳抗张力 lowest ultimate tensile stress
最底点 nadir
最底元素 bottommost element
最东的 easternmost
最陡坡度 angle of steepest slope; steepest gradient
最陡上升 steepest ascent
最陡上升抽样 steepest ascent sampling
最陡上升法 steepest ascent method; steepest ascent procedure
最陡下降 steepest descent
最陡下降法 method of steepest descent; steepest descent method; steepest gradient method
最陡下降近似法 steepest descent approximation
最陡下降算法 steepest descent algorithm
最陡下降向量 steepest descent vector
最短摆 minimum pendulum
最短曝光时间 minimum exposure time
最短波长 minimal wave length; wavelength minimum
最短测地线 shortest geodesic
最短超车视距 passing minimum sight distance
最短程 minimal path
最短程线的 geodesic
最短初始绿(灯) <全感应信号的> minimum initial green
最短初始绿灯时间 minimum initial green period
最短传输时间 minimum transmission time
最短的停机时间 minimum down time
最短等待时间编码 minimal-latency coding; minimum access coding; minimum delay coding; minimum latency coding
最短工期合同 minimal time contract
最短航程 beeline
最短航线 orthodrome
最短间隔时间 <公交前车离站与后车进站之间的> bus-stop clearance time; clearance time at bus stop
最短距离 beed line; beeline; geodetic distance; minimum distance; shortest distance
最短距离解码器 minimum-distance decoder
最短距离码 minimum distance code
最短距离译码 minimum-distance decoding
最短距离译码器 minimum-distance decoder
最短可觉时间 least perceptible duration
最短历时 minimum duration
最短路程 minimum path
最短路径 beeline; minimal path; minimum path; shortest path; shortest route
最短路径法 critical path method; minimal path method
最短路径蔓 minimum path vine
最短路径树 minimum path tree
最短路径算法 shortest-path algorithm
最短路径问题 shortest path problem; shortest route problem
最短路线 critical path
最短路线计算机 quickest route computer
最短绿 <半感应信号> minimum green
最短绿灯时间 minimum green period; minimum green time
最短盘车时间 minimum turning time
最短摄影基线长度 shortest photobaseline
最短渗流路线 shortest path of percolation
最短时程 brachistochrone; least time path; minimum-time path
最短时间 crash time; minimum duration; shortest time
最短时间操舵 time optimal steering
最短时间定律 law of least time
最短时间估计 <统筹方法中,若一切事项都出乎意料地顺利,则活动在这段时间之内完成> optimistic time estimate
最短时间控制 minimal time control; minimum-time control
最短时间路径树 minimum-time path tree
最短时间原理 least-time principle
最短事件内时间 minimum interevent time
最短视距 minimum sight distance
最短试验时间 minimum test duration
最短寿命 minimum life
最短停产时间 minimal down time
最短停车视距 non-passing minimum sight distance
最短停止距离 short stopping distance
最短通达距离 minimum service range
最短通路算法 minimum path-length algorithm
最短线 line of shortest length; geodetic line; geodesic line <沿地球面的>
最短线的 geodesic; geodetic
最短线式结构 geodetic construction
最短信号距离码 minimum distance code
最短主臂 basic jib
最短转弯路线 minimum turning path
最多 maximum[复 maximums/maxima]
最多车速 modal speed
最多车速段 pace
最多风向 dominant wind direction; most frequent wind direction
最多滚子数 maximum number of rollers
最多居住人数 maximum occupancy
最多票数 <美> plurality
最多球数 maximum number of balls
最恶劣的条件 extreme condition; extremely hard condition
最恶劣使用条件 extreme operating condition
最繁忙时间 peak hour
最符合曲线 best fitting curve
最概然速度 most probable velocity
最概然速率 most probable speed
最概然误差 most probable error
最概然值 most probable value
最干旱期 driest period
最干硬稠度 <即最小坍落度> lowest slump; stiffest consistency
最高 maximum[复 maximums/maxima]
最高安全工作压力 maximum safe working pressure
最高安全速度 maximum safe speed; maximum safe velocity
最高班进尺 maximum shift footage
最高报价 highest bid; highest quotation; highest tender
最高泵压 maximum pump pressure
最高变速比 top gear
最高标 highest bid
最高标价 highest bid
最高波 highest wave
最高波峰 highest wave crest
最高操作温度 maximum allowable operating temperature
最高层 topmost stor(e)y; top story; uppermost stor(e)y
最高层管理者 chief executive
最高产量 maximizing production potential; maximum output; maximum production; peak; peak output; peak production; peak yield
最高产量与允许产量之差 shut in production
最高产率 maximum output
最高潮 culmination; grand climax; highest tide; maximum tide
最高潮水位 highest tide level
最高潮位 highest high water(level); highest tide level; maximum tide level; maximum water level
最高潮位和最低潮位之间的中值 mid-extreme tide
最高潮位线 high tide limit; high water line; sea mark
最高车速 highest speed
最高成本 cost ceiling
最高持续产量 maximum substainable yield
最高持续出力 maximum continuous output
最高齿加速度 top gear acceleration
最高齿轮速比 top gear ratio
最高出价 best bid; highest bid
最高出力 maximum output
最高处 topmost
最高垂直分解力 limiting vertical resolution
最高磁化率 maximum susceptibility
最高磁通密度 peak flux density
最高淬火硬度 maximum quenching hardness
最高存活时间 maximum survival time
最高存量管理法 maximum stock control
最高大气浓度 maximum atmospheric concentration
最高档 stratosphere
最高的 highest; supreme; tallest; top; uppermost; utmost; paramount
最高的估计 outside estimate
最高低潮位 highest low tide level; highest low water
最高低水位 highest low water
最高地方税征收额 maximum rating
最高地下水位 phreatic high
最高递价 highest bid
最高点 acme; climax; culminating point; highest elevation; noon tide; noontime; tiptop; top point
最高点统计 peak
最高点自动保持系统 peak-holding system
最高电流 maximum current
最高电容 maximum capacity
最高电位 maximum potential
最高电压 ceiling voltage; maximum tension; maximum voltage; overvoltage
最高电压值 peak value
最高电阻 maximum resistance
最高电阻截止频率 maximum resistive cutoff frequencies
最高调制频率 maximum modulating frequency; maximum modulation frequency
最高定额 maximum rate
最高度光泽 full gloss
最高端 most significant end
最高额 maximum amount
最高额定速度 maximum specified speed
最高额发行法 maximum issue method
最高额限制 maximum amount limitations
最高发射效率 maximum transmitting efficiency
最高发展 crest development
最高法院 supreme court
最高法院法官 <美> justice
最高帆 kite
最高防洪运行水位 maximum flood control operating level
最高沸点混合物 maximum boiling-point mixture
最高费用 ultimate cost; ultimate expenses
最高峰(值) climax; peak-peak; sharp peak
最高辐射 maximum radiation
最高负荷 maximum load(ing); peak load
最高负荷操作 peak load operation
最高负荷指示器 maximum load indicator
最高负载指示器 maximum load indi-

cator
最高干球温度 maximum dry bulb temperature
最高纲领 maximum program(me)
最高高潮 highest high water(level); high tide high water; extreme high water
最高高潮痕 tide mark; water mark
最高高潮水位 highest high water (level)
最高高潮位 highest high tide level; highest high water (level); high high water level; maximum high water(level); extreme high water
最高高程 maximum level
最高高水位 highest high water(level); high tide high water; maximum high water(level)
最高工资 ceiling on wage
最高工作电压 maximum working voltage
最高工作频率 maximum operating frequency
最高工作水位 maximum operating level
最高工作速度 full operating speed
最高工作纬度 maximum altitude
最高工作温度 maximum service temperature; maximum working temperature
最高工作压力 maximum operating pressure; maximum working pressure
最高公因式【数】 highest common factor
最高公因数 highest common factor
最高公约式【数】 highest common divisor
最高公约数 highest common divisor
最高功率 maximum power
最高功率输出 maximum power output
最高功率因数 highest power factor
最高共沸混合物 maximum azeotropic mixture
最高估计 outside estimate
最高故障率 maximum failure rate
最高观测频率 highest observed frequency
最高管理阶层 top management
最高含氧量 maximum amount of oxygen
最高航速 emergency speed; highest speed
最高和最低(差别)的运价率 maximum and minimum rates
最高和最低电压差 voltage spread
最高和最低(限度) ceiling and floor
最高荷载 peak load
最高洪峰 highest flood peak
最高洪水期 highest flood period
最高洪水水位 peak flood level
最高洪水位 extreme high water; highest ever-known water level; highest flood level; highest high water(level); maximum flood level
最高洪水位线 maximum flood line
最高汇率 ceiling exchange rate
最高混浊区 maximum turbidity zone; turbidity maximum zone
最高火焰温度 peak flame temperature
最高级 top class
最高级的 de luxe; first degree
最高级光泽 super-gloss
最高级会议 summit conference
最高极限 upper limit
最高极限温升 maximum temperature-rise
最高记录 all-time high
最高记录磁平 maximum recording level
最高记录的 maximum-recorded
最高记录水位 highest recorded(water) stage; highest recorded(water) level
最高纪录年份 peak year
最高剂量 maximum dose level
最高价【化】 absolute valence[valency]; highest valence; maximum valence[valency]; maxivalence
最高价订价法 skimming-the-cream price
最高价格 ceiling price; maximum price; price ceiling; top price; ceiling rate
最高价格法 highest price method
最高价竞买人 highest bidder
最高价投标人 highest bidder
最高键控频率 maximum keying frequency
最高阶段 maximum stage
最高接收效率 maximum receiving efficiency
最高经济收益 maximum economic(al) yield
最高静水头 maximum static head
最高聚集人数 maximum number of passengers in peak hours
最高绝热火焰温度 maximum adiabatic flame temperature
最高开动钻机数 maximum amount of operating rigs
最高可采灰分 admissible maximum of ash content
最高可采油指数 maximum producible oil index
最高可及频率 highest-probable frequency
最高可能的频率 highest possible frequency
最高可听频率 upper-frequency limit of audibility
最高可用负载 maximum useful load
最高可用频率 maximum usable frequency
最高可用压缩比 highest useful compression ratio
最高库存量 maximum quantity of stores; maximum stock
最高拉伸倍率 maximum draw ratio
最高力矩 peak moment
最高利润点 profit maximization
最高连续出力 maximum continuous output
最高亮度 maximum brightness
最高灵敏度 maximum sensitivity; ultimate sensibility
最高流冰水位 highest ice floe water level
最高流动点 upper pour point
最高流量 maximum flow
最高流量水平 highest flow level
最高流速 peak flow rate
最高率 maximum rate
最高内在水分 moisture-colding capacity
最高能量 highest energy; maximum energy
最高年产量 peak annual output
最高排出压力 maximum delivery pressure; maximum discharge pressure; maximum output pressure
最高排油压力 maximum discharge pressure
最高赔偿限额 limit of liability
最高配筋率 maximum reinforcement ratio
最高频率 highest frequency
最高平均价格 maximum average price
最高平均气温 maximum mean temperature
最高平均温度 maximum average temperature; maximum mean temperature
最高破波带 zone of highest breakers
最高期 maximum phase
最高倾点 upper pour point
最高权威 last word
最高燃气温度 maximum gas temperature
最高燃烧温度 peak firing temperature
最高燃烧压力 maximum combustion pressure
最高燃速混合物 maximum speed mixture
最高日供水量 maximum daily water supply output; maximum water daily output; water output on maximum day
最高日耗量 maximum daily consumption
最高日剂量 maximum daily dose
最高日进尺 maximum daily footage
最高日射温度 solar maximum temperature
最高日用水量 maximum daily consumption of water
最高荣誉的 blue ribbon
最高容许标准 maximum allowable standard
最高容许的胶质含量 gum tolerance
最高容许量 maximum permissible dose
最高容许浓度 maximum allowable concentration
最高容许速度 maximum permissible speed; permissible maximum speed
最高容许温度 maximum safety temperature
最高容许压力 maximum allowable pressure
最高容许沾染水平 maximum permissible contamination level
最高熔化率 maximum melting rate
最高沙槛高程 maximum sill elevation
最高闪点指数 max flash temperature index
最高上诉法院 court of last resort
最高(上)涌浪 extreme upsurge; highest upsurge
最高上游水位 maximum headwater level
最高烧成温度 peak firing temperature
最高设计电压 maximum design voltage
最高设计压力 extreme design pressure; maximum design pressure
最高升水位 top surge water level
最高生产率 maximum efficiency; maximum productivity
最高生产能力 maximum capacity
最高生产能量成本 capacity cost
最高生长期 maximum growth period
最高生长温度 maximum growth temperature
最高声压级 peak sound pressure level
最高失真 peak distortion
最高使用极限 application limit
最高使用温度 maximum operation temperature; maximum service temperature
最高使用压力 extreme service pressure; maximum operation pressure; maximum working pressure
最高收益 highest yield
最高输出 peak output
最高输出电压 maximum output voltage
最高输出功率 maximum power output
最高输出量 maximum output
最高输入电压 maximum input voltage
最高数位 most significant position
最高水库水位 highest reservoir level
最高水平 top level; topmost level
最高水头 maximum head
最高水位 crest stage; highest stage; highest water level; highest water table; high high water level; maximum level; maximum stage; maximum surface elevation; maximum water level; peak stage; top water level <挡水结构后面水体的>
最高水位标志 high water mark
最高水位出现时间 time of maximum water level
最高水位计 crest-stage indicator
最高水位记录器 crest-stage marker
最高水位时超高 flood freeboard; net freeboard
最高水位水尺 crest-stage ga(u)ge
最高水位线 highest water mark
最高水位预报 crest-stage forecast(ing)
最高水位指示器 crest-stage indicator; high water alarm
最高税率 maximum tariff
最高瞬间燃料变动率 maximum instantaneous fuel change
最高速度 full speed; highest speed; maximum speed; maximum velocity; peak speed; top gear; top speed; top velocity; peak velocity
最高速度试验 maximum speed test
最高速率 maximum speed; top speed
最高碎波带 zone of highest breakers; zone of highest breaks
最高梯级 last step
最高天文潮 highest astronomical tide
最高速度调节器 maximum speed governor
最高通过瞬间电流 peak let-through current
最高通航水位 highest navigable flood level; highest navigable stage
最高通航水位保证率 guaranteed rate of the highest navigable stage; guaranteed rate of the highest navigable water level
最高图像电流频率 maximum frequency of picture current
最高图像频率 maximum picture frequency
最高退磁温度 the highest demagnetizing temperature
最高位 highest order; most significant bit
最高位数位 high order digit
最高位置 extreme higher position
最高温度 ceiling temperature; extreme temperature; highest temperature; maximum temperature; peak temperature
最高温度表 maximum thermometer
最高温度计 maximum thermometer
最高温度期 period at maximum temperature; period of maximum temperature
最高温度期限 maximum temperature period
最高温升 maximum temperature-rise; peak temperature rise
最高污染物水平 maximum polluted level
最高污染物水平指标 maximum polluted level goal
最高无烟火焰高度 maximum flame height without smoking
最高吸入压力 maximum suction pres-

sure
最高系统电压 maximum system voltage
最高限 maximum breaker
最高限度 ceiling;superior limit
最高限断路器 maximum breaker
最高限额 ceiling amount;ceiling limitation;maximum limit;numeric(al) ceiling;top limit;upper limit
最高限价 ceiling price;price ceiling
最高限价担保合同 guaranteed maximum cost contract
最高限值 ceiling value
最高限制 maximum limit
最高消化效率 maximum nitrification efficiency
最高小时交通量 maximum hourly traffic volume
最高效率 maximum efficiency; peak efficiency; top efficiency; top performance
最高效率曲线 maximum efficiency curve
最高信道频带利用率 maximum data rate-to-band-width ratio
最高行驶速度测定 maximum travel(l)ing speed measurement
最高行走速度 maximum travel speed
最高性能 maximum performance
最高性能水平 peak performance
最高需量 maximum demand
最高需量记录装置 maximum recording attachment
最高需量控制(装置) demand control
最高需水量 maximum water demand
最高序列 highest order
最高蓄水位 top water level
最高循环温度 maximum gas temperature
最高压 maximal pressure
最高压级 ultor
最高压力 extreme pressure;maximum pressure; peak pressure; top pressure
最高压力调节器 maximum pressure governor
最高压力指示器 peak pressure indicator
最高岩芯采取率 maximum percentage recovery of core
最高岩芯收获率 maximum core recovery
最高要求 peak demand
最高一层房屋 top flight
最高一级踏步竖板 <楼梯的> top riser
最高一列舷板 saxboard
最高音 treble
最高硬度试验 maximum hardness test
最高壅水位 maximum water level; top water level
最高涌浪 maximum surge;top surge
最高用水量 maximum water consumption;maximum water rate
最高油位 maximum oil level
最高有效频率 the highest effective frequency
最高有效数(字) most significant digit
最高有效水量 field capacity
最高有效位 highest significant position; most significant bit; most significant position
最高有效位字符 most significant character
最高有效温度 maximum effective temperature
最高有效压缩比 highest useful compression ratio
最高月份价格 peak monthly price
最高月进尺 maximum monthly footage
最高月效率 maximum monthly efficiency
最高允许(工程)造价 fixed limit of construction cost
最高允许工作压力 maximum allowable working pressure
最高允许利率 usury ceiling
最高允许速度 maximum authorized speed; maximum permissible speed;maximum permissive speed; maximum permitted speed
最高允许温度 maximum permissible temperature
最高允许压力 maximum working pressure
最高允许值 threshold limit value
最高运行水位 full supply level;maximum operating water level
最高运行速度 maximum operating speed; maximum running speed; maximum service speed
最高运行温度 maximum operating temperature
最高载荷 top load
最高载重 peak load
最高责任 maximum liability
最高障碍物容许限额 dominant obstacle allowance
最高真空度 maximum vacuum
最高振幅 crest amplitude; peak amplitude
最高振型 highest mode
最高蒸发量 maximum evapo(u)rative capacity
最高蒸压温度期限 <指养护期限> maximum temperature period
最高蒸煮压力 maximum cooking pressure
最高帧频 maximum picture frequency
最高值 crest value;peak(ing) value
最高值曲线 peaky curve
最高值指示器 maximum value indicator
最高质量 extra best best;top-notch
最高质量或光泽最纯 <钻石或珍珠等宝石> first-water
最高致死温度 fatal high temperature
最高致死氧浓度 maximum lethal oxygen concentration
最高转化率 maximum conversion
最高转数 maximum number of revolutions
最高转数调速器 maximum speed regulator
最高转速 maximum revolution speed; maximum speed of revolution
最高装载量 peak load
最高资金需要额 maximum financial requirement
最高自记温度表 maximum recording thermometer
最高自然光时 peak natural light hours
最高总效率 maximum overall efficiency
最高租价 ceiling rent
最高最低税则 maximum and minimum tariff
最高最低温度表 maximum(-and)-minimum thermometer
最高最低温度计 maximum(-and)-minimum thermometer
最关键路线 most critical path
最光亮 super-gloss
最好不过 leave nothing to be desired
最好承担 <包工合同> best efforts
最好的 first-class; blue ribbon; optimum;tiptop
最好的措施 best bet
最好的效果 best result
最好的质量 best quality
最好可成交价格 best obtainable price
最好能见度 excellent visibility
最好上升角 best climbing angle
最好调节 best setting
最好温度 optimum temperature
最好性能 top performance
最合适的订购数 best order quantity
最合适人口数 optimum population
最合算的 cost-optimal
最后饱和 final saturation
最后保留体积 final retention volume
最后报告 final report
最后报价 last bid
最后表层涂料 final coat plaster
最后捕收槽 final save-all cell
最后布置 final layout
最后步骤 last step
最后部分 declining
最后裁决 award in final;final award; final ruling
最后采用的坡度线 final grade line
最后残渣 final residuum
最后草图 final scheme
最后测试 last test
最后测试步 last test step
最后产率 final yield
最后产品 end product;final goods;final product
最后产物 end product
最后偿还期 final maturities
最后沉淀池 final sedimentation tank
最后沉积池 final sedimentation tank
最后沉降 final settlement
最后成本 final cost
最后成果的 end product
最后成品精整 final product finishing
最后成为 eventuate
最后乘次优惠 last ride bonus
最后乘子 last multiplier
最后程序 final process;final program(me)
最后澄清 final clarification; principal defecation
最后澄清槽 final settling tank
最后澄清池 final clarification tank
最后持水量 final retention
最后齿轮减速 final gear reduction
最后充填物【地】 mesostasis
最后冲刺 <学习或工作接近终了> end spurt
最后冲洗 final rinse
最后出口边 last exit edge
最后储藏区 ultimate storage area
最后处理 final treatment
最后处置 ultimate disposal
最后船运通知 final shipping instruction
最后存量 closing inventory
最后带间 tail bay
最后贷款人 lender of last resort
最后单件上涂料 final individual coat
最后单元 last location
最后当前状态 last current state
最后到达港 last port
最后的 conclusive; definitive; eventual; final; last; nett; rear most; ultimate
最后的补救办法 last resort
最后的但不是最不重要的 last but not the least
最后的焊道 final pass; final run; finish cover pass
最后的焊珠 finish cover pass
最后的抹灰 final coat exterior plaster
最后的收报局 last office of destination
最后的巡查 jerquer
最后的钻杆 finishing steel
最后定线 final location
最后定线测量 final location survey
最后定型 final shaping
最后冻结温度 final temperature of freezing
最后读出 end readout
最后读数 final reading
最后端 rearmost end
最后端的 fear most
最后断裂 eventual failure
最后断面 final profile
最后方案 extreme scenario
最后方向修正量 final deflection
最后分级 final sizing
最后分类 final clarification
最后分离池 final separating tank
最后分析 ultimate analysis
最后符号 last symbol
最后腐蚀 final etching
最后付款 payment by results
最后付款期限 deadline of payment
最后付款证明书 last pay certificate
最后付清 final payment
最后负担 final incidence
最后负荷 final load
最后负进位 final negative carry
最后干燥 final drying
最后高程 <土建项目> new or finished
最后工序 final operation;finishing work
最后工作 final work
最后公路 final common path
最后估价 final valuation; final value estimate;definitive estimate
最后股息 final dividend
最后固化 final cure
最后刮平 <平地机的> final blading
最后贯入度 final penetration;final set
最后归并阶段 final merge phase
最后规定级别 finish grade
最后规划 final planning
最后滚边机 final mo(u)lder
最后过滤 after-filtration; final filtration
最后过滤器 post-filter;ultimate filter
最后焊层 cap weld
最后焊接 final welding
最后号数 last number(s)
最后合金料 finishing material
最后合拢 final closure
最后合同费用 final contract cost
最后核对图像 final check picture
最后核准 final approval
最后荷载 final load
最后红利 final dividend
最后宏指令 last macro instruction
最后厚度 final thickness
最后恢复 final recovery
最后回报 final return
最后汇报 final return
最后基准线 last scratch
最后激励时间 final actuation time
最后几击平均贯入量 average set of final blows in pile driving
最后计算值 latest entry
最后技术 end-of-pipe technology
最后加工 final working;finishing touch;finishing work
最后加工地带 <路面> finishing belt
最后加工工作 mandrel work
最后加工精度 accuracy of finish
最后加力 final stressing
最后加速阶段 final stage of acceleration
最后价格 final price
最后价值量 ultimate quantity of value
最后检查 final check(ing); final inspection; final test; finish turn inspection
最后检验 final check(ing); final examination; final inspection; final

proofing;final test
最后交割通知日 last notice day
最后交货日期 end delivery date
最后交接试航 final acceptance trial
最后交易日 last trading day
最后阶段 concluding stage;final stage
最后接触开关 final contact switch
最后节点 end event
最后节间 tail bay
最后结果 bottom line;final product;final result;net result;end product
最后结果处理 final data reduction
最后结果分析 end point analysis
最后结果规范 <只规定最后成品质量的规范> end-result specification
最后结局 final outcome
最后结论 ultimatum
最后结算 final statement
最后结账 final closing
最后截止日期 final limited date
最后进场 final;final approach
最后进价法 last invoice price
最后精度 resultant accuracy
最后精加工 final finish
最后精炼 final purification;final refining
最后精选摇床 finishing table
最后精整 finishing tough
最后精制 final refining
最后净化 after purification;eventual purification;final purification
最后决策 final decision
最后决定 final decision;final judgment
最后决议书 final act
最后竣工 final completion
最后竣工阶段 final completion stage
最后竣工证书 final completion certificate
最后卡片 last card
最后开间 tail bay
最后开挖深度 final cut;final dredged depth
最后开挖线 final cut
最后控制 ultimate control
最后冷却器 after-cooler
最后离港定位 taking a departure
最后力矩 final moment
最后粒子 final particle
最后连接 final connection
最后流出物 final effluent
最后馏分 end cuts;last cut;tail fraction;tail oil
最后轮廓线 final profile
最后面的 back-most;rear most
最后停泊港 port of last call
最后抹灰罩面 finish plaster
最后抹面 putty coat
最后目的港 final port of designation
最后碾磨工序 finish grinding
最后碾压 finish rolling
最后排样 final composing
最后判定 terminal decision
最后判决 definitive judg(e)ment;final decree
最后抛光 glossy finish(ing)
最后平地机工序 final blading
最后平整 final grading;final level(l)ing;finish grading
最后破坏 eventual failure
最后期限 deadline;deadline date;last date
最后切断 final cut-off;final cutting
最后清理工作 final cleanup
最后清扫工作 final cleanup
最后区 area postrema
最后取样单位 ultimate sampling unit
最后确定 finalize[finalise]
最后确定的保额 closed line
最后日期 final date
最后如期完工时间 latest event occurrence time
最后撤布 final application
最后筛分 final sizing
最后扇段 last sector
最后上的漆 finish varnishing
最后上蜡 final varnishing
最后上升 final ascent
最后设计 final design;final project
最后申报 final return
最后生存年金 last-survivor annuity
最后生的 ultimogenitary
最后湿度 final humidity
最后湿量 final moisture content
最后实际赔偿数 ultimate net loss
最后实际损失 ultimate net loss
最后事件发生时间 latest event occurrence time
最后试验 final test
最后手段 last resort
最后输出(结果)final output
最后输出值 final output
最后数字码 final digit code
最后刷白 finish plaster
最后提到的 last named
最后条款 final articles;final clause;final provisions
最后调节 final adjustment
最后调整 final adjustment
最后通牒 ultimatum
最后通过阶段 last pass phase
最后通路 final common path
最后涂层 final closing;final coat(ing)
最后涂的水泥层 final coating cement;wash cement
最后退卷 last-off
最后拖光 <混凝土路面> final belting
最后外层涂浆 final coat external plaster
最后弯矩 final moment
最后完成日期 latest finish date
最后完工的工作 copestone
最后完工日期 latest finish date
最后维修 final repair
最后位置 rearmost position
最后温度 final temperature
最后文本 definitive text
最后文件 final act
最后吸气 final gettering
最后吸水量 final retention
最后下标 last subscript
最后限制信号 <驼峰编组场> last limit signal
最后项 last term
最后消费价格 final consumer price
最后消费者 ultimate consumer
最后消费者市场 ultimate consumer market
最后消隐 final blanking
最后效用 terminal utility
最后校定 post-edition
最后校验 terminal check
最后校样 final proof;foundry proof <打纸型制版前的>
最后协定 final agreement
最后协议 final agreement
最后卸货港 final port of discharge
最后行尾技术 end-of-point technology
最后形状 final form;final shape
最后性能 terminal feature
最后修饰 finish
最后修整 finishing touch
最后修整磨光工序 finish grinding
最后修琢 finishing chip
最后旋回圈半径 final diameter
最后压力 end pressure;final pressure
最后研磨 finishing grind(ing)
最后验收 final (inspection and) acceptance
最后阳极 final anode
最后一笔付款数额大的分期贷款 balloon-payment load
最后一笔金额特大的分期付款法 balloon repayment
最后一锤下沉量 final set
最后一次延长绿灯 last call
最后一档 last step
最后一道工序 finish
最后一道抹灰层 final plaster coat
最后一道清管【给】final pigging;final scraping
最后一道细灰泥 <湿壁画的> intonaco
最后一段锤击数 final blow count
最后一行 last column
最后一级过滤层 <多级过滤系统的> final filter
最后一着 last resort
最后移位数 final carry digit
最后议定书 final protocol
最后应力 final stress
最后用户 end-user
最后(再)浮选 final recleaner flo(a)tation
最后再筛分 finish rescreening
最后增压 final boost
最后蒸发 final evapo(u)ration
最后整平 finish grading
最后整形 final shape
最后整型 final shaping;final grading <路基>;finish grading <路基>
最后正式答复 definitive answer
最后执行指令 last executed instruction
最后值 ultimate value
最后制品 end product
最后致死地带 zone of eventual death
最后终端符 last terminal
最后重量 final weight
最后装配 final assembly
最后组装 final assembly
最厚氧化层厚度 maximum oxide thickness
最坏布局检查 worst status check
最坏的材面 worst face
最坏情况 worst-case
最坏情况法 worst-case method
最坏情况分析 worst-case analysis
最坏情况界线 worst-case bound
最坏情况设计 worst-case design
最坏情况特性 worst-case performance
最坏情况循环 worst-case loop
最坏情况噪声 worst-case noise
最坏情况噪声模式 worst-case noise pattern
最坏情况噪音 worst-case noise
最坏条件 worst-case condition
最坏状态检验 worst status check
最惠国待遇 most favo(u)red nation treatment
最惠国待遇条款 most favo(u)red nation treatment clause
最惠国税率 most favo(u)red nation tariff
最惠国条款 most favo(u)red nation clause
最惠特许条款 most favo(u)red licence clause
最或然船位 most probable position;probable position
最或然改正函数 most probable correction function
最或然率 most probable ratio
最或然数 most probable number
最或然修正函数 most probable correction function
最或然值 best possible value
最或是值 most probable value
最基本的 bottom-most
最极目标 ultimate objective
最急转弯 sharpest turn;sharpness turn
最佳 pH 值 optimum pH
最佳 pH 值上下限 optimum pH range
最佳报价 best offer
最佳爆炸高度 optimum height of burst
最佳爆炸时刻 optimum time of burst
最佳逼近 best approximation;optimal approximate;optimal approximation
最佳比例 optimum proportioning
最佳编码 optimum code;optimum coding
最佳编码程序 optimally coded program(me)
最佳标价 best bid
最佳标准 optimality criterion
最佳标准重力 optimum standard gravity
最佳波导管长度 optimum guide length
最佳波段 optimum band
最佳波形 optimum waveform
最佳剥比 optimum stripping ratio
最佳剥采比 optimum stripping ratio
最佳采样密度 optimum sampling density
最佳参数 optimal parameter;optimatical parameter;optimum parameter
最佳参数选择 optimal parameter selection
最佳操作 optimal performance;optimum performance
最佳操作程序 process optimization program(me)
最佳操作密度 optimum running density
最佳策略 optimal policy;optimal strategy
最佳掺和量比 optimum proportion
最佳长度 optimum length
最佳常数控制 optimal constant control
最佳车 optimized vehicle
最佳车流量 optimal car flow
最佳成本 optimum cost
最佳程序 optimum program(me);optimal programming
最佳程序设计 optimum programming
最佳尺寸 optimum size
最佳抽气次数 air exhaustion optimal number
最佳抽取 optimum extraction
最佳出力 optimum output
最佳处治 optimum cure
最佳传输电流 optimum transmitted current
最佳传输频率 frequency of optimum traffic;optimum transmission frequency
最佳船舶航线 optimum ship routing
最佳纯预测器 optimum pure predictor
最佳存货 optimum inventory
最佳代码 optimum code
最佳导引 optic(al)guidance
最佳的 best;optima;optimum
最佳的齿倾角 optimum tine angle
最佳的出行 choice trip
最佳的方位间距 optimum splitting of azimuth
最佳等时线 the optimum isochron
最佳地沥青含量 optimum asphalt content
最佳地面高程 optimum ground elevation
最佳地区 hundred-percent location
最佳点火定时自调系统 optimizer system
最佳点态控制 optimal pointwise control

最佳电路 optimal circuit
最佳电平 optimizing level
最佳电位差 optimum potential difference
最佳电压 optimum voltage
最佳调节 best setting
最佳调谐点 optimum tuning point
最佳调整 optimum setting
最佳叠加 optimum stack
最佳订购单 optimum quantity order
最佳订购点 optimum order point
最佳订购量 optimum reorder quantity
最佳订货点 optimum order point
最佳订货量 economic order(ing) quantity
最佳断面 most advantageous section
最佳对半检索树 optimal binary search tree
最佳对称主点 principal point of optimum symmetry
最佳多项式 optimum polynomial
最佳多项式逼近【数】 best polynomial approximation
最佳二叉查找树 optimal binary search tree
最佳反射率 optimum reflectivity
最佳反应 optimum response
最佳方案 best decision;optimal decision;optimal design;optimum alternative;optimum scheme;optimum solution;preferred plan
最佳方法 best way;optimum method
最佳防治理论 optimal control theory
最佳费用 optimum cost
最佳分辨带宽 optimum resolution bandwidth
最佳分辨力 optimum resolving power
最佳分辨率 optimum resolution
最佳分布 optimal distribution;optimum distribution
最佳分级制 optimal ramification
最佳分解力 optimum resolution
最佳分离系数 optimum separation factor
最佳分配 optimal allocation;optimum allocation
最佳服务 optimum quantity service
最佳符合法 optimum fitting method
最佳幅度响应 optimum amplitude response
最佳负载 optimum load(ing)
最佳负载电阻 optimum load resistance
最佳覆盖厚度 optimum depth
最佳赶工计划 optimum crash program(me)
最佳刚度 optimum rigidity
最佳高度 optimum height
最佳工厂规模 optimal plant size
最佳工况 optimum condition;optimum operation
最佳工况调节装置 optimizer control
最佳工期 optimum duration
最佳工艺特性 optimum process performance
最佳工作边坡 optimum operating slope
最佳工作点 best operating point;optimum operating point
最佳工作电压 optimum operating voltage
最佳工作方式 optimum mode of operation
最佳工作能力 optimum workability
最佳工作频率 frequency optimum traffic;optimum operating frequency;optimum traffic frequency;optimum working frequency
最佳工作区域 optimum working range
最佳工作曲线 optimum-curve for operation
最佳工作条件 optimal operation condition
最佳工作温度 optimum working temperature
最佳工作效率 optimum working efficiency
最佳公式 optimum formula
最佳共鸣 optimum resonance
最佳共振 optimum resonance
最佳钩车间隔 optimum spacing of cuts
最佳估计 best approximation;optimum estimation;optimal estimation
最佳估计量 optimal estimator;best estimator
最佳估计算法 optimal estimation algorithm
最佳估计值 best estimate
最佳估算 optimal estimate
最佳估算值 optimum estimation value
最佳管理实践 best management practice
最佳光阑 optimum diaphragm
最佳规避机动 optimum evasive maneuver
最佳规划 preferred plan
最佳规划法 optimum programming
最佳规模集成 optimum scale integration
最佳轨道 optimal trajectory;optimum orbit
最佳轨道上升 optimal zoom climb
最佳过渡 optimum transfer
最佳含砂率 optimum fine aggregate percentage
最佳含水层 optimum water bearing bed
最佳含水量 optimum moisture content;optimum water content
最佳含油量 optimum oil content
最佳汉明码 optimal Hamming code
最佳航道尺度 optimum channel dimension
最佳航线拟定 optimum(track ship-) routing
最佳航线选定 optimum(track ship-) routing
最佳航向 optimum heading
最佳航向指示 recommended course indicator
最佳航向指示器 recommended course indicator
最佳耗量 optimum consumption
最佳耗水量 optimum consumptive [consumption] use
最佳耗用量 optimum consumptive [consumption] use
最佳和易性 optimum workability
最佳河湾 optimum bend
最佳河湾半径 optimum bend radii
最佳荷载 optimal load
最佳厚度 optimum thickness
最佳滑翔角 optimum gliding angle
最佳化 optimalize;optimization;optimize;optimizing
最佳化的 optimized
最佳化定理 optimization theorem
最佳化方法论 optimization methodology
最佳化计算 optimization calculation
最佳化技术 optimization technique
最佳化阶段 optimization phase
最佳化控制方式 optimizing control mode
最佳化器 optimizer
最佳化算法 optimization algorithm
最佳化系统 optimization system
最佳化属性 optimization attribute
最佳化准则 criterion of optimality
最佳化钻探 optimized drilling
最佳化钻探服务 optimized drilling service
最佳环境 optimum environment
最佳环境质量 optimum environmental quality
最佳回流比率 optimum reflux ratio
最佳混合料 optimum mix
最佳混合气强度 optimum mixing strength
最佳混合物 optimum mixture
最佳混响 optimum reverberation
最佳混响时间 optimal reverberation time;optimum reverberation time
最佳级配 optimal gradation;optimum gradation
最佳集装箱货 prime containerizable cargo
最佳挤压凹模角 optimum extrusion angle of die
最佳记录电流 optimum recording current
最佳继电器随动系统 optimum relay servomechanism
最佳继电伺服机构 optimum relay servomechanism
最佳加权叠加 optimum weighted stack
最佳加溶温度 optimum temperature of solubilization
最佳加速状态 optimal acceleration regime
最佳间距 optimum spacing
最佳检测滤波器 optimum detecting filter
最佳检索法 optimal searching
最佳降深(抽水) optimum drawdown
最佳交混回响 optimum reverberation
最佳焦点 best focus;optimum focus;pinpointed focus
最佳角度 optimum angle
最佳接触角 optimum contact angle
最佳接收法 optimum reception
最佳接收机 optimum receiver
最佳结构 optimum structure
最佳结合料用量 optimum binder content
最佳截击航向 optimum intercept course; optimum interception course
最佳解 optimal solution
最佳解法 best solution;optimum solution
最佳解离点 optimum liberation point
最佳解图 optimal solution graph
最佳进场航线 optimum approach path
最佳进料点 optimum feed location
最佳进气口外形 optimum intake contour
最佳晶格 optimum lattice
最佳距离 optimum distance
最佳聚积 optimum stacking
最佳聚焦 optimum focusing
最佳聚束 optimal bunching;optimum bunching
最佳决策 optimizing decision making
最佳决策法则 optimum decision rule
最佳决策系统 optimum decision system
最佳决定律 optimal decision rule
最佳决断系统 optimum decision system
最佳开发 optimum development
最佳开发方案 optimum development plan
最佳开关函数 optimum switching function
最佳开关面 optimal switch surface
最佳开关线路 optimum switching line
最佳抗风蚀清漆 spar varnish
最佳可靠性 optimum reliability
最佳可行技术 best available technology
最佳可用技术 best available technology
最佳空气供应量 optimum air supply
最佳空气燃料比 air rich
最佳空中目标选定传感器 optimum aerial target sensor
最佳控制 optimal control;optimalizing control;optimized control;optimizing control;optimum control
最佳控制的 optimally controlled
最佳控制计算法 optimal control algorithm
最佳控制理论 optimum control theory;theory of optimal control
最佳控制律 optimal control law
最佳控制器 optimizer;optimizing controller;optimum controller
最佳控制系统 optimal control system;optimizing control system
最佳库存量 optimal inventory
最佳跨长 optimum span length
最佳宽带垂直叠加 optimum wide band vertical stack
最佳宽深比 optimum width-depth ratio
最佳沥青含量 optimum asphalt content
最佳粒度 optimum size
最佳列车交会方案 optimum crossing of trains
最佳灵敏度 optimum sensitivity
最佳流量 optimum flow
最佳流速 most advantage velocity;optimum flow rate;optimum velocity
最佳滤波 optimum filtering
最佳滤波器 optimum filter
最佳路径 optimal path
最佳路况 optimum utilization
最佳路线 optimal routing;optimum route
最佳码 optimum code
最佳密度 optimum density;optimum concentration【交】
最佳密度范围 optimum density range
最佳磨矿粒度 optimum grind
最佳磨矿速度 optimum milling rate
最佳能见度 excellent visibility
最佳能力 optimum capacity
最佳拟合 optimal fitting;optimum fit;best fitting
最佳拟合大圆 best-fit large circle
最佳拟合多项式 best fitting polynomial
最佳拟合方程 best-fit equation
最佳拟合环带轴 best-fit girdle axis
最佳拟合面 optimal fitting plane
最佳拟合曲线 best fitting curve
最佳拟合线 line of best fit
最佳拟合小圆 best-fit small circle
最佳耦合 optimal coupling;optimum coupling
最佳判定 optimal decision
最佳判定规则 optimum decision rule
最佳判据 optimality criterion
最佳判决定则 optimum rule
最佳配(钢)筋 optimum reinforcement
最佳配合 best fit;best match(ing);optimum match(ing)
最佳配合比 optimum mix;optimum proportioning
最佳配(合)比设计 optimum mix design
最佳配筋率 economic ratio of reinforcement
最佳配置 optimum allocation
最佳喷油定时自调系统 optimum oil-spraying timing self-system;optimizer system
最佳喷嘴 optimum nozzle

最佳批量 optimum lot size
最佳疲劳强度 optimum fatigue strength
最佳匹配 best match(ing);optimum match(ing)
最佳偏磁 optimum bias
最佳偏压 optimal bias
最佳偏压测定 determination of optimum bias
最佳偏置 just bias;optimum bias
最佳漂移角 optimum drift angle
最佳频率 optimal frequency; optimum frequency
最佳频率调谐 optimal frequency tuning
最佳期望值 optimum expectation
最佳启动 optimum start-up
最佳启停控制器 optimum start/stop programmer
最佳气候 optimum climate; weather window <浇注混凝土的>; window <浇筑混凝土的>
最佳气体流速 optimum gas velocity
最佳气味 excellent fume
最佳切入速度 optimum penetration speed
最佳切削速度 optimum cutting speed
最佳切削条件 optimum cutting condition
最佳倾角 optimum angle of incidence
最佳清晰度 optimum resolution
最佳情况 optimum
最佳曲线 best curve; optimal fitting curve
最佳取向 preferential orientation; preferred orientation
最佳权分配 most favo(u)rable distribution of weight
最佳燃烧参数 optimum gas parameter
最佳人工散布 optimum artificial dispersion
最佳人口 optimum population
最佳人口密度 optimum population density
最佳人选 optiman
最佳容量 optimum capacity
最佳栅格 optimum lattice
最佳商业票据 prime commercial paper
最佳上升航线 optimum climb path
最佳上升曲线 synergic curve
最佳设计(法) optimal design; optimum design
最佳深度 optimum depth
最佳深度范围 best depth range
最佳生产或订购的一次批量 economic lot size;optimum lot size
最佳生长条件 optimal growing condition; optimum growing condition; the best growing condition
最佳生育季节 appropriate season of giving birth
最佳湿度 optimum humidity; optimum moisture content
最佳石灰石 cairn's lodge
最佳时差 optimising offset
最佳实际气体流速 optimum practical gas velocity
最佳实践方法 best practicable means
最佳实用技术 best practical technology
最佳实用可行技术 optimum practicable technology currently available
最佳使用年限 optimum useful life
最佳使用状况 optimum utilization
最佳视距 optimum viewing distance
最佳收益网络 optimum profitability network
最佳输出 optimum output
最佳输入值给定器 optimalizing input drive
最佳束 optimum beam
最佳数 optimum number
最佳数据 optimum data
最佳衰减 optimized attenuation
最佳水分 optimum moisture
最佳水力断面 best hydraulic cross-section; optimum hydraulic cross-section
最佳水力横断面 optimum hydraulic cross-section
最佳水深 optimum depth
最佳水位变动区间 interval of optimal water level change
最佳松弛因子 optimum relaxation factor
最佳速度 optimum speed; optimum velocity
最佳速率 <在一定道路上通过量可以达到最高值的平均速率> optimum speed
最佳随动系统 best servo
最佳随机化代换 optimal randomized replacement
最佳随机控制 optimal stochastic control
最佳特性 optimal characteristics; optimum performance
最佳特性曲线 optimum performance curve
最佳提取 optimum extraction
最佳提升速度 optimum hoisting speed
最佳天线阵电流 optimum array current
最佳条件 optimality condition; optimum condition;top condition
最佳条件选择 optimization
最佳停车动作 optimum stopping performance
最佳停堆 optimal reactor shutdown
最佳通风 optimal ventilation
最佳通信[讯]频率 optimum traffic frequency
最佳通信[讯]使用频率 frequency of optimum traffic
最佳通行税率 best prevailing tariff rate
最佳投标 best bid
最佳突发错纠正码 optimum burst correcting code
最佳图像编码器 optimum image coder
最佳图像质量 optimum picture quality
最佳推算程序 optimal routing program(me)
最佳退磁场 the best demagnetizing field
最佳退磁温度 the best demagnetizing temperature
最佳挖掘高度 optimum height of cut
最佳挖掘深度 optimum depth of cut
最佳外形 optimum configuration
最佳弯曲半径 optimum bend radius
最佳弯曲原理 optimal crooked principle
最佳位置 optimum position
最佳温度 optimum temperature
最佳温度梯度 optimum temperature gradation
最佳稳定度 optimum stability
最佳无偏检验 best unbiased test
最佳无滞后过滤器 optimum zero-lag filter
最佳无滞后滤波器 optimum zero-lag filter
最佳误差检测码 optimum error detection code
最佳系统控制 optimum system control
最佳系统函数 optimum system function
最佳线路分析 best route analysis
最佳线性测量 optimum linear measurement
最佳线性过滤 optimum linear filtering
最佳线性滤波 optimum linear filtering
最佳线性拟合 best linear fit
最佳线性无偏估计 best linear unbiased estimate
最佳线性系统 optimum linear system
最佳线性预测 optimum linear prediction
最佳相似法 method of maximum likelihood
最佳相位 optimum phase
最佳消耗量 optimum consumptive [consumption] use
最佳消落 optimum drawdown
最佳效率 best efficiency;optimum efficiency
最佳效率点 best efficiency point
最佳协调 optimal coordination
最佳协调解 best compromise solution
最佳斜率 the optimum slope
最佳谐振 optimum resonance
最佳泄降 optimum drawdown
最佳泄降量 optimum drawdown
最佳泄降深度 optimum drawdown
最佳信号 optimum signal
最佳信号噪声比 optimum signal to noise ratio
最佳信息值 optimal information value
最佳信噪比 optimum signal to noise ratio
最佳形状 optimum shape
最佳形状透镜 best form lens
最佳性方程 optimality equation
最佳性分析 optimality analysis
最佳性能 optimal performance; optimum performance; top performance
最佳性能参数 optimum performance parameter
最佳性能曲线 best performance curve
最佳性原理 principle of optimality
最佳性准则 optimality criterion
最佳絮凝点 optimum point of coagulation
最佳选交通信号灯相模式 optimum traffic light pattern
最佳选择 optimal choice
最佳选择分集 optimal selection diversity
最佳压实 optimum compaction
最佳压缩比 optimum compression ratio
最佳养护 optimum cure
最佳一致逼近 minimax approximation
最佳译码 optimal decoding
最佳应力 optimal stress; optimum stress
最佳用途 highest and best use
最佳(优)参数选择 optimization of parameters
最佳优先搜索 best-first search
最佳有效控制技术 best available control technology
最佳预报值 best predictor
最佳预测 optimum prediction
最佳预臭氧化剂量 optimum pre-ozonation dose
最佳圆柱形柱 optimum cylindrical column
最佳运输工具【交】 best mode
最佳运行条件 optimal operation condition
最佳载频 optimum carrier frequency
最佳增益 optimum gain
最佳增益频率 optimum-gain frequency
最佳照度 optimum illumination
最佳照明高度 optimum burning altitude
最佳震源深度 optimal focal depth
最佳值 best value; optimal value; optimum value
最佳质量 best in quality; optimum quality
最佳致密作用 optimum compaction
最佳周期 optimum cycle
最佳周转港 best turnaround port
最佳轴载 optimum axle load
最佳柱长 optimum column length
最佳柱温 optimum column temperature
最佳柱效(能) optimum column efficiency
最佳砖块 body brick
最佳装球量 optimum ball charge
最佳装药 optimum charge
最佳状态 optimum condition; optimum performance
最佳状态的 groovy
最佳状态的输入给定器 optimalizing input drive
最佳状态上升 optimum climb
最佳状态选择器 optimalizing input drive
最佳准则 optimum criterion
最佳自动控制 self-optimizing control
最佳自控 self-optimizing control
最佳自调系统 optimizer system
最佳综合政策 optimal policy mix
最佳纵断面线 optimum profile
最佳阻抗 optimum impedance
最佳阻尼 optimal damping; optimum damping
最佳阻尼参数 optimum damping parameter
最佳组成设计 optimal constituent design
最佳作业水准 optimal activity level
最简的 minimal
最简分式 fraction in lowest terms
最简明的核查 foolproof verification
最简形式 minimum form
最接近道路 closest approach
最接近的 proximal; proximate
最接近的附近地区 immediate vicinity
最接近点 closest approach
最接近点距离 distance to closest point of approach
最接近发盘 closest offer
最接近瞬间 closest instant
最接近状态 closest approach
最捷径 minimum-time path
最紧密堆积 most dense pile; closest packing
最紧迫阿尔发水平检定 most stringent level alpha test
最紧迫检验 most stringent tests
最近背景 immediate background
最近成本 recent cost
最近单元 constructional unit
最近的 red hot; up to the-minute
最近的逼近 nearest approach
最近的估计指出 recent estimates indicate
最近点 closest point
最近渡越点 cross-over
最近发行的运价表 latest issue of tariff
最近方位 shortest direction
最近分数式 closest rational approximation
最近会遇时间 time to closest point of approach
最近距离 closest range; minimum range
最近邻 nearest neighbo(u)r

最近邻分析法 nearest neighbo(u)r analysis

最近邻近似法 nearest neighbo(u)r approximation

最近邻能 nearest neighbo(u)r energy

最近邻频率 nearest neighbo(u)r frequency

最近邻区次策规则 nearest neighbo(u)r decision rule

最近邻顺序分析 nearest neighbo(u)ring sequence analysis

最近邻相互作用 nearest neighbo(u)r interaction

最近邻域分类 nearest neighbo(u)r classification

最近邻域模式分类法 nearest neighbo(u)r pattern classification

最近末端的 endmost

最近拍摄距离 minimum photographic(al) distance

最近市场报告 latest market report

最近似的 best approximated

最近似值 most probable number; optimal approximate value

最近有理逼近 closest rational approximation

最近最多使用的条目 most-recently-used entry

最近最少使用法 least recently used [LRU] method

最经济的 cost-optimal

最经济的采收率 maximum economic(al) recovery

最经济的导线 most economic conductor

最经济订货批量 economic order(ing) quantity

最经济断面 most economical section

最经济负载 most economic load

最经济管理法 value engineering

最经济航程 most economical range

最经济坡度 <以增加的工程费和节约的行车费为对比而得出的> most economical grade

最精采的场面 highlight

最精确聚焦 sharpest focus

最可几大小 most frequent size

最可几的 most probable

最可几端距 most probable end-to-end distance

最可几速度 most probable velocity

最可几值 most probable value

最可能的 most probable

最可能极值 most probable extreme value

最可能渗径 preferential path

最可能时间 most likely time

最可能时间估计 <统筹方法中,对于活动所需要耗费的时间的一个最切合实际的估计> most likely time estimate

最可能数 most probable number

最可能数法 most probable number method

最可能数值 most probable number value

最可能数指数 most probable number index

最可能速度 most probable velocity

最可能位置 most probable position

最可能误差 most probable error

最可能值 most probable value

最可行技术 best practical technology

最枯水期 driest period

最快存取 minimum access; minimum latency

最快存取编程 minimum access programming

最快存取编码 minimal access coding; minimum access coding; minimum delay coding; minimum latency coding

最快存取程序 minimal access program(me); minimal-latency program(me); minimum latency routine

最快存取程序设计 minimal-latency programming

最快存取码 minimum access code; minimum latency code

最快存取时间 minimum access time

最快存取子程序 minimal-latency subroutine

最快取数 minimum access

最快取数编码 minimal-latency coding; minimum access coding; minimum delay coding; minimum latency coding

最快取数程序 minimal-latency routine; minimum access routine

最快取数程序设计 minimal access programming

最快扫描 fastest sweep

最快速下降线 line of quickest descent

最快随机存取程序设计 minimum random access programming

最快下降法 method of steepest descent

最快线路 quickest route

最快装卸速度 utmost dispatch[despatch]

最宽沸程范围 widest boiling point range

最困难的国家 hard core country

最乐观时间 <即最短时间> most optimistic time

最冷期等水温线 isocryme

最冷月份 the coldest month

最冷月份室外平均大气压 outside mean atmospheric pressure in coldest month

最冷月份室外平均温度 outside mean temperature in coldest month

最冷月平均温度 mean temperature in coldest month

最冷月室外平均温度 outdoor mean temperature in coldest month

最里面部 innermost

最里面的 inmost

最理想的 optimal

最廉价交易 dead bargain

最亮信号 highland signal; highlight signal

最劣度 pessimum

最劣情况下应用的器件 worst-case device

最劣天气 minimum weather

最劣质水 worst-case water

最令人满意的 optimal

最满意定价 penetrating pricing

最密集值 thickest value

最密装填 closest packing

最明部分 extreme highlight

最明亮的部分 highlight

最末端 extreme end

最末端的 extreme

最末股息 final dividend

最末号码重拨 last number redial

最末路由 final route

最末中继线群 final trunk group

最末组件 final assembly

最内部的 inmost; innermost

最内层 innermost layer

最内层的 innermost

最内层归约结构 innermost reduction architecture

最内层计算规则 innermost computation rule

最内层循环 innermost loop

最内层循环语句 innermost loop statement

最内层柱面 innermost cylinder

最内地震线 innermost isoseismal line

最南端的 southern most

最难行车【铁】 worst rolling car

最平幅度近似 maximally flat amplitude approximation

最平滑逼近 smoothest approximation

最平时延逼近 maximally flat delay approximation

最平特性带通网络 maximally flat bandpass network

最普通的 modal

最普通金属 most common metal

最前部前列 forefront

最前部前面 forefront

最前部(前线) forefront

最前端的 front most

最前方的工作面 forefield

最前肋骨 foremost frame

最前面 forefront

最前线 forefront

最强地震区 meizoseismal area

最强浪向 direction of most severe wave attack

最强烈灯光 maximum dazzle

最强烈活动时期 climax phase of fault activity

最强通风 maximum draft; maximum draught

最强信号 peak signal

最强氧化剂 strongest oxidant; strongest oxidizing agent

最强震的 meizoseismal; meizoseismic

最强震度力的 meizoseismic

最轻磅皮 extreme light weight hide

最轻的元素 the lightest element

最轻度的 first degree

最轻荷载 lightest load

最轻(重量)设计 minimum weight design

最清晰表面 surface of best definition

最清晰平面 plane of best definition

最确数 most probable number

最热点温度容差 hottest-spot temperature allowance

最热月平均温度 mean temperature in hottest month

最弱边 the weakest side

最弱边相对中误差 relative mean square error of the weakest side

最弱边中误差 mean square error of the weakest side

最弱点 the weakest point

最弱点点位中误差 mean square error of position of the weakest position

最弱点高程中误差 mean square error of height of the weakest point

最弱剪切面 weakest shear plane

最上 tiptop

最上部地壳 uppermost crust

最上层 stratosphere; surface layer; uppermost layer

最上层船桥 flying bridge

最上层甲板 hurricane deck

最上层屋面板 cap sheet

最上的 supreme; topmost; uppermost

最上等的 first rate; super-fine

最上顶线 highest nuchal line; linea nuchae suprema

最上帆 trinket

最上公倍数 smallest common multiple

最上连续甲板 uppermost continuous deck

最上面的 uppermost

最上面的地层 uppermost layer

最上全通甲板 uppermost complete deck

最上统长甲板 uppermost full length deck

最上一块门芯板 frieze panel

最少 minimum[复 minima/minimauls]

最少费用方案 least cost option

最少费用分析 cost-minimization analysis

最少耕作法机具 minimum tillage machine

最少配筋率 minimum reinforcement ratio

最少级的级联 minimum stage cascade

最少价格影响 least price distortion

最少孔隙含量原则 principle of minimum voids content

最少量 shred

最少绿灯时间 minimum green period; minimum period

最少绿化空地 minimum landscaped open space

最少母料 mini-master batch

最少破坏原则 principle of least disruption

最少时间 <感应车辆存在后绿灯再延续一段的> unit extension

最少水泥含量 minimum cement content

最少显著差 least significant difference

最少有效锌 the lowest available zinc

最少余气 minimal residual air

最少运费确定规则 overflow rule

最少制造量 minimum manufacturing quantity

最深部分 innermost part

最深层电子 innermost electron

最深处 innermost; penetralia <尤指房屋的内院,高宇的内殿>

最深处的 inmost

最深的 bottom-most; innermost

最深方向 depth direction

最深分舱载重线 deepest subdivision loadline

最深谷底线【地】 thalweg; valley line

最深海底带 abyssal benthic zone

最深阴影 inmost shadow

最省时飞行路线 minimal flight path

最湿稳定稠度 wettest stable consistency[consistence]

最时新花样 latest fashion flower

最适 pH 值 optimum pH

最适比例 optimal proportions

最适比例带 equivalence zone; optimal proportion zone; zone of optimal proportion

最适比率 optimum ratio

最适产量 optimum yield

最适长度 optimal length

最适承载能力 optimal carrying capacity

最适承载容量 optimal carrying capacity

最适持久产量 optimum sustainable yield

最适处置 optimum cure

最适当出口税 optimal export tax

最适当的工厂规模 optimum plant size

最适当的权分配 most favo(u)rable distribution of weight

最适当的装球量 optimum ball charge

最适当的资金结构 optimal capital structure

最适当的最终研磨 optimum grind

最适当的最终研磨粒度 optimum grind size

最适当规模 optimum scale

最适当加价率 optimal rate of mark

最适当价格政策 optimal price policy

最适当生产向量 optimal production vector

最适当温度 optimum temperature

最适的 optimal
最适的变式 optimum variant
最适的变形 optimum variant
最适度的平衡 optimum balance
最适度含水量 optimum moisture content
最适度利用 optimum utilization
最适度下的 suboptimal
最适幅度 optimum range
最适负荷 optimal load
最适干燥条件 optimum drying condition
最适关税 optimum tariff
最适光强度较差 optimum light intensity range
最适含水量 optimum moisture content;optimum water content
最适含水率 optimum moisture percentage
最适利率 optimum rate of interest
最适量 optimal dose
最适量存货 optimal inventory
最适量现金余额 optimal cash balance
最适硫化 optimum cure
最适率 optimal ratio
最适密度 optimum density
最适面积 optimum plot size
最适捻度 optimum twist
最适浓度 optimum concentration
最适气候 optimum climate
最适强度 optimal intensity
最适曲线 optimum curve
最适人口数 optimum population
最适生长温度 optimum growth temperature
最适湿度 optimum humidity
最适寿命 optimal life
最适水分 optimum moisture; optimum watering
最适条件 optimum; optimum condition
最适条件带 zone of optimum condition
最适条件区 zone of optimum condition
最适温差 optimum temperature difference
最适温度 optimal temperature
最适温度分布图 optimum temperature profile
最适线性预测法 optimum linear prediction
最适线性预测值 optimum linear prediction
最适压 optimum pressure
最适盐度 optimal salinity
最适宜的 optimal;optimum
最适宜的混合物 mixture optimum
最适宜的投资 optimal investment
最适宜的政策 optimal policy
最适宜点 optimum
最适宜回流 optimum reflux
最适宜情况 optimum condition
最适宜速度 optimum velocity
最适宜条件 optimum
最适宜温差 optimum temperature difference
最适宜温度 optimum temperature
最适应的电场强度 most suitable field intensity
最适应的曲线 curve of best fit
最适用工艺 best available technology
最适用性 optimum
最适渔获量 optimum catch
最适种群 optimum population
最适资本结构 optimum leverage
最舒适角度 optimum comfort angle
最舒适温度 optimum temperature
最顺进路 most favo(u)rable route
最速存取时间 minimum access time
最速降线 brachistochrone;least time path;minimum-time path
最速上升 steepest ascent
最速上升抽样 steepest ascent sampling
最速上升法 steepest ascent method;steepest ascent procedure
最速下降法 gradient method;saddle-point method;steepest descent method;steepest gradient method
最速下降线 line of quickest descent
最外边的受拉纤维 extreme tension fiber
最外边的纤维 extreme fiber[fibre]
最外部反射层 outermost reflector
最外侧大梁 <桥的> outside girder
最外侧股道 marginal track
最外层 outermost layer;outermost shell
最外层表面 outermost surface
最外层的 outermost;outmost
最外层计算规则 outmost computation rule
最外层纤维 outermost fiber[fibre]
最外方的 outermost;outmost
最外轨道 outermost orbit
最外面 outermost
最外面的 outmost
最外面涂层 top coat
最外切削点 outmost cutting point
最外挖掘点 outmost cutting point
最外纤维 extreme fiber[fibre]
最外(缘)纤维应力 extreme fiber stress
最完全解理【地】most perfect cleavage
最晚封冻日 latest complete freezing
最晚解冻日 latest breakup;latest final ice clearance
最晚开工日期 latest start date
最晚开工时间 latest start date;latest starting time
最晚开始时间 latest start date;latest starting time
最晚完成时间 latest finish date;latest finish time
最晚完工时间 latest finish date;latest finish time
最晚终冰日 latest final ice clearance
最危险的滑动圆弧 <土坡的> most dangerous slip circle
最危险滑动面位置 slip-plane position of up-to-danger
最危险滑动中心位置 sliding center position of up-to-danger
最微小的距离 hairbreadth
最稀站网 minimum network
最细锉 super-fine file
最下部 foot
最下层 bottom-most;orlop
最下层地下室 <多层地下室中的最下层> subcellar
最下船侧纵通角 orlop stringer angle bar
最下的 undermost
最下甲板 orlop
最下甲板横梁板 orlop beam plate
最下甲板梁 orlop beam
最先进的事物 last word
最先进技术水平 technologic(al) state-of-the-art
最先污染物 priority pollutant
最显著的 predominant
最显著(引人注意)的地位 foreground
最显著(优先)地位 foreground
最现代化的 go-go
最小 minimum[复 minima/minimauls]
最小 χ^2 估计【数】minimum chi-squared estimate
最小安全半径 minimum safe(ty) radius
最小安全防护层 minimum protective layer in full-scale
最小安全距离 minimal safe distance;minimum safe distance
最小安全稳定性 minimum safety stability
最小安全系数 minimum factor of safety;minimum safety factor
最小安装调节距离 minimum installation allowance
最小半径 least radius;minimum radius
最小半径曲线 curve of minimum radius
最小包络曲线 minimum envelope curve
最小保证流量 lowest assured discharge;lowest guaranteed discharge
最小保证天然径流 lowest assured natural stream-flow
最小爆破断面 <隧洞工程中的> minimum excavation line
最小崩矿层厚线 burden line
最小比降 minimum grade
最小闭合高度 <模具的> minimum shut height
最小闭合邻域 smallest closed neighbo(u)rhood
最小闭开拓 minimal closed extension
最小壁厚 minimum wall thickness
最小变形圆锥投影图 minimum error conic(al) projection
最小辨视阈 minimum cognoscible
最小并矢展开 minimal dyadic expansion
最小波长 minimum wavelength
最小波浪历时 minimum wave duration
最小捕获速度 minimum capture velocity
最小不动点 least fixed-point
最小彩色分辨度 minimum colo(u)r acuity
最小操作误差 least operator bias
最小测程 minimum ranging distance
最小策动点函数 minimum-driving-point function
最小层次树 minimum height tree
最小长度 minimum length
最小超车视距 minimum passing sight distance;safe passing sight distance
最小超高度 minimum superelevation
最小车间距 minimum heading;minimum headway
最小车头间距 minimum space headway
最小车头时距 minimum(-time) headway
最小沉降量 minimum settlement
最小衬砌厚度 minimum thickness of tunnel(l)ing
最小成本 least cost
最小成本的投入组合 least cost input combination
最小成本法 least cost method
最小成本网络流程 minimal-cost network-flow
最小承载力 minimum bearing capacity
最小承载性能 minimum bearing property
最小吃水 minimum draught
最小持水量 least water-holding capacity;minimum water-holding capacity
最小尺寸 finest size;inferior limit;lower limit;lowest limit;minimum dimension;minimum size
最小尺寸问题 minimum size problem
最小尺寸钻杆 smallest size drill rods
最小尺度 least dimension
最小齿距 minimum pitch
最小充分统计量 minimal sufficient statistic
最小出力 minimum output
最小储藏效应 minimum storage effect
最小传导时间 minimum conduction time
最小传动比 fastest ratio
最小传热阻 minimum resistance of heat transfer
最小船舶长度 length of the smallest vessel
最小椽子间隔 minimum rafter spacing
最小椽子间距 minimum rafter spacing
最小吹送距离 minimum fetch
最小存取编码 minimal access coding
最小存取程序设计 minimum access programming
最小代格路径 minimal cost path
最小代价函数估计 minimum-cost estimate
最小代价网络 minimum cost network
最小代数群 smallest algebraic group
最小单位包装 unit package
最小导纳频率 minimum admittance frequency
最小的 lowest;minimal
最小的比较劣势 least comparative disadvantage
最小的地方行政区 commune
最小的可靠产量 minimum reliable yield
最小的空铅 hair space
最小的最大失误 mini-maxi regret
最小等数时间编码 minimum latency coding
最小等数时间程序设计 minimum latency programming
最小滴定量 titer[titre]
最小抵抗力 least resistance;minimal resistance
最小抵抗面 plane of weakness;weakness plane
最小抵抗线 least resistance line;line of least resistance;line of weakness;minimum burden;burden <岩石爆破的>
最小地块 minimum lot
最小地震 minimum earthquake
最小地震区 minimum seismic area
最小点 minimal point
最小点火电流 minimum firing current
最小电离速度 minimum ionizing speed
最小电流 valley current
最小电流断路器 minimum current circuit breaker
最小电流梯度 minimal current gradient
最小电压 minimum voltage
最小电阻 minimum resistance
最小定点 least fixed-point
最小定律 minimum law
最小动力负荷 minimum live load
最小动力条件 minimum dynamic(al) condition
最小动作电流 minimum working current
最小动作励磁 minimum working excitation
最小毒性剂量 minimal toxic dose;minimum toxic dose
最小读数 least count;least reading <按照仪器分度可能测读的最小值>
最小断面 minimum cross-section
最小对岸距离 minimum fetch
最小对集 minimum matching
最小多项式 minimal polynomial;minimum polynomial
最小多余度码 minimum redundancy
最小二乘逼近法 least square approximation

最小二乘多项拟合 least square polynomial fit
最小二乘法 least square method; least square procedure; least squares; method of least squares; method of minimum squares
最小二乘法逼近 approximation by least squares
最小二乘法程序 least square procedure
最小二乘法计算 least square calculation
最小二乘法解 least square solution
最小二乘法理论 least square theory
最小二乘法滤波 least square filtering
最小二乘法拟合 least square fit(ting)
最小二乘法配置 least square collocation
最小二乘法平差 adjustment by the method of least squares; least square adjustment
最小二乘法速度滤波 minimum least squares velocity filter
最小二乘法线 line of best fit
最小二乘法相关 least square correlation
最小二乘法选配 least square fit(ting)
最小二乘法原理 principle of least squares
最小二乘方 least square
最小二乘方逼近 least square approximation; least square error approximation
最小二乘方的权 weight in least square
最小二乘方分析 least-square analysis
最小二乘方估算 least squares prediction
最小二乘方计算 least squares calculation
最小二乘方拟合【数】 least square fit
最小二乘方线 line of least squares
最小二乘方准则 least square criterion
最小二乘估计 least square estimation
最小二乘估值 least square estimate
最小二乘回归 least square regression
最小二乘解 least square solution
最小二乘滤波 least squares filtering
最小二乘律 law of least square
最小二乘判别准则 least square criterion
最小二乘推估 least square prediction
最小二乘推估法 least square estimation process
最小二乘圆滑 the least square smoothing
最小发电水池 minimum power pool
最小反射率 minimum reflectivity
最小反射体 minimal reflector
最小反应量 minimal reacting dose
最小反应浓度 minimum response concentration
最小反应转速 <流速仪的> minimum speed of response
最小范围 minimum range
最小方差 minimum variance
最小方差法 varimin method
最小方差估计 minimum variance estimate
最小方差估值 minimum variance estimate
最小方差近似 minimum-squared-error approximation
最小方差控制 minimum variance control
最小方差量化 minimum variance quantizing
最小方差(判别)准则 minimum variance criterion
最小方差直线拟合 least squares straight line fit
最小仿射 least affine
最小仿射倍式 least affine multiple
最小费用 minimal fee; minimum fee
最小费用点 least cost point
最小费用法 least cost method
最小费用分析 least cost analysis
最小费用解 least cost solution
最小费用路由 least cost route
最小费用模型 least cost model
最小分辨角 angle of minimum resolution; limiting resolving angle
最小分辨率 minimum resolution
最小分辨限度 minimum resolution
最小风区(距离) minimum fetch
最小风险估计 minimum risk estimation
最小风险浓度 minimum risk concentration
最小峰宽 minimum peak width
最小服务 minimum service head
最小服务距离 minimum service range
最小负荷 minimum load; valley; period of minimum consumption【交】; valley consumption【交】
最小附着力 minimum adhesion
最小覆盖 minimal cover
最小覆土厚度 minimum soil cover thickness
最小改善路网 do minimum network
最小改善状况 do minimum
最小干扰 minimum disturbance
最小干扰频率 minimum disturbing frequency
最小干容重 minimum dry density
最小干围 minimum girth
最小干舷吃水 minimum freeboard draught
最小干舷(高度) minimum freeboard
最小钢筋比 critical steel ratio
最小工业厚度 minimum economic thickness
最小工作场半径 minimum good field radius
最小工作电流 minimum operating current; minimum working current
最小工作负荷 minimum live load
最小工作概率值 minimum working probit
最小公倍数 least common multiple; lowest common multiple; smallest common multiple
最小公分母 least common denominator; lowest common denominator
最小功 least work; minimum work
最小功变形原理 principle of least work of deformation
最小功法 method of least work
最小功耗 minimum power consumption
最小功理论 least-work theory
最小功率放大器 minimum watt amplifier
最小功率因数 minimum power factor
最小功(能)定理 theorem of least work; least-work theorem
最小功原理 <解超静定结构的一种方法> principle of least work; least-work principle; theorem of least work
最小固体量 least amount of solid
最小拐点 minimum yield point
最小管道管径 minimum sewer size
最小贯入度 minimum penetration
最小灌注速率 minimum irrigation rate
最小海拔高度 minimum height above sea level
最小含筋量 minimum reinforcement content
最小含水量 minimum moisture content
最小含水率 minimum moisture content
最小含氧层 minimum oxygen layer; oxygen minimum layer
最小旱天流量 minimum dry weather flow
最小焊接强度 minimum joint strength
最小航高【航测】 minimum flight height
最小耗水量 minimum water consumption
最小和 minimal sum
最小河道流量 minimum in stream flow
最小河道流量算法 minimum in stream flow algorithm
最小河流功率 least stream power; minimum stream power
最小荷载 minimum load
最小厚度 minimum depth; minimum ga(u)ge; minimum thickness; waist <钢筋混凝土楼梯板的>
最小弧距 minimum radial distance
最小化控制方式 minimizing control mode
最小化问题 minimization problem
最小化学势能 minimum chemical potential energy
最小环境流量 minimum environment(al) discharge
最小回波 minimum echo
最小回流 minimum reflux; minimum return flow
最小回流比 minimum reflux ratio
最小回流级联 minimum reflux cascade
最小回转半径 least radius of gyration; minimum turning radius
最小回转曲线 minimum turning curve
最小回转圆 minimum turning circle
最小混凝土保护层 minimum concrete cover
最小活荷载 minimum live load
最小或是值 most probable value
最小击穿电压 down voltage minimum break
最小积分平方误差 least-integral-square error
最小吉布斯能总量 minimum total Gibbs energy
最小极限 least limit; lower limit
最小极限尺寸 minimum limit of size
最小极限应力 limiting minimum stress
最小计数 least count
最小加法器 minimum adder
最小间隔 minimum interval; minimum separation; minimum spacing
最小间隔码 minimum distance code
最小间距 minimum distance; minimum spacing
最小间隙 minimum clearance; minimum pause; least play【机】
最小检测差 minimum detectable difference
最小检测范围 minimum detection limit
最小检测量 minimum detectable amount; minimum detectable quantity; minimum detection amount; minimum detection quantity
最小检测能力 minimum detectable activity; minimum detectable power
最小检测浓度 minimal detectable concentration; minimum detectable concentration
最小检测水平 minimum detectable level
最小检测温差 minimum detectable temperature difference
最小检测信号 minimum detectable signal
最小检测值 minimum detectable value
最小检出(放射)活性 minimum detectable activity
最小检出量 minimum detectable amount; minimum detection amount
最小检出水平 minimum detection level
最小检出限 minimum detection limit
最小建筑费用 least cost to-build
最小交会角 minimum intersection angle
最小接触角 minimal contact angle
最小截面 minimum section
最小解 minimal solution
最小金额 minimum amount
最小经济坡降 minimum economical descending grade
最小精密因数 least precise factor
最小净空 minimum clearance
最小净面积 minimum clear floor space
最小径流量 lowest streamflow; minimum runoff; minimum streamflow
最小局部实现 minimal partial realization
最小距离 least distance; minimum distance
最小距离调节 minimum range adjustment
最小距离法 minimum distance method
最小距离分类(法) minimum distance classification; least distance classification
最小距离分类器 minimum distance classifier
最小距离分压器 minimum range potentiometer
最小距离决策规则 minimum distance decision rule
最小距离码 minimum distance code
最小距离误差校正 minimum distance error-correcting
最小距离选择 minimum range selection
最小距离准则 minimum distance criterion
最小绝对误差投影 absolute minimum error projection
最小觉察到的效应浓度 lowest observed effect concentration
最小觉察到的效应水平 lowest observed effect level
最小觉察到的有害效应水平 lowest observed adverse effect level
最小均方速度 least mean square velocity
最小均方误差 least mean square error
最小均方最佳化 least mean square optimization
最小卡方 minimum χ^2
最小开机转矩 <内燃机的> breakaway torque
最小开挖线 minimum excavation line; pay line; payment line
最小抗滑线 line of least resistance
最小抗压强度 minimum compression strength; minimum compressive strength
最小科研卫星 minimum scientific satellite
最小可辨别信号 minimum discernible signal
最小可辨彩色差 just perceptible visual colo(u)r difference

Z

最小可辨差别 least perceptible difference
最小可辨差量 minimum perceptible difference
最小可辨差异 just perceptible difference
最小可辨亮度差 minimum perceptible brightness difference
最小可辨色度差 least perceptible chromaticity difference; minimum perceptible chromaticity difference
最小可辨温差 minimum readable temperature differential
最小可辨噪声 just perceptible noise; minimum perceptible noise
最小可采宽度 minimum minable width
最小可测信号功率 minimum detectable signal power
最小可分辨角 minimum resolvable angle
最小可分域 least splitting field
最小可检查臭味浓度 minimum detectable threshold odor concentration
最小可检测回声级 minimum detectable echo level
最小可检测亮度 minimum detectable brightness
最小可检测流量密度 minimum detectable flux density
最小可检测温度 minimum detectable temperature
最小可检测信号 minimum detectable signal
最小可检出放射性强度 minimum detectable activity
最小可见损害作用水平 minimum observed adverse effect level
最小可接受空档 minimum acceptable gap
最小可觉差 least perceptible difference
最小可靠天然径流 lowest assured natural stream-flow
最小可能饱和浓度 the minimum-possible saturation concentration
最小可能运行时间 shortest possible journey time
最小可能值 the smallest possible value
最小可迁集 smallest transitive set
最小可探测温差 minimum detectable temperature difference
最小可提取性百分率 lowest percentage extractability
最小可听度 minimum audibility; threshold of audibility
最小可听声压 minimum audible sound pressure
最小可闻场 minimum audible field
最小可闻区域 minimum audible field
最小可用信号 minimum useful signal
最小可住房间大小 minimum habitable room size
最小可住房间高度 minimum habitable room height
最小刻度 minimum scale
最小刻度值 minimum scale value
最小空隙混合料 minimum-void mix
最小孔径 minimum value aperture
最小孔隙比 minimum void ratio; void ratio in densest state
最小孔隙度 minimum porosity
最小孔隙混合料 minimum-void mix
最小孔隙率 minimum porosity
最小控制 minimum control
最小枯季径流 minimum dry weather flow
最小枯水流量 lowest ever-known discharge; minimum dry weather flow
最小库容 minimum storage
最小宽度 minimum width
最小拉力强度 minimum tensile strength
最小离地间隙 minimum ground clearance; ground clearance
最小离地距离 minimum ground clearance
最小离地空隙 <叉车> minimum gap above the floor
最小理论级数 minimum number of theoretical stage
最小理想 smallest ideal
最小力矩 least moment; minimum moment
最小历时 minimum duration
最小立方体受压强度 minimum cube strength
最小连接强度 minimum joint strength
最小链 minimal chain
最小梁高 minimum depth
最小亮度 minimum brightness
最小量 minimal dose; minimum [复 minima/minimauls]; minimum amount; particle
最小量程 minimum range
最小量程分压器 minimum range potentiometer
最小量耕作法 minimum till
最小裂变弹头 minimum-fission warhead
最小临界点 lowest critical point
最小临界热通量比 minimum critical heat flux ratio
最小临界值 lowest critical value
最小临界质量 minimum critical mass
最小临界质量实验 minimum-critical-mass experiment
最小流量 extreme small discharge; lowest discharge; minimum discharge; minimum flow
最小流量泵 minimal flow pump
最小流量控制 minimum-flow control
最小流速 minimum current velocity; minimum velocity
最小流态化 minimum fluidized voidage
最小楼板面积要求 minimum floor area requirement
最小炉膛尺寸 minimum fire box dimension
最小路径 minimal path
最小路径覆盖 minimum path cover
最小绿灯时间 minimum green period
最小落潮流(速度) lesser ebb; minimum ebb
最小马力 minimum horse power
最小脉冲持续时间 minimum pulse duration
最小脉冲时间 minimum pulse
最小铆钉间距 minimum rivet spacing
最小密度 minimum density
最小面 minimal face; minimum face
最小面积 minimum area
最小敏感限 contrast threshold
最小模 minimum modulus
最小模原理 minimum modulus principle
最小摩擦 minimized friction
最小磨削直径 minimum grinding diameter
最小内部尺寸 minimum internal dimension
最小挠曲半径 minimum bend(ing) radius
最小挠曲度 minimum deflection
最小能见度 minimum visibility
最小能力 minimum capacity
最小能量 minimising energy; minimum energy
最小能量比原理 minimum energy ratio principle
最小能量定理 theorem of minimum energy
最小能量法 minimizing energy method; minimum energy method
最小能量理论 least energy theory; minimum energy theory
最小能量时的共振 lowest resonance
最小能量梯度 minimum energy gradient
最小能量线 minimum energy line
最小能量原理 least energy principle; minimum energy principle; principle of minimum energy
最小能原理 principle of minimum energy
最小能最法 minimizing energy
最小年洪水量 minimum annual flood
最小年降水量 minimum annual precipitation
最小年径流量 minimum annual runoff
最小浓度 minimum concentration
最小排放量 minimum discharge
最小排污浓度测定值 lowest rejected concentration tested
最小配筋率 critical steel ratio; minimum ratio of reinforcement; minimum steel ratio
最小配置 minimal configuration
最小偏差 minimum deflection; minimum deviation
最小偏差法 minimum deflection method; minimum deviation method
最小偏差角法则 minimum deviation law
最小偏向光栅 minimum deviation grating
最小偏向角 angle of minimum deviation; minimum deviation
最小偏斜 minimum deflection
最小偏移距 minimum offset
最小偏转 minimum deflection
最小频移键控 minimum shift keying
最小平方 least square
最小平方法 least square method; method of least squares; method of minimum squares
最小平方反滤波 least square inverse filtering
最小平方反褶积 least squares deconvolution
最小平方近似 least square approximation
最小平方滤波 least squares filtering
最小平方线性关系 least square linear relationship
最小平方映射技术 least square mapping technique
最小平均B位形 minimum average B configuration
最小平均询问次数 minimal mean number of inquiry
最小平均致突变剂量 minimal average dosage for mutation
最小平曲线半径 minimum radius of horizontal curve
最小平曲线长度 minimum horizontal curve length
最小平挖半径 <挖掘机> minimum level radius
最小坡度 minimum grade; minimum gradient; minimum slope
最小坡段长度 minimum length of grade section
最小坡降 minimum fall; minimum grade
最小破坏强度 minimum failure strength
最小破坏应变 minimum failure strength
最小剖面 minimum section
最小期限 minimum period
最小齐次理想 smallest homogeneous ideal
最小起爆药量 minimum initiating charge; minimum priming charge
最小起步转矩 breakaway torque
最小起点电容 minimum capacity
最小起动扭矩 pull-up torque
最小汽车旅馆单元 minimum hotel-motel units
最小铅字 minikin
最小强度 intensity minimum; minimum intensity; minimum strength
最小强度线 line of weakness
最小墙厚度 minimum wall thickness
最小切割布局法 min-cut placement method
最小切削长度 minimum cut length; minimum length cut
最小倾翻负荷 minimum tipping load
最小倾翻力矩 minimum capsizing moment
最小晴天流量 minimum dry weather flow
最小求距角 minimum distance angle
最小曲线半径 minimum curve radius; minimum radius of curve; smallest curve radius
最小屈服点 minimum yield point
最小屈服抗挤强度 minimum yield collapsing strength
最小权路径选择 minimum weight routing
最小权值路由选择 minimum weight routing
最小燃油限止螺钉 minimum fuel limit screw
最小热量 minimum heat
最小热损失 minimum heat loss
最小热阻 minimum thermal resistance
最小人体有害剂量 minimum human infectious dose
最小日供气量 minimum day sendout
最小日照间距 minimum sunlighting spacing
最小容量 minimum capacity; primary capacitance <可变电容器的>
最小容许半径 minimum allowable radius; smallest permissible radius
最小容许尺寸 minimum admissible dimension; minimum permissible dimension
最小容许轨距 minimum permissible ga(u)ge
最小容许流量 minimum acceptable flow
最小容许流速 minimum permissible velocity
最小容许排水量 minimum acceptable discharge
最小容许溶解氧 minimum allowable dissolved oxygen concentration
最小容许值 minimum acceptance value
最小溶度点 point of minimum solubility
最小溶解氧浓度 minimum dissolved oxygen concentration
最小熔断电流 minimum blowing current
最小冗余原理 principle of minimal redundancy
最小入土长度 <桩的> minimum embedment length
最小入土深度 <桩的> minimum embedment depth
最小润湿速度 minimum wetting rate
最小撒布厚度 minimum spreading

Z

thickness
最小熵 minimum entropy
最小熵产生定理 theorem of minimum entropy production
最小熵反褶积 minimum entropy deconvolution
最小上界【数】least upper bound;supremum
最小设计流速 minimum designing velocity
最小设计坡降 minimum designing gradient
最小设计重量 minimum design weight
最小设置高度 minimum setting height
最小社会影响 minimize social impact
最小射程 minimum range
最小射角 minimum elevation
最小深度 least depth;minimum depth
最小渗透率 minimum permeability
最小生成树 minimum spanning tree
最小声速级 level of minimum sound velocity
最小声响 minimum sound
最小剩余 least residue
最小湿周边 least wetted perimeter
最小时程 least time path;minimum-time path
最小时钟脉冲持续时间 minimum clock-pulse duration
最小时钟频率 minimum clock frequency
最小识别单位 atom
最小实现 minimal realization
最小势能 minimum potential energy
最小势能原理 principle of minimum potential energy
最小势能原理法 method of minimum potential
最小视角 minimal visual angle
最小视距三角形 minimum sight triangle
最小视觉亮度差 minimum perceptible brightness difference
最小室外空气量 minimum outside air
最小竖曲线半径 minimum vertical curve radius
最小竖曲线长度 minimum length of vertical curve
最小水池 minimum pool
最小水力蕴藏量 minimum potential of water power
最小水流功率 minimum stream power
最小水平曲线半径 minimum horizontal curve radius
最小水深 minimum depth
最小水头 minimum head;minimum water head
最小水头损失 minimum head loss
最小水压力 minimum water pressure
最小水阻力 minimum water resistance
最小速度 minimum speed;minimum velocity
最小速率 minimum speed
最小塑性抗挤强度 minimum plastic collapsing strength
最小损耗匹配 minimum-loss matching
最小损耗衰减器 minimum-loss attenuator;minimum-loss pad
最小损失防火理论 minimum-damage-fire-control theory
最小塌落度 lowest slump
最小坍落度 stiffest consistency
最小弹塑性抗挤强度 minimum elastic-plastic collapsing strength
最小弹性抗挤强度 minimum elastic collapsing strength
最小探测距离 minimum detectable range
最小套节深度 minimum depth of socket
最小体积 minimum volume
最小田间持水量 least field capacity;minimum field capacity
最小填土高度 minimum fill height;minimum height of fill
最小条件 minimal condition
最小调和优函数 least harmonic majorante
最小调焦距离 shortest focusing distance
最小跳步 minimum hop
最小跳步通路 minimum hop path
最小听压 minimum audible pressure
最小通风量 minimum ventilation rate
最小通航水深 least navigable depth;least navigation depth;minimum navigable depth;minimum navigation depth
最小通信[讯]连接量 minimum tie set
最小凸多角形 minimal convex polygon
最小凸集 minimum convex set
最小推力 minimum thrust
最小推力自动首向调整 autoheading for minimum thrust
最小退潮流 lesser ebb;minimum ebb
最小挖宽【疏】minimum cut(ting) width;minimum dredging width
最小挖深 minimum dredging depth
最小外接圆法 minimum circumscribed circle method
最小外形尺寸 minimum contour size
最小弯矩 minimum moment
最小弯曲半径 minimum bend(ing) radius
最小完备类 minimal complete class
最小完全防护层 minimum full protection layer
最小维持流量 minimum maintained flow
最小位能定理 theorem of minimum potential energy
最小位能定律 law of minimum potential energy
最小位能原理 law of minimum potential energy;principle of minimum potential energy
最小位置 minimum position
最小屋顶坡度 minimum roof slope
最小误差 least error;minimum error
最小误差译码 minimum error decoding
最小下限 smallest limit
最小(显示)变化 <最短绿灯时间完了后的显示变化> minimum change
最小显著差(数) least significant difference
最小显著差异 least significant difference
最小显著数 least significant difference
最小线间距 minimum distance between centers of tracks
最小线路通过能力 minimum line capacity;minimum track capacity
最小限 irreducible minimum
最小限度 interior limit;lower limit;minimize;minimum limit
最小限度的 minimal
最小限度活动的原则 principle of minimal action
最小限界 minimum clearance
最小限制应力 limiting minimum stress
最小相角 minimum phase angle
最小相位 minimum phase
最小相位滤波器 minimum phase filter
最小相位系统 minimum phase system
最小相位因子 minimum phase operator
最小相移 minimum phase
最小相移键控 minimum phase shift keying
最小相移网络 minimum phase-shift network
最小相移系统 minimum phase shift system
最小响应 minimum response
最小项 minterm
最小像差圆 least circle of aberration
最小消光角 least extinction angle
最小新风量 minimum fresh air requirement
最小信号法 minimum signal method
最小信号接收 minimum reception
最小信息组长度 minimum block length
最小行车间隔 minimum running interval
最小行宽 minimum width
最小行政区 minimum administration district
最小型站 minimum-sized station
最小性 minimality
最小性能要求 minimum performance requirement
最小需光度 minimum light-intensity
最小需水量 minimum water requirement
最小需氧量 minimum oxygen requirement
最小需要风量 minimum air requirement
最小需要量 minimum demand
最小旋回圈 minimum turning circle
最小压力稳定器 minimum pressure governor
最小压力线 line of least pressure
最小压实力 minimum compactive effort
最小压缩轴 axis of least compression
最小烟囱尺寸 minimum stack dimension
最小烟囱断面 minimum stack area
最小延迟 minimum delay
最小延迟编码 minimum delay coding
最小延迟的程序设计 minimum delay programming
最小延迟电路 minimum delay circuit
最小延迟条件 minimum delay condition
最小延迟整形 minimum-lag shaping
最小仰角 minimum angle of elevation;minimum elevation
最小样方面积 minimum quadrat area
最小要求浓度 minimum required concentration
最小夜间比 minimum night ratio
最小液流量 minimum liquid rate
最小液气比 minimum liquid-gas ratio
最小一般可感 smallest generally felt
最小异常值 minimum anomaly value
最小抑制剂量 minimal inhibiting dose;minimal inhibitory dose
最小抑制浓度 minimal inhibitory concentration
最小因素 minimum factor
最小应变 minimum strain;minor strain
最小应变能原则 principle of minimum strain-energy
最小应变轴 least strain axis;minimum strain axis
最小应变轴方位 orientation of minimum strain axis
最小应力 minimum stress;minor stress
最小应力原理 principle of minimum stress
最小游隙 minimum play
最小有效高度 minimum effective height
最小有效剂量 least effective dose;minimum effective dose;minimal effective dose
最小有效量 least effective dose
最小有效流量 minimum effective liquid rate
最小有效数 least significant digit
最小有效位 least significant bit
最小有效字符 least significant character
最小有影响浓度 minimal effective concentration;minimum effective concentration
最小余能原理 principle of minimum complementary energy
最小预付保险费 minimum and deposit premium
最小预紧密封比压 minimum pretightening sealing load
最小预应力 minimum prestressing
最小缘石支距 minimum curb offset
最小远交 least affine
最小约束原理 principle of least constraint
最小允许半径 minimum allowable radius
最小允许电流 minimum acceptable current
最小允许含量 minimum permissible content
最小允许量 minimum acceptable discharge
最小允许流量 minimum acceptable flow
最小允许流速 minimum permissible velocity
最小允许强度系数 minimum allowable strength factor
最小运费 minimum freight
最小运输宽度 minimum shipping width
最小运行方式 minimum operation mode
最小运转时间 minimum operating time
最小噪声 minimum noise
最小站间隔 minimum station interval
最小张紧调节距离 minimum takeup allowance
最小涨潮流 lesser flood;minimum flood
最小涨潮速度 minimum flood velocity
最小照度 minimal illumination
最小整数 smallest positive integral
最小正规子群 minimal normal subgroup
最小正整数解 minimal positive integral solution
最小直读 least readout
最小直径 minimum diameter
最小值 minimal value;minimum[复 minima/minimauls];minimum value
最小(值)定律 law of minimum
最小值法 minimization
最小值原理 minimum principle
最小指令间隔 minimum command interval
最小指数 minimal index
最小制动位 minimum reduction position

Z

最小致电离速率 minimum ionizing speed
最小致死剂量 least fatal dose; minimum lethal dose
最小致死浓度 minimum lethal concentration
最小致死氧浓度 minimal lethal oxygen concentration
最小重复天线阵 minimum-redundancy array
最小重力势能 minimum gravitational potential energy
最小重量 minimum weight
最小重量百分率 minimum percentage by weight passing
最小重量设计 least weight design
最小主应变 minimum principal strain
最小主应力 least principal stress; minimum principal stress; minor principal stress
最小主应力方位角 direction of the minimum principal stress
最小主应力倾角 dip of the minimum principal stress
最小主应力值 minimum principal stress value
最小主应力轴 axes of minor stress
最小住房尺寸要求 minimum house size requirement
最小转换键控 minimum shift keying
最小转数 minimum revolution
最小转弯半径 smallest turning radius; minimum radius of turning; minimum turning radius; clearance radius
最小转弯轨迹 minimum turning path
最小转弯圆圈 minimum turning circle
最小装入量 minimum fill
最小状态 minimum rating
最小子 A 模 smallest sub-A-module
最小子环 smallest subring
最小子空间 smallest subspace
最小自动机 minimal automaton
最小总资源量 minimum total resource
最小纵坡 minimum longitudinal grade; minimum longitudinal gradient
最小纵坡度 minimum grade
最小阻抗频率 minimum impedance frequency
最小阻抗线 line of least resistance
最小阻力 least resistance
最小阻力冲角 angle of incidence of minimum drag
最小阻力攻角 angle of attack of minimum drag
最小阻力区 zone of least resistance
最小阻力系数 minimum drag coefficient
最小阻力线 line of least resistance
最小最大法 minimum maximum basis
最小最大后悔值法 smallest of largest regret value method
最小最大应力比 minimum to maximum stress ratio
最小最大原则 minimax principle
最小作业宽度 minimum operating width
最小作用距离 minimum action range
最小作用量 least action; minimal effective dose
最小作用量原理 least action principle; principle of least action
最小作用水平 minimal effect level
最小作用原理 principle of least action
最小座距 minimum seat spacing
最新版本技术标准 latest technical standard
最新产品 last word; latest off spring
最新成就 last word
最新出版物 new publication
最新出品的 brand-new
最新的 ultra-modern; up-to-date style
最新的产品 up-to-date product
最新的电话号码本 up-to-date directory
最新地图 current map; up-to-date map
最新动态 hot news
最新动态介绍的 <美> refresher
最新动态介绍课 <美> refresher course
最新发明 last word
最新工艺 latest technology; updating technique; updating technology
最新估算 current estimate
最新光导摄像管摄像机系统 advanced vidicon camera system
最新技术 updating technique
最新进展 recent development
最新精简的技术标准 streamlined specification
最新卡片 last card
最新科学技术 state-of-the-art technology
最新科学技术成就 most-up-to-date science and technology
最新流行的式样 time of day
最新流行品 last cry; last cry first-out
最新期刊号 current number
最新情报 up-to-date information
最新生产型飞机 ultimate production aircraft
最新式 up-to-date type
最新式差 ultra-modern ship
最新式的 down-to-date; go-go; up to the-minute; ultra-modern; up-to-date
最新式样 up-to-the-minute styling
最新信息 up-to-date information
最新研究报告集 advanced research monograph
最新样式 new look
最新资料 latest information; recent information; update; update data
最新自动编图系统 advanced automatic compilation system
最汹水位 most turbulent flow stage
最严重的 first degree
最一般的合一 most general unifier
最宜场阳极温度 optimal field-anode temperature
最宜价格产量政策 price-output policy
最宜进料位置 optimum feed location
最宜温度 optimal temperature; optimum temperature
最易带出的元素 most active element
最易通过的渗径 preferential path
最易行车(车辆)【铁】 easiest rolling car
最引入注意的位置 foreground
最硬的 rugged
最优 inferior to none
最优泵量 optimum flow rate
最优逼近 best approximation
最优逼近多项式 best-fit polynomial approximation
最优编码 optimum code
最优薄锡层 best cokes
最优薄锡层镀锡薄钢板 best cokes
最优材面 hest face
最优采购量 optimum purchase
最优参数 optimum parameter
最优操作点 optimum operating point
最优策略 optimal policy; optimal strategy; optimum strategy
最优插值 best interpolation; optimum interpolation
最优承租人 prime tenant
最优程序设计 optimum programming
最优稠度 optimal consistency
最优存取 minimum access
最优(代)码 optimum code
最优贷款利率 top priority lending rate
最优的 optimal; optimalizing; optimum
最优点 optimum point
最优调度 optimal scheduling
最优调度策略 optimal scheduling strategy
最优动力特性 optimum dynamic(al) characteristic
最优堆积 optimum packing
最优钝化 optimal desensitization
最优发动机燃料 ultimate motor fuel
最优法 optimization
最优法处理过程 optimization procedure
最优反应器设计 optimum reactor design
最优方案 optimal design; optimum alternative; optimum design; optimum scheme; optimization program(me)
最优方案的性质 property of optimal plan
最优方法 best practice; optimum method
最优分割 optimum division
最优分割法 optimal segmentation method
最优分配 allocation optimum; optimum allocation
最优负荷因数 optimum load factor
最优负荷因素 optimum load factor
最优干度 dry of optimum
最优刚度分布 optimum rigidity distribution
最优耕作 optimum till
最优工作时间 optimum work time
最优攻角 optimum angle of attack
最优估计 optimal estimation; optimum estimate; optimum estimation
最优估算 optimal estimate
最优灌溉 full-delta irrigation
最优灌溉需水量 optimum irrigation requirement
最优归并模式 optimum merging pattern
最优规划(方案) optimum programming
最优规模 optimum size
最优过程 optimal process
最优含水量 optimum moisture content; optimum water content
最优含水率 optimum water content
最优耗量 optimum consumptive [consumption] use
最优河湾 optimum bend
最优厚的条件 softest possible terms
最优化 optimalize; optimization; optimize; optimizing; optimum
最优化程序 optimum procedure
最优化处理过程 optimization procedure
最优化的边缘条件 marginal conditions of optimization
最优化的应用 application of optimization
最优化法 optimizing method
最优化方法 optimization approach; optimization method; optimization procedure
最优化分析 optimality analysis
最优化计算 optimization calculation
最优化技术 optimization technique
最优化技术体系 optimization technique system
最优化进出口商品结构 optimal structure of imports and exports
最优化控制 optimum control
最优化理论 mathematic(al) programming; optimization theory
最优化模式 optimization mode
最优化模型 optimization model
最优化设计 optimization design
最优化水平【数】 level of optimization
最优化算法 optimization algorithm
最优化特性 optimization characteristics
最优化性能 optimization performance
最优化研究 optimization study
最优化原理 principle of optimality
最优化原则 optimization criterion; principle of optimality
最优化钻井程序 optimized drilling program(me)
最优化钻探程序 optimized drilling program(me)
最优惠贷款率 best lending rate
最优惠短期贷款利率 prime lending rate
最优惠国 most-favo(u)red nation
最优惠国待遇 most favo(u)red nation treatment
最优惠国条款 most favo(u)red nation clause
最优惠利率 optimum rate of interest; prime rate; prime rate of interest
最优惠条件 best terms
最优基本解 optimal basic solution
最优基本可行解 optimal basic feasible solution; optimum general feasible solution
最优级材料 research class
最优级(木材) supreme
最优级配 optimum gradation
最优集合 optimal set
最优加价率 optimal mark-up rate
最优加权函数 optimum weight function
最优价格 best price
最优检索算法 optimal searching algorithm
最优奖获得者 grand champion
最优结构 optimal structure; optimum structure
最优结构设计 optimal structural design; optimum structural design
最优解 optimal solution; optimum solution
最优近似 best-fit approximation; optimal approximation
最优决策 optimal decision
最优开发 optimum development
最优开发计划 optimum development plan
最优抗震设计 optimum seismic design
最优颗粒堆积 optimum particle packing
最优可行解 optimal feasible solution
最优控制 optimal control; optimizing control; optimum control
最优控制法 method in optimal control; method of optimal control
最优控制方程 optimal control equation
最优控制理论 optimal control theory; optimizing control theory
最优控制问题 optimal control problem
最优控制系统 optimal control system; optimizing control system

最优块体设计 optimal block design
最优良的 first line
最优滤波 optimal filter(ing)
最优路径 optimal path
最优路线法 critical path diagram; critical path method
最优论文 best-paper
最优满足法 best-fit method
最优密度 optimum density
最优模型 optimal model
最优内插【数】best interpolation; optimum interpolation
最优能力 optimum capacity
最优凝聚点 optimum point of coagulation
最优排放水平 optimal level of emission
最优排量 optimum flow rate
最优配合比 optimum mix; optimum program(me); optimum proportioning
最优配置 optimal allocation; optimum allocation
最优平滑滤波器 optimal smoother
最优平滑器 optimal smoother
最优普氏密度 <土的> Proctor optimum density; optimum Proctor density
最优期望值 optimum expectation
最优汽车汽油 ultimate motor fuel
最优强度 optimum strength
最优强度分布 optimum strength distribution
最优区组设计 optimal block design
最优曲线 optimum curve
最优确定调度 optimal deterministic schedule
最优上升曲线 synergic curve
最优设计 optimal design; optimum design
最优射线 optimal ray
最优生长条件 optimal growing condition; optimum growing condition
最优湿度 wet of optimum
最优湿度值 optimum moisture level; optimum moisture value
最优实用手段 best practicable means
最优势期的 phyloephebic
最优适合 best fit
最优适应的 optimal-adaptive
最优输出 optimum output
最优水力断面 best hydraulic cross-section
最优松弛参数 optimum relaxation parameter
最优搜索过程 optimal search procedure
最优碳化含水量 optimal moisture for carbonation
最优梯度法 optimum gradient method
最优条件 optimal condition; optimality condition
最优调整 optimal adjustment; optimum setting
最优投量 optimum dosage
最优土法 optimum soil process
最优推算 optimal prediction
最优完成时间 optimal finish time
最优网络 optimum network
最优微差时间间隔 optimum short-delay time interval
最优卫星配置 optimum satellite configuration
最优温度序列 optimum temperature sequence
最优系统 optimal system
最优先 top priority
最优先处理 foreground processing
最优先的 prepreference
最优先等级的通信[讯] high priority communication
最优先股 prepreference share
最优先投资项目 high priority investment project
最优先中断【计】override interrupt
最优现金结存额 optimal cash balance
最优线路法 critical path method
最优线性估计 optimum linear estimate
最优线性规划 optimum linear program(me)
最优线性拟合 best linear fit
最优线性无偏估算子 best linear unbiased estimator
最优线性系统 optimum linear system
最优线性预报值 optimal linear predictor
最优线性预测法 optimum linear prediction
最优线性预测值 optimum linear prediction
最优效率 optimum efficiency
最优效率点运行 point-of-best-efficiency operation
最优斜量法 optimum gradient method
最优性 optimality
最优性标准 optimality criterion; optimality standard
最优性方程 optimality equation
最优性能 optimal performance; top performance
最优性条件 optimality condition
最优性原理 optimality principle; principle of optimality
最优秀 first-water
最优(选)点插值 optimum interval interpolation
最优选择 optimal choice
最优压实度 optimal compaction
最优营业规模 optimal size of business
最优预测 optimum prediction
最优在库量 optimal inventory
最优政策 optimal policy
最优值 figure of merit; optimality level; optimal value; optimum value
最优质的 top-quality
最优质设备 top-quality equipment
最优轴载 optimum axle load
最优逐次超松弛 optimum successive overrelaxation
最优转速 optimum rotary speed
最优转速选择器 optimum table-speed selector
最优装配 optimal assembly
最优自动控制 self-optimizing control
最优纵断面线 optimum profile
最优组合 optimum combination
最优钻速 optimum penetration rate
最优钻头水马力 optimum bit hydraulic horse power
最优钻压 optimum bit weight
最优(最适)安排 optimum arrangement
最有价值的 most worthy
最有可能的售价 most probable selling price
最有利的 optimum
最有利方向 most favo(u)rable direction
最有利信号 most favo(u)rable signal
最有破坏性 most destructive
最有效比特 most significant bit
最右边车道 right hand lane
最右边的 rightmost
最右边行车道 right hand lane
最右端 low-order end
最右判读 rightmost interpretation
最远的 endmost; outermost; outmost; utmost
最远点 apog(ee)
最远距离 extreme range
最远行走距离 furthest distance of travel
最远轴距【机】distance between the extremes of any of two more consecutive axles
最早初冰日 earliest first appearance of ice
最早到达时间 earliest arrival time
最早封冻日 earliest complete freezing
最早结束时间 earliest finish time
最早结束时刻 early finish date
最早解冻日 earliest breakup
最早开工 earliest start
最早开工日期 earliest start date
最早开工时间 earliest start time
最早开始 earliest start
最早开始时间 earliest start time; the earliest starting time
最早可能日期 earliest possible date
最早生长的植物 pioneer plant
最早完成时间 earliest finish time
最早项目完成时间 earliest event occurrence time
最早预计日期 <统筹方法中,预期事项发生最早日期> earliest expected date
最早终冰日 earliest final ice clearance
最值得的 most worthy
最终摆动挤压整平装置 final oscillating extrusion finisher
最终报表 final statement
最终报告 close out report; completion report; final report; final statement
最终被控变量 ultimately controlled variable
最终闭括号 final closing bracket
最终边 final
最终边帮台阶 final wall bench
最终边坡设计 ultimate slope design
最终不知 final ignorance
最终财富所有者 ultimate wealth owner
最终财务成果 final financial result
最终采收率 ultimate recovery
最终彩色图像 final colo(u)r picture
最终残渣 final residuum
最终测井解释成果图 epilog
最终测量 completion survey; final survey
最终查勘 final survey
最终产量 ultimate output; ultimate production; ultimate yield
最终产率 ultimate yield
最终产品 end item; termination product; end product
最终产品的 end product
最终产品法 <控制所做沥青混凝土质量的一种合同方法,即不问选用何种设备,只控制最后成品质量的一种方法> end product approach
最终产品规范 end-product specification
最终产品规格 end-product specification
最终产物的 end product
最终沉淀槽 final sedimentation tank; final settling tank
最终沉淀池 final clarifier; final sedimentation tank; final settling basin; final settling tank; final tank
最终沉淀池的沉淀物 final sedimentation
最终沉淀作用 final sedimentation
最终沉降池 final settling basin
最终沉降(量) final settlement; final settling; final subsidence; ultimate settlement; ultimate subsidence
最终沉陷量 final settlement; final settling; final subsidence; ultimate settlement; ultimate subsidence
最终成本 ultimate cost
最终成分 end composition
最终成果 final result
最终成果解释 end product interpretation
最终成品 final product
最终成品断面 finished product section
最终承载力 ultimate loading capacity
最终乘积 final product
最终澄清池 final clarifier
最终弛垂 final sag
最终尺寸 final size; finishing size; resulting dimension
最终齿轮减速【机】final gear reduction
最终充电电流 finishing rate
最终出流物 final effluent
最终出水 final effluent
最终储量 ultimate reserves
最终处理 ultimate disposal
最终处理厂 terminal treatment plant
最终处理过程 final treatment process
最终处置 final disposal; ultimate disposal
最终传动【机】final drive
最终传动机构 final drive gear
最终传动减速【机】final drive reduction
最终传动外侧 outer section of final drive
最终传动小齿轮【机】final drive pinion
最终传动圆柱直齿轮 final drive spur gear
最终传送 fast pass
最终纯损 ultimate net loss
最终存量 closing inventory
最终答案 final result
最终大圆航向 final great-circle course
最终到达港 final destination; ultimate destination
最终的 extreme; final; ultimate
最终的精整加工 final finishing
最终等时线 final isochrone
最终地段墙 final plat
最终地形 ultimate landform
最终地址 final address
最终地质报告 final geologic(al) report
最终电流 ultimate current
最终定线 final location
最终读出 end reading; end readout
最终读数 final reading
最终断面 finished section
最终对准 final alignment
最终发票 final invoice
最终发运计划 final delivery schedule
最终反应 final reaction
最终方案 final plan; final project
最终放大率 final magnification
最终放矿 final draw
最终费用 terminal expenses
最终费用概算 final cost estimate
最终分布 final distribution; ultimate distribution
最终分号 final semicolon
最终分离器 final separator
最终分配 final distribution; ultimate distribution
最终分析 ultimate analysis
最终封井压力 final shut-in pressure
最终付款 final payment

最终付款凭证 final certificate
最终付款申请初稿 draft final statement
最终付款证书 final payment certificate;final certificate for payment
最终干燥 terminal drying
最终工程报告 final engineering report
最终(工程)质量规定 end product specifications
最终供应 final supply
最终购买者 ultimate purchaser
最终购入价格法 method of price of last purchase
最终估价测量 final estimate survey
最终股利 final dividend
最终鼓风 final blast
最终固化度 ultimate set
最终贯入度 final penetration
最终轨道 last turn
最终过滤器 ultimate filter
最终含水饱和度 final water saturation
最终含水量 final moisture content
最终航向【航空】final heading
最终合同报告 final contract report
最终和数 final sum
最终后验分布 final posterior distribution
最终厚度 finished thickness
最终环境报告书 final environmental statement
最终环境鉴定 final environmental statement
最终环境影响报告书 final environmental impact statement
最终环境影响报告书附录 final environmental impact statement supplement
最终环境影响鉴定 final environmental impact statement
最终环境影响鉴定附录 final environmental impact statement supplement
最终挥发速率 final evapo(u)ration rate
最终回采率 ultimate recovery
最终回填料 final backfill
最终回旋直径 final turning diameter
最终毁坏性破裂 final catastrophic failure
最终混合料 ultimate mixture
最终货品 final goods
最终集装箱装载图 complete stowage plan
最终计量 final measurement
最终计算长度 final ga(u)ge length
最终继续行 final continuation line
最终加工 final processing
最终加工精度 accuracy of finish
最终加热 final heating
最终价 ultimate cost
最终价格 final price
最终价值 final cost
最终减速齿轮 final reduction gear
最终减速器壳 final drive casing
最终减速总成 final reduction assembly
最终检查 final examination;final inspection
最终检验 final inspection
最终检验合格证书 final examination certificate
最终阶段 final stage
最终节点 finish node
最终结果 end product;end result;final outcome;final result;net result
最终结果分析 end point analysis
最终解决防治问题 the final solution to pest control
最终进场 final approach
最终精加工 final finishing
最终精矿 final concentrate;finished concentrate
最终精选 final clearing
最终精制 final refining;polishing
最终净化操作【给】polishing operation
最终净化处理 polishing treatment;final treatment
最终净化滤池 polished filter
最终净化塘 polished pond
最终净化装置 final purification plant;final purifier
最终净化作用 eventual purification;final purification
最终静水位 final static level
最终决算 final decision
最终开采量 ultimate recovery
最终开航 final sailing
最终开挖标高 final dredged level
最终开挖线 final excavation limit
最终勘探报告 final exploratory report
最终颗粒大小 final grain size;final particle size
最终可溶解浓度 final soluble concentration
最终孔隙比 final void ratio
最终控股公司 ultimate holding company
最终控制驱动 final controlling drive
最终控制拖动 final controlling drive
最终库存 final stock;final store
最终冷凝器 final condenser
最终冷却 final cooling
最终冷却带 finish cooling zonal final cooling zone
最终利润 final return
最终利益 ultimate interest
最终粒度 final grain size;final particle size
最终链路 final link
最终流程 final flowsheet
最终流出物 final effluent
最终隆起量 final heave
最终露天矿边界 final open-pit boundary;final pit boundary
最终滤清器 final filter
最终路由 final route
最终瞄准 final laying;terminal guidance
最终磨细炉渣 finish grinding
最终目标 end-all
最终目的地 ultimate destination
最终耐用性 <即终止使用时的耐用性> terminal serviceability
最终耐用性指数 final serviceability index;terminal serviceability index
最终配时(方案)final setting
最终膨胀 final bulking
最终频率 final frequency
最终品质证明书 final certificate;final certificate of quality
最终平衡浓度 final equilibrium concentration
最终平衡态 final equilibrium state
最终平均年生长量 final mean annual increment
最终平面布置图 final layout
最终评价 final evaluation
最终坡度 final grade;finish grade
最终破裂强度 ultimate strength of rupture
最终破碎 final crushing;finished crushing
最终剖面 final profile
最终普查报告 final prospecting report
最终启航 final sailing
最终气化温度 final vapo(u)rization temperature
最终弃置 final disposal
最终潜在可采储量 ultimate potential recovery
最终强度 final strength
最终强度极限 ultimate load
最终清洗 final cleaning
最终任务 actual order
最终容积 final volume
最终溶液 final solution
最终入渗能力 final infiltration capacity
最终入渗水量 final infiltration capacity
最终筛 finishing screen
最终筛分 final sizing
最终烧成 final firing
最终设计 final design;ultimate design
最终设计审查会 final design review meeting
最终设计说明书 description of final design
最终设计图 final design drawing;final scheme
最终设计修正 adjustment of final design
最终深度 ultimate depth
最终渗水量 final infiltration capacity
最终升运器 final elevator
最终生化需氧量 biochemical oxygen demand ultimate;ultimate biochemical oxygen demand
最终施工费用 final construction cost
最终湿度 final moisture content
最终实际损失 ultimate net loss
最终实际自留 ultimate net retention
最终使用能力指数 final serviceability index
最终使用温度 end-use temperature;ultimate use temperature
最终使用需求 end-use demand
最终试验 final test
最终收缩度值 final contraction value
最终寿命 ultimate life
最终受控变量 ultimately controlled variable;final controlled variable
最终受控条件 final controlled condition
最终受益人 final beneficiary
最终书面报告 final letter report
最终疏浚标高 final dredged level
最终输出 final output
最终数位码 final digit code
最终水分 final moisture;final water
最终水灰比 final water-cement ratio
最终水头高度 final height of water head
最终死亡 ultimate death
最终速度 final speed;final velocity
最终速度模型 end velocity model
最终坍岸带宽度 final width of bank ruin of reservoir
最终特性 final response
最终条件 end condition
最终条例 final act
最终投保单 closing declaration
最终图件 final map
最终图样 final drawing
最终土壤水分 final soil moisture
最终挖成标高 final dredged level
最终外径尺寸 finished outside diameter ga(u)ge
最终完工 final completion
最终(完工)计量时间 period of final measurement
最终尾矿 end discard;final tailings;true tailings
最终位移 final movement
最终位置 extreme position;final location;final position
最终温度 final temperature;finishing temperature
最终文字材料 final literal material
最终稳定温度 ultimate stable temperature
最终无载垂度 final unloaded sag
最终下沉 final subsidence
最终显示 resulting display
最终消费 final consumption
最终消费部门 final consumption sector
最终消费需求 final consumption demand
最终消费者 end-user;final consumer;ultimate consumer
最终销售 final sale
最终效用 final utility
最终效用价格理论 final utility theory of value
最终卸载升运器 final delivery elevator
最终形变热处理 final thermomechanical treatment
最终性能 final performance
最终修饰 final finishing
最终修整 final finishing
最终需求 final demand;ultimate demand
最终需氧量 limited oxygen demand;ultimate oxygen demand
最终需要量 ultimate demand
最终压力 final pressure
最终压力分布 final pressure distribution
最终验收 acceptance examination;final acceptance
最终验收试验 final acceptance test
最终验收线 final acceptance line
最终验收证 final acceptance certificate
最终饮用水水质 final drinking water quality
最终应力(值)final stress
最终硬度 final hardness
最终用户 end-user;ultimate consumer
最终用途 end point usage;end use
最终用途竞争 end-use competition
最终油样 final sample
最终淤积量 ultimate deposition
最终余震 final aftershock
最终预报误差 final prediction error
最终预算 definition estimate
最终预应力 final prestress(ing force)
最终预应力值 final prestress value
最终再精选 final recleaner flo(a)tation
最终再筛分 finish rescreening
最终在野外使用的机械 end point field usage
最终张拉 final stressing
最终账目 final account
最终整平 fine grading
最终证据 conclusive evidence
最终证书 final certificate
最终支付证书 final certificate of payment
最终直径 final diameter
最终值 end value
最终制品 final product
最终质量 final mass
最终贮存 terminal storage
最终贮存量 ultimate storage
最终转差率 final slip
最终转动箱 case of final drive
最终装机(容量)ultimate installation
最终状况 final condition
最终状态 final state;terminal status
最终状态集 set of final state

最终总沉降量 final total settlement
最重车轮荷载 heaviest wheel load
最重工作鞋 extreme service shoe
最重件重量 heaviest piece weight
最重要变量规则 most important variable rule
最重要的 first line
最重要的实际价值 the most important practical benefit
最重要的是 above all
最重要的四种环境因素 four of the most important environmental factors
最重要优先项目 highest priority
最重轴载 heaviest axle
最主要的 supreme;uppermost
最左边车道 left hand lane
最左单元 left-most cell
最左导出 left-most derivation
最左端 high order end
最左树 left-most tree
最左位 left-most bit

罪犯 culprit;offender;outlaw

醉度计 < 查探酒醉程度的仪器 > Alcometer

醉汉 < 美 > rum

尊重的需求 esteem need

遵导波 guided wave

遵守 abide by;conform;observe
遵守安全规程 safety conscious
遵守程序 procedure to follow
遵守的水准 level of compliance
遵守法规调查 < 如交通规则 > law observance study
遵守法律 observe laws
遵守规定 procedure to follow
遵守规章 abidance by rules
遵守国际惯例 follow the international practice
遵守合同 abide by the contract;keep contract;key the contract
遵守技术规范 adherence to specifications
遵守交通规则调查 law observance study
遵守进度表 maintain a schedule
遵守礼节 observe the proprieties
遵守诺言 keep one's words
遵守施工进度计划 keep to program(me);keep to schedule
遵守条例 abidance by rules
遵守协议 abide by the agreement
遵守者 observer
遵循 follow(ing)
遵循规定的限度 adhere to assigned limits
遵照 conform;in compliance with
遵照处置 disposal compliance
遵照联邦规则 federal regulations compliance
遵照议程 adherence to the agenda
遵照州规则 abide by state regulations
遵重经济规律 recognize economic law

左岸 < 面向下游 > left(side)bank

左岸标志 port hand mark
左岸浮标 port-hand buoy
左岸航标 port hand mark; portside aids
左岸支流 left bank tributary
左半球 left hemisphere
左半轴 left half axle
左半轴壳 left half axle housing
左伴随 left adjoint
左边 first member; left; left-hand edge;left-hand side;port aide
左边岸 left marginal bank
左边部分 left-hand member
左边车道 < 多车道道路的 > left-lane
左边储物舱 port locker
左边的 left-hand
左边的数字 left-hand digit
左边对齐 left alignment
左边驾驶汽车 left-hand control car
左边零 left-hand zero
左边有危险信标 port-hand danger beacon
左边与右边 port and starboard
左边缘瓦 left verge tile
左表 left-handed watch
左表布局 left list layout
左波形括号 left curly bracket
左部 left part
左部表 left part list
左侧 left side;offside < 对右行制交通体系而言 >
左侧变量 left-hand variable
左侧标 left-hand mark
左侧操纵的道岔【铁】 left-hand points
左侧操作 left-hand operation
左侧操作机床 left-hand machine
左侧超车 off-side overtaking
左侧道岔 left-hand turnout
左侧的 left;leftward
左侧断层 left-lateral fault; sinistral fault
左侧对齐的 left justification
左侧分配器盖 left-distributor cover
左侧浮标 port-hand buoy
左侧航标 port hand mark; portside aids
左侧合适的 left justification
左侧记入 left entry
左侧驾驶台 left-hand drive
左侧进站 left-hand stone
左侧履带 left track
左侧螺旋桨 left screw
左侧螺旋线 left-hand helix
左侧面 left surface
左侧排气主口 left main exhaust port
左侧式机 left-hand machine
左侧式转辙机 left-hand point machine;left-hand switch machine
左侧输送机回动盖 left conveyor reverse cover
左侧数 left-component
左侧踏板 left-hand pedal
左侧调整 left justified
左侧位 left-lateral position
左侧卧位 left-lateral position
左侧信号 left-hand signal
左侧信号机【铁】left-hand signal
左侧行车 left-hand drive; left-hand operation; left-hand running; left-track drive
左侧行车的交通 left-hand traffic
左侧行驶 left driving
左侧型破碎机 left-hand crusher
左侧型摇床 left-hand table
左侧圆角阳角砖 left-hand round edge external corner
左侧缘 left border
左侧褶皱 sinistral fold
左侧制分量 left-hand component
左侧装置 left-hand unit
左叉臂 left half fork arm
左车道 left-lane
左车停车 port engine stop
左乘 premultiplication
左乘法 left-handed multiplication
左传动机心 left-hand movement
左搓钢丝绳 Lang's left hand lay
左搓绳 backhanded rope; back laid rope; left-handed rope; left laid rope
左搓绳股 Spanish fox
左导数 derivative on the left; left-hand derivative
左到右分析 left-right parse
左到右分析算法 left-to-right parser
左到右扫描 left-to-right scan
左的 sinistral
左递归 left recursion
左递归定义 left recursive definition
左递归方式 left recursive way
左递归规则 left recursive rule
左读数 back reading
左渡线 left-hand crossing
左端 first member;left-hand end;left margin
左端标记 left end marker
左端回指连线 left-hand side back link
左端墙 sidewall left
左端终止集 left-most terminal set
左对齐文本 left justified text
左对位 left justify
左舵 left rudder;port rudder;port the helm
左舵多一点点 port more
左舵规则 port helm rule
左舵一点点 port a little
左方错位式交叉口 left-hand staggered junction
左方括号 left square bracket
左方连续 left-hand continuity
左方起重机 port crane
左方上导数 left-hand upper derivative
左方下导数 left-hand lower derivate
左方向盘车辆 left-hand control car; left-hand drive car
左扶手楼梯 left-hand stairway
左股钢轨 left rail
左柜 port tank
左焊法 forehand welding; forward welding
左后方【船】port quarter
左后轮 left rear wheel
左滑 left slip
左换向离合器 left reverse clutch
左极限 left limit;limit on the left
左角 left angle
左角自底向上 left corner bottom-up
左绞车 port winch
左截断 left truncation
左进一 ahead one port
左经度 left longitude
左卷向量 sinistral vector
左开窗 left-handed window
左开道岔 left(-hand) turnout; left-turn switch;straight left-hand turnout
左开关 left-hand(ed)opening
左开门 left-hand(ed)door
左扣可卸式安全接头 safety joint with left-hand release
左括号 left bracket;left parent
左利手 left-handedness
左轮迹 left wheel path
左轮手枪 revolver
左螺纹扳牙 die for left hand thread
左螺旋 left-hand helicity; left-hand screw;left-hand spin;left screw
左满舵 hard a-port
左锚 port anchor
左锚绞车 port anchor winch
左锚链 port chain cable
左面 left-hand side
左面的 left-hand;sinistral
左面通行标志 walk on left sign
左挠的 sinistrorse
左捻 < 即 Z 捻 > left(-hand)lay;left-laid
左捻钢丝绳 cable-laid rope; S-lay wire rope;left lay rope < 反时针方向捻成的钢丝绳 >
左扭钢轨 left-hand twist rail
左扭转 left-handed twist
左偏离摄影术 left-averted photography
左偏手性 left-handedness
左平移 left translation
左坡角 left slope angle
左前 left anterior
左前方【船】port bow
左前方电气设备安装 left front electrical equipment mounting
左前降支 left anterior descending branch
左切刀 left-cut tool;left-hand tool
左切铰刀 left-hand rotation reamer
左三角翼 triangular panel
左山墙瓦 left gable tile
左深舱 < 油、水 > port deep tank
左式 levoform
左式破碎机 left-hand crusher
左式曲轴 left-hand crankshaft
左视图 left elevation;left side view;left view
左手 left-hand
左手参考系统 left-handed reference frame
左手测流三通 left-hand side outlet tee
左手测流肘管 left-hand side outlet elbow
左手的 left-hand; left-handed; sinistrorse
左手定律 left-hand rule
左手定则 left-hand(ed)rule
左手方向的 left-handed
左手螺纹 left-hand thread
左手门锁 left-hand lock
左手偏置刀架 left-hand off-set tool holder
左手三指定则 left-hand three finger rule
左手外开门 left-hand reverse door
左手系 left-hand(ed)system
左手织机 left-hand loom
左手坐标系 left-handed coordinate system
左首锚 port bower;second bower
左顺捻 left lay
左丝扣钻杆 left-hand-thread rod
左踏板 left-hand pedal
左台车总成 left under carriage assembly
左调整 left justify
左图廓 left-hand border
左推断层 left-lateral fault; sinistral fault
左推移 left translation
左瓦 left tile
左微商 derivative on the left
左无序记录 left-out-of-order record
左舷 < 船的左帮 > larboard; port; port beam;portside
左舷标志 port aids;port hand mark
左舷侧 portside
左舷侧浮箱 portside pontoon
左舷储物舱 port locker
左舷船岸安全间距 port side(red) bank clearance
左舷的 portside
左舷灯 port lamp;port lantern;port

(side) light;red side light
左舷灯玻璃 port light glasses
左舷吊杆柱 port king post
左舷发动机 port engine
左舷方 port hand
左舷方向 port hand
左舷方向的 larboard
左舷风道 port air duct
左舷浮标 port-hand buoy
左舷柜 port tank
左舷锅炉 port boiler
左舷航标 portside aids
左舷和右舷 port and starboard
左舷后部 port aft
左舷回波 port echo
左舷锚 port anchor;port bower
左舷内 port inboard
左舷前 port forward
左舷倾侧 list to port;port list
左舷识别灯 portside light
左舷首锚 port bower
左舷受风的帆船 port tack yacht
左舷通风干路 port air duct
左舷外 port outboard
左舷/右舷靠 berthing port/starboard side
左舷正横 port beam
左舷值班 port watch
左舷主机 port main engine
左舷主锚 port bower
左舷主柱 port king post
左线 down track;left line
左线性形式产生式 left linear form production
左向 left-hand
左向标志 left-hand designation
左向道岔 left-hand switch
左向发条轮 left-hand fusee
左向焊 forehand welding; forward welding
左向焊法 left-hand welding method; leftward welding; Swiss welding method
左向铰链 left-hand hinge
左向螺旋线 left-hand(ed) helix
左向偏振 left-hand polarization
左向曲线 left-hand curve
左向绕组 left-hand winding
左向弯曲基本轨 left-hand bent stock rail
左向斜板 leftward skew slab
左向旋转 left-hand rotation;left-hand turn(ing)
左向旋转螺旋 left-hand auger
左向旋转绳索 left swing line
左向转弯 left-hand turning
左像片 left-hand photograph
左削侧面粗车刀 left-hand side rough turning tool
左削侧面粗刨刀 left-hand side rough shaping tool
左削侧刨刀 left-hand side tool
左削粗车刀 left-hand rough turning tool
左削刀具 left-hand tools
左削刀头 left-hand tool bit
左削刀头车刀 left-hand knife turning tool
左削刀头刨刀 left-hand knife shaping tool
左削平面车刀 left-hand facing tool
左斜体注记 backward-sloping lettering
左行断层 left-lateral fault; sinistral fault
左行离距 left-handed separation
左行逆向滑动断层 left-reverse-slip fault
左行扭动 sinistral torsion
左行正向滑动断层 left-normal-slip fault
左行直移 sinistral translation
左行走向滑动断层 left-lateral strike-slip fault
左形 left-handed form
左形移动 left-handed separation
左旋 counter-clockwise rotation; laevorotation; left-handed rotation; left-turn;levogyration;levorotation
左旋错动 sinistral dislocation
左旋错位断层 sinistral-separation fault
左旋的 left-handed; levo; levorotatory;sinistral;sinistrorse
左旋定则 left-handed rule; left-handed system
左旋断层 left-hand fault; left-lateral fault;sinistral fault
左旋断层作用 sinistral faulting
左旋阀 left-hand opening valve
左旋横推运动 sinistral transcurrent movement
左旋机 left-hand machine
左旋极化 left-handed polarization
左旋极化波 left-handed polarized wave
左旋极限 left-hand limit
左旋螺钉 left-hand screw
左旋螺杆 left-handed screw
左旋螺母 left-handed nut
左旋螺纹 left-hand thread; left-handed screw; left-hand screw thread; minus screw
左旋螺纹螺母 left-hand nut
左旋螺纹丝锥 left-hand thread tap
左旋螺旋桨 left-hand airscrew; left-handed propeller
左旋螺旋伞齿轮 left-hand spiral bevel gear
左旋螺旋线 left-handed twist
左旋偏振 left-handed polarization
左旋平移断层 sinistral wrench fault
左旋平移运动 sinistral translation movement
左旋石英 laevorotatory quartz; left-handed quartz
左旋水晶 laevorotatory quartz; left-handed quartz
左旋糖 fructose;levulose
左旋梯形螺纹 reverse-buttress thread
左旋推进器船 left-handed vessel
左旋位移 sinistral displacement
左旋有孔虫丰度 abundance of counter clock wise foram
左旋有孔虫与右旋有孔虫相对丰度 relative abundance of counterclockwise foram and clockwise foram
左旋褶皱 sinistral fold
左旋转撑脚 left-turn leg
左旋转轮 left-hand runner
左旋转支柱 left-turn leg
左旋转子 left-hand runner
左旋锥形特性 left-hand taper
左旋走滑断层 sinistral strike-slip fault
左旋坐标系 left-handed coordinate system
左循环问题 left-recursion problem
左移加法器 left-hand adder
左移(位) left shift;shift left
左翼轨 left-hand wing rail
左油箱 nearside tank
左右 left-right
左右摆动六分仪 rocking the sextant; swinging the arc
左右摆动式指针 right-left needle
左右不对称 left-right asymmetry
左右侧管 azimuth by-pass(ing)
左右侧后视镜 LH/RH rear view mirror
左右错位式交叉(口) left-right staggered junction
左右颠倒的 inverted left-to-right
左右调整 left-right control
左右对称 bilateral symmetry
左右对映现象 enanfiomorphy
左右翻转的 reversed left to right
左右方零件互换性 right-to-left interchangeability
左右方位指示器 right-left bearing indicator
左右方向 left-right indicator; side-to-side direction
左右方向指导信号 right-left signal
左右方向指针 left-right needle
左右钢轨互换 rail transposing
左右航迹指示仪 left-right track indicator
左右互换性 right-to-left interchangeability
左右肩调整垫 fork arm shim
左右交叉螺纹 cross screw
左右交错 in staggered pattern
左右交换捻 reverse lay
左右两侧对称的 medianly zygomorphic
左、右轮独立传动系统 dual-power train;split power train
左右轮距 wheel ga(u)ge
左右轮距 tread
左右螺纹的 right-and-left threaded
左右偶联管 rigging-and-left coupling
左右偏差 lateral deviation; left-right deviation
左右偏斜角 <推土机> right left angle
左右倾侧角 roll angle
左右倾斜角 <推土机> tilt angle
左右视差【测】 horizontal parallax
左右视差滑尺 horizontal parallax slide
左右视差较 horizontal parallax difference
左右视差螺旋 horizontal parallax screw
左右手程序图 right-and-left-hand chart
左右水平 track cross level
左右通航标 bifurcation mark; middle ground mark
左右通航浮标 bifurcation buoy; middle ground buoy
左右纹螺母 rigging-and-left nut; right-and-left nut
左右舷 port and starboard
左右舷角九十度 broad on the beam
左右舷角四十五度 broad on/off the bow
左右舷角一百三十五度 broad on the quarter
左右相反形 congruent form
左右向螺杆夹 rigging-and-left-hand-screw clip
左右向指针 left-right needle
左右旋螺栓 right-and-left screw
左右摇摆 yawing
左右移动的幕 <剧院舞台> travel-(l)er
左右异向的 heterochiral
左右运动 side-to-side movement
左右振动 side-to-side vibrations
左右制导 left-right guidance
左右转车辆蔽藏线道 bay-turn
左右转车辆蔽藏线道或车转蔽藏道 pocket turn
左御 left-hand drive
左御车 left-hand control car; left-hand drive car
左缘 left border;left margin
左缘支 left marginal branch
左照片 left-hand photograph
左褶曲 sinistral fold
左褶皱 left fold
左支承 left support
左支架 left support
左制动臂 left brake arm
左制动踏板 left brake pedal
左置换 left substitutability
左主锅炉 port main boiler
左主机 port engine
左转 left-hand corner;left-handed rotation;left-hand turn
左转变 left-turn
左转磁电机 left-hand magneto
左转的 sinistrorsal
左转动 counter-clockwise motion
左转发动机 left-hand(ed) engine
左转角 angle to left
左转力矩 left-handed moment
左转楼梯 left stair(case)
左转绿箭头灯 green turn arrow
左转螺旋 left-handed screw
左转扭绞 left-hand lay
左转驶出行驶 left-off movement
左转(通)风机 left-hand rotating fan
左转弯 left-hand turn; turn left; turn to port
左转(弯)车道 <交叉口处的> left-turn lane
左转弯导向线 left-turn guide line
左转弯的 left-turn
左转弯线 port swing line
左转弯楔形衬砌环 left tapered plate
左转弯专用道 left-turn slot
左转弯走 left-wheel
左转向梯形臂 left steering trapezoid connection
左转向线 port swing line
左转行车 left-turning traffic
左转匝道 left-turn ramp
左转站立 left-turn stand
左转支座 left-turn stand
左转专用车道 left-turn lane
左子树 left subtree
左座驾驶 left-hand drive; left-track drive
左座驾驶车辆 left-hand control car; left-hand drive car
左座转向 left-hand steering

佐

佐贝尔滤波器 Zobel filter
佐川造山旋回【地】 Sakawa orogeny cycle
佐川造山运动【地】 Sakawa orogeny
佐迪阿克电阻合金 Zodiac
佐迪阿克铜镍锌合金 Zodiac
佐尔治统 <早寒武世> Georgian series
佐硅钛钠石 zorite
佐剂 builder
佐利特耐热合金 zorite
佐治亚 <美国州名> Georgia
佐治亚数字式错台仪 <美>【地】 Georgia digital faultmeter

作

作180°弯折 bending through 180°; doubling on itself
作凹处 recess
作凹口 notch
作罢 retreat
作保证人 guarantee
作壁画的 frescoed
作标记 marking
作槽舌 grooving and tonguing
作衬垫的石棉毡 inodorous felt
作成断面 grading
作成对角线 diagonalization
作成拱 arch
作成拱形 vault
作成麻面 ballast-surfaced

作成三份 triplicate
作成图表 diagrammatize
作成细粒 beading
作成一定断面的 graded
作出标记 identifying
作出波纹 gauffrage
作出带有端面的深槽 cup out
作出断面的 cross-sectioned
作出浮花 gauffrage
作出结论 draw a conclusion; finalize [finalise]
作出决定的过程 decision process
作出判定 decision-making
作出判断 enter a judgement; exercise judgment
作存储器信息变换 file conversion
作答 respond
作大修改 up-end
作的功 workdone
作滴嗒声 tick
作滴嗒响的东西 ticker
作抵押 <以房屋> on mortgage
作抵押品的地产 estate in ga(u)ge
作丁零声 jingle
作动流体 working fluid
作动器 actuator
作动筒 actuating cylinder
作动液 working fluid
作动油 working oil
作短途旅行 hop
作对称的布褶或布卷花饰的雕刻版 linenfold
作恶者 perpetrator
作法 facture
作翻新改造 retrofit
作方 squaring
作房屋纵断面图 sciagraph
作非常不规则的振荡 <电子> squeg
作废 abatement; abating; abort; become annulled; blank out; cancel; cancellation; defeasance; delete; desuetude; disannul; null and void; nullification; nullify; repeal; vacate
作废的 invalid
作废符(号) ignore character
作废键 cancel character
作废码 cancel code; reject code
作废通告 notice of rescission
作废图章 cancelling stamp
作废信号 cancel signal
作废信息 cancel message
作废烟道 dead flue
作废之物 obsolete
作废支票 voided check
作废字符 cancel character; delete character
作浮雕 emboss
作浮凸饰 emboss
作覆瓦状 imbricate
作功的水 power water
作功率 rate of doing work
作功能力 capacity for work
作功因素 workdone factor
作功指数 work index
作拱 imbow
作管子标记 marking pipe
作光扫描 photoscan
作过于微细的剖析 split hairs
作过运行时间试验的 time-tested
作好准备 preparedness
作恒值线法 contouring
作户外运动 sport
作会议主席 preside
作基础的木材 groundsel; ground sill
作棘齿 ratching
作几何学图形 geometrize
作记号 marking
作家 author; penman
作价价值 trade-in value
作价剩余资产 reevaluate the remaining assets
作茧 cocooning
作茧喷涂法 cocoonization
作检查的机器人 robot inspector
作交易 barter
作截面图 sectioning
作决定 decision-making
作"卡嗒"声 chatter
作刻痕 nick
作坑 pit
作矿藏略图 block out ore reserves
作框架用 enframe
作垄 earth up; ridging
作垄备灌 ridging for irrigation
作垄犁体 bedding bottom; ridger body
作垄器安装梁 bedding bar
作炉嘴 nosing
作螺旋接合 make a screwed joint
作面支架 foreset
作砰然声 thud
作品 brain child; workmanship
作坡泻水的【建】weathered
作畦埂器 ridge dammer
作企口(槽舌接合) grooving and tonguing
作汽车驾驶试验的环路 ride and handling loop
作铅管细工 plumb
作轻敲声 click
作曲器 composertron
作三角测量 triangulate
作散工 char
作适当的代替 make suitable substitution
作手工者 manipulator
作台工 bench work
作特性曲线 run a curve
作填料层的石头 blinding stone
作投影图 sciagraph
作凸饰 emboss
作凸缘 collaring; flange(up)
作突出的堞眼于胸墙 machicolate
作图 build-up; chart-making; description; drafting; plotting
作图表示 graphic(al) representation
作图插值法 graphic(al) interpolation
作图的 constructive
作图的密度系数 density factor of mapping
作图点 original plotting point
作图法 diagraphy; graphic (al) construction; graphing method; mapping method
作图内插法 graphic(al) interpolation
作图器 diagraph
作图线 construction line
作图准确度 accuracy factor of mapping
作陀螺运动 circumgyration
作万一准备 prepare for the worst
作为标志的语句 logo
作为抵押品的资产 assets held as collateral
作为典型 set an example
作为发电机的感应电动机 inductor motor
作为肥料 as fertilizer sources
作为肥料使用 use in fertilizer
作为肥料使用的有机质 organic matter used as fertilizer
作为分期付款 on account; pay on account
作为腐烂物的来源 to act as a source of decaying material
作为红利发行的股票 bonus issue
作为或不作为 act or omission
作为见习 on probation
作为面层的混凝土砌块 facing block
作为模范 set an example
作为试验样品的电动机 model motor
作为条件来要求 stipulate
作为有效荷载的仪器 payload instrument
作围堰 coffering
作伪 falsification
作物 crop plant
作物病害 crop pest
作物布局 crop disposition
作物残茬 crop residue
作物残体 crop residue
作物产量 crop yield
作物倒茬顺序 cropping sequence
作物地 cropland
作物调查 agronomical survey
作物覆盖 crop cover
作物耕作机 crop-tillage machine
作物耕作制 cropping system
作物灌溉需水量 crop irrigation requirement
作物耗水量 consumptive use of crop; consumptive use of water; crop water requirement
作物截留量 interception by crops
作物净耗水量 net consumptive use; net consumptive use by crop
作物类型 agroindustrial-type; agrotype
作物轮作 crap rotation
作物轮作系数 crop-rotation factor
作物区 crop tract
作物伤害 crop damage
作物生产力 crop-producing power
作物生长期 crop growing period
作物收成 crop return
作物收获指数 cropping index
作物损坏 crop failure
作物天气 crop-weather
作物条分法 crop stripping
作物图 crop map
作物-土壤水分应力 plant-soil moisture stress
作物微气象学 crop micrometeorology
作物形态学特性 <遥控> plant morphologic characteristic
作物需水量 crop needs; crop requirement; crop water requirement
作物学 agronomy
作物栽培 plant culture
作息时间 working and recreating time
作息制度 regime(n); work-rest program(me)
作息制度卫生 hygiene of regimen
作型芯 coring
作修正 make allowance
作序言 preamble; prolog(ue)
作穴机 dibbling machine
作样品 on sample
作业 handling; handling operation; operation; performance; procedure; processing; task; working
作业安排操作 job shop operation
作业安排调度 job shop scheduling
作业安排定序问题 job shop sequencing problem
作业安排模拟 job shop simulation
作业安排模拟器 job shop simulator
作业安全 safety of operation; operational security
作业安全设备 operational safety equipment
作业案例 job case
作业班 gang; operating crew; operation crew; work team
作业班长 foreman
作业班制 working shift
作业班组 gang; operating crew
作业半径 working radius
作业保障 job security
作业编号 job number; operation number
作业变更 conversion
作业标 working post; working sign
作业标志 operation sign; working signal
作业标准 operation standard; performance standard; work standard
作业表 worksheet
作业步(骤) job step
作业步控制 job step control
作业步控制分程序 job step control block
作业步任务 job step task
作业参数 job parameter
作业草图 field sketch
作业测验 performance test
作业场所 operational field
作业车辆 truck
作业成本 activity-based cost(ing); activity cost; cost of activity; cost of operation; operation cost
作业成本法 activity-based costing method; comparative statement of operation cost
作业成本计算 activity costing
作业成本控制 operational cost control
作业程序 operating procedure; job program(me); operation procedure; operation schedule; working sequence
作业程序方式 job program(me) mode
作业程序手册 procedure manual
作业程序图 flow chart
作业程序图表 program(me) of operations
作业程序细则 detailed operating procedure
作业池 process tank
作业持续时间 activity duration; duration of activity; duration of operation
作业处理 job processing
作业处理部件 job processing unit
作业处理监督(程序) job processing monitor
作业处理监督系统 job processing monitor system
作业处理监督终止 job processing monitor termination
作业处理控制 job processing control
作业处理控制程序 job processing control program(me)
作业处理命令 job action command
作业处理能力 job treating throughput
作业处理字 job processing word
作业船 maneuver boat; work boat
作业窗 operation opening
作业次序网络图 precedence network
作业待命 operational readiness
作业单位 work unit
作业的 operational
作业的中止 job abort
作业灯 working lamp; work light
作业登记 job log-in
作业登记卡 job card
作业地带 operating zone; work(ing) area; work(ing) place
作业地带空气流速 air velocity at work area
作业地带温度 temperature at work area; temperature at work place
作业地面 operating surface
作业地区 scene of operation
作业地图 working map
作业点 setting
作业调度 job scheduling
作业调度程序 job processor; job scheduler; job scheduling routine

作业调度程序任务 job scheduler task
作业调度器 job scheduler
作业调度算法 job scheduling algorithm
作业调入系统 job entry system
作业定额 activity quota;job rate
作业定额核定 job rating
作业定时记录 job logging
作业定序模块 job-sequencing module
作业定序系统 job sequencing system
作业定义 job definition
作业段 job step
作业段控制块 job step control block
作业堆积 job stacking
作业堆栈 job stack
作业堆栈系统 job-stack system
作业队 crew;working crew;working gang;working party
作业法 working system
作业范围 operating envelope;work area;working range
作业方法 job method;job practice;method of operation;operational approach;operational considerations;operational procedure
作业方式 operating type
作业房屋 <绘画、摄影、播音、演奏等用的> studio house
作业废水 working wastewater
作业费 working cost
作业费用总账 abstract ledger
作业分额 drifts
作业分解 job-splitting
作业分解结构 <把最终目标按统筹方法分解为分支> work breakdown structure
作业分类 job class;operation type
作业分析 job analysis;operation analysis
作业服务程序 job service program(me)
作业服务前导 job service prefix
作业工程 job engineering
作业估算法 <直接费的一种估算方法> operational rate estimating
作业管理 job management;operation management;word management
作业管理程序 job handling routine;job management program(me)
作业规程 working specification
作业规范 job specification
作业规则 operating rule
作业号码 job number
作业和转移分开的司机室 separate operator and transport station
作业划区 location of industry
作业环境 environment of work;working environment
作业环境照明 task-ambient lighting
作业会计 activity accounting
作业混合 taskmix
作业混合法 job mixing
作业机具 fighter;working implement
作业机械化 mechanization of operation
作业基准 job reference
作业计划 operating program(me);operating schedule;operational plan(ning);operational program(me);operative plan;plan of operation;schedule of operations;scheme of operation;working plan
作业计划表 operation schedule
作业计划单 job route sheet
作业计划制度 operational planning system
作业记录 operating record;timesheet
作业记录簿 charge book
作业记入 job entry
作业间 operating room
作业间歇 intermittence in operation
作业监督程序 job monitor
作业检查 operational checkout
作业箭头 activity arrow
作业脚手架 working scaffold(ing)
作业接口 job interface
作业结束 end of job
作业结束程序 end-of-job routine
作业进程 job process
作业进度 job scheduling
作业进度标志 work-in-progression sign
作业进度表 operation schedule
作业进度统计 schedule statistics
作业进展略图 progress sketch
作业经济性 operating economy
作业决策 operating decision
作业卡片 job card
作业卡片组 job deck
作业开始时间 starting time of activity;starting time of operation
作业控制 job control;job management
作业控制表 job control table
作业控制程序 job control program(me)
作业控制块 job control block
作业控制流 job control stream
作业控制命令 job control command
作业控制设备 job control facility
作业控制通信[讯] job control communication
作业控制信息 job control information
作业控制语句 job control statement
作业控制语言 job control language
作业库 job library
作业扩展 job enlargement
作业类别 job class
作业利用率 operational availability coefficient
作业利用系数 operational availability coefficient
作业量 quantity of work;work load
作业料仓 service bin
作业流 job stream;production current
作业流产 abort
作业流程 flow of operations;work flow;working operation
作业流程控制 job flow control
作业流程时间 job flow-time
作业流程图 job flow chart;operation process chart;control diagram;flow chart
作业流输入 job stream input
作业路面系统 operational pavement system
作业面 point of collection;side;working face
作业面积 operational area;working area
作业面宽度 operating width
作业名 job name
作业命令单 command sheet
作业目标 performance target
作业内容 job content
作业能力 work capacity;working ability
作业排队 job queue
作业排队入口 word queue entry
作业配电板 operational distribution panel
作业配合 job mix
作业票 working bill;work sheet
作业平台 working platform
作业起始 job initiation
作业前导 job prefix
作业前试验 preoperation(al) test
作业请求 job request
作业请求控制程序 job request control routine
作业请求选择 job request selection
作业区 change;target area;working area;working district;work zone
作业区泊位 terminal berth
作业区面积 terminal area
作业区平面布置 terminal layout
作业区设施 terminal facility
作业区生产率 terminal productivity
作业取样 performance sampling
作业全程 complete operation
作业人员 operating staff;working personnel
作业日期 job date
作业入口子系统 job entry subsystem
作业设计 job design
作业时的定额 rating under working condition
作业时的故障 field failure
作业时间 operating time;operation time;productive time;up-time;working hours;working time
作业时间表 operating scheme
作业时间数 work time
作业时间损失系数 coefficient of time lost
作业时间限 job time limit
作业时全开发动机油门 full throttle operation
作业试验 operating test;operation(al) test(ing)
作业室 operating room;studio
作业手册 operating manual
作业输出队列 job output queue
作业输出流 job output stream
作业输出排队 job output queue
作业输出前导 job output prefix
作业输出设备 job output device
作业输出文件 job output file
作业输出装置 job output device
作业输入 job entry
作业输入流 job input stream
作业输入排队 job input queue
作业输入文件 job input file
作业输入装置 job input device
作业输入子系统 job entry subsystem
作业数据 job-data
作业数据表 job data sheet
作业水文学 operational hydrology
作业顺序 job order;sequence of operations
作业顺序单 job order sheet
作业说明书 job description;job specification
作业搜索 job research
作业算账程序 job accounting routine
作业损失时间的扣除 deduction for time lost
作业台 working bench
作业特征调查 job diagnostic survey
作业特征曲线 operating characteristic curve
作业条件 field condition;operational condition
作业调换 operational change over
作业通知单 job ticket;work notice;work order sheet
作业图 flow process diagram;flow sheet;operating chart;working diagram;working drawing;working map
作业图表 schedule of operations;working schedule
作业吞吐能量 job throughput
作业完成时间 activity finish time
作业完成指数 performance index
作业网络 activity network
作业温度 processing temperature;working point
作业温度范围 range of working temperature;working range of temperature
作业文件控制分程序 job file control block
作业文件控制块 job file control block
作业文件索引 job file index
作业系统 operating system;system of operation
作业细则 instruction manual;operating instruction
作业现场 on-the-job
作业线 activity line;operating line;operation rack;productive line;service line;working line;gang
作业线能力 line capacity;line carrying capacity
作业相关图 activity relation chart
作业箱 jumper locker
作业项目 activity;event in operation
作业小时数 hours of operation
作业效率 operating efficiency
作业效应 action effect
作业形式 operation type
作业性 workability
作业性规划 operational plan(ning)
作业性能 operational function
作业序列 job sequence
作业选择 job selection
作业循环 operation cycle;working cycle
作业循环时间 operation cycle time
作业研究 job study;operation research;work study
作业用门 service door
作业用竖井施工法 shaft adit
作业优先权 job priority
作业与维修部门 operation and maintenance department
作业语句 job statement
作业预报 operational forecasting
作业预先权 job priority
作业员 operator;tower man <驼峰楼>
作业照明 task illumination
作业照明灯 working light
作业知识 working knowledge
作业职能 operation function
作业指标 performance indicator
作业指令 handling command
作业制动器 service brake
作业制动系统 service braking system
作业中校正 on-the-job calibration
作业中心 focal point of working
作业终端 job-oriented terminal
作业终止信号 end-of-work signal
作业钟 on-line clock
作业重量 <机械的> working weight
作业周期 duty cycle;operating period;operational cycle;operational period;working cycle
作业注册 job logging
作业注销 job log-out
作业转换 operational change over
作业装配区 job pack area
作业装入 job load
作业装入特性 job load characterization
作业装填区 job pack area
作业装填区队列 job pack area queue
作业状况 job status
作业总高度 overall operating height
作业总时间 total operating time
作业租约 operating lease
作业组 hand crew;side
作业组织 work organization
作业最迟开始时间 latest starting time of activity
作业最大冰层厚度 maximum ice thickness to operate
作业最大波高 maximum wave height to operate

作业最大横流 maximum cross current to operate
作业最大水深 maximum depth of water to operate
作业最大涌高 maximum swell height to operate
作业最小水深 minimum depth of water to operate
作业最早开始时间 earliest starting time of activity
作业最早完成时间 earliest finish time of activity
作一百八十度折叠 doubling on itself
作议定书 protocol
作用 effect; function; influence; purpose; role
作用半径 action radius; amplitude; effect radius; operating radius; operating range; radius of action; radius of operation; reach range; steam circle; working radius
作用泵 process pump
作用变量 action variable; actuating variable; influencing variable
作用变数 action variable
作用标准值 characterisation value of action; characteristic value of action
作用参数 operational factor; operational parameter
作用层 active layer
作用长度 action length; active length
作用程度 degree of action
作用齿轮 operating gear
作用冲击波 applied shock
作用传递函数 actuating transfer function
作用锤击能量 applied driving energy
作用代表值 representative value of actions
作用带区 application zone
作用导阀 application pilot valve
作用的 acting
作用点 acting point; action point; application point; point of action; point of application; point of attack; working point
作用电极 active electrode
作用电流 action current; actuating current
作用电路 application circuit
作用动力 dynamic(al) force applied
作用端 < 杠杆的 > working end
作用阀 application valve
作用法 action method
作用反作用定律 action-reaction law
作用范围 acceptance range; action range; action space; coverage; radius of action; range coverage; range of action; sphere of action; sphere of function; sphere of influence
作用范围顺序校核方式 sequence check mode in action range
作用方框图 functional diagram
作用方式 mode of action
作用方式变化 mode change
作用分路 working branch
作用分项系数 action sub-coefficient; partial safety factor for action
作用风缸 application cylinder; application reservoir
作用符号 functional character
作用杆 actuating rod; operating strut
作用功率 speed of power
作用功能图 functional diagram
作用(光)点 action spot
作用光谱 action spectrum
作用过度 overaction
作用和反作用 interplay
作用荷载 applied load(ing); imposed load; service load; useful load; work(ing) load
作用恒载 service dead load
作用弧 arc of action
作用滑移方向 operative glide direction
作用滑移面 operative glide plane
作用活度 activity
作用活荷载 service live load
作用活塞 application piston
作用活塞扁销 application piston cotter
作用机构 actuation gear; operating gear
作用机理 effect mechanism; function mechanism; mechanism of action
作用机制 active mechanism
作用基 functional group
作用集 active set
作用集中点 < 采矿 > focal point of subsidence
作用剂 agent
作用角 angle of action; angle of application
作用节距 action pitch
作用界限 action limit
作用距离 acceptance range; coverage; decisive range; full operating range; operating distance; operating radius; operating range; operational range; operative range; range coverage; range of action
作用距离连续接通技术 sequential range gating technique
作用距离预报 range prediction
作用控制阀 application control valve
作用力 acting force; action; action force; application force; applied force; apply force; efficient; working force
作用力臂 lever arm
作用力矩 applied moment
作用力示意图 force diagram; reciprocal diagram
作用力与反作用力 action and reaction
作用力中心 center[centre] of effort
作用量 actuating quantity
作用量积分 action integral
作用量原理 action principle
作用量子 quantum of action
作用面 acting face; acting surface; pressure surface; working face
作用面积 area of extraction; area of influence; operating area
作用能 interaction energy
作用频遇值 frequent value of action
作用期 action period
作用期限 activity duration
作用气室 brake chamber
作用腔 functional chamber
作用情况指数 performance index
作用区(域) active region; sphere of action; zone of action; action area; active zone
作用设计 functional design
作用设计值 design value of an action
作用升力 active lift
作用时间 acting time; action time; actuation duration; actuation time; time of operation
作用室 application chamber; functional chamber
作用室管 application chamber pipe
作用输入 action entry
作用水平 action level
作用水头 acting head; working head
作用顺序 sequence of operations
作用速率 speed of action
作用特征 action token
作用筒 ram
作用图 symbolic circuit
作用推力 applied thrust
作用位置系数 < 机场跑道设计的 > place coefficient
作用温度 operative temperature
作用物 reactant
作用物质 working substance
作用系统 actuating system
作用线 action line; active line; line of action; line of application; operating line
作用相 reacting phase
作用相界 reacting interphase
作用效应 action effect; effects of actions
作用效应基本组合 fundamental combination for action effects
作用效应偶然组合 accidental combination for action effects
作用效应系数 coefficient of action effects
作用效应组合 combination for action effects
作用斜度 obliquity of action
作用信号 actuating signal
作用信号比 actuating signal ratio
作用形式 type of action
作用压力 actuating pressure; applied pressure; working pressure
作用压力水头 working pressure head
作用因素 influencing factor
作用应力 applied stress
作用于 act on
作用于拱顶的集中荷载 point load at the crown
作用于柱的力矩 moment at column
作用与反作用相等原理 principle of action and reaction
作用域 scope
作用域的属性 scope attribute
作用元件 actuating unit; functional element
作用在挡土墙上土压力 earth pressure on retaining wall
作用在挂钩上的横力 side draft on the drawbar
作用在机体上力矩 moment on body
作用在基槽挡土板上的土压力 earth pressure sheeted excavation
作用张力 working tension
作用支柱 actuating strut
作用直径 functional diameter
作用指示符 role indicator
作用指示器 role indicator
作用质量 service quality
作用中心 action center[centre]; center[centre] of action; center[centre] of effort
作用重量 working weight
作用周期 action cycle; action period
作用轴 operating shaft
作用轴套 operating shaft bush
作用轴蜗轮 operating shaft worm gear
作用轴蜗轮盒 operating shaft worm gear box
作用准永久值 quasi-permanent value of actions
作用组合 combination of actions
作用组合系数 coefficient for combination value of actions
作用组合值 combination value of actions
作用组合值系数 coefficient for combination value of actions
作宇宙航行 astrogate
作栅栏材料 cribbing
作战海图 combat chart
作战领航图 operational navigation chart
作战区域 operational zone
作战图 battle map; combat map; operation map; war map
作战要求 operational requirement
作者的说明 auctorial comment
作者原图 author's draft map; compilation manuscript
作证 attestation; bear witness; give evidence; testify; verification; vouch
作证者 deponent
作之字形转弯 zig
作纸型用的纸 flong
作中间产品的石膏板 plasterboard for secondary processing
作专业评判 expertise[expertize]
作资信调查 make credit inquiry

坐 perch

坐板 seating slab
坐崩(滑崩) landslide
坐便器 pedestal pan; sitting W.C. pan; water closet
坐标 coordinate; grid reference
坐标北 grid north; northing
坐标闭合差 closure error of coordinate; coordinate closure; position misclosure
坐标变化平差法 adjustment method of variation of coordinates
坐标变换 coordinate conversion; coordinate transformation; transformation of coordinates
坐标变换板 conversion board
坐标变换参数 coordinate conversion parameter
坐标变换测绘板 conversion board
坐标变换法 method of variation of coordinates; variation of coordinate method
坐标变换矩阵 coordinated transformation matrix
坐标变换器 coordinate converter
坐标变形法 coordinate transformation
坐标变形试验法 < 金属冷加工过程的 > photogrid
坐标标度 coordinate scale
坐标表 list of coordinates
坐标测定器 < 从模拟数据中测定 X-Y 坐标的仪器 > X-Y scaler
坐标测量 measurement of coordinates
坐标测量机 coordinate measuring machine
坐标测量仪 coordinate measuring apparatus; coordinate measuring instrument
坐标测设 coordinate setting
坐标差 coordinate difference
坐标尺 coordinatometer; rectangular coordinatometer < 直角的 >
坐标尺寸 coordinate dimension; coordinating dimension
坐标尺寸确定法 coordinate dimensioning
坐标传递 transfer of coordinates
坐标存储器 coordinate storage; coordinate store
坐标存取列 coordinate access array
坐标带 coordinate zone; grid zone
坐标导数积分控制 derivative proportional-integral control
坐标地带 gore
坐标点 coordinate point; fiducial mark
坐标电位计 coordinate type potentiometer
坐标电位器 coordinate potentiometer
坐标定位 coordinate setting; fix by

position lines
坐标定位装置 coordinate location device
坐标读出 coordinate readout
坐标读数 coordinate reading; coordinate readout
坐标读数器 coordinatograph
坐标对 coordinate pair
坐标法 coordinate method
坐标反算 inverse computation of coordinates
坐标方程 coordinate equation
坐标方格 grid square
坐标方位角 bearing angle; coordinate azimuth; grid azimuth; plane coordinate azimuth; grid bearing
坐标方位角法 <用于方向平差> method of bearings
坐标方位仪 coordinator
坐标分析器 coordinate analyser[analyzer]
坐标复接存储电路 coordinate multiplex storage circuit
坐标复接电路 coordinate multiplexing circuit
坐标格网 abac; base grid; coordinate frame; coordinate grid; grid reference
坐标格网尺 coordinate grid scale; romer
坐标格网点 grid point
坐标格网距离 grid distance
坐标格网模片 overlapping grid template
坐标格网线 coordinate line; grid line
坐标格网原点 grid origin
坐标格网值 grid value
坐标工作台 coordinate setting table; coordinate table
坐标函数 coordinate function; position function
坐标换算 coordinate transformation
坐标换算器 coordinate conversion calculator
坐标绘图机 coordinate plotter
坐标集 coordinate set
坐标几何 coordinate geometry
坐标计数器 coordinate counter
坐标计算 computation of coordinate
坐标计算中心 coordinating calculating center[centre]
坐标记录 coordinate record(ing); recording of coordinates
坐标记录器 coordinate recorder
坐标记录仪 coordinate printer; coordinate registrator
坐标记录装置 coordinate recording unit
坐标架 frame of axes
坐标检索法 coordinate indexing
坐标检验机 coordinate inspection machine
坐标角 coordinate angle
坐标经纬 transit square
坐标经纬仪 jig transit
坐标卡集【数】 atlas
坐标控制图 control sheet
坐标量测 coordinate measurement; measurement of coordinates
坐标量测仪 comparator; coordinate comparator; coordinate measuring apparatus; coordinate measuring instrument
坐标量度仪 coordinate comparator; coordinate measuring instrument
坐标量角器 coordinate protractor
坐标邻域 coordinate neighbo(u)rhood
坐标零点 zero of coordinate system
坐标流 stream of coordinate
坐标轮换法 cycle coordinate method
坐标面 coordinate surface; coordinating plane; plane of coordinates
坐标磨床 jig grinder; jig grinding machine
坐标磨削 jig grinding
坐标逆读 coordinate readback
坐标平差【测】 coordinate adjustment; adjustment by coordinates; adjustment by variation of coordinates
坐标平面 coordinate plane
坐标曲率线 coordinate line of curvature
坐标曲面 coordinate curved surface; coordinate surface
坐标曲线 coordinate curve
坐标确定 determination of coordinate
坐标摄动 coordinate perturbation
坐标时 coordinate time
坐标视差 parallax of coordinates
坐标视差位移 parallactic displacement of coordinates
坐标数据 coordinate data; object data
坐标数据发送机 coordinate data transmitter
坐标数据接收机 coordinate data receiver
坐标数字串 coordinate string
坐标数字化仪 coordinate digitizer
坐标数字化转换器 coordinate digitizer
坐标算子 coordinate operator
坐标镗床 coordinate setting boring machine; jig borer; jig-boring machine; jig mill
坐标镗床显微镜 jig-borer microscope
坐标镗削 jig boring
坐标条件 coordinate condition
坐标条件自由项 free term of coordination condition
坐标投影屏 charted screen
坐标网 abac; coordinate grid; coordinate net; grid; grid system; layout grid; network of coordinates
坐标网板 grid board; grid plate; grid sheet
坐标网尺度常数 grid scale constant
坐标网代码 map code
坐标网格 coordinate grid; grid coordinate system; grid space; reference grid
坐标网格玻璃板 grid glass plate
坐标网格尺 grid scale
坐标网格带编号 grid zone designation
坐标网格法 coordinate grid method
坐标网格距离 grid distance
坐标网格模片 grid template
坐标网格值 graticule value
坐标网航空图 grid air navigation chart
坐标网交点 grid intersection
坐标网距 grid length
坐标网说明 coordinate note; grid note
坐标网延长线 graticule tick
坐标网延伸短线 grid tick; numbered tick
坐标网原图 grid drawing
坐标网注记 grid letter
坐标网子午带 meridianal zone of grid system
坐标网纵线 grid meridian
坐标位置 coordinate position; grid location; grid position
坐标误差 error of coordinates; gisement
坐标系(统) system of coordinates; frame of axes; set of coordinates; system of axes; reference frame; coordinate system; grid system
坐标系统变换 coordinate system conversion
坐标系统变换器 coordinate system converter
坐标系转换 transfer of axes
坐标显示 coordinate display
坐标线 datum line; grid line
坐标向量 coordinate vector
坐标象限角 grid bearing
坐标性变化 coordinated variation
坐标序列 coordinate sequence
坐标选择器开关 coordinate selector switch
坐标寻址存储器 coordinate-addressed memory
坐标仪 coelosphere; coordinatometer
坐标仪量测坐标 comparator coordinates
坐标仪像片盘 stage of comparator
坐标与高程 coordinate and elevation
坐标原点 grid origin; origin of coordinates; true origin; zero of coordinate system; zero point; orifice of coordinates
坐标增量 increment of coordinate; increment of coordination
坐标增量闭合差 closing error in coordinate increment; closure error in latitude and departure
坐标增量表 table of increase of coordinates; traverse table
坐标增量配赋法 transit rule
坐标展点仪 flat-bed coordinatograph; coordinatograph
坐标展点仪铅笔 travel(l)ing pencil
坐标正算 direct computation of coordinates
坐标值 coordinate figure; coordinate value
坐标纸 coordinate paper; coordinate system paper; cross-section paper; graphic (al) paper; graph paper; plotting paper; profile paper; scale paper; squared paper
坐标制 coordinated system; crossbar system; system of coordinates
坐标制图机 coordinate graph
坐标制图器 coordinatograph
坐标(制图)仪 coordinatograph
坐标中(均方)误差 mean square error of coordinate
坐标重复系统 coordinate repeating system
坐标轴 axis of coordinates; axis of reference; coordinate axis; reference axis
坐标轴变换 transfer of axes
坐标轴的平移 translation of axes
坐标轴的转动 rotation of coordinate axis
坐标轴配置 arrangement of axes; arrangement of coordinate axis
坐标轴平移 translation of coordinate axis
坐标轴束 sheaf of coordinate axes
坐标轴位移 displacement of coordinate axis
坐标轴系 system of axis
坐标轴线 agonic line; axis of coordinates
坐标轴旋转 rotation of axes
坐标轴转换 translation of coordinate axis
坐标属性 coordinate attribute
坐标转换 coordinate conversion; coordinate transfer
坐标转换计算机 coordinate conversion computer
坐标转换误差 coordinate conversion error
坐标转换装置【测】 coordinate system converter
坐标状态线 coordinate status line
坐标自动记录器 trace assessor
坐标自动记录装置 automatic coordinate recording device
坐标纵线 north-and-south line
坐标纵线偏角 convergence of grid bearing; Gauss meridianal convergence; grid convergence; meridian(al) convergence; gisement
坐标纵线中点 north-south middle point
坐标钻床 coordinate setting drilling machine
坐车中转 transshipment by lying a part of goods unloaded in a part-load wagon
坐船出去 launch
坐床式圆筒直立堤 seated-cylinder vertical breakwater
坐等捕食者 sit and wait predator
坐凳栏杆 seat rail
坐垫 pier cap
坐高 sitting height
坐高尺 sitting length scale
坐格网带编号 grid zone designation
坐骨的 sciatic
坐骨神经痛 sciatica
坐浆 bedding course mortar; bed mortar; mortar bed
坐浆面 bedding face
坐礁 go aground; run on rocks; stranding
坐落在 situated
坐泥泊位 mud berth
坐起室 withdrawing room
坐球面的平均烛光 mean hemi-spheric(al) candle power
坐式便盆 pedestal pan
坐式便器 commode-type toilet
坐式便桶 commode-type closet
坐式(抽水)马桶 pedestal closet
坐式巨像 seated colossus
坐式庞大塑像 seated colossal statue
坐式人像 seated figure
坐式塑像 seated statue
坐式浴缸 seat bath tub
坐滩 pile
坐卧两用车 coach-sleeper; couchette car; sleeper coach
坐卧两用电力轨道车 dual-purpose electric(al) rail-car
坐卧两用沙发 davenport bed <美>; sofa bed; sofa sleeper
坐卧两用室 bed-sitting room
坐卧两用椅 chair-bed
坐坞龙骨 docking keel
坐席定员 seating capacity
坐席区 seating section; seating space
坐雪橇 ski
坐椅间膝部空间 knee room
坐椅头枕 head rest
坐浴 hip bath; seat bath
坐浴盆 site bath; sitz bath
坐浴浴盆 bidet; demi-bath; demi-tub; hip bath; sits bath; tub
坐在车上观看的露天电影院 drive-in
坐在汽车里看戏的电影院 drive-in
坐姿身高 sitting height
坐姿视高 eye height
坐姿膝高 knee height

柞

柞蚕丝(织物) Tussah(wild) silk
柞树 Mongolian oak

唑 azole

座 foundation; hander; mounting; nest; strapple

座板 bearer; bed piece; bed plate; chair plate; hearth block; masonry plate; mounting plate; riser plate; saddling; seat board; shoe plate; stay plate
座板基础 bed-plate foundation
座板锚碇 anchoring of bedplate
座板升降结 bosum; chair hitch
座包 stationary ladle
座包式降落伞 seat-pack parachute
座保护钢板 column guard
座表面 seating face
座舱 cab(in); cabinet; capsule cabin; cockpit; seating accommodation
座舱安全带 seat belt
座舱等级 class fare basis
座舱顶灯 dome light
座舱盖 canopy
座舱盖开启作动筒 canopy actuating cylinder
座舱盖抛放 hood jettison
座舱盖手控柄 canopy manual operating handle
座舱高度 cabin altitude
座舱加温器 cabin heater
座舱控制 cab control
座舱式起重机 cabin type crane
座舱压差 cabin differential pressure
座舱仪表板 flying panel
座舱增压器 cabin blower; cabin supercharger
座舱增压式飞机 pressurized aircraft
座舱照明 cabin lightning
座舱走道 companion way
座撑 seat stay
座床 bunk
座的修整 <如阀座等> reseating
座底 foot stall
座底槽板 seat bottom channel
座底木 <水闸或桥梁的> mud sill
座底式平台 bottom-supported platform; submersible platform
座底式钻机 submersible drilling rig
座底式钻井平台 bottom supporting drilling platform; submersible drilling platform
座底式钻井装置 sit-on-bottom drilling rig
座垫 chair; seat cushion; seating washer; seat pad
座垫底板 seat cushion board
座垫框 cushion frame
座垫弹簧 cushion(ing) spring
座垫套 webbing
座垫填充材料 seat filler material
座垫镶边 seat cushion edge
座垫毡条边 felt edge
座墩底脚 pedestal foot(ing)
座阀 seated valve; seating valve
座阀配流泵 seated valve pump
座封封隔器 setting the packer
座封装置 <泵的> seating and sealing unit
座盖 fixton pad; flap; seat lid <便桶上的>
座果 setting
座合压力 <气门等的> seating force
座环 fixed guide wheel; guide vane ring; seating ring; socket ring; speed ring; stand ring; stay ring; stay vane ring; supporting ring; vane intake ring <贯流式水轮机的>
座环的下环 bottom ring of stay ring
座环碟形边 stay ring transition piece
座环过渡段 stay ring transition piece
座环内圈 <贯流式水轮机> inner stay cone; inner gate barrel
座环外圈 <贯流式水轮机> outer stay cone
座簧 marshall springs
座基上挑檐 cornice of pedestal
座架 bed frame; bedstead; mount; pillow block bearing; seat frame; stock; mounting <系统的>
座架刚度调节器 driver's weight adjustment
座架加油装置 stock oiler
座架木墩 timber column bent
座架木排架 timber frame bent
座架木排架底木 timber sill of frame bent
座架排架混凝土底座 concrete pedestal of frame bent
座架排架木柱 timber column for frame bent
座架式显微镜 stand microscope
座架式压铆机 pedestal-type squeezer
座坚壳属 <拉> Rosellinia
座间通道 seatway
座角钢 seat angle; shelf angle
座脚 seat leg
座铰 abutment hinge; articulated gate shoe
座接法 seat connection
座壳孢属 <拉> Aschersonia
座口磨削机 seat grinder
座块 saddle
座梁 <桥的> seat beam
座落 lie
座落在…… land in
座面 seat surface
座面阀 miter valve
座面快速磨合 rapid seating
座囊菌属 <拉> Systremma
座盘 <古希腊柱的> attic base
座盘孢属 <拉> Discula
座盘菌属 <拉> Stromatinia
座盘饰 tore; torus[复 tori]
座盘饰凸圆线脚 ba(s)ton
座桥看台 seat stand
座圈 ball race; bezel; housing washer; pedestal ring; race; raceway; retainer; seat retainer seat ring; stay ring; support
座圈密封装置 race seal
座圈缘 race lip
座石 bedding stone; bedstone; socle; zacco; zoccola[zoccolo]; zocle <雕像、石柱等的>; padstone
座式 stock engine
座式贯入装置 seating drive
座式换向阀 seated directional valve; seating directional valve
座式继电器 desk-type relay; shelf-type relay
座式取暖器 block heater
座式旋转液压锤 pedestal swing hydraulic hammer
座式轴承 pedestal bearing
座式子钟 desk type secondary clock; pedestal-type secondary clock
座台 seat stand
座谈会 colloquium; forum[复 forums/fora]; panel discussion; rap session; seminar; symposium
座套 seat cover; socket sleeve
座/天 set-day
座艇 barge
座位 locus; place; seat(ing)
座位安排 arrangement of seats; seating diagram
座位安全带总成 seat belt assembly
座位布置 seat layout
座位布置图 <汽车车身内的> seating diagram
座位的空间 seating room
座位定额 seating capacity
座位端墙 seat end wall
座位间距 seat spacing
座位靠背 seat backrest
座位量 seating
座位排列 tier of seats
座位配置 seating configuration
座位配置方案 seating layout
座位坡度 seating rake
座位起降器 seat riser
座位前端活动盖板 cover flaps at the seat's front
座位容量 seat(ing) capacity
座位数 number of seats; seated number; seat(ing) capacity
座位数量 seating capacity
座位调节装置 seat adjuster
座位外形 surround section
座位形状 surround section
座位预订表示器 seat reservation indicator
座位预订系统 seat reservation system
座位预约 seat reservation
座位支架 seat frame
座位总成 seat assembly
座卧两用床 bunk
座卧两用铺位 bunk
座坞应力 docking stress
座坞肘板 <中内龙骨上的> docking bracket
座席按钮 position switch
座席拨号 position dialing
座席拨号盘 operator dial
座席灯 position lamp
座席电路 operator's circuit
座席分配 allocation of seats
座席分区 <古希腊楔形或梯形的> kerkis
座席构架 seat framework
座席计次器 peg count register; position register
座席记发器 operator's register
座席继电器组 position relay set
座席间中继线 interposition trunk
座席开关 position switch
座席客车 day coach
座席利用系数 coefficient of occupation of seats
座席容量 seating capacity
座席塞绳 <交换机的> drope cord
座席设备 equipment of position
座席伸展部分 <设有腿靠> seat extension
座席线路 position link
座席线路监控员 position link controller
座席引示灯 position pilot lamp
座席用户占线继电器 position busy relay
座席预订系统 seat reservation system
座席占用系数 coefficient of occupation of seats; seat occupancy coefficient
座席中继线 position trunk
座席钟 position clock
座席装饰 upholstery of seats
座下发动机 engine under seat
座下温暖器 under seat heater
座销 seat pin
座椅 chair; seat
座椅安全带 seat harness
座椅安装铁 seat tapping plate
座椅臂 seat arm; seat end arm; striker arm
座椅臂盖 seat arm cap
座椅臂靠 seat arm rest
座椅臂靠板 chair arm plate
座椅布置间隔 seating space
座椅侧支架 seat pedestal
座椅导轨 seat guide rail; seat rail
座椅底板 stuffing bottom plate
座椅底座 seat base
座椅电梯 chair lift
座椅垫子 matting
座椅端板 seat end
座椅端板框 seat end frame
座椅分离机构 seat separation equipment
座椅隔臂 seat division
座椅横挡 seat rail
座椅滑轨 seat slide
座椅滑架垂直移动机构 seat carriage vertical release
座椅加温控制 seat warmer control
座椅加温器 seat warmer
座椅间隔 seat pitch
座椅间距 seat spacing
座椅看台 seat stand
座椅靠背 backrest; seat back; seat end arm
座椅靠背板 backrest
座椅靠背软垫 seat back cushion
座椅靠背总成 backrest assembly
座椅框 seat frame
座椅框挡铁 seat frame tie plate
座椅框架 seated frame
座椅拉手 seat pull
座椅取暖器 seat heater
座椅伸腿空间 foot room
座椅升降机 seat lift
座椅升降机构 seat riser
座椅水平移动板 seat horizontal release
座椅弹簧 seat spring
座椅弹射 seat fire
座椅弹射起爆器 seat-ejection initiator; seat initiator
座椅弹射器 seat catapult
座椅套 seat webbing
座椅调节导轨 seat guide; seat track
座椅调整机构 seat regulator
座椅调整装置 seat adjuster
座椅铁锁 seat lock
座椅头靠端 seat head end
座椅腿 seat leg; seat leg
座椅镶条 car seat moulding; seat moulding
座椅枕垫 head rest
座椅支柱 pedestal
座椅座 seat support
座罩 seat cover
座支架 seat support
座柱 seat pillar
座砖 brick cup; nozzle seating block; seating brick

做 180°弯折 bending through 180 degrees

做凹口 notching
做包工的人 jobber
做边条 edge banding
做不到的 impossible
做裁缝 tailor
做槽材料 troughing
做成包裹 packet
做成超高 super-elevate
做成肥料 to make into fertilizer
做成复式车行道 dualling
做成复线 dualling
做成偏心的 weighted at one side
做成穹棱 groining
做成球状 pellet; pelletize
做成台阶 terracing
做成图表 diagrammatization; diagrammatize
做成网状的 reticulate

做成斜边 bevel(l)ing
做成斜坡 ramp
做成一式四份 quadruplicate
做成圆拱形 concamerate
做成直角 square up
做窗格材料 staff bead
做得过分 overshooting
做法 modus operandi
做复制品 replicate
做工好的 well-made
做功 apply work
做功系数 workdone factor
做功循环 expander cycle; work cycle; work extraction cycle
做雇工 work for hire
做过动态平衡的 dynamically balanced
做好 clear off
做好出海准备 secure for sea
做好干旱准备 drought preparedness
做好居住准备 ready for occupation
做好准备 stand-by
做基础 founding
做记号 marking out
做交叉拱 groining
做交易 transact
做结论 draw a conclusion
做井圈 curbing
做空白试验 running a blank
做苦工 moil
做框架 framing
做零活的人 jobber
做垄筑埂 ridging
做路标 blaze the rail
做路缘 curbing
做路缘石的材料 curbing material
做码头工人 stevedore
做模型用黏[粘]土 model(l)ing clay
做木工 carpenter
做木排路 < 美 > corduroy
做试验 do experiment; put to test; run an experiment
做榫舌等 tongue
做替工 subbing
做投机买卖 speculate
做完工作付款 paid in arrears
做围篱 paling
做围圈 boxing up
做线条 parg(et)ing
做镶嵌细工用的有色玻璃 smalto
做学徒 apprenticeship
做样盒 mocking up
做一笔生意 do a stroke of business
做预报 preparation of forecast
做圆 round(ing) off
做栅栏 paling
做栅栏的尖条板 sharp planks for paling
做栅栏用桩 paling

阿拉伯数字、西文字母开头的词语

0-4-0 式调车机车【铁】four-wheel switcher
0-4-4 式水柜机车 Forney four-coupled locomotive
0-6-0 式调车机车 six-wheel switcher
0-6-4 式水柜机车 Forney six-coupled locomotive
0-8-0 式调车机车 eigh t-wheel switcher
0-8-0 式蒸汽机车 eigh t-coupled locomotive
0-10-0 式调车机车 ten-wheel switcher
1/2 砖 half bat
1/3 倍频程 1/3 octave
1:25000 比例尺目标镶嵌图 target centered series 25 mosaic
1-8 萜二醇 terpine
1 的 n 次原根 primitive nth root of unity
1 输出信号 one output signal
1 输出与部分输出比 one output to partial output ratio
1 型极值分布 type 1 extreme-value distribution
1 型裂纹 <张开型或拉伸型裂纹> mode one crack[opening or tensile crack]
1 状态 one condition;one state
2×4×1 木板 two-four-one timber
2/4 线终端盘 two/four wire termination
2-4-0 式蒸汽机车【铁】four coupled locomotive
2-4-2 式蒸汽机车 Columbia type locomotive
2-6-0 式蒸汽机车 Mogul locomotive
2-6-2 式蒸汽机车 Prairie locomotive
2-6-4 式蒸汽机车 double-ended locomotive
2-8-0 式蒸汽机车 consolidation locomotive
2-8-2 式蒸汽机车 Mikado locomotive
2-8-4 式蒸汽机车 Berkshire locomotive
2-10-0 式蒸汽机车 decapod locomotive
2-10-2 式蒸汽机车 Santa Fe type locomotive
2-10-4 式蒸汽机车 Texas type locomotive
2C 较差 discrepancy between twice collimation errors
2 型极值分布 type 2 extreme-value distribution
3/4 砖 three-quarter bat
3/4 砖的丁砖 three-quarter header
3P 计划 pollution prevention pays
3 翼钻头 3-wing bit
3 英寸长大钉 tenpenny (nail)
4-4-0 式蒸汽机车【铁】American type locomotive;eight-wheel locomotive
4-4-2 式蒸汽机车 Atlantic locomotive
4-4-4 式蒸汽机车 Reading type locomotive
4-6-0 式蒸汽机车 ten-wheel locomotive
4-6-2 式蒸汽机车 Pacific type locomotive
4-6-4 式水柜机车 Baitic type locomotive;Hudson type locomotive
4-8-0 式蒸汽机车 twelve-wheel locomotive
4-8-2 式蒸汽机车 Mohawk type locomotive;mountain type locomotive
4-8-4 式蒸汽机车 Northern type locomotive; Pocono type locomotive; confederation locomotive
4-10-0 式蒸汽机车 Mastoden locomotive
4-10-2 式蒸汽机车 Southern Pacific locomotive;Union Pacific locomotive
4S 施工法 <即清理、清除、整理、整顿> four-S construction
4 线载波系统 4-wire carrier system
7/8 英寸企口板 sheeting
8 字形结(节) figure-8 knot
8 字形砂胶试体 mortar briquet(te)s
8 字形试块 briquet(te)
9 补码 nine's complement complement on nine
9 位行供给 nine-edge leading
9 位行输送 nine-edge leading
12 缸发动机 12-cylinder engine
12 信道群 12-channel group
14 英寸×7 英寸板条 narrow ladies
21 式增音机 twenty-one type repeater
21 式中继器 twenty-one type repeater
22 式增音机 twenty-two type repeater
22 式中继器 twenty-two type repeater
24 路组合设备 24-circuit combining equipment
24 小时(的) round-the-clock
24 小时卫星 fixed satellite; geopause satellite; geostationary satellite; motionless satellite; stationary satellite; synchronous satellite
28 英寸×23 英寸的纸张 elephant
33.5 度屋面坡度 third pitch 1:3
40 英寸×26.5 英寸的纸张 double element
60 路超群 60-channel supergroup
90°转弯 right-angled turn
100 英尺弧长所含的圆心角 <美国> degree of curvature; degree of curve
180°换向 opposite change
180°半环形管 return bend
180°转弯 U-turn
180°弯头 return bend
625-line system standard 625 行制标准 <电视>
6500 箱集装箱船 6500-box container ship

A.A 布罗德斯基分类 A. A. Brodisky classification
A/D 转换 analog(ue)/digit
ABC 分类法 ABC system
ABC 分析法 ABC analysis method
ABC 净化法 <也称古古诺法> ABC process
ABC 污水化学沉淀法 ABC process
ABC 污水净化三级过程 ABC process of sewage purification
ABD 型制动阀 ABD brake valve
ABS 树脂 acrylonitrile-butadiene-styrene
AB 式梁桥 AB type bridge
AB 型制动 AB brake
A-D 变换器 analog(ue)-digital converter
AGE 方式【机】AGE system
AGT 车辆 automated guideway transit vehicle
AGT 电车(系统) automated guided transit
API 比重计 American Petroleum Institute[API] hydrometer
API 比重计标度 American Petroleum Institute[API] hydrometer scale
API 比重指数 API degree
API 法 <桩工> API method
API 污水隔油池 American Petroleum Institute[API] oil interceptor
ARPA 计算机网 Advanced Research Projects Agency net
A 波段 A-band
A 玻璃 A glass
A 层(土) horizon A
A 层位 horizon A
A 电池组 A-battery
A 电源 A-power supply
A 级 grade A
A 级绝缘 class A insulation
A 加权测声网络 A-weighted network
A 加权声功率级 A-weighted sound power level
A 架坝 A-frame dam
A 架堰 A-frame weir
A 类部分预应力混凝土 type A partial prestressed concrete
A 类钢 A-type steel
A 类饰面 type A finish
A 镍铬耐热钢 A metal
A 清道证 <准许列车通过表示停车地点的证明> clearance form A
A 声级 A sound level; A-weighed sound pressure level
A 声压级 A-scale
A 式俯冲盆地 A-subduction basin
A 式水平仪 A-level
A 水泥石 <水泥熟料中的矿物成分> alite
A 台 A board;A-position;A station
A 台话务员 A-telephonist
A 线理【地】A-type lineation
A 形材料架 A-frame rack
A 形电杆 A fixture
A 形电缆杆 A cable pole
A 形丁坝 A-frame groin; A frame groyne
A 形分线杆 pole strutted in pyramidal form
A 形俯冲带【地】A-type subduction zone
A 形杆 A-pole
A 形杆式起重机 jinniwink
A 形钢架式摇臂起重机 A-framed derrick
A 形钢筋定位件 A-spacer
A 形构架 A-frame
A 形构架岸边集装箱起重机 A frame portainer crane
A 形构架(活动)坝 A-frame dam
A 形构架(活动)堰 A-frame weir
A 形构架锚桩 A-frame pile anchor
A 形构架木坝 A-frame (type of) timber dam
A 形构架桥塔 A-frame tower
A 形桁架 A-truss
A 形架杆 A-frame
A 形架起重机 <电动的,用于跨装或跨卸运货汽车> A frame crane
A 形架转臂起重机 A-frame derrick
A 形井架 A-type mast
A 形框架 A-frame
A 形框架结构 A-frame design
A 形框架设计 A-frame design
A 形陆上蒸发器 class A land evapo(u)ration pan;class A land pan
A 形门架集装箱起重机(装卸桥) A portainer
A 形木构架坝 A-frame timber dam
A 形起重架 A-frame
A 形砌块 <一端开口,一端封闭,中间有一肋的空心砌块> A-block
A 形轻便井架 A-mast
A 形曲线代替层参数 parameter of substitution layer for A-type curve
A 形燃具 type A appliance
A 形饰面 type A finish
A 形索塔 A-frame tower
A 形塔 A-shaped tower
A 形塔架 <桥塔> A-shaped pylon
A 形塔式起重机 jinniwink
A 形桅杆 double mast;full-view mast
A 形蒸发器 class A evapo(u)ration pan
A 形支柱 A fixture
A 形柱 A-pole
A 形转臂起重机 A-derrick
A 选择器 A-digit selector;A-selector
A 盐 alite
A 站 station A
A 状态 <即空号状态或称起脉冲状态> A-condition
A 字形的 A-line
A 族 subgroup A
A 组电池 A-battery
B.A.苏林分类 B.A. Sulin classification
BAT 形孔隙水压力计 BAT piezometer
BAT 形孔隙压力探头 BAT pore-pressure probe
B.B.R.V 体系 <瑞士 Birkenmaier、Brandestini、Rös、Vogt 所发明的预应力钢丝镦头锚及其张拉系统的简称> B.B.R.V. system
BCH 分组循环码 BCH code
B-H 曲线 B-H curve
BJ5 型集装箱 BJ5 type container
BOD 负荷 biochemical oxygen demand[BOD] load(ing)
BOD 自动测定记录仪 BOD automatic recorder
BOG 总管 BOG header
Burbank 试验 <测定岩石试样相对磨耗度> Burbank test
B 层 horizon B
B 层土壤 eluvial horizon
B 电池 anode battery
B 电路 B-circuit
B 电源 B power supply
B 构造岩 B-tectonite
B 黄铜 B-brass
B 级 grade B
B 级门 <防火标准> B-label(l)ed door
B 阶酚醛树脂 resite stage B
B 类部分预应力混凝土 type B partial prestressed concrete
B 类钢 B-type steel
B 类饰面 type B finish
B 声压级 B-scale
B 树 B-tree
B 水泥石 belite
B 台 B board;B-position;B station

B 线理【地】B-type lineation
B 形燃具 type B appliance
B 选择器 B-selector
B 岩组构造岩 B-tectonite
B 样条函数 B-spline
B 指令 B-instruction
B 族 subgroup B
B 座席 B-position
C14 测定工具 carbon dating; radiocarbon dating
CBR 法 <设计柔性路面厚度的一种方法> CBR method
CBR 法的怀俄明州修正法 <设计柔性路面厚度的一种方法> wrought adaptation of CBR Method
CBR 试验的应力贯入度曲线 stress-penetration curve for CBR test
CGS 单位制 centimeter-gram-second system of units
CS 型阿斯卡尼亚重力仪 CS Askania gravimeter
C 波 coupled wave
C 层 horizon C
C 层土的亚层 Ccs-Horizon
C 层土壤 <第三层土壤> C-horizon
C 常数 C-constant
C 电源 C-power supply
C 级 grade C
C 接帚 private wiper
C 类钢 C-type steel
C 类饰面 type C finish
C 偏压 C bias
C 曲线 time-temperature-transformation diagram
C 水泥 celite
C 系数 <立体测图系的总效率-航高可准确绘出的等高距> C factor [contour factor]
C 线 C-wire;C-line <土塑性图的>
C 形垫圈 C-washer
C 形端架 C end frame
C 形轭 C-type yoke
C 形反曲线 sigmoid
C 形防裂钉 C-iron
C 形或 S 形旋纹酒桶状家具脚 Flemish foot
C 形夹具 C-clamp
C 形夹钳 C-clamp
C 形夹子 C-clamp
C 形框架 gap frame
C 形梁架 C-frame
C 形梁枢轴 <推土机> C-frame pivot
C 形龙骨 C-stud profile
C 形起重机 C-type crane
C 形燃具 type C appliance
C 形网络 C network
C 形涡卷装饰 <家具> C-scroll
C 形支承 C-leg
C 形支杆 C-leg
C 形支架 C-leg
C 形支腿 C-leg
C 形支柱 C-leg
C 选择器 C-digit selector
C 盐 celite
C 因数 C factor[contour factor]
C 因素 <航高与等高距离之比> C factor[contour factor]
C 因子 C factor[contour factor]
C 语言 C language
D/A 转换 digit/analog(ue)
D-H 曲线 discharge head curve
D 层(土) horizon D
D 等年变率图 isoporic chart of D
D 等值线图 isogonic chart
D 电离层 <离地球表面最低的电离层> D-layer;Chapman layer
D 级 grade D
D 级木材 D-grade wood
D 节理【地】D-joint
D 类饰面 type D finish
D 裂缝 D-line crack
D 算法 D-algorithm
D 维熵 d-dimensional entropy
D 形瓣 D-trap
D 形柄手铲 D-handle shovel
D 形存水弯 D-trap
D 形阀 D-value
D 形隔气具 D-trap
D 形裂纹 D crack(ing)
D 形密集裂纹 D(-line) crack(ing)
D 形塔楼 D-shaped tower
D 形炸药 dunnite;explosive D
D 形中空护舷 D hollow fender
D 指数 d-exponent
E1 地震作用 earthquake action E1
EFT 法 <一种冲积土灌浆法> EFT process
e-logp 压缩曲线 e-logp curve
ETL 机器人 <一种双臂多关节机械手,用以钉箱、搬箱、拉锯等> ETL robot
E 玻璃 E glass
E 电离层 Kennely-Heaviside layer
E 级绝缘 class E insulation
E 类饰面 type E finish
E 面 T 形接头 E-plane T junction
E 面弯头 <波导管> E-plane bend
E 区 E-region
E 双晶纹 E-twin lamellae
E 形标尺分划 field graduation
E 形波 E mode
E 形玻璃 electric(al) glass
E 形水尺 E-type staff ga(u)ge
FAA 因子设计 <用于实验设计> FAA [Federal Aviation Agency] factorial design
F 波段吸收 F-band absorption
F 电离层 Appleton layer
F 分布 F distribution; variance-ratio distribution
F 格式 <数据装置记录的一种格式,其中逻辑记录是等长的> F format
F 含锌硬铝 F alloy
F 数 f-number; focal ratio; stop number
F 形接头 <铁烟道的> canister elbow
F 形头汽缸 F head cylinder
F 指数 <光圈的> F stop
GBM 工程 highway standardized and beautified project
GIS 基图 GIS[geographic information system] base map
G-M 法 quasi-slab method
GPS 接收机 GPS receiver
GSSI 脉冲雷达系统 <地球物理勘探用> GSSI impulse radar system
G 铝合金 G alloy
G 容限 g-tolerance
g 系数 g factor
G 形夹【机】G type cramp
G 形夹具 G-clamp
G 形螺旋夹钳 G screw clamp
H155 铬镍钴耐热钢 H155 alloy
H-E 图 hea T-content entropy chart
H.I.普洛特尼科夫分类 N.I. Protnikov classification
H-S 车辆荷载 <美国设计公路桥的一种标准汽车荷载> H-S loadings
H 波 H-wave;hydrodynamic(al) wave
H 参数 H parameter
H 层 H-horizon
H 杆 H pole
H 光灯 fluorescent lighting
H 环 H-ring
H 级绝缘材料 H class insulating material
H 面 T 型接头 H-plane T junction
H 面弯头 <波导管> H-plane bend
H 形波 H mode wave
H 形(测流)槽 H flume
H 形(的) aitch
H 形车架 H frame
H 形撑杆 H-stretcher
H 形等温线 H-shaped isotherm
H 形电(线)杆 H pole; H fixture; H frame
H 形电杆的线路 H-fixture line; H-pole line
H 形垫条 H-type gasket
H 形断面钢柱 steel H-column
H 形钢 H-bar; H-section steel; wide flange steel
H 形钢承重桩 steel H bearing pile
H 形钢打桩导向架 H-beam lead
H 形钢夹 H-clip
H 形钢桩 H-pile;H-section steel pile;H-shaped steel pile; steel H bearing pile;steel H-pile;steel H-section pile
H 形合页 parliament hinge
H 形铰链 H-hinge;parliament hinge
H 形截面 H-section
H 形截面电枢 H-armature
H 形截面衔铁 H-armature
H 形空心砌块 open-end block
H 形框架 H frame
H 形梁 H-beam; H-girder; H-shaped beam;wide flange beam
H 形梁钢 H-beam steel
H 形滤波器 H-filter
H 形铝支架条 <胶合板连接用> ply-clip
H 形锚具 H-anchorage
H 形锚桩 H-section anchor pile
H 形密封垫 H-type gasket
H 形密封条 H-type gasket
H 形圈 H-ring
H 形全焊接钢构架 all-welded steel H frame
H 形四端网络 H-network
H 形通路 H-shaped forehearth
H 形橡胶护舷材 H-type rubber fender
H 形支架 H frame
H 形支柱 H fixture
H 形柱 H-post
H 形桩 H-beam pile;H-column;H-section pile;H-pile
H 形钻头 H-bit
H 桩的锯齿形铸钢桩尖 hard-bite
H 字杆 H-post
H 字截面 H-section
ICOS 工法 <意大利 ICOS 公司开发的一种地下连续墙施工方法> ICOS method
IC 卡 identification card;integrate circuit card
IP 电话 IP-phone
IP 协议 internet protocol
I-V 图 heat-content volume chart
I 型裂纹 <张开型或拉伸型裂纹> mode one crack[opening or tensile crack]
J 积分法 method of J-integral
J 形槽 single-J-groove
J 形吊钩 J hook
J 形钩 J-hook
J 形密封材料 J seal
J 形密封结构 J seal
J 形坡口 J-groove
J 形弯管 J-bend
J 形止水结构 J seal
K-S 方式 K-S system
K-S 检验法【道】Kolmogorov-Smirnov test
K 波段 <约 10000 兆赫> K band
K 阶乘联络 K-factorial connection
K 矩阵 K-matrix
K 蒙乃尔合金 K Monel
K 式锚 K-type anchorage
K 系数 K factor
K 形撑架 K-bracing
K 形桁架 K-truss;truss with sub-ties
K 形桁架桥 K-truss bridge
K 形连接杆 K-bracing
K 形联结杆 K-bracing
K 形路口 K-crossing
K 形膨胀水泥 K-type expansive cement
K 形坡口 double bevel; double bevel groove
K 形三通阀 K triple valve
K 形坍落度试验 K-slump test
K 形支撑框架 K-braced frame
K 型膨胀水泥 <硫铝酸盐型> type K expansive cement
K 型曲线代替层参数 parameter of substitution layer for K-type curve
K 因素 K factor
K 因子 K factor
K 值 <表示路基或基层承载力的指标> K-value
LASH 门吊载驳船型驳船 LASH-type barge
L-CNG 加气站 L-CNG filling station
LC 振荡器 LC oscillator
LNG 和 L-CNG 加气站 LNG and L-CNG filling station
L 波 <地震时一种在地表传播的表面波> L wave
L 波段 <390 ~ 1550 兆赫兹> L-band
L 波段辐射计 L-band radiometer
L 波段雷达 L-band radar
L 杆 triangular pole
L 节理【地】L joint
L 形岸壁 L-wall
L 形暗扶壁墙 L-shaped counterfort wall
L 形坝座 L-abutment
L 形板 angle plate
L 形侧沟 curb and gutter
L 形大梁 L girder
L 形挡土墙 L-shaped retaining wall;L type retaining wall
L 形的 dog leg
L 形吊墙 L-shaped suspended wall
L 形丁坝 L-shaped groin;L-head spur dike;L groyne
L 形顶撑 L-shore
L 形顶撑端头 L-head
L 形端墙 L-endwall
L 形短管 branch ell
L 形断面 angle section;L-section
L 形阀门 angle threshold
L 形防波堤 L-shaped breakwater
L 形钢 L(-shaped) beam
L 形钢筋 <用来把一些木材保持在一起> dog anchor
L 形钢筋混凝土挡土墙 retaining wall L-shape reinforced concrete
L 形供水短管 service ell
L 形构件 L-shaped unit
L 形管接头 street ell;union ell
L 形贯流式水轮机 L-tubular turbine
L 形焊接 angle weld(ing)
L 形检修(用)延长部分 service ell
L 形角钢 L-section
L 形角铁截面 angle section
L 形接合板 angle fishplate;angle type joint bar
L 形结构 L-shaped structure
L 形截面 L-shaped section
L 形块吊耳 drop ell
L 形块式码头 L-block type wharf
L 形块体 L-shaped block
L 形框架 L-frame;L-shaped frame
L 形梁 angle beam;ell-beam;L-bar;L-beam;toe beam
L 形滤波器 L-type filter
L 形路缘 L-shaped curb;L-type curb
L 形螺纹管接头 service ell
L 形码头岸壁 L-shaped quay wall
L 形模壳 L-shaped form
L 形耐火砖分隔 L-block separation

L 形墙 angle type wall
L 形墙段 L-panel
L 形桥台 L-abutment;L-type abutment
L 形曲线 dog leg
L 形人孔 turning manhole
L 形视车镜 < 观察车速用的仪器 > enoscope
L 形四端网络 L-network
L 形踏步 L-shaped step
L 形天线 L-antenna
L 形突堤码头 L-head pier
L 形温度计 angle thermometer
L 形型材 L-section
L 形悬臂墙 L-shaped cantilever wall
L 形鱼尾板 angle joint bar
L 形凿 bent chisel
L 形闸阀 angle gate valve
L 形栈桥 L-headed jetty
L 形栈桥式码头 L-shaped pier
L 形直角尺 try square
L 形柱 < 预制混凝土框架 > L-column
L 形砖 L-shaped brick
L 形装载 L type loading
M 齿锯 M-tooth(ed) saw
M 答应 M response
M 系列钻头 M-series bit
M 响应 < 电传打字端机对查询或地址选择的答应 > M response
M 形齿锯 M-tooth(ed) saw
M 形齿钻头 M-tooth bit
M 形导出滤波器 derived M type filter
M 形公路 M type highway
M 形梁 M(-shaped) beam
M 形膨胀水泥 < 硅酸盐形 > type M expansive cement
M 形推演式滤波器 derived M type filter
M 形屋顶 ridge and valley roof; trough roof
M 形屋面 M-roof
M 形褶皱 M-shaped fold
M 形正交信号的决择 resolution of M-orthogonal signals
n 百分数概率洪水 n percent chance flood
n 倍 n-fold
n 倍的 nth
n 步决策问题 n-stage decision problem
n 重 n-fold
n 次变分 n-th variation
n 次超静定结构 structure indeterminate to the nth degree
n 次导数 nth derivative
n 次的 nth
n 次多项式校正 nth polynomial correction
n 次方 nth power
n 次方程 equation of nth degree; equation of nth order
n 次方根 nth root
n 次幂 nth power
n 次幂信道 nth power channel
n 次群 nth-order group
N 地址 N address
N 地址指令 N-address instruction
N 地址指令格式 N-address instruction format
n 个变数立方体 n-variable cube
n 沟道耗尽型 MOS 场效应管 n-channel depletion MOS field-effect transistor
n 股电缆线 cable of n wires
N 级地址 N-level address
N 级逻辑 N-level logic
n 阶 n^{th} order
n 阶的 n^{th}
n 阶分布 n^{th}-order distribution
n 阶行列式 determinant of the nth order
n 进制补码 compilation on n
n 进制反码 compilation on n-1
n 立方体 n-cube
n 年内可靠产水量 n-year dependable yield
n 年内可靠供水量 n-year dependable yield
n 年一遇洪峰流量 n-year peak discharge
n 年一遇洪水 n-year flood
n 年一遇枯水 n-year low water
n 年最大洪水 n-year peak flow
n 生产阶段的地层压力 formation pressure at the n production stage
N 式格构大梁 N-girder
n 维空间 n-dimensional space
n 维立方体 n-dimensional cube
n 位谓词 n-place predicate
n 线旋转式寻线机 n-point single-motion finder
n 心电缆 n-conductor cable
n 心塞绳 n-conductor cord
N 形半导体 N-type semi-conductor
N 形电杆 N-pole
N 形沟道 N-channel
N 形桁架 N-truss
N 形回弹仪 N-hammer
N 形支柱 N fixture
n 元代码 n-ary code
n 元关系 n-ary relation
n 元组 n-tuple
N 值 < 贯入试验的打击次数 > N-value
N 中取 M 码 M-out-of-N code
OD 调查 origin-destination[OD] study;origin-destination[OD] survey
OD 交通量 OD traffic volume
O 形(密封)圈 O-ring
O 形环 O-ring
O 形环的挡环 backup ring
O 形环接合 O-ring joint
O 形环圈 O-ring
O 形密封环槽 O-ring groove
O 形密封圈 O-ring seal;O-seal ring
O 形圈柱塞阀 O-ring spool valve
O 形橡胶圈 O-rubber ring
PCA 道路平整度仪 PCA meter
PCA 平方和 < 一种 PCA 平整度指标 > PCA sum-of-squares
PDCA 循环法 plan-do-check-action management
PE 管 polyethylene pipe
pH 计 acidimeter[acidometer];pH acidimeter;pH meter
P-E 图 pressure-enthalpy chart
pH 值 hydrogen ion exponent;pH value;potential of hydrogen
pH 值测定仪 pH meter
pH 值记录仪 pH-recorder
pH 值控制 pH control
pH 值指示计 pH-indicator
PM 电车(系统) automated guided transit;people mover
p-n-i-p 型晶体管 p-n-i-p transistor
PRT 交通系统 personal rapid transit [PRT] system
PSK 精密坐标仪 < 德国制造 > PSK stereocomparator
PVC 管 polyvinyl chloride[PVC] pipe [piping];PVC tube[tubing]
P-V 图 pressure-volume diagram
PZ 法【道】PZ method
P 波 longitudinal wave;primary seismic wave;primary wave;P wave < 地震波 >
P 偶图 P-diagram
P 形半导体 P-type semi-conductor
P 形存水弯 P-trap
P 形公路 P-type highway
P 形回弹仪 P-hammer
P 形晶体管 P-type transistor
P 形橡皮止水 music-note type rubber seal
P 值数据切除值 cut value of P-value data
Q-FLEX 型液化天然气运输船 Q-FLEX LNG vessel
Q-MAX 型液化天然气运输船 Q-MAX LNG vessel
Q 倍增器. Q-multiplier
Q 表 Q-meter
Q 波 Q wave
Q 波段 Q-band
Q 单位 Q-unit
Q 点 Q-point
Q 电子 Q-electron
Q 电子管 Q tube
Q 共轭元 Q-conjugate
Q 管 Q tube
Q 节理【地】Q-joint
Q 进制多项式 q-ary polynomial
Q 开关 < 激光术语 > Q-switch
Q 频带 Q-band
Q 旗 Q-flag
Q 式分析 Q-pattern analysis
Q 天线 Q-aerial;Q-antenna
Q 突变 Q-switch
Q 突变技术 Q-switching technique
Q 系数 Q-factor
Q 现象 < 制动机不能缓解 > Q-phenomenon
Q 形聚集分析 Q-mode cluster analysis
Q 形曲线代替层参数 parameter of substitution layer for Q-type curve
Q 形因子分析 Q-mode factor analysis
Q 值 Q-value
Q 值反演滤波 Q-inverse filtering wave
RAS 技术 < 为可靠性、有效使用率和维修效率等技术的总称 > RAS technique
RTF 型三向联结构造 RTF triple junction structure
R 波 < 地震时一种在地表传播的表面波 > R wave;Rayleigh's wave
R 控制图 R chart
R 年一遇波浪 R-year wave
R 年一遇大风 R-year wind
R 年一遇洪水 R-year flood
R 式分析 R-pattern analysis
R 形曲线 R curve
R 形聚集分析 R-mode cluster analysis
R 形因子分析 R-mode factor analysis
S.A.舒卡列夫分类 S. A. Shukarlev classification
SBS 热塑性橡胶【化】styrene-butadiene-styrene block copolymer
SF 形悬浮预热器 SF suspension preheater
SHRP 计划 < 美国 > strategic highway research program(me)[SHRP]
SHRP 沥青研究计划 SHRP asphalt research program(me)
S 波 distortional wave; second (ary body) wave; shear wave; S-wave; secondary seismic wave
S 波段 < 约 3000 兆赫 > S-band
S 玻璃 S glass
S 等价 S-equivalence
S 等值原则 principle of S-equivalence
S 构造岩 S-tectonite
S 过程线 S-hydrograph
S 节理【地】S-joint
S 脉型 sigmoidal vein
S 面 S-surface
S 区曲线 curve of S zone
S 曲线 S-curve; S-hydrograph; time-temperature-transformation diagram
S 弯头 double bend
S 线 sleeve wire;sline;S-wire
S 形 sigmoid
S 形凹凸榫 hook rebate
S 形凹凸线饰 chinbeak mo(u)lding
S 形边垫片 ogee washer
S 形边垫圈 ogee washer
S 形槽口 S-rabbet
S 形长椅 visavis
S 形超高顺坡 S-shaped superelevation ramp
S 形存水弯 S-trap
S 形的 sigmate;sigmoidal;S-shaped
S 形低坝 ogee weir
S 形垫片 ogee washer
S 形钉 steel dog;timber dog
S 形顶堰 ogee crested weir
S 形反射 sigmoid reflection configuration
S 形防裂钉 S-iron
S 形工程进度曲线 S-curve
S 形拱 ogee arch
S 形管 double elbow pipe
S 形(家具)腿 ogee bracket foot
S 形浇口 swan-neck swan
S 形铰 S-hinge
S 形接缝 hook joint
S 形裂纹 S-crack
S 形路线 switchback
S 形膨胀水泥 type S expansive cement
S 形曲颈管 S-trap
S 形曲线 sigmoid curve; ogee curve; S-curve;serpentine
S 形曲线函数 S-shaped curve function
S 形曲线理论 sigmoid curve theory
S 形曲线设备 swan-neck
S 形熔线 fusestat
S 形生长线 sigmoid growth curve
S 形失真 sigmoid distortion
S 形饰 ogee decoration
S 形(双头死)扳手 S(-type) spanner
S 形水轮机 S turbine
S 形水文曲线 S-curve hydrograph
S 形顺坡 S-shaped ramp
S 形铁 S-iron
S 形瓦 S-tile
S 形弯 gooseneck
S 形弯管 double (offset U) bend; gooseneck pipe;S-bend pipe
S 形弯曲 reverse bend
S 形弯曲河段 full meander
S 形弯头 S-bend
S 形屋顶 ogee roof
S 形屋顶雨水沟 ogee roof gutter
S 形下水管 S-trap
S 形衔铁 S armature
S 形线 gula
S 形卸料小车 travel(l)ing tripper
S 形堰 ogee weir
S 形移动 S-shaped motion
S 形溢洪道 ogee spillway
S 形褶皱【地】S-shaped fold
S 形砖 S-tile
S 形(装饰)线脚 ogee mo(u)lding
S 岩组构造 S-tectonite
S 转换管 S transfer tube
TM 场波 transverse magnetic field wave
TRRL 梁式平整度仪 TRRL beam
T 带 T zone
T 等价 T-equivalence
t 对流 counterjet
T 方形架 T-square frame
t 分布【数】t-distribution
T 光圈 T-stop
T 级张拉千斤顶 < 张拉单跟钢丝束 > T-range jack
T 检验 T-test
T 节滤波器 T-section filter
T-截面格栅 half-joist

T 式接合 double square junction
T 试验台 triple valve test rack
T 头键 T-head key
T 线 T-wire
T 形把手 T-handle
T 形板 T-panel; T-slab
T 形布置 perpendicular layout
T 形材 T-section
T 形操纵杆 T-handle
T 形槽 T-slot
T 形槽铣刀 T-slot milling cutter
T 形槽用螺栓 T-head bolt
T 形测平板 bending board
T 形长尾铰链 T-long tail hinge
T 形尺 T-ruler; T-square (ruler)
T 形除水阀 T trap
T 形大梁 T(-section) girder; T-shaped girder
T 形大楼 T-block
T 形带 T trap
T 形单元 T-unit
T 形弹簧吊架 T-hanger
T 形挡土墙 T-shaped retaining wall
T 形刀架 T-rest
T 形的 T-shaped; T-type
T 形吊耳 drop tee
T 形丁坝 hammerhead groin
T 形断面 T cross-section
T 形刚构桥 rigid T-frame bridge; T-shaped rigid frame bridge; T-type rigid-frame bridge
T 形刚架桥 T-shaped rigid frame bridge
T 形钢 structural T; T-bar; T(-section) steel
T 形钢暗榫 T-steel dowel
T 形钢板 T-steel plate
T 形钢棒 T-bar
T 形钢杆 T-steel bar
T 形钢轨 T-rail
T 形钢加强的混凝土楼板 Homan and Rodger's floor
T 形钢筋 rail steel reinforcement
T 形钢筋混凝土挡土墙 T-shape reinforced concrete retaining wall
T 形格栅 T-grill; T-joist
T 形根(叶片) T root
T 形公路 T-type highway
T 形拱座 T-abutment
T 形构架 T-frame
T 形构件中段截面 mid-section of T member;stalk
T 形拐肘 T-crank;three-arm crank
T 形管 T-branch (pipe); T-pipe; T-tube
T 形管接头 service tee;union tee
T 形管节 skew tee
T 形管配件 T-pipe fitting
T 形规划线 T-ga(u)ge
T 形轨 T-trail
T 形焊(缝) T-weld
T 形焊接 T-welding
T 形焊接头 jump weld
T 形环 T-ring
T 形混凝土板桩 T-shaped concrete sheet pile
T 形建筑单元 T-building unit
T 形建筑构件 T-building member; T-component
T 形交叉(道) T-type intersection
T 形交叉口 T junction
T 形铰合钢带 tee-hinge strap
T 形铰合构件 tee-hinge member
T 形铰合连接 tee-hinge junction
T 形铰合面板 tee-hinge panel
T 形铰合皮带 tee-hinge strap
T 形铰合墙板 tee-hinge panel
T 形铰链 cross-garnet (hinge);hook-and-band hinge; T-hinge
T 形接板 T-strap
T 形接缝 T-junction
T 形接合 T joint
T 形接口 T joint
T 形接头 T-connection; T-joint; T-junction; T-piece
T 形接头连接板桩 T-connection sheet pile
T 形接头上斜面焊接 bevel weld in a T-joint
T 形接头贴角焊 T-fillet welding
T 形节理【地】T-joint
T 形截面 T shaped section; T-section
T 形截面的 T-shaped
T 形截面格栅 T-half-joist
T 形截面有效宽度 effective width of T section
T 形距离尺 range rake
T 形块 T-piece
T 形块料 T-block
T 形立体交叉 T-grade separation
T 形连接 T-connection
T 形连接自耦变压器 T-connected autotransformer
T 形梁 double T; T-beam; T-iron; T-shaped beam
T 形梁板构造 T-beam and slab construction
T 形梁板结构 T-beam and slab construction
T 形梁的主筋 rib reinforcement
T 形梁钢筋 joist bar
T 形梁桁架 T-beam girder
T 形梁节间 T-beam floor
T 形梁肋 T-beam rib
T 形梁楼板 T-beam floor
T 形梁桥 T-beam bridge
T 形梁支座 joist chair
T 形梁柱脚 T-beam footing
T 形料道 T-shaped forehearth
T 形楼板 T-floor slab
T 形滤波器 T-filter
T 形铝杆 alumin(i)um T bar
T 形螺栓 T(-type) bolt
T 形码头 hammerhead pier; T-shaped pier
T 形码头的突堤 T- headed jetty
T 形码头或 L 形突码头 T-head pier or L-shape pier
T 形门铰 garnet hinge
T 形面砖 T tile
T 形配件 T-fitting
T 形片 T-piece
T 形墙 T-wall
T 形墙段 T-panel
T 形桥接滤波器 bridged-T network
T 形桥梁 slab-and-beam bridge
T 形桥面板 T-floor slab
T 形桥台 stub abutment; T-abutment
T 形曲拐 T-crank
T 形十字架 crux commissa;Tau cross
T 形石油码头 hammerhead oil pier
T 形手扳钳 T-wrench
T 形手柄 T-handle
T 形顺岸码头 T-type marginal wharf
T 形四端网络 T-network
T 形套筒 T-junction box
T 形天线 T-shaped antenna; T-type antenna
T 形条缝送风口 T-bar slot air diffuser
T 形铁 T-bar; T-iron
T 形铁棒 T-bar
T 形铁件 T-iron
T 形铁角 three-way strap
T 形头部 T head
T 形头部顶撑 T shore
T 形头部突堤码头 T-head pier
T 形头的 T-headed
T 形头螺栓 T-headed bolt
T 形头气动缸 T-head cylinder
T 形头汽缸 T-head cylinder
T 形凸缘 flange tee
T 形突堤码头 T-jetty
T 形图 <一种计划协调的方法> T-map
T 形托架 T-bracket
T 形瓦 T tile
T 形弯折 bent angle tee
T 形网络 T-circuit
T 形物 tee
T 形弦杆 T-chord
T 形斜角规 T-bevel
T 形型钢 T-profile; T-steel bar
T 形悬臂墙 cantilever T-wall
T 形旋臂起重机 <俚语> hammer-head
T 形异径管接头 reducing tee
T 形预制建筑构件 T-building component
T 形元件 tee
T 形支撑 T-head; T-shore
T 形支撑架 T-support
T 形支管 T-fitting
T 形支架 half-set
T 形支柱 T-trust
T 形柱头 T-head
T 形装载 T-type loading
T 形着陆标志 landing T
T 形组成部分 T-building component
T 形组合角钢 back-to-back angles
T 指数 T-number
T 字布 landing T
T 字槽螺丝 T-slot screw
T 字管 T-bend
T 字管接头 T-branch
T 字接头 T joint;triple joint
T 字手柄形 T-handle
T 字铁 T-iron
T 字头形 tee-head type
T 字形【物】tau
T 字形接头 T-bend
T 字形三岔路口 T-junction
T 字形柱 T-head post
U 格式 <数据装置的一种格式,其中数据组不规定长度> U format
U 形 U-shape;horseshoe
U 形岸墩 U-abutment
U 形扳手 U spanner
U 形泵 U pump
U 形冰蚀谷 U(-shaped) glaciated valley
U 形玻璃管风压表 U-tube draft ga(u)ge
U 形补偿器 expansion loop
U 形操纵台 U(-shaped) control console
U 形槽口 U-notch
U 形插塞 U-link
U 形插销连接系统 clevis and pin connection
U 形铲刀式推土机 U-blade bulldozer
U 形铲推土机 U-dozer
U 形承载槽 carrying channel
U 形磁铁 horseshoe shape magnet (iron)
U 形存水弯 hairpin bend; horseshoe bend;running trap;U-trap
U 形搭接钢带 stirrup strap
U 形刀板推土机 U blade dozer
U 形导向架 box lead
U 形的 U-form;U-shaped;U-type
U 形垫圈 U-washer
U 形吊钩 U-hanger
U 形叠加褶皱 U-type superposed folds
U 形钉 staple;U-staple;wire staple;iron dog
U 形钉锤 staple hammer
U 形洞道 siphon
U 形断面 U(-shaped) section
U 形断面盘根 U-packing
U 形杆 banjo bar
U 形钢 channel steel;trough iron;U-steel
U 形钢板桩 U-section steel pile
U 形钢箍 double leg stirrup
U 形钢筋箍 U-stirrup
U 形钢梁 channel beam
U 形拱座 U-abutment
U 形钩 clevice[clevis](pin);shackle;U-hook
U 形钩眼圈 clevis eye
U 形箍 stirrup
U 形箍筋 hairpin;U(-shaped) stirrup
U 形谷 trough valley; U (-shaped) valley
U 形管 gooseneck;hair tube;open return bend;return offset;tube turn;U-form tube;U(-shaped) pipe[piping];U(-shaped) tube[tubing]
U 形管夹 U-clamp
U 形管曝气 U-tube aeration
U 形管曝气装置 U-tube aeration system
U 形管热交换器 U-tube heat exchanger
U 形管式给水加热器 U-tube type feed water heater
U 形管束 hairpin tube bundle
U 形管水银压力计 U-type mercury manometer
U 形管压力计 liquid level manometer;U-tube ga(u)ge;U-tube manometer;U-tube type manometer
U 形管真空计 U-type-vacuum ga(u)ge
U 形河谷 U(-shaped) valley
U 形横截面 open box section
U 形湖 oxbow lake
U 形环 norman
U 形环密封件 U cup packing
U 形混凝土过梁 U-block
U 形混凝土膨胀剂 U type concrete expansion admixture
U 形活塞泵 <往复式泵的一种型式> U piston pump
U 形火焰炉膛 U-flame furnace
U 形加强架 clevice[clevis]
U 形夹铁 cramp iron
U 形夹销 clevis pin
U 形夹(子) clevice[clevis](bracket);U-link
U 形检测管 <检测管系是否漏气设施> U ga(u)ge
U 形检漏器 U ga(u)ge
U 形浆槽 oarlock
U 形浆架 oarlock
U 形胶带输送机 U profile belt conveyer[conveyor]
U 形接头 U(-shaped) joint
U 形截面皮碗 U-cup
U 形紧固件 strap bolt
U 形抗拉试验 U-tension test
U 形控制盘 U(-shaped) control panel
U 形控制台 U(-shaped) control console
U 形捆扎钢条 <木结构的> U-strap
U 形冷却器 hairpin cooler
U 形连接 <指电焊槽焊> U-type joint
U 形连接环 U-link
U 形连接件 U-tie
U 形链 U-link
U 形梁 U(-shaped) beam
U 形龙骨 channel stud;U(-shaped) stud
U 形楼梯斜梁 channel string
U 形螺栓 anchor loop;bulldog grip;norman;U(-shaped) bolt
U 形螺栓柱插 stake pocket strap;stake pocket U-bolt

U 形螺栓铸件 U-bolt casting
U 形螺旋式钻头 U(-shaped) rotary bit
U 形铆顶棍 banjo bar
U 形密度计 U tube densitometer
U 形黏[粘]度计 channel viscosimeter
U 形牛颈弯 oxbow
U 形排水沟 U-drain
U 形盘管 hairpin coil
U 形膨胀弯管 U-expansion bend
U 形皮碗 U cup packing;U-leathering ring
U 形坡口 U-groove
U 形起重机 U-type crane
U 形砌块 U-block
U 形牵引钩 shackle link
U 形桥台 U(-shaped) abutment
U 形曲线 hairpin curve
U 形伸缩(弯)管 expansion U-bend
U 形伸缩器 U(-shaped) expansion loop
U 形深槽 undercut U groove
U 形水管卡 U-iron
U 形顺岸码头 U-type marginal wharf
U 形天线 U-antenna
U 形调整器 U-type compensator
U 形铁 U-iron
U 形铁箍 fastener
U 形铁条 <木结合用> U-strap
U 形铁芯 U-core
U 形铁支柱 channel strut
U 形通风管道 U-duct
U 形突堤码头 U-pier
U 形推土板 U blade
U 形托 <煤水车的> cylinder lever guide
U 形托架 crevice bracket
U 形弯(头) hairpin bend;U-bend
U 形弯钩 <钢筋的> U(-shaped) bend
U 形弯管 U-bend;U-trap
U 形弯曲 oxbow
U 形弯头 close return bend; hairpin turn;U-turn
U 形涡流 U-vortex
U 形物 horseshoe
U 形系墙铁 U-tie
U 形楔子 <固定模板用> hairpin
U 形压板 U(-shaped) clamp
U 形压差计 U-type barometer
U 形压力表 U-ga(u)ge
U 形压力计 U-ga(u)ge; U-type manometer
U 形烟道 U-duct
U 形窑 hairpin furnace
U 形预埋件 hairpin
U 形胀缩器 expansion U-bend
U 形蒸发器 U-type evaporator
U 形支腿 U-end frame
U 形支柱 clevice[clevis]
U 形直立钢筋 <支承上层钢筋网的> standee
U 形柱 supporting channel
U 形转弯 hairpin bend;U-turn
U 形转折 U-turn
V.E.米赫也夫鉴定表 V. E. Michev discrimination table
VB 仪 V-B consistometer
VR 式沟 VR ditch
VSL 锚 <瑞士 Losinger 公司创制,用钢锥锲将若干根钢丝锲紧于锚圈内周,利用锚圈内螺纹可进行多次张拉> VSL anchorage
VSL 体系 <预应力混凝土用的一种锥形锚具体系> VSL system
VST 体系 <预应力混凝土用的一种锥形锚具> VST system
V 波段 V-band
V 答应 V response
V 带角度 angle of the V-belt
V 格式 <数据装置的一种格式,其中数据组长度是可变的,并有一个组长指示符> V format
V 环式换向器 V-ring type commutator
V 频带 V-band
V 式八汽缸 eight in V
V 响应 <电传打字端机对征询或地址选择的答应> V response
V 心 V centre[center]
V 形凹槽 V-depression;V-notch
V 形凹槽板 notch board
V 形凹口 jag;V-notch
V 形八缸发动机 V-8-engine
V 形板条 V(-shaped) batten
V 形边沟 V-drain;V(-shaped) ditch
V 形波纹板片 V-shape corrugated plate
V 形槽 V flume;V(-shaped) trough; V-slot
V 形槽焊接 V-groove weld
V 形槽胶合板 V-grooved and tonguing;V-grooved plywood
V 形槽口 V-notch
V 形槽口板 notch plate
V 形槽口量水堰 V-notched weir
V 形槽口溢流闸板 notch plate
V 形槽轮 V pulley
V 形槽墙板 V-grooved wall panel
V 形槽饰 V channel decoration
V 形槽堰 V-notch weir
V 形槽闸 V-notch lock
V 形柴油机 V engine type diesel engine
V 形铲 V-sweep
V 形铲刀 V-shaped
V 形铲土法 <推土机> V-cut
V 形车轮夹 <自行车停车场> V(-shaped) wheel grip
V 形齿齿轮 herringbone gear;V-gear
V 形传动皮带 wedge-shaped belt
V 形雌雄榫 V-grooving and tonguing
V 形带 V-type belt
V 形带钢 V-strap
V 形刀片 V blade
V 形导槽 inverted V guidance;plain V slide;Vslide
V 形导轨 inverted V (slide) way;V-guide way
V 形道路交叉 V junction
V 形道路枢纽 V junction
V 形的 V-shaped;V-type
V 形低(气)压 V(-shaped) depression
V 形底 V-bottom
V 形底斗仓 V-bottom bucket
V 形电桥 V-bridge
V 形吊顶板 V-line ceiling panel
V 形吊墙板 V-line wall panel
V 形钉 staple
V 形斗底铲斗 V-bottom bucket
V 形斗链式提升机 V-bucket carrier
V 形斗式输送机 V-bucket conveyer [conveyor]
V 形墩 V(-shaped) pier
V 形发动机 V(-type) engine;V-type motor
V 形伐树刀 V-type shearing blade
V 形伐树机 V-tree cutter
V 形分级箱 V-box
V 形风扇皮带 V-type fan belt
V 形缝 V(-shaped) joint
V 形钢筋 V(-shaped) reinforcement
V 形钢索 V-rope
V 形割法 V-cut
V 形隔膜泵 V-type diaphragm pump
V 形工具 V-tool
V 形工具勾缝 V tooled joint
V 形拱 V-arch; inverted-V corbel bracket <唐朝的一种斗拱>
V 形勾缝 V joint pointing;V(-shaped) joint;V-tooled joint
V 形沟 V gutter;V-gully;V(-shaped) ditch
V 形沟渠 V-drain;V(-shaped) trench
V 形沟渠扫污机 V-crowder
V 形谷 V-valley
V 形刮板 V blade
V 形刮片刮泥机 V-blade scraper
V 形管 V(-shaped) pipe
V 形管蒸发器 herringbone evaporator;V-coil evaporator
V 形焊缝 V-weld
V 形焊接 cleft weld
V 形航路 V(-shaped) airway
V 形河谷 V(-shaped) valley
V 形护舷 V(-shaped) fender
V 形滑槽 inverted V slip
V 形滑座 V slide seat
V 形混合机 twin-cylinder mixer; twin-shell blender;V-type mixer
V 形混合器 V-blender;V crack
V 形极板 V plate
V 形加劲肋钢丝网条板 V-stiffened wire lath
V 形夹 V-clamp
V 形交叉 V junction
V 形胶带传动 V-belt drive
V 形角 V-angle
V 形接法 <三相交流电源> open-delta connection;V-connection
V 形接缝 V joint(ing)
V 形接合 V-joint
V 形接口 toe-jointing
V 形接头 V junction; V(-shaped) joint(ing)
V 形截面 arris section
V 形截面雨水槽 V-gutter
V 形紧固夹板 forked strap;two-way strap
V 形开口 open V
V 形开挖 V-cut
V 形空心砖 V-brick
V 形孔口阀 V-port valve
V 形块 clevis;V-block
V 形拉线 V stay
V 形连接 open-delta;V-type joint;V-connection
V 形裂缝 V-crack
V 形螺纹 triangular thread;V(-type) thread
V 形螺旋 V-thread screw
V 形镘 <装在滑模摊铺机后面的> V-float
V 形锚铁 government anchor
V 形门 V-door
V 形耙土机 V-type ripper
V 形耙土器 V-type scarifier
V 形排水槽 miter[mitre] drain; V-drain;V-gutter
V 形排水工程 V-drainage
V 形排水沟 V-drain
V 形排水系统 herringbone drain-(age) system
V 形排泄口 V-let
V 形皮带 cone belt; pulley conical disk;V belt
V 形皮带轮 V-grooved pulley
V 形皮带无级变速传动装置 adjustable V-belt drive
V 形片 V-strip
V 形偏光透镜 half toric sector;ho T-strip
V 形平整器 V-leveler
V 形坡口 single V-groove
V 形企口 V-grooving and tonguing
V 形汽车式雪犁 V-type truck snow plough
V 形汽缸 V-type cylinder
V 形汽缸排列风压机 angle compressor
V 形汽缸体 V-block
V 形砌缝 mason's joint
V 形砌沟 angle of V;V-gutter
V 形桥 V-type bridge
V 形切割 V-cut
V 形切口 chev(e)ron notch; Izod notch;V-notch
V 形切口试件 specimen with V-notch
V 形切土角 V-cut angle
V 形曲线 V-curve
V 形缺口 Izod notch; jag; V(-type) notch
V 形缺口冲击试件 V-notch Charpy specimen
V 形缺口垫板 slewing V-block
V 形缺口堰 triangular notch weir;V-notch weir
V 形沙丘 chev(e)ron dune
V 形绳索 V-rope
V 形十二汽缸发动机 twin-six motor
V 形双犁式扫雪机 V-type snow plough
V 形水槽 V flume;V-gutter
V 形水库 gorge type reservoir
V 形水泥砂浆垫块 cement V washer
V 形水平勾缝 V joint
V 形松土机 V-type ripper
V 形松土器 V-type scarifier
V 形塔轮 V-type step pulley
V 形掏槽 V-cut
V 形掏槽炮眼 V-cut hole
V 形掏心 V-cut
V 形条 V-strip
V 形条板压路机 V-cleated roller
V 形铁 V-block
V 形铁条 V(-shaped) iron bar <木结合的>;V-strap
V 形铜片 copper V-strip
V 形挖泥机 V-crowder
V 形尾部 butterfly tail
V 形涡纹 crotch swirl
V 形屋顶 double lean-to roof; double pent;V(-shaped) roof
V 形屋面 V-roof
V 形线饰 V-mo(u)lding
V 形箱 V-box
V 形镶板 V-jointed panel;V-panel
V 形橡胶防舷材 V-typerubberfender
V 形消解 V-resolution
V 形小槽 V-groove
V 形卸料器 V-stripper
V 形雪铧犁 V-type snow plough
V 形雪犁 <双犁式扫雪机> V-plough
V 形压缩机 V(-type) compressor
V 形檐槽【建】 arris gutter
V 形檐沟【建】 arris gutter
V 形堰 notch weir
V 形翼摊铺器 V-blade distributor
V 形凿 V(-shaped) chisel
V 形折板 V-beam sheeting
V 形折射透镜 half toric sector;ho T-strip
V 形折射仪 V-type refractometer
V 形整平机 V-leveler
V 形支架 V-rest
V 形支座 V support
V 形轴封 V-packing
V 形柱 V(-shaped) column
V 形抓地板压路机 V-cleated roller
V 形装载 V-type loading
V 形装载槽 angle loading chute
V 形锥轮 V-type step pulley
V 值 V-value
WZ 碳化钛烧结合金 WZ alloy
W 系数 W coefficient
W 形车站雨棚 W-shaped canopy of station
W 形叠加褶皱【地】 W-type superposed folds
W 形谷 W-shaped valley

W 形桁架 W-truss
W 形护栏 W-type guardrail
W 形接头 double steps
W 形膨胀水泥 W-type expansion joint
W 指数 W-index
X-Y 绘图机 X-Y plotter
X-Y 绘图仪 coordinatograph
X-Y 轴标绘器 X-Y plotter
X-Y 轴数字转换器 X-Y digitizer
X-Y 坐标记录仪 <将图形上任一点位置转换成数字坐标的装置> X-Y recorder
X 板 X-plate
X 波 X-wave
X 波段参量放大器 X-band parametric amplifier
X 波段二极管移相器 X-band diode phase shifter
X 波段功率放大器 X-band power amplifier
X 波段雷达信标 X-band radar beacon
X 波段脉冲发射机 X-band pulse transmitter
X 波段平面阵 X-band planar array
X 波段三极管振荡器 X-band triode oscillator
X 波段铁氧体调制器 X-band ferrite modulator
X 波段微波源 X-band microwave source
X 波段无源阵 X-band passive array
X 波段限幅衰减器 X-band limiter attenuator
X 波段行波放大器 X-band travelling wave amplifier
X 波段移相器 X-band phase shifter
X 步进记录器 step X register
X 穿孔 X punch
X 带 X zone
X 单位 <一种旧波长单位，$=10^{-10}$毫米> X-unit
X 等年变率图 isoporic chart of X
X 等值线图 isodynamic(al) chart of X
X 方向 X-direction
X 方向流速分量 X-velocity component
X 方向缩放比例系数 X zoom scale factor
X 非效率 <资源分配> X-inefficiency
X 分量 X-component
X 光 roentgen rays;X-ray
X 光部分 X-ray department
X 光测厚计 Measuray
X 光防护门 X-ray shield door
X 光分光计 X-ray spectrometer
X 光分析 X-ray analysis
X 光辐射 X-radiation
X 光管 X-ray tube
X 光机 X-ray apparatus; X-ray machine;X-ray unit
X 光检查 X-ray examination;X-ray inspection
X 光检查装置 panoramic projector
X 光片 roentgenogram;X-ray film
X 光摄影 shadowgraphy
X 光室 X-ray room
X 光探伤 X-ray fault detection;X-ray test
X 光探伤器 X-ray flaw detector
X 光透视 fluoroscope;X-ray examination
X 光透视检查 fluoroscopy
X 光线 Roentgen ray;X-ray
X 光形貌法 X-ray topography
X 光衍射 X-ray diffraction
X 光验测 X-ray test
X 光应力量测 X-ray stress measurement
X 光硬度计 qualimeter
X 光照片 exograph; radiograph; shadowgraph
X 光照相术 radiography
X 光诊断 Roentgen diagnosis; X-ray diagnosis
X 光治疗 X-ray therapy
X 检验 X-test
X 截割晶体 normal cut crystal;X-cut crystal;zero angle cut crystal
X 截距 X-intercept
X 理论 <美国 D.McGregor 提出，认为管理的对策是指导和控制> X theory
X 铝合金 X alloy
X 频带 <频率范围为 5200 ~ 11000MHz> X-band
X 期 X-stage
X 切割 x-cut
X 切割晶片 X-cut crystal
X 射线 Roentgen ray; X-radiation; X-ray
X 射线比浊法 X-ray turbidimetry
X 射线测厚仪 X-ray thickness ga(u)ge
X 射线测角仪 X-ray goniometer
X 射线测量密度法 X-ray absorption method
X 射线沉降分析仪 X-ray sedimentometer
X 射线穿透计 radiochrometer
X 射线传感器 X-ray sensor
X 射线单色仪 X-ray monochromator
X 射线的边界辐射 grenz ray
X 射线底片 X-ray film
X 射线地形测量术 X-ray topography
X 射线定量分析 X-ray quantitative analysis
X 射线发射 X-ray emission
X 射线发射表 X-ray emission ga(u)ge
X 射线发射分析 X-ray emission analysis
X 射线发射谱 X-ray emission spectrum
X 射线法 X-ray method
X 射线反射形貌法 Berg-Barrett method
X 射线防护玻璃 X-ray protective glass
X 射线防护屏 X-ray shield
X 射线放射疗法 X-ray therapy
X 射线废物 X-ray waste
X 射线分光计 X-ray spectrometer
X 射线分析 X-ray analysis;X-raying
X 射线分析法 X-ray analysis method; X-ray testing
X 射线分析仪 X-ray analyser[analyzer]
X 射线粉晶数据 X-ray powder crystal data
X 射线粉晶衍射法 X-ray powdered crystal diffraction method
X 射线粉末法 X-ray powder method
X 射线粉末衍射 X-ray powder diffraction
X 射线辐射 X-ray radiation
X 射线辐射强度 X-ray intensity
X 射线干涉仪 X-ray interferometer
X 射线高温照相机 X-ray high temperature camera
X 射线管 X-ray tube
X 射线光电子光谱学 X-ray photoelectron spectroscopy
X 射线光谱化学分析 X-ray spectrochemical analysis
X 射线激光器 X-ray laser
X 射线激射器 xaser
X 射线计 radiationmeter
X 射线检测仪 X-ray instrumentation
X 射线检查 X-ray examination;X-ray inspection
X 射线检验 radiographic(al) test;X-ray examination;X-ray test
X 射线鉴定 X-ray identification
X 射线结构 x-ray structure
X 射线结构测角法 X-ray texture goniometry method
X 射线结构分析 X-ray structural analysis
X 射线结构分析仪器 X-ray structural analysis instrument
X 射线结晶学 X-ray crystallography
X 射线金相学 X-ray metallography
X 射线晶体密度 X-ray crystal density
X 射线晶体照相法 X-ray crystal diagram method
X 射线密度探测仪 X-ray density probe
X 射线能谱分析 X-ray energy spectrum analysis
X 射线频谱学 X-ray spectroscopy
X 射线谱 X-ray spectrum
X 射线强度计 skiameter;X-ray intensity meter
X 射线取出角 X-ray take out angle
X 射线全息摄影 X-ray holography
X 射线扫描 x-ray scanning
X 射线色散 X-ray dispersion
X 射线烧伤 X-ray burn
X 射线设备 X-ray installation
X 射线摄谱仪 X-ray spectrograph
X 射线摄影测量 X-ray photogrammetry
X 射线摄影术 roentgenography
X 射线术 sciography
X 射线束 pencil of X rays;X-ray beam
X 射线探测器死时间校正 dead time correction of X-ray detector
X 射线探伤 X-ray detection of defects;X-ray inspection
X 射线探伤法 X-ray defectoscopy[defectoscope]
X 射线探伤器 X-ray flaw detector
X 射线探伤仪 X-ray flaw detector
X 射线透视法 radioscopy; roentgenoscopy
X 射线透视检查法 radioscopy
X 射线透视镜 skiascope
X 射线透视术 skiascopy
X 射线图 X-ray diagram
X 射线图案 X-ray pattern
X 射线图式 X-rayogram
X 射线图像 X-ray image
X 射线望远镜 X-ray telescope
X 射线微量分析仪 X-ray microanalyzer
X 射线违禁品检查仪 inspectoscope
X 射线物相鉴定 X-ray mineral phase identification
X 射线吸收 X-ray adsorption
X 射线吸收光谱法 X-ray absorption spectrometry
X 射线吸收限光谱法 X-ray absorption edge spectrometry
X 射线显微分析 X-ray microanalysis
X 射线显微分析法 X-ray microanalysis method
X 射线显微术 X-ray microscopy
X 射线显微照相术 microradiography
X 射线显微照像术 X-ray micrography
X 射线相定量分析 X-ray phase quantitative analysis
X 射线相分析 X-ray phase analysis
X 射线型金刚石钻机 X-ray drill
X 射线型岩芯 X-ray core
X 射线学 roentgenology
X 射线岩组分析 X-ray rock fabric analysis
X 射线衍射 X-ray diffraction
X 射线衍射分布象 X-ray diffraction distribution image
X 射线衍射分析 X-ray diffraction analysis
X 射线衍射图 X-ray diffractogram
X 射线衍射图案 X-ray diffraction pattern
X 射线衍射图像 X-ray diffraction pattern
X 射线衍射图形 X-ray diffraction pattern
X 射线衍射系统 X-ray diffraction system
X 射线衍射仪 X-ray diffractometer
X 射线应力测定法 X-ray method for stress measurement
X 射线应力分析 X-ray stress analysis
X 射线应力仪 X-ray structuring goniometer
X 射线应用 X-ray application; X-ray use
X 射线荧光 X-ray fluorescence
X 射线荧光分析 X-ray fluorescence [fluorescent] analysis
X 射线荧光分析技术 analysis technique of X-ray fluorescence
X 射线荧光光谱 fluorescent X ray spectrum
X 射线荧光光谱法 X-ray fluorescence spectrometry
X 射线荧光光谱分析 fluorescent X-ray spectrographic analysis
X 射线荧光光谱仪 X-ray fluorescence [fluorescent] spectrometer
X 射线荧光屏摄影术 fluorography
X 射线荧光谱仪 X-ray spectrometer
X 射线荧光吸收 X-ray fluorescence absorption
X 射线硬度测定仪 radiochrometer
X 射线在晶体中的衍射 X-ray diffraction in crystals
X 射线照片 roentgenogram; roentgenograph; Roentgen photograph; sciagram; sciagraph; scotograph; X-rayogram; X-ray pattern; X-ray photograph
X 射线照射 X-ray illumination
X 射线照相法 radiographic(al) method
X 射线照相机 X-ray diffraction camera
X 射线照相技术 radiographic(al) technique
X 射线照相检验 X-ray photo detection
X 射线照相试验 radiographic(al) testing
X 射线照相术 sciagraphy; X-radiography
X 射线装置 X-ray generator
X 射线组构测角仪 X-ray structuring goniometer
X 射线组织分析 X-ray difference method
X 视差 horizontal parallax;X-parallax
X 水平偏转板 X-plate
X 危险性测量器 X-ray hazard meter
X 线量计 skiameter
X 效率 <资源分配> X-efficiency
X 形 X-shape;X-type;X-frame
X 形车架 X(-type) frame
X 形撑 X-shaped bracing
X 形的 double vee
X 形电桥 X-bridge
X 形对焊 double V-butt welding
X 形对接焊 double V-butt weld
X 形构架 X-frame
X 形管 X-tube
X 形横梁 X-crossing member
X 形交叉 decussation
X 形接合 X-joint
X 形梁 X-type girder
X 形构件 X-type member
X 形滤波器 lattice-type filter
X 形坡口 X-type groove <熔缝的>; double V groove
X 形钎头 X-bit;X-chisel
X 形桥墩 X-shaped pier
X 形十字架 crux decussate
X 形弹簧 X-spring

X 形天线 X-antenna
X 形网络 lattice-type network;X-network
X 形支撑 X-brace[bracing]
X 形滞火石膏墙板 type X gypsum wallboard
X 形柱条 X-brace[bracing]
X 形钻头 X-bit;X-chisel
X 荧光测井仪 X-fluorescence logging meter
X 荧光测量装置 apparatus of X-fluorescence survey
X 荧光灯测井 X-ray fluoremetry log
X 荧光灯测井曲线 X-ray fluoremetry log curve
X 荧光灯测井仪 X-ray fluoremetry logger
X 荧光井中测量 X-fluorescence survey in borehole
X 荧光实验室分析 X-fluorescence survey in laboratory
X 荧光现场测量 X-fluorescence survey in field
X 正向极性亚带 X normal polarity subzone
X 轴 abscissa axis;axis of abscissa;X-axis
X 轴方向 X-direction
X 轴方向改正 X-correction
X 轴方向偏移 X-offset
X 轴方向位移 X-displacement
X 轴方向移动 X-displacement; X-movement
X 轴方向运动 X-motion
X 轴分量 X-component
X 轴线 X-line
X 坐标 X-coordinate
yo-yo 钻探法 < 钢绳冲击钻探法 > yo-yo drilling
Y-Y 接法 star-star connection
Y-Y 接线 star-star connection
Y-Y 形接线 Y-Y connection
Y-Z 变换钮 Y-Z interchange knob
Y-Δ 接法 star-delta connection
Y-Δ 起动 star-delta starting
Y-Δ 起动继电器 star-delta starting relay
Y-Δ 起动器 star-delta starter;Y-delta starter
Y-Δ 形连接法 Y-delta connection
Y 边导前 Y-edge leading
Y 步进记录器 step Y register
Y 参数 Y-parameter
Y 叉木制薄片 crotch veneer
Y 穿孔 Y punch
Y 带 Y zone
Y 等年变率图 isoporic chart of Y
Y 等值线图 isodynamic(al) chart of Y
Y 电压 Y-voltage
Y 方位角 Y-azimuth
Y 辐射 gamma radiation
Y 高强度耐热铝合金 Y alloy
Y 格网线 Y-grid
Y 合金 Y-alloy
Y 截割 Y-cut;parallel cut
Y 截割晶体 Y-cut crystal
Y 截距 Y-intercept
Y 理论 < 美国 D.McGregor 提出,认为管理的对策是确定目标和诱导,与 X 理论相反 > Y theory
Y 平行截割晶体 Y-parallel cut crystal
Y 钎头 double arc bit
Y 曲折接法 star-zigzag connection
Y 视差 Y-parallax
Y 天线 Y-aerial
Y 通轻质多孔砌块 Y-tong block
Y 线 Y-track
Y 信道 Y-channel
Y 形 Y-type
Y 形叉管 Y branch
Y 形岔线 Y-track
Y 形衬垫 Y-type gasket
Y 形错位式交叉 < 中间形成一个三角形 > crossing at triangle
Y 形道路枢纽 Y-junction
Y 形的 Y-shaped
Y 形电流 star current
Y 形阀 Y(-shaped) valve
Y 形分叉管 Y-type branch
Y 形割法 Y-cut
Y 形谷 Y(-shaped) valley
Y 形管 Y-bend;Y(-shaped) pipe;Y-tube
Y 形管接头 branching fitting;Y-fitting
Y 形轨 Y-track
Y 形轨道 wrought track;Y-track
Y 形交叉 fork junction;Y-junction
Y 形交叉口 Y-crossing; Y-intersection;Y junction
Y 形接法 star connection; Y-connection;Y-joint;Y-section
Y 形接法的 Y-connected
Y 形接头 Y-connection; Y-joint; Y-section;Y(-shaped) fitting
Y 形接线【电】 star connection
Y 形结线 Y-connection
Y 形截距 Y-intercept
Y 形截面 Y-section
Y 形框架 Y-frame
Y 形拉线 Y-stay
Y 形肋 Y(-shaped) rib
Y 形立体交叉 Y-grade separation
Y 形连接【电】 star connection; Y-connection
Y 形连接点 Y-junction
Y 形锚具 Y-type anchorage
Y 形密封垫 Y-type gasket
Y 形密封条 Y-type gasket
Y 形排水管 Y-drain
Y 形匹配 Y-matching
Y 形平面 Y-plan
Y 形砌块 Y(-shaped) building
Y 形桥墩 Y(-shaped) pier
Y 形桥塔 Y(-shaped) pylon
Y 形切割 Y-cut
Y 形清扫设备 Y-clean-out
Y 形球阀 Y globe
Y 形曲线 Y-curve
Y 形三通 Y-type T
Y 形三通管接 pitch T
Y 形三通管接头 Y-branch;Y-joint;Y-lateral
Y 形设计 Y-pattern design;Y(-shaped) design
Y 形水准仪 Y-level
Y 形弯头 Y-bend
Y 形线 Y track
Y 形线路连接器 pigtail
Y 形翼 Y-wing
Y 形支管 Y-branch;Y(-shaped) bend
Y 形支管配件 Y-branch fittings
Y 形支架 Y bearing
Y 形支柱 Y-strut
Y 形肘管 Y-piece
Y 值 < 即饱和度 > Y-value
Y 轴 ordinate axis;Y-axis
Y 轴方向 Y-direction
Y 轴方向偏移 Y-offset
Y 轴方向位移 Y-displacement
Y 轴方向移动 Y-movement
Y 轴线 Y-line
Y 坐标 Y-coordinate
Y 坐标细分划盘 fine index for Y
Y 坐标轴 Y-axis
ZJ1031 型集装箱 ZJ1031 type container
ZSM 型石英弹簧重力仪 ZSM quartz-spring gravimeter
Z 变换 Z-transform
Z 变换值 Z-transform value
Z 的垂直梯度值 vertical gradient of Z
Z 的水平梯度值 horizontal gradient of Z
Z 等年变率图 isoporic chart of Z
Z 等值线图 isodynamic(al) chart of Z
Z 缓冲区绘制 Z-buffer rendering
Z 缓冲区几何 Z-buffer geometry
Z 缓冲区投影 Z-buffer projection
Z 截距 Z-intercept
Z 截式晶体 z-cut crystal
Z 绝缘计数器 Z-absolute counter
Z 肋钢丝网 Z-rib metal lath
Z 理论 < 日本 W.G.Ouchi 提出,认为劳资关系并非对立,其积极性可以溶为一体 > Z theory
Z 铝基轴承合金 z-alloy
Z 捻度 Z-twist
Z 切割 z-cut
Z 形板桩 Z-type piling section;Z-type piling bar
Z 形插孔 < 印制电路的 > accordion
Z 形铲尖 Z-bit
Z 形车站【铁】 switchback station
Z 形的 zigzag
Z 形断面 Z-section
Z 形泛水 Z flashing
Z 形分岔 Z-shape splitting
Z 形缝隙 zigzag crack
Z 形钢 Z bar steel;Z-steel
Z 形钢板桩 Z-piling (bar);Z-section steel sheet pile
Z 形钢筋垫 Z-stool
Z 形钢钎 Z steel bit
Z 形钢丝 Z-shaped steel wire
Z 形钢丝网 Z-rib metal lath
Z 形钢柱 Z-bar column
Z 形格栅 Z grate
Z 形刮板式混料机 Z-blade mixer
Z 形桁架 Z-truss
Z 形角钢 Z-angle
Z 形角铁 Z-angle
Z 形搅拌器 thimble
Z 形截面 Z-section
Z 形截面构件 Z-section member
Z 形肋布置 Z-rib pattern
Z 形连杆 Z-link
Z 形连杆系 Z-linkage
Z 形梁 Z-beam
Z 形檩条 Z purlin(e)
Z 形路 zigzag route
Z 形埋铁 Z-tie
Z 形剖面钢丝 Z-profile wire
Z 形曲柄 Z-crank
Z 形刃口 Z-bit
Z 形双拔钉机 Z-shaped double nail puller
Z 形铁 Z-iron
Z 形铁道路线【铁】 switchback
Z 形铁钎 Z iron rod
Z 形图案 zigzag pattern
Z 形图式 zigzag pattern
Z 形图像 zigzag pattern
Z 形托臂 Z-corbel
Z 形弯管 offset bend
Z 形系墙铁 Z-tie
Z 形衔铁 Z armature
Z 形线 zigzag form;zigzag line
Z 形线路【铁】 switchback
Z 形型钢 Z-steel section
Z 形型式 Z-rib pattern
Z 形窑 zigzag kiln
Z 形褶皱 Z-shaped fold
Z 形桩 Z-(configuration) pile
Z 形装置 Z-linkage
Z 形钻头 Z (drill) bit
Z 值图 Z plot
Z 轴(线) Z-axis
Z 轴方向 Z-direction
Z 轴方向运动 Z-motion
Z 轴调制 Z axis modulation
Z 珠光体可锻铸铁 Z metal
Z 字板桩 Z-type piling bar
Z 字钢 Z-bar;Z-beam;Z-iron;Z-steel
Z 字钢板桩 Z-type piling bar
Z 字形 zigzag
Z 字形钢板桩 Z-piling bar
Z 字形连接 < 变压器中的一种绕组连接方式 > zigzag connection
Z 坐标 Z-axis

α 固溶体 alpha solid solution

α 射线【物】 alpha ray
β 蜕变 β-transformation
β 硅酸二钙 larnite
β 函数 beta function
β 回授网络 beta network
β 玫瑰砷钙石 β-roselite
Δ-Y 接法 delta-star connection
Δ-Δ 接法 delta-delta connection
Δ-Δ 接线 delta-delta connection
δ 调谐 delta tune
δ 分布【数】 delta distribution
δ 函数 delta function
Δ 接法 delta connection
Δ 接线 delta connection
δ 粒子 delta ray
δ 匹配转接 delta-matching
δ 射线 delta ray
δ 铁 delta-iron
Δ 网络 delta network
δ 形的 deltoid(al)
Δ 形结线 delta connection
Δ 形天线 delta-type antenna
Δ 形阻抗匹配天线 delta-match impedance antenna
δ 桩 < 带预制桩尖夯扩桩 > delta pile
θ 角 theta angle
λ 点 lambda point
λ 线 lambda line
ξ 电位 zeta potential
π 电子 pi-electron
π 定理 < 量纲分析用 >【数】 pi-theorem
π 函数 pi function
π 介子 pi-meson;pion
π 形的 double tee
π 形地板 double T floor (slab)
π 形钢 double T-iron
π 形(混凝土)板 double-tee slab
π 形混凝土板地板 double T floor(ing) system
π 形节 pi section
π 形滤波器 pi-type filter
π 形四端网络 pi-network
π 形网络 pi-network
ρ 介子 < 一种质量大于电子 1400 倍的极不稳定的基本粒子 > rho meson
φ 角 phi rotation
φ 粒度分级标准 phi grade scale
ψ 形四通 pitcher crossing
ω-φ-K 旋转角 omega-phi-kappa rotation angle
Ω 网络 omega network
μ 介子素 < μ 介子与电子组合成的耦合系统 > muonium
μ 空间 mu space
μ 子素 muonium
η 粒子 < 不带电,不自旋,质量为电子的 1074 倍,并迅速衰变为介子或伽马射线 >【物】 eta particle
χ^2(方) chi-square
χ^2 分布 chi-square distribution; χ^2-distribution
χ^2 检验 chi-square test
χ^2 置信区间 chi-square confidence interval
χ^2 准则 chi-square criterion

参 考 文 献

[1] 罗新华. 英汉土木工程大词典[M]. 北京:人民交通出版社,2014.

[2] 中国社会科学院语言研究所词典编辑室. 现代汉语词典[M]. 6 版. 北京:商务印书馆,2012.

[3] 罗新华. 汉英港湾工程大词典[M]. 北京:人民交通出版社,2000.

[4] 中国社会科学院语言研究所. 新华字典[M]. 11 版. 北京:商务印书馆,2011.

[5] 方天中. 英汉港口工程词典[M]. 北京:人民交通出版社,2006.

[6] 张泽祯. 英汉水利水电技术词典[M]. 3 版. 北京:中国水利水电出版社,2005.

[7] 全国石油天然气标准技术委员会液化天然气分技术委员会. 液化天然气词汇:SY/T 6936—2013[S]. 北京:石油工业出版社,2014.

[8] 科学出版社名词室. 英汉建筑工程词汇[M]. 北京:科学出版社,2005.

[9] 全国科学技术名词审定委员会. 土木工程名词[M]. 北京:科学出版社,2003.

[10] 顾兴蓥. 新编氧化空调暖通制冷技术词典[M]. 北京:人民交通出版社,1999.

[11] 本词典编写组. 日英汉土木建筑词典[M]. 北京:. 中国建筑工业出版社,1996.

[12] 邱民. 英汉航海・航运・船舶大词典[M]. 北京:人民大学出版社, 1995.

[13] 洪庆余. 现代英汉水利水电科技词典[M]. 武汉: 武汉出版社, 1990.

[14] 谢凯成. 英汉・汉英涂料技术词典[M]. 北京:化学工业出版社, 1999.

[15] 瑞典 BENGT B. BROMS. 英汉对照图示基础工程学[M]. 史佩栋,编译. 北京: 人民交通出版社, 2005.

[16] 宁滨. 英汉汉英铁路词典[M]. 北京:中国铁道出版社,2005.

[17] 张文健. 汉英英汉地铁轻轨词汇[M]. 成都:西南交通大学出版社,2003.

[18] 本词汇编辑组. 英汉测绘词汇[M]. 北京:测绘出版社,1985.

[19] 王业俊,许保玖. 英汉给排水辞典[M]. 北京:中国建筑工业出版社,1989.

[20] 陈明扬. 英汉船舶机电词典[M]. 北京:人民交通出版社, 1996.

[21] 李若珊,等. 新编英汉科技常用词汇[M]. 北京:中国宇航出版社, 1994.

[22] 陆谷孙. 英汉大词典[M]. 缩印本. 上海:上海译文出版社,1993.

[23] 夏行时. 新英汉建筑工程词典[M]. 北京:中国建筑工业出版社,1999.

[24] 李开运. 英汉建筑工程大辞典[M]. 南京:河海大学出版社,1989.

[25] 王同亿英汉科技词天[M]. 北京:中国环境科学出版社,1987.

[26] 本词汇编订组. 新英汉机械工程词汇[M]. 北京:科学出版社,2008.

[27] 马家驹,李骝. 英汉铁路综合词典[M]. 北京:中国铁道出版社,1992.

[28] 潘钟林. 英汉起重装卸机械词典[M]. 北京:人民交通出版社, 1983.

[29] 李轸,朱和. 英汉石油化工词典[M]. 北京:化学工业出版社, 1997.

[30] 交通部基建管理司. 英汉水运工程词典[M]. 北京:人民交通出版社, 1997.

[31] 余昌菊,等. 英汉电工词汇[M]. 2 版. 北京:科学出版社,1987.

[32] 麦世基,等. 英汉工程机械词汇[M]. 哈尔滨:黑龙江科学技术出版社,1992.

[33] 本词典编辑组. 英汉地质词典[M]. 北京:地质出版社,1983.

[34] 马怀平,邵伯岐. 最新汉英审计・会计・金融大辞典[M]. 北京:中国审计出版社,1993.

[35] 周湘寅. 实用汉英机电词典[M]. 北京:人民交通出版社,1994.

[36] 戎培康. 英汉 - 汉英建材工业大词典[M]. 北京:中国建材工业出版社,2000.

[37] 张人琦. 汉英建筑工程词典[M]. 北京:中国建筑工业出版社,1993.

[38] 赵祖康,徐以枋. 汉英土木建筑工程词典[M]. 北京:人民交通出版社,1997.

[39] 赵祖康,等. 英汉道路工程词汇[M]. 4 版. 北京:人民交通出版社,2001.

[40] 李育才. 英汉建筑装饰工程词典[M]. 北京:中国建筑工业出版社,1997.

[41] 本词典编委会. 新英汉建筑工程词典[M]. 北京:中国建筑工业出版社,1991.

[42] 科学出版社名词室. 汉英生物学词汇[M]. 北京:科学出版社,1998.

[43] 科学出版社名词室. 汉英化学化工词汇[M]. 北京:科学出版社,2007.

[44] 吴钰. 英汉铁路工务工程词汇[M]. 北京:中国铁道出版社,2003.

[45] 辞海编辑委员会. 辞海[M]. 缩印本. 上海:上海辞书出版社,1989.

[46] 林鸿慈. 英汉港口航道工程词典[M]. 2 版. 北京:人民交通出版社,1997.

[47] 白英彩. 英汉计算机技术大词典[M]. 上海:上海交通大学出版社,1997.

[48] 中国土木工程学会土力学及基础工程学会. 土力学及基础工程名词(汉英及英汉对照[M]. 2 版. 北京:中国建筑工业出版社,1991.

[49] 李泽成,等. 英汉/汉英航海词典[M]. 北京:人民交通出版社,1998.

[50] 林宗元. 岩土工程勘察设计手册[M]. 沈阳:辽宁科学技术出版社,1996.

[51] 本词汇编辑组. 英汉现代科学技术词汇[M]. 上海:上海科技出版社,1982.

[52] FG Bell with specialist contributors. *Ground Engineer's Reference Book* [M]. ,Butterworths and Co (publishers) Ltd ,1987.

[53] BritishStandard(BS).

[54] American Society for Testing and Materials(ASTM).

[55] 中华人民共和国住房与城乡建设部. 岩土工程勘察术语标准:JGJ/T 84—2015[S]. 北京:中国建筑工业出版社,2015.

[56] 全国暖通空调及净化设备标准化技术委员会. 采暖、通风、空调净化设备术语:GB/T 16803—1997[S]. 北京:中国标准出版社, 1997.

[57] 中华人民共和国住房和城乡建设部. 供暖通风与空气调节术语标准:GB/T 50155—2015[S]. 北京:中国建筑工业出版社, 2015.

[58] 全国地理信息标准化技术委员会. 测绘基本术语:GB/T 14911—2008[S]. 北京:中国标准出版社, 2008.

[59] 中华人民共和国建设部. 城市规划基本术语标准:GB/T 50280—98[S]. 北京:中国计划出版社, 1999.

[60] 中华人民共和国住房和城乡建设部. 城市轨道交通工程基本术语标准:GB/T 50833—2012[S]. 北京:中国建筑工业出版社, 2012.

[61] 全国地理信息标准化技术委员会. 大地测量术语:GB/T 17159—2009 [S]. 北京:中国标准出版社, 2009.

[62] 中华人民共和国交通部. 道路工程术语标准:GBJ 124—88[S]. 北京:中国计划出版社, 1989.

[63] 全国地理信息标准化技术委员会. 地图学术语:GB/T 16820—2009 [S]. 北京:中国标准出版社, 2009.

[64] 全国国土资源标准化技术委员会. 地质矿产勘查测绘术语:GB/T 17228—1998 [S]. 北京:中国标准出版社, 1998.

[65] 中华人民共和国建设部. 电力工程基本术语标准:GB/T 50297—2006[S]. 北京:中国计划出版社, 2006.

[66] 中华人民共和国住房和城乡建设部. 房地产业基本术语标准:JGJ/T 30—2015[S]. 北京:中国建筑工业出版社,2016.

[67] 中华人民共和国交通运输部. 港口工程基本术语标准:GB/T 50186—2015 [S]. 北京:中国计划出版社, 2014.

[68] 中华人民共和国交通运输部. 港口装卸工属具术语:JT/T 392—2013[S]. 北京:人民交通出版社, 2013.

[69] 中华人民共和国住房和城乡建设部. 给水排水工程基本术语标准:GB/T 50125—2010[S]. 北京:中国计划出版社, 2010.

[70] 中华人民共和国住房和城乡建设部. 工程测量基本术语标准:GB/T 50228—2011[S]. 北京:中国计划出版社, 2012.

[71] 全国海洋船标准化技术委员会. 工程船术语:GB/T 8843—2002[S]. 北京:中国标准出版社, 2003.

[72] 全国海洋船标准化技术委员会. 海洋调查船术语:GB/T 7391—2002[S]. 北京:中国标准出版社, 2004.

[73] 中华人民共和国交通部. 航道工程基本术语标准:JTJ/T 204—1996[S]. 北京:人民交通出版社, 1997.

[74] 中华人民共和国住房和城乡建设部. 建材工程术语标准: GB/T 50731—2011[S]. 北京:中国计划出版社, 2012.

[75] 中华人民共和国住房和城乡建设部. 建筑材料术语标准:JGJ/T 191—2009[S]. 北京:中国建筑工业出版社, 2010.

[76] 中华人民共和国住房和城乡建设部. 工程结构设计基本术语标准:GB/T 50083—2014[S]. 北京:中国 建筑工业出版社, 2015.

[77] 中华人民共和国住房和城乡建设部. 岩土工程勘察术语标准:JGJ/T 84—2015[S]. 北京:中国 建筑工业出版社,2015.

[78] 中华人民共和国住房和城乡建设部. 建筑照明术语标准:JGJ/T 119—2008[S]. 北京:中国建筑工业出版社, 2009.

[79] 陈宽基. 科技标准术语词典:第一卷 综合[M]. 北京:中国标准出版, 1995.

[80] 中华人民共和国住房和城乡建设部,国家质量监督检验检疫局. 民用建筑设计术语标准:GB/T 50504—2009 [S]. 北京:中国建筑工业出版社, 2009.

[81] 中华人民共和国国家质量监督检验检疫总局,中国国家标准化管理委员会. 摄影测量与遥感术语: GB/T 14950—2009 [S]. 北京:中国标准出版社, 2009.

[82] 中华人民共和国工业和信息化部. 石油化工配管工程术语:SH/T 3051—2014 [S]. 北京:建材工业出版社,2015.

[83] 孙跃东. 实用英汉汉英土木工程词汇与术语[M]. 北京:人民交通出版社, 2005.

[84] 中华人民共和国住房和城乡建设部. 水文基本术语和符号标准:GB/T 50095—2014 [S]. 北京:中国计划出版社,2015.

[85] 中华人民共和国住房和城乡建设部. 铁路工程基本术语标准:GB/T 50262—2013 [S]. 北京: 中国计划出版社,2014.

[86] 铁道部标准计量研究所. 铁路隧道术语:GB/T 16566—1996[S]. 北京:中国标准出版社, 1997.

[87] 中华人民共和国住房和城乡建设部. 岩土工程基本术语标准:GB/T 50279—2014[S]. 北京:中国计划出版社, 2015.

[88] 全国阀门标准化技术委员会. 蒸汽疏水阀 术语、标志、结构长度:GB/T 12250—2005[S]. 北京:中国标准出版社, 2006.

[89] 全国质量管理和质量保证标准化技术委员会. 质量管理体系 基础和术语: GB/T 19000—2008[S]. 北京:中国标准出版社, 2009.

[90] 本书编委会. 工程地质手册[M]. 4 版. 北京:中国建筑工业出版社,2007.